MOUNTAIN VIEW PUBLIC LIBRARY

1000608741

For reference

Not to be taken from the room.

Jane's
ALL THE WORLD'S AIRCRAFT

Editor-in-Chief: Paul Jackson MRAeS
Deputy Editor: Kenneth Munson AMRAeS
Assistant Editor: Lindsay Peacock

Ninety-second year of issue
2001-2002

Total Number of Entries 1,604 New and Updated Entries 1,528
Total Number of Images 2,299 New Images 940

Bookmark Jane's homepage on
http://www.janes.com

Jane's award-winning web site provides you with continuously updated news and information. As well as extracts from our world renowned magazines, you can browse the online catalogue, visit the Press Centre, discover the origins of Jane's, use the extensive glossary, download our screen saver and much more.

Jane's now offers powerful electronic solutions to meet the rapid changes in your information requirements. All our data, analysis and imagery is available on CD-ROM or via a new secure web service – Jane's Online at www.janesonline.com.

Tailored electronic delivery can be provided through Jane's Data Services. Contact an information consultant at any of our international offices to find out how Jane's can change the way you work or e-mail us at

info@janes.co.uk *or* info@janes.com

ISBN 0 7106 2307 0
"Jane's" is a registered trade mark

Copyright © 2001 by Jane's Information Group Limited, Sentinel House, 163 Brighton Road, Coulsdon, Surrey CR5 2YH, UK

In the USA and its dependencies
Jane's Information Group Inc, 1340 Braddock Place, Suite 300, Alexandria, Virginia 22314-1651, USA

The KFIR 2000

MULTI-ROLE FIGHTER: FIERCE AND FLEXIBLE

Discover how a proven fighter has been modernized to be a winner once again. Featuring the latest generation of avionics and a newly configured glass cockpit, the KFIR 2000 incorporates the unmatched experience and know-how of the Israeli Air Force as well as of IAI's LAHAV – the world leader in fighter upgrading.

See how the ace fighter – the KFIR 2000 can meet your toughest challenges. Contact LAHAV today.

The KFIR 2000 offers:

- Advanced tactical awareness and capabilities
- State-of-the-art radar
- Beyond visual range capability
- Missionized cockpit
- Precision navigation
- Modern HOTAS based cockpit
- Accurate weapon delivery

The Power to Empower Your Fleet

IAI LAHAV DIVISION
Military Aircraft Group
ISRAEL AIRCRAFT INDUSTRIES LTD

www.iai.co.il

Israel Tel: 972-3-935-3163/8198, Fax: 972-3-935-3687 • **U.S.A.** Tel: 1-703-875-3777, Fax: 1-703-875-3740
Europe Tel: 33-1-46.40.47.47, Fax: 33-1-46.40.47.48.

Contents

This Edition has been compiled by:

Susan Bushell	AIRCRAFT: GENERAL AVIATION (part) and AIRLINERS (part)
Bill Gunston, OBE, FRAeS	GLOSSARY; AERO-ENGINES TABLES
Paul Jackson, MRAeS	AIRCRAFT: FRANCE (part), INTERNATIONAL (part), RUSSIAN FEDERATION (civil), UKRAINE (civil), UNITED KINGDOM (part) and UZBEKISTAN
Mike Jerram	AIRCRAFT: GENERAL AVIATION (part) and AIRLINERS (part)
Jon Lake	AIRCRAFT: INTERNATIONAL (part), RUSSIAN FEDERATION (military) and UKRAINE (military)
Kenneth Munson, AMRAeS, ARHistS	AIRCRAFT: CIVIL AND MILITARY (part); LIGHTER THAN AIR
Lindsay Peacock	AIRCRAFT: UNITED STATES OF AMERICA (military)

Users' Charter .... [6]

How to use *Jane's All the World's Aircraft* .... [8]

Alphabetical list of advertisers .... [12]

Foreword .... [15]

Type Classification .... [22]

First Flights .... [24]

Aerospace Calendar .... [26]

Official Records .... [30]

International aircraft registration prefixes .... [33]

Glossary .... [35]

Aircraft

Argentina	1
Australia	2
Austria	10
Belgium	13
Bosnia-Herzegovina	14
Brazil	15
Canada	27
Chile	60
China, People's Republic	61
Colombia	85
Czech Republic	86
Ethiopia	108
Finland	108
France	108
Georgia	142
Germany	142
Greece	167
India	168
Indonesia	178
International Programmes	181

ASTRA SPX & GALAXY MULTI-MISSION AIRCRAFT

MISSIONIZED FOR VALUE

From Israel Aircraft Industries, the ultimate platform solutions for all your multi-mission requirements. Whether you choose the **Astra SPX** or the high volume **Galaxy**, these agile, high speed exceptionally robust platforms feature state-of-the-art avionics and wing hard points. With low DOC, long inspection intervals and a maintenance program supporting over 1,000 flight hours a year, it's clear why you get more value. With the **Astra SPX's** IFR range of more than 2,900NM and the **Galaxy's** more than 3,600NM, your fleet is always ready for long range, high and low altitude, long endurance missions.

Add a powerhouse of capabilities to your fleet, and get more value out of every mission.

- Land/maritime patrol & surveillance (SAR/ISAR)
- Target tow: dual targets on wing hard points
- Aerial reconnaissance & photography
- ELINT, COMINT, EW, C^3I
- Airborne ECM training & target simulation
- Special transport (secure communications, self-protection suite)
- Search & Rescue
- MEDEVAC
- Aircrew training

IAI COMMERCIAL AIRCRAFT GROUP
Israel Aircraft Industries Ltd

Israel Tel: (972)3-935-3565, Fax: (972)3-935-8071 • **U.S.A** Tel: (1)703-875-3727, Fax: (1)703-875-3760
Europe Tel: (33)1-46-40-47-47, Fax: (33)1-46-40-47-48 • E-mail: mzilber@comgrp.iai.co.il • www.iai.co.il

CONTENTS

Iran	257
Israel	260
Italy	266
Japan	292
Korea, South	303
Malaysia	307
Netherlands	309
New Zealand	311
Pakistan	312
Philippines	314
Poland	314
Romania	337
Russian Federation	342
Singapore	472
Slovak Republic	472
South Africa	473
Spain	475
Sri Lanka	478
Sweden	478
Switzerland	481
Taiwan	492
Turkey	494
Ukraine	495
United Kingdom	510
United States of America	539
Uzbekistan	796
Federal Republic of Yugoslavia	796

Lighter than Air

Canada	797
China, People's Republic	799
Czech Republic	799
France	799
Germany	799
Korea, South	802
Netherlands	802
Russian Federation	803
United Kingdom	805
United States of America	809

Air-Launched Missiles	817
Aero-Engines	823
Indexes	837

Users' Charter

This publication is brought to you by Jane's Information Group, a global company with more than 100 years of innovation and an unrivalled reputation for impartiality, accuracy and authority.

Our collection and output of information and images is not dictated by any political or commercial affiliation. Our reportage is undertaken without fear of, or favour from, any government, alliance, state or corporation.

We publish information that is collected overtly from unclassified sources, although much could be regarded as extremely sensitive or not publicly accessible.

Our validation and analysis aims to eradicate misinformation or disinformation as well as factual errors; our objective is always to produce the most accurate and authoritative data.

In the event of any significant inaccuracies, we undertake to draw these to the readers' attention to preserve the highly valued relationship of trust and credibility with our customers worldwide.

If you believe that these policies have been breached by this title, you are invited to contact the editor.

A copy of Jane's Information Group's Code of Conduct for its editorial teams is available from the publisher.

INVESTOR IN PEOPLE

Avanti P180

BEST in CLASS

SPEED COMFORT

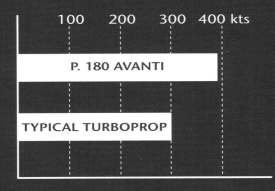

CERTIFIED CEILING: 41,000 ft

PIAGGIO AERO INDUSTRIES

via Cibrario, 4 - 16154 Genova - Italy
telephone +39.010.6481.1 - fax +39.010.6520160
e-mail: Marketing@piaggioaero.it
www.piaggioaero.it

a.d. Giulia Majolino

How to use *Jane's All the World's Aircraft*

Described on the following pages are all known powered aircraft, of which details have been received, currently in, or anticipating, commercial production in all countries of the World, apart from rapidly dismantled, ultralight recreational machines. Exceptions to these criteria are made in the cases of 'one off' aircraft of technical interest – for example those undertaking pioneering work with NASA or designed as a practical exercise by aeronautical universities. Many other of the World's aircraft remaining in service, but no longer being built, will be found in *Jane's Aircraft Upgrades*.

Entries in this book are in alphabetical order (a) by country (including International) and then (b) by manufacturer's name. In the cases of manufacturers producing a diversity of aircraft, those of obvious military potential are presented first. However, it should be noted that a few of the larger constructors are divided into operating divisions, within which individual aircraft types are arranged. In the case of the multinational EADS, still in process of consolidation, aircraft descriptions are retained in the French, German and Spanish sections, alphabetically under 'E'.

Company entries begin with a brief introduction, including postal address, telephone/fax numbers, e-mail and website addresses and the names of some significant executives. The last-mentioned listing is by no means exhaustive and further details will be found in *Jane's International ABC Aerospace Directory* and *Jane's International Defence Directory*. Subcontractors and suppliers to the aerospace industry appear in *Jane's Aircraft Component Manufacturers*, which also provides rapid references to the subcontractors involved in certain prominent aircraft programmes.

For ease of access to information, entries on individual types of aircraft are subdivided under the following headings:

TYPE: A brief description of the aircraft's function. Further explanation of type descriptions appears later in these introductory pages.
PROGRAMME: A record of key events in an aircraft's production history.
CURRENT VERSIONS: Where applicable, details of available models (marks) and cross-reference to earlier versions now out of production.
CUSTOMERS: Present total on order and produced, often in tabular form, for military aircraft and those civil types for which such a list would not be of prohibitive length.
COSTS: Price per unit or programme price, plus any other disclosed information on R&D expenditure.
DESIGN FEATURES: Where appropriate, opens with a broad statement of design objectives and the means by which they were achieved. This is followed by details such as aerofoil section and helicopter rotor speeds.
FLYING CONTROLS: Here is described the method of controlling the aircraft, it being assumed that the reader has a basic understanding of the function of ailerons, flaps, rudders, trim tabs and the other conventional manoeuvring surfaces (refer to 'conventional and manual' in the Glossary). Descriptions are concerned with the method by which the controls are operated (manual/powered) and appropriate control inputs determined (autopilot/fly-by-wire, for example).
STRUCTURE: Configuration, materials and any special manufacturing methods; details of subcontractors or partners producing significant elements of the airframe.
LANDING GEAR: Includes tyre sizes and pressures for wheeled aircraft, as well as ground turning circle. Braking parachutes, where fitted.
POWER PLANT: Number and power of engines; helicopter transmission ratings; fuel capacity. Brief additional details are provided in the Aero-Engines listing towards the end of the book; fuller descriptions of turbine power plants are in *Jane's Aero-Engines*.
ACCOMMODATION: Seating arrangements, access, environmental control and, for transport aircraft, cargo loading capacity; type of ejection seat, if fitted.
SYSTEMS: Power generation provisions, de-/anti-icing equipment, pressurisation/air conditioning and similar equipment.
AVIONICS: The entry is subdivided into communications, radar, flight aids, instruments, mission equipment (mostly military or law-enforcement) and self-defence (military).
EQUIPMENT: Cargo-handling aids, spraying/firefighting apparatus, lighting, ballistic recovery parachutes and similar items.
ARMAMENT: Fixed and air-dropped/launched weapons listed by the manufacturer as actual or potential armament. Not all items may have been cleared for carriage and not all operators will use those which have. Refer also to the Missiles listing, where appropriate.
DIMENSIONS, EXTERNAL: Includes door sizes and certain ground clearances.
DIMENSIONS, INTERNAL: Includes areas and volumes where relevant.
AREAS: Wings, fixed tail surfaces and control surfaces.
WEIGHTS AND LOADINGS: As supplied by the manufacturer; individual aircraft may vary.
PERFORMANCE: Observations as above; all speeds assumed TAS unless stated otherwise.
OPERATIONAL NOISE LEVELS: Internationally recognised measurements of landing and take-off sound at airports.

Measurements of all types are given in both SI (metric) and Imperial units, the more common conversion factors for which are in the Glossary. Performance details are quoted in good faith and certain critical conversions 'rounded' to give a margin of safety, although *Jane's* does not purport to be an alternative to the manufacturer's operating notes.

Following the main section on aeroplanes and helicopters are others detailing Lighter than Air craft, Air-Launched Missiles and Aero-Engines, the first also arranged alphabetically by country. Lighter than Air covers commercially and militarily operated airships, and vehicles of special technical interest, but not recreational balloons. Air-Launched Missiles is a rapid reference which seeks not to duplicate the separate and vastly more detailed *Jane's Air-Launched Weapons*, indicating instead how the potential of aircraft described is augmented by the weapons they carry. Aero-Engines are similarly arranged in alphabetical order of manufacturer, irrespective of their country of origin, within the classifications of piston, turboprop, turboshaft and jet engine.

At the end of the book, Indexes are arranged to speed reference to both current and past aircraft. The first index, on white paper, deals with aircraft in this volume only and is followed by a second, on grey-tinted pages, covering the past 10 editions of the book. Further listings are given for those types of aircraft no longer included in *Jane's*, such as sailplanes, hang gliders and homebuilts constructed from sets of plans. Lighter than Air and the former comprehensive engines section have their separate indexes, but the Air-Launched Missiles and Aero-Engines listings are self-indexing, with multiple cross-references.

Each entry for a company or aircraft is annotated to indicate whether or not it has been altered since the last edition was published. Three descriptions are employed, as under:

● *VERIFIED* The editor has made a detailed examination of the entry's content and checked its relevance and accuracy for publication in the new edition to the best of his ability.

● *UPDATED* During the verification process, significant changes to content have been made to reflect the latest position known to *Jane's* at the time of publication.

● *NEW ENTRY* Information on new equipment and/or systems appearing for the first time in the title.

The most recent drawings, photographs and diagrams are tagged with the date on which they first appeared in *Jane's All the World's Aircraft*. Most are followed by a seven digit number for ease of identification by the *Jane's* image library. General arrangement drawings which have been modified are re-dated, although the original illustrator's name is retained in the caption.

Total Number of Entries 1,604 New and Updated Entries 1,528
Total Number of Images 2,299 New Images 940

All rights reserved. No part of this publication may be reproduced, stored in retrieval systems or transmitted in any form or by any means, electronic, mechanical, photocopying, recording or otherwise, without the prior written permission of the Publishers.
Licences, particularly for use of the data in databases or local area networks are available on application to the Publishers.
Infringements of any of the above rights will be liable to prosecution under UK or US civil or criminal law.

Copyright enquiries
Contact: Keith Faulkner, Tel/Fax: +44 (0) 1342 305032, e-mail: keith.faulkner@janes.co.uk

British Library Cataloguing-in-Publication Data.
A catalogue record for this book is available from the British Library.

Printed and bound in Great Britain by Bath Press, Bath and Glasgow

DISCLAIMER While all reasonable care has been taken in the compilation and accuracy of the material in this publication, the publishers, authors and editors shall not be liable for any loss, injury or damage suffered by any person, corporation or other entity acting or refraining from action as a result of material in, or omitted from, this publication.

EDITORIAL AND ADMINISTRATION

Publishing Director: Alan Condron, e-mail: Alan.Condron @janes.co.uk

Managing Editor: Simon Michell, e-mail: Simon.Michell@janes.co.uk

Content Production Manager: Anita Slade, e-mail: Anita.Slade@janes.co.uk

Application Support Manager: Ruth Simmance, e-mail: Ruth.Simmance@janes.co.uk

Content Editing Manager: Jo Fenwick, e-mail: Jo.Fenwick@janes.co.uk

Pre-Press Manager: Christopher Morris, e-mail: Christopher.Morris@janes.co.uk

Team Leaders: Sharon Marshall, e-mail: Sharon.Marshall@janes.co.uk
Neil Grace, e-mail: Neil.Grace@janes.co.uk

Database Editors: Diana Burns, e-mail: Diana.Burns@janes.co.uk
Tracy Johnson, e-mail: Tracy.Johnson@janes.co.uk

Jane's Information Group Limited, Sentinel House, 163 Brighton Road, Coulsdon, Surrey CR5 2YH, UK
Tel: (+44 20) 87 00 37 00   Fax: (+44 20) 87 00 37 88
e-mail: jawa@janes.co.uk

SALES OFFICE

Send enquiries to: *Julie Reeder – Group Sales Manager*
Jane's Information Group Limited, Sentinel House, 163 Brighton Road, Coulsdon, Surrey CR5 2YH, UK
Tel: (+44 20) 87 00 37 00   Fax: (+44 20) 87 63 10 06
e-mail: info@janes.co.uk

Send USA enquiries to: *Robert Loughman – Vice-President Product Sales*
Jane's Information Group Inc, 1340 Braddock Place, Suite 300, Alexandria, Virginia 22314-1651, USA
Tel: (+1 703) 683 37 00   Fax: (+1 703) 836 00 29   Telex: 6819193
Tel: (+1 800) 824 07 68   Fax: (+1 800) 836 02 97
e-mail: order@janes.com

Send Asia enquiries to: *David Fisher – Group Sales Manager*
Jane's Information Group Asia, 60 Albert Street, #15-01 Albert Complex, Singapore 189969
Tel: (+65) 331 62 80   Fax: (+65) 336 99 21
e-mail: info@janes.com.sg

Send Australia/New Zealand enquiries to: *David Moden – Business Manager*
Jane's Information Group, PO Box 3502, Rozelle Delivery Centre, New South Wales 2039, Australia
Tel: (+61 2) 85 87 79 00   Fax: (+61 2) 85 87 79 01
e-mail: info@janes.thomson.com.au

ADVERTISEMENT SALES OFFICES

Australia: *Richard West* (UK Head Office)

Austria: *Dr Uwe Wehrstedt* (See Germany)

Benelux: *Steven Soffe* (UK Head Office)

Channel Islands: *Steven Soffe* (UK Head Office)

Eastern Europe: *Dr Uwe Wehrstedt* (See Germany)

Egypt: *Steven Soffe* (UK Head Office)

France: *Patrice Février*
BP 418, 35 avenue MacMahon,
F-75824 Paris Cedex 17, France
Tel: (+33 1) 45 72 33 11   Fax: (+33 1) 45 72 17 95
e-mail: patrice.fevrier@wanadoo.fr

Germany: MCW *Dr Uwe Wehrstedt*
Hagenbrite 9, D-06463 Ermsleben, Germany
Tel: (+49 34) 74 36 20 91 Fax: (+49 34) 74 36 20 90

Greece: *Steven Soffe* (UK Head Office)

India: *Joni Beeden* (UK Head Office)

Ireland: *Joni Beeden* (UK Head Office)

Israel: *Oreet Ben-Yaacov*
Oreet International Media, 15 Kinneret Street,
IL-51201 Bene Berak, Israel
Tel: (+972 3) 570 65 27   Fax: (+972 3) 570 65 26
e-mail: oreet@oreet-marcom.com

Italy and Switzerland: Ediconsult Internazionale Srl
Piazza Fontane Marose 3, I-16123 Genoa, Italy
Tel: (+39 010) 58 36 84   Fax: (+39 010) 56 65 78
e-mail: ediconsult@iol.it

Korea, South: *Young Seoh Chinn*
JES Media International, 2nd Floor ANA Building,
2571 Myungil-Dong, Kangdong-Gu, Seoul 134 070, South Korea
Tel: (+82 2) 481 34 11   Fax: (+82 2) 481 34 14
e-mail: jesmedia@unitel.co.kr

Middle East: *Steven Soffe* (UK Head Office)

Pakistan: *Joni Beeden* (UK Head Office)

Rest of the World: *Joni Beeden* (UK Head Office)

Russian Federation and Associated States (CIS): *Simon Kay*
33 St John's Street, Crowthorne, Berkshire RG45 7NQ, UK
Tel: (+44 1344) 77 71 23   Fax: (+44 1344) 77 58 85
e-mail: crowkay@msn.com

Saudi Arabia: *Steven Soffe* (UK Head Office)

Scandinavia: *Gillian Thompson*
The Falsten Partnership, PO Box 21175,
London N16 6ZG, UK
Tel: (+44 20) 88 06 23 01   Fax: (+ 44 20) 88 06 81 37
e-mail: falsten@dial.pipex.com

Singapore: *Richard West* (UK Head Office)

South Africa: *Stephen Judge* (UK Head Office)

Spain: *Macarena Fernandez de Grado*
Via Exclusivas SL, Viriato 69SC, E-28010 Madrid, Spain
Tel: (+34 91) 448 76 22   Fax: (+34 91) 446 01 98
e-mail: amd@varenga.com

Thailand: *Joni Beeden* (UK Head Office)

Turkey: *Richard West* (UK Head Office)

United Arab Emirates: *Steven Soffe* (UK Head Office)

UK – London: *Joni Beeden* (UK Head Office)

UK – South, South East, South West, North West, Wales and West Scotland:
Steven Soffe (UK Head Office)

UK – North East, East, East Scotland: *Joni Beeden* (UK Head Office)

Senior Key Accounts Manager: *Richard West*
Jane's Information Group Limited, Sentinel House, 163 Brighton Road,
Coulsdon, Surrey CR5 2YH, UK
Tel: (+44 1892) 72 55 80 Fax: (+44 1892) 72 55 81
e-mail: richard.west@janes.co.uk

Senior Advertisement Sales Executive: *Stephen Judge*
Jane's Information Group Limited, Sentinel House, 163 Brighton Road,
Coulsdon, Surrey CR5 2YH, UK
Tel: (+44 20) 87 00 38 53   Fax: (+44 20) 87 00 37 44
e-mail: stephen.judge@janes.co.uk

Advertising Sales Executive: *Joni Beeden*
Jane's Information Group Limited, Sentinel House, 163 Brighton Road,
Coulsdon, Surrey CR5 2YH, UK
Tel: (+44 20) 87 00 39 63   Fax: (+44 20) 87 00 37 44
e-mail: joni.beeden@janes.co.uk

Advertising Sales Executive: *Steven Soffe*
Jane's Information Group Limited, Sentinel House, 163 Brighton Road,
Coulsdon, Surrey CR5 2YH, UK
Tel: (+44 20) 87 00 39 43   Fax: (+44 20) 87 00 37 44
e-mail: steven.soffe@janes.co.uk

USA and Canada: *Ronald R Lichtinger III,* Advertising Sales Director
Jane's Information Group Inc, 1340 Braddock Place, Suite 300,
Alexandria, Virginia 22314-1651, USA
Tel: (+1 703) 683 37 00   Fax: (+1 703) 836 55 37
e-mail: lichtinger@janes.com

North Eastern USA and East Canada: *Harry Carter*
Jane's Information Group Inc, 1340 Braddock Place, Suite 300,
Alexandria, Virginia 22314-1651, USA
Tel: (+1 703) 683 37 00   Fax: (+1 703) 836 55 37
e-mail: carter@janes.com

South Eastern USA: *Kristin D Schulze*
5370 Eastbay Drive, Suite 104,
Clearwater, Florida 33764, USA
Tel: (+1 727) 524 77 41   Fax: (+1 727) 524 75 62
e-mail: kristin@intnet.net

Western USA and West Canada: *Richard L Ayer*
127 Avenida del Mar, Suite 2A,
San Clemente, California 92672, USA
Tel: (+1 949) 366 84 55   Fax: (+1 949) 366 92 89
e-mail: ayercomm@earthlink.net

Administration:
USA and Canada: *Maureen Nute*
Jane's Information Group Inc, 1340 Braddock Place, Suite 300,
Alexandria, Virginia 22314-1651, USA
Tel: (+1 703) 683 37 00   Fax: (+1 703) 836 00 29
e-mail: nute@janes.com

UK and Rest of World: *Sue Tucker*
Jane's Information Group Limited, Sentinel House, 163 Brighton Road,
Coulsdon, Surrey CR5 2YH, UK
Tel: (+44 20) 87 00 37 42   Fax: (+44 20) 87 00 38 59
e-mail: sue.tucker@janes.co.uk

Air family of titles

Jane's 3-D
Electronic reference guide with rotational 3-D images of 120 of the world's most significant fighter/attack aircraft. Available on CD-ROM. Features 4 views of each aircraft, information on systems and weapons as well as side by side comparison with another aircraft plus up to five photographs.

Jane's Aero-Engines
Provides information on civil and military engines that are currently in production or still in service throughout the world. Reviews the market trends and examines engine specifications including programme history and technical capabilities.

Jane's Aircraft Component Manufacturers
This extensive resource analyses each sector, such as brakes and engine nacelles, in terms of market size and share, giving vital information on the capabilities of individual companies. Find out who is selling what to whom, where the market opportunities lie and what technical advances are being made.

Jane's Aircraft Upgrades
The companion reference to Jane's All the World's Aircraft. Details information on civil and military aircraft no longer in production, but still in service, including technical descriptions of landing gear, accommodations, systems and avionics. Also includes aircraft modernisation and performance enhancement packages.

Jane's Air-Launched Weapons
With details of over 580 individual air-launched weapons, you are kept up-to-date with the latest developments throughout the world. Find out how each weapon works, when it entered service, who purchased it and which aircraft are cleared to carry which weapons.

Jane's All the World's Aircraft
Expertly details more than 1000 civil and military aircraft currently being produced or under development, providing you with the ability to evaluate competitors, identify potential buyers, locate possible business partners and examine aircraft equipment.

Jane's Avionics
Find detailed information on the avionic equipment in military and civilian aircraft and helicopters in this extensive guide to avionics. Stay up-to-date with the latest developments and new production lines. Discover how large scale integration is changing the way systems are designed and employed worldwide.

Jane's Combat Aircraft Camouflage
Database providing national camouflage markings for combat aircraft (fixed-wing and helicopters). Aircraft types covered include fighter, attack, bomber, cargo, patrol and EW. The database covers all major air arms and contains separate patterns from over 350 different aircraft types. Each aircraft is depicted in four-view line drawings. Files can be supplied in .jpg and .eps formats on a user-friendly interface or in a customised format according to requirements.

Jane's Helicopter Markets and Systems
The most comprehensive resource on the world's manned and unmanned helicopters and engines in use, in production, under development or being upgraded — in civilian and military markets.

Jane's Space Directory
Profiles hundreds of space programmes and their different technologies enabling you to identify thousands of different commercial and defence applications. Review key objectives, developments and technical specifications plus receive listings of suppliers and manufacturers.

Jane's Unmanned Aerial Vehicles and Targets
With details of over 140 UAVs, 100 aerial targets and 180 subsystems, this regularly updated publication is the most comprehensive of its kind. Each entry details the manufacturer – complete with contact information and the civil and military organisations using the aircraft.

Jane's World Air Forces
The premier intelligence source on air forces, naval and army aviation and paramilitary air arms around the globe. Profiles squadrons, reporting structures and inventories including make, role and exact model.

Other Jane's titles

Magazines
Jane's Airport Review
Jane's Asian Infrastructure Monthly
Jane's Defence Industry
Jane's Defence Upgrades
Jane's Defence Weekly
Jane's Foreign Report
Jane's Intelligence Digest
Jane's Intelligence Review
Jane's International Defense Review
Jane's International Police Review
Jane's Islamic Affairs Analyst
Jane's Missiles and Rockets
Jane's Navy International
Jane's Terrorism and Security Monitor
Jane's Transport Finance
Police Review

Security
Jane's Chem-Bio Handbook
Jane's Chemical-Biological Defense Guidebook
Jane's Counter Terrorism
Jane's Facilities Handbook
Jane's Intelligence Watch Report
Jane's Sentinel Security Assessments
Jane's Terrorism Watch Report
Jane's World Insurgency and Terrorism

Transport
Jane's Air Traffic Control
Jane's Airports and Handling Agents
Jane's Airports, Equipment and Services
Jane's High-Speed Marine Transportation
Jane's Road Traffic Management and ITS
Jane's Urban Transport Systems
Jane's World Airlines
Jane's World Railways

Industry
Jane's International ABC Aerospace Directory
Jane's International Defence Directory
Jane's World Defence Industry

Systems
Jane's C^{4}I Systems
Jane's Electronic Mission Aircraft
Jane's Electro-Optic Systems
Jane's Military Communications
Jane's Radar and Electronic Warfare Systems
Jane's Simulation and Training Systems
Jane's Strategic Weapon Systems

Land
Jane's Ammunition Handbook
Jane's Armour and Artillery
Jane's Armour and Artillery Upgrades
Jane's Explosive Ordnance Disposal
Jane's Infantry Weapons
Jane's Land-Based Air Defence
Jane's Military Biographies
Jane's Military Vehicles and Logistics
Jane's Mines and Mine Clearance
Jane's Nuclear, Biological and Chemical Defence
Jane's Personal Combat Equipment
Jane's Police and Security Equipment
Jane's World Armies

Sea
Jane's Amphibious Warfare Capabilities
Jane's Exclusive Economic Zones
Jane's Fighting Ships
Jane's Marine Propulsion
Jane's Merchant Ships
Jane's Naval Construction and Retrofit Markets
Jane's Naval Weapon Systems
Jane's Survey Vessels
Jane's Underwater Technology
Jane's Underwater Warfare Systems

For more information on any of the above products, please contact one of our sales offices listed below:

Europe, Middle East & Africa
Jane's Information Group
Sentinel House
163 Brighton Road
Coulsdon, Surrey, CR5 2YH, UK
Tel: (+44 20) 87 00 37 00
Fax: (+44 20) 87 63 10 06

The Americas
Jane's Information Group
1340 Braddock Place
Suite 300, Alexandria
Virginia 22314-1657, USA
Tel: (+1 703) 683 37 00
Fax: (+1 703) 836 02 97

Asia
Jane's Information Group
60 Albert Street
15-01 Albert Complex
Singapore 189969
Tel: (+65) 331 62 80
Fax: (+65) 336 99 21

Australia
Jane's Information Group
PO Box 3502
Rozelle Delivery Centre
New South Wales 2039, Australia
Tel: (+61 2) 85 87 79 00
Fax: (+61 2) 85 87 79 01

USA West Coast
Jane's Information Group
201 East Sandpointe Avenue
Suite 370, Santa Ana
California 92707, USA
Tel: (+1 714) 850 05 85
Fax: (+1 714) 850 06 06

Alphabetical list of advertisers

B

British Aerospace Defence
Warwick House. Farnborough Aerospace Centre, Hampshire GU14 6YU, UK .... *Bookmark*

I

IAI LAHAV Division
Israel Aircraft Industries Ltd, Ben Gurion International Airport, IL-70100, Israel .... [2]

IAI Commercial Aircraft Group
Israel Aircraft Industries Ltd, Ben Gurion International Airport, IL-70100, Israel .... [4]

P

Piaggio Aero Industries
Via Cibrario 4, I-16154 Genova, Italy .... [7]

Information Services & Solutions

Jane's is the leading unclassified information provider for military, government and commercial organisations worldwide, in the fields of defence, geopolitics, transportation and law enforcement.

We are dedicated to providing the information our customers need, in the formats and frequency they require. Read on to find out how Jane's information in electronic format can provide you with the best way to access the information you require.

Jane's Online

Search across the complete portfolio of Jane's Information, via the Internet

Created for the professional seeking specific detailed information, this user-friendly service can be customised to suit your ever-changing information needs. Search across any combination of titles to retrieve the information you need quickly and easily. You set the query — Jane's Online finds the answer!

Key benefits of Jane's Online include:
- the most up to date information available from Jane's
- accessible anytime, anywhere
- saves time — research can be carried out quickly and easily
- archives enable you to compare how specifics have changed over time
- accurate analysis at your fingertips
- site licences available
- user-friendly interface
- high-quality images linked to text

Check out this site today: **www.janesonline.com**

Jane's CD-ROM Libraries

Quickly pinpoint the information you require from Jane's

Choose from nine powerful CD-ROM libraries for quick and easy access to the defence, geopolitical, space, transportation and law enforcement information you need. Take full advantage of the information groupings and purchase the entire library.

Libraries available:
Jane's Air Systems Library
Jane's Defence Equipment Library
Jane's Defence Magazines Library
Jane's Geopolitical Library
Jane's Land and Systems Library
Jane's Market Intelligence Library
Jane's Police and Security Library
Jane's Sea and Systems Library
Jane's Transport Library

Key benefits of Jane's CD-ROM include:
- quick and easy access to Jane's information and graphics
- easy-to-use Windows interface with powerful search capabilities
- online glossary and synonym searching
- search across all the titles on each disc, even if you do not subscribe to them, to determine whether you would like to add them to your library
- export and print out text or graphics
- quarterly updates
- full networking capability
- supported by an experienced technical team

Jane's Data Service

Jane's information on your intranet or controlled military network

Jane's Data Service brings together more than 200 sources of near-realtime and technical reference information serving defence, intelligence, space, transportation and law enforcement professionals. By making Jane's data (HTML) and images (JPEG) available for integration behind Intranet environments or closed networks, this unique service offers you a way to receive information that is updated frequently and works in tune with your organisation.
We can also offer a complete management service where Jane's hosts the information and server for you. The most secure way to access Jane's Information.

Jane's Consultancy

A service as individual as your needs

Whether it is research on your competitors' markets, in-depth analysis or customised content that you require, Jane's Consultancy can offer you a tailored, highly confidential personal service to help you achieve your objectives. However large or small your requirement, contact us in confidence for a free proposal and quotation.

Jane's Consultancy will bring you a variety of benefits:
- expert personnel in a wide variety of disciplines
- a global and well-established information network
- total confidentiality
- objective analysis
- Jane's reputation for accuracy, authority and impartiality

The information you require, delivered in a format to suit your needs.

The Joint Strike Fighter competition in the USA reaches its climax this year, following maiden flights by both contenders during 2000. Boeing's X-32A is seen here approaching Edwards AFB at the conclusion of its first sortie, on 18 September 2000

FOREWORD: The Year of the Giants

"It's always best on these occasions to do what the mob do." "But suppose there are two mobs?" suggested Mr Snodgrass. "Shout with the largest," replied Mr Pickwick. (Charles Dickens, The Pickwick Papers)

Three decades ago, in 1970, a young aviation enthusiast — unwittingly researching the foreword for this book, but who, more properly, should have been attending to other duties — witnessed the delivery to London/Heathrow of BOAC's first Boeing 747. This aerial leviathan, towering above the Boeing 707s and, even, VC10s, made a profound impression upon aviation *cognoscenti* and laymen, alike; and the popular epithet of 'Jumbo Jet' reinforced the impression that, surely, this was as big as they got. Well, not quite. A few months later, the same keen spotter was outside RAF Lakenheath to welcome the first Lockheed C-5A Galaxy into the UK. A little bigger and a little heavier than a 747, said the *Jane's* of the time; and there was no doubt in the popular mind that the USA had overtaken the USSR in creation of a new genus of monster carriers for both civil and military aviation.

That was not the view from Kiev, now in Ukraine, where Antonov had designed the mighty An-22 Antheus military turboprop which had taken the Paris Air Show by storm in 1965. Twenty years later, also at Le Bourget, that same reporter witnessed the public debut of the four-jet An-124 Ruslan as it taxied into the static park to be thronged by a crowd apparently consisting more of agents of Western intelligence bureaux than of the aviation press. However, even this first glimpse of such a mighty beast was overshadowed four years later as the six-engined An-225 Mriya broke low cloud in the Paris circuit on its arrival for the Show, complete with Buran space shuttle riding pickaback. Now, that's what you call *big*.

Having been progressively stripped of spare parts in the decade since the collapse of the USSR, the unique Mriya is being refurbished with the intention of once again towering over the Paris flight line as this book is published. Indeed, that will be an appropriate moment for the aircraft to confirm that its commercial time has come, because the same event will afford Airbus Industrie the twofold opportunity of parading its newly acquired status as an integrated company and presenting more details of the recently launched, massive airliner we are now learning to call the A380. Truly, this is the year of the giants.

Even when A380 takes its first bow as a flying prototype, it will be slightly smaller and lighter than a fully laden Mriya yet, in one important respect, it will be infinitely larger. The ample Antonov came into being during an age in which money was no object for priority programmes of the Soviet State; its unfortunate recent history, though it mirrored that of its parent, was not directly the cause of it. Conversely, A380 must make its own way in the world market; its commercial failure could presage catastrophe for even a company as currently successful as Airbus.

With the minimum 50-order threshold broken, Airbus signalled full throttle for its former A3XX project in December 2000. Designations A350, A360 and A370 were bypassed so that the centre digit might reflect the two-deck cross-section selected to convey a nominal 555 passengers between continents in a degree of luxury last witnessed during the golden age of the airship. Even though the A380's launch coincided with high-profile media reporting of concerns about the health dangers of international flights in cramped airliners, Airbus itself concedes that its artist's impressions of shops, bars and gymnasia come under the heading of 'options', not 'standard fittings and fixtures'. Thus are we gently awoken from blissful dreams of sumptuousness.

Despite some of its press conferences giving the impression that Airbus's prime function is to manufacture anti-American rhetoric, rather than airliners, it cannot be denied that Boeing's cautious — even timid — rejoinder merely added strength to the Europeans' charge that the 747X was hardly a worthy challenger. Notwithstanding plans to certify it as an all-new design, Boeing was relying on the view that adding a few extra fuselage frames and some other tweaks to 40-year old technology would see off the rival to its long-established pre-eminence in airliner conception and manufacture. Airbus avows that an all-new design can be the only solution to the demands of tomorrow's mass air travel market, whatever the risks inherent in launching a new project with a tight timetable.

Judging from the different levels of enthusiasm shown for the A380 and 747X, airlines seemed inclined to shout with the largest mob, and favour the untried A380 to the solid, dependable 747X. And not just the airlines; to gain the support which would have been an essential prerequisite to programme launch, Boeing had to convince Japanese aerospace companies, including those bruised by an earlier false start on a further stretched Jumbo, that the 747X would have presented a viable commercial proposition. Patently, as we now know, neither demand nor backing was available in sufficient measure to warrant Boeing matching Airbus's US$11 billion programme with a less adventurous US$4 billion gamble of its own. However, as befits a world-class combatant, Boeing did not slink from the battlefield under the cover of a smokescreen of excuses and recriminations. On 29 March it astounded the professional and popular media alike by boldly announcing its Sonic Cruiser programme for a futuristic-looking, 225-seat twin-jet. The 747X has been quietly forgotten, save that some of its features may see the light of day in a more moderate venture which could become the 747-500X.

Has Boeing blinked? While many in Europe would like to think so, future historians of air travel may opine that the US company has cannily taken the opposite fork in the road. In the A380, Airbus provides the potential for cheaper transport through the economies of scale; others may be prepared to pay a little extra to get there more rapidly. Accommodating the same number of passengers as the 767-300, the Sonic Cruiser lives up to its name by flying almost M0.2 faster than its stable-mate (and, indeed, the current generation of airliners). Concorde has proved a premium can be charged for a means of transport which leaves the *hoi polloi* behind; flying at up to M0.98, the Sonic Cruiser will provide this cachet without encountering too many of the technical obstacles to supersonic flight.

Furthermore, Boeing's action has averted the potentially damaging situation of the two major aircraft producers becoming embroiled in a bitter sales war within a

Airbus has launched the A380, opening a new page in the annals of air transport. Spacious interiors are available, if desired by the individual airline, and within a decade, a stretched version could be carrying 1,000 passengers on shorter sectors
2001/0062413

FOREWORD

Unveiled in late March, the Sonic Cruiser is intended to raise airliner operating speeds by a quarter, securing an uncontested market sector for Boeing   *2001*/0110061

mega-airliner marketplace too small to sustain both. Perhaps in 20 years, both firms will have both categories of aircraft available; for now, "Airbus is bigger, but Boeing is faster" seems a sensible compromise. It may not be publicly acknowledgeable on the east side of the Atlantic, but Boeing could have rendered a great service to the European aerospace industry by backing off from a fight. That Airbus provides 61 per cent of EADS' turnover illustrates the size of the business risk, should A380 fail.

Some commentators may already have decided that this will be its fate. British popular media coverage of the launch of Europe's most ambitious aviation project since Concorde was far from universally enthusiastic, being peppered with accusations of this "political" project being "A great white elephant" and prophesies that "In the end, it will be the European taxpayers, including us, who suffer." It is difficult to refute such sentiment when the politics of European federalisation become intertwined with purely business decisions between national aerospace industries.

Britain and France managed to produce Concorde while being at loggerheads over the former's proposed EC membership, although further pursuance of this analogy does not necessarily bring into mind the phoenix-like revival of that aircraft's fortunes after its enforced grounding in the wake of the catastrophic events of July 2000. A melancholy document filed in the editorial archives three decades ago proudly lists the planned delivery positions of the first 74 production Concordes before fuel crises and environmental objections brought the production lines to an abrupt halt at No. 14. Then, it was an earlier generation of taxpayers which settled the outstanding bills.

Further haunting the darker recesses of the collective aerospace mind are the promotional computer images of an A380 standing beside a small fleet of London buses for emphasis of scale — a pose recalling childhood picture books placing inhabitants of the Jurassic era in readily imaginable proportion against a trusty, red, double-decker. With good fortune and good management — *their* relative proportions yet to be determined — Airbus's monster will not turn out to be a dinosaur.

Slight turbulence; no serious injuries

Stepping back to view the airliner scene as a whole, the aerospace industry is now beginning its recovery from the recession that hardly anybody noticed. Deliveries of new airliners by the two world leaders dipped some 12 per cent last year, almost entirely the consequence of Boeing slowing down the 747 line and terminating both the MD-11 and MD-90. With a bulging order book, Airbus was able to continue boosting production rates in spite of the market cycle having reached one of its periodic lows, giving the impression that it was catching up on its American rival. Euphoria, generated in Toulouse by Airbus's winning of the 1999 order 'war', will dissipate as Boeing pulls away once more in the production race, but neither firm has much to concern it in view of the current world airliner backlog of 3,250. The late J C Bamford once declared, "A large backlog is an admission of failure", although it must be conceded that the mechanical excavator market may differ from that of airliners.

Following protracted negotiations among its member companies, Airbus became an integrated company this year. It has a busy time ahead: flying the super-stretched A340-600; building the super-shrunk A318; and floating the planned A330-500. Next year's *Jane's* may well report launching of the enlarged Boeing 747-500X, compressed 717-100X and longer-legged 757-200X, in addition to first deliveries of the fourth Next-Generation 737, the -900. Now that the MD-90 and MD-11 are no more, the 717 is the sole survivor of the McDonnell Douglas range of airliners originating in Long Beach. If it is to remain that way, it could do with more than the 21 orders it received last year, particularly as Boeing's 737-600 would pick up many contracts from airlines thwarted by its disappearance from the market.

Russia has some latitude for cautious celebration, having ended 2000 with a 20 per cent increase in sales of commercial aircraft of all classes. Unfortunately, this is from such a pitifully low base that the overall state of the aerospace industry is little changed from previous years. An estimated 14 airliners, 55 helicopters and 80 medium and light aircraft were supplied in 2000, including an Antonov An-124 heavy freighter from restarted production. New Year certification of the Tupolev Tu-214 (Kazan-built Tu-204) and an imminent first flight of the Tu-234 are further small boosts for the airliner makers, offset by the decision not to produce more Tu-154s after the present stock of 'white tails' is sold. (This could be some time. Justifying the opinion of former editor John W R Taylor that "the Russians never throw anything away", we have just discovered a handful of unsold Ilyushin Il-62 'Classics' still standing outside the factory, patiently awaiting buyers more than a decade after the type was deleted from this annual.)

Perhaps of more significance is the move towards a union of Russia's large and medium commercial aircraft producers: Ilyushin, Tupolev and Yakovlev. Last December, the trio announced plans to form a holding company, taking with them the all-important manufacturing plants Aviastar and KAPO (both allied to Tupolev); VAPO (in the Ilyushin camp); and SAZ and Sokol (both Yakovlev associates). Such a grouping would be a first step towards restoring a world-class position to a Russian civil aircraft industry crippled by funding shortages. Cases in need of immediate financial assistance would include the stagnating Il-96M programme and Tupolev's Tu-334. The latter is being assisted through its certification programme by RSK 'MiG'/Voronin Production Centre, but progress has been slow and the existence of only one prototype was shown to be an Achilles' heel when it was damaged (fortunately slightly) in a heavy landing late last year. A second prototype will not become available until next January, the programme finally getting into top gear on completion of the third, in July 2002.

A sombre official report released late last year implies that Russia has already lost all

The 600th Bombardier de Havilland Dash 8, delivered in March, is an undoubted token of success, yet orders for turboprop-powered regional airliners are rapidly shrinking   *2001*/0062417

hope of a presence in the European and US civil markets and will become entirely uncompetitive within a maximum of five years. Industry ills are blamed in no small measure on the rush to privatise in the early 1990s, resulting in foreign investors gaining substantial interests in aerospace companies and then failing to deliver the promised investment. In the cases of Aviacor and Aviastar production plants, privatisation was reckless and, perhaps, illegal, says the Russian parliamentary study.

While the report and its conclusions may be accused of hankering after the more certain life which State control provided, it is undeniable that it describes a decade of unprecedented collapse. During its death throes of 1991, the USSR still produced 150 civil transports, 300 civil helicopters, 620 military aeroplanes and 390 military helicopters; in 1999, a mere nine civil and 21 military aircraft were built — and the home air force could afford none of them. Whether the recommendations of increased State control, taxation and regulation can remedy the situation within the stated five-year period is unlikely.

In charting the precipitate fall of Russia's 26 per cent share of the world airliner market from 1992, the report may make a political point for home consumption; the indecent haste with which Eastern European airlines unloaded their Ilyushins and Tupolevs as soon as the Kremlin's ties were loosened makes an economic one: short of military conquest, the export market can only be regained through a long-term policy of providing affordable quality, as Western manufacturers must do. As for the internal market, only a state-subsidised Aeroflot functioning as a scarcely disguised military reserve could again support the vast fleet of under-utilised transports which could be found languishing on airport ramps throughout the old USSR. Those days have gone.

Regional props in need of a prop

Last year's Foreword noted that in 1999, for the first time, jet deliveries gained ascendancy over turboprops in the regional airliner market. The proportion is now nearly three-quarters in favour of jets (300:120) and showing no signs of stabilisation. With the regional airliner market having grown at an annual rate of almost 40 per cent recently, the outcome is a current backlog of around 1,400 aircraft, of which 90 per cent are jets.

Such figures leave ATR, in particular, in an exposed position, as it has only one product in the window. This year began with the company planning an ATR 42/72 relaunch and beginning a programme of integration of facilities to improve its efficiency, but not immediately to the full extent now practised by Airbus. Reorganisation is intended to retain profitability at production rates as low as 1.5 per month — or somewhat less than, say, Embraer's current 16 regional jets per month and its plans to achieve 20 after a new 'green field' production site becomes operational in 2002.

When Embraer and Bombardier are not accusing each other of accepting improper subsidies from their respective governments, the latter offers the Dash 8Q as another significant player in the turboprop regional field, yet orders for its extra long Q400 might be arriving more slowly than envisaged when the programme was launched in 1995. To this shrinking market comes the Antonov An-140, now in production in Iran, as well as in Ukraine and (in desultory fashion) Russia, yet unlikely to have negated too many Western manufacturers' sales prospects in these regions.

For all the impressive backlogs quoted by Bombardier (480), Embraer (600) and Fairchild Dornier (200), the regional jet stable has suffered a recent round of culling, as 'obituaries', in this issue, of the embryo 100-seat Bombardier BRJ-X and 44-place Fairchild Dornier 428JET testify. Fairchild's decision, while fiscally correct in order to prevent over-reaching, both removes a product in the fastest growing (30 to 50 seat) sector of the RJ market and leaves the company with a hole in its product range.

No such reticence inhibits China, whose new 70-seater project was launched last November. Fearless in the face of world competition, it will be built by a recently reorganised aerospace industry. We also report on the Alliance StarLiner regional jet project which intends to employ the design and manufacturing expertise of the Sukhoi group to produce a family of low-cost RJs that can be assembled in the US at a competitive cost. This innovative project has just suffered an uncharacteristic setback — along the lines of the samovar calling the kettle black — following Sukhoi's accusation that it was the Americans who had not properly structured and funded their business.

The current edition also welcomes into the air the Bombardier CRJ900 and Embraer ERJ-140, both of them developments of existing aircraft. Belatedly, the Avro RJX should have joined them before this is read. Orders for this upgraded variant stood at the beginning of the year at "two firm and six options", inclusion of the latter always being a sign that the manufacturer was rather hoping to have been able to present more substantial figures. Less understandable is the apparent need of more successful RJ makers to lump orders and options together, even double-counting in the instance of one order transferred to a related variant.

More money for the military?

Entries in the US section this year record maiden flights by the two contenders for the JSF crown. Boeing's X-32 and the Lockheed Martin X-35 are each hoping to secure for their companies orders for an eventual 3,000 aircraft in three distinct versions: land-based, carrier-based and STOVL. Short, but intensive test programmes have been flown and the Pentagon's decision may, or may not, be known by the end of the year. 'Fly-off' is a misnomer for the independent assessment of two technology demonstrators which has been taking place, yet it must be remarked that the USA is the last nation capable of funding two separate aircraft up to flying status in order to select one for production.

The United Kingdom's pretensions in this field ended nearly half a century ago — and, even then, the RAF forgot it was supposed to choose between the Vulcan and Victor. Russia may come to count the MiG 1.44 and Sukhoi S-37 Berkut as its last such contest, even though, had the USSR survived, derivatives of both might now have been in service. France's 'industry of prototypes' was curtailed due to financial restrictions at the same time as its British neighbour, while others never aspired to this twin-track approach. Some managed well without that luxury of choice — others didn't; and for every Saab Viggen there was a HAL Marut.

It may seem unkind in the next breath to Marut to mention HAL's current military venture, the LCA but, despite the success of the latter's maiden flight in January, the fact remains that the programme is now so seriously behind time that India has been forced to obtain a licence from Sukhoi to build 180 part-substitute Su-30s. LCA still has to fly with its indigenous Kaveri turbofan — and Kaveri has still to fly in *any* aircraft (a Tupolev Tu-16 testbed is standing by and the new engine might just get into an LCA in three years' time) — so completion of a fully functioning combat aircraft is still in the middle distance, even if the politicians do not lose patience.

China scored what is variously viewed as a coup or an own goal in January when a

India's LCA has finally slipped the surly bonds of computer software and achieved its maiden flight, albeit on a borrowed engine. The first new warplane of 2001 still has a long and perilous development mission to fly before arriving at its first IAF squadron
2001/0073898

FOREWORD

picture of its new J-10 fighter was posted on the Internet and hastily withdrawn. This apparently unauthorised action may have been prompted by the LCA's first aerial venture a few days earlier and it was certainly useful in confirming that the aircraft *does* exist and looks almost exactly like the artist's impression, by the late Keith Fretwell, *Jane's* has carried since 1997. A contemporary of Eurofighter, it still appears to be some way short of entering squadron service.

In Europe, with the first production Eurofighter Typhoon now taking shape, attention in the military sphere shifts to the next major development programme: the A400M transport. Following an interminable conception, the programme now has its formal commitments (even including a solitary aircraft for the non-existent 'Luxembourg Air Force') and has gained, at the very last moment before launch, a power plant. Conventional wisdom in the aerospace industry would have it that an engine takes longer to develop than an airframe, thereby putting in jeopardy the target dates of first flight in the second half of 2005 and initial deliveries in mid-2007.

However, that view dates from the days when life was simpler. The A400M design has been evolving since the Manchester division of Hawker Siddeley began looking at the question of a 'Hercules replacement' nearly 30 years ago, while the chosen TP400 engine is an amalgam of two well-proven designs: a Snecma M88 core married to Rolls-Royce Trent architecture. So, what exactly is new, and what is tried and tested? Some answers may be found in the Foreword to the 2006-07 edition of *Jane's*.

Many other matters affecting world military aviation will be decided by the new regime installed in the White House in January. Republicans are expected to be good news for the defence industries, but that does not mean extra shares all round for everyone. Bell and Boeing could be the first high-profile discoverers of that fact, following two recent fatal crashes of V-22 Ospreys, full production authorisation on 'hold' and the coming to power of an administration which has long harboured doubts as to the tiltrotor's viability.

Also suffering consequences of the administrative transition is the Lockheed Martin F-22 Raptor. Winter weather delayed by just over a month its achievement of capability milestones which would loosen the purse strings for production. Then, in February, just as compliance was demonstrated, all urgency evaporated from the official programme pending completion of a strategic review by the new Washington incumbents. This seems to suit the USAF, which would like to delay production for a year in order to implement cost-saving measures in the production process.

New Orleans and all that jazz

General aviation is now riding the crest of a wave. In terms of numbers delivered, totals are as high as two decades ago, before the liabilities issue temporarily forced manufacturers out of the light aircraft market. Sales of twin-props and business jets have increased just recently and are now expected to fall back slightly until the end of the current decade. Downward curving lines on the graph might plunge more steeply than anticipated if early warnings of a slowing of the US economy prove to have foundation.

Deliveries by the world's (mostly US) leading private and business aircraft manufacturers totalled 3,000 in the last calendar year, compared with 2,650 in the immediately previous period and 2,300 in 1998. Cessna, Piper and Raytheon stride ahead, but so do the likes of Aviat, Commander, Mooney and newcomer Cirrus, whose SR20 enjoyed its first year of high-rate production and still boasts a healthy order book.

The annual National Business Aircraft Association Convention — held in New Orleans for 2000 — is one of the most reliable market indicators. Its big announcements included unveiling of the Hawker 450 (eight seats, filling the product range between Premier 1 and Horizon) and two new versions of existing designs: Dassault Falcon 2000EX and Gulfstream GV-SP. Notwithstanding the surprise of the last-mentioned announcement, resulting from Gulfstream's year-long disinformation campaign suggesting a GIV upgrade was in prospect, this was confirmation that the executive jet industry is throttling back to cruise power for the next few years. In 1998, it will be recalled, four new programmes were announced by Cessna alone.

Embraer could contain itself only until the eve of the Farnborough Air Show (newly positioned to July) to take the wraps off its Legacy programme. In spite of a new name, the Legacy is but an ERJ-135 with VVIP trimmings, hardly qualifying as a new design. Early this year came rumours of Dassault planning to release details of its next business jet at Paris in June, apparently a clean-sheet-of-paper project to replace the Falcon 900 with a long-range aircraft capable of Mach 0.9.

Out and about for the first time over the past year are three of those Cessnas from 1998: the CJ1, CJ2 and Encore. Raytheon's protracted Premier I is about to join them, while maiden flights by the Bombardier Continental and Hawker Horizon should have taken place shortly before *Jane's* was published. Among the big boys, the Boeing Business Jet 2 is now a reality following its first 'green' delivery in March 2001, responding to a demand that even Boeing admits is greater than anticipated for this longer-fuselage 737 variant.

Patent applications continue to provide brief insights to American plans for a supersonic business jet which continues to be studied jointly by Lockheed Martin and Grumman, with involvement by DARPA and NASA. The two government agencies justify their participation because of the new technology to be explored, not the least of which is the prospect of a supersonic aircraft which produces a minimal supersonic boom. Northrop Grumman has already been awarded a DARPA contract to study a quiet supersonic platform (QSP) for both military and civil applications. QSP guidelines are a 45,360 kg (100,000 lb) aircraft carrying a 9,072 kg (20,000 lb) payload over 6,000 n miles (11,112 km; 6,904 miles) at Mach 2.4, yet which still meets Stage 3 noise regulations (the stumbling block for Dassault's recent aspirations).

Some revolutionary shapes are being explored as a means to beat the boom or, at least, diminish it to what is presumed to be a publicly acceptable 2.44 kg/m^2 (0.5 lb/sq ft) shockwave but, before proceeding too far down this avenue, those involved would be advised to consult with Raytheon, whose Starship programme is said to have floundered because of conservative businessmen's attitudes to its futuristic shape. Perhaps also having suffered this prejudice, Piaggio's graceful, but distinctly unconventional, Avanti turboprop twin makes a welcome return to (moderate) quantity production, after a five-year hiatus.

First flying photographs appear in this edition of the Ibis Aerospace Ae 270, a 10-seat utility turboprop which made its initial, brief appearance in the 1990-91 edition of *Jane's* and has been described in detail (originally under Aero in the Czech Republic section) since 1993. Hailed as the first new Czech commercial aircraft for 50 years, it has had an unconsciously long gestation, matched only by the length of the list of potential commercial rivals. Swift

Raytheon announced the Hawker 450 during the 2000 NBAA Convention — the only all-new business jet of the year

FOREWORD

certification and aggressive marketing are now called for, backed, as in the past, by financial support from Taiwan.

It must be conceded that there is an unfortunate and recent parallel to this international liaison, from which the American Ayres company is only just recovering. Having embarked on the ambitious Loadmaster parcels carrier project, Ayres extended itself further in taking a 93 per cent share in Czech-based Let Kunovice in 1998, only to be dragged down when the state-owned Konsolidacni Banka filed for bankruptcy. Boardroom changes, including a lower-profile role for the redoubtable Fred Ayres, has been the price paid for its rescue under Chapter 11 protection.

An additional — not to mention innovative — competitor in the Ae 270's market niche is the Intracom GM-17 Feniks, making its debut in this edition. Feniks's conventional features are reassuringly familiar; no small wonder when the penny drops that this is a Piper Navajo, whose twin Lycoming pistons have been swept from the wings and replaced by a single Czech-built M 601 turboprop in the nose. Out of Russia by way of a sales office in Switzerland, the aircraft is intended to form the basis of a family of GMs ranging from amphibians to armed trainers, all with maximum commonality.

Private flying, too, can sometimes seem to be trapped in a different time. One of light aviation's anomalies is the backwards leap in propulsion technology made each time a pilot gets out of his car and climbs into an aeroplane. Steps are, finally, being taken to address that situation, evidenced by recent certification of the SMA SR305 (formerly MR230) flat-four turbocharged diesel, rated at 230 cv (169 kW; 227 hp) and offering increased fuel economy, notable noise reduction, a 3,000-hour life and 300-hour TBO. Early applications of what will become a family of SMA engines are in the forthcoming EADS Socata range of new lightplanes. In the USA, Piper is expected to announce a new technology infusion for its piston-engined range in the near future, but Cessna, apart from revealing a turbocharger and some cosmetic changes for the Skylane in 2000, seems content to be edging back to the days of yore, when every year brought a new model with a fresh set of minor upgrades.

EADS Eurocopter surprised the aerospace world in February when it revealed the EC 130. This much modified Ecureuil managed to complete its certification programme without attracting the attention of the media
2001/0062415

Last year's Foreword made patriotic mention of the Farnborough F1 aerial taxi, and this year we can point to fresh life being breathed into the Speedtwin sportplane and the launching of a certified variant of Ivan Shaw's extremely popular Europa. As the Liberty XL-2, the last-mentioned has the potential for large-scale production — in the United States, if not its homeland. The demise of the British lightplane industry has been lamented since Beagle was placed in receivership in 1969; its resurgence could be germinating in the small North Yorkshire village of Kirkbymoorside, where both Slingsby (whose Firefly has now lost some of its commercial sparkle) and Europa are based.

Helicopters and helium

One surprise addition to the ranks of helicopters paraded in this issue is the Eurocopter EC 130. A re-engineered version of the AS 350 Ecureuil, most distinctively sporting a Fenestron, plus windscreen and skids from the EC 120, its two prototypes managed to complete their certification programmes at Marignane without the world's aviation press being even unofficially alerted. Perhaps this airfield would be a suitable (and cheaper) venue for the US to test its 'black' programmes. 'EC' designations are gradually taking over at Eurocopter, it will be noted, with December having seen the 'Plus' Super Puma and military Cougar variant becoming the EC 225/725. Whatever they may be called, latest results show they and their stablemates have, creditably, captured half the civil helicopter market.

Bell/Agusta now has a flying product to its name in the wake of the maiden flight in February of the AB139, a 15-seat utility helicopter which may come to replace the Bell 412, despite being officially its complement. Before the end of the year, the same partnership plans to have its civilian tiltrotor, the BA609, in the air. Little has been heard of this in the past 12 months, notably in the area of updates to the orders situation, so it may be presumed that the initial market enthusiasm for this novel form of air transport has waned somewhat.

In the current edition, a new general arrangement drawing records the not insignificant changes which Sikorsky has made to the S-92, partly in response to the wishes of potential customers. Its rear end has been extensively re-profiled, following a lead set by the Boeing Sikorsky RAH-66 Comanche, suggesting a previously unremarked trend in rearranging the tail feathers of vertically mobile avians. Having been announced in March 1992 and formally launched in 1995, the almost leisurely S-92 appears to be running to a Russian timetable, but Sikorsky is fortunate that so, too, is the Euromil Mi-38.

A further area of this book which seems to have returned to an earlier age of glory is Lighter than Air. When the Boeing 747 was

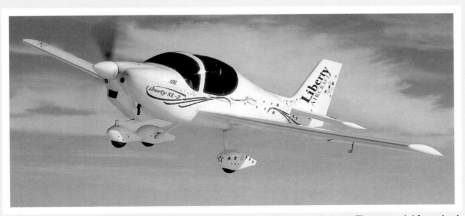

Liberty's XL-2 is a factory-built, certified version of the popular Europa kitplane. The potential for sales is large, so large that a production facility for this British design has been established in the United States
2001/0062414

FOREWORD

entering service, the 'Airships' section struggled to fill a single page, although it did have the distinction of being compiled by The Lord Ventry, an aristocrat who practised what he preached. Now, the greatly enlarged review of current commercial gas-filled aircraft is a dynamic element of this book, with regular comings and goings — admittedly, a high proportion of them unfortunately going before they have properly come, such as the three 'millennium' airship projects, all of which failed to materialise.

Notable among this year's new offerings (and thoroughly in concert with our theme of giants of the air) is the ATG SkyCat, a British craft combining airship and air cushion vehicle, which had been kept well away from prying eyes (including *Jane's*) until its formal launch shortly before the most recent Farnborough Show. Little extra can be added to the CargoLifter entry beyond that which the company — which receives EC funding — makes available in its general publicity material and on its website. We are in the bizarre position of having more illustrative material concerning the construction of its shed near Berlin than on the small-scale prototype that has been flying for 18 months, all pleas for pictures and drawings of the latter having fallen on deaf ears. In view of doubts expressed in the technical press concerning viability of the CargoLifter concept, a more open approach would be in the best interests of the company and taxpayers.

Sadly, more and more companies seem to be relying too heavily on a website to convey their message to the world. *Jane's* regularly visits these to check for new information and, while some are well designed, a high proportion of them is found to be lacking in

Large aircraft do not have to be aeroplanes. Making its first appearance in this annual is the ATG SkyCat hybrid airship which will, when it is built, dwarf the USAF's massive C-5 Galaxy airlifter   *2001*/0073147

some way or other. The most valuable are those which include a 'press' section giving access to archived news releases and high-resolution imagery, these being mostly the sites of the larger companies. Many others appear to be orientated towards the casual surfer or, on the lightplane side, be an unabashed sales pitch to potential customers. It is disturbing how many such sites contradict themselves or proffer data which is at variance with printed material from the same source. Others attempt to provide a description of their product while omitting such vital information as dimensions or empty weight and offering illustrative material which is of poor quality.

Availability of the World Wide Web at one time briefly seemed to offer the prospect of easy and immediate acquisition of information which *Jane's* could present to its subscribers with minimal effort and delay. A more studied assessment, in the light of experience, now suggests the WWW is a mixed blessing; it makes more data available at the touch of a button, yet that which is gained is not necessarily one jot more reliable than any other uncorroborated fact gleaned from another source. To employ an aeronautical analogy, a website is but the equivalent of the automatic terminal information service (ATIS) which blindly broadcasts an airport's basic information to visiting pilots on a closed loop of tape; efficient operation still demands a human air traffic controller (the aerospace company's information officer) to complete the service on a case-by-case basis. Thus it is likely to remain for the foreseeable future.

Paul Jackson
Pulham Market, Norfolk, June 2001

FOREWORD

Notes and acknowledgements

In addition to describing the world's aviation programmes, and detailing 12 months of their progress — or otherwise — each issue of this annual attempts to improve on the overall quality of information presented to the reader. Principal among the changes in the present volume has been a reappraisal of the 'TYPE:' descriptions that begin each aircraft entry, our intention being to bring a greater measure of standardisation in order to expedite and assist comparison of aircraft with similar configurations and roles.

The task proved to be simpler to conceive than to execute. What is important when attempting to describe an aircraft's purpose in no more than half a dozen words? Accepting that no two researchers would be seeking exactly the same set of parameters at least eased the editorial conscience after arbitrary decisions were made, although we have been deliberately inconsistent between certain broad categories in order to bring to the fore that information which might be considered important in one class of aircraft and less so in another. A related consideration has been how many aircraft fall within each category, so that a researcher might be presented with a not overwhelming number of aircraft types within which to further refine his or her search.

For examples, we may begin with the plethora of light aircraft seating between one and six occupants. We have supposed that a reader scanning this category would be keenly interested in the number of seats; category of licence required by the pilot (ultralight, single engine land or twin rating); and whether or not he could, or would have to, build the aircraft himself. Conversely, the number of regional jets in, or anticipating, production is sufficiently small to delegate to the reader the task of separating two- and four-engine types and arranging seating capacity in order, if those parameters are an important consideration for the exercise in hand. Where all lightplanes are singles, unless stated otherwise, we have assumed the baseline business jet to be a twin, and remarked only on single- and tri-engined members of the genus.

Classifications adopted have been determined with due reference to the manufacturer's description but, as the saying has it, 'the editor's decision is final and no correspondence can be entered into.' Differences from the maker's brochure have usually involved placing an aircraft in a lower category than claimed by its designer, but this has always been in the interests of keeping like aircraft together. For example, what some enthusiastic sales managers describe as 'basic trainers' would be regarded by air forces merely as introductory types to the high-powered turboprops which lead pilots on to jet trainers. Similarly, we have declined to describe as five-seaters those aircraft which fall in this category only when three of their occupants are children.

In attempting to present concise descriptions, we have been trapped by the constraints of uniformity and sometimes led into the mires of convoluted language, perhaps better read than spoken. Perversely, in the realm of business jets, we have discarded the multitude of strata (what, exactly, is "super mid-size"?) devised by the industry and adopted broader categories. Some may take issue with what *Jane's* defines as 'long range'; simply, as a European-based publication we have examined the needs of a businessman wishing to reach the US eastern seaboard from a European capital city. After that, New York to Los Angeles doesn't seem far.

Type descriptions do not take into account the possibilities for adapting an aircraft for other duties; that information will be found among the detail of the description, usually under 'CURRENT VERSIONS'. Even under this self-imposed rigour, we have employed some 140 separate descriptions of aeroplane in 11 broad categories, plus 33 more for rotary-wing and eight for lighter than air machines.

Elsewhere in the aircraft section, we have brought more consistency to aircraft tyre sizes, or at least as much as the several classification systems in current use will allow. Manufacturers of these accessories are sometimes less than uniform in the manner in which they present their own designations, so *Jane's* has made a few arbitrary decisions on their behalf.

Some entries have moved this year. The coming into being of EADS has been responsible for some such changes, because these initials are now used as a prefix to old, or revised company names. Consistent with our policy of describing matters as they are, and not how they are supposed to be, the individual companies which formed EADS remain in their national sections, for they will continue to exist as separate legal entities for some time yet. EADS commands its own short entry in the International section, but only for the purpose of describing the 'big picture'. A similar strategy is adopted for AgustaWestland, another new international company consisting of two smaller (by no means insignificant) firms which seem to be carrying on in much the same manner as before.

Like a dictionary, it is *Jane's* function to reflect the use of words and designations, not prescribe it. That does not mean, of course, that we will give equal status to incorrect nomenclature, however widely employed. Aircraft names will change to EADS parentage only after the company unequivocally makes that adoption. Conversely, we shall refrain from quoting the 'BAe Systems', 'BAES' and 'British Aerospace Systems' which have regularly appeared elsewhere and detracted from that company's sought-after corporate image, as a consequence of its adoption of a 'word mark' universally incompatible with media house styles.

This year, the compiling team has added 56 new companies and 97 new aircraft to these pages, the eventual contents (after removal of terminated and moribund programmes and companies) turning out as 553 and 1,051, respectively. Concerning those which also appeared in last year's book, 89 per cent of the company entries and 98 per cent of the aircraft descriptions have been updated.

Illustrations in the Aircraft and Lighter than Air sections now total 2,055. Of the 1,031 which depict complete aircraft (as opposed to illustrating details of potential interest), 64 per cent are new since the last edition. There are also 261 close-ups, 77 artist's impressions, 22 cutaway drawings, 153 diagrams and 511 general arrangement drawings, of which 52 are new and 54 modified in the past 12 months.

Such a prodigious rate of change would be impossible without the greatly appreciated assistance received by the editor from the compiling team of Sue Bushell, Mike Jerram, Jon Lake, Kenneth Munson and Lindsay Peacock plus, of course, technical artist James Goulding and proofreader-cum-backstop Maurice Allward. Engine data and the glossary came from the author of far larger treatises on both, Bill Gunston. Splendid, annotated cutaway drawings have been supplied by Mike Badrocke.

Professional colleagues have been most supportive in supplying additional photographs and data, their number including Peter J Cooper, John Fricker, Pierre Gaillard, Jean-Louis Gaynecoetche, Yefim Gordon, Robert Hewson, Geoffrey P Jones, Ryszard Jaxa-Malachowski, Arnold Nayler, Jim Thorn (Australian Aviation), Mike Vines (Photo Link), Simon Watson and Sebastian Zacharias.

Nor must we forget those in the aircraft industry who have responded to requests to update their entries in this book. The constraints of managing such data gathering for a company producing over 40 yearbooks means that contact cannot be as personal as we would prefer, with the consequence that we are unable to be as specific (and insistent) with our demands for information as in less hectic times. Organisers of air shows around the world have made *Jane's* welcome and sometimes afforded additional privileges which have added to the completeness of this volume — most notably during a rewarding editorial foray to Avalon, in Australia. (Others — curiously, in countries claiming to be keen to promote trade with the West — have been lamentably incommunicative, forcing the cancellation of plans for attendance.)

At the *Jane's* HQ, further back-up has come from managing editor Simon Michell; the compositing team of Lynette Murphy, Hayley Austin and Jack Brenchley, led by Chris Morris; Tracy Johnson, database editor, Christine Varndell, typesetter; and Diana Burns, our indefatigable and irreplaceable production editor. Heartfelt thanks to all.

Type Classifications

To assist comparison between different makes of aircraft undertaking broadly similar roles, the 'TYPE' classifications which introduce each aircraft description have been standardised and arranged into 13 classes. Of these, seven are of a sufficiently interrelated nature to benefit from cross-comparison as one group. The others demand separate consideration, as detailed below.

Classification is according to *Jane's* own criteria; in some instances this may differ from a manufacturer's description. Exact capacities and engine types and numbers for large aircraft are omitted in these shorthand descriptions. However, as size diminishes, such aspects assume greater importance and, accordingly, are quoted.

Except where the type description immediately and obviously precludes it (for example, 'Strategic transport'), aircraft are assumed to be monoplanes powered by a single piston engine and propeller. Multiple engines, turboprop and jet propulsion are specifically mentioned. In deference to common usage, 'jet' includes turbojets and turbofans, as more properly described in the 'POWER PLANT' paragraph.

All aircraft are Class A (or equivalent) certified/certifiable and factory-assembled; where clarification is necessary, the term 'lightplane' is employed. Ultralights and/or aircraft built from kits are specifically noted as such. However, in view of widely differing national legislation on ultralights (from prohibited, up to 540 kg; 1,190 lb maximum take-off weight) it must be noted that the term has several interpretations; the most liberal is used in this book.

The more versatile makes of aircraft cannot be given justice in deliberately short type descriptions. Readers should be aware that the ability to undertake surveillance, geological survey, VIP transport, water bombing and many other duties can be easily bestowed on medium transports and some other types of aircraft. Crop sprayers can be reconfigured in moments for firefighting and oil pollution dispersal. The full story will be found under the 'CURRENT VERSIONS' heading.

Class 1: Bomber and surveillance
These are military or paramilitary aircraft of widely differing size and performance.

```
Strategic Bomber
Strategic/maritime reconnaissance bomber
Medium/maritime reconnaissance bomber
Maritime reconnaissance four-jet
Maritime surveillance twin-jet
Maritime surveillance twin-turboprop
Maritime surveillance turboprop
Airborne early-warning and control system
Airborne ground surveillance system
Multisensor surveillance twin-turboprop
Multisensor surveillance lightplane
```

Class 2: Fighter and trainer
All fighters are jet-powered; trainers are generally single-engined. A few trainers have civil applications, but most are military.

```
Air superiority fighter
Multirole fighter
Attack fighter
Light fighter
Light fighter/advanced jet trainer
Advanced jet trainer
Advanced jet trainer/light attack jet
Basic jet trainer
Basic jet trainer/light attack jet
Basic jet trainer/light transport
Basic turboprop trainer/attack lightplane
Basic turboprop trainer
Basic prop trainer/attack lightplane
Basic prop trainer
```

Class 3: Miscellaneous and/or government
Aircraft of diverse or multiple duties employed generally, but not exclusively, by the state.

```
Missile defence system
Multirole twin-jet
Trials support platform
Space vehicle launch platform
Aerospace craft
High-altitude platform
```

Class 4: Transport
Generally of a military nature, often with rear loading ramps. The larger aircraft are usually, but not exclusively, jet-powered. Light transports are those not exceeding 5,670 kg (12,500 lb).

Class 5: Airliner and freighter
Civilian passenger and cargo aircraft. Jet power is implied for all large airliners; number of engines given only for medium-size aircraft; regional jets are all twins.

Class 6: Business
The established configuration of a business jet being a twin, only single- and tri-jet configurations are specifically noted. Detailed strata devised by participants in the business jet market are not reproduced here. Instead, the main category is subdivided into small business jets with accommodation for six or fewer passengers; long-range (sub-airliner size) business jets with the 4,000 n mile (7,400 km; 4,600 mile) range necessary to fly from the US eastern seaboard to Western Europe; and the central core of twin-jets with upwards of seven seats.

Class 7: Utility
Restricted to single- and twin-engine passenger or passenger/light freight aircraft with accommodation, typically, for four or six persons. An intermediate category bridging commercial and private ownership. Crop sprayers are included.

Class 8: Amphibian
All such aircraft (including the occasional flying-boat) are included here for ease of reference, even those assembled from kits or classified as ultralights.

Class 9: Lightplane (factory built)
Single-prop aircraft seating up to five persons and available only in complete form.

Class 10: Utility kitbuilt
This class reflects the growing number of kitbuilt aircraft intended for more than recreational use. The largest has 10 seats and turboprop power.

The contents of Classes 4 to 10 are presented in a single table on the opposite page. Approximate equivalence between classes is indicated by an entry on the same line, but in a different column. For example, a military strategic transport is in the same capacity bracket as a civil outsize freighter; or the prospective purchaser of a business jet may wish to be reminded that a business jet amphibian is also available. Large and high-performance aircraft are in the table's upper left areas; single-seat lightplanes in the lower right. A few lightplanes are optimised for easy conversion to ultralight category.

Class 11: Kitplanes and/or ultralights
Single-prop machines intended for private ownership. 'Kitbuilt' aircraft are in the lightplane category, unless described as ultralights (and flyable as such in at least some countries). Some aircraft are available only in complete form, while kit manufacturers may offer the option of fly-away aircraft to those requiring the aeroplane, but not the assembly work. These are indicated as follows:
Ultralight Available complete, factory built only
Ultralight kitbuilt Available as a kit only
Ultralight/kitbuilt Option of fly-away or home assembly
Note also that some aircraft are optimised for conversion between ultralight and motor glider categories.

Class 12: Rotary wing
Includes tiltrotors, autogyros and lifting platforms. Light utility helicopters are those not exceeding 5,000 kg (11,023 lb) maximum take-off weight.

Series built	Kitbuilt and ultralight
Attack helicopter	
Armed observation helicopter	
Naval combat helicopter	
Amphibious helicopter	
Light observation helicopter	
Heavy-lift helicopter	
Medium-lift helicopter	
Light-lift helicopter	
Medium transport helicopter	
Multirole medium helicopter	
Medium utility helicopter	
Light utility helicopter	
Multirole light helicopter	
AEW helicopter	
Four-seat helicopter	Four-seat helicopter kitbuilt
Three-seat helicopter	
Two-seat helicopter	Two-seat helicopter kitbuilt
	Two-seat ultralight helicopter kitbuilt
	Sport helicopter kitbuilt
Light helicopter trainer	
	Single-seat helicopter kitbuilt
Autogyro	
	Two-seat autogyro kitbuilt
	Two-seat autogyro ultralight
	Single-/two-seat autogyro kitbuilt
	Single-seat autogyro ultralight
New-concept rotorcraft	
Lifting vehicle	
Two-seat lifting vehicle	
Light utility tiltrotor	
Multimission tiltrotor	
Convertiplane	

Class 13: Lighter than air
All are (at least broadly) cylindrical dirigibles, unless otherwise specified.

```
Helium rigid
Helium semi-rigid
Helium semi-rigid hybrid
Helium non-rigid
Helium non-rigid special shape
Helium non-rigid hybrid
Hybrid air vehicle
Hot air non-rigid
```

Single-seat	Tandem-seat (two-seat)	Side-by-side (two-seat)
kitbuilt	kitbuilt	kitbuilt
	lightplane/kitbuilt	lightplane/kitbuilt
sportplane kitbuilt	sportplane kitbuilt	performance kitbuilt
	turboprop sportplane kitbuilt	
	turboprop sportplane/kitbuilt	
ultralight/motor glider		lightplane/motor glider kitbuilt
ultralight	ultralight	ultralight
ultralight kitbuilt	ultralight kitbuilt	ultralight kitbuilt
ultralight/kitbuilt	ultralight/kitbuilt	ultralight/kitbuilt
	biplane kitbuilt	
ultralight biplane kitbuilt	ultralight biplane kitbuilt	
	ultralight motor glider	
kitbuilt twin	ultralight twin	
agricultural ultralight	agricultural ultralight	
agricultural ultralight kitbuilt		

TYPE CLASSIFICATIONS

COMPARISON: CLASSES 4 to 10

Transport	Airliner/freighter	Business	Utility	Amphibian (all)	Lightplane (factory-built)	Utility kitbuilt
	Supersonic airliner Outsize freighter	Supersonic business jet				
Strategic transport						
	High-capacity airliner					
	Wide-bodied airliner					
	New concept airliner					
Tanker-transport	Four-jet freighter					
Tanker						
Medium transport/ multirole				Four-turboprop amphibian		
	Twin-jet transport					
	Twin-jet airliner	Long-range business jet				
	Twin-jet freighter					
	Regional jet airliner	Large business jet		Twin-jet amphibian		
		Long-range business tri-jet				
		Business tri-jet				
		Business jet		Business jet amphibian		
		Light business jet			Two-seat jet sportplane	
		Business mono-jet				
	Twin-turboprop airliner					
Twin-turboprop transport	Twin-turboprop freighter					
Twin-turboprop light transport		Business twin-turboprop	Utility turboprop twin	Twin-turboprop amphibian		
		Business twin-prop	Light utility twin-prop transport			
			Six-seat utility twin	Six-seat utility twin-prop amphibian		
			Four-seat utility twin			Two-seat kitbuilt twin
			Turboprop utility transport			
		Business turboprop	Light utility turboprop		Five-seat lightplane/ turboprop	
			Light utility turboprop biplane			
		Business prop	Light utility transport			Utility kitbuilt Utility/kitbuilt
			Light utility biplane			Utility biplane kitbuilt
			Six-seat utility transport	Six-seat amphibian Six-seat amphibian/ kitbuilt		Six-seat kitbuilt
				Five-seat amphibian	Five-seat lightplane	Four-/six-seat kitbuilt
				Four-seat amphibian	Four-seat lightplane	
				Four-seat amphibiam kitbuilt		
						Four-seat performance kitbuilt
				Two-/four seat amphibian kitbuilt		Four-seat kitbuilt
				Three-seat flying boat kitbuilt	Three-seat lightplane	Three-seat kitbuilt
						Three-seat biplane kitbuilt
					Two-/three-seat lightplane	
					Primary prop trainer/ sportplane	
					Aerobatic two-seat sportplane	
					Aerobatic two-seat lightplane	
				Two-seat amphibian	Two-seat lightplane	
					Two-seat lightplane/ motor glider	
				Two-seat amphibian kitbuilt		
					Motor glider	
					Two-seat lightplane ultralight	
				Two-seat amphibian ultralight kitbuilt		
					Aerobatic two-seat biplane	
					Aerobatic single-seat sportplane	
					Single-seat sportplane	
					Glider tug	
			Agricultural sprayer			

First Flights

Some of the first flights made during the period 1 January 2000 to 31 January 2001

Lockheed Martin's X-35A contender for the JSF contest made its initial flight on 24 October 2000

Brazil
Embraer ERJ-140 (converted prototype ERJ-145) (PT-ZJA)	27 Jun 00

Canada
Found Aircraft Canada FBA-2C1 Bush Hawk, first production (C-GDWS)	2 Mar 00

China
Evektor EV-97 Eurostar, first Chinese-assembled	24 Oct 00
NAMC K-8E Karakorum, first for Egyptian Air Force	5 Jun 00

Czech Republic
ATEC Zephyr Solo (OK-FUH 01)	29 Oct 00
Wolfsberg-Evektor Raven 257 (OK-RAV)	28 Jul 00

France
CAP 222, first production (F-WWMZ)	15 May 00
Dyn'Aéro MCR4S (F-WWUZ)	14 Jun 00
Reims Aviation F 406 Surpolmar, first for Hellenic Coast Guard (F-WWSS)	21 Aug 00
Socata TB 20 GT Trinidad, first production (N163GT)	.. Mar 00

Germany

AkaFlieg München Mü30 Schlacro (D-EKDF)	16 Jun 00
Aquila A210 Adler (D-EQUI)	5 Mar 00

India
ADA LCA (KH2001)	4 Jan 01

International
Airbus A310-300F, first for Federal Express	30 Jan 01
Airbus A340-300 testbed, first flight with (one) Rolls-Royce Trent 500 engine (F-WWAI)	20 Jun 00
EH Industries EH 101, first AW 320 (CH-149) Cormorant for Canadian Forces (14901)	25 Feb 00
Eurocopter EC 725 Cougar Mk II+ (F-ZVLR)	30 Nov 00
Eurocopter Tigre/Tiger, first preproduction HAP version (PS1)	21 Dec 00

Eurocopter/Kawasaki EC 145 (BK 117C-2), first Japanese-built	15 Mar 00
Eurocopter/Kawasaki EC 145 (BK 117C-2), third (second European) prototype	14 Apr 00
Eurocopter/Kawasaki EC 145 (BK 117 C-2), fourth (third European) prototype (F-WQEP)	27 Nov 00
Ibis Aerospace Ae 270 Ibis (OK-EMA)	25 Jul 00

SEPECAT Jaguar GR. Mk 3, first flight with (one) Adour Mk 106 engine (XX108) — 7 Jun 00

Italy
LMATTS (Lockheed Martin/Alenia) C-27J Spartan, first production (I-FBAX)	12 May 00
LMATTS (Lockheed Martin/Alenia) C-27J Spartan, third prototype (MM62127)	8 Sep 00

Russian Federation
Ilyushin Il-103SKh	29 Mar 00
MiG 1-44 (01)	29 Feb 00
MiG-29 Sniper conversion (32367)	5 May 00
Mil Mi-24PN	.. Sep 00
Sukhoi Su-30MKI, first full-specification prototype	26 Nov 00
Tupolev Tu-234 (Tu-204-300), first production (RA-64026)	8 Jul 00

Turkey
TAI ZIU	26 Jun 00

Ukraine
Aeroprakt-22, first UK-assembled SLAC Foxbat (G-FBAT)	12 Aug 00

UK
ATG SkyKitten	28 Jun 00
BAE Systems Hawk Mk 115, first for NFTC programme (ZJ669)	.. May 00

BAE Systems Hawk Mk 127, first Australian-assembled (A27-010)	12 May 00
Southdown Aero Services 'Colditz Cock' replica (BGA 4732)	2 Feb 00
Westland WAH-64D Apache AH. Mk 1, first British-assembled production aircraft (ZJ172)	18 Jul 00

Now to be built by Letov, the prototype Raven 257 flew on 28 July as a product of the Wolfsberg-Evektor partnership *(Sebastian Zacharias)*

FIRST FLIGHTS

USA

Adam Aircraft Industries M-309 (N309A)	21 Mar 00
American Champion 7ACA Champ (N82AC)	00
Bell AH-1Z SuperCobra (162549)	7 Dec 00
Boeing 720 testbed (N720PW/C-FWXI), first flight with (one) Pratt & Whitney PW6000 engine	21 Aug 00
Boeing 737-900 (N737X)	3 Aug 00
Boeing 777-200ER (F-GSPL), first flight with GE90-94B engines	14 Jun 00
Boeing YAL-1A (747) (00-0001)	6 Jan 00
Boeing C-40A Clipper (converted 737-700C) (165829)	14 Apr 00
Boeing X-32A JSF prototype	18 Sep 00
Boeing Sikorsky YRAH-66 Comanche prototype, first flight with redesigned tail (94-0327)	18 Dec 00
Cessna 525A Citation CJ2, first production (N13CJ)	.. Feb 00
Explorer Aircraft Explorer 500T (converted prototype VH-ONA)	9 Jun 00
Groen Brothers Jet Hawk 4T (N402GB)	12 Jul 00

Kaman SH-2G(NZ) Super Seasprite, first new-built for Royal New Zealand Navy (N352KA/ NZ01)	2 Aug 00
Lancair Columbia 400	..00
Lockheed Martin C-130J Hercules, first for Italian Air Force (N4099R)	11 Feb 00
Lockheed Martin F-22A Raptor, third EMD aircraft (91-4003)	6 Mar 00
Lockheed Martin F-22A Raptor, fourth EMD aircraft, first with full avionics suite (91-4004)	15 Nov 00
Lockheed Martin F-22A Raptor, fifth EMD aircraft (91-4005)	5 Jan 01
Lockheed Martin KC-130J Hercules, first for US Marine Corps (165735)	9 Jun 00
Lockheed Martin X-35A JSF prototype (301)	24 Oct 00
Lockheed Martin X-35C JSF prototype (300)	16 Dec 00
Lockheed Martin/NASA F-16/AFTI, first flight as JSF fly-by-wire testbed (75-0750)	24 Oct 00
Luscombe Model 11E Spartan (N707BM), first production	.. Mar 00
Piper Malibu Meridian, first production (N375RD)	30 Jun 00
Pulsar Super Pulsar 100 (N601SP)	.. Aug 00
Rans S-18 Stinger II (N24052)	9 Sep 00
Sikorsky CH-60S Knighthawk, first production (165742)	27 Jan 00
Sikorsky S-70A-28D Black Hawk, first 'glass cockpit' version for Turkish Army	29 Mar 00
Sikorsky S-70B-28 Seahawk, first for Turkish Navy (N6227)	18 Jan 01
Sino-Swearingen SJ30-2 (N138BF)	30 Nov 00
Van's RV-9A, second (first definitive) prototype (N129RV)	15 Jun 00

Ibis Aerospace flew its Ae 270 light utility turboprop in the Czech Republic on 25 July 2000

Aerospace Calendar
Some significant aerospace events April 2000 to January 2001

The first two of a planned 72 Aero L 159 light combat aircraft were delivered to the Czech Air Force on 10 April 2000 *(Jan Kouba/Aero)*

2000	Military aviation	Civil aviation
3 Apr		Britten-Norman placed in receivership
10 Apr	First two Aero L 159s handed over to Czech Air Force	
13 Apr		Dornier Luftfahrt GmbH renamed Fairchild Dornier Aerospace
14 Apr	Finnmeccanica and EADS agreed funding of joint venture European Military Aircraft Company (EMAC)	
17 Apr		100th production Cessna 560XL Citation Excel business jet aircraft rolled out
26 Apr		FAA certification awarded to Cessna 560 Citation Encore business jet
		Britten-Norman crisis averted when new buyer is found

28 Apr	4,000th Lockheed Martin F-16 Fighting Falcon delivered (an F-16C for the Egyptian Air Force)	
Apr	Boeing revealed 737 Multimission Maritime Aircraft (MMMA) proposal as possible replacement for US Navy P-3 Orion	Aero-Design & Development Ltd announced launch of CityHawk VTOL 'flying car' project
12 May	Initial flight of first new-build LMATTS C-27J Spartan	
	Maiden flight of first BAE Systems Hawk assembled in Australia	

2000	Military aviation	Civil aviation
15 May		Initial production CAP 222 made first flight
16 May	Meteor medium-range air-to-air missile selected for use by RAF Eurofighter Typhoons	
	United Kingdom announced intent to purchase 25 Airbus Military A400M Future Large Aircraft for Royal Air Force	
	United Kingdom announced plan to lease four Boeing C-17A Globemaster III strategic transport aircraft for Royal Air Force from 2001	
	First Boeing AH-64D Apache for Royal Netherlands Air Force arrived at Gilze-Rijen	

25 May		Roll-out of 500th Bombardier Challenger at Dorval, Canada
1 Jun		Mil unveiled mock-up of Mi-52 Snegir (Bullfinch) light helicopter

AEROSPACE CALENDAR

2000	Military aviation	Civil aviation
6 Jun		Alliance Aircraft Corporation of USA launched StarLiner regional jet family at Berlin Air Show
9 Jun	France and Germany announced intention to buy 50 and 73 examples, respectively, of the Airbus Military A400M Future Large Aircraft First flight of Lockheed Martin KC-130J Hercules tanker version	First flight of Explorer Explorer 500T turboprop conversion
14 Jun		First flight of Dyn'Aero MCR4S four-seat lightplane
15 Jun	US Navy signed multiyear procurement contract covering acquisition of 222 Boeing F/A-18E/F Super Hornets during FY00 to FY04	First flight of second, definitive, prototype of Van's RV-9A
16 Jun		First flight of AkaFlieg München Mü30 Schlacro
26 Jun		First flight of prototype TAI ZIU agricultural aircraft
28 Jun	Assembly began of first Boeing C-17A Globemaster III to be leased by the RAF	
27 Jun		First flight of prototype Embraer ERJ-140
29 Jun	Lockheed Martin Argentina announced intent to restart production of IA 63 Pampa for Argentine Air Force	
30 Jun		First flight of first production Piper PA-46-500TP Malibu Meridian
Jun		Announcement of order to build prototype Sukhoi Su-38L agricultural aircraft US lightplane manufacturer Akrotech laid off workforce and suspended operations 500th RotorWay International Exec 162F kitbuilt helicopter delivered
4 Jul	First BAE Systems CT-155 (Hawk Mk 115) for NATO flying training arrived in Canada	
5 Jul	Initial flight of NAMC K-8E (Egyptian version) jet trainer	
8 Jul		Maiden flight of Tupolev Tu-234 version of Tu-204 airliner

2001/0110004

2000	Military aviation	Civil aviation
10 Jul	European Aeronautic, Defence and Space Company NV (EADS) officially formed	
11 Jul	First Lockheed Martin C-130J Hercules for Italy rolled out at Marietta, Georgia Popular name 'Clipper' assigned to US Navy's C-40A (Boeing 737-700C)	
14 Jul		Metal cutting began for first pre-series Embraer ERJ-170

2001/0105061

2000	Military aviation	Civil aviation
17 Jul	First delivery of Beech/Pilatus PC-9 Mk II to Greek Air Force	First production-conforming prototype Sino Swearingen SJ30-2 rolled out
18 Jul	Initial flight of first UK-assembled GKN Westland WAH-64D Apache Longbow at Yeovil	
19 Jul	French Navy received its first Rafale M	
20 Jul		FAA certification and authorisation for 180-minute ETOPS awarded to Boeing 767-400

2001/0105062

2000	Military aviation	Civil aviation
24 Jul		Simultaneous roll-out of 400th Gulfstream IV and 100th Gulfstrem V business jets JAA certification awarded to Boeing 767-400 Embraer Legacy business jet announced at Farnborough International Emirates became first airline to commit to Airbus A3XX (now A380) purchase, in advance of industrial launch. Air France follows immediately
25 Jul	First launch of AIM-9M Sidewinder AAM by Lockheed Martin F-22A Raptor	First flight of Ibis Aerospace Ae 270 Ibis
	Aermacchi announced M-346 Westernised version of Yak-130 advanced trainer	Embraer ERJ-145XR announced at Farnborough International with launch order for 75 aircraft from Continental Express Seabird signed agreement with Evektor for components manufacture of Seeker observation aircraft in Czech Republic
26 Jul	Name AgustaWestland announced as new title for merged helicopter divisions of Finnmeccanica and GKN	Kolb Kolbra ultralight announced and publicly displayed at Oshkosh

AEROSPACE CALENDAR

2000	Military aviation	Civil aviation
27 Jul	Formal announcement of acceptance of Airbus Military A400M Future Large Aircraft by seven participating member states	
28 Jul		First flight of Wolfsberg-Evektor Raven 257 utility aircraft
Jul	Egypt announced intention to upgrade 35 Boeing AH-64D Apache combat helicopters to AH-64D Apache Longbow standard, with initial delivery scheduled for July 2003	Mil announced plan to proceed with development of Mi-60 three-seat helicopter
2 Aug	First flight of first Kaman SH-2G Super Seasprite for New Zealand	Maiden flight of Boeing's latest 737, the -900
7 Aug	Sukhoi Su-35UB makes first flight	

2000	Military aviation	Civil aviation
28 Sep		FAA certification awarded to Schweizer 333 light helicopter
29 Sep	Second multiyear procurement contract valued at US$2.3 billion for 269 AH-64D Apache Longbow combat helicopters awarded to Boeing	First delivery of Cessna 560 Citation Encore business jet

8 Aug		Fairchild Dornier announced abandonment of 428JET airliner programme
11 Aug		First delivery (to Delta Airlines) of -400 series Boeing 767
24 Aug		IPTN adopted new name as PT Dirgantara Indonesia
25 Aug	Final first-generation Boeing Hornet (an F/A-18D) delivered to US Marine Corps	
29 Aug		OMF-160 Symphony awarded German type certification
Aug	First Lockheed Martin C-130J Hercules handed over to Italy and flown to Pisa	Explorer Explorer 500T made public debut at EAA AirVenture at Oshkosh Maverick TJ-1500 Twinjet 1500 made public debut at EAA AirVenture at Oshkosh XAC MA-60 entered revenue-earning service
1 Sep		Bombardier delivered the 500th Canadair Challenger business jet
4 Sep	UK finalised contract for purchase of four Boeing C-17A Globemaster IIIs	
9 Sep		Rans S-18 Stinger II flown for the first time
13 Sep		Boeing announced launch of -400ERX version of 767 airliner
18 Sep	First flight of first prototype Boeing X-32A Joint Strike Fighter	First metal cut for Fairchild Dornier 728JET
Sep	Final assembly of first production Eurofighter Typhoon Begun at Warton, UK, following arrival of German-built centre fuselage on 6 September and Italian rear fuselage shortly after KAI and Lockheed Martin establish T-50 International Company to market the T-50 aircraft outside South Korea	
3 Oct		200th Bombardier Aerospace Learjet 31A delivered (to Falcon Air Services of Phoenix, Arizona); coincidentally, 100th Learjet 45 also delivered in same month
9 Oct		Announcements on eve of NBAA Convention at New Orleans revealed new Hawker 450 business jet programme and plans for upgrades of existing types, in the shape of Dassault Falcon 2000X and Gulfstream V-SP
24 Oct	First flight of Lockheed Martin X-35A Joint Strike Fighter demonstrator First launch of AIM-120 AMRAAM by Lockheed Martin F-22A Raptor	
25 Oct		Scaled Composites 281 Proteus set first three world altitude records; two more records followed during next two days Diamond DA 40 Diamond Star gains JAR 23 certification
2 Nov	First EMD Lockheed Martin F-22A Raptor ended flying career with transfer to Wright-Patterson AFB, Ohio, for series of live fire trials to assess durability	
6 Nov	F-7MF version of CAC J-7 and GAIC FTC-2000 revealed at Zhuhai Air Show, China	
7 Nov		Official launch of AVIC I NRJ new regional jet programme
15 Nov	First flight of fourth EMD Lockheed Martin F-22A Raptor	
21 Nov	Lockheed Martin X-35A JSF contender exceeded Mach 1	Wolfsberg reassigned prospective production of Raven 257 light freighter from Evektor to Letov
27 Nov	First flight of third European-built Eurocopter/Kawasaki EC 145	

20 Sep		Prototype -600 version of Airbus A340 rolled off assembly line to begin engine installation and outfitting
27 Sep		FAA certification awarded to Piper PA-46-500TP Malibu Meridian

AEROSPACE CALENDAR

2000	Military aviation	Civil aviation
30 Nov	Maiden flight of Eurocopter Cougar Mk II+	First flight of production-conforming prototype Sino Swearingen SJ30-2 business jet
Nov		First customer delivery of Cessna 525A Citation CJ2 business jet
		First customer delivery of Piper PA-46-500TP Malibu Meridian
1 Dec	First flight of upgraded Bell AH-1Z SuperCobra	
14 Dec	WAH-64D Apache AH. Mk 1 gained IOC with British Army on delivery of ninth aircraft	
16 Dec	US Navy-standard Lockheed Martin X-35C made maiden flight	
18 Dec	Boeing Sikorsky RAH-66 Comanche returned to flight testing after fitment of a modified design of empennage	

2000	Military aviation	Civil aviation
19 Dec	Boeing's X-32A JSF contender undertook first aerial refuelling (and exceeded Mach 1 two days later)	Airbus board announced industrial launch of A3XX ultra-large airliner programme, simultaneously changing its designation to A380
	Australia signed a contract with Boeing for four 737 AEW&C surveillance aircraft (plus three options) to satisfy the Project Wedgetail requirement.	
28 Dec	India signed US$3 billion contract for local manufacture by HAL of 180 Sukhoi Su-30MKI fighters	
Dec	Turkey announced intent to purchase six Boeing 737 AEW&C surveillance aircraft	
2001		
4 Jan	First flight of Hindustan Aeronautics-built Light Combat Aircraft	
5 Jan	Fifth EMD Lockheed Martin F-22A Raptor flew for the first time	
16 Jan	Formal UK Release To Service of WAH-64D Apache AH. Mk 1	
18 Jan	First flight of first Sikorsky S-70B Seahawk for Turkish Navy	
23 Jan		First flight of PAC Fletcher FU24 conversion with Ford V-8 engine

Scaled Composites' Proteus set five world altitude records during October 2000

Official Records

(Corrected to 1 January 2001)

Two airship records set by the Graf Zeppelin in 1928 have yet to be bettered

Records are as verified by the Fédération Aéronautique Internationale (FAI). The full range of categories is:

Class A Free balloons	Class G Parachuting	Class O Hang gliders and paragliders
Class B Airships	Class H VTOL aeroplanes	Class P Aerospacecraft
Class C Aeroplanes	Class I Human-powered aircraft	Class R Microlights
Class D Gliders and motor gliders	Class K Spacecraft	Class S Space models
Class E Rotorcraft	Class M Tilting wing/engine aircraft	
Class F Aeromodelling	Class N STOL aeroplanes	

ABSOLUTE WORLD RECORDS

CLASS A (Free Balloons)

Four records are classed as Absolute World Records for free balloons by the FAI, as follows:

Duration (Switzerland)
  Bertrand Piccard (Switzerland) and Brian Jones (UK) in the Cameron R-650 combination hot air/helium balloon HB-BRA *Breitling Orbiter 3*, from Château d'Oex, Switzerland, to near Dâkhla, Egypt, between 1 and 21 March 1999. 477 hours 47 minutes.

Distance (Switzerland)
  Bertrand Piccard (Switzerland) and Brian Jones (UK) in the Cameron R-650 combination hot air/helium balloon HB-BRA *Breitling Orbiter 3*, from Château d'Oex, Switzerland to near Dâkhla, Egypt, between 1 and 21 March 1999. 22,037.8 n miles (40,814.0 km; 25,361.34 miles).

Altitude (USA)
  Cdr M D Ross and Lt Cdr V A Prather in Winzen Research ONXR290WRP-1 *Lee Lewis Memorial*, from USS *Antietam*, Gulf of Mexico, on 4 May 1961. 34,668 m (113,740 ft).

Shortest time around the world (Switzerland)
  Bertrand Piccard (Switzerland) and Brian Jones (UK) in the Cameron R-650 combination hot air/helium balloon HB-BRA *Breitling Orbiter 3*, from and to meridian 9° 27′ West, over Mauritania, between 4 and 20 March 1999. 370 hours 24 minutes.

Grob Egrett-1 N14ES holds the altitude record for turboprop aeroplanes

OFFICIAL RECORDS

CLASS B (Airships)

Four records are classed as Absolute World Records for airships by the FAI, as follows:

Duration (Germany)
Kapt Hugo Eckener and crew in Zeppelin D-LZ127 *Graf Zeppelin* from Lakehurst, New Jersey, to Friedrichshafen, Germany, between 29 October and 1 November 1928. 71 hours 7 minutes.

Distance in a straight line (Germany)
Kapt Hugo Eckener and crew in Zeppelin D-LZ127 *Graf Zeppelin* from Lakehurst, New Jersey, to Friedrichshafen, Germany, between 29 October and 1 November 1928. 3,447.35 n miles (6,384.50 km; 3,967.14 miles).

Altitude (USA)
Brian Boland in Boland Rover A-2 N9029Q at Luxembourg 5 August 1988. 5,059 m (16,598 ft).

Speed in a straight line (USA)
Paul Woessner and Coy Foster in Thunder & Colt GA-42 G-BRHK, from Dallas to Marion, Texas, 24 October 1990. 41.85 kt (77.50 km/h; 48.16 mph).

CLASS C (Aeroplanes)

Seven records are classed as Absolute World Records for aeroplanes by the FAI, as follows:

Distance in a straight line (USA), and **Distance in a closed circuit** (USA)
Richard 'Dick' Rutan and Jeana Yeager in the Voyager (N269VA), between 14 and 23 December 1986. Circumnavigation of the World, starting and finishing at Edwards AFB, California. 21,712.816 n miles (40,212.139 km; 24,986.664 miles).

Altitude (USSR)
Alexander Fedotov in an E-266M (MiG-25) on 31 August 1977. 37,650 m (123,523 ft).

Altitude in sustained horizontal flight (USA)
Captain Robert C Helt and Major Larry A Elliott, USAF, in an unidentified Lockheed SR-71A on 28 July 1976 at Beale AFB, California. 25,929.031 m (85,069 ft).

An-124 SSSR-82009 holds the jet landplanes record for distance in a closed circuit *(Paul Jackson/Jane's)*

Altitude, after launch from a 'mother-plane' (USA)
Major R White, USAF, in the North American X-15A-3 (56-6672) on 17 July 1962, at Edwards AFB, California. 95,935.99 m (314,750 ft).

Speed on a straight course (USA)
Captain Eldon W Joersz and Major George T Morgan Jr, USAF, in an unidentified Lockheed SR-71A on 28 July 1976 over a 15/25 km course at Beale AFB, California. 1,905.81 kt (3,529.56 km/h; 2,193.17 mph).

Speed in a closed circuit (USA)
Major Adolphus H Bledsoe Jr and Major John T Fuller, USAF, in Lockheed SR-71A 64-17958, on 27 July 1976, over a 1,000 km closed circuit from Beale AFB, California. 1,818.154 kt (3,367.221 km/h; 2,092.294 mph).

WORLD CLASS RECORDS

Following are details of some of the more important world class records confirmed by the FAI:

CLASS C, GROUP 1 (Landplanes with piston engines):

Great circle distance without landing and **Distance in a closed circuit**
See Absolute World Records.

Altitude (Italy)
Mario Pezzi, in a Caproni Ca 161*bis*, on 22 October 1938. 17,083 m (56,046 ft).

Speed in a straight line (USA)
Lyle Shelton in the modified Grumman F8F Bearcat *Rare Bear* (N777L), with a 2,833 kW (3,800 hp) Wright R-3350 engine, on 21 August 1989, over a 3 km course at Las Vegas, New Mexico. 459.10 kt (850.25 km/h; 528.33 mph).

CLASS C, GROUP 2 (Landplanes with turboprop engines):

Great circle distance without landing (USA)
Lt Col E L Allison and crew in a Lockheed HC-130H Hercules (65-0972), on 20 February 1972. 7,587.99 n miles (14,052.95 km; 8,732.098 miles).

Distance in a closed circuit (USA)
Cdr Philip R Hite and crew in the Lockheed RP-3D Orion (158227), on 4 November 1972. 5,455.46 n miles (10,103.51 km; 6,278.03 miles).

Altitude (USA)
Einar Enevoldson in Grob Egrett-1, N14ES, on 1 September 1988, at Majors Field, Greenville, Texas. 16,329 m (53,573 ft).

Speed on a straight course (USA)
Cdr Donald H Lilienthal and crew in a Lockheed P-3C Orion, over a 15/25 km course on 27 January 1971. 435.26 kt (806.10 km/h; 500.89 mph).

Speed in a closed circuit (USSR)
Ivan Sukhomlin and crew in Tupolev Tu-114 SSSR-76459, on 9 April 1960, carrying a 25,000 kg payload over a 5,000 km circuit. 473.66 kt (877.212 km/h; 545.07 mph).

CLASS C, GROUP 3 (Landplanes with jet engines):

Great circle distance without landing (USA)
Major Clyde P Evely, USAF, in a Boeing B-52H Stratofortress, between 10 and 11 January 1962, from Okinawa to Madrid, Spain. 10,890.27 n miles (20,168.78 km; 12,532.3 miles).

Distance in a closed circuit (USSR)
Vladimir Terski and crew in Antonov An-124 SSSR-82009, between 6 and 7 May 1987. Moscow-Astrakhan-Tashkent-Lake Baikal-Petropavlovsk-Chukot Peninsula-Murmansk-Zhdanov-Moscow. 10,880.625 n miles (20,150.921 km; 12,521.201 miles).

Altitude, speed on straight course and **speed in 1,000 km closed circuit**
See Absolute World Records.

Speed over a 3 km course at restricted altitude (USA)
Darryl Greenamyer in the modified Red Baron F-104RB Starfighter (N104RB), on 24 October 1977, at Mud Lake, Tonopah, Nevada. 858.77 kt (1,590.45 km/h; 988.26 mph).

Speed in a 100 km closed circuit (USSR)
Alexander Fedotov in an unidentified Mikoyan E-266 (MiG-25) on 8 April 1973. 1,406.641 kt (2,605.1 km/h; 1,618.734 mph).

Speed in a 500 km closed circuit (USSR)
M Komarov in a Mikoyan E-266 (MiG-25), on 5 October 1967, near Moscow. 1,609.88 kt (2,981.5 km/h; 1,852.62 mph).

Speed around the world (Eastbound; without aerial refuelling) (France)
Michel Dupont and Claude Hetru in Concorde F-BTSD on 16 August 1995, 704.68 kt (1,305.93 km/h; 811.46 mph). (Note: Exceeds record for circumnavigation with aerial refuelling).

Greatest load carried to a height of 2,000 m (USSR)
Alexander V Galunenko in Antonov An-225 Mriya SSSR-82060, on 22 March 1989. 508,200 kg (1,120,370 lb).

CLASS C.2, ALL GROUPS (Seaplanes):

Great circle distance without landing (UK)
Capt D C T Bennett and First Officer I Harvey, in the Short-Mayo *Mercury*, G-ADHJ, between 6 and 8 October 1938, from Dundee, Scotland, to the Orange River, South Africa. 5,211.66 n miles (9,652 km; 5,997.5 miles).

Altitude (USSR)
Georgi Buryanov and crew of two in Beriev M-10 '40', on 9 September 1961, over the Sea of Azov. 14,962 m (49,088 ft).

Speed on a straight course (USSR)
Nikolai Andrievsky and crew of two in Beriev M-10 '40', on 7 August 1961, at Joukovski-Petrovskoe, over a 15/25 km course. 492.44 kt (912 km/h; 566.69 mph).

OFFICIAL RECORDS

CLASS E.1 (Helicopters):

Great circle distance without landing (USA)
R G Ferry in a Hughes YOH-6A, between 6 and 7 April 1966, from Culver City, California, to Ormond Beach, Florida. 1,923.08 n miles (3,561.55 km; 2,213 miles).

Altitude (France)
Jean Boulet in an Aerospatiale SA 315B Lama on 21 June 1972, at Istres, France. 12,442 m (40,820 ft).

Speed on a straight course (UK)
Trevor Egginton and Derek Clews in Westland Lynx G-LYNX, on 11 August 1986, over a 15/25 km course, at Glastonbury, Somerset. 216.45 kt (400.87 km/h; 249.09 mph).

Speed in a 100 km closed circuit (USSR)
Boris Galitsky and crew of five in a Mil Mi-6, on 26 August 1964, near Moscow. 183.67 kt (340.15 km/h; 211.36 mph).

Speed in a 500 km closed circuit (USA)
Thomas Doyle in Sikorsky S-76A N5445J, at West Palm Beach, Florida, on 8 February 1982. 186.68 kt (345.74 km/h; 214.83 mph).

CLASS E.3 (Autogyros):

Altitude without payload (USA)
William B Clem III, in homebuilt Rotorflight Dynamics Dominator N36MR, on 17 April 1998, at Wauchula, Florida. 7,456 m (24,462 ft).

Great circle distance without landing (UK)
Wing Cdr K H Wallis, in Wallis WA-116/F G-ATHM, from Lydd Airport, Kent, to Wick, Scotland, on 28 September 1975. 471.79 n miles (874.32 km; 543.29 miles).

Distance in a closed circuit (UK)
Wing Cdr K H Wallis, in Wallis WA-116/F/S G-BLIK, on 5 August 1988, at Waterbeach Airfield, Cambridge. 541.44 n miles (1,002.75 km; 623.08 miles).

Speed on a straight course (UK)
Wing Cdr K H Wallis, in a Wallis WA-116/F/S G-BLIK, over a 3 km course, at RAF Marham, Norfolk, on 18 September 1986. 104.5 kt (193.6 km/h; 120.3 mph).

International aircraft registration prefixes

The following two tables list (by alphanumeric order of prefix and alphabetical order of country, respectively) the markings which identify the national registrations of civil aircraft. Initiated in 1919, the system is now administered by the International Civil Aviation Organisation, although the tables include some *de facto* prefixes which do not conform to ICAO rules. Mongolia, for example, is now standardising on JU-, replacing both MT- and BNMAU-. Not all allocated markings are in current use, examples including Gibraltar and Vatican City. Brazil used only PP- and PT- from its allocation until PR- was brought into use in April 1999.

Certified civil aircraft registrations generally comprise five letters of the Roman alphabet, of which the first one or two identify the country of registry and are usually followed by a hyphen. Number suffixes (with or without additional letters) are permitted as an alternative, the most commonly seen being those of the USA and former constituents of the USSR. Letter/number prefixes were introduced in 1948, while some countries which follow the national identifier with numbers omit the hyphen. In the case of some small members of the British Commonwealth, one or two letters after the hyphen are required to complete the national identity, leaving scope for as few as 26 aircraft on the national register.

Within national authorities, rules for allocation vary considerably. Not all letters of the alphabet need be employed. Owners may be permitted to request personal suffixes outside the strict alphabetical progression of the register, or certain classes of aircraft (by weight band or number of engines, for example) might be assigned specific groups of registrations. Civil aircraft flown under permit, sailplanes, ultralights, microlights and aircraft undergoing manufacturer's testing are in some countries allocated registration suffixes of an entirely different character (for example, numbers instead of letters).

It should be borne in mind that some suffix batches are reserved for military use. These call-signs may – notably in Africa – be applied externally to aircraft. Equally, some countries have a second, internal civil registration system, not mentioned here, for aircraft which do not normally cross international borders. Examples include UK sailplanes and French microlights. New Zealand-registered aircraft not intending overseas flight are permitted to operate without the national prefix.

Registration prefixes: by prefix

Prefix	Country	Prefix	Country	Prefix	Country
AP-	Pakistan	J3-	Grenada	VP-A	Anguilla
A2-	Botswana	J5-	Guinea-Bissau	VP-B	Bermuda
A3-	Tonga	J6-	St Lucia	VP-C	Cayman Islands
A4O-	Oman	J7-	Dominica	VP-F	Falkland Islands
A5-	Bhutan	J8-	St Vincent & Grenadines	VP-G	Gibraltar
A6-	United Arab Emirates	LN-	Norway	VP-LM	Montserrat
A7-	Qatar	LQ-, LV-	Argentina	VP-LV	British Virgin Islands
A9C-	Bahrain	LX-	Luxembourg	VQ-H	St Helena & Ascension
B-	China	LY-	Lithuania	VQ-T	Turks & Caicos Islands
(B-H	Hong Kong)	LZ-	Bulgaria	VT-	India
(B-M	Macao)	N	USA	V2-	Antigua & Barbuda
B-	Taiwan	OB-	Peru	V3-	Belize
C-, CF-	Canada	OD-	Lebanon	V4-	St Kitts & Nevis
CC-	Chile	OE-	Austria	V5-	Namibia
CN-	Morocco	OH-	Finland	V6-	Micronesia
CP-	Bolivia	OK-	Czech Republic	V7-	Marshall Islands
CR-, CS-	Portugal	OM-	Slovak Republic	V8-	Brunei
CU-	Cuba	OO-, OQ-	Belgium	XA- to XC-	Mexico
CX-	Uruguay	OY-	Denmark	XT-	Burkina Faso
C2-	Nauru	P-	Korea, North	XU-	Cambodia
C3-	Andorra	PH-	Netherlands	XY-, XZ-	Myanmar
C5-	Gambia	PJ-	Netherlands Antilles	YA-	Afghanistan
C6-	Bahamas	PK-	Indonesia	YI-	Iraq
C9-	Mozambique	PP- to PT-	Brazil	YJ-	Vanuatu
D-	Germany	PZ-	Suriname	YK-	Syria
DQ-	Fiji	P2-	Papua New Guinea	YL-	Latvia
D2-	Angola	P4-	Aruba	YN-	Nicaragua
D4-	Cape Verde	RA-, RF-	Russia	YR-	Romania
D6-	Comoros	RDPL-	Laos	YS-	El Salvador
EC-	Spain	RP-	Philippines	YU-	Yugoslavia
EI-, EJ-	Ireland	SE-	Sweden	YV-	Venezuela
EK-	Armenia	SP-	Poland	Z-	Zimbabwe
EL-	Liberia	ST-	Sudan	ZA-	Albania
EP-	Iran	SU-	Egypt	ZK- to ZM-	New Zealand
ER-	Moldova	SU-YA	Palestine	ZP-	Paraguay
ES-	Estonia	SX-	Greece	ZS- to ZU-	South Africa
ET-	Ethiopia	S2-	Bangladesh	Z3-	Macedonia
EW-	Belarus	S5-	Slovenia	3A-	Monaco
EX-	Kyrgyzstan	S7-	Seychelles	3B-	Mauritius
EY-	Tajikistan	S9-	São Tomé e Príncipe	3C-	Equatorial Guinea
EZ-	Turkmenistan	TC-	Turkey	3D-	Swaziland
E3-	Eritrea	TF-	Iceland	3X-	Guinea
F-	France and colonies	TG-	Guatemala	4K-	Azerbaijan
G-	United Kingdom	TI-	Costa Rica	4L-	Georgia
HA-	Hungary	TJ-	Cameroon	4R-	Sri Lanka
HB-	Switzerland & Liechtenstein	TL-	Central African Republic	4X-	Israel
HC-	Ecuador	TN-	Congo (Republic)	5A-	Libya
HH-	Haiti	TR-	Gabon	5B-	Cyprus
HI-	Dominican Republic	TS-	Tunisia	5H-	Tanzania
HK-	Colombia	TT-	Chad	5N-	Nigeria
HL-	Korea, South	TU-	Côte d'Ivoire	5R-	Madagascar
HP-	Panama	TY-	Benin	5T-	Mauritania
HR-	Honduras	TZ-	Mali	5U-	Niger
HS-	Thailand	T2-	Tuvalu	5V-	Togo
HV-	Vatican City	T3-	Kiribati	5W-	Samoa
HZ-	Saudi Arabia	T7-	San Marino	5X-	Uganda
H4-	Solomon Islands	T9-	Bosnia-Herzegovina	5Y-	Kenya
I-	Italy	UK-	Uzbekistan	60-	Somalia
JA-	Japan	UN-	Kazakhstan	6V-, 6W-	Senegal
JU-	Mongolia	UR-	Ukraine	6Y-	Jamaica
JY-	Jordan	VH-	Australia	7O-	Yemen
J2-	Djibouti	VN-	Vietnam	7P-	Lesotho
				7Q-	Malawi
				7T-	Algeria
				8P-	Barbados
				8Q-	Maldives
				8R-	Guyana
				9A-	Croatia
				9G-	Ghana
				9H-	Malta
				9J-	Zambia
				9K-	Kuwait
				9L-	Sierra Leone
				9M-	Malaysia
				9N-	Nepal
				9Q-	Congo (Democratic Republic)
				9U-	Burundi
				9V-	Singapore
				9XR-	Rwanda
				9Y-	Trinidad & Tobago

The apparently dual personality of this Junkers Ju 52/3m g8e derives from the fact that the German authorities permit it to represent a pre-Second World War aircraft of the same type (D-AQUI), although post-war regulations require it to be registered in the D-C series reserved for aircraft with gross weights between 5.7 and 14 tonnes *(Paul Jackson/Jane's)*   2001/0105016

INTERNATIONAL AIRCRAFT REGISTRATION PREFIXES

Registration prefixes: by country

Afghanistan	YA-
Albania	ZA-
Algeria	7T-
Andorra	C3-
Angola	D2-
Anguilla	VP-LA
Antigua & Barbuda	V2-
Argentina	LQ-, LV-
Armenia	EK-
Aruba	P4-
Australia	VH-
Austria	OE-
Azerbaijan	4K-
Bahamas	C6-
Bahrain	A9C-
Bangladesh	S2-
Barbados	8P-
Belarus	EW-
Belgium	OO-, OQ-
Belize	V3-
Benin	TY-
Bermuda	VP-B
Bhutan	A5-
Bolivia	CP-
Bosnia-Herzegovina	T9-
Botswana	A2-
Brazil	PP- to PT-
Brunei	V8-*
Bulgaria	LZ-
Burkina Faso	XT-
Burundi	9U-
Cambodia	XU-
Cameroon	TJ-
Canada	C-, CF-
Cape Verde	D4-
Cayman Islands	VP-C
Central African Republic	TL-
Chad	TT-
Chile	CC-
China	B-
(Hong Kong)	B-H)
(Macao)	B-M)
Colombia	HK-
Comoros	D6-
Congo (Democratic Republic)	9Q-
Congo (Republic)	TN-
Costa Rica	TI-
Côte d'Ivoire	TU-
Croatia	9A-
Cuba	CU-
Cyprus	5B-
Czech Republic	OK-
Denmark	OY-
Djibouti	J2-
Dominica	J7-
Dominican Republic	HI-
Ecuador	HC-
Egypt	SU-
Equatorial Guinea	3C-
Eritrea	E3-
Estonia	ES-
Ethiopia	ET-
Falkland Islands	VP-F
Fiji	DQ-
Finland	OH-
France and colonies	F-
Gabon	TR-
Gambia	C5-
Georgia	4L-
Germany	D-
Ghana	9G-
Gibraltar	VP-G
Greece	SX-
Grenada	J3-
Guatemala	TG-
Guinea	3X-
Guinea-Bissau	J5-
Guyana	8R-
Haiti	HH-
Honduras	HR-
Hungary	HA-
Iceland	TF-
India	VT-
Indonesia	PK-
Iran	EP-
Iraq	YI-
Ireland	EI-, EJ-
Israel	4X-
Italy	I-
Jamaica	6Y-
Japan	JA
Jordan	JY-
Kazakhstan	UN-
Kenya	5Y-
Kiribati	T3-
Korea, North	P-
Korea, South	HL-
Kuwait	9K-
Kyrgyzstan	EX-
Laos	RDPL-
Latvia	YL-
Lebanon	OD-
Lesotho	7P-
Liberia	EL-
Libya	5A-
Liechtenstein	HB-
Lithuania	LY-
Luxembourg	LX-
Macedonia	Z3-
Madagascar	5R-
Malawi	7Q-
Malaysia	9M-
Maldives	8Q-
Mali	TZ-
Malta	9H-
Marshall Islands	V7-
Mauritania	5T-
Mauritius	3B-
Mexico	XA- to XC-
Micronesia	V6-
Moldova	ER-
Monaco	3A-
Mongolia	JU-
Montserrat	VP-LM
Morocco	CN-
Mozambique	C9-
Myanmar	XY-, XZ-
Namibia	V5-
Nauru	C2-
Nepal	9N-
Netherlands	PH-
Netherlands Antilles	PJ-
New Zealand	ZK- to ZM-
Nicaragua	YN-
Niger	5U-
Nigeria	5N
Norway	LN-
Oman	A4O-
Pakistan	AP-
Palestine	SU-YA
Panama	HP-
Papua New Guinea	P2-
Paraguay	ZP-
Peru	OB-
Philippines	RP-
Poland	SP-
Portugal	CR-, CS-
Qatar	A7-
Romania	YR-
Russian Federation	RA-, RF-
Rwanda	9XR-
El Salvador	YS-
Samoa	5W-
San Marino	T7-
São Tomé e Princípe	S9-
Saudi Arabia	HZ-
Senegal	6V-, 6W-
Seychelles	S7-
Sierra Leone	9L-
Singapore	9V-
Slovak Republic	OM-
Slovenia	S5-
Solomon Islands	H4-
Somalia	6O-
South Africa	ZS- to ZU-
Spain	EC-
Sri Lanka	4R-
St Helena & Ascension	VQ-H
St Kitts & Nevis	V4-
St Lucia	J6-
St Vincent & Grenadines	J8-
Sudan	ST-
Suriname	PZ-
Swaziland	3D-
Sweden	SE-
Switzerland	HB-
Syria	YK-
Taiwan	B-
Tajikistan	EY-
Tanzania	5H-
Thailand	HS-
Togo	5V-
Tonga	A3-
Trinidad & Tobago	9Y-
Tunisia	TS-
Turkey	TC-
Turkmenistan	EZ-
Turks & Caicos	VQ-T
Tuvalu	T2-
Uganda	5X-
Ukraine	UR-
United Arab Emirates	A6-
United Kingdom	G-
United States	N
Uruguay	CX-
Uzbekistan	UK-
Vanuatu	YJ-
Vatican City	HV-
Venezuela	YV-
Vietnam	VN-
Virgin Islands	VP-L
Yemen	7O-
Yugoslavia	YU-
Zambia	9J-
Zimbabwe	Z-

'NEEL before G-ESUS. A pair of RotorWay Execs demonstrates the enhanced effect which two personal registrations can generate as the result of coincidental parking *(Paul Jackson/Jane's)* 2001/0105017

Czech (OK-) ultralights, such as the ATEC Zephyr, all include U as the fourth letter and differ from Class A registered aircraft in having a sequential number as part of the registration 2001/0105066

Glossary of aerospace terms in this book

Afterburning: Temporary thrust augmentation by burning additional fuel in the jetpipe, demonstrated by a Sukhoi Su-37

AAM Air-to-air missile.
AAR Air-to-air refuelling.
AATH Automatic approach to hover.
AB Aktiebolag (Swedish company constitution).
absolute ceiling Greatest altitude attainable by aircraft in level flight.
AC Alternating current.
ACC Air Combat Command (US).
ACE Actuator control electronics.
ACLS Automatic carrier landing system.
ACMI Air combat manoeuvring instrumentation.
ACN Aircraft classification number (ICAO system for aircraft pavements).
ADC Air data computer.
ADF Medium-frequency automatic direction-finding (equipment).
ADG Accessory-drive generator.
ADI Attitude/director indicator.
adjustable pitch Propeller with blades that can be adjusted in pitch by an engineer on the ground. Compare variable pitch.
aerofoil Any solid body so shaped that, as a fluid medium (air or hot gas) moves past it, it experiences a useful force perpendicular to the direction of relative motion; thus, a wing generates lift, while a turbine blade generates torque on a shaft.
aeroplane (N America, airplane) Heavier-than-air aircraft with propulsion and a wing that does not rotate in order to generate lift.
AEW Airborne early warning.
AFB Air Force Base (USAF).
AFCS Automatic flight control system.
AFRC Air Force Reserve Command (US).
AFRP Aramid fibre-reinforced plastics.
afterburning Temporarily augmenting the thrust of a turbofan or turbojet by burning additional fuel in the jetpipe.
AGM Air-to-ground missile.
Ah Ampère-hours.
AHRS Attitude/heading reference system.
AIDS Airborne integrated data system.
airbrake Passive device extended from aircraft to increase drag. Most common form is hinged flap(s) or plate(s), mounted in locations where operation causes no significant deterioration in stability and control at any attainable airspeed.
aircraft All manmade vehicles for off-surface navigation within the atmosphere, including helicopters and balloons. For practical purposes, air-cushion vehicles and wing-in-ground-effect vehicles are excluded from the classification.
airship Power-driven lighter-than-air aircraft. Traditional classes are: blimp, a small non-rigid; non-rigid, in which envelope is essentially devoid of rigid members and maintains shape by inflation pressure; semi-rigid, non-rigid with strong axial keel acting as beam to support load; and rigid, in which envelope is itself stiff in local bending or supported within or around rigid framework.
airstair Retractable stairway built into aircraft.
AIS Advanced instrumentation subsystem.
ALARM Air-launched anti-radiation missile.
ALCM Air-launched cruise missile.
Allithium Aluminium-lithium alloy.
ALLTV All-light level television.
AM Amplitude modulation.
AMC Air Mobility Command (US).
AMRAAM Advanced Medium-Range AAM.
ANG Air National Guard (US).
anhedral Downward slope of wing or tailplane from root to tip.
anti-balance tab Hinged surface on trailing-edge of stabilator and operating in same direction, so as to dampen its movement.
ANVIS Aviator's night vision system.
AO Aktsionernoye Obshchestvo (Co Ltd; Russian company constitution).
AoA Angle of attack (see 'attack' below).
AOOT Aktsionernoye Obshchestvo Oktrytogo Tipa (Russian company constitution).
approach noise Measured 1 n mile from downwind end of runway with aircraft passing overhead at 113 m (370 ft).
APR Auxiliary power reserve.
APU Auxiliary power unit (part of aircraft).
ARINC Aeronautical Radio Inc, US company whose electronic box sizes (racking sizes) are the international standard.
ARM Anti-radiation missile.
ArNG Army National Guard (US).
ASE (1) Automatic stabilisation equipment; (2) Aircraft survivability equipment.
ASI Airspeed indicator.
ASM Air-to-surface missile.
aspect ratio Measure of wing (or other aerofoil) slenderness seen in plan view, usually defined as the square of the span divided by gross area.
ASPJ Advanced self-protection jammer.
AST Air Staff Target (UK).
ASTOVL Advanced STOVL.
ASUW Anti-surface unit warfare.
ASV Anti-surface vessel.
ASW Anti-submarine warfare.
ATC Air traffic control.
ATDS Airborne tactical data system.
ATHS Airborne target handover (US, handoff) system.
ATPCS Automatic take-off power control system.
ATR Airline Transport Radio ARINC 404 black box racking standards.
attack, angle of (alpha) Angle at which airstream meets aerofoil (angle between mean chord and free-stream direction). Not to be confused with angle of incidence (which see).
augmented Boosted by afterburning (turbofan).

autogyro Rotary-wing aircraft propelled by a propeller (or other thrusting device) and lifted by a freely running autorotating rotor.
AUW All-up weight (term meaning total weight of aircraft under defined conditions, or at a specific time during flight). Not to be confused with MTOW (which see).
avionics Aviation electronics.
AWACS Airborne warning and control system (aircraft category).
axisymmetric intakes Twin, circular engine air intakes mounted astride the spinner of New Piper light aircraft.
axisymmetric nozzle Circular jet-engine nozzle capable of unrestricted vectoring movement (within a cone specified by mechanical limitation) to enhance aircraft manoeuvrability.

ballistic parachute Emergency recovery parachute installed in (generally light) aircraft and capable of supporting both machine and occupants.
band See radar frequency.
bar Non-SI unit of pressure adopted by this yearbook pending wider acceptance of Pa. 1 bar = 10^5 Pa. ISA pressure at S/L is 1013.2 mb or just over 1 bar. ICAO has standardised hectopascal for atmospheric pressure, in which ISA S/L pressure is 101.32 hPa.
basic operating weight MTOW minus payload (thus, including crew, fuel and oil, bar stocks, cutlery and so on).
bearingless rotor Rotor in which flapping, lead/lag and pitch change movements are provided by the flexibility of the structural material and not by bearings. No rotor is truly rigid.

GLOSSARY

BITE Built-in test equipment.
bladder tank Fuel (or other fluid) tank of flexible material.
BLC Boundary-layer control.
bleed air Hot high-pressure air extracted from gas turbine engine compressor or combustor and taken through valves and pipes to perform useful work such as pressurisation, driving machinery or anti-icing by heating surfaces.
blown flap Flap across which bleed air is discharged at high (often supersonic) speed to prevent flow breakaway.
BOW Basic operating weight (which see).
BPR Bypass ratio.
BTU Non-SI energy unit (British Thermal Unit) = 0.9478 J.
bulk cargo All cargo not packed in containers or on pallets.
bus Busbar, main terminal in electrical system to which battery or generator power is supplied.
BV Besloten Vennootschap (Netherlands company constitution).
BVR Beyond visual range.
BWB Blended wing/body.
bypass ratio Air flow through fan duct (not passing through core) divided by air flow through core.
byte Group of bits of information forming unit in computer processing.

C³ Command, control and communications.
CAA Civil Aviation Authority (UK).
cabane Structure, usually of braced struts, to support load above fuselage or wing. May carry parasol wing, engine nacelle or upper wing of most biplanes.
cabin altitude Height above S/L at which ambient pressure is same as inside cabin.

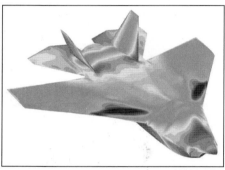
2001/0106241

CAD/CAM Computer-assisted design/computer-assisted manufacture.
CAHI Central Aero and Hydrodynamics Institute of the Russian Federation; also transliterated as TsAGI.
canards Foreplanes, fixed or controllable aerodynamic surfaces ahead of CG.
capacity The volume swept out on each stroke by the pistons of a piston engine. It is expressed in cc (cubic centimetres) for small engines and in litres (1 litre = 1,000 cc) for larger ones. Also known as displacement or swept volume.
carbon fibre Fine filament of carbon/graphite used as strength element in composites.
CAS (1) Calibrated airspeed, ASI calibrated to allow for air compressibility according to ISA S/L; (2) close air support.
casevac Casualty evacuation.
CATIA Computer-aided three-dimensional interactive analysis; Anglicised form of French CAD proprietary system (*Conception assistée tridimensionelle interactive d'applications*).
CBU Cluster bomb unit.
CCV Control-configured vehicle.
CEAM Centre d'Expériences Aériennes Militaires.
CEAT Centre d'Essais Aéronautiques de Toulouse.
Ceconite Manmade covering material for light aircraft; trade name.
CEO Chief executive officer.
CEV Centre d'Essais en Vol.
CFE Conventional Forces in Europe.
CFRP Carbon fibre-reinforced plastics.
CG Centre of gravity.
chaff Thin slivers of radar-reflective material cut to length appropriate to wavelengths of hostile radars and scattered in bundles to protect friendly aircraft.
chord Distance from leading-edge to trailing-edge measured parallel to longitudinal axis.
CIS Commonwealth of Independent [ex-USSR States]. See also RFAS.
CKD Component knocked down, for assembly elsewhere.
clean (1) In flight configuration with landing gear, flaps, slats and so on retracted; (2) Without any optional external stores.
c/n Construction (or constructor's) number; manufacturer's serial number.

C of A Certificate of Airworthiness; awarded to each individual aircraft (compare Type Certificate).
COIN Counter-insurgency.
collective pitch Controls pitch of all blades of helicopter main rotor in unison.
combi Civil aircraft carrying both freight and passengers on main deck.
comint Communications intelligence.
composite Material made of two constituents, such as filaments or short whiskers plus adhesive forming binding matrix.
constant-speed Variable-pitch propeller governed by a CSU so that its rotational speed is held constant.
contrarotating Propellers on same axis turning in opposite directions (compare C/R).
conventional and manual Aeroplane manoeuvring surfaces mechanically linked to pilot's hand and foot controls, unassisted (except, optionally, by aerodynamic or mass balances) and comprising ailerons on the outboard wing, rudder(s) to the rear of fixed tailfins(s) and elevators to the rear of a fixed (but optionally, incidence angle trimmable) tailplane. The description optionally includes leading-edge slats, flaps on inboard trailing-edges and trim tabs, all of which are mentioned separately, if installed. Ailerons which droop in unison with flaps (and thus are not the primary means of lowering stalling speed) are regarded as conventional. Control systems not conforming to the above – in that they have foreplanes, one or more all-moving tail surfaces, or flaperons, and those with mechanical/electronic assistance or interception of the pilot's movements – are described in appropriate detail.
convertible Transport aircraft able to be equipped to carry passengers or cargo, but not both simultaneously.
COO Chief operating officer.
core Gas generator portion of turbofan comprising compressor(s), combustion chamber and turbine(s).
C/R Counter-rotating; propellers of multi-engined aircraft turning in opposite directions on different axes (compare contrarotating).
CRT Cathode-ray tube.
cruising speed Flight speed on less than full engine power; maximum is normally at 75%, if not otherwise specified, but some manufacturers use higher throttle settings.
CSAS Command and stability augmentation system (part of AFCS).
CSD Constant-speed drive (output shaft speed held steady, no matter how input may vary). Used to enable an alternator to generate AC at a fixed frequency.
CTOL Conventional take-off and landing (compare V/STOL).
CVR Cockpit voice recorder.
cyclic pitch Controls variation of pitch as helicopter rotor blade makes each revolution.

Dacron Artificial fabric for light aircraft covering; trade name.
DADC Digital air data computer.
DADS Digital air data system.
DA/FD Digital autopilot/flight director.
DARPA Defense Advanced Research Projects Agency (US) (briefly ARPA before February 1996).
databus Electronic highway for passing digital data between aircraft sensors and system processors, usually MIL-STD-1553B or ARINC 419 (one-way) and 619 (two-way) systems.
dB Decibel.
DC Direct current.
DECU Digital engine (or electronic) control unit.
dem/val Demonstration/validation.
DERA Defence Evaluation and Research Agency (UK). Divided, 2001, as QinetiQ and Defence Science and Technology Laboratory.
derated Engine restricted to power less than potential maximum (usually such engine is flat rated, which see).
design weight Different authorities have different definitions; weight chosen as typical of mission but usually much less than MTOW.
DF Direction-finder, or direction-finding.
DGAC Direction Générale à l'Aviation Civile.
dihedral Upward slope of wing from root to tip.
disposable load Sum of masses that can be loaded or unloaded, including payload, crew, removable equipment, usable fuel and other consumables; MTOW minus OWE.
DME UHF distance-measuring equipment; gives slant distance to a beacon; DME element of Tacan.
DoD Department of Defense (US).
dog-tooth A sharp discontinuity in the leading-edge of a wing or tail surface resulting from an increase in chord (see also sawtooth).
Doppler Short for Doppler radar – radar using fact that received frequency is a function of relative velocity between transmitter or reflecting surface and receiver; used for measuring speed over ground or for detecting aircraft or moving vehicles against static ground or sea.
double-slotted flap One having an auxiliary aerofoil ahead of main surface to increase maximum lift.

EAA Experimental Aircraft Association (divided into local branches called Chapters).
EAS Equivalent airspeed, RAS minus correction for compressibility.
ECCM Electronic counter-countermeasures.
ECM Electronic countermeasures.
ECS Environmental control system.
EEZ Economic exclusion (or exclusive-economic) zone.
EFIS Electronic flight instrument(ation) system, in which large multifunction CRT displays replace traditional instruments.
EGT Exhaust gas temperature.
ehp Equivalent horsepower, measure of propulsive power of turboprop made up of shp plus addition due to residual thrust from jet.
EICAS Engine indication (and) crew alerting system.
ekW Equivalent kilowatts, SI measure of propulsive power of turboprop (see ehp).
elevon Wing trailing-edge control surface combining functions of aileron and elevator.
ELF Extreme low frequency.
elint electronics intelligence.
ELT Emergency locator transmitter, to help rescuers home on to a disabled or crashed aircraft.
EMD Engineering and manufacturing development.
EMP Electromagnetic pulse of nuclear or electronic origin.
EO Electro-optical.
EPNL Effective perceived noise level.
EPNdB Effective perceived noise decibel, SI unit of EPNL.
EPU Emergency power unit (part of aircraft, not used for propulsion).
ERU Ejector release unit.
ESM (1) Electronic surveillance (or support) measures; (2) Electronic signal monitoring.
ETOPS Extended-range twin (engine) operations (thus sometimes given as EROPS), routeing not more than a given flight time (120, 180 or 240 minutes) from a usable alternative airfield.
EW Electronic warfare.

FAA Federal Aviation Administration.
FAC Forward air control (or controller).
factored Multiplied by an agreed number to take account of extreme adverse conditions, errors, design deficiencies or other inaccuracies.
FADEC Full-authority digital engine (or electronic) control.
FAI Fédération Aéronautique Internationale.
fail-operational System which continues to function after any single fault has occurred.
fail-safe Structure or system which survives failure (in case of system, may no longer function normally).
FAR Federal Aviation Regulations.
FAR Pt 23 Defines the airworthiness of private and air taxi aeroplanes of 5,670 kg (12,500 lb) MTOW and below.
FAR Pt 25 Defines the airworthiness of public transport aeroplanes exceeding 5,670 kg (12,500 lb) MTOW.
FBL Fly-by-light (which see).
FBW Fly-by-wire (which see).
FCS Flight control system.
FDR Flight data recorder (which see).
FDS Flight director system.
feathering Setting propeller blades at pitch aligned with slipstream to minimise drag.
fence A chordwise projection on the surface of a wing, used to modify the distribution of pressure.
Fenestron Helicopter tail rotor with many slender blades rotating in short duct (registered name).
ferry range Extreme safe range with zero payload.
FFAR Folding-fin (or free-flight) aircraft rocket.
field length Measure of distance needed to land and/or take off; many different measures for particular purposes, each precisely defined.
fixed-pitch Propeller with blades fixed to the hub.
flap A surface carried on the leading- or trailing-edge of a wing and able to move relative to it. The simplest leading-edge flap and so-called plain (trailing-edge) flap is formed by hinging the entire edge of the wing. The Krueger is a leading-edge flap forming part of the wing undersurface, swung down and forwards on arms to give a bluff leading-edge. A split flap is formed by hinging only the undersurface of the trailing-edge. A slotted flap is a hinged trailing-edge which moves aft as well as down on tracks to leave a narrow slot ahead of it; hence double- and triple-slotted. A Fowler flap is a complete auxiliary aerofoil mounted on tracks under a fixed trailing-edge; initially it moves aft, to emerge behind the fixed part of the wing, and at the end of its travel it rotates down. A Gouge flap has an upper surface forming part of a cylinder, and rotates (on rails or brackets) about that cylinder's centre.
flaperon Wing trailing-edge surface combining functions of flap and aileron.
flat-four Piston engine having four horizontally opposed cylinders; thus, flat-twin, flat-six and so on.
flat rated Propulsion engine capable of giving full thrust or power for take-off at an airfield well above S/L and/or at high ambient temperature (thus, probably derated at S/L).

GLOSSARY

flight data recorder Crash-protected recorder of dynamic/static pressure, air temperature, control-surface and slat/flap positions, 3-axis accelerations, engine parameters and possibly other variables.
FLIR Forward-looking infra-red.
fly-by-light Flight control system in which signals pass between computers and actuators along fibre-optic leads.
fly-by-wire Flight control system with electrical signalling, without mechanical interconnection between cockpit flying controls and control surfaces.
FM Frequency modulation.
FMCS Flight management computer system.
FMS (1) Foreign military sales (US DoD); (2) Flight management system.
FOD Foreign-object damage.
footprint (1) A precisely delineated boundary on the surface around an airfield, inside which the perceived noise of an aircraft exceeds a specified level during take-off and/or landing; (2) Dispersion of weapon or submunition impact points.
foreplanes Pivoted canard surfaces forming part of the primary flight control system with authority in pitch and possibly also in roll. See also canards.
FOV Field of view.
Fowler flap See flap.
FRADU Fleet Requirements and Air Direction Unit (UK).
frequency See radar frequency.
frequency agile (frequency hopping) Making a transmission harder to detect by switching automatically to a succession of frequencies.
FSD Full-scale development.
FSED Full-scale engineering development.
FY Fiscal year; in US government affairs, runs from 1 October to 30 September (FY02 begins 1 October 2001); in Japan, from 1 April (FY13 or FY01 began 1 April 2001).

g Acceleration due to mean Earth gravity, that is of a body in free-fall; or acceleration due to rapid change of direction of flight path.
gallons Non-SI measure; 1 Imp gallon (UK) = 4.546 litres, 1 US gallon = 3.785 litres.
GFRP Glass fibre-reinforced plastics.
'glass cockpit' Cockpit in which dial instruments are replaced by multifunction electronic displays.
glass fibre Spun molten glass; see GFRP.
glide slope Element giving vertical (height) guidance in ILS.
glove (1) Fixed portion of wing inboard of variable sweep wing; (2) additional aerofoil profile added around normal wing for test purposes.
GmbH Gesellschaft mit beschränkter Haftpflicht (or Haftung) (German company constitution).
GPS Global Positioning System, US military/civil satellite-based precision navaid.
GPU Ground power unit (not part of aircraft).
GPWS Ground-proximity warning system.
green aircraft Aircraft flyable but unpainted, unfurnished and basically equipped.
gross wing area See wing area.
gunship Aircraft designed for battlefield attack; helicopter gunships normally use slim body carrying pilot and weapon operator only.
GUP Gosudarstvennoye Unitarnoye Predpriyatie (Russian State Unitary Enterprise).

h Hour(s).
handed Rotating in opposite directions.
hardened Protected as far as possible against nuclear explosion.
hardpoint Reinforced part of aircraft to which external load can be attached, for example weapon or tank pylon.
HDU Hose-drum unit.
head-down display On the cockpit instrument panel (as distinct from a HUD).
head-level display Immediately below HUD.
hectopascal Unit of pressure (hPa) equal to 100 Pascals.
helicopter Rotary-wing aircraft both lifted and propelled by one or more power-driven rotors turning about substantially vertical axes.
HF High frequency.
HIFR Helicopter in-flight refuelling.
HIRF High-intensity radiated field(s).
HMD Helmet-mounted display; hence HMS = sight.
HOCAC Hands on cyclic and collective.
homebuilt Aircraft built/assembled from plans or kits.
hot and high Adverse combination of airfield height and high ambient temperature, which lengthens required take-off distance.
HOTAS Hands on throttle and stick.
hot refuelling Replenishment of fuel while engine(s) running.
hovering ceiling Ceiling of helicopter (corresponding to air density at which maximum rate of climb is zero), either IGE or OGE.
HP High pressure (HPC, compressor; HPT, turbine).
hp Horsepower, non-SI unit of power.
HSI Horizontal situation indicator.
HUD Head-up display (bright numbers and symbols projected on pilot's aiming sight glass and focused on infinity so that pilot can simultaneously read display and look ahead). The term is increasingly rendered in the USA as ''heads up'', which is incorrect.
Hz Hertz, cycles per second.

IAS Indicated airspeed, airspeed indicator reading corrected for instrument error.
IATA International Air Transport Association.
ICAO International Civil Aviation Organisation.
IDG Integrated-drive generator.
IFF Identification friend or foe.

2001/0051034

IFR (1) Instrument flight rules (compare VFR); (2) in-flight refuelling.
IGE In ground effect: helicopter performance with theoretical flat horizontal surface just below it (for example mountain).
IIR Imaging infra-red.
ILS Instrument landing system.
Imperial gallon 1.20095 US gallons; 4.546 litres.
IMS Integrated multiplex system.
INAS Integrated nav/attack system.
Inc Incorporated (company constitution).
incidence The angle at which the wing is set in relation to the fore/aft axis. Often wrongly used to mean angle of attack (which see).
inertial navigation Measuring all accelerations imparted to a vehicle and, by integrating these with respect to time, calculating speed at every instant (in all three planes) and, by integrating a second time, calculating total change of position in relation to starting point.
INS Inertial navigation system.
integral construction Machined from solid instead of assembled from separate parts.
integral tank Fuel (or other liquid) tank formed by sealing part of structure.
intercom Wired telephone system for communication within aircraft.
inverter Electric or electronic device for inverting (reversing polarity of) alternate waves in AC power to produce DC.
IOC Initial operational capability.
IR Infra-red.
IRCM Infra-red countermeasures.
IRLS Infra-red linescan (builds TV-type picture showing cool and hot regions as contrasting shades).
IRS Inertial reference system.
IRST Infra-red search and track.
ISA International Standard Atmosphere (1013.25 mb, 1,225 g/m^3 and 15°C at mean sea level; lapse rate 1.98°C per 1,000 ft up to −56.5°C at 36,090 ft).

J Joule, SI unit of energy.
JAA Joint Aviation Authorities.
JAR Joint Aviation Requirements, agreed by all major EC countries (JAR 25 equivalent to FAR Pt 25).
JAR-VLA JAR classification for Very Light Aircraft (MTOW limit of 750 kg; 1,653 lb).
JASDF Japan Air Self-Defence Force.
JATO Jet-assisted take-off (actually means rocket-assisted).
JCAB Japan Civil Airworthiness Board.
JDA Japan Defence Agency.
JGSDF Japan Ground Self-Defence Force.
JMSA Japan Maritime Safety Agency.
JMSDF Japan Maritime Self-Defence Force.
joined wing Tandem wing layout in which forward and aft wings are swept so that the outer sections meet.
joule SI unit of energy, = 1 Nm = 1 Ws.
JPATS Joint Primary Aircraft Training System (Raytheon T-6A Texan II).
JSC Joint stock company.
JSF Joint Strike Fighter.
J-STARS US Air Force/Navy Joint Surveillance and Target Attack Radar System in Northrop Grumman E-8C.
JTIDS Joint Tactical Information Distribution System (NATO Link 16).

kbit One thousand bits of memory.
Kevlar Aramid fibre used as basis of high-strength composites material.
kg Kilogramme (2.20462 lb).
kitbuilt Prefabricated aircraft for amateur assembly.
KK Kabushiki Kaisha (Japanese company constitution).
km/h Kilometres per hour.
kN KiloNewtons (N×10^3). See N.
knot 1 n mile per hour (1.852 km/h; 1.15078 mph).

Krueger flap Hinges down and then forward from below the leading-edge.
kVA Kilovolt-ampères.
kW Kilowatt, SI measure of all forms of power (not just electrical).

LAMPS Light airborne multipurpose system.
LANTIRN Low-altitude navigation and targeting infra-red, night.
LAPES Low-altitude parachute extraction system.
LBA Luftfahrtbundesamt (German civil aviation authority).
lb Pound, non-SI unit of weight: 0.453592 kg.
lb st Pounds of static thrust.
LCD Liquid crystal display, used for showing instrument information.
LCN Load classification number, measure of 'flotation' of aircraft landing gear linking aircraft weight, weight distribution, tyre numbers, pressures and disposition.
LED Light-emitting diode.
lift dumper Spoiler designed to open on landing to reduce lift and thus increase effectiveness of wheel braking.
LINS Laser inertial navigation system.
litre SI unit of volume (0.264177 US gallon; 0.219975 Imp gallon).
LLTV Low-light TV (thus, LLLTV, low-light level); see ALLTV.
LO Low-observables, which see (stealth).
load factor (1) Percentage of maximum payload; (2) design factor (g limit) for airframe.
LOC Localiser (which see).
localiser Element giving steering guidance in ILS.
LOH Light observation helicopter.
loiter Fly for maximum endurance, at much less than normal cruise speed.
longerons Principal fore-and-aft structural members (for example in fuselage).
Loran Long-range navigation; family of hyperbolic navaids based on ground radio emissions, now mainly Loran C.
LOROP Long-range oblique photography.
LOS Line of sight.

2000/0084658

low-observables Materials, structures and techniques designed to minimise aircraft signatures of all kinds.
lox Liquid oxygen.
LP Low pressure (LPC, compressor; LPT, turbine).
LRIP Low-rate initial production.
LRMTS Laser ranger and marked-target seeker.
LRU Line-replaceable unit.
Ltd Limited (company constitution).
LV Low volume (cropspraying intensity).

m Metre(s), SI unit of length (3.28084 feet).
M or Mach number The ratio of the speed of a body to the speed of sound (340 m; 1,116 ft/s in air at 15°C) under the same ambient conditions.
MAD Magnetic anomaly detector.
mass balance Mass attached to flight control surface, typically ahead of hinge axis, internally or externally, to reduce or eliminate coupling with airframe flutter modes.
mass flow Mass of air passing per second (usually at T-O, S/L).
MAWS Missile-approach warning system.
mb Millibars, bar × 10^{-3}.
Mbyte One million (10^6) bytes.
MCM Mine countermeasures.
medevac Medical evacuation.
MFD Multifunction (electronic) display.
MHz Megahertz: 1 million (10^6) Hertz.
microlight See ultralight.
MIDS Multifunction information distribution system.
MLS Microwave landing system.
MLU Mid-life update.
MLW Maximum landing weight.
mm Millimetres, metres × 10^{-3}.
M$_{MO}$ Maximum operating Mach number.
MMS Mast-mounted sight.
MoD Ministry of Defence.
monocoque Structure with strength in outer shell, devoid of internal bracing (semi-monocoque, with some internal supporting structure).
MoU Memorandum of Understanding.
MPA Maritime patrol aircraft.
mph Miles per hour.
MSIP Multistaged improvement program (US).
MTBF Mean time between failures.

GLOSSARY

MTBR Mean time between removals.
MTI Moving-target indication (radar).
MTOW Maximum take-off weight (minus taxi/run-up fuel).
MYP Multiyear procurement (US).
MZFW Maximum zero-fuel weight.

N Newton, SI unit of force, = 0.22480455 lb force.
NACES Navy aircrew common ejection seat (US).
NAS Naval Air Station (US).
NASA National Aeronautics and Space Administration (US).
NASC Naval Air Systems Command (also several other aerospace meanings) (US).
NATC Naval Air Training Command or Test Center (also several other aerospace meanings) (US).
NATO North Atlantic Treaty Organisation.
NBAA National Business Aircraft Association (US).
NBC Nuclear, biological, chemical (warfare).
NDT Non-destructive testing.
Newton See N.
NFO Naval flight officer; second crew member in US Navy aircraft; compare WSO.
Ni/Cd Nickel/cadmium.
Nib Forward-pointing extension at inner end of fixed glove on VG aircraft or leading-edge root extension on light aircraft.
n mile nautical mile, 1.852 km, 1.15078 miles.
NOE Nap-of-the-Earth (low flying in military aircraft, using natural cover of hills, trees and so on).
NOS Night observation surveillance.
NV Naamloze Vennootschap (Belgian/Netherlands company constitution).
NVG Night vision goggles.
NVS Noise vibration suppression.

OAO Otkrytolye Aktsionernoye Obshchestvo (JSC; Russian company constitution).
OAT Outside air temperature.
OBIGGS Onboard inert gas generating system.
OBOGS Onboard oxygen generating system.
OCU (1) Operational Conversion Unit; (2) operational capabilities upgrade.
OEI One engine inoperative.
OEU Operational Evaluation Unit.
offset Workshare granted to a customer nation to offset the cost of an imported system.
OGE Out of ground effect; helicopter hovering, far above nearest surface.
OKB Opytnyi Konstruktorskoye Byuro (Russian experimental design bureau).
Omega Long-range hyperbolic radio navaid.
omni Generalised word meaning equal in all directions (as in omnirange, omniflash beacon).
on condition maintenance According to condition rather than at fixed intervals.
OOO Obshchestvo Ogranichennoye Otvetstvennostyu (Russian company constitution).
opeval Operational evaluation.
OTH Over-the-horizon (OTHT adds targeting).
OTPI On-top position indicator (indicates overhead of submarine in ASW).
OWE Operating weight empty. MTOW minus payload, usable fuel and oil and other consumables (thus, includes crew).

PA system Public or passenger address.
pallet (1) for freight, rigid platform for handling by forklift or conveyor; (2) for missile, interface mounting and electronics box outside aircraft.
Pascal SI unit of pressure =1 Nm^{-2} (one Newton per square metre).
payload Disposable load generating revenue (passengers, cargo, mail and other paid items); in military aircraft, loosely used to mean total load carried of weapons, cargo or other mission equipment.
Performance Aircraft capabilities after S/L take-off at MTOW in ISA with normal full tankage, except as otherwise specified, and with landing data at MLW (where different).
PFCS Primary flight computer system.
PGM Precision-guided munition.

2000/0084663

phased array Radar in which the beam is scanned electronically in one or both axes without moving the antenna.

Pirate Passive infra-red airborne tracking equipment.
PLA Prelaunch activities.
plane A lifting surface (for example wing, tailplane).
plc Public limited company (company constitution).
plug door Door larger than its frame in pressurised fuselage, either opening inwards or arranged to retract parts before opening outwards.
plume The region of hot air and gas emitted by a helicopter jetpipe.
ply Indication (ply rating) of tyre strength in a specific application; not necessarily the actual number of carcass plies in the tyre.
pneumatic de-icing Covered with flexible surfaces alternately pumped up and deflated to throw off ice.
port Left side, looking forward.
power-by-wire Using electric power alone (not electro-hydraulic) to drive control surfaces and perform other mechanical tasks.
power loading Aircraft weight (usually MTOW) divided by total propulsive power or thrust at T-O. For helicopters, based on transmission rating rather than total engine power.
power train A complete mechanical drive system, for example the sequence of gearwheels, clutches and shafts transmitting power from one or more engines to the rotors of a helicopter.
PPV Pre-production verification.
prepreg Glass fibre cloth or rovings pre-impregnated with resin to simplify layup.
pressure fuelling Fuelling via a leakproof connection through which fuel passes at high rate under pressure.
primary flight controls Those used to control trajectory of aircraft (thus, not trimmers, tabs, flaps, slats, airbrakes or lift dumpers, and so on).
primary flight display Single screen bearing all data for aircraft flight-path control.
propfan A family of new-technology propellers characterised by multiple scimitar-shaped blades with thin sharp-edged profile. Single and contrarotating examples promise to extend propeller efficiency up to an aircraft Mach number of about 0.8.
proprotor Large propeller, tilting for forward or vertical flight.
PT Pesawat Terbang (Indonesian company constitution).
Pty Proprietary (company constitution).
pulse Doppler Radar sending out pulses and measuring frequency-shift to detect returns only from moving target(s) seen against background clutter.
pylon Structure linking aircraft to external load (engine nacelle, drop tank, bomb, and so on).

radar frequency Operating bands of airborne radars are given according to frequency. That part of the electromagnetic spectrum appropriate to above-surface short-range communication and radar (but not OTH) used in aviation is given in the adjacent table with an approximate cross-reference to previously used wavelength bands.

Frequency Band	Frequency	Wavelength	Wave Band
HF	10-30 MHz	30-10 m	HF
VHF	30-100 MHz	10-3 m	VHF
A	100-300 MHz	3-1 m	VHF
B	300-500 MHz	100-60 cm	UHF
C	0.5-1 GHz	60-30 cm	UHF
D	1-2 GHz	30-15 cm	L
E	2-3 GHz	15-10 cm	S
F	3-4 GHz	10-7.5 cm	S
G	4-6 GHz	7.5-5 cm	C
H	6-8 GHz	5-3.75 cm	C
I	8-10 GHz	3.75-3 cm	X
J	10-20 GHz	30-15 mm	X/Ku
K	20-40 GHz	15-7.5 mm	K/Ka
L	40-60 GHz	7.5-5 mm	nil
M	60-100 GHz	5-3 mm	nil

radius The approximate distance an aircraft can fly from base and return without intermediate landing.
RAI Registro Aeronautico Italiano (Italian civil aviation authority).
RAM Radar absorbent material.
ram pressure Increased pressure in forward-facing aircraft inlet, generated by converting relative kinetic energy to pressure.
ramp weight Maximum weight at start of flight (MTOW plus taxi/run-up fuel).
range Too many definitions to list, but essentially the distance an aircraft can fly (or is permitted to fly) with specified load and usually while making allowance for specified additional manoeuvres (diversions, standoff, go-around and so on).
RANSAC Range surveillance aircraft.
RAS Rectified airspeed, IAS corrected for position error.
raster Generation of large-area display, for example TV screen, by close-spaced horizontal lines scanned either alternately or in sequence.

RAT Ram air turbine.
rating Any of several values of thrust or shaft power which an engine is qualified (usually also guaranteed) to develop under specified conditions.
RCS Radar cross-section; apparent size of echo.
redundant Provided with spare capacity or data channels and thus able to survive failures.
reversion Ability to switch to manual control following failure of a powered system.
RFAS Russian Federation and Associated States (CIS).
RFP Request(s) for proposals.
rigid rotor See bearingless rotor.
RMI Radio magnetic indicator; combines compass and navaid bearings.
R/Nav Calculates position, distance and time from groups of airways beacons.
RON Research octane number of fuel.
roving Multiple strands of fibre, as in a rope (but usually not twisted).
rpm Revolutions per minute.
RPV Remotely piloted vehicle (pilot in other aircraft or on ground); contrast UAV.
RSA Réseau du Sport de l'Air.

2000/0084660

ruddervators Flying control surfaces, usually a V tail, that control both yaw and pitch attitude.
RWR Radar warning receiver.

s Second(s).
SA Société Anonyme (France, Romania), Sociedad Anónima (Brazil, Spain) or Spółka Akcyjna (Poland) (company constitution).
safe-life A term denoting that a component has proved by testing that it can be expected to continue to function safely for a precisely defined period before replacement.
SAM Surface-to-air missile.
SAR (1) Search and rescue; (2) synthetic aperture radar.
SAS Stability augmentation system.
satcom Satellite communications.
sawtooth Same as dog-tooth.
SCAS Stability and control augmentation system.
Sdn Bhd Sendirian Berhad (Malaysian company constitution).
SEAD Suppression of enemy air defence(s).
second-source Production of identical item by second factory or company.
semi-active Homing on to radiation reflected from target illuminated by radar or laser energy beamed from elsewhere (for example, from launch aircraft).
sensitive altimeter Altitude indicator of mechanical type, having acute sensitivity.
service ceiling Usually height equivalent to air density at which maximum attainable rate of climb is 100 ft/min.
servo A device which acts as a relay, usually augmenting the pilot's efforts to move a control surface, or the like.
SFAR Special Federal Aviation Regulation(s).
sfc Specific fuel consumption (which see).
shaft Connection between gas-turbine and compressor or other driven unit. Two-shaft engine has second shaft, rotating at different speed, surrounding the first (thus, HP surrounds inner LP or fanshaft).
shipment One item or consignment delivered (by any means of transport) to customer.
shp Shaft horsepower, measure of power transmitted via rotating shaft.
shroud Many meanings, including: (1) a fixed circular duct surrounding a fan or propfan; (2) a ring formed by lateral projections on a rotor (for example fan) blade (part-span or at the tip); (3) a portion of a wing or other fixed aerofoil projecting aft over the leading-edge of a hinged or otherwise movable surface such as a flap, aileron or elevator.
sideline noise EPNdB measure of aircraft landing and taking off, at point 0.25 n mile (2- or 3-engined) or 0.35 n mile (4-engined) from runway centreline.
sidestick Control column in the form of a short handgrip beside the pilot.
SIF Selective identification facility.
sigint Signals intelligence.
signature Characteristic 'fingerprint' of all acoustic or electromagnetic radiation (radar, IR, and so on).
single-aisle Passenger cabin has seats on each side of a single aisle along or near the centre.
single-shaft Gas-turbine in which all compressors and turbines are fixed to common shaft.
S/L Sea level.
SLAR Side-looking airborne radar.

GLOSSARY

slat Auxiliary curved or mini-aerofoil surface designed to prevent flow breakaway from a wing or tail. On a tail leading-edge it may be fixed, leaving a narrow slot. On a wing it is almost always retractable, normally flush with the wing profile but extended (under power or by aerodynamic lift) to leave a narrow slot for take-off, low-speed loiter or landing.

slot, slotted See slat.

snap-down Air-to-air interception of low-flying aircraft by AAM fired from fighter at a higher altitude.

soft target Not armoured or hardened.

SONAR, sonar Sound navigation and ranging.

SpA Società per Azioni (Italian company constitution).

specific fuel consumption Rate at which fuel is consumed divided by power or thrust developed, and thus a measure of engine efficiency. For jet engines (air-breathing, not rockets) unit is mg/Ns, milligrams per Newton-second; for shaft engines unit is μg/J, micrograms (millionths of a gram) per Joule (SI unit of work or energy).

spoiler Plank-like surface normally recessed into top of wing, hinged up under power to reduce (spoil) lift and increase drag. Used asymmetrically for lateral control.

spoileron Small spoiler augmenting ailerons.

sportplane Light aircraft design in which performance takes precedence over utility.

SRL Società Reponsibilita Limitata (Italian company constitution).

Sp. z o.o Spółka z ograniczoną odpowiedzialnością (Polish company constututio)

SSB Single-sideband (radio).

SSR Secondary surveillance radar.

SST Supersonic transport.

st Static thrust.

stabilator One-piece, all-moving horizontal tail, combining functions of horizontal stabiliser and elevator.

stabiliser Fin (thus, horizontal stabiliser = tailplane).

stall Sudden near-total loss of lift of a wing because AoA has exceeded a critical value.

stall strips Sharp-edged strips on wing leading-edge to induce stall to initiate at that point.

stalling speed Airspeed at which aircraft stalls at $1g$.

starboard Right side, looking forward.

static inverter Solid-state (not rotary machine) inverter of alternating wave-form to produce DC from AC.

STC Supplementary Type Certificate.

stealth See low-observables.

stick-pusher Stall-protection device that forces pilot's control column forward as stalling angle of attack is neared.

stick-shaker Stall-warning device that noisily shakes pilot's control column as stalling angle of attack is neared.

STOL Short take-off and landing. (Several definitions, stipulating allowable horizontal distance to clear screen height of 35 or 50 ft or various SI measures.)

store Object carried as part of payload on external attachment (for example bomb, drop tank).

STOVL Short take-off, vertical landing.

strobe light High-intensity flashing beacon.

supercritical wing Wing of relatively deep, flat-topped profile generating lift right across upper surface instead of concentrated close behind leading-edge.

sweepback Backwards inclination of wing or other aerofoil, seen from above, measured relative to fuselage or other reference axis, usually measured at quarter-chord (25 per cent) or at leading-edge.

t Tonne, 1 Megagram, 1,000 kg.

tab Small auxiliary surface hinged (flight-adjustable) or attached in a fixed position (ground-adjustable) to trailing-edge of control surface for trimming, balancing (reducing hinge moment: force needed to operate main surface) or in other way assisting pilot. Compare anti-balance tab.

tabbed flap Fitted with narrow-chord tab along trailing-edge which deflects to greater angle than main surface.

Tacan Tactical air navigation, UHF navaid giving bearing and distance to ground beacons; distance element (see DME) can be paired with civil VOR.

TACCO Tactical commander, ASW aircraft.

taileron Left and right tailplanes used as primary control surfaces in both pitch and roll.

tailplane Horizontal stabiliser; main horizontal tail surface, originally fixed and carrying hinged elevator(s) but today often a single 'slab' serving as control surface (see stabiliser, stabilator).

TANS Tactical air navigation system; Decca Navigator or Doppler-based computer, control and display unit.

TAS True airspeed, EAS corrected for density (often very large factor) appropriate to aircraft altitude.

TBO Time between overhauls.

t/c ratio Ratio of the thickness (aerodynamic depth) of a wing or other surface to its chord, both measured at the same place parallel to the fore-and-aft axis.

TCAS Traffic-alert and collision-avoidance system.

Tercom Terrain-comparison (or contour-matching), navigation aid which compares relief of terrain with profile stored in memory.

TFR Terrain-following radar (for low-level attack).

thickness Depth of wing or other aerofoil; maximum perpendicular distance between upper and lower surfaces.

thrust vectoring Rotation of a vehicle's thrust axis to control its trajectory or support its weight.

2001/0105049

TIALD Thermal imaging and laser designation (pod).

tiltrotor Aircraft with fixed wing and rotors that tilt up for hovering and forward for fast flight.

T-O Take-off.

T-O noise EPNdB measure of aircraft taking off, at point directly under flight path 3.5 n miles from brakes-release.

TOGW Take-off gross weight (not necessarily MTOW).

ton Imperial (long) ton = 1.016 t or 2,240 lb, US (short) ton = 0.9072 t or 2,000 lb.

track Distance between centres of contact areas of main landing wheels measured left/right across aircraft (with bogies, distance between centres of contact areas of each bogie).

transceiver Radio transmitter/receiver.

transformer-rectifier Device for converting AC to DC at a different voltage.

transponder Radio transmitter triggered automatically by a particular received signal, as in secondary surveillance radar (SSR).

TRU Transformer/rectifier unit.

TsENTROSPAS (in Russian Federation) Ministry for Civil Defence, Emergencies and Elimination of the Consequences of Natural Disasters.

turbofan Gas-turbine jet engine generating most thrust by a large-diameter cowled fan, with small part added by jet from core.

turbojet Simplest form of gas turbine comprising compressor, combustion chamber, turbine and propulsive nozzle.

turboprop Gas turbine in which as much energy as possible is taken from gas jet and used to drive reduction gearbox and propeller.

turboshaft Gas turbine in which as much energy as possible is taken from gas jet and used to drive high-speed shaft (which in turn drives external load such as helicopter transmission).

twist Progressive change of angle of incidence of a wing, rotor blade or other aerofoil from root to tip.

Type Certificate Airworthiness licence granted to enable a manufacturer to produce and market a specified type of aircraft (compare C of A).

tyre sizes Five systems of classification are in current use; Type I, consisting of a single figure indicating nominal diameter in inches, is obsolete. See adjacent table and also 'ply'.

UAV Unmanned (or uninhabited) aerial vehicle; contrast RPV.

UHCA Ultra-high capacity airliner.

UHF Ultra-high frequency.

ultralight Light aircraft with parameters below specified national limits, qualifying for less rigorous licensing; also known as microlight. See table.

ULV Ultra-low volume (cropspraying intensity).

unfactored Performance level expected of average pilot, in average aircraft, without additional safety factors.

upper surface blowing Turbofan jet expelled over upper surface of wing to increase lift.

usable fuel Total mass of fuel consumable in flight, usually 95 to 98 per cent of system capacity.

useful load Usable fuel and other consumables plus payload.

US gallon 0.83267 Imperial gallon; 3.785 litres.

UV Ultra-violet.

variable geometry Capable of grossly changing shape in flight, especially by varying sweep of wings.

variable pitch Propeller with its blades held in rotary bearings in the hub, so that pitch (of all blades in unison) can be altered in flight. See constant speed; compare adjustable pitch.

V_D Maximum permitted diving speed.

VDU Video (or visual) display unit.

vectored Capable of being pointed in different directions.

vertrep Vertical replenishment.

VFR Visual flight rules.

VHF Very high frequency.

VLF Very low frequency (area-coverage navaid).

V_{MO} Maximum permitted operating flight speed (IAS, EAS or CAS must be specified).

VMS Vehicle management system.

V_{NE} Never-exceed speed (aerodynamic or structural limit).

VOR VHF omnidirectional range (network of VHF radio beacons each providing to/from bearing).

vortex generators Small blades attached to wing and tail surfaces to energise local air flow and improve control.

vortillon Short-chord fence (particularly on MD-80 series) ahead of and below leading-edge.

VSI Vertical speed (climb/descent) indicator.

V/STOL Vertical/short take-off and landing.

washout Inbuilt twist of wing or rotor blade reducing angle of incidence towards the tip.

Watt SI unit of power, equal to 1 Js^{-1} (one Joule per second).

WDNS Weapon delivery and navigation system.

wet Housing fuel; wet wing often has extra connotation of integral tankage. Wet pylon can accommodate external fuel tank.

wheelbase Minimum distance from nosewheel or tailwheel (centre of contact area) to line joining mainwheels (centres of contact areas).

wide-body Passenger aircraft with cabin wide enough to have two longitudinal aisles between seats.

wing area Total projected area of clean wing (no projecting flaps, slats and so on) including all control surfaces and area of fuselage bounded by leading- and trailing-edges projected to centreline (inapplicable to slender-delta aircraft with extremely large leading-edge sweep angle). Described in *Jane's* as gross wing area; net area excludes projected areas of fuselage, nacelles, and so on.

2000/0084659

wing loading Aircraft weight (usually MTOW) divided by wing area.

Tyre classification systems				
Classification name	Example	Nominal diameter	Nominal section width	Nominal rim diameter
Type III	8.50-10		8½ in	10 in
Type VII	49×17	49 in	17 in	
Three Part	49×19.0-20	49 in	19.0 in	20 in
Radial	32×8.8R16	32 in	8.8 in	16 in
Metric	670×210-12	670 mm	210 mm	12 in

Example microlight/ultralight maxima						
Country	Name	Seat(s)	Empty weight	MTOW	Fuel	Vso
Australia	Ultralight	1-2		540 kg		
France	ULM[1]	1		300 kg[2]		35 kt[3]
	ULM[1]	2		450 kg[2]		35 kt[3]
UK	Microlight	1-2		390 kg	50 l	
	SLA[4]	1-2		450 kg		35 kt
USA	Ultralight	1	254 lb		5 USg	

[1] Ultra léger motorisé; criteria accepted by several European countries, including Germany
[2] Plus 10 per cent for seaplanes and amphibians
[3] Plus 5 per cent for seaplanes and amphibians
[4] Small light aeroplane; interim category, pending redefinition of microlight

GLOSSARY

winglet Small auxiliary aerofoil, usually sharply upturned and often sweptback, at tip of wing.

WSO Weapon(s) system(s) officer.

yoke Pilot's flight control interface for pitch and roll axes in the form of a stick (control column) to the top of which is laterally pivoted a pair of handgrips in the form of a Y.

ZAO Zakrytoe Aktsionernoye Obshchestvo (Russian company constitution).

zero-fuel weight MTOW minus usable fuel and other consumables, in most aircraft imposing severest stress on wing and defining limit on payload.

zero/zero seat Ejection seat designed for use even at zero speed on ground.

ZFW Zero-fuel weight.

μg Microgrammes, grammes $\times 10^{-6}$.

Zero/zero seat being tested from a ground rig

AIRCRAFT

ARGENTINA

LAVIASA

LATIN AMERICANA de AVIACION SA
Mendoza
e-mail: laviasa@ba.net
Web: http://www.laviasa.com.ar

On 15 April 1998, New Piper Aircraft Inc sold the PA-25 Pawnee type certificate to LAVIASA, which then began manufacturing, or remanufacturing, this agricultural sprayer at Mendoza under the local name of **Puelche**; designation PA-25-235/-260 is retained. Prototype LV-X229 was subsequently registered LV-ZOE and sold to Agro Aerea Norte SA at Villa Angela, and at least seven more had been produced by late 1999.

A description of the Pawnee last appeared in the 1982-83 *Jane's*. Differences from original include use of 4130 steel and Alclad 2024 in construction; increased fuel totalling 182 litres (48.0 US gallons; 40.0 Imp gallons); and empty weights of 690 kg (1,323 lb) for PA-25-235, 698 kg (1,540 lb) for -260 with fixed-pitch propeller and 706 kg (1,556 lb) for -260 with constant-speed propeller.

UPDATED

LMAASA

LOCKHEED MARTIN AIRCRAFT ARGENTINA SA
Avenida Fuerza Aérea Argentina Km 5500, CP5010 Córdoba
Tel: (+54 351) 466 87 33
Fax: (+54 351) 466 87 34
e-mail: jkaras@powernet.com.ar
PRESIDENT: James Taylor
OPERATIONS GENERAL MANAGER, ENGINES AND AIRCRAFT: Howard Atwood
BUSINESS DEVELOPMENT DIRECTOR: Bernard Kelleher

Original FMA (Military Aircraft Factory) came into operation 10 October 1927 as central organisation for aeronautical research and production; underwent several name changes (see 1987-88 and earlier *Jane's*) before reverting 1968 to original title as component of Area de Material Córdoba (AMC) of Argentine Air Force. FMA converted into a joint stock company from April 1992, Air Force buying 30 per cent of the shares to establish itself as holding and management authority; control of FMA handed over to Planning Secretary of MoD on 20 December 1993. MoD and Lockheed Aircraft Argentina SA signed concession agreement on 15 December 1994, allocating management of FMA to Lockheed Martin, now Lockheed Martin Aircraft Argentina SA, from 1 July 1995.

Principal activities are aircraft design, manufacture, upgrading maintenance and repair, current tasks including maintenance and modification of A-4AR Fighting Hawks, C-130 Hercules, IA 58 Pucarás and IA 63 Pampas and manufacture of a further batch of 12 Pampas. Further data in *Jane's Aircraft Upgrades*.

Laboratories, factories and other aeronautical division buildings occupy total covered area of 253,000 m² (2,723,300 sq ft); had workforce of more than 1,000 in 2000. Córdoba facility also accommodates Centro de Ensayos en Vuelo (Flight Test Centre), a separate division also controlled by Argentine Air Force, at which all aircraft produced in Argentina undergo certification testing.

UPDATED

LMAASA IA 63 PAMPA NG
TYPE: Basic jet trainer/light attack jet.
PROGRAMME: First-generation Pampa initiated by Fuerza Aérea Argentina (FAA) 1979, eventual configuration being selected over six other designs early 1980, with Dornier of Germany providing technical assistance (including manufacture of prototypes' wings and tailplanes); two static/fatigue test airframes and three flying prototypes built (first flight, by EX-01, 6 October 1984); first flight of production Pampa October 1987; Fourteen of initial batch of 18 (including three prototypes) delivered to Argentine Air Force (survivors currently serving I Escuadron of 4 Grupo de Caza at El Plumerillo, Mendoza) from 1988; expected follow-on order for 46 did not materialise, and Argentine Navy requirement for 12 in abeyance, but one new aircraft (E-816), assembled from existing and new components, delivered to Fuerza Aérea Argentina on 28 September 1999. Resumption of production announced 29 June 2000 for a further batch of 12, with deliveries beginning in 2003. US$230 million contract also includes upgrade of 12 existing Pampas. Argentine Navy interest renewed in 2000, with initial batch of between eight and 12 in prospect.

'New-generation' Pampa NG revealed late 1997 and offered to Argentine Air Force and Navy. No orders had been reported by late 2000.
CURRENT VERSIONS: **Pampa:** Standard Argentine Air Force version; last described in 1998-99 *Jane's*.
Naval version: Strengthened landing gear, uprated engine and some changed avionics.
Pampa 2000 International: Version of first-generation Pampa proposed for, but later eliminated from, USAF/USN JPATS trainer programme. Details in 1997-98 *Jane's* and earlier editions.
Pampa NG A: Proposed advanced trainer with updated avionics.
Pampa NG B: Proposed combat-capable version; uprated engine and avionics.
Following description applies to Pampa NG:
DESIGN FEATURES: Enhanced close support capabilities and extended range. New engine, underwing fuel tanks, digital nav/attack system and new avionics. Intended for cost-effective pilot training in mission management techniques, advanced fighter lead-in training and extended-range anti-drug patrol missions. High degree of commonality with the original IA 63 Pampa, providing a customised low life cycle cost fleet.

Non-swept shoulder-mounted wings and anhedral tailplane, sweptback fin and rudder; single engine with twin lateral air intakes. Wing section Dornier DoA-7/-8 advanced transonic; thickness/chord ratio 14.5 per cent at root, 12.5 per cent at tip; anhedral 3°.
FLYING CONTROLS: Conventional hydraulically powered ailerons, rudder, all-moving tailplane, single-slotted Fowler flaps, and door-type airbrake on each side of upper rear fuselage; primary surfaces have Liebherr tandem actuators and electromechanical trim.
STRUCTURE: Conventional all-metal semi-monocoque/stressed skin; two-spar wing box forms integral fuel tank.

'First-generation' IA 63 Pampa tandem-seat trainer *(Argentine Air Force)* 2000/0048258

Pampa NG B has reinforced structure for +7/–3 *g* load factors in clean configuration. Tailored fatigue spectrum based on low-altitude ground attack missions providing up to 10,000 flying hours based on MIL-A-8860 A criteria. Modified fuselage nosecone, tailcone and fin-tip.
LANDING GEAR: SHL (Israel) retractable tricycle type, with hydraulic extension/retraction and emergency free-fall extension. Oleo-pneumatic shock-absorbers. Single Messier-Bugatti wheel on each unit with Goodrich (main) or Continental (nose) low-pressure tyre; nosewheel offset 10 cm (3.9 in) to starboard. Tyre sizes 6.50-10 (10 ply) on mainwheels, 380×150 (4/6 ply rating) on nosewheel, with respective pressures of 6.55 bars (95 lb/sq in) and 4.00 bars (58 lb/sq in). Nosewheel retracts rearward, mainwheels inward into underside of engine air intake trunks. Messier-Bugatti mainwheel hydraulic disc brakes incorporate anti-skid device; nosewheel steering (±47°). Gear designed for operation from unprepared surfaces. Pampa NG B has reinforced gear, including new shock-absorbers, tyres and more powerful brakes.
POWER PLANT: One 15.57 kN (3,500 lb st) Honeywell TFE731-2-2N turbofan (Pampa NG A) or 18.9 kN (4,250 lb st) TFE731-40R (Pampa NG B) installed in rear fuselage. Single-point pressure refuelling, plus gravity point in upper surface of each wing. Standard internal fuel capacity of 968 litres (255 US gallons; 213 Imp gallons) in integral wing tank of 550 litres (145 US gallons; 121 Imp gallons) and 418 litre (110 US gallon; 92.0 Imp gallon) flexible fuselage tank with a negative *g* chamber permitting up to 10 seconds of inverted flight. Additional 415 litres (109 US gallons; 91.0 Imp gallons) can be carried in auxiliary tanks inside outer wing panels, to give a maximum internal capacity of 1,383 litres (364 US gallons; 304 Imp gallons). Provision on centre underwing stations for two external drop tanks, each of 317 litres (83.7 US gallons; 69.7 Imp gallons) usable capacity is optional on Pampa NG A. Total fuel capacity 2,017 litres (533 US gallons; 444 Imp gallons).
ACCOMMODATION: Tandem, rear seat elevated, on UPC (Stencel) S-III-S3IA63 zero/zero ejection seats. Ejection procedure can be preselected for separate single ejections, or for both seats to be fired from front or rear cockpit. Dual controls standard. One-piece wraparound windscreen. One-piece canopy, with internal screen, is hinged at rear and opens upward. Entire accommodation pressurised and air conditioned.
SYSTEMS: Honeywell environmental control system, maximum differential 0.30 bar (4.4 lb/sq in), supplied by high- or low-pressure engine bleed air, provides a 1,980 m (6,500 ft) cockpit environment up to flight level 5,730 m (18,800 ft) and also provides ram air for negative *g* system and canopy seal. Oxygen system supplied by 10 litre (0.35 cu ft) lox converter. Engine air intakes anti-iced by engine bleed air.

Two independent hydraulic systems, each at pressure of 207 bars (3,000 lb/sq in), each supplied by engine-driven pump. Each system incorporates a bootstrap reservoir pressurised at 4 bars (58 lb/sq in). No. 1 system, with flow rate of 16 litres (4.2 US gallons; 3.5 Imp gallons)/min, actuates primary flight controls, airbrakes, landing gear and wheel brakes; No. 2 system, with flow rate of 8 litres (2.1 US gallons; 1.75 Imp gallons)/min, actuates primary flight controls, wing flaps, emergency and parking brakes, and nosewheel steering. Honeywell ram air turbine provides emergency hydraulic power for No. 2 system if

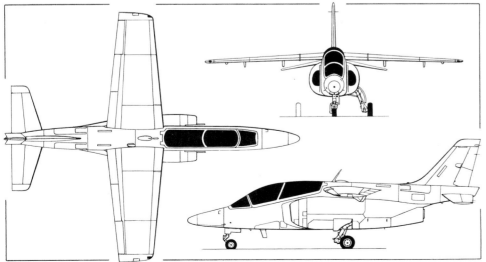

LMAASA IA 63 Pampa NG two-seat basic and advanced jet trainer *(Dennis Punnett/Jane's)*

engine shuts down in flight and pressure in this system drops below minimum.

Electrical system (28 V DC) supplied by Lear Siegler 400 A 11.5 kW engine-driven starter/generator; secondary supply (115/26 V AC power at 400 Hz) from two Flite-Tronics 450 VA static inverters and two SAFT 27 Ah Ni/Cd batteries. Thirty minutes emergency electrical power available in case of in-flight engine shutdown.

AVIONICS: Common avionics core for both Pampa NG variants.

Comms: Both variants have two VHF/UHF transceivers, intercom and optional IFF or ATC transponder.

Flight: Both variants have two VOR/ILS with marker beacon receiver, AHRS, HSI, ADF, altitude indicator, ELT and GPS; optional DME or Tacan.

Instrumentation: HUD (front cockpit), video repeater (rear cockpit), weapon aiming computer in both variants; multifunction displays in NG B.

Mission: Weapon management and inertial navigation systems in both variants. Radar altimeter, laser rangefinder, radar warning system, video camera, HOTAS controls and chaff/flare dispensers in NG B.

ARMAMENT: Pampa NG B has seven stations for external stores, stressed for 450 kg (1,014 lb) on centre fuselage and each inboard underwing station, 300 kg (661 lb) on each central underwing station and 170 kg (375 lb) on each outboard underwing station, all at +5.5/−2 g. Several external stores configurations including provision for AAMs and ASMs.

DIMENSIONS, EXTERNAL:
Wing span	9.69 m (31 ft 9½ in)
Wing aspect ratio	6.0
Length overall	10.90 m (35 ft 9¼ in)
Height overall	4.29 m (14 ft 1 in)
Tailplane span	4.58 m (15 ft 0⅓ in)
Wheel track	2.66 m (8 ft 8¾ in)
Wheelbase	4.42 m (14 ft 6 in)

AREAS:
Wings, gross	15.63 m² (168.3 sq ft)
Ailerons (total)	0.89 m² (9.58 sq ft)
Trailing-edge flaps (total)	2.93 m² (31.54 sq ft)
Fin	1.86 m² (20.02 sq ft)
Rudder	0.655 m² (7.05 sq ft)
Tailplane	4.35 m² (46.87 sq ft)

WEIGHTS AND LOADINGS (estimated, Pampa NG B):
Weight empty	2,980 kg (6,570 lb)
Max fuel weight:	
(A) fuselage	327 kg (721 lb)
(B) wings, internal inboard	430 kg (948 lb)
(C) wings, internal outboard	325 kg (717 lb)
underwing drop tanks	496 kg (1,093 lb)
Max external stores load with normal internal fuel	2,000 kg (4,409 lb)
Max T-O weight: clean	
normal internal fuel (A + B)	3,907 kg (8,613 lb)
max internal fuel (A + B + C)	4,232 kg (9,330 lb)
with external stores	6,000 kg (13,227 lb)
Typical landing weight (50% normal internal fuel)	3,530 kg (7,782 lb)
Max wing loading:	
at clean T-O weight:	
normal internal fuel (A + B)	250 kg/m² (51.20 lb/sq ft)
max internal fuel (A + B + C)	271 kg/m² (55.50 lb/sq ft)
at max T-O weight with external stores and normal internal fuel (A + B)	384 kg/m² (78.64 lb/sq ft)
Max power loading:	
at clean T-O weight:	
normal internal fuel (A + B)	207 kg/kN (2.03 lb/lb st)
max internal fuel (A + B + C)	224 kg/kN (2.20 lb/lb st)
at max T-O weight with external stores and normal internal fuel (A + B)	317 kg/kN (3.11 lb/lb st)

PERFORMANCE (estimated at clean T-O weight with normal internal fuel, except where indicated):
Max level speed at S/L	445 kt (825 km/h; 512 mph)
Approach speed at S/L, 50% normal internal fuel, flaps down	110 kt (204 km/h; 127 mph)
Stalling speed at S/L, 50% normal internal fuel:	
flaps up	104 kt (193 km/h; 120 mph)
flaps down	84 kt (156 km/h; 97 mph)
Max rate of climb: at S/L	1,920 m (6,300 ft)/min
at 4,575 m (15,000 ft)	1,440 m (4,724 ft)/min
Service ceiling	12,900 m (42,320 ft)
T-O run, clean	420 m (1,380 ft)
Landing run, clean, 50% normal internal fuel	480 m (1,575 ft)

Radius of action:
air-to-air (hi-hi), T-O weight of 4,387 kg (9,672 lb) with 477 kg (1,052 lb) external load, 5 min allowance for dogfight, normal internal fuel, 30 min reserves 260 n miles (481 km, 299 miles)
air-to-ground (hi-lo-lo-hi), 30 n miles dash out/in, T-O weight of 5,615 kg (12,379 lb) with 1,410 kg (3,109 lb) external load, max internal fuel, 5 min allowance for weapon delivery, plus reserves 270 n miles (500 km, 310 miles)

Range at 320 kt (593 km/h; 368 mph) at 10,670 m (35,000 ft), ISA with reserves:
normal internal fuel 725 n miles (1,342 km, 834 miles)
max internal fuel 1,090 n miles (2,018 km; 1,254 miles)

Endurance at 270 kt (500 km/h; 312 mph) at 9,450 m (31,000 ft), ISA with reserves:
normal internal fuel	3 h 10 min
max internal fuel	4 h 25 min
max total fuel	5 h 55 min
g limits: clean	+7/−3
with external stores	+5.5/−2

UPDATED

AUSTRALIA

AEROSPORT

AEROSPORT PTY LTD
PO Box 630, Oakbank, South Australia 5243
Tel: (+61 8) 83 88 43 49 and 83 88 47 47
Fax: (+61 8) 83 88 46 44
e-mail: supapup@academy.net.au
Web: http://www.users.setnet.com.au
MANAGING DIRECTOR: John Cotton

Early versions (Mks 1, 2 and 3) of SupaPup hand-built at Hahndorf, South Australia, until cost of compliance with local airworthiness directives forced closure of company. Aerosport redesigned aircraft as Mk 4 in 1992 and launched production line for kits.

UPDATED

AEROSPORT SUPAPUP Mk 4
TYPE: Single-seat kitbuilt.
PROGRAMME: Aerosport developed and produced three earlier versions of the SupaPup from the mid-1980s onwards; SupaPup Mk 4 first flown in early 1998.
COSTS: Kit, less engine A$17,000 (2000).

DESIGN FEATURES: Strut-braced, high-wing cabin monoplane; wings fold rearwards for storage or transportation without disconnection of fuel lines or control runs. Quoted build time 300 hours.
Wing dihedral 2°.
FLYING CONTROLS: Conventional and manual.
STRUCTURE: Welded steel tube fuselage covered in Stitts fabric; glass fibre engine cowling and wingtips.
LANDING GEAR: Tailwheel type; fixed.
POWER PLANT: One 34.0 kW (45.6 hp) Rotax 503 UL-1V two-cylinder two-stroke or 40.3 kW (54 hp) Jabiru 1600 four-cylinder four-stroke. Jabiru 2200 and Rotax 582 and 618 also suitable. Fuel contained in wing tanks, combined capacity 54 litres (14.3 US gallons; 11.9 Imp gallons).
EQUIPMENT: Optional cargo pod under development.

DIMENSIONS, EXTERNAL:
Wing span	7.925 m (26 ft 0 in)
Length overall	5.70 m (18 ft 8½ in)

DIMENSIONS, INTERNAL:
Cabin max width	0.58 m (1 ft 10¾ in)

AREAS:
Wings, gross	9.50 m² (102.3 sq ft)

WEIGHTS AND LOADINGS:
Weight empty	180 to 200 kg (397 to 441 lb)

Aerosport SupaPup single-seat kitbuilt 2001/0087863

Baggage capacity	20 kg (44 lb)
Max T-O weight	340 kg (750 lb)

PERFORMANCE:
Never-exceed speed (V_NE)	130 kt (240 km/h; 149 mph)
Max level speed	100 kt (185 km/h; 115 mph)
Cruising speed	95 kt (176 km/h; 109 mph)
Stalling speed	35 kt (65 km/h; 41 mph)
Max rate of climb at S/L	335 m (1,100 ft)/min
T-O to 15 m (50 ft)	200 m (660 ft)
Endurance with 10% reserves	4 h 30 min

UPDATED

BAE

BAE SYSTEMS AUSTRALIA LIMITED
14 Park Way, Technology Park, Mawson Lakes, South Australia 5095
Tel: (+61 8) 82 90 88 88
Fax: (+61 8) 82 90 88 00
HAWK PROJECT DIRECTOR: Nigel Whitehead

SUBSIDIARY COMPANY:
Hunter Aerospace Corporation Pty Ltd
94 Park Avenue, Kotara, New South Wales 2289
Tel: (+61 2) 49 52 69 00
Fax: (+61 2) 49 57 06 68

Already established as a defence equipment manufacturer, BAE has been selected to assemble and, thereafter, maintain (at least until 2006, under present contract) Hawk Mk 127 lead-in fighter trainers for the RAAF. The company's 4,070 m² (43,800 sq ft) Lead-in Fighter Support Facility was officially opened at Williamtown RAAFB on 15 April 1999 and in late September (although officially celebrated on 28 October) received kit for first (10th RAAF overall) of 21 aircraft it is assembling from UK-supplied components. This first flew (A27-010) on 12 May 2000.

A further 12 Hawks delivered complete, of which first (A27-001) flew at Warton, UK, on 16 December 1999 and was delivered to Australia by airfreight on 6 March 2000.

Australian Hawk A27-010 making its maiden flight at Williamtown in May 2000 2001/0067903

Direct deliveries began on 24 August 2000 with departure of two aircraft from UK. Assembly is delegated to BAE-owned Hunter Aerospace Corporation Pty Ltd. Associated contractors include Airflite, Hawker de Havilland and Qantas, with major component manufacture for tailplanes, weapon pylons, airbrakes and flaps being in Melbourne, Australia. Local involvement accounts for some A$400 million of the A$850 million programme cost. BAE Hawk is described in the United Kingdom section. Last Australian Hawk will be delivered in July 2001.

UPDATED

EAGLE

EAGLE AIRCRAFT PTY LTD

Lot 700, Cockburn Road, Henderson, Western Australia 6166
POSTAL ADDRESS: PO Box 586, Fremantle, Western Australia 6959
Tel: (+61 8) 94 14 11 74
Fax: (+61 8) 94 14 11 75
e-mail: sales@eagleair.com.au
CEO: Nor Manshor Gafar
MARKETING MANAGER: Ron Scherpenzeel
MARKETING EXECUTIVE: Barbara Moyser

US OFFICE:
Eagle Aircraft North America Inc
1900 Hillcrest Street, Orlando, Florida 32803, USA
Tel: (+1 407) 894 53 36
Fax: (+1 407) 894 53 90
Web: http://www.eagleairusa.com

UPDATED

EAGLE AIRCRAFT EAGLE 150

TYPE: Two-seat lightplane.
PROGRAMME: Launched 1981 with objective of producing first all-composites light aircraft in Australia; single-seat POC (proof-of-concept) aircraft now displayed at Power House Museum, Sydney; construction of two-seat preproduction prototype Eagle X started fourth quarter 1987; first flight (VH-XEG) second quarter 1988, with 58 kW (78 hp) Aeropower engine, replaced later by 74.5 kW (100 hp) Continental O-200; 200 hour test programme, meeting all original design criteria, completed by October 1988; plans for 1989 production start aborted (see 1991-92 *Jane's*); component manufacture began March 1991; production prototype (VH-XEP) made first flight 6 November 1992; weight-restricted certification by Australian CASA 21 September 1993; European JAR-VLA certificate not awarded at that time and aircraft consequently redesigned with changes including IO-240-B engine; increased span and chord on foreplane and mainplane flaps; redesigned and repositioned wing cuffs, with leading-edge extension outboard; vortex generators and repositioned horizontal stabiliser.

Series production in Australia, as Eagle X-TS, launched August 1993; first flight (VH-AHH) 23 October 1993 and first customer delivery December 1993 (VH-FPO to Department of Conservation and Land Management in Western Australia). Initial production in Australia only, but using some components manufactured by Eagle Aircraft (Malaysia) of which Eagle Aircraft Pty Ltd is a wholly owned subsidiary. A company announcement in April 1998 indicated that it would be 'at least five years' before there is a full Malaysian assembly line.

First Series 150 (converted production prototype VH-XEP) rolled out August 1997; certified on 13 November 1997 by CASA (Civil Aviation Safety Authority of Australia) to JAR-VLA standards at MTOW of 640 kg (1,411 lb); this certification also valid in Malaysia. FAA certification achieved 11 February 1999, followed by New Zealand approval in mid-1999; JAA certification expected in 2000.

CURRENT VERSIONS: **Series 100:** Retrospective designation of former Eagle X-TS: production completed. Total of 10 built, of which five converted or undergoing conversion to Series 150 by late April 1998. Described in Malaysia section in 1998-1999 *Jane's*.

Series 150: *As described.* Current production version, designated **150A** with IO-240-A engine and **150B** with IO-240-B. Engines have same maximum power of 93.2 kW (125 hp), but B version has increased mid-range power of almost 29 per cent to 69.4 kW (93 hp) and is 3.7 per cent quieter than A model. Available in Basic, Sports and Training variants. US promotion devoted to 150B version only.

Executive: Luxury version with leather upholstery, moving-map GPS and night VFR avionics package.

RPV: Remotely piloted surveillance version, to be developed by BAE Systems of California, which will fit sensors and develop and integrate RPV equipment.

CUSTOMERS: Ten Series 100/Eagle X-TS built, including two for Malaysia (c/n 0003 and 0005) and one for John Roncz in USA (c/n 0010): five conversions (c/n 0002/3/5/7/8) to Series 150.

Total of 25 Series 150s sold by April 2000, comprising: Aerostaff (one), Australian Flying Training School (one), DTIL New Zealand (one), Guernsey Aviation USA (two), HGL Aero (four), Horizon Airways (one), International Aviation and Travel Academy USA (two), Malacca Flying Club Malaysia (three), Perak Aero Club Malaysia (one), Phoenix Aviation (two), Royal Queensland Aero Club (two), Tanjung Flying Club (one), Troy Aviation (one) and Victoria Civil Aviation Academy (three).

COSTS: Basic Series 150 A\$150,000 (US\$104,900), Sports US\$119,900, Training US\$121,400 (all 1999). Basic Eagle 150B US\$121,200, excluding avionics and wheel fairings (2000). Total maintenance cost US\$9.98 per hour over 2,000 hours' utilisation (2000).

DESIGN FEATURES: Intended primarily for *ab initio* training, recreational flying and surveillance. 'Tri-surface' configuration, with high-mounted mainplane, large low-mounted foreplane, and tailplane. Stall strips on foreplane ensure that it stalls before the mainplane.

Mainplane of tailored Roncz aerofoil section, thickness/chord ratio 16 per cent; no sweep; no dihedral.

FLYING CONTROLS: Conventional and manual. Slotted ailerons on mainplane, elevators on tailplane, and rudder, all with normal manual/mechanical actuation. Pushrods on ailerons and elevators, cables on rudder. Electric pitch trim (tab in starboard elevator), manual roll trim; rudder tab. Electrically actuated single-slotted flaps on foreplane (full span) and mainplane (part span). Leading-edge stall strips, vortex generators and fences on mainplane, stall strips on foreplane. Flap and air flow control systems designed to achieve low stalling speed with relatively high wing loading, to provide good ride quality in turbulence.

STRUCTURE: Except for metal engine mounts and flight control rods, Model 150 built entirely of composites, with 90 per cent of structure bonded/assembled by Composites Technology Research Malaysia (CTRM). Wings, fuselage and all control surfaces are Nomex honeycomb or high-density foams, sandwiched between multiple layers of carbon fibre; Kevlar reinforcement around wing leading-edges and shoulder of mainplane; cockpit is impact-resistant capsule of multilayered Kevlar and carbon fibre; carbon fibre spars. Entire structure uses Eagle-designed vinylester resins for strength/longevity/impact resistance and to minimise 'wet environment' problems inherent in standard epoxies.

LANDING GEAR: Non-retractable tricycle type, with glass fibre/epoxy self-sprung main legs and oleo nose leg. Cleveland 5.00-5 wheel on each unit, with optional speed fairings. Cleveland hydraulic single-disc brakes on mainwheels. Castoring nosewheel, size 11 × 4.00. Minimum ground turning radius 5.21 m (17 ft 1 in). Brakes applied by pressure on rudder pedals. Twin landing lights in nosewheel fairing.

POWER PLANT: One 93.2 kW (125 hp) Teledyne Continental IO-240-A or IO-240-B7B flat-four engine (see Current Versions above); respectively driving a McCauley IA135BRM7054 or IA135CRM7057 two-blade fixed-pitch metal propeller. Fuel capacity 100 litres (26.4 US gallons; 22.0 Imp gallons), of which 97 litres (25.6 US gallons; 21.3 Imp gallons) are usable.

ACCOMMODATION: Two seats side by side with Y-shape control column operable from either seat. Bubble canopy hinged at front and opens upward. Hat shelf and two baggage compartments.

SYSTEMS: Hydraulic system for manual brake actuation only; 12 V DC electrical system with 60 A alternator.

AVIONICS: *Comms:* Honeywell KY 97A VHF radio; Sigtronics SPA 400 intercom; Honeywell KT 76A transponder; E-01 ELT. Optional Garmin GTX 320 or GTX 327 transponder in sports and training packages.

Flight: Garmin GNS 250XL GPS in Sports model package, moving-map GPS in Executive package, ADF in Training package.

Instrumentation: Sports and Training packages add turn co-ordinator, artificial horizon, directional gyro, clock and flight hour meter.

EQUIPMENT: Anti-collision beacon in Training model package.

DIMENSIONS, EXTERNAL:
Wing span	7.16 m (23 ft 6 in)
Wing chord, constant	0.74 m (2 ft 5 in)
Wing aspect ratio	9.9
Foreplane span	4.88 m (16 ft 0 in)
Foreplane chord, constant	0.74 m (2 ft 5 in)
Length overall	6.45 m (21 ft 2 in)
Height overall	2.31 m (7ft 7 in)
Tailplane span	3.30 m (10 ft 10 in)
Wheel track	1.93 m (6 ft 4 in)
Wheelbase	1.40 m (4 ft 7 in)
Propeller diameter	1.78 m (5 ft 10 in)

DIMENSIONS, INTERNAL:
Cabin: Length	1.37 m (4 ft 6 in)
Max width	1.07 m (3 ft 6 in)
Max height	0.86 m (2 ft 10 in)

AREAS:
Wings, gross	5.20 m^2 (56.0 sq ft)
Foreplanes, gross	3.62 m^2 (39.0 sq ft)
Wing flaps (total)	0.90 m^2 (9.68 sq ft)
Foreplane flaps (total)	0.90 m^2 (9.68 sq ft)
Tailplane	1.49 m^2 (16.00 sq ft)

WEIGHTS AND LOADINGS:
Weight empty	429 kg (946 lb)

Eagle 150 instrument panel

Eagle Series 150 (Teledyne Continental IO-240 flat-four engine) *(Paul Jackson)*

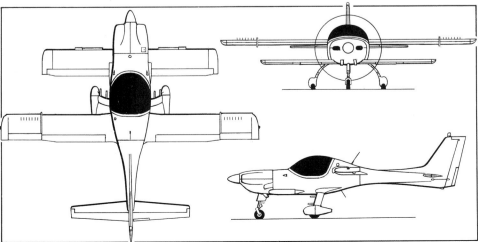

Eagle 150 two-seat light aircraft *(Mike Keep/Jane's)*

AUSTRALIA: AIRCRAFT—EAGLE to GA

Baggage capacity: shelf		9 kg (20 lb)
compartments (total)		36 kg (80 lb)
Max T-O weight		648 kg (1,430 lb)
Max wing/foreplane loading		73.5 kg/m² (15.05 lb/sq ft)
Max power loading		6.96 kg/kW (11.44 lb/hp)

PERFORMANCE:
Max level speed at S/L   130 kt (241 km/h; 150 mph)
Cruising speed at 75% power at 660 m (2,000 ft)   125 kt (232 km/h; 144 mph)
Stalling speed at S/L: flaps up   52 kt (97 km/h; 60 mph)
    flaps down   43 kt (80 km/h; 50 mph)
Max rate of climb at S/L   322 m (1,055 ft)/min
Service ceiling   4,575 m (15,000 ft)
T-O to 15 m (50 ft)   349 m (1,145 ft)
Landing from 15 m (50 ft)   365 m (1,200 ft)
Range with max fuel   520 n miles (963 km; 598 miles)
Endurance at 60% power   5 h
g limits: JAR-VLA   +3.8/−1.9
    ultimate   +8.55/−4.27

UPDATED

EXPLORER

EXPLORER AIRCRAFT CORPORATION PTY LTD

Explorer Aircraft Corporation Pty Ltd (formerly Aeronautical Engineers Australia Research Pty Ltd), which has developed the Explorer utility aircraft, has relocated to Denver, Colorado, USA. See entry in the US section.

UPDATED

GA

GIPPSLAND AERONAUTICS PTY LTD

Latrobe Valley Airfield, PO Box 881, Morwell, Victoria 3840
Tel: (+61 3) 51 74 30 86
Fax: (+61 3) 51 74 09 56
DIRECTORS:
    George Morgan
    Peter Furlong
CHIEF DESIGNER: Colin Nicholson

Involved since 1971 in design and modification programmes for wide range of aircraft, from wooden homebuilts to pressurised turboprops. Conversion of five Piper Pawnees to two-seat configuration in second half of 1980s led eventually to a totally new design, the GA-200 cropsprayer. Since 1997, Gippsland has been seeking partners in Asia and South America for licensed production of the GA-200. Second Gippsland aircraft, GA-8 Airvan, first flew in 1995. First deliveries, originally due in 1999, will follow certification in late 2000.

UPDATED

GIPPSLAND GA-200 FATMAN

TYPE: Agricultural sprayer.
PROGRAMME: Precursor was two-seat conversion of five Piper PA-25-235 Pawnees to -235/A8 or -235/A9 Fatman standard in 1986-90. Prototype GA-200 (VH-BCE) registered 1991; this and second aircraft used airframes from damaged Pawnees; No 3 (VH-SKG) was first new-build, in 1992. Designation derived from hopper capacity in US gallons. Full Australian CAA certification in both Normal and Agricultural categories (to CAO 101.16 and 101.22 and FAR Pt 23 Amendment 23.36) awarded 1 March 1991; FAR Pt 23 certification awarded 15 October 1997 in Restricted category. Certification also achieved in Brazil, Canada and elsewhere. Production at Traralgon, Victoria, since 1993.

Higher gross weight version powered by engines in the 186 to 224 kW (250 to 300 hp) class and with 800 to 1,060 litre (211 to 280 US gallon; 176 to 233 Imp gallon) hopper, available from 1998.

CURRENT VERSIONS: **GA-200**: Initial version. Hopper capacity 776 litres (205 US gallons; 170 Imp gallons). Replaced by GA-200C.
  GA-200 Ag-trainer: Training version, with dual controls, dual rudder pedals and smaller hopper.
  GA-200B: Certified in Brazil; first, from US assembly, exported 1999.
  GA-200C: In early 1998, 21st production (23rd overall) aircraft completed as prototype GA-200C; all subsequent aircraft to this standard. US certification pending in mid-2000.
Description applies to GA-200C.

Gippsland GA-200 Fatman agricultural aircraft

CUSTOMERS: Total of about 50 orders received by October 1997, of which 41 (including prototypes) registered by February 2000; exports to China (nine), New Zealand (seven, beginning 1994), South Africa (one, in March 1999) and USA (five, assembled locally from Australian-manufactured airframe components and US engines, wheels, brakes, instruments and avionics, plus at least two more, exported in early 2000).
COSTS: A$224,600 (1995).
DESIGN FEATURES: Purpose-designed cropsprayer. Braced low wing, with large integral hopper forward of cockpit; crash-resistant, corrosion-proofed structure; gap-sealed ailerons; detachable wingtips. Although initially based on Piper Pawnee, GA-200 in current form is substantially different. Wing dihedral 7° from roots. Flaps and ailerons non-handed.
FLYING CONTROLS: Conventional and manual. Single-slotted trailing-edge wing flaps can be deployed to tighten turning radius during agricultural operations; T-O setting 15°, maximum 38°. Interconnect system applies bias to elevator trim spring when flaps extended, to avoid pitch trim changes.
STRUCTURE: Fuselage of welded SAE 4130 chromoly steel tube with removable metal side panels; wings (braced by overwing inverted V strut of aluminium each side) and tail surfaces conventional all-metal, but wing spars constructed from sheet metal to expedite repairs; wingtips detachable.
LANDING GEAR: Non-retractable, with 6 in diameter Cleveland P/N 40-84A mainwheels mounted on tubular steel side Vs; rubber cord shock-absorption with hydraulic dampers and Cleveland hydraulic disc brakes. Mainwheel tyres size 8.50 × 6 (6 ply). Scott 3200 steerable/castoring tailwheel, mounted on multileaf flat springs.
POWER PLANT: One 224 kW (300 hp) Textron Lycoming IO-540-K1A5 flat-six engine, driving an HC-C2YR-1BF/F8475R metal propeller. Fuel in integral tank in each wing, combined usable capacity 200 litres (52.8 US gallons; 44.0 Imp gallons), plus small 14 litre (3.7 US gallon; 3.1 Imp gallon) header tank in upper front fuselage. Oil capacity 11.4 litres (3.0 US gallons; 2.5 Imp gallons).
ACCOMMODATION: Two energy-absorbing seats, side by side (right-hand seat for loader/driver); dual controls and second set of rudder pedals for right-hand seat in Ag-trainer. Four-point 25 g restraint harness(es). Cockpit doors open upward, each with bubble window to improve shoulder room and outward view.
SYSTEMS: 14 V 55 A automotive alternator and automotive or R-35 aviation battery for electrical power; 50 A circuit breaker switch serves as master switch.
AVIONICS: *Instrumentation:* Basic VFR instruments only.
EQUIPMENT: One 1,045 litre (276 US gallon; 230 Imp gallon) capacity hopper in forward fuselage (approximately 76 litres; 20.0 US gallons; 16.7 Imp gallons less in Ag-trainer). Multirole door/hopper outlet, eliminating need to change outlet when changing between solids and liquids, can also be used for fire bombing or laying of fire retardants. Spreader vanes can be added to increase swath width. 100 W landing light in each wingtip; 28 V night working lights system (two retractable underwing 600 W lights, powered by separate 28 V 55 A alternator) also available.
DIMENSIONS, EXTERNAL:
  Wing span, standard version   11.985 m (39 ft 3¾ in)

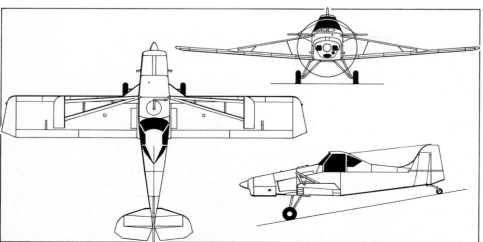

Gippsland GA-200 Fatman in standard two-seat form *(Mike Keep/Jane's)*

Wing chord, constant	1.60 m (5 ft 3 in)
Wing aspect ratio	7.3
Length overall (flying attitude)	7.48 m (24 ft 6½ in)
Height (static) over cockpit canopy	2.33 m (7 ft 7¾ in)
Tailplane span	2.90 m (9 ft 6¼ in)
Wheel track	2.335 m (7 ft 8 in)
Propeller diameter	2.13 m (7 ft 0 in)

AREAS:
Wings, gross 19.60 m² (211.0 sq ft)

WEIGHTS AND LOADINGS (GA-200C, agricultural):
Operating weight empty 770 kg (1,698 lb)
Max T-O weight 1,995 kg (4,400 lb)
Max wing loading 101.8 kg/m² (20.85 lb/sq ft)
Max power loading 8.92 kg/kW (14.67 lb/hp)

PERFORMANCE (GA-200C, agricultural):
Long-range cruising speed (clean)
  110 kt (204 km/h; 127 mph)
Stalling speed at 1,656 kg (3,650 lb)
  flaps up 57 kt (106 km/h; 66 mph) IAS
  flaps down 51 kt (95 km/h; 59 mph) IAS
Max rate of climb (clean) at S/L 149 m (490 ft)/min

UPDATED

Second prototype Gippsland GA-8 Airvan *(Paul Jackson/Jane's)*

GIPPSLAND GA-8 AIRVAN

TYPE: Light utility transport.

PROGRAMME: Design completed and prototype construction started early 1994; first flight 3 March 1995 (aircraft unregistered, but marked 'GA-8' on the fin); 186 kW (250 hp) Textron Lycoming O-540; prototype destroyed 7 February 1996 during spinning trials; second airframe completed to major component stage for static testing; third airframe (also unmarked; 'AIRVAN' on fin; VH-ZGI not worn) was second flying prototype, first flown in August 1996 with an interim 224 kW (300 hp) Textron Lycoming IO-540K-1A5 engine and three-blade propeller, pending installation of intended (but temporarily discontinued) production standard IO-580 in 1997; total of 350 hours flight testing completed by two prototypes by November 1998; second prototype re-registered VH-XGA in January 1999; provisional type certification achieved in early 1999, with full certification to FAR Pt 23 then scheduled for third quarter of 1999, but not achieved until 18 December 2000. First delivery (VH-RYT) to Fraser Island Air 22 December 2000.

CURRENT VERSIONS: **GA-8 Airvan:** Initial production version; IO-540 engine. To be recertified subsequently with IO-580. Hot-and-high subvariant under consideration would have turbocharged engine.

Turboprop: Stretched, 11-seat, version; Rolls-Royce 250-C20 engine with Soloy gearbox. Proposed 1999.

CUSTOMERS: Eight sold by February 2001, including exports to Canada and South Africa.

COSTS: A$500,000 (1998).

DESIGN FEATURES: Strut-braced, high-wing monoplane with sweptback vertical tail and fixed tricycle landing gear; designed to operate from unprepared strips. Fin and rudder modified, and ventral finlet added on second prototype, late 1996.
  Wing aerofoil modified V-35; dihedral 2°; incidence 2°; twist 1.6°.

FLYING CONTROLS: Conventional and manual. Trimmable tailplane; plain flaps, maximum deflection 55°.

STRUCTURE: Light alloy; two-spar wing based on GA-200 Fatman unit; entire structure designed for easy manufacture, maintenance and repair.

LANDING GEAR: Non-retractable tricycle type; single-piece tubular spring main gear, steel spring/oleo nose leg; Cleveland hydraulic brakes; mainwheels 8.50-6; nosewheel 6.00-6. Wipline float installation under development.

POWER PLANT: One 224 kW (300 hp) Textron Lycoming IO-540 flat-six engine driving a three-blade constant-speed propeller. Fuel capacity 340 litres (89.8 US gallons; 74.8 Imp gallons) in two wing tanks, of which 335 litres (88.5 US gallons; 73.7 Imp gallons) usable. Oil capacity 10 litres (2.6 US gallons; 2.2 Imp gallons).

ACCOMMODATION: Pilot and up to seven passengers or equivalent cargo; crashworthy seats. Forward-opening door each side of flight deck; flight openable forward-sliding cargo door on port side aft of wing. Large side windows serve as emergency exits.

AVIONICS: *Flight:* Standard VOR, ADF, ILS, DME, GPS.

EQUIPMENT: Quick-change passenger/cargo kit.

DIMENSIONS, EXTERNAL:
Wing span	12.37 m (40 ft 7 in)
Wing chord, constant	1.60 m (5 ft 3 in)
Wing aspect ratio	7.9
Length overall	8.79 m (28 ft 10 in)
Fuselage max width	1.37 m (4 ft 6 in)
Height overall	2.82 m (9 ft 3 in)
Tailplane span	4.11 m (13 ft 6 in)
Wheel track	2.79 m (9 ft 2 in)
Wheelbase	2.31 m (7 ft 7 in)
Cargo/passenger door: Height	1.09 m (3 ft 7 in)
Width	1.07 m (3 ft 6 in)
Height to sill	0.89 m (2 ft 11 in)

DIMENSIONS, INTERNAL:
Cabin: Length, incl flight deck	4.01 m (13 ft 2 in)
Max width	1.27 m (4 ft 2 in)
Max height	1.19 m (3 ft 11 in)
Floor area: incl flight deck	5.02 m² (54.0 sq ft)
excl flight deck	3.44 m² (37.0 sq ft)
Volume, incl flight deck	5.1 m³ (180 cu ft)

AREAS:
Wings, gross	19.32 m² (208.0 sq ft)
Ailerons (total)	0.80 m² (8.60 sq ft)
Trailing-edge flaps (total)	0.80 m² (8.60 sq ft)
Fin	1.35 m² (14.50 sq ft)
Rudder	0.73 m² (7.90 sq ft)
Tailplane	2.41 m² (25.90 sq ft)
Elevators (total)	1.76 m² (18.90 sq ft)

WEIGHTS AND LOADINGS:
Weight empty	862 kg (1,900 lb)
Max fuel	270 kg (595 lb)
Max T-O and landing weight	1,814 kg (4,000 lb)
Max wing loading	93.9 kg/m² (19.23 lb/sq ft)
Max power loading:	
first prototype	9.74 kg/kW (16.00 lb/hp)
production	8.11 kg/kW (13.33 lb/hp)

PERFORMANCE:
Never-exceed speed (V_{NE})	185 kt (342 km/h; 212 mph)
Max cruising speed at 1,525 m (5,000 ft)	130 kt (241 km/h; 150 mph)
Econ cruising speed at 1,525 m (5,000 ft)	120 kt (222 km/h; 138 mph)
Stalling speed: flaps up	59 kt (109 km/h; 68 mph)
flaps down	53 kt (98 km/h; 61 mph)
Max rate of climb at S/L	228 m (750 ft)/min
Service ceiling, estimated	6,100 m (20,000 ft)
T-O run	305 m (1,000 ft)
T-O to 15 m (50 ft)	457 m (1,500 ft)
Landing from 15 m (50 ft)	490 m (1,600 ft)
Landing run, estimated	305 m (1,000 ft)
Range: with max fuel	650 n miles (1,203 km; 748 miles)
with max payload	100 n miles (185 km; 115 miles)

UPDATED

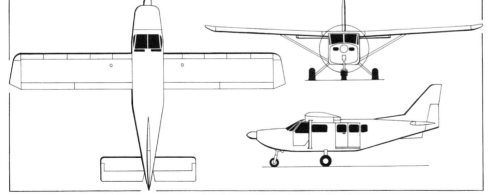

Gippsland GA-8 Airvan (Textron Lycoming IO-540 engine) *(Paul Jackson/Jane's)*

GOAIR

GOAIR PRODUCTS

Hangar 675, Drover Road, Bankstown Airport, New South Wales 2200
Tel: (+61 2) 97 96 34 26
Fax: (+61 2) 97 91 03 54

MANAGING DIRECTOR: Phil Goard

Company machines, welds and manufactures aircraft components. During the 1990s it designed, built and flight-tested the Goair Trainer, production of which is reported to be imminent.

UPDATED

GOAIR GT-1 TRAINER

TYPE: Two-seat lightplane.

PROGRAMME: Four years in design; public debut (VH-BBR, then unflown) at Avalon Air Show, March 1995; first flight July 1995; some 120 hours of flight testing completed by November 1998; first production aircraft was expected to

Prototype Goair Trainer at Avalon in March 1995 *(Gerard Frawley/Australian Aviation)*

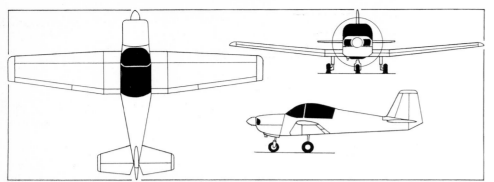

Goair Trainer (118 hp Textron Lycoming O-235) in prototype configuration *(Paul Jackson/Jane's)*

fly in early 1999, but not entered on Australian civil aircraft register until January 2001; further four under construction by late 2000; intended for certification in VLA category.
COSTS: US$70,000 (2001).
DESIGN FEATURES: Conventional low-wing, fixed landing gear monoplane; design goals strength, simplicity, low costs, docile stall characteristics and ease of maintenance. Production aircraft differs from prototype in having cabin 10 cm (4 in) wider; larger ailerons and flaps; and leaf spring main landing gear legs.
FLYING CONTROLS: Conventional and manual.
STRUCTURE: All metal.
LANDING GEAR: Non-retractable tricycle type.
POWER PLANT: One 88 kW (118 hp) Textron Lycoming O-235 flat-four engine, driving a two-blade metal propeller. Fuel capacity 130 litres (34.3 US gallons; 28.6 Imp gallons.)
ACCOMMODATION: Two persons, side by side under rearward-sliding bubble canopy; dual controls standard; seats adjust fore and aft. Baggage compartment behind seats.

DIMENSIONS, EXTERNAL:
Wing span: prototype 8.53 m (28 ft 0 in)
 production 8.69 m (28 ft 6 in)
Wing chord at root 1.52 m (5 ft 0 in)
Wing aspect ratio: prototype 6.9
 production 7.2
Length overall: prototype 6.25 m (20 ft 6 in)
 production 7.16 m (23 ft 6 in)
Height overall: prototype 2.03 m (6 ft 8 in)
 production 2.29 m (7 ft 6 in)
DIMENSIONS, INTERNAL:
Cabin max width 1.14 m (3 ft 9 in)
Baggage compartment volume 0.17 m³ (6.0 cu ft)
AREAS:
Wings, gross 10.50 m² (113.0 sq ft)
WEIGHTS AND LOADINGS:
Weight empty 408 kg (900 lb)
Max T-O weight 748 kg (1,650 lb)
Max wing loading 71.3 kg/m² (14.60 lb/sq ft)
Max power loading 8.51 kg/kW (13.98 lb/hp)
PERFORMANCE:
Max level speed 115 kt (213 km/h; 132 mph)
Cruising speed 100 kt (185 km/h; 115 mph)
Stalling speed 45 kt (84 km/h; 52 mph)
Max rate of climb at S/L 244 m (800 ft)/min
T-O and landing run 138 m (450 ft)
Endurance 5 h

UPDATED

HUGHES

HOWARD HUGHES ENGINEERING PTY LTD
PO Box 89, Lot 8, Southern Cross Drive, Ballina, New South Wales 2478
Tel: (+61 2) 66 86 86 58
Fax: (+61 2) 66 86 83 43
e-mail: alw@spot.com.au
Web: http://www.lightwing.com.au
MANAGING DIRECTOR: Howie Hughes

The Hughes Australian Light Wing series has developed from GR-582 via GA-55 (see 1994-95 *Jane's*) to GR-912 and the associated Sport 2000. Most production, undertaken at Ballina Airport, is for local use and either full certification or registry under AUF rules. Proposed Chinese manufacture of the GR-582 has not taken place (see SFLC in 2000-01 and earlier editions).

A further design, the **Australian Light Wing 180**, with 180 kt (333 km/h; 207 mph) cruising speed, will be launched as a kitbuilt in 2001.

UPDATED

HUGHES AUSTRALIAN LIGHT WING GR-582
TYPE: Side-by-side ultralight/kitbuilt.
PROGRAMME: Prototype Australian Light Wing (ALW) made first flight June 1986; many sold conforming to CAO 95-25 requirements; also produced under homebuilt or CAO 101.28 regulations in non-retractable tailwheel and twin-float configurations. Type certificate issued 1998.
DESIGN FEATURES: Modified Clark Y wing section.
FLYING CONTROLS: Conventional and manual.
STRUCTURE: Strut-braced metal wings; welded steel tube fuselage and tail; all fabric covered except for glass fibre engine cowling.
LANDING GEAR: Non-retractable tailwheel type; optional amphibious floats with retractable mainwheels and non-retractable tailwheels (latter doubling as water rudders); ski gear also optional.
POWER PLANT: One 47.8 kW (64.1 hp) Rotax 582 UL flat-twin engine, with 2.58:1 reduction gear. Fuel capacity 60 litres (15.9 US gallons; 13.2 Imp gallons).

Hughes GR-912-T Sport 2000

Hughes GR-912 floatplane version

DIMENSIONS, EXTERNAL:
Wing span 9.70 m (31 ft 10 in)
Length overall 5.80 m (19 ft 0½ in)
Height overall 1.90 m (6 ft 2¼ in)
AREAS:
Wings, gross 14.55 m² (156.6 sq ft)
WEIGHTS AND LOADINGS:
Weight empty 240 kg (529 lb)
Max T-O weight 480 kg (1,058 lb)
PERFORMANCE:
Max level speed at 915 m (3,000 ft)
 90 kt (167 km/h; 104 mph)
Max cruising speed 75 kt (139 km/h; 92 mph)
Stalling speed, power off 35 kt (65 km/h; 41 mph)
Max rate of climb at S/L 305 m (1,000 ft)/min
T-O run 31 m (100 ft)
Landing run 92 m (300 ft)
Range 300 n miles (555 km; 345 miles)
g limits +6/−4

UPDATED

HUGHES AUSTRALIAN LIGHT WING GR-912
TYPE: Side-by-side ultralight/kitbuilt.
PROGRAMME: Uprated version of the GR-582. Type certificate issued 25 September 1998.
CURRENT VERSIONS: **GR-912**: Tailwheel version. *As described.*
 GR-912-T: Tricycle landing gear version; marketing designation Sport 2000; see following entry.
CUSTOMERS: 150 aircraft (all models) delivered by 1996 (latest information); most registered under AUF rules.

Claimed 29 per cent share of Australian two-seat light trainer market with GR-912.
COSTS: A$85,000 factory built (2000).
DESIGN FEATURES: Modified Clark Y wing section.
FLYING CONTROLS: Conventional and manual. Optional flaps.
STRUCTURE: Strut-braced metal wings; welded steel tube fuselage and tail; tail surfaces are wire-braced; composites and polyfibre covering.
LANDING GEAR: Non-retractable tailwheel type; optional amphibious floats with retractable mainwheels and non-retractable tailwheels (latter doubling as water rudders).
POWER PLANT: One 59.6 kW (79.9 hp) Rotax 912 UL flat-four driving a two- or three-blade propeller. Fuel capacity 60 litres (15.8 US gallons; 13.2 Imp gallons).
DIMENSIONS, EXTERNAL:
Wing span 9.60 m (31 ft 6 in)
Length overall 5.70 m (18 ft 8½ in)
AREAS:
Wings, gross 14.86 m² (160.0 sq ft)
WEIGHTS AND LOADINGS:
Weight empty 295 kg (650 lb)
Max T-O weight 520 kg (1,146 lb)
PERFORMANCE:
Max level speed 95 kt (176 km/h; 109 mph)
Max cruising speed 80 kt (148 km/h; 92 mph)
Loiter speed 50 kt (93 km/h; 58 mph)
Stalling speed, power off 35 kt (65 km/h; 41 mph)
Max rate of climb at S/L 244 m (800 ft)/min
Range with max fuel 216 n miles (400 km; 248 miles)

UPDATED

HUGHES AUSTRALIAN LIGHT WING SPORT 2000
TYPE: Side-by-side ultralight/kitbuilt.
PROGRAMME: Developed version of GR-912 (which see); certified in Australian Primary category. Type certificate issued September 1998.
COSTS: A$86,500 factory built; A$45,900 kit, including alternative 59.6 kW (79.9 hp) Werner engine (2000).
FLYING CONTROLS: Conventional and manual. Horn-balanced tail surfaces; elevator pushrod-operated; aileron and rudder cable-operated.
STRUCTURE: Welded steel tube fuselage; riveted aluminium alloy wings; mainly fabric covered, with non-structural glass fibre fairings.
LANDING GEAR: Non-retractable tricycle type.

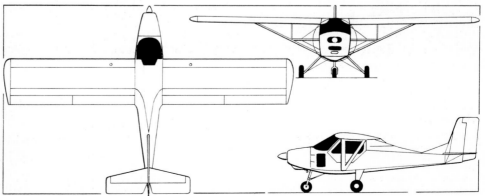

Hughes GR-912 tailwheel version

Hughes Australian Light Wing Sport 2000 (Rotax 912 UL flat-four) *(James Goulding/Jane's)*

HUGHES to MILLICER—AIRCRAFT: AUSTRALIA

POWER PLANT: One 59.6 kW (79.9 hp) Rotax 912 UL flat-four driving a two-blade fixed-pitch wooden propeller. Fuel capacity 62 litres (16.4 US gallons; 13.6 Imp gallons).
DIMENSIONS, EXTERNAL:
Wing span 9.50 m (31 ft 2 in)
Length overall 5.60 m (18 ft 4½ in)
Height overall 2.20 m (7 ft 2½ in)
AREAS:
Wings, gross 15.14 m² (163.0 sq ft)
WEIGHTS AND LOADINGS:
Weight empty 300 kg (661 lb)
Max T-O weight 480 kg (1,058 lb)
PERFORMANCE:
Cruising speed at 75% power 72 kt (133 km/h; 83 mph)
Stalling speed 35 kt (65 km/h; 41 mph)
T-O to 15 m (50 ft) 330 m (1,085 ft)
Range with max fuel 300 n miles (555 km; 345 miles)
g limits +6/−4
UPDATED

HUGHES POCKET ROCKET
TYPE: Tandem seat ultralight/kitbuilt.
PROGRAMME: Offered as PR-582 or PR-618 with Rotax 582 or 618 two-stroke engines, or with Rotax 912 four-stroke.
DESIGN FEATURES: Narrow fuselage version of earlier ALW designs. Quoted kit build time 400 hours.
Description generally as for ALW except the following.
DIMENSIONS, EXTERNAL:
Wing span 7.30 m (23 ft 11½ in)
Length overall 5.60 m (18 ft 4½ in)
AREAS:
Wings, gross 13.47 m² (145.0 sq ft)
PERFORMANCE (Rotax 582):
Cruising speed 80 kt (148 km/h; 92 mph)
Stalling speed 40 kt (75 km/h; 46 mph)
NEW ENTRY

JABIRU

JABIRU AIRCRAFT PTY LTD
PO Box 5186, Bundaberg West, Queensland 4670
Tel: (+61 7) 41 55 17 78
Fax: (+61 7) 41 55 26 69
e-mail: jabiru@tpgi.com.au
Web: http://www.jabiru.net.au
JOINT MANAGING DIRECTOR: Phil Ainsworth

EUROPEAN AGENT:
ST Aviation Ltd, Technology House, High Street, Downham Market, Norfolk PE38 9HH, UK
Tel: (+44 1366) 38 55 58
Fax: (+44 1366) 38 55 59
Web: http://www.jabiru.co.uk

Jabiru formed 1988 by Rodney Stiff and Phil Ainsworth to produce Jabiru LSA 55/2K, but branched into engine design and manufacture when Italian-built KFM engine, which powered this variant, was withdrawn from market. Now produces 1,600 and 2,200 cc flat-four and 3,300 cc flat-six engines for its own and other manufacturers' aircraft. A factory-built SLA version of Jabiru, the UL 450, was launched in 2000. The company and aircraft are named for Australia's sole indigenous stork. Production averages 100 aircraft and 360 engines per year.
UPDATED

JABIRU JABIRU
TYPE: Side-by-side ultralight/kitbuilt.
PROGRAMME: Design of original Jabiru LSA, started 1987; prototype (first of two, VH-JCX and VH-JQX) made first flight late 1989; first customer delivery April 1991; certified to CAO 101.55 on 1 October 1991; 20 built with KFM engine before change to 44.7 kW (60 hp) Jabiru 1600 flat-four; kitplane construction certification achieved in Australia, New Zealand, South Korea, UK and USA by early 1997; marketed in US as amateur-built kit, achieving FAR Pt 21.191 (g) certification on 8 February 1996; marketing office opened at Aiken, South Carolina, in anticipation of substantial US sales of at least 100 per year. Total of 54 built with 1600 engine between April 1993 and March 1996. More powerful ST version introduced in 1994. Joint venture announced in 1998 with CDE Aviation Company of Sri Lanka, subsidiary of Lionair, to establish Jabiru Asia for manufacture of Jabiru ST3 for Asian markets, especially India. Jabiru factory at Koggala, in southern Sri Lanka, will have production capacity for 40 aircraft per year, using engines and avionics imported from Australia, and airframes manufactured locally.
CURRENT VERSIONS: **Jabiru LSA:** Light Sport Aircraft; factory-built; registered under AUF (Australian Ultralight Federation) rules. 'Short span; short fuselage' version:

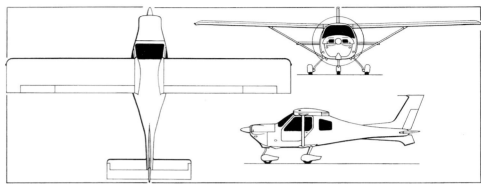
Jabiru ST two-seat ultralight aircraft *(Mike Keep/Jane's)* 0010202

span 8.03 m (26 ft 4 in), length 5.03 m (16 ft 6 in), wing area 7.90 m² (85.0 sq ft).
Jabiru ST: Certified version; in factory production from mid-1994; first was 50th production Jabiru. In mid-1998, the 17th production ST was registered VH-JRU to Jabiru Aircraft as the first **ST3** short span and fuselage.
Jabiru SK: Quick-build kit version of ST (components are manufactured alongside production aircraft); first was 66th production Jabiru; quoted home-build time is 600 hours. First UK kit delivered June 1996; registered G-OJAB for agency, ST Aviation, and first flown 21 October 1996.
Jabiru SP: Combines longer fuselage of UL 450 with short-span wings of SK. MTOW 470 kg (1,036 lb) with Jabiru 2200 engine or 500 kg (1,102 lb) with 3300 engine. V$_{NE}$ 132 kt (244 km/h; 151 mph).
Jabiru SP-T: Tailwheel version of SP. Enlarged rudder; mainwheel legs moved forward; steerable tailwheel and differential braking. Prototype built by Peter Kayne of Narrogin, Western Australia.
Jabiru SP 480: Proposed version of SP for JAR-VLA certification, with maximum take-off weight of 480 kg (1,058 lb); under development in late 1999.
Jabiru UL 450: Factory-built. SLA (Small Light Aeroplane to UK's BCAR Section S), launched 2000 and meeting 36 kt (67 km/h; 42 mph) stalling speed requirement. Long span and fuselage. Australian certification near completion by June 2000; UK certification due shortly thereafter.
Four-seat Jabiru: Prototype, with stretched centre section rolled-out February 2001. Remains in ultralight category when flown with two occupants.
CUSTOMERS: Production of complete and kit aircraft totalled some 450 by early 2001. Sold to 16 countries.

COSTS: £36,000, factory-built UL 450 (2000).
DESIGN FEATURES: Designed to Australian CAO 101.55 and British BCAR Section S, with reference to FAR Pt 23 and JAR-VLA. Ultralight or certified aircraft, suitable for factory or amateur assembly. Unswept wing (braced) and tailplane, swept fin; dorsal and ventral fins; all flying surfaces square-tipped; wings detachable for storage and transportation.
Wing section NASA 4412; drooped wingtips; dihedral 1° 15′; incidence 2° 30′; no twist.
FLYING CONTROLS: Conventional and manual. In-flight adjustable pitch trim; wide-span slotted flaps.
STRUCTURE: All-GFRP except metal wing struts, nosewheel assembly and engine mount.
LANDING GEAR: Non-retractable tricycle type, with 15° steerable nosewheel coupled to rudder. Cantilever self-sprung GFRP mainwheel legs; rubber block nosewheel shock-absorption; wheel fairings standard over 4.00-6 or 5.00-6 tyres; Bigfoot 6.00-6 tyres optional. Nosewheel 2.60-4, 4.00-4 or 5.00-6; tailwheel 250×50.
POWER PLANT: One 59.7 kW (80 hp) Jabiru 2200A/J flat-four engine, driving a two-blade fixed-pitch wooden propeller. Fuel capacity 65 litres (17.2 US gallons; 14.3 Imp gallons).
SYSTEMS: 12 V DC electrical system with 100 W capacity; 20 Ah battery.
AVIONICS: *Comms:* VHF antenna (inside fin) and intercom standard.
Instrumentation: Basic VFR; vacuum instruments optional.
Data below refer to Jabiru UL 450.
DIMENSIONS, EXTERNAL:
Wing span 9.40 m (30 ft 10 in)
Length overall 5.64 m (18 ft 6 in)
Height overall 2.01 m (6 ft 7 in)
DIMENSIONS, INTERNAL:
Cockpit max width at shoulder 1.07 m (3 ft 6 in)
AREAS:
Wings, gross 9.29 m² (100.0 sq ft)
WEIGHTS AND LOADINGS:
Weight empty, equipped 242 kg (534 lb)
Max T-O weight 450 kg (992 lb)
PERFORMANCE:
Never-exceed speed (V$_{NE}$) 116 kt (215 km/h; 133 mph)
Max level speed 110 kt (204 km/h; 127 mph)
Cruising speed at 75% power
 100 kt (185 km/h; 115 mph)
Stalling speed, flaps down 32 kt (60 km/h; 37 mph)
Max rate of climb at S/L 305 m (1,000 ft)/min
T-O run 100 m (330 ft)
Landing run 160 m (525 ft)
Range with max fuel 450 n miles (833 km; 517 miles)
UPDATED

Jabiru UL two-seat kitbuilt ultralight *(Paul Jackson/Jane's)* 2001/0092043

MILLICER

MILLICER AIRCRAFT INDUSTRIES
PO Box 1242, Sale, Victoria 3850
Tel: (+61 3) 51 44 76 67
Fax: (+61 3) 51 44 74 43
e-mail: infomai@netspace.net.au
Web: http://www.milliceraircraftindustries.com.au
CHAIRMAN: Chris Hamilton

CEO AND MANAGING DIRECTOR: David Rees
DIRECTOR AND ENGINEERING MANAGER: Bob MacGillivray
SALES MANAGER: John Preece
MARKETING AND PUBLIC RELATIONS: Robyn MacGillivray

This company, formed in 1995, bears the name of the late Dr Henry Millicer, designer of the Airtourer, and planned to return that aircraft to production.

Millicer Aircraft Industries acquired, from the Airtourer Co-operative Group, production rights for the former Victa Airtourer two-seat trainer/tourer, some 250 of which were built in various models in Australia and New Zealand during the 1960s, together with more than 100 military CT4 trainer derivatives (which see under PAC in New Zealand section). Original wooden aircraft designed in 1953 and first flown in 1959; all-metal version followed in 1961.

AUSTRALIA: AIRCRAFT—MILLICER

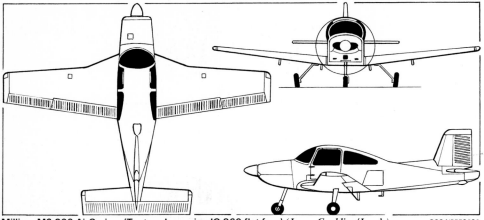

Millicer M9-200 AirCruiser (Textron Lycoming IO-360 flat-four) *(James Goulding/Jane's)*

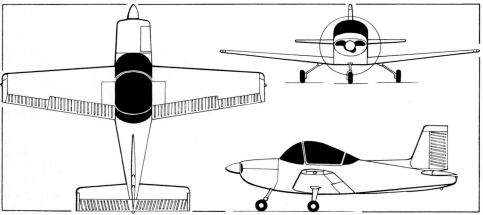

Millicer M10 AirTourer *(James Goulding/Jane's)*

In mid-2000 Millicer was negotiating a 60 per cent interest in a joint venture company in China which would manufacture AirTourer and AirCruiser components, as well as complete airframes for the Chinese market. However, company was placed in administrator's hands in late 2000.

UPDATED

MILLICER M9-200 AIRCRUISER

TYPE: Four-seat lightplane.
PROGRAMME: Sole prototype of original Aircruiser 210CS (VH-MVR) built in 1966. Design rights acquired by Millicer Aircraft Industries in 1996. First flight of revised M9-200 AirCruiser was scheduled for September 2000, with second prototype expected to fly in December 2000, but programme suspended and neither aircraft had flown by early 2001. Retractable landing gear and six-seat derivatives under consideration in 2000.
COSTS: Estimated A$225,000 (1999).
DESIGN FEATURES: Generally as for AirTourer, but with large wingroot glove.
 Changes from original Aircruiser include 149 kW (200 hp) Textron Lycoming IO-360-C1A engine replacing the original's 157 kW (210) hp Teledyne Continental power plant; ergonomically designed interior, choice of sticks or control wheels and IFR/GPS avionics.
 Wing dihedral 6° 4′.
FLYING CONTROLS: As AirTourer; flap deflections 0, 15 and 30°.
STRUCTURE: Mostly light alloy.
LANDING GEAR: Non-retractable tricycle type.
POWER PLANT: One 149 kW (200 hp) Textron Lycoming IO-360-C1A driving a Hartzell two-blade constant-speed propeller. Usable fuel capacity 200 litres (52.8 US gallons; 44.0 Imp gallons). Oil capacity 7.6 litres (2.0 US gallons; 1.07 Imp gallons).
ACCOMMODATION: Four persons in enclosed cabin; single forward-opening door on port side. Baggage compartment aft of rear seats with external access door on port side.
AVIONICS: IFR, with GPS standard.
DIMENSIONS, EXTERNAL:
Wing span	7.92 m (26 ft 0 in)
Wing mean aerodynamic chord	1.51 m (4 ft 11½ in)
Wing aspect ratio	5.2
Length overall	7.06 m (23 ft 2 in)
Height overall	2.59 m (8 ft 6 in)
Wheel track	2.97 m (9 ft 9 in)

AREAS:
Wings, gross	11.98 m² (129.0 sq ft)

WEIGHTS AND LOADINGS (estimated):
Operating weight empty	680 kg (1,500 lb)
Max T-O weight	1,143 kg (2,520 lb)
Max baggage	54 kg (120 lb)
Max wing loading	95.4 kg/m² (19.53 lb/sq ft)
Max power loading	7.67 kg/kW (12.60 lb/hp)

PERFORMANCE (estimated):
Never-exceed speed (VNE)	190 kt (351 km/h; 218 mph)
Normal operating speed	152 kt (282 km/h; 175 mph)
Cruising speed at 75% power at 1,525 m (5,000 ft):	145 kt (269 km/h; 167 mph)
Stalling speed: flaps up	57 kt (106 km/h; 66 mph)
flaps down	47 kt (87 km/h; 55 mph)
Max rate of climb at S/L	305 m (1,000 ft)/min
Range with max fuel	697 n miles (1,290 km; 802 miles)

UPDATED

MILLICER M10 AIRTOURER

TYPE: Aerobatic two-seat lightplane.
PROGRAMME: Proof-of-concept prototype Airtourer 160 (VH-BWG) first flown 20 August 1998; public debut at Aviex '98 at Sydney-Bankstown Airport 19 November 1998; certification to FAR Pt 23 Aerobatic category, and series production, was imminent in 2000.
CURRENT VERSIONS: Offered in two versions, with different engine powers, both approved for aerobatics: **M10-140 AirTourer 140** and **M10-160 AirTourer 160**.
CUSTOMERS: Orders for 47 held by mid-2000.
COSTS: AirTourer 160 with VFR nav/com and GPS A$210,000 (1999).
DESIGN FEATURES: Conventional low wing, mid-mounted tailplane configuration; wing tapered; leading edges of tail surfaces slightly swept.
 Changes from original Airtourer include a strengthened airframe with simplified fuselage structure for ease of manufacture; new cowling design; relocation of fuel from a fuselage bladder tank to integral tanks in the wings, revised landing gear and brakes; modified wingtips incorporating landing lights; electric flaps, with original Airtourer's centre flap deleted; electric elevator trim; revised control circuits; redesigned cockpit which retains the Airtourer's characteristic central control column, but with seats moved rearwards by 51 mm (2 in); and adjustable rudder pedals.
FLYING CONTROLS: Conventional and manual; full-span horn-balanced elevator with electric pitch trim; ground-adjustable tab on horn-balanced rudder.
STRUCTURE: Mostly light alloy. Fluted skins on control surfaces. Flaps; maximum deflection 12°.
LANDING GEAR: Non-retractable tricycle type with cantilever main units; Parker Aerospace wheels and brakes.
POWER PLANT: *AirTourer 140:* One 104 kW (140 hp) Textron Lycoming O-320-E2A engine driving a Sensenich fixed-pitch propeller.
 AirTourer 160: One 119 kW (160 hp) IO-320-D1A driving a Hartzell constant-speed propeller.
 Fuel in integral wing tanks, maximum capacity 170 litres (44.9 US gallons; 37.4 Imp gallons).
ACCOMMODATION: Two persons, side by side in enclosed cockpit with rearward-sliding canopy; central control coloum can be operated from either seat.

DIMENSIONS, EXTERNAL:
Wing span:	7.92 m (26 ft 0 in)
Wing aspect ratio	5.6
Length overall	6.71 m (22 ft 0 in)
Height overall	2.29 m (7 ft 6 in)

AREAS:
Wings, gross	11.15 m² (120.0 sq ft)

WEIGHTS AND LOADINGS (A: AirTourer 140, B: AirTourer 160):
Operating weight empty: A	549 kg (1,210 lb)
B	581 kg (1,280 lb)
Max T-O weight: A, B Normal	861 kg (1,900 lb)
A, B Aerobatic	816 kg (1,800 lb)
Max baggage: A, B	45 kg (100 lb)
Max wing loading: A, B	77.3 kg/m² (15.83 lb/sq ft)
Max power loading: A	8.26 kg/kW (13.57 lb/hp)
B	7.23 kg/kW (11.875 lb/hp)

PERFORMANCE:
Never-exceed speed (VNE):	
A, B	203 kt (376 km/h; 233 mph)
Cruising speed at 75% power at 2,135 m (7,000 ft):	
A	115 kt (213 km/h; 132 mph)
B	130 kt (241 km/h; 150 mph)
Stalling speed, A, B:	
flaps up	55 kt (102 km/h; 64 mph)
flaps down	47 kt (87 km/h; 55 mph)
Max rate of climb at S/L: A	259 m (850 ft)/min
B	320 m (1,050 ft)/min
g limits, Aerobatic: A, B	+6/−3

UPDATED

Prototype Millicer M10 AirTourer, following long-abandoned convention with 'circled P' experimental markings

SEABIRD

SEABIRD AVIATION AUSTRALIA PTY LTD
Hervey Bay Airport, PO Box 618, Hervey Bay, Queensland 4655
Tel: (+61 7) 41 25 31 44
Fax: (+61 7) 41 25 31 23
e-mail: seabird@mlink.com.au
PRODUCTION DIRECTOR: Peter Adams

Seabird Aviation was founded in 1983 to develop the Seeker observation aircraft. Despite having built Seekers for certification, Seabird's business plan was always to franchise Seeker production capability, rather than sell individual aircraft; entire manufacturing equipment for a Seeker production line can be forwarded in two standard-size ISO shipping containers. On 25 July 2000, Seabird signed an agreement with Evektor-Aerotechnik of the Czech Republic (which see) leading to the formation of a joint venture company, Evektor-Seabird Pty Ltd, to manufacture and market the Seeker.

NEW ENTRY

SEABIRD SB7L-360 SEEKER

TYPE: Multisensor surveillance lightplane.
PROGRAMME: Design began January 1985, construction January 1988; first flight of first (SB5N) prototype (VH-SBI, c/n 001, with Norton rotary engine) 1 October 1989, and of SB5E second prototype (VH-SBU, c/n 003, with Emdair engine) 11 January 1991 (c/n 002 was structural test airframe); fourth aircraft (c/n 004, VH-ZIG) is SB7L Seeker prototype, making first flight 6 June 1991; definitive SB7L-360A (VH-DPT; first flight early 1993) has Textron Lycoming O-360 engine for higher performance and received Australian CAA certification to FAR Pt 23 on 24 January 1994. Operational testing of Seeker included use as an airborne surveillance platform in October 1995 trials by the Australian Army. Components for an initial production batch of five aircraft completed by December 1996. Project then frozen and franchised producer sought.
CUSTOMERS: Orders for 12 received by July 2000.
DESIGN FEATURES: Braced high-wing monoplane with pod and boom fuselage, extensively glazed cabin and slightly sweptback vertical tail. Ventral and auxiliary fins added early 1993. Primary applications are observation/reconnaissance, offering helicopter-like view from cockpit and good low-speed handling and loiter capabilities; training and agricultural use also foreseen.
Wing section NACA 63_2-215 (modified); wedge tips; twist 3°, incidence 4° at root, 1° at tip; dihedral 2° 30′ from roots.
FLYING CONTROLS: Conventional and manual. Rod-actuated slotted ailerons, one-piece horn-balanced elevator (with trim tab) and horn-balanced rudder; slotted flaps on wing trailing-edge; fixed incidence tailplane. Inboard stall strips; vortex generators on wing upper surface near leading-edge and lower surface forward of ailerons; elevator bias trim.
STRUCTURE: Front fuselage mainly of 4130 chromoly steel tube with Kevlar non-load-bearing skin; aluminium alloy tubular tailboom; wings and tail unit conventional aluminium alloy stressed skin structures, former with single bracing strut and jury strut each side.
LANDING GEAR: Non-retractable, with Cleveland 8.00-6 mainwheels and fairings on cantilever spring steel legs; fully castoring Scott 8 in tailwheel with oil/nitrogen oleo strut and 210×65 McCreary tyre. Mainwheel tyre pressure 1.38 bars (20 lb/sq in); Cleveland disc brakes. Float gear to be developed.
POWER PLANT: One Textron Lycoming O-360-B2C flat-four engine, derated to 125 kW (168 hp) from 134 kW (180 hp) with lower compression ratio of 7.5:1 for Mogas fuel. Low propeller tip speed for minimal noise; Bishton BB177 two-blade fixed-pitch wood/composites pusher propeller. Fuel in two 96 litre (25.4 US gallon; 21.1 Imp gallon) integral wing tanks, each with overwing gravity filling point. At maximum T-O weight of 897 kg (1,977 lb) with two 77 kg (170 lb) occupants, maximum fuel load 152 litres (40.2 US gallons; 33.4 Imp gallons); higher MTOW of 925 kg (2,040 lb) would permit this to be increased to 182 litres (48.0 US gallons; 40.0 Imp gallons). Provision for auxiliary fuel tanks on underwing hardpoints. Oil capacity 7 litres (1.9 US gallons; 1.5 Imp gallons).
ACCOMMODATION: Side-by-side seats, adjustable fore and aft, for pilot and co-pilot or observer/passenger in extensively glazed cabin. Dual controls and four-point inertia reel seatbelts standard. Split (upward/downward-hinged) door each side, both removable. Ram air cabin ventilation. Space for 43 kg (95 lb) of baggage aft of seats.
SYSTEMS: 28 V electrical system, with 70 A alternator and 18 Ah Gill G-25 battery, for engine start, instruments and lighting.
AVIONICS: *Comms:* Honeywell KY 97A VHF com, KT 76A transponder, AR 850 encoder and Sigtronics SPA 400 intercom standard.
Flight: Honeywell nav/com, ADF, second com and ELT optional.
EQUIPMENT: Hardpoint for 60 kg (132 lb) of external stores beneath each wing. Quick-change photo/survey modules, stretcher or 100 litre (26.4 US gallon; 22 Imp gallon) spraytank optional in place of right-hand seat.

DIMENSIONS, EXTERNAL:
Wing span	11.07 m (36 ft 4 in)
Wing chord, constant	1.22 m (4 ft 0 in)
Wing aspect ratio	9.4
Length overall	7.01 m (23 ft 0 in)
Fuselage: max width	1.14 m (3 ft 9 in)
Height overall, propeller vertical	2.49 m (8 ft 2 in)
Tailplane span	2.90 m (9 ft 6 in)
Wheel track	2.03 m (6 ft 8 in)
Wheelbase	4.75 m (15 ft 7 in)
Propeller diameter	1.77 m (5 ft 9¾ in)
Cabin doors (two, each): Height	0.83 m (2 ft 8¾ in)
Width	0.98 m (3 ft 2½ in)
Height to sill	0.98 m (3 ft 2½ in)

DIMENSIONS, INTERNAL:
Cabin, incl baggage space:	
Length	2.21 m (7 ft 3 in)
Max width	1.12 m (3 ft 8 in)
Max height	1.09 m (3 ft 7 in)
Baggage compartment volume	0.42 m³ (15.0 cu ft)

AREAS:
Wings, gross	13.05 m² (140.5 sq ft)

WEIGHTS AND LOADINGS:
Basic weight empty	604 kg (1,332 lb)
Max fuel	115 kg (254 lb)
Max T-O weight	896 kg (1,977 lb)
Max wing loading	68.7 kg/m² (14.07 lb/sq ft)
Max power loading	7.16 kg/kW (11.77 lb/hp)

PERFORMANCE:
Never-exceed speed (V_{NE})	129 kt (238 km/h; 148 mph) CAS
Cruising speed at 75% power	112 kt (207 km/h; 129 mph)
Minimum patrol speed	65 kt (120 km/h; 75 mph) CAS
Stalling speed, flaps down	50 kt (93 km/h; 58 mph) IAS
Rate of climb: at S/L	288 m (944 ft)/min
at 1,830 m (6,000 ft)	173 m (567 ft)/min
Service ceiling	4,648 m (15,250 ft)
T-O run	265 m (870 ft)/min
Landing run	199 m (654 ft)
Range with reserves: at min patrol speed	476 n miles (881 km; 547 miles)
at 65% power	470 n miles (870 km; 540 miles)
Endurance, with reserves: at min patrol speed	7 h 15 min
at 65% power	4 h 30 min

NEW ENTRY

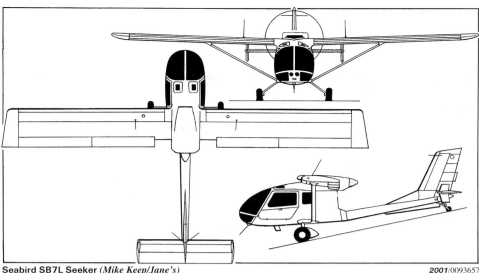

Seabird SB7L Seeker *(Mike Keep/Jane's)*
2001/0093657

SLEPCEV

NESTOR SLEPCEV
Details of Slepcev's replica Storch now appear in a separate entry.

UPDATED

STORCH

STORCH AVIATION AUSTRALIA PTY LTD
113 Koree Island Road, Beechwood, New South Wales 2446
Tel/Fax: (+61 2) 65 85 64 58
Fax: (+61 2) 65 85 66 22
e-mail: slepcev@nor.com.au
Web: http://www.storch.com.au
DESIGNER: Nestor Slepcev

Mr Slepcev designed, built and flew two earlier aircraft before producing the Storch. In August 1996, he landed a Storch Mk 4 on Gran Sasso, Italy, in re-enactment of the rescue of Benito Mussolini, 53 years earlier.

NEW ENTRY

STORCH AVIATION STORCH SS Mk 4

TYPE: Two-seat lightplane/kitbuilt.
PROGRAMME: First flight 1991, then with single seat and powered by 59.6 kW (79.9 hp) Rotax 912 engine; two-seat version first flown 1994; JAR-VLA certification achieved in Australia, 14 October 1999.
CURRENT VERSIONS: **Storch 'Ultra':** Ultralight; Australian Primary Category; MTOW 544 kg (1,200 lb).
Storch 'Muster': Utility version; envisaged uses include cattle mustering.
Storch 'Criquet': Powered by Rotec Fireball 7 seven-cylinder radial engine of 82.0 kW (110 hp), thus resembling Morane-Saulnier MS 502 Criquet. Prototype VH-AJH public debut at Avalon Air Show, February 2001.
CUSTOMERS: More than 80 sold by June 2000 (45 from factory and 35 kitbuilt versions) to customers in 19 countries in Africa, Australasia, Europe, North America and Pacific Rim.
COSTS: US$58,000, factory-complete (2000).
DESIGN FEATURES: Three-quarters scale version of Second World War German Fieseler Fi 156 Storch STOL liaison and observation aircraft. Wings are removable for transport or storage. Quoted build time for kitbuilt version approximately 600 hours.
FLYING CONTROLS: Conventional and manual. Actuation by pushrods and cables; ground-adjustable tab on each aileron. Flaps and full span leading-edge slats. Horn-balanced tail surfaces. Electronic flight-adjustable tailplane incidence for pitch trim in Ultralight; VLA has fixed tailplane with trim tab in port elevator.
STRUCTURE: Welded 4130 chromoly steel tube fabric-covered fuselage and tail surfaces; metal engine cowling. Strut-braced metal wings with stamped aluminium ribs and D-section leading-edge; latest versions aluminium-covered, except fabric flaps and ailerons. Composites wheel fairings.
LANDING GEAR: Tailwheel type; fixed. Original compression spring suspension replaced by bungee. Mainwheel size 8.00×6-6; Maule tailwheel, diameter 15 cm (6 in); hydraulic disc brakes. Float installation under development in late 1999.
POWER PLANT: Standard factory-built VLA version has one 73.5 kW (98.6 hp) Rotax 912 ULS flat-four driving two-blade Bruce de Chastel wooden propeller; 84.6 kW

Storch Aviation Storch three-quarters scale VLA version of Fieseler Storch *(Paul Jackson/Jane's)*

Storch 'Criquet' power plant *(Paul Jackson/Jane's)*

(113.4 hp) turbocharged Rotax 914 optional. Fuel in two wing tanks, combined capacity 76 litres (20.1 US gallons; 16.7 Imp gallons); belly tank, capacity 100 litres (26.4 US gallons; 22.0 Imp gallons) optional.

ACCOMMODATION: Two, in tandem, with dual controls.

DIMENSIONS, EXTERNAL:
Wing span	10.00 m (32 ft 9¾ in)
Length overall	6.80 m (22 ft 3¾ in)
Propeller diameter	1.88 m (6 ft 2 in)

DIMENSIONS, INTERNAL:
Cabin: Length	1.50 m (4 ft 11 in)
Max width	0.80 m (2 ft 7½ in)
Max height	1.00 m (3 ft 3¼ in)

AREAS:
Wings, gross	16.00 m² (172.2 sq ft)

WEIGHTS AND LOADINGS (Storch 'Ultra'):
Weight empty	345 kg (761 lb)
Max T-O weight	544 kg (1,200 lb)

PERFORMANCE:
Max level speed	78 kt (144 km/h; 90 mph)
Cruising speed	70 kt (130 km/h; 81 mph)
Stalling speed, flaps down, power on	22 kt (41 km/h; 26 mph)
Max rate of climb at S/L	213 m (700 ft)/min
Service ceiling	4,575 m (15,000 ft)
T-O and landing run	9 to 15 m (30 to 50 ft)
Endurance: standard fuel	3 h 0 min
optional fuel	7 h 0 min
g limits	+6/−3

UPDATED

AUSTRIA

C.CRAFT

C.CRAFT WOZLMAYER GmbH
Hietzinger Hauptstrasse 99, A-1130 Wien
Tel: (+43 1) 879 29 78
e-mail: c.craft@netway.at
SALES MANAGER: Wilhelm Leitner

UPDATED

C-CRAFT MARTLET

No reports have been received concerning completion of the prototype Martlet utility lightplane, last described in the 2000-01 edition.

UPDATED

DIAMOND

DIAMOND AIRCRAFT INDUSTRIES GmbH
N A Otto-Strasse 5, A-2700 Wiener Neustadt
Tel: (+43 2622) 267 00
Fax: (+43 2622) 267 80
e-mail: sales@diamond-ac-ind.co.at
Web: http://www.diamond.at
PRESIDENT: Christian Dries
MANAGING DIRECTOR: Wolfgang Grumeth
CHIEF DESIGNER: Martin Volck
SALES DIRECTOR: Michael Feinig

Company formed 1981 in Carinthia, southern Austria, as Hoffman Flugzeugbau-Friesach GmbH; re-formed after August 1984 bankruptcy as Hoffman Aircraft Ltd; relocated to new facility at Wiener Neustadt 1987; name HOAC Austria Flugzeugwerk Wiener Neustadt GmbH adopted 1990 after 1989 management buyout; renamed Diamond Aircraft Industries GmbH March 1996.

Factory expanded to 5,200 m² (56,000 sq ft) of floor space in 1996 to accommodate wider model range, increase production levels and provide new research and maintenance facilities. Workforce 118. Current products include the HK 36 Super Dimona/Katana Xtreme and refurbished DV 20 Katana 100. Diamond Aircraft Corporation produces the two-seat DA 20 in Canada (which see); the four-seat DA 40 Diamond Star became available in 2000.

UPDATED

Diamond HK 36TTC motor glider wearing its North American name *(Paul Jackson/Jane's)*

DIAMOND HK 36
European marketing name: Super Dimona
North American marketing name: Katana Xtreme

TYPE: Two-seat lightplane/motor glider.

PROGRAMME: Original Hoffman H-36 Dimona (Diamond) was designed by Austrian-German team, first flight by first of three prototypes taking place in Germany 9 October 1980; production, in Austria, started May 1981; Dimona Mk II introduced May 1985 (smaller wingtip fairings, modified cowling, better propeller pitch control, stronger main gear, sprung steerable tailwheel); production of H-36 totalled over 275 by 1989.

Development of HK 36R Super Dimona (considerably redesigned by Dieter Kohler (hence K in designation) started March 1987; first flight, with 67.1 kW (90 hp) Limbach L 2400 engine, October 1989, and one or two more completed with Limbach engines, but Rotax 912 adopted as standard power plant. Production began April 1990 and Austrian certification to JAR 22 awarded 15 May 1990. Manufacture totalled 114 by 1995, when supplanted by current versions. 'Short-wing' LF 2000 was proof-of-concept for DA 20 Katana; TS version introduced 1995. In 1997, export Dimonas for the North American market were renamed Katana Xtreme.

CURRENT VERSIONS: **HK 36TS**: *Turbo/Schleppe*: Turbocharged/Tug. Introduced 1995; prototypes OE-9415 and OE-9416; fitted with winglets; maximum rate of climb at S/L increased to 246 m (807 ft)/min; certified for glider- and banner-towing; capable of towing gliders of up to 370 kg (816 lb) maximum T-O weight. One further TS built in 1997.

HK 36TC: Tricycle landing gear version of the above. Some 20 TS/TC built in 1996, four in 1998 and eight for Indian Air Force in early 2000. Additionally, **Eco** Super Dimona HB-2335, operated on weather research missions by Met Air of Switzerland, fitted with two underwing sensor pods.

DIAMOND—AIRCRAFT: AUSTRIA

HK 36TTS: Introduced 1996; 85.8 kW (115 hp) Rotax 914F-3 engine; carbon fibre main spar; increased useful load; capable of towing gliders of up to 600 kg (1,323 lb) maximum T-O weight. Four built in late 1997.

HK 36TTC: Tricycle landing gear and Rotax 914. Majority of production is this version. **Eco** version of TTC registered in Canada (C-GETC), May 2000.

CUSTOMERS: Two prototypes and over 165 of T/TT series built by January 2000. Previous versions given under Programme.

COSTS: TS DM208,000, TC DM211,000, TTS DM226,900, TTC DM228,000 (2000).

DESIGN FEATURES: Conventional motor glider, with T tail and low/mid wing. Wings can be detached and folded back alongside fuselage for transportation and storage. Improvements of T/TT series (from 1996) include carbon fibre main spar. Compared with H-36 Dimona, HK 36 has modified inboard leading-edge, sweptback wingtips and Schempp-Hirth (instead of DFS) type airbrakes; better access to engine and control system via completely removable cowlings; improved spin characteristics; stronger main landing gear and improved tailwheel springing; larger canopy with new mechanism; wingroot fairings and rear fuselage redesigned.

Wortmann wing sections: FX-63-137 at root, FX-71 at tip fairings

FLYING CONTROLS: Conventional and manual. Longitudinal stability improved in Super Dimona over original by increasing elevator chord and adding trim tab; airbrakes in wing upper surface.

STRUCTURE: Mainly of GFRP; carbon fibre wing spar and main bulkhead. GFRP mainwheel legs and fairings.

LANDING GEAR: Non-retractable tailwheel type standard. Cantilever main legs with 6 in wheels; sprung, steerable tailwheel. Removable fairings on main legs and wheels. Tricycle gear optional, with free-castoring nosewheel, ±30°.

POWER PLANT: TS and TC have one 59.6 kW (79.9 hp) Rotax 912A-3 flat-four with 2.27:1 reduction drive to an MT composites, two-blade, three-position constant-speed and feathering propeller. TTS and TTC have one 84.6 kW (113.4 hp) Rotax 914F-3. Fuel tank in fuselage, capacity 77 litres (20.3 US gallons; 16.9 Imp gallons).

ACCOMMODATION: Two seats side by side. Cockpit canopy hinged at rear to open upward. Baggage space aft of seats.

DIMENSIONS, EXTERNAL:
Wing span	16.33 m (53 ft 7 in)
Wing aspect ratio	17.4
Length overall	7.28 m (23 ft 10½ in)
Width, wings folded	2.20 m (7 ft 2½ in)
Height over tail	2.40 m (7 ft 10½ in)

AREAS:
Wings, gross	15.30 m² (164.7 sq ft)

WEIGHTS AND LOADINGS:
Weight empty: TS	540 kg (1,190 lb)
TC	555 kg (1,224 lb)
TTS	545 kg (1,202 lb)
TTC	560 kg (1,235 lb)
Max T-O weight: all	770 kg (1,697 lb)
Max towing load: TS, TC	370 kg (815 lb)
TTS, TTC	600 kg (1,322 lb)
Max wing loading: all	50.3 kg/m² (10.31 lb/sq ft)
Max power loading: TS, TC	12.92 kg/kW (21.23 lb/hp)
TTS, TTC	9.10 kg/kW (14.95 lb/hp)

PERFORMANCE, POWERED:
Never-exceed speed (V_NE): all	141 kt (261 km/h; 162 mph)
Max cruising speed: TS	111 kt (205 km/h; 127 mph)
TC	108 kt (200 km/h; 124 mph)
TTS	121 kt (225 km/h; 140 mph)
TTC	119 kt (220 km/h; 137 mph)
Cruising speed at 65% power:	
TS	92 kt (170 km/h; 106 mph)
TC	90 kt (167 km/h; 104 mph)
TTS	108 kt (200 km/h; 124 mph)
TTC	106 kt (196 km/h; 122 mph)
Max rate of climb at S/L: TS	246 m (807 ft)/min
TC	240 m (787 ft)/min
TTS, TTC	324 m (1,063 ft)/min

Seventh prototype/pre-series DA 40 Diamond Star, fitted with revised ventral strake and rudder (Paul Jackson/Jane's) 2001/0092045

Instrument panel of the Diamond DA 40 Diamond Star (Paul Jackson/Jane's) 2001/0092046

T-O run: TS	161 m (528 ft)
TC	201 m (659 ft)
TTS	134 m (440 ft)
TTC	182 m (600 ft)
T-O to 15 m (50 ft): TS	306 m (1,005 ft)
TC	338 m (1,110 ft)
TTS	251 m (825 ft)
TTC	274 m (900 ft)
g limits: all	+5.3/−2.65

PERFORMANCE, UNPOWERED:
Best glide ratio: TS, TTS	28
TC, TTC	27
Min sink rate: TS, TTS	1.14 m (3.74 ft)/s
TC, TTC	1.17 m (3.84 ft)/s

UPDATED

DIAMOND DV 20 KATANA and KATANA 100
See Canadian section under DA 20.

DIAMOND DA 40-180 DIAMOND STAR

TYPE: Four-seat lightplane.

PROGRAMME: Formally launched 23 April 1997 at the Aero 97 show at Friedrichshafen, Germany, when mockup displayed; initially retained name of Katana; proof-of-concept prototype DA 40-V1 (OE-VPC) first flew on 5 November 1997, powered by a Rotax 914 engine; a Teledyne Continental IO-240-engined DA 40-V2 (OE-VPE) followed shortly thereafter; neither event was publicised; a third prototype, DA 40-V3, with Textron Lycoming IO-360 engine, flew in late 1998; fourth, near-production standard, prototype (OE-VPM) joined the test programme in mid-1999 followed by three more prototypes comprising DA40-P5 (c/n 40005/OE-VPQ), -P6 (40006/OE-VPB), -P7 (40007/OE-VPW), these having enlarged ventral strake and cutaway rudder. First production aircraft is c/n 40008. Lycoming-powered variant is production version, for which JAR/FAR 23 certification for VFR operation was expected in August 2000, followed by IFR approval at the end of the year. Target production rate three per week by May 2001, with total of 150 scheduled for 2001 and 300 in 2002.

CUSTOMERS: Total of 300 orders and options by July 2000.

COSTS: DM265,000 VFR glider towing equipped; DM270,000 with 'Night VFR' package; DM318,000 IFR equipped (all excluding VAT, 1999).

DESIGN FEATURES: Four-seat development of DA 20-C1; to be certified to FAR/JAR 23 or equivalents in Europe, North America, Russian Federation, South Africa and Turkey. DA 20-C1 wings attach to new, wider centre-section to increase span; downturned tips to horizontal tail; single strake/tail bumper. Enlarged cabin increases fuselage length by 0.67 m (2 ft 2¼ in).

Wing section is Diamond-modified Wortmann FX 63-137/20; washout 1°.

FLYING CONTROLS: Conventional and manual. Ailerons and elevators operated via pushrods with low-friction bearings, rudder by cables; manual pitch trim standard, electric pitch trim optional via control stick-mounted switch; ground-adjustable trim tab on rudder; electrically actuated slotted flaps.

STRUCTURE: Fuselage, with integral vertical fin, is mostly of GFRP construction with local CFRP reinforcement in high-stress areas, comprising two half-shells bonded together. Two-spar wings with GFRP/PVC foam/GFRP sandwich skins; horizontal tail and rudder surfaces similarly covered.

LANDING GEAR: Fixed tricycle type, with cantilever spring steel leg on each main unit and elastomeric suspension on nose leg. Mainwheel size 6.00-6. Steering is provided by differential braking of mainwheel and friction-damped castoring nosewheel. Speed fairings on all three wheels.

POWER PLANT: One 134 kW (180 hp) Textron Lycoming IO-360-M1A flat-four with Lasar electronic ignition and self-adapting inlet cooling, driving an MTV-12-B/180-17 three-blade constant-speed propeller. Fuel in two wing tanks, standard capacity 155 litres (40.9 US gallons; 34.1 Imp gallons); optional 200 litres (52.8 US gallons; 44.0 Imp gallons).

ACCOMMODATION: Four persons in two side-by-side pairs. Single-piece canopy lifts up and forwards for access to front seats; rear occupants board through an upward-opening door to port. Cabin has 26 g composites seats with three-point automatic safety harnesses and roll-over bar. Baggage compartment behind rear seats. Rigid baggage compartment extension for long objects such as skis.

SYSTEMS: 28 V electrical system includes 70 A alternator; 28 V 35 Ah battery.

AVIONICS: *Comms:* Standard VFR package includes Honeywell KX 125 nav/com, KT 76A Mode C transponder and four-position voice-activated intercom. Optional IFR package adds KX 155A nav/com in exchange for KX 125, KT 76C transponder in exchange for KT 76A, KR 87 ADF and PMA 7000 audio selector panel. ELT optional.

Flight: Optional 'Night VFR' package includes a KI 208 VOR indicator. Optional IFR package includes Garmin GNS 430 nav/com/GPS, GI 106 VOR/LOC/GS indicator with GPS interface, KN 62A DME, and S-Tec System 30 two-axis autopilot including Garmin MD 41.

Instrumentation: Standard VFR package includes ASI, VSI, altimeter, compass, tachometer, engine hours meter, manifold pressure, oil pressure/temperature, EGT/CHT, fuel quantity/flow and OAT gauges, ammeter and voltmeter. 'Night VFR' package adds a turn co-ordinator, directional gyro, artificial horizon and chronometer. IFR package adds a second altimeter and alternate static port.

EQUIPMENT: Landing light, taxying light, position lights, instrument lighting, overhead cabin light and pitot/static heat all standard. Optional equipment includes glider towing kit, with retractor winch.

DIMENSIONS, EXTERNAL:
Wing span	12.00 m (39 ft 4½ in)

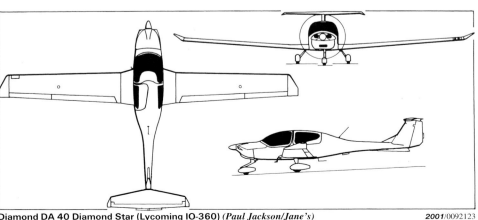

Diamond DA 40 Diamond Star (Lycoming IO-360) (Paul Jackson/Jane's) 2001/0092123

AUSTRIA: AIRCRAFT—DIAMOND to HB FLUGTECHNIK

Wing aspect ratio	10.7
Length overall	8.00 m (26 ft 3 in)
Height overall	2.00 m (6 ft 6¾ in)
DIMENSIONS, INTERNAL:	
Baggage compartment: Length	1.08 m (3 ft 6½ in)
Max width	0.35 m (1 ft 1¾ in)
Max height	0.80 m (2 ft 7½ in)
AREAS:	
Wings, gross	13.50 m² (145.3 sq ft)
WEIGHTS AND LOADINGS:	
Weight empty	700 kg (1,543 lb)
Max T-O weight	1,150 kg (2,535 lb)
Max wing loading	85.2 kg/m² (17.45 lb/sq ft)
Max power loading	8.58 kg/kW (14.10 lb/hp)
PERFORMANCE:	
Max level speed	155 kt (287 km/h; 178 mph)
Cruising speed at 75% power at 1,981 m (6,500 ft)	147 kt (272 km/h; 169 mph)
Econ cruising speed at 50% power at 3,050 m (10,000 ft)	120 kt (222 km/h; 138 mph)
Stalling speed	49 kt (91 km/h; 57 mph)
Max rate of climb: at S/L	350 m (1,150 ft)/min
at 3,050 m (10,000 ft)	168 m (550 ft)/min
T-O run to 15 m (50 ft)	350 m (1,150 ft)
Range, 45 min reserves:	
standard fuel	594 n miles (1,100 km; 683 miles)
optional fuel	799 n miles (1,480 km; 919 miles)

UPDATED

DIAMOND FUTURE DEVELOPMENTS

In early 2000, Diamond Aircraft revealed that it is developing a six-seat version of the DA 40 Diamond Star; a pressurised single-engine design; and a four/six-seat light twin. No further details are known.

NEW ENTRY

HB FLUGTECHNIK

HB FLUGTECHNIK GmbH

Dr Adolf Scharf Strasse 42, Postfach 74, A-4053 Haid-Ansfelden
Tel: (+43 7229) 791 04 and 791 17
Fax: (+43 7229) 791 04 15 and 791 17 15
e-mail: office@hb-aviation.com
Web: http://www.hb-aviation.com
MANAGING DIRECTOR: Ing Heino Brditschka
TECHNICAL DIRECTOR: Georg Passenbrunner

DISTRIBUTOR:
Aero-Service Thüringen GmbH, Mühlberger Strasse 13, D-99869, Wandersleben, Germany
Tel: (+49 36202) 823 95
Fax: (+49 36202) 823 86
e-mail: info@aeroserv.de
Web: http://www.aeroserv.de
DIRECTOR: Karl-Heinz Maletschek

Company name indicates Heino Brditschka, designer of motor gliders, including the HB-23 and HB-202, last described in the 1991-92 *Jane's*. Both had the unconventional mid-fuselage propeller, to which HB Flugtechnik has returned for its most recent project, the Tornado. HB provides a wide variety of other aviation services, including overhaul and repair, aerial photography and advertising, flying training, aircraft charter and support of amateur constructors. By 2000, distribution of HB-207 kits had been assigned to Aero-Service Thüringen.

UPDATED

HB FLUGTECHNIK HB-207 ALFA

TYPE: Side-by-side kitbuilt.
PROGRAMME: First flight (HB-207RG OE-CHC) 14 March 1995. Second prototype (HB-207 OE-CAA) made public debut, unmarked and then unflown, at AERO 97 at Friedrichshafen April 1997.
CUSTOMERS: Total 63 sold and 17 completed in Austria, France, Germany and Switzerland by mid-2000.
CURRENT VERSIONS: **HB-207:** Fixed landing gear version.
HB-207RG: Retractable landing gear version.
COSTS: Standard aircraft fast-build kit, no engine, propeller or avionics Sch399,750 (1996).
DESIGN FEATURES: Quoted build time 1,000 hours. Conforms to JAR-VLA requirements for amateur assembly. Constant chord low wings with raked tips; slightly sweptback fin and rudder; one-piece elevator aft of vertical tail. Airframe features guide derigging and rigging for road transport or storage.
FLYING CONTROLS: Conventional and manual. Ailerons and elevator actuated by pushrods; rudder by cables. Large central trim tab in elevator. Manual (optionally electric) operation for flaps.
STRUCTURE: Mainly metal primary structure with GFRP skin; moving surfaces have Ceconite covering.
LANDING GEAR: HB-207RG has retractable tricycle type, with size 4.00-4 tyre on each unit. Main units have brakes and rubber-in-compression shock-absorption, and retract inward; nosewheel retracts rearward. HB-207 has fixed tricycle type with streamlined fairings and spats on each unit.
POWER PLANT: Standard engine is an 82.0 kW (110 hp) VW-Porsche HB-2400 G/2 flat-four based on that of Porsche 911 sports car; alternatives include 59.6 kW (79.9 hp) Rotax 912 A, and Rotax 914, or Limbach or Textron Lycoming engines up to 74.6 kW (100 hp). Geared drive to choice of propellers: two-blade wooden fixed-pitch, three-blade adjustable- or variable-pitch, or (as on prototypes) five-blade adjustable- or variable-pitch. Fuel tank in each wing, combined capacity 100 litres (26.4 US gallons; 22.0 Imp gallons).
ACCOMMODATION: Rearward-sliding fully transparent canopy.
SYSTEMS: Electrical system: 200 A 15 Ah (optionally 20 Ah) battery.
AVIONICS: *Comms:* Honeywell 760-channel KX 155 transceiver, KT 76A transponder and ELT optional.
Flight: Honeywell KX 125 VOR with KI 208 indicator optional.
Instrumentation: VFR flight and engine transmission instrumentation standard.

HB-207 Alfa two-seat lightplane (Paul Jackson/Jane's) 2001/0092047

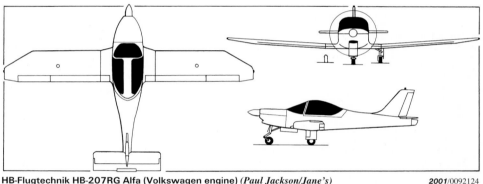

HB-Flugtechnik HB-207RG Alfa (Volkswagen engine) (Paul Jackson/Jane's) 2001/0092124

DIMENSIONS, EXTERNAL:	
Wing span	9.00 m (29 ft 6¼ in)
Wing aspect ratio	8.5
Length overall	5.95 m (19 ft 6¼ in)
Height overall	1.95 m (6 ft 4¼ in)
DIMENSIONS, INTERNAL:	
Cabin: Length	1.94 m (6 ft 4½ in)
Max width	1.10 m (3 ft 7¼ in)
Max height	0.97 m (3 ft 2¼ in)
AREAS:	
Wings, gross	9.50 m² (102.3 sq ft)
Vertical tail surfaces (total)	0.65 m² (7.00 sq ft)
Horizontal tail surfaces (total)	1.68 m² (18.08 sq ft)
WEIGHTS AND LOADINGS (U: Utility, N: Normal category):	
Weight empty: U, N	450 kg (992 lb)

Cockpit of the HB-207 Alfa 1999/0044444

Unflown prototype HB-204 Tornado displayed at Aero '99 at Friedrichshafen (Paul Jackson/Jane's) 2000/0064424

Max T-O weight: U	640 kg (1,411 lb)
N	700 kg (1,543 lb)
Max wing loading: U	67.4 kg/m² (13.80 lb/sq ft)
N	73.7 kg/m² (15.09 lb/sq ft)
Max power loading:	
VW-Porsche HB-2400 G/2	8.54 kg/kW (14.03 lb/hp)
Rotax 912A	11.75 kg/kW (19.31 lb/hp)

PERFORMANCE (HB-2400 engine, U and N as above):
Never-exceed speed (VNE): U, N	166 kt (309 km/h; 192 mph)
Cruising speed: U	132 kt (245 km/h; 152 mph)
N	130 kt (240 km/h; 149 mph)
Stalling speed: flaps up: U	51 kt (93 km/h; 58 mph)
N	53 kt (98 km/h; 61 mph)
30° flap: U	46 kt (84 km/h; 53 mph)
N	47 kt (87 km/h; 55 mph)
Max rate of climb at S/L: U	330 m (1,082 ft)/min
N	300 m (984 ft)/min
T-O run: U	190 m (625 ft)
N	210 m (690 ft)
T-O to 15 m (50 ft): U	300 m (985 ft)
N	320 m (1,050 ft)
Range: U, N	647 n miles (1,200 km; 745 miles)
g limits: U	+4.4/−2.2
N	+3.8/−1.9

UPDATED

HB FLUGTECHNIK HB-204 TORNADO

TYPE: Two-seat lightplane.
PROGRAMME: Developed from previous, unbuilt, design; unflown airframe, based on components from first two aircraft, exhibited at Aero '99 at Friedrichshafen in April 1999; first flight then expected in late 1999 or early 2000 but postponed to second half of 2000.
CUSTOMERS: Two ordered (by one customer) by April 1999. No further orders by mid-2000.
DESIGN FEATURES: Unswept wing with wingtip pods housing landing lights; cheek engine air intakes aft of cabin; pusher propeller mounted at mid-fuselage, rotating in triangular-shaped cutout.
FLYING CONTROLS: Conventional and mechanical. Trim tab in starboard elevator; horn-balanced rudder; two-section Fowler-type flaps.
STRUCTURE: Primarily composites, with welded steel tube fuselage frame and alloy control surfaces.
LANDING GEAR: Retractable tricycle type; mainwheels retract inwards, nosewheel forwards.
POWER PLANT: One 147 kW (197 hp) Porsche Carrera aircooled flat-four, driving a mid-mounted HB five-blade, constant-speed pusher propeller at up to 2,000 rpm, via an extension shaft.
ACCOMMODATION: Two in tandem under one-piece sideways-opening canopy with fixed windscreen.

DIMENSIONS, EXTERNAL:
Wing span	7.75 m (25 ft 5 in)
Wing aspect ratio	6.8
Length overall	6.60 m (21 ft 7¾ in)

AREAS:
Wings, gross	8.85 m² (95.3 sq ft)

WEIGHTS AND LOADINGS:
Max T-O weight	750 kg (1,653 lb)
Max wing loading	84.7 kg/m² (17.36 lb/sq ft)
Max power loading	5.03 kg/kW (8.27 lb/hp)

PERFORMANCE (estimated):
Never-exceed speed (VNE)	217 kt (403 km/h; 250 mph)
Cruising speed	194 kt (360 km/h; 224 mph)
Stalling speed: flaps up	60 kt (110 km/h; 69 mph)
flaps down	53 kt (98 km/h; 61 mph)
Max rate of climb at S/L	945 m (3,100 ft)/min
T-O to 15 m (50 ft)	390 m (1,280 ft)
g limits	+6/−3

UPDATED

BELGIUM

LAMBERT

LAMBERT AIRCRAFT ENGINEERING BVBA
Lupinestraat 5, B-8500 Kortrijk
Tel/Fax: (+32 56) 21 33 47
DIRECTOR: Filip Lambert

In late 1995 Filip Lambert, a student at the College of Aeronautics, Cranfield, UK, was named one of three joint winners of the Royal Aeronautical Society Light Aircraft Design Competition with his Mission M212-100, described below. The prototype is being fabricated in Belgium, assembled at Brooklands and will undertake test flying at Cranfield. No decision has yet been announced as to where production will take place.

UPDATED

LAMBERT MISSION M212

TYPE: Side-by-side lightplane/kitbuilt.
PROGRAMME: Design started November 1992; construction of proof-of-concept prototype began at Geluveld, Belgium, January 1996; structural test programme completed 23 December 1998; first flight at Cranfield, UK, scheduled for late 1999 but not reported by mid-2000, although prototype was registered G-XFLY in February 2000.
CURRENT VERSIONS: Family of five aircraft planned around common core design, including 2+2, aerobatic and four-seat tourer versions, all derived with minimum of modifications to basic two-seat aircraft. Initial versions will be **M212-100** with two seats (as prototype), and **M212-200**, with 2+2 seating. Available factory-built or as kit.
CUSTOMERS: Options on 96 kits held by end of November 1998, from customers in the Benelux countries, France and the UK.
COSTS: Kit, including engine, propeller and VFR avionics, £38,500 (1997).
DESIGN FEATURES: Designed using CAD and wind tunnel research facilities; will meet FAR Pt 23 and JAR 23 certification criteria. Low-wing monoplane; wings, which have moderate taper and upturned tips, can be folded in under 5 minutes by one person without tools. Quoted kit build time 600 hours.
FLYING CONTROLS: Conventional and manual. Cable and pushrod actuation. Single-slotted flaps.
STRUCTURE: All-composites; designed for ease of assembly, robustness, low maintenance and long life.
LANDING GEAR: Fixed tricycle type. Mainwheels size 6.00-6, nosewheel 5.00-5 or 6.00-6; Cleveland brakes. Steerable nosewheel, maximum steering angle ±50°. Potential for development of tailwheel and retractable versions.
POWER PLANT: One 112 kW (150 hp) Zoche ZO-01A radial four-cylinder, two-stroke diesel engine with multifuel capability, driving a three-blade constant-speed propeller. Fuel capacity 100 litres (26.4 US gallons; 22.0 Imp gallons). In developed form as four-seat tourer, suitable for engines in 172 to 194 kW (230 to 260 hp) power range.
ACCOMMODATION: Two persons, side by side, or four in 2+2 configuration, under one-piece, forward-hinged canopy. Seats and rudder pedals adjustable; dual controls standard.
All data are provisional.

DIMENSIONS, EXTERNAL:
Wing span	9.80 m (32 ft 1¾ in)
Wing aspect ratio	8.0
Length overall	7.40 m (24 ft 3½ in)
Height overall	2.90 m (9 ft 6 in)
Tailplane span	3.20 m (10 ft 6 in)
Wheel track	2.60 m (8 ft 6¼ in)

DIMENSIONS, INTERNAL:
Cabin: Length	2.50 m (8 ft 2½ in)
Max width	1.12 m (3 ft 8 in)
Max height	1.25 m (4 ft 1¼ in)
Baggage compartment volume:	
M212-100	0.85 m³ (30.0 cu ft)
M212-200	0.14 m³ (5.0 cu ft)

AREAS:
Wings, gross	12.00 m² (129.2 sq ft)
Ailerons (total)	0.60 m² (6.46 sq ft)
Flaps (total)	1.08 m² (11.62 sq ft)
Elevators (total)	2.65 m² (28.52 sq ft)

WEIGHTS AND LOADINGS (A: M212-100, B: M212-200):
Weight empty: A	460 kg (1,014 lb)
B	485 kg (1,069 lb)
Max T-O weight: A	750 kg (1,653 lb)
B	900 kg (1,984 lb)
Max wing loading: A	62.5 kg/m² (12.80 lb/sq ft)
B	75.0 kg/m² (15.36 lb/sq ft)
Max power loading: A	6.70 kg/kW (11.00 lb/hp)
B	8.04 kg/kW (13.20 lb/hp)

PERFORMANCE: Never-exceed speed (VNE):
A	183 kt (338 km/h; 210 mph)
B	183 kt (338 km/h; 210 mph)
Max level speed at S/L: A	138 kt (256 km/h; 159 mph)
B	136 kt (252 km/h; 157 mph)
Max cruising speed at 2,440 m (8,000 ft):	
at 75% power: A	134 kt (248 km/h; 154 mph)
B	131 kt (243 km/h; 151 mph)
at 60% power: A	121 kt (224 km/h; 139 mph)
B	118 kt (219 km/h; 136 mph)
Stalling speed, flaps down: A	45 kt (84 km/h; 52 mph)
B	49 kt (91 km/h; 57 mph)
Max rate of climb at S/L: A	378 m (1,240 ft)/min
B	293 m (960 ft)/min
T-O run: A	175 m (575 ft)
B	225 m (740 ft)
Range with max fuel: A, B:	
at 75% power	640 n miles (1,185 km; 736 miles)
at 60% power	750 n miles (1,389 km; 863 miles)

UPDATED

Model of the Lambert Mission M212-100
(Paul Jackson/Jane's)

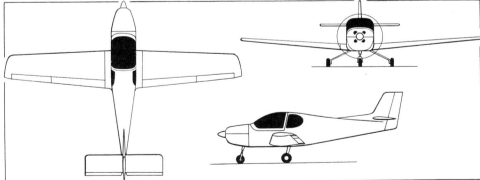

Lambert Mission M212-100 (Zoche ZO-01A diesel engine) *(James Goulding/Jane's)*   1999/0044443

MASQUITO

MASQUITO AIRCRAFT n.v.
Reigersbaan 31, B-1760 Roosdaal-Strijtem
Tel: (+32 54) 34 30 08
Fax: (+32 54) 34 30 09
e-mail: info@masquito.be
Web: http://www.masquito.be

FOUNDERS: Paul Maschelein
Stefaan Maschelein
DIRECTOR: John Pescod

This company was established to produce the Masquito M80 ultralight helicopter, two prototypes of which are now flying.

UPDATED

MASQUITO M80

TYPE: Two-seat ultralight helicopter kitbuilt.
PROGRAMME: Design started November 1994; construction of prototype began December 1995; first flight (G-MASZ) May 1996; 25 hours of test flying completed by April 1997. Prototype reconfigured as M80 with Jabiru engine and began ground trials in late 1997. Two preproduction prototypes (G-MASX and G-MASY) registered June 1998

Masquito two-seat ultralight helicopter *(Paul Jackson/Jane's)* 2001/0092048

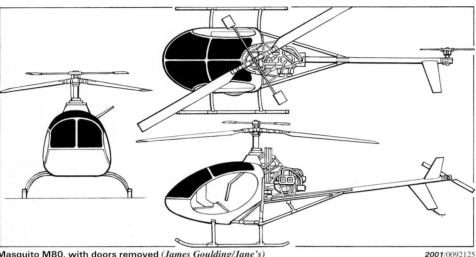

Masquito M80, with doors removed *(James Goulding/Jane's)* 2001/0092125

in M80 configuration; one displayed unflown at the PFA International Air Rally at Cranfield in July 1999, before first flight in February 2000. UK certification will be to new BCAR-VLH requirements.

CURRENT VERSIONS: **M58**: Prototype only; initially with 47.7 kW (64 hp) Rotax 582 engine.
M80: Intended production version; Jabiru engine. *As described.*

COSTS: Kit, approximately €46,000 or US$50,000 (2000), including engine and basic instruments.
DESIGN FEATURES: Conforms to FAR Pt 27 crashworthiness requirements. Two-blade, teetering main rotor and two-blade tail rotor; main rotor blades of modified Wortmann S aerofoil section with 13 per cent thickness/chord ratio and 8° 15′ linear twist. Hiller servo rotor. Main rotor speed 750 rpm; 2 kg (4.4 lb) lead weight in each tip to increase inertia and improve stability at the hover. Tail rotor 3,400 rpm. Ventral fin offset 2°. Composites 'virtual hinge' main rotor head with elastomeric collective pitch thrust bearings. Prototype had Rotax gearbox for primary reduction and Gates powerdrive toothed aramid fibre belts for secondary reduction; production version employs toothed drivebelt, centrifugal clutch and flywheel as primary transmission, with flexible driveshaft for tail rotor.
FLYING CONTROLS: Conventional and manual, with Hiller servo rotor for cyclic pitch control.
STRUCTURE: Primary structure of L-section 6061 alloy, riveted and glued; tailboom of 8.9 cm (3½ in) diameter aluminium tube. Main rotor blades have unidirectional composites spar with PVC foam trailing-edge core and bidirectional composites skin.
LANDING GEAR: Two fixed skids of 6061 alloy mounted on unidirectional glass fibre supports.
POWER PLANT: One 59.7 kW (80 hp) Jabiru 2200 flat-four; transmission rating in excess of 74.6 kW (100 hp). Fuel capacity 62 litres (16.4 US gallons; 13.6 Imp gallons).

DIMENSIONS, EXTERNAL:
Main rotor diameter	5.00 m (16 ft 4¾ in)
Tail rotor diameter	1.00 m (3 ft 3¼ in)
Length: overall, rotors turning	5.96 m (19 ft 6¾ in)
fuselage	4.85 m (15 ft 11 in)
Height overall	2.25 m (7 ft 4½ in)
Width of pod	1.30 m (4 ft 3¼ in)
Skid track	1.70 m (5 ft 7 in)

AREAS:
Main rotor disc	16.62 m² (178.90 sq ft)
Tail rotor disc	0.79 m² (8.45 sq ft)

WEIGHTS AND LOADINGS:
Weight empty, equipped	215 kg (474 lb)
Max T-O and landing weight	430 kg (948 lb)

PERFORMANCE (estimated):
Never-exceed speed (V_{NE})	97 kt (180 km/h; 112 mph)
Max cruising speed	80 kt (148 km/h; 92 mph)
Max rate of climb at S/L	335 m (1,100 ft)/min
Hovering ceiling IGE	3,050 m (10,000 ft)
Range with max internal fuel, no reserves	324 n miles (600 km; 372 miles)
Endurance	4 h

UPDATED

WOLFSBERG

WOLFSBERG AIRCRAFT CORPORATION NV
Woudstraat 23, B-3600 Genk
Tel: (+32 89) 38 08 31
Fax: (+32 89) 38 61 41
DIRECTOR AND CHIEF DESIGNER: Alec N Clark

Formerly Triloader Aircraft Corporation NV (see entry in 1996-97 edition), which intended to develop the Clark-Norman Triloader, described in the UK section of the 1996-97 *Jane's*. Wolfsberg Aircraft Corporation NV is now developing the Raven 257 small multirole aircraft. The programme was previously managed by Wolfsberg-Evektor in the Czech Republic; the prototype, which flew on 28 July 2000, was built by Evektor-Aerotechnik in Kunovice, with assistance from Letov Air. However, on 21 November 2000 it was announced that development and production of the Raven 257 had been assigned to Letov, under which heading it is described in the Czech section.

UPDATED

BOSNIA-HERZEGOVINA

SOKO

SOKO AIR MOSTAR
Rodoc bb, 88000 Mostar
Tel: (+387 88) 35 00 80 and 35 00 70
Fax: (+387 88) 35 01 34 and 35 00 81
DIRECTOR: Miroslav Coric

Soko (Falcon) Aircraft Industry founded in former Yugoslavia on 14 October 1950; initially subcontractor to existing aircraft industry; prime contractor on Type 522 and G-2A Galeb trainers, plus Jastreb attack aircraft; manufactured Kraguj COIN type, G-4 Super Galeb, Eurocopter Gazelle, (including subassemblies and complete airframes for French assembly line) and undertook final assembly of Orao attack aircraft. Also produced aircraft ground and test equipment, rocket pods and bomb racks. Soko privatised March 1991, 3,000 of its 3,500 workers having held 20 per cent of US$110 million capital.

Independence of Bosnia-Herzegovina from former Yugoslavia recognised by European Community 7 April 1992. Successor state of Yugoslavia transferred many Soko activities to Utva (which see); Mostar factory ceased production 1 May 1992; some partly complete Super Galebs abandoned on production line; infrastructure damaged by civil war; numerous subcontract activities, for Embraer, Dassault, de Havilland Canada, Tupolev and others, were abandoned.

Soko reverted to peacetime activities in 1996, initially with a two-seat sport aircraft. As yet, however, this does not appear to have entered production.

UPDATED

Soko 2 in original configuration with small winglets *(Soko/M Majersky)* 0010209

SOKO 2

TYPE: Tandem-seat ultralight.
PROGRAMME: Design began November 1992; first flight of prototype 16 November 1996. Certification stage was reached by mid-1998, following implementation of several minor modifications during early 1997, but was delayed because Bosnia-Herzegovina has no state airworthiness authority. First flight of floatplane prototype also postponed for this reason. Croatian airworthiness certificate and registration (9A-USM) awarded to prototype 002 in late July 1999; undertook promotional tour of Croatia in August and September 1999.
DESIGN FEATURES: To meet JAR-VLA criteria. High-wing (with winglets) pod-and-boom configuration, with T tail and pusher propeller. Modifications following early tests included enlarged winglets, redesigned engine cowling, addition of engine silencer, wider seats and a ballistic parachute.
FLYING CONTROLS: Conventional and manual. All-moving tailplane, with anti-balance tab, mounted on top of fin. No flaps.

STRUCTURE: Fuselage of light alloy semi-monocoque structure. Fabric-covered, single-spar, braced light alloy wing with tubular steel oblique auxiliary spar. Wings fold backwards for storage. Single-spar light alloy fin.
Wing section NACA 23018, dihedral 1°, incidence 14° at root, 12° 30′ at tip.

LANDING GEAR: Fixed tailwheel type. Oblique cantilever main gear struts hinged to fuselage with Soko rubber-in-compression shock-absorbers and auxiliary drag brace-struts. Steerable solid rubber tailwheel mounted on cantilever steel leaf-sprung arm attached to tailboom. Mainwheel tyres 5.00-5, pressure 2.50 bars (36 lb/sq in). Hydraulic disc brakes operated by lever on control column. Optional glass fibre floats.

POWER PLANT: One 59.7 kW (80 hp) Limbach L-2000 EC1 four-cylinder four-stroke driving a two-blade, fixed-pitch, wooden pusher propeller. Fuel tank forms leading-edge of wing centre-section; capacity 50 litres (13.2 US gallons; 11 Imp gallons).

SYSTEMS: 12 V DC electrical system powered by engine-driven alternator. Optional landing and taxying lights. Self-contained hydraulic system for brakes.

AVIONICS: *Comms:* Honeywell KX 99 VHF com; intercom.
Flight: Garmin ADV-100 GPS.
Instrumentation: Full IFR optional.

EQUIPMENT: Optional BRS 5 parachute recovery system.

DIMENSIONS, EXTERNAL:
Wing span	9.70 m (31 ft 10 in)
Wing chord, constant	1.50 m (4 ft 11 in)
Wing aspect ratio	6.5
Width, wings folded	2.40 m (7 ft 10½ in)
Length overall	7.90 m (25 ft 11 in)
Height: overall	2.00 m (6 ft 6¾ in)
wings folded	2.60 m (8 ft 6¼ in)

AREAS:
Wings, gross	14.55 m² (156.6 sq ft)

WEIGHTS AND LOADINGS (A: landplane, B: floatplane):
Weight empty: A	322 kg (710 lb)
B	377 kg (831 lb)
Max T-O and landing weight: A	535 kg (1,179 lb)
B	590 kg (1,300 lb)

PERFORMANCE (landplane unless otherwise stated):
Never-exceed speed (V_{NE})	99 kt (185 km/h; 114 mph)
Max operating speed (V_{MO})	78 kt (145 km/h; 90 mph)
Max cruising speed	76 kt (140 km/h; 87 mph)
Stalling speed, power off	40 kt (75 km/h; 47 mph)
Max rate of climb at S/L	276 m (905 ft)/min
Service ceiling	5,520 m (18,120 ft)
T-O run	100 m (330 ft)
T-O to 15 m (50 ft)	185 m (610 ft)
T-O run from water (floatplane at 525 kg; 1,157 lb)	300 m (985 ft)
Landing from 15 m (50 ft)	255 m (840 ft)
Landing run	100 m (330 ft)
Range with max fuel, no reserves	270 n miles (500 km; 310 miles)
Endurance	5 h 30 min
g limits	+4.4/−2.2

UPDATED

BRAZIL

AEROMOT

AEROMOT INDÚSTRIA MECANICO-METALURGICA LTDA

Caixa Postale 8031, Avenida das Indústrias 1210, 90200-290 Porto Alegre, RS
Tel: (+55 51) 371 16 44
Fax: (+55 51) 371 16 55
e-mail: industria@aeromot.com.br
Web: http://www.ximango.com and http://www.aeromot.com
PRESIDENT: Claudio B Viana
MANAGING DIRECTOR: João Claudio Jotz
COMMERCIAL DIRECTOR: Vitor J P Neves
MARKETING CONTACT: Patricia Munstock

Aeromot Indústria is part of Aeromot Group with Aeromot Aeronaves e Motores SA (parent company, founded 1967) and Aeroeletrônica Indústria de Componentes Aviónicos SA (established 1981); former sells aircraft and spares and provides maintenance; latter certifies and manufactures avionic equipment for civil and military aircraft, including 11 items for Embraer Tucano and 13 for Italian-Brazilian AMX.

Aeromot Indústria initially designed, certified and built seats for aircraft built by Embraer; it later produced several structural parts for Embraer aircraft, and designed and certified seats for Airbus, Boeing, Fokker and McDonnell Douglas (now Boeing) commercial transports. In 1985 it purchased assets of Fournier motor glider factory in France, including sole manufacturing rights for RF-10; has since incorporated several improvements to basic model, including a turbocharged engine and a trainer variant.

Factory has shop floor area of 3,100 m² (33,375 sq ft) and workforce of approximately 150. Future plans include eventual manufacture under licence of high-performance foreign sailplanes and of all-composites four-seat light aircraft.

UPDATED

AEROMOT XIMANGO

TYPE: Motor glider.

PROGRAMME: Brazilian production version of French Aérostructure (Fournier) RF-10; first flight (French prototype) 6 March 1981; see Sailplanes section of 1990-91 *Jane's* for French production history. All production rights sold to Aeromot July 1985; Brazilian CTA certification of AMT-100 granted 5 June 1986 and French (DGAC) 10 October 1990; AMT-200, developed by Aeromot, made first flight July 1992; was certified 3 February 1993 in Brazil, on 29 December 1993 in USA (to FAR Pt 22), during 1995 in UK and on 24 November 1998 in Canada. Certification of AMT-200 also achieved in Australia, Colombia, France, Germany, Japan, Netherlands and New Zealand.

CURRENT VERSIONS: **AMT-100 Ximango:** Initial production version; powered by one 59.7 kW (80 hp) Limbach L 2000 EO1 flat-four engine driving a Hoffmann HO-V62RL/160BT two-blade three-position variable-pitch propeller. Total 43 built, 1988-93.

AMT-100P and -100R Ximango: Military/police (100P) and observation (100R) versions. Two side windows below canopy; underfuselage pod for 100 kg (220 lb) of weapons and avionics or surveillance equipment. In service with Brazilian military police and other law enforcement agencies.

AMT-200 Super Ximango: Generally similar to AMT-100 except for Rotax 912A power plant. Main production variant since 1995. Improvements introduced in 1997 include redesigned instrument panel; new canopy locking mechanism; fully enclosing main landing gear doors; tailwheel fairing; and metal fuel tanks.

AMT-200S Super Ximango S: Version intended for pilot training and soaring. Generally similar to AMT-200 except for Rotax 912 S4 engine and optional removable winglets. Prototype first flew in September 1999; deliveries from December 1999, to customers in Australia (one), South Africa (one), UK (two) and USA (three).

Detailed description applies to the AMT-200 except where indicated.

AMT-300 Turbo Ximango Shark: Prototype (PT-ZAM, converted from an early AMT-100) flew July 1997; based on AMT-200, but with Rotax 914F turbocharged engine, redesigned cockpit, new engine cowling and winglets. Brazilian certification achieved 31 March 1999, followed by FAA approval 19 July 1999; deliveries began in May 1999 with two aircraft for US market (PT-PRS and PT-PRT) delivered to Ximango US at Spruce Creek, Florida.

AMT-300R Reboque: Glider tug version of AMT-300.

AMT-600 Guri: Trainer and sport aerobatic version. Described separately.

CUSTOMERS: Initial order for 100 from Brazilian Civil Aeronautical Department; also produced for military/paramilitary roles such as observation, patrol and counter-insurgency. Total of 130 AMT-100s, AMT-200s and AMT-300s produced by mid-2000, including deliveries to customers in Argentina (two), Australia (seven), Belgium (one), Brazil (63), Canada (one), Colombia (one), France (seven), Germany (two), Japan (four), UK (six) and USA (27). AMT-300 production began at No. 106, but continued in parallel with AMT-200.

COSTS: US$115,000 basic for AMT-200S; US$125,000 basic for AMT-300 (both 2000).

DESIGN FEATURES: Typical motor glider; low, tapered wing with fixed incidence T tail. Wings detachable for transportation and storage; outboard half of wing can be folded inward without disconnecting aileron controls.
Wing section NACA 64₃-618; dihedral 2° 30′.

FLYING CONTROLS: Conventional and manual. Schempp-Hirth airbrakes in wing upper surface.

STRUCTURE: All-GFRP except for carbon fibre main spar and light alloy airbrakes.

LANDING GEAR: Mechanically retractable mainwheels (tyre size 330×130), with hydraulic suspension and Cleveland hydraulic disc brakes with differential action; steerable tailwheel with size 210×65 tyre.

POWER PLANT: One 59.6 kW (79.9 hp) Rotax 912A flat-four with Hoffmann HO-V62R/170FA propeller. Fuel in two main tanks in wings, combined capacity 90 litres (23.8 US gallons; 19.75 Imp gallons). AMT-200S has one 73.5 kW (98.6 hp) Rotax 914 S4, driving a three-position variable-pitch Hoffmann propeller. AMT-300 has one 84.6 kW (113.4 hp) Rotax 914 F3 turbocharged flat-four driving an MT/MTV-21 constant-speed feathering propeller; fuel capacity 90 litres (23.8 US gallons; 19.8 Imp gallons).

ACCOMMODATION: Two seats side by side. One-piece canopy hinged at rear to open upward. Dual controls standard.

SYSTEMS: Electric starter and 12 V 30 A alternator.

DIMENSIONS, EXTERNAL:
Wing span: 200, 200S	17.47 m (57 ft 3¾ in)
300	17.70 m (58 ft 0¾ in)
Wing aspect ratio: 200, 200S	16.3
300	16.7
Width, wings folded	10.15 m (33 ft 3½ in)
Length overall: 200, 200S	8.05 m (26 ft 5 in)
300	8.08 m (26 ft 6 in)
Height overall: 200, 200S, 300	1.93 m (6 ft 4 in)
Propeller diameter: 200, 200S	1.70 m (5 ft 7 in)
300	1.65 m (5 ft 5 in)

DIMENSIONS, INTERNAL:
Cockpit: Length	1.85 m (6 ft 0¾ in)
Max width	1.15 m (3 ft 9¼ in)
Max height	1.00 m (3 ft 3¼ in)

AREAS:
Wings, gross: 200, 200S	18.70 m² (201.3 sq ft)
300	18.75 m² (201.8 sq ft)
Ailerons (total)	2.72 m² (29.28 sq ft)
Airbrakes (total)	1.16 m² (12.49 sq ft)
Fin	1.44 m² (15.50 sq ft)
Rudder	0.69 m² (7.43 sq ft)
Tailplane	1.95 m² (20.99 sq ft)
Elevator	0.95 m² (10.23 sq ft)

WEIGHTS AND LOADINGS:
Weight empty: 200	620 kg (1,367 lb)
200S	625 kg (1,378 lb)
300	630 kg (1,389 lb)
300R	620 kg (1,367 lb)
Max T-O weight: 200, 200S, 300	850 kg (1,874 lb)
Max baggage: 300	10 kg (22 lb)
Max wing loading: 200, 200S	45.5 kg/m² (9.31 lb/sq ft)
300	45.3 kg/m² (9.28 lb/sq ft)
Max power loading: 200	14.26 kg/kW (23.42 lb/hp)
200S	11.56 kg/kW (19.00 lb/hp)
300	9.92 kg/kW (16.29 lb/hp)

PERFORMANCE, POWERED:
Never-exceed speed (V_{NE}):	
200, 200S, 300	133 kt (245 km/h; 153 mph)
Max cruising speed: 200	110 kt (205 km/h; 127 mph)
200S	119 kt (220 km/h; 137 mph)
300	124 kt (230 km/h; 143 mph)
Normal cruising speed: 300	122 kt (225 km/h; 140 mph)
Stalling speed: 300	42 kt (78 km/h; 48 mph)
Max rate of climb at S/L: 200	156 m (512 ft)/min
200S	180 m (591 ft)/min
300	198 m (650 ft)/min

Winglet-equipped Aeromot AMT-200S Super Ximango S two-seat motor glider *(Paul Jackson/Jane's)*

Service ceiling: 200, 200S more than 4,900 m (16,076 ft)
 300 more than 8,700 m (28,543 ft)
T-O run, hard runway: 200 226 m (745 ft)
 200S 210 m (689 ft)
 300 174 m (570 ft)
T-O to 15 m (50 ft), hard runway:
 200 305 m (1,001 ft)
 300 296 m (970 ft)
Landing from 15 m (50 ft) less than 300 m (984 ft)
Landing run less than 150 m (492 ft)
Range with max fuel:
 200: best power 594 n miles (1,100 km; 683 miles)
  best economy 756 n miles (1,400 km; 870 miles)
 200S: best power 575 n miles (1,065 km; 661 miles)
  best economy 783 n miles (1,450 km; 901 miles)
 300: best power 548 n miles (1,015 km; 630 miles)
  best economy 810 n miles (1,500 km; 932 miles)
Max endurance:
 200: best power 5 h 30 min
  best economy 12 h
 200S: best power 4 h 54 min
  best economy 12 h
 300: best power 4 h 30 min
  best economy 12 h
g limits: 200, 200S, 300 +5.3/−2.65
PERFORMANCE, UNPOWERED:
Best glide ratio at 58 kt (108 km/h; 67 mph):
 200, 200S 31.1
 300 32.1
Min rate of sink at 52 kt (97 km/h; 60 mph):
 200, 200S 0.93 m (3.05 ft)/s
 300 0.96 m (3.15 ft)/s
Stalling speed: 200, 200S, 300 42 kt (78 km/h; 48 mph)
UPDATED

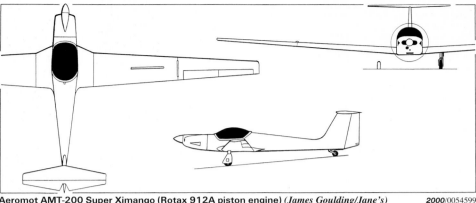

Aeromot AMT-200 Super Ximango (Rotax 912A piston engine) *(James Goulding/Jane's)*

AEROMOT AMT-600 GURI
English name: Boy
TYPE: Aerobatic two-seat lightplane.
PROGRAMME: Design began July 1998; prototype (PP-XBS) first flown 13 July 1999; total of more than 110 flight hours logged by mid-2000, with Brazilian CTA and FAA certification scheduled before end of 2000, but not reported by January 2001.
CURRENT VERSIONS: **AMT-600 Guri**: Initial production version, as described.
Data below are provisional, and refer to prototype.
 Advanced Trainer: Projected derivative with 134.2 kW (180 hp) Lycoming engine, retractable landing gear and IFR avionics.
CUSTOMERS: Brazilian state aero clubs seen as initial target market.
DESIGN FEATURES: Generally as Ximango, but shortened wings (of same aerofoil section) and tricycle landing gear. Wings can be removed for storage.
FLYING CONTROLS: Conventional and manual; ailerons and elevator operated by pushrods, rudder by cables; ground-adjustable trim tab on rudder and starboard aileron. Control surface maximum deflections: elevator +26/−21°, ailerons +15/−27°, rudder 21°, flaps 60°.
STRUCTURE: Primarily wet-moulded GFRP and carbon fibre; ailerons and flaps have alloy spars, GFRP/foam composites ribs and alloy skins.
LANDING GEAR: Non-retractable tricycle type, with hydropneumatic trailing-link suspension on main units and rubber-in-compression suspension on nose leg. Oldi wheels and brakes. Mainwheel tyre size 6.00×6, nosewheel 5.00×5.
POWER PLANT: One 85.8 kW (115 hp) Textron Lycoming O-235-NBR flat-four, driving a two-blade propeller. Fuel contained in two wing tanks, with filler port in upper surface of each wing.
ACCOMMODATION: Two persons, side by side, under single-piece, upward- and rearward-hinged canopy. Dual controls standard. Boarding step on each side forward of wing on prototype will be relocated aft of wing on production aircraft.
AVIONICS: VFR avionics, including Garmin GPS 100; IFR avionics optional.

DIMENSIONS, EXTERNAL:
Wing span 10.50 m (34 ft 5½ in)
Wing aspect ratio 8.0
Length overall 8.07 m (26 ft 5¾ in)
Height overall 2.65 m (8 ft 8¼ in)
Tailplane span 2.60 m (8 ft 6¼ in)
Aileron span 1.99 m (6 ft 6½ in)

Prototype Aeromot AMT-600 Guri

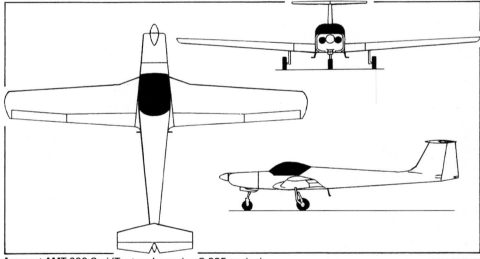

Aeromot AMT-600 Guri (Textron Lycoming O-235 engine)

Flap span 2.44 m (8 ft 0 in)
Wheel track 2.74 m (9 ft 0 in)
Wheelbase 2.11 m (6 ft 11 in)
DIMENSIONS, INTERNAL:
Cabin max width 0.86 m (2 ft 10 in)
AREAS:
Wings, gross 13.79 m² (148.4 sq ft)
WEIGHTS AND LOADINGS:
Weight empty 675 kg (1,488 lb)
Max T-O weight 900 kg (1,985 lb)
Max wing loading 65.26 kg/m² (13.37 lb/sq ft)
Max power loading 10.51 kg/kW (17.26 lb/hp)
PERFORMANCE:
Never-exceed speed (V_{NE}) 156 kt (289 km/h; 180 mph)
Cruising speed 119 kt (220 km/h; 137 mph)
Stalling speed 44 kt (82 km/h; 51 mph)
Max rate of climb at S/L 183 m (600 ft)/min
Service ceiling 4,875 m (16,000 ft)
Max operating altitude 5,490 m (18,000 ft)
T-O run 180 m (591 ft)
T-O to 15 m (50 ft) 290 m (951 ft)
Max range 441 n miles (817 km; 508 miles)
Endurance 4 h 30 min
NEW ENTRY

EDRA HELICENTRO

EDRA HELICENTRO, PEÇAS e MANUTENÇÃO LTDA
Estrada Estadual SP-191 Km 87.5, 13537-000 Ipeúna, São Paulo
Tel: (+55 19) 576 60 00
Fax: (+55 19) 576 13 92
e-mail: mnd@edraescola.com.br

NORTH AMERICAN DISTRIBUTOR:
Amphibian Airplanes of Canada Ltd
Box 1100, Squamish, British Columbia, V0N 3G0, Canada
Tel: (+1 604) 898 53 27
Fax: (+1 604) 898 20 09

Following cessation of manufacture by the Billie company in France, the Pétrel amphibian is being produced in Brazil for local and export markets.
UPDATED

EDRA HELICENTRO PATURI
English name: Masked Duck
Export marketing name: SeaStar
TYPE: Two-seat amphibian/kitbuilt.
PROGRAMME: Derived from Claude Tisserand's Hydroplum II, prototype of which first flew 1 November 1986; SMAN bought design rights in 1987; construction of Pétrel prototype began January 1989; first flight July 1989. At least 70 built in France; rights passed to Billie Aero Marine. Production transferred to Brazil by 1999.

CUSTOMERS: Total of 135 produced by Edra by late 1999. Production rate five kits and one factory-built aircraft per month in 1999.
COSTS: Kit, less engine, US$26,000; fly-away US$43,000 (both 2000).
DESIGN FEATURES: Pusher-engined biplane; road-towable on custom-designed trailer with quoted assembly/disassembly time of 10 minutes.
Equal-span, constant-chord wings, upper unit mounted on cabane; V-type interplane struts, with diagonal strut brace to fuselage from upper plane. Single-step hull. Boom-mounted empennage, wire-braced. Floats at mid-span of lower plane. (French-built Pétrels have shorter lower span with floats at tips.)
Wing section NACA 2412; dihedral 2° 13′ on upper wings, 3° 26′ on lower; sweepback 4° upper, 2° lower.
FLYING CONTROLS: Conventional and manual. Ailerons on upper wing only. No flaps. Actuation by wires (rudder) and pushrods.
STRUCTURE: Moulded monocoque single-step hull of epoxy/carbon fibre foam with carbon fibre tailboom; wings have 2014-T6 aluminium alloy tubular main spar with PVC foam ribs and glass fibre/epoxy leading-edge and tips covered with fabric and braced by single diagonal strut and V interplane struts of 6061-T6 aluminium. Wings fold for ground transportation. Tail surfaces of glass fibre spars and PVC foam ribs, with fabric and glass fibre covering, stainless steel wire braced.
LANDING GEAR: Retractable tricycle type; main units retract upwards into hull and undersurface of lower wing; nosewheel retracts upwards and forwards, tyre remaining partially exposed to serve as docking bumper when operating on water. Nosewheel maximum steering angle 80°. Hydraulic disc brakes on main units.
POWER PLANT: One 59.6 kW (79.9 hp) Rotax 912 UL four-stroke piston engine driving an Airplast 175 three-blade pusher propeller. Optional in-flight adjustable pitch propeller with electric or manual control. Fuel capacity 50 litres (13.2 US gallons; 11.0 Imp gallons), of which 45 litres (11.9 US gallons; 9.9 Imp gallons) usable.
ACCOMMODATION: Two, side by side in open or enclosed cockpit, with windscreen or single-piece forward-hinging canopy. Dual controls.

DIMENSIONS, EXTERNAL:
Wing span (both)	8.25 m (27 ft 8 in)
Length overall	6.47 m (21 ft 2¾ in)
Height overall	2.26 m (7 ft 5 in)
Wheel track	1.78 m (5 ft 10 in)
Wheelbase	2.10 m (6 ft 10¾ in)
Propeller diameter	1.65 m (5 ft 5 in)

DIMENSIONS, INTERNAL:
Cabin max width	1.13 m (3 ft 8½ in)

AREAS:
Wings, gross	16.50 m² (177.6 sq ft)

EDRA Helicentro Paturi, marketed abroad as the SeaStar *(Paul Jackson/Jane's)* 2000/0084581

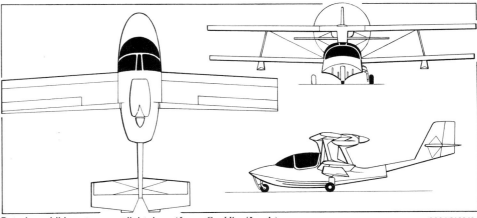
Paturi amphibious two-seat lightplane *(James Goulding/Jane's)* 2001/0100404

WEIGHTS AND LOADINGS:
Weight empty	300 kg (661 lb)
Max T-O weight	600 kg (1,323 lb)
Max wing loading	36.4 kg/m² (7.45 lb/sq ft)
Max power loading	10.08 kg/kW (16.56 lb/hp)

PERFORMANCE:
Never-exceed speed (V_{NE})	98 kt (180 km/h; 112 mph)
Max level speed	89 kt (165 km/h; 102 mph)
Max cruising speed	78 kt (145 km/h; 90 mph)
Manoeuvring speed	70 kt (130 km/h; 81 mph)
Stalling speed	35 kt (65 km/h; 41 mph)
Max rate of climb at S/L	244 m (800 ft)/min
T-O run: on land	200 m (660 ft)
on water	250 m (820 ft)
Landing run: on land	250 m (820 ft)
on water	150 m (500 ft)
Endurance	6 h 0 min
g limits	+4/−2

UPDATED

EMBRAER

EMPRESA BRASILEIRA DE AERONÁUTICA SA

Av Brig Faria Lima 2170, Caixa Postal 343, 12227-901 São José dos Campos, SP
Tel: (+55 12) 345 26 71 and 13 11
Fax: (+55 12) 345 16 32 and 24 11
Web: http://www.embraer.com
Telex: 1233589 EBAE BR
PRESIDENT AND CEO: Maurício Novis Botelho
EXECUTIVE VICE-PRESIDENT, INDUSTRIAL: Satoshi Yokota
EXECUTIVE VICE-PRESIDENT, COMMERCIAL: Frederico Pinheiro Fleury Curado
EXECUTIVE VICE-PRESIDENT, PLANNING AND ORGANISATIONAL DEVELOPMENT: Horácio Aragonés Forjaz
EXECUTIVE VICE-PRESIDENT, FINANCE: Antonio Luis Pizarro Manso
VICE-PRESIDENT, COMMERCIAL (Military Marketing): Romualdo Monteiro de Barros
VICE-PRESIDENT, CORPORATE JETS: Sam Hill
PRESS OFFICER: Marcia Benevides

Created 19 August 1969, Embraer began operating on 2 January 1970. It has a 250,000 m² (2,691,000 sq ft) factory area; global workforce was almost 11,000 in January 2001. Neiva (which see) became a subsidiary in March 1980. Embraer and subsidiaries have delivered nearly 5,500 aircraft.

Privatisation of Embraer was undertaken on 7 December 1994 with the auction of 55.4 per cent of voting stock. A consortium led by Bozano Simonsen bank acquired 45.44 per cent of auctioned stock, assumed a controlling interest in the company, and provided an extra US$36 million in capitalisation. A further 10 per cent of voting stock was offered to the public within 60 days, with Embraer employees and the Brazilian government retaining 10 and 18.4 per cent respectively. In October 1999 a consortium of French aerospace companies, comprising Aerospatiale Matra (now EADS France), Dassault, SNECMA and Thomson-CSF (now Thales), acquired 20 per cent of Embraer's voting shares. Embraer also embarked on a joint venture with Liebherr International of Germany to establish Embraer-Liebherr Equipamentos do Brasil SA (ELEB) to create additional opportunities for the company's landing gear and hydraulics components business. Embraer's controlling group – with 60 per cent of voting shares – comprises pension funds Previ and Sistel, plus investment house Companhia Bonzano, Simonsen.

Principal current own-design manufacturing programmes are EMB-120 Brasilia commuter transport, ERJ-135/ERJ-140/ERJ-145 and ERJ-170/190 regional jet families and EMB-312/EMB-314 Tucano turboprop military trainer. Subsidiary Neiva (which see) manufactures EMB-202 Ipanema agricultural aircraft and, under licence from Piper, PA-32-301 Saratoga (as EMB-720D Minuano) and PA-34-220T (as EMB-810D Seneca IV).

In subcontract field, first deliveries made in 1994 of wingtips and vertical fin fairings for Boeing 777 under 1991 contract.

Embraer is a risk-sharing partner in the Sikorsky S-92 Helibus programme; under a June 1995 contract valued at US$170 million, it will supply 730 sets of S-92 fuel sponsons. In 1995 Embraer signed a co-operation agreement with PZL Warszawa-Okecie of Poland under which Embraer subsidiary Neiva SA would market PZL light aircraft in Brazil and PZL subsidiary Skypol would acquire Brasilias for its freight and charter services. This appears to have lapsed.

Embraer delivered 71 aircraft comprising 23 ERJ-135s and 48 ERJ-145s, in the first half of 2000, when production rate was 14 per month, scheduled to rise to 16 per month by December 2000. Delivered 300th regional jet on 25 August 2000.

UPDATED

EMBRAER DELIVERIES
(at January 2001)

Type	Deliveries
EMB-110 Bandeirante	469
EMB-111 Bandeirante Patrulha	31
EMB-120 Brasilia	352
EMB-121 Xingu	105
ERJ-135	61
ERJ-145	293
EMB-200/201/202 Ipanema	826
EMB-312 Tucano	650
EMB-326GB Xavante	182
AMX	56
Light aircraft (incl Neiva)	2,469
Total	**5,494**

EMBRAER EMB-312 TUCANO

Detailed description last appeared in the 2000-01 *Jane's*.

UPDATED

EMBRAER EMB-314 SUPER TUCANO

English name: Super Toucan
Brazilian Air Force designations: A-29 and AT-29
TYPE: Basic turboprop trainer/attack lightplane.
PROGRAMME: Design of original EMB-312 Tucano started January 1978. Ministry of Aeronautics contract received 6 December that year for two flying prototypes and two static/fatigue test airframes; first prototype (Brazilian Air Force serial number 1300) made first flight 16 August 1980, second (1301) on 10 December 1980; third to fly (PP-ZDK, on 16 August 1982) was to production standard. Total of 650 built by 1998; further details in 2000-01 and previous *Jane's* and in *Jane's Aircraft Upgrades*.

Development of EMB-314 began January 1991; announced (as EMB-312H) at Paris Air Show June 1991; Embraer development aircraft PT-ZTW (c/n 312161, previously used as prototype for TPE331 powered Tucano adopted by Royal Air Force) modified as Tucano H proof-of-concept (POC) prototype, making first flight in this form 9 September 1991. This aircraft toured US Air Force/Navy bases August and September 1992 as preliminary to Super Tucano entry in JPATS competition; Embraer teamed with Northrop May 1992 to bid Super Tucano for JPATS, but was unsuccessful. Provisional Brazilian type certification granted August 1994 after 500 hour, 396 sortie test and certification programme.

BRAZIL: AIRCRAFT—EMBRAER

CURRENT VERSIONS: **EMB-314**: Two EMB-312H prototypes (PT[later PP]-ZTV, c/n 312454, first flight 15 May 1993, and PP-ZTF, c/n 312455, first flight 14 October 1993) tailored to US JPATS requirements as EMB-312HJ. Designation subsequently changed to EMB-314 to reflect extensive modifications to structure and systems.

ALX (EMB-314M): Brazilian Air Force (FAB) version, for border patrol missions under its SIVAM (*SIstema de Vigilancia da AMazonia*) programme. FAB finalised specification in early 1994; trials to validate projected flight characteristics, using POC aircraft and both Super Tucano prototypes, completed 1994. US$50 million development contract signed 18 August 1995 for two prototypes (one single-seat) to be modified from Super Tucano prototypes. The first flew in May 1996 and is being used for external stores compatibility and handling qualities testing; the second, which flew in early 1997, for testing the advanced weapons systems.

FAB ordered 99 ALXs, of which 50 will be two-seat **AT-29**, 30 of these replacing AT-26 Xavante with 2°/5° Grupo of Training Command at Natal AFB, remainder configured for night intruder role and expected to serve with 1°/3° Grupo at Boa Vista and 2°/3° Grupo at Porto Velho; 49 will be single-seat **A-29**. Prototype 5700 (the former EMB-314 PP-ZTV), designated **YA-29**, rolled out 28 May 1999; IOC in May 2001. Elbit selected December 1996 to suply mission avionics, including ventral FLIR Systems AN-AAQ-22 turret, GPS/INS, radalt, Mode S transponder, DME, ILS, ADF, VOR, RWR, MAWS and chaff/flare dispenser. Export variants of both versions will be offered for border patrol/COIN missions and for basic/advanced pilot training.

DESIGN FEATURES: Meets requirements of FAR Pt 23 Appendix A, and MIL and CAA Section K specifications. Low-mounted wings, stepped cockpits in tandem, fully aerobatic. Small fillet forward of tailplane root each side.

EMB-314 differs from EMB-312 mainly in having more powerful engine, reprofiled wing and plugs of 0.37 m (1 ft 2½ in) forward and 1.00 m (3 ft 3¼ in) aft of cockpit to accommodate longer engine and retain CG and stability. Other changes include strengthened airframe for higher g loads and longer fatigue life; ventral strakes; five weapons hardpoints; NVG-compatible 'glass cockpit' with HOTAS controls. Able to cover whole primary and half of advanced pilot training syllabus, and fly precision weapons delivery and target towing missions.

Wing section NACA 63-415 at root; NACA 63A-212 at tip.

FLYING CONTROLS: Conventional and manual. Primary surfaces internally balanced; electrically actuated trim tab in, and small geared tab on, each Frise aileron; electromechanically actuated spring tab in rudder and port elevator. Electrically actuated single-slotted Fowler flaps on wing trailing-edges. Fixed incidence tailplane. Ventral airbrake.

STRUCTURE: Conventional all-metal construction from 2024 series aluminium alloys; continuous three-spar wing box forms integral fuel tankage. Steel flap tracks.

LANDING GEAR: Hydraulically retractable tricycle type, with single wheel and Piper oleo-pneumatic shock-absorber on each unit. Accumulator for emergency extension in the event of hydraulic system failure. Hydraulic steering for nose unit. Rearward-retracting steerable nose unit; main units retract inward into wings. Parker Hannifin 40-130 mainwheels, Oldi-DI-1.555-02-OL nosewheel. Tyre sizes 6.50-10 (8 ply) tubeless on mainwheels, 5.00-5 (6 ply) tubeless on nosewheel. Tyre pressures (±0.21 bar; 3 lb/sq in in each case) are 5.17 bars (75 lb/sq in) on mainwheels, 4.48 bars (65 lb/sq in) on nosewheel. Parker Hannifin 30-95A hydraulic mainwheel brakes.

POWER PLANT: *ALX*: One 1,193 kW (1,600 shp) Pratt & Whitney Canada PT6A-68-1 turboprop, driving a Hartzell five-blade constant-speed fully feathering reversible-pitch propeller.

EMB-314: One 969 kW (1,300 shp) PT6A-68A.

Single-lever combined control for engine throttling and propeller pitch adjustment. Two integral fuel tanks in each wing, total capacity 694 litres (183.3 US gallons; 152.7 Imp gallons). Fuel tanks lined with anti-detonation plastics foam. Optional self-sealing tank in rear cockpit. Single-point pressure refuelling. Fuel system allows nominally for up to 30 seconds of inverted flight. Provision for two underwing ferry fuel tanks, total capacity 660 litres (174.4 US gallons; 145 Imp gallons). Four 'wet' hardpoints on ALX.

ACCOMMODATION: Two in tandem, on Martin-Baker Mk 10 LCX zero-zero ejection seats, in air conditioned and pressurised cockpit. One-piece fully transparent vacuum-formed canopy, opening sideways to starboard, with internal and external jettison provisions. Rear seat elevated 25 cm (9.9 in). Dual controls standard. Baggage compartment in rear fuselage, with access via door on port side.

SYSTEMS: Freon air cycle conditioning system, with engine-driven compressor. Single hydraulic system, consisting basically of (a) control unit, including reservoir with usable capacity of 1.9 litres (0.5 US gallon; 0.42 Imp gallon); (b) an engine-driven pump with nominal pressure of 131 bars (1,900 lb/sq in) and nominal flow rate of 4.6 litres (1.22 US gallons; 1.01 Imp gallons)/min at 3,800 rpm; (c) landing gear and gear door actuators; (d) filter; (e) shutoff valve; and (f) hydraulic fluid to MIL-H-5606. Under normal operation, hydraulic system actuates landing gear extension/retraction and control of gear doors. Landing gear extension can be performed under emergency operation; emergency retraction also possible during landing and T-O with engine running. Reservoir and system are suitable for aerobatics. No pneumatic system. 28 V DC electrical power provided by a 6 kW starter/generator, 26 Ah battery and, for 115 V and 26 V AC power at 400 Hz, a 250 VA inverter. Onboard oxygen generation system. Canopy and propeller de-icing.

AVIONICS: *Comms:* Standard Rockwell Collins equipment.

Instrumentation: ALX has two 152 × 203 mm (6 × 8 in) MFDs.

EQUIPMENT: Landing light in each wing leading-edge; taxying lights on nosewheel unit. Optional Kevlar cockpit armour.

ARMAMENT: Provision for a variety of ordnance including Giat NC621 20 mm cannon pod on centreline hardpoint, one 12 mm machine gun and 200 rounds mounted in each wing, Mk 81/82 bombs, MAA-1 Piranha AAMs, BLG-252 cluster bombs or LGBs on underwing stations.

Prototype YA-29 Super Tucano 5700 first flew in 1999 after conversion from EMB-314 *2000/0084583*

DIMENSIONS, EXTERNAL:
Wing span	11.14 m (36 ft 6½ in)
Wing chord: at root	2.30 m (7 ft 6½ in)
at tip	1.07 m (3 ft 6⅛ in)
Wing aspect ratio	6.4
Length overall	11.42 m (37 ft 5¾ in)
Fuselage: Length (excl rudder)	10.53 m (34 ft 6½ in)
Max depth	1.86 m (6 ft 1¼ in)
Height overall (static)	3.90 m (12 ft 9½ in)
Tailplane span	4.66 m (15 ft 3½ in)
Wheel track	3.76 m (12 ft 4 in)
Wheelbase	3.36 m (11 ft 0¼ in)
Propeller ground clearance (static)	0.345 m (1 ft 1½ in)
Baggage compartment door:	
Height	0.60 m (1 ft 11½ in)
Width	0.54 m (1 ft 9¼ in)
Height to sill	1.25 m (4 ft 1¼ in)

DIMENSIONS, INTERNAL:
Cockpits: Combined length	2.90 m (9 ft 6⅛ in)
Max height	1.55 m (5 ft 1 in)
Max width	0.85 m (2 ft 9½ in)
Baggage compartment volume	0.17 m³ (6.0 cu ft)

AREAS:
Wings, gross	19.40 m² (208.8 sq ft)
Ailerons (total)	1.97 m² (21.20 sq ft)
Trailing-edge flaps (total)	2.58 m² (27.77 sq ft)
Fin, incl dorsal fin	2.29 m² (24.65 sq ft)
Rudder, incl tab	1.38 m² (14.85 sq ft)
Tailplane, incl fillets	4.77 m² (51.34 sq ft)
Elevators, incl tab	2.00 m² (21.53 sq ft)

WEIGHTS AND LOADINGS (EMB-314, except where indicated):
Basic weight empty	2,420 kg (5,335 lb)
Max internal fuel load (usable)	538 kg (1,186 lb)
Max T-O weight: EMB-314, clean	3,190 kg (7,033 lb)
ALX	3,600 kg (7,936 lb)
Max ramp weight	3,210 kg (7,077 lb)
Max zero-fuel weight: EMB-314	2,670 kg (5,886 lb)
ALX	3,150 kg (6,945 lb)
Max wing loading	164.4 kg/m² (33.68 lb/sq ft)
Max power loading	3.29 kg/kW (5.41 lb/shp)

PERFORMANCE (EMB-314, at max clean T-O weight except where indicated):
Max level speed: EMB-314 at 6,100 m (20,000 ft)	301 kt (557 km/h; 346 mph)
ALX with external stores	245 kt (454 km/h; 282 mph)
Max cruising speed at 6,100 m (20,000 ft)	286 kt (530 km/h; 329 mph)
Econ cruising speed at 6,100 m (20,000 ft)	228 kt (422 km/h; 262 mph)
Stalling speed, power off:	
flaps and landing gear up	85 kt (157 km/h; 98 mph) EAS
flaps and landing gear down	78 kt (145 km/h; 90 mph) EAS
Max rate of climb at S/L	895 m (2,925 ft)/min
Service ceiling	10,670 m (35,000 ft)
T-O run	350 m (1,150 ft)
T-O to 15 m (50 ft)	550 m (1,805 ft)
Landing from 15 m (50 ft)	860 m (2,820 ft)
Landing run	550 m (1,805 ft)
Range at 9,150 m (30,000 ft) with max fuel, 30 min reserves	847 n miles (1,568 km; 974 miles)
Ferry range at 7,620 m (25,000 ft) with underwing tanks and 30 min reserves	1,495 n miles (2,768 km; 1,720 miles)
Endurance on internal fuel at econ cruising speed at 7,620 m (25,000 ft), 30 min reserves	6 h 30 min
g limits: fully Aerobatic category at 2,770 kg (6,107 lb)	+7/−3.5
at 2,770 kg (6,107 lb) with external stores	+4/−2.2

UPDATED

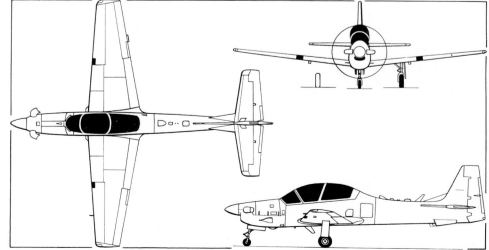

AT-29 advanced trainer version of Embraer Super Tucano (*James Goulding/Jane's*) *2000/0056316*

EMBRAER EMB-120 BRASILIA
Brazilian Air Force designation: VC-97

TYPE: Twin-turboprop airliner.

PROGRAMME: Design began September 1979; three flying prototypes, two static/fatigue test aircraft and one pre-series demonstrator built (first flight, by PT-ZBA, 27 July 1983); Brazilian CTA certification with original PW115 engines on 10 May 1985 followed by FAA (FAR Pt 25) type approval 9 July 1985, British/French/German approval 1986 and Australian in April 1990; deliveries began June 1985 (to Atlantic Southeast Airlines, USA, entering service that October); first order for corporate version (United Technologies Corporation, USA) received August 1986, delivered following month.

From October 1986 (c/n 120028), all Brasilias delivered incorporate composites materials equivalent to 10 per cent of aircraft basic empty weight; also since late 1986 has been available in hot-and-high version (certified 26 August 1986) with PW118A engines which maintain maximum output up to ISA + 15°C at S/L (first customer Skywest of USA); on 4 January 1989, first prototype began flight trials with Honeywell TPE331-12B turboprop on port side of rear fuselage, as testbed for engine installation of subsequently cancelled Embraer/FMA CBA-123; 300th Brasilia delivered 4 September 1995; more than five million flying hours by September 1998, at which time Brasilias had carried more than 60 million passengers.

Extended-range EMB-120ER (now standard version) announced June 1991, certified by CTA February 1992.

CURRENT VERSIONS: **EMB-120**: Launch production version, with 1,118.5 kW (1,500 shp) PW115 engines and Hamilton Sundstrand 14RF four-blade propellers (replaced at early stage by higher output PW115s of 1,193 kW; 1,600 shp). Details in 1996-97 and earlier *Jane's*.

EMB-120RT (Reduced Take-off): Initial production version; 1,342 kW (1,800 shp) PW118 engines for better field performance; most early aircraft now retrofitted to RT standard. Also, from late 1986, in hot/high version with PW118As (see Programme above). See 1996-97 and earlier editions for details.

EMB-120FC: 'Full Cargo' conversion of EMB-120RT developed in USA by IASG and approved by Embraer. Prototype (N453UE) converted June 2000 and exhibited at Farnborough in following month. Cargo volume 31.0 m³ (1,095 cu ft); maximum payload 3,700 kg (8,157 lb); empty weight 7,200 kg (15,873 lb); maximum T-O weight 12,000 kg (26,455 lb). Initial 10 kits ordered by North South Airways.

EMB-120 Combi: Mixed configuration version of ER with quick-release seats, 9 g movable rear bulkhead and cargo restraint net; retains toilet and galley; typical capacity 30 passengers and 700 kg (1,543 lb) of cargo or 19 passengers and 1,100 kg (2,425 lb) cargo.

EMB-120QC (Quick-Change): Convertible in 40 minutes from 30-passenger layout to 3,500 kg (7,716 lb) all-cargo configuration with floor and sidewall protection, fire detection system, smoke curtain aft of flight deck, 9 g movable rear bulkhead and cargo restraint net; conversion from cargo to passenger interior takes 50 minutes. First customer (14 May 1993 delivery), Total Linhas Aereas of Brazil.

EMB-120ER Brasilia Advanced: First delivery to Skywest Airlines in May 1993; standard production version from 1994, previously referred to as EMB-120X or Improved Brasilia. Additional range obtained by increasing maximum T-O weight without major structural change; this also allows for increased standard passenger-plus-baggage weight (97.5 kg; 215 lb per person instead of 91 kg; 200 lb) and obligatory fitting of TCAS and flight data recorder. Incorporates several modifications and some redesign aimed at maximising both passenger comfort and dispatchability while simultaneously reducing operational and maintenance costs. These include redesigned, interchangeable leading-edges for all flying surfaces; deletion of the fin de-icing boot; improved door seals and redesigned interior panel joints to reduce passenger cabin noise; more comfortable Sicma-built pilot seats; redesigned flight deck door; and new-design overhead bins with wider doors and increased capacity. Other features include new passenger cabin lighting; redesigned flight deck and cabin ventilation systems; new windscreen frame fairings to improve resistance to condensation; new flight deck sun visors; improved flap system; and an increase in baggage/cargo compartment capacity to 700 kg (1,543 lb).

Existing Brasilias can be retrofitted to ER standard (first two customers for retrofit Delta Air Transport and Luxair); initial production deliveries (two) to Great Lakes Aviation in USA, December 1994.

Detailed description applies to EMB-120ER except where indicated.

EMB-120 Cargo: All-cargo version of ER with 4,000 kg (8,818 lb) payload capacity; floor and sidewall protection, fire detection system, smoke curtain between flight deck and cargo cabin, and cargo restraint net.

EMB-120K: Proposed maritime surveillance or ASW version for Brazilian Air Force; would be equipped with in-flight refuelling, giving 15 hour endurance; under study in 1999.

VC-97: VIP transport model for Brazilian Air Force's 6° Esquadrao de Transporte Aérea and Grupo de Transporte Especial, both at Brasilia.

CUSTOMERS: Embraer delivered 352nd (including prototypes) aircraft to SkyWest in September 1999. Production plans called for eight per year in 2000 and 2001, but no more produced during following 12 months. Total of 32 operators in 13 countries include Atlantic Southeast (with current fleet of 62), Comair (29), Continental Express (32) and SkyWest (92), all in USA.

DESIGN FEATURES: Low-mounted unswept wings, circular-section pressurised fuselage, all-swept T tail unit. Fixed incidence tailplane.

Wing section NACA 23018 (modified) at root, NACA 23012 at tip; incidence 2°; at 66 per cent chord, wings have 6° 30′ dihedral.

FLYING CONTROLS: Conventional and assisted. Internally balanced ailerons, and horn-balanced elevators, actuated mechanically (ailerons by dual irreversible actuators); serially hinged two-segment rudder actuated hydraulically by Bertea CSD unit; trim tabs in ailerons (one port, two starboard) and each elevator. Hydraulically actuated, electrically controlled, double-slotted Fowler trailing-edge flap inboard and outboard of each engine nacelle, with small plain flap beneath each nacelle; no slats, slots, spoilers or airbrakes. Small fence on each outer wing between outer flap and aileron; twin ventral strakes under rear fuselage.

STRUCTURE: Kevlar-reinforced glass fibre for wing and tailplane leading-edges and tips, wingroot fairings, dorsal fin, fuselage nosecone and (when no APU fitted) tailcone; Fowler flaps of carbon fibre. Remainder conventional semi-monocoque/stressed skin structure of 2024/7050/7475 aluminium alloys with chemically milled skins. Fuselage pressurised between flat bulkhead forward of flight deck and hemispherical bulkhead aft of baggage compartment, and meets damage tolerance requirements of FAR Pt 25 (Transport category) up to Amendment 25-54. Wing is single continuous three-spar fail-safe structure, attached to underfuselage frames; tail surfaces also three-spar.

LANDING GEAR: Retractable tricycle type, with Goodrich twin wheels and oleo-pneumatic shock-absorber on each unit (main units 12 in, nose unit 8 in). Hydraulic actuation; all units retract forward (main units into engine nacelles). Hydraulically powered nosewheel steering. Mainwheel tyres 24×7.25-12 (12 ply) tubeless; nose tyre 18×5.5 (8 ply) tubeless; pressure 6.90 to 7.58 bars (100 to 110 lb/sq in) on main units, 4.14 to 4.83 bars (60 to 70 lb/sq in) on nose unit. Goodrich carbon brakes standard (steel optional). Hydro Aire anti-skid system standard; autobrake optional. Minimum ground turning radius 15.76 m (51 ft 8½ in). Nosewheel guard optional for operation from unpaved surfaces.

POWER PLANT: Two Pratt & Whitney Canada PW118 or PW118A turboprops, each rated at 1,342 kW (1,800 shp) for T-O and maximum continuous power, and driving a Hamilton Sundstrand 14RF-9 four-blade constant-speed reversible-pitch autofeathering propeller with glass fibre blades containing aluminium spars. Fuel in two-cell 1,670 litre (441 US gallon; 367.2 Imp gallon) integral tank in each wing; total capacity 3,340 litres (882 US gallons; 734.4 Imp gallons), of which 3,308 litres (874 US gallons; 728 Imp gallons) are usable. Single-point pressure refuelling (beneath outer starboard wing), plus gravity point in upper surface of each wing. Oil capacity 9 litres (2.4 US gallons; 2.0 Imp gallons).

ACCOMMODATION: Two-pilot flight deck. Main cabin accommodates attendant and 30 passengers in three-abreast seating at 79 cm (31 in) pitch, with overhead lockable baggage racks, in pressurised and air conditioned environment. Passenger seats of carbon fibre and Kevlar, floor and partitions of carbon fibre and Nomex sandwich, side panels and ceiling of glass fibre/Kevlar/Nomex/carbon fibre sandwich. Provisions for wardrobe, galley and toilet. Quick-change interior available optionally (first customer, Total Linhas Aéreas in 1993), for 30 passengers or 3,500 kg (7,716 lb) of cargo. Downward-opening main passenger door, with airstairs, forward of wing on port side. Type II emergency exit on starboard side at rear. Overwing Type III emergency exit on each side. Pressurised baggage compartment aft of passenger cabin, with large door on port side.

Also available with all-cargo interior; executive or military transport interior; or in mixed-traffic version with 24 or 26 passengers (toilet omitted in latter case), and 900 kg (1,984 lb) of cargo in enlarged rear baggage compartment.

SYSTEMS: Honeywell air conditioning/pressurisation system (differential 0.48 bar; 7 lb/sq in), with dual packs of recirculation equipment. Duplicated hydraulic systems (pressure 207 bars; 3,000 lb/sq in), each powered by an engine-driven pump, for landing gear, flap, rudder and brake actuation, and nosewheel steering. Emergency standby electric pumps on each system, plus single standby hand pump, for landing gear extension. Main electrical power supplied by two 28 V 400 A DC starter/generators; two 28 V 100 A DC auxiliary brushless generators for secondary and/or emergency power; one 24 V 40 Ah

First Embraer EMB-120FC Brasilia freighter conversion *(Paul Jackson/Jane's)*

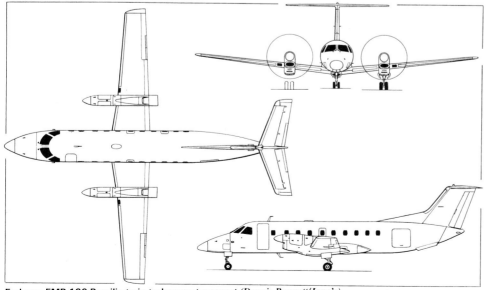

Embraer EMB-120 Brasilia twin-turboprop transport *(Dennis Punnett/Jane's)*

BRAZIL: AIRCRAFT—EMBRAER

Ni/Cd battery for assisted starting and emergency power. Main and standby 450 VA static inverters for 26/115 V AC power at 400 Hz. Single high-pressure (127.5 bars; 1,850 lb/sq in) oxygen cylinder for crew; individual chemical oxygen generators for passengers. Pneumatic de-icing for wing and tail leading-edges; electrically heated windscreens, propellers and pitot tubes; bleed air de-icing of engine air intakes. Optional Honeywell GTCP36-150(A) APU in tailcone, for electrical and pneumatic power supply.

AVIONICS: Rockwell Collins Pro Line II digital package as core system.

Comms: Dual VHF-22 com transceivers, one TDR-90 transponder, Dorne & Margolin DMELT-81 emergency locator transmitter, Fairchild voice recorder, dual Avtech audio/interphones and Avtech PA and cabin interphone all standard. Third VHF com, second transponder, Motorola Selcal and flight entertainment music all optional. Alternative Honeywell avionics package available to special order.

Radar: Rockwell Collins WXR-270 weather radar standard; WXR-300 optional.

Flight: Dual VIR-32 VHF nav receivers, one ADF-60A, DME-41 and CLT-22/32/62/92 control heads. Optional single/dual JET RNS-8000 3D or Racal Avionics RN 5000 nav and second DME. Optional equipment includes APS-65 digital autopilot, single/dual FCS-65 digital flight director, single/dual BAE Systems Canada CMA-771 Alpha VLF/Omega, MLS, GPWS, flight recorder.

Instrumentation: Dual AHRS-85 digital strapdown AHRS, dual ADI-84, dual EHSI-74, dual RMI-36 and JET standby altitude indicator standard. Optional equipment includes dual EFIS-86 electronic flight instrumentation systems, MFD-85 multifunction display, dual ALT-55 radio altimeters and altitude alerter/preselect.

DIMENSIONS, EXTERNAL:
Wing span	19.78 m (64 ft 10¾ in)
Wing chord: at root	2.81 m (9 ft 2¾ in)
at tip	1.40 m (4 ft 7 in)
Wing aspect ratio	9.9
Length overall	20.07 m (65 ft 10¼ in)
Fuselage: Length	18.73 m (61 ft 5½ in)
Max diameter	2.28 m (7 ft 5¾ in)
Height overall	6.35 m (20 ft 10 in)
Elevator span	6.94 m (22 ft 9¼ in)
Wheel track (c/l of shock-struts)	6.58 m (21 ft 7 in)
Wheelbase	6.97 m (22 ft 10½ in)
Propeller diameter	3.20 m (10 ft 6 in)
Propeller ground clearance (min)	0.52 m (1 ft 8½ in)
Passenger door (fwd, port): Height	1.70 m (5 ft 7 in)
Width	0.77 m (2 ft 6¼ in)
Height to sill	1.47 m (4 ft 10 in)
Cargo door (rear, port): Height	1.36 m (4 ft 5½ in)
Width	1.30 m (4 ft 3¼ in)
Height to sill	1.67 m (5 ft 5¾ in)
Emergency exit (rear, stbd): Height	1.37 m (4 ft 6 in)
Width	0.51 m (1 ft 8 in)
Height to sill	1.56 m (5 ft 1½ in)
Emergency exits (overwing, each):	
Height	0.91 m (3 ft 0 in)
Width	0.51 m (1 ft 8 in)
Emergency exits (flight deck side windows, each):	
Min height	0.48 m (1 ft 7 in)
Min width	0.51 m (1 ft 8 in)

DIMENSIONS, INTERNAL:
Cabin, excl flight deck and baggage compartment:	
Length	9.38 m (30 ft 9¼ in)
Max width	2.10 m (6 ft 10¾ in)
Max height	1.76 m (5 ft 9¼ in)
Floor area	15.0 m² (161 sq ft)
Volume	27.4 m³ (968 cu ft)
Rear baggage compartment volume:	
30-passenger version	6.4 m³ (226 cu ft)
all-cargo version	2.7 m³ (95 cu ft)
passenger/cargo version	11.0 m³ (388 cu ft)
Cabin, incl flight deck and baggage compartment:	
total volume	approx 41.8 m³ (1,476 cu ft)
Max available cabin volume (all-cargo version)	
	31.1 m³ (1,098 cu ft)

AREAS:
Wings, gross	39.43 m² (424.4 sq ft)
Ailerons (total)	2.88 m² (31.00 sq ft)
Trailing-edge flaps (total)	3.23 m² (34.77 sq ft)
Fin, incl dorsal fin	5.74 m² (61.78 sq ft)
Rudder	2.59 m² (27.88 sq ft)
Tailplane	6.10 m² (65.66 sq ft)
Elevators, (total, incl tabs)	3.90 m² (41.98 sq ft)

WEIGHTS AND LOADINGS:
Weight empty, equipped*	7,150 kg (15,763 lb)
Operating weight empty	7,560 kg (16,667 lb)
Max usable fuel	2,660 kg (5,864 lb)
Max payload*	3,340 kg (7,363 lb)
Max T-O weight	11,990 kg (26,433 lb)
Max ramp weight	12,080 kg (26,632 lb)
Max landing weight	11,700 kg (25,794 lb)
Max zero-fuel weight	10,900 kg (24,030 lb)
Max wing loading	304.1 kg/m² (62.28 lb/sq ft)
Max power loading	4.47 kg/kW (7.34 lb/shp)

*4 kg (8.8 lb) payload decrease/empty weight increase with PW118A engines

PERFORMANCE:
Max operating speed (V_{MO})	272 kt (504 km/h; 313 mph) EAS
Max level speed at 6,100 m (20,000 ft)	
	327 kt (606 km/h; 377 mph)
Max cruising speed at 7,620 m (25,000 ft)	
PW118	300 kt (555 km/h; 345 mph)
PW118A	313 kt (580 km/h; 360 mph)
Long-range cruising speed at 7,620 m (25,000 ft)	
	270 kt (500 km/h; 311 mph)
Stalling speed, power off:	
flaps up	120 kt (223 km/h; 138 mph) CAS
flaps down	89 kt (165 km/h; 103 mph) CAS
Max rate of climb at S/L	762 m (2,500 ft)/min
Rate of climb at S/L, OEI	168 m (550 ft)/min
Service ceiling: PW118	8,840 m (29,000 ft)
PW118A	9,750 m (32,000 ft)
Service ceiling, OEI, AUW of 11,200 kg (24,690 lb):	
PW118 or PW118A	4,600 m (15,100 ft)
FAR Pt 25 T-O field length	1,550 m (5,085 ft)
FAR Pt 135 landing field length, MLW at S/L	
	1,390 m (4,560 ft)
Range at 9,150 m (30,000 ft), reserves for 100 n mile (185 km; 115 mile) diversion and 45 min hold:	
with max (30-passenger) payload (2,721 kg; 6,000 lb):	
PW118	850 n miles (1,575 km; 979 miles)
PW118A	800 n miles (1,482 km; 921 miles)
with max fuel:	
PW118, 20 passengers (1,785 kg; 3,935 lb payload)	
	1,629 n miles (3,017 km; 1,874 miles)
PW118A, 20 passengers (1,785 kg; 3,935 lb payload)	
	1,570 n miles (2,908 km; 1,806 miles)

OPERATIONAL NOISE LEVELS (FAR Pt 36, BCAR-N and ICAO Annex 16):
T-O	81.2 EPNdB
Approach	92.3 EPNdB
Sideline	83.5 EPNdB

UPDATED

EMBRAER ERJ-135 and ERJ-140
Follow entry for ERJ-145, of which they are variants.

EMBRAER ERJ-145
Brazilian Air Force designations: R-99A and R-99B

TYPE: Regional jet airliner.

PROGRAMME: Development plans revealed 12 June 1989, aimed at first flight late 1991 and first deliveries mid-1993; programme delayed by company cutbacks, complete redesign of wing and other changes, as described in 2000-01 and earlier *Jane's*.

First metal cut for prototype, and tooling fabrication, in second quarter 1993; assembly of prototype (PT-ZJA) began October 1994; fuselage sections mated January 1995; first flight 11 August 1995 ahead of formal roll-out and 'official' first flight a week later; first of three preseries aircraft (PT-ZJB) first flown 17 November 1995; second ('ZJC) flew on 14 February and third ('ZJD) 2 April 1996; FAA and Brazilian CTA certification (to FAR/JAR 25, FAR Pt 36, ICAO Annex 16 and FAR Pt 121) achieved 16 December 1996. Single prototype and three preseries aircraft undertook a 1,600 hour, 13 month development flight testing and certification programme. Deliveries began on 19 December 1996 with two aircraft (N15925 and N15926) to US launch customer Continental Express. Designation changed from EMB-145 to ERJ-145 in October 1997 to reflect 'Regional Jet' terminology. Certified by the aviation authorities of 27 countries by September 1998.

In addition to its suitability for executive transport and corporate shuttle roles, Embraer foresees military potential for the ERJ-145 as a tanker aircraft for small combat units, and as an AEW, elint, comint, sigint or battlefield surveillance platform. See Current Versions.

ERJ-145 G-EMBP was delivered to British Airways on 25 August 2000 as the 300th Embraer regional jet

EMBRAER—AIRCRAFT: BRAZIL

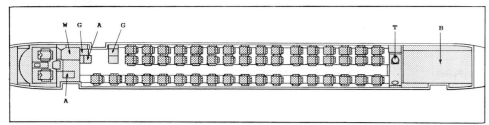

Standard 50-passenger layout in ERJ-145 *(James Goulding/Jane's)*
A: attendant's seat, B: baggage, G: galley, T: toilet, W: wardrobe

Two static test airframes: first (802) completed trials on 30 August 1996; second (803) began 10 year programme in December 1996.

CURRENT VERSIONS: **ERJ-145ER** (Extended Range): Initial version.

ERJ-145LR (Long Range): Introduced in 1998 for full passenger payload range of 1,640 n miles (3,037 km; 1,887 miles); increases in fuel capacity and all operating weights; and uprated AE 3007A1 turbofans providing 15 per cent more thermodynamic power, but flat rated to 33.1 kN (7,430 lb st), for improved climb and hot weather cruise performance.

ERJ-145XR (Extra Long Range): Announced at Farnborough International Air Show 25 July 2000 with launch order for 75 from Continental Express; features winglets, and uprated AE 3007A1E turbofans providing lower specific fuel consumption, improved hot-and-high operation and higher single-engine ceiling; range 2,000 n miles (3,704 km; 2,301 miles). First flight in first quarter 2001, deliveries from second quarter 2002.

ERJ-135: Short-fuselage, 37-seat version; described separately.

ERJ-140: Mid-size, 44-seat version; described separately.

EMB-145AEW&C: Formerly EMB-145SA; Brazilian Air Force designation R-99A. Airborne early warning and remote sensing version of ERJ-145LR developed for Brazilian government's *SIstema de Vigilancia de AMazonia* (SIVAM) programme for which Raytheon is prime contractor; initial requirement for five; contract signature March 1997; deliveries for completion by May 2002. Selected on 15 December 1998 by Greek Air Force for its four-aircraft AEW requirement, with delivery from 2002; contract, signed 1 July 1999, valued at US$500 million. Announced late 1996 and features a strengthened fuselage, ventral strakes, more powerful APU, increased fuel capacity (three extra tanks at extreme rear of cabin, plus jettison capability), extending endurance to more than 8 hours; enhanced electrical system, five seats for relief crew and four (with provision for additional two) operators' consoles, including tactical co-ordinator. Flight crew of two.

Mission systems comprise civil version of Ericsson PS-890 Erieye side-looking airborne radar with antenna housed in long overfuselage fairing, optimised for lower-speed targets typically encountered in border incursions; five Erieye systems purchased at cost of US$145 million in 1997 for installation in EMB-145SAs, with first system delivery scheduled for 1999. Onboard command and control system, and BAE Systems North America Comms-Non Comms system. Radar is pulse Doppler type, operating in E/F-band, offering coverage from very low level up to about 25,000 m (82,000 ft) and at ranges exceeding 162 n miles (300 km; 186 miles). Datalink; GPS; secure communications. Second airframe (PP-XSA/6702) was first to fly, on 22 May 1999, ahead of formal roll-out on 28 May. Service entry with 2°/6°Grupo at Anapolis AFB scheduled for June 2001, following systems integration by Raytheon in USA.

Greek selection of Erieye-equipped EMB-145 announced 1 July 1999; four aircraft; deliveries due from 2002.

EMB-145RS: Brazilian Air Force designation R-99B. Remote sensing version, of which three ordered for FAB's SIVAM programme for delivery commencing first quarter 2001. Similar to AEW variant, and with ventral strakes, but different mission systems for primary roles in natural resources exploitation, environmental and river pollution control, economic activities, ground occupation monitoring and illegal activities surveillance. Main sensor is version of Canadian MacDonald Dettwiler IRIS (Integrated Radar Imaging System) synthetic aperture radar, installed in underfuselage bulge with auxiliary antennas beneath wingroots, operating in D-band interferometric mode and capable of generating 3-D imagery. Other main sensors include Star Safire FLIR mounted behind nosewheel bay, Daedalus ultraviolet/visible/infra-red linescanner and BAE Systems North America Comms-Non Comms system. Roll-out (PP-XRT/6751) November 1999, with first delivery to 2°/6° Grupo at Anapolis AFB was due in January 2001 and final delivery in May 2002.

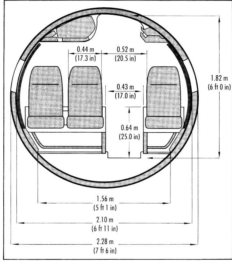

ERJ-145 fuselage cross-section

EMB-145AGS: Airborne ground sensor version, under study during 2000; equipped with a mission package comprising Airborne Platform Subsystem (APS), Airborne Mission Equipment Subsystem (AMES) and Ground Exploitation Station Subsystem (GESS) including HF, UHF, VHF, ELINT and IMINT equipment, providing a self-deployable and cost-effective surface reconnaissance system.

EMB-145MP and **EMB-145MP/ASW**: Maritime patrol and anti-submarine warfare versions, under development during 2000; equipped with surveillance radar with multiple target track-while-scan mode, autodetection, FLIR interface, digital map, incorporated tactical aids, SAR/ISAR mode allowing real-time imaging, adaptive processing for different sea states, and simultaneous side and range views; high-altitude and resolution FLIR; ESM suite; COMINT/ELINT; MAD; IFF/SSR and acoustics.

CUSTOMERS: Total of 549 ordered and 294 delivered by 31 January 2001; further 316 options. Three hundredth ERJ delivery was an ERJ-145 to British Airways on 25 August 2000. Deliveries in 2000 were 112. See table.

EMBRAER ERJ-145 PRODUCTION
(at May 2000)

Operator	Orders	Options	Delivered	Backlog
Alitalia Express	6	15	1	5
American Eagle	50	17	50	0
British Midland	10	10	5	5
Brymon Airways	7	14	3	4
Cirrus	1	0	1	0
City Airlines/ Skyways	4	11	4	0
Continental Express	161	39	64	97
Crossair	25	15	4	21
ERA	2	0	2	0
Flandre Air	2	0	0	2
KLM Excel	3	2	1	2
Lot Polish	6	6	4	2
Luxair	9	2	7	2
Manx Airlines	23	2	15	8
Mesa Airlines	36	64	2	34
Portugalia	8	0	8	0
Proteus	8	0	2	6
Regional Airlines	15	0	9	6
Rheintalflug	2	6	1	1
Rio-Sul	15	15	15	0
Trans States Airlines	9	0	9	0
Wexford Management	45	45	7	38
Totals	**447**	**263**	**214**	**233**

Note: Additional 102 firm orders received by 31 December 2000 comprising Air Caraibes (two), Air Moldova (two), Alitalia Express (two), American Eagle (six), Axon Airlines (four), Continental Express (64), Flandre Air (one), Lot Polish (10), Proteus (two), Rheintalflug (one), Sichuan (five) and Trans States Airlines (three).

COSTS: Estimated development costs US$300 million; unit cost US$15.5 million (1997). AEW version (PS-890 radar) US$50 million to US$60 million (1996).

Following description applies to ERJ-145ER except where indicated.

DESIGN FEATURES: Stretched EMB-120 Brasilia fuselage (with tailcone adapted for rear-mounted engine installation), allied to new-design wing with Embraer supercritical section; CBA-123 nose and cabin; T tailplane.
Wing sweepback 22° 43' 48" at quarter-chord.

FLYING CONTROLS: Conventional and assisted. Ailerons and two-section rudder hydraulically actuated, with artificial feel; mechanically actuated elevator with automatic and spring tab. Four-segment in-flight and ground spoilers; two pairs of electrically actuated double-slotted flaps.

STRUCTURE: Fuselage as for Brasilia; two-spar wing with integral fuel tanks, plus auxiliary third spar supporting landing gear; T tail unit with aluminium main boxes; wing and tailplane leading-edges aluminium, fin leading-edge composites sandwich. Gamesa (Spain) builds wings, wing/body fairings, main landing gear doors and engine nacelles; rear fuselage section 1, including engine pylons and passenger/service/baggage doors, plus centre-fuselage section 1, including doors, by Sonaca (Belgium); fin, tailplane and elevators by ENAER (Chile); engine nacelles and thrust reversers by International Nacelle Systems; nose radome by Norton; passenger cabin and baggage compartment interiors by C & D Interiors (USA). Structure designed for an economical service life of 60,000 flights.

LANDING GEAR: Twin-wheel main legs retract inward into wing/fuselage fairings; twin-wheel nose unit retracts forward. EDE/Liebherr landing gear system, with EDE responsible for whole system and Liebherr for development and production of nose unit. Goodrich wheels and carbon brakes. Tyre sizes 30×9.5-14 (16 ply) tubeless

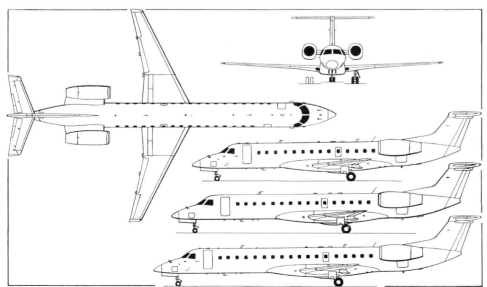

ERJ-145 rear-engined twin-turbofan regional airliner with additional side views of ERJ-135 (top) and ERJ-140 (centre) *(Mike Keep/Jane's)*

BRAZIL: Aircraft—EMBRAER

Second (first to fly) Embraer EMB-145AEW&C surveillance aircraft with dorsal Erieye surveillance radar antenna

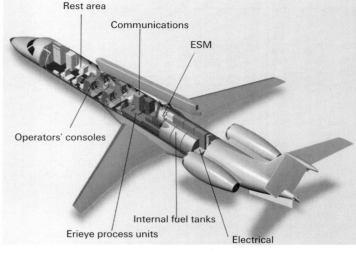

Main internal features of the EMB-145AEW&C

(main), 19.5×6.75-8 (8 ply) tubeless (nose); tyre pressure 8.60 to 9.00 bars (125 to 130 lb/sq in). Minimum ground turning radius at nosewheel 12.51 m (41 ft 0 in). Minimum turning circle 29.22 m (95 ft 10½ in).

POWER PLANT: Two turbofans pylon-mounted on rear cone of fuselage: ERJ-145ER has two 31.3 kN (7,040 lb st), FADEC-equipped Rolls-Royce AE 3007As; and ERJ-145LR has two AE 3007A1s flat rated at 33.1 kN (7,430 lb st), which are also optional on the ERJ-145ER, both with FADEC. ERJ-145XR has two AE 3007A1Es. For fuel, see Weights and Loadings. Parker Hannifin fuel system. Clamshell-type thrust reversers optional.

ACCOMMODATION: Two pilots, flight observer and cabin attendant. Standard accommodation for 50 passengers, three-abreast at seat pitch of 79 cm (31 in). Carry-on baggage wardrobe, galley and cabin attendant's seat at front of cabin; toilet and main baggage compartment at rear of cabin. Cabinet plus overhead bins carry-on baggage capacity 358 kg (789 lb); underseat capacity 450 kg (992 lb); main baggage compartment capacity 1,200 kg (2,646 lb). Additional baggage cabinet or galley capacity can be provided by removing one or two single forward passenger seats. Outward-opening plug-type door, incorporating airstair, at front on port side, identical to that of EMB-120; upward-sliding baggage door at rear on port side; sideways-opening service door at front on starboard side; inward-opening emergency exit above wing on each side. Entire accommodation, including baggage compartments, pressurised and air conditioned.

SYSTEMS: Liebherr Aerospace pressurisation system (maximum differential 0.54 bar; 7.8 lb/sq in) maintains 2,440 m (8,000 ft) cabin altitude to 11,275 m (37,000 ft). Hamilton Sundstrand air conditioning and bleed air systems (wing and tailplane leading-edges and engine intakes anti-iced by engine bleed air); electric anti-icing system for windscreen and static and pitot tubes and sensors. Lucas electrical power generation system. APIC 18.6 kW (25 shp) APS-500 APU. Honeywell air turbine starter. Parker Hannifin flight control and steering systems. Hydro-Aire brake-by-wire control system. EROS oxygen system.

AVIONICS: Honeywell Primus 1000 as core system.
 Comms: Dual Primus II radios and radio management units.
 Radar: Primus 1000 colour weather radar.
 Flight: Dual digital air data computers, dual AHRS, TCAS and GPWS standard. FMS/GPS optional. Flight Dynamics HUD selected April 1998 for certification in 2000, providing Cat. III landing capability.
 Instrumentation: EFIS panel comprising five 280 × 180 mm (11 × 7 in) displays, two PFDs, two MFDs and IECAS.

DIMENSIONS, EXTERNAL:
Wing span	20.04 m (65 ft 9 in)
Wing chord: at root	4.09 m (13 ft 5 in)
at tip	1.04 m (3 ft 5 in)
Wing aspect ratio	7.9
Length overall	29.87 m (98 ft 0 in)
Fuselage: Length	27.93 m (91 ft 7½ in)
Max diameter	2.28 m (7 ft 5¾ in)
Height overall	6.76 m (22 ft 2 in)
Tailplane span	7.55 m (24 ft 9 in)
Wheel track (c/l of shock-struts)	4.10 m (13 ft 5½ in)
Wheelbase	14.45 m (47 ft 5 in)
Passenger door (fwd, port): Height	1.70 m (5 ft 7 in)
Width	0.71 m (2 ft 4 in)
Height to sill (max)	1.63 m (5 ft 4 in)
Baggage door (rear, port): Height	1.12 m (3 ft 8 in)
Width	1.00 m (3 ft 3¼ in)
Height to sill (max)	1.76 m (5 ft 9¼ in)
Service door (rear, stbd): Height	1.42 m (4 ft 8 in)
Width	0.62 m (2 ft 0½ in)
Height to sill	1.60 m (5 ft 3 in)
Emergency exits (overwing, each):	
Height	0.92 m (3 ft 0¼ in)
Width	0.51 m (1 ft 8 in)

DIMENSIONS, INTERNAL:
Cabin (excl flight deck and baggage compartment, incl toilet):	
Length	16.49 m (54 ft 1¼ in)
Max width	2.10 m (6 ft 10¾ in)
Max height	1.83 m (6 ft 0 in)
Max aisle width	0.52 m (1 ft 8½ in)
Floor area	25.7 m² (277 sq ft)
Volume	53.0 m³ (1,872 cu ft)
Baggage compartment: Length	3.26 m (10 ft 8¼ in)
Baggage volume:	
wardrobe and stowage compartment	1.4 m³ (49 cu ft)
overhead bins	1.9 m³ (67 cu ft)
underseat	2.3 m³ (80 cu ft)
baggage compartment	9.2 m³ (325 cu ft)

AREAS:
Wings, gross	51.18 m² (550.9 sq ft)
Ailerons (total)	1.70 m² (18.30 sq ft)
Trailing-edge flaps (total)	8.36 m² (89.99 sq ft)
Spoilers (total)	2.32 m² (24.97 sq ft)
Fin	5.07 m² (54.57 sq ft)
Rudder	2.13 m² (22.93 sq ft)
Tailplane	11.20 m² (120.55 sq ft)
Elevators (total, incl tabs)	3.34 m² (35.95 sq ft)

WEIGHTS AND LOADINGS:
Operating weight empty:	
ERJ-145ER	11,700 kg (25,794 lb)
ERJ-145LR	11,800 kg (26,015 lb)
ERJ-145XR	12,934 kg (28,515 lb)
Max fuel: ERJ-145ER	4,173 kg (9,200 lb)
ERJ-145LR	5,187 kg (11,435 lb)
Max payload: ERJ-145ER	5,400 kg (11,905 lb)
ERJ-145LR	6,100 kg (13,448 lb)
ERJ-145XR	5,566 kg (12,271 lb)
Max T-O weight: ERJ-145ER	20,600 kg (45,415 lb)
ERJ-145LR	22,000 kg (48,500 lb)
ERJ-145XR	23,990 kg (52,889 lb)
Max ramp weight: ERJ-145ER	20,700 kg (45,635 lb)
ERJ-145LR	22,100 kg (48,721 lb)
Max landing weight: ERJ-145ER	18,700 kg (41,226 lb)
ERJ-145LR	19,300 kg (42,549 lb)
ERJ-145XR	20,000 kg (44,092 lb)
Max zero-fuel weight:	
ERJ-145ER	17,100 kg (37,698 lb)
ERJ-145LR	17,900 kg (39,463 lb)
ERJ-145XR	18,500 kg (40,785 lb)
Max wing loading:	
ERJ-145ER	402.5 kg/m² (82.44 lb/sq ft)
ERJ-145LR	429.8 kg/m² (88.04 lb/sq ft)
Max power loading:	
ERJ-145ER	329 kg/kN (3.23 lb/lb st)
ERJ-145LR	333 kg/kN (3.26 lb/lb st)

PERFORMANCE:
High cruising speed: all	M0.78 (450 kt; 833 km/h; 518 mph)
Time to climb to 10,670 m (35,000 ft)	20 min
Service ceiling: both	11,275 m (37,000 ft)
Service ceiling, OEI: both	6,100 m (20,000 ft)
FAR T-O field length at S/L: ERJ-145ER/LR	1,970 m (6,465 ft)
ERJ-145XR	2,430 m (7,972 ft)
FAR landing field length, S/L, at typical landing weight:	
ERJ-145ER/LR	1,300 m (4,265 ft)
ERJ-145XR	1,440 m (4,724 ft)
Range, 50 passengers, 100 n miles (185 km; 115 mile) diversion, 45 min reserves	
ERJ-145ER	1,600 n miles (2,963 km; 1,841 miles)
ERJ-145XR	2,000 n miles (3,704 km; 2,301 miles)

UPDATED

EMBRAER ERJ-135

TYPE: Regional jet airliner.

PROGRAMME: Launched 16 September 1997; two preseries ERJ-145s (001/PT-ZJA and 002/PT-ZJC) modified to create two prototype ERJ-135s; roll-out (PT-ZJA) 12 May 1998, followed by first flight 4 July 1998; public debut at Farnborough Air Show September 1998; second aircraft (PT-ZJC), flown 24 September 1998, for systems testing before conversion to production standard in March 1999. Brazilian Centro Técnico Aeroespacial (CTA) certification achieved in June 1999; FAA certification 16 July 1999; JAA approval was expected in October 1999. First delivery 23 July 1999 to Continental Express; other early aircraft to American Eagle. Deliveries have been 16 in 1999 and 45 in 2000; total increased to 66 by 31 January 2001.
 Programme based on estimates of 500 sales.

CURRENT VERSIONS: **ERJ-135:** Regional airliner, *as described*.
 Legacy: Corporate version; described separately.

CUSTOMERS: Total of 146 firm commercial orders (plus 61 options) by 31 December 2000, comprising American Eagle (40), British Midland (four), City Airlines AB (two), Continental Express (50), Flandre Air (seven), Proteus (three), Regional Airlines France (five), Regional Air Lines Morocco (five), and South African Airlink (30). Additionally, one VIP-configured ERJ-135LR handed over to Greek Air Force on 7 January 2000. Deliveries have been 16 in 1999 and 45 in 2000; total increased to 66 by 31 January 2001.

COSTS: Development cost US$100 million, of which 40 per cent provided by risk-sharing partners. Unit cost US$11.8 million.

DESIGN FEATURES: Shares 96 per cent commonality with ERJ-145 including engines, wings, tail surfaces, flight deck and main systems; fuselage shortened by 3.53 m (11 ft 7 in) by removal of two frames (4.84 m; 15 ft 10½ in ahead of wing and 3.07 m; 10 ft 0¾ in at rear) and substitution of two shorter frames (2.85 m; 9 ft 4¼ in and 1.53 m; 5 ft 0¼ in, respectively).

FLYING CONTROLS: As for ERJ-145.
STRUCTURE: As for ERJ-145.
LANDING GEAR: As for ERJ-145.
POWER PLANT: ERJ-135ER has two 31.3 kN (7,040 lb st) Rolls-Royce AE 3007A turbofans. ERJ-135LR has two Rolls-Royce AE 3007A1 turbofans.
ACCOMMODATION: Standard accommodation for 37 passengers in three-abreast configuration.
SYSTEMS: As for ERJ-145.
DIMENSIONS, EXTERNAL: As for ERJ-145 except:
Length: overall	26.33 m (86 ft 4½ in)
fuselage	24.39 m (80 ft 0¼ in)
Wheelbase	12.43 m (40 ft 9¼ in)

The prototype remote sensing EMB-145RS, with belly-mounted IRIS radar

EMBRAER—AIRCRAFT: BRAZIL

Second prototype Embraer ERJ-135 *(Paul Jackson/Jane's)*

Flight deck of Embraer ERJ-135

DIMENSIONS, INTERNAL:
  Cabin (excl flight deck and baggage compartment, incl
    toilet): Length   12.95 m (42 ft 5¼ in)
  Baggage compartment: Length   3.34 m (10 ft 11½ in)
  Baggage volume:
    wardrobe and stowage compartment 1.0 m³ (35 cu ft)
    overhead bins   1.4 m³ (49 cu ft)
    underseat   1.7 m³ (60 cu ft)
  Galley volume   0.99 m³ (35 cu ft)
WEIGHTS AND LOADINGS:
  Operating weight empty:
    ERJ-135ER   11,200 kg (24,692 lb)
    ERJ-135LR   11,300 kg (24,912 lb)
  Max fuel: ERJ-135ER   4,173 kg (9,200 lb)
    ERJ-135LR   5,187 kg (11,435 lb)
  Max payload: ERJ-135ER   4,400 kg (9,700 lb)
    ERJ-135LR   4,700 kg (10,362 lb)
  Max T-O weight: ERJ-135ER   19,000 kg (41,888 lb)
    ERJ-135LR   20,000 kg (44,092 lb)
  Max landing weight:
    ERJ-135ER, ERJ-135LR   18,500 kg (40,785 lb)
  Max ramp weight: ERJ-135ER   19,100 kg (42,108 lb)
    ERJ-135LR   20,100 kg (44,313 lb)
  Max zero-fuel weight: ERJ-135ER 15,600 kg (34,392 lb)
    ERJ-135LR   16,000 kg (35,274 lb)
  Max wing loading:
    ERJ-135ER   371.2 kg/m² (76.04 lb/sq ft)
    ERJ-135LR   390.8 kg/m² (80.04 lb/sq ft)
  Max power loading:
    ERJ-135ER   304 kg/kN (2.98 lb/lb st)
    ERJ-135LR   319 kg/kN (3.13 lb/lb st)
PERFORMANCE (estimated):
  Max cruising speed   M0.78 (450 kt; 833 km/h; 518 mph)
  Time to 10,670 m (35,000 ft)   20 min
  Service ceiling   11,275 m (37,000 ft)
  T-O field length at S/L   1,700 m (5,577 ft)
  Landing field length, S/L, at typical landing weight
                                       1,360 m (4,460 ft)
  Range with 37 passengers, 100 n mile (185 km;
    115 mile) diversion, 45 min reserves
                          1,700 n miles (3,148 km; 1,956 miles)

UPDATED

EMBRAER LEGACY

TYPE: Business jet.
PROGRAMME: Announced on eve of Farnborough International Air Show 23 July 2000; first flight of prototype, converted from second ERJ-135 prototype (PT-ZJC), then scheduled for February 2001; FAA/JAA certification expected in third quarter 2001; first deliveries September 2001.
CURRENT VERSIONS: Offered in **Executive**, **Corporate Shuttle** and **Government VIP** variants.
CUSTOMERS: Total of 31 orders and 31 options by October 2000; launch customer Swift Aviation of Phoenix, Arizona, ordered 25, with 25 options in July 2000; other announced customers include the Greek Air Force, which has ordered one in executive configuration; market estimated at 240 aircraft over a ten-year period.
COSTS: Corporate Shuttle US$15.5 million; Executive US$19 million (2000).

Description for ERJ-135 applies also to the Legacy, except the following.
DESIGN FEATURES: Compared with ERJ-135 has belly and aft fuel tanks in extended underwing fairing, plus winglets.
STRUCTURE: Airframe manufactured as ERJ-135 and modified to Legacy configuration; Embraer performs interior completion, using components supplied by Nordam.
POWER PLANT: Legacy Executive has two 33.0 kN (7,426 lb st) Rolls-Royce AE 3007A1P turbofans; Legacy Corporate Shuttle has 31.4 kN (7,057 lb st) AE 3007A1/3s.
ACCOMMODATION: Typical accommodation for 10 passengers in Executive configuration, or 20 in two-abreast arrangement at 91 cm (36 in) seat pitch in Corporate Shuttle layout.
AVIONICS: Honeywell Primus 1000 as core system.
WEIGHTS AND LOADINGS (A: Executive, B: Corporate Shuttle):
  Basic operating weight: A   12,990 kg (28,638 lb)
    B   12,510 kg (27,580 lb)
  Max fuel: A   8,030 kg (17,703 lb)
    B   5,135 kg (11,321 lb)
  Max payload: A   3,010 kg (6,636 lb)
    B   3,490 kg (7,694 lb)
  Payload with max fuel: A   970 kg (2,138 lb)
    B   2,355 kg (5,192 lb)
  Fuel with max payload: A   5,990 kg (13,206 lb)
    B   4,000 kg (8,818 lb)
  Max T-O weight: A   21,990 kg (48,479 lb)
    B   20,000 kg (44,092 lb)
  Max landing weight: A, B   18,500 kg (40,785 lb)
  Max ramp weight: A   22,060 kg (48,634 lb)
    B   20,070 kg (44,247 lb)
  Max zero-fuel weight: A, B   16,000 kg (35,274 lb)
  Max wing loading: A   429.7 kg/m² (88.00 lb/sq ft)
    B   390.8 kg/m² (88.04 lb/sq ft)
  Max power loading: A   332 kg/kN (3.26 lb/lb st)
    B   318 kg/kN (3.12 lb/lb st)
PERFORMANCE (estimated):
  Cruising speed at 11,885 m (39,000 ft):
    A, B   M 0.78 (447 kt; 828 km/h; 514 mph)
  T-O field length: A   1,924 m (6,312 ft)
    B   1,708 m (5,604 ft)
  Landing field length: A, B   1,360 m (4,462 ft)
  Range with NBAA IFR reserves:
    A with 10 passengers
                   3,200 n miles (5,926 km; 3,682 miles)
    B with 20 passengers
                   1,834 n miles (3,396 km; 2,110 miles)

NEW ENTRY

EMBRAER ERJ-140

TYPE: Regional jet airliner.
PROGRAMME: Launched 30 September 1999 at the European Regional Airline Association annual meeting in Paris; first flight of prototype, modified from prototype ERJ-135 c/n 801/PT-ZJA, 27 June 2000; public debut at Farnborough International Air Show July 2000; Brazilian CTA and FAA certification then anticipated first quarter 2001 with service entry shortly thereafter.
CUSTOMERS: Launch customer American Eagle announced order for 66 (plus 31 options) on 27 September 2000 and will receive a further 64 through transfer of EMB-135 orders beginning with August 2001 deliveries. No further orders by 31 December 2000, but 139 firm and 25 options by 1 February 2001, when American Eagle exercised some of its options.
COSTS: US$15.2 million (1999).

Prototype Embraer ERJ-140 making its international debut at Farnborough in July 2000 *(Paul Jackson/Jane's)*

Computer-generated image of Embraer Legacy twin-turbofan business aircraft

BRAZIL: AIRCRAFT—EMBRAER

Descriptions for the ERJ-135 and ERJ-145 apply also to the ERJ-140 except as follows:

DESIGN FEATURES: Shares 98 per cent commonality with ERJ-135/145, including engines, wings, tail surfaces, flight deck and main systems; ERJ-135 fuselage stretched by 2.30 m (7 ft 6½ in) by removal of two frames (2.85 m; 9 ft 4¼ in ahead of wing and 1.35 m; 4 ft 5 in at rear) and substitution of two longer frames (3.94 m; 12 ft 11 in and 2.56 m; 8 ft 4¾ in respectively).

ACCOMMODATION: Standard accommodation for 44 passengers, three-abreast at seat pitch of 39 cm (31 in). Wardrobe/carry-on baggage cabinet and galley at front of cabin, lavatory at rear.

DIMENSIONS, EXTERNAL:
Length: overall 28.45 m (93 ft 4 in)
fuselage 26.58 m (87 ft 2½ in)
Wheelbase 13.53 m (44 ft 4¾ in)

DIMENSIONS, INTERNAL:
Baggage volume:
Wardrobe and stowage compartment
0.93 m³ (32.84 cu ft)

WEIGHTS AND LOADINGS:
Basic operating weight 11,740 kg (25,882 lb)
Max fuel 5,187 kg (11,435 lb)
Max T-O weight 21,100 kg (46,517 lb)
Max landing weight 18,700 kg (41,226 lb)
Max ramp weight 21,200 kg (46,738 lb)
Max zero-fuel weight 17,100 kg (37,699 lb)
Max wing loading 412.3 kg/m² (84.44 lb/sq ft)
Max power loading (AE 3007A1)
319 kg/kN (3.13 lb/lb st)

PERFORMANCE (estimated, AE 3007A1):
T-O field length 1,970 m (6,463 ft)
Landing field length 1,340 m (4,396 ft)
Range with 44 passengers at long-range cruising speed, 100 n mile (185 km; 115 mile) diversion and 45 min reserves 1,630 n miles (3,018 km; 1,875 miles)
UPDATED

EMBRAER ERJ-170 and ERJ-190

TYPE: Regional jet airliner.

PROGRAMME: Announced (officially 'pre-launch') in February 1999; engine selection May 1999; first orders announced 14 June 1999; risk-sharing partners (see 'Structure', below) revealed at European Regional Airline Association annual meeting in Paris, 30 September 1999. Joint definition phase leading to ERJ-170 design freeze, completed in April 2000; first metal cut for first of six preseries ERJ-170s 14 July 2000; ERJ-170 first flight scheduled for late 2001, with six-aircraft, 1,500- to 1,800-hour flight test programme, culminating in certification in fourth quarter 2002; initial production rate up to two per month in 2002, rising to three per month in 2003 and four per month in 2004, with maximum capacity of six per month. Final assembly of the first of two preseries ERJ-190-200s will begin in late 2002; first flight in third quarter of 2003; first delivery (to Crossair) in second quarter 2004; first of two preseries ERJ-170-100s will fly in 2004 with certification and deliveries in mid-2005.

CURRENT VERSIONS: **ERJ-170**: Baseline version with 70 seats; available in Standard and Long-Range versions; first delivery December 2002.

ERJ-190-100: Longer version with 6.25 m (20 ft 6 in) fuselage stretch to accommodate 98 passengers; wing span increased by 2.56 m (8 ft 4¾ in); GE CF34-8E-10 engines; strengthened landing gear; available in Standard and Long-Range versions.

ERJ-190-200: Further stretch by 2.41 m (7 ft 11 in) to accommodate 108 passengers; available in Standard and Long-Range versions; first delivery June 2004.

Corporate: Proposed variant of ERJ-170 Long-Range with additional fuel tanks in baggage compartment area to extend range to more than 4,000 n miles (7,408 km; 4,603 miles).

CUSTOMERS: See table. Launch customers Crossair and Regional Airlines of France announced at Paris Air Show 14 June 1999; Crossair ordered 30 ERJ-170s and 30 ERJ-190-200s, with options on a further 100 ERJ-170/190s, for delivery (ERJ-170) from December 2002; Regional Airlines ordered 10 ERJ-170s, with options on a further five; General Electric Capital Aviation Services (GECAS) ordered six ERJ-170s, with 100 options on ERJ-170s and ERJ-190s, on 7 June 2000.

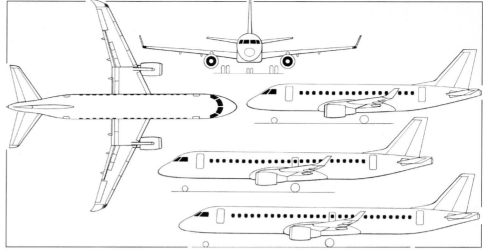

Provisional arrangement of ERJ-170 regional jet, with additional side views of ERJ-190-100 (centre) and ERJ-190-200 (bottom) *(James Goulding/Jane's)*
2001/0100394

EMBRAER ERJ-170/190 ORDERS
(at 31 December 2000)

Customer	Type	Orders
Crossair	170	30
	190	30
GECAS	170	50
Regional Airlines	170	10
Total		**120**

COSTS: ERJ-170 US$21 million; ERJ-190, US$24 million (both 1999). Development US$600 million for ERJ-170, plus US$150 million for ERJ-190.

DESIGN FEATURES: Design goals include low weight, simplicity of operation, high reliability, ease and economy of maintenance and ability to operate from same airports as ERJ-135/145. Low-wing airliner of conventional appearance, with podded engine below each wing and (on ERJ-170) winglets; airframe designed for an economic life of 60,000 to 80,000 cycles.

FLYING CONTROLS: Fly-by-wire. Ailerons, rudder and all-moving tailplane. Double-slotted flaps, five-section leading-edge slats and five-section spoilers on each wing.

STRUCTURE: Embraer is responsible for radome, forward fuselage, centre fuselage II, wing-to-fuselage fairing and final assembly; risk-sharing partners are C & D Interiors (cabin interior); Gamesa (rear fuselage and horizontal and vertical tail surfaces); General Electric (power plant and nacelles); Hamilton Sundstrand (tailcone, APU and air management and electrical systems); Honeywell (avionics); Kawasaki (wing stub, fixed leading- and trailing-edge assemblies, flaps, spoilers and control surfaces); Latécoère (centre fuselage I and III); Liebherr (landing gear); Parker Hannifin (hydraulic, flight control and fuel systems); and Sonaca (wing slats).

LANDING GEAR: Retractable tricycle type. Twin wheels on each unit.

POWER PLANT: ERJ-170 has two 62.28 kN (14,000 lb st) General Electric CF34-8E turbofans; ERJ-190 has two 82.29 kN (18,500 lb st) GE 34-8E-10 turbofans; both engines have FADEC.

ACCOMMODATION: Total of 70 (ERJ-170), 98 (ERJ-190-100) or 108 (ERJ-190-200) passengers, four abreast at 81 cm (32 in) seat pitch.

AVIONICS: Honeywell Primus Epic five-tube EFIS.

DIMENSIONS, EXTERNAL (all versions, except where stated):
Wing span: ERJ-170 over winglets 26.00 m (85 ft 3½ in)
ERJ-190-100, -200 28.56 m (93 ft 8½ in)
Length overall: ERJ-170 29.90 m (98 ft 1¼ in)
ERJ-190-100 36.15 m (118 ft 7¼ in)
ERJ-190-200 38.56 m (126 ft 6 in)
Height overall: ERJ 170 9.67 m (31 ft 8¾ in)
ERJ-190-100, -200 10.48 m (34 ft 4½ in)
Fuselage depth 3.35 m (11 ft 0 in)

DIMENSIONS, INTERNAL:
Cabin (excl flight deck):
Length: ERJ-170 19.37 m (63 ft 6½ in)
ERJ-190-100 25.39 m (83 ft 3½ in)
ERJ-190-200 27.83 m (91 ft 3¾ in)
Max width 2.74 m (9 ft 0 in)
Max height 2.00 m (6 ft 6¾ in)

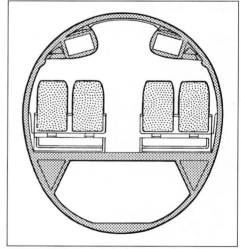

ERJ-170/190 fuselage cross-section
(James Goulding/Jane's)
2001/0100408

Artist's impression of Embraer ERJ-170 70-seat regional airliner wearing winglets added to design in 2000
2001/0100407

Artist's impression of Embraer ERJ-190-100
2000/0062958

EMBRAER to KOVACS—AIRCRAFT: BRAZIL

WEIGHTS AND LOADINGS:
Basic operating weight:
ERJ-170	20,150 kg (44,423 lb)
ERJ-190-100	26,200 kg (57,761 lb)
ERJ-190-200	27,100 kg (59,745 lb)
Max payload: ERJ-170	9,100 kg (20,062 lb)
ERJ-190-100	12,400 kg (27,337 lb)
ERJ-190-200	12,700 kg (27,999 lb)
Max fuel: ERJ-170	9,470 kg (20,878 lb)
ERJ-190-100, -200	13,000 kg (28,660 lb)

Max T-O weight:
ERJ-170 Standard	35,450 kg (78,153 lb)
ERJ-170 Long-Range	36,850 kg (81,240 lb)
ERJ-190-100 Standard	45,990 kg (101,390 lb)
ERJ-190-100 Long-Range	48,500 kg (106,924 lb)
ERJ-190-200 Standard	46,990 kg (103,595 lb)
ERJ-190-200 Long-Range	48,990 kg (108,004 lb)
Max landing weight: ERJ-170	32,450 kg (71,540 lb)
ERJ-190-100	42,500 kg (93,696 lb)
ERJ-190-200	44,500 kg (98,106 lb)

Max power loading:
ERJ-170 Standard	284 kg/kN (2.79 lb/lb st)
ERJ-170 Long-Range	296 kg/kN (2.90 lb/lb st)
ERJ-190-100 Standard	279 kg/kN (2.74 lb/lb st)
ERJ-190-100 Long-Range	295 kg/kN (2.89 lb/lb st)
ERJ-190-200 Standard	285 kg/kN (2.80 lb/lb st)
ERJ-190-200 Long-Range	298 kg/kN (2.92 lb/lb st)

PERFORMANCE (estimated):
Max cruising speed M0.80 (470 kt; 870 km/h; 541 mph)
Max speed below 3,050 m (10,000 ft):
ERJ-170 300 kt (556 km/h; 345 mph)
T-O field length: ERJ-170 1,676 m (5,500 ft)
ERJ-190-100 1,829 m (6,000 ft)
ERJ-190-200 1,920 m (6,300 ft)
Landing field length at S/L, ISA, at typical landing weight:
ERJ-170 1,219 m (4,000 ft)
ERJ-190-100 1,280 m (4,200 ft)
ERJ-190-200 1,326 m (4,350 ft)

Range with max passengers at long-range cruising speed:
ERJ-190-200 Standard
1,400 n miles (2,592 km; 1,611 miles)
ERJ-170, ERJ-190-100 Standard, ERJ-190-200 Long-Range 1,800 n miles (3,333 km; 2,071 miles)
ERJ-170 Long-Range
2,100 n miles (3,889 km; 2,416 miles)
ERJ-190-100 Long-Range
2,300 n miles (4,259 km; 2,646 miles)
UPDATED

OTHER AIRCRAFT
Refer to International section for **AMX**, **AMX-T** and other variants. Details of the **EMB-202 Ipanema** and **Piper** aircraft built under licence granted to Embraer can be found under the Neiva entry in this section and in the US section of this and earlier editions of *Jane's*.
VERIFIED

HELIBRAS
HELICÓPTEROS DO BRASIL SA
(Subsidiary of Eurocopter)
Caixa Postal 184, Dist Industrial, Av Santos Dumont, 37500-000 Itajubá, MG
Tel: (+55 35) 623 20 00
Fax: (+55 35) 623 20 01
e-mail: informática@helibras.com.br
Telex: 35 4268 HL BR

COMMERCIAL DEPARTMENT:
Avenida Paulista, 1499 10° Andar CJ 1006, CEP 01311-928
Tel: (+55 11) 251 17 22

Fax: (+55 11) 283 29 78
e-mail: helibras@mandic.com.br
SUPERINTENDENT: Michael Meylan
COMMERCIAL MANAGER: Patrick De La Revelière
MARKETING: Luis Henrique Testa

Formed 1978; owned jointly by Grupo Bueninvest (30 per cent), MGI Participações (25 per cent) and Eurocopter France (45 per cent). First assembly hall inaugurated 28 March 1980; facility occupies 12,000 m² (129,200 sq ft); new facilities opened at São Paulo in second quarter of 1998 for maintenance, spares, training, sales and marketing.

Assembly, marketing and overhaul, under Eurocopter licence, of single- and twin-engined Ecureuil/Fennec helicopters and twin-engined Dauphin/Panther; markets in Brazil and overhauls civil twin-engined Eurocopter AS 350/355 Esquilo (Fennec), EC 135, AS 332 Super Puma; and Eurocopter/Kawasaki BK 117C-1 (EC 145) (see International section).

Total of 300 helicopters sold by December 1997 (including 30 in 1997) to more than 90 customers, including over 110 to Brazilian Army, Navy and Air Force; nearly 10 per cent of overall total exported. Supply of new EC 120B Colibri to local owners began in 1999. Details of local Esquilo versions last appeared in 2000-01 *Jane's*.
UPDATED

KOVACS
JOSEPH KOVACS
Rua Maria de Lourdes Barbosa 32, 12244-690 São José dos Campos, SP
Tel: (+55 12) 97 19 78 17
e-mail: avkovacs@vol.com.br
DESIGNER: Joseph Kovács
MARKETING: André Kovács

Joseph Kovács was responsible for design of the Neiva Regente and Universal, more recently having been similarly involved with the Embraer Tucano. He is now engaged in the design of private aircraft.
UPDATED

KOVACS K-51 PEREGRINO
English name: Peregrine
TYPE: Aerobatic two-seat sportplane.
PROGRAMME: Prototype (PP-XLI) first flew 28 November 1998, at which time a commercial manufacturer was being sought to launch production; kitbuilt version planned.
DESIGN FEATURES: Conventional low-wing design, similar to Embraer Tucano, but (initially, at least) with tailwheel. Intended for civil or military training, including aerobatics, and with provision for higher-powered engines; suitable for glider towing.

Wing section NACA 64_{15} A-313 ½ (a = 0.8) at root and NACA 64A-212 (modified) at tip for optimum performance in inverted flight and spins. Dihedral 5°; incidence 1° at root and 0° at tip.

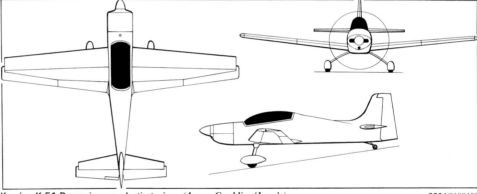

Kovács K-51 Peregrino aerobatic trainer *(James Goulding/Jane's)*

FLYING CONTROLS: Conventional and manual. All manoeuvring surfaces horn balanced. Fully balanced slotted ailerons interconnected with flaps to enhance aerobatic performance; slotted flaps adopt slight reflex angle for cruising flight. Trim tab in port elevator.
STRUCTURE: Wood, metal and composites. Wing has single-piece wood box spar with trellis ribs, rear auxiliary spar, plywood covering on lower surface and 10 mm (0.4 in) glass fibre honeycomb on upper surface; wooden ailerons and flaps covered with 3 mm (⅛ in) glass fibre/Nomex honeycomb sheet. Centre fuselage of welded steel tube with semi-structural plywood skin panels; two-cell semi-monocoque wooden empennage with plywood covering and integral fin. Rudder and elevators covered with 3 mm (⅛ in) glass fibre/Nomex honeycomb.
LANDING GEAR: Tailwheel type; fixed. Machined, conically tapered cantilever spring steel main gear legs constructed of five tubes of different lengths; glass fibre speed fairings. Cleveland mainwheels and hydraulic brakes; tyre size 5.00-5. Steerable R & K tailwheel, size 225×6. Tricycle landing gear optional.
POWER PLANT: Prototype has one 149 kW (200 hp) Textron Lycoming IO-360-C1C flat-four, driving a McCauley two-blade constant-speed metal propeller. Airframe designed to accommodate engines in 119 to 224 kW (160 to 300 hp) class. Fuel for aerobatic flying in 92 litre (24.3 US gallon; 20.2 Imp gallon) tank in forward fuselage; non-aerobatic tank beneath baggage compartment, capacity 112 litres (29.6 US gallons; 24.6 Imp gallons). Total fuel 204 litres (53.8 US gallons; 44.9 Imp gallons). Fuel and oil systems modified for long periods of inverted flight.
ACCOMMODATION: Two persons, in tandem, under starboard-hinged, single-piece canopy. Adjustable seats and dual controls; rear seat, raised 15 cm (6 in), is occupied when flown solo. Baggage compartment behind rear seat.
SYSTEMS: Electrical system 28 V DC.
AVIONICS: *Comms:* Honeywell KT 76A transponder.
Flight: Garmin GNC 250 GPS.
Instrumentation: Standard VFR.
DIMENSIONS, EXTERNAL:
Wing span	8.50 m (27 ft 10¾ in)
Wing chord: at root	1.70 m (5 ft 7 in)
at tip	0.84 m (2 ft 9 in)
Wing aspect ratio	6.5
Length overall	7.03 m (23 ft 0¾ in)
Height overall	2.05 m (6 ft 8¾ in)
Tailplane span	3.20 m (10 ft 6 in)

The Kovács K-51 Peregrino photographed during its maiden flight

Wheel track	2.20 m (7 ft 2½ in)	Max T-O weight: A	840 kg (1,852 lb)	Max rate of climb at S/L: A	600 m (1,969 ft)/min	
Wheelbase	4.38 m (14 ft 4½ in)	U	920 kg (2,028 lb)	U	415 m (1,362 ft)/min	
Propeller diameter	1.93 m (6 ft 4 in)	Max wing loading: A	76.1 kg/m² (15.59 lb/sq ft)	Service ceiling: A	8,052 m (26,417 ft)	
DIMENSIONS, EXTERNAL:		U	83.3 kg/m² (17.07 lb/sq ft)	U	7,076 m (23,215 ft)	
Cockpit: Length	2.13 m (7 ft 0 in)	Max power loading: A	5.64 kg/kW (9.26 lb/hp)	T-O run: A	158 m (520 ft)	
Max width	0.73 m (2 ft 4¾ in)	U	6.17 kg/kW (10.14 lb/hp)	U	247 m (810 ft)	
Baggage compartment volume	0.20 m³ (7.0 cu ft)	PERFORMANCE (A, U as above, 119 kW; 160 hp engine):		T-O to 15 m (50 ft): A	219 m (720 ft)	
AREAS:		Never-exceed speed (V$_{NE}$):		U	328 m (1,080 ft)	
Wings, gross	11.04 m² (118.8 sq ft)	A, U	200 kt (370 km/h; 230 mph)	Landing from 15 m (50 ft): A	425 m (1,395 ft)	
Ailerons (total)	1.08 m² (11.63 sq ft)	Max level speed at S/L: A	177 kt (328 km/h; 204 mph)	U	475 m (1,560 ft)	
Trailing-edge flaps (total)	1.16 m² (12.49 sq ft)	U	175 kt (324 km/h; 201 mph)	Range at max cruising speed and altitude, no reserves:		
Fin	1.22 m² (13.13 sq ft)	Max cruising speed at 75% power at 2,400 m (8,000 ft):		A	496 n miles (920 km; 571 miles)	
Rudder	0.65 m² (7.00 sq ft)	A	172 kt (319 km/h; 197 mph)	U	1,041 n miles (1,927 km; 1,198 miles)	
Tailplane	2.40 m² (25.83 sq ft)	U	170 kt (315 km/h; 196 mph)	g limits	+6/−3	
Elevators, incl tab	0.95 m² (10.23 sq ft)	Stalling speed: flaps up: A	55 kt (102 km/h; 64 mph)		UPDATED	
WEIGHTS AND LOADINGS (A: Aerobatic, U: Utility with 149 kW; 200 hp engine):		U	60 kt (112 km/h; 69 mph)			
		flaps down: A	50 kt (93 km/h; 58 mph)			
Weight empty, equipped: A, U	600 kg (1,323 lb)	U	54 kt (100 km/h; 63 mph)			

NEIVA

INDUSTRIA AERONAUTICA NEIVA SA (Subsidiary of Embraer)

Caixa Postale 1011, Avenida Alcides Cagliari 2281, 18608-900 Botucatu-SP
Tel: (+55 14) 68 02 20 00
Fax: (+55 14) 68 21 21 10
e-mail: neiva@laser.com.br
PRESIDENT AND DIRECTOR: Mauricio Novis Botelho
MANAGING DIRECTOR: Paolo Urbanavicius
SALES MANAGERS: Luiz Fabiano
 Zaccarelli Cunha

Formed October 1953 by José Carlos de Barros Neiva; produced Paulistinha, Regente, Universal and related lightplanes of own design. Became wholly owned Embraer subsidiary 10 March 1980; factory area 18,550 m² (199,700 sq ft), workforce 227. Substantial participation in Embraer general aviation programmes, including complete production of EMB-711ST Corisco Turbo, EMB-710 Carioca, EMB-720D Minuano (of which some 300 delivered), EMB-721 Sertanejo and EMB-810D Seneca IV (licence-built Piper models); also responsible since 1980 for total production of EMB-202 and now discontinued EMB-201A Ipanema. Current production comprises EMB-202 and, at low rate of two/three per year, EMB-810D.

UPDATED

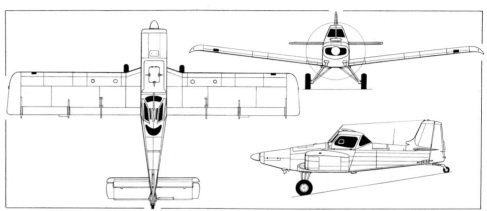

Neiva EMB-202 Ipanema (one Textron Lycoming IO-540) *(Dennis Punnett/Jane's)*

NEIVA EMB-202 IPANEMA

TYPE: Agricultural sprayer.
PROGRAMME: Embraer design; first flight (EMB-200 prototype PP-ZIP) 30 July 1970; certified 14 December 1971; see 1977-78 and earlier editions for EMB-200/200A (49/24 built), EMB-201 (200 built) and EMB-201R (three built); and 1992-93 edition for EMB-201A (402 built). Manufacture transferred to Neiva in 1980; many detail improvements (listed in 1991-92 and earlier *Jane's*) introduced late 1988.
CURRENT VERSIONS: **EMB-202:** Current model from 1992; generally similar to EMB-201A except for increased wing span, larger hopper and choice of Lycoming or Continental engine. First delivery 2 October 1992.
CUSTOMERS: 678 of earlier models built (see Programme). Production of all variants exceeds 800; registration allocated to 811th in January 2000.
DESIGN FEATURES: Designed to Brazilian Ministry of Agriculture specifications. Typical agricultural aircraft features, including hopper ahead of pilot and high cockpit for good view over nose; low-mounted unbraced wings, with cambered leading-edges (detachable) and tips; rectangular-section fuselage; slightly sweptback vertical tail.

Wing section NACA 23015 (modified); incidence 3°; dihedral 7° from roots.
FLYING CONTROLS: Conventional and manual. Frise ailerons; trim tab in starboard elevator, ground-adjustable tab on rudder. Slotted flaps on wing trailing-edges.
STRUCTURE: All-metal safe-life fuselage frame of welded 4130 steel tube, with removable skin panels of 2024 aluminium alloy and glass fibre and specially treated against chemical corrosion. All-metal wings (single-spar) and tail surfaces (two-spar).
LANDING GEAR: Non-retractable mainwheels and tailwheel, with rubber shock-absorbers in main units. Tailwheel has tapered spring shock-absorber. Mainwheels and tyres size 8.50-10. Tailwheel diameter 250 mm (10 in). Tyre pressures: main, 2.07 to 2.41 bars (30 to 35 lb/sq in); tailwheel, 3.79 bars (55 lb/sq in). Hydraulic disc brakes on mainwheels.
POWER PLANT: One 224 kW (300 hp) Textron Lycoming IO-540-K1J5D flat-six engine, driving a Hartzell two-blade (optionally three-blade) constant-speed metal propeller. Optional engine is Teledyne Continental 224 kW (300 hp) IO-550-D with McCauley two-blade constant-speed propeller. Integral fuel tanks in each wing leading-edge, with total capacity of 292 litres (77.1 US gallons; 64.2 Imp gallons), of which 264 litres (69.75 US gallons; 58.1 Imp gallons) are usable. Refuelling point on top of each tank. Oil capacity 12 litres (3.2 US gallons; 2.6 Imp gallons).
ACCOMMODATION: Single horizontally/vertically adjustable seat in fully enclosed cabin with bottom-hinged, triple-lock, jettisonable window/door on each side and two overhead windows. Ventilation airscoop at top of canopy front edge; second airscoop on fin leading-edge pressurises interior of rear fuselage. Inertial shoulder harness standard.
SYSTEMS: 28 V DC electrical system supplied by 24 Ah batteries and a Bosch KI 28 V 35 A alternator. Power receptacle for external battery (AN-2552-3A type) on port side of forward fuselage.
AVIONICS: *Comms:* Optional portable VHF transceiver.
 Flight: Optional portable Garmin GPS.
EQUIPMENT: Hopper for agricultural chemicals, capacity 950 litres (251 US gallons; 209 Imp gallons) liquid or 750 kg (1,653 lb) dry. Dusting system below centre of fuselage. Spraybooms or Micronair atomisers aft of or above wing trailing-edges respectively. Options include ram air pressure generator for use with liquid spray system; improved lightweight spraybooms; smaller/lighter Micronair AU5000 rotary atomisers; and trapezoidal spreader with adjustable inlet to improve application of dry chemicals.

DIMENSIONS, EXTERNAL:
Wing span	11.69 m (38 ft 4¼ in)
Wing chord, constant	1.71 m (5 ft 7½ in)
Wing aspect ratio	6.9
Length overall (tail up)	7.43 m (24 ft 4½ in)
Height overall (tail down)	2.20 m (7 ft 2½ in)
Fuselage: Max width	0.93 m (3 ft 0½ in)
Tailplane span	3.66 m (12 ft 0 in)
Wheel track	2.20 m (7 ft 2½ in)
Wheelbase	5.20 m (17 ft 7¼ in)
Propeller diameter:	
Hartzell two- and three-blade	2.13 m (7 ft 0 in)
McCauley two-blade	2.18 m (7 ft 2 in)

Neiva EMB-202 Ipanema single-seat agricultural aircraft

Neiva-built EMB-810

DIMENSIONS, INTERNAL:
  Cockpit: Length   1.20 m (3 ft 11¼ in)
    Max width   0.85 m (2 ft 9½ in)
    Max height   1.34 m (4 ft 4¾ in)
AREAS:
  Wings, gross   19.94 m² (214.6 sq ft)
  Ailerons (total)   1.60 m² (17.22 sq ft)
  Trailing-edge flaps (total)   2.30 m² (24.76 sq ft)
  Fin   0.58 m² (6.24 sq ft)
  Rudder   0.63 m² (6.78 sq ft)
  Tailplane   3.17 m² (34.12 sq ft)
  Elevators (total, incl tab)   1.50 m² (16.15 sq ft)
WEIGHTS AND LOADINGS (N: Normal, R: Restricted category):
  Weight empty: N, R   1,020 kg (2,249 lb)
  Max payload: N, R   741 kg (1,633 lb)
  Max T-O and landing weight: N 1,550 kg (3,417 lb)
    R   1,800 kg (3,968 lb)
  Max wing loading: N   77.7 kg/m² (15.92 lb/sq ft)
    R   90.3 kg/m² (18.49 lb/sq ft)
  Max power loading: N   6.92 kg/kW (11.39 lb/hp)
    R   8.03 kg/kW (13.23 lb/hp)
PERFORMANCE (clean):
  Never-exceed speed (VNE):
    N   147 kt (272 km/h; 169 mph)
  Max level speed at S/L:
    N   124 kt (230 km/h; 143 mph)
    R   121 kt (225 km/h; 140 mph)
  Max cruising speed at 75% power at 1,830 m (6,000 ft):
    N   115 kt (213 km/h; 132 mph)
    R   111 kt (206 km/h; 128 mph)
  Stalling speed, power off (N):
    flaps up   56 kt (103 km/h; 64 mph)
    8° flap   54 kt (100 km/h; 62 mph)
    30° flap   50 kt (92 km/h; 57 mph)
  Stalling speed, power off (R):
    flaps up   60 kt (110 km/h; 68 mph)
    8° flap   58 kt (107 km/h; 66 mph)
    30° flap   53 kt (99 km/h; 61 mph)
  Max rate of climb at S/L, 8° flap:
    N   283 m (930 ft)/min
    R   201 m (660 ft)/min
  Service ceiling, 8° flap: R   3,470 m (11,380 ft)
  T-O run at S/L, asphalt runway:
    N   200 m (655 ft)
    R   354 m (1,160 ft)
  T-O to 15 m (50 ft), conditions as above:
    N   332 m (1,090 ft)
    R   564 m (1,850 ft)
  Landing from 15 m (50 ft) at S/L, 30° flap, asphalt
    runway: N   412 m (1,355 ft)
    R   476 m (1,565 ft)
  Landing run, conditions as above: N 153 m (505 ft)
    R   170 m (560 ft)
  Range at 1,830 m (6,000 ft), no reserves:
    N   506 n miles (938 km; 583 miles)
    R   474 n miles (878 km; 545 miles)

UPDATED

NEIVA EMB-810D
TYPE: Six-seat utility twin-prop.
PROGRAMME: Licence-built Piper PA-34 Seneca IV, which see in the US section of the 1996-97 *Jane's* for a detailed description; approximate data (Seneca V) in this edition.
CUSTOMERS: Some 874 produced in Brazil by late 1999; recent customers include the state governments of Paraiba and Roraima, Bradesco Leasing and Eletrica Bragantina.

UPDATED

PZL

PZL-MIELEC DO BRASIL FABRICA D'AVIÕES LTDA
Anapolis

PZL Mielec (now Polish Aircraft Factory) of Poland opened a joint venture assembly plant in Brazil on 1 July 1997 and planned to begin delivering locally assembled M-18 Dromader agricultural aircraft during 1998. Refer to Embraer entry and Polish section of 2000-01 *Jane's* for further details. By mid-2000, there was no evidence of Dromaders appearing on the Brazilian civil aircraft register, and the programme is assumed to have lapsed.

UPDATED

CANADA

ALBERTA

ALBERTA AEROSPACE CORPORATION
Suite 804, 7015 Macleod Trail South, Calgary, Alberta T2H 2K6
Tel: (+1 403) 255 28 10 or (+1 877) FJ44JET
Fax: (+1 403) 255 26 49
e-mail: sales@phoenixfanjet.com
Web: http://www.phoenixfanjet.com
PRESIDENT AND CEO: C Raymond Johnson
VICE-PRESIDENT: Jacques 'Red' Dewaelheyns
SALES DIRECTOR: Tom Heath

EUROPEAN SALES OFFICE:
  6 Avenue des Allies, B-6000 Charleroi, Belgium
Tel: (+32) 71 30 29 15
e-mail: europe@phoenixfanjet.com
Web: http://www.phoenixfanjet.com

Alberta Aerospace Corporation was incorporated on 5 December 1995, having acquired manufacturing rights to the Promavia Jet Squalus F1300 jet trainer, last described fully in the Belgian section of the 1995-96 *Jane's*. The company is developing the aircraft as the Phoenix FanJet, but was delayed by legal issues (surrounding the bankruptcy of Promavia) which were resolved in July 1999. Acquisition was eventually achieved in November 1999, when all remaining assets were secured from the Belgian authorities; type certification is due in mid-2001. Manufacture will be at former Claresholm RCAF base, Alberta.

UPDATED

ALBERTA AEROSPACE PHOENIX FANJET
TYPE: Basic jet trainer/light transport.
PROGRAMME: Development of Promavia Jet Squalus, which was designed by Stelio Frati and originally flew in November 1960 as Procaer F.400 Cobra; see Promavia entry in Belgian section of 1995-96 and earlier editions of *Jane's*. Two prototypes (one of which, OO-JET, was unflown) bought in February 1996 and shipped to Canada; flying prototype (OO-SQA) re-registered N112SQ in March 1996 and re-engined with a Williams FJ44-1 replacing its original AlliedSignal TFE109-1; two-phase modification programme planned to bring the aircraft to production standard. In Phase I, which began in September 1997 on N112SQ, wing span was due to be increased by 1.22 m (4 ft 0 in) in order to reduce stalling speed to the 61 kt (113 km/h; 71 mph) maximum required by FAA rules for FAR Pt 23 certification of single-engined aircraft; Martin-Baker Mk 11 lightweight ejection seats replaced with airliner type of flight deck seats; and Honeywell Primus 1000 EFIS panel installed, with the aim of adding approval for aerobatics, measured take-off and high-altitude capability, and Cat. II landings to the original type certificate.

Phase II will involve converting OO-JET to four-seat configuration, adding pressurisation, air conditioning and de-icing, and relocating all fuel to tanks in the wings.

Phoenix FanJet in its two-seat form

Prototype had flown over 700 hours by mid-2000. FAA/JAA certification of two-seat version scheduled for mid-2001, with certification of four-seat version following in early 2002. Manufacture of components will be contracted out to Canadian and US companies, with final assembly at Alberta Aerospace's Claresholm plant, maximum production capacity of which is estimated at 10 per month.
CURRENT VERSIONS: **SigmaJet:** Unpressurised two-seat version aimed at the airline pilot training market or private high-performance sportplane; designated **Trainer** and **Sport**, respectively.
  MagnaJet: Pressurised four-seat version aimed at owner-flown personal transport market.
CUSTOMERS: Eight orders (including five trainers) and 42 expressions of interest by mid-1998. Launch order for three four-seat aircraft placed by Earth Search Sciences of McCall, Idaho. Sales projections for Trainer of 12 in first year, increasing to 24 by fourth year; for transport, 18 in second year, increasing to 45 in fifth.
COSTS: SigmaJet US$1.575 million (plus US$650,000 for optional EFIS cockpit); MagnaJet US$1.875 million (1999) including type-rating and training. Estimated operating cost less than US$250 per hour.
DESIGN FEATURES: Low-wing layout, with bifurcated intake at wingroots; slightly swept wing leading-edges and tailplane; sweptback fin.

Retouched photograph showing the intended four-seat conversion of Phoenix FanJet

Wings of supercritical section (thickness/chord ratio 13 per cent constant); incidence 1° at root, −1° 45′ at tip; dihedral 6° from roots.
FLYING CONTROLS: Conventional and manual. Frise differential ailerons; hydraulic actuation for wing trailing-edge flaps and two-piece underfuselage airbrake; ailerons each have servo tab, starboard elevator electrically operated trim tab; fixed incidence tailplane.
STRUCTURE: Composites for fairings and some non-structural components, otherwise basically metal throughout (semi-monocoque/flush riveted stressed skin); large quick-disconnect panel in lower rear fuselage for rapid engine access/removal.
LANDING GEAR: Retractable tricycle type, with single wheel and oleo-pneumatic shock-absorber on each unit. Mainwheels retract inward, nosewheel rearward. Hydraulic actuation, with built-in emergency system. Main gear of trailing-arm type. Nosewheel steerable ±18°. Mainwheels and tyres size 6.00-6, nosewheel 5.00-5.
POWER PLANT: *Phoenix SigmaJet:* One 8.45 kN (1,900 lb st) Williams FJ44-1A turbofan; derated to 7.12 kN (1,600 lb st). Fuel, total capacity 700 litres (185 US gallons; 154 Imp gallons), contained in semi-integral fuselage tank.
  Phoenix MagnaJet: One 8.45 kN (1,900 lb st) Williams FJ44-1A turbofan. Fuel, total capacity 1,060 litres (280 US gallons; 233 Imp gallons), contained in wing tanks.
ACCOMMODATION: Two or four persons (one or two side-by-side pairs), according to version. Four-seat version pressurised to equivalent of 3,050 m (10,000 ft) at 7,620 m (25,000 ft) and air conditioned. No emergency escape systems.
SYSTEMS: In four-seat configuration, cabin will be pressurised to 0.32 bar (4.67 lb/sq in), maintaining a 3,050 m (10,000 ft) altitude to 7,620 m (25,000 ft).
AVIONICS: Optional three-tube EFIS with Honeywell Primus 1000 as core system.
The following data are provisional.
DIMENSIONS, EXTERNAL:
  Wing span   10.26 m (33 ft 8 in)
  Wing chord at root   1.90 m (6 ft 2¾ in)
  Length of fuselage   9.36 m (30 ft 8½ in)

28   CANADA: AIRCRAFT—ALBERTA to BELL

Height overall	3.60 m (11 ft 7¼ in)	Max wing loading: A	153.7 kg/m² (31.48 lb/sq ft)	T-O to 15 m (50 ft): A	701 m (2,300 ft)
Tailplane span	3.80 m (12 ft 5½ in)	B	189.9 kg/m² (38.89 lb/sq ft)	B	915 m (3,000 ft)
Wheel track	3.59 m (11 ft 9¼ in)	Max power loading: A	325 kg/kN (3.19 lb/lb st)	Landing from 15 m (50 ft): A	731 m (2,400 ft)
Wheelbase	3.58 m (11 ft 9 in)	B	338 kg/kN (3.32 lb/lb st)	B	915 m (3,000 ft)

DIMENSIONS, INTERNAL (Phoenix FanJet):
   Cabin: Length   2.44 m (8 ft 0 in)
     Max width   1.22 m (4 ft 0 in)
     Max height   1.22 m (4 ft 0 in)

AREAS:
   Wings, gross   15.05 m² (162.0 sq ft)

WEIGHTS AND LOADINGS (A: SigmaJet, B: MagnaJet):
   Weight empty: A   1,546 kg (3,408 lb)
     B   1,727 kg (3,808 lb)
   Max T-O weight: A   2,313 kg (5,100 lb)
     B   2,857 kg (6,300 lb)

PERFORMANCE (estimated, A and B as above):
   Max cruising speed at 7,620 m (25,000 ft):
     A   311 kt (576 km/h; 358 mph)
     B   345 kt (639 km/h; 397 mph)
   Stalling speed in landing configuration:
     A, B   61 kt (113 km/h; 71 mph)
   Max rate of climb at S/L: A   719 m (2,360 ft)/min
     B   801 m (2,630 ft)/min
   Service ceiling: A   10,210 m (33,500 ft)
     B   10,670 m (35,000 ft)
   Max certified altitude   7,620 m (25,000 ft)

Range with max fuel and reserves:
   A   715 n miles (1,324 km; 822 miles)
   B   1,179 n miles (2,183 km; 1,356 miles)
g limits: A   +6/−3
   B   +3.6/−1.7

UPDATED

AWT

ADVANCED WING TECHNOLOGIES

1080-885 West Georgia Street, Vancouver, British Columbia V6C 3E8
Tel: (+1 604) 685 92 66
Fax: (+1 604) 685 65 01

Plans to build the El Gavilán **EL-1 Gavilán 358** in Canada had been abandoned by mid-2000, although the Colombian manufacturer anticipated securing an alternative North American partner later in that year.

UPDATED

BELL

BELL HELICOPTER TEXTRON CANADA (Division of Textron Canada Ltd)

12800 Rue de l'Avenir, Mirabel, Quebec J7J 1R4
Tel: (+1 514) 437 34 00
Fax: (+1 514) 437 60 10
e-mail: bhtchr@bellhelicopter.textron.com
PRESIDENT: Bob MacDonald

Memorandum of Understanding to start helicopter industry in Canada signed 7 October 1983; 38,900 m² (418,725 sq ft) factory opened late 1985 and employs 1,700 people; US civil production of 206B JetRanger transferred to Canada in 1986, 206L LongRanger by early 1987; then 212 in 1988 and 412 in 1989. The 230 was introduced and certified in Canada in 1992, but has been replaced by the 430. Newest products are the 407 and 427. About half of each helicopter made in Canada (dynamic systems supplied by Bell Fort Worth); Total of 233 commercial helicopters delivered in 1997, followed by 197 in 1998, 146 in 1999 and 145 in 2000.

Bell helicopter 'families' do not follow a chrono-numerical pattern; accordingly, entries in this section are grouped in the following order:
  Bell 206, 407 and 427
  Bell 212 and 412
  Bell 430

UPDATED

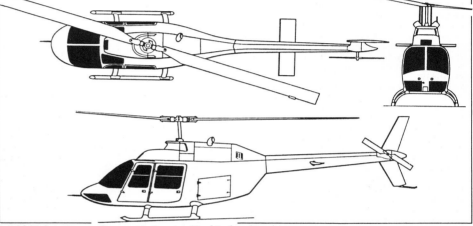

Bell 206B-3 JetRanger five-seat helicopter *(Paul Jackson/Jane's)*

BELL 206B-3 JETRANGER III
US Army designation: TH-67 Creek
TYPE: Light utility helicopter.
PROGRAMME: Model 206 originally offered (unsuccessfully) to US Army as OH-4, having first flown on 8 December 1962. Civil version flew 10 January 1966, after extensive redesign, and entered service as 206A JetRanger; followed by 206A-1 and 206B JetRanger II. Delivery of current 206B-3 JetRanger III began Summer 1977; transferred to Mirabel, Canada, 1986. Also built in Italy by Agusta. Upgraded version, under consideration in 1999, would feature improved engine, transmission and dynamic components, with target selling price under US$900,000.
CURRENT VERSIONS: **206B-3 JetRanger III:** Current civil production version.
Following description refers to JetRanger III.
   TH-67 Creek (Bell designation TH-206): Selected March 1993 as US Army NTH (New Training Helicopter) choice to replace UH-1 at pilot training school, Fort Rucker, Alabama (223 Aviation Regiment at Cairns AAF and 212 Aviation Regiment at Lowe AHP). Instructor and pupil in front seats; plans abandoned for second pupil to be seated at the rear, observing flight instruments by closed-circuit TV screen mounted on back of right-hand front seat. Powered by Rolls-Royce (formerly Allison) 250-C20JN engine. First batch included nine cockpit procedures trainers outfitted by Frasca International; three configurations: VFR, IFR, and VFR with IFR provision.
   TH-67 features also include dual controls, crashworthy seats, five-point seat restraints, heavy-duty battery, particle separator, bleed air heater, heavy-duty skid shoes and enlarged instrument panel. IFR version additionally possesses force trim system, auxiliary electrical system and is FAA certified for dual pilot operation. Also supplied to Taiwan Army, based at Gwe-Ren.
CUSTOMERS: TH-206 (TH-67 Creek) declared winner of US Army NTH competition March 1993; initial 102 ordered in IFR configuration; second batch of 35 VFR helicopters ordered February 1994; deliveries began 15 October 1993 with N67001 and N67014 (all TH-67s have civilian registrations); 45 aircraft (and six procedures trainers) delivered in time for first training course to open 5 May 1994; initial orders 137 TH-67s and nine cockpit procedures trainers; all delivered by late 1995. Further orders placed on behalf of Taiwanese Army, which received 30 TH-67s in 1997-99. Deliveries in 1999 included six to the Bulgarian Air Force.
   Over 7,725 JetRangers produced by Bell and licensees, including 4,400 Bell 206Bs and 2,200 military OH-58 series; and 900 in Italy. Bell Canada delivered 28 206Bs in 1999 and 14 in 2000.
COSTS: US Army initial NTH contract US$84.9 million. Civilian 206B-3, typically equipped, US$730,000 with C20J engine or US$765,000 with C20R (1999). Average direct operating cost US$191 per hour (1997).
DESIGN FEATURES: Typical light utility helicopter with skid landing gear, high-mounted tailboom and horizontal stabiliser. Two-blade teetering main rotor with preconed and underslung bearings; blades retained by grip, pitch change bearing and torsion-tension strap assembly; two-blade tail rotor; main rotor rpm 374 to 394.
FLYING CONTROLS: Hydraulic fully powered cyclic and collective controls and foot-powered tail rotor control; tailplane with highly cambered inverted aerofoil section and stall strip produces appropriate nose-up and nose-down attitude during climb and descent; optional autostabiliser, autopilot and IFR systems.
STRUCTURE: Conventional light alloy structure with two floor beams and bonded honeycomb sandwich floor; transmission mounted on two beams and deck joined to floor by three fuselage frames; main rotor blades have

Bell 206B JetRanger III (Rolls-Royce 250 turboshaft) *(Paul Jackson/Jane's)*   2001/0103571

BELL—AIRCRAFT: CANADA

Bell 206L LongRanger IV general purpose helicopter *(Paul Jackson/Jane's)*

extruded aluminium D-section leading-edge with honeycomb core behind, covered by bonded skin; tail rotor blades have bonded skin without honeycomb core.

LANDING GEAR: Aluminium alloy tubular skids bolted to extruded cross-tubes. Tubular steel skid on ventral fin to protect tail rotor in tail-down landing. Special high skid gear (0.25 m; 10 in greater ground clearance) available for use in areas with high brush. Pontoons or stowed floats, capable of in-flight inflation, available as optional kits.

POWER PLANT: One Rolls-Royce 250-C20J turboshaft, rated at 313 kW (420 shp) for T-O; 276 kW (370 shp) max continuous. Optionally, one Rolls-Royce 250-C20R/4 rated at 336 kW (450 shp) for T-O. Transmission rating 236 kW (317 shp) for T-O; 201 kW (270 shp) continuous. Rupture resistant fuel tank below and behind rear passenger seat, usable capacity 344 litres (91.0 US gallons; 75.75 Imp gallons). Refuelling point on starboard side of fuselage, aft of cabin. Oil capacity 5.68 litres (1.5 US gallons; 1.25 Imp gallons).

ACCOMMODATION: Two seats side by side in front and three-seat rear bench. Dual controls optional. Two forward-hinged doors on each side, made of formed aluminium alloy with transparent panels (bulged on rear pair). Baggage compartment aft of rear seats, capacity 113 kg (250 lb), with external door on port side.

SYSTEMS: Hydraulic system, pressure 41.5 bars (600 lb/sq in), for cyclic, collective and directional controls. Maximum flow rate 7.57 litres (2.0 US gallons; 1.65 Imp gallons)/min. Open reservoir. Electrical supply from 150 A starter/generator. One 28 V 17 Ah Ni/Cd battery; auxiliary 13 Ah battery optional. Optional ECS.

AVIONICS: *Comms:* VHF communications, intercom and speaker system.
Flight: VOR, ADF, DME and R/Nav optional.

EQUIPMENT: Standard equipment includes cabin fire extinguisher, first aid kit, door locks, night lighting, and dynamic flapping restraints. Optional items include clock, engine hour meter, turn and slip indicator, custom seating, internal stretcher kit, rescue hoist, cabin heater, camera access door, high-intensity night lights, engine fire detection system, and external cargo hook of 680 kg (1,500 lb) capacity.

DIMENSIONS, EXTERNAL:
Main rotor diameter	10.16 m (33 ft 4 in)
Main rotor blade chord	0.33 m (1 ft 1 in)
Tail rotor diameter	1.65 m (5 ft 5 in)
Tail rotor blade chord	0.12 m (4¾ in)
Distance between rotor centres	5.96 m (19 ft 6½ in)
Length: overall, rotors turning	11.82 m (38 ft 9½ in)
fuselage, incl tailskid	9.50 m (31 ft 2 in)
Height: over tailfin	2.54 m (8 ft 4 in)
to top of rotor head:	
standard skids	2.89 m (9 ft 6 in)
high skids	3.17 m (10 ft 4¾ in)
emergency floats	3.20 m (10 ft 6 in)
Stabiliser span	1.97 m (6 ft 5¾ in)
Width over landing gear:	
standard skids	1.95 m (6 ft 4¾ in)
high skids	2.07 m (6 ft 9½ in)
emergency floats	2.68 m (8 ft 9½ in)
Forward cabin doors (each): Height	1.02 m (3 ft 4 in)
Width	0.61 m (2 ft 0 in)
Rear cabin doors (each): Height	1.02 m (3 ft 4 in)
Width	0.91 m (3 ft 0 in)

DIMENSIONS, INTERNAL:
Cabin: Length of seating area	2.13 m (7 ft 0 in)
Max width	1.27 m (4 ft 2 in)
Max height	1.28 m (4 ft 3 in)
Volume	1.1 m³ (40 cu ft)
Baggage compartment:	
Max width	1.10 m (3 ft 7¼ in)
Max height	0.55 m (1 ft 9½ in)
Max length	0.94 m (3 ft 1¼ in)
Volume	approx 0.45 m³ (16 cu ft)

AREAS:
Main rotor blades (each)	1.68 m² (18.05 sq ft)
Tail rotor blades (each)	0.11 m² (1.18 sq ft)
Main rotor disc	81.07 m² (872.7 sq ft)
Tail rotor disc	2.14 m² (23.05 sq ft)
Stabiliser	0.90 m² (9.65 sq ft)

WEIGHTS AND LOADINGS:
Weight empty: standard civil	772 kg (1,702 lb)
TH-67: VFR	852 kg (1,879 lb)
IFR	911 kg (2,009 lb)
Operating weight, TH-67: VFR	1,369 kg (3,019 lb)
IFR	1,428 kg (3,149 lb)
Max payload: internal	635 kg (1,400 lb)
external	680 kg (1,500 lb)
Max T-O weight: internal load	1,451 kg (3,200 lb)
external load	1,519 kg (3,350 lb)
Max disc loading:	
internal load	17.9 kg/m² (3.67 lb/sq ft)
external load	18.7 kg/m² (3.84 lb/sq ft)

Transmission loading at max T-O weight and power:
internal load	6.15 kg/kW (10.09 lb/shp)
external load	6.43 kg/kW (10.57 lb/shp)

PERFORMANCE (at internal load max T-O weight, ISA):
Never-exceed speed (V$_{NE}$) at S/L	122 kt (225 km/h; 140 mph)
Max and econ cruising speed at S/L	115 kt (214 km/h; 133 mph)
Max rate of climb at S/L	390 m (1,280 ft)/min
Vertical rate of climb at S/L	91 m (300 ft)/min
Service ceiling	4,115 m (13,500 ft)
Hovering ceiling: IGE	4,025 m (13,200 ft)
OGE	1,615 m (5,300 ft)
Range with max fuel, no reserves:	
at S/L	374 n miles (692 km; 430 miles)
at 1,525 m (5,000 ft)	395 n miles (732 km; 455 miles)
Range, TH-67 VFR with normal 314 litres (83.0 US gallons; 69.0 Imp gallons) fuel	327 n miles (605 km; 376 miles)
Endurance	4 h 30 min

OPERATIONAL NOISE LEVELS (FAR Pt 36):
T-O	88.7 EPNdB
Flyover	85.4 EPNdB
Approach	90.6 EPNdB

UPDATED

BELL 206L-4 LONGRANGER IV

TYPE: Light utility helicopter.

PROGRAMME: Stretched JetRanger (see previous entry). LongRanger announced 25 September 1973; first flight 11 September 1974; initial versions were 206L and 206L-1 LongRanger II, of which 790 built; limited production (17) of 206L-2; replaced by 206L-3 LongRanger III in 1982, of which 612 built, production having transferred to Canada in January 1987.

CURRENT VERSIONS: **LongRanger IV:** Announced March 1992 as new current standard model; transmission uprated to absorb 365 kW (490 shp) instead of 324 kW (435 shp) from same engine; gross weight raised from 1,882 kg (4,150 lb) to 2,018 kg (4,450 lb); certified late 1992, delivered from December 1992.

TwinRanger: Twin-engined version; model **206LT**; 13 built on LongRanger IV line between 1993 and 1997. No longer available; see 1998-99 *Jane's* for description.

Gemini ST: Twin-engined rebuild of LongRanger III/IV developed by Tridair and Soloy in USA: see *Jane's Aircraft Upgrades*.

CUSTOMERS: Over 1,680 LongRangers produced by late 2000, including 265 LongRanger IVs and 206LTs. Total of 12 delivered in 1999 and 27 in 2000. Recent customers include the Procuraduria General de Republica of Mexico, which ordered 24 in October 1999 for delivery between December 1999 and September 2000.

COSTS: US$1.12 million, typically equipped (1999). Average direct operating cost US$293 per hour (1997).

DESIGN FEATURES: As JetRanger, but cabin length increased to make room for club seating and extra window; Bell Noda-Matic transmission to reduce vibration; vertical stabilisers added to horizontal tail surfaces. Improvements introduced on LongRanger II include new freewheel unit, modified shafting and increased-thrust tail rotor; main rotor rpm 394; rotor brake optional.

FLYING CONTROLS: As JetRanger, but with endplate fins on tailplane; single-pilot IFR with Rockwell Collins AP-107H autopilot; optional SFENA autopilot with stabilisation and holds for heading, height and approach.

STRUCTURE: As JetRanger.

LANDING GEAR: As JetRanger.

POWER PLANT: One 485 kW (650 shp) Rolls-Royce 250-C30P turboshaft (maximum continuous rating 415 kW; 557 shp). Transmission rated at 365 kW (490 shp) for take-off, with a continuous rating of 276 kW (370 shp); 340 kW (456 shp) transmission optional. Rupture resistant fuel system, comprising three interconnected cells, total usable capacity 419 litres (111 US gallons; 92.0 Imp gallons).

ACCOMMODATION: Redesigned rear cabin, more spacious than JetRanger. With a crew of two, standard cabin layout accommodates five passengers in two canted rearward-facing seats and three forward-facing seats. Optional DeLuxe and Custom DeLuxe interiors with fabric, fabric/vinyl, all vinyl or fabric/leather seats. Port forward passenger seat has folding back to allow loading of a 2.44 × 0.91 × 0.30 m (8 × 3 × 1 ft) container, making possible carriage of such items as survey equipment, skis and other long components. Double doors on port side of cabin provide opening 1.55 m (5 ft 1 in) wide, for straight-in loading of stretcher patients or utility cargo; in ambulance or rescue role two stretcher patients and two ambulatory patients/attendants can be carried. Dual controls optional.

SYSTEMS: Hydraulic system; 28 V DC electrical power from 180 A starter/generator and 17 Ah battery. Engine bleed air ECS optional.

AVIONICS: *Comms:* Optional Honeywell suite includes dual nav/com and transponder.
Flight: ADF, DME and marker beacon receiver. Honeywell R/Nav, radio altimeter and encoding altimeter optional.

EQUIPMENT: Optional kits include emergency flotation gear, 907 kg (2,000 lb) cargo hook, rescue hoist, Nightsun searchlight (requires high skid gear).

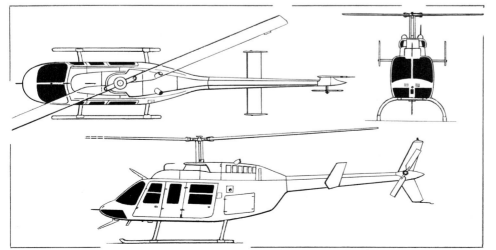
Bell 206L-4 LongRanger IV general purpose light helicopter *(Dennis Punnett/Jane's)*

CANADA: AIRCRAFT—BELL

DIMENSIONS, EXTERNAL:
Main rotor diameter	11.28 m (37 ft 0 in)
Main rotor blade chord	0.33 m (1 ft 1 in)
Tail rotor diameter	1.65 m (5 ft 5 in)
Tail rotor blade chord	0.135 m (5¼ in)
Length: overall, rotors turning	13.02 m (42 ft 8½ in)
fuselage, incl tailskid	9.82 m (32 ft 2½ in)
Height: over tailfin	3.12 m (10 ft 2¾ in)
to top of rotor head	3.14 m (10 ft 3¾ in)
Fuselage: Max width	1.32 m (4 ft 4 in)
Stabiliser span	1.98 m (6 ft 6 in)
Width over skids	2.34 m (7 ft 8 in)
Forward cabin doors (each): Height	1.04 m (3 ft 5 in)
Max width	0.61 m (2 ft 0 in)
Centre cabin door (port): Height	1.04 m (3 ft 5 in)
Width	0.64 m (2 ft 1¼ in)
Rear cabin doors (each): Height	1.04 m (3 ft 5 in)
Width	0.91 m (3 ft 0 in)
Baggage door: Height	0.55 m (1 ft 9½ in)
Width	0.94 m (3 ft 1¼ in)

DIMENSIONS, INTERNAL:
Cabin: Length	2.74 m (9 ft 0 in)
Max width and height	as JetRanger
Volume	2.4 m³ (83 cu ft)
Cabin cargo volume (all passenger seats removed)	2.83 m³ (100 cu ft)
Baggage compartment volume	0.45 m³ (16.0 cu ft)

AREAS:
Main rotor disc	99.89 m² (1,075.2 sq ft)
Tail rotor disc	2.13 m² (22.97 sq ft)

WEIGHTS AND LOADINGS:
Weight empty, standard	1,047 kg (2,309 lb)
Max external load	907 kg (2,000 lb)
Max T-O weight: normal	2,018 kg (4,450 lb)
external load	2,064 kg (4,550 lb)
Max disc loading: normal	20.7 kg/m² (4.23 lb/sq ft)
external load	19.3 kg/m² (3.95 lb/sq ft)
Transmission loading at max T-O weight and power:	
internal load	5.52 kg/kW (9.08 lb/shp)
external load	5.65 kg/kW (9.29 lb/shp)

PERFORMANCE (at max normal T-O weight, ISA):
Never-exceed speed (V_{NE}):	
at S/L	130 kt (241 km/h; 150 mph)
at 1,525 m (5,000 ft)	133 kt (246 km/h; 153 mph)
Max cruising speed: at S/L	110 kt (204 km/h; 127 mph)
at 1,525 m (5,000 ft)	111 kt (206 km/h; 128 mph)
Max rate of climb at S/L	408 m (1,340 ft)/min
Service ceiling at max cruise power	3,050 m (10,000 ft)
Hovering ceiling: IGE	3,050 m (10,000 ft)
OGE	1,980 m (6,500 ft)
Range with max fuel, no reserves:	
at S/L	324 n miles (600 km; 372 miles)
at 1,525 m (5,000 ft)	357 n miles (661 km; 411 miles)
Endurance	3 h 42 min

OPERATIONAL NOISE LEVELS (FAR Pt 36):
T-O	88.4 EPNdB
Flyover	85.2 EPNdB
Approach	90.7 EPNdB

UPDATED

Bell 407 seven-seat light helicopter in Malaysian ownership (*Paul Jackson/Jane's*) 2001/0103573

Bell 407 panel, showing standard and optional instruments 0010221

BELL 407

TYPE: Light utility helicopter.
PROGRAMME: Design definition launched in 1993 as Bell Light Helicopter to supplement, and eventually replace, JetRanger and LongRanger; concept demonstrator Model 407 (N407LR) first flown 21 April 1994 (standard Bell 206L-3 modified with tailboom and dynamic system of military OH-58D, plus sidewall fairings to simulate broader fuselage); programme first revealed at Heli-Expo '95, Las Vegas, January 1995. Two prototype/preproduction 407s (C-GFOS and C-FORS) first flown on 29 June and 13 July 1995, respectively; first production airframe (C-FWQY/N407BT) flown 10 November 1995. Transport Canada certification 9 February 1996, with FAA certification following on 23 February; first customer delivery at Heli-Expo '96, Dallas, in February. MoU of June 1996 provided for licensed assembly and marketing by Dirgantara (formerly IPTN) of Indonesia.
CUSTOMERS: Launch customers Petroleum Helicopters, Niagara Helicopters and Greenland Air. Total of 62 delivered in 1999. The 450th production Bell 407 registered in September 2000. Total of 62 delivered in 2000. Recent customers include the Procuraduria General de Republica of Mexico, which has taken delivery of three for anti-drug operations, and Rocky Mountain Helicopters, which has ordered five for its EMS services.
COSTS: US$1.37 million (1999) flyaway. Average direct operating cost US$307 per hour (1999). Development cost estimated as US$50 million, of which US$9 million provided by Canadian government.
DESIGN FEATURES: As for JetRanger and LongRanger. Based on Bell 206L-4 LongRanger fuselage with cabin widened by 17.8 cm (7 in); 35 per cent larger cabin window area; Litton LCD active matrix cockpit displays; all-composites four-blade main rotor based on that of OH-58D, with soft-mounted pylon isolation system; single Rolls-Royce 250-C47 turboshaft; enlarged vertical stabilisers. 'Ring-fin' shrouded rotor tested in 1995 as possible future option. Rotor speed 413 rpm.
STRUCTURE: Generally as for LongRanger, but with carbon fibre tailboom and flush-fitting cabin door windows and skylights.
LANDING GEAR: LongRanger unit modified to match new rotor's ground-resonance characteristics, and with rear cross-tube pivoted in centre between outboard damping pads.
POWER PLANT: One Rolls-Royce 250-C47B turboshaft rated at 606 kW (813 shp) for T-O, 523 kW (701 shp) maximum continuous; transmission rating 503 kW (674 shp) for T-O, 470 kW (630 shp) continuous operation. Single-channel FADEC standard. Standard usable fuel capacity 484 litres (128 US gallons; 106.6 Imp gallons); optional auxiliary fuel tank in aft baggage compartment, usable capacity 75 litres (20.0 US gallons; 16.5 Imp gallons). 'Quiet Cruise' system, in development in early 1997, reduces flyover noise level by using FADEC to reduce rotor rpm to 90 to 93 per cent in the cruise, with accompanying reduction in V_{NE} to 110 kt (204 km/h; 126 mph).
ACCOMMODATION: With a crew of two, standard cabin layout accommodates five passengers in two rearward-facing seats with centre armrest/console, and three forward-facing seats, all fabric-covered. Optional utility cabin has vinyl-covered seats; corporate interior features two extra-wide forward-facing seats in the outboard positions and an occasional-use seat in the centre.
AVIONICS: Optional Honeywell avionics suite.

DIMENSIONS, EXTERNAL:
Main rotor diameter	10.67 m (35 ft 0 in)
Main rotor blade chord	0.27 m (10¾ in)
Tail rotor diameter	1.65 m (5 ft 5 in)
Tail rotor blade chord	0.16 m (6½ in)
Length: overall, rotors turning	12.74 m (41 ft 9½ in)
rotor in X configuration*	11.16 m (36 ft 7¼ in)
fuselage	10.58 m (34 ft 8½ in)
Stabiliser span over endplate fins	2.22 m (7 ft 3½ in)
Height: over tailfin	3.10 m (10 ft 2 in)
overall: low skids	3.60 m (11 ft 9½ in)
high skids	3.81 m (12 ft 6 in)
Width over skids	2.29 m (7 ft 6 in)
Rear cabin door:	
Width: port	1.55 m (5 ft 1 in)
starboard	0.91 m (3 ft 0 in)
Height to sill (standard skids)	0.51 m (1 ft 8 in)

*Blades at 45° to fuselage centreline

DIMENSIONS, INTERNAL:
Cabin: Max width	1.37 m (4 ft 6 in)
Max height	1.00 m (3 ft 3¼ in)
Baggage compartment volume	0.45 m³ (16.0 cu ft)

AREAS:
Main rotor disc	89.38 m² (962.1 sq ft)
Tail rotor disc	2.08 m² (22.34 sq ft)

WEIGHTS AND LOADINGS:
Weight empty, equipped	1,187 kg (2,617 lb)
Max payload	1,089 kg (2,402 lb)
Max hook capacity	1,200 kg (2,646 lb)
Max T-O weight: internal load:	
standard	2,268 kg (5,000 lb)
optional	2,381 kg (5,250 lb)
external load	2,721 kg (6,000 lb)
Max disc loading	27.9 kg/m² (5.72 lb/sq ft)
Transmission loading at max T-O weight and power:	
internal load: standard	4.52 kg/kW (7.42 lb/shp)
optional	4.74 kg/kW (7.79 lb/shp)
external load	5.42 kg/kW (8.90 lb/shp)
Max disc loading: internal load:	
standard	25.4 kg/m² (5.20 lb/sq ft)
optional	26.6 kg/m² (5.46 lb/sq ft)
external load	30.4 kg/m² (6.24 lb/sq ft)

PERFORMANCE (at internal load MTOW, ISA):
Never-exceed speed (V_{NE})	140 kt (259 km/h; 161 mph)
Max cruising speed:	
at S/L	128 kt (237 km/h; 147 mph)
at 1,220 m (4,000 ft)	131 kt (243 km/h; 151 mph)
Long-range cruising speed:	
at S/L	112 kt (207 km/h; 129 mph)
at 1,220 m (4,000 ft)	115 kt (213 km/h; 132 mph)
Max certified T-O height	5,180 m (17,000 ft)
Max certified altitude	6,100 m (20,000 ft)
Hovering ceiling: IGE	3,720 m (12,200 ft)
OGE	3,170 m (10,400 ft)
Max range	330 n miles (611 km; 379 miles)
Endurance	3 h 42 min

UPDATED

BELL 427

TYPE: Light utility helicopter.
PROGRAMME: Launched as New Light Twin (NLT) in February 1996 on signature of collaborative agreement with Samsung Aerospace Industries of South Korea. Prototype assembly began early 1997; first flight (C-

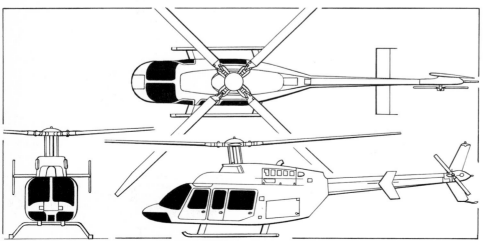

Bell 407 seven-seat light helicopter (*James Goulding/Jane's*) 2001/0103545

BELL—AIRCRAFT: CANADA

Second production Bell 427 twin-turbine utility helicopter *(Paul Jackson/Jane's)* 2000/0062710

GBLL) 11 December 1997; second prototype (C-FCSS) completed February 1998; two prototypes undertook flight test programme, gaining Transport Canada certification on 19 November 1999; FAA VFR certification achieved in January 2000, followed by Dual Pilot IFR (DPIFR) Category A certification on 24 May 2000. First production aircraft (C-GDEJ) flown June 1998; compared with prototypes, production 427 has longer exhausts and revised upper surface contours. Alternative FAR Pt 29 version, with increased T-O weight, will also be available.

CURRENT VERSIONS: **Bell 427:** *As described.*

SB427: Proposed South Korean-assembled version; see SSB entry in that national section.

Super Kiowa: Proposed armed scout offered by Bell.

CUSTOMERS: Orders for 22 placed during the mockup's first public display at Farnborough Air Show 1996; 85 on order by May 2000 from 50 customers, including five sold by Samsung to CitiAir in South Korea. Deliveries began 2000 and totalled 10 by October 2000. Expected to be offered to South African Air Force as Alouette III replacement from 2003, possibly with local assembly. MoU signed in 1999 with Elbit Systems Inc for joint pre-design phase of military light helicopter using Bell 427 as baseline platform. Total of five delivered during 2000.

COSTS: Projected unit cost of US$2.2 million in 1999 and US$2.3 million in 2000 (both IFR). VFR version available from 2000 at US$2.675 million (dual pilot, without AFCS) or US$2.8 million (single pilot, with AFCS). Direct operating cost US$438 per hour (1999).

DESIGN FEATURES: Similar in appearance to Bell 407, with cabin stretch of 33 cm (13 in), but is all-new design, incorporating twin-engine safety margins. Flight dynamics based on four-blade rotor system of Bell OH-58D Kiowa (see US section), allied to tail rotor of Bell 407; folding main blades. Main rotor rpm 395; tail rotor rpm 2,375. Purpose-designed 'flat pack' main transmission, with direct input from both engines, has only four gear meshes to simplify design and operation. Transmission attached to airframe by four liquid-inertia vibration eliminators. First Bell helicopter designed entirely with use of computer (Dassault CATIA programme).

FLYING CONTROLS: Sextant AFDS 95-1 AFCS, with two- to four-axis autopilot computer, flight director computer and associated equipment, to be certified by 2000.

STRUCTURE: Generally as for Bell 206, but extensive use of carbon/epoxy composites reduces airframe parts count by some 33 per cent. Cabin floor and roof are flat panels for ease of manufacture; minimal use of curved panels elsewhere. Composites main and tail rotors; main blades have nickel-plated stainless steel leading-edges. Soft-in-plane hub of main rotor employs a composites flexbeam yoke and elastomeric joints, eliminating lubrication and maintenance requirements. Brake and main rotor blade folding optional. Composites cabins and rolled aluminium tailbooms built by Samsung; assembly in Canada, except for sales to Korea and China; Hexcel honeycomb as stiffener.

LANDING GEAR: Twin skids with dynamically tuned cross tubes to reduce ground resonance. Low skids standard; optional high skids and emergency floats.

POWER PLANT: Two Pratt & Whitney Canada PW207D turboshafts with FADEC, each rated at 529 kW (710 shp) for T-O (5 minutes) or 466 kW (625 shp) maximum continuous; OEI ratings 611 kW (820 shp) for 30 seconds, 582 kW (780 shp) for 2 minutes, 559 kW (750 shp) for 30 minutes or 529 kW (710 shp) maximum continuous. Twin-engine transmission rating, T-O and maximum continuous, 597 kW (800 shp). OEI transmission rating, 485 kW (650 shp) for 30 seconds; 451 kW (605 shp) for 2 minutes; 343 kW (460 shp) maximum continuous.

Fuel contained in three crash-resistant tanks; two forward, one aft, total usable capacity 770 litres (203.5 US gallons; 169 Imp gallons). One forward fuel tank can be removed in EMS configurations to provide additional stretcher space in cabin or to permit stretcher to extend into port side of cockpit.

ACCOMMODATION: Standard accommodation is for two crew in cockpit, on 20 g energy-attenuating seats, and six passengers in cabin on two rows of three seats in club configuration (all-forward-facing seats optional); all seats equipped with inertia-reel shoulder harnesses. Alternative configurations include corporate club-four seating with refreshment/entertainment console between each pair of seats. Optional EMS interiors provide for carriage of one or two stretchers with up to two medical attendants, affording either full patient or head-only in-flight access, with single- or two-person crew. In cargo configuration, with all passenger seats removed, an optional removable flat cargo floor can be installed, equipped with integral tie-downs. Two forward-hinged doors each side; cabin doors, both sides, are forward-hinged, but port unit can be replaced by optional rearward-sliding door for cargo handling. External door on starboard side to rear baggage hold.

SYSTEMS: Hydraulic system, operating pressure 86 bars (1,250 lb/sq in), provides boost power for main and tail rotor controls; 28 V DC electrical power from 17 Ah Ni/Cd battery and two engine-mounted 17 Ah starter/generators; 28 Ah battery and 200 A starter/generator optional. Air conditioning optional.

AVIONICS: Rogerson-Kratos NeoAV two-screen LCD integrated instrument display system (IIDS) for monitoring engine instruments, fuel quantity, hydraulic and electrical systems and weight and balance functions. Rogerson Kratos NeoAV EFIS optional, featuring GPS interface, area navigation map, non-precision approach capability and weather radar display. Optional Honeywell nav/com avionics suite to customer's choice.

EQUIPMENT: Optional kits include engine air particle separator, cargo hook, cargo floor, sliding cabin door, external rescue hoist, NightSun searchlight, EMS installation and wire strike protection system.

DIMENSIONS, EXTERNAL:
Main rotor diameter	11.28 m (37 ft 0 in)
Main rotor blade chord	0.27 m (10½ in)
Tail rotor diameter	1.73 m (5 ft 8 in)
Tail rotor blade chord	0.18 m (7¼ in)
Length: overall, rotors turning	13.07 m (42 ft 10¾ in)
overall, rotor in X configuration*	11.55 m (37 ft 10¾ in)
fuselage, incl tailskid	10.94 m (35 ft 10¾ in)
Height over tailfin	3.49 m (11 ft 5¼ in)
Ground clearance: low skids	0.41 m (1 ft 4¼ in)
high skids	0.67 m (2 ft 2½ in)
Width: overall, rotor in X configuration*	8.24 m (27 ft 0½ in)
over skids	2.36 m (7 ft 9 in)
Cabin door: Height	1.07 m (3 ft 6 in)
Width	1.24 m (4 ft 0¾ in)

Blades at 45° to fuselage centreline

DIMENSIONS, INTERNAL:
Cabin: Max height	1.30 m (4 ft 3 in)
Baggage hold volume	0.76 m³ (27.0 cu ft)

AREAS:
Main rotor disc	99.89 m² (1,075.2 sq ft)
Tail rotor disc	2.34 m² (25.2 sq ft)

WEIGHTS AND LOADINGS (provisional):
Weight empty	1,743 kg (3,842 lb)
Max payload	1,021 kg (2,250 lb)
Max baggage weight	113 kg (250 lb)
Max T-O weight: internal load	2,721 kg (6,000 lb)
external load	2,948 kg (6,500 lb)
Cargo hook capacity	1,361 kg (3,000 lb)
Max cargo floor loading	366.2 kg/m² (75 lb/sq ft)
Max disc loading	29.52 kg/m² (6.05 lb/sq ft)
Transmission loading at max T-O weight and power:	
internal load	4.56 kg/kW (7.50 lb/shp)
external load	4.95 kg/kW (8.13 lb/shp)

PERFORMANCE (estimated, PW206D engines, at internal load max T-O weight, ISA):
Max cruising speed at S/L	140 kt (259 km/h; 161 mph)
Econ cruising speed at S/L	133 kt (246 km/h; 153 mph)
Service ceiling:	
max continuous power	5,547 m (18,200 ft)
OEI (30 min)	3,825 m (12,550 ft)
Hovering ceiling: IGE	4,938 m (16,200 ft)
OGE	4,237 m (13,900 ft)
Range with max fuel, econ cruising speed, no reserves	394 n miles (730 km; 453 miles)
Endurance	4 h 0 min

UPDATED

BELL 212 TWIN TWO-TWELVE

US military designation: UH-1N
Canadian Forces designation: CH-135
Israel Defence Force name: Anafa (Heron)

There has been no known production of this twin-turbine utility helicopter since 1998. Full description last appeared in 1998-99 *Jane's*; abbreviated data in 2000-01 edition.

UPDATED

BELL 412

Canadian Forces designation: CH-146 Griffon
UK armed forces designation: Griffin HT. Mk 1

TYPE: Multirole medium helicopter.
PROGRAMME: Original 412 announced 8 September 1978 (see earlier editions of *Jane's*); FAR Pt 29 VFR approval

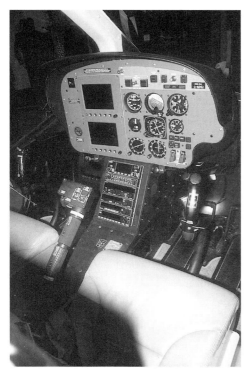

Bell 427 instrument panel *(Paul Jackson/Jane's)*
2000/0062711

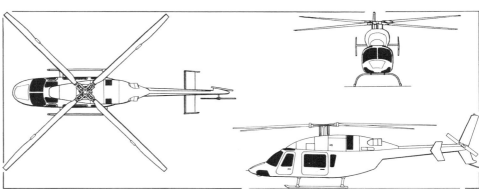

Provisional general arrangement of Bell 427 *(Paul Jackson/Jane's)* 2000/0062757

received 9 January 1981, IFR 13 February 1981; 213 built in USA; production (SP version) transferred to Canada February 1989; first delivery (civil) 18 January 1981. Production licences obtained by IPTN of Indonesia and Agusta of Italy (which see).

CURRENT VERSIONS: **412SP**: Special Performance version with increased maximum T-O weight, new seating options and 55 per cent greater standard fuel capacity. Superseded by 412HP early 1991. Details in 1991-92 *Jane's*.

Military 412: Announced by Bell June 1986; fitted with Lucas Aerospace chin turret and Honeywell Head Tracker helmet sight similar to that in AH-1S; turret carries 875 rounds, weighs 188 kg (414 lb) and can be removed in under 30 minutes; firing arcs ±110° in azimuth, +15° and −45° in elevation; other armament includes twin dual FN Herstal 7.62 mm gun pods, single FN Herstal 0.50 in pod, pods of seven or nineteen 2.75 in rockets, M240E1 pintle-mounted door guns, FN Herstal four-round 70 mm rocket launcher and a 0.50 in gun or two Giat M621 20 mm cannon pods.

412HP: Improved transmission giving better OGE hover; FAR Pt 29 certification 5 February 1991, first delivery (c/n 36020) later that month.

412EP (Enhanced Performance): PT6T-3D engine, dual digital automatic flight control system (DDAFCS), three-axis in basic aircraft but customer option for four-axis and EFIS. Category A certification was imminent in late 1998. Also customer option for SAR fit. Now standard current model.

Detailed description applies to Bell 412EP.

412CF (CH-146) Griffon: Canadian Forces C$700 million contract for 100 CH-146s (modified Bell 412EP) placed in 1992. Duties include armed support, troop/cargo transport, medevac, ASW, SAR and patrol; first flight (146000) 30 April 1994; deliveries began 14 October 1994; completed early 1998. Generally as commercial Bell 412EP except for avionics and mission equipment; see 1998-99 and earlier *Jane's*. Empty weight 3,402 kg (7,500 lb); maximum weight as civil version.

412EP Sentinel: First of two modified by Heli-Dyne Systems with quick-change ASV and ASW mission packages delivered to Ecuadorean Navy late 1998. ASV equipment comprises Honeywell RDR-1500B chin radar, Wescam sensor turret and possibly Penguin Mk 2 Mod 7 ASMs; ASW fit is L3 Ocean Systems AN/AQS-18A dipping sonar and Raytheon Mk 46 torpedo.

412 Plus: Study announced early 1999; MTOW increased to 5,647 kg (12,450 lb), uprated PT6C engines, new dynamic components, Rogerson-Kratos avionics; target selling price US$4 million to US$4.5 million (1999). Launch decision was expected by end of 1999.

NBell-412: Indonesia's IPTN (which see) has licence to produce up to 100 Model 412SPs.

CUSTOMERS: Some 580 Bell 412s of all versions built in North America by September 2000, including 26 delivered in 1999.

Military deliveries include Venezuelan Air Force (two), Botswana Defence Force (three), Public Security Flying Wing of Bahrain Defence Force (two), Sri Lankan armed forces (four), Nigerian Police Air Wing (two), Mexican government (two VIP transports), South Korean Coast Guard (one), Honduras (10), Royal Norwegian Air Force (19, of which 18 assembled by Helikopter Service, Stavanger, to replace UH-1Bs of 339 Squadron at Bardufoss and 720 Squadron at Rygge). Three 412EPs delivered to Slovenian Territorial Forces in 1995, for border patrol and rescue duties; four ordered by Philippine Air Force late in 1996, comprising two for VVIP transport and two SAR; first of nine 412EPs entered service in April 1997 with civilian-operated Defence Helicopter Flying School at RAF Shawbury, UK, within which they constitute No. 60 (Reserve) Squadron, RAF. Four in SAR/utility fit delivered to Venezuelan Navy, 1999.

Bell 412EP civilian demonstrator *(Paul Jackson/Jane's)*

COSTS: Bell 412EP, VFR-equipped US$4.895 million (1999); Bell 412EP, IFR-equipped US$5.12 million (1999). Average direct operating cost US$746 per hour (1997).

DESIGN FEATURES: Four-blade main rotor with blades retained within central metal star fitting by single elastomeric bearings; shorter rotor mast than 212; blades can be folded; rotor brake standard; two-blade tail rotor; main rotor rpm 314.

FLYING CONTROLS: Fully powered hydraulic controls; gyroscopic stabiliser bar above main rotor; automatic tailplane incidence control.

STRUCTURE: Generally of conventional light metal. Main rotor blade spar unidirectional glass fibre with 45° wound torque casing of glass fibre cloth; Nomex rear section core with trailing-edge of unidirectional glass fibre; leading-edge protected by titanium abrasion strip and replaceable stainless steel cap at tip; lightning protection mesh embedded; provision for electric de-icing heater elements; main rotor hub of steel and light alloy; all-metal tail rotor.

LANDING GEAR: High skid, emergency pop-out float or non-retractable tricycle gear optional. Spats optional for last-named.

POWER PLANT: Pratt & Whitney Canada PT6T-3D Turbo Twin-Pac, rated at 1,342 kW (1,800 shp) for T-O and 1,193 kW (1,600 shp) maximum continuous. OEI ratings 850 kW (1,140 shp) for 2½ minutes, or 723 kW (970 shp) for 30 minutes. Transmission rating 1,022 kW (1,370 shp) for T-O, 828 kW (1,110 shp) maximum continuous; OEI rating 850 kW (1,140 shp). Optional 30 kW (40 shp) for accessory drives from main gearbox.

Seven interconnected rupture-resistant fuel cells, with automatic shutoff valves (breakaway fittings), have a combined usable capacity of 1,249 litres (330 US gallons; 275 Imp gallons). Two 76 or 310.5 litre (20.0 or 82.0 US gallon; 16.7 or 68.3 Imp gallon) auxiliary fuel tanks, in any combination, can increase maximum total capacity to 1,870 litres (494 US gallons; 411.6 Imp gallons). Single-point refuelling on starboard side of cabin.

ACCOMMODATION: Pilot and up to 14 passengers: one in front port seat and 13 in cabin. Dual controls optional. Accommodation heated and ventilated.

SYSTEMS: Dual hydraulic systems, pressure 69 bars (1,000 lb/sq in) each. 28 V DC electrical system supplied by two completely independent 450 VA inverters. 40 Ah Ni/Cd battery.

AVIONICS: *Comms:* Optional IFR avionics include dual Honeywell Gold Crown III.
Radar: Weather radar optional.
Flight: Dual KNR 660A VOR/LOC/RMI receivers, KDF 800 ADF, KMD 700A DME, KXP 750A transponder and KGM 690 marker beacon/glide slope receiver. Optional Honeywell AFCS.

EQUIPMENT: Optional equipment includes cargo sling and rescue hoist.

DIMENSIONS, EXTERNAL:
Main rotor diameter	14.02 m (46 ft 0 in)
Main rotor blade chord: at root	0.40 m (1 ft 4 in)
at tip	0.22 m (8½ in)
Tail rotor diameter	2.62 m (8 ft 7 in)
Tail rotor blade chord	0.29 m (11½ in)
Length: overall, rotors turning	17.12 m (56 ft 2 in)
fuselage, excl rotors	12.70 m (41 ft 8 in)
Height: to top of rotor head	3.48 m (11 ft 5 in)
overall, tail rotor turning	4.57 m (15 ft 0 in)
Stabiliser: span	2.87 m (9 ft 5 in)
chord	0.79 m (2 ft 7 in)
Width over skids	2.84 m (9 ft 4 in)
Door sizes	as Bell 212

DIMENSIONS, INTERNAL:
Baggage compartment volume	0.79 m³ (28.0 cu ft)

AREAS:
Main rotor disc	154.40 m² (1,661.9 sq ft)
Tail rotor disc	5.38 m² (57.86 sq ft)

WEIGHTS AND LOADINGS:
Weight empty, standard equipped	3,079 kg (6,789 lb)
Max external hook load	2,041 kg (4,500 lb)
Max T-O and landing weight, internal or external load	5,397 kg (11,900 lb)
Max disc loading	35.0 kg/m² (7.16 lb/sq ft)
Transmission loading at max T-O weight and power	5.29 kg/kW (8.69 lb/shp)

PERFORMANCE:
Never-exceed speed (V$_{NE}$)	140 kt (259 km/h; 161 mph)
Max cruising speed: at S/L	122 kt (226 km/h; 140 mph)
at 1,525 m (5,000 ft)	124 kt (230 km/h; 143 mph)
Long-range cruising speed at 1,525 m (5,000 ft)	130 kt (241 km/h; 150 mph)
Service ceiling, OEI, 30 min power rating	1,920 m (6,300 ft)
Hovering ceiling: IGE	3,110 m (10,200 ft)
OGE	1,585 m (5,200 ft)
Range at 1,525 m (5,000 ft), long-range cruising speed, standard fuel, no reserves	402 n miles (744 km; 462 miles)
Endurance	3 h 42 min

OPERATIONAL NOISE LEVELS (FAR Pt 36) (412 EP):
T-O	92.8 EPNdB
Flyover	93.4 EPNdB
Approach	95.6 EPNdB

UPDATED

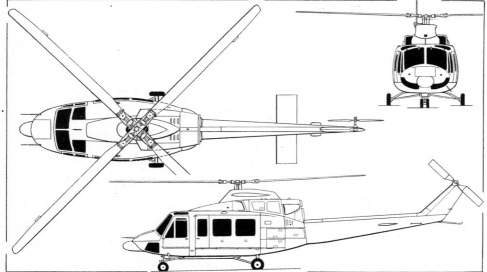

Bell 412EP twin-turboshaft utility helicopter *(Dennis Punnett/Jane's)*

BELL 430

TYPE: Light utility helicopter.

PROGRAMME: Preliminary design 1991; four-blade rotor, higher powered and stretched variant of Bell 230; programme launched February 1992; two prototypes, modified from Bell 230 airframes; first prototype (C-GBLL; wheel-equipped) flown 25 October 1994; second prototype (C-GEXP; skid-equipped, with complete avionics suite) flown 19 December 1994; first flight of production 430 (C-GRND) in 1995; deliveries began 25 June 1996 after Canadian type approval on 23 February. Single-pilot IFR and Category A certification anticipated early 1999. MoU of June 1996 with Dirgantara (formerly IPTN) for licensed assembly and marketing in Indonesia.

Second production aircraft (N4300) circumnavigated the world in a record time of 17 days 6 hours 14 minutes, landing back at Fairoaks, UK, on 3 September 1996.

CUSTOMERS: First delivery 25 June 1996, when sixth production aircraft (N6282X) handed over to IPTN of Indonesia in eight-seat executive configuration. Thirteen delivered in 1996; followed by eight, 15 and 18 in 1997-99, and 11 in 2000. Total of 71 registered by September 2000.

Bell 430 cockpit test-rig, showing Rogerson-Kratos IIDS display in centre

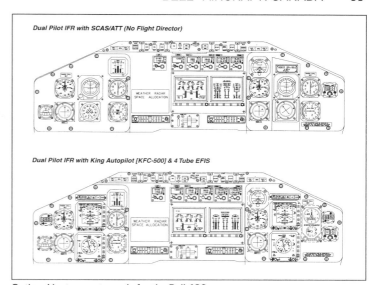

Optional instrument panels for the Bell 430

COSTS: Programme US$18 million, 35 per cent financed by Canadian Defence Industry Productivity Program (DIPP) and repayable as royalty on each sale. Unit costs (1999) US$4.09 million, dual pilot, no autopilot; US$4.25 million, single pilot and autopilot; US$4.295 million, dual pilot and autopilot; US$4.395 million, single/dual pilot and autopilot; all versions equipped for IFR. Average direct operating cost US$565 per hour (1997).

DESIGN FEATURES: Optimised for high cruising speed with (retractable) wheel landing gear, although traditional skids optional; inclined towards executive transport market. Bell 230 fuselage lengthened by 0.46 m (1 ft 6 in) plug; Bell 680 all-composites four-blade bearingless, hingeless main rotor; approximately 10 per cent power increase over Bell 230; uprated transmission; and optional EFIS. Short-span sponson each side of fuselage houses mainwheel units and fuel tanks, and serves as work platform.

FLYING CONTROLS: Fully powered hydraulic, with elastomeric pitch change and flapping bearings; fixed tailplane with leading-edge slats and endplate fins; strakes under sponsons; single-pilot IFR system without auto-stabilisation.

STRUCTURE: Semi-monocoque fuselage of light alloy, with limited use of light alloy honeycomb panels. Fail-safe structure in critical areas. One-piece nosecone tilts forward and down for access to avionics and equipment bay. Short span cantilever sponson set low on each side of fuselage, serving as main landing gear housings, fuel tanks and work platforms. Section NACA 0035. Dihedral 3° 12′. Incidence 5°. Sweepback at quarter-chord 3° 30′.

Fixed vertical fin in sweptback upper and lower sections. Tailplane, with slotted leading-edge and endplate fins, mounted midway along rear fuselage. Small skid below ventral fin for protection in tail-down landing. Four-blade main rotor with stainless steel spars and leading-edges, Nomex honeycomb trailing-edge with glass fibre skin, and glass fibre safety straps; tail rotor blades stainless steel. Rotors shaft-driven through gearbox with two spiral bevel reductions and one planetary reduction. Main blade and hub life, 10,000 hours.

LANDING GEAR: Tubular skid type on Utility version. Executive version has hydraulically retractable tricycle gear, single mainwheels retracting forward into sponsons; forward-retracting nosewheel fully castoring and self-centring; hydraulic disc brakes on main units. Mainwheel tyre size 18×5.5, nosewheel 5.00-5. Emergency floats optional.

POWER PLANT: Two Rolls-Royce 250-C40B turboshafts, each rated at 584 kW (783 shp) for T-O and 521 kW (699 shp) maximum continuous. OEI ratings 701 kW (940 shp) for 30 seconds, 656 kW (880 shp) for 2 minutes, 623 kW (835 shp) for 30 minutes and 602 kW (808 shp) continuous. Chandler Evans FADEC. Transmission rating 779 kW (1,045 shp) for 5 minutes for T-O, 737 kW (989 shp) maximum continuous. Power train TBO, 5,000 hours. Usable fuel capacity 935 litres (247 US gallons; 206 Imp gallons) in skid version, 710 litres (187.5 US gallons; 156 Imp gallons) in wheeled version; provision in both versions for 182 litre (48 US gallon; 40 Imp gallon) auxiliary tank. Fuel system is rupture resistant, with self-sealing breakaway fittings.

ACCOMMODATION: Standard layout has forward-facing seats for nine persons (two-two-two-three) including pilot. Options include 10-seat layout (two-two-three-three); eight-seat executive (rear six in club layout), six-seat executive (rear four in club layout with console between each pair); and five- and four-seat executive with one or two refreshment cabinets; seat pitches vary between 86 cm (34 in) and 91 cm (36 in). Pilots on crashworthy (energy attenuating) seats, which are optional for passengers. Customised emergency medical service (EMS) versions also available, configured for pilot-only operation plus one or two pivotable stretchers and four or three medical attendants/sitting casualties respectively. Two forward-opening doors each side; EMS version has optional stretcher door between forward and rear doors on port side. Entire interior ram air ventilated and soundproofed. Dual controls optional.

SYSTEMS: Dual hydraulic systems (dual for main rotor collective and cyclic, single for tail rotor). Dual 28 V DC electrical systems, powered by two 30 V 200 A engine-mounted starter/generators (derated to 180 A) and a 24 V 28 Ah Ni/Cd battery.

AVIONICS: *Comms:* Honeywell Gold Crown III.

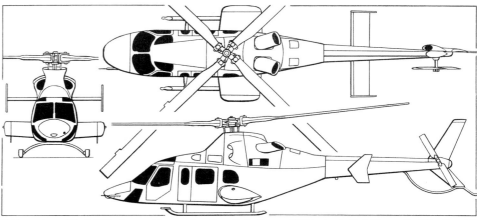

Skid version of Bell 430 nine-seat helicopter (two Rolls-Royce 250 turboshafts) *(James Goulding/Jane's)*

Flight: Honeywell KFC 500 AFCS. GPS optional.
Instrumentation: Rogerson-Kratos LCD integrated instrument display system (IIDS) comprising two active matrix LCDs displaying engine and system parameters; optional Rogerson-Kratos EFIS.

EQUIPMENT: Standard equipment includes rotor and cargo tiedowns, ground handling wheels for skid version, retractable 450 W search/landing light. Options include dual controls, auxiliary fuel tankage, force/feel trim system, more comprehensive nav/com avionics, 272 kg (600 lb) capacity rescue hoist, 1,587 kg (3,500 lb) capacity cargo hook, emergency flotation gear, heated windscreen, particle separator and snow baffles.

DIMENSIONS, EXTERNAL:
Main rotor diameter	12.80 m (42 ft 0 in)
Main rotor blade chord	0.36 m (1 ft 2¼ in)
Tail rotor diameter	2.10 m (6 ft 10½ in)
Tail rotor blade chord	0.25 m (10 in)

Bell 430, showing landing gear extended *(Paul Jackson/Jane's)*

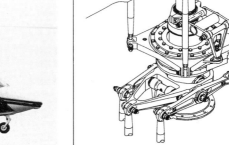

Rotor head of the Bell 430

CANADA: AIRCRAFT—BELL to BOMBARDIER

Length: fuselage (incl tailskid)	13.44 m (44 ft 1¼ in)	Max width	1.27 m (4 ft 2 in)	Long-range cruising speed:	
overall, rotors turning	15.30 m (50 ft 2½ in)	Max height	1.45 m (4 ft 9 in)	A	131 kt (243 km/h; 151 mph)
Fuselage max width over sponsons	3.45 m (11 ft 4 in)	Volume	4.5 m³ (158 cu ft)	B	128 kt (237 km/h; 147 mph)
Height to top of rotor head:		Baggage compartment volume	1.0 m³ (37 cu ft)	Service ceiling: A, B	5,590 m (18,340 ft)
standard skids	4.03 m (13 ft 2½ in)	AREAS:		Hovering ceiling, IGE: A, B	3,565 m (11,700 ft)
high skids	4.24 m (13 ft 10¾ in)	Main rotor disc	128.71 m² (1,385.4 sq ft)	OGE: A, B	1,890 m (6,200 ft)
wheels	3.72 m (12 ft 2½ in)	Tail rotor disc	3.45 m² (37.12 sq ft)	Max range, standard fuel, no reserves:	
Skid track	2.54 m (8 ft 4 in)	WEIGHTS AND LOADINGS (A: wheeled version; B: skid version):		A	275 n miles (509 km; 316 miles)
Wheel track	2.78 m (9 ft 1½ in)	Weight empty: A	2,423 kg (5,342 lb)	B	353 n miles (653 km; 406 miles)
Wheelbase	4.17 m (13 ft 8¼ in)	B	2,407 kg (5,308 lb)	Max endurance	3 h 35 min
Passenger doors: forward:		Max external load: A, B	1,587 kg (3,500 lb)	OPERATIONAL NOISE LEVELS (FAR Pt 36, Stage 2):	
Height	1.34 m (4 ft 4¾ in)	Max T-O weight, all conditions	4,218 kg (9,300 lb)	T-O	92.4 EPNdB
Width	0.88 m (2 ft 10¾ in)	Max disc loading	32.8 kg/m² (6.71 lb/sq ft)	Sideline	91.6 EPNdB
aft:		Transmission loading at max T-O weight and power		Approach	93.8 EPNdB
Height	1.22 m (4 ft 0 in)		5.42 kg/kW (8.90 lb/shp)		**UPDATED**
Width	0.91 m (3 ft 0 in)	PERFORMANCE:			
Baggage door: Height	0.60 m (1 ft 11½ in)	Never-exceed speed (V$_{NE}$):			
Width	0.85 m (2 ft 9½ in)	A, B	150 kt (277 km/h; 172 mph)		
DIMENSIONS, INTERNAL:		Max level speed: A	139 kt (257 km/h; 160 mph)		
Cabin: Length, excl cockpit	2.87 m (9 ft 5 in)	B	135 kt (250 km/h; 155 mph)		

BELL/AGUSTA AB139
A description of this joint venture helicopter appears under the Bell/Agusta entry in the International section.

BLUE YONDER
BLUE YONDER AVIATION
Box 12, Suite 9, RR5, Calgary, Alberta T2P 2G6
Tel: (+1 403) 936 57 67
Fax: (+1 403) 936 51 08
e-mail: mailto:ezflyer@sprint.ca
PRESIDENT: Wayne Winters

Blue Yonder designed the E-Z Flyer in 1991; built under licence by Merlin in USA until bankruptcy of December 1995; production tooling transferred to Canada and manufacture resumed; Aerocomp (which see) is US agent, although Blue Yonder responsible for global sales.
UPDATED

BLUE YONDER E-Z FLYER
TYPE: Tandem-seat kitbuilt.
PROGRAMME: First kitbuilt E-Zs delivered by Merlin mid-1995. Programme transferred to Blue Yonder in 1998.
CUSTOMERS: Registration of 25th aircraft was effected in October 1999. Total 31 kits sold by June 2000.
DESIGN FEATURES: Open tube fuselage married to high-mounted Aerocomp Merlin GT wing and tail sections.
FLYING CONTROLS: Manual. Non-drooping Junkers-type ailerons.
STRUCTURE: Fuselage is welded 4130 chromoly. Wings are aluminium D-cell construction with aluminium and Styrofoam constructed ribs; Stits Poly-Fiber covering.
LANDING GEAR: Conventional tricycle. Hegar rims with brakes; solid spring steel legs; 18 in tundra tyres on mainwheels.
POWER PLANT: One 37.0 kW (49.3 hp) Rotax 503 or 47.8 kW (64.1 hp) Rotax 582 engine, driving a two- or three-blade pusher propeller. Fuel capacity 34 litres (9.0 US gallons; 7.5 Imp gallons).
DIMENSIONS, EXTERNAL:
Wing span 7.92 m (26 ft 0 in)

Blue Yonder E-Z Flyer two-seat kitbuilt

Length overall	6.40 m (21 ft 0 in)	PERFORMANCE:	
Height overall	1.83 m (6 ft 0 in)	Max cruising speed	56 kt (105 km/h; 65 mph)
AREAS:		Stalling speed	35 kt (65 km/h; 40 mph)
Wings, gross, incl extended tips	15.33 m² (165.0 sq ft)	Max rate of climb at S/L	137 m (450 ft)/min
WEIGHTS AND LOADINGS:		T-O to 15 m (50 ft)	99 m (325 ft)
Weight empty	213 kg (469 lb)	Landing from 15 m (50 ft)	61 m (200 ft)
Max T-O weight	607 kg (1,340 lb)		**UPDATED**

BOMBARDIER
BOMBARDIER AEROSPACE
400 Chemin de la Côte Vertu West, Dorval, Québec H4S 1Y9
Tel: (+1 514) 855 50 00
Fax: (+1 514) 855 79 03
Web: http://www.aero.bombardier.com

DESIGN DIVISIONS:
Canadair: follows Bombardier entry
de Havilland: follows Canadair entry
Learjet: see US section
Shorts: see 1996-97 edition and *Jane's Aircraft Upgrades*

PRESIDENT AND COO: Michael S Graff
VICE-PRESIDENT, OPERATIONS: Ken Brunrle
VICE-PRESIDENT, FINANCE: James Stewart
VICE-PRESIDENT, ENGINEERING AND PRODUCTION DEVELOPMENT: John Holding
PRESIDENT, REGIONAL AIRCRAFT: Steve Ridolfi
PRESIDENT, BUSINESS AIRCRAFT: Pierre Beaudoin
PRESIDENT, AMPHIBIOUS AIRCRAFT: Thomas Appleton

Bombardier Inc, a diversified Canadian corporation with 56,000 employees, formed Bombardier Aerospace in 1986, subsequently combining design and manufacturing activities of Canadair (1986), de Havilland (1992), Learjet (1990) and Shorts (1989). Sales, marketing and support are conducted by the Amphibious Aircraft, Regional Aircraft and Business Aircraft units.
Bombardier Aerospace also designs and manufactures components for Airbus and Boeing.
Bombardier Aerospace's revenue for the year ended 31 January 2000 totalled C$8.1 billion.
UPDATED

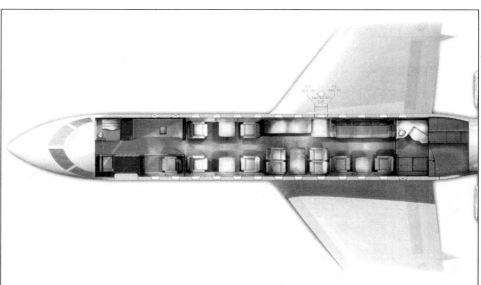

Typical Global Express cabin layout

BOMBARDIER BD-700 GLOBAL EXPRESS
TYPE: Long-range business jet.
PROGRAMME: Announced 28 October 1991 at NBAA Convention; full-scale cabin mockup exhibited at NBAA Convention September 1992; conceptual design started early 1993. Programme launched 20 December 1993; high-speed configuration frozen June 1994; low-speed configuration established August 1994.
Ground test programme using static test airframe c/n 0001 began August 1996; prototype C-FBGX (engineering designation BD-700-1A10) rolled out 26 August 1996; first flight 13 October 1996; public debut at NBAA

Bombardier Global Express long-range business jet

Convention at Orlando, Florida, November 1996; prototype and three other aircraft undertook 2,000-hour, 18-month flight test programme based at Bombardier's flight test centre in Wichita, Kansas; second aircraft, (C-FHGX), which is used for systems evaluation and testing, flew 3 February 1997; third (C-FJGX), which is used for avionics and autopilot testing, 22 April 1997; fourth (C-FKGX), first flown 8 September 1997 and the first to be fully outfitted, was used for function and reliability testing.

Transport Canada certification 31 July 1998; FAA certification 13 November 1998; JAA certification 7 May 1999; German LBA certification 26 May 1999; first customer delivery of completed aircraft 8 July 1999 to AirFlite Inc of Long Beach, California, which operates the aircraft on behalf of Toyota Motor Sales USA; 50th 'green' airframe delivered to Montreal completion centre 7 June 2000. Total of 18 aircraft in customer service and 9 'green' airframes delivered to completion centres by 9 October 2000, by which date in-service aircraft had accumulated 7,239 flying hours in 2,992 flights with a despatch rate of 98.15 per cent.

CURRENT VERSIONS: **Corporate Transport:** *As described.*
Airliner: Market studies were being conducted in early 1999 for a modified version of the Global Express, seating 12 to 16 business class passengers in a three-abreast layout, for use on scheduled long-haul passenger flights between secondary airports.
Global Express-ASTOR: On 15 June 1999 Raytheon Systems Limited was chosen as the preferred bidder for the UK Ministry of Defence's Airborne Stand-Off Radar (ASTOR) programme, for which five Global Express airframes will be modified by Bombardier's Shorts division to provide the airborne platform for radar and communications systems. Contract value £800 million (1999). Preliminary Design Review (PDR) scheduled for completion in March 2001; 200-hour flight test programme of an aerodynamically representative airframe expected to begin in early 2001, leading to Critical Design Review (CDR) and design freeze in November 2001. In-service date is 2004, with aircraft based at RAF Waddington, Lincolnshire.

Artist's impression of Global Express in ASTOR configuration

CUSTOMERS: More than 120 firm orders by October 2000. Announced customers include Bombardier's Flexjet fractional ownership programme, which has ordered 22 for delivery from 2000, the Royal Malaysian Air Force, which has taken delivery of one for VIP duties; Dogus Air of Turkey, which has ordered one for delivery in 2001, Ford Europe, which operates two in corporate shuttle configuration and the Japanese Civil Aviation Bureau (JCAB), which has ordered two for flight inspection and airways calibration duties. Estimated market for 500 to 800 long-range business jets over 15 years; Bombardier anticipates capturing 50 per cent of the market, breaking even at approximately 100; target production rate 34 per year; total of 35 delivered in 2000.
COSTS: Development costs C$800 million; half carried by Bombardier, balance by risk-sharing partners. Unit cost approximately US$40.66 million outfitted (2000).
DESIGN FEATURES: Design goal was longest possible range at highest speed from short runway with 99.5 per cent despatch reliability; wide-body fuselage, combining Challenger cabin cross-section with cabin length of Regional Jet; all-new, 'third-generation supercritical' wings with leading-edge slats and winglets.

Wing sweep 35° at quarter-chord, thickness/chord ratio 11 per cent, dihedral 2° 30′, root incidence 2° 30′. Wing, high-lift devices and wing/fuselage interface and area-ruled rear fuselage/engine pylon junction contours developed with extensive use of computational fluid dynamics (CFD). Rear-mounted engines. Sweptback T tail with 38° sweep and 5° anhedral on tailplane, 45° sweep on fin.
FLYING CONTROLS: Conventional and mechanical. Fully powered primary flying controls with variable artifical feel and emergency back-up via ram air turbine following triple hydraulic failure; dual sidestick controllers; duplicated cable runs with automatic disconnect in the event of control surface jamming; dual power control units on ailerons (maximum deflections +26.5/–23° and elevators (maximum deflections +24/–19°), triple units on rudder (maximum deflection 37° left/right). Eight-section (total) leading-edge slats (maximum deflection 20°) and six-section (total) single-slotted Fowler flaps (maximum deflection 30°) are signalled by dual electronic control units and operated by dual-motor power units connected by rigid driveshafts to ball-screw actuators. Electrically signalled, hydraulically actuated multifunction spoilers (four per side, outboard, operating differentially to assist ailerons and improve roll response, and symmetrically for speed brake or lift dump functions) and ground spoilers (two per side, inboard), maximum deflection +40°. Horizontal stabiliser incidence adjustable for pitch trim (+13/–2°) via dual-channel electrically driven screw actuator; roll trim accomplished by electric trim actuator located at aileron feel simulator unit; yaw trim accomplished by electric trim actuator at summing unit in fin. Dual yaw damper stability augmentation system and stick shaker/pusher stall protection system standard.
STRUCTURE: Semi-monocoque fuselage with chemically milled C-188 A1 aluminium alloy skin riveted over alloy frames and stringers to form damage-tolerant structure; main two-spar torsion box wing structure mostly of alloy construction, with machined alloy spars and ribs and polyurethane-coated machined alloy skin panels; two-spar winglets of mixed alloy/composites construction; multispar fin is alloy; ailerons, flaps, spoilers, rudder, two-spar tailplane, elevators, wing/fuselage fairings, flap track fairings, main landing gear bay, upper and lower engine nacelle doors and cabin floor panels are of composites construction.

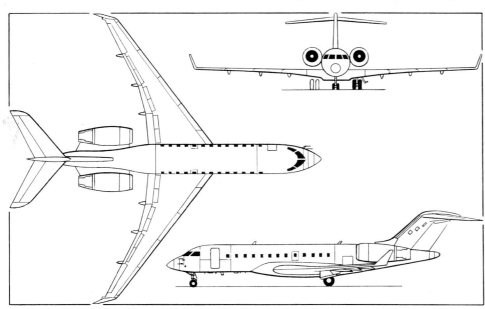

Bombardier BD-700 Global Express (*James Goulding/Jane's*)

CANADA: AIRCRAFT—BOMBARDIER

Global Express flight deck

Global Express cabin interior

Bombardier's Canadair division is design authority and manufactures nose section; de Havilland manufactures rear fuselage, engine pylons and vertical stabiliser and is responsible for final assembly at Downsview; Mitsubishi supplies wings and centre fuselage; Short Brothers designed and manufactures forward fuselage, engine nacelles, horizontal stabiliser and other composites components; Bombardier is responsible for interior completions at its Montreal and Tucson, Arizona, facilities. Other participants in the programme are: Honeywell Aerospace (APU), Ametek Aerospace (data acquisition unit, engine vibration monitoring system, fuel flow transmitters and engine thermocouples), Rolls-Royce Deutschland (power plant), Hella (lighting systems), Honeywell (avionics), Liebherr-Aerospace Toulouse (air management system), Lucas Aerospace (electrical systems), Messier-Dowty International (landing gear), Parker Bertea Aerospace (flight controls and fuel and hydraulic systems), Raytheon E-Systems (pitch feel systems), Sextant Avionique (flight control system) and Hamilton Sundstrand (slat/flap actuation system and ram air turbine).

LANDING GEAR: Hydraulically retractable tricycle type with Messier-Dowty oleo-pneumatic shock-absorber and twin wheels on each unit; main units retract inward into wing, nosewheel forwards. Goodyear tyres, mainwheels tyre size H38×12.0-19 (20 ply) tubeless, maximum pressure 11.45 bars (166 lb/sq in); nosewheel tyres 21×7.25-10 (12 ply) tubeless (deflector-type), maximum pressure 9.93 bars (144 lb/sq in). Carbon brakes with dual Goodrich/HydroAire hydraulic digital brake-by-wire/modulated anti-skid system providing pilot-selectable, three-level autobrake capability. Steerable nosewheel, maximum steering angle ±75°; minimum ground turning radius 20.73 m (68 ft 0 in).

POWER PLANT: Two rear-mounted 65.6 kN (14,750 lb st) Rolls-Royce Deutschland BR710A2-20 turbofans, flat rated to ISA + 20°C, with FADEC. International Nacelle Systems (Shorts/Hurel-Dubois joint venture) hydraulically actuated two-petal target-type thrust reversers.

Fuel contained in two integral wing tanks, each of 8,479 litres (2,240 US gallons; 1,865 Imp gallons) capacity, centre-section tank, capacity 6,117 litres (1,616 US gallons; 1,346 Imp gallons), and auxiliary tank in aft fuselage, capacity 1,234 litres (326 US gallons; 271 Imp gallons), giving total standard capacity of 24,310 litres (6,422 US gallons; 5,347 Imp gallons). Fuel from centre-section and auxiliary tanks is transferred to wing tanks from where two AC main pumps and DC back-up pump feed to engines; automatic fuel management system balances quantities in port and starboard wing tanks.

Gravity and pressure refuelling; single-point pressure fuelling/defuelling coupling in starboard wing/fuselage fairing. Oil capacity 20 litres (5.3 US gallons; 4.4 Imp gallons) with oil replenishment tank, capacity 5.7 litres (1.5 US gallons; 1.2 Imp gallons) permitting remote oil servicing from the cockpit.

ACCOMMODATION: Crew of three or four (including cabin attendant) and eight to 19 passengers depending on interior fit. Customised cabin interior according to customer requirements. Typical arrangement comprises three-compartment cabin with lavatory at rear, crew rest area, galley, small lavatory and wardrobe forward, and provision for 'office in the sky', stateroom or conference area. Flight-accessible baggage compartment at rear of cabin with external plug-type door forward of port engine intake. Accommodation is heated, air conditioned and pressurised; predicted cabin noise level 52 dB. Thirteen cabin windows per side, each 40 cm (1 ft 3¾ in) high × 27.4 cm (10¾ in) wide; one window over wing on starboard side doubles as plug-type emergency exit. Electrically operated airstair door at front of cabin on port side.

SYSTEMS: Integrated air management system by Liebherr-Aerospace Toulouse provides engine bleed, wing anti-ice, air conditioning, cabin pressurisation and avionics ventilation. Digitally controlled dual cooling pack system with ozone converters and bleed air filters provides cabin air circulation at standard rate 1.81 m³ (64 cu ft)/min/person, maximum rate 2.29 m³ (81 cu ft)/min/person with crew-selectable 100 per cent fresh air or recirculation and three air sources for cabin pressure control; maximum pressure differential 0.66 bar (9.64 lb/sq in) maintains a 1,525 m (5,000 ft) cabin altitude to 12,500 m (41,000 ft) and a 2,200 m (7,220 ft) cabin altitude to maximum operating altitude of 15,545 m (51,000 ft). Engine bleed air anti-icing for wing leading-edge fixed surfaces and slats; tail surfaces unprotected; bleed management system automatically switches between low- and high-pressure compressor air to improve engine efficiency. Oxygen system comprises four 1,417 litre (50 cu ft) oxygen cylinders pressurised to 127.6 bars (1,850 lb/sq in) for passenger and crew use.

Lucas/Leach electrical power generation and distribution system comprises two 40 kVA variable frequency generators on each engine, supplying primary 115/200 V three-phase AC electrical power at 324 to 596 Hz; alternative AC power provided by 45 kVA APU-mounted generator and emergency power by 9 kVA air-driven generator, the latter automatically deployed in the event of power loss; electrical management system automatically performs priority-based load-shedding and reconfiguration in event of failure. Four 150 A TRUs convert AC to 28 V DC; emergency DC provided by 25 Ah and 42 Ah low-maintenance Ni/Cd batteries. Provision for external AC and DC power connection. Triple logic controlled AC power centre performs primary AC power distribution and high-power secondary distribution via solid-state switches and 'smart'-contactors; low-power AC distributed through thermal circuit breakers in the cockpit. Triple logic controlled DC power centre provides non-interruptible primary DC power distribution, emergency bus supplies and normal DC supplies to four secondary power distribution assemblies (SPDAs) throughout the aircraft to provide remote logic controlled power to all DC loads. Two CDUs in the cockpit allow for remote sensing/setting and resetting of circuit breakers.

Tailcone-mounted Honeywell RE220(GX) APU provides electrical power (45 kVA ground; 40 kVA flight), as well as bleed air and main engine starting; APU is certified for operation up to 13,715 m (45,000 ft), in-flight starting up to 11,280 m (37,000 ft) and engine starting up to 9,145 m (30,000 ft).

Triple-redundant hydraulic systems at pressure of 207 bars (3,000 lb/sq in), with bootstrap reservoirs.

Walter Kidde Aerospace integrated aircraft fire detection and extinguishing system provides continuous fire detection monitoring in engine nacelles, APU compartment, main landing gear bays and cabin; dual extinguishers provide two-shot fire suppression in main engine and APU bays.

AVIONICS: Honeywell Primus 2000XP as core system.
Comms: Dual VHF (third optional); dual Rockwell Collins HF; dual transponders; dual radio management systems; Coltech five-channel Selcal; Honeywell digital FDR and CVR; ELT; satcom optional, with antenna mounted in fin cap.
Radar: Colour weather radar with dual controllers.
Flight: Dual flight management systems (third optional) with dual Cat. II autopilots and triple digital air data computers providing fail-safe AFCS; triple laser gyro inertial reference systems; GPS with option for second sensor; ADF; VOR/ILS; DME; TCAS II; Honeywell EGPWS with terrain database integrated into Primus 2000XP system for EFIS display.
Instrumentation: Dual EFIS comprising six 203 × 178 mm (8 × 7 in) CRT multifunction displays, for PFD and EICAS functions; dual Rockwell Collins digital radio altimeter; combined standby airspeed/altimeter, standby artificial horizon and stowable standby heading indicator. Thales HUD with Cat. II landing capability and lightning sensor system option.

DIMENSIONS, EXTERNAL:
Wing span over winglets	28.65 m (94 ft 0 in)
Wing chord: at root	6.43 m (21 ft 1 in)
at tip	1.24 m (4 ft 1 in)
Wing aspect ratio	8.6
Length: overall	30.30 m (99 ft 5 in)
fuselage	26.31 m (86 ft 4 in)
Diameter of fuselage, constant	2.69 m (8 ft 10 in)
Height overall	7.57 m (24 ft 10 in)
Tailplane span	9.68 m (31 ft 9 in)
Wheel track (c/l of shock-absorbers)	4.06 m (13 ft 4 in)
Wheelbase	12.78 m (41 ft 11 in)
Passenger door: Height	1.83 m (6 ft 0 in)
Width:	0.91 m (3 ft 0 in)
Baggage door: Height	0.84 m (2 ft 9 in)
Width	1.09 m (3 ft 7 in)
Emergency exit: Height	0.99 m (3 ft 3 in)
Width	0.51 m (1 ft 8 in)

DIMENSIONS, INTERNAL:
Cabin (excl flight deck): Length	14.73 m (48 ft 4 in)
Width at floor	2.11 m (6 ft 11 in)
Max width	2.49 m (8 ft 2 in)

Cutaway drawing of ASTOR cabin

BOMBARDIER—AIRCRAFT: CANADA

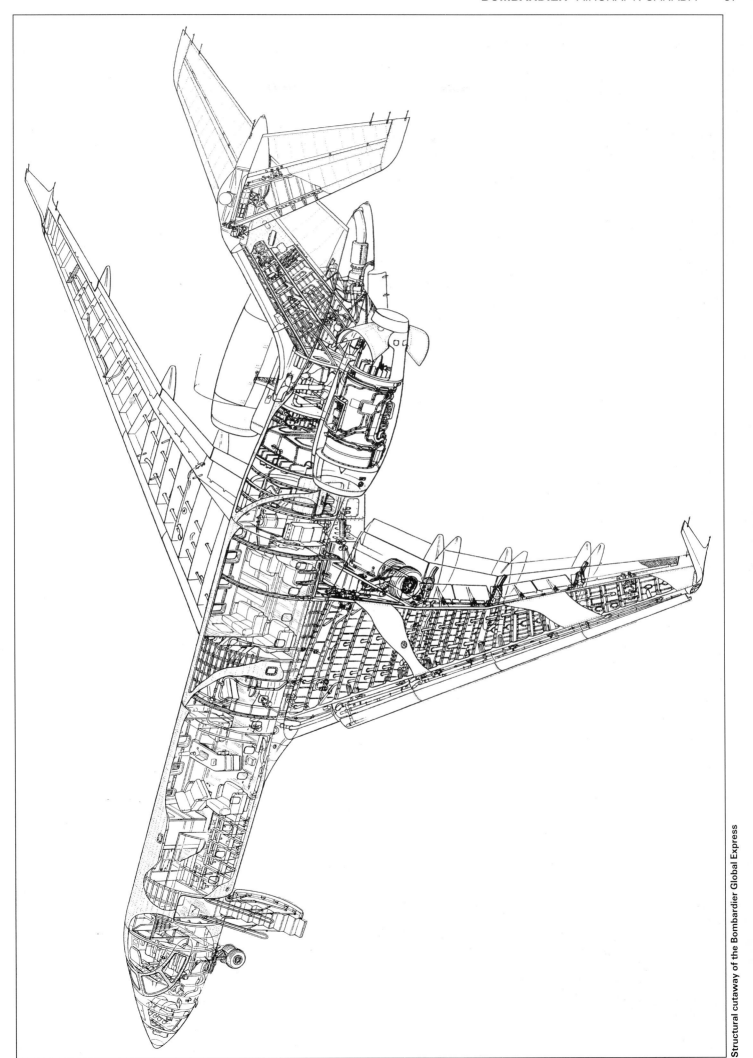

Structural cutaway of the Bombardier Global Express

Max height	1.90 m (6 ft 3 in)
Floor area	31.1 m² (335 sq ft)
Volume, incl baggage compartment	60.6 m³ (2,140 cu ft)

AREAS:
Wings, basic	94.95 m² (1,022.0 sq ft)
Horizontal tail surfaces (total)	22.76 m² (245.0 sq ft)
Vertical tail surfaces (total)	17.28 m² (186.0 sq ft)

WEIGHTS AND LOADINGS (provisional):
Operating weight empty	22,816 kg (50,300 lb)
Max payload	2,585 kg (5,700 lb)
Payload with max fuel	725 kg (1,600 lb)
Max fuel weight	19,663 kg (43,350 lb)
Max ramp weight: standard	43,205 kg (95,250 lb)
optional	43,658 kg (96,250 lb)
Max T-O weight: standard	43,091 kg (95,000 lb)
optional	43,545 kg (96,000 lb)
Max landing weight	35,652 kg (78,600 lb)
Max zero-fuel weight	25,401 kg (56,000 lb)
Max wing loading: standard	453.8 kg/m² (92.95 lb/sq ft)
optional	458.6 kg/m² (93.93 lb/sq ft)
Max power loading: standard	328 kg/kN (3.22 lb/lb st)
optional	332 kg/kN (3.25 lb/lb st)

PERFORMANCE (at max T-O weight, ISA, except where indicated):
Max level speed (V$_{MO}$): S/L to 2,440 m (8,000 ft)	300 kt (555 km/h; 345 mph) CAS
2,240 m (8,000 ft) to 9,410 m (30,870 ft)	340 kt (629 km/h; 391 mph) CAS
above 9,410 m (30,870 ft)	M0.89
High cruising speed	M0.88 (505 kt; 935 km/h; 581 mph)
Normal cruising speed	M0.85 (488 kt; 904 km/h; 562 mph)
Long-range cruising speed at 13,715 m (45,000 ft)	M0.80 (459 kt; 850 km/h; 528 mph)
Max rate of climb at S/L	1,097 m (3,600 ft)/min
Initial cruising altitude	13,105 m (43,000 ft)
Time to climb to initial cruising altitude	30 min
Max certified altitude	15,545 m (51,000 ft)
T-O to 11 m (35 ft)	1,774 m (5,820 ft)
Landing from 15 m (50 ft) at max landing weight	814 m (2,670 ft)
Runway LCN	55

Range with max fuel and eight passengers, NBAA IFR reserves:
at M0.85	6,010 n miles (11,130 km; 6,916 miles)
at M0.87	5,276 n miles (9,771 km; 6,071 miles)

Range with max payload:
at M0.85	5,187 n miles (9,606 km; 5,969 miles)
at M0.87	4,596 n miles (8,511 km; 5,289 miles)
Design g limit	+2.5

OPERATIONAL NOISE LEVELS:
T-O	82.1 EPNdB
Approach	89.8 EPNdB
Sideline	88.7 EPNdB

UPDATED

BOMBARDIER BRJ-X

Bombardier announced in mid-2000 that it was suspending development of the proposed BRJ-X twin-turbofan regional airliner following market evaluation studies. A description, artist's impression and three-view can be found in the 2000-2001 edition.

UPDATED

BOMBARDIER BD-100 CONTINENTAL

TYPE: Business jet.
PROGRAMME: Design study, then known as 'Bombardier Model 70', revealed at the Paris Air Show in June 1997; formally announced at NBAA Convention at Las Vegas 18 October 1998; launched at Paris Air Show 13 June 1999; first metal cut 21 October 1999 following completion of joint definition phase; AS907 engine first flown 29 January 2000, engine certification expected in March 2001; wing/fuselage mating of first aircraft (c/n 20001) achieved 19 November 2000; first flight scheduled for mid-2001; five aircraft test and certification programme scheduled to culminate in Transport Canada and FAA certification in third quarter of 2002; first 'green' delivery late 2002; JAA certification in early 2003.

CUSTOMERS: Two orders signed at time of launch, by customers in Germany and United Arab Emirates; total of 100 firm orders received by 9 October 2000, including 25 for Bombardier's Flexjet fractional ownership programme. Bombardier anticipates gaining 30 per cent of estimated 1,200-aircraft market in this class by 2012, with fractional ownership operations especially targeted.

COSTS: Development cost C$500 million (1998); break-even estimated at 300th aircraft. Unit cost US$14.25 million typically equipped (1998). Direct operating cost estimated at US$770 per hour.

DESIGN FEATURES: Design goals included coast-to-coast range across USA with eight passengers in cabin with stand-up headroom and take-off field length less than 1,525 m (5,000 ft). General configuration is as shown in the accompanying illustrations; supercritical wing with winglets, sweepback 27° at quarter-chord.

FLYING CONTROLS: Conventional and assisted. Elevators and rudder hydraulically actuated via cables and pulleys, with manual reversion; ailerons manually actuated by cables, pulleys and pushrods; two-segment rudder; variable incidence tailplane for pitch trim; one fly-by-wire spoiler (outboard) and one lift dumper (inboard) per wing; no-leading-edge high-lift devices; yaw damper standard.

STRUCTURE: Primarily light alloy, with composites for some non-structural fairings; fuselage of semi-monocoque construction with frames and stringers; two-spar wing. Programme suppliers include: AIDC Taiwan (rear fuselage and tail unit); Canadair (cockpit, forward fuselage and primary flight controls); GKN Westland (engine nacelles); Hawker-de Havilland Australia (tailcone and APU installation kit); Honeywell (power plant and APU); Hurel-Dubois (thrust reversers); Intertechnique (fuel system); Liebherr Aerospace-Toulouse (environmental control and anti-icing systems); Liebherr Aerospace Germany (flap control system); Messier-Dowty (landing gear); Mitsubishi Heavy Industries (wing); Moog (secondary flight controls); Parker Aerospace (hydraulic system); Rockwell Collins (avionics), and Shorts (centre fuselage). Final assembly will be at Bombardier's Learjet facility in Wichita, with interior completion in Tucson.

LANDING GEAR: Hydraulically retractable tricycle type by Messier-Dowty, with two wheels on each unit; main units retract inwards, nosewheel forwards. Steerable nosewheel, maximum deflection ±65°. Mainwheel tyre size 26.5×8.0-14, nosewheel tyre size 18×5.5-8. Goodrich carbon composites multiple disc brakes. Turning radius 17.68 m (58 ft 0 in).

POWER PLANT: Two Honeywell AS907 turbofans, each with thermodynamic rating of 35.81 kN (8,050 lb st), flat-rated to 28.91 kN (6,500 lb st) with APR at ISA+15°C. All fuel contained in two integral wing tanks, combined capacity 7,684 litres (2,030 US gallons; 1,690 Imp gallons). Gravity fuelling point in top of each wing, near leading-edge, plus single-point pressure refuelling/defuelling. Thrust reversers standard.

ACCOMMODATION: Two crew flight deck; cabin, with flat floor, accommodates eight passengers in standard 'double club' arrangement on tracking, swivelling and reclining seats; work table, 110 V power point and telephone jack at each club seating position; galley to rear of flight deck; lavatory, and flight-accessible baggage compartment at rear of cabin. Cabin door on port side aft of flight deck; single emergency exit over starboard wing. Cabin and baggage compartment are pressurised and air conditioned.

SYSTEMS: Two independent hydraulic systems with one engine-driven pump and one DC motor pump per system, pressure 206.8 bars (3,000 lb/sq in), plus one auxiliary system powered by an accumulator. Pressurisation system, differential 0.60 bars (8.78 lb/sq in), with auxiliary system providing pressurisation up to 10,670 m (35,000 ft). 28 V DC electrical system comprises three 400 Ah DC brushless generators and two Ni/Cd batteries which provide power for APU starting, inflight emergency power and ground power. Oxygen system, capacity to suit customer requirements, with demand-type masks for crew and drop-down masks for passengers.

Engine bleed air automatically controlled anti-icing for wing leading-edges and nacelle lips; electrically anti-iced windscreen; heated pitot probes. Honeywell 36-150B APU, will be certified for operation up to 11,280 m (37,000 ft) and in-flight starting to 9,150 m (30,000 ft).

AVIONICS: Rockwell Collins Pro Line 21 as core system.
Radar: Rockwell Collins weather radar standard.
Flight: Dual VOR/ILS, FMS, air data system, AHRS, ADF, DME, GPS, EGPWS, TCAS II, radar altimeter.
Instrumentation: Four-tube EFIS with two-tube EICAS.

DIMENSIONS, EXTERNAL:
Wing span over winglets	19.46 m (63 ft 10 in)
Length overall	20.93 m (68 ft 8 in)
Height overall	6.17 m (20 ft 3 in)
Fuselage max diameter	2.34 m (7 ft 8 in)
Tailplane span	8.45 m (27 ft 8½ in)
Wheel track	3.20 m (10 ft 6 in)
Wheelbase	8.47 m (27 ft 9½ in)
Passenger door: Height	1.89 m (6 ft 2½ in)
Width	0.76 m (2 ft 6 in)
Baggage door: Height	0.76 m (2 ft 6 in)
Width	0.61 m (2 ft 0 in)
Height to sill	1.61 m (5 ft 3½ in)

Computer-generated image of Bombardier BD-100 Continental

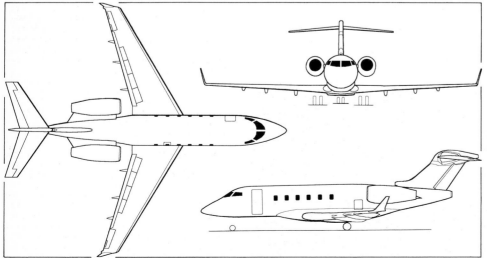

Provisional general arrangement of the Bombardier Continental (Paul Jackson/Jane's)

BOMBARDIER (Canadair)—AIRCRAFT: CANADA

Bombardier Continental flight deck *2000*/0044516

Bombardier Continental cabin mockup *1999*/0044517

Emergency exit: Height	0.91 m (3 ft 0 in)
Width	0.52 m (1 ft 8½ in)
DIMENSIONS, INTERNAL:	
Cabin (excl cockpit):	
Length	7.23 m (23 ft 8½ in)
Width: at centreline	2.18 m (7 ft 2 in)
at floor	1.55 m (5 ft 1 in)
Max height	1.85 m (6 ft 1 in)
Floor area	13.5 m² (146 sq ft)
Volume	24.4 m³ (860 cu ft)
Baggage compartment volume	3.0 m³ (106 cu ft)
AREAS:	
Wings, gross	48.49 m² (522.0 sq ft)
Ailerons, total	0.93 m² (10.00 sq ft)
Trailing-edge flaps, total	7.56 m² (81.40 sq ft)
Spoilers, total	3.62 m² (39.00 sq ft)
Rudder, incl tabs	1.89 m² (20.40 sq ft)
Tailplane	3.92 m² (42.20 sq ft)
Elevators	3.81 m² (41.00 sq ft)
WEIGHTS AND LOADINGS (provisional):	
Operating weight empty	10,138 kg (22,350 lb)
Outfitting allowance	1,315 kg (2,900 lb)
Max payload	1,360 kg (3,000 lb)
Max fuel	6,214 kg (13,700 lb)
Max T-O weight	17,010 kg (37,500 lb)
Max landing weight	15,308 kg (33,750 lb)
Max zero-fuel weight	11,498 kg (25,350 lb)
Payload with max fuel	725 kg (1,600 lb)
Max wing loading	350.8 kg/m² (71.84 lb/sq ft)
Max power loading	294 kg/kN (2.88 lb/lb st)

First Bombardier Continental, following assembly at the company's Wichita facility *2001*/0062409

PERFORMANCE (estimated):	
Max level speed	476 kt (882 km/h; 548 mph)
High cruising speed	M0.82 or 470 kt (870 km/h; 541 mph)
Normal cruising speed	M0.80 or 459 kt (850 km/h; 528 mph)
Max rate of climb at S/L	1,097 m (3,600 ft)/min
Rate of climb at S/L, OEI	205 m (673 ft)/min
Initial cruising altitude	12,500 m (41,000 ft)
Max certified ceiling	13,715 m (45,000 ft)
T-O balanced field length	1,509 m (4,950 ft)
Landing run at max landing weight	808 m (2,650 ft)
Range with eight passengers, NBAA IFR reserves	3,100 n miles (5,741 km; 3,567 miles)

UPDATED

BOMBARDIER AEROSPACE CANADAIR OPERATIONS

400 Chemin de la Côte Vertu West, Dorval, Québec H4S 1Y9
POSTAL ADDRESS: PO Box 6087, Station Centreville, Montréal, Québec H3C 3G9
Tel: (+1 514) 855 50 00
Fax: (+1 514) 855 79 03
Web: http://www.aero.bombardier.com
VICE-PRESIDENT AND GENERAL MANAGER, OPERATIONS, CANADAIR: Serge Perron
VICE-PRESIDENT, ST LAURENT PLANT, OPERATIONS, CANADAIR: Jean Séguin
VICE-PRESIDENT, DORVAL PLANT, OPERATIONS, CANADAIR: Réal Gervais
INFORMATION OFFICER: Anne-Marie Laroche

Acquired by Bombardier Inc 23 December, 1986, Canadair has manufactured more than 5,000 aircraft since 1944.

Dorval plant, adjacent to Montréal International Airport, comprises 78,975 m² (850,080 sq ft) of floor space for manufacture and assembly of the Challenger and Regional Jet. The Canadair 415 is assembled in a 4,750 m² (51,100 sq ft) plant at North Bay, Ontario. Parts, components and spare parts for various aircraft, including Challenger, Regional Jet, Regional Jet Series 700, Global Express and Canadair 415, are manufactured at St Laurent plant, 248,620 m² (2,676,110 sq ft) in Québec. Structural components for other aircraft builders, such as Boeing and Aerospatiale Matra, are also manufactured in this plant.

Global Express is completed at the 2,820 m² (30,350 sq ft) Bombardier Completion Center, Montréal, beside the Bombardier Aerospace headquarters at Dorval. Total workforce in the Montréal area is more than 11,500.

UPDATED

CANADAIR CL-600 CHALLENGER
Canadian Forces designations: CC-144, CC-144B and CE-144A

TYPE: Business jet.
PROGRAMME: First flight of first of three prototypes (C-GCGR-X) 8 November 1978; first flight production Challenger 600 with AlliedSignal ALF 502L-2 turbofans 21 September 1979; first customer delivery 30 December 1980; first flight Challenger 601 with GE CF34s 10 April 1982; first 601-1A delivered 6 May 1983; first 601-3A 6 May 1987 and first 601-3A/ER 19 May 1989; first 601-3R 14 July 1993; first 604 25 January 1996. Challenger certified for operation in 40 countries by 1998. By June 2000 more than 500 Challengers had been delivered and flown 1.75 million hours. 500th Challenger rolled out 'green' 25 May 2000 and handed over (as N816CC) 1 September 2000.

CURRENT VERSIONS: **Challenger 600:** Total 84 built after certification in 1980 (76 since retrofitted with winglets); 12 delivered to Canadian Department of National Defence as CC-144 (three) and CE-144A (three) (see 1989-90 and earlier *Jane's*), plus three for coastal patrol, two for general transport and one test aircraft. Production completed with final delivery on 22 June 1983.

Challenger 601-1A: First production version to have CF34 engines (see 1990-91 *Jane's*); first flight 17 September 1982. Deliveries (66, including four CC-144Bs) between 6 May 1983 and 29 May 1987.

Challenger 601-3A: Version with 'glass' cockpit and CF34-3A engines; first flight 28 September 1986; Canadian and US certification 21 and 30 April 1987; also certified for Cat. II and in 22 other countries; improvements include CF34-3A engines flat rated to 21°C, and fully integrated digital flight guidance and flight management systems. Total of 134 delivered between 6 May 1987 and 29 October 1993.

Canadair Challenger 604 of the Royal Danish Air Force *(Paul Jackson/Jane's)* *2001*/0105012

CANADA: AIRCRAFT—BOMBARDIER (Canadair)

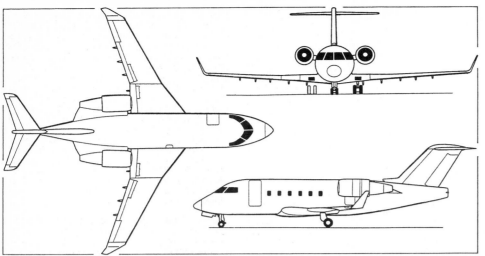

Canadair Challenger 604 *(Paul Jackson/Jane's)*

Challenger 601-3R: Extended-range option available on new 601-3As since 1989 (c/n 5135 and onwards) and as retrofit to 601-1As and 601-3As; range increased to 3,585 n miles (6,639 km; 4,125 miles) with NBAA IFR reserves; first flight 8 November 1988; Canadian certification 16 March 1989; tail fairing replaced with conformal tailcone fuel tank which extends fuselage length by 46 cm (1 ft 6 in) and adds 118 kg (260 lb) to operating weight empty; maximum ramp weight increased by 680 kg (1,500 lb). Optional gross weight increase of 227 kg (500 lb). Total of 92 modification kits supplied between March 1989 and October 1993. Challenger 601-3ER, incorporating extended-range modifications, CF34-3A1 engines and 20,457 kg (45,100 lb) max T-O weight, was standard production version from 14 July 1993 (first delivery); 59 new-build aircraft delivered by early 1996; no further production.

Challenger 604: Has range of 4,077 n miles (7,550 km; 4,691 miles) at M0.74 and is powered by General Electric CF34-3B engines each rated at 38.8 kN (8,729 lb st) T-O power at ISA + 15°C. Prototype (C-FTBZ) modified on the production line from a Challenger 601-3R; engineering designation CL-600-2B16; first flight (with CF34-3A engines) 18 September 1994; first flight with definitive CF34-3B engines 17 March 1995. Exploits systems developed in Regional Jet programme. Rockwell Collins Pro Line 4 EFIS; extra 1,242 litres (328 US gallons; 273 Imp gallons) of fuel in aft equipment bay, forward fuselage tank and tail tank. Automatic aft-CG control to reduce trim drag for longer range. New landing gear, carbon brakes and anti-skid system; strengthened tail unit; new wing-to-fuselage and underbelly fairings. Maximum T-O weight 21,863 kg (48,200 lb). Transport Canada certification achieved 20 September 1995; FAA certification 2 November 1995; 100th delivery to a customer was made in mid-1999.

From early 2001 Challenger 604s will be delivered with upgraded Rockwell Collins Pro Line 4 avionics aimed at reducing pilot workload and making the aircraft more compatible with future air traffic environments. Features will include three-dimensional FMS mapping capability; automatic look-up for thrust settings, take-off, approach and landing, weight limits, runway lengths and climb gradients; automatic computation of cruising speeds for maximum range and maximum speed; improved waypoint reporting on North Atlantic routes, automatic generation by FMS of waypoints for fixed search patterns, expanded FDR functions, and extension of autothrottle functions for take-off and landing. The avionics upgrade will also be available for retrofit to earlier Challenger 604s.

Detailed description applies to Challenger 604.

CUSTOMERS: See under individual headings in Current Versions. Total of 506 Challengers of all versions delivered (including to completion centres) by August 2000, including 153 Challenger 604s. Recent customers include the Royal Danish Air Force, which ordered one in January 1998, plus two options, for delivery from mid-1999; the Korean National Maritime Police, which ordered one in 1999 for delivery in second quarter 2001 for maritime surveillance and SAR duties, with total requirement for up to five; the Royal Jordanian Air Force, which has ordered two in VIP configuration for delivery during 2000 and the Australian government, which ordered three on 16 August 2000 for delivery in 2001 to Qantas Airways, which will operate them on behalf of the Royal Australian Air Force for transport of senior government officials. Annual deliveries have included 33 in 1997, 36 in 1998, 40 in 1999, and 38 in 2000.

COSTS: Unit cost (604), US$22.5 million, typically equipped (2000).

DESIGN FEATURES: Advanced wing section; quarter-chord sweep 25°; thickness/chord ratio 14 per cent at root, 12 per cent at leading-edge sweep break and 10 per cent at tip; dihedral 2° 33′; incidence at root 3° 30′; fuselage circular cross-section, pressurised.

FLYING CONTROLS: Conventional, fully powered hydraulic controls; electrically actuated variable incidence tailplane; two-segment spoilers (outboard airbrake panels, inboard lift dumpers); two-segment double-slotted flaps.

STRUCTURE: Two-spar wing torsion box; chemically milled fuselage skin panels with riveted frames and stringers form damage-tolerant structure; multispar fin and tailplane.

LANDING GEAR: Hydraulically retractable tricycle type, with twin wheels and Dowty oleo-pneumatic shock-absorber on each unit. Mainwheels retract inward into wing centre-section, nose unit forward. Nose unit steerable and self-centring. Mainwheels have H27×8.5-14 (16 ply) tubeless tyres, pressure 12.07 bars (175 lb/sq in); nosewheels have Goodrich 18×4.4 (12 ply) tubeless (deflector type) tyres, pressure 10.00 bars (145 lb/sq in). ABS (Aircraft Braking Systems) hydraulically operated multiple-disc carbon brakes with fully modulated anti-skid system. Minimum ground turning radius 12.19 m (40 ft 0 in).

POWER PLANT: Two General Electric CF34-3B1 turbofans, each rated at 41.0 kN (9,220 lb st) with automatic power reserve, or 38.8 kN (8,729 lb st) without APR, pylon-mounted on rear fuselage and fitted with cascade-type fan-air thrust reversers. Nacelles and thrust reversers by Shorts. Integral fuel tank in centre-section, capacity 2,839 litres (750 US gallons; 624 Imp gallons), one in each wing (each 2,725 litres; 720 US gallons; 600 Imp gallons) and auxiliary tanks (combined capacity 1,181 litres; 312 US gallons; 260 Imp gallons) beneath cabin floor. Saddle tanks, total capacity 999 litres (264 US gallons; 220 Imp gallons); tank in tailcone, capacity 745 litres (197 US gallons; 164 Imp gallons). Total fuel capacity 11,214 litres (2,963 US gallons; 2,468 Imp gallons). Pressure and gravity fuelling and defuelling. Oil capacity 13.6 litres (3.6 US gallons; 3.0 Imp gallons).

ACCOMMODATION: Two-pilot flight deck with dual controls. Blind-flying instrumentation standard. Cabin interiors to customer's specifications; maximum of 19 passenger seats and three crew approved. Typical installations include lavatory, buffet, bar and wardrobe. Medevac version can carry up to seven stretcher patients, infant incubator, full complement of medical staff and comprehensive intensive care equipment. Baggage compartment, with own loading door, accessible in flight. Downward-opening, power-assisted door on port side, forward of wing. Overwing emergency exit on starboard side. Entire accommodation heated, pressurised and air conditioned. Optional extended cabin interior increases cabin length by 0.51 m (1 ft 8 in) and provides two additional cabin windows by removing rear closet and moving the lavatory and baggage compartment bulkheads, with corresponding 0.20 m³ (7 cu ft) reductions in baggage capacity. Ultra Electronics active noise vibration control (ANVC) system optional.

SYSTEMS: Honeywell pressurisation and air conditioning systems, maximum pressure differential 0.63 bar (9.1 lb/sq in). Three independent hydraulic systems, each of 207 bars (3,000 lb/sq in). No. 1 system powers flight controls (via servo-actuators positioned by cables and pushrods); No. 2 system for flight controls and brakes; No. 3 system for flight controls, landing gear extension/retraction, brakes and nosewheel steering. Nos. 1 and 2 systems each powered by an engine-driven pump, supplemented by an AC electric pump; No. 3 system by two AC pumps. Two 30 kVA engine-driven generators supply primary 115/200 V three-phase AC electric power at 400 Hz. Four transformer-rectifiers to convert AC power to 28 V DC; one primary 24 V 17 Ah Ni/Cd battery and one auxiliary 24 V 43 Ah battery. Alternative primary power provided by APU and/or an air-driven generator, latter deployed automatically in flight if engine-driven generators and APU are inoperative. Stall warning system, with stick shakers and stick pusher. Honeywell GTCP-100E gas-turbine APU for engine start, ground air conditioning and other services. Electric anti-icing of windscreen, flight deck side windows and pitot heads; Hamilton Sundstrand bleed air anti-icing of wing leading-edges, engine intake cowls and guide vanes. Gaseous oxygen system, pressure 127.5 bars (1,850 lb/sq in). Continuous-element fire detectors in each engine nacelle, APU and main landing gear bays; two-shot extinguishing system for engines, single-shot system for APU.

AVIONICS: Rockwell Collins Pro Line 4 nav/com.

Comms: Dual VHF; dual ATC transponders; dual HF; cockpit voice recorder.

Radar: Rockwell Collins TWR-854 colour digital weather radar with turbulence detection.

Flight: Dual VHF nav with provision for third; dual DME; dual ADF; dual Litton TN-101 laser inertial reference systems (LIRS) with full provision for third; dual flight management system with provision for third; digital automatic flight control system, with dual-channel

Interior of a Challenger 604

Canadair Challenger 604 flight deck with Rockwell Collins Pro Line 4 integrated avionics

autopilot and flight director; Mach trim and auto trim; dual digital air data system. Flight Dynamics HGS 2150 HUD received FAA approval on 4 April 2000 and is optional. Space provisions for flight data recorder, ELT, dual GPS, EGPWS, AFIS, TCAS and autothrottle.

Instrumentation: Rockwell Collins digital avionics include Pro Line 4 six-tube EFIS with 184 × 184 mm (7¼ × 7¼ in) CRT displays which include two-tube EICAS display (MFD); standby instruments (artificial horizon, airspeed indicator, compass and altimeter). Systems certified for Cat.II operations.

EQUIPMENT (Medevac version): Includes cardiopulmonary resuscitation unit; physio control lifepack comprising heart defibrillator, ECG and cardioscope; ophthalmoscope; respirators and resuscitators; infant monitor; X-ray viewer; cardiostimulator; foetal heart monitor; and anti-shock suit.

DIMENSIONS, EXTERNAL:
Wing span over winglets	19.61 m (64 ft 4 in)
Wing chord: at root	3.99 m (13 ft 1 in)
at tip	1.27 m (4 ft 2 in)
Wing aspect ratio (excl winglets)	8.0
Length overall	20.85 m (68 ft 5 in)
Fuselage: Max diameter	2.69 m (8 ft 10 in)
Length	18.77 m (61 ft 7 in)
Height overall	6.30 m (20 ft 8 in)
Tailplane span	6.20 m (20 ft 4 in)
Wheel track (c/l of shock-struts)	3.18 m (10 ft 5 in)
Wheelbase	7.99 m (26 ft 2½ in)
Passenger door (port, fwd): Height	1.78 m (5 ft 10 in)
Width	0.94 m (3 ft 1 in)
Height to sill	1.63 m (5 ft 4 in)
Baggage door (port, rear): Height	0.84 m (2 ft 9 in)
Width	0.71 m (2 ft 4 in)
Height to sill	1.73 m (5 ft 8 in)
Overwing emergency exit (stbd):	
Height	0.91 m (3 ft 0 in)
Width	0.51 m (1 ft 8 in)

DIMENSIONS, INTERNAL:
Cabin: Length, incl galley, toilet and baggage area, excl
flight deck	8.66 m (28 ft 5 in)
Max width	2.49 m (8 ft 2 in)
Width at floor level	2.18 m (7 ft 2 in)
Max height	1.85 m (6 ft 1 in)
Floor area	18.8 m² (202 sq ft)
Volume	32.6 m³ (1,150 cu ft)

AREAS:
Wings, gross (excl winglets)	48.31 m² (520.0 sq ft)
Ailerons (total)	1.39 m² (15.0 sq ft)
Trailing-edge flaps (total)	7.80 m² (84.0 sq ft)
Fin	9.18 m² (98.8 sq ft)
Rudder	2.03 m² (21.9 sq ft)
Tailplane	6.45 m² (69.4 sq ft)
Elevators (total)	2.15 m² (23.1 sq ft)

WEIGHTS AND LOADINGS:
Manufacturer's weight empty	9,806 kg (21,620 lb)
Operating weight empty	12,079 kg (26,630 lb)
Max fuel	9,072 kg (20,000 lb)
Max payload	2,435 kg (5,370 lb)
Payload with max fuel: standard	485 kg (1,070 lb)
optional	757 kg (1,670 lb)
Max T-O weight: standard	21,591 kg (47,600 lb)
optional	21,863 kg (48,200 lb)
Max ramp weight: standard	21,636 kg (47,700 lb)
optional	21,908 kg (48,300 lb)
Max landing weight	17,236 kg (38,000 lb)
Max zero-fuel weight	14,515 kg (32,000 lb)
Max wing loading:	
standard	446.9 kg/m² (91.54 lb/sq ft)
optional	452.6 kg/m² (92.69 lb/sq ft)
Max power loading:	
standard	263 kg/kN (2.58 lb/lb st)
optional	267 kg/kN (2.61 lb/lb st)

PERFORMANCE (at standard max T-O weight, except where indicated):
High speed cruising speed	M0.82 (470 kt; 870 km/h; 541 mph)
Normal cruising speed	M0.80 (459 kt; 851 km/h; 529 mph)
Long-range cruising speed	M0.74 (425 kt; 787 km/h; 489 mph)
Time to initial cruising altitude	21 min
Initial cruising altitude	11,430 m (37,500 ft)
Max certified altitude	12,500 m (41,000 ft)
Service ceiling, OEI: at mid-cruise weight 17,373 kg	
(38,300 lb)	6,920 m (22,700 ft)
at max T-O weight	5,170 m (16,960 ft)
Balanced T-O field length (ISA at S/L)	1,737 m (5,700 ft)
Landing distance at S/L at max landing weight	846 m (2,775 ft)

Range with max fuel and five passengers, NBAA IFR reserves (200 n mile; 370 km; 230 mile alternate):
long-range cruising speed	4,077 n miles (7,550 km; 4,691 miles)
normal cruising speed	3,769 n miles (6,980 km; 4,337 miles)
Design g limit	+2.5

OPERATIONAL NOISE LEVELS:
T-O	80.9 EPNdB
Sideline	86.2 EPNdB
Approach	90.3 EPNdB

UPDATED

BOMBARDIER CRJ100 and CRJ200

TYPE: Regional jet airliner.

PROGRAMME: Design studies began in third quarter of 1987; basic configuration frozen June 1988; engineering designation CL-600-2B19; formal programme go-ahead given 31 March 1989; extended-range CRJ100ER announced September 1990. Three development aircraft built (c/n 7001-7003), plus static test airframe (c/n 7991) and forward fuselage test article (7992); first flight of 7001 (C-FCRJ) 10 May 1991; 7002 (C-FNRJ) first flew 2 August 1991 and 7003 on 17 November 1991; all three in 1,400 hour flight test programme in Wichita, USA. CF34-3A1 engine obtained its US type certificate 24 July 1991. Transport Canada type approval (CRJ100 and CRJ100ER) 31 July 1992. Japanese Civil Aviation Bureau certification 23 May 2000.

First delivery aircraft (c/n 7004) flew 4 July 1992, and to Lufthansa CityLine of Germany (as D-ARJA) 29 October 1992; European JAA and US FAA certification 14 and 21 January 1993 respectively; long-range CRJ100LR certified 29 April 1994; CRJ200 with CF34-3B1 engines announced in 1995. Production rate six per month by late 1998, scheduled to rise to 7.5 per month by third quarter 1999. Total fleet time at May 2000 was 2,652,604 flight hours and 2,371,774 cycles; 200th aircraft delivered (to Lufthansa) 24 October 1997; 300th to Atlantic Coast Airlines in April 1999. 400th to Delta Connection/SkyWest in July 2000. Production of CRJ100/200 running at 9.5 per month in 2000, rising to 12.5 per month by late 2001, with annual targets of 135 in 2001 and 150 in 2003.

CURRENT VERSIONS: **CRJ100**: Original standard aircraft.

CRJ100ER: Replaced by CRJ200ER.

CRJ100LR: Announced March 1994; launch customer, Lauda Air of Austria; replaced by CRJ200LR.

CRJ200: Standard aircraft; designed to carry 50 passengers over 985 n mile (1,824 km; 1,133 mile) range CF34-3B1 engines with 2.8 per cent lower specific fuel consumption than CF34-3A1 of CRJ100, increasing initial cruise altitude by 213 m (700 ft), cruising speed by 2.5 kt (4.5 km/h; 3 mph), and range typically by 1.5 per cent; Class C baggage compartment as standard. First delivery, to Tyrolean Airways as OE-LCF, 15 January 1996. Further improvements in development for introduction on CRJ200 variants during early 1996 included 3 kt (5.5 km/h; 3.5 mph) reduction in V_2 speed to provide 91 m (300 ft) reduction in T-O run at maximum T-O weight; 1 kt (1.8 km/h; 1.2 mph) reduction in V_{REF} to provide 15 m (50 ft) reduction in landing run at typical landing weights; new 8° flap setting to improve second-segment climb performance; and GPS integrated with an upgraded FMS.

CRJ200ER: Extended-range capability with optional increase in maximum T-O weight to 23,133 kg (51,000 lb) and optional additional fuel capacity, for range of 1,645 n miles (3,046 km; 1,893 miles).

CRJ200LR: Longer-range version of CRJ200ER (more than 2,005 n miles; 3,713 km; 2,307 miles); maximum T-O weight increased by 907 kg (2,000 lb) to 24,040 kg (53,000 lb).

CRJ200B, CRJ200B ER and CRJ200B LR: As above, but with optional hot-and-high CF34-3B1 engines providing normal T-O thrust up to ISA+7.8°C (ISA+6.1°C for standard engines), and APR thrust up to ISA+15°C (ISA+6.1°C for standard engines).

CRJ700: Described separately.

Corporate Jetliner: Company shuttle version with more spacious cabin accommodation for 18 to 30 passengers. One delivered June 1993 to Xerox Corporation. Five ordered by the People's Republic of China in January 1997; operated on behalf of PRC government by China United Airlines crews; contract value C$116 million, including outfitting, pilot and maintenance staff training and spares.

Special Edition (SE): Corporate version developed in consultation with launch customer TAG Aeronautics Ltd to meet requirement for non-stop flights, London to Jeddah or equivalent, with three crew and five passengers; or between Middle East city pairs with 15 passengers. First flown 26 May 1995 and formally announced at Paris Air Show in the following month; first delivery (N877SE) to TAG during Dubai International Aerospace Show in November 1995; second TAG aircraft delivered November 1997. Accommodation for up to 19 passengers in customised cabin; additional 1,814 kg (4,000 lb) of fuel carried in two auxiliary tanks behind main cabin, extending range to more than 3,000 n miles (5,556 km; 3,452 miles); maximum T-O weight 24,040 kg (53,000 lb); first aircraft powered by standard CF34-3A1 turbofans, but subsequent examples are equipped with CF34-3B1s increasing range to 3,120 n miles (5,778 km; 3,590 miles); Rockwell Collins Pro Line 4 avionics as on RJ, but with third FMS, third VHF, dual Collins HF and Selcal. Manufactured to special order only.

CUSTOMERS: See table.

CRJ100/200 PRODUCTION
(at 1 December 2000)

Customer (Srs)	Orders	Delivered	Backlog
Adria Airways (200)	4	4	
Air Canada (100)	24	24	
(200)	2	2	
Air Dolomiti (200)	3		3
Air Littoral (100)	19	19	
Air Nostrum (200)	21	6	15
Air Wisconsin (200)	9	8	1
Atlantic Coast Airlines (200)	96	36	60
Atlantic Southeast Airlines (200)	45	40	5
Brit Air (100)	20	20	
British European (200)	4	3	1
Cimber Air (200)	2	2	
COMAIR (100)	110	104	6
DAC Air (200)	2	2	
Delta Connection (200)	79		79
GECAS (200)	15		15
J-Air (200)	4	1	3
Kendall/Ansett (200)	12	8	4
Lauda Air (100)	8	8	
Lufthansa (100)	45	42	3
Maersk (200)	11	10	1
Mesa (200)	32	32	
Midway (200)	26	23	3
Northwest Airlines (200)	54	8	46
Palestinian (200)	2		2
Saeaga (200)	1	1	
Shandong Airlines (200)	5	2	3
Shanghai Airlines (200)	3	1	2
SkyWest (100)	10	10	
(200)	55	4	51
South African Express (200)	6	6	
Southern Winds (200)	2	2	
The Fair Inc (200)	2		2
Tyrolean (200)	15	10	5
Yunnan (200)	6		6
Undisclosed (200)	2		2
Corporate variants	13	13	
Totals	**769**	**451**	**318**

COSTS: Programme development costs C$275 million. Atlantic Coast Airlines order for 10 CRJ200ERs valued at US$200 million (September 1998).

DESIGN FEATURES: Evolved from Challenger (which see), designed expressly for regional airline operating environment. Advanced transonic wing design, with

Special Edition of TAG Aeronautics Ltd

CANADA: AIRCRAFT—BOMBARDIER (Canadair)

400th Bombardier CRJ, delivered to Delta Connection/SkyWest in July 2000

winglets for high-speed operations; fuel-efficient GE turbofans; options include higher design weights, additional fuel capacity, more comprehensive avionics, and maximum certified altitude raised to 12,500 m (41,000 ft).

Wings, designed with computational fluid dynamics (CFD), have 13.2 per cent (root) and 10 per cent (tip) thickness/chord ratios, 2° 20′ dihedral, 3° 25′ root incidence and 24° 45′ quarter-chord sweepback.

FLYING CONTROLS: Conventional and power-assisted. Primary controls with cables and push/pull rods for multiple redundancy; hydraulically actuated ailerons, elevators and rudder with at least two hydraulic power control unit actuators per surface (three on rudder and elevator); ailerons and elevators fitted with flutter dampers (dual on elevators); rudder with dual-channel control yaw damping; artificial feel and electric trim for roll and yaw; electronically controlled, variable incidence T tailplane for pitch trim and electronically controlled artificial pitch feel. Double-slotted electromechanical flaps with electronically controlled Datron electric motors; BAE fly-by-wire spoiler and spoileron system, four spoilers each side, with inner two functioning as ground spoilers, outer two comprising one flight spoiler and one spoileron, both also providing lift dumping on touchdown. Avionics suite includes engine indication and crew alerting system (EICAS).

STRUCTURE: Semi-monocoque fuselage is damage tolerant FAR/JAR 25 certified airframe with chemically milled skins; flat pressure bulkheads forward of flight deck and aft of baggage compartment; extensive use of advanced composites in secondary structures (passenger compartment floor, wing/fuselage fairings, nacelle doors, wing access door covers, winglets, tailcone, avionics access doors and landing gear doors); comprehensive anti-corrosion treatment and drainage. Wing is one-piece unit mounted to underside of fuselage; two-spar box joined by ribs, covered top and bottom with integrally stiffened skin panels (three upper and three lower each side) for smooth flow; machined or built-up spars and shearweb-type ribs. Short Brothers (UK) manufactures fuselage central section, fore and aft fuselage plugs, wing flaps, ailerons, spoilerons and inboard spoilers.

LANDING GEAR: Hydraulically retractable tricycle type, manufactured by Dowty. Inward-retracting main units each have 15 in Aircraft Braking System (ABS) wheels with H29×9.0-15 (16 ply) Goodyear tubeless tyres, pressure 11.17 bars (162 lb/sq in) unladen. Nose unit has 18×4.4 (12 ply) tyres (deflector-type) and Dowty Canada steer-by-wire steering; unladen tyre pressure 8.62 bars (125 lb/sq in). Aircraft Braking System steel multidisc brakes and fully modulated Hydro Aire Mk III anti-skid system. Minimum taxiway width for 180° turn (with 3.35 m; 11 ft 0 in safety margin) is 22.86 m (75 ft 0 in).

POWER PLANT: Two General Electric CF34-3B1 turbofans, each rated at 41.0 kN (9,220 lb st) with APR and 38.8 kN (8,729 lb st) without. Nacelles produced by Short Brothers. Pneumatically actuated thrust reversers. Fuel in two integral wing tanks, combined capacity 5,300 litres (1,400 US gallons; 1,166 Imp gallons); increasable to 8,080 litres (2,135 US gallons; 1,778 Imp gallons) with optional centre-wing tank. Pressure refuelling point in starboard leading-edge wingroot; transfer rate 474 litres (125 US gallons; 104 Imp gallons)/min at 3.45 bars (50 lb/sq in); two gravity points on starboard wing (one for centre tank) and one on port wing.

ACCOMMODATION: Two-pilot flight deck; one or two cabin attendants. Main cabin seats up to 50 passengers in standard configuration, four-abreast at 79 cm (31 in) pitch, with centre aisle; maximum capacity 52 seats. Various configurations, from 15 to 50 seats, available for corporate version. Downward-opening front passenger door with integral airstairs on port side; plug-type forward emergency exit/service door opposite on starboard side (inoperative on SE). Inward-opening baggage door on port side at rear. Overwing Type III emergency exit each side (port side door inoperative on SE). Entire accommodation pressurised, including rear baggage compartment.

SYSTEMS: Cabin pressurisation and air conditioning system (maximum differential 0.57 bar; 8.3 lb/sq in). Primary flight control systems powered by hydraulic servo-actuators with distinct, alternate paths cable and pushrod systems. Electric trim and dual yaw dampers. Three fully independent 207 bar (3,000 lb/sq in) hydraulic systems. Three-phase 115 V AC electrical primary power at 400 Hz supplied by two 30 kVA engine-driven generators; alternative power provided by APU and air-driven generator. Conversion to 28 V DC by five transformer-rectifier units. Main (Ni/Cd) battery 17 Ah, APU battery 43 Ah. Honeywell GTCP 36-150 (RJ) APU and two-pack air conditioning system in rear of fuselage. Wing leading-edges and engine intake cowls anti-iced by engine bleed air. Electric anti-icing of windscreen and cockpit side windows, pitot heads, air data vanes, static sources and sensors. Ice detection system standard.

AVIONICS: *Comms:* Dual VHF nav/com radios. Options include HF radio, single Selcal and 8.33 kHz VHF.
Radar: Rockwell Collins digital weather radar system; split-scan weather radar and radar with turbulence mode optional.
Flight: Dual flight management systems optional. GPWS, windshear detection system and TCAS. EGPWS optional in place of GPWS. Lockheed Martin Fairchild flight data recorder. Dual FMS 4200 and dual IRS in Corporate Jetliner and Special Edition.
Instrumentation: Rockwell Collins Pro Line 4 integrated all-digital suite, including dual primary flight displays, dual multifunction displays, dual EICAS, dual AFCS, dual AHRS, dual air data system and Cat. II capability with Cat. IIIa optional using head-up guidance system. Dual inertial reference system optional in lieu of AHRS. Flight Dynamics Inc HGS 2100 HUD approved by Transport Canada November 1995, permitting Cat. IIIa operation.

DIMENSIONS, EXTERNAL: As for Challenger 604 except:

Wing span over winglets	21.21 m (69 ft 7 in)
Wing chord: at fuselage c/l	5.13 m (16 ft 10 in)
at tip	1.27 m (4 ft 2 in)
Wing aspect ratio (excl winglets)	8.9
Length: overall	26.77 m (87 ft 10 in)
fuselage	24.38 m (80 ft 0 in)
Height overall	6.22 m (20 ft 5 in)
Fuselage max diameter	2.69 m (8 ft 10 in)
Wheel track	3.17 m (10 ft 5 in)
Wheelbase	11.41 m (37 ft 5 in)
Passenger/crew door (fwd, port):	
Height	1.78 m (5 ft 10 in)
Width	0.91 m (3 ft 0 in)
Service door (stbd, fwd): Height	1.22 m (4 ft 0 in)
Width	0.61 m (2 ft 0 in)
Height to sill (crew/service)	1.63 m (5 ft 4 in)
Baggage door (port, rear): Height	1.09 m (3 ft 7 in)
Width	0.84 m (2 ft 9 in)
Height to sill	1.63 m (5 ft 4 in)
Emergency exit (overwing, stbd):	
Height	0.96 m (3 ft 2 in)
Width	0.51 m (1 ft 8 in)
Turning circle	22.86 m (75 ft 0 in)

DIMENSIONS, INTERNAL: As for Challenger 604 except:

Cabin (incl baggage compartment, excl flight deck):	
Length	14.76 m (48 ft 5 in)
Max height	1.85 m (6 ft 1 in)
Width: at centreline	2.57 m (8 ft 5 in)
at floor level	2.18 m (7 ft 2 in)
Floor area: except SE	32.1 m² (346 sq ft)
SE	30.3 m² (326 sq ft)
Volume: except SE	57.1 m³ (2,015 cu ft)
SE	53.8 m³ (1,900 cu ft)
Stowage volume: main (rear) baggage compartment	
	9.0 m³ (318 cu ft)
wardrobes/bins/underseat (total)	4.7 m³ (165 cu ft)

AREAS:

Wings: gross (excl winglets)	54.54 m² (587.1 sq ft)
net	48.35 m² (520.4 sq ft)
Ailerons (total)	1.93 m² (20.8 sq ft)
Trailing-edge flaps (total)	10.60 m² (114.1 sq ft)
Spoilers (total)	2.26 m² (24.3 sq ft)
Winglets (total)	1.38 m² (14.9 sq ft)
Fin	9.18 m² (98.8 sq ft)
Rudder	2.03 m² (21.9 sq ft)
Tailplane	9.44 m² (101.6 sq ft)
Elevators (total)	2.84 m² (30.52 sq ft)

WEIGHTS AND LOADINGS:

Manufacturer's weight empty:	
200	13,236 kg (29,180 lb)
200ER, 200LR	13,243 kg (29,195 lb)
Corporate Jetliner	11,703 kg (25,800 lb)
Operating weight empty: 200	13,730 kg (30,270 lb)
200ER, 200LR	13,835 kg (30,500 lb)
Corporate Jetliner	14,424 kg (31,800 lb)
SE	15,377 kg (33,900 lb)
Max payload (structural): 200	5,411 kg (11,930 lb)
200ER, 200LR	6,124 kg (13,500 lb)
Corporate Jetliner	5,533 kg (12,200 lb)
SE	2,476 kg (5,460 lb)

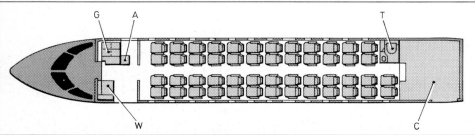

Bombardier CRJ standard 79 cm (31 in) pitch 50-seat layout *(Mike Keep/Jane's)*
A: attendant's seat, C: cargo, G: galley, T: toilet, W: wardrobe

Max fuel: 200   4,254 kg (9,380 lb)
  200ER, 200LR, Corporate Jetliner
       6,489 kg (14,305 lb)
  SE   8,303 kg (18,305 lb)
Payload with max fuel: 200   3,651 kg (8,050 lb)
  200ER   2,923 kg (6,445 lb)
  200LR   3,831 kg (8,445 lb)
  Corporate Jetliner   2,333 kg (5,145 lb)
  SE   410 kg (905 lb)
Max T-O weight: 200   21,523 kg (47,450 lb)
  200ER, Corporate Jetliner   23,133 kg (51,000 lb)
  200LR, Corporate Jetliner (optional), SE
       24,040 kg (53,000 lb)
Max ramp weight: 200   21,636 kg (47,700 lb)
  200ER   23,246 kg (51,250 lb)
  200LR, SE   24,154 kg (53,250 lb)
Max zero-fuel weight: SE   17,917 kg (39,500 lb)
  200   19,141 kg (42,200 lb)
  200ER, 200LR, Corporate Jetliner
       19,958 kg (44,000 lb)
Max landing weight: 200   20,275 kg (44,700 lb)
  200ER, 200LR, Corporate Jetliner, SE
       21,319 kg (47,000 lb)
Max wing loading: 200   394.6 kg/m² (80.82 lb/sq ft)
  200ER   424.1 kg/m² (86.87 lb/sq ft)
  200LR Corporate Jetliner (optional), SE
       440.8 kg/m² (90.27 lb/sq ft)
Max power loading (APR rating):
  200   263 kg/kN (2.57 lb/lb st)
  200ER, Corporate Jetliner   282 kg/kN (2.77 lb/lb st)
  200LR, Corporate Jetliner (optional), SE
       293 kg/kN (2.87 lb/lb st)

PERFORMANCE:
Max operating speed:
  above 9,570 m (31,400 ft)   M0.85
  below 7,740 m (25,400 ft) 335 kt (621 km/h; 386 mph)
High-speed cruising speed: CRJ200 at 11,275 m
  (37,000 ft)   M0.81 or 464 kt (859 km/h; 534 mph)
  Corporate Jetliner, SE
       M0.80 or 459 kt (850 km/h; 528 mph)
Normal cruising speed: CRJ200 at 11,275 m (37,000 ft)
       M0.74 or 424 kt (785 km/h; 488 mph)
  Corporate Jetliner, SE
       M0.77 or 442 kt (819 km/h; 509 mph)
Long-range cruising speed, Corporate Jetliner, SE
       M0.74 or 424 kt (785 km/h; 488 mph)
Approach speed, 45° flap, AUW of 19,504 kg (43,000 lb)
       135 kt (250 km/h; 155 mph)
Max rate of climb at 457 m (1,500 ft), 250 kt CAS/M0.74
  climb schedule: 200   1,128 m (3,700 ft)/min
  200LR   1,036 m (3,400 ft)/min
  SE   1,034 m (3,395 ft/min)
Max certified altitude   12,500 m (41,000 ft)
FAR T-O field length at S/L, ISA:
  200   1,527 m (5,010 ft)
  200ER   1,768 m (5,800 ft)
  200LR   1,917 m (6,290 ft)
  Corporate Jetliner   1,765 m (5,790 ft)
  SE   1,918 m (6,295 ft)
FAR landing field length at S/L, ISA, at max landing
  weight: 200   1,423 m (4,670 ft)
  200ER, 200LR   1,478 m (4,850 ft)
  Corporate Jetliner, SE   887 m (2,910 ft)
Range with max payload at long-range cruising speed,
  FAR Pt 121 reserves:
  200   965 n miles (1,787 km; 1,110 miles)
  200ER   1,645 n miles (3,046 km; 1,893 miles)
  200LR   2,005 n miles (3,713 km; 2,307 miles)
  Corporate (30 seats), NBAA IFR reserves
       2,017 n miles (3,735 km; 2,321 miles)
  SE with 3,674 kg (8,100 lb) payload
       1,541 n miles (2,853 km; 1,773 miles)
Range with max fuel:
  Corporate Jetliner
       2,250 n miles (4,167 km; 2,589 miles)
  SE   3,120 n miles (5,778 km; 3,590 miles)

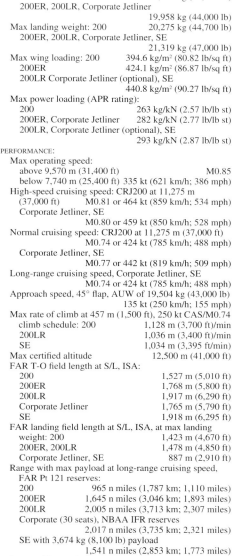

Bombardier CRJ200 (two General Electric CF34-3B1 turbofans)

OPERATIONAL NOISE LEVELS: (CRJ200, FAR Pt 36):
  CRJ200:
    T-O   89 EPNdB
    Approach   98 EPNdB
    Sideline   94 EPNdB
  Corporate Jetliner, SE:
    T-O   79 EPNdB
    Approach   92 EPNdB
    Sideline   82 EPNdB

UPDATED

BOMBARDIER CRJ700

TYPE: Regional jet airliner.
PROGRAMME: Design and market evaluation began in 1995 in consultation with 15-member advisory panel of airline operators; stretched derivative of Regional Jet; originally designated CRJ-X; engineering designation CL-600-2C10. GE CF34-8C1 engines selected February 1995; low-speed wind tunnel testing began early 1995 at Institute of Aeronautical Research, Ottawa; high-speed wind tunnel testing began November 1995 at Rockwell facilities in California. More than 650 hours of wind tunnel testing completed by September 1996. Bombardier Board gave approval for launch on 21 January 1997, at which time orders included four from launch customer Brit Air of France. Aerodynamic configuration frozen 14 March 1997; design frozen 17 July 1998. New version of CF34 engine first ground run in February 1998, followed by flight tests on GE-owned Boeing 747 testbed in first quarter of 1999 and certification in November 1999. First flight (C-FRJX c/n 10001) 27 May 1999; official roll-out 28 May 1999; public debut (c/n 10004) at Farnborough International Air Show 23 July 2000; Transport Canada certification achieved 22 December 2000; first delivery, to Brit Air of France, in February 2001.

Four flying aircraft and two static test airframes participated in the certification programme; one static test airframe was used for Complete Aircraft Static Test (CAST), while the other underwent Durability and Damage Tolerance Test (DDTT), and was subjected to the equivalent of 160,000 flight cycles. Test programme was conducted at the Bombardier Flight Test Centre at Wichita, Kansas, totalling 1,600 flight hours on four aircraft: C-FRJX (handling and performance evaluation), 10002/C-FJFC (systems), 10003/C-FBKA (avionics) and 10004/C-FCRJ, the first to be fully furnished (function and reliability testing), which initially flew on 16 December 1999.

CURRENT VERSIONS: **CRJ700**: 68-seat version in standard and extended range (ER) weight options.
  CRJ701: 70-seat version in standard and extended range (ER) weight options.
  CRJ702: 72- to 78-seat version in standard and extended range (ER) weight options.
  CRJ900: further stretch, described separately.
CUSTOMERS: Announced customers at January 2001 were American Eagle (25), Atlantic Southeast Airlines (12), Brit Air (12), Comair (20), Delta Connection (25), GECAS (25), Horizon Air (30), Lufthansa (20), Maersk (5), and Shandong Airlines (10). Total 184 firm orders.
COSTS: Development cost C$645 million, of which C$440 million provided by Bombardier and balance by risk-sharing partners. Unit cost US$24.7 million (1998). Break-even point is at 200th of 400 aircraft production run.
DESIGN FEATURES: Commonality (including crew training and type rating) with CRJ100/200; direct operating cost per seat-mile to be 20 per cent lower than that of CRJ100/200 and lowest of any airliner in its class. Fuselage stretched 4.72 m (15 ft 6 in) by plugs fore and aft of centre-section to seat 70 passengers; rear pressure bulkhead moved aft by 1.29 m (4 ft 3 in); cabin 6.02 m (19 ft 9 in) longer; APU moved to tailcone; cabin floor lowered by 2.5 cm (1 in) and ceiling raised by 1.3 cm (½ in) to provide 1.89 m (6 ft 2¼ in) headroom; cabin windows raised 11.5 cm (4½ in); new underfloor baggage compartment, volume 3.09 m³ (109 cu ft) to facilitate ramp check-in; wing span increased by 1.83 m (6 ft 0 in) by wingroot plug; wing leading-edge extended and equipped with high-lift devices; larger horizontal tail surfaces; new pitch control system; main landing gear lengthened; new wheels, tyres, brakes; air conditioning and anti-icing systems upgraded; new engines; new underfloor baggage compartment on forward port side; overhead stowage bins redesigned.
FLYING CONTROLS: Sextant and Menasco primary and secondary flight control system; flying controls are actuated via a dual network of cables, pulleys and pushrods which operate hydraulic power drive units.
STRUCTURE: Programme participants include Avcorp (vertical and horizontal stabilisers); Bombardier Canadair operations (wing, cockpit, rudder and doors, electrical system, primary flight controls, plus final assembly and interior completion), C & D Interiors (cabin interior), General Electric (power plant), GKN Westland (tailcone and doors), Hella Aerospace GmbH (lighting system), Honeywell (APU), Intertechnique (fuel system), Liebherr Aerospace Toulouse (air management system), Menasco Aerospace (landing gear), Mitsubishi Heavy Industries (aft

Fourth development CRJ700 in the colours of French carrier Brit Air

44 CANADA: AIRCRAFT—**BOMBARDIER** (Canadair)

fuselage), Parker Abex (hydraulic system), Rockwell Collins (avionics), Sextant Avionique (flight control system), Shorts (nacelles and thrust reversers) and Hamilton Sundstrand (flaps, leading-edge slats and electrical system).

LANDING GEAR: Hydraulically retractable tricycle type by Menasco with twin wheels on each unit. Mainwheel tyre size H36×12.0-8, pressure 10.55 bars (153 lb/sq in); nosewheel tyre size H20.5×6.75-10, pressure 9.24 bars (134 lb/sq in).

POWER PLANT: Two General Electric CF34-8C1 turbofans with dual-channel FADEC. Engine rating 56.4 kN (12,670 lb st), or 61.3 kN (13,790 lb st) with automatic power reserve, flat rated to ISA + 15°C. Intertechnique fuel management system with Ratier-Figeac controls. Fuel capacity 11,488 litres (3,035 US gallons; 2,527 Imp gallons).

ACCOMMODATION: Two-pilot flight deck. Main cabin seats 70 passengers, four-abreast at 79 cm (31 in) pitch. Baggage compartment and lavatory at rear of cabin; underfloor baggage compartment; various combinations of galleys, wardrobes and lavatory at front of cabin according to seating capacity. Passenger door, forward emergency exit/service door, overwing emergency exits and baggage door as for Regional Jet.

SYSTEMS: Hamilton Sundstrand electrical generation system comprising two 40 kVA integrated drive generators; tailcone-mounted Honeywell APU approved for operation up to 12,500 m (41,000 ft); Liebherr air management system; Intertechnique fuel system; Walter Kidde fire detection system; Goodrich portable water system; Hella lighting system; Parker/Abex Hydraulics hydraulic systems, and Teleflex anti-icing system.

AVIONICS: *Radar:* Rockwell Collins digital weather radar.
Flight: Rockwell Collins AHRS and TCAS.
Instrumentation: Rockwell Collins Pro Line 4 EFIS with six 127 × 178 mm (5 × 7 in) CRT displays, including dual PFD, dual MFD and dual EICAS; Flight Dynamics HGS 2000 head-up guidance system. Autopilot, FMS and centralised avionics maintenance functions are provided by integrated avionics processing system (IAPS). Windshear detection and recovery system standard. Equipped for Cat. II landings.

Following data are provisional.

DIMENSIONS, EXTERNAL:
Wing span	23.24 m (76 ft 3 in)
Wing aspect ratio	7.4
Length overall	32.51 m (106 ft 8 in)
Max diameter of fuselage	2.69 m (8 ft 10 in)
Height overall	7.57 m (24 ft 10 in)
Tailplane span	8.53 m (28 ft 0 in)
Turning circle	22.86 m (75 ft 0 in)

Passenger door (port, forward):
Height	1.78 m (5 ft 10 in)
Width	0.91 m (3 ft 0 in)
Height to sill	1.73 m (5 ft 8 in)

Baggage door (port, aft):
Height	0.84 m (2 ft 9 in)
Width	1.09 m (3 ft 7 in)
Height to sill	2.31 m (7 ft 7 in)

Baggage door (port, forward):
Height	0.51 m (1 ft 8 in)
Width	1.07 m (3 ft 6 in)
Height to sill	1.28 m (4 ft 2½ in)

Service door (starboard, forward):
Height	1.22 m (4 ft 0 in)
Width	0.61 m (2 ft 0 in)
Height to sill	1.73 m (5 ft 8 in)

DIMENSIONS, INTERNAL:
Cabin (excl flight deck):
Length	20.78 m (68 ft 2 in)
Max width: at centreline	2.57 m (8 ft 5 in)
at floor level	2.13 m (7 ft 0 in)
Max height	1.89 m (6 ft 2¼ in)
Floor area, excl cockpit	42.8 m² (461 sq ft)
Volume	75.95 m³ (2,682 cu ft)
Baggage volume (total)	23.3 m³ (824 cu ft)

AREAS:
Wings, net	68.63 m² (738.7 sq ft)
Horizontal tail surfaces (total)	20.74 m² (223.3 sq ft)
Vertical tail surfaces (total)	13.36 m² (143.8 sq ft)

WEIGHTS AND LOADINGS:
Operating weight empty	19,269 kg (42,480 lb)
Max payload	8,528 kg (18,800 lb)
Payload with max fuel: standard	4,364 kg (9,620 lb)
ER	5,384 kg (11,870 lb)
Max fuel weight	9,017 kg (19,880 lb)
Max T-O weight: standard	32,999 kg (72,750 lb)
ER	34,019 kg (75,000 lb)
Max ramp weight: standard	33,112 kg (73,000 lb)
ER	34,132 kg (75,250 lb)
Max landing weight	30,390 kg (67,000 lb)
Max zero-fuel weight	28,259 kg (62,300 lb)
Max wing loading: standard	480.8 kg/m² (98.48 lb/sq ft)
ER	495.7 kg/m² (101.53 lb/sq ft)
Max power loading: standard	293 kg/kN (2.87 lb/lb st)
ER	302 kg/kN (2.96 lb/lb st)

PERFORMANCE:
Cruising Mach No.:
high speed	0.825 (470 kt; 870 km/h; 541 mph)
normal	0.77 (442 kt; 818 km/h; 508 mph)
Max certified altitude	12,500 m (41,000 ft)

T-O field length at S/L, ISA:
standard	1,564 m (5,130 ft)
ER	1,648 m (5,406 ft)

Landing field length at S/L, ISA, at max landing weight:
standard and ER	1,478 m (4,850 ft)

Range, 70 passengers, at M0.77, IFR reserves:
standard	1,687 n miles (3,124 km; 1,941 miles)
ER	1,984 n miles (3,674 km; 2,283 miles)

OPERATIONAL NOISE LEVELS (FAR Pt 36):
T-O at max T-O weight	89 EPNdB
Approach at max landing weight	98 EPNdB
Sideline	94 EPNdB

UPDATED

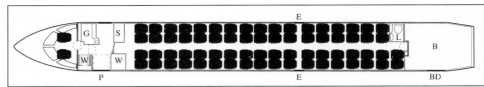

Typical CRJ700 seating plan for 70 passengers at 79 cm (31 in) pitch, with 15.7 m³ (555 cu ft) of baggage
B: baggage, BD: baggage door, E: emergency exit, G: galley, L: lavatory, P: entrance, S: stowage, W: wardrobe

BOMBARDIER CRJ900

TYPE: Regional jet airliner.

PROGRAMME: Announced October 1999; stretched derivative of CRJ700; wind tunnel testing began November 1999; partner/supplier selection finalised second quarter 2000; interior mockup completed March 2000; formal launch at Farnborough International Air Show 24 July 2000; prototype, modified from CRJ700 prototype C-FRJX, with fuselage plugs but retaining CRJ700 wings, landing gear and engines, was flown for the first time on 21 February 2001; assembly of first production aircraft, c/n 15001, scheduled to begin in first quarter 2001; certification third quarter 2002; customer deliveries from first quarter 2003.

CURRENT VERSIONS: **CRJ900**: Standard version.
CRJ900ER: Extended-range version.
CRJ900ER European: As 900ER but with maximum T-O weight limited to 36,995 kg (81,560 lb) to minimise weight-related charges when operating in European airspace.

COSTS: Development cost C$200 million; unit cost C$29 million (2000).

CUSTOMERS: Firm orders for 14, plus 48 options, by 1 December 2000, comprising: Air Nostrum (eight options), Brit Air (four firm, eight options), GE Capital Aviation Services (10 firm, 20 options), and Tyrolean Airways (12

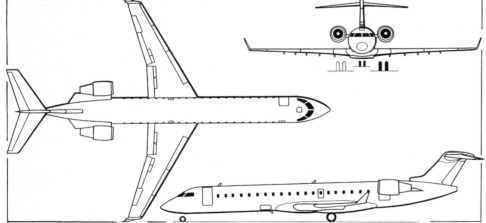

Bombardier CRJ700 regional jet airliner *(Paul Jackson/Jane's)* 2001/0105027

The Bombardier CRJ700 flight deck is based on that of the CRJ100/200

Cabin mockup of CRJ700

Jane's All the World's Aircraft 2001-2002 www.janes.com

BOMBARDIER (Canadair)—AIRCRAFT: CANADA

Prototype Bombardier CRJ900 during its maiden flight on 21 February 2001   *2001/0062439*

options). Estimated market for 800 aircraft in CRJ900 class over 20-year period.

DESIGN FEATURES: Compared to CRJ700, has fuselage stretched – by means of 2.29 m (7 ft 6 in) plug forward of centre-section and 1.57 m (5 ft 2 in) plug aft of centre-section – to accommodate 86-90 passengers in four-abreast configuration with fore and aft lavatories; 5 to 10 per cent higher thrust engines; strengthened main landing gear with upgraded wheels and brakes; strengthened wing; two additional overwing emergency exits; increased volume in forward underfloor baggage hold, and an additional underfloor baggage door and aft service door on starboard side. Common crew qualification with CRJ200/700 series.

STRUCTURE: Programme partners generally as for CRJ700 except: Gamesa (vertical and horizontal stabilisers), Hamilton Sundstrand (flaps), and Shorts (mid-fuselage section). Final assembly and completion by Bombardier at new CRJ700/900 facility at Montreal-Mirabel Airport, scheduled to open in second quarter 2001.

POWER PLANT: Two General Electric CF34-8C5 turbofans. Engine rating 58.4 kN (13,123 lb st) or 63.4 kN (14,255 lb st) with automatic power reserve, flat rated to ISA +15°C. Fuel capacity 11,148 litres (2,945 US gallons; 2,452 Imp gallons).

ACCOMMODATION: Standard dual-class accommodation for 86 passengers in four-abreast configuration at 79 cm (31 in) seat pitch with fore and aft lavatories and forward galley; alternative configurations include high density with accommodation for 90 passengers at 79 cm (31 in) seat pitch, dual-class with 15 business class seats three-abreast at 86 cm (34 in) seat pitch in forward section and 60 economy class four-abreast at 79 cm (31 in) seat pitch at rear, and dual class with 55 business class seats four-abreast at 84 cm (33 in) seat pitch in forward section and 24 economy class at 79 cm (31 in) seat pitch at rear. Standard additional floor beam facilitates offset seat rail for three-abreast seating throughout cabin.

AVIONICS: As for CRJ700.

Following data are provisional.

DIMENSIONS, EXTERNAL: As for CRJ700 except

Wing span	23.24 m (76 ft 3 in)
Length overall	36.19 m (118 ft 9 in)
Height overall	7.49 m (24 ft 7 in)
Wheelbase	14.73 m (48 ft 4 in)
Aft service door (starboard):	
Height	1.22 m (4 ft 0 in)
Width	0.61 m (2 ft 0 in)
Height to sill	1.73 m (5 ft 8 in)
Emergency exits (four, overwing):	
Height	0.91 m (3 ft 0 in)
Width	0.51 m (1 ft 8 in)

DIMENSIONS, INTERNAL:

Cabin (excl flight deck):	
Baggage volume: checked	16.81 m³ (593.5 cu ft)
total	25.57 m³ (903 cu ft)

WEIGHTS AND LOADINGS:

Operating weight empty	21,546 kg (47,500 lb)
Max payload	10,206 kg (22,500 lb)
Payload with max fuel: 900	6,065 kg (13,370 lb)
900ER	7,135 kg (15,730 lb)
Max usable fuel	8,949 kg (19,730 lb)
Max T-O weight: 900	36,514 kg (80,500 lb)
900ER	37,421 kg (82,500 lb)
900ER European	36,995 kg (81,560 lb)
Max landing weight	33,339 kg (73,500 lb)

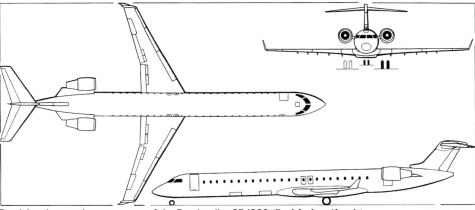

Provisional general arrangement of the Bombardier CRJ900 *(Paul Jackson/Jane's)*   *2001/0105028*

Max zero-fuel weight	31,751 kg (70,000 lb)
Max wing loading: 900	532.1 kg/m² (108.98 lb/sq ft)
900ER	545.3 kg/m² (111.68 lb/sq ft)
900ER European	539.1 kg/m² (110.41 lb/sq ft)
Max power loading: 900	313 kg/kN (3.07 lb/lb st)
900ER	320 kg/kN (3.14 lb/lb st)
900ER European	317 kg/kN (3.11 lb/lb st)

PERFORMANCE:

Cruising Mach No:	
high speed	0.81 (464 kt; 859 km/h; 534 mph)
for max range	0.77 (442 kt; 819 km/h; 509 mph)
Max certified altitude: 900, 900ER	12,500 m (41,000 ft)
Service ceiling, OEI: 900	4,968 m (16,300 ft)
900ER	4,755 m (15,600 ft)
FAR T-O field length:	
900	1,878 m (6,160 ft)
900ER	1,970 m (6,462 ft)
FAR landing field length at S/L, ISA, at max landing weight:	
900, 900ER	1,565 m (5,136 ft)
Range, 86 passengers, at M0.77:	
900	1,498 n miles (2,774 km; 1,723 miles)
900ER	1,732 n miles (3,207 km; 1,993 miles)

UPDATED

CANADAIR 415 SUPERSCOOPER

TYPE: Twin-turboprop amphibian.

PROGRAMME: Introduced as product follow-on to piston-engined CL-215; given new designation Canadair 415 in 1991 to distinguish new production turboprop model from CL-215T retrofit, but engineering designation retained and new-build turboprop versions as CL-215-6B11. Launched officially 16 October 1991 with firm orders from France and (August 1992) Quebec; first flight (C-GSCT) 6 December 1993, although preceded by initial CL-215T conversion (C-FASE), flown on 8 June 1989. Canadian certification 24 June 1994 in Restricted and Utility categories, FAA approval 14 October 1994 in Restricted category. RAI (Italy) approval 27 October 1994 in Restricted category. Fleet had achieved 34,027 flying hours and 126,318 scooping sorties by November 1999. Final assembly relocated to North Bay, Ontario, in November 1998.

CURRENT VERSIONS: Standard **415 SuperScooper** and first production units are in **firefighting** configuration; **415M** can be modified for **maritime, SAR** and **special missions**.

415GR: Ordered by Greece, January 1999; increased weights; boat handling and cargo hoisting provisions. Two (plus an optional third) will be configured for combat search-and-rescue (C-SAR) role, for which, during 2000, SAAB Nyge Aero of Sweden was awarded a three-year contract to install MSS 5000 mission equipment including SLAR, wing-mounted FLIR, nose-mounted Honeywell 1400C weather/search radar, digital cameras, autopilot and provision for Have Quick secure radios and rescue beacon receivers. The aircraft will also have an enlarged cargo door to facilitate deployment of an inflatable rescue boat. The Hellenic Air Force C-SAR aircraft will be based at Elefsis AB.

CUSTOMERS: See table. First French Canadair 415 delivered to CEV experimental unit 8 February 1995, but trials revealed need for modifications and acceptance delayed until 13 June 1995; deliveries completed June 1997. Ontario provincial government announced order for nine on 2 April 1998. Total 53 built by December 2000.

CANADAIR 415 CUSTOMERS
(at July 2000)

Customer	Qty	First aircraft	Delivered
Canada: Ontario govmt	9	C-GAO1	29 Apr 98
Quebec govmt	8	C-GQBA	Jul 95
Croatia[1]	3	9A-CAG	Feb 97
France: Sécurité Civile	12	F-ZBFS	8 Feb 95
Greek govmt[3]	10	2039	Jan 99
Italy: SISAM[2]	15	I-DPCD	27 Jan 95
Total	**57**		

[1] Requirement for six
[2] Societa Italiana Servizi Aerei Mediterranei
[3] Plus five options

COSTS: Approximately US$23 million (2000).

DESIGN FEATURES: Retains well-proven basic airframe of piston-engined CL-215 (thick wing with zero dihedral and 2° incidence; row of vortex generators on each wing outboard of fence; long stall strip inboard of starboard fence; leading-edge strakes and fences beside engine nacelles; water scoops behind planing step; anti-spray channels in planing bottom chine) but incorporates upgrading modifications and improvements including higher operating weights for increased firefighting productivity; pressure refuelling; wing endplates for lateral stability; finlets and tailplane/fin bullet to recover longitudinal and directional stability affected by relocated

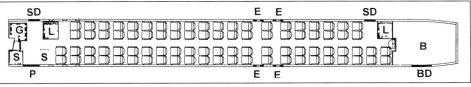

Bombardier CRJ900 seating 86 passengers
B: baggage, BD: baggage door, E: Type III exit, G: galley, L: lavatory, P: Type I passenger door, S: Stowage, SD: Type I service door   *2001/0105026*

Canadair 415 SuperScooper (two P&WC PW123 turboprops)

thrust line, increased power and new propellers; powered rudder, ailerons and elevators; new electrical system; new 'glass' cockpit with air conditioning; enlarged four-tank firefighting drop system.

FLYING CONTROLS: Conventional and power-assisted. Hydraulically actuated ailerons, elevators and rudder, standard; manual reversion in event of hydraulic failure; geared tab in each aileron, spring tab in rudder and each elevator, plus trim tab in port aileron and port elevator. Hydraulically operated single-slotted flaps, each supported by four external hinges.

STRUCTURE: No-dihedral, no-twist high wing, of constant chord; one-piece structure with two conventional spars, extruded spanwise stringers and interspar ribs and aluminium alloy skins. All-metal, fail-safe, single-step boat-hull fuselage with numerous watertight compartments. Tail surfaces of aluminium alloy sheet and extrusions, with honeycomb panels on control surfaces.

LANDING GEAR: Hydraulically retractable tricycle type. Self-centring twin-wheel nose unit retracts rearward into hull and is fully enclosed by conformal doors. Nosewheel steering standard. Main gear legs retract into wells in sides of hull. Plate mounted on each main gear assembly encloses bottom of wheel well. Mainwheel tyres 15.00-16 (16 ply) tubeless, pressure 5.31 bars (77 lb/sq in); nosewheel tyres 6.50-10 (10 ply) tubed, pressure 6.55 bars (95 lb/sq in). Hydraulic disc brakes. Non-retractable stabilising floats, each carried near wingtip on pylon cantilevered from wing box structure, with breakaway provision.

POWER PLANT: Two 1,775 kW (2,380 shp) Pratt & Whitney Canada PW123AF turboprops, on damage-tolerant mounts capable of withstanding a breach of the compressor/turbine casing, each driving a Hamilton Sundstrand 14SF-19 four-blade constant-speed fully feathering reversible-pitch propeller. Two fuel tanks, each of eight identical flexible cells, in wing spar box, with total usable capacity of 5,796 litres (1,531 US gallons; 1,275 Imp gallons). Single-point pressure refuelling (rear fuselage, starboard side), plus gravity points in wing upper surface.

ACCOMMODATION: Normal crew of two side by side on flight deck, with dual controls. Additional station in maritime patrol/SAR versions for third cockpit member, mission specialist and two observers. For water bomber cabin installation, see Equipment paragraph. Combi layout offers cargo at front, full firefighting capability, plus 11 seats at rear. Other quick-change interiors available for utility/paratroop (up to 14 troop-type folding canvas seats in cabin) or other special missions according to customer's requirements. Flush doors to main cabin on port side of fuselage forward and aft of wings. Optional aft cargo door, height 1.33 m (4 ft 4½ in), width 1.46 m (4 ft 9½ in). Emergency exit on starboard side aft of wing trailing-edge. Crew emergency hatch in flight deck roof on starboard side. Mooring hatch in upper surface of nose. Provision for additional cabin windows.

SYSTEMS: Vapour cycle air conditioning system and combustion heater. Hydraulic system, pressure 207 bars (3,000 lb/sq in), utilises two engine-driven pumps (maximum flow rate 45.5 litres; 12 US gallons; 10 Imp gallons/min) to actuate nosewheel steering, landing gear, flaps, water drop doors, pickup probes, flight controls, main gear unlocking and wheel brakes. Hydraulic fluid (MIL-H-83282) in air/oil reservoir slightly pressurised by engine bleed air. Electrically driven third pump provides hydraulic power for emergency actuation of landing gear and brakes and closure of water doors. Electrical system includes two 800 VA 115 V 400 Hz static inverters, two 28 V 400 A DC engine-driven starter/generators and two 40 Ah Ni/Cd batteries. Pneumatic/electric intake de-icing system; airframe ice protection system optional.

AVIONICS: Dual Honeywell Primus 2 digital integrated VHF nav/com.

Comms: Global VHF/UHF/AM/FM and Rockwell Collins HF radios with central control heads, ELT and dual transponders.

Radar: Search/weather radar optional.

Flight: Dual ADF, VOR/ILS, marker beacon receivers and single DME. Optional Garmin GPS and Goodrich Stormscope.

Instrumentation: Honeywell EDZ-605 EFIS with three-tube Integrated Instrument Display System for EADI and EHSI; dual Litef/Honeywell AHRS, dual air data computers, Honeywell radio altimeter.

EQUIPMENT (firefighter): Four integral water tanks in main fuselage compartment, near CG (combined capacity 6,137 litres; 1,621 US gallons; 1,350 Imp gallons), plus eight inward-facing seats in forward cabin. Tanks filled by two hydraulically actuated scoops aft of hull step, fillable also on ground by hose adaptor on each side of fuselage. Four independently openable water doors in hull bottom. Onboard foam concentrate reservoirs (capacity 680 kg; 1,500 lb) and mixing system. Improved drop pattern and drop door sequencing compared with CL-215. Optional spray kit can be coupled with firefighting tanks for large-scale spraying of oil dispersants and insecticides. In a typical firefighting mission, with a water source 6 n miles (11 km; 7 miles) from the fire, aircraft can remain on station for 3 hours, dropping 55,267 litres (14,600 US gallons; 12,157 Imp gallons)/h. Water tanks can be scoop-filled completely (ISA at S/L, zero wind) in 12 seconds over a water distance of 1,341 m (4,400 ft); partial water loads can be scooped on smaller bodies of water. Minimum safe water depth for scooping is only 1.40 m (4 ft 7 in).

EQUIPMENT (other versions): Stretcher kits, passenger or troop seats, cargo tiedowns, searchlight and other equipment according to mission and customer requirements. Provision for underwing pylon attachment points for variety of stores. Canadair 415 can be equipped with maritime surveillance radar and electro-optical sensors, precision navigation and communications equipment and autopilot.

DIMENSIONS, EXTERNAL:

Wing span	28.63 m (93 ft 11 in)
Wing chord, constant	3.54 m (11 ft 7½ in)
Wing aspect ratio	8.2
Length overall	19.82 m (65 ft 0½ in)
Beam (max)	2.59 m (8 ft 6 in)
Length/beam ratio	7.5
Height overall: on land	8.98 m (29 ft 5½ in)
on water	6.88 m (22 ft 7 in)
Draught: wheels up	1.12 m (3 ft 8 in)
wheels down	2.03 m (6 ft 8 in)
Tailplane span	10.97 m (36 ft 0 in)
Wheel track	5.28 m (17 ft 4 in)
Wheelbase	7.23 m (23 ft 9 in)
Propeller diameter	3.97 m (13 ft 0¼ in)
Propeller fuselage clearance	0.59 m (1 ft 11¼ in)
Propeller water clearance	1.30 m (4 ft 3¼ in)
Propeller ground clearance	2.77 m (9 ft 1 in)
Forward door: Height*	1.37 m (4 ft 6 in)
Width	1.03 m (3 ft 4 in)
Height to sill	1.68 m (5 ft 6 in)
Rear door: Height	1.12 m (3 ft 8 in)
Width	1.02 m (3 ft 4 in)
Height to sill	1.83 m (6 ft 0 in)
Water drop door: Length	1.60 m (5 ft 3 in)
Width	0.81 m (2 ft 8 in)
Emergency exit: Height	0.91 m (3 ft 0 in)
Width	0.51 m (1 ft 8 in)

*incl 25 cm (10 in) removable sill

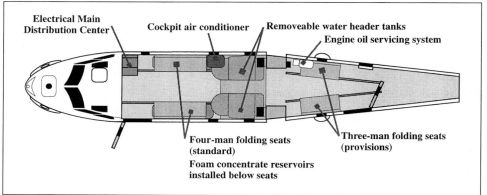

Internal layout of Canadair 415 SuperScooper amphibian in firefighting configuration

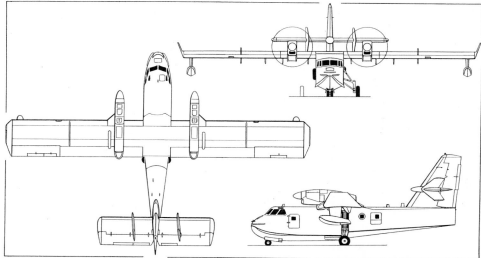

Canadair 415 SuperScooper twin-turboprop general purpose amphibian (Dennis Punnett/Jane's)

DIMENSIONS, INTERNAL:
  Cabin, excl flight deck: Length   9.38 m (30 ft 9½ in)
    Max width   2.39 m (7 ft 10 in)
    Max height   1.90 m (6 ft 3 in)
    Floor area   19.7 m² (212 sq ft)
    Volume   35.6 m³ (1,257 cu ft)
AREAS:
  Wings, gross   100.33 m² (1,080.0 sq ft)
  Ailerons (total)   8.05 m² (86.60 sq ft)
  Flaps (total)   22.39 m² (241.00 sq ft)
  Fin   11.22 m² (120.75 sq ft)
  Rudder, incl tabs   6.02 m² (64.75 sq ft)
  Tailplane   20.55 m² (221.20 sq ft)
  Elevators (total, incl tabs)   7.88 m² (84.80 sq ft)
WEIGHTS AND LOADINGS (A: firefighter, B: utility, land or water based):
  Typical operating weight empty:
    A   12,882 kg (28,400 lb)
    B   12,733 kg (28,072 lb)
  Max internal fuel weight: A, B   4,649 kg (10,250 lb)
  Max payload: A (disposable)   6,123 kg (13,500 lb)
    B   3,777 kg (8,328 lb)
  Max ramp weight: A (land)   19,958 kg (44,000 lb)
    A (water), B   17,236 kg (38,000 lb)
  Max T-O weight: A (land)   19,890 kg (43,850 lb)
    A (water), B (land and water)   17,168 kg (37,850 lb)
  Max touchdown weight for water scooping:
    A   16,420 kg (36,200 lb)
  Max flying weight after water scooping:
    A   21,319 kg (47,000 lb)
  Max landing weight:
    A, B (land and water)   16,783 kg (37,000 lb)
  Max zero-fuel weight: A   19,504 kg (43,000 lb)
    B   16,511 kg (36,400 lb)
  Max wing loading:
    A (after scoop)   212.5 kg/m² (43.52 lb/sq ft)
    A (land)   198.2 kg/m² (40.60 lb/sq ft)
    B (land and water)   171.1 kg/m² (35.05 lb/sq ft)
  Max power loading:
    A (after scoop)   6.01 kg/kW (9.87 lb/shp)
    A (land)   5.60 kg/kW (9.21 lb/shp)
    B (land and water)   4.84 kg/kW (7.95 lb/shp)
PERFORMANCE (at weights shown):
  Max cruising speed at 1,525 m (5,000 ft)
                                        197 kt (365 km/h; 227 mph) TAS
  Long-range cruising speed at 3,050 m (10,000 ft), AUW
    of 14,741 kg (32,500 lb)   145 kt (269 km/h; 167 mph)

Canadair 415 SuperScooper flight deck with Honeywell integrated instrument display system

Patrol speed at S/L, AUW of 14,741 kg (32,500 lb)
                                        130 kt (241 km/h; 150 mph)
Stalling speed: 15° flap, AUW of 21,319 kg (47,000 lb)
                                        80 kt (149 km/h; 93 mph)
Max rate of climb at S/L, AUW of 21,319 kg (47,000 lb)
                                        396 m (1,300 ft)/min
T-O distance at S/L, ISA:
  land, AUW of 19,890 kg (43,850 lb)   844 m (2,770 ft)
  land, AUW of 17,168 kg (37,850 lb)   799 m (2,620 ft)
  water, AUW of 17,168 kg (37,850 lb)   814 m (2,670 ft)

Landing distance at S/L, ISA:
  land, AUW of 16,783 kg (37,000 lb)   674 m (2,210 ft)
  water, AUW of 16,783 kg (37,000 lb)   665 m (2,180 ft)
Scooping distance at S/L, ISA (incl safe clearance
  heights)   1,341 m (4,400 ft)
Ferry range with 499 kg (1,100 lb) payload
                                        1,310 n miles (2,426 km; 1,507 miles)
Design g limits (15° flap)   +3.25/−1

UPDATED

BOMBARDIER AEROSPACE DE HAVILLAND

Garratt Boulevard, Downsview, Ontario M3K 1Y5
Tel: (+1 416) 633 73 10
Fax: (+1 416) 375 45 46
Telex: 0622128
VICE-PRESIDENT, CANADIAN OPERATIONS: Serge Perron
VICE-PRESIDENT AND GENERAL MANAGER, OPERATIONS:
  Alain Dugas
VICE-PRESIDENT, DASH 8 SERIES 100/200/300 OPERATIONS:
  Rob Comand
VICE-PRESIDENT, DASH 8 SERIES 400 OPERATIONS: Bruce Vannus
VICE-PRESIDENT, GLOBAL EXPRESS OPERATIONS:
  Trevor Anderson
VICE-PRESIDENT, FINANCE: Colin Fernie
VICE-PRESIDENT, HUMAN RESOURCES: Bernard Cormier
MANAGER, PUBLIC RELATIONS: Colin Fisher

Established 1928 as The de Havilland Aircraft of Canada Ltd, subsidiary of The de Havilland Aircraft Company Ltd, both absorbed 1961 by Hawker Siddeley Group; ownership transferred to Canadian government 26 June 1974; purchased by Boeing Company 31 January 1986 and made a division of Boeing of Canada Ltd; Boeing's intention to sell announced July 1990.

Sale to Bombardier Inc (51 per cent) and government of Ontario (49 per cent), signed 22 January 1992, supported by help from Ontario and federal governments, with Canadian Export Development Corporation to provide sales financing for Dash 8; all government support conditionally repayable. Bombardier acquired remaining 49 per cent of de Havilland in January 1997.

De Havilland manufactures Dash 8Q series, including spare parts and components, and also some components of the Bombardier Global Express; performs final assembly of Dash 8Q and Global Express; designed and builds wings for Learjet 45.

VERIFIED

DHC-8 DASH 8 Q100 and Q200

TYPE: Twin-turboprop airliner.
PROGRAMME: Launched 1980; first flight of first prototype (C-GDNK) 20 June 1983, second prototype (C-GGMP) 26 October 1983, third 22 December 1983; fourth aircraft, first with production P&WC PW120 engines, 3 April 1984.

Certified to Canadian DoT, FAR Pts 25 and 36 and SFAR No. 27 on 28 September 1984, followed by FAA type approval; also certified in Australia, Austria, Brazil, Cameroon, China, Colombia, Germany, Ireland, Italy, Maldives, Netherlands, New Zealand, Norway, Papua New Guinea, South Africa, Taiwan, UK and United Arab Emirates. First delivery (NorOntair) 23 October 1984 followed by service entry in December; 500th Dash 8 (a Series 200, N355PH) delivered 21 November 1997 to Horizon Air. The Dash 8 fleet exceeded 10 million flight hours on 21 November 2000; 600th of type delivered 6 March 2001 (see Q300).

CURRENT VERSIONS: **Dash 8 Q:** Redesigned interior and noise and vibration suppression system (NVS) standard from second quarter of 1996, reducing cabin noise levels by 12 dB; all current production aircraft are equipped with NVS and known as Dash 8 Qs. Further Noise Technology Demonstration Program, under way in mid-1999, is expected to result in additional vibration/noise damping in fuselage and floor areas. See below for specific versions.

Dash 8 Series 100: Initial version, with choice of PW120A or PW121 engines; described in 1992-93 and earlier editions.

Dash 8 Series 100A: Introduced 1990; PW120A (or optional PW121) engines and restyled interior with 6.35 cm (2.5 in) more headroom in aisle; first delivery to Pennsylvania Airlines July 1990. Specifications in 1997-98 and earlier *Jane's*.

Dash 8 Series 100B: Improved version from 1992; PW121 engines enhance airfield and climb performance. Specifications in 1997-98 and earlier *Jane's*.

Dash 8 Q100: Introduced 1998; PW 120A or optional PW 121 in basic weight version, PW 121 standard in high gross weight (HGW) version.

Dash 8 Series 200A: Increased speed/payload version of Series 100A with PW 123C engines. Transport Canada certification March 1995; first delivery 19 April 1995. Specifications in 1998-99 and earlier editions.

Dash 8 Series 200B: As 200A, but with PW123D engines for full power at higher ambient temperatures. Specifications in 1998-99 and earlier editions.

Dash 8 Q200: Introduced in 1998; increased speed/payload version of Q100; increased OEI capability and greater commonality with Series 300. Same airframe as Q100, but PW 123C/D engines give 30 kt (56 km/h; 35 mph) increase in cruising speed, allowing airlines to increase frequencies or operational radius. PW 123D engine offers full power at higher ambient temperatures for improved hot-and-high airfield performance.

Detailed description applies to Q200 except where indicated.

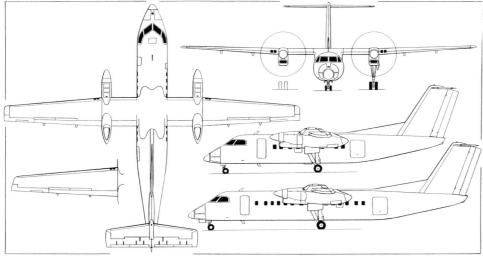

DHC-8 Dash 8 Q100, with additional side view (bottom) and wingtip of Q300 *(Dennis Punnett/Jane's)*

CANADA: AIRCRAFT—BOMBARDIER (de Havilland)

DHC-8 Dash 8 Q200

Dash 8M and **Dash 8 Q300 and Q400**; described separately.

CUSTOMERS: Total 393 firm orders for civil and military (Dash 8M) Series 100/100A/100B/Q100 and 200A/B/Q200 by 1 December 2000, of which 387 delivered. Recent customers include Mexico's Secretaria de Marina, which has ordered one Q200, with one option, for delivery in September 2001, and Nagasaki Airways, which has ordered two Q200s for delivery in the first and fourth quarters of 2001.

COSTS: Unit cost approximately US$13 million (1998).

DESIGN FEATURES: T tail, swept fin with large dorsal fin. Wing has constant chord inboard section and tapered outer panels; thickness/chord ratio 18 per cent at root, 13 per cent at tip; dihedral 2° 30′ on outer panels; inboard leading-edges drooped; 3° washout at wingtips.

FLYING CONTROLS: Conventional and power-assisted. Fixed tailplane; horn-balanced elevator with four tabs; mechanically actuated horn-balanced ailerons with inset tabs; hydraulically actuated roll spoilers/lift dumpers forward of each outer flap; two-segment serially hinged rudder, hydraulically actuated; yaw damper; stall strips on leading-edges outboard of engines; two-section slotted Fowler flaps. Digital AFCS.

STRUCTURE: Fuselage near-circular section, flush riveted and pressurised; adhesive bonded stringers and cutout reinforcements; wing leading-edge, radome, nose bay, wing/fuselage and wingtip fairings, dorsal fin, fin leading-edge, fin/tailplane fairings, tailplane leading-edges, elevator tips, flap shrouds, flap trailing-edges and other components of Kevlar and Nomex; wing has tip-to-tip torsion box. Wheel doors of Kevlar and other composites.

LANDING GEAR: Retractable tricycle type, by Dowty Aerospace, with twin wheels on each unit. Steer-by-wire nose unit retracts forward, main units rearward into engine nacelles. Goodrich mainwheels and brakes; Hydro-Aire Mk 3 anti-skid system. Tyres 26.5×8.0-13 (12 ply) or high flotation H31×9.75-13 (10 ply) on main units; 18×5.5 (10 ply) or 22×6.50-10 (12 ply) on nose unit. Standard tyre pressures: main 9.03 bars (131 lb/sq in), nose 5.52 bars (80 lb/sq in). Low-pressure tyres optional, pressure 5.31 bars (77 lb/sq in) on main units, 3.31 bars (48 lb/sq in) on nose unit.

POWER PLANT: *Q100*: Two 1,491 kW (2,000 shp) Pratt & Whitney Canada PW 120A turboprops, driving Hamilton Sundstrand 14SF-7 four-blade constant-speed fully feathering aluminium/glass fibre propellers with reverse pitch, standard on basic weight version. Two 1,603 kW (2,150 shp) PW 121s optional on this version and standard on High Gross Weight version.

Q200: Two 1,603 kW (2,150 shp) PW 123C/D turboprops, driving Hamilton Sundstrand 14SF-23 propellers; PW 123C is flat-rated for full power at up to 26°C; PW 123D maintains same power at up to 45°C.

Standard usable fuel capacity (in-wing tanks) 3,160 litres (835 US gallons; 695 Imp gallons); optional auxiliary tank system increases this to 5,700 litres (1,506 US gallons; 1,254 Imp gallons). Pressure refuelling point in rear of starboard engine nacelle; overwing gravity point in each outer wing panel. Oil capacity 21 litres (5.5 US gallons; 4.6 Imp gallons) per engine.

ACCOMMODATION: Crew of two on flight deck, plus cabin attendant. Dual controls standard. Standard commuter layout provides four-abreast seating, with central aisle, for 37 passengers at 79 cm (31 in) pitch, plus buffet, lavatory and large rear baggage compartment. Wardrobe at front of passenger cabin, in addition to overhead lockers and underseat stowage, provides additional carry-on baggage capacity. Alternative 39-passenger, passenger/cargo or corporate layouts available at customer's option. Movable bulkhead to facilitate conversion to mixed traffic. Port side airstair door at front for crew and passengers; large inward-opening port side door aft of wing for cargo loading. Emergency exit each side, in line with wing leading-edge, and opposite passenger door on starboard side. Entire accommodation pressurised and air conditioned.

SYSTEMS: Pressurisation system with maximum differential 0.38 bar (5.5 lb/sq in). Normal hydraulic installation comprises two independent systems, each having an engine-driven variable displacement pump and an electrically driven standby pump; accumulator and hand pump for emergency use. Electrical system DC power provided by two starter/generators, two transformer-rectifier units, and two Ni/Cd batteries. Variable frequency AC power provided by two engine-driven AC generators and three static inverters. Ground power receptacles in port side of nose (DC) and rear of starboard nacelle (AC). Rubber-boot de-icing of wing, tailplane and fin leading-edges and nacelle intakes by pneumatic system; electric de-icing of propeller blades, pitot static ports, stall warning transducer, engine intake adaptor and elevator horn leading-edge. APU standard in corporate version. Simmonds fuel monitoring system.

AVIONICS: Honeywell Gold Crown III com/nav; Rockwell Collins equipment alternative option.

Comms: KTR 908 VHF com and KXP 756 transponder. Avtech audio integrating system. Telephonics PA system.

Radar: Primus P660 colour weather radar.

Flight: KNR 806 ADF, KDM 706A DME. Honeywell SPZ-8000 dual-channel digital AFCS with integrated fail-operational flight director/autopilot system, dual digital air data system. Optional FMS and GPS.

Instrumentation: Honeywell EFIS. Hughes/Flight Dynamics HGS 2000 head-up guidance system for Cat. IIIa operation optional.

DIMENSIONS, EXTERNAL:
Wing span	25.91 m (85 ft 0 in)
Wing aspect ratio	12.4
Length overall	22.25 m (73 ft 0 in)
Fuselage: Max diameter	2.69 m (8 ft 10 in)
Height overall	7.49 m (24 ft 7 in)
Elevator span	7.92 m (26 ft 0 in)
Wheel track (c/l of shock-struts)	7.87 m (25 ft 10 in)
Wheelbase	7.95 m (26 ft 1 in)
Propeller diameter	3.96 m (13 ft 0 in)
Propeller ground clearance	0.94 m (3 ft 1 in)
Propeller fuselage clearance	0.76 m (2 ft 6 in)
Passenger/crew door (fwd, port):	
Height	1.66 m (5 ft 5½ in)
Width	0.76 m (2 ft 6 in)
Height to sill	1.09 m (3 ft 7 in)
Baggage door (rear, port): Height	1.52 m (5 ft 0 in)
Width	1.27 m (4 ft 2 in)
Height to sill	1.09 m (3 ft 7 in)

DIMENSIONS, INTERNAL:
Cabin (excl flight deck): Length	9.14 m (30 ft 0 in)
Max width	2.49 m (8 ft 2 in)
Width at floor	2.03 m (6 ft 8 in)
Max height	1.96 m (6 ft 5 in)
Floor area	23.60 m² (254.0 sq ft)
Volume	37.6 m³ (1,328 cu ft)
Baggage compartment volume	8.5 m³ (300 cu ft)

AREAS:
Wings, gross	54.35 m² (585.0 sq ft)
Ailerons (total)	1.12 m² (12.1 sq ft)
Fin	9.81 m² (105.6 sq ft)
Rudder	4.31 m² (46.4 sq ft)
Tailplane	8.97 m² (96.5 sq ft)
Elevators (total)	4.97 m² (53.5 sq ft)

WEIGHTS AND LOADINGS:
Operating weight empty: Q100	10,406 kg (22,941 lb)
Q200	10,486 kg (23,117 lb)
Max usable fuel: standard	2,576 kg (5,678 lb)
optional	4,647 kg (10,244 lb)
Max payload: Q100	4,109 kg (9,059 lb)
Q200	4,211 kg (9,283 lb)
Payload with max fuel: Q100	3,484 kg (7,681 lb)
Q200	3,404 kg (7,505 lb)
Max T-O weight: Q100	15,649 kg (34,500 lb)
Q100 HGW, Q200	16,465 kg (36,300 lb)
Max landing weight: Q100	15,377 kg (33,900 lb)
Q200	15,649 kg (34,500 lb)
Max zero-fuel weight:	
Q100, Q200 JAA	14,515 kg (32,000 lb)
Q200	14,696 kg (32,400 lb)
Max wing loading: Q100	287.9 kg/m² (58.97 lb/sq ft)
Q100 HGW, Q200	303.0 kg/m² (62.05 lb/sq ft)
Max power loading: Q100	5.25 kg/kW (8.63 lb/shp)
Q100 HGW, Q200	5.14 kg/kW (8.44 lb/shp)

PERFORMANCE (at 95% standard MTOW, except where indicated):
Max cruising speed: Q100	265 kt (491 km/h; 305 mph)
Q100 HGW	270 kt (500 km/h; 311 mph)
Q200	290 kt (537 km/h; 334 mph)
Stalling speed, flaps down: all	72 kt (134 km/h; 83 mph)
Max rate of climb at S/L: all	450 m (1,475 ft)/min
Max certified altitude: all	7,620 m (25,000 ft)
Service ceiling: Q100	4,503 m (14,775 ft)
Q100 HGW	5,105 m (16,750 ft)
Q200	4,938 m (16,200 ft)
FAR Pt 25 T-O field length:	
Q100 (PW 121)	991 m (3,250 ft)
Q200	1,000 m (3,280 ft)
FAR Pt 25 landing field length at max landing weight:	
Q100	785 m (2,575 ft)
Q200	780 m (2,560 ft)
Range with 37 passengers:	
Q100	1,020 n miles (1,889 km; 1,173 miles)
Q200	925 n miles (1,713 km; 1,064 miles)

OPERATIONAL NOISE LEVELS: (FAR Pt 36 Stage 3 and ICAO Annex 16):
T-O	80.5 EPNdB
Sideline	85.6 EPNdB
Approach	94.8 EPNdB

UPDATED

DHC-8 DASH 8 Q300

TYPE: Twin-turboprop airliner.

PROGRAMME: Dash 8 Series 300 announced mid-1985 as stretch of Series 200; launched March 1986; first flight (modified Series 100 prototype C-GDNK) 15 May 1987; Canadian DoT certification 14 February 1989; first delivery (Time Air) 27 February 1989; FAA type approval 8 June 1989; now also certified in Argentina, Australia, Austria, Brazil, China, Colombia, Germany, Ireland, Italy, Jordan, Malaysia, Maldives, Netherlands, Norway, South Africa, Taiwan, Thailand and Zambia. Low-noise Dash 8 Q (see Series 200 for details) became standard version from 1996. A Q300 delivered to Air Nippon on 6 March 2001 was the 600th Dash 8.

CURRENT VERSIONS: **Series 300, 300A, 300B and 300E**: Initial versions, described in 1998-99 and earlier editions.

Q300: Introduced 1998; differs from Q200 in having extended wingtips; 3.43 m (11 ft 3 in) two-plug fuselage extension giving standard seating for 50 at 81 cm (32 in) pitch or 56 at 74 cm (29 in) pitch, plus second cabin attendant; also larger galley, galley service door, additional wardrobe, larger lavatory, dual air conditioning packs and optional Turbomach T-40 APU; powered by 1,775 kW (2,380 shp) P&WC PW123s driving Hamilton Sundstrand 14SF-23 four-blade propellers standard in basic version; 1,864 kW (2,500 shp) PW 123Bs optional in basic and standard in high gross weight (HGW) versions provide increased mechanical power for improved take-off performance in low and cold conditions; optional 1,775 kW (2,380 shp) PW 123Es provide 5 per cent increase in thermodynamic power up to 40°C (96°F) for improved hot-and-high performance; fuel capacity as Q200; tyre pressures increased (mainwheels 6.69 bars; 97 lb/sq in, nosewheels 4.14 bars; 60 lb/sq in). NVS system standard on all aircraft produced from second quarter 1996.

CUSTOMERS: Firm orders for 199 by 1 December 2000, of which 166 then delivered. Recent customers include

Dash 8 Q300 of Austrian operator, Rheintalflug *(Paul Jackson/Jane's)*

Sextant five-screen suite of the Dash 8 Q400

Representative flight deck of the Dash 8 Series 100/200/300

Eastern Australia Airlines (two), Air Senegal (one), and UNI Airways (one).

COSTS: Air Senegal order valued at US$14.3 million (2000).

DIMENSIONS, EXTERNAL: As for Q200 except:
Wing span	27.43 m (90 ft 0 in)
Wing aspect ratio	13.4
Length overall	25.68 m (84 ft 3 in)
Wheelbase	10.01 m (32 ft 10 in)

DIMENSIONS, INTERNAL: As for Q200 except:
Cabin (excl flight deck): Length	12.65 m (41 ft 6 in)
Floor area	30.57 m² (329.0 sq ft)
Volume	52.0 m³ (1,838 cu ft)
Baggage compartment volume:	
with 50 passengers	9.1 m³ (320 cu ft)
with 56 passengers	7.9 m³ (280 cu ft)

AREAS:
Wings, gross	56.21 m² (605.0 sq ft)
Ailerons (total)	1.87 m² (20.18 sq ft)
Tail surfaces	as for Series 100A/200A

WEIGHTS AND LOADINGS:
Operating weight empty: basic	11,709 kg (25,814 lb)
HGW	11,791 kg (25,995 lb)
Max usable fuel: standard	2,576 kg (5,679 lb)
optional	4,646 kg (10,243 lb)
Max payload: basic	5,166 kg (11,386 lb)
HGW	6,126 kg (13,505 lb)
Payload with max fuel	5,138 kg (11,327 lb)
Max T-O weight: basic	18,642 kg (41,100 lb)
HGW	19,504 kg (43,000 lb)
Max landing weight: basic	18,144 kg (40,000 lb)
HGW	19,050 kg (42,000 lb)
Max zero-fuel weight: basic	16,873 kg (37,200 lb)
HGW	17,917 kg (39,500 lb)
Max wing loading: basic	331.7 kg/m² (67.93 lb/sq ft)
HGW	347.0 kg/m² (71.07 lb/sq ft)
Max power loading: basic	5.25 kg/kW (8.63 lb/shp)
HGW	5.49 kg/kW (9.03 lb/shp)

PERFORMANCE (at 95% standard MTOW except where indicated):
Max cruising speed: basic	287 kt (532 km/h; 330 mph)
HGW	285 kt (528 km/h; 328 mph)
Stalling speed, flaps down	77 kt (141 km/h; 88 mph)
Max rate of climb at S/L	549 m (1,800 ft)/min
Rate of climb at S/L, OEI	137 m (450 ft)/min
Max certified altitude	7,620 m (25,000 ft)
Service ceiling, OEI: basic	4,145 m (13,600 ft)
HGW	3,734 m (12,250 ft)
FAR Pt 25 T-O field length at S/L, ISA, 15° flap:	
basic	1,097 m (3,600 ft)
HGW	1,180 m (3,870 ft)
FAR Pt 25 landing field length at max landing weight:	
basic	1,010 m (3,315 ft)
HGW	1,040 m (3,415 ft)
Range, HGW with 50 passengers:	
standard fuel	841 n miles (1,557 km; 967 miles)
auxiliary fuel	1,098 n miles (2,033 km; 1,263 miles)

OPERATIONAL NOISE LEVELS: (FAR Pt 36):
T-O	79.5 EPNdB
Sideline	87 EPNdB
Approach	93.3 EPNdB

UPDATED

DHC-8 DASH 8 Q400

TYPE: Twin-turboprop airliner.

PROGRAMME: Launched June 1995 as stretch of Series 300; component manufacture began December 1996; engine first test flown in January 1997, mounted on nose of Pratt & Whitney Canada Boeing 720B testbed; roll-out 21 November 1997, first flight (C-FJJA) 31 January 1998; five aircraft participated in the 1,900 hour, 1,400 sortie flight test programme, based at the Bombardier Flight Test Centre in Wichita, Kansas, (see 2000-01 edition), leading to Transport Canada CAR 525 Amendment 86 certification on 14 June 1999, JAA approval in December 1999 and FAA FAR Pt 25 approval on 8 February 2000. First delivery (OY-KCA) to SAS Commuter 20 January 2000, followed by service entry 7 February on Copenhagen, Denmark, to Poznan, Poland route. Q400 fleet flight time exceeded 4,000 hours on 17 July 2000, including 2,212 flight hours and 3,543 flights by the six aircraft then in service with SAS Commuter. Mitsubishi joined programme as risk-sharing partner in October 1995, with responsibility for design, development and manufacture of fuselage and tail sections.

CUSTOMERS: Total of 68 firm orders by 1 December 2000, at which time 17 had been delivered; see table. Launch customer was Great China Airlines (now UNI Airways), which ordered six (since reduced to three) in February 1996. US launch customer Horizon Air, which ordered 15 with 15 options, in June 1999, took delivery of its first aircraft on 8 January 2001. First for Changan handed over 27 October 2000.

DASH 8Q-400 PRODUCTION
(at 1 December 2000)

Customer	Ordered	Delivered	Backlog
Augsburg Airways	5		5
British European	4		4
Changan Airlines	3	2	1
Horizon Air	15		15
SAS Commuter	28	11	17
Tyrolean	6	4	2
UNI Airways	6		6
Widerøe	1		
Totals	68	17	51

COSTS: June 1999 order for 15 aircraft for Horizon Air valued at US$321 million.

DESIGN FEATURES: Compared with Q300, fuselage stretched 6.83 m (22 ft 5 in) to seat up to 78. Other new features include engines; propeller; tapered inboard wing section increasing propeller/fuselage clearance to 1.09 m (3 ft 7 in); revised wing/fuselage fairings; ailerons, elevators and fin cap fairing; landing gear; baggage/service doors; upgraded avionics. Airframe designed for crack-free life of 40,000 flight hours/80,000 flight cycles and an economic life of 80,000 flight hours/160,000 flight cycles. Common type rating with Q100/200/300.

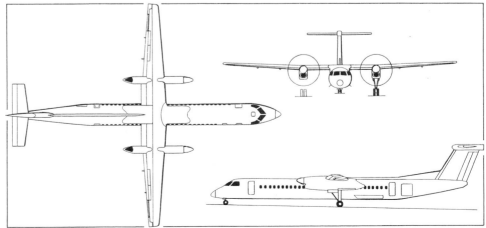

General arrangement of the Dash 8 Q400 (Paul Jackson/Jane's)

CANADA: AIRCRAFT—BOMBARDIER (de Havilland)

Dash 8 Q400 in de Havilland house colours

STRUCTURE: Generally as for Q100 and Q200. De Havilland Canada manufactures cockpit section and wing and performs final assembly; other participants in programme include AlliedSignal (electrical power system); Goodrich (brakes); Dowty Aerospace (propellers); Menasco (landing gear); Micro Technica (flap system); Mitsubishi (fuselage and tail sections); Parker Bertea Aerospace (hydraulics and fuel system); Pratt & Whitney Canada (engines); Sextant Avionique (avionics system); and Shorts (engine nacelles).

POWER PLANT: Two Pratt & Whitney Canada PW150A turboprops with FADEC, each flat rated to 3,781 kW (5,071 shp) at 37.4°C, driving Dowty R408 six-blade, slow-turning (1,020 rpm for T-O; 850 rpm cruise) composites propellers giving 15 per cent reduction in T-O rpm and 6 to 19 per cent reduction in cruise rpm over Q100/300. Fuel capacity 6,708 litres (1,772 US gallons; 1,475 Imp gallons).

ACCOMMODATION: Variety of cabin configurations providing four-abreast seating, with central aisle, for 70 passengers at 84 cm (33 in) pitch. 74 passengers at 79 cm (31 in) pitch, 78 passengers at 76 cm (30 in), or two-class layout for 10 business class passengers at 86 cm (34 in) pitch and 62 economy class passengers at 79 cm (31 in); NVS system standard.

AVIONICS: Prime contractor, Sextant Avionique.
 Comms: Dual VHF nav/com and mode S transponder; solid-state cockpit voice recorder; ELT antenna.
 Radar: Weather radar.
 Flight: GPWS, radar altimeter, ADF and DME. Provision for FMS, GPS and TCAS II.
 Instrumentation: EFIS employs five 152×203 mm (6×8 in) LCDs. Optional Flight Dynamics holographic guidance system.

DIMENSIONS, EXTERNAL:
Wing span	28.42 m (93 ft 3 in)
Wing aspect ratio	12.8
Length overall	32.84 m (107 ft 9 in)
Fuselage max diameter	2.69 m (8 ft 10 in)
Height overall	8.36 m (27 ft 5 in)
Tailplane span	9.27 m (30 ft 5 in)
Wheel track	8.79 m (28 ft 10 in)
Wheelbase	13.94 m (45 ft 9 in)
Propeller diameter	4.11 m (13 ft 6 in)
Propeller ground clearance	0.97 m (3 ft 2 in)
Propeller fuselage clearance	1.10 m (3 ft 7¼ in)
Passenger/crew door (port, fwd):	
Height	1.66 m (5 ft 5½ in)
Width	0.76 m (2 ft 6 in)
Height to sill	1.17 m (3 ft 10 in)
Passenger/crew door (port, rear):	
Height	1.73 m (5 ft 8 in)
Width	0.71 m (2 ft 4 in)
Height to sill	1.55 m (5 ft 1 in)
Baggage door (port, rear):	
Height	1.55 m (5 ft 1 in)
Width	1.27 m (4 ft 2 in)
Height to sill	1.55 m (5 ft 1 in)
Baggage door (stbd, fwd):	
Height	1.37 m (4 ft 6 in)
Width	0.61 m (2 ft 0 in)
Height to sill	1.17 m (3 ft 10 in)
Service door (stbd, aft):	
Height	1.42 m (4 ft 8 in)
Width	0.69 m (2 ft 3 in)
Height to sill	1.55 m (5 ft 1 in)

DIMENSIONS, INTERNAL:
Cabin (excl flight deck):	
Length	18.80 m (61 ft 8 in)
Max width	2.49 m (8 ft 2 in)
Max height	1.95 m (6 ft 5 in)
Floor area	43.6 m² (469 sq ft)
Volume	77.6 m³ (2,740 cu ft)
Baggage volume:	
forward	2.58 m³ (91.0 cu ft)
aft	11.6 m³ (411 cu ft)
total	14.2 m³ (502 cu ft)

AREAS:
Wings, gross	63.08 m² (679.0 sq ft)
Ailerons (total)	1.87 m² (20.18 sq ft)
Vertical tail surfaces (total)	14.12 m² (152.0 sq ft)
Horizontal tail surfaces (total)	16.72 m² (180.0 sq ft)

WEIGHTS AND LOADINGS:
Operating weight empty	17,108 kg (37,717 lb)
Max payload	8,747 kg (19,283 lb)
Payload with max fuel	6,831 kg (15,059 lb)
Max T-O weight: basic	27,329 kg (60,250 lb)
intermediate	27,995 kg (61,720 lb)
HGW	29,256 kg (64,500 lb)
Max landing weight	28,009 kg (61,750 lb)
Max zero-fuel weight	25,855 kg (57,000 lb)
Max wing loading: basic	433.2 kg/m² (88.73 lb/sq ft)
intermediate	443.8 kg/m² (90.90 lb/sq ft)
HGW	463.8 kg/m² (94.99 lb/sq ft)
Max power loading: basic	3.62 kg/kW (5.94 lb/shp)
intermediate	3.70 kg/kW (6.09 lb/shp)
HGW	3.87 kg/kW (6.36 lb/shp)

PERFORMANCE (estimated, at 95% of basic MTOW, except where indicated):
Max cruising speed	350 kt (648 km/h; 403 mph)
Max certified altitude: standard	7,620 m (25,000 ft)
optional	8,230 m (27,000 ft)
Service ceiling, OEI:	
standard	6,250 m (20,500 ft)
optional	5,790 m (19,000 ft)
FAR T-O field length at S/L, MTOW	1,193 m (3,915 ft)
FAR landing field length at S/L, MLW	1,287 m (4,221 ft)
Max range, 70 passengers, IFR reserves	1,360 n miles (2,518 km; 1,565 miles)

OPERATIONAL NOISE LEVELS (estimated, FAR Pt 36):
T-O	78.3 EPNdB
Sideline	84 EPNdB
Approach	94.8 EPNdB

DASH 8 PRODUCTION
(at 1 December 2000)

Customer	Orders	Delivered	Backlog
Abu Dhabi Aviation	2	2	
AGES/Worldwide	6	6	
Air Atlantic	15	15	
AirBC	18	18	
Air Creebec	1	1	
Air Dolomiti	3	3	
Air Maldives	1	1	
Air Manitoba	1	1	
Air Nippon	3		3
Air Niugini	2	2	
Air Nostrum	29	–	29
Air Nova	6	6	
Air Ontario	36	36	
Air Wisconsin	12	12	
Alberta government	1	1	
ALM	2	2	
Amakusa	1	1	
America West	12	12	
Ansett New Zealand	2	2	
Augsburg Airways	14	9	5
Australian Airlines	4	4	
Aviaco/Furlong	4	4	
Avline	3	3	
Bahamasair	3	3	
BPX Colombia	1	1	
British European	11	7	4
Brymon Airways	10	10	
BWIA	3	3	
CCAir	2	2	
Changan Airlines	3	2	1
City Express	4	4	
Contact Air	6	6	
DAC Air Romania	1	1	
DND Canada	6	6	
Eastern Australia	1	1	
Eastern Metro Express	8	8	
Elveden Investments	4	4	
Fairways Corp	1	1	
GPA Jetprop	30	30	
Hamburg Airlines	4	4	
Horizon Air	64	49	15
Interot	3	3	
LADS	1	1	
Liat	8	8	
MarkAir	2	2	
Mesa Air	12	12	
Mexican Navy	1	–	1
Midroc Aviation	2	2	
Mobil Oil	1	1	
Nagasaki Airways	2		2
National Jet	7	7	
Norfolk Airlines	1	1	
norOntair	2	2	
Northwest Airlines	25	25	
Norwegian CAA	1	1	
Palestinian	2	2	
Qantas	2	2	
Rheintalflug	6	6	
Royal Wings	1	1	
Ryukyu Air Commuter	4	3	1
Saega Airlines	2	2	
SAS Commuter	28	11	17
Saudi Aramco	3	3	
Schreiner Airways	3	3	
South African Express	12	12	
Talair	2	2	
TAVAJ	2	2	
Time Air	14	14	
Transport Canada	2	2	
Tyrolean	36	34	2
UNI Airways	26	19	7
USAF/Sierra	2	2	
US Airways*	75	75	
Widerøe	20	19	1
Zhejiang Airlines	2	2	
Undisclosed	8	6	
Totals	**660**	**570**	**90**

Notes: Operators with leased Dash 8 aircraft include, but are not necessarily restricted to, the following: Air Alliance, ERA, Lloyd Aviation, Lufthansa CityLine, Sabena, TABA.
* US Airways operators include Piedmont Airlines and Allegheny Commuter.
UNI Airways received initial aircraft when named Great China Airlines.
British European received initial aircraft when named Jersey European Airways.

Dash 8 Q400 in 74-seat configuration
A: airstair/Type I exit, A2: passenger door/Type I exit, B: baggage, BD: baggage door, F: folding seat, G: galley, L: lavatory, S: service door/Type I exit, TI/TII: Type I/Type II exits, W: wardrobe

SUMMARY

	Delivered	Backlog	Total
Series 100	298	1	299
Series 200	89	5	94
Series 300	166	33	199
Series 400	17	51	68
pugTotals	570	90	660

UPDATED

DHC-8 DASH 8M and TRITON
Canadian Forces designations: CC-142 and CT-142
US Air Force designation: E-9A
TYPE: Multisensor surveillance twin-turboprop.
CURRENT VERSIONS: **CC-142 and CT-142**: Passenger/cargo transport and navigation trainer versions of Dash 8M-100 respectively; both have long-range fuel tanks, low-pressure tyres, high-strength floors and mission avionics; Canadian Department of National Defence operates two CC-142s and four CT-142s, latter with mapping radar in extended nose.

E-9A: US Air Force Dash 8-100 as missile range control aircraft; relays telemetry, voice and drone and fighter control data and simultaneously observes range with radar; avionics include large electronically steered phased-array radar in fuselage side and AN/APS-128D surveillance radar in ventral dome.

Aerial Survey/Coastal Patrol: LADS Corporation of Adelaide, Australia, operates a Series 200 equipped with a Visions System Group laser airborne depth sounder (LADS) for hydrographic survey of shallow coastal waters. Canadian Coast Guard aircraft are equipped with environmental monitoring equipment installed in fairing on starboard side of fuselage.

Surveillance Australia of Adelaide took delivery in mid-2000 of first of two Q200s that it operates for maritime patrol on behalf of the Australian Customs Service's Coastwatch. The Q200s, which supplement three earlier Series 200s Surveillance Australia has operated since 1996, were modified for the maritime patrol role by Field Aviation Inc of Toronto, which installed a belly-mounted Raytheon SV1022 search radar, Wescam 16DS turret containing FLIR and daylight TV camera, large observation window inserts at each of the mid-fuselage emergency exits, and a systems operator's console in the main cabin. In Coastwatch configuration the Q200 can transit 300 n miles (556 km;

CT-142 navigation trainer of No. 402 Squadron, Canadian Forces *(Paul Jackson/Jane's)*

345 miles) at maximum cruising speed, fly a low-level search track of about 1,000 n miles (1,852 km; 1,150 miles) covering some 260,000 km² (100,390 sq miles) per mission, and return to base with adequate fuel reserves.

Triton: Maritime surveillance version; promotion began in 1997 of the Triton 300, based on the Series 300 airframe, but with weapon pylons on the lower fuselage and provision for ventral radar, MAD tail and wingtip ESM.

CUSTOMERS: Norwegian CAA (one Dash 8M-100), Canadian Department of Transport (two Dash 8M-100), Canadian Forces (two CC-142, four CT-142, all with No. 402 Squadron at Winnipeg), USAF (two E-9A at Tyndall AFB, Florida); all 11 included in Series 100/100A delivery total.
Data follow for Triton 300 where different from Dash 8 Series 300.
WEIGHTS AND LOADINGS:
Max fuel weight	4,647 kg (10,245 lb)
Max T-O weight	19,504 kg (43,000 lb)
Max ramp weight	19,595 kg (43,200 lb)
Max landing weight	19,051 kg (42,000 lb)
Max zero-fuel weight	17,916 kg (39,500 lb)

PERFORMANCE:
Econ cruising speed	240 kt (444 km/h; 276 mph)
Patrol speed	135 kt (250 km/h; 155 mph)
Endurance	9 h

UPDATED

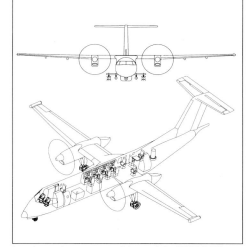

Interior and weapon positions of the DHC-8 Triton 300 maritime patrol aircraft

---

CAG

CANADIAN AEROSPACE GROUP INTERNATIONAL
5353 John Lucas Drive, Burlington, Ontario L7L 6A8
Tel: (+1 905) 331 03 55
Fax: (+1 905) 331 03 56
e-mail: pnelso2@ibm.net
Web: http://www.canadianaerospacegroup.com
CHAIRMAN: Joseph Rotolo
PRESIDENT: Philip A Nelson
DIRECTOR, FLIGHT OPERATIONS: Wayne Thompson
DIRECTOR, MARKETING: Thomas Bunker

Under a joint venture announced in the early months of 1998, CAG subsidiary Panda Aircraft of North Bay, Ontario, will receive bare HAMC Y-12 (IV) (see Chinese section) airframes shipped from China and will equip them with 559 kW (750 shp) PT6A-34 engines, instrumentation and cabin furnishing. The aircraft will be marketed for approximately US$3 million, under the export name **Twin Panda** and the fact that landing gear will also be installed in Canada lends strength to the intention to offer a floatplane version as well as the standard wheeled landing gear.

The initial agreement covers 50 Y-12s in the first three years with options for up to 150 more later.

The first airframe was due to be delivered to Panda in the second quarter of 1999 and the first two Twin Pandas (for an undisclosed customer) were to be ready for delivery, subject to recertification, by the end of the year. However, the programme was rescheduled, providing for one basic Y-12, plus a demonstrator, to be shipped to Canada in early 2000, with delivery of the first (of five) to South African operator Interdoc Aerospace in mid-year. No confirmation of these deliveries has been received. Deliveries of the PT6A-34-powered version were targeted to begin in January 2001 to Air Group of Brazil, which has ordered 30. The Brazilian contract is reportedly worth C$94.5 million, and will spread deliveries over a two-year period.

Under a second agreement, announced in April 1998, CAG obtained for its shareholder company, Can-Aero Leasing, the rights to assemble and complete at St Hubert, Quebec, (including new engine, landing gear and avionics installations) up to 240 NAMC N-5A agricultural aircraft, now known as the **Hongdu AG Dragon**; Can-Aero was to form AG Dragon Aircraft Corporation for this purpose. In return, NAMC has the option of producing in China the Windeagle light aircraft and/or the **Monitor Jet**, a turbofan trainer based on the Bede BD-10 but entirely redesigned by a CAG team and also offered as a UAV with SL Ventures guidance system. In December 1999, coproduction of the Monitor Jet was offered to Portugal as part of Sikorsky's bid to sell that country S-92 Helibus helicopters.

Windeagle Aircraft Corporation of Canada acquired from Composite Aircraft Corporation of Texas, USA, all rights to the Windecker Eagle all-composites light aircraft, last described (under Composite Aircraft Company) in the 1984-85 edition of *Jane's*. CAG planned to resume production of the aircraft, renamed **Windeagle** and powered by a 212 kW (285 hp) Teledyne Continental IO-520C engine, in 1997, at a price of US$198,000. Confirmation of new production had not been forthcoming by late 2000. Proposed future developments include a turboprop version powered by a 335 kW (450 shp) Pratt and Whitney Canada PT6A and flight tests with a 447 kW (600 hp) Orenda V-8 engine. By mid-1999 CAG had received orders for 41 Windeagles, including 15 turboprop models for Interdoc Aerospace of South Africa, valued at US$9 million, since when no further details have been received.

CAG has also acquired rights to the Timberwolf, a Windeagle derivative optimised for military training. This was last described in the 1995-96 *Jane's* under the Sino-Canadian partnership Venga/Baoshan.

UPDATED

CANADA AIR RV

CANADA AIR RV
Hangar 11, 11760 109th Street, Edmonton, Alberta T5G 2T8
Tel: (+1 780) 479 23 46
Fax: (+1 780) 479 10 02
POSTAL ADDRESS: PO Box 34013, Kingsway Mall, Edmonton, Alberta T5G 3G4

The Griffin is this company's sole current product; in late 2000, Canada Air RV was reportedly attempting to raise finance to continue production.

NEW ENTRY

CANADA AIR RV GRIFFIN
TYPE: Side-by-side kitbuilt.
CURRENT VERSIONS: **Griffin**: Baseline version, *as described*.
Griffin Mk III: Lower-powered version.
CUSTOMERS: Eight flying by mid-2000.
COSTS: Kit US$15,874 excluding engine (2000).
DESIGN FEATURES: Good all-round visibility enhanced by instrument panel positioned low in cockpit.
FLYING CONTROLS: Manual. Full-span three-position flaperons deflect to 20° and also reflex slightly for cruise drag reduction. Electrically operated full-span trim on port elevator; ground-adjustable tab on starboard wing.
STRUCTURE: Metal and fabric construction. Single-strut laminar flow high-aspect ratio wing. Optional winglets. Low-mounted tailplane.
LANDING GEAR: Choice of tricycle or tailwheel layout; float/ski attachment points standard. Sprung steel mainwheel legs. Tricycle model has oleo-pneumatic nosewheel suspension. Tailwheel version has steerable tailwheel. Optional speed fairings.
POWER PLANT: *Griffin*: One 119 kW (160 hp) Textron Lycoming O-320 flat-four engine, driving a two-blade fixed-pitch wooden propeller in demonstrator; engines in range 48.5 to 119 kW (65 to 160 hp) can be fitted; Griffin Mk III has 74.6 kW (100 hp) CAM 100 piston engine; mounts for Subaru engines under development. Fuel capacity 140 litres (37.0 US gallons; 30.8 Imp gallons).
ACCOMMODATION: Pilot and passenger in side-by-side seating. Upward-opening door on each side of fuselage.
DIMENSIONS, EXTERNAL:
Wing span	10.82 m (35 ft 6 in)
Wing aspect ratio	11.0
Length overall	6.25 m (20 ft 6 in)
Height overall	2.44 m (8 ft 0 in)

DIMENSIONS, INTERNAL:
Cabin max width	1.19 m (3 ft 11 in)

AREAS:
Wings, gross	10.68 m² (115.0 sq ft)

WEIGHTS AND LOADINGS (Griffin with 119 kW; 160 hp engine, Griffin Mk III with 74.6 kW; 100 hp engine):
Weight empty	381 kg (840 lb)
Max T-O weight: Griffin	787 kg (1,735 lb)
Griffin Mk III	680 kg (1,500 lb)

CANADA: AIRCRAFT—CANADA AIR RV to DIAMOND

Max wing loading: Griffin 73.7 kg/m² (15.09 lb/sq ft)
Griffin Mk III 63.7 kg/m² (13.04 lb/sq ft)
Max power loading: Griffin 6.60 kg/kW (10.84 lb/hp)
Griffin Mk III 9.13 kg/kW (15.00 lb/hp)
PERFORMANCE (Griffin with 119 kW; 160 hp engine; Griffin Mk III with 74.6 kW; 100 hp engine):

Max operating speed (V_{MO}):
Griffin 169 kt (313 km/h; 195 mph)
Griffin Mk III 115 kt (214 km/h; 133 mph)
Max cruising speed at 75% power:
Griffin 130 kt (241 km/h; 150 mph)
Griffin Mk III 105 kt (195 km/h; 121 mph)
Stalling speed: both 40 kt (73 km/h; 45 mph)

Max rate of climb at S/L 259 m (850 ft)/min
Service ceiling 5,485 m (18,000 ft)
T-O run 76 m (250 ft)
Landing run 137 m (450 ft)
Range with max fuel 828 n miles (1,533 km; 952 miles)

NEW ENTRY

CUSTOM FLIGHT

CUSTOM FLIGHT COMPONENTS LTD
RR1, Perkinsfield, Ontario L0L 2J0
Tel: (+1 705) 526 96 26
Fax: (+1 705) 526 25 29
PRESIDENT: Morgan Williams

Custom Flight Components produces the North Star and more basic Bright Star two-seat kitplanes, both modelled on the Piper Super Cub. The North Star wing can also be bought as a separate kit and retrofitted to Piper Super Cub and Aeronca Champ and Citabria aircraft. Completions continue at the rate of two or three per year.

VERIFIED

CUSTOM FLIGHT NORTH STAR
TYPE: Tandem-seat kitbuilt.
PROGRAMME: Prototype North Star (C-FGYD) first flew in 1991 and made its public debut that year. Offered in kit form from 1992 onwards.
CURRENT VERSIONS: **North Star**: Standard version, *as described below*.
Bright Star: Basic version with no flaps, single one-piece door, leaf spring steel main landing gear and fuel capacity of 98 litres (26 US gallons; 21.6 Imp gallons); introduced in 1996; optional tricycle or tailwheel versions. Accepts engines from 74.6 kW to 112 kW (100 to 150 hp).
CUSTOMERS: Production aircraft Nos. 7, 8 and 16 completed in Ontario by D E Blackburn, A King and R M Harkness in 1998. Nos. 3 and 15, built by W Smith, Ontario, and J Y Gratton, Quebec, followed in the first half of 1999. Others produced in USA, including No. 12 in late 1999.
COSTS: North Star: approximately C$30,000; Bright Star: C$20,000 (1997).
DESIGN FEATURES: Strut-braced mainplanes; wire-braced tailplanes; raked wingtips. Easy maintenance/inspection covers. Quoted build time 1,000 hours.
FLYING CONTROLS: Conventional and manual. Four-position flaps.
STRUCTURE: Welded 4130 chromoly tube, fabric-covered fuselage. Tubes injected with oil as anti-corrosion measure. Wing of prototype constructed of wood, but kit aircraft have I-beam spar and aluminium ribs. Aerofoil USA 35B.
LANDING GEAR: Tailwheel configuration (fixed) with coil-sprung suspension. Steerable tailwheel. Optional floats and skis. Mainwheel tyres 8.50-6 in; tailwheel 2.80-4 in.

Prototype Custom Flight North Star two-seat lightplane

POWER PLANT: One 112 kW (150 hp) Textron Lycoming O-320-E3D flat-four driving a McCauley two-blade fixed-pitch propeller. Fuel capacity 197 litres (52.0 US gallons; 43.3 Imp gallons) in two wing tanks.
ACCOMMODATION: Pilot and passenger in tandem within enclosed cockpit. Door on each side – upper half Perspex and lower half fabric-covered steel tubing. Baggage compartment reached through separate door on starboard side of fuselage.
DIMENSIONS, EXTERNAL:
Wing span: North Star 11.07 m (36 ft 4 in)
Bright Star 10.97 m (36 ft 0 in)
Wing chord, average 1.60 m (5 ft 3 in)
Length overall: North Star 6.86 m (22 ft 6 in)
Bright Star 7.16 m (23 ft 6 in)
Height overall: North Star 2.08 m (6 ft 10 in)
Bright Star 2.29 m (7 ft 6 in)
Propeller diameter 2.08 m (6 ft 10 in)
AREAS:
Wings, gross: North Star 17.72 m² (190.8 sq ft)
Bright Star 17.65 m² (190.0 sq ft)
WEIGHTS AND LOADINGS:
Weight empty: North Star 531 kg (1,170 lb)
Bright Star 454 kg (1,000 lb)
Baggage capacity 41 kg (90 lb)
Max T-O weight: North Star 997 kg (2,200 lb)
Bright Star 816 kg (1,610 lb)
Max wing loading: North Star 56.3 kg/m² (11.53 lb/sq ft)
Bright Star 46.3 kg/m² (9.47 lb/sq ft)

Max power loading (112 kW; 150 hp engine):
North Star 8.93 kg/kW (14.67 lb/hp)
Bright Star 7.30 kg/kW (12.00 lb/hp)
PERFORMANCE:
Max level speed 122 kt (225 km/h; 140 mph)
Max cruising speed:
North Star 100 kt (185 km/h; 115 mph)
Bright Star 87 kt (161 km/h; 100 mph)
Normal cruising speed:
North Star 85 kt (158 km/h; 98 mph)
Stalling speed, power off:
flaps up 39 kt (73 km/h; 45 mph)
flaps down 35 kt (65 km/h; 41 mph)
Stalling speed, full power:
flaps up 35 kt (65 km/h; 41 mph)
flaps down 26 kt (49 km/h; 30 mph)
Max rate of climb at S/L:
North Star 335 m (1,100 ft)/min
Bright Star 305 m (1,000 ft)/min
Service ceiling 4,880 m (16,000 ft)
T-O run: North Star 55 m (180 ft)
Bright Star 69 m (225 ft)
Landing run: North Star 61 m (200 ft)
Bright Star 84 m (275 ft)
Range 521 n miles (965 km; 600 miles)

UPDATED

DIAMOND

DIAMOND AIRCRAFT CORPORATION
1560 Crumlin Sideroad, London, Ontario N5V 1S2
Tel: (+1 519) 457 40 00
Fax: (+1 519) 457 40 21
e-mail: sales@diamondair.com
Web: http://www.diamondair.com
CEO: Christian Dries
MARKETING DIRECTOR: Jeff Owen
TECHNICAL DIRECTOR: Peter Maurer

Diamond Aircraft Corporation established June 1994 at London, Ontario, to build DV 20 Katana light aircraft.

An experimental design, the DA 20-B1 (C-GKAT), was completed in June 1996, followed by the DA 20-C1 prototype in February 1997. Series production of the DA 20-C1 is currently under way, and the company has announced plans to manufacture the DA 40 Diamond Star (which see in Austrian section) in Canada. Workforce reduced from about 220 to 60 in March 2000, with reduction in production rate of DA 20-C1s, pending reorganisation of company.

UPDATED

DIAMOND DA 20-A1 KATANA
English name: Samurai Sword
TYPE: Two-seat lightplane.
PROGRAMME: See following entry for early details of Katana programme. Initial Canadian aircraft (C-FSQN) first flight 29 June 1994. First recipient, in March 1995, was Missouri State University; large orders include 42 for Spartan Flight School of Tulsa, Oklahoma. Diamond had built 331 Katana A1s in Canada up to October 1998, when manufacture of the variant was suspended. At least 20 were supplied to Diamond in Austria in 1998, being placed in storage to meet future European orders for Rotax-powered aircraft.
CURRENT VERSIONS: **DA 20-A1 Katana**: New-build version.

DV 20-A1 converted to Katana 100 *(Paul Jackson/Jane's)* 2001/0089474

Katana 100: Rebuild and upgrade offered by Diamond in Austria from 1999 and Diamond Aircraft in Canada from late 2000. Refitted with 74.6 kW (100 hp) Rotax 912S engine. Prototype OE-VPP (DV 20).
Further details generally as in following entry on modified C1 variant, but detailed description of -A1 in Austrian section of 1997-98 and earlier editions. Specific -A1 details below.
DESIGN FEATURES: Incidence 3°.
POWER PLANT: One 59.6 kW (79.9 hp) Rotax 912A-3 flat-four engine; 2.27 reduction gear to two-blade constant-speed Hoffmann HO-V72F-M/S propeller with composites blades. Fuel capacity 79 litres (20.9 US gallons; 17.4 Imp gallons), normal; 55 litres (14.5 US gallons; 12.1 Imp gallons), optional. Fuel grades 100LL or mogas (95 octane leaded/unleaded).
WEIGHTS AND LOADINGS (DA 20-A1 and Katana 100, where different):
Weight empty: A1 494 kg (1,090 lb)
100 520 kg (1,146 lb)
Max fuel 57 kg (125 lb)
Max T-O weight 730 kg (1,610 lb)
Max wing loading 62.9 kg/m² (12.89 lb/sq ft)
Max power loading: A1 12.25 kg/kW (20.13 lb/hp)
100 9.79 kg/kW (16.08 lb/hp)

PERFORMANCE:
Never-exceed speed (V_{NE}):
A1 161 kt (298 km/h; 185 mph)
100 157 kt (291 km/h; 180 mph)
Cruising speed at 75% power:
A1 119 kt (220 km/h; 137 mph)
100 129 kt (239 km/h; 149 mph)
Stalling speed: flaps up 41 kt (76 km/h; 48 mph)
flaps down 37 kt (69 km/h; 43 mph)
Max rate of climb at S/L: A1 207 m (680 ft)/min
100 288 m (945 ft)/min
Service ceiling: A1 4,270 m (14,000 ft)
100 more than 4,000 m (13,120 ft)
T-O to 15 m (50 ft): A1 488 m (1,600 ft)
100 244 m (800 ft)
Landing from 15 m (50 ft): A1 454 m (1,490 ft)
Range with max fuel:
A1 523 n miles (968 km; 601 miles)
100, with 45 min reserves
500 n miles (927 km; 576 miles)
g limits +4.4/−2.2
OPERATIONAL NOISE LEVELS:
T-O 65.2 EPNdB

UPDATED

DIAMOND DA 20-C1 KATANA

English name: Samurai Sword

TYPE: Two-seat lightplane.

PROGRAMME: Original Austrian-built Katana revealed following first flight on 16 March 1991 of proof-of-concept LF 2000 (OE-VPX); followed by LF 2 Katana predecessor (OE-CPU, first flight December 1991); DV 20 prototype (OE-AKL, first flight 17 December 1992) developed from this by Dries and Volck and was representative of production aircraft; Austrian and German certification 26 April 1993; JAR-VLA certification received later in 1993; now also certified by Canada, UK and USA (FAA). Second production line opened in Canada, initially for the DA 20-A1; European-built aircraft designated DV 20. European manufacture of the DV 20 ended in mid-1996.

The current C1 version, introducing minor airframe improvements, was announced at Aero '97 at Friedrichshafen, shortly after first flight of the Canadian prototype (C-FDVA); production of this version began in early 1998; exports of -C1s to USA began June 1998 with seventh production aircraft.

CURRENT VERSIONS: **DA 20-C1:** *As described*. The 100th C1 (C-GFLR) registered September 1999, although only further ten had been built by second quarter of 2000. Most to USA, but some to Canada and one each to New Zealand (in March 1999) and UK (July 1999).

DA 20-A2: Intended to have DA 20-C's improved airframe, but retain DA 20-A's 59.7 kW (80 hp) Rotax 912 flat-four engine. Not launched.

CUSTOMERS: Total of 600 Katanas of all versions delivered by early 2000 from Austrian and Canadian production. European production of DV 20s terminated with the 160th aircraft in July 1996.

DESIGN FEATURES: Conventional, mid-wing, T-tail monoplane of composites construction.

Wing section Wortmann FX-63-137/20; dihedral 4° at root, 27° at tip; sweepback 1°; incidence, 2° 30'. Fin sweepback 57°; tailplane incidence −2°.

FLYING CONTROLS: Conventional and manual. Ailerons and elevators are operated by push-pull tubes, rudder by cables; electrically actuated spring bias provides elevator trimming; electrically actuated flaps.

STRUCTURE: Fuselage, with integral vertical fin, is mostly of GFRP construction with local CFRP reinforcement in high-stress areas, comprising two half-shells which are bonded together and incorporate transverse bulkheads in the cabin/centre-section area and three ring bulkheads in the tailcone. Wings comprise I-section spar of GFRP, with ribs providing mounting surfaces for control tubes and bellcranks; wing skins are of GFRP/PVC foam/GFRP sandwich construction, as are the rudder and horizontal tail surfaces.

LANDING GEAR: Fixed tricycle type, with cantilever self-sprung aluminium leg on each main unit and elastomeric suspension on nose leg; latter offset to starboard by 8 cm (3 in). Steering is provided by differential braking of the mainwheels and friction-damped castoring nosewheel. Optional speed fairings on all three wheels.

POWER PLANT: One 93 kW (125 hp) Teledyne Continental IO-240-B flat-four driving a Hoffman HO-14HM-175-15 wood and composites propeller. Fuel capacity 95 litres (25.0 US gallons; 20.8 Imp gallons); oil capacity 5.7 litres (1.5 US gallons; 1.25 Imp gallons).

ACCOMMODATION: Two seats side by side, with baggage space to rear; leather seats optional. Canopy hinged at rear to open upward.

SYSTEMS: Electrical system includes 12 V 20 Ah battery.

AVIONICS: *Comms:* Honeywell KX 125 com/nav with KT 76A transponder standard; other avionics to customer's choice.

Flight: Optional GPS nav.

Instrumentation: VSI, turn co-ordinator, turn and slip indicator, directional gyro and artificial horizon optional.

EQUIPMENT: Options include fire extinguisher, first aid kit, baggage harness and landing/position lights.

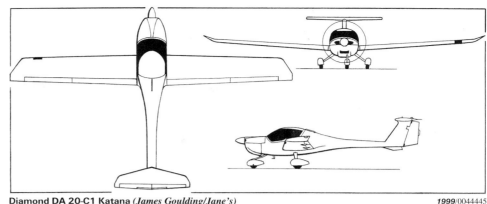

Diamond DA 20-C1 Katana (James Goulding/Jane's)   1999/0044445

DIMENSIONS, EXTERNAL:
Wing span	10.84 m (35 ft 6¾ in)
Length overall	7.165 m (23 ft 6 in)
Height overall	2.18 m (7 ft 2 in)
Tailplane span	2.65 m (8 ft 8¼ in)
Wheel track	1.90 m (6 ft 2¾ in)
Wheelbase	1.72 m (5 ft 7¾ in)
Propeller diameter	1.75 m (5 ft 9 in)

AREAS:
Wings, gross	11.60 m² (124.9 sq ft)

WEIGHTS AND LOADINGS:
Weight empty	529 kg (1,166 lb)
Max T-O weight: JAA	750 kg (1,653 lb)
FAA	780 kg (1,720 lb)
Max wing loading (JAA)	64.6 kg/m² (13.23 lb/sq ft)
Max power loading (JAA)	8.03 kg/kW (13.20 lb/hp)

PERFORMANCE:
Never-exceed speed (V_{NE})	164 kt (303 km/h; 188 mph)
Cruising speed at 75% power at 1,830 m (6,000 ft)	132 kt (244 km/h; 152 mph)
Manoeuvring speed	106 kt (196 km/h; 122 mph)
Stalling speed: flaps up	42 kt (78 km/h; 49 mph)
flaps down	34 kt (63 km/h; 40 mph)
Max rate of climb at S/L	337 m (1,105 ft)/min
Service ceiling	4,000 m (13,120 ft)
T-O run	291 m (955 ft)
T-O to 15 m (50 ft)	385 m (1,265 ft)
Landing from 15 m (50 ft)	377 m (1,235 ft)
Landing run	168 m (550 ft)

UPDATED

Diamond DA 20 Katana registered in the USA (Paul Jackson/Jane's)   2001/0089475

DIAMOND MPX

TYPE: Multisensor surveillance lightplane.

PROGRAMME: Developed by Diamond Aircraft Corporation in association with Wescam Inc of Ontario; multimission utility aircraft version of Diamond HK 36TC Katana Extreme, described under the Diamond Aircraft Industries GmbH entry in Austrian section. MPX incorporates hardpoint under each wing at approximately mid-span for quick-release sensor pods. Internal military-specification electronics racks can accommodate up to 91 kg (200 lb) of equipment, including gyrostabilised video surveillance cameras (such as Wescam Model 16 ENG broadcast camera system) and IR sensors, with provision for onboard data- and image collection, or downlink telemetry. An independent 28 V 40 A DC power supply is provided for surveillance equipment, and a cockpit equipment bay, capacity 0.4 m³ (14 cu ft), accommodates rack-mounted electronics or instruments for in-flight monitoring. Standard fuel capacity 109.8 litres (29 US gallons; 24.1 Imp gallons) provides endurance up to 7 hours. Prototype C-GEPR; FAA certification granted 8 April 1999.

UPDATED

DJ

DJ AIRCRAFTS INTERNATIONAL

7594 Tessier Brossard, Québec J4W 2H5
Tel: (+1 514) 466 11 51

This company's DJ2 was due to fly in 1998. However, by mid-2000 no confirmation had been received of it having done so.

UPDATED

DJ AIRCRAFTS DJ 2

TYPE: Side-by-side kitbuilt.

PROGRAMME: Prototype reported under construction by 1998.

COSTS: Kit US$25,000 (1997).

DESIGN FEATURES: Composites construction. Mid-wing, pusher propeller and tricycle landing gear. Quoted build time 300 hours.

POWER PLANT: Suitable for rotary engines in the 59.7 to 112 kW (80 to 150 hp) range. Prototype has 112 kW (150 hp) engine. Fuel capacity 114 litres (30 US gallons; 25 Imp gallons).

ACCOMMODATION: Pilot and passenger beneath one-piece Perspex canopy.

DIMENSIONS, EXTERNAL:
Wing span	5.67 m (18 ft 7¼ in)

AREAS:
Wings, gross	9.75 m² (105.0 sq ft)

WEIGHTS AND LOADINGS (with 112 kW; 150 hp engine):
Weight empty	272 kg (600 lb)
Max T-O weight	544 kg (1,200 lb)

PERFORMANCE (estimated, with 112 kW; 150 hp engine):
Max level speed	191 kt (354 km/h; 220 mph)
Cruising speed	174 kt (322 km/h; 200 mph)
Stalling speed	39 kt (71 km/h; 44 mph)
Max rate of climb at S/L	640 m (2,100 ft)/min
T-O run	61 m (200 ft)
Landing run	183 m (600 ft)
Max range	1,529 n miles (2,832 km; 1,760 miles)

VERIFIED

EUROCOPTER CANADA

EUROCOPTER CANADA LTD
(Subsidiary of Eurocopter SA)

HEAD OFFICE: PO Box 250, 1100 Gilmore Road, Fort Erie, Ontario L2A 5M9
Tel: (+1 905) 871 77 72
Fax: (+1 905) 871 33 20 and 35 99
e-mail: euro@eurocopter.ca
Web: http://www.eurocopter.ca

PRESIDENT AND CEO: Olivier Francou
EXECUTIVE VICE-PRESIDENT AND COO: Serge Panabiere
DIRECTOR, CUSTOMER SUPPORT AND SALES: Gordon Kay
MANAGER, MARKETING SERVICES AND CUSTOMER RELATIONS: Diane Bourdeau

Started as MBB Helicopter Canada on 30 March 1984, Eurocopter Canada Limited (ECL) now employs 150 people at its 9,290 m² (100,000 sq ft) facility at Fort Erie. The company holds the world product mandate and design authority for the BO 105 LS A-3 and provides products and services for the complete Eurocopter range (which see in International section), including the new EC 135 and EC 120B Colibri. ECL's production department is certified for AQA P1, JAR 145 and FAR Pt 45 requirements; the engineering department makes use of the CATIA design system and is responsible for certification of modifications and repairs for the Eurocopter range. Pilot and engineer training is also carried out by the company and offered as part of the purchase package. Warehouses at Fort Erie, Montreal

CANADA: AIRCRAFT—EUROCOPTER to FOUND

and Vancouver provide parts support for over 250 Eurocopter helicopters at present operational in Canada and also supply approximately 20 new and used helicopters per year. Six used and eight new (one BO 105 LS plus seven assembled from European production) helicopters were supplied in 1999.

ECL introduced a new seven-seat interior for the AS 350 series in 1996, allowing the pilot to operate from either front seat.

Current activities also involve assembly of the EC 120 B Colibri, AS 350/355 Ecureuil and EC 135. Turnover in 1998 was US$44 million and 1999 was US$62 million. After production at Donauwörth ceased, Eurocopter Canada became sole manufacturer of BO 105 LS A-3.

UPDATED

EUROCOPTER BO 105 LS

TYPE: Light utility helicopter.
PROGRAMME: Version of European-built BO 105; first flight of this version 23 October 1981; German LBA certification July 1984; extended for landing and take-off at 6,100 m (20,000 ft) April 1985; extended again 7 July 1986 to cover BO 105 LS A-3; FAA and Canadian DoT certification followed.
CURRENT VERSIONS: **BO 105 LS A-3**: Production version; first delivery February 1987.
CUSTOMERS: Chile, Japan, Mexico, Nigeria, Peru, South Africa and USA. Total 54 built by January 1999, including two in 1994, one each in 1995 and 1996, and six in following two years, plus one in 1999 for US Drug Enforcement Agency.
COSTS: US$2.04 million (1999).
DESIGN FEATURES: Four-blade main rotor with rigid titanium hub; only articulation is roller bearings for blade pitch change; all-composites blades of NACA 23012 section with drooped leading-edge and reflexed trailing-edge; 8° linear twist; pendulous vibration damper near each blade root; main rotor brake standard; two blades can be folded; two-blade semi-rigid tail rotor.
FLYING CONTROLS: Fully powered controls through hydraulic actuation pack mounted on transmission casing; very high control power and aerobatic capability, including claimed sustained inverted 1 g flight; IFR certification without autostabiliser.
STRUCTURE: Light alloy structure with GFRP main and tail rotor blades and top fairing.
LANDING GEAR: Skid type, with aluminium alloy cross-tubes designed for energy absorption by plastic deformation in event of heavy landing. Inflatable emergency floats can be attached to skids.
POWER PLANT: Two Rolls-Royce 250-C28C turboshafts, each rated at 410 kW (550 shp) for 2½ minutes, and with 5 minute T-O and maximum continuous power ratings of 373 kW (500 shp) and 368 kW (493 shp) respectively. Main transmission, Type ZF-FS 112, rated for independent restricted input of 310 kW (416 shp) per engine at T-O power or 294 kW (394 shp) per engine for maximum continuous operation; or single-engine restricted input of 368 kW (493 shp) at maximum continuous power, or 410 kW (550 shp) for 2½ minutes at T-O power. Bladder fuel tanks under cabin floor, usable capacity 570 litres (150.6 US gallons; 125.3 Imp gallons); fuelling point on port side of cabin; optional auxiliary tanks in freight compartment. Oil capacity 4.5 litres (1.2 US gallons; 1.0 Imp gallon) per engine; gearbox 11.6 litres (3.06 US gallons; 2.55 Imp gallons).

The 52nd production Eurocopter Canada BO 105 LS prior to delivery in August 1998   1999/0041218

ACCOMMODATION: Pilot and co-pilot or passenger on individual longitudinally adjustable front seats with four-point harnesses; optional dual controls. Bench seat at rear for three or four persons, removable for cargo and stretcher carrying; cargo space behind rear seats, plus additional 20 kg (44 lb) in baggage compartment. EMS versions available; entire rear fuselage aft of seats and under power plant available as freight and baggage space, with straight-in access through two clamshell doors at rear; two standard stretchers side by side in ambulance role; one forward-opening hinged and jettisonable door and one sliding door on each side of cabin; ram air and electrical ventilation system; heating system optional.
SYSTEMS: Dual hydraulic systems, pressure 103.5 bars (1,500 lb/sq in), for powered main rotor controls; system flow rate 6.2 litres (1.64 US gallons; 1.36 Imp gallons)/min; bootstrap/fluid reservoir, pressurised at 1.70 bars (25 lb/sq in); electrical system powered by two 150 A 28 V DC starter/generators and a 24 V 25 Ah Ni/Cd battery; external power socket. Stability augmentation system standard; optional bleed air anti-icing.
AVIONICS: Wide variety of avionics available including weather radar, Doppler and GPS navigation, Honeywell RDR 1500B 360° search radar, FLIR, TV broadcast and microwave datalink.
EQUIPMENT: Standard equipment includes heated pitot, tiedown rings in cargo compartment, cabin and cargo compartment dome lights, position lights and collision warning lights; options include dual controls, heating system, windscreen wiper, rescue winch, landing light, searchlight, externally mounted loudspeaker, fuel dump valve, external load hook, settling protectors, snow skids, manual main rotor blade folding, emergency floats, sand filters, firefighting kit, weapons fittings, mast-mounted sight and a wide range of EMS and VIP arrangements. NVG-compatible cockpit, first used operationally on 5 February 1999 by Rocky Mountain Helicopters.
DIMENSIONS, EXTERNAL:
Main rotor diameter	9.84 m (32 ft 3½ in)
Tail rotor diameter	1.90 m (6 ft 2¾ in)
Main rotor blade chord	0.27 m (10¾ in)
Tail rotor blade chord	0.18 m (7 in)
Distance between rotor centres	5.95 m (19 ft 6¼ in)
Length: incl main and tail rotors	11.86 m (38 ft 11 in)
excl rotors	8.81 m (28 ft 11 in)
fuselage pod	4.55 m (14 ft 11 in)
Height to top of main rotor head	3.02 m (9 ft 11 in)
Width over skids: unladen	2.53 m (8 ft 3½ in)
laden	2.58 m (8 ft 5½ in)
Rear-loading doors: Height	0.64 m (2 ft 1 in)
Width	1.40 m (4 ft 7 in)

DIMENSIONS, INTERNAL:
Cabin, incl cargo compartment:	
Max width	1.40 m (4 ft 7 in)
Max height	1.25 m (4 ft 1 in)
Volume	3.1 m³ (109 cu ft)
Cargo compartment: Length	1.85 m (6 ft 0¾ in)
Max width	1.20 m (3 ft 11¼ in)
Max height	0.57 m (1 ft 10½ in)
Floor area	2.25 m² (24.2 sq ft)
Volume	1.27 m³ (45 cu ft)

AREAS:
Main rotor disc	76.05 m² (818.6 sq ft)
Tail rotor disc	2.835 m² (30.5 sq ft)

WEIGHTS AND LOADINGS:
Weight empty, basic	1,430 kg (3,152 lb)
Standard fuel	456 kg (1,005 lb)
Max internal payload	637 kg (1,404 lb)
Max T-O weight	2,600 kg (5,732 lb)
Max disc loading	34.2 kg/m² (7.00 lb/sq ft)
Transmission loading at max T-O weight and power	4.19 kg/kW (6.89 lb/shp)

PERFORMANCE (at T-O weight of 2,400 kg; 5,291 lb, ISA):
Never-exceed speed (VNE) at S/L	145 kt (270 km/h; 167 mph)
Max cruising speed at S/L	131 kt (243 km/h; 151 mph)
Max forward rate of climb at S/L	551 m (1,810 ft)/min
Max vertical rate of climb at S/L	427 m (1,400 ft)/min
Max operating altitude	6,100 m (20,000 ft)
Service ceiling, OEI	1,800 m (5,900 ft)
Hovering ceiling: IGE	3,500 m (11,480 ft)
OGE	2,550 m (8,370 ft)
Range at S/L: standard fuel, max internal payload, no reserves	278 n miles (515 km; 320 miles)
with overload fuel	475 n miles (880 km; 547 miles)

UPDATED

FOUND

FOUND AIRCRAFT CANADA INC

RR# 2, Site 12, Box 10, Georgian Bay Airport, Parry Sound, Ontario P2A 2W8
Tel: (+1 705) 378 05 30
Fax: (+1 705) 378 05 94
e-mail: found@zeuter.com
Web: http://www.foundair.com

Found Aircraft Canada has reinstated production of the FBA-2 Bush Hawk, beginning deliveries in mid-2000.

UPDATED

FOUND FBA-2C BUSH HAWK

TYPE: Six-seat utility transport.
PROGRAMME: Found Aircraft Canada restarted production of updated version of Found FBA-2C, which first flew on 9 May 1962 and was last described in the 1967-68 *Jane's*. Production numbered 27, plus five improved Centenniel 100s (1969-70 edition) before original company ceased trading. Prototype FBA-2C1 (C-FSVD), converted from an existing aircraft with IO-540-D engine, flew at Georgian Bay Airport, Ontario, in November 1996 and completed test flying, on wheels and floats, during 1998 and 1999. New production was due to start in July 1997 but programme slipped; preproduction aircraft C-GDWO (c/n 28; IO-540-L engine) made first flight October 1998 and certified 5 March 1999; was followed by C-GDWQ in 1999; initial production Bush Hawk C-GDWS (c/n 30) flew 2 March 2000. FAA certificate A7EA reinstated 10 March 2000, covering both FBA-2C and FBA-2C1.
CURRENT VERSIONS: **FBA-2C Bush Hawk-260:** 194 kW (260 hp) version.
Description and data apply to Bush Hawk-260 unless otherwise stated.
FBA-2C1 Bush Hawk-300: 224 kW (300 hp) version.
CUSTOMERS: Eleven on order by end-1999; customers include Reliance Airways, Northwest Territories; Nakina Outpost Camps; and Air Service, Ontario. Found anticipates production to be initially one per month, rising to 50 per year. Second production aircraft (fourth overall) under construction in June 2000.
COSTS: Estimated C$165,000-C$190,000 (2000).
DESIGN FEATURES: Fully corrosion-proofed, all-metal construction for operation in exceptionally poor weather over all terrain.
High wing with constant chord to mid-span; outer panels unswept, with sharply swept forward trailing-edges.

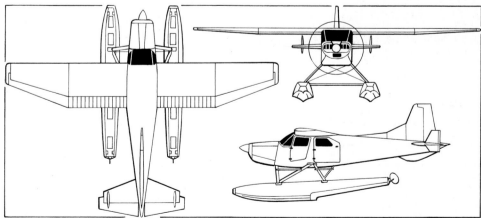

Found FBA-2C Bush Hawk six-seat utility transport (*James Goulding/Jane's*)   2001/0099158

First production Found FBA-2C Bush Hawk-260 *2001/0087862*

Sweptback fin with two-stage fillet; mid-mounted tailplane. Floatplane directional stability enhanced by underfin and auxiliary fins above and below tailplane at three-quarters span.
FLYING CONTROLS: Conventional and manual. Horn-balanced rudder and elevators. Flaps.
STRUCTURE: Aluminium-covered steel tube front fuselage and semi-monocoque rear fuselage. Deep-section, low-drag, all-aluminium wing.
LANDING GEAR: Fixed. Prototype, preproduction and first production aircraft have tailwheel configuration; all following have tricycle layout. Sprung steel main legs with urethane shock-absorbers. Mainwheels 8.00-6, but 66 cm (26 in) 8.50-10 and 76 cm (30 in) Tundra tyres being evaluated. Other options include Edo Straight 2960 or Aerocet 3500 floats and Federal C-3200 retractable skis.
POWER PLANT: One 194 kW (260 hp) Textron Lycoming IO-540-D4A5 or 224 kW (300 hp) IO-540-L1C flat-six engine, driving a Hartzell three-blade propeller. Fuel in two wing tanks, total capacity 379 litres (100 US gallons; 83.3 Imp gallons).
ACCOMMODATION: Pilot and co-pilot, plus four passengers in cabin, with crew entrance through forward-hinged door on each side of fuselage. Rear hammock-style seat rolls up for storage in cabin roof when used in freight role; optional individual folding seats available. Rear cabin typically holds four 204 litre (54 US gallon; 45 Imp gallon) drums and has large, forward-hinged door on each side; door sills are flush with floor; doors swing 180° to allow easy loading. Alternatively, one stretcher and attendant in rear cabin. Camera hatch can be fitted in cabin floor.

DIMENSIONS, EXTERNAL:
Wing span	10.97 m (36 ft 0 in)
Wing aspect ratio	7.2
Length overall	7.77 m (25 ft 6 in)
Propeller diameter	2.13 m (7 ft 0 in)
Cabin door: Width	0.67 m (2 ft 2½ in)

DIMENSIONS, INTERNAL:
Cabin: Length	3.51 m (11 ft 6 in)
Volume (incl baggage compartment)	3.40 m³ (120 cu ft)

AREAS:
Wings, gross	16.72 m² (180.0 sq ft)

WEIGHTS AND LOADINGS (A: wheels, B: floats):
Max T-O weight: A, skis	1,451 kg (3,200 lb)
B	1,492 kg (3,290 lb)
Max wing loading: A	86.8 kg/m² (17.78 lb/sq ft)
B	89.2 kg/m² (18.28 lb/sq ft)
Max power loading: A	7.49 kg/kW (12.31 lb/hp)
B	7.70 kg/kW (12.65 lb/hp)

PERFORMANCE (A, B, as above):
Max operating speed: A	140 kt (259 km/h; 161 mph)
B	117 kt (217 km/h; 135 mph)
Cruising speed: at 75% power:	
A	128 kt (237 km/h; 147 mph)
B	112 kt (208 km/h; 129 mph)
at 60% power:	
A	114 kt (211 km/h; 131 mph)
B	97 kt (180 km/h; 111 mph)
Max rate of climb at S/L:	
A	293 m (960 ft)/min
B	256 m (840 ft)/min

UPDATED

FREEDOM LITE

FREEDOM LITE INC
General Delivery, New Hamburg, Ontario N0B 2G0
Tel: (+1 519) 662 29 80
e-mail: skywatch@wcl.on.ca
CEO: Rob McIntosh

Freedom Lite began production of the Skywatch SS-11 in 1996.

UPDATED

Freedom Lite Skywatch SS-11 two-seat ultralight *1999/0044474*

FREEDOM LITE SKYWATCH SS-11
TYPE: Tandem-seat ultralight kitbuilt.
PROGRAMME: Based on the Beaver RX-650 (see 1992-93 edition of *Jane's*) which first flew 1991. First flight of SS-11, mid-1995.
CURRENT VERSIONS: Standard version *as described*; 47.8 kW (64.1 hp) Rotax 582 and 55.0 kW (73.8 hp) Rotax 618 engines optional.
CUSTOMERS: Nine completed by mid-2000.
COSTS: Standard kits: with Rotax 503 US$18,980 (1999); with Rotax 582 US$19,976 (1997).
DESIGN FEATURES: Simple, pod-and-boom configuration, with braced wing and tailplane, plus pusher propeller. Optional folding of wings and horizontal tail surfaces. Parts manufacture 90 per cent computer numeric controlled (CNC).
FLYING CONTROLS: Conventional and manual. No aerodynamic balances.
STRUCTURE: Metal airframe with composites fuselage fairing and fabric covering on flying surfaces.
LANDING GEAR: Fixed tricycle type; steerable nosewheel. Optional 4.57 m (15 ft 0 in) composites amphibian floats.
POWER PLANT: One 37.0 kW (49.6 hp) Rotax 503 UL-2V. Standard fuel capacity 34 litres (9.0 US gallons; 7.5 Imp gallons); optional 51 litres (13.5 US gallons; 11.2 Imp gallons).

DIMENSIONS, EXTERNAL:
Wing span	10.06 m (33 ft 0 in)
Length overall, wings folded	6.55 m (21 ft 6 in)
Height overall, wings folded	2.18 m (7 ft 2 in)

AREAS:
Wings, gross	14.40 m² (155.0 sq ft)

WEIGHTS AND LOADINGS:
Weight empty	190 kg (420 lb)
Max T-O weight	430 kg (950 lb)

PERFORMANCE:
Cruising speed	61-78 kt (113-145 km/h; 70-90 mph)
Stalling speed	26 kt (47 km/h; 29 mph)
Max rate of climb at S/L	238 m (780 ft)/min
T-O run	61 m (200 ft)
Landing run	77 m (250 ft)
Range	217 n miles (402 km; 250 miles)

UPDATED

MURPHY

MURPHY AIRCRAFT MANUFACTURING LTD
Unit 1, 8155 Aitken Road, Chilliwack, British Columbia V2R 4H5
Tel: (+1 604) 792 58 55
Fax: (+1 604) 792 70 06
e-mail: mursales@murphyair.com
Web: http://www.murphyair.com
PRESIDENT: Darryl Murphy
SALES MANAGER: Colleen Dyck
MARKETING MANAGER: Dave Walker

In addition to the Renegade II and Renegade Spirit sport biplanes, Murphy has developed the three-seat Rebel and Rebel Elite which can be built as a homebuilt or ARV, or as an ultralight in some countries. The four- to six-seat Super Rebel was added to the range in 1996.

Formed in 1985, the company moved from initial 745 m² (8,000 sq ft) plant to new 4,925 m² (53,000 sq ft) premises in 1995 to accommodate increase in business.

UPDATED

MURPHY RENEGADE II
TYPE: Tandem-seat ultralight biplane kitbuilt.
PROGRAMME: Prototype Renegade II made first flight May 1985; first kits became available September that year. Ready assembled aircraft, plans and kits are available.
CUSTOMERS: By mid-2000, 350 Renegade IIs and Renegade Spirits had been completed and flown.
COSTS: Quick-build kit US$12,995; full materials kit US$8,695; partial parts kit US$3,330; plans US$275 (2000 prices).
DESIGN FEATURES: Sporting biplane of classic configuration. Conforms to TP101.41 ultralight regulations in Canada and some European countries. Optional wingtip extensions (for countries such as UK requiring lower wing loadings) increase gross area to 15.83 m² (170.4 sq ft). Optional extended wings and rounded tips; optional floats and parachute.

NACA 23012 wing section; 10° sweepback on upper wings; 3° dihedral on non-swept lower wings. Angle of incidence for both wings of 4°.

Kits in four standards ranging from partial parts kit to 300/400 working hour quick-build kit with all parts premanufactured plus partly assembled fuselage, wings and engine mounting.
STRUCTURE: Fabric-covered aluminium tubing; rear fuselage top-decking of preformed glass fibre; glass fibre wingtips optional; wire-braced tail assembly.
LANDING GEAR: Tailwheel type; fixed. Tailwheel steerable. Bungee cord suspension system.
POWER PLANT: One 34.0 kW (45.6 hp) Rotax 503 UL-1V two-cylinder two-stroke engine standard; 37.0 kW (49.6 hp) twin-carburettor Rotax 503 UL-2V optional. Alternative engines up to 59.6 kW (80 hp) can be fitted, including Rotax 582 and Teledyne Continental O-65, O-75 and O-85. Fuel capacity 53 litres (14.0 US gallons; 11.6 Imp gallons).

DIMENSIONS, EXTERNAL:
Wing span (normal): upper	6.48 m (21 ft 3 in)
lower	6.05 m (19 ft 10 in)
Length overall	5.61 m (18 ft 5 in)
Height overall	2.08 m (6 ft 10 in)

AREAS:
Wings, gross	14.29 or 15.83 m² (153.8 or 170.4 sq ft)

CANADA: AIRCRAFT—MURPHY

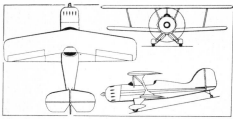

Murphy Renegade Spirit

WEIGHTS AND LOADINGS:
 Weight empty   170-236 kg (375-520 lb)
 Max T-O weight, Rotax 503   431 kg (950 lb)
PERFORMANCE (Rotax 503 engine and two crew):
 Max level speed at 305 m (1,000 ft)
    70 kt (129 km/h; 80 mph)
 Cruising speed at 75% power   56 kt (105 km/h; 65 mph)
 Stalling speed, power off   33 kt (61 km/h; 38 mph)
 Max rate of climb at S/L   152 m (500 ft)/min
 T-O and landing run   92 m (300 ft)
 Range at 75% power   197 n miles (365 km; 227 miles)

UPDATED

MURPHY RENEGADE SPIRIT

TYPE: Tandem-seat ultralight biplane kitbuilt.
PROGRAMME: First flight (C-IJLW) 6 May 1987; kit production began a month later.
DESIGN FEATURES: Variant of Renegade II. Alternative power plants; internal and external refinements.
COSTS: Fast-build kit US$12,995; materials kit US$8,995; plans US$275 (2000 prices).
POWER PLANT: One 47.8 kW (64.1 hp) Rotax 582 in-line engine standard, in radial-type cowling; 59.7 kW (80 hp) Rotax 912 flat-four or Alvis rotary engine optional.
DIMENSIONS, EXTERNAL: As for Renegade II
AREAS: As for Renegade II
WEIGHTS AND LOADINGS:
 Weight empty   190-227 kg (420-500 lb)
 Max T-O weight   431 kg (950 lb)
PERFORMANCE (two crew and Rotax 582 engine):
As for Renegade II except:
 Max level speed   78 kt (145 km/h; 90 mph)
 Cruising speed at 75% power   63 kt (116 km/h; 72 mph)
 Stalling speed   35 kt (64 km/h; 40 mph)
 Max rate of climb at S/L   152 m (500 ft)/min
 T-O and landing run   107 m (350 ft)
 Range at 75% power   206 n miles (381 km; 237 miles)

UPDATED

MURPHY REBEL

TYPE: Three-seat kitbuilt.
PROGRAMME: Construction of prototype started October 1989; first flight 1990; building of production aircraft/kits began in October 1990.
CURRENT VERSIONS: **Rebel**: Standard version, *as described*.
 Rebel Elite: Modified tricycle landing gear version, with metal-covered split flaps and enlarged fin; O-320 or optional 134 kW (180 hp) Lycoming O-360 engine; and increased maximum T-O weight to 816 kg (1,800 lb). First flight (C-FWSF) 29 February 1996.
 Super Rebel and Maverick: Described separately.
CUSTOMERS: Total of 700 complete aircraft or kits (including 82 Elites) ordered by 1 January 2000; over 400 flying by late 1999.
COSTS: Rebel kit without engine US$13,995; Elite kit without engine US$18,700 tailwheel, US$19,775 tricycle (2000). Available in three sub-component kits.
DESIGN FEATURES: High-wing, strut-braced cabin monoplane for recreational, training, cross-country, border patrol and other roles; designed to conform to FAR Pt 23 and JAR (Utility category); STOL performance. Wings detachable and horizontal tail folds upward for transportation and storage. Quoted build time 800 to 1,000 hours. NACA 4415 (modified) wing section.
FLYING CONTROLS: Conventional and manual. Flaps. Elite has split flaps, metal-covered ailerons and cantilevered horizontal stabiliser.
STRUCTURE: Aluminium alloy, semi-monocoque fuselage; flaperons fabric covered; glass fibre wingtips. Tailplane also strut-braced. Rebel Elite wing leading-edge 60 per cent thicker than Rebel.
LANDING GEAR: Non-retractable tailwheel type with mainwheel brakes. Optional tricycle gear (Rebel Elite), skis, or Murphy 1800 straight or wheeled or Montana 2100 wheeled floats. 6.00-6 wheels with 18 in high-profile tyres are standard. Bungee suspension system is standard; optional aluminium spring gear.
POWER PLANT: One 59.6 kW (79.9 hp) Rotax 912 UL, or 86.5 kW (116 hp) Textron Lycoming O-235-N2C engine; 2.27:1 reduction gear with Rotax engine and direct drive with O-235. Elite has 119 kW (160 hp) Lycoming O-320; or optional 134 kW (180 hp) Lycoming O-360; ground-adjustable wooden two-blade propeller. Standard fuel capacity 167 litres (44.0 US gallons; 36.6 Imp gallons) in wing tanks; optional extra 220 litre (58.0 US gallon; 48.3 Imp gallon) tank.
ACCOMMODATION: Pilot and two passengers.
DIMENSIONS, EXTERNAL:
 Wing span: Rebel   9.14 m (30 ft 0 in)
  Rebel Elite   9.25 m (30 ft 4 in)
 Wing chord, constant   1.52 m (5 ft 0 in)
 Wing aspect ratio   6.0
 Length overall: Rebel   6.50 m (21 ft 4 in)
  Rebel Elite   6.78 m (22 ft 3 in)
 Fuselage max width   1.12 m (3 ft 8 in)
 Height overall: Rebel, Elite (tailwheel)   2.03 m (6 ft 8 in)
  Elite (tricycle)   2.39 m (7 ft 10 in)
 Tailplane span   2.79 m (9 ft 2 in)
 Wheel track: Rebel   2.18 m (7 ft 2 in)
  Elite   2.32 m (7 ft 7½ in)
 Wheelbase: Rebel, Elite (tailwheel)   4.93 m (16 ft 2 in)
  Elite (tricycle)   1.70 m (5 ft 7 in)
 Propeller diameter   1.78 m (5 ft 10 in)
AREAS:
 Wings, gross: Rebel   13.94 m² (150.0 sq ft)
  Elite   14.12 m² (152.0 sq ft)
WEIGHTS AND LOADINGS (A: Rotax 912, B: O-235, E: Elite, O-360):
 Weight empty: A   295-317 kg (650-700 lb)
  B   374-408 kg (825-900 lb)
  E   445 kg (980 lb)
 Baggage capacity   45.5-68 kg (100-150 lb)
 Max T-O weight: A   657 kg (1,450 lb)
  B   748 kg (1,650 lb)
  E   816 kg (1,800 lb)
 Max wing loading: A   47.2 kg/m² (9.67 lb/sq ft)
  B   53.7 kg/m² (11.00 lb/sq ft)
  E   57.8 kg/m² (11.84 lb/sq ft)
 Max power loading: A   11.03 kg/kW (18.13 lb/hp)
  B   8.66 kg/kW (14.22 lb/hp)
  E   6.09 kg/kW (10.00 lb/hp)

Murphy Renegade Spirit (nearest) and Renegade II

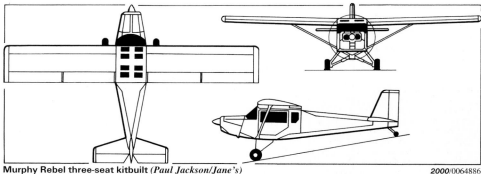

Murphy Rebel three-seat kitbuilt *(Paul Jackson/Jane's)*

Murphy Rebel with standard bungee suspension *(Paul Jackson /Jane's)*

PERFORMANCE (A, B, E as above):
Never-exceed speed (V_{NE}):
A	124 kt (230 km/h; 143 mph)
B	131 kt (243 km/h; 151 mph)
E	136 kt (252 km/h; 157 mph)

Max level speed: A 87 kt (162 km/h; 100 mph)
B 108 kt (201 km/h; 125 mph)
E 126 kt (233 km/h; 145 mph)
Cruising speed at 65% power:
A 69 kt (129 km/h; 80 mph)
B 91 kt (169 km/h; 105 mph)
E 115 kt (212 km/h; 132 mph)
Stalling speed, power off, flaps up:
A 35 kt (65 km/h; 40 mph)
B 39 kt (71 km/h; 44 mph)
E 40 kt (74 km/h; 46 mph)
Stalling speed, power off, flaps down:
A 32 kt (58 km/h; 36 mph)
B 35 kt (65 km/h; 40 mph)
E 37 kt (68 km/h; 42 mph)
Max rate of climb at S/L: A 164 m (500 ft)/min
B 244 m (800 ft)/min
E 457 m (1,500 ft)/min
Service ceiling: B 3,960 m (13,000 ft)
T-O run: A 137 m (450 ft)
B 122 m (400 ft)
E 143 m (470 ft)
T-O to 15 m (50 ft): A, B 244 m (800 ft)
Landing run: A 92 m (300 ft)
B, E 122 m (400 ft)
Range with max fuel:
A 764 n miles (1,416 km; 880 miles)
B 691 n miles (1,281 km; 796 miles)
E 593 n miles (1,099 km; 683 miles)
Endurance: B 7 h 36 min
E 5 h 12 min
g limits +3.8/−2.5
(+5.7/−3.8 ultimate)
UPDATED

MURPHY SR 2500 SUPER REBEL

TYPE: Six-seat kitbuilt.
PROGRAMME: Announced April 1995; prototype (C-GKSR) first flew November 1995; parts shipments began in late 1995 with tail sections. Tricycle version first flew May 1997.
CUSTOMERS: Over 130 aircraft sold.
COSTS: Basic kit less engine, propeller, instruments and avionics: tailwheel version US$22,995; tricycle version US$24,315 (2000).
DESIGN FEATURES: Development of Murphy Rebel (which see). All-metal construction. Quoted build time 1,400 to 1,800 hours. Fast-build kits have quoted build time of 800-1,000 hours.
FLYING CONTROLS: Conventional and manual. Electric trim and three-stage flaps; ailerons deflect 20° up and 12° down; ailerons droop 10° when 40° of flap deployed.
STRUCTURE: Wing is identical in shape to that of Rebel and Elite; aerofoil modified NACA 4415. Three-spar wing; leading-edge of 0.032 sheet aluminium.
LANDING GEAR: Tailwheel configuration is standard (SR 2500TD); tricycle also available. Optional hardpoints enable rapid conversion. Main and nose tyres 8.00-6 in; tailwheel has 9 in tyre. Dual brakes. Optional floats.
POWER PLANT: Prototype has 186 kW (250 hp) Textron Lycoming O-540-A4A5 driving two-blade constant-speed Hartzell propeller, but the design accommodates engines from 134 to 224 kW (180 to 300 hp). Fuel in two wing tanks, total capacity 227 litres (60.0 US gallons; 50.0 Imp gallons). Optional larger tanks increase capacity to 303 litres (80.0 US gallons; 66.6 Imp gallons).
ACCOMMODATION: Pilot and up to five passengers in two pairs of side-by-side seats plus optional jump seat; dual controls. Rear bench seat can be removed to provide cargo capacity. Baggage compartment is reached by large cargo door on port side of fuselage, behind passenger door.
DIMENSIONS, EXTERNAL:
Wing span 11.07 m (36 ft 4 in)
Wing chord, constant 1.52 m (5 ft 0 in)
Wing aspect ratio 7.3
Length overall 8.46 m (27 ft 9 in)
Fuselage max width 1.12 m (3 ft 8 in)
Height overall: tailwheel 1.98 m (6 ft 6 in)
tricycle 2.67 m (8 ft 9 in)
Tailplane span 3.30 m (10 ft 10 in)
Wheel track 2.29 m (7 ft 6 in)
Wheelbase: tailwheel 5.94 m (19 ft 6 in)
tricycle 2.03 m (6 ft 8 in)
Propeller diameter 2.13 m (7 ft 0 in)
Passenger door: Height 1.07 m (3 ft 6 in)
Width 1.07 m (3 ft 6 in)
Baggage door: Height 0.83 m (2 ft 8½ in)
Width 0.93 m (3 ft 0½ in)
DIMENSIONS, INTERNAL:
Cabin: Max width 1.12 m (3 ft 8 in)
Height 1.30 m (4 ft 3 in)
Length 2.62 m (8 ft 7 in)
AREAS:
Wings, gross 16.91 m² (182.0 sq ft)
Ailerons (total) 2.37 m² (25.50 sq ft)
Flaps (total) 2.09 m² (22.50 sq ft)
WEIGHTS AND LOADINGS (186 kW; 250 hp engine):
Weight empty 726 kg (1,600 lb)
Baggage capacity 113 kg (250 lb)
Max T-O weight 1,361 kg (3,000 lb)
Max wing loading 80.5 kg/m² (16.48 lb/sq ft)
Max power loading 7.30 kg/kW (12.00 lb/hp)
PERFORMANCE (186 kW; 250 hp engine):
Never-exceed speed (V_{NE}) 153 kt (284 km/h; 177 mph)
Max level speed 139 kt (257 km/h; 160 mph)
Max cruising speed at 75% power
135 kt (249 km/h; 155 mph)
Stalling speed, power off:
flaps up 46 kt (84 km/h; 52 mph)
flaps down 40 kt (74 km/h; 46 mph)
Max rate of climb at S/L 335 m (1,100 ft)/min
Service ceiling 4,575 m (15,000 ft)
T-O run 183 m (600 ft)
Landing run 152 m (500 ft)
Range: with max standard fuel
537 n miles (996 km; 619 miles)
with optional larger tanks
729 n miles (1,350 km; 839 miles)
Endurance 4 h 15 min
g limits (ultimate) +5.7/−3.8
UPDATED

MURPHY MAVERICK

TYPE: Side-by-side ultralight kitbuilt.
PROGRAMME: Original design shelved in favour of Rebel, but revived in 1994 to meet Japanese ultralight restrictions; prototype/demonstrator G-MYSS used to gain UK BCAR Section S type approval in 1995.
CUSTOMERS: 34 completed and flown by late 1999; at which time 119 kits had been sold.
COSTS: Kit: US$11,850 without engine (2000).
DESIGN FEATURES: Derivative of Rebel, with 40 per cent commonality of parts, including optional wingtip extensions.
FLYING CONTROLS: As Rebel, except cable-operated ailerons only instead of full-span flaperons.
STRUCTURE: Similar to Rebel, but with weight-saving features. Wings omit full-span stringers and are fabric covered; glass fibre engine cowling; tail surfaces also fabric covered.
LANDING GEAR: As Rebel.
POWER PLANT: One two-cylinder two-stroke engine (39.5 kW; 53 hp Rotax 503 DC or 47.8 kW; 64.1 hp Rotax 582 UL); GSC two-blade propeller. Testing with Hpower HKS-700E four-stroke engine completed in November 1999. Fuel capacity 18.9 litres (5.0 US gallons; 4.2 Imp gallons) standard, 53 litres (14.0 US gallons; 11.7 Imp gallons) in optional wing tanks.
DIMENSIONS, EXTERNAL:
Wing span: standard 8.97 m (29 ft 5 in)
with extended tips 9.88 m (32 ft 5 in)
Length overall, tail up 6.30 m (20 ft 8 in)

Murphy Maverick two-seat ultralight

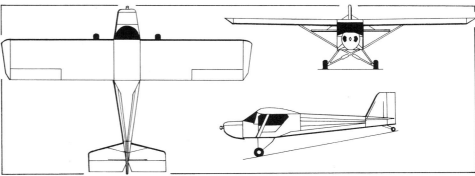
Murphy Maverick two-seat ultralight *(Paul Jackson/Jane's)*

Murphy SR 2500 Super Rebel four-seat lightplane

AREAS:
  Wings, gross: standard   13.66 m² (147.0 sq ft)
    with extended tips   15.05 m² (162.0 sq ft)
WEIGHTS AND LOADINGS (standard span and Rotax 503 DC engine):
  Weight empty   179-191 kg (395-420 lb)
  Max T-O weight   431 kg (950 lb)
PERFORMANCE (standard span and Rotax 503 DC engine):
  Max level speed   82 kt (153 km/h; 95 mph)
  Cruising speed at 75% power 70 kt (129 km/h; 80 mph)
  Stalling speed   28 kt (52 km/h; 32 mph)
  Max rate of climb at S/L 183 m (600 ft)/min
  T-O run   46 m (150 ft)
  Landing run   61 m (200 ft)
Range with max fuel at 75% power
                            243 n miles (450 km; 280 miles)
g limits   +5.7/-3.8
UPDATED

PAXMAN
PAXMAN'S NORTHERN LITE AEROCRAFT
PO Box 1155, Glenwood, Alberta T0K 2R0
Tel: (+1 403) 626 30 94
Fax: (+1 403) 626 34 90

UPDATED

PAXMAN VIPER
TYPE: Side-by-side kitbuilt.
PROGRAMME: Prototype first flew 1994.
CUSTOMERS: Two flying by mid-2000.
COSTS: Kit US$17,500 (2000); with engine and instruments. Plans only, US$300.
DESIGN FEATURES: Low wing monoplane; constant chord wings; broad, sweptback fin. Wood and fabric construction; quoted 600 hour build time. Wings detach for storage.
FLYING CONTROLS: Conventional and manual. Flaps.
STRUCTURE: Steel tube covered with wood and fabric; glass fibre engine cowling.
LANDING GEAR: Tailwheel type; fixed. Sprung steel main legs; brakes on mainwheels.
ACCOMMODATION: Dual controls. Dual gull-wing doors; baggage compartment behind seats.
POWER PLANT: One 74.6 kW (100 hp) Subaru EA81; alternative engines between 48.5 and 85.8 kW (65 and 115 hp) can be fitted. Fuel capacity 64 litres (17.0 US gallons; 14.2 Imp gallons).
DIMENSIONS, EXTERNAL:
  Wing span   7.92 m (26 ft 0 in)
  Wing aspect ratio   6.6
  Length overall   5.94 m (19 ft 6 in)
  Height overall   1.88 m (6 ft 2 in)
  Wheel track   2.13 m (7 ft 0 in)
DIMENSIONS, INTERNAL:
  Cabin: Length   1.09 m (3 ft 7 in)
    Max width   1.04 m (3 ft 5 in)
    Max height   1.02 m (3 ft 4 in)
AREAS:
  Wings, gross   9.48 m² (102.0 sq ft)
WEIGHTS AND LOADINGS (74.6 kW; 100 hp engine):
  Weight empty   295 kg (650 lb)
  Baggage capacity   22 kg (50 lb)
  Max T-O weight   589 kg (1,300 lb)
  Max wing loading   62.2 kg/m² (12.75 lb/sq ft)
  Max power loading   7.91 kg/kW (13.00 lb/hp)
PERFORMANCE:
  Never-exceed speed (V_NE) 126 kt (233 km/h; 145 mph)
  Max level speed   113 kt (209 km/h; 130 mph)
  Cruising speed   100 kt (185 km/h; 115 mph)
  Stalling speed   33 kt (62 km/h; 38 mph)
  Max rate of climb at S/L 457 m (1,500 ft)/min
  T-O run   91 m (300 ft)
  Landing run   122 m (400 ft)
  Range   477 n miles (885 km; 550 miles)
g limits   +4.4/-2.2
UPDATED

RAF
ROTARY AIR FORCE INC
PO Box 1236, 1107-9th Street W, Kindersley, Saskatchewan S0L 1S0
Tel: (+1 306) 463 60 30
Fax: (+1 306) 463 60 32
e-mail: raf2000@sk.sympatico.ca
Web: http://www.raf2000.com
SALES MANAGER: Sim Besse

Incorporated in 1987, Rotary Air Force employs 16 people. For details of the single-seat RAF 1000/GT, see the 1992-93 edition of *Jane's*.

UPDATED

RAF 2000
TYPE: Two-seat autogyro kitbuilt.
CURRENT VERSIONS: **2000 STD-SE**: Basic version, *as described*.
  2000 GTX-SE: Top-of-range model; kit includes rotor brake, heater, dual controls and adjustable pitch and roll trim assembly.
CUSTOMERS: RAF autogyros sold in Argentina, Australia, Austria, Brazil, Canada, Chile, China, Ecuador, Germany, Greece, Hungary, Ireland, Italy, Japan, Kazakhstan, Mexico, Netherlands, New Caledonia, New Zealand, Norway, Portugal, Puerto Rico, Russia, South Africa, Spain, Thailand, UK and USA. At least 430 believed completed and flown by end 1999.
COSTS: 2000 STD-SE kit price US$20,615, including engine; 2000 GTX-SE at US$22,500 (2000); both prices include 10 hours of dual instruction.
DESIGN FEATURES: Conventional autogyro. Suitable for training, crop-spraying, power line inspection and aerial photography. Quoted build time 150-250 hours.
STRUCTURE: Composites RAF rotor blade features aluminium spar and foam filler. Patented 5 × 10 cm (2 × 4 in) rigid rotor mast.
LANDING GEAR: Tricycle configuration; fixed; optional speed fairings.
POWER PLANT: One 97 kW (130 hp) Subaru EJ22 16-valve four-cylinder liquid-cooled engine driving a three-blade Warp Drive composites propeller through RAF cog belt reduction gear, ratio 2.10:1. Fuel capacity 87 litres (23.0 US gallons; 19.2 Imp gallons) of premium unleaded Mogas, of which 79 litres (21.0 US gallons; 17.5 Imp gallons) are usable. Fuel consumption 18.2 litres (4.8 US gallons; 4.0 Imp gallons)/h at 75 per cent power.
ACCOMMODATION: Pilot and passenger side by side in enclosed cabin.
SYSTEMS: 35A alternator, electric starter.
DIMENSIONS, EXTERNAL:
  Rotor diameter   9.14 m (30 ft 0 in)

RAF 2000 GTX-SE autogyro *(Paul Jackson/Jane's)*

  Rotor blade chord   0.22 m (8½ in)
  Fuselage: Length   4.11 m (13 ft 6 in)
    Max width   1.08 m (3 ft 6½ in)
  Width over wheels   1.59 m (5 ft 2½ in)
  Height overall: STD-SE   2.50 m (8 ft 2½ in)
    GTX-SE   2.58 m (8 ft 5½ in)
  Propeller diameter   1.73 m (5 ft 8 in)
  Wheel track   1.55 m (5 ft 1 in)
  Wheelbase   3.91 m (12 ft 10 in)
AREAS:
  Rotor disc   65.67 m² (706.9 sq ft)
WEIGHTS AND LOADINGS:
  Weight empty: STD-SE   345 kg (760 lb)
    GTX-SE   331 kg (730 lb)
  Max payload: STD-SE   261 kg (575 lb)
    GTX-SE   317 kg (700 lb)
  Max T-O weight   698 kg (1,540 lb)
  Max disc loading   10.6 kg/m² (2.18 lb/sq ft)
  Max power loading   7.21 kg/kW (11.85 lb/hp)
PERFORMANCE (two occupants):
  Max operating speed (V_MO) 87 kt (161 km/h; 100 mph)
  Cruising speed: STD-SE   61 kt (113 km/h; 70 mph)
    GTX-SE   65 kt (121 km/h; 75 mph)
  Min flying speed   17-26 kt (32-48 km/h; 20-30 mph)
  Unstick speed:
    STD-SE   26-35 kt (48-64 km/h; 30-40 mph)
    GTX-SE   39-43 kt (72-80 km/h; 45-50 mph)
  Max rate of climb at S/L: STD-SE 213 m (700 ft)/min
    GTX-SE   305 m (1,000 ft)/min
  Service ceiling   3,050 m (10,000 ft)
  T-O run   23-107 m (75-350 ft)
  Landing run   3 m (10 ft)
  Range with max fuel:
    STD-SE   278 n miles (515 km; 320 miles)
    GTX-SE   182 n miles (338 km; 210 miles)
  Endurance (with 30 min reserves): STD-SE   4 h 0 min
    GTX-SE   3 h 0 min
UPDATED

ST JUST
ST JUST AVIATION INC
1310 Gay-Lussac, Boucherville, Québec J4B 7G4
Tel: (+1 450) 641 86 86
Fax: (+1 450) 641 84 21

e-mail: st-justaviation@sympatico.ca
Web: http://www3.sympatico.ca/st-justaviation
PARTNERS: Simon Lavoie
  François Desagné
  Réal Blouin
  Pierre Tremblay

Original Avionnerie du Lac St John formed in 1991 to provide parts for Cessna 185 owners after that company removed the aircraft from its product line; later progressed to manufacturing kits of the Cyclone. Sold to St Just Aviation in 1997.

NEW ENTRY

ST JUST SUPER CYCLONE

TYPE: Four-seat kitbuilt.
PROGRAMME: Introduced in 1997; based on reverse engineering of the Cessna 180/185 (which see in *Jane's Aircraft Upgrades*).
CUSTOMERS: Total 45 sold and 17 flying by mid-2000.
COSTS: Basic kit US$39,900 (2001). Fast-build kits US$3,000 extra.
DESIGN FEATURES: Compared to Cessna 180/185, Super Cyclone has span increased by addition of 30 cm (1 ft) at each wingroot, which increases flap length, lowers stalling speed, enhances stability and increases maximum T-O weight. Quoted build time 2,000 hours.
FLYING CONTROLS: Conventional and manual. Flaps.
STRUCTURE: Single-strut braced, two-spar wing with constant-chord centre-section and tapered outer sections. Wing and empennage constructed of 2024 T-3 aluminium. Large fin fillet. Low-mounted tailplane.
LANDING GEAR: Tailwheel type, with sprung steel main legs; Federal wheel/skis and floats can be fitted. Tricycle version under development.
POWER PLANT: One 224 kW (300 hp) Teledyne Continental IO-520, flat-six engine, driving a three-blade, constant-speed McCauley propeller. Optional alternative engines between 149 and 261 kW (200 and 350 hp). Fuel capacity 341 litres (90.0 US gallons; 75.0 Imp gallons), of which 318 litres (84.0 US gallons; 70.0 Imp gallons) are usable.
ACCOMMODATION: Pilot and three passengers in two pairs of seats with upward-hinged door on each side of cabin. Additional baggage door on port side of fuselage behind cabin. Compared to Cessna 185, has a third cabin window each side.
AVIONICS: Full IFR panel available.
DIMENSIONS, EXTERNAL:
Wing span 11.58 m (38 ft 0 in)
Wing aspect ratio 7.2
Length overall 7.92 m (26 ft 0 in)
Height overall 2.41 m (7 ft 11 in)
Propeller diameter 2.18 m (7 ft 2 in)
DIMENSIONS, INTERNAL:
Cabin max width 1.07 m (3 ft 6 in)
AREAS:
Wings, gross 17.74 m² (191.0 sq ft)
WEIGHTS AND LOADINGS:
Weight empty 839 kg (1,850 lb)
Baggage capacity 68 kg (150 lb)
Max payload 454 kg (1,000 lb)
Max T-O weight 1,587 kg (3,500 lb)
Max wing loading 89.5 kg/m² (18.32 lb/sq ft)
Max power loading 7.10 kg/kW (11.67 lb/hp)
PERFORMANCE:
Max operating speed 152 kt (281 km/h; 175 mph)
Max cruising speed 143 kt (266 km/h; 165 mph)
Stalling speed: flaps up 37 kt (68 km/h; 42 mph)
 flaps down 33 kt (62 km/h; 38 mph)
Max rate of climb at S/L 488 m (1,600 ft)/min
Service ceiling 6,096 m (20,000 ft)
T-O run 76 m (250 ft)
Landing run 122 m (400 ft)
Range with max fuel 791 n miles (1,466 km; 911 miles)
g limits +3.8/−1.5
NEW ENTRY

SUPER-CHIPMUNK

SUPER-CHIPMUNK INC
109, Route 201, St Louis-de-Gonzague, Québec J0S 1T0
Tel: (+1 450) 373 07 06
Fax: (+1 450) 373 31 44
e-mail: info@super-chipmunk.com
Web: http://www.super-chipmunk.com
PRESIDENT: Gilles Leger

This company was launched to manufacture a modern reproduction of the de Havilland Canada DHC-1 Chipmunk, externally similar rear of the firewall, but with all-new structural design. Modification kits for existing Chipmunks are available.
UPDATED

SUPER-CHIPMUNK SUPER-CHIPMUNK

TYPE: Aerobatic, two-seat lightplane kitbuilt.
PROGRAMME: Original Chipmunk described in *Jane's Aircraft Upgrades*. Earlier Super Chipmunk modification of surplus military aircraft, with Textron Lycoming IO-360 engine, achieved only limited sales.
Prototype of current Super-Chipmunk (C-GLSC), with similar power plant, constructed in 1999 from DHC-1 wings and rear fuselage, but with steel tube centre section providing additional accommodation for pilots of above average height and/or girth. Debut at Sun 'n' Fun, Lakeland, Florida, April 2000, when all-new fuselage of third aircraft also exhibited. Complete airframe will be available for assembly from kit.
COSTS: Kit US$33,900 (2000).
DESIGN FEATURES: Primary trainer adapted for civilian sporting use. Low wing and mid-mounted tailplane, both tapered; elliptical fin/rudder. Super-Chipmunk features new internal structure and Canadian DHC-1 standard single-piece canopy and lack of strakes ahead of tailplane.
FLYING CONTROLS: Conventional and manual. Horn-balanced rudder and elevators. Flaps. Trim tabs on both elevators.
STRUCTURE: Metal throughout, except fabric-covered rudder, ailerons, elevator and flaps.
LANDING GEAR: Tailwheel type; fixed. Mainwheel brakes.
POWER PLANT: One 149 kW (200 hp) Textron Lycoming IO-360-C flat-six driving a two-blade propeller. Alternative engines between 112 and 224 kW (150 and 300 hp) with fixed-pitch or constant-speed propellers. Fuel capacity 114 litres (30.0 US gallons; 25.0 Imp gallons) normal, 189 litres (50.0 US gallons; 41.7 Imp gallons)

Prototype Super-Chipmunk reproduction of de Havilland DHC-1 *(Paul Jackson/Jane's)* 2000/0084582

optional. Oil capacity 9.5 litres (2.5 US gallons; 2.0 Imp gallons).
ACCOMMODATION: Two in tandem, beneath single-piece, rearward-sliding canopy with fixed windscreen. Single cockpit.
DIMENSIONS, EXTERNAL:
Wing span 10.57 m (34 ft 8 in)
Length overall 7.75 m (25 ft 5 in)
Height overall 2.13 m (7 ft 0 in)
Tailplane span 3.63 m (11 ft 11 in)
Wheel track 2.84 m (9 ft 4 in)
Wheelbase 5.23 m (17 ft 2 in)
DIMENSIONS, INTERNAL:
Cockpit: Max height 1.40 m (4 ft 7 in)
Width: max 0.91 m (3 ft 0 in)
 at shoulders 0.69 m (2 ft 3 in)
AREAS:
Wings, gross 16.07 m² (173.0 sq ft)
Ailerons (total) 1.29 m² (13.90 sq ft)
Flaps (total) 2.04 m² (22.00 sq ft)
Fin 0.55 m² (5.90 sq ft)
Rudder 0.63 m² (6.80 sq ft)
Tailplane 1.58 m² (17.00 sq ft)
Elevators (total) 1.41 m² (15.20 sq ft)
WEIGHTS AND LOADINGS:
Weight empty 693 kg (1,527 lb)
Baggage capacity 40 kg (90 lb)
Max T-O weight 1,088 kg (2,400 lb)
Max wing loading 8.44 kg/m² (13.87 lb/sq ft)
Max power loading 58.6 kg/m² (12.00 lb/hp)
PERFORMANCE:
Never-exceed speed (V_{NE}) 173 kt (321 km/h; 200 mph)
Max level speed 156 kt (290 km/h; 180 mph)
Cruising speed at 75% power
 139 kt (257 km/h; 160 mph)
Max rate of climb at S/L 610 m (2,000 ft)/min
Service ceiling 5,245 m (17,200 ft)
T-O to 15 m (50 ft) 204 m (670 ft)
Landing from 15 m (50 ft) 283 m (930 ft)
Max range with optional fuel
 521 n miles (965 km; 600 miles)
g limits +9/−6
UPDATED

WINDEAGLE

WINDEAGLE AIRCRAFT CORPORATION
Brief details of the Windeagle light aircraft appear under the CAG heading in this section.
VERIFIED

ZENAIR

ZENAIR LTD
Huronia Airport, Midland, Ontario L4R 4K8
Tel: (+1 705) 526 28 71
Fax: (+1 705) 526 80 22
e-mail: zenair@zenair.com
Web: http://www.zenair.com
PRESIDENT AND DESIGNER: Christophe Heintz
VICE-PRESIDENT: Mathieu Heintz
GENERAL MANAGER: Bruce Barker

Founded in 1974, Zenair designs and develops light aircraft. Christophe Heintz was formerly chief engineer of Avions Pierre Robin in France. In 1992 Zénith Aircraft Company was formed to manufacture and market all models except the CH 2000 in the USA (which see), and in September 1996 reached full production status. In June 1999, Zenith announced that it would transfer production of the CH 2000 to a new, purpose-built 2,800 m² (30,000 sq ft) factory at Eastman, Georgia; Aircraft Manufacturing and Development Company (which see in US section) was formed to produce the aircraft.

Zenair has licensed representatives in Africa, South America, Belgium, Czech Republic (see entry for CZAW), France, Germany, Israel, Italy, Spain and the UK.
Zenith Aircraft Company (Mexico, Missouri, USA) licenses kit production of Zenair designs. In addition to light aircraft, company constructs metal floats and amphibious floats.
UPDATED

CHILE

ENAER
EMPRESA NACIONAL DE AERONÁUTICA DE CHILE
Avenida José Miguel Carrera 11087, El Bosque, Santiago
Tel: (+56 2) 383 18 73 and 383 18 63
Fax: (+56 2) 05 28 26 99
e-mail: enaercom@interaccess.cl
Web: http://www.enaer.cl
PRESIDENT: Gen Patricio Rios
MANAGING DIRECTOR: Brig Gen Alfredo Guzmán
TECHNICAL DIRECTOR: Rodolfo Pinto
COMMERCIAL DIRECTOR: Florencio Dublé
SALES DIRECTOR: Manuel Vargas
MARKETING MANAGER: Felipe Contardo
COMMUNICATIONS MANAGER: Cecilia Sola

State-owned company, formed 1984 from IndAer industrial organisation set up 1980 by Chilean Air Force (FACh); aircraft manufacture started 1979 with assembly of 27 Piper PA-28 Dakota lightplanes for Chilean Air Force and flying clubs. Covered plant area 47,000 m² (505,900 sq ft); workforce of 1,200 in 1997 (latest figure supplied). Sales turnover for 1999 was more than US$50 million.

Recent main programmes have included Pantera upgrade of FACh Mirage 50s, Tigre upgrade of FACh F-5Es and co-production of CASA C-101 as A-36 Halcón. Has modified, with IAI assistance, a Chilean Air Force Boeing 707-320B to combined tanker/AEW configuration with similar nose and fuselage antenna fairings to those of original FACh Phalcon plus hose and drogue aerial refuelling system. Currently undertaking sole-source fin/tailplane manufacture of ERJ-145/135 as Embraer risk-sharing partner, and subcontract manufacture of part of tail unit of Airtech CN-235; is also to manufacture components for Dassault Falcon 900 and Falcon 2000. Has recently collaborated with Snow Aviation of USA in developing a cockpit avionics upgrade package for Cessna T-37.

Own-design programmes have comprised T-35 piston-engined Pillán (recently returned to production), T-35DT Turbo Pillán military trainer (1996-97 and earlier *Jane's*) and Ñamcu light aircraft. Last-named is now produced in Netherlands as the Euro-ENAER EE-10 Eaglet (which see); ENAER is South American distributor for this aircraft.

UPDATED

ENAER (ECH-51) T-35 PILLÁN
English name: Devil
Spanish Air Force designation: E.26 Tamiz (Grader)
TYPE: Basic prop trainer.
PROGRAMME: First two prototypes (first flight 6 March 1981) developed by Piper; followed by three Piper kits for ENAER assembly (first flight, by FACh s/n 101, 30 January 1982); then slight redesign, replacing all-moving tailplane by electrically trimmable tailplane with conventional elevator, increasing rudder mass balance and deepening canopy; ENAER series production started September 1984; first flight of production T-35A, 28 December 1984; deliveries to FACh began 31 July 1985. Supply of ENAER T-35C kits to CASA started 27 December 1985 (first flight 12 May 1986), completed September 1987; T-35D deliveries to Panama began 1988; first flight of single-seat T-35S made 5 March 1988; T-35A/B deliveries to FACh completed by second quarter of 1990; original export deliveries completed 1991. Production resumed in late 1990s to build eight for Dominican Republic, which were delivered 8 October 1999 (four) and 24 January 2000 (four). Further Dominican contract for four announced in May 2000; these included in planned production of at least 12 in 2000, to meet renewed interest from Latin American countries.
CURRENT VERSIONS: **T-35A:** Primary trainer version for Chilean Air Force. In service with Escuela de Aviación 'Capitán Avalos' at El Bosque AB, Santiago.
Detailed description applies to the above version except where indicated.
T-35B: Chilean Air Force instrument trainer, with more comprehenive instrumentation.
T-35C: Primary trainer for Spanish Air Force (designation E.26 Tamiz), assembled by CASA from ENAER kits.
T-35D: Instrument trainer for Panama and Paraguay.
T-35S: Single-seat prototype (CC-PZB); illustrated in 1991-92 *Jane's*.
T-35DT Turbo Pillán: Turboprop version; described in 1996-97 and earlier editions. Prototype set Class C1c point-to-point speed record of 209.8 kt (388.5 km/h; 241.4 mph) on 10 March 2000. Reported interest expressed by two potential customers, but no orders announced by May 2000.
Pillán 2000: Version proposed in 1998, retaining the existing T-35 fuselage, tail unit and piston engine, combined with a lighter weight but increased span wing designed by the Russian companies Technoavia and Tyazhpromexport. Plans also included the option of a turboprop version. No go-ahead for this version had been announced at the time of going to press.
CUSTOMERS: Total of 111 (excluding prototypes and trials aircraft) built by 1991 for air forces of Chile (34 T-35A, 14 T-35B), Panama (10 T-35D), Paraguay (13 T-35D) and Spain (40 T-35C/E.26). Surplus FACh aircraft sold subsequently to El Salvador (five) and Guatemala (five). Production restarted in response to November 1998 order for eight from Dominican Republic; resumed again in 2000 after follow-on order for four from same customer. During the FIDAE Air Show in February 2000, ENAER reported purchase agreements by three Latin American countries, worth US$8.5 million, for 12 to 14 Pilláns.
COSTS: "Less than US$1 million", T-35DT (2000).
DESIGN FEATURES: Low-risk, low-cost and international marketability achieved by reliance on Piper for development, some structural components and FAA compliance. Based on Piper Cherokee series (utilising many components of PA-28 Dakota and PA-32 Saratoga); cleared to FAR Pt 23 (Aerobatic category) and military standards for basic, intermediate and instrument flying training. Low-wing, tandem-seat design; sweptback vertical tail, non-swept horizontal surfaces.
Wing section NACA 65_2-415 on constant chord inboard panels. NACA 65_2-415 (modified) at tips; incidence 2° at root, −0° 30′ at tip; dihedral 7° from roots.
FLYING CONTROLS: Conventional and manual. Mass-balanced elevators and rudder; single-slotted wing flaps, aileron trim tab (port) and tailplane/elevator trim all electrically actuated; variable incidence tailplane.
STRUCTURE: Main structure of aluminium alloy and steel, with riveted skins, except for glass fibre engine cowling, wingtips and tailplane tips. Single-spar fail-safe wings, with components mainly from PA-28-236 Dakota (leading-edges) and PA-32R-301 Saratoga (trailing-edges), modified for shorter span; vertical tail virtually identical with Dakota; tailplane uses some standard components from Dakota and PA-31 (Navajo/Cheyenne); tailcone from Cherokee components, modified for narrower fuselage.
LANDING GEAR: Hydraulically retractable tricycle type, with single wheel on each unit. Main units retract inward, steerable nosewheel rearward. Piper oleo-pneumatic shock-absorber in each unit. Emergency free-fall extension. Cleveland mainwheels and McCreary tyres size 6.00-6 (8 ply), nosewheel and tyre size 5.00-5 (6 ply). Tyre pressures: 2.62 bars (38 lb/sq in) on mainwheels, 2.41 bars (35 lb/sq in) on nosewheel. Single-disc air-cooled hydraulic brake on each mainwheel. Parking brake. Minimum ground turning radius 6.20 m (20 ft 4 in).
POWER PLANT: One 224 kW (300 hp) Textron Lycoming IO-540-K1K5 flat-six engine, driving a Hartzell HC-C3YR-4BF/FC7663R three-blade constant-speed metal propeller. Fuel in two integral aluminium tanks in wing leading-edges, total capacity 291.5 litres (77.0 US gallons; 64.1 Imp gallons), of which 278 litres (73.4 US gallons; 61.1 Imp gallons) are usable. Overwing gravity refuelling point on each wing. Oil capacity 11.4 litres (3.0 US gallons; 2.5 Imp gallons). Fuel and oil systems permit up to 40 seconds of inverted flight.
ACCOMMODATION: Two vertically adjustable seats, with seat belts and shoulder harnesses, in tandem beneath one-piece transparent jettisonable canopy which opens sideways to starboard. One-piece acrylic windscreen and one-piece window in glass fibre fairing aft of canopy. Rear (instructor's) seat 22 cm (8.7 in) higher. Dual controls standard. Baggage compartment aft of rear cockpit, with external access on port side. Cockpits ventilated; cockpit heating and canopy demisting by engine bleed air.
SYSTEMS: Electrically operated hydraulic system, at 124 bars (1,800 lb/sq in) pressure for landing gear retraction and 44.8 bars (650 lb/sq in) for gear extension; separate system at 20.7 bars (300 lb/sq in) for wheel brakes. Electrical system is 24 V DC, powered by 28 V 70 A engine-driven Prestolite alternator and 24 V 15.5 Ah lead-acid battery, with inverter for AC power at 400 Hz to operate RMIs and attitude indicators. External power socket. No oxygen or de-icing provisions.
AVIONICS: Standard basic avionics by Rockwell Collins; optional items by Honeywell.
Comms: Two VHF-251 com transceivers, two AMR-350 audio selector panels, TDR-950 transponder and Isocom interphone standard; KX 165 VHF/VOR, KMA 24H audio selector and KT 76A transponder optional.
Flight: VIR-351 VOR with IND-350A, ADF-650A with IND-650 indicator and TOR-950 IFF standard; options include KR 87 ADF with KA 44B indicator, KN 63 DME with KDI 572 or IND 450 indicator, KDI 573 DME remote indicator, KR 21 or KML 351 marker beacon receiver, KA 40 marker beacon remote, KI 525A HSI and KNI 582 RMI.

ENAER Pillán in service with the Chilean Air Force   2000/0062607

DIMENSIONS, EXTERNAL:
Wing span	8.84 m (29 ft 0 in)
Wing aspect ratio	5.7
Length overall	8.00 m (26 ft 3 in)
Height overall	2.64 m (8 ft 8 in)
Tailplane span	3.05 m (10 ft 0 in)
Wheel track	3.02 m (9 ft 11 in)
Wheelbase	2.09 m (6 ft 10¼ in)
Propeller diameter	1.93 m (6 ft 4 in)

DIMENSIONS, INTERNAL:
Cockpit: Length	3.24 m (10 ft 7½ in)
Max width	1.04 m (3 ft 5 in)
Max height	1.48 m (4 ft 10¼ in)

AREAS:
Wings, gross	13.69 m² (146.3 sq ft)

WEIGHTS AND LOADINGS:
Weight empty, equipped	930 kg (2,050 lb)
Fuel weight	210 kg (462 lb)
Max aerobatic T-O weight	1,315 kg (2,900 lb)
Max T-O and landing weight	1,338 kg (2,950 lb)
Max wing loading	97.73 kg/m² (20.03 lb/sq ft)
Max power loading	5.98 kg/kW (9.83 lb/hp)

PERFORMANCE:
Never-exceed speed (V_{NE})	241 kt (446 km/h; 277 mph)
Max level speed at S/L	168 kt (311 km/h; 193 mph)
Cruising speed:	
at 75% power at 2,680 m (8,800 ft)	144 kt (266 km/h; 166 mph) IAS
at 55% power at 5,120 m (16,800 ft)	138 kt (255 km/h; 159 mph) IAS
Stalling speed: flaps up	67 kt (125 km/h; 78 mph)
flaps down	62 kt (115 km/h; 72 mph)
Max rate of climb at S/L	465 m (1,525 ft)/min
Time to: 1,830 m (6,000 ft)	4 min 42 s
3,050 m (10,000 ft)	8 min 48 s
Service ceiling	5,840 m (19,160 ft)
Absolute ceiling	6,250 m (20,500 ft)
T-O run	287 m (940 ft)
T-O to 15 m (50 ft)	494 m (1,620 ft)
Landing from 15 m (50 ft)	509 m (1,670 ft)
Landing run	238 m (780 ft)
Range with 45 min reserves:	
at 75% power at 2,440 m (8,000 ft)	590 n miles (1,093 km; 679 miles)
at 55% power at 3,660 m (12,000 ft)	650 n miles (1,204 km; 748 miles)
Range, no reserves:	
at 75% power at 2,440 m (8,000 ft)	680 n miles (1,260 km; 783 miles)
at 55% power at 3,660 m (12,000 ft)	735 n miles (1,362 km; 846 miles)
Endurance at S/L: at 75% power	4 h 24 min
at 55% power	5 h 36 min
g limits	+6/−3

UPDATED

CHINA, PEOPLE'S REPUBLIC

AVIC

CHINA AVIATION INDUSTRY CORPORATION (Zhongguo Hangkong Gongye Zonggongsi)
67 Jiao Nan Street (PO Box 33), Beijing 100009
Tel: (+86 10) 64 09 31 14
Fax: (+86 10) 64 01 36 48
PRESIDENTS:
 Liu Gaozhuo (AVIC I)
 Zhang Yanzhong (AVIC II)
VICE-PRESIDENTS:
 Yang Yuzhong (AVIC I)
 Liang Zhenhe (AVIC II)
DIRECTORS-GENERAL OF MARKETING:
 Tang Xiaoping (AVIC I)
 Cui Degang (AVIC II)

INTERNATIONAL MARKETING:
CATIC (Zhongguo Hangkong Jishu Jinchukou Gongsi: China National Aero-Technology Import and Export Corporation)
CATIC Plaza, 8 Beichen East Street, Chaoyang District, Beijing 100101
Tel: (+86 10) 64 94 03 86
Fax: (+86 10) 64 94 03 85
PRESIDENT: Yang Chunshu
DIRECTOR, PUBLIC RELATIONS: Bi Jianfa

Present Chinese aviation industry created in 1951 and has since manufactured some 14,000 aircraft (including more than 10,000 military), more than 50,000 aero-engines and 10,000 air-to-air and tactical missiles. Some 700 aircraft have been exported, approximately 10 per cent of them civil types.

Former Ministry of Aero-Space Industry abolished 1993 and AVIC created on 26 June 1993 as economic entity to develop market economy and expand international collaboration in aviation programmes. CATIC Group formed 26 August 1993, with CATIC (founded January 1979) as its core company, to be responsible for import and export of aero and non-aero products, subcontract work and joint ventures.

Xian, Chengdu, Shanghai, Shenyang, Harbin and other factories carry out subcontract work on Airbus A300/310/318/320; ATR 42; Avro RJ; Boeing 737/747/757 and MD-80/90 series; de Havilland Dash 8Q; and Canadair CL-415. Licensed manufacture of Sukhoi Su-27s is undertaken at Shenyang. Co-development with UK/France/Italy and Singapore of AE-100 regional airliner abandoned in 1998 and replaced by New Regional Jet project for 58/76-seat aircraft (see NRJ entry in this section). Most factories also have wide range of non-aerospace products.

Total workforce of aerospace industry was reduced to about 500,000 in 1998, when about 34,000 workers were made redundant and some 14,000 others transferred to non-aerospace activities. AVIC's President announced plans at Airshow China in November 1998 to restructure aircraft industry into two "competing and co-operating" groups in the near future. Draft organisation plan submitted to State Council in February 1999, resulting in division of AVIC into two separate companies (AVIC I and II) with effect from July 1999. Initial workforces are 281,000 and 220,000, although further reductions are expected during the next few years. AVIC I comprises 104 enterprises and 31 research institutes; corresponding figures for AVIC II are 79 and three. Principal organisations in the new structure, and their major programmes, are as follows. AVIC I plans to separate its military programmes in 2003; AVIC II likely to follow a similar path.

AVIC I
Beijing Aviation Simulator
Chengdu Aircraft Industry Group (J-7/F-7, FC-1 and J-10)
China Air-to-Air Missile Research Institute
Guizhou Aviation Industry Group (JJ-7/FT-7; jet engines; missiles)
NRJ programme
Shanghai Aviation Industry Group (airliners)
Shenyang Aircraft Industry Group (J-8/F-8 and J-11/Su-27; civil subcontracts)
Xian Aero-Engine (WP8 and WS9)
Xian Aircraft Industry Group (JH-7/FBC-1, Y-7 and MA 60)

AVIC II
Changhe Aircraft (Z-11)
Chengdu Engine (WP7 and WP13)
China Helicopter Research and Development Institute (Z-10)
Dongan Engine (WJ5)
Harbin Aircraft Industry Group (Hafei Z-9 and Y-12)
Hongdu Aviation Industry Group (NAMC Q-5/A-5, K-8 and N-5)
Shaanxi Aircraft (Y-8)
Shijiazhuang Aircraft Manufacturing (Y-5)
South Aero-Engine (HS5, WJ6 and WZ8)
UPDATED

BKLAIC

BEIJING KEYUAN LIGHT AIRCRAFT INDUSTRIAL COMPANY LTD
7 Keyuan South Road, Zhongguanchun, Haidian, Beijing 100080
Tel: (+86 10) 62 57 29 22, 62 57 28 22 and 62 55 79 43
Fax: (+86 10) 62 57 29 22
CHAIRMAN: Yuan Yong Min
VICE GENERAL MANAGER: Fangjun Dong

This company produces, as the AD-200, the tandem two-seat version of the AD-100 light aircraft described under the NLA heading in this section. It is also responsible for the Zhong Hua hot-air airship (see Lighter Than Air section, 2000-01 and previous editions).
UPDATED

BKLAIC AD-200
English name: Blue Eagle
TYPE: Two-seat lightplane.
PROGRAMME: Design started September 1987; prototype construction began June 1988; conforms to Chinese (CCAR) equivalents of FAR Pts 21 and 23. Generally similar to AD-100T (see under NLA in this section) except for tandem seating and increased dimensions. First flight 1 September 1988; CAAC type and production certificates awarded 26 December 1995. Initial AD-200A version, with 32.1 kW (43 hp) Rotax 447 engine, no longer produced; details in 1998-99 *Jane's*.
CURRENT VERSIONS: **AD-200B:** Current version, first flown March 1996. More powerful engine and increased payload.
 AD-200N: Preceded AD-200B (first flight October 1995), to which it is similar except for fitment of agricultural equipment: 120 kg (265 lb) pesticide tank in rear cockpit, plus underwing spraybars.
CUSTOMERS: Total of 30 ordered (all versions), of which 20 delivered, by December 1997 (latest figure received).
COSTS: AD-200B US$25,000; AD-200N US$28,000 (1997).
Following description applies to both versions except where indicated.
DESIGN FEATURES: Mid-mounted wings with endplate fins and horn-balanced rudders; foreplanes; pusher engine.
 Wing has aerofoil section 748 with thickness/chord ratio of 15 per cent, 15° sweep on leading-edge, 1° dihedral and 0° incidence; no twist.
FLYING CONTROLS: Manually operated ailerons, foreplane elevators and twin rudders.
STRUCTURE: Glass fibre and wood.
LANDING GEAR: Fixed tricycle type with self-sprung shock-absorption. Mainwheel tyres 373×120, pressure 3.00 to 3.40 bars (43.5 to 49 lb/sq in); nosewheel tyre 299×92, pressure 2.00 to 2.50 bars (29 to 36 lb/sq in).
POWER PLANT: One 47.8 kW (64.1 hp) Rotax 582 UL-2V two-cylinder two-stroke engine, driving a two-blade fixed-pitch pusher propeller. Fuel capacity 49 litres (12.9 US gallons; 10.8 Imp gallons), of which 46 litres (12.2 US gallons; 10.1 Imp gallons) are usable. Refuelling point on top of fuselage.
ACCOMMODATION: Separate cockpit canopies, opening sideways to starboard.
SYSTEMS: 12 V 16 Ah battery.
AVIONICS: *Comms:* Radio optional.
 Instrumentation: ASI, VSI, altimeter and compass.

DIMENSIONS, EXTERNAL:
Wing span	9.41 m (30 ft 10½ in)
Wing chord: at root	1.45 m (4 ft 9 in)
at tip	0.93 m (3 ft 0½ in)
Wing aspect ratio	7.8
Foreplane span	3.80 m (12 ft 5½ in)
Length overall	6.29 m (20 ft 7¾ in)
Fuselage: Length	4.14 m (13 ft 7 in)
Max width	0.98 m (3 ft 2½ in)
Height overall	2.01 m (6 ft 7¼ in)
Wheel track	1.64 m (5 ft 4½ in)
Wheelbase	2.44 m (8 ft 0 in)
Propeller diameter	1.73 m (5 ft 8 in)
Propeller ground clearance	0.48 m (1 ft 7 in)

DIMENSIONS, INTERNAL:
Cockpits: Total length	2.08 m (6 ft 10 in)
Max width: front	0.68 m (2 ft 2¾ in)
rear	0.59 m (1 ft 11¼ in)
Max height: front	1.05 m (3 ft 5¼ in)
rear	1.12 m (3 ft 8 in)

AREAS:
Wings, gross	11.29 m² (121.5 sq ft)
Foreplanes, gross	3.04 m² (32.7 sq ft)
Ailerons (total)	0.39 m² (4.20 sq ft)
Fins (total)	1.30 m² (13.99 sq ft)
Rudders (total)	0.44 m² (4.74 sq ft)
Elevators (total)	0.53 m² (5.70 sq ft)

WEIGHTS AND LOADINGS:
Weight empty, equipped: 200B	285 kg (628 lb)
200N	335 kg (739 lb)
Max fuel weight	40 kg (88 lb)
Max payload:	180 kg (397 lb)
Max T-O and landing weight	550 kg (1,212 lb)
Max wing loading	48.7 kg/m² (9.98 lb/sq ft)
Max power loading	11.53 kg/kW (18.94 lb/hp)

PERFORMANCE:
Never-exceed speed (V$_{NE}$)	85 kt (158 km/h; 98 mph)
Max level speed	81 kt (150 km/h; 93 mph)
Max cruising speed	76 kt (140 km/h; 87 mph)
Econ cruising speed	57 kt (105 km/h; 65 mph)
Crop-spraying speed:	
200N	59-65 kt (110-120 km/h; 68-75 mph)
Stalling speed	37 kt (68 km/h; 43 mph)
Max rate of climb at S/L: 200B	210 m (689 ft)/min
200N	more than 180 m (591 ft)/min
Service ceiling	3,000 m (9,840 ft)
T-O and landing run	less than 100 m (330 ft)
T-O to 15 m (50 ft)	278 m (915 ft)
Landing from 15 m (50 ft)	280 m (920 ft)
Crop-spraying swath width: 200N	22-30 m (72-98 ft)
Max range	216 n miles (400 km; 248 miles)
g limits	+3.8/-1.9

UPDATED

Beijing Keyuan AD-200B tandem two-seat floatplane *(Photo Link)*

The AD-200N with underwing spraybars *(Sebastian Zacharias)*

BUAA

BEIJING UNIVERSITY OF AERONAUTICS AND ASTRONAUTICS
37 Xue Yuan Road, Haidian, Beijing 100083
Tel: (+86 10) 82 31 75 91
Fax: (+86 10) 82 31 75 90
e-mail: webmaster@buaa.edu.cn

Beijing University's Aeronautics and Astronautics department has designed several microlight aircraft during the past two decades, under the series name Mifeng (Bee). Its most recent known manned aircraft project is the M-16 small autogyro, which made its international debut in November 2000.

UPDATED

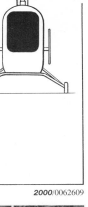

BUAA MIFENG M-16
English name: Bee
TYPE: Single-seat autogyro ultralight.
PROGRAMME: Prototype completed August 1998. Seen publicly at Airshow China, Zhuhai, November 2000.
DESIGN FEATURES: Twin two-blade contrarotating rotors; open-sided cabin pod with one-piece flush windscreen; open framework fuselage; tailplane with twin endplate fins and rudders.
FLYING CONTROLS: Manually actuated.
STRUCTURE: Main framework of steel tube.
LANDING GEAR: Twin-skid type.
POWER PLANT: One 59.7 kW (80 hp) Rotax 912A four-cylinder four-stroke engine; two-blade pusher propeller.
DIMENSIONS, EXTERNAL:
  Rotor diameter, each   5.00 m (16 ft 4¾ in)
  Fuselage length   3.60 m (11 ft 9¾ in)
  Height overall   2.20 m (7 ft 2½ in)
AREAS:
  Rotor discs, each   19.63 m² (211.3 sq ft)
WEIGHTS AND LOADINGS:
  Weight empty   220 kg (485 lb)
  Max T-O weight   350 kg (771 lb)
PERFORMANCE:
  Max level speed   71 kt (132 km/h; 82 mph)
  Max cruising speed   54 kt (100 km/h; 62 mph)
  Econ cruising speed   32 kt (60 km/h; 37 mph)
  Max rate of climb at S/L   432 m (1,417 ft)/min
  Service ceiling   3,000 m (9,840 ft)
  Endurance   1 h 36 min

UPDATED

BUAA Mifeng M-16 single-seat light autogyro *(James Goulding/Jane's)*

Mifeng M-16 displayed at Zhuhai, November 2000 *(Photo Link)*

CAC

CHENGDU AIRCRAFT INDUSTRIAL GROUP (Subsidiary of AVIC I)
PO Box 800, Chengdu, Sichuan 610092
Tel: (+86 28) 740 10 33
Fax: (+86 28) 740 49 84
e-mail: cacoa@mail.cac.com.cn
Web: http://www.cac.com.cn
PRESIDENT: Yang Baoshu
GENERAL MANAGER: Yang Tingkuo
DEPUTY GENERAL MANAGER: Li Shaoming
DIRECTOR, INTERNATIONAL CO-OPERATION DIVISION:
  Wang Yinggong

Major centre for fighter development and production, founded 1958; has since built over 2,000 fighters of more than 10 models or variants; current facility occupies site area of 462 ha (1,142 acres) and had 1995 workforce (latest figure known) of nearly 20,000. Production continues mainly to concern J-7/F-7 fighter series (several models); limited batch production of JJ-5/FT-5 fighter trainer (see 1994-95 and earlier editions) has ended. New fighters are currently under development (see J-10 and FC-1 entries which follow).

Subcontract work includes passenger doors for the Airbus A320. Non-aerospace products previously accounted for about 10 per cent of current output, but were targeted to reach 40 per cent by 2000.

UPDATED

CAC (MIKOYAN) J-7
Chinese name: Jianjiji-7 (Fighter aircraft 7)
Westernised designation: F-7
TYPE: Multirole fighter.
PROGRAMME: Soviet licence to manufacture MiG-21F-13 and its R-11F-300 engine granted 1961, when some pattern aircraft and CKD kits delivered, but technical documentation not completed; assembly of first J-7 using Chinese-made components began early 1964; original plan for Chengdu and Guizhou to become main airframe/engine production centres, backed up by Shenyang until these were fully productive, delayed by cultural revolution.

First flight of Shenyang-built J-7, 17 January 1966; Chengdu production of J-7 I began June 1967 (first flight 16 June 1969); development of J-7 II began 1975, followed by first flight 30 December 1978 and production approval September 1979; development of F-7M and J-7 III started 1981; J-7 III first flight 26 April 1984; F-7M (first flight 31 August 1983) revealed publicly October 1984, production go-ahead December 1984, named Airguard early 1986; first F-7P deliveries to Pakistan 1988; first F-7MPs to Pakistan mid-1989; F-7MG public debut November 1996; J-7FS revealed September 1998 and F-7MF November 2000.

CURRENT VERSIONS (domestic): **J-7**: Initial Shenyang licence version using Chinese-made components; few only.
  J-7 I: Initial Chengdu version (1967), with variable intake shock cone and second 30 mm gun; not accepted in large numbers.
  J-7 II: Modified and improved J-7 I, with WP7B turbojet of increased thrust (43.2 kN; 9,700 lb st dry, 59.8 kN; 13,450 lb st with afterburning); 720 litre (190 US gallon; 158 Imp gallon) centreline drop tank for increased range; brake-chute relocated at base of rudder to improve landing performance and shorten run; rear-hinged canopy, jettisoned before ejection seat deploys; new Chengdu Type II seat operable at zero height and speeds down to 135 kt (250 mph; 155 mph); and new Lanzhou compass system. Small batch production (typically, 14 in 1989), notwithstanding advent of J-7 III and J-7E. Still the major PLAAF variant.
  J-7 III: Chinese equivalent of MiG-21MF, with blown flaps and all-weather, day/night capability. Main improvements are change to WP13 engine with greater power; additional fuel in deeper dorsal spine; JL-7 (J-band) interception radar, with correspondingly larger nose intake and centrebody radome; sideways-opening (to starboard) canopy, with centrally located rearview mirror; improved HTY-4 low-speed/zero height ejection seat; more advanced fire-control system; twin-barrel 23 mm gun under fuselage (with HK-03D optical gunsight); broader-chord vertical tail surfaces, incorporating antennas for LJ-2 omnidirectional RWR in hemispherical fairing each side at base of rudder; increased weapon/stores capability (four underwing stations), similar to that of F-7M; and new or additional avionics (which see). Joint development by Chengdu and Guizhou (GAIC); entered PLA Air Force and Navy service from 1992; production continuing in 1996 (latest report).
  J-7E: Upgraded version of J-7 II with modified, double-delta wing, retaining existing leading-edge sweep angle of 57° inboard but reduced sweep of only 42° outboard; span increased by 1.17 m (3 ft 10 in) and area by 1.88 m² (20.2 sq ft), giving 8.17 per cent more wing area; four underwing stations instead of two, outer pair each plumbed for 480 litre (127 US gallon; 106 Imp gallon) drop tank; new WP7F version of WP7 engine, rated at 44.1 kN (9,921 lb st) dry and 63.7 kN (14,330 lb st) with afterburning; armament generally as listed for F-7M, but capability

Canards and rectangular chin intake characterise the F-7MF *(Photo Link)*

extended to include PL-8 (Python 3) air-to-air missiles; *g* limits of 8 (up to M0.8) and 6.5 (above M0.8); avionics include head-up display and air data computer. Believed to have made first flight in April 1990 and entered service 1993. In production.

J-7EB: Version of J-7E equipping PLA Air Force 'August 1st' aerobatic team (nine in 1998 display season); fitted with smoke canisters for display purposes.

J-7FS: Technology demonstrator, modified from standard J-7 II, with new chin-mounted intake and central splitter plate under reconfigured ogival nosecone; more powerful (73.4 to 78.3 kN; 16,502 to 17,604 lb thrust class with afterburning) Liyang (LMC) WP13F II turbojet. Began 22 month flight test programme on 8 June 1998. Enlarged nose avionics bay able to accept 60 cm (23.6 in) diameter multimode pulse Doppler radar believed to be under development in China. Development work has formed basis for new F-7MF export variant (which see). Planned future changes include wing modifications based on J-7E/F-7MG double-delta configuration.

JJ-7: Tandem two-seat operational trainer, based on J-7 II and MiG-21US; developed at Guizhou and described under GAIC entry.

CURRENT VERSIONS (export): **F-7A:** Export counterpart of J-7 I, supplied to Albania and Tanzania.

F-7B: Export version of J-7 II, with R550 Magic missile capability; supplied to Egypt and Iraq in 1982-83 and also to Sudan. Some supplied to Air Force of Zimbabwe (known locally as F-7 II) appear to be of this version.

F-7BS: Hybrid version supplied to Sri Lanka 1991: has F-7B fuselage/tail and Chinese avionics (no HUD), combined with four-pylon wings of F-7M. Equips No. 5 Squadron of SLAF. Zimbabwe also has some four-pylon aircraft with its No. 5 Squadron; these known locally as the F-7 IIN.

F-7M Airguard: Upgraded export version, developed from J-7 II; new avionics imported from May 1979 included Marconi HUDWAC (head-up display and weapon aiming computer); new ranging radar, air data computer, radar altimeter and IFF; more secure com radio; improved electrical power generation system for the new avionics; two additional underwing stores points; improved WP7B(BM) engine; birdproof windscreen; strengthened landing gear; ability to carry PL-7 air-to-air missiles; nose probe relocated from beneath intake to top lip of intake, offset to starboard. Exported to Iran and Myanmar. Reported in early 1999 that Zimbabwe planned to purchase 12, but no order yet confirmed.

Description applies to F-7M version, except where indicated.

F-7MB: Variant of F-7M; mentioned in 1996 F-7MG brochure. Customer believed to be Bangladesh (16); one planned to be equipped with reconnaissance pod.

F-7N: Variant of F-7M; mentioned in 1996 F-7MG brochure, but no details given. Possibly an alternative designation for Zimbabwe F-7 IIN (see F-7BS paragraph above).

F-7P Airguard: Variant of F-7M (briefly called Skybolt), embodying 24 modifications to meet specific requirements of Pakistan Air Force, including ability to carry four air-to-air missiles (Sidewinders) instead of two and fitment of Martin-Baker Mk 10L ejection seat. Delivered 1988-91. Planned upgrade with improved (±20° compared with ±10° scan) version of Grifo 7 radar.

F-7MP: Further modified variant of F-7P; improved cockpit layout and navigation system incorporating

'Red 139', the Chengdu J-7FS technology demonstrator prototype
(Li Yong/China Aviation News) 1999/0044054

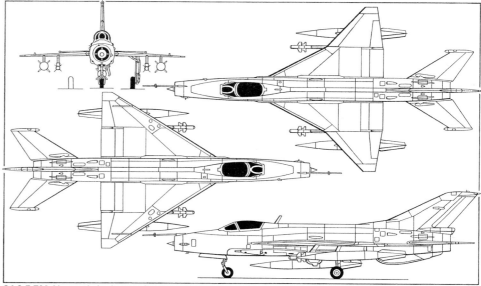

CAC F-7M Airguard single-seat fighter and close support aircraft; upper plan view shows modified outer wings of J-7E and F-7MG *(Mike Keep/Jane's)*

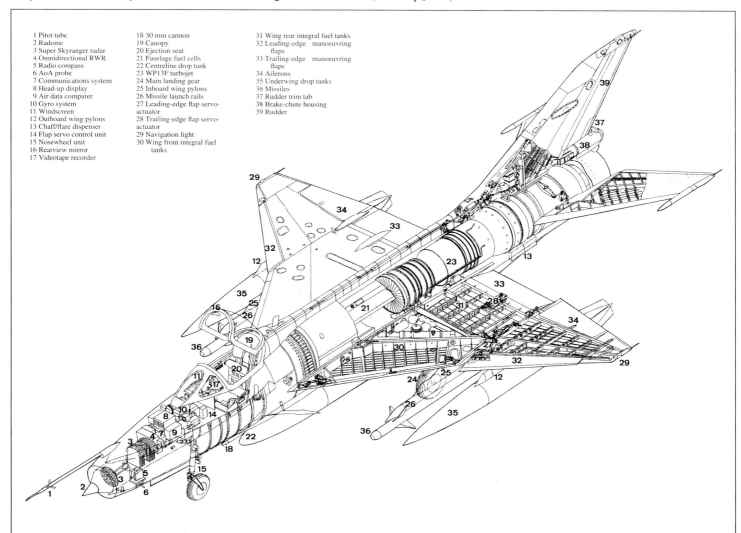

1 Pitot tube
2 Radome
3 Super Skyranger radar
4 Omnidirectional RWR
5 Radio compass
6 AoA probe
7 Communications system
8 Head-up display
9 Air data computer
10 Gyro system
11 Windscreen
12 Outboard wing pylons
13 Chaff/flare dispenser
14 Flap servo control unit
15 Nosewheel unit
16 Rearview mirror
17 Videotape recorder
18 30 mm cannon
19 Canopy
20 Ejection seat
21 Fuselage fuel cells
22 Centreline drop tank
23 WP13F turbojet
24 Main landing gear
25 Inboard wing pylons
26 Missile launch rails
27 Leading-edge flap servo-actuator
28 Trailing-edge flap servo-actuator
29 Navigation light
30 Wing front integral fuel tanks
31 Wing rear integral fuel tanks
32 Leading-edge manoeuvring flaps
33 Trailing-edge manoeuvring flaps
34 Ailerons
35 Underwing drop tanks
36 Missiles
37 Rudder trim tab
38 Brake-chute housing
39 Rudder

CAC F-7MG, cutaway drawing key

CHINA: AIRCRAFT—CAC

Cockpit layout of the F-7MG

F-7MG weapon options

Stores	O-Board	I-Board	C-Line	I-Board	O-Board
2 x PL-7 (or R550) Missiles		○		○	
2 x AIM-9P Missiles		○		○	
4 x AIM-9P Missiles	○	○		○	○
4 x AIM-9P Missiles + 500L Tank	○	○	●	○	○
2 x AIM-9P Missiles + 3 x 500L Tanks	●	○	●	○	●
2 x PL-7 (or R550) Missiles + 3 x 500L Tanks	●	○	●	○	●
4 x HF-7C Rockets Pods	○	○		○	○
2 x HF-7C Rockets Pods + 3 x 500L Tanks	●	○	●	○	●
4 x HF-7C Rockets Pods + 800L Tank	○	○	●	○	○
2 x 500 kg Bombs		○		○	
2 x 500 kg Bombs + 800L Tank		○	●	○	
4 x MK-82 Bombs	○	○		○	○
2 x MK-82 Bombs + 3 x 500L Tanks	●	○	●	○	●
4 x MK-82 Bombs + 800L Tank	○	○	●	○	○
2 x MK-82 Bombs + 2 x HF-7C Rockets Pods + 800L Tank	○	○	●	○	○
2 x 500L Tanks + 800L Tank	●		●		●

Rockwell Collins AN/ARN-147 VOR/ILS receiver, AN/ARN-149 ADF and Pro Line II digital DME-42. Avionics (contract for up to 100 sets) delivered to China from early 1989. FIAR Grifo 7 fire-control radar (range of more than 30 n miles; 55 km; 34 miles) for F-7P and MP ordered 1993, to replace Marconi Skyranger; flight trials began May 1996 and completed in 1997.

F7-3: Export designation of J-7 III. No orders known.

F-7MF: Latest known variant; debut at Airshow China, November 2000; further development of J-7FS and F-7MG for export market. Larger 'solid' nose; shorter, rectangular intake located farther back under nose; small, shoulder-mounted canards just forward of wingroot leading-edge; WP13F engine; 1553B databus; avionics to include 43 n mile (80 km; 50 mile) range pulse Doppler radar, single HUD and dual HDDs; 3,000 kg (6,614 lb) external stores load.

Wind tunnel testing completed; first flight targeted for late 2001/early 2002. Performance expectations include M1.8 top speed, 16,000 m (52,500 ft) ceiling, 650 m (2,135 ft) T-O run and 1,403 n mile (2,600 km; 1,615 mile) ferry range.

F-7MG: Improved version of F-7M (G suffix indicates *gai:* modified), combining double-delta wings of J-7E with upgraded avionics and other changes including uprated (WP13F) engine and leading/trailing-edge manoeuvring flaps. Said to have 45 per cent better manoeuvrability than F-7M. Public debut (aircraft 0142 and 0144) at China Air Show, Zhuhai, November 1996; Bangladesh and Pakistan seen as potential launch customers; latter negotiating for 30 to 40 in late 2000 to offset delays in FC-1 programme. If purchased, would have PAF designation **F-7PG**.

FT-7: Export designation of JJ-7 two-seat trainer.

CUSTOMERS: Several hundred built for Chinese air forces; nearly 400 exported to Albania (12 F-7A), Bangladesh (16 F-7MB), Egypt (approximately 90 F-7B?), Iran (18 or more F-7M), Iraq (approximately 90 F-7B?), Myanmar (24 F-7M), Pakistan (20 F-7P and 100 F-7MP, all designated F-7P by PAF), Sri Lanka (four F-7BS), Sudan (22 F-7B), Tanzania (16 F-7A) and Zimbabwe (approximately 12 F-7B/F-7IIN variants). Pakistan Air Force squadrons are No. 2 at Masroor, Nos. 18 and 20 at Rafiqui and No. 25 at Mianwali.

DESIGN FEATURES: Typical mid-1950s design of fighter, incorporating diminutive delta wing (double-delta on J-7E/EB and F-7MG), with clipped tips to mid-mounted wings, plus all-moving horizontal tail; circular-section fuselage with dorsal spine; nose intake with conical centrebody; swept tail, with large vertical surfaces and ventral fin.

Wing anhedral 2° from roots; incidence 0°; thickness/chord ratio approximately 5 per cent at root, 4.2 per cent at tip; quarter-chord sweepback 49° 6′ 36″ (reduced on J-7E/F-7MG outer panels); no wing leading-edge camber.

FLYING CONTROLS: Manual operation, with autostabilisation in pitch and roll; hydraulically boosted inset ailerons; plain trailing-edge flaps, actuated hydraulically; forward-hinged door-type airbrake each side of underfuselage below wing leading-edge; third, forward-hinged airbrake under fuselage forward of ventral fin; airbrakes actuated hydraulically; hydraulically boosted rudder and all-moving, trimmable tailplane. Leading/trailing-edge manoeuvring flaps on F-7MG.

STRUCTURE: All-metal; wings have two primary spars and auxiliary spar; semi-monocoque fuselage, with spine housing control pushrods, avionics, single-point refuelling cap and fuel tank; blister fairings on fuselage above and below each wing to accommodate retracted mainwheels.

LANDING GEAR: Inward-retracting mainwheels, with 600×200 tyres (pressure 11.50 bars; 167 lb/sq in) and LS-16 disc brakes; forward-retracting nosewheel, with 500×180 tyre (pressure 7.00 bars; 102 lb/sq in) and LS-15 double-acting brake. Nosewheel steerable ±47°. Minimum ground turning radius 7.04 m (23 ft 1¼ in). Tail braking parachute at base of vertical tail.

POWER PLANT: *F-7M:* One LMC (Liyang) WP7B(BM) turbojet (43.2 kN; 9,700 lb st dry, 59.8 kN; 13,448 lb st with afterburning).

J-7 III: LMC WP13 turbojet (40.2 kN; 9,039 lb st dry, 64.7 kN; 14,550 lb st with afterburning).

F-7MG: WP13F (44.1 kN; 9,921 lb st dry, 64.7 kN; 14,550 lb st with afterburning).

Total internal fuel capacity 2,385 litres (630 US gallons; 524.5 Imp gallons), contained in six flexible tanks in fuselage and two integral tanks in each wing. Provision for carrying a 500 or 800 litre (132 or 211 US gallon; 110 or 176 Imp gallon) centreline drop tank, and/or a 500 litre drop tank on each outboard underwing pylon. Maximum internal/external fuel capacity 4,185 litres (1,105 US gallons; 920.5 Imp gallons).

ACCOMMODATION: Pilot only, on CAC zero-height/low-speed ejection seat operable between 70 and 459 kt (130 and 850 km/h; 81 and 528 mph) IAS. Martin-Baker Mk 10L seat in F-7P/MP. One-piece canopy, hinged at rear to open upward. J-7 III/F-7-3 canopy opens sideways to starboard.

SYSTEMS: Improved electrical system in F-7M, using three static inverters, to cater for additional avionics. Jianghuai YX-3 oxygen system.

AVIONICS: *Comms:* BAE Systems AD 3400 UHF/VHF multifunction com, Chinese Type 602 IFF transponder; Type 605A ('Odd Rods' type) IFF in J-7 III.

Radar: BAE Systems Type 226 Skyranger ranging radar in F-7M; FIAR Grifo 7 in F-7P/MP; Chinese JL-7 fire-control radar in J-7 III. BAE Systems Super Skyranger in F-7MG (look-down, shoot-down and track-while-scan capability).

Flight: Navigation function of BAE Systems HUDWAC includes approach mode. WL-7 radio compass, XS-6A marker beacon receiver, Type 0101 HR A/2 radar altimeter and BAE Systems air data computer in F-7M. Beijing Aeronautical Instruments Factory KJ-11 twin-channel autopilot and FJ-1 flight data recorder in J-7 III. F-7MG suite includes VOR/DME/INS and Tacan.

Instrumentation: BAE Systems Type 956 HUDWAC (head-up display and weapon aiming computer) in F-7M provides pilot with displays for instrument flying, with air-to-air and air-to-ground weapon aiming symbols integrated with flight-instrument symbology. It can store 32 weapon parameter functions, allowing for both current and future weapon variants. In air-to-air combat its four modes (missiles, conventional gunnery, snapshot gunnery, dogfight) and standby aiming reticle allow for all eventualities. VCR and infra-red cockpit lighting in F-7MG, for which licence-built Russian helmet sight, slaved to PL-9 AAM, is also in production.

Self-defence: Skyranger ECCM in F-7M. Chinese LJ-2 RWR and GT-4 ECM jammer in J-7 III.

ARMAMENT (F-7M): Two 30 mm Type 30-1 belt-fed cannon, with 60 rds/gun, in fairings under front fuselage just forward of wingroot leading-edges. Two hardpoints under

CAC J-7 III of the Chinese PLA Air Force taking off for a mission

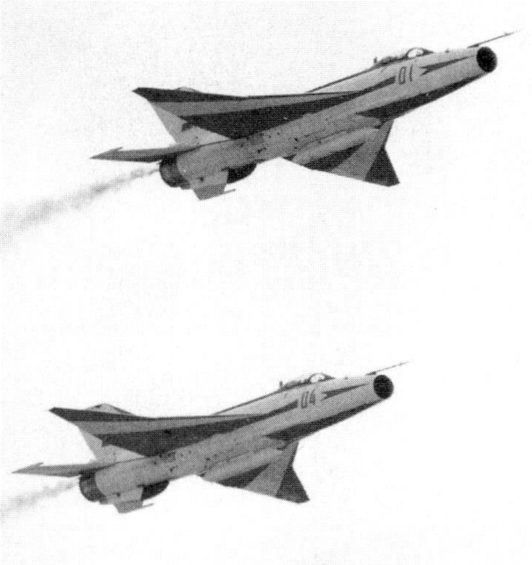

CAC J-7EBs of the PLAAF aerobatic team ***(Photo Link)*** ***1999***/0044482

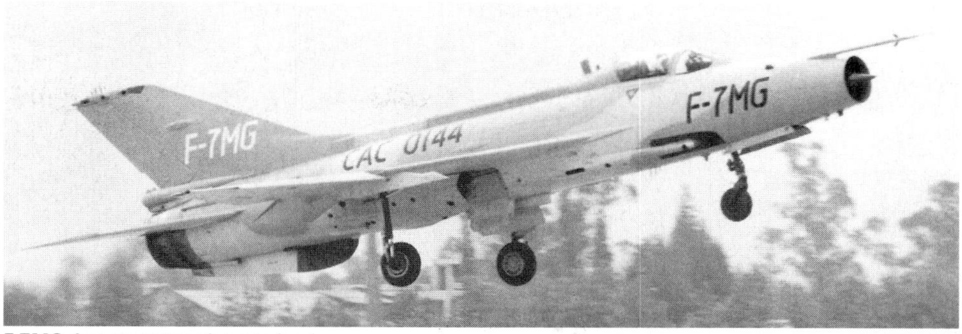

F-7MG demonstrator about to land ***2001***/0106477

each wing, of which outer ones are wet for carriage of drop tanks. Centreline pylon used for drop tank only. Each inboard pylon capable of carrying a PL-2, -2A, -5B, -7 or -8 (Python 3) missile (and PL-9 on F-7MG) or, at customer's option, an R550 Magic; one 18-tube pod of Type 57-2 (57 mm) air-to-air and air-to-ground rockets; one Type 90-1 (90 mm) seven-tube pod of air-to-ground rockets; or a 50, 150, 250 or 500 kg bomb. Each outboard pylon can carry one of above rocket pods, a 50 or 150 kg bomb, or a 500 litre drop tank.

ARMAMENT (J-7 III): One 23 mm Type 23-3 twin-barrel gun in ventral pack. Five external stores stations can carry two to four PL-2 or PL-5B air-launched missiles; two or four Qingan HF-16B 12-round launchers for Type 57-2 or seven-round pods of Type 90-1 rockets; or two 500 kg, four 250 kg or ten 100 kg bombs, in various combinations with 500 litre (one centreline and/or one under each wing) or 800 litre (underfuselage station only) drop tanks.

DIMENSIONS, EXTERNAL:
Wing span: except J-7E/F-7MG 7.15 m (23 ft 5½ in)
J-7E/F-7MG 8.32 m (27 ft 3½ in)
Wing chord: at root 5.51 m (18 ft 0¾ in)
at tip (except J-7E/F-7MG) 0.46 m (1 ft 6¼ in)
Wing aspect ratio: except J-7E/F-7MG 2.2
J-7E/F-7MG 2.8
Length overall: excl nose probe 13.945 m (45 ft 9 in)
incl nose probe 14.885 m (48 ft 10 in)
Fuselage: Length 12.175 m (39 ft 11½ in)
Max diameter 1.34 m (4 ft 4¾ in)
Height overall 4.105 m (13 ft 5½ in)
Tailplane span 3.74 m (12 ft 3¼ in)
Wheel track 2.69 m (8 ft 10 in)
Wheelbase 4.805 m (15 ft 9¼ in)

AREAS:
Wings, gross: except J-7E/F-7MG 23.00 m² (247.6 sq ft)
J-7E/F-7MG 24.88 m² (267.8 sq ft)
Ailerons (total): except J-7E/F-7MG 1.18 m² (12.70 sq ft)
Trailing-edge flaps (total) 1.87 m² (20.13 sq ft)
Fin 3.48 m² (37.46 sq ft)
Rudder 0.97 m² (10.44 sq ft)
Tailplane 3.94 m² (42.41 sq ft)

WEIGHTS AND LOADINGS:
Weight empty: F-7M 5,275 kg (11,629 lb)
F-7MG 5,292 kg (11,667 lb)
Normal T-O weight with two PL-2 or PL-7 air-to-air missiles: F-7M 7,531 kg (16,603 lb)
J-7 III 8,150 kg (17,967 lb)
F-7MG 7,540 kg (16,623 lb)
Max T-O weight: F-7MG 9,100 kg (20,062 lb)
Wing loading at normal T-O weight:
F-7M 327.4 kg/m² (67.06 lb/sq ft)
J-7 III 354.3 kg/m² (72.56 lb/sq ft)
Max wing loading: F-7MG 365.8 kg/m² (74.91 lb/sq ft)
Power loading at normal T-O weight:
F-7M, J-7 III 126 kg/kN (1.23 lb/lb st)
Max power loading: F-7MG 141 kg/kN (1.38 lb/lb st)

PERFORMANCE (F-7M at normal T-O weight with two PL-2 or PL-7 air-to-air missiles, except where indicated):
Never-exceed speed (VNE) above 12,500 m (41,010 ft) M2.35 (1,346 kt; 2,495 km/h; 1,550 mph)
Max level speed between 12,500 and 18,500 m (41,010-60,700 ft) M2.05 (1,175 kt; 2,175 km/h; 1,350 mph)
Unstick speed 167-178 kt (310-330 km/h; 193-205 mph)
Touchdown speed 162-173 kt (300-320 km/h; 186-199 mph)
Max rate of climb at S/L 10,800 m (35,435 ft)/min
Acceleration from M0.9 to M1.2 at 5,000 m (16,400 ft) 35 s
Max sustained turn rate: M0.7 at S/L 14.7°/s
M0.8 at 5,000 m (16,400 ft) 9.5°/s
Service ceiling 18,200 m (59,720 ft)
Absolute ceiling 18,700 m (61,360 ft)
T-O run 700-950 m (2,300-3,120 ft)
Landing run with brake-chute 600-900 m (1,970-2,955 ft)
Typical mission profiles:
combat air patrol at 11,000 m (36,080 ft) with two air-to-air missiles and three 500 litre drop tanks, incl 5 min combat 45 min
long-range interception at 11,000 m (36,080 ft) at 351 n miles (650 km; 404 miles) from base, incl M1.5 dash and 5 min combat, stores as above
hi-lo-hi interdiction radius, out and back at 11,000 m (36,080 ft), with three 500 litre drop tanks and two 150 kg bombs 324 n miles (600 km; 373 miles)
lo-lo-lo close air support radius with four rocket pods, no external tanks 200 n miles (370 km; 230 miles)
Range: two PL-7 missiles and three 500 litre drop tanks 939 n miles (1,740 km; 1,081 miles)
self-ferry with one 800 litre and two 500 litre drop tanks, no missiles 1,203 n miles (2,230 km; 1,385 miles)
g limit +8

PERFORMANCE (J-7 III at normal T-O weight):
Max operating Mach No. (MMO) 2.1
Unstick speed with afterburning 173 kt (320 km/h; 199 mph)
Touchdown speed with flap blowing 146 kt (270 km/h; 168 mph)
Min level flight speed 140 kt (260 km/h; 162 mph)
Max rate of climb at S/L 9,000 m (29,525 ft)/min
Service ceiling 18,000 m (59,060 ft)
Acceleration from M1.2 to M1.9 at 13,000 m (42,660 ft) 3 min 27 s
Air turning radius at 5,000 m (16,400 ft) at M1.2 5,093 m (16,710 ft)
T-O run with afterburning 800 m (2,625 ft)
Landing run with flap blowing, drag-chute and brakes deployed 550 m (1,805 ft)
Range: on internal fuel 518 n miles (960 km; 596 miles)
with 800 litre belly tank 701 n miles (1,300 km; 807 miles)
with 800 litre belly tank and two 500 litre underwing tanks 1,025 n miles (1,900 km; 1,180 miles)
g limits: up to M0.8 +8.5
above M0.8 +7

PERFORMANCE (F-7MG):
Max operating Mach No. (MMO) 2.0
Max level speed 648 kt (1,200 km/h; 745 mph) IAS
Min level speed 114 kt (210 km/h; 131 mph) IAS
Max rate of climb at S/L 11,700 m (38,386 ft)/min
Max instantaneous turn rate 25.2°/s
Sustained turn rate: at 1,000 m (3,280 ft) 16°/s
at 5,000 m (16,400 ft) 11°/s
at 8,000 m (26,240 ft) 8°/s
Service ceiling 17,500 m (57,420 ft)
Theoretical ceiling 18,000 m (59,060 ft)
T-O run, with afterburning, and landing run 600-700 m (1,970-2,300 ft)
Operational radius:
air superiority (hi-hi-hi) with two AIM-9P AAMs and three 500 litre drop tanks, incl 5 min combat with afterburner 459 n miles (850 km; 528 miles)
air-to-ground attack (lo-lo-hi) with two Mk 82 bombs and two 500 litre drop tanks 297 n miles (550 km; 342 miles)
Ferry range 1,187 n miles (2,200 km; 1,367 miles)
g limits +8/–3

UPDATED

CAC FC-1

TYPE: Attack fighter.

PROGRAMME: Fighter China (FC) programme launched 1991 following cancellation of US participation in development of Chengdu Super-7 (see 1995-96 *Jane's*), which it replaces. Some design assistance from MiG OKB, possibly based on (then-designated MiG-33) mid-1980s project for single-engined variant of the MiG-29. (Sources at MiG experimental bureau quoted as saying that FC-1 was designed there to a military specification as Izd (*Izdeliye*: article) 33 and now offered for Chinese production following cancellation of Russian requirement.)

Two static test airframes were due to begin testing 1996, though not yet known to have done so; first flight expected in 2001. Avionics competition still open in early 2001; contenders include SAGEM and Thales. Choice of a multimode pulse Doppler radar apparently still lies between Phazotron of Russia (Kopyo), FIAR of Italy (Grifo 7), and Thales (RC 400). Selection had not been made by early 2001, due to Western misgivings about technology transfer; China may look elsewhere if clearance is much further delayed.

Chinese domestic requirement for the FC-1 is said to be for several hundred aircraft, with Pakistan wanting about 150, but a Pakistani order is conditional upon PLAAF commitment, which was still awaited in early 2001. A letter of intent for joint development by China and Pakistan was reportedly signed in February 1998, followed by signature of a joint development and production agreement in June 1999.

CUSTOMERS: Envisaged for air forces of China and Pakistan initially, but costed to be competitive in wider export market. Seen as potential replacement for Shenyang J-6, Chengdu J-7, Nanchang Q-5, Northrop F-5 and Dassault Mirage III/5.

COSTS: CATIC forecast (1995) of unit cost below US$15 million seems optimistic.

DESIGN FEATURES: Agile light fighter. Mid-mounted delta wing with narrow wingroot strakes at leading-edge; leading-edge manoeuvring flaps; single turbofan engine; side-mounted twin intakes, with splitter plates; large intake trunks provide space for considerable internal fuel capacity. Large main fin with dorsal fairing; two smaller, uncanted ventral fins.

FLYING CONTROLS: Conventional hydraulic servo-operated control of ailerons, rudder and all-moving tailplane initially, with single analogue fly-by-wire system for back-up; provision for FBW to become primary system later. Trailing-edge and leading-edge flaps.

STRUCTURE: Primary structure conventional aluminium and steel alloy semi-monocoque. Some components may be manufactured in Pakistan.

LANDING GEAR: Retractable tricycle type, with single wheel and oleo shock-absorber on each unit. Mainwheels retract upward into engine intake trunks; nosewheel retracts rearward.

POWER PLANT: One Klimov RD-93 (RD-33 derivative) turbofan (81.4 kN; 18,300 lb st with afterburning), possibly to be licence built by Liyang Machinery Corporation (LMC) for production aircraft. Could have alternative Western engine at customer's option. Substantial internal fuel capacity. Provision for external fuel tanks.

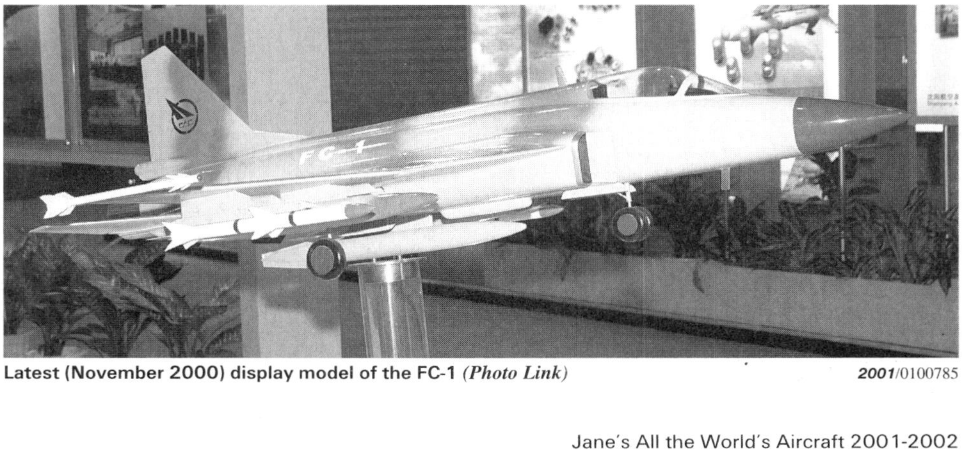

Latest (November 2000) display model of the FC-1 ***(Photo Link)*** ***2001***/0100785

CHINA: AIRCRAFT—CAC to CHAIC

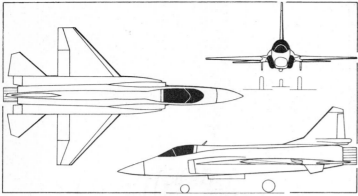

Provisional drawing of the CAC FC-1 multirole fighter *(Paul Jackson/Jane's)*

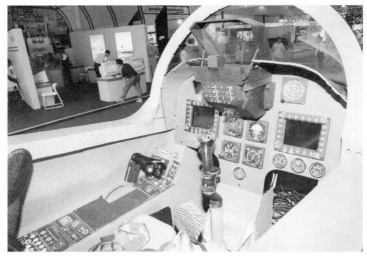

Cockpit mockup of the projected FC-1 fighter, 1998 *(Photo Link)*

ACCOMMODATION: Single seat (Martin-Baker zero/zero Mk 10 in any aircraft for Pakistan) under one-piece canopy. Two-seat training versions also planned.
AVIONICS: Still to be selected in early 2001 (see Programme above). Cockpit mockup exhibited at China Air Show '96 featured BAE (Marconi) HUD and dual head-down MFDs.
ARMAMENT: Underfuselage centreline station for 23 mm GSh-23-2 twin-barrel cannon or other store; two attachments under each wing and one at each wingtip. Weapons expected to include PL-7 and/or PL-10 AAMs, ASMs, bombs, gun and rocket pods, or other stores.

DIMENSIONS, EXTERNAL:
Wing span, to c/l of AAMs	9.00 m (29 ft 6¼ in)
Length overall	14.00 m (45 ft 11¼ in)
Height overall	5.10 m (16 ft 8¾ in)
Wheel track	2.30 m (7 ft 6½ in)
Wheelbase	5.14 m (16 ft 10¼ in)

WEIGHTS AND LOADINGS:
Max external stores load	3,800 kg (8,378 lb)
T-O weight: normal (with wingtip AAMs only)	9,100 kg (20,062 lb)
max	12,700 kg (27,998 lb)
Max power loading	156 kg/kN (1.53 lb/lb st)

PERFORMANCE (estimated):
Max level speed at altitude, clean	M1.6
Service ceiling	16,500 m (54,140 ft)
T-O run	500 m (1,640 ft)
Landing run	700 m (2,300 ft)
Combat radius:	
fighter	648 n miles (1,200 km; 745 miles)
ground attack	378 n miles (700 km; 435 miles)
Max range on internal fuel	864 n miles (1,600 km; 994 miles)
Max ferry range	1,619 n miles (3,000 km; 1,864 miles)
g limit	+8

UPDATED

CAC J-10
Chinese name: Jianjiji-10 (Fighter Aircraft 10)
Westernised designation: F-10

TYPE: Multirole fighter.
PROGRAMME: Reports of the existence of this new Chinese fighter began to emerge in October 1994, following the detection by a US intelligence satellite of a prototype at Chengdu. Said to be in the weight and performance class of the Eurofighter 2000 and Dassault Rafale, the J-10 bears a close external resemblance to the cancelled IAI Lavi (see 1988-89 *Jane's*), despite Israeli government statements that fears of unauthorised technology transfer are unfounded. A photograph, released in 1996 by the People's Liberation Army, of a wind tunnel model of the J-10 showed it to be outwardly identical to the Lavi in all essential respects, apart from slightly raised foreplanes and the addition of wingroot trailing-edge extensions.

This was confirmed in January 2001 when an unauthorised source posted a photograph of prototype '01' on the Worldwide Web (although it was rapidly removed). The photograph showed the accompanying artist's impression to be accurate in all respects, except that fin chord on the actual aircraft is some 20 per cent greater from root to tip.

In late 1995, Russian sources indicated that the first flight, then expected during the early part of 1996, would be powered by a single 122.6 kN (27,560 lb st) Saturn/Lyulka AL-31F turbofan. Russian avionics manufacturer Phazotron offered its Zhuk (Beetle) radar (already selected for the F-8 IIM upgrade programme) and the more capable Zhemchug (Pearls) and RP-35 as alternatives to the Elta

First prototype of the J-10, as seen briefly on the internet in January 2001 *(via Yihong Zhang)*

EL/M-2035 or a derivative. Delivery of first Zhemchug was due late 1999.

Taxi and ejection seat trials began in late 1997; first flight took place on 24 March 1998. According to China-based sources, 10 AL-31F engines had been imported by then and four J-10 prototypes completed. An unofficial CAC source in mid-1999 stated that two prototypes were then flying, with four others undergoing static test, still being assembled or only just completed. Chinese sources at Zhuhai in November 2000 stated that between five and eight J-10s then built and more than 140 test flights made. Service entry should be in about 2005 and some may be deployed in the two aircraft carriers that are due to be built by then.

Reports have suggested a PLAAF requirement for up to 300 J-10s, but achievement of this may depend upon China's progress with licensed manufacture of the Sukhoi Su-27 (see entry for Shenyang (SAC) J-11).
Following estimated data have not been confirmed.

DIMENSIONS, EXTERNAL:
Wing span	8.8 m (28.9 ft)
Length overall	14.6 m (47.9 ft)
Height overall	4.8 m (15.75 ft)

AREAS:
Wings, gross	33 m² (355.2 sq ft)

WEIGHTS AND LOADINGS:
Weight empty	6,940 kg (15,300 lb)
Max T-O weight: clean	10,000 kg (22,046 lb)
with max external stores	18,400 kg (40,565 lb)

PERFORMANCE:
Max level speed	M1.85
Range	1,000 n miles (1,850 km; 1,150 miles)

UPDATED

Artist's impression of the J-10 fighter, allegedly based on the IAI Lavi *(Keith Fretwell/Jane's)*

CHAIC

CHANGHE AIRCRAFT INDUSTRIES CORPORATION (Changhe Feiji Gongye Gongsi) (Subsidiary of AVIC II)
PO Box 109, Jingdezhen, Jiangxi 333002
Tel: (+86 798) 844 20 19 and 20 23
Fax: (+86 798) 844 29 40 and 14 60
PRESIDENT: Yang Jinhuai

CHAIC (formerly Changhe Aircraft Factory), built on a 234 ha (578 acre) site at Jingdezhen, began producing coaches and commercial road vehicles in 1974. These and other automotive products still account for most of output, but batch-produced helicopters have included the Z-8 (1998-99 *Jane's*) and Z-11. Reports from the November 2000 Airshow China indicated that CHAIC was in talks with Pratt & Whitney Canada concerning a possible re-engining programme for the Z-8, limited production and/or upgrading of which may be continuing. CHAIC had a 1999 workforce of some 7,100.

CHAIC is responsible for manufacture of the tailcone, vertical fin and horizontal stabiliser of the Sikorsky S-92 helicopter (see US section). The tail for the first S-92 was delivered to Sikorsky in May 1997.

UPDATED

CHAIC Z-11
Chinese name: Zhishengji-11 (Vertical take-off aircraft 11)

TYPE: Light utility helicopter.
PROGRAMME: First details officially released at China Air Show in Zhuhai November 1996, together with photographs showing one or two Z-11s in flight. Quoted first flight date of 22 December 1994 thought to refer to Chinese re-engined modification of two second-hand ex-US AStars (Ecureuils) acquired earlier that year. In early 1997, a Chinese government agency announced that the Z-11 had flown for the first time on 26 December 1996.

The Z-11 appears identical externally to Eurocopter AS 350B Ecureuil (see International section) except for nose contours, but Eurocopter (which sold eight Ecureuils to China in 1996) declines to comment. No further evidence of the Z-11's provenance had emerged by early

Changhe Z-11 in military camouflage

2001. Domestic certification was said to be imminent at that time.
CURRENT VERSIONS: Suitable for pilot training, police/security patrol, reconnaissance, coastguard, geological survey, rescue and forestry protection.
CUSTOMERS: Eight said to have been completed by late 1996, of which four had been delivered. Customers not identified, but all photographs released before late 1998 showed Z-11 in military camouflage with PLA insignia. One report from Aviation Expo China in October 1997 stated that the PLA had ordered 20 military Z-11s.
POWER PLANT: One 510 kW (684 shp) WZ8D turboshaft (licence-built Turbomeca Arriel 1D), reportedly produced by Liming engine factory.

WEIGHTS AND LOADINGS:
Weight empty 1,256 kg (2,769 lb)
Max T-O weight 2,200 kg (4,850 lb)
PERFORMANCE:
Max level speed 150 kt (278 km/h; 172 mph)
Cruising speed 130 kt (240 km/h; 149 mph)
Hovering ceiling IGE 5,250 m (17,225 ft)
Max range 322 n miles (598 km; 371 miles)
Endurance 3 h 42 min

UPDATED

Civil Z-11 on display at Zhuhai in November 2000 *(Photo Link)*

CHRDI

CHINESE HELICOPTER RESEARCH AND DEVELOPMENT INSTITUTE (Zhongguo Zhishengji Sheji Yanjiuso) (Subsidiary of AVIC II)
Jingdezhen, Jiangxi

This institute (workforce approximately 2,000) is the overall design authority for Chinese indigenous helicopter programmes. Few details yet available of Z-10; other projects may include 8 to 10 tonne heavy-lift and, longer-term, a dedicated attack helicopter.

NEW ENTRY

CHRDI Z-10
Chinese name: Zhishengji-10 (Vertical take-off aircraft 10)
TYPE: Light utility helicopter.
PROGRAMME: Thought to have been initiated in about 1994; Nos. 602 and 608 Institutes reported to be taking part in airframe design. Eurocopter became partner in programme 15 May 1997 with US$70 million to US$80 million, nine-year contract to assist in developing rotor system; joined 22 March 1999 by Agusta with contract (approximately Lit50 billion; US$30 million) to be responsible for transmission system and vibration analysis. Reports suggest that three transmission sets will be built in Italy and four, collaboratively, in China. In size and weight class of Bell 412 and Sikorsky S-76 (MTOW variously reported as 5, 5.5 or 6 tonnes; 11,023 to 13,227 lb), with capacity for up to 14 passengers or troops. First flight thought to be targeted for 2006. Eventual production likely to be entrusted to Hafei or Changhe or, possibly, both.
DESIGN FEATURES: None yet officially confirmed, but intention of both military and civil use suggests that such features as composites construction, crashworthy airframe and seats, rotor system ballistic tolerance, engine run-dry capacity and ability to carry slung loads are likely.
POWER PLANT: Reported requirement is for two 969 kW (1,300 shp) class turboshafts. Talks under way in late 2000 with Turbomeca and Pratt & Whitney Canada and, possibly, other potential suppliers; possibility of indigenous engines not ruled out.

NEW ENTRY

GAIC

GUIZHOU AVIATION INDUSTRY GROUP (Subsidiary of AVIC I)
110 Jinjiang Road, Xiaohe District, Guiyang, Guizhou 550009
Tel: (+86 851) 380 80 94
Fax: (+86 851) 380 80 84
e-mail: gaic@public2.gy.gz.cn
Web: http://www.gaic.com.cn
CHAIRMAN: Zhou Wancheng
PRESIDENT: Zhang Jun

The Guizhou Aviation Industry Group incorporates many enterprises, factories and institutes engaged in various aerospace and non-aerospace activities; aerospace workforce is about 6,000. Aviation programmes include JJ-7/FT-7 fighter trainer, two series of turbojets, air-to-air missiles and rocket launchers, plus participation in Chengdu (CAC, which see) production of single-seat J-7/F-7. GAIC also manufactures maintenance jigs and tools for the Airbus airliner family.

UPDATED

The FT-7P stretched version of the GAIC FT-7 *(Sebastian Zacharias)*

GAIC JJ-7
Chinese name: Jianjiji Jiaolianji-7 (Fighter training aircraft 7)
Westernised designation: FT-7
TYPE: Advanced jet trainer.
PROGRAMME: Launched October 1982; first metal cut April 1985; first flight 5 July 1985; series production began February 1986; production FT-7 first flight December 1987; FT-7P first flight 9 November 1990 and received MAS production approval 13 May 1992. Improved version of FT-7P with lengthened fuselage, internal 30 mm cannon and increased avionics volume in production 1996.
CURRENT VERSIONS: JJ-7: Basic domestic version, based on single-seat J-7 II and MiG-21US.
FT-7: Export version of JJ-7. Two FT-7Bs supplied to Zimbabwe in 1991 are known locally as the FT-7BZ.
Detailed description applies to FT-7 except where indicated.
FT-7P: Stretched version of FT-7 for Pakistan: lengthened by insertion of 0.60 m (1 ft 11½ in) fuselage plug behind rear cockpit, allowing increased fuel load (for 25 per cent improvement in operational range) and fitment of one internal cannon; AoA sensor on port side of nose. Imported fire-control system; HUD and air data computer; improved instrument layout; two underwing pylons each side.
CUSTOMERS: PLA Air Force (JJ-7). Export deliveries to Bangladesh Air Force (seven FT-7B); Myanmar AF (six FT-7); Pakistan AF (15 FT-7P); Sri Lanka AF (one FT-7); Zimbabwe AF (two FT-7B). One more FT-7B reportedly due for delivery to Bangladesh in 2000.
DESIGN FEATURES: Generally as J-7/F-7 except for twin canopies opening sideways to starboard (rear one with retractable periscope), twin ventral strakes of modified

Front cockpit of the GAIC FT-7

Rear cockpit of the GAIC FT-7

Production line-up of standard JJ-7s 2001/0106475

shape, and removable saddleback fuel tank aft of second cockpit. Can provide full training syllabus for all J-7/F-7 versions, plus most of that necessary for Shenyang J-8.
FLYING CONTROLS: As J-7/F-7.
STRUCTURE: As J-7/F-7.
LANDING GEAR: As J-7/F-7.
POWER PLANT: As F-7M. Internal fuel tank arrangement as for F-7M, but fuselage and wing total capacities are 1,880 and 560 litres respectively (496.5 and 148 US gallons; 413.5 and 123 Imp gallons), giving total capacity of 2,440 litres (644.5 US gallons; 536.5 Imp gallons). Internal capacity increased to 2,800 litres (740 US gallons; 616 Imp gallons) in FT-7P. External drop tanks as for F-7M.
ACCOMMODATION: Second cockpit in tandem, with dual controls. Canopies open sideways to starboard.
SYSTEMS: Cockpits pressurised and air conditioned (maximum differential 0.30 bar; 4.4 lb/sq in). Hydraulic system pressure 207 bars (3,000 lb/sq in), flow rates 36 litres (9.51 US gallons; 7.92 Imp gallons)/min in main system, 4 litres (1.06 US gallons; 0.88 Imp gallon)/min in standby system. Pneumatic system pressure 49 bars (711 lb/sq in). Electrical power provided by QF-12C 12 kW engine-driven starter/generator; 400 A static inverters for 28.5 V DC power. YX-1 oxygen system for crew. De-icing/anti-icing standard.
AVIONICS: *Comms:* Include CT-3M VHF com transceiver. Other (unidentified) Chinese nav/com avionics designated Type 222, Type 262, JT-2A and J7L.
Flight: WL-7 radio compass and XS-6A marker beacon receiver. FJ-1 flight data recorder.
ARMAMENT: Single underwing pylon each side (two on FT-7P) for such stores as a PL-2B air-to-air missile, an HF-5A 18-round launcher for 57 mm rockets, or a bomb of up to 250 kg size; can also be fitted with Type 23-3 twin-barrel 23 mm gun in underbelly pack and HK-03E optical gunsight. Gun mounted internally in FT-7P.
DIMENSIONS, EXTERNAL: As F-7M except:
  Length overall, incl probe: FT-7 14.87 m (48 ft 9½ in)
    FT-7P 15.47 m (50 ft 9 in)

WEIGHTS AND LOADINGS:
  Weight empty, equipped: FT-7 5,519 kg (12,167 lb)
    FT-7P 5,330 kg (11,750 lb)
  Max fuel weight: internal: FT-7 1,891 kg (4,169 lb)
    FT-7P approx 2,170 kg (4,784 lb)
    external 558 kg (1,230 lb)
  Max external stores load: FT-7 1,187 kg (2,617 lb)
  Normal T-O weight with two PL-2 air-to-air missiles:
    FT-7 7,590 kg (16,733 lb)
  Max T-O weight: FT-7 8,555 kg (18,860 lb)
    FT-7P 9,550 kg (21,054 lb)
  Max landing weight: FT-7 6,096 kg (13,439 lb)
  Max zero-fuel weight: FT-7 7,300 kg (16,094 lb)
  Max wing loading: FT-7 372.0 kg/m² (76.18 lb/sq ft)
    FT-7P 415.2 kg/m² (85.04 lb/sq ft)
  Max power loading: FT-7 143 kg/kN (1.40 lb/lb st)
    FT-7P 160 kg/kN (1.56 lb/lb st)
PERFORMANCE (FT-7):
  Never-exceed speed (V_{NE}) above 12,500 m (41,000 ft)
    M2.35 (1,346 kt; 2,495 km/h; 1,550 mph)
  Max level speed above 12,500 m (41,000 ft)
    M2.05 (1,175 kt; 2,175 km/h; 1,350 mph)
  Max cruising speed at 11,000 m (36,080 ft)
    545 kt (1,010 km/h; 627 mph)
  Econ cruising speed at 11,000 m (36,080 ft)
    516 kt (956 km/h; 594 mph)
  Unstick speed 170-181 kt (315-335 km/h; 196-208 mph)
  Touchdown speed
    165-175 kt (305-325 km/h; 189-202 mph)
  Stalling speed, flaps down 135 kt (250 km/h; 156 mph)
  Max rate of climb at S/L 9,300 m (30,510 ft)/min
  Service ceiling 17,300 m (56,760 ft)
  Absolute ceiling 17,700 m (58,080 ft)
  T-O run 827 m (2,715 ft)
  T-O to 15 m (50 ft) 931 m (3,055 ft)
  Landing from 15 m (50 ft) 1,368 m (4,490 ft)
  Landing run 1,060 m (3,480 ft)
  Range at 11,000 m (36,080 ft):
    max internal fuel (clean)
      545 n miles (1,010 km; 627 miles)
    max internal and external fuel
      788 n miles (1,459 km; 906 miles)
    max external stores 708 n miles (1,313 km; 816 miles)
  g limit with two PL-2B missiles +7

UPDATED

GAIC FTC-2000
TYPE: Advanced jet trainer.
PROGRAMME: Revealed, in model form, at Airshow China, November 2000. Proposed as lead-in fighter trainer for CAC FC-1 (which see) and J-8 IV/F-8D. Wind tunnel testing completed; prototype could fly by 2003.
COSTS: Predicted at US$2.4 million, excluding radar.
DESIGN FEATURES: Upgraded derivative of JJ-7/FT-7, having J-7E wings, lateral intakes and 'solid' nose to permit installation of fire-control radar. Airframe otherwise as JJ-7/FT-7.
POWER PLANT: One 59.8 kN (13,448 lb st) LMC WP7B (BM) or 64.7 kN (14,550 lb st) WP13B turbojet. In-flight refuelling probe on nose, common with that of J-8 IV/F-8D. Underfuselage and inboard underwing stations 'wet' for carriage of drop-tanks.
ARMAMENT: Stores attachments under fuselage (one) and wings (two each side).
WEIGHTS AND LOADINGS:
  Normal max T-O weight 7,800 kg (17,196 lb)
PERFORMANCE (estimated):
  Max level speed M1.8
  Unstick speed 135 kt (250 km/h; 155 mph)
  Touchdown speed 124 kt (230 km/h; 143 mph)
  Min flying speed 114 kt (210 km/h; 131 mph)
  T-O and landing distance 430 m (1,410 ft)
  Operational endurance 2 h

NEW ENTRY

Model of the GAIC FTC-2000 *(Photo Link)* 2001/0100783

GEAC
GUIZHOU EVEKTOR AIRCRAFT CORPORATION
Guizhou

This company was created in June 2000, as a joint venture between GAIC (which see) and EV-AT of the Czech Republic, to co-manufacture, market and support the lightplanes and ultralights of the latter concern. Launch activity is the assembly of the EV-97 Eurostar from CKD kits, the first of which was imported in August 2000 and made its maiden flight on 24 October that year. Co-production of the Wolfsberg-Letov Raven 257 is also planned.

NEW ENTRY

HAI
HAFEI AVIATION INDUSTRY COMPANY (Subsidiary of AVIC II)
Flat 34, Centre Complex, Harbin High-Tech Development Zone, Harbin, Heilongjiang 150066
Tel: (+86 451) 650 11 22 and 653 07 36
Fax: (+86 451) 652 01 44
e-mail: hai@hafei.com
Web: http://www.hafei.com
PRESIDENT: Cui Xuewen
GENERAL MANAGER: Wang Bin
EXECUTIVE DEPUTY GENERAL MANAGER: Xu Zhanbin

Established as Harbin Aircraft Manufacturing Corporation (Harbin Feiji Zhizao Gongsi: HAMC) 1952, subsequently producing H-5 light bomber (Soviet-designed Il-28) and Z-5 helicopter (Soviet-designed Mi-4) in large numbers, as well as smaller quantities of Chinese-designed SH-5 flying-boat and Y-11 agricultural light twin (see earlier editions of *Jane's* for details). Now core company of Harbin Aircraft Industry Group (HAIG, which see) under AVIC II. Occupies 514 ha (1,270 acre) site, including 350,000 m² (3,767,400 sq ft) of workshop space. Workforce in 1998 numbered approximately 18,000.

Currently producing own-design Y-12 utility light twin and licence-manufacturing Eurocopter Dauphin 2 as Z-9. Subcontract work includes doors for Avro RJ series, and Dauphin doors for Eurocopter.

HAI is partnered by Eurocopter and Singapore Technologies Aerospace in the Eurocopter EC 120 Colibri programme, as described in the International section; more than 200 EC-120 fuselages delivered to Eurocopter by late 2000; production rate 10 fuselages per month.

UPDATED

HAI (EUROCOPTER) Z-9 HAITUN
Chinese name: Zhishengji-9 (Vertical take-off aircraft 9)
English name: Dolphin
TYPE: Multirole light helicopter.
PROGRAMME: Licence-built Eurocopter AS 365N Dauphin 2 (which see in International section). Licence agreement (Aerospatiale/CATIC) signed 2 July 1980; first (French-built) example made initial acceptance flight in China 6 February 1982; Chinese parts manufacture began 1986; initial agreed batch of 50, last of which delivered January 1992. Production continuing under May 1988 domestic contract, with much increased local manufacture (72.2 per cent of airframe and 91 per cent of engine).
CURRENT VERSIONS: **Z-9:** Initial Chinese version, equivalent to French AS 365N. Details in 1994-95 and earlier *Jane's*.

HAI—AIRCRAFT: CHINA 69

Camouflaged WZ-9 Haitun equipped with two seven-round rocket pods and roof-mounted sight
2001/0106479

Red and white Z-9A used for Arctic scientific survey
2001/0106478

Service ceiling: Z-9A/B	6,000 m (19,680 ft)
WZ-9	4,940 m (16,200 ft)
Hovering ceiling: IGE	2,150 m (7,050 ft)
OGE	1,150 m (3,770 ft)

Max range at 135 kt (250 km/h; 155 mph) normal cruising speed, no reserves:
standard tanks:
Z-9A/B	464 n miles (860 km; 534 miles)
WZ-9	358 n miles (664 km; 412 miles)

with auxiliary tank:
Z-9A/B	539 n miles (1,000 km; 621 miles)

UPDATED

Z-9A: Later aircraft in initial 50, to AS 365N1 standard and with increased proportion of locally manufactured components.

Z-9B: Follow-on production version, to AS 365N2 standard, with much increased local manufacture (see Programme); also known as Z-9A-100. First flight 16 January 1992; flight test programme completed 20 November 1992 after almost 200 flight hours (408 flights); Chinese type approval received 30 December 1992. Total of 25 ordered by December 1997, of which 22 had then been delivered; further seven ordered in 1998 and possibly more subsequently. Modified Fenestron with 11 widerchord, all-composites blades instead of 13 metal blades as in AS 365N1. Principal PLA version for SAR, artillery direction, EW, troop transport (accommodates eight), communications and other duties.

Data apply to Z-9A and B except where indicated.

Z-9C: Version for PLA Naval Air Force, for deployment aboard certain classes of destroyers and frigates; in service by late 2000. Believed to be equivalent to Arriel 2-engined Eurocopter AS 565 Panther, but equipped with Thales HS-12 dipping sonar; armament includes torpedoes or TV-guided C-701 anti-surface vessel missiles.

WZ-9: Armed (*wuzhuang*) version (also reported as Z-9W); eight Norinco HJ-8 (*Hongjian*: Red Arrow) wire-guided anti-tank missiles, twin gun/or rocket pods and gyrostabilised, roof-mounted optical sight. First flight thought to have been in early 1989; in production. Export designation **Z-9G**, available with or without roof-mounted sight.

CUSTOMERS: In service with CAAC and Chinese armed services (People's Navy Aviation and Army Aviation each thought to have more than 25). Entered service with two PLA army groups January and February 1988 (Beijing and Shenyang Military Regions respectively). People's Navy Aviation believed to use Z-9B for commando transport as well as shipboard communications. Ten based at former Royal Air Force airfield at Sek Kong in third quarter of 1997, following 1 July handover of Hong Kong by UK.

Civil models used for various duties including offshore oil rig support and air ambulance (four stretchers/two seats or two stretchers/five seats). In 1992, Flying Dragon Aviation received two (late production Z-9As, augmenting an Aerospatiale Dauphin) which are operated on behalf of the Ministry of Forestry. One Z-9A deployed to Arctic regions in mid-2000, on scientific survey ship.

STRUCTURE: Transmission manufactured by Dongan Engine Manufacturing Company at Harbin, hubs and tail rotor blades by Baoding Propeller Factory.

POWER PLANT: Arriel 1C and 1C1 turboshafts produced by SAEC at Zhuzhou as WZ8 and WZ8A; fuel capacity 1,140 litres (301 US gallons; 251 Imp gallons). Option for 180 litre (47.5 US gallon; 39.6 Imp gallon) auxiliary tank.

ARMAMENT (WZ-9): Up to eight HJ-8 or HJ-8E ATMs (range 1.6 n miles; 3 km; 1.9 miles); twin 12.7 mm or 23 mm gun pods; or two pods of 57 or 90 mm rockets. Possible other weapons include TY-90 IR-guided AAM (already test-flown on Z-9 in 1998; range of 3.2 n miles; 6 km; 3.7 miles) and C-701 TV-guided anti-ship missile (8.1 n miles; 15 km; 9.3 miles).

WEIGHTS AND LOADINGS:
Weight empty, equipped	2,050 kg (4,519 lb)
Max payload	2,038 kg (4,493 lb)
Max load on cargo sling	1,600 kg (3,527 lb)
Max T-O weight, internal or external load	4,100 kg (9,039 lb)

PERFORMANCE:
Max level speed	170 kt (315 km/h; 195 mph)
Max cruising speed at S/L	151 kt (280 km/h; 174 mph)

Max vertical rate of climb at S/L:
Z-9A/B	252 m (827 ft)/min
WZ-9	210 m (689 ft)/min

Max forward rate of climb at S/L:
Z-9A/B	396 m (1,299 ft)/min
WZ-9	468 m (1,535 ft)/min

HAI 'Z-X'

HAMC is understood to be undertaking a design study for a new, twin-engined light helicopter in the 5,000 to 5,500 kg (11,023 to 12,125 lb) weight class, possibly derived from the Z-9. It is not yet clear whether this work is related to the Z-10 programme described under the CHRDI heading, although this seems likely.

UPDATED

HAI Y-12

Chinese name: Yunshuji-12 (Transport aircraft 12)
Export marketing name: Twin Panda
TYPE: Twin-turboprop light transport.
PROGRAMME: Initiated as uprated version of Y-12 (I) (1987-88 and earlier *Jane's*); first flight 16 August 1984. Y-12 (IV) received production approval September 1995.
CURRENT VERSIONS: **Y-12 (I):** Initial version (first flight 14 July 1982), with PT6A-11 engines; three prototypes and approximately 30 production examples built; described in 1987-88 *Jane's*.

Y-12 (II): Current production version; higher rated engines, no leading-edge slats and smaller ventral fin. Certified by CAAC 25 December 1985 and UK CAA (BCAR Section K) 20 June 1990.

Detailed description applies to Y-12 (II) except where indicated.

Y-12 (IV): Improved version (first flight 30 August 1993). Sweptback wingtips; modifications to control surface actuation, main gear and brakes; redesigned seating for 18 to 19 passengers; starboard side rear baggage door; maximum payload and maximum T-O weight increased. Further changes include Rockwell Collins or Honeywell com/nav for both VFR and IFR, plus optional colour weather radar, GPS, Omega navigation, wing and tail de-icing, and oxygen system. Domestic certification received 3 July 1994 and FAR Pt 23 approval on 26 March 1995.

Twin Panda: Version of Y-12 (IV) completed and marketed by Canadian Aerospace Group (CAG, which see) and its Panda Aircraft Company subsidiary, primarily as a replacement for the DHC-6 Twin Otter. More powerful PT6A-34 engines, strengthened landing gear, upgraded avionics. Harbin/CAG agreement was signed on 31 March 1998, covering a 10-year period and potentially 200 Y-12s (initial 50, plus 150 options) for the North American market. At November 2000, CAG had firm customer orders for 15 (including seven as floatplanes) and had signed order for further 10 from HAI. First two aircraft due for delivery to CAG by end of 2000.

Y-12E: Next-generation version for Chinese market, under development. To have 559 kW (750 shp) PT6A-135A engines (derated to 462 kW; 620 shp), retractable landing gear, strengthened structure, reduced noise levels and improved avionics. Announced at Airshow China, November 2000; launch customer Sichuan Airlines.

Future versions: Plans under consideration include stretched version and one with pressurised cabin.

CUSTOMERS: Over 120 built by Harbin by November 2000; see table. Reported hiatus in production due to priority helicopter order.

Torpedo-armed Z-9C of PLA Naval Aviation
2001/0106480

Y-12 (II) DELIVERIES
(at December 2000)

Country	Operator	Quantity
Domestic		
China	AVIC (No. 630 Institute)	1
	China General Aviation	5
	China Southwest Airlines	4
	Flying Dragon Aviation	9
	Guizhou Aviation Corporation	1
	State Oceanic Administration	1
Subtotal		**21**
Export*		
Bangladesh	Aero Bengal Airlines	2
Cambodia	Air Force	2
Egypt	Air Force	2
Eritrea	Air Force	3
	Red Sea Air†	1
Fiji	Fiji Air	4
Iran	Air Force	9
Kenya	Air Force	6
Kiribati	Air Kiribati	1
Laos	Lao Aviation	7
Malaysia	Berjaya Air Charter	2
Mauritania	Air Force	2
Mongolia	Mongolian Airlines	6
Namibia	Defence Force	2
Nepal	Nepal Airways	5
Pakistan	Air Force	2
	Army	2
Peru	Air Force	6
	National Police	3
Sri Lanka	Air Force	9
Tanzania	Air Force	2
Zambia	Air Force	4
Total		**102**

* Had risen to 94 by November 2000
† Unclear if aircraft is ex-military

COSTS: Batch of 15 PT6-engined aircraft valued at US$58 million (1999).
DESIGN FEATURES: Designed to standards of FAR Pts 23 and 135 Appendix A and developed to improve upon modest payload/range of piston-engined Y-11. Constant-chord high braced wings, with small stub-wings at cabin floor level supporting mainwheel units; basically rectangular-section fuselage, upswept at rear; non-swept tail surfaces; large dorsal fin; ventral fin under tailcone.
Wing section LS(1)-0417; thickness/chord ratio 17 per cent; dihedral 1° 41′; incidence 4°.
FLYING CONTROLS: Conventional and manual. Drooping ailerons, horn-balanced elevators and rudder; trim tab in starboard aileron, rudder and each elevator; electrically actuated two-segment double-slotted flaps on each wing trailing-edge.
STRUCTURE: Conventional all-metal structure with two-spar fail-safe wings and stressed skin fuselage; Ziqiang-2 resin bonding on 70 per cent of wing structure and 40 per cent of fuselage; integral fuel tankage in wing spar box; bracing strut from each stub-wing out to approximately one-third span.
LANDING GEAR: Non-retractable tricycle type, with oleo-pneumatic shock-absorber in each unit. Single-wheel main units, attached to underside of stub-wings. Single non-steerable nosewheel. Mainwheel tyres size 640×230, pressure 5.50 bars (80 lb/sq in); nosewheel tyre size 480×200, pressure 3.50 bars (51 lb/sq in). Hydraulic brakes. Minimum ground turning radius 16.75 m (54 ft 11½ in).

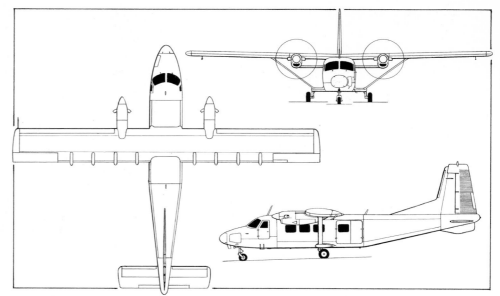

HAI Y-12 (II) twin-turboprop STOL general purpose transport *(Dennis Punnett/Jane's)*

Y-12 (II) in Peruvian Air Force markings *(Peter R Foster)* 2000/0062961

POWER PLANT: Two Pratt & Whitney Canada PT6A-27 turboprops, each derated to 462 kW (620 shp) and driving a Hartzell HC-B3TN-3B/T10173B-3 three-blade constant-speed reversible-pitch propeller. All fuel in tanks in wing spar box, total capacity 1,616 litres (427 US gallons; 355.5 Imp gallons), with overwing gravity filling point each side. Domestic WJ9 turboprop (442 to 520 kW; 593 to 697 shp) under development for Y-12 and other aircraft.
ACCOMMODATION: Crew of two on flight deck, with access via forward-opening door on port side. Four-way adjustable crew seats. Dual controls. Main cabin can accommodate up to 17 passengers in commuter configuration, in three-abreast layout (with aisle), at seat pitch of 75 cm (30 in). Alternative layouts for up to 15 parachutists, or all-cargo configuration with 11 tiedown rings. Passenger/cargo double door on port side at rear, rear half of which opens outward and forward half inward; foldout steps in passenger entrance. Emergency exits on each side at front of cabin and opposite passenger door on starboard side at rear. Baggage compartments in nose and at rear of passenger cabin, for 100 kg (220 lb) and 260 kg (573 lb) respectively.
SYSTEMS: Hamilton Sundstrand R70-3WG environmental control system. Hydraulic system (operating pressure 118 to 147 bars; 1,711 to 2,132 lb/sq in) for mainwheel brakes. Two 6 kW DC starter/generators, two 600 VA 400 Hz single-phase static inverters and one 43 Ah Ni/Cd battery for electrical power. Goodrich Type 29S-7D 5178 anti-icing system optional for wing, tailplane and fin leading-edges. Oxygen system optional.
AVIONICS: *Comms:* VHF-251 and HF-230 radio, AUD-251A intercom and TDR-950 transponder.
Radar: Honeywell 1400C, RDS-81 or RDS-82 weather radar optional.
Flight: ALT-50 radio altimeter, DME-451, ADF-650A, MKR-350 marker beacon receiver, KLN 90 GPS receiver, VIR-351 VOR, GLS-350 glide slope receiver and PN-101 pictorial navigation display. Doppler navigation with satellite responder (Y-12 (IV), Omega nav) optional (for example, in mineral detection role).
Instrumentation: 20025-11324 airspeed indicator, 510-8D10 horizon, 101420-11934 encoding altimeter, 30230-11101 vertical speed indicator, 9551-BN541 bank indicator, LC-2 magnetic compass, ZWH-1 outside air

Y-12 flight deck 2000/0085019

The Y-12 (IV) demonstrates its extended span and sweptback wingtips *(Photo Link)* 1999/0044488

temperature indicator, and ZEY-1 flap position indicator; dual engine torquemeters, interturbine temperature indicators, gas generator tachometers, oil temperature and pressure indicators, fuel pressure and quantity indicators; 309W clock; and XDH-10B warning light box.

EQUIPMENT: Hopper for 1,200 litres (317 US gallons; 264 Imp gallons) of dry or liquid chemical in agricultural version. Appropriate specialised equipment for firefighting, geophysical survey (for example, long, kinked sensor tailboom) and other missions.

DIMENSIONS, EXTERNAL:
Wing span: Y-12 (II)	17.235 m (56 ft 6½ in)
Y-12 (IV)	19.20 m (63 ft 0 in)
Wing chord, constant	2.00 m (6 ft 6¾ in)
Wing aspect ratio: Y-12 (II)	8.7
Length overall	14.86 m (48 ft 9 in)
Height overall	5.575 m (18 ft 3½ in)
Elevator span	5.365 m (17 ft 7¼ in)
Wheel track	3.61 m (11 ft 10¼ in)
Wheelbase	4.70 m (15 ft 5 in)
Propeller diameter	2.49 m (8 ft 2 in)
Propeller ground clearance	1.325 m (4 ft 4¼ in)
Distance between propeller centres	4.94 m (16 ft 2½ in)
Fuselage ground clearance	0.65 m (2 ft 1½ in)
Crew door: Height	1.12 m (3 ft 8 in)
Width	0.65 m (2 ft 1½ in)
Passenger/cargo door: Height	1.38 m (4 ft 6¼ in)
Width (passenger door only)	0.65 m (2 ft 1½ in)
Width (double door)	1.45 m (4 ft 9 in)
Emergency exits (three, each):	
Height	0.68 m (2 ft 2¾ in)
Width	0.68 m (2 ft 2¾ in)
Baggage door (nose, port):	
Max height	0.56 m (1 ft 10 in)
Width	0.75 m (2 ft 5½ in)

DIMENSIONS, INTERNAL:
Cabin, excl flight deck and rear baggage compartment:
Length	4.82 m (15 ft 9¾ in)
Max width	1.46 m (4 ft 9½ in)
Max height	1.70 m (5 ft 7 in)
Floor area	7.0 m² (75.8 sq ft)
Volume	12.9 m³ (455 cu ft)
Baggage compartment volume:	
nose	0.77 m³ (27.2 cu ft)
rear	1.9 m³ (67 cu ft)

AREAS:
Wings, gross: Y-12 (II)	34.27 m² (368.9 sq ft)
Ailerons (total, incl tab)	2.88 m² (31.00 sq ft)
Trailing-edge flaps (total)	6.00 m² (64.58 sq ft)
Fin, incl dorsal fin	2.24 m² (24.11 sq ft)
Rudder, incl tab	3.34 m² (35.95 sq ft)
Tailplane	3.10 m² (33.37 sq ft)
Elevators (total, incl tabs)	4.06 m² (43.70 sq ft)

Y-12 (IV) decorated with the national flags of the Y-12's export customers (*Photo Link*) 2001/0100797

WEIGHTS AND LOADINGS:
Max usable fuel weight	1,230 kg (2,712 lb)
Max payload: Y-12 (II)	1,700 kg (3,748 lb)
Y-12 (IV)	1,984 kg (4,374 lb)
Max T-O weight: Y-12 (II)	5,300 kg (11,684 lb)
Y-12 (IV)	5,670 kg (12,500 lb)
Max ramp weight: Y-12 (II)	5,330 kg (11,750 lb)
Y-12 (IV)	5,700 kg (12,566 lb)
Max landing weight: Y-12 (II)	5,300 kg (11,684 lb)
Y-12 (IV)	5,400 kg (11,905 lb)
Max zero-fuel weight: Y-12 (II)	4,900 kg (10,803 lb)
Y-12 (IV)	5,188 kg (11,438 lb)
Max cabin floor loading (cargo)	750 kg/m² (154 lb/sq ft)
Max wing loading: Y-12 (II)	145.9 kg/m² (29.90 lb/sq ft)
Max power loading: Y-12 (II)	5.74 kg/kW (9.42 lb/shp)
Y-12 (IV)	6.14 kg/kW (10.08 lb/shp)

PERFORMANCE (Y-12 (II)):
Max operating speed (VMO) at 3,000 m (9,840 ft)	177 kt (328 km/h; 204 mph)
Max cruising speed at 3,000 m (9,840 ft)	157 kt (292 km/h; 181 mph)
Econ cruising speed at 3,000 m (9,840 ft)	135 kt (250 km/h; 155 mph)
Max rate of climb at S/L	486 m (1,594 ft)/min
Rate of climb at S/L, OEI	84 m (275 ft)/min
Service ceiling	7,000 m (22,960 ft)
Service ceiling, OEI, 15 m (50 ft)/min rate of climb, max continuous power	3,000 m (9,840 ft)
T-O run, 15° flap: normal	340 m (1,115 ft)
STOL	230 m (755 ft)
T-O to 15 m (50 ft), 15° flap: normal	425 m (1,395 ft)
STOL	370 m (1,215 ft)
Landing from 15 m (50 ft), 20° flap: with braking and propeller reversal: normal	480 m (1,575 ft)
STOL	370 m (1,215 ft)
with brakes only: normal	620 m (2,035 ft)
STOL	200 m (656 ft)
Landing run, 20° flap:	
with braking and propeller reversal	200 m (660 ft)
with brakes only	340 m (1,120 ft)
Range at 135 kt (250 km/h; 155 mph) at 3,000 m (9,840 ft) with max fuel, 45 min reserves	723 n miles (1,340 km; 832 miles)
Endurance, conditions as above	5 h 12 min

PERFORMANCE (Y-12 (IV)):
Max operating speed (VMO) at 3,000 m (9,840 ft)	162 kt (300 km/h; 186 mph) CAS
Econ cruising speed at 3,000 m (9,840 ft)	140 kt (260 km/h; 162 mph)
Max rate of climb at S/L	468 m (1,535 ft)/min
Rate of climb at S/L, OEI	90 m (295 ft)/min
Service ceiling	7,000 m (22,960 ft)
T-O run	370 m (1,215 ft)
T-O to 15 m (50 ft)	490 m (1,610 ft)
Landing from 15 m (50 ft)	630 m (2,070 ft)
Landing run	340 m (1,120 ft)
Max range, 45 min reserves	707 n miles (1,310 km; 814 miles)
Max endurance, 45 min reserves	5 h 15 min

UPDATED

HAIG (Harbin)

HARBIN AIRCRAFT INDUSTRY GROUP (Subsidiary of AVIC II)
15 Youxie Street, Pingfang, PO Box 201-29, Harbin, Heilongjiang 150066

Tel: (+86 451) 650 11 22
Fax: (+86 451) 650 22 73
PRESIDENT AND GENERAL MANAGER: Cui Xuewen
VICE-PRESIDENT: Xu Zhanbin
PUBLIC RELATIONS: Chen Xiaoyi

HAIG at Harbin is the parent organisation of Hafei Aviation Industry Company (HAI, which see).

NEW ENTRY

HAIG (Hongdu)

HONGDU AVIATION INDUSTRY GROUP (Subsidiary of AVIC II)
Xinxiqiao, PO Box 5001, Nanchang, Jiangxi 330024
Tel: (+86 791) 846 84 01 and 846 84 02
Fax: (+86 791) 845 14 91

Hongdu is the parent organisation of Nanchang Aircraft Manufacturing Company (NAMC, which see).

NEW ENTRY

NAI

NANJING AERONAUTICAL INSTITUTE
29 Yu Dao Jie, Nanjing, Jiangsu 210016
Tel: (+86 25) 489 15 14 and 489 27 76
Fax: (+86 25) 489 15 14
e-mail: nhwrjj@public1.ptt.js.cn

NAI (formerly NUAA) was responsible for design and part-production of the AD-100 and FT-3000 light aircraft described in the following entries; design of the tandem-seat AD-200 (see under BKLAIC in this section); and production of the Chang Kong 1 pilotless aerial target (see *Jane's Unmanned Aerial Vehicles and Targets*).

NEW ENTRY

NAI AD-100

TYPE: Single-/two-seat ultralight.
PROGRAMME: Designed by NUAA (now NAI); a tandem-seat version is produced by Beijing Keyuan (BKLAIC, which see) as the AD-200.

CURRENT VERSIONS: **AD-100**: Single-seat landplane version.
Following description applies to AD-100.
AD-100S: Single-seat twin-float version of AD-100. Further details in 1997-98 *Jane's*.
AD-100T: Side-by-side two-seat landplane version of AD-100; 38.3 kW (51.6 hp) Rotax 462, three-blade propeller and 45 litres (11.9 US gallons; 9.9 Imp gallons) fuel. Further details in 1997-98 *Jane's*. One aircraft, No. 012, sold to US owner (as N168Y) in 1996.
CUSTOMERS: Total production of 30 by 1999 (latest figure known).
DESIGN FEATURES: High-mounted sweptback wings with endplate fins and rudders; nose-mounted canards; small fuselage pod.
FLYING CONTROLS: Manual.
STRUCTURE: Mainly of GFRP composites.
LANDING GEAR: Non-retractable tricycle type.
POWER PLANT: One 20.1 kW (27 hp) Rotax 277 engine, with two-blade pusher propeller. Fuel capacity 19 litres (5.0 US gallons; 4.2 Imp gallons).
ACCOMMODATION: Single seat in fully enclosed cockpit.
DIMENSIONS, EXTERNAL:
Wing span	8.68 m (28 ft 5¾ in)
Foreplane span	3.50 m (11 ft 5¾ in)
Length overall	5.02 m (16 ft 5½ in)
Height overall	1.79 m (5 ft 10½ in)

AREAS:
Wings, gross	10.63 m² (114.4 sq ft)
Foreplanes, gross	2.79 m² (30.0 sq ft)

WEIGHTS AND LOADINGS:
Weight empty	150 kg (331 lb)
Max T-O weight	290 kg (639 lb)

PERFORMANCE, POWERED:
Max level speed	75 kt (140 km/h; 87 mph)
Max cruising speed	48 kt (88 km/h; 55 mph)
Max rate of climb at S/L	180 m (591 ft)/min
T-O run	50 m (165 ft)
Landing run	60 m (200 ft)
Range at best cruising speed	189 n miles (350 km; 217 miles)
Endurance	4 h 0 min

PERFORMANCE, UNPOWERED:
Best glide ratio	13

UPDATED

AD-100, produced by Nanjing Aeronautical Institute *(Kenneth Munson)*

Floatplane version of the NAI FT-300 three-seat lightplane

NAI FT-300

TYPE: Three-seat lightplane.
PROGRAMME: Development started June 1993; first flights June 1994 (wheeled landing gear) and November 1994 (float gear). In production.
DESIGN FEATURES: Scaled-up development of AD-100/200 (which see).
FLYING CONTROLS: Manual.
LANDING GEAR: Non-retractable tricycle type; interchangeable with twin-float gear.
POWER PLANT: One 59.7 kW (80 hp) piston engine; three-blade pusher propeller.
ACCOMMODATION: Fully enclosed cabin; dual controls.
DIMENSIONS, EXTERNAL:
Wing span   9.295 m (30 ft 6 in)
Foreplane span   4.50 m (14 ft 9¼ in)
Length overall   5.56 m (18 ft 3 in)
Landing gear track (c/l of mainwheels and floats)
                                 1.50 m (4 ft 11 in)
WEIGHTS AND LOADINGS (A: landplane; B: floatplane):
Weight empty: A   300 kg (661 lb)
       B   350 kg (772 lb)
Max T-O weight: A, B   620 kg (1,366 lb)
Max power loading: A, B   10.39 kg/kW (17.08 lb/hp)
PERFORMANCE (A and B as above):
Max level speed   81 kt (150 km/h; 93 mph)
Min flying speed   36 kt (65 km/h; 41 mph)
Service ceiling   2,000 m (6,560 ft)
T-O run: A   150 m (495 ft)
       B   400 m (1,315 ft)
Landing run: A, B   150 m (495 ft)
Max range   270 n miles (500 km; 310 miles)
Max endurance   5 h
UPDATED

NAMC

NANCHANG AIRCRAFT MANUFACTURING COMPANY (Subsidiary of AVIC II)

PO Box 5001-506, Nanchang, Jiangxi 330024
Tel: (+86 791) 846 84 01 and 84 02
Fax: (+86 791) 845 14 91
PRESIDENT: Jiang Liang
INFORMATION: Feng Jinghua

Nanchang Aircraft Manufacturing Company (NAMC), which became a core unit of the Hongdu Aviation Industry Group (HAIG) in 1998, was created in 1951; it built 379 CJ-5s (licence Soviet Yak-18s) between 1954 and 1958, and in 1960s shared in large production programme for J-6 fighter (Chinese development of MiG-19); also built (1957-68) 727 Y-5 (Chinese An-2) biplanes (see under SAMC in this section and under NAMC in 1991-92 *Jane's*). Current programmes are Q-5/A-5 attack derivative of J-6, K-8 jet trainer and N-5A agricultural aircraft. Agreements with and via Canadian Aerospace Group (see CAG entry) in 1998 provided for North American assembly and marketing of N-5A in return for option to build CAG Windeagle and Monitor Jet in China; by mid-1999, the N-5A programme appeared to be significantly in arrears.

The company occupies a 500 ha (1,235 acre) site, with 10,000 m² (107,600 sq ft) of covered space, and has a workforce of about 20,000; it delivered its 4,000th aircraft in 1993. About 80 per cent of its activities are non-aerospace.
UPDATED

NAMC K-8 KARAKORUM 8

TYPE: Basic jet trainer/light attack jet.
PROGRAMME: Launched publicly (as L-8) at 1987 Paris Air Show as proposed export aircraft to be developed jointly with international partner. Subsequently proposed to be co-developed with Pakistan as partner (25 per cent share); aircraft then redesignated K-8 and named after mountain range forming part of China/Pakistan border. Pakistan decided 1994 against own assembly line, but agreed to reconsider if subcontracting share should later increase to 45 per cent.

Manufacture of four prototypes started January 1989; three flying prototypes: 001 (first flight 21 November 1990), 003 (first flight 18 October 1991) and 004; 002 is static and fatigue test aircraft.

First preproduction aircraft (1001/L8 320101) used as demonstrator. Preproduction batch of six for Pakistan Air Force evaluation (ordered 9 April 1994) handed over in China on 21 September 1994 and delivered to PAF on 10 November that year; PAF evaluation (approximately 1,200 hours) completed in August 1995; aircraft reported subsequently in use at Air Academy, Risalpur. China importing Progress AI-25 turbofans for use in domestic version (see under Power Plant), designated K-8J. One aircraft in use since 1997 by China Flight Test Establishment as variable stability testbed (K-8VSA) with fly-by-wire flight control system.
CURRENT VERSIONS: **K-8**: Initial version with TFE731 engine; as described.
K-8E: TFE731-powered version for Egyptian Air Force, 80 of which ordered under US$345 million contract signed on 27 December 1999. First 10 Chinese-built; Arab Organisation for Industrialisation (AOI) will then assemble batches of 15 and 10 from CKD kits of medium-sized and smaller components respectively before progressing to 90 per cent local manufacture of final 45. Total of 33 items newly selected, developed or upgraded for this version include instrument panels and consoles; com/nav systems; fire-control system; fuel system; environmental control system; hydraulic system; and landing gear. K-8E flown for first time on 5 July 2000.
K-8J: Reported designation of Chinese domestic version when fitted with AI-25 engine.
K-8VSA: In-flight variable stability testbed (serial number 320203), in use 1998 by China Flight Test Establishment (CFTE). Equipped with digital fly-by-wire AFCS, sidestick controller and data acquisition system.
CUSTOMERS: Total of 12 preproduction K-8s (six each to PLA Air Force and Pakistan Air Force) delivered by end of 1996. Original joint venture agreement reportedly involved up to 75 for Pakistan Air Force, but Pakistan Secretary for Defence Production quoted in early 1996 as saying up to 100 needed eventually to replace Cessna T-37; however, SLEP for T-37 has postponed main K-8 requirement to about 2005. Chinese PLAAF originally had

K-8 displayed at Paris Air Show, June 1999 *(Paul Jackson/Jane's)*

K-8 front cockpit

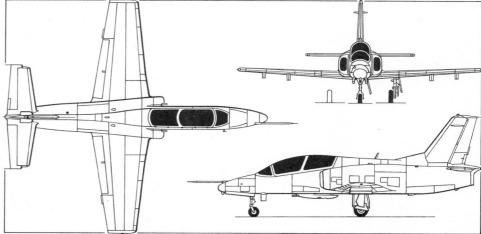
K-8 Karakorum 8 jet trainer and light attack aircraft *(Mike Keep/Jane's)*

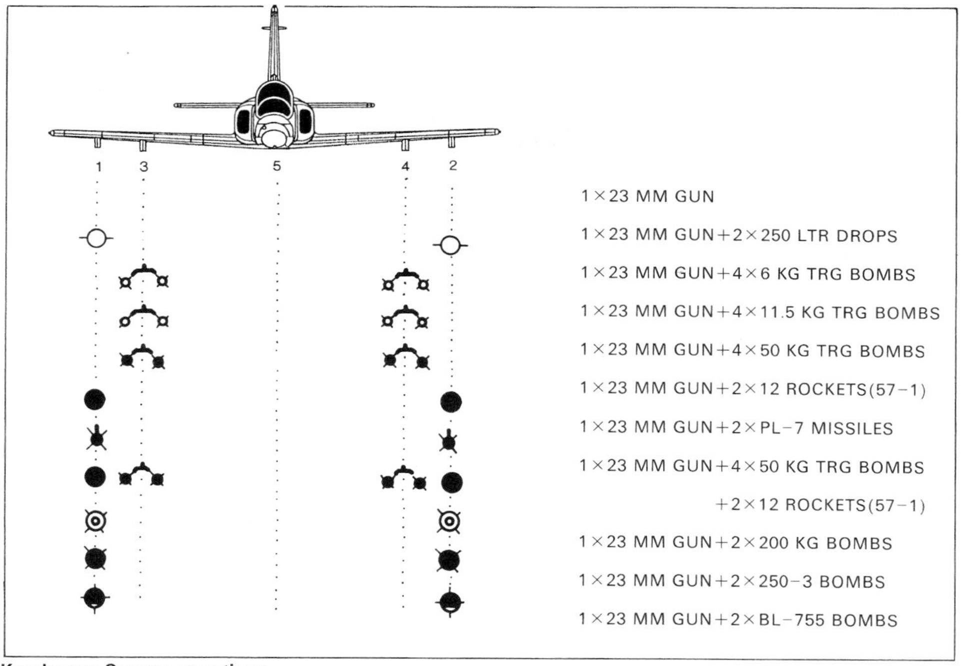

Karakorum 8 weapon options

requirement for several hundred K-8Js, of which 25/30 delivered by late 1999, but now widely reported to have abandoned plans for full adoption. Myanmar Air Force ordered 12, of which last three delivered in September 1999. Zambian Air Force eight, and Namibian Air Force four, also delivered in 1999. Deliveries to Sri Lanka expected in mid-2001. Egyptian Air Force has ordered 80 of K-8E version to replace Aero L-29. Interest also reported from Bangladesh, Cambodia, Eritrea, Laos and Thailand. Production totalled 50 by June 1999, according to test pilot of K-8 displayed at Paris Air Show.

K-8 EXPORT ORDERS
(at January 2001)

Country	Qty
Egypt	80
Myanmar	12[1]
Namibia	4
Pakistan	6
Sri Lanka	6
Zambia	8[2]
Total	**116**

[1] Reported total requirement for 30
[2] Reported option for further 8

COSTS: US$3 million to US$3.5 million flyaway (1996) with TFE731 and Western avionics.

DESIGN FEATURES: Intended for full basic flying training plus parts of primary and advanced syllabi, but capable also of light ground attack missions. Sweptback vertical/non-swept horizontal tail surfaces.

Tapered, non-swept low wings, with NACA 64A-114 root and NACA 64A-412 tip sections; 2° incidence at root, 3° dihedral from roots.

FLYING CONTROLS: Conventional and power-assisted; ailerons have hydraulic boost and artificial feel; variable incidence tailplane; electrically operated trim tab in rudder and port elevator. Two-position Fowler flaps, and split airbrake under fuselage just aft of mainwheel doors, are hydraulically actuated.

STRUCTURE: All-metal damage-tolerant main structure; ailerons of honeycomb, fin and rudder of composites. PAC share (initially only tailplane and elevators) increased to include fin, rudder, rear fuselage and engine cowling/access panels. First Pakistan-built subassemblies delivered to China in mid-1997.

LANDING GEAR: Hydraulically retractable tricycle type, with single wheel and oleo-pneumatic shock-absorber on each unit. Main units retract inward into underside of fuselage; nosewheel, which has hydraulic steering, retracts forward. Mainwheel tyres size 561×169, pressure 6.90 bars (100 lb/sq in). Chinese hydraulic disc brakes. Anti-skid units. Minimum ground turning radius 6.69 m (21 ft 11½ in).

POWER PLANT: Including prototypes, all except K-8J have one 16.01 kN (3,600 lb st) Honeywell TFE731-2A-2A turbofan with Lucas Aerospace FADEC, mounted in rear fuselage. Production K-8J for China powered by 16.87 kN (3,792 lb st) ZMKB Progress AI-25TLK turbofan, ordered in 1997; initial batch of 58 imported.

Fuel in two flexible tanks in fuselage and one integral tank in wing centre-section, combined capacity 1,000 litres (264 US gallons; 220 Imp gallons); single refuelling point in fuselage. Provision for carrying one 250 litre (66.0 US gallon; 55.0 Imp gallon) drop tank on outboard pylon under each wing. Oil capacity 4 kg (8.8 lb).

ACCOMMODATION: Instructor and pupil in tandem, on Martin-Baker Mk 10L zero/zero ejection seats; rear seat elevated 28 cm (11 in). One-piece wraparound windscreen; two-piece canopy opens sideways to starboard. Cockpits pressurised and air conditioned.

SYSTEMS: Honeywell ECS 51833 air conditioning and pressurisation system, with maximum differential of 0.27 bar (3.9 lb/sq in). Hydraulic system, pressure 207 bars (3,000 lb/sq in), for operation of landing gear extension/retraction, wing flaps, airbrake, aileron boost, nosewheel steering and wheel brakes. Flow rate 15 litres (3.96 US gallons; 3.30 Imp gallons)/min, with air pressurised reservoir, plus emergency back-up hydraulic system. Abex AP09V-8-01 pump. Electrical systems 28.5 V DC (primary) and 24 V DC (auxiliary), with 115/26 V single-phase AC and 36 V three-phase AC available from Ni/Cd battery and static inverter, both at 400 Hz. Liquid oxygen system for occupants. Demisting of cockpit transparencies.

AVIONICS: *Comms:* Honeywell VHF and Rockwell Collins EFIS-86 system selected for first 100 aircraft, incorporating CRT primary flight and navigation displays for each crew member plus dual display processing units and selector panels for tandem operation. Magnavox AN/ARC-164 UHF radios.

Flight: Honeywell KNR 634A VOR/ILS with marker beacon receiver, ADF, Type 265 radio altimeter, AHRS and air data computer and KTU-709 Tacan; WL-7 radio compass.

Instrumentation: Rockwell Collins EFIS-86T. Blind-flying instrumentation standard. Standby flight instruments include ASI, rate of climb indicator, barometric altimeter, emergency horizon and standby compass.

ARMAMENT: (optional): One 23 mm gun pod under centre-fuselage; self-computing optical gunsight in cockpit, plus gun camera. Two external stores points under each wing. Twin ejector racks on inboard stations can carry total of four 6, 11.5 or 50 kg practice bombs; single-store outboard stations can each carry a PL-7 air-to-air missile, a 12-round pod of 57 mm rockets, a 200 kg, 250 kg or BL755 bomb, or a drop fuel tank.

DIMENSIONS, EXTERNAL:
- Wing span: 9.63 m (31 ft 7¼ in)
- Wing aspect ratio: 5.4
- Length overall, incl nose pitot: 11.60 m (38 ft 0¾ in)
- Height overall: 4.21 m (13 ft 9¾ in)
- Elevator span: 4.20 m (13 ft 9½ in)
- Wheel track: 2.54 m (8 ft 4 in)
- Wheelbase: 4.44 m (14 ft 6¾ in)

AREAS:
- Wings, gross: 17.02 m^2 (183.2 sq ft)
- Ailerons (total, incl tab): 1.095 m^2 (11.80 sq ft)
- Trailing-edge flaps (total): 2.65 m^2 (28.55 sq ft)
- Fin: 1.975 m^2 (21.27 sq ft)
- Rudder, incl tab: 1.115 m^2 (12.02 sq ft)
- Tailplane: 3.035 m^2 (32.69 sq ft)
- Elevators (total, incl tab): 1.32 m^2 (14.21 sq ft)

WEIGHTS AND LOADINGS:
- Weight empty, equipped: 2,757 kg (6,078 lb)
- Max fuel: internal: 780 kg (1,720 lb)
  - external (two drop tanks): 388 kg (855 lb)
- Max external stores load: 943 kg (2,080 lb)
- T-O weight clean: 3,700 kg (8,157 lb)
- Max T-O weight with external stores: 4,400 kg (9,550 lb)
- Max wing loading: 254.5 kg/m^2 (52.13 lb/sq ft)
- Max power loading (TFE731): 271 kg/kN (2.65 lb/lb st)

PERFORMANCE: (at clean T-O weight):
- Never-exceed speed (V_{NE}): 512 kt (950 km/h; 590 mph) IAS
- Max level speed at S/L: 432 kt (800 km/h; 497 mph)
- Unstick speed: 100 kt (185 km/h; 115 mph)
- Approach speed: 108 kt (200 km/h; 124 mph)
- Touchdown speed, 35° flap: 86 kt (160 km/h; 99 mph)
- Stalling speed, 35° flap: 81 kt (150 km/h; 94 mph)
- Max rate of climb at S/L: 1,800 m (5,905 ft)/min
- Service ceiling: 13,600 m (44,620 ft)
- T-O run: 440 m (1,445 ft)
- T-O to 15 m (50 ft): 600 m (1,970 ft)
- Landing from 15 m (50 ft): 518 m (1,700 ft)
- Landing run: 530 m (1,740 ft)
- Range:
  - max internal fuel: 842 n miles (1,560 km; 969 miles)
  - max internal/external fuel: 1,155 n miles (2,140 km; 1,329 miles)
- Endurance: max internal fuel: 3 h 12 min
  - max internal/external fuel: 4 h 12 min
- *g* limits: +7.33/−3

UPDATED

NAMC Q-5

Chinese name: Qiangjiji-5 (Attack aircraft 5)
Westernised designation: A-5
NATO reporting name: Fantan

TYPE: Attack fighter.

PROGRAMME: Derivative of J-6 fighter (see 1989-90 and earlier *Jane's* for details of changes), originating August 1958 as Shenyang design proposal; responsibility assigned to Nanchang; prototype programme cancelled 1961, but kept alive by small team and resumed officially 1963; first flight 4 June 1965; preliminary design certificate awarded and preproduction batch authorised late 1965, but further modifications (to fuel, armament, hydraulic and other systems) found necessary, leading to flight test of two much modified prototypes from October 1969; series production approved at end of 1969, deliveries beginning 1970.

Improved Q-5 I proposed 1976, flight tested late 1980 and certified for production 20 October 1981, by which time (April 1981) Pakistan had placed order for A-5C modified export version; first A-5C deliveries January 1983, completed January 1984; domestic Q-5 IA, incorporating many of A-5C improvements, certified January 1985.

Upgrade programmes involving Western avionics started in 1986 with France (Q-5K Kong Yun) and Italy (A-5M), but Kong Yun programme terminated 1990

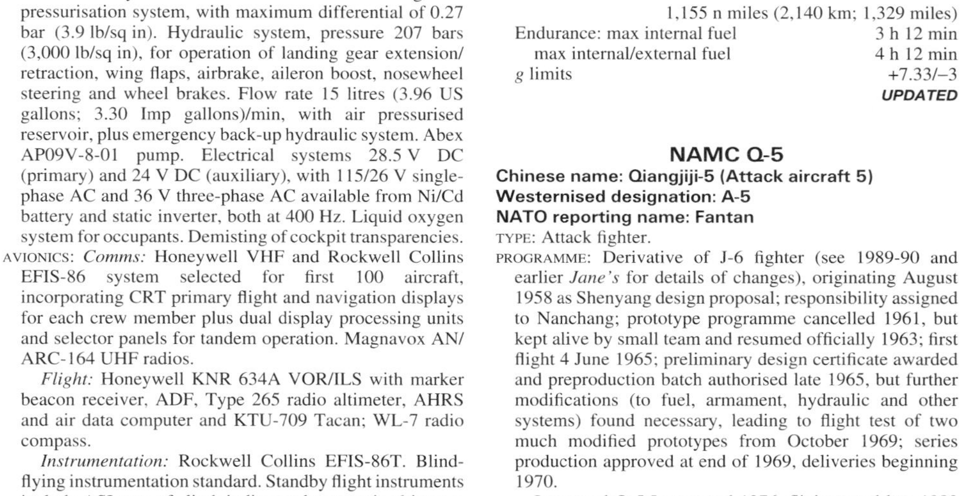

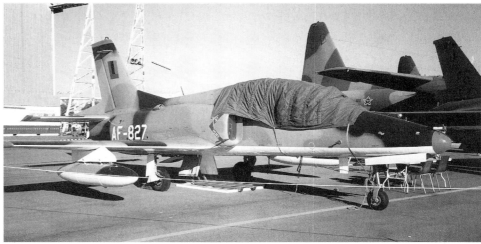

K-8 in the insignia of the Zambian Air Force *(Craig Hoyle/Jane's)* ***2001***/0105032

Cockpit of A-5C

Latest version of the long-serving Q-5 to be revealed is the laser-equipped A-5D

(details in 1990-91 *Jane's*) and A-5M (1996-97 and earlier editions) also abandoned. More recently, Russian Phazotron Komar pulse Doppler radar offered as possible upgrade. Production now probably for attrition replacement only.

CURRENT VERSIONS: **Q-5:** Initial production version, with internal fuselage bay approximately 4.00 m (13 ft 1½ in) long for two 250 kg or 500 kg bombs, two underfuselage attachments adjacent bay for two similar bombs, and two stores pylons beneath each wing; Series 6 WP6 turbojets; brake-chute in tailcone, between upper and lower pen-nib fairings. Some adapted for nuclear weapon delivery tests in early 1970s.

Q-5 I: Extended payload/range version, with internal bomb bay blanked off and space used to enlarge main fuselage fuel tank and add a flexible tank; underfuselage stores points increased to four; improved series WP6 engines; modified landing gear; brake-chute relocated under base of rudder; improved Type I rocket ejection seat; HF/SSB transceiver added. Some aircraft, adapted for PLA Naval Air Force to carry two underfuselage torpedoes, reportedly have Doppler-type nose radar and 20 m (66 ft) sea-skimming capability with C-801 (YJ-8) anti-shipping missiles.

Q-5 IA: Improved Q-5 I, with additional underwing hardpoint each side (increasing stores load by 500 kg; 1,102 lb), new gun/bomb sighting systems, pressure refuelling, and added warning/countermeasures systems.

Q-5 II: As Q-5 IA, but fitted (or retrofitted) with radar warning receiver.

A-5C: Export version for Pakistan Air Force (and later customers), involving 32 modifications from Q-5 I, notably upgraded avionics, Martin-Baker Mk 10 zero/zero seat, and adaptation of hardpoints for 356 mm (14 in) lugs compatible with Sidewinder missiles and other PAF weapons; three prototypes preceded production programme; in service with Nos. 7, 16 and 26 Squadrons of PAF, by whom designated **A-5-III.** Ordered also by Bangladesh and Myanmar.

Description applies to Q-5 IA and A-5C except where indicated.

A-5D: First seen in company brochure in 2000. Upgraded from A-5C to carry laser-guided weapons; small undernose fairing for ALR-1 laser range-finder, a second and slightly larger underfuselage bulge just aft of the cockpit, and what may be AoA sensors on either side of the nose.

CUSTOMERS: China (PLA Air Force and Navy). Nearly 1,000 (all versions) built to date, including 140 for export to Bangladesh (24 A-5C ordered, of which about 16 reported in service by early 1999, with No. 21 Squadron), North Korea (40 Q-5 IA), Myanmar (24 A-5C) and Pakistan (52 A-5-III). Phase-out of PLA aircraft has reportedly begun.

DESIGN FEATURES: Mid-mounted sweptback wings with deep, full-chord fence on each upper surface at mid-span; air intake on each side of fuselage abreast of cockpit; twin jetpipes side by side at rear with upper and lower pen-nib fairings aft of nozzles; fuselage has area-ruled 'waist'; rear fuselage detachable aft of wing trailing-edge for engine access; dorsal spine fairing; shallow ventral strake under each jetpipe; all-swept tail surfaces.

Wings have 52° 30′ sweep at quarter-chord, 0° incidence and 4° anhedral from roots; tailplane has 6° 30′ anhedral.

FLYING CONTROLS: Internally balanced ailerons and fully powered slab tailplane; mechanically actuated mass-balanced rudder; hydraulically actuated Gouge flaps on inboard trailing-edges; electrically operated trim tab in port aileron and rudder; forward-hinged, hydraulically actuated door-type airbrake under centre of fuselage, forward of bomb attachment points; anti-flutter weight on each tailplane tip. Aileron deflection 18° 30′ up/down; tailplane 12° 30′ up/30° down; rudder 25° left/right.

STRUCTURE: Conventional all-metal stressed skin structure. Multispar wings have three-point attachment to fuselage; fuselage built in forward and rear portions.

LANDING GEAR: Hydraulically retractable wide-track tricycle type, with single wheel and oleo-pneumatic shock-absorber on each unit. Main units retract inward into wings, non-steerable nosewheel forward into fuselage, rotating through 87° to lie flat in gear bay. Mainwheels have size 830×205 tubeless tyres and disc brakes; nosewheel tyre size 595×230. Tail braking parachute, deployed when aircraft is 1 m (3.3 ft) above the ground, in bullet fairing beneath rudder.

POWER PLANT: Two LM (Liming) WP6 turbojets, each rated at 25.5 kN (5,730 lb st) dry and 31.9 kN (7,165 lb st) with afterburning, mounted side by side in rear of fuselage. Improved WP6A engines (29.4 kN; 6,615 lb st dry and 39.7 kN; 8,930 lb st with afterburning) available optionally. Lateral air intake, with small splitter plate, for each engine. Hydraulically actuated nozzles.

Internal fuel in three forward and two rear fuselage tanks with combined capacity of 3,648 litres (964 US gallons; 802.5 Imp gallons). Later PLAAF aircraft fitted from about 1998 with in-flight refuelling probes for use with Xian H-6 (Tu-16) bombers converted as tankers. Provision for carrying a 760 litre (201 US gallon; 167 Imp gallon) drop tank on each centre underwing pylon, to give maximum internal/external fuel capacity of 5,168 litres (1,366 US gallons; 1,136.5 Imp gallons). When centre wing stations are occupied by bombs, a 400 litre (105.7 US gallon; 88.0 Imp gallon) drop tank can be carried instead on each outboard underwing pylon.

ACCOMMODATION: Pilot only, under one-piece jettisonable canopy which is hinged at rear and opens upward. Downward view over nose, in level flight, is 13° 30′. Low-speed seat allows for safe ejection within speed range of 135 to 458 kt (250 to 850 km/h; 155 to 528 mph) at zero height or above. Pakistan's A-5s have Martin-Baker Mk 10 zero/zero seats. Armour plating in some areas of cockpit to protect pilot from anti-aircraft gunfire. Cockpit pressurised and air conditioned.

SYSTEMS: Dual air conditioning systems, one for cockpit environment and one for avionics cooling. Two independent hydraulic systems, each operating at pressure of 207 bars (3,000 lb/sq in). Primary system actuates landing gear extension and retraction, flaps, airbrake and afterburner nozzles; auxiliary system supplies power for aileron and all-moving tailplane boosters. Emergency system, operating pressure 108 bars (1,570 lb/sq in), for actuation of main landing gear. Electrical system (28 V DC) powered by two 6 kW engine-driven starter/generators, with two inverters for 115 V single-phase and 36 V three-phase AC power at 400 Hz.

AVIONICS: Space provision in nose and centre-fuselage for additional or updated avionics.

Comms: CT-3 VHF transceiver; YD-3 IFF ('Odd Rods' type aerials under nose on Q-5s, replaced on some A-5Cs by single blade antenna).

Flight: WL-7 radio compass; WG-4 low-altitude radio altimeter; LTC-2 horizon gyro; XS-6 marker beacon receiver.

Instrumentation: SH-1J or ABS1A optical sight originally, for level and dive bombing, or for air-to-ground rocket launching. Latest aircraft reportedly have HUD, ballistic computer and ALR-1 laser range-finder/designator.

Self-defence: Type 930 RWR (antenna in fin-tip); provision for ECM pod on each centre wing station.

EQUIPMENT: Combat camera in small teardrop fairing on starboard side of nose (not on export models). Landing light under fuselage, forward of nosewheel bay and offset to port; taxying light on nosewheel leg.

ARMAMENT: Internal armament consists of one 23 mm cannon (Norinco Type 23-2K), with 100 rounds, in each wingroot. Ten attachment points normally for external stores: two pairs in tandem under centre of fuselage, and three under each wing (one inboard and two outboard of mainwheel leg).

Fuselage stations can each carry a 250 kg bomb (Chinese 250-2 or 250-3, US Mk 82 or Snakeye, French Durandal, or similar). Inboard wing stations can carry 6 kg or 25 lb practice bombs, or a pod containing eight Chinese 57-2 (57 mm), seven 68 mm, or seven Norinco 90-1 (90 mm) or four 130-1 (130 mm) rockets. Centre wing stations can carry a 500 kg or 750 lb bomb, a BL755 600 lb cluster bomb, a Chinese 250-2 or -3 bomb, US Mk 82 or Snakeye, French Durandal, or similar, or a Chinese C-801 anti-shipping missile. Normal bomb carrying capacity is 1,000 kg (2,205 lb), maximum capacity 2,000 kg (4,410 lb). Outboard wing stations can each be occupied by air-to-air missiles such as PL-2, PL-2B, PL-7, AIM-9 Sidewinder and R550 Magic.

Within overall maximum T-O weight, all stores mentioned can be carried provided that CG shift remains within allowable operating range of 31 to 39 per cent of mean aerodynamic chord; more than 22 external stores configurations possible. Some early aircraft in Chinese service modified to carry a single 5 to 20 kT nuclear bomb.

DIMENSIONS, EXTERNAL:
Wing span: Q-5 IA	9.68 m (31 ft 9 in)
A-5C	9.70 m (31 ft 10 in)
Wing chord (mean aerodynamic)	3.095 m (10 ft 2 in)
Wing aspect ratio	3.4
Length overall:	
incl nose probe: Q-5 IA	15.65 m (51 ft 4¼ in)
A-5C	16.77 m (55 ft 0¼ in)
excl nose probe: A-5C	15.695 m (51 ft 6 in)
Height overall: Q-5 IA	4.335 m (14 ft 2¾ in)
A-5C	4.515 m (14 ft 9¾ in)
Wheel track	4.40 m (14 ft 5¼ in)
Wheelbase	4.01 m (13 ft 2 in)

AREAS (Q-5 IA and A-5C):
Wings, gross	27.95 m² (300.9 sq ft)
Ailerons (total)	3.30 m² (35.52 sq ft)
Airbrake	1.775 m² (19.08 sq ft)
Vertical tail surfaces (total)	4.64 m² (49.94 sq ft)
Horizontal tail surfaces:	
movable	5.00 m² (53.82 sq ft)
total, incl projected fuselage area	8.62 m² (92.78 sq ft)
Ventral strakes (total)	1.60 m² (17.22 sq ft)

WEIGHTS AND LOADINGS:
Weight empty: Q-5 IA	6,375 kg (14,054 lb)
A-5C	6,638 kg (14,634 lb)
Fuel: max internal	2,827 kg (6,232 lb)
two 400 litre drop tanks	620 kg (1,367 lb)
two 760 litre drop tanks	1,178 kg (2,597 lb)
max internal/external	4,005 kg (8,829 lb)

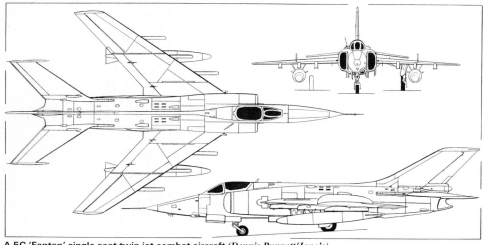

A-5C 'Fantan' single-seat twin-jet combat aircraft *(Dennis Punnett/Jane's)*

Max external stores load	2,000 kg (4,410 lb)
Max T-O weight:	
clean: Q-5 IA	9,486 kg (20,913 lb)
A-5C	9,715 kg (21,418 lb)
with max external stores:	
Q-5 IA	11,830 kg (26,080 lb)
A-5C	12,000 kg (26,455 lb)
Max wing loading:	
clean: Q-5 IA	339.4 kg/m² (69.51 lb/sq ft)
A-5C	340.0 kg/m² (69.58 lb/sq ft)
with max external stores:	
Q-5 IA	423.3 kg/m² (86.69 lb/sq ft)
A-5C	429.3 kg/m² (87.93 lb/sq ft)
Max power loading (WP6 engines):	
clean: Q-5 IA	149 kg/kN (1.46 lb/lb st)
A-5C	152 kg/kN (1.49 lb/lb st)
with max external stores:	
Q-5 IA	186 kg/kN (1.82 lb/lb st)
A-5C	188 kg/kN (1.85 lb/lb st)

PERFORMANCE (A-5C at max clean T-O weight, with afterburning, except where indicated):
Max limiting Mach No. 1.205
Max level speed: at 11,000 m (36,080 ft)
  M1.12 (643 kt; 1,190 km/h; 740 mph)
  at S/L 658 kt (1,220 km/h; 758 mph)
Unstick speed:
  clean, 15° flap 162 kt (300 km/h; 186 mph)
  with max external stores, 25° flap
    178 kt (330 km/h; 205 mph)
Touchdown speed: 25° flap, brake-chute deployed
    150-165 kt (278-307 km/h; 172-191 mph)
Max rate of climb: at S/L 8,880 m (29,134 ft)/min
  at 5,000 m (16,400 ft)
    4,980-6,180 m (16,340-20,275 ft)/min
Service ceiling 15,850 m (52,000 ft)
T-O run:
  clean, 15° flap 700-750 m (2,300-2,460 ft)
  with max external stores, 25° flap 1,250 m (4,100 ft)
Landing run:
  25° flap, brake-chute deployed 1,060 m (3,480 ft)
Combat radius with max external stores, afterburners off:
  lo-lo-lo (500 m; 1,640 ft)
    216 n miles (400 km; 248 miles)
  hi-lo-hi (8,000/500/8,000 m; 26,250/1,640/26,250 ft)
    324 n miles (600 km; 373 miles)
Range at 11,000 m (36,090 ft) with max internal and external fuel, afterburners off
    982 n miles (1,820 km; 1,130 miles)
g limits:
  with full load of bombs and/or drop tanks +5
  with drop tanks empty +6.5
  clean +7.5

UPDATED

NAMC CJ-6A
Chinese name: Chuji Jiaolianji-6A (Basic training aircraft 6A)
Westernised designation: PT-6A

TYPE: Basic prop trainer.
PROGRAMME: Design initiated at Shenyang in second half of 1957 as Chinese-engineered successor to CJ-5 (licence Yak-18: see 1991-92 and earlier *Jane's*); first flight of first prototype (108 kW; 145 hp Mikulin M-11ER engine) 27 August 1958, but trials disappointing; modified version with 194 kW (260 hp) Ivchenko AI-14R made first flight 18 July 1960. Responsibility subsequently transferred to Nanchang, further redesign preceding first flight of production-standard prototype 15 October 1961; production go-ahead for aircraft January 1962, for HS6 engine (Chinese AI-14R) June 1962. No longer produced; see 2000-01 and earlier *Jane's* for description.
CURRENT VERSIONS: **CJ-6A:** Standard version from December 1965, with uprated HS6A engine.
CJ-6B: Armed version: 10 built 1964-66.

NAMC CJ-6A primary trainer *(Paul Jackson/Jane's)*

Haiyan A: Prototype of civil agricultural version, described in 1991-92 and earlier *Jane's*. Programme superseded by N-5A (which see).
CUSTOMERS: Total of 1,796 (all versions) built by end of 1986 (or perhaps earlier), mainly for PLA Air Force but including well over 200 for foreign military customers. Exported to Albania (20), Bangladesh (36), Cambodia, North Korea, Sri Lanka (10), Tanzania and Zambia (10). Large numbers of ex-PLA aircraft disposed of to civilian owners in Australia (from 1991), UK, USA and elsewhere. Main production at factory No. 320 in 26 batches up to 1970 and further 20 or so batches from about 1977 to 1984; however factory No. 512 built some 300 in five batches, 1970 to 1976.

UPDATED

NAMC N-5A
Chinese name: Nongye Feiji 5 (Agricultural aircraft 5)

TYPE: Agricultural sprayer.
PROGRAMME: Design began November 1987; first details revealed at Farnborough Air Show September 1988; first of three prototypes (B-501L) made first flight 26 December 1989; CAAC type certificate awarded July 1992 and production certificate in June 1995; production rate in 1997 (latest information received) reported as 10 per year, but according to CAG (see Canadian section) production suspended after 14th aircraft.
CURRENT VERSIONS: **N-5A:** Standard version, *as described*.
AG Dragon: Version for North American customers, assembled and marketed by AG Dragon Aircraft Corporation at St Hubert, Quebec, Canada, from 1999; offered with choice of piston or turboprop engine. MoU, signed with parent Can-Aero Leasing on 2 April 1998, provides for export of up to 240 N-5As for North American market over five years. Canadian company now seeking alternative to IO-720 engine. None had been registered in Canada by late 2000.
CUSTOMERS: At least seven ordered by Chinese agricultural agencies or farms by end of 1993; others leased from manufacturer during that year. Operators by early 1997 included Longken General Aviation (five). Estimated domestic market for more than 300 as Y-5 replacement.
DESIGN FEATURES: Designed to meet Chinese (CCAR) and US (FAR Pt 23) Normal category requirements, for specialised farming and forestry applications; crash-resistant forward fuselage with quickly removable side panels; cable cutter on windscreen, with deflector cable from this to tip of fin.
Wings mainly constant chord, of LS(1)-0417 Mod section (17 per cent thickness/chord ratio), but inboard leading-edge gloves swept back 18°; dihedral 4° 30′ from roots; incidence 2°. Tailplane has inverted aerofoil section.
FLYING CONTROLS: Conventional and manual. Differential ailerons. Ground-adjustable tab on rudder and each aileron; electrically actuated trim tab in starboard elevator. Single-slotted trailing-edge flaps, also actuated electrically.
STRUCTURE: All-metal, with two-spar thin-wall wings, two-spar fin and tailplane, welded alloy steel tube forward fuselage and riveted duralumin rear fuselage; lower fuselage skin panels are of stainless steel and non-removable. Entire structure anodised before assembly and finished in polyurethane enamel paint.
LANDING GEAR: Non-retractable tricycle type, with single wheel and oleo-pneumatic shock-absorber on each unit. Nose unit has a telescopic strut, a size 400×150 tyre, pressure 2.50 bars (36 lb/sq in), and is fitted with a shimmy damper and wire cutter. Main gear legs are of trailing-link type, with tyres size 500×200, pressure 3.00 bars (44 lb/sq in). Hydraulic mainwheel disc brakes and parking brake.
POWER PLANT: One 298 kW (400 hp) Textron Lycoming IO-720-D1B flat-eight engine, driving a Hartzell HC-C3YR-1RF/F8475R three-blade variable-pitch metal propeller. All fuel in wing tanks with combined capacity of 315 litres (83.2 US gallons; 69.3 Imp gallons). Gravity fuelling point in upper surface of each wing at root. Oil capacity 14 kg (31 lb).
ACCOMMODATION: Tandem seats and inertia reel safety harnesses for pilot and, when required, a loader/mechanic, under hard-top framed canopy with all-round field of view. Downward-opening window/door on each side. Cockpit semi-sealed with ram air ventilation, slightly pressurised to prevent chemical ingress. Cockpit heating optional. Windscreen washer, wiper and demister standard. Deflector cable from top of windscreen cable cutter to tip of fin.
SYSTEMS: Hydraulic system for brakes only. No pneumatic system. Electrical system powered by Prestolite 28 V 100 A AC generator and 30 Ah battery.
AVIONICS: *Comms:* Honeywell KY 96A VHF com transceiver standard; KHF 950 HF/SSB com transceiver optional.
Flight: Options include Type 263 radio altimeter; LC-2 magnetic compass; WL-7A radio compass; and XS-6B marker beacon receiver. Stall warning system.
EQUIPMENT: Glass fibre honeycomb hopper, for liquid or dry chemicals, forward of cockpit, with quick-dump system permitting release of all contents within 5 seconds. Solid or spray dispersal system, as appropriate. Dispersal of liquids, powered by fan-driven pump, is via Y-type filter and 60-nozzle spraybars and is suitable for high-, medium- or low-volume application. Wire cutters on main landing gear and in front of cockpit canopy.

DIMENSIONS, EXTERNAL:
Wing span	13.42 m (44 ft 0¼ in)
Wing chord: at root	2.32 m (7 ft 7½ in)
constant portion	1.875 m (6 ft 2 in)

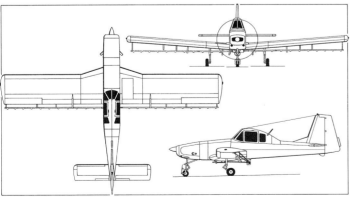

Nanchang N-5A agricultural monoplane *(Mike Keep/Jane's)*

The single/two-seat N-5A crop-sprayer/duster *(Photo Link)*

Wing aspect ratio	6.8	WEIGHTS AND LOADINGS:		Min banking turn radius: A	145 m (480 ft)	
Length overall	10.485 m (34 ft 4⅞ in)	Weight empty	1,328 kg (2,928 lb)	B	140 m (460 ft)	
Height overall	3.78 m (12 ft 4¼ in)	Fuel weight: normal	85 kg (187 lb)	Service ceiling: A	3,750 m (12,300 ft)	
Fuselage: Max width	1.01 m (3 ft 3¾ in)	max	233 kg (513 lb)	B	4,280 m (14,040 ft)	
Max depth	1.735 m (5 ft 8¼ in)	Payload: normal	760 kg (1,675 lb)	T-O run: A	303 m (995 ft)	
Tailplane span	4.59 m (15 ft 0¾ in)	max	960 kg (2,116 lb)	B	296 m (975 ft)	
Wheel track	3.53 m (11 ft 7 in)	Max T-O weight: normal	2,250 kg (4,960 lb)	T-O to 15 m (50 ft): A	569 m (1,870 ft)	
Wheelbase	2.715 m (8 ft 10¾ in)	overload	2,450 kg (5,401 lb)	B	553 m (1,815 ft)	
Propeller diameter	2.185 m (7 ft 2 in)	Max wing loading: normal	86.5 kg/m² (17.72 lb/sq ft)	Landing from 15 m (50 ft): A	373 m (1,225 ft)	
DIMENSIONS, INTERNAL:		overload	94.2 kg/m² (19.30 lb/sq ft)	B	379 m (1,245 ft)	
Cockpit: Length	2.29 m (7 ft 6¼ in)	Max power loading: normal	7.55 kg/kW (12.40 lb/hp)	Landing run: A	246 m (810 ft)	
Max width	1.00 m (3 ft 3¼ in)	overload	8.22 kg/kW (13.50 lb/hp)	B	252 m (830 ft)	
Max height	1.26 m (4 ft 1½ in)	PERFORMANCE (A: with, B: without dispersal equipment):		Normal range with max payload, 45 min reserves:		
Hopper volume	1.2 m³ (42 cu ft)	Max level speed: A	111 kt (205 km/h; 127 mph)	A	135 n miles (250 km; 155 miles)	
AREAS:		B	118 kt (220 km/h; 136 mph)	B	152 n miles (282 km; 175 miles)	
Wings, gross	26.00 m² (279.9 sq ft)	Normal operating speed:		Ferry range with max fuel:		
Ailerons (total)	2.08 m² (22.39 sq ft)	A, B	81-92 kt (150-170 km/h; 93-105 mph)	A	528 n miles (979 km; 608 miles)	
Trailing-edge flaps (total)	4.06 m² (43.70 sq ft)	Stalling speed:		Endurance with max payload, 45 min reserves:		
Fin	2.28 m² (24.54 sq ft)	A, B, flaps up	57 kt (105 km/h; 66 mph)	A	1 h 48 min	
Rudder, incl tab	1.57 m² (16.90 sq ft)	A, B, flaps down	47 kt (86 km/h; 54 mph)	B	1 h 56 min	
Tailplane	4.68 m² (50.38 sq ft)	Max rate of climb at S/L: A	257 m (845 ft)/min	Endurance (self-ferry) with max fuel	5 h 45 min	
Elevators (total, incl tabs)	2.20 m² (23.68 sq ft)	B	281 m (922 ft)/min		**UPDATED**	

NGA

NANJING GROEN AVIATION INDUSTRIAL LTD

140 Guangzhou Road, Suite 22D, Nanjing, Jiangsu 210024
Tel: (+86 25) 360 50 73
Fax: (+86 25) 360 49 53

On 22 July 1996, Shanghai Energy and Chemicals Corporation (SECC) signed a letter of intent with Groen Brothers Aviation Inc of the USA (which see), committing to buy 200 of the latter company's H2X Hawk III three-seat autogyros with which to form an air taxi service in the Shanghai district. Parts and components would be supplied by the US company for final assembly in "an existing aircraft manufacturing plant" in Shanghai. In 1998, Groen Brothers announced new commitments from SECC for a further 200 Hawks, comprising 100 each of the five-seat Hawk V and eight-seat Hawk VIII.

Nanjing Groen Aviation Industrial Ltd (NGA) was registered as a Chinese subsidiary of Groen Brothers in 1996; this plant, too, will import Hawk components from the USA for final assembly and sale in China, once the aircraft have received FAA certification. This was still under way in early 2001.

UPDATED

NLA

NANJING LIGHT AIRCRAFT INCORPORATED COMPANY

Room 102, Building D, 29 Yudao Street, Nanjing, Jiangsu 210016
Tel: (+86 25) 489 16 36 and 36 55
Fax: {+86 25) 489 16 37
e-mail: njqf@pub.jlonline.com
PRESIDENT: Wo Dingzhu

NLA was created in 1998 by Nanjing University of Aeronautics and Astronautics (NUAA, now Nanjing Aeronautical Institute), the municipality of Nanjing, the provincial government of Jiangsu and other business interests. It is responsible for the five-seat AC-500, which was planned to fly for the first time in late 2000.

UPDATED

NLA AC-500 AIRCAR

TYPE: Five-seat lightplane.
PROGRAMME: Revealed at Airshow China in November 1998. Prototype planned to fly in late 2000; scheduled to become available in 2002 after FAR Pt 23 certification.
CUSTOMERS: Initial order from Yuanda Air Conditioning Company.
COSTS: Estimated US$190,000 (1998).
DESIGN FEATURES: General appearance similar to that of Piper and Socata types in same weight/performance category. Constant-chord, low-mounted wings with upturned tips; sweptback vertical tail and small underfin.
FLYING CONTROLS: Conventional and manual; flaps fitted. One-piece horn-balanced elevator; ground-adjustable tab on rudder.
STRUCTURE: All-metal main structure; some non-load-bearing components manufactured from composites.
LANDING GEAR: Non-retractable tricycle type; single wheel on each unit.
POWER PLANT: One 194 kW (260 hp) Textron Lycoming IO-540 flat-six engine, driving a three-blade propeller.
ACCOMMODATION: Pilot and co-pilot or passenger in front, with two rearward-facing seats behind and a forward-facing single seat aft of these. Crew door each side of cockpit, plus passenger door aft of wing on port side.

DIMENSIONS, EXTERNAL:
Wing span 10.20 m (33 ft 5½ in)
Length overall 8.14 m (26 ft 8½ in)
Fuselage max width 1.20 m (3 ft 11¼ in)
Height overall 3.00 m (9 ft 10 in)
Wheel track 2.80 m (9 ft 2¼ in)
WEIGHTS AND LOADINGS:
Max baggage weight 65 kg (143 lb)
Max T-O weight 1,540 kg (3,395 lb)
Max power loading 7.95 kg/kW (13.06 lb/hp)
PERFORMANCE (estimated):
Max level speed 135 kt (250 km/h; 155 mph)
Stalling speed 60 kt (110 km/h; 69 mph)
Service ceiling 3,000 m (9,840 ft)
Range 432 n miles (800 km; 497 miles)
Endurance 4 h

UPDATED

Prototype of the five-seat NLA AC-500 *(Photo Link)* 2001/0100793

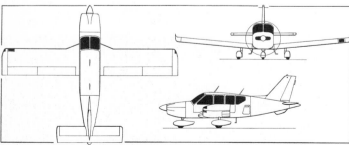

NLA AC-500 light aircraft *(James Goulding/Jane's)* 2000/0062612

Instrument panel of the AC-500 *(Photo Link)* 2001/0100800

NRJ

NEW REGIONAL JET PROGRAMME MANAGEMENT COMPANY (Subsidiary of AVIC I)

PARTICIPATING COMPANIES:
Shanghai Aircraft Research Institute (SARI)
Shanghai Aircraft Industrial Company (SAIC)
Xian Aircraft Design and Research Institute (XADRI)
Xian Aircraft Company (XAC)
MANAGING DIRECTOR: Gao Dacheng (President, XAC)
TECHNICAL DIRECTOR: Wu Xingshi (Chief Designer, SARI)

Six Chinese organisations (see 2000-01 *Jane's*) formed a consortium in 1998 to study the prospects of launching, with risk-sharing foreign industrial partners, a programme for a new regional jet in the 70-seat class. Initial concepts were focused on 58-seat (NRJ 58) and 76-seat (NRJ 76) variants.

On 7 November 2000, the NRJ was officially launched by AVIC I as a US$700 million programme under the auspices of a newly created New Regional Jet Programme Management Company. This will be responsible for defining airframe configuration; defining and subcontracting workshares; overall budget allocation; schedule and quality control; final assembly, certification, marketing and customer support. The 50/70-seat, twin-turbofan design would have rear-mounted engines and a T tail. Power plant and avionics are expected to be of Western origin, with Rolls-Royce Deutschland (BR710 or derivative), Pratt & Whitney Canada (Advanced Technology Fan Integrator) and Honeywell (avionics and mechanical systems) already having expressed interest.

Following data are provisional:
WEIGHTS AND LOADINGS:
Max T-O weight 26,080-31,200 kg (57,496-68,784 lb)
PERFORMANCE:
Range
1,200-2,000 n miles (2,222-3,704 km; 1,380-2,301 miles)
UPDATED

SAC

SHAANXI AIRCRAFT COMPANY (Subsidiary of AVIC II)

PO Box 35, Chenggu, Shaanxi 723213
Tel: (+86 916) 21 56
Fax: (+86 916) 220 21 58
e-mail: sac@public.hanzhong.sn.cn
PRESIDENT: Wang Wenfeng
MARKETING MANAGER: Li Yousen

Founded early 1970s; occupies a 204 ha (504 acre) site and had 1994 workforce (latest received information) of about 10,000; covered workspace includes largest final assembly building in China. Main aircraft programme, recently afforded increased priority, is Y-8 transport; non-aerospace products include 36-seat coaches and small trucks.
UPDATED

SAC Y-8

Chinese name: Yunshuji-8 (Transport aircraft 8)
TYPE: Medium transport/multirole.
PROGRAMME: Redesign, as Chinese development of Antonov An-12B, started at Xian March 1969; first flight of first (Xian-built) prototype 25 December 1974, followed by second (c/n 001802, first built by SAC) 29 December 1975; production go-ahead given January 1980. Pressurised Y-8C made first flight 17 December 1990. Production rate was approximately five per year during first half of 1990s; new (mainly freighter) designations introduced with Y-8F100 from 1997; apparently now seen to have more future in cargo and military roles, following reported difficulties with pressurisation system.
CURRENT VERSIONS: **Y-8**: Prototype and baseline military transport.
Y-8A: Helicopter carrier. Main cabin height increased by 120 mm (4.72 in) by deleting internal gantry; downward-opening rear ramp/door, as in Y-8C. Deliveries began 1987. In service.
Y-8B: Mainly unpressurised civil transport. First deliveries 1986; CAAC certification 1993. Military equipment deleted; empty weight reduced by 1,720 kg (3,792 lb); some avionics differ. In service.
Y-8C: Fully pressurised version for civil and military applications, developed with Lockheed collaboration. Changes include redesigned (downward-opening ramp type) cargo loading door and main landing gear; handling system for standard freight containers and pallets; improved com/nav/ATC equipment, air conditioning and oxygen systems; additional emergency exits. First flight 17 December 1990 (SAC 182, converted from first Shaanxi prototype); CAAC certification 1993. Five delivered by January 1994 (latest figure received); apparently superseded by Y-8F200/F400.
Y-8D: Export version, with main avionics by Rockwell Collins and Litton. Latest versions designated **Y-8D II**. Deliveries began 1987 (Y-8D) and 1992 (Y-8D II); eight delivered by early 1997; none reported since then.
Y-8E: Drone carrier version for Chang Hong 1 (Long Rainbow) UAV. Forward pressure cabin accommodates drone controller's console; carrier/launch trapeze for one UAV under each outer wing panel. First flight 1989; first deliveries later that year, replacing obsolete Tu-4s. In service.
Y-8F: Livestock carrier, with cages to hold up to 500 sheep or goats. First flight early 1990; first deliveries later that year; CAAC type approval January 1994. In service.
Y-8F100: Cargo version for China Postal Airlines (three); delivered 1997 with WJ6A engines. Apparently now the baseline unpressurised version.
Y-8F200: Pressurised version of F100, certified by CAAC in late 1997. Increased internal passenger/cargo volume.
Y-8F300: Freighter. Generally as Y-8F100, but with 'solid' nose and reduced (three-person) flight crew.
Y-8F400: Pressurised freighter. Generally as Y-8F200, but with 'solid' nose and reduced (three-person) flight crew. Reported in 2000 to have entered production.
Y-8F600: Further Westernised freighter version, with Pratt & Whitney Canada PW150 turboprops and Dowty six-blade composites propellers, Honeywell Primus Epic avionics suite, three-person flight crew; otherwise generally similar to Y-8F400. Two under construction late 2000; first flight expected in 2001.
Y-8H: Aerial survey version. No details yet available.
Y-8X: Maritime patrol prototype (B-4101), with Western com/nav, radar, surveillance and search equipment; larger chin radome. Received type approval in September 1984; no further examples confirmed, although some reports suggest PLA Navy has eight. Details in earlier editions.
'Y-8 AEW': At least one aircraft fitted with Racal Skymaster AEW radar, up to eight of which ordered under US$66 million contract in 1996. Reportedly first flew November 1998. Radar, installed in bulbous nose fairing, has 360° scan and more than 200 n miles (370 km; 230 miles) range. (Other sources have suggested radar may be BAE Systems Argus 2000.) Used in recent PLA Navy exercises, reportedly transmitting information via datalink to ship-based Z-9 helicopters. Chinese designation not yet known.

Detailed description applies to F100/200 unless otherwise indicated.

CUSTOMERS: Total of about 75, including exports, reportedly delivered (of which 50 can be confirmed) by December 2000. In service in China with commercial operators, China Postal Service (three) and PLA Air Force; eight military exports (Y-8D) to air forces of Myanmar (four), Sri Lanka (three) and Sudan (one). First two Sri Lankan aircraft modified locally for use as bombers (both since lost). Two (including at least one Y-8F100) leased to BonAir of Iran from May 1998.
DESIGN FEATURES: High-mounted wing; circular-section fuselage (forward section and tail turret pressurised), upswept at rear; angular tail surfaces with large dorsal fin. More pointed nose transparencies than An-12, probably from Chinese H-6 (Tu-16) production.
Wing sections C-5-18 at root, C-3-16 at rib 15 and C-3-14 at tip, final two digits indicating thickness/chord ratio; incidence 4°; 1° dihedral on intermediate panels, 4° anhedral on outboard panels; 6° 50′ sweepback at quarter-chord; fixed-incidence tailplane.
FLYING CONTROLS: Conventional and manual. Aerodynamically balanced differential ailerons, elevators and rudder, each of which has inset trim tab; two-segment, hydraulically actuated double-slotted Fowler flaps on each wing trailing-edge; comb-shaped spoilers forward of flaps.
STRUCTURE: All-metal (aluminium alloy) conventional semi-monocoque/stressed skin; wings, tailplane and fin are all

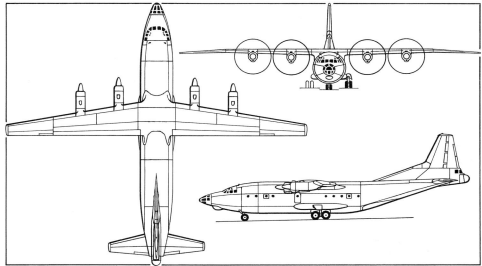

SAC Y-8 four-turboprop multipurpose transport *(Mike Keep/Jane's)*

Model of the solid-nosed Y-8F400 freighter *(Photo Link)*

Canadian turboprops and six-blade propellers characterise the Y-8F600 *(Photo Link)*

Shaanxi Y-8F200, showing lowered rear-loading ramp *(Photo Link)* ***2001***/0100798

two-spar box structures; landing gear and all hydraulic components manufactured by Shaanxi Aero-Hydraulic Component Factory (SAHCF).

LANDING GEAR: Hydraulically retractable tricycle type, with Shaanxi (SAHCF) nitrogen/oil shock-struts on all units. Four-wheel main bogie on each side retracts inward and upward into blister on side of fuselage. Twin-wheel nose unit, hydraulically steerable to ±35°, retracts rearward. Mainwheel tyres size 1,050×300, pressure 28.40 bars (412 lb/sq in); nosewheel tyres size 900×300, pressure 16.70 bars (242 lb/sq in). Hydraulic disc brakes and Xingping inertial anti-skid sensor. Minimum ground turning radius 13.75 m (45 ft 1½ in).

POWER PLANT: Four 3,126 kW (4,192 shp) SAEC (Zhuzhou) WJ6A turboprops, each driving a Baoding four-blade J17-G13 constant-speed fully feathering propeller. All fuel (F100/F300) in two integral tanks and 29 bag-type tanks in wings (20,102 litres; 5,310.5 US gallons; 4,422 Imp gallons) and fuselage (10,075 litres; 2,661.5 US gallons; 2,216 Imp gallons), giving total capacity of 30,177 litres (7,971.5 US gallons; 6,638 Imp gallons). Reduced fuel load in F200/F400 (see under Weights and Loadings). Refuelling points in starboard side of fuselage (between frames 14 and 15), mainwheel fairing, and in wing upper surface.

ACCOMMODATION: Flight crew of five (pilot, co-pilot, navigator, engineer and radio operator) in F100/F200; three crew only in F300/F400. Forward portion of fuselage in F100/F300 is pressurised, and can accommodate up to 14 passengers in addition to crew. Cargo compartment (between frames 13 and 43) is unpressurised in these version. Pressurised volume increased in F200/F400; effective length of cargo hold extended internally by 2.00 m (6 ft 6¾ in). Maximum accommodation for up to 96 troops; or 80 paratroops; or up to 92 casualties with three medical attendants; or two 'Liberation' army trucks, plus Jeep-sized vehicle on loading ramp.

Short hold freighter versions (Y-8F100 and Y-8F300 can accommodate optimum eight 2.24 × 1.37 m (88 × 54 in) plus four 2.24 × 2.74 m (88 × 108 in) standard cargo pallets or 17 LD3 containers. Larger (Y-8F200 and Y-8F400) hold can accept optimum four 2.24 × 3.18 m (88 × 125 in) plus four 2.44 × 3.18 m (96 × 125 in) pallets or 19 LD3s. Individual cargo items of up to 7,400 kg (16,315 lb) can be airdropped.

Crew door and two emergency exits in forward fuselage. Three additional emergency exits in cargo compartment, access to which is via a large rear-loading ramp/door in underside of rear fuselage.

SYSTEMS: Forward fuselage of F100/F200 pressurised to maintain a differential of 0.20 bar (2.8 lb/sq in) at altitudes above 4,300 m (14,100 ft). (F300/F400 fully pressurised.) Two independent hydraulic systems, with operating pressures of 152 bars (2,200 lb/sq in) (port) and 147 bars (2,130 lb/sq in) (starboard), plus hand and electrical standby pumps, for actuation of landing gear extension/retraction, nosewheel steering, flaps, brakes and rear ramp/door. Electrical DC power (28.5 V) supplied by eight 12 kW generators, an 18 kW (24 hp) Xian Aero Engine Company APU (mainly for engine starting) and four 28 Ah batteries. Four 12 kVA alternators provide 115 V AC power at 400 Hz. Gaseous oxygen system for crew. Electric de-icing of windscreen, propellers and fin/tailplane leading-edges; hot air de-icing for wing leading-edges. WDZ-1 APU.

AVIONICS (Y-8F300/400): *Comms:* Rockwell Collins VHF-42B and HF-9000 com radios and TDR-94 ATC transponder.

Radar: Rockwell Collins TWR-850 colour weather radar.

Flight: Universal Avionics UNS-1K flight management system; Rockwell Collins VOR-432, DME-442, AHS-85E AHRS and EFIS-86E; Honeywell ED-55 flight data recorder.

Instrumentation: Honeywell Mk V nav display.

EQUIPMENT: Electric winch, tow, and 2,300 kg (5,070 lb) capacity hoist for cargo loading and unloading. Roller system for containerised or palletised cargo handling.

DIMENSIONS, EXTERNAL (all versions, except where indicated):

Wing span	38.00 m (124 ft 8 in)
Wing chord: at root	4.73 m (15 ft 6¼ in)
at tip	1.69 m (5 ft 6½ in)
mean aerodynamic	3.45 m (11 ft 3¾ in)
Wing aspect ratio	11.9
Length overall: F100, F200	34.02 m (111 ft 7¼ in)
F300, F400	32.93 m (108 ft 0½ in)
Fuselage:	
Max diameter of circular section	4.10 m (13 ft 5½ in)
Height overall	11.16 m (36 ft 7½ in)
Tailplane span	12.195 m (40 ft 0¼ in)
Wheel track (c/l of shock-struts)	4.92 m (16 ft 1¾ in)
Wheelbase (c/l of main bogies)	9.575 m (31 ft 5 in)
Propeller diameter	4.50 m (14 ft 9¼ in)
Propeller ground clearance	1.89 m (6 ft 2½ in)
Crew door: Height	1.455 m (4 ft 9¼ in)
Width	0.80 m (2 ft 7½ in)
Rear-loading hatch: Length	7.67 m (25 ft 2 in)
Width: min	2.65 m (8 ft 8¼ in)
max	3.10 m (10 ft 2 in)
Emergency exits (each): Height	0.55 m (1 ft 9¾ in)
Width	0.60 m (1 ft 11½ in)

DIMENSIONS, INTERNAL:

Cabin (incl flight deck, galley and toilet):	
Length: F100, F300	13.70 m (44 ft 11¼ in)
F200, F400	15.70 m (51 ft 6 in)
Width: min	3.00 m (9 ft 10 in)
max	3.50 m (11 ft 5¾ in)
Height: min	2.20 m (7 ft 2½ in)
max	2.60 m (8 ft 6½ in)
Floor area: F100, F300	55.0 m² (592 sq ft)
Volume: F100, F300	123.3 m³ (4,354 cu ft)
F200, F400	137.6 m³ (4,859 cu ft)

AREAS:

Wings, gross	121.86 m² (1,311.7 sq ft)
Ailerons (total)	7.84 m² (84.39 sq ft)
Trailing-edge flaps (total)	26.91 m² (289.66 sq ft)
Rudder	6.535 m² (70.34 sq ft)
Tailplane	27.05 m² (291.16 sq ft)
Elevators (total)	7.10 m² (76.42 sq ft)

WEIGHTS AND LOADINGS:

Weight empty, equipped:	
F100	34,500 kg (76,060 lb)
F200	34,760 kg (76,635 lb)
Max fuel load: F100, F300	22,909 kg (50,505 lb)
F200, F400	14,566 kg (32,115 lb)
Max payload:	
containerised: all versions	15,000 kg (33,069 lb)
bulk cargo: F100, F200	20,000 kg (44,090 lb)
Max airdroppable cargo: total	13,200 kg (29,100 lb)
single piece	7,400 kg (16,315 lb)
Max T-O weight	61,000 kg (134,480 lb)
Max ramp weight	61,500 kg (135,585 lb)
Max landing weight	58,000 kg (127,870 lb)
Max zero-fuel weight:	
F100, F300	55,278 kg (121,865 lb)
F200, F400	55,538 kg (122,440 lb)
Max wing loading	500.6 kg/m² (102.53 lb/sq ft)
Max power loading	4.88 kg/kW (8.02 lb/shp)

PERFORMANCE:

Max level speed at 7,000 m (22,960 ft):	
F100, F300	357 kt (662 km/h; 411 mph)
F200, F400	345 kt (640 km/h; 397 mph)
Max cruising speed at 8,000 m (26,250 ft):	
all versions	297 kt (550 km/h; 342 mph)
Econ cruising speed at 8,000 m (26,250 ft)	
all versions	286 kt (530 km/h; 329 mph)
Unstick speed	129 kt (238 km/h; 148 mph)
Touchdown speed at MLW	130 kt (240 km/h; 150 mph)
Max rate of climb at S/L	473 m (1,552 ft)/min
Rate of climb at S/L, OEI	231 m (758 ft)/min
Service ceiling, AUW of 51,000 kg (112,435 lb):	
F100, F300	10,400 m (34,120 ft)
F200, F400	10,050 m (32,970 ft)
Service ceiling, OEI, AUW of 51,000 kg (112,435 lb)	8,100 m (26,580 ft)
Runway CAN	15
T-O run : all versions	1,270 m (4,170 ft)
FAR T-O field length	1,900 m (6,235 ft)
T-O to 15 m (50 ft)	3,007 m (9,870 ft)
Landing from 15 m (50 ft) at MLW	2,174 m (7,135 ft)
FAR landing field length	1,650 m (5,415 ft)
Landing run at MLW: all versions	1,050 m (3,445 ft)
Range with max payload:	
F100	687 n miles (1,273 km; 791 miles)
Range with max fuel:	
F100	3,032 n miles (5,615 km; 3,489 miles)
F200, F400	1,857 n miles (3,440 km; 2,137 miles)
F300	2,861 n miles (5,300 km; 3,293 miles)

UPDATED

SAC

SHENYANG AIRCRAFT CORPORATION (Subsidiary of AVIC I)

1 Lingbei Street, Huanggu District, Shenyang, Liaoning 110034
Tel: (+86 24) 86 89 66 80 and 86 59 92 05
Fax: (+86 24) 86 89 66 89
Web: http://www.sac.com.cn
PRESIDENT: Li Fangyong

Pioneer Chinese fighter design centre; built 767 examples of J-5 (licence MiG-17F) from 1956-59 and from 1963 was major producer of several thousand of J-6 series (reverse engineered MiG-19), including 634 JJ-6 tandem-seat fighter trainers; initiated development and early production of J-7 (see under CAC). On 29 June 1994, SAC became core enterprise in newly formed Shenyang Aircraft Industry Group (SAIG). Now occupies site area of more than 800 ha (1,976 acres) and has workforce of 30,000; only some 30 per cent of current activities are in aerospace.

Principal recent programme has been J-8 II fighter, currently the subject of an engine and avionics upgrade programme (F-8 IIM). Now also engaged in major assembly/co-production programme for Sukhoi Su-27 variants for PLA Air Force.

Aerospace subcontract manufacture includes cargo doors for Boeing 757 and baggage/service/emergency exit doors for de Havilland Dash 8Q (100th Dash 8 door delivered 12 June 1992), wing ribs and emergency exits for Airbus A319 and A320, tailcone/landing gear door/pylon components for Lockheed Martin C-130, rear fuselage and tail components for Boeing 737, and other machined parts for BAE, Boeing and EADS Deutschland. Collaboration with Hellenic Aerospace Industry Ltd announced in February 1997 on formation of Shenyang Hellenic Aircraft Repair Company.

UPDATED

SAC J-8 II

Chinese name: Jianjiji-8 (Fighter aircraft 8)
Westernised designation: F-8
NATO reporting name: Finback-B

TYPE: Multirole fighter.

PROGRAMME: Development of original J-8 started 1964; first flight 5 July 1969; initial production authorised July 1979 and ended 1987 after approximately 100 J-8s and J-8 Is; see 2000-01 and earlier *Jane's* for history of these versions. Present configuration dates from 12 June 1984 with maiden flight of first of four prototypes of much redesigned J-8 II. Peace Pearl programme (1990-91 *Jane's*), to upgrade J-8 II with Western avionics, embargoed by US government mid-1989 and cancelled by China 1990; alternative domestic (F-8 IIM) upgrade programme now in progress (see separate entry). Trials with some parts of structure covered in Xikai SF18 radar-absorbent material were reported in early 1999.

CURRENT VERSIONS: **J-8 II** ('Finback-B'): All-weather dual-role version (high-altitude interceptor and ground attack), some 70 per cent redesigned compared with J-8 I. Main configuration change is to 'solid' nose and twin lateral air intakes, providing more nose space for fire-control radar and other avionics, plus increased air flow for more powerful WP13A II turbojets. In production and service, with at least 24 built by 1993, but manufactured in small economic batches rather than continuous production.

Detailed description applies to J-8 II.

J-8 II ACT: Active Control Technology (fly-by-wire) testbed, which first flew 29 December 1996 and completed its 49th and last sortie on 21 September 1999; shown in model form at Airshow China in November 2000. Full-authority quad-redundant, three-axis digital AFCS; small canards mounted high on air intakes to induce instability; two 1553B-standard flight computers with databus interface; integrated servo actuators for all moving control surfaces.

J-8 IID: Designation of J-8 II modified for in-flight refuelling. Non-retractable probe on starboard side, below cockpit. In service with PLAAF and Naval Air Force (approximately 30 each).

F-8 II: Originally proposed export version: WP13B engines (uprated by 4 per cent to 68.7 kN; 15,430 lb st with afterburning), pulse Doppler look-down radar, digital avionics (including HUD and two HDDs) with 1553B databus, leading-edge flaps, in-flight refuelling, seven

The J-8 I remains in service, though no longer in production

stations for 4,500 kg (9,921 lb) stores load, maximum speed M2.2. Programme modified to become present F-8 IIM.

F-8 IIM: Upgraded version of F-8 II; described separately.

CUSTOMERS: PLA Air Force operates regiments within the 1st Air Army/Shenyang Military Region (J-8 I/II), 7th AA/Guangzhou MR (J-8 II), 8th AA/Nanjing MR (J-8 II) and 10th AA/Beijing MR (J-8 I/II). People's Naval Air Force reported to have about 30 J-8 IIDs in service in 2000, following conversion of 10 H-6 (Tu-16) bombers into aerial refuelling tankers.

DESIGN FEATURES: Extension of late 1950s Soviet heavy fighter theory. Thin-section, mid-mounted delta wings and all-sweptback tail surfaces; fuselage has area rule 'waisting', detachable rear portion for engine access, and dorsal spine fairing. Large ventral fin under rear fuselage, main portion of which folds sideways to starboard during take-off and landing, provides additional directional stability; small fence on each wing upper surface near tip; small airscoops at foot of fin leading-edge and at top of fuselage each side, above tailplane. Sweepback 60° on wing and tailplane leading-edges; wings have slight anhedral.

FLYING CONTROLS: Hydraulically boosted ailerons, rudder and low-set all-moving tailplane; two-segment single-slotted flaps on each wing trailing-edge, inboard of aileron; four door-type underfuselage airbrakes, one under each engine air intake trunk and one immediately aft of each mainwheel well.

STRUCTURE: Conventional aluminium alloy semi-monocoque/stressed skin construction, with high-tensile steel for high load-bearing areas of wings and fuselage and titanium in high-temperature fuselage areas; ailerons, rudder and rear portion of tailplane are of aluminium honeycomb with sheet aluminium skin; dielectric skins on nosecone, tip of main fin, and on non-folding portion of ventral fin leading-edge.

LANDING GEAR: Hydraulically retractable tricycle type, with single wheel and oleo-pneumatic shock-absorber on each unit. Steerable nose unit retracts forward, main units inward into centre-fuselage; mainwheels turn to stow vertically inside fuselage, resulting in slight overwing bulge. Brake-chute in bullet fairing at base of rudder.

POWER PLANT: Two LMC (Liyang) WP13A II turbojets, each rated at 42.7 kN (9,590 lb st) dry and 65.9 kN (14,815 lb st) with afterburning, mounted side by side in rear fuselage with pen-nib fairing above and between exhaust nozzles. Lateral, non-swept air intakes, with automatically regulated ramp angle and large splitter plates similar in shape to those of MiG-23. Internal fuel capacity (four integral wing tanks plus fuselage tanks) approximately 5,400 litres (1,426 US gallons; 1,188 Imp gallons). Single-point pressure refuelling. Provision for auxiliary fuel tanks on fuselage centreline and each outboard underwing pylon. J-8 IID retrofitted with probe for in-flight refuelling from Xian H-6 (Tu-16) bombers converted as aerial tankers.

ACCOMMODATION: Pilot only, on zero/zero ejection seat under one-piece canopy hinged at rear and opening upward. Cockpit pressurised, heated and air conditioned. Heated windscreen.

SYSTEMS: Two simple air-cycle environmental control systems, one for cockpit heating and air conditioning and one for radar cooling; cooling air bled from engine compressor. Two 207 bar (3,000 lb/sq in) independent hydraulic systems (main utility system plus one for flight control surfaces boost), powered by engine-driven pumps. Primary electrical power (28.5 V DC) from two 12 kW engine-driven starter/generators, with two 6 kVA alternators for 115/200 V three-phase AC at 400 Hz. Pneumatic bottles for emergency landing gear extension. Pop-out ram air emergency turbine under fuselage.

AVIONICS: *Comms:* VHF/UHF and HF/SSB radios; 'Odd Rods'-type IFF.
Radar: Monopulse radar in nose.
Flight: ILS, Tacan, marker beacon receiver, radio compass, radar altimeter, autopilot.
Mission: Gyro gunsight and gun camera.
Self-Defence: RWR (antenna in fin-tip); chaff/flare dispensers in tailcone.

Enlarged avionics bays in nose and fuselage provide room for modernised fire-control system and other upgraded avionics.

ARMAMENT: One 23 mm Type 23-3 twin-barrel cannon, with 200 rounds, in underfuselage pack immediately aft of nosewheel doors. Seven external stations (one under fuselage and three under each wing) for a variety of stores which can include PL-2B IR air-to-air missiles, PL-7 medium-range semi-active radar homing air-to-air missiles, Qingan HF-16B 12-round pods of 57 mm Type 57-2 unguided air-to-air rockets, launchers for 90 mm air-to-surface rockets, bombs, or (centreline and outboard underwing stations only) auxiliary fuel tanks.

DIMENSIONS, EXTERNAL:
Wing span	9.345 m (30 ft 8 in)
Wing aspect ratio	2.1
Length overall: incl nose probe	21.59 m (70 ft 10 in)
excl nose probe	21.39 m (70 ft 2¼ in)
Height overall	5.41 m (17 ft 9 in)
Wheel track	3.74 m (12 ft 3¼ in)
Wheelbase	7.335 m (24 ft 0¾ in)

AREAS:
Wings, gross	42.20 m² (454.2 sq ft)

WEIGHTS AND LOADINGS:
Weight empty	9,820 kg (21,649 lb)
Normal T-O weight	14,300 kg (31,526 lb)
Max T-O weight	17,800 kg (39,242 lb)
Wing loading:	
at normal T-O weight	338.9 kg/m² (69.40 lb/sq ft)
at max T-O weight	421.8 kg/m² (86.39 lb/sq ft)
Power loading:	
at normal T-O weight	108 kg/kN (1.06 lb/lb st)
at max T-O weight	135 kg/kN (1.32 lb/lb st)

PERFORMANCE:
Design max operating Mach No. (M_{MO})	2.2
Design max level speed	701 kt (1,300 km/h; 808 mph) IAS
Unstick speed	175 kt (325 km/h; 202 mph)
Touchdown speed	156 kt (290 km/h; 180 mph)
Max rate of climb at S/L	12,000 m (39,370 ft)/min
Acceleration from M0.6 to M1.25 at 5,000 m (16,400 ft)	54 s
Service ceiling	20,200 m (66,275 ft)
T-O run with afterburning	670 m (2,200 ft)
Landing run, brake-chute deployed	1,000 m (3,280 ft)
Combat radius	432 n miles (800 km; 497 miles)

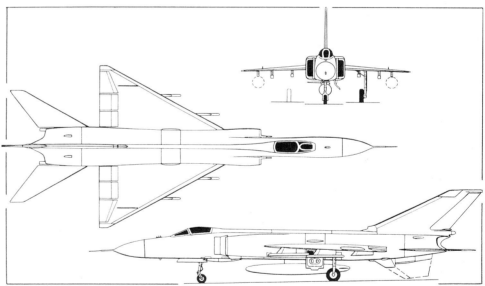

J-8 II version of the 'Finback' twin-jet fighter *(Dennis Punnett/Jane's)*

PLA Air Force squadron line-up of J-8 IID fighters

Model of J-8 II ACT testbed with underfuselage targeting pod and wing-mounted AAMs *(Photo Link)*

Ferry range	1,188 n miles (2,200 km; 1,367 miles)
g limit in sustained turn at M0.9 at 5,000 m (16,400 ft)	+4.83

UPDATED

SAC F-8 IIM

TYPE: Multirole fighter.

PROGRAMME: Upgraded J-8 II; outgrowth of earlier proposals for F-8 II export version. Developed jointly by SAC and Shenyang Aircraft Research Institute (SARI); first flight 31 March 1996, two years after delivery of drawings; flight testing of aircraft and WP13B engine completed 19 January 1998. Second F-8 IIM completed by late 1998. Additional proposed upgrades revealed by then included No. 607 Institute Blue Sky low-altitude navigation pod; Southwest China Electronic Equipment Research Institute KG 300G airborne self-protection jammer pod; No. 613 Institute FLIR/laser targeting pod; and a triple redundant digital fly-by-wire flight control system. These would presumably be similar to the systems quoted for the J-8 II ACT.

CUSTOMERS: Developed primarily for export, but may also be adopted as upgrade for in-service PLAAF J-8 IIs.

DESIGN FEATURES: Main differences from J-8 II are more powerful engine, improved avionics and modernised cockpit with HOTAS controls.

POWER PLANT: Two LMC (Liyang) WP13B turbojets, each rated at 47.1 kN (10,582 lb st) dry and 68.7 kN (15,432 lb st) with afterburning.

SYSTEMS: Electrical system includes two 15 kVA starter/generators.

AVIONICS: *Comms:* Advanced com/nav radios; IFF.
Radar: Phazotron Zhuk-8 II multifunction, look-up/look-down pulse Doppler radar, with 38 n mile (70 km; 43.5 mile) detection range for approaching targets and 21.5 n mile (40 km; 25 mile) range for receding targets (both targets assumed to have 3 m²; 32.3 sq ft radar cross-section). N010 Zhuk-27 being flight tested by late 1998, presumably in second aircraft.
Flight: Include Tacan and datalink; locally developed HUD and INS/GPS navigation; MFDs; HOTAS controls; ARINC 429 databus.
Self-defence: Omnidirectional RWR; rear hemisphere noise jammer against threat (including pulse Doppler) radars; chaff/flare dispenser.

ARMAMENT: Up to six Chinese PL-5 or PL-9 short-range AAMs on underwing stations, or two Russian R-27R1 (AA-10 'Alamo') medium-range AAMs; up to four seven-round pods of 90 mm Type 90-1 rockets; up to 10 anti-runway bombs or 10 Type 250-III or 250-IV low-drag bombs (four under wings and six under fuselage); or eight anti-tank bombs; or five 500 kg low-drag bombs. Internal cannon as for J-8 II.

DIMENSIONS, EXTERNAL, AND AREAS: As for J-8 II

WEIGHTS AND LOADINGS:
Weight empty	10,371 kg (22,864 lb)
Normal fuel load	4,200 kg (9,259 lb)
Max external stores load	4,500 kg (9,921 lb)
Normal T-O weight	15,288 kg (33,704 lb)
Max T-O weight	18,879 kg (41,621 lb)
Max wing loading	447.4 kg/m² (91.63 lb/sq ft)
Max power loading	138 kg/kN (1.35 lb/lb st)

PERFORMANCE:
Max operating Mach No. (M$_{MO}$)	2.2
Unstick speed	179 kt (330 km/h; 206 mph)
Touchdown speed	162 kt (300 km/h; 186 mph)
Level acceleration:	
M0.7 to M1.0 at 1,000 m (3,280 ft)	21 s
M0.6 to M1.25 at 5,000 m (16,400 ft)	55 s
Max rate of climb at M0.9:	
at 1,000 m (3,280 ft)	13,440 m (44,094 ft)/min
at 5,000 m (16,400 ft)	9,600 m (31,496 ft)/min
Service ceiling	18,000 m (59,060 ft)
T-O run with afterburning	630 m (2,070 ft)
Landing run with brake-chute	900 m (2,955 ft)

Typical mission radius:
air-to-air interception, out and back at M0.8:
at 500 m (1,640 ft) with 3 min combat
189 n miles (350 km; 217 miles)
at 11,000 m (36,080 ft) with 5 min combat
540 n miles (1,000 km; 621 miles)
combat air patrol, incl 10 min patrol and 5 min combat, out at M0.8 at 5,000 m (16,400 ft), back at M0.8 at 11,000 m (36,080 ft)
324 n miles (600 km; 372 miles)
air-to-ground attack, out and back at M0.8 at 10,000 m (32,800 ft), incl 5 min combat
486 n miles (900 km; 559 miles)
Ferry range 1,026 n miles (1,900 km; 1,180 miles)
g limits during sustained turn at M0.9:
at 1,000 m (3,280 ft)	+6.9
at 5,000 m (16,400 ft)	+4.7

UPDATED

SAC (SUKHOI Su-27) J-11

TYPE: Air superiority fighter.

PROGRAMME: In February 1996, the Russian military sales organisation Rosvooruzheniye announced a contract under which China would be licensed to manufacture the Sukhoi Su-27 'Flanker' (up to 50 per year; total of 200 in all) at Shenyang.

An initial batch of 26 Russian-built Su-27s, delivered from 1992, comprised 20 Su-27SKs ('Flanker-B') and six two-seat Su-27UBK combat trainers ('Flanker-C'). These were followed in 1996 by a further 24 (18 and six, respectively). Agreement for an additional 50 to 60, for 1997-98 delivery, was reached in August 1997. This was preceded in February 1997 by Russian licence for Chinese manufacture at Shenyang, initially in the form of CKD kit assembly; in 1998 KnAAPO delivered the first two kits of a reported batch of 15; both of these made their first flight in December 1998, and about half of batch reportedly completed by end of 2000. However, it was also reported in mid-2000 that substandard work had caused Russian technicians to rebuild first two aircraft, necessitating import, from late 2000 onwards, of 28 additional Russian-built two-seat Su-27UBKs to offset shortfall in Chinese production. (Six or seven Su-27s were planned to be assembled annually during 1999-2001, increasing to 15 per year from 2002.) Chinese-built Su-27s are designated **J-11** (single-seat) and **JJ-11** (two-seat). Later production will be of upgraded Su-30 variant. Description in Russian section.

UPDATED

F-8 IIM in attack mode, with bombs, rocket pods and air-to-air missiles

The modernised cockpit of the F-8 IIM

The first F-8 IIM retracting its landing gear after take-off

SAIC

SHANGHAI AVIATION INDUSTRY GROUP
(Subsidiary of AVIC I)

2668 Zhongshan North Road, Shanghai 200063
Tel: (+86 21) 62 57 33 51
Fax: (+86 21) 62 57 33 50
e-mail: saiccn@sh163a.sta.net.cn
Web: http://www.saic-china.com
CHAIRMAN: Li Wanxin
VICE-PRESIDENT: Liu Qianyou

Shanghai Aircraft Manufacturing Factory (SAMF) created 1951; now part of Shanghai Aviation Industry Group; occupies site area of 135.5 ha (334.8 acres); SAMF workforce 4,630 in late 2000. Has produced main and landing gear doors for McDonnell Douglas (now Boeing) MD-80 series since 1979; also produces cargo and service doors, avionics access doors and tailplanes. Was due to begin delivering tailplanes for Boeing 737 NG in 1999. See 1994-95 and earlier *Jane's* for details of recent manufacturing programme for MD-82/83. Reportedly in talks with Boeing in 1999 in attempt to avoid extensive redundancies seen as inevitable result of termination of MD-90-30T Trunkliner programme. Also builds Q-2 ultralight.

SAIC and Shanghai Aircraft Research Institute (SARI) are taking part in studies for a new Chinese regional jet transport. All known details appear under the NRJ heading in this section.

UPDATED

SAMC

SHIJIAZHUANG AIRCRAFT MANUFACTURING CORPORATION
(Subsidiary of AVIC II)

PO Box 164, 25 Beihuanxilu, Shijiazhuang, Hebei 050062
Tel: (+86 311) 777 52 83
Fax: (+86 311) 775 29 93
GENERAL MANAGER: Cheng Bingyou
CHIEF ENGINEER: Zhang Zengshou

SAMC has produced Chinese Y-5 versions of the An-2 general purpose biplane since 1970, concentrating since 1989 on Y-5B and C customised agricultural, forestry, tourist and parachutist versions; other products have included W-5 and W-6 ultralight series (see 1992-93 and earlier *Jane's*). With the effective cessation of Polish An-2 production, SAMC is the only company now building new examples of this long-serving biplane.

Occupies 460,000 m² (4,951,400 sq ft) site, including over 100,000 m² (1,076,400 sq ft) of covered space. Workforce of 4,000 in 1998 included some 700 engineers and technicians. Became part of Xian Aircraft Industrial Group July 1992.

UPDATED

SAMC Y-5

Chinese name: Yunshuji-5 (Transport aircraft 5)

TYPE: Light utility biplane.

PROGRAMME: Antonov An-2 has been built under licence in China since 1957, chiefly at Nanchang (727 produced up to 1968) and latterly by SAMC; production of Y-5N civil transport and general purpose version ended in 1986, at which time 221 had been completed. Y-5B dedicated agricultural and forestry version made first flight 2 June 1989; now in batch production; first nine Y-5Bs produced in 1990; total of 83 Y-5B/Cs produced by beginning of 1998. SAMC now only source of current production of this aircraft.

CURRENT VERSIONS: **Y-5B:** Dedicated agricultural and forestry version; certified to CCAR 23 Chinese equivalent of FAR Pt 23.

Following description applies to Y-5B, except where indicated.

Y5B-100: Designation of version displayed at Zhuhai in November 1998, fitted with wing 'tipsails' of the Y-5C and registered B-8448.

Y-5B(K): Tourist version, first flown 1993; certified to CCAR 23.

Y-5B(D): Multipurpose (agri-forest or tourist) version, first flown 1995; certified to CCAR 23.

Y-5C: Parachutist version, with wingtip vanes ('tipsails'); first flown 1996. Ordered by PLA Air Force.

CUSTOMERS: Operators of Y-5B have included Xinjiang Airlines (eight) and Jiangnan General Aviation (three); at least 36 ordered by CASC (China Aviation Supplies Corporation) since 1993; 24 Y-5C ordered by PLA Air Force in 1996.

DESIGN FEATURES: Unequal span, single-bay biplane; braced wings and tail; fuselage circular section forward, rectangular in cabin section, oval in tail section; fin integral with rear fuselage. RPS wing section, thickness/chord ratio 14 per cent (constant); dihedral, both wings, approximately 2° 48′.

FLYING CONTROLS: Conventional and manual. Differential ailerons, elevators and rudder, using cables and push/pull rods; electric trim tab in port aileron, rudder and port elevator; full-span automatic leading-edge slots on upper wings; electrically actuated slotted trailing-edge flaps on both wings. Wingtip vanes on some aircraft (see Current Versions) for additional control at low speeds.

STRUCTURE: All-metal, with fabric covering on wings aft of main spar and on tailplane. Y-5B specially treated to resist corrosion; cabin doors sealed against chemical ingress; empty weight reduced.

LANDING GEAR: Non-retractable split axle type, with long-stroke oleo-pneumatic shock-absorbers. Mainwheel tyres size 800×260, pressure 2.25 bars (33 lb/sq in). Pneumatic shoe brakes on main units. Fully castoring and self-centring tailwheel, size 470×210, with electropneumatic lock. For rough field operation shock-absorbers can be charged from compressed air cylinder installed in rear fuselage. Interchangeable ski landing gear available optionally.

POWER PLANT: One 735 kW (985 hp) PZL Kalisz ASz-62IR-16 or SAEC (Zhuzhou) HS5 nine-cylinder radial engine, respectively driving an AW-2 or Baoding J12B-G15 four-blade variable-pitch propeller. Fuel capacity 1,240 litres (328 US gallons; 273 Imp gallons).

ACCOMMODATION: Flight crew of one or two; dual controls; seats for 12 tourist class passengers (Y-5N) or 10 parachutists (Y-5C). Emergency exit on starboard side at rear. Cabin heating and ventilation improved by new ECS.

AVIONICS: *Comms:* Honeywell KHF 950 HF and KY 196 VHF radios and KMA 24 audio control panel; some electrical and other instrument installations also improved.
Flight: GPS 155.

EQUIPMENT: Large hopper/tank with emergency jettison of contents; high flow rate, wind-driven pump; sprayers with various nozzle sizes, depending upon spray volume required.

DIMENSIONS, EXTERNAL:
Wing span: upper	18.18 m (59 ft 7¾ in)
lower	14.24 m (46 ft 8½ in)
Wing aspect ratio: upper	7.6
lower	7.3
Length overall, tail down	12.40 m (40 ft 8¼ in)
Height overall, tail down	4.01 m (13 ft 2 in)
Wheel track	3.36 m (11 ft 0¼ in)
Wheelbase	8.19 m (26 ft 10½ in)
Propeller diameter: AW-2	3.60 m (11 ft 9¾ in)
J12-G15	3.40 m (11 ft 2 in)
Cargo door (port): Mean height	1.55 m (5 ft 1 in)
Mean width	1.39 m (4 ft 6¾ in)

DIMENSIONS, INTERNAL:
Cargo compartment: Length	4.10 m (13 ft 5½ in)
Max width	1.60 m (5 ft 3 in)
Max height	1.80 m (5 ft 10¾ in)

AREAS:
Wings, gross: upper	43.54 m² (468.7 sq ft)
lower	27.98 m² (301.2 sq ft)

WEIGHTS AND LOADINGS (A: Y-5B with dry chemical spreader, B: Y-5B with liquid spray system, C: Y-5C parachutist version):
Max payload: A, B, C	1,500 kg (3,307 lb)
Max T-O weight: A, B, C	5,250 kg (11,574 lb)
Max wing loading: A, B, C	73.4 kg/m² (15.03 lb/sq ft)
Max power loading: A, B, C	7.15 kg/kW (11.75 lb/hp)

PERFORMANCE (A, B and C as for Weights):
Max level speed at S/L:	
A	110 kt (205 km/h; 127 mph)
B	108 kt (200 km/h; 124 mph)
C	129 kt (239 km/h; 148 mph)
Max level speed at 1,700 m (5,575 ft):	
A	119 kt (220 km/h; 137 mph)
B	116 kt (215 km/h; 133 mph)
C	138 kt (256 km/h; 159 mph)
Operating speed: A, B	86 kt (160 km/h; 99 mph)
Stalling speed: A, B, C	46 kt (85 km/h; 53 mph)
Max rate of climb at S/L: A	120 m (394 ft)/min
B	114 m (374 ft)/min
C	213 m (699 ft)/min
Rate of climb at 1,600 m (5,250 ft):	
A	133 m (436 ft)/min
B	123 m (404 ft)/min
C	225 m (738 ft)/min
Service ceiling: A, B	4,500 m (14,760 ft)
C	5,000 m (16,400 ft)
Air turning radius: A, B, C	350 m (1,150 ft)
T-O run: A	170 m (560 ft)
B	180 m (595 ft)
C	150 m (495 ft)
Landing run: A	160 m (525 ft)
B	157 m (515 ft)
C	170 m (560 ft)
Range at S/L with fuel load of 670 litres (177 US gallons; 147 Imp gallons):	
A, B	456 n miles (845 km; 525 miles)
C	475 n miles (880 km; 546 miles)
Endurance, conditions as above: A, B	5 h 36 min
Swath width (A, B): HV, LV	40-50 m (130-165 ft)
ULV	60 m (200 ft)

UPDATED

Latest Chinese military version of the An-2 is the 'tipsailed' Y-5C for the PLA Air Force

Y-5B(K) civil transport *(Sebastian Zacharias)*

SFLC

SHANGHAI FEITENG LIGHTPLANE COMPANY
Pudong, Shanghai

SFLC is a joint venture between Australia and Shanghai Aircraft Manufacturing Factory (SAMF, which see). It is co-producing the Austflight ULA Drifter and Hughes Australian Light Wing GR-582 ultralights. Initial plans for the Drifter were for Shanghai to produce 100 in its first year of the venture, increasing to 160 in the second year. It is not known whether this was achieved.

VERIFIED

SFLC SB-582
TYPE: Side-by-side ultralight.
PROGRAMME: Licence-built Hughes Australian Light Wing GR-582 (which see in Australian section for a description). Manufacturing agreement 1996; first Shanghai-assembled SB-582 (for Shanghai-Ballina) shipped back to Australia on 5 July 1996. No subsequent details known.

UPDATED

Australian Drifter ultralight built by Shanghai Feiteng

SSLF

SHENYANG SAILPLANE AND LIGHTPLANE FACTORY
17 Shen-Liao East Road, Tiexi District, Shenyang, Liaoning 110021
Tel: (+86 24) 589 42 17 and 01 87
Fax: (+86 24) 589 23 97
GENERAL MANAGER: Lin Jia Ru
CHIEF ENGINEER: Li Ji Junx

Details of SSLF's earlier HU-1 Seagull and HU-2B last appeared in the 1996-97 and 1999-2000 editions respectively. Latest known products are the HU-2C and HU-2D.

UPDATED

SSLF HU-2C PETREL 650C
TYPE: Two/three-seat lightplane.
PROGRAMME: Development of HU-2B Petrel 650B (see 1998-99 *Jane's*). First flown 26 August 1997 and five ordered by end of that year; awaiting Chinese certification in late 1997, but no confirmation of receipt by early 2001. Described and illustrated in 2000-01 *Jane's*.

UPDATED

SSLF HU-2D PETREL 650D
Four-seat lightplane.
PROGRAMME: First flown 1996; revealed at China Air Show in November of that year. No further details received since then. See 2000-01 edition for description.
DESIGN FEATURES: Derivative of HU-2B/C series with extra seat and more powerful engine.

UPDATED

WHGAC

WUHAN HELICOPTER GENERAL AVIATION CORPORATION
2 Gaoxiong Road, Wuhan City, Hubei 430015
Tel: (+86 27) 579 21 87
Fax: (+86 27) 577 69 24
CHAIRMAN AND PRESIDENT: Liu Bichao

Established August 1993 to co-produce Enstrom F28F, 280FX, TH-28 and 480 helicopters (see US section); assembled first two (280FX and TH-28) from CKD kits October 1993; delivered to Wuhan Public Security Bureau and PLA respectively; at least one more US-built TH-28 supplied to China in May 1997. Two Enstrom 480s delivered for reassembly in June 1998. WHGAC also operates an Enstrom 480 as demonstrator and for charter work. New factory completed on a 150 ha (370.7 acre) site and began operating in 1998; expected to have eventual capability for completing up to 100 helicopters per year.

UPDATED

Wuhan Enstrom TH-28 in Chinese markings
1999/0044499

XAC

XIAN AIRCRAFT COMPANY (Subsidiary of AVIC I)
PO Box 140-84, Xian, Shaanxi 710089
Tel: (+86 29) 684 56 65
Fax: (+86 29) 620 37 07
e-mail: XACMKT@public.xa.sn.cn
PRESIDENT OF XIAN AIRCRAFT INDUSTRIAL GROUP, AND GENERAL MANAGER OF XAC: Gao Dacheng
VICE-PRESIDENT, XAC: Nie Zhongliang
AIRCRAFT MARKETING MANAGER: Wang Zhigang

Aircraft factory established at Xian 1958; current XAC is core enterprise of Xian Aircraft Industrial Group (XAIG); it has covered area of some 300 ha (741.5 acres) and 1996 workforce of over 21,000, of whom about 90 per cent are engaged in aircraft production. Earlier programmes have included licence production of Tu-16 twin-jet bomber (as H-6: see 1991-92 and earlier *Jane's*); six H-6s converted to aerial tanker role in 1998, for J-8 II fighter and Q-5 attack aircraft, and first seen publicly during 1 October 1999 flypast to celebrate 50 years of Communist rule in China. Major current programmes concern JH-7 attack aircraft and Y-7/MA 60 transport series of An-24/26 derivatives. XAIG and Xian Aircraft Design and Research Institute (XADRI) are participants in the regional jet study programme described under the NRJ heading in this section.

Subcontract work includes glass fibre header tanks, water float pylons, ailerons and various doors for Bombardier CL-415 amphibian; fins and tailplanes for Boeing 737/757 (439 fins and 104 tailplanes delivered by February 1996); panel assemblies for Raytheon Beech 1900D (from late 1997); and various components for Airbus Industrie. In 1997, subcontracting for ATR (which began with ATR 42 wingtips in 1986) was extended to include ATR 72 rear fuselage sections. The long-term possibility of licensed production of the latter aircraft in China was continuing to be discussed in 2000.

UPDATED

XAC H-6
Chinese name: Hongzhaji-6 (Bomber aircraft 6)
Westernised designation: B-6

TYPE: Bomber (see 1991-92 and earlier *Jane's*) and aerial tanker.
PROGRAMME: Need for aerial refuelling tanker to support Q-5 and more modern fighter and attack types such as JH-7, J-8 IID and Su-27/J-11 resulted in programme to convert H-6 (Tu-16 'Badger') bombers for this role. Programme said to have started in late 1980s, with first flight 1990, although example not seen until October 1999 flypast over Beijing during 50th anniversary celebrations. Tanker is fitted with two underwing hose/drogue pods, but said to carry only limited amount (less than 10,000 kg; 22,046 lb) of transferable fuel. Correct Chinese designation of tanker is thought to be **H-6U**, although HY-6 and JHU-6 have also been reported. At least 10 conversions are believed to have been made, with more in prospect.

NEW ENTRY

XAC JH-7
Chinese name: Jianjiji Hongzhaji-7 (Fighter-Bomber aircraft 7)
Westernised designation: FBC-1 (formerly B-7)
Export name: Flying Leopard

TYPE: Attack fighter.
PROGRAMME: Revealed publicly September 1988 as model at Farnborough International Air Show; first of four or five prototypes said to have been rolled out during previous month; first flight 14 December 1988 and first supersonic flight 17 November 1989; service entry originally scheduled for 1992-93, but delayed; first seen openly in TV broadcast of October 1995 naval exercise.

First public appearance (by third prototype '083', wearing Flight Test Establishment colours) was made in November 1998 flypast at Airshow China in Zhuhai. This coincided with announcement of new export designation and name, and statement that aircraft was ''being redesigned'' for export market. This, plus age of programme and heavy Chinese commitment to Su-27/Su-30 manufacture, make it seem doubtful whether domestic orders will materialise.
CURRENT VERSIONS: **JH-7**: Domestic version.
Following description applies to JH-7 except where indicated.
JH-7A: A US-based source reports that No. 603 Institute still undertaking improvements that include No. 613 Institute FLIR/laser targeting pod and No. 607 Institute Blue Sky low-altitude navigation pod; LETRI JL-10A radar; digital fly-by-wire controls; a health monitoring system; INS/GPS; new databus; two additional underwing hardpoints; one-piece wraparound windscreen;

A preproduction JH-7 about to land
2001/0105277

XAC—AIRCRAFT: CHINA

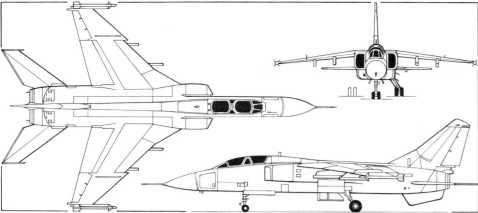

XAC JH-7 interdictor *(James Goulding/Jane's)*

and integration of additional weapons such as PGMs and Kh-31 (AS-7 'Krypton') standoff missiles.

FBC-1 Flying Leopard: Designation (from 1998) of proposed export version; would have customer-defined radar, avionics and armament.

CUSTOMERS: Earlier (1997) Chinese reports, of up to 24 in service with PLA Naval Air Force in operational evaluation role in mid-1997, seem to have been exaggerated (stated at 1998 Airshow China that only seven prototype/preproduction aircraft then completed). Latest best estimate (early 2001) is of 20 aircraft, including prototypes, of which 18 confirmed by known serial numbers to be operated by two squadrons of unknown regiment; this is compatible with known number (50) of Spey engines so far purchased. PLA Air Force and Naval Air Force reportedly remain officially interested, but series production not yet ordered for domestic service.

DESIGN FEATURES: In same role and configuration class as Russian Sukhoi Su-24 'Fencer'. High-mounted wings with compound sweepback, dog-tooth leading-edges and 7° anhedral; twin turbofans, with lateral air intakes; all-swept tail surfaces, comprising large main fin, single small ventral fin and low-set all-moving tailplane with anti-flutter weights at tips; small overwing fence at approximately two-thirds span. Wings have 47° 30′ sweepback on leading-edges. Quarter-chord sweep angles approximately 45° on fin, 55° on tailplane.

STRUCTURE: Conventional all-metal, except for dielectric panels. Plans to fit SF18 radar-absorbent material revealed in early 1999.

LANDING GEAR: Retractable tricycle type, with twin wheels on each unit. Trailing-link main units retract inward, nose unit rearward.

POWER PLANT: All aircraft so far built are powered by two Xian WS9 turbofans (licence Rolls-Royce Spey Mk 202: each 91.2 kN; 20,515 lb st with afterburning), despite persistent reports that their power is inadequate and maintenance is troublesome. Centreline and outboard underwing stations plumbed for auxiliary fuel tanks.

Original intention believed to be to power production JH-7 with LM (Liming) WS6 turbofans of 71.1 kN (15,990 lb st) dry rating (122.1 kN; 27,445 lb st with afterburning). Suggested in mid-1997 that Xian was considering licensed production of Russian alternative such as Saturn/Lyulka AL-31F. SNECMA M53-P2 has also been considered. However, Chinese industry sources in late 2000 stated that plans to re-engine the aircraft have been abandoned; at that time, negotiations were reportedly under way to acquire additional Speys.

ACCOMMODATION: Crew of two in tandem (rear seat elevated). HTY-4 ejection seats, operable at speeds from zero to 540 kt (1,000 km/h; 621 mph) and altitudes from S/L to 20,000 m (65,600 ft). Individual canopies, hinged at rear and opening upward.

AVIONICS (FBC-1): *Comms:* Short wave and ultra short wave transceivers of Italian origin.
Radar: LETRI JL-10A Shen Ying J-band pulse Doppler fire-control radar (search range 43 n miles; 80 km; 50 miles and tracking range 22 n miles; 40 km; 25 miles in look-up mode, or 29 n miles; 54 km; 34 miles and 17 n miles; 32 km; 20 miles, respectively, in look-down mode). Scans ±60° in azimuth and can track four targets simultaneously.
Flight: Automatic flight control system; GPS/INS navigation.
Instrumentation: Includes No. 603 Institute HUD and two other MFDs.
Mission: AFCS linked to fire-control system; No. 613 Institute helmet-mounted sight.
Self-defence: Chaff/flare dispenser at base of vertical tail; omnidirectional RWR; active and passive ECM.

ARMAMENT: Twin-barrel 23 mm gun, with 200 rounds, in starboard side of lower fuselage, just forward of mainwheel bay; fuselage centreline stores station, plus two under each wing and rail for PL-5B, PL-7 or similar close-range air-to-air missile at each wingtip. Typical underwing load for maritime attack, two C-801 or C-802K sea-skimming anti-ship missiles (inboard) and two drop tanks (outboard). Other potential weapons may include C-701 TV-guided anti-ship missiles and 500 kg LGBs.

DIMENSIONS, EXTERNAL:
Wing span	12.705 m (41 ft 8¼ in)
Wing aspect ratio	3.1
Length overall: excl probe	21.025 m (68 ft 11¾ in)
incl probe	22.325 m (73 ft 3 in)
Fuselage length (excl probe)	19.105 m (62 ft 8¼ in)
Height overall	6.575 m (21 ft 6¾ in)
Tailplane span	7.39 m (24 ft 3 in)
Wheel track	3.06 m (10 ft 0½ in)
Wheelbase	7.805 m (25 ft 7¼ in)

AREAS:
Wings, gross	52.30 m² (563.0 sq ft)

WEIGHTS AND LOADINGS:
Max fuel weight	10,050 kg (22,156 lb)
Max external stores load	6,500 kg (14,330 lb)
Max T-O weight	28,475 kg (62,776 lb)
Max landing weight	21,130 kg (46,583 lb)
Max wing loading	544.5 kg/m² (111.51 lb/sq ft)
Max power loading (WS9 engines)	156 kg/kN (1.53 lb/lb st)

PERFORMANCE:
Max level speed at 11,000 m (36,080 ft) (clean)	M1.7 (975 kt; 808 km/h; 1,122 mph)
Max operating speed	653 kt (1,210 km/h; 751 mph) IAS
Cruising speed	M0.80-0.85 (459-487 kt; 850-903 km/h; 528-561 mph)
Service ceiling (clean)	15,600 m (51,180 ft)
T-O run	920 m (3,020 ft)
Landing run	1,050 m (3,445 ft)
Combat radius	891 n miles (1,650 km; 1,025 miles)
Ferry range	1,970 n miles (3,650 km; 2,268 miles)
g limit	+7

UPDATED

XAC MA-60

TYPE: Twin-turboprop airliner.

PROGRAMME: Model of proposed 56-seat Y7-200 (PW127 turboprop) variant first shown at Asian Aerospace, Singapore, February 2000. First flight reportedly 12 March 2000; revenue service began August 2000; public debut (Airshow China) November 2000. To replace earlier Y-7 variants, which from 2001 will no longer meet amended Chinese airworthiness requirements. XAC production capacity 12 to 15 a year; domestic operators will lease from central leasing agency SFLC, which will be responsible for sole-source marketing (first arrangement of this kind in China).

CURRENT VERSIONS: **MA-60:** Initial production version; *as described*. Current life-cycle of 25,000 landings and 30,000 flight hours.
Improved MA-60: Improvements planned for introduction by 2002 include reduction in empty weight by about 400 kg (882 lb); increase in 'hot-and-high' MTOW of about 900 kg (1,984 lb); increase in OEI ceiling; 162 n mile (300 km; 186 miles) increase in range; airframe drag reduction; 'glass cockpit' avionics upgrade.
MA-MPA (Fearless Albatross): Projected maritime patrol version (previously Y7-200BF); nose-mounted search radar; raked wingtips; auxiliary fuel tanks scabbed to fuselage sides to extend range to estimated 1,350 n miles (2,500 km; 1,553 miles); four underwing stations for air-to-air and anti-ship missiles. Shown as model at Airshow China, November 2000.

CUSTOMERS: Sole-source domestic marketing by newly formed Shenzhen Finance Leasing Corporation which, on 7 November 2000, signed contract for delivery of 60 MA-60s over five-year period. Launch operator Sichuan Airlines (five ordered) introduced first leased example (B-3426) into revenue service August 2000; other early operators reported to include Changan, China Northern and Wuhan Airlines. Most of first 60 expected to be of Improved MA-60 standard.

DESIGN FEATURES: Further upgrade of Y-7 series, mainly through Western engines and avionics and improved passenger comfort; incorporates best features of earlier Y7-200A and B programmes.

POWER PLANT: Two 2,051 kW (2,750 shp) Pratt & Whitney Canada PW127J turboprops, each driving a Hamilton Sundstrand 247F-3 four-blade 'scimitar' propeller.

ACCOMMODATION: Flight crew of three, plus cabin attendant(s); 56 to 60 passengers.

SYSTEMS: Honeywell APU.

AVIONICS: Rockwell Collins EFIS and navigation suite, including APS-85 AFCS; GPS nav, weather radar, ADF, TCAS, air data system and radio altimeter; assembled in-country by Leihua Electronic Technology Research Institute (LETRI).

DIMENSIONS, EXTERNAL:
Wing span	29.20 m (95 ft 9½ in)
Wing aspect ratio	11.4

First in-service example of the MA-60, in Sichuan Airlines livery *(Photo Link)*

Length overall	24.71 m (81 ft 0¾ in)
Height overall	8.85 m (29 ft 0½ in)
Wheel track (c/l of shock-struts)	7.90 m (25 ft 11 in)
Wheelbase	9.565 m (31 ft 4½ in)

AREAS:
Wings, gross	74.98 m² (807.1 sq ft)

WEIGHTS AND LOADINGS:
Operating weight empty	13,700 kg (30,203 lb)
Max fuel	4,030 kg (8,885 lb)
Max payload	5,500 kg (12,125 lb)
Max T-O weight	21,800 kg (48,060 lb)
Max ramp weight	21,950 kg (48,391 lb)
Max landing weight	21,200 kg (46,738 lb)
Max zero-fuel weight	19,500 kg (42,990 lb)
Max wing loading	290.7 kg/m² (59.55 lb/sq ft)
Max power loading	5.32 kg/kW (8.74 lb/shp)

PERFORMANCE:
Max cruising speed	272 kt (504 km/h; 313 mph)
Econ cruising speed at 6,000 m (19,680 ft)	248 kt (460 km/h; 286 mph)
Max operating altitude	7,600 m (24,940 ft)
T-O run	724 m (2,375 ft)
T-O to 15 m (50 ft)	901 m (2,955 ft)
T-O balanced field length	1,150 m (3,775 ft)

Range at 6,000 m (19,680 ft) with max payload and 1,540 kg (3,395 lb) fuel, reserves for 100 n mile (185 km; 115 mile) diversion and 45 min hold
864 n miles (1,600 km; 994 miles)
Range with max fuel
1,322 n miles (2,450 km; 1,522 miles)

UPDATED

Model of the projected 'Fearless Albatross' variant of the MA-60 *(Photo Link)*

XAC Y-7
Chinese name: Yunshuji-7 (Transport aircraft 7)
TYPE: Twin-turboprop airliner.
PROGRAMME: Reverse engineering of 48/52-passenger Antonov An-24 began in 1966; three prototypes (first flight 25 December 1970) and two static test airframes completed; Chinese C of A awarded 1980; preproduction Y-7 made public debut 17 April 1982; put into operation (CAAC Shanghai Administration) 24 January 1984; small batch production began 1984; first scheduled passenger services with CAAC, 29 April 1986; Y7-100 prototype conversion by HAECO of Hong Kong in 1985 (first flight of production Y7-100 late 1985), followed by domestic certification 23 January 1986. More than half a million hours flown by mid-1996, and 18 million passengers carried. Production appears to have ended, and phase-out of early aircraft has begun.

CURRENT VERSIONS: **Y-7:** Initial production version; 16 built; two became prototypes for Y-7H (c/n 02702) and Y7-100 (c/n 03702). Full description in 1988-89 *Jane's*. Many retrofitted with winglet modification of Y7-100.

Y-7 Freighter: Civil cargo version of Y-7, first flown late 1989; five delivered to domestic operators in 1992 (first one on 24 June).

Y7-100: Improved, main production version, meeting BCAR standards. Winglets, new three-person (instead of five) flight deck, all-new cabin interior with 52 reclining seats, windscreen de-icing, new HF/VHF com, new nav equipment, addition of oxygen/air data/environmental control systems. One Y-7 (B-3499, c/n 03702) converted as prototype 1985 by Hong Kong Aircraft Engineering Company (HAECO). Subvariants **Y7-100C1/C2/C3** have five-person crew and avionics and equipment to customer's requirements.

Detailed description applies to standard Y7-100, except where indicated.

Y7-100 Maritime: One former China Eastern aircraft (B-3489), equipped with unspecified surveillance equipment, noted in coastguard service at Yamliang, 1998.

Y7-200A: Improved Y7-100: 2,051 kW (2,750 shp) Pratt & Whitney Canada PW127C turboprops, Hamilton Sundstrand 247F-3 four-blade propellers and Rockwell Collins EFIS 85/86 avionics; Honeywell GTCP-150CY APU; empty weight reduced; winglets deleted; fuselage lengthened to accommodate 56 (standard) or a maximum of 60 passengers (seat pitch 78 cm; 31 in and 72 cm; 28 in respectively); two-person flight deck (plus observer's seat) with improved field of view. Three prototypes built; first flight (B-570L, c/n 007A01) 26 December 1993; Chinese CAAC 25 type certificate awarded in 1998. First production aircraft (B-3720) noted in Changan Airlines colours in 1999, but other reported orders (2000-01 *Jane's*) have apparently not materialised, and FAA certification still not obtained by early 2001. Programme superseded by MA-60 (which see). Details in 2000-01 and earlier editions.

Y7-200B: Improved Y7-100 for domestic market, with Dongan WJ5A IG improved turboprops, new four-blade propellers, Rockwell Collins EFIS 85B14 avionics, modified wing leading-edges, higher maximum lift coefficient, ground spoilers added, improved stall characteristics, lower fuel consumption; winglets deleted; overall length increased by 0.74 m (2 ft 5¼ in) and tailplane span by 1 per cent; empty weight reduced by 500 kg (1,102 lb); Cat. II automatic landing capability. First flight (B-528L) 23 November 1990; same aircraft test-flown on 25 December 1999 with two large (1,400 litre; 370 US gallon; 308 Imp gallon) underwing auxiliary fuel tanks. Details in 2000-01 and earlier editions. Superseded by MA-60 (which see).

Y7-200BF: Maritime patrol project; now redesignated MA-MPA and described in MA-60 entry.

Y-7E: Hot/high version with different (more powerful?) APU; first flight 5 July 1994. Nothing further known.

Y7H and Y7H-500: Military and civil cargo versions, derived from Antonov An-26; Only three prototypes known.

MA-60: Described separately.

CUSTOMERS: More than 105 Y-7s and Y7-100s delivered to Chinese domestic airlines (see table) and armed services by early 1997, including PLA Naval Air Force (two PLANAF aircraft converted as aircrew trainers in 1995, with new radar and displays). Mauritanian Air Force received one (5T-MAF) in 1995, but no new aircraft reported since that time.

Y-7 COMMERCIAL OPERATORS
(at mid-2000)

Country	Operator	Y-7	Y7-100
China	Air China		4
	Changan Airlines		7
	China Eastern Airlines		4
	China Northern Airlines		10
	Guizhou Airlines	1	6
	Nanjing Airlines		2
	Shaanxi Aviation		3
	Sichuan Airlines		5
	Wuhan Airlines		7
	Zhongyuan Airlines		2
Laos	Lao Aviation		
Totals		**1**	**52**

Note: Further 10 Y7-100s operated by CAAC Flying College

DESIGN FEATURES: Non-swept high-mounted wings, with 2° 12′ 12″ anhedral on tapered outer panels; wingtip winglets standard on Y7-100, retrofitted on many Y-7; basically circular-section fuselage; sweptback fin and rudder; 9° dihedral on tailplane; twin ventral fins under tailcone. Wing sweepback 6° 50′ at quarter-chord of outer panels; incidence 3°.

LANDING GEAR: Retractable tricycle type with twin wheels on all units. Hydraulic actuation, with emergency gravity extension. All units retract forward. Mainwheels are size 900×300, tyre pressure 5.39 to 5.88 bars (78 to 85 lb/sq in); nosewheels size 700×250, tyre pressure 3.92 bars (57 lb/sq in). (Mainwheel tyre pressures variable to cater for different types of runway.) Disc brakes on mainwheels; hydraulically steerable (±45°) and castoring nosewheel unit.

POWER PLANT: Two DEMC (Dongan) WJ5A I turboprops, each rated at 2,080 kW (2,790 shp) for T-O and 1,976 kW (2,650 shp) at ISA + 23°C; Baoding J16-G10A four-blade constant-speed fully feathering propellers. Fuel in integral wing tanks immediately outboard of nacelles, and four bag-type tanks in centre-section, total capacity 5,550 litres (1,466 US gallons; 1,220 Imp gallons). Provision for four additional tanks in centre-section. Pressure refuelling point in starboard engine nacelle; gravity fuelling point above each tank.

One 8.83 kN (1,984 lb st) RU 19A-300 APU in starboard engine nacelle for engine starting, to improve take-off and in-flight performance, and to reduce stability and handling problems if one turboprop fails in flight.

ACCOMMODATION: Crew of three on flight deck (five in Y7-100C1/C2/C3), plus one or two cabin attendants. Standard layout has four-abreast seating, with centre aisle, for 52 passengers in air conditioned, soundproofed (by Tracor) and pressurised cabin. Galley (by Lermer) and lavatory at rear on starboard side. Baggage compartments forward and aft of passenger cabin, plus overhead stowage bins in cabin. Alternative layouts available for 48 or 50 passengers, or 20-passenger executive interior. Passenger airstair door on port side at rear of cabin. Emergency exit each side at front of cabin. Doors to forward and rear baggage compartments on starboard side. All doors open inward.

DIMENSIONS, EXTERNAL:
Wing span (over winglets)	29.67 m (97 ft 4 in)
Wing aspect ratio	11.7

Prototype of the PW127C-engined Y7-200A, now superseded by the MA-60 *(Photo Link)*

Length overall	24.22 m (79 ft 5½ in)
Height overall	8.85 m (29 ft 0½ in)
Fuselage: Max width	2.90 m (9 ft 6¼ in)
Max depth	2.50 m (8 ft 2½ in)
Wheel track (c/l of shock-struts)	7.90 m (25 ft 11 in)
Wheelbase	7.90 m (25 ft 11 in)
Propeller diameter	3.90 m (12 ft 9½ in)

DIMENSIONS, INTERNAL:
Cabin: Length, incl flight deck	10.50 m (34 ft 5½ in)
Max width	2.76 m (9 ft 0¾ in)
Max height	1.90 m (6 ft 2¾ in)
Volume	56.0 m³ (1,978 cu ft)
Baggage compartment volume:	
fwd	4.5 m³ (159 cu ft)
rear	6.7 m³ (237 cu ft)

AREAS:
Wings, gross	75.26 m² (810.1 sq ft)

WEIGHTS AND LOADINGS:
Operating weight empty	14,988 kg (33,042 lb)
Max fuel	4,790 kg (10,560 lb)
Max payload	5,500 kg (12,125 lb)
Max T-O and landing weight	21,800 kg (48,060 lb)
Max zero-fuel weight	19,655 kg (43,332 lb)
Max wing loading	289.7 kg/m² (59.34 lb/sq ft)
Max power loading	5.24 kg/kW (8.61 lb/shp)

PERFORMANCE:
Max level speed	271 kt (503 km/h; 313 mph)
Max cruising speed at 6,000 m (19,680 ft)	257 kt (476 km/h; 296 mph)
Econ cruising speed at 6,000 m (19,680 ft)	228 kt (423 km/h; 263 mph)
Max rate of climb at S/L, AUW of 21,000 kg (46,297 lb)	458 m (1,504 ft)/min
Service ceiling, AUW of 21,000 kg (46,297 lb)	8,750 m (28,700 ft)
Service ceiling, OEI, AUW of 19,000 kg (41,887 lb)	3,850 m (12,640 ft)

T-O run at S/L, FAR Pt 25, AUW of 21,000 kg
(46,297 lb): ISA 546 m (1,795 ft)
ISA + 20°C 1,398 m (4,590 ft)
Landing run, AUW of 21,000 kg (46,297 lb)
620 m (2,035 ft)
Range: max (52-passenger) payload
491 n miles (910 km; 565 miles)
max standard fuel
1,070 n miles (1,982 km; 1,231 miles)
standard and auxiliary fuel
1,296 n miles (2,400 km; 1,491 miles)
UPDATED

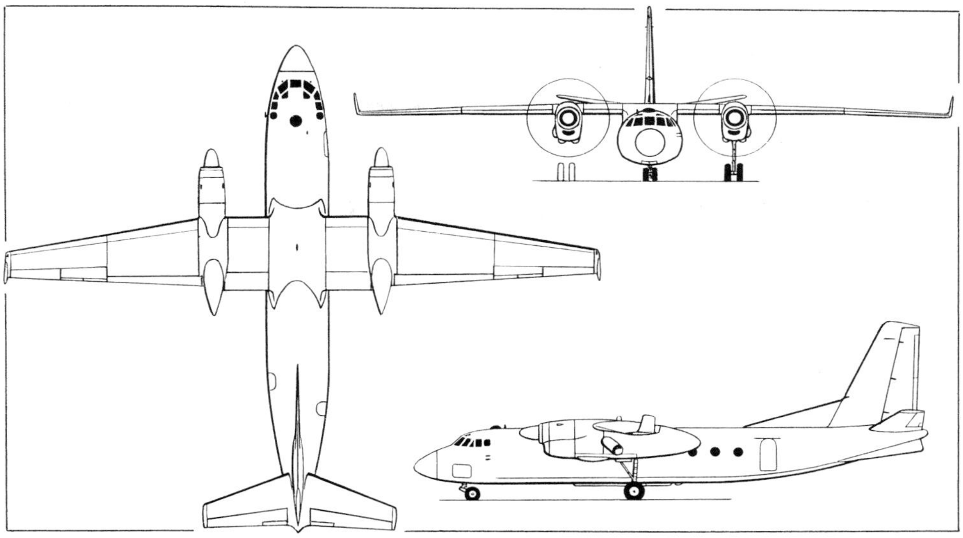

XAC Y7-100 twin-turboprop transport *(Dennis Punnett/Jane's)*

XAC Y7H
Details of this twin-turboprop airliner and cargo transport appeared in the 2000-01 and earlier editions.
UPDATED

COLOMBIA

GAVILAN

EL GAVILAN SA
Carrera 3 No 56-19, Apartado Aéreo 180112, Santa Fé de Bogotá DC
Tel: (+57 1) 676 50 01
Fax: (+57 1) 676 06 50
e-mail: sales@elgavilan.com
Web: http://www.elgavilan.com
GENERAL MANAGER: James S Leaver
TECHNICAL DIRECTOR: Eric C Leaver
OPERATIONS AND PRODUCTION DIRECTOR: Jaime E Silva
MARKETING MANAGER: Omar Porras Pineda
ENGINEERING MANAGER: David Fernando Muñoz Galeano

Formerly Aero Mercantil (see 1991-92 and earlier *Jane's*), associated with Piper Aircraft Corporation as dealer and then distributor since 1952; sold its shares in AICSA and in 1991 formed new Gavilán (Sparrowhawk) company to pursue development and manufacture of own-design EL-1 utility aircraft.

Advanced Wing Technologies (AWT) of Canada (which see) had planned to build Gavilán 358 under late 1999 agreement, but this had been abandoned by mid-2000, at which time alternative North American production facilities being sought.
UPDATED

GAVILAN EL-1 GAVILAN 358 and 508T
English name: Sparrowhawk
TYPE: Light utility transport.
PROGRAMME: Launched March 1986; sponsored by Leaver Group, of which El Gavilián SA is a member; construction began May 1987; first flight (HK-3500-Z) 27 April 1990; after 50 hours' flying, fuselage stretched 0.305 m (1 ft 0 in) immediately forward of main spar and gross weight increased by 136 kg (300 lb); new first flight 7 November 1990; wing incidence increased 1991 to improve cruising speed, and passenger windows reshaped; damage from emergency landing in USA on 1 February 1992 delayed progress and company decided to obtain FAA (FAR Pt 23, Amendment 45) certification from outset; this gained by second prototype, built by General Aviation Technical Services in USA; funding arranged December 1994 for FAA certification. Second aircraft (N358EL/HK-4120-Z) flown (in USA) 29 May 1996 and gained FAR Pt 23 (to Amendment 46) certification 26 May 1998. First production aircraft flew on 7 August 1997; used as North American demonstrator. Potential North American assembly site being sought in late 2000.
CURRENT VERSIONS: **Gavilán 358**: Piston-engined standard passenger, parachutist or military version.
Following description applies to Gavilán 358.
Gavilán 508T: Turboprop version, announced May 1998; under development; choice of P&WC PT6A-34 or Walter M 601E-11 engine. Programme scheduled to start in mid-2001, deliveries from mid-2002.
CUSTOMERS: Nine confirmed orders by November 1998, including four (delivered 29 July to 30 September 1998) for Colombian Air Force (FAC), which has option on eight more. Eleven Gavilán 358s delivered by September 2000, including FAC (four), Colombian Navy (one military, delivered 23 December 1999), Colombian commercial operators (two) and Tikal Jets of Guatemala (one). Several more orders being negotiated at that time. Planned production rate of four per month.
COSTS: Standard Gavilán 358, VFR equipped US$379,0 fully equipped US$393,000 (2000).
DESIGN FEATURES: Rugged utility transport wit door and wide track landing gear, optimi from short and unprepared airstrips.
Constant chord, unswept brac section; thickness/chord ratio incidence 5° 30′; washou
FLYING CONTROLS: Conve cables). Mass-balanced horn-balanced; spring bias

Second prototype Gavilán 358, following its re-registration in Colombia

Mini-tractor stowed aboard a Gavilán 358
1999/0044503

Gavilán 358 instrument panel *1999/0044502*

starboard elevator; manual single-slotted flaps (settings 15, 30 and 40°) on offset hinges actuated by handle on central console. Autopilot planned.

STRUCTURE: All-metal, mainly of 2024-T3 aluminium alloy sheet; fuselage frame of 4130N steel tube skinned with 2024-T3; two-spar wing with single strut each side; two-spar fin and tailplane. Entire airframe extensively corrosion-proofed.

LANDING GEAR: Non-retractable tricycle type, with elastomeric shock-absorption and single wheel on each unit. Steel tube mainwheel legs; trailing-link nose unit, with fully castoring nosewheel. McGreary tyres, sizes 7.00×6 (8 ply) main and 7.00×6 (6 ply) nose; pressures 3.59 bars (52 lb/sq in) and 2.41 bars (35 lb/sq in) respectively. Cleveland hydraulic mainwheel brakes.

POWER PLANT: One Textron Lycoming TIO-540-W2A flat-six engine (261 kW; 350 hp at 2,600 rpm), driving a three-blade constant-speed Hartzell propeller. Three rubber fuel cells in each wing, combined capacity 405 litres (107 US gallons; 89.2 Imp gallons), of which 394 litres (104 US gallons; 86.7 Imp gallons) usable. Refuelling point in top of each tank. Gravity feed system with auxiliary electric pump and header tanks. Oil capacity 11.4 litres (3.0 US gallons; 2.5 Imp gallons).

ACCOMMODATION: Pilot and co-pilot or one passenger at front; dual control wheels. All transparencies of flat Plexiglas. Two quickly removable rows of three seats to rear of pilots, facing each other. Door at front on each side, plus larger double door at rear on port side. Accommodation for four stretchers and one attendant in air ambulance version; up to eight jumpers in parachutist version.

SYSTEMS: Engine-mounted vacuum pump for gyro instruments, driven at 0.31 bar (4.5 lb/sq in). Electrical power from 28 V 70 A engine-driven alternator and 24 V 18 Ah battery. Hydraulic system for brakes only. Gaseous oxygen system optional.

AVIONICS: *Comms:* Honeywell KT 76A transponder and AK-350 encoder.
Radar: Colombian Air Force aircraft to be equipped with Honeywell RDR-2000 weather radar in wing leading-edge radome.
Flight: Honeywell KLX 135A nav/com/GPS.
Instrumentation: Standard VFR.

EQUIPMENT: Standard cargo net and cargo tiedowns in cargo version. Optional cargo kit and air ambulance kit for passenger versions.

ARMAMENT: Colombian Navy aircraft equipped with two 7.62 mm SS-77 machine guns.

DIMENSIONS, EXTERNAL:
Wing span	12.80 m (42 ft 0 in)
Wing chord, constant	1.55 m (5 ft 1 in)
Wing aspect ratio	7.8
Length overall	9.53 m (31 ft 3 in)
Fuselage max width	1.42 m (4 ft 8 in)
Height overall	3.73 m (12 ft 3 in)
Tailplane span	4.68 m (15 ft 4¼ in)
Wheel track	3.35 m (11 ft 0 in)
Propeller diameter	2.18 m (7 ft 2 in)
Propeller ground clearance	0.17 m (6¾ in)
Crew doors (two, each):	
Height	1.17 m (3 ft 10 in)
Max width	0.86 m (2 ft 10 in)
Height to sill	1.02 m (3 ft 4 in)
Passenger/cargo door: Height	1.17 m (3 ft 10 in)
Width	1.24 m (4 ft 1 in)
Height to sill	1.02 m (3 ft 4 in)

DIMENSIONS, INTERNAL:
Cabin: Length	3.40 m (11 ft 2 in)
Max width	1.37 m (4 ft 6 in)
Max height	1.37 m (4 ft 6 in)
Floor area: excl flight deck	4.09 m² (44.0 sq ft)
incl flight deck	4.70 m² (50.6 sq ft)
Volume: excl flight deck	4.0 m³ (140 cu ft)
incl flight deck	6.4 m³ (226 cu ft)
Baggage compartment volume	0.48 m³ (17.0 cu ft)

AREAS:
Wings, gross	18.95 m² (204.0 sq ft)
Ailerons (total)	0.97 m² (10.40 sq ft)
Trailing-edge flaps (total)	1.12 m² (12.00 sq ft)
Fin	1.48 m² (15.91 sq ft)
Rudder	0.91 m² (9.81 sq ft)
Tailplane	1.51 m² (16.30 sq ft)
Elevators (total, incl tab)	0.90 m² (9.70 sq ft)

WEIGHTS AND LOADINGS:
Weight empty, standard equipped	1,270 kg (2,800 lb)
Max fuel	291 kg (642 lb)
Max payload with 30 min fuel at 75% power	488 kg (1,076 lb)
Max T-O and landing weight	2,041 kg (4,500 lb)
Max wing loading	107.7 kg/m² (22.06 lb/sq ft)
Max power loading	7.82 kg/kW (12.86 lb/hp)

PERFORMANCE:
Never-exceed speed (V_{NE})	155 kt (287 km/h; 178 mph) IAS
Max level speed at 4,575 m (15,000 ft)	140 kt (259 km/h; 161 mph)
Cruising speed at 3,050 m (10,000 ft):	
at 75% power	135 kt (250 km/h; 155 mph)
at 65% power	130 kt (241 km/h; 150 mph)
at 55% power	126 kt (233 km/h; 145 mph)
Stalling speed, power off:	
flaps up	65 kt (121 km/h; 75 mph) IAS
40° flap	58 kt (108 km/h; 67 mph) IAS
Max rate of climb at S/L	244 m (800 ft)/min
Max certified altitude	3,180 m (12,500 ft)
Service ceiling	6,860 m (22,500 ft)
T-O run, 15° flap	375 m (1,230 ft)
T-O to 15 m (50 ft)	595 m (1,955 ft)
Landing from 15 m (50 ft)	500 m (1,640 ft)
Landing run	327 m (1,075 ft)
Range, 45 min reserves:	
at 75% power	770 n miles (1,426 km; 886 miles)
at 65% power	820 n miles (1,518 km; 943 miles)
at 55% power	945 n miles (1,750 km; 1,087 miles)

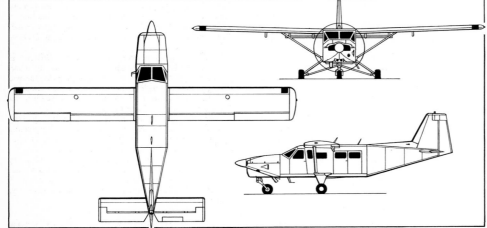

Gavilán 358 seven-passenger utility transport *(Mike Keep/Jane's)* *1999/0044532*

UPDATED

CZECH REPUBLIC

AERO

AERO HOLDING AS
Beranových 130, CZ-199 04 Praha 9-Letňany
Tel: (+420 2) 88 40 65 and 88 27 47
Fax: (+420 2) 88 65 81, 88 40 65 and 88 27 47
DIRECTOR GENERAL: Ing František Petrašek
EXECUTIVE DIRECTOR, TECHNICAL AND COMMERCIAL STRATEGY:
 Ing Jan Bartoň

This joint stockholding management organisation replaced the state-owned Aero Concern of Czechoslovak Aerospace Industry on 1 December 1990; airframe, engine and equipment factories and research centres became limited companies on 1 January 1991. The joint stock company was partly privatised in 1992, basically changing the current role of Aero Holding from an operational holding to a financial one (100 per cent ownership of 10 subsidiaries); full privatisation of Aero Holding was approved by the Czech government in June [199]3. Activities accordingly now comprise organisation, co-[ordin]ation and financing of research, development, [produc]tion and sale of aircraft and other aviation products.
[Share] holdings in wholly owned subsidiaries have been [reduced] since September 1993 by sales to various foreign [domes]tic partners. Čenkovské Strojírní, Teset Semily and Microtechna Holešovice subsidiaries have been liquidated; Moravan (Zlin) sold in July 1994 to Investiční Poštovní Banka; major part of Aero Holding shares in Aero Vodochody, Letov and VZLU were recapitalised and transferred to the new state bank, Konsolidační Banka, and the commercial banks Investiční Poštovní Banka and Československá Obchodní Banka. Major share (93.3 per cent) in Let Kunovice acquired in August 1998 by Ayres Corporation of USA.

Principal (61.8 per cent) shareholder in Aero Holding AS in third quarter of 1999 was the Czech National Property Fund (FNM).
 Subsidiaries at that time were as follows:
  Aero Trade Prague
  AXL Semily
  VZLU (Aeronautical Research and
   Test Institute) Prague
  Walter (Motorlet) Prague

ASSOCIATED COMPANIES:
Aero Trade AS
(address as for Aero Holding AS)
Tel: (+420 2) 88 25 09
Fax: (+420 2) 859 01 16
e-mail: slezak@czcom.cz

DIRECTOR GENERAL: Ing Jiří Holub
MARKETING DIRECTOR: Jaroslav Hošek

Joint stock company, formed by Aero Holding April 1993; responsibilities include sale and leasing, and sale of spare parts, for all aircraft and other aeronautical products manufactured by the Czech aeronautical industry.

Omnipol AS
Nekázanka 11, CZ-112 21 Prague 1
Tel: (+420 2) 24 01 11 11
Fax: (+420 2) 24 01 22 41
e-mail: Omnipol@gts.cz
Web: http://www.omnipol.cz
DIRECTOR GENERAL: Michal Hon
MARKETING DIRECTOR: Peter Jacina
PUBLICITY MANAGER: Zdenek Burian

Joint stock company specialising in export and import of all kinds of goods including aircraft, aeronautical products and defence equipment. No capital connection with Aero Holding AS.

UPDATED

AERO

AERO VODOCHODY AS

CZ-250 70 Odolena Voda
Tel: (+420 2) 688 00 41
Fax: (+420 2) 82 31 72 and 687 25 05
e-mail: mr@aero.cz
Web: http://www.aero.cz
PRESIDENT: Petr Hora
CEO: Vladimír Nývlt
VICE-PRESIDENTS:
 Igor Stratil (Finance)
 Zdeněk Prokop (Marketing and Sales)
 Jiří Fidranský (Research and Development)
 Adam Straňák
DIRECTORS:
 Viktor Kučera (Military Aircraft Programmes)
 Miloš Vališ (Ae 270 Programme)
 František Bílek (Aerostructures)
 Martin Paloda (Marketing)
 Václav Pavlíček (Services)
 Dušan Milowsky (Commercial)
 Petr Dub (Technical)
CHIEF DESIGNERS:
 Zdeněk Stucklík (Military Programmes)
 Josef Jironč (Civil Programmes)

Established on 1 July 1953, Aero produced (with Let) 3,665 L-29 Delfin jet trainers between 1961 and 1974.

The company underwent a change of ownership as the consequence of a restructuring programme approved on 31 July 1996, when it had yet to recover fully from debts occurring in 1990-91. It is now a joint-stock company in which the majority shareholder is the Czech government, through shares held by Letka AS (35.66 per cent) and the Konsolidační Banka (29.00 per cent). From 17 August 1998, a further 35.29 per cent was acquired by Boeing Česká sro, a joint venture comprising Boeing and CSA Czech Airlines, owned 90 per cent by Boeing and 10 per cent by CSA. The remaining 0.05 per cent of Vodochody is owned by other shareholders. Workforce in July 2000 was 1,974.

In May 1998, Aero Vodochody became 100 per cent owner of Technometra Radotin, producer of landing gears and other components for the aircraft of Aero and Let Kunovice. It has also acquired 42.6 per cent of the Prague company Letov, which produces wings, rear fuselages and tail assemblies for Aero's jet trainers; 10 per cent of Aero Media; and 5.3 per cent of Aero Trade.

With effect from 1 October 1999, the company was reorganised into three main business units: Military Aircraft, Commercial Aircraft and Aerostructures. Current aircraft programmes are, respectively, the L-39/139/59/159 family of jet trainers described below; the Ae 270 Ibis civil transport, in a 50/50 venture with AIDC of Taiwan (see International section); and airframe components for Bombardier de Havilland (Dash 8Q tailplanes) and Boeing Commercial Airplane Group (747/757).

A US$230 million deal with Sikorsky was announced in June 2000, under which Aero Vodochody will undertake fabrication and subassembly of S-76 helicopter airframes (but not their dynamic systems), assisted by Fischer Advanced Composite Components of Austria as subcontractor. The agreement calls for delivery of 15 airframes per year over a seven-year period.

UPDATED

AERO L-39 and L-139 ALBATROS

TYPE: Basic jet trainer/light attack jet.
PROGRAMME: First flight 4 November 1968; 10 preproduction aircraft from 1971; selected 1972 to succeed L-29 as standard trainer for USSR, Czechoslovakia and East Germany and production started same year; service trials in USSR and Czechoslovakia 1973; entered service with Czechoslovak Air Force 1974. Worldwide fleet has accumulated more than 4 million flying hours. Modular upgrades being offered for all L-39 variants.
CURRENT VERSIONS: **L-39 C**: Initial pilot trainer, with two underwing stations; details in 1994-95 and earlier *Jane's*. Most recent export customer (Yemen) ordered 12 in February 1999; delivered September to December same year. Eight Czech Air Force aircraft recently refurbished with new wings, front and rear fuselages and other changes under contracts placed in June and November 1999, for one and seven respectively; first aircraft (0113) redelivered 15 December 1999, remainder by May 2000.
 L-39 V: Target towing version for Czech and former East German use; eight only.
 L-39 ZO (Z = *Zbrojni*: armed): Reinforced wings with four underwing stations; first flight (X-09) 25 August 1975. Production completed; weight, performance and other data in 1991-92 and earlier *Jane's*. Two to Ghana in 1999.
 L-39 ZA: Ground attack and reconnaissance version of ZO; four underwing stations and centreline gun pod; reinforced wings and landing gear; prototypes (X-10 and X-11) flown 29 September 1976 and 16 May 1977; production continuing in 2000 for possible future orders. Customer versions completed with Western avionics (HUD, mission computer, Honeywell avionics and navigation equipment) as **L-39 ZA/MP** (for multipurpose). Version for Thailand has Elbit avionics and is designated **ZA/ART**; deliveries to RTAF began in 1993. Latest customer is Lithuania (two delivered October 1998).
Detailed description applies to ZA version except where indicated.
 L-39 MS: Developed version, similar to L-59 and described under that entry.
 L-59: New advanced training version with more powerful engine and improved avionics; described separately.
 L-139 Albatros 2000: Trainer, powered by 18.15 kN (4,080 lb st) TFE731-4-1T turbofan under preliminary Aero/AlliedSignal agreement signed June 1991; Flight Visions HUD and Honeywell avionics; VS-2A zero/zero seats. First flight (5501) 8 May 1993. None yet ordered; possible upgrade with some L 159 avionics being considered.
CUSTOMERS: Total of 2,935 L-39/59s built (excluding five prototypes) by December 1999; see table. Eight former Czech Air Force L-39ZAs refurbished by Israel Aircraft Industries for Cambodian Air Force; deliveries by IAI began in late 1996/early 1997.
DESIGN FEATURES: Tandem two-seater with ejection seats and pressurisation; fixed tip tanks also contain navigation/landing lights; rear fuselage and tail, attached by five bolts, allow easy removal for access to engine; tapered wing has NACA 64A012 Mod.5 section; quarter-chord sweepback 1° 45'; dihedral 2° 30' from roots; incidence 2°.
FLYING CONTROLS: Conventional and manual, by pushrods. Electrically actuated trim tab in each elevator; balance tab in rudder; mass-balanced ailerons with balance tabs (port tab actuated electrically for trimming); two airbrake panels under fuselage just below wing leading-edge, operated by single hydraulic jack, extend automatically as speed approaches M0.8; double-slotted flaps extended by rods from single hydraulic jack, retracting automatically as airspeed reaches 167 kt (310 km/h; 193 mph).
STRUCTURE: One-piece all-metal stressed skin wing with main and auxiliary spars; four-point wing attachment; fuselage ahead of engine bay has three sections: nose containing avionics/battery/antennas/air and oxygen bottles/nosewheel unit; next section forming pressurised cockpit; third section containing air intakes/fuel tanks/engine bay.
LANDING GEAR: Retractable tricycle type, with single wheel and oleo-pneumatic shock-absorber on each unit; designed for touchdown sink rate of 3.4 m (11.15 ft)/s at AUW of 4,600 kg (10,141 lb). Retraction and extension operated hydraulically, with electrical control. All wheel well doors close automatically after wheels are lowered, to prevent ingress of dirt and debris. Mainwheels retract inward into wings (with automatic braking during retraction), nosewheel forward into fuselage. K28 mainwheels with 610×215 tyres and K27 nosewheel with 450×165 tyre. Hydraulic disc brakes and anti-skid units on mainwheels; shimmy damper on nosewheel leg. Minimum ground turning radius (about nosewheel) 2.50 m (8 ft 2 in). L-39 ZA can operate from grass strips with bearing strength of 6 kg/cm^2 (85 lb/sq in).
POWER PLANT: One 16.87 kN (3,792 lb st) Progress AI-25 TL turbofan in rear fuselage, with semi-circular lateral air intake, with splitter plate, on each side of fuselage above wing centre-section.
 Fuel in five rubber main bag tanks aft of cockpits, with combined capacity of 1,055 litres (279 US gallons; 232 Imp gallons), and two 100 litre (26.5 US gallon; 22.0 Imp gallon) non-jettisonable wingtip tanks. Total internal fuel capacity 1,255 litres (332 US gallons; 276 Imp gallons). Gravity refuelling points on top of fuselage and on each tip tank. Provision for two 150 or 350 litre (39.6 or 92.5 US gallon; 33.0 or 77.0 Imp gallon) drop tanks on inboard underwing pylons, increasing total overall fuel capacity to a maximum of 1,955 litres (517 US gallons; 430 Imp gallons). Fuel system permits up to 20 seconds of inverted flight.
ACCOMMODATION: Crew of two in tandem, on Czech VS-1-BRI rocket-assisted ejection seats, operable at zero height and at speeds down to 81 kt (150 km/h; 94 mph); individual canopies hinge sideways to starboard and are jettisonable. Rear seat elevated. One-piece windscreen hinges forward for access to front instrument panel. Internal transparency between cockpits. Dual controls standard.
SYSTEMS: Cockpits pressurised (standard differential 0.23 bar; 3.3 lb/sq in, maximum overpressure 0.29 bar; 4.2 lb/sq in) and air conditioned, using engine bleed air and cooling unit. Air conditioning system provides automatic temperature control from 10 to 25°C at ambient air temperatures from −55 to +45°C.
 Main and standby interconnected hydraulic systems, main system having variable flow pump with operating pressure of 147 bars (2,133 lb/sq in) for actuation of landing gear, flaps, airbrakes, ram air turbine and (at 34.3 bars; 500 lb/sq in pressure) wheel brakes. Emergency system, for all of above except airbrakes, incorporates three accumulators. Pneumatic canopy seals supplied by a 2 litre (0.07 cu ft) compressed air bottle in nose (pressure 147 bars; 2,133 lb/sq in).

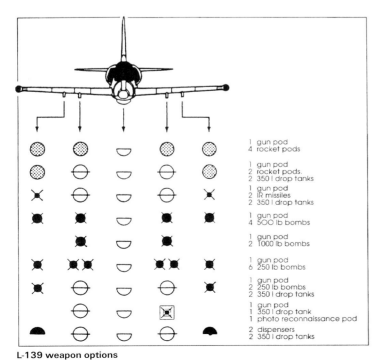

Aero Vodochody's L-139 demonstrator

Ex-military L-39 C registered in Estonia and based in the UK
(Paul Jackson/Jane's)

L-139 weapon options

L-39 and L-59 CUSTOMERS

Country	L-39 C	L-39 V	L-39 Z0	L-39 ZA	L-39 MS	L-59	Cum Total	First Delivery
Afghanistan	26						26	1977
Algeria	7			32			39	1987
Bangladesh				8			8	1995
Bulgaria				36			36	1986
Cuba	30						30	1982
Czechoslovakia	33	8[2]		30	5		76[1]	1972
Egypt						48	48	1993
Ethiopia	26						26	1983
Germany (East)			52[3]				52	1977
Ghana			2				2	1999
Iraq	22		59				81	1975
Libya			181[4]				181	1978
Lithuania				2[9]			2	1998
Nigeria				24[5]			24	1986
Romania				32			32	1981
Syria			55	44			99	1980
Thailand				40			40	1993
Tunisia						12	12	1995
USA	5						5[7]	1996
USSR	2,080						2,080[8]	1974
Vietnam	24						24	1980
Yemen	12						12	1999
Totals	**2,265**	**8**	**349**	**248**	**5**	**60**	**2,935**[6]	

Notes:
[1] Aircraft delivered to former Czechoslovakia divided between Czech and Slovak (nine Cs, two Vs and nine ZAs) air forces in 1992
[2] Two transferred to East Germany
[3] 20 transferred to Hungary
[4] 10 transferred to Egypt
[5] 1992 order for further 27 not delivered
[6] Excluding five prototypes
[7] Private ownership; several others acquired second-hand
[8] Final delivery 25 January 1991; subsequently redistributed to Azerbaijan, Kazakhstan, Kyrgizia, Lithuania, Russia and Ukraine
[9] From stock (built 1992)

Electrical system (27 V DC) powered by 7.5 kVA engine-driven generator; if primary generator fails, V 910 ram air turbine extends automatically into airstream and generates up to 3 kVA emergency power for essential services. 12 V 28 Ah SAM 28 lead-acid or Marathon Ni/Cd battery for standby power and APU starting. Two 800 VA static inverters (first for radio equipment, ice warning lights, engine vibration measurement and air conditioning, second for navigation and landing systems, IFF and air-to-air missiles) provide 115 V single-phase AC power at 400 Hz. Second circuit incorporates 500 VA rotary inverter and 40 VA static inverter for 36 V three-phase AC power, also at 400 Hz.

Saphir 5 APU and SV-25 turbine for engine starting. Air intakes and windscreen anti-iced by engine bleed air; anti-icing normally sensor-activated automatically, but manual standby system also provided. Six-bottle oxygen system for crew, pressure 147 bars (2,133 lb/sq in).

AVIONICS (L-39 ZA/ART): *Comms:* ARC-186 VHF (30 to 78 MHz/FM and 108 to 151.75 MHz/AM) and ARC-164 VHF (220 to 399.975 MHz) radios; KXP 756 transponder; IFF. Orbit crew intercom.

Flight: KNR 634A VOR; KTU 709 Tacan; RV-5M radar altimeter.

Instrumentation: Elbit WDNS (weapon delivery and navigation system) with HUD and video camera in front cockpit and monitor in rear cockpit.

ARMAMENT (L-39 ZA/ART): Underfuselage pod below front cockpit, housing a single 23 mm GSh-23 two-barrel gun; ammunition (maximum 150 rounds) housed in fuselage above gun pod. Gun/rocket/missile firing and weapon release controls in front cockpit only. Four underwing hardpoints, inboard pair each stressed for up to 500 kg (1,102 lb) and outer pair for up to 250 kg (551 lb) each; maximum underwing stores load 1,000 kg (2,205 lb). Non-jettisonable pylons, each comprising an MD3-57D stores rack. Typical underwing stores can include various combinations of bombs (two of up to 500 kg or four of up to 250 kg); four rocket launchers for 2.75 in FFAR or CRV7 rockets; AIM-9 air-to-air missiles (outboard stations only); two 150 or 350 litre drop tanks (see under Power Plant) (inboard stations only); or two training dispensers. See 1993-94 and earlier editions for L-39 ZA armament.

DIMENSIONS, EXTERNAL:
Wing span, incl tip tanks	9.46 m (31 ft 0½ in)
Wing chord (mean)	2.15 m (7 ft 0½ in)
Wing aspect ratio, incl tip tanks	4.8
Length overall	12.13 m (39 ft 9½ in)
Height overall	4.77 m (15 ft 7¾ in)
Tailplane span	4.40 m (14 ft 5 in)
Wheel track	2.44 m (8 ft 0 in)
Wheelbase	4.39 m (14 ft 4¾ in)

AREAS:
Wings, gross	18.80 m² (202.4 sq ft)
Ailerons (total)	1.69 m² (18.19 sq ft)
Trailing-edge flaps (total)	2.68 m² (28.89 sq ft)
Airbrakes (total)	0.50 m² (5.38 sq ft)
Fin	2.59 m² (27.88 sq ft)
Rudder, incl tab	0.91 m² (9.80 sq ft)
Tailplane	3.93 m² (42.30 sq ft)
Elevators (total, incl tabs)	1.14 m² (12.27 sq ft)

WEIGHTS AND LOADINGS (L-39 ZA):
Weight empty, equipped	3,565 kg (7,859 lb)
Fuel load: fuselage tanks	824 kg (1,816 lb)
wingtip tanks	156 kg (344 lb)
Max external stores load	1,290 kg (2,844 lb)
T-O weight clean	4,635 kg (10,218 lb)
Max T-O weight	5,600 kg (12,346 lb)
Max wing loading	297.9 kg/m² (61.01 lb/sq ft)
Max power loading	332 kg/kN (3.25 lb/lb st)

WEIGHTS AND LOADINGS (L-139):
Basic weight empty	3,460 kg (7,628 lb)
Fuel weights (fuselage and wingtip tanks)	as for L-39 ZA
Max external fuel (four 350 litre tanks)	1,088 kg (2,398 lb)
Max underwing stores	1,500 kg (3,307 lb)
Max ramp weight: clean	4,600 kg (10,141 lb)
with external stores	6,000 kg (13,227 lb)
T-O weight clean	4,550 kg (10,031 lb)
Max landing weight	4,800 kg (10,582 lb)
Wing loading at clean T-O weight	241.9 kg/m² (49.55 lb/sq ft)
Power loading at clean T-O weight	251 kg/kN (2.46 lb/lb st)

PERFORMANCE (L-39 ZA at max T-O weight, except where indicated):
Max limiting Mach No.	0.80
Max level speed: at S/L	329 kt (610 km/h; 379 mph)
at 5,000 m (16,400 ft)	340 kt (630 km/h; 391 mph)
Stalling speed	103 kt (190 km/h; 118 mph)
Max rate of climb at S/L	810 m (2,657 ft)/min
Time to 5,000 m (16,400 ft)	10 min
Service ceiling	7,500 m (24,600 ft)
T-O run (concrete)	970 m (3,185 ft)
Landing run (concrete)	800 m (2,625 ft)
g limits:	
operational: at 4,200 kg (9,259 lb) AUW	+8/−4
at 5,500 kg (12,125 lb) AUW	+5.2/−2.6
ultimate, at 4,200 kg (9,259 lb) AUW	+12

PERFORMANCE (L-139 at clean T-O weight, except where indicated):
Never-exceed speed (V_{NE})	491 kt (910 km/h; 565 mph)
Max level speed at 6,100 m (20,000 ft)	410 kt (760 km/h; 472 mph)
Touchdown speed	99 kt (183 km/h; 114 mph)
Stalling speed, flaps down	87 kt (160 km/h; 100 mph)
Max rate of climb at S/L	1,344 m (4,409 ft)/min
Service ceiling	12,000 m (39,360 ft)
T-O run	540 m (1,775 ft)
Landing run	560 m (1,840 ft)
Radius of action with gun, two drop tanks and two Mk 83 (1,000 lb) bombs:	
lo-lo-lo	291 n miles (539 km; 335 miles)
hi-lo-hi	401 n miles (743 km; 461 miles)
Max range on internal fuel, 10% reserves	730 n miles (1,352 km; 840 miles)
Max endurance, conditions as above	3 h 19 min
g limits	+8/−4

UPDATED

Front cockpit of the L-139

Rear cockpit of the L-139

AERO L-39 MS and L-59
Tunisian Air Force name: Fennec

TYPE: Advanced jet trainer/light attack jet.

PROGRAMME: Early development under designation L-39 MS; versions of L-39 with more powerful engine and improved avionics. First flight in definitive configuration (X-22 prototype OK-184) 30 September 1986; two more prototypes (X-24, X-25) flown 26 June and 6 October 1987; first flight of production L-59, 1 October 1989; first flight of first L-59 E made on 22 April 1992; deliveries of L-59 E began 29 January 1993 (two aircraft) and were completed in early 1994. Digital engine control system for Egyptian Air Force L-59s was under development by AlliedSignal (now Honeywell) in 1997.

CURRENT VERSIONS: **L-39 MS:** First version of L-59; five delivered to Czechoslovak Air Force (three in 1991, two in 1992); sixth (0001) retained for trials.

L-59 E: Production two-seat version for Egyptian Air Force.

Detailed description applies to L-59 E, except where indicated.

L-59 T: Version for Tunisian Air Force; deliveries completed in early 1996.

L 159: Light ground attack/lead-in fighter trainer version of L-59; described separately.

AERO—AIRCRAFT: CZECH REPUBLIC

CUSTOMERS: See table. Czechoslovak Air Force (five L-39 MS, of which three retained by Czech Republic and two transferred to Slovak Air Force); Egyptian Air Force (48 L-59 E): Tunisian Air Force (12 L-59 T). Total of 65 L-39 MS/L-59 built by January 1998. No further orders by October 2000.

COSTS: Egyptian order reportedly worth US$204 million.

DESIGN FEATURES: Main changes are a reinforced fuselage, new and more powerful engine, upgraded avionics, more pointed nose, powered controls and enlarged tip tanks.

FLYING CONTROLS: Generally as for L-39, except that ailerons and elevators have Czech-designed irreversible power actuators and no tabs.

STRUCTURE: Generally as for L-39, except for reinforced wings and fuselage.

LANDING GEAR: Czech-designed oleo-pneumatic shock-absorption; K36 mainwheels (610×215) and K37 nosewheel (460×180). Mainwheel tyre pressures 6.00 bars (87 lb/sq in) on clean aircraft, 8.00 bars (116 lb/sq in) on combat-equipped version; corresponding nosewheel tyre pressures are 3.50 bars (51 lb/sq in) and 4.50 bars (65 lb/sq in). Six-piston, air-cooled hydraulic disc brakes on mainwheels, with electronic anti-skid units.

POWER PLANT: One 21.57 kN (4,850 lb st) PS/ZMK DV-2 turbofan. Internal fuel in fuselage tanks (total 1,077 litres; 284.5 US gallons; 237 Imp gallons) and two 230 litre (60.8 US gallon; 50.6 Imp gallon) permanent wingtip tanks. Total usable internal fuel 1,537 litres (406 US gallons; 338 Imp gallons). Provision for four underwing 150 or 350 litre (39.6 or 92.5 US gallon; 33.0 or 77.0 Imp gallon) drop tanks. Fuel and oil systems permit up to 30 seconds of inverted flight.

ACCOMMODATION: Crew of two in tandem on Czech VS-2A zero/zero ejection seats. One-piece canopy, hinged at rear and opening upward hydraulically.

SYSTEMS: Cockpits pressurised (maximum overpressure 0.30 bar; 4.4 lb/sq in) and air conditioned, using engine bleed air (25 litres/min; 0.88 cu ft/min) and cooling unit. Automatic temperature control from 15 to 30°C. Hydraulic system comprises first and second subsystems each with engine-driven variable flow pump with operating pressure of 150 bars (2,175 lb/sq in), maximum flow rate 25 litres (6.6 US gallons; 5.5 Imp gallons)/min. Emergency hydraulic pump for second subsystem driven by APU. Main (9 kW) and standby (6 kW) generators for electrical power, plus 25 Ah Ni/Cd battery. Gaseous oxygen system for crew. Saphir 5M APU for engine starting and drive of standby hydraulic pump and generator.

AVIONICS: *Comms:* LPR 80 VHF/UHF radio with intercom; LUN 3524 standby radio; Honeywell KXP 756 transponder.
Flight: Honeywell KNS 660 flight management system includes KNR 634 VOR, KTU 709 Tacan, KDF 806 ADF, KRA 405 radar altimeter, KLN 670 GPS, KAH 460 AHRS and KAD 480 air data system.
Instrumentation: EFS 40 EFIS; Flight Visions FV-2000 HUD and mission computer, with video camera in front cockpit and monitor in rear cockpit.

ARMAMENT: Single twin-barrel 23 mm GSh-23 gun in underfuselage pod below front cockpit; ammunition (150 rounds) housed in fuselage. Four underwing hardpoints, inner ones each with 500 kg (1,102 lb) capacity, outer ones each 250 kg (551 lb) capacity. Suitable for underwing stores of former Soviet types, including bombs of up to 500 kg size and UB-16-57M (57 mm) rocket launchers, or Western equivalents such as Expal BR-250 and Mk 82 bombs, FFAR and CRV7 rockets and Sidewinder AAMs.

DIMENSIONS, EXTERNAL: As for L-39 except:
Wing span, incl tip tanks   9.54 m (31 ft 3½ in)

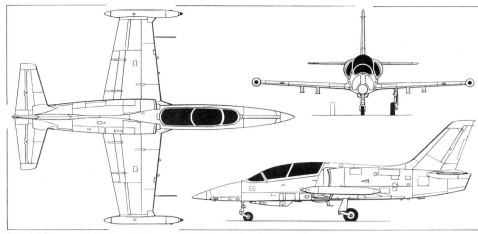

Aero L-59 two-seat basic and advanced jet trainer *(Dennis Punnett/Jane's)*

Wing chord: at root   2.80 m (9 ft 2¼ in)
at tip   1.40 m (4 ft 7 in)
Length overall   12.20 m (40 ft 0¼ in)
AREAS: As for L-39 except:
Tailplane   4.15 m² (44.64 sq ft)
WEIGHTS AND LOADINGS:
Weight empty:
trainer, incl GSh-23 gun   4,030 kg (8,885 lb)
Max fuel weight:
internal, incl wingtip tanks   1,200 kg (2,645 lb)
external (four 350 litre drop tanks)   1,088 kg (2,398 lb)
Max T-O weight: trainer, clean   5,560 kg (12,257 lb)
with external stores*   7,000 kg (15,432 lb)
Max landing weight*   6,000 kg (13,228 lb)
Max wing loading:
clean   286.7 kg/m² (58.72 lb/sq ft)
at 7,000 kg (15,432 lb) max T-O weight
   372.3 kg/m² (76.26 lb/sq ft)
Max power loading:
clean   250 kg/kN (2.45 lb/lb st)
at 7,000 kg (15,432 lb) max T-O weight
   324 kg/kN (3.18 lb/lb st)
Data for concrete runway. On unpaved surface, MTOW 6,000 kg (13,228 lb), MLW 5,700 kg (12,566 lb)
PERFORMANCE (A: at max trainer clean T-O weight, B: at 6,613 kg; 14,579 lb T-O weight with gun pod, two rocket pods and two 350 litre drop tanks):
Max limiting Mach No.   0.82
Never-exceed speed (V_{NE}) at S/L:
A   496 kt (920 km/h; 571 mph)
Max level speed at 5,000 m (16,400 ft):
A   472 kt (875 km/h; 544 mph)
B   452 kt (837 km/h; 520 mph)
Stalling speed, flaps up: A   117 kt (216 km/h; 135 mph)
B   129 kt (238 km/h; 148 mph)
Stalling speed, flaps down:
A   99 kt (182 km/h; 114 mph)
B   110 kt (203 km/h; 127 mph)
Max rate of climb at S/L: A   1,500 m (4,921 ft)/min
B   840 m (2,756 ft)/min
Time to 5,000 m (16,400 ft): A   3 min 30 s
B   5 min
Service ceiling: A   11,730 m (38,480 ft)
B   9,860 m (32,340 ft)
T-O distance (concrete runway): A   640 m (2,100 ft)
B   1,080 m (3,545 ft)
T-O distance (unpaved runway): A   860 m (2,825 ft)

Landing distance at 5,500 kg (12,125 lb) weight:
A   720 m (2,365 ft)
B   765 m (2,510 ft)
Range at 5,000 m (16,400 ft), 5% reserves:
A   653 n miles (1,210 km; 752 miles)
B   796 n miles (1,475 km; 917 miles)
Max load factor in turn at S/L: A   3.42 *g*
B   2.25 *g*
g limits:
up to 5,560 kg (12,257 lb)   +8/−4
above 5,560 kg (12,257 lb)   +6/−3

UPDATED

AERO L 159

TYPE: Light fighter/advanced jet trainer.
PROGRAMME: Design started late 1992; F124 turbofan selected as power plant mid-1994; Rockwell North American (now Boeing) awarded avionics contract late 1994. Development approved by Czech government 7 April 1995, followed by production commitment later the same month and 72-aircraft production contract 4 July 1997. Two prototypes (plus two airframes for static and dynamic test) built. First (two-seat) prototype/engine demonstrator (5831) rolled out 12 June and made first flight 2 August 1997.

Second (single-seat) prototype (5832), in Czech Air Force configuration (first flight 18 August 1998) named ALCA (Advanced Light Combat Aircraft); is first with full avionics and is avionics/weapons trials aircraft. Official first flight of 5832 was 20 August; public debut at Hradec Králové 29 August 1998. Gun firing tests in Czech Republic followed by air-launched weapon trials in Norway in April and May 1999; nav/attack system trials August/September 1999. The two prototypes had flown 1,120 hours (506 by single-seater, 614 by two-seater) by 25 August 2000.

First flight of production L 159A (6001) 20 October 1999. Deliveries to CzAF due to have begun late 1999 (five aircraft, followed by 16 in 2000, 26 in 2001 and final 25 in 2002); however, first two (6001 and 6002) not handed over (to 4 Tactical Fighter Base at Caslav) until 10 April 2000, and deliveries of remainder may be stretched out to offset demands on defence budget. Contract under government criticism in late 2000 due to deteriorating exchange rate between Czech crown and US dollar, previous government having priced contract in latter currency. One proposal being considered is to offer early export availability before completion of CzAF deliveries.

CURRENT VERSIONS: **L 159A ALCA:** Single-seater.
Detailed description applies to L 159A except where indicated.
L 159B: Tandem-seat prototype. Expected to be alternative production version for Czech Air Force (part of 72-aircraft order); also available for export as lead-in fighter trainer.

CUSTOMERS: Czech Air Force order for 72 L 159As valued at approximately US$1 billion. Production lots 1 and 2 comprise five and 10 aircraft respectively. Primary roles will be close air support, counter-insurgency, anti-ship missions, tactical reconnaissance, point air defence, air defence against slow/low-flying targets and border patrol. Prospect of exports to other East European and Baltic countries, including Polish Navy, to whom a Maverick-armed version was offered in late 2000.

DESIGN FEATURES: Configuration generally as L-39/L-59 except for larger, redesigned nose to accommodate radar; longer fuselage; single-seat armoured cockpit; HOTAS controls; Western avionics; 297 kg (655 lb) additional fuel tank in lieu of second seat; permanent wingtip fuel tanks; seven external stores stations.

Wing section NACA 64A-012; sweepback 6° 30′ on leading-edges; dihedral 2° 30′; incidence 2°; twist 0°.

FLYING CONTROLS: Conventional primary surfaces; ailerons and elevators actuated hydraulically, rudder mechanically with electrically operated trim tab. Artificial feel standard.

Aero L-59 demonstrator *(Paul Jackson/Jane's)*

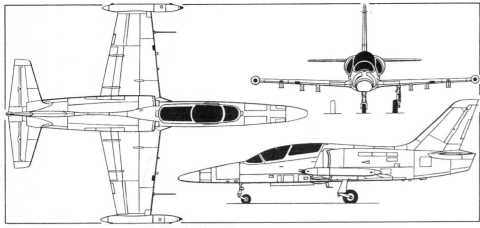

Aero L 159 light attack aircraft *(Paul Jackson/Jane's)*

Instrument panel of the L 159 *(Paul Jackson/Jane's)*
2000/0062963

Twin underfuselage airbrakes and double-slotted trailing-edge flaps, also actuated hydraulically. Maximum flap deflection 44°.

STRUCTURE: Based on that of L-59, but substantially modified.

LANDING GEAR: Hydraulically retractable tricycle type, with single wheel and oleo-pneumatic shock-absorber on each unit. Main units retract inward into wings, nosewheel forward. Moravan K36 wheels and 610×215 tyres (pressure 6.00 to 8.00 bars; 87 to 116 lb/sq in) on main units; nose unit has Moravan K37 wheel and 460×180 tyre (pressure 3.52 to 4.48 bars; 51 to 65 lb/sq in). Moravan K52 air-cooled hydraulic brakes with anti-skid system. Nose unit hydraulically steerable ±59°. Minimum ground turning radius 5.70 m (18 ft 8½ in).

POWER PLANT: One 28.2 kN (6,330 lb st) Honeywell/ITEC F124-GA-100 turbofan with dual-redundant FADEC. Internal fuel of L 159A (total capacity 1,980 litres; 523 US gallons; 436 Imp gallons) is contained in six fuselage and two wingtip tanks and protected by OBIGGS (onboard inert gas generation system); L 159B omits one 350 litre (92.5 US gallon; 77.0 Imp gallon) fuselage tank, internal capacity thus 1,630 litres (431 US gallons; 359 Imp gallons) and does not have OBIGGS. Up to four external tanks can be carried underwing: two 525 litre (139 US gallon; 115.5 Imp gallon) tanks on inboard stations, two 350 litre (92.5 US gallon; 77 Imp gallon) tanks on inboard or centre stations. Pressure fuelling point in port side of lower front fuselage; gravity points in upper fuselage aft of canopy and in each wingtip tank. In-flight refuelling capability under development for L 159B.

ACCOMMODATION: Single VS-2B ejection seat in armoured cockpit. Canopy opens sideways manually in L 159A, upwards hydraulically in L 159B.

SYSTEMS: PBS simple cycle environmental control system, using engine bleed air; Honeywell cockpit pressure control system (maximum differential 0.28 bar; 4.1 lb/sq in). Dual hydraulic systems, each at pressure of 210 bars (3,045 lb/sq in), actuating ailerons, elevators, flaps, airbrakes, landing gear, wheel brakes and nosewheel steering. System maximum flow rate 35 litres (9.25 US gallons; 7.7 Imp gallons)/min. Hydraulic tank pressurised by nitrogen at 2.50 bars (36.3 lb/sq in). Pneumatic system (1.80 bars; 26.1 lb/sq in) for canopy sealing.

Hamilton Sundstrand 40 kVA integrated drive generator provides 200/115 V three-phase main AC power at 400 Hz; 28 V DC power from 6 kW VUES LUN-2134 back-up generator and Varta/Aero 25 Ah battery. Intertechnique OBOGS oxygen system. Windscreen and air intakes de-iced by engine bleed air. PBS Safir 5F APU, operable up to 9,000 m (29,520 ft) altitude, for engine starting and emergency hydraulic or electrical power.

AVIONICS: Integration of complete avionics suite by Boeing.
 Comms: Two Rockwell Collins ARC-210 UHF/VHF radios; Honeywell AN/APX-100 IFF.
 Radar: FIAR Grifo L multimode pulse Doppler radar.
 Flight: Honeywell KNR 634A VOR, KDM 706A DME and air data computer; Honeywell H-746G ring laser gyro INS with embedded GPS.
 Instrumentation: Flight Visions FV-3000 HUD; dual Honeywell 102 × 102 mm (4 × 4 in) liquid crystal colour head-down MFDs; Astronautics mechanical ADI and HSI. Optional helmet-mounted display/sight and NVG compatible lighting.
 Mission: Dual Flight Visions 6400-100 mission computers with HUD symbology generator; provision for reconnaissance and jamming pod. Hamilton Sundstrand weapon delivery system. Dynamic Control Corporation stores management system.
 Self-defence: BAE Systems Sky Guardian 200 RWR; Vinten Vicon 78 Series 455 countermeasures dispenser.

ARMAMENT: No built-in armament. Seven external stores stations: one under fuselage and three under each wing. Fuselage station stressed for 300 kg (661 lb) load; inboard, centre and outboard wing pylons stressed for 550, 320 and 150 kg (1,213, 705 and 331 lb) each respectively. Primary armament comprises AIM-9 Sidewinder AAMs, AGM-65 Maverick ASMs, and SUU-20 or CRV7 unguided rocket pods. Has also been trialled with Brimstone missile and TIALD targeting pod. For typical weapon loads, see accompanying illustration.

DIMENSIONS, EXTERNAL:
Wing span: excl tip tanks	8.70 m (28 ft 6½ in)
Incl tip tanks	9.54 m (31 ft 3½ in)
Wing chord: at root	2.80 m (9 ft 2¼ in)
at tip	1.40 m (4 ft 7 in)
Wing aspect ratio, incl tip tanks	4.8
Length: overall	12.72 m (41 ft 8¾ in)
fuselage	12.59 m (41 ft 3¾ in)
Height overall	4.77 m (15 ft 7¾ in)
Tailplane span	4.70 m (15 ft 5 in)
Wheel track	2.42 m (7 ft 11¼ in)
Wheelbase	4.56 m (14 ft 11½ in)

AREAS:
Wings, gross	18.80 m² (202.4 sq ft)
Ailerons (total)	1.69 m² (18.19 sq ft)
Trailing-edge flaps (total)	2.68 m² (28.85 sq ft)
Fin	2.59 m² (27.88 sq ft)
Rudder, incl tab	0.91 m² (9.80 sq ft)
Tailplane	3.90 m² (41.98 sq ft)
Elevators (total)	1.40 m² (15.07 sq ft)

WEIGHTS AND LOADINGS:
Weight empty, equipped	4,320 kg (9,524 lb)
Max fuel weight:	
internal, L 159A	1,596 kg (3,519 lb)
internal, L 159B	1,273 kg (2,806 lb)
external, both (two 500 litre and two 350 litre)	1,377 kg (3,036 lb)
Max external stores load	2,340 kg (5,159 lb)
Max T-O and ramp weight	8,000 kg (17,637 lb)
Max landing weight	6,000 kg (13,228 lb)
Max zero-fuel weight	6,810 kg (15,013 lb)
Max wing loading	425.5 kg/m² (87.15 lb/sq ft)
Power loading, L 159A clean with 50% internal fuel	284 kg/kN (2.79 lb/lb st)

PERFORMANCE (C: clean, S: with 1,470 kg; 3,241 lb of external stores, except where indicated):
Never-exceed speed (V$_{NE}$)	518 kt (960 km/h; 596 mph)
Max level speed at S/L: C	505 kt (936 km/h; 581 mph)
S	468 kt (867 km/h; 539 mph)
*Stalling speed, engine idling, flaps and gear down	100 kt (185 km/h; 115 mph)
Max rate of climb at S/L: C	3,726 m (12,220 ft)/min
S	2,280 m (7,480 ft)/min
Service ceiling	13,200 m (43,300 ft)
T-O run: C	470 m (1,545 ft)
S	853 m (2,800 ft)

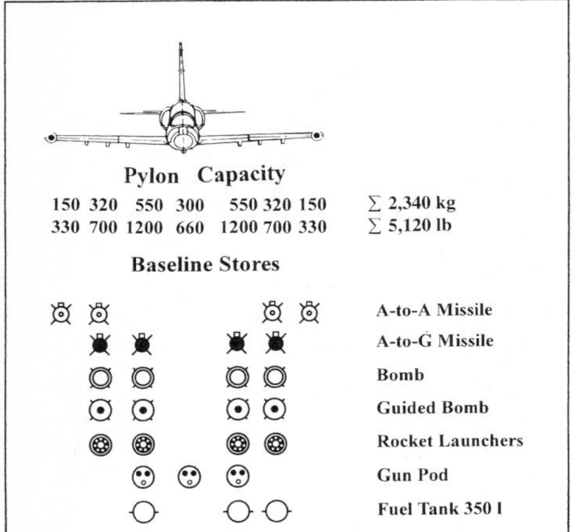

Pylon Capacity

| 150 | 320 | 550 | 300 | 550 | 320 | 150 | Σ 2,340 kg |
| 330 | 700 | 1200 | 660 | 1200 | 700 | 330 | Σ 5,120 lb |

Baseline Stores

					Symbol
⊗	⊗			⊗	⊗	A-to-A Missile
■	■			■	■	A-to-G Missile
○	○		○	○		Bomb
◉	◉		◉	◉		Guided Bomb
✳	✳		✳	✳		Rocket Launchers
	⊙	⊙	⊙			Gun Pod
	○		○			Fuel Tank 350 l
		○				Fuel Tank 525 l
	●		●			Cluster Bomb
◐	◐			◐	◐	Training Bomb/Rocket Pod

Growth Stores

⊗				⊗	A-to-A Medium Range Missile
	⊙				EO Pod
	ECM				ECM Pod
	⊙				Recce Pod

L 159 weapon options 2000/0085020

First prototype after repainting in tactical camouflage *(Paul Jackson/Jane's)*

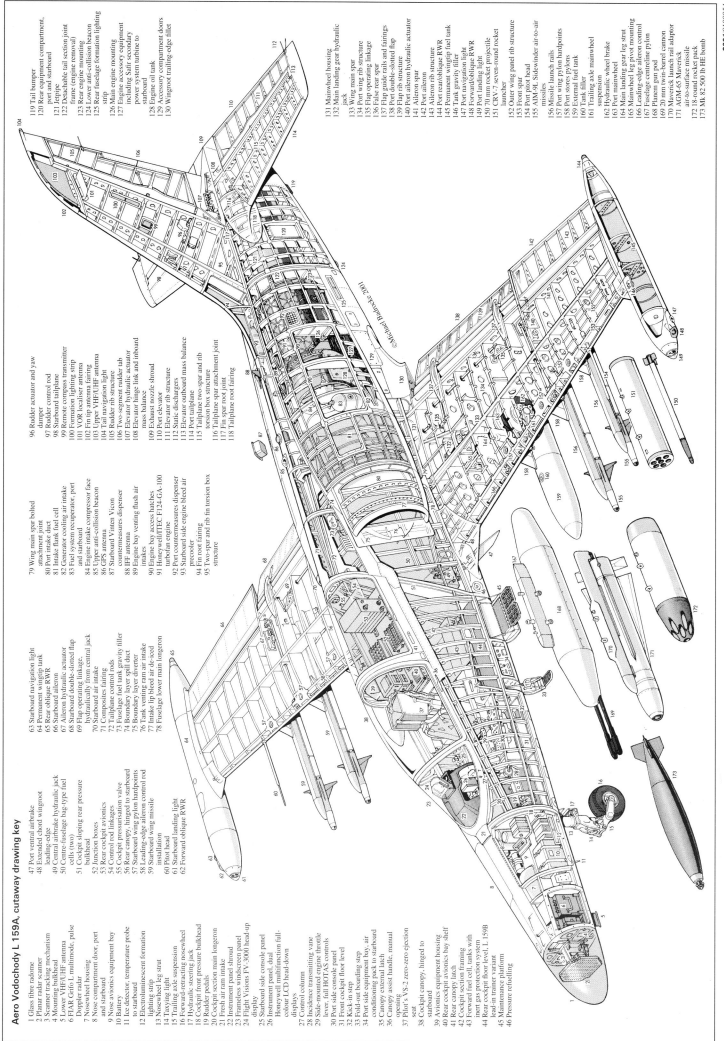

T-O to 15 m (50 ft): C	720 m (2,365 ft)
S	1,276 m (4,185 ft)
Landing from 15 m (50 ft): C	992 m (3,255 ft)
S	816 m (2,675 ft)
Landing run: C	628 m (2,060 ft)
S	816 m (2,675 ft)

Typical mission radius (A: L 159A, B: L 159B):
  with gun pod, two Mk 82 bombs, two AIM-9 missiles
    and two 525 litre drop tanks:
    A, lo-lo-lo   270 n miles (500 km; 310 miles)
    A, hi-lo-hi   380 n miles (705 km; 438 miles)
    B, lo-lo-lo   227 n miles (420 km; 261 miles)
    B, hi-lo-hi   305 n miles (565 km; 351 miles)
  with gun pod, two Mk 82 bombs and two 525 litre
    drop tanks:
    A, lo-lo-lo   299 n miles (555 km; 345 miles)
    A, hi-lo-hi   421 n miles (780 km; 484 miles)
    B, lo-lo-lo   251 n miles (465 km; 289 miles)
    B, hi-lo-hi   335 n miles (620 km; 385 miles)
  with two 525 litre drop tanks only:
    A, lo-lo-lo   318 n miles (590 km; 366 miles)
    A, hi-lo-hi   521 n miles (965 km; 599 miles)
    B, lo-lo-lo   264 n miles (490 km; 304 miles)
    B, hi-lo-hi   432 n miles (800 km; 497 miles)
Range, reserves of 10% of internal fuel:
  L 159A, max internal fuel only
                   848 n miles (1,570 km: 975 miles)
  L 159B, max internal fuel only
                   659 n miles (1,220 km; 758 miles)
  L 159A, max internal and external fuel
                1,366 n miles (2,530 km; 1,572 miles)
  L 159B, max internal and external fuel
                1,242 n miles (2,300 km; 1,429 miles)
  g limits   +8/−4
*At 5,500 kg (12,125 lb) AUW

UPDATED

Ae 270 IBIS
This aircraft is described under the Ibis Aerospace heading in the International section.

First production L 159A in Czech Air Force markings

Second (first single-seat) L 159 prototype 5832

ATEC
ATEC v.o.s.
Opolanská 171, CZ-289 07 Libice nad Cidlinou
Tel: (+420 324) 773 71
Fax: (+420 324) 30 02
e-mail: sales@atecvos.cz
Web: http://www.atecvos.cz
DIRECTOR: Jan Franěk

UPDATED

ATEC ZEPHYR
TYPE: Side-by-side ultralight.
PROGRAMME: Prototype (OK-CUA 01) built 1997.
CURRENT VERSIONS: **Zephyr Standard:** Original version, with 10.60 m (34 ft 9¼ in) wing span and 59.6 kW (79.9 hp) Rotax 912 UL engine. Other details in 2000-01 *Jane's*.
  Zephyr 2000: Shorter wing span and Rotax 912 ULS engine; otherwise generally as Zephyr Standard.

Description below applies to Zephyr 2000.
  Zephyr Solo: Single-seat version. Prototype (OK-FUH 01) made initial flight 29 October 2000.
CUSTOMERS: Total 58 built and sold by December 2000. Exported to Canada, Germany and Netherlands.
COSTS: Zephyr Standard: DM86,000; Zephyr 2000: DM96,000 (2000).
DESIGN FEATURES: Low-wing monoplane with upturned wingtips; T tail.
  Laminar flow UA-2 wing aerofoil section.

ATEC Zephyr 2000 two-seat ultralight

FLYING CONTROLS: Conventional and manual; one-piece elevator with central spring tab. Plain flaps (settings of 15, 30 and 50°). Dual controls.
STRUCTURE: Fuselage has welded steel tube central framework and wooden bulkheads. Composites fuselage shell, wingtips, wing and tailplane leading-edges, rudder, and nosewheel and mainwheel legs. Wings, ailerons, flaps and elevator fabric-covered.
LANDING GEAR: Fixed, tricycle; cantilever mainwheel legs. Castoring nosewheel. Tyres sizes (400×100 (main), 300×100 (nose) Hydraulic brakes. Speed fairings on all three wheels.
POWER PLANT: One 59.6 kW (79.9 hp) Rotax 912 UL or 73.5 kW (98.6 hp) Rotax 912 ULS flat-four engine; FITI two- or three-blade ground-adjustable propeller. Fuel capacity 60 litres (15.9 US gallons; 13.2 Imp gallons) standard, 80 litres (21.1 US gallons; 17.6 Imp gallons) optional.

DIMENSIONS, EXTERNAL:
Wing span 9.60 m (31 ft 6 in)
Length overall 6.20 m (20 ft 4 in)
Height overall 1.90 m (6 ft 2¾ in)
AREAS:
Wings, gross 10.10 m² (108.7 sq ft)
WEIGHTS AND LOADINGS:
Weight empty 280 kg (617 lb)
Max T-O weight 450 kg (992 lb)
PERFORMANCE (912 ULS engine):
Never-exceed speed (V_NE) 148 kt (275 km/h; 170 mph)
Max level speed 135 kt (250 km/h; 155 mph)
Econ cruising speed 97 kt (180 km/h; 112 mph)
Stalling speed, flaps down 36 kt (65 km/h; 41 mph)
Max rate of climb at S/L 390 m (1,280 ft)/min
T-O run 120 m (395 ft)
Landing run 200 m (660 ft)
Range: standard fuel 432 n miles (800 km; 497 miles)
   optional fuel 594 n miles (1,100 km; 683 miles)

UPDATED

CZAW

CZECH AIRCRAFT WORKS s.r.o.

Luční 1824, CZ-686 02 Staré Město
Tel: (+420 632) 54 34 56
Fax: (+420 632) 54 36 92
e-mail: aircraft@mbox.vol.cz
Web: http://www.zenithair.com/dlr/czaw
CEO: Chip W Erwin

In 1997 Zenair Ltd (see Canadian section) formed a joint venture with CZAW for manufacture in the Czech Republic of its STOL CH 701 and Zodiac CH 601 series of light aircraft (described under the AMD and Zenith headings in the US section). The STOL CH 801 and CH 2000 are now also available from CZAW.

CZAW offers the CH 601, 701 and 801 in three different stages of completion: as a factory-jigged, fast-build kit; and in full flyaway form with Rotax 912 engine, avionics, instruments and two-tone paint finish. Development was completed in late 1998 of a modified CH 701 for ULV agricultural applications. A Zodiac floatplane, flown by CZAW's CEO, won the 1999 Schneider Cup seaplane race at Desenzano del Garda, Italy, on 19 September 1999.

CZAW produces a complete line of wheeled and non-wheeled aircraft floats, with displacements ranging from 250 to 1,000 kg (551 to 2,205 lb), suitable for such aircraft as the GlaStar, Kitfox, Zenair and Van's types. Stoddard-Hamilton 'jump-start' wings for the GlaStar kitplane are also being manufactured.

In addition, CZAW is preparing to manufacture the Luscombe 8F under an agreement with Renaissance Aircraft LLC and Zenair Ltd. The first preproduction airframe was being assembled at the Zenair factory in Canada, with workers from CZAW, in November 1998, but had not appeared by early 2000.

UPDATED

CZAW (ZENITH) CH 701-AG STOL

TYPE: Agricultural sprayer.
PROGRAMME: Developed by CZAW following cessation of Zenith production of a dedicated agricultural version.
COSTS: US$50,000 (1998), standard; US$48,500 to early purchasers.

CZAW-built Zodiac CH 601 OK-DUR 99, winner of the 1999 Schneider Cup *(ULM Europe)* 2001/0089711

DESIGN FEATURES: STOL performance, plus ULV crop treatment system optimised for low cost and precise application.
Differences from basic CH 701 (see US section) as follows:
EQUIPMENT: Spray Miser ULV crop treatment system, comprising 113.6 litre (30 US gallon; 25 Imp gallon) belly tank and 12 wind-driven rotary atomisers with controlled droplet application. Hydraulic nozzles produce droplet sizes varying from 1 to 500 microns in size. Electrically powered pump (12 V DC), with remotely controllable on/off switch, can deliver liquid at up to 30.3 litres (8 US gallons; 6.7 Imp gallons)/min. Tank has emergency dump facility.

WEIGHTS AND LOADINGS:
Weight empty, equipped 270 kg (595 lb)
Max T-O weight 450 kg (992 lb)
PERFORMANCE:
Application speed range
   39-65 kt (72-121 km/h; 45-75 mph)
Max rate of climb at S/L 305 m (1,000 ft)/min
Swath width 9.1-30.5 m (30-100 ft)
Range 217 n miles (402 km; 250 miles)

VERIFIED

Czech Aircraft Works CH 701-AG STOL two-seat cropsprayer 1999/0044633

CH 701-AG STOL with CZAW amphibious floats 1999/0044634

DELTA

DELTA SYSTEM-AIR AS

Bratři Štefanů 101, CZ-500 03 Hradec Králové
Tel/Fax: (+420 49) 541 14 06
e-mail: office@dsa.cz
Web: http://www.dsa.cz
MANAGER, FLIGHT OPERATIONS: Radek Dlouhy

UPDATED

DELTA PEGASS

There has been no recent news of the Pegass ultralight, last described in the 1999-2000 *Jane's*, although at least one is available for renting from Delta's diverse fleet of aircraft.

VERIFIED

EV-AT

EVEKTOR-AEROTECHNIK AS
Kunovice-Letiště, CZ-686 04 Kunovice
Tel: (+420 632) 53 73 29
Fax: (+420 632) 53 79 10
e-mail: marketing@evektor.cz
Web: http://www.evektor.cz
CHAIRMAN: Jaroslav Ružička
CEO: Marián Mečiar
COMMERCIAL DIRECTOR: Milan Bříštěla
MARKETING MANAGER: Jan Fridrich

The Evektor aircraft design office (workforce 150 in 1999) was formed in 1991, a major part of its initial work consisting of contribution to the designs of the L 159 ALCA and Ae 270 Ibis under contract to Aero Vodochody. On 2 February 1996 it acquired a 100 per cent shareholding in Aerotechnik CZ, and in 1996-97 the two companies (now a single joint-stock company) developed and delivered preseries and test front fuselage components of the L 159 to Aero.

Aerotechnik was formed in 1970, originally as an enterprise of the former Czechoslovak defence organisation Svazarm. Its first aircraft programmes involved single- and two-seat rotorcraft. Two facilities, at Kunovice and Moravská Třebová, have a combined floor space of 7,340 m² (79,000 sq ft). The combined Evektor-Aerotechnik workforce totalled 350 in mid-2000.

In recent years Aerotechnik produced kits for the Pottier P 220 S Koala all-metal lightplane, powered by its own Mikron III engine. This design was further developed via the P 220 UL into the EV-97 Eurostar (which see) with a Rotax 912 UL engine, and became the first aeroplane to bear the Evektor name. A VLA version of the Eurostar is now under development.

Evektor-Aerotechnik assisted in early stages of the Wolfsberg Raven 257 programme described in the Letov Air entry, including assembly of the prototype. A separate joint venture was formed in October 1999 with GAIC of China (which see), leading to establishment in June 2000 of Guizhou Evektor Aircraft Corporation for local production, sales and support of EV-AT aircraft in that country. The first EV-97 CKD kit for Chinese assembly was delivered in August 2000. This arrangement supersedes that previously agreed for Guizhou to become a licensed production centre for the Raven 257. On 26 July 2000, Evektor-Aerotechnik signed an agreement with Seabird Aviation Australia to manufacture and market the latter's Seeker SB7L-360 light observation aircraft (which see).

Other activities include development and production of Mikron III AE and III B piston engines; hot-air balloons and other commercial inflatables; spares and components for light and ultralight aircraft; and the overhaul and re-engining of L-13SV Vivat and L-60 Brigadýr aircraft. The P 220 S Koala (1999-2000 *Jane's*) is no longer produced.

UPDATED

First EV-AT EV-97 Eurostar built in the UK *(Paul Jackson/Jane's)*

EV-AT EV-97 EUROSTAR

TYPE: Side-by-side ultralight/kitbuilt.
PROGRAMME: Developed as P 220 UL from Pottier P 220 S Koala; extensive redesign started in September 1996; prototype (OK-CUR 97) made first flight May 1997; first customer delivery October 1997. Certified as ultralight in Czech Republic, Germany and Slovak Republic. First UK-built example (G-NIDG, by distributor Cosmik Aviation) flown in April 2000; local assembly in China by Guizhou Evektor Aircraft Corporation (GEAC, which see); first flight by Chinese-assembled EV-97, 24 October 2000.
CURRENT VERSIONS: **Eurostar 2000**: Introduced late 1999. Increased fuselage width (4 cm; 1½ in); modified wingtips and increased span; larger ailerons; steerable nosewheel; flush, oval nose air intake. Available factory-built or as kit.
CUSTOMERS: More than 100 (all versions) ordered, including 36 Model 2000, by July 2000; production rate then three to four per month; customers in Czech Republic, Germany, Italy, Slovak Republic, UK and USA.
COSTS: DM104,226 flyaway, including 16 per cent tax (1999); Chinese version approximately RMB700,000 Yuan (2000).
Following description applies to Model 2000.
DESIGN FEATURES: Low-wing monoplane; modified NACA 4415 aerofoil section; dihedral 3°. Wing folding optional.
FLYING CONTROLS: Conventional and manual. Four-position (0/15/30/50°) split flaps. Elevator, starboard aileron and rudder trim tabs.
STRUCTURE: Mostly of riveted metal; wingtips and mainwheel legs in GFRP.
LANDING GEAR: Fixed tricycle with hydraulic mainwheel disc brakes and steerable nosewheel. Wheel sizes 350×140 (main), 300×100 (nose). Optional wheel fairings or mudguards.
POWER PLANT: One 59.6 kW (79.9 hp) Rotax 912 UL flat-four engine standard, with V 230 C two-blade fixed-pitch wooden propeller. Fuel capacity 50 litres (13.2 US gallons; 11.0 Imp gallons) standard, optionally increasable to 70.0 or 100 litres (18.5 or 26.4 US gallons; 15.4 or 22.0 Imp gallons).
AVIONICS: Honeywell KY 97A com radio and VFR instrumentation standard; ELT and other avionics at customer's option.
EQUIPMENT: Optional ballistic parachute, wing-folding mechanism and aero-tow hook.

DIMENSIONS, EXTERNAL:
Wing span	8.10 m (26 ft 7 in)
Length overall	5.98 m (19 ft 7½ in)
Height overall	2.34 m (7 ft 8¼ in)

DIMENSIONS, INTERNAL:
Cabin max width	1.00 m (3 ft 3¼ in)

AREAS:
Wings, gross	9.84 m² (105.9 sq ft)

WEIGHTS AND LOADINGS:
Weight empty	262 kg (578 lb)
*Max T-O weight	450 kg (992 lb)

**480 kg (1,058 lb) permitted in Slovak Republic, Germany and USA*

PERFORMANCE (at 450 kg; 992 lb MTOW, Rotax 912 UL engine):
Max level speed	121 kt (225 km/h; 139 mph) IAS
Max cruising speed	97 kt (180 km/h; 112 mph) IAS
Stalling speed, flaps down	36 kt (65 km/h; 41 mph) CAS
Max rate of climb at S/L	330 m (1,083 ft)/min
Service ceiling	5,000 m (16,400 ft)
T-O run	145 m (475 ft)
Landing run	200 m (660 ft)
Max range, standard fuel	351 n miles (650 km; 403 miles)
g limits	+6/−3

OPERATIONAL NOISE LEVELS:
To German LSL Chapter 6	60 dBA

UPDATED

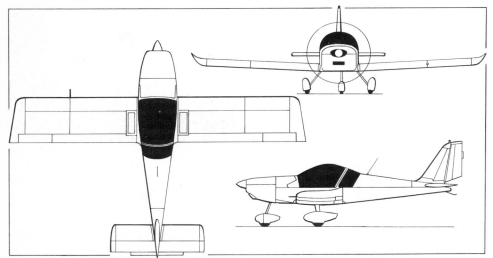

EV-AT EV-97 Eurostar 2000 two-seat ultralight *(James Goulding/Jane's)*

EV-AT EUROSTAR VLA

TYPE: Two-seat lightplane.
PROGRAMME: Revealed 1999; due to fly late 2000, with public debut at Friedrichshafen in April 2001, certification late 2001 and first deliveries early 2002.
DESIGN FEATURES: Heavier/more powerful version of EV-97 Model 2000, conforming to JAR-VLA and BCAR Section S.
Description as for Eurostar Model 2000 except as follows:
POWER PLANT: One 73.5 kW (98.6 hp) Rotax 912 ULS flat-four engine; three-blade variable-pitch propeller. Standard fuel capacity 70 litres (18.5 US gallons; 15.4 Imp gallons).
ACCOMMODATION: Two seats side by side; dual controls. Front-hinged canopy. Baggage space aft of seats.
DIMENSIONS AND AREAS: As for Eurostar Model 2000.

WEIGHTS AND LOADINGS:
Weight empty	320 kg (705 lb)
Max T-O weight	550 kg (1,212 lb)
Max wing loading	55.9 kg/m² (11.45 lb/sq ft)
Max power loading	7.48 kg/kW (12.29 lb/hp)

PERFORMANCE (estimated):
Never-exceed speed (V_{NE})	145 kt (270 km/h; 167 mph) IAS
Max level speed	129 kt (240 km/h; 149 mph) IAS
Max cruising speed	108 kt (200 km/h; 124 mph) IAS
Stalling speed, flaps down	44 kt (80 km/h; 50 mph) CAS
Max rate of climb at S/L	330 m (1,083 ft)/min
Service ceiling	5,000 m (16,400 ft)
T-O run	130 m (425 ft)
T-O to 15 m (50 ft)	260 m (855 ft)
Landing from 15 m (50 ft)	480 m (1,575 ft)
Landing run	240 m (790 ft)
g limits	+5.7/−2.25

NEW ENTRY

First Chinese-assembled Eurostar at Airshow China, Zhuhai, in November 2000 *(Photo Link)*

FANTASY AIR

FANTASY AIR GROUP — PROXIMEX s.r.o.
ulica Krbu 35, CZ-100 00 Prague 10
Tel: (+420 2) 74 77 25 57 and 74 77 23 94
Fax: (+420 2) 74 77 33 43
e-mail: tech@proximex.cz
Web: http://www.proximex.cz

Fantasy Air established at Písek 1995; initial product (see 1999-2000 *Jane's*) launched as Cora, in Legato and Allegro versions. By mid-2000 was standardising on five variants of Allegro, as described in following entry.

UPDATED

FANTASY AIR ALLEGRO

TYPE: Side-by-side ultralight/kitbuilt.
PROGRAMME: Certified in Czech Republic; public debut (Cora), Friedrichshafen Air Show, May 1997.
CURRENT VERSIONS: **Allegro ST**: Baseline version; 47.8 kW (64.1 hp) Rotax 582 UL.
Allegro 200: As ST, but with 55.0 kW (73.8 hp) Rotax 618 UL.
Allegro 2000: Kit version (of 200 ?), introduced from 1 March 2000.
Allegro SW: Short-span version. New SM 701 aerofoil section wing and 59.6 kW (79.9 hp) Rotax 912 UL. Four to France in September 1999, where certification nearing completion at end of that year. Kit version available from 1 March 2000.
Arius F: Standard-span wing (SM 701 section) and 73.5 kW (98.6 hp) Rotax 912 ULS. Two to Finland in August 1999, where this version to be certified at 520 kg (1,146 lb) MTOW to permit flying with twin-float gear. First flight 5 November 1999.
CUSTOMERS: Total 20 (including a few Cora Legatos) flying in Czech Republic by May 1997 (latest figure known).
DESIGN FEATURES: Braced high-wing monoplane; UA-2 aerofoil section on ST and 200, improved SM 701 on SW and Arius; T tail. Ground-adjustable rudder tab.
FLYING CONTROLS: Conventional and manual. Electrically actuated flaperons.
STRUCTURE: Fabric-covered metal wings with welded steel tube centre-section and full-span metal flaperons; laminated glass fibre fuselage and integral fin; all-metal rudder, tailplane and one-piece elevator.
LANDING GEAR: Fixed tricycle with steel tube legs, rubber-block shock-absorbers, mainwheel brakes and steerable nosewheel; wheel spats. Full Lotus twin-float gear to be certified for Arius F.
POWER PLANT: One Rotax engine (details under Current Versions); three-blade propeller.
EQUIPMENT: Recovery parachute optional.
DIMENSIONS, EXTERNAL (A: ST, B: 200, C: SW, D: Arius):
 Wing span: A, B, D 10.60 m (34 ft 9¼ in)
  C 9.63 m (31 ft 7¼ in)
 Length overall: all 6.10 m (20 ft 0¼ in)
 Height overall: all 1.95 m (6 ft 4¾ in)
AREAS (A-D as above):
 Wings, gross: A 11.80 m² (127.0 sq ft)
  B, D 11.40 m² (122.7 sq ft)
  C 11.50 m² (123.8 sq ft)
WEIGHTS AND LOADINGS (A-D as above):
 Weight empty, depending on version
  265-292 kg (584-644 lb)
 Max T-O weight: A, B, C 450 kg (992 lb)
  D 520 kg (1,146 lb)
PERFORMANCE (A-D as above):
 Never-exceed speed (V_{NE}):
  A 87 kt (162 km/h; 100 mph)
  B 108 kt (200 km/h; 124 mph)
  C 102 kt (190 km/h; 118 mph)
  D 110 kt (205 km/h; 127 mph)
 Max cruising speed:
  A 59 kt (110 km/h; 68 mph)
  B, C 76 kt (140 km/h; 87 mph)
  D 81 kt (150 km/h; 93 mph)
 Stalling speed: A 34 kt (63 km/h; 40 mph)
  B, C, D 36 kt (65 km/h; 41 mph)
 Max rate of climb at S/L: A 180 m (591 ft)/min
  B 240 m (787 ft)/min
  C, D 330 m (1,083 ft)/min
 Range, depending on version:
  243-324 n miles (450-600 km; 279-372 miles)

UPDATED

Fantasy Air Cora Allegro *(Paul Jackson/Jane's)* 1999/0044636

INTERPLANE

INTERPLANE spol s.r.o.
Letiště Zbraslavice, CZ-285 21 Zbraslavice
Tel/Fax: (+420 327) 59 13 81
e-mail: InPlane@mira.cz
Web: InterplaneAircraft.com
SALES MANAGER AND EXECUTIVE: Andrea Greplová
DIRECTOR OF MANUFACTURING: Radek Hurt

NORTH AND SOUTH AMERICAN SALES:
 Interplane USA
 39440 South Avenue, Zephyrhills, Florida 33540, USA
 Tel: (+1 813) 782 79 00
 Fax: (+1 813) 788 66 00
 e-mail: InPlaneUSA@aol.com
 PRESIDENT: Ben Dawson

Interplane was created as an ultralight manufacturer in 1992; design and product development is by Gryf Development company at Hodonin. Earlier variations between versions led the company to rationalise its Skyboy and Griffon product line in 1999. There are now two basic versions of the Skyboy: the EX (standard version for European customers and US and Australian kitbuilders) and the UL (to US FAR Pt 103 ultralight standards). Earlier Griffons are now replaced by the UL 103 version for the US market. Griffons and Skyboys now flying in Australia, Belgium, France, Germany, Luxembourg, South Africa, Spain, Sweden, Turkey and USA, as well as in Czech and Slovak Republics.

Long-term projects include **P6 Diblík** single-seat biplane (wing span 6.00 m; 19 ft 8¼ in, area 13.00 m²; 139.9 sq ft; MTOW 270 kg; 595 lb), **P15 Classic** two-seat ultralight (wing span 9.50 m; 31 ft 2 in, 73.5 kW; 98.6 hp Rotax 912 ULS, 81 kt; 150 km/h; 93 mph cruising speed) and **P16 Nemesis** improved version of Skyboy.

UPDATED

INTERPLANE GRIFFON 103

TYPE: Single-seat ultralight/kitbuilt.
PROGRAMME: First flight by Gryf ULM-1 prototype (OK-WUC 01) made on 17 March 1989. Early versions detailed in 1999-2000 *Jane's*. Available factory-built or (with or without power plant and instruments) as kit. Conforms to FAR Pt 103; Griffon 103 introduced 1999.
CUSTOMERS: Approximately 15 to 20 (including kitbuilt examples) completed by late 1999.
DESIGN FEATURES: Generally as for Skyboy (which see).
FLYING CONTROLS: Conventional and manual. Differential ailerons.
STRUCTURE: Generally similar to that of Skyboy.
LANDING GEAR: Non-retractable tricycle type; castoring nosewheel, with optional mechanical brakes.
POWER PLANT: One 37.0 kW (49.6 hp) Rotax 503 UL-2V or 31.0 kW (41.6 hp) Rotax 447 UL-2V two-cylinder two-stroke in-line engine; Junkers ProFly three-blade pusher propeller. Fuel capacity 18.9 litres (5.0 US gallons; 4.2 Imp gallons).
DIMENSIONS, EXTERNAL:
 Wing span 9.00 m (29 ft 6¼ in)
 Length overall 5.73 m (18 ft 9½ in)
 Height overall 2.20 m (7 ft 2½ in)
AREAS:
 Wings, gross 13.96 m² (150.3 sq ft)
WEIGHTS AND LOADINGS (Rotax 447):
 Weight empty 115 kg (254 lb)
 Max T-O weight 235 kg (518 lb)
PERFORMANCE (Rotax 447):
 Max level speed 55 kt (102 km/h; 63 mph)
 Econ cruising speed 49 kt (90 km/h; 56 mph)
 Stalling speed 25 kt (45 km/h; 28 mph)
 Max rate of climb at S/L 210 m (689 ft)/min

UPDATED

Interplane Griffon single-seat ultralight 2001/0092118

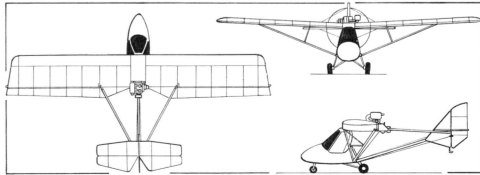

Current version of Interplane Griffon, fitted with 9.00 m (29 ft 6¼ in) wings *(James Goulding/Jane's)* 2001/0092146

INTERPLANE SKYBOY

TYPE: Side-by-side ultralight/kitbuilt.
PROGRAMME: Prototype designed and flown 1993; entered production 1994; conforms to Czech UL-2 and German BFU-DULV standards. Certified in Czech Republic (1994), France (1996), Slovak Republic and Germany (excluding noise tests, 1999). Current EX and UL versions introduced in 1999. New canopy design permits flight without doors for hot-weather flying.
CURRENT VERSIONS: **Skyboy:** Initial version, with Rotax 503 or 582 engine as standard; no longer produced. Details in 1998-99 *Jane's*.
 Skyboy S: Reinforced version for higher MTOW and speeds; Rotax 582 engine and choice of wing spans; no longer produced. Details in 1999-2000 *Jane's*.
 Skyboy EX: Standard version from 1999; for European customers; also for customers in Australia and USA, where it is classified in the FAA Experimental category.
 Skyboy UL: Specially lightened (empty weight 225 kg; 496 lb) to meet US FAR Pt 103 ultralight trainer regulations.
CUSTOMERS: Approximately 40 (all versions, including prototypes) built by late 1999.
Description applies to Skyboy EX.
DESIGN FEATURES: High, constant-chord wing, boom-mounted empennage and pusher propeller. Available factory-built or (with or without power plant and instruments) as kit. Wing struts and tail surfaces can be folded for transportation and storage.
 Wing aerofoil section NACA 4412; dihedral 1° 30′; incidence 4° 30′. Leading-edges sweptforward 2°.
FLYING CONTROLS: Conventional and manual. Tabs on rudder and each elevator; no flaps.
STRUCTURE: Aluminium tube and glass fibre; fabric covering.
LANDING GEAR: Non-retractable tricycle type; castoring nosewheel; mechanical or hydraulic brakes. Ski gear optional; float gear available in 2000.
POWER PLANT: One 47.8 kW (64.1 hp) Rotax 582 UL two-cylinder in-line two-stroke engine standard; 59.6 kW (79.9 hp) Rotax 912 UL flat-four; Rotax 462, 503 or 618; Hirth 2706; or Verner SVS 1400 optional. Wide choice of two- to six-blade wooden or laminate pusher propellers (three-blade Junkers ProFly standard). Fuel capacity 42 litres (11.1 US gallons; 9.2 Imp gallons).
EQUIPMENT: Optional Magnum 450 ballistic recovery parachute.

DIMENSIONS, EXTERNAL:
 Wing span 9.50 m (31 ft 2 in)
 Length overall 6.37 m (20 ft 10¾ in)
 Height overall 2.13 m (6 ft 11¾ in)
DIMENSIONS, INTERNAL:
 Cabin max width 1.20 m (3 ft 11¼ in)
AREAS:
 Wings, gross 13.50 m² (145.3 sq ft)
WEIGHTS AND LOADINGS:
 Weight empty, equipped 265 kg (584 lb)
 Max T-O weight 450 kg (992 lb)
PERFORMANCE (Rotax 582):
 Max level speed 76 kt (140 km/h; 87 mph)
 Econ cruising speed 49 kt (90 km/h; 56 mph)
 Stalling speed 34 kt (63 km/h; 40 mph)
 T-O run 150 m (495 ft)
 Landing run 100 m (330 ft)
 Range with max fuel 162 n miles (300 km; 186 miles)
 UPDATED

INTERPLANE STARBOY 912

Promotion of this side-by-side ultralight has been discontinued. Data and a general arrangement drawing last appeared in the 2000-01 *Jane's*.
UPDATED

Interplane Skyboy ultralight *(Pierre Gaillard)*

Skyboy EX new cabin design, shown at Sun 'n' Fun, 1999

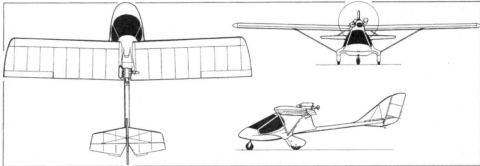
Interplane Skyboy (Rotax 582 two-stroke) *(Paul Jackson/Jane's)*

KAPPA

KAPPA 77 AS
Brtnická 21, CZ-586 01 Jihlava
Tel: (+420 66) 730 81 21
Fax: (+420 66) 730 81 22
e-mail: info@kappa77.cz
Web: http://www.kappa77.cz and http://www.ultralight.cz
MANAGING DIRECTOR: Ing Miroslav Navrátil
CHIEF DESIGNER: Doc Ing Antonin Pištěk
MARKETING DIRECTOR: Vit Tausch
PUBLIC RELATIONS MANAGER: Ing Tomáš Navrátil

Established 1991, initially for machine manufacture; founders previously employed by Aero. Design of KP-2U began 1995; production started 1998.
UPDATED

KAPPA KP-2U SOVA
English name: Owl
TYPE: Two-seat lightplane.
PROGRAMME: Designed by Kappa company and Institute of Aerospace Engineering at Brno Technical University. Prototype (OK-BUU 23) rolled out and first flown 26 May 1996.
CUSTOMERS: Total of 46 sold by 1 November 2000; customers in Chile, Czech Republic, Ecuador, Italy, Netherlands, Slovak Republic, Spain and USA. Marketed in USA by Interplane (which see).
COSTS: US$53,900 flyaway with Rotax 912 (2000).
DESIGN FEATURES: Low-wing monoplane conforming to BFU 95 regulations; JAR-VLA version being developed.
 Wing sections NASA GA(W)-1 at root and GA(W)-2 at tip; upturned wingtips. Sweptback vertical tail.
FLYING CONTROLS: Conventional and manual. Electrically operated trim tab in port elevator; Fowler flaps.
STRUCTURE: All-metal.

KP-2U Sova ultralight *(Paul Jackson/Jane's)*

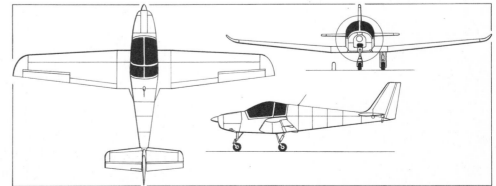
Kappa KP-2U Sova two-seat ultralight *(James Goulding/Jane's)*

LANDING GEAR: Electrically retractable tricycle type. Steerable nosewheel; mechanically operated brakes. Fixed gear optional.
POWER PLANT: One 59.6 kW (79.9 hp) Rotax 912 UL engine, with 4:1 reduction drive to a two- or three-blade propeller; can alternatively accommodate Rotax 582 UL, 618 UL, 912 ULS or 914. Fuel capacity (two wing tanks) 64 litres (16.9 US gallons; 14.1 Imp gallons).
ACCOMMODATION: Two seats side by side under forward-hinged canopy. Dual controls. Space behind seats for 14 kg (31 lb) of baggage.

DIMENSIONS, EXTERNAL:
Wing span 9.90 m (32 ft 5¾ in)
Length overall 7.18 m (23 ft 6¾ in)
Height overall 2.60 m (8 ft 6¼ in)
AREAS:
Wings, gross 11.85 m² (127.6 sq ft)
WEIGHTS AND LOADINGS:
Weight empty 282 kg (622 lb)
Max T-O weight 450 kg (992 lb)
PERFORMANCE (Rotax 912 UL engine):
Max level speed 129 kt (240 km/h; 149 mph)

Cruising speed at 75% power
 108 kt (200 km/h; 124 mph)
Stalling speed, flaps down 26 kt (48 km/h; 30 mph)
Max rate of climb at S/L 390 m (1,279 ft)/min
T-O run 90 m (295 ft)
Range with max fuel 540 n miles (1,000 km; 621 miles)
UPDATED

LET

LET AS, KUNOVICE

PO Box 1177, Uherské Hradiště, CZ-686 04 Kunovice
Tel: (+420 632) 56 41 24 and 56 41 42
Fax: (+420 632) 56 41 02 and 56 41 42
e-mail: let@let.cz
Web: http://www.let.cz
ADMINISTRATOR: Zlatava Davidova
TECHNICAL DIRECTOR: Zdenek Hradsky
COMMERCIAL DIRECTOR: Lubor Smericka
MANUFACTURING DIRECTOR: Pavel Zabrana
CHIEF DESIGNER: Miroslav Pěsák

Established at Kunovice in 1936 as a repair shop for Ba 33, B-534, Junkers W 34 and Arado Ar 96B aircraft. Construction of new plant started in 1950, producing Yak-11 trainer as C-11; subsequently involved in programmes for Aero Ae 45/145, L-200 Morava, L-29 Delfin, L-13 Blanik sailplane, and Zlin Z-37. Ayres Aviation Holdings Inc, an affiliate of Ayres Corporation of the USA (which see), acquired a 93.3 per cent holding in Let AS in August 1998. However, in mid-2000, Ayres funding was suspended, all 1,400 employees on forced leave and production halted due to expiry of credit arrangement with largest creditor, Konsolidacni Banka, which filed a bankruptcy suit against Let in September 2000. Search for new investors continuing at that time, including talks with BAE Systems and Israel Aircraft Industries. Former could place offset work with Let if Czech Air Force orders Gripen; latter may view involvement in Loadmaster as alternative to its own Airtruck programme.

Let has focused recently on upgrading the L-410/420; repairing, modernising and selling used L-410s, especially low-time aircraft from the former USSR; continuing the L-610G/Ayres 7000 certification programme; and attracting subcontract work. Let was made responsible for all wings and tail units for the Ayres LM-200 Loadmaster, as well as fuselages for Loadmasters assembled in the Czech Republic, and had delivered some components to USA before mid-2000 stoppage. First Loadmaster deliveries by Let, for European operations by Federal Express and others, were originally due in March 2001, at a rate of one per month. By then, funding only to maintain Let's contribution to Ayres Loadmaster programme had apparently been obtained, and L 610 programme had been put up for sale.
UPDATED

LET L 410UVP-E and L 420

TYPE: Twin-turboprop transport.
PROGRAMME: First flight 16 April 1969; entered production 1970; details of prototypes and early versions in 1980-81 and earlier *Jane's;* early production, excluding three XL 410 prototypes and three static test airframes, totalled 139 (29 L 410A series in 1970-75 and 110 L 410M series in 1973-78, respectively with PT6A-27 and M 601 engines). L 410UVP first flight 1 November 1977 and entered production 1979; in 1980 became first non-Soviet aircraft to gain Soviet NLGS-2 certification; UVP production (502 built, plus five prototypes) ended late 1985.

Present UVP-E version first flew (OK-120) 30 December 1984, received NLGS-2 certification March 1986; 1,000th aircraft of L 410 family delivered (to Aeroflot) 28 November 1990. First flight of L 420 (OK-150/OK-XYA) 10 November 1993; this aircraft registered in USA to Ayres Aircraft Holdings, December 1998. UVP-E versions (including three static test) built by 1997 included at least 40 unsold or cancelled by former USSR; production then suspended. Approximately 25 of these late-production aircraft were virtually sales-ready at the time of Let's acquisition by Ayres in August 1998.

Let L 410UVP-E general purpose transport *(Paul Jackson/Jane's)* 2000/0062725

CURRENT VERSIONS: **L 410UVP-E:** Standard current model.
Description applies to L 410UVP-E, except where indicated.
L 410 Special Missions: One aircraft modified by BAT Systems Inc, Giddings, Texas, with FLIR chin turret, telephoto video, radar and elint/sigint receivers, plus four monitoring consoles in cabin. Project name Dark Sentinel.
L 420: Improved L 410UVP-E20 with M 601F engines, detail changes and Western avionics. One prototype (OK-XYA) completed 1993; to USA as demonstrator (N420Y). FAA certification to FAR Pt 23 Amendment 41 was received in March 1998. Five ordered by Pan Pacific Airways in June 1999, with options on a further 15. For delivery (one per month) from third quarter 1999.
L 430: Under study as projected stretch of L 410UVP-E to be powered by P&WC PT6 turboprops; increased payload, speed and range and higher MTOW.
CUSTOMERS: L 410 production (all versions, including unsold airframes) is 1,100; see table in 2000-01 *Jane's* for early models, which totalled 649 including prototypes. UVP-E production exceeds 440. Exports to former Soviet Union comprised 303 military and 559 civil, not including 44 of latter not delivered. Customers are identified by suffixes between E2 and E20, with further subdivisions such as E8A, E10A, E19A and E20D. Currently available versions are E9 and E20. Numerous ex-Soviet aircraft have been sold to civil operators in Africa, Asia and Latin America. Pan Pacific Airways of USA ordered five L 420s in 1999, and holds option on 15 more.
DESIGN FEATURES: General purpose medium transport, optimised for operation by Aeroflot from remote airfields and as military communications aircraft by former Soviet forces. Compared with UVP, the UVP-E has baggage compartment and lavatory moved aft to accommodate four more passengers in same fuselage length; wings reinforced to carry wingtip tanks increasing maximum range by 40 per cent; maximum flap deflection increased; spoiler setting 72° on ground; new vacuum-sintered oil cooler; oil/fuel heat exchanger on each engine firewall to avoid use of additives; engine fire bottles moved to port rear wing/fuselage fairing; separate engine and propeller indicators for each engine; portable oxygen in cabin and improved PA system; fire extinguishing system in nose baggage compartment; operating ambient temperature range −50 to +50°C; design life 20,000 hours and 20,000 cycles.

Wing section NACA 63A418 at root, 63A412 at tip; dihedral 1° 45′; incidence 2° at root, −0° 30′ at tip; twist −2° 48′; sweepback 0° at quarter-chord; tailplane dihedral 7°. Fixed tailplane; dorsal and ventral fins.
FLYING CONTROLS: Conventional and manual. Ailerons and elevators actuated by pushrods; rudder by rods and cables. Trim tab in port aileron and geared tab in rudder actuated electromechanically, elevator trim tab by cables; pop-up bank control surfaces ahead of ailerons rise automatically after engine failure to reduce lift on side of running engine. Hydraulic actuation of two-segment double-slotted flaps and ground spoilers/lift dumpers ahead of flaps. Autopilot optional in L 420.
STRUCTURE: All-metal, two-spar torsion box wing with chemically milled skins; four wing attachment points; one-piece tailplane; fabric-covered elevators and rudder.
LANDING GEAR: Retractable tricycle type, with single wheel on each unit. Hydraulic retraction, nosewheel forward, mainwheels inward to lie flat in fairing on each side of fuselage. Technometra Radotin (L 410) or Teset Semily (L 420) oleo-pneumatic shock-absorbers. Non-braking nosewheel, with servo-assisted 50° steering, fitted with 9.00×6 (10 ply) tubeless tyre (550×225 on L 420), pressure 4.50 bars (65 lb/sq in). Nosewheel is also steerable by rudder pedals. Mainwheels fitted with 12.50×10 or 29×11.00-10 (10 ply) tubeless tyres (720×310 on L 420), pressure 4.50 bars (65 lb/sq in). All wheels manufactured by Moravan Otrokovice; tyres by Barumtech Zlin. Moravan Otrokovice K38-3200.00 hydraulic three-disc brakes, parking brake and inertial anti-skid units on mainwheels. Minimum ground turning radius 13.40 m (43 ft 11½ in). Metal ski landing gear, with plastics undersurface, optional.
POWER PLANT: Two 559 kW (750 shp) Walter M 601E turboprops in L 410 (580 kW; 778 shp M 601F in L 420), each driving an Avia Hamilton Sundstrand V510 five-blade constant-speed reversible-pitch metal propeller with manual and automatic feathering and Beta control.

Eight bag fuel tanks in wings, total capacity 1,290 litres (341 US gallons; 284 Imp gallons), plus additional optional 200 litres (52.8 US gallons; 44.0 Imp gallons) of fuel in each wingtip tank. Pressure and gravity refuelling. Fuel system operable after failure of electrical system. Total oil capacity (including oil in cooler) 22 litres (5.8 US gallons; 4.8 Imp gallons). Water tank capacity (for injection into compressor) 11 litres (2.9 US gallons; 2.4 Imp gallons).

Port rear freight door of L 410 and (illustrated) L 420 *(Paul Jackson/Jane's)* 2000/0084002

Prototype of the Let L 420, now flying as N420Y *(Paul Jackson/Jane's)* 2000/0084001

CZECH REPUBLIC: AIRCRAFT—LET

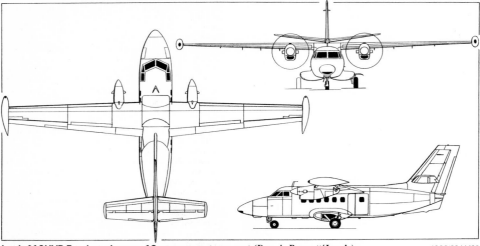

Let L 410UVP-E twin-turboprop 19-passenger transport *(Dennis Punnett/Jane's)* 1999/0044639

ACCOMMODATION: Crew of one or two on flight deck, with dual controls. Standard accommodation in main cabin for 19 passengers, with pairs of adjustable seats on starboard side of aisle and single seats opposite, all at 76 cm (30 in) pitch. Baggage compartment (at rear, accessible externally and from cabin), toilet and wardrobe standard. Cabin heated by engine bleed air. Alternative layouts include all-cargo; ambulance, accommodating six stretchers, five sitting patients and a medical attendant; accommodation for 18 parachutists and a dispatcher/instructor; firefighting configuration, carrying 16 firefighters and a pilot/observer. All-cargo version has protective floor covering, crash nets on each side of cabin, and tiedown provisions; floor is at truckbed height. Aircraft can also be equipped for aerial photography or for calibration of ground navigation aids.

Double upward-opening doors aft on port side, with stowable steps; right-hand door serves as passenger entrance and exit. Both doors open for cargo loading, and can be removed for paratroop training missions. Rearward-opening door, forward on starboard side, serves as emergency exit. Additional emergency exit under wing on each side in L 420.

SYSTEMS: No APU or pressurisation systems. Enviro Systems 125-0400 series vapour cycle air conditioning system in L 420. Duplicated hydraulic systems, main system (pressure 144 bars; 2,088 lb/sq in) actuating landing gear, flaps, spoilers, automatic pitch trim surfaces, mainwheel brakes, nosewheel steering and windscreen wipers. Emergency system for landing gear extension, flap actuation, mainwheel brakes and parking brake.

Electrical system (28 V DC) supplied by two 5.6 kW starter/generators, connected for autonomous starting, plus two 24 V 25 Ah batteries for emergency power. Two LUN 2450 inverters for three-phase AC power (200 V/115 Hz) and two PC 250 inverters for 115/26 V AC at 400 Hz. Two 3.7 kVA alternators supply power for windscreen heating and propeller de-icing. Port alternator provides for windscreen heating, starboard one for propeller blade de-icing. Two static inverters provide three-phase 36 V/400 Hz AC. Two 115 V/400 Hz inverters. One three-phase 36 V/400 Hz static inverter for standby horizon. De-icing for windscreen and propeller blades (electrical) and lower intakes (bleed air); anti-icing flaps inside each nacelle. Gumotex-Břeclav pneumatic de-icing of wing and tail leading-edges.

Two portable oxygen breathing sets on flight deck and two in passenger cabin. Fire extinguishing system for engines and nose baggage compartment.

AVIONICS: Standard Honeywell avionics provide for flight in IMC conditions, with all basic instruments duplicated and three artificial horizons.

Comms: KTR 908 with KFS 598 VHF (two), KHF 950 with KCU 951 HF and KMA 24H-70 intercom (two); KXP 756 transponders (two); Pointer 3000 ELT; A-100A cockpit voice recorder.
Radar: RDS 81 or RDR 2000 weather radar.
Flight: KNR 634A VOR/ILS/marker beacon receiver (two), Loran; KDF 806 ADF (two), KDM 706A DME (two) and KRA 405 radar altimeter; KFC 325 autopilot optional in L 420.
Instrumentation: KPI 553A HSI (two), KNI 582 RMI (two), KI 204 and KI 207 CDIs, KDI 573 DME indicators (two), KDI 415 radar altimeter indicators (two); BUR-1-26 or Fairchild F1000 SSF flight director recorder; rate of climb indicators; LUN 1215 turn and bank indicator.

EQUIPMENT: Cockpit, instrument and passenger cabin lights, navigation lights, three landing lights in nose (each with two levels of light intensity), cockpit and cabin fire extinguishers, windscreen wipers and electrically heated windscreen standard.

DIMENSIONS, EXTERNAL:
Wing span: over tip tanks	19.98 m (65 ft 6½ in)
excl tip tanks	19.48 m (63 ft 11 in)
Wing chord: at root	2.535 m (8 ft 3½ in)
at tip	1.12 m (3 ft 8 in)
Wing aspect ratio	10.9
Length overall	14.425 m (47 ft 3¾ in)
Fuselage: Length	13.205 m (43 ft 3¾ in)
Max width	2.08 m (6 ft 10 in)
Max depth	2.10 m (6 ft 10¾ in)
Height overall	5.83 m (19 ft 1½ in)
Tailplane span	6.79 m (22 ft 3¼ in)
Wheel track	3.65 m (11 ft 11½ in)
Wheelbase	3.665 m (12 ft 0½ in)
Propeller diameter	2.30 m (7 ft 6½ in)
Propeller ground clearance	1.25 m (4 ft 1¼ in)
Distance between propeller centres	4.815 m (15 ft 9¾ in)
Passenger/cargo door (port, rear):	
Height	1.46 m (4 ft 9½ in)
Width overall	1.25 m (4 ft 1¼ in)
Width (passenger door only)	0.80 m (2 ft 7½ in)
Height to sill	0.79 m (2 ft 7 in)
Baggage compartment door (L 420, stbd, rear):	
Height	1.12 m (3 ft 8 in)
Width	0.56 m (1 ft 10 in)
Height to sill	0.96 m (3 ft 1¾ in)
Emergency exit (stbd, fwd):	
Height	0.97 m (3 ft 2¼ in)
Width	0.665 m (2 ft 2¼ in)
Height to sill	0.87 m (2 ft 10¼ in)
Emergency exits (L 420, beneath wing, each):	
Height	0.73 m (2 ft 4¾ in)
Width	0.51 m (1 ft 8 in)
Height to sill	1.38 m (4 ft 6¼ in)

DIMENSIONS, INTERNAL:
Cabin, excl flight deck: Length	6.345 m (20 ft 10 in)
Max width	1.92 m (6 ft 3½ in)
Max height	1.66 m (5 ft 5¼ in)
Aisle width at 0.4 m (1 ft 3¾ in) above cabin floor	0.34 m (1 ft 1½ in)
Floor area	10.0 m² (108 sq ft)
Volume	17.9 m³ (632 cu ft)
Baggage compartment volume:	
nose: L 410UVP-E	0.60 m³ (21.2 cu ft)
L 420	0.76 m³ (26.8 cu ft)
rear: 19 passengers	0.77 m³ (27.2 cu ft)
17 passengers	1.4 m³ (48 cu ft)

AREAS:
Wings, gross	34.86 m² (375.2 sq ft)
Ailerons (total)	2.88 m² (31.00 sq ft)
Automatic bank control tabs (total)	0.49 m² (5.27 sq ft)
Trailing-edge flaps (total)	5.895 m² (63.45 sq ft)
Spoilers (total)	0.88 m² (9.47 sq ft)
Fin	4.595 m² (49.51 sq ft)
Rudder, incl tab	2.70 m² (29.06 sq ft)
Tailplane	6.535 m² (70.29 sq ft)
Elevators, incl tabs	3.025 m² (32.61 sq ft)

WEIGHTS AND LOADINGS (L 410UVP-E):
Weight empty	3,960-4,020 kg (8,730-8,863 lb)
Operating weight empty	4,120-4,150 kg (9,083-9,149 lb)
Max usable fuel	1,300 kg (2,866 lb)
Max payload	1,710 kg (3,770 lb)
Max ramp weight	6,620 kg (14,595 lb)
Max T-O weight	6,600 kg (14,550 lb)
Max zero-fuel weight	5,870 kg (12,941 lb)
Max landing weight	6,400 kg (14,110 lb)
Max wing loading	189.3 kg/m² (38.78 lb/sq ft)
Max power loading	5.90 kg/kW (9.70 lb/shp)

WEIGHTS AND LOADINGS (L 420): As L 410UVP-E except:
Basic weight empty	4,065 kg (8,962 lb)
Operating weight empty	4,225 kg (9,314 lb)
Max zero-fuel weight	5,900 kg (13,007 lb)
Max power loading	5.69 kg/kW (9.35 lb/shp)

PERFORMANCE (L 410UVP-E):
Max operating speed (V_{MO})	185 kt (343 km/h; 213 mph) EAS
Max cruising speed at 4,200 m (13,780 ft)	210 kt (388 km/h; 241 mph)
Econ cruising speed at 4,200 m (13,780 ft)	197 kt (365 km/h; 227 mph)
Stalling speed: flaps up	86 kt (158 km/h; 99 mph) EAS
flaps down	67 kt (123 km/h; 77 mph) EAS
Max rate of climb at S/L	396 m (1,300 ft)/min
Rate of climb at S/L, OEI	100 m (328 ft)/min
Service ceiling: practical	6,460 m (21,200 ft)
theoretical	7,460 m (24,475 ft)
Service ceiling, OEI: practical	2,900 m (9,525 ft)
theoretical	3,800 m (12,475 ft)
T-O run	455 m (1,495 ft)
T-O to 10.7 m (35 ft)	565 m (1,855 ft)
Landing from 15 m (50 ft) at MLW	840 m (2,755 ft)
Landing run at MLW	309 m (1,015 ft)
Range at 4,200 m (13,780 ft), max cruising speed, 45 min reserves:	
with max payload	362 n miles (670 km; 416 miles)
with max fuel	711 n miles (1,317 km; 818 miles)

PERFORMANCE (L 420):
Max operating speed (V_{MO})	205 kt (380 km/h; 236 mph) EAS
Max cruising speed at 4,200 m (13,780 ft)	210 kt (388 km/h; 241 mph)
Econ cruising speed at 3,050 m (10,000 ft)	197 kt (365 km/h; 227 mph)
Stalling speed: flaps up	86 kt (158 km/h; 99 mph) CAS
flaps down	67 kt (123 km/h; 77 mph) CAS
Max rate of climb at S/L	420 m (1,378 ft)/min
Rate of climb at S/L, OEI	108 m (354 ft)/min
Service ceiling (30.5 m; 100 ft/min climb)	7,410 m (24,300 ft)

Let L 410 modified by BAT Systems with FLIR and other sensors 2001/0093317

Interior of BAT Systems Dark Sentinel L 410 2001/0093318

LET—AIRCRAFT: CZECH REPUBLIC

Flight deck of the L 410UVP-E 1999/0041251

Service ceiling, OEI (15 m; 50 ft/min climb)	
	3,870 m (12,700 ft)
T-O run	440 m (1,445 ft)
T-O to 10.7 m (35 ft)	500 m (1,640 ft)
T-O to 10.7 m (35 ft), OEI	830 m (2,725 ft)
Accelerate/stop distance	1,050 m (3,445 ft)
Landing from 15 m (50 ft) at MLW	840 m (2,755 ft)
Landing run	305 m (1,000 ft)

Range at 4,200 m (13,780 ft), max cruising speed, 45 min reserves:
  with max payload 302 n miles (560 km; 348 miles)
  with max fuel (excl tip tanks)
    507 n miles (940 km; 584 miles)
  with max fuel (incl tip tanks) and 1,015 kg (2,237 lb) payload 733 n miles (1,358 km; 843 miles)

UPDATED

LET L 610 G and AYRES 7000

TYPE: Twin-turboprop transport.
PROGRAMME: First flight (X01/OK-130) 28 December 1988; five development aircraft (X02/OK-WZA, X03/OK-024, X05/OK-134, OK-XZA and OK-136), plus one for static test (X04) and one demonstrator (OK-CZD). Two versions originally planned: L 610 M with 1,358 kW (1,822 shp) Walter M 602 turboprops and L 610 G with General Electric CT7s; contract 18 January 1991 for General Electric to provide CT7-9Ds for L 610 G (first two delivered shortly afterward); first flight of this version (OK-XZA) 18 December 1992. Programme for L 610 M (see 1996-97 *Jane's*) abandoned 1996. In November 1999, Swiss marketing agency Airlines Partners SA sold two aircraft to Burundi, using marketing name Ayres 7000. Certification to FAR Pt 25 expected about nine months after contract signature by first US customer. Twelve production aircraft in various stages of completion by June 2000.

CURRENT VERSIONS: **L 610 G**: General Electric CT7-9D turboprops, four-blade propellers and Rockwell Collins digital avionics including EFIS, weather radar and autopilot. First prototype undergoing Czech and US certification testing; second prototype (OK-CZD) completed and was exhibited at Farnborough Air Show in September 1998; third (first series production) aircraft under construction at that time.

Ayres 7000: Production version of L 610 G. Described by Ayres as 'face-lift' upgrade of L 610 G. Marketing, sales and leasing by Airlines Partners SA, Switzerland. Czech-built 'green' aircraft to be shipped to Switzerland for interior furnishing. Standard civil version is 40-passenger regional transport.

Following description applies to Ayres 7000 except where indicated.

Ayres 7000F: Freighter version, lacking passenger cabin windows of 7000 and having instead a large, square cargo door in port side of rear fuselage.

Ayres 7000M: Military version of 7000F. Can be configured as troop, personnel or cargo transport; paratroop carrier; or for medevac/ambulance roles.

CUSTOMERS: Seven, including two production L 610 Gs, flown by mid-2000. Options for 16 from CSA and other (unidentified) operators. Pan Pacific Airways (USA) announced in June 1999 that it planned to acquire two. Two Ayres 7000 ordered in November 1999 by City Connexion of Burundi.

COSTS: US$8.75 million (2000).

DESIGN FEATURES: Intended to meet FAR Pt 25 requirements; high wing for propeller ground clearance and pannier-mounted landing gear to reduce mainwheel leg length and simplify loading/unloading; high-mounted tailplane for optimum control authority.

Wing sections MS(1)-0318D at root, MS(1)-0312 at tip; thickness/chord ratios 18.29 (root) and 12 per cent (tip); dihedral 2°; incidence 3° 8′ 38″ at root, 0° at tip; quarter-chord sweepback 1°.

FLYING CONTROLS: Conventional and manual. Ailerons are horn balanced. Rudder trim tab and both aileron trim tabs actuated by electromechanical strut. Elevator trim tabs actuated mechanically by screw-nut mechanism.

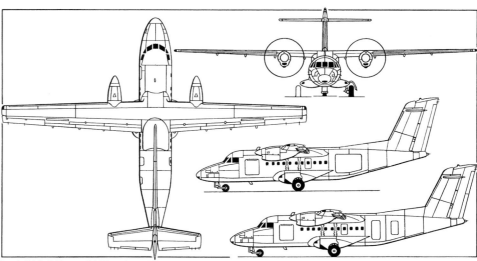

Let L 610 G/Ayres 7000 twin-turboprop regional transport with additional side view (upper) of 7000F and 7000M *(Mike Keep/Jane's)* 2001/0093853

Second prototype Let L 610 G, marked as an Ayres 7000 *(Paul Jackson/Jane's)* 2001/0092508

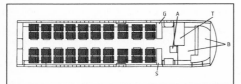

Ayres 7000 standard 40-seat layout
A: attendant's seat, B: baggage, G: galley, S: storage, T: lavatory

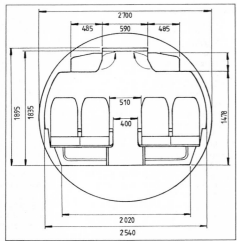

Cabin cross-section of the Ayres 7000; dimensions in millimetres

Flight deck of the L 610 G

Automatic spring tab in rudder. Electrohydraulically actuated single-slotted Fowler flaps. Ground spoilers. Lateral control spoilers deflected proportionally to aileron deflection. Electrically actuated gust lock.

STRUCTURE: All-metal, fail-safe stressed skin structure; circular-section fuselage between flight deck and tail; wing contains high-grade aluminium and high-strength steel; honeycomb spoiler panels.

LANDING GEAR: Retractable tricycle type, with single wheel on each unit. Hydraulic actuation, mainwheels retracting inward to lie flat, without doors, in fairing each side of fuselage; nosewheel retracts forward. Oleo-pneumatic shock-absorber in each unit. Mainwheels are type K 34-3100-7, with 15.00-16 (16 ply) or 1050×390-480 tubeless tyres; type K 35-1100-7 nosewheel has a 720×310-10 or 29×11.00-10 (10 ply) tubeless tyre. Hydraulic disc brakes and electronically controlled anti-skid units. Minimum ground turning radius 18.33 m (60 ft 1¾ in) about wingtip, 7.83 m (25 ft 8¼ in) about nosewheel.

POWER PLANT: Two 1,305 kW (1,750 shp) General Electric CT7-9D turboprops, each driving a Hamilton Sundstrand HS-14RF-23 four-blade fully feathering metal propeller with reversible pitch. Fuel in two integral wing tanks, combined usable capacity 3,420 litres (903.5 US gallons; 752 Imp gallons). Two 578 litre (153 US gallon; 127 Imp gallon) integral reserve tanks in 7000F and 7000M, raising total usable capacity to 4,576 litres (1,209 US gallons; 1,007 Imp gallons). Pressure refuelling point in fuselage, gravity points in wings. Oil capacity 30 litres (7.9 US gallons; 6.6 Imp gallons).

ACCOMMODATION: Crew of two on flight deck, plus one cabin attendant. Standard accommodation for 40 passengers, four-abreast at seat pitch of 76 cm (30 in). Galley, two wardrobes, toilet, freight and baggage compartment, all located at rear of cabin. Alternative mixed (passenger/cargo) and all-cargo layouts available; latter can accommodate six 1.63 × 2.08 × 1.57 m (64 × 82 × 62 in) containers or pallets in main cabin plus a further 200 kg (441 lb) in aft bulk cargo area at rear of main cargo hold. Military version has capacity for 30 fully equipped troops, 30 paratroops plus jumpmaster, 19 stretchers or five critically ill patients plus medical attendants, or up to 5,000 kg (11,023 lb) of cargo plus a loadmaster.

Passenger door in Ayres 7000 at rear of fuselage, freight door at front, both opening outward on port side. Outward-opening service door on starboard side, opposite passenger door, serving also as emergency exit; outward-opening emergency exit beneath wing on each side. Large cargo door aft of wing on port side in 7000F and 7000M. Paratroop version has inward-opening/sliding door aft of wing on port side. Entire accommodation pressurised and air conditioned.

SYSTEMS: Dual Hamilton Sundstrand R 79-3W engine bleed air air conditioning systems. Nord Micro digital, fully automatic pressurisation system gives 0.36 bar (5.2 lb/sq in) differential at flight level of 7,200 m (23,625 ft) and a cabin altitude of 2,400 m (7,875 ft). Duplicated hydraulic systems (one main and one standby), operating at pressure of 210 bars (3,045 lb/sq in). Optional 20 kW Safir 5 K/G APU in tailcone, for engine starting and auxiliary on-ground and in-flight power.

Electrical system powered by two 115/200 V 25 kVA variable frequency AC generators, plus a third 8 kVA 115/200 V three-phase AC generator driven by APU. System also includes two 115 V 400 Hz inverters (each 1.5 kVA), two 27 V DC transformer-rectifiers (each 4.5 kW), and a 25 Ah Ni/Cd battery for APU starting and auxiliary power supply. Gumotex Břeclav or Goodrich pneumatic de-icing boots on wing and tail unit leading-edges; ACT electric anti-icing system for engine inlets; electric de-icing of propeller blade roots, windscreen, pitot static system and horn balances. Oxygen system for crew and four passengers.

AVIONICS: Rockwell Collins Pro Line II suite and EFIS-86 standard.

Comms: Dual 760-channel VHF; single HF (optional); ATC transponder; intercom/PA system; cockpit voice recorder.

Radar: WXR-350 weather radar.

Flight: APS-65 autopilot; AHS-85 AHRS; dual ILS with two LOC/glide slope receivers and two marker beacon receivers; single or dual ADF; dual compasses; single or dual radio altimeters; navigation computer; flight data recorder; Cat. II approach aids.

Instrumentation: Five-tube EFIS-86 with EADI and HSI for each crew member and central MFD; weather radar data can be displayed on HSI and/or MFD.

DIMENSIONS, EXTERNAL:
Wing span	25.60 m (84 ft 0 in)
Wing chord: at root	2.92 m (9 ft 7 in)
at tip	1.46 m (4 ft 9½ in)
Wing aspect ratio	11.7
Length overall	21.72 m (71 ft 3 in)
Fuselage: Length	20.53 m (67 ft 4¼ in)
Max diameter	2.70 m (8 ft 10¼ in)
Height overall	8.19 m (26 ft 10½ in)
Tailplane span	7.91 m (25 ft 11½ in)
Wheel track	4.59 m (15 ft 0¾ in)
Wheelbase	6.61 m (21 ft 8¼ in)
Propeller diameter	3.35 m (11 ft 0 in)
Propeller fuselage clearance	0.66 m (2 ft 2 in)
Propeller ground clearance	1.71 m (5 ft 7¼ in)
Distance between propeller centres	7.00 m (22 ft 11½ in)
Passenger door (7000): Height	1.625 m (5 ft 4 in)
Width	0.76 m (2 ft 6 in)
Height to sill	1.45 m (4 ft 9 in)
Freight door (7000): Height	1.30 m (4 ft 3¼ in)
Width	1.25 m (4 ft 1¼ in)
Height to sill	1.45 m (4 ft 9 in)
Service door (7000): Height	1.29 m (4 ft 2¾ in)
Width	0.61 m (2 ft 0 in)
Rear cargo door (7000F, 7000M):	
Height	2.03 m (6 ft 8 in)
Width	1.68 m (5 ft 6¼ in)
Height to sill	1.50 m (4 ft 11 in)
Emergency exits, (all underwing, each):	
Height	0.915 m (3 ft 0 in)
Width	0.515 m (1 ft 8¼ in)

DIMENSIONS, INTERNAL:
Cabin (excl flight deck): Total length	11.10 m (36 ft 5 in)
Main cargo hold length (7000F, 7000M)	9.70 m (31 ft 10 in)
Passenger cabin: Max length	7.84 m (25 ft 8¾ in)
Max width	2.54 m (8 ft 4 in)
Width at floor	2.02 m (6 ft 7½ in)
Aisle width	0.51 m (1 ft 8 in)
Max height	1.835 m (6 ft 0¼ in)
Floor area	22.4 m² (241 sq ft)
Volume	34.2 m³ (1,208 cu ft)
Wardrobe volume (total)	1.0 m³ (35 cu ft)
Overhead bins volume (total)	1.975 m³ (69.75 cu ft)
Baggage/freight hold volume (total)	6.4 m³ (226 cu ft)
Aft bulk cargo volume (7000F, 7000M)	1.9 m³ (67 cu ft)
Available volume for freight/cargo	44.7 m³ (1,579 cu ft)

AREAS:
Wings, gross	56.0 m² (602.8 sq ft)
Ailerons (total)	3.27 m² (35.20 sq ft)
Trailing-edge flaps (total)	11.29 m² (121.52 sq ft)
Spoilers (total)	3.54 m² (38.10 sq ft)
Fin	8.46 m² (91.06 sq ft)
Rudder, incl tabs	6.84 m² (73.63 sq ft)
Tailplane	8.07 m² (86.86 sq ft)
Elevators (total, incl tabs)	5.43 m² (58.45 sq ft)

WEIGHTS AND LOADINGS (A: 7000, F: 7000F, M: 7000M):
Operating weight empty: A	9,860 kg (21,738 lb)
F, M	8,310 kg (18,320 lb)
Max fuel: all	2,700 kg (5,952 lb)
Max payload: A	4,240 kg (9,348 lb)
F, M	5,000 kg (11,023 lb)
Max ramp weight: A	15,150 kg (33,400 lb)
F, M	15,140 kg (33,378 lb)
Max T-O weight	15,100 kg (33,289 lb)
Max landing weight	14,800 kg (32,628 lb)
Max zero-fuel weight	13,800 kg (30,424 lb)
Max wing loading	269.6 kg/m² (55.23 lb/sq ft)
Max power loading	5.79 kg/kW (9.51 lb/shp)

PERFORMANCE:
Max operating Mach No. (M_{MO}) above 4,875 m (16,000 ft)	0.443
Max operating speed (V_{MO}) up to 4,875 m (16,000 ft)	213 kt (396 km/h; 246 mph) IAS
Max cruising speed at 6,100 m (20,000 ft)	238 kt (440 km/h; 273 mph)
Long-range cruising speed at 7,315 m (24,000 ft)	205 kt (380 km/h; 236 mph)
Min paratroop deployment speed (7000M)	85 kt (157 km/h; 98 mph) IAS
Approach speed	99 kt (184 km/h; 114 mph) IAS
Stalling speed: flaps up	98 kt (182 km/h; 113 mph) EAS
flaps down	76 kt (141 km/h; 88 mph) EAS
Max rate of climb at S/L	498 m (1,634 ft)/min
Rate of climb at S/L, OEI	132 m (433 ft)/min
Max operating altitude	7,315 m (24,000 ft)
Service ceiling, OEI	4,500 m (14,760 ft)
FAR/JAR 25 required T-O field length at S/L:	
ISA	1,050 m (3,445 ft)
ISA + 15°C	1,140 m (3,740 ft)
FAR/JAR 25 required landing field length at MLW, at S/L:	
with propeller reversal	1,030 m (3,380 ft)
without propeller reversal	1,140 m (3,740 ft)
Range, reserves for 45 min hold and 100 n mile (185 km; 115 mile) diversion:	
with 40 passengers	680 n miles (1,259 km; 782 miles)
with max fuel	1,281 n miles (2,373 km; 1,475 miles)

OPERATIONAL NOISE LEVELS:
Comply with FAR Pt 36, Annexe 16, Section 5

UPDATED

LET-MONT

LET-MONT s.r.o.
Vikyrovice 226, CZ-787 01 Sumperk
Tel: (+420 649) 21 40 85 and 21 40 81
Fax: (+420 649) 21 40 85 and 21 38 77
DIRECTORS:
 O Klier
 H Podešva

VERIFIED

LET-MONT TULAK and PIPER UL
English name: Rambler
TYPE: Side-by-side ultralight/kitbuilt (Tulák); tandem-seat ultralight (Piper UL).
PROGRAMME: Sixth production aircraft delivered to Poland in 1999.
CURRENT VERSIONS: **Tulák:** Side-by-side seats; factory-built or kit.
 Piper UL: Tandem seats; otherwise as Tulák. No kit version. German name **Tandem Tulák**.
COSTS: DM67,804 including tax, factory-built, with Rotax 503 (1999).
Data apply to both models except where indicated.
DESIGN FEATURES: High, constant-chord wing braced by V struts; optional tapered or downturned wingtips. Wings fold back for storage and transportation. Resembles Piper designs of 1940s/1950s.
 Clark Y wing section, thickness/chord ratio 12.5 per cent.
FLYING CONTROLS: Conventional and manual. Trim tab in each elevator. Plain flaps.
STRUCTURE: Fabric-covered metal tube fuselage; two-spar wooden wing; engine cowling metal; some composites.
LANDING GEAR: Tailwheel type; fixed. Steerable tailwheel; cable-operated mainwheel brakes. Mainwheel spats optional.
POWER PLANT: One 37.0 kW (49.6 hp) Rotax 503 UL-2V engine; Sport Prop three-blade ground-adjustable pitch propeller. Alternative engines up to 67 kW (90 hp) may be installed. Fuel capacity 50 litres (13.2 US gallons; 11.0 Imp gallons).
DIMENSIONS, EXTERNAL:

Wing span: tapered tips	9.60 m (31 ft 6 in)
downturned tips	9.40 m (30 ft 10 in)
Width, wings folded	3.10 m (10 ft 2 in)
Length overall	5.60 m (8 ft 4½ in)
Height overall	2.00 m (6 ft 6¾ in)

DIMENSIONS, INTERNAL:

Cabin max width: Tulák	1.14 m (3 ft 9 in)
Piper UL	0.70 m (2 ft 3½ in)

AREAS:

Wings, gross	13.00 m^2 (139.9 sq ft)

WEIGHTS AND LOADINGS:

Weight empty	240-280 kg (529-617 lb)
Max T-O weight	450 kg (992 lb)

PERFORMANCE:

Max cruising speed	59 kt (110 km/h; 68 mph)
Stalling speed, flaps down	33 kt (60 km/h; 38 mph)
Max rate of climb at S/L	180 m (590 ft)/min
T-O run	65 m (215 ft)
Landing run	60 m (200 ft)
g limits	+6/−4

UPDATED

Let-Mont Tulák two-seat ultralight

LETOV AIR

LETECKA TOVARNA s.r.o.
Beranových 65, CZ-199 02 Praha 9-Letňany
Tel: (+420 2) 82 56 51
Fax: (+420 2) 858 71 75
DIRECTOR GENERAL: Jan Šimúnek

Control of Letov Air is now in the hands of Wolfsberg Aircraft Corporation NV (which see in Belgian section). Letov produces the rear fuselage and tail surfaces for the Aero L 159 ALCA, in addition to the ultralights described below, but gained a major programme when it was assigned design and production responsibility for the Raven 257 cargo aircraft in an announcement made on 21 November 2000.
 Letov's production facilities cover 18,000 m^2 (193,750 sq ft).

UPDATED

WOLFSBERG-LETOV RAVEN 257
TYPE: Light utility twin-prop transport.
PROGRAMME: Market study commenced June 1996; general design began June 1997, and design of major structures in January 1998; manufacture of prototype by Evektor-Aerotechnik began at Kunovice, Czech Republic, in June 1998, first flight (OK-RAV) 28 July 2000; total of 4 hours flight testing completed by September 2000. Six aircraft (prototype, structural test and four pre-production airframes) will be used in the flight test and certification programme, leading to FAR Pt 23 approval in April 2002; production at Letňany was scheduled to have begun in June 2001.
CURRENT VERSIONS: **Baseline:** 224 kW (300 hp) engines. *As described.*
 Turbo: Turbocharged TSIO-550 engines, each 224 kW (350 hp); MTOW 3,000 kg (6,613 lb); maximum level speed 157 kt (291 km/h; 181 mph).
COSTS: Approximately US$850,000 (September 2000).
DESIGN FEATURES: High-wing, twin-boom configuration with square section fuselage and high-set tailplane. Design goals include ability to operate from short unprepared strips, simple systems and easy field maintenance for operation in less-developed areas, combined with low acquisition and operating costs; seen as modern replacement for aircraft in the B-N Islander and Piper Aztec class, and as a 'step up' for operators of single-engined utility aircraft such as the Cessna 206, with applications in passenger, cargo, medevac and airdrop roles.
 NACA 23018 section wings, with 2° 30′ dihedral; 3° incidence; NACA 0012 section horizontal and vertical tail surfaces.
FLYING CONTROLS: Conventional and manual, via push-pull rods. Control surfaces statically balanced. Two-section elevator with independent controls, trim tab on port side and balance tab on starboard side, maximum deflections +27/−22°, aileron deflections +25/−15°; rudders maximum deflection ±25°, with trim tab on starboard rudder. Four-section electrically actuated flaps, T-O setting 20°, landing 48°.
STRUCTURE: Conventional, mostly riveted metal, with glass fibre wingtips, nosecone, cargo hatch and engine cowlings.
LANDING GEAR: Non-retractable tricycle type, based on, and mostly interchangeable with, that of B-N Islander with twin wheels on each main unit and single nosewheel; oleo strut suspension; mainwheel legs braced to bottom of fuselage. Mainwheel tyre size 7.00-6; nosewheel 6.00-6; Cleveland 40-90F brakes on mainwheels; 40-76D on nosewheel.
POWER PLANT: Two 224 kW (300 hp) Teledyne Continental IO-550-N8B flat-six engines, driving two-blade Hartzell HC-C2YF-2CUF/FC8475K-6/SM8 composites propellers. Fuel contained in two integral tanks in wing centre-section, combined capacity 536 litres (141.5 US gallons; 118 Imp gallons), with filler port in upper surface of each wing.
ACCOMMODATION: Two-crew cockpit (although aircraft will be certified for single-pilot operation); up to seven passengers. Crew door on port side; sliding semi-bulkhead separates flight deck from cabin/cargo area, which has flat floor and quick-change features, with passenger door at rear on port side and upward-hinged rear cargo door. In cargo configuration, fuselage can accommodate ATA D and IATA 8, 12 and 17-type containers, 304 × 139 cm pallets, 1.2 × 1.0 m Europallets and bulk cargo under

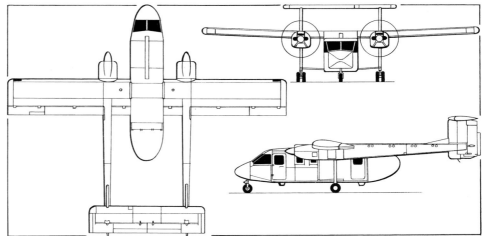
Wolfsberg-Letov Raven 257 utility aircraft (*James Goulding/Jane's*)

First flight of Wolfsberg-Letov Raven 257 at Kunovice, 28 July 2000

Wolfsberg-Letov Raven 257 shortly after roll-out

Letov ST-4 Azték (Rotax 582 engine) *1999/0044647*

LK-2M Sluka single-seat ultralight *(Paul Jackson/Jane's)* *2001/0092098*

ST-4 Azték cockpit *2000/0062597*

netting. Baggage door on port side, emergency exit on starboard side.
SYSTEMS: 24 V 74 A electrical system includes two engine-driven alternators and 24 V 19 Ah accumulator.
AVIONICS: Dual VHF nav/com, ADF, transponder, GPS.

DIMENSIONS, EXTERNAL:
Wing span	14.00 m (45 ft 11¼ in)
Wing chord, constant	1.85 m (6 ft 0¾ in)
Wing aspect ratio	7.6
Length overall	12.29 m (40 ft 3¾ in)
Height overall	3.99 m (13 ft 1 in)
Tailplane span	5.80 m (19 ft 0¼ in)
Wheeltrack	4.00 m (13 ft 1½ in)
Wheelbase	4.30 m (14 ft 1¼ in)
Propeller diameter	1.98 m (6 ft 6 in)

DIMENSIONS, INTERNAL:
Cabin, excl flight deck: Length	3.91 m (12 ft 10 in)
Max width	1.47 m (4 ft 10 in)
Max height	1.44 m (4 ft 8¾ in)
Volume	8.30 m³ (293 cu ft)

AREAS:
Wings, gross	25.86 m² (278.3 sq ft)
Ailerons (total)	1.48 m² (15.93 sq ft)
Trailing-edge flaps (total): inner	1.06 m² (11.41 sq ft)
outer	1.60 m² (17.22 sq ft)
Tailplane	4.54 m² (48.87 sq ft)
Elevator, incl tabs	3.36 m² (36.17 sq ft)
Fin (total)	3.84 m² (41.33 sq ft)
Rudders (total)	2.36 m² (25.40 sq ft)

WEIGHTS AND LOADINGS:
Weight empty	1,636 kg (3,607 lb)
Max T-O weight	2,700 kg (5,952 lb)
Max zero-fuel weight	2,630 kg (5,798 lb)
Max wing loading	104.4 kg/m² (21.38 lb/sq ft)
Max power loading	6.03 kg/kW (9.90 lb/hp)

PERFORMANCE (estimated):
Max level speed	145 kt (268 km/h; 167 mph)
Cruising speed, 75% power	138 kt (255 km/h; 158 mph)
Econ cruising speed	130 kt (240 km/h; 149 mph)
Stalling speed	53 kt (98 km/h; 61 mph)
Max rate of climb at S/L	444 m (1,457 ft)/min
Rate of climb at S/L, OEI	108 m (354 ft)/min
Range with max fuel	715 n miles (1,324 km; 822 miles)

UPDATED

LETOV LK-2M SLUKA

TYPE: Single-seat ultralight/kitbuilt.
PROGRAMME: First flight 1987; 70 built by late 1998 (latest information).
DESIGN FEATURES: Strut-braced high-wing monoplane. Wings fold.
STRUCTURE: Fabric-covered light alloy tubing; GFRP nacelle.
POWER PLANT: One 31.0 kW (41.6 hp) Rotax 447 UL-2V engine. Fuel capacity 34 litres (9.0 US gallons; 7.5 Imp gallons).

DIMENSIONS, EXTERNAL:
Wing span	9.20 m (30 ft 2¼ in)
Length overall	5.10 m (16 ft 8¾ in)
Height overall	2.05 m (6 ft 8¾ in)

AREAS:
Wings, gross	13.34 m² (143.6 sq ft)

WEIGHTS AND LOADINGS:
Weight empty	150 kg (331 lb)
Max T-O weight	270 kg (595 lb)

PERFORMANCE:
Never-exceed speed (V_{NE})	67 kt (125 km/h; 77 mph)
Cruising speed	46 kt (85 km/h; 53 mph)
Stalling speed	27 kt (49 km/h; 31 mph)
Max rate of climb at S/L	270 m (885 ft)/min
T-O run	60 m (200 ft)
Landing from 15 m (50 ft)	100 m (330 ft)
Range, 10 min reserves	172 n miles (320 km; 199 miles)

UPDATED

LETOV ST-4 AZTEK

English name: Aztec
TYPE: Side-by-side ultralight/kitbuilt.
PROGRAMME: Prototype first flew in 1995. Designed for agricultural work, pilot training or surveillance. Short-span version optional. Kit- or factory-built.
CUSTOMERS: Total 21 built by late 1998 (latest information).
DESIGN FEATURES: High, constant-chord wing with V strut bracing. Pod-and-boom configuration, with high-mounted empennage. Wings fold for transport and storage. Shorter-span version available optionally.
STRUCTURE: Steel tube and aluminium; GFRP skin.
LANDING GEAR: Non-retractable tricycle type; wheel spats.
POWER PLANT: One 47.8 kW (64.1 hp) Rotax 582 UL two-cylinder two-stroke standard. Alternatives include Rotax 447/503/618, Mid West MWAE 50R or Walter M 202. Junkers composites propeller (two to six blades). Fuel capacity 45.5 litres (12.0 US gallons; 10.0 Imp gallons) standard; larger tank (60 litres; 15.8 US gallons; 13.2 Imp gallons) optional.

DIMENSIONS, EXTERNAL:
Wing span: normal	10.40 m (34 ft 1½ in)
optional	9.60 m (31 ft 6 in)
Length overall	5.90 m (19 ft 4¼ in)
Height overall	2.25 m (7 ft 4½ in)

AREAS:
Wings, gross	15.30 m² (164.7 sq ft)

WEIGHTS AND LOADINGS:
Weight empty	220 kg (485 lb)
Max T-O weight	450 kg (992 lb)

PERFORMANCE:
Max level speed: normal span	70 kt (130 km/h; 81 mph)
optional span	72 kt (145 km/h; 90 mph)
Cruising speed: normal span	54 kt (100 km/h; 62 mph)
optional span	59 kt (110 km/h; 68 mph)
Stalling speed	27 kt (50 km/h; 32 mph)
Max rate of climb at S/L	240 m (787 ft)/min
T-O run	60 m (200 ft)
Range	183 n miles (340 km; 211 miles)

UPDATED

S-WING

S-WING spol. s.r.o.
Letiště Hosin, CZ-373 41 Hluboká nad Vltavou
Tel/Fax: (+420 38) 72 21 65

VERIFIED

S-WING SWING

TYPE: Side-by-side ultralight.
PROGRAMME: Czech Republic customers only by mid-1997. All-composites wings planned for introduction later. No recent reports of sales.
DESIGN FEATURES: Wings and tailplane foldable or detachable for storage and transportation. Nosewheel version has wings braced by single strut each side, and unbraced tailplane. Tailwheel version has A wing bracing struts and inverted V tailplane braces pivoting on hinge line.
FLYING CONTROLS: Conventional and manual ailerons and rudder, without tabs; all-moving tailplane with full-span, electrically operated trim tab. Three-position plain flaps. Dual controls.
STRUCTURE: Mainly of composites; wings of GFRP sandwich with rigid leading-edge and fabric covering; control surfaces similar.
LANDING GEAR: Fixed tricycle type with steerable nosewheel; wheel spats. Tailwheel version, with mainwheels farther forward, also available. Mechanical drum brakes standard, hydraulic disc brakes optional.
POWER PLANT: One 59.6 kW (79.9 hp) Rotax 912 UL flat-four engine; Junkers ProFly 1700 Maxi ground-adjustable propeller with two composites blades; alternative propellers optional. Can also be powered by 47.8 kW (64.1 hp) Rotax 582 UL. Fuel capacity 75 litres (19.8 US gallons; 16.5 Imp gallons).
EQUIPMENT: USH 520 recovery parachute.

DIMENSIONS, EXTERNAL:
Wing span	10.20 m (33 ft 5½ in)

Length overall	6.29 m (20 ft 7½ in)
Height overall	2.40 m (7 ft 10½ in)
DIMENSIONS, INTERNAL:	
Cabin max width	1.31 m (4 ft 3½ in)
AREAS:	
Wings, gross	12.90 m² (138.9 sq ft)
WEIGHTS AND LOADINGS:	
Weight empty	276 kg (608 lb)
Max T-O weight	450 kg (992 lb)
PERFORMANCE (Rotax 912 UL engine):	
Max level speed	108 kt (200 km/h; 124 mph)
Max cruising speed at 85% power	97 kt (180 km/h; 112 mph)
Stalling speed	33 kt (60 km/h; 38 mph)
Max rate of climb at S/L	420 m (1,378 ft)/min
T-O run	70 m (230 ft)
Landing run	140 m (460 ft)
Max range	540 n miles (1,000 km; 621 miles)

UPDATED

Swing two-seat ultralight (Rotax 912 UL engine)
(Paul Jackson/Jane's)

TL

TL ULTRALIGHT
Dobrovského 734, CZ-500 02 Hradec Králové
Tel: (+420 49) 61 17 53 and 61 89 10
Fax: (+420 49) 61 33 78
DIRECTOR: Ing Ivan Lelák

VERIFIED

TL-32 TYPHOON
TYPE: Side-by-side ultralight.
PROGRAMME: Certified and in production. At least four on Czech register in 1999.
POWER PLANT: One 37.0 kW (49.6 hp) Rotax 503 UL-2V, 38.2 kW (51.3 hp) Rotax 462 or 47.8 kW (64.1 hp) Rotax 582 UL two-cylinder two-stroke in-line engine. Kremen two-blade or Ivoprop three-blade propeller.

DIMENSIONS, EXTERNAL:	
Wing span	10.70 m (35 ft 1¼ in)
Length overall	5.90 m (19 ft 4¼ in)
Height overall	2.45 m (8 ft 0½ in)
AREAS:	
Wings, gross	15.20 m² (163.6 sq ft)
WEIGHTS AND LOADINGS:	
Weight empty	200 kg (441 lb)
Max T-O weight	400 kg (882 lb)
PERFORMANCE:	
Max level speed	59 kt (110 km/h; 68 mph)
Cruising speed	49 kt (90 km/h; 56 mph)
Stalling speed	25 kt (45 km/h; 28 mph)
Range	162 n miles (300 km; 186 miles)

UPDATED

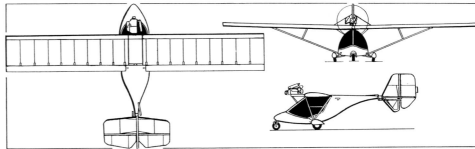

TL-32 Typhoon ultralight *(Paul Jackson/Jane's)*

TL-96 STAR
TYPE: Side-by-side ultralight.
PROGRAMME: Prototype first flew November 1997; has German certification.
CUSTOMERS: At least 34 built by late 1999. Several registered in Netherlands.
COSTS: DM108,000, including tax (basic aircraft, 1999).
DESIGN FEATURES: Conventional, low-wing monoplane with constant-chord wings and tailplane, plus sweptback fin. MS 313 wing aerofoil section.
FLYING CONTROLS: Manual. All-moving tailplane with anti-balance tab. Plain flaps.
STRUCTURE: All-composites (GFRP and CFRP).
LANDING GEAR: Fixed tricycle type; wheel speed fairings. Cable brakes.
POWER PLANT: One 59.6 kW (79.9 hp) Rotax 912 UL-DCDI or 84.6 kW (113.4 hp) Rotax 914 flat-four engine; three-blade Albastar propeller. Optional 67.1 kW (90 hp) Aero Prag AP-45 flat-four and two-blade wooden propeller. Fuel capacity (RON 95) 50 litres (13.2 US gallons; 11.0 Imp gallons).

DIMENSIONS, EXTERNAL:	
Wing span	9.00 m (29 ft 6¼ in)
Length overall	5.50 m (18 ft 0½ in)
Height overall	2.15 m (7 ft 0¾ in)
DIMENSIONS, INTERNAL:	
Cabin max width	1.15 m (3 ft 9¼ in)
AREAS:	
Wings, gross	12.10 m² (130.2 sq ft)
WEIGHTS AND LOADINGS:	
Weight empty	290 kg (639 lb)
Max T-O weight	450 kg (992 lb)
PERFORMANCE, POWERED (Rotax 912):	
Max level speed	140 kt (260 km/h; 161 mph)
Max cruising speed	127 kt (235 km/h; 146 mph)
Stalling speed	34 kt (63 km/h; 40 mph)
Max rate of climb at S/L	276 m (906 ft)/min
PERFORMANCE, UNPOWERED:	
Best glide ratio	16

UPDATED

TL-96 Star two-seat ultralight *(Paul Jackson/Jane's)*

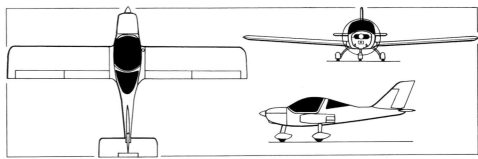

The two-seat TL-96 Star *(Paul Jackson/Jane's)*

TL-132 CONDOR and TL-232 CONDOR PLUS
TYPE: Side-by-side (TL-132) or tandem-seat (TL-232) ultralight.
PROGRAMME: TL-132 first flown 1993; entered production, and TL-232 introduced, in 1994.
CURRENT VERSIONS: **TL-132**: With 37.0 kW (49.6 hp) Rotax 503 UL-2V engine.
TL-232: More powerful engine, cut-down rear fuselage and additional glazing for rear cockpit. Known as **Power Condor** with 73.5 kW (98.6 hp) engine.
CUSTOMERS: Total of 35 TL-132 and six TL-232 built by February 1996 (latest information provided). Customers in Czech Republic, Germany, Netherlands, Poland and Sweden.
Detailed description applies to TL-232 except where indicated.
DESIGN FEATURES: Constant-chord, high wing with V-strut bracing; mid-mounted tailplane; sweptback fin with long fillet.
FLYING CONTROLS: Conventional and manual. In-flight-adjustable tab (ground-adjustable on TL-132) on starboard elevator. Flaps fitted. Dual controls.

Tandem-seat, tricycle-gear TL-232 Condor Plus

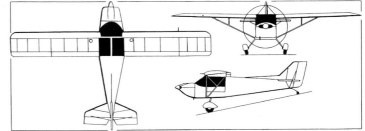

Side-by-side seat, tailwheel TL-132 Condor *(Paul Jackson/Jane's)*

STRUCTURE: Fabric-covered metal. Composites rear top-decking in TL-232.
LANDING GEAR: Non-retractable. Choice of tricycle or tailwheel configuration. Cable brakes. Leg and wheel speed fairings optional.
POWER PLANT: One 47.8 kW (64.1 hp) Rotax 582 UL-DCDI two-cylinder two-stroke in-line engine in basic TL-232, driving a two-blade propeller. Optional alternatives include 59.6 kW (79.9 hp) Rotax 912 UL-DCDI or 73.5 kW (98.6 hp) Rotax 912 ULS flat-four with Ivoprop three-blade composites propeller. Fuel capacity 50 litres (13.2 US gallons; 11.0 Imp gallons) standard, 70 litres (18.5 US gallons; 15.4 Imp gallons) optional.
DIMENSIONS, EXTERNAL:
Wing span 10.60 m (34 ft 9¼ in)
Length overall: tailwheel approx 6.50 m (21 ft 4 in)
tricycle 6.08 m (19 ft 11¼ in)
Height overall: tailwheel approx 1.85 m (6 ft 1 in)
tricycle 2.30 m (7 ft 6½ in)
AREAS:
Wings, gross 14.84 m² (159.7 sq ft)
WEIGHTS AND LOADINGS:
Weight empty 265-280 kg (584-617 lb)
Max T-O weight 450 kg (992 lb)
PERFORMANCE (TL-232 with Rotax 912 UL-DCDI engine):
Max level speed 97 kt (180 km/h; 111 mph)
Cruising speed 81 kt (150 km/h; 93 mph)
Stalling speed 31 kt (57 km/h; 36 mph)
Max rate of climb at S/L 276 m (906 ft)/min

UPDATED

TL-532 FRESH
TYPE: Side-by-side ultralight.
PROGRAMME: First flight 7 March 1995. No recent news.
POWER PLANT: One 47.8 kW (64.1 hp) two-cylinder Rotax 582 UL in-line engine.
DIMENSIONS, EXTERNAL:
Wing span 8.80 m (28 ft 10½ in)
WEIGHTS AND LOADINGS:
Weight empty 220 kg (485 lb)
Max T-O weight 450 kg (992 lb)
PERFORMANCE:
Max level speed 118 kt (220 km/h; 136 mph)
Cruising speed 76 kt (140 km/h; 87 mph)
Stalling speed 36 kt (65 km/h; 41 mph)

UPDATED

UNIS

UNIS OBCHODNI spol. s.r.o.
Jundrovská 33, CZ-624 00 Brno
Tel: (+420 5) 41 51 51 11
Fax: (+420 5) 41 21 03 61
e-mail: ljaros@unis.cz
MANAGING DIRECTOR: Karel Kuhnl
DESIGNER: Ing Jan Namisnak
MARKETING: Libor Jaroš

UPDATED

UNIS NA 40 BONGO
TYPE: Two-seat helicopter.
PROGRAMME: Design and construction started 1996; first flight early 1997; public debut at Brno International Machinery Fair 1997 by first prototype OK-CIU; two further prototypes completed by mid-1998. UNIS plans Normal category certification to FAR Pts 27, 33, 34 and 36. Series production expected to start by late 1999 but not confirmed; up to 50 per year envisaged.
DESIGN FEATURES: Two-blade teetering rotor; no tail rotor; pod and boom fuselage with inverted Y tail unit.
FLYING CONTROLS: Conventional cyclic and collective controls. Ducted air anti-torque system.
STRUCTURE: Mainly composites, including rotor blades; laminated, vacuum-formed elastomeric rotor head.
LANDING GEAR: Tubular twin-skid gear with ground handling wheels. Inflatable permanent or emergency floats optional.
POWER PLANT: Two 50 to 70 kW (67 to 94 shp) PBS Velká Bíteš TE 50B turboshafts, with FADEC and dual ignition. Engines mounted side by side behind cockpit; transmission via combining gearbox. Single self-sealing fuel tank beneath engines, capacity 184 litres (48.6 US gallons; 40.5 Imp gallons).
ACCOMMODATION: Adjustable, foldable and anatomically shaped seats for pilot and passenger. Baggage space aft of seats; provision for additional, aerodynamic baggage container under fuselage. Gull-wing window/door each side, hinged on centreline and opening upward.
SYSTEMS: Dual 27 V DC electrical systems; external power receptacle.
AVIONICS: *Instrumentation*: VFR standard.

DIMENSIONS, EXTERNAL:
Rotor diameter 7.40 m (24 ft 3¼ in)
Fuselage: Length 5.99 m (19 ft 7¾ in)
Max width 1.22 m (4 ft 0 in)
Height overall 2.54 m (8 ft 4 in)
Skid track 1.76 m (5 ft 9¼ in)
DIMENSIONS, INTERNAL:
Baggage space (approx) 0.08 m³ (2.8 cu ft)
AREAS:
Rotor disc 43.01 m² (462.9 sq ft)
WEIGHTS AND LOADINGS:
Weight empty, equipped 330 kg (728 lb)
Max load on external sling 240 kg (529 lb)
Max T-O weight: internal load 650 kg (1,433 lb)
with external sling load 720 kg (1,587 lb)
Max disc loading:
internal load 15.11 kg/m² (3.09 lb/sq ft)
with external sling load 16.74 kg/m² (3.43 lb/sq ft)
PERFORMANCE:
Never-exceed speed (V_{NE}) 151 kt (280 km/h; 174 mph)
Max level speed 135 kt (250 km/h; 155 mph)
Cruising speed 124 kt (230 km/h; 143 mph)
Max rate of climb at S/L 540 m (1,772 ft)/min
Hovering ceiling IGE 4,000 m (13,120 ft)
Max range 378 n miles (700 km; 435 miles)

UPDATED

UNIS NA 40 Bongo two-seat light utility helicopter

Rotor head details of the NA 40 Bongo

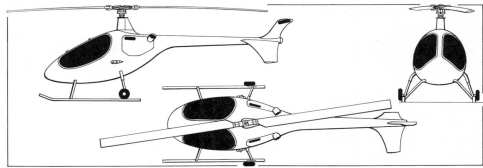

General arrangement of the UNIS NA 40 Bongo light helicopter *(James Goulding/Jane's)*

URBAN

URBAN-AIR s.r.o.
CZ-561 16 Libchavy 83
Tel/Fax: (+420 465) 58 21 53
e-mail: urbanair@libchavy.cz
Web: http://www.libchavy.cz/~urbanair
DIRECTOR: Pavel Urban
SALES MANAGER: Ing Milos Mladek

This company produces the UFM 10 Samba and three versions of the Lambáda ultralight at its three factories in the area of Usti nad Orlici.

UPDATED

URBAN UFM 11 and UFM 13 LAMBÁDA
TYPE: Side-by-side ultralight/kitbuilt.
PROGRAMME: Design started June 1994; first flight (UFM 11 prototype) 23 May 1996; production began October 1996; first customer delivery October 1997.
CURRENT VERSIONS: **UFM 11**: Standard version. See separate entry for low-wing **UFM 10** variant.
UFM 13: Extended-span version. Also available as sailplane **(UFM 13W)** with single mainwheel.
UFM 15: Motor glider version.
CUSTOMERS: Total 20 ordered by late 1998, of which nine (six UFM 11 and three UFM 13) then completed. Exported to Australia, Austria, Germany and elsewhere. Three exported to Ireland in 2000, identified as UFM 11 Moira Lambada.
COSTS: Flyaway, excluding VAT, DM98,000 (2000).
Data apply to UFM 11 and UFM 13 except where indicated.
DESIGN FEATURES: Conforms to JAR-VLA. Forward-swept wings with cranked leading-edge and laminar profile (aerofoil section SM 701); incidence 3°. Wings and horizontal tail detach for storage and transportation.
FLYING CONTROLS: Conventional (manual) flaperons (UFM 13) or ailerons (UFM 11), rudder and one-piece elevator; split flaps on UFM 11, upper surface Schempp-Hirth spoilers on UFM 13.
STRUCTURE: Laminated glass fibre and carbon fibre, with CFRP wing spar.
LANDING GEAR: Fixed tricycle or tailwheel type; hydraulic brakes. Mainwheel tyres 4.00-8 in.

POWER PLANT: One 52.2 kW (70 hp) HKS 700E four-stroke or 31.0 kW (41.6 hp) Rotax 447 UL-2V two-stroke and two-blade Aero Sport carbon fibre, fixed-pitch propeller standard; 73.5 kW (98.6 hp) Rotax 912 ULS four-stroke and Kremen SR 200 three-blade propeller optional in UFM 11. BSAB 430 engine in prospect. Fuel capacity 40 litres (10.6 US gallons; 8.8 Imp gallons) standard, 50 litres (13.2 US gallons; 11.0 Imp gallons) optional with Rotax 912.

DIMENSIONS, EXTERNAL:
Wing span: 11	11.80 m (38 ft 8½ in)
13	13.00 m (42 ft 7¾ in)
15	15.00 m (49 ft 2½ in)
Length overall: 11, 13	6.60 m (21 ft 7¾ in)
Height overall: 11, 13	1.85 m (6 ft 0¾ in)

DIMENSIONS, INTERNAL:
Cabin max width: 11, 13	1.06 m (3 ft 5¾ in)

AREAS:
Wings, gross: 11	10.80 m² (116.25 sq ft)
13	12.16 m² (130.9 sq ft)

WEIGHTS AND LOADINGS:
Weight empty: 11, 13 (HKS 700E)	255 kg (562 lb)
11 (Rotax 447)	240 kg (529 lb)
13 (Rotax 447)	245 kg (540 lb)
15	290 kg (639 lb)
Max T-O weight: all versions	450 kg (992 lb)

PERFORMANCE, POWERED (HKS 700E engine):
Max cruising speed: 11	92 kt (170 km/h; 105 mph)
13	86 kt (160 km/h; 99 mph)
15	81 kt (150 km/h; 93 mph)
Stalling speed: 11	36 kt (65 km/h; 41 mph)
13	35 kt (63 km/h; 40 mph)
Max rate of climb at S/L: 11, 13	210 m (689 ft)/min
T-O run: 11	72 m (240 ft)
13	70 m (230 ft)
Range: 11, 13	432 n miles (800 km, 497 miles)

PERFORMANCE, UNPOWERED:
Best glide ratio: 11	20
13	23
15	28
Min rate of sink: 11	1.30 m (4.27 ft)/s
13	1.10 m (3.61 ft)/s

UPDATED

Urban UFM 13 S Lambáda built in Germany to Experimental rules by Prof Uwe Wiltin and Ing Rudi Hackel *(Paul Jackson/Jane's)* 2001/0092099

URBAN UFM 10 SAMBA

TYPE: Side-by-side ultralight/kitbuilt.
PROGRAMME: Previously known as Speed Lambáda. Prototype OK-EUU 38 flying by 1999; first production aircraft (OK-EUU 39) flew 21 July 1999.
Details as for UFM 11 except as follows:
DESIGN FEATURES: Short-span version of UFM 11 (which see); low-mounted wing and tailplane; shorter wheelbase.
COSTS: DM96,000 to DM98,500, including engine (2000).
FLYING CONTROLS: Conventional and manual. Split flaps.
POWER PLANT: One 59.6 kW (79.9 hp) Rotax 912 UL or 73.5 kW (98.6 hp) Rotax 912 ULS flat-four, driving two-blade, ground-adjustable pitch propeller. Fuel capacity 50 litres (13.2 US gallons; 11.0 Imp gallons).

DIMENSIONS, EXTERNAL:
Wing span	10.00 m (32 ft 9¾ in)
Length overall	5.90 m (19 ft 4¼ in)
Height overall	1.60 m (5 ft 3 in)

WEIGHTS AND LOADINGS:
Weight empty	270 kg (595 lb)
Max T-O weight	450 kg (992 lb)

PERFORMANCE, POWERED (estimated):
Max cruising speed	108 kt (200 km/h; 124 mph)
Stalling speed: flaps up	38 kt (70 km/h; 44 mph)
flaps down	35 kt (64 km/h; 40 mph)
T-O run	75 m (250 ft)

PERFORMANCE, UNPOWERED (estimated):
Best glide ratio	19
Min rate of sink	1.50 m (4.92 ft)/s

UPDATED

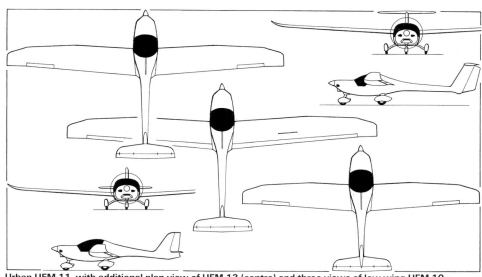

Urban UFM 11, with additional plan view of UFM 13 (centre) and three views of low-wing UFM 10 (bottom) 2001/0092147

First production Urban UFM 10 Samba *(Paul Jackson/Jane's)* 2001/0092100

WOLFSBERG-EVEKTOR

WOLFSBERG-EVEKTOR s.r.o.
Kunovice-Letiště, CZ-686 04 Kunovice
Tel: (+420 632) 53 75 01
Fax: (+420 632) 53 79 09
e-mail: wacz@evektor.cz

This company was formed in early 1999 to manage the programme for the Wolfsberg Raven 257 utility transport described in the Belgian section. Wings and horizontal tail of the prototype (OK-RAV), which made its maiden flight on 28 July 2000, were manufactured by Letov Air of Prague and landing gear by GAIC of China; prototype completion was undertaken by EV-AT (which see). On 21 November 2000, however, Wolfsberg Aircraft Corporation NV (see Belgian section) announced cancellation of Evektor agreement and reassignment of Raven's design and production to Letov (see this section).

UPDATED

ZLIN

MORAVAN AEROPLANES INC (A division of Moravan Inc)
Letiště 1578, CZ-765 81 Otrokovice
Tel: (+420 67) 767 20 00 and 767 20 01
Fax: (+420 67) 792 21 48
e-mail: moravan@moravan.cz
Web: http://www.moravan.cz
PRESIDENT: Libor Soska
VICE-PRESIDENT: Patrik Joachimczyk

MANAGING DIRECTOR: Marek Dolansky
TECHNICAL DIRECTOR: Vaclav Krizek
SALES DIRECTOR: Daniel Sabat
MARKETING DIRECTOR: Viktor Slampa

Formed 18 September 1934 as Zlinská Letecká Akciová Společnost (Zlin Aviation Company Ltd) in Zlin; manufacture of Zlin aircraft started 1933 by Masarykova Letecká Liga (Masaryk League of Aviation); total of approximately 650 aircraft, including gliders and military transports, manufactured by end of Second World War.

Company then renamed Moravan; postwar aircraft production totalled 3,128 by 3 December 1996, including 1,495 of the Z 26 family (162 Z 26, 170 Z 126, 366 Z 226, 436 Z 326, 331 Z 526 and 30 Z 726) built between 1949 and 1975 and 698 Z 42/43/142 (see 1999-2000 edition for details) produced between 1970 and 1995. Algerian Z 43s (five supplied in 1991) designated **Safir 43**; Algerian Z 142s (17 complete and 29 kits in 1987-91) designated **Firnas 142**. Only 25 aircraft (11 Z 143Ls and 14 Z 242Ls) had been produced by July 1997, when the factory suffered serious flood damage, causing production to be suspended. It was

CZECH REPUBLIC: AIRCRAFT—ZLIN

Zlin Z 142 C aerobatic light trainer/tourer 2000/0064437

resumed in 1998. Workforce was approximately 1,250 at end of 1997.

In June 1999, Moravan announced the sale of 100 Z 242s to Egypt, equivalent to a two-year backlog of work at then-current production rates. This order then represented approximately half of Moravan's backlog value of US$40 million, but was later cancelled. A threefold increase in production was planned in 2000, compared with 1999. Aerolease, a joint venture with Dutch financial support, was formed in 2000 to lease Zlin aircraft to European flying clubs and flying training schools; it was expected to come into operation in June 2000.

Moravan also manufactures aircraft equipment.

UPDATED

ZLIN Z 142 C

TYPE: Primary prop trainer/sportplane.
PROGRAMME: Design of original Z 142 begun late 1977/early 1978, prototype (OK-078) construction April 1978; first flight 29 December 1978; FAR Pt 23 certification (Aerobatic, Utility and Normal) 1980; deliveries from April 1980; Canadian DoT certification 26 November 1991. Z 142 C received Czech certification on 18 July 1991; also certified in Australia (5 March 1996), Canada (4 August 1993) and Slovak Republic (18 June 1991). Production ended in 1995, but aircraft still being promoted in 1999 and available to order; standard delivery time, four months after contract signature. Full description in 1996-97 *Jane's*; shortened version follows.
CURRENT VERSIONS: **Z 142 C**: Current version, certified in A (aerobatic trainer), U (general trainer) and N (touring) categories. *Description applies to this version.*
 Z 142 CAF: Military version; eight delivered to Czech Air Force.
 Z 242 L: Lycoming-powered development of Z 142; described separately.
CUSTOMERS: Total 378 (all versions) built by 1995, including 10 Z 142 C for Canada/USA and eight Z 142 CAF for Czech Air Force.
COSTS: Standard aircraft US$119,860 (VFR) or US$139,860 (IFR) (1999).
DESIGN FEATURES: Development of Zlin Z 142 M (1980-81 *Jane's*); basic, advanced and aerobatic training, glider towing and (with appropriate equipment) night flying and IFR training.
 Wing section NACA $63_2 416.5$; dihedral 6° from roots; sweepforward 4° 20′ at quarter-chord.
POWER PLANT: One LOM M 337 AK inverted six-cylinder air-cooled in-line engine (157 kW; 210 hp at 2,750 rpm), with supercharger and low-pressure injection pump, driving a two-blade Avia V 500 A constant-speed metal propeller. Fuel tanks in each wing leading-edge, with combined capacity of 120 litres (31.7 US gallons; 26.4 Imp gallons). Normal category version has auxiliary 50 litre (13.2 US gallon; 11.0 Imp gallon) tank at each wingtip, increasing usable fuel capacity to 220 litres (58.1 US gallons; 48.4 Imp gallons). Fuel and oil systems permit inverted flying for up to 1½ minutes. Oil capacity 12 litres (3.2 US gallons; 2.6 Imp gallons).
ACCOMMODATION: As for Z 242 L.
DIMENSIONS, EXTERNAL:
 Wing span 9.16 m (30 ft 0½ in)
 Length overall 7.33 m (24 ft 0½ in)
 Height overall 2.75 m (9 ft 0¼ in)
 Wheel track 2.33 m (7 ft 7¾ in)
 Wheelbase 1.66 m (5 ft 5¼ in)
 Propeller diameter 2.00 m (6 ft 6¾ in)
DIMENSIONS, INTERNAL:
 Cockpit max width 1.12 m (3 ft 8 in)
AREAS:
 Wings, gross 13.15 m² (141.5 sq ft)
WEIGHTS AND LOADINGS (A: Aerobatic, U: Utility, N: Normal category):
 Basic weight empty (all versions) 730 kg (1,609 lb)
 Max T-O weight: A 970 kg (2,138 lb)
  U 1,020 kg (2,248 lb)
  N 1,090 kg (2,403 lb)
 Max landing weight: A 970 kg (2,138 lb)
  U 1,020 kg (2,248 lb)
  N 1,050 kg (2,315 lb)
 Max wing loading: A 73.8 kg/m² (15.11 lb/sq ft)
  U 77.6 kg/m² (15.89 lb/sq ft)
  N 82.9 kg/m² (16.98 lb/sq ft)
 Max power loading: A 6.19 kg/kW (10.17 lb/hp)
  U 6.51 kg/kW (10.69 lb/hp)
  N 6.96 kg/kW (11.43 lb/hp)
PERFORMANCE (at relevant max T-O weight):
 Never-exceed speed (V_{NE}) (all versions):
  at S/L 170 kt (316 km/h; 196 mph)
  at 500 m (1,640 ft) 174 kt (323 km/h; 200 mph)
 Max level speed at S/L:
  A, U 125 kt (232 km/h; 144 mph)
  N 123 kt (228 km/h; 142 mph)
 Max cruising speed at S/L at 75% power:
  A, U 112 kt (207 km/h; 129 mph)
  N 109 kt (202 km/h; 126 mph)
 Max cruising speed at 2,000 m (6,560 ft):
  A 115 kt (213 km/h; 132 mph)
  U 113 kt (210 km/h; 130 mph)
  N 111 kt (206 km/h; 128 mph)
 Econ cruising speed at S/L at 65% power:
  A, U 104 kt (193 km/h; 120 mph)
  N 100 kt (186 km/h; 116 mph)
 Econ cruising speed at 500 m (1,640 ft):
  A, U 97 kt (179 km/h; 111 mph)
  N 92 kt (171 km/h; 106 mph)
 Stalling speed at S/L, flaps up:
  A 61 kt (113 km/h; 71 mph)
  U 63 kt (115 km/h; 72 mph)
  N 65 kt (120 km/h; 75 mph)
 Stalling speed at S/L, T-O flap setting:
  A 60 kt (110 km/h; 69 mph)
  U 61 kt (112 km/h; 70 mph)
  N 63 kt (116 km/h; 73 mph)
 Stalling speed at S/L, flaps down:
  A 53 kt (98 km/h; 61 mph)
  N 56 kt (103 km/h; 64 mph)
 Max rate of climb at S/L, ISA:
  A 312 m (1,023 ft)/min
  U 388 m (945 ft)/min
  N 252 m (827 ft)/min
 Service ceiling: A 5,000 m (16,400 ft)
  N 4,200 m (14,180 ft)
 T-O run: A 231 m (760 ft)
  U 236 m (775 ft)
  N 252 m (830 ft)
 T-O to 15 m (50 ft): A 510 m (1,675 ft)
  U 560 m (1,840 ft)
  N 620 m (2,035 ft)
 Landing from 15 m (50 ft): A 425 m (1,395 ft)
  U 440 m (1,445 ft)
  N 460 m (1,510 ft)
 Landing run: A 190 m (625 ft)
  U 200 m (660 ft)
  N 220 m (725 ft)
 Range at econ cruising speed:
  A, U 316 n miles (586 km; 364 miles)
  N 522 n miles (967 km; 601 miles)
 g limits: A +6/−3.5
  U +6/−3
  N +3.8/−1.5

UPDATED

ZLIN Z 242 L

TYPE: Primary prop trainer/sportplane.
PROGRAMME: Lycoming-powered version of Z 142 (see 1997-98 and earlier *Jane's*). Design started December 1988; first flight (OK-076) 14 February 1990; second prototype (SE-KMS) followed, both converted on production line from Z 142; Czech certification to FAR Pt 23 (A, U and N categories) 22 March 1992; now certified in Argentina, Australia, Austria, Belgium, Canada, Israel, Luxembourg, Slovenia, Sweden, UK and USA.
CURRENT VERSIONS: **Z 242 L**: Aerobatics-capable trainer. *Description applies to this version.*
 Z 242 LA: Lower-powered, non-aerobatic version for primary, day/night and IFR training and glider/banner towing; *g* limits +4.4/−1.76. Prototype (OK-ANA) on test early 1996. No subsequent information received. All known details in 1998-99 edition.
CUSTOMERS: Total of 96 delivered by 21 November 2000, for customers in Argentina (one), Australia (one), Canada (15), Czech Republic (four), Israel (three), Macedonia (four), Peru (Air Force, 18), Slovenia (Defence Force, seven), Sweden (two), UK (three) and USA (38). Egyptian Air Force order for 100 announced at Paris Air Show, June 1999, but cancelled in favour of Grob G 115 in 2000. Israel has shown interest in up to 25.
DESIGN FEATURES: Conventional low-wing monoplane. Constant-chord horizontal surfaces with wingroot glove. Sweptback fin with fillet. Changes from Z 142, apart from new engine, include redesigned (and shorter) engine cowling and front fuselage, wing incidence, 0° sweep, wingroot glove, redesigned wing- and tailplane tips, redesigned fuel system and updated instruments. Spin recovery strake each side of cowling.
 Wing section NACA $63_2 416.5$.
FLYING CONTROLS: Conventional and manual. Slotted, mass-balanced surfaces used for ailerons and flaps; horn-balanced elevator with trim tab; ground-adjustable tabs in ailerons and rudder; ailerons and elevator operated by rods, rudder by cables.
STRUCTURE: Mainly metal; wing has main and auxiliary spars; duralumin skins, fluted on control surfaces; metal engine

Zlin Z 242 L (Textron Lycoming AEIO-360 engine) *(Paul Jackson/Jane's)* 2001/0092101

cowlings; centre-fuselage is steel tube cage with composites skin panels.
LANDING GEAR: Non-retractable tricycle type, with nosewheel offset 13 cm (5.1 in) to port. Oleo-pneumatic nosewheel shock-absorber. Mainwheels carried on flat spring steel legs. Nosewheel steered (±38°) by rudder pedals. Mainwheels and Barum tyres size 420×150 or Goodyear 6.00-6.5, pressure 1.90 bars (28 lb/sq in); nosewheel and Barum tyre size 350×135 or Goodyear 5.00-5, pressure 2.50 bars (36 lb/sq in). Hydraulic disc brakes on mainwheels can be operated from either seat. Parking brake standard.
POWER PLANT: One Textron Lycoming AEIO-360-A1B6 flat-four engine (149 kW; 200 hp at 2,700 rpm) driving a Mühlbauer MTV-9-B-C/C-188-18a three-blade constant-speed wood/composites propeller or Hartzell HC-C3YR-4BF/FC 6890 three-blade constant-speed metal propeller. Fuel capacity 120 litres (31.7 US gallons; 26.4 Imp gallons); Normal category version wingtip tanks 55 litres (14.5 US gallons; 12.1 Imp gallons) each, bringing usable capacity to 230 litres (60.7 US gallons; 50.6 Imp gallons). Inverted flight limited to 1 minute. Oil capacity 8 litres (2.1 US gallons; 1.8 Imp gallons).
ACCOMMODATION: Side-by-side seats for two persons, instructor's seat to port. Both seats adjustable and permit use of back-type parachutes. Baggage space aft of seats. Cabin and windscreen heating and ventilation standard. Forward-sliding cockpit canopy. Dual controls standard.
SYSTEMS: Electrical system includes 1.6 kW 28 V engine-driven generator and 24 V 19 Ah Gill battery. External power source can be used for engine starting.
AVIONICS: To customer's specification, usually from Honeywell Silver Crown or Garmin range.
EQUIPMENT: Standard equipment includes EGT gauge, fuel flow indicator, g meter and anti-collision beacon. Optional items include cockpit/instrument/cabin lights, landing/taxying lights, anti-collision lights, and towing gear for gliders of up to 500 kg (1,102 lb) weight.

DIMENSIONS, EXTERNAL:
Wing span	9.34 m (30 ft 7¾ in)
Wing chord, constant portion	1.50 m (4 ft 11 in)
Wing aspect ratio	6.3
Length overall	6.94 m (22 ft 9¼ in)
Height overall	2.95 m (9 ft 8¼ in)
Elevator span	3.20 m (10 ft 6 in)
Wheel track	2.33 m (7 ft 7¾ in)
Wheelbase	1.76 m (5 ft 9¼ in)
Propeller diameter (MTV)	1.88 m (6 ft 2 in)
Propeller ground clearance (MTV)	0.33 m (1 ft 1 in)

DIMENSIONS, INTERNAL:
Cabin: Length	1.80 m (5 ft 10¾ in)
Max width	1.12 m (3 ft 8 in)
Max height	1.20 m (3 ft 11¼ in)
Baggage space	0.20 m³ (7.1 cu ft)

AREAS:
Wings, gross	13.86 m² (149.2 sq ft)
Ailerons (total)	1.41 m² (15.18 sq ft)
Trailing-edge flaps (total)	1.41 m² (15.18 sq ft)
Fin	0.54 m² (5.81 sq ft)
Rudder, incl tab	0.81 m² (8.72 sq ft)
Tailplane	1.23 m² (13.24 sq ft)
Elevator, incl tabs	1.36 m² (14.64 sq ft)

WEIGHTS AND LOADINGS (A: Aerobatic, U: Utility, N: Normal category):
Basic weight empty: A, U, N	730 kg (1,609 lb)
Max T-O weight: A	970 kg (2,138 lb)
U	1,020 kg (2,248 lb)
N	1,090 kg (2,403 lb)
Max landing weight: A	970 kg (2,138 lb)
U	1,020 kg (2,248 lb)
N	1,050 kg (2,315 lb)
Max wing loading: A	70.3 kg/m² (14.40 lb/sq ft)
U	73.6 kg/m² (15.07 lb/sq ft)
N	78.6 kg/m² (16.11 lb/sq ft)
Max power loading: A	6.50 kg/kW (10.68 lb/hp)
U	6.84 kg/kW (11.24 lb/hp)
N	7.31 kg/kW (12.01 lb/hp)

PERFORMANCE:
Never-exceed speed (V$_{NE}$): all versions	170 kt (315 km/h; 195 mph)
Max level speed at S/L: A	127 kt (236 km/h; 146 mph)
U	125 kt (233 km/h; 144 mph)
N	124 kt (231 km/h; 143 mph)
Max cruising speed at 2,000 m (6,560 ft) at 75% power: A	123 kt (227 km/h; 141 mph)
U	121 kt (225 km/h; 140 mph)
N	120 kt (223 km/h; 139 mph)
Stalling speed at S/L: flaps up: A	60 kt (111 km/h; 69 mph)
U	62 kt (114 km/h; 71 mph)
N	64 kt (118 km/h; 74 mph)
T-O flap setting: A	57 kt (105 km/h; 66 mph)
U	58 kt (107 km/h; 67 mph)
N	60 kt (111 km/h; 69 mph)
flaps down: A	51 kt (94 km/h; 59 mph)
U	53 kt (97 km/h; 61 mph)
N	54 kt (100 km/h; 63 mph)
Max rate of climb at S/L: A	336 m (1,102 ft)/min
U	300 m (984 ft)/min
N	270 m (886 ft)/min
T-O run: A	210 m (690 ft)
U	233 m (765 ft)
N	266 m (875 ft)
T-O to 15 m (50 ft): A	450 m (1,480 ft)
U	495 m (1,625 ft)
N	565 m (1,855 ft)
Landing from 15 m (50 ft): A	500 m (1,645 ft)
U	525 m (1,725 ft)
N	540 m (1,775 ft)
Range with max fuel: A, U	267 n miles (495 km; 308 miles)
N	570 n miles (1,056 km; 656 miles)
g limits: A	+6/−3.5
U	+5/−3
N	+3.8/−1.5

UPDATED

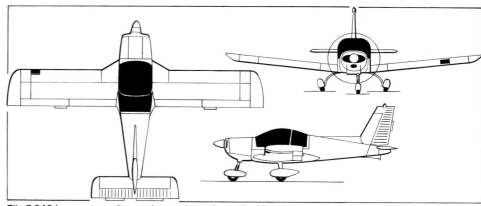

Zlin Z 242 L two-seat trainer and sportplane (*James Goulding/Jane's*) 0010599

ZLIN Z 143
Czech Air Force designation: Z 143 MAF
TYPE: Four-seat lightplane.
PROGRAMME: Launched May 1991, with prototype construction beginning mid-year and first flight (Z 143 L prototype OK-074) 24 April 1992. Czech certification of this version 10 June 1994; now also certified in Austria, Bulgaria, Germany, Slovak Republic, Switzerland and USA.
CURRENT VERSIONS: **Z 143 L**: Initial version, with Textron Lycoming engine.
Detailed description applies to Z 143 L.
Z 143 LSi: Version with 194 kW (260 hp) Textron Lycoming fuel-injection engine. Under development.
CUSTOMERS: Total of 37 delivered by 1 December 2000 to customers in Argentina (one), Austria (one), Bulgaria (two), Czech Republic (Air Force, four), Egypt (five), Germany (eight), Israel (two), Slovenia (two), Switzerland (four), Thailand (two) and USA (six).
COSTS: Standard aircraft US$149,000 (VFR), US$169,000 (IFR) (1999).

DESIGN FEATURES: Generally as for Z 242. For basic aerobatics (except inverted flight), glider towing and (with equipment) night and IFR training. Non-swept wings, of similar planform to Z 242 L but with NACA 63$_2$416.5 wing section and 6° dihedral.
FLYING CONTROLS: As for Z 242 L.
STRUCTURE: As for Z 242 L.
LANDING GEAR: As for Z 242 L.
POWER PLANT: One Textron Lycoming O-540-J3A5 flat-six engine (175 kW; 235 hp at 2,400 rpm); Mühlbauer MTV-9-B/195-45a three-blade variable-pitch propeller. Fuel capacity as for Z 242 L.
ACCOMMODATION: Four seats in 2 + 2 configuration, with pilot in front left-hand seat. All seats have four-point safety belts. Baggage compartment (capacity 60 kg; 132 lb) aft of rear seats, with external door on port side. One-piece forward-sliding canopy. Dual controls standard. Cabin and windscreen heating and ventilation standard.
SYSTEMS: As for Z 242 L.
AVIONICS: As for Z 242 L.

DIMENSIONS, EXTERNAL:
Wing span	10.14 m (33 ft 3¼ in)
Wing chord, mean aerodynamic	1.49 m (4 ft 10¾ in)
Wing aspect ratio	7.0
Length overall	7.58 m (24 ft 10½ in)
Height overall	2.91 m (9 ft 6½ in)
Elevator span	3.01 m (9 ft 10½ in)
Wheel track	2.44 m (8 ft 0 in)
Wheelbase	1.75 m (5 ft 9 in)
Propeller diameter	1.95 m (6 ft 4¾ in)
Propeller ground clearance	0.28 m (11 in)

DIMENSIONS, INTERNAL:
Baggage compartment volume	0.25 m³ (8.83 cu ft)

AREAS: As for Z 242 L except:
Wings, gross	14.78 m² (159.1 sq ft)

WEIGHTS AND LOADINGS (U: Utility, N: Normal category):
Weight empty, equipped: U, N	840 kg (1,852 lb)
Max T-O weight: U	1,080 kg (2,381 lb)
N	1,350 kg (2,976 lb)

Z 143 L instrument panel

Zlin Z 242 L instrument panel

Zlin Z 143 L operated in USA (*Paul Jackson/Jane's*)

Max landing weight: U	1,080 kg (2,381 lb)				
N	1,280 kg (2,822 lb)				
Max wing loading: U	73.1 kg/m² (14.97 lb/sq ft)				
N	91.4 kg/m² (18.72 lb/sq ft)				
Max power loading: U	6.16 kg/kW (10.13 lb/hp)				
N	7.70 kg/kW (12.66 lb/hp)				

PERFORMANCE:
Never-exceed speed (VNE):
  U, N   170 kt (315 km/h; 195 mph)
Max level speed: U   144 kt (267 km/h; 166 mph)
  N   141 kt (262 km/h; 163 mph)
Max cruising speed at 75% power:
  U   127 kt (235 km/h; 146 mph)
  N   125 kt (232 km/h; 144 mph)
Econ cruising speed at 65% power:
  U   116 kt (216 km/h; 134 mph)
  N   113 kt (209 km/h; 130 mph)
Stalling speed:
  flaps up: U   58 kt (108 km/h; 67 mph)
  N   63 kt (117 km/h; 73 mph)
  T-O flap setting: U   54 kt (100 km/h; 63 mph)
  N   58 kt (107 km/h; 67 mph)
  flaps down: U   54 kt (100 km/h; 63 mph)
  N   60 kt (110 km/h; 69 mph)
Max rate of climb at S/L: U   444 m (1,457 ft)/min
  N   294 m (964 ft)/min
Service ceiling: U   5,700 m (18,700 ft)
  N   4,170 m (13,680 ft)
T-O run: U   170 m (560 ft)
  N   295 m (970 ft)
T-O to 15 m (50 ft): U   450 m (1,480 ft)
  N   640 m (2,100 ft)
Landing from 15 m (50 ft): U   590 m (1,935 ft)
  N   765 m (2,510 ft)
Landing run: U   305 m (1,000 ft)
  N   380 m (1,250 ft)
Range at 3,050 m (10,000 ft) at 60% power:
  U   307 n miles (568 km; 353 miles)
  N   708 n miles (1,311 km; 814 miles)
g limits: U   +4.4/−1.76
  N   +3.8/−1.52

UPDATED

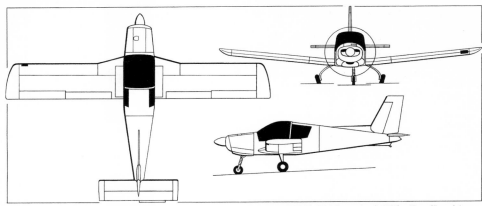

General arrangement of the Zlin Z 143 L (Textron Lycoming O-540-J3A5 engine) *(Mike Keep/Jane's)*

ETHIOPIA

EAL
ETHIOPIAN AIRLINES ENTERPRISE
PO Box 1755, Addis Ababa
Tel: (+251 1) 61 22 22
Fax: (+251 1) 61 14 74
e-mail: eal@telecom.net.ett
Telex: 21012 ETHAIR ADDIS
CEO: Bisrat Nigatu
AIRCRAFT DIVISION MANAGER, AGRO AIRCRAFT MANUFACTURING:
  Tilahun Kassa

Ag-Cat Corporation Ag-Cat Super B Turbine assembled and part-manufactured by EAL under name **Eshet** for Ethiopian use and export; first aircraft (ET-AIY) rolled out 20 December 1986; same as former Ag-Cat Corporation version (see US section of 1997-98 *Jane's*). Power plant is one 559 kW (750 shp) PT6A-34AG turboprop driving 2.69 m (8 ft 10 in) Hartzell HC-B3TN-3D three-blade propeller; optional engine is 507 kW (680 shp) PT6A-15AG; maximum fuel load 435 litres (115 US gallons; 95.75 Imp gallons) and oil capacity 10.6 litres (2.8 US gallons; 2.3 Imp gallons).

EAL has sole rights to build, market and service Ag-Cat throughout Africa except Algeria, Tunisia and South Africa; proportion of local manufacture 35 per cent since 1995. By December 1998 (latest information received), EAL had delivered 12 Ag-Cats; had two available for delivery; and three more in process of assembly.

UPDATED

Ethiopian Airlines licence-built version of the Ag-Cat Super B Turbine

FINLAND

FINAVITEC
PATRIA FINAVITEC OY
FIN-35600 Halli
Tel: (+358 3) 580 91
Fax: (+358 3) 580 96 67
e-mail: finavitec@patria.fi
Web: http://www.patria.fi
CORPORATE VICE-PRESIDENT AND CEO: Veijo Vartiainen
VICE-PRESIDENTS: Aarne Nieminen (Maintenance)
  Veli-Matti Erkkilä (Structures)
  Jukka Holkeri
  Esko Kurki
  Pauli Juuti (Systems)
COMMUNICATIONS MANAGER: Raili Saarinen

Finavitec (previously known as Valmet Aviation Industries – part of the Valmet group) and its predecessors have built 30 types of aircraft, including 19 of Finnish design, since 1922; current activities include production of aircraft and aero-engine parts; and maintenance, overhaul, repair and modification of aircraft and accessories, aero-engines and accessories, marine diesel engines, accessories and gears. Manufacture of Valmet L-90 TP Redigo ended in 1995, but the programme was acquired by Aermacchi of Italy (which see) as the M-290TP. In August 2000, Finavitec subsidiary Finavicomp delivered last of 57 Boeing F-18C Hornets to Finnish Air Force from CKD kits; also co-produced fuselage dorsal and side skin panels. Approval by Boeing to manufacture composites dorsal covers for the F/A-18E/F was given in April 1998. Finavitec additionally manufactures rear and centre fuselage sections for the Embraer ERJ-135/145 under subcontract to Sonaca of Belgium, to which first shipsets were delivered in second half of 1998.

Systems Division (electrical ground support and test systems, surveillance systems and space electronics) created 1 November 1997 by acquisition of Finnyards Ltd Electronics of Tampere; workforce 1,000 in late 1998 (including 320 in Finavicomp). Other facilities at Kuorevesi and Linnavuori. State-owned shares in Finavitec (50.1 per cent) were transferred to Patria Industries Oy, a 100 per cent state-owned company. MoUs signed in December 1999 by Finavitec and Airbus Industrie for former to take part in conceptual definition of A3XX and to explore other possible industrial co-operation.

UPDATED

FRANCE

AAT
AMEUR AVIATION TECHNOLOGIE
Lotissement Pasqualini, Baléone, F-20167 Sarolla Carcopino
Tel: (+33 4) 95 20 77 14
Fax: (+33 4) 95 20 76 12
CHIEF DESIGNER: Boussad Ameur

Ameur Aviation Technologie is based on the island of Corsica, where it produces the Baljims kitplane, previously known as the Balbuzard.

UPDATED

AAT BALJIMS 1A
TYPE: Side-by-side kitbuilt.
PROGRAMME: Prototype, then known as Balbuzard, first flew in 1995, powered by a 59.7 kW (80 hp) Rotax 912A. Second prototype (F-PARA) followed in July 1996 and features many changes to original design, including O-235 engine; lower positioned wing with increased span and

AAT Baljims 1A two-seat kitplane *(Pierre Gaillard)*

winglets; and redesigned landing gear of increased height. Name changed to Baljims in July 1998, when that designation applied to F-PTCD (c/n 01).
CUSTOMERS: First kit aircraft registered 1998.
DESIGN FEATURES: Pusher layout with V tail. Air intake above the cockpit. Wings removable for storage.
FLYING CONTROLS: Manual. Ailerons and combined elevators and rudders.
STRUCTURE: Generally of composites.
LANDING GEAR: Mainwheels retract inwards; nosewheel rearwards. Hydraulic brakes on mainwheels.
POWER PLANT: One 88.0 kW (118 hp) Textron Lycoming O-235 flat-four driving a two-blade fixed-pitch wooden propeller.
ACCOMMODATION: Pilot and passenger in side-by-side seating under one-piece blown Perspex canopy. Baggage locker.

DIMENSIONS, EXTERNAL:	
Wing span	7.90 m (25 ft 11 in)
Wing chord, mean	0.80 m (2 ft 7½ in)
Wing aspect ratio	10.1
Length overall	5.78 m (18 ft 11½ in)
Height overall	2.36 m (7 ft 9 in)
Tailplane span	2.30 m (7 ft 6½ in)
Wheel track	2.10 m (6 ft 10¾ in)
Wheelbase	1.63 m (5 ft 4¼ in)
AREAS:	
Wings, gross	6.20 m² (66.7 sq ft)
WEIGHTS AND LOADINGS:	
Weight empty	395 kg (871 lb)
Baggage capacity	45 kg (99 lb)
Max fuel weight	87 kg (192 lb)
Max T-O weight	700 kg (1,543 lb)
Max wing loading	112.9 kg/m² (23.12 lb/sq ft)
Max power loading	7.95 kg/kW (13.07 lb/hp)
PERFORMANCE:	
Max operating speed (V_{MO})	199 kt (370 km/h; 229 mph)
Cruising speed at 80% power	194 kt (360 km/h; 224 mph)
T-O speed	67 kt (125 km/h; 78 mph)
Stalling speed, power off	57 kt (105 km/h; 66 mph)
Max rate of climb at S/L	420 m (1,380 ft)/min
Range, 50 min reserves, 15°C	800 n miles (1,482 km; 920 miles)
Endurance, 50 min reserves	4 h 30 min
g limits	+4.4/−2.2

VERIFIED

AEROKUHLMANN

AEROKUHLMANN
Aérodrome de Cerny, F-91590 Cerny
Tel: (+33 1) 69 90 17 80
Fax: (+33 1) 69 23 34 78
e-mail: aerokuhlmann@magic.fr
Web: http://www.chez.com/scub
DIRECTOR: Hervé Kuhlmann
COMMERCIAL DIRECTOR: Daniel Feldzer

AeroKuhlmann's first product, built in the company's 600 m² (6,450 sq ft) factory on La Ferté-Alais airfield, near Cerny, is the SCUB.

UPDATED

AEROKUHLMANN SCUB
TYPE: Tandem-seat kitbuilt.
PROGRAMME: Prototype (91-KJ) first flew 5 May 1996. Certified to JAR-VLA standard in France.
CURRENT VERSIONS: **Ultralight:** Optional 450 kg (992 lb) MTOW in Europe or 544 kg (1,200 lb) in USA.
Certified: MTOW 598 kg (1,320 lb).
Description applies to certified version, except where otherwise stated.
CUSTOMERS: Production totalled 17 by June 1999.
COSTS: FFr290,000, basically equipped, plus VAT (1997). Floats FFr36,000 extra.
DESIGN FEATURES: Braced high-wing monoplane, suitable for wide range of tasks including surveillance, mapping, cargo, medical and agricultural duties. Wings and tail fold for storage. Airframe design life of 10,000 hours.
Wing supercritical profile NASA L51 0417 mod; incidence 4° 30′, dihedral 2°.
FLYING CONTROLS: Conventional and manual. Horn-balanced rudder. Elevator movement +24° to −24°, rudder movement +21° to −21°, aileron movement +8° to −15°. Flight-adjustable trim tab on starboard elevator. No flaps.
STRUCTURE: Dacron-covered chromium-molybdenum tubular steel fuselage. Wing main spar of three-ply birch with carbon fibre spar cap, glass fibre ribs and plywood-covered leading-edge, all covered with Dacron.
LANDING GEAR: Tailwheel type, non-retractable; drum brakes. Mainwheels 8.00-6. Optional float (Stéphane Morosini/Hydro Aero Concept two-step) or ski gear can be installed in quoted time of 2 hours; additional pick-up point for both forward of main legs. Tundra tyres can be fitted for rough field operations.
POWER PLANT: One 62.5 kW (83.8 hp) JPX 4TX75A flat-four with dual ignition, driving a ULX two-bladed wooden propeller. Fuel capacity 110 litres (29.0 US gallons; 24.2 Imp gallons). Oil capacity 11.4 litres (3.0 US gallons; 2.5 Imp gallons).
ACCOMMODATION: Pilot and passenger in tandem, with dual controls.
SYSTEMS: All electric systems corrosion proof on floatplane version.
AVIONICS: *Flight:* GPS optional.
Instrumentation: ASI, VSI, compass, altimeter, oil temperature, oil pressure, RPM, fuel gauges.
EQUIPMENT: For surveillance, video cameras can be fitted beneath wings; four Micronair 7000 spray atomisers can be fitted on underwing bars, with a 150 litre (40 US gallon; 33 Imp gallon) tank fitted under fuselage.

AeroKuhlmann SCUB lightplane equipped with Micronair spraying equipment *(Paul Jackson/Jane's)*
2001/0089476

DIMENSIONS, EXTERNAL:	
Wing span	11.00 m (36 ft 1 in)
Wing aspect ratio	8.5
Width, wings folded	1.83 m (6 ft 0 in)
Length overall	7.01 m (23 ft 0 in)
Height: overall	1.83 m (6 ft 0 in)
Folded	2.29 m (7 ft 6 in)
Tailplane span	2.74 m (9 ft 0 in)
Tailplane chord	0.71 m (2 ft 4 in)
AREAS:	
Wings, gross	14.30 m² (153.9 sq ft)
Ailerons (total)	0.72 m² (7.80 sq ft)
Fin	0.37 m² (4.00 sq ft)
Rudder, incl tab	0.56 m² (6.00 sq ft)
Tailplane	1.21 m² (13.00 sq ft)
Elevators, incl tab	1.21 m² (13.00 sq ft)
WEIGHTS AND LOADINGS:	
Weight empty: landplane	250 kg (551 lb)
floatplane	354 kg (780 lb)
Max T-O weight: landplane	450 kg (992 lb)
floatplane	500 kg (1,102 lb)
Max wing loading: landplane	31.5 kg/m² (6.45 lb/sq ft)
floatplane	35.0 kg/m² (7.16 lb/sq ft)
Max power loading: landplane	7.20 kg/kW (11.83 lb/hp)
floatplane	8.00 kg/kW (13.14 lb/hp)
PERFORMANCE, POWERED:	
Never-exceed speed (V_{NE})	91 kt (169 km/h; 105 mph)
Max operating speed	67 kt (124 km/h; 77 mph)
Stalling speed	33 kt (60 km/h; 38 mph)
Max rate of climb at S/L	240 m (787 ft)/min
Service ceiling	4,570 m (15,000 ft)
T-O and landing run: on land	55 m (180 ft)
on water	91 m (300 ft)
T-O to 15 m (50 ft)	137 m (450 ft)
Landing from 15 m (50 ft)	201 m (660 ft)
Range with max fuel	543 n miles (1,005 km; 625 miles)
Endurance	7 h 0 min
g limits	+6.6/−3.3
PERFORMANCE, UNPOWERED:	
Best glide ratio at 49 kt (90 km/h; 56 mph)	15

UPDATED

APEX

APEX INTERNATIONAL
3 rue Troyon, F-75017 Paris
Tel: (+33 1) 40 72 61 10
Fax: (+33 1) 40 72 62 98
Web: http://www.capaviation.com/site/gb/apex
CHAIRMAN AND CEO: Guy Pellissier

Three French light aircraft manufacturers are included in the holding company Apex International (formerly known as Aéronautique Service); products of BUL, CAP and Robin are described under those headings later in this section; related companies are detailed below. In total, the group had 170 employees in 1999.

Constructions Aéronautiques de Bourgogne (CAB): Construction of BUL Zùlù, CAP 10 and CAP 222 at Darois, and Robin aircraft at Bernay. President: Patrice Serignac.

CAP Industries: Construction of CAP 232 and after-sales service of all CAP aircraft (including former Mudry CAP designs) at Bernay. Director-General: Eric Willaert.

Aviation Service Center (ASCO): After-sales service and support. Director-General: Ludovic Paul.

Group administrative services are provided by **Aéronautique Financements & Services.**

UPDATED

ASSOCIATION ZEPHYR

ASSOCIATION ZEPHYR
34 rue Victor Hugo, F-92300 Levallois-Perret
Tel/Fax: (+33 1) 41 06 69 88
e-mail: Zephyr@estaca.fr

Association Zéphyr was founded in 1994 by student engineers of the Ecole Supérieure des Techniques Aérotechniques et de Construction Automobile (ESTACA), who have designed the Alizé light aircraft, intended for amateur construction.

UPDATED

ASSOCIATION ZEPHYR ALIZE
English name: Tradewind
TYPE: Two-seat lightplane.
PROGRAMME: Design started 1996; construction of prototype began January 1998; first flight scheduled for 2000, but not reported by September of that year.
DESIGN FEATURES: Low wing; swept fin and rudder. Wing section NACA NLF 0215F.
STRUCTURE: All-wood airframe.
LANDING GEAR: Retractable tricycle type.
POWER PLANT: One 119 kW (160 hp) Textron Lycoming IO-320-B1E, driving an Evra propeller in prototype, with future provision for SMA MR 200 of 149 kW (200 hp). Fuel in two tanks, combined capacity 140 litres (37.0 US gallons; 30.8 Imp gallons).
ACCOMMODATION: Two persons, side by side, under bubble canopy.
DIMENSIONS, EXTERNAL:
  Wing span (without wingtips) 7.53 m (25 ft 8½ in)
  Wing aspect ratio 7.0
  Length overall 6.40 m (21 ft 0 in)
DIMENSIONS, INTERNAL:
  Cabin max width 1.10 m (3 ft 7¼ in)
  Baggage volume 0.04 m³ (1.41 cu ft)
AREAS:
  Wings, gross 8.10 m² (87.2 sq ft)
WEIGHTS AND LOADINGS:
  Weight empty 440 kg (970 lb)
  Max T-O weight 720 kg (1,587 lb)
  Max wing loading 88.8 kg/m² (18.21 lb/sq ft)
  Max power loading (prototype) 6.04 kg/kW (9.92 lb/hp)

Computer graphic of the Association Zéphyr Alizé *1999/0044674*

PERFORMANCE (estimated):
  Cruising speed 173 kt (320 km/h; 199 mph)
  Landing speed 59 kt (110 km/h; 68 mph)
  Max rate of climb at S/L 390 m (1,280 ft)/min
  Range 648 n miles (1,200 km; 745 miles)
  g limits +4.4/−2.2

UPDATED

AVIAKIT

AVIAKIT
rue Patures, F-10120 Pouange
Tel: (+33 3) 25 41 71 11

Development and refinement of Aviakit's Hermes continues.

UPDATED

AVIAKIT HERMES
TYPE: Side-by-side ultralight kitbuilt.
COSTS: Basic kit FFr98,000 (1996); engine, propeller and instruments extra.
DESIGN FEATURES: Low-wing tourer with high-mounted canopy and centre fuselage of sharply reduced cross-section aft of cockpit; strut-braced, mid-mounted tailplane; small underfin.
NACA 23015 wing section with 5° dihedral. Quoted build time is 200 hours.
FLYING CONTROLS: Conventional and manual. Large ground-adjustable tab added to rudder by 1998.
STRUCTURE: Fuselage is carbon-epoxy construction. Wings are wood/composites with carbon spar; strut-supported mid-mounted tailplane.
LANDING GEAR: Fixed tricycle type with speed fairings.

Aviakit Hermes two-seat kitplane *(Pierre Gaillard)*   *1999/0044653*

POWER PLANT: One two-cylinder 47.8 kW (64.1 hp) Rotax 582 or four-cylinder 59.6 kW (79.9 hp) Rotax 912 UL, according to customer's specification. Fuel capacity 60 litres (15.9 US gallons; 13.2 Imp gallons).
DIMENSIONS, EXTERNAL:
  Wing span 10.00 m (32 ft 9¾ in)
  Length 7.01 m (23 ft 0 in)
AREAS:
  Wings, gross 15.70 m² (169.0 sq ft)
WEIGHTS AND LOADINGS:
  Weight empty: Rotax 582 215 kg (474 lb)
  Rotax 912 UL 225 kg (496 lb)
  Max T-O weight 450 kg (992 lb)
PERFORMANCE:
  Max operating speed (V_MO) 100 kt (185 km/h; 115 mph)
  Cruising speed 90 kt (166 km/h; 103 mph)
  Stalling speed 33 kt (60 km/h; 38 mph)

UPDATED

BUL

BUL AVIATION
1 bis route de Troyes, F-21121 Darois
Tel: (+33 3) 80 44 20 75
Fax: (+33 3) 80 44 20 69 or 80 35 60 80
e-mail: bulaero@wanadoo.fr
PRESIDENT: Philippe Corne
CONTACT: David Morin

BUL (Bourgogne Ultra Léger) is one of three light aircraft manufacturers owned by Apex International (formerly Aéronautique Service), its specialist area being ultralights. Details of the Zùlù, its first product, were announced in 1998.

UPDATED

BUL ZÙLÙ
TYPE: Side-by-side ultralight.
PROGRAMME: Design began August 1996 and programme launched two months later; prototype construction began November 1996; first flight 15 March 1998; French certification achieved July 1998. Assembly of first production aircraft began October 1998 for first flight in August 1999, but no confirmation 12 months later that it had done so.
CUSTOMERS: First order placed 5 October 1998; four on order by December 1998 (latest information received). Planned production 100 per year.
COSTS: FFr375,000, incl VAT (1999).
DESIGN FEATURES: Based on Robin ATL (1991-92 *Jane's*), but with new wing and V tail replaced by conventional surfaces. Intended for easy and safe operation in a variety of roles, rapid removal of wings for road transport and certification in ULM category.
Wing section NACA 43015. Leading-edge sweep 4° 30′ 15″; thickness/chord ratio 15 per cent; dihedral 3° 40′; incidence 3° 24′; no twist.
FLYING CONTROLS: Manually actuated. Ailerons; rudder; all-moving horizontal tail with anti-balance tabs and electric trim. Flaps.
STRUCTURE: Composites throughout.
LANDING GEAR: Fixed, tricycle type, with elastomeric suspension and hydraulic disc brakes. Floats, skis, balloon tyres and speed fairings optional (available from September 1999).

BUL Zùlù ultralight (Rotax 912 piston engine)   *2001/0093660*

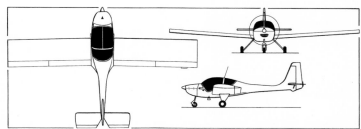

BUL Zùlù two-seat ultralight *(Paul Jackson/Jane's)*   *2000/0075927*

POWER PLANT: One 59.6 kW (79.9 hp) Rotax 912 UL flat-four driving a Duc adjustable-pitch propeller. Fuel capacity 70 litres (18.5 US gallons; 15.4 Imp gallons).
ACCOMMODATION: Two persons, side by side, beneath forward-hinged canopy. Dual controls.
EQUIPMENT: Ballistic parachute and anti-collision light standard.
DIMENSIONS, EXTERNAL:
Wing span   10.00 m (32 ft 9¾ in)
Length overall   6.60 m (21 ft 8 in)
Height overall   2.35 m (7 ft 8½ in)
AREAS:
Wings, gross   15.20 m² (163.6 sq ft)
WEIGHTS AND LOADINGS:
Weight empty   272 kg (600 lb)
Max T-O weight   472 kg (1,040 lb)
PERFORMANCE:
Max level speed   113 kt (210 km/h; 130 mph)
Econ cruising speed   86 kt (160 km/h; 99 mph)
Stalling speed, flaps down   32 kt (58 km/h; 36 mph)
Max rate of climb at S/L   300 m (984 ft)/min
T-O run   60 m (200 ft)
Landing run   45 m (150 ft)
Range with max fuel   432 n miles (800 km; 497 miles)
UPDATED

CAP

CAP AVIATION

9 rue de l'Aviation, F-21121 Dijon-Darois
Tel: (+33 3) 80 35 65 10
Fax: (+33 3) 80 35 65 15
e-mail: akrotech@altavista.net
Web: http://www.capaviation.com
PRESIDENT, CEO AND HEAD OF DESIGN: Dominique Roland
SALES MANAGER, EUROPE: Alain Ruelloux

CAP Aviation (known as Akrotech Europe until January 1999) is owned by holding company Apex International (formerly known as Aéronautique Service – AES) (which see), also responsible for the Robin and BUL concerns. It formed in 1997, initially to market the Giles G-202 two-seat aerobatic aircraft, now known as the CAP 222.

Apex took over production of the Mudry series of aerobatic aircraft on 12 May 1997 following that company's bankruptcy. All aerobatic aircraft produced by Apex will be marketed by CAP Aviation: the CAP 232 built at Bernay and CAP 10 and CAP 222 at Darois. During 1998, 12 CAPs of various models were completed.
UPDATED

CAP 10 B two-seat aerobatic light aircraft *(Paul Jackson/Jane's)*

CAP 10 B

TYPE: Aerobatic two-seat sportplane.
PROGRAMME: Derived from Piel CP-30/CP-301 Emeraude, first flown 19 June 1954 and built by amateur constructors as well as by 11 commercial plants, under various names. Super Emeraude modified with taller fin and 119 kW (160 hp) engine as CP-100 prototype, flown 13 August 1966; with broader chord rudder and 134 kW (180 hp) engine, became prototype CAP 10 B, first flight (F-WOPX) 22 August 1968; further three preseries aircraft; certified 4 September 1970; FAA certification for day and night VFR 1974. Produced by Mudry until 1996.
CURRENT VERSIONS: **CAP 10 B:** Main production version.
Description refers to CAP 10 B.
  CAP 10 C: Announced in early 1999; features carbon fibre wing spar for reduced weight and increased speed and roll rate. Static test of wing undertaken 27 April 2000; prototype first flight then scheduled for November 2000.
CUSTOMERS: Total 284 built to mid-1999, of which 279 completed by Mudry before 1996 bankruptcy. Most recent five produced in temporary factory at Bernay and include exports to UK (three) and USA (one). Production then paused, pending launch of CAP 10 C. Previous major users, French Air Force (56) and Mexican Air Force (20) began disposals in 1994-95. Two more recent aircraft operated by 208 Squadron of South Korean Air Force; French Navy has eight.
COSTS: FFr750,000 plus tax (1999); US price US$147,000 (1999).
DESIGN FEATURES: Simple, sporting lightplane with low, basically elliptical wing, mid-mounted tailplane and curved rudder.
  Wing section NACA 23012; dihedral 5°; incidence 0°.
FLYING CONTROLS: Conventional and manual. Slotted ailerons; trim tabs on both elevators; balance tab on rudder; tailplane incidence adjustable on ground; plain flaps.
STRUCTURE: All-spruce single-spar wing torsion box; rear auxiliary spar; okoumé ply skin on wing and fuselage with polyester fabric covering; double skin on forward fuselage; fin spar integral with fuselage and tailplane. New wing with carbon fibre main spar and the ailerons of the CAP 231 under development in late 1997.
LANDING GEAR: Tailwheel type, non-retractable. Mainwheel legs of light alloy, with ERAM type 9 270 C oleo-pneumatic shock-absorbers. Single wheel on each main unit, tyre size 380×150. Solid tailwheel tyre, size 6 × 200. Tailwheel is steerable by rudder linkage but can be disengaged for ground manoeuvring. Hydraulically actuated mainwheel disc brakes (controllable from port seat) and parking brake. Streamline fairings on mainwheels and legs.
POWER PLANT: One 134 kW (180 hp) Textron Lycoming AEIO-360-B2F flat-four engine, driving a Hoffman two-blade fixed-pitch wooden propeller. Standard fuel tank aft of engine fireproof bulkhead, capacity 72 litres (19.0 US gallons; 15.8 Imp gallons). Optional auxiliary tank, capacity 75 litres (19.8 US gallons; 16.5 Imp gallons), beneath baggage compartment. Inverted fuel and oil (Aviat/Christen) systems permit continuous inverted flight.
ACCOMMODATION: Side-by-side adjustable seats for two persons, with provision for back parachutes, under rearward-sliding and jettisonable moulded transparent canopy. Special aerobatic shoulder harness standard. Space for 20 kg (44 lb) of baggage aft of seats in training and touring models.
SYSTEMS: Electrical system includes Delco-Rémy 40 A engine-driven alternator and STECO ET24 Ni/Cd battery.
AVIONICS: Honeywell avionics standard.
DIMENSIONS, EXTERNAL:
Wing span   8.08 m (26 ft 6 in)
Wing aspect ratio   6.0
Length overall   7.16 m (23 ft 6 in)
Height overall   2.55 m (8 ft 4½ in)
Tailplane span   2.90 m (9 ft 6 in)
Wheel track   2.06 m (6 ft 9 in)
DIMENSIONS, INTERNAL:
Cockpit max width   1.05 m (3 ft 5½ in)
AREAS:
Wings, gross   10.85 m² (116.8 sq ft)
Ailerons (total)   0.79 m² (8.50 sq ft)
Vertical tail surfaces (total)   1.32 m² (14.25 sq ft)
Horizontal tail surfaces (total)   1.86 m² (20.02 sq ft)
WEIGHTS AND LOADINGS (A: Aerobatic, U: Utility):
Weight empty, equipped: A, U   549 kg (1,210 lb)
Fuel weight: A   54 kg (119 lb)
  U   108 kg (238 lb)
Max T-O weight: A   760 kg (1,675 lb)
  U   830 kg (1,830 lb)
Max wing loading: A   70.0 kg/m² (14.35 lb/sq ft)
  U   76.5 kg/m² (15.67 lb/sq ft)
Max power loading: A   5.66 kg/kW (9.31 lb/hp)
  U   6.19 kg/kW (10.16 lb/hp)
PERFORMANCE:
Never-exceed speed (V_{NE})   183 kt (340 km/h; 211 mph)
Max level speed at S/L   148 kt (274 km/h; 170 mph)
Max cruising speed at 75% power
                                                135 kt (250 km/h; 155 mph)
Stalling speed: flaps up   53 kt (99 km/h; 61 mph) IAS
  flaps down   43 kt (80 km/h; 50 mph) IAS
Max rate of climb at S/L   488 m (1,600 ft)/min
Service ceiling   5,000 m (16,400 ft)
T-O run   350 m (1,150 ft)
T-O to 15 m (50 ft)   450 m (1,480 ft)
Landing from 15 m (50 ft)   600 m (1,970 ft)
Landing run   360 m (1,185 ft)
Range with max fuel   540 n miles (1,000 km; 621 miles)
g limits   +6/–4.5
UPDATED

CAP 222

TYPE: Aerobatic two-seat sportplane.
PROGRAMME: Conceived as European, factory-built version of US Experimental category Akrotech Giles G-202 with JAR 23 certification.
  Three G-202 kits imported for development, initially designated G222.
  The first European-built CAP 222 (F-WWMY) flew on 12 June 1997; made its static debut at Le Bourget two days later; and performed its first public air display on 18 June. Second and third followed by January 1999. Failure to meet JAR 23 certification criteria for wing structural strength demanded redesign which was completed in April 1999.
  First production CAP 222 used for static tests; second made first flight on 15 May 2000 (F-WWMZ) and initial public appearance at Aerofair, North Weald, UK on 2 June 2000. To CEV for certification trials, which were expected to be complete by early 2001.
COSTS: US$220,000 (2000).
Data generally as for Akrotech Giles G-202 (US section) except that below.
POWER PLANT: One 149 kW (200 hp) Textron Lycoming AEIO-360-A1E driving an MT three-blade, constant-

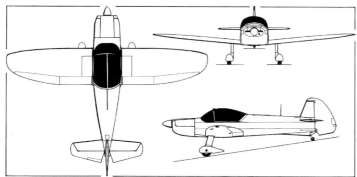
CAP 10 B (134 kW; 180 hp AEIO-360 engine) *(Paul Jackson/Jane's)*

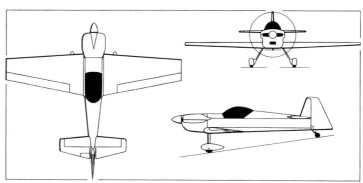
CAP Aviation CAP 232 *(Paul Jackson/Jane's)*

First (flying) production CAP 222, shortly after its maiden flight on 15 May 2000 *(Paul Jackson/Jane's)*

speed propeller. Fuel in main 68 litre (18.0 US gallon; 15.0 Imp gallon) fuselage tank and optional 155 litre (41.0 US gallon; 34.2 Imp gallon) wing tank.

WEIGHTS AND LOADINGS:
Weight empty   454 kg (1,000 lb)

PERFORMANCE:
Manoeuvre speed	180 kt (333 km/h; 207 mph)
Stalling speed	57 kt (106 km/h; 66 mph)
Rate of roll	500°/s
g limits	±10

NEW ENTRY

CAP 232

TYPE: Aerobatic single-seat sportplane.

PROGRAMME: Prototype (F-WZCH) first flew 7 July 1994; French certification and first sales March 1995. First place in 2000 World Aerobatic Championship in France was taken by Eric Vazeille in CAP 232 F-GKDX; Catherine Maunoury was fifth (women's first) in F-GXCM.

CUSTOMERS: Total 27 built by August 2000. Mainly French operators (including two to French Air Force), but others sold to Australia (one, 1996), Switzerland (one), UK (two, 1998) and USA (two, 1998 and four, 2000). Moroccan Air Force placed order for nine in mid-2000 to re-equip Marche Verte aerobatic team.

COSTS: FFr1.25 million plus tax (1999); US price US$250,000 (1999).

DESIGN FEATURES: Optimised for world-class aerobatic competition. Similar to, and improvement on, CAP 231 EX. Strong, carbon fibre wing of broad root chord, sharply tapered for high roll rates; tapered tail surfaces.

FLYING CONTROLS: Conventional and manual. Electrically actuated elevator tab; elevator servo tab to reduce stick forces. Almost full-span ailerons for high roll rates.

STRUCTURE: Two-spar carbon fibre wings; 10 ribs.

LANDING GEAR: Fixed tailwheel type; mainwheel tyres 5.00-5.

POWER PLANT: One 224 kW (300 hp) Textron Lycoming AEIO-540-L1B5 flat-six engine, driving a constant-speed Mühlbauer MTV-9-BC-C200-15 or MTV-14-BC-C190-17 four-blade propeller. Total fuel capacity 189.3 litres (50.0 US gallons; 41.6 Imp gallons) in one fuselage tank and two wing tanks. Fuel and oil system designed for prolonged inverted flight.

ACCOMMODATION: Single seat under rear-hinged canopy. Baggage space behind pilot.

SYSTEMS: Electrical system includes engine-driven alternator and 12 V 60 Ah battery.

AVIONICS: Customer specified.
Comms: Radio and transponder optional.
Flight: GPS optional.

EQUIPMENT: Sighting frame can be attached to wingtip for judging exact verticals in competition aerobatics.

DIMENSIONS, EXTERNAL:
Wing span	7.39 m (24 ft 3 in)
Wing chord: at root	1.83 m (6 ft 0 in)
at tip	0.91 m (3 ft 0 in)
Wing aspect ratio	5.4
Length overall	6.76 m (22 ft 2 in)
Tailplane span	2.74 m (9 ft 0 in)
Wheel track	1.75 m (5 ft 9 in)
Propeller diameter	1.90 m (6 ft 2¾ in)
Propeller ground clearance	0.30 m (1 ft 0 in)

AREAS:
Wings, gross	10.15 m² (109.3 sq ft)
Ailerons (total)	1.00 m² (10.76 sq ft)
Fin	0.55 m² (5.92 sq ft)
Rudder	0.77 m² (8.29 sq ft)
Tailplane	1.02 m² (10.99 sq ft)
Elevators	1.11 m² (11.93 sq ft)

WEIGHTS AND LOADINGS (A: Aerobatic, N: Normal category):
Weight empty	585 kg (1,290 lb)
Max payload	227 kg (500 lb)
Fuel weight	123 kg (270 lb)
Max T-O and landing weight: A	730 kg (1,610 lb)
N	821 kg (1,810 lb)
Max wing loading	80.9 kg/m² (16.56 lb/sq ft)
Max power loading	3.67 kg/kW (6.03 lb/hp)

PERFORMANCE:
Never-exceed speed (V_{NE})	219 kt (405 km/h; 252 mph)
Max level speed	189 kt (349 km/h; 217 mph)
Max cruising speed at 75% power	180 kt (333 km/h; 207 mph)
Manoeuvre speed (V_A)	169 kt (314 km/h; 195 mph)
Econ cruising speed	145 kt (269 km/h; 167 mph)
Stalling speed, power off	57 kt (105 km/h; 65 mph)
Max rate of climb at S/L	1,003 m (3,290 ft)/min
Rate of roll at manoeuvre speed	420°/s
Service ceiling	4,575 m (15,000 ft)
T-O run	150 m (495 ft)
T-O to 15 m (50 ft)	180 m (595 ft)
Landing from 15 m (50 ft)	450 m (1,480 ft)
Range with max fuel, 45% power	650 n miles (1,203 km; 748 miles)
g limits	±10

UPDATED

CAP 232 F-GXCM, flown by 2000 women's aerobatic champion Catherine Maunoury

CHEREAU

ARNO CHEREAU AERONAUTIQUE
La Tiercerie, F-37110 Morand
Tel: (+33 2) 47 29 66 66
Fax: (+33 2) 47 56 96 03

Arno Chereau Aéronautique manufactures the J.300 Series 2 ultralight in its 3,000 m² (32,300 sq ft) factory southwest of Paris.

VERIFIED

CHEREAU J.300 SERIES 2

TYPE: Tandem-seat ultralight kitbuilt.

CURRENT VERSIONS: **Civilian version** is used for private flying, cropspraying and aerial photography. **Military version** features larger tyres and a full instrument panel including artificial horizon, UHF radio, GPS and transponder. Float-equipped aircraft are offered as an option.

CUSTOMERS: Sales include two to Mauritania. Two evaluated by the French Army in 1996; selected to replace 21 Mignet Balerits in current service.

COSTS: From FFr190,000 (1996).

DESIGN FEATURES: Braced, high-wing monoplane with mid-mounted, braced tailplane; constant chord wing with tapered tips.

FLYING CONTROLS: Conventional and manual. Horn-balanced rudder.

STRUCTURE: Alloy tube airframe; wood and fabric covering.

LANDING GEAR: Tailwheel type; fixed. Mainwheel speed fairings optional; hydraulic brakes.

POWER PLANT: One 59.6 kW (79.9 hp) Rotax 912 flat-four driving a three-blade propeller is standard; options include engines from BMW, JPX and Limbach.

DIMENSIONS, EXTERNAL:
Wing span	10.20 m (33 ft 5½ in)
Length overall	5.26 m (17 ft 3 in)

AREAS:
Wings, gross	17.60 m² (189.4 sq ft)

WEIGHTS AND LOADINGS:
Weight empty	174 kg (384 lb)
Max T-O weight	450 kg (992 lb)

Chereau J.300 ultralight *(Geoffrey P Jones)*

PERFORMANCE:
Max operating speed (V_{MO})	102 kt (190 km/h; 118 mph)
Econ cruising speed	65 kt (120 km/h; 75 mph)
Stalling speed	32 kt (58 km/h; 37 mph)
Max rate of climb at S/L	288 m (945 ft)/min
T-O run	90 m (295 ft)
Range	875 n miles (1,620 km; 1,007 miles)
Endurance	10 h

UPDATED

DASSAULT

DASSAULT AVIATION
9 Rond-Point Champs Elysées-Marcel Dassault, F-75008 Paris
Tel: (+33 1) 53 76 93 00
Fax: (+33 1) 53 76 93 20

COMMUNICATIONS OFFICE: 27 rue du Professeur Pauchet, F-92420 Vaucresson
Tel: (+33 1) 47 95 86 87
Fax: (+33 1) 47 95 87 40
Web: http://www.dassault-aviation.fr

WORKS: F-92214 St Cloud, F-95100 Argenteuil,
F-33127 Martignas,
F-33701 Bordeaux-Mérignac,
F-33260 Cazaux, F-64205 Biarritz-Parme,
F-13801 Istres, F-74370 Argonay,
F-59472 Lille-Seclin, F-86580 Poitiers
HONORARY CHAIRMAN: Serge Dassault
CHAIRMAN AND CEO: Charles Edelstenne
VICE-CHAIRMAN: Bruno Revellin-Falcoz
SENIOR EXECUTIVE VICE-PRESIDENT, OPERATIONS: Christian Decaix
SENIOR EXECUTIVE VICE-PRESIDENT, INTERNATIONAL: Pierre Chouzenoux
VICE-PRESIDENT, COMMUNICATIONS: Gérard David

Former Avions Marcel Dassault-Breguet Aviation formed from merger of Dassault and Breguet aircraft companies in December 1971; French government acquired 20 per cent of stock in January 1979, raised to 45.76 per cent in November 1981; present company name adopted June 1990. Of government shareholding, 10.75 per cent held double voting rights so that French government had majority (55 per cent) control. On 17 September 1992, Dassault joined with Aerospatiale in state-owned joint holding company, SOGEPA (which see), which has 35 per cent holding in Dassault Aviation; Dassault and Aerospatiale then pooled resources and co-ordinated R&D strategy, although each remained separate entity.

French government intention to merge Aerospatiale and Dassault Aviation implemented after State relinquished control of Aerospatiale; agreement, ratified by shareholders on 23 December 1998, allocated government's shares in Dassault Aviation to Aerospatiale, but without double voting rights. Aerospatiale later became Aerospatiale Matra and then incorporated into EADS. Last-mentioned affirmed continued stake in Dassault on 27 June 2000 and this endorsed by shareholders on 4 September 2000.

Related measures include separation from Dassault Aviation of Dassault Systèmes, this becoming Dassault Participations, owned 45.76 per cent by French government, 49.93 per cent by Groupe Industriel Marcel Dassault and 4.31 per cent privately. Dassault intention to separate civil and military businesses, with Falcon business jet activities concentrated at Biarritz, Martignas and Seclin, plus Mérignac for flight test, vetoed by Aerospatiale Matra on 30 June 1999. However, *de facto* partition implemented on 1 January 2000, on "internal and informal basis".

Employees total some 8,710 at 10 industrial sites. Business divided approximately 30 per cent military and 70 per cent civil and space. Revenue in 1999 financial year derived 68 per cent from civil business, while exports were 68 per cent of the total. Sales in 1999 totalled FFr18.9 billion (US$2.9 billion); orders placed in 1999 for 72 business jets and 28 Rafales, exports accounting for 58 per cent of total and civil business for 55 per cent. In 2000, Dassault sold 90 Falcon business jets and planned delivery in 2001 of 73. It was expected to unveil a new business jet project at the 2001 Paris Air Show.

Dassault assembles and tests its civil and military aircraft in its own factories, but operates wide network of subcontractors. Other products include flight control system components, maintenance and support equipment and CAD/CAM software (CATIA). Related companies include Dassault Falcon Jet Corporation and SOGITEC. All are subordinate to Groupe Industriel Marcel Dassault, previously including Dassault Electronique, which became part of Thomson-CSF Detexis on 1 January 1999.

Dassault Aviation shared in Atlantique programme with Belgium, Germany, Italy Netherlands and UK; SEPECAT (Jaguar) with UK; and Alpha Jet with Dornier. Offset manufacture of Dassault aircraft components arranged with Belgium, Egypt, Greece and Spain. European Aerosystems Ltd (which see in International section) established with BAE as a joint venture military aircraft company.

Dassault has produced some 1,370 executive jets for service in 64 countries and received orders for over 2,700 Mirages of all types. The 7,000th Dassault fuselage (a Mirage 2000) was completed at the Argenteuil plant on 21 May 1996. Dassault aircraft have been sold in 73 countries and accumulated 15 million flying hours. In 1999, some 2,500 Dassault military aircraft were flying in nearly 40 countries.

UPDATED

Dassault Mirage 2000N strike aircraft of Escadron de Chasse 2/4 'La Fayette' *(Paul Jackson/Jane's)*
2001/0093664

DASSAULT MIRAGE 2000
Indian Air Force name: Vajra (Divine Thunder)
TYPE: Multirole fighter.
PROGRAMME: Selected as main French Air Force combat aircraft 18 December 1975; first of four single-seat prototypes flew 10 March 1979, followed by two-seat version on 11 October 1980; initially developed as interceptor with SNECMA M53 power plant and Thomson-CSF (now Thales) RDM multimode pulse Doppler radar; M53-5 in early production aircraft succeeded by M53-P2; fitted with RDI radar from 38th French Air Force 2000C onwards; first flight of production 2000C 20 November 1982; first flight of production two-seat 2000B, 7 October 1983; first unit, EC 1/2 'Cigognes', formed at Dijon 2 July 1984. Subsequently developed for strike/attack as 2000N/D.

Second-generation Mirage 2000-5 first flown as prototype 24 October 1990; initial production aircraft flown in October 1995 and type qualification granted (for SF1 French production standard) by DGA procurement agency on 13 June 1997; initial export delivery in May 1997, followed by first to French Air Force in December 1997.

Potential engine upgrade revealed in early 2000; M53 PX3 under development to provide additional 8 to 10 per cent thrust.

CURRENT VERSIONS: **2000B** (*Biplace*: two-seat). Trainer counterpart of 2000C; Nos. 501 to 514 (Series) S3 with RDM radar and M53-5 power plant; Nos. 515 to 520, S4 with RDI J1-1 radar and M53-5, but Nos. 516 and 517 retrofitted with RDM; No. 521 S4-2 with RDI J2-4 and M53-5; No. 522 also S4-2, but M53-P2; Nos. 523 to 530 S5 with RDI J3-13 and M53-P2. Final delivery in December 1994. See also Mirage 2000DA below. One or two Mirage 2000Bs distributed to each 2000C operating squadron, but most operated by OCU, EC 2/5 at Orange, which assumed task from EC 2/2 at Dijon on 1 July 1998. S3 aircraft have been upgraded with RDI radar.

One aircraft, No. 504, converted to **2000BOB** (*Banc Optronique Biplace:* two-seat electro-optics testbed); flown 28 June 1989 after modification by CEV at Brétigny-sur-Orge; trials of Rubis FLIR, VEH-3020 holographic HUD, night vision goggles and other electro-optical systems, mainly in connection with the Dassault Rafale programme. Fitted with OBOGS.

2000C: (*Chasse:* fighter). Standard interceptor; Nos. 1 to 37 built as series S1, S2 and S3 with RDM radar and M53-5 power plants; since upgraded to S3 radar standard. Loosely called Mirage **2000RDM**; Mirage 2000B and C collectively known as Mirage **2000DA** (*Défense Aérienne*). Later aircraft (loosely **2000RDI**) have RDI radar and M53-P2 power plants: Nos. 38 to 48 Series S4, delivered from July 1987 and later upgraded to S4-1; Nos. 49 to 63 S4-1; Nos. 64 to 74 S4-2; Nos. 75 to 124 Series S5, delivered between late 1990 and June 1995. Equipment standards of Mirage 2000B/Cs are **S3**, incapable of

Retouched photograph depicting Dassault Mirage 2000-9 in UAE colours *(Aviaplans/F Robineau)*
2001/0093652

114 FRANCE: AIRCRAFT—DASSAULT

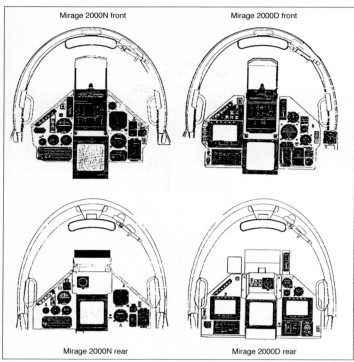

Comparison of strike/attack Mirage 2000 cockpits

Mirage 2000-5 potential weapons options (excluding two internal 30 mm cannon and chaff/flare system). Hardpoint maximum loads are given in kilogrammes; (1) internal guns in single-seat versions only

0010578

Internal:
Two DEFA 554 30 mm (1) guns
EM and IR decoys

300	1830	400	400	1400	400	400	1830	300	
		●	●		●	●			Mica active air-to-air missile
●								●	Magic IR missile
	●						●		Underwing extra fuel tank
				●					Fuselage extra fuel tank
	●●	●	●	●●	●	●	●●		500 lb/250 kg bomb
		●		●		●			2,000 lb/1,000 kg laser guided bomb
		●	●	●●	●	●			Durandal bomb
				18					BAP 100 penetration bomb
	●	●	●	●	●	●	●		Beluga grenade dispenser
				●					F 2 practice bomb launcher
●	●						●	●	F 4 (18) 68 mm rocket launcher
			●		●				E.O. or I.R. targeting pod
	●						●		AS 30 laser guided missile
				●					Apache missile
	●						●		Anti-radiation missile
							●		AM 39 anti-ship missile
					●				Flir pod navigation
				●					Twin gun pod
				●					Recce pod
				●					Offensive or Intelligence ECM pod
				●					Buddy refuelling pod

launching Matra BAe 530D (530F only); **S4** RDI J1-1 radar; **S4-1** retrofit of all S4s with improved J1-2; **S4-2** further radar upgrade to J2-4; **S4-2A** retrofit of all S4-1 and S4-2 aircraft (Nos. 38 to 74, 515 and 518 to 522) with HOTAS-type throttle and improved J2-5 radar; **S5** definitive standard with J3-13 radar and, from No. 93 onwards, Spirale chaff/flare dispenser; and **S5-2C** retrofit introduced 1995 (aircraft of EC 1/12) providing better anti-jamming protection for RDI radar. Conversion completed of 37 2000B/Cs to Mirage 2000-5F (which see). Withdrawal of S3 began in January 1998 for upgrading with RDI radar and final aircraft returned to service (at Orange) in March 1999; EC 2/2 gained late production (RDI) aircraft, with which it became operational in the air defence role on 1 August 1998.

Detailed description applies to 2000C except where indicated.

2000D: Two-seat conventional attack version of 2000N, lacking ASMP missile interface and nose pitot but with HOTAS controls, additional display screens for both crew members, Antilope 5-3D terrain-following/terrain-reference radar, GPS and improved (ICMS Mk 2) ECM; functions of pilot and navigator more clearly demarcated. First flight (D01, ex-N01) 19 February 1991; second prototype, D02 (ex-N02), flown 24 February 1992; first 2000D (No. 601) delivered CEAM Mont-de-Marsan for trials 9 April 1993; first squadron, EC 1/3 'Navarre', achieved limited IOC (six aircraft only) 29 July 1993 at CEAM. Combat debut was 5 September 1995, launching AS 30L ASMs in Bosnia. Deliveries in 1999 totalled 11; final delivery was due in May 2001.

Initial six aircraft built to 'interim baseline' configuration known as **R1N1L**, with ability to launch laser-guided weapons and Magic missiles only. Further few, designated **R1N1**, had slightly expanded weapon options. Later **R1** aircraft have full range of current French Air Force armament (BAP 100, BAT 120, Belouga, 68 mm rockets, AS 30L, GBU-12 (Mk 82), BGL 1000, Magic 2 and PDLCT designator). PDLCT-S (*Synergie*), first employed over Yugoslavia in March 1999, is modified version with improved imagery contrast. **R2** standard, from 2000-01 (including retrofit of all earlier aircraft), introduces APACHE standoff weapons dispenser, Eclair fully integrated self-defence suite and Atlis II laser designator pods (from Jaguar force). First two R2 conversions completed early 2000 at Nancy; further 68 following at three/four per month. Planned **R3** version, with SCALP SOM and a reconnaissance pod, was abandoned in June 1996, but had been reinstated by 1999, envisaging carriage of a new, modular AASM SOM, JTIDS and MIDS datalinks for full NATO interoperability and service entry in 2003-05. All interim 2000Ds were modified to full R1 standard by June 1995.

2000E: Multirole fighter for export; M53-P2 power plant throughout. Details in Customers paragraph. Baseline version for India, Egypt and Peru, differences from 2000C including RDM radar with CW illumination for Super 530D AAM; two main computers, with expanded memory; ULISS 52 INS; improved ECM (integrated system with VCM-65 display or, alternatively, Remora and Caiman pods); VE-130 HUD; VMC-180 head-down; and expanded weapon options. Abu Dhabi and Greek 2000Es have extra computing power, further armament options and improved self-defence (SAMET system for Abu Dhabi; ICMS Mk 1 for Greece).

2000ED: Two-seat trainer counterpart of 2000E.

2000N: Two-seat low-altitude penetration version to deliver ASMP nuclear standoff missile; two prototypes; first flight 3 February 1983; one preseries aircraft (No. 301) built at Istres and first flown 3 March 1986; first 24 production aircraft (Nos. 302-325) were 2000N-K1 with ASMP capability only; from July 1988 remaining aircraft, designated 2000N-K2, have full conventional and ASMP capability; production ended 1993; some K1s (initially of squadrons 3/4 and part of 2/4) retrofitted with partial air-to-ground capability for conventional attack, all having been modified by late 1998. Equipment includes Antilope 5 terrain-following radar, two SAGEM inertial platforms, two improved AHV-12 radio altimeters, colour head-down CRT, two Magic self-defence missiles, and ICMS (integrated countermeasures system) comprising Sabre jamming system, Serval RWR and Spirale automatic chaff/flare dispenser system.

2000N' (N Prime): Initial designation of 2000D.

2000R: Single-seat day/night reconnaissance export version of 2000E but with normal nose; various sensor pods possible (see Avionics paragraph).

2000-3: Private venture upgrade, begun 1986, with Rafale-type multifunction (five-CRT) cockpit displays known as APSI (advanced pilot system interface); prototype BY1/F-ZJTB (ex-No. B01) flew 10 March 1988. Later received RDY radar.

2000-4: Private venture integration of Matra BAe MICA AAM. First guided flight against target drone, 9 January 1992.

2000-5: Multirole upgrade incorporating -3 and -4 improvements, plus Thomson-CSF Detexis RDY radar and new central processing unit, Sextant VEH 3020 HUD and ICMS Mk 2 countermeasures; laser-guided bombs and ASMs, or APACHE standoff dispenser in air-to-ground role. First Mirage flight of RDY radar in BY1 (later numbered BY2) May 1988; first flight of full 2000-5 (two-seat) 24 October 1990 (same aircraft; initially no serial number; later reverted to B01); first single-seater, 01 (conversion of trials aircraft CY1) flown 27 April 1991.

Export orders from Taiwan and Qatar; air-to-air firing trials with MICA completed July 1996; air-to-ground trials completed May 1997; operational use (export; Taiwan) from June 1997.

FFr4,600 million (US$830 million) conversion programme announced November 1992 for upgrading 37 French Air Force 2000Cs (envisaged as 34 S4-2As and three S5s) mainly from EC 2/5 and 3/5 at Orange to **2000-5F** for continued service; contract awarded 25 November 1993; only one reworked aircraft funded in 1994 defence budget; 10 more in 1995 funding; and 23 in 1996. Delivery schedule (not achieved) was one in 1997; 11 in 1998; 22 in 1999; and three in 2000; one lost on 16 March 1999. Rework (at Argenteuil, with reassembly at Bordeaux) required six months per aircraft and involved complete dismantling and return of fuselages to Argenteuil. RDI radars from these upgraded aircraft retrofitted in early production (RDM) Mirage 2000Cs.

Two-seat trainer Mirage 2000B in the markings of OCU, EC 2/5 *(Paul Jackson/Jane's)* 2000/0062720

Prototypes were conversions of Nos. 51 and 77 (latter an S5); both modified at Istres; initial aircraft reflown 26 February 1996; second followed two months later.

First 'production' conversion, No. 38 handed over 30 December 1997 at Istres test centre before delivery in April 1998 to CEAM, Mont-de-Marsan, to begin conversion of pilots of EC 1/2 from Dijon. IOC at Dijon (12 aircraft) 31 March 1999; FOC 4 February 2000; re-equipment of EC 2/2 began late 1999; total 30 delivered by February 2000, including 18 in 1999.

Identification features are additional (LAM) 'bullet' antenna on fin leading-edge, absence of nose pitot and four horizontal antenna bands on radome. Normal external configuration is two MICA and two Magic AAMs, plus two 2,000 litre (528 US gallon; 440 Imp gallon) and one 1,300 litre (343 US gallon; 286 Imp gallon) external tanks. Optional configuration (Standard Kilo) is four MICAs and two Magics. MICA delivered (first 25 rounds) from August 1999. Early 2000-5F deliveries are to **SF1** standard; **SF2** projected with GPS, provision for helmet-mounted sight and ability to carry unspecified long-range identification aid.

2000-5 Mk 2: Announced early 1999; version of 2000-9 purchased initially by Greece. Features of 2000-9 and 2000-5 Mk 2 include modular avionics, laser gyro INS, upgraded ECM, expanded aircraft-missile datalink, Damoclès laser-designation pod (known as Shehab on 2000-9), Nahar navigational FLIR in Damoclès pylon, upgraded version of RDY radar (multitarget air-to-sea search and track, high-resolution DBS mapping mode and search and track of mobile land targets), increased MTOW of 17,500 kg (38,580 lb), OBOGS (in development), new multichannel recording system, new rear cockpit colour display repeater, helmet-mounted sight, autopilot terrain-following system based on digital terrain file and display of digital map on head-down screen. Can carry six MICA AAMs in addition to air-to-surface weapons.

2000-8: Mirage 2000EAD/RAD/DAD supplied to Abu Dhabi/UAE from 1989.

2000-9: Version of 2000-5 for United Arab Emirates incorporating long-range air-to-ground capability with weapons including Black Shahine and Hakim. Configuration includes M53-P2 engines, RDY-7 radar with synthetic aperture and beam-sharpening modes, Sextant Totem 3000 laser INS, upgraded air conditioning system and BAE Systems Terprom digital terrain system. First 2000-9s due to undertake flight testing and proving (three aircraft) at Istres between late 2000 and late 2001; first flight (two-seat version) 14 December 2000; initial single-seat 2000-9 flew 25 January 2001. Further features as for 2000-5 Mk 2.

2000S: Export attack version of 2000D; promotion discontinued.

CUSTOMERS: See table for rapid reference. **France** required seven prototypes and 372 production aircraft; reduced in late 1991 to 318 by abandonment of final 24 2000Cs and 30 2000Ds and transfer of some single-seat aircraft to trainer contract; no Mirages funded in 1992 or 1993, but one 2000B and 14 2000Cs cancelled, then re-ordered in 1994 defence budget as 2000Ds. French orders subsequently amended to 30 2000Bs, 124 2000Cs, 75 2000Ns and 86 2000Ds (total 315); last was due in May 2001. Mirage 2000C equipped three squadrons of EC 2 at Dijon (1984-86), three of EC 5 at Orange (1988-90) and two of EC 12 at Cambrai.

Mirage 2000N deliveries began at Luxeuil 30 March 1988; EC 1/4 'Dauphine' formed 1 July 1988, now with full dual-role K2 series aircraft, followed by two more squadrons. Pending 2000D, EC 2/3 'Champagne' operational at Nancy 1 September 1991 with 2000N in conventional role. EC 1/3 'Navarre' fully operational with 2000D from 31 March 1994; EC 3/3 'Ardennes' began re-equipment, July 1994; EC 2/3 began converting to 2000D

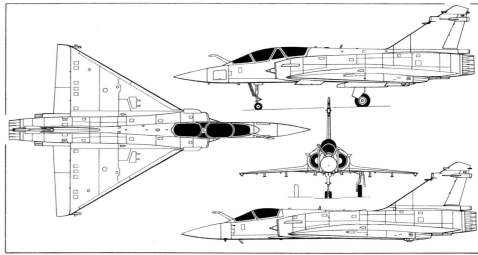

Dassault Mirage 2000-5DDA, with added side view (bottom) of Mirage 2000-5Ei *(Dennis Punnett/Jane's)*

FRENCH MIRAGE 2000 SQUADRONS

Squadron	Base	Type	Commissioned	Remarks
EC 1/2 'Cicognes'	Dijon	2000C (RDM)	2 Jul 84	Temp stood down Jan 98[4]
		2000-5F	31 Mar 99	Operational 4 Feb 00
EC 2/2 'Cote d'Or'	Dijon	2000B/C (RDM)	27 Jul 86	OCU until Jan 98
		2000C	1 Aug 98	
		2000-5F	2000	
EC 3/2 'Alsace'	Dijon	2000C (RDM)	Mar 86	Disbanded 31 Jul 93
EC 1/3 'Navarre'	Nancy	2000D	1 Apr 94[1]	
EC 2/3 'Champagne'	Nancy	2000N	30 Aug 91	Conventional attack only
		2000D	24 Apr 97[2]	Converted 1996-98
EC 3/3 'Ardennes'	Nancy	2000D	21 Aug 95	
EC 1/4 'Dauphine'	Luxeuil	2000N	1 Jul 88	
EC 2/4 'La Fayette'	Luxeuil	2000N	1 Jul 89	
EC 3/4 'Limousin'	Istres	2000N	1 Jul 90	
EC 1/5 'Vendée'	Orange	2000C	1 Apr 89	
EC 2/5 'Ile de France'	Orange	2000C	1 Apr 90	
		2000B/C	1 Jul 98	OCU
EC 3/5 'Comtat Venaissin'	Orange	2000C	3 Sep 90[3]	Disbanded 31 Aug 97
EC 1/12 'Cambrésis'	Cambrai	2000C	1 Jul 92	
EC 2/12 'Picardie'	Cambrai	2000C	1 Sep 93	

[1] Limited IOC 29 Jul 93
[2] When only partly equipped with 2000D
[3] First delivery
[4] And equipped with Alpha Jets to maintain pilot currency

Prototype Dassault Mirage 2000B posing as a 2000-5 Mk 2, with four MICA AAMs under the fuselage, two Magic 2s outboard underwing and two LGBs on the centreline, complete with laser designation pod *(Aviaplans/F Robineau)*

Mirage 2000-5F of second operating squadron EC 2/2 'Cote d'Or' wearing new grey colour scheme *(Michel Fournier)* ***2001***/0093663

(ex-2000N) in mid-1996, completing conversion in 1999. Mirage 2000N depot servicing required every 900 hours or 36 months, whichever sooner.

Abu Dhabi (United Arab Emirates) ordered 18 aircraft on 16 May 1983 and took up 18 options in 1985 for a total of 22 2000EADs, eight 2000RADs and six 2000DADs; deliveries delayed from 1986 to 1989 by provision for US weapons such as Sidewinder; deliveries to Maqatra/Al Dhafra completed November 1990 for Nos. 1 and 2 Shaheen (Warrior) Squadrons. Abu Dhabi 2000RADs carry COR2 multicamera pod, but alternatives include Raphaël-type SLAR 2000 or Harold pods; second 18 have Elettronica ELT/158 threat warning receivers and ELT/558 self-protection jammers; all with Spirale chaff/flare system. Self-defence suite code-named SAMET. Deliveries 7 November 1989 to November 1990. Weapons include BAE Systems PGM Hakim ASM.

Follow-on batch authorised in late 1996; confirmed 16 December 1997 as 30 new 2000-9s and upgrade to this standard of 33 earlier aircraft; contract signed 18 November 1998; total value estimated as US$3.4 billion for aircraft, or US$5.5 billion, including weapons and support. Weapons include Black Shahine version of Storm Shadow/SCALP SOM, IR and active radar MICA AAMs; ICMS Mk 3, incorporating Spirale and Eclair ECM systems; DDM missile approach warning. Shehab target designator pod (export version of Damoclès/PDLCT-S).

Egypt ordered 16 2000EMs and four BMs in December 1981; deliveries 30 June 1986 to January 1988; based at Berigat.

Greece ordered 36 2000EGs and four 2000BGs on 20 July 1985; handed over from 21 March 1988 and delivered from 27 April 1988 for 331 'Aegeas' and 332 'Geraki' Mire Pandos Kairou within 114 Pterix Mahis at Tanagra; deliveries suspended October 1989 at 28th aircraft; resumed 1992 and completed 18 November 1992. RDM3 radar; Spirale chaff/flare system installed, as part of ICMS self-defence suite; full ICMS system operational from August 1995. Some upgraded to 2000EG-SG3 standard with provision for AM 39 Exocet anti-ship missile. Further 15 Mirage 2000-5 Mk 2s confirmed 30 April 1999 and formally ordered in August 2000, together with upgrade to this standard of 10 earlier aircraft. Deliveries from 2001, with equipment including Thomson-CSF Damoclès laser designation pods and ICMS Mk 3. New weapons include MICA AAM (Standard EG 51) and SCALP EG ASM (EG 52). Option held on further three new aircraft and five upgrades.

India first ordered 36 2000Hs and four THs in October 1982; 26 Hs and four THs temporarily powered by M53-5; final 10 by M53-P2 from outset; first flight of 2000H (KF-101) 21 September 1984; first flight of 2000TH (KT-201) early 1985; No. 7 IAF Squadron 'Battle Axe' formed at Gwalior AB 29 June 1985, coincident with first arrivals in India. Named *Vajra* (Divine Thunder); second Indian order for six Hs and three THs signed March 1986 and delivered April 1987 to October 1988 to complete No. 1 'Tigers' Squadron. Third order placed 19 September 2000 for four 2000Hs and six 2000THs to be delivered in 2003/04. Despite grey-and-blue air defence camouflage, unconfirmed report claimed Indian Mirages optimised for attack, with Antilope 5 radar and twin INS. Role (at least) confirmed by 1996 selection (subject to operational trials) of the Rafael Litening laser-designator pod for IAF Mirage 2000s and Jaguars. By 1993 IAF aircraft appearing in experimental brown-and-green low-level colours, with Spirale chaff dispensers. By 1998, 38 remaining 2000Hs had been upgraded by HAL at Bangalore with local flare dispensers and were preparing for receipt of LGBs.

Peru ordered 24 2000Ps and two 2000DPs in December 1982, but reduced this to 10 2000Ps and two 2000DPs; first 2000DP handed over 7 June 1985; deliveries to Peru from December 1986; Escuadron de Caza-Bombardeo 412 of Grupo Aéreo de Caza 4 at La Joya inaugurated 14 August 1987.

Qatar ordered 12 Mirage 2000-5s under contract 'Falcon' on 31 July 1994, together with MICA and Magic 2 AAMs and Black Pearl variant of APACHE standoff weapon. Versions are 2000-5EDA (nine; single-seat) and 2000-5DDA (three; two-seat). Equipment includes GPS, ICMS Mk 2 and provision for Spirale. First flight late 1995; first three handed over at Bordeaux on 8 September 1997; four delivered to Qatar 18 December 1997; four more on 1 April 1998.

Taiwan ordered 60 Mirage 2000-5s with M53-P2 power plants on 17 November 1992. First export sale for -5; first flight late 1995; first handed over on 9 May 1996 for training in France; initial five arrived in Taiwan as sea freight on 5 May 1997; equips 41, 42 and 48 Squadrons of 2nd Tactical Fighter Wing at Hsinchu; IOC of initial squadron, November 1997. Final eight delivered late October 1998, at which time second squadron declared operational. Versions are 2000-5Ei (48 aircraft; single-seat) and 2000-5Di (12 aircraft; two-seat). Requirement revealed in 1997 for an additional 60. Armament includes MICA and Magic AAMs.

Additionally, Jordan ordered 10 2000EJs and two 2000DJs on 22 April 1988; all were cancelled in August 1991.

Total 599 firm orders (excluding seven prototypes and six company-owned trials and demonstrator aircraft) by 1 January 2001 (315 French; 284 exports); by late 2000 545 had been delivered, including 229 exports. France had received all 124 Cs, 30 Bs, and 75 Ns by 31 December 1995, only 2000Ds being delivered after that date (approximately 74 by mid-2000). Provisional agreement signed with Pakistan in January 1992 for 44 Mirage 2000Es; under further discussion in 1994-95, quantity having been revised to 36, including nine two-seat aircraft, all to 2000-5 standard. No further progress made.

Demonstration and evaluation sorties by a Mirage 2000-5 two-seat version given to Czech, Hungarian and Polish air forces in August 1996. Possible Polish final assembly by PZL Mielec is covered by an MoU announced in June 1997.

MIRAGE 2000 ORDERS

Customer	Qty	Version	First aircraft
Abu Dhabi/UAE[1]	6	2000DAD	701
	8	2000RAD	711
	22	2000EAD	731
	30	2000-9	—
Egypt	16	2000EM	101
	4	2000BM	201
France	30	2000B	501
	124[2]	2000C	1
	86	2000D	601
	75	2000N	301
Greece	36[3]	2000EG	210
	4	2000BG	201
	15	2000-5 Mk 2	—
India	46	2000H	KF101
	13	2000TH	KF201
Peru	10	2000P	050
	2	2000DP	193
Qatar	9	2000-5EDA	QA90
	3	2000-5DDA	QA86
Taiwan	48	2000-5Ei	2001
	12	2000-5Di	2051
Subtotal	**599**		
Prototypes	7		
Development	6		
Total	**612**		

[1] 33 being upgraded to 2000-9DAD/RAD/EAD
[2] 37 upgraded to 2000-5F
[3] 10 to be upgraded to 2000-5 Mk 2

COSTS: Programme unit cost: Taiwan FFr 333 million (1997).

DESIGN FEATURES: Multirole combat aircraft; low-set, thin delta wing for high internal volume and low wave drag (as in Mirage III/5), but with delta's disadvantages in manoeuvrability and landing/take-off requirements offset by relaxed stability and leading-edge slats. Area-ruled fuselage.

Wing has cambered section, 58° leading-edge sweep and moderately blended root employing Karman fairings. Cleared for 9 *g* and 270°/s roll at sub- and supersonic speed carrying four air-to-air missiles.

FLYING CONTROLS: Full fly-by-wire control with SFENA autopilot; two-section elevons on wing move up 16° and down 25°; inner leading-edge slat sections droop up to 17° 30′ and outer sections up to 30°; fixed strakes on intake ducts create vortices at high angles of attack that help to correct yaw excursions; small airbrakes above and below wings.

STRUCTURE: Multispar metal wing; elevons have carbon fibre skins with AG5 light alloy honeycomb cores; carbon fibre/light alloy honeycomb panel covers avionics bay; most of fin and all rudder skinned with boron/epoxy/carbon; rudder has light alloy honeycomb core.

LANDING GEAR: Retractable tricycle type by Messier-Bugatti, with twin nosewheels; single wheel on each main unit. Hydraulic retraction, nosewheels rearward, main units inward. Oleo-pneumatic shock-absorbers. Electro-hydraulic nosewheel steering (±45°). Manual disconnect permits nosewheel unit to castor through 360° for ground towing. Light alloy wheels and tubeless tyres, size 360×135-6 or 360×135R6, pressure 8.00 bars (116 lb/sq in) on nosewheels; size 750×230R15, pressure 15.00 bars (217 lb/sq in) on mainwheels. Messier-Bugatti hydraulically actuated polycrystalline graphite disc brakes on mainwheels, with anti-skid units. Compartment in lower rear fuselage for brake parachute, arrester hook or chaff/flare dispenser.

POWER PLANT: One SNECMA M53-P2 turbofan, rated at 64.3 kN (14,462 lb st) dry and 95.1 kN (21,385 lb st) with afterburning. Alternative M53-P20, rated at 98.1 kN (22,046 lb st), is no longer offered. Movable half-cone centrebody in each air intake.

Internal wing fuel tank capacity 1,480 litres (391 US gallons; 326 Imp gallons); fuselage tank capacity 2,498 litres (660 US gallons; 549 Imp gallons) in single-seat aircraft, 2,424 litres (640 US gallons; 533 Imp gallons) in two-seat aircraft. Total internal fuel capacity 3,978 litres (1,050 US gallons; 875 Imp gallons) in 2000C and E, 3,904 litres (1,030 US gallons; 859 Imp gallons) in 2000B, N, D and S. Provision for one jettisonable 1,300 litre (343 US gallon; 286 Imp gallon) RPL-522 96 kg (212 lb) fuel tank under centre of fuselage, and a 1,700 litre (449 US gallon; 374 Imp gallon) RPL-501/502 210 kg (463 lb) drop tank under each wing. Total internal/external fuel capacity 8,678 litres (2,291 US gallons; 1,909 Imp gallons) in 2000C and E, 8,604 litres (2,271 US gallons; 1,892 Imp gallons) in 2000B.

Detachable flight refuelling probe forward of cockpit on starboard side. (Availability of in-flight refuelling on export aircraft not disclosed, although probes fitted to Abu Dhabi's 2000RADs.) Dassault type 541/542 tanks of 2,000 litres (528 US gallons; 440 Imp gallons) are available for the 2000-5, 2000N, D and S wing attachments (and optional on 2000B/C), empty weight 240 kg (529 lb) each, increasing internal/external fuel to 9,204 litres (2,430 US gallons; 2,025 Imp gallons).

ACCOMMODATION: One or two occupants (see Current Versions) on Hispano-Suiza licence-built Martin-Baker Mk 10Q zero/zero ejection seat(s), in air conditioned and pressurised cockpit. Pilot-initiated automatic ejection in two-seat aircraft; 500 microseconds delay between departures. Canopy/ies hinged at rear to open upward and, on Mirage 2000D, covered in gold film to reduce radar signature.

SYSTEMS: ABG-Semca air conditioning and pressurisation system. Two independent hydraulic systems, pressure 280 bars (4,000 lb/sq in) each, to actuate flying control servo units, landing gear and brakes. Hydraulic flow rate 110 litres (29.1 US gallons; 24.2 Imp gallons)/min. Electrical system includes two Auxilec 20110 air-cooled 20 kVA 400 Hz constant frequency alternators (25 kVA in Mirage 2000D and 2000-5), two Bronzavia DC transformers, a SAFT 40 Ah battery and ATEI static inverter. Eros oxygen system.

AVIONICS: *Comms:* Thomson-CSF Communications ERA-7000 V/UHF com transceiver, ERA-7200 UHF (with optional Have Quick II) or SCP 5000 secure voice com; Thomson-CSF Communications NRAI-7A/NRAI-11 IFF transponder/interrogator (SC10/IDEE 1 on 2000-5).

Radar: Thomson-CSF Detexis RDM multimode radar or RDI pulse Doppler radar, each with operating range of 54 n miles (100 km; 62 miles). (Mirage 2000N/D have Thomson-CSF Detexis Antilope terrain-following radar for automatic flight down to 61 m (200 ft) at speeds not exceeding 600 kt (1,112 km/h; 691 mph); Antilope 5TC in 2000N includes altitude-contrast updating of navigation system; Antilope 5-3C in 2000D has full terrain-reference navigation facility.) Thomson-CSF Detexis RDY multimode, multitarget radar in 2000-5 has the ability to detect 24 targets while tracking eight.

Flight: SOCRAT 8900 solid-state VOR/ILS and IO-300-A marker beacon receiver, Thomson-CSF Communications radio altimeter (AHV-6 in 2000B and C, AHV-9 in export aircraft, two AHV-12 in 2000N and AHV-17 in 2000-5). Thomson-CSF Communications NC12 or Deltac Tacan. SAGEM Uliss 52 inertial platform (52D in 2000C and B; 52E for export; and two 52P in 2000N/D; 52ES in 2000-5; plus integrated GPS in 2000D and 2000-5). Sextant Totem 3000 RLG in 2000-9. Thomson-CSF Detexis Type 2084 central digital computer and Digibus digital databus (2084-XR in 2000D; XRI3 in 2000-5). Sextant AP 605 autopilot (606 in 2000N, 607 in 2000D, 608 in 2000-5). Sextant MMR (multimode receiver) in 2000-9, combining ILS, microwave landing and differential GPS. Undisclosed operator ordered BASE (BAE Systems) Terprom GPWS in 1996 for retrofit by Orbital Sciences Corporation.

Instrumentation: Sextant TMV 980 data display system (VE 130 head-up and VMC 180 head-down) (two ICC 55 head-down in 2000N/D). Mirage 2000-5 has Sextant Comète multidisplay system and VEM 130 HUD with Thomson-CSF Optronique recording camera. Provision optional in 2000-5 and -9 for Sextant Topsight E helmet-mounted sight/display system.

Mission: Mirage 2000-5 has LAM (*liaison avion-missile*) 'bullet' antenna on fin leading-edge for mid-course guidance of MICA AAMs. Sensors of strike/attack and export versions include 570 kg (1,257 lb) Thomson-CSF Detexis Raphaël SLAR 2000 pod, 400 kg (882 lb) Thomson-CSF Optronique COR2 multicamera pod or 680 kg (1,499 lb) Dassault AA-3-38 Harold long-range oblique photographic (Lorop) pod; 110 kg (243 lb) Thomson-CSF/TRT/Intertechnique Rubis FLIR pod; Thomson-CSF Optronique Atlis laser designator and marked target seeker (in pod on forward starboard underfuselage station); 340 kg (750 lb) Thomson-CSF Optronique PDLCT day/night (TV/thermal imaging) laser designator pod on Mirage 2000D (or CLDP/Atlis II on export aircraft)(2000D squadrons also being issued with Atlis II designation pods – being refurbished Atlis systems from Jaguar force), while PDLCT-S available from mid-1999; one 550 kg (1,213 lb) Thomson-CSF TMV 004 (CT51J) Caiman offensive or intelligence ECM pod and one 400 kg (882 lb) Thomson-CSF Astac elint (interferometer) pod. Export PDLCT-S known as Damoclès (and Shehab in UAE).

Self-defence: Systems in 2000C and 2000N include Thomson-CSF Detexis Serval radar warning receiver (antennas at each wingtip and on trailing-edge of fin, near tip, plus VCM-65 cockpit display); Thomson-CSF Detexis Caméléon (2000N), Caméléon C2 (2000D) or Sabre (2000C) jammer at base of fin (detector on fin leading-edge); and Matra BAe Spirale, comprising chaff dispensers in Karman fairings at wing trailing-edge/fuselage intersection and flares in lower rear fuselage. French Air Force DDM (*Détecteur Départ Missile*) missile plume detector requirement satisfied by 1994 purchase of SAGEM SAMIR system for 1995 fitment in rear of Magic launch rails (2000D/N first, but also to 2000Cs patrolling Bosnia. Spirale fitted to 2000N-K2; retrofitted to 2000N-K1 and installed on 2000Cs from No. 93; earlier 2000Cs have Eclair system (Alkan LL5062 chaff and flare launcher) in place of braking parachute, lacking automatic operation. Spirale on some export 2000Es.

Upgrade planned of 2000E with ICMS Mk 1 (integrated countermeasures system; as in 2000EG) comprising RWR and SHR, Thomson-CSF high-band jammer (leading-edge of fin and bullet fairing at base of rudder) and Spirale; automated ICMS Mk 3 of Mirage 2000-5 adds receiver/processor in nose to detect missile command links; extra pair of antennas near top of fin and additional DF antennas scabbed to existing wingtip pods (fin and secondary wingtip antennas also on Greek Mirage 2000s); ABD2000 export version of Sabre in some Mirage 2000Es. External equipment can include two 182 kg (401 lb) Thomson-CSF DB 3141/3163 Remora self-defence ECM pods.

EQUIPMENT: Optional 250 kg (551 lb) Intertechnique 231-300 buddy-type in-flight refuelling pod.

ARMAMENT: Two 30 mm DEFA 554 guns in 2000C, 2000E and single-seat 2000-5 (not fitted in B, D or N), with 125 rds/gun. Nine attachments for external stores, five under fuselage and two under each wing. On 2000-5, fuselage centreline stressed for 1,400 kg (3,086 lb) loads; other four fuselage points for 400 kg (882 lb) each; inner wing pylons for 1,830 kg (4,034 lb) each; and outboard wing points for 300 kg (661 lb) each.

Typical interception weapons comprise two 275 kg (606 lb) Super 530D or (if RDM radar not modified with target illuminator) 250 kg (551 lb) 530F missiles (inboard) and two 90 kg (198 lb) 550 Magic or Magic 2 missiles (outboard) under wings. Alternatively, each of four underwing hardpoints can carry a Magic. MICA AAM (110 kg; 243 lb) optional on Mirage 2000-5. Primary weapon for 2000N is 900 kg (1,984 lb) ASMP tactical nuclear missile mounted on LM-770 centreline pylon.

In air-to-surface role, the Mirage 2000 can carry up to 6,300 kg (13,890 lb) of external stores, including Matra BAe 250 kg retarded bombs or 32.5 kg (72 lb) TDA BAP 100 anti-runway bombs; 16 Matra BAe Durandal 219 kg (483 lb) penetration bombs; one or two 990 kg (2,183 lb) Matra BAe BGL 1000 laser-guided bombs; five or six 305 kg (672 lb) Matra BAe Belouga cluster bombs or 400 kg (882 lb) TDA BM 400 modular bombs; one Rafaut F2 practice bomb launcher; US Mk 20, Mk 82, GBU-10 and GBU-12 bombs; two 520 kg (1,146 lb) AS 30L, Armat anti-radar, or 655 kg (1,444 lb) AM 39 Exocet anti-ship, air-to-surface missiles; four 185 kg (408 lb) Matra BAe LR F4 rocket launchers, each with eighteen 68 mm rockets; two packs of 100 mm rockets; or a 765 kg (1,687 lb) Dassault CC 630 gun pod, containing two 30 mm guns and total 600 rounds of ammunition.

Mirage 2000D (and export 2000-5) to receive 1,200 kg (2,646 lb) APACHE-AP standoff (216 n mile; 400 km; 249 mile) weapons dispenser. APACHE-SCALP (*Système de Croisière Autonome à Longue Portée*) selected December 1994 to satisfy APTGD (*Arme de Précision Tirée à Grande Distance*) requirement for 216 to 324 n mile (400 to 600 km; 249 to 374 mile) range stealthy cruise missile; service entry in 2002. For air defence weapon training, a Cubic Corporation AIS (airborne instrumentation subsystem) pod, externally resembling a Magic missile, can replace Magic on launch rail, enabling pilot to simulate a firing without carrying actual missile.

DIMENSIONS, EXTERNAL:
Wing span	9.13 m (29 ft 11½ in)
Wing aspect ratio	2.0
Length overall: 2000C, E	14.36 m (47 ft 1¼ in)
-5*	14.65 m (48 ft 0¾ in)
2000B, N*	14.55 m (47 ft 9 in)
Height overall: 2000C, E, -5	5.20 m (17 ft 0¾ in)
2000B, N, D	5.15 m (16 ft 10¾ in)
Wheel track	3.40 m (11 ft 1¾ in)
Wheelbase	5.00 m (16 ft 4¾ in)

*2000D, S and -5 versions lack nose pitot

AREAS:
Wings, gross	41.0 m² (441.3 sq ft)

WEIGHTS AND LOADINGS:
Weight empty: 2000C, E, -5	7,500 kg (16,534 lb)
2000B, N, D	7,600 kg (16,755 lb)
Max internal fuel: 2000C, -5	3,160 kg (6,967 lb)
2000B, N, D	3,100 kg (6,834 lb)
Max external fuel	4,100 kg (9,039 lb)
Max external stores load	6,300 kg (13,890 lb)
Combat weight: 2000-5	9,500 kg (20,944 lb)
T-O weight clean: 2000C, E, -5	10,860 kg (23,940 lb)
2000B, N, D	10,960 kg (24,165 lb)
Max T-O weight: except -5, -9:	
normal	14,000 kg (30,864 lb)
overload	17,000 kg (37,480 lb)
-5, -9	17,500 kg (38,580 lb)
Max landing weight	16,700 kg (36,817 lb)
Max wing loading	414.6 kg/m² (84.92 lb/sq ft)
Max power loading (M53-P2)	179 kg/kN (1.75 lb/lb st)

PERFORMANCE (M53-P2 power plant):
Max level speed: at height	M2.2
at S/L	M1.2
Max continuous speed: 2000C, E	M2.2
2000N, D	M1.4
Min speed in stable flight	100 kt (185 km/h; 115 mph)
Authorised min flight speed	zero
Approach speed	140 kt (259 km/h; 161 mph)
Landing speed	125 kt (232 km/h; 144 mph)
Max rate of climb at S/L:	
M53-P2	17,060 m (56,000 ft)/min
M53-P20	18,290 m (60,000 ft)/min
Time to 11,000 m (36,080 ft) and M1.8:	
2000-5	approx 5 min
Time from brake release to intercept target flying at M3.0 at 24,400 m (80,000 ft)	less than 5 min
Service ceiling: 2000C	16,460 m (54,000 ft)
2000-5	18,290 m (60,000 ft)
Range: hi-hi-hi	1,000 n miles (1,852 km; 1,151 miles)
interdiction, hi-lo-hi	800 n miles (1,480 km; 920 miles)
attack, hi-lo-hi	650 n miles (1,205 km; 748 miles)
attack, lo-lo-lo	500 n miles (925 km; 575 miles)
with one 1,300 litre and two 1,700 litre drop tanks	1,800 n miles (3,333 km; 2,071 miles)
Operational loiter (2000-5) at M0.8 at 7,620 m (25,000 ft) with three external tanks, four MICA and two Magic AAMs	2 h 30 min
Operational range (2000-5) for 5 min combat at M0.8 at 9,145 m (30,000 ft) with four MICA and two Magic AAMs, tanks jettisoned	780 n miles (1,445 km; 898 miles)
g limits:	+9.0/–3.2 normal
	+11.0/–4.5 ultimate

UPDATED

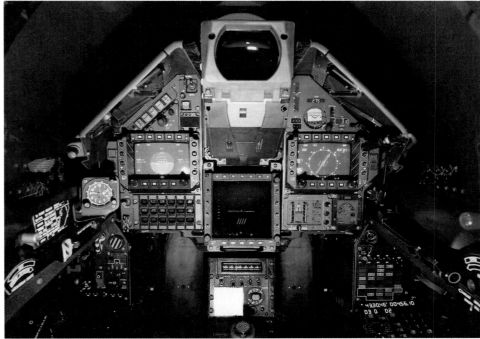
Mirage 2000-5 cockpit

DASSAULT RAFALE

English name: Squall

TYPE: Multirole fighter.

PROGRAMME: Ordered (as *Avion de Combat Tactique; ACT*) to replace French Air Force Jaguars and (as *Avion de Combat Marine; ACM*) Navy Crusaders and Super Etendards; for early development history, see 1990-91 and earlier *Jane's*; first flight of Rafale A prototype (F-ZJRE) 4 July 1986; first flight with SNECMA M88 replacing one GE F404, 27

Dassault Rafale B01, the first two-seat aircraft, equipped with two APACHE SOMs and three external fuel tanks (*Aviaplans/F Robineau*)

FRANCE: AIRCRAFT—DASSAULT

February 1990 (was 461st flight overall); 867th and final sortie, 24 January 1994. ACE International (*Avion de Combat Européen*) GIE set up in 1987 by Dassault Aviation, SNECMA, Thomson-CSF and Dassault Electronique, partly to attract international partners; none found. Four preproduction aircraft, as described under Current Versions (specific). Production launch officially authorised, 23 December 1992 (and 31 December 1992 for M88-2 power plant). First Rafale B and Rafale M ordered 26 March 1993; first production aircraft (Rafale B No. 301) flew 24 November 1998 and made 'inaugural' (official) first flight 4 December; to CEV at Istres, early 1999, for development of F2 production standard. First production Rafale M (No. 1) flew at Bordeaux 7 July 1999. Rafale had by then accumulated over 4,000 sorties and 3,450 hours, including 50 sorties by No 301 and 12 by No 1. Third production aircraft (No 302) flew late 1999. First Rafale C to be delivered for service use mid-2002.

Test airframe, in Rafale M configuration, delivered to CEAT at Toulouse for ground trials 10 December 1991. Between 17 December 1991 and 2 March 1993, completed 10,000 simulated flights, including 3,000 catapult take-offs and 3,000 deck landings. Rafale structural validation achieved 15 December 1993.

French Air Force preference switched to operational two-seat (pilot and WSO) derivative of Rafale C in 1991; announced 1992 that 60 per cent of procurement to be two-seat, although 16 aircraft deleted from requirements at this time. Procurement target further reduced, as detailed in Customers paragraph. Two-seat version of naval Rafale announced September 2000.

Funding constraints and French government demands for cost reductions resulted in suspension of Rafale programme in November 1995 and temporary blocking of most 1996 funds. Plans were simultaneously abandoned for three progressively more sophisticated service standards of Rafale (*Standard Utilisateur* 0, 1 and 2) and replaced by a common French military standard and an export parallel, although three basic software standards (F1 to F3) will phase-in operational capabilities, as described under Current Versions (general). Work on production Rafales temporarily halted in April 1996. On 22 January 1997, Dassault and French defence ministry agreed on 48-aircraft multiyear procurement (1997-2002) in return for a 10 per cent cost reduction, effectively relaunching the programme. This initiative lapsed with the change of government after June 1997, but reinstated in January 1999, with firm orders for 28, plus 20 options, covering deliveries between 2002 and 2007.

First production aircraft completed late 1998; second and third (No. 302 and M1) delivered to CEV trials unit late 1999/early 2000; delivery of M2 to Aéronavale in September 2000; IOC (naval version) in June 2001 (six aircraft); first naval squadron complete (10 Rafales) by June 2002; Rafale B No. 303 will be first for air force, in late 2002; first air force squadron equipped with 20 aircraft during 2005; 80th air force aircraft due in 2010, 140th in 2015; final (294th) French delivery due in 2023.

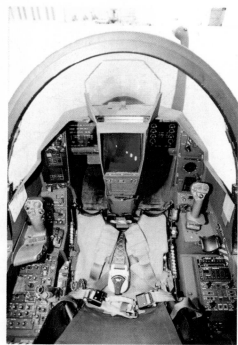

Rafale cockpit *(Dassault/Michel Isaac)* 0010616

Associated programmes include Thomson-CSF electronic scanning RBE2 (*Radar à Bayalage Electronique deux plans*) multimode radar, ordered November 1989; test flights begun in Falcon 20 No. 104, 10 July 1992; first RBE2 flight in Rafale 7 July 1993 (B01); first production RBE2 flew on 16 October 1997 in a Falcon 20 before being refitted in Rafale B01 from November 1997. Development authorised in early 1999 of upgraded RBE2 version (full air-to-ground weapons capability) for 2003 delivery and installation in F2 standard Rafales.

Thomson-CSF/Aerospatiale Matra defensive aids package named Spectra (*Système pour la Protection Electronique Contre Tous les Rayonnements Adverses*); wholly internal IR detection, laser warning, electromagnetic detection, missile approach warning, jamming and chaff/flare launching; nine prototypes ordered; total weight 250 kg (551 lb); Spectra trials begun on Mirage 2000 in 1992, while full suite installed in Falcon 20 No. 252 by Dassault at Istres between December 1992 and September 1994 before flight trials; Spectra flown in Rafale M02 on 20 September 1996 at launch of integration programme at CEV Istres. Development contract awarded 1991 for Thomson-CSF Optronique/SAGEM OSF (*Optronique Secteur Frontal*) with IRST, FLIR and laser range-finder in two modules ahead of windscreen; surveillance, tracking and lock-on by port module; target identification, analysis and optical identification by starboard module; combined output in pilot's head-level display; initially tested in Mirage 2000BOB (which see).

M88 engine, which has flown only on Rafales, achieved type certification on 22 March 1996. First of initial production batch of 42 M88-2 engines delivered by SNECMA 30 December 1996. Integration of Matra BAe Dynamics MICA AAM completed 5 July 2000, after 27 launches.

Rafale offered for export; Greek evaluation in January 2000; model displayed February 2000 carried two conformal fuel tanks, each of 1,250 litres (330 US gallons; 275 Imp gallons), increasing range with two SCALP missiles to 1,000 n miles (1,852 km; 1,150 miles), hi-lo-hi. Rafale B No. 302 equipped with supplementary software to enable demonstration of LGB capability.

CURRENT VERSIONS (general): **Rafale B:** Originally planned two-seat, dual-control version for French Air Force; weight initially envisaged as 350 kg (772 lb) more than Rafale C; 3 to 5 per cent higher cost than Rafale C. Being developed into fully operational variant for either pilot/WSO or single-pilot combat capability. Serial numbers begin at 301. First two assigned to CEV.

Rafale C: Single-seat combat version for French Air Force. Serial numbers begin at 101. Deliveries due in 2002.

Rafale D: Original configuration from which production versions derived; now '*Rafale Discret*' (stealthy) generic name for French Air Force versions.

Rafale M: Single-seat carrierborne fighter; serial numbers begin at 1. Navalisation weight penalty, 610 kg (1,345 lb); take-off weight from existing French carrier *Foch* limited to 16,500 kg (36,376 lb); has 80 per cent structural and equipment commonality with Rafale C, 95 per cent systems commonality. Navy's financial share of French programme cut in 1991 from 25 to 20 per cent.

Initial operational software standard for Rafale M, designated F1, will permit air defence missions against multiple targets using Magic and radar-homing version of MICA; self-defence provided by Spectra system. In 2002, F1.1 will add IR-guided MICA and MIDS datalink for communication with Northrop Grumman E-2C Hawkeye AEW aircraft. F2 will apply to both naval and air force Rafales delivered from late 2004 and combine F1/F1.1 with air-to-ground radar modes and the ability to launch IR-guided MICA, SCALP and AASM weapons as well as OSF electro-optics suite and MIDS datalink; operational in 2006. Funding for F2 development granted 31 December 1998; first stage of F2, known as E12F, is OSF, SCALP and standard air-to-surface ordnance. Finally, in 2008 (operational 2009), F3 standard (see also below) provides full capabilities to naval and air force Rafales, including air-to-sea attack, ANF and ASMP-A weapons, refuelling and reconnaissance pods, and Topsight E pilot's helmet. Unspecified and unfunded F4 envisaged for 2010, but early naval aircraft will all have been upgraded to F3 by 2008. By early 2001, F4 configuration under preliminary study for second decade of 21st Century.

Rafale M01, the first naval prototype, launching a MICA AAM *(Matra BAe Dynamics)* 2001/0073275

DASSAULT—AIRCRAFT: FRANCE

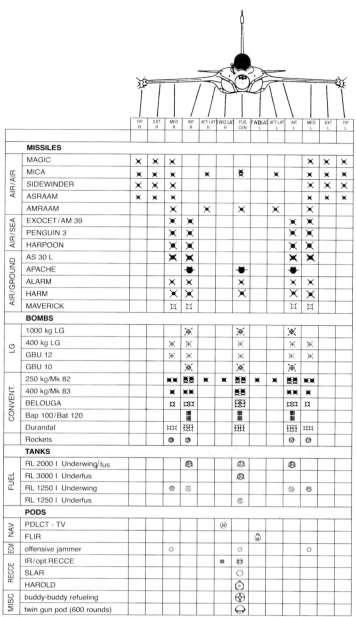

Rafale weapon options

'Fastback' conformal fuel tanks on a Rafale model *(Paul Jackson/Jane's)*

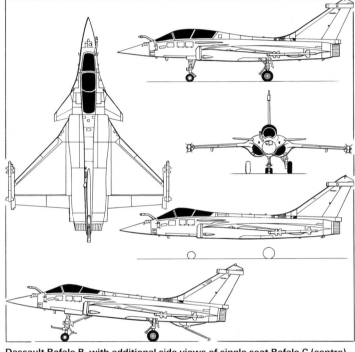

Dassault Rafale B, with additional side views of single-seat Rafale C (centre) and navalised Rafale M (bottom) *(James Goulding/Jane's)*

Service familiarisation began 4 December 2000 when Nos. 2 and 3 delivered to Landivisiau naval air base.

Rafale: Two-seat naval version; announced September 2000; prototype, No. 15, for trials in 2005; requirement for 35 to 40 aircraft within overall Navy purchase; first production aircraft to be No. 21, in 2006; service entry 2005. Cost 5 per cent more than single-seat M; 200 kg (441 lb) heavier; 85 per cent commonality with three previous versions; development cost FFr1,500 billion (€228 million); deletion of internal canon replaces 200 litres (52.8 US gallons; 44.0 Imp gallons) of fuel lost with addition of second seat.

'Rafale R': Initial studies launched by procurement agency, DGA, late 1997, into stealthy sensor pod which would allow Rafales to replace Mirage F-1CRs and naval Super Etendards in reconnaissance role.

'Rafale Mk 2': Export version, under active consideration by 2000, featuring active antenna radar and M88-3B engines of 88.3 kN (19,850 lb st) each. Available from 2006.

CURRENT VERSIONS (specific): **C01:** Single-seat Rafale C prototype, C01/F-ZWVR, ordered 21 April 1988; flown 19 May 1991; officially flight tested at CEV in October 1991, two months ahead of schedule; 100th sortie 12 May 1992. First Rafale gun firing, 5 March 1993; first Magic 2 AAM launch, 26 March 1993. Continued high AoA trials in 2000, having exceeded 30°.

Second Rafale C order (C02) not placed; abandoned 1991.

M01: First navalised prototype, F-ZWVM, ordered 6 December 1988; flown 12 December 1991. Assigned to structural qualification, FCS and aerodynamic trials. Catapult trials ashore at US Naval Air Warfare Center, Patuxent River, and Lakehurst, 13 July/23 August 1992; second series of US trials 15 January/18 February 1993 followed by deck trials on *Foch*; first deck landing 19 April 1993; first deck launch 20 April 1993, although first take-offs with 'jump strut' nosewheel leg began in following month; third US trials series 18 November/16 December 1993, carrying external loads; fourth series, October/December 1995, including dummy deck launch at maximum weight of 22,300 kg (49,163 lb). On 8 June 1995, M01 made the first launch of a MICA AAM against an aerial target acquired by RBE2 radar fitted in a Rafale. Employed on development trials for F2 production standard. Total 438 hours in 723 sorties by October 1997. In storage late 1999, awaiting next assignment.

B01: Two-seat dual-control trainer Rafale B prototype, B01/F-ZWVS, ordered 19 July 1989 as first with RBE2 radar and Spectra defensive systems; first flight 30 April 1993; first flight with RBE2 7 July 1993. Longest Rafale sortie, Istres to Dubai, November 1995: approximately 3,020 n miles (5,600 km; 3,475 miles) in 6 hours 30 minutes with three aerial refuellings (including one precautionary). Heavy configuration trials (23,400 kg; 51,588 lb, including two APACHEs) completed in February 1997. Total 990 sorties by mid-1999.

M02: Second naval prototype, ordered 4 July 1990; first flight 8 November 1993; assigned to operational and maintenance testing aboard ship and navigation/weapons trials. Joint carrier trials with M01 aboard *Foch* (second series) 27 January/4 February 1994. Third series of deck trials (M02 only) aboard *Foch* begun 17 October 1994 for three weeks (total 28 launches including two at night); included maintenance, electromagnetic compatibility, RBE2 radar and Spectra ECM tests. Fitted with model of OSF in late 1994, for vibration tests. Flew Singapore to Istres (6,300 n miles; 11,668 km; 7,250 miles) in under 15 hours, February 1996. First launch of Magic 2 AAM at moving target, 4 April 1996. Total 297 hours in 348 sorties by October 1997. With B01, employed by 1998 on development of F2 and F3 avionics standards. On 6 July 1999 became only second jet fighter (following Super Etendard) to land on new carrier FS *Charles de Gaulle*. Continued Spectra trials in 2000.

CUSTOMERS: Anticipated worldwide market for 500 aircraft in addition to originally planned 250 for French Air Force (225 Cs and 25 Bs) and 86 French Navy; former service announced revised requirement for 234, comprising 95 Rafale Cs and 139 two-seat (pilot and WSO) combat versions, in 1992. Defence economies in 1996 included reduction of requirements to 212 B/Cs and 60 Ms. Naval deliveries began with No. 2 to CEPA trials unit on 19 July 2000, followed by No.3 in September 2000; Landivisiau-based 12 Flottille to achieve IOC with 10 aircraft in February 2002 for interceptor duties aboard *Charles de Gaulle* (commissioned 1999); balance replaces Super Etendards, 11 Flottille being first recipient, in 2005. Navy to receive 10 aircraft in F1 configuration, 15 F2s and 35 F3s, last in 2012, completing third squadron, all at Landivisiau. Air Force deliveries originally planned for 1996-2009, including first 20 in interim configuration; two year postponement announced 1992; further slips in development funding delayed first receipts to 2002 and IOC to 2005. Only Air Force's first three aircraft 301, 302 and 101 (two Bs and one C) to F1 standard.

Official authorisation to launch production given 23 December 1992. Initial production contract in 1993 defence budget (formally awarded 26 March 1993), comprising one aircraft each for Air Force and Navy. Total 13 by 1996, while 1997-2002 plan envisaged 33 B/Cs and 15 Ms (total 48) to be ordered (and two Bs and 12 Ms delivered), followed by orders for 15 B/Cs per year from 2003. In early 1997, French defence procurement agency, DGA, agreed with Dassault a multiyear procurement of the 48 aircraft in return for a 10 per cent cost reduction, but this later suspended. Authorisation to order the first 13 Rafales was only granted in May 1997, however. At the same time, separate plans were being formulated for acceleration into service of the first 10 air force Rafales to equip an export-promoting and operational trials (half) squadron, but those plans also soon abandoned. Eventually, go-ahead given on 14 January 1999 for 48 aircraft, including 20 options to be confirmed in 2000. Export versions of naval variant were available to potential customers from 1999 onwards. China expressed interest in 1996-97. By 2015 French Air Force combat arm is expected to comprise 140 Rafales in front-line units.

RAFALE PROCUREMENT

Year	Rafale C	Rafale B	Rafale M/BM	Total
1993		1	1	2
1994		1	2	3
1995			5	5
1996		1	2	3
1997				
1998				
1999†	7	14	7	28
Subtotals	7	17	17	41
2001*	5	7	8	20
Totals	12	24	25	61

† start of F2 production version; earlier aircraft are F1; 62nd and subsequent will be F3
* to be confirmed; options placed in 1999
Note: Orders as announced. Third Rafale B was subsequently exchanged with first production Rafale C.

RAFALE M DECK TRIALS

Campaign	Dates	LAND Location	Aircraft	Dates	SEA Location	Aircraft
First	13 Jul – 23 Aug 1992	Patuxent & Lakehurst	M01	19 Apr – 7 May 1993	FS *Foch*	M01
Second	15 Jan – 18 Feb 1993	Lakehurst	M01	27 Jan – 7 Feb 1994 and	FS *Foch*	M01, M02
				11 Apr – 3 May 1994	FS *Foch*	M01, M02
Third	18 Nov – 16 Dec 1993	Lakehurst	M01	17 Oct – Nov 1994	FS *Foch*	M02
Fourth	30 Oct – 7 Dec 1995	Lakehurst	M01	11 Sep – 15 Sep 1995	FS *Foch*	M02
Fifth	n/a			19 Jan – 6 Feb 1998	FS *Foch*	M02
Sixth	n/a			6 Jul – 22 Jul 1999	FS *Charles de Gaulle*	M02

COSTS: Programme estimated at FFr155 billion (1991), including FFr40 billion for R&D; revised to FFr178 billion in 1993, FFr198.4 billion in 1995 and FFr202.37 billion in 1996. Last-mentioned total comprises FFr48.62 billion (of which 25 per cent paid by industry) for development, FFr17.583 billion for industrialisation, FFr76.25 billion for 234 Rafale B/Cs, FFr20.89 billion for 60 Rafale Ms, FFr37.812 billion for spares and FFr1.215 billion for simulators. Early 1997 agreement on 10 per cent cost reduction resulted in flyaway price falling to FFr282 million for a Rafale C, FFr299 million for Rafale B and FFr315 million for Rafale M. Total of FFr30 billion spent by 1995. Second production order for eight aircraft (1994/1995 authorisations) estimated at FFr1.5 billion, excluding engines, radar and weapons system. In 1998, however, new cost estimate for 294 aircraft was FFr320 billion as a consequence of programme delays. Cost (1999) of 48 aircraft given as FFr17.2 billion.

DESIGN FEATURES: Multirole combat aircraft, rivalling Eurofighter Typhoon. Minimum weight and volume structure to hold costs to minimum; thin, mid-mounted delta wing with moving canard; individual fixed, kidney-shaped intakes without shock cones. HOTAS controls, with sidestick controller on starboard console and small-travel throttle lever.
Wing leading-edge sweepback approximately 48°.

FLYING CONTROLS: Fully fly-by-wire controls with fully modulated two-section leading-edge slats and two elevons per wing; canard incidence automatically increased to 20° when landing gear lowered; airbrake panels in top of fuselage beside leading-edge of fin. Specification includes 30° AoA in stable flight.

STRUCTURE: Most of wing components made of carbon fibre including elevons; slats in titanium; wingroot and tip fairings Kevlar; canard made mainly by superplastic forming and diffusion bonding of titanium; fuselage 50 per cent carbon fibre; fuselage side skins of aluminium-lithium alloy; wheel and engine doors carbon fibre; fin made primarily of carbon fibre with aluminium honeycomb core in rudder. Composites account for 25 per cent by weight of structure and 20 per cent of surface area; weight saving directly attributable to composites is 300 kg (661 lb) – equivalent to a 1 tonne reduction in empty weight.

LANDING GEAR: Hydraulically retractable tricycle type supplied by Messier-Dowty, with single 790×275-15 (20 ply) or 790×275R15 mainwheels and twin, hydraulically steerable, 360×135-6 or 360×135R6 nosewheels. All wheels retract forward. Designed for impact at vertical speed of 3 m (10 ft)/s, or 6.5 m (21 ft)/s in naval version, without flare-out. Rafale M has same mainwheels; but 520×140R10.5 nosewheels. Messier-Bugatti carbon brakes on all three units, controlled by fly-by-wire system.
Rafale M has 'jump strut' nosewheel leg which releases energy stored in shock-absorber at end of deck take-off run, changing aircraft's attitude for climb-out without need for ski-jump ramp. 'Jump strut' advantage equivalent to 9 kt (16 km/h; 10 mph) or 900 kg (1,984 lb) extra weapon load; not to be used aboard carrier *Foch*, which to have 1° 30′ ramp giving 20 kt (37 km/h; 23 mph) or 2,000 kg (4,409 lb) advantage. Dowty Aerospace Yakima holdback fitting. Naval nosewheel steerable ±70°; or almost 360° under tow. Hydraulic (Rafale M) or tension-stored (Rafale B/C) arrester hook. Landing gear management (braking and steering) by Thomson-CSF computer.

POWER PLANT: Two SNECMA M88-2 augmented turbofans, each rated at 48.7 kN (10,950 lb st) dry and 72.9 kN (16,400 lb st) with afterburning. Stage 4 standard engines in production Rafales. M88-3 of 88.3 kN (19,840 lb st) maximum rating offered as alternative, subject to programme launch. Internal tanks in single-seat versions for approximately 5,700 litres (1,506 US gallons; 1,254 Imp gallons) of fuel. Fuel system by Lucas Air Equipement, Lebozec and Zenith Aviation; equipment by Intertechnique. Five 'wet' hardpoints: centreline, two inboard wing and two centre wing; all able to accommodate a 1,250 litre (330 US gallon; 275 Imp gallon) external tank; alternative 2,000 litre (528 US gallon; 440 Imp gallon) centreline and inboard; alternative 3,000 litre (793 US gallon; 660 Imp gallon) tank on centreline. Max external fuel capacity 9,500 litres (2,510 US gallons; 2,090 Imp gallons). Total internal and external fuel (single-seat versions) approximately 15,200 litres (4,015 US gallons; 3,344 Imp gallons); slightly less in two-seat version. Pressure refuelling in 7 minutes, or 4 minutes for internal tanks only. Fixed (detachable) in-flight refuelling probe on all versions.

ACCOMMODATION: Pilot only, on SEMMB (Martin-Baker) Mk 16 zero/zero ejection seat, reclined at angle of 29°. One-piece Sully Produits Spéciaux blister windscreen/canopy, hinged to open sideways to starboard. Canopy gold-coated to reduce radar reflection.

SYSTEMS: Technofan cockpit air conditioning system; Cryotechnologies avionics cooling system. Dual hydraulic circuits, pressure 350 bars (5,075 lb/sq in), each with two Messier-Bugatti pumps and Bronzavia ancillaries. Auxilec electrical system, with two 30/40 kVA Auxilec variable frequency alternators. Triplex digital plus one dual analogue fly-by-wire flight control system, integrated with engine controls and linked with weapons system. Air Liquide OBOGS; EROS oxygen system; L'Hotellier fire detection system; Microturbo APU.

AVIONICS: Provision for more than 780 kg (1,720 lb) of avionics equipment and racks.
Comms: EAS V/UHF and Thomson-CSF Communications TRA 6032 SATURN UHF radios; Thomson-CSF Communications TSB 2500 transponder/interrogator. TEAM intercom; Sextant Avionique voice-activated radio controls and voice alarm warning system. Chelton aerials.
Radar: GIE Radar (Thomson-CSF Detexis) RBE2 lookdown/shoot-down radar, able to track up to eight targets simultaneously, with automatic threat assessment and allocation of priority.
Flight: Thomson-CSF Communications TLS-2020 integrated ILS/MLS, VOR/DME; SAGEM Sigma 95N (RL-90) RLG INS (SAGEM Telemir interface with carrier's navigation on Rafale M); Thomson-CSF Communications NC 12E Tacan interrogator; Thomson-CSF Communications AHV-17 radio altimeter and SFIM/Thomson-CSF ESPAR static memory flight recorder. Sextant Avionique GPS.
Instrumentation: Digital display of fuel, engine, hydraulic, electrical, oxygen and other systems information on two 127 × 127 mm (5 × 5 in) lateral multifunction touch-sensitive colour LCD displays by Sextant. Third cockpit screen is 20 × 20° head-level tactical navigation/sensor display. Sextant Avionique CTH3022 wide-angle, holographic HUD (30 × 22° field of view) incorporating Signaal USFA OTA-1320 CCD camera and recorder. Sextant/Intertechnique Topsight E helmet-mounted sight.
Mission: Thomson-CSF Optronique/SAGEM OSF electro-optical sensors. MIDS (Multifunctional Information Distribution System) datalink (equivalent to JTIDS/Link 16). Various reconnaissance, ECM, FLIR and laser designation pods, including Thomson-CSF Optronique Damoclès target designator. Terrain-following system cleared (1999) to 152 m (500 ft); eventual 30 m (100 ft) over land and 15 m (50 ft) over water is envisaged.
Self-defence: Spectra radar warning and ECM suite by Thomson-CSF Detexis and Aerospatiale Matra. Thomson TTD Optronique DAL (*Détecteur d'Alerte Laser*) system.

EQUIPMENT: Integral, electrically operated, folding ladder in Rafale M.

ARMAMENT: One 30 mm Giat DEFA 791B cannon in side of starboard engine duct (except naval two-seat). Fourteen external stores attachments: two on fuselage centreline, two beneath engine intakes, two astride rear fuselage, six under wings and two at wingtips. Forward centreline position deleted on Rafale M. Normal external load 6,000 kg (13,228 lb); maximum permissible, 9,500 kg (20,944 lb); see weapon options table. In strike role, one Aerospatiale Matra ASMP standoff nuclear weapon. In interception role, up to eight MICA AAMs (with IR or active homing) and two underwing fuel tanks; or six MICAs and three external fuel tanks. In air-to-ground role, typically sixteen 227 kg (500 lb) bombs, two MICAs and two 1,250 litre (330 US gallon; 275 Imp gallon) tanks; or two APACHE standoff weapon dispensers, two MICAs and three tanks; or FLIR pod, Atlis laser designator pod, two 1,000 kg (2,205 lb) laser-guided bombs, two AS.30L laser ASMs, four MICAs and single tank. In anti-ship role, two Exocet sea-skimming missiles, four MICAs and two external fuel tanks. Future weapons will include AASM (*Armement Air-Sol Modulaire*) powered LGB.

DIMENSIONS, EXTERNAL:
Wing span, incl wingtip missiles 10.80 m (35 ft 5¼ in)
Wing aspect ratio 2.6

First production Rafale, No. B 301, flew on 24 November 1998

DASSAULT—AIRCRAFT: FRANCE

Dassault Rafale B, cutaway drawing key

1. Kevlar composites radome
2. RBE2 electronically scanned multimode radar scanner
3. Fixed (detachable) in-flight refuelling probe
4. OSF IR scanner/tracker
5. OSF passive visual sight (LLTV)
6. OSF electronics module
7. Air flow sensors, pitch and yaw
8. Temperature probe
9. Radar equipment module
10. Dynamic pressure probe
11. Cockpit front pressure bulkhead
12. Instrument panel shroud
13. Rudder pedals
14. Canopy emergency release
15. Electroluminescent formation lighting strip
16. Nose landing gear bay
17. Port side console
18. Engine throttle lever with HOTAS controls
19. Elbow rest
20. Sidestick controller
21. Pilot's wide-angle holographic HUD
22. Head-level tactical navigation indicator (HSI)
23. Lateral multifunction touch-sensitive colour LCD displays
24. Systems displays
25. Standby horizontal situation indicator (HSI)
26. Frameless windscreen panel
27. Canopy open position
28. Thomson-CSF Atlis laser designator pod
29. Atlis mounting pylon
30. Pilot's rearview mirrors
31. Pilot's helmet with integrated sight display
32. Cockpit canopy, hinged to starboard, incorporating miniature detonating cord (MDC) emergency breaker
33. Pilot's SEMMB (licence-built Martin-Baker) Mk 16 zero/zero ejection seat
34. Forward fuselage/cockpit section carbon fibre structure
35. Lateral equipment bays, port and starboard
36. Nose landing gear pivot mounting
37. Lower UHF antenna
38. Taxying lights
39. Nosewheel hydraulic steering jacks
40. Twin nosewheels, forward-retracting
41. Hydraulic retraction jack
42. Port engine air intake
43. Boundary layer splitter plate
44. Ventral intake suction relief door
45. Port forward-oblique Spectra ECM antenna
46. Spectra RWR antenna
47. Onboard oxygen generation system (OBOGS)
48. Rear instrument display console
49. Blast screen between cockpits
50. Canopy centre arch
51. Lateral avionics equipment bays, port and starboard
52. Canard hydraulic actuator
53. Canard hinge mounting
54. Environmental control system (ECS) equipment bay
55. Heat exchanger exhaust
56. Cockpit rear bulkhead
57. Second Pilot/Weapons System Officer's (P2/WSO) SEMMB Mk 16 ejection seat
58. Starboard canard
59. Carbon fibre structure with honeycomb core
60. Starboard navigation light
61. Dorsal avionics equipment bay
62. Centre fuselage aluminium-lithium structure
63. Intake ducting
64. Fuselage integral fuel tanks
65. Port fuselage main longeron
66. ADF flush antenna
67. Dorsal spine fairing containing systems ducting
68. Anti-collision beacon
69. Starboard fuselage integral fuel tankage
70. Kevlar composites wing/fuselage fairing panels
71. Starboard wing integral fuel tank
72. Wing pylon hardpoints
73. Leading-edge slat hydraulic jacks and position transmitters
74. Slat guide rails
75. Starboard two-segment automatic leading-edge slats
76. Starboard external fuel tank
77. Giat DEFA 791B 30 mm cannon, mounted beneath starboard wingroot
78. Forward RWR antenna
79. Wingtip fixed missile pylon
80. Matra BAe MICA AAM, IR variant
81. Rear RWR antenna
82. Starboard outboard elevon
83. Elevon hydraulic actuator
84. Wing carbon fibre skin panelling
85. Inboard elevon
86. Fuselage aluminium-lithium skin panelling, carbon fibre ventral engine bay access panels
87. APU intake grilles
88. Microturbo APU
89. Wing panel attachment machined fuselage main frames
90. Engine compressor intake with variable guide vanes
91. SNECMA M88-2 afterburning turbofan
92. Forward engine mounting
93. APU exhaust
94. Carbon fibre engine bypass duct
95. Rear engine mounting
96. Fin attachment main frames
97. Fin root bolted attachment fittings
98. Rudder hydraulic actuator
99. Carbon fibre multispar fin structure
100. Carbon fibre leading-edge
101. Flight control system airflow sensor
102. Formation lighting strip
103. VOR localiser antenna
104. Forward ECM transmitting antenna
105. Spectra ECM equipment housing
106. Fin tip antenna fairing
107. V/UHF antenna
108. Rear position light
109. Aft ECM transmitting antenna
110. Rudder
111. Carbon fibre rudder skin panelling
112. Extended wingroot trailing-edge fairing
113. Carbon fibre elevon skin panels
114. Brake parachute housing
115. Variable area afterburner nozzle shroud plates
116. Nozzle actuators (total five)
117. Afterburner ducting
118. Formation lighting strip
119. Chaff/decoy launcher
120. Extended wingroot trailing-edge fairing
121. Flight control system equipment
122. Wing rear spar attachment joint
123. Engine accessory equipment
124. Engine oil tank
125. Inboard elevon hydraulic actuator
126. Stored energy (spring loaded) emergency runway arrester hook
127. Port inboard elevon
128. Carbon fibre elevon skin panels
129. Aluminium honeycomb core structure
130. Elevon hydraulic actuator in ventral fairing
131. Port outboard elevon
132. Port rear RWR antenna
133. Matra BAe MICA AAM, active radar variant
134. Matra BAe Magic short-range IR AAM
135. Forward RWR antenna
136. Port wingtip missile pylon
137. Wing outboard missile pylon
138. Outer pylon hardpoint
139. Leading-edge slat guide rails and hydraulic jacks
140. Port automatic leading-edge slat segments, diffusion-bonded superplastic-formed titanium structure
141. External fuel tank
142. Port intermediate wing pylon
143. Leading-edge spar
144. Intermediate pylon hardpoint
145. Titanium wing ribs
146. Carbon fibre multispar wing panel structure
147. Port wing integral fuel tankage
148. Inboard pylon hardpoint
149. Rear fuselage ventral Matra BAe MICA missile pylon
150. Wing panel titanium bolted attachment fittings
151. Hydraulic reservoir and accumulator, port and starboard dual system
152. Airframe-mounted auxiliary equipment gearbox, port and starboard (interconnected)
153. Main landing gear leg pivot mounting
154. Hydraulic retraction jack
155. Leg rotating link; wheel lies flat beneath intake duct
156. Mainwheel shock-absorber strut
157. Port mainwheel
158. Torque scissor links
159. Inboard wing pylon
160. Mainwheel leg breaker strut
161. Port navigation light
162. Landing light
163. Front spar/fuselage attachment joint
164. Electrically driven standby hydraulic pump
165. Blended wing/fuselage chine
166. Port canard
167. Position of cannon muzzle aperture on starboard fuselage
168. TRT/Intertechnique Rubis FLIR pod beneath port intake duct
169. Matra BAe APACHE standoff weapons dispenser
170. Folding wing panels
171. APACHE jettisonable intake fairing
172. Matra BAe BGL 1000 laser-guided bomb

© Michael Badrocke 1998

122   FRANCE: AIRCRAFT—DASSAULT

Length overall	15.27 m (50 ft 1¼ in)
Height overall (Rafale D)	5.34 m (17 ft 6¼ in)

AREAS:
Wings, gross	45.70 m² (491.9 sq ft)

WEIGHTS AND LOADINGS (estimated):
Basic weight empty, equipped:	
Rafale B	10,000 kg (22,047 lb)
Rafale D	9,060 kg (19,973 lb)
Rafale M	9,670 kg (21,319 lb)
External load: normal	6,000 kg (13,228 lb)
max	9,500 kg (20,944 lb)
Max fuel weight: internal:	
single seat	4,500 kg (9,921 lb)
two seat	4,300 kg (9,524 lb)
external	7,500 kg (16,535 lb)
Max T-O weight: early production	19,500 kg (42,990 lb)
subsequent production	22,500 kg (49,604 lb)
developed version	24,500 kg (54,013 lb)
Max landing weight	22,500 kg (49,605 lb)
Max wing loading:	
initial version	426.7 kg/m² (87.39 lb/sq ft)
developed version	536.1 kg/m² (109.80 lb/sq ft)
Max power loading:	
initial version	134 kg/kN (1.31 lb/lb st)
developed version (M88-3)	141 kg/kN (1.38 lb/lb st)

PERFORMANCE (estimated):
Max level speed: at altitude	M1.8
at low level	750 kt (1,390 km/h; 864 mph)
Approach speed	120 kt (223 km/h; 139 mph)
Max rate of climb at S/L	approx 18,290 m (60,000 ft)/min
Roll rate	270°/s
Max instantaneous turn rate	up to 30°/s
Service ceiling	16,765 m (55,000 ft)
T-O distance: air defence	400 m (1,315 ft)
attack	600 m (1,970 ft)
Landing distance	450 m (1,480 ft)
Radius of action: low-level penetration with 12 × 250 kg bombs, four MICA AAMs and 4,000 litres (1,056 US gallons; 880 Imp gallons) of external fuel in three tanks	570 n miles (1,055 km; 655 miles)
air-to-air, long-range with eight MICA AAMs and 6,000 litres (1,585 US gallons; 1,320 Imp gallons) of external fuel in four tanks, 12,200 m (40,000 ft) transit	950 n miles (1,759 km; 1,093 miles)
Operational loiter	up to 3 h
g limits	+9.0/−3.2

UPDATED

Modified image showing proposed Dassault Atlantic 3 *(2e Maître Gisselbrecht)*   2000/0062694

DASSAULT ATLANTIC 3

TYPE: Maritime surveillance twin-turboprop.

PROGRAMME: Former Breguet Br 1150 Atlantic, designed to NATO specification, built by SECBAT consortium; first flight 21 October 1961; last of four prototype and 87 production aircraft completed July 1974 (see *Jane's Aircraft Upgrades*). Dassault Atlantique 2 (ATL2 or *Atlantique Nouvelle Génération*) development began September 1978; first flight Atlantic 1 converted to ATL2 prototype 8 May 1981; second converted prototype 26 March 1982; first flight production aircraft 19 October 1988. Final (28th) production aircraft delivered 15 January 1998. Atlantics of first two generations have flown some 1 million hours.

Promotion continues of third-generation version.

CURRENT VERSIONS: **Atlantic 3 (ATL3):** Developed version, first proposed 1988; renewed promotion in 1995 for UK's Nimrod replacement, but withdrawn from competition in January 1997; known until 2000 as Atlantique 3. Features include Rolls-Royce AE 2100H turboprop engines with six-blade propellers; optional refuelling probe; all-new mission system avionics; improved armament with provision for four self-protection AAMs; and two-crew EFIS flight deck. Capabilities bestowed by 15 per cent lower fuel consumption and flat-rating of engines to ISA + 20°C (68°F) include 10 to 12 hours on-task at 800 n miles (1,482 km; 920 miles) from base. Currently being offered (in competition with Lockheed Martin P-3 Orion) to meet anticipated German (12 aircraft) and Italian (14) requirements for delivery from 2007, and is offered to French Navy as MLU for Atlantique 2.

CUSTOMERS: None. However, some 40 Atlantics and 28 Atlantique 2s remain in service in France, Germany, Italy and Pakistan.

DESIGN FEATURES: Aerodynamically conventional mid-wing twin-turboprop aircraft with double-bubble fuselage; retractable radome; weapons bay and ejectable stores stowage in lower fuselage section; avionics pods at tips of wings and fin; underwing stores stations for anti-ship or self-defence missiles.

Wing section NASA 64 series; incidence 3°; dihedral 6° outboard of engines; sweepback 9° on leading-edge.

FLYING CONTROLS: All fully powered by tandem hydraulic power units; no trim tabs; ailerons assisted by three spoilers on each wing, some of which also act as airbrakes; fixed tailplane; three segments of slotted flaps on each wing.

STRUCTURE: Employs refined honeycomb panel bonding technique to improve corrosion resistance and maintainability of original structure; wing torsion box, pressure cabin walls, flaps and large doors made of bonded honeycomb panels; weapon bay and nose landing gear doors slide over external skin to open.

LANDING GEAR: Retractable tricycle type, with twin wheels on each unit. Hydraulic retraction, nosewheels rearward, main units forward into engine nacelles. Tyre size 39×13-20 (22 ply) tubeless on mainwheels, pressure 12.00 bars (170 lb/sq in); 26×7.75-13 or 26×8.0-13 tubeless on nosewheels, pressure 6.50 bars (94 lb/sq in).

POWER PLANT: Two Rolls-Royce AE 2100H turboprops, each driving a six-blade Dowty 945 composites propeller. Four pressure-refuelled integral fuel tanks in wings, with total capacity of 23,120 litres (6,108 US gallons; 5,085 Imp gallons).

ACCOMMODATION: Normal flight crew of two, plus eight/ten systems operators and observers. Rest/relief crew compartment in centre of fuselage. Primary access via extending airstair door in bottom of rear fuselage. Emergency exits above and below flight deck and on each side, above wing trailing-edge.

SYSTEMS (Atlantique 2): Air conditioning system supplied by two compressors driven by gearboxes. Heat exchangers and bootstrap system for cabin temperature control. Duplicated hydraulic system, pressure 186 bars (2,700 lb/sq in), to operate flying controls, landing gear, flaps, weapons bay doors and retractable radome. Hydraulic flow rate 17.85 litres (4.7 US gallons; 3.9 Imp gallons)/min.

Three basic electrical systems: variable frequency three-phase 115/200 V AC system, with two 60/80 kVA Auxilec alternators and modernised control and protection equipment; fixed frequency three-phase 115/200 V 400 Hz AC system, with four 15 kVA generators, two on each engine; 28 V DC system, with four 6 kW transformer-

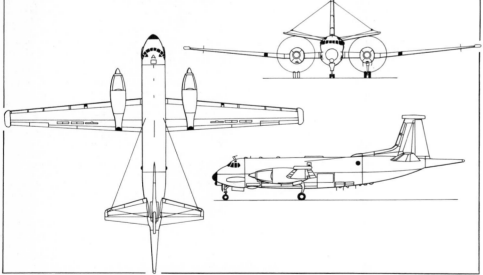

Projected Dassault Atlantic 3 maritime patrol aircraft *(Paul Jackson/Jane's)*   1999/0044513

Computer-generated image of Atlantic 3 interior equipped with seven starboard-facing consoles *(DCN International)*   2000/0062753

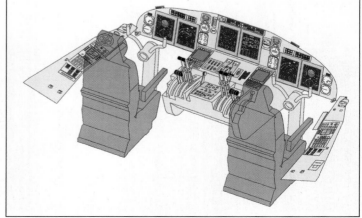

Two-crew flight deck proposed for the Atlantic 3   0011130

Jane's All the World's Aircraft 2001-2002   www.janes.com

rectifiers supplied from variable frequency AC system, and one 40 Ah battery. One 60 kVA emergency AC generator, driven at constant speed by APU.

Individual oxygen bottles for emergency use. Air Equipement/Kléber-Colombes pneumatic de-icing system on wing and tail leading-edges. Electric anti-icing for engine air intake lips, propeller blades and spinners. Turbomeca/ABG-Semca Astadyne gas-turbine APU for engine starting, electrical supply, and air conditioning in emergency on ground.

AVIONICS: New mission system, employing many components already developed for other programmes, will replace Atlantique 2 sensors.

Radar: Will be capable of SAR, ISAR and profile imagery. Manufacture by DASA, Thomson-CSF and FIAR.

Mission: ESM able to detect short emissions in a severe environment and correlate with threat library. Long-range, day/night electro-optic sensors for IR bands 1 and 2 and LLLTV. Acoustic processor with 64 channels; MAD; two digital cameras; integrated self-protection system including towed decoy, jammer, directional IRCM, MAWS and LWS; Have Quick II and SATURN standard V/UHF communications; satcom and NATO datalinks.

EQUIPMENT: More than 200 sonobuoys in compartment aft of weapons bay.

ARMAMENT: Main weapons bay in unpressurised lower fuselage can accommodate all NATO standard bombs, eight depth charges, eight Mk 46 homing torpedoes, seven Eurotorp MU90 Impact advanced torpedoes, four AGM-84D Harpoon or two AM 39 Exocet or ANF air-to-surface missiles. Four underwing attachments for a combined total of 3,500 kg (7,716 lb) of external stores, including two or four air-to-surface and air-to-air missiles or pods.

Following data are provisional.
DIMENSIONS, EXTERNAL:
Wing span: incl wingtip pods	37.46 m (122 ft 10¾ in)
excl wingtip pods	36.30 m (119 ft 1¼ in)
Wing aspect ratio	11.7
Length: overall	31.71 m (104 ft 0 in)
with refuelling probe (ATL3)	33.12 m (108 ft 8 in)
Height overall (30 tonnes)	10.89 m (35 ft 8¾ in)
Fuselage max depth	4.00 m (13 ft 1½ in)
Tailplane span	12.31 m (40 ft 4½ in)
Wheel track (c/l of shock-struts)	9.00 m (29 ft 6¼ in)
Wheelbase	9.40 m (30 ft 10 in)
Distance between propeller centres	9.00 m (29 ft 6¼ in)

DIMENSIONS, INTERNAL:
Cabin, incl rest compartment, galley, toilet, aft observers' stations: Length	18.50 m (60 ft 8½ in)
Max width	2.90 m (9 ft 6¼ in)
Max height	2.00 m (6 ft 6¾ in)
Floor area	55.0 m² (592 sq ft)
Volume	92.0 m³ (3,250 cu ft)
Main weapons bay: Length	9.00 m (29 ft 6¼ in)
Width	2.60 m (8 ft 6¼ in)
Depth	1.00 m (3 ft 3¼ in)
Volume	27.0 m³ (953 cu ft)

AREAS:
Wings, gross	120.34 m² (1,295.3 sq ft)
Ailerons (total)	5.26 m² (56.62 sq ft)
Flaps (total)	26.42 m² (284.38 sq ft)
Spoilers (total)	1.66 m² (17.87 sq ft)
Fin, incl dorsal fin	10.68 m² (114.96 sq ft)
Rudder	5.96 m² (64.15 sq ft)
Tailplane	24.20 m² (260.49 sq ft)
Elevators (total)	8.30 m² (89.34 sq ft)

WEIGHTS AND LOADINGS:
Weight empty, equipped, standard mission	25,700 kg (56,659 lb)
Max weapon load: internal	6,500 kg (14,330 lb)
external	3,500 kg (7,716 lb)
Max fuel	18,500 kg (40,785 lb)
Max T-O weight	46,200 kg (101,850 lb)

PERFORMANCE (at T-O weight of 45,000 kg; 99,200 lb except where indicated):
Max Mach No.	0.73
Max level speed	350 kt (648 km/h; 402 mph)
Max cruising speed at 7,620 m (25,000 ft)	310 kt (574 km/h; 357 mph)
Normal patrol speed, S/L to 1,525 m (5,000 ft)	170 kt (315 km/h; 195 mph)

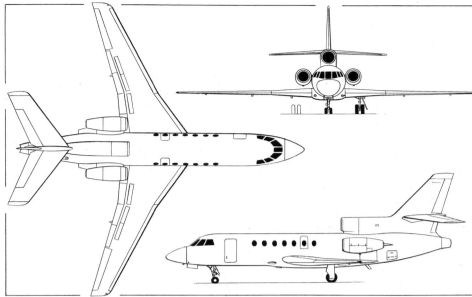

Dassault Falcon 50 long-range three-turbofan business transport *(Dennis Punnett/Jane's)*

T-O balanced runway length at MTOW	2,300 m (7,545 ft)
T-O to 10.7 m (35 ft)	1,840 m (6,040 ft)
Landing from 15 m (50 ft)	1,500 m (4,925 ft)
Ferry range, ICAO reserves	4,800 n miles (8,889 km; 5,523 miles)

Time on station:
at 1,000 n miles (1,852 km; 1,150 miles) from base	8 h
at 1,500 n miles (2,778 km; 1,726 miles) from base	5 h
Endurance	more than 18 h

UPDATED

DASSAULT FALCON 50EX
Spanish Air Force designation: T.16
TYPE: Business tri-jet.
PROGRAMME: First flight of prototype Falcon 50 (F-WAMD) 7 November 1976; second prototype 18 February 1978; first (and only) preproduction 13 June 1978. French certification 27 February 1979; FAA certification 7 March 1979; deliveries began July 1979. Available in sigint version from 1994.

Falcon 50EX announced 26 April 1995. Long-range variant; uprated turbofans provide a 7 per cent improvement in fuel consumption and improved initial cruising altitude of 12,500 m (41,000 ft). Production of initial batch of 40 Falcon 50EX airframes began 1995. First flight (F-WOND, c/n 251) 10 April 1996; new avionics suite installed in Falcon 50 c/n 252 for integration flight testing leading to avionics certification by FAA on 15 July 1996; DGAC certification achieved 15 November 1996, followed by FAA approval 20 December; initial production aircraft (F-WWHA, c/n 253) first flight October 1996; flown in 'green' condition Le Bourget, Paris, to Teterboro, New Jersey, 16 November, thence to Orlando, Florida, for US debut at National Business Aircraft Association Convention; first customer delivery (to Volkswagen of Germany) February 1997. Estimated market for 150 to 200 of EX version over unspecified period.

CURRENT VERSIONS: **Falcon 50:** Previous version, now superseded by Falcon 50EX.
Falcon 50EX: *As described.*
Falcon 50 Surmar: Maritime surveillance version. Order for French Navy announced 12 November 1996 covering four aircraft (compared with five in original June 1995 statement) at estimated total cost of FFr750 million. Sensors include a Thomson-CSF Ocean Master 100 search radar and Thomson-TTD Chlio FLIR. Three mission specialist/console stations, two at front of cabin, one at rear; airdrop door; provision for carriage of up to eight 25-person airdroppable liferafts; two observation windows.

Capabilities include 4 hours on station at 400 n miles (740 km; 460 miles) from base, cruising at 200 kt (370 km/h; 230 mph) at 915 m (3,000 ft).

First flight (No. 36/F-ZWTA) November 1998; first delivery in January 2000; IOC for first two aircraft mid-2000 with Flottille 24 at Lann-Bihoué; remaining two scheduled for delivery in 2001.

Falcon 50 Sigint: Model of a potential signals intelligence Falcon 50 displayed at Dubai Air Show in November 1997.

CUSTOMERS: Total 300 Falcon 50s and 50EXs sold between July 1979 and October 1999; 300th scheduled for delivery in late 2000, following interior completion. Adopted by governments of Burundi, Djibouti, France, Iraq, Italy, Jordan, Libya, Morocco, Portugal, Rwanda, South Africa, Spain, Sudan and Yugoslavia; three of Italian Air Force convertible for medevac. Production rate of 1.5 per month in mid-2000.

COSTS: US$17.25 million (1998).

DESIGN FEATURES: Three-engine layout permits overflight of oceans and desert areas within public transport regulations. Sharply waisted rear fuselage and engine pod designed by computational fluid dynamics; wing has compound leading-edge sweep (24° 50' to 29° at quarter-chord) and optimised section.

FLYING CONTROLS: Fully powered controls with pushrods, dual-barrel hydraulic actuators and artificial feel; variable incidence anhedral tailplane with dual electrical actuation by screwjack; drooped leading-edge inboard and slats outboard; double-slotted flaps; three two-position airbrake/spoiler panels on each wing.

STRUCTURE: All-metal, circular-section fuselage with rear baggage compartment inside pressure cabin; wing boxes are integral fuel tanks bolted to carry-through box. Latécoère, Potez, Reims Aviation, SEFCA and Socata are subcontractors on the programme; other suppliers include Dassault Equipements (flight controls, flaps, slats and airbrakes), Liebherr-Aerospace Toulouse (engine bleed air system); SARMA (flight control actuating rods) and Sully Produits Spéciaux (windscreen panels).

LANDING GEAR: Retractable tricycle type by Messier-Dowty, with twin wheels on each unit. Hydraulic retraction, main units inward, nosewheels forward. Nosewheels steerable ±60° for taxying, ±180° for towing. ABS wheels, brakes and braking system. Mainwheel tyres size 26×6.6 (14 ply) or 26×6.6R14 tubeless, pressure 14.34 bars (208 lb/sq in). Nosewheel tyres size 14.5×5.5-6 (14 ply) tubeless, pressure 8.96 bars (130 lb/sq in). Four-disc brakes designed for 400 landings with normal energy braking. Minimum ground turning radius (about nosewheels) 13.54 m (44 ft 5 in).

POWER PLANT: Three Honeywell TFE731-40 turbofans, each rated at 16.46 kN (3,700 lb st) at ISA + 17°C in Falcon 50EX. Two engines pod-mounted on sides of rear fuselage, third attached by two top mounts. Thrust reverser on centre engine. Fuel in integral tanks, with capacity of 5,787 litres (1,529 US gallons; 1,273 Imp gallons) in wings and 2,976 litres (786 US gallons; 655 Imp gallons) in fuselage tanks. Total fuel capacity 8,763 litres (2,315 US gallons; 1,928 Imp gallons). Single-point pressure fuelling. Intertechnique fuel distribution and gauges.

ACCOMMODATION: Standard accommodation for two crew and nine passengers. In typical arrangement, cabin is divided into a forward section with four armchairs and two fold-out tables, and a rear section with three-seat sofa (convertible into a single bed) and two armchairs separated by a fold-out table; two galleys at forward end of cabin, toilet at rear. Alternative layouts to customer choice, with accommodation for a maximum of 19 passengers. Cabin and rear baggage compartment are pressurised and air

Falcon 50 Surmar of French Navy (note enlarged observation window) *(Paul Jackson/Jane's)*

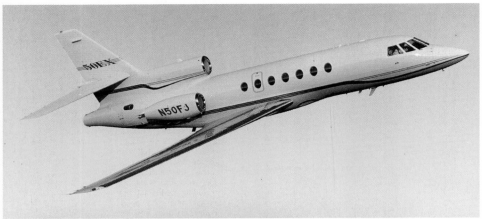

Dassault Falcon 50EX long-range business jet *(Paul Bowen/Dassault)* 2001/0093651

conditioned; Barber-Colman temperature controls. Access is by separate door on port side.

SYSTEMS: Air conditioning system utilises bleed air from all three engines or APU. Maximum pressure differential 0.61 bar (8.8 lb/sq in). Pressurisation maintains a maximum cabin altitude of 2,440 m (8,000 ft) to a flight altitude of 14,935 m (49,000 ft). Two independent Messier-Bugatti and Vickers-Sterer hydraulic systems, pressure 207 bars (3,000 lb/sq in), with three engine-driven pumps and one emergency electric pump, actuate primary flying controls, flaps, slats, landing gear, wheel brakes, airbrakes and nosewheel steering. Plain reservoir, pressurised by bleed air at 1.47 bars (21 lb/sq in). 28 V DC electrical system, with a 9 kW 28 V DC Auxilec starter/generator on each engine and two 23 Ah batteries. Labinal electrical harnesses. Wing leading-edge, centre engine S-duct and engine nacelles have engine bleed air anti-icing. Automatic emergency oxygen system. Honeywell 36-150 APU standard.

AVIONICS: *Comms:* Dual Rockwell Collins VHF and mode S transponders; dual Rockwell Collins HF-9000 HF transceivers with Selcal; Teledyne Controls MagnaStar or Honeywell Flitefone 800 radiotelephones; cockpit voice recorder and ELT standard.

Radar: Rockwell Collins TWR 850 Doppler turbulence detection weather radar.

Flight: Dual Rockwell Collins VIR-432 VOR, ADF-462 and DME-442. Rockwell Collins APS-4000 autopilot, ADC-850C air data systems, dual Honeywell FHS-6100 with integrated GPS receiver, Honeywell EGPWS and dual Honeywell Laseref III laser gyro inertial reference systems standard. Dual Universal UNS-1C FMS optional, replacing GNS-XES.

Instrumentation: Rockwell Collins Pro Line 4 (EFIS-4000) four-tube EFIS; Sextant Avionique three-tube LCD engine indicating electronic display (EIED).

DIMENSIONS, EXTERNAL:
Wing span	18.86 m (61 ft 10½ in)
Wing chord (mean)	2.84 m (9 ft 3¾ in)
Wing aspect ratio	7.6
Length: overall	18.52 m (60 ft 9¼ in)
fuselage	17.66 m (57 ft 11 in)
Height overall	6.98 m (22 ft 10¾ in)
Tailplane span	7.74 m (25 ft 4¾ in)
Wheel track	3.98 m (13 ft 0¾ in)
Wheelbase	7.24 m (23 ft 9 in)
Passenger door: Height	1.52 m (4 ft 11¾ in)
Width	0.80 m (2 ft 7½ in)
Height to sill	1.30 m (4 ft 3¼ in)

Emergency exits (each side, over wing):
Height	0.92 m (3 ft 0¼ in)
Width	0.51 m (1 ft 8 in)
Baggage door: Height	0.73 m (2 ft 4¾ in)
Width	0.99 m (3 ft 3 in)

DIMENSIONS, INTERNAL:
Cabin, incl forward baggage space and rear toilet:
Length	7.16 m (23 ft 6 in)
Max width	1.86 m (6 ft 1¼ in)
Max height	1.80 m (5 ft 10¾ in)
Volume	20.2 m³ (712 cu ft)
Baggage space	0.71 m³ (25.0 cu ft)
Baggage compartment (rear)	3.26 m³ (115 cu ft)

AREAS:
Wings, gross	46.83 m² (504.1 sq ft)
Horizontal tail surfaces (total)	13.35 m² (143.69 sq ft)
Vertical tail surfaces (total)	9.82 m² (105.70 sq ft)

WEIGHTS AND LOADINGS:
Weight empty, equipped	9,603 kg (21,170 lb)
Basic operating weight	9,888 kg (21,800 lb)
Baggage capacity (rear)	1,000 kg (2,205 lb)
Max payload: normal	1,710 kg (3,770 lb)
with max fuel	1,170 kg (2,579 lb)
Max fuel	7,040 kg (15,520 lb)
Max T-O weight: standard	18,007 kg (39,700 lb)
optional	18,497 kg (40,780 lb)
Max ramp weight: standard	18,098 kg (39,900 lb)
optional	18,497 kg (40,780 lb)
Max zero-fuel weight	11,600 kg (25,570 lb)
Max landing weight	16,200 kg (35,715 lb)
Max wing loading	384.5 kg/m² (78.75 lb/sq ft)
Max power loading	365 kg/kN (3.58 lb/lb st)

PERFORMANCE:
Max operating Mach No. (M_{MO})	0.86
Max operating speed (V_{MO}):	
at S/L	350 kt (648 km/h; 402 mph) IAS
at 7,225 m (23,700 ft)	370 kt (685 km/h; 425 mph) IAS
Max cruising speed	M0.85 or 487 kt (902 km/h; 560 mph)
Normal cruising speed	M0.80 or 459 kt (850 km/h; 528 mph)
Long-range cruising speed at 10,670 m (35,000 ft)	M0.75 (430 kt; 797 km/h; 495 mph)
Approach speed, eight passengers and NBAA IFR reserves	107 kt (198 km/h; 123 mph)
Initial cruising altitude	12,497 m (41,000 ft)
Max certified altitude	14,935 m (49,000 ft)
FAR Pt 25 balanced T-O field length, S/L, ISA with eight passengers and max fuel	1,437 m (4,715 ft)
FAR Pt 91 landing distance with eight passengers and NBAA IFR reserves:	
at max landing weight	666 m (2,185 ft)
Range with eight passengers and NBAA IFR reserves:	
at M0.80	3,025 n miles (5,602 km; 3,481 miles)
at M0.75	3,285 n miles (6,083 km; 3,780 miles)

OPERATIONAL NOISE LEVELS:
T-O	83.8 EPNdB
Approach	95.2 EPNdB
Sideline	92.0 EPNdB

UPDATED

DASSAULT FALCON 900B/C and 900EX
Spanish Air Force designation: T.18

TYPE: Long-range business tri-jet.

PROGRAMME: Falcon 900 announced 27 May 1983; first flight of prototype (F-GIDE *Spirit of Lafayette*) 21 September 1984, second aircraft (F-GFJC) 30 August 1985; flew non-stop 4,305 n miles (7,973 km; 4,954 miles) Paris to Little Rock, Arkansas, September 1985; returned Teterboro, New Jersey, to Istres, France, at M0.84; French and US certification March 1986, including status close to FAR Pts 25 and 55 for damage tolerance of entire airframe.

Prototype Falcon 900 in use as testbed (first flight 12 April 1994) for new laminar flow wing section intended to provide significant reductions in drag. New section installed as sleeve on inner wing and designed to demonstrate hybrid laminar flow with boundary layer suction via seven channels in laser-drilled titanium skin over some 10 per cent chord of upper wing surface. Following test programme, modified aircraft has returned to Dassault Falcon Service to validate laminar flow capability under normal commercial operating conditions.

CURRENT VERSIONS: **Falcon 900B:** French and UK certification received end 1991; complies with FAR Pt 36 Stage III and ICAO 16 noise requirements; approved for Cat. II approaches, for operations from unpaved fields; re-engined with TFE731-5BR-1C turbofans, to give 5.5 per cent power increase; initial cruising altitude 11,855 m (39,000 ft) and NBAA IFR range increased by 100 n miles (185 km; 115 miles); retrofit offered to existing operators.

Detailed description applies to Falcon 900B, except where indicated.

Falcon 900C: Announced 26 June 1998. Combines airframe, engines and cabin of 900B with Honeywell Primus 2000 avionics of 900EX, but without autothrottles; 900B F-WWFP/F-GRDP (c/n 169) served as prototype for certification; first flown 17 December 1998; first deliveries in 2000, replacing 900B on production line. French DGAC certification achieved 15 June 1999, with FAA certification following on 26 August 1999.

Falcon 900EX: Long-range development of 900B, announced October 1994. Re-engined with 22.24 kN (5,000 lb st) (ISA+17°C) Honeywell TFE731-60 turbofans, to give 5.8 per cent increase in retained thrust at 12,200 m (40,000 ft) and more than 8 per cent improvement in cruise specific fuel consumption. Engine nacelles, pylons, thrust reversers and portions of centre engine S-duct redesigned; maximum fuel capacity increased to 11,865 litres (3,134 US gallons; 2,610 Imp gallons) by increasing capacity of centre-fuselage tank to 591 kg (1,303 lb) and addition of tank in rear fuselage, capacity 240 kg (530 lb).

Upgraded standard avionics comprise fully integrated Honeywell Primus 2000 suite with five-tube 20 × 17.75 cm (8 × 7 in) colour EFIS; one engine instrument display; three IC-800 integrated avionics

Falcon 50EX flight deck is based on that of Falcon 2000 *(Bud Shannon/Dassault)* 2001/0100520

Typical cabin layout of a Falcon 900EX *(Bud Shannon/Dassault)* 2001/0100519

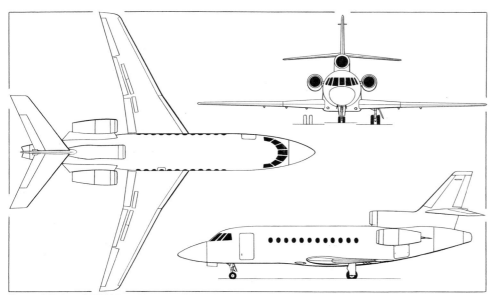

Dassault Falcon 900B (three Honeywell TFE731 turbofans) *(Dennis Punnett/Jane's)*

computers; dual FMZ-2000 flight management systems with a third optional; dual fail-operational autopilots; T-O to landing autothrottle; Honeywell EGPWS; dual Laseref III inertial reference systems with third optional; Primus colour weather radar; optional single or dual 12-channel GPS, multichannel satcom, communications management unit (CMU) and Flight Dynamics HGS-2850 head-up display.

Risk-sharing partners, representing 20 per cent of total development investment, are Honeywell (engines and primary avionics), SABCA (centre engine intake cowlings), Hellenic Aircraft Industries (rear fuselage fuel tank), Latécoère (T5 fuselage section and engine pylons), and Alenia (nacelles and centre engine thrust reverser).

Prototype (F-WREX) rolled out 13 March 1995; first flight 1 June 1995; flew Luton, England, to Las Vegas, Nevada, non-stop on 24 September 1995, completing the 4,700 n mile (8,704 km; 5,409 mile) flight in 11 hours 40 minutes including 30 minutes hold for air traffic delays; DGAC certification 31 May 1996; FAA approval granted 19 July after 350-hour flight test programme; first customer delivery to Anheuser-Busch Companies Inc (N200L) 1 November 1996; production aircraft delivered to Little Rock, Arkansas, for outfitting. Prototype completed demonstration tour of China in early November 1996, then flew Sydney, Australia, to Orlando, Florida, with single fuel stop at Maui, Hawaii, covering a total ground distance of 8,635 n miles (15,992 km; 9,937 miles) in an airborne time of 17 hours 57 minutes.

Japan MSA: Two Falcon 900s for long-range maritime surveillance entered service with the Japan Maritime Safety Agency September 1989; US search radar, special communications radio, operations control station, U-125A-style search windows and drop hatch for sonobuoys, markers and flares.

CUSTOMERS: Total 186 Falcon 900s and 74 Falcon 900EXs delivered to customers or completion centres by August 2000. Government/VIP versions operated by Algeria, Australia, Equatorial Guinea, France, Gabon, Italy, Malaysia, Nigeria, Russia, Saudi Arabia, Spain, Syria and United Arab Emirates. Production totalled 20 in 1998 and 24 in 1999.

COSTS: Standard equipped 900B US$26.2 million (1998); 900EX US$29.4 million (1998).

DESIGN FEATURES: Larger cross-section and cabin length than Falcon 50; added economy and further power increase of engines achieved by mixer compound nozzle tailpipe, mixing cold and hot flows.

Wing adapted from Falcon 50 but increased span and area, and optimised for M0.84 cruise; compound leading-edge sweep (24° 50′ to 29° at quarter-chord); dihedral 0° 30′.

FLYING CONTROLS: Fully powered flying controls with artificial feel and variable-incidence tailplane as for Falcon 50; full-span slats and double-slotted Fowler flaps; three-position airbrakes.

STRUCTURE: Design and manufacture computer assisted; damage-tolerant structure; extensive use of carbon fibre and aramid (Kevlar); Kevlar radome, wingroot fairings and tailcone; secondary rear cabin pressure bulkhead allows access to baggage in flight and additional protection against pressure loss. Nosewheel doors of Kevlar; mainwheel doors of carbon fibre. Kevlar air intake trunk for centre engine, and rear cowling for side engines. Carbon fibre central cowling around all three engines.

New horizontal tail surface featuring cast titanium central box with resin transfer moulding composite spars and carbon fibre skin panels, resulting in a 13.6 kg (30 lb) reduction in weight, certified by the DGAC in December 1999 and introduced as standard on Falcon 900C and 900EX from December 2000 deliveries.

LANDING GEAR: Retractable tricycle type by Messier-Bugatti, with twin wheels on each unit. Hydraulic retraction, main units inward, nosewheels forward. Oleo-pneumatic shock-absorbers. Mainwheels fitted with Michelin radial tyres size 29×7.7-15, pressure 13.30 bars (193 lb/sq in). Nosewheel tyres size 17.5×5.75R8, pressure 9.80 bars (142 lb/sq in). Hydraulic nosewheel steering (±60° for taxying, ±180° for towing). Messier-Bugatti triple-disc carbon brakes and anti-skid system. Minimum ground turning radius (about nosewheels) 13.54 m (44 ft 5 in).

POWER PLANT: Three Honeywell TFE731-5BR-1C turbofans, each rated at 21.13 kN (4,750 lb st) at ISA + 10°C. Thrust reverser on centre engine. Fuel in two integral tanks in wings, centre-section tank, and two tanks under floor of forward and rear fuselage. Total fuel capacity 10,825 litres (2,860 US gallons; 2,381 Imp gallons).

ACCOMMODATION: Type III emergency exit on starboard side of cabin permits wide range of layouts for up to 19 passengers. Flight deck for two pilots, with central jumpseat. Flight deck separated from cabin by door, with crew wardrobe and baggage locker on either side. Galley at front of main cabin, on starboard side opposite main cabin door. Passenger area is divided into three lounges. Forward zone has four 'sleeping' swivel chairs in facing pairs with tables. Centre zone is dining area, with two double seats facing a transverse table. On starboard side, storage cabinet contains foldaway bench, allowing five to six persons to be seated around table, while leaving emergency exit clear. In rear zone, inward-facing three-seat settee on starboard side converts into a bed. On port side, two armchairs are separated by a table. At rear of cabin, a door leads to toilet compartment, on starboard side, and a second structural plug door to large rear baggage area. Baggage door is electrically actuated.

Other interior configurations available. Alternative eight-passenger configuration has bedroom at rear and three personnel seats in forward zone. A 15-passenger layout divides a VIP area at rear from six (three-abreast) chairs forward; full fuel can still be carried with 15 passengers. The 18-passenger scheme has four rows of three-abreast airline-type seats forward, and VIP lounge with two chairs and settee aft. Many optional items, including additional windows, front toilet unit, video system with one or more monitors, 'Airshow 200' navigation display system, compact disc deck, aft cabin partition, one or two couches in aft cabin convertible to bed (s), storage cabinet in baggage hold, aft longitudinal table, individual listening devices for passengers, lifejackets and rafts.

SYSTEMS: Air conditioning system uses engine bleed air or air from Honeywell GTCP36-150 APU installed in rear fuselage. Softair pressurisation system, with maximum differential of 0.64 bar (9.3 lb/sq in), maintains sea level cabin environment to height of 7,620 m (25,000 ft), cabin equivalent of 2,440 m (8,000 ft) at 15,550 m (51,000 ft). Cold air supply by single oversize air cycle unit. Two independent hydraulic systems, pressure 207 bars (3,000 lb/sq in), with three engine-driven pumps and one emergency electric pump, actuate primary flying controls, flaps, slats, landing gear retraction, wheel brakes, airbrakes, nosewheel steering and thrust reverser. Bootstrap hydraulic reservoirs. DC electrical system supplied by three 9 kW 28 V Auxilec starter/generators and two 23 Ah batteries. Heated bleed air anti-icing of wing leading-edges, intakes and centre engine duct; electrically heated windscreens. Eros (SFIM/Intertechnique) oxygen system.

AVIONICS: *Comms:* Rockwell Collins Pro Line II avionics (Pro Line 4 in 900C and 900EX) including dual TDR 94D transponders; dual VHF 22A; dual Honeywell KHF 950 HF com. Dual VHF 422C and dual Rockwell Collins HF-9000 in 900C and 900EX.

Radar: Honeywell Primus 870 colour weather radar.

Flight: Dual Honeywell FMZ 800 flight management system, associated with two AZ 810 air data computers and two Honeywell Laseref II ring laser inertial reference systems. Rockwell Collins dual VIR 32 VOR/ILS/marker beacon receiver; dual ADF 60B; dual DME 42; and

Flight deck of Falcon 900EX

Honeywell RT 300 radio altimeter. Dual bidirectional Honeywell ASCB digital databus operating in conjunction with dual SPZ 8000 flight director/autopilot.

Falcon 900C and 900EX have Honeywell FMZ-2000 FMS with two AZ-840 micro air data systems and two Laseref III LINS; Rockwell Collins dual VIR-432 VOR/ILS marker receiver, dual ADF-462 and DME-442; Honeywell AA-300 radar altimeter and IC-800 autopilot; and Honeywell EGPWS.

Instrumentation: Honeywell EDZ 820 five-tube EFIS. Falcon 900C and 900EX have Primus 2000 system.

DIMENSIONS, EXTERNAL:
Wing span	19.33 m (63 ft 5 in)
Wing chord: at root	4.08 m (13 ft 4¾ in)
at tip	1.12 m (3 ft 8 in)
Wing aspect ratio	7.6
Length overall	20.21 m (66 ft 3¾ in)
Fuselage: Max diameter	2.50 m (8 ft 2½ in)
Height overall	7.55 m (24 ft 9¼ in)
Tailplane span	7.74 m (25 ft 4¾ in)
Wheel track	4.45 m (14 ft 7¼ in)
Wheelbase	7.90 m (25 ft 11 in)
Passenger door: Height	1.72 m (5 ft 7¾ in)
Width	0.80 m (2 ft 7½ in)
Height to sill	1.64 m (5 ft 4½ in)
Emergency exit (overwing, stbd):	
Height	0.92 m (3 ft 0¼ in)
Width	0.53 m (1 ft 8¾ in)
Baggage door:	
Height	0.75 m (2 ft 5½ in)
Width	0.95 m (3 ft 1½ in)

DIMENSIONS, INTERNAL:
Cabin, excl flight deck, incl toilet and baggage	
compartments: Length	10.11 m (33 ft 2 in)
Max width	2.34 m (7 ft 8¼ in)
Width at floor	1.91 m (6 ft 3¼ in)
Max height	1.88 m (6 ft 2 in)
Volume	35.8 m³ (1,264 cu ft)
Rear baggage compartment volume	3.6 m³ (127 cu ft)
Flight deck volume	3.8 m³ (132 cu ft)

AREAS:
Wings, gross	49.00 m² (527.4 sq ft)
Vertical tail surfaces (total)	9.82 m² (105.7 sq ft)
Horizontal tail surfaces (total)	13.35 m² (143.7 sq ft)

WEIGHTS AND LOADINGS:
Weight empty, equipped (typical):	
900B	10,455 kg (23,049 lb)
900EX	10,829 kg (23,875 lb)
Operating weight empty: 900B	10,545 kg (23,248 lb)
900C	10,832 kg (23,880 lb)
900EX	11,204 kg (24,700 lb)
Max payload: 900B	2,185 kg (4,817 lb)
900C	1,969 kg (4,341 lb)
900EX	2,796 kg (6,164 lb)
Payload with max fuel: 900B	1,385 kg (3,053 lb)
900C	1,204 kg (2,654 lb)
900EX	1,270 kg (2,800 lb)
Max fuel: 900B	8,690 kg (19,158 lb)
900C	8,693 kg (19,165 lb)
900EX	9,525 kg (21,000 lb)
Max ramp weight:	
900B, 900C standard	20,729 kg (45,700 lb)
900C optional	21,183 kg (46,700 lb)
900EX standard	22,000 kg (48,500 lb)
900EX optional	22,317 kg (49,200 lb)
Max T-O weight:	
900B, 900C standard	20,640 kg (45,500 lb)
900C optional	21,092 kg (46,500 lb)
900EX standard	21,909 kg (48,300 lb)
900EX optional	22,226 kg (49,000 lb)
Max landing weight:	
900B, 900C, 900EX	19,050 kg (42,000 lb)
Normal landing weight, eight passengers and fuel	
reserves: 900B	12,250 kg (27,000 lb)
900C	12,494 kg (27,545 lb)
900EX	12,846 kg (28,321 lb)
Max zero-fuel weight: 900B	12,800 kg (28,220 lb)
900EX	14,000 kg (30,865 lb)
Max wing loading:	
900B, 900C standard	421.2 kg/m² (86.27 lb/sq ft)
900C optional	430.5 kg/m² (88.17 lb/sq ft)
900EX standard	447.1 kg/m² (91.58 lb/sq ft)
900EX optional	453.6 kg/m² (92.91 lb/sq ft)
Max power loading:	
900B, 900C standard	326 kg/kN (3.19 lb/lb st)
900C optional	333 kg/kN (3.26 lb/lb st)
900EX standard	328 kg/kN (3.22 lb/lb st)
900EX optional	333 kg/kN (3.27 lb/lb st)

PERFORMANCE (at AUW of 12,250 kg; 27,000 lb, except where indicated):
Max operating speed (V_{MO}): 900, 900B, 900C, 900EX:	
at S/L	M0.87 (350 kt; 648 km/h; 403 mph IAS)
between 3,050-7,620 m (10,000-25,000 ft)	
	M0.84 (370 kt; 685 km/h; 425 mph IAS)
Max cruising speed:	
900C, 900EX	M0.84 or 481 kt (891 km/h; 554 mph)
Normal cruising speed:	
900C, 900EX	M0.80 or 459 kt (850 km/h; 528 mph)
Long-range cruising speed:	
900C, 900EX	M0.75 or 430 kt (796 km/h; 495 mph)
Approach speed, eight passengers and NBAA IFR fuel	
reserves: 900B	108 kt (200 km/h; 125 mph)
900C	107 kt (199 km/h; 124 mph)
900EX	109 kt (202 km/h; 126 mph)
Stalling speed:	
900B, 900EX, clean	106 kt (196 km/h; 122 mph)
900B, 900EX, landing configuration	
	85 kt (158 km/h; 98 mph)
Initial cruising altitude:	
900C, 900EX	11,885 m (39,000 ft)
Max certified altitude:	
900B, 900C, 900EX	15,550 m (51,000 ft)
FAR Pt 25 balanced T-O field length at standard max	
T-O weight: 900B	1,426 m (4,680 ft)
900C	1,504 m (4,935 ft)
900EX	1,590 m (5,215 ft)
FAR Pt 121 landing distance:	
with eight passengers and NBAA IFR reserves:	
900B	1,189 m (3,900 ft)
900C	716 m (2,350 ft)
900EX	724 m (2,375 ft)
at max landing weight:	
900B, 900EX	1,793 m (5,880 ft)
Range with NBAA IFR reserves:	
900C with five passengers at M0.75	
	4,000 n miles (7,408 km; 4,603 miles)
900EX with eight passengers:	
at M0.84	3,810 n miles (7,056 km; 4,384 miles)
at M0.80	4,335 n miles (8,028 km; 4,988 miles)
at M0.75	4,500 n miles (8,334 km; 5,178 miles)

OPERATIONAL NOISE LEVELS:
T-O: 900C, 900EX	79.8 EPNdB
Approach: 900C	91.7 EPNdB
900EX	92.3 EPNdB
Sideline: 900C	91.2 EPNdB
900EX	90.5 EPNdB

UPDATED

Dassault Falcon 900C business jet *(Paul Bowen/Dassault)*

DASSAULT FALCON 2000

TYPE: Business jet.

PROGRAMME: Announced Paris Air Show 1989 as Falcon X; follow-on to Falcon 20/200; launched as Falcon 2000 on 4 October 1990, following first orders; Alenia joined as 25 per cent risk-sharing partner February 1991, with responsibility for rear fuselage section and engine nacelles; selection of CFE738 engine announced 2 April 1990; first flight (F-WNAV) 4 March 1993; one prototype only; third airframe (F-WWFA) was second to fly, on 10 May 1994; ferried 'green' to US subsidiary Dassault Falcon Jet Corporation at Little Rock, Arkansas, 13 July 1994 for completion as US demonstrator, and set two world speed records 31 October to 1 November 1994 flying Los Angeles (Chino) to Bangor, Maine, in 4 hours 36 minutes 27 seconds and Bangor to Paris in 5 hours 26 minutes 12 seconds; Dassault demonstrator, second airframe (F-WNEW), flew 11 July 1994.

JAA certification to JAR 25 obtained 30 November 1994, at which time five aircraft (prototype, two demonstrators and two customer aircraft) had accumulated 1,055.5 flight hours; FAR Pt 25 and FAR Pt 36 Stage 3 noise levels certification 2 February 1995; first delivery (F-WNEW/ZS-NNF to South African customer) 16 February 1995. Approved to operate into London City Airport during 1996. Commonwealth of Independent States certification 14 April 1997. HGS-2850 HUD first flown in prototype 8 December 1995; initial JAA certification

Dassault Falcon 2000 US demonstrator *(Paul Bowen/Dassault)*

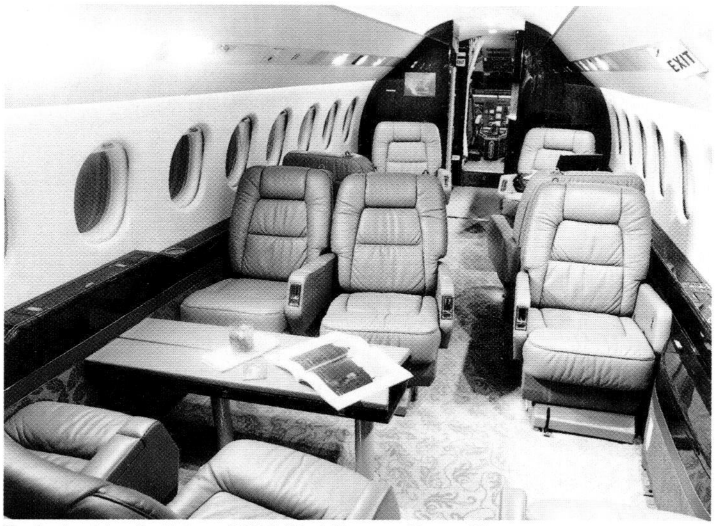

Typical cabin of a Falcon 2000 ***(Bud Shannon/Dassault)*** ***2001***/0093649

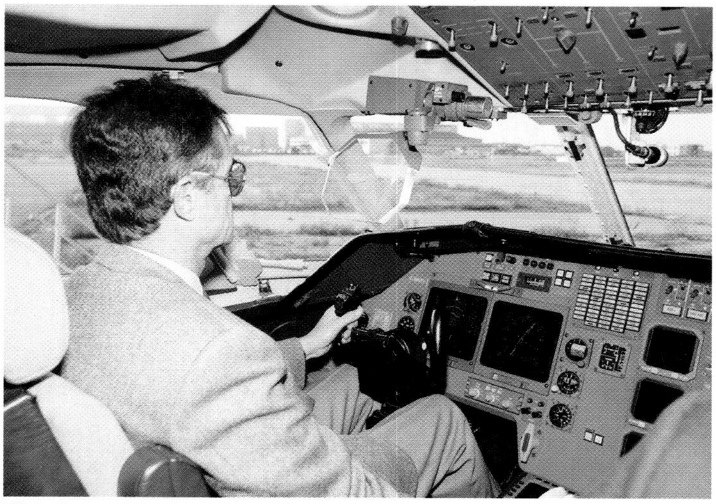

Flight Dynamics HGS-2850 head-up display installed in prototype Falcon 2000
0010621

granted 30 August 1996, followed by approval for low-visibility take-offs (down to 91 m; 300 ft RVR) and hand-flown Cat. II and Cat. IIIa instrument approach approved by FAA 30 July 1997. Full JAA certification 16 December 1997; FAA certification 18 May 1998.

CURRENT VERSIONS: **Falcon 2000:** Initial production version. *As described.*

Falcon 2000EX: Extended-range version. Development began October 1999; announced at the NBAA Convention in New Orleans, 10 October 2000. First flight scheduled for the fourth quarter of 2001, with certification in the third quarter of 2002 and customer deliveries of outfitted aircraft in early 2003. Falcon 2000EX will be manufactured in parallel with Falcon 2000.

New features include two Pratt & Whitney Canada PW3087C turbofans, with FADEC, each rated at 31.1 kN (7,000 lb st) for take-off; Nordam advanced single pivot (ASP) thrust reversers; revised fuel system with 1,730 kg (3,814 lb) increase in fuel capacity; strengthened main landing gear with heavy duty braking system and Falcon 900EX nose landing gear; and new bleed air system. The first 20 to 30 Falcon 2000EXs will retain the current Collins Pro Line 4 avionics suite, but later aircraft will be equipped with Dassault EASy flight deck with four-tube flat panel displays based on those of the Honeywell Primus Epic modular avionics system.

Preliminary specifications include: Basic operating weight 10,505 kg (23,160 lb); max payload 2,966 kg (6,539 lb); max fuel weight 7,244 kg (15,970 lb); max ramp weight 18,552 kg (40,900 lb); max T-O weight 18,461 kg (40,700 lb); max landing weight 17,373 kg (38,300 lb); max zero-fuel weight 13,472 kg (29,700 lb); max operating Mach No. (MMO) 0.85-0.86; normal cruising speed Mach 0.80; time to climb to 12,495 m (41,000 ft) 21 min; T-O balanced field length 1,756 m (5,760 ft); landing run 802 m (2,635 ft); range with six passengers, NBAA IFR reserves at Mach 0.80 3,800 n miles (7,037 km; 4,373 miles).

CUSTOMERS: Total of 160 ordered by 27 September 1999, of which 100 then delivered. Production totalled 20 in 1997, 15 in 1998 and 34 in 1999. By August 2000, 128 had been delivered to customers or completion centres. Total of 66 ordered by Executive Jet International for its NetJets and NetJets Middle East fractional ownership programmes in Europe, Middle East and USA, of which 15 were in operation by October 2000. Recent recipients include Volkswagen motorcar company. Royal Australian Air Force ordered three in 2000 to replace five leased Falcon 900s. Total market estimated at more than 300 in 10 years.

COSTS: Basic price 2000 US$19.5 million (1998); 2000EX US$24 million (2000).

DESIGN FEATURES: Same fuselage cross-section as Falcon 900, but 1.98 m (6 ft 6 in) shorter. Falcon 900 wing with modified leading-edge and no inboard slats; sweepback at quarter-chord 24° 50′ to 29°.

STRUCTURE: Largely as for Falcon 900. Rear fuselage, including engine pods and pylons, by Alenia and Piaggio; thrust reversers by Dee Howard. New horizontal tail surface featuring cast titanium central box with resin transfer moulding composite spars and carbon fibre skin panels, resulting in a 13.6 kg (30 lb) reduction in weight, certified by the DGAC in December 1999 and introduced as standard from December 2000 deliveries.

LANDING GEAR: Retractable tricycle type; mainwheels retract inwards, nosewheel forwards. Main tyres 26×6.6R14 (14 ply) tubeless; nose 14.5×5.5R6 tubeless.

POWER PLANT: Two GE/Honeywell CFE738-1-1B turbofans, each rated at 26.3 kN (5,918 lb st) at ISA + 15°C. Usable fuel capacity 6,865 litres (1,814 US gallons; 1,510 Imp gallons). Dee Howard clamshell-type thrust reversers certified by FAA and JAA during 1995.

ACCOMMODATION: Up to 19 passengers and two flight crew; standard passenger accommodation is four seats in forward lounge and four seats and a two-person sofa in aft lounge. Pressurised, flight accessible baggage compartment at rear of cabin.

AVIONICS: Rockwell Collins Pro Line 4.

Comms: Dual com/nav; dual TRD-94D Mode-S transponders; dual RTU-4420 radio tuning units; Rockwell Collins HF-9000 transceiver.

Radar: Rockwell Collins WXR-840 colour weather radar standard; TWR-850 Doppler weather radar optional.

Flight: Rockwell Collins FMS-6100 flight management system and Cat. II autopilot standard. Rockwell Collins dual DME-442, dual ADF-462; dual AHRS and dual digital air data computers linked to dual-channel integrated avionics processor system (IAPS). Dual Honeywell Laseref III, Honeywell EGPWS, IRS, second FMS (with GPS), flight data recorder, traffic alert and collision avoidance system (TCAS) and GPWS all optional.

Instrumentation: Rockwell Collins Pro Line 4 four-tube EFIS-4000. Sextant Avionique three-tube engine indicating electronic display (EIED); Flight Dynamics HGS-2850 HUD.

Mission: Optional satcom antenna in fintip pod.

DIMENSIONS, EXTERNAL:

Wing span	19.33 m (63 ft 5 in)
Wing chord (mean)	2.89 m (9 ft 5¾ in)
Wing aspect ratio	7.6
Length overall	20.22 m (66 ft 4 in)
Height overall	7.06 m (23 ft 2 in)
Wheel track	4.45 m (14 ft 7¼ in)
Wheelbase	7.39 m (24 ft 3 in)
Passenger door: Height	1.72 m (5 ft 7¾ in)
Width	0.80 m (2 ft 7½ in)
Emergency exits (Type III): Height	0.92 m (3 ft 0¼ in)
Width	0.53 m (1 ft 8¾ in)
Baggage door: Height	0.775 m (2 ft 6½ in)
Width	0.75 m (2 ft 5½ in)

DIMENSIONS, INTERNAL:

Cabin: Length	7.98 m (26 ft 2¼ in)
Max width	2.34 m (7 ft 8¼ in)
Max height	1.88 m (6 ft 2 in)
Volume	29.0 m^3 (1,024 cu ft)
Baggage volume: standard	3.8 m^3 (134 cu ft)
optional	5.4 m^3 (191 cu ft)

AREAS:

Wings, gross	49.02 m^2 (527.6 sq ft)

WEIGHTS AND LOADINGS:

Weight empty, equipped	9,405 kg (20,735 lb)
Basic operating weight	9,730 kg (21,450 lb)
Max payload	3,270 kg (7,210 lb)
Max fuel weight	5,513 kg (12,154 lb)
Max ramp weight: standard	16,329 kg (36,000 lb)
optional	16,647 kg (36,700 lb)
Max T-O weight: standard	16,238 kg (35,800 lb)
optional	16,556 kg (36,500 lb)
Max landing weight	14,970 kg (33,000 lb)
Max zero/fuel weight	13,000 kg (28,660 lb)
Max wing loading: standard	331.3 kg/m^2 (67.85 lb/sq ft)
optional	337.8 kg/m^2 (69.18 lb/sq ft)
Max power loading: standard	308 kg/kN (3.02 lb/lb st)
optional	314 kg/kN (3.08 lb/lb st)

PERFORMANCE:

Max operating Mach No. (MMO)	0.85-0.87
Max operating speed (VMO)	350-370 kt (648-685 km/h; 403-425 mph) IAS
Max cruising speed	M0.84 or 481 kt (891 km/h; 554 mph)
Normal cruising speed	M0.80 or 459 kt (850 km/h; 528 mph)
Long-range cruising speed	M0.75 or 430 kt (796 km/h; 495 mph)
Approach speed	109 kt (202 km/h; 125 mph)
Initial cruising altitude: at M0.80	12,497 m (41,000 ft)
at M0.75	13,106 m (43,000 ft)
Max certified altitude	14,330 m (47,000 ft)
FAR Pt 25 balanced T-O field length, S/L, ISA:	
with eight passengers and max fuel	1,597 m (5,240 ft)
at max T-O weight	1,658 m (5,440 ft)
Landing field length:	
with eight passengers and NBAA IFR reserves	1,295 m (4,250 ft)
at max landing weight	1,591 m (5,220 ft)
Range with eight passengers and NBAA IFR reserves:	
at M0.75	3,120 n miles (5,778 km; 3,590 miles)
at M0.80	3,000 n miles (5,556 km; 3,452 miles)
g limits	+2.64/−1

OPERATIONAL NOISE LEVELS:

T-O	79.4 EPNdB
Approach	93.1 EPNdB
Sideline	86.4 EPNdB

UPDATED

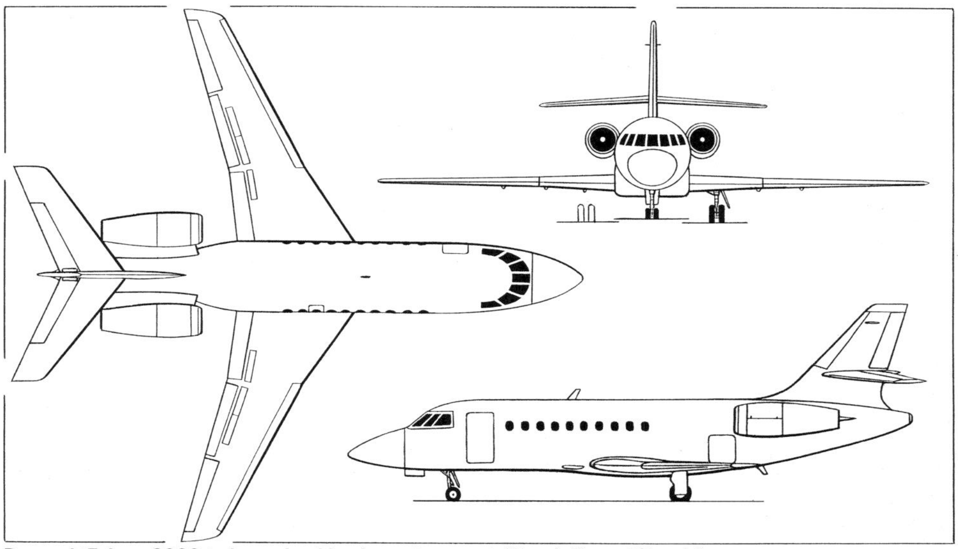

Dassault Falcon 2000 twin-engined business transport ***(Dennis Punnett/Jane's)***

DYN'AERO

SOCIETE DYN'AERO
11 rue de l'Aviation, F-21121 Darois
Tel: (+33 3) 80 35 60 62
Fax: (+33 3) 80 35 60 63
CHAIRMAN: Christophe Robin
DESIGN ASSOCIATE: Michel Colomban

Dyn'Aéro formed September 1992 by Christophe Robin, son of the founder of Avions Pierre Robin; his initials (CR) appear in aircraft designations. Initial product was CR100, flown two weeks before formation of company. Two new designs derived from CR100 announced in March 1995: CR110 and CR120; simultaneously, MCR01 also announced. The new MCR01 Club flew in June 1998 and a four-seat version in June 2000. Production facilities at Darois aerodrome comprise 2,450 m² (26,375 sq ft) of workshops, storage and offices.

UPDATED

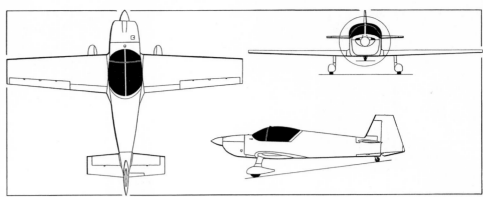

General arrangement of the Dyn'Aéro CR100 *(Paul Jackson/Jane's)*

DYN'AERO CR100, CR110 and CR120

TYPE: Aerobatic, two-seat lightplane/kitbuilt.
PROGRAMME: Prototype CR100 (F-PDYN) first flew 27 August 1992; first five CR100s flew total of 600 hours of aerobatics in first year of operations. Available factory-built or as a kit.
CURRENT VERSIONS: **CR100:** Baseline version.
Description applies to CR100, except where otherwise stated.
 CR110: Uprated version with 149 kW (200 hp) Textron Lycoming engine.
 CR120: High-agility version. First flight (F-PPCR) September 1996; second aircraft (F-PTCR) flying by 1998. Engine as CR110, but airframe all carbon fibre; longer span ailerons (2.70 m; 8 ft 10¼ in), no flaps, shorter wingspan (7.80 m; 25 ft 7 in), 270°/s rate of roll at 140 kt (260 km/h; 162 mph) compared with CR100's 195°/s, MTV-12-B-C/C-183-17e two-blade propeller, 85 litres (22.4 US gallons; 18.7 Imp gallons) aerobatic fuel in starboard wingroot and 105 litres (27.7 US gallons; 23.1 Imp gallons) in port wingroot, tyre pressure 2.00 bars (29.0 lb/sq in). Aircraft are registered under CR100 designation.
CUSTOMERS: Two CR100s ordered March 1995 by l'Equipe de Voltige de l'Armée de l'Air at Salon de Provence, of which first (No. 22 'CJ') delivered 21 July 1995; however, both withdrawn by 1997 and sold in 2000, one to UK owner Cole Aviation. Total of 35 CR100/110/120 kits sold and 12 flying, by mid-1999.
COSTS: FFr400,000 as a kit without engine; about FFr600,000 assembled, with engine and instruments.
DESIGN FEATURES: Single-engined low-wing monoplane of conventional layout.
 Aerofoil section NACA 21015/12 modified; dihedral 0°.
FLYING CONTROLS: Conventional and manual. Elevator tab for pitch trim. Ground-adjustable tab on ailerons on rudder. Plain flaps.
STRUCTURE: Wood and fabric, but including carbon fibre main wing spar.
LANDING GEAR: Tailwheel type; fixed. Oleo-pneumatic shock-absorption. Speed fairings on mainwheel legs. Disc brakes. Mainwheel tyre size 380×150; pressure 2.30 bars (33.0 lb/sq in).
POWER PLANT: One 134 kW (180 hp) Textron Lycoming AEIO-360-B2F flat-four engine, driving Arplast Dyn'Aéro CR100 two-blade propeller. Compatible with alternative engines of 119 kW (160 hp) and above. Fixed-pitch or constant-speed propellers can be fitted. Fuel capacity for aerobatics 85 litres (22.4 US gallons; 18.7 Imp gallons), of which 82 litres (21.7 US gallons; 18.0 Imp gallons) are usable; max fuel capacity 160 litres (42.3 US gallons; 35.2 Imp gallons). Oil capacity 7.6 litres (2.0 US gallons; 1.7 Imp gallons).
ACCOMMODATION: Two seats, side by side, with dual controls. Fixed windscreen and rearward-sliding canopy.
SYSTEMS: Electrical system: 12 V 30 Ah battery.
AVIONICS: Customer specified.
DIMENSIONS, EXTERNAL:

Wing span	8.50 m (27 ft 10½ in)
Wing aspect ratio	6.8
Length overall	7.10 m (23 ft 3½ in)

Dyn'Aéro CR120 (marked as CR100) *(Paul Jackson/Jane's)*

Tailplane span	2.80 m (9 ft 2¼ in)
Wheel track	2.37 m (7 ft 9¼ in)
Propeller diameter	1.90 m (6 ft 3 in)
AREAS:	
Wings, gross	10.63 m² (114.4 sq ft)
Ailerons (total)	1.215 m² (13.08 sq ft)
Trailing-edge flaps (total)	1.21 m² (13.02 sq ft)
Vertical tail surfaces (total)	1.71 m² (18.40 sq ft)
Horizontal tail surfaces (total)	2.20 m² (23.68 sq ft)
WEIGHTS AND LOADINGS:	
Weight empty	550 kg (1,213 lb)
Max T-O weight: Aerobatic	760 kg (1,676 lb)
Normal	850 kg (1,873 lb)
Max wing loading: Aerobatic	71.5 kg/m² (14.64 lb/sq ft)
Normal	80.0 kg/m² (16.38 lb/sq ft)
Max power loading: Aerobatic	5.66 kg/kW (9.31 lb/hp)
Normal	6.34 kg/kW (10.41 lb/hp)
PERFORMANCE:	
Never-exceed speed (V_{NE})	205 kt (379 km/h; 235 mph)
Max level speed	171 kt (317 km/h; 197 mph)
Max cruising speed	154 kt (285 km/h; 177 mph)
Manoeuvring speed (V_A)	140 kt (260 km/h; 161 mph)
Max rate of climb at S/L	480 m (1,575 ft)/min
T-O run	153 m (500 ft)
T-O to 15 m (50 ft)	351 m (1,150 ft)
g limits	+8/−6

UPDATED

DYN'AERO MCR01

TYPE: Side-by-side kitbuilt.
PROGRAMME: Composites version of the metal MC100 Ban-Bi designed by Michel Colomban. MCR identifies Michel Colomban/Christophe Robin. Prototype (F-PECH) flew 1995.
CURRENT VERSIONS: **MCR01 VLA:** Certified version meeting JAR-VLA requirements; **VLA 912** or **VLA 914**, according to engine.
 Ban Bi: US name for VLA version marketed and assembled by American Ghiles Aircraft Inc, 522 East Washington Street, Orlando, Florida 32802-3666; Tel (+1 407) 839 05 81; Fax (+1 407) 872 32 33; Web: http://www.wicghilesbanbi.com.
 MCR01 Club: Version of MCR01 VLA 912 intended for flight training. First flight (F-PCLB) June 1998. Certification to JAR-VLA due in 2001.
 MCR01 ULM: Ultralight version meeting FAI ULM requirements; 59.7 kW (80 hp) engine. Canadian **MCR01 ULC** has 480 kg (1,058 lb) MTOW.
 Motor Glider: Under development. 'Winggrid' wingtips developed in conjunction with La Roche Consulting of Switzerland; 59.7 kW (80 hp) engine, best glide ratio 30:1.
 MCR M: Tailwheel version.
Description applies to all versions except where indicated.
CUSTOMERS: Total of 170 kits sold, and 80 flying by June 2000. Most constructed in France, but others in Netherlands, UK and USA.
COSTS: Kits, less engine: VLA FFr165,000, ULM FFr170,000 (1997).
DESIGN FEATURES: Single-engined low-wing monoplane with T tail; design goals included extremely low structural weight, fast kit production and high performance in comfort and safety.
FLYING CONTROLS: Manual. All-moving tailplane with trim tab, which is full span on ULM version; electrically actuated flaps and pitch trim; ULM version has full-span double-slotted flaperons; VLA has flaperons; Club has separate flaps and ailerons.
STRUCTURE: Wing and control surfaces built on carbon fibre spars and cellular ribs, to which preformed aluminium skins are glued in partial vacuum; monocoque carbon fibre fuselage; quoted build time of 500 to 1,000 hours from three modular kits without special tools. Airframe disassembles for storage or road-towing, with recommended disassembly/reassembly time of 4 minutes, single-handed.
LANDING GEAR: Fixed tricycle type with speed fairings on all three units; nosewheel steerable. Large wheels optional for rough field operation. Tailwheel configuration optional on MCR01 club and MCR01 ULC models.
POWER PLANT: One 59.6 kW (79.9 hp) Rotax 912 UL; 63.4 kW (85 hp) JPX 4T75; or 73.5 kW (98.6 hp) normally aspirated Rotax 912 ULS or turbocharged Rotax 914 flat-four driving a two-blade fixed-pitch propeller optimised for cruise; three-blade constant-speed propeller optional. Standard fuel capacity 88 litres (23.2 US gallons; 19.4 Imp gallons) in all versions; optional additional 90 litre (23.75 US gallon; 19.8 Imp gallon) tank in Club and ULM or 264 litre (69.7 US gallon; 58.1 Imp gallon) tank in VLA.
AVIONICS: To customer's choice.
EQUIPMENT: Optional road-towing trailer/storage hangar. BRS ballistic parachute recovery system optional on ULM.
DIMENSIONS, EXTERNAL:

Wing span: VLA	6.63 m (21 ft 9 in)
Club	6.92 m (22 ft 8½ in)
ULM	8.28 m (27 ft 2 in)
Motor Glider	9.80 m (32 ft 1¾ in)
Wing aspect ratio: VLA	8.5
Club	7.0
ULM	8.6
Length overall: all	5.66 m (18 ft 6¾ in)
DIMENSIONS, INTERNAL:	
Cabin max width: all	1.12 m (3 ft 8 in)
AREAS:	
Wings, gross: VLA	5.20 m² (56.0 sq ft)
Club	6.45 m² (69.4 sq ft)
ULM	7.95 m² (85.6 sq ft)
Motor Glider	9.35 m² (100.6 sq ft)

Dyn'Aéro MCR01 Club, demonstrating double-slotted flaps, plus ailerons *(Paul Jackson/Jane's)*

WEIGHTS AND LOADINGS:
Weight empty, equipped: VLA 220 kg (485 lb)
    Club 235 kg (518 lb)
    ULM 250 kg (551 lb)
    Motor Glider 262 kg (578 lb)
Max T-O weight: VLA, ULM 450 kg (992 lb)
    Club 490 kg (1,080 lb)
PERFORMANCE (59.7 kW; 80 hp engine):
Never-exceed speed (V_{NE}): normal:
    VLA 172 kt (320 km/h; 198 mph)
    Club, ULM 162 kt (300 km/h; 186 mph)
    with BRS: all 145 kt (270 km/h; 167 mph)
Max level speed: VLA 912 163 kt (302 km/h; 188 mph)
    VLA 914 197 kt (364 km/h; 226 mph)
    Club 151 kt (281 km/h; 174 mph)
    ULM 150 kt (278 km/h; 172 mph)
    Motor Glider 135 kt (250 km/h; 155 mph)
Max cruising speed at 75% power:
  at 2,440 m (8,000 ft):
    VLA 912 158 kt (292 km/h; 181 mph)
    Club, ULM 146 kt (271 km/h; 168 mph)
    Motor Glider 130 kt (240 km/h; 149 mph)
  at 3,350 m (11,000 ft):
    VLA 914 178 kt (329 km/h; 204 mph)
Stalling speed, flaps down:
    VLA 47 kt (87 km/h; 54 mph)
    Club 43 kt (79 km/h; 49 mph)
    ULM 35 kt (64 km/h; 40 mph)
    Motor Glider 30 kt (55 km/h; 35 mph)
T-O run: VLA 285 m (935 ft)
    Club 300 m (985 ft)
    ULM 195 m (640 ft)
T-O to 15 m (50 ft): VLA, Club 424 m (1,395 ft)
    ULM 298 m (980 ft)
Range at econ cruising speed and altitude:
  with normal fuel:
    VLA 894 n miles (1,657 km; 1,029 miles)
    Club 829 n miles (1,536 km; 954 miles)
    ULM 834 n miles (1,545 km; 960 miles)
  with max optional fuel:
    VLA 3,452 n miles (6,394 km; 3,973 miles)
    Club 1,623 n miles (3,006 km; 1,867 miles)
    ULM 1,611 n miles (2,985 km; 1,854 miles)
g limits: all +4/−2

UPDATED

DYN'AERO MCR4S

TYPE: Four-seat kitbuilt.
PROGRAMME: Mockup first shown at Aero '99, Friedrichshafen, April 1999. Kits available from late 1999. Prototype (FWWUZ) first flew 14 June 2000 and displayed at PFA International Air Rally, Cranfield, UK, 23-25 June 2000.
CURRENT VERSIONS: **Three-seat**: With 59.6 kW (79.9 hp) Rotax 912 UL or JPX 4TX75 engine; 640 kg (1,411 lb) MTOW.
Four-seat: With 73.5 kW (98.6 hp) normally aspirated engine and fixed-pitch propeller; 750 kg (1,653 lb) MTOW.
Four-seat Performance: 84.6 kW (113.4 hp) Rotax 914 UL turbocharged engine and constant-speed propeller; optimised for cruise at 3,050 m (10,000 ft).
COSTS: Kit FFr285,000, less engine; FFr400,000, FFr445,000 or FFr485,000 with Rotax 912 ULS plus fixed propeller, 912 ULS plus c/s propeller or 914 plus c/s propeller (1999).
DESIGN FEATURES: Three/four-seat version of MCR01, having 75 per cent commonality. Dihedral 3°.
Description generally as for MCR01, except that below.
FLYING CONTROLS: Ailerons, plus two-segment flaps. Aileron max deflection +20°/−10°. Tailplane (all moving) max deflection +5°/−10°. Double-slotted flaps, deflections 0°, 17° and 45°.
POWER PLANT: See Current Versions. Fuel in two wing tanks; total capacity 120 litres (31.7 US gallons; 26.4 Imp gallons).
ACCOMMODATION: Four persons, in two side-by-side pairs. Forward-hinged, two-piece canopy/windscreen, plus two fixed side windows.
EQUIPMENT: Ballistic parachute optional.
DIMENSIONS, EXTERNAL:
    Wing span 8.72 m (28 ft 7¼ in)
    Wing chord (mean) 0.96 m (3 ft 1¾ in)
    Wing aspect ratio 9.2
    Length overall 9.20 m (30 ft 2¼ in)
    Height overall 1.90 m (6 ft 2¾ in)
    Tailplane span 2.50 m (8 ft 2½ in)
DIMENSIONS, INTERNAL:
    Cabin max width 1.20 m (3 ft 11¼ in)
AREAS:
    Wings, gross 8.30 m² (89.3 sq ft)
WEIGHTS AND LOADINGS:
    Weight: empty 300 kg (661 lb)
        equipped 350 kg (772 lb)
    Max T-O weight 640-750 kg (1,411-1,653 lb)
PERFORMANCE (Rotax 912, 912 ULS and 914 UL engines):
    Never-exceed speed (V_{NE}) 172 kt (320 km/h; 198 mph)
    Max level speed: 912 143 kt (265 km/h; 165 mph)
        912 ULS 156 kt (288 km/h; 179 mph)
        914 UL 173 kt (320 km/h; 199 mph)
    Manoeuvring speed 117 kt (217 km/h; 134 mph)
    Max cruising speed: 912 136 kt (252 km/h; 157 mph)
        912 ULS 150 kt (277 km/h; 172 mph)
        914 UL 155 kt (287 km/h; 178 mph)
    Econ cruising speed: 912 131 kt (242 km/h; 150 mph)
        912 ULS, 914 UL 146 kt (270 km/h; 168 mph)
    Stalling speed 45 kt (82 km/h; 51 mph)
    T-O run with c/s prop: 912 317 m (1,040 ft)
        912 ULS 203 m (670 ft)
        914 UL 153 m (505 ft)
    T-O to 15 m (50 ft) with c/s prop: 912 455 m (1,495 ft)
        912 ULS 295 m (970 ft)
        914 UL 221 m (725 ft)
    Range with max fuel:
        912 1,034 n miles (1,916 km; 1,190 miles)
        912 ULS, 914 UL 921 n miles (1,707 km; 1,060 miles)
    g limits +3.8/−1.5

UPDATED

Dyn'Aero MCR01 ULM, with additional side view (lower) and half-plan view of VLA, plus scrap side view of Club *(Paul Jackson/Jane's)* 2001/0092148

Dyn'Aero MCR01 built in the UK *(Paul Jackson/Jane's)* 2001/0092050

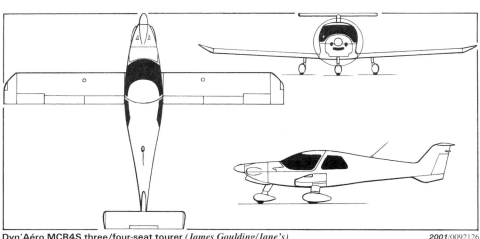

Dyn'Aéro MCR4S three/four-seat tourer *(James Goulding/Jane's)* 2001/0092126

Prototype MCR4S at its first public showing, Cranfield, June 2000 *(Paul Jackson/Jane's)* 2001/0092051

'Winggrid' wingtips to be fitted to the Motor Glider version of Dyn'Aero MCR01 *(Paul Jackson/Jane's)* 2000/0085617

EADS FRANCE

AEROSPATIALE MATRA

37 boulevard de Montmorency, F-75781 Paris Cedex 16
Tel: (+33 1) 42 24 24 24
Fax: (+33 1) 42 24 26 19 and 45 24 54 14
Web: http://www.aeromatra.com
CHAIRMAN: Jean-Luc Lagardère
CEO: Philippe Camus
GROUP MANAGING DIRECTOR, AERONAUTICS:
  Jean-François Bigay
VICE-PRESIDENT, CORPORATE COMMUNICATIONS: Pierre Bayle
PRESS OFFICER: Roland Sanguinetti

Aerospatiale formed 1 January 1970 by French government-directed merger of Sud-Aviation, Nord-Aviation and SEREB. Revised organisation from 1 January 1995 consisted of three major businesses, plus unchanged Missiles Division.

Aerospatiale participated with BAe (now BAE Systems) and DASA in formulating a blueprint for a restructured European aerospace industry which was presented to the French, UK and German governments in March 1998. Preparatory steps included July 1998 agreement to sell up to 33 per cent of Aerospatiale to Lagardère group on 1 January 1999 and reduce French government's share below 50 per cent by sales to staff and outside investors. On 1 January 2000, the French government's share stood at 47.7 per cent.

In 1999, Aerospatiale recorded orders of FFr100 billion and backlog of FFr292 billion. Sales for the year totalled FFr84.6 billion and for the first three months of 2000 reached €2,747 million. Employees in June 2000 totalled 56,500.

In preparation for the integration of European aerospace industries, Aerospatiale formed separate subsidiary companies to undertake certain activities in the aircraft and space & defence sectors; these were aligned with some existing semi-autonomous companies of the Aerospatiale Group. In 1998, the French government transferred to Aerospatiale its 45.76 per cent controlling interest in Dassault Aviation (which see). On 23 July 1998 the French government announced Aerospatiale's merger with the Lagardère group, owners of Matra Hautes Technologies, and the flotation of approximately 20 per cent of the new company on the Paris stock market. The new entity, comprising Aerospatiale and Matra HT, was to have become fully operational on 1 January 1999, but negotiations were only completed in late February; merger was completed on 6 May 1999 and floated on the Paris stock market on 4 June 1999.

On 14 October 1999, DaimlerChrysler, Lagardère and the French government announced an agreement to merge Aerospatiale Matra and DASA into the European Aeronautic, Defence and Space company (EADS) as described in the International section.

Aerospatiale Matra has 37.9 per cent share in Airbus Industrie (see International section); main Airbus assembly plant is at Aerospatiale base at Toulouse. However, in 2000, Airbus became a self-managing, entirely separate corporate entity. Likewise, the ATR regional transport business (see International section) is 50 per cent owned and is also to be separated. Share of Dassault Aviation is 45.76 per cent, and of Eurocopter (International section) 70 per cent; wholly owned subsidiaries are Socata light aircraft company (this section) and Sogerma aircraft support company (*Jane's Aircraft Upgrades*).

Refer also to International section for details of Aerospatiale Matra involvement in UHCA/VLCT large transport and Supersonic Airliner studies. Including Dassault, Aerospatiale Matra activities are devoted 34 per cent to airliners with over 100 seats, 14 per cent to regional airliners, 6 per cent to combat aircraft, 14 per cent to helicopters, 12 per cent to space, 13.5 per cent to missiles and 6.5 per cent to telecommunications and technology.

In January 2001, Aerospatiale Matra remained a legal entity, despite operating under the EADS name since that company formed on 10 July 2000. Components of the French company were then as below.

Aéronautique
  Aerospatiale Matra Airbus
    including Corse Composite Aéronautiques
  Aerospatiale Matra ATR
  Eurocopter
Sector Aéronautique
  Airbus Industrie (GIE)
  ATR (GIE)
  Sogerma
    including REVIMA and SECA
  Socata
    including Aéroport Tarbes
Systèmes, Services et Télécommunications
  Pole C4ISR
    comprising MS&I, APIC, ISTAR and Fleximage
  Pole Services et Tests
    comprising APSYS, Sycomore, CRIS and SG2I
  Poles Conception et Fabrication Industrielle
  Pole Internet et Services Opérateur
    comprising MCN Sat Holding, Matra Grolier Network and Multicoms
  Poles Télécommunications
    comprising Matra Nortel Communication
Affaires Militaires
  Aerospatiale Matra Missiles
    comprising Celerg, Euromissile, EMDG, Eurosam, SERAT, ASB and GDI Simulation
  Matra BAe Dynamics
    comprising Alkan and Matra Electronique
Affaires Spatiales
  Matra Marconi Space
  Aerospatiale Matra Lanceurs
    comprising Starsem, Cryospace, Nucletudes, CILAS and SODERN
Co-ordination Stratégique
Activités Nouvelles
  Gamaero
  Composites Atlantic
Lagardère International

UPDATED

EADS SOCATA

SOCATA GROUP AEROSPATIALE MATRA

Le Terminal Bât. 413, Zone d'Aviation d'Affaires, F-93352 Le Bourget Cedex
Tel: (+33 1) 49 34 69 69
Fax: (+33 1) 49 34 69 71
Telex: SOCATA 520 828 F
Web: http://www.socata.com

WORKS AND AFTER-SALES SERVICE:
Aérodrome de Tarbes-Ossun-Lourdes, BP 930, F-65009 Tarbes Cedex
Tel: (+33 5) 62 41 76 00
Fax: (+33 5) 62 41 76 54
PRESIDENT: Stéphane Bernard
CHAIRMAN AND CEO: Philippe Debrun
SALES DIRECTOR: Christophe van den Broek
TECHNICAL DIRECTOR: Dominique Deschamps
INFORMATION AND COMMUNICATION: Philippe de Segovia

US OPERATING AND SERVICE FACILITY:
Socata Aircraft, North Perry Airport, 7501 Pembroke Road, Pembroke Pines, Florida 33023
Tel: (+1 954) 893 14 00
Fax: (+1 954) 893 14 02

Formed 1966 as a subsidiary of Aerospatiale responsible for light aircraft (is acronym of Société de Constructions d'Avions de Tourisme et d'Affaires). By mid-2000, Socata had produced over 2,000 TB series aircraft, excluding the TBM 700. In 1999, sold and delivered 40 TBs; sold 16 and delivered 21 TBM 700s. Reduced deliveries of 45 TBs and 33 TBMs in 2000; target for 2001 is 100+ and 33, respectively. Revenue for 2000 was expected to total US$160 million.

Also makes components for Airbus A300/320/330/340, Lockheed Martin C-130, ATR 42/72, Dassault Falcons, and Eurocopter Super Puma, Dauphin, Ecureuil and satellite structures. Covered floor area 57,000 m² (613,450 sq ft); workforce 750.

Following formation of EADS on 10 July 2000, Socata is a subsidiary of that company, although immediately subordinate to Aerospatiale Matra pending full integration of EADS members.

UPDATED

SOCATA, TB 9C TAMPICO CLUB, TB 9 SPRINT GT, TB 10 TOBAGO GT and TB 200 TOBAGO XL GT

TYPE: Four-seat lightplane.
PROGRAMME: Design launched 1975; first flight of original TB 10 (F-WZJP), powered by 119 kW (160 hp) Textron Lycoming O-320, 23 February 1977; second prototype powered by 134 kW (180 hp) Textron Lycoming.
CURRENT VERSIONS: **TB 9C Tampico Club:** Four-seater with 119 kW (160 hp) Textron Lycoming O-320-D2A and Sensenich fixed-pitch propeller; superseded Tampico FP and CS in 1989 (see 1988-89 *Jane's*); fuel capacity 158 litres (41.75 US gallons; 34.75 Imp gallons) of which 152 litres (40.2 US gallons; 33.4 Imp gallons) are usable; non-retractable landing gear; first flight 9 March 1979; French certification 27 September 1979. Specification data refer to Club version.

TB 9 Sprint: Launched in September 1997; features trailing-link main landing gear with speed fairings and new propeller; cruising speed increased by 10 kt (18 km/h; 11 mph); first deliveries early 1998.

TB 10 Tobago: With 134 kW (180 hp) Textron Lycoming O-360-A1AD, rear bench seat for optional fifth occupant and non-retractable landing gear. French certification 26 April 1979; FAA certification 27 November 1985.

TB 10 Tobago Privilège: Limited edition version of Tobago, for French market only; launched March 1999.

Detailed description applies to TB 10 Tobago, except where indicated.

TB 200 Tobago XL: Optional five-seater with 149 kW (200 hp) Textron Lycoming IO-360-A1B6; otherwise generally as TB 10; first flight 27 March 1991; French certification 30 October 1991.

TB 9 Tampico Sprint GT, TB 10 Tobago GT and TB 200 Tobago XL GT: In June 1999, Socata announced its Nouvelle Génération (New Generation) series of single piston-engined models. The title Generation Two (GT) was subsequently adopted. These feature aerodynamic and other improvements that may subsequently be incorporated in the Socata Morane diesel-engined light aircraft described elsewhere in this entry. Modifications include a curved dorsal fairing for the fin, upturned wingtips, raised cabin roofline with new single-piece carbon fibre/honeycomb roof, revised window pillar design, flush-mounted windows, larger baggage door, redesigned interior and new fuel filler door. A TB 10 Tobago NG (F-GNHG) was displayed at the 1999 Paris Air Show.
CUSTOMERS: Total of 468 Tampicos and Tampico Clubs in service by June 1999 with customers in Africa and Middle East, Asia-Pacific, Australasia, Europe, and North America. Total of 637 Tobagos and 87 Tobago XLs delivered by June 1999, to customers in Africa and Middle East, Australasia, Europe, North America and elsewhere. Deliveries in 2000 totalled 45. Launch order for six TB 200 GTs, plus six options, placed by Aeronautical Academy of Europe.
DESIGN FEATURES: Conventional light tourer and trainer.
  Wing section RA 16.3C3; thickness/chord ratio 16 per cent; dihedral 4° 30′.
FLYING CONTROLS: Manual, with pushrod actuation for ailerons, rudder and all-moving tailplane; cable-actuated anti-balance tab on tailplane, plus ground-adjustable tabs on ailerons and rudder; strakes on lower edges of fuselage just aft of wing control turbulence under rear fuselage; electrically actuated flaps.
STRUCTURE: Conventional light alloy; single-spar wing; GFRP tips and engine cowlings. Triple anti-corrosion protection.

Socata TB 9C Tampico Club

2000/0064447

Socata TB 10 Tobago five-seat tourer *(Paul Jackson/Jane's)*

LANDING GEAR: Non-retractable tricycle type, with steerable nosewheel. Oleo-pneumatic shock-absorber in all three units; trailing-link suspension main landing gear legs introduced on all fixed-gear TB series from 1997. Mainwheel tyres size 6.00-6 (6 ply), pressure 2.30 bars (33 lb/sq in) on TB 10/200; 15×6.00-6 (6 ply) on TB 9/20/21. Nosewheel 5.00-5 on TB 10/200 and 5.00-4 on TB 9/20/21. Glass fibre wheel fairings on all three units. Hydraulic disc brakes. Parking brake.

POWER PLANT: One 134 kW (180 hp) Textron Lycoming O-360-A1AD flat-four engine, driving a Hartzell two-blade constant-speed propeller. Two integral fuel tanks in wing leading-edges; total capacity 210 litres (55.5 US gallons; 46.2 Imp gallons), of which 204 litres (53.9 US gallons; 44.9 Imp gallons) are usable. Oil capacity 8 litres (2.1 US gallons; 1.8 Imp gallons).

ACCOMMODATION: Four or five seats in enclosed cabin, with dual controls. Separate, adjustable front seats with inertia reel seat belts. Two separate rear seats or removable three-place bench seat with safety belts. Sharply inclined low-drag windscreen. Access via upward-hinged window/doors of glass fibre. Baggage compartment aft of cabin, with external door on port side. Cabin carpeted, soundproofed, heated and ventilated. Windscreen defrosting standard.

SYSTEMS: Electrical system includes 28 V 70 A alternator and 24 V 10 A battery. Hydraulic system for brakes only.

AVIONICS: Honeywell Silver Crown avionics suite to customer choice, including KLN 90 GPS.
Instrumentation: Typical IFR package includes dual altimeters; dual true airspeed indicators; dual artificial horizons; vertical speed indicator, electric turn co-ordinator; directional gyro, tachometer; oil temperature, oil pressure, fuel pressure, fuel quantity, manifold pressure, CHT/EGT and OAT gauges; ammeter, voltmeter and compass.

EQUIPMENT: Includes armrests for all seats, map pockets, anti-glare visors, stall warning indicator, tiedown fittings and towbar, landing and navigation lights, four individual cabin lights and instrument panel lighting.

DIMENSIONS, EXTERNAL (TB 9, TB 9C, TB 10 and TB 200):
Wing span	9.77 m (32 ft 0¾ in)
Wing chord, constant	1.22 m (4 ft 0 in)
Wing aspect ratio	8.0
Length overall	7.70 m (25 ft 3 in)
Height overall	3.02 m (9 ft 11 in)
Tailplane span	3.20 m (10 ft 6 in)
Wheel track	2.33 m (7 ft 7¾ in)
Wheelbase	1.96 m (6 ft 5 in)
Propeller diameter	1.88 m (6 ft 2 in)
Propeller ground clearance	0.10 m (4 in)
Cabin doors (each): Width	0.90 m (2 ft 11½ in)
Height	0.76 m (2 ft 6 in)
Baggage door: Width	0.64 m (2 ft 1¼ in)
Max height	0.44 m (1 ft 5¼ in)

DIMENSIONS, INTERNAL (TB 9, TB 9C, TB 10 and TB 200):
Cabin: Length:	
firewall to rear bulkhead	2.53 m (8 ft 3½ in)
panel to rear bulkhead	2.00 m (6 ft 6¾ in)
Max width: at rear seats	1.28 m (4 ft 2¼ in)
at front seats	1.15 m (3 ft 9¼ in)
Max height	1.12 m (3 ft 8 in)

AREAS (TB 9, TB 9C, TB 10 and TB 200):
Wings, gross	11.90 m² (128.1 sq ft)
Ailerons (total)	0.91 m² (9.80 sq ft)
Trailing-edge flaps (total)	3.72 m² (40.04 sq ft)
Fin	0.88 m² (9.47 sq ft)
Rudder	0.63 m² (6.78 sq ft)
Horizontal tail surfaces (total)	2.56 m² (27.56 sq ft)

WEIGHTS AND LOADINGS:
Weight empty, with unusable fuel and oil:	
TB 9	647 kg (1,426 lb)
TB 10	700 kg (1,543 lb)
TB 200	715 kg (1,576 lb)
Baggage: all	65 kg (143 lb)
Max T-O weight: TB 9	1,060 kg (2,336 lb)
TB 10, TB 200	1,150 kg (2,535 lb)
Max wing loading: TB 9	89.1 kg/m² (18.25 lb/sq ft)
TB 10, TB 200	96.6 kg/m² (19.79 lb/sq ft)
Max power loading: TB 9	8.91 kg/kW (14.64 lb/hp)
TB 10	8.57 kg/kW (14.08 lb/hp)
TB 200	7.72 kg/kW (12.68 lb/hp)

PERFORMANCE:
Max level speed: TB 9	122 kt (226 km/h; 140 mph)
TB 10	133 kt (247 km/h; 153 mph)
TB 200	140 kt (259 km/h; 161 mph)
Max cruising speed at 75% power:	
TB 9 at 1,830 m (6,000 ft)	107 kt (198 km/h; 123 mph)
TB 10 at 2,590 m (8,500 ft)	127 kt (235 km/h; 146 mph)
TB 200 at 2,590 m (8,500 ft)	130 kt (240 km/h; 149 mph)
Econ cruising speed at 60% power:	
TB 9 at 1,830 m (6,000 ft)	105 kt (194 km/h; 121 mph)
TB 10 at 2,590 m (8,500 ft)	108 kt (200 km/h; 124 mph)
TB 200 at 2,590 m (8,500 ft)	115 kt (213 km/h; 132 mph)
Stalling speed:	
flaps up: TB 9	58 kt (107 km/h; 67 mph)
TB 10	61 kt (112 km/h; 70 mph)
flaps down: TB 9	48 kt (89 km/h; 56 mph)
TB 10	52 kt (97 km/h; 60 mph)
TB 200	53 kt (98 km/h; 61 mph)
Max rate of climb at S/L: TB 9	225 m (738 ft)/min
TB 10	240 m (790 ft)/min
TB 200	305 m (1,000 ft)/min
Max certified altitude: TB 9	3,350 m (11,000 ft)
TB 10	3,965 m (13,000 ft)
TB 200	4,875 m (16,000 ft)
T-O run: TB 9	340 m (1,115 ft)
TB 10	325 m (1,070 ft)
TB 200	290 m (955 ft)
T-O to 15 m (50 ft):	
TB 9 at T-O weight of 1,060 kg (2,337 lb)	520 m (1,706 ft)
TB 10 at T-O weight of 1,150 kg (2,535 lb)	505 m (1,657 ft)
TB 200 at T-O weight of 1,150 kg (2,535 lb)	475 m (1,558 ft)
Landing from 15 m (50 ft): TB 9	420 m (1,378 ft)
TB 10	460 m (1,509 ft)
TB 200	449 m (1,473 ft)
Landing run: TB 9	195 m (640 ft)
TB 10	190 m (625 ft)
Max range: TB 9	556 n miles (1,029 km; 639 miles)
TB 10	697 n miles (1,290 km; 802 miles)
TB 200	637 n miles (1,179 km; 733 miles)

UPDATED

SOCATA TB 20 TRINIDAD GT and TB 21 TRINIDAD GT TURBO

Israel Defence Force name: Pashosh (Lark)

TYPE: Four-seat lightplane.

PROGRAMME: First flight TB 20 (F-WDBA) 14 November 1980; French certification 18 December 1981; FAA certification 27 January 1984; first delivery (F-WDBB) 23 March 1982; first flight TB 21, 24 August 1984; French certification 23 May 1985; FAA certification 5 March 1986.

CURRENT VERSIONS: **TB 20 Trinidad GT**: At the Paris Air Show in June 1999, Socata announced its Nouvelle Génération (New Generation) series of single piston-engined aircraft, the prototype of which, TB 20 NG F-WWRG, had first flown on 21 April 1999. The title Generation Two (GT) was subsequently adopted. The TB 20 GT received DGAC certification on 31 January 2000. The GT series was formally launched at Tarbes on 2 February 2000 with roll-out of the first production TB 20 GT, which became US demonstrator (N163GT c/n 2000) and exhibited at Sun 'n' Fun, April 2000.

GT variants feature aerodynamic and other improvements that will be incorporated in the Socata Morane diesel-engined series described below. Identifying features include upturned wingtip, similar to those on the TBM 700 turboprop; a curved dorsal fillet; raised cabin roofline with new single-piece carbon fibre/honeycomb roof; revised cabin window/pillar design with flush-mounted windows; plus redesigned interior, larger baggage door, new fuel filler door, retractable footstep and optional three-blade Hartzell propeller.

Details of earlier TB 20 Trinidad, TB 20 Trinidad Excellence, TB 20 C and TB 21 Trinidad TC variants (of which some 700 built) may be found in the 2000-01 and previous editions of *Jane's*.

TB 21 Trinidad GT Turbo: As above, but with computer-controlled turbocharger.

CUSTOMERS: At time of launch of GT series, the Aeronautical Academy of Europe in Evora, Portugal, took options on four TB 20 GTS. By mid-2000, TB 200 GTs had been delivered to customers in France, UK and USA from planned first year production of 72. Orders in 2000 included 69 from New Avex, distributor for US West Coast; recent customers include the Nigerian College of Aviation Technology (NCAT) of Zaria, Kaduna State, which ordered 10 Trinidad GTs for delivery beginning mid-2001.

DESIGN FEATURES: Mainly as for Tobago; dihedral 6° 30′.

FLYING CONTROLS: As Tobago, but rudder trim and flap preselector added.

STRUCTURE: Largely as Tobago.

LANDING GEAR: Hydraulically retractable tricycle type, with single wheel on each unit. Free-fall emergency extension. Steerable nosewheel retracts rearward. Main units, with trailing-link suspension, retract inward into fuselage. Hydraulic disc brakes. Parking brake.

POWER PLANT: *TB 20 GT:* One 186 kW (250 hp) Textron Lycoming IO-540-C4D5D flat-six, driving a two-blade Hartzell constant-speed propeller.

TB 21 GT: One 186 kW (250 hp) Textron Lycoming TIO-540-AB1AD turbocharged flat-six. Two-blade Hartzell constant-speed propeller; three-blade propeller optional.

Fuel in two integral wing tanks, total capacity 336 litres (88.8 US gallons; 73.9 Imp gallons), of which 326 litres (86.1 US gallons; 71.7 Imp gallons) usable. Oil capacity 12.6 litres (3.3 US gallons; 2.8 Imp gallons).

ACCOMMODATION: Generally as for TB 9 Sprint GT, TB 10 Tobago GT and TB 200 XL GT; new headrests for all seats introduced on GT variants; rear seat can be removed for carrying 250 kg (551 lb) of cargo.

SYSTEMS: Electrical system comprises 28 V 70 A alternator and 24 V 10 Ah battery.

AVIONICS: TB 20/21 GT basic navigation package comprises single nav/com and VOR receiver. IFR packages and EFIS optional.

DIMENSIONS, EXTERNAL: As for TB 10, except:
Wing span	9.85 m (32 ft 1¾ in)
Wing aspect ratio	8.2
Length overall	7.75 m (25 ft 5 in)
Height overall	2.85 m (9 ft 4¼ in)
Tailplane span	3.68 m (12 ft 0¾ in)
Wheel track	2.17 m (7 ft 1½ in)

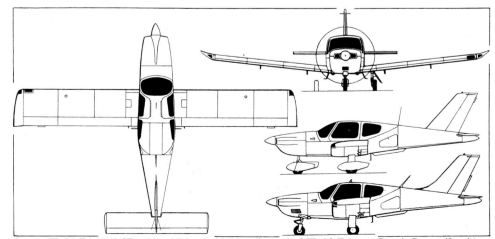

Socata TB 21 Trinidad GT, with additional side view (upper) of TB 10 Tobago *(Dennis Punnett/Jane's)*

Early production EADS Socata TB 21 Trinidad GT Turbo

Wheelbase	1.91 m (6 ft 3¼ in)
Propeller diameter	2.03 m (6 ft 8 in)

DIMENSIONS, INTERNAL (TB 20 GT):
Cabin: Width 1.28 m (4 ft 2½ in)
    Max height 1.17 m (3 ft 10 in)
AREAS: As for TB 10, except:
    Horizontal tail surfaces (total) 3.06 m² (32.94 sq ft)
WEIGHTS AND LOADINGS:
    Operating weight empty: TB 20 GT 859 kg (1,894 lb)
        TB 21 GT 867 kg (1,911 lb)
    Max baggage: TB 20 GT, TB 21 GT 65 kg (143 lb)
    Max T-O and landing weight: TB 20 GT, TB 21 GT
        1,400 kg (3,086 lb)
    Max wing loading:
        TB 20 GT, TB 21 GT 117.6 kg/m² (24.10 lb/sq ft)
    Max power loading:
        TB 20 GT, TB 21 GT 7.51 kg/kW (12.35 lb/hp)
PERFORMANCE:
    Cruising speed at 75% power:
        TB 20 GT at 1,981 m (6,500 ft)
            163 kt (302 km/h; 188 mph)
        TB 21 GT at 7,620 m (25,000 ft)
            190 kt (352 km/h; 219 mph)
    Econ cruising speed at 65% power:
        TB 20 GT at 1,981 m (6,500 ft)
            157 kt (291 km/h; 181 mph)
        TB 21 GT at 7,620 m (25,000 ft)
            169 kt (313 km/h; 194 mph)
    Rate of climb at S/L: TB 20 GT 366 m (1,200 ft)/min
        TB 21 GT 343 m (1,126 ft)/min
        TB 21 GT at 5,180 m (17,000 ft) 244 m (800 ft)/min
    Certified ceiling: TB 20 GT 6,100 m (20,000 ft)
        TB 21 GT 7,620 m (25,000 ft)
    T-O to 15 m (50 ft): TB 20 GT 635 m (2,085 ft)
        TB 21 GT 595 m (1,955 ft)
    Landing from 15 m (50 ft): TB 20 GT 555 m (1,825 ft)
        TB 21 GT 540 m (1,775 ft)
    Max range:
        TB 20 -GT 1,110 n miles (2,055 km; 1,277 miles)
        TB 21 GT 1,035 n miles (1,916 km; 1,191 miles)

UPDATED

SOCATA MORANE LIGHT AIRCRAFT

In January 1997 Aerospatiale and Renault Sport formed a jointly owned subsidiary, Société de Motorisations Aéronautiques (SMA), to develop a new generation of light aircraft piston engines, under the marketing name Morane Renault. Their first application is a new range of Socata Morane aircraft, based on the current piston-engined TB series. Three engines were planned, sharing common core components; 134 to 149 kW (180 to 200 hp), 186 kW (250 hp) and 224 kW (300 hp), the last-mentioned two with reduction gearing to bring propeller speed down to 2,000 rpm. The 224 kW (300 hp) engine will also be certified for aerobatic use. First to be certified, however, is a 169 kW (227 hp) variant, the SR305.

The engines are four-cylinder, horizontally opposed, turbocharged diesels with 19:1 compression ratio, designed to run on jet fuel. They feature single-lever power control, electronic management and integral recording and checking systems, and are around 10 dB quieter and more fuel-efficient than existing aero-engines, with estimated operating costs of 30 to 40 per cent less per hour than for comparable light aircraft power plants. The engines will be capable of maintaining 75 per cent of rated power up to 7,620 m (25,000 ft); TBO will be 3,000 hours.

Two prototype engines, then designated MR 200, began ground running trials in mid-1997. A single TB 20 Trinidad served as flying testbed for all three engines; this aircraft was rolled out on 24 January 1998 and first flew (F-WWRS) with an MR 200 on 3 March 1998.

At the Paris Air Show in June 1997, Socata announced a range of light aircraft to be powered by the Morane Renault engines. The aircraft also revive the Morane title of the former Morane-Saulnier company. Three types will be available initially; brief details of the Morane range are given below, where these differ from corresponding TB-series models.

MS 200 FG: Previously MS 180. Socata TB 10 Tobago-derived fixed landing gear, four/five-seat tourer fitted with 134 kW (180 hp) MR 200 engine with direct drive to three-blade constant-speed Hartzell propeller, diameter 1.88 m (6 ft 2 in).

MS 200 RG: Previously MS 250. Socata TB 20 Trinidad-derived retractable landing gear tourer with 186 kW (250 hp) engine with geared drive to three-blade, constant-speed Hartzell propeller, diameter 2.03 m (6 ft 8 in). Mockup displayed at Paris Air Show in June 1997. Prototype, initially to be powered by a conventional Textron Lycoming IO-540 engine, under construction by late 1998.

MS 300 Epsilon II: Initially named Sabre; Socata TB 30 Epsilon-derived two-seat fully aerobatic trainer with 224 kW (300 hp) MR 300 engine with geared drive to three-blade constant-speed Hartzell propeller, diameter 1.98 m (6 ft 6 in). Empty weight 928 kg (2,046 lb). 'Prototype' (actually Epsilon demonstrator, No. 3, in non-flying condition as 'F-EMKZ' displayed at Paris Air Show in June 1999; certification scheduled for mid-2001; available for delivery 24 months from launch order. The Epsilon was last described in the 1990-91 edition and appears in the current *Jane's Aircraft Upgrades*.

PERFORMANCE (estimated):
    Cruising speed:
        200 FG at 3,050 m (10,000 ft)
            134 kt (248 km/h; 154 mph)
        200 RG at 4,575 m (15,000 ft)
            189 kt (350 km/h; 217 mph)
        300 at S/L 206 kt (382 km/h; 237 mph)
        300 at 3,050 m (10,000 ft) 223 kt (413 km/h; 267 mph)
    Stalling speed, landing configuration
        300 63 kt (117 km/h; 73 mph)
    Max rate of climb at S/L:
        200 FG more than 229 m (750 ft)/min
        200 RG more than 335 m (1,100 ft)/min
    Rate of climb at 3,050 m (10,000 ft):
        300 567 m (1,860 ft)/min
    Service ceiling: 200 FG 5,180 m (17,000 ft)
        200 RG 7,010 m (23,000 ft)
    T-O to 15 m (50 ft): 200 FG 505 m (1,660 ft)
        200 RG 655 m (2,150 ft)
        300 590 m (1,940 ft)

Artist's impression of the Socata MS 200 RG

MR 250 engine testbed Trinidad, first flown on 3 March 1998 *(Paul Jackson/Jane's)*

Landing from 15 m (50 ft):
200 FG 460 m (1,510 ft)
200 RG 533 m (1,750 ft)
300 660 m (2,165 ft)
Max range: 200 FG 996 n miles (1,844 km; 1,146 miles)
200 RG 1,433 n miles (2,654 km; 1,649 miles)
300 at 222 kt (411 km/h; 255 mph) at 4,575 m
(15,000 ft) 830 n miles (1,537 km; 955 miles)
Endurance at 6,100 m (20,000 ft) at max cruising speed:
200 RG 8 h 10 min

UPDATED

SOCATA TBM 700

TYPE: Business turboprop.
PROGRAMME: Three prototypes built: first flights 14 July 1988 (F-WTBM), 3 August 1989 (F-WKPG) and 11 October 1989 (F-WKDL); French certification received 31 January 1990; FAR Pt 23 type approval awarded 28 August 1990; first delivery 21 December 1990; Canadian public transport certification 1993. Wider, single-piece, upward opening door replaced horizontally split door, port side, rear, from late 1998 and is standard. Aircraft used for freighting have optional crew door, port, forward.
CURRENT VERSIONS: **TBM 700:** In addition to basic transport, Socata offers multimission versions, including military medevac, target towing, ECM, freight, maritime patrol, law enforcement, navaid calibration and vertical photography versions. Prototype of last-mentioned (F-GLBF) exhibited at the 1996 Farnborough Air Show.
TBM 700B: Large door version, introduced in 2000, with 1.19 m (3 ft 10¾ in) × 1.07 m (3 ft 6 in) cabin door providing easier access to main cabin and baggage compartments and enhancing multimission capability. First three aircraft handed over to French Army on 28 June 2000.
TBM 700C: Cargo version initially designated TBM 700C, with port side cargo door and separate port side cockpit door; reinforced metal cargo floor with tiedown points in rails; maximum cargo capacity 825 kg (1,819 lb); volume 3.5 m³ (124 cu ft); maximum T-O weight increased to 3,300 kg (7,275 lb); quick conversion to six/seven-seat passenger configuration; announced June 1995; French certification 23 November 1998. Launch customer Air Open Sky of France, which began TBM 700 freight operations in November 1999.
CUSTOMERS: Total 160 delivered by late 2000, to customers in Asia-Pacific, Europe and North America. Deliveries to French Air Force began 27 May 1992 with first of initial six for liaison duties with Groupe Aérien d'Entraînement et de Liaison (GAEL) and ETE 43; further six supplied 1993-94 (also for ETE 41 and ETE 44, plus CEAM); total Air Force requirement is 22 (of which 17 funded by mid-1998); two officially handed over to French Army 13 January 1995 for 3 GHL at Rennes, followed by six more, last three of which delivered (in TBM 700B configuration) 28 June 2000. Two delivered to Indonesian Civil Aviation Academy at Curug in September 1996. Production targets rising from 21 in 1999 to 28-30 in 2000 and 30-35 in 2001.
COSTS: US$2.5 million, typically equipped (1999); direct operating cost US$230 per hour (2000).
DESIGN FEATURES: High-speed, long-range, single-turboprop business transport. Low wing; twin strakes under rear fuselage; sweptback fin (with dorsal fin) and mass balanced rudder; non-swept tailplane with mass balanced elevators.
Wing of Aerospatiale RA 16-43 root section with 6° 30′ dihedral from roots.
FLYING CONTROLS: Conventional and manual. Pushrod and cable actuated with electric trim tabs in port aileron, rudder and each elevator; 'scaled-down ATR' single-slotted Fowler flaps, also electrically actuated, along 71 per cent of each wing trailing-edge; slotted spoiler forward of each flap at outer end, linked mechanically to aileron; yaw damper.

TBM 700B (large door version) in French Army service *(Paul Jackson/Jane's)*

Instrument panel of Socata TBM 700

STRUCTURE: Mainly of light alloy and steel except for control surfaces, flaps, most of tailplane and fin of Nomex honeycomb bonded to metal sheet; wing leading-edges and landing gear doors GFRP/CFRP; tailcone and wingtips GFRP; two-spar torsion box forms integral fuel tank in each wing.
LANDING GEAR: Hydraulically retractable tricycle type, with emergency manual operation. Inward-retracting main units of trailing-link type; rearward-retracting steerable nosewheel (±28°). Main tyres 18×5.5 (8 ply) tubeless; nose 5.00-5 (10 ply) tubeless. Parker hydraulic disc brakes. Minimum ground turning radius (based on nosewheel) 23.98 m (78 ft 8 in).
POWER PLANT: One 1,178 kW (1,580 shp) Pratt & Whitney Canada PT6A-64 turboprop, flat rated at 522 kW (700 shp), driving a Hartzell HC-E4N-3/E9083S(K) four-blade constant-speed fully feathering reversible-pitch metal propeller. Fuel in integral tank in each wing, combined capacity 1,100 litres (290.5 US gallons; 242 Imp gallons), of which 1,066 litres (282 US gallons; 234 Imp gallons) usable. Gravity filling point in top of each tank. Oil capacity 12 litres (3.2 US gallons; 2.6 Imp gallons).
ACCOMMODATION: Adjustable seats for one or two pilots at front. Dual controls standard. Four seats in club layout aft of these, with centre aisle, or five seats in high-density layout. Large upward-opening door on port side aft of wing; overwing emergency exit on starboard side. Oxygen system, comprising three under-seat bottles with individual emergency oxygen mask for each passenger and masks with integral microphones for crew. Pressurised baggage compartment at rear of cabin, with internal access only; additional unpressurised compartment in nose, between engine and firewall, with external access via door on port side. Optional crew door, port side, front. Reinforced metal floor with tie-down points optional.
SYSTEMS: Engine bleed air pressurisation (to 0.43 bar; 6.2 lb/sq in) and air cycle (optionally Freon) air conditioning. Hydraulic system for landing gear only. Electrical system powered by two 28 V 200 A engine-driven starter/generators (one main, one standby) and a 28 V 40 Ah Ni/Cd battery. Pneumatic rubber-boot de-icing of wing/tailplane/fin leading-edges. Propeller blades anti-iced electrically, engine inlets by exhaust air. Electric anti-icing and hot air demisting of windscreen.
AVIONICS: Honeywell Silver Crown digital IFR package.
Comms: KY 196 VHF transceiver; KT 79 transponder; KMA 24H interphone.
Radar: Honeywell RDR-2000 weather radar optional.
Flight: KX 165 nav/com; KNS 80 R/Nav; KR 21 marker beacon receiver; KR 87 ADF; KFC 275 autopilot with KAS 297C altitude preselect/alerter. Optional GPS.
Instrumentation: Optional three-screen EFIS.

DIMENSIONS, EXTERNAL:
Wing span 12.68 m (41 ft 7¼ in)
Wing chord, mean aerodynamic 1.51 m (4 ft 11½ in)
Wing aspect ratio 8.9
Length overall 10.645 m (34 ft 11 in)
Height overall 4.35 m (14 ft 3¼ in)
Tailplane span 4.99 m (16 ft 4½ in)
Wheel track 3.87 m (12 ft 8¼ in)
Wheelbase 2.91 m (9 ft 6½ in)
Propeller diameter 2.31 m (7 ft 7 in)
Crew door (optional): Height 1.02 m (3 ft 4 in)
  Width 0.78 m (2 ft 6¾ in)
Cabin door: Height 1.19 m (3 ft 10¾ in)
  Width 1.08 m (3 ft 6½ in)
DIMENSIONS, INTERNAL:
Cabin: Length 4.05 m (13 ft 3½ in)
  Max width 1.21 m (3 ft 11¾ in)
  Max height 1.22 m (4 ft 0 in)
Baggage compartment volume:
  front 0.25 m³ (8.8 cu ft)
  rear 0.90 m³ (31.8 cu ft)
AREAS:
Wings, gross 18.00 m² (193.75 sq ft)
Vertical tail surfaces (total) 2.56 m² (27.55 sq ft)
Horizontal tail surfaces (total) 4.76 m² (51.24 sq ft)
WEIGHTS AND LOADINGS:
Weight empty, equipped 1,860 kg (4,101 lb)
Fuel weight (usable) 866 kg (1,910 lb)
Baggage: front 50 kg (110 lb)
  rear 100 kg (220 lb)
Max T-O weight 2,984 kg (6,578 lb)
Max ramp weight 3,000 kg (6,613 lb)
Max landing weight 2,835 kg (6,250 lb)
Max wing loading 165.8 kg/m² (33.95 lb/sq ft)
Max power loading 5.72 kg/kW (9.48 lb/shp)
PERFORMANCE:
Max cruising speed at 7,925 m (26,000 ft)
  300 kt (555 km/h; 345 mph)
Econ cruising speed at 9,140 m (30,000 ft)
  243 kt (450 km/h; 280 mph)
Stalling speed, flaps and landing gear down
  61 kt (113 km/h; 71 mph)
Max rate of climb at S/L 725 m (2,380 ft)/min
Max certified altitude 9,140 m (30,000 ft)
T-O to 15 m (50 ft) 650 m (2,135 ft)
Landing from 15 m (50 ft):
  with reverse pitch 500 m (1,640 ft)
  without reverse pitch 650 m (2,135 ft)
Range, ISA, 45 min reserves:
  with max payload:
    at max cruising speed
      900 n miles (1,666 km; 1,035 miles)
    at econ cruising speed
      1,079 n miles (2,000 km; 1,242 miles)
  with max fuel:
    at max cruising speed
      1,350 n miles (2,500 km; 1,553 miles)
    at econ cruising speed
      1,550 n miles (2,870 km; 1,783 miles)
g limits +3.8/−1.5

UPDATED

SOCATA TB 360 TANGARA

Socata's offer of a production partnership for this business twin-prop has not been taken up and the programme remains dormant. Details last appeared in the 2000-01 edition.

UPDATED

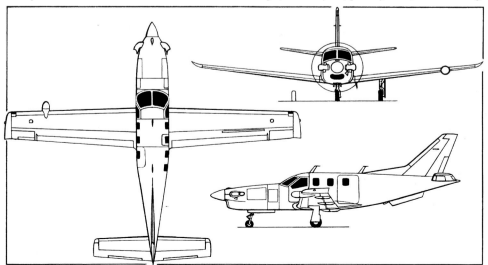

Socata TBM 700 business and multirole aircraft *(Dennis Punnett/Jane's)*

EURAVIAL

EURAVIAL

Euravial was established to produce the René Fournier-designed RF-47 under licence, but closed in late 1999 having encountered intractable bureaucratic obstacles in obtaining JAA certification for the aircraft. In late 2000, M Fournier was seeking another manufacturer for the RF-47, a description, photograph and three-view of which last appeared in the 2000-01 edition.

UPDATED

FOURNIER

AVIONS FOURNIER
René Fournier, F-37270 Athée sur Cher
Tel: (+33 2) 47 50 68 30
Fax: (+33 2) 47 50 24 22

Earlier designs by René Fournier remain in production elsewhere, as described in other sections of this book: RF-6 adopted and converted as T67 Firefly by Slingsby in UK; RF-9 intended for resumed production by ABS of Switzerland; and RF-10 as Ximango by Aeromot in Brazil. In June 2000, M Fournier was seeking a manufacturer for the RF-47 following the closure of Euravial, which see in this section. A description, photograph and three-view drawing of the RF-47 last appeared under the Euravial entry in the 2000-01 edition.

NEW ENTRY

HUMBERT

HUMBERT AVIATION
rue du Menil, F-88160 Ramonchamps
Tel: (+33 3) 29 25 05 75
Fax: (+33 3) 29 25 98 97
DESIGNER: Jacques Humbert

NEW ENTRY

HUMBERT TETRAS
TYPE: Side-by-side ultralight kitbuilt.
PROGRAMME: Prototype unveiled at 1992 RSA Rally at Moulins.
COSTS: Kit, with Rotax 912 UL engine, FFr217,779 (1999).
DESIGN FEATURES: High wing with single bracing strut on each side; strut-braced tailplane. Wing section NACA 23012. Quoted build time 150 to 250 hours.
FLYING CONTROLS: Conventional and manual. Three-position plain flaps, deflections 0, 15 and 45°.
STRUCTURE: Steel tube fuselage with Dacron covering; composites wing; light alloy ailerons and flaps.
LANDING GEAR: Non-retractable tailwheel type with bungee cord suspension on main units.
POWER PLANT: One 53.7 kW (72 hp) Humbert-Volkswagen HW 2000 or 59.6 kW (79.9 hp) Rotax 912 UL flat-four, driving a two-blade composite propeller. Other options available. Fuel in two wing tanks, combined capacity 60 litres (15.9 US gallons; 13.2 Imp gallons).
DIMENSIONS, EXTERNAL:
Wing span	10.10 m (33 ft 1¾ in)
Length overall	6.50 m (21 ft 4 in)

Humbert Tetras two-seat ultralight (Paul Jackson/Jane's)

Height overall	2.06 m (6 ft 9 in)
Propeller diameter	1.71 m (5 ft 7¼ in)
AREAS:	
Wings, gross	15.70 m² (169.0 sq ft)
WEIGHTS AND LOADINGS:	
Weight empty	260 kg (573 lb)
Max T-O weight	450 kg (992 lb)
PERFORMANCE (Rotax 912 UL):	
Max level speed	105 kt (194 km/h; 121 mph)
Cruising speed at 75% power at 3,050 m (10,000 ft)	101 kt (187 km/h; 116 mph)
Stalling speed, flaps down	27 kt (50 km/h; 32 mph)
Max rate of climb at S/L	302 m (990 ft)/min
T-O run	50 to 80 m (164 to 262 ft)
g limits	+6/-3

NEW ENTRY

ISSOIRE

ISSOIRE AVIATION
ZA la Béchade, Aérodrome Issoire-Le Broc, F-63501 Issoire
Tel: (+33 4) 73 89 01 54
Fax: (+33 4) 73 89 54 59
e-mail: iav@issoire-aviation.com
Web: issoire-aviation.com
PRESIDENT AND CEO: Philippe Moniot
DIRECTOR: Laurent Bourdier
SALES MANAGER: Yves Moyne-Bressand

Issoire Aviation was established in 1978 as a successor to Wassmer Aviation. Its subsidiary, Rex Composites, is responsible for carbon fibre airframes. Current product is the APM-20/21 lightplane.

UPDATED

ISSOIRE APM-20 LIONCEAU and APM-21 LION
English name: Lion Cub
TYPE: Two-seat lightplane.
PROGRAMME: Design (as Moniot APM-20) started 1992 by Les Industries de Composites d'Auvergne Réunités (ICAR); prototype (F-WWMP) exhibited at Paris Air Show 1995 before first flight on 21 November 1995; third (second flying) aircraft (F-WWXX) exhibited statically at Paris Air Show, June 1997, fitted with JPX flat-four engine, which is not offered on production aircraft. No. 4 (also F-WWXX) exhibited at Paris in June 1999. Certified to JAR-VLA 17 May 1999. First all-carbon fibre, single-engine aircraft to gain JAR-VLA certification. First production batch of five APM-20s under construction in mid-2000 with two scheduled for delivery by end of that year. APM-21 will follow at unspecified date.
CURRENT VERSIONS: **APM-20:** Primary trainer *As described.*
  APM-21 Lion: Aerial work version powered by 73.5 kW (98.6 hp) Rotax 912 ULS engine driving a two-blade, constant-speed Hoffmann propeller. Prototype (F-WWMP, modified from prototype Lionceau) demonstrated at Paris Air Show in June 1999.
  Four-seat: Under development; no major structural changes.
CUSTOMERS: By mid-2000 Issoire Aviation had received 20 firm orders. Initial production rate planned of two per month, possibly rising to 50 per year. Break-even point estimated at 200 aircraft.
COSTS: FFr620,000, minimally equipped or FFr700,000 with standard equipment (2000). Operating cost FFr350 per hour (1998).
DESIGN FEATURES: Low wing, NACA 63618 aerofoil, thickness/chord ratio 18 per cent, dihedral 3°, incidence 2°, twist 1°.
FLYING CONTROLS: Conventional and manual. Spring elevator tab for pitch trim. Electrically operated Fowler flaps to about two-thirds span.
STRUCTURE: Carbon fibre/epoxy.
LANDING GEAR: Fixed tricycle type with spats; oleo-sprung steerable nose leg, composites main legs. Mainwheels and nosewheel diameter 330 mm; maximum pressure 2.35 bars (34 lb/sq in).
POWER PLANT: One 59.6 kW (79.9 hp) Rotax 912 UL four-cylinder four-stroke driving a two-blade fixed-pitch propeller. Fuel capacity 68 litres (18 US gallons; 15 Imp gallons), of which 65 litres (17.2 US gallons; 14.3 Imp gallons) are usable. Refuelling point on port side of fuselage. Oil capacity 4 litres (1.1 US gallons; 0.9 Imp gallon).
ACCOMMODATION: Two, side by side under rearward-sliding tinted canopy. Dual controls standard. Baggage compartment at rear of seats.
AVIONICS: *Instrumentation:* VFR panel standard.

Issoire APM-20 Lionceau (Rotax 912 UL engine) (Paul Jackson/Jane's)

DIMENSIONS, EXTERNAL:
Wing span	8.66 m (28 ft 5 in)
Wing chord: at root	1.10 m (3 ft 7¼ in)
at tip	0.84 m (2 ft 9 in)
Wing aspect ratio	7.9
Length overall	6.60 m (21 ft 7¾ in)
Max width of fuselage	1.15 m (3 ft 9¼ in)
Height overall	2.40 m (7 ft 10½ in)
Tailplane span	2.50 m (8 ft 2½ in)
Wheel track	1.92 m (6 ft 3½ in)
Propeller diameter	1.64 m (5 ft 4½ in)
Propeller ground clearance	0.31 m (1 ft 0¼ in)
AREAS:	
Wings, gross	9.50 m² (102.3 sq ft)

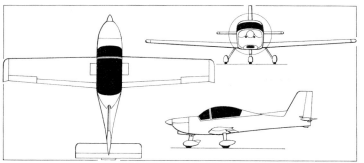

APM-20 Lionceau (Rotax 912 UL engine) *(James Goulding/Jane's)*   2000/0064443

Prototype Issoire APM-21 Lion all-composites two-seat trainer *(Paul Jackson/Jane's)*   2000/0064440

Ailerons (total)	0.50 m² (5.38 sq ft)	Max T-O weight	620 kg (1,366 lb)	Stalling speed: flaps up	54 kt (100 km/h; 63 mph)
Trailing-edge flaps (total)	1.40 m² (15.07 sq ft)	Max wing loading	65.3 kg/m² (13.37 lb/sq ft)	25° flap	44 kt (80 km/h; 50 mph)
Fin	1.80 m² (19.37 sq ft)	Max power loading	10.39 kg/kW (17.06 lb/hp)	Max rate of climb at S/L	204 m (669 ft)/min
Tailplane	2.00 m² (21.53 sq ft)	PERFORMANCE:		T-O run	240 m (787 ft)
Elevator, incl tab	0.50 m² (5.38 sq ft)	Never-exceed speed (V$_{NE}$)	135 kt (250 km/h; 155 mph)	T-O to 15 m (50 ft)	460 m (1,509 ft)
WEIGHTS AND LOADINGS:		Cruising speed: at S/L	113 kt (210 km/h; 130 mph)	Landing run	150 m (492 ft)
Weight empty	380 kg (838 lb)	at 1,676 m (5,500 ft)	124 kt (230 km/h; 143 mph)	Endurance	4 to 5 h
Baggage capacity	20 kg (44 lb)	Econ cruising speed	94 kt (175 km/h; 109 mph)		**UPDATED**
Max fuel	49 kg (108 lb)	Manoeuvring speed	108 kt (200 km/h; 124 mph)		

JURCA

MARCEL JURCA
3 Allées des Bordes, F-94430 Chennevières-sur-Marne
Tel/Fax: (+33 1) 45 94 01 38

M Marcel Jurca has designed a series of high-performance light aircraft and replicas of Second World War fighters. Most have wood or metal tube fuselages; stress analysis is verified by the ESTACA university (see Association Zéphyr entry in this section).

Jurca aircraft (all available to amateur builders) are MJ2 Tempête; MJ3 Dart; MJ4 Shadow; MJ5 Sirocco; MJ51 Sperocco; MJ53 Autan; and MJ55 Biso (Provence wind) two-seat, fully aerobatic trainer which employs NACA 21012 section wing of MJ7/MJ77, 134 kW (180 hp) Textron Lycoming engine and first flew (F-PJJR) in October 1997.

Replicas comprise MJ7 Gnatsum two-third size Mustang Replica; MJ77 Gnatsum three-quarter size, two-seat Mustang (early- and late-style canopies); MJ8 three-quarter size Fw 190 Replica; MJ9 three-quarter size Messerschmitt Bf 109 Replica (one under construction in Canada); MJ10 three-quarter size Spitfire Replica; MJ12 three-quarter size P-40 Replica; MJ100 full-size Spitfire Replica (prototype first flown 14 October 1994; two others under construction at Prescott, Arizona, with 1,044 kW; 1,480 hp Allison piston engines; one with Rolls-Royce Merlin Mk 58 at Annemasse, France); MJ90 full-size Messerschmitt Bf 109 Replica (410 kW; 550 hp Double Ranger V12-770C-1 engine); MJ80 full-size Focke-Wulf Fw 190 Replica (first under construction in Germany with 895 kW; 1,200 hp Pratt & Whitney R-1830 radial engine); MJ70 full-size P-51D Mustang Replica (under development by January 1996, will be powered by 1,044 kW; 1,400 hp Allison engine).

Under development in 2000 are the MJ11 single-seat, 119 kW (160 hp) biplane based on the Bücker Bü 133 Jungmeister and MJ16 racer, with sportplane derivative.

Some 150 MJ aircraft have been registered and flown and a further 20 are under construction. Leading types are MJ2 (46 built) and MJ5 (70).

UPDATED

MJ12 replica at three-quarter scale of Curtis P-40, built by Russel Moynagh at Santa Ana, California   2001/0100412

MJ2 NAP version of Tempête *(Nominal 6m, Ailes Pliables)* with folding wing, built by Dominique Corbaye of Belgium   2001/0100411

JURCA MJ5 SIROCCO

TYPE: Two-seat lightplane.
PROGRAMME: Prototype first flight 3 August 1962, powered by 78.5 kW (105 hp) Potez 4 E-20 engine; factory-built model subsequently awarded type certificate in Utility category. Version for amateur construction (70 built by January 2000) is generally similar to factory-built version.

DESIGN FEATURES: Tandem two-seat development of MJ2 Tempete. Versions available with fixed or retractable gear; retractable tailwheel also available. Aerobatic version with 149 kW (200 hp) engine has strengthened wing spar.
POWER PLANT: Engines available include the Textron Lycoming series of the following powers: 85.75 kW (115 hp), 112 kW (150 hp), 119 kW (160 hp), 134 kW (180 hp) and 149 kW (200 hp).

Details which follow refer to a Sirocco with a 149 kW (200 hp) Textron Lycoming IO-360 engine.
DIMENSIONS, EXTERNAL:
Wing span 7.00 m (23 ft 0 in)
Wing aspect ratio 4.9
Length overall 6.15 m (20 ft 2 in)
Height overall, tail up:
 with modified rudder 2.60 m (8 ft 6¼ in)
 standard rudder 2.80 m (9 ft 2¼ in)

MJ8 three-quarter scale Focke-Wulf Fw 190A-5 (224 kW; 300 hp Lycoming) built by David Spencer of Martinsville, Virginia   2000/0084369

Jurca MJ77 Gnatsum three-quarter scale Mustang replica built by Jim Riechert at Owaso, Michigan (388 kW; 520 hp Ranger V-770C-1 piston engine) *(Pierre Gaillard)*   2001/0100410

136   FRANCE: AIRCRAFT—JURCA to POTTIER

AREAS:
Wings, gross   10.00 m² (107.6 sq ft)
WEIGHTS AND LOADINGS:
Weight empty   565 kg (1,246 lb)
Max T-O weight   850 kg (1,874 lb)
Max wing loading   85.5 kg/m² (17.51 lb/sq ft)
Max power loading   5.74 kg/kW (9.43 lb/hp)
PERFORMANCE:
Max level speed   162 kt (300 km/h; 186 mph)
Cruising speed   140 kt (260 km/h; 162 mph)
Stalling speed   59 kt (110 km/h; 68 mph)
Max rate of climb at S/L   838 m (2,750 ft)/min
Service ceiling   6,000 m (19,680 ft)
T-O run   280 m (920 ft)
Landing run   500 m (1,640 ft)
Endurance   3 h 30 min
g limits   +6/−3

UPDATED

Jurca MJ5 Sirocco, with 149 kW (200 hp) Lycoming, built by Daniel Lioult

NOIN

NOIN AERONAUTIQUE
R N 85, Chateauvieux, F-05130 Tallard
Tel/Fax: (+33 4) 92 54 15 04 and 00 99
DIRECTOR: Claude Noin

The Choucas is the second design produced by Claude Noin, following the Sirius. Noin's father was a noted soaring pilot of the 1950s and was first to cross the Alps by glider.

VERIFIED

NOIN CHOUCAS
English name: Jackdaw

TYPE: Side-by-side ultralight kitbuilt.
PROGRAMME: Design started 1994 by Claude Noin and Charly Baum; construction of prototype started 1995; first flight February 1996; construction of production prototype started 1996, first flight at that time scheduled for March 1997, but had not been notified to *Jane's* by late 2000. Marketed as basic kit for advanced builders, quoted build time 1,200 hours; advanced kit with main components prefabricated, quoted build time 600 hours; or ready-to-fly. A special version for physically handicapped pilots is to be launched, if there is sufficient demand.
CUSTOMERS: Three ordered by January 1997 (latest information).
COSTS: Basic kit (glass fibre) FFr44,500 (carbon fibre) FFr49,600; advanced kit FFr169,000; ready-to-fly FFr298,000 (all excluding VAT; 1997).
DESIGN FEATURES: Tail-less design, stated to be spinproof; design aimed at meeting ultralight category certification requirements while matching performance of modern gliders and conventional three-axis ultralight aircraft. Dihedral 4°; zero twist.
FLYING CONTROLS: Unconventional flying wing; pitch control by elevators in inboard wing trailing-edge; ailerons outboard; large rudder. Trim tabs on elevators; spoilers/airbrakes on upper surface of wing.
STRUCTURE: Standard fuselage moulded in half-shells of carbon fibre/epoxy; glass fibre/epoxy fuselage available at lower cost for advanced builders, but with weight penalty; composites wing structure with carbon fibre main spar, glass fibre/epoxy-stiffened Styrofoam ribs and birch plywood D-section leading-edge, covered with heat-shrink Dacron; moulded carbon fibre/epoxy ailerons; Lexan/polycarbonate canopy with carbon fibre frame.
LANDING GEAR: Tailwheel type, fixed; streamline fairings on main legs and wheels; drum brakes.
POWER PLANT: One 37.0 kW (49.6 hp) Rotax 503 UL-2V with 2.58:1 reduction gear driving a two-blade ground-adjustable DUC 20 carbon fibre propeller. Fuel capacity 36 litres (9.5 US gallons; 7.9 Imp gallons), of which 35 litres (9.25 US gallons; 7.7 Imp gallons) are usable.
ACCOMMODATION: Two, side by side beneath one-piece upward-hinging canopy; baggage shelf to rear of seats.
SYSTEMS: 14 V DC alternator provides electrical power.
AVIONICS (factory-built aircraft): *Comms:* VHF comm and intercom.
Instrumentation: ASI, altimeter, variometer, slip indicator, compass, tachometer, fuel gauge, chronometer, CHT and EGT gauges and voltmeter.
EQUIPMENT: BRS 5 ballistic parachute recovery system optional.
DIMENSIONS, EXTERNAL:
Wing span   14.35 m (47 ft 1 in)
Length overall   5.35 m (17 ft 6½ in)
Height overall   1.95 m (6 ft 4¾ in)
Propeller diameter   1.60 m (5 ft 3 in)
DIMENSIONS, INTERNAL:
Cabin max width   0.97 m (3 ft 2 in)

Prototype Noin Choucas (Rotax 503 piston engine)

AREAS:
Wings, gross   21.30 m² (229.3 sq ft)
WEIGHTS AND LOADINGS:
Weight empty   260-290 kg (573-639 lb)
Max T-O weight:
FAI ultralight certification   450 kg (992 lb)
motor glider certification   510 kg (1,124 lb)
PERFORMANCE, POWERED:
Cruising speed at 75% power   70 kt (130 km/h; 81 mph)
Stalling speed   33 kt (60 km/h; 38 mph)
Max rate of climb at S/L   180 m (590 ft)/min
T-O run   100 m (330 ft)
Landing run   150 m (495 ft)
Range at 75% power, no reserves
                           175 n miles (325 km; 201 miles)
PERFORMANCE, UNPOWERED:
Best glide ratio   22
Min rate of sink   1.00 m (3.28 ft)/s

UPDATED

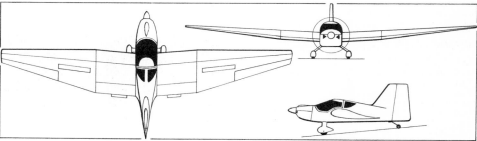

Noin Choucas tail-less light aircraft (*James Goulding/Jane's*)

POTTIER

AVIONS POTTIER
4 rue de Poissy, F-78130 Les Mureaux
Tel: (+33 1) 30 99 13 85
Fax: (+33 1) 34 92 97 26
DIRECTOR: Jean Pottier

The prolific Jean Pottier has designed numerous light aircraft for amateur construction or assembly. Most recent is the P 200 series, although the V-tail P 300 Ara will be added to the range.

VERIFIED

POTTIER P 200 SERIES

TYPE: Four-seat kitbuilt.
PROGRAMME: First aircraft in P 200 series was single-seat P 210 S Coati with tailwheel-type landing gear. This does not appear to have entered production, but several two- to four-seat kitbuilt versions have appeared subsequently.
CURRENT VERSIONS: **P 220 S Koala:** Side-by-side two-seat version; 55.9 kW (75 hp) VW/Limbach engine standard. At least 20 kits delivered by early 1996; six flying and 20 building reported by early 1997. Modified version with Walter Mikron engine formerly produced in Czech Republic by Evektor-Aerotechnik (which see) and now marketed in Poland as the AT-3 by Aero Sp z.o.o. (see appropriate entry). Rotax-, Lycoming- and Continental-engined versions built in Czech Republic by Evektor-Aerotechnik and in Italy by SG Aviation (both which see).
P 230 S Panda: Three-seater; 74.6 kW (100 hp) Continental engine standard. Prototype first flew 1990; at least 30 kits delivered by early 1996; eight said to be flying and 36 building by early 1997. Reportedly also available as **P230i** ultralight with 63.4 kW (85 hp) Rotax 912 engine.
P 240 S Saiga: Four-seat version (2 + 2); 134 kW (180 hp) Textron Lycoming engine. One said to be under construction in 1997. No specification data received for this version, and may no longer be current.
P 250 S Xerus: Two-seater; essentially tandem-seat version of Koala, but some dimensions differ. Four reportedly under construction in 1997.
P 270 S Amster: Four-seater (2 + 2); similar to P 240 S but with 112 kW (150 hp) Textron Lycoming engine. Five reported under construction in 1997.
CUSTOMERS: Sold to customers in Czech Republic, Finland, France, Germany, Netherlands, Poland and elsewhere.
DESIGN FEATURES: Typical low-wing monoplane with sweptback vertical tail; wing aerofoil section NACA 4415 on all versions.
FLYING CONTROLS: Conventional and manual. Aileron deflection +20/−15° on all versions; all-moving tailplane with automatic balance/trim tab. All versions have flaps (settings 0, 10 and 30°).
STRUCTURE: Mainly metal; some components in composites.
LANDING GEAR: All versions have non-retractable tricycle gear. Mainwheel/nosewheel sizes 330 and 270 mm respectively on two-seaters, 350 and 300 mm on three- and four-seaters. Hydraulic mainwheel brakes. Wheel speed fairings optional.
POWER PLANT: See under Current Versions above.
ACCOMMODATION: See under Current Versions above. Dual controls standard. One-piece canopy on P 250 S, two-piece on other versions.

POTTIER to REIMS AVIATION—AIRCRAFT: FRANCE

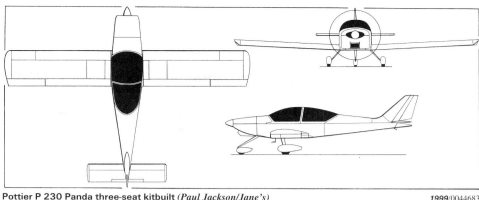

Pottier P 230 Panda three-seat kitbuilt *(Paul Jackson/Jane's)*

DIMENSIONS, EXTERNAL:
Wing span: 220	6.50 m (21 ft 4 in)
230, 270	8.10 m (26 ft 7 in)
250	6.40 m (21 ft 0 in)
Wing chord, constant: all	1.25 m (4 ft 1¼ in)
Wing aspect ratio: 220	5.3
230, 270	6.6
250	5.2
Length overall: 220	5.54 m (18 ft 2 in)
230	6.35 m (20 ft 10 in)
250	5.82 m (19 ft 1¼ in)
270	6.45 m (21 ft 2 in)
Height overall: 220, 250	1.95 m (6 ft 4¾ in)
230	2.05 m (6 ft 8¾ in)
270	2.15 m (7 ft 0¾ in)
Tailplane span: 220, 230, 250	2.60 m (8 ft 6½ in)
270	3.00 m (9 ft 10 in)
Wheel track: 220	1.70 m (5 ft 7 in)
230, 270	1.75 m (5 ft 9 in)
250	1.40 m (4 ft 7 in)
Wheelbase: 220	1.32 m (4 ft 4 in)
230, 270	1.45 m (4 ft 9 in)
250	1.55 m (5 ft 1 in)
Propeller diameter: 220	1.44 m (4 ft 8¾ in)
230	1.72 m (5 ft 7¾ in)
250	1.50 m (4 ft 11 in)

DIMENSIONS, INTERNAL:
Cabin: Length: 220	1.64 m (5 ft 4½ in)
230, 250	2.00 m (6 ft 6¾ in)
270	2.10 m (6 ft 10¾ in)
Max width: 220	1.04 m (3 ft 5 in)
230, 270	1.10 m (3 ft 7¼ in)
250	0.70 m (2 ft 3½ in)
Max height: 220	1.05 m (3 ft 5¼ in)
230, 270	1.10 m (3 ft 7¼ in)
250	1.00 m (3 ft 3¼ in)

AREAS:
Wings, gross: 220	8.00 m² (86.1 sq ft)
230, 270	10.00 m² (107.6 sq ft)
250	7.90 m² (85.0 sq ft)
Ailerons (total): 220, 250	0.40 m² (4.31 sq ft)
230, 270	0.60 m² (6.46 sq ft)
Flaps (total): 220, 250	0.90 m² (9.69 sq ft)
230, 270	1.10 m² (11.84 sq ft)
Vertical tail surfaces (total):	
220, 230, 250	0.80 m² (8.61 sq ft)
270	1.00 m² (10.76 sq ft)
Horizontal tail surfaces (total):	
220, 230, 250	1.80 m² (19.38 sq ft)
270	2.10 m² (22.60 sq ft)

WEIGHTS AND LOADINGS:
Weight empty: 220	275 kg (606 lb)
230	380 kg (838 lb)
250	240 kg (529 lb)
270	450 kg (992 lb)
Max T-O weight: 220	500 kg (1,102 lb)
230	700 kg (1,543 lb)
250	470 kg (1,036 lb)
270	870 kg (1,918 lb)
Max wing loading: 220	62.5 kg/m² (12.80 lb/sq ft)
230	70.0 kg/m² (14.34 lb/sq ft)
250	59.5 kg/m² (12.19 lb/sq ft)
270	87.0 kg/m² (17.82 lb/sq ft)
Max power loading: 220	8.95 kg/kW (14.70 lb/hp)
230	9.39 kg/kW (15.43 lb/hp)
250	9.70 kg/kW (15.94 lb/hp)
270	7.78 kg/kW (12.78 lb/hp)

PERFORMANCE:
Never-exceed speed (V_{NE}):	
220, 230, 250	135 kt (250 km/h; 155 mph)
270	156 kt (290 km/h; 180 mph)
Max level speed: 220	113 kt (210 km/h; 130 mph)
230	121 kt (225 km/h; 140 mph)
250	119 kt (220 km/h; 137 mph)
270	146 kt (270 km/h; 168 mph)
Max cruising speed: 220	103 kt (190 km/h; 118 mph)
230	113 kt (210 km/h; 130 mph)
250	108 kt (200 km/h; 124 mph)
270	130 kt (240 km/h; 149 mph)
Stalling speed, flaps up:	
220, 230	49 kt (90 km/h; 56 mph)
250	48 kt (88 km/h; 55 mph)
270	54 kt (100 km/h; 63 mph)
Stalling speed, flaps down:	
220, 250	44 kt (80 km/h; 50 mph)
270	49 kt (90 km/h; 56 mph)
Rate of climb at S/L: 220	210 m (689 ft)/min
230	240 m (787 ft)/min
250	222 m (728 ft)/min
270	396 m (1,299 ft)/min
T-O run: 220	200 m (660 ft)
230	210 m (690 ft)
250	180 m (590 ft)
270	330 m (1,085 ft)
Landing run: 220	220 m (725 ft)
230	240 m (790 ft)
250	200 m (660 ft)
270	370 m (1,215 ft)
Range with max fuel:	
220, 270	378 n miles (700 km; 435 miles)
230	405 n miles (750 km; 466 miles)
250	351 n miles (650 km; 403 miles)
g limit: all	+5.7

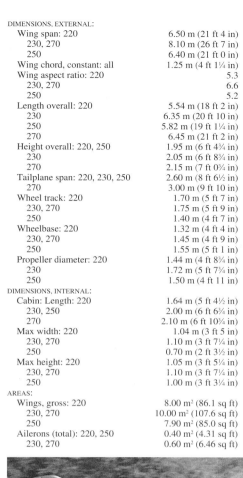

French homebuilt Pottier P 230 Panda *(Paul Jackson/Jane's)*

UPDATED

REIMS AVIATION

REIMS AVIATION SA
Aérodrome de Reims-Prunay, BP 2745, F-51062 Reims Cedex
Tel: (+33 3) 26 48 46 84
Fax: (+33 3) 26 49 18 57
e-mail: reims.aviation@reims-aviation.fr
Web: http://www.reims-aviation.fr
Telex: REMAVIA 830754 F
CEO: Jean-Paul Chaufour
VICE-PRESIDENT, MARKETING AND SALES: Gildas Illien
AREA SALES MANAGERS: Laurent Mesmin, Mathieu Quoi
EXTERNAL RELATIONS: Max Boirame

Originally Avions Max Holste, founded 1933; Cessna acquired 49 per cent share in February 1960; name then changed to Reims Aviation, which licensed to manufacture Cessna aircraft for sale in Europe, Africa and Asia, but stopped making small piston-engined types when Cessna did (see below); Reims had built 6,345 aircraft of all types by late 1998. Reims developed twin-turboprop F 406 Caravan II (see below); but Cessna sold its share in Reims Aviation to Compagnie Française Chaufour Investissement (CFCI) in early 1989. Remains French distributor for Cessna 172R Skyhawk, 182S Skylane and 206 Stationair.

Reims makes components for Dassault Falcons and Mirage 2000, ATR 42/72, Airbus A300/A310/A320 series/A330/A340 and Embraer ERJ-145. Other activity is maintenance of general aviation aircraft via its Reims Aviation Maintenance et Service (RAMS) subsidiary. At the Paris Air Show in June 1999, Reims announced plans to build a new maintenance facility for aircraft in the ATR and Boeing 737 class at the former NATO reserve air base at Vatry. The company offers a total rebuild of Cessna F 152 light aircraft, effectively to 'zero life' standard. Interest was shown by the company in 1999 in becoming European agent for the Antonov An-140 twin-prop airliner (which see).

Workforce was 360 in late 1998; office and factory floor space covers 28,300 m² (304,500 sq ft).

UPDATED

Reims F 406 Vigilant comint/imint intelligence-gathering aircraft

REIMS F 406 CARAVAN II

TYPE: Utility turboprop twin.
PROGRAMME: Announced mid-1982; first flight (F-WZLT) 22 September 1983; French certification 21 December 1984, FAA later; first flight production F 406 (F-ZBEO) 20 April 1985.
CURRENT VERSIONS: **F 406 Caravan II**: Initial production version, available in passenger, freight (with underbelly cargo pod), medevac, skydiving, aerial survey, training, navaid calibration and target towing variants.
F406 NG (New Generation): Upgraded version announced in October 2000. Features include 473.5 kW (635 shp) PT6A engines driving four-blade propellers; reduced structural weight; 200 kg (441 lb) increase in maximum take-off weight; quick-release access panels for

easier maintenance; new lightweight instrument panel with liquid crystal displays; and redesigned cabin interior by Air Esthetic featuring improved acoustic and thermal insulation, new ergonomically designed seats with integral harnesses for improved comfort on long missions and integral window blinds. Take-off, climb and OEI rate of climb performance are improved, and endurance increased by up to one hour.

Additionally, six versions of the basic **Maritime Patrol** configuration, each with a ventral 360° radar:

Vigilant: Surveillance version (land or sea) with mission equipment to suit customer requirement, including (in latest configuration) Thomson-CSF Airborne Maritime Situation Control System (AMASCOS) which comprises belly-mounted Telephonics AN/APS-143, Texas Instruments 1022 or Thomson-CSF/DASA Ocean Master 100/200 360° surveillance radars, chin-mounted FLIR turret for FSI Ultra, Inframetrics Mk III, SAGEM Hesis, Star Safire, Thomson Chlio or Wescam FLIR, mission management and communications systems and single operator's console in Srs 100 form; or with additional console for electronic surveillance operator in Srs 200. BAE Systems Seaspray 2000 radar for Scottish Fisheries; Texas Instruments AN/APS-134 radar for Australian Customs.

Vigilant Frontier: Border patrol and anti-drug surveillance version with mission equipment including surveillance radar (Northrop Grumman APG-66) in anti-drug configuration, FLIR and datalink.

Vigilant Polmar II: Pollution surveillance (maritime police) version with SAGEM Cyclope 2000 IR linescanner and Ericsson SLAR.

Vigilant Polmar III: As Polmar II, but based on F406 NG airframe.

Vigilant Surmar: Surveillance (maritime) version with Texas Instruments AN/APS-134 or Telephonics AN/APS-143 radar and Litton night vision system. In 1996 Reims developed underwing hardpoints for the F 406 which enable it to carry light weapons such as machine gun pods or rocket launchers, or camera pods or SAR liferafts.

Vigilant Comint/Imint: Dedicated communications and imaging intelligence version of the Caravan II, developed jointly with Thomson-CSF Communications, first shown at Paris '97. Equipped with Thomson TCC SAS airborne Elint system, an operator's console with MS Windows NT-based workstation and an array of direction-finding antennas housed in a ventral radome. Is also first Vigilant with a pair of underwing pylons for carriage of stores which can include gun or rocket pods and airdroppable SAR equipment.

CUSTOMERS: The 86th production aircraft (excluding prototype) and first of the second production batch was delivered in early 2000. F 406 is operated by 45 customers in 25 countries including ALH (Netherlands), Acrop Mad (Salvador), Aerovia (Spain), Aerotuy (Venezuela), Air Atlantic (Netherlands), Air Guyane (French Guyana), Air St Martin (Guadeloupe), Aircraft Ing (Kenya), Arcus Air Logistic (Germany), Aviazur (New Caledonia), Blue Bird Aviation (Kenya), Comair (South Africa), DDF (Zimbabwe), Direct Flight (UK), Excel Air (Malta), Fehlhaber Flugdienst (Germany), French Army (two used for target towing), French Customs Service (see below), Grupo America (Salvador), Hellenic Ministry of Merchant Marine (see below), Hellras (Kenya), Kensoma Ltd (UK), Luchtvaartmij (Netherlands), Namibian Fisheries, National Jet Systems (Australia, see below), NDG Aviation (USA), Overflight Support (South Africa), Polarwing (Finland), Republic of Korea Navy (see below), SAL (Angola), Scan Equipment (Norway), Sembawang (Singapore), Seychelles Coast Guard, Somacvram (Madagascar), Span Aviation (India), TCS (Pakistan), TAAG (Angola), Tanzanian Air, Tawakal Airlines, TCS (Saudi Arabia), Trackmark (Kenya), US Customs Service, US Navy and Wing Airline (Belgium).

French Customs Service received eight in Polmar II configuration, and took delivery in 1995 of three in Surmar configuration with one further aircraft, fitted with Honeywell 1500 radar and Hesis FLIR, supplied in late 1999. One in Polmar III configuration, ordered in late 2000 for delivery by 2002. First Surmar (No. 74, although registered as a Vigilant) handed over 11 January 1995, followed by Nos. 75 and 77 on 12 September 1995, the latter two replacing Cessna 404s in Martinique, French Antilles; National Jet Systems Australia ordered three Vigilants for delivery early 1996 for Surveillance Australia programme (aircraft Nos. 76, 78 and 79) and Scottish Fisheries has four Vigilants for fisheries patrol, the latest delivered in April 1998.

Recent customers for the F 406 include the Republic of Korea Navy, which ordered five in target tug configuration, the first delivered on 23 November 1998 and the last in mid-June 1999, total contract value being US$24 million (1997); and the Hellenic Ministry of Merchant Marine, which, in June 1999, ordered two for delivery in late 2000 as Surmar variants and additional one in November 1999 in Polmar guise; the first Surmar for this contract made its maiden flight on 21 August 2000. Total 94 sold by mid-1999. An agreement with the Brazilian government signed in April 1998 provides for the purchase of five Caravan IIs, comprising one Vigilant and four standard transports to be delivered between 1999 and 2001.

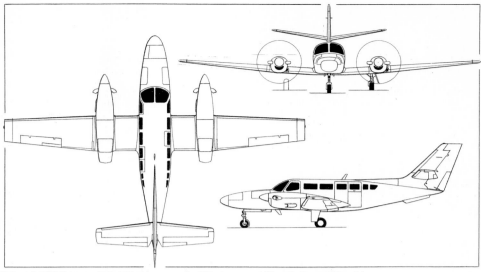

Reims F 406 Caravan II light business and utility transport *(Dennis Punnett/Jane's)*

Daihyaku Shoji of Japan ordered one F406 NG during 2000, for operation by an undisclosed local customer.

COSTS: Standard commuter aircraft US$2.1 million (2000). Maritime patrol versions typically US$3 million to US$4 million with mission equipment. Korean Navy contract for five aircraft valued at US$24 million, including technician training and one-year technical support (1997). Greek contract for two Surmars valued at FFr90 million (1999); single Polmar FFr133 million (1999). Direct operating cost US$338 per hour (1997).

DESIGN FEATURES: Extrapolated from Cessna Conquest airframe; wing section NACA 23018 at root and 23012 at tip; dihedral 3° 30' on centre-section, 4° 55' on outer panels; twist −3°; incidence 2° at root; fin offset 1° to port; tailplane dihedral 9°; cabin not pressurised.

FLYING CONTROLS: Conventional and manual; Trim tabs in elevators, port aileron and rudder; hydraulically operated Fowler flaps.

STRUCTURE: Conventional light metal with three-spar fail-safe wing centre-section to SFAR 41C; two-spar outer wings.

LANDING GEAR: Hydraulically retractable tricycle type with single wheel on each unit. Mainwheel tyre size 22×7.75-10, nosewheel 6.00-6. Main units retract inward into wing, nosewheel rearward. Emergency extension by means of a 138 bar (2,000 lb/sq in) rechargeable nitrogen bottle. Cessna oleo-pneumatic shock-absorbers. Main units of articulated (trailing link) type. Single-disc hydraulic brakes. Parking brake.

POWER PLANT: Two Pratt & Whitney Canada PT6A-112 turboprops (each 373 kW; 500 shp), each driving a McCauley 9910535-2 three-blade reversible-pitch and automatically feathering metal propeller. Fuel capacity 1,823 litres (481 US gallons; 401 Imp gallons) of which 1,798 litres (475 US gallons; 395.5 Imp gallons) usable. Oil capacity 17.2 litres (4.5 US gallons; 3.8 Imp gallons).

ACCOMMODATION: Crew of two and up to 12 passengers, in pairs facing forward, with centre aisle, except at rear of cabin in 12/14-seat versions. Alternative basic configurations for six VIP passengers in reclining seats in business version, and for operation in mixed passenger/freight role. Business version has partition between cabin and flight deck, and toilet on starboard side at rear. Split main door immediately aft of wing, on port side, with built-in airstair in downward-hinged lower portion. Optional cargo door forward of this door to provide single large opening. Overwing emergency exit on each side. Passenger seats removable for cargo carrying, or for conversion to ambulance, air photography, maritime surveillance and other specialised roles. Baggage compartments in nose, with three doors, at rear of cabin and in rear of each engine nacelle. Ventral cargo pod optional.

SYSTEMS: Freon air conditioning system of 17,500 BTU capacity, plus engine bleed air and electric boost heating. Electrical system includes 28 V 250 A starter/generator on each engine and 39 Ah Ni/Cd battery. Hydraulic system, pressure 120 bars (1,750 lb/sq in), for operation of landing gear. Separate hydraulic system for brakes. Optional Goodrich pneumatic de-icing of wings and tail unit, and electric windscreen de-icing.

AVIONICS: Standard Honeywell Silver Crown; Gold Crown optional.
Comms: Dual Honeywell transceivers.
Radar: Honeywell RDR 2000 weather radar optional. Maritime surveillance radar as described under Current Versions.
Flight: Dual ADF and marker beacon receiver. Autopilot optional.
Instrumentation: Provision for equipment to FAR Pt 135A standards, including dual controls and instrumentation for co-pilot.

EQUIPMENT: Optional cargo interior includes heavy-duty sidewalls, utility floorboards, cabin floodlighting and cargo restraint nets. Optional pylons under each wing for carriage of stores including gun or rocket pods and airdroppable SAR equipment.

DIMENSIONS, EXTERNAL:
Wing span	15.08 m (49 ft 5½ in)
Wing aspect ratio	9.7
Length overall	11.89 m (39 ft 0¼ in)
Height overall	4.01 m (13 ft 2 in)
Tailplane span	5.87 m (19 ft 3 in)
Wheel track	4.28 m (14 ft 0½ in)
Wheelbase	3.81 m (12 ft 6 in)
Propeller diameter	2.36 m (7 ft 9 in)
Cabin door: Height	1.27 m (4 ft 2 in)
Width	0.58 m (1 ft 10¾ in)
Cargo double door (optional):	
Total width	1.24 m (4 ft 1 in)

DIMENSIONS, INTERNAL:
Cabin (incl flight deck): Length	5.71 m (18 ft 8¾ in)
Max width	1.42 m (4 ft 8 in)
Max height	1.31 m (4 ft 3¼ in)
Min height (at rear)	1.21 m (3 ft 11½ in)
Width of aisle	0.29 m (11½ in)
Volume	8.6 m³ (305 cu ft)
Baggage compartment (nose):	
Length	2.00 m (6 ft 6¾ in)
Volume	0.74 m³ (26.0 cu ft)
Nacelle lockers: Length	1.55 m (5 ft 1 in)
Width	0.73 m (2 ft 4¾ in)
Baggage volume: total, internal	2.2 m³ (79 cu ft)
incl cargo pod	3.5 m³ (124 cu ft)

AREAS:
Wings, gross	23.48 m² (252.8 sq ft)
Ailerons (total)	1.36 m² (14.64 sq ft)
Trailing-edge flaps	3.98 m² (42.84 sq ft)
Fin	4.05 m² (43.59 sq ft)
Rudder, incl tab	1.50 m² (16.15 sq ft)
Tailplane	5.81 m² (62.54 sq ft)
Elevators, incl tabs	1.66 m² (17.87 sq ft)

WEIGHTS AND LOADINGS:
Standard empty weight	2,283 kg (5,033 lb)
Max payload	2,219 kg (4,892 lb)
Max fuel	1,444 kg (3,183 lb)
Max ramp weight	4,502 kg (9,925 lb)
Max T-O and landing weight	4,468 kg (9,850 lb)
Max zero-fuel weight	3,856 kg (8,500 lb)
Max wing loading	190.3 kg/m² (38.97 lb/sq ft)
Max power loading	5.99 kg/kW (9.85 lb/shp)

PERFORMANCE:
Max operating Mach No. (M_{MO})	0.52
Max operating speed (V_{MO})	229 kt (424 km/h; 263 mph) IAS
Max cruising speed	246 kt (455 km/h; 283 mph)
Econ cruising speed	200 kt (370 km/h; 230 mph)
Stalling speed:	
wheels and flaps up	94 kt (174 km/h; 108 mph) IAS
wheels and flaps down	81 kt (150 km/h; 93 mph) IAS

Reims F406 flight deck

REIMS AVIATION to ROBIN—AIRCRAFT: FRANCE

Max rate of climb at S/L	564 m (1,850 ft)/min
Rate of climb at S/L, OEI	121 m (397 ft)/min
Service ceiling	9,145 m (30,000 ft)
Service ceiling, OEI	4,935 m (16,200 ft)
T-O run	526 m (1,725 ft)
T-O to 15 m (50 ft)	803 m (2,635 ft)
Landing from 15 m (50 ft), without reverse pitch	674 m (2,215 ft)
Range with max fuel, at max cruising speed, 45 min reserves	1,153 n miles (2,135 km; 1,327 miles)
g limits	+3.6/−1.44

OPERATIONAL NOISE LEVELS:
Flyover   72.0 EPNdB

UPDATED

REIMS FUTURE CARGO AIRCRAFT

At the Paris Air Show in June 1999, Reims CEO Jean-Paul Chaufour revealed plans to develop, or produce under licence, a new cargo aircraft with a 50 to 60 m³ (1,766 to 2,119 cu ft), 4,000 to 5,000 kg (8,818 to 11,023 lb) cargo hold and range of 378 to 1,080 n miles (700 to 2,000 km; 435 to 1,243 miles). The proposed aircraft would be manufactured at Vatry.

VERIFIED

ROBIN

ROBIN AVIATION

1 route de Troyes, F-21121 Darois
Tel: (+33 3) 80 44 20 50
Fax: (+33 3) 80 35 60 80
Telex: 350 818 ROBIN F
NON-EXECUTIVE CHAIRMAN: Philippe Corne
CEO AND GENERAL MANAGER: Guy Pellissier
SALES MANAGER, EUROPE: Pierre Pelletier

Formed October 1957 as Centre Est Aéronautique; name changed to Avions Pierre Robin 1969; acquired July 1988 by Compagnie Française Chaufour Investissement (CFCI) and incorporated into Aéronautique Service group with Robin SA (after-sales support company of Dijon Val-Suzon); Pierre Robin left company 1990. Total of 3,497 aircraft produced by December 1998, including 48 (and seven – of 35 on order – Sperwer UAVs for SAGEM) in 1998. Factory area 11,500 m² (123,785 sq ft); workforce 90.

Former Robin subdivisions, Constructions Aéronautiques de Bourgogne and Aéronautique Finance et Services, are now separate companies. The first-mentioned is responsible for the construction of Robin aircraft, as explained in the entry for Aéronautique Service.

UPDATED

ROBIN HR 200/120B and 160

TYPE: Primary prop trainer/sportplane.
PROGRAMME: New production version, since 1993, of original Robin HR 200/120B, which had first flown in 1971 and was built through much of 1970s (see 1977-78 *Jane's*); incorporates minor modifications.
CURRENT VERSIONS: **Robin HR 200/120B:** *As described.*
Robin HR 200/160: Power plant as Robin 2160, but with prominent exhaust below cabin and retaining Robin 200's rudder. Available from 1999.
CUSTOMERS: Total of 180 built between 1971 and late 2000, including 109 in first series. Combined 310 HR 200/R 2160s built by late 2000.
COSTS: FFr420,550 with Package 1 avionics; FFr497,785 with Package 2.
DESIGN FEATURES: Typical Robin low-wing, tricycle landing gear tourer/trainer. Constant-chord, unswept wings and tailplane, latter all-moving. Relaunched version has new instrument panel, adjustable seats, new engine cowlings and propeller spinner, and improved anti-corrosion treatment.
Wing section NACA 64A515 (mod); dihedral 6° 18′ from roots; incidence 6°; no sweepback.
FLYING CONTROLS: Manual. Conventional rudder and cable-actuated Frise-type ailerons. One-piece all-moving tailplane with trim and anti-balance tabs; electrically actuated trailing-edge slotted flaps.
STRUCTURE: All-metal; aluminium alloy stressed skin and ribs.
LANDING GEAR: Non-retractable tricycle type, with single wheel on each unit; nosewheel leg offset to starboard, steered by rudder bar; streamline leg and wheel fairings; damped tailskid; hydraulic disc brakes and parking brake. All three wheels and tyres are size 380×150.
POWER PLANT: One 88.0 kW (118 hp) Textron Lycoming O-235-L2A flat-four engine, driving a two-blade fixed-pitch propeller. Fuel capacity 120 litres (31.7 US gallons; 26.4 Imp gallons). Auxiliary tanks optional.
ACCOMMODATION: Pilot and passenger side by side under forward-sliding jettisonable canopy with anti-glare tint; dual stick controls; dual left-hand throttles; dual toe brakes.
SYSTEMS: Cabin ventilated and heated, with windscreen defrosting standard. Electrical system includes 12 V 32 Ah battery, 12 V 50 A alternator and starter.
AVIONICS: Customer selection; following are available.
Comms: Honeywell KY 97 or KX 155-38 transceivers; SPA 400 intercom; dual PTT buttons; KT 6A transponder with ACK 30 height encoder.
Flight: KI 208 VOR, KY 97 or KX 155-38 nav/com. Flight hour recorder.
Instrumentation: Vacuum-driven gyro horizon and direction indicator (engine-driven pump), electric turn and slip indicator, magnetic compass, rate of climb indicator and rest of blind-flying instruments.
EQUIPMENT: Navigation lights, strobe lights, landing light.
DIMENSIONS, EXTERNAL:

Wing span	8.33 m (27 ft 4 in)
Wing chord, constant	1.50 m (4 ft 11 in)
Wing aspect ratio	5.6
Length overall	6.35 m (20 ft 10 in)
Height overall	1.94 m (6 ft 4½ in)
Tailplane span	2.64 m (8 ft 8 in)
Wheel track	2.88 m (9 ft 5½ in)
Wheelbase	1.465 m (4 ft 9½ in)

DIMENSIONS, INTERNAL:

Cabin max width	1.07 m (3 ft 6 in)

AREAS:

Wings, gross	12.50 m² (134.5 sq ft)
Ailerons, total	1.06 m² (11.41 sq ft)
Trailing-edge flaps, total	1.34 m² (14.42 sq ft)
Elevators, incl tabs	2.03 m² (21.85 sq ft)

WEIGHTS AND LOADINGS:

Weight empty	525 kg (1,157 lb)
Baggage capacity	35 kg (77 lb)
Max T-O weight	780 kg (1,719 lb)
Max wing loading	62.4 kg/m² (12.78 lb/sq ft)
Max power loading	8.86 kg/kW (14.57 lb/hp)

Robin HR 200/160 (one Textron Lycoming flat-four) *(Paul Jackson/Jane's)*   2001/0093691

Instrument panel of the Robin HR 200

PERFORMANCE:

Cruising speed:	
at 75% power	115 kt (213 km/h; 132 mph)
at 65% power	105 kt (194 km/h; 121 mph)
Stalling speed	52 kt (96 km/h; 60 mph)
Max rate of climb at S/L	235 m (770 ft)/min
Service ceiling	3,900 m (12,800 ft)
T-O and landing run	230 m (755 ft)
T-O to 15 m (50 ft)	510 m (1,675 ft)
Landing from 15 m (50 ft)	445 m (1,460 ft)
Range	566 n miles (1,050 km; 652 miles)
Endurance	4 h 35 min

UPDATED

ROBIN R 2160

TYPE: Two-seat lightplane.
PROGRAMME: Certified in France in mid-1978 and in USA (FAR Pt 23 Aerobatic and Utility category) 15 November 1982. Some aircraft assembled in Canada (1983-85). Production then ceased, but restarted in France January 1994.
CURRENT VERSIONS: **R 2160 D:** Baseline version, *as described below.*
R 2160i: Certified 8 July 1998, with 119 kW (160 hp) Textron Lycoming AEIO-320-D2B fuel-injected flat-four. Three delivered by end of 1998, including two to CATC training college in Thailand.
R 2120U: One trials aircraft, F-WZZX, built early 2000. No further details revealed.
CUSTOMERS: Total of some 130 sold in Europe, Australia, Canada and USA by late 2000, including 97 in first series.
DESIGN FEATURES: Generally as Robin HR 200, but with extended rudder chord, underfin and other detail changes.
Wing section NACA 23015; dihedral 6° 20′; incidence 3° at root.
FLYING CONTROLS: Manual. Conventional fully balanced slotted ailerons and horn-balanced rudder without tabs. All-moving tailplane with anti-balance tabs each side. Slotted flaps.
STRUCTURE: Aluminium alloy wing spars and skinning; semi-monocoque fuselage.
LANDING GEAR: Non-retractable tricycle type with fairing, oleo-pneumatic shock-absorber and tyre (380 × 150 mm) on each leg. Cleveland disc brakes on mainwheels. Nosewheel steering through rudder bar.

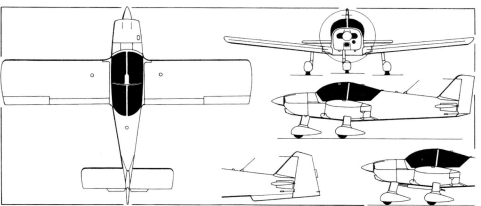

Robin HR 200/120B two-seat trainer, with additional scrap views of HR 200/160 forward fuselage and R 2160 empennage *(Paul Jackson/Jane's)*   2000/0062966

FRANCE: AIRCRAFT—ROBIN

Robin R 2160i (Textron Lycoming 0-320 flat-four) *(Paul Jackson/Jane's)*

POWER PLANT: One 119 kW (160 hp) Textron Lycoming O-320-D2A flat-four engine, driving a Sensenich 74DM6S5-2-64 two-blade fixed-pitch propeller. Provisional design studies completed for installation of 134 to 149 kW (180 to 200 hp) engine. Christen inverted oil system standard. Fuselage fuel tank, capacity 120 litres (31.7 US gallons; 26.4 Imp gallons). If operated in Utility category, optional fuel tank of 160 litres (42.3 US gallons; 35.2 Imp gallons).
ACCOMMODATION: Two seats side by side.
SYSTEMS: 12 V 24 Ah electrical system.
AVIONICS: *Flight:* VOR, ADF, ILS, GPS and other items, at customer's choice.
Instrumentation: Blind-flying panel optional.

DIMENSIONS, EXTERNAL:
Wing span	8.33 m (26 ft 4 in)
Wing chord, constant	1.56 m (5 ft 1¼ in)
Wing aspect ratio	5.3
Length overall	7.10 m (23 ft 3½ in)
Height overall	2.13 m (7 ft 0 in)
Tailplane span	3.03 m (9 ft 11¼ in)
Wheel track	2.91 m (9 ft 6½ in)
Wheelbase	1.44 m (4 ft 8½ in)
Propeller diameter	1.88 m (6 ft 2 in)

DIMENSIONS, INTERNAL:
Cabin: Max width	1.07 m (3 ft 6 in)
Height (seat cushion to canopy):	
R 2160	0.92 m (3 ft 0¼ in)

AREAS:
Wings, gross	13.01 m² (140.0 sq ft)

WEIGHTS AND LOADINGS:
Weight empty	550 kg (1,212 lb)
Max T-O weight: Aerobatic	800 kg (1,764 lb)
Utility	900 kg (1,984 lb)
Max baggage weight	40 kg (88 lb)
Max wing loading: Aerobatic	61.5 kg/m² (12.59 lb/sq ft)
Utility	69.2 kg/m² (14.17 lb/sq ft)
Max power loading: Aerobatic	6.71 kg/kW (11.02 lb/hp)
Utility	7.55 kg/kW (12.40 lb/hp)

PERFORMANCE (O-320 engine at Aerobatic max T-O weight):
Never-exceed speed (V_{NE})	180 kt (333 km/h; 207 mph)
Max level speed	138 kt (256 km/h; 159 mph)
Cruising speed:	
at 75% power	131 kt (241 km/h; 151 mph)
at 65% power	120 kt (222 km/h; 138 mph)
Stalling speed: flaps up	52 kt (96 km/h; 60 mph)
flaps down	46 kt (86 km/h; 53 mph)
Max rate of climb at S/L	314 m (1,030 ft)/min
Service ceiling (30.5 m; 100 ft/min rate of climb)	4,575 m (15,000 ft)
T-O and landing run	230 m (755 ft)
T-O to 15 m (50 ft)	410 m (1,345 ft)
Landing from 15 m (50 ft)	425 m (1,395 ft)
Range at 75% power:	
standard fuel	363 n miles (672 km; 418 miles)
optional fuel	484 n miles (896 km; 557 miles)
g limits	+6/−3

UPDATED

ROBIN DR 400 DAUPHIN
English name: Dolphin

TYPE: Four-seat lightplane.
PROGRAMME: First flight original DR 400 Petit Prince 15 May 1972; French and UK certification 1977; DR 400 Dauphin introduced 1979; improvements introduced 1988 and 1993.
CURRENT VERSIONS: **DR 400/120 Dauphin 2+2:** Production version with 83.5 kW (112 hp) engine, to carry two adults and two children.
  DR 400/125i: Version of Dauphin 2+2 with fuel-injected 93.2 kW (125 hp) Teledyne Continental IO-240 driving three-bladed Mühlbauer constant-speed propeller, diameter 1.70 m (5 ft 7 in); prototype (F-WNNK) first flown 1995 and made its public debut at the Paris Air Show in June 1995; engine/propeller combination improves take-off, climb and cruising performance and reduces noise levels; one delivered in 1996. Weights and performance data in 2000-01 and earlier editions. No evidence of production.
  DR 400/140B Dauphin 4: Full four-seater with 119 kW (160 hp) engine.
  DR 400/160 and **DR 400/180:** Described separately.
CUSTOMERS: Some 1,775 of DR 400 series built by late 2000, including some 30 registered in that year.
DESIGN FEATURES: Generally as for HR 200.
  Wing section NACA 23013.5 modified with leading-edge droop; centre panels parallel chord, slight twist; outer panels tapered with dihedral 14°; twist −6°.
FLYING CONTROLS: Manual. Conventional ailerons and rudder. All-moving tailplane with trimmable anti-balance tab on each side; plain flaps.
STRUCTURE: All-wood; single box spar with ribs threaded over box; plywood-covered leading-edge box; fabric covering elsewhere; fuselage plywood-covered; flaps all-metal and interchangeable; ailerons interchangeable.
LANDING GEAR: Non-retractable tricycle type, with oleo-pneumatic shock-absorbers and hydraulically actuated disc brakes. All three wheels and tyres are size 380×150, pressure 1.57 bars (23 lb/sq in) on nose unit, 1.77 bars (26 lb/sq in) on main units. Nosewheel steerable via rudder bar. Fairings over all three legs and wheels. Tailskid with damper. Toe brakes and parking brake.
POWER PLANT: *Dauphin 2+2:* One 83.5 kW (112 hp) Textron Lycoming O-235-L2A flat-four engine, driving a Sensenich 72 CKS 6-0-56 two-blade fixed-pitch metal propeller, or Hoffmann two-blade wooden propeller.
  Dauphin 4: One Textron Lycoming O-320-D2A flat-four engine developing 104 kW (140 hp) at 2,300 rpm and 119 kW (160 hp) at 2,700 rpm.
  Both versions have fuel tank in fuselage, capacity 109 litres (28.8 US gallons; 24.0 Imp gallons) with filler port on left side; Dauphin 4 has optional 51 litre (13.5 US gallon; 11.2 Imp gallon) auxiliary tank, with filler port on right side. Oil capacity 5.7 litres (1.5 US gallons; 1.25 Imp gallons).
ACCOMMODATION: Enclosed cabin, with seats for three or four persons. Maximum weight of 154 kg (340 lb) on front pair and 136 kg (300 lb), including baggage, at rear in Dauphin 2+2. Additional 55 kg (121 lb) of disposable load in Dauphin 4. Access via forward-sliding jettisonable transparent canopy. Dual controls standard. Cabin heated and ventilated. Baggage compartment with internal access.
SYSTEMS: Standard equipment includes a 12 V 50 A alternator, 12 V 32 Ah battery, electric starter, audible stall warning and windscreen de-icing.
AVIONICS: Radio, blind-flying equipment, and navigation, landing and anti-collision lights, to customer's requirements.

DIMENSIONS, EXTERNAL:
Wing span	8.72 m (28 ft 7¼ in)
Wing chord:	
centre-section, constant	1.71 m (5 ft 7½ in)
at tip	0.90 m (2 ft 11½ in)
Wing aspect ratio	5.6
Length overall	6.96 m (22 ft 10 in)
Height overall	2.23 m (7 ft 3¾ in)
Tailplane span	3.20 m (10 ft 6 in)
Wheel track	2.60 m (8 ft 6¼ in)
Wheelbase	5.20 m (17 ft 0¾ in)
Propeller diameter	1.78 m (5 ft 10 in)

DIMENSIONS, INTERNAL:
Cabin: Length	1.62 m (5 ft 3¾ in)
Max width	1.10 m (3 ft 7¼ in)
Max height	1.23 m (4 ft 0½ in)
Baggage volume	0.39 m³ (13.8 cu ft)

AREAS:
Wings, gross	13.60 m² (146.4 sq ft)
Ailerons, total	1.15 m² (12.38 sq ft)
Flaps, total	0.70 m² (7.53 sq ft)
Fin	0.61 m² (6.57 sq ft)
Rudder	0.63 m² (6.78 sq ft)
Horizontal tail surfaces, total	2.88 m² (31.00 sq ft)

WEIGHTS AND LOADINGS:
Weight empty, equipped: 2+2	550 kg (1,212 lb)
4	580 kg (1,279 lb)
Max baggage: 2+2, 4	40 kg (88 lb)
Max T-O and landing weight: 2+2	900 kg (1,984 lb)
4	1,000 kg (2,205 lb)
Max wing loading: 2+2	66.2 kg/m² (13.56 lb/sq ft)
4	73.5 kg/m² (15.05 lb/sq ft)
Max power loading: 2+2	10.78 kg/kW (17.71 lb/hp)
4	8.38 kg/kW (13.78 lb/hp)

PERFORMANCE:
Never-exceed speed (V_{NE}):	
2+2, 4	166 kt (308 km/h; 191 mph)
Max level speed at S/L:	
2+2	130 kt (241 km/h; 150 mph)
4	143 kt (265 km/h; 165 mph)
Max cruising speed: 2+2	116 kt (215 km/h; 133 mph)
4	117 kt (216 km/h; 134 mph)
Stalling speed, flaps down:	
2+2	45 kt (82 km/h; 51 mph)
4	47 kt (87 km/h; 54 mph)
Max rate of climb at S/L: 2+2	183 m (600 ft)/min
4	264 m (865 ft)/min
Service ceiling: 2+2	3,660 m (12,000 ft)
4	4,265 m (14,000 ft)
T-O run: 2+2	235 m (775 ft)
4	245 m (805 ft)
T-O to 15 m (50 ft): 2+2	535 m (1,755 ft)
4	485 m (1,595 ft)
Landing from 15 m (50 ft): 2+2	460 m (1,510 ft)
4	470 m (1,545 ft)
Landing run: 2+2	200 m (660 ft)
4	220 m (725 ft)
Range with standard fuel at max cruising speed, no reserves: 2 + 2	500 n miles (926 km; 575 miles)
4	464 n miles (860 km; 534 miles)
Range, Dauphin 4 with optional fuel at max cruising speed, no reserves	740 n miles (1,370 km; 851 miles)

UPDATED

ROBIN DR 400/160 MAJOR

TYPE: Four-seat lightplane.
PROGRAMME: First flight of original DR 400 Chevalier 29 June 1972; certified France and UK same year; Major introduced 1980.
CUSTOMERS: See main DR 400 entry; total 140 delivered by December 1998.
DESIGN FEATURES: Main differences from Dauphin (see preceding entry) are additional rear cabin window on each side, external baggage compartment door on port side and extended wingroot leading-edges to house additional fuel tanks.
Differences from Dauphin listed below:
POWER PLANT: One 119 kW (160 hp) Textron Lycoming O-320-D flat-four engine, driving a Sensenich two-blade metal fixed-pitch propeller. Fuel tank in fuselage, capacity 110 litres (29.1 US gallons; 24.2 Imp gallons), and two tanks in wingroot leading-edges, giving total capacity of

Robin DR 400/140 Dauphin 4 *(Paul Jackson/Jane's)*

190 litres (50.2 US gallons; 41.8 Imp gallons), of which 182 litres (48.1 US gallons; 40.0 Imp gallons) are usable. Provision for auxiliary tank, raising total capacity to 240 litres (63.4 US gallons; 52.8 Imp gallons). Oil capacity 7.6 litres (2.0 US gallons; 1.7 Imp gallons).

ACCOMMODATION: Seating for four persons, on adjustable front seats (maximum load 154 kg; 340 lb total) and rear bench seat (maximum load 154 kg; 340 lb total). Forward-sliding transparent canopy, but port rear window/door provides outside access to baggage area. Up to 40 kg (88 lb) of baggage can be stowed aft of rear seats when four occupants are carried.

DIMENSIONS, EXTERNAL:
Propeller diameter	1.83 m (6 ft 0 in)
Baggage door: Height	0.47 m (1 ft 6½ in)
Width	0.55 m (1 ft 9½ in)

AREAS:
Wings, gross	14.20 m² (152.8 sq ft)

WEIGHTS AND LOADINGS:
Weight empty, equipped	598 kg (1,318 lb)
Max T-O and landing weight	1,050 kg (2,315 lb)
Max wing loading	74.0 kg/m² (15.15 lb/sq ft)
Max power loading	8.81 kg/kW (14.47 lb/hp)

PERFORMANCE:
Never-exceed speed (V_{NE})	166 kt (308 km/h; 191 mph)
Max level speed at S/L	146 kt (271 km/h; 168 mph)
Max cruising speed at 75% power at 2,440 m (8,000 ft)	132 kt (245 km/h; 152 mph)
Econ cruising speed at 65% power at 3,200 m (10,500 ft)	130 kt (241 km/h; 150 mph)
Stalling speed: flaps up	56 kt (103 km/h; 64 mph)
flaps down	50 kt (93 km/h; 58 mph)
Max rate of climb at S/L	255 m (836 ft)/min
Service ceiling	4,115 m (13,500 ft)
T-O run	295 m (970 ft)
T-O to 15 m (50 ft)	590 m (1,940 ft)
Landing from 15 m (50 ft)	545 m (1,790 ft)
Landing run	250 m (820 ft)
Range at econ cruising speed, no reserves:	
standard fuel	825 n miles (1,530 km; 950 miles)
optional fuel	1,026 n miles (1,900 km; 1,180 miles)

UPDATED

Robin DR 400/160, showing additional rear transparencies and port rear door to baggage area *(Paul Jackson/Jane's)*

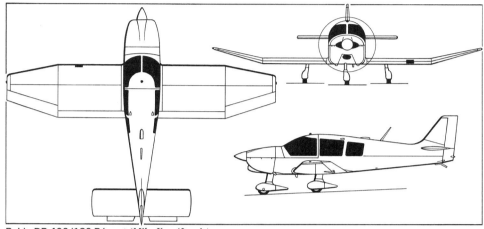

Robin DR 400/180 Régent *(Mike Keep/Jane's)*

ROBIN DR 400/180 REGENT

TYPE: Four-seat lightplane.
PROGRAMME: First flight 27 March 1972; certified 10 May 1972.
CUSTOMERS: See main DR 400 entry; total 363 delivered by early 1999.
DESIGN FEATURES: Generally as for DR 400 series; two-seat rear bench.

Differences from DR 400/160 listed below:

POWER PLANT: One 134 kW (180 hp) Textron Lycoming O-360-A flat-four engine. Fuel tankage unchanged.
ACCOMMODATION: Basically as for DR 400/160. Baggage capacity 60 kg (132 lb).

DIMENSIONS, EXTERNAL:
Propeller diameter	1.93 m (6 ft 4 in)

WEIGHTS AND LOADINGS:
Weight empty, equipped	610 kg (1,345 lb)
Max T-O and landing weight	1,100 kg (2,425 lb)
Max wing loading	77.48 kg/m² (15.87 lb/sq ft)
Max power loading	8.21 kg/kW (13.47 lb/hp)

PERFORMANCE (at max T-O weight):
Max level speed at S/L	150 kt (278 km/h; 173 mph)
Max cruising speed at 75% power at 2,285 m (7,500 ft)	140 kt (260 km/h; 162 mph)
Econ cruising speed at 60% power at 3,660 m (12,000 ft)	132 kt (245 km/h; 152 mph)
Stalling speed: flaps up	57 kt (105 km/h; 65 mph)
flaps down	52 kt (95 km/h; 59 mph)
Max rate of climb at S/L	252 m (825 ft)/min

Robin DR 400/180 Régent (Textron Lycoming O-360-A engine) *(Paul Jackson/Jane's)*

Service ceiling	4,720 m (15,475 ft)	Range, no reserves:	
T-O run	315 m (1,035 ft)	standard fuel	785 n miles (1,453 km; 903 miles)
T-O to 15 m (50 ft)	610 m (2,000 ft)	optional fuel	975 n miles (1,805 km; 1,122 miles)
Landing from 15 m (50 ft)	530 m (1,740 ft)		
Landing run	249 m (820 ft)		

UPDATED

Robin Remo 180 glider tug *(Paul Jackson/Jane's)*

Robin DR 500/200i Président standard IFR instrument panel *(Paul Jackson/Jane's)*

ROBIN DR 400/180R REMO 180
TYPE: Glider tug.
PROGRAMME: First flight and certification 1972 as DR 400/180R (*Remorqueur*, abbreviated Remo); flown 1985 with Porsche PFM 3200 engine as DR 400RP or Remo 212; became first Porsche-powered aircraft to be certified; Remo 212 production ceased 1990 after 29 had been built (details in 1991-92 *Jane's*).
CUSTOMERS: See main DR 400 entry; total 311 Remos delivered by December 1998.
DESIGN FEATURES: Same as Régent, except no external baggage door, and baggage compartment covered in transparent Plexiglas to maximise rearward view; towing hook under tail; Dauphin wing (13.60 m²; 146.4 sq ft) without extended wingroot leading-edges.
Differences from DR 400/180 listed below:
POWER PLANT: One 134 kW (180 hp) Textron Lycoming O-360-A flat-four engine, driving (for glider towing) a Sensenich 76 EM 8S5 058 or Hoffmann HO-27-HM-180/138 two-blade, fixed-pitch propeller. For touring, a coarser pitch Sensenich 76 EM 8S5 064 propeller of same diameter can be fitted. Fuel 109 litres (28.8 US gallons; 24.0 Imp gallons) normal; additional 51 litre (13.5 US gallon; 11.2 Imp gallon tank optional).
WEIGHTS AND LOADINGS:
Weight empty, equipped 592 kg (1,305 lb)
Max T-O and landing weight 1,000 kg (2,205 lb)
Max wing loading 73.5 kg/m² (15.05 lb/sq ft)
Max power loading 7.46 kg/kW (12.25 lb/hp)
PERFORMANCE (A: at max T-O weight, B: with 600 kg; 1,323 lb glider and max normal fuel):
Max level speed 146 kt (270 km/h; 168 mph)
Cruising speed at 70% power at 2,440 m (8,000 ft)
124 kt (230 km/h; 143 mph)
Stalling speed, flaps down: A 47 kt (87 km/h; 54 mph)
B 45 kt (84 km/h; 52 mph)
Max rate of climb at S/L: A 336 m (1,100 ft)/min
B 210 m (690 ft)/min
Service ceiling 6,100 m (20,000 ft)

Robin DR 500/200i Président *(Paul Jackson/Jane's)* 2001/0093695

T-O run: A 205 m (675 ft)
B 300 m (985 ft)
T-O to 15 m (50 ft): A 400 m (1,315 ft)
B 535 m (1,755 ft)
Landing from 15 m (50 ft) 470 m (1,545 ft)
Range, no reserves:
max normal fuel 426 n miles (788 km; 490 miles)
with supplementary fuel
610 n miles (1,129 km; 702 miles)
UPDATED

ROBIN DR 500 PRESIDENT
TYPE: Four-seat lightplane.
PROGRAMME: Prototype (F-WZZY), then known as DR 400/200i, first flew 5 June 1997; revealed at Paris Air Show, 13 June. Renamed as above but certified as DR 400/500.
CURRENT VERSIONS: **DR 500/200i Président**: As described.
DR 500 Super Régent: 180 hp engine, lacks fuel injection. None built.
CUSTOMERS: Deliveries began in 1998 with eight aircraft. Total 25 built by mid-2000.
DESIGN FEATURES: Uprated version of DR 400/180 Régent, including that version's 14.20 m² (152.8 sq ft) wing; 149 kW (200 hp) Textron Lycoming IO-360-A1B6 engine driving a Hartzell F7 666A-2 constant-speed propeller; width of cabin increased by 10 cm (4 in) and headroom by same amount; IFR instrumentation as standard; Honeywell avionics, including KAP 140 autopilot. Fuel capacity increased. Flaps electrically operated.
Differences from DR 400/180 listed below:
LANDING GEAR: Nosewheel size 15×6.00-5, maximum pressure 1.8 bars (26 lb/sq in); mainwheel size 330×150, pressure 2.0 bars (29 lb/sq in).
POWER PLANT: One 149 kW (200 hp) Textron Lycoming IO-360-A1B6 driving a Hartzell two-blade, constant-speed metal propeller. Fuel contained in four tanks: two, each of capacity 40 litres (10.6 US gallons; 8.8 Imp gallons), in wingroots; main tank, capacity 105 litres (27.7 US gallons; 23.1 Imp gallons), below cabin floor; and fourth tank, capacity 90 litres (23.8 US gallons; 19.8 Imp gallons), below baggage compartment. Total capacity of 275 litres (72.6 US gallons; 60.5 Imp gallons).
ACCOMMODATION: Four adults, or two plus three children. Baggage shelf at rear of seats; upward-opening rear window on port side doubles as baggage door.
DIMENSIONS, EXTERNAL:
Length overall 7.05 m (23 ft 1½ in)
Propeller diameter 1.93 m (6 ft 4 in)
DIMENSIONS, INTERNAL:
Cabin max width 1.20 m (3 ft 11¼ in)
WEIGHTS AND LOADINGS:
Weight empty 650 kg (1,433 lb)
Baggage capacity 60 kg (132 lb)
Max T-O weight 1,150 kg (2,535 lb)
Max wing loading 81.0 kg/m² (16.59 lb/sq ft)
Max power loading 7.72 kg/kW (12.68 lb/hp)
PERFORMANCE:
Max level speed 147 kt (272 km/h; 169 mph)
Cruising speed 143 kt (265 km/h; 165 mph)
Max rate of climb at S/L 305 m (1,000 ft)/min
T-O to 15 m (50 ft) 450 m (1,476 ft)
Range, no reserves:
at 75% power 995 n miles (1,842 km; 1,145 miles)
at 65% power 1,090 n miles (2,018 km; 1,254 miles)
UPDATED

ROBIN R 1180 AIGLON
The mooted relaunch of Aiglon production has not been confirmed. Details last appeared in the 2000-01 edition.
UPDATED

SOGEPA
SOCIETE DE GESTION DE PARTICIPATIONS AERONAUTIQUES
Details of this holding company's earlier operations last appeared in the 2000-01 *Jane's*. It is currently used to manage the French government's 50 per cent interest (with Lagardère and private investors contributing the remainder) in France's holding in EADS.
UPDATED

GEORGIA

TASA
TBILISI AVIATION STATE ASSOCIATION
ulitsa Khmelnitski 181B, 380036 Tbilisi
Tel: (+995 32) 98 53 90 and 70 88 38
Fax: (+995 32) 96 43 69 and 98 25 51
GENERAL DIRECTOR: Pantiko Tordia

TASA began aircraft production in 1941, with the LaGG-3 fighter, and progressed through La-5, Yak-3, Yak-15, Yak-17/17UTI, Yak-23, MiG-15 and MiG-17 to building 1,677 MiG-21U/UMs (for which it offers an upgrade programme) between 1957 and 1984. The plant also began Su-25/25U production in 1978 and has built 875 basic single-seat versions, including 50 Su-25BM target-tugs. In total, TASA has built over 8,070 manned aircraft, 2,555 UAVs and over 36,000 AAMs (30,000 R-60/AA-8s and 6,000 R-73/AA-11s). Current manufacture is of the Su-25UB two-seat trainer and its Su-25T (Su-39) derivative. Production is also under way of a batch of 20 Yak-58 light transports.
Privatisation plans for TASA are proceeding and the reorganised company will be known as Tbilisi Aerospace Manufacturing. The 250,000 m² (2,691,000 sq ft) plant employed 15,000 personnel at its peak, but currently has a workforce of about 4,000. A 1999 alliance with Kelowna Flightcraft of Canada (see *Jane's Aircraft Upgrades*) provides for TASA to assist with production of materials for the Convair 5800 conversion programme and eventually undertake the entire programme, if demand warrants. Re-engining of Yak-40s and avionics upgrades for Su-25s are also under consideration by the partnership.
UPDATED

GERMANY

AEROSTYLE
AEROSTYLE ULTRALEICHT FLUGZEUGE GmbH
Norderreeg 2, D-25852 Bordelum
Tel: (+49 4671) 93 13 93
Fax: (49 4671) 93 13 94
e-mail: rmagnussen@t-online.de
DIRECTOR: Ralf Magnussen

Production is now under way of the company's first product, the Breezer.
UPDATED

AEROSTYLE BREEZER
TYPE: Side-by-side ultralight.
PROGRAMME: Prototype (D-MOOV) exhibited unflown at Aero '99 at Friedrichshafen in April 1999; first flight was made in December 1999. Noise tests under way in mid-2000; German certification was then anticipated in August or September 2000.
CUSTOMERS: Two complete aircraft and three kits sold by mid-2000, representing capacity for that year.
COSTS: €43,500 excluding taxes (2000).
DESIGN FEATURES: Low, constant-chord wing, with wingtips upturned at rear. Sweptback fin. Wing aerofoil section NACA 4414.

AEROSTYLE to AIR LIGHT—AIRCRAFT: GERMANY

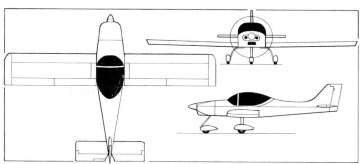

Aerostyle Breezer all-metal two-seat ultralight aircraft *(James Goulding/Jane's)*

Prototype Aerostyle Breezer

FLYING CONTROLS: Conventional and manual. Horn-balanced elevators and rudder; trim tab in port elevator; half-span Fowler flaps.
STRUCTURE: Mainly riveted aluminium, with glass fibre for some non-structural fairings.
LANDING GEAR: Non-retractable tricycle type with composites cantilever main gear legs and composites trailing link nose leg.
POWER PLANT: One 67.1 to 74.6 kW (90 to 100 hp) fuel-injected BMW Take Off engine driving a three-blade ground-adjustable Neuform propeller. Fuel capacity 55 litres (14.5 US gallons; 12.1 Imp gallons).

DIMENSIONS, EXTERNAL:
Wing span 6.40 m (21 ft 0 in)
Length overall 8.80 m (28 ft 10½ in)
Height overall 2.15 m (7 ft 0¾ in)
AREAS:
Wings, gross 12.00 m² (129.2 sq ft)
WEIGHTS AND LOADINGS (estimated):
Weight empty 278 kg (613 lb)
Max T-O weight 450 kg (992 lb)
PERFORMANCE (estimated, A: 67.1 kW/90 hp engine; B: 74.6 kW/100 hp engine):
Never-exceed speed (V_{NE}):
  A, B 132 kt (245 km/h; 152 mph)
Max level speed: A 116 kt (215 km/h; 134 mph)
  B 121 kt (225 km/h; 140 mph)
Cruising speed: A 108 kt (200 km/h; 124 mph)
  B 113 kt (210 km/h; 130 mph)
Stalling speed: A, B 32 kt (58 km/h; 36 mph)
Max rate of climb at S/L 300 m (984 ft)/min
T-O and landing run 120 m (395 ft)
g limits +4/−2

UPDATED

AIR LIGHT

AIR LIGHT GmbH
Flugplatzstrasse 18, D-97437 Hassfurt
Tel: (+49 9521) 61 83 93
Fax: (+49 9521) 61 83 95
Web: http://www.air-light.de

The Romanian-developed WT01 Wild Thing lightplane is marketed by Air Light in Germany. A nosewheel version was introduced in 1999.

UPDATED

AIR LIGHT WILD THING

TYPE: Side-by-side ultralight/kitbuilt.
PROGRAMME: Developed by Aerostar (which see) in Romania; prototype registered YR-6107. Launched at Friedrichshafen Air Show, April 1997; certification August 1997. Initial production aircraft had Jabiru 2200 engines; alternatives available from 1998.
CURRENT VERSIONS: **WT01**: Baseline version; tailwheel.
  WT02: Nosewheel version introduced early 1999 (D-MWTE, 50th production aircraft, as demonstrator).
CUSTOMERS: Total of 50 built by April 1999.
COSTS: Quick-build kit, excluding engine, DM39,900 (2000). Power plant DM17,000 to DM26,000 extra. Factory-built DM77,500 (Jabiru 2200) to DM87,500 (Rotax 912 ULS), according to engine (2000). Nosewheel DM3,590 extra.
DESIGN FEATURES: Bears strong outward resemblance to Murphy Rebel (which see in Canadian section). Built by Aerostar in Romania for distribution out of Germany. Wings fold alongside fuselage for storage or transportation on aluminium trailer.

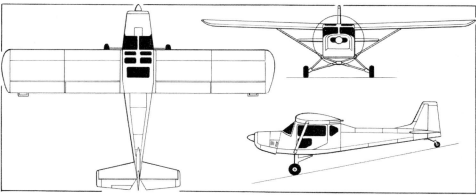

WT01 Air Light WT01 Wild Thing *(James Goulding/Jane's)*

FLYING CONTROLS: Conventional and manual elevators and ailerons pushrod-operated, rudder cable-operated. Horn-balanced tail surfaces; trim tab in port elevator; ground-adjustable tab on rudder. Flaps.
STRUCTURE: Main structure is metal monocoque, with fabric-covered ailerons and flaps.
LANDING GEAR: Non-retractable tailwheel type with steerable tailwheel via spring links to rudder horn. Mainwheel size 8.00-6, tailwheel size 200×50. Cable-operated disc brakes. Nosewheel (with repositioned mainwheel legs) optional.
POWER PLANT: One 59.7 kW (80 hp) Jabiru 2200, 59.6 kW (79.9 hp) Rotax 912 UL, 78.3 kW (105 hp) Mid West Hawk, 89.5 kW (120 hp) Jabiru 3300 or 73.5 kW (98.6 hp) Rotax 912 ULS driving a Junkers or Kremen three-blade wooden or carbon fibre propeller. Fuel contained in two wing tanks, combined capacity 80 litres (21.1 US gallons; 17.6 Imp gallons).
AVIONICS: Becker, Filser and Icom avionics to customer's choice.
EQUIPMENT: Junkers Magnum speed parachute emergency rescue system optional.
DIMENSIONS, EXTERNAL:
Wing span 9.20 m (30 ft 2¼ in)
Length overall 6.50 m (21 ft 4 in)
Height overall 1.92 m (6 ft 3½ in)
AREAS:
Wings, gross 13.94 m² (150.0 sq ft)
WEIGHTS AND LOADINGS:
Weight empty (typical) 300 kg (661 lb)
Max T-O weight 450 kg (992 lb)
PERFORMANCE (Jabiru 2200 engine):
Max level speed 86 kt (160 km/h; 99 mph)
Cruising speed 76 kt (140 km/h; 87 mph)
Stalling speed 32 kt (58 km/h; 36 mph)
Max rate of climb at S/L 152 m (500 ft)/min
T-O run 98 m (320 ft)
Landing run 44 m (145 ft)
Range with max fuel 432 n miles (800 km; 497 miles)

UPDATED

Air Light WT01 Wild Thing fitted with Mid West Hawk engine and Kremen wooden propeller *(Paul Jackson/Jane's)*

WT02 nosewheel version of Air Light Wild Thing, introduced in 1999 *(Paul Jackson/Jane's)*

AKAFLIEG MÜNCHEN

AKAFLIEG MÜNCHEN eV
Arcisstrasse 21, D-80333 München
Tel/Fax: (+49 89) 28 61 11
e-mail: akaflieg@lrz.tu-muenchen.de
Web: http://www.akaflieg.vo.tu-muenchen.de

Akademische Fliegergruppe München (Munich Academy Flying Group) formed in 1924 to provide students with theoretical and practical knowlege of aircraft design and construction; now affiliated to Munich Technical University. Over 20 types have been built; most have been sailplanes, but the latest venture is a powered design.

UPDATED

AKAFLIEG MÜNCHEN Mü30 SCHLACRO
TYPE: Aerobatic, two-seat sportplane.
PROGRAMME: Work began in 1985; originally designed for Porsche PFM 3200 engine, development of which was abandoned in 1992; name derived from glider tug (*schlepp*) and aerobatic (*acrobatic*); prototype (V-1/D-EKDF) displayed, then unflown, at Berlin in May 1998; first engine runs conducted on 8 October 1999 followed by first flight 16 June 2000.
DESIGN FEATURES: Low-wing monoplane.
FLYING CONTROLS: Conventional and manual. Plain flaps. Trim tabs in rudder and starboard elevator only.
STRUCTURE: Metal tube fuselage frame with carbon fibre and Kevlar skinning; all-carbon fibre single-spar wing; ailerons, rudder and elevator of aramid, carbon fibre and glass fibre composites.
LANDING GEAR: Steel cantilever sprung mainwheels, tyre size 15×6.0-6, with speed fairings; steerable tailwheel.
POWER PLANT: One 224 kW (300 hp) Textron Lycoming AEIO-540-L1B5D flat-six driving a Mühlbauer MTV-14-B-C/C190-17 four-blade, constant-speed propeller. Christensen inverted fuel and oil system. Fuel capacity 300 litres (79.2 US gallons; 66.0 Imp gallons).
ACCOMMODATION: Two in tandem beneath separate, single-piece canopies.

DIMENSIONS, EXTERNAL:
Wing span	8.82 m (28 ft 11¼ in)
Wing aspect ratio	6.5
Length overall, fuselage horizontal	7.36 m (24 ft 1¾ in)
Height overall, tail down	2.50 m (8 ft 2½ in)
Propeller diameter	1.90 m (6 ft 2¾ in)

AREAS:
Wings, gross	11.96 m² (128.7 sq ft)
Elevators	2.44 m² (26.26 sq ft)

WEIGHTS AND LOADINGS:
Weight empty	730 kg (1,609 lb)
Max T-O weight: Aerobatic	850 kg (1,873 lb)
Utility	1,050 kg (2,314 lb)
Max wing loading: Aerobatic	71.1 kg/m² (14.56 lb/sq ft)
Utility	87.8 kg/m² (17.98 lb/sq ft)
Max power loading: Aerobatic	3.80 kg/kW (6.24 lb/hp)
Utility	4.69 kg/kW (7.71 lb/hp)

PERFORMANCE (estimated, at Utility MTOW):
Max level speed	194 kt (360 km/h; 224 mph)
Cruising speed at 75% power at 4,000 m (13,120 ft)	178 kt (330 km/h; 205 mph)
Max rate of climb at S/L	840 m (2,756 ft)/min
T-O to 15 m (50 ft)	220 m (722 ft)
Range	647 n miles (1,200 km; 745 miles)
g limits	
at 850 kg (1,873 lb)	9.7
at 1,050 kg (2,314 lb)	8.0

UPDATED

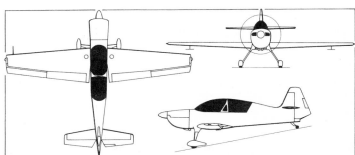

General arrangement of the AkaFlieg München Mü30 Schlacro (Paul Jackson/Jane's) 0022157

AkaFlieg München Mü30 Schlacro
2001/0099162

AQUILA

AQUILA TECHNISCHE ENTWICKLUNGEN GmbH
Dorfstrasse 47, D-14959 Schönhagen
Tel: (+49 33) 73 11 46 97
Fax: (+49 33) 73 11 46 98
e-mail: info@aquila-aero.com
Web: http://www.aquila-aero.com
DIRECTORS: Peter Grundhoff
 Alfred Schmiderer
 Markus Wagner
MARKETING MANAGER: Siegfried Dörfler

Company formed August 1996 to develop A210; employed 12 staff at Schönhagen aerodrome by April 2000 and expected workforce increase to 20 during 2001, following mid-2000 commissioning of new production hangar; capacity 50 aircraft per year.

UPDATED

AQUILA A210
TYPE: Two-seat lightplane.
PROGRAMME: Launched 1995; design, using CAD/CAM techniques, began 1997, assisted by Berlin Technical University; prototype (D-EQUI) displayed at Aero '99, Friedrichshafen, in April 1999; first flight 5 March 2000; initially flew without winglets; certification to JAR-VLA then anticipated in third quarter 2000.
COSTS: DM210,000, including tax, for first 20 production aircraft (2000).
DESIGN FEATURES: Design goals included crashworthy cabin structure, forgiving handling characteristics and low operating costs. Low wing with sharply tapered fuselage aft of cabin and sweptback fin; high aspect ratio wing with upturned tips; laminar flow wing section Horstmann/Quast HQ-42 modified.
FLYING CONTROLS: Conventional and manual. Actuation via pushrods for elevators and ailerons; cables for rudder. Externally hinged, single-slotted Fowler flaps, deflections 0, 20 and 35°.
STRUCTURE: Composites, with skins formed in CNC-milled moulds. Cabin of vinyl-carbon fibre (adopted from F1 racing car technology) for additional crash protection.
LANDING GEAR: Non-retractable tricycle type; spring steel main legs; steerable nose leg with rubber-in-compression suspension; all wheels size 5.00-5; speed fairings on all wheels; hydraulic brakes; combined ventral fillet/tailskid.
POWER PLANT: One 73.5 kW (98.6 hp) Rotax 912 ULS flat-four engine driving a Hoffmann HO-V352F/170 FQ +10 two-blade, constant-speed propeller. Fuel in two wing tanks, total capacity 120 litres (31.7 US gallons; 26.4 Imp gallons). Fuel filler in upper surface of each wing.
ACCOMMODATION: Two persons side by side under forward- and upward-opening windscreen/canopy; baggage door on port side of aft wing. Accommodation heated and ventilated.
SYSTEMS: 14 V electrical system, including 40 Ah alternator and 12 V battery.
AVIONICS: *Comms:* Honeywell KX 125 nav/com; KT 76A transponder; ELT; and PS Engineering PM 501 intercom all standard. Optional Honeywell KX 155A nav/com; KT 76C and Garmin GMA 340 and GTX 320 transponders; and PS Engineering PM 6000/6000M audio/intercom/marker.
 Flight: Honeywell KI 209 VOR/LOC/GS. Option of KLN 89B, KLN 94, Garmin GNC 250XL or GNS 430 GPS, with/without Skyforce Skymap III colour moving map display.
 Instrumentation: Standard VFR.
EQUIPMENT: Navigation and anti-collision lights and tiedown fittings standard.

DIMENSIONS, EXTERNAL:
Wing span, excl winglets	10.30 m (33 ft 9½ in)
Wing aspect ratio	10.1
Length overall	7.30 m (23 ft 11½ in)
Height overall	2.30 m (7 ft 6½ in)

DIMENSIONS, INTERNAL:
Cabin max width	1.15 m (3 ft 9¼ in)
Baggage volume	0.50 m³ (17.66 cu ft)

AREAS:
Wings, gross	10.50 m² (113.0 sq ft)

Prototype Aquila A210 on its maiden flight
2001/0092130

WEIGHTS AND LOADINGS:
  Weight empty   490 kg (1,080 lb)
  Baggage capacity   35 kg (77 lb)
  Max T-O weight   750 kg (1,653 lb)
  Max wing loading   71.4 kg/m² (14.63 lb/sq ft)
  Max power loading   10.20 kg/kW (16.76 lb/hp)
PERFORMANCE (estimated):
  Max level speed   164 kt (305 km/h; 189 mph)
  Cruising speed at 75% power
                                     130 kt (240 km/h; 149 mph)
  Stalling speed: flaps up   56 kt (102 km/h; 64 mph)
            flaps down   44 kt (81 km/h; 51 mph)
  Max rate of climb at S/L   288 m (945 ft)/min
  Service ceiling   5,000 m (16,400 ft)
  T-O run   160 m (525 ft)
  T-O to 15 m (50 ft)   340 m (1,115 ft)
  Landing from 15 m (50 ft)   320 m (1,050 ft)
  Range with max fuel at 1,525 m (5,000 ft):
    at 75% power   745 n miles (1,380 km; 857 miles)
    at 55% power   999 n miles (1,850 km; 1,149 miles)
                                         UPDATED

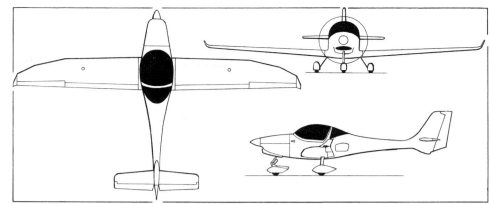

Aquila A210 all-composites light aircraft (Rotax 912 ULS flat-four) *(James Goulding/Jane's)* 2001/0092131

B&F

B&F TECHNIK VERTRIEBS GmbH
Am neuen Rheinhafen 10, D-67346 Speyer
Tel: (+49 62) 327 20 76
Fax: (+49 62) 327 20 78

B&F continues to market the original FK 9 ultralight in parallel to a new version in composites. Latest products are the nostalgic FK 12 and FK 14 Polaris composites ultralights. The FK series is designed by Dipl Ing Otto Funk and Peter Funk.

VERIFIED

B&F FK 9 MARK 3
TYPE: Side-by-side ultralight.
PROGRAMME: The earlier versions of B&F's FK 9 (see 1990-91 *Jane's*) were built at Krosno, Poland, employing metal tube fuselages and composites wings, both fabric-covered. The Mark 3 all-composites version was announced in 1997; 85 per cent of airframe is manufactured in Poland, for completion in Germany. Launched at Friedrichshafen Air Show, April 1997; UK sales promotion from July 1998. Following trials in earlier FK 9 (D-MCFK), 40.5 kW (54.2 hp) Suprex three-cylinder engine, developed from that of DaimlerChrysler Smart car, offered as alternative power plant from 2000.
CUSTOMERS: Total of 140 FK 9s of all series built by April 1999.
COSTS: DM87,500 including tax (1998).
STRUCTURE: Composites wing and fuselage; metal tailplane; fabric-covered ailerons, slotted flaps and elevators.
LANDING GEAR: Fixed tricycle type with speed fairings.
POWER PLANT: One 59.6 kW (79.9 hp) Rotax 912 UL driving a Junkers three-blade carbon fibre propeller. Max fuel capacity 65 litres (17.2 US gallons; 14.3 Imp gallons).
DIMENSIONS, EXTERNAL:
  Wing span   9.85 m (32 ft 3¾ in)
  Length overall   5.85 m (19 ft 2¼ in)
WEIGHTS AND LOADINGS:
  Weight empty   275 kg (606 lb)
  Max T-O weight   450 kg (992 lb)
PERFORMANCE:
  Cruising speed:
    at 75% power   103 kt (190 km/h; 118 mph)
    at 65% power   97 kt (180 km/h; 112 mph)
  Stalling speed   36 kt (65 km/h; 41 mph)
  Range   540 n miles (1,000 km; 621 miles)
                                         UPDATED

B&F FK 12 Comet ultralight aerobatic biplane *(Pierre Gaillard)* 2001/0092129

Production B&F FK 14 Polaris visiting PFA International Rally, Cranfield, UK, June 2000 *(Paul Jackson/Jane's)* 2001/0092055

B&F FK 12 COMET
TYPE: Aerobatic, tandem-seat ultralight biplane.
PROGRAMME: Design, by Peter Funk, started late 1994; construction of prototype began 1995; first flight (D-MPLI) March 1997; first flight of production aircraft May 1997; certification proceeding in 1999.
CUSTOMERS: Total of 40 sold by April 1999.
COSTS: Kit DM43,000, ready-to-fly DM89,000, including taxes (1998).
DESIGN FEATURES: Wings, of laminar aerofoil section, have sweepback, and fold back for storage.
FLYING CONTROLS: Conventional and manual rudder and elevator; full-span flaperons on upper and lower wings.
STRUCTURE: Welded steel tube forward fuselage with riveted aluminium tube rear section; composites and GFRP laminate skinning; wings have carbon fibre main spar and leading-edge with aluminium secondary spar and flaperons, rear section fabric-covered.
LANDING GEAR: Tailwheel type; non-retractable.
POWER PLANT: One 59.6 kW (79.9 hp) Rotax 912 UL, 73.5 kW (98.6 hp) Rotax 912 ULS or turbocharged 84.6 kW (113.4 hp) Rotax 914 driving a two-blade wooden MT propeller; three-blade CFRP Junkers propeller optional. Fuel capacity 44 litres (11.6 US gallons; 9.7 Imp gallons).
DIMENSIONS, EXTERNAL:
  Wing span   6.70 m (21 ft 11¾ in)
  Length overall   5.30 m (17 ft 4¾ in)
  Height overall   1.98 m (6 ft 6 in)
AREAS:
  Wings, gross   13.40 m² (144.2 sq ft)
WEIGHTS AND LOADINGS:
  Weight empty   240 kg (529 lb)
  Max T-O weight   450 kg (992 lb)
PERFORMANCE:
  Max level speed   100 kt (185 km/h; 115 mph)
  Cruising speed   89 kt (165 km/h; 103 mph)
  Stalling speed, flaperons down   36 kt (65 km/h; 41 mph)

Demonstrator B&F FK 9 Mark 3 two-seat ultralight *(Paul Jackson/Jane's)* 1999/0044695

Max rate of climb at S/L	305 m (1,000 ft)/min
T-O run	120 m (394 ft)
Range with max fuel	243 n miles (450 km; 279 miles)

UPDATED

B&F FK 14 POLARIS

TYPE: Side-by-side ultralight.
PROGRAMME: Development began early 1998; prototype (D-MVFK), then unflown, exhibited at Aero '99 at Friedrichshafen in April 1999. In production by 2000.
COSTS: With Rotax 912, DM119,480; with Rotax 912 ULS, DM123,540; both including tax (1999).
DESIGN FEATURES: Low wing with upturned wingtips incorporating navigation lights; wing section designed with assistance of Stuttgart University wind tunnel.
FLYING CONTROLS: Conventional and manual; two-section flaps extend on rails; trim tab in each elevator; horn-balanced rudder.
STRUCTURE: Mainly composites, with metal ailerons, tailplane and first section of flaps; elevators, rudder and second section of flaps fabric-covered.
LANDING GEAR: Non-retractable tricycle type; composites cantilever main legs; steerable nosewheel; brakes; speed fairing on all wheels.
POWER PLANT: One 59.6 kW (79.9 hp) Rotax 912 UL or 73.5 kW (98.6 hp) Rotax 912 ULS driving a Junkers three-blade, ground-adjustable, carbon fibre propeller. Fuel contained in single tank behind cockpit, capacity 42 litres (11.1 US gallons; 9.2 Imp gallons).
EQUIPMENT: BRS 5 emergency parachute recovery system optional.

DIMENSIONS, EXTERNAL:
Wing span	8.90 m (29 ft 2½ in)
Length overall	5.35 m (17 ft 6½ in)
Height overall	1.88 m (6 ft 2 in)

WEIGHTS AND LOADINGS:
Weight empty	260 kg (573 lb)
Max T-O weight	450 kg (992 lb)

PERFORMANCE (estimated):
Never-exceed speed (VNE)	156 kt (290 km/h; 180 mph)
Min flying speed	35 kt (65 km/h; 41 mph)

UPDATED

BUSSARD

BUSSARD DESIGN GmbH

Marie-Curiestrasse 6, D-85055 Ingolstadt
Tel: (+49 841) 901 44 60
Fax: (+49 841) 901 44 62
e-mail: info@bussard-design.de
DEVELOPMENT DIRECTOR: Georg Heinrich

A design services subcontractor to the aerospace and automotive industries, Bussard plans to enter lightplane manufacturing with its own concept, the Raptor.

NEW ENTRY

BUSSARD RAPTOR

TYPE: Side-by-side ultralight.
PROGRAMME: Scale model shown at ILA, Berlin, June 2000, at which time tooling being assembled for prototype. Debut planned at Aero '01, Friedrichshafen, April 2001. Available factory-built only. Alternative version offered with 55.1 kW (74 hp) Wankel 814Tgi and 113 kt (210 km/h; 130 mph) cruising speed.
DESIGN FEATURES: Pod-and-boom configuration with pusher propeller; high wing with constant chord on inner three-quarters and two-stage sweepback on outer panels. Mid-mounted tailplane and sweptback fin.
FLYING CONTROLS: Conventional and manual.
STRUCTURE: Aluminium fuselage structure with carbon fibre sandwich covering, except Kevlar reinforcement around cockpit. Carbon fibre wings.
LANDING GEAR: Tricycle type; fixed. Steerable nosewheel. Speed fairings on all wheels.
POWER PLANT: One digitally controlled 84.6 kW (113.4 hp) Rotax 914 F flat-four driving a pusher propeller. Fuel capacity 70 litres (18.5 US gallons; 15.4 Imp gallons).

DIMENSIONS, EXTERNAL:
Wing span	8.50 m (27 ft 10¾ in)
Length overall	6.20 m (20 ft 4 in)

AREAS:
Wings, gross	10.50 m² (113.0 sq ft)

WEIGHTS AND LOADINGS:
Weight empty	200 kg (441 lb)
Baggage capacity	20 kg (44 lb)
Max T-O weight	450 kg (992 lb)

PERFORMANCE (estimated):
Max level speed	135 kt (250 km/h; 155 mph)

Model of Bussard Raptor shown at Berlin, June 2000 *(Paul Jackson/Jane's)* 2001/0092053

Normal cruising speed	124 kt (230 km/h; 143 mph)
Stalling speed	36 kt (65 km/h; 41 mph)
Range	Up to 540 n miles (1,000 km; 621 miles)

NEW ENTRY

DORNIER

DORNIER LUFTFAHRT GmbH

Renamed Fairchild Dornier (which see) on 13 April 2000.

UPDATED

EADS DEUTSCHLAND

EADS DEUTSCHLAND GmbH

PO Box 801109, D-81663 München
Tel: (+49 89) 60 70
Fax: (+49 89) 60 72 64 81
Web: http://www.eads-nv.com
PRESIDENT AND CEO: Rainer Hertrich
HEADS OF BUSINESS UNITS:
 Werner Heinzmann (Defence and Civil Systems)
 Dr Gustav Humbert (Commercial Aircraft)
 Josef Kind (Space Infrastructure)
 Karl-Heinz Hartmann (Military Aircraft)
SENIOR VICE-PRESIDENT, CORPORATE PLANNING AND TECHNOLOGY: Dr Thomas Enders
SENIOR VICE-PRESIDENT, COMMUNICATIONS: Christian Poppe
VICE-PRESIDENT, PRESS AND INFORMATION: Rainer Ohler

Former Deutsche Aerospace (DASA), established 19 May 1989, was renamed Daimler-Benz Aerospace on 1 January 1995 and DaimlerChrysler Aerospace on 17 November 1998 following merger of Daimler-Benz AG and Chrysler Corporation (but retained DASA initials); was the aircraft, defence, space and propulsion systems arm of the Daimler-Chrysler Group; integrated Dornier, former Messerschmitt-Bölkow-Blohm (MBB), MTU (Motoren- und Turbinen-Union München) and former Telefunken SystemTechnik (TST); Hamburg company operated as DaimlerChrysler Aerospace Airbus, a subsidiary of DaimlerChrysler Aerospace.

On 14 October 1999, DASA and France's Aerospatiale Matra signed agreement for merger as basis of EADS (see International section), CASA of Spain following suit on 2 December 1999. EADS formed 10 July 2000.

Workforce 46,107 at end of 1999, including 16,754 employed on civil aeroplanes, 6,308 on military aeroplanes, 4,052 on helicopters and 6,875 on aero-engines, but scheduled to have been reduced by 880 by end of 2000, mostly within Defence and Civil Systems Business Unit. Revenues in 1999 totalled €9,191 million, producing operating profit of €730 million. Company organised into six business units, as detailed below.

Efforts to become an international competitor and equal partner in international programmes have included MTU general co-operation agreement with Pratt & Whitney, formation of Eurocopter with Aerospatiale, and senior membership of Airbus consortium through DaimlerChrysler Aerospace Airbus. An MoU was signed in September 1993 covering co-operation between DASA and two all-Russian materials institutes on aluminium-lithium alloys.

Former East German Elbe Flugzeugwerke incorporated into the then Deutsche Aerospace Airbus and Flugzeugwerke Ludwigsfelde into DASA Propulsion Group during 1991. In September 1995, DASA and Alenia of Italy established a working group to determine and evaluate areas of mutual interest with a view to establishing a strategic alliance through integration of some programmes, notably high-capacity transports, regional transports and space, possibly also involving transfer of some production work to Italy.

Commercial Aircraft and Helicopters Business Unit consists of Airbus division and Elbe Flugzeugwerke; activities include Airbus family and helicopters (last-mentioned detailed under Eurocopter in International section). Participation in Airbus consortium is 37.9 per cent; holds 20 per cent of EADS Dornier.

Military Aircraft Business Unit concentrates on development of Eurofighter Typhoon, as described in International section and support of Tornado (see *Jane's Aircraft Upgrades*). The Mako advanced trainer and light attack fighter is under development.

Space Infrastructure Business Unit concentrates on European orbital programmes, including the Ariane 5, Columbus Orbital facility (a manned space module to be docked with the International Space Station) and the Eureca free-flying retrievable research carrier.

Satellite Systems Business Unit comprises EADS Dornier Satellitensysteme GmbH, producing satellite systems, payloads and subsystems for a variety of civil, security and defence applications.

Defence and Civil Systems Business Unit products include anti-tank, anti-ship and anti-aircraft missiles, dispenser systems, UAVs, radar, and systems for training, environmental protection and other high-technology civil applications.

Aero-Engines Business Unit propulsion projects include repair and overhaul of large civil and military engines, and joint development and production with Rolls-Royce, IAE, Pratt & Whitney and Turbomeca of such engines as CF6, RB199, EJ200, V2500, PW300, PW2000, PW4000, M138 and MTR 390. Is bidding T700/T6E for equipment of German NH 90 helicopters.

Details follow of aircraft business units and their products.

UPDATED

EADS M (Military Aircraft)

PO Box 801160, D-81663 München
Tel: (+49 89) 60 70
Fax: (+49 89) 60 72 64 81
CHAIRMAN: Rainer Hertrich
DIRECTOR, EUROFIGHTER AND TRAINING AIRCRAFT:
 Erwin Obermeier
DIRECTOR, SUPPORT PROGRAMMES AND TORNADO:
 Rüdiger Sanitz
HEAD OF COMMUNICATION: Wolfdietrich Hoeveler

Major activities of Group include light fighter/trainer aircraft, airborne reconnaissance systems, aircraft armament, disarmament verification systems (the former PRISMA system and BICES), simulation and training systems, and research into advanced aircraft systems, materials and manufacturing technologies. Division also makes Airbus subassemblies. Augsburg, Manching and Ottobrunn plants employed 5,750 personnel in 2000; EADS M employees then totalled 7,400 in all countries.

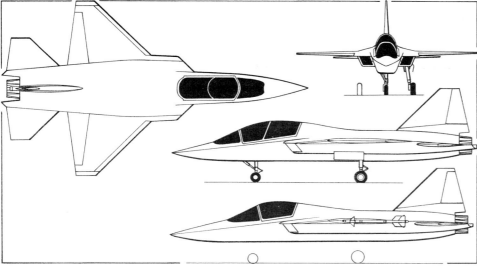

Provisional general arrangement of the Mako-AT advanced trainer, with additional side view (bottom) of single-seat Mako-LCA attack aircraft *(James Goulding/Jane's)* 2000/0062399

Mockup of the EADS Mako-AT 2001/0092110

Refer to International section for Military Aircraft Business Unit participation in the **Eurofighter Typhoon** programme, plus *Jane's Aircraft Upgrades* for the DASA **F-4F Phantom ICE** (improved combat effectiveness) programme and **Panavia Tornado**. Will participate in Europatrol (International section) programme. In partnership with US Navy and Boeing, is undertaking further series of trials of Rockwell/DASA X-31 vectored thrust testbed, 2000-02, with particular reference to carrier-type short landings. Unit offers upgrades for MiG-29 under agreement signed with MAPO in June 1999; completed radar upgrade of NATO E-3A AWACS fleet in 2000.

Spanish (CASA; which see) inputs to EADS M include support of EF-18 Hornet, AV-8B Harrier, Mirage F1, P-3 Orion, C-101 Aviojet, F-5B Freedom Fighter and E.26 Tamiz.

UPDATED

EADS MAKO
English name: Shark
TYPE: Advanced jet trainer/light attack jet.
PROGRAMME: Joint Aermacchi/DASA AT-2000 programme launched in April 1989, Aermacchi withdrew in 1994; DASA maintained the programme in hope of a joint development venture with South Korea. Following the latter's expressed preference in 1996 for a partnership with Lockheed Martin to satisfy its KTX-2 requirement, DASA began seeking an alternative collaborator in addition to Denel in South Africa, which was to provide avionics. Hyundai of South Korea expressed interest in the venture in early 1997, intending to build non-composites components of the centre and rear fuselage and tail. Design revealed in October 1996, after completion of wind tunnel tests and when radar cross-section trials in hand. Feasibility study completed by DASA in late 1997. Named Mako in 1998, by which time was foreseen as a 'modular design concept' with different versions for air-to-air, air-to-ground and training missions.

Predefinition phase began in early 1998. Full definition began by mid-1999, following a decision by the DASA board; full-scale mockup shown at 1999 Paris Air Show. Once the definition phase is completed, development will only be launched if DASA can find risk-sharing partners; the company failed to bring its original target partners (South Africa and South Korea) into the programme. There is no home-market (Luftwaffe) interest in the aircraft. Efforts are now concentrating on trainer market opportunities in Greece, Spain and the UAE. Full-scale development still officially expected from 'late 2000', although programme was progressing only slowly in mid-2000 due to former DASA reorganisation within newly formed EADS and potential rivalry in Europe from Aermacchi M-346/Yak-130. First of three prototypes to fly in 2003; production from 2007; first deliveries from 2008.

CURRENT VERSIONS: **Mako-AT:** Advanced trainer. Original baseline advanced two-seat trainer, available with or without radar and internal gun. Capable of in-flight weapons system simulation, with generation of synthetic threats and targets. Also suitable for some combat roles, including ground attack and reconnaissance.

Mako-LCA: Light combat aircraft. Single-seat, dedicated combat version for air defence or reconnaissance. Now being promoted as primary variant, with emphasis switching from the advanced trainer to the light combat roles. EADS perceives a world market for 2,500 to 3,000 light combat aircraft between 2007 and 2027.

Unmanned: UCAV derivative being studied.

CUSTOMERS: Predicted 800 sales by 2030, of which 30 per cent would be two-seat version. Entered (1999) in Brazilian F-X BR fighter replacement competition (to replace Mirage III and F-5E); also being promoted in Spain. MoU signed with UAE November 1999. Requires 300-350 orders to break even. Initial wave of international orders anticipated in 2008-2012, with second period of high interest foreseen in 2018.

COSTS: Estimated development cost US$1,300 million; unit cost US$15 million to US$22 million.

DESIGN FEATURES: Modular design for ease of meeting different roles; simple structure permits low production cost. Conventional configuration described as 'transonic layout with supersonic performance'. Advanced trainer version is capable of combat missions, employing LO technology and thrust vectoring. LO features include chined forward section, wing/forebody blending and 'caret' engine air intakes to give an RCS of 1 m² (10.8 sq ft) at 24 n miles (44 km; 28 miles) range. EFCS adapted from Rockwell/DASA X-31 research aircraft.

Wing leading-edge sweep 45°.

FLYING CONTROLS: Reprogrammable quadruplex digital FBW FCS with 'carefree handling'. Large, single-section flaperons; all-moving tailplane (tailerons); inset rudder; and full-span wing leading-edge slats. Four door-type airbrakes adjacent to jetpipe.

STRUCTURE: Centre-fuselage of aluminium; wing skins, engine air inlets and empennage both of carbon fibre.

LANDING GEAR: Retractable tricycle type; single wheel on each leg. Based on Saab JAS 39 Gripen.

Mako-AT mockup cockpit *(Paul Jackson/Jane's)* 2000/0081737

POWER PLANT: One Eurojet EJ200 turbofan; installed derated to 75.0 kN (16,860 lb st) with afterburning; optionally with thrust vectoring (provision for which is made in the FCS and airframe structure); or normally dry rated (approximately 60 kN; 13,490 lb st); or with full afterburning (approximately 90 kN; 20,250 lb st), according to customer's requirement. General Electric F404-402, F414 and SNECMA M88-2 or -3 under consideration as alternative power plants. FCS incorporates provision for vectoring nozzle. Single-seater carries 3,150 kg (6,945 lb) of internal fuel; trainer has 2,840 kg (6,261 lb).

ACCOMMODATION: One or two pilots, according to version; layout and instrumentation generally as in Eurofighter Typhoon. Single-piece fixed windscreen hinges forward for avionics access; blown canopy, opening to starboard. Trainer has stepped cockpits, with 15° view over nose and 40° each side; seats inclined rearwards at 18°.

SYSTEMS: OBOGS. APU.

AVIONICS: *Radar:* Optional 'AN/APG-67 class' multimode radar; other contenders include FIAR Grifo, Thomson-CSF RD-400 and BAE Systems Bluehawk.

Instrumentation: Three MFDs in each cockpit, reflecting modern combat aircraft (such as Eurofighter) environment. Modular concept to allow easy upgrade or tailoring to individual customer requirements. Expected to include provision for FLIR, HMS and similar equipment.

ARMAMENT: Optional Mauser BK 27 mm internal gun and seven low RCS external stores hardpoints: one at each wingtip, two under each wing and one below centreline. Centreline and inboard wing pylons stressed to 1,350 kg (2,976 lb); outboard to 675 kg (1,488 lb); wingtip for AAM only.

All data are provisional.

DIMENSIONS, EXTERNAL:
Wing span	8.25 m (27 ft 0¾ in)
Wing aspect ratio	2.6
Length overall	13.75 m (45 ft)
Height overall	4.5 m (15 ft)
Wheel track	2.4 m (8 ft)
Wheelbase	4.8 m (16 ft)

AREAS:
Wings, gross	26.70 m² (287.4 sq ft)

WEIGHTS AND LOADINGS:
Weight empty: LCA	6,200 kg (13,669 lb)
AT	5,800 kg (12,787 lb)
Max weapon load	4,500 kg (9,220 lb)
Max fuel weight	3,300 kg (7,275 lb)
T-O weight: normal: LCA	9,400 kg (20,723 lb)
AT	8,100 kg (17,857 lb)
max: LCA, AT	13,000 kg (28,660 lb)
Wing loading at normal TOW:	
LCA	352.1 kg/m² (72.11 lb/sq ft)
AT	303.4 kg/m² (62.14 lb/sq ft)
Power loading (full afterburning) at normal TOW:	
LCA	104 kg/kN (1.02 lb/lb st)
AT	90 kg/kN (0.88 lb/lb st)

PERFORMANCE (estimated):
Max level speed	M1.5
Service ceiling	14,400 m (47,240 ft)
T-O run	450 m (1,480 ft)
Landing run	750 m (2,460 ft)
Ferry range	2,000 n miles (3,704 km; 2,301 miles)
g limits	+9/−3

UPDATED

EADS DEUTSCHLAND AIRBUS

PO Box 950109, Kreetslag 10, D-21111 Hamburg
Tel: (+49 40) 743 70
Fax: (+49 40) 743 44 22
Telex: 21950-0 DA D
WORKS: Hamburg, Bremen, Buxtehude, Nordenham, Varel, Stade, Laupheim, Dresden
PRESIDENT AND CEO: Gustav Humbert
PRESS AND INFORMATION: Rolf Brandt

This division, formerly DaimlerChrysler Aerospace Airbus, undertakes about one-third of the development and manufacturing work on the Airbus airliner family and represents 90 per cent of EADS Deutschland's civil aeroplanes business. Through the organisation Airspares and acting on behalf of Airbus Industrie, it is responsible for provision of spare parts worldwide for the Airbus fleet.

The company produces fuselage sections and vertical tails and is main cabin furnishing centre for Airbus family (except for A330/A340); fits all movable wing parts to wing torsion boxes produced by BAE Systems. A319, A319CJ and A321 are assembled by EADS Deutschland Airbus (241 complete aircraft delivered in 2000); A318 will follow; freighter conversion is carried out by EADS EFW, formerly Elbe Flugzeugwerke GmbH, in Dresden.

Refer to the International section for EADS Deutschland Airbus participation in the **SATIC Airbus Super Transporter, Very Large Commercial Transport/Ultra-High Capacity Transport (VLCT/UHCA), Airbus Military Company A400M FLA** and **Supersonic Airliner Studies** projects, plus *Jane's Aircraft Upgrades* for the **Transall C-160** life extension programme.

UPDATED

EXTRA

EXTRA-FLUGZEUGBAU GmbH

Flugplatz Dinslaken, Schwarze Heide 21, D-46569 Hünxe
Tel: (+49 2858) 913 70
Fax: (+49 2858) 91 37 30
MANAGING DIRECTOR: Walter Extra
INFORMATION CONTACT: Bruno Van Waeyenberghe

EXTRA 400 EUROPEAN DISTRIBUTION:
Diamond Aircraft Service GmbH
  Flughafen Siegerland, Werfthalle 1, D-57299 Burbach
Tel: (+49 2736) 442 80
Fax: (+49 2736) 44 28 50
e-mail: info@diamond-aircraft.de
Web: http://www.diamond-aircraft.de

Extra produces aerobatic sportplanes and a turboprop business aircraft. Workforce is approximately 130.

UPDATED

EXTRA 200

TYPE: Aerobatic, two-seat sportplane.
PROGRAMME: Prototype (D-ETEL) first flown 2 April 1996 and almost immediately shipped to USA for demonstrations and public debut at EAA Sun 'n' Fun in Lakeland, Florida, later that month; second prototype and principal flight test aircraft (D-EAZG) exhibited at ILA, Berlin, May 1996; German certification scheduled for 15 June 1996, followed by FAA certification (initially in Experimental/Exhibition Category). Engineering designation **Extra 300/200**.
CUSTOMERS: Prototype and five production aircraft delivered in 1996; 15 in 1997; five in 1998; and further two in mid-1999, increasing total to 28; none further known by mid-2000. Exports to USA account for 85 per cent of production.
COSTS: About DM300,000 (1998).
Data generally as for Extra 300, except that below:
POWER PLANT: One 149 kW (200 hp) Textron Lycoming AEIO-360-A1E flat-four engine driving a two-blade Mühlbauer MT constant-speed propeller; three-blade optional. Standard fuel capacity 116 litres (30.6 US gallons; 25.5 Imp gallons), of which 114 litres (30.0 US gallons; 25.0 Imp gallons) are usable; optional long-range tanks increase total capacity to 154 litres (40.6 US gallons; 33.8 Imp gallons), of which 151 litres (40.0 US gallons; 33.3 Imp gallons) are usable.

DIMENSIONS, EXTERNAL:
Wing span	7.50 m (24 ft 7¼ in)
Wing aspect ratio	5.4
Length overall	6.81 m (22 ft 4 in)
Height overall	2.56 m (8 ft 4¾ in)

AREAS:
Wings, gross	10.40 m² (111.9 sq ft)

WEIGHTS AND LOADINGS:
Weight empty	544 kg (1,199 lb)
Max T-O weight: Aerobatic, solo	608 kg (1,340 lb)
Aerobatic, two-seat	838 kg (1,848 lb)
Normal	868 kg (1,914 lb)
Max wing loading:	
Aerobatic, solo	57.75 kg/m² (11.97 lb/sq ft)
Aerobatic, two-seat	80.6 kg/m² (16.50 lb/sq ft)
Normal	83.5 kg/m² (17.09 lb/sq ft)
Max power loading:	
Aerobatic, solo	4.08 kg/kW (6.70 lb/hp)
Aerobatic, two-seat	5.62 kg/kW (9.24 lb/hp)
Normal	5.82 kg/kW (9.57 lb/hp)

PERFORMANCE (N: at Normal category T-O weight, SA: at solo Aerobatic weight):
Never-exceed speed (VNE):	
N	220 kt (407 km/h; 253 mph)
Manoeuvring speed	158 kt (293 km/h; 182 mph)
Cruising speed at 75% power:	
N	150 kt (278 km/h; 173 mph)
Stalling speed: N	56 kt (104 km/h; 64 mph)
SA	52 kt (96 km/h; 60 mph)
Max rate of climb at S/L: N	488 m (1,600 ft)/min
SA	686 m (2,250 ft)/min
Service ceiling: N	4,575 m (15,000 ft)
Range with single pilot at 75% power at 2,440 m (8,000 ft), 45 min reserves:	
N: standard fuel	450 n miles (833 km; 517 miles)
optional fuel	600 n miles (1,111 km; 690 miles)
g limits: SA	±10

UPDATED

EXTRA 300 and 330

TYPE: Aerobatic, two-seat sportplane.
PROGRAMME: Design began January 1987; first flight (D-EAEW) 6 May 1988; LBA certification 16 May 1990; certified FAR Pt 23 Amendment 33 Normal and Aerobatic categories for USA and Europe; production started October 1988.
CURRENT VERSIONS: **Extra 300**: Baseline, mid-wing, two-seat aircraft.
Details apply to Extra 300 except where indicated.
Extra 300L: Low-wing, two-seat version with other structural changes; low wing enhances visibility during take-off and landing and eases installation of full IFR instrumentation; new ailerons give slightly higher roll rate (approximately 400°/s) and improved snap-rolling capability, both erect and inverted; shorter fuselage with all-composites side and belly access panels; cockpit interior modified for more comfort, with new reclined carbon fibre seats with optional leather trim; first aircraft (D-EUWH) delivered November 1994.
Extra 300S: Single-seat version of 300 with same power plant; wing shortened by 50 cm (19½ in) and more powerful ailerons; first flight (D-ESEW) 4 March 1992; certified March 1992; US, FAA and French certification.
Extra 330L: Strengthened version of 300L with CFRP fuselage panels, fin and rudder; wider-chord rudder and elevators, all with horn balances; 246 kW (330 hp) Textron Lycoming AEIO-580 engine. Prototype (D-ESEW), modified from prototype 300S first flown January 1998 and since converted to 330L. First production aircraft (D-EPET) flew 11 May 1998 and delivered to Hungarian aerobatic pilot Peter Besenyei; normally flown with single-seat cockpit canopy; first public performance at North Weald, UK, 29 May 1998.
Extra 330LX: Two-seat version of 330; prototype D-EDGE first flown early 1999.
CUSTOMERS: Total 66 (including prototype) two-seat original ('mid-wing') EA 300s delivered by April 1997; additional one to USA in November 1998. Further 28 EA 300Ss built by August 1995, plus one in 1997 and one in 1999. EA 300L production totalled 110 by March 2000. EA 330L/LX production three (including prototype). Almost all for customers in USA, although recent sales included six for Romania and one to South Africa..
COSTS: Basic price DM340,000 (January 1996); 330, DM350,000 (1998).
DESIGN FEATURES: Designed for unlimited competition aerobatics; tapered, square-tipped mid-mounted wing with 4° leading-edge sweepback; aerofoil symmetrical MA-15S at root, MA-12S at tip; no twist or dihedral.
FLYING CONTROLS: Conventional and manual. Rod and cable operated; nearly full-span ailerons, assisted by spade-type suspended tabs; trim tab in starboard elevator.
STRUCTURE: Fuselage (excluding tail surfaces) of steel tube frame with part aluminium, part fabric covering; wing spars of carbon composites and shells of carbon composite sandwich; tail surfaces of carbon spars and glass fibre shells.
LANDING GEAR: Fixed cantilever composites arch main gear with single polyamid-faired wheels; Cleveland brakes; leaf-sprung steerable tailwheel.
POWER PLANT: One Textron Lycoming AEIO-540-L1B5 flat-six engine giving 224 kW (300 hp), driving Mühlbauer MTV-9-B-C/C 200-15 three-blade constant-speed propeller (four-blade MTV-14 optional). With either standard or optional propeller and Gomolzig Type 3 silencer, Extra 300s meet German and US noise limits. Christen Industries inverted oil system. Fuel capacity: 300L standard capacity of 171 litres (45.1 US gallons; 37.6 Imp gallons), of which 169 litres (44.6 US gallons; 37.1 Imp gallons) usable; optional long-range tanks raise total fuel capacity to 208 litres (55.1 US gallons; 45.9 Imp gallons), of which 206 litres (54.5 US gallons; 45.4 Imp gallons) usable. 300S standard fuel capacity 160 litres (42.3 US gallons; 35.2 Imp gallons).
ACCOMMODATION: Pilot and co-pilot/passenger in tandem under single-piece canopy opening to starboard; additional transparencies in lower sides of cockpit.
SYSTEMS: 12 V generator and battery.
AVIONICS: *Comms:* Becker AR 3201 VHF radio standard.

Extra 300L operated in USA *(Paul Jackson/Jane's)*

Extra 200 aerobatic/training aircraft *(Paul Jackson/Jane's)*

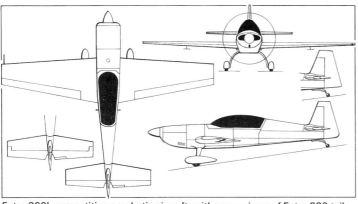

Extra 300L competition aerobatic aircraft, with scrap views of Extra 330 tail surfaces *(James Goulding/Jane's)* 2000/0064457

D-EDGE, the first Extra 330LX *(Paul Jackson/Jane's)* 2001/0092057

DIMENSIONS, EXTERNAL:
  Wing span: 300   8.00 m (26 ft 3 in)
    300L   7.70 m (25 ft 3 in)
    300S   7.50 m (24 ft 7¼ in)
  Wing chord:
    at root: 300, 300S   1.85 m (6 ft 0¾ in)
    at tip: 300   0.83 m (2 ft 8¾ in)
    300S   0.93 m (3 ft 0½ in)
  Wing aspect ratio: 300   6.0
    300S   5.4
  Length overall: 300   7.12 m (23 ft 4¼ in)
    300L   6.94 m (22 ft 9¼ in)
    300S   6.65 m (21 ft 9¾ in)
  Height overall: 300, 300S   2.62 m (8 ft 7¼ in)
  Tailplane span: 300   3.20 m (10 ft 6 in)
  Wheel track: 300   1.80 m (5 ft 10¾ in)
  Propeller diameter: MTV-9   2.00 m (6 ft 6¾ in)
    MTV-14   1.90 m (6 ft 2¾ in)
AREAS:
  Wings, gross: 300, 300L   10.70 m² (115.2 sq ft)
    300S   10.44 m² (112.4 sq ft)
  Ailerons, total: 300, 300L   1.71 m² (18.41 sq ft)
WEIGHTS AND LOADINGS:
  Weight empty: 300   630 kg (1,389 lb)
    300L   653 kg (1,440 lb)
    300S   609 kg (1,343 lb)
  Max T-O weight:
    300, Aerobatic, solo and 300S   821 kg (1,810 lb)
    300, Aerobatic, two-seat   870 kg (1,918 lb)
    300, 300L, 300S, Normal   950 kg (2,095 lb)
  Max wing loading:
    300, Aerobatic, solo   76.5 kg/m² (15.67 lb/sq ft)
    300, 300L, 300S, Normal   88.8 kg/m² (18.19 lb/sq ft)
  Max power loading:
    300, Aerobatic, solo   3.72 kg/kW (6.12 lb/hp)

Flight deck of the Extra 400 *(Paul Jackson/Jane's)* 2000/0064463

    300, Normal   4.24 kg/kW (6.98 lb/hp)
    300S   3.66 kg/kW (6.02 lb/hp)
PERFORMANCE:
  Never-exceed speed (V_{NE}):
    300L, 300S   220 kt (407 km/h; 253 mph)
  Max level speed: 300S   185 kt (343 km/h; 213 mph)
  Max manoeuvring speed: 300L, 300S
     158 kt (293 km/h; 182 mph)
  Cruising speed, 45% power at 2,440 m (8,000 ft):
    300L   170 kt (315 km/h; 196 mph)
  Stalling speed:
    300L, 300S at solo Aerobatic max T-O weight
     55 kt (102 km/h; 64 mph)
    300L, 300S at Normal max T-O weight
     60 kt (111 km/h; 69 mph)
  Max rate of climb at S/L:
    300L, 300S   975 m (3,200 ft)/min
  Service ceiling: 300L, 300S   4,875 m (16,000 ft)
  T-O run   199 m (655 ft)
  T-O to 15 m (50 ft)   545 m (1,790 ft)
  Range at 65% power cruising speed at 2,440 m
    (8,000 ft), 45 min reserves:
    300S   440 n miles (814 km; 506 miles)
  Range at 45% power cruising speed, at 2,440 m
    (8,000 ft) 45 min reserves:
    300L, standard fuel   415 n miles (768 km; 477 miles)
    300L, long-range fuel   510 n miles (944 km; 586 miles)
  Endurance: 300S   2h 30 mins
  g limits: Aerobatic, solo   ±10
    Two-seat Aerobatic   ±8

UPDATED

EXTRA 400

TYPE: Business turboprop.
PROGRAMME: Announced February 1993; designed in collaboration with Delft University; first flight (D-EBEW) 4 April 1996; initial LBA certification to FAR Pt 23 received 23 April 1997, to be followed by LBA/FAA certification for pressurised IFR operation and flight into known icing conditions. A turboprop version was being considered by early 1997. Second prototype (D-EGBU) flew April 1998, featuring downturned wingtips and ventral stake, both of which had been added to first aircraft before its retirement in 1997. This aircraft lost on delivery to first customer, 21 August 1998. Third and fourth aircraft exhibited at Aero '99, Friedrichshafen, April 1999. Target production rate 60 per year.
CUSTOMERS: Two prototypes and four production aircraft flying by mid-2000.
COSTS: Typically equipped, US$998,500 (2000).
DESIGN FEATURES: Cantilever high-wing, T tail layout; spacious cabin; optimised for low internal and external noise.
FLYING CONTROLS: Conventional and manual. Frise ailerons; two-section Fowler flaps approximately two-thirds of span, deflection 0, 15 and 30°; horn-balanced elevators; trim tab in starboard elevator; ground-adjustable tab on rudder. Interconnected aileron and rudder controls, with override.
STRUCTURE: Composites throughout. Airframe consists of skin with integral longerons and frames. Wing built on

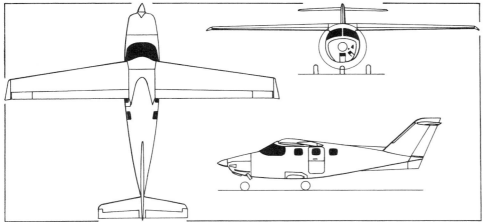

Provisional general arrangement of Extra 400 *(Paul Jackson/Jane's)* 2001/0092132

Fourth production Extra 400, fitted with wingtip radar pod *(Paul Jackson/Jane's)* 2001/0092059

Extra 400 main landing gear *(Paul Jackson/Jane's)* 2000/0064462

double front spar and rear spar, interconnected by ribs; tailplane has front and rear spar. Skins generally of carbon fibre facings and honeycomb; longerons, frames, spars and ribs of carbon fibre with foam core; wing leading-edge of glass fibre with honeycomb and galass fibre ribs. Service life 25 years/20,000 hours/30,000 cycles/22,500 pressurisation cycles.

LANDING GEAR: Hydraulically retractable tricycle type; designed and manufactured by Gomolzig; single wheels throughout; main units retract forward into fuselage sides; nosewheel retracts rearwards. Mainwheel size 15×6.0-6, nosewheel size 5.00-5. Minimum turning radius 20.4 m (67 ft). Independent hydraulic brakes on mainwheels. Nosewheel steerable ±30°.

POWER PLANT: One 261 kW (350 hp) Teledyne Continental Voyager TSIOL-550-C six-cylinder liquid-cooled engine, driving a four-blade, constant-speed Mühlbauer MTV-14-D/195-30a hydraulic propeller; three-blade propeller optional. Fuel in two integral wing tanks, combined capacity 468 litres (123.6 US gallons; 102.9 Imp gallons). Oil capacity 11.4 litres (3.0 US gallons; 2.5 Imp gallons).

ACCOMMODATION: Pilot and five passengers in pressurised cabin; rear seats in club configuration; single door on port side is horizontally split, with airstair in bottom half; top half hinges upwards. Emergency exit starboard side, opposite door.

SYSTEMS: Electrical system 24 V, with direct-driven 100 A alternator, belt-driven 85 A alternator and 24 V battery in tailcone. External power receptacle in tailcone. Pressurisation system, maximum differential 0.38 bar (5.5 lb/sq in). Wing and empennage leading-edges de-iced by Teflon inflatable boots.

AVIONICS: Honeywell package as standard, including two-tube EFIS with EFIS 40 electronic EHSI. Dual Garmin GNS 430 package with Litef LCR 92 laser gyro AHRS and Meggitt System 55 autopilot optional.

Radar: Optional Honeywell RDR 2000 colour weather radar, housed in streamlined pod on port wingtip.

Instrumentation: Becker LCD engine monitoring system standard.

DIMENSIONS, EXTERNAL:
Wing span	11.50 m (37 ft 8¾ in)
Wing aspect ratio	9.3
Length overall	9.57 m (31 ft 4¾ in)
Max width of fuselage	1.46 m (4 ft 9½ in)
Height overall	3.09 m (10 ft 1¾ in)
Tailplane span	3.80 m (12 ft 5½ in)
Wheel track	2.20 m (7 ft 2½ in)
Wheelbase	2.56 m (8 ft 4¾ in)
Propeller diameter	1.95 m (6 ft 4¾ in)
Cabin door:	
Width	0.68 m (2 ft 2¾ in)
Height	1.15 m (3 ft 9¼ in)

DIMENSIONS, INTERNAL:
Cabin (between forward and aft bulkheads):
Length	4.13 m (13 ft 6½ in)
Max width	1.39 m (4 ft 6¾ in)
Max height	1.24 m (4 ft 0¾ in)

AREAS:
Wings, gross	14.26 m² (153.5 sq ft)

WEIGHTS AND LOADINGS:
Basic operating weight empty	1,389 kg (3,062 lb)
Useful load	611 kg (1,347 lb)
Max T-O and landing weight	1,999 kg (4,407 lb)
Max zero-fuel weight	1,772 kg (3,907 lb)
Max wing loading	140.2 kg/m² (28.71 lb/sq ft)
Max power loading	7.66 kg/kW (12.58 lb/hp)

PERFORMANCE (estimated, at mid-cruise weight):
Max level speed	259 kt (480 km/h; 298 mph)
Manoeuvring speed	156 kt (289 km/h; 179 mph)
Max cruising speed	235 kt (435 km/h; 270 mph)
Stalling speed in landing configuration	59 kt (110 km/h; 68 mph)
Max rate of climb at S/L	427 m (1,400 ft)/min
Max certified altitude	7,620 m (25,000 ft)
T-O run*	440 m (1,444 ft)
Landing run*	270 m (886 ft)
Range at 55% power	1,160 n miles (2,148 km; 1,335 miles)
g limits	+4.0/−1.6

* At 1,795 kg (3,957 lb)

UPDATED

FAIRCHILD DORNIER

FAIRCHILD DORNIER

Flugplatz Oberpfaffenhofen, PO Box 1103, D-82230 Wessling
Tel: 49 (81) 533 00 20 25
Fax: 49 (81) 53 30 20 07
Web: http://www.fairchilddornier.com
CHAIRMAN OF THE BOARD: Charles P Pieper
CEO: Louis F Harrington
COO: John Wolf
PRESIDENT, CORPORATE AIRCRAFT DIVISION: Dean Rush
EXECUTIVE VICE-PRESIDENT: Barry Eccleston
VICE-PRESIDENTS:
 Kent Hollinger (Technical Support)
 Duncan Koerbel (928JET Programme)
 Robert Stangarone (Communications)

SALES OFFICE:
Worldgate Plaza IV, 8th Floor, 12801 Worldgate Drive, Herndon, Virginia 20170-4381, USA
Tel: (+1 703) 375 3600

Dornier GmbH, formerly Dornier-Metallbauten, formed 1922 by late Professor Claude Dornier; Daimler-Benz AG acquired majority holding in 1985, later assumed by Daimler-Benz Aerospace AG and subsequently by DaimlerChrysler Aerospace AG.

In June 1996, Fairchild Aerospace of San Antonio, Texas, acquired an 80 per cent shareholding and took over most of Dornier's operations, including manufacture of the 328 regional aircraft, Airbus subcontract work, life extension programme for German armed forces' Bell UH-1D helicopters, business and commuter aircraft service and support operations and international logistics activities. Support of NATO AWACS aircraft, previously undertaken by Dornier Luftfahrt GmbH, was not included in the acquisition. Company acquired by an investment fund led by Clayton, Dubilier & Rice and Allianz Capital Partners in April 2000 and renamed Fairchild Dornier on 13 April 2000. Former Fairchild Aerospace (see US section of 2000-01 *Jane's*) transferred to Germany, apart from US sales element.

Current activities at Fairchild Dornier Oberpfaffenhofen facility include manufacture, sales and support of the Dornier 328JET; development of the Fairchild Aerospace 528JET, 728JET and 928JET jets; support of the Dornier 328 and Fairchild Metro range of twin-turboprop aircraft; and support of German military Bell UH-1Ds. German production of Dornier 228 ended in 1999; HAL in India to become sole source; description of aircraft can be found under that heading. Orders and options totalled 542 (half of them firm) in January 2001.

NEW ENTRY

DORNIER 228

See under HAL in the Indian section for a description.

FAIRCHILD DORNIER 328JET

TYPE: Regional jet airliner.
PROGRAMME: Development of Dornier 328 turboprop twin begun in mid-1980s, but then suspended; relaunched 3 August 1988; rolled out 13 October 1991; first flight (D-CHIC) 6 December 1991; first flight of first production 328-100 (D-CITI) 23 January 1993. JAA 25 certification 15 October 1993; FAA certification 10 November 1993; first delivery, to Air Engiadina, 21 October 1993.

Fairchild Dornier 328JET twin-turbofan regional transport *(Paul Jackson/Jane's)*

Production (including three prototypes) totalled 111; final aircraft (OE-LKC) delivered to Air Alps Aviation, Austria, 13 October 1999.

Twin turbofan version announced 5 February 1997 and formally launched at Paris Air Show in June 1997; marketed as 328JET, engineering designation being 328-300; high-speed wind tunnel tests, confirming cruise performance, and ground vibration tests completed by November 1997; prototype D-BJET, converted from second prototype 328 turboprop (D-CATI), rolled out 6 December 1997; first flight 20 January 1998; public debut 4 February 1998; three further prototypes built on the production line; second (D-BWAL) dedicated to performance certification testing, flew 20 May 1998; third (D-BEJR), used for avionics certification, flew 10 July 1998; fourth (D-BALL), used for function and reliability testing and simulated airline operation trials, flew on 15 October 1998; JAA certification was achieved on 8 July 1999 after completion of 1,560 flight hours in 950 sorties; FAA certification 15 July 1999.

First delivery in July 1999 to Skyway Airlines (N351SK); total of 15 delivered in 1999. By 21 September 2000 the in-service fleet of 34 aircraft had logged more than 30,000 departures, with a 99.46 per cent flight completion rate; high-time aircraft, operated by Hainan Airlines, had then logged 1,992 departures.

CURRENT VERSIONS: **328JET**: Regional airliner, *as described*; 32 to 34 passengers at 79 cm (31 in) seat pitch, and up to 34 in high-density **Corporate Shuttle** configuration.

Freighter: Proposed all-cargo version with large door in aft fuselage for loading of containers or palletised freight.

Envoy 3: Corporate version, launched at the 1997 National Business Aviation Association Convention in Dallas, Texas; accommodation for up to 19 passengers; typical executive accommodation for 12 to 14 passengers with forward service galley and wardrobe and aft galley and toilet; optional long-range fuel tank extends maximum range to 2,000 n miles (3,704 km; 2,302 miles).

CUSTOMERS: Total of 232 orders and options by 2 January 2001, at which time 40 were in service. Announced customers include: Atlantic Coast Airlines (62, first handed over 27 April 2000; includes 30 for ACJet); EuroCityline (nine, plus six options); Gandalf Airlines, Italy (four, plus three options, delivered from November 1999); Hainan Airlines, China (19, plus 20 options, from December 1999); Johnson Controls (one, in quick-change corporate configuration, delivered in early 2000); Midwest Express (five, plus 10 options); Minerva (two); Modern Air/Zweibrücken (10); MTM Aviation (one); Ozark Airlines (two, plus three options from September 1999); Proteus Airlines (six); Shell Petroleum Development Company of Nigeria (three, in corporate configuration, from August 1999); Senator Aviation Charter (one from September 1999); Skyway Airlines (five, plus 10 options); St Thomas Energy products (one, in December 1999); Tyrolean Jet Service (two, from September 1999); and Wanair Tahiti (one, from April 2000). One leased to Air Vallee from May 2000.

COSTS: Regional airliner US$11.9 million; Corporate US$12.9 million, outfitted (both 1999). Airliner direct operating costs estimated at US$3.16 per n mile, based on 32 passengers and 300 n mile sectors.

DESIGN FEATURES: Combines basic TNT supercritical wing of Dornier 228 with new pressurised fuselage from NRT (Neue Rumpf Technologien) programme; internal volume designed to give passengers more seat width than in a Boeing 727 or 737 and stand-up headroom in aisle. Jet retains high commonality with turboprop model; major changes include

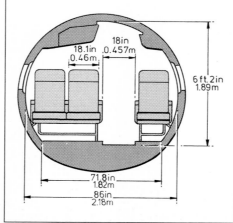

Cabin cross-section of Fairchild Dornier 328JET *(Mike Keep/Jane's)*

FAIRCHILD DORNIER—AIRCRAFT: GERMANY

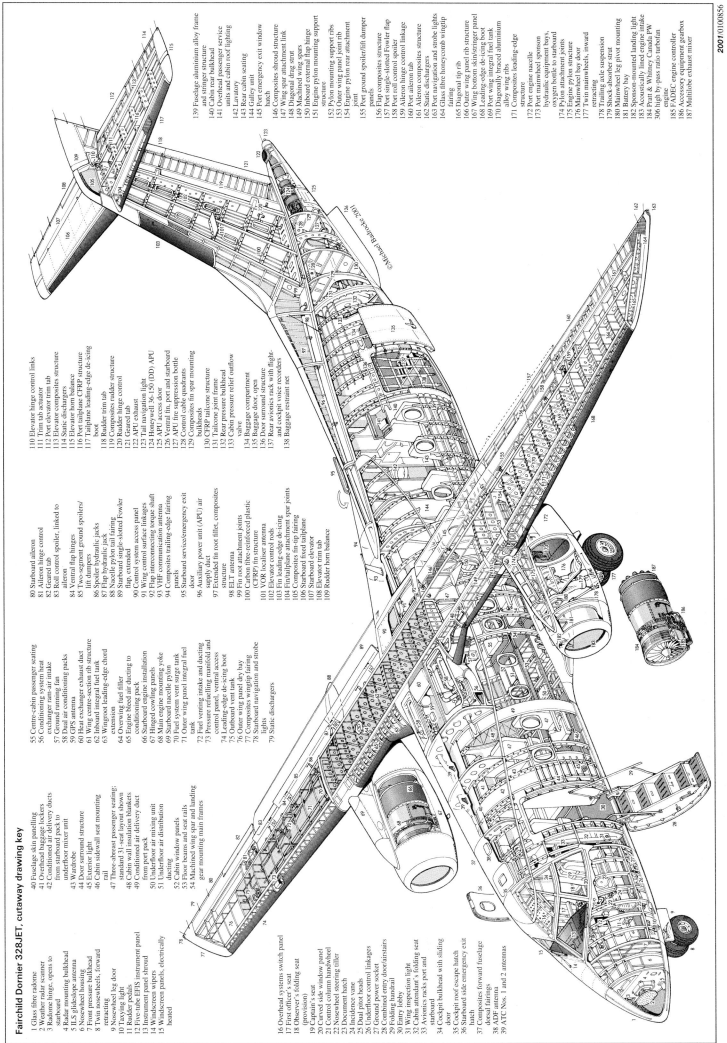

Fairchild Dornier 328JET, cutaway drawing key

1 Glass fibre radome
2 Weather radar scanner
3 Radome hinge, opens to starboard
4 Radar mounting bulkhead
5 ILS glideslope antenna
6 Nosewheel housing
7 Front pressure bulkhead
8 Twin nosewheels, forward retracting
9 Taxiing light
10 Nosewheel leg door
11 Rudder pedals
12 Five-tube EFIS instrument panel
13 Instrument panel shroud
14 Windscreen wipers
15 Windscreen panels, electrically heated
16 Overhead systems switch panel
17 First officer's seat
18 Observer's folding seat (provision)
19 Captain's seat
20 Curved side window panel
21 Control column handwheel
22 Nosewheel steering tiller
23 Document hatch
24 Entry lobby
25 Dual pitot heads
26 Underfloor control linkages
27 Ground power socket
28 Combined entry door/airstairs
29 Folding handrail
30 Entry lobby
31 Wing inspection light
32 Cabin attendant's folding seat
33 Avionics racks port and starboard
34 Cockpit bulkhead with sliding door
35 Cockpit roof escape hatch
36 Starboard side emergency exit hatch
37 Composites forward fuselage dorsal fairings
38 ADF antenna
39 ATC Nos. 1 and 2 antennas
40 Fuselage skin panelling
41 Overhead baggage lockers
42 Conditioned air delivery ducts from starboard pack to underfloor mixer unit
43 Wardrobe
44 Door surround structure
45 Exterior light
46 Cabin sidewall seat mounting rail
47 Three-abreast passenger seating; standard 31-seat layout shown
48 Cabin wall insulation blankets
49 Conditioned air delivery duct from port pack
50 Underfloor air mixing unit
51 Underfloor air distribution ducting
52 Cabin window panels
53 Floor beams and seat rails
54 Machined wing spar and landing gear mounting main frames
55 Centre-cabin passenger seating
56 Conditioning system heat exchanger ram-air intake
57 Ground running fan
58 Dual air conditioning packs
59 GPS antenna
60 Heat exchanger exhaust duct
61 Wing centre-section rib structure
62 Inboard integral fuel tank
63 Wingroot leading-edge chord extension
64 Overwing fuel filler
65 Engine bleed air ducting to conditioning pack
66 Starboard engine installation
67 Hinged cowling panels
68 Main engine mounting yoke
69 Starboard nacelle pylon
70 Fuel system vent surge tank
71 Outer wing panel integral fuel tank
72 Fuel venting intake and ducting
73 Pressure refuelling manifold and control panel, ventral access
74 Leading-edge de-icing boot
75 Outboard vent tank
76 Outer wing panel dry bay
77 Composites wingtip fairing
78 Starboard navigation and strobe lights
79 Static dischargers
80 Starboard aileron
81 Aileron hinge control
82 Geared tab
83 Roll control spoiler, linked to aileron
84 Ventral flap hinges
85 Two-segment ground spoilers/lift dumpers
86 Spoiler hydraulic jacks
87 Flap hydraulic jack
88 Nacelle pylon tail fairing
89 Starboard single-slotted Fowler flap, extended
90 Control system access panel
91 Wing control surface linkages
92 Flap interconnecting torque shaft
93 VHF communication antenna
94 Composites trailing-edge fairing panels
95 Starboard service/emergency exit door
96 Auxiliary power unit (APU) air supply duct
97 Extended fin root fillet, composites structure
98 ELT antenna
99 Fin root attachment joints
100 Carbon fibre-reinforced plastic (CFRP) fin structure
101 VOR localiser antenna
102 Elevator control rods
103 Fin leading-edge de-icing
104 Fin/tailplane attachment spar joints
105 Composites fin-tip fairing
106 Starboard fixed tailplane
107 Starboard elevator
108 Elevator trim tab
109 Rudder horn balance
110 Elevator hinge control links
111 Trim tab actuator
112 Port elevator trim tab
113 Elevator composites structure
114 Static dischargers
115 Elevator horn balance
116 Port tailplane CFRP structure
117 Tailplane leading-edge de-icing boot
118 Rudder trim tab
119 Composites rudder structure
120 Rudder hinge control
121 Geared tab
122 APU exhaust
123 Tail navigation light
124 Honeywell 36-150 (DD) APU
125 APU access door
126 Ventral fin, port and starboard
127 APU fire suppression bottle
128 Control cable quadrants
129 Composites fin spar mounting bulkheads
130 CFRP tailcone structure
131 Tailcone joint frame
132 Rear pressure bulkhead
133 Cabin pressure relief outflow valve
134 Baggage compartment
135 Baggage door, open
136 Door surround structure
137 Rear avionics rack with flight- and cockpit voice recorders
138 Baggage restraint net
139 Fuselage aluminium alloy frame and stringer structure
140 Cabin rear bulkhead
141 Overhead passenger service units and cabin roof lighting
142 Lavatory
143 Rear cabin seating
144 Galley unit
145 Port emergency exit window hatch
146 Composites shroud structure
147 Wing spar attachment link
148 Diagonal drag strut
149 Machined wing spars
150 Inboard external flap hinge
151 Engine pylon mounting support structure
152 Pylon mounting support structure
153 Outer wing panel joint rib
154 Engine pylon rear attachment joint
155 Port ground spoiler/lift dumper panels
156 Flap composites structure
157 Port single-slotted Fowler flap
158 Port roll control spoiler
159 Aileron hinge control linkage
160 Port aileron tab
161 Aileron composites structure
162 Static dischargers
163 Port navigation and strobe lights
164 Glass fibre honeycomb wingtip fairing
165 Diagonal tip rib
166 Outer wing panel rib structure
167 Wing bottom skin/stringer panel
168 Leading-edge de-icing boot
169 Engine pylon structure
170 Port wing integral fuel tank
171 Composites leading-edge structure
172 Port engine nacelle
173 Port mainwheel sponson hydraulic equipment bays, oxygen bottle to starboard
174 Pylon attachment joints
175 Engine pylon structure
176 Mainwheel bay door
177 Twin mainwheels, inward retracting
178 Trailing axle suspension
179 Shock-absorber strut
180 Mainwheel leg pivot mounting
181 Battery bay
182 Sponson-mounted landing light
183 Acoustically lined engine intake
184 Pratt & Whitney Canada PW 306 high by-pass ratio turbofan engine
185 FADEC engine controller
186 Accessory equipment gearbox
187 Multilobe exhaust mixer

GERMANY: AIRCRAFT—FAIRCHILD DORNIER

Fairchild Dornier/Envoy 3 flight deck *(Paul Jackson/Jane's)* 0084563

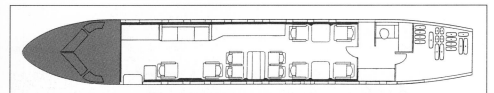

Typical Fairchild Dornier Envoy 3 internal arrangement 1999/0044702

new pylon/nacelle and centre wing attachments, strengthened landing gear and brakes, APU as standard, and modifications to Honeywell Primus 2000 avionics software, FADEC and environmental control systems.

FLYING CONTROLS: Conventional and manual. Optional lateral control (one) and ground (two) spoilers ahead of each aileron; horn-balanced elevators; trim tab in each elevator and rudder; single-slotted Fowler flaps.

STRUCTURE: Wing mainly light alloy structure; entire rear fuselage and tail surfaces of CFRP, except dorsal fin made of Kevlar/CFRP sandwich and aluminium alloy tailplane leading-edge; Kevlar/CFRP sandwich also used for wing trailing-edge structure, nosecone, tailcone and for long wing/fuselage fairing housing system components outside pressure hull; cabin doors of superplastic formed aluminium alloy; engine nacelles of superplastic formed titanium and carbon composites.

OGMA of Portugal manufactures fuselage shells that are assembled into complete fuselages by risk-sharing partner Aermacchi in Italy, which also manufactures flight deck structure; engine nacelles and doors by Westland Aerostructures; wings and tail surface components milled by DASA at Augsburg and Munich; rear fuselage and tail surfaces by Fairchild Dornier; final assembly at Oberpfaffenhofen.

LANDING GEAR: ERAM (with SHL of Israel) retractable tricycle type, with twin Honeywell wheels on each unit; nose unit retracts forward, main units into Kevlar/CFRP sandwich unpressurised fairings on fuselage sides. Main tyres size 24×7.7 (14 ply), nose tyre 19.5×6.75-8 (10 ply); tyre pressures 4.40 bars (64 lb/sq in) on nose unit, 8.00 bars (116 lb/sq in) on main units; alternative 25.5×8.75-10 (14 ply) flotation main tyres; Honeywell brakes.

POWER PLANT: Two 26.9 kN (6,050 lb st) Pratt & Whitney Canada PW 306/9 turbofans with FADEC, pod-mounted under wings.

ACCOMMODATION: Flight crew of two and cabin attendant. Main cabin seats 32 to 34 passengers, three-abreast at 79 cm (31 in) or 76 cm (30 in) pitch, with single aisle; galley to rear of passenger seats; lavatory at rear of cabin, baggage compartment between passenger cabin and rear pressure bulkhead, with external access via baggage door in port side; crew/passenger airstair door at front on port side, with Type III emergency exit opposite; Type III emergency exit on port side at rear of cabin, with service door Type II exit at rear on starboard side.

SYSTEMS: Air conditioning and pressurisation systems standard (maximum differential 0.47 bar; 6.75 lb/sq in), with GKN Westland ECS. Hydraulic and two independent AC/DC electrical systems housed in main landing gear fairings. Tailcone-mounted Honeywell 36-150 (DD) APU as standard.

AVIONICS: Honeywell Primus 2000 suite.
Comms: Dual Primus II integrated radio system and Mode S transponder standard. HF comms optional.
Radar: Primus 650 weather radar standard; Primus 880 weather radar optional.
Flight: AFCS, AHRS, dual integrated avionics computer, dual digital air data reference unit, TCAS, GPWS with windshear detection system standard. Optional GPS with GPS approach and head-up guidance system allowing Cat.II landings.

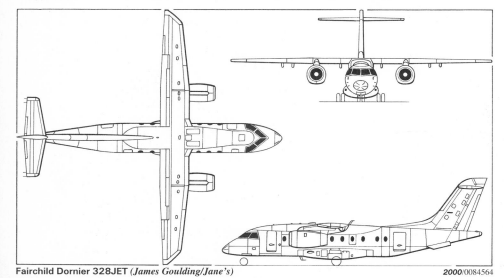

Fairchild Dornier 328JET *(James Goulding/Jane's)* 2000/0084564

Computer-generated image of Fairchild Dornier Envoy 3 business aircraft 2000/0084565

Instrumentation: Five-tube EFIS with EICAS.

DIMENSIONS, EXTERNAL:
Wing span	20.98 m (68 ft 10 in)
Wing aspect ratio	11.0
Length overall	21.28 m (69 ft 9¾ in)
Fuselage: Length	20.92 m (68 ft 7¾ in)
Max width	2.415 m (7 ft 11 in)
Max depth	2.425 m (7 ft 11½ in)
Height overall	7.24 m (23 ft 9 in)
Elevator span	6.70 m (21 ft 11¾ in)
Wheel track (c/l of shock-struts)	3.22 m (10 ft 6¾ in)
Wheelbase	7.42 m (24 ft 4¼ in)
Passenger door (fwd, port): Height	1.70 m (5 ft 7 in)
Width	0.70 m (2 ft 3½ in)
Service door (rear, stbd): Height	1.25 m (4 ft 1¼ in)
Width	0.51 m (1 ft 8 in)
Baggage door (rear, port): Height	1.40 m (4 ft 7 in)
Width	0.92 m (3 ft 0¼ in)

DIMENSIONS, INTERNAL:
Cabin (excl flight deck): Length	10.33 m (33 ft 10¾ in)
Max width	2.18 m (7 ft 2 in)
Width at floor	1.83 m (6 ft 0 in)
Max height in aisle	1.89 m (6 ft 2½ in)
Baggage hold volume	6.4 m³ (226 cu ft)

AREAS:
Wings, gross	40.00 m² (430.6 sq ft)
Ailerons (total)	2.42 m² (26.00 sq ft)
Trailing-edge flaps (total)	7.61 m² (81.90 sq ft)
Fin	11.06 m² (119.00 sq ft)
Rudder	3.92 m² (42.20 sq ft)
Tailplane	9.03 m² (97.20 sq ft)
Elevators	3.08 m² (33.20 sq ft)

WEIGHTS AND LOADINGS (A: standard, B: high gross weight):
Operating weight empty: A	9,344 kg (20,600 lb)
B	4,394 kg (20,710 lb)
Max fuel: A, B	3,646 kg (8,039 lb)
Max payload: A	3,266 kg (7,200 lb)
B	3,676 kg (8,104 lb)
Max T-O weight: A	15,200 kg (33,510 lb)
B	15,660 kg (34,524 lb)
Max landing weight: A	14,090 kg (31,063 lb)
B	14,390 kg (31,724 lb)
Max ramp weight: A	15,350 kg (33,841 lb)
B	15,780 kg (34,789 lb)
Max zero-fuel weight: A	12,610 kg (27,800 lb)
B	13,070 kg (28,814 lb)
Max wing loading: A	380.0 kg/m² (77.83 lb/sq ft)
B	391.5 kg/m² (80.18 lb/sq ft)
Max power loading: A	282.0 kg/kN (2.77 lb/lb st)
B	291.0 kg/kN (2.85 lb/lb st)

PERFORMANCE:
Max cruising speed, M_{MO} at 7,620 m (25,000 ft), ISA +10°:
A, B	405 kt (750 km/h; 466 mph)
Service ceiling: A, B	10,670 m (35,000 ft)
Service ceiling, OEI: A, B	7,528 m (24,700 ft)
T-O required field length: A	1,359 m (4,460 ft)
B	1,422 m (4,665 ft)
Landing required field length: A	1,291 m (4,235 ft)
B	1,387 m (4,550 ft)

Range with max payload, at 9,450 m (31,000 ft), 45 min reserves and allowance for 100 n mile (185 km; 115 mile) diversion:
A	740 n miles (1,370 km; 851 miles)
B	900 n miles (1,666 km; 1,035 miles)

UPDATED

FAIRCHILD DORNIER 428JET

TYPE: Regional jet airliner.

PROGRAMME: Fairchild Dornier management announced on 8 August 2000 that no business case could be foreseen for an aircraft of this size in the company portfolio and that, accordingly, the programme was being discontinued. See 2000-01 edition for description and illustrations.

UPDATED

FAIRCHILD DORNIER 528JET, 728JET and 928JET

TYPE: Regional jet airliner.

PROGRAMME: Baseline 728JET initially designated X28JET; design study, to Lufthansa requirements, announced at the Dubai Air Show in October 1997; launched at ILA Berlin Aerospace Show 19 May 1998 in five-abreast configuration; General Electric CF34 engine selection announced 3 August 1998; baseline configuration frozen

FAIRCHILD DORNIER—AIRCRAFT: GERMANY

Computer-generated image of Fairchild Dornier 728JET in the colours of launch customer Lufthansa

December 1999; first metal cut by SABCA (Belgium) on 18 September 2000; design freeze December 2000; final assembly to begin October 2001; first flight of 728JET due 2002; five aircraft (including first production machine) to fly 1,800-hour test programme, complementing 'iron bird' engineering rig, which commissioned in November 2000; service entry mid-2003. Initial production target (all versions) 100 per year.

CURRENT VERSIONS: **728JET**: As described.

528JET: Projected shortened version with accommodation for 55 to 63 passengers in single or mixed class layouts; derated CF34-8D3 engines; service entry scheduled for mid-2004.

928JET: Growth version with accommodation for 85 to 105 passengers in single or mixed-class layouts; 4.10 m (13 ft 5½ in) fuselage stretch in centre-section; 1.95 m (6 ft 4¾ in) increase in wing span; 82.3 kN (18,500 lb st GE CF34-10D engines; strengthened main landing gear; uprated environmental control system; maximum T-O weight 47,600 kg (104,940 lb); maximum range 1,925 n miles (3,565 km; 2,215 miles). Orders in June 2000 comprised four firm and two options for launch customer Bavaria International Aircraft Leasing GmbH.

1128JET: Possible further stretched version with 110 seats, first mooted in 1998.

Envoy 5: Proposed corporate version of 528JET, not formally launched by early 2000.

Envoy 7: Corporate version of 728JET, with increased range; launched at NBAA Convention in Las Vegas, October 1998; launch customer Flight Options Inc announced order for 25 for its fractional ownership programme at the Paris Air Show on 14 June 1999, to be delivered at the rate of five per year from late 2002; full-scale cabin mockup exhibited at NBAA Convention at Atlanta, Georgia, in October 1999. Lufthansa Technik AG of Hamburg, Germany, appointed European completion centre on 10 October 2000, and scheduled to deliver first aircraft in early 2004 to Safadi Holdings of Beirut; total orders stood at 28 by 1 January 2001.

Envoy 9: Corporate version of 928JET; not formally launched by late 2000.

CUSTOMERS: Total of 281 orders and options by 1 January 2001; launch (728) customer Lufthansa CityLine ordered 60, with 60 options, in April 1999, for delivery between 2003 and 2006. Other announced customers include GE Capital Aviation Services Inc (GECAS), which ordered 50, with up to 100 options, in June 2000, for delivery from 2003; Bavaria International Aircraft Leasing GmbH, which ordered two, with two options, in June 2000; and Sol Air, which has ordered two.

COSTS: Programme US$1.1 billion for all versions. 728JET programme US$800 million; 528JET and 928JET US$150 million each. Unit cost: 728JET US$27 million, Envoy 7 US$30.5 million. Lufthansa order for 60 valued at US$1.6 billion; Flight Options order for 25 Envoy 7s valued at more than US$760 million (all 1999).

DESIGN FEATURES: Conventional twin-jet airliner with low-mounted wings carrying podded engines; sweptback fin and mid-mounted tailplane.

FLYING CONTROLS: Digital fly-by-wire. Ailerons, rudder and all-moving tailplane. Four-section slats on each wing leading-edge; three-section spoilers on each wing upper surface. Flaps.

STRUCTURE: Airframe participants include: Honeywell (avionics, environmental control system and APU); Goodrich (wheels, tyres, brakes, nosewheel steering system and fuel system); SABCA (cockpit and rear fuselage); CASA/EADS (wings and tail surfaces); General Electric (power plant); Hurel-Dubois (engine nacelles); Parker Aerospace (hydraulic system and primary FCS), and Hamilton Sundstrand (electrical system secondary FCS and ram air turbine).

LANDING GEAR: Retractable tricycle type with twin wheels on each unit.

POWER PLANT: Two General Electric CF34-8D3 turbofans, each rated at 60.37 kN (13,572 lb st) with APR, pod-mounted below wings. Standard fuel capacity 12,000 litres (3,170 US gallons; 2,640 Imp gallons).

ACCOMMODATION: Flight deck crew of two with two flight attendants and 68 to 78 passengers; in typical single-class configuration, cabin accommodates 75 passengers, five-abreast, at 84 cm (33 in) seat pitch, or in high-density layout, 85 passengers, five-abreast at 76 cm (30 in) seat pitch with single 48 cm (19 in) aisle; alternative two-class configurations accommodate 60 passengers in five-abreast seating at 84 cm (33 in) seat pitch and eight or ten in four-abreast seating at 96 cm (38 in) pitch at forward end of cabin; galley, lavatory and wardrobe forward, galley and toilet aft. Front crew/passenger airstair door and aft emergency exit on port side; forward emergency exit and aft service door/emergency exit and aft baggage door on starboard side. Two underfloor baggage compartments with doors, port side, forward and aft of wing; upper baggage compartment at rear of cabin. Cabin pressurised, maximum differential 0.57 bar (8.3 lb/sq in), maintaining sea level pressure to 6,431 m (21,000 ft)..

SYSTEMS: Honeywell RE220FD APU.
AVIONICS: Honeywell Epic avionics suite.
Comms: Dual Primus II integrated radio system with VHF com and Mode S transponder standard.

Cabin mockup of Fairchild Dornier Envoy 7 business jet

Computer-generated image of Fairchild Dornier 928JET 85 to 105-passenger airliner

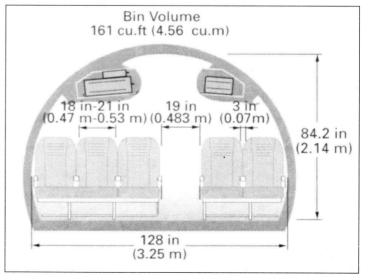

Cabin cross-section of Fairchild Dornier 728JET series

154　GERMANY: AIRCRAFT—**FAIRCHILD DORNIER to FLÄMING AIR**

Radar: Primus 660 weather radar.
Flight: VOR, DME, ADF, ILS standard, with integrated sensor suite containing inertial measurement unit, dual air data modules, AFCS and GPS sensor modules; TCAS 2000; EGPWS; windshear detection system.
Instrumentation: Six-tube EFIS with LCD PFDs, MFDs and EICAS.
The following data are provisional and apply to 728.

DIMENSIONS, EXTERNAL:
Wing span	27.13 m (89 ft 0 in)
Length overall	27.05 m (88 ft 9 in)
Height overall	9.045 m (29 ft 8 in)
Wheel track	4.85 m (15 ft 11 in)
Baggage doors: Height to sill	1.22 m (4 ft 0 in)

DIMENSIONS, INTERNAL:
Cabin (excl rear upper baggage area):
Length	16.13 m (52 ft 11 in)
Max width	3.25 m (10 ft 8 in)
Max height	2.14 m (7 ft 0¼ in)
Rear upper baggage area: Length	1.27 m (4 ft 2 in)
Volume: Forward under floor	6.34 m³ (224 cu ft)
Rear under floor	8.04 m³ (284 cu ft)
Overhead bins (total)	4.56 m³ (161 cu ft)

AREAS:
Wings, gross	75.00 m² (807.3 sq ft)

WEIGHTS AND LOADINGS:
Max fuel weight	10,100 kg (22,267 lb)
Max ramp weight	35,400 kg (78,043 lb)
Max T-O weight	35,300 kg (77,823 lb)
Max landing weight	32,420 kg (71,474 lb)
Max zero-fuel weight	30,300 kg (66.800 lb)
Max wing loading	470.67 kg/m² (96.40 lb/sq ft)
Max power loading	292 kg/kN (2.87 lb/lb st)

PERFORMANCE:
Max cruising speed at 11,280 m (37,000 ft)	M0.81
Max certified altitude	12,495 m (41,000 ft)
T-O required field length	1,527 m (5,010 ft)
Landing required field length	1,282 m (4,206 ft)

Range: 728JET with 70 passengers
　1,800 n miles (3,333 km; 2,071 miles)
Envoy 7 with 16 to 19 passengers, cruising at M0.80 at 12,495 m (41,000 ft)
　4,200 n miles (7,778 km; 4,833 miles)
UPDATED

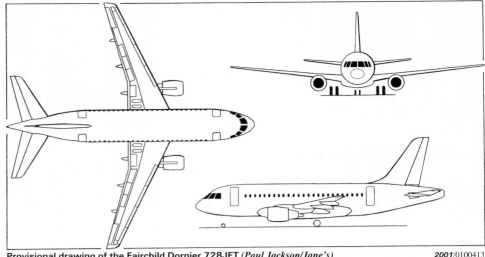
Provisional drawing of the Fairchild Dornier 728JET *(Paul Jackson/Jane's)*　2001/0100413

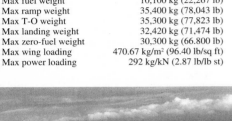
Computer-generated image of Fairchild Dornier 528JET twin-turbofan airliner　2000/0084573

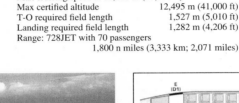
Alternative seating arrangements for Fairchild Dornier 728JET　2001/0104359

FLÄMING AIR

FLÄMING AIR GmbH
Flugplatz Oehna-Zellendorf, D-14913 Zellendorf
Tel: (+49 33742) 603 30
Fax: (+49 33742) 60 33 10
e-mail: flugplatzoehna@t-online.de
Web: http://www.flaemingair.de

In addition to operating a flying school and air charter business, Fläming Air imports and sells Eastern European aircraft to the German market. These include the ATEC Zephyr, Urban Lambáda/Samba and the Podeswa Trener Baby, which is marketed under the Fläming Air name.
NEW ENTRY

FLÄMING AIR TRENER BABY
TYPE: Tandem-seat ultralight/kitbuilt.
PROGRAMME: Designed and built as ultralight prototype by Peter Podeswa; scale replica of Czechoslovak Zlin 126 Trener (member of Zlin 26 family, which was produced between late 1940s and mid-1970s). Production version built by son, Tomas Podeswa at Hranice; certified and distributed by Fläming Air; prototype (OK-FUD 01) exhibited at ILA, Berlin, June 2000. Available from second half of 2000.
CURRENT VERSIONS: **Ultralight:** Wooden wings; MTOW limited to 450 kg (992 lb). Also available as kit, meeting '51 per cent rule'.
　Experimental: Metal wings; 520 kg (1,146 lb) MTOW; kitbuilt.
COSTS: DM90,000, factory-built (2000).
DESIGN FEATURES: Low wing with sweptback leading-edge and unswept trailing-edge; mid-mounted, braced tailplane. Replica of Zlin 126 aerobatic trainer to 80 per cent scale, except full-scale cockpit.
FLYING CONTROLS: Conventional and manual. Split flaps. Flight-adjustable tabs on both elevators; ground-adjustable tab on rudder and starboard aileron.
STRUCTURE: Steel tube fuselage with fabric covering, except for sheet metal upper surface and engine cowling; fabric-covered empennage. Wings wood or metal, according to version, with aluminium leading-edge and fabric covering. Composites nose fairing.
LANDING GEAR: Tailwheel type; fixed. Mainwheels 5.00-5; tailwheel 200×50. Hydraulic brakes.
POWER PLANT: One 55.1 kW (73.9 hp) Walter Mikron III four-cylinder, in-line piston engine driving a Kremen two-blade, electronically controlled, constant-speed propeller. (VZLÚ two-blade, wooden propeller on prototype.) Fuel capacity 35 litres (9.2 US gallons; 7.7 Imp gallons).
AVIONICS: To customer's specification.

DIMENSIONS, EXTERNAL:
Wing span	9.20 m (30 ft 2¼ in)
Length overall	6.20 m (20 ft 4 in)
Height overall	1.80 m (5 ft 10¾ in)

WEIGHTS AND LOADINGS (UL: Ultralight, E: Experimental):
Weight empty: UL	285 kg (628 lb)
E	320 kg (705 lb)
Max T-O weight: UL	450 kg (992 lb)
E	520 kg (1,146 lb)

PERFORMANCE (UL, E, as above):
Never-exceed speed (V_NE):	
UL	145 kt (270 km/h; 167 mph)
E	162 kt (300 km/h; 186 mph)
Cruising speed at 75 % power	
	108 kt (200 km/h; 124 mph)
Stalling speed: UL	36 kt (65 km/h; 41 mph)
E	38 kt (70 km/h; 44 mph)
g limits	+6/−3

NEW ENTRY

Prototype Trener Baby scale replica of Zlin 126 *(Paul Jackson/Jane's)*
2001/0092060

FLIGHT DESIGN

FLIGHT DESIGN GmbH
Sielminger Strasse 65, D-70771 Leinfelden-Echterdingen
Tel: (+49 711) 90 28 70
Fax: (+49 711) 902 87 99
e-mail: flightdesign@t-online.de

PRODUCTION CENTRE:
Ost-Vest Konsalting SP
ulítsa Neftyanikov 15-A, 325000 Kherson, Ukraine
Tel: (+38 55) 223 51 35
Fax: (+38 55) 229 90 32
e-mail: owc@tlc.kherson.ua
DIRECTORS: Andrei Ya Prtygin, Aleksandr A Bondar

Flight Design collaborated with ASO Flugsport of Bottrop in producing the CT. Fabrication of parts is at a Ukrainian factory producing composites to German specifications, with ASO responsible for certification. Several detail changes were introduced in 1999 to meet certification requirements.

UPDATED

FLIGHT DESIGN CT

TYPE: Two-seat lightplane/kitbuilt.
PROGRAMME: Certified in 1997, having flown some 300 hours.
CURRENT VERSIONS: **CTM:** Microlight/Ultralight version; 450 kg (992 lb) or 540 kg (1,190 lb) MTOW, according to country of residence.
CTX: Rapid-assembly kitbuilt; Experimental category.
COSTS: Airframe only, less engine and propeller, DM72,300; complete, with Rotax 912 UL-2, propeller and BRS UL 1050 ballistic parachute recovery system, DM110,440, both including tax (1999). Operating cost DM51 per hour for 1,000 hours per year utilisation (1999).
DESIGN FEATURES: Optimised for safe handling, speed and range. High-mounted, cantilever, constant chord wing, sharply waisted fuselage and sweptback fin with large underfin.
Wing dihedral 1° 30′.
FLYING CONTROLS: Conventional ailerons and horn-balanced rudder, manually actuated; all-moving tailplane with anti-balance tab. Electrically actuated plain flaps, deflection −8° and +30°; internal (external on early aircraft) mass balances for ailerons and tailplane.
STRUCTURE: Principally of composites, including carbon/glass fibre and carbon/aramid sandwich construction.
LANDING GEAR: Fixed tricycle type with swept forward aluminium tube main and nose legs and steerable nosewheel. Optional speed fairings. Hydraulic mainwheel brakes.
POWER PLANT: One 59.6 kW (79.9 hp) Rotax 912 UL-2 or 73.5 kW (98.6 hp) Rotax 912 ULS flat-four, driving a two-blade constant-speed or fixed-pitch Neuform propeller. Fuel in two wing leading-edge tanks combined capacity 110 litres (29.0 US gallons; 24.2 Imp gallons).
ACCOMMODATION: Two persons, side by side; dual controls; upward-hinged door for each occupant. Baggage space behind seats, with external door on each side.
AVIONICS: Optional avionics include Becker Radio AR 4201 com, Honeywell KX 125 nav/com, KT 76A Mode A/C transponder, Garmin GNC 300 XL nav/com/GPS with Jeppesen database, Garmin Pilot III GPS, Flightcom headsets and ELT.
Instrumentation: Winter variometer and Rotax FlyDat engine management system optional.
EQUIPMENT: Optional equipment includes cabin comfort pack; anti-collision and position lights; and Junkers High Speed or BRS UL 1050 ballistic parachute recovery systems.

DIMENSIONS, EXTERNAL:
Wing span	9.31 m (30 ft 6½ in)
Wing chord, constant	1.47 m (4 ft 9¾ in)
Wing aspect ratio	8.0
Length overall	6.22 m (20 ft 5 in)
Height overall	2.165 m (7 ft 1¼ in)
Tailplane span	2.38 m (7 ft 9¾ in)
Wheelbase	1.56 m (5 ft 1½ in)
Propeller diameter	1.73 m (5 ft 8 in)

DIMENSIONS, INTERNAL:
Cabin max width	1.24 m (4 ft 0¾ in)

AREAS:
Wings, gross	10.80 m² (116.3 sq ft)
Ailerons (total)	0.98 m² (10.55 sq ft)
Flaps (total)	1.51 m² (16.25 sq ft)

WEIGHTS AND LOADINGS:
Weight empty: CTM	265 kg (584 lb)
CTX	310 kg (683 lb)
Max T-O weight: CTM	450 kg (992 lb)
	or 540 kg (1,190 lb)
CTX	600 kg (1,322 lb)
Max baggage weight	60 kg (132 lb)
Max wing loading: CTX	55.6 kg/m² (11.38 lb/sq ft)
Max power loading, CTX:	
912 UL-2 engine	10.06 kg/kW (16.53 lb/hp)
912 ULS engine	8.05 kg/kW (13.22 lb/hp)

PERFORMANCE (CTM with Rotax 912 ULS and variable-pitch propeller, except where indicated):
Never-exceed speed (VNE)	167 kt (310 km/h; 192 mph)
Max level speed	159 kt (295 km/h; 183 mph)
Cruising speed at 75% power:	
912 UL-2	130 kt (240 km/h; 149 mph)
912 ULS	146 kt (270 km/h; 168 mph)
Stalling speed	36 kt (65 km/h; 41 mph)
T-O run	90 m (295 ft)
T-O to 15 m (50 ft)	130 m (430 ft)
Range with max fuel, 30 min reserves	810 n miles (1,500 km; 932 miles)
g limits: at 450 kg (992 lb) MTOW	+6
at 600 kg (1,322 lb) MTOW	+4.5

UPDATED

Flight Design CTM ultralight *(Paul Jackson/Jane's)*

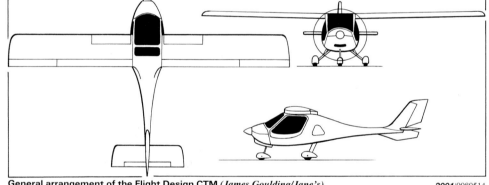
General arrangement of the Flight Design CTM *(James Goulding/Jane's)*

FLIGHT TEAM

FLIGHT TEAM ULTRALEICHTFLUGZEUGE und FLUGSCHULE
Hauptstrasse 63, D-97258 Ippesheim
POSTAL ADDRESS: Lessingstrasse 6, D-97072 Würzburg
Tel: (+49 9339) 12 97 or (+49 171) 770 54 92
Fax: (+49 9339) 998 51
e-mail: flight.team@t-online.de
Web: http://www.home.t-online.de/home/flight.team
DIRECTOR: Peter Götzner

Flight Team markets the Sinus ultralight and motor glider that was developed by Pipistrel in Slovenia with support from the Oskar Ursinus Association (named for a 1920s aviation pioneer, being the Deutscher Verein zur Förderung des Eigenbaus von Luftfahrtgerät; German Association for Promotion of Personal Aviation).

UPDATED

FLIGHT TEAM SINUS

TYPE: Two-seat lightplane/motor glider.
PROGRAMME: Design begun 1994. Prototype (S5-NBP) built and certified in Slovenia. Promotion continues, but no known production.
COSTS: Standard equipped DM71,990, including tax (2000).
DESIGN FEATURES: Pod-and-boom fuselage with high cantilever wing (except on prototype, which was strut-braced), upturned wingtips, and strut-braced T tail. Optional folding propeller for improved engine-off performance.
Aerofoil section IMD 029-b.
FLYING CONTROLS: Conventional elevator and rudder, manually operated. Flight-adjustable trim tab in starboard elevator. Flaperons on inboard 90 per cent of wing trailing-edge, deflections −4°, +8°, +15°. Schempp-Hirth-type spoilers in upper wing, one per side.

Folding propeller of the Flight Team Sinus *(Paul Jackson/Jane's)*

Prototype Sinus, fitted with non-folding propeller *(Paul Jackson/Jane's)*

STRUCTURE: Principally of glass fibre, carbon fibre and Kevlar. Folding wings optional.
LANDING GEAR: Non-retractable tailwheel type with speed fairings on mainwheels.
POWER PLANT: One 37.0 kW (49.6 hp) Rotax 503 UL two-cylinder two-stroke driving a two-blade Pipistrel FB 2 ground-adjustable propeller; optionally propeller folds rearwards into cowling recesses. Fuel in two tanks, combined capacity 60 litres (15.9 US gallons; 13.2 Imp gallons).
ACCOMMODATION: Two persons, side by side in enclosed cockpit, each with upward-opening door.
EQUIPMENT: USH or BRS 1050 ballistic parachute systems optional.

DIMENSIONS, EXTERNAL:	
Wing span	14.80 m (48 ft 6¾ in)
Length overall	6.40 m (21 ft 0 in)
Height overall	1.70 m (5 ft 7 in)
AREAS:	
Wings, gross	12.80 m² (137.80 sq ft)
WEIGHTS AND LOADINGS:	
Weight empty	260 kg (573 lb)
Max T-O weight	450 kg (992 lb)
PERFORMANCE, POWERED:	
Never-exceed speed (V_NE)	124 kt (230 km/h; 143 mph)
Max level speed	100 kt (186 km/h; 116 mph)
Cruising speed	92 kt (170 km/h; 106 mph)
Stalling speed: flaps up	36 kt (66 km/h; 41 mph)
flaps down	32 kt (59 km/h; 37 mph)
Max rate of climb at S/L	228 m (748 ft)/min
T-O run	118 m (387 ft)
Range	513 n miles (950 km; 590 miles)
g limits	+4/−2
PERFORMANCE, UNPOWERED:	
Best glide ratio: fixed propeller	22
folded propeller	27
Min rate of sink at 49 kt (90 km/h; 56 mph):	
fixed propeller	1.24 m (4.1 ft)/s
folded propeller	1.02 m (3.3 ft)/s

UPDATED

FLUG WERKE

FLUG WERKE GmbH
Priel 5/a, D-85408 Gammelsdorf
Tel: (+49 87) 668 44
Fax: (+49 87) 66 13 51
DIRECTORS: Claus Colling
Hans Günther Wildmoder

Specialising in the reproduction of classic German aircraft, Flug Werke plans to augment its Fw 190 project with resumed manufacture of the Messerschmitt Bf 109E and Arado Ar 96 (with Chevrolet V-12 engine). A Heinkel He 112 replica is also planned. New 600 m² (6,450 sq ft) hangar inaugurated 31 March 2000.

UPDATED

FLUG WERKE Fw 190A-8/N
TYPE: Single-seat sportplane.
PROGRAMME: Flug Werke GmbH is assembling a batch of 12 Focke-Wulf Fw 190A-8 Second World War fighters, their /N designation indicating *nachbau*: replica. Components are being manufactured in seven countries (principally in Romania by Aerostar) from original drawings, resulting in airframes claimed to be 98 per cent faithful to the original. Power plant is a 1,400 kW (1,877 hp) ASh-82T 14-cylinder twin-row radial engine, a Russian version of the original BMW 801D. The first aircraft was completed in early 2000; second (D-FWKC) shown statically, April 2000, but first flight unreported by late 2000. Three Fw 190s will be sold in Europe, the remaining nine going to the USA, South Africa and Australia; eight had found firm buyers by April 2000. Price is DM990,000 or US$495,000.

First new production Flug Werke/Focke-Wulf Fw 190A-8/N (990001) nearing completion *(Paul Jackson/Jane's)*

UPDATED

GROB

BURKHART GROB LUFT- UND RAUMFAHRT GmbH & Co KG (Associated with Grob-Werke GmbH & Co KG)
Lettenbachstrasse 9, D-86874 Tussenhausen-Mattsies
Tel: (+49 8268) 99 80
Fax: (+49 8268) 99 81 14
e-mail: grob_aerospace.info@t-online.de
Web: http://www.grob.de
PRESIDENT AND CEO: Dr hc Dipl Ing Burkhart Grob
VICE-PRESIDENTS:
Dipl Ing Klaus-Harald Fischer
Dipl Ing Roland Rischer

Company founded 1928; began aviation activities in 1971, since when has built more than 3,500 aircraft; name changed in 1988 from Burkhart Grob Flugzeugbau (formed 1974) to Burkhart Grob Luft- und Raumfahrt GmbH with light and heavy sections; light section produces sailplanes and the G 115 powered aircraft; heavy section (see 1997-98 *Jane's*) supports Egrett and Strato 2C and deals with space activities in support of Weltraum-Institut Berlin. During 1999-2001 was delivering 99 G 115s for RAF University Air Squadrons.

UPDATED

GROB G 115B, C and D
Royal Navy (Fleet Air Arm) name: Heron
TYPE: Aerobatic two-seat sportplane.
PROGRAMME: First flight November 1985; first flight of second prototype second quarter of 1986 with taller fin and rudder and relocated tailplane; LBA certification to FAR Pt 23 on 31 March 1987; British certification February 1988; later gained full public transport certification and German spinning clearance; production suspended briefly in 1988 after 88 aircraft, resumed in September 1989, then terminated August 1990 after total production of 107 G 115/115As. Power plants and equipment updated and designations changed late 1992 for 1993 product line; total of 200 built by December 1999, comprising 107 early versions, 12 G 115TAs, single G 115B and 85 G 115C/Ds. Descriptions, specifications and photographs of the Grob G 115B, G115C1/C2/C2 IFR Trainer, G115D and G115D2 may be found in the 1999-2000 and earlier editions of *Jane's*.
CURRENT VERSIONS: **G 115E Tutor** and **G 115EG**: Described separately.

UPDATED

Grob G 115E Tutors wearing the markings of East Lowlands (nearest), Aberdeen & St Andrews, Glasgow, Oxford and London UASs *(Paul Jackson/Jane's)*

GROB G 115E and G 115EG
Royal Air Force name: Tutor
TYPE: Aerobatic two-seat sportplane.
PROGRAMME: Earlier versions described in previous entry.
CURRENT VERSIONS: **G 115E**: Selected in June 1998 for UK Ministry of Defence requirement to replace BAe Bulldogs in RAF University Air Squadrons (UAS) and Air Cadet Air Experience Flights, plus a squadron of the Central Flying School (CFS). Total of 99 ordered for operation at 13 sites in civil markings under contract managed by Vosper Thornycroft Aerospace (formerly Bombardier Defence Services). Development aircraft (D-ERAF) first flown early 1999; first production aircraft (D-EUKB/G-BYUA) and three others delivered to UK 15 July 1999; official handover of first five aircraft (G-BYUB/C/D/E/F) at RAF Cranwell, Lincolnshire, 13 September 1999, for CFS; deliveries to Cambridge UAS began 14 September 1999; final delivery scheduled for September 2001.
Description applies to G 115E RAF version.
G 115EG: Version based on G115E for the Egyptian Air Force, which ordered 74 in May 2000 for service at the Egyptian Air Force Academy at Bilbays Air Base. First aircraft formally handed over in Germany 9 October 2000, and scheduled to be delivered to Egypt as part of first batch

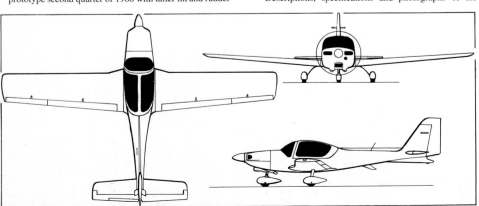

General arrangement of Grob G 115E Tutor *(James Goulding/Jane's)*

Grob Tutor instrument panel *(Paul Jackson/Jane's)*

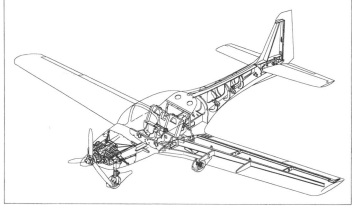
Structural details of Grob G 115E Tutor

of eight in December; deliveries scheduled to continue at the rate of four per month until mid-2001, then increasing to eight per month until contract completion in February 2002.

DESIGN FEATURES: Low-wing monoplane of composites construction optimised for flight training. Wing trailing edge moderately swept forward; slightly tapered horizontal tail surfaces; sweptback fin. G 115E based on earlier G 115s, but with fuselage entirely of carbon fibre, saving more than 40 kg (88 lb) and thus allowing aerobatics to be performed at MTOW.

Wing section Eppler E696; dihedral 5°; incidence +2° at root, −2° at tip; tailplane section NACA 64010.

FLYING CONTROLS: Conventional and manual, via pushrods. Horn-balanced elevators with mechanically actuated trim tab in left unit; elevator deflection +34/−20°; mass-balanced ailerons with ground-adjustable trim tabs, aileron deflection +20° 25′/−18° 25′; horn-balanced rudder with ground adjustable tab, rudder deflection ±30°. Electrically actuated plain flaps, deflection 0, 15 and 60°; intermediate 45° setting optional.

STRUCTURE: Entirely of CFRP; semi-monocoque fuselage has rigid CFRP shell formed in two vertically split halves with frames and web members and integral fin; wing of CFRP/honeycomb sandwich with I-beam main spar and carbon fibre roving spar caps; CFRP/honeycomb horizontal tail surfaces of two-spar construction; rudder, ailerons and flaps are GFRP/rigid foam sandwich. Airframe is protected from lightning strike damage by means of aluminium fibres embedded in the outer layer of carbon fabric and copper mesh in the wing surfaces above and below the fuel tanks, all bonded to metal ground strips. Airframe fatigue life 12,000 hours.

LANDING GEAR: Non-retractable tricycle type with GFRP wheel fairings; cantilever spring steel suspension on main units, gas damping on nose unit; mainwheel tyre size 6.00-6, nosewheel tyre size 5.00-5; nosewheel steering via rudder pedals, maximum deflection ±9° or ±47° with use of differential braking; hydraulic disc brakes.

POWER PLANT: One 134 kW (180 hp) Textron Lycoming AEIO-360-B1F flat-four driving a three-blade wood/composites constant-speed Hoffmann HO-V343K-V/183GY propeller. Christen inverted oil system. Fuel in two integral wing tanks, combined capacity 150 litres (39.6 US gallons; 33.0 Imp gallons) with collector tank, capacity 5.4 litres (1.4 US gallons; 1.2 Imp gallons) for up to three minutes' inverted flight; fuel filler port in top of each wing.

ACCOMMODATION: Two seats side by side, under one-piece, rearward-sliding framed canopy; dual controls (sticks) standard; seats can accommodate backpack parachutes and have five-piece harnesses; baggage space behind seats. Cockpit is heated and ventilated.

SYSTEMS: Electrical system comprises 28 V DC 35 Ah engine-driven generator and 24 V high-capacity battery for engine starting and to provide 30 minutes' emergency power in the event of generator or main bus failure.

AVIONICS: Basic aircraft supplied without avionics. Optional minimum avionics pack includes Honeywell KX 155A navcom/GPS, KI 203 VOR/LOC indicator and KT 76A transponder. Typical customer-specified Honeywell avionics fit for military trainer detailed below.

Comms: KTR 909 UHF transceiver with KFS 599A control unit; dual KX 155A with KI 204 indicator; KT 76A transponder; KMA 26 audio control panel; ELT; PTT switches on control columns and NATO-standard quick-release sockets for headsets.

Flight: KCS55A with KG 102A slaved gyro unit; KI 525A HSI; KA 51B slaving control unit with KMT 112 magnetic azimuth transmitter; KN 63 DME with KDI 572 indicator; KCS 55A slaved remote gyrocompass; KG 102A slaved directional gyro; Filser LX500TR differential GPS.

Instrumentation: Standard ASI, VSI, turn and slip indicator, attitude gyro, directional gyro, tachometer, magnetic compass, CHT/fuel pressure gauge, manifold pressure/fuel flow gauge, oil pressure/temperature gauge, OAT/EGT gauge, fuel quantity gauge, voltage indicator, g meter and clock.

EQUIPMENT: Standard equipment includes navigation, anti-collision, landing and instrument panel lights; map light on flexible arm between seats optional.

DIMENSIONS, EXTERNAL:
Wing span	10.00 m (32 ft 9¾ in)
Wing chord: at root	1.43 m (4 ft 8¼ in)
at tip	0.94 m (3 ft 1 in)
Wing aspect ratio	8.2
Length overall	7.79 m (25 ft 6¾ in)
Height overall	2.82 m (9 ft 3 in)
Tailplane span	3.50 m (11 ft 5¾ in)
Wheel track	2.56 m (8 ft 4¾ in)
Wheelbase	1.51 m (4 ft 11½ in)
Propeller diameter	1.83 m (6 ft 0 in)
Propeller ground clearance	0.19 m (0 ft 7½ in)

DIMENSIONS, INTERNAL:
Cabin (incl baggage area):
Length	2.14 m (7 ft 0¼ in)
Max width	1.04 m (3 ft 5 in)
Max height	1.13 m (3 ft 8½ in)
Baggage volume	0.22 m³ (7.8 cu ft)

AREAS:
Wings, gross	12.21 m² (131.4 sq ft)
Ailerons (total)	1.12 m² (12.06 sq ft)
Fin	1.05 m² (11.30 sq ft)
Rudder	0.64 m² (6.89 sq ft)
Tailplane	1.86 m² (20.02 sq ft)
Elevators (total)	0.86 m² (9.26 sq ft)

WEIGHTS AND LOADINGS:
Weight empty, basic	670 kg (1,477 lb)
Baggage capacity	55 kg (121 lb)
Max fuel	103 kg (227 lb)
Max T-O weight	990 kg (2,182 lb)
Max wing loading	81.08 kg/m² (16.61 lb/sq ft)
Max power loading	7.38 kg/kW (12.12 lb/hp)

PERFORMANCE:
Never-exceed speed (V$_{NE}$)	184 kt (341 km/h; 212 mph)
Max level speed	135 kt (250 km/h; 155 mph)
Long-range cruising speed	96 kt (178 km/h; 110 mph)
Stalling speed: flaps up	53 kt (99 km/h; 61 mph)
60° flap	49 kt (91 km/h; 57 mph)
Max rate of climb at S/L	277 m (909 ft)/min
Max operating altitude	6,095 m (20,000 ft)
T-O run	248 m (815 ft)
T-O to 15 m (50 ft)	462 m (1,515 ft)
Landing from 15 m (50 ft)	457 m (1,500 ft)
Landing run	171 m (560 ft)

Range, with 45 min reserves:
75% power at 1,220 m (4,000 ft)
446 n miles (826 km; 513 miles)
55% power at 2,440 m (8,000 ft)
569 n miles (1,053 km; 654 miles)
45% power at 2,440 m (8,000 ft)
600 n miles (1,111 km; 690 miles)
g limits +6/−3

UPDATED

GROB G 115TA ACRO

There has been no production of this G 115 variant since 1997. Details last appeared in the 2000-01 *Jane's*.

UPDATED

GROB G 120A

TYPE: Aerobatic two-seat lightplane.

PROGRAMME: First flight (D-ELHU; unannounced) 2000; development of G 115 series to meet modern airline pilot and military training requirements, having advanced features including EFIS. With retractable landing gear and 194 kW (260 hp) Textron Lycoming O-540 flat-six engine, G 120A achieves 172 kt (319 km/h; 198 mph) level speed and 365 m (1,200 ft)/min initial climb rate.

Certification and first deliveries due September 2001; announced January 2001, when Lufthansa ordered three (plus four options) for its US training operation at Goodyear, near Phoenix, Arizona, USA.

NEW ENTRY

HEPP

AUGUST HEPP

Schienenhof 4, D-88427 Bad Schussenried
Tel: (+49 73) 921 81 84
Fax: (+49 73) 92 15 04 24

No recent news has been received of the Hepp Surprise kitbuilt ultralight, prototype of which was nearing completion in 1998. A description and photograph last appeared in the 2000-01 edition.

UPDATED

HK

HK AIRCRAFT TECHNOLOGIE AG

Am Flugplatz, D-88512 Hohentengen
Tel: (+49 7572) 71 20 13 and 71 20 15
Fax: (+49 7572) 71 20 14

VERIFIED

HK WEGA

English name: Vega
TYPE: Two-seat lightplane.
PROGRAMME: Prototype (D-ETHK) first flew on 9 April 1998; designation Wega Ex (experimental) applied to initial aircraft, which had completed about 75 hours of flight testing by spring 1999. JAR-VLA certification scheduled for end of 1999. No further reports received.

COSTS: DM150,000 (1998) flyaway, with standard instrumentation.

DESIGN FEATURES: German production version of the Pulsar Aircraft Corporation (Aero Designs) Turbo Pulsar kitbuilt, certified to JAR-VLA. Production version has cockpit width increased by 10 cm (4 in) to 1.06 m (3 ft 5¾ in).

FLYING CONTROLS: Conventional and manual. Electrically actuated pitch trim via tab on port elevator; ground-adjustable tab on starboard aileron; electrically actuated flaps.

STRUCTURE: Principally GFRP with carbon fibre or aluminium reinforcement in high stress areas and foam cores. Wing skins of composites.

LANDING GEAR: Fixed tricycle type with speed fairing on each unit.

POWER PLANT: One 84.6 kW (113.4 hp) Rotax 914 turbocharged flat-four driving a three-blade ground-adjustable Hoffmann HO-V383F propeller. Fuel in two tanks, total capacity 104 litres (27.5 US gallons; 22.9 Imp gallons).

ACCOMMODATION: Two, side by side in reclined bucket seats; forward-sliding, one-piece canopy. Central control column.

AVIONICS: *Comms:* Nav/com and transponder standard.

DIMENSIONS, EXTERNAL:
Wing span	7.62 m (25 ft 0 in)
Wing aspect ratio	7.8
Length overall	6.10 m (20 ft 0 in)
Height overall	2.07 m (6 ft 9½ in)

AREAS:
Wings, gross	7.43 m² (80.0 sq ft)

WEIGHTS AND LOADINGS:
Weight empty	320 kg (705 lb)
Max T-O weight	620 kg (1,367 lb)
Max wing loading	83.4 kg/m² (17.09 lb/sq ft)
Max power loading	7.34 kg/kW (12.05 lb/hp)

PERFORMANCE:
Never-exceed speed (V_{NE})	180 kt (333 km/h; 207 mph)
Max cruising speed	170 kt (315 km/h; 196 mph)
Stalling speed	45 kt (84 km/h; 52 mph)
Max rate of climb at S/L	760 m (2,500 ft)/min
Service ceiling	6,100 m (20,000 ft)
T-O run	350 m (1,150 ft)
Landing run	380 m (1,250 ft)
Range with max fuel	809 n miles (1,500 km; 932 miles)
g limits	+3.8/−2

UPDATED

Prototype HK Wega, German series production version of the Pulsar Aircraft Turbo Pulsar *(Paul Jackson/Jane's)*

IKARUS

IKARUS DEUTSCHLAND (Comco Ikarus Gerätebau GmbH)

Am Flugplatz 11, Flugplatz Mengen, D-88367 Hohentengen
Tel: (+49 7572) 50 81
Fax: (+49 7572) 33 09
e-mail: comco-ikarus.de

Undertakes sales and distribution of the C 42 ultralight in Europe. Ikarus also produces the C 22 open cockpit ultralight, described in the 1992-93 *Jane's*. Over 1,500 Ikarus aircraft have been sold.

VERIFIED

IKARUS C 42

TYPE: Side-by-side ultralight/kitbuilt.
PROGRAMME: Simultaneously launched in Europe (Friedrichshafen Air Show) and North America (Sun 'n' Fun) in second quarter of 1997. European deliveries began immediately. On 10 May 1998 an Ikarus C 42 set a new ultralight world speed record over a closed circuit of 54 n miles (100 km; 62 miles) at 90.4 kt (167.5 km/h; 104.1 mph), and on 13 May 1998 established another world record over a 270 n mile (500 km; 311 mile) closed circuit at 71.78 kt (132.93 km/h; 82.60 mph). Upgraded version, introduced April 1999, offers optional winglets of GFRP and further option of GFRP ailerons, flaps, elevators and rudder.
CURRENT VERSIONS: **Ikarus C 42 Standard:** A*s described.*
C 42 Competition: Kitbuilt version, available from April 1999.
CUSTOMERS: First production C 42 to Netherlands owner (PH-2Y5) mid-1997.
COSTS: Kit, including engine £20,026, factory built £23,359 both excluding VAT (2000).
DESIGN FEATURES: Strut-braced high-wing monoplane; strut-braced tail unit. Wings fold for storage.
NACA 2412 wing section.
FLYING CONTROLS: Conventional and manual. Ailerons and elevators pushrod-operated, rudder cable-operated. Plain flaps standard; electrically actuated flaps optional; flight adjustable trim tab on port elevator.
STRUCTURE: Glass fibre fuselage assembled around full-length aluminium boom; tubular aluminium wings and tail surfaces, covered with Mylar/polyester laminate (GT-foil). Glass fibre-reinforced epoxy resin fairings; stainless steel fittings; Makrolon transparencies. Optional GFRP winglets.
LANDING GEAR: Fixed tricycle type with oleo-pneumatic suspension and speed fairings on all three units; hydraulic brakes on main units.
POWER PLANT: One 59.6 kW (79.9 hp) Rotax 912 UL flat-four driving a two-blade CFK carbon fibre propeller via 1:2.27 reduction gearing; Warp Drive three-blade ground-adjustable propeller optional. Standard fuel capacity 50 litres (13.2 US gallons; 11 Imp gallons). 100 litres (26.4 US gallons; 22 Imp gallons) optional.
EQUIPMENT: Optional BRS 50L4 or Magnum Speed S ballistic parachute, and ELT.

Standard version of Ikarus C 42 *(Paul Jackson/Jane's)*

DIMENSIONS, EXTERNAL:
Wing span: standard	9.45 m (31 ft 0 in)
winglet option	9.50 m (31 ft 2 in)
Length overall	6.20 m (20 ft 4 in)
Height overall	2.34 m (7 ft 8¼ in)

DIMENSIONS, INTERNAL:
Cabin max width	1.22 m (4 ft 0 in)

AREAS:
Wings, gross	12.50 m² (134.5 sq ft)

WEIGHTS AND LOADINGS:
Weight empty: Standard	265 kg (584 lb)
Competition	251 kg (553 lb)
Max T-O weight, all	450 kg (992 lb)

PERFORMANCE:
Cruising speed at 75% power	91 kt (169 km/h; 105 mph)
Stalling speed	34 kt (63 km/h; 40 mph)
Max rate of climb: solo	396 m (1,300 ft)/min
dual	299 m (980 ft)/min
T-O run	79 m (258 ft)
Range with standard fuel	347 n miles (643 km; 400 miles)

UPDATED

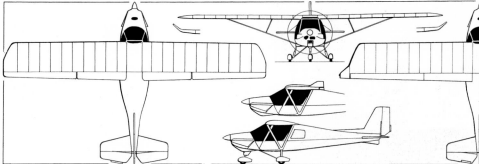

Ikarus C 42 with scrap views of winglet-equipped version *(James Goulding/Jane's)*

KAISER

KAISER FLUGZEUGBAU GmbH

Flugplatz Schönhagen, D-14959 Schönhagen
Tel: (+49 3045) 02 28 58 or (+49 3329) 629 40
Fax: (+49 3329) 629 41
Web: http://www.Kaiser-Flugzeugbau.de
DIRECTOR: Dipl Ing Jörg Kaiser

WORKS:
Wojskowe Zaklady Lotnicze Nr 3
PL-085-21 Deblin-3, Poland
Tel: (+48 81) 883 01 22
Fax: (+48 81) 883 02 48

Kaiser's first product, the Magic, had its debut as a static exhibit at the Berlin Air Show in May 1998. The partly complete Whisper prototype was shown at Aero '99, Friedrichshafen, in April 1999. By mid-2000, neither had flown.

UPDATED

KAISER MAGIC

TYPE: Aerobatic, two-seat sportplane.
PROGRAMME: Design begun in 1995; prototype almost complete when exhibited at Berlin, May 1998, but maiden flight delayed pending static testing of new structural methods. Registered SP-PCM in early 2000; official authorisation for flight cancelled by Polish authorities; remained unflown in June 2000. First four aircraft, all Franklin-engined, had been expected to be completed in 1999. Initial certification to be in Poland, followed by Germany.
CURRENT VERSIONS: **Four-cylinder:** See Power Plant.
Six-cylinder: As prototype; see Power Plant.
COSTS: DM138,000 (four-cylinder) or DM168,000 (six-cylinder) (2000).
DESIGN FEATURES: Fully aerobatic, classic, single-bay biplane with N-type interplane struts and wire-braced empennage. Extra airframe strength achieved by reinforcement of strategic metal tube joints with composites cord binding. German design. Short span optimises roll rate, which approximately 360°/s.

FLYING CONTROLS: Conventional and manual. Ailerons on both wings, interconnected by rods. No flaps.
STRUCTURE: Aluminium tube fuselage with Ceconite covering, except for composites engine cowling, cockpit rim and wing/fuselage fairings. Aluminium-covered ailerons and wing leading-edges; Ceconite elevator and rudder. Glass fibre cantilever mainwheel legs and tailwheel arm. Upper wing on aluminium tube cabane ahead of cockpit; N interplane bracing strut each side, plus bracing wires inboard; fin and tailplane mutually wire-braced. Polish construction by Military Aircraft Works No. 3 at Deblin; available factory-built only.
LANDING GEAR: Tailwheel type; fixed, with speed fairings on mainwheel legs; main tyres 15 × 6.00-6; solid, steerable tailwheel. Hydraulic mainwheel brakes.
POWER PLANT: One Polish (PZL)-built Franklin horizontally opposed piston engine: 93.2 kW (125 hp) four-cylinder 4A-235B4 or 153 kW (205 hp) six-cylinder 6A-350C1R. Three-blade Mühlbauer MTV-9-D-C/C188-18a hydraulically operated constant-speed propeller for six-cylinder version; type for four-cylinder version not yet selected. Fuel tank ahead of cockpit; capacity 100 litres (26.4 US gallons; 22.0 Imp gallons) with six-cylinder engine or approx 160 litres (42.2 US gallons; 35.2 Imp gallons) with four-cylinder engine.
ACCOMMODATION: Two persons in tandem, with dual controls, under single-piece acrylic canopy in six-cylinder version; in open cockpit in four-cylinder version.
AVIONICS: *Flight:* GPS standard.
Instrumentation: Front cockpit has basic instruments only.
EQUIPMENT: Navigation lights standard.
DIMENSIONS, EXTERNAL:
Wing span 6.80 m (22 ft 3¾ in)
Length overall 6.50 m (21 ft 4 in)
Height overall 2.20 m (7 ft 2½ in)
AREAS:
Wings, gross 15.40 m² (165.8 sq ft)
WEIGHTS AND LOADINGS (A: four-cylinder, B: six-cylinder):
Weight empty: A 450 kg (992 lb)
B 500 kg (1,102 lb)
Max T-O weight: A, B 750 kg (1,653 lb)
Max wing loading: A, B 48.7 kg/m² (9.97 lb/sq ft)
Max power loading: A 8.05 kg/kW (13.22 lb/hp)
B 4.91 kg/kW (8.06 lb/hp)

Kaiser Magic aerobatic biplane wearing Polish registration in June 2000 *(Paul Jackson/Jane's)* 2001/0089479

PERFORMANCE (estimated; A, B as above):
Never-exceed speed (V$_{NE}$): A, B
 189 kt (350 km/h; 217 mph)
Max rate of climb at S/L: A 600 m (1,969 ft)/min
B 900 m (2,953 ft)/min
Range: A 432 n miles (800 km; 497 miles)
g limits: B +6/−3
UPDATED

KAISER WHISPER
TYPE: Side-by-side ultralight kitbuilt.
PROGRAMME: Exhibited incomplete at Aero '99 (April 1999) and ILA, Berlin, June 2000. First flight planned for late 2000.
COSTS: Ready to fly, DM49,900; kit, less engine, instruments and parachute recovery system, DM24,500 (both 2000, including tax).
DESIGN FEATURES: Pod-and-boom fuselage. Wings braced by V-type struts; mid-mounted tailplane braced to lower fuselage by parallel struts.
FLYING CONTROLS: Conventional and manual. Two-section Junkers-type flaps, plus separate ailerons. Dual controls.
STRUCTURE: Principally riveted aluminium; Ceconite covering and glass fibre pod and control surfaces.
LANDING GEAR: Tailwheel type; fixed. Carbon fibre spring main gear; speed fairings on mainwheels; floats optional.
POWER PLANT: One 59 kW (79 hp) Hirth F30 flat-four in prototype; production version with BMW horizontally opposed, two-cylinder piston engine; both driving a Junkers three-blade, carbon fibre, ground-adjustable pitch pusher propeller turning at 2,200 rpm, or a ducted pusher propeller. Suitable for other engines in 60 to 89 kW (80 to 120 hp) class.
SYSTEMS: Ballistic parachute recovery system.
DIMENSIONS, EXTERNAL:
Wing span 9.00 m (29 ft 6¼ in)
Length overall 6.60 m (21 ft 7¾ in)
Height overall 2.60 m (8 ft 6¼ in)
AREAS:
Wings, gross 12.90 m² (138.9 sq ft)
WEIGHTS AND LOADINGS:
Weight empty 230 kg (507 lb)
PERFORMANCE (estimated):
Never-exceed speed (V$_{NE}$) 102 kt (190 km/h; 118 mph)
Cruising speed 76 kt (140 km/h; 87 mph)
Max rate of climb at S/L 360 m (1,180 ft)/min
UPDATED

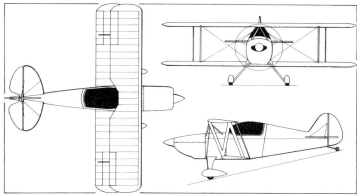

General arrangement of the Kaiser Magic *(Paul Jackson/Jane's)* 0022156

Partly complete prototype Kaiser Whisper *(Paul Jackson/Jane's)* 2001/0089480

MK
MK HELICOPTER GmbH
Hans-Böcklerstrasse 30, D-65468 Trebur-Astheim
Tel: (+49 6147) 91 91 28
Fax: (+49 6147) 91 91 29
e-mail: MK-helicopter-Engineering@t-online.de

DIRECTORS: Uwe Mathes
 Jens Klüver

The MK company was founded on 1 May 1991 to develop, certify and market an economic, environmentally friendly light helicopter.
VERIFIED

MK HELICOPTER MK2
TYPE: Two-seat helicopter.
PROGRAMME: Development, as the Ultralight Mk 2, began 1990; prototype construction started 1995; now aimed at JAR 27 certification. First flight planned for mid-1998, but had not been reported to *Jane's* by June 2000.
COSTS: Target price US$150,000 (1997).

Mockup of the MK Helicopter MK2
0010528

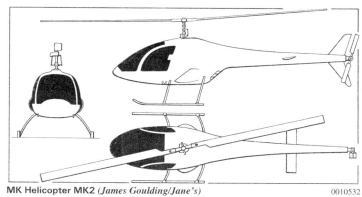

MK Helicopter MK2 *(James Goulding/Jane's)* 0010532

DESIGN FEATURES: Two-blade semi-articulated main rotor; blades comprise glass/carbon-reinforced plastic (GCRP) skin over Rohacell foam core. Rotor brake standard. Two-blade tail rotor on port side; rotor blades of aluminium extrusions. Engine drives two-stage transmission through polymer belts and centrifugal clutch. Helicopter designed by finite element method (FEM) analysis; prototype built using Unigraphics CAD system.
STRUCTURE: Engine, transmission, rotor controls and landing gear attached to aluminium chassis made from two high-strength alloy plates; fuselage of CFRP sandwich with Nomex core. Plexiglas windows.
LANDING GEAR: Skid type with strengthened aluminium tubes.
POWER PLANT: Upgraded version of Rotax 912 four-stroke water- and air-cooled engine with turbocharger, developing 93 kW (125 hp) at 5,500 rpm and 107 kW (144 hp) at 5,800 rpm maximum continuous. Fuel contained in central twin-compartmented tank, usable capacity 80 litres (21.1 US gallons; 17.6 Imp gallons).
ACCOMMODATION: Two persons, side by side in enclosed cabin.

DIMENSIONS, EXTERNAL:
Main rotor diameter	8.06 m (26 ft 5¼ in)
Tail rotor diameter	1.50 m (4 ft 11 in)
Fuselage: Length	6.99 m (22 ft 11¼ in)
Max width	1.32 m (4 ft 4 in)
Height overall	2.73 m (8 ft 11½ in)

AREAS:
Main rotor disc	51.02 m² (549.2 sq ft)
Tail rotor disc	1.77 m² (19.05 sq ft)

WEIGHTS AND LOADINGS:
Weight empty, equipped	388 kg (855 lb)
Max T-O weight	650 kg (1,433 lb)
Max disc loading	12.7 kg/m² (2.61 lb/sq ft)
Max power loading	6.06 kg/kW (9.95 lb/hp)

PERFORMANCE (estimated):
Never-exceed speed (V$_{NE}$)	108 kt (200 km/h; 124 mph)
Max cruising speed at S/L	92 kt (170 km/h; 106 mph)
Max rate of climb at S/L	480 m (1,575 ft)/min
Service ceiling	4,500 m (14,765 ft)
Hovering ceiling: IGE	3,000 m (9,840 ft)
OGE	1,500 m (4,920 ft)
Range with max fuel at S/L	323 n miles (600 km; 372 miles)

UPDATED

MYLIUS

MYLIUS FLUGZEUGWERK GmbH & Co KG
Am Tower 10, D-54634 Bitburg
Tel: (+49 6561) 950 50
Fax: (+49 6561) 95 05 50
e-mail: verwaltung@mylius-flugzeugwerke.de
Web: http://www.mylius-flugzeugwerke.de
MANAGING DIRECTOR: Albert Mylius
FINANCIAL DIRECTOR: A Ruf
PRODUCTION DIRECTOR: Wilfried Stiefel

Mylius Flugzeugwerk was founded in 1996 by Albert Mylius, son of Hermann Mylius whose MHK-101 design was the proof-of-concept prototype for the MBB (Bölkow) BO 209 Monsun light aircraft, last described in the 1973-74 *Jane's*. The company is developing a new range of light aircraft, based on Hermann Mylius' designs and similar in appearance to the Monsun, which employs modular, interchangeable structures for manufacturing efficiency, minimal spares stockage and low maintenance down-time. All models will be certified initially to JAA standards and later to FAR Pt 23.

UPDATED

Prototype Mylius MY-103/200 Mistral Basic Trainer *(Paul Jackson/Jane's)*

MYLIUS MY-103

TYPE: Aerobatic, two-seat sportplane.
PROGRAMME: Two-place development of MY-102 (see 1999-2000 *Jane's*); previously named Mistral. Prototype (MY-103/200 D-ETMY) exhibited unflown at Aero '97 at Friedrichshafen in April 1997; first flight 23 May 1998; five preproduction aircraft in progress in April 1999, but no report of their completion by mid-2000.
CURRENT VERSIONS: **MY-103/180 Standard:** Lycoming IO-360 engine.
MY-103/200 Basic Trainer: Commercial airline or military pilot trainer, also suitable for glider/banner towing; Lycoming AEIO-360 engine; enlarged canopy for helmet clearance in accordance with MIL-STD 1333B.
DESIGN FEATURES: Intended as an affordable aerobatic aircraft, trainer and glider/banner tug for private or club use. Design goals for this and MY-104 include high commonality of parts, responsive controls, stable and predictable handling, high payload/empty weight ratio and low noise emissions. Low-wing monoplane. Tapered, wing with end caps; swept fin with dorsal strake.
FLYING CONTROLS: Conventional ailerons and rudder, manually actuated; all-moving tailplane. Ailerons, rudder and electrically actuated flaps are pushrod-operated; tailplane cable-operated. Ailerons and flaps are each of approximately half-span; anti-servo trim tab on tailplane; external mass balance on each side of rudder.
STRUCTURE: All-metal, apart from some carbon fibre/glass fibre in non-structural components. Wing has single-piece main spar with D-section nose skin forming part of the load-bearing structure; flaps and ailerons hinged to rear auxiliary spar. Wing attached to fuselage by two main and one auxiliary bolts. Rectangular cross-section fuselage of stringer/aluminium skin construction with stainless steel firewall.

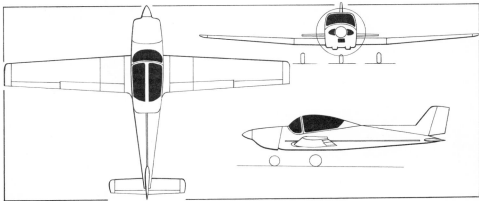
Mylius MY-104/200 four-seat tourer *(James Goulding/Jane's)*

LANDING GEAR: Tricycle type; spring steel main legs are non-retractable; nosewheel steerable and retracts rearwards; streamlined fairings on mainwheels. Mainwheel tyre size 6.00-5; nosewheel 5.00-5. Disc brakes.
POWER PLANT: *MY-103/180 Standard* has one 134.2 kW (180 hp) Textron Lycoming IO-360 flat-four driving a Mühlbauer MTV four-blade constant-speed propeller.
MY-103/200 Basic Trainer has one 149 kW (200 hp) Textron Lycoming AEIO-360 flat-four driving a Mühlbauer MTV-22-B/C174-08 four-blade constant-speed propeller.
Fuel in two wing leading-edge tanks, total capacity 197 litres (52.0 US gallons; 43.0 Imp gallons). Inverted engine lubricating system optional.
ACCOMMODATION: Two, side by side under rearward-sliding large bubble canopy with fixed windscreen. Dual controls standard.
SYSTEMS: Electrical system includes 24 V 13.6 Ah battery.
AVIONICS: VFR or IFR avionics to customer's choice, installed in modular instrument panel facilitating maintenance and replacement.

DIMENSIONS, EXTERNAL:
Wing span	8.65 m (28 ft 4½ in)
Wing aspect ratio	7.2
Length overall	6.50 m (21 ft 4 in)
Height overall	2.34 m (7 ft 8¼ in)

AREAS:
Wings, gross	10.31 m² (111.0 sq ft)

WEIGHTS AND LOADINGS (A: MY-103/180; B: MY-103/200):
Weight empty: A	583 kg (1,285 lb)
B	600 kg (1,323 lb)
Max fuel weight: A, B	142 kg (312 lb)
Payload with max fuel: A	227 kg (500 lb)
B	210 kg (463 lb)
Max T-O weight (Normal category):	
A	950 kg (2,095 lb)
B	930 kg (2,051 lb)
Max wing loading: A	92.15 kg/m² (18.87 lb/sq ft)
B	90.21 kg/m² (18.48 lb/sq ft)
Max power loading: A	7.08 kg/kW (11.64 lb/hp)
B	6.24 kg/kW (10.25 lb/hp)

PERFORMANCE (A, B as above):
Never-exceed speed (V$_{NE}$):	
A, B	216 kt (400 km/h; 249 mph)
Normal cruising speed: A	138 kt (256 km/h; 159 mph)
B	148 kt (274 km/h; 170 mph)
Manoeuvring speed: A, B	146 kt (270 km/h; 168 mph)
Landing speed: A, B	57 kt (106 km/h; 66 mph)
Max rate of climb at S/L: A	312 m (1,024 ft)/min
B	432 m (1,417 ft)/min
Service ceiling: A	5,490 m (18,000 ft)
B	6,100 m (20,000 ft)
Range with max fuel, 15 min reserves:	
A	837 n miles (1,550 km; 963 miles)
B	665 n miles (1,231 km; 765 miles)

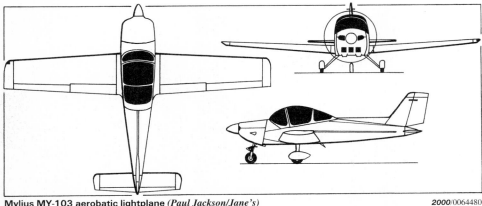
Mylius MY-103 aerobatic lightplane *(Paul Jackson/Jane's)*

Model of the Mylius MY-104/200
(Paul Jackson/Jane's) 2000/0064475

Endurance at 65% power: A	6 h 18 min
B	4 h 30 min
g limits: A, B	+6/−4.5

UPDATED

MYLIUS MY-104/200

TYPE: Four-seat lightplane.
PROGRAMME: First flight was planned in 2000, but has been postponed. Registration D-ETMJ reserved in 1998.
Description as for MY-103 except as follows:
DESIGN FEATURES: Wing span increased by means of new centre-section with dihedral only on outer panels; lengthened fuselage and fully retractable landing gear.
FLYING CONTROLS: Flaps have increased span.
LANDING GEAR: Retractable tricycle type; electric actuation; main legs retract inwards, nosewheel rearwards.
POWER PLANT: One 149 kW (200 hp) Textron Lycoming IO-360 flat-four driving a Mühlbauer MTV three-blade constant-speed propeller. Fuel in two wing tanks, total capacity 276 litres (73 US gallons; 61 Imp gallons).
ACCOMMODATION: Four in two pairs, under large rearward-sliding bubble canopy with fixed windscreen. Cockpit is heated and ventilated.
DIMENSIONS, EXTERNAL:

Wing span	10.40 m (34 ft 1¼ in)
Wing aspect ratio	8.0
Length overall	7.50 m (24 ft 7¼ in)
Height overall	2.38 m (7 ft 9¾ in)
AREAS:	
Wings, gross	13.47 m² (145.0 sq ft)
WEIGHTS AND LOADINGS:	
Weight empty	679 kg (1,498 lb)
Max fuel	200 kg (441 lb)
Payload with max fuel	318 kg (700 lb)
Max T-O weight (Normal category)	1,258 kg (2,774 lb)
Max wing loading	93.4 kg/m² (19.13 lb/sq ft)
Max power loading	8.44 kg/kW (13.87 lb/hp)
PERFORMANCE (estimated):	
Never-exceed speed (V$_{NE}$)	216 kt (400 km/h; 248 mph)
Normal cruising speed	148 kt (274 km/h; 170 mph)
Manoeuvring speed	135 kt (250 km/h; 155 mph)
Landing speed	57 kt (107 km/h; 66 mph)
Max landing crosswind limit	22 kt (41 km/h; 25 mph)
Max rate of climb at S/L	348 m (1,142 ft)/min
Service ceiling	5,490 m (18,000 ft)
Range with max fuel, 15 min reserves	986 n miles (1,826 km; 1,135 miles)
Endurance at 65% power	7 h 0 min
g limits, Utility	+4.4/−2.4

UPDATED

NITSCHE

NITSCHE FLUGZEUGBAU GmbH
Streichenweg 21, D-83246 Unterwössen
Tel: (+49 8641) 69 00 26
Fax: (+49 8641) 69 00 27
e-mail: samburo@t-online.de

This company's Samburo motor glider has now entered production, although retrofits of earlier AVo 68-S variants to current standard are also being undertaken. Additionally, Nitsche also assembles Romanian-fabricated Pretty Flight lightplanes for PC-Flight Flugzeugbau (which see).

UPDATED

NITSCHE AVo 68-R SAMBURO

TYPE: Motor glider.
PROGRAMME: Earlier AVo 68-S programme last described in 1979-80 edition. New AVo 68-R variant, using converted AVo 68-S D-KFGN (c/n 024) as prototype, first flew March 1995, retaining original landing gear of single mainwheel. Second prototype converted from AVo 68-S D-KALO (029), incorporating new twin leg landing gear design. First production AVo 68-R D-KSAM, with turbocharged and intercooled Rotax 914 engine, lengthened and widened fuselage and new canopy design, exhibited unflown at Aero '99 at Friedrichshafen in April 1999; first flight shortly thereafter. Second aircraft (D-KWEN, with Rotax 912) under construction in 2000.
CURRENT VERSIONS: **AVo 68-R100:** With Rotax 912 FS engine
AVo 68-R115: With Rotax 914 engine
CUSTOMERS: Earlier AVo 68v and AVo 68-S production totalled 29. Orders for five of new version reported by early 1997.
COSTS: Standard aircraft with Rotax 912 A4 engine, DM155,000 plus tax (1999).
DESIGN FEATURES: Mid-wing configuration with wingtip fences and moderate sweepback on all leading-edges. Improvements over AVo 68-S include more powerful engine, wider cockpit, glider towing capability, conventional aeroplane landing gear and modernised avionics. Wing folding optional. Maintenance checks at 100 hour intervals.
FLYING CONTROLS: Conventional and manual. Adjustable tab on starboard elevator and plain (door-type) spoilers in upper wings. No flaps.
STRUCTURE: Steel fuselage frame covered with glass fibre. Wings of wood with glass fibre covering to two-thirds chord on inboard panels and whole of outboard fixed surface; polyester covering on remaining wing surface, ailerons and elevator and rudder.
LANDING GEAR: Non-retractable tailwheel type with composites cantilever main legs and castoring tailwheel which unlocks on application of differential hydraulic braking; speed fairings standard. Monowheel configuration with outrigger under each wing optional.
POWER PLANT: One Rotax 912 A3/A4 rated at 59.6 kW (79.9 hp), driving a two-blade fixed-pitch Hoffmann propeller. Rotax 912 FS3/FS4 rated at 73.5 kW (98.6 hp) for take-off and 69.8 kW (93.6 hp) maximum continuous, and Rotax 914 F3 rated at 84.6 kW (113.4 hp) for take-off and 73.5 kW (98.6 hp) maximum continuous, and Hoffmann HO-V352F-S2 constant-speed feathering propeller, all optional. Throttle connected to variable pitch control in one combined lever to replicate fixed-pitch propeller motor gliding handling. Fuel contained in two wing tanks plus header tank, total capacity 80 litres (21.1 US gallons; 17.6 Imp gallons).
ACCOMMODATION: Two persons, side by side under rearward-sliding canopy. Baggage compartment behind seats.
AVIONICS: *Comms:* Optional equipment includes Filser ATR 720A, Becker AR 4201 or Dittel FSG 71M or FSG 90 coms; Honeywell KY 125 nav/com; Honeywell KT 76A or Becker ATC 2000 transponder; and ACK Technologies ELT.

First production AVo 68-R Samburo *(Paul Jackson/Jane's)* 2001/0092064

Instrumentation: Winter 6 FMS ASI, 4 FGH 10 altimeter, 5 StV variometer, compass, ammeter, oil temperature and pressure gauges and CHT gauge all standard. Optional instruments include second altimeter, RC Allen or SFENA 703 artificial horizon, gyrocompass and Elba electronic fuel indicator.
EQUIPMENT: Standard equipment includes ergonomic seats with back supports, four-point seat harnesses, carburettor heat, cockpit heating and ventilation, rudder-mounted antenna and two-colour external paint scheme. Options include hydraulic brake for single-wheel landing gear, glider-towing kit, shift lock for airbrakes, landing and position lights and additional external generator.
DIMENSIONS, EXTERNAL:

Wing span	16.68 m (54 ft 8¾ in)
Wing aspect ratio	13.4
Width, wings folded (option)	10.00 m (32 ft 9¾ in)
Length overall	8.05 m (26 ft 5 in)
Height overall	1.78 m (5 ft 10 in)
Tailplane span	3.47 m (11 ft 4½ in)
Wheel track	1.15 m (3 ft 9¼ in)
AREAS:	
Wings, gross	20.70 m² (222.81 sq ft)
WEIGHTS AND LOADINGS (A: Rotax 912 A3/A4; B: Rotax 912 FS3/FS4; C: Rotax 914 F3):	
Weight empty	530 kg (1,168 lb)
Max T-O weight	730 kg (1,609 lb)
Max wing loading	35.3 kg/m² (7.22 lb/sq ft)
Max power loading: A	12.26 kg/kW (20.14 lb/hp)
B	9.93 kg/kW (16.32 lb/hp)
C	8.64 kg/kW (14.19 lb/hp)
PERFORMANCE, POWERED:	
Never-exceed speed (V$_{NE}$)	116 kt (215 km/h; 133 mph)
Cruising speed, 59.6 kW (79.9 hp) engine at 75% power	97 kt (180 km/h; 112 mph)
Stalling speed	28 kt (51 km/h; 32 mph)
Max rate of climb at S/L: A	216 m (709 ft)/min
B, estimated	246 m (807 ft)/min
C, estimated	276 m (906 ft)/min
T-O run: A	100 m (328 ft)
B, estimated	86 m (282 ft)
C, estimated	78 m (256 ft)
T-O to 15 m (50 ft): A	220 m (722 ft)
B, estimated	183 m (600 ft)
C, estimated	165 m (541 ft)
PERFORMANCE, UNPOWERED:	
Best glide ratio, at 51 kt (94 km/h; 59 mph)	27
Min rate of sink: single wheel	1.03 m (3.38 ft)/s
twin wheel	1.07 m (3.51 ft)/s

UPDATED

Nitsche AVo 68-R Samburo motor glider *(James Goulding/Jane's)* 2001/0092133

OMF

OSTMECHLENBURGISCHE FLUGZEUGBAU GmbH

Flughafenstrasse, D-17039 Trollenhagen
Tel: (+49 395) 42 56 00
Fax: (+49 395) 425 60 02
e-mail: info@omf-aircraft.com
Web: http://www.omf-aircraft.com
PROJECT DIRECTOR: Tim Wright

OMF acquired aerodynamic technology for the Stoddard-Hamilton GlaStar (which see in US section) from Arlington Aircraft Developments in March 1998. Company's 1,500 m² (16,146 sq ft) factory on Trollenhagen airfield at Neubrandenburg was opened on 10 May 2000. In June 2000, the company had 58 employees; it hopes to expand to 280 within three years of product certification.

NEW ENTRY

OMF-160 Symphony, on show at Berlin in June 2000 *(Paul Jackson/Jane's)* 2001/0093678

OMF-160 SYMPHONY

TYPE: Two-seat lightplane.
PROGRAMME: Redesign of GlaStar to JAR 23 certification standards began in May 1998. First prototype (D-ETCW, c/n P1) was essentially unchanged and first flew May 1999; second prototype D-EMVP (c/n 0001), with modifications to meet JAR 23, first flown 7 October 1999; third aircraft (D-ENVG) had not flown by June 2000, when it made type's public debut at Berlin Air Show. Deliveries due to have begun in late 2000. LBA evaluation began mid-July 2000 and certification and production certificate both awarded 29 August 2000.
CURRENT VERSIONS: Base model is two-seater; three- and four-seat versions are planned, plus diesel-powered and IFR-capable variants.
CUSTOMERS: Two flying by mid-2000; by October 2000, deposits held for 37 aircraft and production positions reserved for further 52, mainly by North American customers; company has a target production rate of 300 aircraft per year.
COSTS: US$105,000 (2000).
DESIGN FEATURES: Intended to be an inexpensive trainer and lightplane. Completely redesigned landing gear and wing; retains only some 10 per cent commonality with GlaStar. Around 60 per cent of aircraft built by OMF; remainder sourced from abroad, including Czech Republic, Poland, France, UK and USA. Service life of 18,000 hours. Braced, unswept high wing of constant chord; can be folded by one person. Tailplane has curved, leading-edge root strakes; tall, sweptback fin.
FLYING CONTROLS: Conventional and manual. Mass-balanced Frise ailerons with three hinges each (compared to GlaStar's two); electrically operated, metal, three-position Fowler flaps (0, 20 and 40°). Horn-balanced, single-piece elevator with anti-balance tab; horn-balanced rudder. Vortex generators added to wings in front of ailerons.
STRUCTURE: Fuselage of GFRP with 4130 steel tube frame strengthening surrounding cockpit sections. Metal wings, struts, flaps, ailerons, tailplane, elevator and rudder; composites wingtips, fin fillet, tailplane strakes and wheel fairings.
LANDING GEAR: Tricycle type; fixed. Castoring nosewheel. Sprung steel mainwheel legs; hydraulic brakes. All wheels 5.00-5.

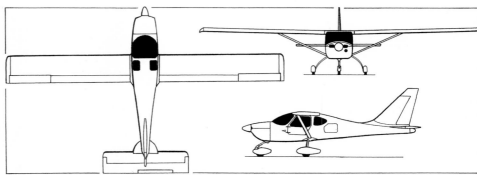

OMF-160 Symphony two-seat lightplane *(Paul Jackson/Jane's)* 2001/0093640

POWER PLANT: One 119 kW (160 hp) Textron Lycoming O-320-D2A flat-four, driving a two-blade fixed-pitch MT propeller. Fuel capacity 114 litres (30.1 US gallons; 25.1 Imp gallons) of which 110 litres (29.1 US gallons; 24.2 Imp gallons) usable.
ACCOMMODATION: Two persons side by side on glass fibre-reinforced composites material seats capable of withstanding 26 *g* crash. Crew door each side; baggage compartment behind crew with separate baggage door on port side.
AVIONICS: Skyforce colour or black and white GPS optional.
DIMENSIONS, EXTERNAL:
Wing span 10.67 m (35 ft 0 in)
Wing chord, constant 1.12 m (3 ft 8 in)
Wing aspect ratio 9.5
Length overall 6.91 m (22 ft 8 in)
Fuselage: max width 1.22 m (4 ft 0 in)
Height overall 2.82 m (9 ft 3 in)
Tailplane span 3.28 m (10 ft 9 in)
Wheelbase 1.73 m (5 ft 8 in)
AREAS:
Wings, gross 11.89 m² (128.0 sq ft)
WEIGHTS AND LOADINGS:
Weight empty 600 kg (1,323 lb)
Max T-O weight 890 kg (1,962 lb)
Max wing loading 74.8 kg/m² (15.33 lb/sq ft)
Max power loading 7.46 kg/kW (12.26 lb/hp)
PERFORMANCE:
Never-exceed speed (V_{NE}) 162 kt (300 km/h; 186 mph)
Max cruising speed at 75% power
 131 kt (243 km/h; 151 mph)
Stalling speed, power off:
 flaps up 55 kt (101 km/h; 63 mph)
 flaps down 46 kt (85 km/h; 53 mph)
Max rate of climb at S/L 365 m (1,198 ft)/min
Service ceiling 5,030 m (16,500 ft)
T-O to 15 m (50 ft) 350 m (1,148 ft)
Landing from 15 m (50 ft) 450 m (1,476 ft)
Range with max fuel 522 n miles (966 km; 600 miles)
g limits +3.8/−1.5

NEW ENTRY

PC-FLIGHT

PC-FLIGHT FLUGZEUGBAU GmbH

Bachweg 3, D-83246 Unterwössen
Tel: (+49 8641) 69 82 55
Fax: (+49 08241) 69 82 56
e-mail: prettyflight@pc-flight.de
Web: http://www.pc-flight.de
PRESIDENT AND DESIGNER: Dipl Ing Calin Gologan

PC-Flight derived its name from the initials of its founders, designer Calin Gologan and test pilot Prof Dr Peter Maderitch. The company's first product, the Pretty Flight, has been built in Romania from 1996 and assembled in Germany by Nitsche Flugzeugbau GmbH. New designs, including low-wing and four-seat versions, will follow.

VERIFIED

PC-FLIGHT PRETTY FLIGHT

TYPE: Side-by-side ultralight.
PROGRAMME: Prototype (D-MNPF) first flew in November 1996; first flight of production aircraft 10 November 1997; German certification achieved 25 September 1998; four prototypes then completed; manufacture of first batch of 10 series production aircraft began in November 1998.
CURRENT VERSIONS: **Ultralight:** MTOW 450 kg (992 lb); *as described.*
 Experimental: Projected version in Experimental category with 550 kg (1,212 lb) MTOW.
COSTS: DM88,888 (1999).

Third prototype PC-Flight Pretty Flight *(Paul Jackson/Jane's)* 2000/0064483

DESIGN FEATURES: Metal structure for high utilisation and reliability; available only in complete form. Designed using company's own CAD programme. High-wing braced monoplane with upturned wingtips, sweptback fin, and tailplane tip fences. Quick-build kit version also available, with quoted build time of 300 hours. Wing section GA(W) PC-1.
FLYING CONTROLS: Manual. Conventional horn-balanced rudder; 75 per cent span flaperons; and all-moving tailplane with anti-balance tabs.

PC-FLIGHT to SCHEIBE—AIRCRAFT: GERMANY 163

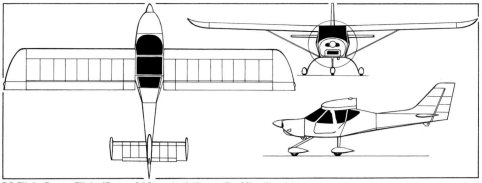

PC-Flight Pretty Flight (Rotax 912 engine) *(James Goulding/Jane's)*   2000/0064484

STRUCTURE: Predominantly metal, with fabric-covered rear wing panels and control surfaces. Production in Romania at Star Tech Impex SRL, with final assembly by Nitsche Flugzeugbau GmbH (which see) in Germany.

LANDING GEAR: Tricycle type, with cantilever spring main legs and optional speed fairings.
POWER PLANT: One 59.6 kW (79.9 hp) Rotax 912 UL flat-four driving a Fiti Speed three-blade fixed-pitch propeller. Fuel in two tanks, combined capacity 100 litres (26.4 US gallons; 22.0 Imp gallons).
EQUIPMENT: BRS 1050 or USH 520 parachute recovery system.

DIMENSIONS, EXTERNAL:
Wing span	10.00 m (32 ft 9¾ in)
Length overall	6.25 m (20 ft 6 in)
Height overall	2.57 m (8 ft 5¼ in)

DIMENSIONS, INTERNAL:
Cabin max width	1.16 m (3 ft 9¾ in)

AREAS:
Wings, gross	11.66 m² (125.5 sq ft)

WEIGHTS AND LOADINGS:
Weight empty, including parachute	290 kg (639 lb)
Max T-O weight	450 kg (992 lb)

PERFORMANCE:
Max level speed	113 kt (210 km/h; 130 mph)
Cruising speed at 75% power	100 kt (185 km/h; 115 mph)
Stalling speed	34 kt (62 km/h; 39 mph)
Max rate of climb at S/L	300 m (984 ft)/min
T-O run	60 m (197 ft)
Range	702 n miles (1,300 km; 807 miles)

UPDATED

REMOS

REMOS AIRCRAFT GmbH
Waldweg 1, D-85283 Eschelbach
Tel: (+49 8442) 96 77 0
Fax: (+49 8442) 96 77 96
e-mail: aircraft@remos.com
Web: http://www.remos.com
PRESIDENT: Lorenz Kreitmayr

Remos formed in 1987; currently markets the G-3 Mirage. Recently discontinued the Gemini Ultra (40 sold). Non-aviation activities include production of automotive components.

UPDATED

REMOS G-3 MIRAGE
TYPE: Side-by-side ultralight.
PROGRAMME: Prototype (D-MRAE) exhibited at Friedrichshafen Air Show, April 1997. First flight 20 September 1997. Production version, shown in April 1999, featured rudder horn balance and minor changes to ailerons and elevator. **Mirage S**, certified 16 August 1999 with 74.6 kW (100 hp) Rotax 912 ULS, has 113 kt (210 km/h; 130 mph) cruising speed. In June 2000, D-MPCJ shown (then unflown) at ILA, Berlin, with 53.7 kW (72 hp) Swiss Auto SAB 430 turbocharged, two-cylinder, four-stroke motorcar engine, which installation weighs 30 kg (66 lb) less than Rotax.
CUSTOMERS: Total of 25 sold by mid-1999. Exports include Argentine police and undisclosed military agency in Romania; one to Netherlands in 1998.

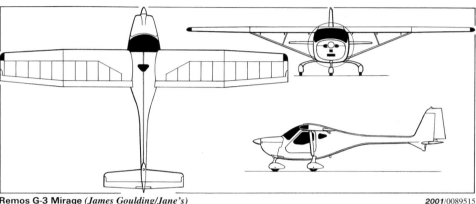

Remos G-3 Mirage *(James Goulding/Jane's)*   2001/0089515

COSTS: DM108,500 (1999).
DESIGN FEATURES: Meets FAR Pt 23. Intended for recreational and instructional use. Waisted rear fuselage, sweptback fin and braced wing with tapered outer panels. Wings fold for transportation and storage.
FLYING CONTROLS: Conventional and manual. Single-piece elevator with flight-adjustable inset tab. Horn-balanced rudder, ailerons and elevator. No rudder tab. Horn-balanced control surfaces on production version. Electrically operated flaps and trim.
STRUCTURE: Composites fuselage; wings of sandwich composites; leading edges and tips composites, remainder fabric-covered; all tail surfaces composites. Tail bumper. GFRP ailerons and flaps. Composites parts produced in Poland; metal parts in Germany.
LANDING GEAR: Tricycle type; fixed. Cantilever main gear with hydraulic drum brakes and integral speed fairings; rubber-in-compression nosewheel leg with speed fairing.
POWER PLANT: One 59.6 kW (79.9 hp) Rotax 912 UL four-stroke engine, driving a GT-2/169.5 two-blade wooden propeller. Fuel capacity: normal 50 litres (13.2 US gallons; 11.0 Imp gallons); optional 70 litres (18.5 US gallons; 15.4 Imp gallons).
AVIONICS: To customer's specification.
EQUIPMENT: Junkers or BRS ballistic parachute.

DIMENSIONS, EXTERNAL:
Wing span	9.80 m (32 ft 1¾ in)
Length overall	6.47 m (21 ft 2¾ in)
Height overall	1.70 m (5 ft 7 in)

AREAS:
Wings, gross	12.04 m² (129.6 sq ft)

WEIGHTS AND LOADINGS:
Weight empty	281 kg (619 lb)
Max T-O weight	450 kg (992 lb)

PERFORMANCE, POWERED:
Never-exceed speed (V_{NE})	121 kt (225 km/h; 139 mph)
Normal cruising speed	105 kt (195 km/h; 121 mph)
Max rate of climb at S/L	348 m (1,141 ft)/min
T-O run	60 m (197 ft)
T-O to 15 m (50 ft)	180 m (590 ft)
Landing from 15 m (50 ft)	200 m (660 ft)
Range	more than 540 n miles (1,000 km; 621 miles)
g limits	+4/−2

PERFORMANCE, UNPOWERED:
Stalling speed, flaps down	34 kt (63 km/h; 40 mph)
Best glide ratio	17
Min rate of sink	1.80 m (5.91 ft)/s

UPDATED

Remos Mirage fitted with SAB 430 motorcar engine *(Paul Jackson/Jane's)*   2001/0089481

SCHEIBE

SCHEIBE FLUGZEUGBAU GmbH
August-Pfaltz-Strasse 23, PO Box 1829, D-85208 Dachau
Tel: (+49 8131) 720 83 and 720 84
Fax: (+49 8131) 73 69 85
e-mail: SFFlugzeug@t-online.de
Web: scheibe-flugzeugbau.de
CO-CHAIRMAN: Matthias Nährlich
CO-CHAIRMAN AND MARKETING DIRECTOR: Werner Hoffman

Scheibe has produced hundreds of gliders since it was founded in 1951 by the late Dipl Ing Egon Scheibe. Current motor glider production comprises the steel tube and wood SF 25C Falke and Rotax Falke; manufacture is also proceeding of the SF 40C ultralight.

UPDATED

SCHEIBE SF 25C FALKE 2000, ROTAX FALKE AND SUPERSCHLEPPER
TYPE: Motor glider.
PROGRAMME: Bergfalke sailplane first flew 5 August 1951; motorised as SF-25A Motor Falke; D-KEDO first flew May 1963. Built in several versions, including under licence by Spartavia-Pützer, Slingsby (T.61) and Loravia (France).
Current SF 25C version first flew (D-KBIK) in March 1971 and gained its type certificate in September 1972. Falke 1700, with 48.5 kW (65 hp) Limbach 1700 engine, has been discontinued. Falke 2100, with Sauer SS2100 engine, tested 1988 (D-KIAC), but only two others built.
CURRENT VERSIONS: **Falke 2000**: Limbach L 2000 EA flat-four engine; monowheel landing gear or conventional tailwheel with two mainwheels available.
Rotax Falke: Prototype D-KIAJ flown 1989. Has essentially the same performance as Falke 2000, but is powered by a water-cooled Rotax 912 A or, from 1999,

Rotax 912 ULS flat-four engine; nosewheel landing gear optional. Most production now of this version; some earlier Falkes have been upgraded with Rotax engines. Tricycle landing gear first tested on Falke 2000 D-KBUG in 1987; deliveries from 1990.

Superschlepper: Glider tug version; has max T-O weight of 600 kg (1,323 lb), in this mode. Rotax 912 A or 912 ULS. Towing trials by D-KIEK in 1998-99.

CUSTOMERS: Over 1,219 Falkes of all types delivered by January 2000, including 658 SF-25Cs by the parent firm. Total 18 registered in 1999, including single exports to Japan and USA. More than 20 Rotax Falkes used as glider tugs.

COSTS: Falke 2000 with Limbach engine and monowheel landing gear, DM143,100; nosewheel landing gear DM5,100 extra; with Rotax 912 A2 engine DM8,600 extra (2000).

DESIGN FEATURES: Docile aerofoil with moderate aspect ratio, designed for safe and simple handling; aircraft can be easily dismantled and transported by trailer. Mid-mounted wing with taper and slight forward sweep; constant chord tailplane; sweptback fin. Various landing gear and engine options.

FLYING CONTROLS: Conventional and manual. Door-type wing spoilers at one-third span; upper wing only. Trim tab on starboard aileron 13 mm (0.5 in) end plates on wing.

STRUCTURE: Steel tube and fabric fuselage; wooden wing and empennage.

LANDING GEAR: Standard landing gear is central monowheel, outriggers under wings, and tailwheel; two-wheel main landing gear with steerable tailwheel, and tricycle landing gear (only with Rotax engine), are optional.

POWER PLANT: Choice of three engines: a 59.6 kW (79.9 hp) Rotax 912 A with water-cooled cylinder heads, running at 5,800 rpm and turning the propeller at 2,500 rpm; a 73.5 kW (98.6 hp) Rotax 912 ULS running at 5,800 rpm and turning the propeller at 2,380 rpm; or a 59.7 kW (80 hp) Limbach L 2000 EA driving the propeller at 3,450 rpm. All engines have electric starter and alternator; propeller feathering optional. Glider tug has MTV-1-A/175-05 constant-speed or MT 175R130-2A fixed-pitch propeller. Standard fuel tankage is 55 litres (14.5 US gallons; 12.0 Imp gallons); optional fuel tankage 80 litres (21.1 US gallons; 17.6 Imp gallons).

ACCOMMODATION: Side-by-side seating under forward-hinged canopy; optional large cockpit opening; instrument panel space for radio.

SYSTEMS: 12 V electrical system.

AVIONICS: Choice of optional nav/com radios.

DIMENSIONS, EXTERNAL:
Wing span	15.30 m (50 ft 2¼ in)
Wing aspect ratio	12.9
Length overall	7.60 m (24 ft 11¼ in)
Height overall: tailwheel	1.68 m (5 ft 6¼ in)
nosewheel	2.50 m (8 ft 2½ in)

AREAS:
Wings, gross	18.20 m² (195.9 sq ft)

WEIGHTS AND LOADINGS (Rotax 912 A):
Manufacturer's weight empty	approx 438 kg (966 lb)
Max T-O weight	650 kg (1,433 lb)
Max wing loading	35.7 kg/m² (7.31 lb/sq ft)
Max power loading	10.90 kg/kW (17.91 lb/hp)

PERFORMANCE, POWERED (L: Limbach, A: Rotax 912 A, S: Rotax 912 ULS, fixed-pitch (FP) or constant-speed (CS) propeller):
Max level speed (all)	102 kt (190 km/h; 118 mph)
Cruising speed: L	81 kt (150 km/h; 93 mph)
A	92 kt (170 km/h; 106 mph)
S	97 kt (180 km/h; 111 mph)
Stalling speed (all)	35 kt (65 km/h; 41 mph)
Max rate of climb at S/L: L	192 m (630 ft)/min
A: FP	240 m (787 ft)/min
CS	288 m (945 ft)/min
S: FP	300 m (984 ft)/min
CS	330 m (1,082 ft)/min
T-O run: L	140 m (460 ft)
A	100 m (330 ft)
S	90 m (295 ft)
Range, standard fuel:	
L, A	378 n miles (700 km; 435 miles)
S	324 n miles (600 km; 372 miles)
Endurance, standard fuel:	
L	4-5 h
A	4 h 30 min
S	4 h

UPDATED

Scheibe SF 25C Falke 2000 with tailwheel landing gear *(Paul Jackson/Jane's)*

SCHEIBE SF 40C ALLROUND

TYPE: Side-by-side ultralight.

PROGRAMME: Prototype **SF 40** first flew May 1994; five of this version built, powered by one 44.7 kW (60 hp) Sauer four-stroke.

SF-40B, first flown 1995, had shorter wingspan but same engine; now superseded. SF 40C introduced in 1996; prototyype D-MSER; further shortened span wing with flaps, plus increased fuel capacity; also suitable for banner and glider towing; type certificate awarded May 1998.

COSTS: DM100,000 plus tax (1998).

DESIGN FEATURES: Wings of constant chord; approximately 5° forward sweep.

FLYING CONTROLS: Conventional and manual. Flaps.

STRUCTURE: Steel tube and fabric fuselage; glass fibre and fabric wing.

LANDING GEAR: Fixed tricycle type.

POWER PLANT: One 59.6 kW (79.9 hp) Rotax 912 flat-four driving a two-blade propeller. Fuel capacity 55 litres (14.5 US gallons; 12.1 Imp gallons).

SYSTEMS: Electrical system: engine-driven alternator, 12 V 30 Ah battery.

DIMENSIONS, EXTERNAL:
Wing span	9.40 m (30 ft 10 in)
Length overall	5.30 m (17 ft 4½ in)
Height overall	1.45 m (4 ft 9 in)

AREAS:
Wings, gross	12.00 m² (129.2 sq ft)

WEIGHTS AND LOADINGS:
Weight empty	280 kg (617 lb)
Max T-O weight: passenger	450 kg (992 lb)
towing	400 kg (882 lb)

PERFORMANCE:
Max level speed	108 kt (200 km/h; 124 mph)
Normal cruising speed	97 kt (180 km/h; 112 mph)
Stalling speed	36 kt (65 km/h; 41 mph)
Max rate of climb at S/L	360 m (1,181 ft)/min
Range with max fuel	378 n miles (700 km; 435 miles)

UPDATED

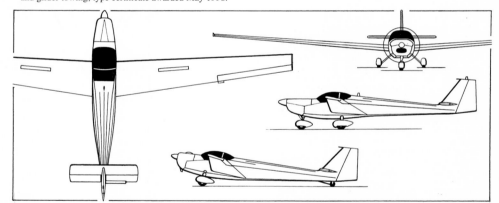

Scheibe SF 25C Falke powered by Rotax 912 A engine, with second side view (lower) of Falke 2000 with tailwheel landing gear *(Mike Keep/Jane's)*

Scheibe SF 40C replaces the original version's Sauer engine by a Rotax 912

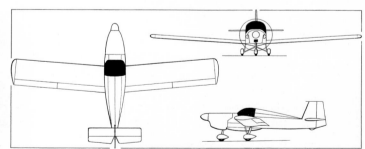

Scheibe SF 40C ultralight *(James Goulding/Jane's)*

STEMME

STEMME GmbH & Co KG

Gustav Meyer Allee 25, D-13355 Berlin (Wedding)
Tel: (+49 30) 463 40 71
Fax: (+49 30) 469 46 49
PRESIDENT: Dr Reiner Stemme
TECHNICAL DIRECTOR: Dipl Ing Gottfried Freudenberger
SALES DIRECTOR: Dipl Ing Sebastian Loewer
MARKETING: Astrid Gaeckler

WORKS:
Strasse F2, Flugplatz, D-15344 Strausberg
Tel: (+49 3341) 361 20
Fax: (+49 3341) 31 11 73
e-mail: S10info@stemme.de
Web: http://www.stemme.de and www.stemme.com

NORTH AMERICAN REPRESENTATIVE:
Stemme USA Inc, 1401 S Brentwood Blvd, Suite 760, Saint Louis, Missouri 63144
Tel: (+1 314) 721 59 04
Fax: (+1 314) 726 51 14
e-mail: info@stemme.com
Web: http://www.stemme.com
PRESIDENT: Barbara Pfifferling

Founded in 1985, Stemme produces components for other manufacturers as well as its own high-performance motor gliders. Flight test and operations moved to Strausberg in 1991; factory area 3,000 m² (32,292 sq ft); 75 employees. 100th aircraft delivered September 1998. Turnover in 1999 was DM7.6 million; expected turnover for 2000 DM8.6 million, rising to DM30 million in 2004; 75 per cent of products are exported. In May 2000 the company announced two new derivatives of the S10: the all-purpose S8 motor glider and S7 sailplane. Production is currently 20 per year.

UPDATED

STEMME S10 and S15

US marketing name: Chrysalis
US Air Force designation: TG-11A

TYPE: Motor glider.

PROGRAMME: Prototype (D-KKST), with Glaser-Dirks DG500 wing, flown 6 July 1986; first flight of second prototype (D-KCHS) with definitive wing 2 June 1987; S10-VC surveillance and observation platform introduced in 1989; designed to JAR 22. Certified in Germany 31 December 1990, in UK 29 October 1991 and in USA, to FAR Pt 21, 8 July 1992. Production of S10 suspended March 1994; resumed late 1994 with S10-V.

CURRENT VERSIONS: **S10**: Original version; 54 produced by March 1994, including prototypes.

Following description applies specifically to S10, but is generally applicable to all variants.

S10-VC: Surveillance/observation conversion with underwing and small wingtip sensor pods for pollution control and resources investigation. One former S10 (D-KGCM/N600PL) used by Greenpeace since May 1994; one former S10-V (PH-1055) operated by Aerial Surveyance den Hollander, Netherlands, since 1995.

S10-V: Variable-pitch propeller version; demonstrator (D-KGCX) completed mid-1994 as conversion of S10 but written off in Tanzania 10 March 1995; one other converted; production version from 1994; certified September 1994; 29 built by end of 1998; none subsequently, but at least six further conversions from S10. (S10-V has engineering designation S14.)

S10-VT: Turbocharged version with 84.6 kW (113.4 hp) Rotax 914 liquid/air-cooled engine driving a variable-pitch propeller with reduction ratio 0.9:1 through toothed gear system. Cruising speed 140 kt (259 km/h; 161 mph); range 696 n miles (1,290 km; 801 miles) with standard fuel; 966 n miles (1,790 km; 1,112 miles) with optional fuel load. Maximum rate of climb at S/L is 240 m (787 ft)/min; service ceiling 9,145 m (30,000 ft). Prototype (D-KGCR) built 1997; LBA certification August 1997, FAA certification September 1997 and CAA certification May 1998; total of more than 36 produced by end of 1999; most to USA. (S10-VT has engineering designation S11.)

S15: Launched at the 1996 Berlin Air Show; has a Rotax 914 engine and a 20.00 m (65 ft 7½ in) wing span; two underwing hardpoints for sensor pods in law-enforcement or scientific research roles suitable for loads up to 65 kg (143 lb) each, including FLIR, night sun searchlights, video cameras, and environmental monitoring equipment being developed by Berlin Technical University's Geography Institute. Prototype (D-ESTE) registered 1996, but no record of completion. S15B, light surveillance version, based on S10-VT, is due to enter service in 2002, followed by S15A advanced surveillance model in 2004; both are being sponsored by Deutsche Bundesstiftung Umwelt. First S10-VT (D-KGCR) shown at Berlin, June 2000, as POC S15.

S-UAV: Unpiloted surveillance version. One S10-V (N600V) currently in use as testbed with Platforms International Corporation, Mojave, California.

CUSTOMERS: Owners in Australia, Austria, Belgium, Brazil, Canada, France, Germany, Mexico, Netherlands, South Africa, Spain, Switzerland, UK and USA. Two delivered in 1995 to 94th Air Training Squadron, USAF Academy, Colorado Springs, Colorado. By mid-2000, 126 S-10 variants (including prototypes) built.

COSTS: S10-VT DM320,000 (2000). S10 and S10-V available to special order.

DESIGN FEATURES: Shoulder-wing, T-tail sailplane with mid-mounted engine and completely retractable propeller. Winglets introduced from 1997. Behaviour tuned for both low and high airspeeds and docility during tight turns; wing not affected by bug accretion or water droplets. Outer wings can be folded by one person for taxying and hangarage; wing sections carried in gantries inside trailer so that one person can fit centre and outer sections straight from trailer to fuselage.

Wing section Horstmann & Quast HQ41/14.35; dihedral 1°.

FLYING CONTROLS: Conventional and manual. Flap/aileron linkage for manoeuvrability and docility at low airspeeds; Schempp-Hirth-type airbrakes in outer ends of centre wing. Six-position flaps.

STRUCTURE: CFRP wings in three sections detachable from fuselage; CFRP rear fuselage and tail mounted to steel tube centre frame carrying wing, engine and landing gear; nose section, of CFRP structure with Kevlar safety lining, mounted at front of centre frame. Engine fully accessible and horizontal firewall separates engine from wing and flying controls.

LANDING GEAR: Tailwheel type; inward-retracting, narrow track mainwheels and fixed tailwheel. Mainwheel tyres 355×122, with electrically actuated, mechanical standby disc brakes. Tailwheel: 210×65 tyre, steerable with rudder, fairing optional on S10 and standard on S10-VT. Larger 369×136 main tyres are optional. Tyre pressure 2.90 bars (42 lb/sq in).

POWER PLANT: One air-cooled 69.4 kW (93 hp) Limbach L 2400 EB1.AD flat-four engine mounted in the central fuselage steel tube frame aft of cockpit; cooling by adjustable ram air intake; 1.90 m (6 ft 2¾ in) CFRP extension shaft in Kevlar tunnel drives folding nose-mounted, two-blade, fixed-pitch propeller through flexible coupling, sliding spline joint and five-belt 1.18:1 reduction gear; engine starting in flight takes 4 seconds; nosecone is moved forward and spring-folded blades emerge through peripheral slot under centrifugal force; centrifugal clutch protects against overspeed, damps starting shocks and allows propeller to slow down independently of engine on shutdown. Two-position variable-pitch propeller on S10-V; inspection interval 200 hours.

Fuel in two 45 litre (11.9 US gallon; 9.9 Imp gallon) fuel tanks in outer ends of centre wing; 60 litre (15.8 US gallon; 13.2 Imp gallon) tanks optional. US market examples have modified fuel system.

ACCOMMODATION: Two pilots side by side; dual controls standard; seats adjustable for position and rake; canopy hinged at forward end and held open by gas struts; heating provided by engine cooling air.

SYSTEMS: 12 V 26 Ah battery and generator; 35 Ah battery optional; full night lighting and landing light optional. Solar cells can be retrofitted into upper cowling to provide 2 A on S10 and is a standard fitting on S10-VT.

AVIONICS: *Comms:* Intercom, VHF radio and transponder optional, as is special cut-down panel for mountain soaring.

DIMENSIONS, EXTERNAL:
Wing span	23.00 m (75 ft 5½ in)
Wing aspect ratio	28.3
Width, wings folded	11.20 m (36 ft 9 in)
Length overall	8.42 m (27 ft 7½ in)
Fuselage max width	1.18 m (3 ft 10½ in)
Height over tailplane	1.80 m (5 ft 10¾ in)
Wheel track	1.15 m (3 ft 9¼ in)
Wheelbase	5.42 m (17 ft 9½ in)
Propeller diameter	1.63 m (5 ft 4¼ in)
Fuselage ground clearance at mainwheels	0.72 m (2 ft 4¼ in)

DIMENSIONS, INTERNAL:
Cockpit: Width	1.16 m (3 ft 9½ in)
Height	0.93 m (3 ft 0½ in)

AREAS:
Wings, gross	18.70 m² (201.3 sq ft)

WEIGHTS AND LOADINGS:
Weight empty	645 kg (1,422 lb)
Max T-O weight	850 kg (1,874 lb)
Max wing loading	45.5 kg/m² (9.31 lb/sq ft)
Max power loading	12.25 kg/kW (20.13 lb/hp)

PERFORMANCE, POWERED:
Never-exceed speed (V_{NE}), smooth air	146 kt (270 km/h; 168 mph)
Manoeuvring speed	97 kt (180 km/h; 112 mph)
Cruising speed:	
fixed-pitch propeller	89 kt (165 km/h; 102 mph)
variable-pitch propeller	121 kt (225 km/h; 139 mph)
Stalling speed	42 kt (78 km/h; 48 mph)
Max rate of climb at S/L:	
fixed-pitch propeller	150 m (492 ft)/min
variable-pitch propeller	210 m (689 ft)/min
T-O run: on concrete	184 m (605 ft)
on grass	300 m (984 ft)
Range: with 90 litres (21 US gallons; 17.6 Imp gallons), variable-pitch propeller, continuous power	863 n miles (1,600 km; 995 miles)
with max fuel and sawtooth climb/glide profile	2,160 n miles (4,000 km; 2,485 miles)

PERFORMANCE, UNPOWERED:
Best glide ratio at 57 kt (106 km/h; 66 mph)	50
Min rate of sink	0.57 m (1.87 ft)/s
g limits	+5.3/−2.65

OPERATIONAL NOISE LEVELS:
To German light aircraft rules	57.3 dBA

UPDATED

Stemme S10-VT two-seat motor glider *(Paul Jackson/Jane's)*

Proof-of-concept Stemme S15 *(Paul Jackson/Jane's)*

Mid-mounted engine of the Stemme S10 *(Paul Jackson/Jane's)*

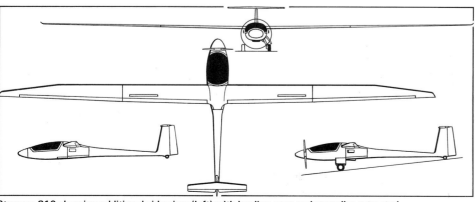
Stemme S10 showing additional side view (left) with landing gear and propeller retracted *(Mike Keep/Jane's)*

STEMME S8

TYPE: Motor glider.
PROGRAMME: Announced at ILA 2000, Berlin, May 2000. First deliveries due early 2002.
CURRENT VERSIONS: S7: Two-seat, high-performance training (20 m) and aerobatic (18 m) glider. Beyond scope of this book.
S8: Two-seat touring/towing motor glider.
DESIGN FEATURES: Based on the Stemme S10 (which see), including mid-mounted engine.
FLYING CONTROLS: As S10.
STRUCTURE: Generally as S10.
LANDING GEAR: Tricycle type; fixed. Speed fairings on all wheels.
POWER PLANT: Touring version will have 73.5 kW (98.6 hp) Rotax 912 ULS flat-four; towing version will have 84.6 kW (113.4 hp) Rotax 914.
ACCOMMODATION: Pilot and passenger/student in side-by-side seating under single-piece forward-hinged canopy.
DIMENSIONS, EXTERNAL (provisional):
Wing span 18.00 m (59 ft 0½ in)
PERFORMANCE, (provisional):
Max operating speed 150 kt (277 km/h; 172 mph)
Service ceiling 3,960 m (13,000 ft)
PERFORMANCE, UNPOWERED (provisional):
Best glide ratio 35

NEW ENTRY

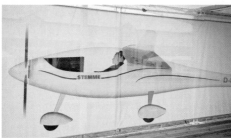
Artist's impression of Stemme S8 *2001/0092084*

TECHNOFLUG

TECHNOFLUG LEICHTFLUGZEUGBAU GmbH

Dr Kurt Steimstrasse 6, D-78713 Schramberg
Tel: (+49 74) 22 84 23
Fax: (+49 74) 22 87 44
e-mail: stz.afl@okay.net
Web: http://www.technoflug.com
CEO: Rolf Schmid
DESIGN ENGINEER: Berthold Karrais
INTERNATIONAL SALES MANAGER: Tom Dietrich

Technoflug is known for its sailplanes and motor gliders; for details of the company's Piccolo ultralight glider (over 120 sold) see 1989-90 *Jane's*. Its newest product, the Carat, was formally announced in 1998.

VERIFIED

TECHNOFLUG TFK-2 CARAT

TYPE: Motor glider.
PROGRAMME: First flown 16 December 1997; flight testing carried out during 1998. Initial production aircraft displayed at Aero '99, Friedrichshafen, in April 1999. Due to have been certified for aerobatics during 1999.
CUSTOMERS: Expected total sales of 350, at around 30 per year, production to start at one or two per month, but none reported built in 1999.
COSTS: DM138,000 (2000) without avionics.
DESIGN FEATURES: Low wing of 21.3 aspect ratio; optional winglets; T tail. Uses wings and horizontal tail (but not water tank) of Schempp-Hirth Discus sailplane, with main spar passing under pilot's knees; wings and tailplane can be detached for storage.
FLYING CONTROLS: Conventional and manual. Schempp-Hirth airbrakes on upper wing surfaces and 'turbulators' below.
STRUCTURE: Generally of glass fibre and carbon fibre composites.
LANDING GEAR: Mainwheels, size 4.00-4 in, retract inwards and forwards electrohydraulically with manual override; fixed steerable 210 × 65 mm tailwheel. Hydraulic disc brakes.
POWER PLANT: One 1,800 cc (109.8 cu in) Sauer E1S conversion of Volkswagen motorcar engine, producing 40 kW (53.6 hp) at max continuous power, driving a 1.40 m (4 ft 7 in) Technoflug KS-F3-1A/140-1 fixed-pitch propeller, blades of which are held forward by gas damping springs located under the spinner when not turning. Other planned engines include 59.6 kW (79.9 hp) Rotax 912. Fuel capacity 45 litres (12.0 US gallons; 10.0 Imp gallons) in single fuel tank situated between cockpit and engine.
AVIONICS: Suggested package includes ATR720A radio, Honeywell transponder and Filser DX50FAI GPS.

DIMENSIONS, EXTERNAL:
Wing span 15.00 m (49 ft 2½ in)
Length overall 6.21 m (20 ft 4½ in)
AREAS:
Wings, gross 10.58 m² (113.9 sq ft)
WEIGHTS AND LOADINGS:
Weight empty 325 kg (717 lb)
Max T-O weight 470 kg (1,036 lb)
PERFORMANCE:
Normal cruising speed at 75% power
 116 kt (214 km/h; 133 mph)
Max rate of climb at S/L 210 m (689 ft)/min
T-O run 226 m (741 ft)
Landing run 137 m (450 ft)
Range with max fuel, 30 min reserves
 486 n miles (900 km; 559 miles)
g limits +5.3/−2.65
PERFORMANCE, UNPOWERED:
Stalling speed 44 kt (80 km/h; 50 mph)
Best glide ratio 35
Min rate of sink 0.75 m (2.46 ft)/s

UPDATED

Prototype Technoflug Carat with propeller retracted *2001/0092143*

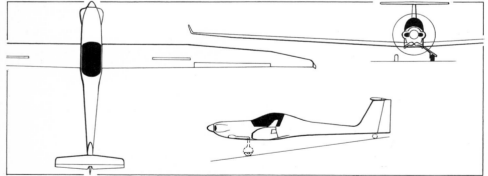
Technoflug Carat (Volkswagen motorcar engine) *(James Goulding/Jane's)* *2001/0092145*

Instrument panel of the Carat *2000/0062971*

USA

ULTRALEICHTFLUGZEUGWERKE SACHSEN-ANHALT

Am Windberg 1, D-39291 Wörmelitz
Tel: (+49 3921) 98 42 90
Fax: (+49 3921) 98 42 90
CHIEF DESIGNER: Fritz Schussler

At the Friedrichshafen Aero 99 show in April 1999, USA launched the SF 45 SA Spirit, a locally modified version of the 3Xtrim lightplane described in the Polish section. Its designation reflects the initials of the designer, the maximum take-off weight and the region in which the company is based. The project failed to develop as rapidly as anticipated, although it remained active in 2000.

UPDATED

USA SF 45 SA SPIRIT

TYPE: Side-by-side ultralight.
PROGRAMME: Upgraded version of Aerotrade Racak 2, of which around 20 built in the Czech Republic, and parallel to certified, Polish-built 3Xtrim (which see). Prototype D-MSBI was due to fly mid-1999, with certification

Prototype USA Spirit at Aero '99, Friedrichshafen, April 1999 *(Paul Jackson/Jane's)* *2001/0093679*

intended soon thereafter. Currently factory built; kit version to be launched at later date.
CUSTOMERS: Four under construction by mid-1999, three of which had been sold.
COSTS: DM87,900 (2000).
DESIGN FEATURES: Braced, high, constant-chord wing and tapered rear fuselage. Extensive use of composites throughout.
FLYING CONTROLS: Conventional and manual. Electric elevator trim. Single-piece, horn-balanced elevator, with trim tab on port side. Flaps.
STRUCTURE: Differs from Racak 2 and 3Xtrim by having all-composites sandwich wing.

LANDING GEAR: Tricycle type; fixed. Rubber-in-compression; nose gear with steerable wheel. Carbon fibre cantilever sprung main gear; hydraulic disc brakes on mainwheels.
POWER PLANT: One 59.6 kW (79.9 hp) Rotax 912 UL flat-four driving a ground-adjustable pitch, two-blade propeller. Kevlar propeller. Fuel tank, capacity 50 litres (13.2 US gallons; 11.0 Imp gallons), behind seats; optional fuel capacity 70 litres (18.5 US gallons; 15.4 Imp gallons). Baggage capacity 18 kg (40 lb).
DIMENSIONS, EXTERNAL:
Wing span 10.00 m (32 ft 9¾ in)
Length overall 6.60 m (21 ft 8 in)
Height overall 2.23 m (7 ft 3¾ in)
AREAS:
Wings, gross 12.50 m² (134.55 sq ft)
WEIGHTS AND LOADINGS:
Weight empty 290 kg (639 lb)
Max T-O weight 450 kg (992 lb)
PERFORMANCE:
Max speed 113 kt (210 km/h; 130 mph)
Cruising speed 100 kt (185 km/h; 115 mph)
Stalling speed 36 kt (65 km/h; 41 mph)
Max rate of climb 204 m (670 ft)/min
Range with max fuel 432 n miles (800 km; 497 miles)
UPDATED

WD

WD FLUGZEUGLEICHTBAU GmbH
Sudetenstrasse 57/2, D-75340 Heubach
Tel: (+49 7173) 92 99 90
Fax: (+49 7173) 92 99 99
e-mail: info@dallach.de
Web: http://www.dallach.de

WD Flugzeugleichtbau markets the Fascination D4 BK, designed by Wolfgang Dallach.
UPDATED

WD D4 BK FASCINATION
TYPE: Side-by-side, ultralight/kitbuilt.
PROGRAMME: Original D4, of steel, wood, fabric and composites construction, first flew in early 1996; public debut (D-MOVE and D-MBWH) at the Réseau du Sport de l'Air rally at Epinal, France, in July 1996. Prototype D4 BK (D-MPMM, carrying spurious registration D-MMMM) exhibited at Aero '99, Friedrichshafen, in April 1999, shortly before maiden flight; US debut at EAA Air Venture at Oshkosh, Wisconsin, in July 2000. JAR-VLA certification scheduled for February 2001.
CURRENT VERSIONS: **D4:** Initial version, last described in 1999-2000 edition.
D4 BK: All-composites version, *as described.* BK indicates *bugrad, kraftstoff:* nosewheel, composites.
CUSTOMERS: Total of 90 D4s sold (including 40 as kits), of which 50 were flying by April 1999, including 15 in Brazil. D4 BK sales totalled five by April 1999.
COSTS: Basic DM139,200; fully equipped DM165,000, including tax (2000). Kit version 15 per cent less.
DESIGN FEATURES: Wings and tail surfaces detach for storage/transport. Quoted kit build time 800 hours.
FLYING CONTROLS: Conventional and manual. Fowler flaps.
STRUCTURE: All composites.
LANDING GEAR: Standard fixed tricycle type, with nosewheel power steering. Retractable optional; main units retract outwards, nosewheel rearwards. Cable-operated disc brakes.
POWER PLANT: One 59.6 kW (79.9 hp) Rotax 912 UL (standard), 73.5 kW (98.6 hp) Rotax 912 ULS (optional) or (on completion of development) 86.8 kW (116.3 hp) DZ 100 flat-four engine driving a Rospeller two-blade electric constant-speed propeller. Fuel capacity 90 litres (23.8 US gallons; 19.8 Imp gallons).

Prototype WD Fascination D4 BK wearing spurious registration *(Paul Jackson/Jane's)* 2001/0092065

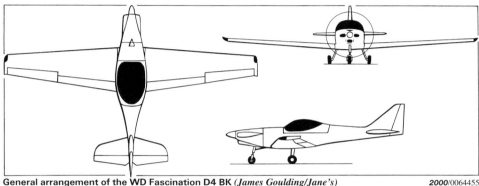
General arrangement of the WD Fascination D4 BK *(James Goulding/Jane's)* 2000/0064455

AVIONICS: GPS and transponder optional.
EQUIPMENT: Ballistic recovery parachute optional.
DIMENSIONS, EXTERNAL:
Wing span 9.00 m (29 ft 6¼ in)
Length overall 6.98 m (22 ft 10¾ in)
Height overall 1.85 m (6 ft 0¾ in)
AREAS:
Wings, gross 12.70 m² (136.7 sq ft)
WEIGHTS AND LOADINGS:
Weight empty 290 kg (639 lb)
Max T-O weight 450 kg (992 lb)
PERFORMANCE (Rotax 912 ULS):
Max level speed 157 kt (290 km/h; 180 mph)
Cruising speed 143 kt (265 km/h; 165 mph)
Stalling speed 35 kt (64 km/h; 40 mph)
Max rate of climb at S/L 390 m (1,280 ft)/min
T-O and landing run 110 m (360 ft)
Range with max fuel 782 n miles (1,450 km; 901 miles)
UPDATED

GREECE

HAT

HELLENIC AERONAUTICAL TECHNOLOGIES
40 Moraiti Street, GR-11525 Athens
Tel/Fax: (+30 1) 677 91 52
GENERAL MANAGER: Anastasios Makrikostas

HAT is a specialist composites materials company. Its initial aircraft, the LS2, is the first Greek design intended for the homebuilt market.
UPDATED

HAT LS2
TYPE: Side-by-side kitbuilt.
PROGRAMME: Design started 1 June 1990; construction of first prototype began 1 February 1993; first flight (SX-LS2) 20 May 1997. Certification sought to JAR 22.
CUSTOMERS: None yet reported.
COSTS: Dr30 million (1998).
DESIGN FEATURES: Low-cost, low-power, simple to build kitbuilt. Designation LS2 indicates landplane, single engine, two-seat. Wings fold for easy storage.
Laminar flow aerofoil; NACA 64₂-415 section at root and 64₁-412 at tip; leading-edge sweep 2° 42′, dihedral 4°, incidence 2°, twist 2°.
FLYING CONTROLS: Conventional and manual. Cable-operated elevator and rudder; pushrods operate ailerons. Sweptback fin with tapered tailplane. Trim tab on elevator, stick force augmentation provided by spring; horn-balanced rudder and elevators. Four-position flaps.
STRUCTURE: Glass fibre throughout, except nose landing gear and engine mount, both of which are 4130 steel. Single-spar wing.
LANDING GEAR: Fixed tricycle type. Original free-castoring (±55°) nosewheel changed to steerable, shock-absorbing unit to improve ground handling. One 4×5.00 Azusa wheel on each unit. Mainwheel tyre pressure 3.45 bars (50 lb/sq in); nosewheel tyre pressure 3.10 bars (45 lb/sq in). Azusa drum brakes on mainwheels; glass fibre speed fairings on all units.
POWER PLANT: Prototype has one 47.7 kW (64 hp) VW motorcar piston engine driving a fixed-pitch wooden two-blade propeller. Fuel capacity 68 litres (18.0 US gallons; 15.0 Imp gallons) in single fuselage tank between cabin and firewall; gravity refuelling point. Oil capacity 2.6 litres (0.7 US gallon; 0.6 Imp gallon).
ACCOMMODATION: Pilot and passenger side by side in enclosed cabin, with upward-hinged gullwing door each side. Baggage compartment behind seats.
SYSTEMS: 20 A, 12 V alternator and 24 Ah battery.
AVIONICS: Basic VFR flight equipped, including VAL Avionics comms.
DIMENSIONS, EXTERNAL:
Wing span 8.20 m (26 ft 10¾ in)
Wing chord: at root 1.20 m (3 ft 11¼ in)
at tip 0.75 m (2 ft 5½ in)
Wing aspect ratio 8.4
Width, wings folded 2.40 m (7 ft 10½ in)
Length overall 5.58 m (18 ft 3¾ in)
Max diameter of fuselage 1.00 m (3 ft 3¼ in)
Height overall 1.96 m (6 ft 5¼ in)
Tailplane span 2.40 m (7 ft 10½ in)
Wheel track 2.00 m (6 ft 6¾ in)
Wheelbase 1.38 m (4 ft 6¼ in)
Propeller diameter 1.32 m (4 ft 4 in)
Propeller ground clearance 0.33 m (1 ft 1 in)
Passenger door: Height 0.48 m (1 ft 7 in)
Width 0.70 m (2 ft 3½ in)
Height to sill 1.05 m (3 ft 5¼ in)
DIMENSIONS, INTERNAL:
Cabin: Length 1.80 m (5 ft 10¾ in)
Max width 0.96 m (3 ft 1¾ in)
Max height 0.90 m (2 ft 11½ in)
Baggage volume 0.60 m³ (21.2 cu ft)
AREAS:
Wings, gross 8.00 m² (86.1 sq ft)
Ailerons (total) 0.53 m² (5.70 sq ft)
Trailing-edge flaps (total) 1.18 m² (12.70 sq ft)
Fin 0.42 m² (4.52 sq ft)
Rudder, incl tab 0.24 m² (2.58 sq ft)
Tailplane 0.88 m² (9.47 sq ft)
Elevators (total, incl tab) 0.5 m² (5.38 sq ft)
WEIGHTS AND LOADINGS:
Weight empty 307 kg (677 lb)
Max fuel weight 50 kg (110 lb)
Max T-O weight 534 kg (1,177 lb)
Max zero-fuel weight 460 kg (1,014 lb)

Max wing loading	66.8 kg/m² (13.67 lb/sq ft)	Econ cruising speed	100 kt (185 km/h; 115 mph)	Landing from 15 m (50 ft)		320 m (1,050 ft)
Max power loading	11.19 kg/kW (18.39 lb/hp)	Stalling speed, power off, flaps down		Landing run		165 m (541 ft)
PERFORMANCE:			46 kt (84 km/h; 52 mph)	Range: with max fuel	564 n miles (1,046 km; 650 miles)	
Never-exceed speed (V$_{NE}$)	169 kt (313 km/h; 195 mph)	Max rate of climb at S/L	213 m (700 ft)/min	with max payload	417 n miles (772 km; 480 miles)	
Max operating speed	126 kt (233 km/h; 145 mph)	T-O run	300 m (980 ft)			UPDATED
Normal cruising speed	109 kt (201 km/h; 125 mph)	T-O to 15 m (50 ft)	480 m (1,575 ft)			

Hellenic Aeronautical Technologies LS2 (VW motorcar engine) *2000/0062384*

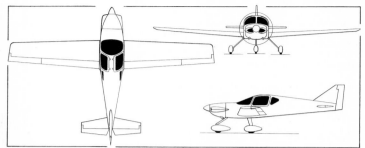
Hellenic Aeronautical Technologies LS2 *(James Goulding/Jane's)* *1999/0044724*

INDIA

ADA
AERONAUTICAL DEVELOPMENT AGENCY
PO Box 1718, Vimanapura Post Office, Bangalore 560 017
Tel: (+91 80) 523 30 60 and 20 83
Fax: (+91 80) 523 84 93 and 44 93
e-mail: info@ada.ernet.in
LCA AND MCA PROGRAMME DIRECTOR: Dr Kota Harinarayana
SENIOR MANAGER, INFORMATION TECHNOLOGY: Dr T N Prakash

The ADA, an autonomous organisation under the Indian Ministry of Defence, was created in 1984 to be responsible for design and development of the next-generation Light Combat Aircraft (LCA); its manufacturing partner is Hindustan Aeronautics Ltd (HAL, which see). Several laboratories of the Defence Research and Development Organisation (DRDO) have also contributed to the LCA programme.

UPDATED

Provisional drawing of the ADA Light Combat Aircraft *(Mike Keep/Jane's)* *1999/0051645*

ADA LIGHT COMBAT AIRCRAFT (LCA)
TYPE: Multirole fighter.
PROGRAMME: Development approved by Indian government in 1983 as MiG-21 replacement; project definition begun second quarter 1987, completed late 1988, with first flight then predicted for April 1990 and service entry in about 1995. However, due to early slippages, basic design not finalised until 1990; construction by HAL started mid-1991. First aircraft (TD1: technology demonstrator; serial number KH2001) was nine months behind revised schedule at roll-out on 17 November 1995; several subsequent postponements of new (June 1996) target date for first flight, which eventually took place on 4 January 2001; TD2 rolled out on 14 August 1998 and now targeted to fly in June 2001. These two aircraft are powered by F404-GE-F2J3 afterburning turbofan, ground runs of which began on 9 April 1998; indigenous Kaveri engine being developed by Gas Turbine Research Establishment (GTRE) Bangalore for next four (originally five) prototype vehicles (PV1 to PV4) and production aircraft. GTRE engine will undertake flight trials in Russia in Tu-16 testbed; bench trials of complete engine had reportedly totalled nearly 1,000 hours by early 2001. Three of the PVs to be single-seaters (including one in naval configuration); the fourth will be a two-seat operational trainer.

Programme delays further exacerbated by May 1998 US embargo on supplies and assistance following India's refusal to abandon nuclear weapons testing, necessitating all-Indian completion of development of integration of digital AFCS and some other systems (LCA is inherently unstable and thus unable to fly without AFCS. However, development of the indigenously developed radar is said to be on schedule; this is to be flight tested initially in a HAL HS 748 and first installed in the PV1 LCA. Work on PV1 and PV2 was under way in late 1998; additional US$125 million released in May 1999 in an attempt to accelerate remaining development.

Latest (February 2001) estimate indicated third (PV1) prototype due to be flying by end of 2001; fourth and fifth (PV2 and 3) in 2002; first two-seater (PV4) in 2003; and

LCA instrument panel *0052370*

series production to begin in 2007, aimed at service entry by 2012.
CURRENT VERSIONS: **LCA:** Single-seat version for Indian Air Force.
  LCA Trainer: Tandem two-seat operational trainer.
  LCA Navy: Single-seat carrierborne version; drooped nose, arrester hook, long-stroke landing gear and movable vortex controls at the wingroot. Design approved early 1999, but still in planning stage.
CUSTOMERS: Indian Air Force (requirement for up to 220).
COSTS: Phase 1 (TD1 and TD2) costs, including engine, estimated at approximately Rs30 billion (US$675 million) by late 2000. Estimated unit cost, based on 220 production total, US$17 million to US$20 million (2001).
DESIGN FEATURES: Tail-less delta planform with relaxed static stability; shoulder-mounted delta wings with compound sweep on leading-edges; large twist from inboard to outboard leading-edges; wing-shielded, side-mounted, fixed-geometry Y-duct air intakes. Advanced materials for minimum structural weight; fly-by-wire and HOTAS controls; high agility; supersonic at all altitudes; wide range of external stores.
FLYING CONTROLS: Two-segment trailing-edge elevons and three-section leading-edge slats; vortex-shedding inboard leading-edges with inboard slats to form vortices over wingroot and fin; airbrake in top of fuselage each side of vertical fin. Quadruplex digital fly-by-wire AFCS.
STRUCTURE: Advanced materials to include aluminium-lithium alloy, carbon composites and titanium alloys; CFRP wings (including elevons), fin and rudder; Kevlar radome. Wings manufactured with one-piece top and bottom skins, bolted on to wing box; majority of spars and ribs in composites. Fin, rudder, elevons, airbrakes and landing gear doors embody co-cured, co-bonded techniques. Carbon fibre composites to be increased from 30 per cent (by weight) in TD1/2 to 45 per cent in PVs, with corresponding reduction of aluminium alloys from 57 to 43 per cent.

Hindustan Aeronautics (HAL) responsible for fuselage, communications system, electrical system, mechanical system LRUs and utility systems management; Aeronautical Development Establishment (ADE) developing flight control system, cockpit displays and display processors; National Aerospace Laboratories (NAL) responsible for development of fin and fabrication of rudder.
LANDING GEAR: Hydraulically retractable tricycle type. Single mainwheels; twin-wheel nose unit. DRDL carbon disc brakes; brake management system. ADRDE brake-chute in fairing at base of rudder.

TD1 first prototype of the LCA during its maiden flight on 4 January 2001 *2001/0073898*

POWER PLANT: One 80.5 kN (18,100 lb st) General Electric F404-GE-F2J3 afterburning turbofan in TD1 and TD2; Indian GTRE GTX-35VS Kaveri turbofan (52.0 kN; 11,700 lb st dry, 80.1 kN; 18,000 lb st with afterburning), with digital engine control unit (KADECU), under development for PV prototypes and production aircraft. Internal fuel in wing and fuselage tanks. Fixed in-flight refuelling probe on starboard side of front fuselage. Provision for up to three 1,200 or five 800 litre (317 or 211 US gallon; 264 or 176 Imp gallon) external fuel tanks.
ACCOMMODATION: Pilot only, on zero/zero ejection seat. Canopy opens sideways to starboard. Development will include two-seat training version.
SYSTEMS: Hydraulic system for powered flying controls, brakes and landing gear; electrical system for fly-by-wire and avionics power supply; environmental control system; lox system; jet fuel starter; aircraft-mounted accessory gearbox; USMS for aircraft system management.
AVIONICS (production version): *Comms:* V/UHF and UHF radios and air-to-air/air-to-ground secure datalink; IFF transponder/interrogator.
Radar: Electronics Research and Development Establishment (ERDE)/HAL multimode radar with multitarget search and track-while-scan and ground mapping functions; coherent pulse Doppler system, with look-up/look-down modes, Doppler beam sharpening and moving target indication.
Flight: RLG-based INS; provision for GPS/INS; radio altimeter.
Instrumentation: NVG-compatible 'glass cockpit'; two Bharat Electronics active matrix colour LCD MFDs; collimated HUD; dedicated LCD 'get-you-home' panel; LCD multifunction keyboard.
Mission: Three multiplexed MIL-STD-1553B digital databusses; 32-bit mission computer operating in Ada, backed up by a second, equally powerful computer; IRST; laser range-finder/designator pod. Provision for reconnaissance, EW or other sensor pods. Helmet-mounted display/sight.
Self-defence: MAWS; RWR; jammer; chaff/flare dispenser.
ARMAMENT: Internally mounted GSh-23 twin-barrel 23 mm gun with 220 rounds. Seven external stores stations (three under each wing and one under fuselage) for wide range of short/medium-range air-to-air missiles, PGMs, air-to-surface (including anti-ship) missiles, unguided rockets, conventional and retarded bombs, and cluster bomb dispensers. Indigenous Astra active radar-guided ASM under development by DRDO. All except outboard underwing stations are wet for carriage of drop tanks.

DIMENSIONS, EXTERNAL:
Wing span	8.20 m (26 ft 10¾ in)
Wing aspect ratio	1.8
Length overall	13.20 m (43 ft 3¾ in)
Height overall	4.40 m (14 ft 5¼ in)
Wheel track	2.20 m (7 ft 2½ in)
Wheelbase	4.34 m (14 ft 2¾ in)

AREAS:
Wings, gross	approx 38.40 m² (413.3 sq ft)

WEIGHTS AND LOADINGS:
Weight empty	approx 5,500 kg (12,125 lb)
Max external stores load	more than 4,000 kg (8,818 lb)
T-O weight (clean)	approx 8,500 kg (18,740 lb)
Wing loading (clean)	approx 221.4 kg/m² (45.35 lb/sq ft)
Power loading (TD1 and 2, clean)	106 kg/kN (1.04 lb/lb st)

PERFORMANCE (estimated):
Max level speed at altitude	M1.8
Service ceiling	above 15,240 m (50,000 ft)
g limits	+9/–3.5

UPDATED

ADA MEDIUM COMBAT AIRCRAFT (MCA)

The ADA is studying an advanced version of the LCA under the above designation, as a potential replacement for Indian Air Force Jaguars and Mirage 2000s from about 2008. According to Defence Research and Development Organisation (DRDO) officials at the Indian Air Show in December 1996, the twin-engined MCA would embody stealth technology, be powered by a non-afterburning variant of the GTRE Kaveri turbofan with thrust vectoring, and have a greater range than the LCA. Other design features, disclosed in late 1998, include the adoption of radar-absorbent materials; twin outward-canted tailfins; and overwing external fuel tanks. Project definition has begun;

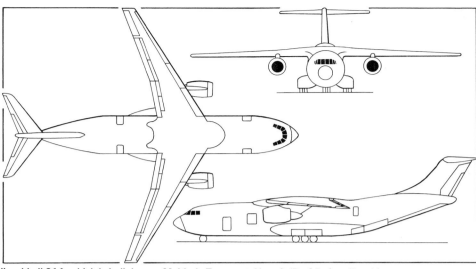

Ilyushin Il-214, which is India's new Multirole Transport Aircraft *(Paul Jackson/Jane's)* 2001/0062411

Model of projected MTA/Il-214 at Aero India, February 2001 *(Craig Hoyle/Jane's)* 2001/0095089

design and development costs have been provisionally estimated at US$2.3 billion in 1996 dollars.

WEIGHTS AND LOADINGS (approx):
T-O weight: clean	12,000 kg (26,455 lb)
max	18,000 kg (39,685 lb)

VERIFIED

ADA MULTIROLE TRANSPORT AIRCRAFT (MTA)

TYPE: Medium transport/multirole.
PROGRAMME: This Indian requirement has become the subject of a joint venture between HAL and Ilyushin (Russian designation Il-214; which see), as the result of an MoU signed in late 2000. Configuration now differs considerably from the original Indian design of 1996, details of which last appeared in the 2000-01 *Jane's*.
CURRENT VERSIONS: **Airliner:** With cabin windows and seating for 100. Russian designation **Il-214-100**.
Tactical transport: Military version with rear-loading ramp; provision for 82 paratroops. Russian designation **Il-214T**.
Commercial freighter: With rear loading ramp.
CUSTOMERS: Required by Indian and Russian air forces; manufacture by HAL and IAPO. Estimated market for 200 to 300. Unit cost US$12 million to US$15 million with PS-90 engines.
COSTS: Development estimated as US$300 million.

POWER PLANT: Two turbofans, each of some 97.9 kN (22,000 lb st). Options include Aviadvigatel PS-90 or Rolls-Royce BR715.

DIMENSIONS, EXTERNAL:
Wing span	30.10 m (98 ft 9 in)
Length overall	33.20 m (108 ft 11 in)
Height overall	10.00 m (32 ft 0¾ in)

DIMENSIONS, INTERNAL:
Cargo hold: Length	13.50 m (44 ft 3½ in)
Max width at floor	3.15 m (10 ft 4 in)
Max height	3.00 m (9 ft 10 in)

WEIGHTS AND LOADINGS:
Max payload	18,500 kg (40,785 lb)
Max fuel weight	13.500 kg (29,762 lb)
Max T-O weight	55,000 kg (121,250 lb)
Max landing weight	49,500 kg (109,125 lb)

PERFORMANCE:
Max level speed	459 kt (850 km/h; 528 mph)
Normal cruising speed	432 kt (800 km/h; 497 mph)
Airdropping speed	135 kt (250 km/h; 155 mph)
Service ceiling	10,975 m (36,000 ft)
T-O to 15 m (50 ft)	1,300 m (4,265 ft)
Landing from 15 m (50 ft)	1,200 m (3,940 ft)
Range: with max payload	1,349 n miles (2,500 km; 1,553 miles)
with 4,500 kg (9,920 lb) payload and auxiliary fuel	3,239 n miles (6,000 km; 3,728 miles)

UPDATED

BHEL

BHARAT HEAVY ELECTRICALS LTD

BHEL House, Sirifort, New Delhi 110 049
Tel: (+91 11) 649 34 37
CHAIRMAN AND MANAGING DIRECTOR: K G Ramachandran

WORKS: Heavy Electrical Equipment Plant, Ranipur, Hardwar PIN 249 403, Uttar Pradesh
Tel: (+91 133) 42 64 59
Fax: (+91 133) 42 64 62 and 42 62 54

EXECUTIVE DIRECTOR: H W Bhatnagar
GENERAL MANAGER, AVIATION: P K Awasthi

BHEL was approved for the manufacture of light aircraft by the Indian Ministry of Industry in March 1991 and by the Indian Directorate General of Civil Aviation in May 1991. Its most recent known product is the Swati light training aircraft.

UPDATED

BHEL LT-IIM SWATI

TYPE: Primary prop trainer/sportplane.
PROGRAMME: Designed and developed by Technical Centre of Directorate General of Civil Aviation (DGCA), India; first flight (VT-XIV, as LT-I) 17 November 1990; technical certificate issued 1991 by Indian Institute of Technology, Kanpur; type certification awarded January 1992; first production batch (four aircraft) delivered March 1993. See 1994-95 *Jane's* for description of this version.

BHEL LT-IIM Swati two-seat club trainer

Design modified by DGCA Technical Centre in 1993 under new designation LT-IIM; prototype was conversion of second production aircraft, VT-STB. Nos. 5 to 7 also converted to LT-IIM; No. 8 and onwards built as such.
Description applies to this version.
CUSTOMERS: Total 14 production aircraft (including LT-Is) completed and delivered by 1 November 1998; further five registered by January 2000. Production aircraft owned by Directorate General of Civil Aviation and assigned to various schools and flying clubs (FC), including Bihar Flying Institute, Bombay FC, Kerala Aviation Training Centre, Madhya Pradesh FC, North India FC, Pinjore FC and Rajasthan State Flying School.
COSTS: Approximately Rs3.8 million (1998).
DESIGN FEATURES: Designed to FAR Pt 23 (Utility and Normal categories) but currently limited to Normal category only at 730 kg (1,609 lb) all-up weight; applications include flying training, sport flying, surveillance, photography and short hauls; capable of stall turns, lazy eights and chandelles.
Wing section NASA GA(W)-1 (constant); dihedral 4°; incidence 2°.
FLYING CONTROLS: Conventional and manual. Actuation by pushrods and cables. Plain flaps; horn-balanced rudder and elevators.
STRUCTURE: Wings constant chord with single box spar of Himalayan spruce, ply and fabric covered; ailerons and flaps similar; fuselage of chromoly steel tube with metal/GFRP engine cowling and skin on forward fuselage; fabric covering aft; GFRP fairings; tail surfaces light metal.
LANDING GEAR: Non-retractable bungee-sprung mainwheels, on side Vs and half-axles, with dual hydraulic brakes. Non-retractable steerable nosewheel. Mainwheel tyres 15×6.00-6, pressure 2.00 bars (29 lb/sq in); nosewheel tyre 5.00-5, pressure 2.07 bars (30 lb/sq in).
POWER PLANT: One 80.5 kW (108 hp) Textron Lycoming O-235-N2C flat-four engine, driving a Hoffmann HO-14-178115 two-blade fixed-pitch composites propeller without spinner. Fuel capacity 90 litres (23.8 US gallons; 19.8 Imp gallons).

ACCOMMODATION: Two seats, with safety harnesses, side by side under one-piece rearward-sliding canopy.
AVIONICS: VFR instrumentation and VHF radio standard.
DIMENSIONS, EXTERNAL:
Wing span	9.20 m (30 ft 2¼ in)
Wing chord, constant	1.30 m (4 ft 3¼ in)
Wing aspect ratio	7.1
Length overall (flying attitude)	7.21 m (23 ft 7¾ in)
Height overall (flying attitude)	2.78 m (9 ft 1½ in)
Tailplane span	2.60 m (8 ft 6½ in)
Wheel track	2.00 m (6 ft 6¾ in)
Wheelbase	1.725 m (5 ft 8 in)
Propeller diameter	1.78 m (5 ft 10 in)

AREAS:
Wings, gross	11.96 m² (128.7 sq ft)
Ailerons (total)	1.10 m² (11.84 sq ft)
Trailing-edge flaps (total)	1.04 m² (11.19 sq ft)
Fin	0.75 m² (8.07 sq ft)
Rudder	0.62 m² (6.67 sq ft)
Tailplane	0.93 m² (10.01 sq ft)
Elevators (total, incl tab)	0.94 m² (10.12 sq ft)

WEIGHTS AND LOADINGS (Normal category):
Weight empty, equipped (typical)	540 kg (1,190 lb)
Max T-O weight	730 kg (1,609 lb)
Max wing loading	61.0 kg/m² (12.50 lb/sq ft)
Max power loading	9.07 kg/kW (14.90 lb/hp)

PERFORMANCE (at Normal category max T-O weight):
Never-exceed speed (V_{NE})	143 kt (265 km/h; 164 mph)
Max level speed	104 kt (193 km/h; 120 mph)
Cruising speed at 75% power	100 kt (185 km/h; 115 mph)
Stalling speed	44 kt (82 km/h; 51 mph)
Max rate of climb at S/L	183 m (600 ft)/min
T-O run	259 m (850 ft)
Landing run	205 m (675 ft)
Range with max fuel	272 n miles (505 km; 313 miles)
Max endurance	3 h 30 min
g limits	+3.8/−1.52

UPDATED

HAL
HINDUSTAN AERONAUTICS LIMITED
CORPORATE OFFICE: PO Box 5150, 15/1 Cubbon Road, Bangalore 560 001
Tel: (+91 80) 286 46 36/7 and 46 39/43
Fax: (+91 80) 286 71 40 and 286 87 58
e-mail: marketing@hal-india.com
Web: http://www.hal-india.com
CHAIRMAN: Dr C G Krishnadas Nair
DIRECTOR, CORPORATE PLANNING AND MARKETING:
  S N Sachindran

GENERAL MANAGER, COMMERCIAL: U B Balachandra
GENERAL MANAGER, MARKETING: D Chowdhury
MANAGER, PUBLIC RELATIONS: Andrew Sunderaj

Formed 1 October 1964; has 14 manufacturing divisions at seven locations (seven at Bangalore, one each at Nasik, Koraput, Hyderabad, Barrackpore, Kanpur, Lucknow and Korwa), plus Design and Development Complex; nine R&D centres located with manufacturing divisions; total workforce 35,000 at end of 1999. Hyderabad Division manufactures avionics for all aircraft produced by HAL, plus air route surveillance and precision approach radars; Lucknow Division produces landing gears and other accessories under licence from manufacturers in France, Russia and the UK; Korwa manufactures inertial navigation and nav/attack systems.

Preliminary agreement with ATR signed in early 1999, under which HAL would initially outfit ATR 42 aircraft for Indian market; could develop later into assembly of that aircraft from CKD kits and later still to full Indian manufacture. Detail negotiations continuing in 2000, at which time coastguard and paramilitary variants were being proposed.

UPDATED

DESIGN AND DEVELOPMENT COMPLEX
PO Box 1789, Bangalore 560 017
Tel: (+91 80) 526 34 57
Fax: (+91 80) 526 30 96
Telex: 845 8083
DIRECTOR: Ashok K Baweja
GENERAL MANAGERS:
  Yogesh Kumar (ARDC)
  K S Sudheendra (Executive Director, RWRDC)

HAL currently has nine Research and Design (R&D) Centres, including one each for fixed-wing aircraft (ARDC) and rotary-wing (RWRDC) programmes. Former's main programmes are LCA (see ADA entry) and HJT-36; latter concerned mainly with ALH. Transport Aircraft R&D Centre was responsible for developing oversize cabin door, integration of Super Marec radar, gun pod, IR/UV scanner and flight data recorder for HAL-built Dornier 228 programme. Earlier designs have included HT-2, Pushpak, Krishak, Basant, Marut, Kiran, Deepak and HTT-34. Jet trainer successor to Kiran was revealed as HJT-36 in 1998.

UPDATED

HAL ADVANCED LIGHT HELICOPTER (ALH)
TYPE: Multirole light helicopter.
PROGRAMME: Agreement signed with MBB (Germany) July 1984 to support design, development and production; design started November 1984; ground test vehicle runs began April 1991; five flying prototypes (two basic, one air force/army, one naval and one civil); PT1 first prototype (Z3182) rolled out 29 June 1992, first flight 20 August and 'official' first flight 30 August 1992; not then fitted with ARIS; PT2 second prototype (Z3183) made its first flight 18 April 1993; PT-A (army/air force prototype Z3268) on 28 May 1994; PT-N (naval prototype), with CTS 800 engines, flew for first time (IN901) on 23 December 1995. Total hours flown, including 'hot-and-high' trials in environments of 50°C (122°F) and more than 6,000 m (19,680 ft), exceeded 1,100 by February 2000; planned programme totals 1,420 hours. Certification was expected by end of 2000.

Naval trials by PT-N conducted in March 1998 aboard aircraft carrier INS *Viraat* and smaller decks of other Indian Navy vessels. May 1998 US trade embargo, imposed following India's refusal to sign nuclear test ban treaty, blocked import of CTS 800 engines (30 ordered) and has delayed planned first flight of civil fifth prototype (VT-XLH) with this engine. Weight reduction programme instigated in mid-1998; RFPs issued later same year for cockpit display system.

Deliveries (four each to Indian Air Force and Army, two each to Navy and Coast Guard) were due to begin in late 1999 but programme has slipped, and only first two (to Army) now due before end of 2000. By the end of 1998, manufacture was well advanced of three preproduction aircraft (PPN-1, PPA-2 and PPA-3: one for each of the three armed services). Initial batch of 30 TM 333-2B2 engines ordered in mid-1999 to power first 12 (including two civil) production ALHs. Further 52 engines ordered mid-2000 to power next 20 ALHs.

Fourth ALH prototype in Indian Navy (PT-N) configuration with landing gear retracted *(Sebastian Zacharias)*
2000/0062688

CURRENT VERSIONS: **Air force/army:** Skid gear, crashworthy fuel tanks, bulletproof supply tanks, IR and flame suppression; night attack capability; roles to include attack and SAR.
  Naval: Retractable tricycle gear, harpoon decklock, foldable tailboom, pressure refuelling; fairings on fuselage sides to house mainwheels, flotation gear and batteries.
  Civil: Roles to include passenger and utility transport, commuter/offshore executive, rescue/emergency medical service and law enforcement. Wheel landing gear and (planned) improved CTS 800 engines. Prototype targeted to fly in 2000; DGCA certification to be followed by FAA/JAA type approval.
  Coast Guard: High commonality with naval version; nose-mounted surveillance radar; roof-mounted FLIR; starboard side, cabin-mounted 7.62 mm machine gun; radar console and operator's seat; liferaft, loudhailer.
CUSTOMERS: Indian government requirement for armed forces and Coast Guard, to replace Chetaks/Cheetahs; letter of intent for 300 (Army 110, Air Force 150, Navy/Coast Guard 40) followed by contract for 100 in late 1996, with

HAL—AIRCRAFT: INDIA   171

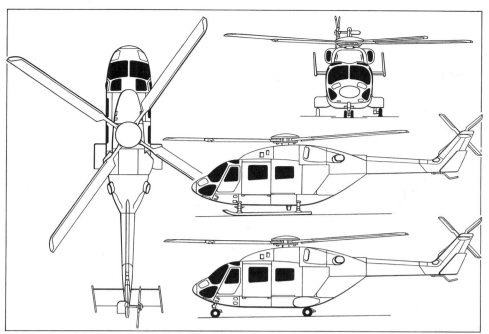

Naval version of the HAL Advanced Light Helicopter, with additional side view (centre) of air force/army variant *(Mike Keep/Jane's)*

deliveries of first 12 (see Programme above) not now expected to begin until late 2000. Second production lot contains 20. HAL predicts total military/civil domestic orders for about 650.

COSTS: Unit price of basic aircraft approximately US$4.5 million (1995). Total programme costs US$170 million by 1997.

DESIGN FEATURES: First modern helicopter of local design and construction. Conventional layout, including high-mounted tailboom to accommodate rear-loading doors. Four-blade hingeless main rotor with advanced aerofoils and sweptback tips; Eurocopter FEL (fibre elastomer) rotor head, with blades held between pair of cruciform CFRP starplates; manual blade folding and rotor brake standard; integrated drive system transmission; four-blade bearingless crossbeam tail rotor on starboard side of fin; fixed tailplane; sweptback endplate fins offset to port; vibration damping by ARIS (anti-resonance isolation system), comprising four isolator elements between main gearbox and fuselage. Folding tailboom on naval variant.

Main rotor blade section DMH 4 (DMH 3 outboard); tail rotor blade section S 102C (S 102E at tip). Rotor speeds 314 rpm (main), 1,564 rpm (tail).

FLYING CONTROLS: Integrated dynamic management by four-axis AFCS (actuators have manual as well as AFCS input); constant-speed rpm control, assisted by collective anticipator (part of FADEC and stability augmentation system acting through AFCS).

STRUCTURE: Main and tail rotor blades and rotor hub glass fibre/carbon fibre; Kevlar nosecone, crew/passenger doors, cowling, upper rear tailboom and most of tail unit; carbon fibre lower rear tailboom and fin centre panels; Kevlar/carbon fibre cockpit section; aluminium alloy sandwich centre cabin and remainder of tailboom.

LANDING GEAR: Non-retractable metal skid gear standard for air force/army version. Hydraulically retractable tricycle

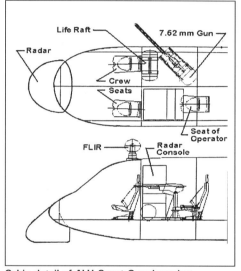

Cabin detail of ALH Coast Guard version
2001/0099548

gear on naval and civil versions, with twin nosewheels and single mainwheels, latter retracting into fairings on fuselage sides which also (on naval version) house flotation gear and batteries; rearward-retracting nose unit; naval ALH has harpoon decklock system. Spring skid under rear of tailboom on all versions, to protect tail rotor.

FPT Industries (UK) Kevlar inflatable flotation bags for prototypes, usable with both skid and wheel gear.

POWER PLANT: First three, and fifth, prototypes each powered by two Turbomeca TM 333-2B turboshafts, with FADEC, rated at 746 kW (1,000 shp) for T-O, 788 kW (1,057 shp) maximum contingency and 663 kW (889 shp) maximum continuous. LHTEC CTS 800-4N (969 kW; 1,300 shp) selected late 1994 as possible alternative power plant; test-flown in fourth prototype, but subsequently embargoed; all ALHs now to have TM 333-2B.

Transmission ratings (two engines) 1,240 kW (1,663 shp) for 5 minutes for T-O and 1,070 kW (1,435 shp) maximum continuous; OEI ratings 800 kW (1,073 shp) for 30 seconds (super contingency), 700 kW (939 shp) for 2½ minutes, 620 kW (831 shp) for 30 minutes and 535 kW (717 shp) maximum continuous. Transmission input from both engines combined through spiral bevel gears to collector gear on stub-shaft. ARIS system gives 6° of freedom damping. Power take-off from main and auxiliary gearboxes for transmission-driven accessories.

Total usable fuel, in self-sealing crashworthy underfloor tanks (three main and two supply), 1,400 litres (370 US gallons; 308 Imp gallons). Pressure refuelling in naval version.

ACCOMMODATION: Flight crew of two, on crashworthy seats in military/naval versions. Main cabin seats 12 persons as standard, 14 in high-density configuration. EMS interior (planned for demonstration in 2000) can accommodate two stretchers and four medical attendants, or four stretchers and two medical personnel. Crew door and rearward-sliding passenger door on each side; clamshell cargo doors at rear of passenger cabin.

SYSTEMS: DC electrical power from two independent subsystems, each with a 6 kW starter/generator, with battery back-up for 15 minutes of emergency operation; AC power, also from two independent subsystems, each with a 5/10 kVA alternator. Three hydraulic systems (pressure 207 bars; 3,000 lb/sq in, maximum flow rate 25 litres; 6.6 US gallons; 5.5 Imp gallons/min): systems 1 and 2 for main and tail rotor flight control actuators, system 3 for landing gear, wheel brakes, decklock harpoon (naval variant) and optional equipment. Oxygen system.

AVIONICS: *Comms:* V/UHF, HF/SSB and standby UHF com radio, IFF and intercom.
Radar: Weather radar optional. Surveillance radar in Coast Guard version.
Flight: SFIM four-axis AFCS, Doppler navigation system, TAS system, ADF, radio altimeter, heading reference standard; Omega nav system optional.
Mission: Roof-mounted FLIR in Coast Guard version.

EQUIPMENT: Depending on mission, can include two to four stretchers, external rescue hoist, liferaft and 1,500 kg (3,307 lb) capacity cargo sling.

ARMAMENT: Cabin-side pylons for two torpedoes/depth charges or four anti-ship missiles on naval variant; on army/air force variant, stub-wings which can be fitted with eight anti-tank guided missiles, four pods of 68 mm or 70 mm rockets or two pairs of air-to-air missiles. Army/air force variant can also be equipped with ventral 20 mm gun turret or sling for carriage of land mines. Cabin-mounted 7.62 mm machine gun in Coast Guard version, firing from starboard side doorway.

DIMENSIONS, EXTERNAL:
Main rotor diameter	13.20 m (43 ft 3¾ in)
Main rotor blade chord: inboard	0.50 m (1 ft 7¾ in)
at tip	0.165 m (6½ in)
Tail rotor diameter	2.55 m (8 ft 4½ in)
Length:	
overall, both rotors turning	15.87 m (52 ft 0¾ in)
fuselage (except Coast Guard version)	13.43 m (44 ft 0¾ in)

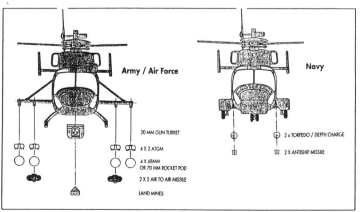

ALH weapon option
0010519

ALH instrument panel
0010517

172　INDIA: AIRCRAFT—HAL

ALH civil commuter interior
2001/0099549

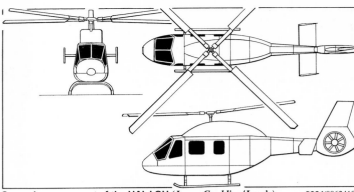

General arrangement of the HAL LOH *(James Goulding/Jane's)*　*2001/0062410*

Height: overall, tail rotor turning:
army/air force version	4.98 m (16 ft 4 in)
naval version	4.91 m (16 ft 1¼ in)
to top of main rotor head	3.93 m (12 ft 10¾ in)
Fuselage: Max width	2.00 m (6 ft 6¾ in)

Width over mainwheel sponsons (naval version)
	3.15 m (10 ft 4 in)
Tail unit span (over fins)	3.19 m (10 ft 5½ in)
Wheel track (naval version)	2.80 m (9 ft 2¼ in)
Wheelbase (naval version)	4.37 m (14 ft 4 in)
Skid track (army/air force version)	2.60 m (8 ft 6¼ in)
Tail rotor ground clearance	2.34 m (7 ft 8¼ in)

DIMENSIONS, INTERNAL:
Cabin, excl flight deck: Max width	1.97 m (6 ft 5½ in)
Max height	1.42 m (4 ft 8 in)
Volume	7.33 m³ (259 cu ft)
Cargo compartment volume	2.16 m³ (76.3 cu ft)

AREAS:
Main rotor disc	136.85 m² (1,473.0 sq ft)
Tail rotor disc	5.11 m² (55.0 sq ft)
Main fin	2.126 m² (22.88 sq ft)
Endplate fins (total)	1.45 m² (15.61 sq ft)
Tailplane	2.40 m² (25.83 sq ft)

WEIGHTS AND LOADINGS (A: army/air force version, B: naval variant):
*Weight empty, equipped: A, B	2,450 kg (5,401 lb)
Max fuel weight: A	1,040 kg (2,293 lb)
B	1,100 kg (2,425 lb)
Max sling load: A	1,000 kg (2,205 lb)
B	1,500 kg (3,307 lb)
Max T-O weight: A	4,500 kg (9,920 lb)
B	5,500 kg (12,125 lb)
Max disc loading: A	32.9 kg/m² (6.74 lb/sq ft)
B	40.2 kg/m² (8.23 lb/sq ft)

Transmission loading at max T-O weight and power:
A	3.63 kg/kW (5.97 lb/shp)
B	4.44 kg/kW (7.29 lb/shp)

*Reduction to 2,250 kg (4,960 lb) planned

PERFORMANCE (at 4,000 kg; 8,818 lb AUW, at S/L, ISA +15°C):
Never-exceed speed (V$_{NE}$)	178 kt (330 km/h; 205 mph)
Max level speed	157 kt (290 km/h; 180 mph)
Max cruising speed	132 kt (245 km/h; 152 mph)
Max rate of climb at S/L	780 m (2,559 ft)/min
Service ceiling	6,000 m (19,680 ft)
Hovering ceiling IGE	above 3,000 m (9,840 ft)

Range: with max fuel, 20 min reserves
	432 n miles (800 km; 497 miles)
with 700 kg (1,543 lb) payload	
	216 n miles (400 km; 249 miles)
Endurance, 20 min reserves	4 h

UPDATED

HAL LOH

TYPE: Light observation helicopter.
PROGRAMME: Announced at Air India Show in December 1998; design study stage at that time. Following brief details received by early 2001.
DESIGN FEATURES: Envisaged as replacement for Cheetah and Chetak; configuration similar to ALH, but smaller and lighter and with ducted tail rotor; bearingless main rotor; construction mainly of composites; 'glass cockpit'; health and usage monitoring system.
POWER PLANT: Single or twin turboshafts.

DIMENSIONS, EXTERNAL:
Main rotor diameter	approx 11.30 m (37 ft 1 in)
Fuselage: Length	10.355 m (33 ft 11¾ in)
Max width	1.60 m (5 ft 3 in)
Height over tailfin	4.185 m (13 ft 8¾ in)
Skid track	2.10 m (6 ft 10¾ in)

WEIGHTS AND LOADINGS (approx):
Max payload	1,500 kg (3,307 lb)
Max T-O weight	3,000 kg (6,613 lb)

PERFORMANCE (estimated):
Hovering ceiling IGE	5,800 m (19,000 ft)

UPDATED

HAL HJT-36

TYPE: Basic jet trainer/light attack jet.
PROGRAMME: Revealed at Singapore Air Show, February 1998. Conceived as successor to IAF HJT-16 Kiran. In design development stage; mockup, with Sextant avionics suite, exhibited at Air India Show in December 1998, differed considerably in appearance from general arrangement released at Singapore. Indian government Rs1.8 billion (US$42 million) contract awarded in mid-1999, covering completion, flight test and certification of two prototypes. First flight targeted for late 2002 and service entry two years later. By early 2000, metal was being cut, although selection of risk-sharing local partners had not been announced by November 2000.
CUSTOMERS: Indian Air Force and Navy as replacement for existing fleet of approximately 170 Kirans. Government approval for 150 to 200 production run.
DESIGN FEATURES: Capable of high-speed training, but with simple handling at low speeds; cockpit layout compatible with current combat aircraft. Conventional, low-wing CAD/CAM design with moderate (18°) wing leading-edge sweepback, slight dihedral and low-mounted tailplane.
FLYING CONTROLS: Conventional and manual, with three-axis trim; horn-balanced elevators.
STRUCTURE: Generally of light alloy, but composites control surfaces. Intended fatigue life of more than 7,500 hours.
LANDING GEAR: Retractable tricycle type. Inward-retracting main units, rearward-retracting nosewheel.
POWER PLANT: One 14.03 kN (3,153 lb st) SNECMA Larzac 04-H20 non-afterburning turbofan in rear fuselage, fed by bifurcated air intake.
ACCOMMODATION: Crew of two in tandem, on lightweight zero/zero ejection seats; rear seat raised. One-piece canopy opens sideways to starboard.
AVIONICS: State-of-the-art, with dual V/UHF comms, GPS, HUD and LCDs. Bidders include Elop (HUD), BAE Systems and Sextant (MFD).
ARMAMENT: One underfuselage and four underwing weapons pylons for bombs, rocket pods and gun pods.

DIMENSIONS, EXTERNAL:
Wing span	9.80 m (32 ft 1¾ in)
Length overall	10.91 m (35 ft 9½ in)
Height overall	4.13 m (13 ft 6½ in)

WEIGHTS AND LOADINGS (approx):
Max external stores load	1,000 kg (2,204 lb)
T-O weight: clean	3,500 kg (7,716 lb)
max	4,500 kg (9,920 lb)

PERFORMANCE (estimated):
Max operating Mach No.	0.80
Max permissible diving speed	
	445 kt (825 km/h; 512 mph)
Service ceiling	12,000 m (39,360 ft)
Endurance	3 h
g limits (ultimate)	+8/−3

UPDATED

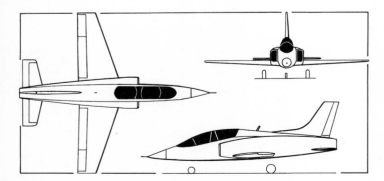

Mockup of HJT-36, intended as HJT-16 Kiran successor　*2000/0084004*

Provisional general arrangement of the HAL HJT-36 jet trainer *(Paul Jackson/Jane's)*
2000/0062685

BANGALORE COMPLEX
Post Bag 1785, Bangalore 560 017
Tel: (+91 80) 526 82 30
Fax: (+91 80) 527 95 64
Telex: 845 2234
MANAGING DIRECTOR: Behari Lal

Aircraft Division
Post Bag 1788, Bangalore 560 017
Tel: (+91 80) 526 89 69
Fax: (+91 80) 526 51 88
Telex: 845 2234 HALM IN
GENERAL MANAGER: K Umamaheswar

Main programmes Jaguar International; composites and metal drop tanks; Dornier 228 landing gears; sheet metal items; forward passenger doors for Airbus A319/A320/A321; overwing emergency exit doors for Boeing 757; pylons for Panavia Tornado.

Helicopter Division
Post Bag 1790, Bangalore 560 017
Tel: (+91 80) 523 86 02
Fax: (+91 80) 523 47 17
Telex: 845 2764
GENERAL MANAGER: S M Kapoor

Manufacture and overhaul of Cheetah, Chetak, Lancer and ALH light helicopters.

Bangalore Complex comprises Aircraft Division, Helicopter Division, Aerospace Division, Engine Division, Overhaul Division, Foundry and Forge Division and Industrial and Marine Gas Turbines Division. Programmes include subcontract work for leading aerospace companies such as BAE, EADS, Boeing, Fairchild Dornier and Latécoère. Aircraft Division will complete 50 of the 66 BAE Hawks intended for the Indian Air Force (eight from CKD kits, 42 by local manufacture, after delivery of 16 UK-built aircraft from BAE.)

Aerospace Division manufactures light alloy structures and assemblies for satellites and launch vehicles. Engine Division manufactures, overhauls and repairs Adour Mk 811, Artouste IIIB and TPE331-5 engines; it also overhauls and repairs Adour Mk 804E, Dart, Gnome, Orpheus and Avon engines. Overhaul of Jaguar, Kiran, Mirage and An-32 aircraft, Cheetah helicopters and Pratt & Whitney and Textron Lycoming piston engines is undertaken by Overhaul Division.

UPDATED

HAL (SEPECAT) JAGUAR INTERNATIONAL
Indian Air Force name: Shamsher (Assault Sword)
TYPE: Attack fighter.
Comprehensive details in Jane's Aircraft Upgrades; following refers generally to Jaguar S, and particularly to Indian version.
PROGRAMME: Forty (including five two-seat) Jaguar Internationals with Adour Mk 804 engines delivered from UK, beginning March 1981; 45 more with Adour Mk 811s assembled in India, making first flight (JS136) 31 March 1982; further 31 manufactured under licence in India (first delivery early 1988); deliveries of further 15, ordered 1993, were due for completion by March 1999 with delivery of last three of this batch. Indian government ordered additional batch of 17 two-seaters in late 1998. These will be equipped with DARIN nav/attack system and employed in a night attack role with laser-guided weapons. Production was due to start in 1999 and deliveries to Indian Air Force to follow in 2001. Agreement with BAe signed mid-1993 to market Jaguar in Oman and to overhaul and repair that country's Jaguars.
CURRENT VERSIONS: **Jaguar B:** Two-seat trainer version; 15 delivered; further 17 on order.
  Jaguar S: Standard single-seat attack version; 106 delivered; further 20 ordered in 2000.
  Maritime Jaguar: Jaguars of IAF No. 6 Squadron, assigned to anti-shipping role, have Thomson-CSF Agave radar, interfaced with DARIN nav/attack system and Sea Eagle anti-shipping missiles; first modified aircraft delivered January 1986; 10 ordered, of which all delivered by end of 1999. To be upgraded by substitution of Elta EL/M-2032 for Agave; flight trials were planned to begin in 1998.
CUSTOMERS: Indian Air Force has received 131, comprising 116 single-seat (including 10 Maritime) and 15 two-seat combat-capable trainers. Further 17 Jaguar Bs (ordered late 1998) in production for delivery from 2001; additional 20 single-seaters ordered in 2000. Basic strike version equips Nos. 5 and 14 Squadrons at Ambala and Nos. 16 and 27 Squadrons at Gorakhpur; anti-shipping version equips No. 6 Squadron at Pune.
DESIGN FEATURES: Purpose-designed attack aircraft. Shoulder-wing monoplane. Anhedral 3°. Sweepback 40° at quarter-chord. Outer leading-edges fitted with slat which also gives effect of extended chord. Tail unit sweepback at quarter-chord 40° on horizontal, 43° on vertical surfaces; tailplane anhedral 10°. Ventral fins beneath rear fuselage.
FLYING CONTROLS: Fairey Hydraulics powered flying controls. No ailerons: control by two-section spoilers, forward of outer flap on each wing, in association (at low speeds) with differential tailplane. Hydraulically operated (by screwjack) full-span double-slotted trailing-edge flaps. Leading-edge slats can be used in combat. All-moving slab-type tailplane, the two halves of which can operate differentially to supplement the spoilers.
STRUCTURE: Wing is an all-metal two-spar torsion box structure; skin machined from solid aluminium alloy, with integral stiffeners. Main portion built as single unit, with three-point attachment to each side of fuselage. Fuselage all-metal, mainly aluminium, built in three main units and making use of panels and, around the cockpit(s), honeycomb panels. Local use of titanium alloy in engine bay area. Two door-type airbrakes under rear fuselage, immediately aft of each mainwheel well. Structure and systems aft of cockpit(s), identical for single-seat and two-seat versions. The tail unit is a cantilever all-metal structure, covered with aluminium alloy sandwich panels. Rudder and outer panels and trailing-edge of tailplane have honeycomb core.
LANDING GEAR: Messier-Hispano-Bugatti retractable tricycle type, all units having Dunlop wheels and low-pressure tyres for rough field operation. Hydraulic retraction, with oleo-pneumatic shock-absorbers. Forward-retracting main units each have twin wheels, tyre size 615×225-10, pressure 5.8 bars (84 lb/sq in). Wheels pivot during retraction to stow horizontally in bottom of fuselage. Single rearward-retracting nosewheel, with tyre size 550×250-6 and pressure 3.9 bars (57 lb/sq in). Twin landing/taxying lights in nosewheel door. Dunlop hydraulic brakes. Anti-skid units and arrester hook standard. Irvin brake parachute of 5.5 m (18 ft 0 in) diameter in fuselage tailcone.
POWER PLANT: Two HAL-built Rolls-Royce Turbomeca Adour Mk 811 turbofans (Phase 3 aircraft onwards), each rated at 25.0 kN (5,620 lb st) dry and 37.4 kN (8,400 lb st) with afterburning. Fixed-geometry air intake on each side of fuselage aft of cockpit. Fuel in six tanks, one in each wing and four in fuselage. Total internal fuel capacity 4,200 litres (1,110 US gallons; 924 Imp gallons). Armour protection for critical fuel system components. Provision for carrying three auxiliary drop tanks, each of 1,200 litres (317 US gallons; 264 Imp gallons) capacity, on fuselage and inboard wing pylons. Provision for in-flight refuelling, with retractable probe forward of cockpit on starboard side.
ACCOMMODATION: Jaguar S has enclosed cockpit for pilot, with rearward-hinged canopy and Martin-Baker Mk 9B zero/zero ejection seat. Jaguar B crew of two in tandem on Martin-Baker Mk 9B Srs II zero/zero ejection seats. Individual rearward-hinged canopies. Rear seat 38 cm (15 in) higher than front seat. Windscreen bulletproof against 7.5 mm rifle fire.
SYSTEMS: Air conditioning and pressurisation systems for cockpit(s) also control temperature in certain equipment bays. Two independent hydraulic systems, powered by two engine-driven pumps. Hydraulic pressure 207 bars (3,000 lb/sq in). First system (port engine) supplies one channel of each actuator for flying controls, hydraulic motors which actuate flaps and slats, landing gear retraction and extension, brakes and anti-skid units. Second system supplies other half of each flying control actuator, two further hydraulic motors actuating slats and flaps, airbrake and landing gear emergency extension jacks, nosewheel steering and wheel brakes. In addition, there is an emergency hydraulic power transfer unit.

Electrical power provided by two 15 kVA AC generators, either of which can sustain functional and operational equipment without load shedding. DC power provided by two 4 kW transformer-rectifiers. Emergency power for essential instruments provided by 15 Ah battery and static inverter. De-icing, rain clearance and demisting

HAL Maritime Jaguar of No. 6 Squadron, Indian Air Force

INDIAN AIR FORCE JAGUARS

Mark/Variant	Serial Nos.	Assembled[1]	Remarks	Qty
GR. Mk 1	JI003 to 018	BAe	Loaned by/returned to RAF[2]	(16)
S(I)	JS101 to 135	BAe	Phase 2; Adour 804, NAVWASS	35
	JS136 to 170	HAL	Phase 3; Adour 811, DARIN	35
	JS171 to 195	HAL	Phase 4; Adour 811, DARIN	25[3]
	JS196 to 206	HAL	Phase 5; Adour 811, DARIN	11
Maritime	JM251 to 256	HAL	Phase 4; Adour 811, DARIN, Agave	6[3]
	JM257 to 260	HAL	Phase 5; Adour 811, DARIN, Agave	4
T. Mk 2	JI001 to 002	BAe	Loaned by/returned to RAF[2]	(2)
B(I)	JT051 to 055	BAe	Phase 2; Adour 804, NAVWASS	5
	JT056 to 065	HAL	Phase 3; Adour 811, DARIN	10
	JT066 *et seq*	HAL	Phase 6; night attack, Adour 811, DARIN	17
Total				**148 (18)**

Notes
[1] All co-built by (SEPECAT) BAe and Dassault
[2] Adour 804 engines and NAVWASS nav/attack avionics; two single-seat lost in Indian service
[3] Phase 4 originally planned as 56 aircraft of entirely local manufacture; amended to 31 kits

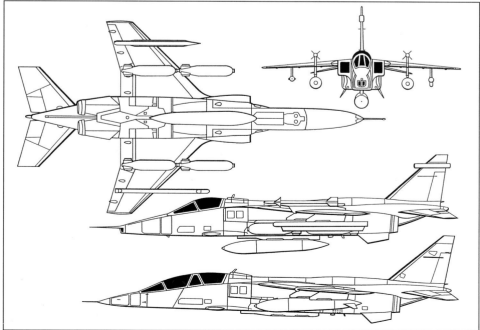

SEPECAT Jaguar single-seat tactical support aircraft and two-seat operational trainer *(Mike Keep/Jane's)*

standard. Liquid oxygen system, which also pressurises pilot's anti-*g* trousers.

Jaguar is fully power-controlled in all three axes and is automatically stabilised as a weapons platform by gyros which sense disturbances and feed appropriate correcting data through a computer to the power control assemblies, in addition to human pilot manoeuvre demands. Power controls are all of duplex tandem arrangement, with mechanical and electrical servo-valves of Fairey platen design. Air-to-air combat capability can be enhanced by inclusion of roll/yaw dampers, to increase lateral stability, and by increasing slat and flap angles.

AVIONICS: HAL-manufactured DARIN (display attack and ranging inertial navigation) nav/attack system initially, incorporating INS, HUDWAC, COMED, interconnected MIL-STD-1553B dual-redundant databus and interfaced with LRMTS. Some ingredients of DARIN now being upgraded (see below).

Comms: Main (20-channel) V/UHF; standby UHF and HF/SSB transceivers; IFF-400 AM transponder.

Radar: Thomson-CSF Agave in Maritime version, interfaced with DARIN system; planned to have been replaced by Elta EL/M-2032 from mid-1998.

Flight: SAGEM ULISS 82 INS; HAL ADF and radar altimeter.

Instrumentation: Smiths HUDWAC (head-up display and weapon aiming computer); BAE Systems COMED 2045 (combined map and electronic display) to be replaced by Sextant digital autopilot and 152 × 152 mm (6 × 6 in) smart multifunction display (SMFD), ordered November 1999.

Mission: Laser ranger and marked target seeker (LRMTS). Provisional selection of Rafael Litening laser designation pod announced February 1996.

Self-defence: RWR; active and passive ECM.

ARMAMENT: Two 30 mm Aden guns in lower fuselage aft of cockpit in single-seater, with 150 rds/gun; single Aden on port side in two-seater. One stores attachment on fuselage centreline and two under each wing. Centreline and inboard wing points can each carry up to 1,134 kg (2,500 lb) of weapons, outboard underwing points up to 567 kg (1,250 lb) each. Typical alternative loads include one air-to-surface missile and two 1,200 litre (317 US gallon; 264 Imp gallon) drop tanks; eight 1,000 lb bombs; various combinations of free-fall, laser-guided, retarded or cluster bombs; overwing R.550 Magic missiles; air-to-surface rockets; or a reconnaissance camera pack. Maritime Jaguars equipped with one or two Sea Eagle anti-shipping missiles.

DIMENSIONS, EXTERNAL:
Wing span	8.69 m (28 ft 6 in)
Wing aspect ratio	3.1
Length overall, incl probe:	
single-seat	16.83 m (55 ft 2½ in)
two-seat	17.53 m (57 ft 6¼ in)
Height overall	4.89 m (16 ft 0½ in)
Wheel track	2.41 m (7 ft 11 in)
Wheelbase	5.69 m (18 ft 8 in)

AREAS:
Wings, gross	24.18 m² (260.3 sq ft)

WEIGHTS AND LOADINGS:
Typical weight empty	7,000 kg (15,432 lb)
Max external stores load (incl overwing)	4,763 kg (10,500 lb)
Normal T-O weight (single-seater, with full internal fuel and ammunition for built-in cannon)	10,954 kg (24,149 lb)
Max T-O weight with external stores	15,700 kg (34,612 lb)
Max wing loading	649.3 kg/m² (132.98 lb/sq ft)
Max power loading	210 kg/kN (2.06 lb/lb st)

PERFORMANCE:
Max level speed at S/L	M0.98 (648 kt; 1,200 km/h; 745 mph)
Max level speed above 6,000 m (19,680 ft)	M1.5 (907 kt; 1,680 km/h; 1,044 mph)
Touchdown speed	115 kt (213 km/h; 132 mph)
Service ceiling	13,715 m (45,000 ft)
T-O run: clean	565 m (1,855 ft)
with four 1,000 lb bombs	880 m (2,890 ft)
with eight 1,000 lb bombs	1,250 m (4,100 ft)
T-O to 15 m (50 ft) with typical tactical load	940 m (3,085 ft)
Landing from 15 m (50 ft) with typical tactical load	785 m (2,575 ft)
Landing run:	
normal weight, with brake-chute	470 m (1,540 ft)
normal weight, without brake-chute	680 m (2,230 ft)
overload weight, with brake-chute	670 m (2,200 ft)
Typical attack radius, internal fuel only:	
hi-lo-hi	460 n miles (852 km; 530 miles)
lo-lo-lo	290 n miles (537 km; 334 miles)
Typical attack radius with external fuel:	
hi-lo-hi	760 n miles (1,408 km; 875 miles)
lo-lo-lo	495 n miles (917 km; 570 miles)
Range with max external fuel	1,400 n miles (2,593 km; 1,611 miles)
g limits	+8.6 (+12 ultimate)/−3

UPDATED

HAL CHEETAH

TYPE: Light utility helicopter.

PROGRAMME: Licence-built Aerospatiale SA 315B Lama. Production by HAL for Indian armed forces started 1972 but now only at very low rate: six built in the three years ending 31 March 1997; none reported since that date but, in early 2000, HAL was predicting continued orders and production for several years and possible employment of Turbomeca TM 333 turboshaft (from ALH programme) to enhance performance. See 1996-97 and earlier *Jane's* for full description; shortened version in 1997-98 edition.

CUSTOMERS: HAL had delivered 244 by 31 March 1997 (latest figure supplied), including two sold to Namibia Defence Force in 1994 and one (No. 244) for regional government of Jammu and Kashmir.

UPDATED

HAL LANCER

TYPE: Armed observation helicopter.

PROGRAMME: Upgraded, counter-insurgency version of Cheetah. Prototype rebuilt from late production aircraft Z2867, unveiled late 1998: cabin has lightweight composites armour protection for pilot, control linkages and fuel tank; bullet-proof flat-plate transparencies; anti-tank missiles or twin gun/rocket pods. Can be rebuild or new production. See *Jane's Aircraft Upgrades* for further details.

CUSTOMERS: Indian Army order for 12, of which first four to be delivered in February/March 2001 and last in September 2001.

ACCOMMODATION: Crew of two, side by side; cleared for single-pilot operation.

ARMAMENT: Tubular metal outrigger each side of cabin, outboard of landing skids, each supporting pod containing one 12.7 mm machine gun and four 70 mm unguided air-to-surface rockets. Pilot's gunsight.

WEIGHTS AND LOADINGS:
Basic weight empty: Cheetah	1,090 kg (2,403 lb)
Lancer, incl weapon pods	1,350 kg (2,976 lb)
Fuel weight	393 kg (866 lb)
Ammunition (500 rds) and six rockets	130 kg (287 lb)
Max T-O weight	1,950 kg (4,299 lb)

PERFORMANCE:
Never-exceed speed (VNE)	113 kt (210 km/h; 130 mph)
Cruising speed at S/L, ISA + 20°C	95 kt (176 km/h; 109 mph)
Max rate of climb at S/L	330 m (1,083 ft)/min
Service ceiling	5,400 m (17,720 ft)
Operational radius, incl 30 min over target and 20 min fuel reserves, ISA + 20°C	78 n miles (145 km; 90 miles)
Endurance	2 h 30 min

UPDATED

HAL CHETAK

TYPE: Light utility helicopter.

PROGRAMME: Licence-built Aerospatiale SA 316B Alouette III. Remains in production only in India, but now at very low rate (seven built in three years ending 31 March 1997; no further examples reported since then but, in early 2000, HAL was expecting further orders to prolong production for several years. Turbomeca TM 333-2B2 turboshaft under consideration as alternative power plant). Shortened description in 1997-98 edition; see 1996-97 and earlier *Jane's* for more detailed version.

CUSTOMERS: HAL output had totalled at least 339 (of which 35 built from French kits) by late 1998, including two sold to Namibia Defence Force in 1994. One due for delivery to Sri Lanka Navy in late 2000.

UPDATED

TRANSPORT DIVISION

PO Box 225, Kanpur 208 008
Tel: (+91 512) 45 03 61
Fax: (+91 512) 45 05 05
Telex: 325 243 HALK IN
GENERAL MANAGER: B K Banerjee

This Division is responsible for manufacture of the Dornier 228, and for overhaul and repair of the Dornier 228, Hawker Siddeley 748, HAL HPT-32 and civil aircraft. Dornier 228 production stated in October 2000 to be continuing, but no changes indicated to 1997 delivery and 1998 production totals. Division also manufacturing prototypes of the NAL Saras (which see).

UPDATED

HAL (DORNIER) 228

TYPE: Twin-turboprop transport.

PROGRAMME: First flight of Dornier 228-100 prototype (D-IFNS) in Germany 28 March 1981; first flight 228-200 prototype (D-ICDO) 9 May 1981; British CAA certification 17 April 1984, FAR Pt 23 and Appendix A 135 11 May 1984, Australian 11 October 1985. For European production versions, see German section of 1996-97 and previous *Jane's*.

Contract for licensed manufacture of up to 150 Dornier 228s in India signed 29 November 1983; one pattern aircraft supplied complete from Germany followed by 16 locally assembled kits; first flight by Kanpur-assembled aircraft (VT-EJN, c/n 1002) 31 January 1986. Full local manufacture from 18th onwards (CG758, first flight 7 March 1991). It was announced in 1996 that production of

HAL—AIRCRAFT: INDIA   175

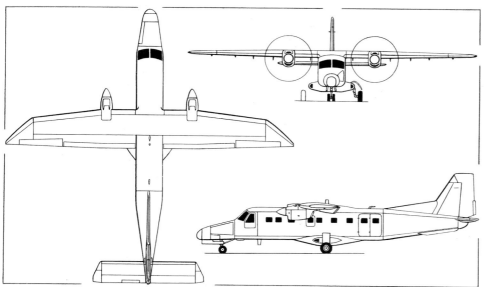

HAL (Dornier) 228 (two TPE331 turboprops) *(Dennis Punnett/Jane's)*

GERMAN/INDIAN DORNIER 228 DELIVERIES
(to March 1997; no further deliveries reported by October 2000)

Country	Operator	Series	Dornier	HAL/CKD	HAL	Total
India	Air Force	201		5	20	25
	Airports Authority	201	1	1		2
	Coast Guard	101	3	4	11	18[1]
	Navy	201	1		4	5[2]
	Oil and Natural Gas Commission	101	1			1
	UB Air	101		1		1
	Vayudoot	201	5	5		10
Mauritius	Coast Guard	101			1	1
Totals			**11**	**16**	**36**	**63**

Notes:
[1] Further 25 on order
[2] Further 10 on order

the 228 would be entirely in India with immediate effect; intention then was for transfer following completion of 241st German production aircraft (c/n 8241); however, in mid-1999, Nos. 243 to 245 (228-212 versions) were on the German production line, although none delivered by late 2000. The original Indian production programme has apparently been at a standstill since first quarter of 1997, but seven orders outstanding for Indian Coast Guard; follow-on orders for further 18 (various customers) awaiting confirmation in late 2000.

CURRENT VERSIONS: **Regional Airliner:** Ten **228-201**s (five each by Dornier and HAL) delivered to Vayudoot.

Maritime Surveillance: Thirty-three **228-101**s ordered for Indian Coast Guard (after three from Germany); CG754 first flight 18 February 1988; 18 delivered by March 1997 (latest figure supplied). Further seven ordered under US$72 million contract in early 2000.

In service with Nos. 744 and 750 Squadrons at Daman and Madras; for coastal, anti-smuggling patrol; 360° Marec 2 search radar under fuselage (replaced in 15 aircraft by Super Marec), Litton Omega navigation system, Matra infra-red/ultraviolet linescanner for pollution detection, search and rescue liferafts, 1 Mcd searchlight, side-mounted loudhailer, marine markers, and provision for two Micronair underwing spraypods to combat oil spills; sliding main cabin door for airdropping six- or 10-man liferafts. Normal crew two pilots, radar operator and observer. Optional armament includes two underwing 7.62 mm twin-gun machine gun pods or air-to-surface missiles.

Anti-ship: Indian Navy has current order for 15 specially equipped Dornier **228-201**s with anti-ship missiles and Super Marec radar (10 radars ordered 1993-94). Deliveries began (IN221) on 24 August 1991; five delivered by March 1996; further four were then due for delivery 1997-98 and six in 1998-99, but appear to still be awaited.

Utility transport: Indian Air Force received 25 **228-201**s; deliveries started (HM667 and HM668) 15 December 1987 to Nos. 41 and 59 Squadrons; carry 21 field-equipped troops on inward-facing folding seats; large cargo double door at rear, port side.

Executive/Air taxi: Various configurations including six- or 10-seat executive or 15-passenger air taxi, with cabin attendant and galley/wardrobe/lavatory; built-in APU for air conditioning and lighting in flight or on ground.

CUSTOMERS: Total 244 German aircraft built (including prototypes and Indian kits) by late 1998. Those supplied to, or built in, India are given in the accompanying table.

DESIGN FEATURES: Optimised for efficient cruising flight. High wing for propeller clearance, despite minimised fuselage ground clearance and pannier-mounted landing gear. Variable sweep wing leading-edge; unswept tailplane; sweptback fin with dorsal fillet.

Special Dornier wing with Do A-5 supercritical aerofoil; 8° leading-edge sweep on outer wing panels; raked tips; no dihedral or anhedral.

FLYING CONTROLS: Conventional and manual. Variable incidence tailplane with actuator switch on aileron wheel; horn-balanced elevators; rudder trim tab; single-slotted Fowler flaps augmented by drooping ailerons; two strakes under rear fuselage for low-speed stability.

STRUCTURE: Two-spar wing box; mainly light alloy structure, but with CFRP wingtips and tips of tailplane and elevators; GFRP nosecone, tips of rudder and fin; Kevlar landing gear fairings and in part of wing ribs; hybrid composites in fin leading-edge; fuselage unpressurised, built in five sections.

LANDING GEAR: Retractable tricycle type, with single mainwheels and twin-wheel nose unit; main units retract inward into fuselage fairings; hydraulically steerable nosewheels retract forward; Goodyear wheels and tyres, 25.5×8.75-10 (650×220-10) (10/12 ply) or 8.50-10 (10 ply) on mainwheels, 6.00-6 (8 ply), on nosewheels; Honeywell carbon brakes on mainwheels.

POWER PLANT: HAL-built Dornier 228-201s have two Honeywell TPE331-5-252D turboprops, each flat rated at 533 kW (715 hp) to ISA +18°C. Primary wing box forms integral fuel tank with total usable capacity 2,386 litres (630 US gallons; 525 Imp gallons); oil capacity per engine 5.9 litres (1.56 US gallons; 1.30 Imp gallons).

ACCOMMODATION: Crew of one or two; pilots' seats adjustable fore and aft; two-abreast seating with central aisle; maximum capacity 19 (more in military versions); flight deck door on port side; combined two-section passenger and freight door, with integral steps, on port side of cabin at rear; one emergency exit on port side of cabin, two on starboard side; baggage compartment at rear of cabin, accessible externally and from cabin; capacity 210 kg (463 lb). Enlarged baggage door optional; additional baggage space in fuselage nose, with separate access; capacity 120 kg (265 lb); modular units using seat rails for rapid changes of role.

SYSTEMS: Entire accommodation heated and ventilated; air conditioning system optional; heating by engine bleed air. Hydraulic system, pressure 207 bars (3,000 lb/sq in), for landing gear, brakes and nosewheel steering; hand pump for emergency landing gear extension. Primary 28 V DC electrical system, supplied by two 28 V 250 A engine-driven starter/generators and two 24 V 25 Ah Ni/Cd batteries; two 350 VA inverters supply 115/26 V 400 Hz AC system. Air intake anti-icing standard; de-icing system optional for wing and tail unit leading-edges, windscreen and propellers.

AVIONICS: *Comms:* Standard avionics include dual Honeywell KY 196 VHF com; KT 76A transponder. Becker audio selector and intercom.

Radar: Weather radar optional.

Flight: Dual Honeywell KN 53 VOR/ILS and KN 72 VOR/LOC converters; KMR 675 marker beacon receiver, dual or single KR 87 ADF; dual or single Aeronetics 7137 RMI; dual or single DME. Autopilot optional to permit single-pilot IFR operation.

Instrumentation: IFR instrumentation standard, comprising dual Honeywell GH14B gyro horizons; dual Honeywell KPI 552 HSIs; dual ASIs; dual altimeters; dual ADIs; dual VSIs. Honeywell five-tube EFIS optional.

EQUIPMENT: Standard equipment includes complete internal and external lighting, hand fire extinguisher, first aid kit, gust control locks and tiedown kit.

DIMENSIONS, EXTERNAL:
Wing span	16.97 m (55 ft 8 in)
Wing aspect ratio	9.0
Length overall	16.56 m (54 ft 4 in)
Height overall	4.86 m (15 ft 11½ in)
Tailplane span	6.45 m (21 ft 2 in)
Wheel track	3.30 m (10 ft 10 in)
Wheelbase	6.29 m (20 ft 7½ in)
Propeller diameter	2.69 m (8 ft 10 in)
Propeller ground clearance	1.08 m (3 ft 6½ in)
Passenger door (port, rear): Height	1.34 m (4 ft 4¾ in)
Width	0.64 m (2 ft 1¼ in)
Height to sill	0.91 m (2 ft 11¾ in)
Freight door (port, rear): Height	1.34 m (4 ft 4¾ in)
Width, incl passenger door	1.28 m (4 ft 2½ in)
Emergency exits (each): Height	0.66 m (2 ft 2 in)
Width	0.48 m (1 ft 7 in)
Baggage door (nose): Height	0.50 m (1 ft 7½ in)
Width	1.20 m (3 ft 11¼ in)
Standard baggage door (rear):	
Height	0.90 m (2 ft 11½ in)
Width	0.53 m (1 ft 9 in)

DIMENSIONS, INTERNAL:
Cabin, excl flight deck and rear baggage compartment:
Length	7.08 m (23 ft 2¾ in)
Max width	1.346 m (4 ft 5 in)
Max height	1.55 m (5 ft 1 in)
Floor area	9.6 m² (103 sq ft)
Volume	14.7 m³ (519.1 cu ft)
Rear baggage compartment volume	2.6 m³ (92 cu ft)
Nose baggage compartment volume	0.89 m³ (31.4 cu ft)

AREAS:
Wings, gross	32.00 m² (344.3 sq ft)
Ailerons (total)	2.71 m² (29.17 sq ft)
Trailing-edge flaps (total)	5.87 m² (63.18 sq ft)
Fin, incl dorsal fin	4.50 m² (48.44 sq ft)
Rudder, incl tab	1.50 m² (16.15 sq ft)
Horizontal tail surfaces (total)	8.33 m² (89.66 sq ft)

Indian Coast Guard HAL Dornier 228-101 *(Simon Watson)*

WEIGHTS AND LOADINGS (Indian-built 228-201):
Operating weight empty 3,687 kg (8,128 lb)
Max T-O weight: civil 6,200 kg (13,668 lb)
military 6,400 kg (14,110 lb)
Max landing weight 5,900 kg (13,007 lb)
Max wing loading: civil 193.8 kg/m² (39.68 lb/sq ft)
military 200.0 kg/m² (40.96 lb/sq ft)
Max power loading: civil 5.82 kg/kW (9.56 lb/shp)
military 6.00 kg/kW (9.87 lb/shp)
PERFORMANCE (228-201):
Never-exceed speed (V_{NE})
255 kt (472 km/h; 293 mph) IAS
Max operating speed (V_{MO})
223 kt (413 km/h; 256 mph) IAS
Max cruising speed: at S/L 222 kt (411 km/h; 255 mph)
at 3,050 m (10,000 ft) 231 kt (428 km/h; 266 mph)
Stalling speed: flaps up 79 kt (147 km/h; 91 mph) IAS
flaps down 63 kt (117 km/h; 73 mph) IAS
Max rate of climb at S/L 582 m (1,909 ft)/min
Service ceiling, 30.5 m (100 ft)/min rate of climb
8,535 m (28,000 ft)
Service ceiling, OEI, 15 m (50 ft)/min rate of climb
3,870 m (12,700 ft)
T-O run 442 m (1,450 ft)
T-O to 15 m (50 ft) 580 m (1,900 ft)
Accelerate/stop distance, with anti-skid 762 m (2,500 ft)
Landing from 15 m (50 ft) at MLW 457 m (1,500 ft)
Range at 3,050 m (10,000 ft) with 19 passengers,
reserves for 50 n mile (93 km; 57 mile) diversion,
45 min hold and 5% fuel remaining:
at max cruising speed
560 n miles (1,038 km; 645 miles)
at max range speed 630 n miles (1,167 km; 725 miles)
Range with 775 kg (1,708 lb) payload, conditions as
above: at max cruising speed
1,160 n miles (2,148 km; 1,335 miles)
at max range speed
1,320 n miles (2,445 km; 1,519 miles)
UPDATED

HAL 100-SEAT TRANSPORT

Indian government start-up funding of US$115 million for a 100-seat transport announced at Indian Air Show in December 1996; discussions for co-development and co-production continuing in 2000. HAL officials stated in February 2000 that India would require 120 such aircraft post-2006.

PERFORMANCE (target):
Cruising speed 459-486 kt (850-900 km/h; 528-559 mph)
Max range
1,080-1,890 n miles (2,000-3,500 km; 1,242-2,174 miles)
UPDATED

MiG COMPLEX

Ojhar Township Post Office, Nasik, Maharashtra 422 207
Tel: (+91 2550) 751 00
Fax: (+91 2550) 758 21
MANAGING DIRECTOR: N R Mohanty
GENERAL MANAGERS:
 D C Das (Nasik Aircraft Division)
 K P Puri (Koraput Engine Division)

This complex comprises the Nasik Aircraft and Koraput Engine Divisions of HAL. Indian MiG-21 production phased out 1986-87 as production of MiG-27M increased. See *Jane's Aircraft Upgrades* for details of Indian Air Force MiG-21 and MiG-27M upgrade programmes.

Licensed manufacture of Sukhoi Su-30MKI is expected to be undertaken by this complex. Agreement signed early October 2000 for local manufacture of up to 140, following earlier order for initial batch of 40 Russian-built Su-30MKIs. Present activities are MiG-21/27 overhaul and repair.
UPDATED

ACCESSORIES COMPLEX

PO Box 215, Lucknow 226 016
Tel: (+91 522) 38 43 47
Fax: (+91 522) 38 00 96
MANAGING DIRECTOR: L M Bhardwaj

Comprises Lucknow (Accessories), Hyderabad (Avionics) and Korwa (Avionics) Divisions.
UPDATED

NAL

NATIONAL AEROSPACE LABORATORIES

PO Bag 1779, Kodihalli, Bangalore 560 017
Tel: (+91 80) 527 05 84 and 526 55 79
Fax: (+91 80) 526 08 62 and 526 77 81
e-mail: nicrom@nalsic.ernet.in
Web: http://www.cmmacs.ernet.in/nal
DIRECTOR: Dr T S Prahlad

Design and development of the Saras 9/14-passenger twin-turboprop business and commuter transport is continuing as an Indian programme, having previously appeared in the International section under the Myasishchev/NAL heading. Government funding of prototypes was approved in June 1999. NAL also developed the Hansa light trainer under an agreement with TAAL (which see for description).
UPDATED

NAL SARAS

English name: Crane

TYPE: Twin-turboprop transport.
PROGRAMME: Originated as six/nine-passenger light transport shown in model form at Moscow Aerospace '90 exhibition; revised design announced by Myasishchev 1993, with name Delphin; NAL India had comparable project; general agreement concluded 1993 to combine Myasishchev and NAL programmes, Russian version being known as M-102 Duet and NAL version as Saras (species of Indian crane). Full-scale mockup shown at 1993 and 1995 Moscow Air Shows. Prototype construction (two in Russia, one in India originally) planned to begin in September 1994, but Indian share of private venture capital not secured until early 1996. Myasishchev proposed TVD-20M turboprop in 1995 as cheaper alternative to PT6A-66. Detailed engineering began in April 1996 following finalisation of design late in previous year; Russia was to be responsible for fuselage, landing gear and flight deck systems but, in 1997, Myasishchev withdrew from the project to proceed with unilateral development of the similar M-202, as described in the Russian Federation section.

India elected to continue with the Saras as a national programme, although some form of collaboration with the former Russian partner was not ruled out. NAL secured private backing of Taneja Aerospace and Kumaran Industries, plus financial support from the Indian Technology Development Board. Hindustan Aeronautics Ltd is also now a major partner in the development programme. Release of further Indian government funding in mid-1999 has enabled work to begin on manufacture (by HAL) of two flying prototypes and a structural test airframe, aimed now at a rescheduled first flight in second half of 2001.
CUSTOMERS: NAL estimates at least 200 Indian orders over 10 years.
COSTS: Design and development cost quoted in February 1996 as US$40 million. Late in 1996, cost re-estimated as US$60 million, including flight test phase. Indian government approved Rs1.3 billion (US$30 million) in June 1999 to enable prototype manufacture to begin.
DESIGN FEATURES: Suitable for operations in hot-and-high conditions (typically, airfield at 2,000 m; 6,560 ft, at up to 45°C), and from semi-prepared runways. Unconventional twin rear pusher propeller configuration. Certification intended to FAR Pt 25. High aspect ratio, low-mounted wings, with straight taper; pressurised circular-section fuselage; T tailplane on sweptback fin; engines pylon-mounted each side of fuselage.

Wing section GA(W)-2 (modified); leading-edge sweep 5°; thickness/chord ratio 15 per cent; 4° dihedral; 2° incidence; 2° twist.
FLYING CONTROLS: Conventional and manual. Tabs on all control surfaces; electrically operated pitch, roll and yaw trim; yaw damper; spoilers and 25 per cent chord, single-slotted Fowler flaps; dual-channel three-axis autopilot. Highly swept ventral fins provide nose-down pitching moment at high AoA.
STRUCTURE: Mixed construction of aluminium alloy (fuselage, engine pylons and fixed portion of wing) and composites (nosecone, wing/body fairing, wing moving surfaces and tips and engine cowlings); designed for 30,000 hour life; fail-safe philosophy for all primary structures and major attachments; incorporates damage tolerance features. Two-spar wing with integrally milled skin riveted to stringers; two-spar flaps with honeycomb sandwich filler; single-spar ailerons with honeycomb filler; single-piece, two-spar metal tailplane and three-spar metal fin.
LANDING GEAR: Hydraulically retractable tricycle type, with single mainwheels and single steerable (±53°) nose unit. Mainwheels retract inward into wingroot fairings, nosewheel forward. Oleo-pneumatic shock-absorbers. Mainwheels 26×8.75, pressure 5.52 bars (80 lb/sq in); nosewheel 17.5×6.3, pressure 3.79 bars (55 lb/sq in); air-cooled carbon brakes. Minimum turning radius 8.10 m (26 ft 7 in) at nosewheel.
POWER PLANT: Two 634 kW (850 shp) Pratt & Whitney Canada PT6A-66 turboprops, pylon-mounted on sides of rear fuselage; Hartzell five-blade, constant-speed pusher propellers, rotating in opposite directions. Integral fuel tank in each wing, combined capacity 1,608 litres (425 US gallons; 354 Imp gallons), of which 1,595 litres (421 US gallons; 351 Imp gallons) are usable. Gravity refuelling point in port wing upper surface; optional single-point pressure refuelling point at starboard wingroot.
ACCOMMODATION: Two-person flight deck with dual controls, but to be certified also for one-pilot operation. Alternative layouts for 14 economy class passengers in single seats at 76 cm (30 in) pitch, with centre aisle, rear lavatory and total of 1.1 m³ (39 cu ft) for baggage in forward and rear compartments; or 18 passengers in commuter layout. Typical executive interior could have eight seats, tables, wardrobe, baggage compartment, galley and lavatory. Ambulance version capable of carrying six stretchers, with seats for two medical attendants, medical supplies storage, baggage compartment and lavatory. Various passenger/cargo arrangements or special mission interiors optional in combi version. Door at front on port side, hinged downward, with integral airstairs; overwing Type III

Model of the NAL Saras *(Simon Watson)*
1999/0044853

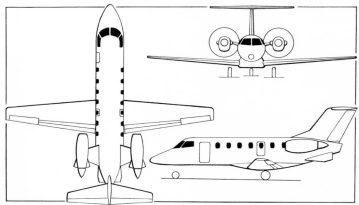

NAL Saras twin-turboprop multipurpose transport *(James Goulding/Jane's)*

emergency exit on each side. Baggage compartment door on port side of tailcone.

SYSTEMS: Bootstrap environmental control system; cabin pressure differential 0.45 bar (6.5 lb/sq in). Hydraulic system (pressure 207 bars; 3,000 lb/sq in) for landing gear actuation, brakes and nosewheel steering; 4 litres (1.06 US gallons; 1.27 Imp gallons)/min electrically powered hydraulic pump; hand pump for emergency lowering of landing gear.

Primary electrical power is 28 V DC, supplied by two 400 A starter/generators; 43 Ah Ni/Cd battery provides emergency power for essential loads for approximately 15 minutes, including three engine restarts. Two solid-state inverters supply 115/26/5 V AC power at 400 Hz for instrumentation; AC power for anti-icing and galley/windscreen heating obtained via alternators. Emergency oxygen system for crew and passengers (two 1,400 litre; 49.4 cu ft bottles), plus two 122 litre (4.3 cu ft) portable bottles and masks for first aid. Engine fire and cabin smoke detection systems. Halon fire extinguishing system. De-icing boots on leading-edges.

AVIONICS: Integrated digital system (ARINC 429 compatible).

Comms: VHF (two), optional HF, intercom/PA system, TCAS II transponder and CVR.

Radar: Weather radar.

Flight: Autopilot (dual-channel, three-axis), ADF, VOR/ILS, marker beacon receiver, DME, radar altimeter, air data sensor and computer, AHRS, flight control computer and flight director standard; GPS, satcom, Selcal and flight management system optional.

Instrumentation: Four-tube EFIS; 'return home' standby instrumentation.

DIMENSIONS, EXTERNAL:

Wing span	14.70 m (48 ft 2¾ in)
Wing chord: at c/l	2.65 m (8 ft 8¼ in)
at tip	0.85 m (2 ft 9½ in)
Wing aspect ratio	8.4
Length overall	15.02 m (49 ft 3¼ in)
Fuselage : Length	13.90 m (45 ft 7¼ in)
Max diameter	1.95 m (6 ft 4¾ in)
Height overall	5.175 m (16 ft 11¾ in)
Tailplane span	6.09 m (19 ft 11¾ in)
Wheel track	3.20 m (10 ft 6 in)
Wheelbase	6.465 m (21 ft 2½ in)
Propeller diameter	2.16 m (7 ft 1 in)
Propeller ground clearance	1.30 m (4 ft 3¼ in)
Distance between propeller centres	3.94 m (12 ft 11 in)
Passenger door (fwd, port): Height	1.48 m (4 ft 10¼ in)
Width	0.75 m (2 ft 5½ in)

DIMENSIONS, INTERNAL:

Cabin: Length	7.74 m (25 ft 4¾ in)
Max width	1.80 m (5 ft 10¾ in)
Aisle width at floor	0.54 m (1 ft 9¼ in)
Max height	1.72 m (5 ft 7¾ in)
Floor area	7.08 m² (76.2 sq ft)
Volume	21.3 m³ (752 cu ft)
Baggage hold volume: fwd, stbd	0.64 m³ (22.6 cu ft)
rear, port	0.46 m³ (16.2 cu ft)

AREAS:

Wings, gross	25.70 m² (276.6 sq ft)
Ailerons (total)	1.335 m² (14.37 sq ft)
Trailing-edge flaps (total)	4.07 m² (43.81 sq ft)
Spoilers (total)	0.65 m² (7.00 sq ft)
Fin	6.38 m² (68.67 sq ft)
Rudder, incl tab	1.815 m² (19.54 sq ft)
Tailplane	7.00 m² (75.35 sq ft)
Elevators, incl tabs	2.05 m² (22.07 sq ft)

WEIGHTS AND LOADINGS:

Weight empty, equipped	4,116 kg (9,074 lb)
Max fuel weight	1,326 kg (2,923 lb)
Max payload	1,232 kg (2,716 lb)
Max T-O and landing weight	6,100 kg (13,448 lb)
Max zero-fuel weight	5,348 kg (11,790 lb)
Max wing loading	237.4 kg/m² (48.61 lb/sq ft)
Max power loading	4.81 kg/kW (7.91 lb/shp)

PERFORMANCE (estimated):

Never-exceed speed (VNE)	372 kt (690 km/h; 428 mph)
Max level speed at 9,000 m (29,520 ft)	324 kt (600 km/h; 373 mph)
Econ cruising speed at 9,000 m (29,520 ft)	270 kt (500 km/h; 311 mph)
Stalling speed at S/L, power off: flaps up	88 kt (163 km/h; 101 mph)
flaps down	72 kt (132 km/h; 82 mph)
Max rate of climb at S/L	740 m (2,427 ft)/min
Rate of climb at S/L, OEI	230 m (754 ft)/min
Max certified altitude	9,000 m (29,525 ft)
Service ceiling	11,500 m (37,720 ft)
Service ceiling, OEI	7,500 m (24,600 ft)
T-O run	430 m (1,410 ft)
T-O field length	570 m (1,870 ft)
Landing field length	605 m (1,985 ft)
Landing run	342 m (1,125 ft)
Runway LCN	10
Range, 30 min reserves: with max fuel	1,511 n miles (2,800 km; 1,739 miles)
with max payload	594 n miles (1,100 km; 683 miles)

UPDATED

TAAL

TANEJA AEROSPACE AND AVIATION LTD

1010 Tenth Floor, Prestige Meridian-1, 29 MG Road, Bangalore 560 001
Tel: (+91 80) 555 06 09 and 555 06 10
Fax: (+91 80) 555 09 55
e-mail: taal@giasbg01.vsnl.net.in
CHAIRMAN: Khushroo Rustumji
MANAGING DIRECTOR: Salil Taneja
VICE-PRESIDENT, MARKETING: Vinod Singel
DIRECTOR, OPERATIONS: A Nanda

TAAL is part of Indian Seamless Metal Tubes Group; has modern plant at Hosur, near Bangalore, with 1,300 m (4,265 ft) captive runway, hangars, laboratories, paint shops and other facilities for aircraft overhaul and manufacture. Entered into technical agreement second quarter 1992 with Aercosmos of Milan to produce Partenavia (now VulcanAir) P.68C/TC, P.68 Observer and AP.68TP-600 Viator light twins in India; for details of aircraft completed by TAAL under this agreement, see the 1998-99 *Jane's*.

No production deliveries of the Hansa light trainer had been reported by October 2000.

UPDATED

NAL/TAAL HANSA-3

TYPE: Primary prop trainer.

PROGRAMME: Developed under agreement with Indian National Aerospace Laboratories (NAL, which see). Known as NALLA: NAL Light Aircraft. Design started May 1989; construction of prototype began December 1991, and this aircraft (VT-XIW) made first flight 17 November 1993. Prototype received Indian DGCA Experimental category type certificate; re-engined 1995 with more powerful IO-240 replacing original 74.6 kW (100 hp) O-200 flat-four. Hansa-3 to be certified initially by DGCA in JAR-VLA (day VFR) category. Deliveries expected to begin in first quarter of 1999, but none reported by October 2000.

CURRENT VERSIONS: **Hansa-2:** Prototype in original configuration; described in 1995-96 *Jane's*.

Hansa-2RE: Prototype re-engined with IO-240; span increased and flaps added; first flight 26 January 1996.

Hansa-3: Production version of Hansa-2RE, first flown (VT-XAL) on 26 November 1996. First flight with Rotax 914 F3 engine, 11 May 1998, made by second prototype VT-XBL. Current version (VT-HNS) has lightning protection for night operations; made first flight 14 May 1999 and received DGCA certification under FAR Pt 23, Amendment 23-42 (JAR-VLA night operation requirements.)

Following description applies to this version.

COSTS: Programme development US$40 million (1997); standard aircraft US$100,000 (1997).

DESIGN FEATURES: Docile handling qualities for *ab initio* training; robust construction and low acquisition/operating cost. Low-wing monoplane with circular-section waisted fuselage; outer wings tapered on leading-edges, with unswept trailing-edge; sweptback fin and rudder; shallow ventral strake; conventional unswept, straight-tapered horizontal tail surfaces.

Laminar-flow wing, with NASA LS(1)-0415 aerofoil section on constant-chord inboard portion, linearly tapered to LS(1)-0413 from there to tip; sweepback 12° 18′ 36″ on outer leading-edges; dihedral 4° from roots; incidence 0°; twist 0° between root and kink; washout –2° between kink and tip.

FLYING CONTROLS: Conventional and manual. Frise ailerons (100 per cent internal mass balance); horn-balanced plain elevators and large rudder, all actuated by pushrods; pitch trim by electrically operated tab in port elevator. Single-slotted Fowler flaps, deflection 20°.

STRUCTURE: Built entirely of composites (CFRP/GFRP reinforced epoxy) with hand-laid, vacuum-bagged, room temperature-cured sandwich shells; components are post-cured before assembly. Conventional rib and three-spar wings, with moulded sandwich shell and PVC foam core; two-spar tail surfaces similar. Fuselage is also a moulded sandwich shell with PVC foam core.

LANDING GEAR: Non-retractable tricycle type, with cantilever steel spring mainwheel legs and steerable (±33°) nosewheel. Cleveland wheels and McCreary tyres (main 6.00-6, nose 5.00-5), pressure 2.07 bars (30 lb/sq in) on all units. Cleveland hydraulic disc brakes. Minimum ground turning radius 2.64 m (8 ft 8 in).

POWER PLANT: One 84.6 kW (113.4 hp at 5,800 rpm) Rotax 914 F3 turbocharged flat-four engine; Hoffmann NOV 352FQ+8 two-blade constant-speed propeller. Single composites fuel tank aft of cockpit, capacity 85 litres (22.5 US gallons; 18.7 Imp gallons). Single gravity refuelling point on top of fuselage, port side. Oil capacity 3 litres (0.8 US gallon; 0.7 Imp gallon).

ACCOMMODATION: Two seats side by side; dual controls standard. Upward-opening gullwing doors. Baggage compartment aft of pilot's seat.

SYSTEMS: Manual (toe-operated) hydraulic mainwheel brakes. Electrical system powered by 14 V, 40 A generator and 12 V, 18 Ah Gill G25M lead-acid battery.

AVIONICS: *Comms:* Honeywell KLX 125 combined VHF/VOR com/nav unit with concealed foil antennas; ACK Technologies E-01 ELT; Clark intercom.

Flight: Optional GPS.

Instrumentation: Conventional VFR (IFR version planned). For day operations: ASI, VSI, altimeter, magnetic compass, slip/skid indicator and OAT gauge. For night operations: artificial horizon, directional gyro, turn co-ordinator, navigation lights, landing lights, anti-collision lights, panel and map lights.

DIMENSIONS, EXTERNAL:

Wing span	10.47 m (34 ft 4¼ in)
Wing chord: at root	1.30 m (4 ft 3¼ in)
at tip	0.80 m (2 ft 7½ in)
Wing aspect ratio	8.8
Length overall	7.66 m (25 ft 1½ in)
Fuselage max width	1.13 m (3 ft 8½ in)
Height overall	2.615 m (8 ft 7 in)
Tailplane span	3.60 m (11 ft 9¾ in)
Wheel track	2.20 m (7 ft 2½ in)
Wheelbase	1.95 m (6 ft 4¾ in)
Propeller diameter	1.70 m (5 ft 7 in)
Propeller ground clearance	0.33 m (1 ft 1 in)

DIMENSIONS, INTERNAL:

Cabin: Length	2.15 m (7 ft 0¾ in)
Max width	1.07 m (3 ft 6¼ in)
Max height	1.00 m (3 ft 3¼ in)
Floor area	0.96 m² (10.3 sq ft)
Volume	1.4 m³ (49.9 cu ft)
Baggage compartment volume	0.24 m³ (8.5 cu ft)

AREAS:

Wings, gross	12.47 m² (134.2 sq ft)
Ailerons (total)	0.97 m² (10.44 sq ft)
Flaps (total)	1.515 m² (16.31 sq ft)
Fin	0.75 m² (8.07 sq ft)

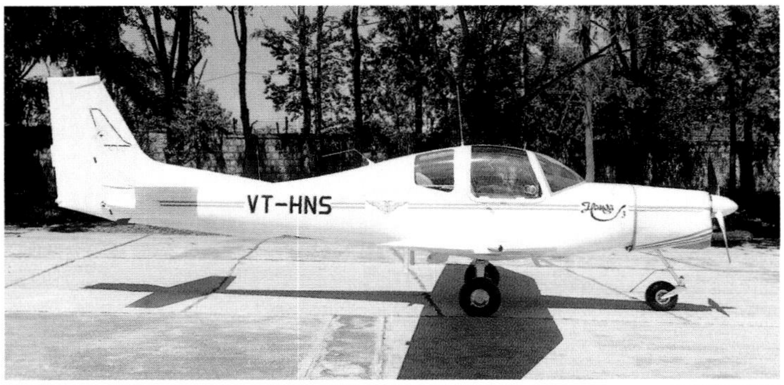

Night-operational version of the NAL/TAAL Hansa-3 civil trainer ***2001***/0097301

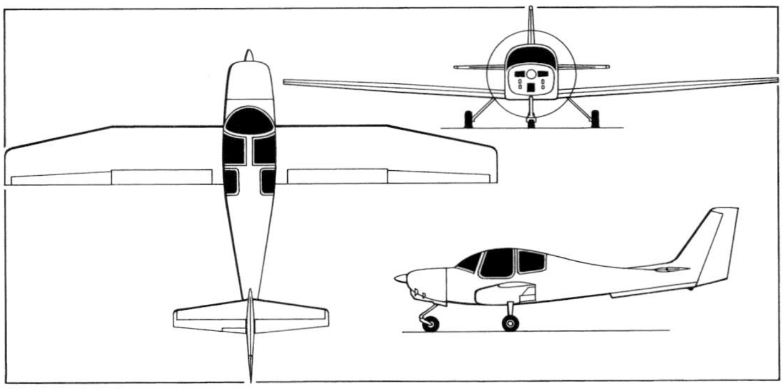

NAL/TAAL Hansa in pre-series configuration ***(Mike Keep/Jane's)***

Rudder	0.70 m² (7.53 sq ft)	Max wing loading	60.1 kg/m² (12.32 lb/sq ft)	T-O to 15 m (50 ft)	450 m (1,480 ft)
Tailplane	1.19 m² (12.81 sq ft)	Max power loading	8.87 kg/kW (14.57 lb/hp)	Landing from 15 m (50 ft)	600 m (1,970 ft)
Elevators (total, incl tab)	0.85 m² (9.15 sq ft)	PERFORMANCE:		Range with max fuel	455 n miles (842 km; 523 miles)

WEIGHTS AND LOADINGS:
Weight empty (night operations) 550 kg (1,213 lb)
Max fuel weight 61.5 kg (135.5 lb)
Max T-O and landing weight 750 kg (1,653 lb)

Max cruising speed at 3,050 m (10,000 ft) 115 kt (213 km/h; 132 mph)
Stalling speed, flaps down 48 kt (89 km/h; 56 mph) IAS
Max rate of climb at S/L 198 m (650 ft)/min

UPDATED

INDONESIA

DIRGANTARA

PT DIRGANTARA INDONESIA (Indonesian Aerospace)

HEAD OFFICE AND WORKS: PO Box 1562 BD, GPM 4th Floor, Jalan Pajajaran 154, Bandung 40174
Tel: (+62 22) 600 25 73
Fax: (+62 22) 603 21 45
e-mail: pub-rel@iptn.co.id
Web: http://www.iptn.co.id
PRESIDENT DIRECTOR AND CEO: Jusman Syafei Djamal
EXECUTIVE VICE-PRESIDENTS:
 Salomo Panjaitan (Operations)
 Ilham Akbar Habibie (Commerce)
 Agung Nugroho (Technology)
 Ending Ridwan (Finance)
CORPORATE PUBLIC RELATIONS OFFICER: Soleh Affandi
PRESS OFFICER: Alex Lim

SALES OFFICE:,
 PO Box 3752 JKT, 14th and 15th Floor, BBD Plaza Building, Jalan Imam Bonjol 61, Jakarta 10310
Tel: (+62 21) 32 22 47
Fax: (+62 21) 310 00 81 and 32 53 19
Telex: 61246 IPTN JK IA

US SUBSIDIARY:
 IPTN North America Inc, 1035 Andover Park West, Suite B, Tukwila, Seattle, Washington 98188-7681
Tel: (+1 206) 575 65 07
Fax: (+1 206) 575 03 18
e-mail: iptn250@aol.com
PRESIDENT: M S Noer

Originally formed by Indonesian government as PT Industri Pesawat Terbang Nurtanio (Nurtanio Aircraft Industry Ltd) 23 August 1976 to centralise all aerospace facilities in one company; renamed PT Industri Pesawat Terbang Nusantara (IPTN) in late 1985. Company restructured in January 2000, with PT Bahana Pengelola Industri Strategis as major shareholder, becoming limited liability company with nine strategic business units, six profit centres, four resource centres and six subsidiaries/corporate functions. New name PT Dirgantara Indonesia (internationally, Indonesian Aerospace, or IAe) inaugurated 24 August 2000 (start of company's 25th year of existence). Workforce in early 2000 was 10,500, including 2,500 engineers and 5,000 technicians and operators, but expected to reduce to a maximum 7,000 by year-end. Site area is 79.3 ha (196 acres) including 600,000 m² (6,458,350 sq ft) covered area.

International Monetary Fund (IMF) bail-out of Indonesia's collapsed economy, beginning in 1997, demanded immediate suspension of further state funding for N-250 and N-2130 programmes. Despite this, N-250 was then expected to survive; alternative support for N-2130 being sought, but no new partners for either venture reported by November 2000. Further estimated investment of US$90 million said to be needed for N-250; discussions with China continuing in late 2000; company reported to be offering entire N-2130 programme for sale.

Fixed-Wing Division co-manufactures CN-235 with CASA (see Airtech in International section), produces CASA C-212 Series 200 under licence, and responsible for indigenous N-250 and N-2130 regional airliners. Six CN-235s delivered to Malaysia and two to Thailand in 1999. Agreement 19 June 1991 with BAe to collaborate on production and assembly of Hawks for Indonesian Air Force.

Helicopter Division is responsible for licensed production of Eurocopter BO 105 and Super Puma (as NBO-105 and NAS-332 respectively), and Bell 412 (as NBell-412). Deliveries resumed in late 1999. Agreement with Bell, June 1996, to include Bell 407 and 430, apears to have lapsed. Production of indigenous helicopters is long-term objective.

Defence Division in Menang Tasikmalaya, West Java (plus smaller plant at Batu Poron, Madura), develops and produces weaponry for IPTN military aircraft.

Space Systems Division main business comprises satellite ground systems and components, application technology, offset and satellite engineering services.

Subcontract work includes production of components for Boeing 737 and 767 and Lockheed Martin F-16. IPTN's 1,400 m² (15,070 sq ft) gas-turbine maintenance centre can maintain, overhaul and repair Rolls-Royce (Allison) 250, P&W JT8D, P&WC PT6T-3/3B, R-R Dart RDa.7, Honeywell LTS101 and TPE331 and General Electric CT7 engines, plus some components of Allison T56 and R-R Tay.

NEW ENTRY

DIRGANTARA N-250

TYPE: Twin-turboprop airliner.
PROGRAMME: Project (Indonesia's first fully indigenously designed transport) announced at Paris Air Show 15 June 1989; engine selected July 1990; first metal cut August 1992; decided mid-1993 to launch with N-250-100 as 64/68-seat sized model instead of originally planned 50/54-seat N-250 (though first prototype remaining configured as 50-seater); this prototype (PA1/PK-XNG *Gatotkoco*) rolled out 10 November 1994; first flight 10 August 1995. Was to be followed by three N-250-100 development aircraft (PA2 *Koconegoro*, PA3 *Krincingwesi* and PA4 *Putut Guritno*) plus two static/fatigue test airframes, with PA3 (systems certification) and PA4 (flight deck certification) to be brought up to production standard later.

Further change of plan in late 1995 made 50-seater an alternative production version after all. A fifth development aircraft (PA5) to this (modified) configuration (see Current Versions) was planned but subsequently cancelled.

PA2 (PK-XNK) originally planned to fly in May 1996, but delayed until 19 December; PA1 had flown 650 hours and PA2 40 hours by December 1997; PA3 and PA4 then expected to fly in second quarter 1998, but suspension of further state funding caused cancellation of PA4 and PA5, as well as slow-down in development programme. Flying programme continued with PA1 and PA2 (some 800 hours flown by late 1999) but, by late 2000, certification still awaited and PA3 still unable to fly until replacement funding obtained. Attempts to secure alternative investment are continuing; if this is secured, part of it will be used to complete PA3 to undertake also the development tasks intended for PA4.

Boeing is assisting (though not financially) in progress towards gaining FAA certification, but plans for second assembly line (for N-270 stretched version) in Mobile, Alabama, USA, under the aegis of American Regional Aircraft Industries (AMRAI, which see), have been frozen.

CURRENT VERSIONS: **N-250-50**: Short-fuselage (50/54-seat) version; represented by PA1 first prototype. Second development aircraft (PA5), planned to differ by having lower-mounted wing box (to reduce weight and drag) and a larger diameter fuselage to increase cabin and baggage volume, cancelled.

N-250-100: Intended initial production version (60 to 68 seats), represented by prototypes PA2 and PA3; certification still not completed by November 2000. Fuselage 1.525 m (5 ft 0 in) longer than N-250 by 0.51 m (1 ft 8 in) insert forward and 1.015 m (3 ft 4 in) aft of wing; shorter engine nacelles. Passenger, cargo and combi versions planned.

Description applies to N-250-100 except where indicated.
N-250-200: Proposed stretched (76-passenger) version.

N-270: Stretched (70-passenger) version of N-250-50, to have been assembled in USA by AMRAI from IPTN kits. Three prototypes were planned.

CUSTOMERS: Orders by November 1997 from Air Venezuela (four), Merpati (15), Sempati (six), Bouraq (five) and Hanze Air (two); options held at that date by Air Venezuela (four), Merpati (85), Sempati (10), Bouraq (57), Columbia Airlines (four), FFV (10), Gulfstream International Airlines (10) and Pakistan International Airlines (15). Total orders and options thus 32 + 195 = 227. Break-even total is 259; estimated Indonesian market is 400, but local orders require reconfirmation in light of recent national financial problems.

DESIGN FEATURES: First fly-by-wire regional turboprop, deriving aerodynamic and operating cost advantages from this technology. Larger fuselage cross-section and longer cabin than CN-235, but no rear-loading ramp; high wing has constant chord centre-section (thickness/chord ratio 17 per cent) and tapered outer panels (13 per cent thickness/chord ratio at tip); sweptback vertical tail surfaces with large dorsal fin; non-swept T tailplane; double-hinged rudder.

Wing is of MS-0317 (modified NASA MD2) (root) and MD3 (tip) aerofoil sections, set at 2° incidence and having 3° dihedral and 3° twist on outboard panels.

FLYING CONTROLS: Lucas/Liebherr fully powered fly-by-wire control of flaps, ailerons, roll spoilers, elevators and rudder; standby fly-by-wire for rudder, mechanically signalled standby for ailerons and elevators. Fixed tailplane. Long-span, fixed-vane double-slotted Fowler flaps, with two-segment spoiler forward of each outer flap. Roll spoilers also function as ground spoilers.

STRUCTURE: Includes composites sandwich for flaps, spoilers, ailerons, elevators, rudder, wing/fin/tailplane tips and leading-edges, wingroot fairings, dorsal fin, tailcone, cowlings, radome and landing gear doors.

Flight deck of the N-250

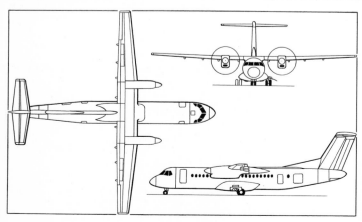

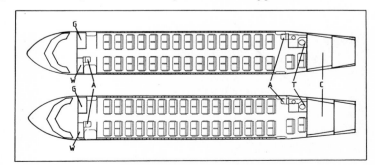

N-250-100 typical 64-seat (top) and 68-seat layouts
A: attendant's seat, C: cargo, G: galley, T: toilet, W: wardrobe

N-250-100 production configuration (*James Goulding/Jane's*)

Formation flight by PA1 (N-250-50 prototype, in lead) and PA2 (N-250-100) *2000*/0088164

LANDING GEAR: Messier-Dowty hydraulically retractable tricycle type, with twin wheels and oleo-pneumatic shock-absorbers on each unit. Main units retract into fairings on fuselage sides, nosewheel forward; nose unit has Lucas 'fly-by-wire' steering (±65°). Mainwheel tyres size 37×11.75-16 (12 ply), pressure 6.96 bars (101 lb/sq in); nosewheel tyres size H21×7.25 (8/12 ply), pressure 4.96 bars (72 lb/sq in). Dunlop steel disc brakes with digital anti-skid system.

POWER PLANT: Two 2,439 kW (3,271 shp) Rolls-Royce AE 2100C turboprops, each driving a Dowty Rotol R384/6123-FX8 six-blade propeller. Engines have Lucas Aerospace FADEC. All fuel in wings, inboard and outboard of nacelles; (see under Weights and Loadings for quantity). EICAS monitoring. Pressure refuelling point in starboard mainwheel fairing, plus overwing gravity point for each tank. Oil capacity 4.8 litres (1.26 US gallons; 1.05 Imp gallons.)

ACCOMMODATION (N-250-100): Flight deck crew of two. Seating (four-abreast with central aisle) for 62 or 64 tourist class passengers at 81 cm (32 in) pitch, 68 at 76 cm (30 in) pitch, or 60 at 81 cm (32 in) with optional increase in cargo. One or two cabin attendants. Storage compartment at front of cabin on port side. Galley at front and lavatory at rear of cabin, both on starboard side. Passenger airstair door at front on port side; service doors at front and rear on starboard side. Type III emergency exit at rear on port side. Large baggage compartment aft of main cabin, with door on starboard side. Additional bulk storage in underfloor compartment, also with external access.

SYSTEMS: Dual Hamilton Sundstrand R90-3WR environmental control systems, pressurised by engine bleed air and APU (maximum pressure differential 0.41 bar; 6.0 lb/sq in). Three hydraulic systems, each at 207 bars (3,000 lb/sq in) pressure. Electrical system has a 40 kVA dual-channel, variable-speed, constant-frequency engine-driven generator to provide 115/200 V three-phase AC power at 400 Hz, with two transformer-rectifier units and two 43 Ah Ni/Cd batteries for 28 V DC normal and emergency power respectively. Rubber boot de-icing of wing, fin and tailplane leading-edges; Westland hot-air anti-icing system with FOD duct for engine intakes. Hamilton Sundstrand APS-1000 APU. Flight Refuelling fuel management system.

AVIONICS: *Comms:* Two VHF 422, one HF 9000 and one TDR 94D (all Rockwell Collins) standard; Avtech intercom and PA system; Antex ELT; Lockheed Martin solid-state CVR. Third VHF 422 and Selcal optional.
Radar: Rockwell Collins WXR 840 weather radar (TWR 850 optional).
Flight: Dual Rockwell Collins AHS 85 AHRS, ADF 462, VOR 432 and ADC 850 air data systems; single Rockwell Collins DME 442 and ALT 55B; Rockwell Collins FCC 4004 AFCS; SFIM solid-state FDR; Hamilton Sundstrand Mk V GPWS. Second DME 442 and ALT 55B, and TCAS, optional.
Instrumentation: Rockwell Collins Pro Line 4 EFIS (four CRTs) and EICAS (one CRT); sixth CRT for GPS optional.

Mockup of the N-2130 flight deck
(Paul Jackson/Jane's) 1999/0044727

DIMENSIONS, EXTERNAL (N-250 and N-250-100, except where indicated):
Wing span	28.00 m (91 ft 10¼ in)
Wing chord: at root	2.80 m (9 ft 2¼ in)
at tip	1.45 m (4 ft 9 in)
Wing aspect ratio	12.1
Length overall: 250-50	26.63 m (87 ft 4½ in)
250-100	28.15 m (92 ft 4¼ in)
Fuselage length: 250-50	25.25 m (82 ft 10 in)
250-100	26.775 m (87 ft 10¼ in)
Fuselage max diameter	2.90 m (9 ft 6¼ in)
Height overall	8.78 m (28 ft 9¾ in)
Tailplane span	9.04 m (29 ft 8 in)
Wheel track (c/l of shock-struts)	4.10 m (13 ft 5½ in)
Wheelbase: 250-50	9.745 m (31 ft 11¾ in)
250-100	10.25 m (33 ft 7½ in)
Propeller diameter	3.81 m (12 ft 6 in)
Propeller/fuselage clearance	0.63 m (2 ft 0¾ in)
Distance between propeller centres	7.67 m (25 ft 2 in)
Passenger door (fwd, port): Height	1.845 m (6 ft 0¾ in)
Width	0.845 m (2 ft 9¼ in)
Height to sill	1.635 m (5 ft 4½ in)
Service doors (fwd and rear, stbd, each):	
Height	1.52 m (4 ft 11¾ in)
Width	0.725 m (2 ft 4½ in)
Height to sill	1.635 m (5 ft 4½ in)
Baggage door (rear, stbd): Height	1.38 m (4 ft 6¼ in)
Width	1.12 m (3 ft 8 in)
Height to sill	1.76 m (5 ft 9¼ in)
Emergency exit (rear, port): Height	0.915 m (3 ft 0 in)
Width	0.61 m (2 ft 0 in)

DIMENSIONS, INTERNAL:
Cabin: Length	13.23 m (43 ft 5 in)
Max width	2.68 m (8 ft 9½ in)
Width at floor	2.41 m (7 ft 10¾ in)
Max height	1.925 m (6 ft 3¾ in)
Main baggage compartment	8.87 m³ (313 cu ft)
Underfloor bulk storage	0.60 m³ (21.2 cu ft)

AREAS:
Wings, gross	65.00 m² (699.7 sq ft)
Vertical tail surfaces (total)	14.72 m² (158.47 sq ft)
Horizontal tail surfaces (total)	16.31 m² (175.56 sq ft)

WEIGHTS AND LOADINGS:
Typical operating weight empty:	
250-50	13,665 kg (30,126 lb)
250-100	15,700 kg (34,612 lb)
Max fuel weight:	
250-50 and 250-100	4,200 kg (9,259 lb)
Max payload: 250-50	6,000 kg (13,227 lb)
250-100	6,200 kg (13,668 lb)
Max ramp weight: 250-50	22,100 kg (48,722 lb)
250-100	24,900 kg (54,895 lb)
Max T-O weight: 250-50	22,000 kg (48,501 lb)
250-100	24,800 kg (54,674 lb)
Max landing weight: 250-50	21,800 kg (48,060 lb)
250-100	24,600 kg (54,233 lb)
Max zero-fuel weight: 250-50	19,665 kg (43,353 lb)
250-100	21,900 kg (48,281 lb)
Max wing loading: 250-50	338.5 kg/m² (69.32 lb/sq ft)
250-100	381.5 kg/m² (78.14 lb/sq ft)
Max power loading: 250-50	4.51 kg/kW (7.41 lb/shp)
250-100	5.09 kg/kW (8.36 lb/shp)

PERFORMANCE (N-250-50, estimated, except where indicated):
Max cruising speed at 6,100 m (20,000 ft), ISA	330 kt (611 km/h; 380 mph)
Econ cruising speed at 6,100 m (20,000 ft)	300 kt (556 km/h; 345 mph)
Stalling speed, power off:	
flaps up	105 kt (195 km/h; 121 mph) EAS
20° flap	90 kt (167 km/h; 104 mph) EAS
Max rate of climb at S/L	564 m (1,850 ft)/min
Rate of climb at S/L, OEI	158 m (520 ft)/min
Service ceiling: 250-50	7,620 m (25,000 ft)
250-100	9,140 m (30,000 ft)
Service ceiling, OEI: 250-50	5,180 m (17,000 ft)
250-100	6,100 m (20,000 ft)
T-O and landing balanced field length at S/L, ISA + 20°C	1,220 m (4,000 ft)
Range with max payload	686 n miles (1,270 km; 789 miles)
Max range with 50 passengers:	
basic	800 n miles (1,481 km; 920 miles)
optional	1,100 n miles (2,037 km; 1,266 miles)

UPDATED

DIRGANTARA N-2130

TYPE: Twin-jet airliner.
PROGRAMME: Initial study (NTP: N-2130 Technology Programme) started October 1994 and continued to March 1997, including 15,000 hours of tunnel testing (half of it transonic). Three versions planned originally (basic 100-seater plus shorter and longer fuselage models), though smallest version (designated I03) was abandoned by early 1997. Ten-year programme anticipated initial customer delivery in June 2006, but acceleration announced February 1996. New timetable then envisaged completion of conceptual design and formal launch in March 1997; start of fabrication and assembly in mid-1998; design 90 per cent complete by early 2001, with subassembly then to

Artist's impression of the baseline Dirgantara N-2130-100 0010512

N-2130 in basic 104/114-seat configuration, with additional side view (bottom) of the projected N-2130-200 *(James Goulding/Jane's)*
0010514

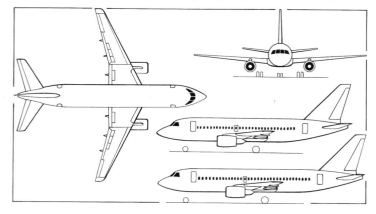

180 INDONESIA: AIRCRAFT—DIRGANTARA

Indonesian-built NBO-105 CBS, equipped for medical evacuation
(Paul Jackson/Jane's) 2000/0064499

NAS-332C Super Puma *(Paul Jackson/Jane's)* 2000/0064498

Indonesian Army NBell-412 *(Paul Jackson/Jane's)*
2000/0064500

begin. Four or five prototypes planned (first roll-out early 2002, first flight mid-2002); certification programme by FAA and Indonesian DGAC; first customer delivery 2004.

Funding for programme raised by local venture capital company named DSTP (*Dua Satu Tiga Puluh:* Indonesian for two, one, three, zero). Following 1997 IMF embargo on state funding for N-2130, IPTN/DSTP in discussions with possible risk-sharing partners (HAL, India, and AIDC, Taiwan, mentioned) in early 1998. Search continued throughout 1998-99; meanwhile, low-priority design development still being undertaken. Company hoped to complete preliminary design phase by end of 2000. In October 2000, IPTN continued to list N-2130 as continuing programme, but only -100 versions then being promoted and entire programme said to be for sale.

CURRENT VERSIONS: **N-2130-100 BGW:** Basic version, with six-abreast seating for 104 passengers in two-class layout or 114 in tourist class with 81 cm (32 in) seat pitch.
Following details refer to this basic version, except where indicated.

N-2130-200: Proposed stretched version, with capacity increased to 122 mixed class or 132 in tourist class layout. See 2000-01 *Jane's* for details.

N-2130-100/200 HGW: Both versions planned to have basic gross weight (BGW) and higher gross weight (HGW) options.

CUSTOMERS: Asia/Pacific market for aircraft of this type estimated at 456, of which IPTN hoped to capture one-third. Initial MoU commitments by Garuda (20) and Sempati (15).

COSTS: Development programme, including all R&D, estimated at US$2,000 million (1995).

DESIGN FEATURES: Envisaged as Asia's first 100-seat regional jet. Conventional design of low-mounted, sweptback wings (25° at quarter-chord) with underwing podded engines; circular-section fuselage; all-swept tail unit. Prominent flap track guides under wing trailing-edge.

FLYING CONTROLS: Digital fly-by-wire flight control system, similar to that of N-250. Conventional ailerons, elevators and rudder. Two-segment trailing-edge flaps; four-segment spoilers/lift dumpers forward of each pair of flaps.

LANDING GEAR: Retractable tricycle type, with twin wheels and oleo-pneumatic shock-absorbers on each unit.

POWER PLANT: Two turbofans, with FADEC, in underwing pods; ratings class 86.1 kN (19,355 lb st) for -100 BGW, 89.3 kN (20,076 lb st) for -100 HGW. Candidates are CFM56-9, Pratt & Whitney PW 6000 and BMW Rolls-Royce BR715-56. Fuel in wing tanks and optional centre-section tank.

ACCOMMODATION: Flight deck crew of two, plus cabin attendant(s). Basic six-abreast seating, with single aisle (see Current Versions). One passenger door forward and one aft of wing on port side, with service door opposite each on starboard side. Emergency exit above wing each side.

DIMENSIONS, EXTERNAL:
Wing span	29.89 m (98 ft 0¾ in)
Wing aspect ratio	8.1
Length overall	31.25 m (102 ft 6¼ in)
Fuselage max diameter	3.95 m (12 ft 11½ in)
Height overall	11.65 m (38 ft 2¾ in)
Tailplane span	12.80 m (42 ft 0 in)
Wheelbase	11.0 m (36 ft 1 in)

DIMENSIONS, INTERNAL:
Cabin, excl flight deck: Length	20.93 m (68 ft 8 in)
Max width	3.71 m (12 ft 2 in)
Aisle width	0.48 m (1 ft 7 in)

AREAS:
Wings, gross	110.00 m² (1,184.0 sq ft)

WEIGHTS AND LOADINGS (design):
Max payload	11,400 kg (25,133 lb)
Max T-O weight: 100 BGW	49,550 kg (109,239 lb)
100 HGW	51,500 kg (113,540 lb)
Max landing weight: 100 BGW	44,595 kg (98,315 lb)
100 HGW	46,350 kg (102,184 lb)
Max wing loading: 100 BGW	450.5 kg/m² (92.26 lb/sq ft)
100 HGW	468.2 kg/m² (95.89 lb/sq ft)
Max power loading: 100 BGW	288 kg/kN (2.82 lb/lb st)
100 HGW	288 kg/kN (2.83 lb/lb st)

PERFORMANCE (design):
Long-range cruising speed	M0.80
Approach speed at MLW	125 kt (232 km/h; 144 mph)
Service ceiling	11,890 m (39,000 ft)
Balanced T-O and landing field length at S/L, ISA + 18°C	1,750 m (5,745 ft)
Range:	
100 BGW	1,200 n miles (2,222 km; 1,380 miles)
100 HGW	1,600 n miles (2,963 km; 1,841 miles)

UPDATED

DIRGANTARA (CASA) NC-212-200

TYPE: Twin-turboprop transport.

PROGRAMME: Original Spanish production of Series 200 ended 1987 (see 1987-88 *Jane's* for details). C-212 manufactured under licence in Indonesia as NC-212 since 1976, IPTN producing 28 NC-212-100s before switching to NC-212-200. Description in 1998-99 *Jane's*.

CUSTOMERS: Total of 95 (28 Srs 100 and 67 Srs 200) produced by December 1997: see 1998-99 *Jane's* for customer details. At least five more completed and five others on production line by October 1998. Outstanding orders at 1 August 1998 totalled 16 Srs 200 (six MPA for Indonesian Navy; six, including one VIP, transports for Indonesian Army; and four for Indonesian Police). Subsequently reported that further production had ceased, and deliveries withheld due to customers' inability to pay for these last aircraft; however, two standard transports were delivered to Indonesian Navy in September 1999. Company was then scheduled to deliver six MPAs to same customer in 2000-2001, but this order said to be under review in early 2000.

UPDATED

DIRGANTARA (EUROCOPTER) NBO-105

TYPE: Multirole light helicopter.

PROGRAMME: Manufactured under licence from MBB (now Eurocopter) as NBO-105 since 1976; only rotors and transmission now supplied from Germany; originally **NBO-105 CB**, but stretched **NBO-105 CBS** available from 101st aircraft onwards; **NBO-105 MPDS** (multipurpose delivery system) can carry 50 to 81 mm unguided rockets, single or twin 0.30 or 0.50 in machine gun pods; and reconnaissance or FLIR pods. Also available in **FAC** (forward air control) version.

CUSTOMERS: At least 121 completed by mid-1996 (partial customer list in previous editions). Three ordered by Indonesian Navy, for over-horizon targeting, in June 1996, equipped with Ocean Master radar and Gemini navigation computer; these largely complete by 1999, and delivery due in 2000. Other completed NBO-105s reportedly remained unsold in mid-1999, but deliveries had restarted (including two to Indonesian Navy) by end of that year.

UPDATED

DIRGANTARA (EUROCOPTER) NAS-332 SUPER PUMA

TYPE: Multirole medium helicopter.

PROGRAMME: Assembly of 11 SA 330J Pumas agreed 1977 and began 1981; switched to AS 332C and L Super Puma in early 1983; available configurations include ASW/MA (anti-submarine warfare/maritime attack). First NAS-332 for Pelita Air Service rolled out 22 April 1983.

CUSTOMERS: Deliveries include four as commando and general purpose transports for Indonesian Navy and two VVIP; see 1997-98 edition for partial customer list. 18 Super Pumas completed or under construction by mid-1996. Sales of seven civil NAS-332s to Iran, reportedly for oil-rig duties, was approved in October 1996 but has been in abeyance (interest reportedly renewed in late 2000); 16 on order for Indonesian Air Force as S-58T replacements. Latter, nearly all completed or in production by April 1998, comprise two VVIP, nine tactical transports and five for combat SAR. Other completed NAS-332s reportedly remained unsold in mid-1999, and none delivered in that year, but five due to be delivered to Indonesian Air Force by end of 2000, and balance (11) will follow.

UPDATED

DIRGANTARA (BELL) NBELL-412

TYPE: Multirole medium helicopter.

PROGRAMME: Licence agreement for Bell 412 (see Canadian section) signed 12 November 1982; covers 100 helicopters; production started 1984; first flight April 1986; NBell-412HP is 40 per cent Indonesian manufactured and employs PT6-3BE engines. FN Herstal **EMA** (external mounting assembly) for 7.62 and 12.7 mm gun pods and 70 mm rocket pods, already certified for Canadian and Italian Agusta-built Bell 412s, qualified for NBell-412 and fitted to several helicopters; **CAS** (close air support) configuration also available.

CUSTOMERS: Production temporarily halted at 27th helicopter, September 1995, restarted three months later, but suspended again as condition of IMF bail-out of Indonesian economy in 1997. None delivered in 1999. One ordered by Indonesian Army in 1996. See 1997-98 edition for partial customer list.

UPDATED

OTHER AIRCRAFT

See under Airtech in International section for details of **CN-235** twin-turboprop military and civil transport.

Agreement for local assembly of **Bell 407** signed 25 June 1996. First IPTN-assembled example, in EMS configuration for MultiPolar of Jakarta, was second of two ordered (first was Canadian-built). No other examples reported; agreement assumed to have been suspended or lapsed.

UPDATED

IPTN

PT. INDUSTRI PESAWAT TERBANG NUSANTARA (Nusantara Aircraft Industries Ltd)

On 24 August 2000, IPTN was renamed Dirgantara Indonesia (which see).

UPDATED

INTERNATIONAL

AGUSTAWESTLAND

AGUSTAWESTLAND
CHAIRMAN: Kevin Smith
CEO: Amedeo Caporaletti
MANAGING DIRECTOR: Richard Case
COO: A Gustapane (Italy)
   A Johnson (UK)
MARKETING MANAGER: G Orsi
PUBLIC RELATIONS MANAGER: Gianluca Grimaldi (Italy)
   David Bath (UK)
PARTICIPATING COMPANIES
   Finmeccanica: see Agusta under Italy.
   GKN: see Westland under UK.

MoU covering joint merger of helicopter divisions signed by Finmeccanica and GKN on 16 April 1998; heads of agreement 18 March 1999; details finalised (subject to regulatory approval) and name AgustaWestland announced 26 July 2000; company declared fully operational 12 February 2000.

Westland also contributed its 50 per cent share in EH 101, GKN's aerospace transmissions business and 50 per cent share in Aviation Training International (joint venture with Boeing to support British Army WAH-64 Apaches). Agusta also contributed half of EH 101, transmissions and aerostructures business and own shares in NH90 (32 per cent) and Bell/Agusta (45 per cent). Joint revenue in 2000 was some US$2,400 million, and order backlog US$8,000 million, placing AgustaWestland second only to Boeing's helicopter activities and ahead of Eurocopter, Bell and Sikorsky. Employees total 10,000

Wholly- and part-owned subsidiaries follow.

Finmeccanica:
   Agusta Aerospace Corp (USA)
   Agusta Aerospace Services BV (Belgium)
   EHI Ltd (50 per cent)
   Bell/Agusta Aerospace Corp (USA) (45 per cent)
   NHI sarl (27 per cent)
   other subsidiaries and holdings

GKN:
   Westland Industrial Products Ltd
   Westland Transmissions Ltd
   EHI Ltd (50 per cent)
   Aviation Training International Ltd (50 per cent)
   other subsidiaries and holdings.

NEW ENTRY

AIRBUS

AIRBUS INDUSTRIE
1 Rond Point Maurice Bellonte, F-31707 Blagnac Cedex, France
Tel: (+33 5) 61 93 33 33
Fax: (+33 5) 61 93 37 92
Web: http://www.airbus.com
CHAIRMAN OF SUPERVISORY BOARD: Manfred Bischoff
CEO: Noël Forgeard
COMPANY SECRETARY: Peter Kleinschmidt
SENIOR VICE-PRESIDENT, CUSTOMER AFFAIRS: John Leahy
SENIOR VICE-PRESIDENT, PROGRAMMES: Gérard Blanc
SENIOR VICE-PRESIDENT, PROCUREMENT: Ray Wilson
SENIOR VICE-PRESIDENT, GOVERNMENT RELATIONS,
   COMMUNICATIONS, EXTERNAL AFFAIRS: Philippe Delmas
SENIOR VICE-PRESIDENT, OPERATIONS: Gustav Humbert
SENIOR VICE-PRESIDENT, ENGINEERING: Alain Garcia
GENERAL MANAGER, PRESS AND INFORMATION SERVICES:
   Barbara Kracht

AIRFRAME PRIME CONTRACTORS:
   EADS: see this section
   BAE Systems Airbus: see under UK

Airbus Industrie set up 18 December 1970 as Groupement d'Intéret Economique (GIE: a company which makes no profits or losses in its own right) and has been profitable since 1990, producing an operating surplus that is shared among its partners. Its first task was to manage development, manufacture, marketing and support of A300; this management now extends to A300-600, A310, A318, A319/ACJ, A320, A321, A330, A340 and A380. Large Airliner Division created in March 1996 to oversee A3XX (later A380) project.

Plans for establishment of a single corporate entity (SCE) overtaken and modified by formation of European Aeronautic, Defense and Space Company (which see). Planned restructuring of the consortium into Airbus Integrated Company (AIC) was revealed in March 2000 and approved by the EC in late 2000; official starting date was 1 January 2001, although completion of formalities was delayed. EADS holds 80 per cent and BAE Systems 20 per cent of AIC, which incorporated in France as an SAS (Société par Actions Simplifiées) following a formal decision on 23 June 2000.

Airbus Industrie responsible for all work by partner companies and has some 3,000 employees, including workers at its spare parts centre in Hamburg, and its US and Chinese subsidiaries; about 41,000 more are directly employed on Airbus work within its partners. Fokker is an associate in A300 and A310 and Belairbus (Belgian consortium) in A310, A320, A330 and A340. Alenia manufactures front fuselage plug for A321.

Airbus Industrie deliveries declined from 163 in 1991 to 157 in 1992, 138 in 1993, 123 in 1994 and rose to 124 in 1995, 126 in 1996, 182 in 1997, 229 in 1998 and 294 in 1999; 277 delivered by 30 November 2000. Orders for 1999 totalled 476 aircraft for 34 customers valued at US$30.5 billion, plus 103 options; orders for 2000 had reached 480 by 30 November. First A340 delivered to Air France in March 1993 was 1,000th Airbus airliner; total orders passed the 2,000 mark in March 1996 (733 A300/310, 959 A319/320/321 and 312 A330/340) and 3,000 in September 1998; 1,500th delivery was an A319 supplied to Lufthansa on 18 February 1997 and 2,000th delivery was an A340-300 to Lufthansa on 18 May 1999. Deliveries totalled 311 in 2000, producing US$17.2 billion turnover. Orders totalled 520, worth US$41.3 billion, not including A380 launch orders. Total order backlog at 1 January 2001 was 1,626.

In early 1997, Airbus initiated a 'build to order only' policy for the A300 and A310 in view of the small backlog then obtaining. In December 1996, Airbus delivered the first aircraft (an A319 to Air Canada) built on a tightened, nine-month production lead time. Production rate of 26 per month in early 2001 due to rise to 38 by 2003.

The Airbus product range, pictured en route to the 1999 Paris Air Show: A300-600ST Beluga (nearest), A340-300, A330-200, A300-600, A310, A321, A320 and ACJ *2000*/0085594

TOTALS OF AIRBUS AIRLINERS
(at 28 February 2001)

	A300	A310	A318	A319	A320	A321	A330	A340	Totals
Firm orders	580	260	161	704	1,452	381	376	309	4,223
Delivered	497	255	0	328	911	180	180	193	2,544
Operating	423	247	0	328	903	180	180	192	2,453

AIRBUS ORDERS and DELIVERIES 2000

Firm orders	2			388			130		
Delivered	8			241			62		

Note: Data for 31 December 2000 was 4,125 orders and 2,499 deliveries

Subsidiaries include Airbus Industrie of North America, Airbus Finance Company (AFC) formed in 1994, Airbus Training Centre (Miami) and Airbus Industrie China. Airbus Industrie's training centre in Toulouse, previously a subsidiary known as Aeroformation, and the main spares centre, Airbus Industrie Materiel Support, formerly Airspares Hamburg, have been integrated within the consortium's Customer Services Directorate; a new, purpose-built mockup

182 INTERNATIONAL: AIRCRAFT—AIRBUS

centre was added to the Toulouse facilities during 1999 and was fully operational by end 2000. A training centre was opened in Miami, Florida, in October 1999.

In March 1994, Airbus signed an agreement with Indonesia's IPTN (now Dirgantara) to assist in flight testing of N-250 turboprop transport (see under Indonesia) up to certification. In July 1996, Airbus Training and Support Centre in Beijing was opened, becoming operational in October 1997.

Agreement of 1995 provides for Airbus Industrie to design, manufacture and market A400M military transport; Airbus Military Company (which see) established for that purpose in January 1999.

Refer also to SATIC (later in this section) for A300-608ST Super Transporter.

UPDATED

AIRBUS A300-600

TYPE: Wide-bodied airliner.

PROGRAMME: Launched 29 May 1969; initial variants were A300B1 (first flight 28 October 1972, service entry November 1974: see 1971-72 *Jane's* for details), A300B2 (first flight June 1973, service entry May 1974) and A300B4 (first flight December 1974, service entry June 1975; 248 built: see 1984-85 and previous editions). A300-600 go-ahead 16 December 1980; first flight (F-WZLR) 8 July 1983; certified (with JT9D-7R4H1 engines) 9 March 1984; first delivery (to Saudia) 26 March 1984.

Improved version with CF6-80C2 engines and other changes (see Current Versions) made first flight 20 March 1985; French certification for Cat. IIIb take-offs and landings 26 March 1985; first delivery of improved version (to Thai Airways) 26 September 1985. Extended-range A300-600R (then known as -600ER) made first flight 9 December 1987, receiving European and FAA certification 10 and 28 March 1988 respectively, deliveries (to American Airlines) beginning 20 April 1988; A300-600 powered by GE CF6-80C2A5 with FADEC granted 180-minute ETOPS April 1994. CIS certification granted May 1996.

CURRENT VERSIONS: **A300-600**: Advanced version of A300B4-200; major A300 version since early 1984. Passenger and freight capacity increased by fitting rear fuselage of A310 with pressure bulkhead moved aft; wings have simple Fowler flaps and increased trailing-edge camber; forward-facing two-person flight deck with EFIS; new digital avionics; new braking control system; new APU; simplified systems; weight saving by use of composites for some secondary structural components; payload/range performance and fuel economy improved by comprehensive drag clean-up. Cargo conversions of A300-600 and earlier A300B4 are offered, see *Jane's Aircraft Upgrades* for details.

Improved version: Introduced 1985. Has CF6-80C2 or PW4000 as engine options, carbon brakes, wingtip fences and 'New World' flight deck; basic equipment of aircraft delivered from late 1991 further improved by incorporating standard options.

Detailed description applies to improved version except where indicated.

A300-600R: Extended-range version of A300-600, differing mainly in having fuel trim tank in tailplane and higher maximum T-O weight.

A300-600 Convertible: Convertible passenger/cargo version, described separately.

A300-600 Freighter: Non-passenger version, described separately.

Airbus Super Transporter: A300-600R conversion as Super Guppy replacement; see under SATIC later in this section.

CUSTOMERS: Total of 580 of all A300 versions ordered, and 497 delivered, by 1 March 2001.

AIRBUS A300 ORDERS
(at 1 March 2001)

Customer	Qty
Air Afrique	3
Air France	23
Air India	3
Air Inter Europe	8
Alitalia	8
American Airlines	35
Amiri Flight	2
Ansett	8
Australian Airlines	5
China Airlines	15
China Eastern Airlines	7
China Northern Airlines	6
China Northwest Airlines	3
CityBird	2
Continental Airlines	3
Cruzeiro	2
Eastern Airlines	34
Egyptair	17
Emirates	5
Federal Express	36
Finnair	2
Garuda	9
Hapag Lloyd	7
Iberia	6
Indian Airlines	10
ILFC	9
Iran Air	8
Japan Air System	29
Japan Fleet Service	2
Korean Air	32
Kuwait Airways	8
Laker	3
LaTur	2
Lufthansa	23
Malaysia Airlines	4
Monarch	4
Olympic	10
Pakistan International Airlines	4
Pan Am	12
Philippine Airlines	5
Polaris Aircraft Leasing	5
Saudi Arabian Airlines	11
SAS	4
Singapore Airlines	8
SOGERMA SOCEA	1
South African Airways	7
Thai Airways	33
Trans European Airways	1
Tunis Air	1
UPS	90
Varig	2
VASP	3
Total	**580**

DESIGN FEATURES: Mid-mounted wings with 10.5 per cent thickness/chord ratio, 28° sweepback at quarter-chord, and (since 1985) tip fences; circular-section pressurised fuselage; all-swept tail unit.

FLYING CONTROLS: Power-assisted. Each wing has three-segment, two-position (T-O/landing) leading-edge slats (no cutout over engine pylon), small Krueger flap at leading-edge wingroot, three cambered tabless flaps on trailing-edge, all-speed aileron between inboard flap and outer pair, and seven spoilers forward of flaps on each wing; flaps occupy 84 per cent of trailing-edge, increasing wing chord by 25 per cent when fully extended; ailerons deflect 9° 2′ downward automatically when flaps are deployed; all 14 spoilers used as lift dumpers: outboard 10 for roll control and inboard 10 as airbrakes; variable incidence tailplane. Ailerons/elevators/rudder fully powered by hydraulic servos (three per surface), controlled mechanically; secondary surfaces (spoilers/flaps/slats) fully powered hydraulically with electrical control, tailplane by two independent hydraulic motors electrically controlled with additional mechanical input; preselection of spoiler/lift dump lever permits automatic extension of lift dumpers on touchdown; flaps and slats have similar drive mechanisms, each powered by twin motors driving ball screwjacks on each surface with built-in protection against asymmetric operation.

STRUCTURE: Two-spar main wing box, integral with fuselage and incorporating fail-safe principles; third spar across inboard sections; semi-monocoque fuselage (frames and open Z-section stringers), with integrally machined skin panels in high-stress areas; primary structure is of high-strength, damage-tolerant aluminium alloy, with steel or titanium for some critical fuselage components, honeycomb panels or selected glass fibre laminates for secondary structures; metal slats, flaps and ailerons. CFRP fins replaced aluminium alloy unit from 1988; secondary structure composites include AFRP for flap track fairings, rear wing/body fairings, cooling air inlet fairings and radome; GFRP for wing upper surface panels above mainwheel bays, fin leading/trailing-edges, fin-tip, fin/fuselage fairings, tailplane trailing-edges, elevator leading-edges, tailplane and elevator tips and elevator actuator access panel; carbon-reinforced GFRP for elevators and rudder; CFRP for spoilers, outer flap deflector doors and fin box; all CFRP moving surfaces have aluminium or titanium trailing-edges. Nosewheel doors and mainwheel leg fairing doors also of CFRP. Nose gear is structurally identical to that of B2/B4/A310; main gear is generally reinforced, with a new hinge arm and a new pitch damper hydraulic and electrical installation. Nacelles have CFRP cowling panels and are subcontracted to Rohr (California); pylon fairings are of AFRP.

Aerospatiale Matra builds nose (including flight deck), lower centre-fuselage, four inboard spoilers, wing/body fairings and engine pylons; Airbus Deutschland builds forward fuselage (flight deck to wing box), upper centre-fuselage, rear fuselage (including tailcone), vertical tail, 10 outboard spoilers and some cabin doors; it also equips wings and installs interiors and seats; BAE Systems (formerly BAe) designed wings and builds wing box; Airbus España manufactures horizontal tail, port and starboard forward passenger doors and mainwheel/nosewheel doors; Fokker produces wingtips, ailerons, flaps, slats and main gear leg fairings. Large, fully equipped and inspected airframe sections airlifted by Beluga to Aerospatiale Matra at Toulouse for assembly and painting, aircraft then being flown to Hamburg for outfitting and returned to Toulouse for customer acceptance.

LANDING GEAR: Hydraulically retractable tricycle type, of Messier-Bugatti design, with Messier-Bugatti/Liebherr/Dowty shock-absorbers and wheels standard; twin-wheel nose unit retracts forward, main units inward into fuselage; free-fall extension; has four-wheel main bogies interchangeable left with right. Standard bogie size is 927 × 1,397 mm (36½ × 55 in); wider bogie of 978 × 1,524 mm (38½ × 60 in) is optional. Mainwheel tyres size 49×17-20 or 49×17.0R20 (30 ply) (standard) or 49×19-20 (30 ply) (wide bogie), with respective pressures of 12.41 and 11.10 bars (180 and 161 lb/sq in). Nosewheel tyres size 40×14 or 40×14.0R16 (22 ply), pressure 9.38 bars (136 lb/sq in). Steering angles 65°/95°. Messier-Bugatti/Liebherr/Dowty hydraulic disc brakes standard on all mainwheels. Normal braking powered by 'green' hydraulic system, controlled electrically through two master valves and monitored by a brake system control box to provide anti-skid protection. Standby braking (powered automatically by 'yellow' hydraulic system if normal 'green' system supply fails) controlled through a dual metering valve; anti-skid protection is ensured through same box as normal system, with emergency pressure supplied to brakes by accumulators charged from 'yellow' system. Automatic braking system optional. Bendix or Goodrich wheels and brakes available optionally. Minimum ground turning radius (effective, aft CG) 22.00 m (72 ft 2¼ in) about nosewheel, 34.75 m (114 ft 0 in) about wingtips.

POWER PLANT: Two turbofans in underwing pods. A300-600 was launched with 249 kN (56,000 lb st) Pratt & Whitney JT9D-7R4H1 and currently available with 249 kN (56,000 lb st) Pratt & Whitney PW4156 or 262 kN (59,000 lb st) General Electric CF6-80C2A1. A300-600R is offered with 274 kN (61,500 lb st) CF6-80C2A5 or 258 kN (58,000 lb st) PW4158. CF6-80C2A5 and PW4158 also available as options on A300-600.

Fuel in two integral tanks in each wing, and fifth integral tank in wing centre-section, giving standard usable capacity of 62,000 litres (16,379 US gallons; 13,638 Imp gallons). Additional 6,150 litre (1,625 US gallon; 1,353 Imp gallon) fuel/trim tank in tailplane (-600R only) increases this total to 68,150 litres (18,004 US gallons; 14,991 Imp gallons). Optional extra fuel cell in aft cargo hold can increase total to 73,000 litres (19,285 US gallons; 16,058 Imp gallons) in -600R. Two standard refuelling

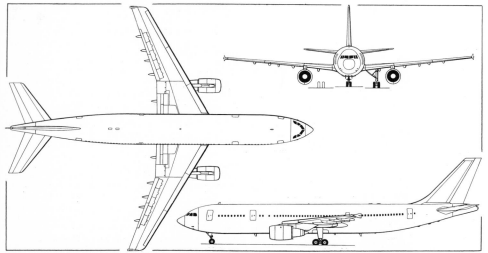

Airbus A300-600R wide-bodied transport (two GE CF6-80C2 turbofans) *(Dennis Punnett/Jane's)*

points beneath starboard wing; similar pair optional under port wing.

ACCOMMODATION: Crew of two on flight deck, plus two observers' seats. Passenger seating in main cabin in six-, seven-, eight- or nine-abreast layout with two aisles; typical mixed class layout has 266 seats (26 first class and 240 economy), six/eight-abreast at 96/86 cm (40/32 in) seat pitch with two galleys and one lavatory forward, one galley and two lavatories at Door 2 position, and one galley and four lavatories at rear; typical economy class layout for 285 passengers eight-abreast at 86 cm (34 in) pitch. Maximum capacity (subject to certification) 361 passengers. Closed overhead baggage lockers on each side (total capacity 10.5 m^3; 370 cu ft) and in double-sided central 'super-bin' installation (total capacity 14.5 m^3; 512 cu ft), giving 0.03 to 0.09 m^3 (1.2 to 3.2 cu ft) per passenger in typical economy layout.

Two outward parallel-opening Type A plug-type passenger doors ahead of wing on each side, and one on each side at rear. Type I emergency exit on each side aft of wing. Underfloor baggage/cargo holds fore and aft of wings, with doors on starboard side; forward hold can accommodate 12 LD3 containers, or four 2.24 × 3.17 m (88 × 125 in) pallets or, optionally, 2.43 × 3.17 m (96 × 125 in) pallets, or engine modules; rear hold can accommodate 10 LD3 containers; additional bulk loading of freight provided for in an extreme rear compartment with usable volume of 17.3 m^3 (611 cu ft); alternatively, rear hold can carry 11 LD3 containers, with bulk cargo capacity reduced to 9.0 m^3 (318 cu ft); bulk cargo compartment can be used to transport livestock. Entire accommodation is pressurised, including freight, baggage and avionics compartments.

SYSTEMS: Air supply for air conditioning system taken from engine bleed and/or APU via two high-pressure points; conditioned air can also be supplied direct to cabin by two low-pressure ground connections; ram air inlet for fresh air ventilation when packs not in use. Pressure control system (maximum differential 0.574 bar; 8.32 lb/sq in) consists of two identical, independent, automatic systems (one active, one standby); automatic switchover from one to other after each flight and in case of active system failure; in each system, pressure controlled by two electric outflow valves, function depending on preprogrammed cabin pressure altitude and rate of change of cabin pressure, aircraft altitude, and preselected landing airfield elevation. Automatic prepressurisation of cabin before take-off, to prevent noticeable pressure fluctuation during take-off. Modular box system provides passenger oxygen to all installation areas.

Hydraulic system comprises three fully independent circuits, operating simultaneously; each system includes reservoir of direct air/fluid contact type, pressurised at 3.52 bars (51 lb/sq in); fire-resistant phosphate ester-type fluid; nominal output flow 136 litres (35.9 US gallons; 30 Imp gallons)/min delivered at 207 bars (3,000 lb/sq in) pressure; 'blue' and 'yellow' systems have one pump each, 'green' system has two pumps. The three circuits provide triplex power for primary flying controls; if any circuit fails, full control of aircraft is retained without any necessity for action by crew. All three circuits supply ailerons, rudder and elevators; 'blue' circuit additionally supplies spoiler 7, spoiler/airbrake 4, airbrake 1, yaw damper and slats; 'green' circuit additionally supplies spoiler 6, flaps, Krueger flaps, slats, landing gear, wheel brakes, steering, tailplane trim, artificial feel, and roll/pitch/yaw autopilot; 'yellow' circuit additionally supplies spoiler 5, spoiler/airbrake 3, airbrake 2, flaps, wheel brakes, cargo doors, artificial feel, yaw damper, tailplane trim, and roll/pitch/yaw autopilot. Ram air turbine pump provides standby hydraulic power should both engines become inoperative.

Main electrical power supplied under normal flight conditions by two integrated drive generators, one on each engine; third (auxiliary) generator, driven by APU, can replace either of main generators, having same electromagnetic components but not constant-speed drive; each generator rated at 90 kVA, with overload ratings of 112.5 kVA for 5 minutes and 150 kVA for 5 seconds; APU generator driven at constant speed through gearbox. Three unregulated transformer-rectifier units (TRUs) supply 28 V DC power. Three 25 Ah Ni/Cd batteries used for emergency supply and APU starting; emergency electrical power taken from main aircraft batteries and emergency static inverter, providing single-phase 115 V 400 Hz output for flight instruments, navigation, communications and lighting when power not available from normal sources.

Hot air anti-icing of engines, engine air intakes, and outer segments of leading-edge slats; electrical heating for anti-icing flight deck front windscreens, demisting flight deck side windows, and for sensors, pitot probes and static ports, and waste water drain masts.

Honeywell 331-250F APU in tailcone, exhausting upward; installation incorporates APU noise attenuation. Self-contained fire protection system, and firewall panels protect main structure from an APU fire. APU provides bleed air to pneumatic system, and drives auxiliary AC generator during ground and in-flight operation; APU drives 90 kVA oilspray-cooled generator, and supplies bleed air for main engine start or air conditioning system. For current deliveries of A300-600, APU has improved relight capability, and can be started throughout flight envelope.

For new A300-600s and -600Rs, two optional modifications offered for compliance with full extended-range twin-engined operations (ETOPS) requirements: hydraulically driven fourth generator and increased cargo hold fire suppression capability. ETOPS kit qualified for aircraft with CF6-80C2 and JT9D-7R series engines and, since mid-1988, for those with PW4000 series.

AVIONICS: *Comms:* Standard communications radios include two VHF, with provision for a third, two HF, two transponders, one Selcal, interphone and passenger address systems, ground crew call system and cockpit voice recorder. Provision for Mode S transponders.

Radar: Weather radar standard, with provision for second.

Flight: Radio navigation avionics include two VOR, two ILS, two DME, one ADF, two marker beacon receivers and two radio altimeters; TCAS and GPWS. Most other avionics are to customer requirements, only those relating to the instrument landing system (Honeywell or Rockwell Collins ILS and Rockwell Collins or TRT radio altimeter) being selected and supplied by the manufacturer. Two Honeywell digital air data computers standard; basic digital AFCS has dual flight control computers (FCCs) for flight director and autopilot functions (for Cat. III automatic landings), single thrust control computer (TCC) for speed and thrust control, and two flight augmentation computers (FACs) to provide yaw damping, electric pitch trim, and flight envelope monitoring and protection. Options include second FCC (for Cat. III automatic landing); second TCC; two flight management computers (FMCs) and two control display units for full flight management system. Basic aircraft also fitted with ARINC 717 data recording system with digital flight data acquisition unit, digital flight data recorder and three-axis linear accelerometer; optional additional level of windshear protection is available. Honeywell Enhanced GPWS available from March 1997.

Instrumentation: Six identical and interchangeable CRT electronic displays (four electronic flight instrument system and two electronic centralised aircraft monitor), plus digitised electromechanical instruments with liquid crystal displays.

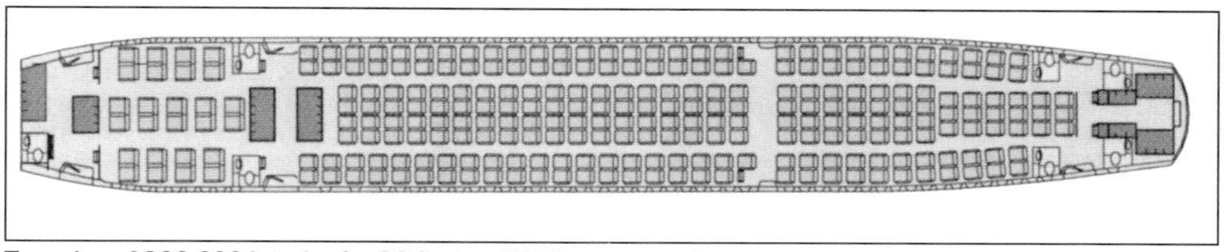

Two-class A300-600 interior for 26 first and 240 economy class passengers ***1999***/0044877

DIMENSIONS, EXTERNAL:
Wing span 44.84 m (147 ft 1 in)
Wing aspect ratio 7.7
Length overall 54.08 m (177 ft 5 in)
Fuselage: Length 53.30 m (174 ft 10½ in)
Max diameter 5.64 m (18 ft 6 in)
Height overall 16.51 m (54 ft 2 in)
Tailplane span 16.26 m (53 ft 4 in)
Wheel track 9.60 m (31 ft 6 in)
Wheelbase (c/l of shock-absorbers) 18.62 m (61 ft 1 in)
Passenger doors (each): Height 1.93 m (6 ft 4 in)
Width 1.07 m (3 ft 6 in)
Height to sill: forward 4.60 m (15 ft 1 in)
centre 4.80 m (15 ft 9 in)
rear 5.50 m (18 ft 0½ in)
Emergency exits (each): Height 1.60 m (5 ft 3 in)
Width 0.61 m (2 ft 0 in)
Height to sill 4.87 m (15 ft 10 in)
Underfloor cargo door (forward):
Height 1.71 m (5 ft 7¼ in)
Width 2.69 m (8 ft 10 in)
Height to sill 3.07 m (10 ft 1 in)
Underfloor cargo door (rear):
Height 1.71 m (5 ft 7¼ in)
Width 1.81 m (5 ft 11¼ in)
Height to sill 3.41 m (11 ft 2¼ in)
Underfloor cargo door (extreme rear):
Height (projected) 0.95 m (3 ft 1 in)
Width 0.95 m (3 ft 1 in)
Height to sill 3.56 m (11 ft 8 in)

DIMENSIONS, INTERNAL:
Cabin, excl flight deck: Length 40.70 m (133 ft 6¼ in)
Max width 5.28 m (17 ft 4 in)
Max height 2.54 m (8 ft 4 in)
Underfloor cargo hold:
Length: forward 10.60 m (34 ft 9¼ in)
rear 7.95 m (26 ft 1 in)
extreme rear 3.40 m (11 ft 2 in)
Max height 1.76 m (5 ft 9 in)
Max width 4.20 m (13 ft 9¼ in)
Underfloor cargo hold volume:
forward 75.1 m^3 (2,652 cu ft)
rear 55.0 m^3 (1,942 cu ft)
extreme rear 17.3 m^3 (611 cu ft)

AREAS:
Wings, gross 260.00 m^2 (2,798.6 sq ft)
All-speed ailerons (total) 7.06 m^2 (75.99 sq ft)
Trailing-edge flaps (total) 47.30 m^2 (509.13 sq ft)
Leading-edge slats (total) 30.30 m^2 (326.15 sq ft)
Krueger flaps (total) 1.115 m^2 (12.00 sq ft)
Spoilers (total) 5.40 m^2 (58.13 sq ft)
Airbrakes (total) 12.59 m^2 (135.52 sq ft)
Fin 45.20 m^2 (486.53 sq ft)
Rudder 13.57 m^2 (146.07 sq ft)
Tailplane 44.80 m^2 (482.22 sq ft)
Elevators (total) 19.20 m^2 (206.67 sq ft)

WEIGHTS AND LOADINGS (A: CF6-80C2A1/A5 engines, B: PW4156/4158 engines, both in 266-seat configuration)*:
Manufacturer's weight empty:
A (600) 79,210 kg (174,630 lb)
A (600R) 80,070 kg (176,525 lb)
B (600) 79,151 kg (174,500 lb)
B (600R) 79,320 kg (174,870 lb)
Operating weight empty:
A (600) 90,115 kg (198,665 lb)
A (600R) 91,040 kg (200,700 lb)
B (600) 90,065 kg (198,565 lb)
B (600R) 90,965 kg (200,550 lb)
Max payload (structural): A (600) 39,885 kg (87,931 lb)
A (600R) 38,962 kg (85,896 lb)
B (600) 39,993 kg (88,169 lb)
B (600R) 39,037 kg (86,061 lb)
Max usable fuel:
600: standard 49,786 kg (109,760 lb)
600R: standard 54,721 kg (120,640 lb)
with optional cargo hold tank
58,618 kg (129,230 lb)
Max T-O weight (A and B):
600 165,000 kg (363,765 lb)
600R (standard) 170,500 kg (375,885 lb)
600R (option) 171,700 kg (378,535 lb)
Max ramp weight (A and B):
600 165,900 kg (365,745 lb)
600R (standard) 171,400 kg (377,870 lb)
600R (option) 172,600 kg (380,520 lb)
Max landing weight (A and B):
600 138,000 kg (304,240 lb)
600R (standard) 140,000 kg (308,645 lb)
Max zero-fuel weight (A and B):
600, 600R (standard) 130,000 kg (286,600 lb)
Max wing loading: 600 634.6 kg/m^2 (129.98 lb/sq ft)
600R (standard) 655.8 kg/m^2 (134.32 lb/sq ft)

**Production aircraft from late 1996 onward. See 1989-90 and previous editions for original versions; 1996-97 and six previous editions for intermediate versions*

PERFORMANCE (A and B as for Weights and Loadings):
Max operating speed (V_{MO}) from S/L to 8,140 m (26,700 ft) 335 kt (621 km/h; 386 mph) CAS

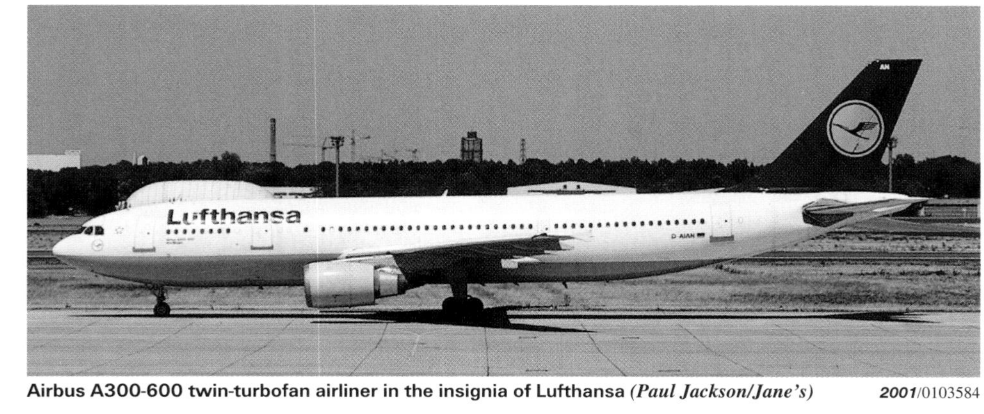

Airbus A300-600 twin-turbofan airliner in the insignia of Lufthansa *(Paul Jackson/Jane's)* ***2001***/0103584

Max operating Mach No. (M$_{MO}$) above 8,140 m
 (26,700 ft) 0.82
Max cruising speed at 7,620 m (25,000 ft)
 480 kt (890 km/h; 553 mph)
Max cruising speed at 9,140 m (30,000 ft)
 M0.82 (484 kt; 897 km/h; 557 mph)
Typical long-range cruising speed at 9,440 m (31,000 ft)
 M0.80 (472 kt; 875 km/h; 543 mph)
Approach speed: 600 135 kt (249 km/h; 155 mph)
 600R 136 kt (251 km/h; 156 mph)
Max operating altitude 12,200 m (40,000 ft)
Runway ACN for flexible runway, category B:
 standard bogie and tyres: 600 56
  600R 59
  600R (option) 60
 optional bogie and tyres: 600 52
  600R 55
  600R (option) 56
T-O field length at S/L, ISA + 15°C:
 600: A 2,378 m (7,800 ft)
  B 2,270 m (7,450 ft)
 600R: A (C2A5 engines) 2,408 m (7,900 ft)
  B (PW4158 engines) 2,362 m (7,750 ft)
Landing field length: 600 1,536 m (5,040 ft)
 600R 1,555 m (5,100 ft)
Range (1996 and subsequent deliveries) at typical airline OWE with 266 passengers and baggage, reserves for 200 n miles (370 km; 230 miles):
 600, GE/PW engines
  3,700 n miles (6,852 km; 4,257 miles)
 600R, GE/PW engines, standard fuel
  4,050 n miles (7,500 km; 4,660 miles)
 600R, GE/PW engines, optional fuel
  4,150 n miles (7,685 km; 4,775 miles)
OPERATIONAL NOISE LEVELS (A300-600R, ICAO Annex 16, Chapter 3):
 T-O: A 91.1 EPNdB (96.3 limit)
  B 92.2 EPNdB (96.3 limit)
 Sideline: A 98.6 EPNdB (99.9 limit)
  B 97.7 EPNdB (99.9 limit)
 Approach: A 99.8 EPNdB (103.3 limit)
  B 101.7 EPNdB (103.3 limit)

UPDATED

AIRBUS A300-600 CONVERTIBLE and A300-600 FREIGHTER

TYPE: Twin-jet freighter.
PROGRAMME: Specialised versions of A300-600. First flight of A300-600F 2 December 1993; certified April 1994 and entered service with Federal Express in same month; A300-600F powered by GE CF6-80C2A5 with FADEC was first A300-600 version to operate with 180-minute ETOPS in May 1994.
CURRENT VERSIONS: **Convertible:** For all-passenger or all-cargo configuration. Typical options include accommodation (in mainly eight-abreast seating) for maximum 375 passengers (subject to certification) on the main deck; or up to twenty 2.24 × 3.17 m (88 × 125 in) pallets; or five 88 × 125 in plus nine 96 × 125 in pallets.
Freighter: For freighting only; no passenger systems provided; various systems options give airlines ability to adapt basic aircraft to specific freight requirements; Airbus offers conversion with port-side forward freight door. Freighter conversions, offered by British Aerospace and EFW, are detailed in *Jane's Aircraft Upgrades*.
CUSTOMERS: Federal Express became A300-600 Freighter launch customer July 1991 with order for 25 and commitments for 50 more, of which 11 confirmed as orders in September 1996; 30 delivered by 31 December 1998. UPS placed order on 9 September 1998 for 30 PW4158 powered A300F4-600R Freighters plus options on a further 45 (later confirmed and a further 15 firm orders, to total 90, with 50 more on option). Deliveries began mid-2000; UPS has an option to convert some of the order to A380s if it wishes. First (of two) for CityBird of Belgium delivered 23 July 1999; production set to continue to at least 2009.
STRUCTURE: Generally similar to A300-600. Main differences are large port-side main deck cargo door, reinforced cabin floor, smoke detection system in main cabin; main deck cargo door is on opposite side to door of forward underfloor hold, allowing simultaneous loading or unloading at all positions.
POWER PLANT: Options as for A300-600R; first example was first Airbus aircraft powered by GE CF6-80C2A5 with FADEC.
DIMENSIONS, EXTERNAL: As A300-600R, plus:
 Main deck cargo door (fwd, port):
  Height (projected) 2.57 m (8 ft 5¼ in)
  Width 3.58 m (11 ft 9 in)
  Height to sill 4.91 m (16 ft 1 in)
DIMENSIONS, INTERNAL:
 Cabin main deck usable for cargo:
  Length 33.45 m (109 ft 9 in)
  Min height 2.01 m (6 ft 7 in)
  Max height:
   ceiling trim panels in place 2.22 m (7 ft 3½ in)
   without ceiling trim panels 2.44 m (8 ft 0 in)
  Volume 192.0-203.0 m³ (6,780-7,169 cu ft)

WEIGHTS AND LOADINGS (basic Convertible. A: with CF6-80C2A5 engines, B: with PW4158 engines):
 Manufacturer's weight empty:
  A, passenger mode 82,555 kg (182,000 lb)
  B, passenger mode 82,470 kg (181,815 lb)
  A, freight mode 80,345 kg (177,130 lb)
  B, freight mode 80,260 kg (176,945 lb)
 Operating weight empty:
  A, passenger mode 93,550 kg (206,240 lb)
  B, passenger mode 93,475 kg (206,075 lb)
  A, freight mode 81,600 kg (179,895 lb)
  B, freight mode 81,525 kg (179,730 lb)
 Max payload (structural):
  A, passenger mode 36,448 kg (80,354 lb)
  B, passenger mode 36,523 kg (80,519 lb)
  A, freight mode 48,400 kg (106,705 lb)
  B, freight mode 48,475 kg (106,870 lb)
 Max T-O weight: A, B 170,500 kg (375,900 lb)
 Max landing weight: A, B 140,040 kg (308,650 lb)
 Max zero-fuel weight: A, B 130,000 kg (286,600 lb)
WEIGHTS AND LOADINGS (basic Freighter variant of -600R):
 Manufacturer's weight empty:
  A 78,335 kg (172,700 lb)
  B 78,250 kg (172,510 lb)
 Operating weight empty: A 79,050 kg (174,275 lb)
  B 78,980 kg (174,120 lb)
 Max payload (structural):
  A, range mode 50,950 kg (112,325 lb)
  B, range mode 51,020 kg (112,480 lb)
  A, payload mode 54,750 kg (120,705 lb)
  B, payload mode 54,820 kg (120,855 lb)
 Max T-O weight: A, B:
  range mode 170,500 kg (375,900 lb)
  payload mode 165,100 kg (363,980 lb)
 Max landing weight: A, B:
  range mode 140,000 kg (308,650 lb)
  payload mode 140,600 kg (309,970 lb)
 Max zero-fuel weight: A, B:
  range mode 130,000 kg (286,600 lb)
  payload mode 133,800 kg (294,980 lb)
PERFORMANCE:
 Range with max (structural) payload, allowances for 30 min hold at 460 m (1,500 ft) and 200 n mile (370 km; 230 mile) diversion: A, B, range mode
  2,650 n miles (4,908 km; 3,050 miles)
  A, B, payload mode
  1,900 n miles (3,519 km; 2,186 miles)

UPDATED

AIRBUS A310
Canadian Forces designation: CC-150 Polaris

TYPE: Wide-bodied airliner.
PROGRAMME: Launched July 1978; first flight (F-WZLH) 3 April 1982; initial French/German certification 11 March 1983; first deliveries (Lufthansa and Swissair) 29 March 1983, entering service 12 and 21 April respectively; JAA Cat. IIIa certification (France/Germany) September 1983; UK certification January 1984; JAA Cat. IIIb November 1984; FAA type approval early 1985. First flight of extended-range A310-300 8 July 1985 (certified with JT9D-7R4E engines 5 December 1985, delivered to launch customer Swissair 17 December); wingtip fences introduced as standard on A310-200 from early 1986 (first delivery: Thai Airways, 7 May); certification/delivery of A310-300 with CF6-80C2 engines April 1986, with PW4152s June 1987. Russian State Aviation Register certification October 1991 (first Western-built aircraft to achieve this status). A310s powered by PW4000 series and CF6-80C2 approved for 180-minute ETOPS. Enhanced GPWS installed March 1997, following certification.
CURRENT VERSIONS: **A310-200:** Basic passenger version. *Detailed description mainly applies to A310-200.*
A310-200C: Convertible version of A310-200; first delivery (Martinair) 29 November 1984.
A310-200F: Conversion of A310 marketed by Elbe Flugzeugwerke GmbH; max payload 40,600 kg (89,508 lb) and max T-O weight 142,000 kg (313,056 lb). Main cargo deck accommodates sixteen 2.24 × 3.17 m (88 × 125 in) pallets plus further three pallets of this size in underfloor hold, in addition to six LD3s; alternatively 14 LD3s can be carried in underfloor hold. See *Jane's Aircraft Upgrades* for further details.
A310-300: Extended-range passenger version, launched March 1983; second member of Airbus family to introduce delta-shaped wingtip fences as standard. Extra range provided by increased basic maximum T-O weight (150,000 kg; 330,695 lb) and greater fuel capacity (higher maximum T-O weights optional); standard extra fuel capacity is in tailplane, allowing in-flight CG control for improved fuel efficiency. For extra long range, one or two ACTs (additional centre tanks) can be installed in part of cargo hold; modification certified November 1987 (first customer Wardair of Canada).
Belgian Air Force obtained two second-hand A310-200s in 1997; Canadian Forces operate six A310-300s, German Luftwaffe seven and French Air Force two; last-named have ETOPS for 180 minutes; Spanish intention to buy two for VVIP use, following modification by EADS CASA, announced December 2000. Details of projected military versions appear under Airbus Military Company at the end of this entry.
CUSTOMERS: Total of 260 sold, and 255 delivered by 1 December 2000.

AIRBUS A310 ORDERS
(at 1 March 2001)

Customer	Qty
Aeroflot	5
Air Afrique	4
Air Algerie	2
Air France	11
Air India	8
Air Niugini	2
Austrian	4
Balair	4
Biman Bangladesh	2
British Caledonian Airways	2
China Eastern Airlines	5
Condor Flugdienst	5
Cyprus Airways	4
Czech Airlines	2
Delta	9
Ecuatoriana	2
Emirates	8
Hapag Lloyd Flug	7
Interflug	3
ILFC	7
Iraqi Airways	5 (not delivered)
Kenya Airways	2
KLM	10
Kuwait Airways	11
Lufthansa	20
Martinair	2
Nigeria Airways	4
Oasis Group	2
Pakistan International Airlines	6
Pan Am	18
Royal Jordanian	6
Royal Thai Air Force	1
Sabena	3
Singapore Airlines	23
Somali Airlines	1
Swissair	9
TAP-Air Portugal	5
Tarom	2
Thai Airways	2
THY	14
Trans European Airways	1
Uzbekistan Airways	1
Wardair	12
Yemenia	2
Undisclosed	2
Total	**260**

DESIGN FEATURES: Retains same fuselage cross-section as A300, but with cabin 11 frames shorter and overall fuselage 13 frames shorter than A300B2/B4-100 and -200; new

UPS is the newest A300 freighter customer with order for 30 placed in 1999; artist's impression shows aircraft in house colours

Hapag-Lloyd Airbus A310-200 retrospectively fitted with winglets *(Paul Jackson/Jane's)*

advanced-technology wings of reduced span and area; new and smaller horizontal tail surfaces; common pylons able to support all types of GE and PW engines offered; advanced digital two-man cockpit; landing gear modified to cater for size and weight changes. Wings have 28° sweepback at quarter-chord, root incidence 5° 3′, dihedral 11° 8′ (inboard) and 4° 3′ (outboard) at trailing-edge, and thickness/chord ratios of 15.2 (root), 11.8 (at trailing-edge kink) and 10.8 per cent (tip).

FLYING CONTROLS: Wing leading-edge movable surfaces as for A300-600; trailing-edges each have single Fowler flap outboard, vaned Fowler flap inboard, lateral control by inboard all-speed aileron and fly-by-wire outboard spoilers, without outboard ailerons; all 14 spoilers used as lift dumpers, inner eight also as airbrakes; fly-by-wire spoiler panels controlled by two independent computer systems with different software to ensure redundancy and operational safety. Tail control surfaces as for A300-600.

STRUCTURE: Mainly of high-strength aluminium alloy except for outer shrouds (structure in place of low-speed ailerons), spoilers, wing leading-edge lower access panels and outer deflector doors, nosewheel doors, mainwheel leg fairing doors, engine cowling panels, elevators and fin box, which are all of CFRP; A310 was first production airliner to have carbon fin box, starting with A310-300 for Swissair in December 1985; flap track fairings, flap access doors, rear wing/body fairings, pylon fairings, nose radome, cooling air inlet fairings and tailplane trailing-edges made of AFRP; wing leading-edge top panels, panel aft of rear spar, upper surface skin panels above mainwheel bays, forward wing/body fairings, glide slope antenna cover, fin leading/trailing-edges, fin and tailplane tips (GFRP); and rudder (CFRP/GFRP). Wing box is two-spar multirib metal structure, with top and bottom load-carrying skins. Undertail bumper beneath rear fuselage, to protect structure against excessive nose-up attitude during T-O and landing.

Manufacturing breakdown differs in detail from that of A300-600: Aerospatiale Matra builds nose section (including flight deck), lower centre-fuselage and wing box, rear wing/body fairings, engine pylons and airbrakes, and is responsible for final assembly; DaimlerChrysler Aerospace Airbus builds forward fuselage, upper centre-fuselage, rear fuselage and associated doors, tailcone, fin and rudder, flaps and spoilers, and fits control surfaces and equipment to main wing structure produced by BAE Systems; CASA's contribution includes horizontal tail surfaces, nose-gear and mainwheel doors, and forward passenger doors; Fokker manufactures main landing gear leg doors, wingtips, all-speed ailerons and flap track fairings; wing leading-edge slats and forward wing/fuselage fairings produced by Belgian Belairbus consortium. An A310 of German Luftwaffe is currently testing a fuselage panel constructed from a new lightweight composites material, Glare, in preparation for its use in A3XX; this is first employment of Glare in a primary structure.

LANDING GEAR: Hydraulically retractable tricycle type. Twin-wheel steerable nose unit (steering angle 65°/95°) as for A300. Main gear by Messier-Bugatti, each bogie having two tandem-mounted twin-wheel units. Retraction as for A300-600. Standard tyre sizes: main, 46×16-20 (28/30 ply), pressure 11.24 bars (163 lb/sq in); nose, 40×14 or 40×14.0R16 (22/24 ply), pressure 9.03 bars (131 lb/sq in). Two options for low-pressure tyres on main units: (1) size 49×17.0R20 (30/32 ply), pressure 9.86 bars (143 lb/sq in); (2) size 49×19-20, pressure 8.89 bars (129 lb/sq in). Messier-Bugatti brakes and anti-skid units standard; Bendix type optional on A310-200. Carbon brakes standard since 1986. Minimum ground turning radius (effective, aft CG) 18.75 m (61 ft 6 in) about nosewheel, 33.00 m (108 ft 3¼ in) about wingtips.

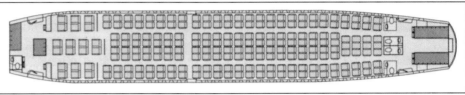

A310 two-class seating for 20 first and 200 economy passengers

POWER PLANT: Launched with two 213.5 kN (48,000 lb st) Pratt & Whitney JT9D-7R4D1 or 222.4 kN (50,000 lb st) General Electric CF6-80A3 turbofans; currently available with 238 kN (53,500 lb st) CF6-80C2A2, or 231 kN (52,000 lb st) Pratt & Whitney PW4152. Available from late 1991 with 262 kN (59,000 lb st) CF6-80C2A8 or 249 kN (56,000 lb st) PW4156A.

Total usable fuel capacity 54,920 litres (14,509 US gallons; 12,081 Imp gallons) in A310-200. Increased to 61,070 litres (16,133 US gallons; 13,434 Imp gallons) in A310-300 by additional fuel in tailplane trim tank. Further 7,200 litres (1,902 US gallons; 1,584 Imp gallons) can be carried in each, of up to two, additional centre tanks (ACT) in forward part of aft cargo hold. Two refuelling points, one beneath each wing outboard of engine.

ACCOMMODATION: Crew of two on flight deck; provision for third and fourth crew seats. Cabin, with six-, seven-, eight- or nine-abreast seating, normally for 210 to 250 passengers, although certified for up to 280; typical two-class layout for 220 passengers (20 first class, six-abreast at 102 cm; 40 in seat pitch, plus 200 economy class mainly eight-abreast at 81 cm; 32 in pitch); maximum capacity for 280 passengers nine-abreast in high-density configuration at pitch of 76 cm (30 in). Standard layout has two galleys and lavatory at forward end of cabin, plus two galleys and four lavatories at rear; depending on customer requirements, second lavatory can be added forward, and lavatories and galleys can be located at forward end at class divider position. Overhead baggage stowage as for A300-600, rising to 0.09 m³ (3.2 cu ft) per passenger in typical economy layout. Four passenger doors, one forward and one aft on each side; oversize Type I emergency exit over wing on each side. Underfloor baggage/cargo holds fore and aft of wings, each with door on starboard side; forward hold accommodates eight LD3 containers or three 2.24 × 3.17 m (88 × 125 in) standard or three 2.44 × 3.17 m (96 × 125 in) optional pallets; rear hold accommodates six LD3 containers, with optional seventh LD3 or LD1 position; LD3 containers can be carried two-abreast, and/or standard pallets installed crosswise.

SYSTEMS: Honeywell 331-250 APU. Air conditioning system, powered by compressed air from engines, APU or a ground supply unit; two separate packs; air is distributed to flight deck, three separate cabin zones, electrical and electronic equipment, avionics bay and bulk cargo compartment; ventilation of forward cargo compartments optional. Pressurisation system has maximum normal differential of 0.57 bar (8.25 lb/sq in). Air supply for wing ice protection, engine starting and thrust reverser system bled from various stages of engine compressors, or supplied by APU or ground supply unit.

Hydraulic system (three fully independent circuits operating at 207 bars (3,000 lb/sq in) details as described for A300-600).

Electrical system, similar to that of A300-600, consists of a three-phase 115/200 V 400 Hz constant frequency AC system and a 28 V DC system; two 90 kVA engine-driven brushless generators for normal single-channel operation, with automatic transfer of busbars in the event of a

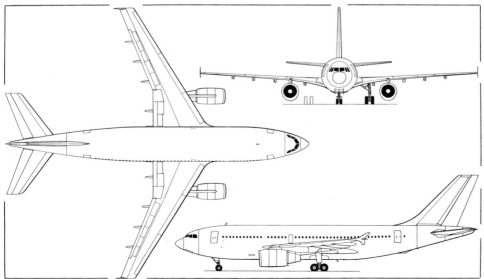

Airbus A310 medium/extended-range airliner *(Dennis Punnett/Jane's)*

Airbus A310-300 of Emirates landing at Dubai *(Paul Jackson/Jane's)* 2001/0103586

generator failure; each has overload rating of 135 kVA for 5 minutes and 180 kVA for 5 seconds; third (identical) AC generator, directly driven at constant speed by APU, can be used during ground operations, and also in flight to compensate for loss of one or both engine-driven generators; current production A310s have APU with improved relight capability, which can be started and operated throughout the flight envelope. Any one generator can provide sufficient power to operate all equipment and systems necessary for indefinite period of safe flight; DC power is generated via three 150 A transformer-rectifiers; three Ni/Cd batteries.

Flight crew oxygen system fed from rechargeable pressure bottle of 2,166 litres (76.5 cu ft) capacity; standard options are second 76.5 cu ft bottle, a 3,256 litre (115 cu ft) bottle, and an external filling connection; emergency oxygen sets for passengers and cabin attendants. Anti-icing of outer wing leading-edge slats and engine air intakes by hot air bled from engines; and of pitot probes, static ports and plates, and sensors, by electric heating.

For current production A310s, an ETOPS modification kit, as for the A300-600, is available.

AVIONICS: As described for A300-600. A310 was first airliner to introduce CRTs with its ECAM system.

DIMENSIONS, EXTERNAL:
Wing span	43.90 m (144 ft 0 in)
Wing chord: at root	8.38 m (27 ft 6 in)
at tip	2.18 m (7 ft 1¾ in)
Wing aspect ratio	8.8
Length overall	46.66 m (153 ft 1 in)
Fuselage: Length	45.13 m (148 ft 0¾ in)
Max diameter	5.64 m (18 ft 6 in)
Height overall	15.80 m (51 ft 10 in)
Tailplane span	16.26 m (53 ft 4¼ in)
Wheel track	9.60 m (31 ft 6 in)
Wheelbase (c/l of shock-absorbers)	15.21 m (49 ft 10¾ in)
Passenger door (forward, port): Height	1.93 m (6 ft 4 in)
Width	1.07 m (3 ft 6 in)
Height to sill at OWE	4.54 m (14 ft 10¾ in)
Passenger door (rear, port): Height	1.93 m (6 ft 4 in)
Width	1.07 m (3 ft 6 in)
Height to sill at OWE	4.85 m (15 ft 11 in)
Servicing doors (forward and rear, stbd)	as corresponding passenger doors
Upper deck cargo door (A310C/F)	as A300-600F
Emergency exits (overwing, port and stbd, each):	
Height	1.39 m (4 ft 6¾ in)
Width	0.67 m (2 ft 2½ in)
Underfloor cargo door (forward):	
Height	1.71 m (5 ft 7¼ in)
Width	2.69 m (8 ft 10 in)
Height to sill at OWE	2.61 m (8 ft 6¾ in)
Underfloor cargo door (rear):	
Height	1.71 m (5 ft 7¼ in)
Width	1.81 m (5 ft 11¼ in)
Height to sill at OWE	2.72 m (8 ft 11 in)
Underfloor cargo door (aft bulk hold):	
Height	0.95 m (3 ft 1½ in)
Width	0.95 m (3 ft 1½ in)
Height to sill at OWE	2.75 m (9 ft 0¼ in)

DIMENSIONS, INTERNAL:
Cabin, excl flight deck: Length	33.25 m (109 ft 1 in)
Max width	5.28 m (17 ft 4 in)
Max height	2.33 m (7 ft 7¾ in)
Volume	210.0 m³ (7,416 cu ft)
Forward cargo hold: Length	7.63 m (25 ft 0½ in)
Max width	4.18 m (13 ft 8½ in)
Height	1.71 m (5 ft 7¼ in)
Volume	50.3 m³ (1,776 cu ft)
Rear cargo hold: Length	5.03 m (16 ft 6¼ in)
Max width	4.17 m (13 ft 8¼ in)
Height	1.67 m (5 ft 5¾ in)
Volume	34.5 m³ (1,218 cu ft)
Aft bulk hold: Volume	17.3 m³ (611 cu ft)
Total overall cargo volume	102.1 m³ (3,605 cu ft)

AREAS:
Wings, gross	219.00 m² (2,357.3 sq ft)
Ailerons (total)	6.86 m² (73.84 sq ft)
Trailing-edge flaps (total)	36.68 m² (394.82 sq ft)
Leading-edge slats (total)	28.54 m² (307.20 sq ft)
Spoilers (total)	7.36 m² (79.22 sq ft)
Airbrakes (total)	6.16 m² (66.31 sq ft)
Vertical and horizontal tail surfaces	as A300-600

WEIGHTS AND LOADINGS (220-seat configuration. C2: CF6-80C2A2 engines, P2: PW4152s, C8: CF6-80C2A8s, P6: PW4156As):
Manufacturer's weight empty:	
200: C2	71,660 kg (157,975 lb)
P2	71,600 kg (157,850 lb)
300: C2	72,140 kg (159,040 lb)
P2	72,080 kg (158,910 lb)
C8	72,525 kg (159,890 lb)
P6	72,455 kg (159,735 lb)
Operating weight empty:	
200: C2	80,140 kg (176,685 lb)
P2	80,125 kg (176,645 lb)
300: C2	81,205 kg (179,025 lb)
P2	81,165 kg (178,940 lb)
C8	81,610 kg (179,920 lb)
P6	81,545 kg (179,775 lb)
Max payload: 200: C2	32,858 kg (72,439 lb)
P2	32,875 kg (72,476 lb)
300: C8	32,388 kg (71,403 lb)
P6	32,456 kg (71,553 lb)
Max usable fuel: 200*	44,100 kg (97,224 lb)
300	49,039 kg (108,110 lb)
Max T-O weight: 200	142,000 kg (313,050 lb)
300	150,000 kg (330,675 lb)
options (300)	153,000 kg (337,300 lb)
	or 157,000 kg (346,125 lb)
	or 164,000 kg (361,550 lb)
Max landing weight: 200, 300	123,000 kg (271,150 lb)
options (200 and 300)	124,000 kg (273,375 lb)
Max zero-fuel weight: 200, 300	113,000 kg (249,120 lb)
options (200 and 300)	114,000 kg (251,325 lb)

optional additional tank in aft cargo hold adds 5,779 kg (12,740 lb) of fuel and increases OWE/reduces max payload by 726 kg (1,600 lb). Two additional tanks add 11,560 kg (25,485 lb) of fuel and increase OWE/reduce max payload by 1,536 kg (3,386 lb)

PERFORMANCE (at basic max T-O weight except where indicated; engines as under Weights and Loadings):
Typical long-range cruising speed at 9,450-12,500 m (31,000-41,000 ft): C2, P2, C8, P6	M0.80
Max operating Mach No. (M$_{MO}$)	0.84
Approach speed at max landing weight:	
C2, P2, C8, P6	135 kt (250 km/h; 155 mph)
T-O field length at S/L, ISA + 15°C:	
200: C2	1,960 m (6,430 ft)
P2	1,890 m (6,200 ft)
300: C2 (at 150 tonne MTOW)	2,410 m (7,910 ft)
P2 (at 150 tonne MTOW)	2,180 m (7,155 ft)
C8 (at 164 tonne MTOW)	2,485 m (8,155 ft)
P6 (at 164 tonne MTOW)	2,360 m (7,745 ft)
Landing field length at S/L, at max landing weight (200 and 300): C2	1,479 m (4,850 ft)
P2	1,555 m (5,100 ft)
Runway ACN for flexible runway, category B:	
standard tyres: 200	43
300	49
optional tyres: 200	41
300	47

Range (1991 and subsequent deliveries) at typical airline OWE with 220 passengers and baggage, international reserves for 200 n mile (370 km; 230 mile) diversion:
200, GE engines	3,600 n miles (6,667 km; 4,142 miles)
200, PW engines	3,650 n miles (6,759 km; 4,200 miles)
300, GE engines	4,300 n miles (7,963 km; 4,948 miles)
300, PW engines	4,350 n miles (8,056 km; 5,005 miles)
300, option, at T-O weight 157,000 kg (346,125 lb):	
GE engines	4,750 n miles (8,797 km; 5,466 miles)
PW engines	4,800 n miles (8,889 km; 5,523 miles)
300, option, at T-O weight 164,000 kg (361,560 lb):	
GE engines	5,150 n miles (9,537 km; 5,926 miles)
PW engines	5,200 n miles (9,630 km; 5,984 miles)

OPERATIONAL NOISE LEVELS (ICAO Annex 16, Chapter 3):
T-O: 200: C2	89.6 EPNdB (95.3 limit)
300: C2	91.2 EPNdB (95.6 limit)
Sideline: 200: C2	96.4 EPNdB (99.2 limit)
300: C2	96.3 EPNdB (99.4 limit)
Approach: 200, 300: C2	98.6 EPNdB (102.9 limit)

UPDATED

AIRBUS A318

Follows entry for the A320, of which it is a derivative.

AIRBUS A319

Follows entry for the A320, of which it is a derivative.

AIRBUS A320

TYPE: Twin-jet airliner.

PROGRAMME: Launched 23 March 1984; four-aircraft development programme (first flight 22 February 1987 by F-WWAI); JAA (UK/French/German/Dutch) certification of A320-100 with CFM56-5 engines, for two-crew operation, awarded 26 February 1988; first deliveries (Air France and British Airways) 28 and 31 March 1988 respectively; JAA certification of A320-200 with CFM56-5s received 8 November 1988, followed by FAA type approval for both models 15 December 1988; certification with V2500 engines (first flown 28 July 1988) received 20 April (JAA) and 6 July 1989 (FAA), deliveries with this power plant (to Adria Airways) beginning 18 May 1989; FAA approved common type rating on A320 and A321 without further training in early 1994; 500th A320 delivered, 20 January 1995, to United Airlines; 1000th member of family (an A319) and 1,001st (A320 for United Airlines) were delivered 15 April 1999.

CURRENT VERSIONS: **A320-100**: Initial version (21 ordered); details in 1987-88 *Jane's*. Superseded by A320-200.

A320-200: Now called simply **A320**. Standard version from third quarter 1988; differs from initial A320 in having wingtip fences, wing centre-section fuel tank and higher maximum T-O weights.

Detailed description applies to A320-200

A320 research: An A320 was used for riblet research 1989-91. In mid-1998 Airbus Industrie began testing an experimental laminar-flow fin on an A320; air is sucked through small holes in the leading-edge to reduce drag and save fuel.

A320 Freighter, Convertible and Quick-Convertible: Freight variants under consideration by Airbus and Hindustan Aeronautics; cargo capacities would be 20,000 kg (44,090 lb), 22,000 kg (48,500 lb) and 30,000 kg (66,140 lb), respectively.

A318: Shortened version of A320; described separately.

A319: Shortened version of A320; described separately.

A321-100 and -200: Stretched versions of A320; described separately.

CUSTOMERS: Total 1,452 sold by 1 March 2001, of which 911 then delivered. Airbus announced in June 2000 that production of single-aisle aircraft would increase to 30 per month by 2002.

Brazil's TAM began receiving Airbus A320s in 2000

2001/0103540

AIRBUS A320 ORDERS
(at 1 March 2001)

Customer	Qty
ACES	8
Adria Airways	5
Aer Lingus	6
Aero Lloyd	4
Aerostar	12
Air 2000	4
Air Canada	28
Air France	27
Air Inter Europe	22
Air Jamaica	4
Air Malta	2
Alitalia	11
All Nippon Airways	25
America West Airlines	21
Ansett Australia	19
Austrian Airlines	13
Boullioun Aviation Services	13
British Airways	31
British Midland Airways	6
Canadian Airlines International	2
Carlisle Aircraft Ltd	3
China Eastern Airlines	10
China Northwest Airlines	13
China Southern Airlines	20
CIT Group	32
Condor Flugdienst	12
Croatia Airlines	2
Cyprus Airways	8
Debis AirFinance	15
Dragonair	7
Edelweiss Air AG	3
Egyptair	7
Finnair	9
Flightlease	9
GATX/CL AIR	18
GATX/Flightlease	12
GB Airways	3
GECAS	76
GPA	51
Gulf Air	14
Iberia	58
Iberworld	2
Indian Airlines	31
Intl Lease Finance Corp	160
jetBlue Airways	33
Kawasaki Leasing Intl	8
Kuwait Airways	3
Kuwait Finance House	4
LAN Chile	25
Lotus Airline	1
LTU	6
Lufthansa	37
Mexicana	20
Midway Airlines	4
Midwest Airlines	2
Monarch Airlines	2
Northwest Airlines	82
Nouvelair	2
ORIX Corporation	24
Philippine Airlines	4
Premiair	6
Qatar Airways (incl Amiri Flight)	7
Royal Jordanian	3
SALE	41
Sabena	3
Shorouk Air	2
Sichuan Airlines	2
Silkair	5
South African Airways	7
Spanair	10
SriLankan Airlines	2
Sudan Airways	1
Swissair	17
Syrian Arab Airlines	6
TACA	32
TAM	21
TAP-Air Portugal	6
TransAsia Airways	5
Tunis Air	12
United Airlines	116
US Airways	50
Zhejiang Airlines	3
Total	**1,452**

DESIGN FEATURES: First subsonic commercial aircraft to have composites for major primary structures, and centralised maintenance system; advanced-technology wings have 25° sweepback at quarter-chord, 5° 6′ 36″ dihedral plus experience from A310 and significant commonality with other Airbus Industrie aircraft where cost-effective; 6° tailplane dihedral.

FLYING CONTROLS: A320 is first subsonic commercial aircraft equipped for fly-by-wire (FBW) control throughout entire normal flight regime, and first to have sidestick controller (one for each pilot) instead of control column and aileron wheel. Thomson-CSF/SFENA digital FBW system features five main computers and operates, via electrical signalling and hydraulic jacks, all primary and secondary flight controls; pilot's pitch and roll commands are applied through sidestick controller via two different types of computer; these have redundant architecture to provide safety levels at least as high as those of mechanical systems they replace; system incorporates flight envelope protection features to a degree that cannot be achieved with conventional mechanical control systems and its computers will not allow aircraft's structural and aerodynamic limitations to be exceeded; even if pilot holds sidestick fully forward, it is impossible to go beyond aircraft's maximum operating speed (V_{MO}) for more than a few seconds; if pilot holds sidestick fully back, aircraft is controlled to an 'alpha floor' angle of attack, a safe airspeed above stall and throttles opened automatically to ensure positive climb. Nor is it possible to exceed g limits while manoeuvring. If a bank angle of more than 30° is commanded with the sidestick, the bank angle is automatically returned to 30° when pressure is released.

Fly-by-wire system controls ailerons, elevators, spoilers, flaps and leading-edge slats; rudder movement and tailplane trim connected to FBW system, but also signalled mechanically when used to provide final back-up pitch and yaw control, which suffices for basic instrument flying. Each wing has five-segment leading-edge slats (one inboard, four outboard of engine pylon), two-segment Fowler trailing-edge flaps, and five-segment spoilers forward of flaps; all 10 spoilers used as lift dumpers, inner six as airbrakes, outer eight and ailerons for roll control and outer four and ailerons for gust alleviation; slat and flap controls by Liebherr and Lucas.

STRUCTURE: Generally similar to A310, but with AFRP for fuselage belly fairing skins; GFRP for fin leading-edge and fin/fuselage fairing; CFRP for wing fixed leading/trailing-edge bottom access panels and deflectors, trailing-edge flaps and flap track fairings, spoilers, ailerons, fin (except leading-edge), rudder, tailplane, elevators, nosewheel/mainwheel doors, and main gear leg fairing doors. A320 was first airliner to go into production with CFRP tailplane.

Aerospatiale Matra builds entire front fuselage (forward of wing leading-edge), cabin rear doors, nosewheel doors, centre wing box and engine pylons, and is responsible for final assembly; centre and rear fuselage, tailcone, wing flaps, fin, rudder and commercial furnishing undertaken by DaimlerChrysler Aerospace Airbus; BAE Systems builds main wings, including ailerons, spoilers and wingtips, and main landing gear leg fairings; Belgian consortium Belairbus produces leading-edge slats; CASA responsible for tailplane, elevators, mainwheel doors, and sheet metal work for parts of rear fuselage; Mitsubishi builds wing root shroud box under BAE Systems subcontract; AVIC I of China provides wing components and signed an MoU in November 2000 to increase its participation, possibly leading to complete wing production by 2007; GKN Aerospace providing cargo door actuators from January 2002. Final assembly undertaken at Toulouse.

LANDING GEAR: Hydraulically retractable tricycle type, with twin wheels and oleo-pneumatic shock-absorber on each unit (four-wheel main-gear bogies, for low-strength runways, optional); Dowty main units retract inward into wing/body fairing; steerable Messier-Bugatti nose unit retracts forward; nosewheel steering angle ±75° (effective turning angle ±70°). Tyre size 46×16 or 46×17.0R20 (30 ply) on main gear and 30×8.8 or 30×8.8-R15 (16 ply) on nose gear; optional tyres for main gear are 49×17 or 49×17R20 or 49×19R20 or 46×16-20 or 49×19-20. Tyres for main-gear bogie option are 915×300R16 or 36×11 or 46×17.0R20. Carbon brakes standard. Minimum width of pavement for 180° turn 23.1 m (75 ft 9½ in).

POWER PLANT: Two 111.2 to 120.1 kN (25,000 to 27,000 lb st) class CFM International CFM56-5A1 turbofans for first aircraft delivery in 1988, with 113.4 kN (25,500 lb st) IAE V2500-A1 engines available for aircraft delivered from May 1989 and 117.9 kN (26,500 lb st) CFM56-5A3 from November 1990; 120.1 kN (27,000 lb st) CFM56-5B4 and 117.9 kN (26,500 lb st) IAE V2527-A5 available from 1994. New aircraft now (1998 onwards) delivered with

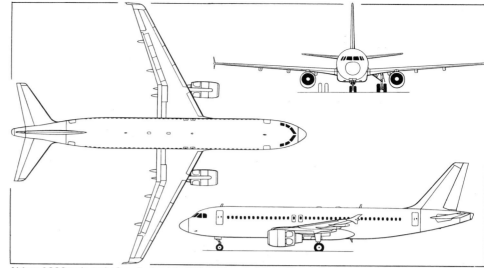

Airbus A320 twin-turbofan single-aisle 150/179-seat airliner (Dennis Punnett/Jane's)

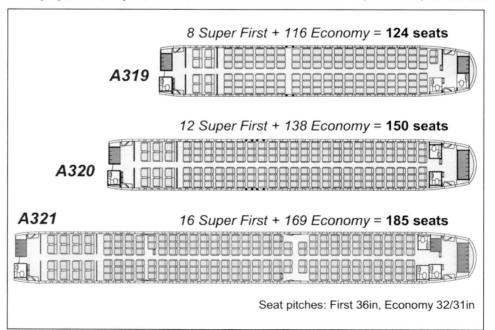

Typical Airbus interiors for the A320 family

Airbus A320 of French operator Star Airlines *(Paul Jackson/Jane's)*

either 120.1 kN (27,000 lb st) CFM56-5B4/P or 117.9 kN (26,500 lb st) IAE V2527E-A5. Nacelles by Rohr Industries; thrust reversers by Hispano-Suiza for CFM56 engines, by IAE for V2500s. Dual-channel FADEC system standard on each engine.

For A320-200, standard fuel capacity in wing and wing centre-section tanks is 23,859 litres (6,303 US gallons; 5,248 Imp gallons); optional additional centreline tank holds 2,900 litres (766 US gallons; 638 Imp gallons); for A320-100, standard fuel capacity without centre-section tank is 15,843 litres (4,185 US gallons; 3,485 Imp gallons).

ACCOMMODATION: Standard crew of two on flight deck, with one (optionally two) forward-facing folding seats for additional crew members; seats for four cabin attendants. Single-aisle main cabin has seating for up to 179 (FAR) or 180 (JAR) passengers, depending upon layout, with locations at front and rear of cabin for galley(s) and lavatory(ies); typical two-class layout has 12 seats four-abreast at 91.5 cm (36 in) pitch in 'super first' and 138 six-abreast at 81 cm (32 in) pitch economy class; alternative 152 six-abreast seats (84 business + 68 economy) at 86 and 78 cm (34 and 31 in) pitch respectively; single-class economy layout could offer 164 seats at 81 cm (32 in) pitch, or up to 179 in high-density configuration. Compared with existing single-aisle aircraft, fuselage cross-section is significantly increased, permitting use of wider triple seats to provide higher standards of passenger comfort; five-abreast business class seating provides standard equal to that offered as first class on major competitive aircraft. In addition, wider aisle permits quicker turnrounds. Overhead stowage space superior to that available on existing aircraft of similar capacity, and provides ample carry-on baggage space; best use of underseat space for baggage is provided by improved seat design and optimised positioning of seat rails.

Passenger doors at front and rear of cabin on port side, forward one having optional integral airstairs; service door opposite each of these on starboard side. Two overwing emergency exits each side. Fuselage double-bubble cross-section provides increased baggage/cargo hold volume and working height, and ability to carry containers derived from standard interline LD3 type. As base is same as that of LD3, all existing wide-body aircraft and ground handling equipment can accept these containers without modification. Forward and rear underfloor baggage/cargo holds, plus overhead lockers; with 164 seats, overhead stowage space per seat is 0.06 m³ (2.0 cu ft). Mechanised cargo loading system will allow up to seven LD3-based containers to be carried in freight holds (three forward and four aft).

SYSTEMS: Liebherr/ABG-Semca air conditioning, Hamilton Sundstrand/Nord-Micro pressurisation, Hamilton Sundstrand electrical system, and Honeywell 36-300 APU, the last named replaced by Honeywell 131-9(A) from September 1998. Sundstrand APS3200 introduced at same time as alternative APU. Primary electrical system powered by two Sundstrand 90 kVA constant frequency generators, providing 115/200 V three-phase AC at 400 Hz; third generator of same type, directly driven at constant speed by APU, can be used during ground operations and, if required, during flight.

AVIONICS: *Flight:* Fully equipped ARINC 700 digital avionics including advanced digital automatic flight control and flight management systems; AFCS integrates functions of SFENA autopilot and Honeywell FMS; Honeywell air data and inertial reference system.

Instrumentation: Each pilot has two Thomson-CSF/VDO electronic flight instrumentation system (EFIS) displays (primary flight display and navigation display); PFD was first on an airliner to incorporate speed, altitude and heading. Between these two pairs of displays are two Thomson-CSF/VDO electronic centralised aircraft monitor (ECAM) displays developed from the ECAM systems on A310 and A300-600; upper display incorporates engine performance and warning, lower display carries warning and system synoptic diagrams.

DIMENSIONS, EXTERNAL:
Wing span	34.09 m (111 ft 10¼ in)
Wing aspect ratio	9.5
Length overall	37.57 m (123 ft 3 in)
Fuselage: Max width	3.96 m (13 ft 0 in)
Max depth	4.14 m (13 ft 7 in)
Height overall	11.76 m (38 ft 7 in)
Tailplane span	12.45 m (40 ft 10 in)
Wheel track (c/l of shock-struts)	7.59 m (24 ft 11 in)
Wheelbase	12.65 m (41 ft 6 in)
Passenger doors (port, forward and rear), each:	
Height	1.85 m (6 ft 1 in)
Width	0.81 m (2 ft 8 in)
Height to sill	3.415 m (11 ft 2½ in)
Service doors (stbd, forward and rear), each	
	as corresponding passenger doors
Overwing emergency exits (two port and two stbd), each:	
Height	1.02 m (3 ft 4¼ in)
Width	0.51 m (1 ft 8 in)
Underfloor baggage/cargo hold doors (stbd, forward and rear), each: Height	1.25 m (4 ft 1¼ in)
Width	1.82 m (5 ft 11½ in)

DIMENSIONS, INTERNAL:
Cabin, excl flight deck: Length	27.50 m (90 ft 2¾ in)
Max width	3.70 m (12 ft 1¾ in)
Max height	2.22 m (7 ft 4 in)
Baggage/cargo hold volume: front	13.3 m³ (469 cu ft)
rear	24.15 m³ (853 cu ft)

AREAS:
Wings, gross	122.60 m² (1,319.7 sq ft)
Ailerons (total)	2.74 m² (29.49 sq ft)
Trailing-edge flaps (total)	21.10 m² (227.12 sq ft)
Leading-edge slats (total)	12.64 m² (136.06 sq ft)
Spoilers (total)	8.64 m² (93.00 sq ft)
Airbrakes (total)	2.35 m² (25.30 sq ft)
Vertical tail surfaces (total)	21.50 m² (231.4 sq ft)
Horizontal tail surfaces (total)	31.00 m² (333.7 sq ft)

WEIGHTS AND LOADINGS (Typical 150-passenger configuration. A: CFM56-5B4/P engines, B: V2527-A5s):
Operating weight empty: A*	42,100 kg (92,815 lb)
B	42,482 kg (93,657 lb)
Max payload: A	18,633 kg (41,079 lb)
B	18,518 kg (40,825 lb)
Max fuel	19,159 kg (42,238 lb)
Max T-O weight: standard	73,500 kg (162,040 lb)
1st option	75,500 kg (166,445 lb)
2nd option	77,000 kg (169,755 lb)
Max ramp weight: standard	73,900 kg (162,920 lb)
1st option	75,900 kg (167,330 lb)
2nd option	77,400 kg (170,635 lb)
Max landing weight: standard	64,500 kg (142,195 lb)
option	66,000 kg (145,505 lb)
Max zero-fuel weight: standard	61,000 kg (134,480 lb)
option	62,500 kg (137,789 lb)
Max wing loading:	
standard	599.5 kg/m² (122.79 lb/sq ft)
1st option	615.8 kg/m² (126.13 lb/sq ft)
2nd option	628.1 kg/m² (128.64 lb/sq ft)
Max power loading (V2527 engines):	
standard	312 kg/kN (3.06 lb/lb st)
1st option	320 kg/kN (3.14 lb/lb st)
2nd option	327 kg/kN (3.20 lb/lb st)

*Options raise OWE to a maximum of 43,000 kg (94,799 lb)

PERFORMANCE (engines A and B as for Weights and Loadings, C: CFM56-5A3/5B4, D: 77,000 kg; 169,756 lb max T-O weight):
Max operating Mach No. (M$_{MO}$)	0.82
Optimum cruise speed	M0.78
Max rate of climb at S/L	152 m (500 ft)/min
Initial cruise altitude	11,280 m (37,000 ft)
Service ceiling	11,890 m (39,000 ft)
Service ceiling, OEI	5,945 m (19,500 ft)
T-O distance at S/L, ISA + 15°C: A	1,960 m (6,430 ft)
B	1,950 m (6,400 ft)
C	2,180 m (7,155 ft)
D	2,250 m (7,385 ft)

Airbus A320 flight deck with sidestick controllers outboard and six-screen EFIS

Landing distance at max landing weight:
    A, B, C, D   1,490 m (4,890 ft)
Runway ACN (flexible runway, category B):
    twin-wheel, standard 45×16R20 tyres   41
    four-wheel bogie option, 36×11-16 Type VII or
        900×300-R16   22
Range with 150 passengers and baggage in two-class
    layout, FAR domestic reserves and 200 n mile
    (370 km; 230 mile) diversion:
    A, standard   2,592 n miles (4,800 km; 2,982 miles)
    B, standard   2,596 n miles (4,807 km; 2,987 miles)
    A, optional   2,800 n miles (5,185 km; 3,222 miles)
    B, optional   2,871 n miles (5,317 km; 3,303 miles)
    2nd option:
        A, C   3,045 n miles (5,639 km; 3,504 miles)
        B   3,065 n miles (5,676 km; 3,527 miles)
OPERATIONAL NOISE LEVELS (ICAO Annex 16, Chapter 3; A, B
    and C as for performance):
    T-O: A   88.0 EPNdB (91.5 limit)
        B   84.8 EPNdB (91.5 limit)
        C   85.4 EPNdB (91.5 limit)
    Sideline: A   94.4 EPNdB (96.8 limit)
        B   92.8 EPNdB (96.8 limit)
        C   93.8 EPNdB (96.8 limit)
    Approach: A   96.4 EPNdB (100.5 limit)
        B   95.9 EPNdB (100.5 limit)
        C   96.0 EPNdB (100.5 limit)
                                         UPDATED

Computer-generated image of Airbus A318

AIRBUS A318

TYPE: Twin-jet airliner.
PROGRAMME: Short-bodied version of A319 (itself a truncated A320). Formally announced at Farnborough Air Show, September 1998, although known to be under consideration (as A319M5) since late 1997 when AVIC/AIA/STPL AE316/AE317 venture became uncertain. Smallest aircraft in Airbus family; programme launched 26 April 1999 with orders, commitments and options for 109 aircraft; launch customer Air France (via ILFC); prototype first flight due first quarter 2002; first deliveries in last quarter 2002. Final assembly at Hamburg, resulting in some A319 production being moved to Toulouse. First metal cut November 2000.
CURRENT VERSIONS: **A318-100**: Baseline version.
CUSTOMERS: Airbus estimates requirement for 2,124 airliners in 70-100 seat and 7,286 in 125-175 seat categories up to 2018. First customer, International Lease Finance Corporation (ILFC), signed MoU for up to 30 aircraft on 17 November 1998, subject to launch of project, followed by firm order 30 April 1999. First conditional airline customer, TWA, signed LoI on 9 December 1998 for 50, confirming 25 by firm order 14 December 1999; other early customers include Egyptair (three on 17 July 1999 – first firm airline order, followed by two further examples), GATX-Flightlease (12), America West (15), British Airways (12) Air China (8), Air France (15), CIT Group (four), Frontier Airlines (five) and GECAS (30). Total orders 161 at 1 March 2001, plus 57 options.
COSTS: Programme cost estimated as US$300 million. Unit price US$36 million (1998).
DESIGN FEATURES: Has 95 per cent commonality with other A320 family members, including A319 wing, pylon and interface. Laser welding (rather than riveting) used on fuselage to reduce costs and weight; first use of this technology on airliner.
FLYING CONTROLS: As A320.
STRUCTURE: As A320, but fuselage 4½ frames (2.39 m; 7 ft 10 in) shorter than A319; 1½ frames removed forward and 3 aft of wing. Fin has tip extension. AVIC of China will have share of production work to offset AE31X project cancellation.
LANDING GEAR: As A320.
POWER PLANT: Two 97.9 kN (22,000 lb st) Pratt & Whitney PW6122 or two 106.8 kN (24,000 lb st) PW6124 turbofans; A318 is launch aircraft for engine family. Alternative power plant is 97.9 kN (22,000 lb st) or 103.6 kN (23,300 lb st) CFM International CFM56-5B, officially announced 4 August 1999. First CFM-powered A318 will enter service mid-2003. Thrust reversing and deflection as A320; fuel tank capacities and locations as A319.
ACCOMMODATION: Two flight crew plus three cabin crew. Typical passenger capacity eight first class and 99 second class with 96/81 cm (38/32 in) seat pitch, or 117 in single-class seating; high-density seating for 129 passengers. Front and rear passenger doors on port side; service door opposite each on starboard side. Overwing emergency exit each side. A318 will not carry containerised freight due to smaller baggage doors (of which size reduced to maintain same engine nacelle clearance for loading vehicles as A319).
SYSTEMS: As A320.
AVIONICS: As A320.
DIMENSIONS, EXTERNAL: As A320 except:
    Length overall   31.45 m (103 ft 2¼ in)
    Height overall   12.56 m (41 ft 2½ in)
    Wheelbase   10.25 m (33 ft 7½ in)
    Baggage doors (each): Height 1.24 m (4 ft 0¾ in)
        Width   1.28 m (4 ft 2½ in)
DIMENSIONS, INTERNAL:
    Cabin: Length   21.38 m (70 ft 1¾ in)
    Baggage hold volume: front 6.51 m³ (230 cu ft)
        rear   14.70 m³ (519 cu ft)
WEIGHTS AND LOADINGS:
    Operating weight empty   39,035 kg (86,057 lb)
    Max payload   13,965 kg (30,788 lb)
    Max T-O weight: normal   59,000 kg (130,075 lb)
        option 1   61,500 kg (135,585 lb)
        option 2   66,000 kg (145,505 lb)
    Max ramp weight: normal   59,400 kg (130,955 lb)
        option 1   61,900 kg (136,465 lb)
    Max landing weight: normal 56,000 kg (123,460 lb)
        option 1   57,500 kg (126,765 lb)
    Max zero-fuel weight: normal 53,000 kg (116,845 lb)
        option 1   54,500 kg (120,150 lb)
    Max wing loading: normal   481.2 kg/m² (98.57 lb/sq ft)
        option 1   501.6 kg/m² (102.74 lb/sq ft)
        option 2   538.3 kg/m² (110.26 lb/sq ft)
    Max power loading, PW6122 engines:
        normal   301 kg/kN (2.96 lb/lb st)
        option 1   313 kg/kN (3.07 lb/lb st)
        option 2   337 kg/kN (3.31 lb/lb st)
PERFORMANCE (estimated):
    Runway ACN at 59,000 kg (130,075 lb): flexible Cat B
        runway   29
    T-O run at normal MTOW 500 n mile (925 km;
        575 mile) mission, elevation 610 m (2,000 ft), ISA
        +15°C: PW6122   1,670 m (5,479 ft)
        CFM56   1,630 m (5,348 ft)
    Landing run at MLW, S/L, ISA:
        PW6122   1,332 m (4,370 ft)
        CFM56   1,355 m (4,446 ft)
    Range with 107 passengers, FAR domestic reserves,
        200 n miles (370 km; 230 miles) diversion, max
        payload:
        normal MTOW:
            PW6122   1,462 n miles (2,707 km; 1,682 miles)
            CFM56   1,455 n miles (2,694 km; 1,674 miles)
        option 1 MTOW:
            PW6122   1,960 n miles (3,630 km; 2,255 miles)
            CFM56   1,980 n miles (3,667 km; 2,278 miles)
        option 2 MTOW:
            PW6122   2,820 n miles (5,222 km; 3,245 miles)
            CFM56   2,880 n miles (5,333 km; 3,314 miles)
OPERATIONAL NOISE LEVELS (estimated):
    T-O with PW6122 at MTOW   79.7 EPNdB
    Sideline   90.4 EPNdB
    Approach   89.7 EPNdB
                                         UPDATED

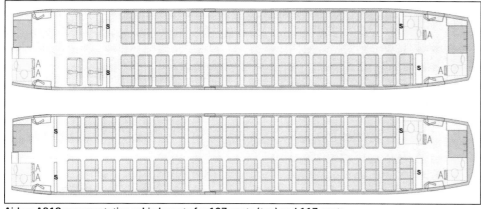

Airbus A318 representative cabin layouts for 107 seats (top) and 117 seats
A: attendant's seat, S: screen

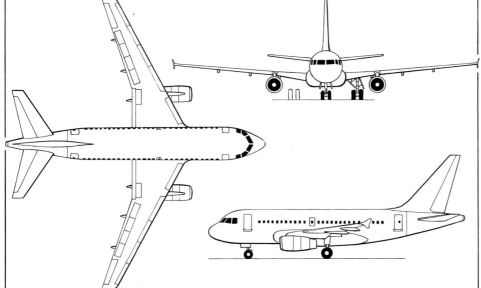

General arrangement of the Airbus A318 (*James Goulding/Jane's*)

AIRBUS A319

TYPE: Twin-jet airliner.
PROGRAMME: Short-fuselage A320. Airbus Board officially authorised start of sales 22 May 1992; programme launched June 1993. Final assembly of first aircraft (F-WWDB, the 546th of the A319/320/321 family) began 23 March 1995; rolled out at Hamburg 24 August; first flight (with CFM56-5A engines) on 29 August; second aircraft (F-WWAS, No. 572) flew (with CFM56-5A engines and commercial interior) 31 October 1995; 650 hour flight test programme resulted in initial certification (CFM56-5B) on 10 April 1996 (CFM56-5A and V2500-A5 followed); first delivery, HB-IPV to ILFC on 25 April 1996 and immediately to Swissair on 30 April, flying first service on 8 May; F-WWAS re-engined with IAE V2524 engines, first flight 22 May 1996 and certified on 18 December 1996; other 1996 first receipts included Air Inter (F-GPMA 21 June); Lufthansa (D-AILA 19 July); and Air Canada (C-FYIY 12 December). JAA 120 minute ETOPS approval granted 14 February 1997 and extended to ACJ in December 2000. 1,000th member of family (an A319 for Air France via ILFC) delivered 15 April 1999.
CURRENT VERSIONS: **A319-100**: Baseline version.
Description applies to A319-100.
  ACJ: Airbus Corporate Jetliner. Announced at 1997 Paris Air Show. Standard aircraft will carry up to 58 passengers over a range of 6,300 n miles (11,667 km; 7,250 miles), cruising at 12,500 m (41,000 ft) at speeds of up to M0.82. Being certified as a commercial airliner to Cat. IIIb landing criteria, and will convert easily to airliner configuration. First customer, announced on 18 December 1997, is Mohamed Abdulmohsin Al Kharafi of Kuwait; 12 other commitments held by December 1998. First aircraft (G-OMAK, 913th of A320 family), type A319-132, first flew 12 November 1998 and delivered to Jet Aviation, Switzerland, for outfitting on 31 December 1998; customer receipt 8 November 1999, operated by Twinjet Aircraft for Al Kharafi.
  Other customers include DaimlerChrysler, Italian Air Force (first of two aircraft delivered 7 March 2000; second in August 2000), Fayair, French Air Force (two) and Aero Service Executive (one). Orders totalled 20 by mid-2000. ACJ under consideration for Royal Australian Air Force Statement of Requirement for two VIP aircraft.
  Set world record non-stop 15 hour 13 minute flight of 6,918 n miles (12,812 km; 7,961 miles) from Santiago to Le Bourget on 16 June 1999. JAR certification (as amendment of A319 certificate) received August 1999.
  Airbus announced in 2000 that production of the ACJ would be restricted to four per year until 2003.
CUSTOMERS: Total of 704 sold, and 328 delivered, by 1 March 2001, including ACJs.

AIRBUS A319 and ACJ ORDERS
(at 1 March 2001)

Customer	Qty
Aer Lingus	4
Aero Services Executive	1
Air Bosna	2
Air Canada	37
Air France	18
Air Inter Europe	9
Air Mauritius	2
Al Kharafi Group	1
Alitalia	12
America West Airlines	26
Boullioun Aviation Services	17
British Airways	39
CIT Group	14
Croatia Airlines	4
DaimlerChrysler Aviation	1
Debis AirFinance	10
Eurowings	6
Finnair	6
French Air Force	2
Frontier Airlines	7
GATX/Flightlease	1
GECAS	25
Government of Italy	2
ILFC	87
Lufthansa	20
Northwest Airlines	74
Qatar Airways	1
SALE	3
Sabena	28
Silkair	4
Swissair	6
TACA	16
TAM	20
TAP-Air Portugal	13
Trans World Airlines	20
Tunis Air	3
United Airlines	76
US Airways	71
Undisclosed	16
Total	**704**

Airbus A319 delivered to China Eastern in 2000 2001/0103541

COSTS: Total development cost estimated at US$275 million, entirely financed by Airbus Industrie. ACJ cost US$35 million (2000), excluding outfitting of between US$4 million and US$10 million.
DESIGN FEATURES: Seven fuselage frames shorter than A320, but otherwise little changed. Seats 124 passengers in typical two-class layout, compared with 150 in A320 and 185 in A321; range of 3,550 n miles (6,574 km; 4,085 miles) said to be the longest in this category of airliner; common pilot type rating with A320 and A321.
FLYING CONTROLS: Same flight deck and flying control system as A320.
STRUCTURE: As A320. Assembled in Germany by DaimlerChrysler Aerospace Airbus, alongside the stretched A321; partner workshares rearranged to maintain overall workshare balance between France and Germany.
LANDING GEAR: As A320, but 46×16 or 46×17.0R20 main tyre options only.
POWER PLANT: Standard: two CFMI CFM56-5A4 or IAE V2522-A5 turbofans, each giving 97.9 kN (22,000 lb st). Options: two CFMI CFM56-5A5, CFM56-5B/P or IAE V2524-A5, each giving 104.5 kN (23,500 lb st). Normal fuel capacity 23,860 litres (6,303 US gallons; 5,248 Imp gallons); one or two additional centreline tanks, each holding 2,900 litres (766 US gallons; 638 Imp gallons), can be fitted.
ACCOMMODATION: Typically 124 passengers (eight in 'super first' class plus 116 economy); maximum 145 passengers in all-economy configuration.

DIMENSIONS, EXTERNAL: As A320 except:
  Length overall   33.84 m (111 ft 0¼ in)
  Wheelbase   11.05 m (36 ft 3 in)
DIMENSIONS, INTERNAL:
  Cabin length (excl flight deck)   23.77 m (77 ft 11¾ in)
WEIGHTS AND LOADINGS:
  Typical operating weight empty:
    standard   40,160 kg (88,537 lb)
    option   41,203 kg (90,837 lb)
  Max payload   16,653 kg (36,714 lb)
  Max T-O weight: standard   64,000 kg (141,095 lb)
    option   75,500 kg (166,450 lb)
  Max landing weight: standard   61,000 kg (134,480 lb)
    option   62,500 kg (137,790 lb)
  Max zero-fuel weight: standard   57,000 kg (125,665 lb)
    option   58,500 kg (128,970 lb)
  Max wing loading:
    standard   522.9 kg/m² (101.09 lb/sq ft)
    option   616.8 kg/m² (126.34 lb/sq ft)
  Max power loading: standard   327 kg/kN (3.21 lb/lb st)
    option   361 kg/kN (3.54 lb/lb st)
PERFORMANCE (A: with CFM56-5A4, CFM56-5B5 or V2522-A5; B: with CFM56-5A5, CFM56-5B6 or V2524-A5):
  Max operating Mach No. (M_{MO})   0.82
  Typical operating Mach No.   0.78
  Max rate of climb   152 m (500 ft)/min
  Service ceiling   11,890 m (39,000 ft)
  Service ceiling, OEI   6,400 m (21,000 ft)

Airbus Corporate Jetliner owned by businessman Mohamed Al Kharafi 2001/0103542

German carrier Eurowings is an operator of the Airbus A319 *(Paul Jackson/Jane's)* 2001/0103588

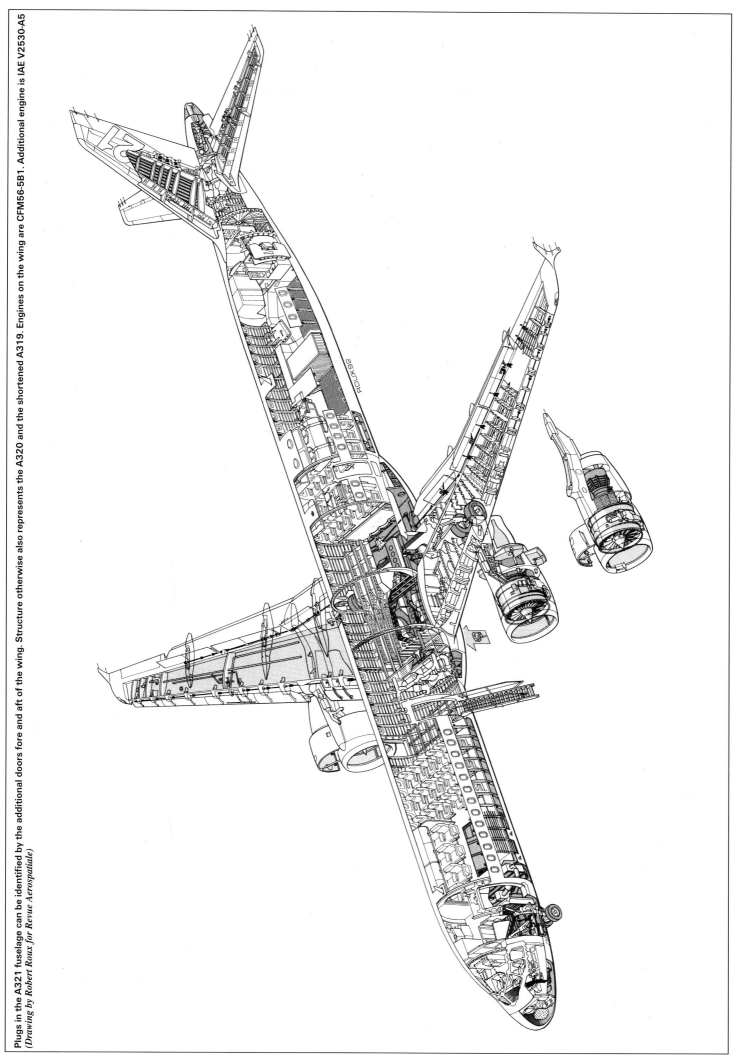

Plugs in the A321 fuselage can be identified by the additional doors fore and aft of the wing. Structure otherwise also represents the A320 and the shortened A319. Engines on the wing are CFM56-5B1. Additional engine is IAE V2530-A5 (Drawing by Robert Roux for Revue Aerospatiale)

192　INTERNATIONAL: AIRCRAFT—AIRBUS

Airbus A321 operated by Virgin Atlantic　　*2001/0103539*

T-O distance, S/L ISA+15°C:
  standard: A　　1,720 m (5,643 ft)
  option: B　　2,640 m (8,661 ft)
Landing distance, S/L ISA: A, B　　1,430 m (4,692 ft)
Range on normal fuel tankage with 124 passengers and baggage, FAR domestic reserves and 200 n mile (370 km; 230 mile) diversion:
  A, at T-O weight 64,000 kg (141,095 lb)
    1,813 n miles (3,357 km; 2,086 miles)
  B, at T-O weight 75,500 kg (166,450 lb)
    3,697 n miles (6,846 km; 4,254 miles)
OPERATIONAL NOISE LEVELS (A, B as above):
  T-O: A　　82.5 EPNdB
      B　　82.4 EPNdB
  Sideline: A　　92.6 EPNdB
      B　　92.2 EPNdB
  Approach: A　　93.9 EPNdB
      B　　94.2 EPNdB
UPDATED

AIRBUS A321

TYPE: Twin-jet airliner.
PROGRAMME: Stretched version of A320. Announced 22 May and launched 24 November 1989; four development aircraft; rolled out 3 March 1993, first flight with V2530 lead engine 11 March 1993 (F-WWIA), second aircraft with alternative CFM56-5B engine in May 1993; V2530-powered version received European JAA certification 17 December 1993; CFM56-5B2-powered version certified by JAA 15 February 1994; CFM56-5B1 certified by JAA 27 May 1994; first delivery (D-AIRA to Lufthansa) 27 January 1994; first service with Lufthansa 18 March 1994. A321 powered by CFM56-5Bs passed cold-weather trials in Kiruna, January 1994; first with alternative engine handed over to Alitalia 18 March 1994. JAA approval for Cat. III automatic landings achieved in December 1994. 120-minute ETOPS granted 29 May 1996. The 100th A321 (Alitalia's I-BIXZ) flew 1 July 1998. A321 also recommended by Northrop Grumman as platform for Joint STARS, competing for NATO's Airborne Ground Surveillance requirement.
CURRENT VERSIONS: **A321-100**: Initial version.
Description applies to A321-100.
  A321-200: Extended-range version, launched April 1995; features reinforced structure, higher-thrust versions of existing engines and optional additional centre tank (ACT), capacity 2,900 litres (766 US gallons; 638 Imp gallons) which increases maximum T-O weight to 89,000 kg (196,210 lb) and range by 350 n miles (648 km; 402 miles). A321-200 expected to have increased market appeal on North American domestic routes; charter routes between northern and southern Europe; and on scheduled routes between Europe and Middle East. First aircraft flew 12 December 1996 at Hamburg; became G-OZBC of Monarch Airlines and delivered 24 April 1997. Higher-MTOW version, with option of second ACT, launched January 1999; first customer Spanair, for delivery September 2000.
  A321CJ: Airbus Industrie reported to be considering a corporate version with additional fuel tanks; no formal plans to launch known by late 2000.
CUSTOMERS: Total of 381 sold and 180 delivered by 1 March 2001.

AIRBUS A321 ORDERS
(at 1 March 2001)

Customer	Qty
Aer Lingus	3
Aero Lloyd	5
Air Canada	12
Air France	8
Air Inter Europe	7
Air Macau	1
Alitalia	23
All Nippon Airways	7
Asiana Airlines	12
Austrian Airlines	7
British Airways	2
British Midland Airways	6
China Northern Airlines	10
CIT Group	8
Debis AirFinance	7
Egyptair	4
Finnair	4
Flightlease	1
GATX/CL AIR	6
GATX/Flightlease	9
GB Airways	3
GECAS	30
Iberia	16
ILFC	78
Leisure Int'l Airways	2
Lufthansa	26
Monarch Airlines	7
Onur Air	2
Sabena	3
SALE	6
Scandinavian Airlines System	12
Sichuan Airlines	2
Spainair	4
Swissair	6
TAP-Air Portugal	2
TransAsia Airways	6
US Airways	34
Total	**381**

DESIGN FEATURES: Compared with A320, A321 has 4.27 m (14 ft 0 in) fuselage plug immediately forward of wing and 2.67 m (8 ft 9 in) plug immediately aft; pairs of wing fuel tanks unified and system simplified; other changes include local structural reinforcement of existing assemblies, slightly extended wing trailing-edge with double-slotted flaps, uprated landing gear and higher T-O weights.
STRUCTURE: As for A320 except for airframe changes noted under Design Features; front fuselage plug by Alenia and rear one by BAE Systems; final assembly and outfitting by DaimlerChrysler Aerospace Airbus at Hamburg.
LANDING GEAR: Uprated, with 22 in wheel rims, 49×18.0-22 or 46×17.0R20 (30 ply) mainwheel tyres and increased energy brakes; wheels and brakes by Aircraft Braking Systems.
POWER PLANT: Standard: two CFM56-5B/P or IAE V2530-A5 turbofans, each rated at 133.4 kN (30,000 lb st). Option:

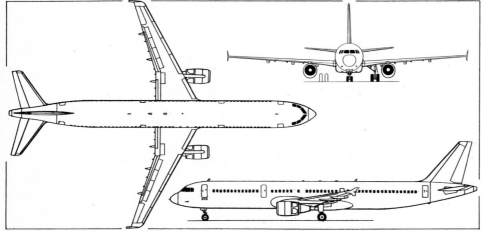

A321 stretched development of the Airbus A320 *(Dennis Punnett/Jane's)*

Aero Lloyd Airbus A321 used as an advertising hoarding by a German sportswear company *(Paul Jackson/Jane's)*　　*2001/0103549*

Jane's All the World's Aircraft 2001-2002　　　　www.janes.com

two CFM56-5B2 engines of 137.9 kN (31,000 lb st); options on A321-200 are two CFM56-5B3 of 142.3 kN (32,000 lb st) or two V2533-A5 of 146.8 kN (33,000 lb st). Normal fuel capacity 23,700 litres (6,261 US gallons; 5,213 Imp gallons); one or two optional 2,900 litre (766 US gallon; 638 Imp gallon) additional centre tanks.

ACCOMMODATION: Typically offers 24 per cent more seats and 40 per cent more hold volume than A320; examples are 185 passengers in two-class layout (16 'super first' class at 91 cm; 36 in seat pitch and 169 economy class at 81 cm; 32 in), or 220 passengers in all-economy high-density configuration. Each fuselage plug incorporates one pair of emergency exits, replacing single overwing pair of A320.

SYSTEMS: Choice of Honeywell 36-300 or APIC APS3200 APU; Honeywell 131-9(A) available from 1998; full commonality with A320 installation.

DIMENSIONS, EXTERNAL:
Wing span	34.09 m (111 ft 10 in)
Length overall	44.51 m (146 ft 0 in)
Height overall	11.81 m (38 ft 9 in)
Wheelbase	16.92 m (55 ft 6¼ in)
Wheel track	7.59 m (24 ft 10¾ in)

Passenger and service doors (port/stbd, forward and rear) as for A320

Emergency exits (forward stbd and rear port/stbd, each):
Height	1.52 m (5 ft 0 in)
Width	0.76 m (2 ft 6 in)

Emergency exit (forward port, usable also as passenger door):
Height	1.85 m (6 ft 1 in)
Width	0.76 m (2 ft 6 in)

DIMENSIONS, INTERNAL: As A320 except:
Cabin, excl flight deck: Length	34.44 m (113 ft 0 in)

Baggage/cargo hold volume:
front	22.81 m³ (806 cu ft)
rear	28.93 m³ (1,022 cu ft)

WEIGHTS AND LOADINGS (A321-100, except where indicated; typical 185-passenger layout. C1: CFM56-5B1, V: V2530-A5, C2: CFM56-5B2, C3: CFM56-5B3, V2: V2533-A5):

Operating weight empty:
C1, C2	48,085 kg (106,010 lb)
V	48,200 kg (106,265 lb)
C3	48,084 kg (106,005 lb)
V2	48,200 kg (106,265 lb)

Max payload: C1, C2	21,416 kg (47,214 lb)
V	21,301 kg (46,960 lb)
Max fuel: C1, V, C2	19,031 kg (41,956 lb)
Max T-O weight: standard	83,000 kg (182,980 lb)
option	85,000 kg (187,390 lb)
A321-200: standard	89,000 kg (196,210 lb)
option	93,000 kg (205,030 lb)

Max landing weight:
A321-100: standard	73,500 kg (162,035 lb)
option	74,500 kg (164,245 lb)
A321-200: standard	75,500 kg (166,450 lb)
option	77,800 kg (171,520 lb)

Max zero-fuel weight:
A321-100: standard	69,500 kg (153,220 lb)
option	70,500 kg (155,430 lb)
A321-200: standard	71,500 kg (157,630 lb)
option	73,800 kg (162,705 lb)

Max wing loading:
standard	678.1 kg/m² (138.89 lb/sq ft)
option	694.4 kg/m² (142.23 lb/sq ft)

Max power loading:
standard	311 kg/kN (3.05 lb/lb st)
option	308 kg/kN (3.02 lb/lb st)

PERFORMANCE:
Max operating Mach No. (M_{MO})	0.82
Typical operating Mach No.	0.78
Max rate of climb at S/L	152 m (500 ft)/min
Service ceiling	11,890 m (39,000 ft)
Service ceiling, OEI: A321-100	5,180 m (17,000 ft)
A321-200	5,000 m (16,400 ft)

T-O distance at max T-O weight, S/L, ISA +15°C:
C1	2,220 m (7,285 ft)
V	2,065 m (6,775 ft)
C2	2,270 m (7,450 ft)
C3	2,330 m (7,645 ft)
V2	2,341 m (7,680 ft)

Landing distance at max landing weight:
C1, V	1,540 m (5,055 ft)
C2	1,530 m (5,020 ft)
C3, V2	1,577 m (5,175 ft)

Runway ACN (flexible runway, category B):
standard	48

Range on normal fuel tankage with 185 passengers and baggage at typical airline OWE, FAR domestic reserves and 200 n mile (370 km; 230 mile) diversion: C1, C2:
standard	2,138 n miles (3,959 km; 2,460 miles)
optional	2,250 n miles (4,167 km; 2,589 miles)
V: standard	2,253 n miles (4,172 km; 2,592 miles)
optional	2,386 n miles (4,418 km; 2,745 miles)

A321-200:
with one ACT	2,700 n miles (5,000 km; 3,107 miles)
with two ACT	3,000 n miles (5,556 km; 3,452 miles)
V2: standard	2,384 n miles (4,415 km; 2,743 miles)
with one ACT	2,714 n miles (5,026 km; 3,123 miles)

OPERATIONAL NOISE LEVELS (ICAO Annex 16, Chapter 3, estimated):
T-O: C1	87.0 EPNdB (92.1 limit)
V	85.4 EPNdB (92.1 limit)
Sideline: C1	96.2 EPNdB (97.2 limit)
V	94.5 EPNdB (97.2 limit)
Approach: C1	96.1 EPNdB (100.9 limit)
V	95.4 EPNdB (100.9 limit)

UPDATED

AIRBUS A330
Follows entry for the A340, to which it is closely related.

AIRBUS A340
TYPE: Wide-bodied airliner.

PROGRAMME: Launched 5 June 1987 as combined programme with A330, differing mainly in number of engines and in engine-related systems; A330 and A340 were first airliners created with 100 per cent computer-aided design (CAD); first six aircraft were four A340-300s and two -200s; first flight 25 October 1991 (A340-300); A340-200 and -300 certified simultaneously by 18 European joint airworthiness authorities (JAA) on 22 December 1992; A340-300 and -200 entered service with Lufthansa January 1993; both received FAA certification 27 May 1993. A340-300 by JAA for GPS satellite navigation January 1994; successful trials conducted at Toulouse in October 1994 using differential global positioning (DGPS) for fully automatic landings, including roll-out. The 200th aircraft from A330/340 assembly line flew in November 1997; 100th A340 delivered to Singapore Airlines in February 1997; 180-minute ETOPS for Rolls-Royce Trent 700 engines awarded May 1996. A340-500 and -600 were launched on 8 December 1997, at an estimated investment cost of US$2.9 billion. 2,000th Airbus was A340-300 for Lufthansa, delivered 18 May 1999. 300th A340 first flown November 1999.

Cathay Pacific took delivery of an A340-300 covered with 700 m² (7,535 sq ft) of plastic film riblets in October 1996 in a trial intended to reduce fuel consumption (average ¾ tonne per medium- or long-range flight) by reducing drag (see illustrations in 1998-99 edition).

CURRENT VERSIONS: **Initial A340-300:** Four-engined higher-capacity version, carrying up to 375 passengers (standard) or 440 (optional) and powered initially by CFM56-5C2 turbofans. Able to carry typical load of 295 passengers over distances of 7,300 n miles (13,519 km; 8,400 miles).

Current A340-300 (engineering designation -300X) powered by 151.2 kN (34,000 lb st) CFM56-5C4 turbofans; maximum T-O weight 271,000 kg (597,450 lb) with optional MTOW of 275,000 kg (606,275 lb); stronger landing gear, and aerodynamic and engine refinements, plus optional additional centre tank (ACT) for increased fuel capacity compared with basic A340-300. First flight (F-WWJH; CFM56-5C4 engines) 25 August 1995; first delivery, to Singapore Airlines, 17 April 1996.

Initial A340-200: Short-fuselage, longer-range version of A340-300, with same initial power plant; seating capacity 420 passengers, but more typically 303 in a two-class or 263 in three-class configuration; first flight (F-WWBA) 1 April 1992; entered service January 1993 with Lufthansa.

Current A340-200: Advanced version (previously referred to as A340-8000) with additional fuel in two tanks in rear cargo hold, strengthened fuselage and wings, CFM56-5C4 engines and 275,000 kg (606,275 lb) maximum T-O weight; first order placed March 1996 by Prince Jefri of Brunei for one VVIP aircraft, which first flew 19 December 1997 and was retained by Airbus during 1998; stored at Schönefeld-Berlin in 'green' condition, in 1999, before delivery to customer.

A340-400: Series discontinued in favour of A340-600; basic details in 1998-99 *Jane's*.

A340-500: Ultra-long-range variant of A340-300 able to carry 313 passengers in three classes across 8,650 n miles (16,020 km; 9,954 miles) and available at the same time as the A340-600. Will be six frames longer than A340-300. Launched at 1997 Paris Air Show. Commonality with existing A330/A340 variants allows cross-crewing and employment of common cargo containers and interiors to minimise training, staffing, provisioning and maintenance costs. Stated to be the world's longest-range airliner; first flight will take place around mid-2001; first delivery (to Air Canada) mid-2002. Development time of A340-500/-600 series reduced by 25 per cent by use of Airbus Concurrent Engineering (ACE) shared CADD-CAD/CAM systems.

Compared with -300, has 0.53 m (1 ft 9 in) fuselage plug ahead of wings and 1.06 m (3 ft 5¾ in) plug to the rear; wings of increased chord (incorporating a third fuselage plug of 1.60 m; 5 ft 3 in) with span stretched to 63.60 m (208 ft 8 in); 31.1° sweepback at quarter-chord; taller vertical stabiliser is married to a new horizontal stabiliser to give increased chord fin with 0.50 m (1 ft 7¾ in) height extension. Max T-O weight 368,000 kg (811,300 lb). Landing gear adaptation involves replacement of central twin-wheel unit with forward-retracting brakable double-bogie unit. Rolls-Royce Trent 553 of 236 kN (53,000 lb st) chosen (non-exclusively) as power plant; this combines fan diameter of Trent 700 with scaled IP and HP compressors and turbines of Trent 800, plus new high-lift LP turbine; test flying of engine began 20 June 2000. Honeywell 331-600 APU chosen for A340-500/-600 in March 1999. A340-500 initial launch commitments received from Air Canada, Emirates, SIA and ILFC. First flight scheduled for December 2001 with service entry during late 2002.

A340-600: Derivative of A340-300 with fuselage stretch, 267 kN (60,000 lb st) class engines, improved aerodynamics and fly-by-wire flight-control system,

A340-200   12 First + 36 Business + 213 Economy = **261 seats**

A340-300   12 First + 42 Business + 241 Economy = **295 seats**

A340-500   12 First + 42 Business + 259 Economy = **313 seats**

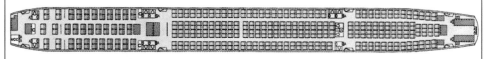

A340-600   12 First + 54 Business + 314 Economy = **380 seats**

Seat pitches: First 62in, Business 40in, Economy 32in

Potential interior arrangements of the Airbus A340   1999/0044886

INTERNATIONAL: Aircraft—AIRBUS

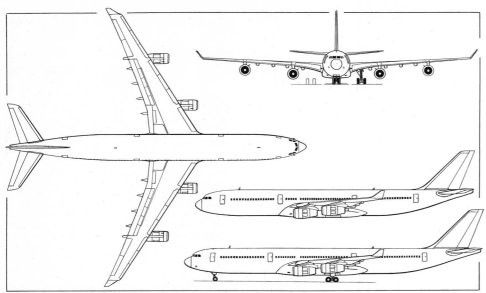

Airbus A340-300 four-turbofan long-range airliner, with additional side view (upper) of A340-200 *(Dennis Punnett/Jane's)*

additional fuel capacity, four-wheel central landing gear and 365,000 kg (804,690 lb) maximum T-O weight; can carry 380 three-class passengers or 485 in all-economy class up to 7,500 n miles (13,890 km; 8,630 miles) and designed as Boeing 747 replacement with significantly lower costs and full commonality with A330/A340 family. Compared with A340-300, has same wing chord/fuselage centre-section, wing span, tailplane and fin modifications as -500, but with further 5.87 m (19 ft 3 in) plug ahead of wing and 3.20 m (10 ft 6 in) to the rear. Airbus and GE Aircraft Engines agreed in April 1996 that latter should be sole power plant source for -600, but this accord dissolved in February 1997 and Rolls-Royce Trent 556 of 249 kN (56,000 lb st) chosen on non-exclusive basis. Launch commitments from Aerolineas Argentinas, Air Canada, DLH, Egyptair, Swissair and Virgin Atlantic; initial firm order placed by Virgin on 15 December 1997 for eight, plus options. First metal cut 27 July 1998. Final assembly began June 2000; airframe completed September 2000 and engines fitted November 2000. First flight due May 2001, with first delivery (to Virgin Atlantic) June 2002.

A340-600F: Proposed freighter version; considered by UPS for order eventually placed for MD-11 freighters; airline has option to convert some of its 60 A300-600Fs to A340-600Fs if required.

CUSTOMERS: Sales at 1 March 2001 totalled 309, of which 193 then delivered. See table.

AIRBUS A340 ORDERS
(at 1 March 2001)

Customer	Qty
Aerolineas Argentinas	6
Air Canada	13
Air China	3
Air France	14
Air Mauritius	3
All Nippon Airways	5
Austrian Airlines	4
Cathay Pacific	11
China Airlines	7
China Eastern Airlines	5
China Southwest Airlines	3
Egyptair	5
Emirates	6
Flightlease	11
GATX/Flightlease	2
Gulf Air	6
Iberia	18
ILFC	28
Kuwait Airways	4
LAN Chile	7
Lufthansa	49
Olympic Airways	4
Philippine Airlines	8
Qatar Airways	1
Sabena	9
SAS	7
Singapore Airlines	22
SriLankan Airlines	3
TAP-Air Portugal	4
Turk Hava Yollari	7
UTA	7
Virgin Atlantic Airways	17
Undisclosed	10
Total	**309**

COSTS: Development cost of A340-500/-600 set at 20 per cent lower than basic A340.

DESIGN FEATURES: A340 capitalises on commonality with A330 (identical wing/cockpit/tail unit and same basic fuselage) to create aircraft for different markets, and also has much in common (for example, existing Airbus widebody fuselage cross-sections, A310/A300-600 fin, advanced versions of A320 cockpit and systems) with rest of Airbus range; FAA has approved cross-crew qualification for A320 series, A330 and A340. New design wing (by BAE Systems), approximately 40 per cent larger than that of A300-600, has 30° sweepback and winglets. A340-500/600 wing 20 per cent larger than basic A340, has increased sweepback of 30° 6′ and 1.60 m (5 ft 3 in) extension to each wingtip.

FLYING CONTROLS: In A330/A340 electronic flight control system (EFCS), roll axis is controlled by two individual outboard ailerons and five outboard spoiler panels on each wing; pitch axis control is by trimmable tailplane and separate left and right elevators; tailplane can also be mechanically controlled from flight deck, but fly-by-wire computer inputs are superimposed; single rudder is directly linked to rudder pedals, with dual yaw damping inputs superimposed. High-lift devices consist of full-span slats, flaps and aileron droop; speed braking and lift dumping by raising all six spoilers on each wing and raising all ailerons. Slats and flaps controlled outside main fly-by-wire complex by duplicated slat and flap control computers (SFCC). See diagram.

Control surface actuation by three hydraulic systems (green, yellow, blue); two powered control units (PCU) at each aileron and elevator are controlled either by primary or secondary flight control computers; single actuators at spoiler panels controlled by primary or secondary flight control computers; dual PCUs for fly-by-wire tailplane trimming, and for centrally located flap and slat actuators; three PCUs at rudder.

Fly-by-wire computers include three flight control primary computers (FCPC) and two flight control secondary computers (FCSC); each computer has two processors with different software; primary and secondary computers have different architecture and hardware; power supplies and signalling lanes are segregated; system provides stall protection, overspeed protection and manoeuvre protection as in the A320, but the A330/340 computer arrangement maintains the protections for longer in the face of failures of sensors and inputs; FCPC and FCSC all operate continuously and provide comparator function to active channels, but only one in active control at any one control surface; reconfiguration logic can provide alternative control after failures; different normal and alternative control laws apply fly-by-wire basically as a g demand in pitch and rate demand in roll, plus complex manoeuvre limitations; if all three inertial systems fail (removing attitude information), system reverts to direct mode in which control surface angle is directly related to sidestick position; ultimate control mode is direct control of rudder and tailplane angle from rudder pedals and manual trim wheel, which is sufficient for accurate basic instrument flight.

Pilots have sidestick controllers and normal rudder pedals; EFIS instrumentation consists of duplicated primary flight displays (PFD), navigation displays (ND) and electronic centralised aircraft monitors (ECAM); three display management computers, with separate EFIS and ECAM channels, can each control all six displays in their four possible formats; flight path control by duplicated flight management and guidance and envelope computers (FMGEC); they control every phase of flight including course, attitude, engine thrust and flight planning using information from GPS and inertial systems; point of no return calculations made automatically for long-range flights in A340/A330. Control system data are collected for maintenance purposes by two flight control data concentrator (FCDC) computers. Honeywell Pegasus flight management system evaluated 1999, before January 2000 certification.

In normal flight, bank angle limited to 33° hands-off (autopilot control) and 67° with full stick displacement; airspeeds limited to 305 kt (565 km/h; 350 mph) and M0.82 under automatic flight control; if stick is held fully forward, the nose is automatically raised when airspeed reaches V_{MO} + 15 kt (28 km/h; 17 mph); if nose is raised, equivalent protections apply; 'alpha max' (13° clean and 19° with flap), slightly below maximum lift coefficient, is the greatest achievable with sidestick; 'alpha floor' is the angle of attack beyond which throttles progressively open to go-around power and airspeed is finally held steady just above stall, even if stick is continuously held back; alpha protection applied at 'normal' and 'hard' modes according to alpha rate; below protection speed (V_{PROT}) of 142 kt (263 km/h; 163 mph) automatic and manual trimming stops; outer ailerons remain centred at over 200 kt (371 km/h; 230 mph); if a spoiler panel fails, the symmetrical opposite panel stops operating; if rudder yaw damping fails, pairs of spoiler panels are used instead; minimum-speed marker on PFD adjusts to changing aircraft configuration, airbrake and pitch rate; fuel automatically transferred between wing and tailplane tanks to minimise trim drag when cruising above 7,620 m (25,000 ft).

Flight decks of the A330 and A340 (illustrated) are almost identical and very similar to the A320. The FAA has approved cross-qualification of pilots

Engines controlled by setting throttle levers to marks on quadrant, such as climb (CLB), maximum continuous and flexible take-off (Flex T-O); digital engine control makes detailed settings appropriate to altitude and temperature.

During landing and take-off, nosewheel steering, by rudder pedals, automatically disengages above 100 kt (185 km/h; 115 mph); demands for more than 4°/s nose-up pitch restrained near ground; maximum airspeeds for flaps and slats signalled on airspeed scale; trim automatically cancels effect of flaps, landing gear and airspeed; ailerons droop with full flap selection and deflect 25° up with spoilers in lift dump after touchdown; thrust reverser failure automatically countered by cancelling symmetrical opposite reverser; voice warning demands throttle closure at 6 m (20 ft) during landing flare; autothrottle disengaged when throttles closed at touchdown; on touch-and-go landing, trim automatically reset for take-off when Flex T-O power selected; engine failure in flight compensated automatically, with wings held level, slight heading drift and spoiler panels sucked down to avoid unnecessary drag.

STRUCTURE: A330 and A340 wings almost identical except latter strengthened in area of outboard engine pylon with appropriate modification of leading-edge slats 4 and 5; main three-spar wing box and leading/trailing-edge ribs and fittings of aluminium alloy, with Al-Li for some secondary structures; steel or titanium slat supports; approximately 13 per cent (by weight) of wings is of CFRP, GFRP or AFRP, including outer flaps and flap track fairings, ailerons, spoilers, leading/trailing-edge fixed surface panels and winglets; common fuselage for all initial versions, except in overall length (A340-300 same size as A330-300; A340-200 and A330-200 respectively eight and ten frames shorter; A340-500 and A340-600 described above; see also dimensions, below); construction generally similar to that of A310 and A300-600 except centre-section to accept new wing; tail unit (common to all versions except A330-200 and A340-500/600, which have larger fin and rudder) utilises same CFRP fin as A300-600 and A310; new tailplane incorporates trim fuel tank and has CFRP outer main boxes bridged by aluminium alloy centre-section.

Work-sharing along lines similar to those for A310 and A300-600, with percentages similar to those held in consortium. Aerospatiale Matra thus responsible for cockpit, engine pylons, part of centre-fuselage, and final assembly and outfitting at Toulouse; BAE Systems (with Textron Aerostructures, USA, as subcontractor) for wings; DaimlerChrysler Aerospace Airbus for most of fuselage, fin and interior; Airbus España for tailplane; Belgian consortium Belairbus for leading-edge slats and slat tracks.

LANDING GEAR: Main (four-wheel bogie) and twin-wheel nose units identical on all A330/340 versions. Main tyres size 54×21.0-23 or 46×17.0R20 or 1400×530R23 (32/36 ply); nose tyres 40.5×15.5-6 or 30×8.8R15 or 1050×395R16 (28 ply). Early A340s have additional twin-wheel auxiliary unit on fuselage centreline amidships, retracting rearward; A340-500 and -600 have four-wheel centre bogie, retracting forward. Goodyear tyres on all units. Landing gear and surrounding structure reinforced for extended-range A340-300.

POWER PLANT: Four 138.8 kN (31,200 lb st) CFM56-5C2 turbofans initially; 144.6 kN (32,500 lb st) CFM56-5C3

Airbus A340-300 in the colours of China Airlines   *2001/0103534*

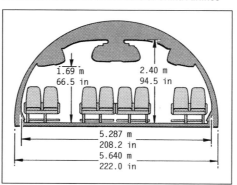

Airbus A330/340 passenger cabin cross-section
(*Mike Keep/Jane's*)   *2001/0106843*

and 151.2 kN (34,000 lb st) CFM56-5C4, Trent 553 and Trent 556 engines for A340-500/-600; CFM56 upgrade announced in July 2000, combining CFM56-5C with core of CFM56-5B/P. Maximum fuel capacity (-200 and -300 until 1996) 138,600 litres (36,614 US gallons; 30,488 Imp gallons); (-200 1996 and subsequent deliveries) 140,640 litres (37,154 US gallons; 30,937 Imp gallons) (-300, 1996 and subsequent deliveries) 141,500 litres (37,381 US gallons; 31,126 Imp gallons); (-213X additional centre tanks) 155,040 litres (40,958 US gallons; 34,105 Imp gallons); 148,700 litres (39,283 US gallons; 32,710 Imp gallons) (A340-300E); (-500) 214,810 litres (56,745 US gallons; 47,251 Imp gallons); (-600) 194,880 litres (51,483 US gallons; 42,869 Imp gallons).

ACCOMMODATION: Crew of two on flight deck (all versions); flight deck can be supplied with humidifier system. Passenger seating typically six-abreast in first class, six-abreast in business class and eight-abreast in economy (nine-abreast optional), all with twin aisles; two-class configurations seat 335 passengers in A340-300, and 303 passengers in A340-200; more typically, a three-class layout seats 295 in A340-300 and 263 in the A340-200. Single-class capacity of A340-300 is 440. Optional lower deck facilities in rear cargo hold of -600, with staircase to main deck, include an area with stand-up headroom and up to six lavatories plus, typically, eight crew rest bunks. A similar facility is offered as an option in the -500 and -600, with up to 10 full-length beds for passengers. A340-600 (except prototype) has two overwing Type III emergency exits in addition to standard eight Type A doors. Underfloor cargo holds house up to 32 LD3 containers or 11 standard 2.24 × 3.17 m (88 × 125 in) pallets in A340-300, 26 LD3s or nine pallets in A340-200, 30 LD3s or 10 pallets in A340-500, and 42 LD3s or 14 pallets in A340-600; front and rear cargo holds have doors wide enough to accept 2.44 × 3.17 m (96 × 125 in) pallets; all versions have a 19.7 m³ (695 cu ft) bulk cargo hold aft of the rear cargo hold.

AVIONICS: Airbus Future Air Navigation System (FANS-A), comprising Smiths Industries digital control and display system married to Honeywell FMS, underwent testing end 1999 and was certified July 2000; can be retrofitted to all A330/A340s.

DIMENSIONS, EXTERNAL:
Wing span: A340-200/300	60.30 m (197 ft 10 in)
A340-500/600	63.70 m (208 ft 11¾ in)
Wing aspect ratio: A340-200/300	10.1
A340-500/600	9.3
Length overall: A340-200	59.42 m (194 ft 11¼ in)
A340-300	63.60 m (208 ft 8 in)
A340-500	67.80 m (222 ft 5¼ in)
A340-600	75.30 m (247 ft 0½ in)
Fuselage: Max diameter (all versions)	5.64 m (18 ft 6 in)
Height overall:	
A340-200/-300	16.83 m (55 ft 2½ in)
A340-500/-600	17.80 m (58 ft 4¾ in)
Wheel track (all versions)	10.69 m (34 ft 5 in)
Wheelbase: A340-200	23.47 m (77 ft 0 in)
A340-300	25.60 m (84 ft 0 in)
A340-500	27.59 m (90 ft 6¼ in)
A340-600	32.89 m (107 ft 11 in)

DIMENSIONS, INTERNAL:
Cabin: Length (excl flight deck):	
A340-200	46.06 m (151 ft 1½ in)
A340-300	50.35 m (165 ft 2¼ in)
A340-500	53.55 m (175 ft 8¼ in)
A340-600	60.95 m (199 ft 11½ in)
Max width	5.28 m (17 ft 4 in)

AREAS:
Wings, gross:	
A340-200/-300	361.60 m² (3,892.2 sq ft)
A340-500/-600	437.00 m² (4,703.8 sq ft)

WEIGHTS AND LOADINGS:
Operating weight empty:	
A340-200 current	129,000 kg (284,395 lb)
A340-300 standard	129,900 kg (286,380 lb)
A340-500	170,400 kg (375,665 lb)
A340-600	177,000 kg (390,220 lb)
Max payload: A340-200 initial	44,000 kg (97,005 lb)
A340-200 current	45,530 kg (100,375 lb)
A340-300 standard	48,100 kg (106,040 lb)
A340-500	43,300 kg (95,460 lb)
A340-600	55,800 kg (123,020 lb)
Max T-O weight:	
A340-200, -300	275,000 kg (606,275 lb)
A340-300	271,000 kg (597,450 lb)
A340-500	368,000 kg (811,300 lb)
A340-600	365,000 kg (804,675 lb)
Max landing weight:	
A340-200 initial	185,000 kg (407,850 lb)
A340-200 current	188,000 kg (414,475 lb)
A340-300 standard	190,000 kg (418,875 lb)

Airbus A340-300 of Singapore Airlines   *2000/0085604*

A340-600 being prepared for engine runs   *2001/0098663*

Airbus A340 wing leading- and trailing-edge moving surfaces are evident in this view *(Paul Jackson/Jane's)*

Double-bogie landing gear of the A340
(Paul Jackson/Jane's)

A340-300 option	192,000 kg (423,275 lb)
A340-500	236,000 kg (520,275 lb)
A340-600	254,000 kg (559,975 lb)

Max zero-fuel weight:
A340-200	173,000 kg (381,400 lb)
A340-300 standard	178,000 kg (392,425 lb)
A340-300 option	181,000 kg (399,025 lb)
A340-500	222,000 kg (489,425 lb)
A340-600	240,000 kg (529,100 lb)

Max wing loading:
A340-200, -300	760.5 kg/m² (155.76 lb/sq ft)
A340-500, -600	835.2 kg/m² (171.07 lb/sq ft)

PERFORMANCE (provisional; definitions as for Weights and Loadings):
Max operating Mach No. (M_{MO})	0.86
Typical operating Mach No.: A340-200/300	0.82
A340-500/600	0.83

Stalling speed:
A340-300, at 267,000 kg (588,625 lb), wheels up:
flaps up	161 kt (299 km/h; 186 mph)
flaps down	133 kt (247 km/h; 153 mph)

T-O at S/L, ISA + 15°C:
A340-200 at 275,000 kg (606,275 lb)
  3,017 m (9,900 ft)
A340-300 at 271,000 kg (597,450 lb)
  3,017 m (9,900 ft)

Range at typical OWE:
A340-200 with 239 passengers and baggage, international allowances and 200 n mile (370 km; 230 mile) diversion
  8,000 n miles (14,816 km; 9,206 miles)
A340-300 with 295 passengers and baggage, allowances as above:
standard	7,200 n miles (13,334 km; 8,285 miles)
optional -300	7,400 n miles (13,704 km; 8,515 miles)

A340-500 with 313 passengers
  8,650 n miles (16,019 km; 9,954 miles)
A340-600 with 380 passengers
  7,500 n miles (13,890 km; 8,630 miles)

OPERATIONAL NOISE LEVELS (A340-300 standard):
T-O, fly-over	95.6 EPNdB
Sideline	96.1 EPNdB
Approach	96.9 EPNdB

UPDATED

AIRBUS A330

TYPE: Wide-bodied airliner.
PROGRAMME: Developed simultaneously with four-engined A340; launched 5 June 1987; first flight (F-WWKA) with GE engines 2 November 1992; first R-R Trent 700-powered A330 flew 31 January 1994; simultaneous European and US certification with initial GE CF6-80E1 engines received 21 October 1993; first delivery (Air Inter) December 1993, entered service January 1994; certification with PW4164/4168 obtained 2 June 1994; A330 powered by GE CF6-80E1 with FADEC granted 120-minute ETOPS approval May 1994; Aer Lingus flew first ETOPS services across the Atlantic in May 1994; A330 powered by PW4164/4168 granted 90-minute ETOPS approval November 1994; certification with R-R Trent achieved 22 December 1994; 90-minute ETOPS approval granted before first delivery, to Cathay Pacific Airways, 24 February 1995. Further extensions of ETOPS to 180 minutes granted on 6 February 1995 (GE engines), 4 August 1995 (P&W) and 17 June 1996 (R-R); 180-minute ETOPS for PW4168A-powered A330-300 granted July 1999.

All main structural and systems information common to both A330 and A340 is listed in A340 entry. Differences in A330 given here.

CURRENT VERSIONS: **A330-300**: Baseline version; seating capacity as for A340-300; 335 passengers in standard two-class, 440 passengers maximum. Payload increase of 7,000 kg (15,432 lb) offered for standard A330-300 in October 1993. Maximum T-O weight increased to 217,000 kg (478,400 lb) from November 1995 to allow typical 335 passengers to be carried 4,850 n miles (8,982 km; 5,581 miles), further increase to 230,000 kg (507,050 lb) with range increased to 5,600 n miles (10,370 km, 6,444 miles) offered in 1997; first aircraft to this standard ordered by Air Canada in 1997, but first in service was a Korean Air example in May 1999. Optional higher maximum T-O weight of 233,000 kg (513,675 lb) available from early 2001; earlier 230,000 kg (507,050 lb) versions can be retrofitted. During August 2000, Airbus was said to be looking at a 240,000 kg (529,100 lb) high gross weight model with increased range; development would proceed after the A330-500 launch.

A330-200: Extended-range version; launched 24 November 1995; 10-frame reduction in fuselage length to 59.00 m (193 ft 6¾ in); maximum T-O weight 230,000 kg (507,050 lb); higher MTOW available from early 2001, as per A330-300. 253 passengers in three classes or 293 passengers in two classes; range with 253 passengers 6,450 n miles (11,945 km; 7,422 miles); engine choice as for A330-300; initial order, for 13, placed in March 1996 by ILFC. First flight (c/n 181, F-WWKA, with CF6-80E1 engines) 13 August 1997 was followed by public debut at Dubai Air Show in November 1997. Canada 3000 first user (on lease from ILFC) with first flight of production aircraft (C-GGWA) 20 January 1998 and delivery on 29 May 1998 following FAA/JAA/Transport Canada certification on 31 March 1998. First direct airline order from Korean Air, received second built in June 1998 (c/n 195, with PW4000 engines) following 4 December 1997 first flight and May 1998 certification. Version with Rolls-Royce Trents (re-engined first prototype) flew 24 June 1998 and following JAR and Transport Canada certification in January 1999 was delivered to Air Transat in February 1999. Estimated development cost US$450 million (1995).

A330-100: Proposed nine-frame reduction of A330-200; abandoned in favour of A330-500; see *Jane's 2000-01* for details.

A330-200F: Proposed freighter version under study as a possible DC-10/MD-11 replacement. Estimated market for 200 aircraft over next 10 years.

A330-500: Proposed eight-frame reduction of A330-200, announced July 2000 and initially internally designated A330-M18; four frames removed from forward fuselage and four from aft to reduce overall length to 54.76 m (179 ft 8 in); will retain one structural design only, but will offer a reduced maximum T-O weight version of 195,000 kg (429,900 lb) to form baseline of possible A300 and A310 replacements, increasing the range to 4,400 n miles (8,148 km; 5,063 miles) with 267 passengers (two-class configuration). Standard 228,000 kg (502,650 lb) maximum T-O weight version allows long-

Airbus A330-200 wearing Emirates' latest colour scheme

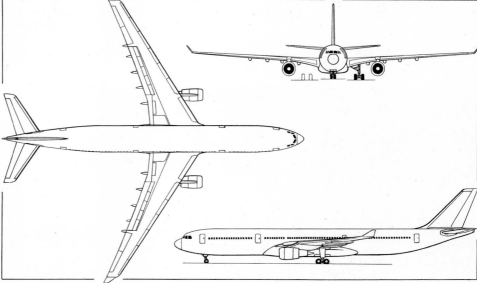

Airbus A330 twin-turbofan airliner, developed in parallel with the A340 *(Dennis Punnett/Jane's)*

AIRBUS—AIRCRAFT: INTERNATIONAL

range applications up to 7,000 n mile (12,964 km; 8,055 mile) range with 224 passengers (three-class configuration). Design requires engines in 320 kN (72,000 lb st) class and will offer the A330 three engine choices: CF6-80-E1A3, PW4168A, Trent 772 for the 228,000 kg version and lower-thrust variants of these engines, 258 kN (58,000 lb st) class, for the 195,000 kg version: CF6-80-E1A5, PW4157, Trent 758. The design was due to be officially launched at the end of 2000 with an in-service date of early 2004, but has been delayed after a less than enthusiastic airline response.

CUSTOMERS: Total of 376 sold, and 180 delivered by 1 March 2001. See table.

Gulf Air began Airbus A330-200 operations in 1999

AIRBUS A330 ORDERS
(at 1 March 2001)

Customer	Qty
Aer Lingus	3
Air Canada	8
Air France	8
Air Inter Europe	12
Airtours Int'l Airways	7
Asiana Airlines	6
Austrian Airlines	3
British Midland Airways	2
Cathay Pacific Airways	20
CIT Group	20
Corsair	2
Dragonair	4
Emirates	21
Flightlease	9
Garuda Indonesia	9
GATX/Flightlease	4
GECAS	17
Gulf Air	6
ILFC	75
Korean Air	19
LTU	5
Malaysia Airlines	10
Monarch Airlines	2
Northwest Airlines	36
Philippine Airlines	8
Qantas	13
Sabena	3
SAS	3
SriLankan Airlines	6
Swissair	4
TAM	5
Thai Airways International	12
US Airways	10
Undisclosed	4
Total	**376**

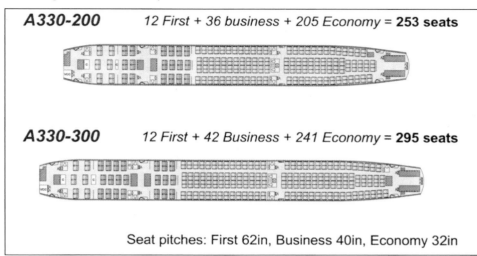

Airbus A330 alternative interiors

POWER PLANT: First deliveries with two 300 kN (67,500 lb st) GE CF6-80E1A4 turbofans; alternative engines, using common pylon and mount, GE CF6-80E1A3, P&W PW4168A or R-R Trent 772B. Longer-range version powered by CF6-80E1A4, PW4168A or Trent 772B. Fuel capacity 97,170 litres (25,670 US gallons; 21,374 Imp gallons) in A330-300; 139,090 litres (36,744 US gallons; 30,596 Imp gallons) in A330-200.

DIMENSIONS, EXTERNAL: As A340 except:
Length overall:
 A330-200 59.00 (193 ft 6¾ in)
 A330-300 63.68 m (208 ft 11 in)
Height overall:
 A330-200 17.89 m (58 ft 8¼ in)
 A330-300 16.84 m (55 ft 3 in)

WEIGHTS AND LOADINGS (A330-300 basic versions; CL: with CF6-80E1A3A4, PL with PW4168A, TL with Trent 772B. A330-200 basic versions; C: with CF6; P: with P&W, T: with Trent):
Operating weight empty:
 CL 127,520 kg (281,125 lb)
 PL 124,855 kg (275,250 lb)
 TL 127,615 kg (281,350 lb)
 C 120,470 kg (265,600 lb)
 P 121,070 kg (266,925 lb)
 T 120,565 kg (265,800 lb)
Max payload:
 CL 48,480 kg (106,880 lb)
 PL 48,145 kg (106,140 lb)
 TL 48,385 kg (106,670 lb)

SriLankan's first Airbus A330-200 was delivered on 26 October 1999

INTERNATIONAL: AIRCRAFT—AIRBUS

C	47,530 kg (104,785 lb)
P	46,930 kg (103,460 lb)
T	47,435 kg (104,575 lb)
Max T-O weight: CL, PL, TL:	
standard	230,000 kg (507,050 lb)
optional	233,000 kg (513,675 lb)
C, P, T	230,000 kg (507,050 lb)
Max landing weight:	
CL, PL, TL: standard	185,000 kg (407,850 lb)
optional	187,000 kg (412,250 lb)
C, P, T	180,000 kg (396,825 lb)
Max zero-fuel weight:	
CL, PL, TL: standard	173,000 kg (381,400 lb)
optional	175,000 kg (385,800 lb)
C, P, T	168,000 kg (370,375 lb)
Max wing loading:	
CL, PL, TL: standard	636.0 kg/m² (130.28 lb/sq ft)
optional	644.4 kg/m² (131.97 lb/sq ft)
C, P, T	633.4 kg/m² (129.74 lb/sq ft)

WEIGHTS AND LOADINGS (A330-200; C: with CF6-80E1A3/A4, P: with PW4168A, T: with Trent 772B):

Typical airline OWE:	
C	120,170 kg (264,925 lb)
P	120,750 kg (266,200 lb)
T	120,265 kg (265,150 lb)
Max T-O weight, all:	
standard	230,000 kg (507,060 lb)
optional	233,000 kg (513,675 lb)
Max landing weight, all	
standard	180,000 kg (396,825 lb)
optional	182,000 kg (401,250 lb)
Max zero-fuel weight, all:	
standard	168,000 kg (370,375 lb)
optional	170,000 kg (374,775 lb)
Max wing loading, all:	
standard	633.4 kg/m² (129.74 lb/sq ft)
optional	644.4 kg/m² (131.97 lb/sq ft)

PERFORMANCE:
Max operating Mach No. (M$_{MO}$) 0.86
Typical operating Mach No. 0.82
T-O run at S/L, ISA + 15°C:
  A330-200 at 230,000 kg (507,050 lb)
    2,650 m (8,695 ft)
  A330-300 at 230,000 kg (507,050 lb)
    2,514 m (8,250 ft)
Range at typical OWE, max passengers, international allowances and 200 n mile (370 km; 230 mile) diversion:
  A330-200 at 230,000 kg (507,050 lb)
    6,450 n miles (11,954 km; 7,422 miles)
  A330-300 at 230,000 kg (507,050 lb)
    5,600 n miles (10,371 km; 6,444 miles)
OPERATIONAL NOISE LEVELS (A330-300: A: 217,000 kg (478,400 lb) with Trent 768; B: 230,000 kg (507,050 lb) with Trent 772):
  T-O, flyover: A 89.8 EPNdB
  B 98.0 EPNdB
  Sideline: A 96.5 EPNdB
  B 101.0 EPNdB
  Approach: A 96.8 EPNdB
  B 104.0 EPNdB

UPDATED

AIRBUS A3XX
Redesignated in December 2000 as A380.

UPDATED

AIRBUS A380
TYPE: High-capacity airliner.
PROGRAMME: Engineering work began in early June 1994; known as A3XX until December 2000; separate A3XX directorate (Large Aircraft Division) formed within Airbus in March 1996; concept definition began April 1996; Airbus expected to allocate up to 40 per cent of development cost to partners; Mitsubishi Heavy Industries reportedly interested in subcontract work. Full-size mockup of fuselage cross-section shown at Paris in 1997. Belairbus signed agreement 17 January 2000 for 49 per cent of A3XX work, believed to include wing leading-edge slats.

Airbus Industrie Supervisory Board authorised programme go-ahead 8 December 1999; commercial launch authorised 23 June 2000; industrial launch and A380 designation confirmed 19 December 2000, by Airbus board, on receipt of required 50th launch order. Final product definition will be completed by end 2001, after which assembly will begin, to be completed by second quarter of 2004. First flight due mid-2004, with certification and entry into service in fourth quarter 2005. Production intended to reach four per month by 2008.

CURRENT VERSIONS: **A380-700**: Short-fuselage, 480-passenger version; intended to fit into the Airbus range between the A340-600 and A380-800. Fuselage shortening achieved by removing sections both fore and aft of the wing. Engine thrust up to 298 kN (67,000 lb) each; fuel capacity as for A380-800.

A380-800: Baseline version; nominal 555 passengers and range of 7,800 n miles (14,445 km; 8,976 miles). Engine thrust 302 kN (67,900 lb st), fuel capacity 325,000 litres (85,858 US gallons; 71,492 Imp gallons). An increased weight **A380HGW** will carry same number of passengers 8,150 n miles (15,093 km; 9,375 miles) using engines of 311 kN (69,915 lb st); fuel capacity unchanged.

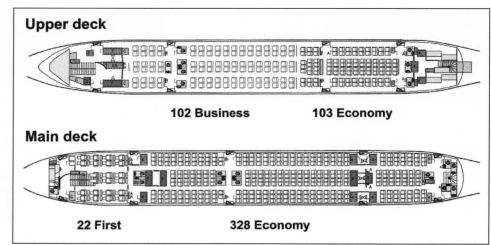
Upper and main decks of projected A380-800 configured for 555 passengers

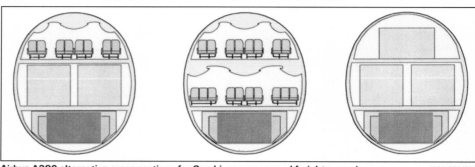
Airbus A380 alternative cross-sections for Combi, passenger and freighter service

Cargo hold capacity for both versions is 12 pallets plus two LD3 containers.

A380-800R: Increased-range version; 555 seats. Engine thrust 333 kN (74,850 lb st) each; fuel capacity 372,000 litres (98,274 US gallons; 81,831 Imp gallons).

A380-800F: All-cargo version, carrying 150,000 kg (330,700 lb) of payload over 5,620 n miles (10,408 km; 6,467 miles). Standard layout is 17 pallets on upper deck, 28 pallets on main deck and 13 pallets in cargo hold/lower deck.

A380-900: Stretched version initially known as A3XX-200; 656 seats in three-class layout and 1,000 at high density. Engines and fuel as -800R.

CUSTOMERS: Potential market for 1,200 passenger and over 300 cargo aircraft in A380 class up to 2020. From 14,000 new passenger airliners required by then, more than 50 per cent of A380 class of aircraft would be required in Asia Pacific region; 20 airlines will take nearly 80 per cent of production. Initial launch customer, on 24 July 2000, was Emirates, which committed to five A380s and two A380-800Fs, plus five options, for delivery from February 2006 (A380) and second quarter of 2008 (A380-800Fs). Air France ordered 10, plus four options, also on 24 July 2000, followed by ILFC (five), Singapore (10 plus 15 options), Qantas (12 plus 12 options) and Virgin Atlantic (six plus six options). Federal Express ordered 10 A380-800Fs on 16 January 2001, for delivery from 2008 onward, to become that model's launch customer. Total orders at 1 January 2001 stood at 50, plus 42 options.

COSTS: Estimated development cost US$10 billion to US$12 billion for whole A380 family. Projected unit cost $217 million (2000) for passenger version; US$233 million for freighter.

DESIGN FEATURES: First version designated A380, is conventional in external appearance, but incorporates new developments in structures, materials, landing gear design and aerodynamics. Dassault CATIA and IBM computer-aided design. Flight deck commonality with earlier Airbuses to permit crew cross-qualification.

Fuselage is vertically orientated oval three-deck arrangement; this 'vertical ovoid' accommodates 10 passengers abreast on main deck and eight-abreast on upper deck, offering greater space per passenger than Boeing 747. Seating ranges from a nominal 555 (22 first and 328 economy on main deck and 102 business and 103 economy on upper deck) in three-class layout to 854 in single-class, high-density Japanese domestic layout. Those ordered by Qantas are 524-seat. Dual-lane boarding stair allows four-aisle boarding and deplaning through main deck.

Lower deck can accommodate shop, bar, restaurant and/or normal range of 36 cargo containers or 12 pallets and 18.4 m³ (650 cu ft) of bulk freight; main deck is large enough to accommodate two 2.44 × 2.44 m (8 ft 0 in × 8 ft 0 in) containers side by side in the freighter version.

Direct operating costs of A380 required to be at least 15 to 20 per cent lower than those of nearest existing competitor.

STRUCTURE: Extensive use of composites for all flaps and spoilers, wing/fuselage fairings, all tail surfaces, tailcone aft of fin leading-edge intersection with fuselage and

Computer-generated image of the twin-deck Airbus A380-800

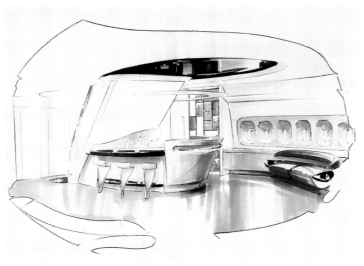
Interior impression of the A380, showing passengers' bar

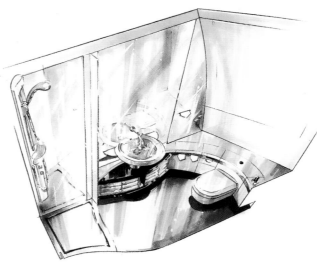
Impression of spacious lavatory aboard A380

engine cowlings and possibly the outer wing. Laser welding will be employed on fuselage to reduce cost and weight. New 'Glare' material being tested on A310 for future use of A380.

Work allocation is Airbus France (St Nazaire): flight deck and centre fuselage; Airbus Deutschland (Hamburg): forward centre fuselage, rear fuselage; Airbus Deutschland (Stade): fin and rudder; Airbus UK (Broughton): wing main panels; Airbus España: wing/fuselage fairings, belly fairing and fixed horizontal tail; Airbus France (Toulouse): engine pylons and final assembly; undetermined: wing leading-edge, flaps, ailerons, winglets, spoilers, elevators and tailcone. Transport of major components undertaken by purpose-built ship from Hamburg via UK and St Nazaire (forward centre fuselage disembarked, joined to flight deck and re-embarked, accompanied by centre fuselage), the aircraft set then transferred to barge at Bordeaux and joined by Spanish-built elements for river transport to within 80 km (50 miles) of Toulouse, completing journey by road. However, both ship- and truck-borne movements are being opposed by local residents.

LANDING GEAR: Messier-Dowty and Goodrich tendering for landing gear contracts. A380 will have six-wheel bogies on main legs.

POWER PLANT: A380 is offered with a choice of Engine Alliance (GE/P&W) GP7200 or Rolls-Royce Trent 900; Airbus Industrie and Rolls-Royce signed an MoU specifying the Trent 900 as favoured power plant in November 1996; is due for certification in 2004; Singapore Airlines has selected R-R power plant, as have Virgin and ILFC.

SYSTEMS: High-pressure hydraulics, using lighter, more compact pipework; variable frequency electrical

Impression of the A380's flight deck

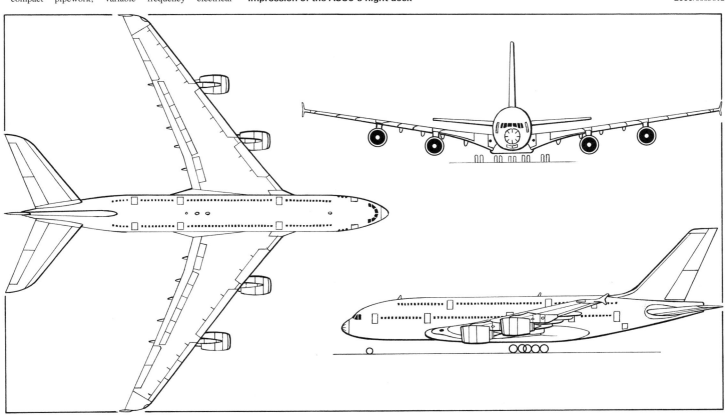
Airbus A380-800 555-seat airliner *(James Goulding/Jane's)*

200 INTERNATIONAL: AIRCRAFT—AIRBUS

generating systems; ethernet system for avionics and communications; and onboard cameras for taxying assistance.

AVIONICS: Cockpit layout, by Aerospatiale Matra, will be compatible with other Airbus family members. Onboard information system (OIS) will integrate databases with operator's in-house software packages. Honeywell terrain guidance and on-ground navigation will be integrated into FMS.

DIMENSIONS, EXTERNAL:
Wing span	79.80 m (261 ft 9¾ in)
Length overall: A380-800	73.00 m (239 ft 6 in)
Height overall	24.10 m (79 ft 0¾ in)
Rear cargo door: A380-800F	
Max width	3.43 m (11 ft 3 in)
Max height	2.54 m (8 ft 4 in)

WEIGHTS AND LOADINGS:
Operating weight empty: A380	275,000 kg (606,275 lb)
A380-800F	248,000 kg (546,750 lb)
Max payload: A380	83,000 kg (182,985 lb)
A380-800F	150,000 kg (330,695 lb)
Max T-O weight: A380	548,000 kg (1,208,125 lb)
A380 HGW	560,000 kg (1,234,575 lb)
A380-800F	583,000 kg (1,285,300 lb)
Max landing weight:	
A380, A380HGW	383,000 kg (844,350 lb)
A380-800F	427,000 kg (941,375 lb)
Max zero-fuel weight:	
A380	358,000 kg (789,250 lb)
A380-800F	399,000 kg (879,650 lb)
Max power loading:	
A380	454 kg/kN (4.45 lb/lb st)
A380HGW	459 kg/kN (4.50 lb/lb st)
A380-800F	483 kg/kN (4.73 lb/lb st)

PERFORMANCE (estimated, Trent 900 engines):
Max operating speed M0.89 (340 kt; 630 km/h; 391 mph)	
Econ cruising Mach No.	0.85
Service ceiling	13,100 m (43,000 ft)
T-O field length, ISA + 15°C	less than 3,353 m (11,000 ft)
Range: A380	7,800 n miles (14,445 km; 8,976 miles)
A380HGW	8,150 n miles (15,094 km; 9,378 miles)
A380-800F	5,620 n miles (10,408 km; 6,467 miles)

NEW ENTRY

AIRBUS ACJ
The Airbus Corporate Jetliner is described under the A319 heading.

AIRBUS INDUSTRIE ASIA (AIA)
AIA was owned 38 per cent by Finmeccanica-Alenia of Italy and 62 per cent by Airbus Industrie, and was formed on 18 June 1997 to undertake European participation in the AVIC/AIA/STPL AE-100 regional airliner as successor to the AI(R)-managed subsidiary Aero International Asia. It was announced on 3 September 1998 that the partners had "jointly concluded that no solid common basis had been found for further developing this new aircraft". Airbus and AVIC simultaneously revealed their intentions to discuss a new project, although this fared little better and the company has now been dissolved.

UPDATED

AIRBUS P305
Proposed in mid-1998 as a replacement programme for the A316/A317, to be undertaken with the Chinese aviation industry. Airliner based on A300/A310 fuselage, but with avionics, engines and FBW controls equivalent to latest Airbus aircraft. Rolls-Royce Trent 500 and a P&W PW4000 derivative were among engine options. Subsequently, Airbus invited AVIC to participate in the A318 programme and production of A320 components, and the P305 project was cancelled.

UPDATED

AIRBUS MILITARY SAS
17 avenue Didier Daurat, F-31707 Blagnac, France
Tel: (+33 5) 62 11 07 82
Fax: (+33 5) 62 11 06 11
Web: http://www.airbusmilitary.com
PRESIDENT: Alain Flourens
INDUSTRIAL DIRECTOR: Michele Priolo
COMMERCIAL DIRECTOR: Richard Thompson
PRODUCTION AND QUALITY CONTROL DIRECTOR: Angel Hurtado
FINANCE AND PROCUREMENT DIRECTOR: Michael Haidinger
HEAD OF MARKETING: David R Jennings
PARTICIPATING COMPANIES:
EADS France
BAE Systems Airbus (UK)
EADS CASA (Spain)
EADS Deutschland (Germany)
Alenia (Italy)
FLABEL (Belgium)
TAI (Turkey)

Airbus Military was legally established, with the above-mentioned companies, in January 1999 as a 'Société aux Actions Simplifiées' as the prospective manufacturer of the Airbus A400M, formerly known as the Future Large Aircraft (FLA). It is also anticipated that AM will promote military versions of the Airbus series of airliners, some of which are already in air force service as transports.

Conceptual work was undertaken by the European FLA Group (Euroflag). Euroflag srl originally formed 17 June 1991, with headquarters in Alenia head office in Rome, to manage European FLA development. Aerospatiale, Alenia, British Aerospace, CASA and Daimler-Benz Aerospace Airbus (DaimlerChrysler from 1998) had equal shares in Euroflag srl; MoUs established 1992 with FLABEL (SABCA, SONACA, ASCO and BARCO) of Belgium, OGMA of Portugal and Turkish Aerospace Industries (TAI) of Turkey to allow integrated participation in FLA programme; BAE and FLABEL were industrial, not national, partners contributing their own funds, although the UK government announced in December 1994 that membership was to be upgraded to national participation.

The partners agreed in September 1994 to industrialise the programme by transferring it to their existing airliner production company; formal announcement was made on 14 June 1995 that Airbus Military Company (AMC) would be established, replacing Euroflag, which then disbanded. AMC would be responsible for FLA design, marketing and production, but was never legally incorporated. With the formation of Airbus Military SAS the five original participating companies are represented by the directors noted above; however, Airbus Industrie is the largest shareholder, while TAI and FLABEL are full risk-sharing partners. Programme management will be through AM making use of Airbus Industrie procedures and industrial infrastructure and will take advantage of technologies developed for Airbus airliners. Participating governments are expected to leave workshare percentage decisions to industry. The projected shares for R&D financing were Germany 25.7, France 17.2, UK 15.5, Italy 15.1, Spain 12.4, Turkey 6.9, Belgium 4.1 and Portugal 3.1.

Computer simulation of A400M loading an APC 2001/0100421

By early 1997, the FLA programme had lost development sponsorship by the principal participating governments, although military commitments remained, subject to the aircraft being produced with commercial funding. Programme was weakened during 1997 by unilateral German negotiations with Ukraine over Antonov An-7X (Westernised An-70). German MoD attempts to involve Russia and Ukraine continued into 1999, but these were not supported by Airbus Industrie, which declined to become the prime contractor and assume the commercial risk of a programme based on the An-70; a study commissioned by the German MoD and carried out by DaimlerChrysler Aerospace reached a similar conclusion in September 1998. By mid-2000 senior German government sources were stressing the need for a European solution, prompting Airtruck to request assurances that its An-7X was still under consideration.

Airbus Military SAS submitted responses to the seven-nation FLA RFP (request for proposals, dated September 1997) on 29 January 1999 and to the competitive Future Transport Aircraft (FTA) RFP issued to Boeing, Airbus and Lockheed Martin by Belgium, France, Spain and the UK on 31 July 1998. Acceptance of A400M was formally announced by all seven members on 27 July 2000.

UPDATED

AIRBUS A400M FUTURE LARGE AIRCRAFT (FLA)
TYPE: Strategic transport.
PROGRAMME: Original FIMA programme (see 1989-90 *Jane's*) replaced April 1989 by five-nation industry MoU to develop four-turbofan, new technology transport to replace C-130 Hercules and C.160 Transall; Independent

Artist's impression of the turboprop-powered Airbus A400M Future Large Aircraft 2001/0089525

European Programme Group (IEPG) defined Outline European Staff Target (OEST) during 1991; initial studies undertaken by Euroflag organisation. Western European Union report in third quarter of 1991 concluded Euroflag FLA should form core of future European military transport capability to support Rapid Reaction Corps; national armament directors of Belgium, France, Germany, Italy, Portugal, Spain and Turkey affirmed support for 12 month prefeasibility study completed by Euroflag in late 1992; UK MoD declined involvement, but retained observer status; UK participation privately maintained by BAe and Shorts (10 per cent of BAe work); European Staff Target and intergovernmental MoU signed by seven nations in 1993; full feasibility programme officially started October 1993, by which time cargo hold width and height increased from original 3.66 m (12 ft 0 in) and 3.55 m (11 ft 7¼ in), respectively; study finished May 1995 and submitted to European defence ministries. Meanwhile, FLA underwent profound change in April 1994 when turbofans deemed incapable of providing desired performance; aircraft recast with four turboprops of new design. Discussion of a 'close association' between Euroflag and Airbus Industrie began third quarter of 1993 and formalised in June 1995.

Launch of the predevelopment phase (PDP) was postponed at least six months from early 1996 as a result of funding uncertainties. Original intention was for PDP to run from 1996 to 1998 and define a comprehensive specification for the aircraft and contractual forms and conditions against which the partner nations would make commitments. Full development and production phase (DPP) scheduled to follow directly on from PDP and terminate with first flight in 2002. Customer deliveries were to begin in 2004.

France announced funding withdrawal from FLA development on 13 May 1996 and UK failed to rejoin the programme later that year, despite intention announced in December 1994 (when the RAF purchased Lockheed Martin C-130J Hercules). However, Germany became first to sign a European Staff Requirement, on 24 July 1996, although having terminated official funding for FLA development in previous month. AMC accordingly announced a 'single-phase' programme in May 1996.

The new programme schedule started in mid-1998 with a set of formal prelaunch activities (PLA), largely funded by industry and lasting 12 months, leading to a fully documented proposal for the purchase of the Airbus A400M. This contains the technical proposal, including the aircraft specification and performance guarantees, and the commercial proposal with firm and fixed prices; a full set of contractual terms and conditions; and a detailed planning of the single-phase programme. Original plans called for a first flight in 2003 and entry into service in 2005, but further slippage has occurred. The proposal was made in the name of Airbus Military SAS (AM), formed in January 1999. Strategic workshares (detailed enough to allow industry to provide the necessary resources to complete the proposal) were agreed at the start of PLA.

February 1999 delivery of the A400M proposal, initiated a 12-month period of negotiation of individual national requirements before planned official launch during early 2000 to meet an ESR now supported by Belgium, France, Germany, Italy, Spain, Turkey and the UK, but not Portugal. The 'PLA + single phase' programme provides industry with an uninterrupted development schedule and strong commitments from governments, while it also meets the customers'

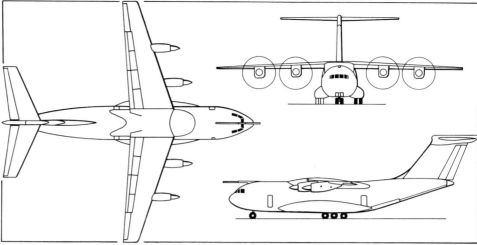

Provisional drawing of the Airbus A400M military transport *(Paul Jackson/Jane's)* 2000/0062400

requirement that industry carries as much of the development risk as possible.

A meeting in March 2000 saw Belgium, France, Italy, Spain and Turkey identify a requirement for 131 A400Ms (37 fewer than expected) while, on 16 May, UK announced its intention to buy 25 aircraft (becoming first to fully commit). France and Germany followed on 9 June 2000. Seven participating nations announced selection of A400M on 27 July, committing to 225, including one for Luxembourg. Portugal announced requirement for four shortly thereafter.

Engine selection delayed until 2000, when six-partner TP400 selected. Flight testing is expected to be at CASA's Seville plant and AMC's Toulouse facility, with certification by a single authority. CASA will have sole production line. Germany committed US$4.4 billion to the programme in November 2000, though this represented 60 per cent of the amount required to guarantee Germany's 73 aircraft and planned 33 per cent workshare. France committed US$2.6 billion at the same time.

A first flight date of 2005 has been tentatively scheduled, with deliveries from 2006/07 to 2016. A timetable of 51 months between contract and flight has been agreed, with an ISD for a common configuration 'logistics' aircraft of 71 months after contract signature.

CURRENT VERSIONS: Primarily for personnel/cargo transport and parachuting of men and equipment; basic aircraft will be capable of rapid conversion to two-point tanker with 25,000 kg (55,115 lb) of transferable fuel after a 3 hour loiter 450 n miles (833 km; 518 miles) from base. Derivatives will include a role-convertible three-point air-refuelling tanker with HDU on rear ramp and possibly surveillance/reconnaissance, long-range maritime patrol and AEW. Three-point tanker could transfer 35,000 kg (77,162 lb) of fuel (in normal tanks plus roll-on/roll-off tanks in cargo bay) after a 3 hour loiter 450 n miles (833 km; 518 miles) from base but is regarded in some circles as being too slow for routine support of fast jets. Proposed maritime patrol version now abandoned. 'K-400M' proposed turbofan-powered tanker derivative to meet 360-aircraft USAF KC-135 replacement requirement, with flying boom systems, redesigned wings and new engines.

CUSTOMERS: Procurement agency is Organisme Conjointe de Co-operation en matière d'Armement (OCCAR) in Bonn, acting for all prospective NATO purchasers. Exports expected, and attempts made in early 1995 to interest Japan; export market estimated at 400 aircraft over 25 years, with A400M to secure 50 per cent share.

Country	Original	Current	First Delivery
Belgium	12	7	2014
France	50	50	2007
Germany	75	73	2008
Italy	44	16	2014
Luxembourg	0	1	2014
Portugal	0	4	n/a
Spain	36	27	2010
Turkey	26	26	2007
UK	45	25	2007
Total	**288**	**229**	

COSTS: Unit cost estimated in 1999 as US$80 million over 300 aircraft production run including amortisation of all R&D costs (estimated as some US$6 billion). In October 1999, AMC offered to 'stake-hold' by providing 20 per cent of development costs. Unit price estimated in 2000 as US$80 million to US$85 million.

DESIGN FEATURES: High-wing, T-tailed aircraft with rough-field landing gear and much larger cabin/hold floor area and cross-section than C-130/C.160, permitting high payload factors with low-density cargo, vehicles or mixed passenger/cargo loads. Use of propellers felt to be essential for adequate thrust-reverse performance for taxying and short landing; for maximising power response; and for minimising FOD vulnerability. Long-range cruising speed of M0.68 up to 11,280 m (37,000 ft). Tactical mission parameters of 150 m (500 ft) AGL in IMC on predetermined route with civil standard of safety. Airbus has noted that its extensive use of new technology gives twice the volume and payload of the C-130J at the same life-cycle cost. Compared with the C-17, the A400M is said to be less than half the price and to have one third of the life-cycle cost. Minimum service life 30,000 hours, including allowance for low-level flight and short-field performance.

Wing sweep 15° at 25 per cent chord; anhedral 2°; taper ratio 0.334; mean aerodynamic chord 5.690 m.

FLYING CONTROLS: Fly-by-wire, hydraulically powered; manually actuated electrohydrostatic back-up for ailerons, elevator and rudder. Four spoilers and two-section flaps on each wing; tailplane trimmable by screw-jack. No slats. Spoilers used for roll control, lift dumping and as speed brakes.

STRUCTURE: Aluminium alloy, with titanium alloy in highly loaded areas (around windscreen, wing/fuselage joint and landing gear anchorage) and glass fibre or carbon fibre for lightly loaded components (landing gear doors and various fairings). Tailplane has aluminium alloy central structural box and two outer composites box structures; elevator primary structure of carbon fibre. Fin has three-spar main box, trailing-edge shroud and single-piece rudder, all primarily of composites, plus metal/composites removable leading-edge. Rudder of carbon fibre, with aluminium, hinge-connecting ribs. Extensive use of composites in wing for skins, stringers, spars and (carbon fibre) moving surfaces; metal for ribs, engine mountings and fuselage pick-ups. Front spar at 15 per cent chord; composites rear spar at 67 per cent chord. Modern design and manufacturing techniques expected to afford major reductions in maintenance man-hour requirements and increases in aircraft availability/survivability.

Flight deck of the Airbus A400M, created by computer imagery 1999/0044888

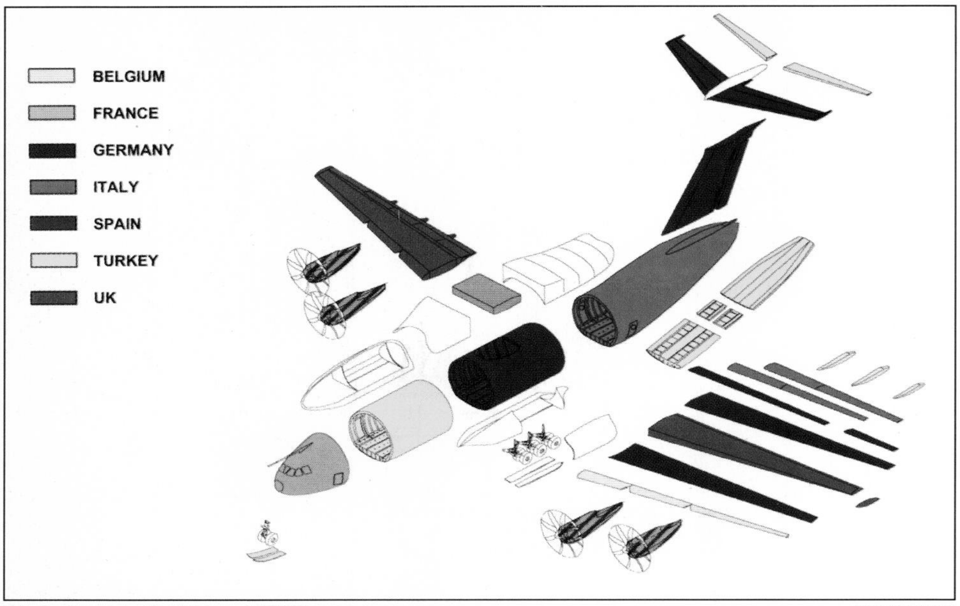

Work shares in the Airbus A400M programme ***2001***/0105703

In 2000, workshares allocated as follows. Aerospatiale Matra: nose/flight deck (with TAI), wing centre box and (with CASA) engine nacelles. Alenia: aft fuselage (with TAI), inboard and outboard flaps and cargo handling system. BAE Systems: overall wing leadership, assembly and equipping; fuel system, ice/rain protection and landing gear. CASA: tailplane (with TAI), engine nacelles (with Aerospatiale Matra) and final assembly. DASA: fuselage team leaders, systems integration, centre fuselage, fin, ailerons, spoilers and wing skins. FLABEL: wing leading-edge and flap tracks. TAI: nosewheel doors and bay; forward centre fuselage; cargo doors and ramp; elevators.

LANDING GEAR: Retractable tricycle type with sufficient 'flotation' for semi-prepared and/or unsurfaced runways. Each main unit has six wheels in tandem pairs, retracting rearwards into fairings on fuselage sides. Each pair of mainwheels has independent, lever-type shock-absorbers. Twin nosewheels retract forwards. Emergency gravity extension of all units. Multidisc carbon brakes on mainwheels can operate differentially to assist steerable nosewheel in ground manoeuvring. Turning radius: landing gear 15 m (50 ft); wingtip 28.6 m (94 ft). Mainwheels can 'kneel' for unloading of large cargoes. Hydraulic strut at rear of each sponson supports and stabilises aircraft during loading and unloading.

POWER PLANT: Initial candidate engines all rated at approximately 6,898 kW (9,250 shp): M138 turboprop offered by Turboprop International SNECMA (33 per cent), MTU (33 per cent), Fiat Avio (22 per cent) and ITP (12 per cent) and based on SNECMA M88-2 core; Rolls-Royce Deutschland offering a turboprop development of the BR715 turbofan, 8,949 kW (12,000 shp) BR700-TP; and Pratt & Whitney Canada proposed a 'Twinpac' version of the existing PW150. The last-mentioned engine was rejected after failing to reach 6,711 kW (9,000 shp) and the required specific fuel consumption. Required engine power, as defined by Airbus, is up to 7,457 kW (10,000 shp).

Eventual choice settled on three-shaft 7,457 to 9,694 kW (10,000 to 13,000 shp) turboprop TP400 developed by Fiat Avio, ITP, MTU, Rolls-Royce, SNECMA and Techspace Aero. Engine couples M88-based core with Rolls-Royce Trent architecture. Final assembly by MTU, which will also supply intermediate- and low-pressure turbines. SNECMA will provide high-pressure compressor and turbine, and the combustion chamber; Rolls-Royce will produce low-pressure compressor and integrate the disparate elements. MTU, SNECMA and Rolls-Royce will each take a 24.8 per cent share. ITP (with 13.6 per cent) will provide casings and dressings; Fiat Avio (8 per cent), propeller gearbox; and Techspace Aero (4 per sent share) accessories. Eight-blade composites curved propellers to be supplied by Dowty Aerospace or Ratier Figeac/Hamilton Sundstrand. Propeller tip speed 229 m (750 ft)/s at T-O; 198 m (650 ft)/s in cruising flight. Composites blades with integral de-icing blankets. Full reversal of pitch permits aircraft to back up 2° slope at MTOW on concrete.

Fuel capacity 64,030 litres (16,915 US gallons; 14,085 Imp gallons) in five tanks (no transfer tank) inside wing box; electric pumps and valves all mounted outside tanks. Detachable in-flight refuelling probe. Provision for inert gas system; wing-mounted refuelling pods; HDU in cargo hold; and additional fuel tanks, totalling 16,250 litres (4,293 US gallons; 3,575 Imp gallons), in fuselage. Pressure refuelling, with gravity back-up.

ACCOMMODATION: Two-man, NVG-compatible flight deck with dual sidestick controllers and additional forward-facing workstation for third 'mission crew member' to assist with tactical and special tasks, when required. View from flight deck exceeds JAR 25 and MIL-STD-850B. Provision for bulletproof flight deck windows, 68 mm (2¾ in) thick, and armour protection around crew's seats. Loadmaster station forward of and overlooking cargo area. Two fixed, screened urinals and fixed hand-basin, starboard, rear. Astrodome for formation surveillance expected. Flight crew rest area with two foldaway bunks.

Two passenger doors forward; two rear. Forward, port, for normal access; forward, starboard, for emergency exit; rear doors for paratroop dropping. Three emergency exits in roof for flight crew and passengers. Cargo door, hinged at aft end, raised hydraulically to hold roof for loading via rear ramp. Closed-circuit TV surveillance of cargo hold, with imagery selectable on flight deck displays.

Cargo floor with 250 tiedown rings stressed to 4,536 kg (10,000 lb) and 60 to 11,340 kg (25,000 lb). Typical loads include Warrior or MRAV armoured transport vehicles; Super Puma or two Tiger helicopters; nine pallets (88 × 108 in military or 88 × 125 in civil); plus 57 troops and second loadmaster on permanent (tip-up) sidewall seats; two 20 ft ISO containers; Patriot SAM system; six Land Rovers, plus trailers; 66 stretchers and 10 medical attendants; or 120 armed troops on sidewall and 62 removable centreline seats. Ramp stressed for 6,000 kg (13,228 lb) loads and has three hydraulically powered toes and 90 tiedowns of 4,536 kg (10,000 lb).

SYSTEMS: FBW FCS derived from Airbus airliners, including sidestick controllers (left hand for captain, right hand for co-pilot, with conventional central power-lever throttle quadrant).

Electrical power provided by four engine-driven generators, each of 75 kVA. Additional power from three-phase generator on APU (90 kVA) in landing gear sponson; three-phase generator on RAT; emergency battery; and external power receptacle. DC power from four 200 A transformer/rectifier units: two feed separated, main DC busbars; one to 'flight essential' busbar and emergency busbar and battery; and one to APU starting system. Two Ni/Cd batteries.

Two hydraulic systems, Blue and Yellow, each operating at 207 bars (3,000 lb/sq in). Blue (driven by Nos. 1 and 4 engines) powers port aileron and elevator, inboard (No. 4) and No. 2 spoiler on each wing, back-up brakes, cargo ramp and ground stabiliser struts. Yellow (Nos. 2 and 3 engines) responsible for starboard aileron and elevator, Nos. 2 and 4 spoilers on each side, landing gear kneeling, brakes and landing gear actuation. Both systems power flaps, rudder and tailplane trim. Each system has 140 litre (37.0 US gallon; 30.8 Imp gallon)/min engine-driven pump and 40 litre (10.6 US gallon; 8.8 Imp gallon)/min alternate current motor pump; power transfer unit between systems for emergency use.

Pneumatic system for air conditioning pressurisation; wing and engine air intake anti-icing; engine starting; and pressurisation of other onboard systems. Two computers control four engine air bleed units. Interior divided into three air conditioning zones; flight deck and two in cargo hold. Cabin pressure altitude 2,440 m (8,000 ft) when flying at 11,280 m (37,000 ft).

AVIONICS: *Comms:* HF, V/UHF, Selcal and optional satcom. Audio management system, cockpit voice recorder and passenger address system. Transponder, ELT and IFF.

Radar: Northrop Grumman AN/APN-241 weather radar with additional ground mapping mode.

Flight: Three inertial platforms with embedded air data systems. VOR, DME, Tacan, multimode receiver (ILS, MLS and GPS), two radar altimeters and GPWS. Optional terrain-reference navigation system.

Instrumentation: Five full-colour MFDs and two HUDs; two multifunction control and display units on centre pedestal, plus one for optional third crew member.

Mission: Secure tactical radios and optional MIDS datalink.

Self-defence: Modular DASS with optional elements including central computer, RWR, MAWS, LWR, chaff/flare dispensers, IR jammers, electronic jammer and towed radar decoy.

EQUIPMENT: Winch at forward end of cargo hold. Optional 5-tonne crane in roof above cargo ramp. Provision for one 1,500 litre (396 US gallon; 330 Imp gallon)/min refuelling pod under each wing and/or 2,250 litre (594 US gallon; 495 Imp gallon)/min HDU in rear of cargo hold.

DIMENSIONS, EXTERNAL:

Wing span	42.40 m (139 ft 1 in)
Wing aspect ratio	8.1
Length overall	42.20 m (138 ft 5½ in)
Height overall	14.73 m (48 ft 4 in)
Wheel track	6.20 m (20 ft 4 in)
Wheelbase	13.60 m (44 ft 7½ in)
Propeller diameter	5.18 m (17 ft 0 in)

DIMENSIONS, INTERNAL:

Hold: Length excl ramp	17.71 m (58 ft 1¼ in)
Width at floor, continuous	4.00 m (13 ft 1½ in)
Height: forward of wing box	3.85 m (12 ft 7½ in)
aft of wing box	4.00 m (13 ft 1½ in)
Floor area: incl ramp	92.4 m² (995 sq ft)
Volume: incl ramp (approx)	356 m³ (12,570 cu ft)
excl ramp (approx)	274 m³ (9,680 cu ft)
Ramp: Length	5.40 m (17ft 8½ in)
Width	4.00 m (13 ft 1½ in)
Cargo door: Length	8.10 m (26 ft 7 in)

AREAS:

Wing area, gross	221.50 m² (2,384.2 sq ft)

WEIGHTS AND LOADINGS (A: logistic operation at max 2.5 *g*, B: logistic at 2.25 *g*, C: tactical operation at 2.5 *g*):

Operating weight, empty*:	66,500 kg (146,605 lb)
Max payload: A	31,500 kg (69,445 lb)
B	37,000 kg (81,570 lb)
C	29,500 kg (65,036 lb)
Max T-O weight: A	126,500 kg (278,885 lb)
B	130,000 kg (286,600 lb)
C	116,500 kg (256,835 lb)
Max landing weight: A, B	114,000 kg (251,325 llb)
C	106,500 kg (234,790 lb)
Max zero-fuel weight: A	98,000 kg (216,055 lb)
B	103,500 kg (228,180 lb)
C	96,000 kg (211,645 lb)

* *including 1,000 kg (2,205 lb) allowance for optional equipment*

PERFORMANCE (estimated):

Max operating speed and Mach No. (VMO/MMO).	M0.72 (300 kt 555 km/h; 345 mph)
Normal cruising Mach No.	0.68
Airdrop speed	130-200 kt (241-370 km/h; 150-230 mph)
Max rate of climb at S/L	1,524 m (5,000 ft)/min
Max certified altitude	11,280 m (37,000 ft)
T-O run at 116,500 kg (256,835 lb)	1,402 m (4,600 ft)
Landing run at 93,500 kg (206,131 lb), max propeller reversal 152 m (500 ft) roll-out	625 m (2,050 ft)

Range with 5% reserves, missed approach, 200 n miles (370 km; 230 miles) diversion and 30 min hold at 445 m (1,500 ft), logistic mission (2.25 *g* load factor):

with 30,000 kg (66,139 lb) payload	2,450 n miles (4,537 km; 2,819 miles)
with 20,000 kg (44,092 lb) payload	3,550 n miles (6,574 km; 4,085 miles)
Ferry range	4,900 n miles (9,074 km; 5,638 miles)

UPDATED

AIRBUS MULTI ROLE TANKER TRANSPORT (MRTT)

TYPE: Tanker-transport.

PROGRAMME: Belgian, Canadian, French, German and Thai air forces already operate A310s variously fitted for VIP, troop and/or freight transport. Airbus has delegated development and marketing of flight refuelling versions to its major partners, using either pre-owned or new aircraft; A340 may also be adapted as tanker later. Demonstrator MRTT produced by conversion of former airline A310-324 N816PA; undertook compatibility trials with RAF aircraft, July 1995. Marketing efforts centre on the A310 version, which is expected to be offered for the RAF's FSTA future tanker aircraft requirement. This will begin replacement of VC10s in 2005, although more than one type of tanker is expected to be obtained, probably by lease rather than purchase. Exclusive marketing rights for MRTT assigned to Raytheon (which will also become Design Authority), June 1999.

CURRENT VERSIONS: Interim MRT (MultiRole Transport) version of A310, without tanker capability, converted by Elbe Flugzeugwerke, Dresden and Lufthansa Technik, Hamburg, for Luftwaffe. Structural strengthening and 3.50 m × 2.50 m (11 ft 5¾ in × 8 ft 2½ in) cargo door added in port forward fuselage. Capacity of up to 214 passengers; or 36 tonnes of cargo/passengers; or 56 stretchers and six intensive care patients in the casevac role. First redelivered from Dresden in June 1999; two of seven Luftwaffe A310s redelivered as MRTs by September 1999. Similarly modified A300 freighters offered in partial satisfaction of RAF's Short Term Strategic Airlifter requirement, before its abandonment in

August 1999. In service, Luftwaffe aircraft proved partially incompatible with military cargo handling equipment and upper deck therefore rarely used for freight; full conversion to tanker configuration now considered unlikely. Tanker-capable versions include: **MRTT 300** powered by GE CF6-50C2; **MRTT 300-600R** powered by GE CF6-80C2A5; **MRTT 310** powered by GE CF6-80C2A2. Customers will also be offered versions with P&W PW4000 turbofan engines.

DESIGN FEATURES: Conversions offer greater refuelling and transport capability than earlier airliners in combination with modern aircraft with better lifetime costs and longer life expectancy. Possible roles include tanker with underwing HDUs and fuselage-mounted boom and/or hose transfer systems and carrying in excess of 77,111 kg (170,000 lb) of fuel; and cargo and personnel transports which can be combined with refuelling, medevac, airborne command post and reconnaissance/airborne warning.

Airbus conversions offer payloads from 35,000 to 50,000 kg (77,161 to 110,231 lb), full payload transatlantic range, long on-station time, combined boom and hosereel transfer capability, standard Airbus forward port-side freight door (projected height 2.57 m; 8 ft 5¼ in, width 3.58 m; 11 ft 9 in); quick-change main deck layout, probe or receptacle fuel receiver capability; commonality with existing airliners and same worldwide support resources, predictable spares requirements and longer remaining airframe life.

About 100 civil operators on all five continents are flying A300/310, and first-generation Airbus airliners are now becoming available on market; military rendezvous and self-protection systems can be fitted; main deck can be converted with palletised seating for up to 270 passengers in under 24 hours; up to 28,000 kg (61,729 lb) of additional fuel can be carried in tanks in underfloor cargo compartments.

DIMENSIONS, EXTERNAL (300: MRTT 300, 600: MRTT 300-600R, 310: MRTT 310):
Length overall, including probe:
310   47.36 m (155 ft 4½ in)

DIMENSIONS, INTERNAL:
Usable cabin length: 300   36.96 m (121 ft 3 in)
  600   40.70 m (133 ft 6 in)
  310   43.90 m (144 ft 0 in)
Cabin height: all   2.28 m (7 ft 5¾ in)
Max cabin width: all   5.29 m (17 ft 4¼ in)
Underfloor freight hold volume:
  300   107.0 m³ (3,778 cu ft)
  600   138.0 m³ (4,873 cu ft)
  310   80.0 m³ (2,825 cu ft)

WEIGHTS AND LOADINGS:
Operating weight empty: 300   88,410 kg (194,910 lb)
  600   89,650 kg (197,645 lb)
  310   80,830 kg (178,200 lb)
Max non-fuel payload: 310   37,000 kg (81,571 lb)
Max normal fuel capacity: 300   48,350 kg (106,595 lb)
  600   53,290 kg (117,485 lb)
  310   47,940 kg (105,690 lb)
Additional fuel: all   28,240 kg (62,258 lb)
Max T-O weight: 300   165,000 kg (363,750 lb)
  600, normal   170,500 kg (375,875 lb)
  600, optional   171,700 kg (378,525 lb)
  310, normal   157,000 kg (346,125 lb)
  310, optional   164,000 kg (361,550 lb)
Max ramp weight: 310   164,900 kg (363,550 lb)
Max landing weight: 310   124,000 kg (273,375 lb)
Max zero-fuel weight: 310   114,000 kg (251,325 lb)

PERFORMANCE:
Refuelling speed:
  all, boom   240-320 kt (444-592 km/h; 276-368 mph)

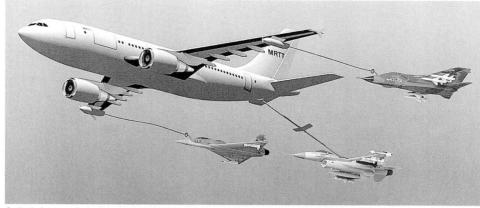
Artist's impression of an Airbus MRTT refuelling a diverse group of NATO aircraft   2000/0062383

One of two Luftwaffe A310-304 MRTs converted to Interim tanker-transport standard, lacking refuelling equipment (Paul Jackson/Jane's)   2001/0089511

  all, hose and drogue   220-320 kt (407-592 km/h; 253-368 mph)
Max range, standard fuel:
  300   3,700 n miles (6,852 km; 4,258 miles)
  600   4,150 n miles (7,685 km; 4,775 miles)
  310   4,800 n miles (8,889 km; 5,523 miles)
Max range, using transferable fuel:
  300, 600   5,400 n miles (10,000 km; 6,214 miles)
  310   7,200 n miles (13,334 km; 8,285 miles)

UPDATED

AIRBUS A310 AEW&C

TYPE: Airborne early warning and control system.
PROGRAMME: A310 selected by Raytheon Systems as a platform for its bid for the Australian AEW&C requirement (Project Wedgetail), competing against Boeing 737 and Lockheed Martin C-130J. Incorporates Elta Phalcon 360° phased-array radar in a fixed dorsal dome; formally announced in January 1997. Australia selected Boeing 737 in July 1999, but A310 AEW&C also offered to Turkey and South Korea. Operating weight empty 80,235 kg (176,890 lb); max T-O weight 164,000 kg (361,550 lb); patrol for over 10 hours on station at 300 n miles (555 km; 345 miles) from base at M0.72 at 8,990 m (29,500 ft).

UPDATED

AIRBUS A321 AGS

TYPE: Airborne ground surveillance system.
PROGRAMME: Northrop Grumman recommended the A321 as its preferred Joint STARS platform (see E-8 entry in US section) to compete for NATO's AGS (airborne ground surveillance) requirement. Considered in 1996 and rejected on cost grounds, the A321 was reconsidered in the light of NATO's unwillingness to accept a Boeing 707-based solution. Plans call for selection in 2002 and IOC, with six aircraft, by 2007. Northrop Grumman's A321 AGS features the US Army's AN/APY-X RTIP (Radar Technology Insertion Program) upgrade developed for the E-8, with a shorter underfuselage electronically scanned antenna array, giving MTI and SAR modes. It would have between 10 and 12 operator stations.

Member countries and participants in industrial working groups for AGS comprise Belgium: Barco; Canada: Computing Devices; Denmark: Terma; France: Aerospatiale Matra and Airbus Industrie; Germany: DaimlerChrysler, Dornier and ESG; Greece: HAI and Intracom; Italy: Alenia/Fiat and IAMCO; Luxembourg: Cargolux and Euro-Composites; Netherlands: Fokker/Stork; Norway: Kongsberg; Portugal: OGMA and TAP Air Portugal; Spain: CASA and Indra; Turkey: Esdas and TAI; and USA: Northrop Grumman.

NEW ENTRY

Airbus A310 AEW&C proposal with non-rotating Elta radar above rear fuselage; configuration may also include a large dorsal air scoop (Paul Jackson/Jane's)   2000/0062385

Cutaway model of Airbus A321 AGS (Paul Jackson/Jane's)   2001/0089512

AIRTECH

AIRCRAFT TECHNOLOGY INDUSTRIES

Web: http://www.casa.es/eng_31125_cn235-vistas
PARTICIPATING COMPANIES:
CASA: see under Spain
Dirgantara: see under Indonesia

Airtech was formed by CASA (now part of EADS) and IPTN (now known as Dirgantara) to develop the CN-235 twin-turboprop transport; design and production is shared 50-50. TAI of Turkey (which see) is assembling and manufacturing some parts for CN-235s for home use and export. CASA unilaterally developed the stretched C-295, described in Spanish section, which it is manufacturing to meet a Spanish military order. That aircraft has been a competitor to the CN-235 in some military competitions.

UPDATED

AIRTECH (CASA/DIRGANTARA) CN-235

Spanish Air Force designations: T.19A and T.19B
TYPE: Twin-turboprop transport.
PROGRAMME: Preliminary design began January 1980, prototype construction May 1981; one prototype completed in each country, with simultaneous roll-outs 10 September 1983; first flights 11 November 1983 (by CASA's ECT-100) and 30 December 1983 (IPTN's PK-

204　INTERNATIONAL: AIRCRAFT—AIRTECH

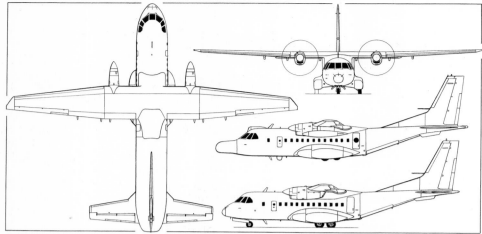

Airtech CN-235 twin-turboprop multipurpose transport, with additional side view (centre) of a representative CN-235 MPA *(Dennis Punnett/Jane's)*

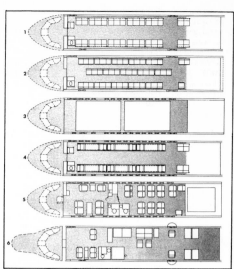

CN-235 M configured for (1) paratroop transport; (2) troop transport; (3) cargo transport on standard pallets; (4) medical evacuation with 18 stretchers and five attendants; (5) VIP communications; and (6) maritime patrol

Flight deck of a CN-235 Series 200

XNC); Spanish and Indonesian certification 20 June 1986; first flight of production aircraft 19 August 1986; FAA type approval (FAR Pts 25 and 121) 3 December 1986; deliveries began 15 December 1986 from IPTN (now Dirgantara) line and 4 February 1987 from CASA; entered service (with Merpati Nusantara Airlines) 1 March 1988; JAR 25 type approval October 1993.

Licence agreement with TAI (see Turkish section) announced January 1990, initially to assemble and later to manufacture locally 50 of 52 ordered; first flight of Turkish-assembled aircraft 24 September 1992; first delivery 13 November 1992; final air force delivery 10 August 1998, but TAI currently producing follow-on batch of nine maritime patrol variants.

In 1995, CASA unilaterally launched development of a stretched CN-235 to accommodate a fourth cargo pallet in conjunction with a strengthened wing, enlarged rudder and P&WC PW127 turboprops. The resultant aircraft, designated C-295, is described in the Spanish section.

CURRENT VERSIONS: **CN-235 Series 10:** Initial production version (15 built by each company), with CT7-7A engines; described in 1986-87 and earlier *Jane's*.

CN-235 Series 100/110: Generally as Series 10, but CT7-9C engines in new composites nacelles; replaced Series 10 in 1988 from 31st production aircraft. Series 100 is Spanish-built and, following JAA certification, was certified by FAA in February 1992. Series 110 is Indonesian-built, with improved electrical, warning and environmental systems to comply with JAR 25;

certification of this version achieved in Europe (JAA), July 1995.

Detailed description applies to the above version except where indicated.

CN-235 Series 200/220: Structural reinforcements to cater for higher operating weights, aerodynamic improvements to wing leading-edges and rudder, reduced field length requirements and much-increased range with maximum payload; Series 200 is Spanish-built and was certified by FAA March 1992. Series 220 is Indonesian-built, with improvements similar to Srs 110; prototype, flown early 1996, is converted from a company development aircraft (PK-XNV, the 20th production aircraft from the Indonesian line); orders include six for Malaysian Air Force, all of which completed to Srs 220 standard (including three in maritime patrol configuration) by early 1998 (34th to 39th Indonesian-built). Revised leading-edge shape led to requirement to requalify pneumatic de-icer boots, delaying initial deliveries.

CN-235 Series 300/330: IPTN originally offered Series 330 **Phoenix** (with new Honeywell avionics, ARL-2002 EW system and 16,800 kg; 37,037 lb MTOW) to Royal Australian Air Force to meet Project Air 5190 tactical airlift requirement, but was forced by financial constraints to withdraw in 1998. Separately, CASA offered its own Series 300 to meet the same specification.

CASA Series 300 under certification in 2000 with an open-systems avionics architecture, based on MIL-STD-1553B and ARINC 429 digital databusses. Full NVG-compatible cockpit; four-dimensional navigation system with avionics suite, including Sextant Avionique Topdeck colour weather radar, radios, solid-state flight data and cockpit voice recorders, enhanced TCAS, enhanced GPWS and four 152 × 203 mm (6 × 8 in) LCDs; twin HUDs and Totem 3000 ring laser gyro INS optional. Other features include in-flight refuelling capability, improved pressurisation (2,440 m; 8,000 ft cabin environment at 7,620 m; 25,000 ft) and provision for optional twin nosewheel installation to provide better soft-field taxying capability.

CN-235 AEW: Proposals were revealed in December 1995 for fitment of an Ericsson Erieye electronically scanned phased-array radar above the fuselage of a CN-235. Initial interest was from the Indonesian Air Force, but primarily in the ocean surveillance role; retrofit of three existing aircraft was considered. Radar, three surveillance operators' positions and associated equipment increase aircraft weight by approximately 2,000 kg (4,409 lb).

CN-235 M: Other military transport versions.

CN-235 MP Persuader and CN-235 MPA: Maritime patrol versions; described separately.

CN-235 QC: Quick-change cargo/passenger version; certified by Spanish DGAC May 1992.

CN-245: Indonesian stretched version; not built. See 1998-99 *Jane's*.

C-295: Refer to CASA in Spanish section.

N2XXM: See CN-245 entry in 1998-99 *Jane's*. Project abandoned.

CUSTOMERS: See table; by October 1999, total orders reported by CASA as 233 (to 34 operators in 24 countries); this reflected suspension of April 1994 commitment by Merpati to buy 16 further Series 220s.

One (s/n 66049) acquired (presumably second-hand) by USAF in 1998. Turkey signed a lease agreement on 16 April 1999 to allow a one-year renewable lease of two Turkish Air Force CN-235s to Jordan. Switzerland briefly leased a Spanish Air Force CN-235 in 1999 to support peacekeeping operations in the former Yugoslavia. Three Merpati aircraft leased to Air Venezuela from May 1999; three leased to Asian Spirit Airlines, Philippines, from March 2000, including two on lease-purchase. One ordered on 1 July 1999 by Macedonian Air Force is

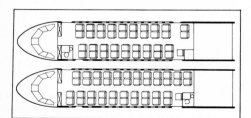

CN-235 in typical configurations for 38 (top) and 44 passengers

Malaysian CN-235, delivered in 1999 with optional enlarged nose *(Paul Jackson/Jane's)*

Jane's All the World's Aircraft 2001-2002

believed to be ex-Turkish. Intention to buy a further two announced by Papua New Guinea in mid-1998. National Jet Systems of Australia interested in two coastal patrol variants; signed MoU for possible acquisition of two, plus five options, February 1998. One CN-235-300 leased by Austrian Ministry of Defence for six months from April 2000. CN-235 is contender in Taiwanese requirement for 18 to 22 light transports, and for the US Army Airborne Common Sensor platform requirement.

COSTS: US$16.9 million (1995) programme unit cost, Malaysia.

DESIGN FEATURES: Optimised for short-haul operations, enabling it to fly four 860 n mile (1,593 km; 990 mile) stage lengths (with reserves) before refuelling and to operate from paved runways or unprepared strips; high-mounted wing; pressurised fuselage (including baggage compartment) of flattened circular cross-section, with upswept rear end incorporating cargo ramp/door; sweptback fin (with dorsal fin) and rudder; low-set non-swept fixed incidence tailplane and elevators; two small ventral fins; vortex generators on rudder and elevator leading-edges; optional extended nose radome.

NACA 65_3-218 aerofoil with no-dihedral/constant chord centre-section; tapered outer panels have 3° dihedral and 3° 51′ 36″ sweepback at quarter-chord.

FLYING CONTROLS: Conventional and manual. Ailerons, elevators and rudder statically and dynamically balanced (duplicated actuation for ailerons); mechanical servo tab and electric trim tab in each aileron, rudder and starboard elevator, trim tab only in port elevator; single-slotted inboard and outboard trailing-edge flaps (each pair interchangeable port/starboard), actuated hydraulically by Dowty irreversible jacks.

STRUCTURE: Conventional semi-monocoque, mainly of aluminium alloys with chemically milled skins; composites (mainly glass fibre or glass fibre/Nomex honeycomb sandwich, with some carbon fibre and Kevlar) for leading/trailing-edges of wing/tail moving surfaces, wing/fuselage and main landing gear fairings, wing/fin/tailplane tips, engine nacelles, ventral fins and nose radome. Propeller blades are of glass fibre, with metal spar and urethane foam core.

CASA builds wing centre-section, inboard flaps, forward and centre fuselage, engine nacelles; Dirgantara builds outer wings, outboard flaps, ailerons, rear fuselage and tail unit; both manufacturers use numerical control machinery extensively. Final assembly line in each country. Part of tail unit built by ENAER Chile under subcontract from CASA. TAI (Turkey) initially assembled under licence before progressing gradually to local manufacture of balance of 50 aircraft for Turkish Air Force.

LANDING GEAR: Messier-Bugatti retractable tricycle type with levered suspension, suitable for operation from semi-prepared runways. Electrically controlled hydraulic extension/retraction, with mechanical back-up for emergency extension. Oleo-pneumatic shock-absorber in each unit. Each main unit comprises two wheels in tandem, retracting rearward into fairing on side of fuselage. Mainwheels semi-exposed when retracted. Single steerable nosewheel (±48°) retracts forward into unpressurised bay under flight deck. Dunlop 28×9.00-12 (12 ply) tubeless mainwheel tyres standard, pressure 5.17 bars (75 lb/sq in) on civil version, 5.58 bars (81 lb/sq in) on military version; low-pressure mainwheel tyres optional, size 11.00-12 (10 ply), pressure 3.45 bars (50 lb/sq in). Dunlop 24×7.7 (10/12 ply) tubeless nosewheel tyre, pressure 5.65 bars (82 lb/sq in) on civil version, 6.07 bars (88 lb/sq in) on military version; optional 8.50×10 (12 ply). Dunlop hydraulic differential disc brakes; Dunlop anti-skid units on main gear. Chilean Army aircraft used in Antarctic have wheel/ski gear. Minimum ground turning radius 9.50 m (31 ft 2 in) about nosewheel, 18.98 m (62 ft 3¼ in) about wingtip.

POWER PLANT: Two General Electric CT7-9C3 turboprops, each flat rated at 1,305 kW (1,750 shp) (S/L, to 41°C) for take-off and 1,394.5 kW (1,870 shp) up to 31°C with automatic power reserve. Hamilton Sundstrand 14RF-21 (14RF-37 in Srs 300) four-blade constant-speed propellers, with full feathering and reverse-pitch capability. Fuel in two 1,042 litre (275 US gallon; 229 Imp gallon) integral main tanks in wing centre-section and two 1,592 litre (421 US gallon; 350 Imp gallon) integral outer-wing auxiliary tanks; total fuel capacity 5,264 litres (1,392 US gallons; 1,158 Imp gallons), of which 5,128 litres (1,355 US gallons; 1,128 Imp gallons) are usable. Single pressure refuelling point in starboard main landing gear fairing; gravity filling point in top of each tank. Propeller braking permits No. 2 engine to be used as on-ground APU. Oil capacity 14 litres (3.7 US gallons; 3.1 Imp gallons).

ACCOMMODATION: Crew of two on flight deck, plus cabin attendant (civil version) or third crew member (military version). Accommodation in commuter version for up to 44 passengers in four-abreast seating, at 76 cm (30 in) pitch, with 22 seats each side of central aisle. Lavatory, galley and overhead luggage bins standard. Pressurised baggage compartment at rear of cabin, aft of movable bulkhead; additional stowage in rear ramp area and in overhead lockers. Can also be equipped as mixed passenger/cargo combi (for example, 19 passengers and two LD3 containers), or for all-cargo operation, with roller loading system, carrying four standard LD3 containers, five LD2s, or two 2.24 × 3.18 m (88 × 125 in) and one 2.24 × 2.03 m (88 × 80 in) pallets; or for military duties, carrying up to 57 fully equipped troops or 46 paratroops. Other options include layouts for aeromedical (24 stretchers and four medical attendants), electronic warfare, geophysical survey or aerial photographic duties.

Main passenger door, outward- and forward-opening with integral stairs, aft of wing on port side, serving also as a Type I emergency exit. Type III emergency exit facing this door on starboard side. Crew/service downward-opening door (forward, starboard) has built-in stairs, and serves also as a Type I emergency exit, or as passenger door in combi version; second Type III exit opposite this door on port side. Wide ventral door/cargo ramp in underside of upswept rear fuselage, for loading of bulky cargo. Accommodation fully air conditioned and pressurised.

CN-235 PRODUCTION
(at 1 June 2000)

Customer	Qty	First order	First aircraft	First delivery	Delivered	Mfr
Civil version:						
Austral (Argentina)	2[15]	19 Dec 1989	LV-VHM	1993	2	CASA
Binter Canarias (Spain)	4[1, 16]	10 Jun 1988	EC-EMO	22 Dec 1988	4	CASA
Binter Mediterraneo (Spain)	4[2]	19 Dec 1989	EC-FAD	4 Sep 1990	3	CASA
Korean Air	1	1997	-	1998	0	CASA
Mandala Airlines (Indonesia)	3	-	-	-	0	Dirgantara
Merpati Nusantara (Indonesia)	15[1]	-	PK-MNA	15 Dec 1986	15	Dirgantara
Military version:						
Abu Dhabi Air Force	7		810	31 Aug 1993	7	Dirgantara
Botswana Defence Force	2[1]	10 Jun 1986	OG-1	21 Dec 1987	2	CASA
Brunei Air Wing	3[3]		-	-	0	Dirgantara
	1		ATU-501	1997	1	Dirgantara
Chilean Army	4	12 Feb 1989	E-216	31 Aug 1989	4	CASA
Colombian Air Force	3	Jul 1997	-	-	0	CASA
Croatian Air Force	5	Oct 1996	-	-	0	TAI
Ecuadorean Army	1	6 Jun 1989	AEE-502	6 Jun 1989	1	CASA
Ecuadorean Navy	1	27 Jul 1988	ANE-204	13 Jun 1989	1	CASA
French Air Force	15[8]	11 Apr 1990	043	28 Feb 1991	15	CASA
Gabon Air Forces	1	26 Feb 1990	TR-KJE	19 Mar 1991	1	CASA
Indonesian armed forces	24[11]		A-2301	12 Jan 1993	7	Dirgantara
Irish Air Corps	1[13]	3 Apr 1991	250	10 Apr 1991	1	CASA
	2[3]	3 Apr 1991	252	8 Dec 1994	2	CASA
Malaysian Air Force	6[7, 9]		M44-01	26 Aug 1999	6	Dirgantara
Moroccan Air Force	7[10]	19 Sep 1989	CNA-MA	27 Sep 1990	7	CASA
Oman Police	2	15 Feb 1992	A40-CU	14 Jan 1993	2	CASA
Panama National Guard	1[1, 12]	19 Mar 1987	SAN-265	13 Sep 1988	1	CASA
Papua New Guinea Defence Force	2	26 Oct 1991	P2-0501	15 Nov 1991	2	CASA
Saudi Air Force	4[1]	5 Feb 1984	118	9 Feb 1987	4	CASA
South African Air Force (ex Bophuthatswana)	1[1]	29 May 1990	8026	6 Jan 1991	1	CASA
South Korean Air Force	12	19 Aug 1992	078	13 Nov 1993	12	CASA
	8[9]	21 Oct 1997	-	2001	0	Dirgantara
Spanish Air Force	2[4]	16 Nov 1988	T.19-01	7 Dec 1988	2	CASA
	18	28 Dec 1990	T.19-03	1 Feb 1991	18	CASA
	4[3]	-	-	-	0	CASA
Thai Ministry of Agriculture and Co-operatives	2[9]	Oct 1996	2221	April 1999	2	Dirgantara
Thai Police	0[2, 14]	Apr 1995	28053	Feb 1996	0	CASA
Turkish Air Force	52[6]	11 Dec 1990	51	25 Jan 1992	52	TAI/CASA
Turkish Navy	6[3]	23 Sep 1998	-	-	0	TAI/CASA
Turkish Coastguard	3[3]	23 Sep 1998	-	-	0	TAI/CASA
UAE Air Force	4[3]	March 1998	-	-	0	Dirgantara
Subtotals	**233**				**175**	
Demo/trials	3[5]		EC-016		3	CASA
	5		PK-XNC		5	Dirgantara
Totals	**241**				**180**	

[1] Series 10
[2] Series 200
[3] Maritime patrol
[4] VIP version
[5] Includes one -100QC; plus one -200QC sold in 1996 to East Texas Aircraft Services Corporation, then Turbo Flight Aviation, March 1998
[6] 50 built in Turkey by TAI
[7] Option on further 12
[8] Including option on seven taken up in February 1996; first six as Srs 100, but upgraded to Srs 200 from 1999
[9] Series 220
[10] Includes one VIP version
[11] Includes six maritime patrol of which three on firm order
[12] To Flight International (USA) 1995
[13] Withdrawn at end of lease, 1995
[14] Former demonstrator
[15] Converted from -100 to Series 200
[16] Withdrawn 1998; two to Luftmeister, South Africa

Note: All are Series 100/110 unless indicated otherwise. Croatian order not counted by CASA, which reports 189 military orders by mid-2000 and implies reduction in Indonesian requirement.

'Rain-making' Airtech CN-235 delivered in 1999 to Thailand's agriculture ministry   2000/0081735

INTERNATIONAL: AIRCRAFT—AIRTECH

SYSTEMS: Hamilton Sundstrand air conditioning system, using engine compressor bleed air. Honeywell electropneumatic pressurisation system (maximum differential 0.25 bar; 3.6 lb/sq in) giving cabin environment of 2,440 m (8,000 ft) up to operating altitude of 5,480 m (18,000 ft). Hydraulic system, operating at nominal pressure of 207 bars (3,000 lb/sq in), comprises two engine-driven, variable displacement axial electric pumps, a self-pressurising standby mechanical pump, and a modular unit incorporating connectors, filters and valves; system is employed for actuation of wing flaps, landing gear extension/retraction, wheel brakes, emergency and parking brakes, nosewheel steering, cargo ramp and door, and propeller braking. Accumulator for back-up braking system.

28 V DC primary electrical system powered by two 400 A Auxilec engine-driven starter/generators, with two 24 V 37 Ah Ni/Cd batteries for engine starting and 30 minutes (minimum) emergency power for essential services. Constant frequency single-phase AC power (115/26 V) provided at 400 Hz by three 600 VA static inverters (two for normal operation plus one standby); two three-phase engine-driven alternators for 115/200 V variable frequency AC power. Fixed oxygen installation for crew of three (single cylinder at 124 bars; 1,800 lb/sq in pressure); three portable units and individual masks for passengers.

Pneumatic boot anti-icing of wing (outboard of engine nacelles), fin and tailplane leading-edges. Electric anti-icing of propellers, engine air intakes, flight deck windscreen, pitot tubes and angle of attack indicators. No APU: starboard engine, with propeller braking, can be used to fulfil this function. Engine fire detection and extinguishing system.

AVIONICS (civil): *Comms:* Two Rockwell Collins VHF-22B com radios, one Avtech DADS crew interphone, Rockwell Collins TDR-90 ATC transponder. Fairchild A-100A cockpit voice recorder, Avtech PACIS PA system. Dorne & Margolin ELT 8-1 emergency transmitter. Optional second TDR-90; optional HF-230 radio.
Radar: Rockwell Collins WXR-300 weather radar.
Flight: Two VIR-32 VOR/ILS/marker beacon receivers; DME-42; ADF-60A; two 332D-11T vertical gyros; two MCS-65 directional gyros; two ADI-85A; two HSI-85; two RMI-36; APS-65 autopilot/flight director; ALT-55B radio altimeter; two 345A-7 rate of turn sensors (all by Rockwell Collins); SFENA H-301 APM standby attitude director indicator; Hamilton Sundstrand Mk II GPWS; and Fairchild/Teledyne flight data recorder. Options include second DME-42 and ADF-60A, Rockwell Collins RNS-325 radar nav, Litton LTN-72R inertial nav or Global GNS-500A Omega navigation system.
Instrumentation: Rockwell Collins EFIS-85B five-tube CRT system standard.

AVIONICS (military) (Indonesian aircraft): *Comms:* Rockwell Collins AN/ARC-182 VHF/UHF; Rockwell Collins HF 9000 HF; IFF.
Flight: Rockwell Collins VIR-32 VHF nav; Litton LTN92 GPS-aided INS; Rockwell Collins DF-206A ADF; Rockwell Collins AN/APS-65F autopilot; GPWS.
Instrumentation: Rockwell Collins EFIS-85B(14) EFIS (four or five screens). IPTN developing cockpit lighting system compatible with night vision goggles.

AVIONICS (military): *Series 300:* Sextant Topdeck suite (see Current Versions) as core system.
Flight: Twin ADU 3000 air data units, GPSs and AHRSs; radar altimeter; TCAS; GPWS; optional Totem 3000 LINS, Cat. II landing capability, MLS and satcom.
Instrumentation: Four 152 × 203 mm (6 × 8 in) LCDs; optional HUDs. Optional electro-optical sensors display imagery on LCDs. NVG compatibility.
Mission: Four-dimensional navigation FMS calculates high-altitude and computed air release points for load-dropping.

EQUIPMENT: Navigation lights, anti-collision strobe lights, 600 W landing light in front end of each main landing gear fairing, taxying lights, ice inspection lights, emergency door lights, flight deck and flight deck emergency lights, cabin and baggage compartment lights, individual passenger reading lights, and instrument panel white lighting, all standard. Hand-type fire extinguishers on flight deck (one) and in passenger cabin (two); smoke detector in baggage compartment.

ARMAMENT (military version): Three attachment points under each wing. Weapons can include Harpoon anti-ship missiles; Indonesian MPA version (which see) can be fitted with two Mk 46 torpedoes or AM 39 Exocet anti-shipping missiles.

DIMENSIONS, EXTERNAL:
Wing span	25.81 m (84 ft 8 in)
Wing chord: at root	3.00 m (9 ft 10 in)
at tip	1.20 m (3 ft 11¼ in)
Wing aspect ratio	10.2
Length overall, standard nose	21.40 m (70 ft 2½ in)
Fuselage: Max width	2.90 m (9 ft 6 in)
Max depth	2.615 m (8 ft 7 in)
Height overall	8.18 m (26 ft 10 in)
Tailplane span	11.00 m (36 ft 1 in)
Wheel track (c/l of mainwheels)	3.90 m (12 ft 9½ in)
Wheelbase	6.92 m (22 ft 8½ in)
Propeller diameter	3.35 m (11 ft 0 in)
Propeller ground clearance	1.66 m (5 ft 5¼ in)
Distance between propeller centres	7.00 m (22 ft 11½ in)
Passenger door (port, rear) and service door (stbd, fwd):	
Height	1.70 m (5 ft 7 in)
Width	0.73 m (2 ft 4¾ in)
Height to sill	1.22 m (4 ft 0 in)
Paratroop doors (port and stbd, rear, each):	
Height	1.75 m (5 ft 9 in)
Width	0.90 m (2 ft 11½ in)
Height to sill	1.22 m (4 ft 0 in)
Ventral upper door (rear): Length	2.365 m (7 ft 9 in)
Width	2.35 m (7 ft 8½ in)
Height to sill	1.22 m (4 ft 0 in)
Ventral ramp/door (rear): Length	3.04 m (9 ft 11¾ in)
Width	2.35 m (7 ft 8½ in)
Height to sill	1.22 m (4 ft 0 in)
Type III emergency exits (port, fwd, and stbd, rear):	
Height	0.91 m (3 ft 0 in)
Width	0.51 m (1 ft 8 in)

DIMENSIONS, INTERNAL (C: CASA-built, N: Dirgantara-built):
Cabin, excl flight deck: Length: C	9.65 m (31 ft 8 in)
N	9.98 m (32 ft 9 in)
Max width	2.70 m (8 ft 10½ in)
Width at floor	2.365 m (7 ft 9 in)
Max height: C	1.88 m (6 ft 2 in)
N	1.90 m (6 ft 2¾ in)
Floor area	22.8 m² (246 sq ft)
Volume: C	43.2 m³ (1,527 cu ft)
N	41.8 m³ (1,475 cu ft)
Baggage compartment volume:	
ramp	5.3 m³ (187 cu ft)
overhead bins	1.7 m³ (59 cu ft)

AREAS:
Wings, gross	59.10 m² (636.1 sq ft)
Ailerons (total, incl tabs)	3.07 m² (33.06 sq ft)
Trailing-edge flaps (total)	10.87 m² (117.00 sq ft)
Fin, incl dorsal fin	11.38 m² (122.49 sq ft)
Rudder, incl tabs	3.32 m² (35.74 sq ft)
Tailplane	25.40 m² (273.40 sq ft)
Elevators (total, incl tabs)	4.25 m² (45.75 sq ft)

WEIGHTS AND LOADINGS:
Operating weight empty:	
passenger version	9,800 kg (21,605 lb)
cargo and military versions	8,800 kg (19,400 lb)
Max fuel	4,230 kg (9,325 lb)
Max payload: passenger version:	
Srs 100	4,000 kg (8,818 lb)
Srs 200	4,300 kg (9,480 lb)
cargo and military versions	6,000 kg (13,227 lb)
Max weapon load (CN-235 M)	3,500 kg (7,716 lb)
Max T-O weight: civil: Srs 100	15,100 kg (33,289 lb)
Srs 200	15,800 kg (34,833 lb)
military: Srs 200	16,000 kg (35,273 lb)
Max ramp weight: civil Srs 200	15,850 kg (34,943 lb)
military: Srs 200	16,500 kg (36,376 lb)
Srs 300	16,500 kg (36,376 lb)
Max landing weight: civil:	
Srs 100	16,500 kg (36,376 lb)
Srs 200	15,400 kg (33,951 lb)
military: Srs 200	15,600 kg (34,392 lb)
Max zero-fuel weight: civil	14,100 kg (31,085 lb)
military	15,400 kg (33,951 lb)
Cabin floor loading: cargo and military versions	
	1,504 kg/m² (308.0 lb/sq ft)
Max wing loading: civil:	
Srs 100	255.5 kg/m² (52.36 lb/sq ft)
Srs 200	267.3 kg/m² (54.75 lb/sq ft)
military	270.7 kg/m² (55.44 lb/sq ft)
Max power loading (without APR): civil:	
Srs 100	5.78 kg/kW (9.51 lb/shp)
Srs 200	6.05 kg/kW (9.95 lb/shp)
military	6.13 kg/kW (10.07 lb/shp)

PERFORMANCE (civil versions):
Max operating speed at 550 m (1,800 ft)	
	240 kt (445 km/h; 276 mph) IAS
Stalling speed at S/L:	
flaps up	100 kt (186 km/h; 116 mph) IAS
flaps down	84 kt (156 km/h; 97 mph) IAS
Max rate of climb at S/L	465 m (1,527 ft)/min
Rate of climb at S/L, OEI	128 m (420 ft)/min
Service ceiling	7,620 m (25,000 ft)
Service ceiling, OEI	3,840 m (12,600 ft)
T-O run: Srs 100	1,217 m (3,995 ft)
Srs 200	1,051 m (3,450 ft)
T-O balanced field length at S/L:	
Srs 100	1,406 m (4,615 ft)
Srs 200	1,275 m (4,185 ft)
Srs 200 at AUW of 14,646 kg (32,290 lb)	
	1,139 m (3,740 ft)
Landing from 15 m (50 ft) at S/L:	
Srs 100	1,276 m (4,190 ft)
Srs 200	670 m (2,200 ft)
Range at 5,485 m (18,000 ft), reserves for 87 n mile (161 km; 100 mile) diversion and 45 min hold:	
Srs 100 with max payload	
	450 n miles (834 km; 518 miles)
Srs 200 with max payload	
	957 n miles (1,773 km; 1,102 miles)
Srs 100 with max fuel	
	2,110 n miles (3,908 km; 2,428 miles)
Srs 200 with max fuel	
	1,974 n miles (3,656 km; 2,272 miles)

PERFORMANCE (CN-235 M):
As for civil versions except:
Max cruising speed at 5,480 m (18,000 ft)	
	228 kt (422 km/h; 262 mph)
Max rate of climb at S/L	579 m (1,900 ft)/min
Rate of climb at S/L, OEI	156 m (512 ft)/min
Service ceiling	6,860 m (22,500 ft)
Service ceiling, OEI	4,040 m (13,260 ft)
T-O to 15 m (50 ft) (CFL): Srs 100	1,290 m (4,235 ft)
Srs 200	1,165 m (3,825 ft)
Srs 300	754 m (2,479 ft)
Landing from 15 m (50 ft): Srs 100	772 m (2,530 ft)
Srs 200	652 m (2,140 ft)
Srs 300	603 m (1,980 ft)
Landing run, with propeller reversal:	
Srs 100	398 m (1,305 ft)
Srs 200	400 m (1,315 ft)
Range at 6,100 m (20,000 ft), long-range cruising speed, reserves for 45 min hold:	
Srs 100 with max payload	
	810 n miles (1,501 km; 932 miles)
Srs 200 with max payload	
	825 n miles (1,528 km; 950 miles)
Srs 100 with 3,550 kg (7,826 lb) payload	
	2,350 n miles (4,352 km; 2,704 miles)
Srs 200 with 3,550 kg (7,826 lb) payload	
	2,400 n miles (4,445 km; 2,762 miles)
Range, no reserves, Srs 300:	
with max fuel	2,700 n miles (5,000 km; 3,107 miles)
with max payload	540 n miles (1,000 km; 621 miles)

OPERATIONAL NOISE LEVELS (civil versions):
T-O	84.0 EPNdB
Approach	87.0 EPNdB
Sideline	86.0 EPNdB

UPDATED

AIRTECH CN-235 MP PERSUADER and CN-235 MPA

TYPE: Maritime patrol versions of CN-235.
CURRENT VERSIONS: **CN-235 MP Persuader:** CASA version; different avionics from Indonesian MPA. In service with Irish Air Corps and ordered by Spain (four) and Turkey (nine: six for Navy, three for Coast Guard, assembled by TAI at Ankara). In mid-1999, Turkey sought proposals from at least seven potential integrators of surveillance systems to provide radar, FLIR and an acoustics suite for naval CN-235s.

CN-235 MPA: Indonesian-developed version; available either with lengthened nose housing radar and IFF; or with normal CN-235 nose, plus belly radar; CN-235 prototype PK-XNC serving as testbed. Maximum T-O weight 15,400 kg (33,951 lb), endurance more than 8 hours. Provision for quick-change configuration for general transport, communications or other duties.

Maritime surveillance CN-235 MP Persuader of the Irish Air Corps *(Paul Jackson/Jane's)*

AIRTECH to AMX—AIRCRAFT: INTERNATIONAL

Required by Indonesian Navy (six included in national order for 24), Indonesian Air Force (three), Brunei (three) and UAE (four ordered March 1998 for 2001 delivery); Abu Dhabi and Pakistan interested in four each. Offered to South Africa, though baseline aircraft may not reach SAAF range requirement.

Indonesia confirmed initial three firm orders in May 2000, when Thomson-CSF selected to supply AMASCOS airborne maritime situation control system, comprising Elettronica ALR-733 RWR, T-CSF Ghlio thermal imager and T-CSF Sextant Gemini navigation computer. Brunei chose Boeing as Argo Systems integrator for its three aircraft in late 1995, specifying individual sensors in October 1996 as AN/AAQ-21 FLIR, Sky Guardian ESM, Cossor 3500 IFF and AN/APS-134 radar, plus two operators' consoles. UAE has also selected AMASCOS 300 with Ocean Master 100 radar and other systems. BAE Australia marketing CN-235 MPA in Asia-Pacific region under September 1997 agreement; BAE also to provide advanced systems development for proposed configurations.

AVIONICS (Persuader): *Radar:* Litton APS-504(V)5.
Mission: FLIR-2000HP undernose-mounted night vision system and Litton AN/ALR-85(V) ESM system, fully integrated via a central tactical processor with reconfigurable consoles.

AVIONICS (CN-235 MPA): *Radar:* BAE Systems Seaspray 4000, or Raytheon AN/APS-134 (LW) or Thomson-CSF Ocean Master 100.

Flight: Litton LN92 ring laser gyro INS; Trimble TNL 7900 Omega/GPS.

Mission: Argo data processing and display system with multifunction consoles. BAE Systems Sky Guardian SG-300, or Argo Systems AR-700 or Litton AN/ALR-93(V)4 ESM. FLIR Systems AN/AAQ-21 Safire or BAE Systems MRT FLIR. Cossor 3500 IFF interrogator. (Trials aircraft originally equipped with APS-504 and Ocean Master; SG-300. Reconfigured by 1994 with AN/APS-134, MRT, AR-700, LN92 and TNL 7900. Further alternatives available at customer's option.)

UPDATED

AMX

AMX INTERNATIONAL
c/o Alenia, Via Faustiniana, I-00131 Rome, Italy
Tel: (+39 06) 522 91
Fax: (+39 06) 52 29 30 94
PRESIDENT: Dott Ing Giovanni Gazzaniga (Alenia)
VICE-PRESIDENTS:
F Grandi (Aermacchi)
R Pesce (Embraer)
PARTICIPATING COMPANIES:
Alenia: see under Italy
Aermacchi: see under Italy
Embraer: see under Brazil

AMX production for the Italian Air Force is now complete, but manufacture continues in Brazil. An Italian mid-life update is planned.

VERIFIED

AMX
Brazilian Air Force designations: A-1 and A-1B
TYPE: Attack fighter.
PROGRAMME: Resulted from June 1977 Italian Air Force specification for small tactical fighter-bomber (see 1987-88 and previous *Jane's* for early background); original Aeritalia/Aermacchi partnership joined by Embraer July 1980; seven single-seat prototypes built (first flight 15 May 1984: further details in 1987-88 and earlier editions); production of first 30 (Italy 21, Brazil nine), and design of two-seater, began mid-1986; first production aircraft rolled out at Turin 29 March 1988, making first flight 11 May; second contract (Italy 59, Brazil 25, including six and three two-seaters respectively) placed 1988.

Deliveries to Italian Air Force (six for Reparto Sperimentale di Volo at Pratica di Mare) began April 1989; production A-1 for Brazilian Air Force (s/n 5500) made first flight 12 August 1989, deliveries (two to Nu A-1 training nucleus at Santa Cruz) following from 17 October 1989; in-flight refuelling test programme completed (by Embraer) August/September 1989; first flight by first (of three) two-seat AMX-T prototypes 14 March 1990 (MM55024), followed by second on 16 July; first flight of Embraer two-seater (serial number 5650), 14 August 1991; third production batch authorised early 1992 (one year late); first two-seater for Brazilian Air Force (5650) delivered 7 May 1992.

Italian single-seater production temporarily halted following delivery on 1 February 1993 of 72nd aircraft (MM7160); resumed late 1994, with both AMX and first

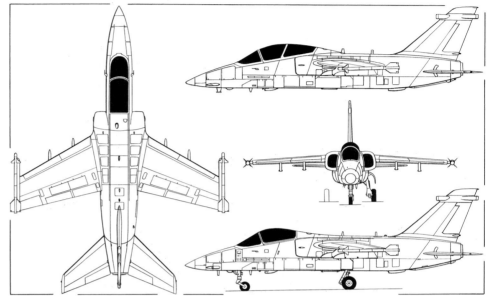

Alenia/Aermacchi/Embraer AMX, for the air forces of Italy and Brazil; upper side view shows two-seater *(Dennis Punnett/Jane's)*

production batch of AMX-T. Final Italian single-seater delivered in 1997; total 110, comprising 74 built by Alenia and 36 by Aermacchi. Batch 4 (35 AMX and 16 AMX-T) and Batch 5 (42 AMX and 9 AMX-T) cancelled. Final Italian two-seater followed in 1998; 26 built: 17 Alenia, nine Aermacchi. Production continued in Brazil, where 50th was delivered on 1 December 1998 and last in 1999.

Istrana- and Amendola-based AMX squadrons flew 252 combat sorties (667 flying hours) during Operation Allied Force against Yugoslavia in 1999, dropping 39 Opher LGBs.

CURRENT VERSIONS: **AMX:** Replaced G91R/Y and some F-104G/S in Italian Air Force (eight squadrons originally planned) and some EMB-326GB Xavante in Brazilian Air Force for close support/interdiction/reconnaissance, sharing counter-air duties with IDS Tornado (Italy) and F-5E/Mirage 50 (Brazil); in service with five Italian Stormi (see table) and 10° and 16° Grupos (Brazil); Brazilian Air Force aircraft (designated **A-1**) differ primarily in avionics and weapon delivery systems, have two 30 mm guns instead of Italian version's single multibarrel 20 mm weapon and are usually fitted with in-flight refuelling probes.

AMX-MLU: Mid-life upgrade configuration for Italian and Brazilian air forces, with new avionics, improved pilot interface, NVG-compatible 'glass cockpit' with three MFDs, new navigation system with embedded GPS, MIL-STD-1760 stores management system, radar altimeter, helmet-mounted display system, FLIR, multimode radar, digital FCS and tactical datalink. Will be cleared with a full range of standoff weapons. First phase has been approved (for Italian AMXs) adding IN/GPS, MIL-STD-1760 SMS, JTIDS datalink and new HDD to the 36 surviving Batch 3 aircraft, at a cost of Lit170 billion (US$84.4 miillion). Work due to begin in 2001. Phase 2 will add the NVG 'glass cockpit', new self-protection suite and new radar. Italy intends to bring all 94 single-seaters to this standard.

Single-seat AMX of the Italian Air Force

Super AMX: Obsolete designation for two-seater which later offered to South Africa the basis of Italian upgrade. Wide-angle HUD, 'glass cockpit', improved HOTAS, GPS, HMD, integrated defensive aids, and new weapons. Also new radar – possibly Scipio already fitted to Brazilian AMX. Studies under way into replacement of Spey engine by non-afterburning 60.0 kN (13,500 lb st) EJ200, derived from Eurofighter Typhoon power plant, giving 20 per cent extra static thrust, 40 per cent extra thrust at Mach 0.8, 40 per cent increase in T-O performance and 20 per cent increase in sustained turn rate. Development of engine flight demonstrator expected, perhaps with first flight during 2000-2001. RFP issued for new digital FCS computer.

Detailed description applies to single-seater except where indicated.

AMX-T: Second cockpit accommodated by removing forward fuselage fuel tank and relocating environmental control system; dual controls, canopy, integration of rear cockpit GEC-Marconi HUD monitor, and oxygen systems, designed/redesigned by Embraer; intended both as operational trainer and, suitably equipped, for such roles as EW, reconnaissance and maritime attack; most Italian AMX-Ts assigned to operational squadrons of 32° Stormo; others assigned, one per squadron, as trainers. Brazilian designation **A-1B**.

AMX-ATA: Eight two-seaters ordered by Venezuela in September 1999 for delivery in the fourth quarter of 2001. The Venezuelan aircraft will have Elbit avionics and may be followed by more AMX-ATA 1s.

AMX-ATA 1: Total of 24 modernised and upgraded trainers still expected to be ordered by Venezuela, based on FAB Block 3 configuration, but with digital cockpit; some ALX-based avionics in package to be supplied and integrated by Elbit. New radar, probably Alenia SCP-01 Scipio.

AMX-ATA 2: Light combat aircraft and advanced trainer for export with AMX-MLU avionics and non-afterburning EJ200 engine.

AMX-E: Planned two-seat EW version intended as escort jammer and SEAD platform using AGM-88 HARM missiles. Flight controls removed from rear cockpit. MM55027 used for development. Definitive AMX-E may have EJ200 engine. Feasibility study complete, further development delayed.

CUSTOMERS: Total of 192 (136 Italy/56 Brazil, including 26 and 11 two-seaters) delivered by March 2000 (see table) to original partners, plus eight two-seaters selected by Venezuela on 9 September 1999 for delivery in 2001. Italy plans to place all 19 remaining Batch 1 aircraft in reserve for possible sale. By 1998, 23 Batch 2 aircraft had been upgraded to 'pre-FOC' (full operational capability) standard and plans were in hand to modify remaining 15 to FOC. These 38 plus all 56 remaining from Batch 3 (FOC), formed baseline fleet of 94, though AMI active inventory quoted as 104 aircraft by May 1999. MLU for AMI AMX described above. Upgrades for remaining aircraft will include Thomson CDLP laser designator and new electro-optical reconnaissance pod. Italian long-term plan is for four squadrons, of which one OCU.

COSTS: US$18.75 million (1999) AMX-T Venezuelan programme unit cost.

DESIGN FEATURES: Intended for high-subsonic/very low-altitude day/night missions, in poor visibility and, if necessary, from poorly equipped or partially damaged runways. Required fatigue life of 16,000 hours being extended to 24,000 hours.
Wing sweepback 31° on leading-edges, 27° 30′ at quarter-chord; thickness/chord ratio 12 per cent.

FLYING CONTROLS: Hydraulically actuated ailerons and elevators; leading-edge slats and Fowler double-slotted trailing-edge flaps (each two-segment on each wing, positions 0, 30 and 41°) actuated electrohydraulically; pair of hydraulically actuated spoilers forward of each flap pair, deployable separately in inboard and outboard pairs; fly-by-wire control of spoilers, rudder and variable incidence tailplane by Alenia/BAE Systems flight control computer; ailerons, elevators, rudder have manual reversion for fly-home capability, even with both hydraulic systems inoperative; spoilers serve also as airbrakes/lift dumpers.

STRUCTURE: Mainly aluminium alloy except for carbon fibre fin and elevators; shoulder-mounted wings, each with three-point attachment to fuselage, have three-spar torsion box with integrally stiffened skins; oval-section semi-monocoque fuselage, with rear portion (including tailplane) detachable for engine access.
Work split gives programme leader Alenia 46.7 per cent (centre-fuselage, nose radome, tail surfaces, ailerons and spoilers); Aermacchi has 23.6 per cent (forward fuselage including gun and avionics integration, canopy, tailcone) and Embraer 29.7 per cent (air intakes, wings, leading-edge slats, flaps, wing pylons, external fuel tanks and reconnaissance pallets); single-sourced production, with final assembly lines in Italy and Brazil.

LANDING GEAR: Hydraulically retractable tricycle type, of Messier-Bugatti levered suspension design, built in Italy by Magnaghi (nose unit) and in France by ERAM (main units). Single wheel and oleo-pneumatic shock-absorber on each unit. Nose unit retracts forward; main units retract forward and inward, turning through approximately 90° to lie almost flat in underside of engine air intake trunks. Nosewheel hydraulically steerable (±6° normal; ±45° with full pedal movement), self-centring, and fitted with anti-shimmy device. Towing travel ±90°. Mainwheel tyres size 670×210-12 (18 ply), pressure 9.65 bars (140 lb/sq in); nosewheel tyre size 18×5.5-8 (10 ply), pressure 10.70 bars (155 lb/sq in). Hydraulic brakes and fully modulated anti-skid system. No brake-chute. Runway arrester hook. Minimum ground turning radius 7.53 m (24 ft 8½ in).

POWER PLANT: One 49.1 kN (11,030 lb st) Rolls-Royce Spey Mk 807 non-afterburning turbofan, built under licence in Italy by Fiat, Piaggio and Alfa Romeo Avio, in association with Companhia Eletro-Mecânica (CELMA) in Brazil. Self-sealing, compartmented, rubber fuselage bag tanks and two integral wing tanks with combined capacity of 3,500 litres (924.6 US gallons; 770 Imp gallons). Brazilian AMX unofficially reported to carry more internal fuel. Auxiliary fuel tanks of up to 1,100 litres (290 US gallons; 242 Imp gallons) capacity can be carried on each inboard underwing pylon, and up to 580 litres (153 US gallons; 128 Imp gallons) on each outboard pylon. Single-point pressure or gravity refuelling of internal and external tanks. Optional in-flight refuelling capability (probe and drogue system).

ACCOMMODATION: Pilot only, on Martin-Baker Mk 10L zero/zero ejection seat; 18° downward view over nose. One-piece wraparound windscreen; one-piece hinged canopy, opening sideways to starboard. Cockpit pressurised and air conditioned. Tandem two-seat combat trainer/special missions version also produced, with Mk 10LY-2 (front) and Mk 10LY-3 (rear) seats.

SYSTEMS: Microtecnica environmental control system (ECS) provides air conditioning of cockpit, avionics and reconnaissance pallets, cockpit pressurisation, air intake and inlet guide vane anti-icing, windscreen demisting and anti-g systems. Duplicated redundant hydraulic systems, driven by engine gearbox, operate at pressure of 207 bars (3,000 lb/sq in); both actuate primary flight control system (aileron, elevators and rudder) and secondary (flap and slat system; No. 1 circuit additionally supplies outboard spoilers, nosewheel steering and gun; No. 2 also supplies inboard spoilers and landing gear actuation. Primary electrical system AC power (115/200 V at fixed frequency of 400 Hz) supplied by two 30 kVA IDG generators, with two transformer-rectifier units for conversion to 28 V DC; 36 Ah Ni/Cd battery for emergency use, to provide power for essential systems in the event of primary and secondary electrical system failure. Aeroeletrônica (Brazil) external power control unit. Fiat FA 150 Argo APU for engine starting. APU-driven electrical generator for ground operation. Liquid oxygen system.

AVIONICS: All avionics/equipment packages pallet-mounted and positioned for rapid access. Modular design and space provisions within aircraft permit retrofitting of alternative avionics.

AMX REQUIREMENTS

	Italy				Brazil				
Batch	AMX	First aircraft	AMX-T	First aircraft	AMX	First aircraft	AMX-T	First aircraft	Total
1	19	MM7089	2	MM55024	8	5500	1	5650	30
2	53	MM7108	6	MM55026	22	5508	3	5651	84
3	38	MM7161	18	MM55034	15	5530	7	5654	78
Subtotal	110		26		45		11		192
4					15		4		19
5					19		0		19
Total	110		26		79		15		230

Notes: Brazil retains option on 23 aircraft from Batches 4 and 5. Excludes seven single-seat prototypes and eight for Venezuela.

ITALIAN AMX UNITS

Squadron	Wing	Base	Type	First aircraft and acceptance
13° Gruppo	32° Stormo	Amendola	AMX	MM7112	Nov 94
			AMX-T[1]	MM55029	by Apr 95
14° Gruppo	2° Stormo	Rivolto[2]	AMX	MM7130	10 Jul 91
			AMX-T[1]	MM55028	24 Nov 94
28° Gruppo[5]	3° Stormo	Villafranca	AMX	MM7129	Jun 93
			AMX-T[1]	MM55029	by Dec 94
101° Gruppo[7]	32° Stormo	Amendola	AMX	MM7091	31 Jul 95
			AMX-T	MM55030	21 Nov 94
103° Gruppo	51° Stormo	Istrana	AMX	MM7098	30 Sep 89
			AMX-T[1]	MM55027	by Apr 95
132° Gruppo	51° Stormo	Istrana[6]	AMX	MM7110	15 Nov 90[4]
			AMX-T[1]	MM55037	4 Dec 95

[1] One aircraft only
[2] At Istrana until February 94; may return
[3] Unit initially re-equipped as 201° Gruppo (until 31 Jul 95)
[4] Initial working up at Istrana
[5] Disbanded 29 Sep 97
[6] Villafranca until March 1999
[7] AMX OCU

BRAZILIAN AMX UNITS

Squadron	Wing	Base	First delivery
1° Esquadrão	16 Grupo de Aviação	Santa Cruz	Formed 7 November 1990
1° Esquadrão	10 Grupo de Aviação	Santa Maria	1996
3° Esquadrão	10 Grupo de Aviação	Santa Maria	1998

An AMX-T of 101° Gruppo 32° Stormo fitted with a refuelling probe *(Paul Jackson/Jane's)* 2001/0092081

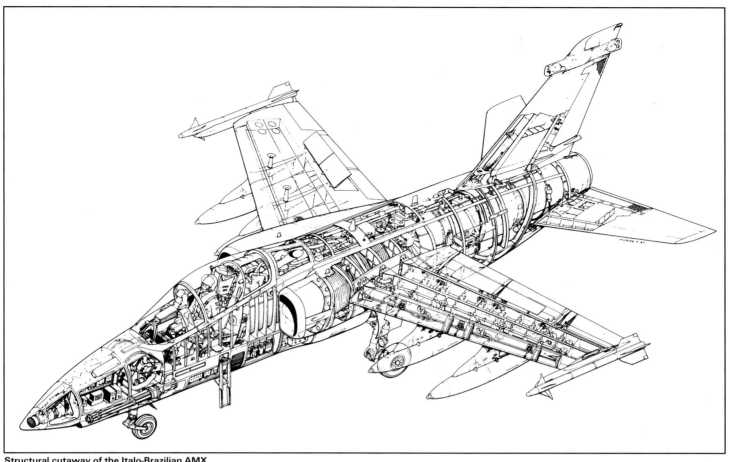

Structural cutaway of the Italo-Brazilian AMX

Comms: UHF and VHF com, and IFF.

Radar: Pointer ranging radar in Italian AMXs is I-band set modified from Elta (Israel) EL/M-2001B and built in Italy by FIAR; Brazilian aircraft have Tecnasa/SMA-built (Alenia) SCP-01 Scipio radar.

Flight: Litton Italia INS, with standby AHRS and Tacan, for Italian Air Force; VOR/ILS for Brazil. Data processing, with Microtecnica air data computer. BAE Systems MED 2067 video monitor display in rear cockpit of two-seater, for use by instructor/navigator as HUD monitor.

Instrumentation: Alenia computer-based weapon aiming and delivery, incorporating radar and Alenia stores management system; digital data displays (OMI/Alenia head-up, Alenia multifunction head-down, and weapons/nav selector). Provision for night vision goggles.

Mission: Italian aircraft of 3° Stormo equipped with Oude Delft Orpheus reconnaissance pods, and was an internal sensor suite being sought for deployment in 2001. This is believed to have been cancelled, but was to have comprised any one of three interchangeable Aeroeletrônica (Brazil) pallet-mounted photographic systems installed internally in forward fuselage, complementing external IR/EO pod on centreline pylon. Each system fully compatible with aircraft, and not affecting operational capability. Camera bay in lower starboard side of fuselage, forward of mainwheel bay.

Self-defence: Elettronica active and passive ECM, including fin-mounted radar warning receiver.

ARMAMENT: One M61A1 multibarrel 20 mm cannon, with 350 rounds, in port side of lower forward fuselage of aircraft for Italian Air Force (one 30 mm DEFA 554 cannon on each side in aircraft for Brazilian Air Force). Single stores attachment point on fuselage centreline, plus two attachments under each wing, and wingtip rails for two AIM-9L Sidewinder or similar IR air-to-air missiles (MAA-1 Piranha on Brazilian aircraft). Fuselage and inboard underwing points each stressed for loads of up to 907 kg (2,000 lb); outboard underwing points stressed for 454 kg (1,000 lb) each. Twin carriers can be fitted to all five stations. Total external stores load 3,800 kg (8,377 lb). Attack weapons can include free-fall or retarded Mk 82/83/84 bombs, laser-guided bombs, cluster bombs, air-to-surface missiles (including area denial, anti-radiation and anti-shipping weapons), electro-optical precision-guided munitions and rocket launchers.

Exocet firing trials conducted 1991; Marte trials 1994; carriage trials of GBU-16 Paveway II LGB on Italian AMX in 1995 and aircraft used Elbit Opher LGB system during Operation Allied Force over Kosovo, May-July 1999.

DIMENSIONS, EXTERNAL:
Wing span:	
excl wingtip missiles and rails	8.875 m (29 ft 1½ in)
over missiles	9.97 m (32 ft 8½ in)
Wing aspect ratio	3.8
Wing taper ratio	0.5
Length: overall	13.23 m (43 ft 5 in)
fuselage	12.55 m (41 ft 2 in)
Height overall	4.55 m (14 ft 11¼ in)
Tailplane span	5.20 m (17 ft 0¾ in)
Wheel track	2.15 m (7 ft 0¾ in)
Wheelbase	4.70 m (15 ft 5 in)

AREAS:
Wings, gross	21.00 m² (226.0 sq ft)
Ailerons (total)	0.88 m² (9.47 sq ft)
Trailing-edge flaps (total)	3.86 m² (41.55 sq ft)
Leading-edge slats (total)	2.07 m² (22.28 sq ft)
Spoilers (total)	1.30 m² (13.99 sq ft)
Fin (exposed)	4.265 m² (45.91 sq ft)
Rudder	0.83 m² (8.93 sq ft)
Tailplane (total exposed)	5.10 m² (54.90 sq ft)
Elevators (total)	1.00 m² (10.76 sq ft)

WEIGHTS AND LOADINGS (all versions):
Operational weight empty	6,730 kg (14,837 lb)
Max fuel weight: internal	2,790 kg (6,151 lb)
external	1,726 kg (3,805 lb)
Max external stores load	3,800 kg (8,377 lb)
T-O weight (clean)	9,694 kg (21,371 lb)
Typical mission T-O weight	10,750 kg (23,700 lb)
Max T-O weight	13,000 kg (28,660 lb)
Normal landing weight	7,000 kg (15,432 lb)
Combat wing loading (clean)	457.1 kg/m² (93.62 lb/sq ft)
Max wing loading	619.1 kg/m² (126.79 lb/sq ft)
Max power loading	265 kg/kN (2.60 lb/lb st)

PERFORMANCE (A at typical mission weight of 10,750 kg; 23,700 lb with 907 kg; 2,000 lb of external stores, B at max T-O weight with 2,721 kg; 6,000 lb of external stores, ISA in both cases):
Max level speed: at S/L	M0.84
at 9,140 m (30,000 ft)	M0.86
Max rate of climb at S/L	3,124 m (10,250 ft)/min
Service ceiling	13,000 m (42,650 ft)
T-O run at S/L: A	631 m (2,070 ft)
B	982 m (3,220 ft)
T-O to 15 m (50 ft) at S/L: B	1,442 m (4,730 ft)
Landing from 15 m (50 ft) at S/L: B	753 m (2,470 ft)
Landing run at S/L	464 m (1,520 ft)
Attack radius, allowances for 5 min combat over target and 10% fuel reserves:	
lo-lo-lo: A	300 n miles (556 km; 345 miles)
B	285 n miles (528 km; 328 miles)
hi-lo-hi: A	480 n miles (889 km; 553 miles)
B	500 n miles (926 km; 576 miles)
Ferry range with two 1,000 litre (264 US gallon; 220 Imp gallon) drop tanks, 10% reserves	
	1,800 n miles (3,333 km; 2,071 miles)
g limits	+7.33/-3

UPDATED

BELL/AGUSTA

BELL/AGUSTA AEROSPACE COMPANY (BAAC)

PO Box 901073, Fort Worth, Texas 76101, USA
Tel: (+1 817) 278 96 00
Fax: (+1 817) 278 97 26
Web: http://www.bellagustaaerospace.com
MANAGING DIRECTOR: Jim Rogers (Bell)
EXECUTIVE MANAGING DIRECTORS:
 Don Barbour (BA609 Programme)
 Antonio Giovannini (AB139 Programme)
PARTICIPATING COMPANIES:
 Bell Helicopter Textron: see under USA
 Agusta: see under Italy

Bell and Agusta announced on 8 September 1998 that they had agreed to establish a joint venture to manage development of two new aircraft: the BA609 tiltrotor, previously a Bell and Boeing programme, and the AB139, a new helicopter announced on the same day. Following approval of both boards, a definitive agreement was signed on 6 November 1998. Bell is the majority shareholder and will undertake final assembly for AB139s delivered to North America. Agusta, which has built Bell helicopters under licence since 1952, will invest and participate in development of the BA609, manufacturing some components and assembling those sold in Europe and certain other parts of the world. Additionally, Agusta is responsible for the AB139's development and certification, with participation by Bell. A military version was revealed in July 2000.

UPDATED

BELL/AGUSTA AB139

TYPE: Light utility helicopter.
PROGRAMME: Announced at Farnborough Air Show, 8 September 1998, as joint venture between Agusta and Bell; full-scale mockup unveiled at Paris Air Show 12 June 1999. Agusta responsible for development, certification to JAR/FAR 29 and transition to production, with participation by Bell on a 75:25 per cent work-share basis; final assembly by Agusta at Vergiate, and by Bell (possibly

at Mirabel, Canada) for American and Pacific Rim customers. First flight 3 February 2001; three aircraft will undertake flight test programme leading to certification and first deliveries in 2002; to complement, rather than replace, Bell 412. Full-scale mockup of AB139 Military unveiled at Farnborough International 2000 in July 2000; development will follow certification of civilian version.

Risk-sharing collaborators include GKN Westland (tail rotor drive train), Honeywell (avionics), Kawasaki (transmission input module), Liebherr Germany (landing gear and air conditioning system), Pratt & Whitney Canada (power plant) and PZL Świdnik (airframe components).

CURRENT VERSIONS: **AB139**: Commercial version, *as described*.

AB139 Military: Multirole military helicopter with provision for armoured crew seats, electronic warfare protection, IR suppressors, two internal pintle-mounted machine guns and easily removable stub-wing weapons supports for gun pods, rocket launchers and AAMs.

CUSTOMERS: Launch customer Bristow Helicopters of UK announced order for two on 26 September 2000; for delivery in 2002. Anticipated market for 900 over 20 years, some 55 per cent for military use; 34 per cent of sales projected in Europe, 23 per cent in Middle East, 18 per cent in Far East, 13 per cent in South America and 12 per cent in North America.

COSTS: Commercial version US$6 million (1999).

DESIGN FEATURES: Design goals include high manoeuvrability and agility, low pilot workload, night/all-weather operation, low acoustic and infra-red emissions and mission flexibility for commercial and military operators. Intended for offshore support, medevac, corporate/VIP transport, SAR and military operations. Able to operate at maximum T-O weight from Class A helipads at 945 m (3,100 ft) at ISA + 20°C. Five-blade, fully articulated, ballistic tolerant main rotor and four-blade tail rotor. Some transmission and rotor elements based on Agusta A129 Mangusta.

FLYING CONTROLS: Four-axis, digital AFCS.

LANDING GEAR: Heavy-duty, retractable tricycle type with twin wheels on nose unit; single wheels on main units, which retract into side sponsons.

POWER PLANT: Two Pratt & Whitney Canada PT6C-67C turboshafts, with FADEC, each rated at 1,252 kW (1,679 shp) for T-O and 1,142 kW (1,531 shp) maximum continuous; OEI ratings 1,289 kW (1,729 shp) for two minutes and 1,252 kW (1,679 shp) maximum continuous. Fuel tanks behind main cabin.

ACCOMMODATION: Up to 15 passengers on crashworthy seats in three rows of five, two forward facing, one rearward facing, in unobstructed cabin with flat floor; flight-accessible baggage compartment at rear of cabin. Alternatively, six stretchers and four attendants in medevac configuration. Plug-type sliding door on each side of cabin, with separate crew doors.

SYSTEMS: Systems duplicated and separated. Main and tail rotor ice protection optional.

AVIONICS: Honeywell Primus Epic as core system. Provision for up to four 203 × 254 mm (8 × 10 in) high-definition colour active matrix liquid crystal displays for MFD, PFD and FLIR/video functions, and four-axis modular digital autopilot with flight director for hands-off operation and SAR modes.

DIMENSIONS, EXTERNAL:
Main rotor diameter	13.80 m (45 ft 3¼ in)
Tail rotor diameter	2.65 m (8 ft 8¼ in)
Length overall, rotors turning	16.65 m (54 ft 7½ in)
Fuselage: Length	13.53 m (44 ft 4¼ in)
Max width: across cabin	2.26 m (7 ft 5 in)
across sponsons	3.20 m (10 ft 6 in)
Height, rotors turning	4.95 m (16 ft 3 in)
Total rotor ground clearance	2.30 m (7 ft 6½ in)
Cabin doors: Height	1.35 m (4 ft 5 in)
Width	1.68 m (5 ft 6 in)

DIMENSIONS, INTERNAL:
Cabin: Length	2.70 m (8 ft 10¼ in)
Max width	2.00 m (6 ft 6¾ in)
Max height	1.42 m (4 ft 8 in)
Volume	8.0 m³ (283 cu ft)
Baggage compartment volume	3.4 m³ (120 cu ft)

AREAS:
Main rotor disc	149.57 m² (1,610.0 sq ft)
Tail rotor disc	5.52 m² (59.37 sq ft)

Bell/Agusta AB139 prototype making its maiden flight

Instrument panel of the AB 139 Military

Artist's impression of Bell/Agusta AB139 Military

Standard 15-seat cabin of the Bell/Agusta AB139 mockup

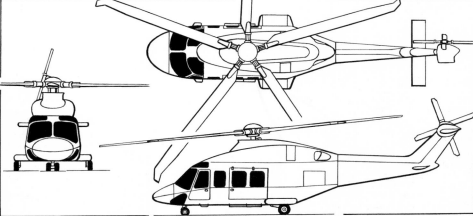
General arrangement of the Bell/Agusta AB139 (*James Goulding/Jane's*)

WEIGHTS AND LOADINGS:
  Useful load   2,500 kg (5,512 lb)
  Max external load   2,750 kg (6,063 lb)
  Max T-O weight   6,000 kg (13,227 lb)
  Max disc loading   40.11 kg/m² (8.22 lb/sq ft)
PERFORMANCE (estimated):
  Never-exceed speed (VNE)   167 kt (309 km/h; 192 mph)
  Max cruising speed   155 kt (287 km/h; 178 mph)
  Max rate of climb at S/L   610 m (2,000 ft)/min
  Service ceiling OEI   2,926 m (9,600 ft)
  Hovering ceiling OGE   3,660 m (12,000 ft)
  Max range, no reserves   400 n miles (740 km; 460 miles)
  Endurance   3 h 54 min
                                              UPDATED

BELL/AGUSTA BA609

TYPE: Light utility tiltrotor.
PROGRAMME: Having earlier joined in partnership to develop the military V-22 tiltrotor, Bell Boeing revealed in February 1996 that studies were in progress for a nine-passenger civil tiltrotor aircraft in the 6,350 kg (14,000 lb) weight class, with the preliminary designation D-600. Subsequently, on 18 November 1996, the two companies announced that a joint venture was being established to design, develop, certify and market a six- to nine-passenger civil tiltrotor as the Bell Boeing 609.

Boeing withdrew as a partner on 1 March 1998 and Bell subsequently teamed with Agusta to develop, produce and market the tiltrotor as the BA609; this arrangement was formally announced at the Farnborough Air Show in September 1998. Agusta is investing and participating in BA609 development and will be responsible for assembly of BA609s sold in Europe and elsewhere.

Preliminary design review completed May 1997. Manufacture of parts for prototypes began at Philadelphia, August 1997. Full-size mockup first exhibited at Paris Air Show, June 1997.

First flight of the BA609 scheduled for August 2001. Four prototypes are being produced for 18-month flight test programme leading to certification under FAR Pt 25 (fixed-wing aircraft) and Pt 29 (helicopters), plus Pt 21.17(b) Special Conditions for unique components. To be capable of single-pilot IFR operation. Initial customer deliveries expected in early 2003.

CURRENT VERSIONS: **BA609**: Initial version *as described*.
HV-609: Multimission version proposed by Bell and Lockheed Martin to satisfy US Coast Guard 'Deepwater' re-equipment programme as potential replacement for Dassault HU-25 Guardian, Eurocopter HH-65 Dolphin and Sikorsky HH-60J Jayhawk. Missions could include drug interdiction and SAR, with 30 to 50 HV-609s possibly being acquired; preliminary studies now under way, with final selection not anticipated before January 2002.
UV-609: Utility version conceived by Bell for US Army combat role, including casualty evacuation, command and control, logistic support and light utility/troop transport tasks. Manufacturer also promoting UV-609 to US Marine Corps as potential training system for V-22 Osprey.
620: Proposed 22-seat version conceived by Bell Boeing team; no recent news received and concept may have lapsed.

CUSTOMERS: Order book opened 2 February 1997 at Heli Expo, with first order placed soon after by unspecified customer. Total of 77 ordered by 42 customers in 18 countries by October 1999; purchasers identified to date include Helitech Pty of Australia; Lider of Brazil; Canadian Helicopter Corporation and Northern Mountain Helicopters Inc of Canada; Petroleum Tiltrotors International of Dubai; Aero-Dienst GmbH of Germany; Mitsui of Japan; United Industries of South Korea; Helikopter Services of Norway; Lloyd's Investments of Poland; Bristow of the UK; Massachusetts Mutual Life Insurance, Austin Jet and Petroleum Helicopters of the USA; and Air Center Helicopters Inc of the US Virgin Islands. Briefing given to US Coast Guard, late 1997, followed by demonstration by XV-15 tiltrotor concept demonstrator aboard Coast Guard Cutter *Mohawk* off Key West, Florida, in May 1999.

COSTS: US$8 million to US$10 million, depending on configuration.

DESIGN FEATURES: Combines the most favourable aspects of helicopter and aeroplane performance in passenger-carrying role. T tail configuration instead of endplate fin layout used by earlier tiltrotor designs (XV-15 and V-22 Osprey); this raises horizontal tailplane above rotor wake to minimise fore and aft pitching moment at transition phase of flight. Size of wing determined by requirement for it to hold all fuel for CG and safety considerations. Composites cross-shafts keep both proprotors turning in event of engine failure. Manual screwjack facility exists whereby the proprotors can be tilted into helicopter mode if cross-shafts fail. Designed using three-dimensional CATIA digital computer design system.

Refer also to Bell Boeing V-22 entry in US section for explanation of tiltrotor concept and for its extension to four-proprotor configuration.

FLYING CONTROLS: BAE Systems triplex digital fly-by-wire flight control system, with Dowty Aerospace actuators. T tail with conventional elevators; no rudder. Two-segment trailing-edge flaperons.

STRUCTURE: Aluminium fuselage structure with composites skinning; composites wing. Production fuselages, including cockpit, cabin and systems installation, will be built by risk-sharing partner Fuji Heavy Industries of Japan, which may also establish a third production line if substantial orders are won from the Japanese government; cabin doors and fuselage tailcone supplied by Kawasaki; wing and nacelles by Bell at Fort Worth, with final assembly line at new Bell tiltrotor manufacturing facility in Amarillo, Texas. Fuselage in three major sections; nose, centre and tail, with fuselage skin incorporating graphite stringers with Japanese Toray composites material. Same used for wing and nacelles, with upper and lower wing surfaces produced as single pieces.

LANDING GEAR: Retractable tricycle type, with twin nosewheels and single wheel on each main unit. Messier-Dowty overseeing design, development and manufacture of integrated landing gear system, including legs, wheels, tyres, brakes, brake control and landing gear control systems.

POWER PLANT: Two 1,447 kW (1,940 shp) Pratt & Whitney Canada PT6C-67A turboshaft engines, installed in tilting nacelles at wingtips, each driving a three-blade proprotor. Nacelle interface units by AMETEK Aerospace Systems.

Artist's impression of the Bell/Agusta BA609 in US Coast Guard HV-609 guise *2001/0093642*

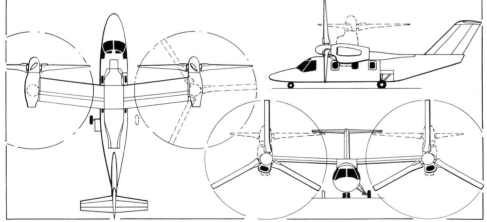
Bell/Agusta BA609 tiltrotor aircraft *(James Goulding/Jane's)* *1999/0044896*

Instrument panel of the Bell/Agusta BA609 mockup *1999/0044897*

Cabin interior of Bell/Agusta BA609 mockup *2000/0079278*

INTERNATIONAL: AIRCRAFT—BELL/AGUSTA to BOEING/BAE

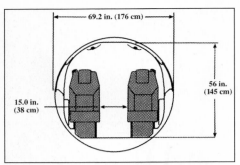

Cabin cross-section of BA609 tiltrotor 2000/0079276

Rockwell Collins EICAS; modified oil system, with dual pumps, to generate sufficient oil pressure when operating in vertical mode; several planetary gears also removed to achieve direct drive 30,000 rpm output. Fuel in integral wing tanks; usable capacity 1,401 litres (370 US gallons; 308 Imp gallons). Provision for auxiliary fuel tanks.

ACCOMMODATION: Crew of two, side by side on flight deck, with dual controls. Maximum of nine passengers in standard aircraft. Crew and passenger door on starboard side, forward of wing. Accommodation pressurised and air conditioned; pressurisation differential 0.38 bars (5.50 lb/sq in). Transparencies by Sully Produits Speciaux of France.

SYSTEMS: Equipped for flight into known icing. Lucas Aerospace DC electrical power systems. Intertechnique brushless electric pumps and motor-operated shut-off valves.

AVIONICS: Rockwell Collins Pro Line 21 package as standard.
Comms: Dual VHF radios; Mode S transponder. Cockpit voice recorder included as standard.
Radar: Optional Rockwell Collins WXR-800 solid-state weather radar.
Flight: Dual VOR/ILS, DME and ADF, with integrated control of sensors by dual Rockwell Collins RTU-4200 radio tuning units. Rockwell Collins ALT-4000 radar altimeter optional. GPS included as standard, along with TCAS and FDR.
Instrumentation: Three 250 × 200 mm (10 × 8 in) active matrix colour LCD adaptive flight displays, including two primary flight displays and one multifunction display. Standby instrument system by Goodrich Aerospace.

Following data are provisional.
DIMENSIONS, EXTERNAL:
Proprotor diameter	7.9 m (26 ft)
Width, rotors turning	18.3 m (60 ft)
Length overall	13.4 m (44 ft)
Max diameter of fuselage	1.75 m (5 ft 9 in)
Height to top of fin	4.6 m (15 ft)
Wheel track	3.0 m (10 ft)
Wheelbase	5.8 m (19 ft)
Distance between proprotor centres	10.0 m (33 ft)
Passenger door: Width	0.76 m (2 ft 6 in)

DIMENSIONS, INTERNAL:
Cabin: Length	5.33 m (17 ft 6 in)
Max width	1.52 m (5 ft 0 in)
Max height	1.42 m (4 ft 8 in)
Baggage hold, volume	1.4 m³ (50 cu ft)

AREAS:
Rotor discs, each	49 m² (530 sq ft)

WEIGHTS AND LOADINGS:
Weight empty	4,765 kg (10,500 lb)
Max T-O weight	7,265 kg (16,000 lb)
Max disc loading	147 kg/m² (30 lb/sq ft)

PERFORMANCE:
Max cruising speed	275 kt (509 km/h; 316 mph)
Max certified altitude	7,620 m (25,000 ft)
Range: normal fuel capacity with 2,500 kg (5,500 lb) payload at 250 kt (463 km/h; 288 mph)	750 n miles (1,389 km; 863 miles)
with auxiliary fuel tanks	1,000 n miles (1,852 km; 1,151 miles)

UPDATED

BOEING/BAE

BOEING/BAE

PARTICIPATING COMPANIES:
Boeing (Military Aircraft and Missile Systems): see under USA
BAE Systems: see under UK

Original McDonnell Douglas Corporation (MDC) and Hawker Siddeley companies initially associated in 1969 through US procurement of BAE Harrier; relationship developed with US Navy selection of BAE Hawk; both types built at St Louis, although new production of the Harrier ended in 1997, leaving only remanufacture and upgrade work on the AV-8B, as described in *Jane's Aircraft Upgrades*. Joint work undertaken on advanced STOVL combat aircraft, with additional participation of Northrop Grumman, until next-stage contracts were awarded to competing designs in November 1996. Boeing took over MDC in August 1997. Production of T-45 Goshawk continues.

VERIFIED

BOEING/BAE T-45 GOSHAWK

TYPE: Advanced jet trainer.
PROGRAMME: Carrier-capable version of BAE Hawk selected 18 November 1981 (from five other candidates) as winner of US Navy VTXTS (now T45TS) competition for undergraduate jet pilot trainer to replace T-2C Buckeye and TA-4J Skyhawk; original plan was for initial 54 'dry' (land-based) T-45Bs followed by 253 carrier-capable 'wet' T-45As; B model eliminated in FY84 in favour of 300 (and then 302) T-45As; FSD phase began October 1984; construction of two prototypes by Douglas began February 1986; funding approved 16 May 1986 for first three production lots (including 60 T-45As and 15 flight simulators during FY88-90); Lot 1 production contract (12 aircraft) awarded 26 January 1988; FSD prototypes made first flights 16 April (162787) and November 1988 (162788); original planned date for first deliveries (October 1989) delayed by further airframe and power plant changes requested by US Navy.

Announced 19 December 1989 that entire T45TS programme to be transferred to McDonnell Aircraft Co at St Louis; modified FSD prototypes made first flights September and October 1990; two Douglas production aircraft (163599 and '600) delivered to NATC Patuxent River, Maryland, on 10 October and 15 November 1990; first carrier landing (162787 on USS *John F Kennedy*) 4 December 1991; first McAir production aircraft (163601) flew at St Louis 16 December 1991 and handed over to USN 23 January 1992. Production one per month in 1995 following successful passing of US DoD Milestone III review on 17 January 1995 authorising full-rate production. Final delivery due in 2005. The planned 187 aircraft will sustain the fleet at its designated level until 2020, but 40 more aircraft would be needed to reach an out-of-service date of 2035 at the end of the Goshawk's airframe life.

Digital/'glass cockpit' developed as 'Cockpit 21'; first to this standard (37th aircraft, 163635) made first flight 19 March 1994; planned production line introduction at 73rd aircraft, to be delivered October 1996, was delayed; flight trials of prototype aboard USS *John C Stennis* conducted April to May 1996 and successfully completed with 30 May approval for 'Cockpit 21' installation from 84th aircraft (165080), which first flew on 21 October 1997 and formally rolled-out on 31 October. Earlier aircraft to be retrofitted (subject to funding) with 'Cockpit 21' between 2002 and 2007, following a validation conversion of 163651 which was undertaken in 1998.

On 16 August 1995, MDC and Rockwell Australia signed a teaming agreement under which the latter would assemble T-45s at its Avalon plant (formerly ASTA) in the event of the T45TS being chosen for the RAAF's 35 to 40 aircraft lead-in fighter competition. ITEC F124 offered as alternative power plant; installed in bailed-back second prototype T-45 162788 and first (and last) flown on 7 October 1996, despite Australian rejection of T-45 on 6 September. F124-engined T-45A LIF (lead-in fighter) for RAAF would have had 10 per cent additional thrust and a maximum T-O weight of 6,185 kg (13,636 lb).

Beginning in 2000, all T-45s are receiving modified engine air intakes as they undergo routine servicing at NAS Kingsville; new intake has the top 'lip' extended forward, giving a raked profile, improving engine air flow at high angles of attack and reducing number of surges and compressor stalls.

CURRENT VERSIONS: **T-45A:** Baseline version.
T-45C: Digital ('Cockpit 21') avionics version with new computer, digital databus, data transfer cartridge and Litton GINA (GPS/INS assembly); first aircraft 165080; otherwise as T-45A; sole external difference is T-45C's GPS antenna on spine, immediately behind canopy.

CUSTOMERS: US Navy: two FSD prototypes and 187 production aircraft currently envisaged, of which 111 delivered up to 1 January 2000; 15 were to follow in 2000. Complete T45TS programme also involves 19 (originally 32) flight simulators (built by Hughes Training Inc); 48 (originally 49) computer-aided instructional devices, one training integration system mainframe, six electronic classrooms, 155 terminals, plus academic materials and contractor-operated logistic support. USN T45TS requirement stipulated 42 per cent size reduction in (TA-4 and T-2) training fleet, 25 per cent fewer flight hours, and 46 per cent fewer personnel.

Four training squadrons (VT) to equip: VT-21 and VT-22 of Training Wing 2 at Kingsville, Texas, in 1992-96; VT-9 and VT-23 (which redesignated VT-7 on 1 October 1999) of TW-1 at Meridian, Mississippi, 1997-2003. TW-2 assigned early aircraft; TW-1 receiving 104 'Cockpit 21' Goshawks (first official handover to VT-23 15 December 1997, aircraft having been received on 10 December). TW-2 will convert to the T-45C as its aircraft undergo conversion. Small numbers of T-45s at both

First T-45C to wear US Marine Corps markings 2001/0100691

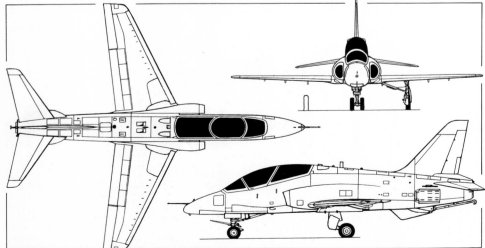

Boeing/BAE T-45C Goshawk tandem-seat basic and advanced trainer with revised intake shape *(Dennis Punnett/Jane's)* 1999/0099603

Boeing/BAE T-45C Goshawk, of Training Wing One launching from USS *John F Kennedy* 2001/0099679

Kingsville and Meridian wear 'Marines' titles to reflect the proportion (about 13 per cent) of USMC student pilots being trained on the type.

VT-21 operational 27 June 1992 for instructor training; four-aircraft operational evaluation begun by VT-21 on 18 October 1993 for one month first phase; student training begun 4 January 1994; first student flight 11 February 1994; first solo 23 March 1994; deployed to Miramar for deck landing course on aircraft carrier, September 1994; course graduated 5 October 1994. Second phase of operational evaluation (advanced tactics/weapons and carrier qualification) on USS *Dwight D Eisenhower* ended with clearance for fleet introduction being recommended on 5 July 1994. Primary sea platform is training carrier USS *John F Kennedy*. VT-23 (now VT-7) of TW-1 began student training in July 1998; VT-9 equips in 2003. T-45 syllabus is 119 sorties (156 hours) plus 10 simulator sessions (95.4 hours). In April 2000, 1,000th student graduated from T45TS course; fleet passed 200,000 hours in March 1999.

US NAVY PROCUREMENT

FY	Lot	Qty	First aircraft
88	1	12	163599
89	2	24	163611
92	3	12	163635
93	4	12	163647
94	5	12	165057
95	6	12	165069*
96	7	12	165081
97	8	12	165457
98	9	15	165469
99	10	15	165484
00	11	15	165499
Subtotal		**153**	
Planned:			
01-03	12-14	34	
Total		**187**	

Notes: Procurement rate slowed by 1994 defence review; was to have involved 24 in FY95 and 30 per year in FY96-00.

Final four batches may be combined into one multiyear procurement (MYP).

* 'Cockpit 21' installed from 12th of this batch (165080).

COSTS: Original US$316 million for 12 aircraft in FY96 procurement augmented by US$6,830,421 for addition of 'Cockpit 21' modifications. FY99 batch of 15 allocated US$342.8 million. Official unit cost US$17.2 million (1999).

DESIGN FEATURES: Carrier-capable adaptation of existing trainer design for cost and risk reduction. Generally as for two-seat BAE Hawk Srs 60 (see UK section), but redesigned (including deeper and longer forward fuselage) and strengthened to accommodate new landing gear and withstand carrier operation (incidentally increasing fatigue life to 14,400 flying hours); twin airbrakes of composites material; fin height increased by 15.2 cm (6 in) and single ventral fin added; rudder modified; tailplane span increased by 10.2 cm (4 in); wingtips squared off; nose tow launch bar added; underfuselage arrester hook, deployable 20° to each side of longitudinal axis. No provision for gun or outboard underwing hardpoints.

T-45C Goshawk 'Cockpit 21' digital avionics

FLYING CONTROLS: Differences from two-seat BAE Hawk include electrically actuated/hydraulically operated full-span wing leading-edge slats (operation limited to landing configuration); aileron/rudder interconnect; two fuselage-side airbrakes instead of one under fuselage and associated autotrim system for horizontal stabiliser when brakes deployed; BAE Systems yaw damper computer and addition of 'smurf' (side-mounted unit root fin), a small curved surface forward of each tailplane leading-edge root, to eliminate pitch-down during low-speed manoeuvres; Dowty actuators for slats and airbrakes.

STRUCTURE: BAE (principal subcontractor) builds wings, centre and rear fuselage, fin, tailplane, windscreen, canopy and flying controls. Intended fatigue life of 14,000 hours. Batch of 54 composites tailplanes being built by Boeing for retrofit to extant T-45s.

LANDING GEAR: Wide-track hydraulically retractable tricycle type, stressed for vertical velocities of 7.28 m (23.9 ft)/s. Single wheel and long-stroke oleo (increased from 33 cm; 13 in of standard Hawk to 63.5 cm; 25 in) on each main unit; twin-wheel steerable nose unit with 40.6 cm (16 in) stroke. Articulated main gear, by AP Precision Hydraulics, is of levered suspension (trailing arm) type with a folding side-stay. Cleveland Pneumatic nose gear, with Sterer digital dual-gain steering system (high gain for carrier deck operations). Nose gear has catapult launch bar and holdback devices. Main units retract inward into wing, forward of front spar; nose unit retracts forward. All wheel doors sequenced to close after gear lowering; inboard mainwheel doors bulged to accommodate larger trailing arm and tyres. Gear emergency lowering by free-fall. Goodrich wheels, tyres and brakes. Mainwheel tyres size 24×7.7-10 (20 ply) tubeless; nosewheels have size 19×5.25-10 (12 ply) tubeless tyres. Tyre pressure (all units) 22.40 bars (325 lb/sq in) for carrier operation; reduced for land operation. Hydraulic multidisc mainwheel brakes with Dunlop adaptive anti-skid system.

POWER PLANT: One 26.00 kN (5,845 lb st) nominal rating Rolls-Royce Turbomeca F405-RR-401 (navalised Adour Mk 871) non-afterburning turbofan; installed rating 24.59 kN (5,527 lb st). (FSD aircraft powered by a 24.24 kN; 5,450 lb st F405-RR-400L, equivalent to Adour Mk 861-49.) Honeywell F124-GA-400 (28.02 kN, 6,300 lb st) assessed as alternative power plant under US Congressional directive; power plant change abandoned January 1994. Plans cancelled for assembly of F405 by Allison (due to have started with the 112th engine in October 1997). Air intakes and engine starting as described for BAE Hawk. Fuel system similar to BAE Hawk, but with revision for carrier operation. Total internal capacity of 1,635 litres (432 US gallons; 360 Imp gallons). Retrofit kits ordered in October 1993 to add 133 litres (35 US gallons; 29 Imp gallons) in engine air intake tanks, but these cancelled in 1994. Provision for carrying one 591 litre (156 US gallon; 130 Imp gallon) drop tank on each underwing pylon.

ACCOMMODATION: Similar to BAE Hawk, except that ejection seats are of Martin-Baker Mk 14 NACES (Navy aircrew common ejection seat) zero/zero rocket-assisted type.

SYSTEMS: Air conditioning and pressurisation systems, using engine bleed air. Duplicated hydraulic systems, each 207 bars (3,000 lb/sq in), for actuation of control jacks, slats, flaps, airbrakes, landing gear, arrester hook and anti-skid wheel brakes. No. 1 system has flow rate of 36.4 litres (9.6 US gallons; 8.0 Imp gallons)/min, No. 2 system a rate of 22.7 litres (6.0 US gallons; 5.0 Imp gallons)/min. Reservoirs nitrogen pressurised at 2.75 to 5.50 bars (40 to 80 lb/sq in). Hydraulic accumulator for emergency operation of wheel brakes. Pop-up Dowty Aerospace ram air turbine in upper rear fuselage provides emergency hydraulic power for flying controls in event of engine or No. 2 pump failure. No pneumatic system. DC electrical power from single brushless generator, with two static inverters to provide AC power and two batteries for standby power. Onboard oxygen generating system (OBOGS).

AVIONICS: Avionics and cockpit displays optimised for carrier-compatible operations.

Comms: Rockwell Collins AN/ARN-182 UHF/VHF, Honeywell APX-100 IFF.

Flight: AN/ARN-144 VOR/ILS by Rockwell Collins, Honeywell AN/APN-194 radio altimeter, Sierra AN/ARN-136A Tacan, BAE Systems yaw damper computer. Digital avionics aircraft (No. 84 onwards) have revised navigation package comprising mission data input device (MDID); Litton LN-100G ring laser gyro and Rockwell Collins five-channel GPS linked by 12-state Kalman filter.

Instrumentation: US Navy AN/USN-2 standard attitude and heading reference system (SAHRS), Smiths Industries Mini-HUD (front cockpit), Racal Acoustics avionics management system, and Teledyne caution/warning system. Digital avionics from 84th production aircraft onwards: two 127 × 127 mm (5 × 5 in) Elbit monochrome multifunction screens in both cockpits, MIL-STD-1553B databus and Smiths HUD.

Mission: Electrodynamics airborne data recorder.

ARMAMENT: No built-in armament, but weapons delivery capability for advanced training is incorporated. Single pylon under each wing for carriage of practice multiple bomb rack, rocket pods or auxiliary fuel tank. Provision also for carrying single stores pod on fuselage centreline. CAI Industries weapon-aiming sight in rear cockpit.

DIMENSIONS, EXTERNAL:

Wing span 9.39 m (30 ft 9¾ in)

Wing chord: at root	2.87 m (9 ft 5 in)	Rudder, incl tab	0.58 m² (6.24 sq ft)	Max level speed at 2,440 m (8,000 ft)	
at tip	0.89 m (2 ft 11 in)	Tailplane	4.43 m² (47.64 sq ft)		543 kt (1,006 km/h; 625 mph)
Wing aspect ratio	5.0	WEIGHTS AND LOADINGS:		Max level Mach No. at 9,150 m (30,000 ft)	0.84
Length, overall, incl nose probe	11.98 m (39 ft 4 in)	Weight empty	4,261 kg (9,394 lb)	Carrier launch speed	121 kt (224 km/h; 139 mph)
Height overall	4.11 m (13 ft 6 in)	Internal fuel: early production	1,312 kg (2,893 lb)	Approach speed (typical)	125 kt (232 km/h; 144 mph)
Tailplane span	4.59 m (15 ft 0¾ in)	enhanced capacity	1,433 kg (3,159 lb)	Max rate of climb at S/L	2,440 m (8,000 ft)/min
Wheel track (c/l of shock-struts)	3.90 m (12 ft 9½ in)	Normal T-O weight	5,783 kg (12,750 lb)	Time to 9,150 m (30,000 ft), clean	7 min 40 s
Wheelbase	4.31 m (14 ft 1¾ in)	Max T-O weight	6,123 kg (13,500 lb)	Service ceiling	12,950 m (42,500 ft)
AREAS:		Max wing loading	346.7 kg/m² (71.02 lb/sq ft)	T-O to 15 m (50 ft)	1,100 m (3,610 ft)
Wings, gross	17.66 m² (190.1 sq ft)	Max installed power loading	249 kg/kN (2.44 lb/lb st)	Landing from 15 m (50 ft)	1,009 m (3,310 ft)
Ailerons (total)	1.05 m² (11.30 sq ft)	PERFORMANCE:		Ferry range	700 n miles (1,296 km; 805 miles)
Trailing-edge flaps (total)	2.50 m² (26.91 sq ft)	Design limit diving speed at 1,000 m (3,280 ft)		g limits	+7.33/–3
Airbrakes (total)	0.88 m² (9.47 sq ft)		575 kt (1,065 km/h; 662 mph)		**UPDATED**
Fin	2.61 m² (28.10 sq ft)	Max true Mach No. in dive	1.04		

DAIMLERCHRYSLER AEROSPACE AIRBUS/TUPOLEV

DAIMLERCHRYSLER AEROSPACE AIRBUS/TUPOLEV

PARTICIPATING COMPANIES:
 DaimlerChrysler Aerospace Airbus: see under Germany
 Tupolev: see under Russian Federation

There have been no known recent developments in this partnership's plans to produce a variant of the A310 airliner powered by cryogenic hydrogen or liquid natural gas. Details last appeared in the 2000-01 *Jane's*.

UPDATED

EADS

EUROPEAN AERONAUTIC, DEFENCE AND SPACE COMPANY NV

Drentestraat 24, NL-1083 HK Amsterdam, Netherlands
Web: http://www.eads-nv.com
PARTICIPATING COMPANIES:
 Aerospatiale Matra: see EADS France in French section
 CASA: see EADS CASA in Spanish section
 DaimlerChrysler Aerospace (DASA): see EADS Deutschland in German section
CHAIRMEN: Jean-Luc Lagardière
 Manfred Bischoff
CHIEF EXECUTIVES: Phillipe Camus
 Rainer Hertrich
COO: Gustav Humbert

Decision to merge Aerospatiale Matra of France and DASA of Germany was announced on 14 October 1999; previously mooted amalgamation of Spain's CASA with DASA was reconfirmed on 2 December 1999, increasing size of prospective company to 87,000 personnel at 90 sites; annual revenues in region of €22.5 billion. Officially formed 10 July 2000, immediately becoming world's third-largest aerospace company. Asset distribution comprises 40 per cent stock flotation; 30 per cent DaimlerChrysler; and 30 per cent owned by holding company, comprising French government via Sogepa (15 per cent), Lagardère SCA (11.1 per cent) and French financial institutions (3.9 per cent). Partners' holdings may vary by acquisition and disposal after July 2003, when CASA's 6.25 per cent share (held by state company Sociedad Estatal de Participaciones Industriales) is sold. Ownership is 34 per cent public, 30 per cent French groups and institutions, 30 per cent DaimlerChrysler and 6 per cent Spanish.

Five major operating divisions of EADS are Airbus (CEO Noël Forgeard), Aeronautics (Dietrich Russell), Military Transport Aircraft (Alberto Fernández), Space (François Auque) and Defence & Civil Systems (Thomas Enders), backed by Strategic Co-ordination (Jean-Louis Gergorin), Marketing (Jean-Paul Gut) and Finance (Axel Arendt) divisions. Aeronautics division comprises Eurocopter and EADS ATR (both in International section of this edition); EADS Socata (French section); EADS Military Aircraft (forming within EADS Deutschland); and EADS Sogerma and EADS EFW for overhaul and maintenance. On 21 December 2000 it was announced that Defence & Civil Systems was first EADS division to begin cross-border integration of assets.

Company controls, Socata and Eurocopter (100 per cent); also holds 80 per cent of Airbus Industrie and 56 per cent of Airbus Military Company; and has significant shares in Dassault (45.8 per cent), Eurofighter (43 per cent), ATR (50 per cent) and missile and space programmes. Related amalgamations in these last-mentioned areas were Matra Marconi Space and DASA Dornier on 18 October 1999 (forming Astrium; with Alenia Spazio joining later); and Matra BAe Dynamics, Aerospatiale Matra Missiles and Alenia Marconi Systems missile division on 20 October 1999.

Corporate headquarters established in Netherlands and EADS is subject to Dutch company law. Activities 76 per cent civil and 24 per cent military. Related venture with Finmeccanica of Italy is EMAC, described later in this section.

UPDATED

Diverse holdings of EADS include 43 per cent of Eurofighter and 26 per cent of Arianespace *(Paul Jackson/Jane's)*

EADS ATR

AVIONS DE TRANSPORT REGIONAL

1 allée Pierre Nadot, F-31712 Blagnac Cedex, France
Tel: (+33 5) 61 21 62 21
Fax: (+33 5) 61 21 63 18
Web: http://www.ataircraft.com
CEO: Antoine Bouvier
SENIOR VICE-PRESIDENT, COMMERCIAL: Alain Brodin
SENIOR VICE-PRESIDENT, SALES AND COMMERCIAL: Ciro Cirillo
SENIOR VICE-PRESIDENT, COMMERCIAL CUSTOMER SUPPORT AND
 ASSET MANAGEMENT: Paolo Revelli-Beaumont
SENIOR VICE-PRESIDENT, CUSTOMER SUPPORT: Ciro Vigorito
COMMUNICATIONS MANAGER: Giancarlo Fre
MEDIA RELATIONS MANAGER: Ann de Crozals

First Aerospatiale/Aeritalia (later Alenia) agreement July 1980; ATR programme started 4 November 1981; Groupement d'Intéret Economique (50:50 joint management company) formally established 5 February 1982 to develop ATR series of transport aircraft. Assembly or licensed production by Xian Aircraft in a new factory at Shenzhen, near Hong Kong, has been discussed, but not pursued; Xian already produces components for ATR. However, HAL of India (which see) anticipates opening a local assembly line if regional airlines place orders for more than 60 ATR 42/72s. First phase of this plan, announced December 1998, and on which preliminary agreement was reached on 1 February 1999, is for HAL to undertake local customisation, beginning with six ATR 42-500s to be acquired by Indian Airlines.

ATR marketing and support office opened in Washington 15 July 1986; ATR airline support centre in Singapore opened 18 November 1988; ATR Training Centre opened 1 July 1989; these functions taken over by AI(R) from 1996, but returned when this joint venture with BAe was dissolved on 1 July 1998 (see 1998-99 edition for final details of AI(R)). Sales total was 365 ATR 42s and 266 ATR 72s by 29 September 2000 to 82 customers; ATR expects to sell 30 new aircraft per year for the next 20 years; it sold 82 in 1999 compared with 61 in 1998. Phased production lines introduced in 1999 have reduced delivery times to three months from one year. The 600th ATR was delivered to Air Dolomiti on 28 April 2000. Sales by December 2000 totalled 365 ATR 42s and 266 ATR 72s, or 631 in all. The company is now planning to reach proposed ICAO Stage 4 noise levels.

ATR's Asset Management arm is responsible for second-hand sales. It sold 34 ATR 42s and 12 ATR 72s in 1999 and 19 ATR 42s and 3 ATR 72s during the first seven months of 2000.

ATR expected to have become a single corporate entity by early 1999 but this has not materialised. It is now earmarked as a division of EADS. Following dissolution of AI(R) in 1998, it revived plans to develop the AirJet 70/84-seat regional airliner, but later in the same year was undertaking talks on a joint venture of similar size with Fairchild Aerospace, involving the 728JET family at which time parallel discussions were in hand with Embraer of Brazil;

EADS ATR—AIRCRAFT: INTERNATIONAL

ATR 42-500 operated by Eurowings of Germany

however, plans to produce jet aircraft were abandoned at the end of 1999, although plans to form an alliance between the two companies continue.

UPDATED

ATR 42

TYPE: Twin-turboprop airliner.

PROGRAMME: Joint launch by Aerospatiale and Aeritalia (now Alenia) in November 1981, following June 1981 selection of P&WC PW120 turboprop as basic power plant; first flights of two prototypes 16 August 1984 (F-WEGA) and 31 October 1984 (F-WEGB); first flight production aircraft 30 April 1985; simultaneous certification to JAR 25 by France and Italy 24 September 1985, followed by USA (FAR Pt 25) 25 October 1985, Germany 12 February 1988, UK 31 October 1989; deliveries began 3 December 1985 to Air Littoral.

Series 500, with 'new look' interior, announced at Paris Air Show 1993; first flight F-WWEZ (c/n 443) 16 September 1994; certification 28 July 1995; first delivery (F-OHFF to Air Dolomiti) 31 October 1995. FAA certification 13 May 1996.

CURRENT VERSIONS: **ATR 42-300:** Initial version; phased out of production in 1996. Two Pratt & Whitney Canada PW120 turboprops, each flat rated at 1,342 kW (1,800 shp) for normal operation and 1,492 kW (2,000 shp) OEI and driving a Hamilton Sundstrand 14SF four-blade constant-speed fully feathering and reversible-pitch propeller. Additional data in 1996-97 *Jane's*.

ATR 42-320: Identical to 42-300 except for optional PW121 engines for improved hot/high performance; OWE increased/payload decreased by 5 kg (11 lb). Phased out in 1996.

ATR 42-400: P&WC PW121A engines with six-blade Hamilton Sundstrand 568F propellers; maximum cruising speed 266 kt (493 km/h; 306 mph); maximum range 825 n miles (1,527 km; 949 miles) with full payload. First flight 12 July 1995 (F-WWEF/OK-AFE); two Srs 420s ordered for CSA, and both delivered 14 March 1996, having received DGAC certification on 27 February. No further civil aircraft.

ATR 42-500: Principal ATR 42 version from 1996. Compared with Series 300, has more powerful engines, reinforced wings to allow greatly increased cruising speed and higher weights; all systems improvements of ATR 72, including flight management computers; cockpit, elevators and fin from ATR 72-210; strengthened landing gear;

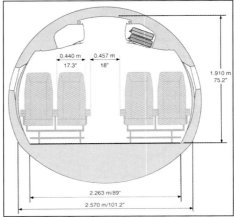

Cross-section of ATR-42/72 cabin

electrically operated main doors; reinforced fuselage and wing centre-section.

Description applies to ATR 42-500, except where indicated.

ATR 42 Cargo QC: Quick-change (1 hour) interior to hold nine containers. Conversion programme includes installation of 2.95 × 1.80 m (9 ft 8 in × 5 ft 11 in) cargo door and modification of cabin into an E-class cargo compartment with strengthening of floor to 400 kg/m² (82 lb/sq ft); total volume for cargo transport 54 m³ (1,907 cu ft); maximum payload 5,800 kg (12,785 lb). Launch customer, DHL Aviation, was due to take delivery of first aircraft before the end of 2000 and undertake retrospective conversion to the new standard during 2001.

ATR 42 F: Military/paramilitary freighter with modified interior, reinforced cabin floor, port-side cargo/airdrop door can be opened in flight; can carry 3,800 kg (8,377 lb) of cargo or 42 passengers over 1,250 n miles (2,315 km; 1,438 miles). One delivered to Gabon 1989.

ATR Calibration: Projected navaid calibration version.

ATR 42L: Projected freighter with lateral cargo door.

ATR 42 Surveyor: Maritime and rescue version; described separately.

CUSTOMERS: Total 365 firm orders by 29 September 2000. Four ordered and 10 delivered in 1998; 14 ordered and 12 delivered in 1999.

COSTS: ATR 42-500 development cost US$50 million; unit price US$13.8 million (2000).

DESIGN FEATURES: Designed to JAR 25/FAR Pt 25; high wing of medium aspect ratio, with constant-chord centre section and tapered outer panels; T-type tail with tapered tailplane and sweptback fin and fillet; pannier-mounted main landing gear.

Wing section Aerospatiale RA-XXX-43 (NACA 43 series derivative); thickness/chord ratio 18 per cent at root, 13 per cent at tip; constant-chord, no-dihedral centre-section with 2° incidence at root; outer panels 3° 6' sweepback at quarter-chord and 2° 30' dihedral.

FLYING CONTROLS: Conventional and manual. Lateral control assisted by single spoiler surface ahead of each outer flap; ailerons each have electrically actuated trim tab; fixed incidence tailplane; horn-balanced rudder and elevators, each with electrically actuated trim tab; two-segment double-slotted flaps on offset hinges with Ratier-Figeac hydraulic actuators.

STRUCTURE: Two-spar fail-safe wings, mainly of aluminium alloy, with leading-edges of Kevlar/Nomex sandwich; wing top skin panels aft of rear spar are of Kevlar/Nomex with carbon reinforcement; flaps and ailerons have aluminium ribs and spars, with skins of carbon fibre/Nomex and carbon/epoxy respectively; fuselage is fail-safe stressed skin, mainly of light alloy except for Kevlar/Nomex sandwich nosecone, tailcone, wing/body fairings, nosewheel doors and main landing gear fairings; fin (attached to rearmost fuselage frame) and tailplane carbon structure; CFRP/Nomex sandwich rudder and elevators; dorsal fin of Kevlar/Nomex and GFRP/Nomex sandwich; engine cowlings of CFRP/Nomex and Kevlar/Nomex sandwich, reinforced with CFRP in nose and underside; propeller blades have metal spars and GFRP/polyurethane skins.

Aerospatiale responsible for design and construction of wings and engine nacelles, flight deck and cabin layout, installation of power plant, flying controls, electrical and de-icing systems, and final assembly and flight testing of civil passenger versions; Alenia builds fuselage and tail unit, installs landing gear, hydraulic system, air conditioning and pressurisation systems. ATR 42/72 manufactured at St Nazaire and Nantes (France), Pomigliano d'Arco and Capodichino (Italy), and assembled in Toulouse.

Manufacturers of the ATR airframe

LANDING GEAR: Hydraulically retractable tricycle type, of Messier-Bugatti/Magnaghi/Nardi trailing-arm design, with twin wheels and oleo-pneumatic shock-absorber on each unit. Nose unit retracts forward, main units inward into fuselage and large underfuselage fairing. Goodrich wheels and multiple-disc mainwheel brakes and Hydro-Aire anti-skid units. Mainwheel tubeless tyres, size 32×8.8R16 (10/12 ply), pressure 8.69 bars (126 lb/sq in) or H34×10.0R16 (14 ply). Nosewheel tubeless tyres, size 450×190-5 (10 ply), pressure 4.34 bars (63 lb/sq in) or 450×190R5 (10 ply). Minimum ground turning radius 17.37 m (57 ft 0 in).

POWER PLANT: Two Pratt & Whitney Canada PW127E turboprops, derated from 2,051 kW (2,750 shp) to 1,610 kW (2,160 shp) normal operation and 1,790 kW (2,400 shp) OEI, giving high power reserve; ATR 72-210 nacelles; six-blade Ratier-Figeac/Hamilton Sundstrand 568F propellers with new electronic control giving faster response and better synchrophasing (as for ATR 72-500 from 1996). Power plant for ATR 42-320/400 is two PW121s, each flat rated at 1,417 kW (1,900 shp) for normal operation and 1,567 kW (2,100 shp) OEI. Propeller brake on starboard engine to enable engine to be used as auxiliary power unit for internal air conditioning.

Fuel in two integral tanks in spar box, total capacity 5,736 litres (1,515 US gallons; 1,262 Imp gallons). Single pressure refuelling point in starboard wing leading-edge. Gravity refuelling points in wing upper surface. Oil capacity 40 litres (10.6 US gallons; 8.8 Imp gallons).

ACCOMMODATION: Crew of two on flight deck; folding seat for observer. Seating for 42 passengers at 84 cm (33 in) pitch; or 46, 48 or 50 passengers at 76 cm (30 in) pitch; compared with Series 300, ATR 42-500 has completely new interior with new ceiling and sidewalls, indirect lighting, more sound damping; call buttons and reading lights relocated; overhead bins lengthened to 2 m (6 ft 6¾ in) to accommodate skis, golf clubs and fishing equipment carried as hand baggage. Baggage volume increased by 40 per cent. Active noise control system, previously offered as an option, is no longer available. ATR 42-500s have structural acoustic treatment comprising reinforcement of seven fuselage frames adjacent propeller plane; dynamic vibration absorbers in this area; and internal aluminium skin damping material forward and aft of wing.

Passenger door, with integral steps, at rear of cabin on port side. Main baggage/cargo compartment between flight deck and passenger cabin, with access from inside cabin and separate loading door on port side; lavatory, galley, wardrobe and seat for cabin attendant at rear of passenger cabin, with service door on starboard side; rear baggage/cargo compartment aft of passenger cabin; additional baggage space provided by overhead bins and underseat stowage. Entire accommodation, including flight deck and baggage/cargo compartments, pressurised and air

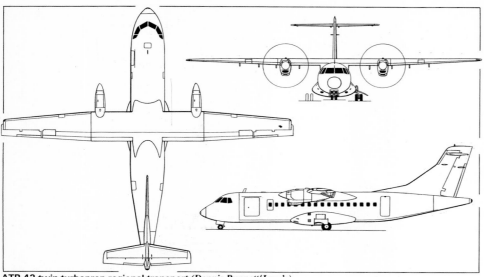

ATR 42 twin-turboprop regional transport *(Dennis Punnett/Jane's)*

conditioned. Emergency exit via rear passenger and service doors, and by window exits on each side at front of cabin.

SYSTEMS: Improved in parallel with development of ATR 72; following refers to baseline aircraft. Honeywell air conditioning and Softair pressurisation systems, utilising engine bleed air. Pressurisation system (nominal differential 0.41 bar; 6.0 lb/sq in) provides cabin altitude of 2,040 m (7,000 ft) at flight altitudes of up to 7,620 m (25,000 ft).

Two independent hydraulic systems, each at pressure of 207 bars (3,000 lb/sq in), driven by electrically operated Abex pump and separated by interconnecting valve controlled from flight deck; system flow rate 7.9 litres (2.09 US gallons; 1.74 Imp gallons)/min; one system actuates wing flaps, spoilers, propeller brake, emergency wheel braking and nosewheel steering; second system for landing gear and normal braking. Kléber-Colombes pneumatic system for de-icing of wing leading-edges, tailplane leading-edges and engine air intakes; noses of aileron and elevator horns have full-time electric anti-icing.

Main electrical system is 28 V DC, supplied by two Auxilec 12 kW engine-driven starter/generators and two Ni/Cd batteries (43 Ah and 15 Ah), with two solid-state static inverters for 115/26 V single-phase AC supply; 115/200 V three-phase supply from two 20 kVA frequency-wild engine-driven alternators for anti-icing of windscreen, flight deck side windows, stall warning and airspeed indicator pitots, propeller blades and control surface horns. Eros/Puritan oxygen system. Instead of APU, starboard propeller braked and engine run to give DC and 400 Hz power, air conditioning and hydraulic pressure.

AVIONICS: Rockwell Collins com/nav equipment. Improved in parallel with development of ATR 72; following refers to baseline aircraft.
 Comms: CVR, PA system.
 Radar: Honeywell P-660 weather radar.
 Flight: Honeywell DFZ 600 AFCS; AZ-800 ADCs; AH 600 AHRS with avionics standard communication bus; Sundstrand GPWS; Loral digital FDR; Collins DME; Honeywell FMZ-800 flight management system and dual GPS receivers installed in four Continental Airlines ATR 42s to allow autonomous approaches.
 Instrumentation: EZ-820 electronic flight instrument system.

DIMENSIONS, EXTERNAL:
Wing span	24.57 m (80 ft 7½ in)
Wing chord: at root	2.57 m (8 ft 5¼ in)
at tip	1.41 m (4 ft 7½ in)
Wing aspect ratio	11.1
Length overall	22.67 m (74 ft 4½ in)
Fuselage: Max width	2.865 m (9 ft 4½ in)
Height overall	7.59 m (24 ft 10¾ in)
Elevator span	7.31 m (23 ft 11¾ in)
Wheel track (c/l of shock-struts)	4.10 m (13 ft 5½ in)
Wheelbase	8.78 m (28 ft 9¾ in)
Propeller diameter	3.93 m (12 ft 10¾ in)
Distance between propeller centres	8.10 m (26 ft 7 in)
Propeller fuselage clearance	0.835 m (2 ft 8¾ in)
Propeller ground clearance	1.10 m (3 ft 7¼ in)
Passenger door (rear, port): Height	1.73 m (5 ft 8 in)
Width	0.64 m (2 ft 1¼ in)
Height to sill (at OWE)	1.375 m (4 ft 6¼ in)
Service door (rear, stbd): Height	1.22 m (4 ft 0 in)
Width	0.61 m (2 ft 0 in)
Height to sill	1.375 m (4 ft 6¼ in)
Cargo/baggage door (fwd, port):	
Height	1.575 m (5 ft 2 in)
Width	1.295 m (4 ft 3 in)
Height to sill (at OWE)	1.15 m (3 ft 9¼ in)
Emergency exits (fwd, each): Height	0.91 m (3 ft 0 in)
Width	0.51 m (1 ft 8 in)

Crew emergency hatch (flight deck roof):
Length	0.51 m (1 ft 8 in)
Width	0.48 m (1 ft 7 in)

DIMENSIONS, INTERNAL:
Cabin: Length (excl flight deck, incl toilet and baggage compartments)	14.72 m (48 ft 3½ in)
Max width	2.57 m (8 ft 5¼ in)
Max width at floor	2.26 m (7 ft 5 in)
Max height	1.90 m (6 ft 2¾ in)
Floor area	31.0 m² (334 sq ft)
Volume	58.0 m³ (2,048 cu ft)
Baggage/cargo compartment volume:	
front (42-46 passengers)	6.0 m³ (212 cu ft)
front (48 passengers)	4.8 m³ (170 cu ft)
front (50 passengers)	3.6 m³ (127 cu ft)
rear	4.8 m³ (170 cu ft)

AREAS:
Wings, gross	54.50 m² (586.6 sq ft)
Ailerons (total)	3.12 m² (33.58 sq ft)
Flaps (total)	11.00 m² (118.40 sq ft)
Spoilers (total)	1.12 m² (12.06 sq ft)
Fin, excl dorsal fin	12.48 m² (134.33 sq ft)
Rudder, incl tab	4.00 m² (43.05 sq ft)
Tailplane	11.73 m² (126.26 sq ft)
Elevators (total, incl tabs)	3.92 m² (42.19 sq ft)

WEIGHTS AND LOADINGS:
Operating weight empty	11,250 kg (24,802 lb)
Max fuel weight	4,500 kg (9,921 lb)
Max payload	5,450 kg (12,015 lb)
Max T-O weight	18,600 kg (41,005 lb)
Max landing weight	18,300 kg (40,345 lb)
Max zero-fuel weight	16,700 kg (36,817 lb)
Max wing loading	341.3 kg/m² (69.90 lb/sq ft)
Max power loading (with power reserve)	4.51 kg/kW (7.42 lb/shp)

PERFORMANCE:
Max cruising speed	300 kt (556 km/h; 345 mph)
T-O distance:	
ISA, S/L	1,165 m (3,825 ft)
ISA +10°C at 915 m (3,000 ft)	1,415 m (4,645 ft)
ISA, S/L for 300 n mile (556 km; 345 mile) stage with 48 passengers	990 m (3,250 ft)
Landing field length:	
S/L at landing weight with max passengers	1,040 m (3,415 ft)
S/L at max landing weight	1,126 m (3,695 ft)
Time to climb to 5,180 m (17,000 ft)	9.9 min
Service ceiling OEI, ISA +10°C, 97% max T-O weight	5,485 m (18,000 ft)
Range with max fuel	1,600 n miles (2,963 km; 1,841 miles)
Max range with 48 passengers	840 n miles (1,555 km; 966 miles)

UPDATED

ATR 42 SURVEYOR

TYPE: Maritime surveillance twin-turboprop.

PROGRAMME: Variant of ATR 42 airliner developed by Alenia. Exhibited in model form at Dubai Air Show, November 1995. Initially designated **SAR 42** and **ATR 42MP**. First airframe modified by Officine Aeronavali's Capodichino plant; maiden flight (CMX62166) in Surveyor configuration, but without equipment installed, 1 February 1999, was also delivery to Alenia at Caselle for systems integration. Italian civil certification received 28 October 1999; first delivery 15 December 1999.

CURRENT VERSIONS: Other missions include economic exclusive zone protection, environmental protection, law enforcement, medical evacuation and VIP/troop/cargo/corporate/humanitarian transport.

CUSTOMERS: Two ordered by Guardia di Finanza (Italian customs service) 1996 and first aircraft (MM62165) delivered that November but in normal transport configuration for training; converted to operational version in 2000; is model ATR 42-400 for first two aircraft; future aircraft; will be based on the 42-500.

Italian Coastguard ordered one plus another option for unarmed version; first delivery was due mid-2000.

Data as ATR 42, except particulars below.

POWER PLANT: Two 1,603 kW (2,150 shp) Pratt & Whitney Canada turboprops driving Hamilton Sundstrand Ratier 568F six-blade propellers.

ACCOMMODATION: ATR 42 has modified cabin with Alenia Difesa MPMS (maritime patrol mission system) comprising two multifunction operator consoles with 48 cm (19 in) display, 25 cm (10 in) sensor display, keyboard, trackball, joystick and colour printer on

First ATR 42 Surveyor, operated by the Italian customs service

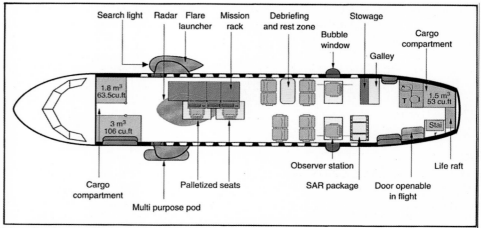

Internal and external features of the ATR 42 Surveyor

starboard side. Provision for future growth of sensor suite. Rest/debrief area towards rear of cabin; galley/lavatory facilities at rear. Observers' stations with bubble windows one each side of fuselage. Rear door modified for in-flight opening; civil-style freight door in forward port side. SAR package behind rest area.

AVIONICS: *Comms:* VHF/UHF transceiver, VHF/FM/HF comms transceiver, transponder/IFF and secure datalink.
  Radar: Weather radar from ATR 42 retained.
  Flight: ADS, VOR, DME, ADF, Tacan, FMS, IRS, INS/GPS, radio altimeter, direction-finder.
  Mission: Alenia Difesa Maritime Patrol Mission System comprises Raytheon SV 2022 360° search radar and EOST 23 surveillance system comprising Wescam turret beneath fuselage containing Galileo FLIR; Spectrolab SX16E searchlight and Thiokol LUU-2B/B flare launcher both mounted in pod on starboard side of forward fuselage; Elettronica ALR-733 ESM. MIL-STD-1553B, RS-422 and ARINC 429 databus system.

ARMAMENT: Optional FN Herstal HPM twin machine gun pod on port side of forward fuselage.

WEIGHTS AND LOADINGS:
Operating empty weight	11,300 kg (24,912 lb)
Max payload	1,100 kg (2,425 lb)
Max T-O weight	18,600 kg (41,005 lb)
Zero-fuel weight: typical	13,200 kg (29,101 lb)
max	16,700 kg (36,817 lb)

PERFORMANCE:
  Endurance on station: at 200 n mile (370 km; 230 mile) radius 8 h
  at 600 n mile (1,111 km; 690 mile) radius 3 h 30 min

UPDATED

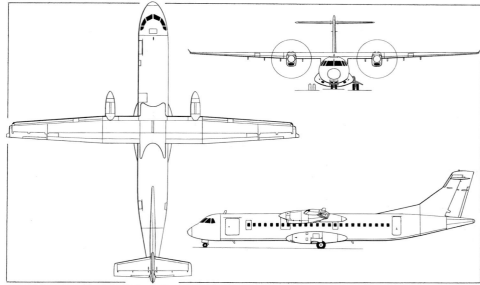

ATR 72-200 (two Pratt & Whitney Canada PW124B turboprops) *(Dennis Punnett/Jane's)*

ATR 72

TYPE: Twin-turboprop airliner.

PROGRAMME: Stretched version of ATR 42; announced at 1985 Paris Air Show; launched 15 January 1986; three development aircraft built: first flights 27 October 1988 (F-WWEY), 20 December 1988 (F-WWEZ, c/n 108) and 18 April 1989 (OH-KRA, c/n 126); French and US certification 25 September and 15 November 1989 respectively; deliveries, to Kar Air of Finland, began 27 October 1989 (OH-KRB).

CURRENT VERSIONS: **ATR 72-200:** Initial production version; two Pratt & Whitney Canada PW124B turboprops, each rated at 1,611 kW (2,160 shp) for normal take-off and 1,790 kW (2,400 shp) with ATPCS; Hamilton Sundstrand 14SF-11 four-blade propellers. Also cargo version, capable of carrying 13 small containers. Now discontinued.

ATR 72-210: Improved hot/high performance version with PW127 engines rated at 1,849 kW (2,480 shp) and Hamilton Sundstrand 247F propellers with composites blades on steel hubs; ATPCS power 2,059 kW (2,760 shp); carries 17 to 19 more passengers than standard ATR 72 in WAT-limited conditions; French and US certification 15 and 18 December 1992, German on 24 February 1993; first delivery December 1992.

ATR 72-500: Launched as ATR 72-210A. Improved hot/high performance version. DGAC certification achieved 14 January 1997.

Description applies to ATR 72-500, except where indicated.

ATR 72 Cargo QC: Planned cargo version similar to ATR 42 Cargo QC; maximum payload will be 8,250 kg (18,188 lb) and total cargo volume 73 m³ (2,578 cu ft).

CUSTOMERS: Total 266 ATR 72s ordered up to 29 September 2000, at which time 254 delivered. Deliveries in 1998 totalled 21 and in 1999 numbered 23.

DESIGN FEATURES: As ATR 42 (which see), but with more power, more fuel, greater wing span/area, and longer fuselage for up to 74 passengers.

FLYING CONTROLS: As for ATR 42 but vortex generators ahead of ailerons and aileron horn balances shielded by wingtip extensions; vortex generators under leading-edge of elevators.

STRUCTURE: Generally as for ATR 42, but new wings outboard of engine nacelles have CFRP front and rear spars, self-stiffening CFRP skin panels and light alloy ribs, resulting in weight saving of 120 kg (265 lb); sweepback on outer panels 2° 18′ at quarter-chord. Trials of an all-composites tail assembly were conducted in 1997 and the structure incorporated in all production aircraft from 1998. The major airframe inspection period for the ATR 72 was increased on 2 October 1997 from 24,000 to 36,000 cycles, with a corresponding reduction in maintenance cost, bringing it in line with the ATR 42 family.

LANDING GEAR: Messier-Hispano-Bugatti units with Dunlop wheels (tyres size H34×10.0R16 (14 ply), pressure 7.86 bars; 114 lb/sq in) and structural carbon brakes; nosewheel tyre as ATR 42. Minimum ground turning radius 19.76 m (64 ft 10 in).

POWER PLANT: *ATR 72-500:* Two Pratt & Whitney Canada PW127F turboprops, each rated at 1,864 kW (2,500 shp) for normal flight and 2,051 kW (2,750 shp) for take-off, driving Hamilton Sundstrand 568F six-blade, all-composites propellers. Fuel capacity 6,337 litres (1,674 US gallons; 1,394 Imp gallons), comprising ATR 42 tanks, plus additional 637 litres (168 US gallons; 140 Imp gallons) in outer wings; pressure refuelling point in starboard main landing gear fairing.

ACCOMMODATION: Basic 68 passengers at 79 cm (31 in) seat pitch; other seating configurations range from 64 seats at 81 cm (32 in) to 72 seats at 76 cm (30 in); plus second cabin attendant's seat. Single baggage compartment at rear of cabin; two at front. Forward door, with a service door opposite on starboard side. Service door on each side at rear, that on port side replaced by a passenger door when cargo door is fitted at front. Two additional emergency exits (one each side); both rear doors also serve as emergency exits. Increased-capacity air conditioning system.

AVIONICS: More advanced avionics of ATR 72-500 have been transferred into ATR 42-500.

DIMENSIONS, EXTERNAL: As ATR 42 except:
Wing span	27.05 m (88 ft 9 in)
Wing chord at tip	1.59 m (5 ft 2½ in)
Wing aspect ratio	12.0
Length overall	27.17 m (89 ft 1¾ in)
Height overall	7.65 m (25 ft 1¼ in)
Wheelbase	10.77 m (35 ft 4 in)
Passenger door (fwd, port): Height	1.575 m (5 ft 2 in)
Width	1.295 m (4 ft 3 in)
Height to sill	1.12 m (3 ft 8 in)

DIMENSIONS, INTERNAL:
Cabin: Length (excl flight deck, incl toilet and baggage compartments)	19.21 m (63 ft 0¼ in)
Cross-section	as for ATR 42
Floor area	41.7 m² (449 sq ft)
Volume	76.0 m³ (2,684 cu ft)

Baggage/cargo compartment volume:
front (68 passengers with front cargo door)	5.8 m³ (205 cu ft)
rear	4.8 m³ (170 cu ft)

AREAS: As ATR 42 except:
Wings, gross	61.0 m² (656.6 sq ft)
Ailerons (total)	3.75 m² (40.36 sq ft)
Flaps (total)	12.28 m² (132.18 sq ft)
Spoilers (total)	1.34 m² (14.42 sq ft)

WEIGHTS AND LOADINGS:
Operating weight empty	12,950 kg (28,550 lb)
Max fuel weight	5,000 kg (11,023 lb)
Max payload: standard	7,050 kg (15,543 lb)
optional	7,350 kg (16,204 lb)
Max T-O weight: standard	22,000 kg (48,501 lb)
optional	22,500 kg (49,604 lb)
Max ramp weight: standard	22,170 kg (48,876 lb)
optional	22,670 kg (49,979 lb)
Max landing weight: standard	21,850 kg (48,171 lb)
optional	22,350 kg (49,273 lb)
Max zero-fuel weight: standard	20,000 kg (44,092 lb)
optional	20,300 kg (44,754 lb)
Max wing loading: standard	360.7 kg/m² (73.87 lb/sq ft)
optional	368.9 kg/m² (75.55 lb/sq ft)
Max power loading: standard	5.36 kg/kW (8.81 lb/shp)
optional	5.49 kg/kW (9.01 lb/shp)

PERFORMANCE:
Max cruising speed: standard	276 kt (511 km/h; 318 mph)
optional	275 kt (509 km/h; 316 mph)
Econ cruising speed at 7,010 m (23,000 ft), at 95% MTOW	248 kt (459 km/h; 285 mph)
Service ceiling	7,620 m (25,000 ft)
Service ceiling, OEI, ISA + 10°C, 97% MTOW	4,330 m (14,200 ft)

T-O balanced field length:
at S/L, ISA: basic	1,223 m (4,015 ft)
optional	1,290 m (4,235 ft)
at 915 m (3,000 ft), ISA +10°C, at T-O weight, 68 passengers, both	1,300 m (4,265 ft)
Landing field length at MLW: basic	1,048 m (3,438 ft)
optional	1,067 m (3,500 ft)

Still air range, reserves for 87 n mile (161 km; 100 mile) diversion and 45 min:
max payload: basic	714 n miles (1,322 km; 821 miles)
optional	890 n miles (1,648 km; 1,024 miles)
max fuel and zero payload	1,956 n miles (3,622 km; 2,251 miles)

UPDATED

ATR-72-500 of Air New Zealand

2000/0099558

EAL

EUROPEAN AEROSYSTEMS LIMITED

Warton Aerodrome, Preston, Lancashire PR4 1AX, UK
Tel: (+44 1772) 63 33 33
Fax: (+44 1772) 63 47 24
CHAIRMAN: Bruno Revellin-Falcoz (Dassault Aviation)
EXECUTIVE DIRECTOR: John Tucker (BAE Systems)

Following from a technology-sharing accord of December 1992, Dassault Aviation and British Aerospace Defence signed an agreement on 31 October 1995, providing for the formation of a joint venture military aircraft company. Intention to establish EAL was announced on 4 September 1998, each participant to hold 50 per cent and contribute three members to directors' board. Initial tasks were to define and undertake research and development projects using integrated teams, formulating technologies and processes applicable both to existing products and to possible technology demonstrators for future fighter aircraft programmes. EAL's duties will further include avionics and weapons integration; methodologies of aircraft development; design and maintenance of weapons systems; and industrial fabrication processes.

UPDATED

EHI

EH INDUSTRIES LIMITED (Subsidiary of AgustaWestland)

Pyramid House, Solartron Road, Farnborough, Hampshire GU14 7PL, UK
Tel: (+44 1252) 37 21 21
Fax: (+44 1252) 38 64 80

EH Industries formed June 1980 by Westland Helicopters and Agusta (50 per cent each), to undertake joint development of new anti-submarine warfare helicopter for Royal Navy and Italian Navy within an integrated programme under which naval, army and civil variants now being produced. Programme handled on behalf of both governments by UK Ministry of Defence; GKN Westland has design leadership for commercial version, Agusta for rear-loading military/utility version; naval version being developed jointly for UK and Italian navies and export. IBM Federal Systems (now known as Lockheed Martin Aerospace Systems Integration Corporation) selected to manage Royal Navy programme in 1991, overseeing and taking responsibility for RN-specific development activity, systems integration and aircraft production and delivery. Canadian programme launched October 1992 and terminated 12 months later, but EH 101 was reselected in December 1997 for purchase as new Canadian SAR helicopter.

Kawasaki of Japan, in conjunction with trading company Okura, signed an agreement in June 1995 for joint marketing and support of the EH 101, this to include local manufacture, if warranted by orders.

A partnership between AgustaWestland, Bell Helicopter Textron and Boeing Canada will promote EH 101 in North American market, for the Canadian maritime helicopter programme and for US requirements to replace HH-60G, MH-53E and VH-3D; total US market for helicopters in this class has been calculated as 280.

UPDATED

Merlin HM. Mk 1 of the Royal Navy

2001/0099602

EH INDUSTRIES EH 101
Royal Navy designation: Merlin HM. Mks 1 and 2
RAF designation: Merlin HC. Mk 3
Canadian Forces name: Cormorant
TYPE: Multirole medium helicopter.
PROGRAMME: Stems from Westland WG 34 (see UK section of 1979-80 *Jane's*), selected in mid-1978 by UK MoD to meet SR(S) 6646 for a Sea King replacement; broadly similar requirement by Italian Navy led to 1980 joint venture with Agusta; subsequent market research confirmed compatibility of basic EH 101 design with commercial payload/range and tactical transport/logistics requirements, resulting in decision to develop naval, civil and military variants based on common airframe.

Nine month project definition phase approved by UK/Italian governments 12 June 1981; full programme go-ahead announced 25 January 1984; design and development contract signed 7 March 1984; selected by Canadian government August 1987; Italian-built iron bird ground test airframe followed by nine preproduction aircraft (PP1-9: see Current Versions); RTM 322 engines selected for RN Merlin June 1990; T700-GE-T6A engines selected for Italian Navy version September 1990; fourth UK/Italian government MoU signed 30 September 1991 (starting industrialisation phase); UK MoD commitment to 44 Merlins 9 October 1991; Canadian order for 50 (35 CH-148 and 15 CH-149) announced 24 July 1992; UK and Italian civil certification of Srs 300 and 500 planned for late 1993, but eventually achieved on 24 November 1994; US FAA approval gained on following day.

Flight testing halted following January 1993 loss of PP2; resumed 24 June 1993. US$1 million study completed for US Marine Corps, 1993, assessing EH 101 as 30-troop or cargo transport as fallback in event of Bell/Boeing MV-22 cancellation. First flight with RTM 322 engines, 6 July 1993; two aircraft to fly total of 260 hours for development. Canadian requirement cut to 43 by deletion of seven CH-148s in August 1993 as cost-saving measure, but incoming government cancelled entire programme on 4 November 1993, despite award of substantial offsets to Canadian firms. UK government confirmed EH 101 as next RAF tactical transport helicopter on 1 December 1993; formal announcement of order for 22 made 9 March 1995; Italian order confirmed October 1995 for 16, plus eight options. First production aircraft, RN01/ZH821 for Royal Navy, first flew 6 December 1995 and officially rolled out on 6 March 1996, although lacking mission avionics. Military Aircraft Release trials began in May 1998, using RN02/ZH822 (first flight 14 January 1997) and the first iteration of the Release was issued in the last quarter of 1998, allowing the start of Intensive Flight Trials with the commissioning of No. 700M Squadron (the Merlin IFTU) at RNAS Culdrose on 1 December 1998, and of RN aircrew training. First Merlin transferred to RN charge was RN05/ZH825 on 12 November 1998; first to civil customer, March 1999; first production EH 101 for Italian Navy flew 6 December 1999; first front-line squadron (824 NAS, Royal Navy) commissioned 2 June 2000 at RNAS Culdrose.

Total EH 101 hours had exceeded 10,000 in August 1999. Intensive flight operations programme (IFOP) by PP7 and PP9 concluded early, in June 2000, after 2,600 flights and 1,200 sorties (2,340 and 2,230 flying hours for each aircraft). During the previous year, the helicopter had demonstrated an MTOW of 15,500 kg (34,170 lb), or 900 kg (1,984 lb) over current limits.

CURRENT VERSIONS (general): **Srs 100/Naval:** Primary roles ASW, ASV, anti-ship surveillance/tracking, amphibious operations and SAR; other roles may include AEW, vertrep and ECM (deception, jamming and missile seduction); designed for fully autonomous all-weather operation from land bases, large and small vessels (including merchant ships) and oil rigs, and specifically from a 3,500 tonne frigate, with dimensions tailored to frigate hangar size. Capabilities include Type 23 frigate launch and recovery in conditions up to Sea State 6 with ship on any heading and windspeed (from any direction) up to 50 kt (93 km/h; 57 mph); endurance and carrying capacity needed to meet expanding maritime tactical requirements of 21st century, including ability to operate distantly for up to 5 hours with state-of-the-art equipment and weapons. See under Italian Navy and Merlin headings below.

Srs 300 Heliliner: Commercial passenger version.
Utility: Army **(Srs 400)** and civil **(Srs 500)** versions with rear ramp; naval **(Srs 200)** version without ramp.
CURRENT VERSIONS (specific): **PP1:** Westland-built first preproduction aircraft; first flight (ZF641) 9 October 1987; completed official test centre review 1991; over 375 hours flown by September 1992, at which time being refitted with T700-GE-T6A engines and 3,878 kW (5,200 shp) main gearbox. To Italy, May 1995 for 21 hour trials programme, following which it was retired from flying after some 700 hours. Used for severe weather operating trials in February 1997, operating from Type 23 frigate HMS *Grafton*. Trials included 500 blade folding cycles in up to Sea State 6, and winds of up to 68 kt (126 km/h; 78 mph).

PP2: Agusta-built second preproduction aircraft; first flight (02) 26 November 1987; deck trials aboard Italian Navy *N Grecale* and *Maestrale* July 1990; was to lead work on achieving new maximum T-O weight of 14,288 kg (31,500 lb), and had been fitted with ACSR (see Design Features), but lost in crash in Italy, 21 January 1993, following a rotor brake fire during noise measurement trials.

PP3: Westland-built; first flight (G-EHIL) 30 September 1988; first civil-configured aircraft; rotor vibration trials and flights with 'soft link' engine mounting struts; icing trials at CFB Shearwater, Canada, for five months from November 1993 (total 23 sorties); continuing programme also includes optimisation of ACSR, and weapon and development trials. Serialled ZH647 for weapons carriage in 1993. Subsequently flew the high-altitude trials programme at up to 4,570 m (15,000 ft). Used for trials to extend MTOW to 15,500 kg (34,170 lb) in support of Cormorant programme. Withdrawn from use for preservation 18 February 1999

PP4: Westland-built; first flight (ZF644) 15 June 1989; general naval variant; overwater navigation equipment trials and AFCS development; flown with RTM 322 engines 6 July 1993, initiating two-year programme of ground running and 160 flight hours. This to have been followed by 50 hours training RN aircrew for operational trials including hot-and-high (military) at Mesa, Colorado, 1995 but aircraft lost after drive train control rod failure on 7 April 1995 after 463 hours in 385 sorties. Tasks assumed by RN01/ZH821.

PP5: Westland-built; dedicated Merlin development aircraft; first flight (ZF649) 24 October 1989; Type 23 frigate interface trials in HMS *Norfolk* (including decklock, recovery handling, weapons handling, tail and blade folding) August 1991; seagoing trials in HMS *Iron Duke* (including overwater sonobuoy release) completed December 1992. Second RTM 322 testbed. Full Merlin avionics by early 1996 for operational trials of datalink, sonar, digital map, colour displays and GPS; also ship-to-helicopter operations limits trials. Used for torpedo trials during mid-1997, making 15 drops of the Sting Ray in Cardigan Bay, operating from DTEO Aberporth. Royal Navy trials, Bahamas, 1999.

PP6: Agusta-built; dedicated Italian Navy development aircraft; first flight (I-RAIA) 26 April 1989; seagoing trials in *Giuseppe Garibaldi* and *Andrea Doria* completed mid-October 1991; rotor downwash trials on behalf of Canadian Forces to assess acceptability for SAR role. Trials of prototype (Mod 1) HELRAS dipping sonar, reserialled MMX605. In storage by mid-2000; 620 hours.

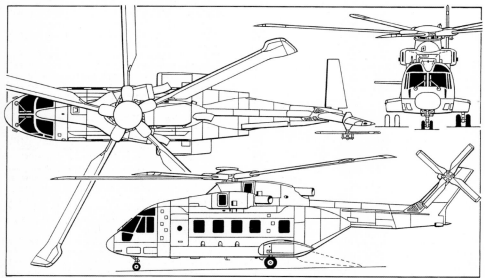

Utility version of the EH 101, showing modified rear fuselage with rear-loading ramp/door *(Mike Keep/Jane's)*

First three RAF Merlin HC. Mk 3s, marking service entry on 27 June 2000

PP7: Agusta-built; military utility development aircraft with rear-loading ramp and horizontal tail surfaces to port as well as starboard; first flight 18 December 1989; low-speed handling trials and tail rotor performance assessment 1992; fitted with ACSR 1992. Registered I-HIOI in November 1993. Severely damaged in attempted run-on landing after tail rotor pitch control failure on 20 August 1996; 450 hours of flight time. Flew again after repair 21 January 1999; re-serialled ZK101 and modified with representative in-flight refuelling probe and FLIR turrets of RAF HC. Mk 3; departed by sea on 12 June 1999 to undertake major demonstration tour along US East Coast, including evaluations by USN, USAF and Coast Guard.

PP8: Westland-built; civil variant; first flight (G-OIOI) 24 April 1990; evaluations of ADF and DME, area nav, electronic instrumentation, civil AFCS and communications equipment; fitted with ACSR. Allocated ZJ116 for trials, 1995. To Brindisi 27 March 1996 for intensive flying trials, with PP9; to Aberdeen, 14 September 1998. Attached to 700M Squadron for one week in May 1999, during which it flew demonstration flights to the Scilly Isles for BIH. Unrefuelled flight of 8 hours 15 minutes, March 2000, simulating 945 n mile (1,750 km; 1,087 mile) SAR mission with 5,500 kg (12,125 lb) of internal fuel. PP8 is fleet leader, with 3,500 flying hours.

PP9: Agusta-built; civil variant; final development aircraft (second with rear ramp); first flight (I-LIOI) 16 January 1991; extensive flight controls survey involving both ground and in-flight trials. Hot-and-high (civil) trials at Mesa, USA, 1995. To Brindisi May 1996, with PP8; to Aberdeen, 14 September 1998. Accommodated 55 people in a simulated SAR/evacuation mission, in support of the Cormorant programme. Had accumulated 2,800 flying hours by July 2000.

Civil Utility: For commercial operators requiring rear-loading facility; represented by PP9. First production **Mk 510** (I-AGWH) built for certification as a 'white tail'; rolled out in Italy 28 May 1997, and flew 17 June. Subsequently used for cold weather trials, operating from Fairbanks, Alaska, at temperatures from −5°C to −32°C (23°F to −26°F) until April 2000, and then for hot-and-high trials in USA. First order in November 1996 for a Mk 510 for Tokyo Metropolitan Police Agency; first flew September 1997; delivered from Italian production line in January 1998 as first to civil customer; fitting-out at Kawasaki Heavy Industries was delayed by bankruptcy of original financiers; eventually handed over at Gifu on 25 March 1999. Named *Ozora Ichigo* (Number One in the Big Sky) in service.

Heliliner: Commercial variant; main certification programme being flown by PP3, with PP8 as demonstrator; intended to offer 360 n mile (666 km; 414 mile) range, with full IFR reserves, carrying 30 passengers and baggage; flight crew of two, provision for cabin attendant, stand-up headroom, airline style seating, overhead baggage stowage, full environmental control, passenger entertainment, and provision for lavatory and galley. Category A VTO performance, capable of offshore/oil rig operations or scheduled flights into city centres at high all-up weights under more rigorous future civil operating rules; rear-loading ramp optional. An executive version is also proposed.

Italian Navy ASW/ASVW/Mk 110: Development aircraft PP4 (basic) and PP6; will operate from both shore bases and aircraft/helicopter carriers against surface and underwater targets with HELRAS dipping sonar and armed with two Marte Mk. 2/S ASMs. First of eight (MMX 84180 '2-01') flew on 6 December 1999.

Italian Navy AEW/Mk 112: Requirement revealed in 1994 for AEW helicopter; system unspecified. Four ordered in 1995. Also known as **ASVW/E** (Anti-Submarine and Vessel Warfare, Enhanced). MM/APS-784-based Eliradar HEW-784 air and surface surveillance radar; radome almost doubled in size (from 1.80 m; 5 ft 10¾ in) to accommodate 3.00 m (9 ft 10 in) antenna, but other avionics similar to Italian Navy Srs 100s.

Italian Navy Utility/Mk 410: Based on rear-ramp Srs 400, but with Officine Galileo GaliFlir FLIR (also specified for other Italian versions), cargo hook, basic avionics and Elettronica ESM/ECM and self-defence suite similar to Italian ASW version. Weather radar. Will serve in the commando support role with Nucleo Elicotteristico per la Lotta Anfibia at Grottaglie. Italian air force also has a requirement for EH 101s to replace two AS-61A-4 VIP helicopters.

Merlin HM. Mk 1: Royal Navy ASW version; EH 101 **Mk 111**; PP4 (basic) and PP5 were development aircraft; in satisfaction of Staff Requirement (Sea) 6646; Loral ASIC is prime contractor for systems integration; will operate from Type 23 frigates, 'Invincible' class carriers, RFAs and other ships, and land bases. Initial production aircraft, RN01/ZH821, first flew 6 December 1995; second, RN02/ZH822, first with mission avionics, flew 14 January 1997. Of first seven production Merlins, RN01 and RN02 initially assigned to operational performance acceptance procedure (OPAP) trials at the fully instrumented Atlantic Underwater Test and Evaluation Center (AUTEC) in Bahamas, 1998-99, and to assistance in sea trials; RN03 to DT&EO, Boscombe Down, in September 1997 for a six-phase military aircraft release trials programme; joined by RN02 for first, pre-IFTU stage of release, which was achieved in November 1998. The fully instrumented RN04 followed in late 1997; RN05-RN08 formed Intensive Flight Trials Unit (No. 700M Squadron) on 1 December 1998; PP5, RN02 and RN03 underwent some operational tests at AUTEC's range in the Bahamas from February 1999 (Operation Pearly King). Subsequent deep water trials were undertaken from Benbecula in 1999. In April 2000, RN12 from Boscombe Down and RN17 from Culdrose undertook extended Ship Helicopter Operational Limit trials aboard the RFA *Argus*, experiencing winds of up to 50 kt (93 km/h; 58 mph) and up to Sea State 10.

RN02, RN03 and RN14 undertook ASuW trials from Benbecula from late June 2000, following a further 30-day series of ASW operational trials off Bahamas, during Operation Trial Wizard in March-April 2000. These trials included the dropping of eight Stingray torpedoes and 800 sonobuoys.

By the end of 1998 the Royal Navy had seven Merlins, two with No. 700M Squadron. No. 700M received its third aircraft on 21 April 1999. 700M evolved from IFTU to OEU during 2001. Training role passed to No. 824 Squadron at Culdrose on 2 June 2000, with No. 814 Squadron forming as the first front-line squadron in the first or second quarter of 2001 for deployment aboard HMS *Ark Royal* after it completes a major refit during the second quarter of 2001 gaining an embarked operational capability in early 2002. The aircraft will then equip a small-ships squadron (No. 829) and a second carrier squadron, No. 820. The Royal Navy received its 25th Merlin in August 2000. An upgrade to the sonar signal and data processing system with COTS hardware and VME architecture is already planned for the HM. Mk 1 from 2001.

Merlin HM. Mk 2: Possible upgrade of Mk 1 to embrace Anti-Surface Warfare (ASuW) role with advanced, next-generation anti-ship missile (FASGW – Future Air-to-Surface Guided Weapon) to give autonomous ASuW capability; to meet SR (Sea Air) 903 under study by 1993, with improved sensors and processing equipment, plus uprated transmission for later versions of RTM 322. For retrofit from 2009; projected follow-on purchase of up to 22 RN Merlins must now be considered unlikely.

Military Utility: Tactical or logistic transport variant (represented by PP7) with rear-loading ramp; able to airlift 6 tons or up to 30 combat-equipped troops. Tail- and rotor-folding Utility version under consideration, with role options including mine countermeasures, towing EDO Mk 106 sled. Version will probably form the basis of the EHI bid to meet the RN Future Amphibious Support Helicopter/Support Amphibious Battlefield Rotorcraft (FASH/SABR) requirement to replace Sea King HC. Mk 4 RAF Pumas, and SAR Sea King HAR. Mk. 3s.

Merlin HC. Mk 3: Bid for RAF contract entered May 1994; revised cockpit layout for low-level operations; provision for pintle-mounted machine guns in side doors. Order for 22 announced 9 March 1995, in satisfaction of Staff Requirement (Air) 440; manufacturer's designation: EH 101 **Mk 411**. Each has provision for rapid installation of chin-mounted FLIR turret and (non-telescopic)

First Canadian Cormorant making its maiden flight *(Oscar Bermardi/EHI)*

Flight deck of Royal Navy Merlin HM. Mk 1   2001/0099601

Interior of RAF Merlin HC. Mk 3   2001/0099600

Civil operation of the EH 101 began with delivery to the Tokyo Metropolitan Police on 25 March 1999   2000/0054392

refuelling probe beneath the nose, offset to starboard (though neither routinely fitted). Uprated engines. Other features include integrated defensive aids system (with Nemesis directional IR countermeasures, AN/AVR-2A(V) laser warning and Sky Guardian 200 RWR), NVG compatibility, crash-attenuating seats for all occupants (two pilots; loadmaster; optional fourth crewman; and 24 troops), active noise-reduction headsets for all passengers, non-folding main rotor blades and tailboom, improved navigation suite (compared with RN version), variable-speed cargo winch and roller conveyor for cargo handling, SAR hoist to starboard and cargo hook for external loads (installation in floor resulting in slightly reduced fuel capacity). Will be capable of carrying long wheelbase Land Rover; an overload of 30 troops can board and strap-in within 2 minutes, and de-plane within 40 seconds. Critical design review completed July 1997. Assembly of first (RAF01/ZJ117) began November 1997; rolled out 25 November 1998; first flew 24 December 1998. First aircraft delivered to DERA Boscombe Down 19 January 2000. Sixth aircraft handed over to Defence Procurement Agency on 27 June 2000, this date thereby becoming official acceptance into RAF service. Fifth and sixth aircraft used for conversion of first RAF instructors at Yeovil in June 2000.

Deliveries due between September 1999 (first to Boscombe Down for DT&EO trials) and October 2001. Initial units, from April 2000, will be Rotary Wing Operational Evaluation and Training Unit (initially at Boscombe Down) and the Operational Conversion Unit (actually a flight of reformed No. 28 Squadron at Benson) with a limited Release-to-Service for day training. Delivery of the first aircraft, due on 6 November 2000, was delayed by a worldwide EH 101 grounding, following the loss of an HM.Mk 1. A new Medium Support Helicopter Aircrew Training Facility with two Merlin simulators opened at Benson on 17 July 2000. Both No. 28 and No. 72 (at Aldergrove) Squadrons will form operational flights in the first quarter of 2002. Allocation is to be 12 to Benson (No. 28 and, eventually, Rotary Wing OEU); six to Aldergrove; and four in-use reserves will be used in support of 16 Air Assault Brigade.

Merlin AEW: UK MoD funded study, submitted August 1998, of possible Sea King AEW. Mk 7 replacement to meet Royal Navy's FOAEW requirement. Ventral radome and stub-wings; latter would permit speeds up to 250 kt (463 km/h; 288 mph).

'CH-X': Offered to USMC; chin gun turret; not adopted.

CH-148 Petrel and **CH-149 Chimo:** Intended Canadian ASW and SAR versions; cancelled after payment of US$353 million in compensation; see 1994-95 and previous *Jane's*.

AW320 Cormorant: Adaptation of civil EH 101 (with ramp) offered to Canada in early 1995 as a CH-113 replacement (in place of the CH-149) for SAR missions;

EH 101 ORDERS

Country	Service	Quantity	Commitment	Mk/Version	First aircraft
Canada	Armed Forces	15	5 Jan 98	AW320 (SAR)	901
Italy	Navy	8	Oct 95	110	MMX84180
	Navy	4	Oct 95	112 (AEW)	
	Navy	4	Oct 95	410 (Util)	
Japan	Tokyo Police	1	Nov 96	510 (SAR)	JA01MP
UK	Navy	44	9 Oct 91	111/HM. Mk 1	ZH821
	RAF	22	9 Mar 95	411/HC. Mk 3	ZJ117

selected late 1997. Deck landing capability; reduced cost achieved by reliance on mainly commercial avionics and by original proposed use of twin hoists in single door, rather than one per side. Since then, the hoist arrangement has been revised, with a primary winch in the starboard door and a secondary in the forward port door. Will have Honeywell RDR-1400 radar, Litton LN-100G embedded laser INS/GPS, FLIR, Spectrolab SX-16 searchlight, crashworthy fuel tanks, Breeze-Eastern primary and secondary hoists and an underslung load hook. Planned to include provision for internal extended range tanks (ERTs) and hover in-flight refuelling from ships under way and wire-strike protection. Selected in preference to Chinook, Cougar and Sikorsky S-70 'Maplehawk'. Total of 15 ordered from Italian production to replace CH-113 Labradors, with deliveries between October 2000 and end of 2002. First (901) flew at Vergiate on 7 March 2000, with an official first flight on 31 May 2000. To equip 442 Squadron (19 Wing) at Comox, 413 Squadron (14 Wing) at Greenwood, 424 Squadron (8 Wing) at Trenton and 103 Squadron (9 Wing) at Gander. Estimated to cost 40 per cent less than original CH-149 with 108 per cent offsets ('industrial regional benefits') totalling C$629 million. The programme remains ahead of schedule and on budget. Associated industrial team includes Bombardier, Bristol Aerospace, Spar Aerospace and CAE, with Canadian Helicopters to provide a leasing and follow-on maintenance option. EHI has demonstrated a 15,500 kg MTOW (900 kg overweight) which would give a 1,080 n mile (2,000 km; 1,242 mile) SAR radius.

Cormorant MHP: Tentative designation for EH 101 subvariant offered by EHI-led AW320 'Team Cormorant' to meet remaining Canadian requirement (Maritime Helicopter Program) for a shipborne ASW Sea King replacement with multifunction radar, sonobuoy processor, dipping sonar IFF, an EO sensor and a datalink. To have a four-man crew and be capable of carrying a stretcher and two passengers or a six-man boarding party; endurance of 2 hours 50 minutes with 30-minute reserves; able to carry at least two Mk 46 torpedoes. Total 28 required for delivery from 2005 under a programme budgeted at C$2.9 billion (US$1.96 billion). To deploy aboard 12 'Halifax' class frigates, four modernised 'Iroquois' class destroyers and two fleet replenishment ships. Boeing joined the team in mid-1999 to supply and integrate the maritime patrol mission system, based on that of the Nimrod MRA. Mk 4.

CUSTOMERS: Orders totalled 98 by January 2000; see table. Total 66 for UK, comprising 44 Royal Navy Merlins and 22 RAF Merlins. Italian Navy ordered 16 (reduced from 36, and then from 24), comprising eight ASW (and ASV) versions, four AEW versions and four marines tactical transports with blade- and tail-folding; option on further eight, (six ASW, two Marines). Westland produced one Merlin in 1995, three in 1997, six in 1998 and 13 in 1999; Agusta built two civil EH 101s in 1997 and delivered first in 1998, the other being retained for trials and demonstration. Long-range utility version for logistic and tactical support role also under consideration by Italy and proposed to US Marine Corps. VVIP version offered to Saudi Arabia in 1995. Portuguese requirement for two fishery and 12 combat SAR helicopters being pursued by EHI. Singapore requires nine to 12 helicopters for its newly ordered 'La Fayette' class frigates. EH 101 is one of three remaining contenders to meet the Nordic Standard Helicopter Programme, which calls for 50 to 100 helicopters for transport, SAR, maritime and shipborne ASW duties. Reportedly under consideration in Ireland, where two or three medium helicopters are required for SAR, and two more for troop transport.

Prototypes assigned a 3,750 hour flight development programme, but had flown 5,000 hours by mid-1999. Additional 5,600 hours flown by PP8 and PP9 from Brindisi (Italy) and from September 1998 Aberdeen (Scotland) in trial between March 1996 and June 2000, to improve reliability and prove extended (2,000 hour) overhaul intervals

COSTS: Royal Navy £1.5 billion (1991) for 44 Merlin; RAF £500 million for 22 Merlin HC. Mk 3s; Canada C$4.4 billion (1992) for 50 CH-148/149, reduced to C$579 million for 15 AW320. Production investment phase (design, tooling, maturity and product support) valued at £200 million (1993). First Italian batch (16) valued at Lit1,250 billion (US$775 million), 1995, of which

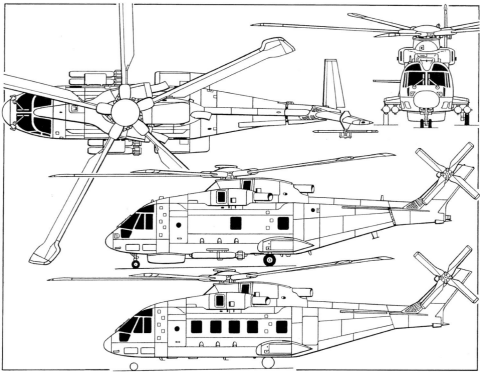

Three-view of the naval EH 101, with additional side view (bottom) of the Heliliner *(Mike Keep/Jane's)*

Westland's share is £150 million. UK official audit of early 1996 reported Merlin HM. Mk 1 as £351 million (36.1 per cent) over budget and 60 months late.

DESIGN FEATURES: Three-engine power margin with long-endurance 120 kt (222 km/h; 138 mph) cruise possible on two engines and good twin-engined hover performance. Fail-safe/ damage-tolerant airframe and rotating components, high system redundancy, onboard monitoring of engines/transmission/avionics/utility systems; airframe/ power plant/rotor and transmission systems/flight controls/ utility systems common to all variants; five-blade main rotor with multiple load path hub and elastomeric bearings; blades of advanced aerofoil section with BERP-derived high-speed tips; four-blade teetering tail rotor; transmission has minimum 30 minutes (60 minutes demonstrated) run-dry capacity; fuselage in four main modules (front and centre ones common to all variants, modified rear fuselage and slimmer tailboom on military utility variant to accommodate rear-loading ramp); automatic power folding of main rotor blades and tail rotor pylon on naval variant, with emergency manual back-up (tail section folds forward/downward, stowing starboard half of tailplane beneath rear fuselage). Folding version of utility (rear ramp) EH 101 has been designed. New active vibration cancelling system ACSR (active control of structural response) reduces vibration by 80 per cent at blade passing frequency.

Airframe overhaul interval 1,000 hours on service entry; eventual target 3,000 hours. Intended service life of 40,000 hours.

FLYING CONTROLS: Dual-redundant digital AFCS.

STRUCTURE: Rotor head of composites surrounding a metal core; composites blades; fuselage mainly aluminium alloy, with bonded honeycomb main panels; composites for such complex shapes as forward fuselage, upper cowling panels, tailfin, tailplane and windscreen. Engine air intakes of Kevlar reinforced with aero-web honeycomb. Tail unit of carbon epoxy and Kevlar epoxy skinned sandwich panels over central skeleton of metal- or foam-cored composites ribs and longerons; Kevlar-Nomex-Kevlar sandwich for leading-edge of tailfin. Single-sourced series production, with final assembly lines in Italy and UK.

LANDING GEAR: Hydraulically retractable tricycle type, with single mainwheels and steerable twin-wheel nose unit, designed and manufactured by AP Precision Hydraulics in association with Officine Meccaniche Aeronautiche. Main units retract into fairings on sides of fuselage. Goodrich wheels, tyres and brakes: main units have size 8.50-10 wheels with 24×7.7 tyres, unladen pressure 6.96 bars (101 lb/sq in); nosewheels have size 19.5×6.75 tyres, unladen pressure 8.83 bars (128 lb/sq in). Twin-mainwheel gear optional for all variants; adopted for all Italian Navy helicopters and RAF Merlin. FPT Industries emergency flotation bags.

POWER PLANT: Three Rolls-Royce Turbomeca RTM 322-01/8 turboshafts in Royal Navy Merlin (maximum contingency rating 1,724 kW; 2,312 shp, T-O rating 1,566 kW; 2,100 shp and maximum continuous rating 1,394 kW; 1,870 shp); RTM 322-02/8 in RAF version, T-O rating 1,670 kW (2,240 shp); General Electric T700-GE-T6A turboshafts in Italian naval variant, and T6A1 in Cormorant, rated at 1,521 kW (2,040 shp) for T-O and 1,327 kW (1,780 shp) maximum continuous. Engines for Italian naval variant will be assembled by Alfa Romeo Avio and Fiat; Cormorant engines assembled in Canada. Commercial and utility variants powered by three General Electric CT7-6 turboshafts (CT7-6A in PP3) with ratings of 1,491 kW (2,000 shp) for take-off and OEI and 1,282 kW (1,718 shp) maximum continuous.

Transmission rated at 4,161 kW (5,580 shp) for T-O, 3,715 kW (4,982 shp) maximum continuous and 2,769 kW (3,713 shp) OEI maximum continuous.

Standard fuel in three tanks, each of 1,074 litres (276 US gallons; 230 Imp gallons) capacity; total 3,222 litres (851 US gallons; 709 Imp gallons). Each tank feeds separate engine, except on selection of emergency cross-feed; self-sealing optional. Additional fourth or fifth tanks (all same size) optional; maximum fuel capacity 5,370 litres (1,417 US gallons; 1,181 Imp gallons). Merlin HC. Mk 3 capacity approximately 4,075 litres (1,077 US gallons; 896 Imp gallons), augmented by optional tank in cargo hold. Computerised fuel management system. Pressure refuelling point on starboard side; maximum transfer rate 682 litres (180 US gallons; 150 Imp gallons)/min; individual gravity refuelling positions on port side. Provision for detachable refuelling probe on RAF Merlins.

ACCOMMODATION: One or two pilots on flight deck (naval version will be capable of single-pilot operation, if required, and RN will operate with pilot observer and crewman; commercial variant will be certified for two-pilot operation). ASW version will normally also carry observer and acoustic systems operator. Italian Navy ASW crew members are designated pilot/mission commander, co-pilot/tactical co-ordinator, and two sensor operators.

AW320 Cormorant will operate with pilot, co-pilot/navigator, flight engineer and two crewmen. Martin-Baker crew seats in naval version, able to withstand 10.7 m (35 ft)/s impact. Socea or Ipeco crew seats in commercial variant. Commercial version able to accommodate 30 passengers four-abreast at approximate seat pitch of 76 cm (30 in), plus cabin attendant, with lavatory, galley and baggage facilities (including overhead bins). Offshore variant offers 'Club 4' grouped seating to facilitate rapid egress through windows in event of ditching. Military variant can accommodate up to 30 (seated) or 45 (non-seated) combat-equipped troops, 16 stretchers plus a medical team, palleted internal loads, or can carry externally slung loads of up to 5,443 kg (12,000 lb). SAR version can seat 28 survivors; or four seated patients, four stretchers and six medics; or 20 fully equipped Arctic rescuers with skis. It has also demonstrated an emergency evacuation capability with 55 in the cabin.

Main passenger door/emergency exit at front on port side with additional emergency exits on starboard side and on each side of cabin at rear, above main landing gear sponson. Large sliding door at mid-cabin position on starboard side, with inset emergency exit. Commercial variant has baggage bay aft of cabin, with external access via door on port side. Cargo loading ramp/door at rear of cabin on military and utility versions. Cabin floor loading 976 kg/m^2 (200 lb/sq ft) on PP1.

SYSTEMS: Hamilton Sundstrand/Microtecnica environmental control system. Dual-redundant integrated hydraulic system, pressurised by three Vickers pumps each supplying fluid at 207 bars (3,000 lb/sq in) nominal working pressure, with flow rates of 55, 59 and 60 litres (14.5, 15.6 and 15.9 US gallons; 12.1, 13.0 and 13.2 Imp gallons)/min respectively. Hydraulic system reservoirs are of the piston load pressurised type, with a nominal pressure of 0.97 bar (14 lb/sq in).

Primary electrical system is 115/200 V three-phase AC, powered by two Lucas brushless, oilspray-cooled 45 kVA generators (90 kVA if Lucas Spraymat blade ice protection system fitted), with one driven by main gearbox and the other by accessory gearbox, plus a third, separately driven standby alternator. APU for main engine air-starting, and to provide electrical power, plus air for ECS, without running main engines or using external power supplies. Lucas Spraymat electric de-icing of main/tail blades standard on naval variant, optional on others; Dunlop electric anti-icing of engine air intakes. Fire detection and suppression systems by Graviner and Walter Kidde respectively.

AVIONICS (military): Integrated systems based on two MIL-STD-1553B multiplex databusses that link basic aircraft management, avionics and mission systems.

Comms: Royal Navy version has BAE Systems communications subsystem including internal voice intercommunication to six positions, plus secure voice transmission via two AD3400 V/UHF radios, UHF and HF, plus M/A-COM Ltd ARI 5983 I-band transponder and Link 11. Racal RA 800 Light secure communications control system for RAF Merlin. Italian Navy equipment,

Heliliner version of EH 101

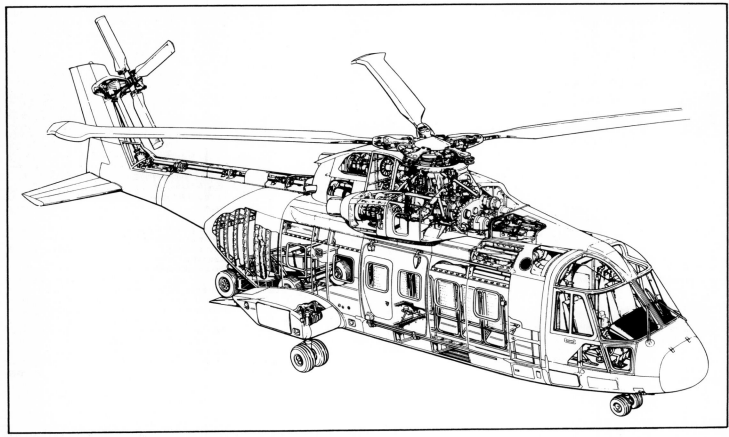

EH 101 cutaway drawing

by Elmer, includes three SRT-651 V/UHF, two SRT-170L HF, TD8503 Link 11, SP-1450 intercom and Italtel Mk 12 IFF. RAF Merlins have ADELT (automatically deployed emergency loader transmitter) containing FDR and CVR and locator beacon, scabbed on to starboard rear fuselage.

Radar: ASW version fitted with 360° search radar (pulse-compressed, frequency-agile BAE Systems Blue Kestrel 5000 in UK's Merlin; Eliradar MM/APS-784 in Italian helicopters). AEW version has Eliradar HEW-784 coherent pulse Doppler, 360° scan radar. Italian Navy utility helicopters have second-hand (from Agusta-Bell 212ASW) Officine Galileo MM/APS-705B search and weather radar which is removable, if demanded by certain missions.

Flight: Smiths Industries OMI SEP 20 dual-redundant digital AFCS is standard, providing fail-operational autostabilisation and four-axis autopilot modes (auto hover, auto transitions to/from hover standard on naval variants, optional on commercial and military variants). AFCS sensors on naval variant include BAE Systems LINS 300 ring laser gyro inertial reference unit (IRU) and Litton Italia LISA-4000 strapdown AHRS; IRU also provides self-contained navigation, with Racal Doppler 91E velocity sensor (Elmer system for Italy); Cossor Electronics GPS receiver selected for Royal Navy variant, Elmer GPS for Italian Navy aircraft. Other avionics on naval variants include Thomson-CSF AHV 16 radar altimeters (two) (Elettronica J-band units for Italian Navy), BAE Systems low-airspeed sensing and air data system, and Alenia/BAE Systems aircraft management computer.

Instrumentation: Litton EFIS with six Smiths multifunction screens.

Mission: On naval variant, main processing element of management system is a dual-redundant aircraft management computer which carries out navigation, control and display management, performance computation and health and usage monitoring of principal systems (engines, drive systems, avionics and utilities); it also controls basic bus. Alenia/Racal cabin mission display unit. Surveillance radar (see above). Underwater detection in Royal Navy Merlin is by active/passive sonobuoys and Thomson Marconi Sonar TMS 118 ADS (active dipping sonar) system, incorporating same company's AQS-903 acoustic processor for LOFAR, DIFAR, VLAD, Barra, DICAS and CAMBS buoys and Thomson Sintra folding lightweight acoustic system for helicopters (FLASH) expandable array, which has 750 m (2,460 ft) of cable and is manoeuvred by a high-speed winch. Merlin has Racal Lightweight Common Control Unit as interface between crew and tactical navigation and communications systems. The HM. Mk 1 has four units (one each for pilot, co-pilot and two operators); HC. Mk 3 has one in the cabin primarily for maintenance management. Royal Navy Merlin also has Racal Orange Reaper ESM, Normalair-Garrett mission recorder and sonobuoy/flare dispenser, Chelton sonobuoy homing system and Ultra processor for Link 11 datalink. The TMS 118 sonar is to be upgraded from 2001 with new, commercial, off-the-shelf processors, adding to speed and capacity. Honeywell HELRAS Mod 2 dipping sonar, Alenia AYK-204 processors (four) and Alenia SL/ALR-735 ESM in Italian ASW version, which also includes two operators' consoles with 350 mm (14 in) displays. ASST (anti-ship surveillance and tracking) version will carry equipment for tactical surveillance and OTH (over the horizon) targeting, to locate and relay to a co-operating frigate the position of a target vessel, and for mid-course guidance of frigate's missiles. On missions involving patrol of an exclusive economic zone it can also, with suitable radar, monitor every hour all surface contacts within area of 77,700 km² (30,000 sq miles); can patrol an EEZ 400 × 200 n miles (740 × 370 km; 460 × 230 miles) twice in one sortie; and can effect boarding and inspection of surface vessels during fishing protection and anti-smuggling missions. FLIR specified for all Italian Navy versions; BAE Systems MST-S FLIR turret, including magnifying mode, can be rapidly installed on RAF Merlins.

Self-defence: RAF Merlins have integrated defensive aids including Raytheon Danbury laser detection, BAE Systems Sky Guardian 2000 RWR, Doppler-based MAWS, Northrop Grumman AN/AAQ-24 Nemesis DIRCM and BAE Systems North America AN/ALE-47 chaff/flare dispensers. Chaff and flare dispensers on all Italian versions, co-ordinated by Elettronica ELT/156X (V2) RWR and BAE Systems RALM/1 laser warner.

AVIONICS (civil): Integrated avionics system of commercial variant based on ARINC 429 data transfer bus.

Comms: Racal intercom system; Rockwell Collins or Honeywell communications system.

Radar: Honeywell weather radar.

Flight: BAE Systems Canada CMA-900 flight management system for fuel flow, fuel quantity and specific range computations; tuning of nav/com radios; interfaces with electronic instrument systems; two-dimensional multisensor navigation; and built-in navigational database with update service. AFCS sensors on commercial variant include two Litton Italia LISA-4000 strapdown AHRS. Standard avionics include Penny and Giles air data system.

Instrumentation: Smiths Industries/OMI electronic instrument system (EIS) providing colour flight instrument, navigation and power systems displays. CMA-900 includes colour CRT display with graphics and alphanumeric capability.

EQUIPMENT: ASW variants have two sonobuoy dispensers, external rescue hoist and Fairey Hydraulics (Merlin) decklock. BAJ Ltd four-float emergency flotation gear.

ARMAMENT (naval and military utility versions): Naval version able to carry up to four homing torpedoes (BAE Systems Sting Ray on Merlin; Mk 46 or Eurotorp MU90 on Italian version) or other weapons, including Mk 11 Mod 3 depth charges. ASV version designed to carry two air-to-surface missiles (Marte Mk 2/S for Italian Navy) and other weapons, for use as appropriate, from strikes against major units using sea-skimming anti-ship missiles to small arms deterrence of smugglers. Armament optional on military utility versions; options include pintle-mounted machine guns in doorway/rear ramp, chin turret for 12.7 mm machine gun and stub-wings for rocket pods.

DIMENSIONS, EXTERNAL (A: naval variant, B: Heliliner, C: military/utility variant):

Main rotor diameter	18.59 m (61 ft 0 in)
Tail rotor diameter	4.01 m (13 ft 2 in)
Length:	
overall, both rotors turning	22.80 m (74 ft 9¾ in)
fuselage	19.53 m (64 ft 1 in)
main rotor and tail pylon folded:	
A	15.75 m (51 ft 8 in)
C	16.00 m (52 ft 6 in)
Width: cabin	2.80 m (9 ft 2¼ in)
fuselage overall (port sponson to starboard tailplane)	5.09 m (16 ft 8¼ in)
over sponsons	4.61 m (15 ft 1½ in)
main rotor and tail pylon folded:	
A	5.20 m (17 ft 0¾ in)
C	5.60 m (18 ft 4½ in)
crew door open	5.32 m (17 ft 5½ in)
Height: overall, both rotors turning	6.62 m (21 ft 8¾ in)
main rotor and tail pylon folded:	
A	5.20 m (17 ft 0¾ in)
C	5.30 m (17 ft 4¾ in)
Tailplane half-span	2.78 m (9 ft 1½ in)
Wheel track	4.55 m (14 ft 11¼ in)
Wheelbase	6.98 m (22 ft 11 in)

Test deployment of the HELRAS sonar array from EH 101 prototype PP6 *(L-3 Communications)*

Passenger door (fwd, port): Height	1.70 m (5 ft 7 in)
Width	0.91 m (3 ft 0 in)
Sliding cargo door (mid-cabin, stbd):	
Height	1.55 m (5 ft 1 in)
Width	1.83 m (6 ft 0 in)
Baggage compartment door (rear, port, B):	
Height	1.38 m (4 ft 6 in)
Width	0.55 m (1 ft 10 in)
Rear-loading ramp/door (rear, military/utility variant):	
Height	1.95 m (6 ft 4¾ in)
Width	2.26 m (7 ft 5 in)
Main rotor ground clearance (turning)	4.70 m (15 ft 5 in)
DIMENSIONS, INTERNAL:	
Cabin:	
Length: A	7.09 m (23 ft 3 in)
C	6.50 m (21 ft 4 in)
Max width	2.49 m (8 ft 2 in)
Width at floor	2.26 m (7 ft 5 in)
Max height: B	1.90 m (6 ft 2¾ in)
Volume: A	29.0 m³ (1,024 cu ft)
B	27.5 m³ (970 cu ft)
Baggage compartment volume (B)	3.8 m³ (135 cu ft)
Rear ramp (C): Length	2.10 m (6 ft 10¾ in)
Width	1.80 m (5 ft 10¾ in)
AREAS:	
Main rotor disc	271.51 m² (2,922.5 sq ft)
Tail rotor disc	12.65 m² (136.2 sq ft)
WEIGHTS AND LOADINGS (A, B, C, as above):	
Operating weight empty (estimated):	
A	10,500 kg (23,149 lb)
B (IFR, offshore equipped)	9,300 kg (20,503 lb)
C	9,350 kg (20,613 lb)
Merlin HC. Mk 3	10,250 kg (22,597 lb)
Max fuel weight (four internal tanks, total):	
A (JP-1)	3,406 kg (7,509 lb)
B, C (JP-4)	3,360 kg (7,408 lb)
Merlin HC. Mk 3	3,200 kg (7,055 lb)
Max fuel weight (five internal tanks, total):	
B, C (JP-4)	4,200 kg (9,259 lb)
Disposable load/payload:	
A (four torpedoes)	960 kg (2,116 lb)
B (30 passengers plus baggage)	2,850 kg (6,283 lb)
C (24 combat-equipped troops)	3,120 kg (6,878 lb)
Max underslung load	4,535 kg (10,000 lb)
Max T-O weight: A, B, C	14,600 kg (32,188 lb)
Max disc loading: A, B, C	53.8 kg/m² (11.01 lb/sq ft)
Transmission loading at max T-O weight and power	
A, B, C	3.51 kg/kW (5.76 lb/shp)
PERFORMANCE:	
Never-exceed speed (V_{NE}) at S/L, ISA	
	167 kt (309 km/h; 192 mph) IAS
Average cruising speed	150 kt (278 km/h; 173 mph)
Best range cruising speed	140 kt (259 km/h; 161 mph)
Best endurance speed	90 kt (167 km/h; 104 mph)
Service ceiling	4,575 m (15,000 ft)
Range (B):	
four tanks, offshore IFR equipped, with reserves	
	610 n miles (1,129 km; 702 miles)
five tanks, offshore IFR equipped, with reserves	
	750 n miles (1,389 km; 863 miles)
Ferry range: C (four tanks plus internal auxiliary tank)	
	1,130 n miles (2,093 km; 1,300 miles)
SAR radius for 26 survivors, with two minute hover per survivor	350 n miles (648 km, 403 miles)
Endurance	5 h
g limit	+3

UPDATED

First production Italian military EH 101 is a naval ASW variant 2001/0099598

EMAC

EUROPEAN MILITARY AIRCRAFT COMPANY

PARTICIPATING COMPANIES
EADS: see this section.
Finmeccanica: see Alenia under Italy.

On 14 April 2000, Finmeccanica and EADS agreed to fund a joint (50:50) venture company with the working title of EMAC and an official starting date of 1 January 2001. Subsequently, this was delayed slightly, to first quarter of 2001. EMAC has a 62.5 per cent share in Eurofighter (which see) and 57.5 per cent share in Panavia (see *Jane's Aircraft Upgrades*).

In all, EMAC employs 17,000 personnel and has total revenues of €2.5 billion. It comprises all civil and military activities of Alenia Aeronautica, plus combat aircraft operations of EADS Deutschland and EADS CASA, plus aircraft spares activities of EADS Deutschland's military division.

UPDATED

EUROCOPTER

EUROCOPTER SAS (Component of EADS)

Aéroport International Marseille-Provence, F-13725 Marignane Cedex, France
Tel: (+33 4) 42 85 85 85
Fax: (+33 4) 42 85 85 00
Web: http://www.eurocopter.com
CHAIRMAN: Jean-François Bigay
DIRECTOR OF COMMUNICATIONS: Xavier Pauporchin
CHIEF, PRESS & INFORMATION: Jean-Louis Espes

Eurocopter SA formed 16 January 1992 by merger of Aerospatiale and MBB (DASA) helicopter divisions. Share capital then held on two levels: Eurocopter Holding owned 60 per cent by Aerospatiale and 40 per cent by DASA; capital of Eurocopter SA held 75 per cent by Eurocopter Holding and 25 per cent by Aerospatiale. However, on 30 May 1997, the separate elements of Eurocopter were merged into a single management structure. Former Eurocopter France, Eurocopter International and Eurocopter Participations became a single entity (Eurocopter France, trading as Eurocopter), with Eurocopter Deutschland (in Germany) as a wholly owned subsidiary. Following formation of EADS, Eurocopter reconstituted on 18 September 2000 as simplified stock company (SAS)

Eurocopter products cover 75 to 80 per cent of the range of helicopters in terms of size and capacity, but company intends to expand this to 95 per cent, particularly with co-operative development of Russian Mil Mi-38 (see Euromil entry in this section).

Employees total 9,650. Eurocopter and its predecessors have produced some 11,000 aircraft for over 1,740 customers in 132 countries. In early 2000, some 8,400 Eurocopter helicopters were active.

Orders received in 2000 for 531 helicopters, comprising 61 Colibris (see Eurocopter/CATIC/ST Aero entry), 137 single-engine and 13 twin-engine Ecureuil/Fennecs, 40 EC 135s, seven BK 117s (see Eurocopter/Kawasaki), 18 Dauphin/Panthers, seven EC 155s, 243 NH 90s and five Super Puma/Cougars (including first EC 225/725s). Deliveries for the same period totalled 289.

New, 1997 structure absorbed overseas elements previously managed by Eurocopter Participations; these include American Eurocopter Corporation (USA), Eurocopter Canada (which see in Canadian section), Eurocopter Service Japan, Eurocopter International Pacific, Samaero (Singapore), Helibrás (Brazil), Eurocopter de Mexico, Lansav (South Africa) and MBB Helicopter Systems (UK). Partly owned are MBB Kutlutas Helikopterleri (Turkey), Euroaircraft Services (Malaysia), MBB Helicopter and Transport (Nigeria), Philippine Helicopter Services, Eurocopter International Belgium, and Eurocopter do Brasil.

Eurocopter France
Address as above
Company headquarters and main production centre at Marignane (Marseille); works at La Courneuve (Paris).

Eurocopter Deutschland
D-81663 Munich, Germany
Tel: (+49 89) 60 00 04
Fax: (+49 89) 60 00 90 33
CHAIRMAN: Dr Siegfried Sobotta
PUBLIC RELATIONS EXECUTIVE: Christina Gotzhein

Industrial concern in charge of production and product support of products originating in German part of Eurocopter. Wholly owned subsidiary.

Launched **Hubschrauber 2010** project early 1995 as four-year study into concepts for civil helicopter to enter service in decade 2010 to 2020; half of required DM120 million (US$80 million) provided by German government; targets include halving of operating costs (partly through 30 per cent reduction in fuel consumption), 20 per cent cut in empty weight, increased cruising speeds up to 190 kt (352 km/h; 219 mph) and noise levels 10 EPNdB below current ICAO limit.

Eurocopter Deutschland, the Federal Ministry of Defence and German Aerospace Research Establishment are jointly providing DM40 million to 50 million funding for development of **Helicopter Simulator for Technology, Operations and Research (HeSTOR)**, based on EC 135, to replace BO 105-based Advanced Technology Testing Helicopter System (ATTHeS) which was lost in an accident on 19 May 1995.

Eurocopter España
EADS-CASA, Avenida de Aragón 404, E-28022 Madrid
PUBLIC RELATIONS EXEUCTIVE: Francisco Salido

Established 28 September 2000, replacing former subsidiary Helicopteros España (HESA). Works at Quatro Vientos responsible for engineering, production and customer support, on par with French and German divisions.

Eurocopter Romania
SC IAR SA, R-2200 Brasov

Agreement to form Eurocopter Romania signed with IAR (which see) on 22 December 2000. Joint venture markets Eurocopter products in Romania; performs subcontract work on support of Puma and Alouette for Eurocopter; and maintains, supports and upgrades Romanian and export Pumas.

UPDATED

EUROCOPTER 665 TIGER/TIGRE

TYPE: Attack helicopter.
PROGRAMME: France and Germany agreed in 1984 to develop a common combat helicopter; Eurocopter Tiger GmbH formed 18 September 1985 to manage development and manufacture for French and German armies; was not a full member of Eurocopter because it was working on a single government contract; executive authority for programme is DFHB (Deutsch Französisches Hubschrauberbüro) in Koblenz; procurement agency is German government BWB (Bundesamt für Wehrtechnik und Beschaffung).

Original 1984 MoU amended 13 November 1987; FSD approved 8 December 1987; main development contract awarded 30 November 1989, when name Tiger (Germany)/Tigre (France) adopted; five development aircraft built, including three unarmed aerodynamic prototypes, used also for core avionics testing (PT1, 2 and 3), one (PT4) in HAP (initially called Gerfaut) configuration and one (PT5) as UHT prototype; PT1 rolled out 4 February 1991; first flight 27 April 1991; fifth prototype flew on 21 February 1996, at which time the first four aircraft had accumulated 1,090 flying hours; further details below; total of 2,500 hours flown by December 1999. Germany confirmed purchase of full 212 required, 1994, having considered cut to 138.

Industrialisation phase brought forward by two years to strengthen export prospects and Franco-German MoU signed 30 June 1995; timetable then was first deliveries in 1999 to France (approximately 10) and for export, but France announced spending moratorium in November 1995, postponing authorisation of further funding commitments until signature of a FFr2.5 billion (DM733.6 million) production investment contract on 20 June 1997. Deliveries then expected in 2001, but further delayed to July 2003 (for HAP; 2011 for HAC) by May 1996 defence plan, which envisaged procurement of only 25 Tigres in 2000-2002 budgets.

INTERNATIONAL: AIRCRAFT—EUROCOPTER

Eurocopter Tigre HAP prototype

In October 1996, Germany announced a 12-month delay in launching Tiger production because of funding constraints. However, the government planned to recoup lost time by accelerating production when eventually begun, maintaining in-service date (ISD) of 2001 and having 50 delivered by 2006, after which the manufacturing tempo would be reduced. However, by 1999, first UHT delivery planned in December 2002. France indicated in early 1997 that it would be prepared to see a single Tiger production line located at Donauwörth in Germany which, combined with other economies, would reduce French expenditure by FFr13.5 billion, but a second assembly line at Marignane has now been confirmed. Production investment agreed June 1997.

On 20 May 1998, France and Germany signed a commitment to order an initial joint batch of 160 Tigers. However, planned late 1998 placing of contracts was delayed by requirement of new German government to conduct a defence review; options included delaying ISD; or reducing numbers; or even cancelling UHT and procuring French HAP version. Production contract was finally signed on 18 June 1999 for the full 160 aircraft; first deliveries in 2002. Production of the first batch of 320 engines (plus 12 spares) began during 2000, and will continue through 2011. June 1999 contract also formalised German contract change from PAH2 to UHT and French change from HAP to HAP-F (*Finalisé*).

Joint team at Marignane is flight testing basic helicopter, updating avionics during trials, and testing HAP variant; similar team at Ottobrunn is qualifying basic avionics, Euromep mission equipment package, and integrating weapons system. Rotor downwash problems resulted in trial forward positioning of horizontal stabiliser; by mid-1994 definitive solution adopted of reversion to original position, but halving area. By January 1998, the design had been frozen, and the development programme was more than 90 per cent complete. By third quarter 2000 there was speculation that France might withdraw from the Trigat missile programme, even as test-firings of the German Trigat-LR were successfully completed.

Eurocopter previously teamed with BAe Defence Dynamics Division for £25 billion UK Army competition for 91 combat helicopters; UK considered full membership of Eurocopter Tiger programme, achieving 100 per cent offset, but Tiger rejected in favour of AH-64A Apache, July 1995.

CURRENT VERSIONS (general): Three versions in two basic configurations with about 80 per cent commonality:
U-Tiger is basis of the UHT and HAC, both with mast-mounted sight and Trigat missiles; **HCP** (*Hélicoptère de Combat Polyvalent*) is basis of the HAP (roof sight and turreted gun).

Tigre HAP: *Hélicoptère d'Appui et de Protection*; name Gerfaut dropped late 1993; escort and fire support version for French Army; armed with 30 mm Giat AM-30781 automatic cannon in undernose turret, with 150 to 450 rounds of ammunition; four Mistral air-to-air missiles and two pods each with twenty-two 68 mm unguided TDA rockets delivering armour-piercing darts, mounted on stub-wings, or 12-round rocket pod instead of each pair of Mistrals, making total of 68 rockets; roof-mounted sight, with TV, FLIR, laser range-finder and direct view optics sensors; image intensifiers integrated in helmets; and extended self-defence system. HAP configuration was approved by late 1998, permitting type qualification in December 2002. Deliveries to begin in 2003; final aircraft due in 2010.

UHT: *Unterstützungshubschrauber Tiger* (previously designated UHU); German Army multirole 'utility' or multirole anti-tank and fire-support helicopter for delivery from 2002; replaces dedicated anti-tank PAH-2 Tiger; type qualification due in December 2002. Underwing pylons for HOT 3 or (from 2006) Trigat missiles, Stinger self-defence missiles, unguided rockets, gun pod and extended self-defence system; mast-mounted TV/FLIR/laser ranger sight for gunner; nose-mounted FLIR for piloting. A mid-life upgrade for the UHT may integrate the Mauser 30 mm gun in a chin turret which traverses ±140° in azimuth and from +20° to –45° in elevation.

Tigre HAC: *Hélicoptère Anti-Char*; anti-tank variant for French Army. Type qualification due in third quarter of 2011; same weapon options (except Mistral AAM in place of Stinger), mast-mounted sight and pilot FLIR system as UHT. A mid-life upgrade to the HCP could see addition of a mast-mounted automatic air surveillance and warning system, in the form of DAV pulse Doppler radar, together with HUMS and an IR jammer.

Export: A basic export version combining features of French and German versions; offered (unsuccessfully) to UK and Netherlands.

Aussie Tiger (HCP): Proposed hybrid Tiger variant to meet Australian Army Air 87 Requirement. Based on French HAP, with undernose Giat 30-781 30 mm cannon, roof-mounted sight and provision for underwing rocket pods, but with added anti-tank capability, initially with HOT missile then (from 2006) with Trigat AC3G. Australia also requires integration of the AGM-114 Hellfire ATM. Maximum mission weight of 6,100 to 6,300 kg (13,448 to 13,889 lb).

HCP Tiger (HCP): Export version based on French Army HAP with the same undernose gun turret and roof-mounted sight, HOT 3 and Trigat missiles; Hellfire optional. Strix roof sight (direct view optics optional) and additional laser designator plus video signal interfacing for Trigat operation. Either Mistral or Stinger air-to-air missiles. A mid-life upgrade to the HCP could see addition of a DAV mast-mounted air surveillance radar (pulse Doppler type) or a mast-mounted MMW radar for automatic ground and air surveillance. No gun pod option.

Tiger T800: LHTEC T800 or CTS800 engines proposed as an engine option for the Tiger. Turkey is sales prospect, as Army has reservations about growth potential of MTR390.

CURRENT VERSIONS (specific): **PT1/F-ZWWW:** Aerodynamic prototype; basic avionics; first flight 27 April 1991. Successively fitted with aerodynamic mockups of mast-mounted and roof-mounted sights, nose-mounted gun and weapon containers. Relegated to ground fatigue testing and static display in early 1996 on completion of flight programme. Flown 502 hours.

PT2/F-ZWWY: HAP aerodynamic configuration; full core avionics; rolled out 9 November 1992; first flight 22 April 1993. Used for radar cross-section and detectability tests. Retrofit with HAP systems completed in November 1996; redesignated PT2R. Mistral launch trials at Landes ranges 14/15 December 1998; technical assessment by French Army at Valence between 17 May and 3 June 1999; rocket qualification, June 1999. Used for HAP version qualification (redesignated PT2R2) at Landes test centre between 4 April and 12 May 2000. Redesignated PT2X to serve as HCP Tiger demonstrator.

PT3/9823: Full core avionics (including navigation and autopilot); first flight (as F-ZWWT) 19 November 1993. Retrofit with UHT systems began in February 1997; redesignated PT3R; Euromep C (see Avionics) from late 1997. HOT launches with mast sight at extreme range in night and smoke conditions, June 1999; hot weather trials at Bateen AB in Abu Dhabi September 1999. Moved back to France for HAC development.

PT4/F-ZWWU: HAP aerodynamic configuration and avionics (including roof sight, HUD and Topowl helmet sight; first Tiger with live weapons system); first flight 15 December 1994. Sighting system trials early 1995; Giat cannon trials (15 ground-based tests) completed at Toulon, April 1995; full testing began at CEV Cazaux, 21 September 1995 and, by late November, had demonstrated airborne cannon firing and launch of Mistral AAM (without seeker); by 1 January 1997 had fired eight Mistrals, 3,000 cannon rounds and 50 rockets; 1997 trials included two more Mistrals, rockets and tests of gun controls. Painted in three-tone disruptive camouflage. Winter trials in Sweden, early 1997 with skid/skis landing gear. Flown 296 hours to 1 December 1997; crashed during night low-level evaluation by Australian Army 17 February 1998.

PT5/9825: Full UHT avionics; first flight 21 February 1996. Undertook German Army weapon trials (Stinger, HOT 2 and 12.7 mm podded gun) in 1997 including the firing of six HOT 2s using Euromep Osiris mast-mounted sight. Retrofitted as PT5R with production-standard weapon system; first flight 8 October 1999.

PT6 and **PT7:** Static test airframes for fatigue and crash-resistance trials.

PS1: Preseries aircraft being built on production tooling; laid down in third quarter of 1998; expected to fly in December 2000. Will validate production methods and planned production configuration.

UHT S01: First true production aircraft; planned to fly on 1 March 2002. To be followed by S02 late 2002. Deliveries December 2002 and April 2003, respectively. S01 will be used for six-month techeval/opeval trials, replacing PT5R.

HAP S01: First production French Tiger, planned to fly during early 2003; delivery to French Army in July 2003.

CUSTOMERS: Original requirement was for 427 (France 75 HAP and 140 HAC, Germany 212 PAH-2); UHU (later UHT) version substituted for PAH-2s in 1993; French order amended by 1994 to 115 HAP and 100 HAC but may be reduced to overall total of 180; Germany committed to

Tiger pilots occupy the front cockpit (left), the gunner located behind

212, of which 112 to be funded between 2001 and 2009; initial commitment of 20 May 1998 confirmed 80 each by France (70 HAP, 10 HAC) and Germany (80 UHT), as agreed by Franco-German Security Council on 9 December 1996. Germany's Tigers are required to equip four 48-aircraft regiments, each supporting an Army division. A joint training school at le Luc is being established as the Ecole Franco-Allemand, or EFA with a Thomson Training and Simulation/STN ATLAS aircrew training system, including six-axis motion simulators and wide-angle visual systems. The planned training course will last 28 weeks for a crew chief, or 19 weeks for a pilot, courses will begin in 2001-2002. Fleet of 14 German and 13 French Tigers will be assigned by 2006, with eleven simulators. In October 2000 there were reports that the German MoD was considering reducing its Tiger buy to 100 helicopters.

Also under consideration in early 2000 by Australia, Poland, Spain and Turkey. Exports of 200 Tigers thought possible.

COSTS: Tiger current development cost, shared equally by France and Germany, reported DM2.2 billion. Production tooling cost FFr2.6 billion (US$500 million) (1996). Unit cost (1996) for UH estimated as US$11 million, including launchers and all government-furnished equipment. Initial batch of 160 assigned FFr21.5 billion, of which FFr13 billion for 80 German helicopters and FFr8.5 billion for 80 French (1998).

DESIGN FEATURES: Robust, tandem-seat design with pylon-mounted armament, representing current combat helicopter style. FEL (fibre elastomer) main rotor designed for simplicity, manoeuvrability and damage tolerance; has infinite life except for inspection of elastomeric elements at more than 2,500 hour intervals; hub consists of titanium centrepiece (including duct for mast-mounted sight) with composites starplates bolted above and below; flap and lead/lag motions of blades allowed by elastic bending of neck region and pitch change by elastic part of elastomeric bearings; lead/lag damping by solid-state viscoelastic damper struts faired into trailing-edge of each blade root; equivalent flapping hinge offset of 10.5 per cent gives high control power; SARIB passive vibration damping system between transmission and airframe; three-blade Spheriflex tail rotor has composite blades with fork roots; built-in ram air engine exhaust suppressors.

FLYING CONTROLS: Fully powered hydraulic flying controls by SAMM/Liebherr; Labinal/Electrométal servo trim; horizontal tail mounted beneath tail rotor; autopilot is part of basic avionics system (see under Avionics heading).

STRUCTURE: 80 per cent CFRP, block and sandwich and Kevlar sandwich; 6 per cent titanium and 11 per cent aluminium; airframe structure protected against lightning and EMP by embedded copper/bronze grid and copper bonding foil; stub-wings of aluminium spars with CFRP ribs and skins; titanium engine deck may be replaced by GFRP; airframe tolerates crash impacts at 10.5 m/s (34.4 ft/s) and meets MIL-STD-1290 crashworthiness standards; titanium main rotor hub centrepiece and tail rotor Spheriflex integral hub/mast; blade spars filament-wound; GFRP, CFRP skins and subsidiary spars and foam filling.

Eurocopter France responsible for transmission, tail rotor, centre-fuselage (including engine installation), avionics, fuel and electrical systems, weight control, maintainability, reliability and survivability; Eurocopter Deutschland for main rotor, flight control and hydraulic systems, front and rear fuselage (including cockpits), prototype assembly, flight characteristics and performance, stress and vibration testing and simulation.

LANDING GEAR: Tailwheel type, non-retractable, with single wheel on each unit. Designed to absorb impacts of up to 6 m (20 ft)/s. Main gear by Messier-Bugatti, tail gear by Liebherr Aerotechnik.

POWER PLANT: Two MTU/Rolls-Royce/Turbomeca MTR 390 modular turboshaft engines mounted side by side above centre-fuselage, divided by armour plate 'keel' (engine first flown in Panther testbed 14 February 1991); power ratings are maximum T-O 958 kW (1,285 shp), super emergency 1,160 kW (1,556 shp), maximum continuous 873 kW (1,171 shp). LHTEC has proposed the T800-801 as a potential alternative power plant for export variants of the Tiger.

Self-sealing crashworthy fuel tanks, with explosion suppression and with non-return valves which minimise leakage in a crash; total capacity 1,360 litres (359 US gallons; 299 Imp gallons). Provision for two external tanks, one on each inboard pylon, each of approximately 350 litres (92 US gallons; 77 Imp gallons) capacity. Gearbox has specified 30 minutes' dry running capability (demonstrated 65 minutes, November 1994).

ACCOMMODATION: Crew of two in tandem, with pilot in front and weapons system operator at rear; full dual controls; both crew members can perform all tasks and weapon operation except that anti-tank missile firing only available to gunner. Armoured, impact-absorbing seats; stepped cockpits, with flat-plate windscreens and slightly curved non-glint transparencies.

SYSTEMS: Redundant hydraulic, electrical and fuel systems. Primary power generation by two 20 kVA alternators; DC power generation by two 300 A 28 V transformer/rectifiers and two 23 Ah Ni/Cd batteries.

Third Tiger prototype (PT3R/9823), the German UHT testbed

Dual redundant AFCS provides four-axis command and stability augmentation. Basic AFCS modes: attitude hold, heading hold. Higher AFCS modes: Heading/acquire/hold, barometric altitude capture/hold, altitude acquire, airspeed hold, vertical speed acquire/hold, nav coupling, radar height hold, Doppler hover hold, line of sight acquisition/ hold. Other AFCS functions: gun recoil force compensation, axis decoupling and tactical mode (follow-up trim on override of break-out forces).

AVIONICS: *Basic or core avionics* common to all three versions include bus/display system, com radio (French and German systems vary), autonomous nav system and radio/Doppler navaids, Thomson-CSF TSC 2000 IFF Mk 12, NH 90-based ECM suite (including laser warning) and AFCS, all connected to and controlled through redundant MIL-STD-1553B data highway.

Flight: Navigation system, by Sextant Avionique, Teldix and DASA, is fully redundant; system contains two Sextant PIXYZ three-axis ring laser gyro units, two air data computers, two magnetic sensors, one Teldix/BAE Canada CMA 2012 Doppler radar, a radio altimeter and GPS; these sensors also provide signals for flight control, information display and guidance; integrated duplex AFCS by Sextant and Nord Micro; AFCS computers produced by Sextant Avionique, VDO-Luft and Litef.

Instrumentation: Colour liquid crystal flight displays showing symbology and imagery (two per cockpit for flight and weapon/systems information) by Sextant and VDO-Luft; each crewman has central control/display unit for inputting all radio, electronic systems and navigation selections; digital map display system by Dornier and VDO-Luft (incorporating NH 90's Eurogrid map generation system); engine and systems data are fed into the databus for in-flight indication and subsequent maintenance analysis. BAE Systems Knighthelm fully integrated day and night helmet ordered for German Tigers; French Tigres have similar Sextant Topowl helmet-mounted sights, with integrated night vision (image intensifiers), FLIR, video and synthetic raster symbology.

Mission: Euromep (European mission equipment package) includes SATEL (Aerospatiale Matra/Pilkington Thorn Optronics/Eltro consortium) Condor 2 pilot vision subsystem (PVS), air-to-air subsystem (Stinger or Mistral), mast-mounted sight and missile subsystem and Euromep management system all connected to separate MIL-STD-1553B avionics data highway. Euromep Standard B avionics first flew February 1995 (PT5); Standard C testing began in late 1997 (PT3R). PVS has 40 × 30° instantaneous field of view (with ±110 × 35° total field of view) thermal imaging sensor steered by helmet position detector giving both crewmen day/night/bad weather vision, flight symbology and air-to-air aiming in helmet-mounted display; mast-mounted sight, gunner sight

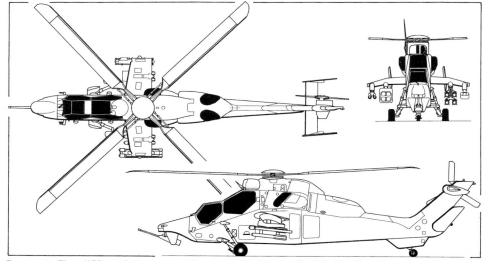

Eurocopter Tigre HCP combat escort and fire support helicopter *(Paul Jackson/Jane's)*

electronics and gunner's head-in target acquisition display and ATGW 3 subsystem connected by separate data highway; HOT 3 missile system also available. Sextant armament control panel and fire-control computer.

HAP combat support mission equipment package includes SFIM/TRT STRIX gyrostabilised roof-mounted sight (with IRCCD IR channel) above rear cockpit; includes direct view optics with folding sight tube, television and IR channels and laser ranger/designator.

Self-defence: DaimlerChrysler Aerospace C-model EW suite (as in NH 90) is one element in HAP's fully integrated avionics suite. This has a DASA laser warning receiver, DASA missile launch warning device *(Lenkflugkörpersysteme)* and Thomson-CSF EW processor and radar warning receiver. Also chaff and flare dispensers (with up to 144 cartridges, sequenced by a Saphir M system). UH is similar, but with option of fitting IR jammer.

ARMAMENT: Tiger has four outboard weapon stations for typical four HOT, four Trigat, two Mistral or Stinger launchers or pods of 12 or 22 rockets; optional 12.7 mm gun pod with 250 rounds or auxiliary fuel tank. HCP (HAP) options all include one 30 mm Giat AM-30781 automatic cannon with up to 450 rounds (traversing from +33 to −30 ° in elevation and through ±90° in azimuth). Possible weapon configurations include (a) HAP: 44 rockets on outer stations, or 24 rockets and four Mistrals, or mix of launchers. (b) HAC: Eight maximum HOT or eight Trigat or HOT/Trigat mix on inner stations, plus four Mistrals on outer stations. (c) UHT: 44 rockets, eight HOT, eight Trigat, or two gun pods on inner stations; four Stingers on outer stations. Combination of auxiliary fuel tank and weapon launcher (rocket, HOT, Trigat, gun pod) on inner stations for extended range missions is possible. (d) HCP (Export): Nose-mounted 30 mm gun with 450 rounds, plus all stores combinations of UHT and HAP (except 12.7 mm gun pods).

DIMENSIONS, EXTERNAL (UT: UHT and HAC, HCP: HAP, where different):

Main rotor diameter	13.00 m (42 ft 7¾ in)
Tail rotor diameter	2.70 m (8 ft 10¼ in)
Length overall, rotors turning	15.80 m (51 ft 10 in)
Length of fuselage: UT	14.08 m (46 ft 2¼ in)
HCP (incl cannon)	15.00 m (49 ft 2½ in)
Height: to top of rotor head	3.83 m (12 ft 6¾ in)
to top of tail rotor disc	4.32 m (14 ft 2 in)
to top of mast sight (UT only)	5.20 m (17 ft 0¾ in)
Width over weapon pylons	4.52 m (14 ft 10 in)
Wheel track	2.38 m (7 ft 9¾ in)
Wheelbase	7.65 m (25 ft 1 in)

AREAS:

Main rotor disc	132.70 m² (1,428.7 sq ft)
Tail rotor disc	5.72 m² (61.63 sq ft)

WEIGHTS AND LOADINGS:

Basic weight empty	3,400 kg (7,496 lb)
Fuel weight: internal	1,080 kg (2,381 lb)
external (two tanks)	555 kg (1,224 lb)
Mission T-O weight	5,300-6,000 kg (11,685-13,227 lb)
Max T-O weight	6,100 kg (13,448 lb)
Main rotor disc loading (max mission T-O weight)	45.2 kg/m² (9.26 lb/sq ft)

PERFORMANCE (UT and HCP, as above):

Never-exceed speed (VNE):	
UHT, HAC	171 kt (298 km/h; 185 mph)
HCP, HAP	175 kt (322 km/h; 200 mph)
Max level speed: UHT	145 kt (269 km/h; 167 mph)
HAP	155 kt (287 km/h; 178 mph)
Cruising speed	124 kt (230 km/h; 143 mph)
Max rate of climb at S/L: UHT	642 m (2,106 ft)/min
HAP	690 m (2,263 ft)/min
Vertical rate of climb at S/L: UHT	312 m (1,023 ft)/min
HAP	384 m (1,259 ft)/min
Hovering ceiling OGE: UHT	3,200 m (10,500 ft)
HAP	3,500 m (11,480 ft)
Range on internal fuel	432 n miles (800 km; 497 miles)
Endurance: operational mission	2 h 50 min
max internal fuel	3 h 25 min

UPDATED

EUROCOPTER AS 332 SUPER PUMA Mk I and AS 532 COUGAR Mk I

Spanish military designations: HD.21 and HT.21
Swedish Air Force designation: Hkp 10

TYPE: Multirole medium helicopter.

PROGRAMME: Early history and versions listed in 1985-86 and earlier *Jane's*; first flight AS 332 Super Puma (F-WZJA) 13 September 1978; six prototypes; deliveries began mid-1981; present version powered by Turbomeca Makila 1A1 introduced 1986; military versions renamed Cougar in 1990; first AS 332L (stretched fuselage) certified to French IFR Cat. II 7 July 1983 and delivered to Lufttransport of Norway; certified for flight into known icing 29 June 1983; FAA Cat. II certification with SFIM CDV 85 P4 four-axis AFCS and FAR Pt 25 Appendix C known icing clearance. Mk I continues in production as long as orders (mainly making good attrition) continue.

Some AS 332/532s built under licence by IPTN in Indonesia and TAI in Turkey; some assembled by CASA in Spain; and some equipped by F+W (now SF) in Switzerland.

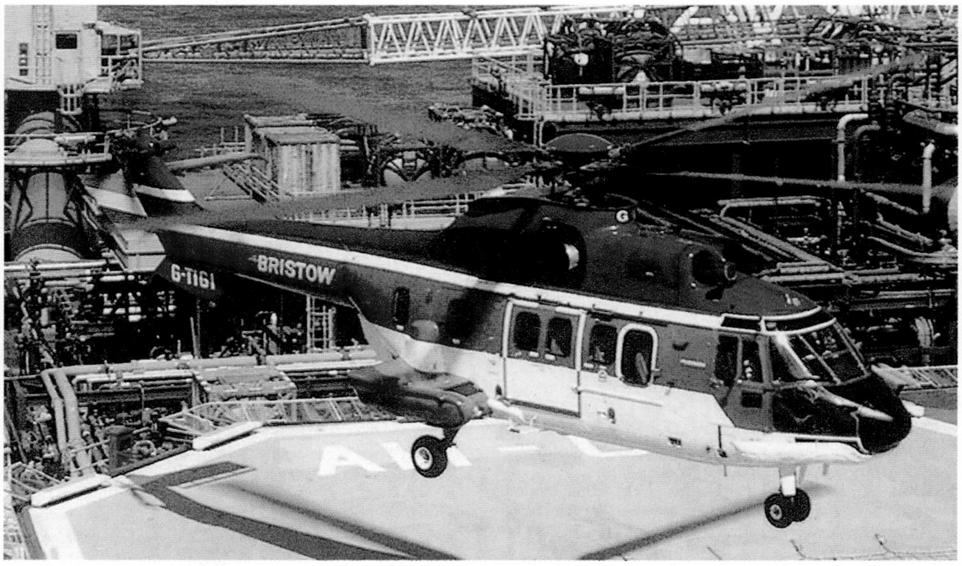

Eurocopter AS 332L Super Puma used for North Sea oil support by Bristow Helicopters ***2001***/0099561

CURRENT VERSIONS: Designation suffixes: U: military unarmed utility, A: armed, S: anti-ship/submarine, C: military court (short) fuselage, L: long fuselage, military or civil.

AS 332L1 Super Puma: Standard Mk I civil version with long fuselage and airline interior for 20 passengers; UK CAA IFR certification 21 April 1992.

AS 532 Cougar 100: Simplified 9 tonne tactical transport; fixed landing gear reduces cruising speed by 5 kt (9 km/h; 6 mph); price reduced by 15 to 20 per cent. External tanks cannot be fitted. Two versions are proposed: **AS 532UB** Short fuselage version and **AS 532UE** stretched fuselage version (first flight due 2000, but not confirmed); only AS 532UE currently marketed.

AS 532UC Cougar: Military Mk I short fuselage unarmed utility; seats up to 21 troops and two crew; cabin floor reinforced for 1,500 kg/m² (307 lb/sq ft). Crashworthy self-sealing fuel tanks standard. Equipment includes IR suppressors, RWR, chaff dispensers, flare launcher and armoured protection for crew and troops.

AS 532UL Cougar: Military Mk I unarmed long fuselage version; cabin lengthened by 0.76 m (2 ft 6 in); extra fuel and two large windows in forward cabin plug; carries up to 25 troops and two crew. Crashworthy self-sealing fuel tanks standard. Equipment as per AS 532UC; sliding forward cabin doors can be fitted to allow carriage of forward-firing 7.62 mm machine guns.

AS 532AC Cougar: Armed AS 532UC (Mk I).

AS 532AL Cougar: Armed AS 532UL (Mk I).

AS 532SC Cougar: Naval short Mk I version with ASW/ASV equipment; folding tail rotor pylon and main rotor blades; deck harpoon landing aid. Crew of two plus up to two operators.

AS 532UL Horizon: Battlefield surveillance radar helicopter; flight trials of small Orphée radar under an AS 330B Puma started 1986; French Army wanted 20 AS 532 Cougar Mk II fitted with radar, dedicated ECM and datalink (Orchidée programme). First flight of full-scale AS 532/Orchidée June 1990; programme cancelled for budgetary reasons August 1990; prototype without datalink flew 24 missions in Gulf War February 1991 in Operation Horus.

Horizon programme (*Hélicoptère d'Observation Radar et d'Investigation sur Zone*) with more effective operational concept and reduced development costs replaced Orchidée; development contract awarded to Eurocopter October 1992 for two (later increased to four) AS 532UL Horizons with same radar capabilities and ECM as Orchidée, but in AS 532UL Cougar Mk I, with standard ECM and longer mission endurance. Antenna rotates at either 2° 24′ or 8°/s; radar range 108 n miles (200 km; 124 miles) with helicopter cruising at 97 kt (180 km/h; 112 mph) at 4,000 m (13,120 ft); normal operational range is 81 n miles (150 km; 93 miles); target speed resolution is 2 m (6.6 ft)/s. Military features include jet efflux deflector/diluters, anti-icing systems, NVG-compatible cockpit lighting, weather radar, encrypted communications, INS/GPS and active/passive countermeasures. First flight Horizon with full radar 8 December 1992; first delivery 24 June 1996; second (plus a ground station) in December 1996; third delivered in 1997 and fourth (plus second ground station) in 1998.

AS 332L2 Super Puma Mk II: Stretched version with Spheriflex rotor heads, new cockpit, enlarged rotors and more powerful Makila 1A2 engines. See separate entry.

AS 532U2/A2 Cougar Mk II: Stretched version with Spheriflex rotor heads, new cockpit, enlarged rotors and more powerful Makila 1A2 engines. See separate entry.

CUSTOMERS: Total 570 AS 332/532s ordered by 81 customers in 47 countries (two-thirds of them military Cougars with 26 air forces, eight armies and five navies in 33 countries) by 1 June 2000; 1999 orders for Super Pumas and Cougars of all marks totalled 22, compared with 43 in 1997 and seven in 1998. 500th helicopter in family (an AS 332L2) was delivered to Bristow Helicopters on 17 June 1999. Some 110 AS 332L1/L2 Super Pumas in civil offshore oil industry support operations; French military orders include six for French Air Force (three for nuclear test facilities in Pacific and three for government VIP flying); French Army (ALAT) has supplemented SA 330 Pumas with AS 532 Cougars; 22 delivered to *Force d'Action Rapide* between late 1988 and end 1991.

Export customers include Abu Dhabi (eight including two VIP of which five to be upgraded with Exocet, sonar and torpedoes under 1995 contract), Brazil (25 including six AS 532SCs, one AS 332M1 and eight ordered January 2000), Cameroon (one), Chile (army, two; navy, four AS 532SCs), China (six), Ecuador (eight including six army), Finland (three for border police), Gabon (two, Presidential Guard), Germany (three, border police), Greece (Ministry of Merchant Marine, four ordered August 1998 and first two delivered 21 December 1999; remaining pair March 2000), Indonesia (built under licence: see IPTN entry); Japan (three, army/VIP), Jordan (eight), South Korea (three army/VIP, one AS 332L1), Mexico (two VIP), Nepal (two, Royal Flight), Nigeria (two), Oman (two, Royal Flight), Panama (one VIP), Saudi Arabia (12 AS 532SCs), Singapore (36), Spain (10 SAR HD.21s, two VIP HT.21s, 18 army tactical transport HT.21s plus 15 AS 532ULs ordered February 1996 and delivered from 1998), Sweden (12 SAR), Switzerland (15 AS 332Ms; plus 12 AS 532ULs ordered December 1998 for delivery from 2000; two to be built by Eurocopter and remainder subcontracted to SF, which see), Togo (one), Turkey (30 AS 532ULs and 20 AS 532ALs in SAR configuration, including 28 assembled by TAI; first Turkish-built AS 532AL handed over 1 May 2000; deliveries to be completed by February 2003; first Turkish-built AS 532UL handed over 31 May 2000), Venezuela (six), Zaïre (one VIP). Competing for Taiwanese UH-1 and Greek UH-1/AB 205 replacement programmes. Bristow Helicopters acquired 31 examples of 19-passenger AS 332L (see 1993-94 *Jane's*) for offshore oil support. See also Super Puma/Cougar Mk II.

COSTS: AS 332L1 US$14.25 million, IFR; direct operating cost US$1,542 per hour (both 1997).

DESIGN FEATURES: Four-blade fully articulated main rotor turning clockwise seen from above; five-blade tail rotor on starboard side of tailboom; engines mounted above cabin have rear drive into main transmission at 23,840 rpm; main rotor turns at 265 rpm, tail rotor at 1,278 rpm; various lateral sponsons available housing partly retracted main landing gear and combinations of additional fuel or pop-out floats; optional air conditioning system housed in casing on port forward flank of cabin; all civil versions certified for IFR category A and B to FAR Pt 29.

FLYING CONTROLS: Dual fully powered hydraulic controls with full-time autostabilisation and yaw damping; machine remains flyable with autostabiliser switched off; cyclic trimming by stick friction adjustment; inverted slot on tailplane to maintain attitude-holding effect at low climb speeds; large ventral fin; saucer fairing on rotor head to smooth wake of hub; four-axis SFIM 155 autopilot standard.

STRUCTURE: Conventional light alloy airframe with some titanium; crashworthy fuel system, impact-absorbing landing gear and other features; main rotor blades of GFRP with CFRP stiffening and Moltoprene filler; elastomeric drag dampers.

LANDING GEAR: Retractable tricycle high energy absorbing design by Messier-Bugatti; all units retract rearward hydraulically, mainwheels into sponsons on sides of fuselage; dual-chamber oleo-pneumatic shock-absorbers; twin-wheel self-centring nose unit, tyre size 7.00-6m (8

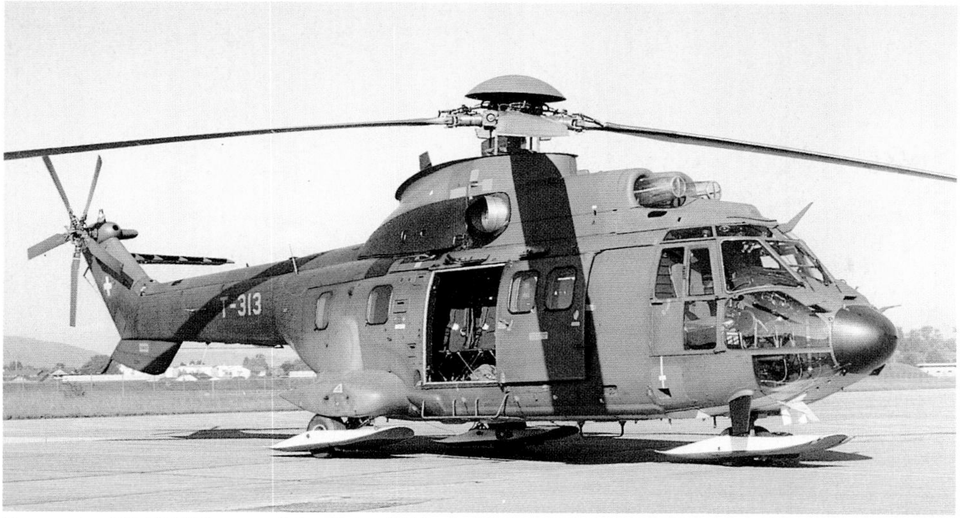

Eurocopter AS 532M Cougar of the Swiss Air Force *(Lindsay Peacock)* ***2001***/0099559

ply) tubeless, pressure 7.00 bars (102 lb/sq in); single wheel on each main unit with tyre size 615×225-10 (12 ply) tubeless or 640×230-10, pressure 9.00 bars (130 lb/sq in); hydraulic differential disc brakes, controlled by foot pedals; lever-operated parking brake; emergency pop-out flotation units can be mounted on main landing gear fairings and forward fuselage.

POWER PLANT: Two Turbomeca Makila 1A1 turboshafts, each with maximum contingency rating of 1,400 kW (1,877 shp) and take-off rating of 1,357 kW (1,820 shp). Maximum continuous rating of 1,184 kW (1,588 shp). Air intakes protected by a grille against ingestion of ice, snow and foreign objects; Centrisep multipurpose intake optional for flight into sandy areas. Transmission rated at 2,238 kW (3,000 shp) for T-O.

AS 532UC/AC have five flexible fuel tanks under cabin floor, with total usable capacity of 1,497 litres (395 US gallons; 329 Imp gallons); AS 532SC has total basic capacity of 2,141 litres (565 US gallons; 471 Imp gallons); AS 332L1/532UE/532UL/532AL have a basic fuel system of six flexible tanks with total capacity of 2,020 litres (533 US gallons; 444 Imp gallons) in the 332 and 2,003 litres (529 US gallons; 440 Imp gallons) in the 532; provision for additional 1,900 litres (502 US gallons; 418 Imp gallons) in four auxiliary ferry tanks installed in cabin; two external auxiliary tanks with total capacity of 650 litres (172 US gallons; 143 Imp gallons) standard on AS 532SC, optional on other versions; AS 532UB Cougar 100 has capacity of 1,533 litres (405 US gallons; 337 Imp gallons). For long-range missions (mainly offshore) in AS 332L1, a special internal auxiliary tank can be fitted in cargo sling well, in addition to the two external tanks, to raise total usable fuel capacity to 2,994 litres (791 US gallons; 658 Imp gallons). Refuelling point on starboard side of cabin; fuel system designed to avoid leakage following a crash; self-sealing tanks standard on military versions, optional on other versions; other options include a fuel dumping system and pressure refuelling.

ACCOMMODATION: One pilot (VFR) or two pilots side by side (IFR) on flight deck, with jump seat for third crew member or paratroop dispatcher; provision for composite light alloy/Kevlar armour for crew protection on military models; door on each side of flight deck and internal doorway connecting flight deck to cabin; dual controls, co-pilot instrumentation and crashworthy flight deck and cabin floors; maximum accommodation for 21 troops in AS 532UC, 20 passengers in AS 332L1, 24 in AS 332L2, and 25 troops in AS 532UL, 20 troops in AS 532UE; interiors available for VIP, air ambulance with six stretchers and 11 seated casualties/attendants; strengthened floor for cargo carrying, with lashing points; jettisonable sliding door on each side of main cabin or port side door with built-in steps and starboard side double door in VIP configuration; removable panel on underside of fuselage, at rear of main cabin, for longer loads; removable door with integral steps for access to baggage racks optional; hatch in floor below main rotor contains hook for slung loads up to 4,500 kg (9,920 lb) on internally mounted cargo sling; cabin and flight deck heated, ventilated and soundproofed; demisting, de-icing, washers and wipers for pilots' windscreens.

SYSTEMS: Two independent hydraulic systems, supplied by self-regulating pumps driven by main gearbox. Each system supplies one set of powered flying controls; left-hand system also supplies autopilot, landing gear, rotor brake and wheel brakes; hydraulically actuated systems can be operated on ground from main gearbox (when a special disconnect system is installed to permit running of port engine with rotors stationary), or by external power through ground power receptacle. Emergency landing gear lowering by standby pump.

Three-phase 200 V AC electrical power supplied by two 20 kVA 400 Hz alternators, driven by port side intermediate shaft from main gearbox and available on ground under same conditions as hydraulic ancillary systems; two 28.5 V DC transformer-rectifiers; main battery used for self-starting and emergency power in flight.

AVIONICS: *Comms:* Optional equipment includes VHF, UHF, tactical HF and HF/SSB radio and intercom.

Radar: Offshore models have nose-mounted radar; search and rescue version has nose-mounted RDR 1400 or chin-mounted 1500 search radar (as on Swedish Hkp 10s); naval ASW and ASV versions can have nose-mounted Thomson-CSF Ocean Master radar, linked to a tactical table in the cabin.

Flight: ADF, VOR/ILS, radio altimeter, GPS, VLF Omega, Decca navigator and flight log, Doppler, SFIM 155 autopilot, with provision for coupling to self-contained navigation and landing systems; search and rescue version has Doppler, and Sextant Avionique Nadir or Decca self-contained navigation system (Nadir Mk 2 in French Army and Greek versions), including navigation computer with SAR patterns, polar indicator, roller map display, hover indicator, route mileage indicator and groundspeed and drift indicator; SFIM CDV 155 autopilot coupler contains automatic nav track including search patterns, transitions and hover; multifunction video display shows radar and route images, SAR patterns and hover indication; Swedish Hkp 10s have Racal RAMS flight management system including BAE Systems AHRS, Decca Doppler and GPS.

Instrumentation: Full IFR instrumentation optional.

Mission: Naval ASW versions have Thomson Marconi HS 312 acoustics suite; upgraded FLASH version currently available.

EQUIPMENT: Optional fixed rescue hoist (capacity 275 kg; 606 lb) starboard side and electrical back-up hoist; equipment for naval missions can include sonar, MAD and sonobuoys.

ARMAMENT (optional): Typical alternatives for army/air force missions are two 20 mm guns or two 7.62 mm machine guns or two 68 mm rocket launchers. Armament for naval missions includes two AM 39 Exocet missiles or two lightweight torpedoes.

DIMENSIONS, EXTERNAL:
Main rotor diameter 15.60 m (51 ft 2¼ in)
Tail rotor diameter 3.05 m (10 ft 0 in)
Main rotor blade chord 0.60 m (1 ft 11½ in)
Length: overall, rotors turning 18.70 m (61 ft 4¼ in)
fuselage, incl tail rotor:
AS 532UC/AC/SC/UB 15.53 m (50 ft 11½ in)
AS 332L1/532AL/UL/UE 16.29 m (53 ft 5½ in)
tail and rotors folded (AS 532SC) 12.95 m (42 ft 5¾ in)
Width, blades folded:
AS 532UB/UC/AC/AL/UL/332L1 3.79 m (12 ft 5¼ in)
AS 532SC 3.86 m (13 ft 3 in)
Width over sponsons:
AS 332L1/AS 532AL/UL/UC/AC 3.38 m (11 ft 1 in)
AS 532SC 3.56 m (11 ft 8¼ in)
Fuselage width 2.00 m (6 ft 6 in)
Height: overall 4.92 m (16 ft 1¾ in)
blades and tail pylon folded:
AS 532UE/UC/AC/SC 4.80 m (15 ft 9 in)
to top of rotor head 4.60 m (15 ft 1 in)
Tailplane half-span 2.11 m (6 ft 11 in)
Wheel track 3.00 m (9 ft 10 in)
Wheelbase: AS 532UC/AC/SC 4.49 m (14 ft 8¾ in)
AS 332L1/532AL/UL/UE 5.25 m (17 ft 2¾ in)
Passenger cabin doors, each: Height 1.35 m (4 ft 5 in)
Width 1.30 m (4 ft 3¼ in)
Floor hatch, rear of cabin: Length 0.98 m (3 ft 2¾ in)
Width 0.70 m (2 ft 3½ in)

DIMENSIONS, INTERNAL:
Cabin: Length: AS 532UC/AC/SC 6.05 m (19 ft 10½ in)
AS 332L1/532AL/UL/UE 6.81 m (22 ft 4 in)
Max width 1.80 m (5 ft 10¾ in)
Max height 1.55 m (5 ft 1 in)
Floor area: AS 532UC/AC/SC 7.80 m² (84.0 sq ft)
AS 332L1/532AL/UL/UE 9.18 m² (98.8 sq ft)
Usable volume:
AS 532UC/AC/SC 11.4 m³ (403 cu ft)
AS 332L1/532AL/UL/UE 13.3 m³ (470 cu ft)

AREAS:
Main rotor disc 191.13 m² (2,057.3 sq ft)
Tail rotor disc 7.31 m² (78.64 sq ft)

WEIGHTS AND LOADINGS:
Weight empty (standard aircraft):
AS 532UC/AC 4,330 kg (9,546 lb)
AS 532SC 4,500 kg (9,920 lb)
AS 332L1/532AL/UL 4,460 kg (9,832 lb)
AS 332UE 4,485 kg (9,888 lb)
Max external payload 4,500 kg (9,920 lb)
Max T-O weight: internal load:
AS 532UC/AC/AL/UB/UL/SC/UB/UE 9,000 kg (19,841 lb)
AS 332L1 8,600 kg (18,960 lb)
external slung load 9,350 kg (20,615 lb)
Max disc loading, external load 48.9 kg/m² (10.02 lb/sq ft)
Transmission loading at max T-O weight and power:
internal load:
AS 532UC/AC/AL/UB/UL/SC 4.03 kg/kW (6.61 lb/shp)
AS 332L1 3.85 kg/kW (6.32 lb/shp)
external load 4.18 kg/kW (6.87 lb/shp)

PERFORMANCE:
Never-exceed speed (VNE) 150 kt (278 km/h; 172 mph)
Cruising speed at S/L:
AS 532UC/AC/AL/UL/SC 139 kt (257 km/h; 160 mph)
AS 532UB/UE 134 kt (249 km/h; 155 mph)
AS 332L1 141 kt (261 km/h; 162 mph)
Max rate of climb at S/L:
AS 532UC/AC/AL/UL 420 m (1,378 ft)/min
AS 532UB/UE 432 m (1,417 ft)/min
AS 532SC 372 m (1,220 ft)/min
AS 332L1 486 m (1,594 ft)/min
Service ceiling: AS 332L1 4,600 m (15,080 ft)
AS 532UC/AC/SC/AL/UL/UB/UE 4,100 m (13,450 ft)
Hovering ceiling IGE:
AS 532AL/UL/AC/UC/SC/UB/UE 2,800 m (9,180 ft)
AS 332L1 3,250 m (10,660 ft)

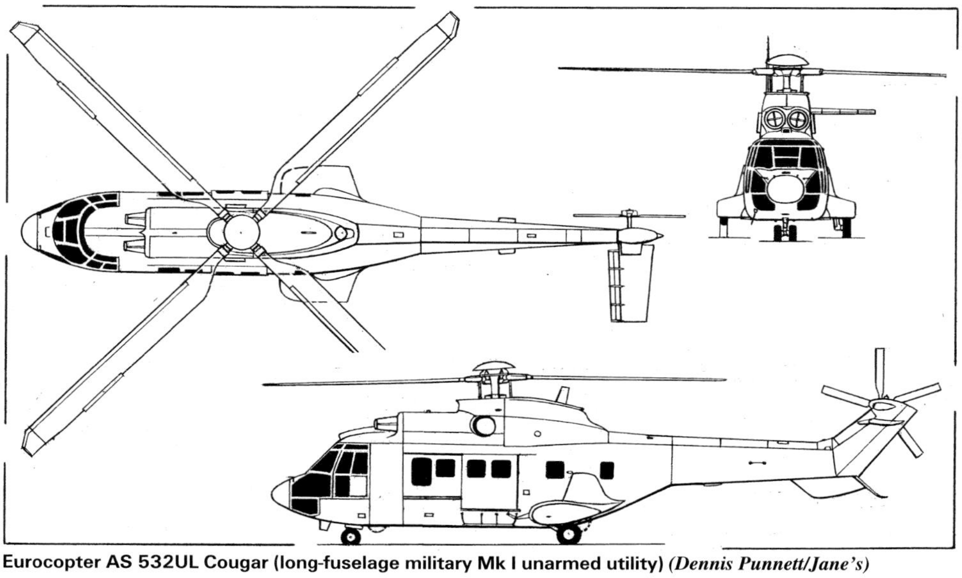

Eurocopter AS 532UL Cougar (long-fuselage military Mk I unarmed utility) ***(Dennis Punnett/Jane's)***

Hovering ceiling OGE:
 AS 532AL/UL/AC/UC/SC/UB/UE 1,650 m (5,420 ft)
 AS 332L1 2,300 m (7,540 ft)
Range at S/L, standard tanks, no reserves:
 AS 332L1 449 n miles (831 km; 516 miles)
 AS 532UC/AC 334 n miles (618 km; 384 miles)
 AS532UB 309 n miles (573 km; 356 miles)
 AS 532SC 492 n miles (911 km; 566 miles)
 AS 532AL/UL 455 n miles (842 km; 523 miles)
 AS 332L1 470 n miles (870 km; 540 miles)
 AS 532UE 416 n miles (770 km; 478 miles)
Range with auxiliary internal tank:
 AS 532UE 486 n miles (900 km; 559 miles)
 AS 532UB 380 n miles (703 km; 437 miles)
Range at S/L with external (two 338 litre) and auxiliary (one 320 litre) tanks, no reserves:
 AS 332L1 659 n miles (1,220 km; 758 miles)
 AS 532UC/AC 528 n miles (977 km; 607 miles)
 AS 532AL/UL 637 n miles (1,179 km; 733 miles)

UPDATED

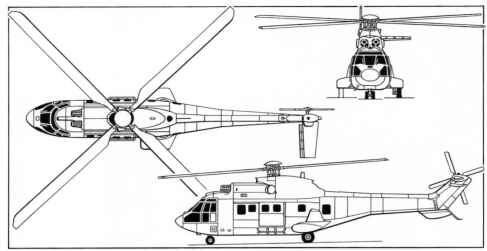

Eurocopter AS 332L2 Super Puma Mk II with Spheriflex main and tail rotor heads *(Mike Keep/Jane's)*

EUROCOPTER AS 332L2 SUPER PUMA Mk II/AS 532 COUGAR Mk II and EC 225 SUPER PUMA Mk II+/EC 725 COUGAR Mk II+

Spanish Army designation: HU.21L

TYPE: Multirole medium helicopter.

PROGRAMME: Derived from Super Puma/Cougar Mk I (which see); first flight of development vehicle 6 February 1987; French certification 2 April 1992; UK BCAR 29 certification 16 November 1992; first delivery August 1993. Earlier AS 332L1 and 532U/A Cougar Mk I remain on production line. Further development is under consideration.

CURRENT VERSIONS: **AS 332L2 Super Puma Mk II:** Current production civil transport; innovations include Spheriflex rotor heads, super-emergency engine rating, EFIS flight deck and built-in health and usage monitoring system, duplex four-axis AFCS, and hydraulically powered standby electrics for 2 hours' operation after complete main generator failure. First delivery 28 August 1993.

AS 332L2 Super Puma Mk II VIP: Eight- to 15-passenger arrangement with attendant and fully equipped galley and lavatory; two four-seat lounges; fine materials and fittings throughout; range with eight passengers and attendant, 635 n miles (1,176 km; 730 miles). Used in 28 countries.

AS 532U2 Cougar Mk II: Unarmed military tactical transport; longest member of Cougar family; carries 29 troops and two-man crew.

AS 532A2 Cougar Mk II: Armed version. First order (for four) by French Air Force for combat SAR, 1995; known as Cougar RESCO (REcherche et Sauvetage en COmbat); prototype (F-ZVMC) undertook maker's trials in 1998; was delivered to CEAM military test unit on 9 September 1999 for one year of evaluation; second aircraft, to determine operational fit, was due to be delivered in 2000. Potential equipment includes RWR, missile approach warner, chaff, flares, rescue hoist, two outboard 20 mm cannon, two door-mounted 12.7 mm machine guns, encrypted Personnel Locator System, third-generation NVG-compatible lighting, air-to-air refuelling capability. Nadir Mk 2 navigation computer, SFIM PA-165 four-axis digital autopilot, VOR, Tacan, DME, GPS, searchlight and FLIR. Alternative maximum T-O weight of 11,200 kg (24,692 lb) permits an 800 n mile (1,482 km; 920 mile), plus 30 min hover and 30 min reserve, unrefuelled radius of action to rescue two persons.

EC 225 Super Puma Mk II+: Growth version, announced June 1998, as competitor to Sikorsky S-92 in offshore support role; has military counterpart in **EC 725 Cougar Mk II+**. Cabin volume increased by 25 per cent to 20.0 m³ (706 cu ft) through increases of 35 cm (13¾ in) in height, 70 cm (2 ft 3½ in) in length and 25 cm (9¾ in) in width. Reinforced gearbox and 1,800 kW (2,414 shp) Makila 1A4 engines with FADEC produce 14 per cent power increase over earlier versions. Five-blade main rotor; diameter unchanged. MTOW increased to 10,400 kg (21,560 lb) for civil and 11,000 kg (24,251 lb) for military version (11,200 kg; 24,691 lb with external load). New integrated flight and display system (IFDS) using four 152 × 203 mm (6 × 8 in) multifunction LCDs plus four-axis autopilot. FAR/JAR-29 certification expected late 2002 and first deliveries to French Air Force, April 2003; sales of 200 anticipated. First prototype (F-ZVLR) made maiden flight 30 November 2000, at which time Eurocopter revealed an order for four from the French Air Force, which requires an eventual 14, including one earlier Mk 2 raised to this standard.

CUSTOMERS: Helikopter Service ordered four AS 332L2 Mk II in May 1992 and a further four in June 1997, latter delivered from 27 May 1998 onwards; RNethAF ordered 17 AS 532U2s October 1993 and deliveries began 1 April 1996 (with 400th AS 332/532) and last was received in October 1997; first of follow-on order for two for Bond Helicopters delivered 27 August 1998. Eventual 14 AS 532A2s required for combat SAR by French Air Force, first of four ordered delivered 9 September 1999; remainder to be delivered in three batches by 2003. Saudi Arabia ordered 12 AS 532A2s in July 1996 for combat SAR, deliveries beginning in 1998; in-flight refuelling tests undertaken during second half of 2000. VIP orders include Samsung (one AS 332L2) and the German government (three AS 532U2s delivered from 4 November 1997 and operated by the Luftwaffe). Three ordered for SAR by Hong Kong Government Flying Service for delivery mid-2001. Total of 46 of both versions in service by June 2000.

COSTS: 17 Netherlands AS 532U2 cost FFr1.4 billion, plus 120 per cent offset and technology offset over 12 years. Spanish order for 15 valued at US$207 million in 1996.

DESIGN FEATURES: Generally as for Mk I, but gross weight increased by 150 kg (331 lb) in 1993; composites plug in rear cabin area accommodates one extra row of seats and moves tail rotor rearwards; Spheriflex main and tail rotor heads with elastomeric spherical bearings and Kevlar retention bands; main rotor with longer blades having parabolic tips; enlarged composites sponsons can contain fuel, liferafts, air conditioning and pop-out floats.

FLYING CONTROLS: Fully powered hydraulically actuated with SFIM 165 four-axis digital AFCS and coupler.

STRUCTURE: Rear fuselage 55 cm (1 ft 9½ in) plug made of composites giving extra row of seats; large windows at rear of cabin; increased use of composites compared with Mk I; fuselage frames strengthened to retain same degree of crashworthiness.

LANDING GEAR: As Super Puma Mk I.

POWER PLANT: Two Turbomeca Makila 1A2 rear drive free turbines; maximum emergency power each (OEI 30 seconds) 1,573 kW (2,109 shp), 12 per cent higher than AS 332 Mk I; intermediate emergency power (OEI 2 minutes) 1,467 kW (1,967 shp); take-off power 1,376 kW (1,845 shp); maximum continuous power 1,236 kW (1,657 shp); protective intake grilles standard; optional Centrisep multipurpose intakes for dusty conditions. Transmission ratings: twin-engined maximum 2,410 kW (3,229 hp), maximum from single engine 1,666 kW (2,232 hp), maximum transient 20 seconds 2,651 kW (3,552 hp), maximum continuous 1,555 kW (2,084 hp); transmission has extended run-dry capability.

Standard fuel tankage under cabin floor 2,020 litres (535 US gallons; 444 Imp gallons); auxiliary tankage includes 324 litres (85.6 US gallons; 71.3 Imp gallons) in cargo hook well; sponson tanks each holding 325 litres (85.9 US gallons; 71.5 Imp gallons); 600 litres (159 US gallons; 132 Imp gallons) in cabin fuel tank; one to five internal ferry tanks each holding 475 litres (126 US gallons; 104.5 Imp gallons). Tankage (self-sealing tanks standard from 1997) AS 532A2 1,896 litres (500 US gallons; 417 Imp gallons); hook well tank 320 litres (84.5 US gallons; 70.4 Imp gallons); 325 litres (85.9 US gallons; 71.5 Imp gallons) in each sponson tank; crashproofing is standard in AS 532 and optional in 332; pressure refuelling optional. Military versions can be fitted with probe for in-flight refuelling.

Spanish Army AS 532UL Cougar *(Paul Jackson/Jane's)*

Super Puma Mk II operated on North Sea oil support by Norsk Helikopter *(Peter J Cooper)* 2001/0099560

ACCOMMODATION: Single pilot in DGAC Category B civil operation; single pilot plus licensed crewman in Category A VFR; two pilots in IFR; military operation one pilot in VFR, two in IFR; civil transport 24 passengers and attendant in airline interior for 220 n miles (408 km; 253 miles) or 19 passengers at 81 cm (32 in) seat pitch for 350 n miles (648 km; 403 miles); military capacity, squad chief plus 28 troops; VIP version for eight to 15 passengers plus attendant; ambulance version holds 12 stretchers and four seated casualties plus attendant.

AVIONICS: *Instrumentation:* Civil and military cockpits have Sextant Avionique four-tube EFIS integrated flight and display system (IFDS) with four 15 × 15 cm (6 × 6 in) screens; subsystems include smart multimode display (SMDS), automatic flight control system (AFCS) and flight data system (FDS); military cockpit compatible with night vision goggles; civil and military IFR system offered, plus military communications radio; SAR system includes radar and FLIR coupled to display system as well as navigation computer with Doppler and GPS sensors; can also be fitted with inertial nav sensor for combat SAR configuration at customer's request; SAR system can include automatic search pattern, transition and hover hold.

Radar: RNethAF AS 532U2s have been retrofitted with Honeywell weather radar.

EQUIPMENT: Pop-out floats, rescue winch (272 kg; 600 lb), external sling. Military equipment (U2/A2) includes IR suppressors, flare/chaff dispensers, RWR, MAWS and armoured protection for crew and troops.

ARMAMENT: Armament choice as for AS 532A Cougar Mk I except for Exocet. Special sliding doors on forward cabin windows allow forward firing 7.62 mm machine guns to be fitted.

DIMENSIONS, EXTERNAL:
Main rotor diameter	16.20 m (53 ft 1½ in)
Tail rotor diameter	3.15 m (10 ft 4 in)
Length overall: rotors turning	19.50 m (63 ft 11 in)
main rotor folded	16.79 m (55 ft 0½ in)
Width: fuselage	2.00 m (6 ft 6 in)
over sponsons	3.38 m (11 ft 1 in)
overall, main rotor folded	3.86 m (12 ft 8 in)
Height: overall, tail rotor turning	4.97 m (16 ft 4 in)
to top of rotor head	4.60 m (15 ft 1 in)
Tailplane span	2.12 m (6 ft 11½ in)
Wheel track	3.00 m (9 ft 10 in)
Wheelbase	5.25 m (17 ft 2¾ in)
Sliding cabin doors, each: Height	1.35 m (4 ft 5 in)
Width	1.30 m (4 ft 3¼ in)
Floor hatch: Length	0.98 m (3 ft 2¾ in)
Width	0.70 m (2 ft 3½ in)

DIMENSIONS, INTERNAL:
Cabin: Max length	7.87 m (25 ft 10 in)
Floor length	6.15 m (20 ft 2¼ in)
Max width	1.80 m (5 ft 10¾ in)
Max height	1.45 m (4 ft 9 in)
Volume	15.0 m³ (530 cu ft)

AREAS:
Main rotor disc	206.12 m² (2,218.7 sq ft)
Tail rotor disc	7.79 m² (83.88 sq ft)

WEIGHTS AND LOADINGS:
Manufacturer's weight empty: L2	4,686 kg (10,331 lb)
U2	4,945 kg (10,902 lb)
A2	5,012 kg (11,050 lb)
Useful load: L2	4,595 kg (10,130 lb)
U2	4,805 kg (10,593 lb)
A2	4,738 kg (10,445 lb)
Standard fuel weight: L2	1,596 kg (3,519 lb)
U2, crashworthy tanks	1,548 kg (3,412 lb)
Max normal T-O weight: L2	9,300 kg (20,502 lb)
U2	9,750 kg (21,495 lb)
A2 Combat SAR	11,200 kg (24,692 lb)
Max slung load: A2/U2	5,000 kg (11,023 lb)
L2	4,500 kg (9,920 lb)
Max flight weight with slung load:	
U2	10,500 kg (23,148 lb)

Max disc loading, normal T-O weight:
L2	45.1 kg/m² (9.24 lb/sq ft)
U2	47.3 kg/m² (9.69 lb/sq ft)
U2, alternate	54.3 kg/m² (11.13 lb/sq ft)

Transmission loading at max T-O weight and power:
L2	3.85 kg/kW (6.34 lb/shp)
U2 with slung load	4.35 kg/kW (7.16 lb/shp)
U2 alternate	4.65 kg/kW (7.64 lb/shp)

PERFORMANCE:
Never-exceed speed (V_{NE}):
L2/U2	170 kt (315 km/h; 195 mph)

Fast cruising speed:
L2	150 kt (277 km/h; 172 mph)
U2	147 kt (273 km/h; 170 mph)

Econ cruising speed:
L2	136 kt (252 km/h; 157 mph)
U2	132 kt (244 km/h; 152 mph)

Rate of climb at 70 kt (130 km/h; 81 mph) at S/L:
L2	441 m (1,447 ft)/min
U2	384 m (1,260 ft)/min

Service ceiling (45.7 m; 150 ft/min climb), ISA:
L2	5,180 m (17,000 ft)
U2	4,100 m (13,450 ft)

Hovering ceiling IGE, normal T-O weight, ISA, T-O power:
L2	3,120 m (10,240 ft)
U2	2,690 m (8,820 ft)

Hovering ceiling OGE, normal T-O weight, ISA, T-O power:
L2	2,250 m (7,380 ft)
U2	1,900 m (6,240 ft)

Hovering ceiling OGE, slung load weight, ISA, T-O power:
L2/U2	960 m (3,150 ft)

Range, no reserves, standard fuel, econ cruise:
L2	447 n miles (828 km; 514 miles)
U2	427 n miles (791 km; 491 miles)

Range, no reserves, max fuel, econ cruise:
L2	656 n miles (1,215 km; 755 miles)
U2	636 n miles (1,178 km; 732 miles)

Endurance, standard fuel, at 70 kt (130 km/h; 81 mph):
L2	4 h 54 min
U2	4 h 20 min

UPDATED

EUROCOPTER AS 350 ECUREUIL/ASTAR and AS 550 FENNEC

Brazilian Air Force designations: CH-50 and TH-5 Esquilo
Brazilian Army designation: HA-1 Esquilo
Brazilian Navy designation: UH-12 Esquilo
UK forces designation: Squirrel HT. Mk 1 and HT. Mk 2

TYPE: Light utility helicopter.

PROGRAMME: First flight (F-WVKH) powered by Lycoming LTS101 turboshaft 27 June 1974; first flight second prototype (F-WVKI) powered by Turbomeca Arriel 1A February 1975; first production version was AS 350B powered by 478 kW (641 shp) Arriel 1B and certified 27 October 1977; LTS101 powered AStar sold only in USA. AS 350BA Ecureuil was powered by 478 kW (641 shp) Turbomeca Arriel 1B and fitted with large main rotor blades of AS 350B2 (see below); maximum T-O weight increased by 150 kg (331 lb); AS 350B upgraded to AS 350BA in field; replaced AS 350B during 1992; French VFR certification 1991; UK and US certifications 1992, Japanese 1993. Production of AS 350BA ended in 1998; see *Jane's* 1999-2000 for details.

Delivery of the 2,000th AS 350/550 (F-WWPZ, later ZJ270 of UK Defence Helicopter Flying School) effected in July 1997. In February 2001, delivery effected of 3,000th of Ecureuil family (including AS 355/555), which also was first EC 130 upgraded version.

Built at Marignane, France, and under licence by Helibrás in Brazil (which see). See also the CHAIC Z-11 entry in the Chinese section.

CURRENT VERSIONS: **AS 350B2 Ecureuil:** Uprated engine and transmission; wide-chord new section main and tail rotor blades originally developed for AS 355 twin; certified 26 April 1989; known as **SuperStar** in North America; supplied to UK Ministry of Defence as **AS 350BB**, designated **Squirrel HT. Mk 1** with normal instrumentation and **Squirrel HT. Mk 2** with provision for pilot's NVGs.

AS 350B3 Ecureuil: Current variant; 632 kW (847 shp) Arriel 2B and wide-chord tail rotor; optimised for high-altitude operation; uprated transmission and digital engine control. First flight (F-WWPB) 4 March 1997; French VFR certification December 1997 after 150 hour test programme; first delivery, to Osterman Helicopter, January 1998; Australian and New Zealand certification achieved late 1998. By 1 June 2000, 164 AS 350B3s had been ordered by 50 customers, of which over 100 were in service in 25 countries; 100th AS 350B3 delivered 14 December 1999 to Japanese customer.

AS 350D AStar: Version of 350B for North American market; improved LTS101 engine, with 6 per cent more power, offered early 1999.

AS 350 Firefighter: Conair system able to pick up water load in 30 seconds while hovering over water; demonstrated 1986; Isolair system certified September 1995.

AS 550C3 Fennec: Light anti-tank version armed with four TOW missiles; has same power plant as AS 550A3 and 2 hour 45 minute endurance.

AS 550A3 Fennec: Light attack version of AS 350B2 with 3-hour endurance; powered by 632 kW (847 shp), Arriel 2B; standard features include taller landing gear, sliding doors; NVG-compatible cockpit; reinforced airframe; provision for armoured seats and engine cowlings; can be armed with single 20 mm cannon or two rocket launchers.

AS 550U3 Fennec: Military utility version with Arriel 2B power plant; capable of performing observation, commando and transport duties.

EC 130: Described separately.

CUSTOMERS: Total 2,314 AS 350s of all types ordered by 70 countries by August 2000. Of these 1,948 single-engined Ecureuils in service at May 2000.

Military customers include Singapore armed forces (10 AS 550B2 and 10 AS 550C2); Australia (RAAF 18 for training; RAN six utility); Danish Army (12 AS 550C2 with ESCO HeliTOW system ordered 1987, delivered 1990); French Army ALAT (originally expressed need for up to 100 to replace Alouette IIs); French CEV (flight test centre), and FBS Ltd, which ordered 38 AS 350BBs, with deliveries beginning November 1996 and service entry in April 1997. Of these, 26 are Squirrel HT. Mk 1s with

Initial French Air Force Cougar RESCO before beginning trials 2000/0048804

INTERNATIONAL: AIRCRAFT—EUROCOPTER

Instrument panel of the IFR-equipped Squirrel HT. Mk 2 *(Paul Jackson/Jane's)*

VFR instrument panel fitted to AS 350B3 Ecureuil *(G Deulin/Eurocopter)*

Defence Helicopter Flying School at Shawbury and 12 are NVG-compatible HT. Mk 2s with 670 Squadron, Army Air Corps, at Middle Wallop. Other orders include 11 for Brazilian Navy for liaison and patrol work; 30 for Brazilian Air Force for training; 36 for Brazilian Army (16 for liaison and 20 for training and firefighting); eight for China (see also CHAIC Z-11) and 40 AS 350B2 options for US Customs Service (augmenting five B2s delivered from 21 October 1998 to supplement 19 earlier AStars). Orders in 1997-99 were 102, 73 and 95.

DESIGN FEATURES: Conventional pod-and-boom design with power plant above cabin; fin, underfin and twin horizontal stabilising surfaces. Starflex bearingless glass fibre main rotor head; all versions now have lifting section composites main rotor blades. Main rotor turns at 394 rpm, tail rotor at 2,086 rpm.

FLYING CONTROLS: Single fully powered controls with accumulators to delay manual reversion following hydraulic failure until airspeed can be reduced; cyclic trim by adjustable stick friction; inverted aerofoil tailplane to adjust pitch attitude in climb, cruise and descent; saucer fairing on rotor head to smooth wake; swept fins above and below tail. Variety of autostabilisers and autopilots available (see Avionics).

STRUCTURE: Main rotor head and much of airframe of glass fibre and aramids; main rotor blades automatically manufactured in composites; self-sealing composites fuel tank in military versions.

LANDING GEAR: Steel tube skid type. Taller version standard on military aircraft. Emergency flotation gear optional.

POWER PLANT: AS 350B2 powered by one 546 kW (732 shp) Turbomeca Arriel 1D1 with transmission rated at 440 kW (590 shp) for T-O; AS 350B3/AS 550C3 powered by one 632 kW (847 shp) Turbomeca Arriel 2B turbine engine controlled by FADEC, with transmission rated at 500 kW (671 shp) for T-O; AS 350D powered by AlliedSignal LTS101 sold only in USA. Plastics fuel tank (self-sealing on AS 550) with capacity of 540 litres (143 US gallons; 119 Imp gallons). Optional 475 litre (125 US gallon; 104 Imp gallon) auxiliary tank in cabin.

ACCOMMODATION: Two individual bucket seats at front of cabin and four-place rear bench standard, optional two-place front bench seat; optional ambulance layout; large forward-hinged door on each side of versions for civil use; optional sliding door at rear of cabin on port side; (sliding doors standard on military version); baggage compartment aft of cabin, with full-width upward-hinged door on starboard side; top of baggage compartment reinforced to provide work platform on each side.

SYSTEMS: Hydraulic system includes four single-body servo units, operating at 40 bars (570 lb/sq in) pressure, and accumulators to delay reversion to manual control; electrical system includes a 4.5 kW engine-driven starter/generator, a 24 V 16 Ah Ni/Cd battery and a ground power receptacle; cabin air conditioning system optional.

AVIONICS: *Comms:* VHF/AM radios, HF/SSB transponder and ICS.
Flight: VOR/ILS, ADF, marker beacon receiver and DME. SFIM PA 85T31 (IFR), GPS; full-colour LCD VEMD (vehicle and engine multifunction display) on AS 350B3 and AS 550C3.

EQUIPMENT: Options include 907 kg (2,000 lb) capacity cargo sling (1,160 kg; 2,557 lb for AS 350B2/550, 1,400 kg; 3,086 lb on AS 350B3), 135 kg (297 lb) capacity electric hoist (204 kg; 450 lb on AS 350B3), a TV camera for aerial filming, SX-16 searchlight, Wescam and FLIR observation systems, and a 735 litre (194 US gallon; 161 Imp gallon) Simplex agricultural spraytank and boom system.

ARMAMENT (AS 550C3): Provision for wide range of weapons, including 20 mm Giat M621 gun, FN Herstal TMP twin 7.62 mm and 12.7 mm machine gun pods, Thomson Brandt 68.12 launchers for twelve 68 mm rockets, Forges de Zeebrugge launchers for seven 2.75 in rockets, and ESCO HeliTOW anti-tank missile system.

DIMENSIONS, EXTERNAL:
Main rotor diameter	10.69 m (35 ft 0¾ in)
Main rotor blade chord	0.35 m (13¾ in)
Tail rotor diameter	1.86 m (6 ft 1¼ in)
Tail rotor blade chord: AS 350D	0.185 m (7¼ in)
AS 350B2/B3/550C3	0.205 m (8 in)
Length: overall, rotors turning	12.94 m (42 ft 5½ in)
fuselage	10.93 m (35 ft 10½ in)
Width: fuselage	1.80 m (5 ft 10¾ in)
overall, blades folded (horizontal stabiliser span)	2.53 m (8 ft 3¾ in)
Height overall:	
AS 350/B2/B3/D	3.14 m (10 ft 3½ in)
AS 550C3	3.34 m (10 ft 11½ in)
Tailplane span	2.53 m (8 ft 3½ in)
Skid track: AS 350/B2/B3/D	2.17 m (7 ft 1½ in)
AS 550C3	2.28 m (7 ft 5¾ in)
Cabin doors (civil versions, standard, each):	
Height	1.15 m (3 ft 9¼ in)
Width	1.10 m (3 ft 7¼ in)

DIMENSIONS, INTERNAL:
Cabin: Length	2.42 m (7 ft 11¼ in)
Width at rear	1.65 m (5 ft 5 in)
Height	1.35 m (4 ft 5 in)
Floor area	2.60 m² (28.0 sq ft)
Baggage compartment volume	1.0 m³ (35 cu ft)

AREAS:
Main rotor disc	89.75 m² (966.1 sq ft)
Tail rotor disc	2.72 m² (29.25 sq ft)

WEIGHTS AND LOADINGS:
Weight empty: AS 350B2	1,171 kg (2,582 lb)
AS 350B3	1,175 kg (2,590 lb)
AS 350D	1,123 kg (2,475 lb)
AS 550C3	1,220 kg (2,689 lb)
Max T-O weight: AS 350B2/B3/550C3	2,250 kg (4,960 lb)
AS 350D	1,950 kg (4,300 lb)
Max weight with slung load:	
AS 350B2	2,500 kg (5,511 lb)
AS 350B3/AS 550C3	2,800 kg (6,173 lb)
AS 350D	2,100 kg (4,630 lb)
Max disc loading:	
AS 350D with slung load	23.4 kg/m² (4.79 lb/sq ft)
AS 350B2/550 clean	25.1 kg/m² (5.13 lb/sq ft)
AS 350B2 with slung load	27.9 kg/m² (5.71 lb/sq ft)
AS 350B3/550C3 with slung load	30.6 kg/m² (6.28 lb/sq ft)
Transmission loading at max T-O weight and power:	
AS 350B2 clean	5.11 kg/kW (8.40 lb/shp)
AS 350B2 with slung load	5.68 kg/kW (9.34 lb/shp)
AS 350B3/550C3 clean	4.50 kg/kW (7.39 lb/shp)
AS 350B3/550C3 with slung load	5.50 kg/kW (9.04 lb/shp)

PERFORMANCE (AS 350D at normal T-O weight, AS 350B2/B3/550 at 2,200 kg; 4,850 lb):
Never-exceed speed (VNE) at S/L:	
AS 350/B2/B3/550	155 kt (287 km/h; 178 mph)
AS 350D	147 kt (272 km/h; 169 mph)
Max cruising speed at S/L:	
AS 350B2	134 kt (248 km/h; 154 mph)
AS 350B3/550C3	140 kt (259 km/h; 161 mph)
AS 350D	124 kt (230 km/h; 143 mph)
Max rate of climb at S/L:	
AS 350B2	534 m (1,752 ft)/min
AS 350B3/550C3	618 m (2,028 ft)/min
Service ceiling: AS 350B2	4,800 m (15,750 ft)
AS 350B3/550C3	5,280 m (17,323 ft)
Hovering ceiling: IGE: AS 350B2	3,200 m (10,500 ft)
AS 350B3/550C3	4,260 m (13,976 ft)
OGE: AS 350B2	2,550 m (8,360 ft)
AS 350B3/550C3	3,630 m (11,909 ft)
Range with max fuel at recommended cruising speed, no reserves:	
AS 350B2	362 n miles (670 km; 416 miles)
AS 350B3/550C3	360 n miles (666 km; 414 miles)

UPDATED

Eurocopter AS 350B3 painted to mark 100th delivery *(G Deulin/Eurocopter)*

EUROCOPTER EC 130B

TYPE: Light utility helicopter.
PROGRAMME: Derived from AS 350 Ecureuil; PT-1 prototype first flew (unannounced) June 1999 and registered F-WQEV in May 2000; second prototype (F-WQES) assisted in 200 hour test programme, including high-altitude testing in Albuquerque, USA, September 2000; JAA certification 14 December 2000; FAA certification 21 December 2000; revealed only in February 2001, when initial production aircraft (also 3,000th of Ecureuil family) handed over to Blue Hawaiian Helicopters at HeliExpo. Four pre-production helicopters delivered to launch customers in first half of 2001; main series production from September 2001.
CURRENT VERSIONS: **EC 130B4**: Initial production version.
COSTS: US$1.6 million, unofficial estimate (2001).
DESIGN FEATURES: As for AS 355B3, but considerably modified external appearance, with windscreen side panels, doors and landing gear from EC 120, plus Fenestron based on EC 135, although symmetrical, carried by redesigned tailboom. Dual hydraulic system from AS 355N. Cabin width increased by 25 cm (10 in) to give additional 23 per cent internal space, accommodating seven or eight seats and enlarged (10 per cent) baggage compartment; optional air conditioning. Retains AS 355B3 engine and transmission, but automatic system (in conjunction with dual-channel FADEC, plus third, digital, backup channel) matches rotor speed to flight conditions for noise reduction; 84.3 EPNdB flyover rating is 7 dB below ICAO limit and 0.5 dB under special Grand Canyon National Park restrictions.
LANDING GEAR: Emergency flotation system available.
POWER PLANT: One 543 kW (728 shp) Turbomeca Arriel 2B1 turboshaft with transmission rated at 632 kW (847 shp) for T-O, with FADEC. Fuel capacity 540 litres (143 US gallons; 119 Imp gallons).
ACCOMMODATION: Pilot and up to seven passengers in transport role; or pilot and one or two stretcher patients plus two medical attendants in casualty evacuation configuration. Dual controls.
SYSTEMS: 4.5 kW 28 V DC starter-generator. 15 Ah battery; 28 V DC cabin outlet.
AVIONICS: *Comms:* Shadin 8800 T altitude encoder; Honeywell KT 76C Mode C transponder and NAT AA95 ICS.
Flight: Two Honeywell KX 165A VHF/VOR/LOC/ glidescope; Garmin GNS 430 GPS; Artex 100 HM ELT; AIM 1110 HSI; Honeywell KI 525 HSI.

DIMENSIONS, EXTERNAL:
Main rotor diameter	10.69 m (35 ft 0¾ in)
Length: overall, rotors turning	12.64 m (41 ft 5½ in)
fuselage	10.68 m (35 ft 0½ in)
Fuselage max width	2.03 m (6 ft 8 in)
Height: to top of fin	3.61 m (11 ft 10¼ in)
to top of rotor head	3.34 m (10 ft 11½ in)
Tailplane span	2.73 m (8 ft 11½ in)
Skid track	2.31 m (7 ft 7 in)

DIMENSIONS, INTERNAL:
Cabin: Length	2.19 m (7 ft 2¼ in)
Max width	1.865 m (6 ft 1½ in)
Max height	1.28 m (4 ft 2½ in)
Volume	3.7 m³ (131 cu ft)

Baggage hold, volume: Front left-hand 0.29 m³ (10.2 cu ft)

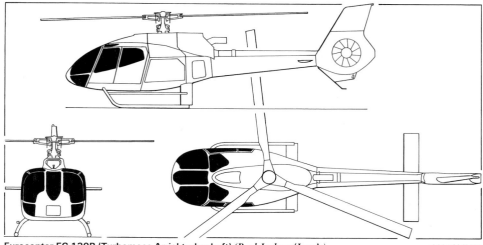

Eurocopter EC 130B (Turbomeca Arriel turboshaft) *(Paul Jackson/Jane's)* 2001/0062408

Front right-hand	0.25 m³ (8.8 cu ft)
Rear	0.57 m³ (20.1 cu ft)

AREAS:
Main rotor disc	89.75 m² (966.1 sq ft)

WEIGHTS AND LOADINGS:
Weight empty	1,360 kg (3,000 lb)
Max T-O weight: internal load	2,400 kg (5,291 lb)
external	2,800 kg (6,173 lb)
Max disc loading:	
internal load	26.74 kg/m² (5.48 lb/sq ft)
external load	31.20 kg/m² (6.39 lb/sq ft)
Max power loading:	
internal load	4.42 kg/kW (7.27 lb/shp)
external load	5.16 kg/kW (8.48 lb/shp)

PERFORMANCE (two occupants):
Never-exceed speed (V_{NE})	155 kt (287 km/h; 178 mph)
Max operating speed	127 kt (235 km/h; 146 mph)
Max rate of climb at S/L	698 m (2,290 ft)/min
Hovering ceiling: IGE	5,875 m (19,280 ft)
OGE	5,320 m (17,460 ft)
Service ceiling	7,010 m (23,000 ft)

OPERATIONAL NOISE LEVELS (ICAO Annex 16 Chap 8):
Average	86.8 EPNdB
Overflight	84.3 EPNdB

NEW ENTRY

EUROCOPTER AS 355 ECUREUIL 2 TWINSTAR and AS 555 FENNEC

Brazilian Air Force designations: CH-55 and VH-55 Esquilo
Brazilian Navy designation: UH-12B Esquilo
Royal Air Force name: Twin Squirrel

TYPE: Light utility helicopter.
PROGRAMME: Twin-engined version of AS 350. First flight of first prototype (F-WZLA) 28 September 1979; details of early production AS 355E/F versions in 1984-85 *Jane's*; AS 355F superseded by AS 355F1 in January 1984; AS 355F2 certified 10 December 1985; AS 355N powered by two Turbomeca TM 319 Arrius 1A certified in France in 1989, UK and USA in 1992; deliveries of this version began early 1992.
CURRENT VERSIONS: Current production AS 355 Ecureuil 2 and AS 555 Fennec are powered by two Turbomeca Arrius1A turboshafts.
AS 355N Ecureuil 2: Current civil production version, adaptable for passengers, cargo, police, ambulance, slung loads, carrying harbour pilots, working on high-tension cables and other missions; Category A OEI performance; known as **TwinStar** in USA.
Details refer to AS 355N, except where stated.
AS 555UN Fennec: Military utility version and French Army ALAT IFR pilot trainer.
AS 555AN Fennec: Armed version; later machines adapted for centreline-mounted 20 mm gun and pylon-mounted rockets.
AS 555MN Fennec: Naval unarmed version. Can carry a 360° chin-mounted radar.
AS 555SN Fennec: Armed naval version for ASW and over-the-horizon targeting, operating from ships of 600 tonnes upwards; armament includes one lightweight homing torpedo or the cannon and rockets of land-based versions; avionics include Honeywell RDR-1500B 360° chin-mounted radar, Sextant Nadir 10 navigation system, Dassault Electronique RDN 85 Doppler and SFIM 85 T31 three-axis autopilot.
CUSTOMERS: Total 840 AS 355/555 ordered by customers in 45 countries by 1 November 1999 including 12 sold to 10 law-enforcement agencies in six countries. Of these, 592 twin-engined Ecureuils in service at May 2000. French Air Force ordered 52 Fennecs, first eight as AS 355F1s powered by Allison 250; flown by 67e Escadre d'Hélicoptères at Villacoublay and other units for communications and, with side-mounted Giat M61 20 mm gun pod, by ETOM 68 in Guyana; delivery of remaining 44 (AS 555AN powered by Arrius) began 19 January 1990; from 24th onwards, provision for centrally mounted 20 mm cannon and T-100 sight, plus Mistral missiles; French Army ordered 18 AS 555UNs for IFR training (delivery from February 1992); Brazilian Air Force has 13

Second prototype Eurocopter EC 130B, derived from AS 350B Ecureuil *(Jérome Deulin/Eurocopter)* 2001/0106462

232　INTERNATIONAL: AIRCRAFT—EUROCOPTER

AS 555s: 11 with armament, designated CH-55, and two VIP transports, designated VH-55; Brazilian Navy acquired 11 UH-12Bs; Brazilian AS 555s assembled by Helibrás in Brazil (which see); four more navies ordered eight AS 555s in 1992, two in 1993, four in 1994 and one in 1995. Two AS 355F1s, leased from OSS of Hayes, UK, supplied to the RAF's No 32 (The Royal) Squadron at Northolt on 1 April 1996 for VIP transport; third subsequently added. Orders in 1997-99 totalled nine, 20 and 10; 1999 deliveries totalled 19, including four for the Jamaica Defence Force.

DESIGN FEATURES: Starflex main rotor; two engine shafts drive into combiner gearbox containing freewheels; main rotor turns at 394 rpm and tail rotor at 2,086 rpm; otherwise substantially as AS 350.

FLYING CONTROLS: Powered, full dual flying controls without manual reversion; trim by adjustable stick friction; inverted aerofoil tailplane; swept fins above and below tailboom.

STRUCTURE: Light metal tailboom and central fuselage structure; thermoformed plastics for cabin structure.

LANDING GEAR: As for AS 350B2/550.

POWER PLANT: Two Turbomeca Arrius 1A turboshafts, each rated at 357 kW (479 shp) for take-off and 302 kW (406 shp) maximum continuous; 388 kW (520 shp) 30 minutes OEI emergency; 407 kW (547 shp) 2.5 minutes OEI emergency; full authority digital engine control (FADEC) allows automatic sequenced starting of both engines, automatic top temperature and torque limiting and preselection of lower limits for practice OEI operation. Transmission rating 511 kW (686 shp) for T-O. Two integral fuel tanks, with total usable capacity of 730 litres (193 US gallons; 160 Imp gallons), in body structure. Optional 475 litre (125 US gallon; 104 Imp gallon) auxiliary tank in cabin.

ACCOMMODATION: As for single-engined Ecureuil and Fennec.

SYSTEMS: As for AS 350B2/550, except two hydraulic pumps, reservoirs and tandem powered flying control units and electric generators.

AVIONICS: Options include a second VHF/AM and radio altimeter; provision for IFR system, SFIM 85 T31 three-axis autopilot and CDV 85 T3 nav coupler.

EQUIPMENT: Casualty installations, TV, FLIR, searchlight and 204 kg (500 lb) capacity winch available. Cargo hook optional; capacity 1,134 kg (2,500 lb). See also Current Version descriptions.

ARMAMENT (AS 555): Optional alternative weapons include TDA or Forges de Zeebrugge rocket packs, Matra or FN Herstal machine gun pods or a Giat M621 20 mm gun. Naval version carries one homing torpedo in ASW role, or SAR winch.

DIMENSIONS, EXTERNAL: As for AS 350B2/550C3, except:
Width: fuselage	1.87 m (6 ft 1½ in)
overall, blades folded	3.05 m (10 ft 0 in)
Tailplane span	3.05 m (10 ft 0 in)

DIMENSIONS, INTERNAL: As for AS 350B2/550C3

WEIGHTS AND LOADINGS:
Weight empty	1,436 kg (3,166 lb)
Max sling load	1,134 kg (2,500 lb)
Max T-O weight:	
internal load	2,600 kg (5,732 lb)
max sling load	2,600 kg (5,732 lb)
Max disc loading:	
internal load	29.0 kg/m² (5.93 lb/sq ft)
external load	29.0 kg/m² (5.93 lb/sq ft)
Transmission loading at max T-O weight and power:	
internal load	5.09 kg/kW (8.36 lb/shp)
external load	5.09 kg/kW (8.36 lb/shp)

PERFORMANCE (ISA):
Never-exceed speed (V_NE)	150 kt (278 km/h; 172 mph)
Max cruising speed at S/L:	
AS 355N	121 kt (224 km/h; 139 mph)
AS 355UN/AN	118 kt (219 km/h; 136 mph)
Econ cruising speed	117 kt (217 km/h; 135 mph)
Max rate of climb at S/L	384 m (1,260 ft)/min
Rate of climb at S/L, OEI	156 m (512 ft)/min
Service ceiling	3,800 m (12,460 ft)
Service ceiling, OEI	1,450 m (4,760 ft)
Hovering ceiling: IGE	2,000 m (6,560 ft)
OGE	750 m (2,460 ft)
Radius, SAR, two survivors	
	70 n miles (129 km; 80.5 miles)
Range with max fuel at S/L, no reserves:	
AS 355N	389 n miles (720 km; 448 miles)
AS 355UN/AN/MN/SN	
	375 n miles (694 km; 431 miles)
Endurance, no reserves:	
one torpedo, or cannon or rocket pods	2 h 20 min
cannon plus rockets	1 h 50 min

UPDATED

EUROCOPTER AS 365N DAUPHIN 2
US Coast Guard designation: HH-65A Dolphin
Israel Defence Force name: Dolpheen

TYPE: Light utility helicopter.

PROGRAMME: Certified for VFR in France November 1989; 500th Dauphin (all versions) delivered in November 1991 was 83rd AS 365N2. Dauphin family had flown 2.7 million flying hours by June 2000.

Eurocopter AS 555SN of the Colombian Navy

Eurocopter AS 355N operated by Malaysian National Police *(Paul Jackson/Jane's)*

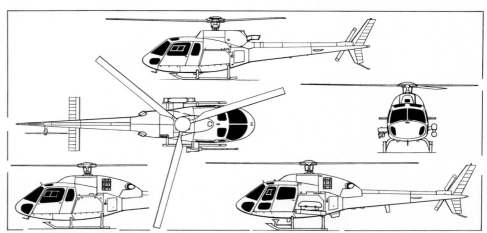
Eurocopter AS 555AN Fennec twin-turbine armed helicopter with side view (top) of single-engined AS 350B3 Ecureuil and scrap view (bottom) of AS 355N civil twin *(Dennis Punnett/Jane's)*

Instrument panel of a naval AS 555MN/SN Fennec *(Patrick Penna/Eurocopter)*

AS 365N2/N3 Dauphin optional instrument panel: EFIS 50, RDR 2000 weather radar and Honeywell avionics *(Patrick Penna/Eurocopter)*

CURRENT VERSIONS: **AS 365N2 Dauphin 2:** Recent baseline version; now produced in parallel with AS 365N3.
Details below refer to N2 version.

AS 365N3 Dauphin: Improved hot-and-high version; technology for N3 and N4 versions developed by experimental DGV (Dauphin *Grande Vitesse*) F-WDFK; DGAC certification in October 1997 and first delivery to Noordzee Helikopter Vlaanderen of Belgium in early 1998. Turbomeca Arriel 2C engines offer maximum T-O weight up to 48°C (118°F), rather than N2's 36°C (97°F). FADEC as standard. Orders had reached 40 by mid-2000.

AS 365N4 Dauphin: Redesignated EC 155; see following entry.

AS 366G1 (HH-65A Dolphin): US Coast Guard version (99 built, plus two ex-trials aircraft bought for evaluation by Israel as Dolpheen); trial installation of 895 kW (1,200 shp) LHTEC T800 turboshaft ordered February 1990, but abandoned November 1991 when LTS101 performance promised to improve; re-engining with French Arriels also declined. For description see 1991-92 and earlier *Jane's*.

AS 565UB Panther: Army version; described separately.

AS 565SB Panther: Naval version; described separately.

Ambulance/EMS: Flight crew of two plus two or four stretchers along cabin sides loaded through rear doors; up to four seats can replace stretchers on one side; doctor sits in middle with equipment; rear doors open through 180° instead of sliding.

Offshore support: Pilot plus up to 12 passengers, autopilot and navigation systems, pop-out floats.

DTV5 (*Démonstrateur Technologie Véhicule*): Demonstrator of new Dauphin variant; new rotor of 13.50 m (44 ft 31/2 in) diameter; Tigre transmission; and two 809 kW (1,085 shp) Turbomeca TM 333-2E1 turboshafts. Conversion of Dauphin N4 prototype, subsequently renamed Dolphin VHS (very high speed).

Dolphin FWB: Testbed for fly-by-wire technology used in NH 90.

CUSTOMERS: Some 677 AS 365/366/565 ordered for civil and military use by 187 customers from 53 countries by mid-2000; totals include 50 produced as Harbin Z-9/9A in China and 101 in HH-65A Dolphin US Coast Guard versions; Chinese production by HAMC (which see) continuing as Z-9A-100. Dauphin and Panther orders totalled 10 in 1999, compared with 28 in 1998, and 16 in 1997. Eurocopter has entered offset agreement with LOT of Slovak Republic to allow modification of AS 365s for East European customers; first sale announced 1998.

DESIGN FEATURES: Progressively tapered rear fuselage with stabilising surfaces comprising horizontal tail with endplate fins, plus central fin; Fenestron anti-torque rotor; engines above cabin.

AS 365N2 and N3 have Starflex main rotor hub with four blades; 11-blade Fenestron; main rotor blades have quick disconnect pins for manual folding; ONERA OA 212 (thickness/chord ratio 12 per cent) at root to OA 207 (thickness/chord ratio 7 per cent) at tip; adjustment tab near tip; leading-edge of tip swept at 45°; main rotor rpm 350; Fenestron rpm 3,665; rotor brake standard. AS 365N3 can be retrospectively fitted with EC 155's five-blade Spheriflex main rotor and 10-blade Fenestron rotor with unevenly spaced blades to reduce vibration.

FLYING CONTROLS: Hydraulic dual fully powered; cyclic trim by adjustable stick friction damper; fixed tailplane, with endplate fins offset 10° to port; IFR systems available.

STRUCTURE: Mainly light alloy; machined frames fore and aft of transmission support; main rotor blades have CFRP spar and skins with Nomex honeycomb filling; Fenestron duct and fin of CFRP and Nomex/Rohacell sandwich; nose and power plant fairings of GFRP/Nomex sandwich; centre and rear fuselage assemblies, flight deck floor, roof, walls and bottom skin of fuel tank bays of light alloy/Nomex sandwich.

LANDING GEAR: Hydraulically retractable tricycle type; twin-wheel self-centring nose unit retracts rearward; single wheel on each main unit; main legs retract into troughs in fuselage without cover doors; all three units have oleo-pneumatic shock-absorber; mainwheel tyres 15×6.00-6 or 380×150-6 (6 ply) tubeless, pressure 8.60 bars (125 lb/sq in); nosewheel tyres size 13×5.00-4, pressure 5.50 bars (80 lb/sq in); hydraulic disc brakes.

POWER PLANT: *AS 365N2:* Two Turbomeca Arriel 1C2 turboshafts, each rated at 551 kW (739 shp) for T-O and 471 kW (631 shp) maximum continuous, mounted side by side aft of transmission with stainless steel firewall between them. Transmission rating 965 kW (1,294 shp) for T-O and maximum continuous.

AS 365N3: Two FADEC-equipped Turbomeca Arriel 2C turboshafts, each rated at 635 kW (851 shp) for T-O, 597 kW (800 shp) maximum continuous and 729 kW (977 shp) for 30 seconds.

Standard fuel in four tanks under cabin floor and fifth tank in bottom of centre-fuselage; total capacity 1,135 litres (300 US gallons; 249.5 Imp gallons); provision for auxiliary tank in baggage compartment, with capacity of 180 litres (47.5 US gallons; 39.5 Imp gallons); or ferry tank in place of rear seats in cabin, capacity 475 litres (125.5 US gallons; 104.5 Imp gallons); refuelling point above landing gear door on port side. Oil capacity 14 litres (3.7 US gallons; 3.1 Imp gallons).

ACCOMMODATION: Standard accommodation for pilot and co-pilot or passenger in front, and two rows of four seats to rear; high-density seating for one pilot and 12 passengers; VIP configurations for four to six persons in addition to pilot; three forward-opening doors on each side; freight hold aft of cabin rear bulkhead, with door on starboard side; cabin heated and ventilated.

SYSTEMS: Air conditioning system optional. Duplicated hydraulic system, pressure 60 bars (870 lb/sq in). Electrical system includes two 4.8 kW starter/generators, one 24 V 27 Ah battery and two 250 VA 115 V 400 Hz inverters.

AVIONICS: *Comms:* Optional transponder.
Radar: Optional weather radar.
Flight: SFIM 155 duplex autopilot; optional VOR, ILS, ADF, DME, GPS and SFIM CDV 85 autopilot coupler.
Instrumentation: Two-pilot IFR instrument panel; optional EFIS.

EQUIPMENT: Includes 1,600 kg (3,525 lb) capacity cargo hook, and 275 kg (606 lb) capacity hoist with 90 m (295 ft) cable.

DIMENSIONS, EXTERNAL (N2 and N3):
Main rotor diameter:	11.94 m (39 ft 2 in)
Fenestron diameter	1.10 m (3 ft 7¼ in)
Main rotor blade chord: basic	0.385 m (1 ft 3¼ in)
outboard of tab	0.405 m (1 ft 4 in)
Length: overall, rotor turning	13.73 m (45 ft 0½ in)
fuselage	11.63 m (38 ft 2 in)

Eurocopter SA 365N2 Dauphin of Noordzee Helikopter Vlaanderen *(Paul Jackson/Jane's)*

234　INTERNATIONAL: AIRCRAFT—EUROCOPTER

Width, rotor blades folded	3.25 m (10 ft 8 in)
Height: to top of rotor head	3.47 m (11 ft 4½ in)
overall (tip of fin)	4.06 m (13 ft 4 in)
Wheel track	1.90 m (6 ft 2¾ in)
Wheelbase	3.64 m (11 ft 11¼ in)
Main cabin door (fwd, each side):	
Height	1.16 m (3 ft 9½ in)
Width	1.14 m (3 ft 9 in)
Main cabin door (rear, each side):	
Height	1.16 m (3 ft 9½ in)
Width	0.87 m (2 ft 10¼ in)
Baggage compartment door (stbd)	
Height	0.51 m (1 ft 8 in)
Width	0.73 m (2 ft 4¾ in)
DIMENSIONS, INTERNAL:	
Cabin: Length	2.30 m (7 ft 6½ in)
Max width	1.92 m (6 ft 3½ in)
Max height	1.16 m (3 ft 9¾ in)
Floor area	4.20 m² (45.2 sq ft)
Volume	5.0 m³ (176 cu ft)
Baggage compartment volume	1.6 m³ (56 cu ft)
AREAS:	
Main rotor disc	111.97 m² (1,205.2 sq ft)
Fenestron disc	0.95 m² (10.23 sq ft)
WEIGHTS AND LOADINGS:	
Weight empty: N2	2,281 kg (5,028 lb)
N3	2,302 kg (5,075 lb)
Max T-O weight: N2	4,250 kg (9,369 lb)
N3	4,300 kg (9,480 lb)
Max disc loading: N2	37.9 kg/m² (7.78 lb/sq ft)
N3	38.4 kg/m² (7.87 lb/sq ft)
Transmission loading at max T-O weight and power:	
N2	4.40 kg/kW (7.24 lb/shp)
PERFORMANCE:	
Never-exceed speed (V_{NE}):	
N2 and N3	155 kt (287 km/h; 178 mph)
Max cruising speed: N2 and N3	
	150 kt (278 km/h; 173 mph)
Econ cruising speed: N2	137 kt (254 km/h; 158 mph)
N3	148 kt (274 km/h; 170 mph)
Max rate of climb at S/L: N2	408 m (1,339 ft)/min
Rate of climb at S/L, OEI: N2	78 m (256 ft)/min
N3	183 m (602 ft)/min
Service ceiling: N2	3,700 m (12,140 ft)
N3	4,700 m (15,420 ft)
Service ceiling, OEI: N2	1,200 m (3,940 ft)
N3	1,870 m (6,140 ft)
Hovering ceiling: IGE: N2	2,000 m (6,560 ft)
N3	2,620 m (8,600 ft)
OGE: N2	1,200 m (3,940 ft)
N3	1,150 m (3,780 ft)
Max range with standard fuel:	
N2	464 n miles (859 km; 534 miles)
N3	440 n miles (815 km; 506 miles)
Endurance with standard fuel: N2	4 h 18 min
N3	4 h 6 min
	UPDATED

EUROCOPTER AS 565 PANTHER (ARMY/AIR FORCE)
Brazilian Army designation: HM-1
TYPE: Multirole light helicopter.
PROGRAMME: Unarmed and armed army/air force versions of (initially) AS 365N2 and AS 365N3 Dauphin 2. First flight of AS 365M Panther (F-WZJV) 29 February 1984; first shown in production form 30 April 1986; armament integration and firing trials completed late 1986. AS 365AB withdrawn from programme; no orders forthcoming.
CURRENT VERSIONS: **AS 565UB:** Utility version replacing earlier AS 565UA from 1997; high-speed assault transport for eight to 10 troops with two crewmen over radius of 215 n miles (400 km; 248 miles); other roles include reconnaissance, aerial command post, electronic warfare, target designation, search and rescue, four-stretcher casevac and slung loads up to 1,600 kg (3,525 lb).
AS 565AB: Replaces AS 565AA; promotion continued in 2000. Armed version; options include two rocket pods

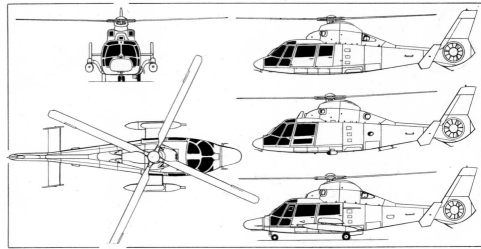

Eurocopter AS 565AA Panther with side views (centre) of AS 366 HH-65A Dolphin and (top) AS 365N2 Dauphin 2 *(Dennis Punnett/Jane's)*

with either 22 rounds of 68 mm calibre or 19 of 2.75 in; or two gun pods, each with 180 rounds.
CUSTOMERS: First order 36 AS 565AAs for Brazilian 1 BAvEx (Army Aviation Branch) at Taubaté delivered from 1989. No further sales reported by Eurocopter. One converted Dauphin 2 (F-ZVLD) is used by France's CEV test establishment for trials of the Trigat anti-tank missile.
DESIGN FEATURES: As AS 365N3, but with greater use of composites materials and greater emphasis on survivability in combat areas; radar and IR signatures reduced by composites and special paints; noise signature low; powered control servos and engine controls armoured; cable cutters; run-dry transmission; crew seats tolerate 20 g; entire basic airframe designed to withstand vertical impact at 7 m (23 ft)/s at maximum T-O weight.
FLYING CONTROLS: As AS 365N3.
POWER PLANT: Two Turbomeca Arriel 2C turboshafts; maximum ratings (each) are contingency 729 kW (977 shp), take-off 635 kW (851 shp) and continuous 597 kW (800 shp); FADEC gives automatically sequenced engine starting, full protection from excess temperature and torque anticipated engine acceleration and selectable torque/temperature limits for training. Fuel system withstands 14 m (46 ft)/s crash and tanks self-seal.
ACCOMMODATION: Crew of two plus up to ten troops or four stretchers plus one doctor/attendant in casualty evacuation role.
AVIONICS: *Comms:* Military communications radios.
Flight: Sextant Avionique self-contained navigation system.
Self-defence: Optional Thomson-CSF TMV 011 Sherloc RWR, IR jammer and chaff/flare dispenser.
EQUIPMENT: Options include 1,600 kg (3,527 lb) capacity cargo sling and 272 kg (600 lb) capacity electric hoist for combat SAR.
DIMENSIONS, EXTERNAL: As for AS 365N3, except:
Length of fuselage　　　　　　　12.08 m (39 ft 7½ in)
WEIGHTS AND LOADINGS:
Weight empty　　　　　　　　　2,305 kg (5,082 lb)
Max sling load　　　　　　　　　1,600 kg (3,527 lb)
Max T-O weight, internal or external load
　　　　　　　　　　　　　　　4,250 kg (9,369 lb)
PERFORMANCE (at average mission weight of 4,000 kg; 8,818 lb):
Never-exceed speed (V_{NE})　155 kt (287 km/h; 178 mph)
Max cruising speed at S/L　　150 kt (278 km/h; 173 mph)
Max rate of climb at S/L　　　468 m (1,535 ft)/min
Hovering ceiling: IGE　　　　　2,600 m (8,540 ft)
　　　　　OGE　　　　　　　2,500 m (8,200 ft)
Range with max standard fuel at S/L
　　　　　　　　　　　443 n miles (820 km; 510 miles)
　　　　　　　　　　　　　　　　　　　UPDATED

EUROCOPTER AS 565 PANTHER (NAVY)
Israel Defence Force name: Atalef (Bat)
TYPE: Multirole light helicopter.
PROGRAMME: Launch order placed by Saudi Arabia 13 October 1980; first flight of modified AS 365N (c/n 5100) as AS 565N prototype 22 February 1982; first flight production AS 365F (c/n 6014) equipped as SAR helicopter on 2 July 1982.
CURRENT VERSIONS: **AS 565MB Panther:** Unarmed naval search and rescue and sea surveillance version; replaced AS 565MA from 1997; rescue hoist, sea search radar, self-contained navigation, automatic hover transition; searchlight and deck-landing harpoon optional.
AS 565SB Panther: Armed ASW and anti-ship version; replaced AS 565SA from 1997; ASW version has two homing torpedoes; four side-mounted AS 15TT radar-guided missiles for anti-ship role; SAR versions have nose-mounted Omera ORB 32 search radar; anti-ship versions have chin-mounted, roll-stabilised Agrion 15 radar. 20 mm side-mounted cannon for self-defence.
AS 565SA Atalef: Ordered by Israel Defence Force, April 1994; total 20 required, of which initial order was for eight; assembled and tested at Grand Prairie, Texas; first flight (MNT01) September 1995. Equipment includes Telephonics search radar in nose radome; FLIR turret port side of cabin; Litton INS and GPS; and Elbit communications and self-defence systems. Deliveries were due from July 1996 for operation from Saar 5 and 4.5 frigates.
CUSTOMERS: Saudi Arabia ordered four SAR/surveillance AS 565MAs and 20 anti-ship AS 565SAs with AS 15TT; Ireland ordered five for SAR with Honeywell RDR 1500 radar, SFIM 155 autopilot, Sextant ONS 200A long-range navigation, Nadir Mk II nav computer, Dassault Electronique Cina B Doppler and five-screen EFIS; these can carry light weapons; French Navy ordered three AS 565F (MA) 'Pédros' in 1988 for aircraft carrier plane guard; 15 more AS 565SAs delivered from 1993; 11, with AS 15TT, ordered by United Arab Emirates 1995, are first AS 565SBs delivered. Icelandic Coast Guard operates single example.
DESIGN FEATURES: Updated version of AS 365F shipborne Panther with enlarged 11-blade Fenestron for control during out-of-wind overwater hover; extended nose to house radar and additional avionics.
FLYING CONTROLS: As AS 365N3, with SFIM CDV 155 flight director and coupler for naval operations.
POWER PLANT: Earlier AS 565MA/SAs have two Turbomeca Arriel 1M1 turboshafts, each giving 558 kW (749 shp) for take-off and 487 kW (653 shp) continuously; current AS 565MB/SBs have Arriel 2C turboshafts, each rated at 635 kW (851 shp) at T-O and 597 kW (800 hp) continuous for better hot-and-high performance; standard fuel capacity 1,135 litres (300 US gallons; 250 Imp gallons); optional 180 litre (47.5 US gallon; 39.6 Imp gallon) auxiliary tank.
ACCOMMODATION: Two-man crew; cabin can hold 10 passengers.
DIMENSIONS, EXTERNAL: As for AS 565 Panther (army/air force), except:
Width over missiles (AS 565SB)　4.20 m (13 ft 9½ in)
WEIGHTS AND LOADINGS:
Weight empty　　　　　　　　　2,305 kg (5,082 lb)
Max sling load　　　　　　　　　1,600 kg (3,527 lb)
Max T-O weight, internal or external load
　　　　　　　　　　　　　　　4,300 kg (9,480 lb)
PERFORMANCE (at average mission weight of 4,000 kg; 8,818 lb):
Never-exceed speed (V_{NE})　160 kt (296 km/h; 184 mph)
Max cruising speed at S/L　　152 kt (281 km/h; 175 mph)
Max rate of climb at S/L　　　480 m (1,575 ft)/min
Hovering ceiling: IGE　　　　　3,300 m (10,820 ft)
　　　　　OGE　　　　　　　2,500 m (8,200 ft)

Early production SA 365N Dauphin employed as a trials support vehicle by France's CEV *(Paul Jackson/Jane's)*　　　　　　　　　　　　　　　　　2001/0099607

Radius of action: anti-shipping, with four missiles, 120 kt
(222 km/h; 138 mph) cruising speed at 915 m
(3,000 ft), ISA −20°C, 30 min reserves
                               135 n miles (250 km; 155 miles)
with two missiles   150 n miles (278 km; 173 miles)
SAR, ISA +20°C, 30 min reserves, carrying six
    survivors   130 n miles (241 km; 150 miles)
Range with max standard fuel at S/L
                               443 n miles (820 km; 510 miles)
Endurance   4 h 30 min
UPDATED

EUROCOPTER EC 155B

TYPE: Light utility helicopter.
PROGRAMME: Programme launched September 1996, as further development of Dauphin 2 (see previous entry). Announced at 1997 Paris Air Show, when known as AS 365N4 Dauphin 2. First flight 17 June 1997 as conversion of DGV testbed F-WDFK (see AS 365N Dauphin 2 entry); 1,000 flying hours achieved by February 1998. New, EC 155 designation revealed at HAI convention, February 1998. First production EC 155 (F-WWOZ) flew 11 March 1998. JAR certification received 9 December 1998, FAR certification was due late 1999. Certification for single pilot in IFR issued 25 January 2000.
CUSTOMERS: First customer, German Border Guard, ordered 13 for delivery between 16 March 1999 and 2000; German Interior Ministry ordered two for Baden-Württemberg regional government in February 2000 for delivery in March 2001; Hong Kong Government Flying Service ordered five for delivery late 2002; firm orders stood at 33 by June 2000, of which eight were placed in 1999.
COSTS: US$5.5 million (1998).
DESIGN FEATURES: Compared to earlier Dauphin models, has bulged sliding cabin doors, redesigned cabin windows and 40 per cent larger cabin area. Five-blade Sphériflex main rotor and 10-blade Fenestron rotor with unevenly spaced blades to reduce noise.
FLYING CONTROLS: As AS 365N.
STRUCTURE: As AS 365N.
LANDING GEAR: As AS 365N.
POWER PLANT: Two FADEC-equipped Turbomeca Arriel 2C1 turboshafts, each rated at 635 kW (851 shp) for T-O, 597 kW (800 hp) max continuous power and 729 kW (977 shp) for 30 seconds. Standard fuel in four tanks under cabin floor and fifth tank in bottom of centre fuselage; total capacity 1,255 litres (332 US gallons; 276 Imp gallons); provision for auxiliary tank in baggage compartment, with capacity for 180 litres (47.6 US gallons; 39.6 Imp gallons); or ferry tank in place of rear seats in cabin, capacity 475 litres (125.5 US gallons; 104.5 Imp gallons); refuelling point above landing gear door on port side. Oil capacity 14 litres (3.7 US gallons; 3.1 Imp gallons).

Eurocopter AS 565SB naval helicopter launching an AS 15TT missile

ACCOMMODATION: Standard accommodation for pilot and co-pilot or passenger in front, and three rows of four seats to rear; high-density seating for one pilot and up to 14 passengers; VIP configurations for nine persons in addition to pilot; up to six stretchers in casevac role; one crew door and one large sliding door on each side; freight hold aft of cabin rear bulkhead, with door on both sides. Option of hinged cabin door on VIP versions.
SYSTEMS: As AS 365N.
AVIONICS: As AS 365N, but optional eighth MFD available (152 × 203 mm; 6 × 8 in) for mission equipment display.
EQUIPMENT: As AS 365N.
DIMENSIONS, EXTERNAL:
Main rotor diameter	12.60 m (41 ft 4 in)
Fenestron diameter	1.10 m (3 ft 7¼ in)
Main rotor blade chord: basic	0.385 m (1 ft 3¼ in)
outboard of tab	0.405 m (1 ft 4 in)
Length: overall, rotor turning	14.30 m (46 ft 11 in)
fuselage	12.73 m (41 ft 9 in)
Width, rotor blades folded	3.48 m (11 ft 5 in)
Height: to top of rotor head	3.64 m (11 ft 11¼ in)
overall (tip of fin)	4.35 m (14 ft 3¼ in)
Wheel track	1.90 m (6 ft 2¾ in)
Wheelbase	3.91 m (12 ft 10 in)
Baggage compartment door (stbd):	
Height	0.51 m (1 ft 8 in)
Width	0.73 m (2 ft 4¾ in)

DIMENSIONS, INTERNAL:
Cabin: Length	2.55 m (8 ft 4½ in)
Max width	2.10 m (6 ft 10¾ in)
Max height	1.34 m (4 ft 4¾ in)
Volume	6.8 m³ (240 cu ft)
Baggage compartment volume	2.0 m³ (71 cu ft)

AREAS:
Main rotor disc	124.69 m² (1,342.1 sq ft)
Fenestron disc	0.95 m² (10.23 sq ft)

WEIGHTS AND LOADINGS:
Weight empty	2,528 kg (5,573 lb)
Max T-O weight	4,800 kg (10,582 lb)
Max disc loading	38.5 kg/m² (7.88 lb/sq ft)
Transmission loading at max T-O weight and power:	
	4.97 kg/kW (8.17 lb/shp)

PERFORMANCE:
Never-exceed speed (V_{NE})	170 kt (314 km/h; 195 mph)
Max cruising speed	145 kt (269 km/h; 167 mph)
Econ cruising speed	143 kt (265 km/h; 165 mph)
Rate of climb at S/L, OEI	306 m (1,004 ft)/min
Service ceiling	5,110 m (16,760 ft)
Hovering ceiling: IGE	1,895 m (6,220 ft)
OGE	865 m (2,840 ft)
Max range with standard fuel	
	448 n miles (830 km; 515 miles)
Endurance with standard fuel	4 h 12 min

UPDATED

EUROCOPTER BO 105

Swedish designation: Hkp 9

TYPE: Multirole light helicopter.
PROGRAMME: First flight of prototype 16 February 1967; LBA certification 15 October 1970; production of 100 BO 105 M (VBH) and 212 BO 105 P (PAH-1) (see 1985-86 *Jane's* and current *Jane's Aircraft Upgrades*) completed 1984. Also assembled by Eurocopter Canada (see Canadian section) and IPTN (Indonesian section). By late 1999 the type had accumulated over five million flight hours; German Army fleet celebrated one million hours on 22 May 2000. European production to order only.
CURRENT VERSIONS: **BO 105 CBS**: Basic model with cabin stretched 0.25 m (10 in) in rear seat area; additional window aft of rear door; FAA certification to SFAR Pt 29-4 early 1983; superseded by CBS-5.
BO 105 CBS-5: Main production model since 1993, formerly known as EC Super Five. Upgraded version of BO 105 CBS-4, derived from German Army PAH-1 programme; announced at Heli-Expo February 1993 and certified late 1993.
Has new main rotor blades with parallel-chord DM-H4 aerofoil to 0.8 radius and tapered DM-H3 to tip; rotor lift increased 150 kg (330 lb); airframe vibration reduced to less than 0.1 g; improved stability. Improvements over CBS-4 also include Cat. A T-O weight in ISA + 20°C increased by 130 kg (287 lb); OGE hover ceiling at maximum T-O weight in ISA +20°C increased by 500 m (1,640 ft); power required to hover OGE at S/L in ISA + 20°C reduced by 26 kW (35 shp); maximum weight in OGE hover, OEI with 2.5 minutes' power in ISA + 20°C, increased by 240 kg (529 lb); service ceiling OEI at maximum weight increased by 900 m (2,950 ft).
Details here apply to BO 105 CBS-5, except where indicated.
BO 105 LSA-3: Hot-and-high variant powered by two Rolls-Royce 250-C28C each rated at 410 kW (550 shp) for 30 seconds; produced exclusively by Eurocopter Canada (which see).
BO 105 LSA-3 Super Lifter: Developed specifically for lifting external loads; maximum T-O weight 2,850 kg (6,283 lb). Transmission rating 620 kW (832 shp) for T-O; 588 kW (788 shp) maximum continuous.
CUSTOMERS: Total 1,387 BO 105s of all models delivered to 42 countries by 1999. Apart from the German Army, major operators include Mexican Navy (12), Spanish Army (70 including 18 armed reconnaissance, 14 observation and 28 anti-tank), German Interior Ministry (BGS) (22), Swedish Army (20), Swedish Air Force (four BO 105 CBS for IFR search and rescue), Netherlands Army (30).
Purchases of CBS-5 include two by Bahrain Navy, 1994, with rescue hoist and nose-mounted 360° radar; two, equipped with rescue hoist, searchlights and auxiliary fuel tanks, to TsENTROSPAS (ministry of the Russian

Bundesgrenzschutz Eurocopter EC 155B, equipped with searchlight and FLIR turret *(Paul Jackson/Jane's)*

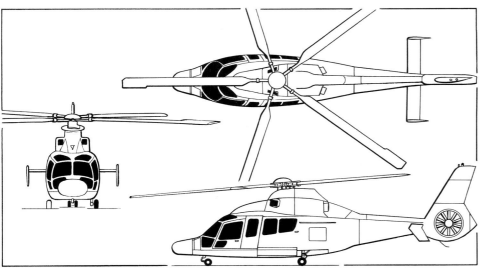

Eurocopter EC 155B *(James Goulding/Jane's)*

INTERNATIONAL: AIRCRAFT—EUROCOPTER

Armed Eurocopter BO 105 CBS-5 for South Korean Army
2000/0085588

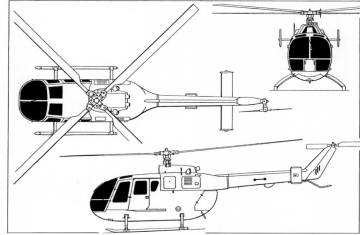

Eurocopter BO 105 CBS-5 multirole light helicopter *(Paul Jackson/Jane's)*

Federation for Civil Defence, Emergencies and Elimination of Consequences of Natural Disasters) delivered May 1995 for operation by MChS (State Unitary Aviation Enterprise). 17 ordered in January 1996 for emergency rescue duties by German Ministry of the Interior, deliveries to BGS Katastrophenschutz beginning immediately and ending early 1997.

In November 1997, South Korean Army selected the CBS-5 for reconnaissance and liaison; initial two delivered end 1999 (first German-built BO 105s produced since 1997); further 10 being assembled by Daewoo from 2000.

DESIGN FEATURES: Traditional pod-and-boom configuration; tail rotor mounted on top of fin; auxiliary horizontal and vertical stabilising surfaces; rear doors for large cargo.

Four-blade main rotor with rigid titanium hub; only articulation is roller bearings for blade pitch change (max 3°); all-composites blades of NACA 23012 section with drooped leading-edge and reflexed trailing-edge; 8° linear twist; pendulous vibration damper near each blade root; main rotor brake standard; two blades can be folded; two-blade semi-rigid tail rotor; main rotor rpm 424, tail rotor rpm 2,220.

FLYING CONTROLS: Fully powered controls through hydraulic actuation pack mounted on transmission casing; very high control power and aerobatic capability, including claimed sustained inverted 1 g flight; IFR certification without autostabiliser.

STRUCTURE: Light alloy structure with GFRP main and tail rotor blades and top fairing. Main production source is Eurocopter Deutschland factory at Donauwörth, but Spanish models assembled by CASA, Indonesian models manufactured and assembled by IPTN and BO 105 LS produced exclusively by Eurocopter Canada (formerly MBB Helicopter Canada) in Ontario (which see).

LANDING GEAR: Skid type, with aluminium alloy cross-tubes designed for energy absorption by plastic deformation in event of heavy landing. Inflatable emergency floats can be attached to skids.

POWER PLANT: Two 313 kW (420 shp) Rolls-Royce 250-C20B turboshafts, each with a maximum continuous rating of 298 kW (400 shp). Transmission rating 566 kW (759 shp) for T-O and maximum continuous. Bladder fuel tanks under cabin floor, usable capacity 570 litres (150.6 US gallons; 125.3 Imp gallons); fuelling point on port side of cabin; optional auxiliary tanks in freight compartment. Oil capacity: engine 12 litres (3.2 US gallons; 2.6 Imp gallons), gearbox 11.6 litres (3.06 US gallons; 2.55 Imp gallons).

ACCOMMODATION: Pilot and co-pilot or passenger on individual longitudinally adjustable front seats with four-point harnesses; optional dual controls. Bench seat at rear for three or four persons, removable for cargo and stretcher carrying; EMS versions available; entire rear fuselage aft of seats and under power plant available as freight and baggage space, with straight-in access through two clamshell doors at rear; two standard stretchers side by side in ambulance role; one forward-opening hinged and jettisonable door and one sliding door on each side of cabin; ram air and electrical ventilation system; heating system optional.

SYSTEMS: Dual hydraulic systems, pressure 103.5 bars (1,500 lb/sq in), for powered main rotor controls; system flow rate 6.2 litres (1.64 US gallons; 1.36 Imp gallons)/min; bootstrap/fluid reservoir, pressurised at 1.70 bars (25 lb/sq in); electrical system powered by two 150 A 28 V DC starter/generators and a 24 V 25 Ah Ni/Cd battery; external power socket.

AVIONICS: Wide variety of avionics available including weather radar, Doppler and GPS navigation, Honeywell RDR 1500B 360° search radar, FLIR, TV broadcast and microwave datalink.

EQUIPMENT: Standard equipment includes pilot's windscreen wiper, 250 W retractable landing light in nose, single-hand starting device, scavenge oil filter, heated pitot, tiedown rings in cargo compartment, cabin and cargo compartment dome lights, position lights and collision warning lights; options include heating system, rescue winch, searchlight, externally mounted loudspeaker, fuel dump valve, external load hook, settling protectors, snow skids, manual main rotor blade folding, emergency floats, sand filters, and a wide range of EMS and VIP arrangements. NVG-compatible cockpit, first used operationally on 5 February 1999 by Rocky Mountain Helicopters.

DIMENSIONS, EXTERNAL:
Main rotor diameter	9.84 m (32 ft 3½ in)
Tail rotor diameter	1.90 m (6 ft 2¾ in)
Main rotor blade chord	0.27 m (10¾ in)
Tail rotor blade chord	0.18 m (7 in)
Distance between rotor centres	5.95 m (19 ft 6¼ in)
Length: incl main and tail rotors	11.86 m (38 ft 11 in)
excl rotors	8.81 m (28 ft 11 in)
fuselage pod	4.55 m (14 ft 11 in)
Fuselage max width	1.58 m (5 ft 2¼ in)
Height to top of main rotor head	3.00 m (9 ft 10 in)
Skid track: unladen	2.53 m (8 ft 3½ in)
laden	2.58 m (8 ft 5½ in)
Rear-loading doors: Height	0.64 m (2 ft 1 in)
Width	1.40 m (4 ft 7 in)

DIMENSIONS, INTERNAL:
Cabin, incl cargo compartment:
Max width	1.40 m (4 ft 7 in)
Max height	1.25 m (4 ft 1 in)
Volume	4.8 m³ (169 cu ft)
Cargo compartment: Length	1.85 m (6 ft 0¾ in)
Max width	1.20 m (3 ft 11¼ in)
Max height	0.57 m (1 ft 10½ in)
Floor area	2.25 m² (24.2 sq ft)
Volume	1.3 m³ (46 cu ft)

AREAS:
Main rotor disc	76.05 m² (818.6 sq ft)
Tail rotor disc	2.835 m² (30.5 sq ft)

WEIGHTS AND LOADINGS:
Weight empty, basic: CBS-5	1,301 kg (2,868 lb)
LSA-3 Super Lifter	1,374 kg (3,029 lb)
Max external load	1,200 kg (2,646 lb)
Standard fuel (usable)	456 kg (1,005 lb)
Max fuel, incl auxiliary tanks	776 kg (1,710 lb)
Max T-O weight: CBS-5:	
internal load	2,500 kg (5,511 lb)
external load	2,600 kg (5,732 lb)
LSA-3 Super Lifter:	
internal load	2,600 kg (5,732 lb)
external load	2,850 kg (6,283 lb)
Max disc loading:	
CBS-5, internal load	32.9 kg/m² (6.74 lb/sq ft)
LSA-3 Super Lifter	34.2 kg/m² (7.00 lb/sq ft)
Transmission loading at max T-O weight and power, internal load:	
CBS-5	4.86 kg/kW (7.98 lb/shp)
LSA-3 Super Lifter	5.05 kg/kW (8.29 lb/shp)

PERFORMANCE:
Never-exceed speed (V_{NE}) at S/L:	
CBS-5	131 kt (242 km/h; 150 mph)
LSA-3 Super Lifter with underslung load	80 kt (148 km/h; 92 mph)
Max cruising speed at S/L:	
CBS-5	132 kt (245 km/h; 152 mph)
Best range speed at S/L	110 kt (204 km/h; 127 mph)
Max rate of climb at S/L, max continuous power	570 m (1,870 ft)/min
Vertical rate of climb at S/L, T-O power	90 m (295 ft)/min
Service ceiling, OEI, 30 min power	1,700 m (5,580 ft)
Max operating altitude:	
at 2,500 kg (5,511 lb)	3,050 m (10,000 ft)
at 2,300 kg (5,070 lb)	5,180 m (17,000 ft)
Hovering ceiling, T-O power: IGE	3,200 m (10,500 ft)
OGE	2,430 m (7,960 ft)
Range with standard fuel and max payload, no reserves:	
at S/L	310 n miles (574 km; 357 miles)
at 1,525 m (5,000 ft)	321 n miles (596 km; 370 miles)
Ferry range with auxiliary tanks, no reserves:	
at S/L	519 n miles (961 km; 597 miles)
at 1,525 m (5,000 ft)	550 n miles (1,020 km; 634 miles)
Endurance with standard fuel and max payload, no reserves: at S/L	3 h 24 min

UPDATED

EUROCOPTER EC 135 and EC 635

TYPE: Light utility helicopter.

PROGRAMME: First flight on 15 October 1988 of technology prototype, D-HBOX, previously known as BO 108, powered by two Rolls-Royce 250-C20R turboshafts with conventional tail rotor; new all-composites bearingless tail rotor tested during 1990; Eurocopter announced in January 1991 that BO 108 was to succeed BO 105; first flight of second prototype (D-HBEC) powered by two Turbomeca TM 319-1B Arrius, 5 June 1991; production main and tail rotors flight tested during 1992 in preparation for certification programme; design revised late 1992 to increase maximum seating to seven; advanced Fenestron adopted.

Two preproduction prototypes D-HECX and D-HECY made first flights respectively 15 February and 16 April 1994 powered by Turbomeca Arrius 2B and P&WC PW206B intended as production alternatives; third preproduction prototype (D-HECZ) made first flight 28 November 1994, powered by Arrius 2B, and subsequently made type's US debut at HeliExpo '95 in Las Vegas in January 1995; total flight time of first three preproduction EC 135s was nearly 1,600 hours by end 1996, by which time all three preproduction prototypes had been retired. VFR certification to JAR 27 16 June 1996 and to FAR Pt 27 with Category A provisions and for both engine options, 31 July 1996; IFR certification awarded jointly by DGAC (France) and LBA (Germany) on 9 December 1998, while that for FAA due mid-2000; LBA (JAA) single pilot IFR certification achieved 2 December 1999; certified in 17 countries by June 2000; first two production aircraft delivered to Deutsche Rettungsflugwacht on 31 July 1996.

CURRENT VERSIONS: **EC 135P1:** Pratt & Whitney engine version. First two (D-HQQQ and D-HYYY; c/ns 0005 and 0006) delivered on 31 July 1996 to Deutsche Rettungsflugwacht.

EC 135T1: Turbomeca engine version. First (0010/N4037A) delivered to USA in November 1996.

EC 135 ACT/FHS: Active control technology and flying helicopter simulator; German fly-by-light trials programme; first flight 1999.

EC 135 Police: Unified mission fit offered by McAlpine Helicopters of UK, 1997; allows simple outfitting with sensors and equipment, according to tasking, using under-fuselage pod; typical equipment, including loudspeakers, searchlights, microwave downlink and multisensor turret, can be fitted externally; TV and video equipment carried internally. Pod reduces maximum speed by 5 kt (9 km/h; 6 mph).

EC 635: Military version; mockup was conversion of first preproduction EC 135 (D-HECX). Offered (unsuccessfully) to South Africa and unveiled at Aerospace Africa Air Show on 28 April 1998. First

Mockup of the Eurocopter EC 635 military helicopter *(Paul Jackson/Jane's)*

Eurocopter EC 135 used by German Army for training *(Herbert Wittmann/Eurocopter)*

customer is Portuguese Army, which ordered nine EC 635T1s on 22 October 1999 for delivery from June 2001.

CUSTOMERS: Announced customers include Bond, UK (15); Deutsche Rettungsflugwacht (German Air Rescue; two); EuropAvia, Switzerland; MFL, France; Helicap of Paris, France (11 EMS versions); Basque Police, Spain; Bavarian Police, Germany (9 with PW206 engines); Sultan of Pahang, Malaysia; Heliair/OAMTC, Austria (five, delivered from 10 August 1997 followed by order for 11 on 27 September 2000 for delivery between December 2000 and early 2002); DAO, Argentina; Laidlaw, UK; Hood, UK; Manichi Press, Japan; Elifriulia, Italy; Proeetus, France; Air Service 51, France; Center for Emergency Medicine of Pittsburgh, Pennsylvania (one); Norsk Luftambulanse (six); Petroleum Helicopters Inc (three); Temsco of Ketchikan, Alaska (one); TexAir of Houston, Texas (one); ADAC, Germany (nine for EMS), Osterman, Sweden (two). Law enforcement users include Mecklenburg-West Pomerania, Germany (two), Saxony Police (one), Abu Dhabi Police (one), Chile Police (one), Greek Police (two), Travis County Police Department, USA (one), plus UK police forces (seven). German Army ordered 15 at cost of DM95 million in August 1997 for delivery in 2000 to replace Alouette II in training role; Border Guard ordered nine (plus two options) in December 1997. Eurocopter expects to win 700 sales out of a world market for 1,350 during the period 1998 to 2007. Total of over 200 on order by 50 customers in September 2000. 100th delivered (to Bavarian Police) on 16 June 1999, some two-thirds with Turbomeca engines; over 150 produced by September 2000; 30 ordered in 1999.

COSTS: Operating cost 25 per cent lower than BO 105; development programme funded by Eurocopter Deutschland and Eurocopter Canada, suppliers, and German Ministries of Economics and Research and Technology; flyaway cost US$2.39 million (1996); programme unit cost DM6.67 million (US$4.4 million), Bavarian police (1996). Portuguese contract for nine valued at €35 million (1999).

DESIGN FEATURES: Designed to FAR Pt 27 including Category A and European JAR 27; pod-and-boom configuration, with Fenestron; forward flight stability by two horizontal and four vertical (fin, underfin and two endplates) surfaces; four-blade FVW bearingless main rotor, single-piece rotor head/mast; rotor rpm are variable; composites blades mounted on controlled flexibility composites arms giving flap, lag and pitch-change freedom; control demands transmitted from rods to root of blade by rigid CFRP pitch cuffs; main rotor blades have DM-H3 and -4 aerofoils with non-linear twist and tapered transonic tips; main rotor axis tilted forward 5°.

Airframe drag 30 per cent lower than BO 105 by clean and compact external shape; cabin height retained by shallow two-stage transmission; vibration reduced by ARIS mounting between transmission and fuselage; all dynamically loaded components to have 3,500 hours MTBR or be maintained on-condition. Fenestron has 10 asymmetrically spaced blades.

Second BO 108 prototype had EFIS-based IFR system; fuselage stretched 15 cm (5.9 in) and interior cabin width extended by 10 cm (3.9 in); main rotor diameter extended to 10.20 m (33 ft 5½ in); for EC 135, tail rotor replaced in 1992 by New Generation Fenestron with 11 fixed flow-straightening vanes in fan efflux designed to avoid momentum losses and improve fan figure of merit; vanes are swept relative to radius and fan has different number of blades to avoid shocks and reduce noise; fan blade tip speed is only 185 m (607 ft)/s; maximum T-O weight increased to 2,720 kg (5,997 lb).

FLYING CONTROLS: Conventional hydraulic fully powered controls with integrated electrical SAS servos; objective is single-pilot IFR with cost-effective stability augmentation. Electric cyclic trim system.

STRUCTURE: Airframe mainly Kevlar/CFRP sandwich composites, except aluminium alloy sidewalls, pod lower module and cabin floor, tailboom and around cargo area; some titanium in engine bay; composites tailplane.

LANDING GEAR: Skid type; ground handling wheels can be fitted.

POWER PLANT: Choice of turboshaft engines. Early Turbomeca-engined aircraft had two Arrius 2B each giving 435 kW (583 shp) at T-O, 417 kW (559 shp) maximum continuous, 473 kW (634 shp) OEI continuous and 519 kW (696 shp) for 2½ minutes with OEI; these have been retrofitted (and new aircraft are now supplied) with two Arrius 2B1 each giving 500 kW (670 shp) at T-O, 425 kW (570 shp) maximum continuous, 560 kW (751 shp) for 2½ minutes with OEI. Alternative power plant is two Pratt & Whitney Canada PW206B each giving 463 kW (621 shp) at T-O, 419 kW (562 shp) maximum continuous, 500 kW (671 shp) OEI continuous and 546 kW (732 shp) for 2½ minutes with OEI. Both types of engine have FADEC. Transmission rating 616 kW (826 shp) maximum T-O, 567 kW (760 shp) maximum continuous, 353 kW (473 shp) OEI continuous, 411 kW (551 shp) for 2½ minutes with OEI.

Fuel capacity 673 litres (178 US gallons; 148 Imp gallons) of which 663 litres (175 US gallons; 146 Imp gallons) are usable; additional long-range tank optional, usable capacity 198.5 litres (52.4 US gallons; 43.7 Imp gallons). Optional self-sealing fuel tanks. Oil capacity 8 litres (2.0 US gallons; 1.75 Imp gallons).

ACCOMMODATION: Pilot(s), plus six or seven passengers on crashproof seats in standard version, or pilot plus four or five passengers in VIP version. Four-point harnesses for front seats; three-point harnesses for remaining seats. Forward-hinged doors for two front occupants; sliding doors for five persons in cabin. Rear of pod has clamshell doors for bulky items/cargo; flights permissible with clamshell doors removed; optional window in each rear door. Unobstructed cabin interior. EMS variant can accommodate one or two pilots with up to two stretcher cases and up to three seated medical staff/attendants.

SYSTEMS: Redundant 28 V DC electrical supply systems to JAR/FAR 27 standards; two 160 A 28 V starter/generators and 24 V 17 Ah Ni/Cd batteries in Arrius 2B variant, two 200 A 28 V starter/generators and 24 V 25 Ah Ni/Cd batteries in PW206B variant. Fully redundant dual hydraulic systems. NATO standard external power connector.

AVIONICS: Honeywell Gold Crown and Sextant Avionique Nouvelle (AN) equipment.
Radar: Provision for integrated weather radar.
Flight: Air data computer; SFIM automatic flight control system (AFCS); GPS. Honeywell Combined solid-state flight data and cockpit voice recorder tested on EC 135 late 1998.
Instrumentation: Liquid crystal dual-screen (Sextant SMD45) vehicle and engine management displays with AN equipment.
Mission: Optional FLIR and NVGs for police and ambulance roles.

EQUIPMENT: Options include cargo hook, external loudspeakers and searchlights, rescue winch (230 kg; 507 lb) with 50 m (164 ft) cable, emergency floats, sand filter, wire-strike protection system and light armour protection.

DIMENSIONS, EXTERNAL:
Main rotor diameter	10.20 m (33 ft 5½ in)
Fenestron diameter	1.00 m (3 ft 3¼ in)
Length: overall, rotor turning	12.16 m (39 ft 10¾ in)
fuselage: incl boom	10.20 m (33 ft 5½ in)
to end of cabin	5.87 m (19 ft 3 in)
Height: overall	3.51 m (11 ft 6 in)
to top of rotor head	3.35 m (11 ft 0 in)
Width: without rotor blades	2.65 m (8 ft 8¼ in)
over skids	2.00 m (6 ft 6¾ in)
fuselage (max)	1.56 m (5 ft 1½ in)
Skid length	3.20 m (10 ft 6 in)
Ground clearance: fuselage	0.40 m (1 ft 3¾ in)
tailboom	0.66 m (2 ft 2 in)

DIMENSIONS, INTERNAL:
Cabin: Length: normal	3.06 m (10 ft 0½ in)
with EMS floor extension	4.11 m (13 ft 5¾ in)
Max width	1.50 m (4 ft 11 in)
Max height	1.26 m (4 ft 1½ in)
Floor area	3.3 m² (36 sq ft)
Volume	3.8 m³ (134.2 cu ft)
Baggage compartment:	
Length	1.05 m (3 ft 5¼ in)
Max width	1.23 m (4 ft 0½ in)
Max height	0.70 m (2 ft 3½ in)
Floor area	1.20 m² (12.9 sq ft)
Volume	1.1 m³ (39 cu ft)

AREAS:
Main rotor disc	81.71 m² (879.5 sq ft)
Fenestron disc	2.84 m² (30.52 sq ft)

WEIGHTS AND LOADINGS:
Weight empty: EC 135	1,490 kg (3,284 lb)
EC 635	1,530 kg (3,373 lb)
Max fuel: standard	536 kg (1,182 lb)
with long-range tank	696 kg (1,534 lb)

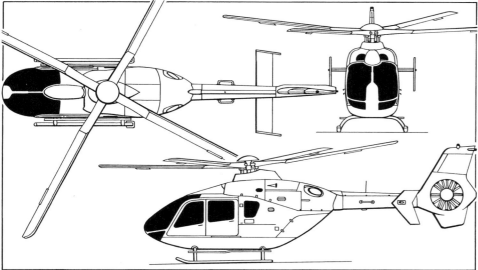
Eurocopter EC 135 seven-seat, twin-engined helicopter *(Mike Keep/Jane's)*

Eurocopter EC 135 light utility helicopter *(Paul Jackson/Jane's)*

Max external load	1,360 kg (2,998 lb)
Max T-O weight: normal	2,835 kg (6,250 lb)
with external load	2,900 kg (6,393 lb)
Max disc loading: normal	34.7 kg/m² (7.11 lb/sq ft)
with external load	35.5 kg/m² (7.27 lb/sq ft)
Transmission loading at max T-O weight and power:	
Normal T-O	4.61 kg/kW (7.57 lb/shp)
T-O with external load	4.71 kg/kW (7.73 lb/shp)

PERFORMANCE (A: Arrius 2B1, B: PW206B, at MTOW):

Never-exceed speed (V_NE):	
A, B	140 kt (259 km/h; 161 mph)
Max cruising speed: A, B	138 kt (256 km/h; 159 mph)
Normal cruising speed: A	127 kt (235 km/h; 146 mph)
B	123 kt (228 km/h; 142 mph)
Max rate of climb at S/L: A, B	457 m (1,500 ft)/min
Rate of climb at S/L, OEI: A, B	61 m (200 ft)/min
Service ceiling: A, B	3,050 m (10,000 ft)
Service ceiling, OEI: A	920 m (3,020 ft)
B	870 m (2,850 ft)
Hovering ceiling, IGE: A, B	1,325 m (5,000 ft)
Hovering ceiling, OGE: A, B	3,110 m (10,200 ft)
Range at S/L, standard fuel:	
A	335 n miles (620 km; 385 miles)
B	340 n miles (629 km; 391 miles)
Ferry range with long-range tank:	
A, B	474 n miles (878 km; 545 miles)
Endurance at S/L, standard fuel: A	3 h 24 min
B	3 h 33 min

UPDATED

EUROCOPTER EC 145

The Eurocopter EC 145 is a development of the Eurocopter/Kawasaki BK 117 and is described following that entry.

EUROCOPTER EC 155

This Dauphin development is described immediately following the Eurocopter AS 565 Panther entry.

EUROCOPTER EC 165

PROGRAMME: Studies continuing for a new helicopter in 4,000 to 6,000 kg (8,818 to 13,228 lb) class, to replace the AS 365N Dauphin 2. The DTV5 programme (see Dauphin 2 entry) will contribute to technology which may be used in the EC 165.

VERIFIED

EUROCOPTER TILT-ROTOR

Eurocopter's efforts in this direction have been merged into the Eurotilt programme, described later in this section.

UPDATED

OTHER AIRCRAFT

Refer elsewhere in this section for **Eurocopter/CATIC/STAe EC 120**, **Eurocopter/Kawasaki BK 117**, **EC 145**, and **Euromil (Eurocopter/Mil/Kazan/Klimov) Mi-38**; Eurocopter also participates in **NH Industries NH 90** international programme. Refer to *Jane's Aircraft Upgrades* for BO 105M PAH-1A1 upgrades.

VERIFIED

EUROCOPTER/CATIC/ST Aero

PARTICIPATING COMPANIES
Eurocopter: see this section
CATIC: see under HAMC in Chinese section
Singapore Technologies Aerospace: see under Singapore

This partnership formed to develop the EC 120 Colibri, production of which is now fully under way. Shares are Eurocopter 61 per cent (and programme leader), CATIC/HAMC 24 per cent and ST Aero 15 per cent.

UPDATED

EC 120 B COLIBRI
English name: Hummingbird
TYPE: Light utility helicopter.
PROGRAMME: Definition phase of original P120L launched 15 February 1990; subsequently redesigned with 500 kg (1,102 lb) lower gross weight and new engine and rotor; development contract signed October 1992; redesignated EC 120, January 1993; design definition completed mid-1993; assembly of first of two prototypes began at Eurocopter France at Marignane in early 1995; first flight (F-WWPA) 9 June 1995; second prototype (F-WWPD) flown 17 July 1996; certification to JAR 27 was achieved on 16 June 1997 (following Arrius 2F engine approval by DGAC on 22 January 1997); FAR Pt 27 certification on 22 January 1998; operations in cold weather certified late 1998; first production EC 120 (c/n 1005) flew 5 December 1997 before delivery to Japanese distributor Nosaki on 23 January 1998 for eventual use (as JA120B) by Nosaki Sangyo, Osaka.

Eurocopter responsible for seats, rotor system, transmission, final assembly, flight test and certification; CATIC (China National Aero-Technology Import & Export Corporation in the form of Harbin Aircraft Manufacturing Corporation) for cabin, landing gear and fuel system; and STAe for tailboom, fins, doors and instrument pedestal. Joint design team working in Eurocopter France at Marignane. Global support system, with 13 product centres, being initiated. Production rate increased to 90 per year in 2000.

CUSTOMERS: Estimated sales about 1,600 to 2,000 total over next 10 years. Recent customer is Spanish Air Force, which ordered 15 in 1999 for basic training; deliveries were to begin in 2000. Orders for 170 by customers from 25 countries by June 2000; 100th aircraft handed over 19 April 2000 to a German customer. Orders for 1999 totalled 36, compared with 45 in 1998 and 68 in 1997.

COSTS: US$840,000 (1999). Spanish contract for 15 valued at €15 million (1999), including training and spares.

DESIGN FEATURES: Pod-and-boom layout; horizontal stabilisers, fin and underfin for directional stability; anti-torque Fenestron. Three-blade main rotor on Spheriflex hub rotating clockwise (nominal rotor speed 406 rpm) integrated with main shaft and transmission; two-stage reduction gear; eight asymmetrically positioned bladed New Generation Fenestron (nominal tail rotor speed 4,567 rpm). Single level cabin floor. Engine mounted to left of main rotor mast to improve balance by counteracting main rotor downwash.

FLYING CONTROLS: Control forces on collective and cyclic reduced by three electrically actuated hydraulic servos operating at 37 bars (537 lb/sq in). Dual controls optional.

STRUCTURE: Composites for main rotor blades; titanium alloy Spheriflex head and shaft made as single composites assembly; metal centre-fuselage; light alloy skid landing gear; crashworthy seats and fuel system. Light alloy tail rotor shaft.

LANDING GEAR: 'Moustache' configuration of fixed skids, having sweptback forward supports and full-width boarding step. Optional 1.50 m (4 ft 11 in) skis for snow operations.

POWER PLANT: One Turbomeca TM 319 Arrius 2F engine selected for first 300 EC 120s; rating 376 kW (504 shp) for T-O, 335 kW (449 shp) max continuous. Transmission

EC 120 cabin layout *(Jérome Deulin/Eurocopter)*

Instrument panel of EC 120 B Colibri

EC 120 B Colibri five-seat light helicopter *(Paul Jackson/Jane's)*

rating 330 kW (442 shp). Fuel capacity 416 litres (109.9 US gallons; 91.5 Imp gallons) in two tanks (one located beneath cabin floor and one above baggage compartment).

ACCOMMODATION: Pilot and four passengers. Luggage compartment below engine on same level as cabin floor, accessible from cabin, by external side door and rear door. Seating conforms to new FAR Pt 27 requirements for 30 g vertical and 18 g horizontal deceleration.

SYSTEMS: VEMD (vehicle and engine multifunction display), fitted as standard, is a fully duplex three-module processing system using glass screens to monitor performance and maintenance requirements. Optional air conditioning.

AVIONICS: *Instrumentation:* Two-screen EFIS.

EQUIPMENT: Optional Aerazur emergency flotation bags, and cable cutters. Optional pilot-controllable swivelling landing light. Optional windscreen wipers.

DIMENSIONS, EXTERNAL:
Main rotor diameter	10.00 m (32 ft 9¾ in)
Main rotor blade chord	0.26 m (10¼ in)
Fenestron diameter	0.75 m (2 ft 5½ in)
Fenestron blade chord	0.06 m (2¼ in)
Length overall, rotors turning	11.52 m (37 ft 9½ in)
Fuselage: Length	9.60 m (31 ft 6 in)
Max width of cabin	1.50 m (4 ft 11 in)
Max height, excl skids:	
to engine cowling	1.67 m (5 ft 5¾ in)
to rotor head	2.52 m (8 ft 3¼ in)
Tailplane span	2.60 m (8 ft 6¼ in)
Height: overall	3.40 m (11 ft 1¾ in)
to top of rotor head	3.08 m (10 ft 1¼ in)
to top of engine cowling	2.23 m (7 ft 3¾ in)
Skid track	2.07 m (6 ft 9½ in)
Height of skids	0.56 m (1 ft 10 in)
Height of boarding step	0.50 m (1 ft 7¾ in)

DIMENSIONS, INTERNAL:
Cabin: Length	2.30 m (7 ft 6½ in)
Max width	1.35 m (4 ft 5¼ in)
Max height	1.25 m (4 ft 1¼ in)
Floor area	3.0 m² (32.29 sq ft)

Baggage compartment volume	0.8 m³ (28.25 cu ft)

AREAS:
Main rotor disc	78.54 m² (845.4 sq ft)
Fenestron disc	0.44 m² (4.73 sq ft)

WEIGHTS AND LOADINGS:
Standard empty weight	960 kg (2,116 lb)
Max sling load	700 kg (1,543 lb)
Max useful load	755 kg (1,664 lb)
Max T-O weight: internal load	1,715 kg (3,781 lb)
external load	1,800 kg (3,968 lb)
Max disc loading:	
internal load	21.8 kg/m² (4.47 lb/sq ft)
external load	22.9 kg/m² (4.69 lb/sq ft)
Transmission loading at max T-O weight and power:	
internal load	5.20 kg/kW (8.54 lb/shp)
external load	5.45 kg/kW (8.96 lb/shp)

PERFORMANCE (at 1,715 kg; 3,781 lb):
Never-exceed speed (V$_{NE}$)	150 kt (278 km/h; 173 mph)
Max cruising speed	122 kt (226 km/h; 140 mph)
Max rate of climb at S/L	396 m (1,300 ft)/min
Service ceiling	5,189 m (17,024 ft)
Hovering ceiling: IGE	2,819 m (9,249 ft)
OGE	2,316 m (7,598 ft)
Range, no reserves	395 n miles (731 km; 454 miles)
Endurance at 65 kt (120 km/h; 75 mph), no reserves	4 h 10 min

UPDATED

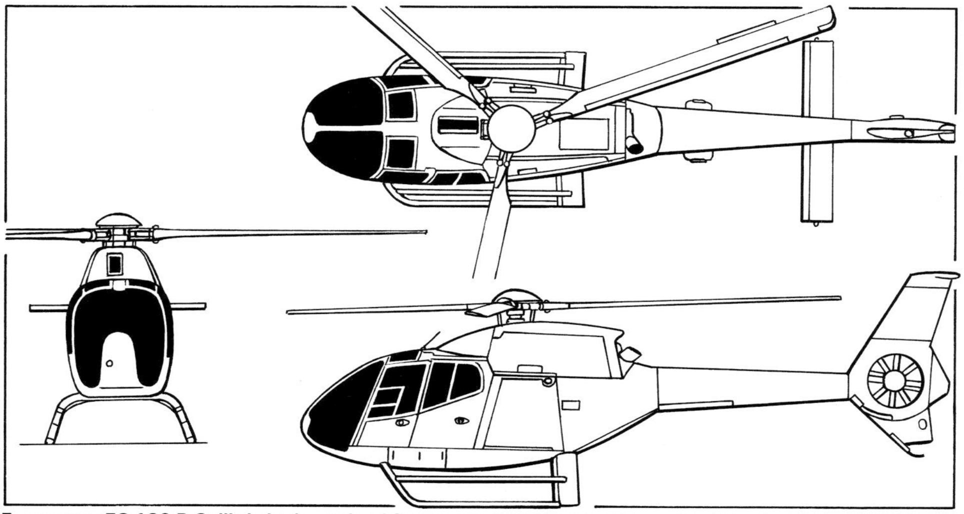

Eurocopter EC 120 B Colibri single-engined five-seat light helicopter *(James Goulding/Jane's)*

EUROCOPTER/KAWASAKI

PARTICIPATING COMPANIES
Eurocopter: see this section
Kawasaki: see under Japan

These two companies formed a partnership on 25 February 1977 to develop the BK 117 multipurpose helicopter. Low-rate production continues in both Germany and Japan. The EC 145/BK 117C-2 development is now being marketed.

UPDATED

EUROCOPTER/KAWASAKI BK 117

TYPE: Light utility helicopter.

PROGRAMME: Developed jointly by partners; four prototypes, first flight 13 June 1979; one preproduction aircraft, first flight 6 March 1981; first flights of production aircraft 24 December 1981 (JQ1001 in Japan) and 23 April 1982 (in Germany); certified in Germany and Japan 9 and 17 December 1982 respectively, followed by US FAA 29 March 1983 (FAR Pt 29, Categories A and B, including Amendments 29-1 to 29-16); deliveries began early 1983.

See 1991-92 and previous editions for earlier A series and B-1. BK 117 fleet reached one million flying hours mid-1999. Low-rate production only.

DASA is developing an all-weather operations system for SAR helicopters involving mast-mounted HeliRadar synthetic aperture radar capable of detecting power lines and similar obstacles with four transmitter/receiver modules on 1.5 m (5 ft) rotating arms; backed by GPS and two 284 × 160 mm (11¼ × 6¼ in) screens for 3-D radar picture and moving map.

240   INTERNATIONAL: AIRCRAFT—EUROCOPTER/KAWASAKI

Eurocopter/Kawasaki BK 117 C-1 demonstrator in decorative colour scheme *(Paul Jackson/Jane's)*

CURRENT VERSIONS: **BK 117 B-2:** Production model from 1992 powered by AlliedSignal LTS 101-750B-1 (with BK 117 C-1 below); German LBA certification 17 January 1992; US FAA certification 7 December 1992; Japanese JCAB certification 18 March 1993; French DGAC certification 15 July 1993; UK CAA certification 5 May 1995; maximum T-O weight increased to 3,350 kg (7,385 lb); payload increased by 150 kg (330 lb). No longer available; last built in 1998; performance figures in 1999-2000 *Jane's*.

BK 117 C-1: Version with Turbomeca Arriel 1E2 engines; first flight (F-WMBB) 6 April 1990; German LBA certification 17 January 1992; US FAA certification 7 December 1992; first delivery (to USA) December 1992. French DGAC certification 15 July 1993. Performance similar to that of BK 117 B-2; better hot-and-high performance. BK 117 C-1 improvement under way to increase OEI performance; includes higher transmission and engine OEI ratings, new tail rotor blades and variable rotor speed to improve tail rotor thrust and reduce external noise, and torque matching system to reduce pilot workload; German LBA certification achieved 28 April 1994, followed by US FAA certification 29 September 1994, Italian RAI certification 24 November 1994, and UK CAA certification 28 July 1995.

Japanese (Kawasaki-built) version which includes both re-engine and above improvements was approved 8 June 1995 by Japanese CAB. A BK 117 C-1 delivered by Kawasaki to Ehime Prefecture in August 1996 has an improved GPS/digital map system with enhanced navigation and ground collision warning by means of terrain direction indicator and voice, in addition to existing system functions such as map display, on which the aircraft's location is indicated.

Details below apply to BK 117 C-1 version.
   BK117C-2/EC145: See separate entry.
   NBK-117: Designation of aircraft built by IPTN (see Indonesian section, 1991-92 *Jane's*) under November 1982 agreement with MBB. Three only; no longer produced.
   BK 117 P5: Advanced technology demonstrator in Japan; fly-by-wire control system; described in 1999-2000 and earlier *Jane's*.

CUSTOMERS: Total of 283 built by MBB/Eurocopter up to late 1999; Kawasaki total was 128 by early 2000. Combined total of 411. Three ordered in 1999, compared with five in 1998 and seven in 1997. Latest customers include NAPOCOR power producer of Philippines (two), Amil of Brazil, Rocky Mountain Helicopters of USA and Italian EMS operators Aeroveneta, Avioriprese, Alidaunia and Helitalia (two).

COSTS: BK 117 B-2 US$2.815 million (1996); BK 117 C-1 US$3.1 million (1996).

DESIGN FEATURES: Pod-and-boom configuration, latter mounted high for cabin access via rear freight doors; tail rotor mounted at fin-tip; large auxiliary fins at tip of each horizontal stabiliser; engines above cabin.

System Bölkow four-blade main rotor head, almost identical to that of BO 105; main rotor blades similar to but larger than those of BO 105, with NACA 23012/23010 (modified) section; optional two-blade folding. Two-blade semi-rigid tail rotor with MBB-S102E performance/noise optimised blade section; rotor rpm 383 (main), 2,169 (tail).

FLYING CONTROLS: Equipped as standard for single-pilot VFR operation; dual controls and dual VFR instrumentation optional; rotor brake and yaw CSAS standard on German-built versions, optional on Kawasaki aircraft; options include IFR instrumentation, two-axis (pitch/roll) CSAS and Honeywell SPZ-7100 dual digital AFCS. Mast moment indicator discourages excessive cyclic control inputs.

STRUCTURE: Main rotor has one-piece titanium hub with pitch-change bearings; fail-safe GFRP blades with stainless steel anti-erosion strip. Tail rotor, mounted on port side of central fin, has GFRP blades of high-impact resistance. Main fuselage pod and tailboom are aluminium alloy with single-curvature sheets and (on fuselage) bonded aluminium sandwich panels; secondary fuselage components are compound curvature shells of Kevlar sandwich. Engine deck, to which tailboom is integrally attached, forms cargo compartment roof and is of titanium adjacent engine bays. Detachable tailcone carries main fin/tail rotor support, and horizontal stabiliser with offset endplate fins.

Eurocopter responsible for rotor systems, tailboom, tail unit, skid landing gear, hydraulic system, engine firewall and cowlings, powered controls and systems integration; Kawasaki for fuselage, transmission, fuel and electrical systems, and standard equipment. Components single-sourced and exchanged for separate assembly lines at Donauwörth and Gifu; some components and accessories interchangeable with those of BO 105 (which see), from which hydraulic-powered control system is also adapted.

LANDING GEAR: Non-retractable tubular skid type, of aluminium construction. Skids are detachable from cross-tubes. Ground handling wheels standard. Emergency flotation gear, settling protectors and snow skids optional.

POWER PLANT: BK 117 B-2 has two Honeywell LTS 101-750B-1 turboshafts, each rated at 410 kW (550 shp) for take-off and maximum continuous power, 441 kW (592 shp) for 30 minutes with OEI. BK 117 C-1 has two Turbomeca Arriel 1E2 turboshafts each rated at 550 kW (738 shp) for take-off, 516 kW (692 shp) maximum continuous and 574 kW (770 shp) for 2½ minutes OEI.

Kawasaki KB 03 main transmission rated at 736 kW (986 shp) for twin-engine T-O, 632 kW (848 shp) maximum continuous; for single-engine operation, 550 kW (738 shp) allowed for 2½ minutes, 404 kW (542 shp) for maximum continuous (see also BK 117 C-1 improvements).

Fuel in four flexible bladder tanks (forward and aft main tanks, with two supply tanks between), in compartments under cabin floor. Two independent fuel feed systems for engines and common main fuel tank. Total standard fuel capacity 697 litres (184 US gallons; 153 Imp gallons), of which 685 litres (181 US gallons; 151 Imp gallons) are usable; single or twin internal auxiliary fuel tanks, each of 200 litres (52.8 US gallons; 44.0 Imp gallons) capacity, and two external auxiliary fuel tanks, each of 150 litres (39.6 US gallons; 33 Imp gallons), optional.

ACCOMMODATION: Pilot and up to six (executive version), seven (Eurocopter standard version) or nine passengers (Kawasaki version). High-density layouts available for up to 10 passengers in addition to pilot. Level floor throughout cockpit, cabin and cargo compartment. Jettisonable forward-hinged door on each side of cockpit, pilot's door having an openable window. Jettisonable rearward-sliding passenger door on each side of cabin, lockable in open position. Fixed steps on each side. Two hinged, clamshell doors at rear of cabin, providing straight-in access to cargo compartment. Rear cabin window on each side. Aircraft can be equipped for offshore support, medical evacuation (one or two stretchers side by side and up to six attendants), firefighting, search and rescue, law enforcement, cargo transport or other operations.

SYSTEMS: Ram air and electrical ventilation system. Fully redundant tandem hydraulic boost system (one operating and one standby), pressure 103.5 bars (1,500 lb/sq in), for flight controls. System flow rate 8.1 litres (2.14 US gallons; 1.78 Imp gallons)/min. Bootstrap/oil reservoir, pressure 1.70 bars (25 lb/sq in). Main DC electrical power from two 150 A 28 V (200 A 28 V for C-1) starter/generators (one on each engine) and a 24 V 25 Ah Ni/Cd battery. AC power provided by inverter; second AC inverter optional; emergency busbar provides direct battery power to essential services, external DC power receptacle.

AVIONICS: *Comms:* VHF-AM/FM, UHF and HF radios to customer's requirements.
   Flight: Long-range navaids optional.
   Radar: Multirole radar optional.
   Instrumentation: Basic instrumentation for single-pilot VFR operation includes airspeed indicator with two electrically heated pitot tubes and static ports, two hydraulic pressure indicators, encoding altimeter, instantaneous vertical speed indicator, 4 in artificial horizon, 3 in standby artificial horizon, HSI, (4 in and 3 in artificial horizons and HSI optional on Kawasaki-built aircraft), gyro magnetic heading system, magnetic compass, ambient air temperature thermometer and clock.

EQUIPMENT: Standard basic equipment includes rotor brake, annunciator panel, two master caution lights, rotor rpm/engine fail warning control unit, fuel quantity indicator and low-level sensor, outside air temperature indicator, engine and transmission oil pressure and temperature indicators, two exhaust temperature indicators, dual torque indicator,

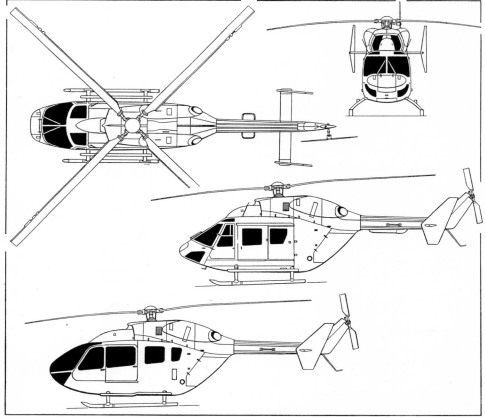

Eurocopter/Kawasaki BK 117 B-2 twin-turboshaft multipurpose helicopter, with additional side view (bottom) of EC 145 version *(Dennis Punnett/Jane's)*

triple tachometer, two N1 tachometers, full internal and external lighting, ground handling wheels, pilot's windscreen wiper, floor covering, interior panelling and sound insulation, ashtrays, map/document case, tiedown rings in cabin and cargo compartment, engine compartment fire warning indicator, engine fire extinguishing system, portable fire extinguisher, first aid kit, and single-colour exterior paint scheme.

Optional equipment includes long-range fuel tanks, high-density seating arrangement, bleed air heating system, crashworthy seats, emergency flotation gear, settling protectors, snow skids, main rotor blade folding kit, dual pilot operation kit, stretcher installation, external cargo hook for 1,500 kg (3,307 lb) maximum load, rescue hoist (90 m; 295 ft cable, maximum load 270 kg; 595 lb, variable winch speed), Spectrolab SX 16 remotely controlled searchlight, external loudspeaker, sand filter and kits for rescue, law enforcement, firefighting and VIP transport.

Active anti-vibration system certified in Japan in mid-1997; available as option or for retrofit.

DIMENSIONS, EXTERNAL:
Main rotor diameter	11.00 m (36 ft 1 in)
Tail rotor diameter	1.96 m (6 ft 5 in)
Tail rotor blade chord	0.22 m (8¾ in)
Length: overall, both rotors turning	13.01 m (42 ft 8¼ in)
fuselage, tail rotor blades vertical	9.93 m (32 ft 7 in)
Fuselage: Max width	1.60 m (5 ft 3 in)
Height: overall, both rotors turning	3.85 m (12 ft 7½ in)
to top of main rotor head	3.36 m (11 ft 0¼ in)
Tailplane span (over endplate fins)	2.71 m (8 ft 10¾ in)
Ground clearance: tail rotor	1.89 m (6 ft 2½ in)
fuselage	0.35 m (1 ft 1¾ in)
Skid track	2.50 m (8 ft 2½ in)

DIMENSIONS, INTERNAL:
Combined cabin and cargo compartment:
Max length	3.02 m (9 ft 11 in)
Width: max	1.59 m (5 ft 2½ in)
min	1.27 m (4 ft 2 in)
Height: max	1.28 m (4 ft 2½ in)
min	0.99 m (3 ft 3 in)
Useful floor area	3.70 m² (39.8 sq ft)
Volume	5.0 m³ (177 cu ft)
Cabin only: Length	2.56 m (8 ft 4¾ in)

AREAS:
Main rotor blades (each)	1.76 m² (18.94 sq ft)
Tail rotor blades (each): B-2	0.098 m² (1.05 sq ft)
C-1	0.108 m² (1.16 sq ft)
Main rotor disc	95.03 m² (1,022.9 sq ft)
Tail rotor disc	3.00 m² (32.24 sq ft)

WEIGHTS AND LOADINGS:
Basic weight empty: B-2	1,745 kg (3,846 lb)
C-1	1,764 kg (3,890 lb)
Fuel: standard usable	558 kg (1,230 lb)
incl first auxiliary tank	718 kg (1,583 lb)
Max T-O weight: internal payload	3,350 kg (7,385 lb)
external payload	3,500 kg (7,716 lb)
Max disc loading:	
internal payload	35.25 kg/m² (7.22 lb/sq ft)
external payload	36.8 kg/m² (7.54 lb/sq ft)
Transmission loading at max T-O weight and power:	
internal payload	4.55 kg/kW (7.48 lb/shp)
external payload	4.76 kg/kW (7.83 lb/shp)

PERFORMANCE (main values for BK 117 C-1; A: at gross weight of 3,000 kg; 6,614 lb, B: at 3,200 kg; 7,055 lb, C: at 3,350 kg; 7,385 lb):
Never-exceed speed (V_{NE}) at S/L:
A	150 kt (277 km/h; 172 mph)
B, C	140 kt (259 km/h; 161 mph)
Max cruising speed: A	135 kt (250 km/h; 155 mph)
B	134 kt (248 km/h; 154 mph)
C	133 kt (246 km/h; 153 mph)
Econ cruising speed: A	127 kt (235 km/h; 146 mph)
B	126 kt (233 km/h; 145 mph)
C	125 kt (231 km/h; 144 mph)
Max rate of climb at S/L: A	655 m (2,150 ft)/min
B	587 m (1,925 ft)/min
C	538 m (1,765 ft)/min
Max certified altitude: A, B, C	4,575 m (15,000 ft)
Service ceiling: A, B	5,480 m (18,000 ft)
C	5,090 m (16,700 ft)
Hovering ceiling IGE (zero wind):	
A	3,690 m (12,100 ft)
B	3,050 m (10,000 ft)
C	2,530 m (8,300 ft)
Hovering ceiling IGE (20 kt; 37 km/h; 23 mph) crosswind: A	3,200 m (10,500 ft)
B	2,530 m (8,300 ft)
C	1,920 m (6,300 ft)
Hovering ceiling OGE: A	3,520 m (11,550 ft)
B	3,000 m (9,840 ft)
C	1,480 m (4,860 ft)
Range: C	292 n miles (540 km; 336 miles)
Endurance: B, C	2 h 50 min

UPDATED

Prototype Eurocopter/Kawasaki EC 145 *(Wolfgang Obrusnik/Eurocopter)*

EUROCOPTER/KAWASAKI EC 145

TYPE: Light utility helicopter.
PROGRAMME: Development started in 1997 as the BK 117C-2 and incorporated EC 135 technology; redesignated end 1999. For BK 117 history see previous entry. First flight of German aircraft (D-HMBL) (unannounced) 12 June 1999; first flight of Japanese prototype 15 March 2000; third prototype joined programme 14 April 2000. Kawasaki builds tail section; Eurocopter responsible for forward section. Certification by LBA and JCAB expected by end 2000 with commercial launch the following year.
CUSTOMERS: Launch customer was French Sécurité Civile, which ordered 32 in December 1997 for delivery between mid-2000 and 2005 to replace Alouette III. Second customer is French Gendarmerie, with firm order for eight placed in 1999.
COSTS: Sécurité Civile contract worth US$170 million. Flyaway cost reported as US$4.9 million (2000).
DESIGN FEATURES: Redesigned tail rotor and forward fuselage and nose section, latter based on EC 135 and providing improved visibility. Rotor blades are same diameter as BK 117, but have EC 135 profile. Optimised for search and rescue and emergency roles.
STRUCTURE: As BK 117.
LANDING GEAR: As BK 117.
POWER PLANT: As BK 117C-1, but fuel capacity increased to 757 litres (200 US gallons; 167 Imp gallons).
ACCOMMODATION: Pilot(s) and up to eight passengers in standard layout; up to 11 passengers in high-density layout; sliding cabin doors.
SYSTEMS: As BK 117.
AVIONICS: Compared to BK 117, EC 145 has entirely redesigned avionics panel. Cockpit is NVG compatible.
EQUIPMENT: As BK 117.

DIMENSIONS, EXTERNAL: As BK 117C-1, except:
Fuselage: Length	10.27 m (33 ft 8¼ in)
Max width	1.70 m (5 ft 7 in)
Tailplane span (over endplate fins)	2.75 m (9 ft 0¼ in)

DIMENSIONS, INTERNAL: As BK 117C-1, except:
Cabin: Length	2.97 m (9 ft 9 in)
Width: max	1.70 m (5 ft 7 in)
min	1.45 m (4 ft 9 in)
Volume	5.9 m³ (207 cu ft)

WEIGHTS AND LOADINGS:
Max T-O weight: internal load	3,500 kg (7,716 lb)
external load	3,650 kg (8,046 lb)
Max disc loading: internal load	36.8 kg/m² (7.54 lb/sq ft)
external load	38.4 kg/m² (7.86 lb/sq ft)
Max power loading:	
internal load	4.76 kg/kW (7.83 lb/shp)
external load	4.97 kg/kW (8.16 lb/shp)

PERFORMANCE (estimated, two occupants):
Max level speed	145 kt (268 km/h; 167 mph)
Service ceiling	4,570 m (15,000 ft)
Range with max fuel	378 n miles (700 km; 435 miles)
Endurance, 30 min reserves	3 h 25 min

UPDATED

EUROFAR

EUROPEAN FUTURE ADVANCED ROTORCRAFT

This programme has been replaced by Eurotilt, as described later in this section.

UPDATED

EUROFIGHTER

EUROFIGHTER JAGDFLUGZEUG GmbH

Am Söldermoos 17, D-85399 Hallbergmoos, München, Germany
Tel: (+49 811) 80 15 55
Fax: (+49 811) 80 15 57
e-mail: eurofighter.pr@ibm.net
Web: http://www.eurofighter.com
CHAIRMAN: Filippo Bagnato
MANAGING DIRECTOR: Bob Haslam
DIRECTOR, MARKETING SALES SUPPORT: Andrew Lewis
VICE-PRESIDENT, COMMUNICATIONS: Ian Bustin

Eurofighter GmbH formed to manage EFA (European Fighter Aircraft) programme June 1986, followed shortly after by Eurojet Turbo GmbH to manage engine programme. Eurofighter GmbH is owned by Alenia (Italy), BAE Systems (UK), CASA (Spain) and DASA (Germany), in the form of a 62.5 per cent holding by EMAC (which see). Eurojet Turbo participants are Fiat Aviazione (Italy), ITP (Spain), MTU-München (Germany) and Rolls-Royce (UK). Radar is provided by the Euroradar consortium of BAE Systems (UK), FIAR (Italy), DASA Sensor Systems (Germany) and ENOSA (Spain). NETMA (NATO Eurofighter and Tornado Management Agency) supervises the programme on behalf of the customer air forces.

All four participating countries agreed on 22 December 1997 to proceed with production and signed the appropriate authorisation on 30 January 1998. The aircraft was formally named Typhoon on 2 September 1998 although, officially, that title is only to be used for marketing outside Europe.

INTERNATIONAL: AIRCRAFT—EUROFIGHTER

Prototype Eurofighter, following installation of EJ200 engines *(Paul Jackson/Jane's)*

On 4 November 1999, four partners announced impending formation of Eurofighter International (EFI) as dedicated sales organisation with target of securing half of available market for 800 combat aircraft over following 30 years. Relations with NETMA remain unchanged. First export commitment issued by Greece on 7 March 2000.

Eurofighter International
London, UK
PRESIDENT: Cesare Gianni
VICE-PRESIDENTS: Manfred Wolff (Campaigns)
 Andrew Lewis (Sales Support)
 Rob New (Contracts and Finance)
 UPDATED

EUROFIGHTER TYPHOON

Spanish Air Force designations: C.16 (single-seat) and CE.16 (two-seat)
TYPE: Multirole fighter.
PROGRAMME: **Politico-industrial** history began with outline staff target for common combat aircraft issued December 1983 by air chiefs of staff of France, Germany, Italy, Spain and UK; initial feasibility study launched July 1984; France withdrew July 1985, shareholdings then being readjusted to 33 per cent each to UK and Germany, 21 per cent Italy and 13 per cent Spain; project definition phase completed September 1986; definitive ESR-D (European Staff Requirement – Development) issued September 1987, giving military requirements in greater detail; definition refinement and risk reduction stage completed December 1987; main engine and weapons system development contracts signed 23 November 1988.

Programme re-examined in 1992 following German demands for substantial cost reduction and studies of alternative proposals, which submitted in October 1992, although none was adopted; Italy and Spain froze EFA work mid-October. Defence ministers' conference of 10 December 1992 relaunched aircraft as Eurofighter 2000, delaying service entry by three years, to 2000, and allowing Germany to incorporate off-shelf avionics (AN/APG-65 radar suggested), lower standard of defensive aids and other deletions to effect 30 per cent price cut. By 1996, however, these downgrades had been abandoned and Eurofighter GmbH was planning for German production almost identical to the common standard.

Additional difficulties resulted from German underfunding and demands for further cost cuts. Political reapportionment of production work-shares negotiated in 1995, following reduction of German requirement. Revised European Staff Requirement – Development signed by four air forces, 21 January 1994 and re-orientation of programme agreed in MoU 4, July 1995. MoU 5, covering work-shares, delayed to 1996 by German claims for 30 per cent, despite 140 aircraft requirement representing only 23 per cent of production. Compromise, agreed January 1996, involves addition of at least 40 (and possibly up to 60) ground attack aircraft to German requirement after 2012 and reduction of UK needs to 232. Work-shares for production phase finally agreed as 30 per cent to Germany, 37 per cent to UK, 19 per cent to Italy and 14 per cent to Spain. However, for first 148 aircraft, agreed 1998, average shares are 30, 36.33, 20 and 13.67 per cent, respectively. Initial production details contained in Quotation 4, submitted in March 1996; this envisaged start of manufacturing in January 1998 and (after minor restructuring) production of three in 2001, 12 in 2002, 37 in 2003, 46 in 2004 and 52 per year thereafter.

Eurofighter GmbH held open the terms of Quotation 4 throughout 1996 as Germany repeatedly postponed a production decision as a consequence of financial constraints. UK was first to declare a firm production commitment, on 2 September 1996, and Spain followed on 21 October 1996, when terms for start-up funding were agreed with industry. All four governments declared support for a production launch on 5 December 1996, although Germany and Italy did not grant funds until 26 November and 9 December 1997, respectively, allowing defence ministers' conference on 22 December 1997 to launch production phase. First metal for a production aircraft was cut at DASA's Augsburg plant in May 1998. Each partner nation is assembling its own aircraft on lines at Manching (Germany), Caselle (Italy), Getafe (Spain) and Warton (UK). Locations for assembly of export aircraft yet to be announced. Assembly of major components for first production aircraft began at Augsburg (Germany) in February 1999 and Caselle in March; final assembly of this aircraft began at Warton in September 2000.

Typhoon was finalist (with F-16 Block 50) in Norwegian competition for 20 fighters to be delivered between 2003 and 2010; letter of request issued to Eurofighter GmbH by Norwegian MoD on 10 February 1999, but programme suspended in 2000. Greek interest announced in February 1999 and followed, on 7 March 2000, by commitment to order 60, plus 30 options.

Engineering and flight-testing programme originally to be based on eight development aircraft (no prototypes apart from BAe EAP – see 1991-92 and earlier *Jane's*); reduced to seven (DA1-7) early 1991, coincident with 11 per cent cut in intended flight test programme to 4,500 hours. Total to be achieved in 2,990 sorties by DA series (635, 575, 430, 420, 385, 315 and 290 for DA1 to DA7, respectively) plus 1,700 sorties by first five production aircraft, which will be fitted with test instrumentation.

No formal roll-out; DA1 and DA2 remained unflown for some 18 months for exhaustive cross-checking of flight control system (FCS). First flight eventually achieved on 27 March 1994, but FCS development resulted in later aircraft flying out of sequence, DA6 being fourth to fly (31 August 1996), by which time earlier aircraft had completed 241 sorties. DA5 flew Eurofighter's 500th sortie (almost 450 hours total airborne time) on 21 October 1997; the 750th sortie (over 630 hours) was flown in June 1998 and the 1,000th in May 1999. Total 1,135 sorties and 931 hours by fourth quarter of 1999; 1,000th hour flown February 2000. Some 90 per cent of flight envelope had been explored by April 1999. Further details of individual aircraft appear below.

Defensive aids subsystem contract awarded to Euro-DASS 13 March 1992, but Germany and Spain initially declined to participate. 'A' version of ECR 90 radar first flew in nose of modified BAe One-Eleven testbed (ZE433) at Bedford, 8 January 1993; 'C' version is first ECR 90 packaged to fit Eurofighter; flown in One-Eleven from July 1996. First development standard radar delivered to DASA in June 1996; flight testing (in DA5) began 24 February 1997. Radar named Captor in 2000; first production unit then due for delivery in February 2001. Electronically scanned version to fly in One-Eleven by 2003, for possible service in 2010.

Major airframe fatigue test (AFT) fuselage at Ottobrunn achieved 6,000 hours in May 1995 and target of 18,000 hours (equivalent to 6,000 hours of service use) on 4 September 1998.

In 1998, BAE undertook a UK Ministry of Defence funded study of a possible short take-off but arrested recovery (STOBAR) Eurofighter which could operate from aircraft carriers. Officially confirmed in early 2000 that navalised Eurofighter was one option under consideration for Royal Navy's Future Carrier-Borne Aircraft (FCBA) requirement.

Thrust vectoring nozzles for EJ200 had undergone 78 hours of bench testing by mid-2000. Study launched by ITP of Spain in 1994, using private funding. Trial system demonstrated 23° 30′ vector angle and 110°/s slew rate; advantages of installation for regular use would include 3 per cent reduction in cruise drag, 7 per cent improvement in sustained turn rate, 7 per cent increase in installed thrust, 25 per cent reduction of take-off run and 3 per cent reduction of mission fuel burn. Flight trials could begin in 2002.

Three-phase introduction of Eurofighter to service begins with PSP 1 standard, suitable only for basic air defence training. PSP 2 configuration to be fully air defence capable, through addition of direct voice input (DVI), digital datalink (MIDS) and MLA, plus limited defensive aids subsystem (DASS). Third phase, PSP 3, gives full 'swing role' capability with air-to-surface weaponry, enhanced situational awareness and improved survivability through incorporation of PIRATE passive tracking system, helmet-mounted display, towed radar decoy and full sensor fusion. Following five instrumented production aircraft, PSP 1 to be achieved in June 2002 on delivery of second two-seater for RAF (production number BT002), which is seventh series production (SPA7). PSP 2 follows in December 2003 with SPA44 (fourth RAF single-seat; BS004); and PSP 3 in April 2005 with SPA115 (20th German single-seat; GS020).

CURRENT VERSIONS (general): **Single-seater:** Standard version.

Two-seater: Combat-capable conversion trainer. Slightly reduced internal fuel capacity.

CURRENT VERSIONS (specific): **DA1/9829:** (DASA-built at Ottobrunn; airframe No. 01; Luftwaffe serial number 9829.) By road to Manching 11 May 1992; first flight, 27 March 1994, using Phase 0 software; planned transfer to Warton for handling and envelope expansion trials (wearing UK serial number ZH586) cancelled; remained at Manching; nine sorties to June 1994, when grounded for

Eurofighter DA7, the final development batch aircraft

EUROFIGHTER—AIRCRAFT: INTERNATIONAL

FCS upgrade to Phase 2; reflown 18 September 1995; first sortie by a German military pilot (Lt Col Heinz Spolgen), March 1996 at start of initial military evaluation phase which completed on 24 April 1996. Total 123 flights by late 1997, when stood down for retrofit with EJ200 series 03Z engines, plus parallel avionics upgrade to 3910 standard and installation of Martin-Baker Mk 16 ejection seats. Upgrade completed in November 1998; returned to flying in third quarter of 1999.

DA2/ZH588: (BAE at Warton; airframe No. 02.) First engine run 30 August 1992, first flight 6 April 1994; assigned to envelope expansion, 'carefree' handling and load trials; nine sorties to June 1994, when stood-down for FCS upgrade; reflown 17 May 1995 with Phase 2 version of flight software and made Eurofighter's world public debut (Paris, 11 June 1995, static) and UK debut (Fairford 22 July 1995, flying). On 9 November 1995, 57th sortie was also first with an RAF pilot (Sqdn Ldr Simon Dyde). Demonstrated 25° AoA in May 1997, followed by radar decoy trials and demonstration of full 'carefree' handling. First M2.0 Eurofighter sortie 23 December 1997; first aerial refuelling (RAF VC10) 14 January 1998. Retrofitted with EJ200 engines, upgraded avionics and Mk 16 ejection seat; reflown late August 1998. Achieved 15,240 m (50,000 ft), April 1999. Prepared for load testing during mid-1999. First to fly with 2B2 software, 7 July 2000, by which time painted black overall.

DA3/MMX602: (Alenia at Turin/Caselle; airframe No. 04.) First with EJ200 power plants (series -01A) for engine trials (originally scheduled from March 1993 but postponed) and gun/weapon release trials. Initial flight 4 June 1995 on Phase 1 software (20° AoA and +6 g limits). Total 53 sorties by 1 September 1996 (all with Phase 1 software). Refitted with EJ200-01C engines in 1996; in-flight engine relight demonstrated December 1996; first sortie with two 1,000 litre underwing tanks 5 December 1997; EJ200-03A engine by early 1998; 1,500 litre tanks, February 1999; M1.6 with two 1,000 litre tanks, March 1999; 200th sortie, November 1999. Initiated ground dropping trials of air-to-surface weapons, 1999.

DA4/ZH590: (BAE; airframe No. 03.) First two-seat and first with full avionics (including ECR 90) for radar development, and 'carefree' handling trials. Rolled out 4 May 1994; first flight 14 March 1997; 14 sorties up to end of first phase of trials on 8 September 1997. First scheduled demonstration of supercruise 20 February 1998; lightning strike trials in test rig at Warton May-June 1998; rear cockpit activated (as on other two-seaters) 19 April 1999; autopilot autothrottle activated 28 April 1999; first flight of helmet-mounted sight 17 June 1999; Eurofighter 1,000th hour flown February 2000 on aircraft's 75th sortie. First to fly with active missile approach warning. First two-seat Eurofighter night flight, May 2000.

DA5/9830: (DASA.) Construction begun 2 November 1992; maiden flight 24 February 1997; first with ECR 90 radar; autopilot and weapons trials. DS-X radar software standard upgraded to DS-C1 in June 1997, at which time first EJ200-03A engines fitted. Reportedly is testbed for latest standard of radar-absorbent materials. Deployed to Rygge, Norway, June 1998 (first visit to prospective customer outside original partner nations), for evaluation. Flown by Norwegian test pilot, 15 December 1998. New standard software (Phase 2B1; also in DA4) flown 1 April 1999, permitting autopilot and autothrottle operation. Radar trials against four simultaneous targets, mid-1999; flew Typhoon's 1,000th sortie 18 May 1999, representing 830 hours. In-flight icing trials, February 2000.

DA6/XCE.16-01: (CASA at Seville.) Second two-seat; performance (including 'carefree' handling), environmental systems, MIDS integration and helmet integration trials. Scheduled (by 1995 reprioritisation) to be fourth to fly; early 1996 date set back by six months for re-checking of Phase 2A software. First flight 31 August 1996; high-temperature trials at Morón, Spain, from 20 July 1998; trials of LCSS (see Systems) cooled aircrew clothing, June 1999; environmental trials (with DA1), including ground tests at Boscombe Down, completed May 2000.

DA7: (Alenia.) Nav/com, performance and weapons integration trials. First flight 27 January 1997. First firing of AIM-9L Sidewinder 15 December 1997; first AIM-120 AMRAAM jettison two days later; first (1,000 litre) external wing tank jettison 17 June 1988.

Further eight ground testing part-airframes and five instrumented production aircraft.

IPA1: (BAE.) Two-seat; first production Eurofighter Typhoon. Production number BTT1. Final assembly from September 2000; delivery in 2001. Defensive aids trials.

IPA2: (Alenia.) Two-seat; air-to-surface weapon and sensor fusion trials.

IPA3: (DASA.) Two-seat; air-to-surface weapon integration.

IPA4: (CASA.) Single-seat; AAM and gun trials; environmental system trials.

IPA5: (BAE.) Single-seat; air-to-air and -surface weapons trials.

CUSTOMERS: Originally declared requirements for 765 (UK and Germany 250 each; Italy 165 and Spain 100). In January 1994, Spain announced firm requirement for 87; Germany revised needs to 180 (including at least 40 fighter-bombers post-2012) under January 1996 work-share agreement. Final total is 620 aircraft, including 1,382 engines; plus options on 90 and 183, respectively (see accompanying table). MoUs 6 and 7 (production and logistical support) of 3 December 1997 and Production Investment contract of 10 December followed on 22 December 1997 by political agreement of all four nations to fund production phase.

Production contract (Supplement 1) for all 620 aircraft (plus 90 options) signed 30 January 1998; Supplement 2 agreement of 18 September 1998 authorised first 148 of these on fixed-price terms, together with 363 engines.

Deliveries to air forces begin in June 2002 with RAF and Italy; first RAF unit to be Operational Evaluation Unit (OEU) at BAE Warton, where first 12 pilots to be converted; in 2004, Coningsby will receive OEU and form OCU, followed by two operational squadrons in 2005 and 2006; two squadrons form at Leeming in 2006 and 2007; three at Leuchars in 2008-2010. First of seven combat squadrons to be air defence tasked; remainder to comprise three air defence, two multirole and one offensive support.

RAF announced in May 2000 that no ammunition would be procured for Eurofighter's cannon, despite installation of Mauser in Tranche 1 aircraft. Decision may be reversed at later stage.

First German wing will be JG 73 at Laage during 2002, followed by JG 74 at Neuberg, JG71 at Wittmund and JG 72 at Hopsten; 135 for air defence and 40 multirole; final German delivery in 2014, but further two squadrons (single- or two-seat to be decided) may be obtained for air-to-ground operations. Italian re-equipment begins with 4° Stormo at Grosseto, other bases being Cameri (53° Stormo) and Trapani (37° Stormo). Other details in accompanying table.

Export orders also being received: Norway entered exploratory discussions in 1997 with a view to acquiring up to 30 aircraft to replace Northrop F-5A/Bs; appointed a liaison officer to NETMA in 1998 and issued formal request for proposal (RFP) on 15 February 1999; suspended programme in early 2000. Greece announced intention to obtain Typhoon on 12 February 1999, government confirmation following in March 2000; local final assembly under consideration. South Korean RFP received June 1999. These and other requests for information summarised in accompanying table. Eurofighter predicts export of 400 Typhoons between 2005 and 2030, worth in excess of £35 billion (2000).

Contract for Eurofighter ground support system signed 24 May 2000.

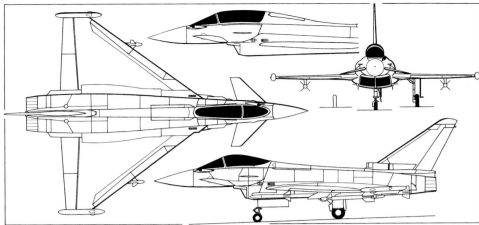
Eurofighter Typhoon with scrap view of two-seat version (*Paul Jackson/Jane's*)

Afterburner take-off by DA5, the second German-built Eurofighter

Spanish-built two-seater, DA6

INTERNATIONAL: AIRCRAFT—EUROFIGHTER

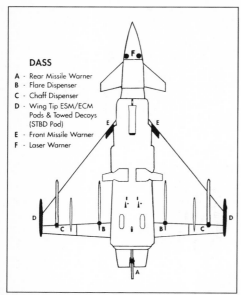

DASS
- A - Rear Missile Warner
- B - Flare Dispenser
- C - Chaff Dispenser
- D - Wing Tip ESM/ECM Pods & Towed Decoys (STBD Pod)
- E - Front Missile Warner
- F - Laser Warner

Configuration of the full Eurofighter defensive aids subsystem *1999/0044917*

EUROFIGHTER TYPHOON REQUIREMENTS BY PARTNER NATIONS

Tranche	Delivery	Germany		Italy		Spain		UK		Totals		
		1s	2s	1s	2s	1s	2s	1s	2s	1s	2s	Comb
1	2001-05	28	16	19	10	12	8	37	18	96	52	148[1,2]
2	2005-10	58	10	43	3	27	6	83	6	211	25	236[3]
3	2010-14	61	7	44	2	33	1	75	13	213	23	236[4]
Subtotals		147	33	106	15	72	15	195	37	520	100	620
		180		121		87		232				
Options		nil		9		16		65				90
Totals		180		130		03		97				710

Notes: 1s is single-seat; 2s is two-seat
[1] Including five instrumented trials aircraft, one for static testing and 37 'IOC' standard aircraft assigned to training
[2] Plus 363 engines and 147 radars
[3] Plus 519 engines
[4] Plus 500 engines

EUROFIGHTER TYPHOON EXPORT ORDERS AND PROSPECTS

Nation	Requirement	Timeframe	Status
Requests for proposals (RFP):			
Greece		contract imminent	Selection confirmed April 1999. EF Office Athens opened July 1999. Intentions confirmed March 2000
South Korea	40-60	decision late 2001 deliveries 2004-2007	EF Office Seoul opened April 1998. RFP received June 1999. Proposal submitted 8 September 1999. Flight evaluation expected early 2000. Final RFP submitted 28 June 2000
Norway	20 plus 10 option	to be determined	EF Office Oslo opened March 1999. RFP received February 1999. Proposal submitted 1 June 1999. Offsets proposal submitted 1 December 1999. Competition suspended 2000
Requests for information (RFI):			
Australia	60	decision after 2000	Discussions under way in 1999
Czech Republic	Possible 36	long-term requirement	RFI received June 1999. RFI response September 1999
Netherlands	possible 100 plus	decision 2005	RFI received June 1999. RFI response October 1999
Poland	60	long-term requirement	RFI received June 1999. RFI response July 1999
Saudi Arabia	possible 50-70	decision after 2000	Early discussions under way in 1999
Singapore	20-40	decision 2000/2001	RFI response submitted late 1999, evaluation under way

COSTS: Estimated £25 to 26.5 million, UK, 1992 unit cost; DM127 million, Germany early 1992, 10 year system price; reduced to DM89 million by late 1992 economies. UK National Audit Office report of early 1996 estimated British share of the development phase to be £1,253 million (43.7 per cent), of which £407 million resulted from restructuring to keep Germany in the programme. Same source reported programme 46 per cent over budget and 36 months late by mid-1997. Total development cost estimated as US$21 billion by 1998. Combined development investment estimated (1996) as DM18 billion (US$12.25 billion); non-recurring production investment estimated (1996) as DM12 billion (US$8.15 billion).

Revised data gives flyaway price of DM75 million to 85 million (US$51 million to US$58 million) and system price of DM150 million to 170 million (US$102 million to US$116 million) at 1996 levels. UK NAO estimated (1996) £15.4 billion for 250 RAF aircraft, including £9.5 billion for production (unit cost £38 million) and this reaffirmed (US$58 million/£37 million) in mid-1998. UK programme cost officially estimated at £16.1 billion (mid-1999).

In January 1997, German government agreed a weapons system price (including logistics support) of DM125.4 million (US$79 million), equivalent to DM23 billion for 180 aircraft. Supplement 2 (148 aircraft) valued at DM14 billion.

DESIGN FEATURES: Agile fighter; subsonic instability exceeds 35 per cent (as achieved by Grumman X-29 research aircraft). Collaborative design by BAE, DASA, Alenia and CASA, incorporating some design and technology (including low detectability) from BAe EAP programme. No official requirement for thrust vectoring (TV), supercruise or high order of 'stealthiness'; however, TV nozzle for Eurofighter is under private development; supercruise was 'inadvertently' demonstrated at high altitude in 1997; and RAF has confirmed that aircraft meets low-observables specification.

Low-wing, low-aspect ratio tailless delta with 53° leading-edge sweepback; underfuselage box with side-by-side engine air intakes, each with fixed upper wedge/ramp and vari-cowl (variable position lower cowl lip) with Dowty actuators.

Intended service life, 6,000 hours or 25 years. Integrated structural health-and-usage monitoring system (first in any combat aircraft) calculates structural fatigue at 20 positions on the airframe 16 times per second during flight. Maintainability features include 10 mmh/fh and single engine change by four engineers in 45 minutes. Operational turn-round by six ground crew in 25 minutes. Germany, and possibly UK, will contract-out maintenance on 'power by the hour' arrangement with private industry.

FLYING CONTROLS: Two-segment automatic slats on wing leading-edges, inboard and outboard flaperons on trailing-edges; all-moving foreplanes below windscreen; rudder; hydraulically actuated airbrake aft of canopy, forming part of dorsal spine; Liebherr primary flight control actuators. Full-authority quadruplex ACT (active control technology) digital fly-by-wire flight control system (team leader DASA; Bodenseewerk and ENOSA flight control computer) combines with mission adaptive configuring and aircraft's instability in pitch to provide required 'carefree' handling, gust alleviation and high sustained manoeuvrability throughout flight envelope; pitch and roll control via foreplane/flaperon ACT to provide artificial longitudinal stability; yaw control via rudder; no manual reversion.

STRUCTURE: Fuselage, wings (including inboard flaperons), wing-fuselage fairings, fin and rudder mainly of CFC (carbon fibre composites) except for foreplanes, outboard flaperons and exhaust nozzle fairings (titanium); nose radome and fin-tip (GFRP); leading-edge slats, wingtip pods, fin base, fin leading-edge, rudder trailing-edge, cockpit side strake and canopy-to-airbrake fairing (aluminium-lithium alloy); and canopy surround (aluminium). CFC constitutes 70 per cent of surface area, with metal 15 per cent, GFRP 12 per cent and other materials 3 per cent. Manufacture includes such advanced techniques as superplastic forming and diffusion bonding; CASA-led joint structures team. On development aircraft (only) BAE responsible for front fuselage, foreplanes, starboard leading-edge slats and flaperons; DASA the centre-fuselage and fin, Alenia the port wing, port leading-edge and flaperons, and stages 2 and 3 of rear fuselage; CASA the rear fuselage stage 1; and CASA/BAE the starboard wing; no duplication of tooling.

Work-share on production aircraft involves BAE for front fuselage, canards, windscreen, canopy, dorsal fairing inboard flaperons, fin and rear fuselage stage 1; DASA for centre-fuselage; Alenia for port wing, outboard flaperons and rear fuselage stages 2 and 3; and CASA for starboard

First two-seat Eurofighter DA4 *2001/0092575*

Eurofighter Typhoon cockpit *2000/0080295*

EUROFIGHTER—AIRCRAFT: INTERNATIONAL

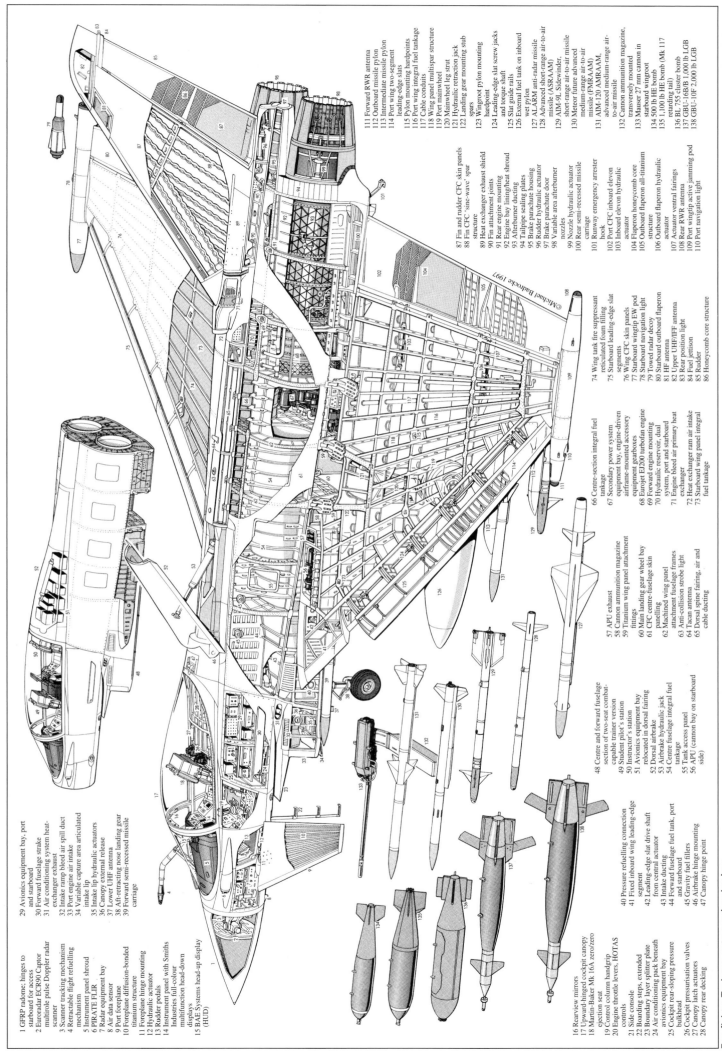

Eurofighter Typhoon, cutaway drawing key

INTERNATIONAL: AIRCRAFT—EUROFIGHTER

In 2000, Eurofighter DA1 was repainted black overall *(Derek Bower)*

wing and leading-edge slats. Assembly at Caselle (Italy), Getafe (Spain), Manching (Germany) and Warton (UK).

LANDING GEAR: Dowty Aerospace retractable tricycle type with SICAMB mainwheels and Magnaghi/OMA nose gear; Dunlop Aviation wheels, brakes and braking system; Ultra Electronics landing gear computer. Single-wheel main units retract inward into fuselage; nosewheel unit forward. Nosewheel steering is subfunction of DFCS. Tyre sizes 28×9.5R15 main; 18×7.75R6 nose. Elektro Metall braking parachute at base of fin.

POWER PLANT: Two Eurojet EJ200 advanced technology turbofans (each of approximately 60 kN; 13,490 lb st dry and 90 kN; 20,250 lb nominal thrust with afterburning), mounted side by side in rear fuselage with ventral intakes. EJ200-01A initially; -1C for early flight tests; -03A first flown June 1997 (DA5); -03B followed in late 1998; and 03Z in December 1999. Staged EJ200 improvements available (but not funded) to 103 kN (23,155 lb st) (designated EJ 230) and then 117 kN (26,300 lb st). MTU FADEC. Lucas Aerospace fuel management system; Aeroquip GmbH fuel ducts; Autoflug sensors; Teldix flowmeters; VDO computer and gauges; Smiths fuel measurement system. First two development aircraft originally powered by two Turbo-Union RB199-122 (Mk 104E) afterburning turbofans (each more than 71.2 kN; 16,000 lb st). Both were retrofitted with EJ200s in 1998.

Internal fuel capacity classified, but believed to total approximately 5,700 litres (1,506 US gallons; 1,254 Imp gallons) in two fuselage tanks and two integral wing tanks. Two-seat trainer lacks forward transfer tank, but partly offsets loss of capacity with auxiliary tank in the enlarged spine. Pressure refuelling point below fuselage, immediately behind air intake. Provision for in-flight refuelling and up to three external fuel tanks: two 1,000 litre (264 US gallon; 220 Imp gallon) or 1,500 litre (396 US gallon; 330 Imp gallon) underwing, plus one 1,000 litre (264 US gallon; 220 Imp gallon) centreline tank. Only the smaller tanks are rated for supersonic flight. In early 1998, BAe was reported to be designing upper fuselage conformal tanks to carry a total of 4,000 kg (8,818 lb) of extra fuel to increase combat radius to 1,500 n miles (2,778 km; 1,726 miles).

ACCOMMODATION: Pilot(s) on Martin-Baker Mk 16A zero/zero ejection seat(s). Single-piece Aerospace Composite windscreen and single-piece, rear-hinged canopy on both versions. Optional liquid-cooled vest for pilot. Helmet-mounted display. Anti-g trousers augmented by pressure breathing system. Two equipment/baggage stowage bays, each 0.01 m³ (0.35 cu ft), above air intakes, port and starboard.

SYSTEMS: Responsibility for systems delegated to Eurofighter GmbH; participants are BAE for electrics, CASA for environmental control and DASA for hydraulics.

Electrical system has Lucas Aerospace as leading supplier and is designed to minimise risk of total power loss by using high level of redundancy and by system positioning. Engine start, systems test, avionics activation and alignment, need no external power. Elements include ZF accessories drive gearbox, two AC and two DC engine-driven Hamilton Sundstrand generators, Honeywell APU in forward port wingroot, Varta battery in corresponding starboard position, two Ferranti Technologies transformer/rectifier units and power converter from same company.

Environmental control system (ECS), with Normalair Garrett as leading supplier, provides conditioned and partly conditioned air to cockpit canopy seal, anti- and de-misting, pilot's anti-g clothing, radar, FLIR, avionics and general equipment. Main oxygen provided by molecular sieve generation system (MSOGS). Precooler (at base of fin), heat exchanger, cold air unit and MSOG all located in aircraft's spine. Liquid cooling subsystem (LCSS) connects aircrew vest to ECS.

Hydraulic system (Magnaghi as leading supplier) comprises two independent circuits supplying power to flight control system (Dowty Boulton Paul actuators), landing gear (including nosewheel and brakes), port and starboard utilities, gun, canopy, airbrake and refuelling probe.

Utilities control system is integrated within overall system architecture and provides for continuous monitoring and fault detection, comprising front computer, fuel computer, secondary power system computer, landing gear computer and maintenance data panel. Integrated monitoring and recording system constantly checks status of all other systems, airframe and engine, to provide rapid, onboard fault diagnosis; functions include crash-survivable memory, bulk storage device, video-voice recording, mission data loading and portable data store, maintenance data panel, portable maintenance data store and air-to-ground relay of data.

AVIONICS: BAE Systems has overall team leadership for avionics development and integration. All avionics, flight control and utilities control systems integrated through STANAG 3910 databus highways with appropriate redundancy levels, using fibre optics and microprocessors. Some functions activated by direct voice input, with 100 word vocabulary.

Comms: Rohde & Schwarz or Elmer VHF/UHF communications, both secure and non-secure. Computing Devices video and voice recorder. DASA and Cossor IFF; Chelton antennas.

Radar: Euroradar ECR 90 Captor multimode pulse Doppler radar.

Flight: Smiths mission data loader and radar altimeter. Litton laser INS; Marconi SpA microwave landing system; ground proximity warning; Tacan; provision for terrain reference navigation; Elmer SpA crash survival memory unit. FCS software updated to Phase 2 in late 1995; Phase 2A 'carefree' handling throughout the subsonic flight envelope, including 25° AoA and over +6 g (cleared by DA2 in January 1998); Phase 2B allows carriage of heavy stores and comprises 2B1 (flown April 1999) for 28° AoA and +7.25 g, and 2B2 for 30-35° and 9 g (first flown 7 July 2000). Phase 3 is IOC standard in 2001 (9 g envelope); Phase 4 covering air-to-surface weapons; and full combat capability Phase 5. DFCS incorporates auto-recovery mode ('panic button') for immediate return to straight-and-level flight in emergency. Ada language, apart from time-critical subroutines in Assembler.

Instrumentation: Special attention given to reducing pilot workload. New cockpit techniques simplify safe and effective operation to limits of flight envelope while monitoring and managing aircraft and its operational systems, and detecting/identifying/attacking desired targets while remaining safe from enemy defences. This achieved through high level of system integration and automation, including HOTAS controls; BAE wide-angle (30° azimuth; 25° elevation) HUD able to display, in addition to other symbology, FLIR pictures from sensor pod mounted externally to port side of cockpit; helmet-mounted sight (HMS), with helmet tracking system and direct voice input (DVI) for appropriate functions; and three Smiths Industries multifunction head-down colour CRT displays (MHDD) and Smiths glareshield standby displays. Dornier, EDS Defence and CANAVA digital map generator; Teldix cockpit interface unit.

Mission: Alenia and Computing Devices nav/attack computers. Eurofirst (FIAR consortium) PIRATE (Passive Infra-Red Airborne Tracking Equipment) port side of windscreen. Secure datalink. RAF aircraft to have optional reconnaissance capability with a long-range electro-optical pod, for which SR(A) 1368 was issued in 1995.

Self-defence: Advanced integrated defensive aids subsystem (DASS), contracted to Euro-DASS consortium, led by BAE Systems, includes Teldix, Computing Devices and Alenia computers, RWR and active jamming pod at each wingtip, laser warning receiver, missile approach warning, Elettronica Aster/GAMESA/CelsiusTech chaff/flare dispenser and towed radar decoys. (Germany initially

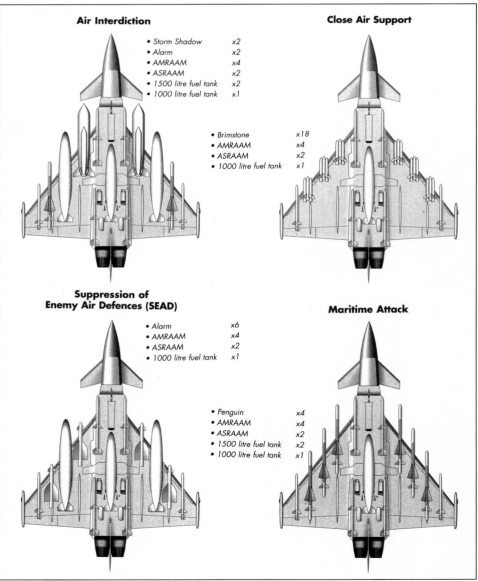

Representative Eurofighter Typhoon weapon loads

did not join Euro-DASS, but is now to adopt standard equipment or alternative locally developed system.) Two BAE Systems radar decoys in the starboard wingtip pod, each on a 100 m (320 ft) fibre optic cable. Italy considering Cross Eye ECM system as alternative to towed decoy. Spain also declined to join Euro-DASS, but is now participating; UK and Spain are only nations to have LWR.
EQUIPMENT: Hella lighting; Logic anti-collision beacons.
ARMAMENT: Total of 13 external stores stations: five (including one wet) under fuselage and four (including one wet) under each wing. Internally mounted 27 mm Mauser gun on starboard side. Interceptor will have mix of medium-range AIM-120 AMRAAM and short-range ASRAAM or IRIS-T air-to-air missiles carried externally; four AIM-120s carried in underfuselage troughs; integration planned of Meteor, following selection for RAF, 16 May 2000, and expression of interest by Italy and Spain. Short-range missiles carried on ML Aviation underwing ejector release units.

Eurofighter will be able to carry considerable load of air-to-surface weapons, including Storm Shadow and Taurus SOMs, Harpoon or Penguin anti-ship missiles, Brimstone anti-armour munition, Future Medium-Range AAM, Extended Range AAM, ALARM ARM, GBU-10/16 and Paveway III laser-guided bombs, BL755 cluster bombs, CRV-7 rocket pods and conventional bombs of up to 2,000 lb size.

Eurofighter DA2 in interdiction fit, with two Taurus, two ALARM, four BVRAAM, two IRIS-T and one 1,000 litre tank

DIMENSIONS, EXTERNAL:
Wing span over ECM pods	10.95 m (35 ft 11 in)
Wing aspect ratio	2.4
Length overall	15.96 m (52 ft 4¼ in)
Height overall	5.28 m (17 ft 4 in)

AREAS:
Wings, gross	50.0 m² (538.2 sq ft)
Foreplanes (total)	2.40 m² (25.83 sq ft)

WEIGHTS AND LOADINGS (approx):
Weight empty, air-to-ground role	10,995 kg (24,240 lb)
Internal fuel load	4,500 kg (9,920 lb)
External stores load (weapons and/or fuel):	
normal	6,500 kg (14,330 lb)
overload	8,000 kg (17,637 lb)
Max T-O weight: normal	21,000 kg (46,297 lb)
overload	23,000 kg (50,706 lb)
Max wing loading	460.0 kg/m² (94.22 lb/sq ft)
Max power loading	128 kg/kN (1.25 lb/lb st)

PERFORMANCE (design):
Max level speed	M2.0
Time to 10,670 m (35,000 ft)/M1.5	up to 2 min 30 s
Runway requirement	700 m (2,300 ft)
T-O run, air combat mission	300 m (985 ft)
Combat radius, single-seat: ground attack, lo-lo-lo	325 n miles (601 km; 374 miles)
ground attack, hi-lo-hi with three LGBs, designator pod and seven AAMs	750 n miles (1,389 km; 863 miles)
air defence with 3 hour CAP	100 n miles (185 km; 115 miles)
air defence with 10 minute loiter	750 n miles (1,389 km; 863 miles)
g limits with full internal fuel and four AIM-120s	+9/−3

UPDATED

EUROMIL

PARTICIPATING COMPANIES:
Eurocopter: see this section
Mil: see under Russian Federation
Kazan: see under Russian Federation
GENERAL DIRECTOR: Vladimir Yablokov

Agreement signed in Moscow 29 September 1994 between Eurocopter, Mil Moscow Helicopter Plant, Kazan Helicopter Plant (KVZ) and Klimov Corporation to form equal share joint venture company, operating under Russian law and covering development and production of 30-passenger Mi-38 helicopter. Eurocopter plans to use Mi-38 to extend its range of products upwards; expects to invest about US$100 million in programme, from total cost (1999) of US$500 million. Each of above-named had one-third share, but risk-sharing partnership expanded to include Sextant Avionique and Pratt & Whitney Canada.

Mil, Kazan and Eurocopter took a board decision in December 1998 to proceed with Mi-38 on new timetable and signed contract in Moscow on 18 August 1999 for completion of demonstrator, as preliminary to manufacture of representative prototype. Final months of 1999 used to establish operational organisation of Euromil, before obtaining design and certification approvals, plus exclusive programme rights from Russian authorities.

UPDATED

EUROMIL Mi-38
TYPE: Medium transport helicopter.
PROGRAMME: Fuselage of demonstrator (plus two static test airframes) had been built by mid-1997, but engine and transmission suppliers had not then been selected. Programme then delayed by Russian financial collapse. P&WC now to supply 2,461 kW (3,300 shp) PW127 turboshafts for demonstrator and production-standard PW127T/S variant for Western exports; Klimov TVA-3000 for CIS operators, subject to completion of development. Revised timetable calls for Mi-38 demonstrator (PT-1) first flight in 2001; four prototypes to follow by 2003; and series production from 2006, following FAR/JAR 29 certification. Sales predicted of 200 in CIS, plus 100 exports. Reports of impending French withdrawal were denied on 16 July 2000.

Mil (see Russian Federation section for Mi-38 description) will handle development, including general design, drawings, component testing and flight testing; Eurocopter to lead in flight deck, avionics system and passenger accommodation and be responsible for preparation for export from Russia; Kazan will manufacture fuselage and rotor blades and undertake final assembly for domestic market. Major Russian subcontractors include Krasny Oktyabr for transmission and Stupino for rotor head.

Model of the projected Mi-38 medium transport helicopter *(Paul Jackson/Jane's)*

UPDATED

EUROPATROL

PARTICIPATING COMPANIES:
Alenia: see under Italy
BAE Systems: see under UK
Dassault: see under France
DaimlerChrysler Aerospace: see under Germany
CASA: see under Spain
Fokker: see under Netherlands

Grouping formed by above manufacturers in mid-1992 to move towards development of a European maritime patrol aircraft and mission system for the 21st century. Initial objectives were to work towards a common requirement; approach governments; and encourage Western European Armaments Group (WEAG) to formulate an Outline European Staff Target (OEST).

Existing, updated or new aircraft are being considered, while the mission system is to be suitable for retrofitting in existing aircraft. BAE Nimrod (which see in UK section), Dassault Atlantic 3 (which see in French section) and Lockheed Martin P-3 Orion (see *Jane's Aircraft Upgrades*) offered as possible existing platforms. Studies continuing at low priority. Germany and Italy have established the MPA2000 group to study requirements for such an aircraft for service from about 2005 to 2007. In late 1999, Germany proposed co-operation agreement with Netherlands, and sought interest from Belgium, Denmark, Norway and other NATO partners in forming a European maritime patrol force.

UPDATED

EUROTILT

EUROPEAN TILTROTOR PROGRAMME
PARTICIPATING COMPANIES:
AgustaWestland: see Italian/UK sections
Eurocopter: see this section

Beginning in 1987, the European Commission sponsored feasibility and definition studies for a tiltrotor transport, as last described under EuroFAR in the 1998-99 *Jane's*. It was hoped by Eurocopter that this ground-work could be incorporated in a forthcoming venture, announced in October 1998, for which other European partners were sought. Eurocopter envisaged a 19/20 seat aircraft with an MTOW of 25,000 kg (55,116 lb) and a service-entry date of around 2015. In early 1999 it undertook a study, sponsored by the French defence ministry, into the military potential of tiltrotors.

By 1999, 33 companies from nine European countries had joined with Eurocopter to apply for EC funding, initially to build a test rig for a proposed 10,000 kg (22,046 lb), 12/19-seat tiltrotor, provisionally known as Eurotilt. Envisaged shares in the project, which could have led to a prototype flying in 2004-05 and entry into service in 2008, included France 36 per cent, Spain 25.7 per cent, Germany 22.3 per cent and Italy 11.6 per cent. Work-shares included Fiat Avio for transmission, CASA for wings and Rolls-Royce Turbomeca for RTM 332 engines. Performance parameters were 300 kt (556 km/h; 345 mph) cruising speed over ranges between 200 and 800 n miles (370 and 1,481 km; 230 and 920 miles) and a service ceiling of 7,620 m (25,000 ft). Development cost (1999) estimated as US$1.05 billion, of which US$87.5 million for ground test rig.

INTERNATIONAL: AIRCRAFT—EUROTILT to IBIS

Eurocopter's concept of a European tiltrotor 2001/0093643
Agusta Erica; main difference to Eurocopter proposal is tilting outboard wing
(James Goulding/Jane's) 2000/0081740

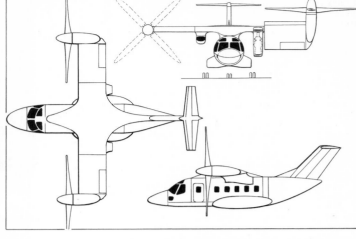

Meanwhile, in July 1999, Agusta (which see) announced the 10-tonne, 20-seat Erica Tiltrotor and was seeking €90 million of funding from EU's 5th Framework research programme to build a ground test vehicle (GTV). Testing was envisaged five years after GTV go-ahead; prototype first flight in further two years. Backing received from 16 companies, including GKN Westland (UK), ZF Luftfahrttechnik (Germany), IAI (Israel), Aermacchi (Italy), NLR (Netherlands), Gamesa (Spain) and Saab (Sweden).

Erica was second-generation tiltrotor, incorporating improvements over BA609 and V-22. Engines mounted inboard, underwing, driving connecting shafts to proprotors at wingtips. Outboard half of wing was to tilt with proprotors, obviating blocking effect of fixed wing and increasing vertical lift by 12 per cent. This improvement would permit smaller proprotor diameter, increasing cruising speed and allowing almost conventional rolling landings, with outboard wing tilted by only 5 to 7° to prevent ground contact.

In October 1999, European Commission rejected separate funding of competing European tiltrotors and urged merger of Erica with Eurotilt.

Common research project, known as 2Gether (second-generation European tilting highly efficient rotorcraft), submitted to European Commission on 31 March 2000 by Eurocopter, Westland and Agusta (since merged as AgustaWestland). Timetable envisages initial €95 million study between 2000 and 2004, followed by demonstrator first flight in 2005 (additional €250 million) and production programme (€1,000 million) leading to series manufacture beginning in 2010. Launch version will have 10-tonne MTOW and carry 20 passengers, but market foreseen for 30-seat and larger tiltrotors from 2020 onwards.

NEW ENTRY

IBIS

IBIS AEROSPACE LTD

PARTICIPATING COMPANIES:
Aero Vodochody: see under Czech Republic
AIDC: see under Taiwan

US SALES OFFICE:
804 Water Street, Kerrville, Texas 78028
Tel: (+1 830) 257 82 00
Fax: (+1 830) 257 82 01
e-mail: ibisaerospace@ktc.com
Web: http://www.ibisaerospace.com
MARKETING DIRECTOR: Jeffrey V Conrad

Ibis Aerospace Ltd is a 50-50 joint venture company created by an agreement of 15 March 1997 under which Aero Vodochody and AIDC are co-developing and producing the Ae 270 Ibis single-turboprop utility aircraft, which made its maiden flight in July 2000. Final assembly is only in the Czech Republic, at least initially.

UPDATED

IBIS AEROSPACE Ae 270 IBIS

TYPE: Light utility turboprop.
PROGRAMME: Announced by Aero Vodochody of Czech Republic in early 1990, originally as L-270; configuration

Main features of the Ibis are well illustrated in this view 2001/0093628

modified 1991; originally planned in two versions (Ae 270 U and Ae 270 MP: see 1993-94 *Jane's*), but revised late 1993 and name Ibis introduced. Chief designer Jan Mikula. Three flying prototypes planned (Nos. 1, 3 and 5), plus one each for static and fatigue testing (Nos. 2 and 4, respectively); to be certified under FAR Pt 23 (Normal category) and be suitable for FAR Pt 135 single-pilot IFR.

Wings for first prototype (Ae 270 P) received from Taiwan August 1999. Roll-out delayed by late arrival of some components, but took place on 10 December 1999; first flight (OK-EMA) 25 July 2000; second (Ae 270 W) prototype due for roll-out in 2000; third to fly with upgraded PT6A-66A, representing Ae 270 HP version disclosed at NBAA Convention in October 2000. Certification scheduled for late 2001 or early 2002.

CURRENT VERSIONS: **Ae 270 P:** Basic pressurised version, with P&WC PT6A-42A engine, Honeywell avionics and retractable gear.

Ae 270 HP: High-performance version, announced 7 October 2000 at NBAA Convention in New Orleans; more powerful PT6A-66A engine offering improved speed, climb and altitude performance and greater fuel efficiency.

Ae 270 W: Basic non-pressurised model (previously called Ae 270 U), with fixed landing gear, Walter M 601F engine and Czech avionics. Plans for this version under review in late 2000.

Ae 270 FP and **FW:** Wheeled-float versions of P and W.

CUSTOMERS: None announced by November 2000.
COSTS: US$1.895 million (2000).
DESIGN FEATURES: Sized between Socata TBM 700 and Pilatus PC-12. High-aspect ratio low wing, circular cabin windows, sweptback vertical tail.

Medium-speed aerofoil section (thickness/chord ratio 17 per cent at root, 12 per cent at tip); leading-edge sweepback 7° 42′; dihedral 6°; incidence 3°; twist 3°.

FLYING CONTROLS: Conventional and manual, augmented by upper-wing roll control spoilers. Elevators trimmed mechanically, ailerons and rudder electromechanically; trim tab in rudder. Wide-span single-slotted Fowler flaps (70 per cent of trailing-edge) actuated hydraulically. Autopilot optional.

STRUCTURE: All-metal semi-monocoque, with fail-safe structural elements in fuselage and two-spar wings. Fuselage of conventionally formed sheet metal bulkheads, stringers and skin; wings (built by AIDC) have formed sheet metal ribs, stringers and chemically milled skin. Production will be in Czech Republic except for wings and

Ae 270 state-of-the-art flight deck with flat panel displays 2001/0100521

landing gear, which will be produced in Taiwan by AIDC; final assembly by Aero Vodochody.

LANDING GEAR: Tricycle type (retractable on Ae 270 P/HP, non-retractable on Ae 270 W), with steerable nosewheel (60° by brakes, 15° by rudder pedals). Inward retraction for mainwheels, rearward for nosewheel. Oleo-pneumatic shock-absorbers in all units; hydraulic disc brakes on mainwheels. Dunlop tubeless tyres: size 6.50-10, pressure 6.90 bars (100 lb/sq in) on mainwheels, size 6.00-6, pressure 3.80 bars (55 lb/sq in) on nosewheel. Minimum ground turning radius (based on nosewheel) 4.00 m (13 ft 1½ in).

POWER PLANT: *Ae 270 P:* One 634 kW (850 shp) (flat rated) Pratt & Whitney Canada PT6A-42A turboprop, driving a Hartzell HC-D4N-3/D9511FK four-blade constant-speed reversible-pitch propeller.

Ae 270 HP: One 634 kW (850 shp) Pratt & Whitney Canada PT6A-66A turboprop.

Ae 270 W: One 580 kW (778 shp) Walter M 601F turboprop (max cruise power 500 kW; 670 shp), with optional water injection system; McCauley 4HFR-34C754/94LA-0 four-blade constant-speed reversible-pitch propeller.

Integral fuel tank in each wing centre-section, combined usable capacity 1,170 litres (309 US gallons; 257 Imp gallons). Gravity refuelling point in top of each wing. Oil capacity 5.7 litres (1.5 US gallons; 1.25 Imp gallons) in Ae 270 P, 7 litres (1.9 US gallons; 1.5 Imp gallons) in Ae 270 W.

ACCOMMODATION: Flight crew of two standard, but to be certified for single-pilot operation. Main cabin suitable for up to eight passengers or 1,200 kg (2,645 lb) of cargo, or combinations of both. Six/seven-seat business or four-seat club layouts permit inclusion of lavatory. Forward-opening crew door at front on port side; upward-opening passenger/cargo door on port side aft of wing; overwing emergency exit on starboard side. Baggage door on starboard side at rear. Cockpit and cabin air conditioning, pressurisation and windscreen heating standard on Ae 270 P/HP. Heating and ventilation standard in Ae 270 W, Honeywell air conditioning optional.

SYSTEMS: Electrical power in both models provided by 28 V 250 A DC engine-driven starter/generator and 24 V 42 Ah lead-acid battery; 28 V DC external power connector. Standby generator optional. Hydraulic system, pressure 150 bars (2,175 lb/sq in), for actuation of flaps, mainwheel brakes and (on Ae 270 P/HP) landing gear extension/retraction; flow rate 11 litres (2.9 US gallons; 2.4 Imp gallons)/min. Landing gear and mainwheel brakes also controllable by separate emergency hand-operated valves and parking brake. Honeywell air conditioning and pressurisation system in Ae 270 P/HP maintains differential of 0.30 bar (4.4 lb/sq in) up to altitude 7,500 m (24,600 ft). Pneumatic (engine bleed air) de-icing of wing and tailplane leading-edges; electric de-icing of windscreens, propeller blades; hot air de-icing of engine air intake; stall warning sensor and pitot tube standard on both models. Emergency oxygen system for crew and passengers.

AVIONICS: Standard flight, navigation and engine instrumentation (VFR or IFR) to comply with FAR Pt 23.
Radar: Weather radar optional.
Flight: IFR package standard for business versions, including VOR/ILS, ADF and GPS nav.
Instrumentation: EFIS optional.

DIMENSIONS, EXTERNAL:
Wing span	13.80 m (45 ft 3¼ in)
Wing chord: at root	1.88 m (6 ft 2 in)
at tip	1.04 m (3 ft 5 in)
Wing aspect ratio	9.1
Length overall	12.24 m (40 ft 1 in)
Fuselage: Length	12.19 m (40 ft 0 in)
Max width	1.60 m (5 ft 3 in)
Max depth	1.75 m (5 ft 9 in)
Height overall	4.79 m (15 ft 8½ in)
Elevator span	5.40 m (17 ft 8½ in)
Wheel track	2.83 m (9 ft 3½ in)
Wheelbase	3.53 m (11 ft 7 in)
Propeller diameter: Hartzell	2.44 m (8 ft 0 in)
McCauley	2.39 m (7 ft 10 in)
Propeller ground clearance:	
Hartzell	0.41 m (1 ft 4¼ in)
McCauley	0.44 m (1 ft 5¼ in)
Passenger/cargo door (port, rear):	
Height	1.25 m (4 ft 1¼ in)
Width	1.25 m (4 ft 1¼ in)
Height to sill	1.30 m (4 ft 3¼ in)
Crew door (port, fwd): Height	1.20 m (3 ft 11¼ in)
Width	0.70 m (2 ft 3½ in)
Height to sill	1.30 m (4 ft 3¼ in)
Emergency exit (stbd, overwing):	
Height	0.71 m (2 ft 4 in)
Width	0.50 m (1 ft 7¾ in)

DIMENSIONS, INTERNAL:
Cabin: Length	4.98 m (16 ft 4 in)
Max width	1.44 m (4 ft 8¾ in)
Max height	1.36 m (4 ft 5½ in)
Volume	7.5 m³ (265 cu ft)

AREAS:
Wings, gross	21.00 m² (226.0 sq ft)
Ailerons (total)	1.00 m² (10.76 sq ft)
Trailing-edge flaps (total)	4.22 m² (45.42 sq ft)
Spoilers (total)	0.325 m² (3.50 sq ft)
Fin, incl dorsal fin	1.96 m² (21.10 sq ft)
Rudder, incl tab	1.11 m² (11.95 sq ft)
Tailplane	2.975 m² (32.02 sq ft)
Elevators (total)	1.885 m² (20.29 sq ft)

WEIGHTS AND LOADINGS:
Weight empty, equipped: P	1,788 kg (3,942 lb)
HP	1,996 kg (4,400 lb)
W	1,700 kg (3,748 lb)
Max fuel weight: P, HP, W	1,034 kg (2,280 lb)
Max payload: P, W	1,200 kg (2,645 lb)

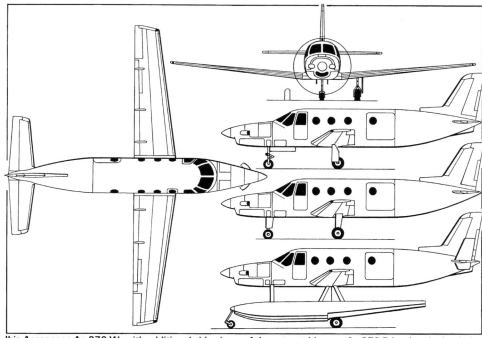
Ibis Aerospace Ae 270 W, with additional side views of the retractable-gear Ae 270 P (top) and wheeled-float Ae 270 FW/FP *(Mike Keep/Jane's)*

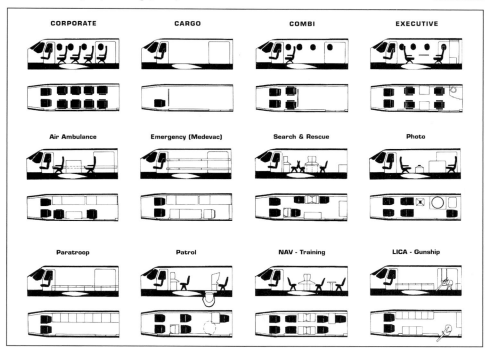
Potential interior configurations of the Ibis utility transport

First prototype Ae 270 P Ibis in flight

Max T-O weight: P, W	3,300 kg (7,275 lb)	HP	260 kt (482 km/h; 299 mph)	HP	397 m (1,300 ft)	
HP	3,515 kg (7,750 lb)	W	172 kt (319 km/h; 198 mph)	W	531 m (1,745 ft)	
Max ramp weight: P, W	3,315 kg (7,308 lb)	at 6,000 m (19,680 ft): P	219 kt (405 km/h; 252 mph)	Landing from 15 m (50 ft): P	853 m (2,800 ft)	
Max landing weight: P, W	3,150 kg (6,944 lb)	W	171 kt (316 km/h; 197 mph)	HP	640 m (2,100 ft)	
Max zero-fuel weight: P	3,065 kg (6,757 lb)	Stalling speed, engine idling:		W	878 m (2,885 ft)	
HP	3,209 kg (7,075 lb)	P, HP, W (flaps up)	80 kt (149 km/h; 93 mph)	Landing run: P	409 m (1,345 ft)	
W	2,977 kg (6,563 lb)	P, HP, W (flaps down)	61 kt (113 km/h; 71 mph)	HP	320 m (1,050 ft)	
Max wing loading: P, W	157.1 kg/m² (32.18 lb/sq ft)	Max rate of climb at S/L: P	492 m (1,614 ft)/min	W	416 m (1,365 ft)	
HP	167.4 kg/m² (34.29 lb/sq ft)	HP	579 m (1,900 ft)/min	Range with max payload, 45 min reserves:		
Max power loading: P	5.21 kg/kW (8.56 lb/shp)	W	326 m (1,069 ft)/min	P	121 n miles (225 km; 140 miles)	
HP	5.55 kg/kW (9.12 lb/shp)	Max certified altitude: HP	7,620 m (25,000 ft)	W	151 n miles (280 km; 174 miles)	
W	5.69 kg/kW (9.35 lb/shp)	Service ceiling: P	9,300 m (30,520 ft)	Range with max fuel, 45 min reserves:		
PERFORMANCE (estimated):		HP	12,190 m (40,000 ft)	P at 8,000 m (26,240 ft)		
Max cruising speed: at S/L:		W	9,200 m (30,180 ft)		1,406 n miles (2,604 km; 1,618 miles)	
P	206 kt (381 km/h; 237 mph)	T-O run: P	266 m (875 ft)	HP at 7,620 m (25,000 ft)		
HP	210 kt (389 km/h; 242 mph)	HP	214 m (700 ft)		1,575 n miles (2,916 km; 1,812 miles)	
W	168 kt (311 km/h; 193 mph)	W	287 m (945 ft)	W at 4,000 m (13,120 ft)		
at 4,000 m (13,120 ft): P	220 kt (408 km/h; 253 mph)	T-O to 15 m (50 ft): P	499 m (1,640 ft)		890 n miles (1,650 km; 1,025 miles)	

UPDATED

INTRACOM

INTRACOM GENERAL MACHINERY SA
Chemin de Faguillon 1, CH-1223 Geneva, Switzerland
Tel: (+41 22) 786 26 57
Fax: (+41 22) 786 27 62
DIRECTOR: Nik Schmidt

Intracom, formed in 1994, is the aircraft design and manufacturing arm of General Machinery SA. It has joined with the Khrunichev company of Russia (which see) to offer the GM-17 light transport in Western markets.

NEW ENTRY

INTRACOM GM-17 FENIKS
English name: Phoenix
TYPE: Light utility turboprop.
PROGRAMME: Designed within Khrunichev bureau in Moscow by Evgeny P Grunin. Considerably modified Piper PA-31P Pressurised Navajo. First flight (RA-01559; the former N38RG) 6 December 2000; 10 sorties and 6 hours flown by February 2001, when temporarily laid-up for minor modifications. Further two prototypes to follow in 2001, using Navajos G-OIEA and SE-IGB as basis. Intracom had previously proposed a collaborative project with Piper, to be known as PA-31TXL, but this was rejected by the US company; projected new production, to begin in 2003, will not use Piper-sourced airframes, therefore. Prototypes converted by Khrunichev; production may be relocated to KnAAPO at Komsomolsk.
GM-17 wing and fuselage are to be used in a series of different, but related aircraft, as described below.
COSTS: US$860,000 (2000).
DESIGN FEATURES: Single-engined turboprop for passenger and light freight transport; suitable for executive ferry, medical evacuation, patrol, geological survey and training. Low-mounted, tapered wing with winglets; mid-positioned, tapered tailplane and sweptback fin.
Wing section P-301; leading-edge sweepback 5°; thickness chord ratio 14 per cent at root, 12 per cent at tip; dihedral 5°; incidence 1°. Tailplane incidence 2°; wingroot nib sweepback 28°.
FLYING CONTROLS: Conventional and manual. Trim tab in starboard aileron, rudder and both elevators. Flaps.
LANDING GEAR: Tricycle type; mainwheels retract inward; nosewheel rearward; single wheel on each unit.

Prototype of Intracom GM-17 Feniks (Phoenix) light transport *2001/0062441*

POWER PLANT: One 560 kW (757 shp) Walter M 601E turboprop, driving a five-blade V510 constant-speed, fully reversing propeller. Three prototypes have M 601D engines and three-blade V508 propellers; TVD-100 turboprop to be offered as alternative on production aircraft. Fuel capacity 1,170 litres (309 US gallons; 257 Imp gallons).
ACCOMMODATION: Two pilots and seven passengers. Single door, port, rear. Emergency exit, starboard.
SYSTEMS: Hydraulic system, pressure 150 bars (2,175 lb/sq in); 27 V electrical system; emergency oxygen. De-icing system on all leading-edges.

AVIONICS: *Comms:* Honeywell KT 76C transponder.
Radar: Optional Honeywell RDS 81 four-colour weather radar.
Flight: Meggitt System 55 autopilot; Chelton GNS-530 Omega/VLF with moving map; Honeywell KR 87 ADF; VOR; ILS.

DIMENSIONS, EXTERNAL:
Wing span	12.73 m (41 ft 9¼ in)
Wing chord: at root	2.28 m (7 ft 5¾ in)
at tip	0.985 m (3 ft 2¾ in)
Wing aspect ratio	7.5
Length: overall	10.98 m (36 ft 0¼ in)
fuselage	10.28 m (33 ft 8¾ in)
Fuselage max width	1.55 m (5 ft 1in)
Height overall	4.08 m (13 ft 4¾ in)
Tailplane span	6.05 m (19 ft 10¼ in)
Wheel track	4.30 m (14 ft 1¼ in)
Wheelbase	2.69 m (8 ft 10 in)
Propeller diameter	2.30 m (7 ft 6½ in)
Propeller ground clearance	0.49 m (1 ft 7¼ in)

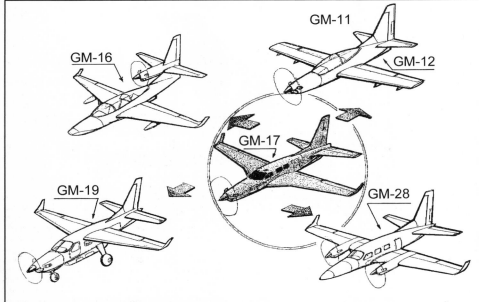

Proposed family of Intracom aircraft *2001/0106484*

Revised nose shape of Navajo following GM-17 conversion *2001/0062440*

INTRACOM to NH INDUSTRIES—AIRCRAFT: INTERNATIONAL

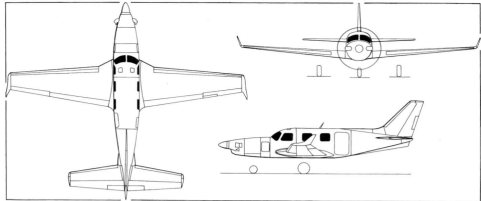

Intracom GM-17 Feniks (Walter M 601 turboprop) *(Paul Jackson/Jane's)* 2001/0093646

Passenger door: Height	1.32 m (4 ft 4 in)
Width	0.78 m (2 ft 6¾ in)
Emergency exit: Height	0.54 m (1 ft 9¼ in)
Width	0.60 m (1 ft 11¾ in)

DIMENSIONS, INTERNAL:
Cabin: Length	4.92 m (16 ft 1¾ in)
Max width	1.41 m (4 ft 7½ in)
Max height	1.45 m (4 ft 9 in)
Baggage hold volume: main	1.9 m³ (67 cu ft)
auxiliary	0.20 m³ (7.1 cu ft)

AREAS:
Wings, gross	21.50 m² (231.4 sq ft)
Ailerons (total)	1.245 m² (13.40 sq ft)
Trailing-edge flaps (total)	3.135 m² (33.74 sq ft)
Winglets (total)	2.86 m² (30.78 sq ft)
Rudder (incl tab)	1.08 m² (11.63 sq ft)
Tailplane	6.08 m² (65.44 sq ft)
Elevators (total)	2.16 m² (23.25 sq ft)

WEIGHTS AND LOADINGS:
Weight empty	1,634 kg (3,602 lb)
Max payload	1,150 kg (2,535 lb)
Max T-O and landing weight	2,960 kg (6,525 lb)
Max zero-fuel weight	2,150 kg (4,740 lb)
Max wing loading	137.7 kg/m² (28.20 lb/sq ft)
Max power loading	5.29 kg/kW (8.68 lb/shp)

PERFORMANCE:
Max level speed at 5,240 m (17,200 ft)
    327 kt (605 km/h; 376 mph)
Max cruising speed at 6,000 m (19,680 ft)
    263 kt (487 km/h; 303 mph)
Max rate of climb at S/L   636 m (2,087 ft)/min
Service ceiling   9,000 m (29,520 ft)
T-O to 15 m (50 ft)   600 m (1,970 ft)
Landing from 15 m (50 ft)   508 m (1,670 ft)
Range: with max fuel
    2,030 n miles (3,760 km; 2,336 miles)
with max payload
    1,349 n miles (2,500 km; 1,553 miles)
with 360 kg (794 lb) payload, 30 min reserves
    1,706 n miles (3,160 km; 1,963 miles)

UPDATED

INTRACOM GM-28

TYPE: Utility turboprop twin.
PROGRAMME: Piper PA-31P with two Walter M 601E engines in usual wing nacelles; added winglets. Proposed conversion of 22 aircraft by 2003, with certification in Experimental category. Dimensions generally as for GM-17.

NEW ENTRY

INTRACOM GM-11 and GM-12

TYPE: Basic turboprop trainer/attack lightplane.
PROGRAMME: Evolution of GM-17 configuration in light attack (GM-11 with four underwing hardpoints) and two-seat trainer (GM-12) forms.

DIMENSIONS, EXTERNAL:
Wing span: GM-11	11.92 m (39 ft 1¼ in)
GM-12	11.74 m (38 ft 6¼ in)
Length overall: GM-11	9.34 m (30 ft 7¾ in)
GM-12	9.89 m (32 ft 5¼ in)
Height overall: GM-11	4.26 m (13 ft 11¾ in)
GM-12	4.40 m (14 ft 5¼ in)
Tailplane span: GM-11	6.045 m (19 ft 10 in)
GM-12	4.30 m (14 ft 1¼ in)
Wheel track: GM-11	4.05 m (13 ft 3½ in)
GM-12	3.85 m (12 ft 7½ in)
Wheelbase	3.72 m (12 ft 2½ in)
Propeller diameter	2.30 m (7 ft 6½ in)

NEW ENTRY

INTRACOM GM-16

TYPE: Six-seat amphibian.
PROGRAMME: Proposed with GM-17 wing mounted mid-fuselage, plus single turboprop forward of fin leading-edge.

DIMENSIONS, EXTERNAL:
Wing span	11.92 m (39 ft 1¼ in)
Length overall	10.94 m (35 ft 1¾ in)
Height: to fintip	4.00 m (13 ft 1½ in)
overall (propeller turning)	4.48 m (14 ft 8½ in)
Tailplane span	6.00 m (19 ft 8¼ in)
Wheel track	4.19 m (13 ft 9 in)
Wheelbase	4.26 m (13 ft 11¾ in)
Propeller diameter	2.50 m (8 ft 2½ in)

NEW ENTRY

INTRACOM GM-19

TYPE: Light utility turboprop.
PROGRAMME: Adaptation of GM-17 as single-engine, six-seat transport, but with high wing and ventral luggage pannier. Stretched version is GM-19T.

DIMENSIONS, EXTERNAL:
Wing span	13.92 m (45 ft 8 in)
Length overall: standard	10.46 m (34 ft 3½ in)
stretched	11.64 m (38 ft 2¼ in)
Height overall	4.84 m (15 ft 10½ in)
Tailplane span	6.045 m (19 ft 10 in)
Wheel track	3.40 m (11 ft 1½ in)
Wheelbase: standard	3.94 m (12 ft 11 in)
stretched	4.40 m (14 ft 5¼ in)
Propeller diameter	2.30 m (7 ft 6½ in)

NEW ENTRY

NH INDUSTRIES

NH INDUSTRIES sarl

Le Quatuor, Batiment C, 42 route de Galice, F-13082 Aix-en-Provence Cedex 02, France
Tel: (+33 4) 42 95 97 00
Fax: (+33 4) 42 95 97 48/49
e-mail: nhi@aix.pacwan.net
Web: http://www.nhindustries.com
GENERAL MANAGER: Philippe Stuckelberger

PARTICIPATING COMPANIES:
Agusta: see under Italy
EADS Eurocopter Deutschland: see this section
EADS Eurocopter France: see this section
Fokker Aerostructures: see under Netherlands

NH Industries established August 1992 to manage design and development of NH 90; in October 1995 workshares were slightly amended to Agusta 28.2 per cent, Eurocopter Deutschland 23.7 per cent, Eurocopter France 41.6 per cent, Fokker Aircraft 6.5 per cent. NFT (Norway) joined in 1994 as risk-sharing partner of Eurocopter France. Netherlands activities transferred to Fokker Aerostructures BV following 1996 bankruptcy of the former Fokker Aircraft. Industrial ownership (and production share) is now EADS 62.5 per cent, Finmeccanica (Agusta) 32 per cent and Fokker Aviation BV 5.5 per cent. NH Industries is prime contractor for design and development, industrialisation and production, logistic support, marketing and sales.

Joint agency NAHEMA (NATO Helicopter Management Agency) formed February 1992 by the four governments, within NATO framework, to manage programme; NAHEMA located alongside NH Industries in Aix.

UPDATED

NH INDUSTRIES NH 90

TYPE: Multirole medium helicopter.
PROGRAMME: Initial studies by NIAG (NATO Industrial Advisory Group) SG14 in 1983-84; September 1985 MoU between defence ministers of France, Germany, Italy, Netherlands and UK (USA and Canada having already withdrawn) preceded 14 month feasibility/predefinition study for new naval/army NH 90 (NATO helicopter for the 1990s); initial design phase approved December 1986; second MoU in September 1987 led to predefinition phase and completion of weapons system definition in 1988; UK withdrew from programme April 1987; German workshare reduced early 1990, Italian participation renegotiated later 1990; French and German launch decision 26 April 1990; two ministerial MoUs in December 1990 and June 1991 cover design and development responsibilities; intercompany agreement signed March 1992; design and development contract signed 1 September 1992; five prototypes and GTV (ground test vehicle) planned, with three prototypes produced in France, TTH in Germany and NFH in Italy; assembly of first prototype began at Marignane, October 1993; first flight 18 December 1995.

Development suspended May 1994; restarted July 1994, but all new-development items subject to rigorous cost examination; preference for off-the-shelf components. Programme doubts dispelled by inclusion of NH 90 in French defence plan adopted in June 1996, which requires first NFH delivery to Aéronavale in 2004 and first TTH to ALAT in 2011, although delivery of the last ALAT aircraft has been delayed to 2025. Bid for production investment and Lot 1 manufacture submitted 5 October 1997; four nations signed political authorisation for first production batch of 154 (amended later to 151) helicopters on 19 May

NH 90 prototype F-ZWTH undergoing trials aboard FNS *Colbert* as MMX612   2000/0062387

1998; pricing proposals submitted by NHI to NAHEMA on 15 July 1998; industrial contract was due in mid-1999 to permit first deliveries in 2003 for Netherlands Navy, but was delayed and latter date rescheduled to 2007.

In September 1999, Germany revealed a cut of 24 in its total order and now plans to receive first TTH in 2004 and NFH in 2007, instead of withdrawing current Westland Lynx in 2003. Italian deliveries (both versions) to begin in April 2004, continuing until 2017-18. NH 90 to be built on parallel production lines in France, Germany and Italy. Netherlands confirmed in November 1998 that its helicopters would be built in Italy; German NFHs will be built in France. Further, in 1998, a basic NH 90 variant was proposed to which 'Kits on Option' could be added to produce specific variants at lower cost than purpose-build.

Rescheduled production contract signature, at Berlin Air Show, 8 June 2000, replaced by MoU committing to 243, plus 55 options and minimum buy of 366 from stated requirements for 595. Production investment and production contract finally signed in Paris on 30 June 2000 on these terms and quoting cost for 298 as €6.6 billion, with industrial participants providing 25 per cent of production investment.

INTERNATIONAL: AIRCRAFT—NH INDUSTRIES

On 6 October 1995, NAHEMA contracted NHI to undertake additional work and national customisation, involving integration of T700 engine to meet Italian requirement; design of rear ramp installation for TTH version; and reinforcement for carriage of stores up to 700 kg (1,543 lb). Associated national requirements include study of command post version; cannon pod; sand filter; radiometer; second VHF/FM radio for TTH; and sonobuoy data relay, Tacan and rear ramp for NFH. Total value of €58.23 million slightly amends national workshares. Italy formally confirmed T700 engine selection in June 2000; Germany and Netherlands simultaneously confirmed RTM322, which earlier selected by France.

First three aircraft had flown more than 420 hours by 31 May 1999, when fourth joined the test programme. Flight envelope expanded by that time to 190 kt (352 km/h; 219 mph), 6,096 m (20,000 ft) altitude and T-O weight of 10,000 kg (22,050 lb). The aircraft has also demonstrated fast rolling landings at 50 kt (93 km/h; 58 mph) and landings on slopes of up to 12°. Total 600 hours by 22 December 1999, when fifth prototype joined programme.

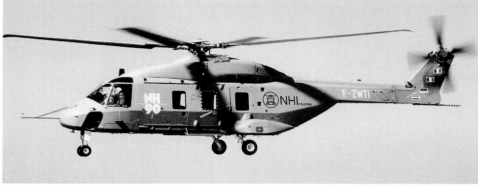

Second NH 90, in naval-style light grey

CURRENT VERSIONS (general): **NFH** (NATO Frigate Helicopter): Naval version, primarily for autonomous ASW and ASVW; additional applications include OTH targeting, vertrep, SAR, transport and anti-air warfare support; designed for all-weather/severe ship motion environment; fully integrated mission system for crew of three (optionally four); ECM, anti-radar and IR protection systems. Italy considering a unique sensor fit from FIAR/Honeywell consortium. Agusta, with its EH 101 and S-61 experience, is responsible for NFH development and integration.

TTH (Tactical Transport Helicopter): Land-based army/air force version, primarily for tactical transport, airmobile operations and SAR with up to 20 armed troops or 2.5 tonnes of supplies; additional applications include tactical support, special EW, airborne command post, VIP transport and training; defensive weapons suite; rear-loading ramp/door, loading winch and rolling strips to be provided for French, Italian and German armies to accommodate light armoured anti-tank missile vehicle; 4-tonne external underslung load hook; high manoeuvrability and survivability for NOE operation near front line. Eurocopter Deutschland is responsible for development of the TTH.

C-SAR: Combat search and rescue version, yet to be fully defined, but probably equipped with refuelling probe, 'bolt-on' armour on lower fuselage, two external fuel tanks, flotation gear, secure datalink, integrated countermeasures, FLIR and four FIM-92 Stinger self-defence AAMs.

'Kits on Option': To accommodate differing budgetary and operational requirements, a baseline configuration has been agreed, to which specific packages could be added. French shipborne utility version combining troop seats and ramp could replace Super Frelon. This version could also form basis of dedicated VIP variant, and even of a civil variant. Eurocopter France is responsible for development of the basic helicopter.

Netherlands NFH: All 20 equipped with common core avionics, FCS, 360° radar, FLIR, ESM, ECM and sonics, but only 14 with full mission system; rest with provision for, but without, LRUs. Proposal to equip six for night special operations/insertion duties. To serve aboard eight M-type frigates, four LCF frigates, two LPDs and two AORs.

CURRENT VERSIONS (specific): **PT1/F-ZWTH**: Prototype; common basic configuration; assembled at Marignane; rolled out 29 September 1995; first flight 18 December 1995 with RTM 322 engines; formal debut (official first flight) 15 February 1996; public debut at Berlin, May 1996; 75 hours airborne by 1 January 1997; tactical (green) colours; with 160 hours logged, PT1 flew to Agusta's Cascina Costa plant on 29 July 1997 for installation of Alfa Romeo Avia/GE T700-T6E engines before an 18-month test programme funded by Italy; first flew as such on 13 March 1998; had amassed 84 hours with these engines by December 1998, including sea trials with French 'La Fayette' class frigate *Courbet*, making 62 deck landings in two days (22/23 July 1998). Flies with Italian serial MMX612. Total 320 hours by 1 January 2000.

PT2/F-ZWTI: Common basic configuration; rolled out November 1996 at Marignane; first flight 19 March 1997; naval (light grey) colours; first with fly-by-wire controls. Initially flew for 5 hours with mechanical back-up flight controls, then first flew with analogue FBW on 2 July 1997; total 20 hours by December 1997; digital FBW installed early 1998 and first flown on 15 May. PT2 also first with automatic electric blade folding intended for NFH, and due to receive IR exhaust suppressors. Total 320 hours by 1 January 2000.

PT3/F-ZWTJ: Common basic configuration; first with core avionics system with AFCS, navigation, communications, IFF and full 'glass' cockpit; retains mechanical control back-up; first flight at Marignane 27 November 1998. Core avionics trials with production-standard LCDs; replaced PT2 in extending flight envelope. Total 45 hours by 1 January 2000.

PT4/9890: TTH configuration; first flight 31 May 1999, at Ottobrunn. First NH 90 to be fitted with C-model prototype of DaimlerChrysler Aerospace electronic warfare suite as one element in a fully integrated avionics suite, with LCD cockpit displays, FLIR, helmet sight, tactical control system, secure communications, and radar (with digital mapping). Being used to clear operation of rear loading ramp. Total 40 hours by 1 January 2000.

PT5/MMX-613: NFH mission system; first flight at Cascina Costa 22 December 1999.

GTV: Ground test vehicle at Cascina Costa. Instrumented to monitor 300 parameters. First run 28 September 1995; 450 hours accumulated by June 1999.

CUSTOMERS: Requirements originally estimated as 726 (France 220, Germany 272, Italy 214 and Netherlands 20); reduced to 647 in July 1996, then to 642 in 1998 and 595 in late 1999. Of 595, 366 covered by firm commitment, of which 298 on order or option; see table. France scheduled eventually to buy 133 TTH for the Armée de l'Air from 2011. Additionally, over 600 export orders are anticipated. NHI has received an RFP for 30 TTH from Oman, and is bidding to meet Singapore requirement for 12 NFH, a Portuguese requirement and Nordic helicopter requirement.

	Firm		Option	
Customer	TTH	NFH	TTH	NFH
France				
Navy		27		
Germany				
Air Force	30		24	
Army	50		30	
Italy				
Air Force			1	
Navy	10	46		
Army	60			
Netherlands				
Navy		20		
Totals	**150**	**93**		**55**
		243		
			298	

Notes: Second contract will be for 68, completing guaranteed purchase of 366. Further 229 required to complete stated needs of four partners. German Air Force total includes 23 for combat SAR

COSTS: Development cost €1,376.15 million (FFr 9.6 billion) (January 1988 values) agreed September 1992, provided by governments and industry: €421.83 million French government plus €161.24 million Eurocopter France; €256.41 million Germany plus €74.59 million Eurocopter Deutschland; €89.44 million Netherlands plus €2.64 million Fokker; and €370 million by Italian government. Additional development, agreed October 1995, adds €58.23 million. By January 1997, an estimated 75 per cent of R&D costs had been disbursed. Projected flyaway prices (1995) are FFr90 million for TTH and FFr145 million for NFH, but further 15 per cent cut demanded by French military procurement office in February 1996. Initial batch of French NFHs were 2000 unit cost of FFr200 million, including spares and tax; NHI quotes FFr115 million, pre-tax. Netherlands paying NLG1.7 billion (US$510 million) giving a unit price of US$25.5 million per aircraft.

DESIGN FEATURES: Compact external dimensions with large cabin (larger than Sikorsky Blackhawk). Low vulnerability/detectability, with aerodynamic and low-observables design, reduced maintenance requirements, and day/night operability within temperature range of −40 to +50°C. Specification requires fewer than 250 failures per 1,000 flight hours; 97.5 per cent mission reliability rate; 87 per cent availability; MTBF better than 4 hours; and under 2.5 mmh/fh (excluding engines). Service life of 10,000 hours over 30 years.

Titanium Spheriflex main rotor hub with elastomeric spherical thrust bearings; four blades with advanced aerofoils and curved tips; main rotor turns anticlockwise at 256.6 rpm with 219 m (719 ft)/s tip speed; bearingless tail rotor with four blades of similar design and construction; tail rotor with cross-beam hub turns at 1,235.4 rpm with 207 m (679 ft)/s tip speed; automatic electronic folding of main rotor blades and tail pylon in NFH; manual blade folding (only) in TTH. Large unobstructed internal volume.

FLYING CONTROLS: Quadruplex fly-by-wire controls eliminate cross-coupling between control axes. NH 90 is the world's first FBW transport helicopter. Flight control system developed by Eurocopter France; optimised for NOE flight and minimal vulnerability to small arms damage.

STRUCTURE: All-composites fuselage; fail-safe design of structure, rotating parts and systems for high safety levels. First set of main and tail rotor blades completed for ground testing May 1994; blades have multibox structure and glass and carbon fibre skin; leading-edge box and roved parts of

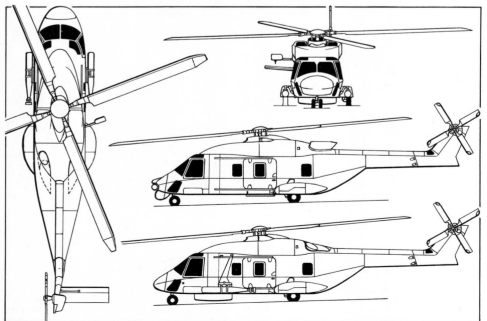

NH Industries NH 90 NFH, with additional side view (upper) of TTH *(Mike Keep/Jane's)*

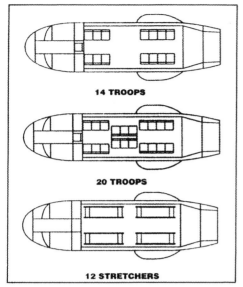

NH 90 TTH seating/carrying options

blade prefabricated to reduce time in main mould; resin used reduces material ageing in damp and hot regions; blades designed and manufactured using CATIA CAD/CAM.

Eurocopter France builds rotor, rear gearbox, auxiliary gearbox, fairings and produces wiring, electrical system and air conditioning. Eurocopter Deutschland builds front and centre fuselage and fuel system. Agusta responsible for rear fuselage, main gearbox, hydraulic system and rear ramp; Fokker for tailboom, sliding doors, landing gear and fairings and intermediate gearbox. Assembly lines in France, Germany (TTH only) and Italy.

LANDING GEAR: Retractable crashworthy tricycle gear by DAF Special Products with twin-wheel nose unit and single-wheel main units; all units retract rearwards; four-balloon emergency flotation gear effective up to Sea State 3. NFH has harpoon-type deck haul-down gear.

POWER PLANT: Two engines; required power from each engine during normal operation at 1,000 m (3,280 ft) in ISA + 15°C is 1,360 kW (1,824 shp) for 30 minutes, 1,253 kW (1,680 shp) maximum continuous, OEI maximum contingency (2½ minutes) 1,484 kW (1,990 shp), emergency at S/L ISA, 1,942 kW (2,604 shp). Engine RFP issued April 1992. Decided on 25 January 1994 that NFH and TTH will be developed and qualified with 1,566 kW (2,100 shp) RTM 322-01/9 made by European consortium of Turbomeca, Rolls-Royce, MTU, Piaggio and Topps; during development, the 1,521 kW (2,040 shp) GE T700-T6E is being qualified by Alfa Romeo and GE to meet Italian military requirements. Engine control by FADEC.

Transmission rating 2,545 kW (3,413 shp) with both engines, 2,050 kW (2,749 shp) for 30 seconds OEI. Modular main gearbox features integrated lubrication system with innovative integral monitoring and diagnostic system and, like the two accessory gearboxes, can run dry for 30 minutes.

Fuel system has eight Uniroyal crash-resistant self-sealing cells and AFG management system. Total internal capacity approx 2,500 litres (660 US gallons; 550 Imp gallons). Provision for pressure refuelling and hover refuelling. Provision for two external tanks (modified from Alpha Jet) each of 750 litres (198 US gallons; 165 Imp gallons).

ACCOMMODATION: Minimum crew one pilot VFR and IFR; NFH crew, pilot, co-pilot/TACCO on flight deck and one system operator in cabin; optionally, two pilots on flight deck and one TACCO, one SENSO in cabin; TTH, two pilots or one pilot and one crewman on flight deck and 14 to 20 equipped troops or one 2 tonne tactical vehicle in cabin. Sogerma composites crew seats withstand 22 g or 11 m (35 ft)/s vertical descent crash loading.

SYSTEMS: Full redundancy for all vital systems; hydraulic system has two Vickers mechanically driven and one electrohydraulic and one electric pumps; redundant flight control hydraulic system has two main separated and independent circuits; independent utilities system provides hydraulic power for landing gear and ancillary operations; electrically driven pump as back up in event of one circuit failing; other circuit permanently supplied in parallel by a utility system pump. Electrical system has two batteries and three 40 kVA Auxilec AC generators (two driven by main gearbox, one by remote accessory gearbox) feeding DC busses through transformer-rectifiers; 70 kW Microturbo Saphyr 100 APU for electrical engine starting, environmental control on ground and emergency use in flight; Kollmorgen emergency DC generator (driven by remote accessory gearbox); fire detection and suppression in engine bays, APU and cabin. Microtecnica air conditioning system (with two identical independent units) operating in three zones: cockpit, cabin (both vapour cycle) and main avionics bays (filtered external air). Anti-icing of main blades (standard on TTH, optional on NFH) by heated leading-edge mat (heating carbon coating and not metal parts). Anti-icing system also on tail rotor blades and horizontal stabiliser, and on air intakes. Ice protection system effective in maximum conditions defined by DEF-STAN 000-970.

AVIONICS: Core avionic system based on dual MIL-STD-1553B data highways (one for the core avionics system, one for the mission system) allows integration of aircraft and mission equipment through several computers. Eurocopter France responsible for basic avionics; Eurocopter Deutschland for TTH avionics; Agusta for NFH mission system.

Comms: Integrated management of communications and identification systems. Two V/UHF; V/UHF DF/homer; HF SSB and Thomson-CSF TSC 2000 IFF. Rohde & Schwarz intercom.

Radar: NFH has Thomson Ocean Master-based 360° track-while-scan surveillance radar with target recognition capability; TTH has weather radar. Helicopter Euro American Radar Team (Elettronica Aster, Siemens, Thomson-CSF, NLR and Raytheon) has offered SeaFalcon to meet NFH radar requirements.

Flight: Twin INS with embedded GPS and Doppler ground speed sensor; twin air data computers; provision for radio altimeter and MLS. SFIM/Alenia flight control computer and Thomson/BAE Systems microwave landing system. Obstacle warning system in TTH.

Instrumentation: NVG compatible. Five 200 mm (8 in) square Sextant LCDs in cockpit of NFH; up to three more in cabin; four in TTH. TTH has digital map. Central warning system, two remote frequency indicators and conventional analogue back-up instruments. Provision for helmet-mounted display and sight.

Mission: NFH: Alenia or Thomson dipping sonar, high focal-length tactical FLIR, MAD, IFF interrogator, Link 11 datalink including ship-to-air umbilical, ESM and telebrief; electronic warfare subsystem. Dedicated video/audio network distributes tactical information to all crew members. TTH: FLIR.

Self-defence: Provision for MAWS, RWR, LWR, chaff/flares and IR jammer. TTH has self-protection suite with passive threat warning system known as EWS (electronic warfare suite) with DASA laser warning receiver, LFK AN/AAR-60 MILDS missile launch warning system and Thomson-CSF EW processor and radar warning receiver. Active elements of self-protection suite include chaff/flare dispensers and IR jammer.

EQUIPMENT: Rosemount windscreen washers/wipers. NFH has 5 ton deck restraint harpoon, rescue hoist and automatic main rotor blade and tail folding.

ARMAMENT: TTH can have area suppression and self-defence armament; NFH to carry air-to-surface missiles weighing up to 700 kg (1,543 lb) and anti-submarine torpedoes (including Murène 90); air-to-air missiles optional. Italian NFH will carry Alenia Marte Mk 2/S.

DIMENSIONS, EXTERNAL:
Main rotor diameter	16.30 m (53 ft 5½ in)
Tail rotor diameter	3.20 m (10 ft 6 in)
Main rotor blade chord	0.65 m (2 ft 1½ in)
Tail rotor blade chord	0.32 m (1 ft 0½ in)
Length: overall, rotors turning	19.56 m (64 ft 2 in)
fuselage	16.14 m (52 ft 11½ in)
folded (NFH)	13.50 m (44 ft 3½ in)
Height: folded (NFH)	4.10 m (13 ft 5½ in)
overall, tail rotor turning	5.44 m (17 ft 10 in)
Width: max	4.52 m (14 ft 10 in)
over mainwheel fairings	3.63 m (11 ft 11 in)
fuselage max	2.60 m (8 ft 6¼ in)
folded (NFH)	3.80 m (12 ft 5½ in)
Tailplane half-span (stbd)	2.55 m (8 ft 4½ in)
Wheel track	3.20 m (10 ft 6 in)
Wheelbase	6.08 m (19 ft 11½ in)
Rear-loading ramp: Length	1.58 m (5 ft 2¼ in)
Width	1.78 m (5 ft 10 in)
Cabin door: Width	1.60 m (5 ft 3 in)
Height	1.50 m (4 ft 11 in)

DIMENSIONS, INTERNAL:
Cabin: Length: excl rear ramp	4.00 m (13 ft 1½ in)
incl rear ramp	4.80 m (15 ft 9 in)
Min usable width	2.00 m (6 ft 6¾ in)
Min usable height	1.58 m (5 ft 2¼ in)
Volume	18.0 m³ (636 cu ft)

AREAS:
Main rotor disc	208.67 m² (2,246.1 sq ft)
Tail rotor disc	8.04 m² (86.57 sq ft)

WEIGHTS AND LOADINGS:
Weight empty, equipped: NFH	6,428 kg (14,171 lb)
Weight empty, basic configuration	5,400 kg (11,905 lb)
Standard fuel: NFH (usable)	2,000 kg (4,409 lb)
Mission payload (both)	2,500 kg (5,512 lb)
Mission T-O weight	10,00 kg (22,046 lb)
Max T-O weight	10,600 kg (23,369 lb)
Max disc loading	50.8 kg/m² (10.40 lb/sq ft)
Transmission loading at max T-O weight and power	4.17 kg/kW (6.85 lb/shp)

PERFORMANCE (estimated; TTH at 1,000 m; 3,280 ft ISA +15°C, NFH at S/L ISA +10°C, except where stated):
Max cruising speed:	
NFH	157 kt (291 km/h; 181 mph)
TTH	160 kt (298 km/h; 185 mph)
Econ cruising speed: NFH	132 kt (244 km/h; 152 mph)
TTH	140 kt (259 km/h; 161 mph)
Max rate of climb: NFH	528 m (1,732 ft)/min
TTH	522 m (1,713 ft)/min
Rate of climb, OEI: NFH	61 m (200 ft)/min
TTH	45 m (148 ft)/min
Hovering ceiling, ISA: IGE: TTH	2,960 m (9,720 ft)
NFH	3,140 m (10,300 ft)
OGE: TTH	2,355 m (7,720 ft)
NFH	2,515 m (8,260 ft)
Radius of action with 2,000 kg (4,409 lb) load, 30 min reserves:	
TTH	160 n miles (296 km; 184 miles)
Mission loiter, 50 n miles (93 km; 58 miles) from base, 20 min reserves: NFH	3 h 20 min
Range: NFH max	490 n miles (907 km; 363 miles)
Ferry: NFH,	655 n miles (1,213 km; 753 miles)
TTH	650 n miles (1,203 km; 748 miles)
Max endurance: NFH	4 h 45 min
TTH	4 h 35 min

UPDATED

Fourth NH 90, 9890, in TTH guise with rear ramp *(Wolfgang Obrusnik/NHI)*

SATIC

SPECIAL AIRCRAFT TRANSPORT INTERNATIONAL COMPANY GIE

9 avenue Georges Guynemer, F-31770 Colomiers
Tel: (+33 5) 61 93 71 51
Fax: (+33 5) 61 93 73 33
Web: http://www.airbustransport.com
PRESIDENT: Heinz Holger Hahn

Formed in 1991, SATIC is a joint venture between DaimlerChrysler Aerospace Airbus and Aerospatiale (each holds 50 per cent), specially created to lead and co-ordinate the Airbus Super Transporter programme. Aircraft completion is undertaken by Sogerma at Toulouse. Although conceived for the transport of Airbus airliner parts, the Super Transporter is increasingly used for commercial operations.

UPDATED

SATIC A300-608ST SUPER TRANSPORTER (BELUGA)

TYPE: Outsize freighter.
PROGRAMME: Announced December 1990 as replacement for turboprop Aero Spacelines Super Guppy transporters (the last of which was retired from Airbus service on 24 October 1997); programme go-ahead November 1991; airframe contractors chosen May 1992; first basic A300-

600 sections delivered to Toulouse third quarter 1993; first nose section delivered August 1993; roll-out 23 June 1994; first flight (F-WAST, later F-GSTA) 13 September 1994; 400 hour test programme leading to DGAC certification 25 October 1995; service entry January 1996; second flew March 1996; third flew 21 April 1997; fourth flew 9 June 1998; fifth due for delivery during 2001. ETOPS qualification for 180 minutes awarded August 1997.

Airbus Transport International offered, unsuccessfully, a new-build A300-600ST (together with four standard A300-600F freighters) to meet the RAF's Short Term Strategic Airlifter requirement. (A second option in the company's bid included two A300-600STs with three standard freighters.) ATI plans to develop an onboard cargo lifter (a folding ramp-type platform) to obviate the need for crane-loading or other ground support equipment. The A300-600ST is one of very few that is able to carry a Boeing CH-47 Chinook without major disassembly (principally, removal of the tail rotor pylon), and can, in fact, carry two Chinooks.

CURRENT VERSIONS: **A300-600ST:** Actual model number 608ST; *as described.*

A340 ST: Design studies under way by 1998 for a transporter of A3XX components; based on A340-600 airliner; possible configurations range from almost standard A340 freighter to Beluga-type conversion; latter would be named **Mega Transporter**. However, by 1999 was examining the possibility of moving A3XX wings attached externally above the fuselage of a conventional A340.

CUSTOMERS: Airbus Industrie ordered five Super Transporters to transport large Airbus subassemblies between group factories. Potential market for up to 20 more has been identified. Airbus Transport International formed in October 1996 to operate the aircraft commercially; first revenue-earning (non-Airbus) load, carried 24 November 1996, was 40 tonne section of Space Station Alpha; four-aircraft fleet will have 800 hours available annually for commercial operations.

COSTS: Programme cost reported as US$1 billion; aircraft price about US$100 million.

DESIGN FEATURES: Internal dimensions determined by size of Airbus airliner major assemblies. Aiframe based on A300-600, with enlarged unpressurised upper fuselage, accessed via upward-hinging door above flight deck; fin raised and tailplane reinforced and fitted with endplate fins; pressurised flight deck set below main deck floor level permits roll-on/roll-off loading of main hold; nose door can be operated in winds up to 30 kt (56 km/h; 35 mph) and turnround times are reduced from several hours to less than 45 minutes. Cargo hold below main floor immediately aft of wing; bulk cargo compartment behind cargo hold. Maximum load cross-section 4.88 × 4.88 m (16 ft 0 in × 16 ft 0 in).

STRUCTURE: SATIC chosen as lead contractor for programme in August 1991 with overall responsibility for programme management and integration of design and manufacturing of 11 structural and 15 systems work packages; participants include: Aerostructure Hamble (main cargo door); CASA (cylindrical upper fuselage shells, tailplane modification and endplates); DaimlerChrysler Aerospace Airbus (structural modifications to vertical tail); Fairchild Dornier (hydraulic systems); Elbe Flugzeugwerke (conical upper fuselage shell); Fokker (water drainage and miscellaneous systems); Labinal (electrical harnesses); Latécoère (linings and insulations and underfloor cockpit section); L'Hoteller (air conditioning system); Ratier-Figeac (mechanical flight controls); and SOGERMA-SOCEA (final assembly).

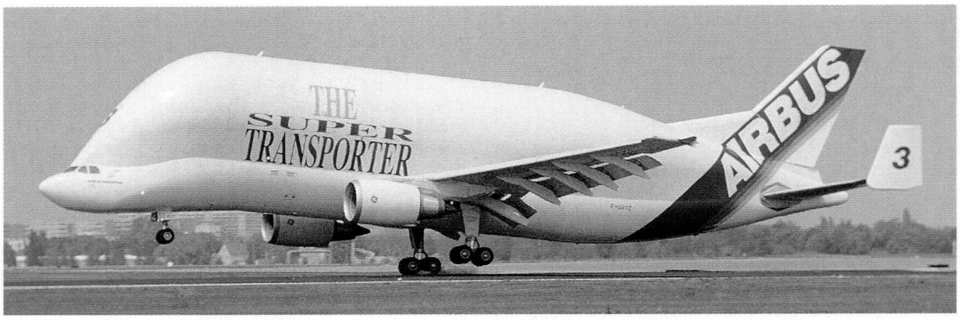

Fourth SATIC A300-608ST Super Transporter *(Paul Jackson/Jane's)* ***2001***/0099613

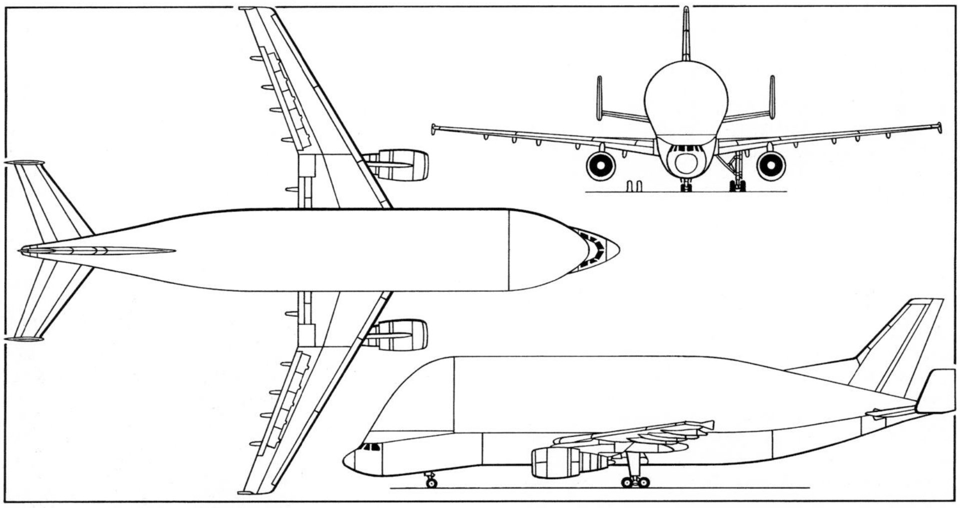
SATIC A300-608ST Super Transporter built to transport Airbus airframe sections *(Mike Keep/Jane's)* 0011276

POWER PLANT: Two General Electric CF6-80C2A8 turbofans (normally fitted to A310) giving 262.4 kN (59,000 lb st) each. Fuel capacity 62,000 litres (16,379 US gallons; 13,638 Imp gallons).

DIMENSIONS, EXTERNAL:

Wing span	44.84 m (147 ft 0 in)
Length overall	56.16 m (184 ft 3 in)
Max width of fuselage	7.31 m (23 ft 11¾ in)
Height: overall	17.24 m (56 ft 6¾ in)
of loading door, when open: at MTOW	16.78 m (55 ft 1 in)
empty	16.98 m (55 ft 8½ in)
to loading door sill: at MTOW	5.01 m (16 ft 5¼ in)
empty	5.20 m (17 ft 0¾ in)
Tailplane span	16.61 m (54 ft 6 in)
Wheeltrack	9.60 m (31 ft 6 in)
Wheelbase	18.60 m (61 ft 0¼ in)

DIMENSIONS, INTERNAL:

Cargo compartment: Length:	
usable	37.70 m (123 ft 8¼ in)
cylindrical portion*	21.34 m (70 ft 0 in)
Width: max	7.10 m (23 ft 3½ in)
at floor	5.105 m (16 ft 9 in)
Height: max	7.10 m (23 ft 3½ in)
Volume	1,400 m³ (49,441 cu ft)
Bulk cargo compartment volume	17.3 m³ (611 cu ft)

**rearwards from freight door hinge*

WEIGHTS AND LOADINGS:

Weight empty	86,500 kg (190,700 lb)
Max payload	47,000 kg (103,617 lb)
Max T-O weight	155,000 kg (341,700 lb)
Max landing weight	140,000 kg (308,645 lb)
Max zero-fuel weight	133,500 kg (294,315 lb)

PERFORMANCE:

Max cruising speed	M0.7 (421 kt; 780 km/h; 484 mph)
Max certified altitude	10,670 m (35,000 ft)
T-O run	1,950 m (6,397 ft)
Landing run	1,176 m (3,858 ft)
Range: with max payload	900 n miles (1,666 km; 1,035 miles)
with 40 tonne payload	1,500 n miles (2,778 km; 1,726 miles)
with 26 tonne payload	2,500 n miles (4,630 km; 2,876 miles)

UPDATED

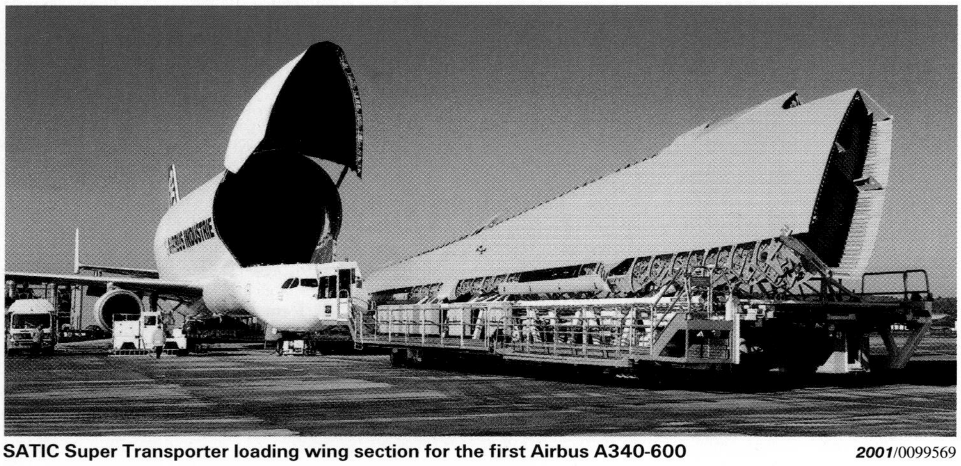
SATIC Super Transporter loading wing section for the first Airbus A340-600 ***2001***/0099569

SEPECAT

SOCIETE EUROPEENNE DE PRODUCTION DE L'AVION E. C. A. T.

PARTICIPATING COMPANIES:

BAE Systems: see under UK

Dassault Aviation: see under France

PRESIDENT: J P Weston (BAE)

VICE-PRESIDENT: R Dubost (Dassault)

Anglo-French company formed May 1966 by Breguet Aviation and British Aircraft Corporation to design and produce Jaguar strike fighter/trainer; production now in India only (see HAL entry); RAF upgrade described here; further data in *Jane's Aircraft Upgrades*.

VERIFIED

SEPECAT JAGUAR

Details of the Jaguar International export version appear in the Indian section (which see); data particular to RAF GR. Mk 1A last appeared under SEPECAT heading in the 1993-94 *Jane's*.

The RAF recently upgraded 11 Jaguars under a UOR (urgent operational requirement) to what was then known as GR. Mk 1B (nine) and T. Mk 2B (two) standard with provision for BAE Systems TIALD target designation pods, MIL-STD-1553B databusses and appropriate cockpit modifications (see *Jane's Aircraft Upgrades*). Trial installation Mk 1B XX748 first flew on 11 January 1995 and was handed over to the RAF on 24 February, the retrofit also involving addition of a Smiths multipurpose colour HDD and a GPS antenna ahead of the windscreen for a FIN 1075G combined INS/GPS. Two Mk 1Bs deployed to Gioia del Colle, Italy, on 27 May 1995, for operational trials over former Yugoslavia alongside No. 6 Squadron detachment. Version subsequently used by all three RAF front-line Jaguar squadrons.

Also in 1995, BAe was awarded a contract to integrate its Terprom ground proximity warning system as part of the

Jaguar 96 upgrade which saw modified aircraft wired for, but not necessarily equipped with, TIALD. Other Jaguar 96 provisions are GEC wide-angle HUD, new hand controller and stick-top, Sky Guardian 200-13PD RWR and similar Granby upgrades as incorporated for the 1991 Gulf War (Mk XII 4 IFF, dual 'Have Quick' frequency-hopping radios and overwing AAM launch rails), plus a colour LCD display to replace the existing moving map (TIALD aircraft only), and video recorders for TIALD pod imagery. First Jaguar 96 (XX738, originally planned as the last Jaguar GR. Mk 1B) flew in January 1996. NVG-compatible cockpit lighting and covert external lighting kits have been procured for fleet-wide incorporation.

Following its formal release to service, due at the end of August 1998 but delayed until mid-1999, the Jaguar 96 was redesignated **GR. Mk 3** (single-seat) and **T. Mk 4** (two-seat). The lack of a full Military Aircraft Release (MAR) did not prevent unrestricted front-line operation of the Jaguar 96, but it was only after MAR that No. 16 (Reserve) Squadron began training students *ab initio* on the GR. Mk 3 and T. Mk 4. Two-seater is not operationally capable, but has old MPCDs from GR. Mk 1B and TIALD provision. First Mk 4, XX150, made its maiden flight on 11 November 1998.

A further upgrade, **Jaguar 97**, is making all single-seat Jaguars capable of reconnaissance and TIALD operations. Additional features include a 152 × 203 mm (6 × 8 in) active

Test Jaguar for Adour Mk 106 engine making its first flight on 7 June 2000 *2001/0034518*

matrix LCD, new digital map/symbol generator, BAE Systems helmet-mounted sight and fully integrated mission planner, electronic FRCs and TAPs giving 'paper-tidy' cockpit. Sole externally obvious change is addition of a head position transmitter in the top of the canopy. Electrical and wiring provision will also be made for ASRAAM. Trial installation (TI) aircraft (XZ399) flew on 4 August 1998 followed by production installation (PI) (XZ116). Designated **GR. Mk 3A** following official release to service on 14 January 2000. First aircraft delivered to No 41(F) squadron 18 January. **T. Mk 4A** designation reserved for as yet unfunded Jaguar 97 version of two-seater.

Interim **Jaguar 96½** features integration of the new mission planner, Tacan and bullseye display in the HUD, GPS height integration, GPS timing, map symbology overlays and some minor weapon aiming display changes. This modification amounts to no more than a software upgrade, and was trialled on XZ369.

In parallel, 82 remaining Jaguars, including some in storage, may be re-engined with Adour Mk 106 power plants, offering reduced operating costs, 10 per cent higher standard ratings and a further 19 per cent combat boost. First trial installation (one Mk 106 only) flown in XX108, 7 June 2000. Other parallel programmes include incorporation of the Series 400 TIALD pod and Vinten Vicon 18 Series 601-GP (1) electro-optical reconnaissance pod. The latter began trials flying in early July 1999 and is replacing the original BAE pod during 2000-2001.

Future upgrades could include improved ECM (with TRD), 'staring' FLIR, IDM, new weapons and the integration of Sky Guardian on the 1553 bus. IDM and ASRAAM have been flown on the DERA 'Nightcat Jaguar' T. Mk 2A XX833. The Jaguar 97 upgrade has also been applied to Omani Jaguars.

VERIFIED

SUPERSONIC AIRLINER STUDIES

Although there is no programme under way to develop a second-generation supersonic transport, possibilities are being studied by several companies and consortia. Cost and the size of the market make it virtually certain that large consortia will form to develop any new SST. Below are the moves so far towards a firm advanced SST programme; supersonic business jets are currently being studied by Sukhoi of Russia and jointly by Gulfstream and Lockheed Martin.

UPDATED

ADVANCED SUPERSONIC AIRLINER

TYPE: Supersonic airliner.
PROGRAMME: Work on second-generation SST begun by Aerospatiale and French national institute ONERA in June 1989; SNECMA and Rolls-Royce began joint engine studies also in 1989; MTU and FiatAvio joined later; Aerospatiale and BAe started joint study in April 1990; SCT (supersonic commercial transport) Group started with Boeing, McDonnell Douglas and DASA Airbus in May 1990; Anglo-French team joined SCT to form Supersonic Commercial Transport International Co-operative Study Group, nicknamed Group of Five; objective to study market and environmental problems; Alenia, Tupolev and Japan Aircraft Development Corporation (Mitsubishi, Kawasaki and Fuji) joined during 1991; P&W and GE working together on engine problems; Phase 1 market, certification, noise and environmental study completed by mid-1993 and Phase 2 launched to study technical and economic feasibility. In a separate move, Gulfstream/Lockheed-Martin is conducting a joint study into supersonic business jets.

NASA launched two High-Speed Research (HSR) programmes. Six year Phase 1 costing US$450 million started in 1990 and examined environmental acceptability. Phase 2 technology programme (US$1.9 billion), started in 1993, is lasting nine years and examining all critical aerodynamic, propulsion and technology areas. Phase 2A, now not to be funded, was to examine building materials and processes. However, the programme ran into noise level problems resulting in NASA extending the projected in-service date for a US HSCT to around 2020 to allow development of engine technology to progress to a point where noise levels would meet Stage IV requirements, that

Japanese design for an SST *2001/0099567*

is, around 10 dB below Stage III requirements, without disproportionate engine weight. NASA initiative of 1999 allocated US$9.6 million to live fire tests (on an SR-71 testbed) of M3.0 pulse-detonation power plant.

As an adjunct to the HSR programme, Northrop Grumman has taken out a patent on a reverse delta-wing layout, incorporating laminar flow control and two pairs of engines in a stacked layout. As yet, Northrop Grumman has not initiated further research into the design.

NASA revealed the Technical Concept Airplane (TCA) in April 1996. TCA is destined to remain a study only, but is judged representative of a viable SST, its evolution being useful in identifying the technologies required for a commercial venture. Status by mid-1997 included mixed-flow turbofans (rather than a variable-cycle engine) fed from axisymmetric inlets; titanium/composites wing structure; fuselage of either the same material or powder metal; wing planform of combined double delta with cranked arrow; crew external vision system and no front visor; wing span 40 m (130 ft); length 98 m (320 ft); 300 seats in three-class layout; cruising speed of M2.4 on four 220 kN (49,500 lb st) engines; 340,200 kg (750,000 lb) maximum T-O weight; and 5,000 n mile (9,260 km; 5,750 mile) range.

US Defense Advanced Research Projects Agency (DARPA) has been allocated US$35 million over two years to develop a high-speed reconnaissance platform using new technologies and layouts, including straight thin wings, which could have commercial applications.

Europe is not providing corresponding funding for SST research, leaving its manufacturers with prospect of subsidiary status in eventual programme. BAE Systems' studies show new SST should have, by comparison with Concorde, 27.5 per cent better cruise lift/drag ratio, 40 per cent lower specific weight and 10.5 per cent lower specific fuel consumption. Maximum T-O weight 317,500 kg (700,000 lb); wing area 780 m² (8,400 sq ft); range with 250 to 300 passengers in three classes: 5,500 n miles (10,175 km; 6,329 miles).

In Japan, National Aerospace Laboratory (NAL, which see) spent 21 per cent of its budget on SST research during late 1990s. Parallel studies are under way by Ministry of International Trade and Industry (MITI), with industrial

Common configuration agreed by Aerospatiale Matra, BAE Systems and DaimlerChrysler Aerospace for the European Supersonic Research Programme (ESRP) advanced supersonic transport *2001/0099640*

teams from Fuji, Ishikawajima-Harima, Kawasaki, Mitsubishi and Nissan. NAL working on eight year, ¥20 billion programme, of which first phase is one-tenth scale, unpowered X-SST-R supersonic glider (11.50 m; 37 ft 8¾ in long, with 4.72 m; 15 ft 5¾ in span cranked-arrow wing) to be ground-launched by rocket in 2001. Same-size X-SST-J demonstrator, with two 4.45 kN (1,000 lb st) class Teledyne (TCAE) J69 turbojets, but still rocket-launched, will fly in 2002. MITI work concentrates on materials and propulsion, including a combined-cycle engine (Hypersonic Transport Propulsion System Research – HYPR) for speeds between M3.0 and M5.0. MITI has also begun funding a five-year follow-on programme to HYPR, which finished in March 1999; the new programme will investigate turbofans for speeds between M2.0 and M3.0.

In May 2000 it was announced that NAL was working in conjunction with the Russian CAHI (TsAGI) on supersonic research, and that Japan was considering the lease of the Tu-144LL testbed.

CURRENT VERSIONS: **Alliance:** Name defines joint BAE Systems/Aerospatiale Matra/ONERA/SNECMA study for Concorde-style 250- to 300-passenger SST with range of 5,935 n miles (11,000 km; 6,835 miles); double delta wing with high-aspect ratio outer panels.

European Supersonic Research Programme (ESRP): Current title of joint European research effort designed to strengthen European position in the face of the US-only HSR programme described above. Aerospatiale Matra, BAE Systems (then British Aerospace) and Daimler-Benz Aerospace (now DaimlerChrysler Aerospace) signed MoU in April 1994 for assessment of technical and economic viability of ESRP, and study of key technological issues such as materials, aerodynamics, systems and power plant integration. In mid-1999, a market for 500 to 1,000 aircraft envisaged, with development costs of around US$15 billion; the consortium plans to produce the first prototype by 2010; Comité d'Orientation sur le Supersonique (COS) inaugurated to investigate problems such as high temperatures and fuel emissions on ESRP, and was to report to the French government (which has commissioned the study) in June 2000; this was delayed to end 2000. Dimensions of ESRP/PERS configuration: fuselage length 89 m (292 ft 0 in), wing span 42 m (138 ft 0 in), take-off weight 340,000 kg (750,000 lb), cruise speed M2.05, range in excess of 5,400 n miles (10,000 km; 6,200 miles).

Tu-244: Tupolev designation for its own SST study as successor to original Tu-144; model first shown at Paris Air Show 1993; 300 seats; range 4,975 n miles (9,200 km; 5,725 miles); maximum T-O weight 350,000 kg (771,625 lb); four engines of 324 kN (72,750 lb st) each.

ASCT: Advanced Supersonic Commercial Transport; generic name for various US and Japanese studies. See also JADC and NAL entries in Japanese section.

HSCT: Boeing (and, previously, McDonnell Douglas) uses this general designation (High-Speed Civil Transport) for a future SST. Details of Douglas HSCT proposal, released shortly before 1997 acquisition by Boeing, included transport of 300 passengers at M2.4 over 5,000 n miles (9,260 km; 5,753 miles); fuel capacity 222,330 litres (59,000 US gallons; 49,128 Imp gallons); weight empty, equipped, 136,985 kg (302,000 lb); maximum zero-fuel weight 167,375 kg (369,000 lb); maximum T-O weight 341,555 kg (753,000 lb); wing span 39 m (128 ft); length overall 102 m (334 ft); and height overall 17 m (56 ft). US research is being assisted by flight trials of a refurbished Tupolev Tu-144 first-generation SST (which see in the Russian Federation Section of 1998-99 and earlier *Jane's*). Talks opened in April 1997 on possibility of further Tupolev participation in US SST; company advising Boeing on improving aerodynamics, strength of fuselage and landing gear. Boeing announced that it was slowing supersonic research in late 1998, while GE and Pratt & Whitney continue research into engine development.

CUSTOMERS: World market originally estimated as US$125-250 billion up to 2025, representing 500 to 1,000 aircraft. However, by 1997, US studies indicated 1,000 to 1,500 aircraft market. No new estimates published since loss of a Concorde in July 2000, although public interest in supersonic air travel maintained despite accident.

DESIGN FEATURES: Generally discussed specifications include cruising speed M2.05 chosen to allow cruising at lower altitudes, where emissions are not so critical; fare surcharge not more than 10 to 20 per cent for three-class layout.

POWER PLANT: Could be four dual-cycle engines meeting normal FAR Pt 36 Stage 3 noise levels; nitrous oxide emission reduced by 80 per cent; dual-cycle concept under study by SNECMA, Rolls-Royce, MTU and FiatAvio is designated Mid Tandem Fan; promises 15 to 20 dB noise reduction at take-off; fuel consumption 0.043 kg/seat/km (0.175 lb/seat/n mile); MTF engine has fan operating in bypass duct mounted between the low- and high-pressure compressor spools. GE has proposed a variable cycle engine.

UPDATED

UHCA/VLCT
ULTRA-HIGH CAPACITY AIRLINER/VERY LARGE COMMERCIAL TRANSPORT

ACTIVE COMPANIES:
 Aerospatiale Matra: see under France
 Airbus Industrie: see in this section
 BAE Systems: see under UK
 Boeing Commercial Airplane: see under USA
 CASA: see under Spain
 DaimlerChrysler Aerospace: see under Germany
 Japan: see Mitsubishi, Kawasaki and Fuji under Japan
 Sukhoi: see under Russia

Only Airbus Industrie has announced firm plans to develop an airliner carrying more than 500 passengers; other companies are progressing more cautiously, especially as their researches have indicated the possibility that less conventional shapes of airliner are better suited for higher capacity. Some recent studies are described and illustrated below.

VERIFIED

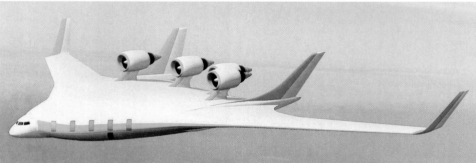
Aerospatiale Matra VHCT (very high capacity transport) study of 1999

ULTRA-HIGH CAPACITY AIRLINER/VERY LARGE COMMERCIAL TRANSPORT

TYPE: New-concept airliner.

PROGRAMME: Airliners for more than 500 passengers; also known as New Large Airplane. Launched in early 1990s, aimed initially at trans-Pacific and southeast Asian region; MoU signed 26 January 1993 between Airbus, Boeing, Aerospatiale (now Aerospatiale Matra), Daimler-Benz (now DaimlerChrysler) Aerospace, British Aerospace (now BAE Systems) and CASA to launch a one year general study of various aspects of UHCA; Airbus Industrie joined programme in first quarter 1994 in an advisory role; programme extended to 1995, beginning to study technical issues. Formal report, 10 July 1995, confirmed technical feasibility, but deferred launch for lack of market.

CURRENT VERSIONS: Each Airbus partner company proposed outline designs, including ASX 500 and 600 from Aerospatiale and A 2000 from Daimler-Benz Aerospace Airbus. Boeing proposals have included double-deck and stretched developments of 747 and the New Large Airplane (NLA), but focused on stretched 747 derivatives by 1996. However, Boeing abandoned initial pair of stretched 747s (-500X and -600X) in January 1997, leaving the field clear for Airbus Industrie's joint proposal, designated A3XX (see entry in this section). Boeing apparently reconstituted the NLA design team in 1997, although not to pursue concepts obtained on acquisition of McDonnell Douglas; it has since confirmed it is studying new concepts under title of Large Airplane Product

Conceptual Megaliner study, 1999-2002, by Germany's Bundesministerium für Wirtschaft und Technologie *(Paul Jackson/Jane's)*

Development, using baseline figures of 421,850 kg (930,000 lb) MTOW and 8,000 n miles (14,800 km; 9,200 miles) range. Stretched 747 again under development (which see), but with lower capacity.

Both Boeing and Airbus UHCA designs have been based on conventional-shaped aircraft carrying 450 to 650 passengers. Although prospects have been mentioned for

CAHI/TsAGI concept of mid-1990s for a flying wing airliner *(Paul Jackson/Jane's)*

Hybrid CAHI/TsAGI design of 1999, featuring circular fuselage cross-section *(Paul Jackson/Jane's)*

later stretching to nearer 1,000 seats, the view has been expressed by an increasing number of designers that the flying wing may offer better characteristics for capacities above 800 passengers. One drawback, however, will be the weight penalty associated with the reinforcement of an irregularly shaped passenger compartment to withstand pressurisation.

During 1995, Airbus revealed a design for a 96 m (315 ft) span flying wing for this number of seats and a parallel study (FW-900) undertaken with the collaboration of Russia's CAHI (TsAGI) for a similar aircraft of 104 m (341 ft) span, carrying 938 persons. These employed four and three engines, respectively, and had total installed power of 1,765 kN (396,780 lb st) and 1,620 kN (364,180 lb st). Aerospatiale Matra continues research into the former, which by mid-1999 had been refined into a 600,000 kg (1,322,775 lb) 1,000-seater, with length of 54 m (177 ft) and wing thickness of 7 m (23 ft). Range is estimated as 6,500 n miles (12,040 km; 7,480 miles) at M0.85. The design has accommodation on two decks, with 800 passengers on upper and 200 on lower. Service entry is envisaged around 2020. Airbus confirmed in 1999 that it is examining new configurations for airliners, but noted that these may also be employed for aircraft with less than 500 seats.

In 1996, shortly before absorption by Boeing, McDonnell Douglas released details of the BWB-1-1 (blended wing body) study giving the most comprehensive analysis yet of the flying wing's advantages for very large transports. For 800 passengers over a 7,000 n mile (12,964 km; 8,055 mile) range, assets included a 15.2 per cent reduction in T-O weight and 12.3 per cent less empty weight; 27.5 per cent reduction in fuel burn; 27.0 per cent reduction in installed thrust; and 20.6 per cent better lift:drag. These are achieved by increasing wing area by 27.9 per cent (reducing loading by 33.6 per cent) and having a 19.2 per cent greater span. Tests of a radio-controlled 5 m (17 ft) span scale BWB-1-1 were undertaken by Stanford University, California, at El Mirage dry lake in July 1997 as part of a US$2.3 million programme to evaluate flight control laws for a flying wing. This, it was hoped, would lead to a US$100 million programme, involving NASA, to fly two 25 per cent scale flying wing airliners by 1999. Service entry could be achieved by 2010. Boeing (assisted by US$1.5 million NASA funding) expects to test a 14.3 per cent scale model of the BWB-1-1 with a 10.70 m (35 ft 1¼ in) span during 2002 to study control and aerodynamic features. In 1999, Boeing and NASA built a carbon fibre wing section, representative of a BWB and weighing 25 per cent less than a comparable metal structure, for static testing.

By 1997, CAHI had progressed to the LK 900 four-engine design (343 kN; 77,160 lb st GE 90, PW4082, Trent 580 or Samara NK-44), still for 938 passengers, which would cruise at M0.8 over a range of 5,400 n miles (10,000 km; 6,210 miles) operating from a 3,200 m

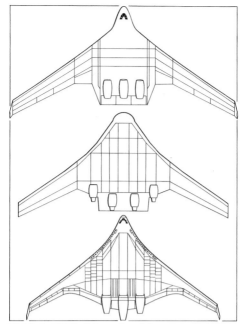

Mid-/late-1990s flying wing airliner proposals from Airbus-CAHI (TsAGI) (top), Airbus and Boeing (McDonnell Douglas) *(James Goulding/Jane's)*

(10,500 ft) runway. Cruise lift-to-drag ratio was estimated as 25 and fuel consumption 30 per cent less than a conventional design. Weight would be 560,000 kg (1,234,600 lb); span 106 m (348 ft); and wing area 1,700 m² (18,300 sq ft). Further research in 1999 turned to wind tunnel tests of a more efficient hybrid blended wing model, reverting to a circular cross-section fuselage and with its podded engines in the more traditional position beneath the wings; would transport 750 passengers over 7,560 n miles (14,000 km; 8,700 miles) and weigh 600,000 kg (1,322,755 lb); lift-to-drag ratio 24.5.

Sukhoi has entered the field with its KR-860 (see Russian section), while Myasishchev included large-capacity airliners in its GP-60 series of design studies, as last described in the 1999-2000 *Jane's*.

CUSTOMERS: Market for large and ultra-large airliners over 500 seats between 1996 and 2015 estimated by Boeing as 470, and by other significant airframe and engine manufacturers as between 500 and 745; however, Airbus Industrie was alone in 1997 in predicting 1,330.

COSTS: Boeing estimated US$7 billion for 747 development to 550 seats; Airbus predicting only US$8 billion for the all-new A3XX, or US$10 billion to 12 billion, for a three-size family; flying wing designs not costed.

DESIGN FEATURES: Design objective is to carry from 500 to 800 passengers over a distance of about 7,000 n miles (12,964 km; 8,055 miles). Engine noise, economy and pollution at the powers envisaged represent the most difficult limitations on the design; wake vortex must not impose burdens on operations of other aircraft; direct operating costs will have to be 15 per cent lower than those of present large airliners.

FLYING CONTROLS: Large size and inertia of the aircraft will demand special flying controls and combinations of control surfaces and very high actuator powers. Shape memory alloys (SMA), being developed for civil and military applications in the USA, hold promise of hingeless control surfaces and lift optimisation through wing twisting at different stages of flight.

STRUCTURE: New materials, such as Glare fibre/metal laminate, will be necessary to control the weight of the UHCA; because the aircraft will be built by a consortium of companies, large assemblies will have to be transported from factory to factory, requiring major structural joints.

*For technical specification of a conventional configuration, 555-seat UHCA/VLCT, see Airbus A3XX; estimated data below refer to the 800-seat **Boeing BWB-1-1** proposal powered by three 275 kN (61,900 lb st) turbofans.*

DIMENSIONS, EXTERNAL:
Wing span: excl winglets	85 m (280 ft)
incl winglets	88 m (289 ft)
Wing aspect ratio	5.1
Length overall	49 m (161 ft)
Height overall	15 m (50 ft)

AREAS:
Wings: trap	728 m² (7,840 sq ft)
gross	1,423 m² (15,325 sq ft)

WEIGHTS AND LOADINGS:
Weight empty	167,750 kg (369,800 lb)
Weight empty, equipped	186,900 kg (412,000 lb)
Max payload	104,800 kg (231,000 lb)
Max fuel weight	122,500 kg (270,000 lb)
Max T-O weight	373,300 kg (823,000 lb)
Max zero-fuel weight	291,650 kg (643,000 lb)
Fuel burn over 7,000 n miles with 800 passengers	96,820 kg (213,450 lb)
Max wing loading	513 kg/m² (105 lb/sq ft)
Max power loading	452 kg/kN (4.4 lb/lb st)

PERFORMANCE:
Normal cruising speed	M0.85
Max approach speed	150 kt (278 km/h; 173 mph) EAS
Initial cruising altitude	10,665 m (35,000 ft)
T-O field length	3,353 m (11,000 ft)
Range with 800 passengers	7,000 n miles (12,964 km; 8,055 miles)

UPDATED

IRAN

AII

AVIATION INDUSTRIES OF IRAN
Km 8, Karaj Old Road (PO Box 14195-111), 15815 Tehran
Tel: (+98 21) 680 92 69 to 680 92 73
Fax: (+98 21) 680 92 68
MANAGING DIRECTOR: Dr S H Pourtakdoust
COMMERCIAL DIRECTOR: A Sardari

AII was formed in April 1993. It is affiliated to the Industrial Development and Renovation Organisation (IDRO) of the Iranian Ministry of Industries to produce, repair and maintain various types of aircraft. It flew its first powered aircraft, the AVA-202 two-seat trainer, in mid-1997; other known projects include the AVA-505 small helicopter and the AVA-404 short-haul transport. A two-seat sailplane, the IR-G1 (now AVA-101) Nasim, made its first flight on 30 May 1996 and received its domestic type certificate in 1997. Most recent project is for an agricultural aircraft designated AVA-303. Design was to have been started in 1999, but no details had been received by October 2000.

UPDATED

AII AVA-202
TYPE: Primary prop trainer/sportplane.
PROGRAMME: Intended for domestic market to avoid dependence on foreign imports; prototype (originally known as IR-02: see 1997-98 *Jane's*) made first flight on 3 June 1997; flight test programme continued through 1998; domestic certification was expected during 1999, but had not been confirmed by October 2000.

DESIGN FEATURES: Similar configuration to Van's RV-6A (US section), but with greater wing span; meets JAR-VLA standards.

NASA aerofoil sections: 63_2-215 (wing), 63_2-015 (horizontal tail) and 63-010 (vertical tail).

FLYING CONTROLS: Conventional and manual. Flaps. Trim tab in port elevator

STRUCTURE: Mainly aluminium alloy; steel alloys and glass fibre in some areas.

LANDING GEAR: Non-retractable tricycle type. Self-sprung cantilever legs; wheel spats.

Prototype AII AVA-202 trainer

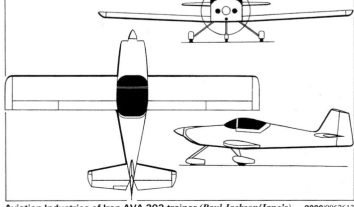

Aviation Industries of Iran AVA-202 trainer *(Paul Jackson/Jane's)*

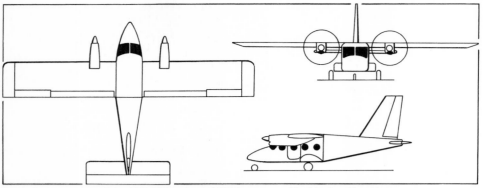

General arrangement of the AVA-404 projected STOL transport *(Paul Jackson/Jane's)*

POWER PLANT: One Textron Lycoming O-320-B2B flat-four engine (119 kW; 160 hp at 2,700 rpm), driving a Sensenich M 74 DM two-blade propeller.
ACCOMMODATION: Two seats side by side. Wraparound windscreen; one-piece rearward-sliding canopy.
DIMENSIONS, EXTERNAL:
Wing span	8.77 m (28 ft 9¼ in)
Wing aspect ratio	7.1
Length overall	6.10 m (20 ft 0 in)
Height overall	2.36 m (7 ft 9 in)

AREAS:
Wings, gross	10.87 m² (117.0 sq ft)
Vertical tail surfaces (total)	1.04 m² (11.19 sq ft)
Horizontal tail surfaces (total)	2.40 m² (25.83 sq ft)

WEIGHTS AND LOADINGS:
Weight empty	449 kg (990 lb)
Max T-O weight	730 kg (1,609 lb)
Max wing loading	67.2 kg/m² (13.75 lb/sq ft)
Max power loading	6.12 kg/kW (10.06 lb/hp)

PERFORMANCE (estimated):
Max level speed at S/L	140 kt (259 km/h; 161 mph)
Service ceiling	6,400 m (21,000 ft)
Range at 3,050 m (10,000 ft) at 65% power	579 n miles (1,073 km; 666 miles)

UPDATED

AII AVA-404

TYPE: Twin-turboprop light transport.
PROGRAMME: In design stage since 1996; no manufacturing timetable announced by October 2000.
DESIGN FEATURES: High-wing monoplane with constant-chord wings and horizontal tail; large, sweptback fin and rudder. Designed to conform to FAR Pts 23 and 135 for VFR operation. Future options to include pressurised fuselage and stretched commuter version.
STRUCTURE: Mainly of composites.
LANDING GEAR: Retractable tricycle type.
POWER PLANT: Two 559 kW (750 shp) Walter M 601E turboprops, each flat rated at 440 kW (590 shp) and driving a three-blade propeller.
ACCOMMODATION: One or two pilots and eight to 11 passengers standard; future commuter, up to 19 passengers.
DIMENSIONS, EXTERNAL:
Wing span	16.80 m (55 ft 1½ in)
Wing aspect ratio	7.0
Length overall	10.40 m (34 ft 1½ in)
Height overall	5.08 m (16 ft 8 in)

DIMENSIONS, INTERNAL:
Cabin: Max width	1.76 m (5 ft 9¼ in)
Max height	1.88 m (6 ft 2 in)

AREAS:
Wings, gross	40.32 m² (434.0 sq ft)

WEIGHTS AND LOADINGS:
Max fuel weight	1,640 kg (3,616 lb)
Max payload	775 kg (1,708 lb)
Max T-O weight	5,080 kg (11,199 lb)
Max wing loading	126.0 kg/m² (25.81 lb/sq ft)
Max power loading	5.78 kg/kW (9.49 lb/shp)

PERFORMANCE (estimated, up to 45°C; 113°F):
Max level speed at S/L	204 kt (378 km/h; 235 mph)
Max cruising speed at 3,050 m (10,000 ft) at 86% power	212 kt (393 km/h; 244 mph)
Stalling speed: flaps up	80 kt (147 km/h; 91 mph)
flaps down	64 kt (118 km/h; 73 mph)
T-O run	412 m (1,350 ft)
Landing run	183 m (600 ft)
Range at 3,050 m (10,000 ft) at 86% power cruising speed	412 n miles (764 km; 475 miles)
Endurance, conditions as above	2 h 24 min
g limits	+3.8/−1.52

UPDATED

AII AVA-505 THUNDER

TYPE: Five-seat helicopter.
PROGRAMME: In design stage by 1996, under former designation of IR-H5; first flight then planned in 1999, but not confirmed by October 2000.
DESIGN FEATURES: Designed for use in hot/high conditions and to have low cost of operation. Two-blade main and tail rotors; pod fuselage; tailboom with mid-mounted horizontal stabiliser; swept upper and lower tailfins. Proposed applications include passenger and light cargo transport; aerial ambulance and disaster relief; mapping, forest protection and cropspraying; urban and road traffic patrol; firefighting and rescue. Able to carry up to 300 kg (661 lb) load on external sling. To conform to FAR Pt 27.
STRUCTURE: Mainly aluminium alloy.
LANDING GEAR: Twin-skid type.
POWER PLANT: One 242 kW (325 hp) piston engine (unidentified), with turbocharger; piston diesel engine optional. Fuel capacity 284 litres (75 US gallons; 62.5 Imp gallons).
ACCOMMODATION: Two seats side by side in front, with bench seat for three more persons at rear. Dual controls optional. Two forward-hinged doors each side. Baggage compartment aft of rear seats, access via door on port side.
DIMENSIONS, EXTERNAL:
Main rotor diameter	10.18 m (33 ft 4¾ in)
Tail rotor diameter	1.65 m (5 ft 5 in)
Fuselage length	8.64 m (28 ft 4¼ in)
Height to top of rotor head	2.47 m (8 ft 1¼ in)
Skid track	1.92 m (6 ft 3½ in)

AREAS:
Main rotor disc	81.39 m² (876.1 sq ft)
Tail rotor disc	2.14 m² (23.02 sq ft)

WEIGHTS AND LOADINGS:
Weight empty	850 kg (1,874 lb)
Max T-O weight: internal load	1,450 kg (3,196 lb)
with sling load	1,500 kg (3,307 lb)
Max disc loading:	
internal load	17.8 kg/m² (3.65 lb/sq ft)
with sling load	18.4 kg/m² (3.77 lb/sq ft)

PERFORMANCE (estimated, at normal internal load MTOW except where indicated):
Never-exceed (VNE) and max level speed	119 kt (220 km/h; 137 mph)
Max cruising speed	113 kt (209 km/h; 130 mph)
Max rate of climb at 4,575 m (15,000 ft):	
ISA	515 m (1,690 ft)/min
ISA + 20°C	506 m (1,660 ft)/min
Service ceiling	5,485 m (18,000 ft)
Hovering ceiling IGE: ISA	5,210 m (17,100 ft)
ISA + 20°C	4,905 m (16,100 ft)
Hovering ceiling OGE at AUW of 1,361 kg (3,000 lb):	
ISA	5,000 m (16,400 ft)
ISA + 20°C	4,695 m (15,400 ft)
Range at S/L, no reserves	415 n miles (770 km; 478 miles)
Endurance at 69 kt (129 km/h; 80 mph) at 1,525 m (5,000 ft), no reserves	6 h

UPDATED

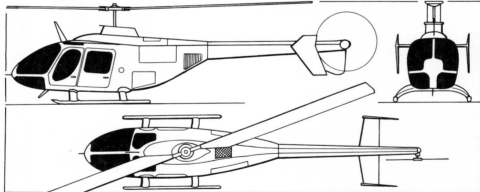

AVA-505 Thunder light helicopter *(Paul Jackson/Jane's)*

DORNA

H F DORNA COMPANY

20 6th Sarvestan Street (PO Box 16315-345), Pasdaran Avenue, Tehran 16619
Tel: (+98 21) 285 48 27
Fax: (+98 21) 284 18 31
e-mail: drna@kavosh.net
Web: http://www.hfdorna.com
MANAGING DIRECTOR: Yaghoob Antesary

Company established March 1989 (see 1991-92 *Jane's*) to specialise in aircraft design and development, and in composites materials technology. It has 3,000 m² (32,300 sq ft) of factory space and had a 1998 workforce (latest figure provided) of 79. Current product is two-seat Blue Bird.

UPDATED

DORNA D139-PT1 BLUE BIRD

TYPE: Two-seat lightplane.
PROGRAMME: Launched 1994; prototype manufacture started mid-1994; first flight was made on 27 July 1998; prototype had accumulated 32 hours' flying in 45 flights by end of November that year. Production was scheduled to begin in February 1999, but not confirmed by October 2000. To be certified by Iranian FAA initially, with full European JAA certification later.
CUSTOMERS: Approximately 30 ordered by 1 December 1998 (latest information received); customer(s) not identified.
COSTS: Programme cost by November 1998 estimated at US$1.5 million excluding R&D; basic aircraft then US$110,000 with two-blade (US$115,000 with three-blade) variable-pitch propeller.
DESIGN FEATURES: Conventional low-wing, fixed-gear lightplane, designed for low-cost, easy-to-fly/maintain operation.
Wings have constant chord, NACA 63-215 aerofoil section, 2° 30′ dihedral from roots, 2° incidence and −2° twist.
FLYING CONTROLS: Conventional and manual. Ailerons and elevators actuated by pushrods; rudder by cables; elevators have electric trim. Plain flaps.
STRUCTURE: All-composites. Two-spar wing, single-spar horizontal tail, monocoque fuselage.
LANDING GEAR: Non-retractable tricycle type, with Goodyear 5.00-5 tyres (6 ply on mainwheels, 4 ply on nosewheel). Cantilever self-sprung steel leg on each unit. Cleveland mainwheel brakes. Composites speed fairings on all three wheels.
POWER PLANT: One 84.6 kW (113.4 hp) Rotax 914 F3 flat-four engine driving an MT Propeller MTV-21-A/175-05 two-blade variable-pitch propeller at 5,800 rpm; three-blade MTV-6-A/172-08 variable-pitch propeller optional. Fuel tank in each wing, combined capacity 128.7 litres (34.0 US gallons; 28.3 Imp gallons); gravity filling point on each tank. Oil capacity 3.0 litres (0.8 US gallon; 0.7 Imp gallon).
ACCOMMODATION: Two seats side by side, with baggage compartment aft of seats. Narrow hardtop roof to cockpit; access via upward-opening window each side, with smaller non-opening window to rear.
SYSTEMS: Electrical system: 12 V 20 A integral AC generator, 12 V 40 A external alternator and 35 Ah battery.
AVIONICS: *Comms:* Radio.
Instrumentation: Conventional VFR.
DIMENSIONS, EXTERNAL:
Wing span	9.30 m (30 ft 6¼ in)
Wing chord, constant	1.17 m (3 ft 10 in)
Wing aspect ratio	7.9
Length overall	6.00 m (19 ft 8¼ in)
Fuselage max width	1.00 m (3 ft 3¼ in)
Height overall	1.96 m (6 ft 5¼ in)
Tailplane span	2.25 m (7 ft 4½ in)
Wheel track	1.59 m (5 ft 2½ in)

Wheelbase	1.50 m (4 ft 11 in)
Propeller diameter	1.70 m (5 ft 7 in)
Propeller ground clearance	0.18 m (7 in)
DIMENSIONS, INTERNAL:	
Cockpit: Max width	1.02 m (3 ft 4¼ in)
Max height	1.02 m (3 ft 4¼ in)
AREAS:	
Wings, gross	10.88 m² (117.1 sq ft)
Ailerons (total)	1.00 m² (10.76 sq ft)
Trailing-edge flaps (total)	1.50 m² (16.15 sq ft)
Fin	0.84 m² (9.04 sq ft)
Rudder	0.40 m² (4.30 sq ft)
Tailplane	1.43 m² (15.39 sq ft)
Elevators (total)	0.50 m² (5.38 sq ft)
WEIGHTS AND LOADINGS:	
Weight empty, equipped	398 kg (877 lb)
Max fuel weight	93 kg (205 lb)
Max T-O and landing weight	658 kg (1,450 lb)
Max wing loading	60.5 kg/m² (12.39 lb/sq ft)
Max power loading	7.78 kg/kW (12.78 lb/hp)
PERFORMANCE (estimated):	
Never-exceed speed (V$_{NE}$)	173 kt (321 km/h; 200 mph)
Max level speed at 3,660 m (12,000 ft)	
	139 kt (257 km/h; 160 mph)
Max cruising speed at 3,660 m (12,000 ft)	
	122 kt (225 km/h; 140 mph)
Econ cruising speed at 3,660 m (12,000 ft)	
	96 kt (177 km/h; 110 mph)
Stalling speed at S/L, flaps down	
	46 kt (84 km/h; 52 mph)
Max rate of climb at S/L	244 m (800 ft)/min
Service ceiling	4,265 m (14,000 ft)
T-O run	214 m (700 ft)
T-O to 15 m (50 ft)	390 m (1,280 ft)
Landing from 15 m (50 ft)	427 m (1,400 ft)
Landing run	244 m (800 ft)
Range with max fuel	521 n miles (965 km; 600 miles)

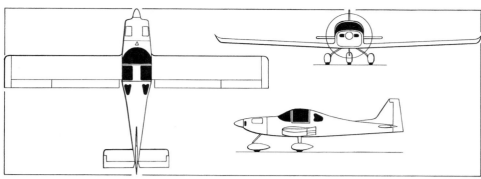

Dorna Blue Bird two-seat light aircraft *(James Goulding/Jane's)* 2000/0062614

UPDATED

FAJR

FAJR AVIATION AND COMPOSITES INDUSTRIES
Tehran

Company capabilities reportedly include composites rotor blades, cockpit transparencies, armoured panels and radar dish antennas, plus non-aerospace products such as hovercraft and speedboats. It is said to have ISO 9001 and JAR 21 ratings.

NEW ENTRY

FAJR-3
English name: Dawn
TYPE: Four-seat lightplane.
PROGRAMME: Design completed 1992; first flight 1994; JAR 23 certification November 2000 after 250 flight hours. Possibly a derivative of two-seat Fajr trainer last described in 1991-92 *Jane's*.
DESIGN FEATURES: Intended for pilot training, reconnaissance and light cargo transportation.
FLYING CONTROLS: Conventional and manual.
STRUCTURE: All-composites.
LANDING GEAR: Retractable tricycle type; outward-retracting mainwheels.
POWER PLANT: One (Western ?) piston engine; three-blade propeller.
ACCOMMODATION: Four seats, including pilot, in side-by-side pairs. Upward-opening gull-wing doors.

DIMENSIONS, EXTERNAL:	
Wing span (estimated)	10.50 m (34 ft 5½ in)
Length overall	8.30 m (27 ft 2¾ in)
Height overall	3.20 m (10 ft 6 in)
AREAS:	
Wings, gross	14.00 m² (150.7 sq ft)
WEIGHTS AND LOADINGS:	
Weight empty	1,100 kg (2,425 lb)
Max T-O weight	1,580 kg (3,483 lb)
Max wing loading	112.8 kg/m² (23.11 lb/sq ft)
PERFORMANCE:	
Max level speed	148 kt (274 km/h; 170 mph)
Service ceiling	5,180 m (17,000 ft)
Range	502 n miles (930 km; 577 miles)

NEW ENTRY

The Fajr-3 four-seat low-wing lightplane *(via Kian Mokhtari)* 2001/0106885

IAIO

IRAN AVIATION INDUSTRIES ORGANISATION
107 Sepahbod Gharani Avenue, Tehran
Tel: (+98 21) 882 50 43 to 882 50 48
Fax: (+98 21) 882 79 05

IAIO is the policy maker, co-ordinator and planner for the manufacturing, overhaul, repair and support of aircraft by Iran's aviation industry. Created to counter problems arising from the Gulf wars and subsequent economic sanctions, its primary concern is in supplying spares for aircraft in service and obtaining the technologies required for manufacturing various types of aircraft for the Iranian armed forces.

Nothing has been heard recently of earlier proposals for a co-production programme involving the Ilyushin Il-114, but talks between Iranian officials and the Tupolev JSC are apparently continuing concerning possible Iranian participation (including local assembly) in the Tu-334 programme.

SUBSIDIARIES:
Iran Aircraft Industries (IACI)
PO Box 14155-1449, Tehran
Tel: (+98 21) 603 56 06
Fax: (+98 21) 600 81 68

Conducts major overhauls of various types of military and civil aircraft, including helicopters, and aero-engines. Established 1970; first Iranian aviation company to obtain ISO 9002 certification. Earlier aircraft projects have included the Fajr (Dawn) piston-engined trainer described briefly in the 1990-91 *Jane's*.

Turbine Engine Manufacturing (TEM)
Subsidiary of IACI. Manufactures engine components at a 100,000 m² (1.08 million sq ft) site with shop floor area of 33,000 m² (355,200 sq ft). Also has well-equipped research laboratories and active design team.

Iran Aircraft Manufacturing Industries Company (IAMI)
PO Box 83145-311, Shahin shahr, Esfahan
Tel: (+98 31) 22 95 06
Fax: (+98 31) 22 76 78
DEPUTY MANAGING DIRECTOR: Gholamreza Barshan
PROGRAMME MANAGER, IRAN 140: Dr Vaziri

Otherwise known as **HESA**, this is a state-owned company controlled jointly by the Iranian Ministry of Defence and Ministry of Industries.

Esfahan factory construction initiated 1975 by Bell Helicopter Textron for production of Bell 214, but suspended 1978 after Islamic revolution when only 11 per cent completed. Reactivated 1983; began piston engine manufacture 1988 and Northrop F-5B overhaul 1989. Helicopter (Bell AH-1J and other) and hovercraft overhaul started 1993; first production piston engine completed in 1996.

Following a February 1996 agreement, a factory was set up, with Ukrainian government assistance, for licensed assembly of the Antonov An-140 twin-turboprop transport. A similar agreement with MAPO organisation, for manufacture of 60 Klimov TV7-117VMA turboprop engines, was announced at the 1997 Paris Air Show and valued at US$180 million. These engines were said to be for installation on 'Ukrainian- and Russian-built' airliners.

Qods Aviation Industries (QAI)
PO Box 13445-873, 4 Km Karaj Road, Tehran
Tel: (+98 21) 450 33 06
Fax: (+98 21) 450 33 07

Established 1984 to design, develop and manufacture surveillance/reconnaissance and attack RPV/UAVs and aerial targets (see *Jane's Unmanned Aerial Vehicles and Targets*). Subsidiary of IAMI/HESA. Also manufactures parachutes for various applications. Factory area 40,000 m² (430,550 sq ft); 1998 workforce approximately 400.

Iranian Aviation Industries (IAI)
MANAGING DIRECTOR: Eng Golestaneh
AERO-ENGINE PROGRAMMES MANAGER: Eng Amiri

Known locally as **SAHA**, this factory was established in the mid-1970s and has since overhauled, maintained and repaired approximately 4,000 aircraft and aero-engines. Like IAMI/HESA, it is controlled jointly by the Ministry of Defence and the Ministry of Industries. It occupies a 45,000 hectare (111,195 acre) site, and had a 1998 workforce of more than 1,000. It is designated to produce the Tu-334 and its D-436 engines if this programme comes to fruition.

Iran Helicopter Support and Renewal Company (IHSRC)
PO Box 13185-1688, Mehrabad Airport Road, Mearadj Avenue, Tehran
Tel: (+98 21) 600 30 31 to 600 30 39
Fax: (+98 21) 95 30 30 and 601 58 62

Established 1969 with assistance of Agusta (Italy) and Bell Helicopter (USA); overhaul and maintenance of Agusta-Bell 205, 212, 214A/B/C; Bell 206/206L JetRanger/LongRanger and AH-1J/JT Cobra series; Boeing CH-47C Chinook; and Sikorsky SH-3D Sea King and RH-53D Sea Stallion. In 1999, it was stated that experience gained with extensive rebuild and refurbishment of AH-1J SeaCobras had given IHSRC (known locally as **PANHA**) the potential for full manufacture, if so required.

UPDATED

IAMI (HESA) AZARAKHSH
English name: Lightning
TYPE: Attack fighter.
PROGRAMME: Development said to have begun in 1986; programme name *Oaj* (Zenith); stated to have passed acceptance tests by April 1997 and completed flight testing in June. Designers' names given as Morteza Sanai and Morteza Satari. Announced by Islamic Republic News

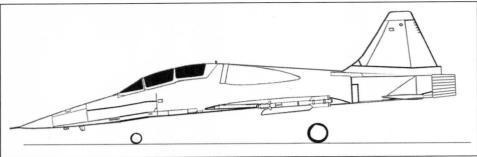

Possible appearance of Azarakhsh fighter, based on latest reports *(Janes Defence Weekly)*   2001/0067681

Agency (IRNA) in late September 1997, quoting Army Chief of Staff Maj Gen Ali Shahbazi, that the aircraft had taken part, earlier that month, in domestic military exercises, during the course of which it had dropped two 113 kg (250 lb) napalm bombs. Announced at that time that production was to begin 'soon', but this now known to be incorrect. In June 1999, Air Force Gen Habibollah Baghai told Iranian journalists that series production had begun, but stated by Lt Gen Iraj Ossareh (IRIAF Deputy Commander) in May 2000 that four aircraft completed at that time, with a fifth under construction. Current plans then to build up to 30 over next three years.

Following details highly provisional and unconfirmed:

DESIGN FEATURES: According to latest (March 2000) Jane's sources, Azarakhsh is a 10 to 15 per cent scaled-up, reverse-engineered Northrop F-5F, re-engined with two Klimov RD-33 turbofans, having shoulder-mounted air intakes.
STRUCTURE: Extensive use of composites reported.
ACCOMMODATION: Two seats in tandem.
AVIONICS: *Radar:* Reported to be locally modified Phazotron N019ME Topaz with BVR capability and enhanced air-to-ground performance.
    Self-defence: Said to have chaff/flare dispensers(s) in fairing beneath rear fuselage.
WEIGHTS AND LOADINGS (approx):
  Weight empty   8,000 kg (17,637 lb)
  Max external stores load   4,000-4,400 kg (8,818-9,700 lb)
PERFORMANCE (estimated):
  Max level speed   M1.6 class
                                                                         UPDATED

IAMI (HESA) SHAHBAZ
English name: Aurora
TYPE: Basic jet trainer.
PROGRAMME: During a military exercise in May 1998, Brig Gen Reza Pardis of the Islamic Republic of Iran Air Force (IRIAF) was quoted by the *Iran Daily* as saying that prototypes of a jet trainer had then completed 120 test flights and that the aircraft was to enter production shortly. IRIAF Gen Habibollah Baghai told Iranian journalists in June 1999 that the trainer was then in production, whereas Lt Gen Ossareh stated in May 2000 that production would start shortly. The aircraft is sometimes referred to by the name Dorna, the programme title under which it was developed. It reportedly resembles the Chinese K-8, but no further details had emerged at the time of going to press.

IAMI (HESA) IRAN 140
TYPE: Twin-turboprop transport.
PROGRAMME: Licence-built Antonov An-140 (see Ukrainian section). Selected by Iran in 1995, against competition from 12 other contenders; initial agreement leading to Iranian manufacture signed February 1996; first stage is assembly from imported CKD kits at rate of up to 10 per year; long-term objective is gradual progress to local parts manufacture; Iranian engineers installed at Antonov from 1997. Contract announced late 1998 for initial quantity of 80 aircraft, for Iranian domestic (civil and military) use only, but possibility of additional production for export to CIS states has been suggested. Assembly of first aircraft began at Esfahan March 1999; maiden flight then scheduled for first half of 2000, but not achieved until 7 February 2001. In May 1999, Iran Electronics and Leninets (St Petersburg) signed contract for Iran 140 avionics development.
CURRENT VERSIONS: Offered for passenger (including VIP) civil transport, aeromedical (ambulance or flying hospital) use, cargo/passenger or parachutist carrying, photogrammetry, geosurvey, environmental research, fisheries protection and pollution control; military versions could include staff or personnel transport, coastal and maritime patrol, radar patrol/missile guidance, elint and crew training. Possibility of future 68-passenger stretched version.
CUSTOMERS: Launch customers reported to be Iran Air and Iran Asseman Airlines, of which latter signed order for 20 in October 1999. Likely also to be procured by Iranian Air Force and Navy (potential roles tactical transport, maritime patrol and EW).
POWER PLANT: Iranian company quotes Klimov/ZMKB Progress TV3-117VMA-SBM1 engines, as in Antonov-built version, but Antonov spokesman indicated in 1998 that Iranian-built aircraft would have an uprated (1,864 kW; 2,500 shp) version of this engine for improved hot/high performance.
                                                                             UPDATED

IAMI (HESA) PARASTU
English name: Swallow
TYPE: Basic prop trainer.
PROGRAMME: First revealed at Mehrabad Air Show in February 1996, when six or seven examples were reported to have been completed; reverse-engineered Raytheon (Beech) F33A Bonanza (see US section of 1997-98 and earlier *Jane's*), with modifications that include performance-improving winglets. Programme initiated September 1987; first flight 5 April 1988. According to an IRNA announcement in September 1997, series production was then due to begin in the near future. However, according to a senior IRIAF spokesman in May 2000, the 11th aircraft was then in manufacture, with full-scale production imminent.
                                                                             UPDATED

IHSRC (PANHA) SHABAVIZ 2-75
English name: Owl
TYPE: Medium utility helicopter.
PROGRAMME: Existence revealed during presentation and flying display in Tehran on 10 December 1998; claimed as "first domestically produced" helicopter, but visibly a Bell 205 except for imported (non-US) engines which were not identified. Is probably an upgraded or reverse-engineered Agusta-Bell 205, of which over 20 supplied to Iran before 1979 Islamic revolution; this appears to be first example completed locally, although the modifications could be applied to large numbers of Bell 212s, 214As and 214Cs, which were also delivered in the 1970s. Details disclosed at launch included payload of 14 passengers or 2,500 kg (5,511 lb), ceiling of 3,800 m (12,460 ft) and unrefuelled range of about 270 n miles (500 km; 310 miles). Intended for both military and civil duties, including an armed version. Significance of 2-75 in designation not apparent. Series production officially launched on 4 May 1999 with opening of a military production plant to build this and the 2061 (which see).
                                                                             UPDATED

IHSRC (PANHA) 2061
TYPE: Five-seat helicopter.
PROGRAMME: Announced, but not seen, at launch of Shabaviz 2-75 in December 1998; reverse-engineered or locally manufactured version of Bell 206 JetRanger, also with non-US power plant. Said to be intended for reconnaissance and training, and also officially began series production on 4 May 1999.
                                                                             UPDATED

IRHSC (PANHA) SHAHED X-5
TYPE: Four-seat helicopter.
PROGRAMME: Sponsored by the Institute of Industrial Research and Development of the Islamic Revolutionary Guard (IRG); reportedly due to have made its first flight in 1997. First production aircraft handed over to IRG 16 September 1999. No descriptive details have yet emerged.
CUSTOMERS: Total of 20 reportedly to be built by end of 2004.
PERFORMANCE (reported):
  Max level speed   97 kt (180 km/h; 112 mph)
  Ceiling   5,200 m (17,060 ft)
                                                                              UPDATED

IRHSC (PANHA) SANJAQAK
English name: Dragonfly
TYPE: Two-seat helicopter.
PROGRAMME: Reported in mid-1997 to be nearing flight test at a state research centre in Esfahan; max T-O weight 450 kg (992 lb). No other details had emerged by October 2000.
                                                                             UPDATED

ISRAEL

AD & D

AERO-DESIGN & DEVELOPMENT LTD
PO Box 565, Gad Finestein Road, Rehovot Industrial Park, IL-76102 Rehovot
Tel: (+972 8) 931 55 33
Fax: (+972 8) 931 55 32
e-mail: ad_d@netvision.net.il
Web: http://www.aero-marine.design-development.com
MANAGING DIRECTORS:
  Dr Raphael Yoeli
  Israel Gottesman

In addition to the Hummingbird described below, AD&D has developed a UAV version known as the Hornet, which began tethered flight testing in January 1999. See *Jane's Unmanned Aerial Vehicles and Targets* for details. A new VTOL 'flying car', the CityHawk, was announced in April 2000.
                                                    UPDATED

AD & D HUMMINGBIRD
TYPE: Lifting vehicle.
PROGRAMME: Launched as private venture, reviving concept explored by US Hiller VZ-1 in the 1950s but incorporating new features. First (tethered) hovering flight 28 August 1997; first free (untethered) flight (4X-BEB) 4 October 1998. Available in kit form from 2000.
COSTS: Estimated US$30,000 (1998).
DESIGN FEATURES: Naturally stable VTOL platform which can be flown by persons with limited piloting experience. Empty weight places it in FAR Pt 103 category. Has full engine redundancy, and can continue flying safely throughout full flight envelope after a single engine failure (initial hovering flights used approximately half of total available power). Air duct has outwards-curved lip on upper surface and diameter:length ratio of approximately 2.75:1. Quoted build time approximately 250 hours.
FLYING CONTROLS: Vertical speed (rise/descent) is controlled by pilot increasing or reducing rotor rpm via a twist-grip throttle with the left hand. Lateral (sideways) and longitudinal (fore/aft) motion is achieved simply by the pilot tilting his body slightly in desired direction. Maximum allowable speed is attained when pilot's body reaches limit imposed by circular railing at waist height. 'Nose' of vehicle can be turned by a small lever at pilot's right hand.
STRUCTURE: Engines are mounted on four sides of a square gearbox which also serves as pedestal supporting occupant. Rotors extend from lower end of gearbox. Engines and propellers are basically off-the-shelf ultralight components. Circular outer duct is of graphite/epoxy construction.
LANDING GEAR: Four non-retractable, free-castoring single wheels, each with shock-absorbing oleo. Can withstand vertical impacts of up to 3.7 m (12 ft)/s without injuring pilot or damaging airframe.
POWER PLANT: Four 16.4 kW (22 hp) Hirth F33-15A single-cylinder piston engines, each with dual ignition, own

Hummingbird hovering in free flight   2000/0080486

carburettor and own fuel line. Engines are coupled through torsional dampers and one-way clutches. These drive, via an AD & D gearbox equipped with a chip detector, two three-blade contrarotating propellers/rotors which are synchronised to obtain practically zero yawing moment at all throttle settings. Alternative engines under consideration. Standard fuel capacity 19 litres (5.0 US gallons; 4.2 Imp gallons).

ACCOMMODATION: Pilot only, standing inside tubular metal frame.

AVIONICS: *Instrumentation:* Battery voltage meter; rotor tachometer; four CHT gauges with an ignition kill switch for each engine; four warning lights (engine out, transmission chips, engine temperature and low battery); master switch; emergency parachute activation handle (optional); and mode selector button for yaw stabilisation augmentation system.

EQUIPMENT: Ballistic parachute recovery system optional.

DIMENSIONS, EXTERNAL:
Diameter	2.20 m (7 ft 2½ in)
Height overall	1.90 m (6 ft 2¾ in)

WEIGHTS AND LOADINGS:
Weight empty	145 kg (320 lb)
Max T-O weight	260 kg (573 lb)

PERFORMANCE:
Never-exceed speed (VNE)	40 kt (74 km/h; 46 mph)
Hovering ceiling: IGE	2,440 m (8,000 ft)
OGE	1,525 m (5,000 ft)
Range, standard fuel	17 n miles (31 km; 19 miles)
Endurance, standard fuel	30 min

UPDATED

AD & D CITYHAWK

TYPE: Two-seat lifting vehicle.

PROGRAMME: Announced April 2000; prototype/demonstrator then under construction. First flight targeted for early 2001.

DESIGN FEATURES: Two-seat extrapolation of Hummingbird concept; T-O/landing area compatible with typical urban parking spaces and garages. Foreseen applications include personal urban transportation, police and traffic patrol, ambulance and urban evacuation, air taxi, and agricultural/environmental monitoring.

FLYING CONTROLS: Manual, from left-hand seat. Differential thrust between fans for pitch and forward speed control; combined thrust of both fans controls vertical speed (rise/descent). Roll and yaw control by cascades of vanes on upper and lower surfaces of each fan.

STRUCTURE: Fans, engines and gearboxes identical with those of Hummingbird.

LANDING GEAR: Tricycle type.

POWER PLANT: Two Hummingbird-type fans, each driven by four 16.4 kW (22 hp) Hirth F33-15A single-cylinder piston engines. Engine redundancy allows vehicle to fly down to a landing even if each fan loses one engine.

ACCOMMODATION: Side-by-side seats for two persons under separate canopies.

EQUIPMENT: Ballistic parachute for emergency recovery.

DIMENSIONS, WEIGHTS AND LOADINGS: Not released

PERFORMANCE (estimated):
Max level speed	80-90 kt (148-166 km/h; 92-103 mph)
Max operating altitude	2,440 m (8,000 ft)
T-O/landing footprint	2.5 × 5.5 m (8.2 × 18 ft)
Endurance	almost 1 h

NEW ENTRY

Hummingbird, complete and in kit form

AD & D HERON

TYPE: Single-seat agricultural sprayer ultralight.

PROGRAMME: Project displayed on website mid-2000, but not in active development by November that year.

DESIGN FEATURES: Pod-and-boom fuselage with inverted V tail unit; unbraced low-wing monoplane with dihedral; optional upper wing. Hopper aft of cabin, capacity 100 litres (26.4 US gallons; 22.0 Imp gallons).

FLYING CONTROLS: Manual.

LANDING GEAR: Tricycle type; fixed.

POWER PLANT: Two 27.6 kW (37 hp) piston engines, pod-mounted above cabin; single two-blade propeller. Fuel capacity 20 litres (5.3 US gallons; 4.4 Imp gallons).

ACCOMMODATION: Pilot only, in enclosed, heated and ventilated cabin with extensive transparencies.

DIMENSIONS, EXTERNAL:
Wing span	12.00 m (39 ft 4½ in)

AREAS:
Wings (lower), gross	17.00 m² (183.0 sq ft)

WEIGHTS AND LOADINGS:
Weight empty	250 kg (551 lb)
Max T-O weight	450 kg (992 lb)

PERFORMANCE (estimated):
Max level speed	78 kt (144 km/h; 90 mph)
Stalling speed	22 kt (41 km/h; 25 mph)
T-O run	70 m (230 ft)
Landing run	40 m (130 ft)
Normal range	52 n miles (96 km; 60 miles)

NEW ENTRY

Artist's impression of twin-fan CityHawk

CityHawk demonstrator under construction

IAI

ISRAEL AIRCRAFT INDUSTRIES LTD
Ben-Gurion International Airport, IL-70100 Tel-Aviv
Tel: (+972 3) 935 31 11
Fax: (+972 3) 971 22 90 and 935 31 31
Web: http://www.iai.co.il
CHAIRMAN: Ori Orr
PRESIDENT AND CEO: Moshe Keret
EXECUTIVE VICE-PRESIDENT: Ovadia Harari
CORPORATE VICE-PRESIDENT: Menham Shmul (General Manager, Military Aircraft Group)
DEPUTY CORPORATE VICE-PRESIDENT: Dr David Harari (Director, Research and Development)
DEPUTY CORPORATE VICE-PRESIDENT, MARKETING: Shimon Eckhaus
DIRECTOR OF CORPORATE INDUSTRIAL SERVICES: Menham Tadmor
DEPUTY CORPORATE VICE-PRESIDENT, COMMUNICATIONS: Doron Suslik

Tel: (+972 3) 935 85 09
Fax: (+972 3) 935 85 12
e-mail: hpaz@hdq.iai.co.il

Founded 1953 as Bedek Aviation; ceased being unit of Ministry of Defence and became government-owned corporation 1967; name changed to Israel Aircraft Industries 1 April 1967; number of divisions reduced from five to four February 1988 (Aircraft, Electronics, Technologies and Bedek Aviation). Further restructuring took place 1994 into

262   ISRAEL: AIRCRAFT—IAI

the present four Groups: Commercial Aircraft, Military Aircraft, Bedek Aviation and Electronics. These business units were expected to become independent subsidiaries as a prelude to full privatisation of IAI, though no such plans yet announced (October 2000). Covered floor space 680,000 m² (7.32 million sq ft); total workforce 14,200 in late 1998.

Approved as repair station and maintenance organisation by Israel Civil Aviation Administration, US Federal Aviation Administration, UK Civil Aviation Authority and Israel Air Force. Products include aircraft of own design; in-house produced airframe systems and avionics; service, upgrading and retrofit packages for civil and military aircraft and helicopters; other activities include space technology, missile and ordnance development, and seaborne and ground equipment.

Revenue for 1999 totalled US$2.0 billion, an increase of 7.2 per cent compared with 1998. Net profit rose 70.7 per cent, from US$41 million to US$70 million. New orders worth US$2.4 billion were taken, a 26 per cent increase over 1998. Backlog at 31 December 1999 stood at an all-time high of US$3.3 billion. The first nine months of 2000 saw a net profit of US$60.9 million (up 26.9 per cent on corresponding period of 1999); revenues for the period were US$1.56 billion, a 7.6 per cent increase. Despite a 1 per cent drop in new contracts value during the first six months of the year, backlog at 30 June 2000 was a new record figure of US$3.76 billion. However, by August 2000, cancellation of such programmes as the Fairchild 428JET and Bedek's Phalcon AEW conversion of a Beriev A-50 for China were among setbacks making significant reductions in workforce seem increasingly likely.

UPDATED

COMMERCIAL AIRCRAFT GROUP
Ben-Gurion International Airport, IL-70100 Tel-Aviv
Tel: (+972 3) 935 33 37 and 935 71 49
Fax: (+972 3) 935 42 26
GENERAL MANAGER: M Boness

Established 1994. Design, development and manufacture of civil aircraft. Made up of three autonomous business divisions (as below), plus Galaxy Aerospace Corporation, the wholly owned Astra and Galaxy series marketing subsidiary in the USA. Galaxy Aerospace, formed November 1996 as a joint venture by IAI and Hyatt Corporation, replaced former Astra Jet Corporation as US sales outlet.

Initial US$80 million agreement with Fairchild Aerospace announced in June 1999, under which IAI was to be responsible for systems engineering and integration, flight testing and certification support for Fairchild Dornier 428JET. Announced subsequently that IAI to manufacture fuselages and undertake final assembly of this aircraft, which it would then ship back 'green' to Dornier for outfitting. However, 428JET programme terminated by Fairchild in August 2000.

Engineering Division
Address as for Commercial Aircraft Group
Tel: (+972 3) 935 81 50 and 971 29 03
Fax: (+972 3) 935 31 30
GENERAL MANAGER: Z Arazi

Analysis, design, development, integrating and testing of aircraft systems and material requirements for aerospace customers.

Production Division
Address as for Commercial Aircraft Group
Tel: (+972 3) 935 31 32 and 971 29 03
Fax: (+972 3) 935 31 30
GENERAL MANAGER: G Cohen

Substructure and components manufacture for domestic and foreign customers. Special capabilities include fabrication of standard and advanced material structures, nacelles and other engine parts (including hot-end components). Development of future civil aircraft, production of Astra SPX and Galaxy business jets, provision of upgrades for IAI Westwind, and product support for Westwind and IAI Arava.

SHL Division
PO Box 190, Industrial Zone, IL-71101 Lod
Tel: (+972 8) 23 91 11 and 22 26 51
Fax: (+972 8) 22 27 92
GENERAL MANAGER: H Porter

Design, development and manufacture of military and civil hydraulic systems and components; aircraft landing gear and shock-absorbers; servohydraulic actuators for flight control systems.

ASTRA AND GALAXY MARKETING:
Galaxy Aerospace Corporation
One Galaxy Way, Alliance Airport, Fort Worth, Texas 76177, USA
Tel: (+1 817) 837 33 33
Fax: (+1 817) 837 38 62
PRESIDENT AND CEO: Brian E Barents
EXECUTIVE VICE-PRESIDENT, SALES AND MARKETING:
  Roger Sperry
VICE-PRESIDENT, CUSTOMER SERVICE: Mike Wuebbling
VICE-PRESIDENT, CORPORATE COMMUNICATIONS: Jeff Miller

Construction of new 13,935 m² (150,000 sq ft) facility began September 1997 and completed in 1998; US$12 million complex includes: worldwide headquarters; marketing and customer support offices; type certificate holder and completion centre for Astra SPX and Galaxy; factory service centre; and a parts distribution facility. 'Green' Astra SPXs are flown from Tel-Aviv to Fort Worth with ferry engines, given factory-new engines and fully outfitted. Similar completion of Galaxy airframes began in first quarter of 1999.

UPDATED

Astra SPX delivered to AIDC, Taiwan, in February 2000 for target towing and other missions   2000/0084392

IAI 1125 ASTRA SPX
US Air Force designation: C-38A
TYPE: Business jet.
PROGRAMME: Descendant of US Aero Commander 1121 Jet Commander (first flight 27 January 1963), acquired by IAI 1967 and developed successively as 1121 Commodore Jet, 1123 and 1124 Westwind and 1124A Westwind 2; sales of 1121/23/24/24A models by Aero Commander and IAI totalled 441 by end of 1987; 1125 model launched at NBAA October 1979, renamed Astra 1981.
Construction of two prototypes and one static/fatigue test aircraft started April 1982; roll-out 1 September 1983; first flight (4X-WIN, c/n 4001) 19 March 1984; first flight 4X-WIA (c/n 4002) August 1984; first flight production Astra (4X-CUA) 20 March 1985; FAR Pts 25 and 36 certification 29 August 1985; first delivery 30 June 1986. Astra SP introduced at NBAA October 1989; first delivery of this version (N60AJ) late 1990.
Astra SPX announced at NBAA 1994 and certified by FAA 8 January 1996. Production rate 12 to 15 per year.
CURRENT VERSIONS: **Astra:** Initial version; 31 production aircraft; see 1992-93 and earlier *Jane's*.
Astra SP: Introduced 1989; superseded by SPX; 37 built; details in 1997-98 *Jane's*
Astra SPX: First flight (4X-WIX) 18 August 1994; production deliveries from early 1996. Winglets and Rockwell Collins Pro Line 4 avionics. Change to more powerful TFE731-40R-200G turbofans with FADEC, hydromechanical fuel control back-up and Dee Howard thrust reversers; increased weights and payload/range performance and shorter T-O run.
Detailed description applies to SPX version.
C-38A: Two SPXs, outfitted by Tracor Flight Systems (now BAE Systems North America) for transport and medevac duties; delivered as C-38As to 201st Airlift Squadron of the US Air National Guard at Andrews AFB, Maryland, in April and May 1998, replacing C-21A Learjets. Contract contains options for an additional two SPXs.
Galaxy: Improved and redesigned version; described separately.
CUSTOMERS: Combined production of Astra and Astra SP totalled 68 (plus two prototypes), of which approximately 80 per cent to US customers.
SPX sales totalled 64 by July 2000, of which 43 delivered to customers or to USA for completion by 1 January 2000; 14 delivered to customers in 1998 and 12 in 1999. Non-US operators for SPX include residents of Bermuda, Brazil, Canada (two), Cayman Islands, Cyprus, Germany and Switzerland; one (B-20001, c/n 119) delivered to Taiwan in April 2000 for use by RoCAF as target tug. Offered to Israeli Navy as replacement for its 20-year-old IAI 1124 Seascan (Westwind) maritime patrol aircraft.
COSTS: US$11 million (1997).
DESIGN FEATURES: Typical sweptwing, rear-engined business jet configuration.
Wing section high-efficiency IAI Sigma 2; leading-edge sweep 34° inboard, 25° outboard; trailing-edge sweep on outer panels; winglets.
FLYING CONTROLS: Conventional and hydraulically powered. Control surfaces operated by pushrods; tailplane incidence controlled by three motors running together to protect against runaway or elevator disconnect; ailerons can be separated in case of jam; spoiler/lift dumper panels ahead of Fowler flaps; horn-balanced rudder with dual actuated trim tab; flaps interconnected with leading-edge slats, both electrically actuated.
STRUCTURE: One-piece, two-spar wing with machined ribs and skin panels, attached by four main and five secondary frames; wing/fuselage fairings, elevator and fin tips and tailcone of GFRP; ailerons, spoilers, inboard leading-edges and wingtips of Kevlar and Nomex honeycomb; nose avionics bay door and nosewheel doors of Kevlar; Kevlar-reinforced nacelle doors and panels; chemically milled fuselage skins; some titanium fittings; heated windscreens of laminated polycarbonate with external glass layer to resist scratching.
LANDING GEAR: SHL hydraulically retractable tricycle type, with oleo-pneumatic shock-absorber and twin wheels on

IAI 1125 Astra SPX twin-turbofan business transport   2000/0062972

IAI—AIRCRAFT: ISRAEL

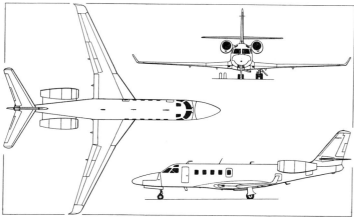

IAI 1125 Astra SPX business transport (two Honeywell TFE731-40R-200G turbofans) *(Dennis Punnett/Jane's)*

Typical executive interior of the Astra SPX

each unit. Trailing-link main units retract inward, nosewheels forward. Tyre sizes 23×7.00-12 (10 ply) (main), 16×4.4 (6 ply) deflector type (nose). Hydraulic extension, retraction and nosewheel steering; hydraulic multidisc anti-skid mainwheel brakes. Compressed nitrogen cylinder provides additional power source for emergency extension.

POWER PLANT: Two 18.90 kN (4,250 lb st) Honeywell TFE731-40R-200G turbofans, with Dee Howard hydraulically actuated target-type thrust reversers and FADEC, pylon-mounted in nacelle on each side of rear fuselage. Standard fuel in integral tank in wing centre-section, two outer-wing tanks, and upper and lower tanks in centre-fuselage (combined usable capacity 4,910 litres; 1,297 US gallons; 1,080 Imp gallons). Additional fuel can be carried in 378.5 litre (100 US gallon; 83.3 Imp gallon) removable auxiliary tank in forward area of baggage compartment. Single pressure refuelling point in lower starboard side of fuselage aft of wing, or single gravity point in upper fuselage, allow refuelling of all tanks from one position. Fuel sequencing automatic.

ACCOMMODATION: Crew of two on flight deck. Dual controls standard. Sliding door between flight deck and cabin. Standard accommodation in pressurised cabin for six persons, two in forward-facing seats at front and four in club layout; galley (port or starboard) at front of cabin, coat closet forward (starboard), toilet at rear. All six seats individually adjustable fore and aft, laterally, and can be swivelled or reclined; all fitted with armrests and headrests. Two wall-mounted foldaway tables between club seat pairs. Coat closet houses stereo tape deck. Maximum accommodation for nine passengers.

New cabin interior (first redesign since 1986, introduced in September 1999, has less intrusive (curved) cabinetry and galley units; restyled headliner and sidewall panels; new seats, tables and lighting. Forward club seating maximises recline angle and personal space for each occupant; lavatory has brighter lighting; more storage space for coats and carry-on baggage.

Plug-type airstair door at front on port side; emergency exit over wing on each side. Heated baggage compartment aft of passenger cabin, with external access. Service compartment in rear fuselage houses aircraft batteries (or optional APU), electrical relay boxes, inverters and miscellaneous equipment. Cabin soundproofing improved compared with Westwind 2.

SYSTEMS: Honeywell environmental control system, using engine bleed air, with normal pressure differential of 0.61 bar (8.8 lb/sq in). Honeywell 36-150(W) APU available optionally. Two independent hydraulic systems, each at pressure of 207 bars (3,000 lb/sq in). Primary system operated by two engine-driven pumps for actuation of anti-skid brakes, landing gear, nosewheel steering, spoilers/lift dumpers and primary control surfaces. Back-up system, operated by electrically driven pump, provides power for emergency/parking brake, primary control surfaces and thrust reversers.

Electrical system comprises two Lucas Aerospace 300 A 28 V DC engine-driven starter/generators, with two 1 kVA single-phase solid-state inverters operating in unison to supply single-phase 115 V AC power at 400 Hz and 26 V AC power for aircraft instruments. Two 24 V Ni/Cd batteries for engine starting and to permit operation of essential flight instruments and emergency equipment. 28 V DC external power receptacle standard.

Pneumatic de-icing of wing leading-edge slats and tailplane leading-edges; thermal anti-icing of engine intakes. Oxygen system for crew (pressure demand) and passengers (drop-down masks) supplied by 1.35 m³ (48 cu ft) cylinder. Two-bottle Freon-type engine fire extinguishing system standard.

AVIONICS: Rockwell Collins Pro Line 4 standard.
Comms: Dual VHF-22A radios and TDR-90 transponders; dual Baker audio systems.
Radar: WXT-250A colour weather radar.
Flight: VIR-32 nav, DME-42, RMI-36, C-14 compass systems and FCS-80 flight director systems (all dual); APS-85 autopilot, ADS-85 air data system, AHS-85 AHRS, VNI-80D vertical nav system, ADF-60A and ALT-50A radio altimeter. Provisions for GNS-1000, GNS-X or UNS-1C flight management system.
Instrumentation: All flight information displayed on four 18.4 cm (7¼ in) square screens. Five-tube EFIS-86C standard.

EQUIPMENT: Standard equipment includes electric windscreen wipers, electric (warm air) windscreen demisting, wing ice inspection lights, landing light in each wingroot, taxying light inboard of each mainwheel door, navigation and strobe lights at wingtips and tailcone, rotating beacons under fuselage and on top of fin, and wing/tailplane static wicks.

DIMENSIONS, EXTERNAL:
Wing span over winglets	16.64 m (54 ft 7 in)
Wing aspect ratio	8.8
Length overall	16.94 m (55 ft 7 in)
Fuselage: Max width	1.57 m (5 ft 2 in)
Max depth	1.91 m (6 ft 3 in)
Height overall	5.54 m (18 ft 2 in)
Tailplane span	6.40 m (21 ft 0 in)
Wheel track (c/l of shock-struts)	2.77 m (9 ft 1 in)
Wheelbase	7.34 m (24 ft 1 in)
Passenger door (fwd, port): Height	1.37 m (4 ft 6 in)
Width	0.66 m (2 ft 2 in)
Overwing emergency exits (each):	
Height	0.69 m (2 ft 3 in)
Width	0.48 m (1 ft 7 in)

DIMENSIONS, INTERNAL:
Cabin: Length: incl flight deck	6.86 m (22 ft 6 in)
excl flight deck	5.21 m (17 ft 1 in)
Max width	1.45 m (4 ft 9 in)
Max height	1.70 m (5 ft 7 in)
Baggage compartment volume	1.6 m³ (55 cu ft)

AREAS:
Wings, gross (excl winglets)	29.41 m² (316.6 sq ft)

WEIGHTS AND LOADINGS:
Basic operating weight empty	6,214 kg (13,700 lb)
Max usable fuel: standard	3,942 kg (8,692 lb)
with long-range tank	4,248 kg (9,365 lb)
Max payload	1,497 kg (3,300 lb)
Max ramp weight	11,249 kg (24,800 lb)
Max T-O weight	11,181 kg (24,650 lb)
Max landing weight	9,389 kg (20,700 lb)
Max zero-fuel weight	7,711 kg (17,000 lb)
Max wing loading	380.3 kg/m² (77.89 lb/sq ft)
Max power loading	296 kg/kN (2.90 lb/lb st)

PERFORMANCE:
Max operating Mach No. (M_{MO})	0.875
Max operating speed (V_{MO}), S/L to 8,230 m (27,000 ft)	353 kt (653 km/h; 406 mph) IAS
Max cruising speed at 10,670 m (35,000 ft), 8,618 kg (19,000 lb) mid-cruise weight	483 kt (895 km/h; 556 mph)
Typical cruising speed	M0.82 (470 kt; 870 km/h; 541 mph)
Econ cruising speed	432 kt (800 km/h; 497 mph)
Max rate of climb at S/L	1,160 m (3,805 ft)/min
Rate of climb at S/L, OEI	411 m (1,348 ft)/min
Max certified altitude	13,715 m (45,000 ft)
FAR Pt 25 T-O balanced field length	1,645 m (5,395 ft)
FAR Pt 25 landing distance at MLW	890 m (2,920 ft)
Range: with eight passengers	2,286 n miles (4,235 km; 2,631 miles)
with four passengers, max fuel and NBAA IFR reserves	2,949 n miles (5,461 km; 3,393 miles)

Flight deck of IAI Astra SPX

UPDATED

IAI 1126 GALAXY

TYPE: Business jet.

PROGRAMME: Initiated as derivative of Astra SP; design (then called Astra IV) finalised late 1992 in anticipation of 1993 launch; co-production with Yakovlev of Russia discussed during early part of 1993; formal announcement of launch as Galaxy, with minor design changes, announced 20 September 1993 just before NBAA Convention in Atlanta, Georgia, USA, followed next day by news that Yakovlev to be risk-sharing partner; other partners to be Rockwell Collins (avionics supplier) and eventual engine manufacturer (P&WC since selected). Additional partner(s) sought 1995: followed by disclosure that replacement for Yakovlev also being sought; Sogerma (France) contracted August 1996 to build production aircraft fuselages and tail units.

Four prototypes (003 and 004 flying plus 001 static and 002 fatigue test); first flight rescheduled for fourth quarter 1996 (later changed to second quarter 1997 and later still to fourth quarter 1997); 003 rolled out 4 September 1997 and made first flight (4X-IGA) 25 December 1997. Static testing of 001 completed September 1998; 002 to begin full-scale fatigue testing on completion of static tests. Second flight test aircraft (4X-IGO, c/n 004) made first flight 21 May 1998; 005 (company demonstrator 4X-IGB/N505GA) first flew on 23 September 1998; combined total of 750 hours in 260 flights by three aircraft by December 1998. Israeli CAA and US FAA certification to FAR Pt 25 Amendment 75 and FAR Pts 34 and 36 awarded 16 December 1998; European JAA certification process started mid-1999. First customer delivery, to TTI of Fort Worth, Texas, made in January 2000. Production rate two per month in mid-2000. Family of variants planned, but no firm details yet released.

CURRENT VERSIONS: Seen as four/eight-passenger business/executive (standard model), with option of alternative interior seating up to 18 passengers for regional transport operation.

CUSTOMERS: Between 45 and 50 commitments by October 1999, of which about 35 per cent from non-US operators in Canada, Europe, Israel, Mexico and South America. First bulk order, for seven, plus option on 15 more, placed mid-1999 by charter operator ILI Aviation of Zurich, Switzerland. Six delivered by 31 July 2000, including one to AvBase Aviation of Cleveland, Ohio, for fractional ownership. IAI estimates break even at 100 sales and potential market for 200. Twenty-first production aircraft delivered to USA in September 2000.

COSTS: Approximately US$18 million flyaway (July 2000); total development programme cost forecast at approximately US$152 million (1993).

DESIGN FEATURES: Designed for transatlantic range (non-stop Paris to New York). Essentially same wing as Astra SPX, except for 34° 30′ inboard leading-edge sweep and addition of Krueger flaps; new wide-body fuselage is longer and has more headroom.

FLYING CONTROLS: All-hydraulic dual actuation except for rudder (manual); aileron movement 10° up/15° down, elevators 27° up/20° down, rudder 20° left/right; tailplane and rudder tab have electric trim. Wings fitted with outboard leading-edge slats (25° fully out), inboard Krueger leading-edge flaps (110°), four-segment upper surface airbrakes/lift dumpers (45° fully up) and Fowler single-slotted inboard and outboard trailing-edge flaps (settings 0, 12, 20 and 40°). Ailerons and elevators can be operated manually in event of hydraulic failure.

STRUCTURE: Generally similar to that described for Astra. Sogerma (France) produced tail units, doors, access panels, tailcones and wing/fuselage fairings for prototypes; beginning with 13th aircraft, will manufacture complete fuselages as well as tail units. Sogerma has subcontracted Romaero of Romania to manufacture fuselage rear sections. Flight Environments cabin flame protection and sound attenuation system.

LANDING GEAR: Hydraulically retractable tricycle type; twin wheels and oleo-pneumatic shock-absorbers on each unit. Nose unit has electrohydraulic steering and retracts forward. Trailing-link mainwheel units retract inward and are equipped with Honeywell multidisc anti-skid carbon brakes.

POWER PLANT: Two FADEC-equipped Pratt & Whitney Canada PW306A turbofans, each flat rated at 26.9 kN (6,040 lb st), pylon-mounted on sides of rear fuselage. Nordam nacelles and thrust reversers. Fuel capacity 7,882 litres (2,080 US gallons; 1,732 Imp gallons).

ACCOMMODATION: One or two pilots; provision for jump-seat for third crew member. Standard club-type seating for four to eight persons in business/executive version; legroom between facing seats allows enough space for full reclining and berthing; large galley with room for refrigerator, microwave oven, coffee maker and storage. Alternative 10-passenger layout has four club seats at front, with a four-place conference group and two on a divan to the rear,

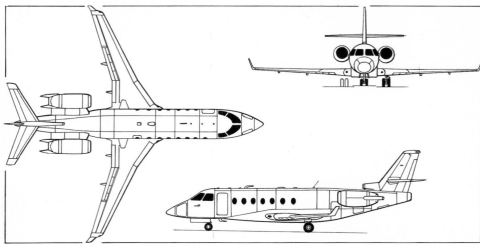

IAI Galaxy twin-turbofan business and commuter transport *(James Goulding/Jane's)*

IAI Galaxy business jet (two PW306A turbofans)

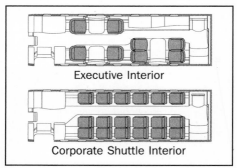

Galaxy alternative executive and corporate shuttle configurations

Family photograph of Astra SPX (left) and Galaxy emphasises the latter's more capacious fuselage

Typical Galaxy business jet interior

Flight deck of IAI Galaxy with Rockwell Collins Pro Line 4 avionics

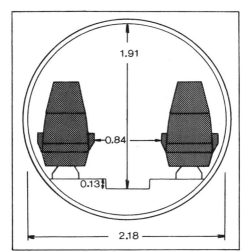
Cross-section of Galaxy cabin (dimensions in metres) *(Paul Jackson/Jane's)*

plus optional lavatory and additional baggage space aft. Three-abreast seating for up to 18 passengers, with single aisle, can be provided in corporate shuttle configuration. Generous baggage compartment in rear fuselage, accessed by external airstair door, can accommodate baggage for all 18 passengers. Entire accommodation, including baggage compartment, is pressurised.

SYSTEMS: Dual hydraulic systems, each 207 bars (3,000 lb/sq in). Electrical system comprises two Lucas Aerospace 28 V 400 Ah engine-driven starter/generators, two 24 V 43 Ah Ni/Cd batteries and a Honeywell GTCP36-150 APU. Pneumatic system for emergency extension of landing gear, actuation of wheel brakes and thrust reversers, and de-icing of wing leading-edges. Cabin pressurisation and air conditioning system, differential 0.61 bar (8.8 lb/sq in). One (optionally two) 2.04 m³ (72 cu ft) oxygen bottles.

AVIONICS: All-digital Rockwell Collins Pro Line 4 suite, including EFIS and EICAS (five 18.4 cm; 7¼ in square screens), standard.
  Comms: Dual VHF and single HF standard; transponder; ELT. Satcom optional.
  Radar: TWR-850 colour weather radar.
  Flight: FCC-4005 flight control system with Cat. II autopilot; VOR/LOC; dual DME; ADF; GPS Nav 1; marker beacon receiver; radio altimeter. Options for second ATC and second radio altimeter, second GPS Nav, TCAS, second ADF and MLS.
  Instrumentation: Goodrich standby instruments.

DIMENSIONS, EXTERNAL:
Wing span	17.70 m (58 ft 1 in)
Length overall	18.97 m (62 ft 3 in)
Wing aspect ratio	9.1
Height overall	6.53 m (21 ft 5 in)
Tailplane span	6.86 m (22 ft 6 in)
Wheel track	3.30 m (10 ft 10 in)
Wheelbase	7.39 m (24 ft 3 in)
Passenger door: Height	1.82 m (5 ft 11½ in)
Width	0.84 m (2 ft 9 in)
Baggage compartment door:	
Height	1.14 m (3 ft 9 in)
Width	0.89 m (2 ft 11 in)

DIMENSIONS, INTERNAL:
Cabin: Length: incl flight deck	9.25 m (30 ft 4 in)
excl flight deck	7.44 m (24 ft 5 in)
Max width	2.18 m (7 ft 2 in)
Max height	1.91 m (6 ft 3 in)
Baggage compartment volume	3.5 m³ (125 cu ft)

AREAS:
Wings, gross	34.28 m² (369.0 sq ft)

WEIGHTS AND LOADINGS:
Basic operating weight empty	8,709 kg (19,200 lb)
Max usable fuel weight	6,804 kg (15,000 lb)
Max payload	2,177 kg (4,800 lb)
Payload with max fuel	365 kg (805 lb)
Max ramp weight	15,876 kg (35,000 lb)
Max T-O weight	15,807 kg (34,850 lb)
Max landing weight	12,701 kg (28,000 lb)
Max zero-fuel weight	10,886 kg (24,000 lb)
Max wing loading	461.1 kg/m² (94.44 lb/sq ft)
Max power loading	294 kg/kN (2.88 lb/lb st)

PERFORMANCE (estimated):
Max operating Mach No. (M_{MO})	0.85
Max operating speed (V_{MO}):	
S/L to 4,575 m (15,000 ft)	310 kt (574 km/h; 356 mph) IAS
4,575-7,620 m (15,000-25,000 ft)	360 kt (667 km/h; 414 mph) IAS
Max cruising speed (V_{MC}) at 9,450 m (31,000 ft), mid-cruise weight of 12,247 kg (27,000 lb)	494 kt (915 km/h; 568 mph)
Typical cruising speed at 11,880 m (39,000 ft)	M0.82 (470 kt; 870 km/h; 541 mph)
Long-range cruising speed	430 kt (797 km/h; 495 mph)
Max certified altitude	13,715 m (45,000 ft)
T-O balanced field length (S/L, ISA)	1,800 m (5,900 ft)
Landing balanced field length at MLW (S/L, ISA)	1,037 m (3,400 ft)
Range with four passengers, NBAA IFR reserves	3,620 n miles (6,704 km; 4,165 miles)

UPDATED

IAI C-5WA AIRTRUCK

TYPE: Twin-turboprop freighter.

PROGRAMME: IAI submission chosen in principle by Federal Express of USA against competition from Saab (Sweden) and Ayres (USA) in mid-1998 as replacement for Fokker Friendship parcel transports. FedEx purchase will be 100, but IAI is seeking further 100 sales to reduce unit price to FedEx-imposed ceiling of US$10 million. Risk-sharing partners also needed to ensure programme launch, but not obtained by October 2000; discussions with potential subcontractors in Europe and Asia during 1998 included Hyundai of South Korea as one possible source of wing. Target first flight in 2001 will not now be met. IAI considering joining Ayres of USA in Loadmaster programme (which see) and abandoning Airtruck.

CUSTOMERS: Unconfirmed reports by late 1999 that FedEx requirement may have been reduced; Global Air (Leasing) Group ordered 50, with further 50 on option, in July 1999.

DESIGN FEATURES: High-wing twin-turboprop with T-tail; optimised for night transport of up to five standard-size cargo containers.

LANDING GEAR: Retractable tricycle type. Main units retract into fairings on fuselage sides.

POWER PLANT: Two 4,098 kW (5,495 shp) class turboprops (to be selected); four-blade propellers.

ACCOMMODATION: Pressurised flight deck, with forward airstair door access on port side. Unpressurised, but temperature-controlled, main cargo compartment has large port-side cargo door aft of wing. Tailcone compartment, also unpressurised, configured for carriage of hazardous materials. Main compartment can accommodate five M1 (2.44 × 2.44 m; 8 × 8 ft) cargo pallets.

SYSTEMS: Redundant fire detection and extinguishing system in tailcone compartment.

DIMENSIONS, EXTERNAL:
Wing span	31.725 m (104 ft 1 in)
Wing aspect ratio	10.8
Length overall	29.985 m (98 ft 4½ in)
Cargo door: Height	2.59 m (8 ft 6 in)
Width	2.74 m (9 ft 0 in)

DIMENSIONS, INTERNAL:
Main cargo compartment: Max height	2.44 m (8 ft 0 in)
Max width	3.18 m (10 ft 5¼ in)
Volume	126 m³ (4,450 cu ft)

AREAS:
Wings, gross	93.00 m² (1,001.0 sq ft)

WEIGHTS AND LOADINGS (design):
Operating weight empty	17,240 kg (38,008 lb)
Max fuel weight	8,300 kg (18,298 lb)
Max payload	12,470 kg (27,492 lb)
Max T-O weight	35,380 kg (78,000 lb)
Max wing loading	380.6 kg/m² (77.96 lb/sq ft)

PERFORMANCE (estimated):
Max cruising speed	300 kt (556 km/h; 345 mph)
Range:	
with max payload	950 n miles (1,759 km; 1,093 miles)
with max fuel and 10,000 kg (22,045 lb) payload	1,500 n miles (2,778 km; 1,726 miles)

UPDATED

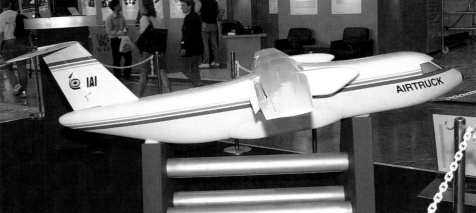
Model of the projected IAI Airtruck *(Photo Link)*

BEDEK AVIATION GROUP
Ben-Gurion International Airport, IL-70100 Tel-Aviv
Tel: (+972 3) 935 89 64
Fax: (+972 3) 971 22 98
GENERAL MANAGER: David Arzi

Group comprises four Divisions: Aircraft, Engines, Components and Ramta. Internationally approved as single-site civil and military airframe, power plant, systems and accessory service and upgrading centre; plants are Matam (aircraft services and infrastructure), Shaham (aircraft maintenance and upgrading), Masham (engine maintenance) and Mashav (components maintenance). Covered floor space 110,000 m² (1,184,000 sq ft). Maintenance subsidiary, Commodore Aviation, based at Miami International Airport, Florida, USA; plans announced in mid-2000 for second facility in US mid-west or on east coast.

More than 25 types of aircraft handled, including Boeing 707/727/737/747/767, former McDonnell Douglas DC-8/-9/-10, Lockheed Martin C-130, Tupolev Tu-134/-154 and Airbus A300/310; more than 30 types of civil and military engine include JT3D, JT8D, JT9D, F100, J79, Atar 9C/9K-50, TFE731, T56, PT6, CFM56, Rolls-Royce 250, T53 and T64; more than 6,000 types of accessory and instruments serviced. Bedek provides total technical support and holds warranty and/or approved service centre appointments for domestic and foreign regulatory agencies, air arms and manufacturers; approving agencies include Israel CAA, Israel Air Force, US military, US FAA, UK CAA and German LBA.

Bedek was responsible for the Phalcon Boeing 707 conversion to AEW configuration for Chile, described in the 1998-99 and earlier editions. Installation of the IAI Elta Phalcon system on a Beriev A-50 destined for the People's Republic of China began in late 1999, following arrival of the aircraft (RA-78740) at Ben-Gurion IAP on 25 October. Contract reportedly included options for up to three similar conversions, but was terminated in mid-2000 under pressure from US government, with installation then only partially completed. Bedek reportedly discussing potential sale of two Phalcon systems to India in August 2000.

UPDATED

MILITARY AIRCRAFT GROUP
Ben-Gurion International Airport, IL-70100 Tel-Aviv
Tel: (+972 3) 935 41 36 and 971 14 71
Fax: (+972 3) 935 34 53
GENERAL MANAGER: Menham Shmul

Design, manufacture and upgrading of manned and unmanned fixed-wing aircraft and helicopters. Made up of three autonomous business divisions, as follows:

Lahav Division
Address as for Military Aircraft Group
Tel: (+972 3) 935 31 63
Fax: (+972 3) 935 36 87
GENERAL MANAGER: David Dagan

Retrofit design, integration and assembly of advanced aircraft. Development and manufacture of customer-specified advanced combat aircraft. Upgrade of MiG series aircraft. Selected mid-2000 to develop avionics upgrade for 22 Spanish Air Force Northrop SF-5Bs.

Mata Helicopters Division
PO Box 27/160, Industrial Zone, IL-91271 Jerusalem
Tel: (+972 2) 84 12 11 and 83 43 13
Fax: (+972 2) 84 12 10 and 84 12 99
GENERAL MANAGER: Ami Davidsohn

Helicopter structures, systems and component maintenance; crash repair; overhaul, modification and upgrading. See 1994-95 edition and *Jane's Aircraft Upgrades* for recent Yasur (Petrel) 2000 upgrade of IDF CH-53Ds. Contract for components and seats for MD Helicopters Explorer.

Malat Division
Address as for Military Aircraft Group
Tel: (+972 3) 935 35 08 and 971 18 78
Fax: (+972 3) 935 41 75
GENERAL MANAGER: R Maor

Concept, design integration and assembly of UAVs, their airframes, onboard components, operational control systems and launch and recovery equipment. (See *Jane's Unmanned Aerial Vehicles and Targets* for current activities.)

UPDATED

ELECTRONICS GROUP
PO Box 105, Industrial Zone, IL-56000 Yehud
Tel: (+972 3) 531 55 55 and 531 40 21
Fax: (+972 3) 536 52 05 and 536 39 75
Telex: 341450 MBT IL
GENERAL MANAGER: S Alkon

Comprises three business divisions (MBT, MLM and Tamam), plus Elta Electronics Industries Ltd. Products include electronics and electro-optical systems and components, space technologies (including SDI environment) and wide range of civil and military hardware and software products and services.

UPDATED

ITALY

AERONAUTICA MACCHI

AERONAUTICA MACCHI SpA
Via Ing Paolo Foresio 1, I-21040 Venegono Superiore (VA)
Tel: (+39 0331) 86 59 12
Fax: (+39 0331) 86 59 10
CHAIRMAN: Dott Fabrizio Foresio

Original Macchi company, founded 1913 in Varese, produced famous line of high-speed flying-boats and seaplanes. Aeronautica Macchi became holding company for Aermacchi SpA (see following entry); SICAMB (airframe and equipment manufacturing, including licensed production of Martin-Baker ejection seats) at Latina; and CIRA. A 25 per cent holding in Aeronautica Macchi was acquired by Aeritalia, now Alenia, in 1983, the balance being held by the Foresio family. In January 1997, Aermacchi acquired the SIAI-Marchetti company from Alenia-Finmeccanica.

Also obtained in the same agreement was Finmeccanica's nacelle manufacturing activities for Airbus airliners and Dassault Falcon business jets, previously the responsibility of Alenia at Turin. Related activities formerly undertaken by SIAI included Airbus A300 and 310 tailcones; overhaul and repair of various types of aircraft (notably C-130 Hercules, DHC-5 Buffalo and Cessna Citation II); participation in national or multinational programmes, producing parts for Eurofighter Typhoon (carbon fibre structures, titanium engine cowlings, ECM pods and weapon pylons), Alenia G222/C-27J (wing and tailerons), Panavia Tornado (wing furniture) and AMX. Macchi is risk-sharing partner in Fairchild Dornier 328JET and 428JET, supplying fuselages for both.

Workforce in 1999 totalled over 2,000, including 1,660 in Aermacchi. Activities in civil field accounted for over 40 per cent of turnover by 1998.

UPDATED

Jet trainers offered by Aeronautica Macchi include the MB-339 (nearest), Yak-130 (alias M-346) and AMX-T
2001/0089522

AERMACCHI

AERMACCHI SpA
(Subsidiary of Aeronautica Macchi SpA)
Via Ing Paolo Foresio 1, PO Box 101, I-21040 Venegono Superiore (VA)
Tel: (+39 0331) 81 31 11
Fax: (+39 0331) 81 31 52
e-mail: aem@aermacchi.it
CHAIRMAN: Dott Fabrizio Foresio
MANAGING DIRECTOR: Dott Ing Giorgio Brazzelli
GENERAL MANAGER: Dott Ing Bruno Cussigh
TECHNICAL DIRECTOR: Dott Ing Massimo Battini
COMMERCIAL MANAGER: Dott Cesare Cozzi

Aermacchi is aircraft manufacturing company of Aeronautica Macchi group; plants at Venegono airfield occupy total area of 274,000 m² (2,949,300 sq ft), including 113,000 m² (1,216,300 sq ft) covered space; flight test centre has covered space of 5,100 m² (54,900 sq ft) in total area of 28,000 m² (301,400 sq ft). Subsidiary companies are Logic (electrical equipment) and Nuova Orione.

Aermacchi active in aerospace ground equipment; produces fuselages for Fairchild Dornier 328JET (100th delivered in 1998) and engine nacelles for Airbus A 318, 319, 320, 321 and 330, Dassault Falcon 900 and Falcon 2000; 500th Airbus nacelle delivered in June 1999. Will build nacelles (in collaboration with Hurel Dubois) for Fairchild 728JET and, possibly, Embraer ERJ-170 families; was one of two Italian assembly centres for AMX (which see in

International section), having produced 36, plus nine AMX-Ts; also has important roles in Eurofighter Typhoon and Yakovlev Yak-130 programmes, offering Westernised version of latter as M-346. In January 1996, Aermacchi obtained production rights to the former Valmet L-90TP Redigo turboprop trainer, which was redesignated M-290TP Redigo. On 1 January 1997, Aermacchi acquired all training activities and programmes of SIAI-Marchetti, including the SF.260 and S.211 (but not SF.600 Canguro) and support of earlier S.205s, S.208s and SM-1019s remaining in service. Production was then transferred from Sesto Calende to the Aermacchi plant at Venegono. Since 1960, Aermacchi trainers have been bought by 39 countries.

UPDATED

AERMACCHI MB-339

TYPE: Basic jet trainer/light attack jet.

PROGRAMME: MB-339A selected by Italian Air Force on 11 February 1975 in preference to MB-338; one static test airframe and two prototypes: first flights (MM588) 12 August 1976 and (MM589) 20 May 1977; first flight of production MB-339A (MM54438) 20 July 1978; initial two deliveries (MM54439 and 54440) on 8 August 1979; first batch of 51 included three used as radio calibration aircraft at Pratica di Mare and 14 PANs for Frecce Tricolori aerobatic team.

MB-339C developed as an uprated MB-339 with a Viper 680 power plant; experimental installation in I-MABX, first flown on 9 June 1983; initial MB-339C first flight (I-AMDA) 17 December 1985; production aircraft (I-TRON) 8 November 1989. The 200th MB-339 (a 339CE for Eritrea) was delivered on 14 April 1997. MB-339s had flown more than 350,000 hours by 1998.

CURRENT VERSIONS: **MB-339A**: Initial military basic/advanced trainer (full details in 1990-91 *Jane's*). Final deliveries (small attrition replacement batch for Italy) in 1995. Main operator is 61° Stormo Scuola Volo Basico Iniziale su Aviogetti at Lecce-Galatina.

Lit110 billion MLU initiated in mid-1999 includes structural modifications on over 90 aircraft to extend service lives from 10,000 to 15,000 hours (from 20 to 30 years). Features of MLU include Litton Italia LISA-FG GPS/INS; new integrated air data transducer; provision for AN/ARC-150(V) Have Quick secure radio; Elmer crash recorder; ELT; instrument panel modifications; radio switches on control column; formation flying lights; higher capacity batteries connected in parallel with starter/generator; provision for target towing; anti-skid brakes; nosewheel splash-guard; repositioned IFF antenna; airborne stress gauge; and improved access, by means of additional removable panels, to wing/fuselage joint for inspection and engine maintenance. Depot maintenance interval will increase to 1,500 hours after MLU. Prototype (MM54453) flown mid-1999; 'production', to begin in 2001, will take between eight and 10 years for whole fleet.

MB-339PAN (*Pattuglia Aerobatica Nazionale*: national aerobatic team): Special version of MB-339A for Italian Air Force; smoke generator added; wingtip tanks deleted. Official debut 27 April 1982. Total 25 built as, or converted to, PAN. Will also be included in MLU.

MB-339AM: Special anti-ship version armed with AOSM Marte Mk 2A missile; avionics, equivalent to MB-339C, include new inertial navigator, Doppler radar, navigation and attack computers, head-up display and multifunction display. Prototype converted from MB-339A; qualification completed January 1995. No known orders.

MB-339B: Powered by 19.57 kN (4,400 lb st) Viper Mk 680-43; larger tip tanks. Prototype (demonstrator I-GROW) modified in 1993 with two LCD EFIS displays and air-to-air refuelling (AAR).

MB-339RM (*Radiomisure*: radio calibration): Three produced for Italian Air Force, 1981; withdrawn and transferred to training as MB-339As.

T-Bird II: One demonstrator modified for US JPATS competition; delivered to Lockheed factory at Marietta, Georgia, on 20 May 1992. Power plant, one Rolls-Royce Viper 680-582 of 17.79 kN (4,000 lb st) with noise reduction kit. Rejected in favour of Pilatus PC-9 (Raytheon T-6A Texan II), June 1995. Continues in use as chase aircraft.

MB-339CB: For Royal New Zealand Air Force; Viper 680 engine. Serves with 14 Squadron at Ohakea in training and light attack roles.

MB-339CD (C Digital): Developed for Italian Air Force advanced/fighter lead-in training. Rolled out 12 April 1996; first flight (conversion of MB-339A MM54544) 24 April 1996. Rolls-Royce Viper Mk 632-43 engine; removable in-flight refuelling probe (first phase of tanker trials completed March 1996 by development MB-339 I-GROW); new Sextant avionic architecture based on a single central mission computer, MIL-STD-1553B digital databus, one ring laser gyro platform with embedded GPS, EFIS cockpits with HUD in each, three identical liquid crystal colour MFDs and HOTAS controls. Other details generally as MB-339CB, but (compared with MB-339A) nose lengthened; elevators modified to give handling characteristics closer to the fast jets to which students will progress; some structural parts and systems modified; and pilots' escape system updated.

Achieved full operational capability with IAF in October 1998 after 148 sorties by prototype (latterly serialled MMX606). In early 2000, IAF was resisting pressure from Aermacchi and Italian industry ministry to order further 15 MB-339CDs.

Detailed description applies to the above version.

Cockpit of MB-339CD/FD

MB-339CE: Variant of MB-339CD; Viper 680 engine. Six ordered by Eritrea on 7 November 1995; deliveries began April 1997.

MB-339FD (Full Digital): Export equivalent of CD, with Viper 680 engine. Offered to Australia in 1994; unsuccessful in competition against the BAe Hawk. Also offered to Brazil and South Africa as Xavante/Impala replacement. Purchase by Venezuela of an initial eight announced 7 July 1998, but decision subsequently reversed and competition reopened.

CUSTOMERS: See table. Fifteen CDs ordered by Italian Air Force; initial delivery of two aircraft on 18 December 1996, followed by further 10 in 1997 and three in 1998 (last in November); first four to Reparto Sperimentale Volo at Pratica di Mare for trials before issue to 61° Stormo; some assigned to base flights at Gioia del Colle and Novara-Cameri for continuation training, including in-flight refuelling.

DESIGN FEATURES: Conventional subsonic jet trainer with tandem, stepped cockpits and low, moderately sweptback wing. Designed to MIL-A-8860A for 10,000 hours service life. Intended maintenance requirement of 5.95 mmh/fh; scheduled maintenance at 150 and 300 hour intervals; IRAN at 1,500 hours (extendable to 1,800 hours).

Wing section NACA 64A-114 (mod) at centreline, 64A-212 (mod) at tip; quarter-chord sweepback 8° 29′.

Aermacchi MB-339CD in service with 61° Stormo (*Paul Jackson/Jane's*)

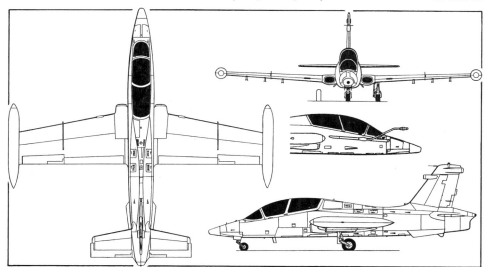

Aermacchi MB-339CD advanced trainer and attack aircraft, with additional starboard side view of optional refuelling probe (*Dennis Punnett/Jane's*)

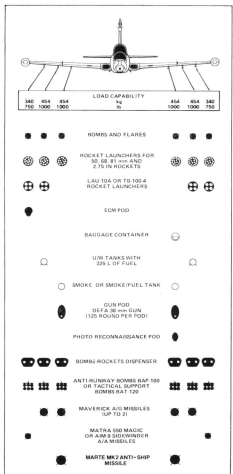

MB-339C weapon options

FLYING CONTROLS: Conventional. Power-assisted ailerons with artificial feel, plus servo tabs to assist emergency manual reversion; elevators and rudder actuated manually by pushrods; electrically controlled servo tab for control assistance and trimming on elevator; hydraulically actuated single-slotted flaps; two ventral strakes under tail; electrohydraulically actuated airbrake panel under forward fuselage; wing fence ahead of aileron inboard edge; both pilots have HUD; rear pilot elevated sufficiently to be able to aim guns and air-to-surface weapons and fly visual approaches (see nav/attack system under Avionics heading).

STRUCTURE: All-metal; stressed skin wings with main and auxiliary spars and spanwise stringers; bolted to fuselage; tip tanks permanently attached; rear fuselage detachable by four bolts for engine access.

LANDING GEAR: Hydraulically retractable tricycle type with oleo-pneumatic shock-absorbers, suitable for operation from semi-prepared runways. Hydraulically steerable nosewheel retracts forward; main units retract outward into wings. Low-pressure mainwheel tubeless tyres size 545 × 175-10 (14 ply rating); nosewheel tubeless tyre size 380 × 150-4 (6 ply rating). Emergency extension system. Hydraulic disc brakes with anti-skid system. Minimum ground turning radius 8.63 m (28 ft 3¾ in).

POWER PLANT: One Rolls-Royce Viper Mk 680-43 turbojet, installed rating 19.31 kN (4,340 lb st). MB-339Cs for Italy retain the MB-339A's 17.79 kN (4,000 lb st) Viper Mk 632-43. Fuel in two-cell rubber fuselage tank, capacity 781 litres (206 US gallons; 172 Imp gallons), and two wingtip tanks with combined capacity of 1,000 litres (264 US gallons; 220 Imp gallons). Total internal usable capacity 1,781 litres (470.5 US gallons; 392 Imp gallons). Self-sealing tanks optional. Single-point pressure refuelling point in port side of fuselage, below wing trailing-edge. Gravity refuelling points on top of fuselage and each tip tank. Provision for two drop tanks, each of 325 litres (86.0 US gallons; 71.5 Imp gallons) usable capacity, on centre underwing stations. Optional refuelling probe on starboard side of cockpits.

ACCOMMODATION: Crew of two in tandem, on Martin-Baker Mk 10LK zero/zero ejection seats in pressurised cockpit. Independent or rear-seat command ejection. Design in accordance with MIL-STD-203F. Rear seat elevated 32.5 cm (1 ft 1 in). Rearview mirror for each occupant. Two-piece moulded transparent canopy, opening sideways to starboard.

SYSTEMS: Pressurisation system maximum differential 0.24 bar (3.5 lb/sq in); cockpit designed for 40,000 pressurisation cycles. Bootstrap-type air conditioning system, also providing air for windscreen and canopy demisting. Hydraulic system, pressure 172.5 bars (2,500 lb/sq in), for actuation of flaps, aileron servos, airbrake, landing gear, wheel brakes and nosewheel steering. Back-up system for wheel brakes and emergency extension of landing gear. Main electrical DC power from one 28 V 9 kW engine-driven starter/generator and one 28 V 6 kW secondary generator; power distribution via five 28 V buses. Two 24 V 22 Ah Ni/Cd batteries for engine starting. Fixed-frequency 115/26 V AC power from two 600 VA single-phase static inverters; provision for additional inverter for three-phase AC. External power receptacle. Low-pressure demand oxygen system, operating at 28 bars (400 lb/sq in). Anti-icing system for engine air intakes.

AVIONICS: Representative fit includes following:
Comms: Honeywell AN/APX-100 IFF.
Flight: BAE 620 kbit navigation computer; RT-1159/A or Rockwell Collins AN/ARN-118(V) Tacan, or Honeywell KDM 706A DME; Rockwell Collins 51RV-4B VOR/ILS and MKI-3 marker beacon receiver; Rockwell Collins ADF-60A ADF/L (or, optionally, DF-301E V/UHF ADF); BAE AD-660 Doppler velocity sensor integrated with Litton LR-80 inertial platform; HOTAS controls. Flight recorder and airborne stress gauge.
Instrumentation: Alenia CRT multifunction display; Alenia/Honeywell HG7505 radar altimeter; Astronautics AN/ARU-50/A attitude director indicator and AN/AQU-13 HSI.
Mission: Kaiser Sabre head-up display and weapon aiming computer; Logic stores management system; provision for FIAR P 0702 laser range-finder; Lockheed Martin Fairchild video camera; photographic pod with four Vinten 70 mm cameras. Optional laser range-finder.
Self-defence: Options include ELT-156 radar warning system; single Elettronica ELT-555 ECM pod combined with AN/ALE-40 chaff/flare dispenser.

ARMAMENT: Up to 1,814 kg (4,000 lb) of external stores on six underwing hardpoints. Four inner hardpoints each stressed for up to 454 kg (1,000 lb) load, and two outer hardpoints each for up to 340 kg (750 lb) load. RNZAF aircraft fitted for AIM-9 Sidewinder and AGM-65 Maverick. Provision on two inner stations for installation of two Macchi gun pods, each containing either a 30 mm DEFA 553 cannon with 120 rounds or a 12.7 mm AN/M-3 machine gun with 350 rounds.

Other typical loads can include two Matra BAe 550 Magic or AIM-9 Sidewinder air-to-air missiles on two outer stations; six general purpose or cluster bombs of appropriate weights; six AN/SUU-11A/A 7.62 mm Minigun pods, each with 1,500 rounds; six Matra launchers, each for eighteen 68 mm rockets; six AN/LAU-68/A or AN/LAU-32G launchers, each for seven 2.75 in rockets; six Aerea AL-25-50 or AL-18-50 launchers, each with twenty-five or eighteen 50 mm rockets respectively; six AL-18-80 launchers, each with twelve 81 mm rockets; four AN/LAU-10/A launchers, each with four 5 in Zuni rockets; four TDA 100-4 launchers, each with four 100 mm TDA rockets; six Bristol Aerospace LAU-5002 launchers for CRV-7 high-velocity rockets; six Aerea BRD bomb/rocket dispensers; six Aermacchi 11B29-003 bomb/flare dispensers; six TDA 14-3-M2 adaptors, each with six BAP 100 anti-runway bombs or BAT 120 tactical support bombs. Marte 2A anti-ship missile completed MB-339 qualification trials in February 1995.

Following data are for MB-339FD, but generally applicable to MB-339CD.

DIMENSIONS, EXTERNAL:
Wing span over tip tanks	11.22 m (36 ft 9¾ in)
Wing aspect ratio	6.5
Length overall	11.24 m (36 ft 10½ in)
Height overall	3.945 m (12 ft 11¼ in)
Elevator span	4.16 m (13 ft 7¾ in)
Wheel track	2.48 m (8 ft 1¾ in)
Wheelbase	4.37 m (14 ft 4 in)

AREAS:
Wings, gross	19.30 m² (207.7 sq ft)
Ailerons (total)	1.33 m² (14.32 sq ft)
Trailing-edge flaps (total)	2.55 m² (27.45 sq ft)
Airbrake	0.52 m² (5.60 sq ft)
Fin	2.21 m² (23.78 sq ft)
Rudder, incl tab	0.68 m² (7.32 sq ft)
Tailplane	3.38 m² (36.38 sq ft)
Elevators (total incl tabs)	0.98 m² (10.55 sq ft)

WEIGHTS AND LOADINGS:
Weight empty	3,334 kg (7,350 lb)
Max weapon load	1,814 kg (4,000 lb)
Max fuel weight: internal	1,430 kg (3,153 lb)
external	522 kg (1,151 lb)
T-O weight, clean	4,950 kg (10,913 lb)
Max T-O weight with external stores	6,350 kg (14,000 lb)
Max wing loading	329.0 kg/m² (67.39 lb/sq ft)
Max power loading	329 kg/kN (3.23 lb/lb st)

PERFORMANCE (at trainer clean T-O weight, except where indicated):
Never-exceed speed (V_{NE})	M0.82 (500 kt; 926 km/h; 575 mph)
Max level speed at S/L:	
clean	490 kt (907 km/h; 564 mph)
with four 500 lb bombs	440 kt (815 km/h; 506 mph)
Max level speed at 9,140 m (30,000 ft)	
	M0.77 (441 kt; 815 km/h; 508 mph)
Max speed for landing gear extension	
	175 kt (324 km/h; 202 mph)
T-O speed	100 kt (185 km/h; 115 mph)
Approach speed over 15 m (50 ft) obstacle	
	102 kt (189 km/h; 117 mph)
Stalling speed, 20% fuel	86 kt (160 km/h; 99 mph)
Max rate of climb at S/L	2,011 m (6,600 ft)/min
Service ceiling	13,715 m (45,000 ft)
T-O run at S/L	540 m (1,775 ft)
Landing run at S/L, 20% fuel	470 m (1,545 ft)
Radius of action, 10% reserves:	
with four 500 lb bombs:	
lo-lo-lo	145 n miles (268 km; 166 miles)
hi-lo-hi	255 n miles (472 km; 293 miles)
with Marte ASM and two underwing tanks:	
lo-lo-lo	230 n miles (426 km; 264 miles)
hi-lo-hi	440 n miles (814 km; 506 miles)
with two 500 lb bombs, two rocket launchers and two gun pods, 20 min loiter and 6 min combat, hi-lo-hi	
	125 n miles (231 km; 143 miles)
Ferry range: clean	950 n miles (1,759 km; 1,039 miles)
with two underwing drop tanks, 10% reserves	
	1,050 n miles (1,944 km; 1,208 miles)
Max endurance with drop tanks	3 h 50 min
g limit	+7.33/−4.0

UPDATED

MB-339 PRODUCTION

Customer	Variant	Qty	First Aircraft	First Delivered
Argentine Navy	AA	10[2]	0761	1980
Eritrean Air Force	CE	6	409[3]	Apr 1997
Ghana Air Force	A	2	G800	1987
		2	G802	1994
Italian Air Force	A[1]	80	MM54438	19 Feb 1979
		21	MM54533	1986
		6	MM55052	1994
	CD	15	MM55062	18 Dec 1996
Malaysian Air Force	AM	13	M34-01	1983
New Zealand Air Force	CB	18	NZ6460	13 Mar 1991
Nigerian Air Force	AN	12	301	Jun 1984
Peruvian Air Force	AP	16	452	Nov 1981
UAE (Dubai) Air Force	A	2	431	24 Mar 1984
		3	433	1987
Subtotal		**206**		
Prototypes		2	MM588	1976
Demonstrator	A	1	I-NEUN	1980
	B	1	I-GROW	1984
	C	1	I-AMDA	1985
Total		**211**		

[1] Also 339PAN and 339RM
[2] One converted to JPATS demonstrator N339L; others lost or withdrawn from service; including at least two believed transferred to UAE
[3] Eritrean aircraft serial numbered in reverse order

AERMACCHI SF-260
Uruguayan Air Force designation: T-260

TYPE: Basic prop trainer/attack lightplane.

PROGRAMME: Originated as F.250 designed by Stelio Frati and made by Aviamilano (see 1965-66 *Jane's*); civil SIAI-Marchetti SF.260A and B detailed in 1980-81 *Jane's*; SF.260C detailed in 1985-86 *Jane's*; SF.260M military trainer detailed in 1984-85 and earlier *Jane's*; still in low-volume production as a military trainer following transfer of design rights to Aermacchi in 1997 as SF-260 (with hyphen).

CURRENT VERSIONS: **SF-260E/F:** Direct-injection (E) and carburetted (F) versions with 100 kg (220 lb) higher aerobatic weight than superseded SF.260D. Certified by RAI on 21 January 1992 and by FAA on 17 August 1994. *Main description applies to SF-260E/F and E/F Warrior.*

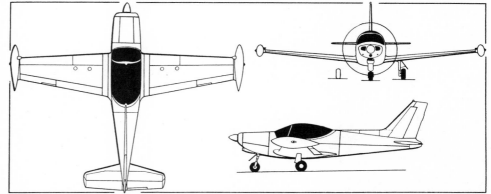
Aermacchi SF-260E/F basic trainer *(Paul Jackson/Jane's)*

SF-260E/F Warrior: Trainer/tactical support version. Two underwing pylons, for up to total 300 kg (661 lb) of external stores, and cockpit stores selection panel. Able to undertake a wide variety of roles, including low-level strike; forward air control; forward air support; armed reconnaissance; and liaison.

SF-260TP: Turboprop-powered version. Described separately.

CUSTOMERS: Total of more than 650 produced (including turboprop version) by SIAI-Marchetti at Sesto Calende for customers in 25 countries; further 49 sold by Aermacchi. By 1999, 26 air arms had flown 1,700,000 hours on SF-260s. Civil operators include Sabena, Royal Air Maroc, and Alitalia. Six SF-260Fs ordered by Air Force of Zimbabwe, early 1997, of which last delivered in October 1998; these were first of type built at Venegono. In June 1998, 12 SF-260Es bought by Venezuelan Air Force, but delivery postponed following reopening of jet trainer competition. Delivery of 13 SF-260Es to Uruguayan Air Force began in September 1999, following first flight on 23 July. Mexico placed order in early 1998, initially for 13, but subsequently 30, of which first six delivered in early 2000.

AERMACCHI SF-260 MILITARY CUSTOMERS

Operator	Variant	Qty	First Aircraft
Belgium	MB	36	ST-01
	D	9	ST-40
Brunei	W	2	129
Burundi	C	3	9U-RUA
Comoros	W	2	
Dubai	TP	6	401
Ethiopia	TP	12	156?
Haiti	TP	6	1271
Ireland	WE	11	222
Italy	260	3	MM91004
	AM	45	MM54418
Libya	W	240*	
Mauritania	E	5†	–
Myanmar	M	20	2001
Mexico	E	30†	–
Philippines	MP	30	601
	WP	19	631
	TP	18	701
Singapore	W	11	151
Somalia	C	6	6O-SBC
	W	6	
Sri Lanka	TP	11	CT121
Thailand	MT	18	15-01
Tunisia	CT	9	W41501
	WT	12	W41401
Turkey	D	40	773
Uganda		6	AF501
Uruguay	E	13†	610
Zaire	MC	12	AT-101
Zambia	MZ	9	AF-507
	TP		AF-
Zimbabwe	C	17	R9900
	W	14	R3260
	F	6†	

* May not all have been delivered
† Built by Aermacchi

Notes: Additional operators are Burkina Faso (aircraft ex-Philippines) and Chad (ex-Libya)

DESIGN FEATURES: Side-by-side, piston-engined aerobatic trainer. Low-mounted wings with forward-swept trailing-edges and tip tanks; sweptback fin.
Wing section NACA 64_1-212 (modified) at root, 64_1-210 (modified) at tip; dihedral 6.5° 20′ from roots.

FLYING CONTROLS: Conventional and manual. Frise ailerons each with servo tab; trim tabs for all three axes; four-way trim button on each stick top; electrically operated slotted flaps; controls operated by dual cables.

STRUCTURE: All-metal stressed skin structure; wing skin formed of butt-jointed panels flush riveted; single main spar and auxiliary rear spar; press-formed ribs; wings bolted together on centreline and attached to fuselage by six bolts.

LANDING GEAR: Electrically retractable tricycle type, with manual emergency actuation. Inward-retracting main gear, of trailing arm type, and rearward-retracting nose unit, each embodying Magnaghi oleo-pneumatic shock-absorber (Type 2/22028 in main units). Nose unit is of leg and fork type, with coaxial shock-absorber and torque strut. Cleveland P/N 3080A mainwheels, with size 6.00-8 tube and tyre (6 ply rating), pressure 2.45 bars (36 lb/sq in). Cleveland P/N 40-77A nosewheel, with size 5.00-5 tube and tyre (8 ply rating), pressure 1.96 bars (28 lb/sq in). Cleveland P/N 3000-500 independent hydraulic single-disc brake on each mainwheel; parking brake. Nosewheel steering (±20°) operated directly by rudder pedals.

POWER PLANT: SF-260F powered by one 194 kW (260 hp) Textron Lycoming O-540-E4A5 flat-six engine. Fuel-injection AEIO-540-D4A5 engine powers SF-260E. Both have a Hartzell HC-C2YK-1BF/8477-8R two-blade constant-speed metal propeller.

Fuel in two light alloy tanks in wings each holding 49.5 litres (13.1 US gallons; 10.9 Imp gallons); and two permanent wingtip tanks each holding 72 litres (19.0 US gallons; 15.85 Imp gallons). Total internal fuel capacity 243 litres (64.2 US gallons; 53.3 Imp gallons), of which 235 litres (62.1 US gallons; 51.7 Imp gallons) are usable. Individual refuelling point on top of each tank. In addition, SF-260W may be fitted with two 80 litre (21.1 US gallon; 17.5 Imp gallon) auxiliary tanks on underwing pylons. Oil capacity (all models) 11.4 litres (3.0 US gallons; 2.5 Imp gallons).

ACCOMMODATION: Two pilots side by side in adjustable seats, with full blind-flying panel on right and reduced panel and radio on left. Third seat in rear. All three seats equipped with lap belts and shoulder harnesses. Baggage compartment aft of rear seat. Upper portion of sliding canopy tinted. Emergency canopy release handle for front seat occupant. Steel tube windscreen frame for protection in the event of an overturn. Air conditioning, oxygen and enlarged canopy optional.

SYSTEMS: Foot-powered hydraulics for mainwheel brakes only. No pneumatic system. 24 V DC electrical system of single-conductor negative earth type, including 70 A Prestolite engine-mounted alternator/rectifier and 24 V 24 Ah battery, for engine starting, flap and landing gear actuation, fuel booster pumps, electronics and lighting. Sealed battery compartment in rear of fuselage on port side. Connection of an external power source automatically disconnects the battery. Heating system for carburettor air intake. Emergency electrical system for extending landing gear if normal electrical actuation fails; provision for mechanical extension in the event of total electrical failure. Cabin heating, and windscreen de-icing and demisting, by heat exchanger using engine exhaust air. Additional manually controlled warm air outlets for general cabin heating.

AVIONICS: *Comms:* Honeywell or Rockwell Collins airways radio with dual com and nav; in SF-260E/F, avionics mounted aft of cabin and battery moved forward.

EQUIPMENT: Military equipment to customer's requirements. External stores can include one or two reconnaissance pods with two 70 mm automatic cameras, or two supply containers or two external tanks. Landing light in nose, below spinner. Two Servaux Samar-Sater 83/0 or 83/1 emergency packs.

ARMAMENT: Warrior has two underwing hardpoints, able to carry external stores on NATO 356 mm (14 in) shackles up to a maximum of 300 kg (661 lb) when flown as a single-seater. Typical alternative loads can include one or two SIAI gun pods, each with one or two 7.62 mm FN machine guns and 500 rounds; two 0.50 Browning machine gun pods; two Aerea AL-8-70 launchers each with eight 2.75 in rockets; two Forges de Zeebrugge LAU-32 launchers each with seven 2.75 in rockets; two Aerea AL-18-50 launchers each with eighteen 2 in rockets; two Aerea AL-8-68 launchers each with eight 68 mm rockets; two Thomson-Brandt 68-7 launchers each with seven 68 mm rockets; two Aerea AL-6-80 launchers each with six 81 mm rockets; two Thiokol LUU-2/B parachute flares; two SAMP EU 32 125 kg general purpose bombs or EU 13 120 kg fragmentation bombs; two SAMP EU 70 50 kg general purpose bombs; Mk 76 11 kg practice bombs; two cartridge throwers for 70 mm multipurpose cartridges, F 725 flares or F 130 smoke cartridges.

DIMENSIONS, EXTERNAL:
Wing span over tip tanks	8.35 m (27 ft 4¾ in)
Wing chord: at root	1.60 m (5 ft 3 in)
mean aerodynamic	1.325 m (4 ft 4¼ in)
at tip	0.785 m (2 ft 7 in)
Wing aspect ratio (excl tip tanks)	6.3
Wing taper ratio	2.2
Length overall	7.10 m (23 ft 3½ in)
Fuselage: Max width	1.10 m (3 ft 7¼ in)
Max depth	1.04 m (3 ft 5 in)

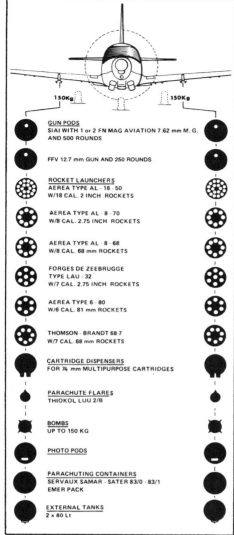

Weapon options for military SF-260E/F or two-pylon configuration for SF-260TP; latter may also carry second pairs of gun pods, parachute flares, bombs or parachuting containers on outboard pylons
0011077

Height overall	2.41 m (7 ft 11 in)
Elevator span	3.01 m (9 ft 10½ in)
Wheel track	2.275 m (7 ft 5½ in)
Wheelbase	1.66 m (5 ft 5¼ in)
Propeller diameter	1.93 m (6 ft 4 in)
Propeller ground clearance	0.32 m (1 ft 0½ in)

DIMENSIONS, INTERNAL:
Cabin: Length	1.66 m (5 ft 5¼ in)
Max width	1.00 m (3 ft 3¼ in)
Height: from seat cushion:	
to canopy	0.98 m (3 ft 2½ in)
to enlarged canopy	1.07 m (3 ft 6 in)
Volume	1.5 m³ (53 cu ft)
Baggage compartment volume	0.18 m³ (6.4 cu ft)

Aermacchi SF-260E with representative armament
2001/0089496

AREAS:	
Wings, gross	10.10 m² (108.7 sq ft)
Ailerons (total, incl tabs)	0.76 m² (8.18 sq ft)
Trailing-edge flaps (total)	1.18 m² (12.70 sq ft)
Fin	0.76 m² (8.18 sq ft)
Dorsal fin	0.16 m² (1.72 sq ft)
Rudder, incl tab	0.60 m² (6.46 sq ft)
Tailplane	1.46 m² (15.70 sq ft)
Elevator, incl tab	0.96 m² (10.30 sq ft)
WEIGHTS AND LOADINGS:	
Manufacturer's basic weight empty	779 kg (1,717 lb)
Max fuel weight: wing internal and tip tanks (all versions)	169 kg (372.5 lb)
underwing tanks	115 kg (254 lb)
Max T-O weight: E/F	1,200 kg (2,645 lb)
Warrior	1,350 kg (2,976 lb)
Max wing loading: E/F	118.8 kg/m² (24.33 lb/sq ft)
Warrior	128.7 kg/m² (26.36 lb/sq ft)
Max power loading: E/F, M	6.18 kg/kW (10.17 lb/hp)
Warrior	6.70 kg/kW (11.02 lb/hp)
PERFORMANCE (at 1,100 kg; 2,425 lb):	
Never-exceed speed (V_{NE})	238 kt (441 km/h; 274 mph)
Max level speed at S/L	182 kt (337 km/h; 209 mph)
Stalling speed, flaps down	59 kt (110 km/h; 68 mph)
Max rate of climb at S/L	548 m (1,800 ft)/min
Service ceiling	5,790 m (19,000 ft)
T-O run	275 m (905 ft)
Landing run	270 m (885 ft)
Range with max internal fuel, two pilots	596 n miles (1,104 km; 686 miles)
Max range, internal and external fuel	800 n miles (1,481 km; 920 miles)
g limits	+6/−3

UPDATED

Aermacchi SF-260TPs of the Sri Lankan Air Force *(Denis Hughes)*   1999/0051081

AERMACCHI SF-260TP

TYPE: Basic turboprop trainer/attack lightplane.
PROGRAMME: Re-engined SF-260. First flight July 1980; airframe virtually unchanged aft of firewall except for inset rudder trim tab and automatic fuel feed system. Italian civil certification received 29 October 1993.

SF-260E/F description applies also to TP, except in the following details:

CUSTOMERS: Some 60 SF.260TPs sold by SIAI-Marchetti to Dubai (seven), Ethiopia (10), Philippines (18 new, plus conversions of SF-260M/W), Sri Lanka (nine) and Zimbabwe (14).
POWER PLANT: One fully aerobatic Rolls-Royce 250-B17D turboprop, flat rated at 261 kW (350 shp) and driving a Hartzell HC-B3TF-7A/T10173-25R three-blade constant-speed fully feathering and reversible-pitch propeller. Electrical anti-icing system for engine air intake optional. Fuel capacity as for SF-260E/F; automatic fuel feed system. Oil capacity 7 litres (1.8 US gallons; 1.5 Imp gallons).
ARMAMENT: Four underwing hardpoints, able to carry external stores on NATO 356 mm (14 in) shackles up to a maximum of 300 kg (661 lb) when flown as single-seater. Outboard pylons rated at 50 kg (110 lb) each; inboard pylons at 150 kg (331 lb) each, or 100 kg (220 lb) when outboard pylons in use. Stores options as for SF-260 Warrior.

DIMENSIONS, EXTERNAL:	
Length overall	7.40 m (24 ft 3¼ in)
WEIGHTS AND LOADINGS:	
Weight empty, equipped	750 kg (1,654 lb)
Max T-O weight: trainer	1,250 kg (2,755 lb)
armed	1,350 kg (2,976 lb)
Max wing loading: trainer	123.7 kg/m² (25.34 lb/sq ft)
Max power loading: trainer	4.79 kg/kW (7.87 lb/shp)
armed	5.18 kg/kW (8.50 lb/shp)
PERFORMANCE (at T-O weight of 1,200 kg; 2,645 lb, ISA):	
Never-exceed speed (V_{NE})	236 kt (437 km/h; 271 mph)
Max level speed at 3,050 m (10,000 ft)	230 kt (426 km/h; 265 mph)
Max cruising speed at 3,050 m (10,000 ft)	216 kt (400 km/h; 248 mph)
Econ cruising speed at 4,575 m (15,000 ft)	170 kt (315 km/h; 195 mph)
Stalling speed at S/L, flaps down, power off	61 kt (113 km/h; 70 mph)
Max rate of climb at S/L	661 m (2,170 ft)/min
Service ceiling	7,500 m (24,600 ft)
T-O run	298 m (980 ft)
Landing run, without reverse pitch	307 m (1,010 ft)
Range at 4,575 m (15,000 ft) with max fuel, 30 min reserves	512 n miles (949 km; 589 miles)
Radius of action, pilot only, 300 kg (661 lb) weapon load, 20 min over target	80 n miles (148 km; 92 miles)

UPDATED

AERMACCHI M-346

TYPE: Advanced jet trainer/light attack jet.
PROGRAMME: Announced 25 July 2000. Westernised version of Yakovlev Yak-130 (which see in Russian section), conforming to requirements of Eurotrainer group. Two prototypes each to be powered by two Honeywell/ITEC F124 turbofans. Roll-out due June 2002, and first flight 12 months later. Other features include Teleavio/Marconi Italia flight control system, Dowty and Microtecnica actuators, Alenia DSAE HUD and main displays/processor, Honeywell GPS/INS and Martin-Baker ejection seats.

Aermacchi shares design rights and has production and modification rights for the Yak-130. Development of the aircraft was transferred to Italy in mid-1998 to prevent further delays resulting from the economic situation in Russia. Some funding provided by Russian Ministry of Industry. Dott Massimo Luchesini is Aeromacchi's AEM/Yak-130 programme director. Italian Air Force expected to certify aircraft and issue service release, although no firm local requirement yet exists.

Cockpit updated from a design derived from MB-339CD and intended to simulate Eurofighter Typhoon, with three active matrix LCD 127×127 mm (5 × 5 in) multifunction displays, local embedded symbol generator and wide-angle raster HUD. Optional two-screen cockpit prepared for customers who wish to simulate JSF environment. Components will be selected according to best price, and not by single-source contracts.

NEW ENTRY

OTHER AIRCRAFT

Production rights to the (former SIAI-Marchetti) S.211 jet trainer and light attack aircraft were transferred to Aermacchi in 1997. Promotion continued during 1999, but no production has taken place since before the sale. An illustrated description last appeared in the 1997-98 *Jane's*. The M-290TP Redigo turboprop trainer was last described in the 1999-2000 *Jane's*.

See International section for Aermacchi/Alenia/Embraer **AMX** attack aircraft.

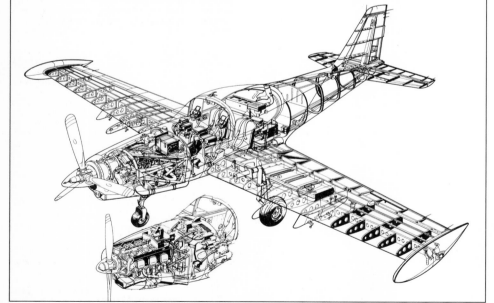

Structural details of the SF-260TP, with scrap view of the piston-engined SF-260E/F   1999/0051082

UPDATED

AGUSTA

AGUSTA (Group IRI Finmeccanica)

Via Giovanni Agusta 520, I-21017 Cascina Costa di Samarate (VA)
Tel: (+39 0331) 22 97 69
Fax: (+39 0331) 22 99 84
Web: http://www.agusta.com
CHAIRMAN AND CEO: Amedeo Caporaletti
VP EXTERNAL RELATIONS AND COMMUNICATIONS:
Gian Luigi Ghezzi

CO-GENERAL MANAGER, MARKETING AND SALES: Giuseppe Orsi
PRESS OFFICER: Gianluca Grimaldi

VERGIATE WORKS:
Via Roma 51, I-21019 Vergiate (VA)
Tel: (+39 0331) 94 01 11
Fax: (+39 0331) 94 79 09

Formed in 1977; the Agusta group (see 1980-81 *Jane's*) completely reorganised from 1 January 1981 under new holding company Agusta; became part of Italian public holding company EFIM, employing nearly 10,000 people in 12 factories in various parts of Italy. Agusta workforce in Italy and abroad was 5,019 in 1999 and turnover was US$800 million. Order book was US$3.4 billion.

As part of recovery of liquidated state-owned EFIM, Agusta group integrated into Finmeccanica on 20 December 1996.

On 16 April 1998, Finmeccanica and the United Kingdom's GKN, owners of Agusta and Westland, respectively, announced signature of an MoU which was intended to lead to an 'alliance of equals' of their helicopter

AGUSTA—AIRCRAFT: ITALY

divisions. This was effected on 26 July 2000, as described under AgustaWestland in the International section. The pooling of design, manufacturing and marketing activities will not require refinancing nor affect previous programme-based agreements with other partners.

In November 1998 Agusta and Bell Helicopter Textron (which see in Canada and United States sections) signed an agreement to form Bell/Agusta Aerospace Company, which will share development, manufacture and marketing of the Bell/Agusta BA609 civil tiltrotor and AB139 utility helicopter, both of which are described in the International section. See also the Eurotilt entry in that section, in which AgustaWestland also involved.

By mid-2000, Agusta had built more than 3,770 helicopters under licence from Bell, Boeing, MD helicopters and Sikorsky.

INTERNATIONAL AFFILIATES:
Agusta Aerospace Corporation
2655 Interplex Drive, Travose, Philadelphia, Pennsylvania 19047, USA
Tel: (+1 215) 281 14 00
Fax: (+1 215) 281 04 40
Telex: 6851181

Agusta Aerospace Service SA
Keiberg Park, 23 Excelsiorlaan, B-1930 Zaventem, Belgium
Tel: (+32 2) 648 55 15
Telex: 63349

Bell/Agusta Aerospace Company
See International section.

EH Industries Ltd
See International section

NH Industries sarl
See International section

UPDATED

CASCINA COSTA WORKS

Original Agusta company established 1907 by Giovanni Agusta; acquired licence for Bell 47 in 1952; first flight of first Agusta example 22 May 1954; later produced Bell 204, 205, 206 and 212; Bell AB412 still in low-volume production; also produced various versions of Sikorsky S-61 under licence and is partner with Westland in EH 101 (see under EHI in International section); participates in NH 90 programme (see International). Own designs include A 109 multirole helicopter, A 119 Koala, A 129 anti-tank helicopter and AB139 joint venture with Bell Helicopter Textron.
UPDATED

AGUSTA A 129 MANGUSTA
English name: Mongoose
TYPE: Attack helicopter.
PROGRAMME: Italian Army specification issued 1972; A 129 given go-ahead March 1978; final form settled 1980; detail design completed 30 November 1982; first flights of five development aircraft 11 September 1983, 1 July and 5 October 1984, 27 May 1985 and 1 March 1986; first delivery October 1990. In 1999, Italian government approved A 129 upgrade, which will raise first 45 army helicopters to multirole standard. Raytheon Stinger, in AAM version, selected in March 2000 to arm Italian A 129s; integration due to be complete by 2002.
CURRENT VERSIONS: **Anti-tank:** Initial Italian Army version; also demonstrated for export *(described below)*. Lot 2 aircraft (from 16th production onward – MM81391) equipped with secure communications, cockpit lighting compatible with third-generation NVGs, IR suppressors, IR jammers, laser warning system, improved main computer software, auxiliary fuel tanks and folding main rotor blades.
Multirole: Italian Army is receiving aircraft Nos. 46 to 60 (MM81421 to 81435) in *combattimento* (combat) configuration; main features are five-blade rotor; transmission uprated to 1,268 kW (1,700 shp); maximum take-off weight increased to 4,600 kg (10,141 lb); Lockheed Martin/Otobreda TM 197B 20 mm cannon in nose turret; and Stinger AAMs. Fifth A 129 prototype (MMX598) converted to this configuration by 1997. Other features are small strake along port side of tailboom; long, tray-shaped ammunition tank on port side of fuselage; revised cockpit layout, including HOCAC controls; updated 'HIRNS Plus' navigation/sighting system; integrated GPS; and uprated defensive aids, comprising BAE RALM-01 laser warning receiver, Elettronica ELT-157 RWR, ECDS chaff/flare dispenser and DASA/BAE passive missile warning (being applied to earlier

Italian Army Agusta A 129 from Lot 2 *(Paul Jackson/Jane's)*

aircraft). Similar upgrade of first 45 Mangustas due for completion by 2006.
Shipborne: Proposed maritime anti-ship version; no orders received by early 2000.
A 129 International: Described separately.
CUSTOMERS: First five of planned 60 for Italian Army anti-tank squadrons (15 Lot 1 and 45 Lot 2) delivered October 1990 after delay of more than a year to allow fitting of Saab/ESCO HeliTOW system with nose-mounted sight; currently operated by 493° Squadrone Elicotteri da Attacco of 49° Gruppo Squadroni at Rimini/Miramare. First Lot 2 A 129, with Honeywell/OMI helicopter IR navigation system (HIRNS), entered service in August 1993; operated by 7th Attack Helicopter Regiment 'Vega' at Casarsa and Army Aviation Centre at Viterbo; 45 in service by end of 1996; last 15 aircraft will be delivered as Multirole version; no known deliveries of these by mid-2000.
DESIGN FEATURES: Fully articulated four-blade main rotor with blades retained by single elastomeric bearing and restrained by hydraulic drag damper and mechanical droop stop; main rotor blade folding on Lot 2 aircraft.
Main transmission has independent oil cooling system; intermediate and tail rotor gearboxes grease lubricated; all designed for at least 30 minutes run dry; accessory gearbox can be run independently on ground without rotor engagement by No. 1 engine engaged by pilot-operated clutch.
FLYING CONTROLS: Full-time dual electronic flight controls, with full manual reversion, provide automatic heading hold, autohover, autopilot modes and autostabiliser modes, all selectable by pilot; gunner in front seat has cyclic side-arm controller, normal collective lever and pedals and has full access to AFCS; electrical inputs from AFCS integrated with hydraulic-powered control units.
STRUCTURE: Composites materials account for 45 per cent of fuselage weight (less engines) and 16.1 per cent of total empty weight; material used for fuselage panels, nosecone, tailboom, tail rotor pylon, engine nacelles, canopy frame and maintenance panels; each blade has CFRP and Nomex main spar, Nomex honeycomb leading- and trailing-edges, composites skins, stainless steel leading-edge abrasion strip and frangible tip; control linkage runs inside driveshaft to reduce radar signature, avoid icing and improve ballistic tolerance; blades tolerant to 12.7 mm hits, possibly also 23 mm; delta-hinged two-blade tail rotor with broad-chord blades for ballistic tolerance. Total 70 per cent of airframe surface is composites. Bulkhead in nose and A frame running up through fuselage to rotor pylon protect crew against roll-over; overall IR suppressing paint; airframe meets MIL-STD-1290 crashworthiness covering vertical velocity changes of 11.2 m (36 ft 9 in)/s and longitudinal changes of 13.1 m (43 ft 0 in)/s.
LANDING GEAR: Non-retractable tailwheel type, with single wheel on each unit. Two-stage hydraulic shock-strut in each main unit designed to withstand normal loads and hard landings at descent rates in excess of 10 m (32.8 ft)/s.

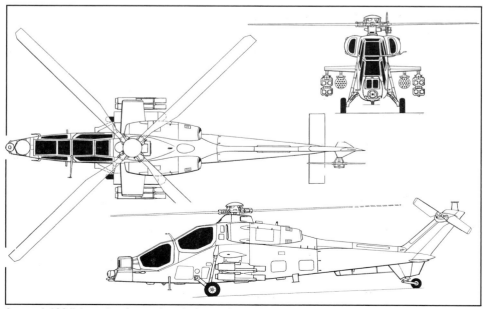

Agusta A 129 light anti-tank, attack and advanced scout helicopter *(Dennis Punnett/Jane's)*

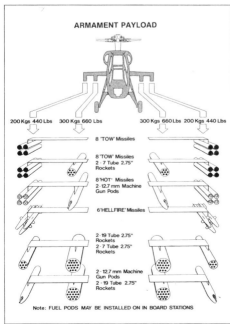

A 129 weapon options

A 129 International armed with Hellfire ATMs, controlled by a Tamam NTS-A sighting system *2001*/0034420

Model of A 129 Scorpion, offered to Australia ***(Paul Jackson/Jane's)*** *2001*/0093629

POWER PLANT: Two Rolls-Royce Gem 1004 turboshafts, each with a T-O rating of 657 kW (881 shp) for 30 minutes; maximum continuous rating of 615 kW (825 shp) for normal twin-engined operation; intermediate contingency rating of 657 kW (881 shp) for 30 minutes; maximum contingency rating of 704 kW (944 shp) for 2½ minutes; and emergency rating (S/L, ISA) of 759 kW (1,018 shp) for 20 seconds. Transmission rating (Lot 2) is 969 kW (1,300 shp) (two engines), 704 kW (944 shp) for single-engined operation, with emergency rating of 759 kW (1,018 shp); power input into transmission is at 27,000 rpm from the RR 1004. Standard transmission rating for Lot 3 (Multirole) aircraft is 1,268 kW (1,700 shp); 1,566 kW (2,100 shp) in emergency. Production engines licence-built in Italy by Piaggio. Fireproof engine compartment, with engines widely spaced to improve survivability from enemy fire.

Two separate fuel systems, with cross-feed capability; interchangeable self-sealing and crash-resistant tanks, self-sealing lines, and digital fuel feed control. Tanks can be foam-filled for fire protection. Single-point pressure refuelling. IR exhaust suppression system (from Lot 2) and low engine noise levels. Separate independent lubrication oil-cooling system for each engine. Provision (Lot 2 aircraft) for auxiliary (self-ferry) fuel tanks on outboard underwing stations.

ACCOMMODATION: Pilot and co-pilot/gunner in separate cockpits in tandem. Elevated rear (pilot's) cockpit. External crew field of view exceeds MIL-STD-850B. Each cockpit has a flat plate low-glint canopy with upward-hinged door panels on starboard side, blow-out port side panel for exit in emergency, and Martin-Baker crashworthy seat with sliding side panels of composites armour. Landing gear design and crashworthy seats reduce impact from 50 *g* to 20 *g* in crash.

SYSTEMS: Hydraulic system includes two main circuits dedicated to flight controls and two independent circuits for rotor and wheel braking. Main system operates at pressure of 207 bars (3,000 lb/sq in) and is fed by two independent power groups integrated and driven mechanically by the main transmission. Tandem actuators are provided for main rotor flight controls. Hydraulic system flow rate 23.6 litres (6.2 US gallons; 5.2 Imp gallons)/min in each main group. Spring-type reservoirs, pressurised at 0.39 bar (5.6 lb/sq in).

AVIONICS: *Comms:* Dual Elmer SRT-651/A U/VHF and single Elmer SRT-170/EB4 HF/SSB radios; Italtel IFF (with Mode 4 encryption unit in Lot 2 aircraft).

Flight: Fully integrated digital multiplex system (IMS) controls navigation, flight management, weapon control, autopilot, monitoring of transmission and engine condition, fuel/hydraulic/electrical systems, caution and warning systems; IMS managed by two Agusta Sistemi/Harris central computers, each capable of operating independently, backed by two interface units which pick up outputs from sensors and avionic equipment and transfer them, via redundant MIL-STD-1553B databusses, to main computers for real-time processing; processed information is presented to pilot and co-pilot/gunner on separate graphic/alphanumeric head-down multifunction displays (MFDs) with standard multifunction keyboards for easy access to information, including area navigation using up to 100 waypoints, weapons status and selection, radio tuning and mode selection, caution and warning, and display of aircraft performance; conventional instruments and dials are provided as back-up; IMS computer can store up to 100 preset frequencies for HF, VHF and UHF radio management; navigation is controlled by navigation computer of IMS coupled to Doppler radar and radar altimeter with low airspeed indicator, normally used for rocket aiming, providing back-up velocity data when the Doppler is beyond limits; synthetic map presentation of waypoints, target areas and dangerous areas is shown on pilot's or co-pilot's MFD; Litton strapdown inertial reference for both flight control and navigation is integrated into the IMS; AFCS provides either three-axis stabilisation or full attitude and heading hold, automatic hover, downward transition to hover or holds for altitude, heading and airspeed or groundspeed and automatic track following.

Instrumentation: Full day/night operational capability, with equipment designed to give both crew members a view outside helicopter irrespective of light conditions; cockpit lighting compatible with night vision goggles.

Mission: Pilot's night vision system (HIRNS: helicopter IR night system) allows nap-of-earth (NOE) flight by night with outside view generated by BAE North America FLIR sensor mounted on a GEC/OMI steerable platform at nose of aircraft and presented to both crewmen through the monocle of the Honeywell integrated helmet and display sighting system (IHADSS), to which it is slaved by helmet position sensors; flight information symbology superimposed on to image, giving true head-up reference. HeliTOW sight gives co-pilot/gunner direct view optics and FLIR, plus laser for ranging; provision for mast-mounted sight (MMS).

Self-defence: Active and passive self-protection systems (ECCM and ECM) standard on Italian Army A 129; Have Quick II frequency-hopping radio in Lot 2; onboard nav/weapon system can connect directly or by datalink with Italian CATRIN C³I combat information system; active and passive electronic warfare systems include Elettronica ELT-554 radar jammer and ELT-156 RWR (ELT-156-05 in Lot 2) radar warning receiver; BAE Italia RALM-101 laser warning receiver; Sanders AN/ALQ-144A IR jammer; provisions for chaff/flare dispenser. Upgraded defensive aids under development for Lot 3 and retrofit, including ELT-156X(V2), DASA/BAE MILDS missile approach warner, MES ECDS chaff/flare dispenser and BAE RALM-1(V2) laser warner.

ARMAMENT: Four attachments beneath stub-wing stressed for loads of 200 kg (441 lb) outboard and 300 kg (661 lb) inboard or 300 kg (661 lb) outboard and 100 kg (220 lb) inboard; outboard stations incorporate articulation which allows pylon to be elevated 2° and depressed 10° to increase missile launch envelope; they are aligned with aircraft automatically, with no need for boresighting. Initial armament of up to eight TOW, ITOW or TOW 2/2A wire-guided anti-tank missiles (two, three or four in carriers suspended from each wingtip station), with Saab/ESCO HeliTOW aiming system; with these can be carried, on inboard stations, either two 7.62, 12.7 or 20 mm gun pods, or two launchers each for seven or 19 air-to-surface rockets. For general attack missions, rocket launchers can be carried on all four stations; Italian Army has specified SNIA-BPD 81 and 70 mm rockets. Multirole and International versions can carry, in addition to the above, a Lockheed Martin/Otobreda TM 197B (standard 350 or optional 500 rounds) or Giat M621 20 mm cannon mounted in a nose turret; and Stinger AAMs. Optional upgrades offered on export versions include an autotracking sight; provision for up to eight Hellfire anti-tank missiles with autonomous laser designation capability; or a mix of four Hellfires and four TOWs. Other armament options include up to eight Hot missiles; AIM-9L Sidewinder, Mistral or Javelin AAMs; or grenade launchers. Lucas 0.50 in self-contained gun turret qualified, but not used by Italian Army.

DIMENSIONS, EXTERNAL:

Main rotor diameter	11.90 m (39 ft 0½ in)
Tail rotor diameter	2.32 m (7 ft 7 in)
Wing span	3.20 m (10 ft 6 in)
Width over TOW pods	3.60 m (11 ft 9¾ in)
Length overall, both rotors turning	14.29 m (46 ft 10½ in)
Fuselage: Length	12.275 m (40 ft 3¼ in)
Max width	0.95 m (3 ft 1½ in)
Height:	
over tailfin, tail rotor horizontal	2.75 m (9 ft 0¼ in)
tail rotor turning	3.315 m (10 ft 10½ in)
to top of rotor head	3.35 m (11 ft 0 in)
Tailplane span	2.50 m (8 ft 2¼ in)
Wheel track	2.23 m (7 ft 3¾ in)
Wheelbase	6.955 m (22 ft 9¾ in)

AREAS:

Main rotor disc	111.20 m² (1,197.0 sq ft)
Tail rotor disc	4.23 m² (45.53 sq ft)

WEIGHTS AND LOADINGS (Italian Army):

Weight empty, equipped	2,529 kg (5,575 lb)
Max internal fuel load	750 kg (1,653 lb)
Max external weapons load	1,200 kg (2,645 lb)
Max T-O weight: standard	4,100 kg (9,039 lb)
multirole version	4,600 kg (10,141 lb)
T-O weight, normal mission	3,950 kg (8,708 lb)
Max disc loading: standard	36.9 kg/m² (7.55 lb/sq ft)
multirole version	41.4 kg/m² (8.47 lb/sq ft)
Transmission loading at max T-O weight and power:	
standard	4.23 kg/kW (6.95 lb/shp)
multirole version	3.63 kg/kW (5.97 lb/shp)

PERFORMANCE (Italian Army, with eight TOW, at mission T-O weight of 3,950 kg (8,708 lb), at 2,000 m (6,560 ft), ISA+20°C, except where indicated):

Dash speed	159 kt (294 km/h; 183 mph)
Max level speed at S/L	135 kt (250 km/h; 155 mph)
Max rate of climb at S/L	612 m (2,008 ft)/min
Service ceiling	4,725 m (15,500 ft)
Hovering ceiling: IGE	3,140 m (10,300 ft)
OGE	1,890 m (6,200 ft)

Basic 2 h 30 min mission profile with eight TOW and 20 min fuel reserves: Fly 54 n miles (100 km; 62 miles) to battle area, mainly in NOE mode, 90 min loiter (incl 45 min hovering), and return to base

Max endurance, no reserves	3 h 5 min
g limits	+3.5/–0.5

UPDATED

AGUSTA A 129 INTERNATIONAL

TYPE: Attack helicopter.

PROGRAMME: Upgraded export version of A 129 (see previous entry) with increased power and military load, plus improved avionics. First flight of prototype powered by two T800 turboshafts October 1988; demonstrated in Arabian Gulf during 1990; first flight with production standard five-blade rotor (converted prototype MMX592, 29002) 9 January 1995; tail rotor diameter slightly increased; T800 gives between 20 and 40 per cent more power than Gem 1004. First firing of AGM-114 Hellfire ATMs, November 1999; Rafael NT-D Dandy ATM integrated and launched in same test series.

CURRENT VERSIONS: **International:** *As described.*

Scorpion: Offered to Australia in 1998 to meet Project Air 87 for 25 to 30 attack helicopters. Based on Multirole version, but some A 129 International features.

A 129DE: Trials aircraft registered (I-INTR) to Agusta in February 1998.

Data generally as for Mangusta, except as below.

DESIGN FEATURES: Five-blade main rotor; pannier on each side of forward fuselage.

POWER PLANT: Two LHTEC T800-LHT-800 turboshafts, each 996 kW (1,335 shp) for twin-engined operation or 1,047 kW (1,404 shp) maximum contingency OEI; transmission rating 1,268 kW (1,700 shp). Optional 33 per cent additional internal fuel.

AVIONICS: *Flight:* INS/GPS as main navigation sensor.

Instrumentation: Two 152 × 203 mm (6 × 8 in) MFDs in each crew position.

Mission: Upgraded observation and targeting sensors (CCD TV and laser range-finder, replacing original TOW telescopic sight), permitting autonomous laser designation and maximum range launch of Hellfire ATMs.

ARMAMENT: Hellfire and/or TOW anti-tank missile, 70 and 81 mm rockets; Lockheed Martin/Otobreda TM 197B 20 mm cannon in nose turret; Stinger or Mistral AAMs.

WEIGHTS AND LOADINGS:

Primary mission T-O weight	4,800 kg (10,582 lb)
Max T-O weight	5,100 kg (11,243 lb)

PERFORMANCE (mixed weapons at mission T-O weight):

Cruising speed	150 kt (278 km/h; 173 mph)
Max rate of climb at S/L	677 m (2,220 ft)/min
Max rate of climb, OEI	274 m (900 ft)/min
Vertical rate of climb at S/L	326 m (1,070 ft)/min
Hovering ceiling: IGE	4,206 m (13,800 ft)
OGE	3,292 m (10,800 ft)
Max range, internal fuel, no reserves	303 n miles (561 km; 348 miles)

UPDATED

AGUSTA A 109

TYPE: Light utility helicopter.
PROGRAMME: First flight of original A 109 4 August 1971; deliveries of A 109A started early 1976; single-pilot IFR certification 20 January 1977; deliveries of uprated A 109 Mk II began September 1981; A 109C certified 1989; first deliveries February 1989; A 109 Max was medevac version; A 109CM and A 109EOA for military use, as described in the 1996-97 and earlier *Jane's*.

First flight of A 109K April 1983; first flight of production representative second aircraft March 1984. A 109K2s of REGA air ambulance fitted with Sextant AFDS 95-1 AFCS, granted FAA single-pilot IFR certification late in 1996.

A 109 Power programme begun in 1993, intending to unify A 109C and A 109K2; initially known as A 109 Unified; launched November 1993; unique A 109D development aircraft with FADEC-equipped Allison 250-C22R9s flew 1 October 1994; prototype A 109E (I-EPWS) flew 8 February 1995. A 109D became second A 109E prototype.

CURRENT VERSIONS: **A 109KM:** Military version of A 109K2; roles include anti-tank/scout, escort, command and control, utility, ECM and SAR/medevac; fixed landing gear; sliding side doors.

A 109KN: Shipboard version with equivalent roles to A 109KM, including anti-ship, over-the-horizon surveillance and targeting and vertical replenishment.

A 109K2: Special civil rescue version first sold to Swiss REGA non-profit rescue service; REGA equipment includes Spectrolab SX-16 searchlight, 1,000 kg (2,204 lb) cargo hook, GPS, Elbit moving map display and single-pilot IFR instrumentation; NVG compatible. Equipped with Sextant AFDS 95-1 AFCS from 1996. Production continues at low rate.

A 109K2 Law Enforcement: Dedicated police version; optional equipment includes 907 kg (2,000 lb) cargo hook, 204 kg (450 lb) capacity variable speed rescue hoist with 50 m (164 ft) of cable, rappeling kit, wire-strike protection, SX-16 searchlight, MA3 retractable light, external loudspeakers, emergency floats, GPS, FM tactical communications, weather radar, LLTV and FLIR.

A 109 Power: Revealed at 1995 Paris Air Show, by which time prototype had accumulated more than 60 hours of flight testing. Engineering designation is **A 109E**. Based on A 109K2 airframe with new, A 129-derived lightweight, low-maintenance titanium main rotor head connected to composites material grips via single elastomeric bearing on each blade; Pratt & Whitney Canada PW206C or Turbomeca Arrius 2K1 engines; new heavy-duty high-clearance landing gear in revised position below fuselage; cabin as for A 109K2, retaining quick-change facility from passenger to EMS operation. First production aircraft (I-PWER) completed late 1995; RAI IFR certification 31 May 1996; FAA IFR certification 26 August 1996. Aircraft with Rogerson IIDS (integrated instrument display system) cockpit (I-EAPW; first production) displayed at 1997 Paris Air Show. PZL-Świdnik of Poland contracted to build fuselages for A 109E between 1996 and 2002 as alternative production source. By 2000, production was some 44 per year.

A 109F: Derivative of A 109E Power, announced early 1999; Turbomeca Arrius 2K1 turboshaft and 'glass cockpit'. Prototype is c/n 11501.

A 109M: Military version of A 109E with PW207C or Arrius 2K2 engine.

CUSTOMERS: Recent customers include the US Coast Guard, which ordered four (with four options) A 109Es in police/utility configuration in April 2000 to equip Helicopter Interdiction Tactical Squadron 10 (HITRON TEN) in Florida for armed anti-drug missions; and Italy's Carabinieri, which took delivery of two A 109Es in 2000. Greek government ordered five A 109Es in EMS configuration in 1999, and Dyfed/Powys Police Authority in the UK took delivery of an A 109E in late 1999. Three A 109K2s delivered to Dubai Police in 1995 and 1996.

Agusta A 109E Power in law enforcement configuration

Switzerland ordered one A 109 Power, which was delivered in May 1998 to the Federal Office for Civil Aviation; 150 orders for A 109Es held by June 2000. Requirement for 40 A 109s announced by South African Air Force in November 1998 of which 30 ordered, with 10 options; these likely to be equipped with IST Dynamics turret mounting a Vektor multiple-calibre gun. Total of more than 600 A 109s ordered by the end of 1999, comprising 336 A 109A series, 79 A 109C series, 41 A 109K2s and some 150 A 109Es. Production of A 109E reached 90 in June 2000.

COSTS: A 109E US$3.0 million (1998). A 109K2 US$3.3 million (1998). Direct operating cost US$386 per hour (1997). US Coast Guard contract for four A 109Es valued at US$18.6 million (2000).

DESIGN FEATURES: Fully articulated four-blade metal main rotor hub with tension/torsion blade attachment and elastomeric bearings; delta-hinged two-blade stainless steel tail rotor; manual blade folding and rotor brake optional. Main blade section NACA 23011 with drooped leading-edge; thickness/chord ratios 11.3 per cent at root, 6 per cent at tip. Tail rotor with Wortmann aerofoil and stainless steel skins; optional rotor brake. Compared with earlier models, A 109K has lengthened cabin to hold two stretchers fore and aft; modified fuel system; and smaller instrument panel.

FLYING CONTROLS: Fully powered hydraulic; IFR system with autopilot available. A 109KM has three-axis stability augmentation/attitude hold system; dual redundant IFR system and four-axis AFCS with flight path computer optional.

LANDING GEAR: A 109K series has non-retractable tricycle type, giving increased clearance between fuselage and ground. A 109E Power has tricycle type, with oleo-pneumatic shock-absorber in each unit. Single mainwheels and self-centring nosewheel castoring ±45°. Hydraulic retraction, nosewheel forward, mainwheels upward into fuselage. Hydraulic emergency extension and locking. Magnaghi disc brakes on mainwheels. All tyres are tubeless, of same size (360×135-6 or 380×135-6 or 14.5×5.5-6, 12/14 ply tubeless) and pressure (5.90 bars; 85 lb/sq in). Tailskid under ventral fin. Emergency pop-out flotation gear and fixed snow skis optional on all models.

POWER PLANT: *A 109K:* Two Turbomeca Ariel 1K1 turboshafts, each rated at 575 kW (771 shp) for 2½ minutes, 550 kW (737 shp) for take-off (30 minutes) and 471 kW (632 shp) maximum continuous power. Engine particle separator optional. Main transmission uprated to 671 kW (900 shp) for take-off and maximum continuous twin-engined operation; single-engine rating is 477 kW (640 shp) for 2½ minutes and 418 kW (560 shp) maximum continuous. Main rotor rpm 384, tail rotor 2,085. Standard usable fuel capacity 750 litres (198 US gallons; 165 Imp gallons), with optional 150 litre (39.6 US gallon; 33.0 Imp gallon) auxiliary tank for EMS operations, or 200 litre (52.8 US gallon; 44.0 Imp gallon) auxiliary tank in the A 109KM. Self-sealing fuel tanks optional. Independent fuel and oil system for each engine.

A 109E: Two Pratt & Whitney Canada PW206C engines, each rated at 477 kW (640 shp) for T-O, 423 kW (567 shp) for twin-engine operation, 546 kW (732 shp) for 2½ minutes OEI contingency rating and 500 kW (670 shp) maximum continuous OEI; or two Turbomeca Arrius 2K1, each rated at 500 kW (670 shp) for T-O, 426 kW (571 shp) for twin-engine operation, 559 kW (750 shp) for 2½ minutes OEI contingencies and 500 kW (670 shp) maximum continuous OEI; transmission rating 671 kW (900 shp) maximum continuous for twin-engine operation, 418 kW (560 shp) maximum continuous and 477 kW (640 shp) for 2½ minutes OEI; FADEC and liquid crystal multifunction displays for engine management; standard fuel capacity (three cells) 605 litres (160 US gallons; 133 Imp gallons); optional capacity 710 litres (187 US gallons; 156 Imp gallons with one extra cell or 870 litres (230 US gallons: 191 Imp gallons) with two extra cells.

SYSTEMS: Military versions have electrical system supplied by two 160 A 28 V DC self-cooled starter/generators and 27 Ah 28 V Ni/Cd battery. Optional AC electrical system comprises two 250 VA or two high-load 600 VA 115/26 V AC 400 Hz static inverters. Optional high-load AC system comprises one 6 kVA alternator and one standby 250 VA solid-state inverter. Dual independent hydraulic systems for flight controls, each capable of operating main actuators in the event of failure of the other system; utility hydraulic system with normal and emergency accumulators for operation of rotor brake, wheel brakes and nosewheel centring.

AVIONICS: A109E has Rockwell Collins ProLine II or Honeywell Silver Crown as core system. Military versions have ergonomic, NVG-compatible flight deck with provision for IFR instrumentation and role/mission dedicated displays.

Comms: VHF/AM, VHF/FM, UHF, HF, three-station intercom, IFF and ELT.

Radar: Colour weather radar optional.

Flight: ADF, VOR/GS/ILS, DME, GPS and VLF-Omega. Optional Rockwell Collins CMS80 cockpit management system with one or more centrally mounted CDUs and automatic target hand-off system compatibility.

Mission: FLIR.

Self-defence: Radar/laser warning receivers and chaff/flare/smoke dispensers.

EQUIPMENT: Optional equipment for military use includes windscreen wipers, rear view mirror, bleed air heater,

Artist's impression of Agusta A 109M in South African markings

Instrument panel of A 109E Power

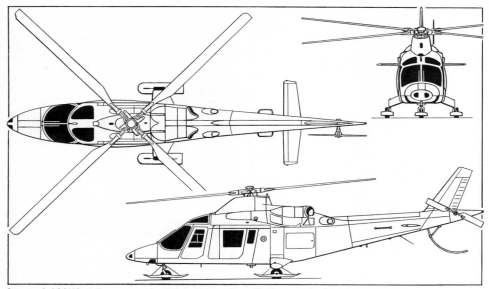

Agusta A 109K2 civil rescue and utility helicopter (two Turbomeca Arriel 1K1 turboshafts) *(Dennis Punnett/Jane's)*

particle separator, engine fire extinguisher, oxygen system, environmental control unit, one- or two-stretcher installation, air ambulance kit, external loudspeaker system, high-intensity searchlight, cargo platform, external cargo hook with maximum capacity 1,000 kg (2,205 lb), rescue hoist maximum capacity 270 kg (595 lb), snow skis and emergency floats.

ARMAMENT: Internal armament comprises pintle-mounted 7.62 mm machine gun and 12.7 mm machine gun in doorway. Provision for carriage of four or eight TOW, TOW 1, TOW 2 or TOW 2A missiles on external lateral pylons, each of 300 kg (661 lb) maximum capacity, with roof-mounted HeliTOW sight or APX M 334 or Helios gyrostabilised sights. Alternatively, pylons can accommodate seven- or 12-tube pods for 2.75 in or 81 mm rockets; rocket/machine gun (RPM) pods each with three 70 mm rockets and a 12.7 mm machine gun with 200 rounds; or machine gun (MG) pods with 12.7 mm gun (and 250 rounds) or 7.62 mm gun. A 109M has 12.7 mm or 20 mm nose-mounted turret gun.

DIMENSIONS, EXTERNAL:
Main rotor diameter	11.00 m (36 ft 1 in)
Tail rotor diameter	2.00 m (6 ft 6¾ in)
Length overall, rotors turning	13.04 m (42 ft 9 in)
Fuselage: Length	11.44 m (37 ft 6 in)
Height over tailfin	3.50 m (11 ft 5¾ in)
Tailplane span	2.88 m (9 ft 5½ in)
Width over mainwheels	2.45 m (8 ft 0½ in)
Wheelbase	3.535 m (11 ft 7¼ in)
Passenger doors (each): Height	1.06 m (3 ft 5¾ in)
Width	1.15 m (3 ft 9¼ in)
Height to sill	0.65 m (2 ft 1½ in)
Baggage door (port, rear): Height	0.51 m (1 ft 8 in)
Width	1.00 m (3 ft 3¼ in)

DIMENSIONS, INTERNAL:
Cabin: Length	2.10 m (6 ft 10¾ in)
Max width: KM, K2	1.59 m (5 ft 2½ in)
E	1.61 m (5 ft 3½ in)
Max height	1.28 m (4 ft 2½ in)
Volume, incl flight deck: K2	4.9 m³ (173 cu ft)
E	5.1 m³ (180 cu ft)
Baggage compartment volume: K2	0.85 m³ (30.0 cu ft)
E	0.95 m³ (33.6 cu ft)

AREAS:
Main rotor disc	95.03 m² (1,022.9 sq ft)
Tail rotor disc	3.14 m² (33.82 sq ft)

WEIGHTS AND LOADINGS:
Weight empty: KM	1,660 kg (3,660 lb)
K2	1,650 kg (3,638 lb)
E	1,570 kg (3,461 lb)
M	1,639 kg (3,614 lb)
Max slung load	1,000 kg (2,204 lb)
Max T-O weight: all	2,850 kg (6,283 lb)
Max T-O weight with slung load: all	3,000 kg (6,613 lb)
Max disc loading: KM, K2, E	30.0 kg/m² (6.14 lb/sq ft)
Transmission loading at max T-O weight and power: all	4.24 kg/kW (6.97 lb/shp)

PERFORMANCE:
Never-exceed speed (VNE):	
KM, K2	152 kt (281 km/h; 174 mph)
E, M	168 kt (311 km/h; 193 mph)
Max cruising speed at S/L, clean:	
KM, K2	143 kt (264 km/h; 164 mph)
E	156 kt (289 km/h; 180 mph)
M	151 kt (280 km/h; 174 mph)
Max rate of climb at S/L: KM, K2	594 m (1,950 ft)/min
E	588 m (1,930 ft)/min
M	588 m (1,929 ft)/min
Rate of climb at S/L, OEI:	
KM, K2, M	274 m (900 ft)/min
Service ceiling: KM, K2	6,100 m (20,000 ft)
E	5,974 m (19,680 ft)
M	5,029 m (16,500 ft)
Service ceiling OEI: KM, K2	3,660 m (12,000 ft)
E, M	3,990 m (13,100 ft)
Hovering ceiling IGE: KM, K2	5,305 m (17,400 ft)
E	5,060 m (16,600 ft)
M	4,328 m (14,200 ft)
Hovering ceiling OGE: KM, K2	3,900 m (12,800 ft)
E	3,597 m (11,800 ft)
M	2,957 m (9,700 ft)
Max range, best height and speed, max optional (five cells) fuel: KM, K2	434 n miles (805 km; 500 miles)
E	521 n miles (964 km; 599 miles)
M	447 n miles (827 km; 514 miles)
Endurance: KM, K2	4 h 0 min
E	5 h 4 min
M	4 h 29 min

UPDATED

AGUSTA A 119 KOALA

TYPE: Light utility helicopter.

PROGRAMME: First flight (I-KOAL) early 1995; public debut at Paris Air Show June 1995; second prototype flew later in 1995; both re-engined with PT6B turboshafts early in 1997; flown more than 500 hours by mid-1999; production target 20 to 25 per year. As part of South African purchase of 40 military A 109s, Denel (which see) will assemble and market the Koala in Africa and provide components for the Italian production line.

CUSTOMERS: Six ordered by Omniflight Helicopters in February 1996; orders and options totalled more than 20 by January 2000 from customers in Australia, Austria, Brazil, UK, USA and Venezuela. Interest shown by Carabinieri (Italian military police) as replacement for AB 206 and Guardia di Finanza (customs) as NH 500 replacement.

COSTS: US$1.85 million (2000).

DESIGN FEATURES: Fully articulated four-blade composites main rotor with titanium hub, composites material grips and elastomeric bearings; two-blade tail rotor.

STRUCTURE: Aluminium alloy fuselage.

LANDING GEAR: Fixed skids.

POWER PLANT: Prototype had 596.5 kW (800 shp) Turbomeca Arriel 1 turboshaft; production version will have Pratt & Whitney Canada PT6B-37A rated at 747 kW (1,002 shp) for T-O, and 650 kW (872 shp) maximum continuous. Transmission rating 671 kW (900 shp) for take-off and maximum continuous. Standard three-cell fuel system, combined capacity 606 litres (160 US gallons; 133 Imp gallons), optional four- and five-cell fuel tanks, combined capacities respectively 721 litres (188 US gallons; 156.5 Imp gallons) and 871 litres (230 US gallons; 191.5 Imp gallons).

ACCOMMODATION: Pilot and passenger in front; six passengers in main cabin in three-abreast club configuration claimed to be 30 per cent larger than that of any current single-turbine light helicopter; flight-accessible baggage compartment in cabin; main baggage compartment in rear fuselage with optional extensions; in EMS configuration can accommodate two stretchers and three medical attendants in main cabin without intrusion into cockpit area. Large sliding doors on each side of main cabin; forward-hinged door to cockpit on both sides. Cabin is heated, air conditioned and soundproofed.

AVIONICS: Honeywell Silver Crown suite standard; integrated instrument display system (IIDS) optional.

Comms: Silver Crown com/nav and ELT standard.

Flight: Three-axis autopilot, flight director, autotrim, radar altimeter, GPS, moving map display.

Instrumentation: NVG compatible.

EQUIPMENT: Optional equipment includes 500 kg (1,102 lb) or 1,000 kg (2,205 lb) fixed cargo hook; 200 kg (441 lb) rescue hoist; snow skids, external emergency floats, particle separator, Spectrolab SX-5 searchlight and FLIR/LLTV camera.

DIMENSIONS, EXTERNAL:
Main rotor diameter	10.83 m (35 ft 6½ in)
Tail rotor diameter	2.00 m (6 ft 6¾ in)
Length overall, rotors turning	13.01 m (42 ft 8¼ in)
Height overall	3.50 m (11 ft 5¾ in)
Fuselage: max width	1.67 m (5 ft 5¾ in)

Agusta A 119 Koala light utility helicopter

AGUSTA—AIRCRAFT: ITALY   275

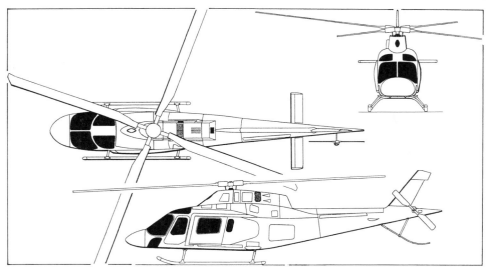

Agusta A 119 Koala (one P&WC PT6B turboshaft) *(Paul Jackson/Jane's)*

Width overall, rotor in 'X'	7.66 m (25 ft 1½ in)
Main rotor tip ground clearance	2.52 m (8 ft 3¼ in)
Tail rotor tip ground clearance	1.29 m (4 ft 2¾ in)
Fuselage ground clearance	0.54 m (1 ft 9¼ in)

DIMENSIONS, INTERNAL:
Cabin: Length	2.10 m (6 ft 10¾ in)
Max width	1.67 m (5 ft 5¾ in)
Max height	1.28 m (4 ft 2½ in)
Floor area	2.60 m² (28.0 sq ft)
Volume, incl flight deck	4.96 m³ (175.2 cu ft)
Baggage compartment: length	2.30 m (7 ft 6½ in)
volume	0.95 m³ (33.5 cu ft)

AREAS:
Main rotor disc	92.12 m² (991.6 sq ft)
Tail rotor disc	3.14 m² (33.82 sq ft)

WEIGHTS AND LOADINGS:
Basic weight empty	1,430 kg (3,153 lb)
Max T-O weight: internal load	2,720 kg (5,996 lb)
external load	3,000 kg (6,613 lb)
Max disc loading:	
internal load	29.53 kg/m² (6.05 lb/sq ft)
external load	32.57 kg/m² (6.67 lb/sq ft)
Transmission loading at max T-O weight and power:	
internal load	4.05 kg/kW (6.65 lb/shp)
external load	4.47 kg/kW (7.35 lb/shp)

PERFORMANCE (max internal load):
Never-exceed speed (V$_{NE}$)	152 kt (281 km/h; 174 mph)
Max speed	144 kt (267 km/h; 166 mph)
Hovering ceiling: IGE	3,353 m (11,000 ft)
OGE	2,682 m (8,800 ft)
Service ceiling	5,700 m (18,700 ft)
Range with max auxiliary fuel, at 1,525 m (5,000 ft), no reserves	552 n miles (1,023 km; 635 miles)
Endurance	5 h 42 min

UPDATED

AGUSTA-BELL 412 and GRIFFON
TYPE: Multirole medium helicopter.
PROGRAMME: First flight of Bell 412SP (in USA) August 1979; deliveries started January 1981; Agusta licensed production of civil version started 1981; first flight of military Griffon August 1982; deliveries began January 1983; Bell 412HP certified 29 June 1990; Bell 412EP (see Canadian section) is a civilian version currently manufactured under licence by Agusta.
CURRENT VERSIONS: **Griffon**: Military derivative developed for direct fire support, scouting, assault transport, equipment transport, SAR and maritime surveillance.
Creso: Battlefield surveillance version; trial installation for Italian Army in AB 412 MM81196 as an element (with Mirach 26 UAVs and other sensors) of the CATRIN C³I system. FIAR Creso E-band MTI radar in a circular radome below the nose, plus Elettronica emitter locators at the nose and on the tailboom. Officine Galileo FLIR turret above pilot's seat. Maximum operational altitude of 1,500 m (4,920 ft) gives radar range of 32 to 38 n miles (60 to 70 km; 37 to 43 miles). First flight of operational system mid-1996; Creso is Italy's contender in the NATO Alliance Ground Surveillance programme.
CUSTOMERS: Those in Italy include Army (24), Carabinieri (34), Civil Protection Service (six), Coast Guard (eight; ultimate requirement for 24), national fire service (14), national forest service (nine) and Guardia di Finanza (16); others include Zimbabwe Air Force (10), Ugandan Army (two), Finnish Coast Guard (four); Royal Netherlands Air Force (three); Dubai Air Force (nine); Ghana Air Force (two, in EMS configuration); Swedish Army (five); and Dubai Police (two); Italian production totalled 240 by mid-2000. Five ordered for SAR by Turkish Coastguard in late 1999.
COSTS: Five Turkish Coastguard, total cost US$52 million (1999).
DESIGN FEATURES: Griffon has reinforced impact-absorbing landing gear, selective armour protection and differences noted below.

POWER PLANT: One Pratt & Whitney Canada PT6T-3D Twin-Pac rated at 1,342 kW (1,800 shp) for T-O and 1,194 kW (1,600 shp) maximum continuous (single-engine ratings 850 kW; 1,140 shp for 2½ minutes and 723 kW; 970 shp continuous). Transmission rating 1,181 kW (1,584 shp) for 5 minutes, 846 kW (1,135 shp) maximum continuous and 850 kW (1,140 shp) single engine. IR emission reduction devices optional. Fuel capacity 1,249 litres (330 US gallons; 275 Imp gallons). Two 75.7 or 341 litre (20.0 or 90.0 US gallon; 16.7 or 74.9 Imp gallon) auxiliary fuel tanks optional; single-point refuelling.
ACCOMMODATION: One or two pilots on flight deck, on energy-absorbing, armour-protected seats. Fourteen crash-attenuating troop seats in main cabin in personnel transport roles, six patients and two medical attendants in ambulance version, or up to 1,814 kg (4,000 lb) of cargo or other equipment. Space for 181 kg (400 lb) of baggage in tailboom. Total of 51 fittings in cabin floor for attachment of seats, stretchers, internal hoist or other special equipment.
SYSTEMS: Generally as for Bell 212/412.
EQUIPMENT: SAR and coastal surveillance versions equipped with 360° panoramic search radar integrated with FLIR and TV; video and still camera systems; datalink; dual digital four-axis AFCS with auto approach to hover, mark on target and search pattern; dual GPS; EFIS-LCD IFR cockpit instrumentation; operator's console in passenger compartment; gyrostabilised binoculars; searchlight and rescue light; electric rescue hoist; liferafts and lightweight emergency floats. BAE MST-S multisensor turret system supplied for undisclosed AB 412EP customer in 1996, to operate in conjunction with Honeywell RDR-1500 maritime surveillance radar.
ARMAMENT: Wide variety of external weapon options for Griffon include swivelling turret for 12.7 mm gun, two 25 mm Oerlikon cannon, four or eight TOW anti-tank missiles, two launchers each with nineteen 2.75 in SNORA or twelve 81 mm rockets, 12.7 mm machine guns (in pods or door-mounted), four air-to-air or air defence suppression missiles, or, for attacking surface vessels, four Sea Skua or similar air-to-surface missiles.
DIMENSIONS: As for Bell 212/412
WEIGHTS AND LOADINGS:
Weight empty, equipped (standard configuration)	2,914 kg (6,425 lb)
Max T-O weight	5,398 kg (11,900 lb)
Transmission loading at max T-O weight and power:	4.57 kg/kW (7.51 lb/shp)

PERFORMANCE:
Never-exceed speed (V$_{NE}$) at S/L	140 kt (259 km/h; 161 mph)
Cruising speed: at S/L	122 kt (226 km/h; 140 mph)
at 1,500 m (4,920 ft)	125 kt (232 km/h; 144 mph)
at 3,000 m (9,840 ft)	123 kt (228 km/h; 142 mph)
Max rate of climb at S/L	542 m (1,780 ft)/min
Rate of climb at S/L, OEI	168 m (551 ft)/min
Service ceiling, 30.5 m (100 ft)/min climb rate	5,395 m (17,700 ft)
Service ceiling, OEI, 30.5 m (100 ft)/min climb rate	2,320 m (7,620 ft)
Hovering ceiling: IGE	3,110 m (10,200 ft)
OGE	1,585 m (5,200 ft)
Range with max standard fuel at appropriate cruising speed (see above), no reserves:	
at S/L	354 n miles (656 km; 407 miles)
at 1,500 m (4,920 ft)	402 n miles (745 km; 463 miles)
at 3,000 m (9,840 ft)	434 n miles (804 km; 500 miles)
Max endurance: at S/L	3 h 36 min
at 1,500 m (4,920 ft)	4 h 12 min

UPDATED

AGUSTA ERICA
Elements of this programme are being incorporated in the Eurotilt project, described in the International section.

UPDATED

Agusta-Bell 412SP of the Royal Netherlands Air Force *(Paul Jackson/Jane's)*   1999/0051072

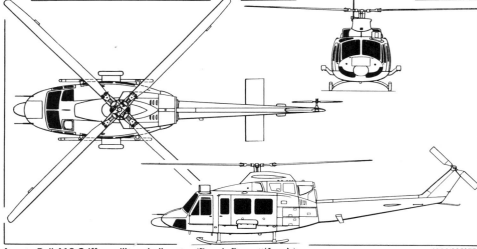

Agusta-Bell 412 Griffon military helicopter *(Dennis Punnett/Jane's)*   1999/0052775

ALENIA

ALENIA AEROSPAZIO (A Finmeccanica company)
Via Giulio Vincenzo Bona 85, I-00156 Roma
Tel: (+39 06) 41 72 31
Fax: (+39 06) 411 44 39
Web: http://www.aleniaerospazio.com
PRESIDENT: Giorgio Zappa
SENIOR VICE-PRESIDENT, HEAD OF STRATEGY AND MILITARY PROGRAMMES: Carmelo Cosentino
SENIOR VICE-PRESIDENT, OPERATIONS: Gianni Cantini
HEAD OF AERONAUTICS DIVISION: Filippo Bagnato
SENIOR VICE-PRESIDENT, TECHNICAL AND PRODUCTION FACILITIES, AERONAUTICS DIVISION: Giuseppe Ragni
MARKETING COMMUNICATIONS OFFICER: Riccardo Rovere

Alenia Aerospazio, a Finmeccanica company, is the leading Italian aerospace designer and manufacturer. Employees total 12,000, engaged on projects and programmes for civil and military applications. Alenia Aerospazio is organised into two Divisions: Aeronautics and Space (Alenia Spazio SpA). On 14 April 2000, it signed a joint venture agreement with EADS to form the European Military Aircraft Company (EMAC).

Aeronautics Division dedicated to full range of activities, from design and production to modification and product support for both military and civil aircraft; most entail collaboration with other aerospace companies.

Division's activities fall into categories of Military Aircraft, Regional Aircraft and Aerostructures. It also operates Officine Aeronavali Venezia, specialising in maintenance, overhaul and modification of commercial and military aircraft.

In military sector, company designs and produces directly, or through international collaborations, combat and transport aircraft such as Tornado, Eurofighter Typhoon, G222, C-27J, A400M and ATR 42 MP Surveyor; was also responsible for updating of the F-104S/ASA-M and TF-104G-M in service with Aeronautica Militare Italiana and has assembled AV-8B Harrier II Plus for the Italian Marina Militare.

In regional aircraft sector, activities include production of the ATR turboprop family developed with Aerospatiale Matra.

In the aerostructures sector, Alenia co-operates with other major aeronautical companies manufacturing structural parts for commercial aircraft such as B767, B777, B717, A321, A300/310, A330/A340, Falcon 900EX and Falcon 2000.

In the modification and maintenance field, Alenia Aerospazio has a full capability of design, production, installation and tests of complex parts and systems.
UPDATED

ALENIA G222
US Air Force designation: C-27A Spartan
In the absence of further orders, production of the T64-engined G222 (2000-01 *Jane's*) has been terminated. Development and marketing of the basically identical, but AE 2100-powered, C-27J continues, as detailed in the following entry.
UPDATED

ALENIA/LOCKHEED MARTIN C-27J SPARTAN
TYPE: Twin-turboprop transport.
PROGRAMME: Conceived in 1960s as jet-powered, V/STOL transport to NATO requirement (NBMR-4), but this version not developed. Original Fiat G222 designed by Giuseppe Gabrielli; two prototypes (lacking the pressurisation standard on later aircraft) flew on 18 July 1970 (MM582) and 22 July 1971 (MM583), both at Turin; MM582 handed over to Italian Air Force for operational evaluation on 21 December 1971. First production G222 (MM62101) flew 23 December 1975; deliveries began 21 November 1976 with single aircraft to Dubai, and to Italy on 21 April 1978. Tenth production aircraft was first to be built at Naples; 27th (March 1979) was 22nd and last built at Turin. Main users in Italy are 2° and 98° Gruppi of the 46° Brigata Aerea at Pisa. Civil category R (equivalent to FAR Pt 25) certification granted to G222SAA by Italian airworthiness authority on 1 April 1997. Several subvariants for specific roles; last detailed in 2000-01 *Jane's*.

Improved version of G222 conceived during 1995 negotiations between Lockheed Martin and Alenia on potential offsets for proposed Italian purchase of C-130J Hercules; initially designated G222J, by reason of having C-130J flight deck features and improved (T64G) versions of the G222's engines with new four-blade propellers. Formally announced as a joint project in February 1996, when commonality with the C-130J was further increased by adoption of the Allison (now Rolls-Royce) AE 2100 as power plant, allied to six-blade propellers. Accordingly redesignated C-27J, to reflect the C-27A version of G222 delivered to the US Air Force. Feasibility phase February to September 1996; definition phase September 1996 to May 1997.

Development and certification costs being shared equally between Alenia Aerospazio and Lockheed Martin; latter responsible for propulsion systems, avionics, worldwide marketing and product support; Alenia for production, flight test and certification; promotion by Lockheed Martin Alenia Tactical Transport Systems. Programme formally launched on 17 June 1997; 'propulsion test' prototype, a converted G222 demonstrator, rolled out at Turin/Caselle on 14 June 1999 and first flew (c/n 4043; I-CERX) 24 September 1999; initial testing completed second quarter of 2000; modified for certification trials and resumed flying on propulsion system, performance and handling evaluation; total 115 hours/61 sorties by July 2000.

Second C-27J and initial new-build aircraft, 4115/I-FBAX, first flew 12 May 2000; first with advanced flight deck and full avionics, new APU and new landing gear; achieved 54 hours/23 sorties in initial two months; will be sold on completion of trials. Third prototype, 4033/MM62127, converted from Italian Air Force G222TCM and will be returned to IAF after completion of trials. C-27J will be certified to civil JAR 25 standards in 750 hour test programme; deliveries from mid-2001, first two batches, totalling 10, being under construction (long lead items) by 1999. Final assembly remains in Italy only.

CURRENT VERSIONS: **Transport:** *As described.*

AEW: Airborne early warning version proposed in 1998, employing Ericsson Erieye system, as fitted to Saab S 100B Argus.

Firefighter: Proposed with 2,200 kg (4,850 lb) mission system and 6,800 kg (14,991 lb) of retardant liquid.

Aerial Sprayer: Proposed with 2,200 kg (4,850 lb) mission system; capable of covering 50 hectares (124 acres) at 150 litres/hectare (16.0 US gallons; 13.0 Imp gallons/acre).

CUSTOMERS: See table. Italian Air Force announced launch order for 12 on 11 November 1999; deliveries between December 2001 and 2004. Up to 500 sales anticipated over a 20 year period, mostly to existing Hercules operators. Lockheed Martin co-markets the aircraft. Greece expressed interest, mid-1998, in 15, plus five options. Sales targets include Argentina, Brazil (20), Switzerland (two) and Taiwan (18 to 22); also promoted for US Army's 40-aircraft Aerial Common Sensor programme and to US

G222 and C-27 CUSTOMERS

Customer	Version	Qty	First aircraft	Delivered
Argentine Army	G222	3	AE260	29 Mar 1977
Congo Air Force	G222	3		on order[4]
Dubai Air Force	G222	1	321	21 Nov 1976
Italy: prototypes	G222	2	MM582	21 Dec 1971
Air Force	G222TCM	40[3]	MM62101	Apr 1978
	C-27J	12		2001
Air Force	G222RM	4	MM62139	Jan 1983
Air Force	G222VS	2	MM62107	1978
SNPC	G222PROCIV	5[6]	MM62145	1987
Libyan Air Force	G222T	20	221	1981
Nigerian Air Force	G222	5	950	Sep 1984
Somali Air Force	G222	2	AM-94	1980
Venezuelan: Army	G222	1[1]	EV-8327	1983
Air Force	G222	6	1258	1984
Thai Air Force	G222	6	60307	2 May 1995
US Air Force	C-27A	10[5]	90-0170	17 Apr 1991
Alenia (demonstrator)	G222/C-27J C-27J	1	I-CERX	1983
		1	I-FBAX	2000
Total		**124[2]**		

Notes:
[1] Transferred to air force
[2] Prototypes and 22 early aircraft built at Turin; remainder at Naples
[3] Including 10 with provision for rapid conversion to 222SAA
[4] Built, but awaiting completion of contractual details
[5] Transferred to US government, with civil registrations, 1999
[6] All transferred to Tunisian Air Force

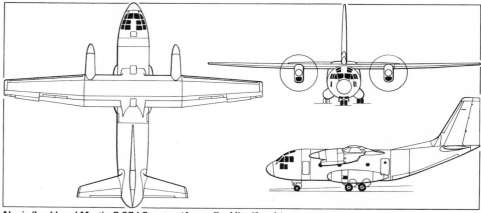

Alenia/Lockheed Martin C-27J Spartan (*James Goulding/Jane's*)   2001/0093635

Upper aspect of C-27J Spartan   2001/0093634

ALENIA—AIRCRAFT: ITALY

Prototype C-27J Spartan, making type's first international trade show appearance at Farnborough in July 2000 *(Paul Jackson/Jane's)*

Army National Guard (40) as replacement for Shorts Sherpas. Production target of 18 per year.

COSTS: US$28 million (2000).

DESIGN FEATURES: Conventional tactical transport configuration of high wing, pannier-mounted main landing gear and upswept rear fuselage with integral loading ramp.

Intended to complement the Lockheed Martin Hercules. Upgraded G222 with new, two-crew flight deck and increased performance. Propulsion system, cargo loading system and many of the avionics and flight controls are adapted from the C-130J Hercules. Compared with the G222, the C-27J is intended to provide increases of 35 per cent in range, 30 per cent cruise ceiling, 15 per cent high-speed cruise and over 200 per cent payload/range/speed; maintainability and reliability are scheduled to increase by 100 per cent for the engine, 275 per cent for propeller, 150 per cent for other systems and 30 per cent for avionics, resulting in a saving of 30 per cent in operating costs (including 5 per cent off fuel).

Wing has max thickness/chord ratio of 15 per cent. Dihedral 2° 30′ on outer panels.

FLYING CONTROLS: Conventional; manually actuated ailerons and elevators; powered rudder. Ailerons each have inset servo tab. Two-section hydraulically actuated spoilers ahead of each outboard flap segment, used also as lift dumpers on landing. Double-slotted flaps extend over 60 per cent of trailing-edge. Spoilers and flaps fully powered by tandem hydraulic actuators. Rudder fully powered by tandem hydraulic actuators. Two tabs in each elevator; no rudder tabs.

STRUCTURE: Wing of aluminium alloy three-spar fail-safe box structure, built in three portions. One-piece constant chord centre-section fits into recess in top of fuselage and is secured by bolts at six main points. Outer panels tapered on leading- and trailing-edges. Upper surface skins are of 7075-T6 alloy, lower surface skins of 2024-T3 alloy. All control surfaces have bonded metal skins with metal honeycomb core.

Pressurised fail-safe fuselage of aluminium alloy stressed skin construction and circular cross-section. Easily removable stiffened floor panels. Cantilever safe-life tail surfaces of aluminium alloy, with sweptback three-spar fin and slightly swept two-spar variable incidence tailplane.

Subcontractors include Aermacchi (outer wings), Piaggio (wing centre-section), Agusta (tail unit), Magnaghi (landing gear) and Aeronavali Venezia (airframe components).

LANDING GEAR: Hydraulically retractable tricycle type, suitable for use from prepared runways, semi-prepared strips or grass fields. Main gear built by APPH; nose gear by Magnaghi. Steerable twin-wheel nose unit retracts forward. Main units, each consisting of two single wheels in tandem, retract into fairings on sides of fuselage. Oleo-pneumatic shock-absorbers. Gear can be lowered by gravity in emergency, the nose unit being aided by aerodynamic action and the main units by the shock-absorbers, which remain compressed in the retracted position. Oleo pressure in shock-absorbers is adjustable to permit variation in height and attitude of cabin floor from ground. Low-pressure tubeless tyres on all units, size 39×13 (14/16 ply) on mainwheels, 29×11.00-12 or 29×11.00-10 (10 ply) on nosewheels. Tyre pressures 4.41 bars (64 lb/sq in) on main units, 3.92 bars (56.88 lb/sq in) on nose unit. Hydraulic multidisc brakes.

POWER PLANT: Two Rolls-Royce AE 2100D2 turboprops, each rated at 3,460 kW (4,640 shp), driving Dowty R391 six-blade composites propellers. Fuel in integral tanks; two in outer wings, combined capacity 6,800 litres (1,796 US gallons; 1,495 Imp gallons); two centre-section tanks, combined capacity 5,200 litres (1,374 US gallons; 1,143 Imp gallons); crossfeed provision to either engine. Total overall fuel capacity 12,000 litres (3,170 US gallons; 2,638 Imp gallons).

ACCOMMODATION: Two-pilot crew on flight deck with third seat; provision for loadmaster or jumpmaster when required. Crew door, port, forward; Type III emergency door, starboard, forward; paratroop door each side, immediately rear of sponsons; rear loading ramp; emergency hatches (three) in roof, above flight deck and level with leading- and trailing edges.

Standard troop transport version has 34 foldaway sidewall seats and 12 stowable seats for 46 fully equipped troops (62 in high density). Paratroop transport can carry between 34 (normal) and 46 (maximum) fully equipped paratroops, and is fitted with 32 sidewall seats, plus eight stowable seats, door jump platforms and static lines. Cargo transport version can accept standard pallets of up to 2.24 m (88 in) wide, and can carry up to 9,000 kg (19,841 lb) of freight. Hydraulically operated rear-loading ramp and upward-opening door in underside of upswept rear fuselage, which can be opened in flight for airdrop operations. Five pallets of up to 1,000 kg (2,205 lb) each can be airdropped from rear opening, or single pallet of up to 5,000 kg (11,023 lb). Paratroop jumps can be made either from this opening or from rear side doors. Medical evacuation accommodation for 36 stretchers and six attendants. Entire accommodation pressurised.

SYSTEMS: Pressurisation system maintains a cabin differential of 0.41 bar (5.97 lb/sq in), giving a 1,200 m (3,940 ft) environment at altitudes up to 6,000 m (19,680 ft). Air conditioning system uses engine bleed air during flight; on ground, it is fed by compressor bleed air from APU to provide cabin heating to a minimum of 18°C. Honeywell 113 kW (152 hp) APU, installed in starboard main landing gear fairing, provides power for engine starting, hydraulic pump and alternator actuation, air conditioning on ground, and all hydraulic and electrical systems necessary for loading and unloading on ground.

Two independent hydraulic systems, each of 207 bars (3,000 lb/sq in) pressure. No. 1 system actuates flaps, spoilers, rudder, wheel brakes and (in emergency only) landing gear extension; No. 2 system actuates flaps, spoilers, rudder, wheel brakes, nosewheel steering, landing gear extension and retraction, rear ramp/door and windscreen wipers. Auxiliary hydraulic system, fed by APU-powered pump, can take over from No. 2 system in flight, if both main systems fail, to operate essential services. In addition, a standby hand pump is provided for emergency use to lower the landing gear and, on the ground, to operate the ramp/door and parking brakes.

Three 45 kVA alternators, one driven by each engine through constant-speed drive units and one by the APU, provide 115/200 V three-phase AC electrical power at 400 Hz. 28 V DC power is supplied from the main AC buses via two transformer-rectifiers, with 24 V 34 Ah Ni/Cd battery and static inverter for standby and emergency power. External AC power socket. Engine intakes anti-iced by electrical/hot air system. Pneumatically inflated de-icing boots on outer wing leading-edges, and fin and tailplane leading-edges, using engine bleed air. Liquid oxygen system for crew and passengers (with cabin wall outlets); this system can be replaced by a gaseous oxygen system if required. Emergency oxygen system available for all occupants in the event of a pressurisation failure.

AVIONICS: *Radar:* Northrop Grumman AN/APN-241.

Instrumentation: Five-screen EFIS based on C-130J flight deck. Radar integrated with moving map display. NVG-compatible.

DIMENSIONS, EXTERNAL:
Wing span	28.70 m (94 ft 2 in)
Wing aspect ratio	10.0
Length overall	22.70 m (74 ft 5½ in)
Height overall: unladen	10.57 m (34 ft 8¼ in)
fully laden	9.70 m (31 ft 10 in)
Fuselage: Max diameter	3.55 m (11 ft 7¾ in)
Tailplane span	12.40 m (40 ft 8¼ in)
Wheel track	3.67 m (12 ft 0½ in)
Wheelbase (to c/l of main units):	
unladen	6.23 m (20 ft 5¼ in)
fully laden	6.40 m (21 ft 0 in)
Propeller diameter	4.11 m (13 ft 6 in)
Distance between propeller centres	9.50 m (31 ft 2 in)
Propeller/fuselage clearance	1.04 m (3 ft 5 in)
Rear-loading ramp/door: Width	2.45 m (8 ft 0½ in)
Height	2.25 m (7 ft 4½ in)
Crew door: Width	0.70 m (2 ft 3½ in)
Height	1.52 m (4 ft 11¾ in)
Emergency door: Width	0.53 m (1 ft 8¾ in)
Height	1.01 m (3 ft 3¾ in)
Paratroop doors: Width	0.91 m (2 ft 11¾ in)
Height	1.92 m (6 ft 3½ in)
Emergency hatch: flight deck:	
Width	0.70 m (2 ft 3½ in)
Length	0.50 m (1 ft 7¾ in)
Cabin (both): Width	0.63 m (2 ft 0¾ in)
Length	0.90 m (2 ft 11½ in)

C-27J's five-screen EFIS

DIMENSIONS, INTERNAL:		
Main cabin: Length	8.58 m (28 ft 1¾ in)	
Width	2.45 m (8 ft 0½ in)	
Height	2.25 m (7 ft 4½ in)	
Floor area: excl ramp	21.0 m² (226 sq ft)	
incl ramp	25.7 m² (276 sq ft)	
Volume	58.0 m³ (2,048 cu ft)	
AREAS:		
Wings, gross	82.00 m² (882.6 sq ft)	
Ailerons (total)	3.65 m² (39.29 sq ft)	
Trailing-edge flaps (total)	18.40 m² (198.06 sq ft)	
Spoilers (total)	1.65 m² (17.76 sq ft)	
Fin (incl dorsal fin)	12.19 m² (131.21 sq ft)	
Rudder	7.02 m² (75.56 sq ft)	
Tailplane	19.09 m² (205.48 sq ft)	
Elevators (total)	4.61 m² (49.62 sq ft)	
WEIGHTS AND LOADINGS:		
Operating weight empty	17,000 kg (37,479 lb)	
Max payload: at 2.5 g	9,000 kg (19,840 lb)	
at 2.25 g	10,225 kg (22,542 lb)	
airdrop	5,000 kg (11.023 lb)	
Max fuel load	9,400 kg (20,725 lb)	
Max T-O weight	31,800 kg (70,106 lb)	
Max landing weight	30,000 kg (66,138 lb)	
Max cargo floor loading	1,500 kg/m² (307.2 lb/sq ft)	
Max wing loading	387.8 kg/m² (79.43 lb/sq ft)	
Max power loading	4.60 kg/kW (7.55 lb/shp)	
PERFORMANCE:		
Max level speed	325 kt (602 km/h; 374 mph)	
Time to 4,570 m (15,000 ft)	7 min	
Initial cruising altitude	8,380 m (27,500 ft)	
Service ceiling	9,145 m (30,000 ft)	
Service ceiling, OEI	3,355 m (11,000 ft)	
T-O run	410 m (1,345 ft)	
T-O to 15 m (50 ft)	640 m (2,100 ft)	
Landing from 15 m (50 ft)	690 m (2,265 ft)	
Landing run	390 m (1,280 ft)	
Range: with max payload	1,160 n miles (2,148 km; 1,334 miles)	
with 6,000 kg (13,228 lb) payload	2,500 n miles (4,630 km; 2,877 miles)	
ferry	3,200 n miles (5,926 km; 3,682 miles)	
Radius of action: with 46 paratroops	1,100 n miles (2,037 km; 1,265 miles)	
with 5,000 kg (11,023 lb) airdrop load	1,215 n miles (2,250 km; 1,398 miles)	
g limit	+3.0	

UPDATED

DRAGON FLY

DRAGON FLY SRL
Via Raffaello 1A, I-22060 Cucciago (CO)
Tel: (+39 031) 72 51 90
Fax: (+39 031) 78 76 42
PRESIDENT: Angelo Castiglioni
VICE-PRESIDENT: Alfredo Castiglioni
MANAGING DIRECTOR AND ADMINISTRATION MANAGER: C Baggioli
COMMERCIAL DIRECTOR: Arnaldo P Ratto
CHIEF PROJECT ENGINEER: Paolo Lucca
OPERATIONS DIRECTOR: Paolo Teso

Company founded in 1993 by twin brothers Angelo and Alfredo Castiglioni specifically to produce the Dragon Fly light helicopter, rights to which were taken over from CRAE Elettromeccanica SpA.

VERIFIED

DRAGON FLY 333
TYPE: Two-seat helicopter.
PROGRAMME: Developed originally by CRAE; design studies and manufacture of single-seat prototype 1985-88; ground and flight testing of this aircraft, 1989-90; two-seat prototype built and tested, 1991-93. Total of two single-seat and three two-seat prototypes, followed by four preseries aircraft; production transferred to new factory from October 1994. Developed and tested by manufacturer to standards approaching FAR Pt 27; initial Italian certification is in ultralight class, but domestic VLR (very light rotorcraft) certification obtained 16 June 1996; JAA and FAA certification was under way by 1998, since when no further information has been received.
CURRENT VERSIONS: **Dragon Fly 333**: *As described.*
Dragon Fly 333AC: Certified version to Italian VLR rules.
Héliot: RPV version developed in association with French companies Etudes et Developpement Techniques (EDT) and CAC Systèmes; launched in June 1996 when prototype displayed at Eurosatory 1996 show at Le Bourget, Paris. Described in *Jane's Unmanned Aerial Vehicles and Targets.*
CUSTOMERS: First delivery in May 1994 to Chinese Civil Protection Volunteers. Total of 80 ordered, of which some 70 delivered, by late 1998; customers in Abu Dhabi, Australia, Belgium, Czech Republic, France, Germany, Italy (37), New Zealand, Portugal and Turkey. First delivery of certified version was in 1997; initial operator was Venice Aero Club.
COSTS: US$105,000 (1997).
DESIGN FEATURES: Two-blade, semi-rigid main rotor and two-blade tail rotor; all blades of NACA 0012 aerofoil section; main rotor nominal speed 500 rpm. Can be road-towed on trailer with main blades folded. Optionally available in kit form.
STRUCTURE: Cabin is welded titanium frame with composites outer shell; aluminium alloy tailboom, rotor blades and landing skids. Full corrosion protection.
LANDING GEAR: Conventional twin-skid type. Emergency floats and skis under development.
POWER PLANT: One Dragon Fly/Hirth F30A26AK four-cylinder two-stroke, rated at 82 kW (110 hp) for T-O, 70.8 kW (95 hp) maximum continuous. Transmission driven through centrifugal clutch and two V-belts. Usable fuel capacity 57 litres (15.0 US gallons; 12.5 Imp gallons).
ACCOMMODATION: Side-by-side seats for two persons. Dual controls standard. Small baggage compartment below seats.
SYSTEMS: 12 V electrical system with engine-driven generator and 12 V 24 Ah battery.
AVIONICS: *Comms:* Provision for transceiver and intercom.
Flight: Optional electric lateral trim.
Instrumentation: Standard VFR.
EQUIPMENT: Medevac and firefighting kits and external load hook under development in 1997.

DIMENSIONS, EXTERNAL:	
Main rotor diameter	6.70 m (21 ft 11¾ in)
Length overall, rotors turning	7.86 m (25 ft 9½ in)
Height to top of rotor head	2.36 m (7 ft 9 in)
DIMENSIONS, INTERNAL:	
Cabin: Length	1.11 m (3 ft 7¾ in)
Max width	1.15 m (3 ft 9¼ in)
Max height	1.13 m (3 ft 8½ in)
AREAS:	
Main rotor disc	35.21 m² (379.0 sq ft)
WEIGHTS AND LOADINGS:	
Weight empty, standard	260 kg (573 lb)
T-O weight: normal	450 kg (992 lb)
max	500 kg (1,102 lb)
PERFORMANCE (two crew, half fuel):	
Max level speed at S/L	72 kt (135 km/h; 83 mph)
Cruising speed at S/L	70 kt (130 km/h; 81 mph)
Max rate of climb at S/L	381 m (1,250 ft)/min
Service ceiling	3,050 m (10,000 ft)
Hovering ceiling: IGE	2,050 m (6,720 ft)
OGE	1,450 m (4,760 ft)
Max range	167 n miles (310 km; 192 miles)
Endurance	2 h 30 min

UPDATED

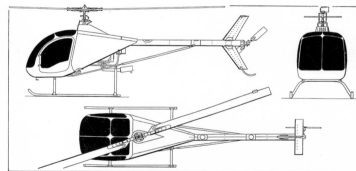

Dragon Fly 333 very light helicopter *(Paul Jackson/Jane's)*

Early production Dragon Fly 333
2000/0044021

ETRURIA

ETRURIA TECHNOLOGY srl
Via Pacinotti, 80, I-55049 Viareggio (LU)
Tel: (+39 0584) 458 81
Fax: (+39 0584) 458 87
Web: http://www.etruria.com
PRESIDENT: Alessandro Mazzoni
CHIEF EXECUTIVE OFFICER: Paolo Margara

RESEARCH AND DEVELOPMENT CENTRE:
Regione Bra 14, I-17033 Garlenda (SV)
Tel: (+39 0182) 58 04 12
Fax: (+39 0182) 58 21 85

The Etruria Vagabond, at present under development, draws on Alessandro Mazzoni's experience with the Piaggio P.180 Avanti.

VERIFIED

ETRURIA VAGABOND
TYPE: Four/five-seat tourer.
PROGRAMME: Work began 1996; announced 1999; no subsequent reports.
CURRENT VERSIONS: **E180**: Baseline model, *as described.*
E250: Proposed turbocharged version. Powered by 186 kW (250 hp) Textron Lycoming TIO-540C flat-six turbocharged engine with longer wing and increased tail area. Data as per E180, except where stated.
COSTS: US$180,000 (2000).
DESIGN FEATURES: Complies with FAR Pt 23. Three lifting surface technology (wing, canard, tailplane) reduces fuel consumption and minimises drag.
FLYING CONTROLS: Flaps on canard and main wing are connected; dual elevator control system.
STRUCTURE: Mixed use of composites and metals throughout. Aluminium fuselage frame bonded to carbon/Kevlar stressed skin. Three-section, two-spar wing is mounted mid-fuselage and has extended laminar flow; composites construction with light alloy flaps and trailing edge. Option to fold wings. Composites canard with 20° forward sweep and 5° anhedral for optimum forward vision; canard flaps of light alloy. Single-piece, constant-section, conventional light alloy tail. Composites tailboom with integral vertical fins.
LANDING GEAR: Hydraulically retractable tricycle type; all units retract into fuselage.
POWER PLANT: One 134 kW (180 hp) Textron Lycoming IO-360B flat-four driving a two-blade, constant-speed, metal pusher propeller, mounted at rear of cabin. Fuel tanks in wing leading-edges, total capacity 200 litres (52.8 US gallons; 44.0 Imp gallons). E250 has extra 100 litre (26.4 US gallon; 22.0 Imp gallon) fuel tank in centre section of main wing torsion box.
ACCOMMODATION: Pilot and up to four passengers in enclosed, crash-resistant cabin; upward-hinging passenger door on port side. Baggage compartment between rear seats and main bulkhead. Dual controls.

ETRURIA to FLY SYNTHESIS—AIRCRAFT: ITALY

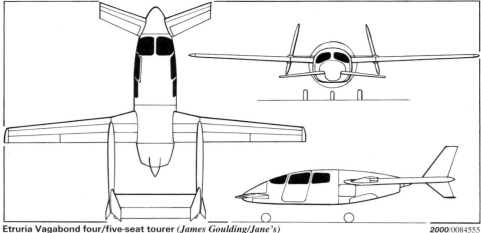

Etruria Vagabond four/five-seat tourer *(James Goulding/Jane's)* 2000/0084555

SYSTEMS: Oxygen supply and air conditioning standard on E250. De-icing boots optional.
DIMENSIONS, EXTERNAL:
Wing span: E180	10.00 m (32 ft 9½ in)
E250	11.00 m (36 ft 1 in)
Wing aspect ratio: E180	12.2
E250	11.0
Length overall	7.60 m (24 ft 11¼ in)
Height overall	2.70 m (8 ft 10¼ in)

AREAS:
Wings, gross: E180	8.20 m² (88.3 sq ft)
E250	11.00 m² (118.4 sq ft)
Foreplanes (total)	2.30 m² (24.76 sq ft)
Tailplane: E180	2.40 m² (25.83 sq ft)
E250	2.70 m² (29.06 sq ft)

WEIGHTS AND LOADINGS:
Weight empty: E180	750 kg (1,653 lb)
E250	850 kg (1,874 lb)
Max baggage capacity: E180	91 kg (200 lb)
Max T-O weight: E180	1,100 kg (2,425 lb)
E250	1,400 kg (3,086 lb)
Max landing weight: E180	1,100 kg (2,425 lb)
E250	1,330 kg (2,932 lb)
Max wing/canard loading: E180	104.8 kg/m² (21.46 lb/sq ft)
E250	105.3 kg/m² (21.56 lb/sq ft)
Max power loading: E180	8.20 kg/kW (13.47 lb/hp)
E250	7.51 kg/kW (12.34 lb/hp)

PERFORMANCE:
Max operating speed: E180	190 kt (352 km/h; 219 mph)
E250	210 kt (389 km/h; 242 mph)
Max cruising speed at 75% power:	
E180	180 kt (333 km/h; 207 mph)
E250	185 kt (343 km/h; 213 mph)
Stalling speed, power off, both:	
flaps up	75 kt (139 km/h; 87 mph)
flaps down	60 kt (111 km/h; 69 mph)
Range, 45 min reserves:	
E180	540 n miles (1,000 km; 621 miles)
E250	999 n miles (1,850 km; 1,149 miles)

UPDATED

EURO ALA

EURO ALA

HEAD OFFICE: C de Vibrata 136, I-64013 Corropoli (Te)
Tel: (+39 0861) 80 80 26
Fax: (+39 0861) 80 83 42
e-mail: euroala@ito.it
PRESIDENT: Alfredo Di Cesare

WORKS: Strada Bonifica, km 17, I-64010 Ancarano (Te)

The Advanced Light Aircraft (ALA) company was established in 1985 and changed name to Euro ALA in 1995.

VERIFIED

EURO ALA JET FOX 97

TYPE: Side-by-side ultralight.
PROGRAMME: Earlier Jet Fox 91 first flown in 1991; 140 sold by 1993. Considerably improved Jet Fox 97 first flown 10 March 1997 and exhibited at Aero '97 at Friedrichshafen the following month; total of 20 to 30 built by early 1999, including remanufactured Jet Fox 91s. Jet Fox series certified in Germany 1993; France and Belgium in 1995.
COSTS: £47,950 with Rotax 582; £55,650 with Rotax 912 (both 1999).

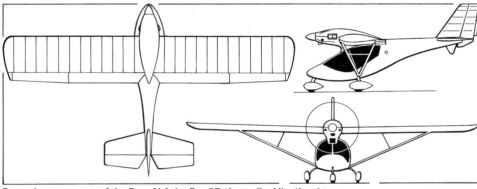

General arrangement of the Euro ALA Jet Fox 97 *(James Goulding/Jane's)* 2000/0054609

DESIGN FEATURES: Considerably refined version of earlier design, but bears structural resemblance to Fox-C22 (1988-89 *Jane's*); fully faired fuselage and engine; flaps. Wings rapidly detachable for transport.
FLYING CONTROLS: Conventional and manual. Flight-adjustable elevator trim.
STRUCTURE: Alloy tube and steel fuselage frame with composites shell; Dacron-covered aluminium wings and tail surfaces; glass fibre and carbon fibre for landing gear.
LANDING GEAR: Fixed tricycle type. Cable-operated disc brakes. Steerable nosewheel. Speed fairings optional.
POWER PLANT: One 47.8 kW (64.1 hp) Rotax 582 UL two-stroke or 59.6 kW (79.9 hp) Rotax 912 UL four-stroke engine, driving a two-blade GT wooden propeller. Fuel capacity 59 litres (15.6 US gallons; 13.0 Imp gallons).
EQUIPMENT: Optional ballistic parachute.
DIMENSIONS, EXTERNAL:
Wing span	9.78 m (32 ft 1 in)
Length overall	5.78 m (18 ft 11½ in)
Height overall	2.80 m (9 ft 2¼ in)

AREAS:
Wings, gross	14.62 m² (157.4 sq ft)

WEIGHTS AND LOADINGS (Rotax 912):
Weight empty, including parachute	290 kg (639 lb)
Max T-O weight	450 kg (992 lb)

PERFORMANCE (Rotax 912):
Max level speed	94 kt (175 km/h; 109 mph)
Cruising speed	81 kt (150 km/h; 93 mph)
Stalling speed	33 kt (60 km/h; 38 mph)
Max rate of climb at S/L	360 m (1,181 ft)/min
T-O run	100 m (330 ft)
Landing run	120 m (395 ft)
Endurance	4 h 30 min
g limits	+6/−3.5

UPDATED

Euro ALA Jet Fox 97 *(Pierre Gaillard)* 2001/0103533

FLY SYNTHESIS

FLY SYNTHESIS SRL

Via Gorizia 63, I-33050 Gonars/Udine
Tel/Fax: (+39 0432) 99 24 82 and 99 35 57
e-mail: flysynth@tin.it
Web: http://www.com pouse.com

The Storch kitbuilt produced by this company also provides the basis of the tailwheel/T tail Flight Team Sinus motor glider and ultralight, described in the German section; this company also markets Storch in Germany. A new design, the Texan, has recently been added to the company's range.

UPDATED

FLY SYNTHESIS STORCH

TYPE: Side-by-side ultralight/kitbuilt.
CURRENT VERSIONS: **CL Pulcino:** Standard version, *as described.*
  CL De Luxe: Heavier (450 kg; 992 lb) version, powered by 47.7 kW (64 hp) Rotax 582 two-stroke.
  CL-J De Luxe: As CL De Luxe, but powered by 59.7 kW (80 hp) Jabiru 2200 four-stroke.
  SS De luxe: As CL De Luxe, but with shorter overall length (5.75 m; 18 ft 10¾ in) and wing span (9.40 m; 30 ft 10 in).
  SS-J De Luxe: Jabiru 2200-powered version of SS De Luxe.
  HS De Luxe: As SS De Luxe, but with shorter (8.71 m; 28 ft 7 in) wing span.
  HS-J De Luxe: Jabiru 2200-powered version of HS De Luxe.
CUSTOMERS: 240 built by end of 1998 (latest figures received).
COSTS: DM69,950 ex-works, Rotax 582 (2000); DM 49,950 kit with Rotax 503 (2000).
DESIGN FEATURES: Laminar flow aerofoil section. Strut-braced wings fold alongside fuselage for storage.
FLYING CONTROLS: Manual. Full-span, Junkers-type ailerons and all-moving tailplane with anti-balance tab.
STRUCTURE: Composites main fuselage shell with welded steel tube frame in cabin area and tubular alloy tailboom;

280 ITALY: AIRCRAFT—FLY SYNTHESIS to GENERAL AVIA

Fly Synthesis Storch two-seat ultralight
1999/0051089

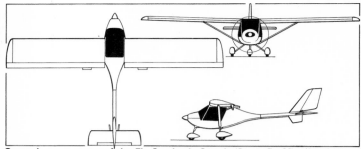
General arrangement of the Fly Synthesis Storch (James Goulding/Jane's)
1999/0051090

light alloy wing with composites skin; all-composites wing optional.
LANDING GEAR: Fixed tricycle type.
POWER PLANT: One 34.0 kW (45.6 hp) Rotax 503 UL-1V two-stroke with reduction gear drive to two- or three-blade propeller. Standard fuel capacity 38 litres (10.0 US gallons; 8.4 Imp gallons); 60 litres (15.9 US gallons; 13.2 Imp gallons) optional.
ACCOMMODATION: Cabin doors open upwards.
EQUIPMENT: BRS ballistic recovery parachute optional.
Data below refer to CL Pulcino.
DIMENSIONS, EXTERNAL:
  Wing span   10.14 m (33 ft 3¼ in)
  Length overall   6.25 m (20 ft 6 in)
  Height overall   2.54 m (8 ft 4 in)
AREAS:
  Wings, gross   13.02 m² (140.15 sq ft)
WEIGHTS AND LOADINGS:
  Weight empty   199-205 kg (439-452 lb)
  Max T-O weight   400-450 kg (881-992 lb)
PERFORMANCE:
  Max level speed   81 kt (150 km/h; 93 mph)
  Cruising speed   67 kt (125 km/h; 78 mph)
  Stalling speed   31-33 kt (57-60 km/h; 36-38 mph)
  Max rate of climb at S/L 150 m (492 ft)/min
  T-O run   105 m (345 ft)
  Landing run   90 m (295 ft)
  Endurance   3 h 18 min
  g limits   +4/−2
UPDATED

FLY SYNTHESIS TEXAN
TYPE: Side-by-side ultralight.
PROGRAMME: First flown 1997.
CUSTOMERS: Three built by end of 1998 (latest information received).
DESIGN FEATURES: Low-wing monoplane. Laminar flow aerofoil section; sweptback fin and rudder.
FLYING CONTROLS: Manual ailerons, rudder and all-moving tailplane. Four-position flaps.
STRUCTURE: Composites throughout.
LANDING GEAR: Fixed tricycle type.
POWER PLANT: One 73.5 kW (98.6 hp) Rotax 912 ULS four-stroke driving a Kremen three-blade propeller. Standard fuel capacity 60 litres (15.9 US gallons; 13.2 Imp gallons).
EQUIPMENT: Optional BRS ballistic recovery parachute system.
DIMENSIONS, EXTERNAL:
  Wing span   8.60 m (28 ft 2½ in)
  Length overall   6.90 m (22 ft 7¾ in)
AREAS:
  Wings, gross   12.04 m² (129.6 sq ft)
WEIGHTS AND LOADINGS:
  Weight empty   285 kg (628 lb)
  Max T-O weight   450 kg (992 lb)
PERFORMANCE:
  Max level speed   121 kt (225 km/h; 139 mph)
  Econ cruising speed   97 kt (180 km/h; 112 mph)
  Stalling speed   34 kt (62 km/h; 39 mph)
  T-O run   135 m (445 ft)
  Landing run   145 m (475 ft)
  Endurance   5 h
  g limits   +4.4/−2.2
UPDATED

FRATI

STELIO FRATI
Via Noe 1, I-20100 Milano
CHIEF DESIGNER: Dott Ing Stelio Frati

In 1998, Dott Ing Frati left General Avia (which see) to resume the activities of a freelance designer – a career successfully practised since 1950. His past designs include the F.8 Falco, F.15 Picchio, F.250 (now manufactured by Aermacchi as the SF-260), SF.600A Canguro (for SIAI-Marchetti, and now VulcanAir) and F.20 Pegaso (see 1981-82 *Jane's*). The F.8 Falco is marketed as a kit aircraft by Sequoia Aircraft Corporation (which see in US section).

Frati is currently working on a new four-seat, all-wood tourer with a 194 kW (260 hp) engine (see illustration); a 'Falco Jet' employing an unspecified Williams turbofan; and a twin-jet commuter airliner with a fuselage possibly based on the Canguro.
UPDATED

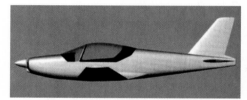
Computer-generated image of new Stelio Frati four-seat tourer
2001/0103604

GENERAL AVIA

GENERAL AVIA COSTRUZIONI AERONAUTICHE SRL
Via U Comandini 38, I-00173 Roma
Tel: (+39 06) 723 16 51
Fax: (+39 06) 723 45 36
PRESIDENT AND MANAGING DIRECTOR: Maurizio Ruggiero
PRODUCTION DIRECTOR: Dott Ing Pasquale De Rosa
MARKETING: Cmdte Alessandro Ghisleni
SALES DIRECTOR: Massimo Piva
PUBLIC RELATIONS: Carla Bielli

WORKS: Viale Roma 25, I-06065 Passignano s T Perugia
Tel: (+39 075) 82 94 44
Fax: (+39 075) 82 94 42

Production of the company's F.22 series has been severely affected by earthquake damage suffered by the Passignano facilities in 1998; low-rate manufacture is continuing, however.
UPDATED

GENERAL AVIA F.22
TYPE: Two-seat lightplane.
PROGRAMME: First flight of prototype F.22-A (I-GEAD) from Orio al Serio (Bergamo) 13 June 1989; certification achieved May 1993; certification of 119 kW (160 hp) F.22-B (prototype I-GEAG) mid-1993, followed by F.22-C (I-GEAH) and F.22-R (I-GEAE, first flown 16 November 1990); certification conducted in association with Aermacchi at Venegono. Further two aircraft as US demonstrators, 1995. All current models had been certified by the RAI, and the F.22-B, F.22-C and F.22-R by the FAA, by late 1997. Series production began in June 1996.
CURRENT VERSIONS: **F.22-A:** Basic version powered by 86.5 kW (116 hp) engine.
  F.22-B: Powered by 119 kW (160 hp) engine; fixed-pitch propeller.
  F.22-B Variabile: As F.22-B but with constant-speed propeller.
  F.22-C: Flying early 1993; originally named Pinguino (penguin); 134 kW (180 hp) engine; constant-speed two-blade metal propeller. Retractable landing gear.
  F.22-R Sprint: As F.22-C, but powered by 119 kW (160 hp) engine.
CUSTOMERS: Total sales by mid-2000 were 43 firm and 50 options, of which 26th (including prototypes) was registered in March 2000; customers in Germany, Italy, Netherlands, New Zealand, UK and USA. In late 1998, Italian Aero Club ordered 12 F.22-Bs for basic training and three F.22-Cs for IFR training; deliveries required between April and December 1999, but by mid-2000 only two F.22-Cs had been received and the delivery schedule was being renegotiated.
COSTS: F.22-A US$98,000, F.22-B US$109,300, F.22-B Variabile US$127,500, F.22-C US$149,500, F.22-R US$143,500, all basically equipped (1997).
DESIGN FEATURES: Classic Frati design; low wing, with moderately swept leading-edges for high cruising speed.
FLYING CONTROLS: Conventional and manual. Elevator trim tab; electrically actuated flaps.
STRUCTURE: All-metal single-spar wing built as one piece; all-metal fuselage. F.22-C and F.22-R are flush-riveted.

General Avia F.22-B Variabile (constant-speed propeller) (Paul Jackson/Jane's)
2000/0064501

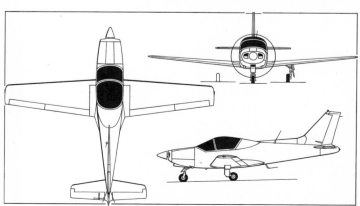
General Avia F.22-R two-seater (Mike Keep/Jane's)

LANDING GEAR: F.22-A and F.22-B have non-retractable tricycle type, with steerable nosewheel; oleo shock-absorbers; faired main legs. F.22-C and F.22-R have retractable landing gear. Wheel/tyre size (all) 5.00-5.
POWER PLANT: F.22-A powered by one 86.5 kW (116 hp) Textron Lycoming O-235-N2C flat-four engine, driving an MT-180-120-2C two-blade wood/composites propeller.
F.22-B powered by 119 kW (160 hp) O-320-D2A driving an MT-180-140 two-blade propeller.
F.22-B Variabile and F.22-R powered by one 119 kW (160 hp) O-320-D1A, driving Hartzell HC-C2YL-1BF two-blade constant-speed propeller.
F.22-C powered by one 134 kW (180 hp) Textron Lycoming O-360-A1A driving Hartzell HC-C2YK-1BF constant-speed metal two-blade propeller.
Fuel capacity 105 litres (28.0 US gallons; 23.0 Imp gallons) in F.22-A, 135 litres (36.0 US gallons; 30.0 Imp gallons) in F.22-B and F.22-B Variabile and 160 litres (42.3 US gallons; 35.2 Imp gallons) in F.22-C and F.22-R; optional long-range tank.
ACCOMMODATION: Two seats side by side: sliding canopy. Dual control columns.
AVIONICS: Honeywell Silver Crown.
 Comms: VHF and transponder.
 Flight: ADF.
DIMENSIONS, EXTERNAL:
 Wing span 8.50 m (27 ft 10¾ in)
 Wing chord: at root 1.59 m (5 ft 2½ in)
  at tip 0.88 m (2 ft 10¾ in)
 Wing aspect ratio 6.7
 Length overall 7.40 m (24 ft 3¼ in)
 Height overall 2.84 m (9 ft 3¾ in)
 Tailplane span 3.00 m (9 ft 10 in)
 Wheel track 2.90 m (9 ft 6¼ in)
 Wheelbase 1.86 m (6 ft 1¼ in)
 Propeller diameter 1.78 m (5 ft 10 in)
AREAS:
 Wings, gross 10.82 m² (116.3 sq ft)
 Fin 0.74 m² (7.97 sq ft)
 Rudder 0.505 m² (5.44 sq ft)
 Tailplane 1.24 m² (13.35 sq ft)
 Elevators (total, incl tab) 1.02 m² (10.98 sq ft)
WEIGHTS AND LOADINGS:
 Weight empty, IFR equipped: F.22-A 510 kg (1,124 lb)
 F.22-B 560 kg (1,235 lb)
 F.22-B Variabile 610 kg (1,345 lb)
 F.22-C 640 kg (1,411 lb)
 F.22-R 620 kg (1,367 lb)
 Max T-O weight:
  Aerobatic: F.22-A 750 kg (1,653 lb)
  F.22-B, F.22-B Variabile 800 kg (1,763 lb)
  F.22-C, F.22-R 850 kg (1,874 lb)
  Utility: F.22-A 780 kg (1,719 lb)
  F.22-B, F.22-B Variabile, F.22-R 850 kg (1,874 lb)
  F.22-C 900 kg (1,984 lb)
 Max wing loading:
  Aerobatic: F.22-A 69.3 kg/m² (14.20 lb/sq ft)
  F.22-B, F.22-B Variabile 73.9 kg/m² (15.14 lb/sq ft)
  F.22-C, F.22-R 78.6 kg/m² (16.09 lb/sq ft)
  Utility: F.22-A 72.1 kg/m² (14.76 lb/sq ft)
  F.22-B, F.22-B Variabile, F.22-R 78.6 kg/m² (16.09 lb/sq ft)
  F.22-C 83.2 kg/m² (17.04 lb/sq ft)
 Max power loading:
  Aerobatic: F.22-A 8.67 kg/kW (14.24 lb/hp)
  F.22-B, F.22-B Variabile 6.71 kg/kW (11.02 lb/hp)
  F.22-C 6.33 kg/kW (10.40 lb/hp)
  F.22-R 7.12 kg/kW (11.70 lb/hp)
  Utility: F.22-A 9.02 kg/kW (14.82 lb/hp)
  F.22-B, F.22-B Variabile, F.22-R 7.12 kg/kW (11.70 lb/hp)
  F.22-C 6.71 kg/kW (11.02 lb/hp)
PERFORMANCE:
 Max level speed: F.22-A 118 kt (220 km/h; 136 mph)
  F.22-B 131 kt (244 km/h; 151 mph)
  F.22-B Variabile 140 kt (260 km/h; 161 mph)
  F.22-C 164 kt (305 km/h; 189 mph)
  F.22-R 158 kt (294 km/h; 182 mph)
 Stalling speed, flaps down:
  F.22-A 51 kt (93 km/h; 58 mph)
  F.22-B, F.22-B Variabile 53 kt (97 km/h; 61 mph)
  F.22-C, F.22-R 54 kt (100 km/h; 63 mph)
 Max rate of climb at S/L: F.22-A 195 m (640 ft)/min
  F.22-B 300 m (984 ft)/min
  F.22-B Variabile 360 m (1,181 ft)/min
  F.22-C, F.22-R 420 m (1,378 ft)/min
 Service ceiling: F.22-A 3,950 m (12,960 ft)
  F.22-B 4,880 m (16,010 ft)
  F.22-B Variabile 5,180 m (16,995 ft)
  F.22-C, F.22-R 5,800 m (19,030 ft)
 T-O run: F.22-A 320 m (1,050 ft)
  F.22-B 260 m (853 ft)
  F.22-B Variabile, F.22-C, F.22-R 220 m (722 ft)
 Landing run: F.22-A 200 m (656 ft)
  F.22-B 220 m (722 ft)
  F.22-B Variabile 230 m (755 ft)
  F.22-C, F.22-R 240 m (787 ft)
 Max range, no reserves:
  F.22-A 405 n miles (750 km; 466 miles)
  F.22-B 496 n miles (920 km; 571 miles)
  F.22-B Variabile 540 n miles (1,000 km; 621 miles)
  F.22-C, F.22-R 594 n miles (1,100 km; 683 miles)
 g limits +6/−3

UPDATED

General Avia F.22-A, first in the UK *(Paul Jackson/Jane's)* 2000/0064502

GENERAL AVIA F.220 AIRONE

All known details of this four-seat tourer last appeared in the 2000-01 edition.

UPDATED

HELISPORT

CH-7 HELISPORT SRL

Strada Traforo del Pino 104, I-10132 Torino
Tel: (+39 011) 899 95 65 and 899 96 18
Fax: (+39 011) 899 96 18 and 898 11 44
Web: http://www.ch-7helicopter.com

Previously known as EliSport, the company markets the CH-7 Angel, a development of the CH-6, produced in 1987 by the Argentine designer Augusto Cicaré. US agent is Lancair (which see).

NEW ENTRY

HELISPORT CH-7 ANGEL

TYPE: Two-seat ultralight helicopter kitbuilt.
PROGRAMME: CH-6 project acquired in 1989 by HeliSport; added new cockpit by sports car designer Marcello Gandini; renamed CH-7 Angel.
CURRENT VERSIONS: **CH-7 Angel 582:** Single-seat version; Rotax 582 two-cylinder engine.
 CH-7 Angel 912: Single-seat version; Rotax 912 four-cylinder engine.
 CH-7 Kompress: Two-seat version; Rotax 914 four-cylinder turbocharged engine and dual controls; developed in 1997.
CUSTOMERS: Some 120 flying by 2000. Owners in Australia, Germany, Italy, South Africa, Taiwan, USA and elsewhere.
COSTS: Basic kit, less engine and tax €45,700 (2001).
DESIGN FEATURES: Small 'penny-farthing' helicopter with two-blade, semi-rigid, teetering main and two-blade rotors; pod-and-boom fuselage; sweptback upper and lower fins, latter with tailskid. Main rotor transmission by single 10 cm (4 in) five-groove Goodyear belt. Quoted build time 200 hours; quick-build kit 85 hours.
 Rotor blade aerofoil sections NACA 8-H-12 (main) and NACA 63-014 (tail); 1.5 kg (3.3 lb) main rotor tip weights; max 520 rpm main, 2,808 rpm tail rotor.
FLYING CONTROLS: Conventional and manual, by pushrods and bell cranks.
STRUCTURE: Welded 4130 steel tube cabin frame, with nitrogen filling; glass fibre cockpit shell and Plexiglas or Lexan canopy; 2024T3 anodised aluminium alloy tailboom and landing skids. Rotor blades of composites (main) and aluminium alloy (tail).
LANDING GEAR: Fixed; twin-skid type.
POWER PLANT: One 47.8 kW (64.1 hp) Rotax 582UL, 59.6 kW (79.9 hp) Rotax 912 UL or 84.6 kW (113.4 hp) Rotax 914 piston engine (see Current Versions). Fuel capacity 40 litres (10.6 US gallons; 8.8 Imp gallons) standard in all versions; additional 19 litre (5.0 US gallon; 4.2 Imp gallon) tank optional in 912 and Kompress.
DIMENSIONS, EXTERNAL:
 Main rotor diameter 6.17 m (20 ft 3 in)
 Tail rotor diameter 1.03 m (3 ft 4½ in)
 Length: fuselage 5.31 m (17 ft 5 in)
  overall, rotors turning 7.41 m (24 ft 4 in)
 Fuselage max width 0.82 m (2 ft 8¼ in)
 Height to top of rotor head 2.31 m (7 ft 7 in)
 Skid track 1.60 m (5 ft 3 in)
DIMENSIONS, INTERNAL:
 Cabin max width 0.76 m (2 ft 6 in)
AREAS:
 Main rotor disc 29.90 m² (321.8 sq ft)
 Tail rotor disc 0.83 m² (8.93 sq ft)
WEIGHTS AND LOADINGS:
 Weight empty: 582 205 kg (452 lb)
  912 215 kg (474 lb)
  Kompress: Europe 245 kg (540 lb)
   USA 281 kg (620 lb)
 Max T-O weight: 582 360 kg (793 lb)
  912 300 kg (661 lb)
  Kompress: Europe 450 kg (992 lb)
   USA 499 kg (1,100 lb)
PERFORMANCE:
 Never-exceed speed (VNE):
  582 80 kt (149 km/h; 92 mph)
  912, Kompress 112 kt (209 km/h; 129 mph)

HeliSport CH-7 Kompress two-seat light helicopter *(Paul Jackson/Jane's)* 2001/0103563

282　ITALY: AIRCRAFT—HELISPORT to III

Max cruising speed: 582	59 kt (110 km/h; 68 mph)	912	2,500 m (8,200 ft)	with optional fuel: Kompress	388 n miles (719 km; 477 miles)
912	81 kt (150 km/h; 93 mph)	Kompress	3,500 m (11,480 ft)	Endurance: 582	3 h
Kompress	86 kt (160 km/h; 99 mph)	OGE: 582, 912	1,500 m (4,920 ft)	912	6 h
Max rate of climb at S/L: 582	600 m (1,968 ft)/min	Kompress	2,500 m (8,200 ft)	Kompress	3 h 40 min
912	300 m (984 ft)/min	Range: with normal fuel:			*NEW ENTRY*
Kompress: Europe	450 m (1,476 ft)/min	582	148 n miles (275 km; 170 miles)		
USA	347 m (1,140 ft)/min	912	486 n miles (900 km; 559 miles)		
Hovering ceiling: IGE: 582	2,100 m (6,900 ft)	Kompress	298 n miles (552 km; 343 miles)		

III

INIZIATIVE INDUSTRIALI ITALIANE SpA
Corso Trieste n 150, I-00198 Rome
Tel: (+39 06) 854 63 41/85 30 14 61/85 30 17 94/855 94 85
Fax: (+39 06) 855 71 62
e-mail: info@skyarrow.com
Web: http://www.skyarrow.com
PRESIDENT AND MANAGING DIRECTOR: Avv Furio Lauri

US AGENT:
Pacific Aerosystem Inc
5760 Chesapeake Court, San Diego, California 92123, USA
Tel: (+1 619) 571 14 41
Fax: (+1 619) 571 08 03

Company formed April 1947 in Trieste, as Meteor SpA. Main R&D and production facility then in Monfalcone, where two- and four-seat aircraft produced for training and sport flying, plus target aircraft and electronic systems for military market. Present company created in 1985; relocated to new facility in Monterotondo 1989; now produces pleasure boats and general aviation aircraft and is developing the Rondine unmanned surveillance system which is based on the Sky Arrow.

VERIFIED

III SKY ARROW (ROTAX 912 VERSIONS)
TYPE: Two-seat lightplane/ultralight.
PROGRAMME: First flight March 1993; certified by FAA in primary aircraft sport category.
CURRENT VERSIONS: Designations refer to maximum T-O weights in kilogrammes or pounds; L and T are landplane, SP is seaplane, C is certified.
450 T: For markets where MTOW for ultralights is 450 kg.
480 T: For markets where MTOW for ultralights is 480 kg. Certified to TP 10141 in Canada.
500 TF: For French market, to meet 500 kg MTOW requirement.
650 LMT: Light Military Trainer; proposed easy to operate, low-cost trainer; provision for belly-mounted IR/video sensors and quick-change amphibious conversion. Under development by late 1997. No known purchases.
650 TC: JAR-VLA certified model, MTOW 650 kg. Prototype 650 T (I-ARRO) flown second quarter of 1995; first 650 TC (I-TREI) built 1996; at least 11 produced for German, Italian and UK owners by mid-1999; none subsequently reported.
650 TCN: FAA-certified model for VFR and night operation. Eight produced by April 1999; none subsequently reported.
650 T and 650 SP: Kitplane versions.
700T Speed Arrow: Prototype (I-RUND) built early 1998; no further data disclosed.
1200 LC: For US market under Primary Category-Sportplane certification rules.
1310 SP: Primary Sport version with increased MTOW.
650 T/914 and 650 A/914: Described separately.
1450 L and **1450 A:** US equivalents of 650, with land and amphibious landing gear (respectively) and 658 kg (1,450 lb) MTOW. Empty weight of 1450 A is 417 kg (920 lb). Marketed by Pacific Aerosystem.
CUSTOMERS: Total of 120 manufactured by December 1997 (latest figure provided), including aircraft for customers in Germany, Ghana, Italy, UK and USA.
COSTS: UK, Sky Arrow 650 TC, £49,000 plus tax (1997). USA, Sky Arrow 650 TCN US$98,500 (1998).

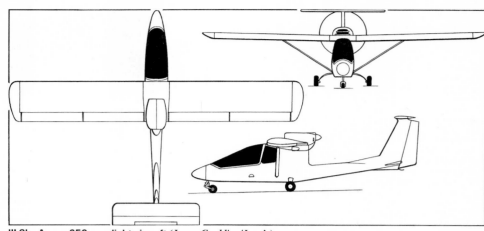

III Sky Arrow 650 very light aircraft *(James Goulding/Jane's)*　0011089

DESIGN FEATURES: Designed with objectives of good view from cockpit, lightness, ease of maintenance, ease of disassembly for transport and storage. Capable of being converted to an amphibian. High-mounted engine behind wing; pusher propeller; low-mounted pod fuselage below propeller carries T tail.
Wing aerofoil Gottingen 398 modified; constant wing chord; dihedral 1° 30′; twist 1°; wings easily detachable at centreline joint.
FLYING CONTROLS: Conventional and manual. Electrically actuated half-span flaps and tabs; centrally mounted elevator tab for trimming. Full dual controls (sidestick and rudder) and two throttles.
STRUCTURE: Airframe, wings, tail unit and landing gear manufactured from carbon sandwich, Kevlar-reinforced, composites material.
LANDING GEAR: Fixed tricycle; hydraulic disc brakes; nosewheel with rubber and spring shock-absorber; twin floats on 1310 SP.
POWER PLANT: One 59.6 kW (79.9 hp) Rotax 912 UL flat-four four-stroke with two-blade fixed-pitch wooden propeller with ground pitch adjustment. Fuel 70 litres (18.5 US gallons; 15.4 Imp gallons).
ACCOMMODATION: Two seats in tandem; sideways-opening canopy; baggage compartments below and behind rear seat.
SYSTEMS: Hydraulic brakes. Electrical system includes 12 V 14 Ah battery; alternator.
AVIONICS: Customer specified. 650 TCN has Honeywell KX 125 nav/com and KT 76A transponder as standard, with second KX 125, KN 62A DME, KR 87 digital ADF and Skyforce moving-map GPS optional.
EQUIPMENT: Optional handicapped pilot kit comprises sidestick controller on port cockpit console for rudder and throttle inputs, enabling the aircraft to be flown by pilots unable to use their legs; baggage compartment accommodates folding wheelchair. Banner/glider towing hook, incorporating LCD micro video camera for monitoring towing operation, optional.

DIMENSIONS, EXTERNAL:
Wing span	9.60 m (31 ft 6 in)
Wing chord	1.40 m (4 ft 7 in)
Wing aspect ratio	6.9
Length overall	7.60 m (24 ft 11¼ in)
Max width of fuselage	0.75 m (2 ft 5½ in)
Height overall: except SPs	2.56 m (8 ft 4¾ in)
SP	2.80 m (9 ft 2½ in)
Tailplane span	2.80 m (9 ft 2¼ in)
Tailplane chord	0.75 m (2 ft 5½ in)
Wheel track	1.75 m (5 ft 9 in)

AREAS:
Wings, gross	13.44 m² (144.7 sq ft)

WEIGHTS AND LOADINGS:
Weight empty: 450 T	289 kg (638 lb)
480 T	303 kg (667 lb)
500 T	319 kg (704 lb)
650 SP	400 kg (882 lb)
650 T, 650 TC	350 kg (772 lb)
1200 LC	322 kg (710 lb)
1310 SP	372 kg (820 lb)
Max T-O weight: 450 T	450 kg (992 lb)
480 T	480 kg (1,058 lb)
500 T	500 kg (1,102 lb)
650 SP, 650 T, 650 TC	650 kg (1,433 lb)
1200 LC	544 kg (1,200 lb)
1310 SP	594 kg (1,310 lb)
Max wing loading (650)	48.3 kg/m² (9.90 lb/sq ft)
Max power loading (650)	10.92 kg/kW (17.93 lb/hp)

PERFORMANCE:
Never-exceed speed (VNE): except 650 series	121 kt (224 km/h; 139 mph)
650 series	148 kt (274 km/h; 170 mph)
Max level speed: 450 T, 480 T, 500 T, 1200 LC	108 kt (200 km/h; 124 mph)
650 SP	90 kt (167 km/h; 104 mph)
650 T, 650 TC	107 kt (198 km/h; 123 mph)
1310 SP	95 kt (176 km/h; 109 mph)
Cruising speed at 75% power at S/L:	
except SPs	90 kt (167 km/h; 104 mph)
SPs	80 kt (148 km/h; 92 mph)
Stalling speed, power off, flaps down:	
450 T, 480 T, 500 T	33 ft (62 km/h; 38 mph)
650 SP, 650 T, 650 TC	37 kt (69 km/h; 43 mph)
1200 LC	34 kt (63 km/h; 40 mph)
1310 SP	35 kt (65 km/h; 41 mph)
Max rate of climb at S/L:	
450 T, 480 T, 500 T	305 m (1,000 ft)/min
650 SP	213 m (700 ft)/min
650 T, 650 TC, 1310 SP	244 m (800 ft)/min
1200 LC	274 m (900 ft)/min
T-O run: 450 T, 480 T, 500 T	110 m (360 ft)
650 SP, 650 T	244 m (800 ft)
650 TC, 1310 SP	183 m (600 ft)
1200 LC	122 m (400 ft)
Landing run: 450 T, 480 T, 500 T	92 m (300 ft)
650 SP	77 m (250 ft)
650 T, 650 TC	153 m (500 ft)
1200 LC	122 m (400 ft)
1310 SP	61 m (200 ft)
Service ceiling: 450 T, 480 T, 500 T, 650 T, 650 TC	3,960 m (13,000 ft)
650 SP	3,050 m (10,000 ft)
1200 LC	3,660 m (12,000 ft)
Range with max fuel, 75% power, 30 min reserves	
except 1310 SP	324 n miles (600 km; 372 miles)
1310 SP	288 n miles (533 km; 331 miles)
Endurance, as above: all	3 h 40 min

UPDATED

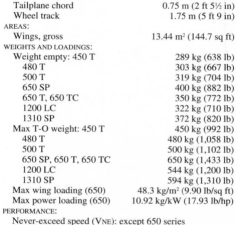

III SKY ARROW (ROTAX 914 VERSIONS)
TYPE: Sports and training aircraft.
PROGRAMME: See previous entry. No reported production.
CURRENT VERSIONS: **650 T/914:** Wheeled version.
650 A/914 Exocet: Amphibian with twin floats.

III Sky Arrow 650 TC certified version *(Paul Jackson/Jane's)*　2001/0103575

III to PIAGGIO—AIRCRAFT: ITALY

POWER PLANT: One 84.6 kW (113.4 hp) Rotax 914 UL flat-four driving a three-blade constant-speed propeller.

WEIGHTS AND LOADINGS (A: 650 A/914, T: 650 T/914):
Weight empty: A		454 kg (1,000 lb)
T		367 kg (810 lb)
Max T-O weight: A, T		650 kg (1,433 lb)
Max wing loading		48.3 kg/m² (9.90 lb/sq ft)
Max power loading		7.69 kg/kW (12.64 lb/hp)

PERFORMANCE:
Never-exceed speed (V$_{NE}$)	148 kt (274 km/h; 170 mph)
Max level speed: at S/L: A	107 kt (198 km/h; 170 mph)
T	120 kt (222 km/h; 138 mph)
at 4,570 m (15,000 ft): A	120 kt (222 km/h; 138 mph)
T	130 kt (241 km/h; 150 mph)
Cruising speed at 75% power: at S/L:	
A	92 kt (170 km/h; 106 mph)
T	108 kt (200 km/h; 124 mph)
at 4,570 m (15,000 ft): A	110 kt (204 km/h; 127 mph)
T	120 kt (222 km/h; 138 mph)
Stalling speed, power off, flaps down:	
A, T	37 kt (69 km/h; 43 mph)
Max rate of climb at S/L: A	335 m (1,100 ft)/min
T	427 m (1,400 ft)/min
T-O run: on land: A	122 m (400 ft)
T	110 m (360 ft)
on water: A	244 m (800 ft)
Landing run: on land: A	214 m (700 ft)
T	153 m (500 ft)
on water: A	77 m (250 ft)
Range with max fuel, 75% power, 30 min reserves:	
A	245 n miles (453 km; 281 miles)
T	270 n miles (500 km; 310 miles)
Endurance: A, T	2 h 15 min
g limits	+5.7/−2.8

UPDATED

III Sky Arrow 650 Exocet *(Keith Wilson/SFB Photographic)* 0064743

MAGNI

MAGNI GYRO
Via della Technica 4/N, I-21044 Cavaria (VA)
Tel: (+39 0331) 21 80 41
Fax: (+39 0331) 21 73 28
e-mail: magnigyro@logic.it
Web: http://www.magnigyro.com

Magni Gyro produces a range of ultralight autogyros, including the open-cockpit single-seat M 18 Spartan and two-seat M 14 Scout and M 16 Tandem Trainer; and the enclosed-cockpit M 19 Shark and M 20 Talon, described below.

UPDATED

MAGNI M 19 SHARK
TYPE: Two-seat autogyro ultralight.
DESIGN FEATURES: Generally as for M 20 Talon.
STRUCTURE: Generally as for M 20 Talon.
LANDING GEAR: Fixed tricycle type.
POWER PLANT: One turbocharged 84.6 kW (113.4 hp) Rotax 914 UL flat-four with electric starter, driving a ground-adjustable pitch, three-blade, carbon fibre pusher propeller. Fuel in integral cockpit seat/tank of epoxy/glass fibre, capacity 74 litres (19.5 US gallons; 16.3 Imp gallons).

DIMENSIONS, EXTERNAL:
Rotor diameter	8.23 m (27 ft 0 in)
Length overall	4.80 m (15 ft 9 in)
Height overall	2.60 m (8 ft 6¼ in)
Propeller diameter	1.70 m (5 ft 7 in)

WEIGHTS AND LOADINGS:
Weight empty	280 kg (617 lb)
Max T-O weight	450 kg (992 lb)

PERFORMANCE:
Max level speed	100 kt (185 km/h; 115 mph)
Cruising speed	78 kt (145 km/h; 90 mph)
Max rate of climb at S/L	300 m (984 ft)/min
Service ceiling	3,500 m (11,480 ft)
T-O run	90 m (295 ft)
Landing run	up to 30 m (100 ft)
Endurance	4 h 30 min

UPDATED

MAGNI M 20 TALON
TYPE: Single-seat autogyro ultralight.
DESIGN FEATURES: Pod and boom fuselage with sweptback tailplane and triple fins; rudder on central fin only.
STRUCTURE: Welded 4130 chromoly main structure with glass fibre fuselage and wheel fairings. Two-blade main rotor of composites construction with mechanical prerotator.
LANDING GEAR: Fixed tricycle type.
POWER PLANT: One 59.6 kW (79.9 hp) Rotax 912 UL flat-four driving a ground-adjustable pitch, three-blade, carbon fibre pusher propeller. Fuel in integral cockpit seat/tank of epoxy/glass fibre, capacity 45 litres (11.9 US gallons; 9.9 Imp gallons).

DIMENSIONS, EXTERNAL:
Rotor diameter	7.62 m (25 ft 0 in)
Length overall	3.92 m (12 ft 10¼ in)
Height overall	2.37 m (7 ft 9¼ in)
Propeller diameter	1.62 m (5 ft 3¾ in)

WEIGHTS AND LOADINGS:
Weight empty	205 kg (452 lb)
Max T-O weight	300 kg (661 lb)

PERFORMANCE:
Max level speed	96 kt (177 km/h; 110 mph)
Cruising speed	78 kt (145 km/h; 90 mph)
Max rate of climb at S/L	300 m (984 ft)/min
Service ceiling	3,500 m (11,480 ft)
T-O run	45 m (150 ft)
Landing run	up to 30 m (100 ft)
Endurance	3 h

UPDATED

Two-seat Magni M 19 Shark *(Kenneth Munson/Jane's)* 2001/0103532

Magni M 20 Talon single-seat ultralight autogyro 2000/0054612

PIAGGIO

PIAGGIO AERO INDUSTRIES SpA
Via Cibrario 4, I-16154 Genova Sestri, Genova
Tel: (+39 010) 648 11
Fax: (+39 010) 60 33 76 (Sales 652 01 60)
e-mail: ge01.piaggio@interbusiness.it

WORKS: Genova Sestri and Finale Ligure (SV)

BRANCH OFFICE: Via A Gramsci 34, I-00197 Rome
CEO: Giuseppe Di Mase
CHAIRMAN: Piero Ferrari
GENERAL MANAGER: Pier Cesare Guenzi
DEPUTY DIRECTOR, MARKETING: Paolo Barbaris

Piaggio began aircraft production at Genoa Sestri 1916 and later extended to Finale Ligure; Rinaldo Piaggio SpA formed 29 February 1964; covered floor area Sestri and Finale Ligure 120,000 m² (1,291,700 sq ft). Early in 1993, Rinaldo Piaggio restructured with Piaggio family retaining 19 per cent of stock; Alenia acquired 31 per cent holding in Piaggio in 1988 (adjusted to 24.5 per cent in 1992, but raised to 30.9 per cent in early 1993) following further restructuring and reorganisation). Industrie Aeronautiche e Meccaniche Rinaldo Piaggio SpA placed in administration in November 1994; following takeover in November 1998 by Turkish state holding company Tushav, which acquired 51 per cent of stock, was renamed Piaggio Aero Industries SpA, with balance of stock held by Italian investment group Aero Trust via Royal Bank of Canada (44 per cent), CSC of Italy (3 per cent) and the Buitoni family (2 per cent). However, Aero Trust share increased to 60 per cent in June 1999, returning company to Italian control and leaving Tushav with 35 per cent, Piero Ferrari with 3 per cent and CSC with 2 per cent. Workforce 1,100 in 1999.

Piaggio also licence-produces engines, including the RRTI RTM 322 (see *Jane's Aero-Engines*) and mobile shelters; manufactures subassemblies of Alenia (Lockheed Martin) G222/C-27J and Dassault Falcon 2000 and overhauls and supports Viper, T53 and T55 power plants.

UPDATED

PIAGGIO P.166
TYPE: Utility turboprop twin.
PROGRAMME: Prototype first flew 26 November 1957; P.166DL2 described in 1978-79 *Jane's* and earlier versions in previous editions; total 113 P.166s of DL2 and earlier versions were built; first flight of DL3 version 3 July 1976 (I-PIAC); Italian and US certification 1978; production continues on demand.
CURRENT VERSIONS: **P.166-DL3SEM:** Most recent production reconnaissance version able to carry chin-mounted radar and underwing FLIR sensor; training, transport, medical and special patrol/observation versions offered.

P.166-DP1: Updated version with Pratt & Whitney Canada PT6A-121 turboprops, Collins avionics suite and updated mission systems; first flight 11 May 1999. Pursuing potential order from Turkish coastguard for up to 16.

Abbreviated description below refers to -DP1; full details in 1990-91 Jane's.

CUSTOMERS: Two DL3 to Alitalia; one DL3 to Transavio; two DL3MP to Somali Air Force; two DL3-Cargo to Somali government; six DL3APH for Italian Air Force communications (303° Gruppo at Rome/Guidonia); (two Alitalia aircraft transferred to Guardia di Finanza customs service); total 13 non-surveillance DL3s, including prototype; further 12 DL3SEM to Capitanerie di Porto coastguard service (four each for 1° Nucleo Volo Capitanerie at Catania; 2° NVC at Luni; and 3° NVC at Pescara) delivered by mid-1990; 10 DL3SEM to Guardia di Finanza between 23 January 1992 and 1994. Total 22 SEMs built by early 1995 and no more by 1999; 35 of all DL3 versions built by 1999. Total of eight DP1s ordered by Italian customs and coastguard; further 28 existing aircraft may be upgraded.

POWER PLANT: Two Pratt & Whitney Canada PT6A-121 turboprops, each rated at 458.6 kW (615 shp) maximum continuous power and driving a Hartzell HC-B3DL/LT10282-9.5 three-blade constant-speed fully feathering metal pusher propeller. Fuel in two 212 litre (56.0 US gallon; 46.5 Imp gallon) outer-wing main tanks, two 323 litre (85.3 US gallon; 71.0 Imp gallon) wingtip tanks, and a 116 litre (30.6 US gallon; 25.5 Imp gallon) fuselage collector tank; total standard internal fuel capacity 1,186 litres (313.3 US gallons; 260.9 Imp gallons). Auxiliary fuel system available optionally, comprising a 232 litre (61.3 US gallon; 51.0 Imp gallon) fuselage tank, transfer pump and controls; with this installed, total usable fuel capacity is increased to 1,418 litres (374.6 US gallons; 312 Imp gallons). Gravity refuelling points in each main tank and tip tank. Provision for two 177 or 284 litre (46.8 or 75.0 US gallon; 39.0 or 62.5 Imp gallon) underwing drop tanks. Single-point pressure refuelling and fuel jettisoning system, permitting partial discharge, optional. Air intakes and propeller blades de-iced by engine exhaust.

ACCOMMODATION: Crew of two on raised flight deck, with dual controls. Aft of flight deck, accommodation consists of a passenger cabin, utility compartment and baggage compartment. Access to flight deck via passenger/cargo double door on port side, forward of wing, or via individual crew door on each side of flight deck. External access to baggage compartment via port side door aft of wing. Passenger cabin extends from rear of flight deck to bulkhead at wing main spar; fitting of passenger carrying, cargo or other interiors facilitated by two continuous rails on cabin floor, permitting considerable flexibility in standard or customised interior layouts. Standard seating for eight passengers, with individual lighting, ventilation and oxygen controls. Flight deck can be separated from passenger cabin by a screen. Door in bulkhead at rear of cabin provides access to utility compartment, in which can be fitted a toilet, bar, or mission equipment for various roles. Entire accommodation heated, ventilated and soundproofed. Emergency exit forward of wing on starboard side. Windscreen hot-air demisting standard. Windscreen wipers, washers and methanol spray de-icing optional.

DIMENSIONS, EXTERNAL:
Wing span over tip tanks	14.69 m (48 ft 2½ in)
Length overall	12.15 m (39 ft 10¼ in)
Height overall	5.00 m (16 ft 5 in)
Cabin door: Height	1.38 m (4 ft 6 in)
Width	1.28 m (4 ft 2 in)

AREAS:
Wings, gross	26.56 m² (285.9 sq ft)

WEIGHTS AND LOADINGS:
Weight empty, equipped	2,850 kg (6,283 lb)
Max fuel	1,130 kg (2,491 lb)
Max T-O weight	4,500 kg (9,920 lb)
Max wing loading	169.4 kg/m² (34.70 lb/sq ft)
Max power loading	4.91 kg/kW (8.07 lb/shp)

PERFORMANCE:
Never-exceed speed (VNE) at 3,050 m (10,000 ft)	220 kt (407 km/h; 253 mph) CAS
Max level and max cruising speed at 3,660 m (12,000 ft)	215 kt (400 km/h; 248 mph)
Max range speed at 3,660 m (12,000 ft)	160 kt (296 km/h; 184 mph)
Stalling speed	75 kt (139 km/h; 87 mph)

UPDATED

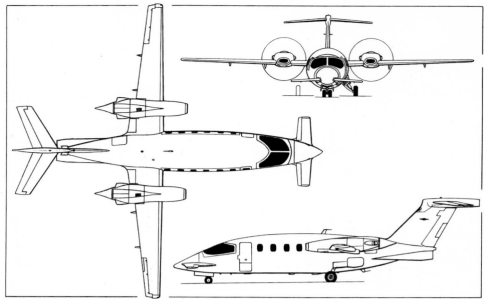

Piaggio P.180 Avanti corporate transport *(Dennis Punnett/Jane's)*

PIAGGIO P.180 AVANTI

TYPE: Business twin-turboprop.

PROGRAMME: Design studies began 1979; launched 1982; Gates Learjet became partner in 1983, but withdrew for economic reasons on 13 January 1986; all existing Learjet P.180 tooling and first three forward fuselages transferred to Piaggio; first flights of two prototypes I-PJAV 23 September 1986 and I-PJAR 14 May 1987; two static test fuselages; first Italian certification 7 March 1990; first flight full production P.180 (I-RAIH/N180BP), 30 May 1990; Italian and US certification 2 October 1990; first customer delivery (N180BP to Robert Pond) 30 September 1990; French certification March 1993. P.180 is certified to Italian RAI 223 and FAR Pt 23 including single pilot, night and day, VFR/IFR and flight into known icing.

CURRENT VERSIONS: Increased gross weight giving higher payload/range decided 1991 and early aircraft retrofitted with minor modifications to allow new weights; weights increased again in 1992. Modified version relaunched in 1997 incorporates changes including fin, rudder and foreplane of aluminium alloy construction and increased fuel capacity. Economy and deluxe interior versions in development in mid-1999.

CUSTOMERS: Total 30 (including prototypes) built by early 1995. 31st followed in early 1999; production plans are five in 1999, 12 in 2000, 21 in 2001 and 26 in 2002. Total 11 in Italian military/paramilitary service; two prototypes; remainder in airline and private ownership in France, Germany, Italy, Spain and USA. Current owners include racing driver Ralf Schumacher. Italian Air Force ordered six for communications; first delivery MM62159 14 May 1993; further five (three for Italian Army; two for SNPC civil protection service) ordered 1994, of which first, MM62167, handed over on 28 July 1997 and delivered 29 August 1997, and second and third delivered March-April 1998, with two SNPC aircraft following in late 1998 and mid-1999. Further 10 ordered for Italian Air Force in July 1997; Turkish military orders for five were under consideration in 1998.

First new aircraft delivery since company reorganisation took place on 2 February 2000 to Aviation Services for Business Aircraft Inc, Portland, Oregon (N680JP). Recent customers include Italian car manufacturer Ferrari, which took delivery of the first of two aircraft (1-FXRB) on 18 May 2000; the Greek Ministry of Health, which has ordered two in EMS configuration; Fox Air of France, which has ordered two; and the Italian Air Force, which has ordered nine, all for delivery during 2000.

COSTS: US$4.3 million (1999).

DESIGN FEATURES: Intended to provide jet-type speeds with turboprop economy. Three-surface control with foreplane and T tail to allow unobstructed cabin with maximum headroom to be placed forward of mid-mounted wing carry-through structure; pusher turboprops aft of cabin and wing reduce cabin noise and propeller vortices on wing, assisting in achievement of 50 per cent laminar flow; midwing avoids root bulges of low-set wings and spar does not pass through cabin; lift from foreplane allows horizontal tail to act as lifting surface and thereby reduce required wing area by 34 per cent.

Laminar flow wing section Piaggio PE 1491 G (mod) at root, PE 1332 G at tip; thickness/chord ratio 13 per cent at root, 14.5 per cent at root; dihedral 2°; sweepback 1° 11' 24"; taper ratio 0.34; foreplane aerofoil Piaggio PE 1300 GN4 unswept; 5° anhedral on foreplane and tailplane; latter sweptback 29° 48' at 25 per cent chord. Tailplane swept 40° at 25 per cent chord.

FLYING CONTROLS: Manual. Aerodynamically and mass-balanced elevators; horn- and mass-balanced rudder. Variable incidence swept tailplane for trim; electrically actuated trim tab in starboard aileron and in rudder; two strakes under tail; electrically actuated outboard Fowler and inboard single-slotted flaps on wing synchronised with single-slotted flaps in foreplanes; three flap positions; dual control circuits; foreplane stalls before wing, providing pitch-down moment. All control surfaces have sealed gaps and are aerodynamically balanced. No dampers or stick-pusher.

STRUCTURE: Airframe 90 per cent aluminium alloy and 10 per cent composites. Fuselage precision stretch-formed in large seamless sections and inner structure matched to precise outer contour in innovative 'outside-in' construction technique; CFRP in high-stress areas; 'Working skin' wing main box, with machined spars and panels, plus integral stiffeners; two main spars; third spar runs from nacelle to fuselage centreline; aluminium leading-edges and aluminium and composites trailing-edges, both connected to main box. Similar construction of forward wing, which connected to lower fuselage at four points and has aluminium alloy leading-edges with electric de-icing blanket and full-depth honeycomb aluminium flaps. Tailplane has two-spar sandwich construction of graphite fabric on Nomex honeycomb core; elevators are single-spar and full-depth aluminium honeycomb with aluminium skins. Fin attached to tailcone bulkheads by four vertical aluminium machined spars; chemically milled aluminium sheet skins. Two-spar rudder, with aluminium alloy skin and carbon fibre and foam core.

Composites parts manufactured by Sikorsky and Edo; wings and tail section produced by Piaggio in Genoa and forward fuselages by Piaggio Aviation in Wichita; final assembly in Genoa.

LANDING GEAR: Dowty Aerospace hydraulically retractable tricycle type, with single-wheel main units and twin-wheel nose unit. Main units retract rearward into sides of fuselage; nose unit retracts forward. Emergency hand pump. Dowty hydraulic shock-absorbers. Electro-

Piaggio P.166-DL3SEM operated by the Italian coastguard 2000

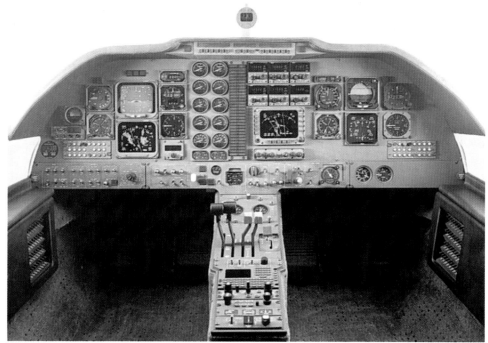

Avanti flight deck   0064740

hydraulic nosewheel steering available through ±20° on take-off and ±50° for taxying. Tyre sizes 6.50-10 (main) and 5.00-5 (nose). Goodrich hydraulic, multidisc carbon brakes.

POWER PLANT: Two 1,107 kW (1,485 shp) Pratt & Whitney Canada PT6A-66 turboprops, flat rated at 634 kW (850 shp), each mounted above wing in all-composites nacelle and driving a Hartzell five-blade constant-speed fully feathering reversible-pitch pusher propeller; propellers handed to counterrotate. Fuel in two fuselage tanks and two wing tanks; total fuel capacity 1,500 litres (396 US gallons; 330 Imp gallons), of which 1,475 litres (388 US gallons; 323 Imp gallons) are usable. Tanks divided into left and right groups (plus primary and secondary collector tanks) which are independent, except during optional pressure refuelling via single-point on starboard centre-fuselage. Gravity refuelling point in upper part of fuselage.

ACCOMMODATION: Crew of one or two on flight deck; certified for single-pilot operation. Seating in main cabin for up to nine passengers, with galley, fully enclosed toilet and coat storage area; choice of nine-passenger high-density or five-seat VIP cabins. Club passenger seats are armchair type, which can be reclined, tracked and swivelled, and locked at any angle. Foldaway tables can be extended between facing club seats. Two-piece wraparound electrically heated windscreen. Rectangular cabin windows, including one emergency exit at front on starboard side. Indirect lighting behind each window ring, plus individual overhead lights. Airstair door at front on port side is horizontally split, upper part forward-opening. Baggage compartment aft of rear pressure bulkhead, with upward-opening door immediately aft of wing on port side. Entire cabin area pressurised and air conditioned. New interior by Sergio Pininfarina under development in mid-1999; total of 15 interior options available.

SYSTEMS: Hamilton Sundstrand R70-3WG three-wheel air-cycle bleed air ECS, with maximum pressure differential of 0.62 bar (9.0 lb/sq in). Single hydraulic system driven by electric motor, with hand pump for emergency back-up, for landing gear (up to 207 bars; 3,000 lb/sq in), brakes and steering (up to 83 bars; 1,200 lb/sq in). Electrical system powered by two 400 A 28 V starter/Lear Siegler generators and 25 V 38 Ah Ni/Cd battery, with triple-redundant essential bus. Two 250 VA static inverters for AC. External power receptacle above port mainwheel well. Scott 1.13 m³ (40 cu ft) oxygen system. Hot air anti-icing of main wing outer and inner leading-edges; electric anti-icing for foreplane and windscreen; rubber boot for engine intakes, with dynamic particle separator; propeller blades de-iced by engine exhaust. Rockwell Collins APS-65A three-axis autopilot.

AVIONICS: *Comms:* Dual Rockwell Collins VHF-22C transceivers; dual Rockwell Collins TDR-90 transponders. Dorne & Margolin DM ELT8 ELT.

Radar: Rockwell Collins WXR-840 weather radar or TWR-850 turbulence-detecting weather radar.

Flight: ADF-462, DME-42, ALT-55B radar altimeter, MCS-65 compass and dual VIR-32 VOR/LOC/MCR (all Rockwell Collins); dual Aeronetics RMI-3337 radio compasses. Rockwell Collins ADS-85 air data system. Optional Universal UNS-1K GPS and second ADF-462.

Instrumentation: Rockwell Collins EFIS-85B system, with two EFD-85 dual colour CRT MFDs for captain and central MFD-85B radar display; EHSI-74 colour display for co-pilot.

Mission: Optional Global/Wulfsberg Flitefone VI in-flight telephone.

DIMENSIONS, EXTERNAL:
Wing span	14.035 m (46 ft 0½ in)
Foreplane span	3.355 m (11 ft 0 in)
Wing chord: at root	1.82 m (5 ft 11¾ in)
at tip	0.62 m (2 ft 0½ in)
Foreplane chord: at root	0.79 m (2 ft 7 in)
at tip	0.55 m (1 ft 9¾ in)
Wing aspect ratio	12.3
Foreplane aspect ratio	5.1
Length overall	14.41 m (47 ft 3½ in)
Fuselage: Length	12.53 m (41 ft 1¼ in)
Max width	1.95 m (6 ft 4¾ in)
Height overall	3.98 m (13 ft 0¾ in)
Tailplane span	4.26 m (13 ft 11¾ in)
Wheel track	2.84 m (9 ft 4 in)
Wheelbase	5.79 m (19 ft 0 in)
Propeller diameter	2.16 m (7 ft 1 in)
Propeller ground clearance	0.795 m (2 ft 7¼ in)
Distance between propeller centres	4.125 m (13 ft 6½ in)
Passenger door (fwd, port): Height	1.345 m (4 ft 5 in)
Width	0.61 m (2 ft 0 in)
Height to sill	0.58 m (1 ft 10¾ in)
Baggage door (rear, port): Height	0.60 m (1 ft 11¾ in)
Width	0.70 m (2 ft 3½ in)
Height to sill	1.38 m (4 ft 6½ in)
Emergency exit (stbd): Height	0.67 m (2 ft 2¼ in)
Width	0.48 m (1 ft 7 in)

DIMENSIONS, INTERNAL:
Passenger cabin (excl flight deck):	
Length	4.42 m (14 ft 6 in)
Max width	1.83 m (6 ft 0 in)
Max height	1.75 m (5 ft 9 in)
Volume	10.6 m³ (375 cu ft)
Flight deck: Length	1.45 m (4 ft 9 in)
Volume	2.3 m² (80 cu ft)
Baggage compartment: Floor length	1.70 m (5 ft 7 in)
Max length	1.95 m (6 ft 4¾ in)
Volume	1.25 m³ (44 cu ft)

AREAS:
Wings, gross	16.00 m² (172.2 sq ft)
Ailerons (total, incl tab)	0.66 m² (7.10 sq ft)
Trailing-edge flaps (total)	1.60 m² (17.23 sq ft)
Foreplane	2.19 m² (23.57 sq ft)
Foreplane flaps (total)	0.58 m² (6.30 sq ft)
Fin	4.73 m² (50.91 sq ft)
Rudder, incl tab	1.05 m² (11.30 sq ft)
Tailplane	3.83 m² (41.23 sq ft)
Elevators (total, incl tabs)	1.24 m² (13.35 sq ft)

WEIGHTS AND LOADINGS:
Weight empty, equipped	3,402 kg (7,500 lb)
Operating weight empty, one pilot	3,479 kg (7,670 lb)
Max usable fuel weight:	
current certification	1,193 kg (2,630 lb)
with max payload	952 kg (2,100 lb)
future certification	1,270 kg (2,800 lb)
Max payload	907 kg (2,000 lb)
Payload with max fuel	667 kg (1,470 lb)
Max T-O weight	5,239 kg (11,550 lb)
Max ramp weight	5,262 kg (11,600 lb)
Max landing weight	4,965 kg (10,945 lb)
Max zero-fuel weight	4,309 kg (9,500 lb)
Max wing loading	327.4 kg/m² (67.07 lb/sq ft)
Max power loading	4.13 kg/kW (6.79 lb/shp)

PERFORMANCE:
Max operating Mach No. (M_{MO})	0.70
Max operating speed (V_{MO})	260 kt (482 km/h; 299 mph) IAS
Max level speed at 8,535 m (28,000 ft)	395 kt (732 km/h; 455 mph)
Manoeuvring speed	199 kt (368 km/h; 229 mph)
Max cruising speed with four passengers at mid-cruise weight:	
at 8,535 m (28,000 ft)	391 kt (724 km/h; 450 mph)
at 11,885 m (39,000 ft)	341 kt (632 km/h; 393 mph)
Stalling speed at max landing weight:	
flaps up	109 kt (202 km/h; 125 mph)
flaps down	93 kt (172 km/h; 107 mph)
Max rate of climb at S/L	899 m (2,950 ft)/min
Rate of climb at S/L, OEI	230 m (755 ft)/min
Max certified altitude	12,500 m (41,000 ft)
Service ceiling	11,885 m (39,000 ft)
Service ceiling, OEI	7,590 m (24,900 ft)
T-O to 15 m (50 ft) ISA, S/L at max T-O weight	869 m (2,850 ft)
Landing from 15 m (50 ft) at max landing weight, no propeller reversal	872 m (2,860 ft)
Range at 11,885 m (39,000 ft), at max range power, one pilot and six passengers: NBAA IFR reserves	1,400 n miles (2,594 km; 1,611 miles)
VFR reserves	1,700 n miles (3,148 km; 1,956 miles)

OPERATIONAL NOISE LEVELS (FAR Pt 36):
Appendix F	76.0 dB(A)
Appendix G	81.8 dB(A)

UPDATED

PIAGGIO AVANTI DEVELOPMENTS

At the 1999 Paris Air Show Piaggio announced that it was undertaking conceptual design work on a family of aircraft based on the Avanti. These include a stretched, 19-passenger regional airliner, and a turbofan-powered business aircraft. The turbofan version would feature a new wing and rear fuselage structure.

VERIFIED

First production Piaggio P.180 Avanti, still operated in the USA by original customer Bob Pond *(Paul Jackson/Jane's)*   2001/0089482

ITALY: AIRCRAFT—RUSALEN & RUSALEN to SG

RUSALEN & RUSALEN

RUSALEN & RUSALEN SdF
Via Nanetti 18, I-33077 Sacile (PN)
Tel: (+39 03 35) 816 88 30
Fax: (+39 04 21) 427 69
e-mail: steni@dacos.it

This company fabricates kits for the Asso lightplane and provides assistance to home constructors. The earlier Asso IV homebuilt continues to be available.

UPDATED

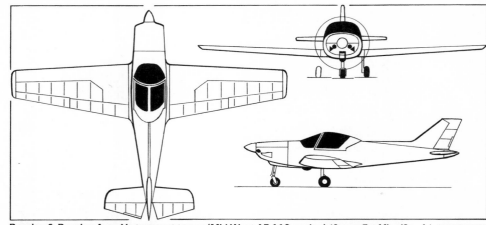

Rusalen & Rusalen Asso Vs two-seat tourer (Mid-West AE 110 engine) *(James Goulding/Jane's)* 2000/0062682

RUSALEN & RUSALEN/AEROSTYLE ASSO Vs

English name: Ace

TYPE: Side-by-side ultralight/kitbuilt.
PROGRAMME: Designed by Aerostyle as Asso V (or Asso 5); 's' suffix indicates 'special' version with new wing section. First flight February 1999 (Italian ultralight category; Mid-West engine); public debut 2 April 1999 and exhibited at Aero '99, Friedrichshafen, later in the same month, at which time total of 11 hours flown. Also offered by Aerostyle in factory-assembled form.
CUSTOMERS: Exports include one to France in 1999.
DESIGN FEATURES: Streamlined, retractable gear, low-wing monoplane, adaptable for Ultralight or Experimental category operation. Tapered wings and horizontal tail surfaces; sweptback fin; upturned wingtips. Large NASA air intakes in cowling for Mid-West-engined version. Asso Vs has biconvex asymmetric wing section. Quoted build time 700 hours for standard kit; 350 hours for fast-build kit.
FLYING CONTROLS: Conventional and manual. Actuation by steel cables. Horn-balanced rudder. Electrically actuated flaps. Trim tabs on rudder and port elevator.
STRUCTURE: Wooden airframe, including wing with single box spar; preformed composites fuselage/fin covering in left and right halves; composites engine cowling and wingtips; Dacron-covered elevators and rudder; plywood-covered ailerons, flaps and tailplane.
LANDING GEAR: Retractable, tricycle type; steerable nosewheel. Mainwheels retract outwards; nosewheel rearwards; electric actuation. No doors, apart from fairing fixed ahead of nosewheel leg. Light alloy wheels with 6-ply tyres.
POWER PLANT: One 78.3 kW (105 hp) Mid-West AE 110 Hawk two-rotor Wankel driving a two-blade, fixed-pitch propeller. Alternative Rotax or Sauer engines. Fuel tanks in each wing.
ACCOMMODATION: Two persons, side by side, beneath rearward-sliding canopy; fixed windscreen. Dual controls. Baggage shelf behind seats.
AVIONICS: To customer's choice.

Asso Vs prototype demonstrated at Aero '99, Friedrichshafen *(Paul Jackson/Jane's)* 2000/0062683

DIMENSIONS, EXTERNAL:
Wing span 7.50 m (24 ft 7¼ in)
Length overall 6.10 m (20 ft 0¼ in)
DIMENSIONS, INTERNAL:
Cockpit, max width 1.06 m (3 ft 5¾ in)
AREAS:
Wings, gross 10.00 m² (107.6 sq ft)
WEIGHTS AND LOADINGS:
Weight empty 285 kg (628 lb)
Max T-O weight (Ultralight) 450 kg (992 lb)
PERFORMANCE (AE 110 engine):
Never-exceed speed (V_{NE}) and max level speed
162 kt (300 km/h; 186 mph)
Cruising speed at 75% power 146 kt (270 km/h; 168 mph)
Econ cruising speed 130 kt (240 km/h; 149 mph)
Stalling speed 33 kt (60 km/h; 38 mph)
T-O and landing run 120 m (395 ft)
Range 971 n miles (1,800 km; 1,118 miles)

UPDATED

SG

SG AVIATION
Via Biancamano, Zona Industriale, I-04016 Sabaudia, Latina
Tel: (+39 0773) 51 52 16
Fax: (+39 0773) 51 14 07
e-mail: sg@storm-sg.it
Web: http://www.storm-sg.it
PRESIDENT: Giovanni Salsedo

NORTH AMERICAN AGENT:
Eden Technologies International LLC
Skylark Airpark, Broad Brook, Connecticut 06016
Tel/Fax: (+1 860) 668 54 74
e-mail: edentech@worldnet.att.net
DIRECTOR: Eric Dixon

In addition to the Storm series of light aircraft (based on an original Pottier design; see French section) and Sea Storm, SG Aviation manufactures spare parts and subassemblies for the Italian Air Force.

UPDATED

SG STORM

TYPE: Side-by-side lightplane/kitbuilt.
PROGRAMME: Design started 11 February 1989 (Storm 280 SI); construction of prototype started 24 November 1990; first flight March 1991; certified to Aero Club Italia ULM category. Production aircraft manufacture began 29 March 1993, first flight July 1993; first delivery August 1993.
CURRENT VERSIONS: Seven versions currently available, both above and below ultralight weight limit: **Storm 280 E**, **280 SI**, **300**, **320 E** and **400 TI** have tricycle landing gear; **Storm 300 Special**, introduced in 1997, has tailwheel. Additionally, the 119 kW (160 hp) **Storm 400 Special** has been developed for the US market. See Power Plant paragraph for details of engines.
Sea Storm: Unrelated design; see separate entry.
CUSTOMERS: By mid-2000 300 Storms ordered, of which 250 were flying. Around 80 per cent are tricycle landing gear versions.
COSTS: Kits: Storm 280 US$17,756; Storm 300 Special US$19,326; Storm 400TI US$20,293. Complete: Storm 280 US$58,500; Storm 300 Special US$68,490; Storm 400TI US$92,650 (All prices 2000).
DESIGN FEATURES: Extensive use of modern materials and technology. All-aluminium fuselage and wings on 280 SI, 300 Special, 320 E and 400 TI. Quoted build time for kits 600 hours.
Wing sections NACA 4415 (Storm 280 SI); GA 3OU-6135 Mod (Storm 300 Special). Thickness/chord ratio 15 per cent; 3° dihedral; 3° incidence.
FLYING CONTROLS: Manual. All-moving tailplane with antibalance tab; no rudder tab. Balanced Frise ailerons on 70 per cent of span. Electric trim controls. Flap settings: 280, 10/20/32°; 12/25/38° on 300 and 400 series, which have Fowler flaps.
LANDING GEAR: Main legs are cantilever sprung. Tricycle versions have tubular steel sprung nosewheel leg steerable ±40°. Tailwheel leg of Storm 300 is metal sprung with a solid tyre. Hydraulic disc brakes on mainwheels. Speed fairings can be fitted. Matco tyres; mainwheels 15×6.00, pressure 2.00 bars (29 lb/sq in); tailwheel (mainwheel in Storm 300 Special) 15 cm (6 in), pressure 1.80 bars (26 lb/sq in).
POWER PLANT: *Storm 280 SI and 280 E:* One 59.6 kW (79.9 hp) Rotax 912 UL or 84.6 kW (113.4 hp) Rotax 914 flat-four four-stroke driving either a two-blade GT166-42 fixed-pitch propeller or a variable-pitch IVO-Prop or AR-Plast propeller; or one 59.7 kW (80 hp) Moto Guzzi driving a two-blade GM fixed-pitch propeller. Fuel capacity for 280, 300 and 320 series 85 litres (22.5 US gallons; 18.7 Imp gallons). Oil capacity 3.5 litres (0.92 US gallon; 0.77 Imp gallon).
Storm 300 and 300 Special: One 84.6 kW (113.4 hp) Rotax 914 turbocharged four-cylinder four-stroke engine driving a two-blade Rospeller constant-speed metal propeller. Mid-West AE 110 Hawk 78.3 kW (105 hp) rotary engine optional. Fuel capacity as Storm 280 E.

A non-standard Storm 300 with nosewheel and three-blade propeller *(Pierre Gaillard)* 2001/0093637

SG—AIRCRAFT: ITALY 287

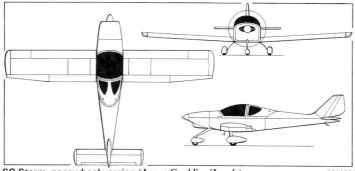

SG Storm, nosewheel version *(James Goulding/Jane's)* 0011094

Storm 300 Special cockpit instruments
0011095

Storm 320 E: One 74.6 kW (100 hp) Teledyne Continental O-200 flat-four driving a two-blade McCauley 1A103/TMC 6948 fixed-pitch propeller.

Storm 400 TI: One 86.5 kW (116 hp) Textron Lycoming O-235-N2C four-cylinder engine driving a two-blade Hartzell HC-C2YL-1BF constant-speed propeller. Fuel capacity 140 litres (37.0 US gallons; 30.8 Imp gallons); optional tanks increase this to 200 litres (52.8 US gallons; 44.0 Imp gallons).

Storm 400 Special: One 119 kW (160 hp) Textron Lycoming O-320 flat-four; engines in the range 82.0 to 119 kW (110 to 160 hp) can be fitted. Fuel capacity as Storm 400 TI.

ACCOMMODATION: Two persons, side by side, with dual controls; single-piece canopy hinged upwards and forwards. Storm 400 TI can accommodate third passenger. Large baggage area behind cockpit.

SYSTEMS: 12 V battery.

AVIONICS: *Instrumentation:* Option of full IFR fit.

DIMENSIONS, EXTERNAL (A: 280 E, B: 300 Special, C: 320 E, D: 400 TI, E: 400 Special):

Wing span: A, C	8.60 m (28 ft 2½ in)
B	7.80 m (25 ft 7 in)
D, E	7.98 m (26 ft 2¼ in)
Wing chord, constant: B, C, D	1.30 m (4 ft 3 in)
Length overall: A, C	6.55 m (21 ft 5¾ in)
B	6.70 m (21 ft 11¾ in)
D, E	6.82 m (22 ft 4½ in)
Height overall: A, C, D, E	2.15 m (7 ft 0¾ in)
B	2.12 m (6 ft 11½ in)

DIMENSIONS, INTERNAL:

Cabin max width: A, B, E	1.10 m (3 ft 7¼ in)

AREAS:

Wings, gross: A, C	11.06 m² (119.0 sq ft)
B	9.98 m² (107.4 sq ft)
D, E	10.45 m² (112.5 sq ft)

WEIGHTS AND LOADINGS:

Weight empty: A	290 kg (640 lb)
B	340 kg (750 lb)
C	320 kg (705 lb)
D, E	400 kg (882 lb)
Max T-O weight: A	520 kg (1,146 lb)
B	570 kg (1,256 lb)
C	520 kg (1,146 lb)
D	720 kg (1,587 lb)
E	750 kg (1,653 lb)
Max wing loading: A	47.0 kg/m² (9.63 lb/sq ft)
B	57.1 kg/m² (11.70 lb/sq ft)
C	47.0 kg/m² (9.63 lb/sq ft)
D	68.9 kg/m² (14.11 lb/sq ft)
E	71.8 kg/m² (14.70 lb/sq ft)
Max power loading: A	8.72 kg/kW (14.34 lb/hp)
B	6.74 kg/kW (11.08 lb/hp)
C	6.98 kg/kW (11.46 lb/hp)
D	8.33 kg/kW (13.68 lb/hp)
E	6.28 kg/kW (10.31 lb/hp)

PERFORMANCE:

Never-exceed speed (V_{NE}):	
A, C, D	136 kt (252 km/h; 156 mph)
B, E	167 kt (310 km/h; 192 mph)
Max operating speed (V_{MO}):	
A	121 kt (225 km/h; 140 mph)
B, C	157 kt (290 km/h; 180 mph)
D	150 kt (278 km/h; 172 mph)
E	160 kt (296 km/h; 184 mph)
Cruising speed at 75% power:	
A	108 kt (200 km/h; 124 mph)
B	143 kt (265 km/h; 165 mph)
E	148 kt (274 km/h; 170 mph)
Stalling speed, power off, flaps down:	
A	31 kt (56 km/h; 35 mph)
B	29 kt (52 km/h; 33 mph)
C	40 kt (74 km/h; 46 mph)
D, E	38 kt (70 km/h; 44 mph)
Max rate of climb at S/L: A	210 m (689 ft)/min
B	300 m (984 ft)/min
C, E	240 m (787 ft)/min
D	330 m (1,083 ft)/min
Service ceiling: A	4,000 m (13,120 ft)
C	4,265 m (14,000 ft)
B, D, E	6,096 m (20,000 ft)
T-O run: A	120 m (395 ft)
B	90 m (295 ft)
C	140 m (460 ft)
D	210 m (690 ft)
E	300 m (985 ft)
Landing run: A, B, C	150 m (495 ft)
D	240 m (790 ft)
E	250 m (820 ft)
Range with max fuel:	
A	648 n miles (1,200 km; 745 miles)
B	718 n miles (1,330 km; 826 miles)
C	621 n miles (1,150 km; 714 miles)
D	453 n miles (839 km; 521 miles)
E	675 n miles (1,250 km; 776 miles)
g limits: A	+5.7/−3
B	+7/−4
C	+4.9/−2.6
D	+3.8/−1.9

UPDATED

SG Aviation Sea Storm amphibian nearing completion in April 2000 *(Paul Jackson/Jane's)* 2000/0085818

SG AVIATION SEA STORM

TYPE: Two/four-seat amphibian kitbuilt.

PROGRAMME: Derived from earlier version, with 59.7 kW (80 hp) engine and metal strut-braced wings, of which some 19 built by 2000. First all-composites Sea Storm under construction in 2000 by SG's North American agent, Eric Dixon.

CURRENT VERSIONS: **Z2**: Two-seat; baseline version.
Z4: Four-seat; increased wing span and power.

COSTS: Z2 US$32,200 (kit), US$69,700 (complete). Z4 US$37,400 (kit), US$83,400 (complete) (2000).

DESIGN FEATURES: High, constant-chord wing, moderately sweptback tailplane and highly sweptback fin; pusher propeller immediately aft of wing trailing-edge; mid-positioned stub-wing with stabilising floats. Quoted kit build time 700 hours.

FLYING CONTROLS: Conventional and manual. Plain flaps.

LANDING GEAR: Optional wheel landing gear attached immediately outboard of wingroots.

POWER PLANT: One pusher piston/rotary engine.
Z2: Option of 73.5 kW (98.6 hp) Rotax 912 ULS or 84.6 kW (113.4 hp) Rotax 914 flat-four or 70.8 kW (95 hp) Mid-West AE 100 rotary driving an adjustable-pitch propeller. Textron Lycoming engines in the 82.0 to 119 kW (110 to 160 hp) range are also suggested. Standard fuel capacity 140 litres (37.0 US gallons; 30.8 Imp gallons), with optional extra 79 litre (20.9 US gallon; 17.4 Imp gallon) tanks available.
Z4: One 157 kW (210 hp) Textron Lycoming O-360; engines recommended in the range 149 to 224 kW (200 to 300 hp). Fuel capacity 160 litres (42.3 US gallons; 35.2 Imp gallons) as standard; optional 61 litre (16.1 US gallon; 13.4 Imp gallon) tanks available.
Idrovario hydraulic, flight-adjustable pitch, two- or three-blade propeller.

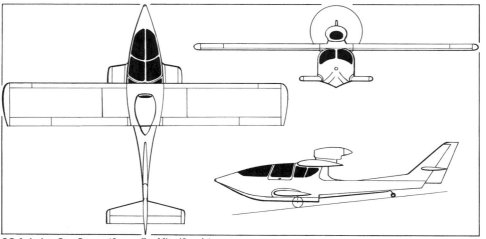

SG Aviation Sea Storm *(James Goulding/Jane's)* 2001/0093636

ACCOMMODATION: Two (side by side) or four persons in enclosed cabin with centreline-hinged door each side.		Max wing loading: A	62.5 kg/m² (12.80 lb/sq ft)	B	4,570 m (15,000 ft)
		B	87.9 kg/m² (18.01 lb/sq ft)	T-O run: on land: A	180 m (591 ft)
All data following (including dimensions) are approximate.		Max power loading:		B	220 m (722 ft)
DIMENSIONS, EXTERNAL (A: Z2; B: Z4):		A (Rotax 912)	11.02 kg/kW (18.11 lb/hp)	on water: A	195 m (640 ft)
Wing span: A	9.60 m (31 ft 6 in)	B (157 kW; 210 hp)	7.64 kg/kW (12.56 lb/hp)	B	336 m (1,100 ft)
B:	10.10 m (33 ft 1¾ in)	PERFORMANCE (A, B, as above):		Landing run: on land: A	200 m (656 ft)
Length overall: A, B	7.44 m (24 ft 5 in)	Never-exceed speed (V$_{NE}$): A		B	250 m (820 ft)
Height overall: A, B	1.83 m (6 ft 0 in)		135 kt (250 km/h; 155 mph)	on water: A	146 m (480 ft)
DIMENSIONS, INTERNAL:		B	162 kt (300 km/h; 186 mph)	B	229 m (750 ft)
Cabin max width: A, B	1.05 m (3 ft 5¼ in)	Max level speed: A	121 kt (225 km/h; 140 mph)	Range, no reserves:	
AREAS:		B	143 kt (265 km/h; 165 mph)	A	917 n miles (1,700 km; 1,056 miles)
Wings, gross: A	12.96 m² (139.5 sq ft)	Cruising speed at 75% power at 2,440 m (8,000 ft):		B	540 n miles (1,000 km; 621 miles)
B	13.65 m² (146.9 sq ft)	A	111 kt (205 km/h; 127 mph)	g limits: A	+4.2/−2.0
WEIGHTS AND LOADINGS (A, B, as above):		B	125 kt (232 km/h; 144 mph)	B	+3.2/−1.8
Weight empty: A	510 kg (1,124 lb)	Stalling speed: flaps up: A	35 kt (64 km/h; 40 mph)		**UPDATED**
B	690 kg (1,521 lb)	B	46 kt (85 km/h; 53 mph)		
Baggage capacity: A	30 kg (66 lb)	flaps down: A	33 kt (61 km/h; 38 mph)		
B	55 kg (121 lb)	B	40 kt (74 km/h; 46 mph)		
Max T-O weight: A	810 kg (1,785 lb)	Max rate of climb at S/L: A	240 m (787 ft)/min		
B	1,200 kg (2,645 lb)	B	210 m (689 ft)/min		
		Service ceiling: A	3,960 m (13,000 ft)		

TECNAM

COSTRUZIONI AERONAUTICHE TECNAM SRL

1a Traversa Via G Pascoli, I-80026 Casoria (Naples)
Tel: (+39 081) 758 32 10, 758 87 51 and 757 41 46
Fax: (+39 081) 758 45 28
e-mail: tecnamca@tin.it
Web: http://www.tecnam.com
MANAGING DIRECTOR: Dott Giovanni Pascale Langer
MARKETING AND SALES DIRECTOR: Paolo Pascale Langer
TECHNICAL DIRECTOR: Prof Luigi Pascale Langer

Company founded 1986, after Pascale brothers were released from the original Pascale company, Partenavia (see now VulcanAir), which had been placed under control of Alenia in 1981. Tecnam manufactures tailplane and other components of ATR 42/72 and parts of A 109, C-27J, SF-260, P.68, plus fuselage panels for Boeing. Workshops qualified to NATO AQA04 and to Italian Aeronautical Register, JAR 21/F. Manufactures P92 Echo ultralight aircraft and P92-J very light aircraft and in 1997 launched production of the P96 and P96-J; P92 SeaSky amphibian added to range in 1998 and P92-S in 1999. Office established in California, 1996, to handle US sales. **UPDATED**

TECNAM P92 ECHO

TYPE: Side-by-side lightplane/kitbuilt.
PROGRAMME: First flight 14 March 1993; production rate by 1996 was six to eight per month. Prototype P92-J (I- TECN) first flown second quarter 1995 and received type certificate 10 November 1995.
CURRENT VERSIONS: **P92:** Standard ultralight version (Italian ULM rules).
P92-J: Detail differences in airframe and equipment to comply with JAR-VLA airworthiness requirements. First delivery (002), to US distributor, second quarter 1996.
P92-S: Added to range in 1999; redesigned rear cabin with additional window; redesigned wing profile and wing-tips, revised engine cowling and windscreen; improvements in performance, *as described.*
P92 2000RG: Retractable landing gear version introduced in 2000. Similar to P92-S, but with shorter (8.74 m; 28 ft 8 in) wing span and reprofiled flaps. Redesigned fuselage, to accommodate retracted gear, is 4 cm (1½ in) wider.
P92 SeaSky: Amphibious floatplane version first flown 1 July 1997; strake below rear fuselage. Full Lotus floats with optional wheels for land operations. Rotax 912

P92-S Echo 100 2000/0062978

Tecnam P92 SeaSky 2000/0062980

ULS engine and increased empty and maximum T-O weights.
Echo 80: Either P92 or P92-S available with standard Rotax 912 UL engine.
Echo 100: Either P92 or P92-S available with optional Rotax 912 ULS engine.
CUSTOMERS: Some 465 delivered by mid-2000; six to Cambodian Air Force (first two September 1994); overseas orders currently account for 30 per cent of production, including many to Israel. P92-J deliveries totalled 12 by mid-1998 (comprising Germany two, Italy five, Netherlands three, New Zealand one and USA one); none further by mid-2000.
COSTS: P92-J, US$80,000 basically equipped (1996).
DESIGN FEATURES: Objectives were lightness, simplicity and accessibility for inspection and servicing. Braced single-spar high wing; aerofoil chosen for good performance at low Reynolds number; untapered with 1° 30′ dihedral; underside of fuselage mostly flat for ease of kit assembly; large one-piece windscreen, inward tapered inboard wing leading-edges, large windows in doors; rearview window (s).
FLYING CONTROLS: Manual. Frise ailerons; all-moving tailplane with anti-balance tab; electrically actuated flaps cover half trailing-edge.
STRUCTURE: Fuselage mainly metal except GFRP lower shell of engine cowling and wing leading-edge; ailerons and part of tailplane fabric covered; GFRP rudder tip; tailplane halves can be unpinned and removed quickly for transport; engine cowling removable by undoing four quick latches to reveal whole power plant. Corrosion protection standard in P92-J and SeaSky.
LANDING GEAR: Non-retractable tricycle type with steel leaf mainlegs attached to bottom of fuselage for easy maintenance; hand-powered hydraulic disc brakes operated together from single lever in cockpit; nos wheel steered from rudder pedals; designed for grass field operation. Wheels 5.00-5 on P92-S; 400-6 on SeaSky. P92 2000RG nosewheel has oleo-pneumatic shock-absorber and retracts rearwards; mainwheels retract inward.
POWER PLANT: One 59.6 kW (79.9 hp) Rotax 912 UL (P92) or 912A (P92-J) or 73.5 kW (98.6 hp) Rotax 912 ULS flat-four engine; integrated reduction gear driving two-blade Bipala wooden propeller: GT-166/146 for Rotax 912 UL; GT-172/164 for Rotax 912 ULS landplane; GT-180/155 for SeaSky.
Limbach 2000 (59.7 kW; 80 hp) and Rotax 582 (47.8 kW; 64.1 hp) two-stroke offered as options for P92; P92 prototype had the latter; Jabiru 2200 under trial in 2000. Fuel capacity 70 litres (18.5 US gallons; 15.4 Imp gallons) in two wing tanks, of which 60 litres (15.9 US gallons; 13.2 Imp gallons) are usable.
ACCOMMODATION: Side-by-side seats with three-point harness; baggage space behind seats. Full dual controls and two throttles. Bulged doors in P92-J give increased cabin width.
SYSTEMS: 14 V 55 A battery; 100 W alternator.
AVIONICS: Customer specified; optional IFR nav/com in P92-J.
DIMENSIONS, EXTERNAL (all versions, except where indicated):
Wing span: P92 2000RG 8.74 m (28 ft 8 in)
P92, P92-S 9.30 m (30 ft 6 in)
P92-J, SeaSky 9.60 m (31 ft 6 in)
Wing chord: P92 2000RG 1.02 m (3 ft 4 in)
Length overall: except P92 2000RG 6.30 m (20 ft 8 in)
P92 2000RG 6.40 m (21 ft 0 in)

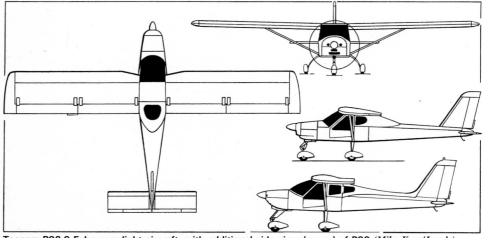

Tecnam P92-S Echo very light aircraft, with additional side view (upper) of P92 *(Mike Keep/Jane's)*
2000/0062979

Tecnam P92 fitted with a Jabiru 2200 engine *(Paul Jackson/Jane's)*   2001/0093680

Height overall: except SeaSky	2.49 m (8 ft 2 in)
SeaSky	2.97 m (9 ft 9 in)
Tailplane span	2.90 m (9 ft 6 in)
Wheel track: except SeaSky and P92 2000RG	
	1.78 m (5 ft 10 in)
P92 2000RG	1.73 m (5 ft 8 in)
Float track (SeaSky)	2.79 m (9 ft 2 in)
Propeller diameter	1.65 m (5 ft 5 in)

DIMENSIONS, INTERNAL:
Cabin max width	1.09 m (3 ft 7 in)

AREAS:
Wings, gross:	
except SeaSky and P92 2000RG	13.20 m² (142.1 sq ft)
P92 2000RG	11.98 m² (129.0 sq ft)
SeaSky	13.40 m² (144.2 sq ft)
Tailplane	1.97 m² (21.20 sq ft)

WEIGHTS AND LOADINGS:
Basic weight empty:	
P92, P92-S, P92 2000RG	281 kg (619 lb)
P92-J	305 kg (672 lb)
SeaSky	330 kg (728 lb)
Max T-O weight:	
ultralight: P92, P92-S, P92 2000RG	450 kg (992 lb)
SeaSky	500 kg (1,102 lb)
certified: P-92, P92-S	550 kg (1,212 lb)
P92-J	535 kg (1,179 lb)
SeaSky	600 kg (1,322 lb)

PERFORMANCE (P92 in Echo 80 form, P92-S in Echo 100 form):
Never-exceed speed (V_{NE}):	
except SeaSky	135 kt (250 km/h; 155 mph)
SeaSky	108 kt (200 km/h; 124 mph)
Max level speed at S/L:	
P92, P92-J	117 kt (218 km/h; 135 mph)
P92-S, P92 2000RG	124 kt (230 km/h; 143 mph)
SeaSky	92 kt (170 km/h; 106 mph)
Cruising speed at 75% power at 2,200 propeller rpm:	
P92, P92-J	100 kt (185 km/h; 115 mph)
P92-S	111 kt (206 km/h; 128 mph)
SeaSky	84 kt (156 km/h; 97 mph)
Stalling speed: flaps up: P92	36 kt (68 km/h; 42 mph)
P92-J	40 kt (74 km/h; 46 mph)
flaps down:	
P92, P92-S, P92 2000RG	33 kt (61 km/h; 38 mph)
P92-J, SeaSky	35 kt (64 km/h; 40 mph)
Max rate of climb at S/L: P92	330 m (1,082 ft)/min
P92-J	262 m (859 ft)/min
P92-S, P92 2000RG	384 m (1,260 ft)/min
SeaSky	244 m (800 ft)/min
Service ceiling: P92, P92-J	4,000 m (13,120 ft)
P92 2000RG	4,510 m (14,800 ft)
P92-S	4,500 m (14,760 ft)
SeaSky	3,500 m (11,480 ft)
T-O run: on land: P92	110 m (360 ft)
P92-J	90 m (295 ft)
P92-S	100 m (330 ft)
P92 2000RG	140 m (460 ft)
SeaSky	120 m (395 ft)
on water: SeaSky	350 m (1,150 ft)
Landing run: on land:	
except P92-J, P92 2000RG, SeaSky	100 m (328 ft)
P92 2000RG	110 m (360 ft)
P92-J	91 m (300 ft)
SeaSky	120 m (395 ft)
on water (SeaSky)	350 m (1,150 ft)
Max range, no reserves:	
except SeaSky	400 n miles (740 km; 460 miles)
SeaSky	300 n miles (555 km; 345 miles)
Endurance: P92	4 h 30 min
P92-J	5 h
Best glide ratio: P92, P92-J	13
g limits: P92-J	+3.8/−1.9
P92-S	+6/−3
SeaSky	+5.3/−2

UPDATED

TECNAM P96 GOLF

TYPE: Side-by-side ultralight.

PROGRAMME: Prototype construction began July 1996; first flight March 1997. Higher-weight (JAR-VLA) P96-J announced second quarter 1997 but no longer available (see 2000-01 and earlier *Jane's*); uprated Golf 100 added in 1999.

CURRENT VERSIONS: **P96 Golf 80**: Ultralight version.
P96 Golf 100: Higher-powered ultralight version.

CUSTOMERS: Some 80 (all versions) produced by mid-1999.

DESIGN FEATURES: Emphasis on reduction of aerodynamic drag without compromising cost. Fuselage cross-sectional area allows easy access for taller occupants by reducing the height of structural carry-through elements. Wing lower surface and fuselage bottom are flush, obviating interference drag and complex fairings. Wing-to-fuselage point optimised by strakes at wingroots. Many parts common to P92 Echo.

FLYING CONTROLS: Manual. Frise ailerons; all-moving tailplane with anti-balance tab; electrically actuated flaps cover half of trailing-edge. Ground-adjustable tab on port aileron and rudder.

STRUCTURE: Fuselage includes a composites spine that follows through to the tailfin and keeps wetted area to a minimum while allowing for good pressure recovery aft of canopy. All-moving horizontal tail reduces drag to a minimum and provides optimum longitudinal stability for stick-free operation. Wing has 5° dihedral.

LANDING GEAR: Fixed tricycle type; hydraulic disc brakes operated from single hand lever in cockpit. Levered suspension nosewheel leg; nosewheel steered by rudder pedals. Mainwheels 5.00-5. Speed fairings.

POWER PLANT: *Golf-80*: One 59.6 kW (79.9 hp) Rotax 912 UL flat-four engine with integrated reduction gear (1:2.27), driving a GT two-blade wooden propeller.
Golf 100: 73.5 kW (98.6 hp) Rotax 912 ULS.
Fuel capacity (both) 70 litres (18.5 US gallons; 15.4 Imp gallons).

ACCOMMODATION: Two seats side by side; baggage space behind seats. Fixed windscreen; canopy slides backwards on guiderails and can be opened with engine on and during flight. Full dual controls and twin throttles.

SYSTEMS: 100 W 12 V alternator and battery.

AVIONICS: To customer's requirements; provision for nav/com radio.

DIMENSIONS, EXTERNAL:
Wing span	8.41 m (27 ft 7 in)
Length overall	6.40 m (21 ft 0 in)
Height overall	2.29 m (7 ft 6 in)
Tailplane span	2.90 m (9 ft 6 in)
Wheel track	1.78 m (5 ft 10 in)
Propeller diameter	1.65 m (5 ft 5 in)

DIMENSIONS, INTERNAL:
Cabin max width	1.13 m (3 ft 8½ in)

AREAS:
Wings, gross	12.20 m² (131.3 sq ft)

WEIGHTS AND LOADINGS:
Weight empty: Golf 80	281 kg (619 lb)
Max T-O weight: both	450 kg (992 lb)

PERFORMANCE:
Never-exceed speed (V_{NE})	140 kt (259 km/h; 161 mph)
Max level speed: Golf 80	121 kt (224 km/h; 139 mph)
Golf 100	130 kt (241 km/h; 150 mph)
Cruising speed at 75% power:	
Golf 80	105 kt (195 km/h; 121 mph)
Golf 100	116 kt (215 km/h; 133 mph)
Stalling speed, power off, flaps down:	
both	33 kt (61 km/h; 38 mph)
Max rate of climb at S/L: Golf 80	271 m (890 ft)/min
Golf 100	360 m (1,180 ft)/min
Service ceiling: Golf 80	4,000 m (13,120 ft)
Golf 100	4,500 m (14,760 ft)
T-O run: Golf 80	110 m (361 ft)
Golf 100	100 m (328 ft)
Landing run: both	100 m (328 ft)
Max range: Golf 80	400 n miles (740 km; 460 miles)
Endurance: Golf 80	4 h 30 min
Best glide ratio: Golf 80	12
g limits: both	+6/−3

UPDATED

Tecnam P96 Golf   2000/0062983

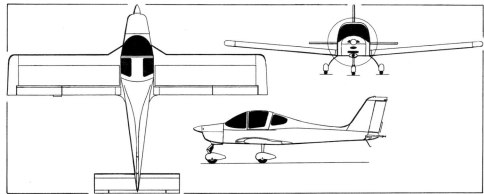

Tecnam P96 Golf two-seat low-wing ultralight aircraft (Rotax 912 piston engine) *(Paul Jackson/Jane's)*   2000/0062982

VULCANAIR

VULCANAIR SpA
Via G Pascoli 7, I-80026 Casoria (NA)
Tel: (+39 081) 591 81 11
Fax: (+39 081) 591 81 72
e-mail: infomarketing@vulcanair.com
Web: http://www.vulcanair.com
PRESIDENT: Carlo de Feo

Bankrupt Partenavia company and Samanta aircraft service company both purchased by Carlo de Feo in 1997; Samanta renamed as VulcanAir in early 1998. VulcanAir returned P.68 series to production in 1999 and continues to supply spares for Partenavia aircraft. Acquired rights to SF.600 Canguro from SIAI-Marchetti on latter's incorporation into Aermacchi in 1997. In June 1999, VulcanAir announced development of the VA 300 twin-diesel-engined light transport derivative of the P.68 and single-engined VA 600W, the latter a variant of the SF.600.

UPDATED

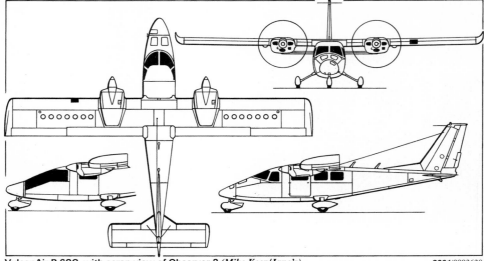

VulcanAir P.68C, with scrap view of Observer 2 *(Mike Keep/Jane's)* 2001/0093638

VULCANAIR P.68C, P.68C-TC, P.68 OBSERVER 2 and OBSERVER 2TC

TYPE: Light utility twin-prop transport.
PROGRAMME: Production of P.68C by Partenavia in Italy started 1978, followed by P.68TC in 1980. Partenavia manufacture ended in 1994, but assets subsequently acquired by VulcanAir. Improvements announced in 1999 include Honeywell Silver Crown avionics and new interiors.
CURRENT VERSIONS: **P.68C**: Basic version.
  P.68C-TC: As P.68C, but with turbocharged engines for better hot-and-high performance.
  P.68 Observer 2: For use by government and specialised services for patrol, surveillance and search; largely transparent nose section with lowered, compact instrument panel; 63 × 46 cm (2 ft 1 in × 1 ft 6 in) underfuselage hatch can carry variety of electro-optical sensors; slightly different equipment from other versions.
  P.68 Observer 2TC: Turbocharged version of Observer 2 for hot/high performance.
CUSTOMERS: Some 403 P.68s of all versions delivered from other sources by end 1998; VulcanAir began production batch of 20 aircraft in 1998; eight Observer 2s sold to Italian police; first delivery (PS-B07) 11 November 1999.
COSTS: US$457,000 (P.68C); US$497,000 (P.68C-TC), both IFR equipped (2000).
DESIGN FEATURES: High wing with NACA 63-3515 aerofoil section and Hoerner tips; dihedral 1°, incidence 1° 3′.
FLYING CONTROLS: Manual. Pushrod and cable actuated, with all-moving tailplane and anti-balance tab; trim tab in rudder; electrically operated single-slotted flaps.
STRUCTURE: Light alloy stressed skin fuselage with frames and longerons; stressed skin two-spar torsion box wing; metal stressed skin tailplane and fin; fuselage/wing fairings mainly GFRP.
LANDING GEAR: Non-retractable, with spring steel main legs; oleo suspension for nosewheel, steered from rudder pedals; mainwheels Cleveland 40-142 with Pirelli 6.00-6 or 7.00-6 (8 ply) tyres; nosewheel Cleveland 40-77B with Goodyear 5.00-5 or 6.00-6 (6 ply) tyre; Cleveland Type 30-61 foot-powered hydraulic disc brakes; streamlined wheel fairings optional. P.68 Observer 2 has larger mainwheel tyres as standard. Minimum ground turning radius 5.70 m (18 ft 8 in).
POWER PLANT: *P.68C:* Two 149 kW (200 hp) Textron Lycoming IO-360-A1B6 flat-four engines, each driving a Hartzell HC-C2YK-2CUF two-blade constant-speed fully feathering propeller.
  P.68C-TC: Two 157 kW (210 hp) Textron Lycoming TIO-360-C1A6D; same propellers as P.68C.
  Fuel capacity 696 litres (184 US gallons; 153 Imp gallons) in integral wing tank, of which 670 litres (177 US gallons; 147 Imp gallons) usable; overwing gravity refuelling. Oil capacity 7.5 litres (2.0 US gallons; 1.7 Imp gallons) for each engine.
ACCOMMODATION: One or two pilots and five or six passengers; cabin has two forward-facing seats in middle and three-seat rear bench; club seating optional; baggage door at rear and pilot door at front on starboard side; passenger door to port in centre cabin; baggage compartment accessible from inside cabin. P.68 Observer 2 has no front starboard door for pilots.

VulcanAir P.68C built by the current manufacturer in 2000 *(Paul Jackson/Jane's)* 2001/0093681

SYSTEMS: Two 24 V 130 Ah alternators and one 24 V 17 Ah battery; Goodrich pneumatic de-icing boots optional; air conditioning optional.
AVIONICS: Choice of VFR or full IFR Honeywell Silver Crown avionics with KFC 150 autopilot.
  Radar: Weather radar optional in Observer 2.
  Mission: Observer 2 can carry FLIR, ATAL video surveillance pod with data downlink and SLAR. Aerial cameras, thermal imager and video cameras can be operated through floor hatch.

DIMENSIONS, EXTERNAL (C: P.68C, TC: P.68C-TC, O: P.68 Observer 2):
Wing span	12.00 m (39 ft 4½ in)
Wing chord, constant	1.55 m (5 ft 1 in)
Wing aspect ratio	7.7
Length overall: C, TC	9.55 m (31 ft 4 in)
O	9.43 m (30 ft 11¼ in)
Height overall	3.40 m (11 ft 1¾ in)
Tailplane span	3.90 m (12 ft 9½ in)
Wheel track	2.40 m (7 ft 10½ in)
Wheelbase	3.65 m (11 ft 11¾ in)
Propeller diameter: all versions	1.83 m (6 ft 0 in)
Propeller ground clearance	0.77 m (2 ft 6¼ in)

DIMENSIONS, INTERNAL:
Cabin: Length	4.05 m (13 ft 3½ in)
Max width	1.16 m (3 ft 9½ in)
Max height	1.20 m (3 ft 11¼ in)
Baggage compartment volume	0.56 m³ (19.8 cu ft)

AREAS:
Wings, gross	18.60 m² (200.2 sq ft)
Ailerons (total)	1.79 m² (19.27 sq ft)
Trailing-edge flaps (total)	2.37 m² (25.51 sq ft)
Fin	1.59 m² (17.11 sq ft)
Rudder, incl tab	0.44 m² (4.74 sq ft)
Tailplane, incl tab	4.41 m² (47.47 sq ft)

WEIGHTS AND LOADINGS:
Weight empty equipped: C, O	1,320 kg (2,910 lb)
TC	1,350 kg (2,976 lb)
Baggage capacity	181 kg (400 lb)
Max T-O weight:	2,084 kg (4,594 lb)
Max ramp weight	2,100 kg (4,630 lb)
Max landing weight	1,980 kg (4,365 lb)
Max zero-fuel weight	1,890 kg (4,167 lb)
Max wing loading	112.0 kg/m² (22.94 lb/sq ft)
Max power loading	6.99 kg/kW (11.49 lb/hp)

PERFORMANCE:
Never-exceed speed (V_{NE}):	
C, TC	193 kt (358 km/h; 222 mph)
O	194 kt (359 km/h; 223 mph)
Max level speed:	
C, O at S/L	173 kt (320 km/h; 199 mph)
TC at S/L	174 kt (322 km/h; 200 mph)
Max cruising speed at 75% power:	
C, O at 2,285 m (7,500 ft)	165 kt (306 km/h; 190 mph)
TC at 3,660 m (12,000 ft)	171 kt (317 km/h; 197 mph)
Stalling speed, power off:	
flaps up	68 kt (126 km/h; 79 mph)
flaps down: TC	57 kt (106 km/h; 66 mph)
Max rate of climb at S/L: C, O	378 m (1,240 ft)/min
TC	438 m (1,437 ft)/min
Max rate of climb, OEI: C, O	64 m (210 ft)/min
TC	78 m (256 ft)/min
Service ceiling: C	5,563 m (18,250 ft)
O	6,000 m (19,685 ft)
TC	6,100 m (20,000 ft)
Service ceiling, OEI: C	1,341 m (4,400 ft)
O	1,750 m (5,740 ft)
TC	2,957 m (9,700 ft)
T-O run: C, O	240 m (790 ft)
TC	230 m (755 ft)
T-O to 15 m (50 ft): C	400 m (1,312 ft)
TC	385 m (1,263 ft)
Landing from 15 m (50 ft): all	600 m (1,970 ft)
Landing run: all	240 m (790 ft)
Range with max payload:	
TC	300 n miles (556 km; 345 miles)
O with auxiliary fuel	590 n miles (1,093 km; 679 miles)
Range with max fuel:	
C	1,525 n miles (2,825 km; 1,755 miles)
O	1,601 n miles (2,965 km; 1,842 miles)
TC	1,403 n miles (2,600 km; 1,615 miles)
Endurance, with max fuel	11 h
g limits	+3.74/−1.50

UPDATED

VULCANAIR AP.68TP 600 VIATOR

TYPE: Utility turboprop twin.
PROGRAMME: First flight 29 March 1985; production started 1989; six built in Italy by Partenavia. Available to special order, but planned to be replaced in the product range by

VulcanAir P.68 Observer 2 *(Paul Jackson/Jane's)* 2001/0093682

VULCANAIR—AIRCRAFT: ITALY

VulcanAir AP.68TP 600 Viator 11-seater

the VA 300. Full description in 2000-01 *Jane's*; shortened version follows.

CUSTOMERS: Total of seven produced before VulcanAir acquisition, including one from CKD kit. (AP.68TP 300 Spartacus, of which 13 built by Partenavia, is also available to special order.)

DESIGN FEATURES: Cantilever, untapered high wing of same span and form as P.68C; aerofoil section NACA 63-3515, with Hoerner wingtips; dihedral 1°, incidence 1° 3′.

FLYING CONTROLS: Conventional and manual. Actuation by pushrods and cables; trim tabs on elevators, rudder and ailerons; elevator has vortex generators under leading-edge; down spring in elevator circuit; stall strips on wing leading-edges; electrically actuated single-slotted flaps.

LANDING GEAR: Retractable tricycle type; main gear retracts hydraulically inward into fuselage fairings, nosewheel forward; Cleveland mainwheels size 40-163EA; Cleveland 40-778 nosewheel; McCreary 6.50-8 mainwheel tyres and 6.00-6 on nosewheel; Cleveland powered hydraulic disc brakes. Minimum ground turning radius 5.45 m (17 ft 11 in).

POWER PLANT: Two Rolls-Royce 250-B17C turboprops, each flat rated at 245 kW (328 shp) and driving a Hartzell HC-B3TF-7A three-blade constant-speed reversible-pitch propeller. One integral fuel tank in each wing and a 38 litre (10 US gallon; 8.35 Imp gallon) tank in each engine nacelle, giving total usable capacity of 840 litres (222 US gallons; 185 Imp gallons).

ACCOMMODATION: One or two pilots, nine or 10 passengers; two doors to port, one for pilot and one for passengers; two doors to starboard, one for co-pilot and one for baggage; baggage compartment variable by using part of cabin; baggage accessible in flight.

AVIONICS: VFR radio and instruments standard; full IFR with weather radar and Honeywell Silver Crown or Rockwell Collins radios optional; observation equipment optional.

DIMENSIONS, EXTERNAL:
Wing span	12.00 m (39 ft 4½ in)
Wing aspect ratio	7.7
Length overall	11.27 m (36 ft 11½ in)
Height overall	3.64 m (11 ft 11 in)
Wheel track	2.17 m (7 ft 1½ in)
Wheelbase	3.51 m (11 ft 6¼ in)
Propeller diameter	2.03 m (6 ft 8 in)

DIMENSIONS, INTERNAL:
Cabin, excl cockpit and baggage compartment:
Length	5.29 m (17 ft 4¼ in)
Max width	1.13 m (3 ft 8½ in)
Max height	1.26 m (4 ft 1½ in)
Floor area	5.75 m² (61.9 sq ft)
Baggage/cargo compartment volume	0.55 m³ (19.4 cu ft)

AREAS:
Wings, gross	18.60 m² (200.2 sq ft)

WEIGHTS AND LOADINGS:
Operating weight empty	1,680 kg (3,704 lb)
Max payload	870 kg (1,918 lb)
Max fuel weight	675 kg (1,488 lb)
Max T-O weight	3,000 kg (6,614 lb)
Max ramp weight	3,025 kg (6,669 lb)
Max landing weight	2,850 kg (6,283 lb)
Max zero-fuel weight	2,550 kg (5,622 lb)
Max wing loading	153.2 kg/m² (31.38 lb/sq ft)
Max power loading	6.13 kg/kW (10.08 lb/shp)

PERFORMANCE:
Max level speed at S/L	200 kt (370 km/h; 230 mph)
Max cruising speed at 3,050 m (10,000 ft)	220 kt (407 km/h; 253 mph)
Econ cruising speed at 3,050 m (10,000 ft)	170 kt (315 km/h; 196 mph)
Stalling speed, power off:	
flaps up	75 kt (139 km/h; 87 mph)
flaps down	69 kt (128 km/h; 80 mph)
Max rate of climb at S/L	472 m (1,550 ft)/min
Rate of climb at S/L, OEI	82 m (270 ft)/min
Service ceiling	7,925 m (26,000 ft)
Service ceiling, OEI	3,475 m (11,400 ft)
T-O run	400 m (1,315 ft)
T-O to 15 m (50 ft)	600 m (1,970 ft)
Landing from 15 m (50 ft)	700 m (2,300 ft)
Landing run	320 m (1,050 ft)
Range:	
with max payload	530 n miles (982 km; 610 miles)
with max fuel	777 n miles (1,440 km; 894 miles)
g limits	+3.54/–1.42

UPDATED

VULCANAIR VA 300

TYPE: Light utility twin-prop transport.

PROGRAMME: Announced at Paris, June 1999, as a concept based on the VulcanAir AP.68TP Viator (which see). Diesel engines planned to be test-flown on a tri-motor configured P.68C during first half of 2000 but apparently not achieved; certification to JAR/FAR 23 planned by 2002.

DESIGN FEATURES: Version of the P.68 series incorporating two diesel engines. Upturned wingtips.

Description generally as for AP.68TP, except that below.

POWER PLANT: Two 224 kW (300 hp) Zoche ZO 02A eight-cylinder, turbocharged diesel engines, each driving a three-blade, constant-speed, fully feathering, reversible-pitch propeller. Fuel capacity 848 litres (224 US gallons; 187 Imp gallons); overwing gravity refuelling.

ACCOMMODATION: Pilot and up to eight passengers.

DIMENSIONS, EXTERNAL:
Wing span	12.00 m (39 ft 4½ in)
Wing chord, constant	1.55 m (5 ft 1 in)
Wing aspect ratio	7.7
Length overall	10.18 m (33 ft 4¾ in)
Height overall	3.63 m (11 ft 11 in)
Tailplane span	4.01 m (13 ft 1¾ in)
Wheel track	3.17 m (10 ft 4¾ in)
Wheelbase	2.37 m (7 ft 9¼ in)
Propeller diameter	1.90 m (6 ft 2¾ in)

DIMENSIONS, INTERNAL:
Cabin (excl flight deck): Length	2.92 m (9 ft 7 in)
Max width	1.32 m (4 ft 4 in)
Max height	1.27 m (4 ft 2 in)
Volume	4.9 m³ (172 cu ft)

AREAS:
Wings, gross	18.60 m² (200.2 sq ft)

WEIGHTS AND LOADINGS (provisional):
Weight empty	1,730 kg (3,814 lb)
Max T-O and landing weight	2,850 kg (6,283 lb)
Max ramp weight	2,875 kg (6,338 lb)
Max zero-fuel weight	2,550 kg (5,622 lb)
Max wing loading	153.2 kg/m² (31.38 lb/sq ft)
Max power loading	6.37 kg/kW (10.47 lb/hp)

PERFORMANCE (estimated):
Max cruising speed at S/L	190 kt (352 km/h; 219 mph)
Normal cruising speed	177 kt (328 km/h; 204 mph)
Stalling speed, flaps down	67 kt (124 km/h; 77 mph)
Max rate of climb at S/L	462 m (1,516 ft)/min
Rate of climb at S/L, OEI	102 m (335 ft)/min
Service ceiling	7,620 m (25,000 ft)
T-O run	394 m (1,295 ft)
T-O to 15 m (50 ft)	600 m (1,970 ft)
Landing from 15 m (50 ft)	650 m (2,135 ft)
Range:	
with max fuel	1,663 n miles (3,080 km; 1,913 miles)
with max payload	718 n miles (1,330 km; 826 miles)

UPDATED

VULCANAIR SF.600A CANGURO

English name: Kangaroo

TYPE: Utility turboprop twin.

PROGRAMME: Prototype F.600 Canguro (I-CANG), designed by Stelio Frati, built by General Avia and then powered by 261 kW (350 hp) Textron Lycoming TIO-540-J flat-six piston engines, made its first flight on 30 December 1978. Described under General Avia heading in the 1979-80 *Jane's*.

Basic production SF.600TP was offered initially by SIAI-Marchetti with Rolls-Royce 250-B17C turboprops and non-retractable landing gear; major options included retractable landing gear and a swing-tail rear fuselage. Late production aircraft had two 335 kW (450 shp) Rolls-Royce 250-B17F engines, in which guise they retrospectively became the SF.600A.

Certification by the RAI was received in second quarter of 1987 in accordance with FAR Pt 23. The initial production batch totalled nine aircraft, and three of these, for the Rome-based air taxi operator Sun Line, were delivered in April 1988. Plans for production in South Korea and Philippines failed to materialise. On purchase of SIAI-Marchetti assets by Aermacchi in 1997, Canguro was excepted and instead passed to VulcanAir, which employed 10th airframe as demonstrator (I-VULA) and also acquired No. 6 (I-CNGT) as part of purchase. First Canguro completed by VulcanAir, No. 11, due to fly in mid-1998 but completion had not been reported by late 2000.

CURRENT VERSIONS: **SF.600A**: Basic design adaptable for passenger/troop or cargo transport, paratroop transport, air ambulance, maritime surveillance, electronic intelligence and agricultural duties. Retractable gear version (not currently offered) is considered especially suitable for maritime surveillance role, equipped with an underfuselage radar system, or FLIR and/or SLAR.

Description applies to SF.600A.

WF 600W: Single-engined version; one 580 kW (778 shp) Walter M 601 F turboprop fitted into nose. Prototype under construction during 1999; first flight was due May 2000 but this date has slipped; certification planned for 2002.

CUSTOMERS: Orders received by VulcanAir for two aircraft; three under construction during 1999. Earlier aircraft still in use include one each with Bum-A (South Korea) and Philippine National Police.

COSTS: US$1.5 million (2000).

Retouched image, showing projected VulcanAir VA 300

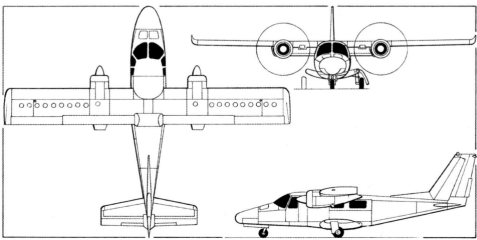

VulcanAir VA 300 (Zoche diesel engines) *(James Goulding/Jane's)*

DESIGN FEATURES: Rugged utility design for operation from short, semi-prepared airstrips.
 Wing section NASA GA(W)-1, with 17 per cent thickness/chord ratio. Dihedral 2°. Incidence (constant) 1° 30′.
FLYING CONTROLS: Conventional and manual. Electrically operated trim tab in port aileron. Trim tabs in rudder (actuated mechanically) and each elevator (electrically/mechanically operated). Electrically operated double-slotted flaps.
STRUCTURE: All-metal riveted structure in aluminium alloy, with stressed skin. Centre-section has main spar and two auxiliary spars; outboard of engines, wings have two spars. All-metal ailerons.
LANDING GEAR: Fixed tricycle gear, of trailing arm type, with oleo-pneumatic shock-absorber in each unit. Twin-wheel main units, mounted on small stub-wings attached to fuselage floor; single steerable nose unit. Mainwheels and tyres size 7.00-6, pressure 2.90 bars (42 lb/sq in); nosewheel and tyre size 6.00-6. Hydraulic disc brakes on main units.
POWER PLANT: Two 335 kW (450 shp) Rolls-Royce 250-B17F1 turboprops, each driving a Hartzell HC-B3TF-7A/T10173-11R three-blade constant-speed fully feathering reversible-pitch propeller. Fuel in four identical outer-wing tanks, total capacity 1,024 litres (270 US gallons; 225 Imp gallons). Self-sealing tanks optional on military versions. Provision for underwing tanks, total capacity 600 litres (158.5 US gallons; 132 Imp gallons). Oil capacity 11.4 litres (3.0 US gallons; 2.5 Imp gallons).
ACCOMMODATION: Pilot and co-pilot or passenger on flight deck. Dual controls standard. Cabin accommodates up to nine passengers at 100 cm (40 in) seat pitch (2-2-2-2-1); six passengers in VIP version, with reclining seats, folding tables, bar and toilet; or 10 paratroops on inward-facing seats; or two stretcher patients and two medical attendants; or freight. Baggage compartment at rear of cabin in standard passenger version; in centre of cabin, opposite toilet, in VIP version; rear compartment used to store folding passenger seats when converted for cargo use. Forward door on port side for crew. Wider, sliding door at rear on port side for passenger and freight loading and paratroop dropping, with smaller emergency door opposite this on starboard side. Cargo version can accept three 1.30 × 1.15 × 1.07 m (51 × 45 × 42 in) containers, or two of size 2.20 × 1.15 × 1.07 m (87 × 45 × 42 in).
SYSTEMS: Standard cabin heating/defrosting system uses engine bleed air; ventilation is provided by ram air; freon air conditioning system optional. Primary electrical system is 28 V DC, powered by two 150 A engine-driven starter/generators, with a 24 V 22 Ah Ni/Cd battery for independent engine starting and emergencies. AC power, 115 V at 400 Hz, is provided when required by a static inverter. Pneumatic de-icing system for wings and tail unit, and electric de-icing of propellers, are optional.
AVIONICS: *Comms:* HF, VHF and UHF.
 Radar: Optional weather radar.
 Flight: ADF, VOR/LOC/ILS and DME.
DIMENSIONS, EXTERNAL:

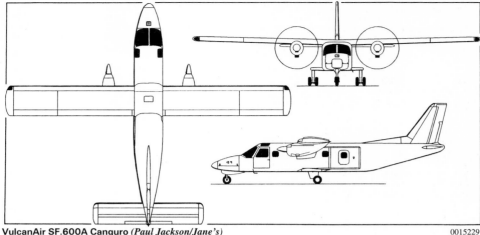

VulcanAir SF.600A Canguro *(Paul Jackson/Jane's)* 0015229

VulcanAir Canguro cockpit 1999/0051098

Demonstrator VulcanAir SF.600A Canguro I-VULA *(Paul Jackson/Jane's)* 2000/0085663

Wing span	15.00 m (49 ft 2½ in)
Wing chord, constant	1.60 m (5 ft 3 in)
Wing aspect ratio	9.4
Length overall	12.21 m (40 ft 0¾ in)
Height overall	4.30 m (14 ft 1¼ in)
Tailplane span	5.89 m (19 ft 4 in)
Wheel track	2.40 m (7 ft 10½ in)
Wheelbase	4.88 m (16 ft 0 in)
Propeller diameter	2.29 m (7 ft 6 in)
Crew door (fwd, port): Height	1.14 m (3 ft 9 in)
Width	0.86 m (2 ft 10 in)
Height to sill	0.90 m (2 ft 11½ in)
Cargo door (rear, port): Height	1.13 m (3 ft 8½ in)
Width	1.49 m (4 ft 10¾ in)
Height to sill	0.90 m (2 ft 11½ in)
DIMENSIONS, INTERNAL:	
Cabin, excl flight deck: Length	5.09 m (16 ft 8½ in)
Width	1.22 m (4 ft 0 in)
Height	1.25 m (4 ft 1¼ in)
Floor area	6.0 m² (65 sq ft)
Volume	7.90 m³ (279 cu ft)
AREAS:	
Wings, gross	24.00 m² (258.3 sq ft)
Fin	1.69 m² (18.19 sq ft)
Rudder	1.35 m² (14.53 sq ft)
Tailplane	3.68 m² (39.61 sq ft)
Elevators (total)	2.76 m² (29.71 sq ft)
WEIGHTS AND LOADINGS:	
Weight empty	2,120 kg (4,674 lb)
Max T-O weight	3,605 kg (7,947 lb)
Max cargo floor loading	400 kg/m² (82 lb/sq ft)
Max wing loading	150.2 kg/m² (30.77 lb/sq ft)
Max power loading	5.37 kg/kW (8.83 lb/shp)
PERFORMANCE:	
Never-exceed speed (V_{NE})	172 kt (320 km/h; 198 mph)
Max cruising speed at 1,525 m (5,000 ft)	165 kt (306 km/h; 190 mph)
Stalling speed, flaps down	67 kt (124 km/h; 77 mph)
Max rate of climb at S/L	387 m (1,270 ft)/min
Rate of climb at S/L, OEI	106 m (350 ft)/min
Service ceiling	6,100 m (20,020 ft)
Service ceiling, OEI	3,200 m (10,500 ft)
T-O run	414 m (1,360 ft)
T-O to 15 m (50 ft)	600 m (1,970 ft)
Landing run	300 m (985 ft)
Range	930 n miles (1,722 km; 1,070 miles)
g limits	+3.5/–1.4

UPDATED

JAPAN

FUJI

FUJI JUKOGYO KABUSHIKI KAISHA (Fuji Heavy Industries Ltd)

Subaru Building, 7-2 1-chome, Nishi-Shinjuku, Shinjuku-ku, Tokyo 160-8316
Tel: (+81 3) 33 47 25 25
Fax: (+81 3) 33 47 25 88
e-mail: okak@sb.hq.subaru-fhi.co.jp
Web: http://www.fhi.co.jp
PRESIDENT: Takeshi Tanaka

Utsunomiya Manufacturing Division
1-11 Yonan 1-chome, Utsunomiya, Tochigi 320-8564
Tel: (+81 28) 684 70 55
Fax: (+81 28) 684 70 71
e-mail: nakagawan@ho.subaru-fhi.co.jp
GENERAL MANAGER: Youichi Sugimura
DEPUTY GENERAL MANAGER: Tsunenori Hoshi

Aerospace Division
SENIOR VICE-PRESIDENT AND GENERAL MANAGER:
 Hiroyuki Nakatsubo
VICE-PRESIDENT AND DEPUTY GENERAL MANAGER:
 Kisaburo Wani
DEPARTMENT GENERAL MANAGERS:
 Kenichiro Usuki (Commercial Marketing and Sales)
 Norihusa Matsuo (Defence Marketing and Sales)
 Shunji Notake (Defence Market Development)
 Takehiko Sarukawa (Administration)

Established 15 July 1953 as successor to Nakajima. Utsunomiya Manufacturing Division occupies 47.7 ha (117.9 acre) site, including 153,000 m² (1,646,880 sq ft) floor area; workforce 2,745 in April 2000.
 Fuji producing Bell UH-1J (see *Jane's Aircraft Upgrades*; 76 completed by 1 October 2000) and recently completed last of 89 AH-1S HueyCobra helicopters; also manufactures wings, tailplanes and canopies for Kawasaki T-4 (which see). Commercial aircraft components produced are spoilers, inboard and outboard ailerons for Boeing 747; outboard flaps for Boeing 757; wing/body fairings and main landing gear doors for Boeing 767 and 777, plus front centre wing box for 777; and complete wing sets for the Raytheon Hawker Horizon business jet (see US section). Company was selected in May 2000 as risk-sharing subcontractor to produce composites fuselages for Bell/Agusta BA609 tiltrotor, with deliveries to start in FY03. Other products include BQM-34AJKai (modified Firebee) and J/AQM-1 target drones and RPH-2 unmanned helicopter (see *Jane's Unmanned Aerial Vehicles and Targets*).
 Fuji has participated in such projects as design of the HOPE-X space shuttle (see NASDA entry), and development of an NAL (which see) aero-spaceplane. Research has also begun on an SST/HST (supersonic/hypersonic transport), including a thermal protection system, heat-resistant structures and composites materials (see under Supersonic Airliner Studies in the International section).

UPDATED

FUJI T-5

TYPE: Turboprop basic trainer.
PROGRAMME: Development of KM-2. Fuji refitted KM-2 with Allison (now Rolls-Royce) 250-B17D turboprop in 1984; first flown 28 June 1984 as KM-2D, later KM-2Kai; JCAB certification (Aerobatic and Utility categories) gained 14 February 1985; ordered March 1987 as replacement for 31 JMSDF KM-2s; first flight of production KM-2Kai 27

April 1988. Deliveries began 30 August 1988; last (37th) aircraft due to have been delivered in March 2000. See 1998-99 *Jane's* for last full description; shortened version in 1999-2000 and 2000-01 editions.

UPDATED

FUJI T-3Kai/T-7

TYPE: Basic turboprop trainer.
PROGRAMME: Modification (Fuji designation KM-2D) of JASDF piston-engined T-3, which it is intended to replace on a one-for-one basis (50 delivered and in service). Selected by JDA (in preference to Pilatus PC-7 Mk IIM) in third quarter 1998 to meet future T-7 requirement; two prototypes included in FY99 budget request, but in late 1998 procurement deferred for one year following bribery allegations against some Fuji executives; Fuji suspended from all contracts until 31 December 1999. JDA reopened competition in August 1999, with bids due in third quarter 2000; T-3Kai selected as basis for T-7, September 2000; 11 T-7s approved in FY01 budget. A T-3Kai prototype has been test flown and awarded JCAB type certificate.
CUSTOMERS: JASDF requirement for approximately 50 T-7s over a 10-year period.
DESIGN FEATURES: Modifications to engine cowling, wings and tail unit compared with T-3.
Wing section NACA 23016.5 at root, NACA 23012 at tip; dihedral 6° from roots; sweepback 0° at quarter-chord.
FLYING CONTROLS: Conventional and manual. Plain ailerons, each with balance tab (port tab controllable for trim); controllable tab in each elevator; anti-balance tab in rudder. Single-slotted flaps.

Fuji T-3, on which the T-3Kai and proposed T-7 are based *(Paul Jackson/Jane's)* 1999/0051099

STRUCTURE: Two-spar all-metal wing; all-metal semi-monocoque fuselage.
LANDING GEAR: Retractable tricycle type with electrical actuation.
POWER PLANT: One 336 kW (450 shp) Rolls-Royce 250-B17F turboprop; three-blade propeller. Two bladder-type fuel tanks in each wing, combined capacity 375 litres (99.0 US gallons; 82.4 Imp gallons).
ACCOMMODATION: Crew of two in tandem. Dual controls standard.
SYSTEMS: Vapour cycle air conditioning system standard.
AVIONICS: Standard avionics include VHF, UHF, ATC transponder, ICS and Tacan.
DIMENSIONS, EXTERNAL:
Wing span	10.04 m (32 ft 11¼ in)
Wing aspect ratio	6.1
Length overall	8.59 m (28 ft 2¼ in)
Height overall	2.96 m (9 ft 8½ in)
Tailplane span	3.71 m (12 ft 2 in)
Propeller diameter	2.12 m (6 ft 11½ in)

AREAS:
Wings, gross	16.50 m² (177.6 sq ft)
Fin	1.28 m² (13.78 sq ft)
Rudder, incl tab	0.66 m² (7.10 sq ft)
Tailplane	2.07 m² (22.28 sq ft)
Elevators (total, incl tabs)	1.39 m² (14.96 sq ft)

WEIGHTS AND LOADINGS:
Max T-O weight	1,585 kg (3,494 lb)
Max wing loading	96.1 kg/m² (19.67 lb/sq ft)
Max power loading	4.73 kg/kW (7.76 lb/shp)

PERFORMANCE (T-3Kai):
Econ cruising speed at 915 m (3,000 ft)	161 kt (298 km/h; 185 mph)
Stalling speed, flaps and gear down	56 kt (104 km/h; 65 mph)
T-O to 15 m (50 ft) at S/L	608 m (1,995 ft)
Landing from 15 m (50 ft) at S/L	566 m (1,860 ft)

UPDATED

FUJI-BELL AH-1S

TYPE: Anti-armour helicopter.
PROGRAMME: Ordered by JGSDF following operational evaluation of two Bell AH-1Es (bought 1977-78) later upgraded to F standard. Fuji selected FY82 as prime contractor for licensed manufacture for JGSDF; first flight 2 July 1984. Final (91st) aircraft delivered March 2000, but AH-1Z (see under Bell in US section) is among candidates for JGSDF's AH-X requirement. Details of AH-1S last appeared in 2000-01 *Janes's*.

UPDATED

JADC

JAPAN AIRCRAFT DEVELOPMENT CORPORATION

Toranomon Daiichi Building, 2-3 Toranomon 1-chome, Minato-ku, Tokyo 105-0001
Tel: (+81 3) 35 03 32 25
Fax: (+81 3) 35 04 03 68
Web: http://www.iijnet.or.jp
CHAIRMAN: Nobuyuki Masuda
VICE-CHAIRMAN: Genhachiro Suzuki
SENIOR MANAGING DIRECTOR: Nobuhiko Tatebe

MANAGING DIRECTORS:
Masaomi Kadoya
Hiroshi Mizuno

Known as CTDC (Civil Transport Development Corporation) from 1972 until 1982, JADC is a consortium established by airframe manufacturers Mitsubishi, Kawasaki and Fuji to promote commercial aircraft business with the support of the Japanese government. Airframe production share (15 per cent of the Boeing 767 and 21 per cent of the 777) is currently managed by JADC's sister organisation, Commercial Aircraft Company (CAC). By 1 January 2000, CAC members had supplied 729 ship-sets of parts for the 767 and 178 for 777.

Current research programmes include supersonic commercial transport studies (see Supersonic Airliner Studies entry in International section and NAL entry in this section); high-temperature composites structures; and an advanced navigation system. Renewed preliminary design studies for an 80/100-passenger aircraft were under way in collaboration with Boeing in late 1999, under the programme name New Small Aircraft (NSA).

UPDATED

JASDF

NIHON KOKU JIEITAI (Japan Air Self-Defence Force)

7-45 Akasaka 9-chome, Minato-ku, Tokyo 107
Tel/Fax: (+81 3) 34 08 52 11

New aircraft programmes currently being formulated by the air force include an air superiority fighter, a primary trainer and a transport.

UPDATED

C-X

The JASDF has for some years required a replacement for the Kawasaki C-1A twin-turbofan transport, with increased range and payload capacity. Desired features include a turbofan power plant and a payload approximately double that of the Lockheed Martin C-130. Tanker capability may be added to the requirement.

Plans announced in late 2000 by the JDA's Technical Research and Development Institute (TRDI) indicate a twin-engined, high-wing aircraft with a rear-loading ramp/door, cruising at M0.8 and featuring a digital AFCS and 'glass' cockpit avionics; flight deck and tail unit are to be common with those of the JMSDF's MPA-X (which see). Between 20 and 50 C-Xs are required, to replace JASDF C-1As, C-130Hs and YS-11s. A two-year design phase will begin in 2003, with prototype/pre-production development following from 2005, engineering and operational testing from 2007 to 2010.

Foreign proposals for the C-X requirement have included the Airbus A310, Boeing C-17 and Lockheed Martin C-130J.

UPDATED

FI-X

Programme for next-generation fighter, to succeed F-15J early in the 21st century. Launched with FY95 allocation of ¥1 billion (US$10.2 million) to IHI to develop new 50 kN (11,240 lb st) class turbofan (XF-7) as power plant, to be test flown in TD-X technology demonstrator. Preliminary TRDI (Japan Defence Agency's Technology Research and Development Institute) design proposal for FI-X showed twin-engined configuration with canards, low aspect ratio tapered wings, twin fins and rudders and thrust-vectoring exhaust nozzles. Construction of up to four prototypes originally expected to begin in FY99 and to include co-cured composites, radar-absorbent materials and digital fly-by-light and engine control systems. Wing span and length provisionally 9.15 m (30 ft) and 13.40 m (44 ft) respectively. Avionics to include conformal radar and IR seeker.

First XF-7 engine was delivered for static testing in June 1998, but FI-X programme has been stretched, and TD-X demonstrator not now expected to fly until 2007.

VERIFIED

PT-X

The FY00 budget approved procurement funding for two new primary trainers for the JASDF. The type to be acquired had not been announced by November 2000, but a leading contender was the Fuji T-3Kai (which see).

UPDATED

JGSDF

NIHON RIKUJYO JIEITAI (Japan Ground Self-Defence Force)

Address and tel/fax number as for JASDF

Two helicopter programmes, for combat and support, are under consideration by Japan's army.

VERIFIED

AH-X

Attack helicopter requirement (approximately 100) to replace JGSDF AH-1S. Apache Longbow thought likely to be too expensive; alternatives could be upgraded AH-1S with more powerful engine, Bell AH-1Z, Eurocopter Tiger, or AH-2 attack development of Kawasaki OH-1. No selection decision made up to November 2000.

UPDATED

LH-X

JGSDF requirement for light transport and reconnaissance helicopter. Fuji and Mitsubishi competing for contract.

UPDATED

JMSDF

NIHON KAIJYO JIEITAI (Japan Maritime Self-Defence Force)

Address and tel/fax number as for JASDF

Naval aviation requires a new maritime patrol aircraft and, against convention, has shown interest in turbofan power.

VERIFIED

MPA-X

JMSDF requirement for approximately 80 maritime patrol/surveillance aircraft to replace Kawasaki (Lockheed Martin) P-3C from about 2010. Provisional initial design parameters envisaged a 50,000 to 70,000 kg (110,230 to 154,325 lb) aircraft powered by four 49 to 67 kN (11,000 to 15,000 lb st) turbofans, able to cruise at M0.85 (500 kt; 926 km/h; 575 mph) at 9,145 m (30,000 ft).

According to JDA's Technical Research and Development Institute (TRDI) in late 2000, MPA-X will be a low-wing design sharing a common flight deck and tail unit with the JASDF's C-X (which see), except that AFCS will be fly-by-light as a safeguard against electromagnetic interference. Like the C-X, it will undergo a two-year (2003 to 2005) design phase, with prototype trials scheduled from 2005 to 2007, but engineering and operational trials a year earlier (2006 to 2009). Radar and other sensors have yet to be specified.

UPDATED

JAPAN: AIRCRAFT—KAC to KAWASAKI

KAC

KANEMATSU AEROSPACE CORPORATION
2-1 Shibaura 1-chome, Minato-ku, Tokyo 105-8005
Tel: (+81 3) 54 40 80 00
Fax: (+81 3) 54 40 65 03
e-mail: pr@kanematsu.co.jp
Web: http://www.kanematsu.co.jp
PRESIDENT AND CEO: Tadashi Kurachi
SALES DIVISION MANAGER: Akiro Ohdoi

Kanematsu (workforce 785 at 31 March 2000) is prime contractor for outfitting Raytheon Hawker 800s (formerly BAe 125 Corporate 800) to JASDF specifications. Full description of standard aircraft under Raytheon heading in the US section; Fuji is systems integrator and responsible for U-125/U-125A maintenance. Kanematsu is also Japanese distributor for Embraer ERJ-145.

UPDATED

U-125/A PROCUREMENT (JASDF)

FY	Ordered U-125	Ordered U-125A	CY	Delivered First Acft	Qty
90	1		92	29-3041	1
91	1		93	39-3042	1
92	1		94	49-3043	1
92		3	95	52-3001	3
93		1	96	62-3004	1
94		1	97	72-3005	2
95		2	98	82-3007	3
96		3	99	92-3010	4
97		4	—	02-3014	—
98		3	—	—	—
99		2	—	—	—
00		2	—	—	—
01		1	—	—	—
Totals	3	22			

KAC (RAYTHEON) HAWKER 800
JASDF designations: U-125 and U-125A
TYPE: Navaid calibration and SAR aircraft.
PROGRAMME: Raytheon Hawker 800 (see US section) selected under JASDF H-X programme to replace Mitsubishi MU-2J and MU-2E in navaid calibration and SAR roles respectively; first U-125 delivered to JASDF 18 December 1992 and first U-125A (52-3003) delivered on 11 December 1994; three U-125As delivered by 31 January 1995 in preparation for formal handover. First six U-125As assembled in UK; seventh aircraft (from Wichita production line) delivered to Kanematsu November 1997; 19th completed in USA by third quarter of 2000.
CURRENT VERSIONS: **U-125:** For navaid flight check role, replacing MU-2J; operated by Flight Check Group at Iruma. Three ordered; all delivered from UK production.
U-125A: Search and rescue version, replacing MU-2E; 360° search radar, FLIR, airdroppable marker flares, liferaft and rescue equipment; first aircraft built is 52-3001. U-125As are operated by detachments of the Air Rescue Wing at Chitose, Hyakuri, Komatsu and Naha, initially going to its Training Squadron at Komaki.
CUSTOMERS: JASDF (three U-125 and 22 U-125A); options for total of 27 U-125As.
COSTS: ¥3.85 billion (2000).
DESIGN FEATURES: U-125A has deep observation 'patio' window each side of fuselage immediately ahead of wing, and dinghy/rescue pack dropping system via pressure door built into lower fuselage which is exposed for operation when landing gear is deployed.
AVIONICS (U-125A): *Radar:* Toshiba-built Raytheon 360° search radar.
Mission: Mitsubishi Electric IR imager in retractable underfuselage turret.
EQUIPMENT (U-125A): Flare and marker buoy dispenser; liferaft.

UPDATED

Raytheon Hawker U-125 navaid calibration aircraft (Peter R Foster)

KAWASAKI

KAWASAKI JUKOGYO KABUSHIKI KAISHA (Kawasaki Heavy Industries Ltd)
Kobe Crystal Tower, 1-3 Higashi-Kawasaki-Cho 1-chome, Chuo-ku, Kobe 650-8680

Aerospace Group/Gas Turbine & Machinery Group
World Trade Center Building, 4-1 Hamamatsu-Cho 2-chome, Minato-ku, Tokyo 105-6116
Tel: (+81 3) 34 35 21 11
Fax: (+81 3) 34 36 30 37
Web: http://www.khi.co.jp/aero/airplane.html
PRESIDENT: Masamoto Tazaki
EXECUTIVE MANAGING DIRECTOR AND AEROSPACE GROUP SENIOR GENERAL MANAGER: Toshiaki Ouchida
MANAGING DIRECTOR AND GAS TURBINE & MACHINERY GROUP SENIOR GENERAL MANAGER: Tadashi Nishimura
INFORMATION: Masayuki Hirata
WORKS: Gifu, Nagoya 1 and 2 (Aerospace Group); Akashi and Seishin (Gas Turbine & Machinery Group)

Kawasaki Aircraft Company has built many US aircraft under licence since 1955; amalgamated with Kawasaki Dockyard Company and Kawasaki Rolling Stock Manufacturing Company to form Kawasaki Heavy Industries Ltd 1 April 1969; Aerospace Group employs some 3,200 people; Gas Turbine & Machinery Group has some 800 aerospace-related employees. Kawasaki has 25 per cent holding in Nippi (which see in 1995-96 and earlier editions).

Kawasaki is currently prime contractor on T-4 programme and for OH-1 observation helicopter; co-developer and co-producer, with Eurocopter, of BK 117 helicopter (see EC 145 entry in International section); is prime contractor for Japanese licence production of CH-47 Chinooks for JGSDF and JASDF; was prime contractor for P-3C variants for JMSDF (now completed); also builds MD Helicopters MD 500 under licence.

Subcontract work includes centre fuselage for Mitsubishi F-2; currently producing forward and centre-fuselage barrel sections and wing ribs for Boeing 767 and 777, plus rear centre wing box and rear pressure bulkhead for 777. Kawasaki also responsible for design and sole-source manufacture of transmission for MD Helicopters Explorer; in 1999 signed agreement with Aerostructures of USA to manufacture cabin doors and tailcones for Bell/Agusta BA609. Nominated as prime contractor for maintenance and support of JASDF E-2C Hawkeyes, E-767 AWACS and C-130 Hercules. Undertaking feasibility studies for JDA for C-X future C-1/C-130YS-11 transport replacement and MPA-X future maritime patrol (P-3C replacement) aircraft (see JASDF and JMSDF entries).

Kawasaki also extensively involved in satellites and launch vehicles; is member of International Aero Engines consortium and produces T53 and T55 engines under licence (see *Jane's Aero-Engines*); overhauls engines; and builds hangars, docks, passenger bridges and other airport equipment.

UPDATED

JASDF Kawasaki T-4 twin-turbofan trainers of the Blue Impulse display team (H Seo/Kawasaki)

KAWASAKI T-4
TYPE: Advanced jet trainer.
PROGRAMME: Kawasaki named prime contractor 4 September 1981 by Japan Defence Agency; T-4 based on Kawasaki KA-851 design, by engineering team led by Kohki Isozaki; basic design studies completed October 1982; funding approved in FY83 and FY84 for four flying prototypes; prototype construction began April 1984; first flight of first XT-4 (56-5601) 29 July 1985; all four delivered between December 1985 and July 1986, preceded by static and fatigue test aircraft.

Production began FY86; first flight of production T-4, 28 June 1988; deliveries started 20 September 1988 to begin replacement of Lockheed T-33A and Fuji T-1A/B. Fuji and Mitsubishi each have 30 per cent share in production programme.
CUSTOMERS: JASDF T-4s used for pilot training, liaison and other duties; total of 212 (including prototypes) ordered by September 2000, of which more than 190 delivered (see table). Used by Nos. 31 and 32 Flying Training Squadrons of 1st Air Wing at Hamamatsu, near Tokyo; and in small numbers by instrument rating/communications flights of

most combat squadrons and regional HQ flights. First delivery to Blue Impulse aerobatic team (11 Squadron/4th Wing) 1994; nine operational for air show season.

COSTS: Unit cost ¥2,251 million (2000).

DESIGN FEATURES: High subsonic manoeuvrability; ability to carry external loads under wings and fuselage; anhedral mid-mounted wings, with extended chord outer panels giving dog-tooth leading-edges; tandem stepped cockpits with dual controls; baggage compartment in centre-fuselage for liaison role.

Supercritical wing section; thickness/chord ratio 10.3 per cent at root, 7.3 per cent at tip; anhedral 7° from roots; incidence 0°; sweepback at quarter-chord 27° 30′.

FLYING CONTROLS: Hydraulically actuated controls; plain ailerons with Teijin powered actuators; all-moving tailplane and rudder use Mitsubishi servo actuators; double-slotted trailing-edge flaps; no tabs; airbrake on each side of rear fuselage.

STRUCTURE: Aluminium alloy wings, with slow crack growth characteristics; CFRP ailerons, fin, rudder and airbrakes; aluminium alloy flaps (with AFRP trailing-edges), tailplane (CFRP trailing-edge) and fuselage with slow crack growth characteristics; titanium used sparingly in critical areas. Kawasaki builds forward fuselage and is responsible for final assembly and flight testing; Fuji builds rear fuselage, wings and tail unit; Mitsubishi builds centre-fuselage and engine air intakes.

LANDING GEAR: Hydraulically retractable tricycle type, with Sumitomo oleo-pneumatic shock-absorber in each unit. Single-wheel main units retract forward and inward; steerable nosewheel retracts forward. Kayaba (Honeywell) mainwheels, tyre size 22×5.5-13.8, pressure 19.31 bars (280 lb/sq in); Kayaba (Honeywell) nosewheel, tyre size 18×4.4-11.6, pressure 12.76 bars (185 lb/sq in). Kayaba (Honeywell) carbon brakes and Sumitomo (Hydro-Aire) anti-skid units on mainwheels. Minimum ground turning radius 9.45 m (31 ft 0 in).

POWER PLANT: Two 16.37 kN (3,680 lb st) Ishikawajima-Harima F3-IHI-30 turbofans, mounted side by side in centre-fuselage. Internal fuel in two 401.25 litre (106 US gallon; 88.3 Imp gallon) wing tanks and two Japanese-built Goodyear rubber bag tanks in fuselage, one of 776 litres (205 US gallons; 170.7 Imp gallons) and one of 662.5 litres (175 US gallons; 145.7 Imp gallons). Total internal capacity 2,241 litres (592 US gallons; 493 Imp gallons). Single pressure refuelling point in outer wall of port engine air intake. Provision to carry one 454 litre (120 US gallon; 100 Imp gallon) ShinMaywa drop tank on each underwing pylon. Oil capacity 5 litres (1.3 US gallons; 1.1 Imp gallons).

ACCOMMODATION: Crew of two in tandem in pressurised and air conditioned cockpit with wraparound windscreen and one-piece sideways (to starboard) opening canopy. Dual controls standard; rear (instructor's) seat elevated 27 cm (10.6 in). UPCO (Stencel) SIIS-3 ejection seats and Teledyne McCormick Selph canopy severance system, licence-built by Daicel Chemical Industries. Baggage compartment in centre of fuselage, with external access via door on port side.

SYSTEMS: Shimadzu bootstrap-type air conditioning and pressurisation system (maximum differential 0.28 bar; 4.0 lb/sq in). Two independent hydraulic systems (one each for flight controls and utilities), each operating at 207 bars (3,000 lb/sq in) and each with separate air/fluid reservoir pressurised at 3.45 bars (50 lb/sq in). Flow rate of each hydraulic system 45 litres (12 US gallons; 10 Imp gallons)/min. No pneumatic system. Electrical system powered by two 9 kW Shinko engine-driven starter/generators. Tokyo Aircraft Instruments onboard oxygen generating system.

AVIONICS: *Comms:* Mitsubishi Electric J/ARC-54 VHF/UHF com, and Nagano JRC J/AIC-103 intercom.

Flight: Nippon Electric J/ARN-66 Tacan, Toshiba J/ARN-69 VOR/ILS, Toyo Communication (Teledyne Electronics) J/APX-106 SIF, Japan Aviation Electronics (Honeywell) J/ASN-3 laser gyro AHRS, Tokyo Keiki (Honeywell) J/ASK-1 air data computer, Tokyo Aircraft Instrument J/ASH-3 VGH recorder and Kanto (Smiths) J/ASH-4 FDR.

Instrumentation: Shimadzu (Kaiser) J/AVQ-1 HUD.

ARMAMENT: No built-in armament.

EQUIPMENT: Two Nippi pylons under each wing for carriage of drop tanks (see Power Plant) or travel pods; one Nippi pylon under fuselage, on which can be carried target towing equipment, ECM/chaff dispenser, travel pod or air sampling pod.

DIMENSIONS, EXTERNAL:
Wing span	9.94 m (32 ft 7½ in)
Wing chord: at root	3.11 m (10 ft 2½ in)
at tip	1.12 m (3 ft 8 in)
Wing aspect ratio	4.7
Length: overall, incl probe	13.00 m (42 ft 8 in)
fuselage	11.96 m (39 ft 3 in)
Height overall	4.60 m (15 ft 1¼ in)
Tailplane span	4.40 m (14 ft 5¼ in)
Wheel track	3.20 m (10 ft 6 in)
Wheelbase	5.10 m (16 ft 9 in)

DIMENSIONS, INTERNAL:
Cockpit: Length	3.20 m (10 ft 6 in)
Max width	0.69 m (2 ft 3 in)
Max height	1.40 m (4 ft 7¼ in)

AREAS:
Wings, gross	21.00 m² (226.1 sq ft)
Ailerons (total)	1.51 m² (16.25 sq ft)
Trailing-edge flaps (total)	2.93 m² (31.54 sq ft)
Fin	3.78 m² (40.69 sq ft)
Rudder	0.91 m² (9.80 sq ft)
Tailplane	6.04 m² (65.02 sq ft)

WEIGHTS AND LOADINGS:
Weight empty	3,840 kg (8,466 lb)
T-O weight, clean	5,730 kg (12,632 lb)
Max design T-O weight	7,500 kg (16,535 lb)
Max wing loading	357.1 kg/m² (73.15 lb/sq ft)
Max power loading	229 kg/kN (2.25 lb/lb st)

PERFORMANCE (in clean configuration. A: at weight of 4,850 kg; 10,692 lb with 50% fuel, B: at T-O weight of 5,730 kg; 12,632 lb):
Max level speed (A): at height	M0.907
at S/L	560 kt (1,038 km/h; 645 mph)
Cruising speed: B	M0.75
Stalling speed: A	90 kt (167 km/h; 104 mph)
Max rate of climb at S/L: B	3,078 m (10,100 ft)/min
Service ceiling: B	14,815 m (48,600 ft)
T-O run, 35°C: B	655 m (2,150 ft)
Landing run, 35°C: B	704 m (2,310 ft)

Range (B) at M0.75 cruising speed at 7,620 m (25,000 ft):
  internal fuel only 700 n miles (1,297 km; 806 miles)
  with two 120 US gallon drop tanks
    900 n miles (1,668 km; 1,036 miles)
g limits +7.33/−3

UPDATED

KAWASAKI T-4 PROCUREMENT AND DELIVERY (JASDF)

	PROCUREMENT				DELIVERY		
FY	Lot	Qty	Cum Total	CY	Qty	First Aircraft	Cum Total
83		3	3				
84		1	4				
85			4	85	2	56-5601	2
86	C-1	12	16	86	2	66-5603	4
87	C-2	20	36	87			4
88	C-3	20	56	88	8	86-5605	12
89	C-4	20	76	89	13	96-5613	25
90	C-5	19	95	90	29	06-5626	54
91	C-6	21	116	91	19	16-5655	73
92	C-7	19	135	92	19	26-5674	92
93	C-8	9	144	93	18	36-5693	110
94	C-9	9	153	94	21	46-5711	131
95	C-10	9	162	95	10	56-5732	141
96	C-11	9	171	96	10	66-5742	151
97	C-12	13	184	97	9	76-5752	160
98	C-13	9	193	98	9	86-5761	169
99	C-14	10	203	99		96-5770	
00	C-15	9	212	00		06-	

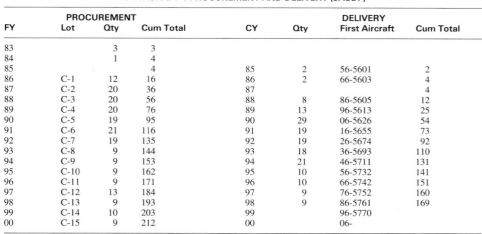

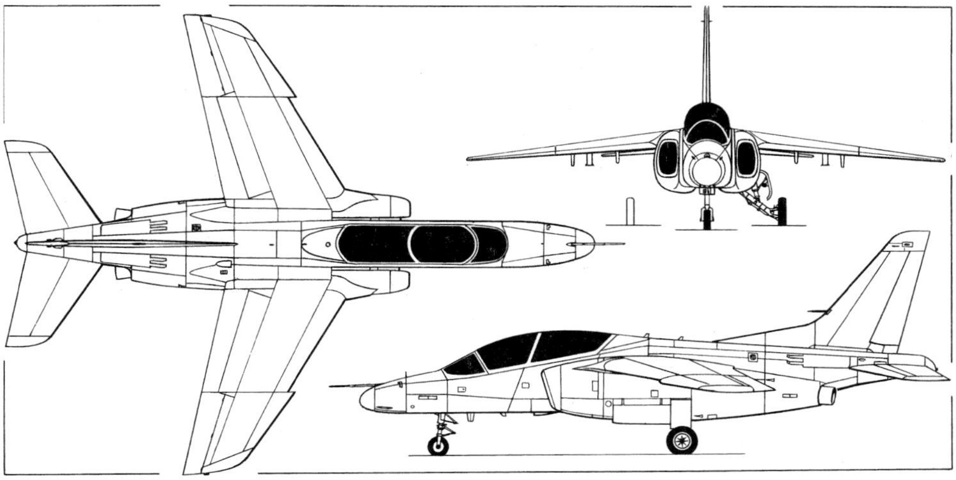

Kawasaki T-4 trainer (two Ishikawajima-Harima F3-IHI-30 turbofans) *(Dennis Punnett/Jane's)*

KAWASAKI OH-1

TYPE: Armed observation helicopter.

PROGRAMME: Developed to replace OH-6Ds of JGSDF. Japan Defence Agency (JDA) awarded ¥2.7 billion (US$22.5 million) in FY92 to cover basic design phase of helicopter then provisionally designated OH-X; RFPs issued by JDA's Technical Research & Development Institute (TRDI) 17 April 1992; Kawasaki selected as prime contractor (60 per cent of programme) 18 September 1992, with Fuji and Mitsubishi (20 per cent each) as partners; Observation Helicopter Engineering Team (OHCET), formed by these three companies, began preliminary design phase 1 October 1992. Mockup made public 2 September 1994 under Japanese name *Kogata Kansoku* (new small observation [helicopter]).

Programme includes six prototypes (four flying, two for ground test); first aircraft (32001) rolled out at Gifu on 15

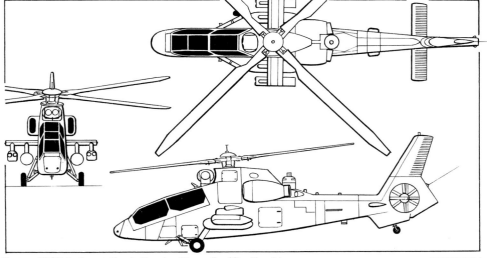

Kawasaki OH-1 observation helicopter *(James Goulding/Jane's)*

JAPAN: AIRCRAFT—KAWASAKI

First production OH-1 in service with the JGSDF Aviation School at Akeno

March 1996 and made first flight 6 August 1996, followed by second prototype on 12 November; OH-1 designation assigned late 1996; first two XOH-1s handed over to JDA on 26 May and 6 June 1997; third flown on 9 January 1997, at which time earlier aircraft had accumulated some 30 and 20 hours, respectively; handed over 24 June 1997; fourth flown on 12 February 1997 and handed over 29 August 1997. Prototypes renumbered by 1999 from 32001-04 to 32601-04.

First three production OH-1s funded FY97 and ordered 1998; first prototype flown with more fuel-efficent TS1-10QT (replacing XTS1-10) engines, 30 March 1998. By early 1999, four prototypes had flown 450 hours and were due to complete further 450 hours by end of 1999, including operational evaluation at Akeno JGSDF base. First production OH-1 (32605) flown July 1999 and handed over to JGSDF at Gifu 24 January 2000; two more due for delivery by 31 March 2000; these three aircraft allocated to JGSDF's Aviation Schools at Akeno (two) and Kasumigaura (one).

CURRENT VERSIONS: **OH-1:** Basic initial production version; *as described.*

Growth versions: Under study with more powerful engines (possibly LHTEC T800 or R-R/Turbomeca/MTU MTR 390) and uprated gearbox. **OH-1Kai** is possible candidate for AH-X requirement (see JGSDF entry) with tentative designation AH-2, armour-plated forward and centre fuselage, upgraded engines and transmission and additional weapons carriage.

CUSTOMERS: See table. Japan Ground Self-Defence Force requirement for 150 to 200.

COSTS: Funding for development, prototypes and flight testing ¥2.5 billion in FY92, ¥10.2 billion in FY93, ¥50.1 billion in FY94 and ¥23.3 billion in FY95. Unit costs of first four production lots ¥1.924 billion (FY97), ¥2.018 billion (FY98), ¥2.229 billion (FY99) and ¥2.075 billion (FY00).

DESIGN FEATURES: Kawasaki hingeless, bearingless and 20 mm ballistic-tolerant four-blade elastomeric main rotor and transmission system; Fenestron-type tail rotor with eight unevenly angled blades (35 and 55°); stub-wings for stores carriage. Active vibration damping system.

FLYING CONTROLS: Integrated AFCS and stability control augmentation system (SCAS).

STRUCTURE: Rotor blades and hub manufactured from GFRP composites; centre-fuselage and engines by Mitsubishi, tail unit/canopy/stub-wings/cowling by Fuji, rest by Kawasaki. Some 37 per cent of airframe (by weight) in GFRP/CFRP.

LANDING GEAR: Non-retractable tailwheel type. Provision for wheel/skis on main units.

POWER PLANT: Twin 662 kW (888 shp) FADEC-equipped Mitsubishi TS1-10QT turboshafts (XTS1-10 originally in prototypes). Possibility of off-the-shelf alternative engines not ruled out. Transmission has 30-minute run-dry capability. Stub-wings can each carry a 235 litre (62.0 US gallon; 52.0 Imp gallon) auxiliary fuel tank.

ACCOMMODATION: Crew of two on tandem armoured seats (pilot in front). Flat-plate cockpit transparencies, upward-opening on starboard side for crew access.

AVIONICS: *Flight:* AFCS with stability augmentation and holding functions; dual HOCAS controls.

Instrumentation: Two Yokogawa Electric large, flat-panel, liquid crystal colour MFDs in each cockpit, linked to a MIL-STD-1553B databus; Shimadzu HUD in front cockpit.

KAWASAKI OH-1 PROCUREMENT AND DELIVERY (JGSDF)

	PROCUREMENT			DELIVERY		
FY	Qty	Cum Total	FY	Qty	First Aircraft	Cum Total
93	2[1]	2				
95	2[1]	4				
97	3	7	97	4	32601	4
98	2	9	99	3	32605	7
99	3	12				
00	4	16				
01	2	18				

Notes:
[1] XOH-1 prototypes

Mission: Kawasaki electrically operated roof-mounted turret combining Fujitsu thermal imager, NEC real-time colour TV camera and NEC laser range-finder/designator; field of regard 110° in azimuth, 40° in elevation.

Self-defence: Spine-mounted IR jammer based on Sanders AN/ALQ-144.

ARMAMENT: Four Toshiba Type 91 (modified) lightweight, short-range, IR-guided AAMs on pylons under stub-wings for self-defence.

Following data all provisional:
DIMENSIONS, EXTERNAL:
Main rotor diameter	11.6 m (38 ft 0¾ in)
Main rotor blade chord, constant portion	0.38 m (1 ft 3 in)
Wing span	3.0 m (9 ft 10 in)
Fuselage: Length	12.0 m (39 ft 4½ in)
Max width	1.0 m (3 ft 3¼ in)
Height: to top of rotor head	3.4 m (11 ft 1¾ in)
over tailfin	3.8 m (12 ft 5½ in)
Tailplane span	3.0 m (9 ft 10 in)

AREAS:
Main rotor disc	105.68 m² (1,137.5 sq ft)

WEIGHTS AND LOADINGS:
Weight empty	2,450 kg (5,401 lb)
Max weapons load	132 kg (291 lb)
T-O weight: design	3,550 kg (7,826 lb)
max	4,000 kg (8,818 lb)
Max disc loading	37.9 kg/m² (7.75 lb/sq ft)

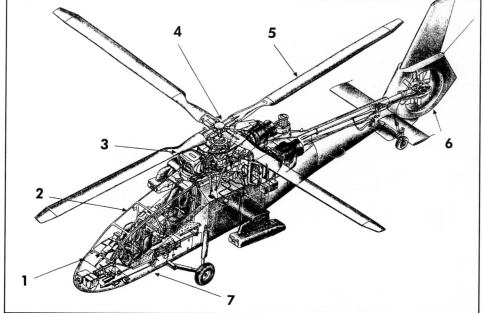

Cutaway drawing of the OH-1 showing (1) integrated cockpits; (2) crashworthy armoured seats; (3) FLIR/TV/LRF sighting system; (4) composites hingeless rotor hub; (5) composites ballistic-tolerant main rotor blades; (6) ducted tail rotor; and (7) AFCS with stability and control augmentation system

PERFORMANCE:
Max level speed 150 kt (277 km/h; 172 mph)
Combat radius 108 n miles (200 km; 124 miles)
Range 297 n miles (550 km; 342 miles)
g limits +3.5/–1

UPDATED

KAWASAKI (LOCKHEED MARTIN) P-3C

Last of 107 licence-built Orions (a UP-3D) delivered on 1 February 2000. See procurement table in 2000-01 *Jane's*; detailed description of Japanese P-3Cs in 1999-2000 and earlier editions. Full P-3 description in *Jane's Aircraft Upgrades*.

UPDATED

KAWASAKI (BOEING) CH-47 CHINOOK

JASDF/JGSDF designations: CH-47J and CH-47JA
TYPE: Medium lift helicopter.
PROGRAMME: FY84 defence budget approved purchase of three Boeing CH-47s, two for JGSDF and one for JASDF; first two built in USA and delivered second quarter 1986; Nos. 3 to 7 delivered as CKD kits for assembly in Japan; Kawasaki granted manufacturing licence for Japanese services' Chinooks; first CH-47Js delivered late 1986.
CURRENT VERSIONS: **CH-47J**: Generally similar to US CH-47D. Total 34 (from all sources) delivered to JGSDF by 1996; procurement continues for JASDF.
CH-47JA: Designation introduced to describe CH-47Js delivered to army aviation from 1996 with weather radar; first (52951) was 35th JGSDF Chinook, 50th Japanese Chinook.
CUSTOMERS: See table. Total 58 (including US-built) delivered by 1 April 2000. JGSDF has requirement for 52, of which 42 in service by 1 April 2000. Operated by 1st and 2nd Squadrons of 1st Helicopter Brigade at Kisarazu (CH-47J), 1st Composite Brigade on Okinawa (CH-47JA) and Aviation School at Akeno.
JASDF 16 CH-47J ordered and delivered by 31 March 2000. Operated by Air Rescue Wing flights at Iruma, Kasuga, Misawa and Naha (Okinawa).
COSTS: Unit cost ¥4.5 billion for CH-47J, ¥5.1 billion for CH-47JA (2000).
AVIONICS: *Flight:* GPS in JGSDF aircraft from 1993; Mitsubishi Precision INS.

KAWASAKI CH-47J/JA PROCUREMENT*

FY	JASDF	First aircraft	JGSDF	First aircraft	Cum Total
84	1	67-4771	2	52901	3
85	1	77-4772	3	52903	7
86	3	87-4773	4	52906	14
87	2	97-4776	4	52910	20
88	3	07-4778	5	52914	28
89	2	27-4481	5	52919	35
90	2	37-4483	5	52924	42
91	1	47-4485	3	52929	46
92			3	52932	49
93			2†	52951	51
94			2†	52953	53
95	1	87-4486	2†	52955	56
96			2†	52957	58
97			2†	52959	60
98			1†	52961	61
99	2	07-4487	2†	52962	65
00	1		2†		68
01	1		2†		71
Totals	20		51		71

* First seven built in USA (two) or assembled from kits
† CH-47JA

UPDATED

Kawasaki-built Boeing CH-47JA Chinook, showing this version's radar nose *(Peter R Foster)* 1999/0051103

KAWASAKI (MDH) MD 500D

JGSDF/JMSDF designations: OH-6D and OH-6DA
TYPE: Light utility helicopter.
PROGRAMME: First flight of initial Kawasaki licence-built Hughes Helicopters (now MD Helicopters) 369D (500D) 2 December 1977; JCAB Normal category certification awarded 20 April 1978. Replacement for JGSDF now in production under designation OH-1 (which see).
CUSTOMERS: See table. JGSDF ordered 193, of which all delivered by FY96; 16 delivered to JMSDF (including one for Antarctic research) by 1999. Nine delivered for civil operation by 1996. One OH-6DA for JMSDF approved in FY00 budget.
COSTS: ¥300 million (2000).

KAWASAKI OH-6D/DA PROCUREMENT

FY	JGSDF	JMSDF	Cum Total
78	10		10
79	12		22
80	10		32
81	8		40
82	6	2	48
83	3	1	52
84	9	2	63
85	7		70
86	12		82
87	12	2	96
88	11		107
89	11	2	120
90	15		135
91	14	1	150
92	13	1	164
93	13		177
94	16		193
95	11	3	207
96			207
97		2*	209
98			209
99			209
00		1*	210
Totals	193	17	210

* OH-6DA

UPDATED

Kawasaki (MDH) OH-6D light helicopter of the JMSDF 0011102

MITSUBISHI

MITSUBISHI JUKOGYO KABUSHIKI KAISHA (Mitsubishi Heavy Industries Ltd)

5-1 Marunouchi 2-chome, Chiyoda-ku, Tokyo 100-8315
Tel: (+81 3) 32 12 31 11 and 32 12 56 41
Fax: (+81 3) 32 12 98 65
e-mail: QQ2100@hq.mhi.co.jp
Web: http://www.mhi.co.jp

NAGOYA AIRCRAFT SYSTEMS WORKS:
10 Oye-cho, Minato-ku, Nagoya 455
PRESIDENT: Takashi Nishioka
EXECUTIVE VICE-PRESIDENT: Kimiyuki Hanada
GENERAL MANAGER, DEFENCE AIRCRAFT AND AERO-ENGINE DEPARTMENT: Tetsuya Otabe

Present Komaki South plant built 1952, adjacent to Nagoya Airport, for assembly, outfitting and flight test. Main Nagoya facility (previously known as Komaki North) is divided into Aerospace Systems Works and Guidance & Propulsion Systems Works; former has 280,900 m² (3,023,575 sq ft) of floor space on a 58.6 ha (144.8 acre) site and had an April 1999 workforce of 3,921; latter occupies a 140.4 ha (346.9 acre) site with 84,530 m² (909,900 sq ft) of covered space and has approximately 1,700 employees. Oye plant manufactures aircraft fuselage components, spacecraft parts and other aero-related equipment; Tobishima plant responsible for aircraft fuselage subassembly, plus assembly and check-out of space systems; fuselage subassembly, final assembly, outfitting, flight test and repair undertaken by Komaki South plant.

Developed MU-2, MU-300 (now Raytheon Beechjet 400, which see in US section), T-2 supersonic trainer and the related close support F-1 for JASDF. Built 167 HSS-2/2A/2B and 18 S-61A helicopters under Sikorsky licence (last aircraft delivered 2 March 1990), and 138 F-4EJ and 199 F-15J/DJ jet fighters under McDonnell Douglas licence (last F-15 delivered December 1999). Is currently prime contractor for F-15J/DJ modernisation programme, F-2 fighter and SH/UH-60J helicopters; has developed own-design MH2000 helicopter; and manufactures centre-fuselage and engines of Kawasaki OH-1 and centre-fuselage and engine intakes for Kawasaki T-4.

Risk-sharing partner in Bombardier Global Express (responsible for wings and centre-fuselage), Dash 8 Q400 (centre and rear fuselage and tail unit), CRJ 700 (rear fuselage) and BD 100 (wings).

Participating in Boeing 767 (rear fuselage panels and cargo doors) and 777 (rear fuselage panels, rear fuselage doors). Subcontract work includes Boeing 737 (inboard flaps), 747 (inboard flaps) and 757-300 (stringers). Undertaking feasibility study for new C-X transport aircraft to replace C-1 and C-130 (see JASDF entry in this section). Responsible for main cabin section of Sikorsky S-92 Helibus medium helicopter.

Aero-engine activities detailed in *Jane's Aero-Engines*; also produces rocket engines; participates in H-I and H-II launchers and Japanese Experimental Module for US space station.

UPDATED

MITSUBISHI F-2

TYPE: Attack fighter.

PROGRAMME: Indigenous design, plus proposals based on F-15, F-16, F/A-18 and Tornado ADV, were originally considered; modified F-16C selected as Japan's FS-X replacement for Mitsubishi F-1 on 21 October 1987; Mitsubishi appointed prime contractor November 1988; initial contracts awarded for airframe design March 1989 and prototype active phased-array radar February 1990; General Electric F110-GE-129 Improved Performance Engine selected 21 December 1990. Programme delayed by questions of development sharing with General Dynamics (now Lockheed Martin Aeronautics Company (LMAC)) and technology transfer to Japan, but agreed at Japan 60 per cent and USA 40 per cent cost sharing (confirmed July 1996); first subcontracts to GD let February 1990 for design and development of rear fuselage, wing, leading-edge flaps, avionics and computer-based test equipment. Active phased-array radar, EW (ECM/ESM), mission computer and inertial reference system being developed using Japanese domestic technology.

Japan totally responsible for programme, including all funding; Japanese airframe subcontractors include Kawasaki and Fuji (see Structure). Programme involves four flying prototypes (two single-seat XF-2A first, then two tandem two-seat XF-2B) and two for static and fatigue test; construction began early 1994, final assembly mid-1994; first prototype (single-seat 63-0001) rolled out 12 January 1995; first flight 7 October 1995, followed by second prototype (63-0002) on 13 December 1995; third prototype (63-0003) flew on 18 April 1996 and fourth (63-0004; first in blue and grey colour scheme) on 24 May 1996. Japanese Cabinet approved 130-aircraft programme and allocated F-2 designation on 15 December 1995. Static test airframe (No. '991') under loading trials by 1995.

First XF-2A handed over to JDA on 22 March 1996, followed by '02 on 26 April 1997 and the two XF-2Bs on 9 August and 20 September 1997; prototypes re-serialled 63-8501/02/03/04 on 1 December 1997. Congressional approval of US participation in September 1996 followed in October by US$75 million Mitsubishi initial production contract to Lockheed Martin. First Lockheed Martin rear fuselage accepted by Mitsubishi in November 1998; first licence-built engines delivered by IHI in early 1999. Discovery in May 1998 of wing cracks, and evidence, when flying with two ASM-2 anti-ship missiles, of severe flutter, necessitated modifications to wingtips and pylon attachments, resulting in nine-month slippage of March 1999 scheduled completion date. Flight and static/fatigue testing rescheduled for completion in December 1999 and deliveries for FY99 (three) and FY00 (eight), but further wing crack problems revealed in mid-1999, causing completion of development testing to be extended to 30 June 2000 and first deliveries rescheduled to third quarter 2000. First production F-2A made maiden flight 12 October 1999 and delivered to JASDF on 25 September 2000. First two production lots (15 F-2As and four F-2Bs) due to be delivered by 31 March 2001.

Development of a dedicated air superiority version is under construction, to replace the F-4EJKai.

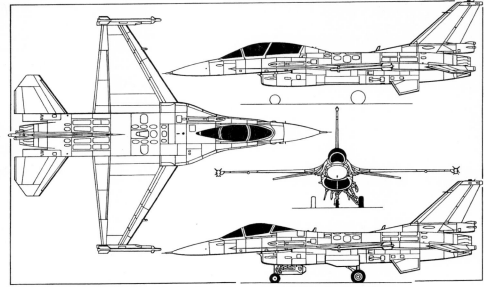

Mitsubishi F-2A, with additional side view (top) of two-seat F-2B *(Mike Keep/Jane's)*

CURRENT VERSIONS: **F-2A**: Single-seat support fighter.
Description applies to single-seat version except where indicated.

F-2B: Combat-capable two-seater.

CUSTOMERS: See table. Total of 57 production aircraft approved by February 2001. JASDF (sole user) originally stated requirement for approximately 72 F-2As to replace F-1s in three support fighter squadrons, plus approximately 50 F-2Bs for OCU and possibly to replace T-2/2A. Current plans are to acquire 130 single/two-seaters (83 + 47, reduced from earlier figure of 141 by cancellation of 11 two-seaters). First recipient originally due to begin conversion in FY99 and to be completely equipped by FY00; however, first delivery (03-8503) not made until 25 September 2000 (formal acceptance 3 October) (six F-2As and eight F-2Bs) delivered by year-end: one (F-2A) temporarily to No. 1 Technical School at Hamamatsu and seven to Rinji F2 Hikotai (temporary F2 Squadron) at Misawa, which became 3 Squadron of 3 Wing at same base when IOC achieved in April 2001. Remainder of first 45 aircraft will be allocated to Matsushima OCU (19 Bs). In 2006, 6 Squadron of 8 Wing at Tsuiki will convert from F-1; followed in 2007 by 8 Squadron/3 Wing at Misawa (which, pending F-2s, has converted from F-1 to F-4EJKai Phantom).

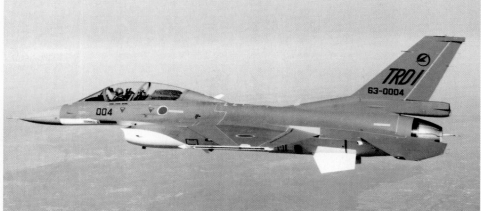

Second Mitsubishi XF-2B two-seat operational trainer

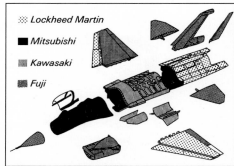

The principal F-2 contractors

Mitsubishi F-2 trials aircraft at TRDI

MITSUBISHI F-2 PROCUREMENT (JASDF)

FY	F-2A	First aircraft	F-2B	First aircraft
92	1[1]	63-8501		
93				
94	1[1]	63-8502	2[1]	63-8503
95				
96	7	03-8503	4	03-8103
97	8	03-8510		
98	2		7	
99	2		6	
00			9	
01	8		4	
Total	29		32	

[1] Prototypes.

COSTS: Total JDA expenditure (1988 to 1995) US$3.27 billion, including ¥75.7 billion (US$575 million) in FY92 to include first prototype and radar development; further ¥96.5 billion (US$804 million) in FY94 provided for three more flying prototypes and two for ground test. First two Mitsubishi contracts to GD totalled US$280.5 million; follow-on contract to Lockheed (5 February 1993) valued at US$74.2 million. FY96 batch of 11 cost US$1,081.58 million (US$98.32 million per unit); FY97 batch of eight cost US$797.47 million (US$99.68 million per unit); FY00 aircraft cost ¥11.8 billion each.

DESIGN FEATURES: Configuration based on Lockheed Martin F-16. New co-cured composites wing of Japanese design, with greater span, root chord and 25 per cent more area than that of F-16; tapered trailing-edge; slightly longer radome and forward fuselage to house new radar and other mission avionics; longer mid-fuselage and shorter jetpipes; increased-span tailplane; addition of brake-chute; adoption of increased performance engine.
Wing leading-edge sweepback 33° 12′; incidence 0°. Tailplane anhedral 8°.

FLYING CONTROLS: Full-span leading-edge flaps and trailing-edge flaperons; all-moving tailplane; rudder; twin, fixed ventral fins, canted outward 15°. Initially planned vertical canards deleted; CCV functions achieved instead by triple-redundant digital fly-by-wire system, developed jointly by Japan Aviation Electronics and Honeywell. Based on earlier Mitsubishi work with T-2 CCV testbed. Available modes include control augmentation, relaxed static stability, manoeuvre load control, decoupled yaw and manoeuvre enhancement. Single analogue back-up flight control system for roll and yaw control.

STRUCTURE: Composites structure wing (except leading-edge flaps), horizontal tail, fin (except leading-edge and base), rudder and landing gear doors; other structures also use advanced materials and structure technology, including Mitsubishi Rayon radar-absorbent material on nose, wing leading-edges and engine intakes; titanium in rear fuselage and tail unit.
Mitsubishi builds forward fuselage and wings; other Japanese airframe companies involved include Fuji (upper wing skins, wing fairings, radome, flaperons, engine air intakes and tail unit) and Kawasaki (fuselage mid-section, mainwheel doors and engine access doors). Lockheed Martin providing rear fuselage, port-side wing boxes, all leading-edge flaps, avionics systems, stores management systems and some test equipment; first LM production rear fuselage accepted 10 November 1998; first wing box delivered March 1999.

LANDING GEAR: Retractable tricycle type, with single wheel on each unit; mainwheels retract inward, nosewheel rearward. Mainwheel tyres size 27.75×8.75R145, pressure 22.06 bars (320 lb/sq in); nosewheel tyre size 18×5.7-8, pressure 20.69 bars (300 lb/sq in). Brake-chute in fairing at base of rudder.

POWER PLANT: One General Electric F110-GE-129 turbofan (131.2 kN; 29,500 lb st with afterburning, licence-built by IHI for production aircraft. Maximum internal fuel capacity 4,637 litres (1,225 US gallons; 1,020 Imp gallons), of which 4,588 litres (1,212 US gallons; 1,009 Imp gallons) are usable; reduced to 3,948 litres (1,043 US gallons; 868.5 Imp gallons), of which 3,903 litres (1,031 US gallons; 858.5 Imp gallons) usable, in F-2B. Maximum external fuel capacity (both) 5,678 litres (1,500 US gallons; 1,249 Imp gallons) (one 1,135.5 litre; 300 US gallon; 249.8 Imp gallon and two 2,271.25 litre; 600 US gallon; 499.6 Imp gallon tanks). No provision for in-flight refuelling.

AVIONICS: *Comms:* Magnavox AN/ARC-164 UHF transceiver; NEC V/UHF transceiver; Hazeltine AIFF; Kokusai Electric HF radio.
Radar: Mitsubishi Electric active phased-array radar.
Flight: Japan Aviation Electronics/Honeywell digital AFCS and laser IRS.
Instrumentation: Yokogawa 127 × 127 mm (5 × 5 in) colour LCD multifunction display and two 102 × 102 mm (4 × 4 in) liquid crystal colour MFDs; Shimadzu holographic HUD.
Mission: Mitsubishi Electric mission computer. Sanders MPS III mission planning system.
Self-defence: Mitsubishi Electric integrated EW system.

ARMAMENT: One internal M61A1 Vulcan 20 mm multibarrel gun in port wingroot, plus 13 external stores stations: Sta 6 on centreline; Sta 1 (port) and 11 at each wingtip; five under each wing (Sta 2, 3, 4L, 4, 5, 6, 8, 8R, 9 and 10); Flight Refuelling common rail launchers, built and installed by Nippi, configured initially for AIM-7F/M Sparrow medium-range air-to-air missiles (Sta 2 and 10); other armament expected to include AIM-9L or Mitsubishi AAM-3 air-to-air missiles (Sta 1, 2, 10 and 11) and ASM-1 and ASM-2 anti-shipping missiles (Sta 3, 4, 8 and 9); 500 lb bombs (Sta 4L and 8R); 340 kg bombs (Sta 4 and 8); CBU-87/B cluster bombs (Sta 4 and 8); and JLAU-3/A or RL-4 rocket launchers (Sta 4 and 8). Centreline and inboard underwing stations (5, 6 and 7) wet for carriage of drop tanks.

DIMENSIONS, EXTERNAL:
Wing span: over missile rails	11.125 m (36 ft 6 in)
excl missile rails	10.80 m (35 ft 5¼ in)
Wing chord: at root	5.27 m (17 ft 3½ in)
at tip	1.185 m (3 ft 10¾ in)

Prototype SH-60JKai shipboard maritime patrol helicopter *(TRDI)*

Wing aspect ratio	3.3
Length overall	15.52 m (50 ft 11 in)
Height overall	4.96 m (16 ft 3¼ in)
Tailplane span	6.045 m (19 ft 10 in)
Wheel track	2.36 m (7 ft 9 in)
Wheelbase	4.05 m (13 ft 3½ in)

AREAS:
Wings, gross	34.84 m² (375.0 sq ft)
Leading-edge flaps (total)	4.70 m² (50.59 sq ft)
Flaperons (total)	3.96 m² (42.63 sq ft)
Fin, incl dorsal fin	4.00 m² (43.06 sq ft)
Rudder	1.08 m² (11.63 sq ft)
Ventral fins (total)	1.50 m² (16.15 sq ft)
Horizontal tail surfaces (total)	7.05 m² (75.89 sq ft)

WEIGHTS AND LOADINGS:
Weight empty: F-2A	9,527 kg (21,003 lb)
F-2B	9,633 kg (21,237 lb)
T-O weight: clean: F-2A	13,459 kg (29,672 lb)
F-2B	13,230 kg (29,167 lb)
with two short-range AAMs:	
F-2A	13,713 kg (30,232 lb)
F-2B	13,412 kg (29,568 lb)
with four short-range and four medium-range AAMs:	
F-2A	15,711 kg (34,637 lb)
F-2B	15,392 kg (33,934 lb)
with two short-range AAMs, six 500 lb bombs and two drop tanks:	
F-2A	19,888 kg (43,845 lb)
F-2B	19,569 kg (43,142 lb)
with two short-range AAMs, four anti-ship missiles and two drop tanks:	
F-2A	20,517 kg (45,232 lb)
F-2B	20,198 kg (44,529 lb)
Design max T-O weight with external stores:	
F-2A, F-2B	22,100 kg (48,722 lb)
Design max landing weight:	
F-2A, F-2B	18,300 kg (40,345 lb)
Design max wing loading	634.3 kg/m² (129.92 lb/sq ft)
Design max power loading	168 kg/kN (1.65 lb/lb st)

PERFORMANCE (F-2A):
Design max level speed: at high altitude	approx M2.0
at low altitude	approx M1.1
Typical combat radius:	
F-2A	more than 450 n miles (833 km; 518 miles)

UPDATED

MITSUBISHI (BOEING) F-15J

TYPE: Air superiority fighter.
PROGRAMME: Two US-built F-15Js followed by eight assembled in Japan from US-supplied CKD kits; first flight of CKD F-15 26 August 1981, delivered 11 December 1981. Last F-15J delivered in 1998 and last F-15DJ on 10 December 1999, but upgrade programme to be defined for flight tests in 2000/02 and subsequent retrofit and likely to include features from F-2 programme (which see) such as radar, EW suite, central computer and possible retrofit with more powerful General Electric F110-GE-129 turbofans. F-15J/DJ description and procurement details in 2000-01 and earlier *Jane's*; see also *Jane's Aircraft Upgrades.*

UPDATED

MITSUBISHI (SIKORSKY) SH-60J

TYPE: Naval combat helicopter.
PROGRAMME: Detail design of S-70B-3 version of Sikorsky SH-60B Seahawk (see US section), to meet JMSDF requirements, started August 1983; Japanese avionics and equipment integrated by Technical Research and Development Institute (TRDI) of JDA. First flight of first of two XSH-60J prototypes (8201), based on imported airframes, 31 August 1987; evaluation by 51st Air Development Squadron of JMSDF at Atsugi completed early 1991; first production SH-60J (8203) flown 10 May 1991, delivered 26 August.

JMSDF No. 121 Squadron Mitsubishi (Sikorsky) SH-60J

Following trials begun by TRDI in 1995 on an all-composites rotor system with redesigned cross-sections, planforms and tips, offering 5 per cent increase in hovering efficiency and 30 per cent vibration reduction, FY97 budget requested funding for one upgraded **SH-60JKai** prototype, with new main rotor hub and blades, active sonar, a tactical data processing and display system, and a laser-based shipboard automatic landing system. This aircraft should have entered flight test in July 2000, but production start planned for FY01 reportedly deferred due to discovery of cracks in main rotor during fatigue tests; development costs said to have totalled some ¥40 billion by mid-2000.

CUSTOMERS: See table. JMSDF requirement for about 100 SH-60Js; 103 ordered, of which more than 65 in service, by 1 April 2001; half of force land-based. Upgrade of nine earlier aircraft also funded in FY00; three new aircraft approved in FY01 to offset delays in SH-60JKai programme. Operated by Nos. 121 and 123 Squadrons (at Tateyama); Nos. 122 and 124 (at Ohmura); and part of No. 211 (training unit at Kanoya); No. 101 Squadron at Tateyama undergoing conversion in 2000.

MITSUBISHI SH/UH-60 PROCUREMENT

FY	SH-60J (JMSDF)	UH-60J (JMSDF)	UH-60JA (JGSDF)	UH-60J (JASDF)	Cum Total
82	1[1]				1
83	1[1]				2
88	12			3[2]	17
89	12	3		2	34
90	11			2	47
91	5	3		4	59
92	7	2		2	70
93	4	2		1	77
94	5	1		2	85
95	6	1	2	2	96
96	6	2	4	1	109
97	7	2	4	3	125
98	7	2	5	2	141
99	9		3	2	155
00	7		3	2	167
01	3	1	2		173
Totals	**103**	**19**	**23**	**28**	**173[3]**

Notes:
[1] Sikorsky-built XSH-60Js
[2] One Sikorsky-built, two Mitsubishi-assembled from CKD kits
[3] Of which 97 delivered by 1 January 2000

COSTS: Unit cost ¥5.07 billion (2000).
POWER PLANT: T700-401C engines manufactured by IHI.
ACCOMMODATION: Crew of three plus five systems operators/observers.
AVIONICS: *Radar:* Japanese HPS-104 search radar.
Flight: Japanese automatic flight management system and ring laser gyro AHRS.
Instrumentation: Japanese controls and displays subsystem, datalink and tactical data processor.
Mission: Japanese HQS-103 sonar; Raytheon AN/ASQ-81D2(V) MAD; Ednac AN/ARR-75 sonobuoy receiver.
Self-defence: General Instruments AN/ALR-66 (VE) RWR; Japanese HLR-108 ESM.
EQUIPMENT: RAST, sonobuoys, rescue hoist and cargo sling.
UPDATED

MITSUBISHI (SIKORSKY) UH-60J

TYPE: Medium transport helicopter.
PROGRAMME: Detail design to Japanese requirements started April 1988; one US-built S-70A-12 imported, followed by two CKD kits for licence assembly (first flight 20 December 1989 at Sikorsky; delivered to JASDF 28 February 1991; second kit aircraft first flew February 1990, delivered to JASDF 29 March 1991). Remainder being built in Japan.

Second prototype Mitsubishi MH2000 twin-turboshaft commercial helicopter 1999/0051108

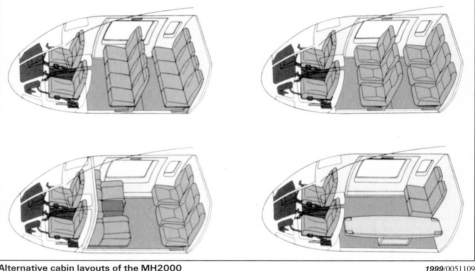
Alternative cabin layouts of the MH2000 1999/0051109

CURRENT VERSIONS: **UH-60J:** Standard search and rescue version for JASDF and JMSDF; can fly one hour search at 250 n miles (463 km; 288 miles) from base.
UH-60JA: JGSDF version; navigation/weather radar and FLIR night and adverse weather vision system. First (43101) completed late 1997; initial deliveries (four) January 1998.
CUSTOMERS: See table. JASDF has total requirement for 46 UH-60Js, JMSDF for 19; total of 21 in service with JASDF and 15 with JMSDF by 1 April 2000. JASDF aircraft operated by detached flights of Air Rescue Wing (HQ: Iruma); JMSDF UH-60Js to Atsugi Rescue Squadron in 1992; subsequently Kanoya, Tokushima, Shimofusa and Hachinohe.
JGSDF plans to procure up to 80 UH-60JAs in a US$2.67 billion programme announced early 1995; procurement began that year; JGSDF will deploy 50 with five district helicopter units alongside UH-1Js, with balance assigned to VIP transport and training; six delivered by 1 April 1999, this remaining as inventory total on 1 April 2000, despite more then awaiting acceptance.
COSTS: Unit costs in FY00 were ¥3.37 billion for UH-60JA and ¥3.6 billion for air force UH-60J.
POWER PLANT: T700-401C turboshafts manufactured by IHI. External long-range fuel tanks.
ACCOMMODATION: Crew of four (JMSDF) or five (JASDF) plus up to 11 other persons. Bubble windows for pilot and on each side at front of main cabin.
AVIONICS: *Radar:* Nose-mounted Japanese search/weather radar.
Flight: Sikorsky self-contained navigation system.
Mission: Turret-mounted FLIR beneath nose.
EQUIPMENT: Lucas rescue hoist (JA), ETS and cargo sling.
UPDATED

MITSUBISHI MH2000

TYPE: Light utility helicopter.
PROGRAMME: Launched in second half of 1995; applications include passenger and business transport, news gathering, law enforcement, search and rescue and emergency medical services. Four development aircraft (two flying, two for ground test). First flight 29 July 1996 (JQ 6003); second prototype (JQ 6004, later JA001M) flew late 1996. JCAB limited certification was awarded June 1997 and full VFR certification on 24 September 1999. First and second prototypes had flown approximately 800 hours (500 + 300) by April 1998. Initial production rate three per year; first production MH2000A (JQ 6005) was handed over to customer (Excel Air Service of Japan) on 1 October 1999. FAA certification anticipated by end of 2001. IFR certification target is 2001; measures being pursued for both internal and external noise reduction. Four of first five (three customer aircraft and one demonstrator) recalled August 2000 when flaws discovered in metal engine covers; sixth MH2000 on production line at that time.
CUSTOMERS: Launch customer is Excel Air Service of Tokyo (one). Two others sold by October 1999, to National Aerospace Laboratory (delivered March 1999 as JA21ME) and a Japanese private customer. Anticipated (mainly Japanese) market for commuter, civil defence, media and cargo use. By July 2000, only three production aircraft had been registered in Japan.
COSTS: Approximately US$3.2 million (1998).
DESIGN FEATURES: Single main rotor configuration (four blades with tapered tips); 10-blade (asymmetrically spaced) Fenestron-type tail rotor. Main gearbox and other drive mechanisms in overhead fairing are located aft of passenger cabin to minimise internal noise and vibration. Mid-mounted tailplane with angular endplate fins.
STRUCTURE: All-composites main and tail rotor blades.
LANDING GEAR: Conventional twin-skid type.

Mitsubishi (Sikorsky) UH-60J rescue helicopter in JASDF markings (*Peter R Foster*) 1999/0051107

POWER PLANT: Two 653 kW (876 shp) Mitsubishi MG5-110 turboshafts, with digital electronic control permitting one-touch changeovers between high-speed and low-noise modes. Crash-resistant fuel tanks aft of passenger cabin, maximum usable capacity 1,132 litres (299 US gallons; 249 Imp gallons).
ACCOMMODATION: Flight crew of two. Main cabin has forward-facing, impact absorbing seats for eight (standard) or six persons; five seats in club layout, with console, in VIP configuration; or two-person SAR team plus stretcher in search and rescue version. Crew door each side at front; large sliding door to main cabin on each side. Baggage compartment aft of fuel tanks, with external access.
AVIONICS: Avionics bay in rear fuselage aft of baggage compartment. Equipment to customer's choice; options to include GPS-based collision avoidance and AFCS.

DIMENSIONS, EXTERNAL:
Main rotor diameter	12.20 m (40 ft 0¼ in)
Length: overall, rotors turning	14.00 m (45 ft 11¼ in)
fuselage	12.20 m (40 ft 0¼ in)
Height over vertical tail	4.10 m (13 ft 5½ in)
Tail unit span	3.10 m (10 ft 2 in)
Skid track	2.70 m (8 ft 10¼ in)
Passenger sliding doors (each):	
Width	1.20 m (3 ft 11¼ in)

DIMENSIONS, INTERNAL:
Baggage compartment volume	2.24 m³ (79 cu ft)

AREAS:
Main rotor disc	116.90 m² (1,258.3 sq ft)

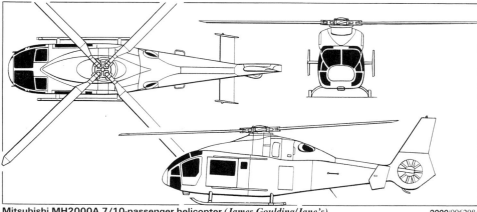

Mitsubishi MH2000A 7/10-passenger helicopter *(James Goulding/Jane's)*

WEIGHTS AND LOADINGS:
Weight empty	2,500 kg (5,512 lb)
Max T-O weight	4,500 kg (9,920 lb)
Max disc loading	38.5 kg/m² (7.88 lb/sq ft)

PERFORMANCE:
Max level speed	151 kt (280 km/h; 174 mph)
Cruising speed	135 kt (250 km/h; 155 mph)
Hovering ceiling IGE	2,700 m (8,860 ft)
Max range at cruising speed above, standard fuel, no reserves	421 n miles (780 km; 484 miles)
Max endurance, conditions as above	4 h

UPDATED

NAL

NATIONAL AEROSPACE LABORATORY
7-44-1 Jindaijihigashi-machi, Chofu City, Tokyo 182
Tel: (+81 422) 47 59 11
Fax: (+81 422) 40 31 21
e-mail: miyakawa@nal.go.jp
Web: http://www.nal.go.jp
DIRECTOR GENERAL: Susuma Toda
DIRECTOR, PLANNING OFFICE: Masaaki Murata
INFORMATION: Tomoko Miyakawa

NAL is a governmental establishment responsible for research and development in aeronautics and space technologies. Its major recent aeronautical activities have included developmental studies for the HOPE-X (H-II Orbiting Plane – Experimental), jointly with NASDA (which see); a future aero-spaceplane (2000-01 *Jane's*); and a supersonic commercial transport (see Supersonic Airliner Studies in the International section).

The NAL plans to develop a series of M2.0 unmanned experimental aircraft to demonstrate such advanced technologies as CFD-based aerodynamic design, lightweight composites structures, and high-performance air intakes. The first, to be tested by 2001, will be a supersonic glider designated **X-SST-R**, to be launched by solid-propellant rocket booster. This will be followed in 2002 by the **X-SST-J**, to be powered by twin turbojets. Further details, and an illustration of the X-SST-J, can be found under the Supersonic Airliner Studies heading in the International section.

NAL operates two research aircraft known as MuPAL (MultiPurpose Aviation Laboratory) for flight systems R&D. MuPAL-α is a modified Dornier 228-200; MuPAL-ε is a Mitsubishi MH2000. The former incorporates a fly-by-wire flight control system, direct lift control flap deflection system, second 'cockpit' in the main cabin, and a data acquisition system with telemetry downlink. Main features of the MuPAL-ε are a variable stability AFCS, cockpit graphic workstation, external view recording system (two video cameras), a hybrid DGPS/INS, main rotor sensors and a data acquisition system.

UPDATED

NAL AERO-SPACEPLANE STUDIES
There have been no recent progress reports for this programme, details of which last appeared in the 2000-01 *Jane's*.

UPDATED

NAL VTOL AIRLINER STUDIES
TYPE: New-concept airliner.
PROGRAMME: Study by NAL, Fuji and Ishikawajima-Harima for medium-range, 100-passenger, VTOL airliner; intended to fly 2010. Wind tunnel tests of preliminary models began in October 1996. No recent programme announcements.
DESIGN FEATURES: Three core engines to drive either six lift fans beside centre-fuselage in aerodynamic wingroot sections or two propulsion fans at tail; rear-mounted sweptback wings with large winglets; sweptback foreplanes; airscoops above rear fuselage.
FLYING CONTROLS: Pitch-axis control by reaction jets or small fans; rudders in winglets.
STRUCTURE: All-composites.

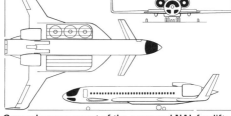

General arrangement of the proposed NAL fan-lift VTOL airliner *(James Goulding/Jane's)*

The following data are provisional and approximate:
DIMENSIONS, EXTERNAL:
Wing span over winglets	24.2 m (79 ft)
Foreplane span	11.0 m (36 ft)
Length overall	30.5 m (100 ft)
Fuselage diameter (cabin section)	3.3 m (11 ft)
Height of winglets	5.0 m (16 ft)
Height overall	7.0 m (23 ft)

AREAS:
Wings, gross	80.0 m² (861 sq ft)
Foreplanes (total)	17.6 m² (189 sq ft)
Winglets (total)	16.0 m² (172 sq ft)

WEIGHTS AND LOADINGS:
Weight empty	24,300 kg (53,600 lb)
Max fuel load	8,000 kg (17,600 lb)
Max payload	10,000 kg (22,000 lb)
Max T-O weight	42,300 kg (93,300 lb)
Max wing/foreplane loading	433 kg/m² (89 lb/sq ft)

PERFORMANCE:
Max cruising speed at 10,670 m (35,000 ft)	M0.80
Range at 10,670 m (35,000 ft):	
at M0.75	1,540 n miles (2,850 km; 1,770 miles)
at M0.80	1,380 n miles (2,560 km; 1,590 miles)

UPDATED

NAL SPACE SHUTTLE
TYPE: Space vehicle launch platform.
PROGRAMME: NAL, jointly with NASDA (which see), started research programme in 1999 with objectives of drastic reduction in space transportation costs and prevention of debris generation in space. Studies consist of three major technology research activities: aerodynamics and flight control, lightweight structures and materials, and reusable rocket engines. Experimental vehicle is planned to verify component technologies and demonstrate reusability.
COSTS: Expenditure approximately US$2 million for technology research in 1999.
DESIGN FEATURES: Fully reusable; horizontal take-off and landing.
POWER PLANT: Liquid oxygen/liquid hydrogen reusable rocket engines, based on LE-5 and LE-7.

UPDATED

NASDA

NATIONAL SPACE DEVELOPMENT AGENCY OF JAPAN
World Trade Center Building, 2-4-1 Hamamatsu-cho, Minato-ku, Tokyo 105-8060
Tel: (+81 3) 34 38 61 11
Fax: (+81 3) 54 02 65 13
Web: http://www.nasda.go.jp

The NASDA HOPE-X (H-II Orbiting Plane – Experimental) programme is for a winged, unmanned, reusable space vehicle intended to serve the International Space Station (ISS) from 2004. After flight demonstration tests, it was to be used to conduct practical missions such as logistics supply to, and recovery from, the ISS and onboard experiments carried in its own cargo bay. It was preceded by three other experimental programmes: OREX (Orbital Re-Entry Experiment), Hyflex (Hypersonic Flight Experiment) and Alflex (Automatic Landing Flight Experiment). Details of Hyflex and Alflex can be found in the 1997-98 *Jane's*. Technologies established by the HOPE-X development

Model of Fuji High-Speed Flight Demonstrator prototype for HOPE-X *(Kenneth Munson)*

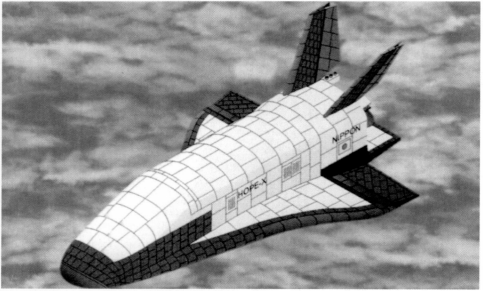

New HOPE-X configuration, as revealed in late 1999

programme were expected to benefit a possible future reusable space transportation system (STS), but in 1999, necessary redesign, coupled with financial restrictions, caused a slippage in the programme.

It was revealed in May 2000 that NASDA had received approval from the government of Kiribati to create a spaceport on Kiritimati (Christmas Island) to support HOPE-X landings. However, two H-II launcher failures in 1999 have cast doubt upon the selection of the uprated H-IIA to launch HOPE-X, leading to a stoppage of work on the latter programme in mid-2000 pending resolution of the launcher situation and consequently a probable further two-year delay in completion of the subscale demonstrators.

UPDATED

NASDA HOPE-X

TYPE: Aerospace craft.
PROGRAMME: HOPE-X (H-II Orbiting Plane – Experimental) is final stage in research programme to develop re-entry technology. Currently in preliminary design phase, with first launch scheduled for early 2000s. Designs proposed by Fuji, Kawasaki and Mitsubishi were combined in single design, as described and illustrated in 1999-2000 *Jane's*. However, this subsequently considered too heavy, and redesign to a lighter configuration began in April 1999, under new ¥9 billion (approximately US$756 million) contract. One-third of this was to fund Fuji to manufacture and fly two quarter-scale High-Speed Flight Demonstrator (HSFD) prototypes (see Current Versions), but first flight of the full-size HOPE-X is now scheduled for 2006, instead of originally planned 2001. New design has conventional twin fins and rudders atop fuselage instead of former integral winglets.

CURRENT VERSIONS: **HSFD 1**: To explore automatic take-off, approach and landing characteristics. Powered by single 44.9 kN (10,100 lb st) Teledyne CAE 382-10 turbojet. Wing span 2.50 m (8 ft 2½ in); length 4.00 m (13 ft 1½ in); height 1.20 m (3 ft 11¼ in); gross weight 735 kg (1,620 lb); ceiling 12,000 m (39,360 ft); maximum speed M0.7.

HSFD 2: Dimensionally as for Demonstrator 1, but gross weight of 500 kg (1,102 lb). To explore transonic flight envelope. Will be air-launched, accelerating to M0.8 to M1.2 in a glide, and be recovered by parafoil. Launch expected to be either from a balloon at about 35,000 m (114,820 ft), or at lower altitude from a piloted aircraft, then using a booster rocket to attain the higher altitude. In February 2000, NASDA was in preliminary agreement with CNES (France) over proposal for latter to assist in programme.

The following approximate and provisional details refer to the redesigned full-size HOPE-X:

DIMENSIONS, EXTERNAL:
Wing span   10 m (32.8 ft)
Fuselage length 16 m (52.5 ft)
WEIGHTS AND LOADINGS:
Weight empty   10,500 kg (23,150 lb)
Payload   3,000-5,000 kg (6,615-11,025 lb)
Max T-O weight 20,000 kg (44,100 lb)

UPDATED

SHINMAYWA

SHINMAYWA KOGYO KABUSHIKI KAISHA (Shinmaywa Industries Ltd)

5-25 Kosone-cho 1-chome, Nishinomiya 663-8122
Tel: (+81 798) 47 03 31
Fax: (+81 798) 41 07 55
CHAIRMAN: Shinji Tamagawa
PRESIDENT: Shiko Saikawa

Aircraft Division
1-1 Ohgi 1-chome, Higashinada-ku, Kobe 652
Tel: (+81 78) 412 64 80
Fax: (+81 78) 435 20 22
EXECUTIVE MANAGING DIRECTOR AND GENERAL MANAGER: Junpei Matsuo
BOARD DIRECTOR AND DEPUTY GENERAL MANAGER: Takaoki Kiho
WORKS: Konan and Tokushima

SALES OFFICE:
Nippon Building, 6-2 Otemachi 2-chome, Chiyoda-ku, Tokyo 100-0004
Tel: (+81 3) 32 45 66 11
Fax: (+81 3) 32 45 66 16
GENERAL MANAGER: Yushi Tanaka

Former Kawanishi Aircraft Company; became Shin Meiwa in 1949 and renamed ShinMaywa Industries in June 1992; major overhaul centre for Japanese and US military and commercial aircraft. Principal activities are production of US-1A for JMSDF, and overhaul work on amphibians. Manufactures external drop tanks for Kawasaki T-4s; tailplanes for Mitsubishi-built SH-60Js; internal cargo handling system for Kawasaki-built CH-47JA; fixed trailing-edges for Boeing 757/767, under subcontract to Vought; and other components for Boeing 767, under subcontract to Mitsubishi; design and manufacture of wing/body fairings for Boeing 777. In 1991, modified five Learjet 36As into U-36A naval fleet training support aircraft for JMSDF; in-service centre for U-36A and Fairchild aircraft.

Continues to study and look for partners to develop Amphibious Air Transport System, which is 30/50-passenger airliner powered by two wing-mounted turbofans with upper surface blowing; range would vary from 500 n miles (926 km; 575 miles) with full payload to 1,200 n miles (2,222 km; 1,381 miles) with full fuel; take-off distance 1,000 m (3,280 ft) on water and 800 m (2,624 ft) on soft ground; cruising speed between 300 and 360 kt (556 and 667 km/h; 345 and 414 mph).

UPDATED

SHINMAYWA US-1A

TYPE: Four-turboprop amphibian.
PROGRAMME: First flown 16 October 1974; first delivery (as US-1) 5 March 1975 (see 1985-86 *Jane's*); all now have T64-IHI-10J engines as US-1As.
CURRENT VERSIONS: **US-1A**: SAR amphibian, developed from PS-1 ASW flying-boat; manufacturer's designation **SS-2A**. *Data apply to US-1A*.

US-1AKai: Proposed upgrade initiated in 1996: FADEC-equipped, 3,356 kW (4,500 shp) Rolls-Royce AE 2100J turboprops, six-blade Dowty R414 propellers, Kawasaki FBW controls, pressurised upper hull, composites wing structure and new avionics similar to those proposed for US-X, including Thomson-CSF/DASA Ocean Master radar. Launch funding sought in FY99 budget, but not obtained; request renewed in FY00. Parts manufacture began in April 2000; first flight targeted for 2003 and service entry for 2009.

Firefighting amphibian: PS-1 modified in 1976 to firefighting configuration, 7,348 kg (16,200 lb) capacity water tank in centre-fuselage aft of step. Since then, one US-1A modified experimentally, with more than 13,608 kg (30,000 lb) tank capacity; tank system developed by Conair of Canada. Development of this version continued in 1995-96 (latest news received).

CUSTOMERS: Total of 18 US-1/1As ordered, of which 17 delivered by 1999; none requested for FY00 but US$283 million requested instead to resume development of US-1AKai. Orders since 1988 have been for attrition replacements and, more recently, due to phase-out of older aircraft. No. 71 SAR Squadron of the JMSDF maintains fleet structure of seven aircraft at Iwakuni and Atsugi bases.

US-1/1A PROCUREMENT (JMSDF)

FY	Qty	Cum Total
72	1	1
73	2	3
77	1	4
78	2	6
79	1	7
80	1	8
82	1	9
83	1	10
84	1	11
88	1	12
92	1	13
93	1	14
95	1	15
96	1	16
97	1	17
99	1	18

COSTS: Unit cost US$67.72 million (1999).
DESIGN FEATURES: Large turboprop-powered amphibian, suitable for maritime patrol and sea rescue missions. Boundary layer control system and extensive flaps for propeller slipstream deflection for very low landing and take-off speeds; low-speed control and stability enhanced by blowing rudder, flaps and elevators, and by use of automatic flight control system (see Flying Controls). Fuselage high length/beam ratio; V-shaped single-step

ShinMaywa US-1A ocean-going search and rescue amphibian *(Dennis Punnett/Jane's)*

ShinMaywa US-1A four-turboprop SAR amphibian *(Katsumi Hinata)* 2001/0093617

planing bottom, with curved spray suppression strakes along sides of nose and spray suppressor slots in lower fuselage sides aft of inboard propeller line; double-deck interior. Large dorsal fin.

FLYING CONTROLS: Automatic flight control system controlling elevators, rudder and outboard flaps. Hydraulically powered ailerons, elevators (with tabs) and rudder, all with feel trim. High-lift devices include outboard leading-edge slats over 17 per cent of wing span and large outer and inner blown trailing-edge flaps deflecting 60° and 80° respectively; outboard flaps can be linked with ailerons; inboard flaps, elevators and rudder blown by BLC system. Two spoilers in front of outer flaps on each wing. Inverted slats on tailplane leading-edge.

STRUCTURE: All-metal; two-spar wing box.

LANDING GEAR: Flying-boat hull, plus hydraulically retractable Sumitomo tricycle landing gear with twin wheels on all units. Steerable nose unit. Oleo-pneumatic shock-absorbers. Main units, which retract rearward into fairings on hull sides, have 40×14-22 tyres, pressure 7.79 bars (113 lb/sq in). Nosewheel tyres size 25×6.75-18, pressure 20.69 bars (300 lb/sq in). Three-rotor hydraulic disc brakes. No anti-skid units. Minimum ground turning radius 18.80 m (61 ft 81/4 in) towed, 21.20 m (69 ft 63/4 in) self-powered.

POWER PLANT: Four 2,535 kW (3,400 ehp) Ishikawajima-built General Electric T64-IHI-10J turboprops, each driving a Sumitomo-built Hamilton Sundstrand 63E60-27 three-blade constant-speed reversible-pitch propeller. Fuel in five wing tanks, with total usable capacity of 11,640 litres (3,075 US gallons; 2,560.5 Imp gallons) and two fuselage tanks (10,849 litres; 2,866 US gallons; 2,386.5 Imp gallons); total usable capacity 22,489 litres (5,941 US gallons; 4,947 Imp gallons). Pressure refuelling point on port side, near bow hatch. Oil capacity 152 litres (40.2 US gallons; 33.4 Imp gallons). Aircraft can be refuelled on open sea, either from surface vessel or from another US-1A with detachable at-sea refuelling equipment.

ACCOMMODATION: Crew of three on flight deck (pilot, co-pilot and flight engineer), plus navigator/radio operator's seat in main cabin. Latter can accommodate up to 20 seated survivors or 12 stretchers, one auxiliary seat and two observers' seats. Sliding rescue door on port side of fuselage, aft of wing.

SYSTEMS: Cabin air conditioning system. Two independent hydraulic systems, each 207 bars (3,000 lb/sq in). No. 1 system actuates ailerons, outboard flaps, spoilers, elevators, rudder and control surface feel; No. 2 system actuates ailerons, inboard and outboard flaps, wing leading-edge slats, elevators, rudder, landing gear extension/retraction and lock/unlock, nosewheel steering, mainwheel brakes and windscreen wipers. Emergency system, also of 207 bars (3,000 lb/sq in), driven by 24 V DC motor, for actuation of inboard flaps, landing gear extension/retraction and lock/unlock, and mainwheel brakes.

Honeywell 85-131J APU provides power for starting main engines and shaft power for 40 kVA emergency AC generator. BLC system includes a C-2 compressor, driven by a 1,119 kW (1,500 shp) Ishikawajima-built General Electric T58-IHI-10-M2 gas turbine, housed in upper centre portion of fuselage, which delivers compressed air at 14 kg (30.9 lb)/s and pressure of 1.86 bars (27 lb/sq in) for ducting to inner and outer flaps, rudder and elevators (US-1AKai has variant of LHTEC CTS800).

Electrical system includes 115/200 V three-phase 400 Hz constant-frequency AC and three transformer-rectifiers to provide 28 V DC. Two 40 kVA AC generators, driven by Nos. 2 and 3 main engines. Emergency 40 kVA AC generator driven by APU. DC emergency power from two 24 V 34 Ah Ni/Cd batteries.

De-icing of wing and tailplane leading-edges. Oxygen system for all crew and stretcher stations. Fire detection and extinguishing systems standard.

AVIONICS: *Comms:* HRC-106, HRC-107 HF and HRC-113 radios; N-CU-58/HRC antenna coupler; HIC-3 interphone; AN/APX-68-NB IFF transponder; RRC-22 emergency transmitter; HGC-102 teletypewriter.

Radar: Raytheon AN/APS-115-2 search radar.

Flight: HRN-101 ADF; AN/ARA-50 UHF/DF; HRN-105B Tacan; HRN-115-1 GPS nav system; HRN-107B-1 VOR/ILS receiver; Honeywell AN/APN-171 (N2) radio altimeter; HPN-101B wave height meter; AN/APN-187C-N Doppler nav; AN/AYK-2 navigation computer; A/A24G-9 TAS transmitter; N-PT-3 dead reckoning plotting board; HRA-5 nav display; and N-ID-66/HRN BDHI.

EQUIPMENT: Marker launcher, 10 marine markers, six green markers, two droppable message cylinders, 10 float lights, pyrotechnic pistol, parachute flares, two flare storage boxes, binoculars, two rescue equipment kits, two droppable liferaft containers, rescue equipment launcher, lifeline pistol, lifeline, three lifebuoys, loudspeaker, hoist unit, rescue platform, lifeboat with outboard motor, camera, and 12 stretchers. Sea anchor in nose compartment. Stretchers can be replaced by troop seats.

DIMENSIONS, EXTERNAL:
Wing span	33.15 m (108 ft 9 in)
Wing chord: at root	5.00 m (16 ft 4¾ in)
at tip	2.39 m (7 ft 10 in)
Wing aspect ratio	8.1
Length overall	33.46 m (109 ft 9¼ in)
Height overall	9.95 m (32 ft 7¾ in)
Tailplane span	12.36 m (40 ft 8½ in)
Wheel track	3.56 m (11 ft 8¼ in)
Wheelbase	8.33 m (27 ft 4 in)
Propeller diameter	4.42 m (14 ft 6 in)

Rescue hatch (port side, rear fuselage):
Height	1.58 m (5 ft 2¼ in)
Width	1.46 m (4 ft 9½ in)

AREAS:
Wings, gross	135.82 m² (1,462.0 sq ft)
Ailerons (total)	6.40 m² (68.90 sq ft)
Inner flaps (total)	9.40 m² (101.18 sq ft)
Outer flaps (total)	14.20 m² (152.85 sq ft)
Leading-edge slats (total)	2.64 m² (28.42 sq ft)
Spoilers (total)	2.10 m² (22.60 sq ft)
Fin	17.56 m² (189.0 sq ft)
Dorsal fin	6.32 m² (68.03 sq ft)
Rudder	7.01 m² (75.50 sq ft)
Tailplane	23.04 m² (248.00 sq ft)
Elevators, incl tabs	8.78 m² (94.50 sq ft)

WEIGHTS AND LOADINGS (search and rescue):
Manufacturer's weight empty	23,300 kg (51,367 lb)
Weight empty, equipped	25,500 kg (56,218 lb)
Usable fuel: JP-4	17,518 kg (38,620 lb)
JP-5	18,397 kg (40,560 lb)
Max oversea operating weight	36,000 kg (79,365 lb)
Max T-O weight: from water	43,000 kg (94,800 lb)
from land	45,000 kg (99,200 lb)
Max wing loading	331.3 kg/m² (67.85 lb/sq ft)
Max power loading	4.24 kg/kW (6.97 lb/ehp)

PERFORMANCE (search and rescue, land T-O. A: at 36,000 kg; 79,365 lb weight, B: at 43,000 kg; 94,800 lb, C: at max T-O weight):
Max level speed: C	276 kt (511 km/h; 318 mph)
Max level speed at 3,050 m (10,000 ft):	
A	282 kt (522 km/h; 325 mph)
Cruising speed at 3,050 m (10,000 ft):	
C	230 kt (426 km/h; 265 mph)
Max rate of climb at S/L: A	713 m (2,340 ft)/min
C	488 m (1,600 ft)/min
Service ceiling: A	8,655 m (28,400 ft)
C	7,195 m (23,600 ft)
T-O to 15 m (50 ft) on land, 30° flap, BLC on (ISA):	
C	670 m (2,200 ft)
T-O distance on water, 40° flap, BLC on (ISA):	
B	735 m (2,410 ft)
Landing from 15 m (50 ft) on land, AUW of 36,000 kg (79,365 lb), 40° flap, BLC on, with reverse pitch (ISA):	
A	810 m (2,655 ft)
Landing distance on water, AUW of 36,000 kg (79,365 lb), 60° flap, BLC on (ISA):	
A	561 m (1,840 ft)
Runway LCN requirement: B	42
Max range at 230 kt (426 km/h; 265 mph) at 3,050 m (10,000 ft)	2,060 n miles (3,815 km; 2,370 miles)

UPDATED

KOREA, SOUTH

DAEWOO

DAEWOO HEAVY INDUSTRIES CO LTD (DHI)

6 Manseog-dong, Dong-gu, Inchon 401702
Tel: (+82 32) 760 11 14
PRESIDENT: H S Choo

Daewoo entered aerospace field in 1984 with a contract to supply F-16 fighter fuselages to General Dynamics of the USA. Since then, it has rapidly expanded its business activities to include commercial and military aircraft and multipurpose telecommunications satellites. Its Aerospace Division became a part of KAI (which see) on 1 October 1999.

Present and recent past contracts include outer wings for the Lockheed Martin P-3C; upper deck frames, walkway panels and wing rib assemblies for the Boeing 747-400; complete wings to BAE Systems for Hawk 100 advanced trainers for the Republic of Korea Air Force; and A319/A320 thrust reversers and A330 engine nacelles for Airbus Industrie. In addition to these commercial programmes, Daewoo has participated in the Korean Fighter Programme, building F-16s for the RoKAF under licence from Lockheed Martin.

Based on the technology and experience gained from these overseas contracts, Daewoo was designated by the South Korean government in 1991 as prime contractor for two major domestic programmes: the KT-1 basic trainer for the RoKAF and the Korean Scout Helicopter (KSH) for the RoK Army.

Daewoo has marketing and co-production agreement with PZL Swidnik for **W-3 Sokoł** helicopter; see Polish section for details.

UPDATED

HYUNDAI

HYUNDAI SPACE AND AIRCRAFT COMPANY (HYSA)

140-2 Gye-dong, Chongro-ku, Hyundai Building, Seoul 110793
Tel: (+82 2) 741 23 01
Fax: (+82 2) 746 10 93

Despite expressing interest in building parts for the DASA Mako (formerly AT-2000) jet trainer/light attack aircraft project described in the German section, Hyundai has so far declined to become a risk-sharing participant.

UPDATED

KAI

KOREA AEROSPACE INDUSTRIES LTD

HEAD OFFICE: 8th Floor, Haedong Insurance Building, 463 3Ka Chungjeong-ro, Seodaemun-ku, Seoul 120709
Tel: (+82 2) 20 01 30 60
Fax: (+82 2) 20 01 30 11
PRESIDENT AND CEO: Lim In Taik
GENERAL MANAGER: Alex Jon

CHANGWON PLANT: 24 Seongju-dong, Changwon-shi, Kyungnam 641120
Tel: (+82 551) 280 65 01
Fax: (+82 551) 285 23 83

SACHON PLANT: 31 Yuchun-ri, Sanam-Myun, Sachon-shi, Kyungnam 356850
Tel: (+82 593) 851 10 00
Fax: (+82 593) 851 10 05

SEOSAN PLANT: 8-2 Kalhyun-ri, Seongyeon-Myun, Seosan-shi, Chungnam 664940
Tel: (+82 455) 661 71 14
Fax: (+82 455) 664 46 12

As part of reform necessary to offset national economic collapse, the new South Korean government, elected in mid-1998, ordered a restructuring of the country's aerospace industry. Initially, this was seen to involve the aerospace activities of Daewoo Heavy Industries (DHI), Hyundai Space and Aircraft Company (HYSA) and Samsung Aerospace (SSA), with Korean Air (which see) possibly to follow. Original consolidation plans of the first three were, however, rejected in December 1998 when their parent companies declined to absorb their aerospace divisions' debts and the government deemed the estimated US$3 billion cost of a full bail-out to be excessive.

A new business plan was prepared during the first half of 1999, aimed at completing the merger by 30 June, but was deferred because Corporate Restructuring Committee (CRC), representing creditor banks, demanded removal of non-aerospace-related assets from valuations. Revised plan again rejected by CRC in third quarter of 1999 on grounds that it was weighted too heavily (more than 75 per cent) in favour of military programmes. Following this, the three companies signified intention on 28 July to proceed with merger first and devise acceptable business plan later. KAI accordingly created officially on 1 October 1999.

Broad proposal is for the three Korean companies to each have 20 per cent holding in KAI, with up to 30 per cent available for foreign investors. Bidders for latter were to have been Lockheed Martin with Aerospatiale Matra (joint bid), Boeing and BAE Systems (joint bid), GEC and DaimlerChrysler Aerospace (DASA). However, due to involvement of Aerospatiale Matra and DASA with formation of EADS (whose Mako 2000 is a potential competitor to KAI T-50), Lockheed Martin bid not made; and GEC merger with BAe resulted in BAE/Boeing bid remaining unopposed. By November 2000 (four months after bid deadline), KAI and BAE/Boeing still unable to resolve some details of proposed agreement, and talks suspended by KAI.

The new KAI is said to have annual sales of US$700 million and a workforce of 3,200. It will assume up to US$600 million of debt from the three Korean constituent companies.

UPDATED

KAI KT-1 WOONG-BEE

English name: Great Flight

TYPE: Turboprop basic trainer.
PROGRAMME: Started February 1988; built under KTX (Korean Trainer Experimental) programme to design by Daewoo and ADD (Agency for Defence Development); construction of first prototype began June 1991; nine prototypes (01-05 flying and 001-004 for static and fatigue test), of which 01, with 410 kW (550 shp) PT6A-25A engine, rolled out November 1991 and made first flight 12 December that year; 02, identical to 01, made first flight 5 February 1993; 03 (with PT6A-62A) made first flight on 10 August 1995; named Woong-Bee in November 1995; 04, which first flew on 10 May 1996, further modified with nose shortened and horizontal tail surfaces remounted lower and farther aft; 03 also modified to this standard. Fifth (preproduction) prototype 05 flew for the first time on 16 March 1998; development and operational testing completed 18 September 1998 after 1,474 flying hours in 1,184 sorties. Series production began 1999; first aircraft handed over to RoKAF at Sachon AB on 7 November 2000.

CURRENT VERSIONS: **KT-1**: Standard trainer, *as described.*
XKO: Forward air control version. Development started in 2000 by modifying aircraft 05 to FAC configuration with weapon management system, HUD, MFD and four underwing hardpoints; first flight in this configuration planned for 2001, with service entry in about 2003.

CUSTOMERS: Approximately 105 required by Republic of Korea Air Force (RoKAF). Contract 9 August 1999 (Swon 600 billion) for 85 KT-1 trainers, to replace Cessna T-37s and T-41s. Follow-on option for 20 XKOs for forward air control duties. Exports planned; Indonesian Air Force order anticipated.

DESIGN FEATURES: Main design objectives 310 kt (574 km/h; 356 mph) maximum speed, 11,280 m (37,000 ft) service ceiling, 950 n mile (1,759 km; 1,093 mile) maximum range at 198 kt (367 km/h; 228 mph) long-range cruising speed.
Unswept low wing with NACA 63-218 (root) and 63-212 (tip) aerofoil sections; tandem cockpits; conventional unswept vertical and horizontal tail surfaces (both with NACA 0012 and NACA 0009 root and tip sections); retractable tricycle landing gear. Curved dorsal fin replaced by original straight-edged one.

FLYING CONTROLS: Primary control surfaces actuated mechanically, with electrically operated trim tab in starboard elevator and port aileron; automatic rudder trim; rudder and one-piece elevator are horn-balanced. Airbrake under centre-fuselage and split flaps on wing trailing-edge are hydraulically operated.

LANDING GEAR: Hydraulically retractable tricycle type, with single wheel and oleo-pneumatic shock-absorber on each unit. Mainwheels retract inward; nosewheel is steerable ±18° and retracts rearward. Parker Hannifin mainwheels with 18×5.5 tyres (10 ply), pressure 9.65 bars (140 lb/sq in). Dunlop nosewheel with 5.00-5 tyre (14 ply), pressure 6.90 bars (100 lb/sq in). Parker Hannifin hydraulic mainwheel brakes.

POWER PLANT: One 708 kW (950 shp) (flat rated) Pratt & Whitney Canada PT6A-62 turboprop, driving a Hartzell HC-E4N-2/E9512CB-1 four-blade constant-speed fully feathering propeller. Fuel in one 275.5 litre (72.8 US gallon; 60.6 Imp gallon) integral tank in each wing, giving total internal fuel capacity of 551 litres (145.6 US gallons; 121.2 Imp gallons). Gravity fuelling point in each wing upper surface. Oil capacity 5.7 litres (1.5 US gallons; 1.25 Imp gallons). Fuel system permits up to 30 seconds of inverted flight.

ACCOMMODATION: Instructor and pupil in tandem cockpits on Martin-Baker Mk 16LF zero/zero ejection seats; rear seat elevated. One-piece canopy with MDC opens sideways to starboard.

SYSTEMS: Honeywell two-wheel bootstrap air conditioning system. Self-pressurised main and emergency hydraulic systems, operating pressure 207 bars (3,000 lb/sq in), flow rate 17.8 litres (4.7 US gallons; 3.9 Imp gallons)/min at 7,650 rpm. Pneumatic back-up system, pressure 3.45 to 4.14 bars (50 to 60 lb/sq in), for landing gear, wheel bay doors, flaps and airbrake. Gaseous oxygen system, capacity 1,367 litres (48.2 cu ft). Electrical power (28 V DC) available from engine-driven starter/generator and built-in battery, or from external power source. Fixed-geometry inertial separator for engine intake anti-icing.

AVIONICS: *Comms:* UHF/VHF radio, IFF, interphone and communication control system.
Flight: VOR/ILS/marker beacon receiver (optional); altitude indicator (including standby) and Tacan.
Instrumentation: EADI, EHSI, AHRS, FCMS, AoA sensor/index and standby magnetic compass.

ARMAMENT (XKO): Two stores stations under each wing for, on KT-1, two LAU-131 seven-round rocket launchers and two machine gun pods; two LAU-131s and two drop tanks; or two gun pods and two drop tanks.

DIMENSIONS, EXTERNAL:
Wing span	10.60 m (34 ft 9¼ in)
Wing aspect ratio	7.0
Length overall	10.26 m (33 ft 8 in)
Height overall	3.67 m (12 ft 0½ in)
Tailplane span	4.16 m (13 ft 7¾ in)
Wheel track	3.54 m (11 ft 7¼ in)
Wheelbase	2.56 m (8 ft 4¾ in)
Propeller diameter	2.30 m (7 ft 11 in)
Propeller ground clearance	0.37 m (1 ft 3 in)

AREAS:
Wings, gross	16.01 m² (172.3 sq ft)
Ailerons (total, incl tabs)	1.11 m² (11.95 sq ft)
Trailing-edge flaps (total)	2.22 m² (23.90 sq ft)
Fin, incl dorsal fin	1.00 m² (10.76 sq ft)
Rudder, incl tab	0.88 m² (9.47 sq ft)
Tailplane	2.32 m² (24.97 sq ft)
Elevators (total, incl tab)	1.19 m² (12.81 sq ft)

WEIGHTS AND LOADINGS:
Weight empty	1,910 kg (4,210 lb)

Fifth (first preproduction) KAI KT-1 two-seat basic trainer *(Peter R Foster)*

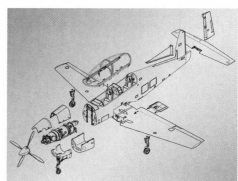

KT-1 major airframe components

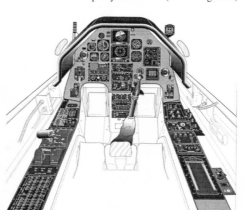

Diagram of KT-1 front cockpit

KAI—AIRCRAFT: KOREA, SOUTH

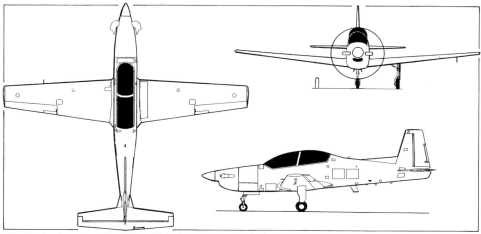

KAI KT-1 basic trainer (P&WC PT6A-62A turboprop engine) *(Paul Jackson/Jane's)*   2000/0062986

Proposed cockpit for the T-50   1999/0051118

Max fuel weight	408 kg (900 lb)
Max T-O and landing weight	2,540 kg (5,600 lb)
Max wing loading	158.7 kg/m² (32.50 lb/sq ft)
Max power loading	3.59 kg/kW (5.89 lb/shp)

PERFORMANCE:
Design diving speed (V_D)	350 kt (648 km/h; 402 mph)
Max level speed at 3,050 m (10,000 ft)	265 kt (490 km/h; 305 mph)
Max operating speed	310 kt (574 km/h; 357 mph)
Stalling speed, flaps down	70 kt (130 km/h; 81 mph)
Max rate of climb at S/L	969 m (3,180 ft)/min
Service ceiling	11,280 m (37,000 ft)
T-O run	250 m (820 ft)
T-O to 15 m (50 ft)	494 m (1,620 ft)
Landing run	397 m (1,300 ft)
Landing from 15 m (50 ft)	727 m (2,385 ft)
Range with max fuel 950 n miles (1,759 km; 1,093 miles)	
g limits: normal operation	+6/–3
ultimate	+7/–3.5

UPDATED

KAI T-50 and A-50 GOLDEN EAGLE

TYPE: Advanced jet trainer/light attack jet.
PROGRAMME: Begun by Samsung Aerospace (SSA) in 1992 under designation KTX-2 (Korean Trainer, Experimental); initial design assistance to Samsung by Lockheed Martin Tactical Aircraft Systems as offset in F-16 Korean Fighter Programme; early design (see 1994-95 *Jane's*) featured shoulder-mounted wings and twin tail unit; revised later to present configuration; basic configuration established mid-1995; full-scale development originally planned to begin in 1997, subject to finding risk-sharing partner; government go-ahead given on 3 July 1997; Samsung/Lockheed Martin agreement September 1997 to continue joint development until 2005; Lockheed Martin Aeronautics at Fort Worth responsible for wings, flight control system and avionics; development phase funded 70 per cent by South Korean government, 17 per cent by Samsung/KAI and 13 per cent by Lockheed Martin.

FSD contract, signed 24 October 1997, calls for two static/fatigue test and four flying prototypes (two T-50A and two T-50B. First prototype entered final assembly January 2001; roll-out targeted for September 2001 and first flight in June 2002); production aircraft to begin manufacture in August 2003. Work split 55 per cent in USA, 44 per cent in South Korea and 1 per cent elsewhere. Preliminary design review (PDR) completed 12 to 16 July 1999; wind tunnel testing completed (4,800 hours) and aerodynamic design frozen November 1999. Critical design review (CDR) passed in August 2000. KTX-2 redesignated T-50 and A-50 in early 2000. T-50 International Company (TFIC) established September 2000 (following July MoU) by KAI and LMAS to market aircraft outside South Korea. Deliveries to RoKAF planned to begin in late 2004 and be completed in 2009.

CURRENT VERSIONS (planned): **T-50A**: Advanced trainer.
T-50B: Lead-in fighter trainer.
A-50: Proposed light combat version.
CUSTOMERS: Initial RoKAF requirement for 94 T-50s (50 T-50s and 44 A-50s), with options for up to 100 more, including further A-50s. Aimed also at F-5 replacement market. Exports (from 2006) estimated potentially at 600 to 800.
COSTS: Development programme cost estimated at US$2,000 million (1995); but re-assessed in 1996 as US$1,500 million, and only US$1,200 million by early 1997. Initial October 1997 FSD contract valued at approximately US$1,270 million. Development phase re-estimated at US$1.8 billion to US$2.1 billion in mid-2000; unit cost US$18 million to US$20 million for trainer, US$20 million to US$22 million for attack version (2000).
DESIGN FEATURES: Mid-mounted, variable camber wings, swept back on leading-edges only; leading-edge root extensions (LERX); all-moving tailplane; sweptback fin leading-edge. Single turbofan engine, with twin side-mounted intakes. KIAT developing fuselage and tail unit.

Avionics to include HUD and colour MFDs. Designed for service life of more than 8,000 hours.
FLYING CONTROLS: Digital fly-by-wire control of elevons, tailplane and rudder. Moog actuators.
STRUCTURE: Lockheed Martin, wings; KAI, fuselage, tail unit and final assembly.
LANDING GEAR: Messier-Dowty KIA retractable tricycle type, with single wheel and oleo-pneumatic shock-absorber on each unit. Mainwheels retract into engine intake trunks, nosewheel forward.
POWER PLANT: One General Electric F404-GE-402 turbofan (78.7 kN; 17,700 lb st with afterburning), equipped with FADEC.
ACCOMMODATION: Crew of two in tandem; stepped cockpits; Martin-Baker ejection seats.
SYSTEMS: Onboard oxygen generating system. Hamilton Sundstrand power generation system. Argo-Tech fuel system.
AVIONICS: *Comms*: UHF/VHF radio; IFF.
Radar: Lockheed Martin AN/APG-67 in T-50B and A-50.
Flight: Digital fly-by-wire flight controls; nav/attack system for fighter lead-in training; ring laser gyro INS; radar altimeter.
Instrumentation: BAE Systems HUD; two 127 mm (5 in) colour MFDs; Honeywell instrumentation displays (eight 76 mm; 3 in) displays, including HSI, attitude indicator, electronic altimeter and Mach speed indicator).
Self-defence: A-50 provision for EW pods and RWR.
ARMAMENT (T-50B and A-50): Internal 20 mm M61 Vulcan cannon with 208 rounds (port LERX). Seven external stations (one on centreline, two under each wing and AAM rail at each wingtip) for AAMs, ASMs, gun pods, rocket pods or bombs. Expected to include AIM-9 Sidewinder and AGM-65 Maverick missiles and Mk 82/83/84 series bombs.

DIMENSIONS, EXTERNAL:
Wing span: over missiles	9.17 m (30 ft 1 in)
excl missiles	9.11 m (29 ft 10½ in)
Length overall	12.98 m (42 ft 7 in)
Height overall	4.78 m (15 ft 8¼ in)

Artist's impression of the T-50/A-50 trainer/light attack aircraft being developed by KAI and Lockheed Martin   2001/0099546

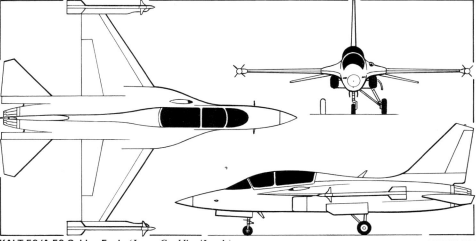

KAI T-50/A-50 Golden Eagle *(James Goulding/Jane's)*   2000/0084394

306 KOREA, SOUTH: AIRCRAFT—KAI to KAL

WEIGHTS AND LOADINGS:
  Weight empty 6,263 kg (13,808 lb)
  Max T-O weight: clean 8,890 kg (19,600 lb)
    with external stores 11,974 kg (26,400 lb)
PERFORMANCE (design):
  Max level speed M1.4
  Max rate of climb at S/L 10,058 m (33,000 ft)/min
  Service ceiling 14,500 m (47,570 ft)
  g limits +8/−3
UPDATED

KAI F-X
TYPE: Multirole fighter (requirement).
PROGRAMME: RoKAF currently considering procurement of up to 40 twin-engined advanced fighters for air-to-air, air-to-ground, air-to-sea and long-range penetration roles; service entry in about 2007; KAI to be programme prime contractor, with expanded local production of selected aircraft. Final RFPs for both direct purchase and licensed production issued in 2000 to Boeing (F-15), Dassault (Rafale), Eurofighter (Typhoon) and Sukhoi (Su-35). Selection process scheduled to result in a decision by late 2001. Flight evaluation of F-15E took place in October 2000.
UPDATED

KAI (LOCKHEED MARTIN) F-16C/D FIGHTING FALCON
TYPE: Multirole fighter.
PROGRAMME: Korean Fighter Programme (KFP) to co-produce 108 of 120 F-16s (80 Block 52D F-16Cs and 40 F-16Ds) announced in Seoul 28 March 1991; Samsung (now KAI) is main Korean contractor, with Daewoo (also now KAI) and Hanjin as main subcontractors; programme is to acquire 12 aircraft and their engines off the shelf, assemble 36 from kits and manufacture remaining 72 locally. First Lockheed-built F-16C handed over 2 December 1994; first kit-assembled F-16C completed late 1995; 49th (first Samsung-built) completed by late 1996; 120th aircraft delivered in April 2000. Decision to produce a further 20 (Block 52 standard) announced by Prime Minister Kim Jong-il on 13 May 1999; approved 2000 as KFP 2; deliveries to start July 2003.
COSTS: US$5.2 billion for 120-aircraft programme; US$663 million (estimated) for additional 20 (1999).
POWER PLANT: One Pratt & Whitney F100-PW-229 turbofan.
AVIONICS: *Mission:* Lockheed Martin LANTIRN pods.
ARMAMENT: Includes AGM-84 Harpoon, AGM-88 HARM and AIM-120 AMRAAM missiles.
UPDATED

KAI (EUROCOPTER) BO 105 CBS-5
TYPE: Armed observation helicopter.
PROGRAMME: Korean Light Helicopter (KLH) programme launched early 1998; BO 105 CBS-5 selected for requirement; first of two aircraft built by Eurocopter and flew (D-HMBT) fourth quarter of 1999; next 10 are being assembled by KAI (Daewoo) from CKD kits, starting late 1999.
CUSTOMERS: Republic of Korea Army (12 ordered so far, but may be increased).
AVIONICS: To include Boeing IR/optical sights and defensive aids systems; for day and night operation, even in extreme environmental conditions.
UPDATED

KAI KMH
TYPE: Multirole light helicopter (requirement).
PROGRAMME: Initiated as part of Republic of Korea Army (RoKA) Force Improvement Plan, with main objective of replacing elderly 500MDs with advanced type usable in combat, light attack, command and control, liaison and passenger-carrying roles. Acquisition method not decided by late 2000. KAI candidate would be SB 427M (which see); see also KAL entry for alternative design submission.
UPDATED

KAI AH-X
TYPE: Attack helicopter (requirement).
PROGRAMME: Aimed at replacing RoKA AH-1 HueyCobra and MD 500TOW Defender helicopters from 2005; to feature improved adverse weather and night operational capabilities. Submissions invited late 1999 and final bids received 20 July 2000. Candidates reported to include Bell (AH-1Z), Boeing (AH-64), Kamov (Ka-52), Mil (Mi-28) and Sikorsky (AUH-60). Selection of acquisition method and winner expected in third quarter of 2001.
UPDATED

KAI-BELL SB 427
TYPE: Light utility helicopter.
PROGRAMME: Licence-assembled Bell 427 (see Canadian section). Canadian type certification obtained in November 1999. Korean-manufactured SB 427s assembled at Sachon. CAAC (China) type certificate gained in April 2000, South Korean certificate in May 2000. KAI production rate one per month until end of 2001, then two per month thereafter.
CURRENT VERSIONS: **SB 427:** Equivalent of standard Canadian Bell 427 (which see).
  SB 427M: Military version proposed by Samsung in 1998 as candidate for Korean Multipurpose Helicopter (KMH) programme (which see) and for local area export. Fuselage-side weapon mountings, mast-mounted target acquisition and designation system (TV/FLIR), optional RWR, uprated transmission, increased rotor blade chord, strengthened skid gear, cockpit armour protection and higher operating weights.
CUSTOMERS: First SB 427 delivered to Broad Air Conditioning Company (China) in 2000; second aircraft due for delivery to Korea National Police in first quarter of 2001. Sales to China totalled 12 by December 2000, including five for Fei Ma Airlines.
Following (provisional) details refer to SB 427M:
POWER PLANT: As Bell 427, but transmission rating increased to approximately 708 kW (950 shp).
ACCOMMODATION: Armour protection for cockpit, cabin and engine cowling.
AVIONICS: Could include RWR and IRCM.
ARMAMENT: Up to four Hellfire or TOW anti-tank missiles, four air-to-air Stingers or 70 mm rocket pods on fuselage sidebars; underfuselage 12.7 mm machine gun pod.
WEIGHTS AND LOADINGS:
  Max T-O weight 3,175 kg (7,000 lb)
UPDATED

KAL

KOREAN AIR LINES CO LTD
KAL Building, 1370 Gonghang-dong, Gangseo-gu, Seoul
Tel: (+82 2) 656 71 14
CHAIRMAN AND COO: Choong-Hoon Cho
PRESIDENT AND CEO: Yi-Taek Shim

Aerospace Division
Address as above
Tel: (+82 2) 656 39 20
Fax: (+82 2) 656 39 17 and 39 18
MANAGING VICE-PRESIDENT: Byung-Sun Lee

KIMHAE PLANT:
103 Daejeo 2-dong, Kangseo-ku, Pusan 618142
Tel: (+82 51) 972 91 11
Fax: (+82 51) 972 62 32
VICE-PRESIDENT: Sang-Mook Suh
GENERAL MANAGER: Yong-An Yoon

Korean Institute of Aeronautical Technology (KIAT),
461 Jonmin-dong, Yusung-ku, Taejon 305390
Tel: (+82 42) 868 61 14
Fax: (+82 42) 868 61 28
VICE-PRESIDENT: Si-Yoong Yoo

Aerospace Division of Korean Air Lines established 1976 to manufacture and develop aircraft; occupies 64.75 ha (160 acre) site at Kimhae, including floor area of 250,000 m² (2.69 million sq ft); 1999 workforce about 1,850. Has overhauled RoKAF aircraft since 1978; programmed depot maintenance of US military aircraft in Pacific area began 1979, including structural repair of F-4s, systems modifications for F-16s, MSIP upgrading of F-15s and overhaul of C-130s. Began assembly in 1981 of first domestically manufactured fighter (Northrop F-5E/F), completing deliveries to RoKAF 1986. Since 1988, KAL has delivered wing components for Boeing 747/777, New-Generation 737 and MD-11 (last-named programme completed July 1999); and fuselage components for MD-80/90 and Airbus A330/A340. Deliveries of 100 nose sections for Boeing 717 started 28 March 1997. KAL has manufactured UH-60P helicopters since 1991, under licence from Sikorsky.

Korean Institute of Aeronautical Technology (KIAT), established as division of KAL in 1978, has grown to become a major Korean aerospace industry R&D centre.

As part of Korean industry development programme from 1988, KAL designed and developed a light aircraft (Chang-Gong 91: see 1995-96 *Jane's*); co-developed (with McDonnell Douglas) the MD 520MK military helicopter derived from the MD 500 (which see under MD Helicopters in US section); is a major member of KFP Korean Fighter Programme (wings and rear fuselages for F-16C/D); and domestically co-developed with Daewoo the KT-1 basic trainer (see KAI entry) for the RoKAF.
UPDATED

KAL (SIKORSKY) UH-60P
TYPE: Multirole medium helicopter.
PROGRAMME: Original licence agreement to build more than 80 UH-60P in Korea (S-70A-18 variant of Sikorsky UH-60L: see US section) to meet domestic military requirements, signed March 1990. Seven US-built aircraft imported 1991, followed by materials including kits and parts for local manufacture; first Korean-built UH-60P made first flight 15 February 1992 and delivered 26 March. Some UH-60Ps now being upgraded for search and rescue, command and control, or special land/sea missions.
CURRENT VERSIONS: **UH-60P:** Generally similar to US Army UH-60L except as noted.
CUSTOMERS: Mainly South Korea Army; some to Navy and Air Force. Initial batch of 81 followed by 57 more; deliveries of second batch began in August 1995; 100th

South Korean UH-60P (Sikorsky S-70A-18) armed with Hellfire ATMs 1999/0051115

Jane's All the World's Aircraft 2001-2002

UH-60P delivered (to Korean Army) October 1996; last delivery was due December 1999; third batch under consideration by South Korean government in 1998, but not funded by October 2000. All known KAL-assembled UH-60s have Sikorsky manufacturer's numbers.
POWER PLANT: Two General Electric T700-GE-701C turboshafts and 2,535 kW (3,400 shp) transmission assembled under licence agreement of March 1990.
ACCOMMODATION: Crew of two plus 11 fully equipped troops.
AVIONICS: *Comms:* Rockwell Collins AN/ARC-217 HF radio.
Flight: Rockwell Collins AR/ARN-147 VOR, AN/ARN-149 ADF and AN/ARN-118 Tacan; Honeywell AN/APN-209 radar altimeter; Kollsman AAU-31/32 absolute altimeter; Sikorsky self-contained navigation system; Rockwell Collins miniature airborne GPS receiver.
Self-defence: Litton AN/APR-39 RWR.
EQUIPMENT: External stores support system (ESSS); stretcher kit; ETS; internal/external hoist; 3,630 kg (8,000 lb) capacity external cargo hook.

UPDATED

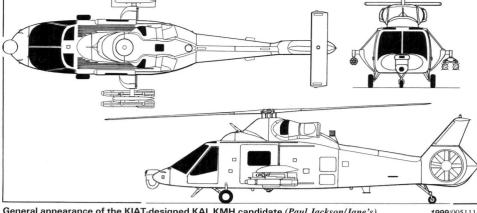

General appearance of the KIAT-designed KAL KMH candidate *(Paul Jackson/Jane's)* 1999/0051114

KAL KMH

TYPE: Multirole light helicopter.
PROGRAMME: Initiated as entrant in Korean Military Helicopter (KMH) competition for armed scout and attack roles, but M now indicating Multipurpose for appeal also to civil operators; design assistance by Sikorsky; mockup displayed and first details released at Seoul Air Show in October 1996. Preliminary design, market survey and solicitation of customer requirements under way in late 1998. No further announcements yet reported, but an alternative design is under development by KAI (which see).
CUSTOMERS: KAL foresees a domestic market for up to 500 military and 150 civil KMHs.
DESIGN FEATURES: Externally similar in some respects to the S-76 or UH-60 Black Hawk, except for rear fuselage (T tail and Fenestron) resembling RAH-66 Comanche (see accompanying illustration).
FLYING CONTROLS: Dual redundant fly-by-light control system.
LANDING GEAR: Non-retractable tailwheel type.
POWER PLANT: Two 671 to 820 kW (900 to 1,100 shp) class turboshafts (type not yet selected).
ACCOMMODATION: Flight crew of two; six seats in main cabin.
AVIONICS: *Mission:* Nose-mounted FLIR sighting system.
Self-defence: Jammer and RWR.
ARMAMENT: Fuselage sidebars for various combinations of Hellfire or Stinger missiles, seven-round rocket launchers, 0.50 in machine gun pods or 454 litre (120 US gallon; 100 Imp gallon) auxiliary fuel tanks.
DIMENSIONS, EXTERNAL: Not provided
WEIGHTS AND LOADINGS:
Max external sling load approx 1,360 kg (3,000 lb)
Max T-O weight approx 3,400 kg (7,500 lb)
PERFORMANCE (design target):
Max level speed 165 kt (305 km/h; 189 mph)
Max rate of climb at S/L 640 m (2,100 ft)/min
Service ceiling 6,100 m (20,000 ft)
Hovering ceiling IGE 4,115 m (13,500 ft)
Range 350 n miles (648 km; 402 miles)

UPDATED

SAMSUNG

SAMSUNG AEROSPACE INDUSTRIES LTD (SSA)

24th Floor, Samsung Life Building, 150-2 ga, Taepyung-ro, Chung-ku, Seoul 100716
Tel: (+82 2) 751 88 07
Fax: (+82 2) 771 65 52
VICE-CHAIRMAN AND CEO: Dae-Won Lee

Second Korean company to enter aerospace; formed August 1977 and began producing jet engines from 1980; moved into commercial aerostructures 1984 and was selected 1986 as prime contractor in licence manufacture of F-16s for RoKAF; also produces wing fixed trailing-edges for Boeing 757/767 and horizontal tail surfaces for Bombardier de Havilland Dash 8. Approved by South Korean government to produce medium helicopters in Korea; agreement with Bell to build major fuselage and tailboom assemblies for 412; initial production began May 1988; plans then called for 93 per cent local manufacture by early 1990s. Also teamed on Bell 427 (see Canadian section), manufacturing cabin and tailboom, and undertaking final assembly of those (designated SB427) for sale in China and South Korea. Signed agreement in October 1996 to produce and market Mil Mi-26 heavy helicopter.

In addition to head office in Seoul, has a 2,100 ha (5,190 acre) aircraft assembly site at Sachon, three plants and an aerospace technology training centre at Changwon, and an R&D centre at Daeduk. Samsung Aerospace became part of KAI (which see) on 1 October 1999.

VERIFIED

MALAYSIA

CTRM

COMPOSITE TECHNOLOGY RESEARCH (MALAYSIA) SDN BHD

Suite 19-14-3, Level 14, UOA Centre, 19 Jalang Pinang, 50450 Kuala Lumpur
Tel: (+60 3) 264 50 74
Fax: (+60 3) 264 50 93
WORKS: Batu Berendam, Malacca
CEO: Dr Chen Lip Keong

CTRM is an investor in Eagle Aircraft of Australia (which see), and since 1997 has manufactured components for the two-seat Eagle 150 at a 24,080 m² (259,200 sq ft) plant on the 200 ha (500 acre) Composite City site, completed in 1996. Following recent increases in orders for the Eagle, CTRM was, in late 2000, collaborating with the Australian Civil Aviation Safety Authority (CASA) to obtain certification of the Malaysian factory as a second-source final assembly centre for the aircraft. At that time, according to Australian press reports, it was assessing the market potential of the Lancair Columbia 300 (see US section), in which it is also an investor, versus a possible four-seat derivative of the Eagle 150.

NEW ENTRY

MEAA

MALAYSIAN EXPERIMENTAL AIRCRAFT ASSOCIATION (Persatuan Pesawa Eksperimental Malaysia)

d/a Muzium TUDM, Pangkalan TUDM KL, Jalan Lapangan Terbang Lama, 50460 Kuala Lumpur
Tel: (+60 3) 245 97 31
Fax: (+60 3) 245 96 14
DIRECTOR: Mej (R) Abdul Razak Bin Omar
CHIEF DESIGNER: En Supli Bin Mat

MEAA formed 1996 to support Rakan Muda Air Recreation Youth Programme, initially with assembly of a Zenith Zodiac CH 601. Association now provides residential pilot training courses using Zodiac, plus Slepcev Storch, Eagle Eagle 150 and Bell 47G helicopter, all located at Sungai Besi. Design has begun of a light aircraft which, it is hoped, will be built and flown by students.

VERIFIED

MEAA RMX-4

TYPE: Four-seat lightplane.
PROGRAMME: Construction of mockup began May 1999; this shown at LIMA '99, Langkawi, in December of same year. Prototype targeted to fly before end of 2000, subject to

Mockup of RMX-4 fuselage (mounted on helicopter skids) on show at LIMA '99, Langkawi, December 1999 *(Paul Jackson/Jane's)* 2000/0062988

receipt of financial support, but not yet reported; certification one year after first flight.

DESIGN FEATURES: First four-seater designed in Malaysia. All design undertaken on computer, providing practical experience for Rakan Muda Aviation Club.
Low-wing monoplane of conventional appearance, operating on diesel fuel. Unswept, constant-chord wings and moderately sweptback tail surfaces.

FLYING CONTROLS: Conventional and manual. Horn-balanced rudder; elevator tips contained within tailplane. Flaps.

STRUCTURE: Aluminium alloy throughout.

LANDING GEAR: Nosewheel type; fixed. Cantilever mainwheel legs; oleo-pneumatic nosewheel leg. Speed fairings on all wheels.

POWER PLANT: One 224 kW (300 hp) horizontally opposed piston engine operating on diesel fuel and driving a constant-speed, composites propeller. Fuel tank in each wing.

ACCOMMODATION: Four persons in side-by-side pairs. Upward-opening door each side. Baggage compartment at rear of cabin.

DIMENSIONS, EXTERNAL:
Wing span	9.50 m (31 ft 2 in)
Length overall	8.00 m (26 ft 3 in)
Fuselage max width	1.14 m (3 ft 8¾ in)
Height overall	2.30 m (7 ft 6½ in)

DIMENSIONS, INTERNAL:
Cabin: Length	1.53 m (5 ft 0¼ in)
Max width	1.08 m (3 ft 6½ in)
Max height	1.18 m (3 ft 10½ in)
Baggage compartment volume	0.50 m (17.7 cu ft)

WEIGHTS AND LOADINGS (estimated):
Weight empty	750 kg (1,653 lb)
Baggage capacity	80 kg (176 lb)
Max fuel weight	400 kg (882 lb)
Max T-O weight	1,350 kg (2,976 lb)
Max power loading	6.04 kg/kW (9.92 lb/hp)

PERFORMANCE (estimated):
Never-exceed speed (V_{NE})	200 kt (370 km/h; 230 mph)
Cruising speed	120 kt (222 km/h; 138 mph)
Max rate of climb at S/L	274 m (900 ft)/min
Service ceiling	4,570 m (15,000 ft)
Range	740 n miles (1,370 km; 851 miles)

UPDATED

SME

SME AVIATION SDN BHD

Lot 1410, Pangkalan TUDM Subang, PO Box 7574, 40718 Shah Alam, Selangor Darul Eshan
Tel: (+60 3) 746 85 77
Fax: (+60 3) 746 85 66
CEO: Lt Col (Retd) Mohamed Tarmizi Ahamd
EXECUTIVE DIRECTOR: Tommy Tay
GENERAL MANAGER: Prabhakaran Nair

US MARKETING:
SME Aero Inc
The Tallahassee Aerospace Center, 3240 Capital Circle South-West, Tallahassee, Florida 32310
Tel: (+1 850) 575 90 02
Fax: (+1 850) 575 43 54
CHAIRMAN: Richard Ledson

SME Aviation Sdn Bhd, established in September 1993, is one of four divisions of SME Technologies Sdn Bhd, a group wholly owned by the Malaysian government which provides products and services for defence, aerospace, plastics and metal-based manufacturing industries. It has manufactured the SME MD3-160 AeroTiga trainer with initial technical support from the aircraft's designer, MDB Flugtechnik of Switzerland, and British Aerospace. Subcontract work includes Hawk wing pylons and (awarded 1999) Avro RJ wing components for BAE Systems. Other contracts, also awarded in late 1999, involve fuselage components for the Alenia/Lockheed Martin C-27J and, for Ulan-Ude Aviation Plant, parts manufacture for the Mi-17 helicopter.

UPDATED

SME MD3-160 AEROTIGA

TYPE: Primary prop trainer/sportplane.

PROGRAMME: Originated by Max Dätwyler in Switzerland in late 1960s, but much redesigned later with view to maximising common-module interchangeability; first flight of MD3-160 (HB-HOH) 12 August 1983; first flight second prototype (HB-HOJ) 1990; Swiss FOCA certification to FAR Pt 23 awarded 22 January 1991; first flight of first preproduction aircraft (HB-HNA) July 1992; FAA type certification to FAR Pts 21 and 23 received July 1992; technology transfer agreement concluded between MDB Flugtechnik AG of Switzerland and SME Group of Malaysia on 30 June 1993. Three preproduction aircraft built in Switzerland, of which two to Malaysia as demonstrator/pattern aircraft and one to USA for SME Aero Inc, a joint venture company of SME Aviation in Tallahassee.

Series production of MD3-160 in Malaysia started in August 1994; first Malaysian aircraft (M42-02, built out of sequence for RMAF) made first flight at Subang on 25 May 1995; this and M42-01 handed over to RMAF on 7 December 1995 (first aircraft of any type wholly constructed in Malaysia). Aircraft awarded type certificate by Malaysian Department of Civil Aviation (DCA) on 29 August 1997; from November 1997 this certification also carries FAA endorsement. Projected US sales (six in 1999 and 12 in 2000) failed to materialise; by December 1999, plans had been postponed for transfer of production to USA; link with European lightplane company then being planned had not materialised by January 2001. Last of 20 for RMAF handed over on 27 January 2000, but final five aircraft were still undergoing modifications at manufacturer in mid-2000.

CURRENT VERSIONS: **MD3-116:** Primary trainer, powered by 85.75 kW (115 hp) Textron Lycoming O-235-N2A flat-four engine; first prototype, fitted with this engine, successfully completed performance and handling test flights third quarter 1991. Not currently offered.

MD3-160: Aerobatic trainer and glider tug, with more powerful Textron Lycoming O-320-D2A engine.

Description applies to MD3-160.

CUSTOMERS: Total of 40 originally ordered by January 1995, comprising 20 each for Indonesian Ministry of Communications' Curug flight training centre and No. 1 Flying Training Centre of Royal Malaysian Air Force at Alor Setar. Indonesian order suspended, due to economic situation in that country, after six built; of these, four transferred to RMAF contract and two delivered to SME Aero in USA as stock in January 1999, although neither had been registered two years later. Total production: five in Switzerland and 22 in Malaysia, all completed by January 2000.

COSTS: Basic standard aircraft approximately US$150,000 (1999).

DESIGN FEATURES: Components interchangeable with each other, and left with right, are: (1) ailerons, flaps, elevators and rudder; (2) wing leading-edges; (3) wing inboard panels; (4) wing outboard panels; (5) wingtips; (6) tailplane halves and fin; (7) elevator and rudder tips; (8) aileron, elevator and rudder tabs.
Wing section NACA 64215-414 (modified); dihedral 5° 30'; incidence 2°.

FLYING CONTROLS: Conventional and manual. Pushrod actuated; elevator and rudder horn-balanced; single-slotted ailerons; trim tabs in each aileron, elevator and rudder; trim controls can be mechanical or electrical; electrically operated flaps, max deflection 48°.

STRUCTURE: All-metal except for glass fibre wingtips, dorsal fin fairing and wheel fairings and CFRP cowling; aluminium honeycomb skins; designed for easy construction; rear fuselage detachable.

LANDING GEAR: Non-retractable tricycle type with steerable nosewheel (30° left, 44° right). Main-gear legs are cantilever steel struts, descending at 45° from fuselage main bulkhead. Nose gear fitted with oleo-pneumatic shock-absorber. Cleveland 6.00-6 mainwheels and 5.00-5 nosewheel with Michelin tyres. Tyre pressure 2.41 bars (35 lb/sq in) on all units. Independent Cleveland hydraulic disc brake on each mainwheel. Wheel fairings on all three wheels.

POWER PLANT: One Textron Lycoming O-320-D2A flat-four engine (119 kW; 160 hp at 2,700 rpm). Sensenich 74DM6S8-0-62 two-blade fixed-pitch metal propeller standard, McCauley 1-C172-AGM-7462 propeller optional. Integral fuel tank in each wing: total capacity 148 litres (39.1 US gallons; 32.6 Imp gallons), of which 144 litres (38.0 US gallons; 31.7 Imp gallons) are usable. Refuelling point in top of each tank. Oil capacity 7.6 litres (2.0 US gallons; 1.7 Imp gallons).

ACCOMMODATION: Side-by-side adjustable seats for pilot and one pupil or passenger. Five-point fixed seat belts.

AeroTiga in RMAF markings *(Paul Jackson/Jane's)* 2000/0062925

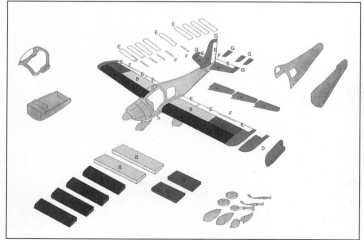

Diagram illustrating the MD3-160's interchangeability of components

Instrument panel of a military MD3-160 AeroTiga *(Paul Jackson/Jane's)* 2000/0062926

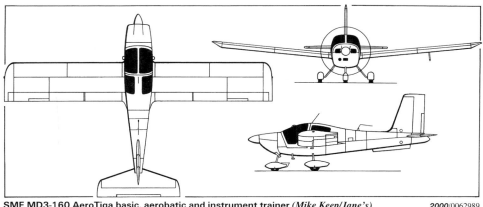

SME MD3-160 AeroTiga basic, aerobatic and instrument trainer *(Mike Keep/Jane's)* 2000/0062989

Forward-sliding jettisonable canopy. Space behind seats for baggage. Dual controls, cabin ventilation and heating standard.
SYSTEMS: Hydraulic system for mainwheel brakes only. Electrical power for engine starting, lighting, instruments and com/nav equipment provided by 28 V 70 A engine-driven alternator and 24 V 30 Ah battery.
AVIONICS: *Comms:* Honeywell VHF radio and blind encoding transponder.
  Flight: Provision for VOR, ADF, GPS, or other items at customer's option.
  Instrumentation: Basic VFR standard; IFR optional.
EQUIPMENT: Equipment for glider towing optional.
DIMENSIONS, EXTERNAL:
  Wing span   10.00 m (32 ft 9¾ in)
  Wing chord, constant   1.50 m (4 ft 11 in)
  Wing aspect ratio   6.7
  Length overall   7.10 m (23 ft 3½ in)
  Fuselage max width   1.13 m (3 ft 8½ in)
  Height overall   2.92 m (9 ft 7 in)
  Tailplane span   3.00 m (9 ft 10 in)
  Wheel track   2.05 m (6 ft 8¾ in)
  Wheelbase   1.55 m (5 ft 1 in)
  Propeller diameter   1.88 m (6 ft 2 in)
  Propeller ground clearance 0.24 m (9½ in)
DIMENSIONS, INTERNAL:
  Cabin, from firewall to rear bulkhead:
    Length   1.30 m (4 ft 3¼ in)
    Max width   1.12 m (3 ft 8 in)
    Max height   1.08 m (3 ft 6½ in)
AREAS:
  Wings, gross   15.00 m² (161.5 sq ft)
  Ailerons (total)   1.13 m² (12.16 sq ft)
  Trailing-edge flaps (total) 1.96 m² (21.10 sq ft)
  Fin   0.89 m² (9.58 sq ft)
  Rudder   0.51 m² (5.49 sq ft)
  Tailplane   1.71 m² (18.41 sq ft)
  Elevators (total)   1.04 m² (11.19 sq ft)
WEIGHTS AND LOADINGS (A: Aerobatic category, U: Utility):
  Weight empty: A, U   670 kg (1,477 lb)
  Baggage capacity   50 kg (110 lb)
  Max fuel weight: A, U   106.5 kg (235 lb)
  Max T-O weight: A   840 kg (1,852 lb)
    U   920 kg (2,028 lb)
  Max landing weight: U   891 kg (1,964 lb)
  Max wing loading: A   56.0 kg/m² (11.47 lb/sq ft)
    U   61.3 kg/m² (12.56 lb/sq ft)
  Max power loading: A   7.05 kg/kW (11.58 lb/hp)
    U   7.72 kg/kW (12.68 lb/hp)
PERFORMANCE (at above A and U max T-O weights except where indicated):
  Never-exceed speed (V_{NE}):
    A   175 kt (324 km/h; 201 mph) IAS
    U   158 kt (292 km/h; 181 mph) IAS
  Max level speed at S/L:
    U   137 kt (253 km/h; 157 mph)
  Cruising speed at 1,525 m (5,000 ft):
    at 75% power: A, U   130 kt (241 km/h; 150 mph) IAS
    at 66% power: A, U   124 kt (230 km/h; 143 mph) IAS
  Max manoeuvring speed (V_M):
    A   121 kt (224 km/h; 139 mph) IAS
    U   110 kt (203 km/h; 126 mph) IAS
  Stalling speed, power off:
    flaps up   56 kt (104 km/h; 65 mph) IAS
    flaps down   47 kt (88 km/h; 55 mph) IAS
  Max rate of climb at S/L: A 300 m (984 ft)/min
  Max certified altitude: A   3,050 m (10,000 ft)
    U   4,875 m (16,000 ft)
  T-O run   165 m (545 ft)
  T-O to 15 m (50 ft)   338 m (1,100 ft)
  Landing from 15 m (50 ft)   308 m (1,010 ft)
  Landing run   135 m (445 ft)
  Range with max fuel, no reserves
                              588 n miles (1,090 km; 677 miles)
  Endurance, 45 min reserves   4 h 45 min
  g limits: A   +6/−3
    U   +4.4/−2.2
                                           UPDATED

NETHERLANDS

EURO-ENAER

EURO-ENAER HOLDING BV
Luchthavenweg 18a, PO Box 3044, NL-1780 GA den Helder
Tel: (+31 223) 67 78 20 and 67 78 22
Fax: (+31 223) 67 78 29
e-mail: sales@euro-enaer.com
Web: http://www.euro-enaer.com
MANAGING DIRECTOR: Michel van Tooren

The company was created in 1997 as a joint venture with ENAER of Chile (which see), to gain JAA and FAA certification for the latter company's Ñamcu all-composites two-seat trainer/tourer; design rights were acquired by Euro-ENAER in 1997. It assembles the aircraft in its approximately 1,000 m² (10,750 sq ft) factory at Lelystad and acts as distributor for the Ñamcu (marketed under its English name, Eaglet) in Europe, Africa and Asia. ENAER will market the aircraft in the rest of the world, except USA.

UPDATED

EURO-ENAER EE 10 EAGLET
TYPE: Aerobatic two-seat lightplane.
PROGRAMME: Launched by ENAER in Chile June 1986, under project name Avión Liviano (light aircraft); prototype construction started February 1987; first flight (CC-PZI) of renamed ECH-02 Ñamcu (Eaglet) April 1989; first flight of second prototype (CC-PZJ) early March 1990; second prototype since lost; fourth (third flying) and final prototype completed subsequently (CC-PZL); re-registered PH-EAG 24 April 1998 and used to test definitive rudder design. Remaining aircraft briefly used by Chilean Air Force.
  Announced in September 1995 that aircraft was to be certified initially to JAR 23, with assembly in Netherlands by Euro-ENAER using composites airframe subassemblies supplied from Chile. Third prototype made first European flight (PH-EAG) on 8 May 1998.
  Euro-ENAER target for FAR/JAR 23 certification was first half of 1999, but this slipped to end 2000. ENAER has also applied for JAR 21 production and design certification. Changes to meet JAR 23 include 119 kW (160 hp) engine as standard; inclusion of resin and fibre in composites materials; redesigned fuel system; and new interior and avionics. Version with 85.8 kW (115 hp) Textron Lycoming O-235-N2C and fixed-pitch propeller may be offered in future.
  Second prototype, incorporating all JAR 23 changes, under construction in early 2000; registration PH-ABG reserved following arrival of kit from Chile in February 2000.

Prototype Euro-ENAER EE 10 Eaglet *(Paul Jackson/Jane's)* 2001/0093683

  Future developments will include IFR-equipped, and fully aerobatic versions, plus possibly a four-seat amphibian.
CUSTOMERS: Planned production for 2000 was 12 aircraft, 12 in 2001 and 24 in 2002, eventually reaching 40 per year.
COSTS: US$160,000 (2000).
DESIGN FEATURES: Limited aerobatic trainer; conforms to JAR 23 (Utility category); low purchase and operating costs.
  Wing section NACA 63_2-415; incidence 3° at root, 0° 30′ at tip; dihedral 5° from roots.
FLYING CONTROLS: Conventional and manual. Actuation by pushrods and cables; plain ailerons, horn-balanced elevators and rudder; plain flaps. Control surface

Euro-ENAER EE 10 Eaglet cockpit layout
1999/0051120

deflections: ailerons +25/−20°; elevators +25/−18°; rudder ±30°; flaps 0, 15 or 30° down.
STRUCTURE: All-composites (glass fibre/foam sandwich), including movable surfaces; semi-monocoque sandwich fuselage with integral wing centre-section; longerons integral with skin; engine and nosewheel leg mounted on steel spaceframe attached to fuselage at four points. Two outer wings bolted to centre-section; integral fuel tanks in leading-edges have aluminium mesh shielding for lightning protection; glass fibre/carbon fibre type C main spar with carbon fibre spar caps. Ailerons, flaps and rudder have two partial-sandwich structural skins; fin has slanting front spar and three ribs, all of sandwich.
LANDING GEAR: Non-retractable tricycle type. Cantilever spring steel main units; steerable and self-centring nose unit, with oleo-pneumatic shock-absorber and shimmy damper. Cleveland wheel and Goodyear tyre on each unit (size 6.00-6 on main units, 5.00-5 on nose unit). Tyre pressure 1.45 bars (21 lb/sq in) on main units; 2.07 bars (30 lb/sq in) on nose unit; Cleveland hydraulic mainwheel disc brakes.
POWER PLANT: One 119 kW (160 hp) Textron Lycoming O-320-D flat-four driving an MT 178 R160-3D two-blade fixed-pitch wooden propeller. Integral fuel tank in each wing leading-edge, combined capacity 150 litres (39.6 US gallons; 33.0 Imp gallons); prototypes' capacity was 100 litres (26.4 US gallons; 22.0 Imp gallons); overwing gravity fuelling point for each tank; engine-driven fuel pump with electrical back-up. Oil capacity 7.6 litres (2.0 US gallons; 1.7 Imp gallons).
ACCOMMODATION: Two seats side by side in fully enclosed cockpit, with headrests and four-piece safety harnesses; seats are electrically adjustable longitudinally. Dual controls. Two independent gull-wing window/doors,

hinged on centreline to open upward; one-piece acrylic windscreen. Space for up to 20 kg (44 lb) of baggage aft of seats. Cockpit heated and ventilated.

SYSTEMS: Hydraulic system for brakes only. Electrical power supplied by Prestolite 12 V 70 A alternator and 12 V 35 Ah battery.

AVIONICS: Honeywell equipment, except where indicated.
Comms: KT 76C transponder with encoding, David Clark Isocom intercom, KMA 24 audio panel and Narco AK-350 blind encoder.
Flight: KX 155 nav/com and KI 209 VOR, KN 64 DME and KR 87 or KI 277 ADF. Garmin 155 TSO GPS.
Instrumentation: VFR flight and engine instrumentation standard; IFR optional.

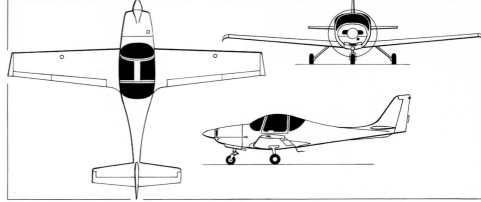

Euro-ENAER EE 10 Eaglet (Textron Lycoming O-320 flat-four) *(James Goulding/Jane's)* 2000/0100523

DIMENSIONS, EXTERNAL:	
Wing span: incl tips	8.70 m (28 ft 6½ in)
excl tips	8.31 m (27 ft 3¼ in)
Wing chord: at root	1.53 m (5 ft 0¼ in)
at tip	0.84 m (2 ft 9 in)
Wing aspect ratio	7.6
Length overall	7.05 m (23 ft 1½ in)
Fuselage max width	1.22 m (4 ft 0 in)
Height overall	2.42 m (7 ft 11¼ in)
Tailplane span	3.00 m (9 ft 10 in)
Wheel track	2.80 m (9 ft 2¼ in)
Wheelbase	1.47 m (4 ft 9¾ in)
Propeller diameter	1.78 m (5 ft 10 in)
DIMENSIONS, INTERNAL:	
Cockpit: Length	1.54 m (5 ft 0½ in)
Max width	1.12 m (3 ft 8 in)
Max height	1.00 m (3 ft 3¼ in)
AREAS:	
Wings gross: incl tips	10.01 m² (107.75 sq ft)
excl tips	9.85 m² (106.0 sq ft)
Ailerons (total)	0.50 m² (5.38 sq ft)
Trailing-edge flaps (total)	0.98 m² (10.55 sq ft)
Fin	0.93 m² (10.01 sq ft)
Rudder	0.41 m² (4.41 sq ft)
Tailplane	2.13 m² (22.93 sq ft)
Elevators (total)	0.84 m² (9.04 sq ft)
WEIGHTS AND LOADINGS:	
Basic weight empty	540 kg (1,190 lb)
Max fuel	100 kg (220 lb)
Max T-O and landing weight	850 kg (1,873 lb)
Max wing loading	84.9 kg/m² (17.39 lb/sq ft)
Max power loading	7.87 kg/kW (12.92 lb/hp)
PERFORMANCE, POWERED:	
Never-exceed speed (V_{NE})	180 kt (333 km/h; 207 mph) IAS
Max level speed at S/L	152 kt (282 km/h; 175 mph)
Max cruising speed	147 kt (272 km/h; 169 mph)
Stalling speed, power off:	
flaps up	62 kt (115 km/h; 72 mph)
flaps down	53 kt (99 km/h; 61 mph)
Max rate of climb at S/L	320 m (1,050 ft)/min
Service ceiling	4,270 m (14,000 ft)
T-O run	305 m (1,000 ft)
T-O to 15 m (50 ft)	412 m (1,355 ft)
Landing from 15 m (50 ft)	376 m (1,235 ft)
Landing run	183 m (600 ft)
Range at 75% power at 2,440 m (8,000 ft) with max fuel, 10% reserves	501 n miles (927 km; 576 miles)
Endurance, conditions as above	4 h 30 min
g limits	+4.4/−2.2
PERFORMANCE, UNPOWERED:	
Best glide ratio	12.5

UPDATED

FOKKER

FOKKER AVIATION BV
PO Box 12222, NL-1100 AE Amsterdam-Zuidoost
Tel: (+31 20) 605 66 66
Fax: (+31 20) 605 70 15
Telex: 11526 FMHS NL
ACTING CEO: E L De Vries
AVIATION MARKETING AND SALES: J Oosterman
OPERATING COMPANIES:
Fokker Services BV, PO Box 3, NL-4630 AA Hoogerheide (aircraft maintenance, repair and overhaul; holder of Fokker aircraft type certificates)
Fokker Elmo BV, Woensdrecht (electrical/electronic systems and components for aviation industry)
Fokker Special Products BV, PO Box 59, NL-7900 AB Hoogeveen (industrial products for civil and military use)

OTHER BUSINESS UNITS:
Fokker Product Support, Schiphol (sale of spares; technical support of Fokker aircraft)

Fokker Aerostructures BV, Papendrecht (manufacture of parts, structural and non-structural lightweight components and subassemblies for aerospace industry)

Failure of several bids to resuscitate the bankrupt aircraft manufacturing arm of Fokker resulted in the last jet airliner, Fokker 70 PH-KZK of Cityhopper, being delivered on 18 April 1997 (the final Fokker 100 preceded it as PT-MRW of TAM Brasil on 21 March 1996); and the final Fokker aircraft of all, Fokker 50 ET-AKS, departing Schiphol for Ethiopian Airlines on 23 May 1997. Recent history of the company and descriptions of the Fokker 50, 60, 70 and 100 last appeared in the 1997-98 edition.

Despite closure of the assembly line, Dutch company **Rekkof Restart** (Rekkof = Fokker spelt backwards), formed early 1998, has been attempting to persuade potential investors and suppliers to support a move to resume production of Fokker 70 and 100. By late 1998 it had backing and orders for up to 20 aircraft and was considering production of up to 24 aircraft per year, with first deliveries from mid-2000. The first 40 would be built by Fokker Services, which currently provides spares and technical support to Fokker products; thereafter the type certificates would be owned jointly by Fokker Services and Rekkof Aircraft. However, by June 1999 this plan had been abandoned.

A second company, **Forward Aircraft,** announced in September 1998 that it was examining prospects for restarting production of the Fokker 50 and 60 for military applications, subject to a minimum of 36 relaunch orders. Design rights to the two turboprops are owned by the Netherlands government. Production, at Groningen/Eelde Airport, would be set at 12 per year, later rising to 20, with deliveries beginning by mid-2000, though this date has now slipped. In mid-1999, Forward was looking for US partners to offer the, by now renamed, Friendship 60 as a contender for the US Army Aerial Common Sensor Program and, in early 2000, was reported to be close to signing a deal with the Royal Netherlands Air Force for three 'Friendship 50s'. By late 2000, nothing had come of this proposal.

UPDATED

MD HELICOPTERS — *see US section*

SSVOBB

STICHTING STUDENTEN VLIEGTUIGONTWIKKELING -BOUW EN -BEHEER (Foundation for Student Aircraft Development, Manufacturing and Operation)
Kluyverweg 1, NL-2629 HS Delft
Tel: (+31 15) 278 10 57 and 90 58
Fax: (+31 15) 278 12 43
e-mail: ssvobb@lr.tudelft.nl
Web: http://www.lr.tudelft.nl/ssvobb
CHAIRMAN: Menno Van Rijn
PUBLIC RELATIONS MANAGER: Jan Smulders

Computer-generated image of the SSVOBB Impuls two-seat light aircraft

Impuls project:
PRODUCTION CO-ORDINATOR: Pieter Segers
ENGINEERING CO-ORDINATOR: Robin in 't Groen

SSVOBB was formed in 1990 by The Society of Aerospace Students 'Leonardo da Vinci' of the Delft University of Technology to co-ordinate the design, manufacture and maintenance of aircraft by students. A replica of the 1937 Lambach HL II aerobatic biplane was constructed between 1989 and 1995, and work is proceeding on an original lightplane design, the Impuls.

UPDATED

SSVOBB IMPULS
English name: Impulse
TYPE: Two-seat lightplane.
PROGRAMME: Preliminary design started early 1994. Registration PH-VXM reserved March 1997. Main fuselage moulds completed early 1999. Main wing moulds were due to be completed late 1999 with wing production due to start early 2000. Certification to FAR Pt 23 being undertaken under supervision of dr ir Middel (general and aerodynamic design); Prof dr ir Theo van Holten (flight performance); Prof dr ir J A Mulder (stability/control), ir F van Dalen (construction); Prof ir R J Zwaan (aero-elasticity); ing K van Woerkom (systems); ir van Badegom (fuel and propulsion systems).

DESIGN FEATURES: Twin-boom layout with pod-type fuselage and high wing of parallel chord; swept fins.
FLYING CONTROLS: Conventional and manual. Actuation by pushrods (ailerons and elevator), with cable and pushrod interconnected twin rudder and nosewheel system. Servo-actuated electric pitch trim and electrically operated single-slotted flaps.
NACA 63₂415 wing section with RA 163CW3 flaps; twin tailbooms joined by NACA 0012 section horizontal stabiliser.
STRUCTURE: All-composites wing with GFRP sandwich skin and spars; GFRP tail unit. Fuselage steel tube frame with glass fibre shell will be used to finalise design of cockpit and control systems.
LANDING GEAR: Non-retractable tricycle type; GFRP cantilever sprung mainwheels. Steerable nosewheel has

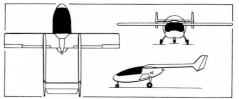

SSVOBB Impuls (MWAE rotary engine) *(Paul Jackson/Jane's)* 1999/0051122

20 cm (8 in) tyre pressurised to 3.14 bars (45.5 lb/sq in); mainwheels have 45 cm (17¾ in) tyres with same pressure. Hydraulically operated brakes on mainwheels.

POWER PLANT: One 74.6 kW (100 hp) Mid-West AE100 rotary engine driving a three-blade fixed-pitch propeller. Fuel capacity 65 litres (17.2 US gallons; 14.3 Imp gallons) in single tank located between cockpit and engine. Oil capacity 3.0 litres (0.8 US gallon; 0.7 Imp gallon).

ACCOMMODATION: Side-by-side seating for pilot and passenger under one-piece, Perspex canopy. Baggage space behind seats.

DIMENSIONS, EXTERNAL:
Wing span	7.75 m (25 ft 5 in)
Wing chord, constant	1.03 m (3 ft 4½ in)
Wing aspect ratio	7.5
Length overall	5.77 m (18 ft 11¼ in)
Fuselage: Length	3.32 m (10 ft 10¾ in)
Max width	1.20 m (3 ft 11¼ in)
Height overall	1.99 m (6 ft 6¼ in)
Tailplane span	2.10 m (6 ft 10¾ in)
Wheel track	2.20 m (7 ft 2½ in)
Wheelbase	1.75 m (5 ft 9 in)
Propeller diameter	1.53 m (5 ft 0¼ in)

AREAS:
Wings, gross	8.00 m² (86.1 sq ft)
Ailerons (total)	0.66 m² (7.10 sq ft)
Flaps (total)	1.06 m² (11.41 sq ft)
Fins (total)	2.00 m² (21.53 sq ft)
Rudders (total)	0.24 m² (2.58 sq ft)
Tailplane	1.51 m² (16.25 sq ft)
Elevators (total)	0.62 m² (6.67 sq ft)

WEIGHTS AND LOADINGS (estimated):
Weight empty	375 kg (827 lb)
Baggage capacity	30 kg (66 lb)
Max T-O weight	575 kg (1,267 lb)
Max wing loading	71.9 kg/m² (14.72 lb/sq ft)
Max power loading	7.71 kg/kW (12.67 lb/hp)

PERFORMANCE (estimated):
Max level speed at S/L	125 kt (232 km/h; 144 mph)
Max cruising speed	108 kt (200 km/h; 124 mph)
Max rate of climb at S/L	246 m (807 ft)/min
T-O to 15 m (50 ft)	340 m (1,115 ft)
Landing from 15 m (50 ft)	430 m (1,410 ft)
Range with max fuel	more than 405 n miles (750 km; 466 miles)
Endurance	5 h

UPDATED

NEW ZEALAND

PAC
PACIFIC AEROSPACE CORPORATION LIMITED
Private Bag HN 3027, Hamilton Airport, Hamilton
Tel: (+64 7) 843 61 44
Fax: (+64 7) 843 61 34
e-mail: pacific@aerospace.co.nz
Web: http://www.aerospace.co.uk
GENERAL MANAGER: Graeme Polley
MANAGING DIRECTOR: Brian Hare
MANAGER, MARKETING: John McWilliam

Pacific Aerospace Corporation formed 1982 following acquisition of assets and undertakings of New Zealand Aerospace Industries; became wholly owned subsidiary of AeroSpace Technologies of Australia (75.1 per cent) and Lockheed Martin, USA (24.9 per cent). In October 1995, Aeromotive, a privately owned New Zealand company, purchased ASTA's 75.1 per cent share. PAC maintains production and support facilities for its own aircraft, the CT4 Airtrainer series, Fletcher FU24 series and Cresco 08-600 and 08-750; by early 2001 some 450 aircraft had been produced. Manufacturing facility also produces items for the Boeing 747/777, Airbus A330/340 and F/A-18 Hornet; 100 employed in the company's 24,155 m² (260,000 sq ft) facility.

UPDATED

PAC FLETCHER FU24-954
TYPE: Agricultural sprayer/light utility transport.
PROGRAMME: First flight of US-built FU24 prototype (then Thorp T-15) 14 June 1954; first flight of first US production aircraft five months later; type certificate granted 22 July 1955; all manufacturing and sales rights transferred to New Zealand 1964; factory refurbishment/upgrade programme offered by PAC for earlier FU24 series aircraft. Total of 286 FU24 series aircraft (of which 257 were -950 versions) built by 1983; further 10 produced by other sources; returned to production in 1989 to build six, including four for Thai Ministry of Agriculture; further batch of five for Syrian Ministry of Agriculture early 1992, of which final aircraft (ZK-FZN) flown June 1992.

Again returned to production in 1995 to provide three more for Syria, although these were not completed. No further sales known by late 2000, although Walter of the Czech Republic confirmed an order for M 601 engines for the FU24 in early 1999 for use in conversions. Super Air of Hamilton flew prototype Ford V-8 conversion on 23 January 2001; this version also equipped with twin mainwheels.

CURRENT VERSIONS: **FU24-954:** Current standard model.
CUSTOMERS: Total 297 by PAC and predecessors comprises 204 for New Zealand customers, 77 for Australia, five each for Pakistan, Syria and Thailand, and one for Turkey. By early 2001, 183 were still in service.

Abbreviated data follow; full details in 1994-95 and earlier Jane's.

DESIGN FEATURES: Primary configuration as agricultural aircraft; with appropriate hopper base, can also be used for firefighting.
Wing section NACA 4415 (constant); 8° dihedral on outer wing panels; incidence 2°.
FLYING CONTROLS: Manual. Horn-balanced ailerons and rudder; all-moving tailplane with full span anti-servo tab; ground-adjustable tab on rudder; single slotted flaps; row of vortex generators forward of each aileron.
STRUCTURE: Conventional light alloy; two-spar wing; cockpit area stressed for 25 *g* impact.
LANDING GEAR: Non-retractable tricycle type; steerable nosewheel. Fletcher air-oil shock struts. Cleveland wheels and hydraulic disc brakes on main units. Goodyear tyres, size 8.50-6 (6 ply); pressures vary between 0.76 and 2.07 bars (11 and 30 lb/sq in).
POWER PLANT: One 298 kW (400 hp) Textron Lycoming IO-720-A1A or A1B flat-eight engine. Fuel capacity 254 litres (67.0 US gallons; 55.8 Imp gallons) normal; 500 litres (132 US gallons; 110 Imp gallons) with optional long-range tanks. Conversion using Walter 601 turbine certified in 2000.
ACCOMMODATION: Two-seat cockpit with sliding canopy. Rear compartment, with port freight door, holds six passengers, equivalent freight or chemical hopper.
EQUIPMENT: 1,211 litre (320 US gallon; 266 Imp gallon) liquid or 1,066 kg (2,350 lb) dry hopper.

DIMENSIONS, EXTERNAL:
Wing span	12.81 m (42 ft 0 in)
Length overall	9.70 m (31 ft 10 in)
Height overall	2.84 m (9 ft 4 in)

AREAS:
Wings, gross	27.31 m² (294.0 sq ft)

WEIGHTS AND LOADINGS:
Weight empty, equipped	1,188 kg (2,620 lb)
Max disposable load: Agricultural	975 kg (2,150 lb)
Max T-O weight: Normal	2,204 kg (4,860 lb)
Agricultural	2,463 kg (5,430 lb)

PERFORMANCE (at Normal max T-O weight):
Never-exceed speed (V_{NE})	143 kt (265 km/h; 165 mph)
Max level speed at S/L	126 kt (233 km/h; 145 mph)
Max cruising speed at 75% power	113 kt (209 km/h; 130 mph)
Max rate of climb at S/L	280 m (920 ft)/min
Service ceiling	4,875 m (16,000 ft)
T-O to 15 m (50 ft)	372 m (1,220 ft)
Landing from 15 m (50 ft)	390 m (1,280 ft)
Range	573 n miles (1,062 km; 660 miles)
Endurance	5 h 6 min

UPDATED

PAC CRESCO
TYPE: Agricultural sprayer/light utility turboprop.
PROGRAMME: Design began 1977; first flight of prototype (ZK-LTP) 28 February 1979; first flight of production aircraft early 1980; entered service January 1982 as Cresco 08-600 powered by 447 kW (599 shp) LTP101-700-1A turboprop. This variant was augmented by Cresco 08-750 (PT6A-34AG version) in November 1992. First flight of PT6A-34AG version (ZK-TMN, c/n 010) 18 November 1992. Work on a wide-body development started in 2000.
CURRENT VERSIONS: **Cresco 08-600:** Initial version, with 447 kW (599 shp) (flat rated) Honeywell LTP 101-700A-1A turboprop. No longer available.
Cresco 08-750: With higher-powered 559 kW (750 shp) Pratt & Whitney Canada PT6A-34AG engine; launched 1992 (first delivery 23 December) as 08-600-34AG.
Cresco 750XL: Wide-body version with fuselage height raised 15 cm (6 in). Passenger compartment 1.57 m (5 ft 2 in) wide, 3.99 m (13 ft 1 in) long, 1.40 m (4 ft 7 in) high; length 11.58 m (38 ft 0 in) remains same. 1.27 × 1.14 m (4 ft 2 in × 3 ft 9 in) cargo door. Optimised for parachute jumping; able to carry 17 jumpers to 4,265 m (14,000 ft) in less than 13 minutes.
CUSTOMERS: Total 28 built by September 2000, of which five delivered in 1997, three in 1998, three in 1999 and three in 2000. One delivered to Kevron Pty Ltd in Australia during 1998 is used for geophysical exploration. Of 1999 deliveries, two were built for parachuting, second for NSW Skydive of Australia. Current operators include New Zealand (seven companies), Australia (two companies), Malaysia and Bangladesh.
DESIGN FEATURES: Turboprop development of FU24, *which see for additional description*; approximately 60 per cent commonality of components with FU24, but markedly different in many other respects; larger hopper.
POWER PLANT: See Current Versions. Hartzell HC-B3TN-3D/T10282 three-blade constant-speed, fully feathering, reversible-pitch metal propeller. Four integral fuel tanks in wing centre-section, total capacity 500 litres (132 US gallons; 110 Imp gallons); ferry configuration capacity 2,400 litres (634 US gallons; 528 Imp gallons) using hopper as auxiliary fuel tank. 750XL fuel capacity 939 litres (248 US gallons; 207 Imp gallons). Two refuelling points in upper surface of each wing. Oil capacity 5.7 litres (1.5 US gallons; 1.25 Imp gallons).

DIMENSIONS, EXTERNAL:
Wing span	12.81 m (42 ft 0 in)
Wing chord, constant	2.13 m (7 ft 0 in)
Wing aspect ratio	6.0
Length overall	11.07 m (36 ft 4 in)
Fuselage max width	1.22 m (4 ft 0 in)
Height overall	3.63 m (11 ft 10¾ in)
Tailplane span	4.62 m (15 ft 2 in)
Wheelbase	2.77 m (9 ft 1¼ in)
Wheel track: 750XL	3.72 m (12 ft 2½ in)
Propeller diameter	2.59 m (8 ft 6 in)
Propeller ground clearance	0.305 m (1 ft 0 in)
Cargo door (port): Height	0.94 m (3 ft 1 in)
Width	0.94 m (3 ft 1 in)
Height to sill	0.91 m (3 ft 0 in)

DIMENSIONS, INTERNAL:
Passenger/cargo compartment (aft of hopper):
Length	3.12 m (10 ft 3 in)
Max width	1.09 m (3 ft 7 in)
Max height	1.24 m (4 ft 1 in)
Floor area	2.79 m² (30.0 sq ft)
Volume: passenger version	3.4 m³ (120 cu ft)
cargo version	3.8 m³ (134 cu ft)
Hopper volume	1.86 m³ (66 cu ft)

AREAS:
Wings, gross	27.31 m² (294.0 sq ft)
Ailerons (total)	1.97 m² (21.18 sq ft)
Trailing-edge flaps (total)	3.05 m² (32.87 sq ft)
Fin	2.16 m² (23.23 sq ft)
Rudder	0.63 m² (6.77 sq ft)
Tailplane	5.09 m² (54.74 sq ft)
Elevator, incl tab	2.32 m² (25.00 sq ft)

WEIGHTS AND LOADINGS:
Weight empty, equipped: 600	1,270 kg (2,800 lb)
750	1,338 kg (2,950 lb)
750XL	1,393 kg (3,072 lb)
Max fuel: 600, 750	435 kg (960 lb)
Typical payload: 750, Normal	1,111 kg (2,450 lb)
750, Agricultural (Restricted)	2,155 kg (4,750 lb)
750XL	2,009 kg (4,428 lb)

Pacific Aerospace Corporation Cresco agricultural aircraft

Max disposable load (fuel + hopper):	
600, Normal	1,578 kg (3,480 lb)
600, Agricultural (Restricted)	1,828 kg (4,030 lb)
750, Normal	1,524 kg (3,360 lb)
750, Agricultural (Restricted)	2,429 kg (5,356 lb)
Max T-O weight: 600, Normal	2,925 kg (6,450 lb)
600, Agricultural (Restricted)	3,175 kg (7,000 lb)
750, Normal	2,925 kg (6,450 lb)
750, Agricultural (Restricted)	3,742 kg (8,250 lb)
750XL	3,402 kg (7,500 lb)
Max landing weight: 600, 750	2,925 kg (6,450 lb)
750XL	3,231 kg (7,125 lb)
Wing loading at Normal max T-O weight:	
600, 750	107.1 kg/m² (21.94 lb/sq ft)
750XL	124.6 kg/m² (25.51 lb/sq ft)
Wing loading at Agricultural (Restricted) max T-O weight:	
600	116.3 kg/m² (23.81 lb/sq ft)
750	137.1 kg/m² (28.08 lb/sq ft)
Power loading at Normal max T-O weight:	
600	6.55 kg/kW (10.77 lb/shp)
750	5.23 kg/kW (8.60 lb/shp)
750XL	6.09 kg/kW (10.00 lb/shp)
Power loading at Agricultural (Restricted) max T-O weight:	
600	7.11 kg/kW (11.69 lb/shp)
750	6.70 kg/kW (11.01 lb/shp)
PERFORMANCE (at max Normal T-O weight, except where indicated):	
Never-exceed speed (V$_{NE}$):	
600, 750	173 kt (320 km/h; 199 mph)
750XL	176 kt (326 km/h; 202 mph)
Max level speed at S/L:	
600	148 kt (274 km/h; 170 mph)
750	152 kt (282 km/h; 175 mph)
750XL	155 kt (287 km/h; 178 mph)
Max cruising speed at 75% power at 305 m (1,000 ft):	
600	133 kt (246 km/h; 153 mph)
750	140 kt (259 km/h; 161 mph)
Stalling speed, flaps down, power off:	
600 at 2,767 kg (6,100 lb) AUW	52 kt (97 km/h; 60 mph)
750	57 kt (106 km/h; 66 mph)
750XL at 3,402 kg (7,500 lb) AUW	59 kt (110 km/h; 68 mph)
Max rate of climb at S/L: 600	379 m (1,245 ft)/min
750	475 m (1,560 ft)/min
750XL	518 m (1,700 ft)/min
Absolute ceiling: 600	5,485 m (18,000 ft)
750	8,230 m (27,000 ft)
750XL	9,140 m (30,000 ft)
T-O run: 600	323 m (1,060 ft)
750	227 m (745 ft)
T-O run at 1,406 kg (3,100 lb) AUW:	
750	45 m (150 ft)
T-O to 15 m (50 ft): 600	436 m (1,430 ft)
750	325 m (1,065 ft)
Landing from 15 m (50 ft): 600	500 m (1,640 ft)
750	427 m (1,400 ft)
Landing run at 1,406 kg (3,100 lb) AUW with propeller pitch reversal: 750	86 m (285 ft)
Range at 75% power with standard fuel, no reserves:	
600 at 3,175 kg (7,000 lb) MTOW	460 n miles (852 km; 529 miles)
750	305 n miles (676 km; 420 miles)
750XL	592 n miles (1,097 km; 682 miles)
Range as above with fuel in hopper:	
750	2,037 n miles (3,774 km; 2,345 miles)
Endurance with standard fuel:	
750 at 75% power	2 h 40 min
750 at 60% power	3 h 0 min
750XL at 75% power	4 h 16 min

UPDATED

PAC AIRTRAINER CT4E

Royal Thai Air Force designation: BF 16

TYPE: Primary prop trainer/sportplane.
PROGRAMME: New Zealand redesign of Australian Victa Aircruiser; first flight 23 February 1972; deliveries began October 1973; 94 (plus two prototypes) built before production ended 1977 (75 CT4A and 19 special military variants); production line reopened 1991 to build 12 civil CT4Bs for Ansett Flying College and six for Thailand; last completed August 1992. Piston-engined CT4A and B last described fully in 1994-95 *Jane's*. First flight of CT4C turboprop prototype (ZK-FXM, converted RAAF CT4B) 21 January 1991.
CURRENT VERSIONS: **T350:** Turboprop version (previously known as CT4C) with 313 kW (420 shp) Rolls-Royce 250-B17D (throttle limited to 261 kW; 350 shp) in lengthened nose. See 1994-95 *Jane's* for description. Certification was expected 1998 in anticipation of Southeast Asian order for 30, but this failed to materialise.
CT4E: Developed version of CT4B with more powerful engine; wing mounted slightly farther forward than on CT4B; first flight (ZK-EUN, converted from RAAF CT4A) 14 December 1991; NZ certification (FAR Pt 23 Amendment 36) 8 May 1992.

Detailed description applies to CT4E.

CUSTOMERS: PAC previously built 75 CT4As and 38 CT4Bs; further production of CT4E began in 1997, of which 13 supplied on 20 year lease to the Royal New Zealand Air Force as replacements for CT4Bs; contract signed 18 August 1998; first (NZ1985) delivered to Pilot Training Squadron at Ohakea in same month; last (NZ1997) on 8 June 1999; by late 2000 the fleet had accumulated 10,000 flying hours. Also used by Red Checkers aerobatic team. Thailand has a requirement for 24 CT4Es, initial order being for 12, deliveries of which began in July 1999 and were completed in June 2000 (final Thai aircraft first flew 11 May 2000). Under consideration for PA-18 replacement by Israel; decision due mid-2001. Total 26 built up to May 2000 temporary suspension of manufacture, one of which is demonstrator.
DESIGN FEATURES: Metal translation of 1953-vintage wooden tourer. Conventional, low-wing lightplane. Tapered wings and constant-chord tailplane. Detachable wingtips to allow fitting of optional wingtip fuel tanks.
Wing section NACA 23012 (modified) at root, NACA 4412 (modified) at top; dihedral 6° 45' at chord line; incidence 3° at root; twist 3°; root chord increased by forward sweep of inboard leading-edges; small fence on outer leading-edges.
FLYING CONTROLS: Conventional and manual. Aerodynamically and mass-balanced bottom-hinged ailerons and statically balanced one-piece elevator actuated by pushrods; rudder by rod and cable linkage; ground-adjustable tab on rudder, electric trim control for elevator and rudder; electrically actuated single-slotted flaps.
STRUCTURE: Light alloy stressed skin except for Kevlar/GFRP wingtips and engine cowling; ailerons and flaps have fluted skins.
LANDING GEAR: Non-retractable tricycle type, with cantilever spring steel main legs; steerable (±25°) nosewheel carried on telescopic strut and oleo shock-absorber. Main units have Dunlop Australia wheels and tubeless tyres size 15×6.00, pressure 1.65 bars (24 lb/sq in); nosewheel has tubeless tyre size 14×4.5, pressure 1.24 bars (18 lb/sq in). Single-disc toe-operated hydraulic brakes, with hand-operated parking lock. Minimum ground turning radius 2.90 m (9 ft 6 in). Landing gear designed to shear before any excess impact loading is transmitted to wing, to minimise structural damage in event of crash landing.

PAC CT4E Airtrainer of the RNZAF overflying its base at Ohakea *(Flt Sgt P Stein/RNZAF)*   1999/0036072

POWER PLANT: One 224 kW (300 hp) Textron Lycoming AEIO-540-L1B5 flat-six engine with inverted oil system; three-blade Hartzell HC-C3YF-4BF/FC7663-2R constant-speed metal propeller. Standard fuel capacity of 204.4 litres (54.0 US gallons; 45.0 Imp gallons) of which 200 litres (52.8 US gallons; 44.0 Imp gallons) are usable; wingtip tanks, each of 77.6 litres (20.5 US gallons; 17.1 Imp gallons) capacity, available optionally. Gravity fuelling point in each wing. Oil capacity 11.4 litres (3.0 US gallons; 2.5 Imp gallons).
ACCOMMODATION: Two seats side by side under hinged, fully transparent jettisonable Perspex canopy. Space to rear for optional third seat or 77 kg (170 lb) of baggage or equipment. Dual controls standard.
SYSTEMS: 28 V DC electrical system; ground power receptacle.

DIMENSIONS, EXTERNAL:	
Wing span	7.92 m (26 ft 0 in)
Wing chord: at root	2.18 m (7 ft 2 in)
at tip	0.94 m (3 ft 1 in)
Wing aspect ratio	5.2
Length overall	7.16 m (23 ft 5¾ in)
Height overall	2.59 m (8 ft 6 in)
Fuselage: Max width	1.12 m (3 ft 8 in)
Max depth	1.40 m (4 ft 7¼ in)
Tailplane span	3.61 m (11 ft 10 in)
Wheel track	2.97 m (9 ft 9 in)
Wheelbase	1.66 m (5 ft 5½ in)
DIMENSIONS, INTERNAL:	
Cabin: Length	2.74 m (9 ft 0 in)
Max width	1.08 m (3 ft 6½ in)
Max height	1.34 m (4 ft 5 in)
AREAS:	
Wings, gross	11.98 m² (129.0 sq ft)
Ailerons (total)	1.07 m² (11.56 sq ft)
Flaps (total)	2.10 m² (22.60 sq ft)
Fin	0.78 m² (8.44 sq ft)
Rudder, incl tab	0.58 m² (6.26 sq ft)
Tailplane	1.46 m² (15.71 sq ft)
Elevator	1.56 m² (16.74 sq ft)
WEIGHTS AND LOADINGS:	
Weight empty, equipped	807 kg (1,780 lb)
Max fuel weight	149 kg (328 lb)
Max T-O weight	1,179 kg (2,600 lb)
Max wing loading	98.38 kg/m² (20.15 lb/sq ft)
Max power loading	5.26 kg/kW (8.67 lb/hp)
PERFORMANCE:	
Never-exceed speed (V$_{NE}$)	209 kt (387 km/h; 240 mph)
Max level speed at S/L	163 kt (302 km/h; 188 mph)
Cruising speed at 2,590 m (8,500 ft) at 75% power	158 kt (293 km/h; 182 mph)
Stalling speed at S/L: flaps up	60 kt (112 km/h; 69 mph)
flaps down	44 kt (82 km/h; 51 mph)
Max rate of climb at S/L, ISA	558 m (1,830 ft)/min
Service ceiling	5,640 m (18,500 ft)
Time to service ceiling	13 min
T-O run at S/L, ISA	183 m (600 ft)
Landing from 15 m (50 ft)	244 m (800 ft)
Landing run	169 m (553 ft)
Range at S/L with max fuel at 75% power, ISA, no reserves	520 n miles (963 km; 599 miles)
Range at 1,830 m (6,000 ft) with max fuel at 45% power, no reserves	520 n miles (963 km; 598 miles)

UPDATED

PAKISTAN

PAC

PAKISTAN AERONAUTICAL COMPLEX

Kamra, District Attock
WORKS: F-6 Rebuild Factory; Mirage Rebuild Factory; Kamra Avionics and Radar Factory; Aircraft Manufacturing Factory (all at Kamra)
Tel: (+92 51) 927 01 11 and 927 01 17
Fax: (+92 51) 927 01 07 and 927 01 00
e-mail: info@pac.org.pk
Web: http://www.pac.org.pk
CHAIRMAN: Air Marshal Pervez A Nawaz

MANAGING DIRECTORS:
    Air Cdre S Sohail H Sadiq (AMF)
    Air Cdre Iftikhar A Gul (F-6RF)
    Air Cdre Basit Raza Abbasi (MRF)
    Air Cdre Saleem Akbar (KARF)
SALES AND MARKETING DIRECTOR:
    Air Cdre Raja Tariq Mahmood
CHIEF TEST PILOT: Wg Cdr Ahmed Faiq

Pakistan Aeronautical Complex (PAC) is the nucleus of the aeronautical industry in Pakistan. Its professional standards and reliability enable domestic customers and foreign agencies alike to undertake joint production with PAC in various fields. PAC Kamra comprises four factories: the Mirage Rebuild Factory (MRF), F-6 Rebuild Factory (F6-RF), Aircraft Manufacturing Factory (AMF) and Kamra Avionics and Radar Factory (KARF). All four are ISO 9000 certified. The main roles and capabilities of these factories are as follows:

Mirage Rebuild Factory (MRF) was established in 1978 to provide in-country maintenance for the Dassault Mirage weapon system. Main mission today is complete overhaul of Mirage III/V aircraft, Atar 09C engines, F100 engine modules (for F-16 aircraft) and all associated aircraft

Instrument panel of the Super Mushshak Agile

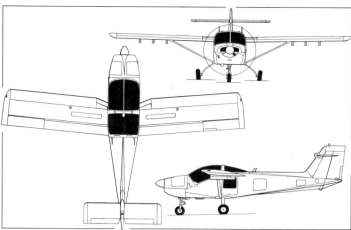
Pakistan Aeronautical Complex Mushshak two/three-seat light aircraft *(Dennis Punnett/Jane's)*

components and engine accessories. The initial task was to overhaul six aircraft and 24 to 36 Atar engines, and their accessories and components, per year. With the induction of more aircraft the task has increased to 12 aircraft and 73 engines per year. Equipped with modern machinery, test equipment, jigs and fixtures, the capabilities include: major structural repair of fuselage and accessories; wing refurbishment; No. 4 rib replacement; fabrication of system pipes; overhaul of ejection seat; repair of 561 subassemblies and components; repair of fuel cells; general overhaul of Atar 09C engine and 65 major subassemblies; life core modification, overhaul and upgrade of F100-200E engine (for F-16); and overhaul of jet fuel starter of F100-200F. Achievements have included recovery of Australian aircraft, avionics upgrading of Mirage III and upgrade of F100-PW-200 engine to F100-PW-200E configuration. MRF has an effective quality system to verify processes and improve product quality. ISO 9000 certification was achieved on 14 September 1995. Over the past two decades MRF has gained valuable experience of overhauling the Atar 09C engine, for which, after ISO 9002 and SNECMA certification, it is now an overhaul subcontractor to SNECMA.

F-6 Rebuild Factory (F6-RF): Although originally established, in 1980, for the F-6 (Chinese version of the MiG-19), the F-6 RF also currently overhauls Chinese origin F-6, FT-6, FT-5, A-5 III, F-7P (variant of MiG-21) and FT-7P aircraft. All the components and instruments of these aircraft are also rebuilt at the factory. Modern technical facilities for aircraft structural repairs and spares manufacture include casting, forging, heat treatment, surface treatment and various machining processes, both conventional and CNC (turning, milling, cutting, copying, welding and EDM die-making). F6-RF also manufactures jettisonable fuel tanks, hardware parts, compression/tension/torsion springs, aluminium and steel washers and clamps, standard harness and rubberised items. It manufactures its own aeronautical grade synthetic rubber and has a moulding and forming facility where a large variety of rubber parts of the aircraft are formed. In addition, the F6-RF has extensive and well-equipped laboratories and a testing facility to meet all aviation standard requirements, including a wide range of inspections ranging from X-rays to non-destructive inspection, geometrical standards (PME), tensile testing (probably the only such facility in the country), hardness testing, flow and pressure measuring, spectrophotometer analysis, chemical testing, rubber testing and salt spray testing.

Aircraft Manufacturing Factory (AMF): In operation since mid-1981, the AMF spearheads the aircraft manufacturing industry in Pakistan. It is the only aircraft manufacturing concern in Pakistan and is at present engaged in the manufacture of the light, robust, basic trainer-cum-surveillance aircraft Mushshak; the Super Mushshak – a variant with a more powerful engine, cockpit air conditioning, electric instruments and several other improvements; the Baaz and Ababeel low-speed target drones for anti-aircraft gunnery training; and development and manufacture, with NAMC of China, of the K-8 (Karakoram) jet trainer aircraft, a few of which are already flying with the Pakistan Air Force. In addition, the AMF has excellent facilities to rebuild and repair the Mushshak at factory level, along with after-sales spares support to Mushshak operators worldwide. AMF also has the skills and capabilities of conventional/CNC machining, sheet metal forming, hand layup, contact suction hot press moulding, stretch moulding, gravity moulding, reaction injection moulding, mould/die and pattern making in woodwork, surface treatment/painting, heat treatment, tig welding (AC and DC), electric arc welding, bronzing, soldering, tube banding, material testing, spectroanalysis, tensile testing, precision measurement chemical analysis, CNS calibration, co-ordinate measurement, mould/dies and fixture designing in tooling facility, CNC copy milling and EDM (spark erosion) milling.

Kamra Avionics Radar Factory (KARF): KARF, the fourth factory of PAC, came into operation in 1987. It is an electronics centre whose activities include the overhaul of ground-based radars and power generators. KARF has proven capabilities for rebuilding of radars, contract and reporting centres (CRC), generators, and manufacture and repair of avionics systems. Currently, it is producing ESM equipment and airborne radar systems. All production activities are carried out in large, spaciously built shops equipped with state-of-the-art machinery. Apart from the production of RWR and Fiar Grifo 7 radar systems and the overhauling of MPDR, CRC and generators, KARF offers services in many fields such as surface mount technology, cable repair and manufacturing inspection/testing of microwave components, testing of gears and synchro bridge, environmental testing, testing hydraulic components, rebuilding/manufacturing of 03 phase air conditioning and general engineering and painting facilities.

UPDATED

PAC (AMF) MUSHSHAK and SUPER MUSHSHAK AGILE
English name: Proficient
TYPE: Two/three-seat lightplane.
PROGRAMME: Fifteen Mushshaks supplied complete from Sweden; 10 then assembled from SKD kits and 82 from CKD kits at Risalpur between 1975 and 1981; completely indigenous production followed, with 180 delivered by December 1996; 113 repaired at Kamra including 100 from 1980 to 1996; engines, instruments, electrical equipment and radios imported; complete airframe manufactured locally. Shahbaz (prototype 86-5147; first flight July 1987; English name Falcon) received US FAR Pt 23 certification 1989. Super Mushshak (see Current Versions) introduced 1997. Production rate (new aircraft and conversion kits) stated to be 24 per year in late 1999.
CURRENT VERSIONS: **Mushshak:** Standard production version; used for training, communications and observation; more than half a million hours flown by late 1999. Pakistan Air Force then upgrading 16 to Super Mushshak; quoted time for conversion: 3 weeks.
Description applies to above version, except where indicated.
Shahbaz: As Mushshak except for 157 kW (210 hp) Teledyne Continental TSIO-360-MB turbocharged engine; four completed by September 1993; no further examples since then. Details in 1997-98 and earlier *Jane's*.
Super Mushshak: Further increase in power by adopting IO-540 flat-six engine. Prototype 95-5385X made first flight 15 August 1996. Revealed at Dubai Air Show in November 1997, at which time five completed. These had flown 800 hours by late 1999, at which time PAC manufacturing conversion kits for PAF.
CUSTOMERS: See table. Total of 15 imported aircraft plus 280 Mushshaks, five Super Mushshaks and four Shahbaz acquired/produced by late 1999. Mushshak deliveries to Iran in 1988-91; six supplied to Syria and three to Oman in 1994. Sale of unspecified number of Super Mushshaks to an unnamed Middle East country was reported to be imminent in late 2000.

MUSHSHAK PRODUCTION
(at November 1997)

Customer	Qty
At Risalpur:	
Pakistan Air Force/Army	92[1]
At Kamra:	
Pakistan Air Force	36[2]
Pakistan Army	101
Pakistan (unknown service)	13
Iran	25
Oman	5
Syria	8
Pakistan Aeronautical Complex	4
Total	**280[3]**

[1] Saab kits (10 SKD, 82 CKD)
[2] First aircraft 83-5116 (numbers 5200-5299 not allocated); equip Nos. 1 and 2 Primary Flying Training Squadrons and three Headquarters Flights
[3] Plus complete aircraft from Saab to PAF

DESIGN FEATURES: Based on Swedish Saab Safari/Supporter, but without latter's armament option. Optimised for military support (originally including light attack) from semi-prepared airstrips; wing position aids ground observation.
Wing thickness/chord ratio 10 per cent; dihedral 1° 30'; incidence 2° 48'; sweepforward 5° from roots.
FLYING CONTROLS: Manual. Mass-balanced ailerons with servo tab in starboard unit; rudder with trim tab; all-moving mass-balanced tailplane with large anti-balance tab. Electrically actuated plain sealed flaps; leading-edge slots.
STRUCTURE: All-metal, except for GFRP tailcone, engine cowling panels, wing strut/landing gear attachment fairings and fin-tip.
LANDING GEAR: Non-retractable tricycle type. Cantilever composites spring main legs. Cleveland 6.00-6 mainwheels and 5.00-5 steerable (±30°) nosewheel. Tyre pressure (all units) 2.07 bars (30 lb/sq in). Cleveland disc brakes on main units.
POWER PLANT: *Mushshak:* One 149 kW (200 hp) Textron Lycoming IO-360-A1B6 flat-four engine, driving a Hartzell HC-C2YK-4F/FC7666A-2 two-blade constant-speed metal propeller.
Super Mushshak: One 194 kW (260 hp) Textron Lycoming IO-540-V4A5 flat-six; Hartzell two-blade (as in

Pakistan Army student pilots undergo their first 50 hours of training on the Mushshak *(Peter R Foster)*

Mushshak) or McCauley B3D3C412-C/G-82NDA-8 three-blade constant-speed propeller of same diameter.

Fuel (both versions) in two integral wing fuel tanks, total capacity 182 litres (48.0 US gallons; 40.0 Imp gallons). Oil capacity (both) 7.5 litres (2.0 US gallons; 1.6 Imp gallons). From 10 to 20 seconds inverted flight (limited by oil system) permitted.

ACCOMMODATION: Side-by-side adjustable seats, with provision for back-type or seat-type parachutes, for two persons beneath fully transparent upward-hinged canopy. Dual controls standard. Space aft of seats for 100 kg (220 lb) of baggage (with external access on port side) or, optionally, a rearward-facing third seat. Upward-hinged door, with window, beneath wing on port side. Cabin heated and ventilated.

SYSTEMS: 28 V DC electrical system (70 A alternator and 70 Ah battery). PAC air conditioning system optional.

AVIONICS: *Comms:* Honeywell KLX 125 and KLX 135 radios; KT 76A/78A ATC transponder; KA 134 audio panel.
Flight: KLX 125 VOR; KLX 135 GPS; KR 87 ADF.
Instrumentation: Provision for full blind-flying instrumentation.

EQUIPMENT: Provision for six underwing attachment points, inner two stressed to carry up to 150 kg (330 lb) each and outer four up to 100 kg (220 lb) each. Options include ULV cropspraying kit, target towing kit, or underwing supply/relief containers.

DIMENSIONS, EXTERNAL:
Wing span	8.85 m (29 ft 0½ in)
Wing chord: at root	1.21 m (3 ft 11¾ in)
outer panels, constant	1.36 m (4 ft 5½ in)
Wing aspect ratio	6.6
Length overall: Mushshak	7.00 m (22 ft 11½ in)
Super Mushshak	7.15 m (23 ft 5½ in)
Height overall	2.60 m (8 ft 6½ in)
Tailplane span	2.80 m (9 ft 2¼ in)
Wheel track	2.20 m (7 ft 2½ in)
Wheelbase: Mushshak	1.61 m (5 ft 3½ in)
Super Mushshak	1.58 m (5 ft 2¼ in)
Propeller diameter	1.88 m (6 ft 2 in)
Cabin door (port): Height	0.78 m (2 ft 6¾ in)
Width	0.52 m (1 ft 8½ in)

DIMENSIONS, INTERNAL:
Cabin: Max width	1.10 m (3 ft 7¼ in)
Max height (from seat cushion)	1.00 m (3 ft 3¼ in)

AREAS:
Wings, gross	11.90 m² (128.1 sq ft)
Ailerons (total)	0.98 m² (10.55 sq ft)
Flaps (total)	1.55 m² (16.68 sq ft)
Fin	0.77 m² (8.29 sq ft)
Rudder, incl tab	0.73 m² (7.86 sq ft)
Tailplane, incl tab	2.10 m² (22.60 sq ft)

WEIGHTS AND LOADINGS (Mushshak and Super Mushshak. A: Aerobatic, U: Utility, N: Normal category):
Weight empty, equipped: Mushshak	646 kg (1,424 lb)
Super Mushshak	760 kg (1,676 lb)
Max external stores load (both)	300 kg (661 lb)
Max T-O weight: A (both)	900 kg (1,984 lb)
U (Mushshak)	1,000 kg (2,205 lb)
U (Super Mushshak)	1,030 kg (2,270 lb)
N (Mushshak)	1,200 kg (2,645 lb)
N (Super Mushshak)	1,250 kg (2,755 lb)
Max wing loading: A (both)	75.6 kg/m² (15.48 lb/sq ft)
U (Mushshak)	84.0 kg/m² (17.20 lb/sq ft)
U (Super Mushshak)	86.6 kg/m² (17.73 lb/sq ft)
N (Mushshak)	100.8 kg/m² (20.65 lb/sq ft)
N (Super Mushshak)	105.0 kg/m² (21.51 lb/sq ft)
Max power loading (Mushshak):	
A	6.04 kg/kW (9.92 lb/hp)
U	6.71 kg/kW (11.02 lb/hp)
N	8.05 kg/kW (13.23 lb/hp)
Max power loading (Super Mushshak):	
A	4.64 kg/kW (7.63 lb/hp)
U	5.31 kg/kW (8.73 lb/hp)
N	6.45 kg/kW (10.60 lb/hp)

PERFORMANCE (Utility category):
Never-exceed speed (V_{NE}): both	196 kt (363 km/h; 226 mph)
Max level speed at S/L:	
Mushshak	128 kt (238 km/h; 148 mph)
Super Mushshak	145 kt (268 km/h; 166 mph)
Cruising speed at S/L at 75% power:	
Mushshak	113 kt (210 km/h; 130 mph)
Super Mushshak	130 kt (240 km/h; 149 mph)
Stalling speed, power off:	
Mushshak, flaps up	60 kt (111 km/h; 69 mph)
Super Mushshak, flaps up	57 kt (105 km/h; 66 mph)
Mushshak, flaps down	54 kt (100 km/h; 63 mph)
Super Mushshak, flaps down	52 kt (96 km/h; 60 mph)
Max rate of climb at S/L:	
Mushshak	312 m (1,024 ft)/min
Super Mushshak	518 m (1,700 ft)/min
Time to 1,830 m (6,000 ft): Mushshak	7 min 30 s
Time to 3,050 m (10,000 ft): Super Mushshak	10 min
Service ceiling: Mushshak	4,800 m (15,740 ft)
Super Mushshak	6,705 m (22,000 ft)
T-O run: Mushshak	150 m (495 ft)
Super Mushshak	183 m (600 ft)
T-O to 15 m (50 ft): Mushshak	305 m (1,000 ft)
Super Mushshak	275 m (900 ft)
Landing from 15 m (50 ft): Mushshak	350 m (1,150 ft)
Super Mushshak	296 m (970 ft)
Landing run: Mushshak	140 m (460 ft)
Super Mushshak	153 m (500 ft)
Max endurance at 65% power at S/L, 10% reserves:	
Mushshak	5 h 10 min
Super Mushshak	4 h 3 min
g limits (both): A	+6/−3
U	+5.4/−2.7
N	+4.8/−2.4

UPDATED

PHILIPPINES

PADC

PHILIPPINE AEROSPACE DEVELOPMENT CORPORATION
PO Box 7395, Domestic Airport Post Office, Lock Box 1300, Domestic Road, Pasay City, Metro Manila
Tel: (+63 2) 832 37 57 and 23 81
Fax: (+63 2) 832 25 68
Telex: 66019 PADC PN
PRESIDENT: Reynato R Jose
SENIOR VICE-PRESIDENT: Josefina Laquindanum

PADC established 1973 as government arm for development of Philippine aviation industry; is now an attached agency of Department of Transportation and Communication (DOTC) and has technical workforce of about 200. Main activities are aircraft manufacturing and assembly; maintenance engineering; aircraft and spare parts sales; service centres for Britten-Norman Islander.

Past programmes include licensed assembly of 67 Islanders (including 22 for Philippine Air Force), 24 Agusta S.211 jet trainers, 16 (of 18) SF-260TP turboprop trainers and 44 BO 105s, as detailed in previous editions of *Jane's*. One of the two Italian-built SF.600 Canguros mentioned in the 1998-99 edition was delivered to the Philippine National Police (PNP); in early 1999 the other was outfitted for the Philippine Coast Guard.

Six Lancair ESs and two Lancair IVs were being assembled from PAI kits at that time for delivery to PNP. PADC's six-seat RPX-Alpha Hummingbird, the first helicopter to be designed and built in the Philippines, was described in the 1998-99 *Jane's*.

PADC has maintenance/repair/overhaul centre for Rolls-Royce 250 series turbine engines and for Textron Lycoming and Teledyne Continental piston engines of up to 298 kW (400 hp). Its Maintenance and Engineering Department undertakes FMS work on 250-C30 engines for Sikorsky helicopters and 250-B17 turboprops and propellers for PAF Nomads.

PADC's 30/70 per cent joint venture with Eurocopter, known as Eurocopter Philippines, offers sales and after-sales service and assembly of Eurocopter helicopters from CKD kits. Two Ecureuils have been supplied to the PNP and one to the Department of Environment and Natural Resources (DENR).

PADCs latest known venture is a partnership with National Airmotive Corporation (NAC) of the USA to establish a centre in the Philippines to support the aviation needs of Southeast Asia (and especially the needs of the Philippine Air Force) regarding overhaul of Rolls-Royce T56/501 and 250 series engines, QEC overhaul and repair, propeller refurbishment and C-130 Hercules airframe components.

UPDATED

PAI

PACIFIC AERONAUTICAL INC
25 First Avenue, Mactan Export Processing Zone, Lapu-Lapu City 6015
Tel: (+63 32) 334 02 83/84/86
Fax: (+63 32) 334 02 85
e-mail: pacaero@skyinet.net
PRESIDENT: Robert C Fair
VICE-PRESIDENTS:
Wilfredo Dela Cruz (Production)
Augusto V Dayrit (Operations)

PAI continues to manufacture kits for Lancair aircraft (see US section), and during 1999 shipped, on average, eight Lancair IV and two Lancair ES kits per month. Planned monthly rate for 2000 was three Lancair IV, two Lancair ES and six fast-build Legacy 2000 kits. Responsibility for the Lancair Columbia 300 lies with Pacific Aeronautical Inc (Malaysia).

Recent orders include two Lancair IVs, two ESs and two Legacy 2000s for the Mexican Navy. The first batch of Legacy 2000 kits was shipped to the USA at the end of June 2000.

UPDATED

POLAND

3XTRIM

ZAKŁADY LOTNICZE 3XTRIM Sp. z o.o.
(3XTRIM Aircraft Factory)
ulica Regera 109, PL-43 382, Bielsko-Biala
Tel: (+48 33) 818 91 21 and 818 92 51
Fax: (+48 33) 818 91 21
e-mail: biuro@3xtrim.pl
Web: http://www.3xtrim.pl
CHIEF DESIGNER: Adam Kurbiel
WORKS: Bielsko-Biala and Szczecin; Warsaw factory planned

Company was formed in 1996 as WNKL A Kurbiel, changing to its present title in 1999, but has its roots in the SZD-PZL-Bielsko which for some 40 years has been the major producer of Polish high-performance sailplanes and gliders. Its main current products are the two-seat lightplanes described in this entry, but glider production is under way and other (non-aerospace) products in composites are also manufactured.

NEW ENTRY

3XTRIM

TYPE: Two-seat lightplane/ultralight.
PROGRAMME: Development of Czech-produced Aerotrade Racek 2, of which some 20 built. VLA version being developed first (prototype SP-PUP made first flight 5 February 2000 and 60 hours flown by May 2000); static/fatigue testing October 1999 to January 2000; JAR-VLA tests completed 16 May 2000. Two prototypes of ultralight version flying by that time. Floatplane and glider tug versions were to follow later in 2000. Modified version marketed in Germany by USA company as SF 45 (which see).

CURRENT VERSIONS: **3Xtrim 550-VLA**: For aero club and flying school (PPL) training; also available with IFR instrumentation as border patrol and reconnaissance aircraft.

3Xtrim 450-UL: Ultralight version. Reduction of MTOW to required 450 kg (992 lb) achieved by wider use of CFRP and omission of some equipment and furnishing (for example, no door stays). Rotax 912 UL engine standard, but can be offered with 912 ULS for higher performance.

COSTS: 550-VLA DM125,000 including VAT; 450-UL DM99,950 including VAT (2000).

DESIGN FEATURES: Typical strut-braced high-wing cabin monoplane. Wing incidence 2° 30′; tailpane incidence −3°.

FLYING CONTROLS: Conventional and manual (pushrod ailerons and elevator, cable-operated rudder). Rudder and one-piece elevator horn-balanced; 450-UL has small, inset, electrically actuated trim tab in port half of elevator; tab in 550-VLA is twice as large and protrudes beyond trailing-edge. Slotted flaps deflect 33° up/15° down.

STRUCTURE: All-composites fuselage (glass and carbon fibre/epoxy cellular construction with plastics foam filling) with integral fin; GFRP sandwich single-spar wing.

LANDING GEAR: Fixed tricycle type, with GFRP, self-sprung cantilever mainwheel legs and castoring (±15°) nosewheel. Mainwheels are size 5.00-5, with 350×135 tyres, pressure 2.50 bars (36 lb/sq in), and have hydraulic disc brakes. Composites twin floats in design stage mid-2000.

POWER PLANT: *550-VLA*: One 73.5 kW (98.6 hp) Rotax 912 ULS flat-four engine, driving an MT 170 R 135-A2 two-blade fixed-pitch propeller (optional also on 450-UL).

450-UL: One 59.6 kW (79.9 hp) Rotax 912 UL and two-blade 3Xtrim 170 propeller with ground-adjustable pitch standard.

Fuel capacity (both) 70 litres (18.5 US gallons; 15.4 Imp gallons) standard; optional larger tank in 550-VLA.

ACCOMMODATION: Two seats side by side; dual controls standard. Fully enclosed, heated and ventilated cabin with upward-opening door on each side. Baggage space behind seats.

SYSTEMS: Hydraulic system for brakes only; 12 V 18 Ah battery for electrical power.

AVIONICS: ICOM A 200 radio and VFR instrumentation standard.

EQUIPMENT: BRS 5 ballistic recovery parachute.

DIMENSIONS, EXTERNAL:
Wing span	10.03 m (32 ft 11 in)
Wing chord: at root	1.30 m (4 ft 3¼ in)
at tip	1.00 m (3 ft 3¼ in)
Wing aspect ratio	8.1
Length overall	6.87 m (22 ft 6½ in)
Fuselage max width	1.18 m (3 ft 10½ in)
Height overall	2.405 m (7 ft 10¾ in)
Tailplane span	2.75 m (9 ft 0¼ in)
Wheel track	1.895 m (6 ft 2½ in)
Wheelbase	1.48 m (4 ft 10¼ in)
Propeller diameter (both)	1.70 m (5 ft 7 in)
Propeller ground clearance	0.195 m (7¾ in)

AREAS:
Wings, gross	12.40 m² (133.5 sq ft)
Horizontal tail surfaces (total)	1.73 m² (18.62 sq ft)

WEIGHTS AND LOADINGS:
Weight empty: A	275 kg (606 lb)
B	315 kg (694 lb)
Max T-O weight: A	450 kg (992 lb)
B	550 kg (1,212 lb)
Max wing loading: A	36.3 kg/m² (7.43 lb/sq ft)
B	44.4 kg/m² (9.08 lb/sq ft)
Max power loading: A	7.55 kg/kW (12.41 lb/hp)
B	7.48 kg/kW (12.29 lb/hp)

PERFORMANCE (A* and B as above):
Never-exceed speed (VNE):	
A	108 kt (200 km/h; 124 mph)
B	118 kt (220 km/h; 136 mph)
Max level speed: A	89 kt (165 km/h; 103 mph)
B	91 kt (170 km/h; 106 mph)
Max cruising speed at 75% power:	
A	82 kt (152 km/h; 94 mph)
B	86 kt (160 km/h; 99 mph)
Stalling speed, flaps up: A	37 kt (68 km/h; 43 mph)
B	38 kt (70 km/h; 44 mph)
Max rate of climb at S/L: A, B	270 m (886 ft)/min
T-O to 15 m (50 ft): A, B	250 m (820 ft)
Range with max standard fuel:	
A, B**	405 n miles (750 km; 466 miles)
g limits: A	+4.0/−2.0
B	+3.8/−1.5

*With standard engine
**B 675 n miles (1,250 km; 776 miles) with larger tank

NEW ENTRY

The 3Xtrim 550-VLA prototype wearing pre-certification registration *(Paul Jackson/Jane's)* 2001/0093672

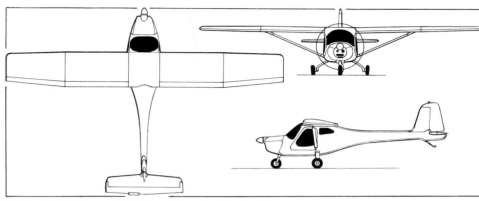
3Xtrim 550-VLA two-seat lightplane *(James Goulding/Jane's)* 2001/0093619

AERO

AERO Sp. z o. o.
Wat Miedzeszyński 844/2, PL-03-942 Warszawa
Tel: (+48 22) 616 20 87
Fax: (+48 22) 617 85 28
DIRECTOR: Ing Tomasz Antoniewicz

Company formed in 1994. Three P.220 S Koala kits were ordered from Aerotechnik (see Czech Republic section) that year, being given Polish designation suffix AT-2 (AT-1 having been a single-seat homebuilt design of some years earlier). These all differed dimensionally and were variously fitted with Walter Mikron or 55.9 kW (75 hp) Limbach engines (first flight 12 December 1995 by c/n 001 SP-PUL, later SP-FUL). Third aircraft subsequently developed with more powerful Limbach and other modifications for production in Poland as AT-3.

VERIFIED

AERO AT-3

TYPE: Two-seat lightplane.
PROGRAMME: Completed initially (c/n 003, SP-PUH) as P.220 S-AT2, but subsequently modified to AT-3 with more powerful engine, revised dimensions and higher weights; given new c/n of 001 and first flown in new configuration 18 January 1998. Received JAR-VLA certification 14 May 1999 and re-registered SP-GUH. First production deliveries were about to be made in late 1999, but no further registrations had been reported by mid-2000.

CUSTOMERS: Orders received at time of certification, though no figure quoted. Potential Polish market for 200 over five-year period.

COSTS: US$50,000 (1999/2000).

Main differences from Czech version are as follows:

DESIGN FEATURES: Changes from AT-2/P.220 S include approximately 1 m² (10.8 sq ft) more wing area, 4 cm (1.6 in) longer cockpit, enlarged fin and rudder, and strengthened landing gear.

FLYING CONTROLS: Manual; ailerons, rudder and all-moving tailplane with anti-balance tab; flaps.

STRUCTURE: Primarily metal, with some skins and fairings of glass fibre.

LANDING GEAR: Non-retractable tricycle type; mainwheel disc brakes; speed fairings on all three wheels.

POWER PLANT: One 67.1 kW (90 hp) Limbach L-2400EB3AC flat-four engine with two-blade Hoffmann HO-11A-150B/108L propeller. Fuel capacity 64 litres (16.9 US gallons; 14.1 Imp gallons).

ACCOMMODATION: Two seats side by side. Upward-opening canopy, hinged at front.

AVIONICS: *Instrumentation:* VFR standard; IFR (plus ADF, directional gyro, attitude indicator and turn co-ordinator) optional.

DIMENSIONS, EXTERNAL:
Wing span	7.32 m (24 ft 0¼ in)
Wing chord, constant	1.27 m (4 ft 2 in)
Wing aspect ratio	5.8
Propeller diameter	1.50 m (4 ft 11 in)
Length overall	5.90 m (19 ft 4¼ in)
Height overall	2.19 m (7 ft 2¼ in)

AREAS:
Wings, gross	9.30 m² (100.1 sq ft)

WEIGHTS AND LOADINGS:
Weight empty	354 kg (780 lb)
Max T-O weight	582 kg (1,283 lb)
Max wing loading	62.6 kg/m² (12.82 lb/sq ft)
Max power loading	8.68 kg/kW (14.26 lb/hp)

PERFORMANCE:
Never-exceed speed (VNE)	155 kt (287 km/h; 178 mph)
Max level speed	105 kt (195 km/h; 121 mph)
Max cruising speed	97 kt (180 km/h; 112 mph)
Max speed for flap extension	87 kt (162 km/h; 100 mph)
Stalling speed	45 kt (83 km/h; 52 mph)
Max rate of climb at S/L	180 m (591 ft)/min
Service ceiling	3,000 m (9,840 ft)
T-O run	240 m (790 ft)
Landing run	180 m (590 ft)
Max range, with reserves	378 n miles (700 km; 435 miles)
Endurance	4 h 30 min
g limits	+3.8/−1.5

UPDATED

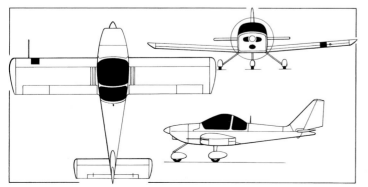

AT-3 prototype two-seat lightplane *(R J Malachowski)* 2001/0093621

Aero (Antoniewski) AT-3 two-seat lightplane *(James Goulding/Jane's)*
1999/0051150

AGROLOT

FUNDACJA AGROLOT (Agrolot Foundation)
aleja Krakowska 110/114, PL-00-971 Warszawa
Tel: (+48 22) 846 00 31 or 846 00 69 extn 586
Fax: (+48 22) 846 27 01
CHAIRMAN: Andrzej Słociński

This foundation (established 1990) is responsible for the PZL-126P Mrówka programme previously managed by PZL Warszawa-Okecie. The prototype made its maiden flight in October 2000, some three years later than originally planned.
UPDATED

AGROLOT PZL-126P MRÓWKA 2001
English name: Ant
TYPE: Agricultural sprayer.
PROGRAMME: See under PZL Warszawa-Okecie in 1993-94 and previous editions for early history. PZL-126 prototype (SP-PMA: see 1995-96 and earlier editions for details) made first flight 20 April 1990; this aircraft was used in 1994 to test a modified 44.7 kW (60 hp) PZL-F2A-120-C1 two-cylinder engine driving a two-blade fixed-pitch propeller. It was then withdrawn for rebuilding as the prototype PZL-126P planned operational version, which was designed and developed at the PZL Warszawa-Okecie factory. Project is funded by Agrolot Foundation; chief designer is Andrzej Słociński. First flight of the PZL-126P (SP-PMB) was first envisaged for 1997, but aircraft not completed until 1998, marked 'Mrówka 2001'; first flight was eventually made on 20 October 2000.
CURRENT VERSIONS: **PZL-126 2001:** Under conversion since 1995. Increased wing span and area, more powerful engine, larger fuel tanks and increased payload compared with PZL-126 prototype. Intended as economical carrier of airborne systems for forestry protection (for example, combating pests, identifying diseased vegetation, detecting/controlling forest fires, spreading non-toxic small-volume selective-action agents for plant protection such as Trichrogramma parasitic wasp eggs laid in eggs of host insect). Other applications include air pollution monitoring and pipeline patrol.
Description applies to PZL-126P 2001.
DESIGN FEATURES: Single-engined low-wing monoplane with rectangular tail surfaces; meets requirements of FAR Pt 23. Possible alternative use as glider tug. Semi-monocoque fuselage with equipment bay in lower portion aft of cockpit, accessible through hatch; upswept rear fuselage facilitates access to this bay. Quick-fastening lock beneath each wingtip for attachment of integral spraypod between main wing and tip. Aircraft can be dismantled quickly for towing on own landing gear by light all-terrain vehicle.
Wing aerofoil section NASA GA(W)-1; dihedral 3°.
FLYING CONTROLS: Manual. Three-section area-increasing flaps and flaperons on each wing, actuated by pushrods and interconnected by single central system located under cockpit floor; elevator actuated by pushrods, rudder by cables from adjustable rudder pedals; trim tab on elevator, ground-adjustable tab on rudder.
STRUCTURE: All-metal except for GFRP/epoxy engine cowling, laminated wingtips and fairings. Single-spar wings and fixed tail surfaces; wings mounted on fuselage mandrels and offset spars connected with single pin; front part of fuselage has two 'tusks' for engine and nosewheel mounting.
LANDING GEAR: Non-retractable tricycle type. Self-springing cantilever mainwheel legs of duralumin; castoring nosewheel with oleo-pneumatic shock-absorber. Wheel and tyre size 350×135 on all three units. Hydraulic differential mainwheel disc brakes.
POWER PLANT: One 74.6 kW (100 hp) Teledyne Continental O-200A flat-four engine, driving a McCauley 1A100MCM6950 two-blade metal propeller. Integral fuel tank in each wing, combined capacity 200 litres (52.8 US gallons; 44.0 Imp gallons).
ACCOMMODATION: Single-seat cockpit, fitted with airbag enabling in-flight seat adjustment. One-piece organic glass moulded canopy, opening sideways to starboard.
SYSTEMS: Hydraulic system for mainwheel brakes only; 24 V electrical system; hand pump to inflate seat cushion.
AVIONICS: *Comms:* Honeywell KX 155 com/nav, KY 96A VHF transceiver and KT 76A transponder.
Flight: GPS 95XL global positioning; stall warning system.
Instrumentation: Prototype has IFR capability, but basic version fitted with VFR instrumentation. Instrument panel, with cover, can be lifted up for access to instruments and front part of cockpit.
EQUIPMENT: Spraypod can be fitted to outer end of each main wing panel, inboard of detachable tip. Pod is an integral unit with all necessary attachments for spraying, attached by quick-fastening locks, electric multiplug socket supplying power to atomisers, and electric valve controlling outflow of liquid from tank. Optional third spray unit can be mounted under fuselage. Up to 35 ha (86.5 acres) can be sprayed in one flight with 70 litres (18.5 US gallons; 15.4 Imp gallons) of pesticide delivered at rate of 2 litres (0.53 US gallon; 0.44 Imp gallon)/ha.
Special biological spreader developed for dosing and spreading eggs of Trichrogramma wasp; eggs are carried in capsules in reel of tape covered with thin paper, spreader holding four such reels. Spreader is powered by electric motor and activated or stopped by push-button on throttle lever; 3 kg (6.6 lb) load of capsules is sufficient, at typical dispersal rate of four capsules every 50 m (164 ft) in rows 50 m apart, to seed an area of 800 ha (1,977 acres). Rotational mounting of spreader allows for full deflection and easy replacement of egg reels.
Aircraft can also carry miniaturised equipment such as photographic, video, thermal or other systems, coupled to satellite navigation system, monitoring and recording pictures or other signals from space.

DIMENSIONS, EXTERNAL:
Wing span: excl pods	7.66 m (25 ft 1½ in)
incl pods	8.46 m (27 ft 9 in)
Wing chord, constant:	
excl extended flaps/flaperons	0.75 m (2 ft 5½ in)
incl extended flaps/flaperons	0.92 m (3 ft 0¼ in)
Wing aspect ratio	8.5
Length overall	5.25 m (17 ft 2¾ in)
Height overall	2.80 m (9 ft 2¼ in)
Tailplane span	2.20 m (7 ft 2½ in)
Wheel track	2.15 m (7 ft 0¾ in)
Wheelbase	1.35 m (4 ft 5¼ in)
Propeller diameter	1.75 m (5 ft 9 in)

AREAS:
Wings, gross: excl pods	6.87 m² (73.9 sq ft)
incl pods	7.27 m² (78.3 sq ft)
Flaperons (total)	0.48 m² (5.17 sq ft)
Trailing-edge flaps (total)	0.97 m² (10.44 sq ft)
Vertical tail surfaces (total)	0.73 m² (7.86 sq ft)
Horizontal tail surfaces (total)	1.08 m² (11.63 sq ft)

WEIGHTS AND LOADINGS:
Max T-O and landing weight	575 kg (1,267 lb)
Max wing loading:	
without pods	83.7 kg/m² (17.14 lb/sq ft)
with pods	79.1 kg/m² (16.20 lb/sq ft)
Max power loading	7.72 kg/kW (12.68 lb/hp)

PERFORMANCE (estimated):
Max level speed	105 kt (195 km/h; 121 mph)
Operating speed	75 kt (140 km/h; 87 mph)
Max rate of climb at S/L	more than 360 m (1,181 ft)/min

UPDATED

The PZL-126P prototype made its first flight on 20 October 2000

IFR instrument panel of the Mrówka prototype *(Andrzej Słociński/Agrolot)*

Mrówka 2001 ventral equipment bay with camera and video camera *(Andrzej Słociński/Agrolot)*

Prototype PZL-126P Mrówka with spraypods fitted *(R J Malachowski)*

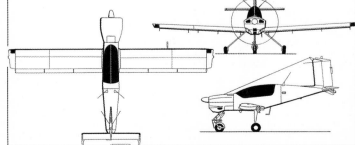
PZL-126P Mrówka light agricultural aircraft *(Paul Jackson/Jane's)*

AVIATION FARM

AVIATION FARM LIMITED
Dobino 12, PL-78-326 Toporzyk
Tel: (+48 961) 626 87
PRESIDENT: Zenona Andruszkiewicz
VICE-PRESIDENT: Andrzej Bieniarz

SALES AGENT:
East European Markets Ltd
ulica Widok 12, PL-00-023 Warszawa
Tel: (+48 22) 827 48 58
Fax: (+48 22) 827 77 88

Mr Bieniarz, a professional pilot of agricultural aircraft, purchased a farm in 1990 after being made redundant. Following early crop-raising activity on the appropriately renamed Aviation Farm, more land was acquired and dairy cattle were introduced, the sale of milk to neighbouring Germany providing the means to erect an aircraft hangar in which he completed his first aircraft, a 176 kg (388 lb) single-seat ultralight motor glider. Rights to the J-5 Marco were acquired from the Alpha company in 1990; its KFM-107 engines, after an abortive two-year attempt to set up licensed manufacture in Poland, were imported from Italy; Czech M 202 was planned to be offered as an alternative, but no reports of company activity have been received since 1996.

UPDATED

AVIATION FARM J-5 MARCO
TYPE: Single-seat ultralight/motor glider.
PROGRAMME: Original J-5 designed in 1983 (first flight 30 October) by Jaroslaw Janowski and marketed in kit form by Marco-Elektronik Co of Lodz (see 1986-87 *Jane's*); Polish certification 1987; later sold by Hewa-Technics of Germany in both kit and ready-assembled forms after JAR 22 certification in 1990 (1990-91 *Jane's*); Polish company relocated to Krakow and renamed Alpha (1992-93 *Jane's*); present company acquired licence from Alpha in 1990, producing in fully assembled form only. JAR 23 certification obtained 1990; JAR-VLA certification was expected to be awarded in 1997, but no confirmation of this had been received by third quarter 2000. See 2000-01 *Jane's* for description and illustration.
A larger version produced by J & AS Aero Design of Lodz (which see) is known as the J6 Fregata.
CURRENT VERSIONS: **Motor glider:** Basic version with monowheel main landing gear; Polish and JAR 22 certification.
Aircraft: JAR 23 version with twin-mainwheel landing gear; certified 1990.

UPDATED

CENZIN

CENZIN FOREIGN TRADE ENTERPRISE CO LTD
ulica Czerniakowska 81/83, PL-00-957 Warszawa
Tel: (+48 22) 841 12 63
Fax: (+48 22) 841 12 66
e-mail: dircom@cenzin.com.pl
Web: http://www.cenzin.com.pl
PRESIDENT: Tadeusz Bednarek
MARKETING MANAGER: Agnes Grzedzinska

Cenzin Foreign Trade Enterprise is the largest Polish export/import company in the defence sector. It specialises in export of Polish-built products for land, air, naval and police forces, and is a major importer of arms and military equipment for the Polish armed forces, special forces and police.

UPDATED

E&K

E&K Sp. z o.o.
ulica Radziwiłłowska 5, PL-20-080 Lublin
Tel: (+48 81) 743 73 57
Fax: (+48 81) 743 66 12
PRESIDENT: Andrzej Izdebski

VERIFIED

E&K SBM-03 KOS
TYPE: Side-by-side ultralight.
PROGRAMME: Single-seat prototype (SP-PEG) first flew August 1998; 55 hours by May 1999; start of certification (in Germany), and first flight of two-seat prototype, expected to follow shortly thereafter, but not reported by third quarter 2000.
FLYING CONTROLS: Conventional and manual. Mass-balanced rudder; flaps.
STRUCTURE: Glass fibre/epoxy construction; semi-monocoque fuselage.
LANDING GEAR: Fixed, with tailwheel. Mainwheels on cantilever, self-sprung leg bows.
POWER PLANT: One 59.6 kW (79.9 hp) Rotax 912 A4 (VLA version) or 912 UL4 (ultralight) flat-four engine; KG-3-150L three-blade propeller. Option of 47.8 kW (64.1 hp) Rotax 582 UL in ultralight. Two fuel tanks aft of cockpit, combined capacity 40 litres (10.6 US gallons; 8.8 Imp gallons).
AVIONICS: *Comms:* Honeywell KY 96A VHF radio.
Flight: Navigation system of customer's choice.
Instrumentation: Standard VFR.
EQUIPMENT: Ballistic parachute in front of cockpit.
DIMENSIONS, EXTERNAL:
Wing span 8.50 m (27 ft 10¾ in)
Length overall 5.95 m (19 ft 6¼ in)
Height overall 1.80 m (5 ft 10¾ in)
AREAS:
Wings, gross 10.20 m² (109.8 sq ft)
WEIGHTS AND LOADINGS:
Weight empty 265 kg (584 lb)
Max T-O weight 450 kg (992 lb)
PERFORMANCE (Rotax 912 A4 engine):
Max level speed 113 kt (210 km/h; 130 mph)
Cruising speed at 75% power 94 kt (175 km/h; 109 mph)
Stalling speed: flaps up 45 kt (83 km/h; 52 mph)
flaps down 39 kt (72 km/h; 45 mph)
Max rate of climb at S/L 540 m (1,771 ft)/min
Range with max fuel, 10% reserves
  367 n miles (680 km; 422 miles)
g limits +3.8/−2.5

UPDATED

SP-PEG, the single-seat Kos prototype *(Paul Jackson/Jane's)* 2000/0062600

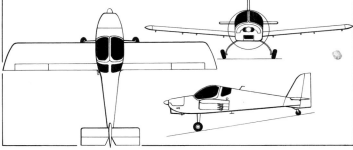
E&K SBM-03 Kos light aircraft *(James Goulding/Jane's)* 2000/0062601

HE-RO'S

HE-RO'S DESIGN LUBLIN
PO Box 106, PL-20-950 Lublin 1
Tel/Fax: (+48 81) 743 47 70
MANAGER AND DESIGNER: H P Rozwadowski

UK ADDRESS:
92 Thorn Court, Manchester M6 5EL
Tel/Fax: (+44 161) 737 93 44

Design work on what is intended to become a related series of aircraft began in 1993; current activities include the HB-2 and a new ultralight/motor glider design, the HGM-2 LadyBird, announced in March 1999. In March 2000, design of the HB-2 was approaching finalisation; market-based partners then being sought for HGM-2 in co-operation with student research teams.
Company's first development partner is Aerokris ZPHU, a Polish plans-based manufacturer of the Zenair CH 601; research undertaken by Lublin Technical University.

UPDATED

HE-RO'S HB-2
TYPE: Two/three-seat lightplane.
PROGRAMME: Design (as HC-2: see 1999-2000 *Jane's*) began 1993; models up to quarter-scale used for proof of concept. Further developed, as HB-2, with deeper fuselage; port-side door to rear seats; and enlarged engine cowling to accommodate higher-rated power plants. First versions available will be the HB-2TR and HB-2CR; others are planned to follow.
CURRENT VERSIONS: **HB-2TR Beta-Trainer:** Two-seat standard, three-seat optional, all with flying controls.
HB-2CR Beta-Cruiser: Three-seat standard, with optional dual controls

HB-2HCR/TR Beta-High Wing Cruiser/Trainer: High-wing development of single-engined HB-2, with higher-rated power plants (up to 149 kW; 200 hp) and fixed landing gear.
HE-2TCR/TR Executive-Twin Cruiser/Trainer: With twin 52.2 to 59.7 kW (70 to 80 hp) engines and retractable landing gear.

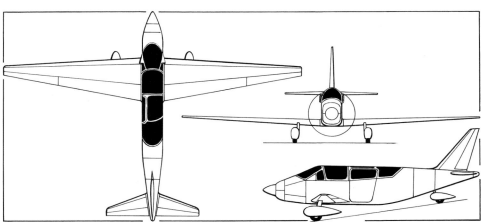
He-Ro's HB-2CR Beta-Cruiser trainer and tourer *(James Goulding/Jane's)* 2000/0085837

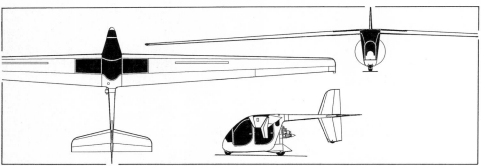

General arrangement of the He-Ro's HGM-2 LadyBird motor glider as envisaged in early 1999

HJ-2R Jet Research: Utilising fuselage core of HB-2 to develop design methodologies for high-subsonic aircraft.

DESIGN FEATURES: Targets of low operating cost and long range. Sharply tapered wings with modest sweepback; tail surfaces tapered and with medium sweepback. Standard versions have 450 kg (992 lb) MTOW; optional versions increased by 40 per cent to 630 kg (1,388 lb).

FLYING CONTROLS: Conventional and manual. Flaps on inboard three-fifths of wing trailing-edge.

STRUCTURE: All-metal.

LANDING GEAR: Fixed, tailwheel type; single wheel and speed fairing on each. Retraction as later option.

POWER PLANT: One 52.2 to 59.7 kW (70 to 80 hp) unidentified piston engine driving a three-blade propeller

ACCOMMODATION: Two or three persons in tandem, according to version; fixed windscreen, with individual canopy for each occupant.

DIMENSIONS, EXTERNAL:
Wing span 9.70 m (31 ft 10 in)
Length overall 6.99 m (22 ft 11¼ in)
DIMENSIONS, INTERNAL:
Cockpit max width 0.64 m (2 ft 1¼ in)
AREAS:
Wings, gross 8.61 m² (92.7 sq ft)

WEIGHTS AND LOADINGS (A: standard, B: optional):
Weight empty: A 220 kg (485 lb)
  B 240 kg (529 lb)
Max payload: A 210 kg (463 lb)
  B 320 kg (705 lb)
Max fuel weight: A 80 kg (176 lb)
  B 160 kg (353 lb)
Max T-O weight: A 450 kg (992 lb)
  B 630 kg (1,388 lb)
Max wing loading: A 52.3 kg/m² (10.70 lb/sq ft)
  B 73.2 kg/m² (14.99 lb/sq ft)
PERFORMANCE (estimated; A and B as above):
Cruising speed 108 kt (200 km/h; 124 mph)
Range: with max payload:
  A 337 n miles (625 km; 388 miles)
  B 1,517 n miles (2,810 km; 1,746 miles)
with max fuel:
  A 1,349 n miles (2,500 km; 1,553 miles)
  B 2,699 n miles (5,000 km; 3,106 miles)

VERIFIED

HE-RO'S HGM-2 LADYBIRD

TYPE: Tandem-seat ultralight/motor glider.

PROGRAMME: Design being finalised in early 1999; development partners then being sought. No subsequent information.

CURRENT VERSIONS: Meets design criteria as both motor glider and ultralight. Optional 'tropical' version without cabin doors and winglets.

DESIGN FEATURES: Intended for club basic training and leisure flying; objectives are high performance and low operating costs.
Cantilever high-mounted wing with approximately 3° anhedral; downward-pointing winglets serve also as supports for self-launching. Very short fuselage. Fins and rudders above and below tailboom.

FLYING CONTROLS: Conventional and manual. Airbrake in upper surface of each wing. Optional flaps on inboard 60 per cent of wing trailing-edge.

STRUCTURE: Metal and composites.

LANDING GEAR: Non-retractable nosewheel and single mainwheel in tandem; nosewheel self-castoring ±14°. Small, non-retractable tailwheel in base of lower fin for protection in event of tail-down landings.

POWER PLANT: Range of small piston engines of 14.9 to 26.1 kW (20 to 35 hp). Two-blade ground-adjustable pitch pusher propeller standard; mechanically or electrically variable-pitch propeller optional.

DIMENSIONS, EXTERNAL:
Wing span, incl winglets 15.00 m (49 ft 2½ in)
Length: fuselage pod 3.99 m (13 ft 1 in)
  overall 4.69 m (15 ft 4¾ in)
Height overall 2.18 m (7 ft 1¾ in)
DIMENSIONS, INTERNAL:
Cabin max width 0.64 m (2 ft 1¼ in)
AREAS:
Wings, gross 11.99 m² (129.1 sq ft)
WEIGHTS AND LOADINGS:
Weight empty 139 kg (306 lb)
Max fuel weight 20 kg (44 lb)
Max T-O weight 299 kg (659 lb)
PERFORMANCE, POWERED (estimated):
Max permissible diving speed
  97 kt (180 km/h; 111 mph)
Cruising speed 65 kt (120 km/h; 75 mph)
PERFORMANCE, UNPOWERED (estimated):
Best glide ratio 30

UPDATED

IL

INSTYTUT LOTNICTWA (Aviation Institute)

aleja Krakowska 110/114, PL-02-256 Warszawa
Tel: (+48 22) 846 00 11 and 846 08 11
Fax: (+48 22) 846 44 32
e-mail: ilot@ilot.edu.pl
GENERAL DIRECTOR: Dr Eng Witold Wisniowski
CHIEF CONSULTANT FOR SCIENTIFIC AND TECHNICAL
  CO-OPERATION: Jerzy Grzegorzewski, MSc Eng

Aircraft Department
Address and fax as above
Tel: (+48 22) 846 01 71
CHIEF DESIGNER, I-23: Dr Eng Alfred Baron

Helicopter Department
Address and fax as above
Tel: (+48 22) 846 57 51
e-mail: romicki@ilot.edu.pl
HEAD OF DEPARTMENT AND IS-2 CHIEF DESIGNER:
  Zbigniew M Romicki, MSc

Founded 1926; directly subordinate to Ministry of Heavy and Machine Building Industry and responsible for most research and development work in Polish aviation industry; conducts scientific research, including investigation of problems associated with low-speed and high-speed aerodynamics, static and fatigue tests, development and testing of aero-engines, flight instruments, space science instrumentation, and other equipment, flight tests, and materials technology; also responsible for construction of aircraft and aero-engines. Current programmes include the I-23 and I-25 light aircraft and the IS-2 small helicopter.

In 1997 IL signed an agreement with Boeing covering possible co-operation in 20 advanced technology areas related to the latter's sales campaign for F-18 Hornets to the Polish Air Force.

UPDATED

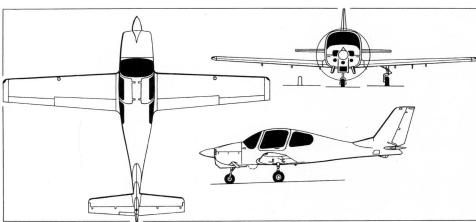

Instytut Lotnictwa I-23 Manager four-seat light aircraft *(Mike Keep/Jane's)*

IL I-23 MANAGER

TYPE: Four-seat lightplane.

PROGRAMME: Launched June 1992; funded by grant from Government Scientific Research Committee; construction started January 1994 of one static test (001) and two flying prototypes; first flight, by 002/SP-PIL, rescheduled for 1998 but delayed until 20 February 1999; original name Ibis changed to Manager 1999; series production planned to begin 2001, followed by certification to Normal category FAR Pt 23.

DESIGN FEATURES: Intended as safe, low-cost, low fuel consumption private owner aircraft. Conventional low-wing, tricycle-gear design with fully enclosed cabin.
Wing has ILL-217 (root) and ILL-213 (tip) aerofoil sections, 2° 30′ leading-edge sweepback, 4° dihedral, 2° 30′ incidence and −1° twist.
Design permits use of any Textron Lycoming flat-four engine from 119 to 149 kW (160 to 200 hp), option for fixed or retractable landing gear, and equipment to customer's specifications.

FLYING CONTROLS: Conventional and manual. Flaps; trim tab in port elevator.

STRUCTURE: Mainly composites (GFRP/CFRP/epoxy), with sandwich skins. First Polish aircraft with prepreg composites.

LANDING GEAR: Electrohydraulically retractable tricycle type, with oleo-pneumatic shock-absorbers; steerable (±25°) nosewheel; mainwheels retract inward, nosewheel rearward, latter remaining half exposed when retracted; Parker wheels and brakes; Goodrich 5.00-5 tyres (main

Prototype Instytut Lotnictwa I-23 Manager (134 kW; 180 hp O-360 engine)

and nose), pressure 3.60 bars (52 lb/sq in) on mainwheels and 3.10 bars (45 lb/sq in) on nosewheel. Non-retractable gear optional.

POWER PLANT: Standard engine is a Textron Lycoming O-360-A1A flat-four (134 kW; 180 hp at 2,700 rpm), driving a Hartzell HC-C2YK-1BF/F766A-4 two-blade constant-speed propeller. Higher- or lower-powered alternatives optional (see Design Features). Integral fuel tank in each wing, combined capacity 191 litres (50.5 US gallons; 42.0 Imp gallons). Gravity fuelling points on each wing. Oil capacity 7.6 litres (2.0 US gallons; 1.7 Imp gallons).

ACCOMMODATION: Four persons (pilot and three passengers) on two rows of seats. Upward-opening door each side. Cabin heated and ventilated.

SYSTEMS: Electrohydraulic system for landing gear actuation; hydraulic system for mainwheel brakes only. Electrical power from 14 V 70 A alternator and 12 V 35 Ah battery. Heated pitot head.

AVIONICS: *Comms:* Honeywell com transceiver and ATC transponder standard.
Flight: Honeywell nav transceiver, ADF, GPS and directional gyro standard; optional full IFR system with KNS 81 integrated nav/RNav, DME and KCS 55A HSI.

DIMENSIONS, EXTERNAL:
Wing span	8.94 m (29 ft 4 in)
Wing chord: at root	1.31 m (4 ft 3½ in)
at tip	0.84 m (2 ft 9 in)
Wing aspect ratio	8.0
Length: overall	7.10 m (23 ft 3½ in)
Fuselage	5.70 m (18 ft 8½ in)
Height overall	2.49 m (8 ft 2 in)
Tailplane span	3.20 m (10 ft 6 in)
Wheel track	2.89 m (9 ft 5¾ in)
Wheelbase	1.65 m (5 ft 5 in)
Propeller diameter	1.83 m (6 ft 0 in)
Propeller ground clearance	0.20 m (7¾ in)

AREAS:
Wings, gross	10.00 m² (107.6 sq ft)
Ailerons (total)	0.73 m² (7.86 sq ft)
Trailing-edge flaps (total)	1.48 m² (15.93 sq ft)
Fin	0.95 m² (10.23 sq ft)
Rudder	0.48 m² (5.17 sq ft)
Tailplane	2.08 m² (22.39 sq ft)
Elevators (total)	0.83 m² (8.93 sq ft)

WEIGHTS AND LOADINGS:
Weight empty, equipped	690 kg (1,521 lb)
Max standard fuel weight	130 kg (287 lb)
Max T-O and landing weight	1,150 kg (2,535 lb)
Max wing loading	115.0 kg/m² (23.55 lb/sq ft)
Max power loading (180 hp)	8.57 kg/kW (14.08 lb/hp)

PERFORMANCE (estimated):
Never-exceed speed (V_{NE})	170 kt (315 km/h; 195 mph)
Max level speed	162 kt (300 km/h; 186 mph)
Max cruising speed at 2,400 m (7,875 ft)	159 kt (295 km/h; 183 mph)
Econ cruising speed at 2,400 m (7,875 ft)	151 kt (280 km/h; 174 mph)
Stalling speed, flaps up, power off	61 kt (113 km/h; 71 mph)
Max rate of climb at S/L	270 m (886 ft)/min
Service ceiling	4,400 m (14,440 ft)
T-O run	250 m (820 ft)
T-O to 15 m (50 ft)	500 m (1,640 ft)
Landing from 15 m (50 ft)	480 m (1,575 ft)
Landing run	300 m (985 ft)
Range with max payload	756 n miles (1,400 km; 870 miles)

UPDATED

IL IS-2

TYPE: Two-seat helicopter.
PROGRAMME: Initiated 1 June 1994; three prototypes planned (first and second for static and ground resonance testing, third for flight test); construction started 1 June 1995; first flight long delayed, but currently planned for second quarter 2001; to be certified to JAR 27.
COSTS: Estimated programme cost US$3.5 million; target unit price US$180,000 (2000).

Computer-generated image of the projected I-25 AS utility lightplane

DESIGN FEATURES: Designed for basic training and pipeline/border patrol, with low cost of operation and high level of safety. Conventional pod and boom configuration, with three-blade main rotor and four-blade Fenestron-type ducted tail rotor. Comparatively low rotation rates of both rotors minimise noise.
Main rotor blades have IL HX4A1 aerofoil section with thickness/chord ratios of 12 per cent at root and 9 per cent at tip.

FLYING CONTROLS: Conventional and manual.

STRUCTURE: Fuselage truss frame of steel tube and aluminium alloy, with glass fibre skin in cockpit area; aluminium alloy semi-monocoque tailboom; glass fibre composites tailfin; all rotor blades of glass fibre/epoxy and carbon fibre composites.

LANDING GEAR: Conventional metal skid type, with small ground wheels.

POWER PLANT: One 134 kW (180 hp) Textron Lycoming O-360 flat-four engine. Fuel in two tanks, total capacity 110 litres (29.1 US gallons; 24.2 Imp gallons).

ACCOMMODATION: Two seats side by side; forward-opening window/door on each side. Baggage space aft of seats. Cabin heated and ventilated.

SYSTEMS: No hydraulic or pneumatic systems; 28 V DC electrical system. De-icing optional.

AVIONICS: *Comms:* Honeywell KT 76A transponder.
Flight: Honeywell KLX 135 GPS receiver and KR 87 ADF.
Instrumentation: Conventional; includes directional gyro and Goodrich attitude gyro.

DIMENSIONS, EXTERNAL:
Main rotor diameter	7.50 m (24 ft 7¼ in)
Main rotor blade chord: at root	0.20 m (7¾ in)
at tip	0.15 m (6 in)
Tail rotor diameter	0.865 m (2 ft 10 in)
Length overall, rotor turning	8.93 m (29 ft 3½ in)
Fuselage: Length	7.07 m (23 ft 2¼ in)
Max width	1.40 m (4 ft 7 in)
Height overall	2.83 m (9 ft 3½ in)
Tailplane span	1.18 m (3 ft 10½ in)

DIMENSIONS, INTERNAL:
Cabin: Length	1.66 m (5 ft 5¼ in)
Max width	1.35 m (4 ft 5¼ in)
Max height	1.35 m (4 ft 5¼ in)
Floor area	1.60 m² (17.2 sq ft)
Volume	2.2 m³ (78 cu ft)

AREAS:
Main rotor disc	44.18 m² (475.5 sq ft)
Tail rotor disc	0.59 m² (6.35 sq ft)
Fin, incl tail rotor cover	1.16 m² (12.49 sq ft)
Tailplane	0.47 m² (5.06 sq ft)

WEIGHTS AND LOADINGS:
Weight empty	550 kg (1,213 lb)
Max internal fuel	82 kg (181 lb)
Max payload	100 kg (220 lb)
Max T-O weight	790 kg (1,741 lb)
Max disc loading	17.9 kg/m² (3.66 lb/sq ft)

Transmission loading at max T-O weight and power 5.89 kg/kW (9.67 lb/hp)

PERFORMANCE (estimated):
Never-exceed speed (V_{NE})	118 kt (220 km/h; 136 mph)
Max level speed at 500 m (1,640 ft)	103 kt (190 km/h; 118 mph)
Max cruising speed at 500 m (1,640 ft)	94 kt (175 km/h; 109 mph)
Econ cruising speed at 500 m (1,640 ft)	84 kt (155 km/h; 96 mph)
Max rate of climb at S/L	414 m (1,358 ft)/min
Hovering ceiling: IGE	1,500 m (4,920 ft)
OGE	1,000 m (3,280 ft)
Max range	216 n miles (400 km; 248 miles)
Max endurance	4 h 5 min

UPDATED

IL I-25 AS

TYPE: Two-seat lightplane.
PROGRAMME: Design started 1 January 1996 and first brief details released 1997. For certification to JAR 23, but present state of progress not disclosed.

DESIGN FEATURES: Braced high-wing monoplane with good view from cabin; suitable for club flying, training, glider towing and patrol.
Constant-chord wings with ILL-417 aerofoil section, 1° dihedral, 2° 30′ incidence and 0° twist.

FLYING CONTROLS: Conventional and manual. Elevators and rudder aerodynamically and horn-balanced; trim tab in one elevator, controlled by trim wheel. Single-slotted flaps, deflecting 15° for take-off and 40° for landing.

STRUCTURE: All-metal primary structure, including 4130 welded steel tube cabin truss; semi-monocoque wings, engine cowling, rear fuselage and tail unit; GFRP wing and tail unit tips.

LANDING GEAR: Non-retractable tricycle type, with cantilever self-sprung mainwheel legs and nose oleo-pneumatic shock-absorption. All three wheels have 360×125 tyres, with pressure of 2.0 bars (29 lb/sq in). Parker Hannifin single-disc hydraulic mainwheel brakes. Nosewheel steerable ±30°.

POWER PLANT: One 119 kW (160 hp) Textron Lycoming O-320-D2A flat-four engine, driving a Sensenich 74DM6S5-0-62 two-blade fixed-pitch metal propeller. Optional 93 kW (125 hp) PZL-Franklin 4A-235-B4 engine. Metal fuel tank in each wing, combined capacity 100 litres (26.4 US gallons; 22.0 Imp gallons). Gravity refuelling point in top of each wing. Oil capacity 7.6 litres (2 US gallons; 1.7 Imp gallons).

ACCOMMODATION: Two persons, side by side in heated and ventilated cabin. Forward-opening door each side. Space for up to 30 kg (66 lb) of baggage aft of seats.

SYSTEMS: 14 V 25 Ah battery.

AVIONICS: Basic VFR instrumentation to JAR 23 standard; IFR and other avionics to customer's requirements.

DIMENSIONS, EXTERNAL:
Wing span	9.82 m (32 ft 2½ in)

Instytut Lotnictwa IS-2 light helicopter prototype on test stand in March 1998 *(W Garbarczyk/IL)*

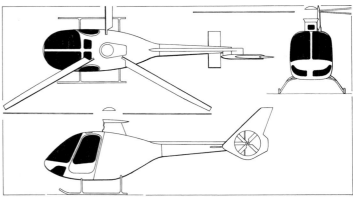

Provisional drawing of Instytut Lotnictwa IS-2 two-seat light helicopter with modified mast and unfinalised rotor head *(Paul Jackson/Jane's)*

POLAND: AIRCRAFT—IL to PZL

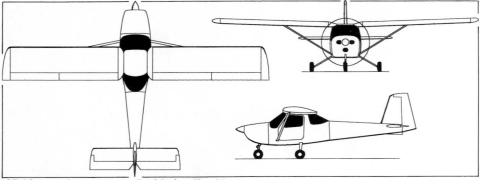

I-25 AS general arrangement *(Paul Jackson/Jane's)* 2001/0093625

Wing chord, constant	1.30 m (4 ft 3¼ in)	Cabin doors (two, each): Height	1.00 m (3 ft 3¼ in)	
Wing aspect ratio	7.8	Width	0.95 m (3 ft 1½ in)	
Length overall	6.92 m (22 ft 8½ in)	DIMENSIONS, INTERNAL:		
Height overall	2.58 m (8 ft 5½ in)	Cabin: Length	1.30 m (4 ft 3¼ in)	
Tailplane span	3.40 m (11 ft 1¾ in)	Max width	1.05 m (3 ft 5¼ in)	
Wheel track	2.56 m (8 ft 4¾ in)	Max height	1.15 m (3 ft 9¼ in)	
Wheelbase	1.58 m (5 ft 2¼ in)	Baggage compartment volume	0.35 m³ (12.4 cu ft)	
Propeller diameter	1.83 m (6 ft 0 in)	AREAS:		
Propeller ground clearance	0.21 m (8¼ in)	Wings, gross	12.30 m² (132.4 sq ft)	

Ailerons (total)	1.54 m² (16.58 sq ft)
Flaps (total)	2.18 m² (23.47 sq ft)
Fin	1.35 m² (14.53 sq ft)
Rudder	0.60 m² (6.46 sq ft)
Tailplane	2.75 m² (29.60 sq ft)
Elevators (total, incl tab)	1.10 m² (11.84 sq ft)
WEIGHTS AND LOADINGS:	
Weight empty	520 kg (1,146 lb)
Max fuel weight	72 kg (159 lb)
Max payload	200 kg (441 lb)
Max T-O weight	775 kg (1,708 lb)
Max wing loading	63.0 kg/m² (12.91 lb/sq ft)
Max power loading	6.50 kg/kW (10.68 lb/hp)
PERFORMANCE (estimated):	
Never-exceed speed (V_{NE})	137 kt (255 km/h; 158 mph)
Max level speed at S/L	111 kt (205 km/h; 127 mph)
Max cruising speed at S/L	100 kt (185 km/h; 115 mph)
Econ cruising speed at S/L	86 kt (160 km/h; 99 mph)
Stalling speed: flaps up	52 kt (95 km/h; 60 mph)
flaps down	44 kt (80 km/h; 50 mph)
Max rate of climb at S/L	270 m (886 ft)/min
Service ceiling	4,000 m (13,120 ft)
T-O to 15 m (50 ft)	450 m (1,480 ft)
Landing from 15 m (50 ft)	400 m (1,315 ft)
Range with max fuel, 30 min reserves	378 n miles (700 km; 435 miles)
Max endurance	4 h 30 min
g limits	+3.8/−1.5

UPDATED

J & AS

J & AS AERO DESIGN Sp. z o.o.
ulica Nowowiejska 2/29, PL-91-061 Łódź
Tel/Fax: (+48 42) 32 35 52
CHAIRMAN: Jan Borowski
DESIGNER, J-6: Jaroslaw Janowski

VERIFIED

J & AS J6 FREGATA
English name: Frigate Bird
TYPE: Single-seat ultralight/motor glider.
PROGRAMME: Larger and more powerful version of J-5 Marco described under Aviation Farm heading in 2000-01 *Jane's*. Prototype SP-P046 (later SP-0046) first flew on 15 December 1993 with a 29 kW (39 hp) Mosler CB-40 engine and with modified Honda power plant on 19 July 1995 but lost in accident, 16 July 1996, by which time second prototype (SP-P049) had been built with conventional (two mainwheels) landing gear. JAR 22 certification was planned, but most recent news was reservation in July 2000 of US registration N105JJ for c/n 005, implying sale also of 003 and 004.
DESIGN FEATURES: Constant-chord, mid-mounted wings with tapered tips. Wertmann FX-67K-170-17 aerofoil section and 1° 30′ diihedral. Pod-and-boom fuselage; pusher engine; V tail.
FLYING CONTROLS: Manual. Twin ruddervators and conventional ailerons. Schempp-Hirth type airbrakes in upper surface of wing.
STRUCTURE: GFRP/epoxy; single-spar wings.
LANDING GEAR: Motor glider has single mainwheel, plus tailwheel, and outrigger wheels slightly inboard of wingtips; tyre sizes 350×135 (main), 120×30 (tailwheel). Ultralight has two cantilever sprung mainwheels, plus tailwheel.
POWER PLANT: One 38.8 kW (52 hp) J & AS 3PZ-800 (Honda BF45A) three-cylinder, liquid-cooled, geared, four-stroke in-line engine; two-blade fixed-pitch wooden pusher propeller. Fuel capacity 60 litres (15.8 US gallons; 13.2 Imp gallons). Oil capacity 2.5 litres (0.7 US gallon; 0.6 Imp gallon).

DIMENSIONS, EXTERNAL:
Wing span 12.55 m (41 ft 2 in)
Length overall 5.03 m (16 ft 6 in)
Height overall 1.58 m (5 ft 2¼ in)
AREAS:
Wings, gross 8.80 m² (94.7 sq ft)
WEIGHTS AND LOADINGS:
Weight empty 245 kg (540 lb)
Max T-O weight 410 kg (903 lb)
PERFORMANCE, POWERED:
Max level speed 103 kt (191 km/h; 118 mph)
Max cruising speed at 75% power
 97 kt (180 km/h; 112 mph)
Stalling speed 38 kt (70 km/h; 44 mph)
Max rate of climb at S/L 396 m (1,299 ft)/min
T-O and landing run 200 m (660 ft)
Range 809 n miles (1,500 km; 932 miles)
PERFORMANCE, UNPOWERED:
Best glide ratio 23
Min rate of sink 1.00 m (3.28 ft)/s

UPDATED

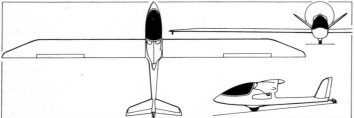

Motor glider version of J6 Fregata *(Paul Jackson/Jane's)* 0011927

J6 Fregata single-seat motor glider *(R J Malachowski)* 1999/0051153

PEZETEL

PEZETEL SA
aleja Stanów Zjednoczonych 61 (PO Box 88), PL-04-028 Warszawa 50
Tel: (+48 22) 810 80 01 and 813 52 85
Fax: (+48 22) 813 23 56
e-mail: board@pezetel.com.pl
GENERAL MANAGER: Andrzej Jesionek
MARKETING DIRECTOR: Witold Kozłowski

Pezetel Foreign Trade Enterprise was for many years responsible for export and import of defence equipment, including fixed- and rotary-wing aircraft, gliders, aero-engines, aircraft parts, flight instruments and other aviation products and services. From it, Pezetel SA was created in May 2000 to specialise in such activities concerning aeronautical products.

UPDATED

PZL

POLSKIE ZAKŁADY LOTNICZE Sp. zo.o.
 (Polish Aviation Factory Ltd)
ulica Wojska Polskiego 3, PL-39-300 Mielec
Tel: (+48 17) 788 63 24
Fax: (+48 17) 788 68 20
e-mail: pzl@ptc.pl
Web: http://www.pzlmielec.pl
PRESIDENT AND MANAGING DIRECTOR: Daniel Romański
COMMERCIAL DIRECTOR: Marian Stachowicz
OPERATIONS DIRECTOR: Tomasz Oleksiak

As the result of a restructuring process carried out by the Polish government, the former WSK-PZL Mielec and its subsidiary PZL Mielec Aircraft Company Ltd (see 1999-2000 *Jane's* for previous history) were subjected to liquidation proceedings in 1998. To replace the parent company, a new business entity under the above title came into operation on 23 October 1998. With its more than 60 years of history, a mid-2000 workforce of about 990, and the production facilities, technology, design and testing capabilities inherited from its predecessor, Polish Aviation Factory Ltd remains the largest and best equipped aircraft manufacturing plant in Poland. It is licensed to operate within the Europark Mielec special economic zone. In July 1999, its production facilities were certified by Poland's General Inspectorate of Civil Aviation under JAR 21 Subsection G, followed in September 1999 by design organisation approval under JAR 21 Subsection A. Quality management system (QMS) certification has been awarded by BAE Systems, and similar approval from Boeing was expected before the end of 1999. In November 1999, the Factory was awaiting ISO 9001 approval.

PZL's current indigenous aircraft programmes comprise the M-18 Dromader, M-20 Mewa, M-26 Iskierka, M-28B Bryza and M-28 Skytruck. Subcontract work continues for BAE (Hawk and Avro RJ), Boeing (757) and AgustaWestland.

UPDATED

PZL (ANTONOV) M-28 BRYZA
English name: Breeze
NATO reporting name: Cash
TYPE: Maritime surveillance twin-turboprop.
PROGRAMME: Developed by Antonov in former USSR as An-28; robust utility transport for service on Aeroflot's shortest routes, particularly those operated by An-2s into places relatively inaccessible to other fixed-wing aircraft; official Soviet flight testing completed 1972; first

Recently delivered Bryza 1R in current Polish Navy all-over grey colour scheme
2001/0098658

Bryza 1R with trial installation of Hartzell five-blade propeller *(R J Malachowski)*   *2001*/0099572

preproduction An-28 (SSSR-19723) originally retained same engines as prototype, but re-registered SSSR-19753 April 1975 when flown with current engines; production assigned to PZL Mielec 1978; temporary Soviet NLGS-2 type certificate awarded 4 October 1978 to second Soviet-built preproduction aircraft.

Polish manufacture started with initial batch of 15; first flight of Polish An-28 (SSSR-28800) 22 July 1984; version received full Soviet type certificate 7 February 1986. Polish production now for domestic military use only; current marketing effort concentrated on M-28 Skytruck derivative (which see). See 1994-95 and earlier *Jane's* for full An-28 description.

CURRENT VERSIONS: **M-28B Bryza 1R**: Originally designated An-28RM Bryza 1RM (*Ratownictwa Morskiego:* maritime reconnaissance). Prototype (SP-PDC) completed 1992 and delivered to Polish Navy in 1993 for evaluation; later upgraded and redelivered (as 0810) March 1999. First production example (serial number 1022) delivered 25 October 1994 and since upgraded; two further examples (1008 and 1017) ordered January 1998 and delivered in March 1999 to No. 3 Naval Air Arm Division at Siemirowice. These were supplied, and earlier aircraft upgraded, with Honeywell Silver Crown avionics. Further four aircraft (including 1114) delivered by end of 2000; eventual requirement reportedly for eight to 12. Upgrade continuing in late 2000 with introduction of Hartzell five-blade propellers; new-production aircraft also to have new (RDR-2000) weather radar and central, single-point refuelling system. Further plans include semi-retractable landing gear, scheduled for maiden flight in second quarter of 2001, and provision for additional fuel tankage.

An-28 Bryza 2RF: Originally designed as elint version, but currently operated in passenger/cargo transport role. One completed (serial number 1007); delivered to Polish Navy November 1996. Honeywell avionics and EFIS displays, Litton INS and autopilot.

An-28TD Bryza 1TD: (*Towaru Desantowy:* cargo airdrop): Transport/paradrop version, designated for paradrop missions. Clamshell rear doors replaced by single door sliding forward under fuselage, similar to that of An-26 (which see). Previously known as An-28B1T. First use may be to transport and drop firefighting teams. Prototype (SP-PDE) converted 1992 and underwent military evaluation as 0723 before return to PZL Mielec by 1996. Two (0404 and 0405) delivered to Polish Navy and one (1003) to Polish Air Force; Navy pair have MSS 5000 maritime surveillance system and are redesignated M-28M (see below)

M-28M: Two aircraft (Polish Navy 0404 and 0405) equipped with Swedish Space Corporation MSS 5000 maritime surveillance system for ecological monitoring;

previously designated An-28TD; 0405 commissioned 28 November 2000, at which time conversion of 0404 under way.

M-28MPA/ASW: Provisional designation of proposed anti-submarine version. Mission subsystems yet to be selected (early 2001); will include FLIR turret in redesigned nose; MAD; EW system; sonobuoy system; and torpedoes.

CUSTOMERS: See 1994-95 and earlier *Jane's*. Total of 166 An-28s of all variants except M-28 built in Poland by October 1999 (plus three in former USSR). Deliveries include seven Bryza 1Rs and two 1TDs to the 3rd Division of the Polish Navy at Siemirowice; plus one Bryza 2RF, also to Navy's 3rd Division. Polish Air Force 13th Transport Regiment at Krakow has one Bryza 1TD (1003, delivered October 1994). Two supplied to Venezuelan civilian operator CIACA in late 1996.

POWER PLANT: Two 716 kW (960 shp) TWD-10B/PZL-10S turboprops, each driving a 2.80 m (9 ft 2¼ in) diameter AW-24AN three-blade or a five-blade Hartzell propeller; first aircraft with latter was 0810, redelivered to Polish Navy on 27 August 2000.

ACCOMMODATION (Bryza 1R): Pilot, co-pilot, engineer and three systems operators.

AVIONICS (Bryza 1R): *Comms:* LS-10 C³ subsystem; Honeywell KT 76A transponder; NATO-compatible IFF.

Radar: ARS-400 360° search and surveillance radar (86 n mile; 160 km; 99 mile range), in underfuselage radome; Honeywell RDR-2000 weather radar in nose.

Flight: Include Honeywell KNS 81S navigation system and KLN 90B GPS receiver.

Mission: Chelton DF 707-1 emergency signals receiver.

EQUIPMENT (Bryza 1R): ACR/RLB-14 radio marker buoys; two 100 kg (220 lb) SAB-100NM illumination flares; five six-seat Mewa-6 dinghies, including two for crew.

WEIGHTS AND LOADINGS (Bryza 1R):
Max fuel weight 1,529 kg (3,371 lb)
Max payload 1,750 kg (3,858 lb)
Max T-O and landing weight 7,000 kg (15,432 lb)

PERFORMANCE (Bryza 1R):
Max cruising speed 189 kt (350 km/h; 217 mph)
Max rate of climb at S/L 480 m (1,575 ft)/min
Service ceiling 6,000 m (19,680 ft)
T-O to 15 m (50 ft) 350 m (1,150 ft)
Landing from 15 m (50 ft) 210 m (690 ft)
Range, 45 min reserves
 484-664 n miles (900-1,230 km; 559-764 miles)

UPDATED

PZL M-28 SKYTRUCK

TYPE: Twin-turboprop transport.

PROGRAMME: Westernised development of the Antonov An-28, with Polish power plant and avionics replaced by P&WC PT6A-65B turboprops, Hartzell five-blade propellers and Honeywell nav/com, weather radar and other equipment. Prototype conversion (SP-PDF) begun early 1991; first flight July 1993; Polish temporary type certificate to FAR Pt 23 Amendment 34 granted March 1994; permanent certificate March 1996, permitting MTOW increase to 7,000 kg (15,432 lb). Certified to FAR Pt 23 Amendment 42. Increased gross weight version (7,500 kg; 16,534 lb), a prototype of which was completed in 1996, offers options that include reinforced landing gear, increased fuel capacity, hydraulically operated rear door and additional emergency exit.

CURRENT VERSIONS: **M-28 Skytruck**: *As described.*

M-28B: Amended current designation of An-28RM Bryza 1RM; described in Bryza entry.

M-28 Skytruck Plus: Provisional designation for stretched version; development concept undergoing redesign in early 2001.

CUSTOMERS: First customer delivery (HK-4066X to Latina de Aviación of Colombia) made in January 1996; firm orders totalled 12 at that time, including six for Venezuelan National Guard (deliveries to which began in December 1996); equip Air Detachment 5 at Caracas (four), AD 3 at Maracaibo (one) and AD 9 at Puerto Agoucho (one). One delivered to Aerogryf in Poland in mid-1996 and one to USA in 1997. Venezuelan National Guard order subsequently (1997) increased to 18, second-batch deliveries beginning mid-1999. Also ordered (unspecified number) by Venezuelan Army; first deliveries April 2000; four received by July 2000. Production totalled 20 by July 2000.

DESIGN FEATURES: Inherited An-28's ability to operate from short, austere airfields; added Western engines and avionics to improve export sales prospects. Braced wings; will not stall because of action of automatic slats. Short

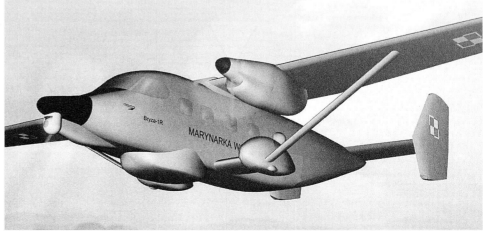

Proposed MPA/ASW version of the Bryza *(PZL via Piotr Butowski)*   *2001*/0034652

PZL M-28 Skytruck from the follow-on batch for Venezuelan National Guard *(R J Malachowski)* 2001/0099573

stub-wing extends from lower fuselage to carry main landing gear and support wing bracing struts, curving forward and downward at front to serve as mudguards; underside of rear fuselage upswept, incorporating clamshell doors for passenger/cargo loading; twin fins and rudders, mounted on inverted-aerofoil, no-dihedral fixed incidence tailplane.

Wing section CAHI (TsAGI) R-II-14; thickness/chord ratio 14 per cent; constant chord, non-swept, no-dihedral centre-section, with 4° incidence; tapered outer panels with 2° dihedral, negative incidence and 2° sweepback at quarter-chord.

FLYING CONTROLS: Manual. Single-slotted mass and aerodynamically balanced ailerons (port aileron has trim tab), designed to droop with large, hydraulically actuated, two-segment double-slotted flaps; electrically actuated trim tabs in elevator have manual back-up; twin rudders each with electrically actuated trim tab; automatic leading-edge slats over full span of wing outer panels; slab-type spoiler forward of each aileron and each outer flap segment at 75 per cent chord; fixed slat under full span of tailplane leading-edge. If an engine fails, patented upper surface spoiler forward of aileron on opposite wing is opened automatically, resulting in wing bearing dead engine dropping only 12° in 5 seconds instead of 30° without spoiler; patented fixed tailplane slat improves handling during high angle of attack climb-out; under icing conditions, if normal anti-icing system fails, ice collects on slat rather than tailplane, to retain controllability.

STRUCTURE: Mostly metal; duralumin ailerons with fabric covering; duralumin slats with CFRP skins; CFRP spoilers and trim tabs. Air intakes lined with epoxy laminate. Two-spar wing with integral fuel tanks between spars.

LANDING GEAR: Non-retractable tricycle type, with single wheel and PZL oleo-pneumatic shock-absorber on each unit. Main units, mounted on small stub-wings, have wide tread balloon tyres, size 720×320, pressure 3.50 bars (51 lb/sq in). Steerable (±50°) and self-centring nosewheel, with size 595×185-280 Stomil (Poland) tyre, pressure 3.50 bars (51 lb/sq in). Multidisc, anti-lock hydraulic brakes on main units, and inertial anti-skid units. Reinforced landing gear and brakes optional. Minimum ground turning radius 16.00 m (52 ft 6 in). Ski gear under development.

POWER PLANT: Two 820 kW (1,100 shp) Pratt & Whitney Canada PT6A-65B turboprops, each driving a Hartzell HC-B5MP-3M10876ASK five-blade propeller. Two main fuel tanks in outer wings, each 660 litres (174 US gallons; 145 Imp gallons); two auxiliaries in wing centre-section, each 290 litres (76.6 US gallons; 63.8 Imp gallons); two auxiliary tanks in outer wings, each 189 litres (49.9 US gallons; 41.6 Imp gallons); total wing fuel thus 2,278 litres (602 US gallons; 501 Imp gallons). Auxiliary tanks gravity-filled; main tanks by pressure or gravity. Additional fuel can be carried in tank in main cabin. Oil capacity 16 litres (4.2 US gallons; 3.5 Imp gallons) per engine.

ACCOMMODATION: Pilot and co-pilot on flight deck, which has bulged side windows and electric anti-icing for windscreens, and is separated from main cabin by bulkhead with connecting door. Dual controls standard. Jettisonable emergency door on each side. Standard cabin layout of passenger version has seats for 18 people, with six single seats on port side and six double seats on starboard side of aisle, at 72 cm (28 in) pitch. All seats easily foldable or removable for carriage of cargo. Aisle width 34.5 cm (13.5 in). Five passenger windows each side of cabin. Seats fold back against walls when aircraft is operated as a freighter or in mixed passenger/cargo role, seat attachments providing cargo tiedown points. Hoist of 500 kg (1,102 lb) capacity able to deposit cargo in forward part of cabin. Optional ventral pannier, increasing baggage/cargo capacity by 300 kg (661 lb). Entire cabin heated, ventilated, soundproofed and, optionally, air conditioned. Outward/downward-opening clamshell doors, under upswept rear fuselage, for passenger and cargo loading. Emergency exit at rear of cabin on each side. Option for hydraulically operated door at rear and additional emergency exit aft of flight deck.

SYSTEMS: No pressurisation or pneumatic systems. Hydraulic system powered by two engine-driven pumps, pressure 147 bars (2,132 lb/sq in), for flap, spoiler and (where fitted) cargo ramp actuation, mainwheel brakes and nosewheel steering, with emergency back-up system for spoiler extension and mainwheel braking. Optional air conditioning system, electrically powered with ground power supply back-up, reduces ambient temperatures of up to 40°C (104°F) by 10 to 12°C (18 to 22°F).

Primary electrical system is three-phase AC, with two 28 V DC 12 kW engine-driven starter/generators; two 200/115 V, 400 Hz, 1,500 VA inverters as secondary supply for windscreen heating. Two 26 Ah Ni/Cd batteries.

Electric anti-icing of flight deck windscreens, propellers, spinners and pitot heads, sufficient for short exposure to icing conditions. Optional pneumatic de-icing of wing, slat, tailplane and fin leading-edges. Oxygen system (for crew plus two passengers) optional. No APU.

AVIONICS: Standard Silver Crown suite by Honeywell.
Comms: KY 196A VHF-AM and KX 165 HF radios; KMA 24-70 audio selector panel and intercom; KT 76A transponder and encoding altimeter. Optional KHF 950 short wave radio.
Radar: Optional RDR-2000 digital weather radar.
Flight: Primary navigation system based on KNS 81 VOR/DME/RNAV with HSI (KPI 552B) indicator; secondary is KX 165-based for VOR/DME. KR 22 marker beacon receiver; twin KR 87 ADF; KCS 55A gyrocompass. Optional KLN 90B GPS and flight data recorder. KFC 325 AFCS and KCP 220 computer.

DIMENSIONS, EXTERNAL:
Wing span	22.06 m (72 ft 4½ in)
Wing chord: at root	2.20 m (7 ft 2½ in)
mean aerodynamic	1.89 m (6 ft 2½ in)
at tip	1.10 m (3 ft 7¼ in)
Wing aspect ratio	12.3
Length overall	13.10 m (42 ft 11¾ in)
Fuselage: Length	12.68 m (41 ft 7¼ in)
Max width	1.90 m (6 ft 2¾ in)
Max depth	2.14 m (7 ft 0¼ in)
Height overall	4.90 m (16 ft 1 in)
Tailplane span	5.14 m (16 ft 10¼ in)
Wheel track	3.405 m (11 ft 2 in)
Wheelbase	4.40 m (14 ft 5¼ in)
Propeller diameter	2.82 m (9 ft 3 in)
Propeller ground clearance	1.06 m (3 ft 5¼ in)
Distance between propeller centres	5.20 m (17 ft 0¾ in)
Rear clamshell doors:	
Length	2.40 m (7 ft 10½ in)
Total width: at top	1.00 m (3 ft 3¼ in)
at sill	1.40 m (4 ft 7 in)
Emergency exits (rear, each):	
Height	0.91 m (3 ft 0 in)
Width	0.51 m (1 ft 8 in)

DIMENSIONS, INTERNAL:
Cabin (excl flight deck): Length	5.26 m (17 ft 3 in)
Max width	1.74 m (5 ft 8½ in)
Max height	1.72 m (5 ft 7¾ in)
Floor area	approx 7.5 m² (80.7 sq ft)
Volume	approx 14.0 m³ (494 cu ft)
Pannier (optional): Height	0.38 m (1 ft 3 in)
Width	1.53 m (5 ft 0¼ in)
Volume	1.35 m³ (47.7 cu ft)

AREAS:
Wings, gross	39.72 m² (427.5 sq ft)
Ailerons (total)	4.33 m² (46.61 sq ft)
Trailing-edge flaps (total)	7.99 m² (86.00 sq ft)
Spoilers (total)	1.67 m² (17.98 sq ft)
Fins (total)	10.00 m² (107.64 sq ft)
Rudders (total, incl tabs)	4.00 m² (43.6 sq ft)
Tailplane	8.85 m² (95.26 sq ft)
Elevators (total, incl tabs)	2.56 m² (27.56 sq ft)

WEIGHTS AND LOADINGS:
Weight empty, equipped	4,060 kg (8,951 lb)
Max fuel load	1,766 kg (3,893 lb)
Max payload	2,000 kg (4,409 lb)
Fuel with max payload	1,227 kg (2,705 lb)
Payload with max fuel	1,460 kg (3.219 lb)
Max T-O and landing weight	7,500 kg (16.534 lb)
Max zero-fuel weight	6,600 kg (14.550 lb)
Max wing loading	188.8 kg/m² (38.67 lb/sq ft)
Max power loading	4.57 kg/kW (7.52 lb/shp)

PERFORMANCE (at 7,000 kg; 15,432 lb MTOW):
Max operating speed (VMO)	191 kt (355 km/h; 220 mph)
Econ cruising speed at 3,000 m (9,840 ft)	146 kt (270 km/h; 168 mph)
Unstick speed	73 kt (135 km/h; 84 mph)
Approach speed	70 kt (130 km/h; 81 mph)
Stalling speed, flaps down	67 kt (123 km/h; 77 mph)
Max rate of climb at S/L	810 m (2,657 ft)/min
Time to 3,000 m (9,840 ft): de-icing off	6 min
de-icing on	9 min
T-O run	255 m (835 ft)
T-O to 10.7 m (35 ft)	325 m (1,065 ft)
Landing from 15 m (50 ft)	560 m (1,835 ft)
Landing run	185 m (605 ft)
Range with max payload, 45 min reserves	755-809 n miles (1,400-1,500 km; 869-932 miles)
g limits	+3/−1

UPDATED

PZL M-28 03/04 SKYTRUCK PLUS

TYPE: Twin-turboprop transport.
PROGRAMME: Conceived in response to interest expressed by potential customers. Ground test prototype, completed in August 1997, uses 10th production An-28 airframe, SP-PDA; first of two flight test prototypes was due to fly in mid-1998 but in early 2001 original concept (see 2000-01 *Jane's*) was being redesigned for eligibility for FAR Pt 25 type certification.

UPDATED

PZL M-18 DROMADER
English name: Dromedary
TYPE: Agricultural sprayer.
PROGRAMME: Designed to meet requirements of FAR Pt 23; prototype (SP-PBW) first flew 27 August 1976; further two

M-28 Skytruck in Venezuelan Army camouflage   2001/0098662

prototypes; M-18 awarded Polish type certificate 27 September 1978; eight preproduction aircraft used for operational trials; later certified in Argentina, Australia, Brazil, Canada, China, former Czechoslovakia, France, former East Germany, Italy, Lithuania, New Zealand, Spain, USA and former Yugoslavia; series production began 1979.

In June 1996 PZL Mielec signed an agreement for the assembly of M-18Bs from Polish CKD kits by a Brazilian local authority, but the project appears not to have been launched.

CURRENT VERSIONS: **M-18**: Initial single-seat agricultural version (see 1988-89 *Jane's*); production ended 1984.

M-18A: Two-seat agricultural version, for operators requiring to transport mechanic/loader to improvised airstrip; production began 1984, following Polish supplementary type certification 14 February 1984; FAA type certificate for M-18 extended to M-18A September 1987; production continued until early 1996. Weight and performance data in 1997-98 and earlier *Jane's*.

M-18AS: Two-seat training version, with smaller hopper to make space for instructor's cockpit in front of trainee pilot; front cockpit installation readily interchangeable with standard hopper of M-18A; first flown 21 March 1988; five built by 1 January 1992; remains available.

M-18B: Standard production version from mid-1996. Improved performance development of M-18A, awarded extension of Polish type certificate 27 January 1994; FAA certification granted 19 December 1995. Elevators have smaller, centrally located trim tabs; spring interconnect between elevators and flaps, and between ailerons and rudder; flaps-down deflection increased from 15 to 30°; maximum payload and overload MTOW increased. Normal category landing run reduced; flaps-down stalling speed reduced; lower control stick force values; enhanced static and dynamic longitudinal stability. Power plant and hopper as for M-18A.

Detailed description applies to M-18B except where indicated.

M-18BS: Two-seat trainer, generally as M-18AS, but using M-18B airframe. Prototype (SP-ZUW, converted from an earlier M-18) first flew in November 1997. Polish type certificate awarded early 1998. Front (instructor's) seat reduces hopper capacity to 900 litres (238 US gallons; 198 Imp gallons). One produced for Iranian operator.

M-18C: Improved M-18B, with same payload but powered by an 895 kW (1,200 hp) PZL Kalisz K-9 engine offering enhanced performance; longer tailwheel leg to improve view over nose; modified wingtips to aid dispersal of chemical. First flown in August 1995, but apparently not developed further.

Turbine Dromader: Turboprop versions with 761 to 875 kW (1,020 to 1,173 shp) P&WC PT6A-45A/B/R or PT6A-65R/AG/AR engines with Hartzell propeller. Developed initially by James Mills Turbines Inc in USA and continued by Turbine Conversions Ltd. Variants with Honeywell 746 kW (1,000 shp) TPE331-10 or T53-L-7 also available from respective STC holders.

CUSTOMERS: Total of more than 700 built (all versions) by end of 1998, of which 88 per cent for export; 28th production batch under way in 2000; sold to operators in Argentina, Australia, Brazil, Bulgaria, Canada, Chile, China, Cuba, former Czechoslovakia and East Germany, Greece (30 for firefighting), Hungary, Iran, Italy, Morocco, Nicaragua, Poland, Portugal, Spain, Swaziland, Trinidad, Turkey, USA (200), Venezuela and former Yugoslavia.

DESIGN FEATURES: Emphasis on crew safety; all parts exposed to chemical contact treated with polyurethane or epoxy enamels, or manufactured of stainless steel; detachable fuselage side panels for airframe inspection and cleaning; braced tailplane.

Wing sections NACA 4416 at root, NACA 4412 at end of centre-section and on outer panels; incidence 6°.

Greek Air Force PZL Dromader equipped as a water-bomber *(Paul Jackson/Jane's)*

FLYING CONTROLS: Conventional and manual. Mass and aerodynamically balanced slotted ailerons with trim tabs, actuated by pushrods; aerodynamically and mass-balanced rudder and horn-balanced elevators with trim tabs, actuated by cables and pushrods respectively; hydraulically actuated two-section trailing-edge slotted flaps. See also Current Versions M-18B paragraph above.

STRUCTURE: All-metal; steel-capped duralumin wing spar; fuselage mainframe of helium-arc welded chromoly steel tube, oiled internally against corrosion; duralumin fuselage side panels and stainless steel bottom covering; corrugated inboard flap and tail unit skins.

LANDING GEAR: Non-retractable tailwheel type. Oleo-pneumatic shock-absorber in each unit. Main units have 800×260 tyres and are fitted with hydraulic disc brakes, parking brake and wire cutters. Fully castoring tailwheel, lockable for take-off and landing, with tyre size 380×150.

POWER PLANT: One PZL Kalisz ASz-62IR nine-cylinder radial air-cooled supercharged engine (731 kW; 980 hp at 2,200 rpm), driving a PZL Warszawa AW-2-30 four-blade constant-speed aluminium propeller. Integral fuel tank in each outer wing panel, combined usable capacity 712 litres (188 US gallons; 157 Imp gallons). Gravity feed header tank in fuselage. For ferrying, additional fuel capacity in chemical hopper. Oil capacity 70 litres (18.5 US gallons; 15.4 Imp gallons).

ACCOMMODATION: Single adjustable seat in fully enclosed, sealed and ventilated cockpit stressed to withstand 40 g impact. Additional cabin located behind cockpit and separated from it by a wall. Latter equipped with rigid, rear-facing seat with protective padding and safety belt, port-side jettisonable door, windows (port and starboard), fire extinguisher and ventilation valve. Communication with pilot provided via window in dividing wall, and by intercom. In M-18AS, standard hopper is replaced by smaller one, permitting installation of bolt-on instructor's cabin. Standard hopper of M-18A and instructor's cockpit of M-18AS interchangeable. Glass fibre cockpit roof and rear fairing, latter with additional small window each side. Front cockpit of M-18AS has more extensive glazing. Adjustable shoulder-type safety harness. Adjustable rudder pedals. Quick-opening door on each side of front cockpit; port door jettisonable.

SYSTEMS: Hydraulic system, pressure 98 to 137 bars (1,421 to 1,987 lb/sq in), for flap actuation, disc brakes and dispersal system. Electrical system powered by 28.5 V 100 A generator, with 24 V 28 Ah lead-acid battery and overvoltage protection relay.

AVIONICS: *Comms*: RS-6102 (Polish-built), Honeywell KX 155 or KY 196A com transceiver.

Instrumentation: KI 208 course indicator, VOR-OBS indicator, artificial horizon with turn and bank indicator, gyrocompass, radio compass and stall warning.

EQUIPMENT: Glass fibre epoxy hopper, with stainless steel tube bracing, forward of cockpit; capacity 2,500 litres (660 US gallons; 550 Imp gallons) of liquid or 2,200 kg (4,850 lb) of dry chemical. Smaller hopper in M-18AS. Deflector cable from cabin roof to fin.

M-18 variants can be fitted optionally with several different types of agricultural and firefighting systems, as follows: spray system with 54/96 nozzles on spraybooms; dusting system with standard, large or extra large spreader; atomising system with 10 atomisers; water bombing installation; fire-bombing installation with foaming agents; and multifunction firefighting system (water bombing or sequential line-drops). Aerial application roles can include seeding, fertilising, weed or pest control, defoliation, forest and bush firefighting, and patrol flights.

Special wingtip lights permit agricultural flights at night, and aircraft can operate in both temperate and tropical climates. Navigation lights, cockpit light, instrument panel lights and two rotating beacons standard. Landing lights, taxying light and night working light optional. Built-in jacking and tiedown points in wings and rear fuselage; towing lugs on main landing gear. Cockpit fire extinguisher and first aid kit.

DIMENSIONS, EXTERNAL:
Wing span	17.70 m (58 ft 0¾ in)
Wing chord, constant	2.29 m (7 ft 6¼ in)
Wing aspect ratio	7.8
Length overall	9.47 m (31 ft 1 in)
Height: over tailfin	3.70 m (12 ft 1¾ in)
overall (flying attitude)	4.60 m (15 ft 1 in)
Tailplane span	5.90 m (19 ft 4¼ in)
Wheel track	3.48 m (11 ft 5 in)
Propeller diameter	3.30 m (10 ft 10 in)
Propeller ground clearance (tail up)	0.23 m (9 in)

DIMENSIONS, INTERNAL:
Hopper volume	2.5 m³ (88 cu ft)

AREAS:
Wings, gross	40.00 m² (430.5 sq ft)
Ailerons (total)	4.33 m² (46.61 sq ft)
Trailing-edge flaps (total)	5.16 m² (55.54 sq ft)
Vertical tail surfaces (total)	3.60 m² (38.75 sq ft)
Horizontal tail surfaces (total)	7.58 m² (81.59 sq ft)

WEIGHTS AND LOADINGS (M-18B):
Max payload	2,200 kg (4,850 lb)
Max fuel weight	510 kg (1,124 lb)
Max T-O weight	5,300 kg (11,684 lb)
Max wing loading	132.5 kg/m² (27.14 lb/sq ft)
Max power loading	7.26 kg/kW (11.92 lb/hp)

PERFORMANCE (M-18B):
Never-exceed speed (V_NE)	124 kt (230 km/h; 142 mph)
Normal operating speed	108 kt (200 km/h; 124 mph)
Stalling speed, flaps down	66 kt (121 km/h; 76 mph)
Max rate of climb at S/L	264 m (866 ft)/min
Service ceiling	6,500 m (21,320 ft)
T-O run	350 m (1,150 ft)
T-O to 15 m (50 ft)	760 m (2,495 ft)
Landing from 15 m (50 ft)	570 m (1,870 ft)
Max range, no reserves:	
normal tankage	523 n miles (970 km; 602 miles)
with hopper fuel	1,079 n miles (2,000 km; 1,242 miles)

UPDATED

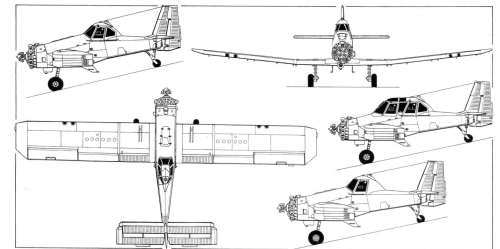

PZL M-18B Dromader, with additional side elevations (right) of M-18BS training version (upper) and original M-18 (lower)

PZL M-20 MEWA

English name: Gull
Export marketing name: Gemini
TYPE: Six-seat utility twin-prop.
PROGRAMME: Developed from PA-34-200T Seneca II light twin under 1977 agreement with Piper Aircraft Corporation, USA; nine Piper kits supplied to Pezetel

PZL M-26 Air Wolf, pictured before delivery to the USA *(R J Malachowski)*

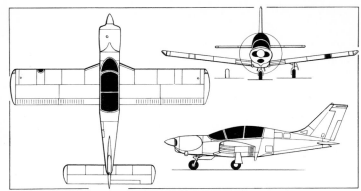

PZL M-26 Iskierka tandem two-seat primary trainer *(James Goulding/Jane's)*

1978-80; adapted to accept PZL-F (Polish Franklin) engines 1978 and made first flight (SP-PKA) 25 July 1979; first four completed as M-20 00 and next five as M-20 01 (first flight, by SP-PKE, 22 September 1982); Polish certification of 00/01 22 September 1983; fifth M-20 01 converted as M-20 02 prototype, making first flight in this form 10 October 1985, but non-availability of production PZL-F engines resulted in switch to Teledyne Continental power plant in current M-20 03 model, which made first flight (SP-DMA, the re-registered prototype) 13 October 1988 and received Polish certification 12 December same year; 03 since certified by Germany (2 October 1991), Australia (3 March 1992) and US FAA (16 July 1993). Last known delivery 1997; see 1999-2000 and earlier *Jane's* for description.

CUSTOMERS: Four M-20 00 and five M-20 01 built 1979-80 and 1983-84 respectively (one 01 converted to M-20 02); at least 13 M-20 03 completed by early 1998; approximately 12 more either in storage awaiting sale or under construction. Two for Polish Interior Ministry delivered 1994-95 to 103rd Air Regiment of Polish Air Force at Warszawa-Bemowo AB. Two registered in Germany, 1995; one delivered to Elektrim SA in Poland, 1996; third to German owner in 1997. No known sales since then, but remains available.

UPDATED

PZL M-26 ISKIERKA
English name: Little Spark
Export marketing name: Air Wolf

TYPE: Primary prop trainer/sportplane.

PROGRAMME: Two versions developed; first flight of first prototype (SP-PIA) with PZL-F engine 15 July 1986; first flight of Textron Lycoming-engined prototype (SP-PIB) 24 June 1987. Second prototype used as US demonstrator until destroyed on 17 August 1995. Polish certification obtained 26 October 1991; production of M-26 01 launched second quarter 1994 with initial 20 aircraft order from US distributor Melex; deliveries to USA began July 1996 with first production aircraft (N2601M). US FAR Pt 23 certification received in second quarter 1998.

CURRENT VERSIONS: **M-26 00:** PZL-F engine and Polish avionics. None yet ordered.

M-26 01: Textron Lycoming AEIO-540-L1B5 engine and Honeywell avionics. In production.

Detailed description applies to this version except where indicated.

CUSTOMERS: Initial order for M-26 01 by Melex USA Inc (20), of which two delivered to Venezuelan National Guard Training Centre at Porlamar, Isla Margarita, in 1998. Selected by Israel Defence Force in 1998 as Piper Super Cub replacement, but no quantity or contract revealed by 2000. Sixth production aircraft to USA early 1999. Seventh production aircraft completed in August 2000; deliveries then comprised four (plus refurbished first prototype) to USA and two to Venezuela.

DESIGN FEATURES: Optimised for civil pilot training and military pilot selection; also suitable as high-performance private aircraft. Selected parts and assemblies of M-20 Mewa used in design of wings, tail unit, landing gear, power plant, and electrical and power systems. Fixed incidence tailplane.

Wing section NACA 65_2-415 (constant chord); 7° dihedral from roots; 2° incidence; leading-edges swept forward at root.

FLYING CONTROLS: Conventional and manual. Frise ailerons, horn-balanced rudder, and elevators with starboard trim tab; single-slotted trailing-edge flaps.

STRUCTURE: Safe-life aluminium alloy.

LANDING GEAR: Retractable tricycle type, actuated hydraulically, with single wheel and oleo strut on each unit. Mainwheels retract inward into wings, nosewheel rearward. Size 6.00-6 wheels on all three units; tyre pressures 3.43 bars (50 lb/sq in) on main units, 2.16 bars (31 lb/sq in) on nose unit. PZL Lucas hydraulic disc brakes on mainwheels. Parking brake.

POWER PLANT: *M-26 00:* One 153 kW (205 hp) PZL-F 6A-350CA flat-six engine, driving a PZL Warszawa-Okecie US 142 three-blade constant-speed propeller, or a two-blade Hartzell BHC-C2YF-2CKUF constant-speed propeller. One 92 litre (24.3 US gallon; 20.2 Imp gallon) fuel tank in each wing leading-edge, plus a 9 litre (2.4 US gallon; 2.0 Imp gallon) fuselage tank, to give total capacity of 193 litres (51.0 US gallons; 42.4 Imp gallons). Gravity fuelling point in top of each wing tank. Oil capacity 10 litres (2.6 US gallons; 2.2 Imp gallons).

M-26 01: One 224 kW (300 hp) Textron Lycoming AEIO-540-L1B5 flat-six engine, driving a Hoffmann HO-V123K-V/200CU-10 or Hartzell HC-C3YR-4BF three-blade constant-speed propeller. Second tank in each wing. Total fuel capacity 369 litres (97.5 US gallons; 81.2 Imp gallons). Gravity fuelling point in top of each outer wing tank. Oil capacity 18 litres (4.75 US gallons; 4.0 Imp gallons).

ACCOMMODATION: Tandem seats for pupil (in front) and instructor, under framed canopy which opens sideways to starboard. Rear seat elevated. Baggage compartment aft of rear seat. Both cockpits heated and ventilated.

SYSTEMS: Two independent hydraulic systems, one operating at 154 bars (2,233 lb/sq in) for landing gear extension/retraction and one at 103 bars (1,494 lb/sq in) for wheel braking. DC electrical power supplied by 24 V alternator (50 A in M-26 00, 100 A in M-26 01) and 25 Ah battery.

AVIONICS: *Comms:* Honeywell KX 165 radio, KT 76A transponder, KMA 24 selector panel and intercom.

Flight: Honeywell KR 22 marker beacon receiver, KR 87 ADF and KCS 55A gyrocompass.

EQUIPMENT: Landing light in port wing leading-edge. Melex offers simulated air combat package of smoke, video guns and laser guns.

ARMAMENT: Up to 120 kg (265 lb) of ordnance on each of two underwing hardpoints; total 240 kg (529 lb).

DIMENSIONS, EXTERNAL:
Wing span	8.60 m (28 ft 2½ in)
Wing chord: at root	1.88 m (6 ft 2 in)
at tip	1.60 m (5 ft 3 in)
Wing aspect ratio	5.3
Length overall	8.29 m (27 ft 2½ in)
Height overall	2.96 m (9 ft 8½ in)
Tailplane span	3.80 m (12 ft 5½ in)
Wheel track	2.93 m (9 ft 7¼ in)
Wheelbase	1.93 m (6 ft 4 in)
Propeller diameter	1.90 m (6 ft 2¾ in)

DIMENSIONS, INTERNAL:
Cockpits: Total length	2.91 m (9 ft 6½ in)
Max width	0.88 m (2 ft 10½ in)
Max height	1.30 m (4 ft 3¼ in)

AREAS:
Wings, gross	14.00 m² (150.7 sq ft)
Ailerons (total)	1.17 m² (12.59 sq ft)
Trailing-edge flaps (total)	1.06 m² (11.41 sq ft)
Fin	1.96 m² (21.10 sq ft)
Rudder	0.89 m² (9.58 sq ft)
Tailplane	3.30 m² (35.52 sq ft)
Elevators (total, incl tab)	1.15 m² (12.38 sq ft)

WEIGHTS AND LOADINGS (M-26 01. A: Aerobatic, U: Utility category):
Weight empty: A, U	1,040 kg (2,292 lb)
Max fuel weight	271 kg (597 lb)
Max payload: A	265 kg (584 lb)
U	350 kg (772 lb)
Max T-O weight: A	1,315 kg (2,899 lb)
U	1,400 kg (3,086 lb)
Max wing loading: A	93.9 kg/m² (19.24 lb/sq ft)
U	100.0 kg/m² (20.48 lb/sq ft)
Max power loading: A	5.88 kg/kW (9.66 lb/hp)
U	6.26 kg/kW (10.29 lb/hp)

PERFORMANCE (M-26 01, A and U as above):
Never-exceed speed (V_{NE}):	
A	207 kt (385 km/h; 239 mph)
U	200 kt (371 km/h; 230 mph)
Max level speed at S/L:	
A	178 kt (330 km/h; 205 mph)
U	173 kt (321 km/h; 199 mph)
Max rate of climb at S/L: A	480 m (1,575 ft)/min
U	360 m (1,181 ft)/min
T-O to 15 m (50 ft): A	520 m (1,710 ft)
U	625 m (2,055 ft)
Landing from 15 m (50 ft): A	636 m (2,090 ft)
U	685 m (2,250 ft)
Max range, 30 min reserves:	
U	761 n miles (1,410 km; 876 miles)
g limits: A	+6/−3
at max T-O weight (U)	+4.4/−1.76

UPDATED

PZL-MIELEC

ZAKŁAD LOTNICZY PZL-MIELEC Sp. z o.o. (PZL-Mielec Aircraft Company Ltd)

ulica Wojska Polskiego 3, PL-39-300 Mielec
Tel: (+48 17) 788 73 00
Fax: (+48 17) 788 72 26
MANAGING DIRECTOR: Marek Woszczyński
COMMERCIAL DIRECTOR: Andrzej B Podsadowski

On 22 March 1999, a petition was filed for the bankruptcy of the PZL-Mielec Aircraft Company Ltd. Approximately 300 people remained to continue outstanding commitments to the Iryda jet trainer programme. Other activities of the former WSK-PZL Mielec SA were simultaneously transferred to the newly formed Polish Aviation Factory Ltd (PZL, which see).

UPDATED

PZL-MIELEC M-93 and M-96 IRYDA
English name: Iridium

TYPE: Two-seat basic and advanced jet trainer.

PROGRAMME: Increased-capability developments of I-22 (first flown 3 March 1985; last described in 1998-99 *Jane's*). Initiated (originally under designation M-92) when fourth I-22 prototype (SP-PWD) refitted with IL K-15 turbojets, making new 'first' flight 22 December 1992; fifth I-22 prototype (SP-PWE), refitted with Viper engines, flew 25 April 1994. These variants, later redesignated M-93K and M-93V respectively, also incorporated other changes; evaluated (as M-93S) with SAGEM Maestro (Modular Avionics Enhancement System Targeted for Retrofit Operations), first test-flown in SP-PWD on 24 May 1994, but Sextant Avionique selected in October 1995 to supply M-96 production avionics (contract signed 26 April 1996). M-93V, M-93S, and projected M-93R (reconnaissance) and M-93M (maritime), were not proceeded with; see earlier editions for details.

M-93K nominated as standard for early aircraft (first deliveries, serialled from 0301 onward, May 1995, plus 0204); however, production halted to permit development of further improved M-96 version; latter's flight testing (nine flights in August 1997) then suspended as a consequence of contractual dispute with Polish Air Force which was still unresolved by late 2000. Future of modernisation programme thus still uncertain, but three aircraft are due to re-enter service in 2001.

CURRENT VERSIONS: **M-93K:** As I-22 but with K-15 engines, redesigned airbrakes, modified systems and equipment, Martin-Baker seats and Western avionics. First flight of production M-93K (SP-PWF/0204), 6 July 1994; state qualification board type certificate January 1995; three more production aircraft (0301 to 0303) delivered to Polish

Air Force by end of 1995; first of next three (0304 to 0306) accepted 20 February 1996, but all held at Mielec for modifications.

Detailed description applies to M-93K except where indicated.

M-96: Aerodynamic improvement of M-93K with LERX, leading-edge wing slats, Fowler flaps, taller fin, increased tailplane span and new Sextant Topflight avionics suite. Changes to aerodynamics intended mainly to reduce approach speed and landing run, and to permit flight at AoA of more than 20°. Modified I-22 (SP-PWD) first flew 21 December 1996, in partial M-96 configuration; second development aircraft, fitted with Topflight avionics, flew April 1997; fully modified M-96 SP-PWG (converted from 0204) first flew 16 August 1997. Six production aircraft (0401 to 0406) ordered on 16 April 1996 to be completed as M-96s.

CUSTOMERS: Four M-93Ks delivered to PAF by end of 1995, plus five I-22s. Six M-96s ordered in 1996, making grand total of five prototypes, five I-22s (four remaining), seven production M-93Ks, six production M-96s and two static airframes, or 25 in all. Assigned to 58 Lotniczy Pulk Szkolny (Air School Regiment) at Deblin-Irena. Eleven (four I-22s and seven M-93Ks) planned to be upgraded to M-96 (103/105/0201/0202 and 0204/0301-0306), but not due to be redelivered to PAF until 2001-2002.

DESIGN FEATURES: Shoulder-wing, high-performance trainer with stepped, tandem cockpits and two engines in panniers (also housing landing gear) on lower sides of fuselage – which layout shared with D-B/D Alpha Jet and LMA Pampa. Service life calculated on basis of 12,500 flying hours or 10,000 take-offs and landings.

Laminar flow aerofoil section (NACA 64A-010 at root, 64A-210 at tip), with 0° incidence and 1° 43′ 48″ geometric twist; sweepback 18° 58′ 12″ on leading-edges, 14° 27′ 36″ at quarter-chord; no sweep on trailing-edges; anhedral from roots 3° on wings, 6° on tailplane. LERX, leading-edge slats, wider-span tailplane and taller fin added on M-96.

FLYING CONTROLS: Mass-balanced, differentially operable ailerons (with hydraulic boost and manual reversion), elevators and rudder all actuated by pushrods; ground-adjustable tab on rudder and each aileron. Hydraulically actuated single-slotted trailing-edge flaps (10° for T-O and 30° for landing), with emergency pneumatic operation (M-96, Fowler flaps); leading-edge slats (M-96 only); variable incidence tailplane (0°/−7.5°) and twin airbrakes (maximum deflection 62°) in upper rear fuselage also actuated hydraulically.

STRUCTURE: All-metal light alloy (aluminium/magnesium/steel) stressed skin, with limited use of glass fibre (for example, fuselage/fin fairing); two-spar wing, built as one unit with centre and inboard portions forming integral fuel tanks. Titanium heatshields in engine bays. Elevator trailing-edges of honeycomb sandwich from third preproduction aircraft onward.

LANDING GEAR: Retractable tricycle type, with oleo-pneumatic shock-absorber, single wheel and tubeless tyre on each unit. Hydraulic extension and retraction: nose unit retracts forward, main units forward and upward into engine nacelles. Auxiliary pneumatic system for emergency extension. Mainwheel tyres size 670×210, pressure 8.83 bars (128 lb/sq in); nosewheel tyre 430×170, pressure 5.78 bars (84 lb/sq in). Nosewheel steerable ±45°. Automatic hydraulic disc brakes on mainwheels; auxiliary mainwheel parking brake serves also as emergency brake. Maximum rate of descent 220 m (722 ft)/min at AUW of 6,600 kg (14,550 lb). SH 21U-1 brake-chute (area 15 m²; 161.5 sq ft, drag coefficient 0.45) in fuselage tailcone. Small tail bumper under rear of fuselage.

POWER PLANT (M-93K and M-96): Two 14.71 kN (3,307 lb st) Instytut Lotnictwa K-15 non-afterburning turbojets, pod-mounted on lower sides of centre-fuselage. Fuel in three integral wing tanks (combined capacity 1,090 litres; 288 US gallons; 240 Imp gallons) and two rubber tanks and two header tanks in fuselage (combined capacity 1,340 litres; 354 US gallons; 295 Imp gallons), to give total internal capacity of 2,430 litres (642 US gallons; 535 Imp gallons). Provision for one 380 litre (100 US gallon; 83.6 Imp gallon) auxiliary fuel tank under each wing.

Fuel system permits up to 30 seconds of inverted flight. Single-point pressure refuelling (at front of port engine nacelle), plus four gravity filling points (two in upper fuselage, one on each wing).

ACCOMMODATION: Pressurised, heated and air conditioned cockpit, with tandem Martin-Baker Mk 10LR zero/zero ejection seats for pupil (in front) and instructor; rear seat elevated 400 mm (15¾ in). For solo flying, pilot occupies front seat. Back-type parachute, oxygen bottle and emergency pack for both occupants. Individual framed canopies, each hinged at rear and opening upward pneumatically. Rearview mirror in front cockpit. Dual controls standard; front cockpit equipped for IFR flying. Windscreen anti-iced by electric heating, supplemented by alcohol spray. Remaining transparencies anti-iced and demisted by hot engine bleed air.

SYSTEMS: Cockpits pressurised (maximum differential 0.19 bar; 2.7 lb/sq in) and air conditioned by engine bleed air. Air from air conditioning system also used to pressurise crew's g suits. Main hydraulic system, pressure 210 bars (3,045 lb/sq in), actuates landing gear extension and retraction, wing flaps, airbrakes, tailplane incidence, brake-chute deployment, differential braking of mainwheels, and nosewheel steering. Auxiliary hydraulic system for aileron control boost. Pneumatic system comprises three separate circuits, each supplied by a nitrogen bottle pressurised at 150 bars (2,175 lb/sq in): one powers emergency extension of wing flaps for landing, one the emergency extension of landing gear; third is for canopy opening, closing and sealing, windscreen fluid anti-icing system, and hydraulic reservoir pressurisation. All three bottles charged simultaneously through a common nozzle.

Electrical system, powered by a pair of 9 kW PR-9 DC starter/generators, supplies 115 V single-phase AC (via two 1 kVA LUN 2458-8 static converters) and 36 V three-phase AC (via two 500 VA PT-500C converters), both at 400 Hz; two 24 V 25 Ah 20NKBN-25 batteries provide DC power in event of a double failure. Each AC voltage supplied by one main converter and one standby, latter automatically assuming full load if a main converter fails. Crew oxygen system, capacity 9 litres (549 cu in). Engine intakes anti-iced by engine bleed air; fire detection and extinguishing system (two Freon bottles in rear fuselage). Electronic control system for gun firing and weapon release.

AVIONICS (M-93K): *Comms:* Unimor RS-6113 VHF/UHF multichannel com radio and SRO-2 IFF.

Flight: ARK-15M ADF; IL RWL-750 (RW-5 in I-22) radio altimeter; IL ORS-2M marker beacon receiver (from third preseries aircraft onward); SOD-57M ILS; S2-3 (SARPP 12MW in I-22) flight data recorder in dorsal fin fillet.

Instrumentation: Blind-flying instrumentation standard.
Mission: Integrated analogue weapon delivery control system.
Self-defence: SPO-1 RWR.

AVIONICS (M-96): Based on Sextant Topflight suite.
Comms: Polish radios.
Flight: Sextant Totem RLG INS with integrated GPS; dual HOTAS controls; Polish radio navigation aids.
Instrumentation: Sextant advanced HUD and SMD54 liquid crystal MFDs; IL altimeter and stores management system; Polish flight data recorder; MIL-STD-1553B databus.

EQUIPMENT: ASP-PFD-I22 gyro gunsight, S-13-100 nose-mounted gun camera and (optionally) SSz-45-1-100-05 firing effect monitor camera. Saturn-2 and Fobos-3 camera pods.

ARMAMENT: One 23 mm GSz-23L twin-barrel gun in ventral pack, with up to 200 rounds in fuselage (normal load 50 rounds on training missions). Four UBP-I-22 underwing multiple stores carriers normally, each stressed for load of up to 500 kg (1,102 lb) (but M-93K maximum external stores load limited to 1,800 kg; 3,968 lb); on these can be carried various bombs of 50, 100, 250 or 500 kg size (alternative MBD2-67U carriers for 50 and 100 kg bombs); from 32 to 128 57 mm S-5 unguided rockets in Mars-2, Mars-4, UB-16-57U or UB-32A-1 launchers; or

M-93/M-96 DEVELOPMENT AND PRODUCTION

c/n	Version	Identity	Remarks
001-01	I-22	SP-PWD	Converted as below
	M-92	SP-PWD	First flight 22 Dec 92, converted as below
	M-93S	SP-PWD	First flight 24 May 94, converted as below
	M-96	SP-PWD	First flight 21 Dec 96 (interim configuration)
	–	–	Cancelled January 00; in storage
002-04	M-93K	SP-PWF	Became 0204; converted as below
	M-96	SP-PWG	First flight 16 Aug 97 (full configuration)
	–	–	Cancelled January 00; in storage
003-01	M-93K	0301	Became SP-PWI 21 Aug 96; in storage by 2000
003-02	M-93K	0302	In storage since 1996
003-03	M-93K	0303	In storage since 1996
003-04	M-93K	0304	In storage since 1996
003-05	M-93K	0305	In storage since 1996
003-06	M-93K	0306	In storage since 1996
004-01	M-96	0401	Stored incomplete since 1998
004-02	M-96	0402	Stored incomplete since 1998
004-03	M-96	0403	Stored incomplete since 1998
004-04	M-96	0404	Stored incomplete since 1998
004-05	M-96	0405	Stored incomplete since 1998
004-06	M-96	0406	Stored incomplete since 1998

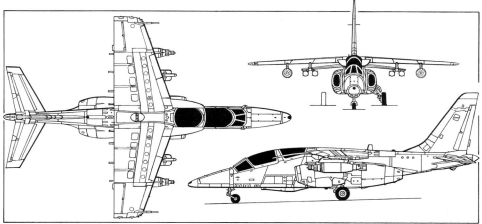

PZL Mielec M-96 Iryda tandem-seat advanced trainer with external armament *(Mike Keep/Jane's)*

M-96 full prototype SP-PWG, displaying to best advantage its taller fin, Fowler flaps, leading-edge slats and LERX

R-60 air-to-air homing missiles on APU60IM launch rails. Alternative underwing loads can include 23 mm UPK-23-250 gun pods, Zeus-1 7.62 mm gun pods, ZK-300 Kisajno cluster bombs, a ZR-8MB bomblet dispenser, 80 mm S-8 rockets in B-8M launchers or other weapons. M-96 able to carry full 2,000 kg (4,409 lb) of stores.

DIMENSIONS, EXTERNAL:
Wing span	9.60 m (31 ft 6 in)
Wing chord: at root	2.90 m (9 ft 6¼ in)
at tip	1.25 m (4 ft 1¼ in)
Wing aspect ratio	4.6
Length overall	13.22 m (43 ft 4½ in)
Height overall: M-93K	4.30 m (14 ft 1¼ in)
M-96	4.80 m (15 ft 9 in)
Tailplane span: M-93K	4.20 m (13 ft 9¼ in)
M-96	4.80 m (15 ft 9 in)
Wheel track	2.71 m (8 ft 10¾ in)
Wheelbase	4.91 m (16 ft 1¼ in)

AREAS:
Wings, gross	19.92 m² (214.4 sq ft)
Ailerons (total)	1.36 m² (14.64 sq ft)
Trailing-edge flaps (total)	3.22 m² (34.66 sq ft)
Airbrakes (total)	0.40 m² (4.31 sq ft)
Fin (M-93K)	2.71 m² (29.17 sq ft)
Rudder, incl tab	0.98 m² (10.55 sq ft)
Tailplane	3.57 m² (38.43 sq ft)
Elevators (total)	1.26 m² (13.56 sq ft)

WEIGHTS AND LOADINGS (M-96 data provisional):
Weight empty, equipped: M-93K	4,650 kg (10,251 lb)
M-96	5,070 kg (11,177 lb)
Max fuel weight: internal	1,895 kg (4,178 lb)
external	608 kg (1,340 lb)
Max external stores load: M-93K	1,800 kg (3,968 lb)
M-96	2,000 kg (4,409 lb)
Max T-O weight: clean: M-93K	6,900 kg (15,211 lb)
with external stores: M-93K	8,600 kg (18,959 lb)
M-96	9,000 kg (19,841 lb)
Max landing weight: M-93K	6,600 kg (14,550 lb)
M-96	7,200 kg (15,873 lb)
Max wing loading:	
M-93K, clean	346.4 kg/m² (70.95 lb/sq ft)
M-93K, with external stores	431.7 kg/m² (88.42 lb/sq ft)
M-96, with external stores	451.8 kg/m² (92.54 lb/sq ft)
Max power loading:	
M-93K, clean	234 kg/kN (2.30 lb/lb st)
M-93K, with external stores	292 kg/kN (2.87 lb/lb st)
M-96, with external stores	306 kg/kN (3.00 lb/lb st)

PERFORMANCE (M-93K at clean T-O weight of 5,900 kg; 13,007 lb except where indicated; M-96 estimated at 5,850 kg; 12,897 lb except where indicated):
Max limiting Mach No.: M-93K	0.80
M-96	0.83
Max level speed at 5,000 m (16,400 ft):	
M-93K	513 kt (950 km/h; 590 mph)
M-96	507 kt (940 km/h; 584 mph)
Stalling speed, flaps down:	
M-93K	110 kt (203 km/h; 127 mph)
M-96	98 kt (180 km/h; 112 mph)
Max rate of climb at S/L:	
M-93K	2,730 m (8,957 ft)/min
M-96	2,520 m (8,268 ft)/min
Rate of climb at S/L, OEI: M-93K	755 m (2,477 ft)/min
M-96	726 m (2,382 ft)/min
Time to climb: M-93K to 5,000 m (16,400 ft)	2 min 30 s
M-96 to 6,000 m (19,680 ft)	3 min 10 s
Service ceiling: M-93K, M-96	13,700 m (44,940 ft)
Service ceiling, OEI: M-93K, M-96	9,000 m (29,520 ft)
T-O run:	
M-93K at max clean T-O weight	670 m (2,200 ft)
M-96 at MLW	670 m (2,200 ft)
T-O to 15 m (50 ft):	
M-93K at max clean T-O weight	1,060 m (3,480 ft)
M-96 at MLW	970 m (3,185 ft)
Landing from 15 m (50 ft) at MLW:	
without brake-chute: M-93K	1,400 m (4,595 ft)
M-96	1,080 m (3,545 ft)
with brake-chute: M-93K	1,150 m (3,775 ft)
M-96	950 m (3,120 ft)
Landing run at MLW:	
without brake-chute: M-93K	670 m (2,200 ft)
M-96	495 m (1,625 ft)
with brake-chute: M-93K	420 m (1,380 ft)
M-96	310 m (1,020 ft)
Tactical radius at 500 m (1,640 ft) at max T-O weight with max external stores:	
M-93K, M-96	135 n miles (250 km; 155 miles)
Range, T-O at 7,200 kg (15,873 lb), no reserves:	
M-96	647 n miles (1,200 km; 745 miles)
Endurance, as above: M-96	2 h 55 min

UPDATED

PZL ŚWIDNIK SA

ZYGMUNTA PULAWSKIEGO-PZL ŚWIDNIK (Zygmunt Pulawski Transport Equipment Manufacturing Centre JSC, Świdnik)

aleja Lotników Polskich 1, PL-21-045 Świdnik k/Lublina
Tel: (+48 81) 751 20 61, 751 35 05 and 468 86 08
Fax: (+48 81) 468 09 19 and 468 09 18
e-mail: hem@pzl.swidnik.pl
Web: http://www.pzl.swidnik.pl
PRESIDENT: Mieczysław Majewski, MSc Eng
MANAGING DIRECTOR: Jan Miroński
COMMERCIAL DIRECTOR: Ryszard Cukierman
DIRECTOR OF RESEARCH AND DEVELOPMENT:
  Ryszard Kochanowski, MSc Eng
MARKETING MANAGER: Andrzej Stachyra

Świdnik factory established 1951; engaged initially in manufacturing components for LiM-1 (MiG-15) jet fighter; began licence production of Soviet Mi-1 helicopter in 1954, building some 1,250 as SM-1s, followed by 450 Świdnik-developed SM-2s; design office formed at factory to work on variants/developments of SM-1 and original projects such as SM-4 Latka. More than 7,200 helicopters of all types built to date, including four pre-series and 5,495 production Mi-2s; in addition, further six Mi-2 static test airframes produced, while eight more incomplete Mi-2s remained in storage in mid-2000.

Świdnik works named after famous pre-war PZL designer, Zygmunt Pulawski; became a joint stock company on 4 January 1991, with Polish Treasury the major shareholder (now 60 per cent); 1998 workforce of about 3,850 reduced to 2,600 by mid-2000. Buyer(s) for company still being sought; decision expected by end of that year. Production concentrates on W-3 Sokół variants, Kania and SW-4. On 15 June 2000, Swidnik was granted AQAP-110 certification, having fully met NATO quality assurance requirements for design, development and production.

Also manufactures PW-5 Smyk and PW-6 composites sailplanes (both recently sold to Egypt) and central wing box for ATR 72; manufacturing fuselages for AgustaWestland A 109 Power between 1996 and 2002; collaborating in development of Bell/Agusta AB139, for which Świdnik builds the fuselage (six due for delivery during 2000). Subsidiaries include Heliseco, which operates over 100 helicopters on agricultural and similar tasks at home and in Egypt, Greece, Oman, Portugal, Spain and elsewhere.

UPDATED

PZL ŚWIDNIK KANIA
English name: Kitty Hawk

TYPE: Light helicopter.
PROGRAMME: Modification of Mi-2 (1994-95 *Jane's*), developed in collaboration with Allison in USA; two prototypes produced by converting Mi-2 airframes; first flight of first prototype (SP-PSA) 3 June 1979; Polish supplementary type certificate to Mi-2 on 1 October 1981; full type certificate as FAR Pt 29 (Transport Category B) day and night VFR multipurpose utility helicopter with Category A engine isolation on 21 February 1986 as considerably improved Kania Model 1. Czech and Slovak certification of this version followed in 1992 and 1994 respectively. No deliveries since 1997, but remains available.

CURRENT VERSIONS: **Kania Model 1:** Intended for passenger transport (with nine-passenger standard, five-passenger executive or customised interiors); cargo transport (internal or slung load); agricultural (LV and ULV spraying/spreading/dusting); medevac (typically, three seats plus two stretchers); training; rescue; and aerial surveillance configurations.

CUSTOMERS: Total of 13 (plus one static test airframe) completed by 1996; none since then; further four incomplete airframes in storage, 2000. Four delivered in 1996 to Polish Ministry of Interior for Border Guard and police use (two each; one of former since lost), two for Helicopter sro of Czech Republic (including one initially supplied to Helicaribe of Venezuela in 1990), two for Cyprus National Guard and one for Air Transport Europa of Slovak Republic. Others (including prototypes) owned by PZL Świdnik.

Full description in 1998-99 and earlier editions; shortened versions in 1999-2000 and 2000-01 editions.

UPDATED

PZL ŚWIDNIK W-3 SOKÓŁ
English name: Falcon

TYPE: Multirole medium helicopter.
PROGRAMME: Developed in second half of 1970s; static/fatigue ground test airframe followed by five flying prototypes, first of which (SP-PSA) made first flight 16 November 1979, and used in subsequent tiedown tests; remaining prototypes embodied changes resulting from tests; manufacturer's flight trials resumed 6 May 1982 with second prototype (SP-PSB); third, fourth and fifth prototypes all made first flights in 1984, on 24 July (SP-PSC), 4 June (SP-PSD) and 26 November (SP-PSE) respectively; certification trials carried out in wide range of operating conditions, including heavy icing and extreme temperatures of −60 and +50°C; certification in Poland 1990, and to Russian NLGW regulations in 1992; production started in 1985 and 20 early aircraft supplied to Aeroflot from 10 August 1988, but subsequently returned to Świdnik for use by Heliseco.

Polish, US and German FAR Pt 29 (VFR) certification of W-3A received 1993; Polish ICAO-standard noise certification 1995. One W-3A2 (c/n 370508, SP-PSL) fitted in early 1998 with Sextant Avionique AFDS 95-1 AFCS (four-axis digital autopilot) with a view to obtaining Polish (GILC) and US (FAA) single-pilot IFR certification.

CURRENT VERSIONS: **W-3 Sokół:** Initial standard civil and military version; production completed.

W-3RL: Two W-3s (0501 and 0502) upgraded for combat SAR role in 1999 as part of Polish qualification for NATO membership; based at Bydgoszcz. New equipment includes Rockwell Collins search radar, Honeywell com radios and intercom, Tacan, GPS, radio compass, IFF, searchlight, rescue hoist, stretchers (two) and auxiliary fuel tanks; standard crew of five.

W-3W: Armed W-3 (W for *Wielozadaniowy*: multipurpose) with starboard-mounted 23 mm GSz-23 twin-barrel gun; Mars-2 launchers for sixteen 57 mm S-5 or 80 mm S-8 unguided rockets, ZR-8 bomblet dispensers, Platan minelaying packs, and six cabin window-mounted AK 47, 5.45 mm Tantal or PKM machine guns. Twenty-one delivered to Polish Ministry of National Defence; in service with Polish Air Force 47 Szkolny Pulk Smiglowcow (Helicopter School Regiment) at Nowe Miasto (five) and Polish Army (16).

W-3A Sokół: Improved version for Western certification; redesign started 1989; first flight 30 July 1992; FAA type approval to FAR Pt 29 received 31 May 1993, German LBA certification 6 December 1993. Dual hydraulic systems, new de-icing system, Western instrumentation. First delivery, to Saxony Police Department, Germany, 20 December 1993. Marketed in the Americas and Pacific Rim by Piasecki Aircraft Corporation, USA.

Detailed description applies to W-3A, except where indicated.

W-3WA: Armed version of W-3A; armament includes GSz-23L gun, Strzala-2 AAMs and Polish Gad fire-control system. Total of 19 delivered to Polish armed forces by

PZL Świdnik W-3 during 1999 firing trials with HOT 3 ATM and SFIM Viviane sighting system above flight deck (*R J Malachowski*)

Instrument panel of the combat SAR W-3RL *(R J Malachowski)*

Rotor head and propulsion package of the W-3A

October 1999: Air Force five, Army 12, Navy two. Three in combat SAR configuration (presumably based on W-3RL) delivered in second quarter 2000 to 7th Cavalry Regiment at Mazowiecki: new comms, NVG-compatible instrumentation, new IFF, classified ESM and armoured crew seats.

W-3AM: Version of W-3A with six inflatable flotation bags. Seven delivered by 1 October 1999.

W-3A2: Sextant Avionique AFCS testbed (see under Programme above).

W-3RM Anakonda: Offshore search and rescue version of W-3; watertight cabin, six inflatable flotation bags, additional window in lower part of each flight deck door. In service with Polish Navy (five delivered, of which two since lost), Petrobaltic (one), Ministry of Interior (one) and for Świdnik trials (one). Latest examples, with US FSI Ultra 4000 FLIR, are designated **W-3WARM** (first aircraft, 360813, delivered August 1998; second, 360815, 15 March 1999), the A indicating American (FAA) certification standard and W indicating armament. Upgrade under test in 2000 includes, a more advanced engine control system blade folding and a deck lock for shipboard deployment.

W-3U Salamandra: Armed prototype (see 1995-96 and earlier *Jane's* for details). Not produced.

S-1RR Procjon: Electronic combat reconnaissance version of W-3; prototype (c/n 310203) developed in 1995 and delivered to Polish Air Force by 1 January 1997. Subsequent aircraft designated **SRR-10**; first of these (c/n 370720) completed in third quarter 1998 and fully equipped in 1999. ECR suite includes two-place console and chaff/flare dispensers; large external antenna housings include one on nose, one on cabin roof and one on starboard side of fuselage; last-named can be rotated downwards for full 360° scan capability.

W-3H: Armed support helicopter, derived from W-3WA; pursued as upgrade under now-abandoned Huzar programme. Local requirement for 96. Avionics and weapons fit decided in favour of Israeli (Elbit avionics and Rafael NT-D ATM) equipment in October 1997, but selection immediately overturned by new Polish government; rival avionics offered by Boeing and Sextant, partnered by Hellfire II and HOT 3, respectively. Programme reinstated October 1998 but foundered due to inability to demonstrate NT-D in Poland within deadline of 30 November, and Israeli deal cancelled by Polish government on 8 December 1998.

September 1998 co-operation agreement between PZL Świdnik and Euromissile led to integration of HOT 3 ATM on Sokół, followed by successful firing tests during demonstration of HOT/Viviane system at Polish firing range in Nowa Deba on 4 March 1999, using Świdnik trials aircraft SP-SUW (c/n 360318). This aircraft utilised in 2000 as prototype to test new main rotor blades with modified leading-edges and increased damage tolerance. Świdnik continues to offer W-3/HOT 3 variant as potential export version, but domestic requirement for W-3H now shelved in favour of a 50-aircraft support role upgrade of existing Sokóls and new open competition for 60 to 70 attack helicopters. Candidates for latter likely to include A 129 International, Apache, Rooivalk, SuperCobra and Tiger; RFP still awaited in third quarter 2000.

W-3WB: Armed prototype: see 1995-96 and earlier *Jane's*. One only; armament deleted and reverted to W-3, 1994.

SW-5: Described separately.

CUSTOMERS: See table. Total of 136 (excluding five flying prototypes) completed by June 2000.

COSTS: US$3.156 million for basic VFR, single-pilot W-3A (1999).

DESIGN FEATURES: Conventional utility helicopter of pod-and-boom layout and with engines above cabin. Four-blade fully articulated main rotor and three-blade tail rotor; main rotor has pendular Salomon-type vibration absorber for smooth flight and low vibration levels. Transmission driven via main, intermediate and tail rotor gearboxes. Tailfin integral with tailboom; fixed incidence horizontal stabiliser, not interconnected with main rotor control system.

Main rotor blades have NACA 23012M aerofoil section and optional manual folding. Rotor brake standard. Rotor rpm 268.5 (main) and 1,342 (tail); main rotor blade tip speed 220.7 m/s (724 ft/s).

FLYING CONTROLS: Three hydraulic boosters for longitudinal, lateral and collective pitch control of main rotor; one booster for tail rotor control. Constant-speed rpm control for continuous operation (manual rpm control also available). Two-axis stability augmentation system with pitch and roll hold. Three- and four-axis AFCS available from late 1994.

STRUCTURE: Rotor blades (main and tail) and single-spar horizontal stabiliser of laminated GFRP impregnated with epoxy resin; tail rotor driveshaft of duralumin tube with splined couplings; duralumin fuselage; GFRP fin trailing-edge.

LANDING GEAR: Non-retractable tricycle type, plus tailskid beneath tailboom. Twin-wheel castoring and self-centring nose unit; single wheel on each main unit. Oleo-pneumatic shock-absorber in each unit. Mainwheel Stomil Poznań tyres size 700×250; nosewheel tyres size 400×140. Tyre pressures 4.90 and 4.40 bars (71 and 64 lb/sq in)

SOKÓŁ OPERATORS
(at 1 October 1999)

Country	Operator	W-3	W-3W	W-3A	W-3WA	W-3AM	W-3RM	S-1RR	Cum Total
Czech Republic	Air Force			11					11
Germany	Saxony Police		2						2
Korea, South	CitiAir			1		2			3
	Daewoo			2[2]		4			6
Myanmar	Air Force	13[3]							13
Nigeria	Okada Air	1							1
Poland	Air Force	7[4]	5		5			1	18
	Army		16		12				28
	Navy	2			2		4[1]		8
	Ministry of Interior	2		1			1		4
	Heliseco	21[5]				1			22
	Petrobaltic						1		1
	Zakopane Mountain Rescue Group			1					1
	Lublin Medical Aviation Centre	1							1
	Daewoo			1					1
	Świdnik trials	4[6]					1		5
Totals		**51**	**21**	**19**	**19**	**7**	**7**	**1**	**125**

Notes: Further deliveries by mid-2000 included one W-3AM to Chong-Nam Fire Department, South Korea
[1] Two others lost; four more reportedly required but all not yet funded; includes two W-3WARM
[2] MoU 13 April 1996 for purchase of 35 W-3A with marketing rights in Asia, but only seven paid for by October 1998 due to economic problems; further deliveries suspended as a result
[3] Including two in VIP configuration
[4] Including one in VIP configuration
[5] Used for firefighting in Spain
[6] Including non-standard versions

PZL Świdnik Sokół converted to W-3RL for combat search and rescue role *(R J Malachowski)*

POLAND: AIRCRAFT—PZL ŚWIDNIK

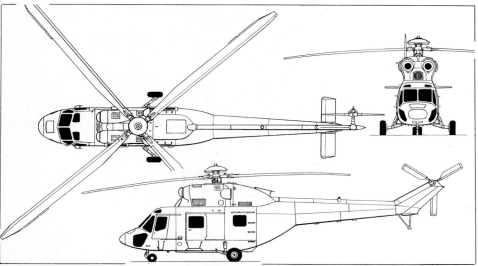

PZL Świdnik W-3A Sokół twin-turboshaft helicopter *(Dennis Punnett/Jane's)*

respectively. Pneumatic disc brakes on mainwheels. Metal ski landing gear optional. Six inflatable flotation bags on Anakonda and W-3AM.

POWER PLANT: Two WSK PZL-Rzeszów PZL-10W turboshafts, each with rating of 671 kW (900 shp) for T-O and emergency ratings of 746 kW (1,000 shp) and 858 kW (1,150 shp) for 30 and 2½ minutes OEI respectively.

Particle separators on engine intakes, and inlet de-icing, standard. Power plant equipped with advanced electronic fuel control system for maintaining rotor speed at pilot-selected value amounting to ±5 per cent of normal rpm, and also for torque sharing as well as for supervising engine limits during start-up and normal or OEI operation. Engines and main rotor gearbox mounted on bed frame, eliminating drive misalignment due to deformations of fuselage. Transmission rating 1,342 kW (1,800 shp) maximum for T-O, 1,163 kW (1,560 shp) maximum continuous and 857 kW (1,150 shp) OEI. Engine input rpm 23,615.

Four bladder fuel tanks beneath cabin floor, with combined capacity of 1,750 litres (462 US gallons; 385 Imp gallons). Auxiliary tank, capacity 1,100 litres (291 US gallons; 242 Imp gallons), optional (not FAA approved). Oil capacity 14 litres (3.7 US gallons; 3.1 Imp gallons) per engine.

ACCOMMODATION: Pilot (port side), and co-pilot or flight engineer, side by side on W-3 flight deck, on adjustable seats with safety belts. W-3A can be flown by single pilot in VFR, with extra passenger in co-pilot seat. Dual controls and dual flight instrumentation optional. Accommodation for 12 passengers in main cabin or up to eight survivors plus two-person rescue crew and doctor in Anakonda SAR version. Seats removable for carriage of internal cargo. Ambulance version can carry four stretcher cases and medical attendant. Baggage space, capacity 180 kg (397 lb), at rear of cabin.

Door with bulged window on each side of flight deck; large sliding door for passenger and/or cargo loading on port side at forward end of cabin; second sliding door at rear of cabin on starboard side. Optically flat windscreens, improving view and enabling wipers to sweep a large area. Accommodation soundproofed, heated (by engine bleed air) and ventilated.

SYSTEMS: Two independent hydraulic systems, working pressure 90 bars (1,300 lb/sq in), for controlling main and tail rotors, unlocking collective pitch control lever, and feeding damper of directional steering system; automatic power changeover if one system fails. Flow rate 11 litres (2.9 US gallons; 2.4 Imp gallons)/min in each system. Vented gravity feed reservoir, at atmospheric pressure. Pneumatic system for actuating hydraulic mainwheel brakes. Electrical system providing 200/115 V three-phase AC power at 400 Hz and 28 V DC power. Electric anti-icing of all rotor blades. Fire detection/extinguishing system. Air conditioning and oxygen systems optional. Neutral gas system optional, for inhibiting fuel vapour explosion.

AVIONICS: Standard VFR and IFR nav/com avionics permit adverse weather operation by day or night. Honeywell avionics standard; alternatives at customer's option.
Comms: Chrom (NATO 'Pin Head') IFF transponder in military versions.
Radar: Honeywell RDS-82 weather radar in W-3A; 5A-813 radar in W-3RM.
Flight: Stability augmentation system standard. AP Decca navigator in W-3RM.
Mission: SPOR search and detection system in W-3RM.
Self-defence: Modified Syrena RWR in military versions initially; to be replaced by Thomson-CSF Detexis EWR-99.

EQUIPMENT: Cargo version equipped with 2,100 kg (4,630 lb) capacity external hook and 150 kg (331 lb) capacity rescue hoist. W-3RM has 270 kg (595 lb) capacity electric hoist; stretchers, two-person rescue basket, rescue belts, liferafts for six people, rope ladder, portable oxygen equipment, electric blankets and vacuum flasks, various types of buoy (light, smoke and radio) and marker, binoculars, flare pistol and searchlights.

ARMAMENT: As described under Current Versions.

DIMENSIONS, EXTERNAL:
Main rotor diameter	15.70 m (51 ft 6 in)
Tail rotor diameter	3.03 m (9 ft 11¼ in)
Main rotor blade chord	0.44 m (1 ft 5¼ in)
Distance between rotor centres	9.50 m (31 ft 2 in)
Length: overall, rotors turning	18.79 m (61 ft 7¾ in)
fuselage	14.21 m (46 ft 7½ in)
Fuselage max width	1.75 m (5 ft 9 in)
Height: to top of rotor head	4.20 m (13 ft 9½ in)
overall, rotors turning	5.135 m (16 ft 10¼ in)
Stabiliser span	3.45 m (11 ft 3¾ in)
Wheel track	3.15 m (10 ft 4 in)
Wheelbase	3.55 m (11 ft 7¾ in)
Passenger/cargo doors:	
Height (each)	1.18 m (3 ft 10½ in)
Width: port	0.94 m (3 ft 1 in)
starboard	1.25 m (4 ft 1¼ in)
Height to sill	0.86 m (2 ft 10 in)

DIMENSIONS, INTERNAL:
Cabin: Length	3.21 m (10 ft 6½ in)
Max width	1.55 m (5 ft 1 in)
Max height	1.38 m (4 ft 6¼ in)
Floor area	4.80 m² (51.7 sq ft)
Volume	6.9 m³ (243 cu ft)

AREAS:
Main rotor disc	193.6 m² (2,083.8 sq ft)
Tail rotor disc	7.21 m² (77.6 sq ft)
Main rotor blades (each)	2.90 m² (31.22 sq ft)
Tail rotor blades (each)	0.28 m² (3.01 sq ft)
Fin	1.00 m² (10.76 sq ft)
Elevator	2.16 m² (23.25 sq ft)

WEIGHTS AND LOADINGS:
Min basic weight empty	3,300 kg (7,275 lb)
Basic operating weight empty (multipurpose versions)	3,850 kg (8,488 lb)
Max fuel weight	1,326 kg (2,923 lb)
Max payload, internal or external	2,100 kg (4,630 lb)
Normal T-O weight	6,100 kg (13,448 lb)
Max T-O weight	6,400 kg (14,110 lb)
Max disc loading	33.1 kg/m² (6.78 lb/sq ft)
Transmission loading at max T-O weight and power	4.77 kg/kW (7.84 lb/shp)

PERFORMANCE (at 5,800 kg; 12,787 lb T-O weight at 500 m; 1,640 ft, ISA, except where indicated):
Never-exceed speed (V_{NE})	140 kt (260 km/h; 161 mph)
Max cruising speed	131 kt (243 km/h; 151 mph)
Touchdown speed for power-off landing	33 kt (60 km/h; 38 mph)
Max rate of climb at S/L	612 m (2,008 ft)/min
Rate of climb at S/L, OEI, 2½ min emergency rating	171 m (561 ft)/min
Vertical rate of climb at S/L:	
at max T-O weight	74 m (243 ft)/min
at normal T-O weight	219 m (718 ft)/min
Service ceiling:	
at normal T-O weight	4,910 m (16,100 ft)
at T-O weight below normal	up to 6,000 m (19,680 ft)
Service ceiling, OEI:	
at 2½ min emergency rating	2,600 m (8,540 ft)
Hovering ceiling at max T-O weight:	
IGE	3,020 m (9,900 ft)
OGE	1,900 m (6,220 ft)
Range: standard fuel, 5% reserves	367 n miles (680 km; 422 miles)
standard fuel, no reserves	402 n miles (745 km; 463 miles)
with auxiliary fuel, no reserves	696 n miles (1,290 km; 801 miles)
with max payload, no reserves	108 n miles (200 km; 124 miles)
Endurance: standard fuel, 5% reserves	3 h 50 min
standard fuel, no reserves	4 h 5 min
with auxiliary fuel, 5% reserves	6 h 41 min
with auxiliary fuel, no reserves	7 h 5 min

PERFORMANCE (W-3RM at typical mission T-O weight of 5,500 kg; 12,125 lb):
Never-exceed speed (V_{NE})	140 kt (260 km/h; 161 mph)
Max cruising speed	131 kt (243 km/h; 151 mph)
Max rate of climb at S/L	672 m (2,205 ft)/min
Service ceiling	5,830 m (19,120 ft)
Hovering ceiling: IGE	4,000 m (13,120 ft)
OGE	2,970 m (9,740 ft)
Max range, no reserves	411 n miles (761 km; 473 miles)
Max endurance, no reserves	4 h 30 min

UPDATED

PZL ŚWIDNIK SW-5

The designation SW-5 is a provisional one allocated to an improved version of the W-3 Sokół under study since at least 1997. Principal changes envisaged include simplified rotor head, wider use of composites, upgraded avionics by Sextant (including four-axis autopilot with AFDS 95-1 flight director) and 746 kW (1,000 shp), FADEC-equipped Pratt & Whitney Canada PT6B-67B engines. No recent news of programme status received.

VERIFIED

PZL ŚWIDNIK SEP

TYPE: Anti-submarine helicopter.
PROGRAMME: Reported in 2000 as new design derived from W-3RM Anakonda; more shipboard-compatible; new-generation avionics and ASW equipment. No other details known at time of going to press.

NEW ENTRY

PZL ŚWIDNIK SW-4

TYPE: Five-seat helicopter.
PROGRAMME: Development began 1985; full-scale mockup completed 1987 (see description and illustrations in 1991-92 and earlier *Jane's*); major redesign undertaken 1989-90, using Allison (now Rolls-Royce) 250 engine in more streamlined fuselage with modified tail unit. Prototype (c/n 600102), rolled out December 1994, is non-flying testbed for ground and equipment tests; '101 is static test airframe, '103 and '104 are flying prototypes. First flight made by 600103 (red overall; later registered SP-PSW) on 26 October 1996 ('official' flight three days later).

Trials in 1997 demonstrated requirement for a new rotor head design, enlarged horizontal stabiliser and more robust hydraulic system. Following 70 hours of test-flying, SP-PSW was grounded in late 1997 for installation of SAMM-designed hydraulic flight control system, with which it was due to return to flying in 1998. Second flying prototype (yellow overall), with improved skids, exhibited statically at Paris Air Show, June 1997; registered as SP-PSZ in

Anakonda in latest W-3WARM configuration *(Paul Jackson/Jane's)*

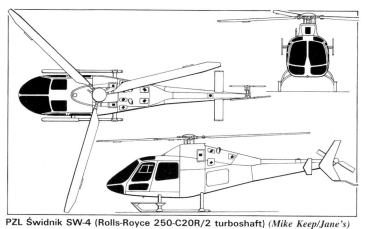
PZL Świdnik SW-4 (Rolls-Royce 250-C20R/2 turboshaft) *(Mike Keep/Jane's)*

First flying prototype of the SW-4 four/five-seat light helicopter *(R J Malachowski)*

October 1998, but had not flown by late 2000. First six production aircraft started during first quarter 1999; customers include Polish Police and Polish MoD (one each).

CURRENT VERSIONS: Will be marketed in two versions:
Economy: With one 336 kW (450 shp) Rolls-Royce 250-C20R engine and 1,600 kg (3,527 lb) normal T-O weight.
High Performance: With one 459 kW (615 shp) Pratt & Whitney Canada PW200/9 engine and 1,800 kg (3,968 lb) normal T-O weight.
Provision has been made for later installation of two engines to fulfil ICAO requirements for OEI operation (including hover).

COSTS: US$690,000 (1999).

Following description applies to first flying prototype:
DESIGN FEATURES: Compact light helicopter, originally envisaged as Eastern Bloc parallel to Eurocopter, Agusta and similar types, but reorientated with Western engines and avionics as rival design. Intended applications include passenger and cargo transport, medevac, border patrol, armed scout and training use. Three-blade main rotor; arrowhead tailfin on port side, with two-blade tail rotor to starboard; narrow tailplane with small endplate fins; skid landing gear.
STRUCTURE: GFRP for approximately 20 per cent of airframe, including all rotor blades; remainder mainly of aluminium alloy.
LANDING GEAR: Skid type; able to accommodate heavy landing sink rate of 3.1 m (10 ft)/s by elastic deflection of cross-tubes.
POWER PLANT: One 336 kW (450 shp) Rolls-Royce 250-C20R/2 turboshaft. Transmission rating 336 kW (450 shp) for T-O, 283 kW (380 shp) maximum continuous; 30 minute run-dry capability. Standard fuel capacity 500 litres (132 US gallons; 110 Imp gallons) in tank below main gearbox.
ACCOMMODATION: Pilot and up to four passengers or one stretcher patient and two medical attendants. One front-hinged and one rearward-sliding door on each side of cabin.
AVIONICS: *Instrumentation:* Honeywell VFR in first prototype; IFR optional in production aircraft.

DIMENSIONS, EXTERNAL:
Main rotor diameter	9.00 m (29 ft 6¼ in)
Tail rotor diameter	1.50 m (4 ft 11 in)
Distance between rotor centres	5.25 m (17 ft 2¾ in)
Length:	
overall, both rotors turning	10.55 m (34 ft 7½ in)
fuselage (incl tailskid)	9.075 m (29 ft 9¼ in)
Max width: cabin	1.515 m (4 ft 11¾ in)
over tail unit endplates	1.83 m (6 ft 0 in)
over skids	2.28 m (7 ft 5¾ in)
Height overall	3.05 m (10 ft 0 in)
Fuselage ground clearance	0.55 m (1 ft 9¾ in)

DIMENSIONS, INTERNAL:
Cabin: Length	2.14 m (7 ft 0¼ in)
Max width	1.42 m (4 ft 8 in)
Max height	1.27 m (4 ft 2 in)
Volume	3.85 m³ (136 cu ft)

AREAS:
Main rotor disc	63.62 m² (684.8 sq ft)
Tail rotor disc	1.77 m² (19.05 sq ft)

WEIGHTS AND LOADINGS:
Weight empty	950 kg (2,094 lb)
Max payload: internal	550 kg (1,213 lb)
on external sling	750 kg (1,653 lb)
Normal T-O weight (internal payload)	1,600 kg (3,527 lb)
Max T-O weight with sling load	1,800 kg (3,968 lb)
Max disc loading	26.7 kg/m² (5.47 lb/sq ft)
Transmission loading at max T-O weight and power	5.07 kg/kW (8.33 lb/shp)

PERFORMANCE (at normal T-O weight, ISA):
Never-exceed speed (V$_{NE}$)	155 kt (288 km/h; 179 mph)
Max level speed	125 kt (232 km/h; 144 mph)
Normal cruising speed	108 kt (200 km/h; 124 mph)
Max rate of climb at S/L	510 m (1,673 ft)/min
Service ceiling: ISA	5,430 m (17,820 ft)
ISA + 20°C	4,780 m (15,680 ft)
Hovering ceiling: IGE: ISA	2,300 m (7,540 ft)
ISA + 20°C	1,500 m (4,920 ft)

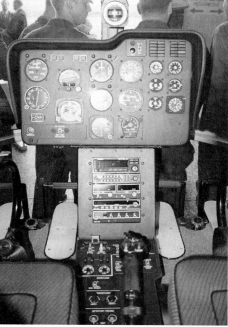
SW-4 instrument panel *(R J Malachowski)*

OGE: ISA	4,820 m (15,820 ft)
ISA + 20°C	4,150 m (13,620 ft)
Max range, standard fuel, no reserves	464 n miles (860 km; 534 miles)
Endurance, as above	5 h 8 min

UPDATED

PZL WARSZAWA-OKĘCIE

PAŃSTWOWE ZAKŁADY LOTNICZE WARSZAWA-OKĘCIE SA (State Aviation Works, Warsaw-Okęcie)
aleja Krakowska 110/114, PL-00-971 Warszawa
Tel: (+48 22) 846 50 61
Fax: (+48 22) 846 27 01
e-mail: wasiucionek@pzl-okecie.pl
Web: http://www.pzl-okecie.com
PRESIDENT AND CEO: Ryszard Leja, MSc
GENERAL DESIGNER: Andrzej Frydrychewicz, Msc Eng
COMMERCIAL MANAGER: Władysław Skorski, MSc Eng

Okęcie factory founded in 1928; responsible for light aircraft development and production, and for design and manufacture of associated agricultural equipment for its own aircraft and those built at other Polish factories; became public stock company, entirely owned by National Treasury, in 1995; by 1 January 2000 had produced 3,777 aircraft since 1945, including 33 in 1995, eight in 1996; 11 in 1997, 20 in 1998 and eight in 1999; workforce in 2000 was 960. MoU signed with Embraer of Brazil in June 1995, under which PZL-104, -106 and -111 marketed via Neiva in Brazil, in exchange for Brasilia order from Warszawa-Okęcie subsidiary Skypol; by October 2000, neither component of the agreement appeared to have been implemented.

Light Aircraft and Agricultural Equipment Pilot Plant
Tel: (+48 22) 846 63 50
e-mail: kubicki@pzl-okecie.com.pl
DIRECTOR: Tomasz Kubicki, MSc Eng

Main function of this PZL Warszawa-Okęcie plant is to develop and perform research tasks, build and flight test prototypes.

UPDATED

PZL WARSZAWA PZL-130 ORLIK
English name: Spotted Eaglet
TYPE: Basic turboprop trainer/attack lightplane.
PROGRAMME: Development of piston-engined Orlik (see 1989-90 *Jane's*) discontinued 1990. Development of turboprop derivative began 1985 and fully described in 1999-2000 and earlier editions.
First deliveries, to Polish Air Force Academy at Deblin, were two PZL-130TMs (c/n 005 and 006) in October 1992, followed by first two PZL-130TBs (c/n 012 and 013) from December 1992. Deliveries suspended in 1995; resumed in early 1998 with manufacture of what were, reportedly, last three TC-1s (043 to 045), but at least two further aircraft (046 and 047) completed late 1999. Fleet now being retrofitted with underwing drop tanks following Polish Air Force dissatisfaction with mission radius capability; first flight in this form in third quarter 1999.
CURRENT VERSIONS: **PZL-130TB:** For Polish Air Force. Differences from TM prototypes listed in 2000-01 *Jane's*. First example (c/n 009) rolled out May 1991, making first flight 17 September 1991 (later converted as TC-1 prototype but lost January 1996). Nine production TBs delivered, of which 021 lost on 30 April 1994. Remainder in service with 1 Eskadra/60 Lotniczy Pulk Szkolny (1st Squadron/60th Air School Regiment) at Radom, all retrofitted to TC-1 standard.
PZL-130TC: Advanced version, with 708 kW (950 shp) P&WC PT6A-62 engine; Honeywell avionics, Martin-

Front cockpit of the PZL-130TC-1 Orlik

POLAND: AIRCRAFT—PZL WARSZAWA-OKĘCIE

Model of 2000 proposal for an upgraded TC-2 version of Orlik *(R J Malachowski)*

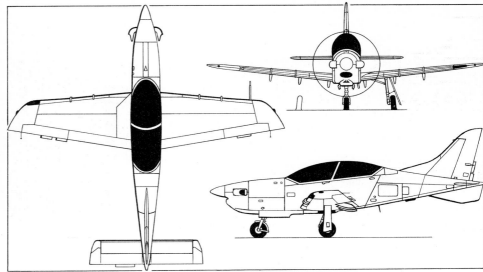

PZL Warszawa-Okęcie PZL-130TC-1 Orlik, showing current Polish Air Force in-service configuration *(Mike Keep/Jane's)*

Baker Mk 11B ejection seats and Flight Visions HUD; c/n 011 completed as prototype/demonstrator SP-PCE and made first flight 2 June 1993; in storage by 1998. Further details in 1998-99 and earlier editions.

PZL-130TC-1: Upgraded version of PZL-130TB, first flown (converted 009/SP-PRF) 9 July 1994. Power plant as for TB, but Martin-Baker Mk 11B zero height/90 kt (167 km/h; 104 mph) seats, Honeywell GPS and multifunction flight data recorder. Modified tailplane and enlarged ventral fin trialled on 037/SP-PCH in May 1996. Total of 23 TC-1s (15 new-build and eight TB conversions) delivered to Polish Air Force by end of 1995. None delivered in 1996 or 1997; three TC-1s built in 1998 and two in 1999. Tail surface modifications accepted by Polish Air Force in September 1997. Further control surface modifications flight-tested on 037 in 1998.

Following description applies to TC-1, except where otherwise indicated; further details in 1999-2000 edition.

PZL-130TC-2: Alternative version; simplified equipment compared with TC, and 559 kW (750 shp) PT6A-25C engine. Weights, loadings and performance data in 1999-2000 *Jane's*. Development programme abandoned and prototype (c/n 014) relegated to static test.

Proposal revealed early 2000 for revised version, still with PT6A-25C engine, but revamped to offer capabilities comparable to US T-6A Texan II. Major changes would include new, extended wingtip 'salmons' of composites; strengthened main wing able to support Alkan 6170 stores carriers for light air-to-ground weapons or 7.62 mm gun pods; landing gear reinforced for higher operating weights; new nav/attack avionics; and HOTAS controls. Could provide basis for upgrade of existing Orlik fleet, but no commitment so far announced.

PZL-140 Orlik 2000: Seven-seat business transport, using some PZL-130TC-1 components; described separately.

CUSTOMERS: See table. By 1995, 23 TC-1s (including eight TB conversions) in Polish Air Force service; situation unchanged in mid-2000. All former TBs in service now upgraded to TC-1s. Four Orliks formed a Polish Air Force aerobatic team in 1998.

FLYING CONTROLS: Conventional and manual. Frise differential ailerons, elevators and rudder all aerodynamically and mass balanced; ailerons actuated by pushrods and torque tube, elevators by rods and cables, rudder by cables; electrically actuated trim tabs on port aileron (two), starboard aileron (one), port elevator (one) and rudder (two). Vortex generators ahead of ailerons on 'TC-2' 037. Three-position double-slotted trailing-edge flaps, also actuated electrically.

New trim system flight tested early 1995, with Lear Astronics computer controlling a movable undertail fin.

LANDING GEAR: Hydraulically retractable tricycle type, all three units retracting into fuselage (mainwheels inward, steerable nosewheel rearward). Nosewheel protrudes slightly when retracted. PZL Warszawa-Okęcie

PZL-130 DEVELOPMENT AND PRODUCTION

c/n	Identity	Version	First flight	Engine	Remarks
001	n/a	130	n/a	n/a	Static test
002	SP-PCA	130	12 Oct 84	M-14 Pm	Piston-engined prototype
003	SP-PCB	130	29 Dec 84	M-14 Pm	Became 130T mockup
004	SP-PCC	130	21 Jan 85	M-14 Pm	Re-engined with PT6A-25A by Airtech as 130T/SP-RCC; flew 13 Jul 86 but lost Jan 87
005	005	130	19 Feb 88	M-14 Pm	Re-engined with M 601E as 130TM; to Deblin Oct 92;
006	006	130	19 Mar 88	K8-AA	Re-engined with M 601E as 130TM; to Deblin Oct 92; preserved at Bemowo
007	007	130TM	12 Jan 89	M 601E	First TM; to Deblin Oct 92; now instructional airframe
008	SP-WCA	130T	15 Mar 90	PT6A-25A	Overseas demonstrator; to Deblin Dec 92
009	SP-PRF	130TB/TC-1	17 Sep 91	M 601T	Comparative trials; lost 25 Jan 96
010	010	130TC	–	–	Static test
011	SP-PCE	130TC	2 Jun 93	PT6A-62	Prototype/demonstrator; now in storage
012, 013	012, 013	130TB/TC-1	21 Aug 92, 2 Mar 93	M 601T	First production TBs; first delivery Dec 92
014	014	130TC-2	–	PT6A-25C	TC-2 prototype (ex TB-1); static test
015-021	015-021	130TB/TC-1	8 May 93	M 601T	To 60 LPS 1993-94; 021 lost 30 Apr 94
022-036	022-036	130TC-1	15 Apr 94	M 601T	To 60 LPS and 23 LES 1994-95
037	037/SP-PCH	'130TC-2'	14 Feb 96	M 601T	First with enlarged ventral fin and squarer tailplane tips; replaces 009
038-042	038-042	130TC-1	24 Jan 96	M 601T	Built early 1996; delivered by early 1998 after modifications
043-045	043-045	130TC-1		M 601T	Completed in 1998
046-047	046-047	130TC-1		M 601T	Completed in 1999

PZL-130TC-1 Orlik turboprop trainer *(Paul Jackson/Jane's)*

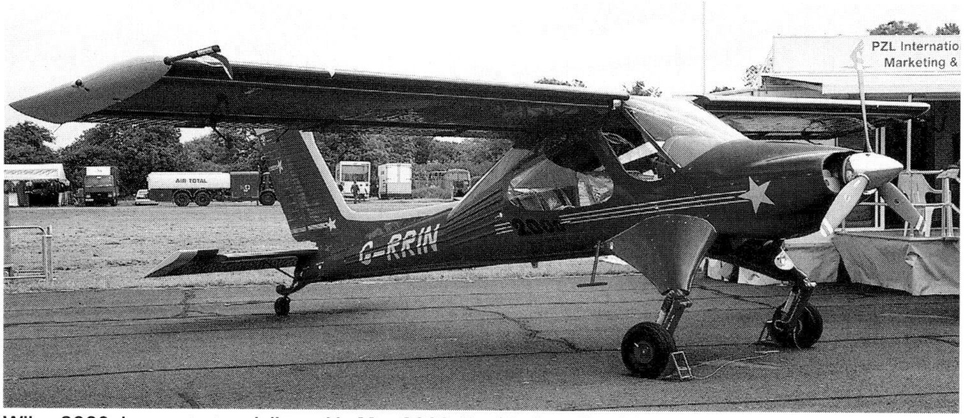

Wilga 2000 demonstrator, delivered in May 2000 *(Paul Jackson/Jane's)* ***2001***/0092510

oleo-pneumatic shock-absorber in each unit. All three wheels same size, with 500×200 tubeless tyres. Differential hydraulic multidisc brakes; parking brakes for main and nosewheels. No anti-skid system.

POWER PLANT: One 560 kW (750 shp) Walter M 601T turboprop, driving a five-blade propeller. Fuel in four integral wing tanks, total usable capacity 540 litres (143 US gallons; 119 Imp gallons). Overwing refuelling point for each tank. Fuel and oil systems permit up to 30 seconds of inverted flight. Retrofit with two 340 litre (89.8 US gallon; 74.8 Imp gallon) underwing drop tanks under way in 1999.

ACCOMMODATION: Tandem seating for pupil (in front) and instructor under one-piece canopy which opens sideways to starboard. Rear seat elevated 65 mm (2.6 in). Polish LFK-K1 ejection seats originally in PZL-130TB (zero height/70 to 323 kt; 130 to 600 km/h; 81 to 373 mph); Martin-Baker Mk 11B zero height/90 kt (167 km/h; 104 mph) seats in TC/TC-1. Baggage space aft of rear seat. Full dual controls standard. Cockpit heating and ventilation; canopy demisting.

ARMAMENT: Three hardpoints under each wing, inboard and centre ones stressed for loads of 160 kg (353 lb) each and outboard stations for 80 kg (176 lb) each. Typical external loads for PZL-130TC-1 include free-fall bombs of up to 100 kg, Tejsy and Mokrzycko bomb canisters, Zeus gun pods each with two 7.62 mm machine guns, launchers for 57 mm or 80 mm rockets, or Strela IR air-to-air missiles; plus S-17 gunsight and S-13 gun camera.

DIMENSIONS, EXTERNAL:	
Wing span	9.00 m (29 ft 6¼ in)
Wing aspect ratio	6.2
Length overall	9.00 m (29 ft 6¼ in)
Height overall *	3.53 m (11 ft 7 in)
Wheel track	3.10 m (10 ft 2 in)
Wheelbase	2.90 m (9 ft 6¼ in)
** Before installation of taller fin*	
AREAS:	
Wings, gross	13.00 m² (139.9 sq ft)
WEIGHTS AND LOADINGS:	
Weight empty	1,600 kg (3,527 lb)
Max external stores load	800 kg (1,764 lb)
Max T-O weight: Aerobatic	2,000 kg (4,409 lb)
Utility	2,700 kg (5,952 lb)
Max wing loading:	
Aerobatic	153.8 kg/m² (31.50 lb/sq ft)
Utility	207.7 kg/m² (42.54 lb/sq ft)
Max power loading: Aerobatic	3.58 kg/kW (5.88 lb/shp)
Utility	4.83 kg/kW (7.94 lb/shp)

PERFORMANCE (at max Aerobatic T-O weight, clean configuration, except where indicated):

Max level speed at S/L	245 kt (454 km/h; 282 mph)
Max level speed at 6,000 m (19,680 ft)	270 kt (501 km/h; 311 mph)
Max rate of climb at S/L	798 m (2,620 ft)/min
Service ceiling	10,060 m (33,000 ft)
T-O run at S/L	222 m (730 ft)
T-O to 15 m (50 ft)	280 m (920 ft)
Range with max fuel	523 n miles (970 km; 602 miles)
Range at 3,000 m (9,840 ft), clean configuration, no reserves	620 n miles (1,150 km; 714 miles)

UPDATED

PZL WARSZAWA PZL-104M WILGA 2000

English name: Oriole

TYPE: Four-seat lightplane.

PROGRAMME: First flight of prototype Wilga 1, 24 April 1962 (see 1968-69 *Jane's*); first flight of improved Wilga 35, 28 July 1967; production of Wilga 35 and 32 began 1968; both received Polish type certificate 31 March 1969 (Wilga 32 described in 1974-75 *Jane's*; see 1975-76 edition for Lipnur Gelatik Indonesian modified Wilga 32); first flight of Wilga 80, 30 May 1979. Both powered by one 194 kW (260 hp) PZL AI-14RA nine-cylinder supercharged radial air-cooled engine. Total of 962 of all earlier versions (except Gelatik) built by 1996, some of which stored and later supplied from stock, including one Wilga 80 in 1999. Further details of Wilga 35 and Wilga 80 variants in 1999-2000 and earlier *Jane's*.

Wilga 2000, improved version of Wilga 35, developed to appeal to Western markets, principally through use of Western engine and avionics. Wing and nose modifications tested on Wilga 35A SP-CSG. First flight of Wilga 2000 prototype (SP-PHG, later -WHG) 21 August 1996; first production (SP-AHV) in mid-1997; FAR Pt 23 certification achieved in 1997. SP-PHG won 1st World Air Games, 1997.

CURRENT VERSIONS: **PZL-104M Wilga 2000**: Standard landplane version; *as described.*

PZL-104MW Wilga 2000 Hydro: Floatplane version. Maximum T-O weight 1,500 kg (3,306 lb); performance generally similar to landplane except T-O run from water 180 m (590 ft) and S/L climb rate 210 m (689 ft)/min.

CUSTOMERS: Ten reported orders for Wilga 2000 by late 1996, including four for Polish National Aeroclub and three for export. First two deliveries in 1997 to Polish Aeroclub and private owner; five built in 1998 and delivered under 1997 contract to Polish Border Guard, first one being handed over 1 June 1998; one built in 1999, delivered to UK as demonstrator in May 2000. Total nine, including prototype, by mid-2000.

COSTS: Approximately US$100,000 (1996).

DESIGN FEATURES: Suitable for wide variety of military, general aviation and flying club duties; STOL performance bestowed by slats, flaps and drooping ailerons, allied to heavy-duty landing gear. High-mounted cantilever wings; braced tail unit; tall landing gear legs. In addition to Western equipment, Wilga 2000 features increased fuel, strengthened wing with integral fuel tanks and shorter mainwheel legs enclosed in fairings. Has 70 per cent structural commonality with original Wilga.

Wing section NACA 2415; dihedral 1°.

FLYING CONTROLS: Conventional and manual. Aerodynamically and mass-balanced slotted ailerons can be drooped to supplement flaps during landing; tab on starboard aileron; aerodynamically, horn- and mass-balanced one-piece elevator and rudder; trim tab in centre of elevator; manually operated slotted flaps; fixed slat on wing leading-edge along full span.

STRUCTURE: All-metal, with beaded skins; single-spar wings, with leading-edge torsion box; fuselage in two portions, forward incorporating main wing spar carry-through structure; rear section is tailcone; cabin floor of metal sandwich, with paper honeycomb core, covered with foam rubber; aluminium tailplane bracing strut.

LANDING GEAR: Non-retractable tailwheel type. Semi-cantilever main legs, of rocker type, have oleo-pneumatic shock-absorbers. Low-pressure tyres size 500×200 on mainwheels. Hydraulic brakes. Steerable tailwheel, tyre size 255×110, carried on rocker frame with oleo-pneumatic shock-absorber. Metal ski landing gear optional. Airtech Canada CAP 3000 twin-float gear on Wilga 2000MW.

POWER PLANT: One 224 kW (300 hp) Textron Lycoming IO-540 flat-six engine, driving a Hartzell F8468A-6R three-blade constant-speed propeller. Fuel in integral wing tanks, total capacity 400 litres (106 US gallons; 88.0 Imp gallons).

ACCOMMODATION: Passenger version accommodates pilot and three passengers, in pairs, with adjustable front seats. Baggage compartment aft of seats, capacity 35 kg (77 lb). Rear seats can be replaced by additional fuel tank for longer-range operation. Upward-opening door on each side of cabin, jettisonable in emergency.

AVIONICS: *Comms:* Honeywell KX 155 main and KY 96A standby transceiver; KT 76A transponder; KMA 24 audio panel; intercom.

Flight: Honeywell KLN 89B GPS receiver and KR 87 radio compass.

Instrumentation: ASI, VSI; ADF, VOR and twin indicators; artificial horizon; directional gyro; magnetic compass; altimeter; stall warning device; signalling and warning panel; clock; rpm tachometer; standard fuel/pressure/temperature gauges and indicators.

Mission: Polish Border Guard aircraft equipped with FLIR.

DIMENSIONS, EXTERNAL:	
Wing span	11.28 m (37 ft 0 in)
Wing chord, constant	1.40 m (4 ft 7 in)
Wing aspect ratio	8.5
Length overall: 2000	8.10 m (26 ft 6¾ in)
2000MW	8.52 m (27 ft 11½ in)
Height overall: 2000	2.96 m (9 ft 8½ in)
2000MW: on water	approx 3.00 m (9 ft 10 in)
on land	3.57 m (11 ft 8½ in)
Tailplane span	3.70 m (12 ft 1¾ in)

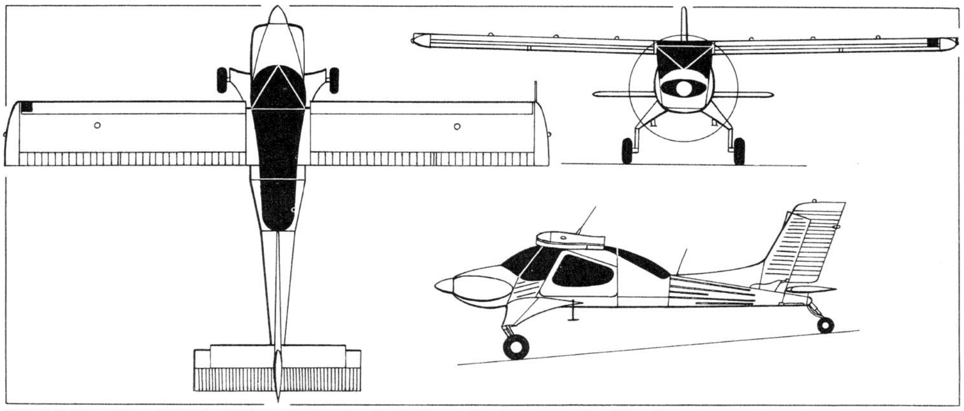

PZL-104M Wilga 2000 (224 kW; 300 hp IO-540 engine) *(James Goulding/Jane's)* ***2001***/0093855

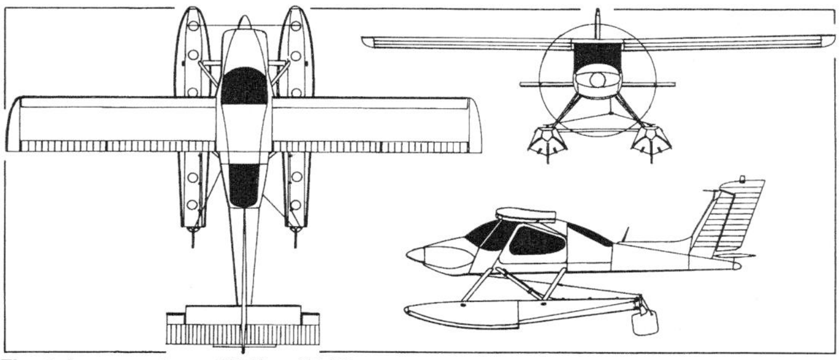

Floatplane version of Wilga 2000 ***2001***/0092544

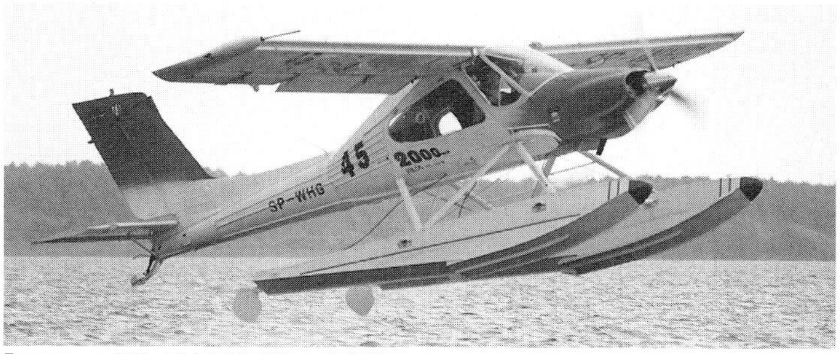

Prototype PZL-104 MW Wilga 2000 Hydro ***2000***/0084610

332 POLAND: AIRCRAFT—PZL WARSZAWA-OKĘCIE

Wheel track	2.75 m (9 ft 0¼ in)
Wheelbase	6.70 m (21 ft 11¾ in)
Distance between float c/l: 2000MW	2.48 m (8 ft 1¾ in)
Propeller diameter	2.03 m (6 ft 8 in)
Passenger doors (each): Height	0.90 m (2 ft 11½ in)
Width	1.50 m (4 ft 11 in)

DIMENSIONS, INTERNAL:

Cabin: Length	2.20 m (7 ft 2½ in)
Max width	1.20 m (3 ft 10 in)
Max height	1.50 m (4 ft 11 in)
Floor area	2.20 m² (23.8 sq ft)
Volume	2.40 m³ (85 cu ft)
Baggage compartment	0.50 m³ (17.5 cu ft)

AREAS:

Wings, gross	15.00 m² (161.5 sq ft)
Tailplane	3.16 m² (34.01 sq ft)
Elevator, incl tab	1.92 m² (20.67 sq ft)

WEIGHTS AND LOADINGS:

Weight empty	900 kg (1,984 lb)
Max T-O weight	1,400 kg (3,086 lb)
Max wing loading	93.3 kg/m² (19.12 lb/sq ft)
Max power loading	6.26 kg/kW (10.29 lb/hp)

PERFORMANCE:

Never-exceed speed (VNE)	131 kt (243 km/h; 151 mph)
Max level speed	112 kt (208 km/h; 129 mph)
Cruising speed at 75% power	103 kt (190 km/h; 118 mph)
Stalling speed: flaps up	58 kt (106 km/h; 66 mph)
flaps down	52 kt (95 km/h; 59 mph)
Max rate of climb at S/L	293 m (961 ft)/min
Service ceiling	4,118 m (13,510 ft)
T-O run (concrete)	272 m (895 ft)
Landing run (concrete)	165 m (545 ft)
Max range	750 n miles (1,390 km; 863 miles)

UPDATED

PZL-106BT Turbo-Kruk agricultural aircraft (Walter M 601D turboprop) *(R J Malachowski)* 2001/0092546

PZL WARSZAWA PZL-106BT TURBO-KRUK
English name: Turbo-Raven

TYPE: Agricultural sprayer.

PROGRAMME: Original piston-engined PZL-106 designed early 1972; first flight (SP-PAS), with IO-720 engine, 17 April 1973; last piston-engined aircraft built in 1990. First flight of turboprop-powered prototype (SP-PAA) 18 September 1985; FAR Pt 23 certification early 1994; features include taller fin and increased payload. A Turbo-Kruk was flown experimentally in 1997 with a 559 kW (750 shp) P&WC PT6A-34AG turboprop, modified cockpit and minor aerodynamic improvements.

CURRENT VERSIONS: **PZL-106BT-601:** Standard production version with Walter M 601D engine.

PZL-106BT-34: Export version with P&WC PT6A-34AG turboprop (559 kW; 750 shp); first flight (SP-PBW, 262nd Kruk built) mid-August 1998. Redesigned engine cowling, improved electrical system and new instrument panel. Certification was expected in 2000. Aimed initially at Latin American market.

Following description applies to current production PZL-106BT-601.

CUSTOMERS: Total of 39 built by mid-2000, including one BT-34 in 1998 and two BT-601s in 1999. Export deliveries include Argentina (21), Egypt, Iran, South Africa (five) and Sudan. Overall total of 270 (all versions) built up to mid-2000, comprising eight prototypes, 143 PZL-106As (of which two converted to -106B) between 1975 and 1981, 79 PZL-106BR/BSs and 42 PZL-106BTs.

DESIGN FEATURES: Typical low-, braced-wing agricultural monoplane with hopper at CG and high cockpit with good forward view.
Wings have upward-cambered tips; braced tailplane. Sweepback 6° at quarter-chord; 6° dihedral.

FLYING CONTROLS: Conventional and manual. Slotted ailerons (each with electrically actuated trim tab); mass-balanced elevators (with port trim tab); rudder has trim tab and anti-servo tab. Full-span, four-segment slats on each wing leading-edge; trailing-edge flaps.

STRUCTURE: Duralumin wings with metal and polyester fabric covering and GFRP tips; flaps and ailerons of duralumin with polyester fabric skins; GFRP/foam core sandwich slats; duralumin V bracing struts. Welded steel tube fuselage, with polyurethane enamel coating and quickly removable light alloy and GFRP panels; structure can be pressure tested for cracks. Duralumin tail unit has metal-skinned fixed surfaces and polyester fabric-covered control surfaces. Structure corrosion resistant, and additionally protected by external finish of polyurethane enamel.

LANDING GEAR: Non-retractable tailwheel type, with oleo-pneumatic shock-absorber in each unit. Mainwheels, with 800×260 low-pressure tyres (2.00 bars; 29 lb/sq in), are each carried on a side V and half-axle. Steerable tailwheel has 350×135 tubeless tyre, pressure 2.50 bars (36 lb/sq in). Pneumatically operated mainwheel hydraulic disc brakes; parking brake.

POWER PLANT: One 544 kW (730 shp) Walter M 601D turboprop; Avia V 508 D three-blade propeller. Fuel in two integral wing tanks, total capacity 560 litres (148 US gallons; 123 Imp gallons); can be increased to total of 950 litres (251 US gallons; 209 Imp gallons) by using hopper as auxiliary fuel tank. Gravity refuelling point on each wing; semi-pressurised refuelling point on starboard side of fuselage.

ACCOMMODATION: Single vertically adjustable seat in enclosed, ventilated and heated cockpit with steel tube overturn structure. Provision for instructor's cockpit with basic dual controls, forward of main cockpit and offset to starboard, for training of pilots in agricultural duties. Optional rearward-facing second seat (for mechanic) to rear. Jettisonable window/door on each side of cabin. Pilot's seat and seat belt designed to resist 40 g impact. Cockpit air conditioning optional.

AVIONICS: *Comms:* VHF com transceiver standard; 720-channel UHF transceiver optional.

EQUIPMENT: Easily removable non-corroding (GFRP) hopper/tank, forward of cockpit, can carry more than 1,000 kg (2,205 lb) (see Weights and Loadings) of dry or liquid chemical, and has maximum capacity of 1,400 litres (370 US gallons; 308 Imp gallons). Enlarged hopper (capacity 1,500 litres; 397 US gallons; 330 Imp gallons) optional.
Turnaround time, with full load of chemical, is approximately 28 seconds. Hopper quick-dump system can release 1,000 kg of chemical in 5 seconds or less. Pneumatically operated intake for loading dry chemicals optional. Distribution system for liquid chemical (jets or atomisers) powered by fan-driven centrifugal pump. Tunnel spreader, with positive on/off action for dry chemicals. For ferry purposes, hopper can be used to carry additional fuel instead of chemical.
When converted into two-seat trainer (see Accommodation), standard hopper can be replaced easily by container with reduced capacity tank for liquid chemical. Steel cable cutter on windscreen and each mainwheel leg; steel deflector cable runs from top of windscreen cable cutter to tip of fin. Windscreen washer and wiper standard. Other equipment includes clock, rearview mirror, second (mechanic's) seat (optional), landing light, anti-collision light, and night working lights (optional).

DIMENSIONS, EXTERNAL:

Wing span	15.00 m (49 ft 2½ in)
Wing chord, constant	2.16 m (7 ft 1 in)
Wing aspect ratio	7.1
Length overall	10.34 m (33 ft 11 in)
Height overall	4.34 m (14 ft 2¾ in)
Tailplane span	5.77 m (18 ft 11¼ in)
Wheel track	3.10 m (10 ft 2 in)
Wheelbase	7.41 m (24 ft 3¾ in)
Propeller diameter	2.50 m (8 ft 2½ in)
Propeller ground clearance (tail down)	0.63 m (2 ft 0¾ in)
Crew doors (each): Height	0.91 m (2 ft 11¾ in)
Width	1.06 m (3 ft 5¾ in)

DIMENSIONS, INTERNAL:

Cabin: Length	1.37 m (4 ft 6 in)
Max width	1.25 m (4 ft 1¼ in)
Max height	1.30 m (4 ft 3¼ in)
Floor area	1.12 m² (12.05 sq ft)

AREAS:

Wings, gross	31.69 m² (341.1 sq ft)
Vertical tail surfaces (total)	3.44 m² (37.03 sq ft)
Horizontal tail surfaces (total)	7.56 m² (81.38 sq ft)

WEIGHTS AND LOADINGS:

Weight empty, equipped	1,750 kg (3,858 lb)
Max chemical payload	1,500 kg (3,307 lb)
Max T-O weight	3,500 kg (7,716 lb)
Max landing weight	3,000 kg (6,614 lb)
Max wing loading	110.4 kg/m² (22.61 lb/sq ft)
Max power loading	6.07 kg/kW (9.97 lb/shp)

PERFORMANCE (at 3,000 kg; 6,614 lb AUW, S/L, ISA. A: clean, B: with tunnel spreader):

Never-exceed speed (VNE)	
A, B	145 kt (270 km/h; 167 mph)
Max level speed: A	130 kt (240 km/h; 149 mph)
Operating speed:	
B	86-97 kt (160-180 km/h; 99-112 mph)
Stalling speed, flaps up, power off:	
A	59 kt (108 km/h; 68 mph)
B	61 kt (112 km/h; 70 mph)
Max rate of climb: A	378 m (1,240 ft)/min
B	330 m (1,083 ft)/min
Service ceiling: A	6,500 m (21,320 ft)
T-O run: A	240 m (790 ft)
B	270 m (885 ft)
Landing run: A	260 m (855 ft)
B	270 m (885 ft)
Swath width: B: atomisers	45 m (148 ft)
spray nozzles	30 m (98 ft)
tunnel spreader	25 m (82 ft)

UPDATED

PZL WARSZAWA PZL-110 KOLIBER
English name: Hummingbird

TYPE: Four-seat lightplane.

PROGRAMME: Licence-built and uprated version of Socata Rallye 100 ST; original (Morane-Saulnier) Rallye flew 10 June 1959; main production in France until 1983.
First PZL-110, modified to receive 86.5 kW (116 hp) PZL-F (Franklin) engine, first flown 18 April 1978; first flight of Koliber 150 prototype (SP-PHA) 27 September 1988; Polish type certificate for 150 awarded January 1989, FAA certificate for 150A February 1995. Koliber 160A introduced in 1998; production deliveries from 1999.

CURRENT VERSIONS: **Koliber:** See 1991-92 and earlier *Jane's* for details of initial **Series I/II/III** with 85.8 kW (115 hp) PZL-Franklin engines; production (total of 40 built) ended December 1989 and included 675 sets of spare fuselage/wing/aileron/flap/control surface assemblies, some of them exported to France.

Koliber 150: Production version from 1990 to 1995, described in 1998-99 and earlier *Jane's*. Higher-powered Textron Lycoming engine (112 kW; 150 hp O-320-E2A)

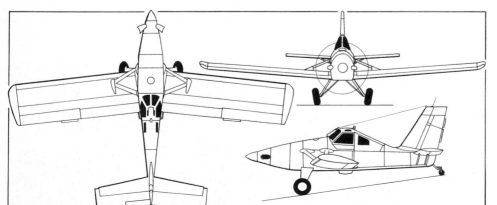

PZL-106BT Turbo-Kruk with a PT6A-34AG turboprop and aerodynamic refinements *(Paul Jackson/Jane's)* 2000/0079002

Jane's All the World's Aircraft 2001-2002 www.janes.com

PZL WARSZAWA-OKĘCIE—AIRCRAFT: POLAND

Instrument panel of a production PZL-110 Koliber 160A *(Paul Jackson/Jane's)*

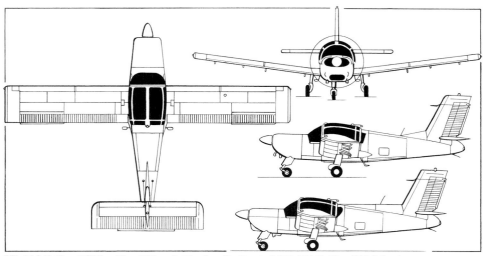

PZL-110 Koliber 160A, with additional side view (bottom) of PZL-111 Koliber 235A Senior *(James Goulding/Jane's)*

and detail improvements. Certified in Australia, Denmark, Germany, Netherlands, Norway, Poland, Sweden and UK.

Koliber 150A: Export version of 150 (known in USA as **Koliber II**); same power plant but with Western avionics, for night/IFR flying; improved electrical system. Performance (see 1998-99 edition) differs slightly from that of Koliber 150. Certified to FAR Pt 23 Amendment 23 in February 1995.

Koliber 160A: Introduced 1998 and is now the only production version. First flight (SP-WGF) 15 April 1998; second prototype (SP-WGG) followed 25 May 1998. First two production aircraft delivered to PZL International at North Weald, UK, in September and October 1999 for onward sale; further three completed in first half of 2000.

Koliber 235A Senior (PZL-111): More powerful version; described separately.

Koliber Junior (PZL-112): Two-seat, new-design primary trainer; described separately.

CUSTOMERS: Koliber 150s and 150As sold in Canada, Denmark, Germany, Netherlands, Sweden, UK and USA. Total of 79 Koliber 110/150/150As built by 1995, of which five held in stock during 1997; no production in 1996-98. Two stock 150As converted to 160A prototypes and delivered to UK as demonstrators in 1998. Five production 160As delivered by mid-2000, including one development aircraft and three supplied to UK distributor.

COSTS: Koliber 160A, US$115,000 (1998). Avionics options US$13,200 extra (Nav I) or US$14,700 (Nav II).

Following description applies to Koliber 160A.

DESIGN FEATURES: Winner of French competition for ideal flying club aircraft (Rallye); side-by-side seating, nosewheel landing gear, leading-edge slats for low-speed safety (minimum tendency to stall) and simple, inexpensive construction. Constant-chord wing, tailplane and fin, last-mentioned with sweepback.
Wing section NACA 63A-416 (modified); dihedral 7° 7′ 30″; incidence 4°.

FLYING CONTROLS: Conventional and manual. Pushrod-actuated, aerodynamically and mass-balanced ailerons, each with ground-adjustable tab; aerodynamically and mass-balanced elevator with central control tab, rudder with ground-adjustable tab; full-span automatic leading-edge slats; electrically actuated Fowler flaps.

STRUCTURE: All-metal; flaps, elevator and rudder have corrugated skins; wing torsion box and trailing-edge segments electrically spot-welded.

LANDING GEAR: Non-retractable tricycle type, with leg fairings and oleo-pneumatic shock-absorption; castoring nosewheel. Tyre sizes 5.00-4 (nose) and 6.00-6 (main) with pressures 1.40 and 1.80 bars (20 and 26 lb/sq in) respectively. Toe-operated hydraulic disc brakes. Parking brake.

POWER PLANT: One 119 kW (160 hp) Textron Lycoming O-320-D2A flat-four engine, driving a Sensenich 74DM6-058 two-blade fixed-pitch metal propeller. Fuel in two metal tanks in wings, with total usable capacity of 161 litres (42.5 US gallons; 35.4 Imp gallons). Refuelling points above wings. Oil capacity 7.6 litres (2.0 US gallons; 1.7 Imp gallons).

ACCOMMODATION: Two fore-and-aft adjustable, side-by-side seats, plus two-person bench seat at rear, under large rearward-sliding canopy. Dual controls. Heating, ventilation and soundproofing standard.

SYSTEMS: 14 V electrical system, with 70 Ah alternator and 70 Ah battery.

AVIONICS: *Comms*: Honeywell KX 155-0042 nav/com VHF; KT 76A transponder; SPA 400 intercom; ELT.

Flight: Honeywell KI 204-0002 VOR/ILS and KMA 24-0002 marker beacon receiver. Audible stall warning system standard. Two optional Honeywell avionics packages: *Nav I* (KLN 89 GPS, KX 155A/glide slope, KR 87 digital ADF and KI 209 VOR/LOC/glide slope indicator): or *Nav II* (KLN 89B GPS/IFR, MD 41-228 GPS nav, KX 155A/glide slope, KR 87 digital ADF and KI 209A VOR/LOC/glide slope with GPS interface).
Instrumentation: Standard VFR; IFR applied for.

EQUIPMENT: Position, anti-collision and landing/taxying lights; instrument panel, cabin and map lights; fixed cabin entrance steps; tiedown rings; fire extinguisher; winterisation kit.

DIMENSIONS, EXTERNAL:
Wing span	9.74 m (31 ft 11½ in)
Wing chord, constant	1.30 m (4 ft 3 in)
Wing aspect ratio	7.5
Length overall	7.37 m (24 ft 2¼ in)
Height overall	2.80 m (9 ft 2¼ in)
Tailplane span	3.67 m (12 ft 0½ in)
Wheel track	2.03 m (6 ft 8 in)
Wheelbase	1.71 m (5 ft 7¼ in)
Propeller diameter	1.88 m (6 ft 2 in)

DIMENSIONS, INTERNAL:
Cabin max width	1.08 m (3 ft 6½ in)

AREAS:
Wings, gross	12.68 m² (136.5 sq ft)
Ailerons (total)	1.56 m² (16.79 sq ft)
Trailing-edge flaps (total)	2.40 m² (25.83 sq ft)
Vertical tail surfaces (total)	1.74 m² (18.73 sq ft)
Horizontal tail surfaces (total)	3.48 m² (37.50 sq ft)

WEIGHTS AND LOADINGS (Normal category):
Weight empty, equipped	607 kg (1,338 lb)
Max fuel weight	116 kg (256 lb)
Max T-O and landing weight	850 kg (1,874 lb)
Max wing loading	67.0 kg/m² (13.72 lb/sq ft)
Max power loading	7.60 kg/kW (12.49 lb/hp)

PERFORMANCE (Normal category):
Never-exceed speed (V_{NE})	136 kt (251 km/h; 156 mph)
Max level speed	119 kt (220 km/h; 137 mph)
Cruising speed at 75% power	105 kt (194 km/h; 121 mph)
Econ cruising speed	75 kt (140 km/h; 87 mph)
Stalling speed: flaps up	47 kt (87 km/h; 55 mph)
Max rate of climb at S/L	210 m (690 ft)/min
Service ceiling	3,500 m (11,480 ft)
T-O run at S/L	168 m (555 ft)
T-O to 15 m (50 ft) at S/L	397 m (1,305 ft)
Landing from 15 m (50 ft) at S/L	320 m (1,050 ft)
Landing run at S/L	137 m (450 ft)
Range with max fuel, no reserves:	
at 75% power	496 n miles (919 km; 571 miles)
at 65% power	518 n miles (960 km; 596 miles)
g limits	+3.8/-1.5

UPDATED

PZL WARSZAWA PZL-111 KOLIBER 235A SENIOR

English name: Hummingbird

TYPE: Four-seat lightplane.

PROGRAMME: Construction of prototype of higher-powered version of now-discontinued Koliber 150A began August 1990; static testing completed March 1993. Prototype (SP-PHI; later -WHI) made its first flight on 14 September 1995. Polish certification 13 December 1996. No production aircraft had emerged by mid-2000. Proposed Koliber 180 and 200, with alternatively rated engines (see 1996-97 *Jane's*), have not yet joined the range.

Description as for Koliber 160A except as follows:

COSTS: US$158,000 (1998) with basic instrumentation. Nav I and II avionics options as for Koliber 160A; glider tow hook option US$2,250.

CURRENT VERSIONS: **Koliber 235A Senior**: With 175 kW (235 hp) Textron Lycoming engine.

DESIGN FEATURES: Numerous small differences from PZL-110 include extended dorsal fin, windows to rear of canopy, modified engine cowling and (in production version) square-topped fin. Intended for basic, navigation and general training; glider towing; sport flying; touring; executive flights; and special missions.

LANDING GEAR: Goodyear 6 ply tyres, size 350×35 (nose) and 6.00-6 Type III (main).

PZL-110 Koliber 160A (Textron Lycoming O-320-D2A engine) *(Paul Jackson/Jane's)*

POLAND: AIRCRAFT—PZL WARSZAWA-OKĘCIE

Prototype PZL-111 Koliber 235A Senior

POWER PLANT: One 175 kW (235 hp) Textron Lycoming O-540-B4B5 flat-six engine; Hartzell HC-C2YK-1BF/F8468A4 two-blade constant-speed metal propeller; three-blade McCauley B3D32C414/G-82NDA-6 propeller also available. Fuel capacity as for Koliber 160A.
ACCOMMODATION: Baggage space aft of seats.
EQUIPMENT: Baggage securing straps; optional glider towing hook.
DIMENSIONS, EXTERNAL: As Koliber 160A except:
Length overall	7.52 m (24 ft 8 in)
Propeller diameter: Hartzell	2.03 m (6 ft 8 in)
McCauley	1.93 m (6 ft 4 in)

WEIGHTS AND LOADINGS (Normal category):
Weight empty	705 kg (1,554 lb)
Max baggage weight	50 kg (110 lb)
Max T-O and landing weight	1,150 kg (2,535 lb)
Max wing loading	89.2 kg/m² (18.27 lb/sq ft)
Max power loading	6.56 kg/kW (10.79 lb/hp)

PERFORMANCE (Normal category):
Max level speed	140 kt (260 km/h; 162 mph)
Cruising speed at 75% power	134 kt (248 km/h; 154 mph)
Stalling speed, flaps up	50 kt (92 km/h; 58 mph)
Max rate of climb at S/L	372 m (1,220 ft)/min
Service ceiling	4,500 m (14,775 ft)
T-O run	135 m (445 ft)
Landing run	171 m (560 ft)
Range with max fuel, no reserves:	
at 75% power	396 n miles (734 km; 456 miles)
at 60% power	517 n miles (958 km; 595 miles)

UPDATED

PZL WARSZAWA PZL-112 KOLIBER JUNIOR

TYPE: Primary prop trainer/sportplane.
PROGRAMME: Designed in collaboration with Warsaw University of Technology; prototype construction began in late 1997. Programme has been delayed, but prototype (SP-PRG) completed by mid-2000. To be certified to FAR/JAR 23 (VFR and IFR), including spinning and other basic aerobatics.
COSTS: Approximately US$40,000 (1997).
DESIGN FEATURES: Low-wing monoplane with sweptback vertical tail and small ventral fin; intended as inexpensive primary trainer for Polish aeroclubs; utilises some components of PZL-110 Koliber. Dihedral 4°.
FLYING CONTROLS: Conventional and manual. Elevator trim tab; rudder and elevators horn-balanced. Fowler flaps.
LANDING GEAR: Non-retractable, trailing-link tricycle type; oleo-pneumatic shock-absorbers. Mainwheel tyres size 15 × 6.00-6 (4 ply rating), pressure 1.80 bars (26 lb/sq in); nosewheel tyre size 5.00 × 4 (6 ply rating), pressure 1.40 bars (20 lb/sq in). Hydraulic disc brakes.
POWER PLANT: One 86.5 kW (116 hp) Textron Lycoming O-235-L2C flat-four engine, driving a Hoffmann HO-14HM-175140 two-blade propeller; Lycoming O-320-D2A, derated to 116 hp, optional. Fuel capacity 105 litres (27.7 US gallons; 23.1 Imp gallons), of which 95 litres (25.1 US gallons; 20.9 Imp gallons) are usable.
ACCOMMODATION: Two seats side by side in fully enclosed cabin. Two-piece gull-wing canopy.
DIMENSIONS, EXTERNAL:
Wing span	8.803 m (28 ft 10½ in)
Wing chord, constant	1.30 m (4 ft 3¼ in)
Wing aspect ratio	6.8
Length overall	6.561 m (21 ft 6¼ in)
Height overall	2.625 m (8 ft 7¼ in)
Tailplane span	3.15 m (10 ft 4 in)
Wheel track	1.87 m (6 ft 1½ in)
Wheelbase	1.513 m (4 ft 11½ in)
Propeller diameter	1.75 m (5 ft 9 in)
Propeller ground clearance	0.12 m (4¾ in)

AREAS:
Wings, gross	11.44 m² (123.1 sq ft)
Ailerons (total)	0.60 m² (6.46 sq ft)
Trailing-edge flaps (total)	2.40 m² (25.83 sq ft)
Fin	0.66 m² (7.10 sq ft)
Rudder	0.50 m² (5.38 sq ft)
Tailplane	1.36 m² (14.64 sq ft)
Elevators (total, incl tab)	1.48 m² (15.93 sq ft)

WEIGHTS AND LOADINGS:
Max T-O and landing weight (Utility category)	712 kg (1,569 lb)
Max wing loading	62.2 kg/m² (12.75 lb/sq ft)
Max power loading	8.24 kg/kW (13.53 lb/hp)

PERFORMANCE (estimated, at Utility max T-O weight):
Never-exceed speed (V_{NE})	141 kt (262 km/h; 162 mph) EAS
Normal operating speed	110 kt (204 km/h; 127 mph) EAS
Approach speed	109 kt (203 km/h; 126 mph) EAS
Max speed for flap extension	71 kt (132 km/h; 82 mph) EAS
g limits	+4.4/−1.8

UPDATED

PZL WARSZAWA PZL-140 ORLIK 2000
English name: Spotted Eaglet

TYPE: Business turboprop.
PROGRAMME: Originated in 1995 design project. For certification to FAR Pt 23 Amendment 45 in Normal category. Prototype said to be under construction, but had not emerged by mid-2000.
COSTS: Estimated unit cost US$1.5 million (1996).
DESIGN FEATURES: Based on, and designed for extensive commonality with, PZL-130TC-1 Orlik military trainer (which see). All-new wingtips (with winglets), cockpit, cabin, rear fuselage and fixed vertical tail; PZL-130TC-1 components comprise forward fuselage, wings (including moving surfaces), rudder, ventral fin, horizontal tail, tailcone and landing gear.
POWER PLANT: Proposed with 559 kW (750 shp) Walter M 601F turboprop and Hamilton Sundstrand-Avia V-510 four-blade propeller, or Pratt & Whitney Canada PT6A-60 (derated to same power) with Hartzell propeller. Fuel capacity 1,150 litres (304 US gallons; 253 Imp gallons).
ACCOMMODATION: Individual seats in pressurised cabin for up to seven persons, including one or two pilots. Alternative layouts for cargo, aeromedical, photogrammetry or other applications. Square, downward-opening main door aft of wing, and emergency exit above leading-edge, both on port side.
AVIONICS: Honeywell KFC 275 autopilot and Silver Crown package.
DIMENSIONS, EXTERNAL:
Wing span	11.50 m (37 ft 8¾ in)
Wing aspect ratio	7.3
Length overall	11.40 m (37 ft 4¾ in)
Height overall	4.20 m (13 ft 9¼ in)
Passenger door: Height	1.25 m (4 ft 1¼ in)
Width	1.25 m (4 ft 1¼ in)

DIMENSIONS, INTERNAL:
Cabin: Length	4.00 m (13 ft 1½ in)
Max width	1.55 m (5 ft 1 in)
Max height	1.45 m (4 ft 9 in)

AREAS:
Wings, gross	18.00 m² (193.75 sq ft)

WEIGHTS AND LOADINGS:
Standard weight empty	2,000 kg (4,409 lb)
Max T-O weight	3,400 kg (7,495 lb)
Max wing loading	188.9 kg/m² (38.69 lb/sq ft)
Max power loading	6.08 kg/kW (9.99 lb/shp)

PZL-112 Koliber Junior prototype *(R J Malachowski)*

Cockpit and instruments of the PZL-112 *(R J Malachowski)*

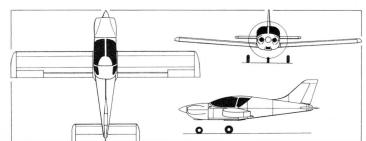

PZL-112 Koliber Junior primary trainer *(Paul Jackson/Jane's)*

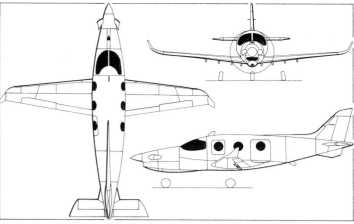

PZL-140 Orlik 2000 small business transport, based on the Orlik turboprop military trainer *(Paul Jackson/Jane's)* 2001/0093856

Model of the PZL-140 Orlik 2000

PERFORMANCE (estimated):
Cruising speed at 5,000 m (16,400 ft)
  243 kt (450 km/h; 279 mph)
Stalling speed 73 kt (135 km/h; 84 mph)
Service ceiling 10,000 m (32,800 ft)
T-O run 650 m (2,135 ft)
T-O to 15 m (50 ft) 1,000 m (3,280 ft)
Landing from 15 m (50 ft) 1,040 m (3,415 ft)
Landing run with propeller reversal 620 m (2,035 ft)
Typical range with seven persons and 140 kg (308 lb) of baggage 1,349 n miles (2,500 km; 1,553 miles)
Max range with five persons and 100 kg (220 lb) of baggage 1,727 n miles (3,200 km; 1,988 miles)
UPDATED

PZL WARSZAWA PZL-240 PELIKAN
English name: Pelican

TYPE: Agricultural sprayer.
PROGRAMME: Designed within framework of larger programme undertaken with co-operation of organisations in Poland (Warsaw University of Technology), Czech Republic (Letecky Instytut), France (Institut Aérotechnique) and UK (Kingston University, Cranfield University and Slingsby Aviation). Design started in early 1996, but changes required by European Commission, to which it was due to be resubmitted in 2000.
Following description applies to original design.

DESIGN FEATURES: Intended for forestry service, for firefighting, crop fertilisation and a wide range of ecological tasks. Braced biplane with sweptback tail surfaces; utilises cockpit, upper rear fuselage, tail unit and two pairs of wings from PZL-106 Kruk to minimise cost; large central hopper/tank forward of cockpit; additional, small belly tank for liquids; small underfin. Dihedral 5° on lower wings only.
FLYING CONTROLS: Conventional and manual. Flaps and ailerons on both upper and lower wings.
LANDING GEAR: Non-retractable tricycle type.
POWER PLANT: One 1,470 kW (1,971 shp) Pratt & Whitney Canada turboprop engine driving a Hamilton Sundstrand five-blade propeller. Standard fuel capacity 1,150 litres (304 US gallons; 253 Imp gallons); provision for 100 litres (26.4 US gallons; 22.0 Imp gallons) of auxiliary fuel in belly tank.
ACCOMMODATION: As for PZL-106.
EQUIPMENT: Central hopper/tank, capacity 6,000 litres (1,585 US gallons; 1,320 Imp gallons); ventral tank, capacity 100 litres (26.4 US gallons; 22.0 Imp gallons), for fire retardant.
DIMENSIONS, EXTERNAL:
Wing span: lower 15.50 m (50 ft 10¼ in)
Length overall 11.00 m (36 ft 1 in)
Height overall 5.10 m (16 ft 8¾ in)
Wheel track 3.60 m (11 ft 9¾ in)
Wheelbase 3.20 m (10 ft 6 in)

AREAS:
Wings, gross 65.00 m² (699.7 sq ft)
Vertical tail surfaces (total) 4.50 m² (48.44 sq ft)
Horizontal tail surfaces (total) 10.00 m² (107.64 sq ft)
WEIGHTS AND LOADINGS:
Operating weight empty 3,000 kg (6,614 lb)
Max T-O weight: Normal category 5,500 kg (12,125 lb)
  Restricted category 7,500 kg (16,534 lb)
Max wing loading:
  Normal category 84.6 kg/m² (17.33 lb/sq ft)
  Restricted category 115.4 kg/m² (23.63 lb/sq ft)
Max power loading:
  Normal category 3.74 kg/kW (6.15 lb/shp)
  Restricted category 5.10 kg/kW (8.38 lb/shp)
PERFORMANCE (estimated):
Max level speed 124 kt (230 km/h; 142 mph)
Operating speed 92 kt (170 km/h; 106 mph)
Stalling speed 44 kt (80 km/h; 50 mph)
Max rate of climb 270 m (886 ft)/min
T-O run on grass 270 m (890 ft)
Landing run 250 m (820 ft)
Range with max fuel 567 n miles (1,050 km; 652 miles)
Endurance:
  with 4,000 kg (8,818 lb) payload 1 h 30 min
  with 3,000 kg (6,614 lb) payload 5 h 0 min
UPDATED

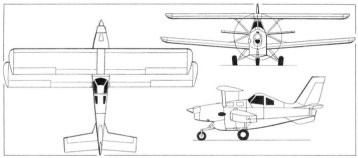

PZL-240 Pelikan turboprop utility biplane *(Paul Jackson/Jane's)* 1999/0051176

Model of the PZL-240 Pelikan *(R J Malachowski)* 2001/0092548

RADWAN

RADWAN LTD
PO Box 23, PL-00-958 Warszawa
Tel: (+48 90) 28 24 26
Fax: (+48 22) 38 15 74

Founded in 1989, Radwan entered the aviation business in 1992, its first commercial venture being the Swift lightplane. However, there has been no recent news of the programme, which was last described in the 2000-01 edition.
UPDATED

SSH

SERWIS SAMOLOTOW HISTORYCZNICH (Historic Aircraft Service)
PL-43 385 Jasienica 829
Tel: (+48 33) 815 34 91
Fax: (+48 33) 815 34 92
e-mail: jungmann@pro.onet.pl
MANAGING DIRECTOR: Janusz Karasiewicz
CUSTOMER SERVICE MANAGER: Julius Zulauf

In addition to producing the Bü 131, SSH is preparing for manufacture of a related aircraft, the Bü 133 Jungmeister. A prototype, to be known as the BU-233P, will be built with the registration SP-PUV, which was reserved in December 1998; however, this had not appeared by mid-2000.
UPDATED

SSH (BÜCKER) T-131 JUNGMANN
English name: Young Man

TYPE: Two-seat biplane.
PROGRAMME: Original German design first flew 1934. Popularity continued after Second World War, with variants built in Czechoslovakia (by Tatra), former West Germany (Canary/Bücker), Spain (CASA) and Switzerland (FFA); combined production exceeded 1,000. SSH T-131 launched in March 1990, based on engineering drawings of Czech version (T and P in designations indicate Tatra and Poland). First of four preseries aircraft flew 8 July 1994; first production aircraft exported to Austria in kit form December 1994 and made maiden flight 7 June 1997.
CURRENT VERSIONS: **T-131P**: Baseline version, with 78.3 kW (105 hp) Walter Minor 4-III four-cylinder inverted in-line engine and Hoffmann HO-40AHM-180 two-blade fixed-pitch metal propeller.
T-131PA: More powerful version, with 103 kW (138 hp) LOM M 332AK four-cylinder, in-line engine.
Description applies to T-131PA.
CUSTOMERS: Three T-131Ps built 1995, of which two exported to Sweden; 13 PAs registered and 12 built by

SSH T-131PA replica of Bücker Jungmann *(Paul Jackson/Jane's)* 2001/0093674

October 2000, including one to Germany and two to Austria.

Development Zl1.2 million (1997); standard T-131PA DM160,000 (2000).

DESIGN FEATURES: Single-bay biplane with interchangeable upper and lower wings, braced by pair of I-struts each side.
Wings have Göttingen aerofoil section (thickness/chord ratio 10.5 per cent) and 11° leading-edge sweepback; dihedral 3° 30′ (upper), 1° 30′ (lower); incidence -1° 30′; twist -1° 30′.

FLYING CONTROLS: Conventional and manual. Ailerons on all four wings; non-balanced rudder; trim tab in each elevator. No flaps.

STRUCTURE: Welded 4130 chromoly steel tube fuselage and tail unit; wooden wing spars (two) and ribs. Mostly Ceconite fabric covering except for metal engine cowling and around cockpit area.

LANDING GEAR: Non-retractable, with sprung 200×50 tailwheel; mainwheels 420×150 on forward-raked legs. Tyre pressures: main 1.50 bars (53 lb/sq in), tail 2.50 bars (88 lb/sq in). Hydraulic brakes.

POWER PLANT: One 104 kW (140 hp) LOM M 332AK four-cylinder in-line engine, driving an MT two-blade fixed-pitch propeller. Fuselage fuel tank, capacity 85 litres (22.5 US gallons; 18.7 Imp gallons), of which 80 litres (21.1 US gallons; 17.6 Imp gallons) are usable. Oil capacity 8.5 litres (2.25 US gallons; 1.9 Imp gallons).

ACCOMMODATION: Two persons in tandem, open cockpits.

SYSTEMS: 24 V DC electrical system (generator and two batteries).

AVIONICS: Basic VFR standard; other avionics to customer's requirements.

DIMENSIONS, EXTERNAL:
Wing span	7.40 m (24 ft 3¼ in)
Wing chord (each, constant)	1.00 m (3 ft 3¼ in)
Wing aspect ratio	4.0
Length overall	6.70 m (21 ft 11¾ in)
Height overall	2.50 m (8 ft 2½ in)
Tailplane span	2.50 m (8 ft 2½ in)
Wheel track	1.75 m (5 ft 9 in)
Wheelbase	4.10 m (13 ft 5½ in)
Propeller diameter	1.88 m (6 ft 2 in)
Propeller ground clearance	0.63 m (2 ft 0¾ in)

AREAS:
Wings, gross	13.70 m² (147.5 sq ft)
Ailerons (four, total)	2.00 m² (21.53 sq ft)
Fin	0.50 m² (5.38 sq ft)
Rudder	0.80 m² (8.61 sq ft)
Tailplane	1.25 m² (13.45 sq ft)
Elevators (total, incl tabs)	1.15 m² (12.38 sq ft)

WEIGHTS AND LOADINGS (approx):
Weight empty	450 kg (992 lb)
Fuel weight	70 kg (154 lb)
Max T-O and landing weight	720 kg (1,587 lb)
Max wing loading	52.5 kg/m² (256.6 lb/sq ft)
Max power loading	6.90 kg/kW (11.33 lb/hp)

PERFORMANCE:
Never-exceed speed (V_{NE})	162 kt (300 km/h; 186 mph)
Max level speed	102 kt (190 km/h; 118 mph)
Max cruising speed at S/L	97 kt (180 km/h; 112 mph)
Econ cruising speed at S/L	86 kt (160 km/h; 99 mph)
Stalling speed, engine idling	46 kt (85 km/h; 53 mph)
Max rate of climb at S/L	210 m (689 ft)/mn
Service ceiling	4,000 m (13,120 ft)
T-O run	220 m (720 ft)
T-O to 15 m (50 ft)	400 m (1,315 ft)
Landing from 15 m (50 ft)	300 m (985 ft)
Landing run	200 m (660 ft)
Max range, 30 min reserves	243 n miles (450 km; 279 miles)

UPDATED

WZL 3

WOJSKOWE ZAKLADY LOTNICZE Nr. 3 (Military Aviation Works No. 3)
PL-08-521 Deblin 3
Tel: (+48 81) 883 01 22
Fax: (+48 81) 883 02 48
Telex: 0642514 WZL PL

Primarily an airframe and engine overhaul plant, WZL 3 began operations in 1945, maintaining the Polikarpov Po-2, and progressed through a variety of Warsaw Pact equipment, its present activities centring on the MiG-21, MiG-23, TS-11 Iskra (including those ordered by India) and An-2. Recently, the plant began subcontract manufacture of ultralight aircraft, including the FF-1 and DEKO-9, the latter marketed in Germany as the Kaiser Magic (which see).

UPDATED

FF-1 ultralight under construction at Deblin
1999/0051180

YALO/AVIATA

ZAKŁAD NAPRAWY I BUDOWY SPRZETU LATAJACEGO YALO SC
Wal Miedzeszyński 646, PL-03-994 Warszawa
Tel/Fax: (+48 2) 671 93 79

MANAGERS:
Janusz Krzywonos
Krzystof Pazura

This consortium's first product, the Gniady, resumed flight trials in late 1997 or early 1998 with a view to certification, although these further tests are being conducted by the Instytut Lotnictwa (which see). Yalo is meanwhile working on the design of two two-seat aircraft, a hydroplane and an ultralight. No details of these projects have been received. The company also built the prototype Radwan Swift (which see in 2000-01 and earlier editions), assembles the Avid Flyer ultralight (see US section) for the domestic market and in 1998 rebuilt two SZD-9bis Bocian sailplanes as Bocian M-2000 motor gliders.

UPDATED

YALO/AVIATA GM-01 GNIADY
English name: Carthorse

TYPE: Glider tug.

PROGRAMME: Prototype (SP-PBN) flew for the first time on 26 April 1995. Initial development costs shared by Yalo and Polish government; additional funding sought to complete certification, but by early 1996 the prototype had been dismantled. Further support has been forthcoming and the prototype was reassembled in 1997, with flight testing continuing in 1999. Production is planned at Deblin.

DESIGN FEATURES: Hybrid design, by Aviata, manufactured by repair and overhaul specialist company Yalo. Utilises power plant, cowling and propeller of PZL Warszawa-Okęcie Wilga 35, wings and tail unit of PZL-110 Koliber, plus Wilga main landing gear, combined with new-design slimmer fuselage.

Capable of towing up to three gliders at a time. Operating costs stated to be half those of comparable aircraft in service, more than offsetting higher fuel consumption of more powerful engine.

FLYING CONTROLS: Conventional and manual. Flaps fitted.

STRUCTURE: All-metal.

Yalo/Aviata GM-01 Gniady glider tug *(R J Malachowski)* 0011950

Cabin/landing gear module of the Yalo-1 two-seat ultralight *(R J Malachowski)*
1999/0051179

POWER PLANT: One 194 kW (260 hp) PZL Kalisz AI-14RA nine-cylinder radial engine; US122000 two-blade propeller. Fuel capacity 105 litres (27.7 US gallons; 23.1 Imp gallons).
ACCOMMODATION: Pilot only, beneath rearward-sliding canopy.
AVIONICS: *Comms:* UHF radio.
Instrumentation: VFR standard.
DIMENSIONS, EXTERNAL:
Wing span	9.74 m (31 ft 11½ in)
Length overall	6.84 m (22 ft 5¼ in)
Height overall	2.61 m (8 ft 6¾ in)

AREAS:
Wings, gross	12.70 m² (136.7 sq ft)

WEIGHTS AND LOADINGS:
Weight empty	740 kg (1,631 lb)
Max T-O weight	905 kg (1,995 lb)
Max wing loading	71.3 kg/m² (14.60 lb/sq ft)
Max power loading	4.67 kg/kW (7.67 lb/hp)

PERFORMANCE:
Never-exceed speed (V_{NE})	128 kt (238 km/h; 147 mph)
Cruising speed	81 kt (150 km/h; 93 mph)
Stalling speed	43 kt (79 km/h; 50 mph)
Max rate of climb at S/L:	
clean	720 m (2,362 ft)/min
with three gliders	300 m (984 ft)/min
Service ceiling	4,000 m (13,120 ft)
T-O run	85 m (280 ft)
Landing run	90 m (295 ft)
Range	151 n miles (280 km; 174 miles)
g limits	+3.8/–1.5

UPDATED

YALO/AVIATA HYDROPLANE

TYPE: Two-seat amphibian.
PROGRAMME: Construction begun at Bemowo by 1997; company withholding details (including name) until after first flight, which had not been announced by mid-2000.
DESIGN FEATURES: Conventional configuration with constant chord unswept wings, highly swept fin, T tail and pusher propeller.
STRUCTURE: Largely of composites.
LANDING GEAR: Single-step hull; stabilising float scabbed to underside of each wing, slightly inboard of tip. Retractable tricycle landing gear, single wheel on each leg; mainwheels retract inwards, nosewheel forwards.

First of two Bocian M-2000 motor gliders produced by Yalo in 1998 *(R J Malachowski)*

Unnamed Yalo/Aviata hydroplane under construction *(R J Malachowski)*

POWER PLANT: One piston engine mounted on metal tube cabane immediately behind cockpit.
ACCOMMODATION: Two persons, side by side, with dual controls.

UPDATED

ROMANIA

IAROM

SC IAROM SA
44A Bulevardul Ficusului, Sector 1, R-71547 Bucuresti
Tel: (+40 1) 232 63 16
Fax: (+40 1) 232 63 46
GENERAL MANAGER: Mihail Toncea
COMMERCIAL AND BUSINESS DEVELOPMENT MANAGER: Antonela Banea
RESEARCH DEPARTMENT MANAGER: Vlad Atanase

IAROM (Industria Aeronautica Romana) is holding company for Romanian aircraft industry, created 1991 to replace former CNIAR (1991-92 and earlier *Jane's*). Aircraft and aero-engine companies and R&D institutes given full autonomy from 1990, when all became joint stock companies as prelude to privatisation. Shareholding was initially divided between State Ownership Fund (70 per cent) and Private Ownership Funds (30 per cent); full privatisation was achieved in 1998.

Main function is to establish general strategy and policy for Romanian aviation industry in fields of research, design, development, production and marketing, including technical support services for Romanian and foreign companies; also acts as agent for foreign companies doing business within Romania.

Tehnoimportexport SA
2 Doamnei Street (PO Box 110), R-1100 Bucuresti
Tel: (+40 1) 312 10 39
Fax: (+40 1) 312 10 38
GENERAL DIRECTOR: Mircea Bortes

Tehnoimportexport deals, on a non-exclusive basis, with sales and purchase of aircraft and related equipment, aircraft marketing and consultancy.

UPDATED

AEROSTAR

SC AEROSTAR SA
9 Condorilor Street, R-5500 Bacau
Tel: (+40 34) 17 50 70
Fax: (+40 34) 17 20 23 and 17 22 59
e-mail: aerostar@aerostar.ro
Web: http://www.aerostar.ro
PRESIDENT AND GENERAL DIRECTOR: Dipl Eng Grigore Filip
DIRECTORS:
Dipl Eng Ovidiu Buhai (Systems Division)
Dipl Eng Petru Placinta (Technology Division)
Dipl Eng Dorin Panfil (Commercial Division)
HEAD OF MARKETING DEPARTMENT: Serban Iosipescu
PUBLIC RELATIONS: Doina Matanie

Factory founded 1953 as URA (later IRAv, then IAv Bacau), originally as repair centre for Romanian Air Force Yak-17/23, MiG-15/17/19/21 and Il-28/H-5 front-line military aircraft and Aero L-29/L-39 jet trainers; built first Romanian prototype of IAR-93 between 1972 and 1974. Five other specialised work sections: (1) landing gears, hydraulic and pneumatic equipment; (2) special production; (3) light aircraft; (4) engines and reduction gears; (5) avionics. Site area 45 ha (111.2 acres), including 25 ha (61.8 acres) of covered workshop space. Workforce in October 2000 totalled 2,800.

Romanian State Ownership Fund (SOF) originally owned 69.99 per cent of Aerostar; 12 per cent was held by SIF Moldova investment organisation and 18 per cent listed on Romanian stock exchange. In preparation for full privatisation, Aerostar was reorganised in 1997 into three semi-autonomous divisions: systems, commercial, and technology support, reporting to a strategic fourth division. Creation of a joint venture (Aerothom Electronics) with Thomson-CSF Communications of France was announced in June 1999; Aerostar owns 60 per cent of this company, formed to supply IFF equipment to Romanian MoD. Aerostar's future as a privatised company was secured on 11 February 2000 by signature of a contract under which a consortium named Aerostar-PAS-IAROM, a grouping of Aerostar management and employees (PAS) with IAROM (which see), acquired the Aerostar stock previously held by the SOF. The new shareholdings are IAROM 65 per cent; SIF 'Moldova' (financial investment company) 11 per cent; PAS-Aerostar 5 per cent; and others 19 per cent. The SOF, however, retains a 'golden share' entitling it to a right of veto in matters affecting the company's defence production capability. Under the terms of the contract, the consortium agrees to take over all Aerostar's obligations and debts; not to dissolve the company for 10 years; and to keep its core business intact for at least five years. Company turnover in 1999 was Lei308.7 billion, of which exports accounted for approximately 58 per cent.

Aerostar factory's main product for many years has been Iak-52 trainer. Company is involved in major Romanian Air Force upgrade (Lancer) programme for 110 MiG-21s with Elbit of Israel; first upgraded Lancer flew on 23 August 1995; see *Jane's Aircraft Upgrades* for details. Now performing similar upgrade (Sniper programme), teamed with Elbit and DASA, and offered for Romanian Air Force MiG-29s; upgrade demonstrator/prototype made first flight 5 May 2000; talks with RSK MiG in progress in third quarter 2000 regarding possibly combining Sniper with latter's MiG-29SMT programme. Manufactures airframes for Air Light WT-01 Wild Thing ultralight (see German section) and components for replicas of Focke-Wulf Fw 190 (see Flug Werke in German section); own-design ultralight, Aerostar 1, under development.

Landing gears and/or hydraulic/pneumatic equipment have been produced for IAR-316B (Alouette III), IAR-330 (Puma), IAR-93, IAR-99, Romaero (BAC) One-Eleven and Iak-52. Engines include M-14P for Iak-52, M-14V26 for Ka-26 and RU-19A-300 APU for An-24; reduction gears include R-26. Avionics factory produces radio altimeters, radio compasses, marker beacon receivers, IFF and other radio/radar items.

UPDATED

AEROSTAR 1

TYPE: Side-by-side ultralight.
PROGRAMME: Illustrated (as project) in 1999 company brochure, but no details given; possibly awaiting outcome of privatisation process before go-ahead decision taken.
DESIGN FEATURES: Conventional low-wing monoplane with nosewheel landing gear; see accompanying illustration.

UPDATED

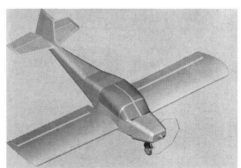

Aerostar 1 ultralight project

AEROSTAR (YAKOVLEV) Iak-52

TYPE: Primary prop trainer/sportplane.
PROGRAMME: Design of Yak-52 in USSR began 1975; series production assigned to Comecon programme (Council for Mutual Economic Assistance), following Romanian-USSR intergovernmental agreement of 1974; designation is transliteration of 'Yak' in Romanian alphabet, but civil aircraft are registered as Yaks.

Construction began 1977 and first Romanian prototype (c/n 780102) made first flight May 1978; series production began 1979; first deliveries (to DOSAAF, USSR) 1975; 1,000th aircraft delivered in 1986 and 1,500th in 1990. After cessation of deliveries to USSR in 1991, production had almost come to a standstill, but interest reawakened by improved Iak-52W (first flight April 1999, debut at Paris Air Show, June 1999). Iak-52W began development 1998 and first flew (12201) March 1999; currently offered as alternative to standard version. Aerostar planning five-year programme of 30 per year; deliveries started August 1999. Tailwheel version announced early 2000.

CURRENT VERSIONS: **Iak-52**: Standard model.
Full description of standard Iak-52 follows, augmented by details of Iak-52W and TW, where different.

Iak-52W: Upgraded ('Westernised') version; introduced 1999. Main changes are increased fuel tankage, three-blade propeller, Western avionics, and metal-skinned control surfaces. In production.

Iak-52TW: Tailwheel version of Iak-52W, revealed early 2000. Production aircraft available from April 2001; US marketing by GeSoCo Industries of Swanton, Vermont, USA.

Yak-54: Armed military version with underwing pylons; designation duplicates that of current tailwheel-equipped aircraft described in Russian section. Three built: one lost in service in Afghanistan; one to Monino museum; third (Yak-54B ES-FYA) currently with AS Strovest in Netherlands.

CUSTOMERS: Estimated 1,815, in 121 production batches, built by 1998, mainly for USSR and Romanian Air Force but also for West European and American civil markets; Hungarian Air Force received 12 in early months of 1994. Production averaged 10 per year between 1992 and 1996 (almost all to USA or Hungary), augmenting large number of military surplus sales. Twelve delivered to Vietnam in 1997. No known deliveries in 1998. Exports of Iak-52W began to USA in mid-1999; one also to Australia in 2000.

COSTS: Iak-52W approximately US$120,000 (1999); Iak-52TW US$125,000 for first 10, US$139,000 thereafter (2000).

DESIGN FEATURES: Tandem-cockpit variant of Yak-50, with unchanged span and length, but with semi-retractable landing gear to reduce damage in wheels-up landing (mainwheels fully retractable in Iak-52TW). Simple, robust, low-wing design with long landing gear legs to accommodate large propeller. Moderately tapered wings and tailplane; curved fin; 'glasshouse' canopy. Current production aircraft lack wingtip fairings. Strengthened wingtips on Iak-52W for optional fitment of tip-tanks. Increased wing span in Iak-52TW.

FLYING CONTROLS: Conventional and manual. Actuation of mass-balanced slotted ailerons by pushrods; mass-balanced elevators by pushrods/cables; and horn-balanced rudder by cables; manually operated trim tab in port elevator; ground-adjustable tab on rudder and each aileron. Pneumatically actuated trailing-edge split flaps.

STRUCTURE: All-metal except for fabric-covered primary control surfaces (these also metal-skinned on Iak-52W). Single-spar wings; modified spar, post-1986 (and some retrofits), permits higher load factors. Iak-52W wingtips strengthened to carry optional tip-tanks.

LANDING GEAR: Semi-retractable tricycle type, with single wheel and oleo-pneumatic shock-absorber on each unit. Iak-52TW mainwheels fully retractable, with flush doors. Cleveland wheels and Western tyres on Iak-52W; Scott steerable tailwheel on TW. Pneumatic actuation, nosewheel retracting rearward, main units forward. All three wheels remain fully exposed to air flow, against undersurface of fuselage and wings respectively, to offer greater safety in event of wheels-up emergency landing. Pneumatic drum (hydraulic disc on Iak-52W) mainwheel brakes, operated differentially from rudder pedals. Non-retractable plastics-coated duralumin skis, with shock-struts, can be fitted in place of wheels for winter operations.

POWER PLANT: One 265 kW (355 hp) Aerostar-built VOKBM (Bakanov) M-14P nine-cylinder air-cooled radial, driving a V-530TA-D35 two-blade constant-speed wooden propeller (three-blade Mühlbauer MTV-9 on Iak-52W and TW). Two aluminium alloy fuel tanks, in wingroots forward of spar; collector tank in fuselage supplies engine during inverted flight.

Total internal fuel capacity (standard Iak-52) 122 litres (32.2 US gallons; 26.8 Imp gallons). Additional integral tank in each wing of Iak-52W and TW, raising capacity to 280 litres (74.0 US gallons; 61.6 Imp gallons). Oil capacity 10 litres (2.6 US gallons; 2.2 Imp gallons). Iak-52TW oil cooler relocated below nose cowling.

ACCOMMODATION: Tandem seats for pupil (at front) and instructor under long 'glasshouse' canopy, with separate rearward-sliding hood over each seat. Hooker Harness seat belts for both occupants. Dual controls standard. Seats and rudder pedals adjustable. Heating and ventilation standard. Optional 0.20 m³ (7.06 cu ft) baggage compartment in standard version, accessible from rear seat; Iak-52W has externally accessible baggage and battery compartments on port side, aft of rear cockpit, plus permanently mounted retractable ladder for cockpit access.

SYSTEMS: Independent main and emergency pneumatic systems, pressure 50 bars (725 lb/sq in), for flap and landing gear actuation, engine starting and brake control. Electrical system (27 V DC) supplied by 3 kW engine-driven generator and (in port wing) 12 V 23 Ah ASAM battery (two 27 V in Iak-52W); two static inverters in fuselage for 36 V AC power at 400 Hz. Oxygen system available optionally.

AVIONICS: *Comms*: Balkan 5 VHF radio and SPU-9 intercom in standard Iak-52; replaced in Iak-52W by Western instruments including Garmin IC-AQ200 com radio and GTX transponder, AmeriKing ELT and altitude encoder, and NAT AA80-20 intercom.
Flight: ARK-15M automatic radio compass, eight-channel ADF and GMK-1A gyrocompass.
Other avionics available at customer's option.

EQUIPMENT: Strobe and navigation lights, recessed landing and taxying lights on Iak-52W; roll bar in Iak-52TW.

DIMENSIONS, EXTERNAL:
Wing span: except 52TW	9.30 m (30 ft 6¼ in)
52TW	9.90 m (32 ft 5¾ in)
Length overall: except 52TW	7.745 m (25 ft 5 in)
52TW	7.70 m (25 ft 3¼ in)
Height overall, propeller turning:	
52	2.70 m (8 ft 10¼ in)
52W, 52TW	2.50 m (8 ft 2½ in)
Wheel track: all	2.70 m (8 ft 10¼ in)
Wheelbase: except 52TW	1.86 m (6 ft 1¼ in)
Propeller diameter: 52	2.40 m (7 ft 10½ in)
52W, 52TW	2.00 m (6 ft 6¾ in)

DIMENSIONS, INTERNAL:
Cockpit: Max width	0.74 m (2 ft 5¼ in)
Max height	1.12 m (3 ft 8 in)

AREAS:
Wings, gross: except 52TW	15.00 m² (161.5 sq ft)
52TW	15.30 m² (164.7 sq ft)
Ailerons (total)	1.98 m² (21.31 sq ft)
Flaps (total)	1.03 m² (11.09 sq ft)
Fin	0.61 m² (6.57 sq ft)
Rudder	0.87 m² (9.36 sq ft)
Tailplane	1.32 m² (14.21 sq ft)
Elevators (total)	1.53 m² (16.47 sq ft)

WEIGHTS AND LOADINGS:
Weight empty, equipped: 52	1,015 kg (2,238 lb)
52TW	950 kg (2,094 lb)
52W	960 kg (2,116 lb)
Max T-O weight: 52	1,305 kg (2,877 lb)
52TW	1,360 kg (2,998 lb)
52W	1,315 kg (2,899 lb)
Max wing loading: 52	87.0 kg/m² (17.82 lb/sq ft)
52TW	88.9 kg/m² (18.21 lb/sq ft)
52W	87.7 kg/m² (17.96 lb/sq ft)
Max power loading: 52	4.93 kg/kW (8.10 lb/hp)
52TW	5.14 kg/kW (8.45 lb/hp)
52W	4.97 kg/kW (8.17 lb/hp)

PERFORMANCE (Iak-52W):
Never-exceed speed (V_{NE})	194 kt (360 km/h; 223 mph)
Max level speed: at S/L	154 kt (285 km/h; 177 mph)
at 1,000 m (3,280 ft)	145 kt (270 km/h; 167 mph)
Econ cruising speed at 1,000 m (3,280 ft)	103 kt (190 km/h; 118 mph)
Landing speed	60 kt (110 km/h; 69 mph)
Stalling speed, flaps down, engine idling	49 kt (90 km/h; 56 mph)
Max rate of climb at S/L	420 m (1,378 ft)/min
Time to 4,000 m (13,120 ft)	15 min
Service ceiling	4,000 m (13,120 ft)
T-O run	170 m (560 ft)
T-O to 15 m (50 ft)	200 m (660 ft)
Landing from 15 m (50 ft)	350 m (1,150 ft)
Landing run	300 m (985 ft)
Range at 500 m (1,640 ft), max fuel, 20 min reserves:	
52	296 n miles (550 km; 341 miles)
52W	648 n miles (1,200 km; 745 miles)
g limits: up to c/n 867215	+5/−3
from c/n 877301	+7/−5

PERFORMANCE (Iak-52TW, estimated):
Max cruising speed	175 kt (325 km/h; 202 mph)
Stalling speed, flaps down	57 kt (105 km/h; 66 mph)
Max rate of climb at S/L	360 m (1,181 ft)/min
Service ceiling	4,000 m (13,120 ft)
T-O run	300 m (985 ft)
Landing run	260 m (855 ft)
Range at 500 m (1,640 ft) at cruising speed, 20 min reserves:	
standard fuel	297 n miles (550 km; 341 miles)
max fuel	648 n miles (1,200 km; 745 miles)
g limits	+7/−5

UPDATED

Third production Aerostar Iak-52W in spurious military markings. Note absence of wingtip fairings *(Paul Jackson/Jane's)*

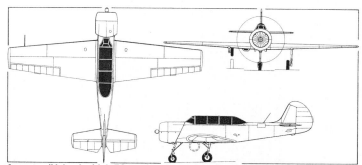

Aerostar (Yakovlev) Iak-52 tandem two-seat primary trainer *(Dennis Punnett/Jane's)*

Front cockpit instrument panel of the upgraded Iak-52W

AVIOANE

AVIOANE CRAIOVA SA
1 Aeroportului Street, PO Box 469, R-1100 Craiova
Tel: (+40 51) 40 20 40
Fax: (+40 51) 40 29 01
e-mail: tudor@acv.ro
Web: http://www.acv.ro
MANAGING DIRECTOR: Tudor Bistreanu
MARKETING MANAGER: Mircea Popa

Founded 1 February 1972 as IAv Craiova, beginning aircraft manufacture with joint Romanian/Yugoslav IAR-93/Orao attack aircraft; changed to present name 29 March 1991; range of products and services for military and civil aviation. Aircraft and equipment design and manufacture, repair and overhaul, life cycle management and integrated logistics support. Granted ISO 9001 status 1998. IAR-99 jet trainer being upgraded under 1998 contract. Site area 1.70 ha (4.20 acres), including 47,500 m² (511,275 sq ft) shop floor area.
UPDATED

Avioane IAR-99 Şoim avionics upgrade prototype *(Paul Jackson/Jane's)*

AVIOANE IAR-99 ŞOIM
English name: Hawk
TYPE: Advanced jet trainer.
PROGRAMME: Design by INAv (Institutul de Aviatie) at Bucharest started 1975; three prototypes, of which S-001 made first flight 21 December 1985; S-002 was static test airframe; S-003 second flying prototype. Initial production order believed for 20, deliveries of which (serial number 701) began 1988; further two said to have been completed for proposed avionics upgrade programme with Jaffe Aircraft of USA in 1991, which failed to materialise; similar venture with IAI about a year later resulted in upgraded IAR-109 Swift prototype (7003), possibly converted from S-003 prototype and flown in November 1993 (see 1995-96 and earlier *Jane's*), but no production of this version ensued. However, contract for 24 upgraded aircraft with Elbit modernised avionics ordered in September 1998, for delivery by 2001; option for further 16 thereafter.
CURRENT VERSIONS: **IAR-99:** Initial production version for Romanian Air Force; in service.
New IAR-99: Upgraded trainer (see Design Features below); Elbit (Israel) teamed with Avioane to produce this version as lead-in fighter trainer for Romanian Air Force and to offer for export. First upgraded aircraft (718; actually constructed 1994) served as development demonstrator; made first flight in new configuration 22 May 1997 and is still (March 2001) highest-serialled aircraft known to have flown, despite reports that first six upgraded Şoims due for delivery by end of 2000. US$21 million contract for 24 aircraft to this standard announced in September 1998 and contains options for further 16 to follow. Provision for eventual retrofit of existing fleet thought to have lapsed.
Following description applies to this version.
CUSTOMERS: Total of 20 (believed S-001 and 003; 701-718) stated to have been delivered to Romanian Air Force by early 1999, of which No. 718 was first to current upgrade standard; about 13 believed still in service; 719 to 721 still on production line, 1998-2000.
COSTS: Upgraded version approximately US$6 million (2000).
DESIGN FEATURES: Typical basic/advanced jet trainer with tandem, stepped cockpits and moderately tapered wings. Provision for armament increases versatility.
Wing section NACA 64₁A-214 (modified) at centreline; 64₁A-212 (modified) at tip; dihedral 3° from roots; quarter-chord sweepback 6° 35'; incidence 1° at root.
Main feature of new Elbit avionics suite is a data transfer system that processes navigational information received via datalink from other aircraft or from a ground station equipped with the same system. This information is presented on a simulated 'virtual radar' display in the cockpit, so that the pilot can be trained in the use of radar without a need for the IAR-99 itself to be fitted with a radar. Upgraded aircraft have a large blade antenna beneath nose on port side.
FLYING CONTROLS: Conventional and partly assisted. Statically balanced ailerons hydraulically actuated with manual reversion; horn-balanced elevators and statically balanced rudder actuated mechanically by push/pull rods; servo tab in port aileron, trim tabs in rudder and each elevator, all operated electrically; ailerons deflect 15° up/15° down, elevators 20° up/10° down, rudder 25° to left and right. Hydraulically actuated single-slotted flaps, deflecting 20° for T-O and 40° for landing, retract gradually when airspeed reaches 162 knots (300 km/h; 186 mph); twin hydraulically actuated airbrakes under rear fuselage.
STRUCTURE: All-metal; aluminium honeycomb ailerons/elevators/rudder; semi-monocoque fuselage includes honeycomb panels for fuel tank compartments; machined wing skin panels form integral fuel tanks.
LANDING GEAR: Retractable tricycle type, with single wheel and oleo-pneumatic shock-absorber on each unit. Mainwheels retract inward, castoring nosewheel forward, all being fully enclosed by doors when retracted. Landing light in port wingroot leading-edge. Mainwheels fitted with tubeless tyres, size 552×164-10, pressure 7.5 bars (109 lb/sq in), and hydraulic disc brakes with anti-skid system. Nosewheel has tubeless tyre size 445×150-6, pressure 4.0 bars (58 lb/sq in).
POWER PLANT: One Turbomecanica Romanian-built Rolls-Royce Viper Mk 632-41M turbojet, rated at 17.79 kN (4,000 lb st). Fuel in two flexible bag tanks in centre-fuselage, capacity 900 litres (238 US gallons; 198 Imp gallons), and four integral tanks between wing spars, combined capacity 470 litres (124 US gallons; 103 Imp gallons). Total internal fuel capacity 1,370 litres (362 US gallons; 301 Imp gallons). Gravity refuelling point on top of fuselage. Provision for two drop tanks, each of 225 litres (59.5 US gallons; 49.5 Imp gallons) capacity, on inboard underwing stations. Maximum internal/external fuel capacity 1,820 litres (481 US gallons; 400 Imp gallons).
ACCOMMODATION: Crew of two in tandem, on Martin-Baker Mk 10L zero/zero ejection seats in pressurised and air conditioned cockpit. Rear seat elevated 35 cm (13.8 in). Dual controls standard. One-piece canopy with internal screen, opening sideways to starboard.
SYSTEMS: Engine compressor bleed air for pressurisation, air conditioning, anti-*g* suit and windscreen anti-icing system, and to pressurise fuel tanks. Hydraulic system, operating at pressure of 206 bars (2,990 lb/sq in), for actuation of landing gear and doors, flaps, airbrakes, ailerons and main-wheel brakes. Emergency hydraulic system for operation of landing gear doors, flaps and wheel brakes. Main electrical system, supplied by 9 kW 28 V DC starter/generator, with 28.5 V 36 Ah Ni/Cd battery, ensures operation of main systems, in case of emergency, and engine starting. Two 750 VA static inverters supply two secondary AC networks: 115 V/400 Hz and 26 V/400 Hz. Oxygen system for two crew for 2 hours 30 minutes.
AVIONICS: Elbit Systems integrated suite, based on MIL-STD-1553B.
Comms: Voice-activated intercom; IFF.
Flight: VOR/ILS; DME; ADF; Litton Italia INS with Trimble GPS nav.

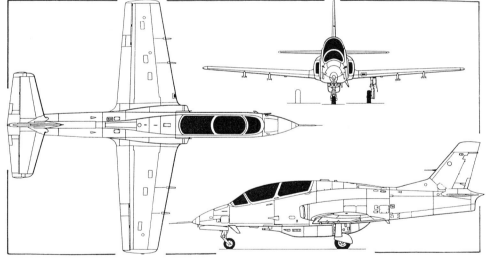

Antenna detail identifies the upgraded IAR-99 *(Dennis Punnett/Jane's)*

IAR-99 modernised front cockpit with up-front control panel

Rear cockpit of the newly upgraded IAR-99

Instrumentation: Modular multirole computer; HOTAS controls; colour CRT MFD in each cockpit; head-up display with up-front control panel (HUD/UFCP) in front cockpit, with aft station HUD monitor (ASHM) in rear (instructor's) cockpit; dual ADIs with HSIs; radar altimeter; Elta data transfer system (DTS) with pilot's virtual radar display; instructor's display and sight helmet (DASH).

Self-defence: Elta RWR and ECM (jammer) pod; chaff/flare dispenser.

ARMAMENT: Removable ventral gun pod containing 23 mm GSh-23 gun with 200 rounds. Gun/rocket firing and weapon release controls, including electrically controlled AA-1F gyroscopic gunsight and AFCT-1 gun camera in front cockpit only. Four underwing hardpoints stressed for loads of 250 kg (551 lb) each. Typical underwing stores can include four 250 kg bombs; four triple carriers each for three 50 kg bombs (or two 100 kg and one 50 kg); four L-16-57 launchers each containing sixteen 57 mm air-to-surface rockets; four L 32-42 launchers each containing thirty-two 42 mm air-to-surface rockets; infra-red air-to-air missiles (inner pylons only); two twin 7.62 mm machine gun pods with 800 rds/pod (inboard pylons only); and auxiliary fuel tanks (see under Power Plant) on inboard pylons.

DIMENSIONS, EXTERNAL:
Wing span	9.85 m (32 ft 3¾ in)
Wing chord: at root	2.305 m (7 ft 6¾ in)
mean	1.965 m (6 ft 5¼ in)
at tip	1.30 m (4 ft 3¼ in)
Wing aspect ratio	5.2
Length overall	11.01 m (36 ft 1½ in)
Height overall	3.90 m (12 ft 9½ in)
Elevator span	4.12 m (13 ft 6¼ in)
Wheel track	2.685 m (8 ft 9¾ in)
Wheelbase	4.38 m (14 ft 4½ in)

AREAS:
Wings, gross	18.71 m² (201.4 sq ft)
Ailerons (total)	1.56 m² (16.79 sq ft)
Flaps (total)	2.54 m² (27.34 sq ft)
Fin, incl dorsal fin	1.92 m² (20.67 sq ft)
Rudder	0.63 m² (6.78 sq ft)
Tailplane	3.125 m² (33.63 sq ft)
Elevators (total)	1.25 m² (13.45 sq ft)

WEIGHTS AND LOADINGS:
Weight empty, equipped	3,200 kg (7,055 lb)
Max fuel weight: internal	1,100 kg (2,425 lb)
external	350 kg (772 lb)
Max T-O weight: trainer	4,400 kg (9,700 lb)
ground attack	5,560 kg (12,258 lb)
Max wing loading: trainer	235.2 kg/m² (48.17 lb/sq ft)
ground attack	297.2 kg/m² (60.86 lb/sq ft)
Max power loading: trainer	247.5 kg/kN (2.42 lb/lb st)
ground attack	312.7 kg/kN (3.06 lb/lb st)

PERFORMANCE:
Max operating Mach number (M_{MO})	0.76
Max level speed at S/L:	
trainer	467 kt (865 km/h; 537 mph)
Max rate of climb at S/L	2,100 m (6,890 ft)/min
Service ceiling	12,900 m (42,325 ft)
Min air turning radius	330 m (1,083 ft)
T-O run: trainer	450 m (1,477 ft)
ground attack	960 m (3,150 ft)
T-O to 15 m (50 ft): trainer	750 m (2,461 ft)
ground attack	1,350 m (4,430 ft)
Landing from 15 m (50 ft): trainer	740 m (2,428 ft)
ground attack	870 m (2,855 ft)
Landing run: trainer	550 m (1,805 ft)
ground attack	600 m (1,969 ft)
Typical combat radius (one pilot, ventral gun, internal fuel only):	
lo-lo-hi, four 16-round rocket pods, AUW 5,000 kg (11,023 lb)	189 n miles (350 km; 217 miles)
hi-lo-hi, two 16-round rocket pods, two 50 kg and four 100 kg bombs, AUW 5,280 kg (11,640 lb)	186 n miles (345 km; 214 miles)
hi-hi-hi, four 250 kg bombs, AUW 5,480 kg (12,081 lb)	208 n miles (385 km; 239 miles)
Max range with internal fuel:	
trainer	593 n miles (1,100 km; 683 miles)
ground attack	522 n miles (967 km; 601 miles)
Max endurance with internal fuel: trainer	2 h 40 min
ground attack	1 h 46 min
g limits	+7/−3.6

UPDATED

IAR

SC IAR SA

1 Aeroportului Street, PO Box 198, R-2200 Brasov
Tel: (+40 68) 15 00 15 and 15 05 55
Fax: (+40 68) 15 13 04 and 15 06 23
e-mail: iar@deuroconsult.ro
Web: http://www.iar.ro
GENERAL DIRECTOR: Dipl Eng Neculai Banea
COMMERCIAL DIRECTOR: Vasile Tanase
MARKETING MANAGER: Stefan Paunescu

Factory, created 1968, continues work begun in 1925 by IAR-Brasov; occupies 136 ha (336 acre) site, including 185,400 m² (1,995,625 sq ft) factory area, and had 2000 workforce of more than 2,300. Current aircraft are Romanian-designed IS-28M2/GR motor glider and IAR-46 very light aircraft; IS-28B2, IS-29D2 and IAR-35 series sailplanes (see 1992-93 and earlier *Jane's*); spares for Puma and Alouette III; aircraft components and equipment. A licence agreement for manufacture and assembly of the Eurocopter AS 350BA Ecureuil and AS 355N Ecureuil 2 (up to 80 total) was signed on 29 May 1996, but no announcement of a starting date had been made by 1 June 2000. At that time, Eurocopter remained only bidder for the 70 per cent state holding in IAR by Romanian State Ownership Fund, Bell Helicopter Textron having abandoned its 1999 bid; bid deadline was later extended to 18 August 2000 due to 'problems regarding legal aspects of privatisation'.

In July 2000, IAR Brasov was named as first partner of ATG (see UK part of Lighter than Air section) in production of latter's new SkyCat 200 airship. First deliveries of major subassemblies are due in early 2001.

For details of Romanian Air Force IAR-330L Puma SOCAT upgrade programme, see *Jane's Aircraft Upgrades*.

UPDATED

IAR IS-28M2/GR

TYPE: Motor glider.
PROGRAMME: Two motor glider versions originally developed from IS-28B2 sailplane, of which tandem-seat IS-28M1 (later redesignated IAR-34) last described in 1981-82 *Jane's*. Principal version was side-by-side IS-28M2; prototype (YR-1013) made first flight 26 June 1976; 50.7 kW (68 hp) engine in earlier aircraft, then 59.7 kW (80 hp) IS-28M2/80 version, certified to JAR 22 in Australia, France, Japan, Norway, Portugal and UK. Current model is IS-28M2/GR.
CURRENT VERSIONS: **IS-28M2/GR**: Upgraded version with Rotax 912 A3 geared engine of 59.7 kW (80 hp) and increased MTOW of 780 kg (1,719 lb). Certified in Romania and Germany under JAR 22 in 1997. Deliveries began 1997.
CUSTOMERS: Earlier models sold in Argentina, Australia, Botswana, Canada, Cyprus, Denmark, France, Germany, Hungary, India, Israel, Japan, Norway, Philippines, Spain, Sweden, Switzerland, Turkey, UK and USA. Total of 80 (all versions) built by mid-1998, of which 45 for export; no known deliveries in 1999 or 2000, but remains available.
COSTS: DM139,900 (1999).
DESIGN FEATURES: Usual sailplane features of high aspect ratio wing and T tail (with dihedral). Highly bulged cockpit and canopy. Wing is low-mounted, with upturned tips.
Wortmann wing sections (root FX-61-163, tip FX-60-126); dihedral 2°; sweep forward 2° 30′ at quarter-chord.
FLYING CONTROLS: Conventional and manual. Two-segment Hütter airbrakes in each wing upper surface; trim tab in each elevator. Plain flaps.
STRUCTURE: Mainly aluminium alloy. Single-spar wing, with metal ribs and skin; all-metal flaps and airbrakes; ailerons, and elevator/rudder trailing-edges, are fabric-covered. Front fuselage has metal longerons and frames with glass fibre fairings and engine cowling panels; centre-fuselage is metal monocoque, rear fuselage metal frames and skin.
LANDING GEAR: Two semi-retractable Tost mainwheels side by side under centre-fuselage, with rubber disc shock-absorbers and mechanical drum brakes; tyre size 340×126. Steerable, non-retractable tailwheel, also with shock-absorber.
POWER PLANT: One 59.6 kW (79.9 hp) Rotax 912 UL-A3 flat-four engine, driving a Hoffmann HO-V352F-S1/S1-170FQ two-blade constant-speed propeller. Single fuel tank aft of cockpit, capacity 55 litres (14.5 US gallons; 12.1 Imp gallons).
ACCOMMODATION: Two seats side by side under rearward-sliding canopy. Dual controls standard.
AVIONICS: *Comms*: Honeywell KT 76A transponder; intercom.
Flight: Honeywell KX 125 nav/com transceiver; Garmin GPS receiver. Honeywell KLN 35A moving map display optional.

DIMENSIONS, EXTERNAL:
Wing span	17.00 m (55 ft 9¼ in)
Wing aspect ratio	15.8
Length overall	7.76 m (25 ft 5½ in)
Height over tail	2.15 m (7 ft 0¾ in)
Wheel track	1.36 m (4 ft 5½ in)
Propeller diameter	1.70 m (5 ft 7 in)

AREAS:
Wings, gross	18.24 m² (196.3 sq ft)

WEIGHTS AND LOADINGS:
Weight empty	580 kg (1,278 lb)
Max T-O weight	780 kg (1,719 lb)
Max wing loading	42.8 kg/m² (8.76 lb/sq ft)
Max power loading	13.07 kg/kW (21.48 lb/hp)

PERFORMANCE, POWERED:
Never-exceed speed (V_{NE})	118 kt (220 km/h; 136 mph)
Max cruising speed, 75% power	108 kt (200 km/h; 124 mph)
Econ cruising speed	57 kt (105 km/h; 65 mph)
Stalling speed, flaps down, power on or off	42 kt (77 km/h; 48 mph)
Max rate of climb at S/L	192 m (630 ft)/min
Service ceiling	6,200 m (20,340 ft)
T-O run (grass)	232 m (765 ft)
T-O to 15 m (50 ft) (grass)	437 m (1,435 ft)
Landing run	90 m (295 ft)
Range with max fuel:	
30 min reserves	302 n miles (560 km; 348 miles)
no reserves	378 n miles (700 km; 435 miles)
g limits	+5.3/−2.65

PERFORMANCE, UNPOWERED:
Max speed: smooth air	102 kt (190 km/h; 118 mph)
rough air	91 kt (170 km/h; 105 mph)
Min rate of sink at 43 kt (80 km/h; 50 mph)	1.15 m (3.77 ft)/s
Best glide ratio at 57 kt (105 km/h; 65 mph)	more than 24

UPDATED

IAR IAR-46

TYPE: Two-seat lightplane.
PROGRAMME: Definition phase and marketing studies started early 1991; detail design began late 1991; development

IAR Brasov IS-28M2/GR motor glider (Rotax 912 engine) *(Paul Jackson/Jane's)*

IAR to ROMAERO—AIRCRAFT: ROMANIA

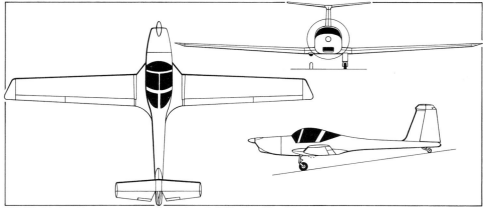

IAR-46 side-by-side two-seat very light aircraft *(Mike Keep/Jane's)*

phase initiated mid-1992; first of two prototypes (YR-1037) shown at Paris Air Show June 1993 and made first flight November that year, followed by flight test evaluation, static test and ground testing for JAR-VLA compliance. Development of 02 second prototype began at end of 1995; this aircraft (YR-BVC) made first flight in April 1997; manufacturer's preliminary flight test programme completed; JAR-VLA certification by Romanian CAA began mid-1998 and received November 1999; FAA certification expected in September 2000. Further three production aircraft built by late 1998, but no customers had been announced by mid-2000.

CURRENT VERSIONS: **IAR-46**: Basic version; *as described*.
 IAR-46S: To be introduced in October 2000; 73.5 kW (98.6 hp) Rotax 912 ULS engine.
DESIGN FEATURES: Low-wing monoplane; GA(W)-1 aerofoil section) with high aspect ratio, raked tips and T tail. Dihedral 2° from centre-section; incidence 4° at root; no twist.
FLYING CONTROLS: Conventional and manual. Actuation by pushrods and cables; trim tab in each elevator. Manually operated plain trailing-edge flaps.
STRUCTURE: All-metal except fabric covering on elevators and rudder and GFRP for non-stressed fairings. Wing has main and auxiliary spars. Fluted aluminium skins on ailerons and flaps.
LANDING GEAR: Retractable (manual/mechanical) single mainwheels with hydraulic shock-absorbers and hydraulic, toe-operated disc brakes; steerable, non-retractable tailwheel with rubber-in-compression shock-absorption. Rearward-retracting 5.00×5 Matco or Cleveland mainwheels, with 360×110 tyres, pressure 3.00 bars (43.5 lb/sq in). Tost 210×65 tailwheel and tyre, pressure 2.50 bars (36 lb/sq in).
POWER PLANT: One 59.6 kW (79.9 hp) Rotax 912 F3 flat-four engine, driving a Hoffmann HO-V352F/170FQ two-blade constant-speed propeller. Fuel in single tank in fuselage, capacity 70 litres (18.5 US gallons, 15.4 Imp gallons).
ACCOMMODATION: Two adjustable backrest seats side by side; dual controls standard; adjustable rudder pedals. Fixed windscreen and rearward-sliding jettisonable canopy. Baggage compartment aft of seats. Cockpit heated and ventilated.
SYSTEMS: Electrical power supplied by 250 W 12 V DC alternator and 12 V 25 Ah battery. Hydraulic system for mainwheel brakes only, with parking brake valve.
AVIONICS: *Instrumentation*: Standard VFR instrumentation to JAR 22.1303 and JAR 22.1305. Options include horizon and directional gyros, turn co-ordinator and Becker nav/com and transponder avionics package.
EQUIPMENT: Anti-collision and position lights standard; ground power receptacle optional.

DIMENSIONS, EXTERNAL:
Wing span	12.05 m (39 ft 6½ in)
Wing chord: at root	1.40 m (4 ft 7 in)
at tip	0.93 m (3 ft 0½ in)
Wing aspect ratio	9.4
Length overall	7.85 m (25 ft 9 in)
Height overall	2.15 m (7 ft 0½ in)
Tailplane span	3.48 m (11 ft 5 in)
Wheel track	1.59 m (5 ft 2½ in)
Wheelbase	4.94 m (16 ft 2½ in)
Propeller diameter	1.70 m (5 ft 7 in)
Propeller ground clearance	0.295 m (11½ in)

DIMENSIONS, INTERNAL:
Cockpit: Length	1.54 m (5 ft 0¾ in)
Max width	1.04 m (3 ft 5 in)
Max height (from seat cushion)	0.85 m (2 ft 9½ in)

AREAS:
Wings, gross	13.87 m² (149.3 sq ft)
Ailerons (total)	0.82 m² (8.83 sq ft)
Trailing-edge flaps (total)	1.36 m² (14.64 sq ft)
Fin	0.60 m² (6.46 sq ft)
Rudder	0.89 m² (9.58 sq ft)
Tailplane	1.64 m² (17.65 sq ft)
Elevators (total, incl tabs)	1.10 m² (11.84 sq ft)

WEIGHTS AND LOADINGS:
Weight empty	550 kg (1,213 lb)
Max fuel weight	53 kg (117 lb)
Max T-O and landing weight	750 kg (1,653 lb)
Max wing loading	54.1 kg/m² (11.08 lb/sq ft)
Max power loading	12.58 kg/kW (20.67 lb/hp)

PERFORMANCE:
Never-exceed speed (V_{NE})	145 kt (270 km/h; 167 mph) CAS
Max cruising speed	97 kt (180 km/h; 112 mph) CAS
Econ cruising speed	81 kt (150 km/h; 93 mph) CAS
Stalling speed at 1,000 m (3,280 ft), engine idling:	
flaps up	48 kt (88 km/h; 55 mph) CAS
flaps down	42 kt (77 km/h; 48 mph) CAS
Max rate of climb at S/L	170 m (558 ft)/min
Service ceiling	4,000 m (13,120 ft)
T-O run	193 m (635 ft)
T-O to 15 m (50 ft)	465 m (1,525 ft)
Landing run	110 m (360 ft)
Range, with reserves	324 n miles (600 km; 372 miles)
g limits	+4.4/−2.2

UPDATED

IAR (EUROCOPTER) AS 350BA ECUREUIL and AS 355N ECUREUIL 2

On 29 May 1996 an agreement was signed for licensed manufacture, subject to French and Romanian government approval, of up to 80 AS 350BA single-engined Ecureuil and twin-engined AS 355N Ecureuil 2 helicopters, described under Eurocopter in the International section. Purchases are anticipated from local civilian and public service operators, while export orders would be handled jointly by Eurocopter and IAR. No announcement of a starting date for this programme had been made by August 2000.

UPDATED

Second prototype IAR-46 *(Paul Jackson/Jane's)*

ROMAERO

SC ROMAERO SA

44 Bulevardul Ficusului, Sector 1 (PO Box 18), R-71544 Bucuresti 1
Tel: (+40 1) 232 37 35 and 232 07 21
Fax: (+40 1) 232 20 82
e-mail: management@romaero.rdsnet.ro
CHAIRMAN AND CEO: Pantelimon Vilceanu
DIRECTOR, SALES AND MARKETING: Anatolie Merling
CHIEF OF PUBLIC RELATIONS: Carmen Gheorghiu

Established 1951 and operated successively under several names (see 1991-92 *Jane's* for details); became commercial company under present name 20 November 1990. Manufactures B-N Islander and Defender 4000; major subassemblies and components for Boeing 737/757; cabin subassemblies and floats for Bombardier (Canadair) CL-415 and auxiliary fuel tanks for Challenger 604; detail parts and subassemblies for BAE Systems (Avro) RJ85 and RJX; tail units for Learjet 45 (subcontracted from Shorts, UK); Astra SPX cabins and Galaxy rear fuselages for IAI; also undertakes maintenance of various types of aircraft, including Islander, BAC One-Eleven, Boeing 707/727 and Antonov transports. Plans to establish an FAA-approved maintenance and repair station for larger aircraft of various types. Romaero's 36.43 ha (90.0 acre) Baneasa Airport site includes a 163,670 m² (1.76 million sq ft) production area; workforce in mid-2000 was 1,280.

The sale of Romaero to Britten-Norman of the UK in January 1999 was aborted when the latter ceased trading three months later, resulting in suspension of Islander production; deliveries, to the newly formed B-N Group, resumed in July 2000. Romaero selected by Integrity Aircraft of New Zealand as assembly centre for its single-turboprop derivative of the former three-piston-engined Britten-Norman Trislander.

UPDATED

ROMAERO (B-N) ISLANDER and DEFENDER

TYPE: Light utility twin-prop transport.
PROGRAMME: First flight of Romanian Islander (assembled from UK kit by former IRMA) 4 August 1969; first Romanian-built flew 18 October 1969. Defender 4000 production started in 1995; first delivery (G-BWPU) 27 September 1997.
CUSTOMERS: Initial commitment to build 215 completed January 1977; total of 509 standard Islanders, six Turbine Islanders and six Defender 4000s delivered to Britten-Norman (now B-N) by 26 September 1998. No further deliveries then until 19 July 2000 (one BN-2B-20).

UPDATED

RUSSIAN FEDERATION

AERO-M

AKTSIONERNOE OBSHCHESTVO AERO-M (Aero-M JSC)
shosse Varshavskoye 132, 113519 Moskva
Tel: (+7 095) 315 34 25
Fax: (+7 095) 315 10 53
GENERAL DIRECTOR: Nikolai I Florov

No progress reports have been received of the A-209 and A-230, both of which were last described in the 1999-2000 edition.

UPDATED

AEROPRAKT

AEROPRAKT OOO (Aeropract JSC)
alleya 9863, 443008 Samara
Tel: (+7 8462) 27 09 65
Fax: (+7 8462) 42 96 13
e-mail: ela@dionis.samtel.ru

Formed in 1974 to design and build gliders, small flying-boats and amphibians, OKB Aeroprakt established a second centre at Kiev in 1986, only to be divided on dissolution of the USSR. Became LM Aeropract Samara, a joint Russo-Finnish venture, in 1991; KB Aeropract in 1993; and Aeropract JSC in 1997.

The Russian Aeropract now adopts a -ct ending in transliteration to differentiate it from the Aeroprakt now in Ukraine (which see). The two companies are no longer connected. Chief designer Igor Vakhrushev was killed in the crash of an Aeropract-23M in June 2000.

UPDATED

AEROPRACT-21M
Export marketing name: Solo

TYPE: Single-seat sport ultralight/kitbuilt.
PROGRAMME: Design started December 1990; construction of A-21 prototype with RMZ-640 Buran engine started March 1991; first flight 14 August 1991; first flight of production A-21M with Rotax 503 engine May 1992; certification to JAR-VLA regulations under way early 1997; demonstrator FLARF-02177 rebuilt by 1999 with new wing, having full-span Junkers flaperons in place of conventional, inset, ailerons and Hirth engine, replacing 37.0 kW (49.6 hp) Rotax.
CUSTOMERS: Five production A-21Ms built by August 1999; two imported into UK 1995.
COSTS: Standard aircraft US$15,000 (1999); aerobatic US$18,000. Kit less engine and avionics US$6,500.
DESIGN FEATURES: Conventional cantilever low-wing monoplane, with constant chord and no sweep; sweptback vertical tail surfaces; high and extensive glazed canopy. Fin fillet added 1996.
Wing section CAHI (TsAGI) R-IIIA-15; thickness/chord ratio 15 per cent; dihedral 5° from roots; incidence 3°; no twist.
FLYING CONTROLS: Manual. Junkers flaperons, horn-balanced rudder and full-span elevator.
STRUCTURE: Single-spar wings and tail surfaces. Airframe entirely glass fibre sandwich construction.
LANDING GEAR: Non-retractable tricycle type; nosewheel steerable ±15.3°; cantilever mainwheel legs. Go-kart tyres; cable brakes on mainwheels. Wheels interchangeable with skis.
POWER PLANT: One 48.5 kW (65 hp) Hirth 2706 piston engine; two/three-blade fixed-pitch propeller. Fuel capacity 36 litres (9.5 US gallons; 7.9 Imp gallons); provision for external tank under fuselage, capacity 32 litres (8.5 US gallons; 7.0 Imp gallons).
DIMENSIONS, EXTERNAL:
Wing span 6.65 m (21 ft 9¾ in)
Length overall 4.73 m (15 ft 6¼ in)
Height overall 1.84 m (6 ft 0½ in)
AREAS:
Wings, gross 5.85 m² (63.0 sq ft)
WEIGHTS AND LOADINGS:
Weight empty 180 kg (397 lb)
Max T-O weight 280 kg (617 lb)
PERFORMANCE:
Max level speed 102 kt (190 km/h; 118 mph)
Cruising speed 81 kt (150 km/h; 93 mph)
Stalling speed 44 kt (80 km/h; 50 mph)
Max rate of climb at S/L 300 m (985 ft)/min
T-O run 100 m (330 ft)
Landing from 15 m (50 ft) 398 m (1,305 ft)
Range:
with internal fuel 116 n miles (215 km; 133 miles)
with max fuel 251 n miles (466 km; 289 miles)
g limits +6/–3

UPDATED

Aeropract-21M Solo single-seat sporting aircraft in 1999, with new wing control surfaces and Hirth engine *(Paul Jackson/Jane's)*

First production Aeropract-23M *(Paul Jackson/Jane's)*

AEROPRACT-23M
Export marketing name: Dragon

TYPE: Tandem-seat lightplane/kitbuilt.
PROGRAMME: Developed from twin-engined A-23; prototype was conversion of A-23 SVS-17033 with a single Rotax 582 engine, 1997. Production prototype, FLARF-02385, which first flew on 5 August 1999 and made public debut at MAKS, Moscow, 12 days later, was of considerably improved detailed design and construction, including T tail. Lost in fatal accident at Samara 24 June 2000.
COSTS: Kit US$12,000, excluding instruments (US$700), landing gear and Rotax 582 engine (US$7,500). Ex-works, complete, with wheels, US$25,000 (1999). Wheel landing gear US$1,000, skis US$250 to US$350; floats US$1,200 to US$1,650.
DESIGN FEATURES: Meets JAR-VLA criteria. Braced, high-wing monoplane with main and jury strut each side; pod-and-boom fuselage; sweptback fin and rudder; unswept wing and horizontal tail surfaces.
Wing section CAHI (TsAGI) R-IIIA-15.5; thickness/chord ratio 15.5 per cent; constant chord; dihedral 1° 30′; incidence 3°; no twist.
FLYING CONTROLS: Manual. Flaperons, almost full-span, have large, suspended auxiliary aerofoils. All-moving tailplane with anti-balance tab and twin mass balances. Full-height rudder with no balances.
STRUCTURE: Composites throughout.
LANDING GEAR: Tricycle type; fixed. Composites cantilever main legs; telescopic, metal leaf-sprung nose leg; metal tail bumper. Option of wheels (with cable-operated mainwheel brakes), floats or skis. Nosewheel steerable ±15°.
POWER PLANT: One 47.8 kW (64.1 hp) Rotax 582 UL two-cylinder, liquid-cooled piston engine driving a three-blade, ground-adjustable pitch propeller. Optional engines include Rotax 912 and VAZ 21083. Fuel tank, capacity 80 litres (21.1 US gallons; 17.6 Imp gallons), behind rear seat, with filler on port side.
ACCOMMODATION: Two persons in tandem beneath two-element, starboard-hinged canopy. Additional, small side windows for rear occupant. Inset boarding step, port side. Dual controls.
AVIONICS: Standard VFR.
EQUIPMENT: Optional ballistic parachute.
DIMENSIONS, EXTERNAL:
Wing span 9.80 m (32 ft 1¾ in)
Length overall (incl nose-probe) 7.50 m (24 ft 7¼ in)
Height overall 2.38 m (7 ft 9¾ in)

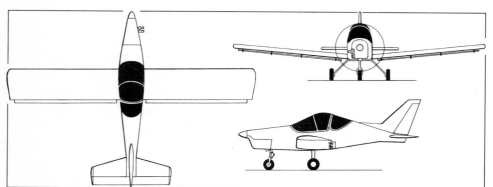

General arrangement of the Aeropract-21M Solo *(James Goulding/Jane's)*

AEROPRAKT—AIRCRAFT: RUSSIAN FEDERATION

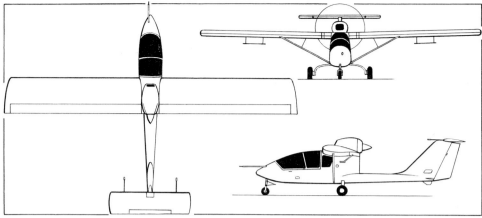

Aeropract-23M in production configuration *(James Goulding/Jane's)* 2000/0062589

WEIGHTS AND LOADINGS:
Weight empty	350 kg (772 lb)
Max T-O weight	600 kg (1,322 lb)
Max power loading (Rotax 582 UL)	12.55 kg/kW (20.62 lb/hp)

PERFORMANCE:
Max level speed	86 kt (160 km/h; 99 mph)
Normal cruising speed	65 kt (120 km/h; 75 mph)
T-O run	100 m (330 ft)
Range	259 n miles (480 km; 298 miles)
g limits	+4/−2

UPDATED

AEROPRACT-25

Export marketing name: Breeze

TYPE: Four-seat amphibian.
PROGRAMME: Designed to FAR Pt 23 and US Primary Category standards; design started January 1992; construction of prototype began 4 January 1993; prototype (FLA RF-02495) first flew 9 November 1994; exhibited at Moscow Air Show '95. Retained for development. To be re-engined with 186 kW (250 hp) Textron Lycoming. No production aircraft reported by mid-2000.
COSTS: Standard aircraft US$120,000 (1997).
DESIGN FEATURES: Conventional light amphibian. Suitable for passenger, ambulance and cargo use without modification. Boat hull; mid-mounted cantilever wings with downturned tips; sweptback fin and rudder, with dorsal fin on shallow slab-sided tailboom; unswept tailplane and elevators mid-mounted on fin; engine pod above fuselage nacelle.
Wing section NACA 63_2417 modified; dihedral 3°, incidence 3°; no twist.
FLYING CONTROLS: Conventional and manual. Outer section of starboard elevator has electric trim; electrically actuated flaps.
STRUCTURE: Single-spar wings and tail surfaces. Entire airframe made of glass fibre sandwich.
LANDING GEAR: Electrically retractable tricycle type; single wheel on each unit. Mainwheels, each on titanium leaf spring, retract inward into bottom of hull; nosewheel retracts forward into nose. Mainwheel tyres 400×150; nosewheel tyre 300×125. Nosewheel steerable ±15°. Wingtips form stabilising floats.
POWER PLANT: One 154 kW (207 hp) LOM M337 six-cylinder air-cooled and supercharged piston engine (132 kW; 177 hp unsupercharged); two-blade fixed-pitch pusher propeller. Optional engines include 157 to 224 kW (210 to 300 hp) Wankel-type rotary engine. Two fuel tanks in wings, total capacity 260 litres (68.7 US gallons; 57.2 Imp gallons); oil capacity 16 litres (4.2 US gallons; 3.5 Imp gallons).
ACCOMMODATION: Four persons in pairs; bench-type rear seat. Two single-piece windscreen/canopy gullwing doors, hinged on centreline. Dual controls.

DIMENSIONS, EXTERNAL:
Wing span	10.60 m (34 ft 9¼ in)
Wing chord, constant	1.35 m (4 ft 5¼ in)
Wing aspect ratio	7.6
Length overall	7.95 m (26 ft 1 in)
Height overall, on wheels	2.96 m (9 ft 8½ in)
Tailplane span	2.92 m (9 ft 7 in)
Wheel track	2.88 m (9 ft 5½ in)
Wheelbase	2.86 m (9 ft 4½ in)
Propeller diameter	1.80 m (5 ft 11 in)
Gullwing doors (each): Width	1.30 m (4 ft 3¼ in)
Height	0.60 m (1 ft 11½ in)

DIMENSIONS, INTERNAL:
Cabin: Length	2.00 m (6 ft 6¾ in)
Max width	1.10 m (3 ft 7¼ in)
Max height	1.10 m (3 ft 7¼ in)
Volume	1.8 m³ (64 cu ft)

AREAS:
Wings, gross	14.70 m² (158.2 sq ft)
Ailerons (total)	1.22 m² (13.13 sq ft)
Trailing-edge flaps (total)	1.42 m² (15.28 sq ft)
Fin	1.43 m² (15.39 sq ft)
Rudder	0.95 m² (10.23 sq ft)
Tailplane	1.46 m² (15.72 sq ft)
Elevators (total)	1.17 m² (12.59 sq ft)

WEIGHTS AND LOADINGS:
Weight empty, equipped	907 kg (2,000 lb)
Max T-O weight	1,200 kg (2,645 lb)
Max fuel weight	190 kg (419 lb)
Max T-O weight	1,255 kg (2,767 lb)
Max wing loading	81.6 kg/m² (16.72 lb/sq ft)
Max power loading	7.78 kg/kW (12.78 lb/hp)

PERFORMANCE:
Never-exceed speed (V_{NE}):	163 kt (302 km/h; 187 mph)
Max level speed	108 kt (200 km/h; 124 mph)
Max cruising speed	97 kt (180 km/h; 112 mph)
Econ cruising speed	81 kt (150 km/h; 93 mph)
Stalling speed: flaps up	60 kt (111 km/h; 69 mph)
40° flap	47 kt (87 km/h; 55 mph)
Landing speed	52 kt (95 km/h; 59 mph)
Max rate of climb at S/L	300 m (985 ft)/min
T-O run: land	290 m (955 ft)
water	400 m (1,315 ft)
Range:	
with max payload	109 n miles (202 km; 125 miles)
with max fuel	526 n miles (975 km; 605 miles)
g limits	+3.8/−2

UPDATED

AEROPRACT-27

TYPE: Side-by-side lightplane/kitbuilt.
PROGRAMME: Prototype (FLARF-02647) first flew (in floatplane configuration) 22 June 1998; public debut at Gelendzhik 98 seaplane show, and in landplane version at MAKS, Moscow, August 1999; evaluated during 1999 and 2000 by AvtoZavod Flying Club in Togliatti.
CURRENT VERSIONS: **A-27**: Baseline version with Hirth engine.
A-27M: Proposed version with 73.5 kW (98.6 hp) Rotax 912 ULS engine.
Description generally as for A-27, except where indicated.
COSTS: Kit US$14,000, excluding instruments (US$700), landing gear and Rotax 912 engine (US$14,000). Ex-works, complete, with wheels, US$35,500 (1999). Wheel landing gear US$1,000; skis US$250 to US$350; floats US$1,200 and US$1,650.

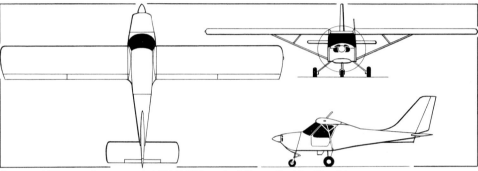

Aeropract-27 general arrangement (Hirth F30 engine) *(James Goulding/Jane's)* 2000/0062588

Aeropract-25 Breeze four-seat light amphibian, with engine cowling removed 2001/0099576

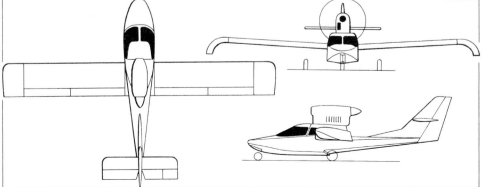

Aeropract-25 multipurpose amphibian *(James Goulding/Jane's)* 0011961

344   RUSSIAN FEDERATION: AIRCRAFT—AEROPRAKT to AEROPROGRESS/ROKS-AERO

DESIGN FEATURES: Braced, high-wing monoplane with single main and two jury struts each side. Sweptback fin and rudder; unswept wings and horizontal tail.
FLYING CONTROLS: Manual. Ailerons, rudder and all-moving tailplane with anti-balance tab. Flaps.
STRUCTURE: Tube metal fuselage with composites skin.
LANDING GEAR: Tricycle type; fixed. Cantilever, metal mainwheel legs; telescopic, leaf-sprung nosewheel leg. Optional floats or skis.
POWER PLANT: One 70.8 kW (95 hp) Hirth F30 flat-four driving a four-blade, ground-adjustable pitch propeller. Fuel capacity 80 litres (21.1 US gallons; 17.6 Imp gallons).
ACCOMMODATION: Two persons, side by side; forward-hinged door each side. Luggage space behind seats.
AVIONICS: Standard VFR.

Data refer to both versions, except where indicated.

DIMENSIONS, EXTERNAL:
Wing span: A27	10.60 m (34 ft 9¼ in)
A-27M	10.00 m (32 ft 9¾ in)
Length overall: A-27	6.30 m (20 ft 8 in)
A-27M	6.50 m (21 ft 4 in)
Height overall	2.22 m (7 ft 3½ in)

WEIGHTS AND LOADINGS:
Weight empty	335 kg (739 lb)
Max T-O weight	600 kg (1,322 lb)

Prototype Aeropract-27 with float landing gear and additional passenger   *2001/0099577*

Max power loading: A-27	8.47 kg/kW (13.92 lb/hp)	A-27M	76-81 kt (140-150 km/h; 87-93 mph)
A-27M	8.05 kg/kW (13.22 lb/hp)	T-O run	100 m (330 ft)
PERFORMANCE:		Range	324 n miles (600 km; 372 miles)
Max level speed	95 kt (175 km/h; 109 mph)		**UPDATED**
Normal cruising speed: A-27	70 kt (130 km/h; 81 mph)		

AEROPROGRESS/ROKS-AERO

AEROPROGRESS/ROKS-AERO

Gosudarstvennoye Kosmichesky Nauchno-Prozivodst-vennoye Tsentr imeni M B Khrunicheva, ulitsa Novozavodskaya 18, 121309 Moskva
Tel: (+7 095) 145 88 60
Fax: (+7 095) 145 94 77
PRESIDENT AND GENERAL DESIGNER: Evgeny P Grunin
DEPUTY GENERAL DESIGNER: Arnold I Andrianov
DEPUTY GENERAL DESIGNER, FOREIGN ECONOMIC RELATIONS: Alexander V Andreev
DESIGN BUREAU MANAGER: Sergei M Zhiganov

Known initially as ROS-Aeroprogress, this organisation was founded in 1990 to design and manufacture utility, commuter, amphibian, aerobatic, agricultural, firefighting, training and attack aircraft, WIG (wing-in-ground-effect) vehicles, replicas and other vehicles. ROKS-Aero Inc is the design bureau of Aeroprogress. Marketing of the T-101 Grach is undertaken by RSK MiG, with production at its Lukhovitsy plant. Divisions of the ROKS-Aero bureau are:

Utility Aircraft Division
DIRECTOR: Yuri I Polavsky
CHIEF DESIGNER: Mikhail M Vasyuk

Amphibian and Special Aircraft Division
DIRECTOR AND CHIEF DESIGNER: Mikhail S Remizov

Business and Touring Aircraft Division
DIRECTOR AND CHIEF DESIGNER: Leonard A Tarasevich

Trainer and Aerobatics Division
DIRECTOR: Valentin A Fomin

Replica Division
DIRECTOR: Mikhail V Korenkov

Projects in the preliminary stage include the T-317 Mukha (Fly), T-435 Pelikan, T-505 Strekoza (Dragonfly), T-515 Shmel (Bumblebee) and T-701 Phoenix.

UPDATED

AEROPROGRESS/ROKS-AERO T-101 GRACH

English name: Rook
TYPE: Light utility turboprop.
PROGRAMME: Design started by Utility Aircraft Division in September 1991, as monoplane successor to Antonov An-2/3 biplane; construction of prototype started April 1992; first flight (FLARF-01466) 7 December 1994; no evidence of reported four additional prototypes, all of which are presumed to have been used for static testing.
Series manufacture by Moscow Aviation Production Organisation (MAPO) initiated January 1993; initial production aircraft shown unpainted at Lukhovitsy factory, August 1999 (see LMZ entry in this section); further seven then substantially complete. One T-101 delivered to Chukota region in July 2000 on six-month lease, representing initial revenue-earning use. A Westernised version, developed via the Aeroprogress ROKS/Aero T-201, was unsuccessfully marketed as the Khrunichev T-201 (which see in 1999-2000 and previous editions).
CURRENT VERSIONS: **T-101:** Basic passenger/cargo transport.
Detailed description applies specifically to T-101.
T-101E: As basic T-101 but with 917 kW (1,230 shp) P&WC PT6A-65AR turboprop, driving Hartzell propeller. Honeywell TPE331-14 turboprop also being considered. Honeywell or Rockwell Collins avionics optional. Length overall 15.30 m (50 ft 2½ in). Maximum T-O weight 5,670 kg (12,500 lb); maximum cargo payload 2,000 kg (4,409 lb); maximum cruising speed 172 kt (320 km/h; 198 mph); T-O run 430 m (1,410 ft), landing run 580 m (1,905 ft); range, 1 hour fuel reserve, 561 n miles (1,040 km; 646 miles).
T-101P: Firefighting version, on non-amphibious floats.
T-101L: As basic T-101 but with ski landing gear.
T-101SKh: Redesigned agricultural aircraft; fuselage and tail unit as basic T-101; strut-braced low wings with considerable dihedral; each mainwheel on strut-braced oleo; spraybars under wings; new cockpit with large flat windscreen and large side windows.
T-101S: Military version of basic T-101; small sweptback winglets; two stores pylons under each wing; small stub-wings added, each with a weapon pylon and wingtip mount for a gun pod or other store.
T-101V: Floatplane version, shown in mockup form at Moscow Air Show in 1992; prototype due to be completed by May 1995, but failed to appear. Described more fully in 1997-98 and previous editions.
Further developments, with tricycle landing gear, increased wing span, more powerful engines and other changes, will be designated **T-102, T-103** and **T-104.**
CUSTOMERS: By January 1995, prospective orders included 50 transports for air forces, 50 agricultural aircraft, 25 for Contingency Ministry, and 150 for airborne forces. However, by 2000 no firm contracts had been announced. Total market predicted at 500 aircraft.

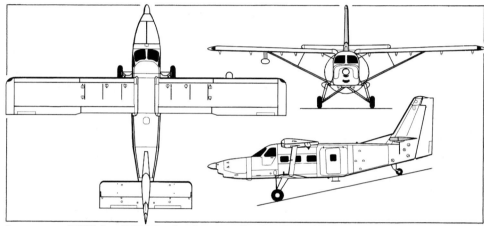

Aeroprogress/ROKS-Aero T-101 Grach turboprop utility aircraft *(Paul Jackson/Jane's)*   *2000/0079324*

COSTS: Approximately US$900,000 (2000).
DESIGN FEATURES: Single-turboprop aircraft for Normal category passenger/cargo transportation and utility applications; suitable to replace Antonov An-2. High-mounted, unswept, constant-chord, braced wing; non-retractable landing gear and unpressurised cabin; STOL capable, with wide CG range; large passenger/freight door; sweptback fin and rudder with dorsal fin; braced constant chord tailplane and elevators.
Wing section CAHI (TsAGI) P-11-14; dihedral 3°; incidence 3° constant.
FLYING CONTROLS: Conventional and manual. Actuation by pushrods and cables. Slotted ailerons, deflection 30° up, 14° down; elevator deflection 42° up, 22.5° down; rudder deflection ±28°. Trim tabs in each aileron, each elevator and rudder. Electrically actuated single-section slotted trailing-edge flap and two-section automatic leading-edge slats on each wing; flap deflection 25° for take-off, 40° for landing.
STRUCTURE: All-metal (aluminium alloy and high-tensile steel) structure; two-spar wings, metal skinned with integral stringers, and ribs; two-spar fin and tailplane; semi-monocoque fuselage, with frames, stringers and stressed skin.
LANDING GEAR: Non-retractable tailwheel type with single wheel on each unit. Main legs of tripod type, with oleo-pneumatic shock-absorption; KT-135D mainwheels, with 720×320 tyres, pressure 3.43 bars (50 lb/sq in); K-392

Aeroprogress/ROKS-Aero T-101 Grach prototype *(Paul Jackson/Jane's)*   *2001/0099617*

Prototype Aeroprogress/ROKS-Aero T-311 Kolibri under construction

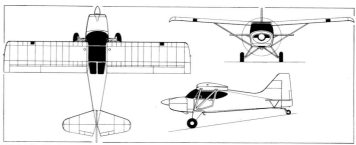

T-311 Kolibri multipurpose lightplane *(James Goulding/Jane's)*

tailwheel, with 380×200 tyres, pressure 3.43 bars (50 lb/sq in); hydraulic brakes and anti-skid units on mainwheels. Skis and floats optional.
POWER PLANT: One 706 kW (947 shp) Mars (Omsk) TVD-10B turboprop, driving AV-24AN three-blade constant-speed propeller with reverse pitch and full feathering. Three fuel tanks in each wing, each 200 litres (52.8 US gallons; 44.0 Imp gallons); total fuel capacity 1,200 litres (317 US gallons; 264 Imp gallons). Oil tank capacity 30 litres (7.9 US gallons; 6.6 Imp gallons).
ACCOMMODATION: Crew of one or two; up to nine passengers or equivalent freight. Forward-opening door each side of flight deck; large upward-opening freight door aft of wing on port side, with integral inward-opening passenger door; door between flight deck and cabin; starboard emergency exit. Cabin ventilated and heated by engine bleed air.
SYSTEMS: Hydraulic system for brakes; maximum flow 4 litres (1.05 US gallons; 0.88 Imp gallon)/min, at 147 bars (2,135 lb/sq in). Three-phase 120/208 V 400 Hz AC electrical system, supplied by 6 kVA 200 A BU6BK brushless alternator, with emergency DC power supply and 24 V 40 Ah Ni/Cd battery. Electric de-icing of propeller blades and spinner; engine air intake de-iced by heated oil.
AVIONICS: *Comms:* R-855A emergency locator beacon and ARB-NK emergency radio buoy.
 Radar: Weather radar pod under starboard wing.
 Flight: Com/nav equipment for VFR and IFR operations by day and night over all terrain; PNP-72-14 nav; A-723 long-range radio nav; ARK-M ADF; GRAN low-altitude radio altimeter; GROM satellite nav; A-611 marker beacon receiver; AP-93 autopilot.
 Instrumentation: Air data system with digital airspeed and altitude indication; VBM-1PB standby altimeter; KC-MC compact compass system; AGB-96 gyro horizon.
DIMENSIONS, EXTERNAL:
Wing span	18.20 m (59 ft 8½ in)
Wing chord: at root	2.40 m (7 ft 10½ in)
at tip	2.45 m (8 ft 0½ in)
Wing aspect ratio	7.6
Length overall	15.06 m (49 ft 5 in)
Fuselage: Max width	1.80 m (5 ft 10¾ in)
Max height	2.52 m (8 ft 3¼ in)
Height overall	4.86 m (15 ft 11¼ in)
Tailplane span	5.82 m (19 ft 1¼ in)
Wheel track	3.36 m (11 ft 0¼ in)
Wheelbase	8.30 m (27 ft 3 in)
Propeller diameter	2.80 m (9 ft 2¼ in)
Propeller ground clearance	0.97 m (3 ft 2¼ in)
Flight deck doors (each): Height	1.20 m (3 ft 11¼ in)
Width: at top	0.43 m (1 ft 4¾ in)
at bottom	0.67 m (2 ft 2¼ in)
Passenger door: Height	1.42 m (4 ft 8 in)
Width	0.81 m (2 ft 7¾ in)
Freight door: Height	1.80 m (5 ft 10¾ in)
Width	1.60 m (5 ft 3 in)
Emergency exit: Height	0.54 m (1 ft 9 in)
Width	0.87 m (2 ft 10¼ in)

DIMENSIONS, INTERNAL:
Cabin: Length	4.50 m (14 ft 9¼ in)
Max width	1.60 m (5 ft 3 in)
Max height	1.85 m (6 ft 0¾ in)
Floor area	6.72 m² (72.3 sq ft)
Volume	12.1 m³ (427 cu ft)
Flight deck/cabin door: Height	1.35 m (4 ft 5 in)
Width	0.51 m (1 ft 8 in)

AREAS:
Wings, gross	43.63 m² (469.6 sq ft)
Ailerons (total)	5.66 m² (60.93 sq ft)
Trailing-edge flaps (total)	4.02 m² (43.27 sq ft)
Leading-edge slats (total)	5.98 m² (64.37 sq ft)
Fin, incl dorsal fin	5.455 m² (58.72 sq ft)
Rudder, incl tab	2.955 m² (31.81 sq ft)
Tailplane	6.34 m² (68.25 sq ft)
Elevator, incl tab	4.10 m² (44.13 sq ft)

WEIGHTS AND LOADINGS:
Weight empty, equipped	3,330 kg (7,342 lb)
Max payload	1,400 kg (3,086 lb)
Max fuel	950 kg (2,095 lb)
Max T-O and landing weight	5,250 kg (11,574 lb)
Max wing loading	120.3 kg/m² (24.65 lb/sq ft)
Max power loading	7.44 kg/kW (12.22 lb/shp)

PERFORMANCE (estimated):
Max level speed at 3,000 m (9,840 ft)	162 kt (300 km/h; 186 mph)
Nominal cruising speed at 3,000 m (9,840 ft)	135 kt (250 km/h; 155 mph)
Service ceiling	4,000 m (13,120 ft)
T-O run	350 m (1,150 ft)
Landing run	200 m (660 ft)
Range:	
with max payload	377 n miles (700 km; 434 miles)
with max fuel	685 n miles (1,270 km; 789 miles)

UPDATED

AEROPROGRESS/ROKS-AERO T-311 KOLIBRI

English name: Hummingbird

TYPE: Four-seat lightplane.
PROGRAMME: Revealed mid-1996, when prototype under construction. No report of completion by late 2000.
CUSTOMERS: Initial order for 100 for the Russian forest protection service was reported in 1996; no further information.
COSTS: Basic production aircraft US$75,000; kitplane US$19,500 (1996).
DESIGN FEATURES: Basic light cargo/passenger transport; adaptable as ambulance with stretcher; for patrol duties with seat for observer and surveillance equipment; and for aerial photography and geological survey, with seats for operators and relevant equipment.
 Conventional high-wing monoplane; V bracing struts each side; high-lift wing section; aileron and flap occupy entire trailing-edge of each wing, except for cambered tip. Constant-chord wing. Tapered tail surfaces, with dorsal fin.
 Wing section USA-35R, thickness/chord ratio 12.5 per cent; dihedral 1°; incidence 2° 48′.
FLYING CONTROLS: Conventional and manual. Actuation via cables; horn-balanced rudder and elevators, each with trim tab; simple flaps.
STRUCTURE: Two-spar duralumin torsion-box wing. Truss-type fuselage of US 4130 or Russian VNS-2 welded steel tubing; welded tail unit; Stits fabric covering.
LANDING GEAR: Non-retractable tailwheel type; single wheel on each unit; oleo spring shock-absorbers; hydraulic disc brakes on mainwheels.
POWER PLANT: One 157 kW (210 hp) Teledyne Continental IO-360-ES flat-six piston engine; McCauley two-blade constant-speed propeller. Alternative engines include Textron Lycoming TO-360-CF. Two fuel tanks in each wing; total capacity 240 litres (63.4 US gallons; 52.8 Imp gallons).
ACCOMMODATION: Up to four persons, in pairs; door on each side. Compartment for up to 300 kg (661 lb) of cargo, with floor and side tiedown fittings. Baggage compartment aft of passenger/cargo cabin.
SYSTEMS: Cabin heated by engine bleed air and ventilated. Engine-driven generator for 27 V DC electrical system; 28 Ah battery.
AVIONICS: Weather radar, autopilot and GPS optional.

DIMENSIONS, EXTERNAL:
Wing span	9.62 m (31 ft 6¾ in)
Wing chord, constant	1.61 m (5 ft 3½ in)
Wing aspect ratio	6.0
Length overall	6.80 m (22 ft 3¾ in)
Fuselage: Max width	1.20 m (3 ft 11¼ in)
Max height	1.30 m (4 ft 3¼ in)
Height overall	2.13 m (7 ft 0 in)
Tailplane span	3.40 m (11 ft 2 in)
Wheel track	1.99 m (6 ft 6¼ in)
Wheelbase	4.97 m (16 ft 3½ in)
Propeller diameter	1.93 m (6 ft 4 in)
Propeller ground clearance	0.48 m (1 ft 7 in)
Cargo/passenger compartment door:	
Height	0.82 m (2 ft 8¼ in)
Max width	1.24 m (4 ft 0¾ in)

DIMENSIONS, INTERNAL:
Cabin: Length	2.60 m (8 ft 6¼ in)
Max width	1.14 m (3 ft 8¾ in)
Max height	1.15 m (3 ft 9¼ in)
Floor area	2.45 m² (26.4 sq ft)
Volume	2.9 m³ (104 cu ft)
Baggage compartment: Volume	0.54 m³ (19.1 cu ft)

AREAS:
Wings, gross	15.30 m² (164.7 sq ft)
Ailerons (total)	1.25 m² (13.45 sq ft)
Trailing-edge flaps (total)	1.84 m² (19.81 sq ft)
Fin	1.33 m² (14.32 sq ft)
Rudder	0.93 m² (10.01 sq ft)
Tailplane	1.49 m² (16.04 sq ft)
Elevators (total)	1.97 m² (21.21 sq ft)

WEIGHTS AND LOADINGS:
Weight empty, equipped	670 kg (1,477 lb)
Normal payload	250 kg (551 lb)
Max fuel	180 kg (397 lb)
Max T-O and landing weight	1,100 kg (2,425 lb)
Max zero-fuel weight	1,015 kg (2,237 lb)
Max wing loading	71.9 kg/m² (14.73 lb/sq ft)
Max power loading	7.03 kg/kW (11.55 lb/hp)

PERFORMANCE (at max T-O weight, estimated):
Max cruising speed at 2,800 m (9,180 ft)	135 kt (250 km/h; 155 mph)
Stalling speed, flaps down	47 kt (87 km/h; 54 mph)
Max rate of climb at S/L	480 m (1,575 ft)/min
Service ceiling	5,000 m (16,400 ft)
T-O run	96 m (315 ft)
T-O to 15 m (50 ft)	197 m (650 ft)
Landing from 15 m (50 ft)	250 m (820 ft)
Landing run	90 m (295 ft)
Range at 2,000 m (6,560 ft), 30 min reserves:	
with max payload	297 n miles (550 km; 341 miles)
with max fuel and 150 kg (330 lb) payload	809 n miles (1,500 km; 932 miles)

UPDATED

AERORIC

AERORIC NAUCHNO-PROIZVODSTVENNOYE PREDPRIYATIE OOO (AeroRIC Scientific-Production Enterprise Ltd)
ulitsa Chaadaeva 1, 603035 Nizhny Novgorod
Tel: (+7 8312) 46 71 27
Fax: (+7 8312) 24 79 66
DIRECTOR AND CHIEF DESIGNER: Victor P Morozov

The AeroRIC design bureau is co-located with the Sokol Aircraft Manufacturing Plant at Nizhny Novgorod.

VERIFIED

AERORIC DINGO

All known details of this light utility aircraft with air cushion landing gear appeared in the 2000-01 and earlier editions of *Jane's*. Completion of the prototype has yet to be achieved.

UPDATED

ALBATROS

TSENTR NAUCHNO-TEKHNICHESKOGO TVORCHESTVA ALBATROS (Albatross Scientific-Technical Works)
ulitsa Tsentral'naya 22/126, 142092 Troitsk, Moskovskaya oblast
Tel/Fax: (+7 096) 751 66 76
CONSTRUCTOR: Sergei V Ignatev
TEST PILOT: Mikhail Y Shevyakov

Original Aviacomplex company established in 1990 to build prototypes of the AS-2 two-seat lightplane. Testing was completed in collaboration with CAHI (TsAGI) and LII Flight Research Institute. Production by Aviacomplex and Myasishchev Engineering Bureau, with participation by other Russian companies. At 1999 Moscow Air Show (MAKS), a production AS-2 was shown under new designation of Albatros AS-3A and first details were revealed of a development, the Sigma-4.

VERIFIED

Albatros AS-3A prototype

ALBATROS AS-2 and AS-3

TYPE: Side-by-side ultralight.
PROGRAMME: Design and prototype construction started 1990; first flight of AS-2 February 1993; prototype exhibited at Moscow Air Show '93 and '95; prototype and production aircraft in '97. Total of three AS-2s built, including **AS-2k** with cockpit side doors. Production AS-2 FLARF-01036 became prototype AS-3A, with metal fuselage and electrically operated flaperons; demonstrated at Moscow, August 1999.
COSTS: US$25,000 (1999).
DESIGN FEATURES: Designed and built to JAR-VLA standards. High-wing monoplane, with single bracing strut each side. Open or enclosed fuselage pod. Cruciform tail unit carried on strut-braced tubular tailboom, extended forward beneath wing and carrying power plant at tip. Sweptback vertical tail surfaces; unswept and strut-braced horizontal tail surfaces. Slight wing dihedral. Easy assembly and disassembly; transportable on trailer.
FLYING CONTROLS: Manual. Full-span flaperons and elevators; full-height rudder. Ground-adjustable tab on port elevator.
STRUCTURE: Aluminium alloy beams, spars, ribs and bracing struts, with steel alloys where high strength required. AS-2 has wing leading-edge and cabin of glass fibre; remainder of wing fabric covered. AS-3 cabin is metal-covered.
LANDING GEAR: Non-retractable tricycle type. Mainwheels carried on strut-braced blade; trailing-link nosewheel. Floats and skis optional.
POWER PLANT: One 47.8 kW (64.1 hp) Rotax 582 UL two-cylinder two-stroke engine; two-blade fixed-pitch propeller. Two fuel tanks, total capacity 120 litres (31.7 US gallons; 26.4 Imp gallons).

DIMENSIONS, EXTERNAL:
Wing span	9.60 m (31 ft 6 in)
Length overall	5.77 m (18 ft 11¼ in)
Height overall	2.61 m (8 ft 6¾ in)

AREAS:
Wings, gross	12.20 m² (131.3 sq ft)

WEIGHTS AND LOADINGS (AS-3):
Weight empty	240 kg (529 lb)
Max T-O weight	420 kg (926 lb)

PERFORMANCE (AS-3):
Max level speed	65 kt (120 km/h; 75 mph)
Max cruising speed	49 kt (90 km/h; 56 mph)
Max rate of climb at S/L	180 m (590 ft)/min
T-O run	60 m (200 ft)
Range with max payload	162 n miles (300 km; 186 miles)

UPDATED

ALBATROS SIGMA-4

TYPE: Four-seat kitbuilt.
PROGRAMME: Design began in third quarter of 1998. Preliminary details announced at MAKS, Moscow, August 1999, when prototype 50 per cent complete.
COSTS: US$32,000 (1999).
DESIGN FEATURES: Four-place development of AS-2. More aerodynamically refined, with streamlined pod, spinner, cowled engine and speed fairings over wheels. Flying boat version also projected.
Wing section GA(W)-1.
STRUCTURE: Metal airframe with glass fibre covering.
LANDING GEAR: Tricycle type; fixed; speed fairings on all wheels. Optional ski and float gear.
POWER PLANT: One 59.6 (79.9 hp) Rotax 912 UL flat-four. Fuel capacity 50 litres (13.2 US gallons; 11.0 Imp gallons).

DIMENSIONS, EXTERNAL:
Wing span	9.80 m (32 ft 1¼ in)

AREAS:
Wings, gross	11.20 m² (120.6 sq ft)

WEIGHTS AND LOADINGS:
Weight empty	280 kg (617 lb)
Max T-O weight	520 kg (1,146 lb)

PERFORMANCE:
Max level speed	105 kt (195 km/h; 121 mph)
Max cruising speed	94 kt (175 km/h; 109 mph)
Landing speed	32 kt (60 km/h; 37 mph)
Max rate of climb at S/L	360 m (1,181 ft)/min
T-O run	65 m (215 ft)
Range	270 n miles (500 km; 310 miles)

UPDATED

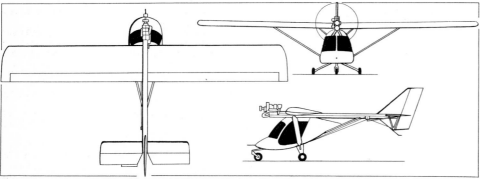

General arrangement of the Albatros AS-3A *(Paul Jackson/Jane's)*

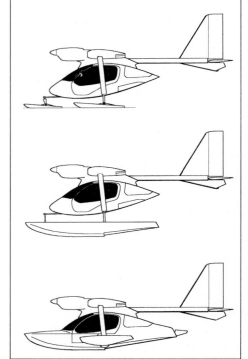

Projected versions of Sigma-4 with ski, twin-float and flying boat landing gear

---

ARSENYEV

ARSENYEVSKOYE AVIATSIONNOYE PROIZVODSTVENNOYE PREDPRIYATIE IMENI N I SAZYKINA (Arsenyev Aviation Production Enterprise 'Progress' named for N I Sazykin)
prospekt Lenina 5, 692335 Arsenyev, Primovsky Kray
Tel: (+7 423) 612 48 97
Fax: (+7 423) 612 61 30
GENERAL DIRECTOR: Vladimir Pechyonkin

Arsenyev plant previously built the Mil Mi-24/25/35 series of combat helicopters, in parallel with Rostvertol, and was also manufacturer of the Yak-55 aerobatic lightplane; a modified version of the last-mentioned, the Technoavia SP-55, was returned to production in 1999, in an initial batch of five, two of which delivered to Technoavia by October 2000.

Arsenyev is responsible currently for the Kamov Ka-50 (including, if ordered, the Ka-50-2 export version) and Mil Mi-34 helicopters, as well as Moskit missiles. It has been assigned the new Kamov Ka-60. Russian government shareholding is 51 per cent.

A government decree of 31 December 1997 authorised the Arsenyev plant to offer the Ka-50 for export and supply spares and support for existing Mi-24/25/35 helicopters. Overseas delivery of P-15U, P-20, P-21 and P-22 cruise missiles is also covered.

UPDATED

AST

AVIASPETSTRANS OAO (Aviaspetstrans JSC)
alleya 230 AST, Zhukovsky-5, 140160 Moskovskoye oblast
Tel: (+7 095) 556 59 93
Fax: (+7 095) 292 65 11
GENERAL DIRECTOR: Valentin A Korchagin

There have been no recent reports of this consortium or its Yamal amphibian aircraft programme. Details of both last appeared in the 2000-01 edition.

UPDATED

AVIA

NAUCHNO- PROIZVODSTVENNOE OBEDINENIE AVIA LTD (Avia Scientific-Production Association JSC)
Studeni proezd 5, 129282 Moskva
Tel: (+7 095) 472 97 89 and 478 55 62
Fax: (+7 095) 479 16 09
e-mail: avia@msk.sitek.net
Web: http://www.windoms.sitek.net/accord 201
GENERAL DIRECTOR: Aleksandr Loshkarev
SALES DIRECTOR: Vadim Salimov

OKB AKKORD (Accord Design Burea)
ulitsa Chaadaeva 12a, 603035 Nizhny Novgorod
Tel: (+7 8312) 46 77 49
e-mail: almaz@vpk.ru (accord)
DIRECTOR: Yury Lakhtachev

EUROPEAN SALES AGENT:
Ground Support Equipment srl
Viale del Vignola 44, I-00196 Roma, Italy
Tel: (+39 06) 322 28 77
Fax: (+39 06) 361 17 15
DIRECTOR: Roberto Salmoni

The Avia company was established on 20 February 1995 to design, manufacture and support light aircraft. Its associated design bureau (itself part of the Sokol plant) is responsible for the Accord range of light twin-engined aircraft. GSE of Italy was appointed as sales agent in 2000.

UPDATED

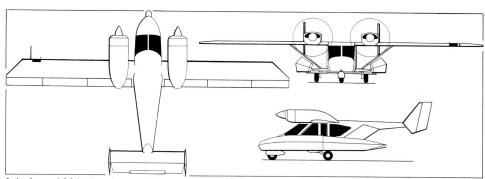

Avia Accord-201, showing position of retracted floats in nose and side views *(Paul Jackson/Jane's)*
2000/0079325

AVIA ACCORD-201

TYPE: Light multipurpose aircraft.
PROGRAMME: Original (lightweight) Accord design began May 1991; first flew 18 April 1994 and was last described in the 1997-98 *Jane's*. Production version is Accord-201; prototype flew 18 August 1997 at Nizhny Novgorod, initially with only VFR avionics and without floats. Demonstrations to prospective customers began December 1997; series of 12 flights from water, using floats, undertaken mid-1998. By early 1999, four more aircraft, including static test specimen, under construction and tooling in place for maximum possible production of 10 per month. Russian AP-23 certification anticipated October 1999, but delayed; new target was first quarter of 2001 (Experimental category certificate, granted mid-2000 with expiry date of 19 June 2001). Total 14 under construction in March 2000.
CURRENT VERSIONS: Offered for a variety of roles, including SAR (**201P**), aerial survey (**201AFS**), ecological monitoring (**201EM**), sigint (**201RC**) and jamming (**201REP**).
CUSTOMERS: First firm order received in May 1998 from Almazy Rossii-Sakha diamond mining company, which ordered two to be used for geomagnetic survey and exploration. Customers in Dubai and Liechtenstein by 2000.
COSTS: Standard aircraft US$298,000 on wheels; US$328,000 on floats. Operating cost US$44 per hour (2000).
DESIGN FEATURES: High-wing monoplane with single bracing strut each side; unswept, high-lift wing section and blown flaps; pod and boom fuselage; cruciform tail surfaces with sweptback fin and rudder; mainwheels at tips of short stub-wings that support bracing struts. Constant-chord main wing panels, with engines on leading-edge; tapered outer panels.
Service life 15,000 hours/15 years. Wing section CAHI (TsAGI) P-II, with 12 per cent thickness/chord ratio at root, CAHI (TsAGI) P-III at tip; tip sweepback 45°; no dihedral, incidence 1°, no twist.
FLYING CONTROLS: Conventional and manual. Ailerons and elevators 100 per cent balanced; twin horn-balanced rudders; all cable actuated. Three-axis electric trim. Two-section slotted flaps each side.

STRUCTURE: All-metal; D16 aluminium alloy and high-strength stainless steel. Fuselage based on welded rectangular-section cabin frame of steel tubing; remainder of airframe conventional light alloy construction.
LANDING GEAR: Non-retractable tricycle type; single wheel on each unit. Shock absorption by specially shaped rubber cushion between strut and wheel suspension lever. Cleveland wheels and brakes; mainwheels type 40-142 with 8.00-6 tyres; nosewheel type 40-140 with 6.00-6 tyres; type 30-127 disc brakes. Minimum turning circle 3.2 m (10 ft 6 in). Maximum nosewheel steering angle ±60°. Optional floats (additional to, not replacing, fixed landing gear) retract upward electrically, outside mainwheels.
POWER PLANT: Two 157 kW (210 hp) Teledyne Continental IO-360-ES7B flat-six engines; Hartzell PHC-H3YF-2UF/FC7453 three-blade, constant-speed, fully feathering propellers. Fuel tank in each engine nacelle; total capacity 750 litres (200 US gallons; 165 Imp gallons). Total oil capacity 15.2 litres (4.0 US gallons; 3.3 Imp gallons).
ACCOMMODATION: Pilot and up to six passengers in 2-3-2 seating, at 80 cm (31.5 in) pitch. Alternatively, pilot and three passengers and cargo, or all-cargo, or one stretcher patient, one or two attendants and medical equipment. Optional dual controls. Two large upward-hinged side doors. Baggage/cargo hold at rear of cabin, with door.
SYSTEMS: SAU-201 autopilot. Webasto Air Top 32 cabin heater. Electrical system includes two 27 V Teledyne N653344 DC generators and one 19 Ah Gill G-247 battery. Elipos-3 pneumatic de-icing system for wing, fin and tailplane leading-edges; Goodrich electric propeller de-icing. Janitrol cabin heater.
AVIONICS: Mainly Honeywell equipment:
Comms: KY 196A VHF transceiver, KLX 135A VHF com transceiver and integral GPS, KHF 950 HF transceiver, KT 76A and CO-94 transponders, KMA 24 audio control console, R-855A1 (Russian) ELT and ET3000.
Radar: Weather radar optional.
Flight: KX 155 digital nav/com, KCS 55A compass system and KR 87 ADF.
INSTRUMENTATION: Conventional.

DIMENSIONS, EXTERNAL:
Wing span	13.75 m (45 ft 1¼ in)
Wing chord, constant	1.30 m (4 ft 3¼ in)
Wing aspect ratio	11.1
Length overall: landplane	8.12 m (26 ft 7¾ in)
amphibian	8.73 m (28 ft 7¾ in)
Height overall: landplane	2.94 m (9 ft 7¾ in)
amphibian	3.455 m (11 ft 4 in)
Span over tailfins	3.66 m (12 ft 0 in)
Wheel track	2.34 m (7 ft 8¼ in)
Float track	2.90 m (9 ft 6¼ in)
Wheelbase	2.845 m (9 ft 4 in)
Propeller diameter	1.93 m (6 ft 4 in)
Propeller ground clearance	1.25 m (4 ft 1¼ in)
Distance between propeller centres	2.80 m (9 ft 2¼ in)
Cabin doors (each): Height	1.10 m (3 ft 7¼ in)
Width: at top	0.75 m (2 ft 5½ in)
at bottom	1.50 m (4 ft 11 in)
Baggage/airdrop door: Height	0.90 m (2 ft 11½ in)
Width: at top	0.85 m (2 ft 9½ in)
at bottom	1.20 m (3 ft 11¼ in)

DIMENSIONS, INTERNAL:
Cabin: Length	3.50 m (11 ft 5¾ in)
Max width	1.31 m (4 ft 3½ in)
Max height	1.20 m (3 ft 11¼ in)
Floor area	3.50 m² (37.7 sq ft)
Volume	4.0 m³ (141 cu ft)
Baggage hold: Volume:	
pilot and six passengers	0.50 m³ (17.7 cu ft)
pilot and four passengers	1.5 m³ (53 cu ft)
pilot and cargo	2.5 m³ (88 cu ft)

AREAS:
Wings, gross	17.00 m² (183.0 sq ft)
Ailerons (total)	1.17 m² (12.59 sq ft)
Trailing-edge flaps (total)	3.23 m² (34.77 sq ft)
Fins (total)	1.36 m² (14.64 sq ft)
Rudders (total)	1.36 m² (14.64 sq ft)
Tailplane	2.18 m² (23.47 sq ft)
Elevators (total)	1.78 m² (19.16 sq ft)

WEIGHTS AND LOADINGS:
Operating weight empty: landplane	1,350 kg (2,976 lb)
amphibian	1,530 kg (3,373 lb)
Max T-O weight	2,200 kg (4,850 lb)
Max wing loading	129.4 kg/m² (26.51 lb/sq ft)
Max power loading	7.03 kg/kW (11.55 lb/hp)

PERFORMANCE (estimated):
Never-exceed speed (V_{NE})	202 kt (375 km/h; 233 mph)
Max level speed at S/L:	
landplane	161 kt (299 km/h; 186 mph)
amphibian	150 kt (277 km/h; 172 mph)
Max cruising speed at 75% power at 1,000 m (3,280 ft):	
landplane	148 kt (275 km/h; 170 mph)
amphibian	138 kt (255 km/h; 158 mph)
Econ cruising speed at 3,000 m (9,840 ft):	
landplane	119 kt (220 km/h; 137 mph)
amphibian	111 kt (205 km/h; 127 mph)
Stalling speed, flaps down, power off:	
landplane	58 kt (106 km/h; 66 mph)
Max rate of climb at S/L:	
landplane	445 m (1,460 ft)/min
amphibian	420 m (1,378 ft)/min
Service ceiling: landplane	6,100 m (20,020 ft)
amphibian	5,800 m (19,020 ft)
T-O run: on land: landplane	240 m (790 ft)
amphibian	250 m (820 ft)
on water	360 m (1,185 ft)
Landing run on land:	
landplane, amphibian	220 m (725 ft)
on water	190 m (625 ft)
Range with max fuel at 3,000 m (9,840 ft):	
landplane	1,592 n miles (2,950 km; 1,833 miles)
amphibian	1,339 n miles (2,480 km; 1,541 miles)
g limits	+3.8/-1.6

UPDATED

Avia Accord-201 landplane version
2001/0099580

Float-equipped Accord-201
2001/0099581

AVIACOR

AVIAKOR MEZHDUNARODNAYA AVIATSIONNAYA KORPORATSIYA OAO (Aviacor International Aviation Corporation JSC)
ulitsa Pskovskaya 32, 443052 Samara
Tel: (+7 8462) 27 03 87 and 27 04 44
Fax: (+7 8462) 27 06 91 and 27 04 77
GENERAL DIRECTOR: Vladimir Belogub
MARKETING DIRECTOR: Sergei Grachev

Founded during the 1930s, the Samara plant (formerly GAZ 18) is now terminating production of the Tupolev Tu-154 three-turbofan transport, the final 12 of which are being completed between 2000 and 2002. It manufactures the Molniya-1 six-seat light aircraft and an air cushion vehicle derivative, and will produce the Tupolev Tu-354. It is also earmarked to build the Ukrainian Antonov An-70 and An-140 in Russia and assist RSK MiG with the Tu-334. By mid-2000, the first An-140 still had not been completed, however. Production plans for An-70 were questioned by Samara local government in 2000. Will manufacture M-12 Kasatik (see following entry). Two new aircraft, known only as the Aist-2 and Aist-4, are reported to be under development. Subsidiary, Aviacor-Service, specialises in repair and maintenance.

In mid-1997 (workforce then approximately 10,000), Aviacor agreed to merge with Tupolev design bureau and Aviastar (which see). Russian government shareholding is 25.5 per cent; controlling interest held by Siberian Aluminium Group since 1998. Undertakes subcontract work for Airbus and Boeing.

UPDATED

AVIAKOR

MEZHDUNARODNAYA AVIATSIONNAYA KORPORATSIYA OAO (Aviacor International Aviation Corporation JSC) OPYTNYI KONSTRUKTORSKOYE BURO FENIX (Phoenix Experimental Design Bureau)
ulitsa Pskovskaya 32, 443052 Samara
Tel: (+7 095) 29 41 51
Fax: (+7 095) 27 06 91
DIRECTOR: Nikolai Masterov

Phoenix, which is part of Aviacor Corporation, first showed its M-12 lightplane at the 1995 Moscow Air Show.

UPDATED

PHOENIX KASATIK M-12
English name: Darling
TYPE: Three-seat multipurpose lightplane.
PROGRAMME: Prototype first flew 1994 and exhibited at Moscow Air Show '95; accumulated over 150 hours of flight testing. Second prototype lost in accident 14 October 1997; replaced by third prototype. Original version, with 25.7 kW (34.5 hp) Vikhr 30 engine, described in 1998-99 edition; heavier version, with Rotax 582s, gained AP-23 certification in 1999 after total of 550 hours. Production deliveries began April 1999 and were intended to continue at rate of 10 per year.
CUSTOMERS: Export orders received from Germany.
COSTS: US$43,000 (1999).
DESIGN FEATURES: Braced high-wing monoplane; single bracing strut each side; pod and boom fuselage; unswept wire-braced tail unit. Can be disassembled easily for transport. Suitable for patrol, mail, ambulance and training duties.
FLYING CONTROLS: Conventional and manual. No aerodynamic balance on ailerons and elevator; balanced rudder.
STRUCTURE: Conventional all-metal semi-monocoque construction.
LANDING GEAR: Non-retractable tricycle type; single wheel on each unit; cantilever mainwheel legs; mainwheel tyre size 300×120; fairings over all three wheels. Optional floats or skis.
POWER PLANT: Two 47.8 kW (64.1 hp) Rotax 582 UL piston engines above wingroot trailing-edge, driving pusher propellers. Fuel capacity 70 litres (18.5 US gallons; 15.4 Imp gallons).
ACCOMMODATION: Pilot at front; two passengers side by side at rear; dual control column between rear seats optional. Extensively glazed cabin for outstanding field of view; rear-hinged fully glazed door each side.
EQUIPMENT: Two landing lights in nose. Ballistic parachute.
DIMENSIONS, EXTERNAL:
Wing span 10.00 m (32 ft 9¾ in)
Length overall 6.20 m (20 ft 4 in)
Height overall 1.95 m (6 ft 4¾ in)
AREAS:
Wings, gross 12.20 m² (131.3 sq ft)
WEIGHTS AND LOADINGS:
Weight empty 360 kg (794 lb)
Max T-O weight 670 kg (1,477 lb)
Max wing loading 54.9 kg/m² (11.25 lb/sq ft)
Max power loading 7.02 kg/kW (11.54 lb/hp)
PERFORMANCE:
Max level speed 103 kt (190 km/h; 118 mph)
Max cruising speed 78 kt (145 km/h; 90 mph)
Service ceiling 3,000 m (9,840 ft)
Range 283 n miles (525 km; 326 miles)
Endurance 3 h 30 min

UPDATED

Phoenix M-12 three-seat light multipurpose aircraft *(Paul Jackson/Jane's)* 2000

AVIASTAR

AVIASTAR, ULYANOVSKY AVIATSIONNYI PROMSHLENNYI KOMPLEKS OAO (Ulyanovsk Aviation Industrial Complex 'Aviastar' JSC)
prospekt Antonova 1, 432072 Ulyanovsk
Tel: (+7 8422) 20 25 75 and 29 10 22
Fax: (+7 8422) 20 35 06 and 29 23 65
Telex: 263816 GORN
DIRECTOR GENERAL: Gennadi I Koradnev
MARKETING DIRECTOR: Aleksandr I Gladkov
EXECUTIVE DIRECTOR: Valery V Savotchenko

Founded in 1976, this 1.5 million m² (16.5 million sq ft) production facility, known as Plant 25, began manufacture of the Antonov An-124 ultra-large airlifter (which see in the Ukrainian section) in 1985, following with the Tupolev Tu-204 transport in 1987 and the related Tu-234 in 1996. In 1996, brief details were given of the Module two-seat light kitplane; no known production; last described in 1999-2000 *Jane's*.

In mid-1997, Aviastar agreed to merge with Tupolev design bureau and Aviacor (which see). Tupolev merger effected 30 June 1999, forming Tupolev OAO with inputs from Russian government (51 per cent), Aviastar (43.6 per cent) and Tupolev Design Bureau (5.4 per cent). Aviacor not included. Also in 1997, Aviastar Asia was formed in Taipei, Taiwan, as a joint venture with Asian interests to promote the Tu-204 in the Far East. Finance is provided by Aviastar Financial International, formed by Aviastar, Perm Motors and Moscow International Bank.

After five-year break, production of An-124 resumed with August 2000 delivery of one to Volga-Dniepr. Tu-204 production rate, previously maximum four per year, accelerated in 2000.

UPDATED

Tupolev Tu-204-200 airliners under construction at the Aviastar plant *(Yefim Gordon)* 2001/0092564

AVIATIKA

KONTSERN AVIATIKA (Aviatika Concern)
Address as for MAPO
Tel: (+7 095) 252 87 87 and 158 44 68
Fax: (+7 095) 250 19 84
e-mail: oskbes@com2com.ru
DEPUTY CHIEF DESIGNER: Nikolay Goryunov

Established on 27 July 1992 as Aviatika JSC by Moscow Industrial Aviation Association named after Dementyev (MIAA), Gromov Flight Research Institute (LII) and Moscow Aviation Institute (MAI). Certification by Aviaregister followed on 17 February 1993; name changed to Aviatika Concern on 20 April 1994. It began exporting its aircraft in 1992, delivering 90 to 12 countries in its first year. Promotion, manufacture and sales are managed by MAPO (which see).

Design of Aviatika aircraft is by MAI, which previously built Kvant aerobatic lightplane, PS-01 and Elf-D UAVs, Elf lightplane (see Dubna entry), Yunior/MAI-890 biplanes and Aviatika-MAI-900 aerobatic single-seater. It has also produced the Aviatika-MAI-920 training glider, with 90 per cent of its components made up of standard -890 parts.

UPDATED

AVIATIKA-MAI-890
TYPE: Single-/two-seat biplane ultralight.
PROGRAMME: Open cockpit Yunior (Junior) first flew July 1987 as single-seater with 5.68 m (18 ft 7½ in) span and 280 kg (617 lb) MTOW. Developed as MAI-89, span 8.11 m (26 ft 7½ in); MTOW 340 kg (750 lb). Production version, MAI-890, flew 1990 and entered series manufacture in 1991, although this halted temporarily in mid-1990s. Manufactured by RSK MiG's Voronin production centre.
CURRENT VERSIONS: **Aviatika-MAI-890:** Single-seater.
Aviatika-MAI-890U: Side-by-side two-seater (*uchebno*: training). First flown August 1991 (Rotax 582

UL). Normally with Rotax 912 UL engine (first flight 16 September 1992). Empty weight 298 kg (657 lb); MTOW 540 kg (1,190 lb); length 5.49 m (18 ft 0¼ in). Certification anticipated in early 2001. Floatplane version completed trials on Lake Shartash, Ekaterinburg, September 2000.

Aviatika-MAI-890SKh: Crop-sprayer version (*selskokhozyaistvenny*: agricultural); marketing name **Farmer**. Chemical tank under engine and spraybars aft of lower wings; weight of agricultural equipment 28 kg (62 lb); chemicals 60 kg (132 lb) maximum with Rotax 582 engine. Certification scheduled for end of 2000. Local market estimated as 1,200 to 1,500.

Aviatika-MAI-890S: Rotax 912 ULS 73.5 kW (98.6 hp) flat-four. Prototype (ZU-BZO) converted in South Africa, 1999, by Aeroserve Pty Ltd.

CUSTOMERS: Total of 300 sold in 17 countries by 2000, including some 15 in crop-spraying configuration. Approximately 30 imported by a South African dealer in 1995; these known locally as **890CSH Mai** and **890U Mai**. Total market for agricultural variant estimated at 800 to 1,000. Contract for 25 aircraft, for undisclosed customer, in negotiation in second quarter of 2000, for delivery in mid-2001.

COSTS: US$25,000 to US$40,000 (2000).

DESIGN FEATURES: Strut- and wire-braced biplane; unspinnable; conventional tail surfaces carried on tubular boom; engine mounted under trailing-edge of upper wing; slight sweepback on all wings; dihedral on lower wings only. Semi-aerobatic and capable of flying in rough air conditions. Designed to JAR-VLA and FAR Pt 23 standards. Open cockpit standard; doors optional, one jettisonable; provision for backpack parachute(s).

FLYING CONTROLS: Conventional and manual. Full-span ailerons on lower wings; large-area rudder and elevators, each with ground-adjustable tab.

STRUCTURE: Aircraft grade aluminium and titanium alloys, alloy steels; fabric covering unaffected by Sun's radiation or atmospheric precipitation.

LANDING GEAR: Non-retractable tricycle type; single wheel on each unit; cantilever spring main legs. Optional skis or floats.

POWER PLANT: One 47.8 kW (64.1 hp) Rotax 582 UL two-cylinder or 59.6 kW (79.9 hp) Rotax 912 UL four-cylinder piston engine; two-blade pusher propeller. Standard fuel 50 litres (13.2 US gallons; 11.0 Imp gallons); provision for 55 litre (14.5 US gallon; 12.1 Imp gallon) auxiliary tank on single-seat version.

EQUIPMENT: Provision for up to 120 kg (265 lb) payload on four attachments, under engine mountings (60 kg; 132 lb), or underbelly (100 kg; 220 lb), or at lower wingtips (each 45 kg; 99 lb), including agricultural dusting and spraygear for Aviatika-MAI-890SKh Farmer. Optional BRS ballistic parachute system.

Following details apply to single-seat 890 with Rotax 912 UL engine.

DIMENSIONS, EXTERNAL:
Wing span, upper	8.11 m (26 ft 7½ in)
Length overall, incl pitot	5.32 m (17 ft 5½ in)
Height to tip of fin	2.30 m (7 ft 6½ in)

AREAS:
Wings, gross, total	14.29 m² (153.8 sq ft)

WEIGHTS AND LOADINGS:
Weight empty	268 kg (591 lb)
Max T-O weight	450 kg (992 lb)

PERFORMANCE:
Max level speed	67 kt (125 km/h; 78 mph)
Cruising speed	54 kt (100 km/h; 62 mph)
Landing speed	35-38 kt (65-70 km/h; 41-44 mph)
Max rate of climb at S/L	300 m (984 ft)/min
T-O run	65 m (215 ft)
Landing run	95 m (315 ft)
Endurance	3 h 7 min
g limits	+5/–2.5

UPDATED

Aviatika-MAI-890 single-seat light biplane — *1999/0051199*

Aviatika-MAI-960 light passenger/freight aircraft, produced by fitting new rear fuselage to the prototype -910 *(Paul Jackson/Jane's)* — *2001/0099579*

AVIATIKA-MAI-890A

TYPE: Single-seat autogyro.

PROGRAMME: Prototype completed 1994; still undertaking flight trials in 1999.

DESIGN FEATURES: Conventional configuration, with fully enclosed cockpit, tail surfaces carried on tubular tailboom; two-blade autorotating rotor; rotor column folds for storage; non-retractable tricycle landing gear. Control movements emulate aeroplane, rather than autogyro, providing safer type conversion for former fixed-wing pilots. Ground-adjustable tab on rudder and each elevator.

POWER PLANT: One 47.8 kW (64.1 hp) Rotax 582 UL piston engine mounted behind top of cabin pod; two-blade pusher propeller. Fuel tank under engine.

DIMENSIONS, EXTERNAL:
Rotor diameter	7.20 m (23 ft 7½ in)
Height overall	2.80 m (9 ft 2¼ in)

WEIGHTS AND LOADINGS:
Weight empty	195 kg (430 lb)
Max T-O weight	360 kg (794 lb)

PERFORMANCE:
Max level speed	72 kt (135 km/h; 83 mph)
Nominal cruising speed	49 kt (90 km/h; 56 mph)
Min level speed	22 kt (40 km/h; 25 mph)
Max rate of climb at S/L	240 m (787 ft)/min
T-O run	30-50 m (100-165 ft)
Landing run	5-6 m (17-20 ft)

UPDATED

AVIATIKA-MAI-900 ACROBAT

In September 2000, Aviatika announced that it had been promised funds from a Lithuanian investor to launch production of the Acrobat lightplane, last described in the 1997-98 *Jane's*. Estimated cost is US$120,000 to US$150,000.

NEW ENTRY

Aviatika-MAI-890A single-seat lightweight autogyro *(Paul Jackson/Jane's)*

AVIATIKA-MAI-910 and 960

TYPE: Two-seat lightplane.

PROGRAMME: Design started and construction of first prototype began 1993; first flight 22 June 1995 in MAI-910 configuration. Promotion began at MAKS '99, Moscow, August 1999.

CURRENT VERSIONS: **Aviatika-MAI-910:** Baseline version with boom-mounted empennage.

Aviatika-MAI-960: Version with conventional rear fuselage, interchangeable with the above in 15 minutes and including ventral strake.

Aviatika-MAI-217: Described separately.

DESIGN FEATURES: Designed to JAR-VLA standards. Braced high-wing monoplane; single bracing strut and jury strut each side; fully enclosed cabin pod; strut- and wire-braced conventional tail unit carried on interchangeable rear fuselage of conventional thickness or on short boom. Wings fold.

Wing section CAHI (TsAGI) Sh-1-14; 1° leading-edge sweepback; thickness/chord ratio 14 per cent; constant chord except for tapered tips; no dihedral; incidence 5°.

FLYING CONTROLS: Conventional and manual. Actuation by cables; long-span flaperon on each wing; ground-adjustable tab on rudder and each elevator.

STRUCTURE: Metal structure with fabric-covered wings and tail surfaces, metal-skinned fuselage. Wings have single tubular spar, diameter 102 mm (4 in), and stamped sheet metal ribs. Materials primarily aluminium alloy, with stainless steel and titanium alloy where appropriate.

LANDING GEAR: Non-retractable tricycle type; single wheel on each unit; cantilever spring legs; mainwheel tyres 400×150 or 300×120; nosewheel tyre 300×120; tyre pressure (all) 2.53 bars (37 lb/sq in). Mechanical brakes with 300×120 tyres; hydraulic brakes with 400×150 tyres. Nosewheel steerable through ±50°. Tailwheel gear optional.

POWER PLANT: One 59.6 kW (79.9 hp) Rotax 912 UL flat-four; Russian two-blade fixed-pitch propeller. Two fuel tanks, each 25 litres (6.6 US gallons; 5.5 Imp gallons); one in each wingroot. Oil capacity 6 litres (1.6 US gallons; 1.3 Imp gallons).

ACCOMMODATION: Two seats side by side in fully enclosed cabin with wide field of view for observation, patrol, search and aerial photography. Passenger seat folds for carriage of stretcher or lengthy freight. Large rearward-sliding window/door each side. Space for baggage or freight behind seats. Cabin heated and ventilated.

SYSTEMS: 12 V AC electrical system; Teledyne battery.

AVIONICS: *Comms:* UHF radio.
Flight: GPS.

DIMENSIONS, EXTERNAL:
Wing span	10.73 m (35 ft 2½ in)

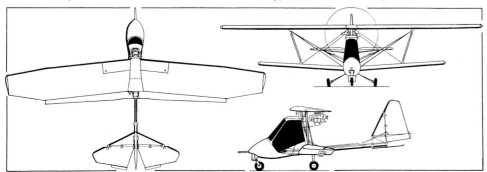
Aviatika-MAI-890 (Rotax piston engine) *(Paul Jackson/Jane's)* — *1999/0051198*

Wing chord: at root	1.10 m (3 ft 7¼ in)
at tip	0.70 m (2 ft 3½ in)
Wing aspect ratio	10.3
Length overall	5.26 m (17 ft 3 in)
Height overall	2.50 m (8 ft 2½ in)
Wheel track	2.20 m (7 ft 2¾ in)
Wheelbase	1.60 m (5 ft 3 in)
Propeller diameter	1.90 m (6 ft 2¾ in)
Propeller ground clearance	0.30 m (11¾ in)
Doors (each): Height	0.87 m (2 ft 10¼ in)
Max width	1.10 m (3 ft 7¼ in)
Baggage door: Length	0.75 m (2 ft 5½ in)
Width	0.70 m (2 ft 3½ in)
DIMENSIONS, INTERNAL:	
Cabin: Max width	1.20 m (3 ft 11¼ in)
Max height	1.08 m (3 ft 6½ in)
Floor area	1.20 m² (12.9 sq ft)
Baggage/freight space: Volume	0.70 m³ (24.7 cu ft)
AREAS:	
Wings, gross	11.20 m² (120.6 sq ft)
WEIGHTS AND LOADINGS:	
Weight empty	380 kg (838 lb)
Max fuel	40 kg (88 lb)
Max T-O weight	620 kg (1,366 lb)
Max wing loading	55.4 kg/m² (11.34 lb/sq ft)
Max power loading	10.40 kg/kW (17.09 lb/hp)
PERFORMANCE:	
Never-exceed speed (VNE)	121 kt (225 km/h; 139 mph)
Max level speed at S/L	84 kt (155 km/h; 96 mph)
Max and econ cruising speed	75 kt (140 km/h; 87 mph)
Stalling speed	41 kt (76 km/h; 48 mph)
Max rate of climb at S/L	210 m (690 ft)/min
Service ceiling	3,000 m (9,840 ft)
T-O run	150 m (495 ft)
T-O to 15 m (50 ft)	300 m (985 ft)
Landing from 15 m (50 ft)	150 m (495 ft)
Range	216 n miles (400 km; 248 miles)
g limits	+4/−2

UPDATED

Aviatika-MAI-910 (Rotax piston engine) *(James Goulding/Jane's)* 0011968

AVIATIKA-MAI-217
TYPE: Two-seat lightplane.
PROGRAMME: Development of MAI-910; production funds promised, and detail design under way, in 2000. Prototype target completion date August 2001. Cost US$60,000 to US$70,000.

NEW ENTRY

AVIATON

AVIATON NAUCHNO-PROIZVODSTVENNAYA AVIATSIONNAYA FIRMA (Aviaton Aviation Scientific-Production Firm)
Balashikhinsky r-on, p/o Chernaya, Mars, 143991 Moskovskoye oblast
Tel: (+7 095) 252 82 21 and 522 90 66
Fax: (+7 095) 191 68 27
GENERAL DIRECTOR: Avtandil Khachapuridze

Aviaton displayed its first product, the Merkury light twin, at the 1997 Moscow Air Show. It purchased a controlling share (91.44 per cent) of the Kutaisi Aircraft Repair Plant in 1998 and will assemble production aircraft there from parts manufactured elsewhere.

UPDATED

AVIATON MERKURY
English name: Mercury
TYPE: Four-seat utility twin.
PROGRAMME: Prototype completed, but apparently not yet flown, by August 1997. Second prototype under construction at Kutaisi by late 2000; third, uprated, prototype being built by RSK 'MiG's' Voronin plant, intended for public debut in August 2001.
CURRENT VERSIONS: **Four-seat:** *As described.*
Six-seat: Third prototype powered by two LOM M337B six-cylinder piston engines, each 176 kW (237 hp); stretched fuselage. Estimated cost (2000) US$360,000.
DESIGN FEATURES: Robust utility design; high wing for ease of access and minimal propeller damage from semi-prepared airstrips.
FLYING CONTROLS: Conventional and manual. Slotted flaps; horn-balanced rudder; flight-adjustable trim tabs in rudder and starboard elevator.
STRUCTURE: Metal throughout; fin and tailplane mutually braced.
LANDING GEAR: Non-retractable tricycle type. Mainwheels on independently hinged, steel tube levers with vertically mounted motorcar-type shock-absorbers; steerable nosewheel. Mainwheel tyres 500×200; nosewheel 400×150.
POWER PLANT: Prototype has two 119 kW (160 hp) Aviadvigatel rotary engines, each driving a three-blade V-532B propeller. Fuel filler on port side, behind cabin.
ACCOMMODATION: Four persons, including two pilots. Large door each side pulls out and slides backwards on internal bearings for access.
EQUIPMENT: Landing light in nosecone.

Aviaton Merkury light transport *(Paul Jackson/Jane's)* 1999/0044527

DIMENSIONS, EXTERNAL:	
Wing span	10.20 m (33 ft 5½ in)
Length overall	6.60 m (21 ft 7¾ in)
Height overall	3.20 m (10 ft 6 in)
WEIGHTS AND LOADINGS:	
Max payload	300 kg (661 lb)
Max T-O weight	1,500 kg (3,306 lb)
Max power loading	6.29 kg/kW (10.33 lb/hp)
PERFORMANCE (estimated):	
Normal cruising speed	162 kt (300 km/h; 186 mph)
T-O run	125 m (410 ft)
Range	809 n miles (1,500 km; 932 miles)

UPDATED

BERIEV

BERIEVA AVIATSIONNYI KOMPANIYA (Beriev Aviation Company)
ploshchad Aviatorov 1, 347923 Taganrog
Tel: (+7 86344) 499 01 and 498 39
Fax: (+7 86344) 414 54
e-mail: info@beriev.com
Web: http://www.beriev.com
Telex: 298207 YAKOR
PRESIDENT AND GENERAL DESIGNER: Gennady S Panatov
HEAD OF MARKETING: Andrei Shishko
HEAD OF INFORMATION: Andrei I Salinkov

Original Beriev design bureau (OKB) founded in October 1934 by Georgy Mikhailovich Beriev (1902-79); except during part of the Second World War, 1942-45, it has been based at Taganrog, in northeast corner of Sea of Azov; since 1948 has been primary centre for Russian seaplane development. Beriev has manufactured more than 20 types of aircraft, most of which have entered series production. In 1990 was redesignated Taganrogsky Aviatsionnyi Nauchno-Tekhnichesky Kompleks imeni G M Berieva (TANTK: Taganrog Aviation Scientific-Technical Complex named for G M Beriev); became part of Sukhoi Aviation – Military Industrial Complex in 1996. As a consequence of this merger, the bureau's products are often manufactured at factories

Beriev A-50 ('Mainstay') 86579 in Russian Air Forces service, before delivery to IAI for conversion to A-50I *(Yefim Gordon)* 2001/0093598

traditionally associated with Sukhoi, the Be-103, for example, at Komsomolsk. On 1 January 1998 adopted style, 'Beriev Aviation Company' for international promotion, retaining TANTK in Russia.

Beriev company now includes the experimental design bureau, experimental production facilities, a flight test complex, economic, financial and logistics support services, with test bases and proving grounds at the Black Sea and Sea of Azov. Its products are experimental prototypes of amphibious aircraft and wing-in-ground-effect (WIG) vehicles, together with test reports and technical documentation for their series production. It undertakes design and development of unconventional aircraft in response to requests for proposals from other companies, testing of aircraft and assemblies in maritime conditions, and training of aircrew and ground personnel for seaplane operation.

UPDATED

BERIEV A-50
NATO reporting name: Mainstay
Russian air force designation: Ilyushin Il-76A-50 or Il-76A

TYPE: Airborne early warning and control system.
PROGRAMME: Began in 1969 with the development of the Shmel (Bumblebee) AWACS system, based on the Grib (Mushroom) radar. This intended to replace Tupolev Tu-126 'Moss', with Liana radar, which was eventually retired in 1984. Liana was ineffective against low-level targets; also extremely dangerous to its crew, several of whom suffered symptoms of radiation sickness.

Tu-126, Tu-142M and Tu-154 evaluated as possible platforms for Shmel before a modified Tu-126, the Tu-56, with D-30KP turbofan engines, was drawn up. This had a double-height rotodome pylon and Boeing 747-type fuselage hump. It was also intended to carry Zaslon radar, together with AA-9 AAMs and a tail gun turret with 30 mm cannon. Eventually the decision was taken to use an existing airframe, and Ilyushin Il-76 was selected.

One Tu-126 airframe delivered to Beriev at Taganrog; used as a testbed for the Shmel AWACS system before conversion of the first Il-76. Initial prototype '10' first flew 19 December 1978. As consequence of slow development, early A-50s understood to have entered service with a modified Liana radar, also confusingly, known as Shmel.

First production A-50 ('30') delivered to Siauliai, Lithuania, on 30 December 1983; officially entered service in 1984, when the 67th OAE DRLO relinquished last of its Tu-126s. Of 28 aircraft, one never completed, and first three retained at Taganrog for trials and further development. Two operated round-the-clock over Black Sea during Operations Desert Shield and Desert Storm (1990-91), monitoring USAF operations from Turkey and keeping watch for 'stray' cruise missiles. Operating unit has moved base and changed designation several times, and has operated a number of detachments. It now has 20 improved A-50M/Us at Ivanovo, not all of them serviceable.

CURRENT VERSIONS: **A-50:** Basic version; all now converted to later standards or withdrawn from use.

A-50U: Announced at Moscow Air Show '95. Generally similar to A-50, but with much improved Vega-M Shmel-II AEW&C radar system in place of Liana. This has a passive mode to detect hostile ECM sources without transmission-induced vulnerability, a computer-based three-dimensional pulse Doppler radar, and a digital MTI subsystem that indicates all moving airborne targets, including their altitude. A-50U is said to have entered service with claimed capabilities equivalent to those of Boeing E-3C Sentry AWACS. Search radius of 124 n miles (230 km; 143 miles) is claimed for targets the size of a MiG-21 fighter, or 216 n miles (400 km; 248 miles) for a ship. Ten mission operators in the main cabin can track 50 targets and guide the interception of up to 10 of them simultaneously. Some sources suggest that the A-50U can simultaneously track 100 targets, or provide 'track information' on 300.

Endurance on internal fuel, at maximum T-O weight is 4 hours at 540 n miles (1,000 km; 621 miles) from base. Datalink range to a C³ point is up to 188 n miles (350 km; 217 miles) by VHF/UHF, up to 1,079 n miles (2,000 km; 1,242 miles) by HF, more by satcom. A flight refuelling probe was standard, but extremely poor flight refuelling capabilities led to removal of probes from many A-50 and A-50M/U aircraft. Sole external distinguishing feature between A-50 and the A-50M/U is a small blister fairing on the lower rear fuselage of the latter aircraft. Normally operates on figure-of-eight course at 10,000 m (33,000 ft), with 54 n miles (100 km; 62 miles) between centres of the two orbits.

Description applies to the above version.

A-50M: Said to apply to aircraft with computer upgrade and Shmel-II AEW&C system. Possible alternative designation for A-50U, or intermediate standard. May be one-off test aircraft.

A-50I (*Izraelsky:* Israeli): Conversion of A-50 by Israel Aircraft Industries. IAI planned to install Elta radar as a potential AEW system to fulfill a US$1 billion contract, to meet an eventual Chinese People's Liberation Army Air Force requirement for up to four aircraft. Prototype (RA-78740, the former Russian Air Forces A-50 '44'),

A-50 OPERATING UNITS

Unit	Base	Notes
67 OAE DRLO	Siauliai (Lithuania)	Redesignated 144 AP
144 AP	Siauliai (Lithuania)	Moved to Berezovka by 1987
144 AP	Berezovka	Moved to Ivanovo by 1999
144 AP	Ivanovo	Current
144 AP (Det)	Klin	No longer current
144 AP (Det)	Pechora-Kamenka	No longer current
144 AP (Det)	Ukurei	No longer current
144 AP (Det)	Vitebsk (Belarus)	No longer current

Notes: AP is Aviatsionnyi Polk (Aviation Regiment); OAE DRLO is (Independent Aviation Squadron, Long-Range Airborne Surveillance)

A-50 PRODUCTION

Code	Serial	Initial operating base		Remarks
		Location	Delivered	
10	09243	Taganrog	-	Prototype; de-modified; scrapped at Chortkov
-	10311	Taganrog	-	Ground trainer at Taganrog
20	30875	Taganrog	30 Mar 81	Ground trainer at Taganrog
30	36059	Siauliai	30 Dec 83	Berezovka; Ivanovo (1999)
46	43258	Siauliai	28 Feb 89	Ukurei; Ivanovo (1999)
38	47379	Siauliai	29 Jun 88	Berezovka; Ivanovo (1999)
34	49460	Ukurei	31 Aug 87	*
46	51498	Taganrog	30 Jun 85	May still be at Taganrog, ex-testbed
39	52537	Siauliai	31 Oct 88	Berezovka;*
47	53577	Vitebsk	9 Dec 86	Taganrog (1999)
33	54618	Siauliai	27 May 87	Berezovka; Ivanovo (1999)
48	58738	Vitebsk	30 Sep 86	Berezovka; Ivanovo (1999)
31	59777	Berezovka	31 Dec 86	Ivanovo (1999)
32	66979	Berezovka	20 May 87	*
49	69057	Ukurei	30 Sep 87	Believed lost
35	73178	Berezovka	31 Dec 87	*
36	75260	Berezovka	25 Jan 88	*
37	76298	Berezovka	29 Jun 88	*
43	79377	Berezovka	31 Aug 89	*
40	81457	Berezovka	29 Dec 88	Ivanovo (1999)
41	83499	Siauliai	29 Dec 88	Vitebsk; Ivanovo (1999)
42	84538	Berezovka	30 Jun 89	Ivanovo (1999)
44	86579	Berezovka	31 Mar 90	*Subsequently earmarked for China†
51	88634	Berezovka	15 Jan 91	Ivanovo (1999)
52	91739	Berezovka	30 Apr 92	Ivanovo (1999)
45	93818	Berezovka	29 Sep 90	Ivanovo (1999)
50	96899	Berezovka	28 Dec 91	Ivanovo (1999)
53	97940	-	-	Believed uncompleted

Notes:
*These believed to be the additional eight aircraft parked, unmarked, at Ivanovo in 1999
†Carried registrations RA-78740 and 4X-AGI

sole aircraft so far funded, arrived at Elta's Ben-Gurion facility on 25 October 1999, with rotodome, but without mission equipment. The installation, completed in late 1999, involved a phased-array radar in a non-rotating radome, in a position similar to that of the Beriev and LII '976' described in the following entry but with three distinct antenna segments, one facing forward and occupying front 120° of circumference, others facing each rear/side segment; this still provides 360° coverage. Radome mounted on new broader, constant-chord struts. New aircraft has prominent ventral fins below tailfin and new trapezoidal wingtips. No ram air intake at base of fin. Additional equipment includes ESM, elint and comint systems similar to those of the Boeing 707-based Phalcon sold by IAI to Chile. Following intense US political and economic pressure, Israel cancelled programme in 2000. China reportedly paid US$240 million for first Phalcon system and US$100 million down-payment on first aircraft. Chinese intention allegedly to evaluate A-50I alongside an Il-76 equipped with BAE Systems' Argus radar. Subsequently China said to have joined India (see below) in discussing lease of Russian A-50s. By October 2000, it was reported that China required three aircraft, and that aircrew had already undergone training in Russia.

A-50Eh (*Ehksport:* export): Proposed to China by Moscow Scientific Research Institute of Instrument

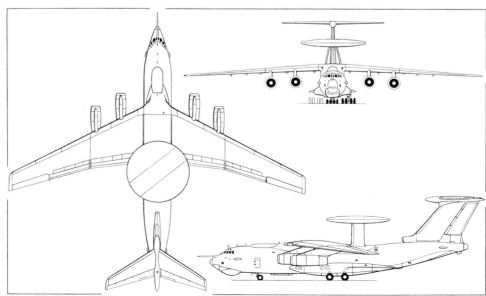

Beriev A-50M AEW&C version of Ilyushin Il-76, known to NATO as 'Mainstay' *(Dennis Punnett/Jane's)*

Engineering (MNIIP) following cancellation of A-50I; cost reportedly US$180 million to US$200 million. Development to take three years; performance data identical to those of A-50U.

CUSTOMERS: About 20 improved A-50M/U aircraft are now based at Ivanovo, and about 16 operate primarily in support of the former IA PVO MiG-29, MiG-31 and Su-27 interceptor squadrons. Beriev reportedly in negotiations with India for lease of three A-50s and in April 2000 a Russian Air Forces A-50 deployed to Chandigarh to participate in an IAF air defence exercise, the aircraft making 10 six-hour flights, funded by India.

COSTS: Given as US$250 million to US$270 million during discussions on India's requirement.

DESIGN FEATURES: Derivative of Il-76MD (which see), with conventionally located rotating 'saucer' radome (diameter 9 m; 29 ft 6 in) on twin de-iced pylons, flight refuelling noseprobe, satellite nav/com antennas in fairing above fuselage forward of wing, new IFF and comprehensive ECM; normal nose glazing around navigator's station replaced by non-transparent fairings; intake for avionics cooling at front of dorsal fin; hemispherical dielectric radome replaces rear gun turret; aerodynamic surface on each outer landing gear fairing, aft of wing trailing-edge, and directly beneath radome, each of which protrudes 2.30 m (7 ft 6½ in).

FLYING CONTROLS: As for Il-76.
STRUCTURE: As for Il-76, except as noted under Design Features.
LANDING GEAR: As for Il-76.
POWER PLANT: As for Il-76. Thrust improvement programme for D-30KP turbofan originally launched specifically for A-50 and Il-78. In-flight refuelling difficult because of severe buffeting induced by rotating radome in tanker's slipstream.
ACCOMMODATION: Five flight crew; 10 systems operators. Rear ramp and cargo provisions deleted.
AVIONICS: Considered capable of detecting and tracking aircraft and cruise missiles flying at low altitude over land and water, and of helping to direct fighter operations over combat areas as well as enhancing air surveillance and defence of RFAS (CIS).
Radar: Weather radar in nose; nav and ground-mapping radar under nose. Shmel AEW&C radar, derived from Tu-126 Liana system, in rotating radome above fuselage.
Flight: Satellite nav/com and satellite datalink to ground stations.
Instrumentation: Colour CRT displays for radar observers.
Self-defence: RWR; flare pack on each side of rear fuselage; wingtip countermeasures pods under development.
EQUIPMENT: LO-82 IR warning receiver for protection against short-range IR-guided air-to-air missiles. New IR launch-warning system detects launch of tactical, medium-range and submarine-launched missiles at ranges up to 540 n miles (1,000 km; 621 miles) with aircraft at 10,000 to 12,000 m (32,800 to 39,370 ft).

WEIGHTS AND LOADINGS:
Weight empty   119,000 kg (262,350 lb)
Max T-O weight   190,000 kg (418,875 lb)

PERFORMANCE:
Max level speed   425 kt (785 km/h; 488 mph)
Service ceiling   10,500 m (34,440 ft)
Range   2,753 n miles (5,100 km; 3,169 miles)

UPDATED

BERIEV Be-32

NATO reporting name: Cuff

TYPE: Twin-turboprop light transport.
PROGRAMME: Development of Be-30, first flown in prototype form on 3 March 1967; eight Be-30s built, but programme terminated when Aeroflot ordered Let L-410As from Czechoslovakia. Hard currency shortage revived programme 1993, with modestly upgraded version known as Be-32; one of original Be-30s (RA-67205) exhibited at 1993 Paris Air Show, as Be-32 demonstrator with original Russian TVD-10B engines; re-engined with PT6A-65B turboprops, RA-67205 first flew as Be-32K on 15 August 1995.

Agreement for licensed production, with PT6As by IAR in Romania (which see), was announced in September 1996, but subsequently denied by IAR. In late 1998, the aircraft was stated to be entering serial production at Taganrog Aviation in expectation of receiving its type certificate. However, in August 1999, flight testing was reportedly still in final stages and production plans remained in preparatory phase; no known developments by late 2000.

Promotion by ZAO Mercury Aviation Scientific and Industrial Company, formed 1999 from TANTK, Taganrog Aviation, Interprofavia and Russian Savings Bank.

CURRENT VERSIONS: **Be-32:** TVD-10B engines; a military version was being promoted in 1998.
Be-32K: PT6 engines. Airliner or business turboprop.
Be-32P: Offered for Border Guards' use (*Pogranichni: Frontier*).

CUSTOMERS: Order for 50 announced by Moscow Airways 1993 (company no longer extant). In March 2000 interest reported from Sverdlovsk regional government (100 to 200), Russian Border Guard (50 to 70), Almazy-Rossii (two) and Gazprom (10). Target for sales is 1,400 between 2000 and 2005, comprising 400 in Russia, 500 to 600 in other CIS countries and 400 abroad.

COSTS: US$2.6 million to 2.8 million (2000).

DESIGN FEATURES: General purpose light transport, optimised for operation by Aeroflot from remote airfields and as military communications aircraft for former Soviet forces. Conventional cantilever high-wing monoplane; three-section wings, with anhedral on outer panels; 42° sweptback vertical tail surfaces; engines at tip of centre-section each side.

Wing section P-20; thickness/chord ratio 18 per cent at centre-section, 14 per cent at tip; 3° twist.

FLYING CONTROLS: Conventional and manual. Trim tabs in port aileron, port elevator and rudder. Double-slotted flaps.
STRUCTURE: All-metal; semi-monocoque fuselage of rectangular section; spars and skin panels of wing torsion box are mechanically and chemically milled profile pressings; detachable bonded leading-edge; half of wings and most of tail unit covered with thin honeycomb panels stiffened with stringers; 70 per cent of fuselage made of adhesive-bonded panels; tips of wings and tail surfaces and wing/fuselage fillets of GFRP.
LANDING GEAR: Tricycle type; single wheel on each unit; nosewheel retracts forward, mainwheels rearward into engine nacelles; tyres size 720×320 on mainwheels, pressure 3.90 bars (57 lb/sq in), 500×150 on nosewheel, pressure 2.95 bars (43 lb/sq in); mainwheel brakes. Optional floats and skis.
POWER PLANT: Two 820 kW (1,100 shp) P&W-Rus PK6A-65B turboprops; three-blade propellers; six integral wing fuel tanks, total capacity 2,250 litres (594 US gallons; 495 Imp gallons).
ACCOMMODATION: Basic seating for two crew and 16 passengers in pairs, seat pitch 73.75 cm (29 in), with centre aisle; other versions include a six-passenger business transport, with a galley; cargo version with cabin 7.10 m (23 ft 3½ in) long, volume 17.0 m³ (600 cu ft), for 1,900 kg (4,190 lb) freight; transport with sidewall seats for 15 paratroops or 17 troops; and ambulance for six stretcher patients, 10 seated casualties and attendant. Carry-on baggage compartment on starboard side, aft of cabin seating, opposite forward-hinged door and airstairs; lavatory to rear.
SYSTEMS: Three-phase AC electrical system of 115/200 V 400 Hz, with two 16 kW GT16P48E alternators; DC supply via VU6SK 25.5 V 12 kW rectifiers. Hot air de-icing system for wing and tail unit leading-edges, engine air intakes and oil cooler; electrothermal system for propeller, spinner and windscreen anti-icing. Cabin and flight deck heated and ventilated. Portable oxygen bottles, masks and smoke protection goggles.
AVIONICS: *Comms:* Com/nav radio, emergency radio, radio beacon buoy.
Radar: Weather radar in nose.
Flight: Automatic flight control system; FON-BP-type air data system.

DIMENSIONS, EXTERNAL:
Wing span   17.00 m (55 ft 9¼ in)
Wing aspect ratio   9.0
Length overall   15.70 m (51 ft 6 in)
Fuselage max width   1.70 m (5 ft 7 in)
Height overall   5.52 m (18 ft 1½ in)
Tailplane span   6.36 m (20 ft 10½ in)
Wheel track   5.20 m (17 ft 0¾ in)
Wheelbase   4.75 m (15 ft 7 in)
Cabin door: Height   1.30 m (4 ft 3 in)
  Width   0.75 m (2 ft 5½ in)
Emergency exits (each): Height   0.92 m (3 ft 0¼ in)
  Width   0.60 m (1 ft 11½ in)

DIMENSIONS, INTERNAL:
Cabin: Length: airliner   6.00 m (19 ft 8¼ in)
  transport   7.10 m (23 ft 3½ in)
Height   1.82 m (5 ft 11¾ in)
Width   1.55 m (5 ft 1 in)
Volume: airliner   13.0 m³ (459 cu ft)
  transport   17.0 m³ (600 cu ft)
Cargo floor area: airliner   3.64 m² (39.2 sq ft)
Baggage hold volume: airliner   1.33 m³ (47.0 cu ft)

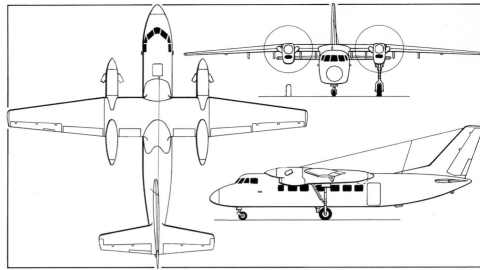
Beriev Be-32K twin-turboprop multipurpose light transport *(Mike Keep/Jane's)* 1999/0051203

Model of proposed military Be-32, retaining original Be-30 power plants *(Yefim Gordon)* 2001/0092520

Beriev Be-32K prototype *(Yefim Gordon)* 2001/0092521

BERIEV—AIRCRAFT: RUSSIAN FEDERATION

AREAS:
Wings, gross 32.00 m² (344.45 sq ft)
Ailerons (total) 2.60 m² (28.00 sq ft)
Trailing-edge flaps (total) 5.80 m² (62.43 sq ft)
Fin 3.14 m² (33.80 sq ft)
Rudder 2.43 m² (26.16 sq ft)
Tailplane 6.28 m² (67.60 sq ft)
Elevators (total) 2.72 m² (29.28 sq ft)
WEIGHTS AND LOADINGS (A: TVD-10B engines, B: PK6A-65B engines):
Weight empty: A 4,760 kg (10,495 lb)
Max payload: A 1,900 kg (4,190 lb)
Max fuel: A 1,700 kg (3,750 lb)
Max T-O weight: A, B 7,300 kg (16,094 lb)
Max landing weight: A, B 6,800 kg (14,991 lb)
Max wing loading: A, B 228.1 kg/m² (46.72 lb/sq ft)
Max power loading: B 4.45 kg/kW (7.32 lb/shp)
PERFORMANCE (estimated):
Never-exceed speed (V_NE):
  A 264 kt (490 km/h; 304 mph)
Max cruising speed at 3,000 m (9,840 ft):
  A 237 kt (440 km/h; 273 mph)
  B 275 kt (510 km/h; 317 mph)
Econ cruising speed at 3,000 m (9,840 ft):
  A 202 kt (375 km/h; 233 mph)
  B 200 kt (370 km/h; 230 mph)
Unstick speed 98 kt (180 km/h; 112 mph)
Landing speed 92 kt (170 km/h; 106 mph)
Min manoeuvring speed: A 79 kt (145 km/h; 90 mph)
Max rate of climb at S/L, 15° flap:
  A 450 m (1,475 ft)/min
Rate of climb at S/L, 15° flap, OEI:
  A 81 m (265 ft)/min
Service ceiling: B 4,200 m (13,780 ft)
T-O 10.7 m (35 ft): A 600 m (1,970 ft)
  B 480 m (1,575 ft)
Landing balanced field length 830 m (2,725 ft)
Landing from 15 m (50 ft): A 620 m (2,035 ft)
  B 500 m (1,640 ft)
Landing run: B 240 m (790 ft)
Range: A, with 17 passengers (1,150 kg; 2,535 lb)
 323 n miles (600 km; 373 miles)
A, with 14 passengers
 518 n miles (960 km; 596 miles)
A, with seven passengers
 944 n miles (1,750 km; 1,087 miles)
B, 30 min reserves, with 1,550 kg (3,417 lb) payload (16 passengers):
 at max cruising speed
 324 n miles (600 km; 372 miles)
 at econ cruising speed
 477 n miles (885 km; 549 miles)
B, business transport, six passengers:
 at max cruising speed
 781 n miles (1,450 km; 901 miles)
 at econ cruising speed
 1,106 n miles (2,050 km; 1,273 miles)
Endurance: B, border patrol mission 5 h 24 min

UPDATED

BERIEV Be-42

Similar to the Beriev A-40, this aircraft project was last described in the 1998-99 edition. Beriev confirmed in 1999 that development is still proceeding, albeit at a slow rate. Prototype 85 per cent complete by September 2000, by which time only Rb19.5 million assigned of Rb307 million required to complete development. Domestic needs are up to eight for military use and up to four for MChS civil protection agency; export interest from Canada, Chile, China, Finland, South Korea and Thailand.

UPDATED

BERIEV Be-103

TYPE: Six-seat utility twin-prop amphibian.
PROGRAMME: Design started 1992; model exhibited and initial data released at Moscow Air Show '92; construction of preproduction batch of four began by KnAAPO at Komsomolsk-on-Amur 1994; prototype (RA-37019), with M-17F engines, displayed statically at Gelendzhik Hydroaviation Show, on the Black Sea, 23 September 1996; first flew at Taganrog on 15 July 1997, but destroyed on its 28th sortie, during practice for Moscow Air Show on 18 August 1997. Second prototype (RA-03002) flew 17 November 1997; made type's first water take-off and landing on 24 April 1998 (aircraft's sixth sortie), but crashed on 30 April 1999 while returning from exhibition at Aero '99, Friedrichshafen; further two structure/static test airframes completed for use at Taganrog by early 1997.

Type certification rescheduled for second half of 1998 and first delivery for 1999, but these dates also not met; development being assisted by three preproduction aircraft, which completed by late 1997, although first of these did not fly until 19 February 1999. Be-103s had flown 180 sorties by August 1999. Further eight production aircraft under construction by September 1998; Be-103 development included, since 1997, in Rostov regional development programme, while funding provided by government's 1999-2000 Light Aircraft Development Programme; KnAAPO announced 20 aircraft under production by second quarter of 1999, and is paying for certification.

Second prototype Beriev Be-103 amphibian *(Paul Jackson/Jane's)*

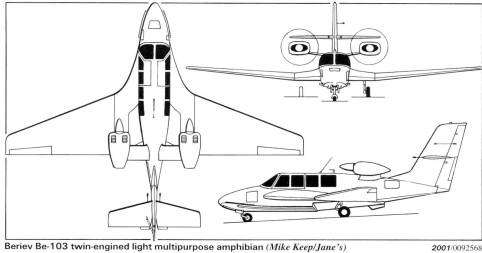
Beriev Be-103 twin-engined light multipurpose amphibian *(Mike Keep/Jane's)*

Marketed by **Takom Avia** consortium, formed by Beriev, KnAAPO and VO Mashinoexport of Moscow. In late 1999, consideration given to establishing final assembly centre in Malaysia, combining Russian airframes and US avionics and engines.
CURRENT VERSIONS: **Be-103** *As described.*
 Single engine: Proposed for Russian Federation market with single 265 kW (355 hp) VOKBM M-14 nine-cylinder radial pylon mounted above the fuselage.
CUSTOMERS: Projected sales of 600 including 230 exports. First order from Russian Border Guards for 20 (out of possible 200 required) for delivery from 1999; Forestry Service requires 30 to 40. First export order announced August 1999 in form of 10 for undisclosed customer.
COSTS: Basic price US$600,000 to US$700,000 (2000), with IO-360 engines. Development costs to 1999 estimated as US$36 million.
DESIGN FEATURES: Low-wing monoplane, with water-displacing wings of moderate sweep with large wingroot extensions; two-step boat hull; no stabilising floats; horizontal strake each side of nose, plus vertical strake ahead of second hull step; wing centre-section increases lift during take-off and landing, partly compensating for absence of flaps. Engine pylon mounted on each side of rear fuselage, aft of wings; engines raised slightly on production version. Sweptback fin and rudder; tailplane mid-set on fin. Designed in compliance with AP-23 and FAR Pt 23 requirements; suitable for passenger and cargo transport, medical evacuation, patrol and ecological monitoring. Operable in wave heights up to 0.5 m (1½ ft) and on water as shallow as 1.25 m (4 ft).

Wing leading-edge sweep 22°; wing section NACA 2412M; dihedral 5° 3' on outer wings; incidence 1°.
FLYING CONTROLS: Manual. All-moving tailplane (with anti-balance tab) and ailerons actuated by pushrods; rudder by cables; two mass balances ahead of tailplane; mass balance (port side only) on rudder, which has flight-adjustable trim tab; electric trim; spring feel in tailplane control; no flaps.
STRUCTURE: All-metal semi-monocoque boat-type fuselage; all-metal single-spar wings. Extensive use of 1446 aluminium-lithium alloy.
LANDING GEAR: Pneumatically retractable tricycle type; single wheel on each unit; mainwheels retract forward into wing centre-section; nosewheel retracts forward; oleo-nitrogen

Beriev Be-103 cockpit *(Yefim Gordon)*

shock-absorbers; brakes on mainwheels; nosewheel tyre size 400×150, mainwheel tyres 7.00-6; pressure 3.00 bars (43 lb/sq in) in all units; disc brakes; self-centring nosewheel steerable ±35°.

POWER PLANT: Two 157 kW (210 hp) Teledyne Continental IO-360-ES4 or 129 kW (173 hp) VOKBM/Bakanov M-17F piston engines, latter driving AV-103 three-blade variable-pitch MTV-12-D-C-F-R-(M)/CFR183-17 tractor propellers, with optional reversible pitch; fuel tank in each wingroot, total capacity 450 litres (119 US gallons; 99.0 Imp gallons).

ACCOMMODATION: Pilot and five passengers in pairs; dual controls optional; large upward-opening door on each side, hinged on centreline; folding seats facilitate entry/loading baggage/freight compartment aft of cabin. Optional ambulance configuration for pilot, one stretcher, three seated persons; all-cargo configuration for items up to 2.00 × 0.70 × 0.70 m (6 ft 6¾ in × 2 ft 3½ in × 2 ft 3½ in) in size; patrol; agricultural and ecological monitoring use.

SYSTEMS: Interior heated and ventilated. Pneumatic system, bottle capacity 6 litres (365 cu in) of air, pressure 49 bars (711 lb/sq in). Electrical system 27 V DC and three-phase 36 V 400 Hz AC, supplied by 3 kW DC generators, rectifiers and 25 Ah battery. Fire suppression system. Optional anti-icing.

AVIONICS: Integrated by Honeywell. VFR standard; IFR optional. Version with Russian avionics may be offered in the future.

Comms: VHF com radio, intercom and emergency locator beacon.

Radar: Optional weather radar.

Flight: Digital flight control and nav equipment; position, angle rate, linear acceleration and angle of attack sensors. Optional autopilot and GPS.

Instrumentation: Conventional.

EQUIPMENT: Landing light in each wing leading-edge.

DIMENSIONS, EXTERNAL:	
Wing span	12.72 m (41 ft 9 in)
Wing chord: at root	6.21 m (20 ft 4¾ in)
at tip	0.83 m (2 ft 9 in)
Wing aspect ratio	6.4
Length: overall	10.65 m (34 ft 11¼ in)
fuselage	9.96 m (32 ft 8 in)
Height overall	3.76 m (12 ft 4 in)
Tailplane span	3.90 m (12 ft 9½ in)
Wheel track	2.91 m (9 ft 6½ in)
Wheelbase	4.055 m (13 ft 3¾ in)
Propeller diameter	1.78 m (5 ft 10 in)
Distance between propeller centres	3.00 m (9 ft 10¼ in)
DIMENSIONS, INTERNAL:	
Cabin: Length	3.65 m (11 ft 11¾ in)
Max width	1.25 m (4 ft 1¼ in)
Max height	1.23 m (4 ft 0½ in)
Baggage compartment: Max width	1.08 m (3 ft 6½ in)
Max height	0.88 m (2 ft 10½ in)
Volume	0.85 m³ (30.0 cu ft)
AREAS:	
Wings, gross	25.10 m² (270.2 sq ft)
Ailerons, total	0.80 m² (8.61 sq ft)
Fin	2.86 m² (30.78 sq ft)
Rudder	1.54 m² (16.58 sq ft)
Tailplane, total	3.68 m² (39.61 sq ft)
WEIGHTS AND LOADINGS:	
Weight empty, equipped	1,540 kg (3,395 lb)
Max payload	385 kg (849 lb)
Max T-O weight	2,270 kg (5,004 lb)
Max wing loading	90.4 kg/m² (18.52 lb/sq ft)
Max power loading	7.25 kg/kW (11.91 lb/hp)
PERFORMANCE (estimated):	
Max level speed	154 kt (285 km/h; 177 mph)
Max cruising speed	140 kt (260 km/h; 162 mph)
Service ceiling	3,000 m (9,840 ft)
T-O run: on land	310 m (1,020 ft)
on water	440 m (1,445 ft)
Landing run: on land	240 m (790 ft)
on water	320 m (1,050 ft)
Range:	
with max payload	270 n miles (500 km; 310 miles)
with max fuel	2,315 n miles (1,250 km; 2,664 miles)

UPDATED

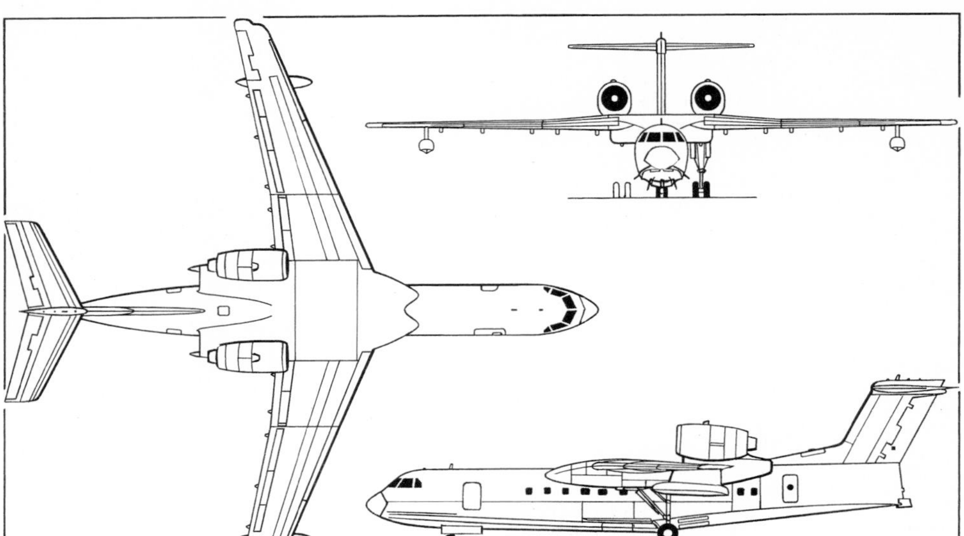

Beriev (Betair) Be-200 civil utility amphibian *(Mike Keep/Jane's)* ***1999***/0051205

BERIEV (BETAIR) Be-200

TYPE: Twin-jet amphibian.

PROGRAMME: Initiated 1989 under design leadership of Alexander Yavkin. Russian government approval for purpose-designed water bomber granted 8 December 1990. Details announced, and model displayed, at 1991 Paris Air Show; full-scale mockup constructed 1991; development by Beta Air (Betair: **Be**riev **ta**ganrog **ir**kutsk) consortium, formed 27 November 1991 from Irkutsk Aircraft Production Association (53.5 per cent), Beriev OKB (20.5 per cent), ILTA Trade Finance SA of Geneva, Switzerland (5 per cent) and the Ukrainian bank Prominvest (15 per cent) plus unidentified partner(s) (6 per cent); included in Civil Aviation Development Programme and state forestry protection programme.

Beriev responsible for development, design and documentation; systems bench-testing; static-, flight- and fatigue-testing of prototypes; certification and design support of serial production. IAPO duties comprise production preparation; manufacture of tooling; production of four prototypes and series aircraft; and spare parts manufacture.

First prototype (001) rolled out 11 September 1996, at Irkutsk; first flight scheduled 1997, but eventually achieved (from land) 24 September 1998; 'official' first flight 17 October 1998; aircraft transferred from Irkutsk to Taganrog (by then with 19 sorties and 26½ flying hours) on 27 April 1999 to begin certification trials, including water drops; exhibited at Paris (by then registered RA-21511) June 1999; first water landings and take-offs, 10 September 1999. Experimental category certification achieved March 2000. Total 80 sorties, including 18 from water, by November 1999, when water operations ceased for winter.

Second flying prototype (003) in firefighting configuration (Be-200ChS). Two test airframes, of which first ferried from Irkutsk to Taganrog by An-124 in March 1995, followed by second in 1997; these, respectively, static test (SI : *staticheskiy ispytaniya*) and fatigue test (RI : *resursnye ispytaniya*). Four production aircraft (0101 to 0104) in hand by late 1998. On original schedule, second prototype was due for delivery in late 1999 to Russian Emergencies Ministry. Initial production and certification supported by Alenia of Italy under US$1.6 million programme financed by European Union. Deliveries of firefighting version scheduled to begin 2000, followed by search and rescue version. By mid-2000, first flight date for second prototype had slipped to late 2000, coincident with Russian certification; international certification predicted in late 2001.

Avionics development and scoop trials for water bomber version being undertaken by the Beriev Be-12P-200/Be-128 testbed since August 1996. TANTK also produced four Be-12Ps (*Pozharny:* firefighting) for concept and tactics development and two Be-12NKh (*Narodno-Khozyaistvennye:* National Economy) general purpose versions, both of which lost in 1994.

CURRENT VERSIONS: **Firefighting;** TsENTROSPAS/MChS designation **Be-200ChS:** Tanks under cabin floor of centre-fuselage, capacity 12 m³ (423 cu ft) water; six tanks in cabin for 1.2 m³ (42 cu ft) liquid chemicals; two retractable water scoops forward of step, two aft; 30 fully equipped smoke jumpers can be carried on seats along sidewalls of cabin, with jump-door at rear of cabin on starboard side; 12 tonnes of water scooped from seas in 14 seconds with waves up to 1.2 m (4 ft).

Fully fuelled, Be-200 can drop total 310,000 kg (683,420 lb) of water in successive flights when airfield to reservoir distance is 108 n miles (200 km; 125 miles) and reservoir to fire zone distance is 5.4 n miles (10 km; 6.2 miles); or 140,000 kg (308,640 lb) when distances are respectively 108 n miles (200 km; 125 miles) and 27 n miles (50 km; 31 miles). Tank emptying time 0.8 to 1.0 second; minimum dropping speed 119 kt (220 km/h; 137 mph). Tanks quickly removable when aircraft carries freight. Flight deck and cargo hold sealed against smoke ingress. ARIA-2000 avionics features include water

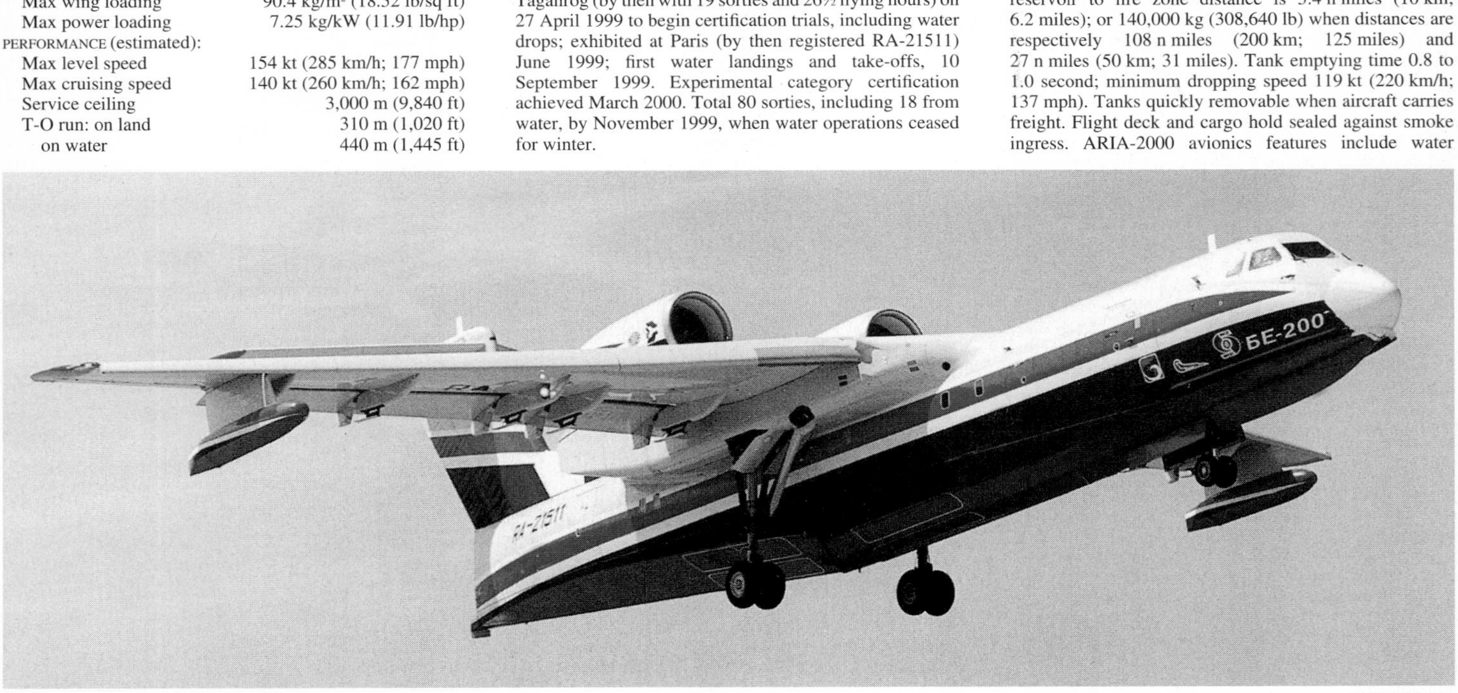

Beriev Be-200 multirole amphibian ***2001***/0092572

source/drop zone track memory, automatic glide slope and digital flight deck/ground fire crew communications.

Passenger: Two flight crew, two cabin attendants, and 72 tourist class passengers four-abreast in pairs, with centre aisle, at seat pitch of 75 cm (29.5 in).

Cargo: Payload 7,500 kg (16,534 lb) in unobstructed cabin 17.00 m (55 ft 9 in) long, 2.6 m (8 ft 6 in) wide and 1.9 m (6 ft 3 in) high.

Ambulance: Two flight crew, seven seated casualties/medical personnel, 30 stretchers in three tiers.

Search and rescue: Sensors and searchlights for detecting survivors, and onboard medical equipment.

Be-200M: Redesignated Be-210 in 1998.

Be-200P: Projected anti-submarine version; combat radius 2,537 n miles (4,700 km; 2,920 miles); endurance 7 hours.

Be-210: Described separately.

CUSTOMERS: Seven ordered (of 20 required) by TsENTROSPAS/MChS (Russian Ministry of Emergency Situations) on 15 January 1997 for firefighting, with further orders to follow for SAR; will equip four SAR stations on Black Sea and Baltic and two in eastern Russia. Russian state forest service/firefighting agency requires up to 54, of which near-term needs total 10 to 15; 50 required by Sakhalin regional administration; five by Irkutsk regional administration. However, at time of maiden flight, only five orders were being claimed as firm. Potential market foreseen for more than 400 by 2010, of which over 60 per cent for export. South Korea evinced interest in 12 of maritime patrol version for police duties during 1998; China considering licensed production at Harbin; Italian interest in 15 reported mid-2000.

COSTS: Basic price US$25 million (2000). Break-even at 46/48th aircraft.

DESIGN FEATURES: First Russian purpose-designed water-bomber. Derived from Beriev A-40, but underwing stabilising floats moved inboard from tips, twin-wheel main landing gear units, and no booster turbojets. Conforms to FAR Pt 25 criteria. Swept wings of moderate aspect ratio, with high-lift devices; single-step hull of high length/beam ratio, which reported to provide world's first variable-rise bottom, providing a considerable improvement in stability and controllability in the water, as well as a reduction in g loads when landing and taking off at sea; small wedge-shape boxes ('hydrodynamic compensators') aft of step aid 'unsticking' from water in wave heights up to 2.2 m (7 ft 2½ in); all-swept T tail; high-mounted engines protected from spray by strakes on each side of nose and by wings; large underwing pod each side of hull, faired into wingroot.

Wing leading-edge sweep 23° 13'; supercritical wing sections, thickness/chord ratio 16 per cent to 11.5 per cent.

FLYING CONTROLS: Conventional. Entire span of each wing trailing-edge occupied by aileron and two-section area-increasing single-slotted flaps; full-span leading-edge slats in three sections each side; five spoiler/lift dumper sections forward of flaps each side.

STRUCTURE: Hull made primarily of high-strength aluminium/lithium alloys; interior of composites; water tanks of ferric alloys of aluminium in firefighting version.

LANDING GEAR: Hydraulically retractable tricycle type. Twin-wheel main units, tyre size 950×300, pressure 9.80 to 10.30 bars (142 to 150 lb/sq in); twin nosewheels, tyre size 620×180, pressure 7.35 to 7.85 bars (106 to 114 lb/sq in). Mainwheels and nosewheels retract rearwards. Water rudder. Ground turning radius 17.4 m (57 ft 1 in). Nosewheel steering angle ±45°.

POWER PLANT: Two ZMKB Progress D-436TP turbofans, each 73.6 kN (16,550 lb st). Rolls-Royce Deutschland BR715s and others under consideration as alternative engines. Fuel system by Pall (USA). Total oil capacity 22 litres (5.8 US gallons; 4.85 Imp gallons).

ACCOMMODATION: Two flight crew; up to 72 tourist class passengers at 75 cm (29.5 in) seat pitch and two attendants; or 10 to 32 first class and business class passengers at up to 102 cm (40 in) seat pitch, with provision for galley, lavatory and baggage stowage. Up to nine freight containers or, typically, six containers and 19 economy class seats. Cargo door, with integral passenger door, starboard side, forward, opens upwards; passenger door port, forward; emergency exits; forward and rear freight/baggage holds. Interior design by AIM Aviation (UK).

SYSTEMS: Barco Display Systems FMS system. All accommodation pressurised. Three hydraulic systems at 207 bars (3,000 lb/sq in); 150 litres (39.6 US gallons; 33.0 Imp gallons) of MGJ-5U fluid; flow rate 70 litres (18.5 US gallons; 15.4 Imp gallons)/min. Capacity of pneumatic system bottles 29 litres (1.02 cu ft). Three-phase 115/220 V 400 Hz AC electrical system; single-phase 115 V 400 Hz AC system; 27 V DC system; supplied by two engine-driven 60 kVA AC generators and three static inverters; three batteries. Gaseous oxygen bottle, pressure 147 bars (2,135 lb/sq in). Provision for de-icing tail unit, slats, engine air intakes and windscreen. TA-12 APU, operable up to 7,000 m (23,000 ft) in starboard wingroot. Propeller-driven emergency generator at base of fin.

AVIONICS: *Radar:* MN-85 weather radar in nose.

Flight: ARIA-200 digital flight and navigation system by American-Russian Integrated Avionics, a joint venture of AlliedSignal (now Honeywell) and Moscow Research Institute of Aircraft Equipment. INS standard.

Instrumentation: Honeywell EFIS displays, with six 152 × 203 mm (6 × 8 in) LCDs.

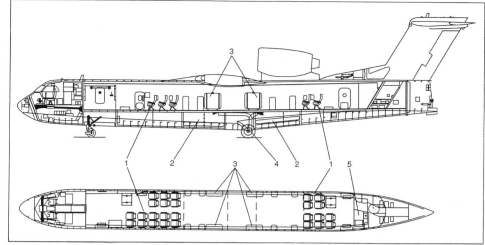

Interior of the firefighting version of Beriev (Betair) Be-200, showing (1) firefighters' seats, (2) water tanks, (3) chemical retardant tanks, (4) water scoop and (5) toilet

DIMENSIONS, EXTERNAL:
Wing span over winglets	32.78 m (107 ft 6½ in)
Wing chord: at root	5.58 m (18 ft 3½ in)
at tip	1.72 m (5 ft 7¾ in)
Wing aspect ratio	9.1
Length overall	32.05 m (105 ft 1¾ in)
Fuselage: Length	29.18 m (95 ft 9 in)
Max diameter	2.86 m (9 ft 4½ in)
Height overall	8.90 m (29 ft 2½ in)
Tailplane span	10.115 m (33 ft 2¼ in)
Wheel track (c/l shock-absorbers)	4.30 m (14 ft 1¼ in)
Wheelbase	11.145 m (36 ft 6¾ in)
Width over stabilising floats	25.60 m (84 ft 0 in)
Passenger doors (each): Height	1.70 m (5 ft 7 in)
Width	0.90 m (2 ft 11½ in)
Cargo door: Height	1.80 m (5 ft 10¾ in)
Width	2.00 m (6 ft 6¾ in)
Emergency exits: Height	1.70 m (5 ft 7 in)
Width	0.90 m (2 ft 11½ in)

DIMENSIONS, INTERNAL:
Cabin, excl flight deck: Length	17.00 m (55 ft 9 in)
Max width: passenger	2.40 m (7 ft 10½ in)
cargo	2.60 m (8 ft 6¼ in)
Max height: passenger	1.80 m (5 ft 10¾ in)
cargo	1.90 m (6 ft 2¾ in)
Floor area	39.0 m² (420 sq ft)
Volume: forward baggage hold	8.8 m³ (310 cu ft)
rear baggage hold	4.5 m³ (159 cu ft)
main cabin and freight/baggage holds, total, cargo configuration	84.0 m³ (2,966 cu ft)

AREAS:
Wings, gross	117.44 m² (1,264.2 sq ft)
Ailerons (total)	3.56 m² (38.32 sq ft)
Flaps (total)	20.43 m² (219.91 sq ft)
Slats (total)	12.61 m² (135.74 sq ft)
Spoilers (total)	4.59 m² (49.41 sq ft)
Fin	12.60 m² (135.63 sq ft)
Rudder	4.60 m² (49.52 sq ft)
Tailplane	17.96 m² (193.33 sq ft)
Elevators (total)	6.96 m² (74.92 sq ft)

WEIGHTS AND LOADINGS:
Max payload	7,500 kg (16,534 lb)
Max fuel weight	12,260 kg (27,025 lb)
Max T-O and ramp weight:	
cargo/passenger	42,000 kg (92,594 lb)
firefighting	37,200 kg (82,011 lb)
Max airborne weight (after water scooping)	43,000 kg (94,800 lb)
Max landing weight, land or water	35,000 kg (77,160 lb)
Max wing loading:	
cargo/passenger	357.6 kg/m² (73.25 lb/sq ft)
firefighting	316.5 kg/m² (64.88 lb/sq ft)
max airborne weight	366.1 kg/m² (74.99 lb/sq ft)

Be-200 releasing water

Max power loading:
cargo/passenger 285 kg/kN (2.80 lb/lb st)
firefighting 253 kg/kN (2.48 lb/lb st)
max airborne weight 292 kg/kN (2.86 lb/lb st)
PERFORMANCE (A: passenger version; B: business; C: cargo/passenger; D: cargo; E: firefighting):
Max Mach No. in level flight: A, B, C, D 0.69
Never-exceed speed (V_{NE}):
A, B, C, D 329 kt (610 km/h; 379 mph) EAS
Max level speed at 7,000 m (22,960 ft):
A, B, C, D 388 kt (720 km/h; 447 mph)
Max cruising speed:
A, C, D at 8,000 m (26,240 ft)
378 kt (700 km/h; 435 mph)
B at 10,000 m (32,800 ft) 378 kt (700 km/h; 435 mph)
E at 3,000 m (9,840 ft) 383 kt (710 km/h; 441 mph)
Econ cruising speed:
A, C, D at 8,000 m (26,240 ft)
297 kt (550 km/h; 342 mph)
B at 10,000 m (32,800 ft) 297 kt (550 km/h; 342 mph)
E at 3,000 m (9,840 ft) 324 kt (600 km/h; 373 mph)
Stalling speed: flaps up 116 kt (215 km/h; 134 mph)
flaps down 84 kt (155 km/h; 97 mph)
Max rate of climb at S/L 840 m (2,755 ft)/min
Rate of climb at S/L, OEI 168 m (550 ft)/min
Service ceiling: A, B, C, D 11,000 m (36,080 ft)
E 8,000 m (26,240 ft)
Service ceiling, OEI 5,500 m (18,040 ft)
T-O to 11 m (35 ft): on land: A, B, C, D 950 m (3,120 ft)
E 700 m (2,300 ft)
on water: A, B, C, D 1,400 m (4,595 ft)
E 1,000 m (3,280 ft)
Landing from 15 m (50 ft): on land: A, B, C, D
800 m (2,625 ft)
E 950 m (3,120 ft)
On water: A, B, C, D 950 m (3,120 ft)
E 1,300 m (4,265 ft)
Balanced runway requirement: all 1,800 m (5,905 ft)
Water scooping distance to 15 m (50 ft)
1,450 m (4,760 ft)
Range:
A with 72 passengers
998 n miles (1,850 km; 1,149 miles)
C, D with 6,500 kg (14,330 lb) freight, 1 h reserves
998 n miles (1,850 km; 1,149 miles)
E, ferry, with 1 h reserves
1,943 n miles (3,600 km; 2,236 miles)
all, with max fuel
2,078 n miles (3,850 km; 2,392 miles)
OPERATIONAL NOISE LEVELS:
T-O 96 EPNdB
Climb 90 EPNdB
Landing 98 EPNdB
UPDATED

BERIEV (BETAIR) Be-210
TYPE: Amphibious airliner.
PROGRAMME: Announced in 1998; development of Be-200 multirole amphibian for airline use in regions of minimal airport infrastructure. Seating for two pilots, two attendants and 72 passengers at 75 cm (29½ in) seat pitch. Other details may be assumed to be similar to Be-200, but airframe strengthened for extra fuel in wing centre-section (42,000 kg; 92,594 lb MTOW) and ferry range increased to 2,483 n miles (4,600 km; 2,858 miles), or 998 n miles (1,850 km; 1,149 miles) with full passenger load. By 2000, Beriev quoting 72-passenger capacity for standard Be-200, implying fusion of two types.
UPDATED

Beriev Be-976 Il-76SKIP range control and surveillance aircraft 76455 at Zhukovsky in 1999 *(Paul Jackson/Jane's)* 2000/0062727

BERIEV Be-976
Alternative designation: Il-76SKIP
TYPE: Trials support platform.
PROGRAMME: Developed via two interim conversions of 'Candids' Il-76 SSSR-86721 'Samolet 676' (Aircraft 676) and Il-76M SSSR-86024 'Samolet 776' (Aircraft 776). These had an L-shaped antenna on each side of the tailfin; long radomes in place of tail turret; and variety of new communications and ECM antennas. Uses included receipt of downlinked telemetry from missile test launches, and limited communications relay.

Aircraft 976, by contrast, designed as dedicated test range radar picket for observing flight tests of aircraft and missiles and for use as range control aircraft, uses same Shmel surveillance radar, satcom and datalink equipment as the basic A-50. Nicknamed *Pogahnka* (toadstool). Its initial 'cover story' was that it was an "airborne laboratory for studying the chemical composition of the atmosphere". Five definitive Be-976s converted before delivery to Zhukovsky test centre (see table), probably intended for use at ranges in what are now independent republics, including Kazakhstan and Turkmenistan. Only one aircraft (76453) shows definite evidence of having flown since 1992, when the 'RA-' registration prefix was introduced, but by 1999, when it was displayed at Zhukovsky, fourth aircraft was marked only as '76455'.

Capabilities are described as remote monitoring of flight test telemetry; monitoring of test aircraft and missile trajectories; radio control of test vehicles; real-time processing and display of test data; rebroadcast of test data via satellite and terrestrial datalinks.
CURRENT VERSIONS: **Be-976/Il-76SKIP:** Designation applied to baseline aircraft. SKIP stands for *Samoletnii Komandno-Izmeritelnii Punkt* (Airborne Control and Measurement System).
Description applies to above version.
Il-76SK: Proposed development (probably to be converted from RA-76453) as Burlak sub-orbital launcher monitoring platform. SK stands for *Spetsialnii Komandnii Samolet* (Special Command Aircraft).
DESIGN FEATURES: Externally similar to A-50 except for retention of navigator's station in nose; Il-76PP-type wingtip avionics pods; new thimble radome in place of tail turret; and non-installation of aerodynamic surfaces that project from A-50's landing gear fairings, RWR, ECM and chaff/flare packs.
AVIONICS: Installations include six-channel telemetry system, capacity 2×10^6 baud, range 540 n miles (1,000 km; 621 miles); 90 channel radio command system; trajectory measuring system, accurate to 30 m (100 ft); information processing and display equipment; and data transmission and relay facilities.
PERFORMANCE:
Endurance 8 h
UPDATED

Be-976 CONVERSIONS

Registration	Delivery	Remarks
SSSR-76452	27 May 87	Withdrawn from service by 1995
SSSR-76453	8 Sep 87	Reregistered RA-76453
SSSR-76454	30 Mar 88	Withdrawn from service by 1995
SSSR-76455	18 May 89	Marked 76455 (only) 1999
SSSR-76456	23 Oct 89	Withdrawn from service by 1995

CHERNOV
OPYTNYI KONSTRUKTORSKOYE BYURO CHERNOV B & M OOO (B & M Chernov Experimental Design Bureau JSC)
ulitsa Elevatornaya 6, korpus 3, 115492 Moskva
Tel: (+7 095) 250 29 91
Fax: (+7 095) 250 21 77

The designs of Boris Chernov have previously appeared under the heading of SKB-1 at Samara (which see). However, in August 1999, the Che-25 amphibian was shown at MAKS '99, credited to Chernov's own design office. It is anticipated that manufacture will be at Samara under the name Hydroplane.
UPDATED

CHERNOV Che-15
TYPE: Side-by-side ultralight/kitbuilt.
PROGRAMME: First flight achieved by 1998. Continues to be promoted by SKB-1 (see SGAU entry).
COSTS: US$12,000, flyaway; kit also available.
DESIGN FEATURES: Open pod for occupants slung below wing/tailboom unit; folding wings; downturned wingtips.
FLYING CONTROLS: Manual. Flaperons along entire trailing-edge of wing.
LANDING GEAR: Tricycle type; fixed. Alternatively skis or twin single-stepped floats.
POWER PLANT: One 47.8 kW (64.1 hp) Rotax 582 UL-2V.
DIMENSIONS, EXTERNAL:
Wing span 9.80 m (32 ft 1¾ in)
Length overall 7.03 m (23 ft 0¾ in)
WEIGHTS AND LOADINGS:
Weight empty 250 kg (551 lb)
Max T-O weight 480 kg (1,058 lb)

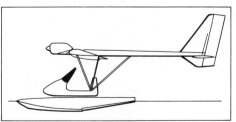

Side elevation of Chernov Che-15 floatplane *(Paul Jackson/Jane's)* 0015088

PERFORMANCE:
Cruising speed 54 kt (100 km/h; 62 mph)
Landing speed 33 kt (60 km/h; 38 mph)
T-O run: land 50 m (165 ft)
water 30 m (100 ft)
Range with max fuel 135 n miles (250 km; 155 miles)
g limits +4/-2
UPDATED

CHERNOV Che-22 CORVETTE
US marketing name: Pelican
TYPE: Three-seat flying-boat/kitbuilt.
PROGRAMME: Developed on basis of four earlier amateur-built, Vikhr-engined Che-20s, winners of many light aircraft competitions; design and construction of prototype started 1988, first flight 1989; construction of production aircraft started 1992, first flight 1993; more than 5,000 hours of test and training flights, and more than 120 amateur pilots trained on prototype and production aircraft

Prototype Che-15 2001/0099582

in first five years. Promotion now undertaken by Refly (which see in this section); launched as Refly Pelican at EAA Air Venture, Oshkosh, 2000.
CURRENT VERSIONS: **Che-22R:** With one Rotax 582 UL two-cylinder two-stroke engine.
Che-22R-2: With two Rotax 503 UL two-cylinder two-stroke engines.
Che-22D: With one Rotax 912 UL four-cylinder four-stroke engine.
Refly Pelican: Generally as for Che-22R.
CUSTOMERS: First preproduction delivery to a forest control organisation 1990. One (Che-22R-2 c/n 006, RP-X1548 of OVOS WLL Co) sold in Philippines. Others to China and Spain. Total 13 built by mid-1999.
COSTS: Standard aircraft US$45,000 (2000); kit US$10,000 (2000) or US$18,000 for fast build, excluding engine.
DESIGN FEATURES: Designed to JAR-VLA, FAR Pt 23 and AP-23 requirements, for maximum simplicity of design and construction; primarily for forestry applications, but suitable for patrol, inspection and ecological monitoring of

CHERNOV—AIRCRAFT: RUSSIAN FEDERATION

hunting and fishing locations, border patrol, training and business operations. Quoted build time 1,500 hours, or 50 hours for fast build kit.

Braced parasol monoplane, with light central cabane and single main bracing strut and bracing wire each side; downturned wingtips serve as stabilising floats; flying boat hull with two steps; unswept strut-braced horizontal tail surfaces mounted on fin of sweptback vertical surfaces; engine(s) mounted on leading-edge of wing centre-section. US demonstrator has fin of reduced height. Unswept constant chord wing; CAHI (TsAGI) P-IIIA section, thickness/chord ratio 15 per cent; incidence 4° 30′; dihedral 1° 30′. Airframe life 3,000 flying hours.

FLYING CONTROLS: Conventional and manual. Full-span slotted flaperons, elevators and horn-balanced rudder. No tabs.

STRUCTURE: Single-spar wing of riveted duralumin; fuselage, tail unit and wingtip floats of vacuum-moulded GFRP sandwich.

LANDING GEAR: Optional tailwheel type; mainwheels, tyre size 350×120, carried on cantilever GFRP spring. GFRP skis optional.

POWER PLANT: One 47.8 kW (64.1 hp) Rotax 582 UL or 59.6 kW (79.9 hp) Rotax 912 UL, or two 37 kW (49.6 hp) Rotax 503 UL-2V or 38 kW (51 hp) Rotax 462 piston engine(s). Two-blade wooden or three-blade GFRP fixed-pitch propeller(s) of Russian manufacture. Two fuel tanks in wing centre-section; total capacity 80 litres (21.1 US gallons; 17.6 Imp gallons) or (Pelican) 100 litres (26.4 US gallons; 22.0 Imp gallons); gravity fuelling.

ACCOMMODATION: Pilot and two passengers under large glazed blister canopy, giving exceptional field of view. Gull-wing canopy/door each side.

SYSTEMS: 12/24 V electrical system.

AVIONICS: *Comms:* VHF radios.
Flight: Garmin GPS.

Che-22R-2 flying-boat (two Rotax 503 UL-2V piston engines)

DIMENSIONS, EXTERNAL:
Wing span	11.70 m (38 ft 4¾ in)
Wing chord, constant	1.40 m (4 ft 7¼ in)
Wing aspect ratio	8.3
Length overall	6.70 m (21 ft 11¾ in)
Height overall	2.20 m (7 ft 2¾ in)
Tailplane span	2.40 m (7 ft 10½ in)
Wheel track	1.80 m (5 ft 10¾ in)
Wheelbase	1.40 m (4 ft 7¼ in)
Propeller diameter	1.50 m (4 ft 11 in)
Canopy doors (each): Height	0.70 m (2 ft 3½ in)
Width	0.80 m (2 ft 7½ in)
Height to sill	0.63 m (2 ft 0¾ in)

DIMENSIONS, INTERNAL:
Cabin: Length	1.50 m (4 ft 11 in)
Max width	1.10 m (3 ft 7¼ in)
Max height	1.20 m (3 ft 11¼ in)

AREAS:
Wings, gross	16.40 m² (176.5 sq ft)
Flaperons (total)	3.40 m² (36.60 sq ft)
Fin	0.55 m² (5.92 sq ft)
Rudder	0.35 m² (3.77 sq ft)
Tailplane	1.68 m² (18.08 sq ft)
Elevators (total)	0.80 m² (8.61 sq ft)

WEIGHTS AND LOADINGS (A: Che-22R, B: Che-22R-2, C: Che-22D):
Weight empty: A	330 kg (728 lb)
B	360 kg (794 lb)
C	340 kg (750 lb)
Max payload: A	190 kg (419 lb)
B	160 kg (353 lb)
C	180 kg (397 lb)
Max fuel: A, B, C	58 kg (128 lb)
Max T-O weight: A, B, C	650 kg (1,433 lb)
Max wing loading: A, B, C	39.6 kg/m² (8.12 lb/sq ft)
Max power loading: A	13.60 kg/kW (22.34 lb/hp)
B	8.78 kg/kW (14.43 lb/hp)
C	10.91 kg/kW (17.92 lb/hp)

PERFORMANCE:
Never-exceed speed (V_{NE}): A, B, C	108 kt (200 km/h; 124 mph)
Max level speed: A	86 kt (160 km/h; 99 mph)
B	91 kt (170 km/h; 105 mph)
C	97 kt (180 km/h; 111 mph)
Max cruising speed: A	64 kt (120 km/h; 74 mph)
B, C	70 kt (130 km/h; 80 mph)
Econ cruising speed: B, C	59 kt (110 km/h; 68 mph)
Stalling speed:	
flaperons up: A, B, C	38 kt (70 km/h; 44 mph)
flaperons down: A, B, C	33 kt (60 km/h; 38 mph)
Max rate of climb at S/L: A	210 m (689 ft)/min
B, C	300 m (985 ft)/min
Rate of climb at S/L, OEI: B	120 m (394 ft)/min
Service ceiling: B	4,000 m (13,120 ft)
T-O run: on land: A	70 m (230 ft)
B	50 m (165 ft)
C	60 m (200 ft)
on water: A	90 m (295 ft)
B	70 m (230 ft)
C	80 m (265 ft)
T-O to 15 m (50 ft): on water: B	150 m (495 ft)
Landing from 15 m (50 ft): on water: B	300 m (990 ft)
Range with standard fuel, 30 min reserves:	
A	296 n miles (550 km; 341 miles)
B	232 n miles (430 km; 267 miles)
C	485 n miles (900 km; 559 miles)
Range with auxiliary fuel:	
A	647 n miles (1,200 km; 745 miles)
B	491 n miles (910 km; 565 miles)
C	971 n miles (1,800 km; 1,118 miles)
g limits	+3.8/−1.5
Permissible wave height	0.5 m (1 ft 7¾ in)

UPDATED

CHERNOV Che-25

TYPE: Four-seat amphibian.

PROGRAMME: Development of Che-22 Corvette. Prototype (FLARF-17039) built by SKB-1 (see SGAU entry in this section) first shown at Gelendzhik Hydro-aviation Show, Russia, September 1996. Aircraft shown at MAKS '99, Moscow, had redesigned empennage with raised tailplane and deletion of rudder's horn balance.

COSTS: US$44,000 (1997).

DESIGN FEATURES: Generally as Che-22. Operable on water in 0.8 m (30 in) swell.

FLYING CONTROLS: Manual, with full-span slotted flaperons. No tabs.

STRUCTURE: Single-spar wing of riveted duralumin; fuselage, tail unit and wingtip floats of vacuum-moulded GFRP sandwich. Wings braced by struts and wires; strut-braced tailplane. Airframe life 3,000 hours.

LANDING GEAR: Optional amphibian configuration; mainwheel legs, attached to fuselage sides, turn 90° forward for retraction; mainwheel tyre size 300×125; tailskid; cable brakes. Tailwheel/water rudder immediately to rear of second hull step.

POWER PLANT: Two 47.8 kW (64.1 hp) Rotax 582 UL-2V two-cylinder piston engines driving two-blade fixed-pitch propellers. Fuel capacity 130 litres (34.3 US gallons; 28.6 Imp gallons). Version planned with Rotax 912 ULS of 73.5 kW (98.6 hp).

ACCOMMODATION: Four persons; dual controls. Gull-wing door each side of cockpit.

DIMENSIONS, EXTERNAL:
Wing span	11.80 m (38 ft 8½ in)
Wing chord, constant	1.40 m (4 ft 7¼ in)
Wing aspect ratio	8.4
Length overall	7.15 m (23 ft 5½ in)
Height overall (on wheels, fuselage horizontal)	2.76 m (9 ft 0¾ in)
Tailplane span	2.80 m (9 ft 2¼ in)
Wheel track (optional)	1.42 m (4 ft 8 in)
Propeller diameter	1.84 m (6 ft 0½ in)
Distance between propeller centres	1.84 m (6 ft 0½ in)

AREAS:
Wings, gross	16.50 m² (177.6 sq ft)

WEIGHTS AND LOADINGS:
Weight empty	510 kg (1,124 lb)
Max payload	300 kg (661 lb)
Max T-O weight	950 kg (2,094 lb)
Max wing loading	57.6 kg/m² (11.79 lb/sq ft)
Max power loading	9.94 kg/kW (16.33 lb/hp)

PERFORMANCE:
Max level speed	97 kt (180 km/h; 112 mph)
Max cruising speed	70 kt (130 km/h; 81 mph)
Unstick speed	38 kt (70 km/h; 44 mph)
Max rate of climb at S/L	300 m (984 ft)/min
Service ceiling	4,000 m (13,120 ft)
T-O run: on land	30 m (100 ft)
on water	50 m (165 ft)
Range	464 n miles (860 km; 534 miles)
g limits	+4/−2

UPDATED

Che-25 four-seat amphibian in its 1999 configuration *(Paul Jackson/Jane's)*

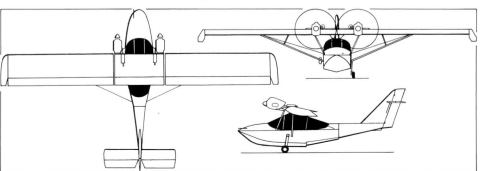

General arrangement of the Che-25 *(Paul Jackson/Jane's)*

RUSSIAN FEDERATION: AIRCRAFT—DUBNA (DMZ) to ELITAR

DUBNA (DMZ)

PROIZVODSTVENNO-TEKHNICHESKY KOMPLEKS DUBNENSKOGO MASHINOSTROITELNOGO ZAVOD AO
(Dubna Machine-Building Plant Production Technical Complex Co Ltd)
ulitsa Zhukovskogo 2, 141980 Dubna, Moskovskaya oblast
Tel: (+7 09621) 514 14, 516 13 and 513 79
Fax: (+7 09621) 235 24
Telex: 206132 TITAN
GENERAL DIRECTOR: Pyotr Lyrschikov

Founded in 1939, the Dubna Plant is an enterprise of the General Department of Aircraft Industry of the State Committee on Defence Industries of the Russian Federation. In the 1960s, it launched airframe production of the MiG-25 fighter and became a major producer of remotely piloted vehicles for Eastern bloc and other countries. As part of its restructuring under the *Konversiya* policy, since 1993, it has produced the Sukhoi Su-29 two-seat aerobatic aircraft and the Dubna series of light aircraft.

UPDATED

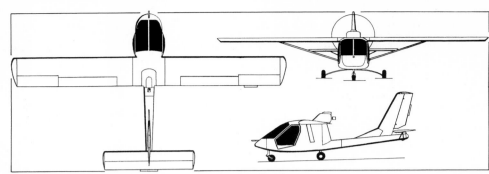

Dubna-2 light aircraft (Rotax 912 engine) *(Paul Jackson/Jane's)* 2001/0092552

DUBNA-1 SHMEL
English name: Bumblebee
TYPE: Single-seat ultralight.
PROGRAMME: Designed by Taifun Experimental Design Bureau, (ulitsa Tchaikovskogo 2, 041980 Dubna) as DMZ-1; resembles mid-wing Elf, flown in 1984 by Moscow Aviation Institute as development of 1978 Elf-D UAV; displayed at Moscow Air Show '95 and '97. Alternative twin-float landing gear shown in 1996.
DESIGN FEATURES: Pod-and-boom fuselage; strut-braced high wing, with single bracing strut each side; sweptback vertical tail surfaces; constant chord horizontal tail surfaces, braced each side by V struts; engine mounted above rear of fuselage pod. Suitable for same duties as DMZ-2.
FLYING CONTROLS: Conventional and manual. Ailerons, single-piece elevator and rudder all have no aerodynamic balance; ground-adjustable tab on starboard aileron, rudder and each side of elevator; large dorsal fin. Slotted flaps.
STRUCTURE: All-metal structure with fabric-covered wings and elevators; semi-monocoque fuselage with tapered tubular tailboom.
LANDING GEAR: As Dubna-2. Optional twin-float landing gear.
POWER PLANT: One Voronezh M-18-01 47 kW (63 hp) piston engine on prototype (later, 47.8 kW; 64.1 hp Rotax 582 UL); three-blade pusher propeller; fuel tank behind cabin; capacity 40 litres (10.5 US gallons; 8.8 Imp gallons).

Dubna-1 Shmel single-seat light aircraft equipped with optional floats *(Sebastian Zacharias)*
1999/0051210

DIMENSIONS, EXTERNAL:	
Wing span	9.92 m (32 ft 6½ in)
Length overall	5.07 m (16 ft 7½ in)
AREAS:	
Wings, gross	10.92 m² (117.5 sq ft)
WEIGHTS AND LOADINGS:	
Weight empty	200 kg (441 lb)
Max T-O weight	450 kg (992 lb)
PERFORMANCE:	
Nominal cruising speed	59 kt (110 km/h; 68 mph)
Landing speed	35 kt (65 km/h; 41 mph)
Max rate of climb at S/L	270 m (885 ft)/min

UPDATED

DUBNA-2 OSA
English name: Wasp
TYPE: Two-seat lightplane.
PROGRAMME: Designed by Taifun Experimental Design Bureau as DMZ-2; prototype (FLARF-01325) displayed at Moscow Air Show '95 and '97 with 59.6 kW (79.9 hp) Rotax 912 engine; promoted in China, 1996. By 1999, when definitive prototype FLARF-01268 shown at Moscow, detail changes included increased wheel track; repositioning of landing gear leg to below fuselage and of wing bracing strut to further forward; and Hirth engine, with additional fuel.
DESIGN FEATURES: As Dubna-1, suitable for passenger, freight and mail transport, cropspraying, forest surveillance, oil/gas pipeline and power line patrol, coastal fishery survey, aerial photography, *ab initio* pilot training and special missions.
FLYING CONTROLS: Conventional and manual. Flight-adjustable tab on elevator; ground-adjustable tabs on starboard aileron and rudder.
STRUCTURE: As Dubna-1.
LANDING GEAR: Non-retractable tricycle type; single wheel on each unit; arched metal leafspring mainwheel legs; trailing-link nosewheel leg. Tyres size 400×150 on mainwheels, 310×135 on nosewheel. Floats and skis optional.
POWER PLANT: One 89.5 kW (120 hp) Hirth F30/4C flat-four driving two-blade pusher propeller. Fuel tank behind cabin, capacity 80 litres (21.1 US gallons; 17.6 Imp gallons).
ACCOMMODATION: Two seats side by side, with dual controls, in fully enclosed, extensively glazed cabin; forward-hinged window/door on each side; space for baggage/freight in compartment behind cabin.

DIMENSIONS, EXTERNAL:	
Wing span	9.92 m (32 ft 6½ in)
Wing aspect ratio	9.0
Length overall	5.97 m (19 ft 7in)
Width over doors	1.25 m (4 ft 1¼ in)
Height overall	2.91 m (9 ft 6½ in)
Tailplane span	3.30 m (10 ft 10 in)
Wheel track	2.40 m (7 ft 10½ in)
Wheelbase	1.90 m (6 ft 2¾ in)
AREAS:	
Wings, gross	10.92 m² (117.5 sq ft)
WEIGHTS AND LOADINGS:	
Weight empty	296 kg (653 lb)
Baggage capacity	30 kg (66 lb)
Max T-O weight	576 kg (1,270 lb)
Max wing loading	52.7 kg/m² (10.81 lb/sq ft)
Max power loading	9.60 kg/kW (15.88 lb/hp)
PERFORMANCE:	
Max level speed	89 kt (165 km/h; 102 mph)
Nominal cruising speed	62 kt (115 km/h; 71 mph)
Landing speed	41 kt (75 km/h; 47 mph)
Unstick speed	46 kt (85 km/h; 53 mph)
Max rate of climb at S/L	240 m (790 ft)/min
Service ceiling	4,000 m (13,120 ft)
T-O run	120 m (395 ft)
Max range	270 n miles (500 km; 310 miles)
g limits	+4.0/−2.5

UPDATED

Production version of Dubna-2 Osa two-seat multipurpose light aircraft *(Paul Jackson/Jane's)* 2001/0092528

ELITAR

EHLITAR OOO
Korp 1, dom 10, ulitsa B. Cheremushkinskaya, 117448 Moskva
Tel: (+7 095) 120 07 07, 155 90 36, 552 79 69

At the August 1999 MAKS air show, Ehlitar (which adopts the Western form, Elitar) displayed the newly flown IE-101 lightplane and released initial details of the Senator.

Members of the Elitar design and production team work for MiG, Molniya and Sukhoi design bureaux and have been responsible for the M-5 Oktyabr and M-9 Marafon ultralights.

VERIFIED

ELITAR IE-101 ELITAR
TYPE: Side-by-side ultralight.
PROGRAMME: Prototype built at Myskova, Moscow; first flown 14 August 1999; public debut at MAKS '99, 17 August 1999.
DESIGN FEATURES: Simple, robust lightplane, meeting FLARF (Russian amateur aviation) airworthiness rules. High, unswept, constant-chord wing with single bracing strut each side. Sweptback fin augmented by underfin. Wings fold for storage.
FLYING CONTROLS: Conventional and manual. Mass-balanced ailerons with ground-adjustable tab to starboard; horn-balanced rudder with ground-adjustable tab; two-piece elevator with flight-adjustable tab to port. Fowler flaps.
STRUCTURE: Fuselage and wing of composites. Metal bracing struts and cantilever mainwheel struts.

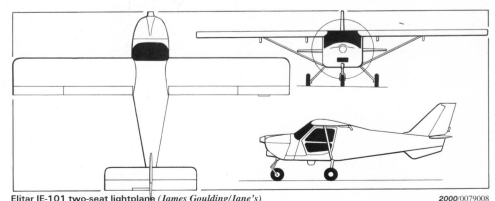

Elitar IE-101 two-seat lightplane *(James Goulding/Jane's)* 2000/0079008

LANDING GEAR: Tricycle type; fixed; speed fairings on each wheel. Hydraulic brakes. Mainwheels 400 × 150 mm. Optional twin-float gear.
POWER PLANT: One 59.6 kW (79.9 hp) Rotax 912 UL flat-four driving a two-blade ground-adjustable pitch propeller. Fuel capacity 90 litres (23.8 US gallons; 19.8 Imp gallons). Optional engines include 85.8 kW (115 hp) Rotax 914 and 89 kW (120 hp) Hirth F30.
ACCOMMODATION: Two persons, side by side, with dual controls. Upward-hinged door each side.
EQUIPMENT: RSV K-500 ballistic parachute.

DIMENSIONS, EXTERNAL:
Wing span	8.70 m (28 ft 6½ in)
Length overall	5.60 m (18 ft 4½ in)
Width, wings folded	2.20 m (7 ft 2½ in)
Height overall	2.35 m (7 ft 8½ in)
Wheel track	2.00 m (6 ft 6¾ in)
Wheelbase	1.40 m (4 ft 7 in)

DIMENSIONS, INTERNAL:
Cabin max width	1.12 m (3 ft 8 in)

AREAS:
Wings, gross	10.00 m² (107.6 sq ft)

WEIGHTS AND LOADINGS:
Weight empty	320 kg (705 lb)
Max T-O weight	530 kg (1,168 lb)

PERFORMANCE:
Max level speed	119 kt (220 km/h; 137 mph)
Max cruising speed	100 kt (185 km/h; 115 mph)
Landing speed	38 kt (69 km/h; 43 mph)
Max rate of climb at S/L	360 m (1,183 ft)/min
T-O run	110 m (360 ft)
Range with max fuel	468 n miles (900 km; 559 miles)
g limits	+4.9/−1.9

UPDATED

ELITAR SENATOR
TYPE: Side-by-side ultralight.
PROGRAMME: Announced, and preliminary details released, August 1999.

DESIGN FEATURES: Low-wing cabin monoplane; constant-chord, cantilever mainplanes; sweptback fin with underfin.
LANDING GEAR: Tricycle type; fixed.
POWER PLANT: One 73.5 kW (98.6 hp) Rotax 912 ULS flat-four driving a two-blade propeller. Optional engines include 89 kW (120 hp) Hirth F30. Fuel capacity 100 litres (26.4 US gallons; 22.0 Imp gallons).

DIMENSIONS, EXTERNAL:
Wing span	7.00 m (22 ft 11½ in)
Length overall	5.40 m (17 ft 8½ in)

AREAS:
Wings, gross	7.70 m² (82.9 sq ft)

WEIGHTS AND LOADINGS:
Weight empty	315 kg (694 lb)
Max T-O weight	525 kg (1,157 lb)

PERFORMANCE:
Max level speed	131 kt (243 km/h; 151 mph)
Max cruising speed	111 kt (205 km/h; 127 mph)
Stalling speed	39 kt (71 km/h; 45 mph)
Max rate of climb at S/L	336 m (1,102 ft)/min
T-O run	114 m (375 ft)
Landing run	130 m (430 ft)
Range with 60% fuel	364 n miles (675 km; 419 miles)

UPDATED

Model of Elitar Senator 2000/0079009

Prototype Elitar Elitar on display at Moscow, August 1999 *(Paul Jackson/Jane's)* 2001/0093699

GIDROPLAN

GIDROPLAN (Hydroplane)
Samara

Amphibian lightplanes designed by the Chernov bureau (which see) have previously been marketed under this name, as last described in 1997-98 *Jane's*. This venture was placed on a firmer footing in late 2000 when Gidroplan invested the equivalent of US$500,000 in acquiring 3,500 m² (37,675 sq ft) of working area from the Samara Instrument Bearing Plant and installing production equipment. Series manufacture of Chernov aircraft was due to begin in early 2001.

NEW ENTRY

IAPO

IRKUTSKOYE AVIATSIONNOYE PROIZVODSTVENNOYE OBEDINENIE OAO (Irkutsk Aviation Production Association JSC)
ulitsa Novatorov 3, 664020 Irkutsk
Tel: (+7 3952) 32 29 09 and 32 29 42
Fax: (+7 3952) 32 29 41 and 32 29 45
e-mail: market@iaia.irk.ru
Web: http://www.iaia.irk.ru
PRESIDENT: Alexei I Fedorov
GENERAL DIRECTOR: Vladimir V Kovalkov
DIRECTOR, MARKETING AND SALES: Yuri Belozerov

Founded on 28 March 1932 and commissioned on 24 August 1934, Irkutsk Aviation Production Association has built some 6,500 aircraft of 16 types from Antonov, Ilyushin, MiG, Pettyakov, Sukhoi, Tupolev and Yakovlev bureaus and supplied them to 21 countries. In recent years, it has manufactured MiG-23UB trainers (1970-85); Su-27UB trainers (from 1986); and is currently responsible for producing the Su-30 fighter (since 1991), Yakovlev Yak-112 lightplane and Beriev (Betair) Be-200 amphibian. IAPO also undertakes Su-30 upgrades. Russian government holding is 14.7 per cent.

In October 2000, IAPO was completing the prototype of a new, three-seat autogyro.

UPDATED

First Indian Sukhoi Su-30 nearing completion at Irkutsk *(Yefim Gordon)* 2001/0092138

ILYUSHIN

MEZHDUNARODNYYI AVIATSIONNYI KOMPANIYA ILYUSHINA (Ilyushin International Aviation Company)

Incorporating aircraft design and production companies, Ilyushin MAK formed April 2000 by government decree. Initial members are Ilyushin Aviation Complex (see below) and VAPO (see later in this section); following negotiations with Uzbekistan, will be expanded to include TAPO (which see) as second production plant; previous Russian and Uzbek government agreement of 1998 covered collaboration in developing and marketing the Il-76 and Il-114.

ILYUSHIN – AVIATSIONNYI KOMPLEKS IMENI S V ILYUSHINA OAO (Aviation Complex named for S V Ilyushin JSC)
Leningradsky prospekt 45G, 125190 Moskva
Tel: (+7 095) 157 35 73
Fax: (+7 095) 212 02 75
e-mail: ilyushin@glasnet.ru
Telex: 411956 Sokol

CHAIRMAN OF THE BOARD OF DIRECTORS AND GENERAL DESIGNER: Genrikh V Novozhilov
GENERAL DIRECTOR AND CEO: Victor V Livanov
HEAD OF INTERNATIONAL RELATIONS AND CHIEF DESIGNER: Igor Ya Katyrev
DEPUTY GENERAL DESIGNER, MARKETING, BUSINESS DEVELOPMENT AND FOREIGN ECONOMIC RELATIONS: Vladimir A Belyakov

Ilyushin OKB is named after Sergei Vladimirovich Ilyushin, who died 9 February 1977, aged 82. Bureau (OKB-240) was

RUSSIAN FEDERATION: AIRCRAFT—ILYUSHIN

founded 1933; has been headed by Genrikh Novozhilov since 1970. About 60,000 aircraft of Ilyushin design have been built. Personel in 2000 totalled about 2,000; by that time, activities were 90 per cent civil orientated.

Funding problems and bank collapses have affected Ilyushin and some of its suppliers, delaying the establishment of an Il-114 leasing arm and the acquisition of Western systems for the Il-96T.

UPDATED

ILYUSHIN Il-18 and Il-20
NATO reporting name: Coot

The UK quarterly *World Air Power Journal* revealed details of the Il-18/-20/-22 radio relay and telemetry family in March 2000. A single Il-22D was converted to serve as the **Il-18SIP** space/missile telemetry aircraft, with receiver in a dorsal hump, and with a downlink in an extended tailcone. Prototype was SSSR-75647, a former Il-18B which is believed to be the same aircraft as a similarly configured machine identified by c/n 188000401 (SSSR-27220). Four new-build production aircraft were produced from 1973 as **Il-20RT**s. These were equipped with an RTS-9 radio telemetry station, BRS-4 telemetry receiver, SYeV-12 time synchronisation system and voice and telegraph communications equipment. Operated from Leninsk, near Baikonur, in Aeroflot colour schemes; replaced by Beriev 976s, then being relegated to use as crew-trainers and transports by the Russian Air Forces.

Three **Il-18D (mod)** aircraft were used by the 235th OAO as communications relay aircraft, replacing the unit's similar **Il-18V**s. These latter were distinguished by two small blade antennas on the spine, whereas the Il-18D had a conduit running forward from the fin fillet. Both types are now used as transports by the 23rd Flight Detachment at Chkalovskaya.

Northern Fleet naval base at Ostrov operates Il-20RT variant for circuit training and transport duties; this aircraft retains large dorsal pannier and abbreviated blunt tail sting. Possibly 'Coot-C' (which reporting name allocated by ASCC to as-yet unknown configuration).

UPDATED

ILYUSHIN Il-18D-36 and Il-22M-11
NATO reporting names: Coot-B and (Coot-C?)

At least 34 new-built 'Coot-B' airborne command post adaptations of the Il-18 transport were delivered in Aeroflot markings (SSSR-758975 to 75928, not necessarily consecutive) to RFAS (CIS) air forces as replacements for the An-12VKP in two basic variants.

First 13 were **Il-18D-36** Bizon (Bison) with very long ventral fairing, and with standard fin-top bullet. The variant was developed by Myasishchev (OKB-23). Two prototypes were converted from Il-18 airliners in 1970. Production aircraft were built 'green' at MMZ 30, then shipped to Zhukhovsky for installation of mission equipment by Myasishchev. A further 21 were new-build **Il-22M-11** Zebra, with shorter ventral fairing and usually with antenna group above fuselage, dominated by single large blade. There were also at least four Il-22M-11s converted from Il-18D airliners. No cabin windows on starboard side of forward fuselage. Ukraine has at least one; another, in Kazakhstan, was converted to a VIP transport with internal mission fittings removed and was subsequently lost. Former Belarusian Il-22M SSSR-75916 sold to Concors of Latvia in 1998 and converted to Il-18D airliner. About six remain in front-line service, with more grounded by funding constraints.

It would be logical to expect variety of external fairings and antennas, differing from one aircraft to another, depending on its specific duties, but all examples photographed recently have common configuration, identified by bullet fairing at fin-tip, shallow container under front fuselage, and many small blade antennas above and below fuselage. SSSR-75915 was Il-18M-15.

'Coot-C' reporting name believed applied to elint platform encountered by Canadian CF-18s during the late 1980s and early 1990s. May be the same as configuration shown on Russian television in early 1999 and illustrated here. This (SSSR-75909) lacked the underfuselage SLAR-type fairing but had a comprehensive antenna group above the fuselage, with large, swept 'hockey stick' antennas over rear part of cabin and another below the centre-section.

UPDATED

Two large spine 'hockey stick' antennas are apparent on this Il-22M-11 Zebra, believed to be a 'Coot-C' which has only the mounting for the normal SLAR fairing – sometimes described as a pair of side-by-side strake-like antennas. *(Yefim Gordon)* 2001/0064894

Ilyushin Il-22M following conversion to passenger-carrying Il-18D in 1998-99 *(Paul Jackson/Jane's)* 2001/0092087

ILYUSHIN Il-20M
NATO reporting name: Coot-A

Ilyushin produced some 21 Il-20M reconnaissance aircraft (the prototype by conversion of existing Il-18D, plus about 20 new-build) for the Soviet/CIS air forces. Designation was *Izdelie* 17 (Article 17). Prototype made its maiden flight on 25 March 1968.

Equipped with Igla-1 phased-array SLAR in a pod below the fuselage, the aircraft also had smaller pods on the sides of the forward fuselage, housing A-87P LOROP cameras and Romb 4 sigint systems. Two large trapezoidal antennas above the fuselage served the Vishnaya comint system, while flush antennas served the Kvadrat 2 sigint system. Crew of 13, including eight operators. Flew mainly from Pushkin. Three aircraft were used by the former Soviet forces in East Germany. About six remain in use, sometimes as transports.

UPDATED

ILYUSHIN Il-24N

The Il-24N is a civil aircraft similar to the Il-20M ('Coot-A') for ice reconnaissance and fishery observation. It retains the large underfuselage side-looking radar (SLAR) fairing, though this contains a Nit'-S1 SLAR and not Igla. All elint equipment is absent and the Il-24N lacks most of the antennas associated with 'Coot-A' configuration. Two late production Il-18Ds were converted to Il-24N standards in 1980. The aircraft were de-converted in 1994 and were returned to passenger duties, initially with Air Transport Office of Zaïre and Daallo of Djibouti. At least one aircraft retains its SLAR fairing.

UPDATED

ILYUSHIN Il-38
NATO reporting name: May

TYPE: Maritime patroller.
PROGRAMME: Developed (as Izdelie 8), initiated by directive dated 18 June 1960 as one element in Vyaz ASW system, intended to counter Polaris missile submarines. Simultaneous Mozhzhevel'nik R&D programme resulted in Il-38's sonobuoy system, Ka-25 sonar and other ASW equipment. Long-range ASW platform element based on an adaptation of the Il-18, but this relegated to a medium-range aircraft with the increasing range of SLBMs, which necessitated the development of the Tu-142 for long-range ASW. Il-18I SSSR-75888 used for fuel system trials for Il-18D and Il-38; Il-18 SSSR-75643 used as radar and systems testbed; Il-38 prototype first flown 28 September 1961, one year ahead of planned schedule; first fully equipped Il-38 flew 1965; State acceptance trials June-December 1965; delivery of first preseries aircraft (10106) October 1967; Il-38 formal commissioning 17 January 1969; 57 built by MAPO (plus one prototype), last of which delivered 22 February 1972; first export order, from India in 1975, fulfilled using refurbished AVMF aircraft.

CURRENT VERSIONS: **May-A:** Main production version. Some aircraft were upgraded (with no change of designation) with a 68-channel Volkhov receiver, new data processing and display equipment and a Bor-1s (APM-73S) MAD – a more modest upgrade than originally envisaged, not incorporating the Korshun STS or a new AFCS. Probably designated Il-38K for export.
Description applies to 'May-A'.

May-B: Has additional teardrop-shape ventral fairing beneath forward weapons bay, as housing for KAS droppable rescue equipment. This can probably be fitted to any Il-38.

Il-38M: One-off conversion as in-flight refuelling receiver, with probe above nose.

Il-38MZ: Aircraft modified as in-flight refuelling receiver with additional tanker capability, using UPAZ-38 refuelling pod under the fuselage.

Il-38P: Alternatively **Il-38R** ('P' may be Cyrillic 'R'). Identified 1996. Some Il-38s upgraded with Emerald ASW system, Bor 15 (APM-73S) MAD and RGB-16 sonobuoys after failed programme to integrate Tu-124M's Korshun ASW suite.

Il-38 IKAR: One Il-38 converted as high-resolution geophysical survey platform by Leninets in the late 1990s.

Sigint Version: Several aircraft with cable antenna running forward from leading-edge to tandem blister fairings on forward fuselage. Shallow box antenna for Vishnya comint system below rear fuselage. Encountered by Swedish Air Force at high altitude over the Baltic.

Upgrades: One aircraft fitted experimentally with SPS-151 and -153 active jammers in 1971-72. Berkut STS upgraded in 1974-75 with ANP-3V automatic navigation device for more accurate ASW pattern flying. Single Il-38 modified by Leninets Holding Company with new Sea Dragon ASW/ASV mission system as prototype for projected fleet upgrade, extending service life by 15 years. Simplified version of Sea Dragon (see Tupolev Tu-142 entry) includes only two full and one partial operator consoles.

Unidentified: At least one Indian Il-38 (IN304) seen at Pushkin with box fairings on sides of nose and on MAD boom root, possibly accommodating ESM antennas or chaff/flare dispensers.

One Russian Navy aircraft, '71', noted at Sherman Grayson Airport, Texas, in 1995, apparently receiving modifications; had returned to Pushkin by 1999.

CUSTOMERS: RFAS (CIS) Naval Aviation. About 36 with Northern Fleet (24th OPLAP-DD at Severomorsk), Pacific Fleet (77th OPLAE at Nikolayevka, with detachment at Yelizovo) and for training with 240th GvSAP at Ostrov. Aircraft with Ukrainian-based detachment not taken over by Ukraine. Five Il-38Ks with INAS 315 at Dabolim, Goa.

Ilyushin Il-20M reconnaissance aircraft with civil registration RA-75923, duplicating an Il-22M *(Yefim Gordon)* 2000/0064891

ILYUSHIN—AIRCRAFT: RUSSIAN FEDERATION

Ilyushin Il-38 (four ZMKB Progress/Ivchenko AI-20M turboprops) *(Paul Jackson/Jane's)*

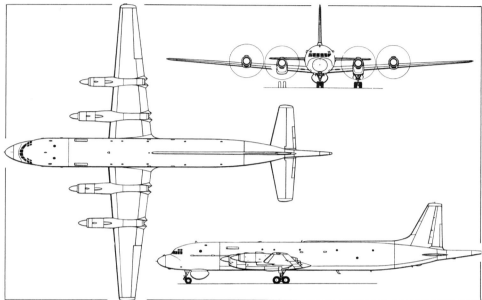

Ilyushin Il-38 anti-submarine/maritime patrol derivative of the Il-18 airliner *(Dennis Punnett/Jane's)*

DESIGN FEATURES: Basic Il-18 four-prop, low-wing airliner, with lengthened fuselage, and wings moved forward approximately 2.75 m (9 ft) to cater for effect of role equipment and stores on CG position resulting in 'Dachshund' nickname; few cabin windows; cooling air duct for avionics on upper port-side fuselage forward of wing leading-edge; large undernose radome; MAD tail sting.
Wing dihedral 3° from roots; mean thickness/chord ratio 14 per cent.

FLYING CONTROLS: Conventional and assisted. Actuation by cables; mass and aerodynamically balanced ailerons with electric trim tabs; hydraulically assisted elevators and rudder, each with electric trim tab; additional rudder spring tab; hydraulically actuated double-slotted trailing-edge flaps. AP-6 Ye autopilot.

STRUCTURE: All-metal; three-spar wing centre-section, two spars in outer wings; circular-section fail-safe semi-monocoque fuselage, with rip-stop doublers around window cutouts, door frames and more heavily loaded skin panels.

LANDING GEAR: Retractable tricycle type, strengthened in comparison with Il-18. Hydraulic actuation; all units retract forward, mainwheels into inboard engine nacelles. Four-wheel bogie main units, with 930×305 tyres and hydraulic brakes. Steerable (±45°) twin-wheel nose unit, with 700×250 tyres and stone guard. Hydraulic brakes. Pneumatic emergency braking.

POWER PLANT: Four ZMKB Progress/Ivchenko AI-20M turboprops, each 3,126 kW (4,190 ehp), with AV-68M four-blade reversible-pitch metal propellers. Multiple bag fuel tanks in centre-section and in inboard panel of each wing, and integral tank in outboard panel; maximum fuel weight 26,500 kg (58,420 lb) in 25 tanks with a maximum capacity of 33,820 litres (8,923 US gallons; 7,440 Imp gallons). Fuel management system automatically maintains centre of gravity. Pressure fuelling through four international standard connections in inner nacelles. Eight-point gravity refuelling. Provision for overwing fuelling. Oil capacity 58.5 litres (15.45 US gallons; 12.85 Imp gallons) per engine. Engines started electrically.

ACCOMMODATION: Crew of seven (or eight), all in pressurised forward compartment. Pilot and co-pilot side by side on flight deck, with dual controls; flight engineer centrally behind them, then navigator/radar operator on port side and communications operator starboard, with seat for eighth crew member (second engineer or instructor); acoustics/MAD operator (port) and tactical commander (starboard) in rearward-facing seats at rear. Forward-hinged ventral entry hatch immediately aft of radome. Dorsal ditching hatch above pressure cabin, another overwing on port side. Flight deck separated from main cabin by pressure bulkhead to reduce hazards following decompression of either. Access to rear fuselage possible when pressure differential eliminated. Main cabin has few windows. Door on starboard side at rear of cabin (location of Il-18 service door). Rear cabin accommodated a crew rest area with a galley, lavatory and folding bunk.

SYSTEMS: Cabin maximum pressure differential 0.49 bar (7.1 lb/sq in). Eight engine-driven generators for 28 V DC and 115 V 400 Hz AC supply. Hydraulic system, pressure 207 bars (3,000 lb/sq in), for landing gear retraction, nosewheel steering, brakes, elevator and rudder actuators, flaps, weapon bay doors and radar antennas. Electrothermal de-icing for wings and tail unit. Wipers and washers on four windscreen panels. TG-16M APU, with exhaust on port side of rear fuselage.

AVIONICS: *Comms:* R-836 Neon and R-847A VHF Peleng HF R-632 decimetre-wave command link, R-802V command link, SPU-7B intercom and MS-61 CVR D-band IFF. Aircraft intercepted since 1996 feature many new antennas, possibly indicating comms upgrade.
Radar: Normal Il-18 navigation/weather radar in nose deleted. J-band 360° 'Berkut' search radar ('Wet Eye') in undernose radome with 120° and 150° sector scanning options.
Flight: RSBN-25 Svod Shoran, DISS-1 Doppler, ARK-II ADF, SP-50 ILS and RV-4 radio altimeter. Also Put'-4B-2K compass.
Mission: 'Berkut' (originally APM-60 then APM-73S) (then Emerald) ASW system, including TsVM-264 digital computer and MAD tailboom. Highly automated with auto-attack and auto-weapons selection.
Self-defence: RWR each side of nose and tailboom. Provision for chaff/flare dispensers each side of nose and tailboom.

ARMAMENT: Two large internal weapons/stores bays, filling much of cabin forward and aft of wing carry-through structure; contents typically 216 RGB-1 non-directional passive sonobuoys, 144 RGB-2 passive sonobuoys and two AT-1 or AT-2 torpedoes (Indian aircraft use AT-1E), 10 PLAB-250-120 depth charges, eight AMD-2-500 mines, or one Ryu-2 nuclear depth bomb. Alternatives include up to three RGB-3 large active/passive sonobuoys. Some aircraft have guidance system for KAB-500PL guided depth charges. UDM-500 mines and GB-100 bombs. Wings too heavily loaded for external missile carriage, though this was requested.

DIMENSIONS, EXTERNAL:
Wing span	37.4 m (122 ft 8¾ in)
Wing chord: at root	5.61 m (18 ft 4¾ in)
at tip	1.87 m (6 ft 1½ in)
Wing aspect ratio	10.0
Length overall	40.1 m (131 ft 7¼ in)
Height overall	10.1 m (33 ft 2 in)
Tailplane span	11.80 m (38 ft 8½ in)
Wheel track	9.00 m (29 ft 6¼ in)
Propeller diameter	4.50 m (14 ft 9¼ in)

AREAS:
Wings, gross	140.0 m² (1,506.9 sq ft)

WEIGHTS AND LOADINGS:
Weight empty	35,500 kg (78,263 lb)
Max weapon load	8,400 kg (18,520 lb)
Max fuel	26,500 kg (58,420 lb)
Max T-O weight	66,000 kg (145,503 lb)
Max wing loading	471.4 kg/m² (96.55 lb/sq ft)
Max power loading	5.28 kg/kW (8.68 lb/ehp)

PERFORMANCE:
Max Mach No.	0.65
Max level speed at 6,400 m (21,000 ft)	390 kt (722 km/h; 448 mph)
Max cruising speed at 8,230 m (27,000 ft)	348 kt (645 km/h; 401 mph)
Patrol speed at 100-1,000 m (330-3,280 ft)	173-216 kt (320-400 km/h; 199-248 mph)
Min flying speed	103 kt (190 km/h; 118 mph)
Min patrol speed	189 kt (350 km/h; 217 mph)
Min patrol height: day	30-60 m (100-200 ft)
night	100 m (330 ft)
Service ceiling	10,000 m (32,800 ft)
T-O run	1,700 m (5,577 ft)
Landing run with propeller reversal	1,070 m (3,510 ft)
Range with max fuel	5,129 n miles (9,500 km; 5,903 miles)
Mission radius	1,187 n miles (2,200 km; 1,367 miles)
Typical mission duration	10-11 h
Patrol endurance with max fuel	12 h

UPDATED

ILYUSHIN Il-76
NATO reporting name: Candid
Indian Air Force name: Gajaraj (King Elephant)
TYPE: Strategic transport.
PROGRAMME: Design began late 1960s, led by G V Novozhilov, to replace turboprop An-12; prototype (SSSR-86712) flew 25 March 1971; Western debut at Paris, June 1971; two prototypes and one static test airframe built at Khodinka (GAZ 30); all subsequent production at TAPOiCh plant (formerly GAZ 34), Tashkent, Uzbekistan; first delivery (SSSR-86600; third production) to Ivanovo on 3 June 1974. Manufacture peaked at 10 per month in 1980s, but rate reduced almost to zero by 1994 and UK aircraft broker, Fortis Aviation, engaged in early 1998 to promote 'white tail' sales, offering PS-90 engine as potential alternative to standard D-30. Russian and Uzbek governments agreed in 1998 to promote Il-76 (and Il-114). In 1999, CFM56 turbofan considered as alternative power plant for stretched

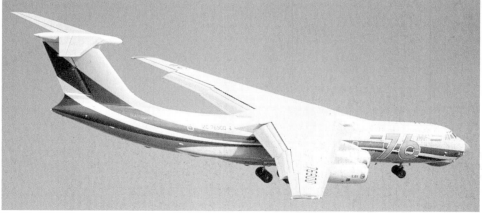

Ilyushin Il-76MF prototype *(Paul Jackson/Jane's)*

Il-76MF and Russian Air Forces announced life-extension programme for older Il-76s.

CURRENT VERSIONS: **Il-76** ('Candid-A'): Initial basic military production version; Aviadvigatel D-30KP turbofans, each 117.7 kN (26,455 lb st). Production completed in 1977.

Il-76T ('Candid-A'): Civil version of Il-76; additional fuel in wing centre-section, above fuselage; maximum T-O weight 170,000 kg (374,785 lb); maximum payload 40,000 kg (88,185 lb); no armament. First aircraft (SSSR-76504; 82nd overall) delivered to Tyumen division of Aeroflot, 20 October 1977.

Il-76M ('Candid-B'): As Il-76T but military; up to 140 troops or 125 paratroops carried as alternative to freight; rear gun turret (not always fitted on export aircraft) containing two 23 mm twin-barrel GSh-23L guns; small ECM fairings (optional on export aircraft) between centre windows at front of navigator's compartment, on each side of front fuselage, and each side of rear fuselage; packs of ninety-six 50 mm IRCM flares on landing gear fairings and/or on sides of rear fuselage of aircraft operating into combat areas. First aircraft (SSSR-86728; 81st overall) delivered to Panevezys 27 August 1977.

Il-76K: Initial cosmonaut (*kosmos*) training version, enabling occupants to experience brief periods of weightlessness. Two aircraft: SSSR-86723, delivered to Chkalovsky 23 July 1977, and SSSR-86729, delivered 29 September 1977. Withdrawn from service.

Il-76TD ('Candid-A'): Unarmed; generally as Il-76T but with strengthened wings and centre-fuselage; Aviadvigatel D-30KP-2 turbofans, maintaining full power to ISA + 23°C against ISA + 15°C for earlier models; maximum T-O weight and payload increased; 10,000 kg (22,046 lb) additional fuel increases maximum fuel range by 648 n miles (1,200 km; 745 miles); upgraded avionics; production began at 271st aircraft (5A-DNC for Libya) but initial delivery was SSSR-76464 (273rd) to Krasnoyarsk division of Aeroflot on 17 May 1982; one specially equipped with seats, soundproofing, buffet kitchen, toilet and working facilities, to carry members of Antarctic expeditions between Maputo, Mozambique, and Molodozhnaya Station, Antarctica (proving flight February 1986 with 94 passengers, 14,000 kg; 30,865 lb of scientific equipment, cargo and baggage containers). Version proposed in early 1990s with CFM56 turbofans, each rated at 138.8 kN (31,200 lb st); range increased 20 to 30 per cent; fuel burn decreased; noise reduced to comply with ICAO Chapter 3 Appendix 16.

Il-76MD ('Candid-B'): Military version; generally as Il-76M but with improvements of Il-76TD. Maximum permissible T-O weight 210,000 kg (462,970 lb) for up to 15 per cent of flights. First production aircraft (SSSR-86871; 251st overall, including prototypes) delivered to Zhukovsky for trials 25 March 1981. One (RA-76753) operated as flying laboratory. Some Il-76MDs converted to TDs by elimination of gun turret and retrofit with TD's navigation equipment.

Detailed description applies to Il-76MD version except where indicated.

Il-76MDK/MDK-2: Adaptation of Il-76MD to enable cosmonauts to experience several tens of seconds of weightlessness during training. Aircraft comprise RA-76766 delivered 31 August 1988; RA-78770 (31 December 1990), RA-78825 (31 March 1991) and RA-78850 (27 April 1991).

Il-76MF: Stretched military version with four Aviadvigatel PS-90AN turbofans, each 156.9 kN (35,275 lb st). Noise and emission characteristics conform with ICAO standards. New flight and navigation equipment includes Kupol III-76MF IFCS and CMCS. Cargo hold lengthened 6.60 m (21 ft 8 in) by two equal-sized plugs, fore and aft of wings; internal volume increased by 400.0 m³ (14,125 cu ft); length overall 53.195 m (174 ft 6¼ in), height overall 14.42 m (47 ft 3¾ in); maximum payload and T-O weight increased;

Il-76TD flown by MChS Rossii *(Paul Jackson/Jane's)*

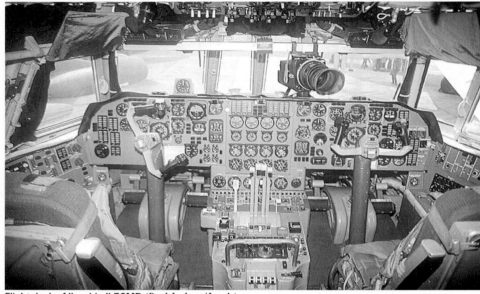

Flight deck of Ilyushin Il-76MD *(Paul Jackson/Jane's)*

alternative payloads include four YAK-10 or YYK-20, or nine YAK-5 or PA-5.6 containers; range, with reserves, with 40,000 kg (88,185 lb) payload 2,805 n miles (5,200 km; 3,230 miles). Two 2.5 tonne travelling cranes in cargo hold; lifting capacity 5 tonnes for a centreline load.

First flight (IS- [later RA-]76900) 1 August 1995; prototype completed official acceptance trials in December 1999, when due to be delivered to Ilyushin's own freight airline, Ilavia, at Zhukovsky; flew in military exercises, June 2000. Built in Tashkent, where 11 in course of assembly by early 2000. Planned Russian Air Forces orders not placed by mid-2000 due to financial problems; requirement is for up to 120; four Il-76MFs were complete by May 2000, apart from engines and avionics.

Il-76MF-100: Stretched version with 151.2 kN (34,000 lb st) CFMI CFM56-5C2 or -5C4 engines; was subject of 1999 study and February 2000 MoU to meet requirements of potential five-aircraft customer, which emerged in April 2000 as Uzbekistan government; aircraft will be used for unspecified 'development work'. CFM56 also being offered as retrofit for standard-length Il-76s.

Il-76TF: Civil version of Il-76MF. MoU of 17 May 2000 requires two complete, but unsold, military Il-76MFs to be fitted at Tashkent with PS-90 engines, stripped of military equipment, and delivered, from early 2001 onwards, to Uzbekistan Airways, which will immediately sublease to East Line freight airline at Moscow-Domodedovo. Variant meets ICAO noise criteria, but FAR Pt 25 certification financially impractical as original Il-76TD was built to NLGS-3 criteria, not Western-compatible AP-25.

Il-76 Aibolit: Mobile operating theatre developed by TAPO at Tashkent; follow-on to military Il-76 Skalpel MT, of which two employed during Afghanistan conflict. (May be variant also designated Il-76TD-S.)

Il-76MDP: Firefighting (*pozharnyye*) conversion of Il-76MD demonstrated first in 1990; up to 44,000 kg (97,000 lb) of water/fire retardant in two cylindrical tanks in hold; discharge, replenishment and draining systems; drop zone aiming devices; up to 384 silver iodide cartridges in dispensers to induce precipitation; able to water-bomb an area of 500 × 100 m (1,640 × 330 ft) — equivalent to saturation of 1.5 litres/m² (4.3 US gallons; 3.6 Imp gallons/sq ft) — or to carry, and parachute when required, 40 fully equipped firefighters; all airborne fire equipment (known as VAP-2, *vylivnoyye aviatsionnye pribor*: dischargeable aviation system; weight 5,000 kg; 11,025 lb) can be installed in standard Il-76, or removed, in 4 hours; tank replenishment time 10 to 15 minutes; discharge time 6 to 7 seconds, with option of successive discharge of tanks to cover 600 × 80 m (1,970 × 260 ft); airspeed during discharge 130 to 215 kt (240 to 400 km/h; 150 to 248 mph) at 80 m (260 ft). Five Il-76TDs employed as water-bombers by MChS Rossii/TsENTROSPAS at Zhukovsky. Used in Greece, 1999.

Il-76LL: Engine testbed conversion, carrying gas turbine of up to 245 kN (55,100 lb st), including turboprops, in place of normal port inner D-30KP; provisions for five test engineers; three of original five Il-76LLs are available, on commercial contract basis, from Gromov Flight Research Institute; engines tested include TV7-117 turboprop, NK-86 and D-18T turbofans and D-236 and D-27 propfans; testing of NK-93 propfan was scheduled to begin 1995, but no confirmation of this; conversion of RA-76751 for PS-90A tests was also abandoned. Maximum vertical *g* +2/–0.3; maximum bank angle 30°; maximum angle of attack 15°; maximum rate of roll 5°/s.

Il-76PP: Version of Il-76MD equipped for ECM (*postanovshchik pomekh*: jammer); not deployed. Landing gear panniers extended forward to slightly ahead of crew door as housing for forward sector dielectric panels of Landysh (lily of the valley) avionics suite. Prototype SSSR-86889 now withdrawn for ground instructional use; three more were converted.

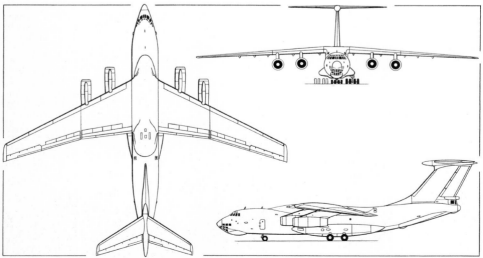

Ilyushin Il-76MD four-turbofan military transport *(Dennis Punnett/Jane's)*

ILYUSHIN Il-76 PRODUCTION

Customer	Il-76	Il-76M	Il-76T	Il-76MD	Il-76TD	Il-78	A-50/976	Il-76MF/TF	Others
Primary Operators									
Prototype/static test	3							1	
Soviet Air Forces	65	112		340		46	32		9
Aeroflot	4		31		108				
Soviet Interior Ministry				11					
Research establishments					3				
MChS Rossii					6				
Algeria (all services)				3	2				
Cuba (all services)				2					
Iraq (all services)	6	12	4	22					
Korea, North (air force)				3					
Libya (all services)		5	3	1	14				
Secondary Operators									
Aeroservice (Kazakhstan)					1				
Air Stan (Russia)					1				
Aviaenergo (Russia)					1				
Avialeasing (Uzbekistan)				2	1				
Cairo Charter					2				
China United Airlines				14					
Gulf Aviation Technical Services					1				
Indian Air Force				17					
Polise Air (Russia)					1				
Sayakhat (Kazakhstan)					3				
Syrianair		2	2						
TASA (Uzbekistan)					1				
Trans Aero Samara (Russia)					1				
Trans Super (Russia)					1				
Turkmenistan Airlines					5				
Uzbekistan Airways					11				
Not known					2				
Undelivered/incomplete					29	2		11	2
Totals (960)	**78**	**131**	**39**	**415**	**194**	**48**	**32**	**12**	**11**

Notes: Variants in 'Others' column comprise one A-50 and one Il-76K from Il-76 series production: and four Il-76MDK, two Il-76VKP, and one each of Il-76RLSBO, Type 576 and Type 1076 from Il-76MD production series

Il-76PS/Il-84: SAR (*poiskovo-spasatelnyi*) version; prototype only; first flight 18 December 1984; cancelled 1989. New version proposed as **Il-76PSD**.

Il-76SKIP: See Beriev '976' entry.

Il-76VKP/Il-82: Airborne communications relay (*Vozdushnyye Komandnyye Punkt*: Airborne Command Post) adaptation of Il-76MD known as version 65C. Two examples (SSSR- [later RA-]76450 and 76451) delivered to Zhukovsky after modifications 22 September and 30 November 1987; first seen 1992; assigned to 8th Special Purpose Aviation Division at Chkalovsky. Detailed description in 1999-2000 and earlier editions.

Specialised variants and developments of Il-76 include transports modified to carry external loads, including Tu-160 tailplane, above fuselage; the AEW&C **A-50** ('Mainstay'), **A-60** airborne laser testbed, and a Chinese AEW version (all described separately under Beriev OKB entry); the **Il-78** ('Midas') flight refuelling tanker (described separately); the AEW&C **Adnan 1** and single-point flight refuelling tanker modified for Iraq.

Further three/four-figure series (possibly allocated by Zhukovsky test centre) used by experimental conversions:

176: Also known as Il-76PP (see below).

576: Unidentified version; one only, c/n 1023412408, built 1992.

676: Telemetry relay platform; Il-76 SSSR-86721 delivered to Zhukovsky 6 July 1977; transferred to Ivanovo in early 1990s.

776: Telemetry replay platform; Il-76M SSSR-86024 delivered to Zhukovsky 20 September 1978; Ivanovo in mid-1990s.

976: See Beriev entry; two aircraft replaced 676 and 776 at Zhukovsky.

1076: Unidentified version; one only, c/n 1033410351, built 1992.

CUSTOMERS: See table. Some 920 built and 40 in various stages of completion by mid-2000. Many early military aircraft now in civilian ownership. Russian air transport force (61st Air Army) has almost 300 Il-76/76M/76MDs, most in Aeroflot markings; air forces of Algeria (eight, including three MD, four TD), Azerbaijan (one M, two MD), Belarus (18 MD, some in civil markings), India (17 MDs, given name Gajaraj), Iran (16 impounded Iraqi MDs and miscellaneous acquisitions), North Korea (three T), Libya (18 M/T/MD), Syria (four M), Ukraine (100) and Yemen (one TD); commercial operators include Aeroflot Russian International (12 TDs), Air Service Ukraine (11), Atlant Soyuz (13), China United (15), Kras Air (11), Transavia Export (22), Uzbekistan Havo Yullari (15) and some 75 smaller operators with less than 10; total approximately 280 in current civilian service. China United fleet acquired since 1991, from new production. Il-76Ms of airlines have no guns in turret; first of two Il-76MDs delivered to Cubana had no tail turret. Chinese Air Force negotiating for 20 Il-76Ms (D-30KR engines) in mid-2000.

COSTS: Il-76MF US$30 million to US$35 million. Unsold Il-76TDs offered at US$20 million each in 1997.

DESIGN FEATURES: Late 1960s requirement was to carry 40 tonnes of freight 2,700 n miles (5,000 km; 3,100 miles) in less than 6 hours, with ability to operate from short unprepared airstrips, in the most difficult weather conditions experienced in Siberia, the north of the Soviet Union and the Far East, while much simpler to service and able to fly much faster than An-12; wings mounted above fuselage to leave interior unobstructed; rear-loading ramp/door; unique landing gear, with two large external fairings for each main gear. All tail surfaces sweptback.

Wing section CAHI (TsAGI) P-151; sweepback 25° at quarter-chord; thickness/chord ratio 13 per cent at root, 10 per cent at tip; anhedral 3° outboard of centre-section; incidence 3° at root, 0° at tip.

FLYING CONTROLS: Conventional, hydraulically boosted; manual operation possible in emergency; mass-balanced ailerons, with balance/trim tabs; two-section triple-slotted trailing-edge flaps over approximately 75 per cent of each semi-span; eight upper-surface spoilers/airbrakes forward of flaps on each wing, four on each inner and outer panel; leading-edge slats over almost entire span, two on each inner panel, three on each outer panel; variable incidence T tailplane; elevators and rudder aerodynamically balanced, each with tab.

STRUCTURE: All-metal; five-piece wing of multispar fail-safe construction, centre-section integral with fuselage; basically circular-section semi-monocoque fail-safe fuselage; underside of upswept rear fuselage made up of two outward-hinged clamshell doors, upward-hinged panel between doors, and downward-hinged ramp.

LANDING GEAR: Hydraulically retractable tricycle type. Steerable nose unit has two pairs of wheels, side by side, with central oleo. Main gear on each side has two units in tandem, each unit with four wheels on single axle. Low-pressure tyres size 1,300×480 on mainwheels, 1,100×330 on nosewheels. Nosewheels retract forward. Main units retract inward into two large ventral fairings under fuselage, with additional large fairing on each side of lower fuselage over actuating gear. During retraction mainwheel axles rotate around leg, so that wheels stow with axles parallel to fuselage axis (that is wheels remain vertical but at 90° to direction of flight). All doors on wheel wells close when gear is down, to prevent fouling of legs by snow, ice, mud and so on. Oleo-pneumatic shock-absorbers. Tyre pressure can be varied in flight from 2.50 to 5.00 bars (36 to 73 lb/sq in) to suit different landing strip conditions. Hydraulic brakes on mainwheels.

POWER PLANT: Four Aviadvigatel D-30KP-2 turbofans, each 117.7 kN (26,455 lb st), in individual underwing pods. Total engine life of 6,500 hours, before scrapping; intermediate overhaul at 3,000 hours. Each pod on large forward-inclined pylon and fitted with clamshell thrust reverser. Integral fuel tanks between spars of inner and outer wing panels. Total fuel capacity 109,480 litres (28,922 US gallons; 24,083 Imp gallons).

ACCOMMODATION: Crew of seven, including two freight handlers. Side-by-side seating for pilot and co-pilot on flight deck. Station for navigator below flight deck in glazed nose. Forward-hinged main cabin door on each side of fuselage forward of wing. Crew emergency escape hatch forward of, and lower than, main door on port side; access via two-piece upward-folding door forming flight deck floor under port rear seat and via door at rear of navigator's compartment. Two windows on each side of hold serve as emergency exits. Hold has reinforced floor of titanium alloys, with folding roller conveyors, and is loaded via rear ramp. Entire accommodation pressurised, permitting carriage of 140 troops or 125 paratroops as alternative to freight.

Advanced mechanical handling systems for containerised and other freight, which can include standard ISO containers, each 12 m (39 ft 4½ in) long, building machinery, heavy crawlers and mobile cranes. Typical loads include six containers measuring either 2.99 × 2.44 × 2.44 m (9 ft 9¾ in × 8 ft × 8 ft) or 2.99 × 2.44 × 1.90 m (9 ft 9¾ in × 8 ft × 6 ft 2¾ in) and with loaded weights of 5,670 kg (12,500 lb) or 5,000 kg (11,025 lb) respectively; or 12 containers measuring 1.46 × 2.44 × 1.90 m (4 ft 9¼ in × 8 ft × 6 ft 2¾ in) and each weighing 2,500 kg (5,511 lb) loaded; or six pallets measuring 2.99 × 2.44 m (9 ft 9¾ in × 8 ft) and each weighing 5,670 kg (12,500 lb); or 12 pallets measuring 1.46 × 2.44 m (4 ft 9¼ in × 8 ft) and each weighing 2,500 kg (5,511 lb). Folding seats along sidewalls in central portion of hold.

Quick configuration changes made by use of modules, each able to accommodate 36 passengers in four-abreast seating, litter patients and medical attendants, or cargo.

Il-76MD in Ukrainian military service (*Paul Jackson/Jane's*)

Three such modules can be carried, each approximately 6.10 m (20 ft) long, 2.44 m (8 ft) wide and 2.44 m (8 ft) high; loaded through rear doors by two overhead travelling cranes, and secured to cabin floor with cargo restraints. Two winches at front of hold, each with capacity of 3,000 kg (6,615 lb). Cranes embody total of four hoists, each with capacity of 2,500 kg (5,511 lb). Ramp can be used as additional hoist, with capacity of up to 30,000 kg (66,140 lb) to facilitate loading of large vehicles and those with caterpillar tracks. Pilot's and co-pilot's windscreens can each be fitted with two wipers, top and bottom.

SYSTEMS: Flight deck only, or entire interior, can be pressurised; maximum differential 0.50 bar (7.25 lb/sq in). Hydraulic system includes servo motors and motors to drive flaps, slats, landing gear and its doors, ramp, rear fuselage clamshell doors and load hoists. Flying control boosters supplied by electric pumps independent of central hydraulic supply. Electrical system includes engine-driven generators, auxiliary generators driven by APU, DC converters and batteries. It powers pumps for flying control system boosters, radio and avionics, and lighting systems.

AVIONICS: *Comms:* R-838 and R-847 radio comms.
Radar: Weather radar in nose; Kupol 3-76 nav and ground mapping radar in undernose radome.
Flight: Full equipment for all-weather operation by day and night, including computer for automatic flight control and automatic landing approach. Il-76M/MD have different navigation avionics from T/TD.

EQUIPMENT: APU in port side landing gear fairing for engine starting and to supply all aircraft systems on ground, making aircraft independent of ground facilities.

DIMENSIONS, EXTERNAL (except Il-76MF):
Wing span	50.50 m (165 ft 8 in)
Wing aspect ratio	8.5
Length: overall	46.60 m (152 ft 10½ in)
fuselage	43.60 m (143 ft 0½ in)
Fuselage: Max diameter	4.80 m (15 ft 9 in)
Height overall	14.76 m (48 ft 5 in)
Tailplane span	17.40 m (57 ft 1 in)
Wheel track	8.16 m (26 ft 9¼ in)
Wheelbase	14.17 m (46 ft 6 in)
Rear-loading aperture: Width	3.40 m (11 ft 1¾ in)
Height	3.40 m (11 ft 1¾ in)
Side doors (each):	
Height	1.90 m (6 ft 2¾ in)
Width	0.86 m (2 ft 9¾ in)

DIMENSIONS, INTERNAL:
Cabin: Length, except Il-76MF:	
excl ramp	20.00 m (65 ft 7½ in)
incl ramp	24.54 m (80 ft 6¼ in)
Length, Il-76MF: excl ramp	26.60 m (87 ft 3¼ in)
incl ramp	31.14 m (102 ft 2 in)
Width at floor	3.45 m (11 ft 3¾ in)
Max height	3.40 m (11 ft 1¾ in)
Volume	235.3 m³ (8,310 cu ft)

AREAS:
Wings, gross	300.0 m² (3,229.2 sq ft)
Ailerons (total)	13.27 m² (142.84 sq ft)
Spoilers (total)	10.86 m² (116.90 sq ft)
Airbrakes (total)	15.80 m² (170.07 sq ft)
Tailplane	45.83 m² (493.31 sq ft)
Elevators (total)	17.17 m² (184.82 sq ft)

WEIGHTS AND LOADINGS:
Operating weight empty: MD	89,000 kg (196,210 lb)
MF	101,000 kg (222,665 lb)
Max payload: T	40,000 kg (88,185 lb)
TD	50,000 kg (110,230 lb)
MD	47,000 kg (103,615 lb)
TD from unprepared surface	33,400 kg (73,633 lb)
MF	52,000 kg (114,640 lb)
Max fuel: T	84,840 kg (187,037 lb)
Max T-O weight: LL, T	170,000 kg (374,785 lb)
MD, TD (normal)	190,000 kg (418,875 lb)
MF	210,000 kg (462,970 lb)
TD from unprepared surface	152,000 kg (335,100 lb)
MD from unprepared surface	157,500 kg (347,225 lb)
Max landing weight: LL	140,000 kg (308,640 lb)
TD	151,500 kg (333,995 lb)
MD	155,000 kg (341,715 lb)
TD on unprepared surface	135,500 kg (298,720 lb)
Permissible axle load (vehicles):	
T	7,500-11,000 kg (16,535-24,250 lb)
Permissible floor loading:	
T	1,450-3,100 kg/m² (297-635 lb/sq ft)
Max wing loading: T	566.7 kg/m² (116.05 lb/sq ft)
TD	633.3 kg/m² (129.72 lb/sq ft)
Max power loading: T	361 kg/kN (3.54 lb/lb st)
TD	404 kg/kN (3.95 lb/lb st)

PERFORMANCE:
Never-exceed speed (V_{NE}): LL	M0.77
Max level speed: LL	323 kt (600 km/h; 372 mph)
T, TD	459 kt (850 km/h; 528 mph)
Cruising speed:	
T, TD	405-432 kt (750-800 km/h; 466-497 mph)
MD, MF	405-420 kt (750-780 km/h; 466-484 mph)
T-O speed: T	114 kt (210 km/h; 131 mph)
Min flight speed: LL	151 kt (280 km/h; 174 mph)
Approach and landing speed:	
T	119-130 kt (220-240 km/h; 137-149 mph)
Normal cruising height:	
T, TD, MD, MF	9,000-12,000 m (29,500-39,370 ft)
Service ceiling: LL	12,000 m (39,380 ft)
Absolute ceiling: T	approx 15,500 m (50,860 ft)
T-O run: T	850 m (2,790 ft)
MD, TD	1,700 m (5,580 ft)
MF	1,000 m (3,280 ft)
T-O to 15 m (50 ft): MF	2,750 m (9,025 ft)
Landing from 15 m (50 ft): MF	2,500 m (8,205 ft)
Landing run: T	450 m (1,475 ft)
MD, TD	900-1,000 m (2,950-3,280 ft)
Required runway length: LL	3,000 m (9,840 ft)
Range with max payload:	
TD	1,970 n miles (3,650 km; 2,265 miles)
MD	2,051 n miles (3,800 km; 2,361 miles)
Max range, with reserves:	
T	3,617 n miles (6,700 km; 4,163 miles)
MD	4,211 n miles (7,800 km; 4,846 miles)
Range with 20,000 kg (44,090 lb) payload:	
MD, TD	3,940 n miles (7,300 km; 4,535 miles)
MF	4,643 n miles (8,600 km; 5,343 miles)

UPDATED

ILYUSHIN Il-78M
NATO reporting name: Midas

TYPE: Tanker.

PROGRAMME: Development began in late 1970s, to replace modified Myasishchev 3MS2 and 3MN2 ('Bison') used previously in this role; based on Il-76MD; prototype (SSSR-76556) first flew 26 June 1983; initial production aircraft (SSSR-76607) to SIBNIA, Novosibirsk, for trials, 19 June 1984; first operational Il-78 entered service with 409 Regiment at Uzin, Ukraine, in 1987, supporting tactical and strategic combat aircraft; some initially operated with one (fuselage) or two (wing) pod(s) only; fully developed Il-78 is three-point tanker.

CURRENT VERSIONS: **Il-78:** Initial version (also known as **Il-78T**, *Transportnyye*: Transport) with up to two cylindrical fuel storage tanks inside cargo hold; convertible into transport by removal of tanks; in addition to tanks in hold, fuel can be transferred from standard tanks in wing torsion box. Later aircraft have fuselage pod in lower position. Able also to refuel aircraft on ground, using conventional hoses. Almost all converted to transports in Ukraine from 1995, serving mostly with BSL Airlines.

Il-78M: Later standard; non-convertible, with two larger tanks in hold; lower structure weight achieved by deletion of cargo-handling equipment; increased MTOW; UPAZ-1M pods with increased delivery rate; and lower-mounted fuselage pod (as late Il-78s). External features include deletion of port crew door. Prototype (SSSR-76701) first flew 7 March 1987 and delivered to Zhukovsky for trials nine days later. First production aircraft (SSSR-78822) delivered to Engels 27 December 1989; last ('36') on 30 November 1991. (Two Il-78s – '32' and '34' – despite having port crew door and, at least in '34', openable cargo door and single removable hold tank, are classified as Il-78Ms in Russian documents, but not regarded as such here.)

Il-78V: Proposed version with Flight Refuelling Ltd Mk 32B refuelling pods.

Il-78MK: Proposed convertible tanker or transport.

CUSTOMERS: Total 36 (including prototype) Il-78s (all except two with civil registrations) and 10 Il-78Ms (all except two in military markings) have been converted; further two incomplete at Tashkent. Of early production aircraft, 23 delivered to Uzin Chepelevka and most passed on to Ukrainian Air Force (409 Aircraft Refuelling Aviation Regiment); of these, 17 sold to BSL Airlines and three remain in military service. Russian Air Forces have 21 with 200 Guards Aircraft Refuelling Aviation Regiment at Engels airbase, Saratov region, comprising nine late production Il-78s and 12 Il-78Ms. Further one at Ivanovo. Initial two of planned six for Indian Air Force were ordered on 27 November 1997. Six ex-Russian aircraft ordered by Algeria; overhauled before delivery by 123 Aviation Repair Depot at Staraya Russia; first four in service at Boufarik by early 2000.

COSTS: US$50 million (reported programme unit price, India, 1998).

DESIGN FEATURES: Development of basic Il-76MD; three-point tanker, with Zvezda UPAZ-1 Sakhalin refuelling pods under outer wings and on port side of rear fuselage; rear turret retained as flight refuelling observation station, without guns; rear-facing radar range-finder built into bottom of upswept rear fuselage. Each UPAZ-1 (*Unifitsirovanny Podvesnoy Agregaht Zaprahvky:* Standardised Suspended Refuelling Unit) has 26 m (85 ft) hose for delivery of fuel at 1,000 litres (264 US gallons; 220 Imp gallons)/min normal rate, or 2,200 litres (581 US gallons; 484 Imp gallons)/min accelerated rate. UPAZ-1M rate increased to 2,340 litres (618 US gallons; 515 Imp gallons)/min.

POWER PLANT: Four Aviadvigatel D-30KP-2 turbofans; each 117.7 kN (26,455 lb st). Fuel jettison system.

ACCOMMODATION: Crew of seven, including refuelling controller in rear turret.

AVIONICS: *Mission:* Kupol navigation system and RSBN Vstrecha (Rendezvous) short-range nav system to permit all-weather day/night mutual detection and approach by receiver aircraft from distances up to 160 n miles (300 km; 185 miles). Systems control the convergence automatically and warn of too-close approach. Refuelling permitted only in direct visibility.

EQUIPMENT: Two electric lights under rear turret illuminate lower fuselage during night refuelling.

DIMENSIONS, EXTERNAL: As Il-76, except:
Distance of refuelling pods from c/l:	
Underwing pods	16.40 m (53 ft 9¾ in)
Rear fuselage pod	3.00 m (9 ft 10¼ in)

WEIGHTS AND LOADINGS:
Weight empty: 78	98,000 kg (216,050 lb)
Fuel load: wing tanks	84,840 kg (187,035 lb)
fuselage tanks: 78	28,000 kg (61,728 lb)
78M	36,000 kg (79,366 lb)
Max T-O weight:	
on concrete runway: 78	190,000 kg (418,875 lb)
78M	210,000 kg (462,950 lb)
on semi-prepared runway with bearing strength less than 6 kg/cm² (85 lb/sq in):	
78, 78M	157,500 kg (347,225 lb)

Final production Il-78M flight refuelling tanker *(Paul Jackson/Jane's)*

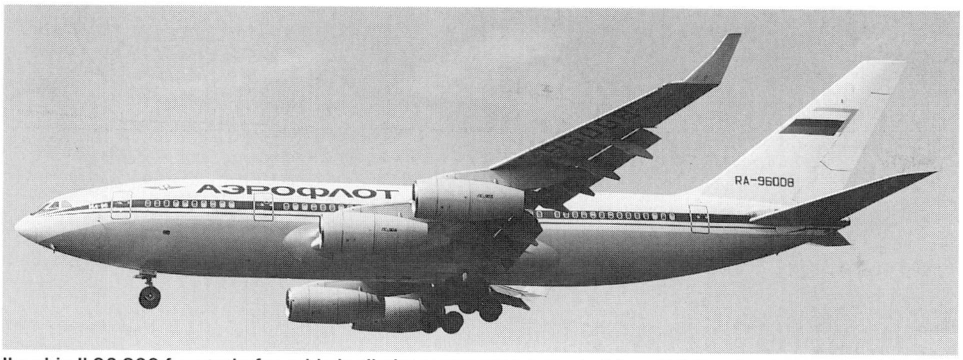

Ilyushin Il-96-300 four-turbofan wide-bodied passenger transport of Aeroflot Russian International Airlines *(Mark Wagner/Flight International)*

Max landing weight: on concrete runway:
78, 78M 151,500 kg (334,000 lb)
on semi-prepared runway:
78, 78M 135,000 kg (297, 625 lb)

PERFORMANCE:
Nominal cruising speed:
78 405 kt (750 km/h; 466 mph)
Refuelling speed:
78 232-318 kt (430-590 km/h; 267-366 mph)
Refuelling height: 78 2,000-9,000 m (6,560-29,525 ft)
T-O run, 78M: concrete runway 2,080 m (6,825 ft)
semi-prepared runway 1,400 m (4,595 ft)
Service ceiling: 78M 12,000 m (39,370 ft)
Refuelling radius, 78:
with 60,000-65,000 kg (132,275-143,300 lb) transfer fuel 540 n miles (1,000 km; 620 miles)
with 32,000-36,000 kg (70,545-79,365 lb) transfer fuel 1,350 n miles (2,500 km; 1,553 miles)

UPDATED

ILYUSHIN Il-96-300

TYPE: Wide-bodied airliner.

PROGRAMME: First of two flying prototypes (SSSR-96000) flew at Khodinka 28 September 1988 and exhibited at Paris in June 1989; second (SSSR-96001) first flew 28 November 1989; further airframe used for static and fatigue testing; all three built at GAZ 30, Khodinka; areas of commonality with Il-86 permitted planned test programme to be reduced to 750 flights totalling 1,200 hours; route proving trials by second production Il-96-300 (SSSR-96005) conducted late 1991; production at VAPO, Voronezh; certification received 29 December 1992.

Following initially poor reliability, Aeroflot (ARIA) serviceability improved, as evidenced by TBO increase from 1,600 to 4,130 hours between 1997 and 1999. PS-90A engine in-flight shutdown had improved to one per 29,000 hours by 1999.

CURRENT VERSIONS: **Il-96-300:** *As described.*

Il-96-300V: Extended-range version on order by Vnukovo Airlines, 1999. Increased fuel for additional 1,200 miles (2,222 km; 1,380 miles).

Il-96M, Il-96T: Described separately.

Il-96PU: One aircraft (RA-96012) built for use of Russian president; VIP interior, additional communications facilities and medical centre.

Il-96-300 Freighter: Design under way by early 2000; development partnership also includes VAPO and Russian banks. To carry 70,000 kg (154,325 lb) payload over 4,859 n miles (9,000 km; 5,592 miles); uprated PS-90 engines each 176.5 kN (39,683 lb st). Estimated unit cost US$30 million (2000).

CUSTOMERS: Aeroflot Russian International Airlines (six delivered by 1996); Domodedovo Airlines (four on order, of which third was delivered 16 April 1999); Russia State Transport Company (three; one delivered 1995; two completed, but not delivered, in 1997); and Atlant Soyuz (received first production aircraft in 1999, following trials use). Total 11 production aircraft delivered: 10th in February 1996; 11th in April 1999. Aeroflot signed lease agreement with Ilyushin Finance in November 1999 for further six, plus sale, refurbishment and leaseback of one of original six; contract involves uncompleted aircraft on production line, permitting deliveries between 2001 and 2003.

COSTS: US$30 million, flyaway (1999).

DESIGN FEATURES: Superficial resemblance to Il-86, but new design, with different engines to overcome performance deficiencies of original Il-86; new structural materials and state-of-the-art technology intended to provide life of 60,000 hours and 12,000 landings; no lower deck passenger entry. Current development aiming at range of 6,475 n miles (12,000 km; 7,450 miles) with 300 passengers.

Conventional wide-bodied airliner, with low wing (and winglets) and four pod-mounted engines.

Supercritical wings, with 30° sweep at quarter-chord; sweepback at quarter-chord 37° 30′ on tailplane, 45° on fin.

FLYING CONTROLS: Triplex fly-by-wire, with manual reversion; each wing trailing-edge occupied by, from root, double-slotted inboard flap, small inboard aileron, two-section single-slotted flaps, and outboard aileron used only as gust damper and to smooth out buffeting; seven-section full-span leading-edge slats on each wing; three airbrakes forward of each inboard trailing-edge flap; six spoilers forward of outer flaps; inboard pair supplement ailerons, others operate as airbrakes and supplementary ailerons; variable incidence tailplane; two-section rudder and elevators, without tabs.

STRUCTURE: Basically all-metal, including new high-purity aluminium alloy, with composites flaps, maindeck floors and underfloor holds of honeycomb and CFRP; inner wings three-spar, outer panels two-spar; each wing has seven machined skin panels, three top surface, four bottom, with integral stiffeners; circular-section semi-monocoque fuselage; leading- and trailing-edges of fin and tailplane of composites. Some components manufactured by PZL Aircraft Factory, Poland.

LANDING GEAR: Retractable four-unit type. Forward-retracting steerable twin-wheel nose unit; three four-wheel bogie main units. Two of latter retract inward into wingroot/fuselage fairings; third is mounted centrally under fuselage, to rear of others, and retracts forward after the bogie has itself pivoted upward 20°. Oleo-pneumatic shock-absorbers. Nosewheel tubeless tyres size 1,260×460; mainwheel tubeless tyres size 1,300×480. Tyre pressure (all) 11.65 bars (169 lb/sq in).

POWER PLANT: Four Aviadvigatel PS-90A turbofans, each 156.9 kN (35,275 lb st), on pylons forward of wing leading-edges. Thrust reversal standard. Integral fuel tanks in wings and fuselage centre-section, total capacity 148,260 litres (39,166 US gallons; 32,613 Imp gallons).

ACCOMMODATION: Pilot, co-pilot and flight engineer; two seats for supplementary crew or observer. Ten or 12 cabin staff. Basic all-tourist configuration has two cabins for 66 and 234 passengers respectively, nine-abreast at 87 cm (34.25 in) seat pitch, separated by buffet counter, video stowage and two lifts from galley on lower deck. Two aisles, each 55 cm (21.65 in) wide. Two toilets and wardrobe at front; six more toilets, a rack for cabin staff's belongings and seats for cabin staff at rear. Seats recline, and are provided with individual tables, ventilation, earphones and attendant call button. Indirect lighting is standard. 235-seat mixed class version has front cabin for 22 first class passengers, six-abreast in pairs, at 102 cm (40 in) seat pitch and with aisles 75.5 cm (29.7 in) wide; centre cabin with 40 business class seats, eight-abreast at 90 cm (35.4 in) seat pitch and with aisles 56.5 cm (22.25 in) wide; and rear cabin for 173 tourist class passengers, basically nine-abreast at 87 cm (34.25 in) seat pitch, with aisle width of 55 cm (21.65 in).

Future Aeroflot standard, defined late 1999, is 12 first class, 21 business class and 180 economy seats, plus complete interior renewal. Refurbishment cost (including some Western avionics) US$1.5 million per aircraft.

Passenger cabin is entered through three doors on port side of upper deck, at front and rear and forward of the wings. Opposite each door, on starboard side, is emergency exit door. Lower deck houses front cargo compartment for six ABK-1.5 (LD3) containers or igloo pallets, central compartment aft of wing for 10 ABK-1.5 containers or pallets, and tapering compartment for general cargo at rear. Three doors on starboard side provide separate access to each compartment. Galley and lifts are between front cargo compartment and wing, with separate door aft of front cargo compartment door.

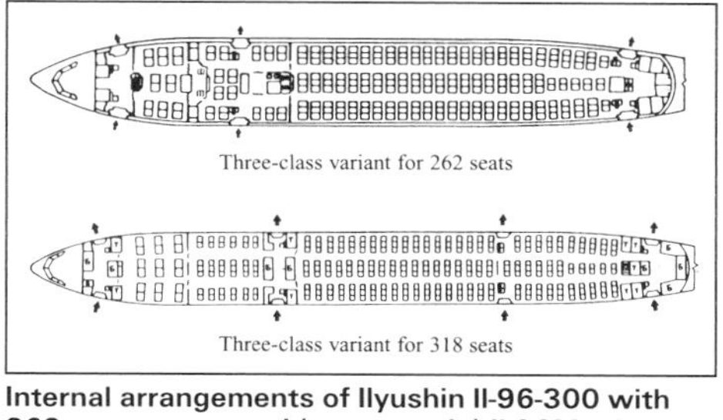

Internal arrangements of Ilyushin Il-96-300 with 262 passengers and (not to scale) Il-96M with 318
0011985

SYSTEMS: Four independent hydraulic systems, using fireproof and explosion-proof fluid, at pressure of 207 bars (3,000 lb/ sq in). APU in tailcone for engine starting and air conditioning on ground.

AVIONICS: *Flight:* Triplex flight control and flight management systems, together with a head-up display, permit fully automatic en route control and operations in ICAO Cat. IIIa minima. Duplex engine and systems monitoring and failure warning systems feed in-flight information to both the flight engineer's station and monitors on the ground. Autothrottle is based on IAS, without angle of attack protection.

Instrumentation: Primary flight information is presented on dual twin-screen colour CRTs, fed by triplex INS, a satellite-based and Omega navigation system and other sensors. Another electronic system provides real-time automatic weight and CG situation data.

DIMENSIONS, EXTERNAL:
Wing span: excl winglets 57.66 m (189 ft 2 in)
over winglets 60.11 m (197 ft 2½ in)
Wing aspect ratio 9.5
Length overall 55.35 m (181 ft 7¼ in)
Fuselage: Length 51.15 m (167 ft 9¾ in)
Max diameter 6.08 m (19 ft 11½ in)
Height overall 17.55 m (57 ft 7 in)
Tailplane span 20.57 m (67 ft 6 in)
Wheel track 10.40 m (34 ft 1½ in)
Wheelbase 20.07 m (65 ft 10 in)
Passenger doors (three): Height 1.83 m (6 ft 0 in)
Width 1.07 m (3 ft 6 in)
Height to sill: Nos. 1 and 2 4.54 m (14 ft 10¾ in)
No. 3 4.80 m (15 ft 9 in)
Emergency exit doors (three):
Height 1.825 m (5 ft 11¾ in)
Width 1.07 m (3 ft 6 in)
Cargo compartment doors (front and centre):
Height 1.825 m (5 ft 11¾ in)
Width 1.78 m (5 ft 10 in)
Height to sill: front 2.34 m (7 ft 8¼ in)
centre 2.48 m (8 ft 1¾ in)
Cargo compartment door (rear):
Height 1.38 m (4 ft 6¼ in)
Width 0.97 m (3 ft 2¼ in)
Height to sill 2.74 m (9 ft 0 in)
Galley door: Height 1.20 m (3 ft 11¼ in)
Width 0.80 m (2 ft 7½ in)

DIMENSIONS, INTERNAL:
Cabins, excl flight deck: Height 2.60 m (8 ft 6¼ in)
Max width approx 5.70 m (18 ft 8½ in)
Volume 350.0 m³ (12,360 cu ft)
Cargo hold volume: front 37.1 m³ (1,310 cu ft)
centre 63.8 m³ (2,253 cu ft)
rear 15.0 m³ (530 cu ft)

AREAS:
Wings, gross 391.60 m² (4,215.1 sq ft)
Vertical tail surfaces (total) 61.00 m² (656.60 sq ft)
Horizontal tail surfaces (total) 96.50 m² (1,038.75 sq ft)

WEIGHTS AND LOADINGS:
Basic operating weight 117,000 kg (257,950 lb)
Max payload 40,000 kg (88,185 lb)
Max fuel 114,900 kg (253,310 lb)
Max T-O weight 216,000 kg (476,200 lb)
Max landing weight 175,000 kg (385,800 lb)
Max zero-fuel weight 157,000 kg (346,125 lb)
Max wing loading 551.6 kg/m² (112.97 lb/sq ft)
Max power loading 344 kg/kN (3.37 lb/lb st)

PERFORMANCE (estimated):
Normal cruising speed at 10,100-12,100 m (33,135-39,700 ft) 459-486 kt (850-900 km/h; 528-559 mph)
Approach speed 140-146 kt (260-270 km/h; 162-168 mph)
Balanced T-O runway length 2,600 m (8,530 ft)
Balanced landing runway length 1,980 m (6,500 ft)
Range, with UASA reserves: with max payload 4,050 n miles (7,500 km; 4,660 miles)
with 30,000 kg (66,140 lb) payload 4,860 n miles (9,000 km; 5,590 miles)
with 15,000 kg (33,070 lb) payload 5,940 n miles (11,000 km; 6,835 miles)

OPERATIONAL NOISE LEVELS: Il-96-300 is designed to conform with ICAO Chapter 3 Annex 16 noise requirements

UPDATED

ILYUSHIN Il-96M and Il-96T

TYPE: Wide-bodied airliner.

PROGRAMME: Projected initially as Il-96-350 stretched version of -300 with Western engines and avionics; designation changed to Il-96M in 1990, when model exhibited at Moscow Air Show '90; then at initial design stage; Pratt & Whitney supplied 10 PW2337 engines for two certification aircraft; conversion of Il-96-300 prototype (RA-96000) to **Il-96MO** (*Opytni:* Test) prototype began early 1992; rolled out at Khodinka 30 March 1993; first flight 6 April 1993; initial order (Aeroflot, 20) placed 3 December 1996.

Planned certification to FAR Pt 25 and ICAO Annex 16 noise levels was to be undertaken in 1996, deliveries beginning the same year from VAPO at Voronezh; however, first production aircraft (an Il-96T, RA-96101) not rolled out until 26 April 1997; first Il-96M was to have flown in mid-1998, but did not do so. Il-96T received

Prototype of increased-capacity Ilyushin Il-96M four-turbofan wide-bodied transport *(Paul Jackson/Jane's)*

First production Ilyushin Il-96T (RA-96101) in ARIA colours *(Paul Jackson/Jane's)*

provisional Russian certification on 31 March 1998 for operation by CIS airlines and FAR Pt 25 approval (after 1,242 flying hours; 593 sorties) on 2 June 1999, clearing way for US Exim Bank funding for purchases of avionics (granted October 2000) and engines. ARIA requirement reconfirmed mid-1998, when Russian banks (National Credit and Vneshtorg) agreed finance for production of airframes. By mid-1999, Western investment included US$50 million from Rockwell Collins and US$80 million from Pratt & Whitney.

CURRENT VERSIONS: **Il-96M:** Basic production version. Deliveries to ARIA scheduled in 1998 (one), 1999 (three), 2000 (five), 2001 (five) and 2002 (three); however, aircraft not certified by mid-2000 and programme has slipped by several years.

Description generally for Il-96M.

Il-96MK: Projected development with Samara NK-93 ducted engines rated at 175 to 195 kN (38,000 to 43,000 lb st), with 17:1 or 18:1 bypass ratio.

Il-96MR: Il-96M proposed with Aviadvigatel PS-90A or Samara NK-92 engines.

Il-96T: Freighter; cargo door 4.85 × 2.87 m (15 ft 11 in × 9 ft 5 in) forward of wing on port side; to carry standard international containers and pallets. First flight (RA-96101) 16 May 1997; initial variant to be certified (2 June 1999) by FAA. Was due to have begun Aeroflot proving flights in April 2000.

Il-96TR: Proposed early 2000; Il-96T powered by uprated PS-90 engines of 176.5 kN (39,683 lb st).

CUSTOMERS: Aeroflot Russian International Airlines (17 Il-96Ms and three Il-96Ts, all leased from National Reserve Bank of Russia – American International Trade Finance consortium; Il-96Ts originally intended for 1997-98 delivery); Transaero (five Il-96Ms, plus six options, ordered June 1997, notionally for delivery from 2001); Volga-Dnepr (four Il-96Ts, plus two options; commitment signed 20 August 1997, originally for 1999 delivery).

COSTS: Basic price US$75 million (1997).

DESIGN FEATURES: Basically as Il-96-300; lengthened fuselage of unchanged cross-section, permitting smaller tailfin; wings identical.

POWER PLANT: Four Pratt & Whitney PW2337 turbofans, each 164.6 kN (37,000 lb st); nacelles supplied by Goodrich, USA. Aviadvigatel PS-90P intended to be eventual primary engine.

ACCOMMODATION: Two flight crew; three passenger arrangements proposed: (1) 18 first class passengers at 152.4 cm (60 in) seat pitch, with 0.77 m (2 ft 6¼ in) aisle; 44 business class at 91.5 cm (36 in) pitch, with 0.625 m (2 ft 0½ in) aisle; 250 tourist class at 86.4 cm (34 in) pitch, with 0.55 m (1 ft 9½ in) aisle; (2) 85 business class and 250

Ilyushin Il-96 cargo door *(Paul Jackson/Jane's)*

tourist class; (3) three tourist class cabins for 124, 162 and 89 passengers. Eight emergency exits. Underfloor hold for 32 standard LD3 containers.

AVIONICS: Rockwell Collins digital EFIS with DO-178A standard software. Designed for Cat. IIIb fully automatic landings.

Comms: Rockwell Collins Srs 700 com/nav radios; Ball Airlink conformal antennas; SAT-900 satellite com.

Radar: Predictive weather radar.

Flight: Triple-redundant Smiths Industries flight management system to ARINC 700 specifications, three inertial platforms with laser gyros, GPWS, Litton LTN-2001 GPS and Glonass satellite nav receivers, Litton Flagship LTN-101 inertial reference centre, Rockwell Collins ASCS (aircraft systems control system).

Instrumentation: Six 203 × 203 mm (8 × 8 in) CRTs; EICAS.

DIMENSIONS, EXTERNAL: As for Il-96-300 except:
Length overall: Il-96M	64.695 m (212 ft 3 in)
Il-96T	63.94 m (209 ft 9¼ in)
Fuselage (Il-96M): Length	60.50 m (198 ft 6 in)
Max diameter	6.08 m (19 ft 11½ in)
Height overall	15.72 m (51 ft 7 in)
Wheel track	10.40 m (34 ft 1½ in)

DIMENSIONS, INTERNAL:
Cabin (incl flight deck; Il-96M):	
Length	49.13 m (161 ft 2¼ in)
Cargo hold (Il-96T): Length	44.24 m (145 ft 1¾ in)

WEIGHTS AND LOADINGS:
Operating weight empty	132,400 kg (291,900 lb)
Max payload: Il-96M	58,000 kg (127,870 lb)
Il-96T	92,000 kg (202,825 lb)
Max T-O weight	270,000 kg (595,225 lb)
Max wing loading	689.5 kg/m² (141.22 lb/sq ft)
Max power loading	410 kg/kN (4.02 lb/lb st)

PERFORMANCE:
Max operating Mach No. (M_{MO})	0.86
Normal cruising speed at 9,000-12,000 m (29,500-39,370 ft):	
Il-96M	448 kt (830 km/h; 516 mph)
Il-96T	459-469 kt (850-870 km/h; 528-540 mph)
T-O run: Il-96M	2,500 m (8,205 ft)
Il-96T	3,350 m (11,000 ft)
Landing run: Il-96M	2,500 m (8,205 ft)
Il-96T	2,400 m (7,875 ft)
Range:	
Il-96M, 30,000 kg; 66,138 lb payload, international rules	6,195 n miles (11,482 km; 7,136 miles)
Il-96T: 92,000 kg (202,825 lb) payload	2,807 n miles (5,200 km; 3,231 miles)
58,000 kg (127,870 lb) payload	5,237 n miles (9,700 km; 6,027 miles)

UPDATED

ILYUSHIN Il-98

Projected twin-engined version of Il-96M/T; originally Il-96MD; under consideration upon 50 firm commitments. Manufacture by MAPO at Lukhovitsy (LMZ). Estimated sales in excess of 500.

CURRENT VERSIONS: **Il-100:** Baseline version; 12 seats.
Il-100A: Stretched version; 18-20 seats.

COSTS: US$1 million to US$1.2 million (2000).

DESIGN FEATURES: Replacement for Antonov An-2 utility biplane, but with 40 per cent reduction in direct operating cost. Rugged, unpressurised and simple to maintain; STOL performance. High-wing, T-tail configuration with unswept rear fuselage, but only port side freight door. Airframe life 60,000 cycles.

WEIGHTS AND LOADINGS:
Max payload	1,500 kg (3,307 lb)

PERFORMANCE:
Cruising speed	200 kt (370 km/h; 230 mph)
T-O run	400 m (1,315 ft)
Landing run	260 m (855 ft)

NEW ENTRY

ILYUSHIN Il-103

TYPE: Four-seat lightplane.

PROGRAMME: Exhibited in model form at Moscow Aerospace '90; programme go-ahead 1990; first flight (RA-10321) 17 May 1994; second prototype (RA-10302) flew 30 January 1995 at LMZ, Lukhovitsy, following August 1993 decision to establish a production line there; preseries batch comprised three flyable and two static test aircraft, former having flown 414 sorties in two years after maiden flight; first production aircraft flew 30 January 1995; by June 1997, total of seven flying and 1,000 hours accumulated. Type certificate by Russian authorities under AP-23 received 15 February 1996; production approval certificates issued 7 March 1998 and 7 April 1998; airworthiness certificate granted 9 July 1997. Amended Russian certification granted 4 December 1997, following addition of GPS, radio compass and AHRS to avionics

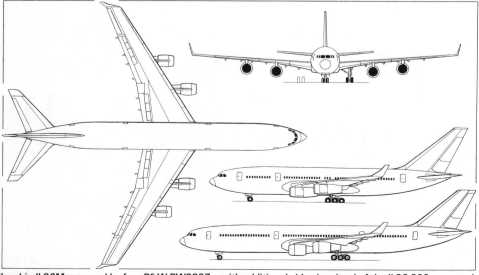

Ilyushin Il-96M powered by four P&W PW2337s, with additional side view (top) of the Il-96-300, powered by four Aviadvigatel PS-90As *(Dennis Punnett/Jane's)*

ILYUSHIN—AIRCRAFT: RUSSIAN FEDERATION

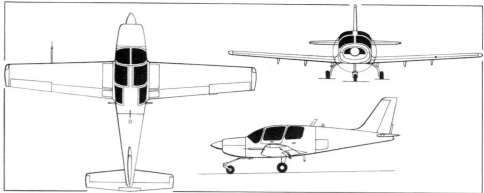

Ilyushin Il-103 two/five-seat multipurpose light aircraft *(James Goulding/Jane's)* 2000/0079319

Instrument panel of the Il-103-10
(Paul Jackson/Jane's) 1999/0008077

suite. FAR Pt 23 certification achieved 9 December 1998 (first by a Russian-built aircraft) after 110-hour/208-sortie programme. Initial export order, six for Peruvian Air Force, agreed June 1999 and finalised in April 2000; however, deliveries had not begun by August 2000, due to lack of payment.

CURRENT VERSIONS: **Il-103-01**: Baseline VFR version for Russian Federation market.

Il-103-10: Export version with fully upgraded avionics suitable for international airways navigation.

Il-103-11: Export version with partly upgraded avionics suitable for local air navigation.

Initial certification is for a max T-O weight of 1,285 kg (2,832 lb), as reflected in weight and performance data below. The Il-103 will later be certified at 1,310 kg (2,880 lb) and, ultimately, 1,460 kg (3,218 lb).

Il-103SKh: Crop-sprayer (*selskokhozyaistvenny*: agricultural) version; first flown 29 March 2000.

CUSTOMERS: Production totalled 25 by May 1999, operators then including Il-Service, Cherepovets Aviation Club, Avialesookhrana (forestry), Special Ecological Aviation Centre, Tatarstan National Flying Club, Civil Aviation Academy and St Petersburg Methodological Centre. Avialine requires fleet of 12 for air taxi service from Moscow-Domodedovo. Federal Fire Service interested in 30. Uzbekistan government announced requirement for 100 in crop-spraying role, late 1999. Russian Federal Air Transport service ordered 30 in 1999 for issue to several flying schools, but total firm orders quoted as 20 (including Peru) in mid-2000. Sales in 2000 included four to GP Bellesavia of Belarus in September for forest fire-fighting; and one to Russian Forestry Service. Ilyushin foresees Russian market for 1,500; plans to produce 100 per year.

COSTS: US$164,000 to 210,000 (2000).

DESIGN FEATURES: Conventional low-wing monoplane, with non-retractable landing gear, originally to meet DOSAAF requirement for 500 military/civil pilot trainers. Designed for daytime VFR flying, in non-icing conditions and ambient air temperature of −35 to +45°C; intended service life 14,000 hours/20,000 landings/15 years.

Wing sweep 0° at quarter-chord, slight dihedral from roots, twist 0°, thickness/chord ratio 16 per cent at root, 15 per cent at tip, taper ratio 1.9.

FLYING CONTROLS: Conventional and manual. Actuation by pushrods, except cable-actuated rudder; horn-balanced rudder and single-piece elevator; single-slotted trailing-edge flaps, 0° or 10° deflection; electrically actuated elevator trim tab. Deflection angles: ailerons +28°/−25°, elevator +20°/−25°, rudder ±25°. Ground-adjustable trim tabs on starboard aileron and rudder.

STRUCTURE: All-metal, basically aluminium alloy, except for titanium firewall frame and wingroot attachments; bonded GFRP wingtips, elevator and rudder tips and elevator tab; and small amounts of magnesium alloys and Al-Li alloys. Single-spar wings, with front and rear false spars, integral fuel tanks and detachable leading-edge, mounted at sides of fuselage. Main spar is riveted beam with extruded caps; false spars stamped from sheet alloy; rolled sheet stringers. Semi-monocoque front/centre fuselage with inbuilt wing carry-through structure; separate rear fuselage; separate tailcone. Longitudinal fuselage members comprise two reinforced spars in cabin floor, two side-ribs of centre wing section and three beams for landing gear attachment. Two-spar, single-piece tailplane attached to fuselage by four bolts; two-spar fin. Detachable wings, fin and tailplane.

LANDING GEAR: Non-retractable tricycle type, with single wheel on each unit. Cantilever spring nose and mainwheel legs of titanium alloy; castoring nosewheel with shimmy damper; oleo-pneumatic nosewheel leg tested on RA-10300 in 1998. Mainwheel tyres size 400×150-115, pressure 3.95 bars (57 lb/sq in) on KT-214-1 wheels; nosewheel tyre size 310×135-99, pressure 3.43 bars (50 lb/sq in) on K-290 wheels. Optional tyres 6.00-6 and 5.00-5, respectively, on Parker 40-75B and 40-77B wheels. Multidisc hydraulic anti-lock brakes on mainwheels, pedal-actuated. Turning radius (outboard wheel) 4.70 m (15 ft 6 in). Floats and skis optional.

POWER PLANT: One 157 kW (210 hp) Teledyne Continental IO-360-ES2B flat-six engine; Hartzell BHC-C2YF-1BF/F8459A-8R metal two-blade variable-pitch propeller. (Manual pitch control on three development aircraft; automatic pitch selection on production aircraft.) Alternative 194 kW (260 hp) Textron Lycoming or similarly rated Teledyne Continental engine under consideration. Fuel and oil systems suitable for inverted flight; two main fuel tanks in wingroots, total capacity 200 litres (52.8 US gallons; 44.0 Imp gallons); supply tank, capacity 3 litres (0.8 US gallon; 0.66 Imp gallon), in fuselage forward of wing front spar carry-through; gravity refuelling points in wingroots.

ACCOMMODATION: Two forward-folding seats side by side at front of cabin; bench seat for two adults or three children at rear; optional control wheel in place of standard stick; optional dual controls; optional front seats for parachutes; space for 220 kg (485 lb) freight with rear bench seat removed; two gull-wing window/doors, hinged on centreline, at front of canopy. Rear windows removable for ground emergency exit. Unrestricted access to baggage hold.

SYSTEMS: Cabin ventilated and heated and windscreen demisted electrically by fan heater. Electrical system 27 V DC, with 1,800 W 60 A generator and 25 Ah battery.

AVIONICS: *Comms:* Yurok VHF radio as standard; P-855A1 optional extra. Il-103-11 equipment includes UBD transponder. Il-103-10 has Honeywell KX 165 nav/com with glide slope receiver (replacing Russian radios) and KT 76A transponder.

Flight: BUR-4 flight data recorder.

Il-103-11 includes MKS-1 compass and Honeywell KR 87 ADF and KLN 89B GPS. Il-103-10 additionally equipped with VOR/ILS, KN 63 DME, KMA 24 marker beacon receiver, KCS 55A compass and KEAA130A encoding altimeter.

DIMENSIONS, EXTERNAL:
Wing span	10.56 m (34 ft 7¾ in)
Wing aspect ratio	7.6
Wing chord: at root	1.825 m (5 ft 11¾ in)
at tip	0.96 m (3 ft 1¾ in)
Length overall	8.00 m (26 ft 3 in)
Fuselage: Length	7.81 m (25 ft 7½ in)
Max height	1.42 m (4 ft 8 in)
Max width	1.40 m (4 ft 7 in)
Height overall	3.135 m (10 ft 3½ in)
Tailplane span	3.90 m (12 ft 9½ in)
Wheel track	2.405 m (7 ft 10¾ in)
Wheelbase	2.045 m (6 ft 8½ in)
Propeller diameter	1.93 m (6 ft 4 in)
Baggage door: Width	0.70 m (2 ft 3½ in)
Height	0.34 m (1 ft 1¼ in)

DIMENSIONS, INTERNAL:
Cabin: Length	2.65 m (8 ft 8¼ in)
Max height	1.27 m (4 ft 2 in)
Max width	1.30 m (4 ft 3 in)

AREAS:
Wings, gross	14.71 m² (158.4 sq ft)
Ailerons (total)	1.137 m² (12.24 sq ft)
Flaps (total)	2.204 m² (23.72 sq ft)
Fin	0.84 m² (9.04 sq ft)
Rudder	0.56 m² (6.03 sq ft)
Tailplane	1.48 m² (15.93 sq ft)
Elevator	1.56 m² (16.79 sq ft)

WEIGHTS AND LOADINGS:
Weight empty	900 kg (1,984 lb)
Baggage capacity	60 kg (132 lb)
Max payload	270 kg (595 lb)
Max fuel	150 kg (330 lb)
Max T-O weight (Utility)	1,285 kg (2,832 lb)
Max wing loading	87.4 kg/m² (17.89 lb/sq ft)
Max power loading	8.21 kg/kW (13.48 lb/hp)

PERFORMANCE:
Never-exceed speed (V_{NE})	183 kt (340 km/h; 211 mph)
Max level speed	119 kt (220 km/h; 137 mph)
Cruising speed	97 kt (180 km/h; 112 mph)
Stalling speed: flaps up	64 kt (117 km/h; 73 mph)
10° flap	60 kt (111 km/h; 69 mph)
Max rate of climb at S/L	190 m (623 ft)/min
Max certified altitude	3,000 m (9,840 ft)
T-O run	380 m (1,250 ft)
Landing run	320 m (1,050 ft)

Max range at cruising speed, pilot and 270 kg (595 lb) payload, 30 min reserves
432 n miles (800 km; 497 miles)

g limits: Utility	+4.4/−1.8
Aerobatic	+6/−3

OPERATIONAL NOISE LEVELS: Designed to conform with GOST 23023-85 and ICAO Annex 16

UPDATED

Prototype Il-103, with ski landing gear in place of experimental oleo-pneumatic nosewheel installation *(Yefim Gordon)* 2000/0079321

Instrument panel of the less comprehensive -11 *(Paul Jackson/Jane's)* 1999/0008078

Ilyushin Il-103 four-seat tourer and trainer *(Yefim Gordon)* 2001/0092069

ILYUSHIN Il-112

TYPE: Twin-turboprop transport.
PROGRAMME: Initiated 1994, with possibility of Russian government funding as alternative to Ukrainian Antonov An-38; design since enlarged and considerably refined; planned manufacture by KAPP at Kumertau and (Il-112V) VAPO at Voronezh.

Revealed July 1999 that Il-112V selected by Russian defence ministry as replacement for Antonov An-26; significant performance improvements to be achieved by redesign begun in same month. Presidential approval granted April 2000.

CURRENT VERSIONS: **Il-112**: Passenger transport, *as described in detail.*

Il-112 business transport: Basically as Il-112 airline version, but cabin divided into three sections, separated by curtains. Front cabin contains four pairs of facing seats, with tables between, and centre aisle, a wardrobe on the port side between the door and flight deck bulkhead, inward-opening door to flight deck, and large baggage compartment on starboard side with outside access.

Central compartment has four armchairs on port side, with tables between, three-seat sofa on starboard side to rear of cabinets; aisle width between armchairs and sofa 86 cm (34 in); wardrobe at rear on each side. Rear section contains main passenger door with airstairs on port side, two buffets, seat for attendant, and access to rear baggage compartment and lavatory.

Il-112T: Freighter with rear-loading ramp; lavatory and seat for loadmaster at front of unobstructed main hold; ramp forms rear wall of hold when stowed. Typical payloads include four 1AK-1.5 (LD3) or five 3AK-1 containers or five PA-1.5 pallets. Airstair door at front of hold on port side; service door opposite; emergency exit on each side in centre of hold.

Il-112V (*voenny:* military): Passenger/freighter for Russian armed forces. Rear ramp; provision for 35 passengers, or 18 stretchers or 34 paratroops. Patrol/surveillance version also to be produced.

DESIGN FEATURES: Based on Il-114 airframe. Conventional high-wing monoplane with T tail; constant-chord wing centre-section, tapered outer panels; sweptback fin and rudder; circular-section pressurised fuselage.

FLYING CONTROLS: Entire trailing-edge of each wing made up of horn-balanced aileron and two-section, area-increasing flaps; two-section spoilers forward of flap on each side of centre-section. Horn-balanced rudder and elevators; tab in rudder.

LANDING GEAR: Retractable tricycle type; twin-wheel nose unit; single wheel on each main unit, retracting into large fairing outside fuselage pressure cell.

POWER PLANT: Two 1,839 kW (2,466 shp) Klimov TV7-117S turboprops.

ACCOMMODATION: Flight crew of two; standard seating for 40 passengers, with 11 pairs of seats on port side and nine on starboard side of 45 cm (17.7 in) wide centre aisle. Baggage/freight compartment with outside access on starboard side at front of cabin. Main airstair passenger door at rear of cabin on port side, with service door opposite; emergency exits at front of cabin on port side, in centre on starboard side. Access to rear baggage compartment and lavatory at rear of cabin; buffet and seat for attendant.

DIMENSIONS, EXTERNAL:
Wing span: except Il-112V	22.55 m (73 ft 11¾ in)
Il-112V	23.45 m (76 ft 11¼ in)
Length: except Il-112V: overall	21.78 m (71 ft 5½ in)
fuselage	20.00 m (65 ft 7½ in)
Il-112V	20.65 m (67 ft 9 in)
Height overall: except Il-112V	7.63 m (25 ft 0¼ in)
Il-112V	7.90 m (25 ft 11 in)
Tailplane span	7.14 m (23 ft 5 in)
Wheel track	3.80 m (12 ft 5½ in)
Passenger door: Height	1.70 m (5 ft 7 in)
Width	0.76 m (2 ft 6 in)
Service door: Height	1.38 m (4 ft 6¼ in)
Width	0.72 m (2 ft 4¼ in)
Forward baggage door: Height	1.30 m (4 ft 3 in)
Width	0.96 m (3 ft 1¾ in)
Emergency exits (each): Height	0.91 m (2 ft 11¾ in)
Width	0.51 m (1 ft 8 in)

WEIGHTS AND LOADINGS (A: passenger transport, B: business transport, C: Il-112V):
Weight empty, equipped: A, B, C	9,100 kg (20,062 lb)
Max payload: A, B	4,000 kg (8,818 lb)
C	6,000 kg (13,227 lb)
Max T-O weight: A	14,530 kg (32,033 lb)
B	13,490 kg (29,740 lb)
C	16,700 kg (36,817 lb)
Max power loading: A	3.95 kg/kW (6.49 lb/shp)
B	4.67 kg/kW (6.03 lb/shp)

PERFORMANCE (estimated):
Nominal cruising speed at 7,600 m (24,950 ft)	
A, B	297-324 kt (550-600 km/h; 342-373 mph)
C	351 kt (650 km/h; 404 mph)
T-O run: A	520 m (1,710 ft)
B	500 m (1,640 ft)
C	650 m (2,135 ft)
T-O balanced field length: A	980 m (3,215 ft)
B	940 m (3,085 ft)
C	1,200 m (3,940 ft)
Landing balanced field length: A, B	1,000 m (3,280 lb)
Landing run: A, B	400 m (1,315 ft)
Range: A	809 n miles (1,500 km; 932 miles)
B	1,943 n miles (3,600 km; 2,236 miles)
C: with 6,000 kg (13,227 lb) payload	540 n miles (1,000 km; 621 miles)
with 2,000 kg (4,409 lb) payload	3,239 n miles (6,000 km; 3,728 miles)

UPDATED

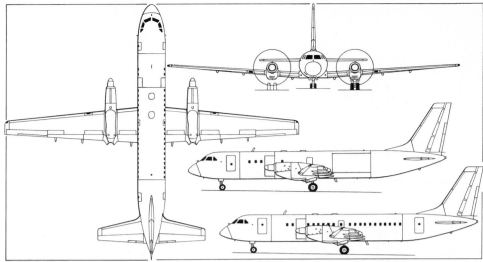

Ilyushin Il-114 short-range passenger transport, with additional side view (upper) of Il-114T freighter *(Mike Keep/Jane's)* 0011989

Prototype Ilyushin Il-114-100 (P&WC PW127H turboprops) *(Paul Jackson/Jane's)* 2001/0092070

ILYUSHIN Il-114

TYPE: Twin-turboprop airliner.
PROGRAMME: Design finalised 1986, as replacement for aircraft in An-24 class; was scheduled to enter service (with Aeroflot's Tashkent division) in 1992. Prototype (SSSR-54000) first flew at Khodinka 29 March 1990; second prototype (SSSR-54001) flew at Khodinka 24 December 1991, but lost in accident on 5 July 1993, resulting in withdrawal of government funding. Series production by TAPO at Tashkent, Uzbekistan, initial aircraft comprising Nos. 0101, 0103 and 0105 for flight development plus 0102 for static tests and 0104 for dynamic tests; first production aircraft flew 7 August 1992. Certification delayed by loss of second prototype and deferred certification of TV7 engines, but finally achieved on 26 April 1997. Negotiations reportedly under way in 1996 for production at Esfahan by Iran Aircraft Manufacturing Company; this venture lapsed upon selection of rival Antonov An-140. Recent development effort centred on -100 version.

In 1998, Russian and Uzbekistani governments agreed to promote Il-114; Ilyushin and Inkombank signed provisional agreement for production funding (although the bank's trading licence was revoked shortly afterwards, resulting in further delays); and Uzbekistan Airlines took delivery of production aircraft, making first commercial flight on 27 August 1998.

CURRENT VERSIONS: **Il-114**: *As described in detail.*

Il-114T: Cargo version developed for Uzbekistan Airlines; port freight door, size 3.25 × 1.715 m (10 ft 8 in × 5 ft 7½ in) in rear fuselage; removable roller floor; cargo attendants' cabin at forward end of freight hold accommodates up to two persons, is smokeproof and variable in size. MTOW and performance as for Il-114 passenger version, except where otherwise indicated. First production example (RA-91005) flew at Tashkent 14 September 1996. Second (UK-91004) converted by 1998.

Il-114-N200S: Rear loading ramp version; none yet built.

Il-114-100: With 2,051 kW (2,750 shp) P&WC PW127H turboprops, Hamilton Sundstrand 586E-7 six-blade propellers, Sextant avionics and new systems; primarily for export; designated **Il-114PC** until late 1997; passenger and cargo (**Il-114-100T**) versions envisaged. Performance generally as Il-114, but with increased range and economy. Weight empty, equipped (including two

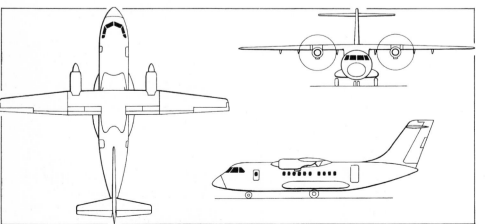

Ilyushin Il-112 twin-turboprop short-range passenger transport *(James Goulding/Jane's)* 2001/0092139

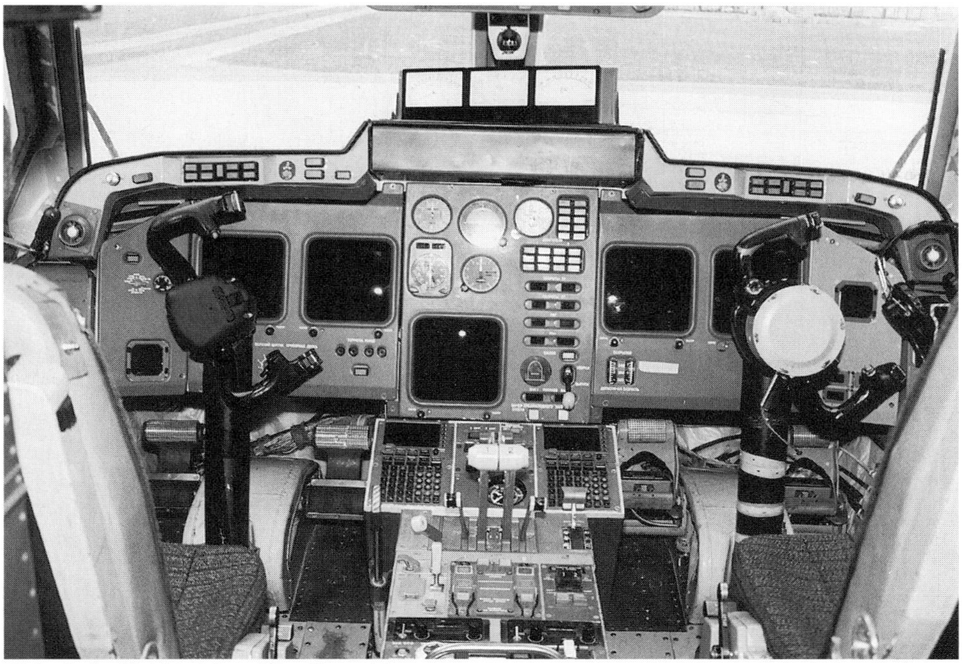

Flight deck of Il-114 airliner *(Piotr Butowski)*

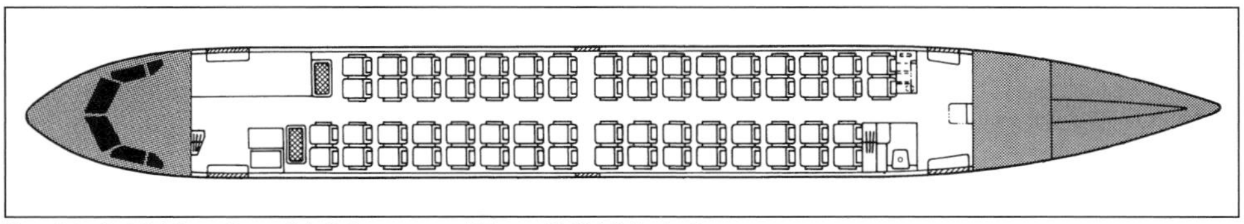

Ilyushin Il-114 interior with 64 seats at 75 cm (29½ in) pitch

crew) 16,100 kg (35,494 lb). Joint venture agreed by Ilyushin and P&WC on 16 June 1997; first flight at Tashkent (UK-91009) 26 January 1999; CIS Interstate Aviation Committee type certificate awarded 27 December 1999.

Il-114M: With TV7M-117 turboprops, increased maximum T-O weight and 7,000 kg (15,430 lb) payload.

Il-114MA: Version of Il-114M with P&WC engines to carry 74 passengers at 324-351 kt (600-650 km/h; 373-404 mph) on 1,079 n mile (2,000 km; 1,242 mile) stages.

Il-114P and Il-114MP: Maritime patrol versions. Described separately.

Il-114FK: Military reconnaissance/cartographic survey version. Described separately.

Il-114PR: Signals intelligence and electronic warfare version; announced October 2000.

Il-140: Described separately.

CUSTOMERS: Uzbekistan Airlines purchased two preseries aircraft in 1994, each flying 300 hours before engine overhaul time expiry and consequent grounding pending establishment of overhaul plant; airline began operating two Il-114Ts on service trials in 1998 and has total requirement for 10 Il-114/Ts, three of which on order by April 2000 for delivery in March and April 2001. Interest expressed by Bulgarian airlines; order for three Il-114-100s from Singaporean purchaser reported early 2000; delivery planned in 2001. Three (excluding prototype) owned by Ilyushin Bureau. Total of 15 flying by January 1998, most awaiting orders.

COSTS: US$10 million (2000).

DESIGN FEATURES: Conventional low-wing monoplane; only fin and rudder swept; slight dihedral on wing centre-section, much increased on outer panels; operation from unpaved runways practical. Service life of 30,000 cycles/30,000 hours/30 years, with overhaul at 6,000 hour intervals.

FLYING CONTROLS: Manual actuation for all except elevator; each wing trailing-edge occupied entirely by aileron, with servo and trim tabs, and hydraulically actuated double-slotted trailing-edge flaps, inboard and outboard of engine nacelle; two airbrakes (inboard) and spoiler (outboard) forward of flaps; spoilers supplement ailerons differentially in event of engine failure during take-off. Elevator control is FBW with back-up cable actuation. Trim and servo tabs in rudder, trim tab in each elevator.

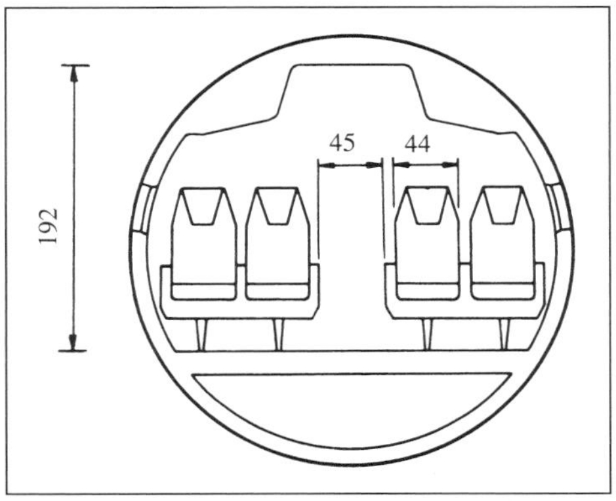

Cross-section of Il-114; dimensions in centimetres

STRUCTURE: Approximately 10 per cent of airframe by weight made of composites; two-spar wings; removable leading-edge on outer panels; circular-section aluminium alloy semi-monocoque fuselage built as five subassemblies; metal tail unit (CFRP tailplane and fin boxes planned for later aircraft).

LANDING GEAR: Retractable tricycle type, with twin wheels on each unit. All retract forward hydraulically; emergency extension by gravity. Oleo-pneumatic shock-absorbers. Tyres size 620×80 on nosewheels, 880×305 on mainwheels. Nosewheels steerable ±55°. Disc brakes on mainwheels. All wheel doors remain closed except during retraction or extension of landing gear.

POWER PLANT: Two 1,839 kW (2,466 shp) Klimov TV7-117S turboprops (with potential to increase to 2,088 kW; 2,800 shp), each driving a low-noise six-blade Stupino SV-34 CFRP propeller. Integral fuel tanks in wings, capacity 8,780 litres (2,319 US gallons; 1,931 Imp gallons).

ACCOMMODATION: Flight crew of two, plus stewardess. Emergency exit window each side of flight deck. Four-abreast seats for 64 passengers in main cabin, at 76 cm (30 in) seat pitch, with central aisle 45 cm (17¾ in) wide. Provision for rearrangement of interior for increased seating, removal of seats for cargo-carrying, and lengthening of fuselage for 70 to 75 passengers. Two passenger doors on port side: airstair door at front of cabin, further door at rear, both opening outward. Galley, cloakroom and lavatory at rear; emergency escape slide by service door on starboard side. Type III emergency exit over each wing. Service doors at front and rear of cabin on starboard side. Baggage compartments forward of cabin on starboard side and to rear of cabin, plus overhead baggage racks. Optional carry-on baggage shelves in lobby by main door at front.

SYSTEMS: Dual-redundant pressurisation and air conditioning system using bleed air from both engines; maximum differential 0.44 bar (6.4 lb/sq in). Two independent hydraulic systems, pressure 207 bars (3,000 lb/sq in), for landing gear actuation, wheel brakes, nosewheel steering, airbrakes and flaps. Three-phase 115/220 V 400 Hz AC electrical system powered by 40 kW alternator on each engine. Secondary 24 V DC system. Wing and tail unit leading-edges de-iced electrically by patented pulse wave system. Electrothermal anti-icing system for propeller blades and windscreen. Engine air intakes de-iced by hot air. APU in tailcone.

AVIONICS: Digital avionics for automatic or manual control by day or night, including automatic approach and landing in limiting weather conditions (ICAO Cat. I and II).

Instrumentation: Two colour CRTs for each pilot for flight and navigation information. Centrally mounted CRT for engine and systems data.

Flight: Barco computer for FMS.

DIMENSIONS, EXTERNAL:

Wing span	30.00 m (98 ft 5¼ in)
Wing aspect ratio	11.0
Length overall	26.875 m (88 ft 2 in)
Fuselage: Length	26.20 m (85 ft 11½ in)
Max diameter	2.86 m (9 ft 4½ in)
Height overall	9.185 m (30 ft 1½ in)
Tailplane span	11.10 m (36 ft 5 in)
Wheel track	8.40 m (27 ft 6½ in)
Wheelbase	9.13 m (29 ft 11½ in)
Propeller diameter	3.60 m (11 ft 9¾ in)
Propeller ground clearance	0.50 m (1 ft 7¾ in)
Propeller fuselage clearance	0.97 m (3 ft 2¼ in)
Passenger doors (each): Height	1.70 m (5 ft 7 in)
Width	0.90 m (2 ft 11¼ in)
Service door (front): Height	1.30 m (4 ft 3¼ in)
Width	0.96 m (3 ft 1¾ in)
Service door (rear): Height	1.38 m (4 ft 6¼ in)
Width	0.72 m (2 ft 4¼ in)
Emergency exit (each): Height	0.91 m (3 ft 0 in)
Width	0.51 m (1 ft 8 in)

DIMENSIONS, INTERNAL:

Length between pressure bulkheads	22.24 m (72 ft 11½ in)
Cabin: Length	18.93 m (62 ft 1¼ in)
Width: max	2.64 m (8 ft 8 in)
at floor	2.28 m (7 ft 5¾ in)
Max height	1.92 m (6 ft 3½ in)
Cargo cabin volume (Il-114T)	76.0 m³ (2,684 cu ft)

AREAS:

Wings, gross	81.90 m² (881.6 sq ft)

WEIGHTS AND LOADINGS:

Operating weight empty	15,000 kg (33,070 lb)
Max payload	6,500 kg (14,330 lb)
Max fuel	6,500 kg (14,330 lb)
Max T-O weight	23,500 kg (51,808 lb)
Max ramp weight	23,600 kg (52,029 lb)
Max wing loading	286.9 kg/m² (58.77 lb/sq ft)
Max power loading	6.39 kg/kW (10.50 lb/shp)

PERFORMANCE:

Max level speed	270 kt (500 km/h; 310 mph)
Cruising speed	254 kt (470 km/h; 292 mph)
Approach speed	100 kt (185 km/h; 115 mph)
Landing speed	87 kt (160 km/h; 100 mph)
Optimum cruising height: Il-114	7,600 m (24,940 ft)
Il-114T	6,000 m (19,680 ft)
T-O run: paved	1,360 m (4,465 ft)
Landing run: paved or unpaved	1,260 m (4,135 ft)
Range, with reserves:	
with 64 passengers	540 n miles (1,000 km; 621 miles)
with 1,500 kg (3,300 lb) payload	2,590 n miles (4,800 km; 2,980 miles)

UPDATED

ILYUSHIN Il-114P

TYPE: Maritime surveillance twin-turboprop.

PROGRAMME: Developed in partnership with Proton-Service scientific research centre, NPO Geophysica, NPO Polyot, VNII Radiotekhnichy 'Skala', NII Systemotekhnichy and NII Pribornoi Avtomatiky; prototype reported to be flying by mid-1996, but no photographs or other description had been made available by October 2000, at which time programme was confirmed as still active.

CURRENT VERSIONS: **Il-114P:** Version for use in ''automated air-to-sea observation systems''.

Il-114MP: Patrol and combat version; *as described.*

DESIGN FEATURES: Generally as Il-114 transport, of which it is a variant, except for slightly changed nose shape to accommodate surveillance radar, and optional MAD tailboom; large freight door at rear of cabin on port side; bubble windows for observation. Suitable for military or paramilitary duties or for use in civil emergencies.

ACCOMMODATION: Forward part of cabin can transport freight pallets or stretcher patients and other casualties; rear section houses Strizh (martin) maritime surveillance systems and two crew workstations.

AVIONICS: *Mission:* Leninets Strizh sensor suite, comprising 360° radar and electro-optical (LLLTV and FLIR) subsystems of Sea Dragon system, plus GPS. Two operator's consoles.

ARMAMENT: Two hardpoints side by side under centre-fuselage and two under each outer wing panel for 23 mm gun pod and other mission stores, including anti-ship missiles, torpedoes, sonobuoys, searchlights and loudspeakers.

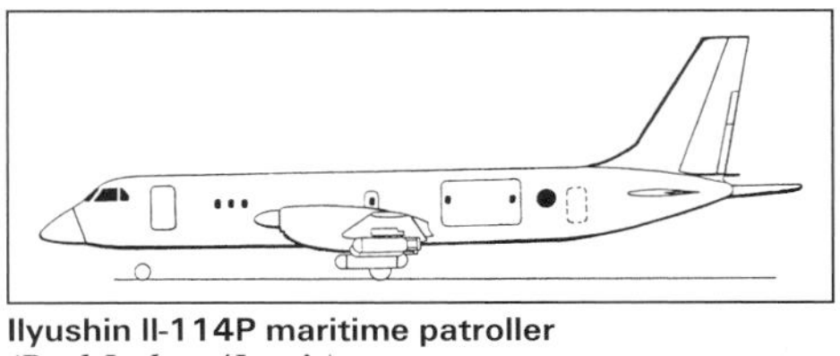

Ilyushin Il-114P maritime patroller *(Paul Jackson/Jane's)* 0011994

RUSSIAN FEDERATION: AIRCRAFT—ILYUSHIN to INTERAVIA

DIMENSIONS, EXTERNAL:
Length without MAD boom:
overall 27.40 m (89 ft 10¾ in)
fuselage 26.725 m (87 ft 8¼ in)
PERFORMANCE (estimated):
Loiter time 162 n miles (300 km; 186 miles) from base
8-10 h
UPDATED

ILYUSHIN Il-114FK
TYPE: Multisensor surveillance twin-turboprop.
PROGRAMME: Announced 1996; confirmed in October 2000 to be still under development.
DESIGN FEATURES: Generally as Il-114 transport, of which it is a variant; elint, survey and cartographic platform; to replace Il-20 and An-30 in Russian military service. Changes include glazed nose; small undernose radome; large observation blister window on side of fuselage below flight deck windows; long container for side-looking airborne radar (SLAR) on port side forward of wingroot. Cabin windows deleted.
DIMENSIONS, EXTERNAL:
Length: overall 27.57 m (90 ft 5½ in)
fuselage 27.06 m (88 ft 9¼ in)
UPDATED

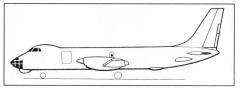

Projected Ilyushin Il-114FK surveillance platform
(Paul Jackson/Jane's) 0011995

ILYUSHIN Il-140
TYPE: Multisensor surveillance twin-turboprop.
PROGRAMME: Two derivatives of Il-114 first revealed to be in project stage in October 2000.
CURRENT VERSIONS: **Il-140**: Tactical air control version with airborne radar.
Il-140M: Patrol, ecological monitoring and maritime SAR version.
NEW ENTRY

ILYUSHIN Il-214
TYPE: Twin-jet airliner.
PROGRAMME: In draft design stage by 2000. Negotiations opened with HAL of India (which see) in March 1999 for possible joint production; protocol signed later in same year. Deliveries to begin within three to five years of funding go-ahead. Included in Russian government's aeronautical development plan for 2001-2015.
CURRENT VERSIONS: **Il-214-100**: Airliner version, with 100 seats.
Il-214T: Transport version, carrying 82 paratroops or cargo.
COSTS: US$12 million to US$15 million with PS-9 engines (2000). Estimated market for 200 to 300.
POWER PLANT: Two turbofan engines: Pratt & Whitney PW6000 or Aviadvigatel PS-9, each rated at up to 97.9 kN (22,000 lb st).
DIMENSIONS, INTERNAL:
Cargo bay: Length 13.50 m (44 ft 3½ in)
Max height 3.00 m (9 ft 10 in)
WEIGHTS AND LOADINGS:
Max payload 15,000 kg (33,069 lb)
Max T-O weight 47,900 kg (105,600 lb)
PERFORMANCE (estimated):
Cruising speed 470 kt (870 km/h; 541 mph)
Range with max payload
1,079 n miles (2,000 km; 1,242 miles)
NEW ENTRY

INTERAVIA

INTERAVIA RADONEZH AO (Interavia Radonezh JSC)
Repikhovo, Sergiev-Posad, Moskovskaya oblast
Tel: (+7 254) 305 04, (+7 095) 258 40 44
Fax: (+7 095) 258 40 50
GENERAL DIRECTOR: Sergei Lebonovich

This company has designed a series of light aircraft, including the multipurpose I-1L, aerobatic I-3 kitplane and Technoavia Finist (which see). Variants of the I-1 and I-3 are produced by other firms, as described in the following entries.
UPDATED

INTERAVIA I-1L
TYPE: Two-seat lightplane.
PROGRAMME: Developed from original I-1 lightplane (which had been intended for production, with VOKBM M-3 engine, by LMZ at Lukhovitsy as the Aviotechnika SL-90 Leshii; see International section, 1997-98 and earlier editions). 'L' in designation signifies change to Textron Lycoming engine. I-1L manufacture at Lukhovitsy began in 1995; promotion by RSK MiG; AP-23 certification awarded December 2000; see also RSK MiG SL-39, which differs only in having LOM M332 engine.
CUSTOMERS: Total 25 built or ordered by mid-1995; no recent information. At least three in USA by 1999.
COSTS: US$86,600 (2000).
DESIGN FEATURES: Basically similar to SL-90 Leshii, designed in 1988 and first flown February 1990; I-1's 82 kW (110 hp) M-3 radial engine replaced by more powerful flat-four engine. Applications include search and rescue, aerial photography, ecological monitoring, agricultural work, forest surveillance and patrol of electricity transmission lines, oil and gas pipelines, with appropriate equipment.
Strut-braced high-wing monoplane with extensively glazed cabin, braced tailplane and sweptback vertical tail surfaces. Airframe life 10,000 hours; 20,000 cycles/15 years.
Constant chord wings with CAHI (TsAGI) P-III-15 section, 3° forward sweep, 1° dihedral and 2° incidence.
FLYING CONTROLS: Conventional and manual. Control surface deflections: aileron +30/−20°, elevator +20/−40° and rudder ±27°. Ground-adjustable tab on each aileron, starboard elevator and rudder. Two fixed balancing surfaces below each aileron. Plain flaps, max deflection 30°. Horn-balanced tail surfaces.
STRUCTURE: Wings have light alloy structure and light alloy skin on centre-section and leading-edge, remainder fabric covered. Light alloy fuselage, with corrugated skin on rear section. All-metal fin and tailplane, with corrugated skin; ailerons, flaps, rudder and elevators fabric covered. Phenolic laminate engine cowling. All aluminium surfaces anodised.
LANDING GEAR: Non-retractable tailwheel type; single wheel on each unit; cantilever steel tube mainwheel legs; steerable tailwheel. Tyres size 400×150 on mainwheels, 200×80 on tailwheel. Cleveland mainwheel brakes.
POWER PLANT: One 112 kW (150 hp) Textron Lycoming O-320-E2A flat-four piston engine; two-blade Mühlbauer MT-180/R/125-3D propeller. Two fuel tanks, combined capacity 120 litres (31.7 US gallons; 26.4 Imp gallons).
ACCOMMODATION: Two seats side by side; upward-hinged window/door each side; baggage/cargo space aft of seats.
AVIONICS: *Comms:* Briz VHF.
DIMENSIONS, EXTERNAL:
Wing span 10.00 m (32 ft 9¾ in)
Wing aspect ratio 7.9
Length overall 6.40 m (21 ft 0 in)
Height overall 1.86 m (6 ft 1¼ in)
Tailplane span 3.30 m (10 ft 10 in)
Wheel track 3.00 m (9 ft 10 in)
Propeller diameter 1.80 m (5 ft 10¾ in)

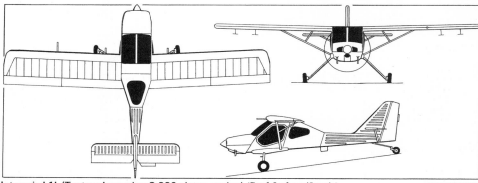

Interavia I-1L (Textron Lycoming O-320 piston engine) *(Paul Jackson/Jane's)* 2001/0089519

DIMENSIONS, INTERNAL:
Cabin: Length 1.10 m (3 ft 7¼ in)
Max width 1.06 m (3 ft 5¼ in)
Max height 1.00 m (3 ft 3¼ in)
Volume 0.7 m³ (24.7 cu ft)
Baggage compartment: Length 0.58 m (1 ft 10¾ in)
Max width 0.56 m (1 ft 10 in)
Max height 0.57 m (1 ft 10½ in)
AREAS:
Wings, gross 12.60 m² (135.6 sq ft)
WEIGHTS AND LOADINGS:
Weight empty, equipped 610 kg (1,345 lb)
Max payload 97 kg (214 lb)
Max fuel weight 90 kg (198 lb)
Max T-O and landing weight 880 kg (1,940 lb)
Max wing loading 69.8 kg/m² (14.30 lb/sq ft)
Max power loading 7.87 kg/kW (12.93 lb/hp)
PERFORMANCE:
Never-exceed speed (VNE) 135 kt (250 km/h; 155 mph)
Max level speed 107 kt (200 km/h; 125 mph)
Nominal cruising speed 92 kt (170 km/h; 106 mph)
T-O and landing speed 62 kt (115 km/h; 71 mph)
Stalling speed 40 kt (73 km/h; 45 mph)
Service ceiling 3,000 m (9,840 ft)
T-O run: paved runway 280 m (920 ft)
unpaved runway 330 m (1,085 ft)
Landing run: paved runway 220 m (725 ft)
unpaved runway 180 m (590 ft)
Range with max fuel 323 n miles (600 km; 372 miles)
g limits +6/−3

Note: *Data published by LMZ gives 820 kg (1,807 lb) MTOW and 242 n mile (450 km; 279 mile) range.*
UPDATED

INTERAVIA I-3
TYPE: Aerobatic two-seat sportplane.
PROGRAMME: Prototypes exhibited at Moscow Air Show in 1993. Production by TMZ at Tushino (which see). Refer also to Technoavia SP-95 in 1999-2000 and previous *Jane's*.
CUSTOMERS: Up to 30 built by late 1998. Four registered to US customers in 1998, following one in 1996 and three in 1997. (Some are documented as Interavia E-3.)
DESIGN FEATURES: Conventional low-wing monoplane, capable of unlimited aerobatics: tapered multispar wings of symmetrical section, without dihedral or anhedral; unswept tail surfaces with pointed rudder tip; clear-view blister canopy over one or two seats. Convertible from two-seater to single-seater in 1 hour by removal of front cabin bay with control panel, installation of top fuselage panel and changing of canopy.

Interavia I-1L two-seat light aircraft *(Paul Jackson/Jane's)* 2001/0089523

FLYING CONTROLS: Manually actuated conventional horn-balanced elevators and horn-balanced rudder; 80 per cent span flaperons. Ground-adjustable tab on rudder; two suspended balance tabs on each flaperon.
STRUCTURE: All-metal; semi-monocoque fuselage; fluted skin on fin, tailplane and all control surfaces.
LANDING GEAR: Tailwheel type, fixed; cantilever mainwheel legs. Mainwheel tyres size 400×150 mm; tailwheel tyre size 200×80 mm.
POWER PLANT: One 265 kW (355 hp) VOKBM M-14P nine-cylinder air-cooled radial engine; three-blade wooden controllable-pitch propeller. Fuel tank in each wingroot; combined capacity 120 litres (31.7 US gallons; 26.4 Imp gallons).
ACCOMMODATION: One or two seats in tandem.
DIMENSIONS, EXTERNAL:
Wing span	8.10 m (26 ft 7 in)
Wing aspect ratio	5.7
Length overall	6.72 m (22 ft 0½ in)
Tailplane span	2.80 m (9 ft 2¼ in)
Propeller diameter	2.40 m (7 ft 10½ in)

AREAS:
Wings, gross	11.54 m² (124.2 sq ft)

WEIGHTS AND LOADINGS:
Weight empty	760 kg (1,675 lb)
Max T-O weight	1,063 kg (2,343 lb)
Max wing loading	92.1 kg/m² (18.86 sq ft)
Max power loading	4.02 kg/kW (6.60 lb/hp)

Interavia I-3 aerobatic aircraft in tandem-seat form *(Paul Jackson/Jane's)*

PERFORMANCE (two seats, except where indicated):
Never-exceed speed (V_{NE})	242 kt (450 km/h; 280 mph)
Max level speed	189 kt (350 km/h; 217 mph)
Max rate of climb at S/L	660 m (2,165 ft)/min
Service ceiling	5,000 m (16,400 ft)
T-O run: paved runway	200 m (656 ft)
unpaved runway	262 m (860 ft)
landing run: paved runway	452 m (1,483 ft)
unpaved runway	477 m (1,565 ft)
Range with max fuel	377 n miles (70 km; 435 miles)
g limits: single seat	+12/–10
two seats	+10/–8

UPDATED

KAMOV

KAMOV OAO (Kamov JSC)

ulitsa 8-go Marta 8, Lyubertsy, 140007 Moskovskaya oblast
Tel: (+7 095) 700 32 04 and 171 37 43
Fax: (+7 095) 700 30 71 and 700 31 10
e-mail: kb@kamov.ru
Web: http://www.kamov.ru
GENERAL DESIGNER: Sergei V Mikhyeyev, PhD
DEPUTY GENERAL DESIGNER: Beniamin A Kasyanikov
CHIEF DESIGNERS:
Boris Gubarev (Ka-115)
Vyacheslav Krygin
Evgeny Pak
Aleksandr Piorzhnikov
Grigory Yakemenko (Ka-50)
Evgeny Sudarev

Formed in 1948, this OKB continues work of Prof Dr Ing Nikolai Ilyich Kamov. All Kamov helicopters in current service have coaxial contrarotating rotors; Ka-60/62, under development, have single main rotor, with anti-torque Fenestron. Latest product to take to the air is the Ka-60, on 24 December 1998, but it was reported in September 2000 that a new transport helicopter, to compete against Western NH90 and S-92, is under development.

Since 1996, Kamov has been a member of the RSK 'MiG' consortium (which see). It is 49 per cent state-owned, and its design bureau and experimental construction plant employed 2,500 people in 1998. There are plans for separation from RSK 'MiG' and integration with those factories producing Kamov helicopters, namely the Arseneyev factory (Ka-50) and the Kumertau factory (Ka-32 family) as well as the Voronezh Mechanical Plant, which produces gearboxes, and the Stupino Metallurgical Plant, which makes rotors and propeller assemblies. The affiliated Aero-Kamov Air Transportation Company (Tel/Fax: (+7 095) 700 31 60) operates a fleet of Ka-32s for diverse tasks, including firefighting, and has helicopters based in Russia, Canada and South Africa.

UPDATED

KAMOV Ka-27 and Ka-28
NATO reporting names: Helix-A and D
TYPE: Naval combat helicopter.
PROGRAMME: Design started 1969 to overcome inability of Ka-25 ('Hormone') to operate dipping sonar at night and in adverse weather; first flight of Ka-25-2 (*Izdelie* D-2 or *Izdelie* 500) prototype (with Ka-25 forward fuselage and lacking slotted fins) 8 August (hovering) and 24 December (full transition) 1973. Trial deployment by pre-production examples aboard *Minsk*, 1978, before completion of full state acceptance tests, December 1978. First open reference in US Department of Defense's 1981 *Soviet Military Power* document, which stated that "Hormone variant" helicopters could be carried in telescoping hangar on 'Sovremenny' class of guided missile destroyers, for ASW missions; photographs of two on stern platform of *Udaloy*, first of new class of ASW guided missile destroyers, taken by Western pilots during Baltic exercises, September 1981, following official service entry on 14 April 1981; at least 16 observed on former 'Kiev' class carrier/cruiser *Novorossiysk* 1983, as stage in continuous replacement of Ka-25s with Ka-27s; 10 Ka-27s and Ka-29s on carrier *Admiral of the Fleet Kuznetsov* when deployed to Adriatic late 1995. Reported 267 manufactured by KAPP.
CURRENT VERSIONS: **Ka-27PL** ('Helix-A'): Basic ASW helicopter with three crew; operational since 1982; normally operated in pairs, one tracking hostile submarine, other dropping depth charges. Later aircraft have TV3-117VK or TV3-117VMA engines. Russian Naval Aviation has 85; Ukraine has six.
Ka-27PL Upgrade (Ka-27PLM): Planned replacement of current Osminog sensor system reportedly will be by a simplified version of Sea Dragon suite to be installed in Tupolev Tu-142 and other aircraft; first reported in 1998 as replacement for abandoned Ka-27M and Ka-27K (Kamerton-1M equipped) upgrades. Upgrade also includes more powerful TV3-117VMA-SB3 engines.
Ka-27PS ('Helix-D'): Search and rescue and plane guard helicopter first flew 8 August 1974; as Ka-27PL, but some operational equipment deleted; internal cabin fuel tanks deleted, but tanks in former weapons bay and external fuel tank each side of cabin, as civil Ka-32; total fuel capacity 3,450 litres (911 US gallons; 759 Imp gallons); LPG-300 electrically powered winch beside port cabin door. Air-droppable dinghy packs in ventral stores bay; racks for marker floats.
Ka-27PK: Proposed version armed with Kh-35 anti-ship missile.
Ka-27PV: Armed version of Ka-27PS, retaining radar (but not sonar/sonics) supplied to Border Troops. (See also Ka-32A7.)
Ka-28 ('Helix-A'): Export version of Ka-27PL, with Izumrud (Emerald) sonobuoy system, using RGB-16 buoys; first flown in 1982. Said to have 1,640 kW (2,200 shp) TV3-117VK turboshafts and 3,680 kg (8,113 lb) of fuel in 12 tanks, including Ka-27PS-type fixed tanks on fuselage sides. Ka-32-type broad cockpit door and bulged window. Higher T-O weight permitted by relaxing 'dip immediately after T-O' requirement. Total fuel capacity 4,760 litres (1,257 US gallons; 1,047 Imp gallons). Indian Ka-28s to be upgraded with Leninets Novella-based Lira systems by the 770th State Aircraft Repair Plant at Sevastopol. May parallel PLM upgrade.
Ka-29 ('Helix-B'): Described separately.
Ka-32 (civil 'Helix-C'): Described separately.

CUSTOMERS: Russian Federation and Associated States (CIS) Naval Aviation (about 125 in Russia, 20 in Ukraine, more elsewhere), China (three Ka-27PL, five Ka-28 to equip two newly-acquired ex-Soviet 'Sovremenny' class destroyers of "up to 12" ordered in 1998 and delivered in late 1999), India (18 Ka-28, including three for training but likely to acquire more for newly acquired Gorshkov), Vietnam (10 Ka-28) and Yugoslavia (two Ka-28).
See also Ka-32 entry for particulars applicable to both types.
DESIGN FEATURES: As Ka-32; compact design and folding rotors enable Ka-27 to stow in shipboard hangars and use deck lifts built for earlier Ka-25.
ACCOMMODATION: Crew of three: pilot, tactical co-ordinator, ASW systems operator.
AVIONICS: Basically as for Ka-32.
Comms: R-832M or R-863 VHF, R-864 HF and secure encrypted maritime tactical datalink system. IFF ('Odd Rods').
Radar: Undernose Osminog (Octopus) 360° search radar linked to PKV-252 navigation system.
Flight: VIAP-1 four-channel autopilot, Privod ILS, and DISS-015 Doppler box under tailboom.
Mission: Ros-V dipping sonar behind clamshell doors at rear of fuselage pod; RGB-N or RGB-NM sonobuoys stowed internally; A-100 Pankhra sonics and APM-73V Bor MAD.
Self-defence: Optional equipment includes RWR on nose and above tailplane; IR jammer ('Hot Brick') at rear of engine bay fairing; chaff/flare dispensers; colour-coded identification flares.
EQUIPMENT: Station-keeping light between ESM radome and jammer.
ARMAMENT: Ventral weapons bay for single AT-1MV torpedo or APR-2 ASW rocket, or two OMAB bombs, or 10 PLAB-250-120, or KAB-250PL depth charges, or up to 36 RGB-NM sonobuoys.
DIMENSIONS, EXTERNAL: As Ka-32

Kamov Ka-28 anti-submarine helicopter ('Helix-A') *(Paul Jackson/Jane's)*

WEIGHTS AND LOADINGS (Ka-28):
Max payload 1,000 kg (2,205 lb)
Max T-O weight 12,000 kg (26,455 lb)
PERFORMANCE (Ka-28):
Max level speed 146 kt (270 km/h; 168 mph)
Max cruising speed 135 kt (250 km/h; 155 mph)
Max rate of climb at S/L 750 m (2,460 ft)/min
Hovering ceiling 5,000 m (16,400 ft)
Radius of action against submarine cruising at up to 40 kt (75 km/h; 47 mph) at depth of 500 m (1,640 ft)
108 n miles (200 km; 124 miles)
Range with max fuel 432 n miles (800 km; 375 miles)

UPDATED

KAMOV Ka-29 and Ka-31
NATO reporting name: Helix-B

TYPE: Medium transport helicopter/AEW helicopter.
PROGRAMME: Developed for AV-MF, following cancellation of proposed joint-service, tandem-rotor, multirole V-50 and its replacement by what became Ka-50, meeting Army requirement only. Ka-252TB prototype (also known as Izdelie D2B or Izdelie 502) first flew 28 July 1976, possibly with Ka-25 nose or original narrow Ka-27/Ka-32 nose. Production at Kumertau (KAPP) from 1984.

Entered service with Northern and Pacific Fleets 1985; photographed on board assault ship *Ivan Rogov* in Mediterranean 1987, thought to be Ka-27B and given NATO reporting name 'Helix-B'; identified as Ka-29 combat transport at Frunze (Khodinka) Air Show, Moscow, August 1989; Ka-31 radar picket version completed initial shipboard trials on aircraft carrier *Admiral of the Fleet Kuznetsov* (then *Tbilisi*) 1990; both versions expected to equip this ship.

CURRENT VERSIONS: **Ka-29TB** ('Helix-B'): Armed derivative of Ka-27 for day/night, VFR and IFR, transport and close support of seaborne assault troops; in-the-field conversion from one role to the other. Non-retractable landing gear and 50 cm (20 in) wider armoured flight deck. Reportedly used by Experimental Combat Group in Chechen War in 1996. Civilianised version will be designated Ka-33.

Detailed description generally as for Ka-32 except as under.

Ka-31 (formerly Ka-29RLD: *radiolokatsyonnogo dozora*: radar picket helicopter): Development began 1980; first flown October 1987; two examples (031 and 032) tested on *Admiral of the Fleet Kuznetsov*; state testing completed in 1996; limited production launched at Kumertau Aircraft Plant, Bashkirskaya, 1999 including first batch of four for Indian Navy, of possible total requirement for 12 Ka-31s. Indian aircraft have 12 channel Kronstadt GPS with Abris digital moving map and a 152 × 203 mm (6 × 8 in) AMLCD screen.

Basic airframe of Ka-27 with broader flight deck; E-801 or E-801E (export) Oko (eye) early warning radar system by Radio Engineering Institute, Nizhny Novgorod, includes large rotating radar antenna (area 6.0 m²; 64.5 sq ft) that stows flat against underfuselage and deploys downward, turning through 90° into vertical plane before starting to rotate at 6 rpm; landing gear retracts upward to prevent interference, nosewheels into long fairings. Once system has been switched on, antenna extended and operation mode selected, data on air targets flying below helicopter's altitude, and on water surface situation, are acquired, evaluated and transmitted automatically to command centre, requiring only two crew (pilot and navigator, latter monitoring – but not operating – the system) in helicopter. Loiter speed 54 to 65 kt (100 to 120 km/h; 62 to 75 mph) at up to 3,500 m (11,480 ft); loiter duration 2 h 30 min. Maximum surveillance radius 54 to 81 n miles (100 to 150 km; 62 to 93 miles) for fighter-size targets, 135 n miles (250 km; 155 miles) for surface vessels; up to 20 targets tracked simultaneously. Antenna can be retracted manually or explosively jettisoned in the event of a forced landing.

Ka-29TB ('Helix-B') combat transport helicopter *(Paul Jackson/Jane's)*

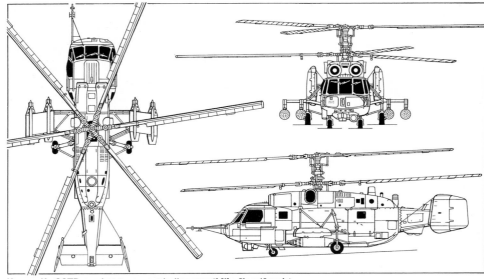

Kamov Ka-29TB combat transport helicopter *(Mike Keep/Jane's)*

Two large panniers starboard side of cabin, fore and aft of main landing gear on helicopter numbered 032 (forward panniers only on 031); starboard airstair-type cabin door, aft of flight deck, divided horizontally into upward- and downward-opening sections, with box fairing in place of window; hatch window deleted above starboard rear pannier; new TA-8Ka APU positioned above rear of engine bay fairing, with slot-type air intake at front of housing, displacing usual ESM and IR jamming pods, gives radar and antenna an independent power supply. Tyre size 620×180 on mainwheels, 480×200 on nosewheels. Tailcone extended by fairing for flight recorder; no armour, gun door, stores pylons or outriggers.

Ka-33: Utility transport. Civilianised version of Ka-29TB shipborne assault transport. Designation revealed at Moscow Air Show in August 1997; no further details released and no known conversions.

CUSTOMERS: Total of 59 built by early 1999 for Russian Federation Naval Aviation (about 45 Ka-29s) and Ukrainian Navy (about 12 Ka-29s); five Ka-31s ordered in 1999 by Indian Navy for delivery in 2001 and basing aboard aircraft carriers and 'Krivak' class destroyers; further five ordered February 2001.

COSTS: Indian Navy batch of four priced at Rs4 billion (US$92 million) (2000).

POWER PLANT: Two Klimov TV3-117VMA turboshafts, each 1,633 kW (2,190 shp). Engines started by APU. Fuel tanks filled with reticulated polyurethane foam for fire suppression.

ACCOMMODATION: Wider flight deck than Ka-27 for two crew; three flat-plate windscreen glazings instead of two-piece curved transparency; 350 kg (772 lb) of armour around cockpit and engines; main cabin port-side door, aft of landing gear, divided horizontally into upward- and downward-opening sections, lower section forming step when open, to facilitate rapid exit of up to 16 assault troops; four stretcher patients, seven seated casualties and medical attendant in ambulance role; internal or slung cargo provisions.

AVIONICS: *Comms:* Two UHF and HF radios.
Radar: Primary radar in port side of nose.
Flight: INS; Doppler box under tailboom; IFF ('Slap Shot').
Mission: Undernose Shturm-V missile guidance and LLTV pods; ESM 'flower pot' above rear of engine bay fairing.
Self-defence: L-166V IR jammer ('Hot Brick'); chaff/flare dispensers.

EQUIPMENT: Station-keeping light between ESM and jammer.
ARMAMENT: Four-barrel Gatling-type GShG-7.62 7.62 mm machine gun, with 1,800 rounds, flexibly mounted behind downward-articulated door on starboard side of nose; four pylons on outriggers, for two four-round packs of 9M114

Second prototype Kamov Ka-31 radar picket helicopter with antenna folded and stowed *(Paul Jackson/Jane's)*

Shturm (AT-6 'Spiral') ASMs and two UV-32-57 57 or B-8V20 80 mm rocket pods. Alternative loads include four rocket packs, two pods each containing a 23 mm gun and 250 rounds, or two ZAB-500 incendiary bombs. Internal weapons bay for torpedo or bombs. Provision for 30 mm Type 2A42 gun above port outrigger, with 250-round ammunition feed from cabin.

DIMENSIONS, EXTERNAL (A: Ka-29, B: Ka-31):
Rotor diameter (each)	15.90 m (52 ft 2 in)
Blade length, aerofoil section (each)	5.45 m (17 ft 10½ in)
Blade chord	0.48 m (1 ft 7 in)
Vertical separation of rotors	1.40 m (4 ft 7 in)
Length overall, excl noseprobe and rotors:	
A	11.30 m (37 ft 1 in)
B	11.25 m (36 ft 11 in)
Height overall: A	5.40 m (17 ft 8½ in)
B	5.60 m (18 ft 4½ in)
Width: between centrelines of outboard pylons	5.65 m (18 ft 6½ in)
over tailfins and centred rudders	3.65 m (12 ft 0 in)
of flight deck	2.20 m (7 ft 2 in)
Mainwheel track	3.50 m (11 ft 6 in)
Nosewheel track: A	1.41 m (4 ft 7½ in)
B	2.41 m (7 ft 11 in)
Wheelbase: A	3.00 m (9 ft 10 in)
B	3.05 m (10 ft 0 in)

AREAS:
Rotor disc (each)	198.50 m² (2,136.6 sq ft)

WEIGHTS AND LOADINGS (A, B as above):
Weight empty: A	5,520 kg (12,170 lb)
Max load: A, internal	2,000 kg (4,409 lb)
A, external	4,000 kg (8,818 lb)
Max combat load: A	1,800 kg (3,968 lb)
Normal T-O weight: A	11,000 kg (24,250 lb)
B	12,500 kg (27,557 lb)
Max T-O weight: A, internal load	11,500 kg (25,353 lb)
B	12,500 kg (27,557 lb)
Max airborne weight:	
A, external slung load	12,600 kg (27,775 lb)

PERFORMANCE (A, B as above):
Max level speed at S/L: A	151 kt (280 km/h; 174 mph)
B	135 kt (250 km/h; 155 mph)
Nominal cruising speed: A	130 kt (240 km/h; 149 mph)
B	119 kt (220 km/h; 137 mph)
Max rate of climb at S/L: A	888 m (2,910 ft)/min
Service ceiling: A	4,300 m (14,100 ft)
B	3,500 m (11,480 ft)
Hovering ceiling OGE: A	3,000 m (9,840 ft)
Combat radius, with six to eight attack runs over target:	
A	54 n miles (100 km; 62 miles)
Range:	
A, max standard fuel	248 n miles (460 km; 285 miles)
B	324 n miles (600 km; 372 miles)
A, ferry	400 n miles (740 km; 460 miles)

UPDATED

KAMOV Ka-32

NATO reporting name: Helix-C

TYPE: Multirole medium helicopter.
PROGRAMME: Development of Ka-27/32 began 1969; first flight of common prototype 1973; first Ka-32 (SSSR-04173) flew 8 October 1980; prototype of utility version shown at Paris Air Show June 1985; new military

Kamov Ka-32A1 emergency helicopter of Moscow fire service *(Paul Jackson/Jane's)*

Flight deck of Kamov Ka-32 utility helicopter *(Paul Jackson/Jane's)*

versions first exhibited at Moscow Air Show '95; Ka-32S and Ka-32T versions in production by KAPP; other conversions by Kamov at Lyubertsy. Klimov VK-3000 turboshaft to be certified in 2001 as alternative power plant.

CURRENT VERSIONS: **Ka-32T** ('Helix-C'): Utility transport (*transportnyi*), ambulance and flying crane; production began in 1987. Limited avionics; for carriage of internal or external freight, and passengers, along airways or over local routes, including support of offshore drilling rigs. Military 'Helix-C' similar; no undernose radome, but with dorsal ESM 'flower pot' and other military equipment. Several seen on board carriers, operating in SAR and planeguard roles. Military version understood to be designated Ka-27 or Ka-27T.

Detailed information applies to Ka-32T and Ka-32S. Ka-32A series generally similar, except as noted.

Ka-32S ('Helix-C'): Shipborne (*sudovoi*) version, intended especially for polar use; in production since 1987. More comprehensive avionics, including autonomous navigation system and Osminog (octopus) undernose radar (search radius 108 n miles; 200 km; 124 miles), for IFR operation from icebreakers in adverse weather and over terrain devoid of landmarks; 300 kg (661 lb) electric load hoist standard; additional external fuel tanks available 1994, strapped on each side at top of cabin; duties include ice patrol, guidance of ships through icefields, unloading and loading ships (up to 30 tonnes an hour, 360 tonnes a day). Simplex carbon fibre/epoxy tank, capacity 1,500 litres (396 US gallons; 330 Imp gallons) or 3,000 litres (792 US gallons; 660 Imp gallons), and 12.0 m (39 ft 4½ in) spraybars can be fitted for maritime anti-pollution work. Spraytime 6 minutes with 1,500 litre tank. In maritime search and rescue role, can loiter for 1 hour anywhere within 260 n miles (480 km; 300 miles) of base, and return carrying four crew and 5,000 kg (11,023 lb) payload. Maximum fuel capacity 2,650 litres (700 US gallons; 583 Imp gallons); weight empty 6,997 kg (15,425 lb); maximum payload 3,300 kg (7,275 lb) internally, 4,600 kg (10,141 lb) externally; maximum level and cruising speeds as Ka-32T.

Ka-32K: Flying crane (*kran*) with retractable gondola for second pilot under cabin. Prototype first flew December 1991; operational testing completed 1992. Supplied to Krasnodar Institute of Civil Aviation.

Ka-32A: Assemblies and systems of basic Ka-32 modified in 1990-93 to meet all requirements of Russian NLG-32-29 and NLG-32-33 and US FAR Pt 29/FAR Pt 33 airworthiness standards in categories A and B. First flight September 1990; Russian type certificates obtained for Ka-32A and its TV3-117VMA engines in June 1993. Production began 1996. Larger tyres. Optional pressure fuelling with reduced fuel capacity. Maximum accommodation for 13 passengers. Advanced avionics available, including Canadian Marconi dual CMA-900 flight management system, with EFIS, AFCS, CMA-2012 Doppler velocity sensor and CMA-3012 GPS sensor. Modification of helicopters to Ka-32A standard started by Kamov 1994.

Ka-32A1: Firefighting version of Ka-32A, first flown 12 January 1994. Equipped with Canadian or Russian variants of 'Bambi bucket', capacity 5,000 litres (1,320 US gallons; 1,100 Imp gallons). Two operated by Moscow fire service, with steerable water cannon and three types of rescue cage, able to lift two, 10 or 20 people from roofs of tall buildings. Other equipment includes searchlights and loudspeakers. Fire service aircraft and others flown by anti-riot police controlled by Aviatika Concern ISC, set up by Moscow city authorities and private investors to develop urban air transport system. Several on lease to South Korean forestry department have Simplex system, including 3,000 litre (793 US gallon; 660 Imp gallon) belly tank which can be refilled in 1½ minutes.

Ka-32A2: Police version used by Moscow Militia, first flown 21 March 1995; seen in camouflage finish (RA-06144) at Moscow Air Show '95. Seats for 11 passengers, two of whom can operate pintle-mounted guns in port-side rear doorway and starboard rear window. Fuel tanks filled with polyurethane foam to prevent explosion after damage or catching fire. Equipped for abseiling from both sides of cabin. Hydraulic hoist; two sets of loudspeakers; L-2AG searchlight under nose. Militia reportedly has 25. Maximum T-O weight 12,700 kg (28,000 lb).

Ka-32A3: Ordered by Russian Ministry of Emergency Situations (MChS) to carry rescue and salvage equipment to disaster areas and evacuate casualties.

Ka-32A7: Armed export version (alternatively known as **Ka-327**) of Russian Border Troops' Ka-27PV developed from military Ka-27PS for frontier and maritime economic zone patrol, with Osminog (octopus) radar and pairs of Kh-25 ASMs, UPK-23-250 pods each

Kamov Ka-32C of the South Korean Maritime Police, fitted with spray equipment

containing a GSh-23L twin-barrel 23 mm gun with 250 rounds, or B-8V-20 pods each with twenty 80 mm S-8 rockets, on four underwing pylons. Displayed – but not yet integrated – with Kh-35 (AS-20 'Kayak') active radar-homing ASMs. Provision for 30 mm Type 2A42 gun above port outrigger. Optional twin searchlights on weapons pylons. Large oblique camera in starboard rear window. Search and rescue equipment standard, with ability to lift up to 10 survivors at a distance of 108 n miles (200 km; 124 miles) from base. Provision for 13 persons in cabin. Maximum T-O weight 11,000 kg (24,250 lb). Maximum level speed 140 kt (260 km/h; 161 mph). First flown 1995.

Ka-32A11BC: Built in accordance with requirements of Transport Canada. FAR Pt 29 certification gained 11 May 1998, but full clearance achieved 26 February 1999, after installation of dual actuators in flight control system; first Russian helicopter to gain Western certification. Two development aircraft delivered to VIH Logging in May 1997; flew 4,000 hours up to February 1999; also used for firefighting; further 15 on order by 1998.

Ka-32A12: Version approved by Aviation Register of Switzerland.

Ka-32M: Under development by Kamov, to increase lifting capability to 7,000 kg (15,432 lb); retrofit with 1,839 kW (2,466 hp) TV3-117VMA-SB3 engines.

CUSTOMERS: Aeroflot and its successors; operators in Bulgaria (32S), Canada (32A), Laos (air force; six Ka-32T), Papua New Guinea (32A), South Africa (32A), Switzerland (32A), Yemen (32S/T). Estimated 132 Ka-32s in civilian use, of which 50 were abroad in 1998, including 31 operating in South Korea with coastguard (32S) and Seoul Fire Brigade (non-radar 32T).

COSTS: US$833,300 for firefighting version (1998).

DESIGN FEATURES: Conceived as completely autonomous 'compact truck', to stow in much the same space as Ka-25 with rotors folded, despite greater power and capability, and to operate independently of ground support equipment; special attention paid to ease of handling with single pilot; overall dimensions minimised by use of coaxial rotors, requiring no tail rotor, and twin fins on short tailboom; upper rotor turns clockwise, lower rotor anti-clockwise; rotor mast tilted forward 3°; twin turbines and APU above cabin, leaving interior uncluttered; lower fuselage sealed for flotation.

FLYING CONTROLS: Dual hydraulically powered flight control systems, without manual reversion; spring stick trim; yaw control by differential collective pitch applied through rudder pedals; mix in collective system maintains constant total rotor thrust during turns, to reduce pilot workload when landing on pitching deck, and to simplify transition to hover and landing; twin rudders intended mainly to improve control in autorotation, but also effective in co-ordinating turns; flight can be maintained on one engine at maximum T-O weight.

STRUCTURE: Titanium and composites used extensively, with particular emphasis on corrosion resistance; fully articulated three-blade coaxial contrarotating rotors have all-composites blades with carbon fibre and glass fibre main spars, pockets (13 per blade) of Kevlar-type material, and filler similar to Nomex; blades have non-symmetrical aerofoil section; each has ground-adjustable tab; each lower blade carries adjustable vibration damper, comprising two dependent weights, on root section, with further vibration dampers in fuselage; tip light on each upper blade; blades fold manually outboard of all control mechanisms, to folded width within track of main landing gear; rotor hub is 50 per cent titanium/50 per cent steel; rotor brake standard; all-metal fuselage; composites tailcone; fixed incidence tailplane, elevators, fins and rudders have aluminium alloy structure, composites skins; fins toe inward approximately 25°; fixed leading-edge slat on each fin prevents air flow over fin stalling in crosswinds or at high yaw angles.

LANDING GEAR: Four-wheel type. Oleo-pneumatic shock-absorbers. Castoring nosewheels. Mainwheel tyres size 600×180 (Ka-32); 620×180, pressure 10.80 bars (156 lb/sq in) (Ka-32A). Nosewheel tyres size 400×150 (Ka-32); 480×200, pressure 5.90 bars (85 lb/sq in) (Ka-32A). Skis optional.

POWER PLANT: Two 1,633 kW (2,190 shp) Klimov TV3-117V (Ka-32) or TV3-117VMA (Ka-32A) turboshafts, with automatic synchronisation system, side by side above cabin, forward of rotor driveshaft. Main gearbox brake standard. Oil cooler fan aft of gearbox. Cowlings hinge downward as maintenance platforms. Fuel in tanks under cabin floor and inside container each side of centre-fuselage; capacity of main tanks 2,180 litres (576 US gallons; 480 Imp gallons); maximum capacity with two underfloor auxiliary tanks 3,450 litres (911 US gallons; 759 Imp gallons). Single-point pressure refuelling behind small forward-hinged door on port side, where bottom of tailboom meets rear of cabin.

ACCOMMODATION: Pilot and navigator side by side on air conditioned flight deck, in adjustable seats. Rearward-sliding jettisonable door with blister window each side. Seat behind navigator, on starboard side, for observer, loadmaster or rescue hoist operator. Alcohol windscreen anti-icing. Direct access to cabin from flight deck. Heated and ventilated main cabin of Ka-32 can accommodate freight or 16 passengers, on three folding seats at rear, six along port sidewall and seven along starboard sidewall (13 passengers in Ka-32A). Lifejackets under seats. Fittings to carry four stretchers. No provisions for lavatory or galley. Pyramid structure can be fitted on floor beneath rotor driveshaft to prevent swinging of external cargo sling loads. Rearward-sliding door aft of main landing gear on port side, with steps below. Emergency exit door opposite. Hatch to avionics compartment on port side of tailboom.

SYSTEMS: Three hydraulic systems: main system supplies servos, mainwheel brakes and hydraulic winch when fitted; standby system supplies only servos after main system failure; auxiliary system supplies brakes after main system failure and adjusts height of helicopter fuselage above ground; it can also be connected to main system for checking all functions on ground. Electrical system includes two independently operating AC generators and two batteries which cut in automatically or manually via inverters after AC generating system failure. After failure of either generator, the other is switched automatically to supply both circuits. Two rectifiers supply DC power. Electrothermal de-icing of entire profiled portion of each blade switches on automatically when helicopter enters icing conditions. Hot air engine intake anti-icing. APU in rear of engine bay fairing on starboard side, for engine starting and to power all essential hydraulic and electrical services on ground, eliminating need for GPU.

AVIONICS: *Flight:* Include electromechanical flight director controlled from autopilot panel, Doppler hover indicator, two HSI and air data computer. Fully coupled three-axis autopilot can provide automatic approach and hover at height of 25 m (82 ft) over landing area, on predetermined course, using Doppler. Radar altimeter. Doppler box under tailboom.

EQUIPMENT: Doors at rear of fuel tank bay provide access to small compartment for auxiliary fuel, or liferafts which eject during descent in emergency, by command from flight deck. Container each side of fuselage, under external fuel containers, for emergency flotation bags, deployed by water contact. Optional rescue hoist, capacity 300 kg (661 lb), between top of door opening and landing gear. Optional external load sling, with automatic release and integral load weighing and stabilisation systems. Firefighting version of Ka-32T demonstrated in 1996.

DIMENSIONS, EXTERNAL:
Rotor diameter (each)	15.90 m (52 ft 2 in)
Length overall: excl rotors	11.27 m (37 ft 11¾ in)
rotors folded	12.25 m (40 ft 2¼ in)
Width, rotors folded	4.00 m (13 ft 1½ in)
Height to top of rotor head	5.45 m (17 ft 10½ in)

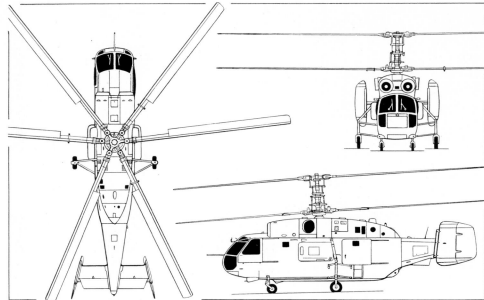
Kamov Ka-32T ('Helix-C') utility helicopter (two Klimov TV3-117V turboshafts) *(Dennis Punnett/Jane's)*

Kamov twin coaxial rotor system on Ka-32, with blades folded *(Paul Jackson/Jane's)*

Model of projected Ka-32A6 passenger transport *(Yefim Gordon)*

KAMOV—AIRCRAFT: RUSSIAN FEDERATION

Production Kamov Ka-50 operated by the Russian Army *(Paul Jackson/Jane's)* 2001/0100475

Wheel track: mainwheels	3.515 m (11 ft 6½ in)
nosewheels	1.41 m (4 ft 7½ in)
Wheelbase	3.03 m (9 ft 11¼ in)
Cabin door: Height	approx 1.20 m (3 ft 11¼ in)
Width	approx 1.20 m (3 ft 11¼ in)

DIMENSIONS, INTERNAL:
Cabin: Length	4.52 m (14 ft 10 in)
Max width	1.30 m (4 ft 3 in)
Max height	1.24 m (4 ft 0¾ in)

AREAS:
Rotor disc (each)	198.50 m² (2,136.6 sq ft)

WEIGHTS AND LOADINGS:
Weight empty	6,610 kg (14,573 lb)
Max payload: internal	3,700 kg (8,157 lb)
external	5,000 kg (11,023 lb)
Normal T-O weight	11,000 kg (24,250 lb)
Max flight weight with slung load	12,700 kg (27,998 lb)

PERFORMANCE (Ka-32):
Max level speed	140 kt (260 km/h; 162 mph)
Max cruising speed	130 kt (240 km/h; 149 mph)
Service ceiling	6,000 m (19,680 ft)
Hovering ceiling OGE	3,500 m (11,480 ft)
Hovering ceiling OGE, OEI	1,705 m (5,600 ft)
Range with max standard fuel	432 n miles (800 km; 497 miles)
Max range with auxiliary fuel	612 n miles (1,135 km; 705 miles)
Endurance with max standard fuel	4 h 30 min
Max endurance with auxiliary fuel	6 h 25 min

UPDATED

KAMOV Ka-50 CHERNAYA AKULA
English name: Black Shark
NATO reporting name: Hokum

TYPE: Attack helicopter.
PROGRAMME: Project launched in December 1977 as **V-80** (*Vertolyet* 80: Helicopter 80); first prototype (010) built by Kamov bureau and hovered at Lyubertsy 17 June 1982 and flew on 23 July, powered by TV3-117V engines; second prototype (011) flew 16 August 1983 with TV3-117VMA engines and mockup of Shkval tracking system, Merkury LLLTV, cannon and K-041 sighting system; both prototypes wore painted 'windows' to simulate fictitious rear cockpits. Initially reported in West in mid-1984, but first photograph did not appear (US Department of Defense's *Soviet Military Power*) until 1989.

First prototype lost in fatal accident on 3 April 1985; replaced by third prototype (012) with Mercury LLLTV system for state comparative test programme against Mil Mi-28, which completed in August 1986. Two preproduction **V-80Sh-1**s (014 and 015) were first to be built at Arsenyev and introduced UV-26 chaff/flare dispensers; second had K-37-800 ejection system and mockup of LLLTV in articulated turret. Ordered into production in December 1987. Further three for continued development work comprised 018 (first flown at Arsenyev 22 May 1991), 020 'Werewolf' and 021 'Black Shark'. (Export marketing name was originally Werewolf, but changed to Black Shark by 1996.) State tests of Ka-50 began in mid-1991 and type was commissioned into Russian Army Aviation in August 1993 for trials at 4th Army Aviation Training Centre, Torzhok. In August 1994, the Ka-50 was included in the Russian Army inventory by Presidential decree, judged winner of the fly-off against Mi-28. The Mi-28 was nominally terminated on 5 October 1994 but the competition continued.

Further army evaluation followed when first two of four production Ka-50s were funded in 1994 and officially accepted on 28 August 1995; third and fourth received in 1996; these four numbered 20 to 23 (prompting preseries 021 to be renumbered 024 to avoid confusion). Arsenyev production was to have increased to one per month during 1997, but this did not occur. The original Ka-50 (and rival Mi-28A) were overtaken by the issue of a revised requirement which emphasised night capability – favouring the two-seat Mi-28. The initial order for 15 Ka-50s was reportedly cancelled in September 1998, with procurement postponed until 2003. Three deployed to Mozdok during 1999 for use in Chechnya.

Klimov VK-3000 turboshaft offered as alternative power plant.

CURRENT VERSIONS: **Ka-50** ('Hokum'): *As described.*

Ka-50N (*Nochnoy*: Nocturnal): Also reported as Ka-50Sh. Night-capable attack version; essentially a single-seat Ka-52. Programme began 1993; originally based on TpSPO-V and Merkury LLLTV systems, which tested on Ka-50 development aircraft. Ka-50N first reported April 1997 as conversion of prototype 018 with Thomson-CSF Victor FLIR turret above the nose and Arbalet (crossbow) mast-mounted radar, plus second TV screen in cockpit; FLIR integrated with Uralskyi Optiko-Mekhanicheskyi Zavod (UOMZ) Shamshit-50 (Laurel-50) electro-optic sighting system, incorporating French IR set. First flight variously reported as 4 March or 5 May 1997; programmed improvements included replacement of PA-4-3 paper moving map with digital equivalent; by August 1997, FLIR turret was repositioned below nose and Arbalet was removed; by mid-1998, had IT-23 CRT display replaced by TV-109, and HUD removed and replaced by Marconi helmet display. Proposed new cockpit shown in September 1998, having two Russkaya Avionika 203 × 152 mm (8 × 6 in) LCDs and central CRT for sensor imagery. Indigenous avionics intended for any local production orders; French systems as interim solution and standard for export. The Republic of Korea Army is reportedly interested in both the Ka-50N and the baseline Ka-50. In 1999, preproduction aircraft 014 was exhibited with a UOMZ GOES sensor turret in place of Shkval.

Ka-50-2: Designation applies to three quite different aircraft. Basic Ka-50-2 is a variant of the Ka-50 single-seater, though the designation is also applied to two twin-seat aircraft; first of these was a version of the Ka-52 Alligator and, as such, is described more fully in that entry. All Ka-50-2s differ from the baseline Ka-52 in retaining attack and anti-tank role using 12 laser beam-riding AT-8 Vikhr ATGMs or 16 Rafael NT-D ATGMs; avionics to be supplied by Israel Aircraft Industries, Lahav Division; 024 used as demonstrator. The basic Ka-50-2 was proposed to China, Finland, India, South Korea, Malaysia, Myanmar, Poland, South Africa, Syria and Turkey.

Second variant of Ka-50-2 is another two-seater, intended to have conventional stepped tandem cockpits; is offered to armed forces which do not accept the single-seat or side-by-side two-seat layouts. A further subvariant of the tandem-seat Ka-50-2, the **Erdoğan** (Turkish for Born Fighter), was proposed to Turkey jointly by Kamov and Israel Aircraft Industries. This would have been fitted with longer-span wings and feature a NATO-compatible Giat 621 turret containing a single 20 mm cannon which would fold down below the belly of the helicopter in flight, for a 360° arc of fire; it would fold to starboard for landing, and could be fired directly forward, even when folded. TV3-117VMA-02 engines. Ten Turkish pilots flew Alligator '061' at Antalya, Turkey, in early 1999 as part of evaluation process; requirement was for 145. Named as second choice when Bell AH-1Z selected.

Ka-52: Two-seat version; described separately.

CUSTOMERS: Four for Russian Army service trials, plus eight flying prototype and preseries helicopters; all delivered. Further 10 ordered in 1997 budget and six in 1998, of which first three were due for delivery before end of 1998; initial helicopter eventually completed in June 1999. One army helicopter lost in accident, 17 June 1998; attributed to rotor clash.

COSTS: Unit price of Ka-50N quoted as between US$12 million and US$15 million in mid-1999.

DESIGN FEATURES: World's first single-seat close support helicopter. Coaxial, contrarotating and widely separated semi-rigid three-blade rotors, with swept blade tip, attached to hub by steel plates; small fuselage cross-section, with nose sensors; flat-screen cockpit, heavily armour protected by combined steel/aluminium armour and spaced aluminium plates, with rearview mirror above windscreen; small sweptback tailfin, with inset rudder and large tab; high-set tailplane on rear fuselage, with endplate auxiliary fins; retractable landing gear; mid-set unswept wings, carrying ECM pods at tips; four underwing weapon pylons; engines above wingroots; high agility for fast, low-flying, close-range attack role; partially dismantled can be air-ferried in Il-76 freighter. Much of fuselage skin formed by large hinged door panels, providing access to interior equipment from ground level.

FLYING CONTROLS: Kamov coaxial design; generally as Ka-32.

STRUCTURE: Fuselage built around steel torsion box beam, of 1.0 m (3 ft 3¼ in) square section. Wing centre-section passes through beam. Cockpit mounted at front of beam, gearbox above and engines to sides. Carbon-based composites materials constitute 35 per cent by weight of structure, including rotors. Approximately 350 kg (770 lb) of armour protects pilot, engines, fuel system and ammunition bay; canopy and windscreen panels are 55 mm (2¼ in) thick bulletproof glass.

LANDING GEAR: Hydraulically retractable tricycle type; twin-wheel steerable nose unit and single mainwheels all semi-exposed when up; all wheels retract rearward; low-pressure tyres.

POWER PLANT: Two 1,633 kW (2,190 shp) Klimov TV3-117VMA turboshafts with VR-80 main reduction gearbox and two PVR-800 intermediate gearboxes, with air intake

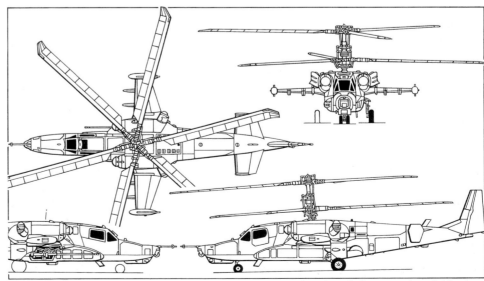

Ka-50 ('Hokum') single-seat combat helicopter with scrap view of gun installation on starboard side *(Mike Keep/Jane's)*

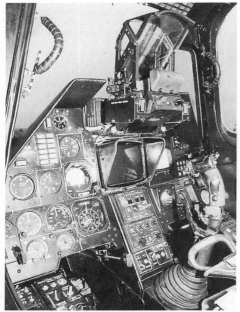

The Ka-50's cockpit has been adapted for use with pilot night-vision goggles despite its old style black-painted panel *(Piotr Butowski)*

Detail view of sensors on the nose of Ka-50N '018' *(Paul Jackson/Jane's)* 0007162

GOES-320 turret in the upper nose position of Ka-50 014 *(Paul Jackson/Jane's)* 2000/0062730

Kamov Ka-50 development aircraft '014' in mid-1999 configuration with GOES and FLIR turrets *(Paul Jackson/Jane's)* 2000/0062731

dust filters and exhaust heat suppressors. Later use of 1,838 kW (2,465 shp) TV3-117VMA-SB3 turboshafts intended. Two primary fuel tanks, filled with reticulated foam, inside fuselage box beam. Total internal capacity approximately 1,800 litres (485 US gallons; 404 Imp gallons). Front tank feeds port engine; rear feeds starboard and APU. Each tank protected by layers of natural rubber. Provision for four 500 litre (132 US gallon; 110 Imp gallon) underwing auxiliary fuel tanks. Transmission remains operable for 30 minutes after oil system failure.

ACCOMMODATION: Double-wall steel armoured cockpit, able to protect pilot from hits by 20 and 23 mm gunfire over ranges as close as 100 m (330 ft). Interior black-painted for use with NVGs. Specially designed Zvezda K-37-800 ejection system, ostensibly for safe ejection at any altitude (actually from 100 m; 330 ft); following explosive separation of rotor blades and opening of cockpit roof, pilot is extracted from cockpit by large rocket; alternatively, he can jettison doors and stores before rolling out of cockpit sideways. Associated equipment includes automatic radio beacon, activated during ejection, inflatable liferaft and NAZ-7M survival kit.

SYSTEMS: All systems configured for operational deployment away from base for up to 12 days without need for maintenance ground equipment; refuelling, avionics and weapon servicing performed from ground level. AI-9V APU for engine starting, and ground supply of hydraulic and electrical power, in top of centre-fuselage. Anti-icing system for engine air intakes, rotors, AoA and yaw sensors; de-icing of windscreen and canopy by liquid spray.

PrPNK Rubikon (L-041) piloting, navigation and sighting system based on five computers: four Orbita BLVM-20-751s for combat and navigation displays and target designation, plus one BCVM-80-30201 for WCS. Incorporates PNK-800 Radian navigation system, with C-061K pitch and heading data, IK-VSP-VI-2 speed and altitude and PA-4-3 automatic position plotting subsystems. Series 3 Tester U3 flight data recorder. Ekran BITE and warning system. KKO-VK-LP oxygen system with 2 litre (0.07 cu ft) supply for 90 minutes. Electrical supply from two 400 kW generators at 115 V 400 Hz three-phase AC; 500 W converter; rectifiers for 27 V DC supply.

AVIONICS: Integrated by NPO Elektro Avtomatika.

Comms: Two R800L1 and one R-868 UHF transceivers, SPU-9 intercom, P-503B headset recorder, Almaz-UP-48 voice warning system and HF com/nav; IFF ('Slap Shot').

Flight: INS; autopilot; Doppler box under tailboom; ARK-22 radio compass; A-036A radio altimeter.

Instrumentation: Conventional instruments; ILS-31 HUD; moving map display (Kronstadt Abris on some aircraft); small IT-23MV CRT beneath HUD, with rubber hood, to display only FLIR and monochrome LLLTV imagery. Pilot has Obzor-800 helmet sight effective within ±60° azimuth and from −20° to +45° elevation; when pilot has target centred on HUD, he pushes button to lock sighting and four-channel digital autopilot into one unit. Displays compatible with OVN-1 Skosok NVGs.

Mission: To reduce pilot workload and introduce a degree of low observability, target location and designation is assigned to other aircraft; equipment behind windows in nose includes I-25IV Shkval-V daylight electro-optical search and auto-tracking system, laser marked target seeker and range-finder; FOV ±35° in azimuth +15° to −80° in elevation. FLIR turret to be added in nose for use with NVGs.

Self-defence: L150 Pastel RWR in tailcone, at rear of each wingtip EW pod and under nose; total of 512 chaff/flare cartridges (in four UV-26 dispensers) in each wingtip pod. L-140 Otklik laser detection system; L-136 Mak IR warning.

ARMAMENT: Four BD3-UV pylons on wings. Up to 80 S-8 80 mm air-to-surface rockets in four underwing B8V20A packs or 20 S-13 122 mm rockets in four B-13L pods; or up to 12 9A4172 Vikhr-M (AT-12) tube-launched laser-guided ASMs with range of 8 to 10 km (5 to 6.2 miles) capable of penetrating 900 mm of reactive armour; or mix of both; Vikhr launched from trainable UPP-800 mounts, which can be depressed to −12°; single-barrel 30 mm

Model of the Ka-50-2 Erdoğan, as proposed to Turkey *(Yefim Gordon)* 2001/0100395

Instrument panel of the prototype Ka-52 *(Yefim Gordon)*

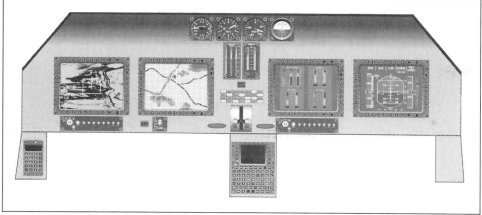
Proposed four-screen EFIS of the Ka-50-2   2000/0062673

2A42 gun on starboard side of fuselage, with up to 470 armour-piercing or high-explosive fragmentation rounds, can be depressed from +3° 30′ to −37° in elevation and traversed from −2° 30′ to +9° in azimuth hydraulically and is kept on target in azimuth by tracker which turns helicopter on its axis; two ammunition boxes in centre-fuselage. Front box contains 240 AP rounds, rear box 230 HE rounds. Selectable rapid (550 to 600 rds/min) or slow (350 rds/min) fire, with bursts of 10 or 20 rounds. Provision for alternative weapons, including UPK-23-250 23 mm gun pods, Igla or R-73 (AA-11 'Archer') AAMs, Kh-25MP (AS-12 'Kegler') ARMs, FAB-500 bombs or dispenser weapons.

DIMENSIONS, EXTERNAL:
Rotor diameter (each)	14.50 m (47 ft 7 in)
Length overall, rotors turning	16.00 m (52 ft 6 in)
Fuselage length, excl noseprobe	14.20 m (46 ft 7 in)
Wing span	7.34 m (24 ft 1 in)
Height overall	4.93 m (16 ft 2 in)
Tailplane span	3.16 m (10 ft 4½ in)
Wheel track: main	2.67 m (8 ft 9 in)
nose	0.34 m (1 ft 1½ in)
Wheelbase	4.19 m (13 ft 9 in)

AREAS:
Rotor disc (each)	165.13 m² (1,777.4 sq ft)

WEIGHTS AND LOADINGS:
Weight empty	7,800 kg (17,196 lb)
Max external stores	3,000 kg (6,610 lb)
Normal T-O weight: Ka-50	9,800 kg (21,605 lb)
Erdoğan	9,800 kg (21,605 lb)
Max T-O weight: Ka-50	10,800 kg (23,810 lb)
Erdoğan	11,300 kg (24,912 lb)

PERFORMANCE:
Max speed:	
in shallow dive	210 kt (390 km/h; 242 mph)
in level flight	162 kt (300 km/h; 186 mph)
in sideways flight	43 kt (80 km/h; 49 mph)
in backward flight	48 kt (90 km/h; 55 mph)
Cruising speed	146 kt (270 km/h; 168 mph)
Vertical rate of climb at 2,500 m (8,200 ft)	600 m (1,970 ft)/min
Service ceiling	5,500 m (18,040 ft)
Hovering ceiling OGE	4,000 m (13,120 ft)
Range: combat	243 n miles (450 km; 279 miles)
with max internal fuel	280 n miles (520 km; 323 miles)
with 4 auxiliary tanks:	
Ka-50	594 n miles (1,100 km; 683 miles)
Erdoğan	626 n miles (1,160 km; 720 miles)
Endurance: standard fuel, 10 min reserves	1 h 40 min
with 2 auxiliary tanks	2 h 50 min
g limit	+3.5

UPDATED

KAMOV Ka-52 ALLIGATOR

TYPE: Combat helicopter.
PROGRAMME: Project revealed at 1995 Paris Air Show; mockup (converted from Ka-50 static test airframe) displayed at Moscow Air Show, August 1995. Prototype (061 c/n 00601, believed converted from centre and rear fuselage of Ka-50 021), with Sextant Avionique avionics, shown to press on 19 November 1996; displayed at Bangalore Air Show, India, from 3 December; first flight 25 June 1997, fitted with flight test nose, lacking sensors; first 'official' flight 1 July. Ka-52 has designation V-80Sh-1.
CURRENT VERSIONS: **Ka-52:** Basic two-seat version of Ka-50 with side-by-side seating. *As described.*
Ka-50-2: Common designation covering three different multirole export variants, all with HOCAS controls, Israeli avionics (including state-of-the-art RWR, MAWS and laser warning) and MIL-STD-1553B/MIL-STD-1760B architecture. Intended roles include anti-armour, armed reconnaissance and observation, CAS, air-to-air combat, helicopter escort, artillery and naval fire correction, training and SAR. One single-seat version based on Ka-50, one based on side-by-side two-seat Ka-52, one proposed tandem two-seater. Last-named forms the basis of Ka-50-2

Erdoğan promoted to Turkey, with long-span wings and folding pylon below fuselage supporting Giat 621 20 mm cannon mounting. When pylon raised (folding to starboard) the cannon can fire straight forward. Provision for 3 tonne capacity external load hook with survivor's cage for SAR. Increased MTOW g loading and maximum speed.

DESIGN FEATURES: Airframe 85 per cent similar to single-seat Ka-50, but front fuselage redesigned to accommodate two crew side by side. Upward-hinged and bulged gull-wing type transparent canopy door over each seat. (Original canopy, built mainly from Ka-50 components, replaced in 1998 by more streamlined unit with upward-hinged doors, but still proofed against 12.7 mm armour-piercing rounds and shrapnel from 23 mm projectiles.) Bottom of nose recessed on starboard side to improve field of fire of 2A42 gun. Flattened nose for avionics led to nickname (since adopted for marketing) 'Alligator'. Ka-52 could be used to detect and designate targets for a formation of Ka-50s, this being referred to by Kamov as 'combat management'.

POWER PLANT: Two Klimov TV3-117VMA turboshafts; each 1,633 kW (2,190 shp). TV3-117VMA-SB3 (1,838 kW; 2,465 shp) in production version.

ACCOMMODATION: Pilot and navigator/weapons operator/pilot or pupil have Zvezda K-37-800 ejection system, for simultaneous emergency escape, similar to that of Ka-50. Full dual controls standard, including two colour and two monochrome SMD 66 multifunction displays.

AVIONICS: MIL-STD 1553B equivalent avionics suite with Russian and Western elements. Integrated by Sextant Avionique, France, supplier of head-down display and NASH (Night Attack System for Helicopters) nav/attack systems, Topowl helmet-mounted displays and image intensifiers.
Radar: FH-01 Arbalet MMW radar, probably by Phazotron, in mast-mounted dome.
Flight: Nadir 10 nav system integrated with GPS, Stratus laser-gyro AHRS and Doppler.
Instrumentation: Arsenal Shchel-V helmet-mounted sight for weapons operator.
Mission: Shamshit-E weapons control system, with TV, laser and radar elements, in ventral ball turret below cockpits. Thomson-CSF FLIR (or optional Russian Khod FLIR), integrated with Rotor electro-optical (TV) sighting system in ball above fuselage aft of canopy. Windows for laser range-finder and IR camera in nose turret.

ARMAMENT: As Ka-50 except only 240 rounds for 2A42 gun. Igla AAMs optional for export.

DIMENSIONS, EXTERNAL (Generally as for Ka-50, except):
Fuselage length, excl noseprobe, Ka-52 plain nose and Ka-50-2	13.53 m (44 ft 4¼ in)

WEIGHTS AND LOADINGS:
Weight empty: 50-2	7,500 kg (16,535 lb)
Max useful load: 50-2	9,700 kg (21,385 lb)
Max fuel weight: internal	1,484 kg (3,271 lb)
external (4 tanks)	1,720 kg (3,792 lb)
Normal T-O weight: 52	10,400 kg (22,925 lb)
50-2	11,300 kg (24,912 lb)
Max T-O weight: 50-2	11,900 kg (26,235 lb)

Prototype Kamov Ka-52 fitted with initial flight test nose, lacking chin sensors   2000/0062677

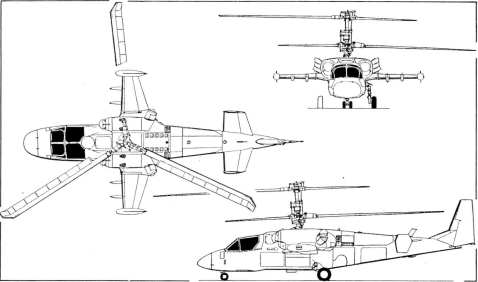
Kamov Ka-52 Alligator combat helicopter in prototype configuration *(Paul Jackson/Jane's)*

378　RUSSIAN FEDERATION: AIRCRAFT—KAMOV

PERFORMANCE (estimated; 52: Ka-52, 50-2: Ka-50-2):
Never-exceed speed (V_{NE}):
  52, 50-2　　　　　　　　189 kt (350 km/h; 217 mph)
Max level speed: 52　　　　162 kt (300 km/h; 186 mph)
  50-2　　　　　　　　　167 kt (309 km/h; 192 mph)
Max speed sideways: 52　　43 kt (80 km/h; 49 mph)
Max speed backward: 52　　48 kt (90 km/h; 55 mph)
Max rate of climb at S/L: 52　480 m (1,575 ft)/min
  50-2　　　　　　　　　966 m (3,170 ft)/min
Vertical rate of climb at S/L: 50-2　792 m (2,600 ft)/min
Hovering ceiling OGE: 52　　3,600 m (11,820 ft)
Service ceiling: 50-2　　　　6,000 m (19,700 ft)
Range: 52: normal　　　　243 n miles (450 km; 279 miles)
  ferry　　　　　　　　647 n miles (1,200 km; 745 miles)
  50-2: ferry　　　　　626 n miles (1,159 km; 720 miles)
Endurance, 20 min reserves, 50-2: normal　　2 h 24 min
  with 2 auxiliary tanks　　　　　　　　3 h 48 min
g limit: 52　　　　　　　　　　　　　　+3
  50-2　　　　　　　　　　　　　　+3.5
UPDATED

KAMOV Ka-60 KASATKA
English name: Killer Whale
TYPE: Medium transport helicopter.
PROGRAMME: Original coaxial rotor, twin tail, single-engined V-60 won Soviet Army lightweight helicopter and Mi-8 replacement competition against heavier, twin-engined Mil Mi-36 in 1982; subsequently, design considerably modified to achieve greater speed through adoption of single main rotor and Fenestron-type of tail rotor with eleven blades. First flight originally due 1993, but programme slowed by funding shortages, and priority changed to promotion of civil variant (see following entry for Ka-62). Ka-60 officially revealed at Lyubertsy on 29 July 1997, when prototype close to completion; first flew (601) 10 December 1998; second sortie 21 December; first official flight 24 December; all were hovering flights; international debut at MAKS '99, Moscow, August 1999; first 'forward flight' 24 December 1999. Conflicting reports quote both Arsenyev and Ulan Ude as prospective production lines, while LMZ was reported in April 2000 to be preparing for production.

CURRENT VERSIONS: Pilot and aircrew training (Ka-60U), utility, shipborne over-the-horizon targeting (Ka-60K) reconnaissance (Ka-60R) anti-tank and anti-helicopter versions proposed; role of Ka-60R, with Shamshit target acquisition system, transferred to Kamov Ka-52.

CUSTOMERS: None. Long-term interest maintained by Russian Army; reportedly evaluated by Iran. Order from Russia expected, initially for rotary-wing pilot training version. Russian military requirements include some 150 Ka-60Us in training role.

COSTS: US$1.7 million, Ka-60U (2000).
DESIGN FEATURES: Generally as for Ka-62 (which see), but with IR- and radar-absorbent coatings, reduced rotor speed, low-IR exhausts.
POWER PLANT: RKBM Rybinsk RD-600V turboshafts, as Ka-62. RRTM RTM322 or GE CT7 available in export versions.
ACCOMMODATION: Up to 16 infantry troops; or six stretchers and three attendants. Pilot (starboard) and co-pilot/gunner (port) side by side. Provision for dual controls; control stick top common with Ka-50/52.
SYSTEMS: All Russian; Western equivalents optional for export.
AVIONICS: As above, including Pastel RWR and Otklik laser warning system.
  Radar: Arbalet MMW antenna in nose.
EQUIPMENT: Cargo hook.
ARMAMENT: One-piece transverse boom through cabin, to rear of doors, optional to provide suspension for total of two B-8V-7 seven-round 80 mm rocket pods, two 7.62 mm or 12.7 mm gun pods, or similar armament.
DIMENSIONS, EXTERNAL: As Ka-62 except:
  Fuselage length　　　　　　　13.465 m (44 ft 2 in)
WEIGHTS AND LOADINGS: Generally as Ka-62 except:
  Max payload: external　　　　2,750 kg (6,062 lb)
UPDATED

KAMOV Ka-62
TYPE: Medium utility helicopter.
PROGRAMME: Funded under Russian programme for development of civil aviation for 2000. Construction of prototype Ka-62 (then known as V-62) began early 1990; had not flown by early 1999; one Ka-60 military version (which see) and two Ka-62s to undertake flight trials; will be certified under Russian AP and FAR Pt 29A/B standards. Republic of Buryatia officially requested Moscow for production rights for UUAP (which see) in February 2000.

CURRENT VERSIONS: **Ka-62**: Basic model for domestic market. *Detailed description applies to the above.*
  Ka-62M: Second and third prototypes; to be certified to Western standards, for sale outside Russian Federation and Associated States (CIS); two 1,212 kW (1,625 shp) General Electric T700/CT7-2D1 turboshafts; five-blade main rotor; avionics to be developed by Aviapribor. Production version is expected to be manufactured by KnAAPO at Komsomolsk.
  Ka-64 Sky Horse: See separate entry.
DESIGN FEATURES: Originated as military transport; all main systems and components duplicated, with main and secondaries routed on opposite sides of airframe; transmission resistant to 12.7 mm bullets; main blades to 23 mm shells; run-dry gearboxes. Advanced technology main rotors with sweptback tips. Production versions will have slower-turning five-blade rotor. Yaw control by 11-blade fan-in-fin. Reverse tricycle landing gear.
STRUCTURE: Composites account for 60 per cent, by weight, of structure, including blades of main rotor; fuselage sides,

Pilot's instruments in the Ka-60 *(Yefim Gordon)*

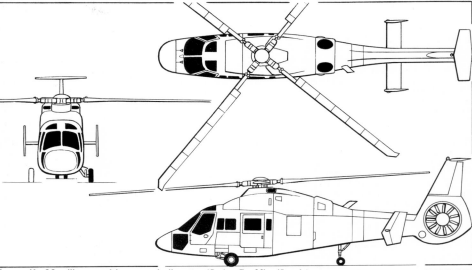
Kamov Ka-60 military multipurpose helicopter *(James Goulding/Jane's)*

Kamov Ka-52 weapon options

Prototype Ka-60 utility helicopter *(Paul Jackson/Jane's)*

KAMOV—AIRCRAFT: RUSSIAN FEDERATION

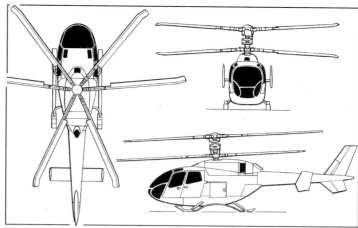

Kamov Ka-115 five/six-seat light helicopter *(James Goulding/Jane's)*

Full-size mockup of Kamov's multirole Ka-115 helicopter
2000/0062671

doors, floor and roof, tailboom, fin, vertical stabilisers, and fan blades of carbon-reinforced Kevlar.
LANDING GEAR: Retractable reverse tricycle type; single KT-217 mainwheels retract inward and upward into bottom of fuselage; twin rear wheels retract forward into tailboom; shock-absorber in each unit. Optional inflatable pontoons for emergency use on water.
POWER PLANT: Basic Ka-62 has two RKBM Rybinsk RD-600V turboshafts, each 956 kW (1,282 shp) max continuous; 1,140 kW (1,529 shp) emergency rating. Fuel tanks under floor, capacity 1,450 litres (383 US gallons; 319 Imp gallons). General Electric T700/CT7-2D1 offered as alternative to RD-600V.
ACCOMMODATION: Crew of one or two, side by side; optional bulkhead divider between flight deck and cabin; up to 14 passengers in four rows; forward-hinged door each side of flight deck; large forward-sliding door and small rearward-hinged door each side of cabin; baggage hold to rear of cabin. VIP configuration to be available, with five to nine seats and refreshment bar.
SYSTEMS: Interior heated and air conditioned. Thermoelectric de-icing system optional. Ivchenko AI-9V APU standard.
AVIONICS: Optional Russian or Western.
EQUIPMENT: Stretchers, hoist above port cabin door, cargo tiedowns, and other items as necessary for variety of roles, including transport of slung freight; air ambulance/operating theatre; search and rescue; patrol of highways, forests, electric power lines, gas and oil pipelines; survey of ice areas; surveillance of territorial waters, economic areas and fisheries; mineral prospecting; and servicing of offshore gas and oil rigs.

DIMENSIONS, EXTERNAL:
Main rotor diameter	13.50 m (44 ft 3½ in)
Tail rotor diameter	1.40 m (4 ft 7 in)
Length overall, rotors turning	15.60 m (51 ft 2¼ in)
Fuselage length	13.465 m (44 ft 2 in)
Height: overall	4.60 m (15 ft 1 in)
to top of rotor head	3.80 m (12 ft 5½ in)
Width over endplate fins	3.00 m (9 ft 10 in)
Wheel track	2.50 m (8 ft 2½ in)
Wheelbase	5.445 m (17 ft 10¼ in)
Cabin doors (each) Height	1.30 m (4 ft 3¼ in)
Width	1.25 m (4 ft 1¼ in)

DIMENSIONS, INTERNAL:
Cabin, excl flight deck: Length	3.40 m (11 ft 2 in)
Width	1.78 m (5 ft 10 in)
Height	1.30 m (4 ft 3¼ in)

Proposed instrument pedestal of the Ka-115 *(Yefim Gordon)* 2000/0084010

AREAS:
Main rotor disc	143.10 m² (1,540.3 sq ft)
Tail rotor disc	1.54 m² (16.57 sq ft)

WEIGHTS AND LOADINGS:
Max payload: internal: 62, 62M	2,000 kg (4,409 lb)
external: 62, 62M	2,500 kg (5,510 lb)
T-O weight: 62, 62M, normal	6,000 kg (13,228 lb)
62, internal load, max	6,250 kg (13,779 lb)
62 external load and 62M, internal load, max	6,500 kg (14,330 lb)
62M, external load, max	6,750 kg (14,880 lb)
Max disc loading: 62	41.9 kg/m² (8.58 lb/sq ft)
62M	47.1 kg/m² (9.66 lb/sq ft)

PERFORMANCE (estimated, at max T-O weight with internal load):
Max level speed: 62, 62M	162 kt (300 km/h; 186 mph)
Cruising speed: 62, 62M	148 kt (275 km/h; 171 mph)
Max rate of climb at S/L: 62	625 m (2,050 ft)/min
62M	690 m (2,263 ft)/min
Rate of climb at S/L, OEI: 62, 62M	123 m (403 ft)/min
Service ceiling: 62	5,150 m (16,900 ft)
62M	5,500 m (18,040 ft)
Hovering ceiling: IGE: 62	2,900 m (9,520 ft)
OGE: 62	2,100 m (6,880 ft)
62M	2,500 m (8,200 ft)
Range with max standard fuel at 2,000 m (6,560 ft), no reserves: 62	332 n miles (615 km; 382 miles)
62M	353 n miles (655 km; 407 miles)
Range with auxiliary fuel at 2,000 m (6,560 ft), no reserves: 62	566 n miles (1,050 km; 652 miles)
62M	553 n miles (1,025 km; 637 miles)

UPDATED

KAMOV Ka-64 SKY HORSE

This development of the Ka-60/62 series is reported to be a joint venture with Agusta of Italy intended for export. Features include a conventional tail rotor, modified passenger cabin, Western avionics and option of General Electric CT7-2DL, LHTEC T800 or RRTI RTM 322 turboshaft engines. Production would be by UUAP at Ulan-Ude.

UPDATED

KAMOV Ka-115 MOSKVICH

English name: Muscovite

TYPE: Light utility helicopter.
PROGRAMME: Announced at Moscow Air Show '95, at which time design had been completed; full-scale mockup first exhibited at Moscow '97; first flight scheduled end 1998, but programme had by then been put in abeyance. In September 1999, Kamov confirmed continued existence of the Ka-115, but declined to predict a new first flight date; this had not occurred by late 2000. Promotion continued in 2000, when mayor of Moscow suggested as initial customer, prompting allocation of type's name; development is included in Russian federal aviation plan 2000-15.
DESIGN FEATURES: Twin coaxial, contrarotating three-blade rotors; no tail rotor. Oval-section cabin; tapered tailboom; sweptback fin with smaller underfin, also swept; tailplane attached to underside of tailboom has small, angular endplate fins. Twin-skid landing gear.
POWER PLANT: One 410 kW (550 shp) P&WC PW 206D turboshaft, with intake particle separator.

ACCOMMODATION: Basic version seats pilot and four or five passengers. Emergency medical service version can accommodate one stretcher, two medical attendants and medical equipment. Side-hinged double doors each side for pilot and passenger access. Internal cargo can be loaded via hatch at rear of cabin and smaller side hatch at floor level each side; larger cargo payload can be carried on external sling. For search and rescue role, cabin door design permits fitment of an external hoist.
SYSTEMS: Hydraulic control system with possible manual monitoring; 27 V DC electrical system; blade de-icing and cabin heating systems.

DIMENSIONS, EXTERNAL:
Rotor diameter (each)	9.50 m (31 ft 2 in)
Length overall, excl rotors	9.20 m (30 ft 2¼ in)
Height to top of rotor head	3.60 m (11 ft 9¾ in)
Skid track	2.00 m (6 ft 6¾ in)

AREAS:
Rotor disc (each)	70.88 m² (763.0 sq ft)

WEIGHTS AND LOADINGS:
Max cargo payload: internal	700 kg (1,543 lb)
on external sling	900 kg (1,984 lb)
Max T-O weight	1,970 kg (4,343 lb)
Max disc loading	13.9 kg/m² (2.85 lb/sq ft)

PERFORMANCE (estimated):
Max level speed	135 kt (250 km/h; 155 mph)
Cruising speed	124 kt (230 km/h; 143 mph)
Max rate of climb at S/L	690 m (2,264 ft)/min
Hovering ceiling OGE	2,350 m (7,700 ft)
Range: with standard fuel	421 n miles (780 km; 484 miles)
with auxiliary tanks	648 n miles (1,200 km; 745 miles)
Endurance, search and rescue within 162 n mile (300 km; 186 mile) radius	4 h

UPDATED

KAMOV Ka-215

TYPE: Light utility helicopter.
PROGRAMME: Twin-engined version of Ka-115 (which see), announced 2000, when cost estimated as more than US$1 million.

NEW ENTRY

KAMOV Ka-226A SERGEI

TYPE: Light utility helicopter.
PROGRAMME: Announced at 1990 Helicopter Association International convention, Dallas, USA. Developed originally for Russian TsENTROSPAS disaster relief ministry, which is providing significant funding. First flight (RA-00199), at Lyubertsy, 3 September 1997; 'official' first flight on following day. Flew total of four sorties by 31 December 1997. Further two prototypes to follow for certification trials; these tests to be completed by end of 1998, but at that time no confirmation available that Nos. 2 and 3 had flown, although a KAPP-built trials aircraft was rolled out at Kumertau on 29 May 1998. FAR Pt 29 Cat A/B certification being sought. Named Sergei in 1999, honouring politician Sergei Shoigu, but programme also guided by Sukhoi General Designer, Sergei Mikheev. By mid-2000 Moscow regional government had provided Rb12 million in development funding and was beginning disbursements under second programme valued at Rb4 million. Initial deliveries due in first half of 2000, but not effected.
Prototypes built jointly by Kamov, Strela, KAPP and Ufa Motors, with final assembly by Strela for production

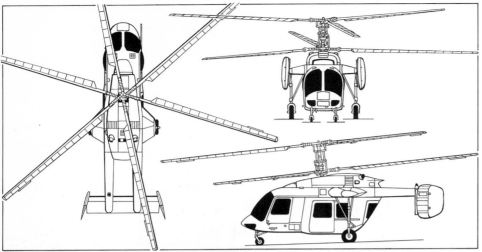

Kamov Ka-226A utility helicopter with production passenger cabin design *(Mike Keep/Jane's)* 2000/0062676

aircraft. Strela scheduled to have delivered five preproduction helicopters to Kamov at Lyubertsy by mid-2000; state ground testing of Ka-226 second prototype completed at Strela's Orenburg plant on 6 March 2000; flight testing of production aircraft then planned for late 2000. KAPP designated second production plant. Motor-Sich of Zaporozhye, Ukraine, negotiating with Kamov in June 2000 to build Ka-226s powered by indigenous ZMKB AI-450 engine. Batch of Rolls-Royce engines ordered by Kamov in July 2000.

CUSTOMERS: Identified requirements include up to 20 for City of Moscow for patrol and medevac; some 250 for MChS Rossii/TsENTROSPAS disaster relief organisation; and up to 75 for Gazprom in gasfield support role. Firm order for 25 reportedly received from TsENTROSPAS by 1997, but quantity had reduced to 10 by 1999; by mid-2000 this quoted as five firm (to be first five production aircraft) and further 15 to be ordered by 2005. Deliveries in 2001 to include three for Moscow and five for Gazprom. By mid-2000, firm orders were about 30 and market being estimated as 400 to 450.

COSTS: US$1.5 million (2000). Development cost Rb108 million (1999).

DESIGN FEATURES: Classic Kamov utility helicopter, featuring interchangeable mission pods. Refined development of Ka-26/126; new rotor system with hingeless hubs and glass fibre/carbon fibre blades; changes to shape of nose, twin tailfins and rudders, and passenger pod; passenger cabin has much larger windows and remains interchangeable with variety of payload modules including agricultural systems with hopper capacity of 1,000 litres (264 US gallons; 220 Imp gallons); new rotor system, interchangeable with standard coaxial system, will become available later.

FLYING CONTROLS: Assisted by irreversible hydraulic actuators. Automatic rotor constant-speed control; conventional four-channel control (longitudinal, lateral, cyclic and differential pitch). Two endplate fins and rudders, toed inward 15°; fixed horizontal stabiliser.

STRUCTURE: Primarily of aluminium alloys, steel alloys and composites sandwich panels of GFRP with honeycomb filler. Rotor blade overhaul interval 2,000 hours; total life 6,000 hours, but to be extended by increments to 18,000 hours.

LANDING GEAR: Non-retractable four-wheel type. Main units at rear, carried by stub-wings. All units embody oleo-pneumatic shock-absorber. Mainwheel tyres size 595×185, pressure 2.50 bars + 0.50 (36.25 lb/sq in + 7.25); forward tyres size 300×125, pressure 3.50 bars + 0.50 (50.75 lb/sq in + 7.25). Forward units of castoring type, without brakes. Rear wheels have pneumatic brakes.

POWER PLANT: Two 335 kW (450 shp) Rolls-Royce 250-C20R/2 turboshafts, side by side aft of rotor mast, with individual driveshafts to rotor gearbox. Two 335 kW (450 shp) Rolls-Royce 250-C20B engines in prototypes. Transmission rating 626 kW (840 shp). Alternatively, two Progress (ZMKB) AI-450 turboshafts, each 335 kN (450 shp).

Standard fuel capacity 770 litres (203 US gallons; 169 Imp gallons), in tanks above and forward of payload module area. Provision for two external tanks, on sides of fuselage, total capacity 320 litres (84.5 US gallons; 70.4 Imp gallons).

ACCOMMODATION: Fully enclosed and lightly pressurised flight deck, with rearward-sliding door each side; normal operation by single pilot; second seat and dual controls optional. Cabin ventilated, and warmed and demisted by air from combustion heater, which also heats passenger cabin when fitted. Space aft of cabin, between main landing gear legs and under transmission, can accommodate variety of interchangeable payloads. Cargo/passenger pod has two bench seats, each accommodating three persons; one bench faces forward, the other, rear; baggage compartment behind rear wall. Seventh passenger beside pilot on flight deck. Provision for cargo sling. Ambulance pod accommodates two stretcher patients, two seated casualties and medical attendant. For agricultural work, chemical hopper (capacity 1,000 litres; 264 US gallons; 220 Imp gallons) and dust spreader or spraybars are fitted in this position, on aircraft's CG. (Flight deck pressurisation protects crew against chemical ingress.) Aircraft can also be operated with either an open platform for hauling freight or hook for slinging loads at end of a cable or in a cargo net.

SYSTEMS: Single hydraulic system, with manual override, for control actuators. Main electrical system 27 V 3 kW DC, with back-up 40 Ah battery; secondary system 36/115 V AC with two static inverters; 115/200 V AC system with 16 kVA generator (6 kVA to power agricultural equipment and rotor anti-icing). Electrothermal rotor blade de-icing; hot air engine air intake anti-icing; alcohol windscreen anti-icing; electrically heated pitot. Pneumatic system for mainwheel brakes, tyre inflation, agricultural equipment control, pressure 39 to 49 bars (570 to 710 lb/sq in). Oxygen system optional.

AVIONICS: Cockpit instrumentation and avionics to customer's choice, including Honeywell equipment for IFR flight.
   Comms: Optional Honeywell KY 196A radio; transponder.
   Flight: Optional Honeywell KN 53 ILS; KR 87 ADF; KLN 90B GPS; and ADF.

EQUIPMENT: Specially equipped payload modules available for variety of roles, including ambulance and agricultural duties.

ARMAMENT: Optional provision for light weapons.

DIMENSIONS, EXTERNAL:
Rotor diameter (each)	13.00 m (42 ft 7¾ in)
Length of fuselage	8.10 m (26 ft 7 in)
Width over stub-wings	3.22 m (10 ft 6¾ in)
Height to top of rotor head	4.15 m (13 ft 7½ in)
Wheel track: nosewheels	0.90 m (2 ft 11½ in)
mainwheels	2.56 m (8 ft 4¾ in)
Wheelbase	3.48 m (11 ft 5 in)
Passenger pod: Length	2.35 m (7 ft 8½ in)
Width	1.40 m (4 ft 7 in)
Height	1.54 m (5 ft 0¾ in)

DIMENSIONS, INTERNAL:
Passenger pod: Length	2.04 m (6 ft 8¼ in)
Width	1.28 m (4 ft 2¼ in)
Height	1.40 m (4 ft 7 in)

AREAS:
Rotor disc (each)	132.70 m² (1,428.4 sq ft)

WEIGHTS AND LOADINGS:
Weight empty	1,952 kg (4,304 lb)
Max payload, internal or slung	1,300 kg (2,865 lb)
Max internal fuel	600 kg (1,322 lb)
Auxiliary fuel	256 kg (564 lb)
T-O weight: normal	3,100 kg (6,835 lb)
max	3,400 kg (7,495 lb)
Transmission loading at max T-O weight and power:	
normal	4.95 kg/kW (8.14 lb/shp)
max	5.43 kg/kW (8.92 lb/shp)

PERFORMANCE (estimated):
Never-exceed speed (V_{NE})	115 kt (214 km/h; 133 mph)
Max level speed	111 kt (205 km/h; 127 mph)
Max cruising speed	105 kt (194 km/h; 120 mph)
Max rate of climb at S/L	610 m (2,000 ft)/min
Rate of climb, OEI	96 m (315 ft)/min
Vertical rate of climb at S/L	168 m (550 ft)/min
Service ceiling	5,700 m (18,700 ft)
Hovering ceiling OGE	2,165 m (7,100 ft)
Range: with max payload	16 n miles (30 km; 18 miles)
with max internal fuel	324 n miles (600 km; 372 miles)
with auxiliary fuel, no reserves	480 n miles (890 km; 552 miles)
Max endurance with internal fuel, no reserves	4 h 42 min

UPDATED

Prototype Kamov Ka-226A displayed at Zhukovsky in August 1999 *(Paul Jackson/Jane's)* 2000/0062723

KAPO

KAZANSKOYE AVIATSIONNOYE PROIZVODSTVENNOYE OBEDINENIE IMENI S P GORBUNOVA (Kazan Aircraft Production Association named for S P Gorbunov)
ulitsa Dementiev 1, 420036 Kazan, Respublika Tatarstan
Tel: (+7 8432) 54 24 32
Fax: (+7 8432) 54 36 93
Telex: 224184 SOKOL

Since 1927, KAPO (formerly GAZ 22) has built more than 18,000 aircraft of 34 types, including the Tu-4, Tu-16, Tu-22, Tu-104 and Il-62. It currently produces the Tupolev Tu-214 and will manufacture the Tu-330 freighter, if it is ordered, and Tu-324 regional airliner. In 2000, KAPO belatedly delivered one Tu-160 strategic bomber, completion of which had been delayed by collapse of the former USSR; a second incomplete airframe may follow, as could two Tu-22Ms, also in stock.

Also in 1999, Kazan delivered one Ilyushin Il-62 airliner (see *Jane's Aircraft Upgrades*) from a stock of five complete and three incomplete airframes held since main production ended in 1993. (Il-62 production thus one static test airframe, four flying prototypes, 94 of first series, 190 complete Il-62Ms, of which three awaiting purchasers, and three partly completed.)

UPDATED

KAPP

KUMERTSKOYE AVIATSIONNOYE PROIZVODSTVENNOYE PREDPRIYATIE (Kumertau Aviation Production Enterprise)
ulitsa Novozarinskaya 15A, 453350 Kumertau, Respublika Bashkortostan
Tel: (+7 34761) 222 53 and 223 00
Fax: (+7 34761) 239 13 and 340 91

GENERAL DIRECTOR: Boris S Malyshev
EXPORT DEPARTMENT: R M Rafikov

KAPP has manufactured helicopters, aeroplanes and related equipment at Kumertau since 1962. Major programmes have involved the Kamov Ka-26 and Ka-27/28/29/32 helicopter series, Myasishchev M-17/M-55, wings for the Tupolev Tu-154M transport, and unmanned air vehicles including the Tu-243 reconnaissance system. Production in 1998 included first of 15 Ka-32A11BCs ordered by Canadian operators.

KAPP is the only manufacturer of carbon fibre/glass fibre rotor blades in the Russian Federation and Associated States (CIS). It has also been chosen to manufacture the Ilyushin Il-112 transport for the Russian Air Forces. Participates in Kamov Ka-226A, and is assigned (after Strela) as second of two production lines.

UPDATED

KAZAN

KAZANSKY VERTOLETNYI ZAVOD AO (Kazan Helicopter Plant JSC)
ulitsa Tetsevskaya 13/30, 420085 Kazan, Respublika Tatarstan
Tel: (+7 8432) 54 45 52
Fax: (+7 8432) 54 52 52
e-mail: market@kazanhelicopter.tatincom.ru
Telex: 224848 AGAT RU
GENERAL DIRECTOR: Aleksandr P Lavrentyev
DEPUTY DIRECTOR: Valery Pashko
ANSAT PROGRAMME DIRECTOR: Valery Dvoeglazov
MARKETING DIRECTOR: Mikhail Tikhonov

Founded 1935 and built 11,000 light aircraft (including Polikarpov Po-2s) up to late 1940s. Since 1951, Kazan (KVZ, formerly GAZ 387) has marketed and built Mil helicopters comprising Mi-1, Mi-4, Mi-8 (from 1965), Mi-14 and Mi-17 (from 1982) series. Exports began in 1956 and now encompass over 60 countries and some 4,000 aircraft; the Mi-17 accounts for 1,200 exports to 30 countries. Total Mi-8/17 production at Kazan exceeds 6,000. Sales over the five years to 1998 were 90 per cent for export, including two Mi-17s to Malaysia in August 1998 for firefighting.

Kazan became a joint stock company in 1994, owned 38 per cent by the state, 62 per cent by its employees. At the 1995 Paris Air Show, it exhibited the Mi-17MD upgraded 40-passenger version of the Mi-17 and a mockup of the first product of its own newly formed (1993) design office, the Ansat light multipurpose helicopter. Modifications and upgrades to Mi-17 have generated new business for Kazan, including Indian contract for 40 in 2000. Russian certification for helicopter design awarded February 1997. A second light helicopter, the Aktai, was revealed at Moscow in August 1997 to be under development, although Kazan is also nominated as potential manufacturer of the Mil Mi-60. Kazan is also a co-founder of Euromil, with Mil Helicopter Plant, Eurocopter and Klimov (see International section), and is manufacturing parts for the Mi-38 prototypes.

UPDATED

KAZAN/MIL Mi-17MD and Mi-17N

TYPE: Medium transport helicopter.
PROGRAMME: See Mil entry for general details; Kazan's Mi-17 development intended to compete with latest Western helicopters (such as Sikorsky S-92, NHI NH-90, EHI EH 101 and Eurocopter AS 332) combining advanced avionics and systems with well-proven airframe and dynamic system. Prototype (RA-70937, converted from Mi-8MTV) displayed at 1995 Paris Air Show with features including widened (from 0.83 m; 2 ft 8¾ in to 1.25 m; 4 ft 1¼ in) forward, port, door; additional starboard door, 1.39 m (4 ft 6¾ in) high and 0.83 m (2 ft 8¾ in) wide; and rear loading via short ramp and two clamshell doors. Further modified, with large, single-piece rear-loading ramp and other changes, for display at ILA '96, Germany, in search and rescue form, designated Mi-17MD Night. By 1997, '70937 had gained a spine-mounted IR jammer and flight deck armour and was stated by Kazan to have the dual designation Mi-17MD/Mi-8MTV-5. At the 1999 Moscow Air Show it had lost its registration and become an 'Mi-8MTV5-1' with the alternative designation Mi-17N (*Noch:* Night). Nose-mounted 8A-813S radar.

Second example (RA-70877) shown at LIMA, Malaysia, November 1999, marked as Mi-17-1V, although with full Mi-17MD airframe modifications, was conversion of prototype Mi-17KF (see following entry), destined for South Korean police force (eventually handed over, January 2000).

Kazan demonstrator RA-70898, rebuilt in 2000 (following accident) with partial 'glass cockpit' supplied by Elbit, GOES FLIR chin turret and additional external fuel tank (900 litres; 238 US gallons; 198 Imp gallons) each side, increasing range to 600 n miles (1,111 km; 690 miles); optimised for offshore oil support and maritime patrol; displayed at Farnborough, July 2000.

CUSTOMERS: Four armed Mi-17MD exported to Rwanda, 1999, with a fifth aircraft in VIP configuration. South Korea (unspecified quantity as Russian debt repayment). Indian Ministry of Defence ordered 40 Mi-17Ms from Kazan in May 2000; deliveries to begin within following few months.

DESIGN FEATURES: Generally similar to latest versions of Mi-17 (which see). Pointed 'Dolphin' nose with increased volume to allow space for radar. Maximum accommodation increased to 40 passengers, or 36 troops in three rows of seats. Hydraulically operated, hinged, built-in ramp forming rear floor of fuselage pod, as fitted to Mil Mi-8MTV-5 (which see). Provision for three stretchers in rear of cabin on starboard side. New cockpit displays. COSPAS/SARSAT equipment, with Glonass antenna on tailboom. Navigation system includes Kurs-MP-70 Shoran with SD-75 VOR/ILS and DME. Extended-range fuel tanks. LLLTV sensor under port side of cockpit; NVGs for crew. Air conditioner/APU intake on starboard side, above cabin door. Flotation system provides buoyancy for at least 30 minutes.

ARMAMENT: Mountings for PKT machine guns in main cabin doors; provision for troops to fire their personal weapons through cabin portholes. Provision for outrigger pylons and compatible with Malyutka-2 or Shturm ATMs.

UPDATED

KAZAN/MIL Mi-17KF KITTIWAKE

TYPE: Medium transport helicopter.
PROGRAMME: Developed jointly by Mil and Kazan, with new avionics and electrical system designed, supplied and integrated by Kelowna Flightcraft Ltd, Canada. Prototype (RA-70877) first flew in Canada on 3 August 1997. Certified to FAR Pt 29 for full IFR operation. Intended as new-build and retrofit configuration. RA-70877 subsequently converted to Mi-17MD at Kazan and sold to LG International for delivery to South Korean police. This proof-of-concept aircraft may be followed by further aircraft for other South Korean paramilitary, military and government agencies.
DESIGN FEATURES: Generally similar to latest versions of Mi-17 (which see).
POWER PLANT: Two 1,434 kW (1,923 shp) Klimov TV3-117MT or 1,545 kW (2,070 shp) TV3-117VM turboshafts. Provision for additional fuel.
ACCOMMODATION: Basic cargo transport, 28-seat passenger or flying hospital configuration complete with operating theatre; other versions to customer's requirements. Optional cargo sling, capacity 5,000 kg (11,020 lb), and hoist, capacity 300 kg (660 lb).
AVIONICS: New six-screen EFIS, integrating data from Honeywell EDZ-756 EFIS with colour displays, Primus 700 colour weather radar, FZ-706 dual flight director system, AA-300 radio altimeter, VG/DG-14 AHRS and Primus II integrated radio system; Transicoil engine

Kazan Mil Mi-17-IV demonstrator RA-70898, fitted with additional external fuel tanks *(Paul Jackson/Jane's)*

Kazan's 'Hip' demonstrator RA-70937 as an Mi-17MD/Mi-8MTV-5 with open rear ramp, ASO-2V flare dispensers, IR suppressor, KO-50 air conditioning pod, flight deck armour and B-8 80 mm rocket pods

382　RUSSIAN FEDERATION: AIRCRAFT—KAZAN

Kelowna Flightcraft instrument panel of the Kazan Mi-17KF
(Michael J Gething/Jane's)

Elbit partial 'glass cockpit' of Mi-17-IV demonstrator RA-70898
(Paul Jackson/Jane's)

instrument system and KGS instrument lighting system. Options include BAE Systems North America Doppler nav system, GPS and NVG-compatible lighting.
EQUIPMENT: Optional flotation system, cargo sling and SAR winch (port door, capacity 300 kg; 661 lb).

WEIGHTS AND LOADINGS:
Max payload: internal　　　　　　4,000 kg (8,818 lb)
　　　　　　external　　　　　　4,500 kg (9,920 lb)
Normal T-O weight　　　　　　11,100 kg (24,470 lb)
Max T-O weight　　　　　　　13,000 kg (28,660 lb)
PERFORMANCE:
Max level speed up to 1,065 m (3,500 ft):
　at normal T-O weight　135 kt (250 km/h; 155 mph)
　at max T-O weight　　124 kt (230 km/h; 143 mph)
Nominal cruising speed up to 1,000 m (3,280 ft):
　at normal T-O weight
　　　　　　119-130 kt (220-240 km/h; 137-150 mph)
　at max T-O weight
　　　　　　111-116 kt (205-215 km/h; 127-134 mph)
Service ceiling:
　at normal T-O weight　　　　6,000 m (19,680 ft)
　at max T-O weight　　　　　4,800 m (15,740 ft)
Hovering ceiling:
　at normal T-O weight　　　　4,000 m (13,120 ft)
Range: with standard fuel at max T-O weight
　　　　　　　　　　324 n miles (600 km; 372 miles)
　with two auxiliary tanks: at normal T-O weight
　　　　　　　　　　567 n miles (1,050 km; 652 miles)
UPDATED

KAZAN ANSAT
English name: Ethereal
TYPE: Light utility helicopter.
PROGRAMME: Design begun at Kazan in 1993; design subcontracts to Kazan State Technical University for structural strength and aerodynamic calculations; Aviacon Scientific and Production Centre for rotor; and Aeromekhanica for transmission. Fuselage mockup exhibited at 1995 Paris Air Show; considerably revised engineering mockup (001) at Paris '97; by August 1998, now marked '01', this had accumulated 10 hours of ground running with engines and rotors. First flight scheduled for late 1997, but initial designated flight trials aircraft (02) exhibited at Farnborough in September 1998, still unflown. First flight (02) was 12 minute hover on 17 August 1999; initial forward flight on 6 October 1999. Trials halted in November 1999, after 4 hours, due to gearbox problems; resumed in second quarter of 2000 with strengthened and redesigned main transmission. Third (second flying) prototype was to have joined the programme in late 1999, but delayed by 12 months. Certification, due by late 2000, similarly postponed to 2001.
CURRENT VERSIONS: Can be optimised for transport (including underslung load), ambulance, SAR, training, patrol and other duties. Provision for attachment of sponsons or tanks to sides of cabin.
CUSTOMERS: Russian Federal Border Service (Federaliya Pogranichiya Sluzhba) order for 100 placed by 1997. Under consideration by Gazprom in 2000 for pipeline inspection. Estimated 600 to 700 sales over 10 years from 2000.
COSTS: US$1.5 million for utility version (2000). Kazan's development expenditure had reached Rb200 million by mid-2000.
DESIGN FEATURES: Meets FAR Pt 29 Category A and Russian AP-29 requirements. Conventional configuration, with high-mounted tailboom carrying horizontal stabiliser and twin fins; power plant above cabin. Bearingless torsion rotor head; four main blades; two-blade tail rotor. Two-stage, VR-23 main rotor reduction gear, ratio 16.4. Rotor brake.
STRUCTURE: Aluminium alloy fuselage; glass fibre main rotor blades, window frames and nosecone.
LANDING GEAR: Twin skids with transverse shock-absorbers; tail bumper to protect anti-torque rotor.
POWER PLANT: Two P&W Rus XRK206S turboshafts, each rated at 477 kW (640 shp) for T-O, 418 kW (560 shp) max continuous, in prototypes. Production version offered with 529 kW (710 shp) PW207s. Fuel in external tanks each side of cabin.
ACCOMMODATION: Ten persons, including one or two pilots; or two stretcher patients and two attendants; or internal or externally slung freight. Two forward-hinged doors each side; cargo bay behind cabin, with rear-facing door.
SYSTEMS: Avionika FBW controls comprise quadruplex electronic system and duplex hydraulic system. Main transmission drives two alternators and two hydraulic fuel pumps. Avionika FBW controls.
AVIONICS: Standard Russian avionics and instruments (apart from prototypes' BAE Systems North America engine parameters display, which to be replaced by Ulyanovsk BISK-A system); full Western avionics fit available as an option.
　Radar: Provision in nosecone.
EQUIPMENT: Optional rescue hoist.
ARMAMENT: Optional external rocket and machine gun pods.

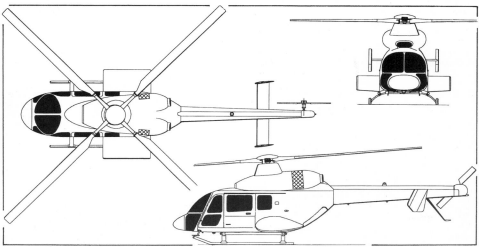

Kazan Ansat light multipurpose helicopter fitted with enlarged sponsons *(Paul Jackson/Jane's)*

Instrument panel of Kazan Ansat light helicopter *(Paul Jackson/Jane's)*

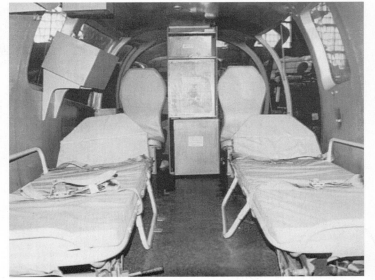

Ansat interior equipped for medical evacuation

Ansat 02 prior to its first flight *(Paul Jackson/Jane's)* 2000/0062724

DIMENSIONS, EXTERNAL:
Main rotor diameter	11.50 m (37 ft 8¾ in)
Tail rotor diameter	2.00 m (6 ft 6¾ in)
Length: overall, rotors turning	13.76 m (45 ft 1¾ in)
fuselage, tail rotor turning	11.54 m (37 ft 10¼ in)
pod	6.95 m (22 ft 9½ in)
Height to top of rotor head	3.40 m (11 ft 1¾ in)
Width: fuselage	1.80 m (5 ft 10¾ in)
over optional sponsons	3.61 m (11 ft 10¼ in)
Skid track	2.50 m (8 ft 2½ in)

DIMENSIONS, INTERNAL:
Cabin: Length	3.50 m (11 ft 5¾ in)
Max width	1.68 m (5 ft 6¼ in)
Max height	1.30 m (4 ft 3¼ in)
Volume	5.3 m³ (187 cu ft)
Cargo bay: Volume	6.5 m³ (230 cu ft)

AREAS:
Main rotor disc	103.87 m² (1,118.0 sq ft)
Tail rotor disc	3.14 m² (33.81 sq ft)

WEIGHTS AND LOADINGS:
Max payload: internal	1,000 kg (2,204 lb)
external	1,300 kg (2,866 lb)
Max T-O weight:	3,300 kg (7,275 lb)
Max disc loading	31.8 kg/m² (6.51 lb/sq ft)

PERFORMANCE (estimated):
Max level speed	151 kt (280 km/h; 174 mph)
Max cruising speed	135 kt (250 km/h; 155 mph)
Max rate of climb at S/L	960 m (3,150 ft)/min
Service ceiling	5,500 m (18,040 ft)
Hovering ceiling OGE	3,000 m (9,840 ft)
Range	321 n miles (595 km; 369 miles)

UPDATED

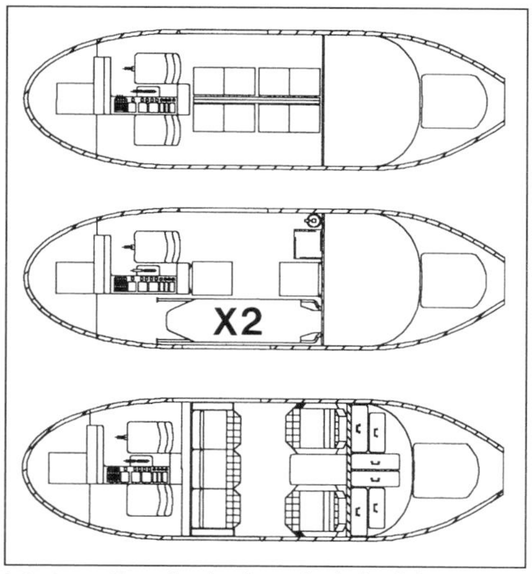

Interior options for the Ansat include eight passengers back-to-back (top); two stretchers and two attendants; and five executives, plus baggage 0015008

KAZAN AKTAI
English name: White Colt

TYPE: Three-seat helicopter.
PROGRAMME: Project revealed at Moscow Air Show, 19 August 1997, with display of mockup. Timetable in August 1999 envisaged initial development between early 1998 and late 1999; preparation for production from early 1999 to third quarter of 2000; prototype construction late 1999 to second quarter 2000; first flight in third quarter of 2000; initial flight trials complete by end 2000; certification trials throughout 2001, ending December; and start of series manufacture in second quarter of 2001.
COSTS: US$280,000 to US$300,000, according to equipment standard (2000). Estimated (2000) market for over 1,000. Development cost (2000) US$10 million.
DESIGN FEATURES: Intended for light transport, patrol, medical evacuation and training. Conventional pod and boom configuration with T tail.
STRUCTURE: Semi-articulated four-blade main rotor; two-blade tail rotor. Extensive use of composites throughout.
LANDING GEAR: Conventional, non-retractable skids on arched support tubes.
POWER PLANT: One 201 kW (270 hp) VAZ-4265 rotary motorcar engine mounted above the cabin and operating on 92/93 grade petrol.
ACCOMMODATION: Pilot and two passengers, side by side, in individual seats; forward-hinged door each side. Twin clamshell doors at rear of pod, below tailboom, provide access to flat freight floor; one casualty stretcher can be loaded through rear doors when passenger seat is removed.

DIMENSIONS, EXTERNAL:
Main rotor diameter	10.00 m (32 ft 9¾ in)
Tail rotor diameter	1.48 m (4 ft 10¼ in)
Fuselage: Length	8.35 m (27 ft 4¾ in)
Max width	1.70 m (5 ft 7 in)
Max height	1.48 m (4 ft 10¼ in)
Height overall	2.69 m (8 ft 10 in)
Tailplane span	1.44 m (4 ft 8¾ in)
Skid track	2.10 m (6 ft 10¾ in)

AREAS:
Main rotor disc	78.54 m² (845.4 sq ft)
Tail rotor disc	1.72 m² (18.52 sq ft)

WEIGHTS AND LOADINGS:
Max payload	300 kg (661 lb)
Max T-O weight: training	900 kg (1,984 lb)
utility	1,150 kg (2,535 lb)
Max disc loading	14.64 kg/m² (3.00 lb/sq ft)

PERFORMANCE (estimated, at training weight, unless otherwise specified):
Max level speed	113 kt (210 km/h; 130 mph)
Normal cruising speed	84 kt (155 km/h; 96 mph)
Service ceiling	4,570 m (15,000 ft)
Hovering ceiling OGE	1,500 m (4,920 ft)
Range:	
with 300 kg (661 lb) payload	54 n miles (100 km; 62 miles)
with 190 kg (419 lb) payload	329 n miles (610 km; 379 miles)
Endurance	7 h

UPDATED

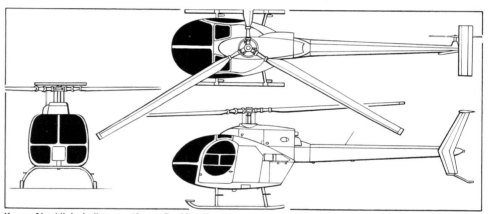
Kazan Aktai light helicopter *(James Goulding/Jane's)* 2000/0062675

Mockup of the Kazan Aktai on show at Moscow, August 1997 *(Paul Jackson/Jane's)* 0008093

KHRUNICHEV
GOSUDARSTVENNYI KOSMICHESKII NAUCHNO-PROIZVODSTVENNYI TSENTR IMENI M V KHRUNICHEVA
(State Research and Production Space Centre named for M V Khrunichev)
ulitsa Novozavodskaya 18, 121309 Moskva
Tel: (+7 095) 145 98 02
Fax: (+7 095) 145 92 03
e-mail: proton@online.ru
Web: http://www.khrunichev.com

GENERAL DIRECTOR: Anatoly I Kiselyov
DEPUTY GENERAL DIRECTORS:
 Alexander V Lebedev
 Anatoly A Kalinin

Aviatsionnoye Otdelenie (Aviation Department)
Tel: (+7 095) 145 93 33
Fax: (+7 095) 145 99 53
e-mail: proton@online.ru
CHIEF DESIGNER: Evgeny P Grunin
DEPUTY CHIEF DESIGNER: Arnold I Andrianov
AVIATION DEPARTMENT MANAGER: Sergei M Zhiganov

RESEARCH AND DEVELOPMENT DIVISION MANAGER:
 Yuri I Polatsky
DESIGN DIVISION MANAGER: Mikhail M Vasyuk

The Aviation Department of this prominent rocket and space vehicle design and manufacturing centre was formed in August 1994 to develop and manufacture general purpose aircraft. It comprises a design bureau, experimental facilities and operational and maintenance services. The main task of the Aviation Department is to develop to the series production stage aircraft designed on the basis of foreign and domestic components, and then to market them worldwide.

Production is being launched of the T-411 Aist, while development efforts are concentrated on the T-440 Mercury turboprop-twin and newly revealed T-517 Farmer agricultural aircraft.

UPDATED

KHRUNICHEV T-21

TYPE: Two-seat lightplane.
PROGRAMME: Announced 2000.
DESIGN FEATURES: Powered, V-strut-braced wing with boom-mounted cruciform empennage and underslung fuselage pod. Intended for training, sport flying, aerial photography/surveillance and crop spraying.
FLYING CONTROLS: Conventional and manual; two-segment Fowler-type flaps occupy some two-thirds of each wing.
LANDING GEAR: Tricycle type; fixed. Single wheel on each unit.
POWER PLANT: One 59.6 kW (79.9 hp) Rotax 912 UL flat-four.
ACCOMMODATION: Two persons, side by side in open cockpit, with windscreen.
DIMENSIONS, EXTERNAL:
Wing span 11.20 m (36 ft 9 in)
Length overall 6.70 m (21 ft 11¾ in)
WEIGHTS AND LOADINGS:
Weight empty 447 kg (985 lb)
Max fuel weight 80 kg (176 lb)
Max T-O weight 700 kg (1,543 lb)
Max power loading 11.74 kg/kW (19.30 lb/hp)
PERFORMANCE:
Cruising speed 70 kt (130 km/h; 81 mph)
T-O run 49 m (161 ft)
T-O to 15 m (50 ft) 150 m (492 ft)
Landing from 15 m (50 ft) 290 m (951 ft)
Landing run 55 m (180 ft)
Max range 432 n miles (800 km; 497 miles)

NEW ENTRY

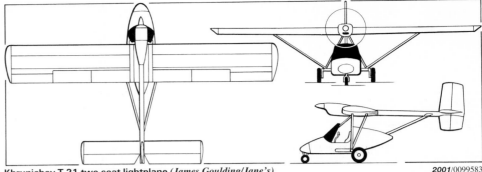
Khrunichev T-21 two-seat lightplane *(James Goulding/Jane's)*

Prototype Khrunichev Aist, powered by a VOKBM M-14X radial *(Paul Jackson/Jane's)*

KHRUNICHEV T-411 AIST

English name: Stork

TYPE: Utility/kitbuilt.
PROGRAMME: Programme started in 1994 when licence acquired from Aeroprogress/ROKS-Aero to produce the T-411 Aist-2 (see 1997-98 and previous editions); Aist-2 design started November 1992; construction of prototype began April 1993; first flight (marked only 'T-411') 10 November 1993 in standard form; 15 March 1994 with ski landing gear. Khrunichev prototype (RA-01585), with several detail changes, first exhibited at Moscow Air Show 1997. Initially designated T-411 Wolverine, which name now transferred to T-421 (which see). North American promotion (in kit form) by Washington Aeroprogress; alternative power plant is Chevrolet eight-cylinder motorcar engine; demonstrator (M-14 engine, and possibly a grounded prototype) 'N01522' based at Boeing Field, Washington, 1998. Manufacture by RSK MiG, whose Voronin plant contracted in March 2000 to build 15 for delivery in 2001.
CURRENT VERSIONS: **T-411:** As described.
T-411A: Agricultural version.
T-411B: Floatplane version with twin floats, for operations in seas with wave height up to 0.35 m (1 ft 1¾ in); length overall 10.00 m (32 ft 9¾ in); maximum take-off weight 1,650 kg (3,637 lb); payload 440 kg (970 lb); cruising speed 113 kt (210 km/h; 130 mph); T-O run 315 m (1,033 ft); T-O to 15 m (50 ft) 475 m (1,558 ft); landing from 15 m (50 ft) 305 m (1,000 ft); landing run 195 m (640 ft); maximum range 607 n miles (1,125 km; 699 miles).
T-411F: As T-411B, but with amphibious floats.
T-421: Described separately.
CUSTOMERS: Orders received from Russian forestry, fisheries and national emergency ministries, Australia, Malaysia and USA from which approximately 70 orders received, including aircraft supplied in kit form.
COSTS: US$130,000 with M-14 engine; US$20,000 more with Lycoming engine included.
DESIGN FEATURES: Conventional strut-braced high-wing monoplane. Constant-chord unswept wings with fixed leading-edge slat, slotted ailerons and single-slotted trailing-edge flaps; V bracing struts. Rectangular-section fuselage. Unswept tail unit with dorsal fin; constant-chord horizontal tail surfaces. Compared with the Aeroprogress Aist-2, the Khrunichev version has slightly stretched airframe and more roomy cabin; different landing gear and power plant.
Wing section NACA 23011; dihedral 2° from root; incidence 3° 30′.
FLYING CONTROLS: Conventional and manual. Actuation by pushrods and cables; electrically controlled trim tab in each elevator; fixed full-span leading-edge slats; slotted flaps.
STRUCTURE: Primary structure of aluminium alloy and alloy steel, part covered with Dacron synthetic fabric. Two-spar wings; metal skin on leading-edge and over fuel tanks at root, Dacron covered between spars. Welded tube truss fuselage, covered with metal sheet and Dacron. Metal tail surfaces, Dacron covered.
LANDING GEAR: Non-retractable tailwheel type; single wheel on each unit, with fairings on mainwheels. Arched cantilever tubular spring main legs. Self-centring steerable tailwheel. Mainwheel tyres 600×180; tailwheel 300×50. Brakes on mainwheels.
POWER PLANT: One 265 kW (355 hp) VOKBM M-14X nine-cylinder radial with Mühlbauer MTV-9 three-blade variable-pitch propeller, or 261 kW (350 hp) Western flat-six (Teledyne Continental TSIO-550 or Textron Lycoming TIO-540) with Hartzell three-blade variable-pitch propeller. Two main and two auxiliary fuel tanks in wingroots, total capacity 340 litres (89.8 US gallons; 74.8 Imp gallons); gravity fuelling.
ACCOMMODATION: Pilot and three or four passengers, on side-by-side front seats and rear bench seat; starboard front seat and rear bench removable, or bench seat foldable, for carrying equivalent freight; convertible into ambulance for one stretcher patient, one seated casualty and medical attendant in addition to pilot. Dual controls standard. Baggage compartment aft of rear seats, with large upward-opening door on port side, used also for loading freight or stretcher. Forward-hinged jettisonable door each side of cabin. Ventilation and heating standard.
SYSTEMS: Pneumatic system for starting engine and operating mainwheel brakes; pressure 50 bars (725 lb/sq in). Electrical system includes GSR-3000M engine-driven 27 V DC generator and 12CAM-28 battery.
AVIONICS: Russian or Western (Honeywell) avionics available. Optional autopilot.
DIMENSIONS, EXTERNAL:
Wing span 13.20 m (43 ft 3¾ in)
Wing aspect ratio 6.7
Length overall 9.405 m (31 ft 0 in)
Height overall 2.60 m (8 ft 6¼ in)
Wheel track 3.13 m (10 ft 3¼ in)
Wheelbase 6.50 m (21 ft 4 in)
Propeller diameter 2.03 m (6 ft 8 in)
Baggage door: Height 0.65 m (2 ft 1½ in)
Width 1.22 m (4 ft 0 in)
DIMENSIONS, INTERNAL:
Cabin: Length 2.94 m (9 ft 7¾ in)
Max width 1.27 m (4 ft 2 in)
Max height 1.30 m (4 ft 3 in)
AREAS:
Wings, gross 24.30 m² (261.6 sq ft)
WEIGHTS AND LOADINGS:
Max payload 720 kg (1,587 lb)
Max fuel weight 255 kg (562 lb)
Max T-O weight 1,600 kg (3,527 lb)
Max wing loading 65.8 kg/m² (13.49 lb/sq ft)
Max power loading 5.96 kg/kW (9.80 lb/hp)
PERFORMANCE:
Max cruising speed 124 kt (230 km/h; 143 mph)
Stalling speed 43 kt (80 km/h; 50 mph)
T-O run 85 m (279 ft)
T-O to 15 m (50 ft) 240 m (790 ft)
Landing from 15 m (50 ft) 250 m (820 ft)
Range with max fuel 864 n miles (1,600 km; 994 miles)

UPDATED

Model of T-411B floatplane

KHRUNICHEV T-420 SKYLARK

TYPE: Light utility twin-prop transport.
PROGRAMME: Design started August 1994; known as Strizh (Martin) until current name adopted in 1999 for piston-engined versions. First flight (T-420 A) was scheduled for 1997, but had not been reported by mid-2000, although promotion continues.
CURRENT VERSIONS: **T-420 A Strizh:** Two 313 kW (420 shp) Rolls-Royce 250-B17C turboprops. Detailed description in 1999-2000 and earlier *Jane's*.
T-420: With two 265 kW (355 hp) VOKBM M-14P piston engines.
T-420 C: With two 261 kW (350 hp) Teledyne Continental TSIO-550-B piston engines.
T-420 CL Strizh: Passenger/cargo aircraft with 'beaver-tail' rear-loading door and fixed landing gear. Described in 1997-98 and previous editions.

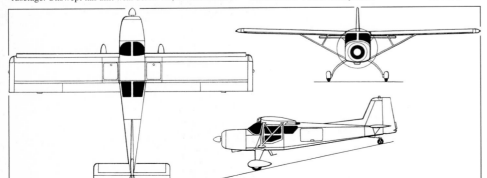
Khrunichev T-411 Aist four/five-seat light aircraft *(James Goulding/Jane's)*

KHRUNICHEV—AIRCRAFT: RUSSIAN FEDERATION

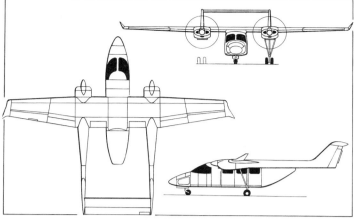

Khrunichev T-420 in model form 2001/0099619

Early design of Khrunichev T-420 A Strizh twin-engined passenger/freight transport *(James Goulding/Jane's)* 2000/0079317

T-420 Skylark: Retractable landing gear and piston engines; promotion began 1999.
Description applies to this version.

DESIGN FEATURES: High-wing monoplane with twin tailbooms, providing easy access to upward-hinged door forming rear of fuselage pod. Unswept wing leading-edge; trailing-edge swept forward outboard of engines; winglets. High-mounted tailplane and single-piece elevator above turbulent air flow from propellers.

FLYING CONTROLS: Conventional and manual. Flight-adjustable trim tabs in port aileron and elevator.

STRUCTURE: All-metal construction; semi-monocoque fuselage pod; two-spar wings with duralumin skin.

LANDING GEAR: Retractable tricycle type; twin wheels on each main unit; oleo-pneumatic shock-absorbers.

POWER PLANT: Two Teledyne Continental flat-six piston engines, each driving three-blade constant-speed propeller. Option of 157 kW (210 hp) IO-360E or 209 kW (280 hp) IO-550. Integral fuel tanks in wings.

ACCOMMODATION: Pilot and one passenger on flight deck, with door each side; five passengers or equivalent freight, in main cabin, standard. Hand baggage hold at rear of cabin; main baggage compartment forward of flight deck. Passenger and freight access via upward-hinged rear door. Heating and ventilation standard.

SYSTEMS: Hydraulic mainwheel brakes. Two 28 V 150 A starter-generators and two voltage regulators in main electrical system; 24 V 29 Ah battery for engine starting and emergency use. Electric anti-icing of propeller blades, pitot tubes and windscreen; pneumatic de-icing of wing and tail unit leading-edges.

DIMENSIONS, EXTERNAL:
Wing span over winglets	13.90 m (45 ft 7¼ in)
Length overall	10.60 m (34 ft 9¼ in)
Height overall	3.10 m (10 ft 2 in)

DIMENSIONS, INTERNAL:
Cabin: Length, incl flight deck	4.50 m (14 ft 9¼ in)
Max width and height	1.25 m (4 ft 1¼ in)

AREAS (approx; for T-420A):
Wings, gross	20.78 m² (223.7 sq ft)
Fins (total)	3.40 m² (36.60 sq ft)
Rudders (total)	1.48 m² (15.93 sq ft)
Tailplane	2.23 m² (24.00 sq ft)
Elevator	1.55 m² (16.68 sq ft)

WEIGHTS AND LOADINGS (A: IO-360, B: IO-550 engine):
Max payload: A	450 kg (992 lb)
B	500 kg (1,102 lb)
Max fuel weight	430 kg (948 lb)
Max T-O weight: A	1,990 kg (4,387 lb)
B	2,130 kg (4,695 lb)
Max power loading: A	6.36 kg/kW (10.45 lb/hp)
B	5.10 kg/kW (8.38 lb/hp)

PERFORMANCE (estimated; A, B as above):
Max cruising speed: A	146 kt (270 km/h; 168 mph)
B	184 kt (340 km/h; 211 mph)
T-O run: A	280 m (920 ft)
B	225 m (740 ft)
Landing run: A	360 m (1,185 ft)
B	375 m (1,230 ft)
Range: with max payload:	
A	378 n miles (700 km; 435 miles)
B	448 n miles (830 km; 515 miles)
with max fuel:	
A	1,133 n miles (2,100 km; 1,304 miles)
B	1,079 n miles (2,000 km; 1,242 miles)

UPDATED

KHRUNICHEV T-421 WOLVERINE

TYPE: Utility/kitbuilt.

PROGRAMME: Westernised (engine and avionics) version of T-411 (which see) optimised for US market; model first shown at Paris, June 1997; promotion continuing in 2000. No known sales.

Data generally as for T-411, except those below.

POWER PLANT: One 261 kW (350 hp) Teledyne Continental TSIO-540 flat-six.

DIMENSIONS, INTERNAL:
Cabin: Length	2.785 m (9 ft 1¾ in)
Max width	1.27 m (4 ft 2 in)
Max height	1.25 m (4 ft 1¼ in)
Baggage compartment: Length	1.30 m (4 ft 3¼ in)
Max width	1.17 m (3 ft 10 in)
Max height	0.735 m (2 ft 5 in)

WEIGHTS AND LOADINGS:
Max payload	720 kg (1,587 lb)
Max fuel weight	255 kg (562 lb)
Max T-O weight	1,600 kg (3,527 lb)
Max power loading	6.13 kg/kW (10.08 lb/hp)

PERFORMANCE (estimated):
Max level speed	151 kt (280 km/h; 174 mph)
Max cruising speed	124 kt (230 km/h; 143 mph)
T-O run	90 m (295 ft)
T-O to 15 m (50 ft)	245 m (804 ft)
Landing from 15 m (50 ft)	250 m (820 ft)
Landing run	136 m (446 ft)
Range with max fuel	864 n miles (1,600 km; 994 miles)

UPDATED

Khrunichev T-421 Wolverine variant of the T-411 Aist *(Paul Jackson/Jane's)* 2001/0099620

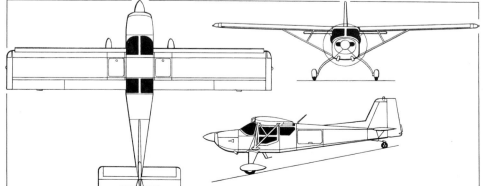

Khrunichev T-421 Wolverine (Teledyne Continental TSIO-550 flat-six) *(James Goulding/Jane's)* 2000/0079314

KHRUNICHEV T-431

Marketing name: Sea Wing

TYPE: Six-seat amphibian.

PROGRAMME: Model first shown at Paris and Moscow, June and August 1999. No published plans for manufacture.

DESIGN FEATURES: Mid-wing configuration; engine mounted on fin for clearance of water spray; high tailplane. Outrigger floats at two-fifths span accommodate mainwheels. Single-step hull.

FLYING CONTROLS: Conventional and manual. Ailerons, each with tab; two-element, horn-balanced rudder; horn-balanced, single-piece elevator with trim tab. Two-element flaps in each wing.

LANDING GEAR: Tricycle type; retractable. Nosewheel retracts forward into fuselage; mainwheels, rearwards into outrigger floats, but protruding at rear.

POWER PLANT: One piston engine driving a three-blade propeller: 224 kW (300 hp) Textron Lycoming flat-six or 265 kW (355 hp) VOKBM M-14P radial.

ACCOMMODATION: Six persons, including one or two pilots. Single-piece windscreen and two-piece, upward-opening (rear-hinged) canopy.

DIMENSIONS, EXTERNAL:
Wing span	15.06 m (49 ft 5 in)
Length overall	10.70 m (35 ft 1¼ in)
Height overall	4.22 m (13 ft 10¼ in)

DIMENSIONS, INTERNAL:
Cabin: Length	3.80 m (12 ft 5½ in)
Max width	1.30 m (4 ft 3¼ in)
Max height	1.25 m (4 ft 1¼ in)

WEIGHTS AND LOADINGS (Lycoming engine):
Max payload	500 kg (1,102 lb)
Max fuel weight	275 kg (606 lb)
Max T-O weight	1,990 kg (4,387 lb)
Max power loading	8.90 kg/kW (14.62 lb/hp)

PERFORMANCE (Lycoming engine):
Max cruising speed	146 kt (270 km/h; 168 mph)
T-O run	290 m (955 ft)
Landing run	260 m (855 ft)
Range with max fuel	912 n miles (1,690 km; 1,050 miles)

UPDATED

RUSSIAN FEDERATION: AIRCRAFT—KHRUNICHEV

Model of the proposed Khrunichev T-431 Sea Wing *(Paul Jackson/Jane's)*

T-433 Flamingo now wears Khrunichev name, but none yet built *(Paul Jackson/Jane's)*

KHRUNICHEV T-433 FLAMINGO

TYPE: Five-seat amphibian.
PROGRAMME: Originated as Aeroprogress/ROKS-Aero T-433, last described in 1999-2000 *Jane's*. Promotion continued under Khrunichev name, but none yet built.

NEW ENTRY

KHRUNICHEV T-440 MERKURY

English name: Mercury
TYPE: Business twin-turboprop.
PROGRAMME: Announced 21 August 1997, when mockup ('RA-44097') displayed at Moscow Air Show. Derived from Aeroprogress/ROKS-Aero T-602 Orel (1995-96 *Jane's*) by way of Khrunichev T-430 Sprinter. Prototype then stated to be due to fly in late 1998. Re-exhibited (as 'RA-44099') at Moscow in August 1999; prototype reported under construction in mid-2000.
COSTS: US$2.1 million (1997).
DESIGN FEATURES: Optimised for long-distance, high-speed executive transport in VMC or IMC conditions from Class 3 airfields (max wheel loading 9 kg/cm^2; 128 lb/sq in). Conventional low-wing monoplane with sweptback tail surfaces; moderate dihedral on wing outboard panels; winglets; pronounced tailplane dihedral.
Aerofoil GA(W)-1; leading-edge sweepback 2°.
FLYING CONTROLS: Conventional and manual. Trim tabs in both ailerons and port elevator. Slotted flaps.
STRUCTURE: Two-spar wing.
LANDING GEAR: Retractable tricycle type; steerable, single nosewheel; twin wheels on each main leg. Mainwheels, size 500×150; all wheels retract rearward.
POWER PLANT: Two 559 kW (750 shp) Pratt & Whitney Canada PT6A-135A turboprops; optionally, two 560 kW (751 shp) Walter M 601 F turboprops. Three- or Hartzell four-blade constant-speed propellers. Fuel in wings, inboard and outboard of engines.
ACCOMMODATION: Alternative interior configurations for between four and eight persons, plus one or two pilots. Single, horizontally split door, hinged top and bottom and with integral stairs, on port side, adjacent to wing trailing-edge. Emergency exit on starboard side, above wing. Lavatory with wash basin and water-heater.
SYSTEMS: Pneumatic system for landing gear actuation. Electrical system 28.5 V DC primary; 115/200 V AC, three-phase, secondary.
AVIONICS: Honeywell equipment, including radar, to customer's specification.
EQUIPMENT: Emergency oxygen bottles for 10 minutes' supply.
DIMENSIONS, EXTERNAL:
Wing span	17.00 m (55 ft 9¼ in)
Wing aspect ratio	11
Length overall	11.66 m (38 ft 3 in)
Height overall	4.54 m (14 ft 10¾ in)
Propeller diameter (4 blade)	2.34 m (7 ft 8¼ in)

DIMENSIONS, INTERNAL:
Cabin: Length	5.80 m (19 ft 0¼ in)
Max width	1.40 m (4 ft 7 in)
Max height	1.35 m (4 ft 5¼ in)
Volume	11.0 m^3 (388 cu ft)

WEIGHTS AND LOADINGS:
Max payload	1,270 kg (2,799 lb)
Max fuel weight	1,600 kg (3,527 lb)
Max T-O weight	4,600 kg (10,141 lb)
Max power loading	4.11 kg/kW (6.76 lb/shp)

PERFORMANCE (estimated):
Max cruising speed	301 kt (557 km/h; 346 mph)
T-O run	260 m (853 ft)
T-O to 15 m (50 ft)	580 m (1,903 ft)
Landing run	410 m (1,345 ft)
Landing from 15 m (50 ft)	530 m (1,739 ft)
Range with max fuel:	
PT6A	2,073 n miles (3,840 km; 2,386 miles)
M 601	1,943 n miles (3,600 km; 2,236 miles)

UPDATED

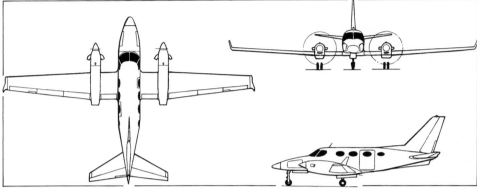

Khrunichev T-440 Merkury in 1999 configuration *(Paul Jackson/Jane's)*

Mockup of the Khrunichev T-440 Merkury *(Paul Jackson/Jane's)*

KHRUNICHEV T-507

Export marketing name: Pegasus
TYPE: Light utility turboprop.
PROGRAMME: Design started 1994 under designation T-417; 261 kW (350 hp) Teledyne Continental TSIO-550 flat-six piston engine (see 1999-2000 edition): prototype under construction by 1996; project terminated and recast as T-507, with turboprop and tailwheel landing gear; assembly of prototype continued during 1999.

Khrunichev T-417 Pegasus light freight transport, now recast as turboprop-powered T-507 *(James Goulding/Jane's)*

Prototype Khrunichev T-507 Pegasus under construction

DESIGN FEATURES: Conventional strut-braced high-wing monoplane, with STOL performance and non-retractable tricycle landing gear (see illustrations).
FLYING CONTROLS: Conventional and manual. Trim tab in rudder and each elevator; single-slotted ailerons and trailing-edge flaps.
STRUCTURE: All-metal. Conventional aluminium alloy wings and tail unit; VNS-2 steel fuselage truss structure, with aluminium alloy skin.
LANDING GEAR: Tailwheel type; fixed. Single wheel on each unit.
POWER PLANT: One 560 kW (751 shp) Walter M 601 turboprop, driving a three-blade propeller; fuel tanks in wingroots.
ACCOMMODATION: Side-by-side seats for two persons on flight deck, with jettisonable forward-hinged door each side. Dual controls standard. Normally flown by pilot only as freighter, but provision for pilot and up to six passengers. Large double freight door on port side of cabin, vertically split to open forward and rearward for unrestricted access. Steps beneath each door. Emergency exit on starboard side of cabin.
SYSTEMS: Interior heated and ventilated. Hydraulic system for flap actuation and wheel brakes. 27 V DC electrical system supplied by engine-driven generator and battery. Pneumatic de-icing system. Engine compartment fire warning and extinguishing system.

No specific data released; following refers to piston-engined version.

DIMENSIONS, EXTERNAL:
Wing span	13.10 m (42 ft 11¾ in)
Wing chord, constant	1.70 m (5 ft 7 in)
Wing aspect ratio	7.7
Length overall	8.95 m (29 ft 4½ in)
Fuselage: Height	1.50 m (4 ft 11 in)
Width	1.40 m (4 ft 7 in)
Height overall	3.60 m (11 ft 9¾ in)
Tailplane span	4.00 m (13 ft 1½ in)
Cabin door: Height	1.15 m (3 ft 9¼ in)
Width	1.25 m (4 ft 1¼ in)

DIMENSIONS, INTERNAL:
Cabin, incl flight deck: Length	3.60 m (11 ft 9¾ in)
Width	1.30 m (4 ft 3 in)

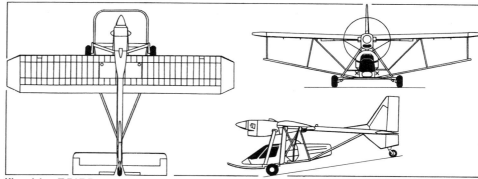

Khrunichev T-517 Farmer agricultural and general-purpose aircraft *(Paul Jackson/Jane's)* 2000/0079010

Height	1.25 m (4 ft 1¼ in)
Volume	5.6 m³ (198 cu ft)

AREAS:
Wings, gross	22.27 m² (239.7 sq ft)
Ailerons (total)	1.52 m² (16.36 sq ft)
Flaps (total)	4.66 m² (50.16 sq ft)
Fin	1.28 m² (13.78 sq ft)
Rudder	0.86 m² (9.26 sq ft)
Tailplane	2.02 m² (21.74 sq ft)
Elevators (total)	2.78 m² (29.92 sq ft)

UPDATED

KHRUNICHEV T-517 FARMER

TYPE: Agricultural sprayer/light utility turboprop.
PROGRAMME: Derived from T-315/T-317 Mukha projects, which envisaged radial engine. Full-size mockup revealed at Arlington, Washington, July 1999. No known production plans.
DESIGN FEATURES: Unconventional 'flying tube platform'; powered wing with boom-mounted empennage and underslung crew and payload pod. Optimised for spraying; high-mounted engine assists dispersal and minimises crop disturbance; open structure minimises lodgement of corrosive chemicals in agricultural or firefighting roles.

Wide track, long wheelbase landing gear and nose-over skid for rough field operations. 'Helicopter-type' visibility from cockpit. Simple structure for minimal cost.
Constant-chord, unswept high wing and tailplane; sweptback fin with large underfin.
FLYING CONTROLS: Manual. Junkers-type flaperons; horn-balanced elevators; both fin leading-edges sweptback; rudder on upper fin only; flight-adjustable trim tab in each elevator.
STRUCTURE: Principally of stainless steel; external bracing of circular- and aerofoil-section tube.
LANDING GEAR: Tailwheel type; fixed. Floats optional. Mainwheels, size 12.50-10, on cantilever tube legs; tailwheel, size 470×210, below underfin.
POWER PLANT: One 485 kW (650 shp) derated turboprop: Walter M 601 F or P&WC PT6A-15AG; three-blade propeller.
ACCOMMODATION: Pilot, only, in extensively glazed cockpit. Forward-sliding canopy/windscreen. Hopper can be replaced by passenger or freight module.
EQUIPMENT: Hopper, capacity 1,600 litres (423 US gallons; 352 Imp gallons), behind cockpit; spraybars, on long suspension arms, below wings; two propeller-driven pumps for spray material. Wire-cutter above windscreen. Two landing lights in port wing leading-edge; one starboard.

DIMENSIONS, EXTERNAL:
Wing span	13.46 m (44 ft 2 in)
Length overall	9.20 m (30 ft 2¼ in)
Height overall	3.34 m (10 ft 11½ in)

AREAS:
Wings, gross	26.92 m² (289.8 sq ft)

WEIGHTS AND LOADINGS:
Max payload	1,200 kg (2,646 lb)
Max fuel weight	520 kg (1,146 lb)
Max T-O weight	3,450 kg (7,605 lb)
Max wing loading	128.2 kg/m² (26.25 lb/sq ft)
Max power loading	7.12 kg/kW (11.70 lb/hp)

PERFORMANCE:
Max cruising speed	157 kt (290 km/h; 180 mph)
Normal spraying speed	70-92 kt (130-170 km/h; 81-105 mph)
Stalling speed	47 kt (87 km/h; 54 mph)
Max rate of climb at S/L	396 m (1,299 ft)/min
T-O run	210 m (690 ft)
Landing run	120 m (395 ft)

Mockup of the Khrunichev T-517 Farmer displayed at Moscow in August 1999 *(Paul Jackson/Jane's)*
2001/0099584

KnAAPO

GOSUDARSTVENNOYE UNITARNOYE PREDPRIYATIE KOMSOMOLSKOE-na-AMURE AVIATSIONNOYE PROIZVODSTVENNOYE OBEDINENIE IMENI Yu A GAGARINA (State Unitary Enterprise Komsomolsk-on-Amur Aircraft Production Association named for Y A Gagarin)
ulitsa Sovetskaya 1, 681018 Komsomolsk-na-Amure
Tel: (+7 421 72) 632 00 and 285 25
Fax: (+7 421 72) 634 51 and 298 51
e-mail: knaapo@kmscom.ru
Web: http://www.ceebd.co.uk/knaapo
Telex: 141149 BURAN
GENERAL DIRECTOR: Viktor I Merkulov
TECHNICAL DIRECTOR: Vyacheslav I Shport
CHIEF, ADVANCED DEVELOPMENTS DEPARTMENT:
Sergei Drobyshev

A major production centre for Sukhoi aircraft at the present time, KnAAPO has manufactured combat aircraft of the single-seat Su-27 series, including the naval Su-33/UB, and the Su-35/37, and is also responsible for the S-80 STOL transport, Beriev Be-103 light amphibian and, when placed in production, Kamov Ka-62M helicopter. Known as GAZ 126 when established in 1934, it has built more than 10,000 aircraft for 20 countries, beginning in May 1936. Test airfield is Dzemgi. Other products include hang-gliders, small aluminium boats, snowmobiles, jetskis and medical

Sukhoi S-80 assembly line at Komsomolsk
2001/0062660

pressure chambers. Operates own transport airline with four An-12s, one An-26, two An-32s, two Il-76s, two Tu-134s and three Mi-8s. The 100 per cent state holding in KnAAPO is to be reduced to 51 per cent. Employees in 1999 totalled 15,000.

UPDATED

LII/PEGAS

LETNO-ISSLEDOVATELSKY INSTITUT IMENI M M GROMOVA, GOSUDARSTVENNYI NAUCHNYI TSENTR (Flight Research Institute named for M M Gromov, State Research Centre)
Zhukovsky-2, 140160 Moskovskoye oblast
Tel: (+7 095) 556 55 44
Fax: (+7 095) 556 53 34
e-mail: vorob@atom.ru
DIRECTOR: Evgeny G Sabadash
MARKETING MANAGER: Andrey Vorobiev

Zhukovsky is the centre of Russian aviation research and testing, dealing in theoretical and practical aspects of flight and their related disciplines. In 1991, the Gromov institute broadened its activities with a tentative venture into light aircraft design attributed to the Pegas bureau. Progress has been slowed by financial constraints, but in January 2001 the programme was confirmed as active.
UPDATED

LII CHIROK
English name: Teal
TYPE: Twin-engined amphibian.
PROGRAMME: Full-size mockup displayed at Moscow Air Show, August 1997. Development was continuing in 2001, with emphasis on wind tunnel trials.
DESIGN FEATURES: Air cushion landing gear for operation from variety of surfaces. Mid-wing monoplane with inverted V tail surfaces on short twin booms and shrouded propellers amidships. No hull step or auxiliary floats; operable in wave heights up to 0.3 m (1 ft). Wing section GA(W)-1; thickness/chord ratio 17 per cent; dihedral 5° on outer panels; incidence 3°.
FLYING CONTROLS: Manual, with full-length ruddervator in each tailfin.
LANDING GEAR: Air cushion generated by 41 kW (55 hp) Voronezh M18-02 piston engine driving two fans, augmented by retractable nosewheel, size 300×125, and tail skid.
POWER PLANT: Two 47.8 kW (64.1 hp) Rotax 582 UL two-cylinder two-stroke engines; optionally, two MWAE 100R rotary engines of 63 kW (85 hp) each; four-blade, ducted propellers.
ACCOMMODATION: Pilot and two rear passengers; baggage space at rear of cabin. Forward-hinged single-piece canopy above pilot; rear-hinged canopy for passengers.

Full-size Chirok mockup; note absence of wheeled landing gear   *2001/0097258*

DIMENSIONS, EXTERNAL:
Wing span	10.01 m (32 ft 10 in)
Wing aspect ratio	10.2
Length overall	5.89 m (19 ft 4 in)
Height overall	2.25 m (7 ft 4½ in)
Tailplane span	2.76 m (9 ft 0½ in)

DIMENSIONS, INTERNAL:
Cabin: Length	2.20 m (7 ft 2½ in)
Max width and height	1.24 m (4 ft 0¾ in)

AREAS:
Wings, gross	9.86 m² (106.1 sq ft)

WEIGHTS AND LOADINGS:
Max payload	210 kg (463 lb)
T-O weight: normal	640 kg (1,411 lb)
max	700 kg (1,543 lb)
Max wing loading	71.0 kg/m² (14.54 lb/sq ft)
Max power loading	7.34 kg/kW (12.05 lb/hp)

PERFORMANCE (estimated):
Max level speed	135 kt (250 km/h; 155 mph)
Cruising speed	108 kt (200 km/h; 124 mph)
T-O speed	51 kt (94 km/h; 59 mph)
Landing speed	38 kt (70 km/h; 44 mph)
Service ceiling	4,000 m (13,120 ft)
T-O run	70-120 m (230-395 ft)
Landing run	70 m (230 ft)
Range at normal T-O weight	216 n miles (400 km; 248 miles)
Ferry range	1,079 n miles (2,000 km; 1,242 miles)
Endurance: normal	2 h
ferry	10 h

UPDATED

LMZ

LUKHOVITSY MASHINOSTROITELNYI ZAVOD (Lukhovitsy Machine-Building Plant)
140500 Lukhovitsy, Moskovskaya oblast
Tel: (+7 096 63) 113 76
Fax: (+7 095) 234 43 13
e-mail: lmz-avia@mtu-net.ru
Web: http://www.ceebd.co.uk/ceebd//lmz.htm or http://www.avicos.ru/lmz or http://www.migavia.ru
DIRECTOR: Vladimir I Nungezer
COMMERCIAL DIRECTOR: Pyotr P Mizin

Founded at Tretyakovo airfield in 1953, this state-owned subsidiary of RSK MiG functioned as a flight test centre until 1968 (and continues to do so for the MiG-29). It has since acquired extensive equipment for manufacturing high-strength aluminium, titanium and composites components in support of RSK MiG production, and also produces a variety of light aircraft including the Ilyushin Il-103 and Interavia I-1L. See under respective bureaux for descriptions. Also participates in manufacture of parts for Sukhoi Su-29 and Su-31 sportplanes. On 7 April 1998 became first (and, currently, only) Russian factory with light aircraft production approval certificate. Assembly is now in hand of the Aeroprogress T-101 Grach. Announced in June 2000 that LMZ will produce Ilyushin Il-100.

Other activities include flying school for commercial or private pilot training on Il-103; freight handling; and production of jetskis and suntanning beds.

UPDATED

Ilyushin Il-103 assembly line at Lukhovitsy
2001/0092553

LVM

PROIZVODSTVENNOYE OBEDINENIE LEGKIE VERTOLETNYI MI OAO (Mi-Light Helicopters Production Association JSC)
alleya 34, 103051 Moskva
Tel/Fax: (+7 095) 200 25 34
GENERAL DIRECTOR: Grigory Bado

LVM, which trades for export as Mi-Light, was formed in April 1993 with responsibility for design, testing, series manufacture, sales, after-sales service, repair, spares provisioning, leasing and general support of the Mi-34 light helicopter, described under the Mil heading. Activities also include the Mi-52 light rotorcraft, although promotion of that product has only recently begun. A further light design, the Mi-60, has recently been revealed.

Participating companies comprise Mil Moscow Helicopter Plant, NIAT – Moscow Aviation Technology Research Institute, Moscow Aviation Industry State Design and Research Institute, St Petersburg Aviation Industry State Design and Research Institute, Perm Motors, Sazykin Arsenyev Aviation Production Plant, Avitek, Aviastar (Ulyanovsk) and Aeromechanica Research and Production Association.

UPDATED

MAPO (VPK MAPO/MIG MAPO) — see RSK 'MiG'

MiG

AVIATSIONNYI NAUCHNO-PROMYSHLENNYI KOMPLEKS MiG (ANPK MiG; MiG Aviation Scientific-Industrial Complex)

a/ya 125299, Leningradskoye shosse 6, Moskva
Tel: (+7 095) 158 18 72
Fax: (+7 095) 943 00 27
ADVISER TO GENERAL DIRECTOR: Rotislav A Belyakov
CHIEF DESIGNER: Lev Shengelaya
GENERAL ENGINEER: Victor M Puzanov
TECHNICAL DIRECTOR: Ivan I Butko
DIRECTOR OF STRATEGIC PLANNING, INFORMATION AND MARKETING: Alexander I Ageev
CHIEF, PUBLIC RELATIONS AND ADVERTISING DEPARTMENT: Svyatoslav Yu Ribas

This bureau originated in GAZ 1 (factory No. 1) founded in 1893. The experimental design bureau, founded by Artem I Mikoyan, dissociated from the factory in 1939 and later became officially the Mikoyan Aviation Scientific-Industrial Complex 'MiG' (the final letter indicating Mikhail I Gurevich). It developed 250 projects and built 120 prototypes. More than 55,000 aircraft of MiG design have been manufactured.

In 1995, the Moscow Aircraft Production Organisation and ANPK MiG were merged into the single state company MAPO 'MiG' (see previous entry). The original 'MiG' acronym continued to be used, though MiG lost its separate legal identity and was prevented from selling its own aircraft. On 4 July 1997 the legal status of MAPO 'MiG' was restored, as reflected by this separate entry. However, some bureau literature now uses a capital I for ANPK MiG and sometimes for designations such as MIG-21 and MIG-23. Following a change of title, the bureau is now subordinate to RSK MiG (which see).

UPDATED

MiG-29

NATO reporting name: Fulcrum
Indian Air Force name: Baaz (Eagle)
TYPE: Multirole fighter.
PROGRAMME: Technical assignment (operational requirement) for PLMI (*perspektiunyi legkiy massouyi istrebitel*), later LPFI (*legkiy perspektivnyi frontovoy istrebitel*: light front-line fighter) issued 1971, with RFP following in 1972, to replace MiG-21, MiG-23, Su-15 and Su-17; initial order placed simultaneously (work on 'fourth-generation fighter' using MiG-29 designation had been under way since 1970). Detail design of 9.11 (and interim 9.11A, planned with S-23 Sapfir radar and other MiG-23 avionics) approved on 15 July 1974; design frozen 1977.

First of five prototypes (numbered 901 to 904 and 908) and nine preseries aircraft (917 to 925) built for factory and State testing (901, Factory index 9.12, with wing area increased by about 11 per cent) flew 6 October 1977; photographed by US satellite, Ramenskoye flight test centre, November 1977 and given interim Western designation 'Ram-L'; second prototype flew June 1978; second and fourth prototypes lost through engine failures; after major design changes (see previous editions of *Jane's*) production began 1982, deliveries to Frontal Aviation (234th GvIAP at Kubinka was the first of 14 regiments) 1983; operational early 1985; first detailed Western study possible after visit of demonstration team to Finland July 1986 and 1988 Farnborough appearance; production completed in 1992 of basic MiG-29 combat aircraft by MAPO, and of MiG-29UB combat trainers by Sokol, for CIS air forces, but very low-rate manufacture may continue sporadically of MiG-29UB, SE and SM for export and to fulfill a Russian requirement for a combined total of about 100 new Su-27s and MiG-29s between 2002 and 2008. There is still a stockpile of unsold aircraft at the LMZ factory airfield, built to meet Soviet orders.

Weapon options of the MiG-29SMT *(Yefim Gordon)*   2001/0100402

Performance demonstrated by series of record flights, 1995. On 26 April Roman Taskayev flew to 27,460 m (90,092 ft) from Akhtubinsk; aircraft powered by RD-33s rated at 81.39 kN (18,298 lb st). On 5 July Oleg Antonovich set C-1h records by carrying 1,000 kg (2,205 lb) payload to 25,150 m (82,513 ft) and 1,066 kg (2,350 lb) payload to 15,000 m (49,212 ft)

CURRENT VERSIONS: **MiG-29** (Factory index 9.12; *Izdelie* 5; 'Fulcrum-A'): Land-based single-seat counter-air tactical fiTcanghter, designed to operate primarily under ground control. Entered production at MMZ No. 30 in 1982. First production series aircraft (about 250) with anti-spin ventral fins (although these already deleted or removed from most prototypes). Some early aircraft also had '*babochka*' (butterfly) gravel-deflector above, and in front of, nosewheels; soon deleted. First 30 aircraft featured composites inlet ducts, cowlings, leading-edge flaps control surfaces, spines and tailfins. D19 aluminium alloy intake ducts and flaps introduced from 31st aircraft for 20 kg (44 lb) weight penalty. Composites cowlings deleted later.

Ventral fins removed on addition of BVP-30-26M chaff/flare dispensers which extend forward from fins on later 'Fulcrum-As'. A-323 Shoran replaced A-312 Radikal NP Shoran on later 9.12s. Control surface deflection range increased and rudder chord extended by 21 per cent from 1984 for improved control at extreme angles of attack. Ailerons set at 5° up in the neutral position to improve spin characteristics. To meet air force requirements, dorsal fins extended forward to house chaff/flare dispensers, so extending keel area without impairing flight stability. Maximum weapon load 2,000 kg (4,409 lb). Some Russian aircraft had nuclear strike capability with 30 kT RN-40 bomb on reinforced port inner pylon. About 840 built by 1991.

MiG-29 (Factory index 9.12A; 'Fulcrum-A'): Version lacking nuclear weapon delivery system, and with OEPrNK-29E electro-optical complex and downgraded IFF, for non-Soviet Warsaw Pact air forces. RLPK-29E (N-019E or N-019EA 'Rubin') radar may also have lacked two modes, giving total of only three. Produced between 1988 and 1991 for Bulgaria, Czechoslovakia, East Germany, Poland and Romania. Most modified to virtual 9.12B configuration (with removal of datalink and IFF) on break-up of USSR and Warsaw Pact. 9.12A/MiG-29A designation originally applied to unbuilt interim aircraft with S-23 radar, cancelled in 1976. (Prototype and preseries aircraft numbered 905 to 907 and 909 to 916.)

MiG-29 (Factory index 9.12B; 'Fulcrum-A'): Downgraded version for non-Warsaw Pact export customers. Lacked Laszlo datalink. Built from 1986. Avionics include N-019EB radar, OEPrNK-29E2 EO complex, L006LM/101 RHAWS and further reduced-capability IFF and ECM.

MiG-29 (DASA upgrade): Aircraft modified to allow operation within NATO air defence structure. MiG Aircraft Product Support GmbH founded 27 July 1993 as joint venture between Rosvooruzheniye, MAPO 'MIG' (now RSK 'MiG') and (then) Daimler-Benz Aerospace for D-level maintenance and modernisation of German MiG-29s. Western IFF, V/UHF, emergency radio, NATO standard units, Tacan and anti-collision lights installed first followed by GPS (on seven aircraft). These two modernisations ICAO I and ICAO II together form Level One. This followed by provision for underwing fuel tanks, extended airframe TBO (from 800 FH or 9 years to 1,300 FH or 15 years) and engine TBO and life (Level Two). Similar upgrade offered to other MiG-29 users, especially in Europe.

Parallel upgrade, with Honeywell KLU-709 Tacan or surplus Luftwaffe AN/ARN-118 (instead of new AN/ARN-153), as well as ILS/VOR and Trimble 2101 I/O Plus GPS, R-865 VHF and new anti-collision lights, being applied to eight Polish MiG-29s (67, 70, 89, 105, 108, 111, 114 and 115) by WZL-2 depot at Bydgoscz, in association with MAPS joint venture. Polish upgraded aircraft have a new F-16A (ADF)-type SB-14 IFF antenna ahead of windscreen and received a new, darker two-tone grey camouflage. Remaining Polish aircraft will probably follow and MAPS has signed a teaming agreement for further work with WZL-2 in 1999, perhaps with further improvements, including 'glass cockpit'. DASA signed MoUs with Bulgaria and Romania in October and November 1999 and opened negotiations with Hungary (including the Danubian Aircraft Company at Tököl) in May 2000. A contract for the upgrade of 14 MiG-29s was signed in August 2000 but was overtaken by a decision to lease F-16s in early 2001. Bulgaria's MiG-29s will also receive a 1,100-hour/nine-year service life extension at the hands of DASA and TEREM, while DASA also has an agreement with Slovakia's LOT. No customers have yet been found for DASA's proposed Level Three upgrade (adding a MIL-STD-1553B databus, LINS/GPS and new IFF, navigation and communications systems) or for the planned Level Four, with a new radar, CDPU, and ECM.

MiG-29 (EADS/Aerostar/Elbit '**Sniper**' upgrade): Following an initial life extension programme on two MiG-29s and two MiG-29UBs, to Level Two standards, Aerostar has teamed with DASA to offer the Romanian Air Force a MiG-29 upgrade which would add most of the

Slovak Air Force MiG-29 'Fulcrum-A' *(Paul Jackson/Jane's)*   2001/0100478

390 RUSSIAN FEDERATION: AIRCRAFT—MiG

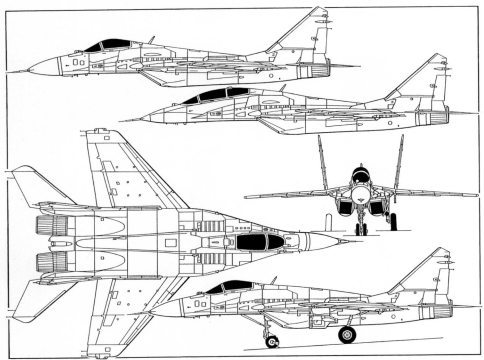

Three-view drawing of MiG-29S 'Fulcrum-C' fighter. Additional side views at top show late-model 'Fulcrum-A' and two-seat 'Fulcrum-B' *(Mike Keep/Jane's)*

same avionics items as were applied to the air arm's MiG-21s under the Lancer upgrade. This would build on the existing EADS NATO compatibility upgrade and life extension, and would use an avionics package supplied and integrated by Elbit, including a modular multirole mission computer linked to a MIL-STD-1553B digital databus, full HOTAS functionality, a second stripped-down Litton Italiana INS integrated with Trimble GPS and a new air data computer, a new Elop wide-angle HUD and two digital colour MFDs with a choice of metric or Imperial units, and with a new up-front controller. New Israeli Elisra SPS-20 RHAWS and compatibility with Western weapons. This upgrade may then be offered to other MiG-29 operators. The prototype (32367, donated by the Romanian Air Force) made its maiden flight on 5 May 2000, and was flown to Berlin for static display at ILA 2000, after completing a 15-flight test programme, on 26 May. Upgrade may be expanded to include an Elta EL/M-2032 pulse Doppler radar, as used by the Lancer-C. A decision by the Romanian Air Force is expected in 2002 or 2003.

MiG-29 (Factory index 9.13; 'Fulcrum-C'): First flown 4 May 1984; generally as 'Fulcrum-A', but introduced deeply curved top to fuselage aft of cockpit to house additional avionics, including L-203BE Gardeniya 1 FU active jammer; internal fuel capacity increased by 75 litres (20.0 US gallons; 17.0 Imp gallons); 1,500 litre (396 US gallon; 330 Imp gallon) underbelly fuel tank optional; weapons load increased to 3,200 kg (7,055 lb). Ammunition shellcase disposal modified to avoid centreline tank, allowing gun-firing with tank in place; auxiliary RWR/ECM antennas added to wingtips, and weapon control system improved. Three prototypes converted from standard 'Batch 4' MiG-29s; trials reportedly began April 1984. About 207 built from 1986 (beginning at serial number 05560). No 'Fulcrum-C' exports, although some were 'inherited' by independent republics on dissolution of USSR.

MiG-29UB (Factory index 9.51; 'Fulcrum-B'): Combat trainer; second K-36DM ejector seat forward of normal cockpit, under continuous canopy, with periscope and HUD repeater for rear occupant; no radar; gun, IRST sensor, laser range-finder and underwing stores pylons retained. Length 17.42 m (57 ft 2 in). Normal T-O weight 14,600 kg (32,187 lb); max speed 1,204 kt (2,230 km/h; 1,386 mph); service ceiling 17,500 m (57,420 ft). Prototype (with ventral fins) first flown 29 April 1981; production began 1982; 197 built by 1991. Phazotron and Sokol offering upgrade with new I-band slotted planar- or phased-array radar.

Detailed description refers to 'Fulcrum-C'.

MiG-29S (Factory index 9.13S; 'Fulcrum-C'): Multistage upgrade of type 9.13; initially with external load increased to 4,000 kg (8,818 lb) and with provision for two 1,150 litre (304 US gallon; 253 Imp gallon) underwing fuel tanks; these later added to some Russian 9.12s and 9.13s. Maximum fuel with external tanks is 8,240 litres (2,177 US gallons; 1,813 Imp gallons) for maximum range of 1,565 n miles (2,900 km; 1,802 miles). AoA operating range increased to 28°; RLPK-29M N019M radar, detection range 54 n miles (100 km; 62 miles) against fighter-size targets; now, with N019ME radar, able to track 10 targets and engage two simultaneously. Offered with third-phase improvements now associated with MiG-29SD, SE and N. Able to carry two pairs of R-27R1, R-27RE1, R-27T1 or R-27TE1 and two to six R-73E AAMs; up to six RVV-AE (R-77; AA-12 'Adder') AAMs after radar upgrade, or up to 4,000 kg (8,820 lb) of bombs; two or four S-24B 240 mm rockets, two or four B-8M1 packs of 20 S-8 80 mm rockets, ZAB-500 napalm tanks or BKF cluster bombs, in addition to standard GSh-301 gun. Some features of trials aircraft first flown 5 May 1984; first true prototype flew 23 December 1990. Maximum T-O weight 19,700 kg (43,430 lb). Two regiments in former East Germany fully equipped with first phase aircraft before Russian withdrawal.

MiG-29S (Factory index 9.12S; 'Fulcrum-A'): Similar to the above; upgrade of 9.12 lacking L203B jammer. Four retained by MiG-MAPO, numbered 333, 555, 777 and 999.

MiG-29SE ('Fulcrum-C'): Current production version of MiG-29S for export; Phazotron NIIR N019ME Topaz radar; six RVV-AE (R-77; AA-12 'Adder') AAMs; optional Western navigation, IFF and radio equipment and English language/Imperial units displays and instruments; downgraded L006LM/108 RHAWS; maximum T-O weight 20,000 kg (44,090 lb); performance data as for MiG-29S.

MiG-29SM (Factory index 9.13M): Current production upgrade of MiG-29S, with added ability to launch ASMs, including two Kh-29T/TE (AS-14 'Kedge') or Kh-31A/P (AS-17 'Krypton'), or four KAB-500KR TV-guided bombs. First version to offer simultaneous dual-target engagement capability. Prototype/demonstrator 17941 first flown 1995, in which year it set C-1h records mentioned earlier. Maximum T-O weight 20,000 kg (44,090 lb). Future enhancements were to include radar with mapping mode. Overtaken by MiG-29SMT.

MiG-29SD ('Fulcrum-A'): Export upgrade of basic MiG-29 (9.12), with most SE improvements, plus provision for 'Dozaprahvka' — inflight refuelling. Prototype (36034 '357') began refuelling trials November 1995. Designated MiG-29N for Malaysia, possibly MiG-29MF or MiG-29FM for Philippines.

'MiG-29N': Malaysian local designation of its new-build version of MiG-29SD; 16 single-seat MiG-29Ns (from production batches 52 and 53) and two two-seat **'MiG-29NUB**s' delivered 1995 for Nos. 17 and 19 Squadrons RMAF. Normal T-O weight 15,000 kg (33,068 lb). Maximum T-O weight 20,000 kg (44,090 lb), maximum weapon load 2,000 kg (4,409 lb). N019ZM radar. All being upgraded at 800 hours servicing to 'full MiG-29N standard', with 3,000 kg (6,614 lb) weapon load; Raytheon IFF; cockpit placards in English; voice warning system, instruments and displays calibrated in feet, knots, feet/minute and nautical miles, in-flight refuelling system using retractable extending probe on port side of front fuselage, AN/ARN-139 Tacan, GPS and ILS. First upgraded aircraft (M43-12) flew on 13 April 1998. Armament includes R-27R1 and R-73E1 AAMs, GSh-301 gun; upgrade adds RVV-AE (R-77; AA-12 'Adder') capability, allied to Phazotron N019ME radar with twin target BVR potential. Performance as MiG-29S except ceiling 18,000 m (59,050 ft). RD-33 engines modified to extend service life.

US$34.4 million, 18-month contract signed 16 October 1997 to upgrade Malaysian aircraft to full SE/SM standards. Payload 4,000 kg (8,818 lb), AAR probe, R-77 capability, radar and systems upgrade incorporated by Aerospace Technologies Systems Corporation at Kuantan between January 1998 and late 1999. First two returned to service on 2 May 1998. Strong interest in further upgrade to MiG-29SMT standards.

MiG-29M (Factory index 9.15; *Izdelie* 5): Advanced, genuinely multirole tactical fighter for control of upper airspace, ground attack and naval high-altitude precision weapons control; increased payload/range and endurance; intended as replacement for basic MiG-29. Preceded by 9.14 with Ryabina (mountain) LLTV targeting pod; 9.14 prototype (07682 '407') first flew 13 February 1985, but became 9.13/MiG-29S development aircraft. Features greatly redesigned airframe; two 86.3 kN (19,400 lb st) Klimov RD-33K turbofans ('land-based' versions of MiG-29K engine); designed for triplex analogue fly-by-wire controls for lateral axis, quadruplex elsewhere, with mechanical back-up to ailerons and rudders (only pitch axis fly-by-wire by late 1996); 'glass' cockpit with two monochrome (green) multifunction CRTs (not push-button, but HOTAS); modifications to extend aft centre of gravity limit for relaxed stability.

First of six prototypes and one static test airframe flown 25 April 1986 with RD-33 engines; first flight with RD-33K engines (previously tested on 921) 26 September 1987; first exhibited at Machulishche airfield, February 1992; flight refuelling trials on standard MiG-29 test aircraft began 16 November 1995, completed January 1996; enlarged engine air intakes with movable lower lip to increase mass flow on take-off. Original FOD doors in air intakes replaced by lighter retractable grids, permitting

Prototype Sniper upgrade of Romanian MiG-26 *(Paul Jackson/Jane's)*   2001/0100477

Malaysian Air Force 'MiG-29N' *(Paul Jackson/Jane's)*   2000/0064947

deletion of overwing louvres and internal ducting in lightweight aluminium-lithium alloy centre-section, providing increased fuel tankage; new intakes tested on 921; total internal fuel capacity 5,700 litres (1,506 US gallons; 1,254 Imp gallons). New wing section, with sharp leading-edge. Increased span ailerons. Bulged wingtips with fore and aft RWRs; more rounded wingtip trailing-edge; larger, sharp-edge and slightly raised LERX; increased-chord horizontal tail surfaces, with dogtooth leading-edge. Bonded aluminium-lithium front fuselage, welded steel behind; nose lengthened by approximately 20 cm (7½ in); 40 mm (1½ in) higher canopy; new IFF and Gardeniya active jammer in dorsal spine, which terminates in 'beaver-tail' structure, containing twin 13 m² (140 sq ft) brake-chutes, that extends beyond jet nozzles; single larger honeycomb composite over-fuselage airbrake. Strengthened landing gear with KT-209 mainwheels. Extensive use of RAM giving claimed '×10' reduction in frontal RCS.

Has 60 per cent lighter Phazotron NIIR N010 Zhuk (RLPK-29M) terrain-following and ground-mapping radar with 680 mm (26.77 in) dish antenna in larger diameter radome, able to track 10 targets and engage four simultaneously over a range of 43 n miles (80 km; 50 miles); (radar first flew in 'Fulcrum-C' '16', *Izdelie* 9.16); new OLS-M longer-range IRST, with added TV channel and laser designator/marked target seeker using common mirror system. TS101 processors with new software. A-331 Shoran. Chaff/flare dispensers relocated in dorsal spine.

Claimed more comfortable to fly, with increased permissible angle of attack (30° during initial tests, subsequently expanded), better manoeuvrability, improved cruise efficiency; eight underwing hardpoints for 4,500 kg (9,920 lb) stores, including four laser-guided Kh-25ML (AS-10 'Karen') or Kh-29L (AS-14 'Kedge'), anti-radiation Kh-25MP (AS-12 'Kegler') and Kh-31A/P (AS-17 'Krypton') or TV-guided Kh-29T (AS-14 'Kedge') ASMs; eight RVV-AE (R-77; AA-12 'Adder') AAMs, R-73E (AA-11 'Archer') AAMs or KAB-500KR 500 kg TV-guided bombs. Rounds for gun reduced to 100. Range on internal fuel 1,079 n miles (2,000 km; 1,242 miles), with three external tanks 1,726 n miles (3,200 km; 1,988 miles). Welded Al-Li structure very expensive and failed to provide promised weight savings. State Acceptance Tests suspended due to funding problems, May 1993. Planned MiG-29UBM trainer (9.61) abandoned. Not ordered for Russian Air Forces at the time, though development was relaunched by MAPO in late 1999.

MiG-29ME: Export version of MiG-29M, with same weapons and equipment but with first-generation WCS based on N-019ME radar of MiG-29SD/SE. Redesignated **MiG-33**, but no longer being promoted; see 1999-2000 edition for details.

MiG-29SMT (Factory index 9.17 MiG-20SMT-I): New version, based on original 9.12/9.13 airframe, and incorporating many of the improvements and capability enhancements planned for the MiG-29M, originally including gridded intakes and still including the dorsal airbrake, but without that variant's new lightweight Al-Li airframe. MiG-29SMTs can be produced on existing production jigs, or by retrofit of in-service aircraft, but the configuration is primarily a retrofit option. 'Fifth generation' avionics system, based around MIL-STD-1553B equivalent databus, building on experience gained with *Izdelie* 9.21 (15125 '211') which tested BTsK-29 digital avionics suite in 1987-88. Two 152 × 203 mm (6 × 8 in) MFI-68 colour LCD MFDs dominate panel, with multifunction control panels incorporating three smaller (96 × 77 mm; 3¾ × 3 in) monochrome LCDs on side consoles. Digital moving map

MiG-29 'Fulcrum-C' of the 'Ukrainian Falcons' aerobatic team *(Paul Jackson/Jane's)* 2000/0064946

Second MiG-29K prototype, showing original wing fold position *(Yefim Gordon)* 2000/0064898

(and possibly Terprom-type elevation database and display) and upgraded HUD. Revised N019MP 'Topaz' radar with synthetic aperture air-to-ground mode and larger 'look' angle (70° in azimuth +50°/–40° in elevation). AoA limit to be raised to 30°.

Last preproduction 'Fulcrum-A' (925 'Blue 25') rebuilt as avionics/cockpit mockup and displayed at Moscow Air Show, August 1997; MiG-29SD 05533 ('Blue 331') rebuilt as avionics/cockpit prototype; first flew as such, 29 November 1997; modified with aerodynamic mockup of revised fuselage spine and flew again on 22 April 1998, having been renumbered 'Blue 405'. The first full-standard 9.17 was a converted MiG-29S/9.12S demonstrator (35400 'Blue 917') which first flew on 14 July 1998 with the new spine fully fitted, and including new fuel tanks.

Revised spine with larger No 1 tank forward and new tank in rear part of fairing. Rear, 'tail' tank protrudes beyond jetpipes and can be retrofitted at unit level. In combination, tanks give 1,475 kg (3,252 lb) (1,000 plus 475 kg; 2,205 plus 1,047 lb) increase in capacity, equivalent to 100 per cent increase in mission radius to 836 n miles (1,550 km; 963 miles) in the air superiority role, or 594 n miles (1,100 km; 683 miles) air-to-ground. Overall range increased to 1,889 n miles (3,500 km; 2,714 miles) or to 3,617 n miles (6,700 km; 4,163 miles) with a single aerial refuelling. Original scheme for SMT added outboard integral wing tanks, tanks in LERXes and in tailbooms, increasing capacity by 2,490 kg (5,490 lb) or 3,170 litres (337 US gallons; 697 Imp gallons).

Aircraft also stated to be compatible with revised, enlarged, 1,800 litre (476 US gallon; 396 Imp gallon) underwing fuel tanks and with bolt-on retractable AAR probe. Warload increased to 5,000 kg (11,023 lb) in air-to-ground role. Installation planned of 98.1 kN (22,050 lb st)

RD-43 (RD-333) engines, possibly later with thrust vectoring. MTOW raised to 21,000 kg (46,297 lb). Service life extended from 4,000 to 6,000 flying hours. Believed aimed primarily at Russian Air Forces as MLU for about 180 existing aircraft but offered for export, including to Ecuador. MIL-STD-1553B bus will simplify integration of Western displays, avionics and weapons. Original Russian plan was for up to 15 upgrades in 1998, building to 40 per year by 2000; however, first air force aircraft (01) completed on 29 December 1998 and demonstrated to customer at Zhukhovsky on 12 January 1999.

MiG-29SMT-II (Factory index 9.17-II): Further upgrades to the SMT are already planned or on offer, under provisional designation MiG-29SMT-II. Improvements include frontal RCS reduction measures, IR signature reduction and further increases in fuel tankage and warload. Fuel capacity to be increased to 5,600 kg (12,346 lb) through installation of new 219 litre (58.0 US gallon; 48.0 Imp gallon) tanks in LERX, replacing auxiliary air intakes and ducts, as in MiG-29M and original SMT scheme. Eight hardpoint wing (either based on MiG-29M, with broad span ailerons, or merely rebuilt standard wing) will allow warload increase to 5,500 kg (12,125 lb). Potential avionics improvements include new radar (N010, Zhuk, Zhuk PH or NIIR Zemchug). Some expect future MiG-29 variants to receive a new engine, the VK-10M, being developed by Klimov for production from 2010, with a thrust of 108 to 113 kN (24,250 to 23,350 lb st). Thrust vectoring may be offered. May use triple redundant digital FBW FCS developed for MiG-29M.

MiG-29SMTK (Factory index 9.17K): Carrierborne variant combining MiG-29SMT features with landing gear, carrier landing system, folding wing, double-slotted flaps, arrester hook and enlarged tailplane of MiG-29K. Believed to have been offered to India and China in association with initial efforts to sell carrier *Admiral Gorshkov*. Latter built to operate STOVL Yak-38, but modifications proposed for STOBAR (short take-off but arrested recovery) operation. Replaced by MiG-29K-2002 and MiG-29K-2008.

MiG-29K (Factory index 9.31; K for *korabelnyy*; ship-based): Maritime version, used for ski-jump take-off and deck landing trials on carrier *Admiral of the Fleet Kuznetsov* (formerly *Tbilisi*), beginning 1 November 1989; two new-build prototypes, using MiG-29M structure; completely redesigned, mainly steel, wing using modified aerofoil section and of increased area (increased span, reduced leading-edge sweep) with double slotted flaps, drooping flaperons and more powerful leading-edge flaps; new spar in front of original wing box; new strengthened centre-section without overwing louvres (see MiG-29M); upward-folding outer wing panels; RD-33K turbofans with 92.2 kN (20,723 lb st) contingency rating for ski-jump take-offs. Fuel capacity reduced to 5,670 litres (1,498 US gallons; 1,247 Imp gallons). First flown (16188 '311') 23 June 1988. (Preceded by MiG-29KVP, conversion of 07687 [preseries aircraft 918 with hook, strengthened landing gear and some carrier landing systems and used for trials at Saki from 1982.) Exhibited at Machulishche airfield, Minsk, February 1992, with typical anti-ship armament of four Kh-31A/P (AS-17 'Krypton') ASMs and four R-73E (AA-11 'Archer') AAMs. Production MiG-29K was intended to use same basic airframe, power

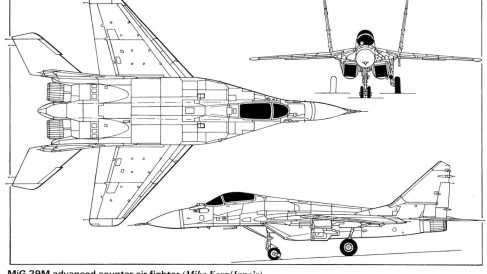

MiG-29M advanced counter-air fighter *(Mike Keep/Jane's)*

Sixth prototype MiG-29M flying with eight underwing stores — AS-17 'Krypton', inboard, and AA-12 'Adder' *(via Yefim Gordon)*

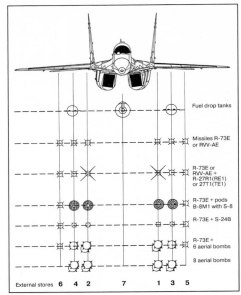

MiG-29SE weapon options

plant, avionics and equipment as MiG-29M, with added wing folding, strengthened landing gear, ±90° nosewheel steering for deck-handling, arrester hook, fully retractable, permanently installed flight refuelling probe, and other naval requirements, including Uzel carrier beacon homing system, SN-K navigation suite with INS-84, Resistor Shoran/ILS, ACLS and new inertial platform and with radar upgraded to RLPK-29IM standards, giving better over-water performance. Ejection seat trajectory laterally inclined 30° so that a deck-level ejection would be into the sea, abeam the carrier, giving extra altitude for parachute to open. Beryoza RHAWS replaced by Pastel, which can function as ELS for Kh-31P ARMs.

State Acceptance Trials suspended due to funding difficulties, early 1992. Further development ended initially when not selected for deployment on *Admiral of the Fleet Kuznetsov*, but resumed at Zhukovsky in September 1996, reportedly in expectation of order from India. (More details in 1993-94 *Jane's*.) First prototype currently grounded; second (27579) returned to flight status in support of MiG-29M programme. Proposed naval **MiG-29KU** (9.62) trainer derivative with two separate stepped cockpits remained unbuilt.

'**MiG-29SMTK**': The original MiG-29SMTK (effectively a MiG-29SMT with the 9.31's folding wing and landing gear), previously offered to India along with the former helicopter carrier *Admiral Gorshkov*, is understood to have been replaced by a new, multirole, carrierborne variant based more closely on the MiG-29K/M, albeit without the expensive Al-Li alloys. With a MIL-STD-1553B-type bus and open systems avionics architecture, the **MiG-29K-2002** is compatible with a wide range of Russian and Western weapons, and may feature the colour displays and GPS-based navigation system of the MiG-29SMT. This variant, possibly designated MiG-29 MTK, is claimed to be able to perform 90 per cent of its missions with a 10 kt (18 km/h; 11 mph) wind-over-deck, even in tropical conditions using new autothrottle. One notable feature of the new aircraft is its much-reduced folded span of 5.80 m (19 ft 0¼ in), achieved by positioning the fold line much closer in to the wing-root, and by adding upward-folding tailplanes. The aircraft can also fold its radome (up and back), reducing overall length to 14.13 m (46 ft 4¼ in). Accordingly, *Admiral Gorshkov* can carry a full air wing of 24 MiG-29Ks (plus six helicopters), or (according to some sources) as many as 30. A projected **MiG-29K-2008** upgrade configuration could add a computer upgrade and additional electro-optic, radar, IIR and recce pods, together with take-off performance improvements. In December 1999, it was reported that India had selected the MiG-29K for use aboard *Admiral Gorshkov*, and an initial order for 50 aircraft (against a total requirement for 60 to 70 aircraft) was expected, with an unknown quantity of Kh-35 anti-ship missiles and Kh-31P ARMs. A refuelling tanker fit has been proposed. Local manufacture by HAL is expected.

MiG-29KUB: Revised carrier-borne two-seat trainer design offered to India, based on MiG-29K-2002 with reduced-span, inboard wing fold and folding tailplanes. Assumed to feature original stepped tandem cockpits of MiG-29KU. Some reports suggest enlarged tailfins with integral fuel tanks, possibly even single centreline tailfin.

MiG-29UBT (Factory index 9.51T; *Izdelie* 30): Private venture programme funded by OKB, consisting of virtual SMT upgrade (increased fuel in enlarged spine, in-flight refuelling capability glass cockpit and enhanced avionics) for UB trainer. This transforms UB into operationally capable two-seat multirole fighter, with potential for Su-24 replacement or for 'pathfinder' use. Expected to be fitted with a millimetre wave terrain-avoidance radar and Osa-2 X-band, phased-array centimetric air-to-air/air-to-ground radar, or with Thomson-CSF RC 400 radar in nose and missile launch and trajectory control system in wingroots. Full dual controls retained in rear cockpit, but augmented by large CRT display screen optimised for display of FLIR or video imagery; WSO would be a rated pilot who could take control if captain incapacitated. Long-serving demonstrator (08134 '304') converted to UBT demonstrator/prototype making type's Western debut at 1998 Farnborough Air Show, after only five test flights in new configuration, having first flown on 25 August 1998. First production conversion 25982 '952', with folding refuelling probe, shown at Paris in June 1999. A further-upgraded aircraft, with four-pylon wings, refuelling probe and OSA-2, is expected to fly in mid-2000.

CUSTOMERS: MAPO production total given as 1,257 by 1997, including 478 to Russia, with 14 prototype/preseries aircraft (four of them built by MiG design bureau) and 197 MiG-29UB trainers. Sokol plant confirms last-mentioned with figure of "about 2000" (2000). Circumstantial evidence suggests about 840 of the 1,257 are 'Fulcrum-A'/9.12 versions and 207 are 9.13s. Some reports imply that small numbers of single-seat MiG-29s also built by Sokol (MMZ No. 21 'Sergo Ordzhonikidze' at Gorkii [Nizhny Novgorod]). Ecuador ordered 10 MiG-29Ss and

1 Oxygen control panel
2 AFCS panel
3 Communications control panel
4 Flaps control panel
5 Throttle lever
6 Aiming control panel
7 AFCS controller
8 Radar control panel
9 Landing gear control valve
10 Canopy manual operating handle
11 Landing lights control panel
12 Landing gear emergency extension handle
13 Pilot approach display
14 Altimeter
15 Airspeed indicator
16 Optical and electronic aiming and navigation control panel
17 AoA and acceleration indicator
18 Emergency braking valve handle

19 Flight director indicator
20 Navigation instruments
21 Heading setting panel for AHRS
22 Pitot static tube selector switch
23 Voltmeter
24 Cabin air temperature selector
25 Braking system pressure gauge
26 Clock
27 Machmeter
28 Vertical speed and turn-and-slip indicator
29 Nosewheel braking handle
30 HUD and control panel
31 Radio altimeter
32 Jamming release system control panel
33 Oxygen supply indicator
34 Hydraulic and pneumatic system pressure indicator
35 Exhaust gas temperature indicator

36 Tachometer
37 Radar screen
38 Fuel quantity and flow meter
39 Air intake ramp position indicator
40 Warning panel
41 EKRAN BITE display
42 Jamming control panel (MiG-29SE)
43 Control stick
44 IFF panel
45 Power plant emergency modes control board
46 Standby compass
47 Annunciators
48 Short-range navigation and landing system control panel
49 Cabin, glass and probe heating control panel
50 Canopy emergency jettison handle
51 Internal and external lighting panel
52 Ventilation selector switch (canopy or pilot)

53 Radio control panel
54 Illumination warning control panel
55 Radio compass controls
56 Lighting panel
57 IFF system panel
58 Guidance system controls
59 Power generating system control board
60 Engine starting control panel
61 Aircraft systems switchboard
62 Aircraft systems control panel
63 Combined armament-control system control panel

EADS/Aerostar/Elbit Sniper upgrade of MiG-29

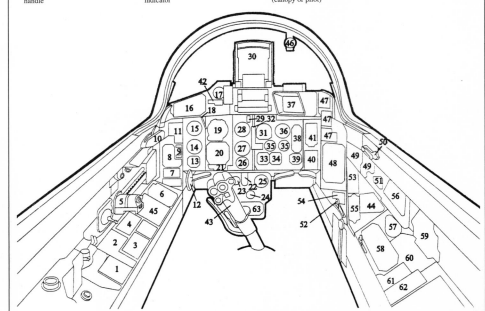

MiG-29SD/SE cockpit

two MiG-29UBs for US$40 million. EADS Deutschland has suggested that 'approximately 1,400' were produced, of which 750 remain in service. Further details in tables.

COSTS: Malaysia reportedly paid US$560 million for its 18 aircraft while Bangladesh paid US$115 to 125 million for eight aircraft, this sum being broken down, according to some reports as US$11 million per aircraft, plus US$27 million for the training/spares/support package.

DESIGN FEATURES: Emphasis from start on high manoeuvrability, to counter US F-15, F-16 and F-18, with target destruction at distances from 200 m (660 ft) to 32 n miles (60 km; 37 miles), and with effective air-to-surface capability. All-swept mid-wing configuration, with wide ogival wing leading-edge root extensions (LERX), 40 per cent of lift provided by lift-generating centre-fuselage, twin tailfins carried on booms outboard of widely spaced engines with wedge intakes; doors in intakes, actuated by extension and compression of nosewheel leg, prevent ingestion of foreign objects during take-off and landing; gap between roof of each intake and skin of wingroot extension for boundary layer bleed.

Fire-control and mission computers link radar with laser range-finder and infra-red search/track sensor, in conjunction with helmet-mounted target designator; radar able to track 10 targets simultaneously; targets can be approached and engaged without emission of detectable radar or radio signals; sustained turn rate much improved over earlier Soviet fighters; thrust/weight ratio better than one; allowable angles of attack at least 70 per cent higher than previous fighters; difficult to get into stable flat spin, reluctant to enter normal spin, recovers as soon as controls released; wing leading-edge sweepback 73° 30′ on LERX, 42° on outer panels; anhedral approximately 3°. Tailfins canted outward 6°; leading-edge sweep 47° 50′ on fins, 50° on horizontal surfaces; anhedral 3° 30′. Fins canted outwards 6°. Engine replacement time 2 hours; preflight preparation 20 minutes. Design flying life 2,500 hours.

FLYING CONTROLS: Conventional. Hydraulically powered surfaces, with SAU-451 three-axis autopilot, ARU-29-2 *g*-feel system and rate dampers; 50S-3M AoA limiter set at 26° in earliest aircraft, with initial service limit of 24° (30° permitted in symmetrical manoeuvres without banking) but can be overriden by pilot by 'pulling through' synthetic stick-stop; automatic 'bank corrector' feeds in rudder to reduce bank at high AoA; computer-controlled four-section leading-edge manoeuvring flaps (maximum downward deflection 20°) over full span of each wing, except tip, and plain trailing-edge flaps (maximum downward deflection 25°); inset ailerons with RP-280A hydraulic actuators; inset rudders with RP-270A actuators and all-moving (+15°/–35° collectively and differentially) horizontal tail surfaces with RP-260A actuators and no tabs; interconnect allows rudders to augment roll rate; mechanical yaw stability augmentation system; hydraulically actuated forward-hinged airbrakes above and below rear fuselage between jetpipes. Pilot may override *g* limiter; few demonstration pilots authorised up to +11 *g*.

STRUCTURE: Approximately 7 per cent of airframe, by weight, of composites; remainder metal, including aluminium-lithium alloy for wing carry-through structure housing fuel tanks; three-spar wings with three 'false spars' two ahead of, one behind, torsion box; 16 stringers and skins reinforced by stringers; trailing-edge wing flaps, ailerons and vertical tail surfaces of carbon fibre honeycomb; approximately 65 per cent of horizontal tail surfaces aluminium alloy, remainder carbon fibre; semi-monocoque all-metal fuselage built around 10 mainframes in three subassemblies, with forward (frames 1 to 3), centre (frames 4 to 7) and rear sections, the latter including the engine bays; fuselage sharply tapered and downswept aft of flat-sided cockpit area, with ogival dielectric nosecone mounting PVD-18 main pitot boom (PVD-7 auxiliary pitot mounted on side of nose); small vortex generator each side of nose helps to overcome early tendency to aileron reversal at angles of attack above 25°; tail surfaces carried on slim booms alongside engine nacelles.

KNOWN RUSSIAN AND SOVIET MiG-29 OPERATORS

Unit	Base	Remarks
4 TsBP I PLS	Lipetsk	Early training unit
8 IAP	Vasilkov	To Ukraine
14 IAP	Zherdevka	
19 GvIAP	Millerovo	
28 GvIAP	Andreapol	
31 Nikopol'sky GvIAP	Zernograd	From Falkenberg/Alt Lönnewitz, Germany
33 IAP	Andreapol	From Wittstock, Germany
35 IAP		From Zerbst, Germany
61 IAP	Kakaydy	To Uzbekistan
62 IAP	Belbek	To Ukraine
67 Attack Aviation Regiment	Mary 2	To Turkmenistan
73 Sevastapolsky GvIAP	Shaykovka	From Köthen
83 GvIAP	Merseburg	To Starokonstantinov, Ukraine
86 IAP	Markuleshty	Transferred to Moldova
114 IAP	Ivano-Frankovsk, Ukraine	From Milovice, Czechoslovakia
116 UtsBP	Astrakhan	
120 IAP	Domna	
160 IIAP	Borisoglebsk	(ex-1080th UATs PLS)
161 IAP, AV-MF	Limanskoye	To Ukraine
176 IAP	Transbaikal district	From Tskhakaya, Georgia
234 Proskurovsky GvIAP	Kubinka	Redesignated 237th GvTsPAT
237 GvTsPAT	Kubinka	
343 IIAP	Sennoy	
642 IAP	Martinovskaya	To Ukraine
715 IAP	Lugovoi	To Kazakhstan
733 IAP	Andreapol	From Putnitz, Germany
787 IAP	Ros, Belarus	From Eberswalde/Finow, Germany
797 UAP	Kuschchevskaya	
871 IAP	Smolensk	
960 IAP	Primorsko Akhtarsk	
968 IISAP	Lipetsk	Was 968th IAP, Nöbitz/Altenburg, Germany
Unidentified UAP	Yeysk	
Combat Training Centre	Mary	
Unidentified IAP	Zherdyovka	Formerly based in Hungary
Unidentified IAP	Orlovka, Khabarovsk	
Unidentified PVO IAP	Privolzhskiy	

Cockpit of first production conversion MiG-29SMT *(Yefim Gordon)* 2000/0064897

MiG-29UB two-seat trainer *(Yefim Gordon)* 2001/0100401

MiG-29 EXPORT CUSTOMERS

Customer	Variant	Source	Delivered/Remaining	ISD
Bangladesh	9.12A	MiG-MAPO	4 of 8 delivered	1999
Belarus	9.12, -13, -51	Soviet/CIS AF	70/58	1991
Bulgaria	9.12A	MiG-MAPO	18/17	1990
Bulgaria	9.51	MiG-MAPO	4/4	1990
Cuba	9.12B	MiG-MAPO	12-14/12?	1990
Cuba	9.51	MiG-MAPO	2/2	1990
Czechoslovakia	9.12A	MiG-MAPO	18/0 to Czech and Slovak	1989
Czechoslovakia	9.51	MiG-MAPO	2/0 to Czech and Slovak	1989
Czech Rep	9.12A	Czechoslovakia	9/0 to Poland	1992
Czech Rep	9.51	Czechoslovakia	1/0 to Poland	1992
East Germany	9.12A	MiG-MAPO	20/19 to Federal Germany	1988
East Germany	9.51	MiG-MAPO	4/0 to Federal Germany	1989
Eritrea	9.12A	Moldova?	c.6/4?	1998
Eritrea	9.12A	MiG-MAPO	c.10 on order	
Germany	9.12A	East German AF	20/19	1990
Germany	9.51	East German AF	4/4	1990
Hungary	9.12B	MiG-MAPO	22/20	1993
Hungary	9.51	MiG-MAPO	6/6	1993
India	9.12B	MiG-MAPO	72/67?	1987
India	9.51	MiG-MAPO	8/?	1987
Iran	9.12B	MiG-MAPO	14/12	1990
Iran	9.12S?	Mig-MAPO	1/1	1997
Iran	9.51	MiG-MAPO	2/2	1990
Iran	9.12B	Iraq	21/21	1991
Iraq	9.12B	MiG-MAPO	36/8 dismantled and stored	1988
Iraq	9.51	MiG-MAPO	6/0	1988
Israel	9.12A	Polish/Hungarian/Romanian loan?	2 or 3, returned	1997
Kazakhstan		Soviet/CIS AF	42/21 (+21 in store)	1991
Korea, North	9.12B, -51	MiG-MAPO	c.25 and 5 UB/?	1988
Malaysia	MiG-29N	MiG-MAPO	16/14	1995
Malaysia	MiG-29NUB	MiG-MAPO	2/2	1995
Moldova	9.12, -13, -51	Soviet/CIS AF	31-34/6 (To Yemen, USA)	1991
Peru	9.12/-13	Belarus	14/4 serviceable	1997
Peru	9.12S	MiG-MAPO	3 ordered in 1998	?
Peru	9.51	Belarus	2/?	1997
Poland	9.12A	MiG-MAPO	9/8	1989
Poland	9.51	MiG-MAPO	3/3	1989
Poland	9.12A	Czech Rep	9/9	1995
Poland	9.51	Czech Rep	1/1	1995
Romania	9.12A	MiG-MAPO	18/16	1989
Romania	9.51	MiG-MAPO	3/2	1989
Russia	9.12, -13, -51	MiG-MAPO	c675/260-460	1983
Slovakia	9.12A	Czechoslovakia	9/9	1992
Slovakia	9.51	Czechoslovakia	1/0	1992
Slovakia	9.12	Russian AF	13/13	1994
Slovakia	9.51	Russian AF	1/1	1995
South Yemen	9.12	Moldova	12/0 (4 returned, 7 lost)	1994
Syria	9.12B	MiG-MAPO	18-42?	1987
Syria	9.51	MiG-MAPO	2-6/?	1988
Turkmenistan	unknown	Soviet/CIS AF	unknown/c.22	1991
Ukraine	9.12, -13, -51	Soviet/CIS	245/62	1991
USA	9.12	Moldova	6/in store	delivered 1997
USA	9.13	Moldova	14/in store	delivered 1997
USA	9.51	Moldova	1/in store	delivered 1997
Uzbekistan	9.13	Soviet/CIS AF	c.36/unknown	1991
Yugoslavia	9.12B	MiG-MAPO	14/c.5-6	1987
Yugoslavia	9.51	MiG-MAPO	2/2	1987

Notes on export users:
Belarus: 61 Fighter Aviation Base, Baranovichi. Russian sources suggest that about 24 of 47 MiG-29s delivered are in service while EADS suggests that 82 are in use, of 80 originally delivered
Bulgaria: 5 Fighter Aviation Base, Burgas/Ravnets. 21-22 remain according to EADS including four trainers and these re-located to Graf Ignatievo during 1999-2000
Cuba: 231° Escuadrón de Caza, 23° Regimiento de Caza, San Julian. EADS gives the number in use as 12
Czechoslovakia: 11 SLP, Zatec
Czech Republic: 2 Squadron, 1 SLP, České Budéjovice
East Germany: 1/ and 2/JG 3 'Vladimir Komarov', Preschen
Germany: JG 3, Preschen (-1991), Erprobungsgeschwader 29, Preschen (1991-1992), 2/JG 73 Laage (1992-)
Hungary: 1 VS 'Puma' and 2 VS 'Dongo' of 59 HVO, Kecskemet (1993-), then 9 HVO. EADS estimates 27 to 28 in use, including six trainers, though only 12 were airworthy at any one time during 2000
India: First export customer. No. 28 Squadron 'First Supersonics', No. 47 Squadron 'Archers', Poona, No. 223 Squadron 'Tridents', Adampur. EADS suggests that 66 remain on charge. Some sources suggest that all single-seaters were 9.12s or 9.12As, last batches almost certainly were
Iran: Project Talle added fixed in-flight refuelling probes, while Tallieh will add new retractable probes. Khorsheed saw the development of indigenous 1,000 and 1,200 litre (264 and 317 US gallon; 220 and 264 Imp gallon) external tanks. EADS gives a total of 35 in use
Iraq: 21 fled to Iran, eight dismantled in 1995, up to six shot down. DASA estimates that 15 remain in use
Israel: 117 Squadron
Japan: Two JASDF test pilots undertook a six-month evaluation of the MiG-29 at Krasnodar during 1998-99
Kazakhstan: 715 IAP, Lugovoi. EADS suggests that 22 are in use. 30 more reportedly to be transferred in compensation for Tu-95 transfers to Russia
Korea, North: Some sources suggest that Korea's MiGs included five or six 9.13s delivered in kit form. Sources differ as to total delivered, from 12, to 20, to 30. EADS estimates that 15 are in service
Malaysia: 17 and 19 Skuadrons, Kuantan. A second batch of 14-18 aircraft is said to be under consideration
Moldova: Inherited 34 MiG-29s from Naval 86 IAP. Total does not tally with known fates. 12 to Yemen (four returned), 21 to USA, '10' to Eritrea. Remaining six said to be 'for sale' in November 2000
Peru: Escuadrón de Caza 611, Grupo Aereo 6, Chiclayo (locally described as MiG-29SE). EADS gives a total of 12 in service
Poland: 1 PLM, Minsk-Mazowiecki
Romania: 57 Regiment de Vinatoure, Mikhail Kogalniceanu. Original batch of 12 plus two UBs augmented by further batches of five and two, including at least one more UB. EADS estimates that 12 are in service, or 18, including 3 trainers
Slovakia: 1 SLP, now 31 Stihaci Letecke Kridlo (Fighter Wing) with 311 and 312 Stihaci Letka (Fighter Squadrons). EADS estimates 12 to 15 in use, including 3 trainers
Syria: A fighter wing at Seikal is believed to operate 42 9.12Bs and six MiG-29UBs, though some reports suggest that only 20 aircraft were delivered. Others suggest that 80 aircraft were delivered, of 150 originally ordered. EADS gives a total of 20 in service)
Turkmenistan: 67 Attack Aviation Regiment, Mary 2
Ukraine: MiG-29 units have included the 8 IAP at Vasilkov, 62 IAP at Belbek, 85 IAP at Starokonstantinov, 114 IAP at Ivano Frankovsk, 161 IAP at Limanskoye, and the 642 IAP Martinovskaya. Ukraine inherited between 216 and 245 MiG-29s (246 according to EADS), including an estimated 155 9.13s. The 62 aircraft given as being in service may not include a large number of aircraft in temporary storage
USA: Air Force Systems Command, Wright-Patterson AFB; two given civil registrations in July 1999 (registered as MiG-29UB, but identities suggest 'Fulcrum-A')
Uzbekistan: 61 IAP, Kakaydy. EADS estimates that 36 are in use
Yemen: EADS suggests that four are still in use
Yugoslavia: 204 LAP at Batajnica, with the 127 LAE. 11 single-seaters remained operational before Operation Allied Force in 1999. About six were destroyed in air combat during the operation, but claims of aircraft destroyed on the ground may refer to the destruction of decoys

First 'production' MiG-29UBT 25982 *(Paul Jackson/Jane's)*

LANDING GEAR: Hydromash retractable tricycle type, with oleo-pneumatic shock-absorbers; single KT-150 wheel on each main unit and twin KT-100 nosewheels. Mainwheels retract forward into wingroots, turning through 90° to lie flat above leg; nosewheels, on trailing-link oleo, retract rearward between engine air intakes. Hydraulic retraction and extension, with mechanical emergency release. Nosewheels steerable ±8° for taxying, T-O and landings, ±31° for slow speed manoeuvring in confined areas (selector in cockpit). Mainwheel tyres size 840×290, pressure 11.75 bars (170 lb/sq in), nosewheel tyres size 570×140, pressure 9.80 bars (142 lb/sq in). Pneumatic steel brakes. Mudguard to rear of nosewheels. Container for 17 m² (183 sq ft) cruciform brake-chute in centre of boat-tail between engine nozzles.

POWER PLANT: Two Klimov/Sarkisov RD-33 turbofans, each 49.4 kN (11,110 lb st) dry and 54.9 to 81.4 kN (12,345 to 18,300 lb st) with afterburning. Engines mounted 4° nose-up, and nacelles toe-in by 1° 30′. Engine ducts canted at approximately 9°, with wedge intakes, sweptback at approximately 35°, under wingroot leading-edge extensions. Multisegment ramp system, including top-hinged forward door (containing a very large number of small holes) inside each intake that closes the duct while aircraft is taking off or landing, to prevent ingestion of foreign objects, ice or snow. Air is then fed to each engine through five louvres in top of each wingroot leading-edge extension and perforations in duct closure door. Doors are opened by extension of nose gear oleo when T-O speed reaches 140 kt (260 km/h; 162 mph) and closed by oleo compression at touchdown. Louvres also have air inlet control function, sometimes asymmetrical, with three lattice spill doors aft of each.

'Fulcrum-A' and 'Fulcrum-C' internal fuel capacities have been quoted differently by the manufacturer on numerous occasions. Basic 'Fulcrum-A' has integral fuel tank formed by torsion box in inboard portion of each wing, capacity 350 litres (92.5 US gallons; 77.0 Imp gallons); four tanks in fuselage, respectively 705 litres (186 US gallons; 155 Imp gallons), 875 litres (231 US gallons; 192.5 Imp gallons), 1,800 litres (476 US gallons; 396 Imp gallons) and 285 litres (75.0 US gallons; 62.5 Imp gallons); total internal fuel capacity 4,365 litres (1,153 US gallons; 960 Imp gallons). In later 'Fulcrum-As', 705 litre tank replaced by one of 780 litres (206 US gallons; 172 Imp gallons) which was replaced in 'Fulcrum-C' by 890 litre (235 US gallon; 196 Imp gallon) tank. Attachment for 1,500 litre (396 US gallon; 330 Imp gallon) non-conformal external fuel tank under fuselage, between ducts. Some (MiG-29S and other) aircraft piped to carry 1,150 litre (304 US gallon; 253 Imp gallon) external tank under each wing. Single-point pressure refuelling through receptacle in port wheel well. Overwing receptacles for manual fuelling. Single GTN-7 turbopumps in tanks 1 and 3, three in tank 2, with fuel jet pumps in wings and tank 3A; system powered by DTsN-80 centrifugal pump. Airscoop for GTDE-117 turboshaft APU, rated at 73 kW (98 eshp) for engine starting, above rear fuselage on port side; exhaust passes through underbelly fuel tank when fitted. In-flight refuelling system, with retractable port-side probe, available as upgrade. 'Fulcrum-A' and 'Fulcrum-B' have KSA-2 gearbox, 'Fulcrum-C' has KSA-3. One Khladon-114V₂ CFC 3-litre fire extinguisher mounted in the spine for use on fires in engine bays or APU/accessory gearbox bay. Actuated automatically by flame sensors.

ACCOMMODATION: Pressurised cockpit enclosed by frames 1 and 2. Pilot only, on 16° rearward-inclined K-36DM Series 2 zero/zero ejection seat, giving −14° view forward over the nose, and under hydraulically actuated, rearward-hinged transparent blister canopy in high-set cockpit. Sharply inclined one-piece curved windscreen of electrically de-iced triple glass. Three internal mirrors provide rearward view.

SYSTEMS: Two independent hydraulic systems powered by NP-103A variable-displacement pumps, driven by the engine accessory gearboxes, pressurised to 207 bars (3,000 lb/sq in), with 80 litres (21 US gallons; 17.5 Imp gallons) fluid. Main system powers one chamber of each control surface actuator, leading-edge and trailing-edge flaps, stick-pusher, artificial feel unit, landing gear extension/retraction, nosewheel steering, intake ramps and APU exhaust door; back-up system powers second chamber of each control surface actuator and stick-pusher, and can be powered by an emergency NS-58 pump.

Electrical system consists of three subsystems: 27-28.5 V DC, 115 V/400 Hz AC (three-phase) and 36 V/400 Hz AC (single-phase). Accessories gearbox drives a 30 kW GSR-ST-12/40A DC generator and a 12 kW GT30NZhCh 12 AC generator. Reserve DC power provided via two 28 V 45 A 15STsS-45B silver-zinc batteries housed in starboard LERX, which can be used for APU starting, if no ground power available. Reserve AC power supplied by 1.5 kW, single-phase/1 kW three-phase PTO-1000/1500M converter. Ground power connectors located on port side of fuselage.

Three separate pneumatic systems, with main system powering the wheel-brakes, canopy, fuel shut-offs and brake parachute actuator and jettison; and emergency system operating mainwheel brakes, and allowing emergency gear extension; final system pressurises hydraulic tanks and avionics bays. Oxygen system totals 16 litres (0.56 cu ft) in four bottles, each charged to 150 kg/cm² (2,133 lb/sq in); provides air/oxygen mix up to 8,000 m (26,240 ft) and pure oxygen thereafter. One bottle of the system is used to ensure reliable engine restarting or for APU start-up on the ground. Pilot's ejection seat incorporates a 0.7 litre (0.02 cu ft) KKO-15LP emergency oxygen bottle.

Air conditioning uses bleed air cooled by heat exhangers and turbocooler. Cockpit temperature maintained at 15 to 25°C (59 to 77°F). System also pressurises pilot's suit, demists screen and cools the gun bay.

Ozh-65 glycol-based liquid cooling system for radar.

AVIONICS: Integrated by NPO Elektroavtomatika with TsVM100.02.02 digital computer.

Comms: R-862 Zhooravl-30 com radio; R-855UM Komar 2M emergency radio and SPU-9 intercom. Index 9.12A version has SRZ-1P interrogator and SRO-1P transponder, forming Parol-2D IFF system, while 9.12B has SRO-2 ('Odd Rods') IFF transponder and SRZ-15 interrogator. Two SO-69 ATC transponder antennas under conformal dielectric fairings in leading-edge of each wingroot extension. ALMAZ-UP cockpit voice recorder and voice warning system. Optional additional V/UHF radio.

Radar: Phazotron RLS RP-29 (N019 Rubin) coherent pulse Doppler look-down/shoot-down engagement radar (NATO 'Slot Back'; able to track 10 targets simultaneously and engage one; search range 38 to 55 n miles; 70 to 102 km; 43 to 63 miles, depending on target size; tracking range 38 n miles; 70 km; 43 miles), target tracking limits 60° up, 38° down, 67° each side, with TS100 digital processor and integrated with OEPrNK-29 targeting/navigation complex including collimated OEPS-29 IRST/laser ranger. Twist cassegrain antenna; some sources suggest N019ME has planar array.

Flight: ARK-19 ADF, A-611 marker beacon receiver, A-037 radar altimeter, A-323 Shoran and ILS. Optional INS, Tacan, VOR/ILS and/or GPS equipment. Tester-

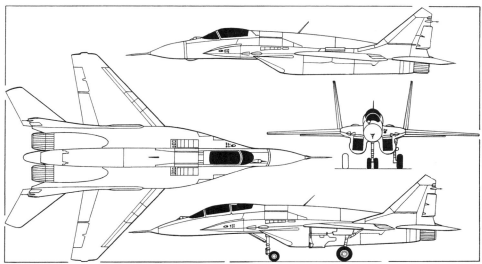

Front cockpit of MiG-29UBT prototype 08134 *(Paul Jackson/Jane's)*

MiG-29SMT, with additional side view (lower) of MiG-29UBT *(James Goulding/Jane's)*

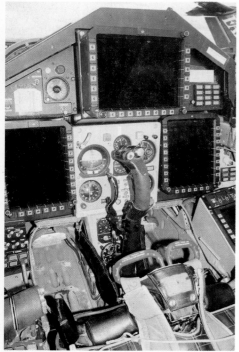

Rear cockpit of MiG-29UBT prototype 08134
(Paul Jackson/Jane's)

UZLK flight data recorder; Ekran-03M BITE. German MiG-29s to receive new Sextant navigation system with upgraded computer, displays and GPS.

Instrumentation: ILS-31 HUD and Shchel-3UM-1 helmet-mounted target designation system for off-axis aiming of air-to-air missiles. SEI-31 integrated display for radar and IRST information controlled by Ts100.02.06 digital computer.

Mission: OEPrNK-29 weapon aiming and navigation system, including NPO Geofizika KOLS (*Izdeliye* 13S) IRST and OEPS-29 electro-optical sight and laser rangefinder (fighter detection range 8 n miles; 15 km; 9.25 miles) forward of windscreen (protected by removable fairing on non-operational flights). Index 9.12 and 9.12A (and 9.13A) have E502-20 datalink, receiving guidance signals and target data from GCT and AWACS.

Self-defence: Sirena 3 SPO-15LM (L006LM Beryoza) 360° radar warning system, with sensors on wingroot extensions, wingtips and port fin. SUVP-29 passive countermeasures control unit; BVP-30-26M dispenser, with 30 PPI-26-1B 26 mm chaff or flare cartridges, in each fin root extension.

EQUIPMENT: FPK-250 taxying light on nosewheel leg; FP-8 or FP-15 landing lights on main landing gear doors.

ARMAMENT: Six close-range, IR-homing R-73 or R-73E (AA-11 'Archer') AAMs, or four R-73/73E and two medium-range radar-guided R-27R1 (AA-10A 'Alamo-A'), on three pylons under each wing; alternative air combat weapons include six close-range R-60T or R-60MK (NATO AA-8 'Aphid') infra-red AAMs, or four R-60T/MK and two R-27R1s. Able to carry 16 OFAB-100 or OFAB-120, eight FAB-250, or four FAB-500 M54 or FAB-500 M62 bombs, KMGU-2 submunitions dispensers, ZB-500 napalm tanks, B-8M1 (20 × 80 mm) rocket packs and 130 mm and 240 mm rockets in attack role. Nuclear weapons carriage by MiG-29 now prohibited by CFE Treaty, but was previously capable of carrying single 30 kT RN-40 tactical store on port inboard pylon. One 30 mm Gryazev/Shipunov GSh-301 (TKB-687/9A4071K) single barrel gun in port wingroot leading-edge extension, with 150 AO-18 rounds.

DIMENSIONS, EXTERNAL:
Wing span	11.36 m (37 ft 3¼ in)
Wing chord: at c/l	5.6 m (18 ft 4½ in)
at tip	1.27 m (4 ft 2 in)
Wing aspect ratio	3.4
Length overall: incl noseprobe	17.32 m (56 ft 10 in)
excl noseprobe	16.28 m (53 ft 5 in)
Length of fuselage, excl noseprobe	14.875 m (48 ft 9¾ in)
Height overall	4.73 m (15 ft 6¼ in)
Tailplane span	7.78 m (25 ft 6¼ in)
Wheel track	3.09 m (10 ft 1¾ in)
Wheelbase	3.645 m (11 ft 11½ in)

AREAS:
Wings, gross	38.00 m² (409.0 sq ft)

WEIGHTS AND LOADINGS (A: MiG-29, B: MiG-29S):
Operating weight empty: A	10,900 kg (24,030 lb)
Max weapon load: A	3,000 kg (6,614 lb)
Max fuel load: A (centreline tank)	4,640 kg (10,230 lb)
B (three tanks)	6,670 kg (14,705 lb)
Normal T-O weight (interceptor):	
A	15,240 kg (33,600 lb)
B	15,300 kg (33,730 lb)
Max T-O weight: A	18,500 kg (40,785 lb)
B	19,700 kg (43,430 lb)
Max wing loading: A	486.8 kg/m² (99.71 lb/sq ft)
B	518.4 kg/m² (106.18 lb/sq ft)
Max power loading: A	114 kg/kN (1.11 lb/lb st)
B	121 kg/kN (1.19 lb/lb st)

PERFORMANCE (A, B as above):
Max level speed, A, B:	
at height	M2.3 (1,320 kt; 2,445 km/h; 1,520 mph)
at S/L	M1.225 (810 kt; 1,500 km/h; 932 mph)
T-O speed: A	119 kt (220 km/h; 137 mph)
B	140-151 kt (260-280 km/h; 162-174 mph)
Acceleration: A, B at 1,000 m (3,280 ft)	
325-595 kt (600-1,100 km/h; 373-683 mph)	13.5 s
595-700 kt (1,100-1,300 km/h; 683-805 mph)	8.7 s
Approach speed: A	140 kt (260 km/h; 162 mph)
Landing speed: A	127 kt (235 km/h; 146 mph)
B	135-140 kt (250-260 km/h; 155-162 mph)
Max rate of climb at S/L: A, B	19,800 m (65,000 ft)/min
Service ceiling: A	17,000 m (55,780 ft)
B	18,000 m (59,060 ft)
T-O run: A, B with afterburning	250 m (820 ft)
B without afterburning	600-700 m (1,970-2,300 ft)
Landing run, with brake-chute:	
A, B	600-700 m (1,970-2,300 ft)
Radius of turn at 3.8 *g*:	
A at 432 kt (800 km/h; 497 mph)	350 m (1,150 ft)
A at 220 kt (408 km/h; 254 mph)	225 m (740 ft)
Range: A with max internal fuel	810 n miles (1,500 km; 932 miles)
B with max internal fuel	772 n miles (1,430 km; 888 miles)
A, B with underbelly auxiliary tank	1,133 n miles (2,100 km; 1,305 miles)
B with three external tanks	1,565 n miles (2,900 km; 1,800 miles)
g limits: above M0.85: A	+7.5/−2.15
below M0.85: A, B	+9/−2.25

UPDATED

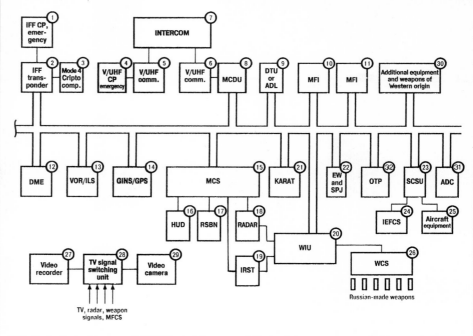

Block diagram of MiG-29SMT cockpit avionics with advanced airborne digital computer system

1* Identification friend-or-foe control panel; emergency IFF
2* IFF transponder with operating modes 1/2/3/C/S
3* Mode 4 Cripto comp
4* V/UHF radio station control panel; emergency control
5, 6 V/UHF radio
7 Intercommunication equipment
8 Multifunctional colour display unit
9 Data entry unit (DTU or ADL)
10, 11 Multifunction indicator (MFI)
12* Tactical navigation system or distance measuring equipment
13* VOR/ILS navigation system
14 Gimbal-less inertial navigation system with embedded global positioning system
15 Multimachine computer system (MCS)
16 Head-up display
17 Short-range navigation system
18 Radar
19 Electro-optical sighting system (IRST)
20 Weapons interface unit
21 Checkout and monitoring system
22 Electronic warfare and countermeasures (EW and SPJ)
23 Signal conditioning and switching unit
24 Integrated electronic flight control system
25 Aircraft equipment
26 Weapons control system
27 Video recorder
28 TV signal switching unit
29* Video camera
30* Other Western equipment
31 Air data computer system
32 Electro-optical targeting pod
* Western equipment

MiG-31

NATO reporting name: Foxhound

TYPE: Air superiority fighter.

PROGRAMME: Designed as a long-range, extended-endurance PVO interceptor to replace the Tu-128 ad MiG-25. Development began in 1967 and the S-155MP avionics complex was ordered for the Ye-155MP interceptor in 1968. First flown, as Ye-155MP (originally Type 83 MiG-25MP) '831', 16 September 1975; second prototype ('832'), with radar, first flew 22 April 1976. Two preproduction aircraft (011 and 012) built by Sokol and flown 13 July and 30 June 1977; six development aircraft (201 to 203 and 301 to 303). Full production (of about 450) started 1979; first of 11 regiments operational 1983, replacing MiG-23s and Su-15s; production has ceased at Sokol Aircraft Building Plant (formerly GAZ-21), Nizhny Novgorod, though the Sokol plant has stated its willingness to reinstate production to meet even small orders. Mikoyan is offering modernisation programme to bring in-service MiG-31s to MiG-31M levels of capability, including integration of R-37 AAM.

CURRENT VERSIONS: **MiG-31** (Type 01; 'Foxhound-A'): Two-seat, all-weather, all-altitude interceptor, able to be guided automatically, and to engage targets, under ground control. Designated MiG-31DZ or MiG-31 01DZ when fitted with AAR probe.

Detailed description applies to the above version.

MiG-31B (Type 01B *Izdeliye* 12): Second production and service variant with improved Zaslon-A radar, ECM and EW equipment and with upgraded R-33S missiles. Replaced 01/01DZ in production in late 1990. Avionics upgrade includes A-723 long-range navigation system, compatible with Loran/Omega and Chaika ground stations.

MiG-31BS (Type 01BS): Designation applied to Type 01/01DZ when converted to MiG-31B standard.

MiG-29SM demonstrator taking-off, with louvres fully open *(Paul Jackson/Jane's)*

MiG—AIRCRAFT: RUSSIAN FEDERATION

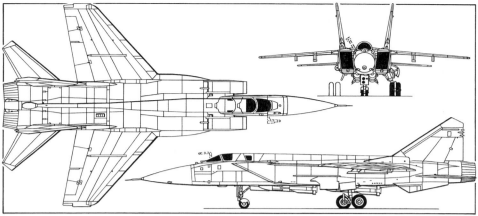

MiG-31M ('Foxhound-B') all-weather interceptor *(Mike Keep/Jane's)*

Front cockpit of MiG-31F/FE/BM demonstrator *(Yefim Gordon)* 2000/0062669

MiG-31E: Export version of basic Type 01. Prototype ('903') first noted 1997; simplified systems, no active jammer, downgraded IFF, radar and DASS. Offered to China, India and other countries.

MiG-31F: Projected multirole interceptor and fighter-bomber using a range of TV-, radar- and laser-guided ASMs. MiG-31F, FE and BM will all have improved air-to-air capability, with a radar upgrade based on technology developed for Zaslon-M of the MiG-31M. This will give a detection capability against ultra-high-speed targets (M6+), and much longer range, as well as better resolution and various new synthetic aperture and real-beam mapping modes for the variants' new air-to-ground role. New versions will also have compatibility with advanced AAMs, including R-77 (AA-12 'Adder') and long-range R-37. Revised cockpit layout includes new HUD, pilot's tactical situation display – MFI-68 152 × 203 mm (6 × 8 in) colour LCD MFD – and three similar MFDs for the navigator. Some sources also suggest upgraded MiG-31s will incorporate structural modifications to increase service life.

MiG-31BM: Designation applied to proposed defence suppression variant based on MiG-31F. Demonstrator (14306 '58'), shown August 1998 with standard R-33S AAMs under fuselage, R-77, Kh-58 and Kh-31P underwing. Possible replacement for MiG-25BM in Russian Air Forces service. Designation now applied to single-role interceptor upgrade with K-37M and K-77M AAMs and new cockpit displays, to be applied to in-service MiG-31B. Two MiG-31BMs built by Sokol, but were inactive at MiG's test airfield in late 2000, awaiting funding.

MiG-31FE: Export version of MiG-31BM or MiG-31F. Alternative designation **MiG-31MF**.

MiG-31LL: Ejection seat testbed (Red '79') used by LII.

MiG-31M (Type 05; possibly 'Foxhound-B'): Improved interceptor, under development since 1984; first prototype ('051') first flew 21 December 1985; first shown publicly February 1992; upgraded engines, with modified nozzles; one-piece rounded windscreen; small side windows only for rear cockpit (though dual controls are fitted); wider and deeper dorsal spine, containing 300 litres (79.3 US gallons; 66.0 Imp gallons) of additional fuel; more rounded wingtips, with flush dielectric areas at front and rear; taller fins with larger, curved root extensions; modified and extended wingroot leading-edge extensions; smaller wing upper-surface fences; all systems upgraded; digital flight controls; multifunction CRT cockpit displays; new multimode Phazotron Zaslon-M phased-array radar, with 1.40 m (55 in) diameter antenna, in 3° 30′ downward-inclined nose; detection range of 360 km (224 miles); retractable flight refuelling probe transferred to starboard side of nose; non-retractable pod with collimated IRST and laser ranger.

No gun; number of fuselage weapon stations increased to six, by addition of two centreline stations with R-37 AAMs in addition to side-mounted R-37s or older R-33s; four new-type underwing pylons for R-77 (AA-12 'Adder') active radar-guided AAMs. First prototype, produced by conversion of MiG-31B '503' lost on 9 August 1991; five or six more prototypes (051 to 057), at least one (057) with cylindrical wingtip ECM/ECCM jammer pods carrying upper and lower winglets. Maximum T-O weight 52,000 kg (114,640 lb), with increased-thrust D-30F6M engines to compensate. Internal fuel 16,350 kg (36,045 lb).

Two **MiG-31D** (Type 07) dedicated anti-satellite models (numbered 071 and 072) were produced and flight-tested in 1986, with ballast instead of radar in nose, a flat fuselage undersurface without recesses, large winglets above and below wingtips, and underwing Vympel ASAT missiles.

MiG-31S: Commercial small satellite launch variant, with Fakel OKB Micron missile capable of delivering a 100 kg (220 lb) payload into a 200 km (124 mile) orbit or a 70 kg (154 lb) payload into a 500 km (311 mile) orbit. The type could also launch the Aerospace Rally System rocket-powered suborbital glider, for astronaut training, upper atmosphere research or space tourism.

MiG-31Eh: *Ehksport* (export) version for China, proposed 2000, and announced at Zhuhai Air Show 6 November 2000.

CUSTOMERS: Russian Federation and Associated States (CIS) Air Forces, including 320 with Russian air defence forces and tactical units; 30 with Kazakhstan. First production aircraft to training unit at Svostleyka; initial operational regiment at Pravdinsk. Others reportedly ordered by China; persistent reports of interest from Iran and Syria.

MiG-31 CURRENT OPERATORS

Unit	Base
Russia	
High Command Air Defence Forces	
148th Combat Training and Conversion Centre	
786th IAP	Pravdinsk
Moscow Air Defence Region	
153rd IAP	Morzhansk
790th IAP	Khotilovo
5th Independent Air Defence Corps	
764th IAP	Bolshoi-Savino
6th Independent Air Defence Corps	
174th GvIAP	Monchegorsk
180th GvIAP	Gromovo
458th IAP	Kotlas
11th Independent Air Defence Corps	
Unidentified IAP	Dolinsk Sokol (Sakhalin)
Unidentified IAP	Yelizovo (Petropavlovsk)
12th Independent Air Defence Corps	
83rd IAP	Rostov-na-Donu
Unidentified Independent Air Defence Corps	
Unidentified IAP	Bratsk (Novosibirsk)
Kazakhstan	
356th IAP	Zhana-Semey (Semipalatinsk)

Notes: IAP is *Istrebitel'nyi Aviatsionny Polk* (Fighter Aviation Regiment)
All units have a small number of MiG-25Us for training
Air Defence Corps are in process of being incorporated into Air Armies and integrated with Regiments formerly assigned to Frontal Aviation

DESIGN FEATURES: Initiated to counter threat of USAF B-52 bombers carrying ALCMs. Basic MiG-25 configuration retained, but very different aircraft, with two seats; strengthened to permit supersonic flight at low altitude; more powerful engines than MiG-25; major requirement increased range, not speed; advanced digital avionics; Zaslon radar was first electronically scanned phased-array type to enter service, enabling MiG-31 to track 10 targets and engage four simultaneously, including targets below and behind its own location; fuselage weapon mountings added.

Wing anhedral 4° from roots; sweepback approximately 41° on leading-edge, 32° at quarter-chord, with small, sharply swept wingroot extensions; all-swept tail surfaces, with twin outward-canted fins and anhedral on horizontal surfaces.

FLYING CONTROLS: Large-span ailerons and flaps; leading-edge slats in four sections on each wing; all-moving horizontal tail surfaces; inset rudders.

MiG-31E prototype, displayed at Moscow in 1999 *(Paul Jackson/Jane's)* 2000/0062733

R-33 (AA-9 'Amos') AAMs beneath the fuselage of the MiG-31F/FE/BM demonstrator
(Paul Jackson/Jane's) 2000/0062732

MiG-31B testbed '368' at Akhtubinsk trials base *(Yefim Gordon)* 2001/0100400

MiG-31F/FE/BM demonstrator '58' pictured shortly after completion *(Yefim Gordon)* 2001/0100399

STRUCTURE: Airframe 49 per cent arc-welded nickel steel, 16 per cent titanium, 33 per cent light alloy; 2 per cent composites, including radome; three-spar wings; no wingtip fairings or mountings; small forward-hinged airbrake under front of each intake trunk; undersurface of centre-fuselage not dished between engine ducts like MiG-25; much enlarged air intakes; jet nozzles extended rearward; shallow fairing extends forward from base of each fin leading-edge; fence above each wing in line with stores pylon.

LANDING GEAR: Retractable tricycle type; offset tandem twin KT-175 wheels with 950×300 tyres on each main unit, retracting forward into air intake trunk, facilitate operation from unprepared ground and gravel; rearward retracting twin nosewheel unit with mudguard and with twin KT-176 wheels with 660×200 tyres.

POWER PLANT: Two Aviadvigatel D-30F6 turbofans, each 93.1 kN (20,930 lb st) dry, 151.9 kN (34,170 lb st) with afterburning; internal fuel capacity 19,940 litres (5,268 US gallons; 4,386 Imp gallons) in seven fuselage tanks, four wing tanks and two fin tanks. Provision for two underwing tanks, each 2,500 litres (660 US gallons; 550 Imp gallons); semi-retractable flight refuelling probe on port side of front fuselage.

ACCOMMODATION: Pilot and weapon systems operator in tandem under individual rearward-hinged canopies; rear canopy has only limited side glazing and blends into shallow dorsal spine fairing which extends to forward edge of jet nozzles. Pilot's ejection seat has heated backrest to allow extended ground alert.

AVIONICS: *Comms:* R-862 UHF, R-864 HF, P-591 voice warning system, SPU-9 intercom; SRO-2P IFF transmitter and SRZ-2P receiver; SO-69 transponder.
Radar: NIIP N007 S-800 SBI-16 (RP-31) Zaslon or Zaslon-A electronically scanned phased-array fire-control radar (NATO 'Flash Dance') in nose; search range of 108 n miles (200 km; 124 miles) in clutter-free forward sector; range in rear sector 48 n miles (90 km; 56 miles); capable of tracking 10 targets and attacking four simultaneously.
Flight: A312 Radikal-NP or A-331 Shoran, A-723 Kvitok-2 Loran. Marshrut long-range and Tropik medium-range nav systems. ARK-19 radio compass, RV-15 radar altimeter, RPM-76 marker beacon receiver.
Mission: In four-aircraft group interception mission, only lead MiG-31 is linked to AK-RLDN automatic guidance network on ground; other three MiG-31s have APD-518 digital datalink to lead aircraft, permitting line-abreast radar sweep of zone 430 to 485 n miles (800 to 900 km; 495 to 560 miles) wide by 140° sector scanning angles. Semi-retractable Type 8TP IR search/track sensor under cockpit; tactical situation display. BAN-75 command link; APD-518 digital air-to-air datalink; Raduga-Bort-MB5U15K air-to-ground tactical datalink; SPO-155L RHAWS; Argon-15 digital computer.

EQUIPMENT: Active IR and electronic countermeasures including UV-3A flare dispensers.

ARMAMENT: Four R-33 or R-33S (NATO AA-9 'Amos') semi-active radar homing long-range AAMs in pairs on AKU ejector pylons under fuselage, plus two R-40T (AA-6 'Acrid') medium-range IR AAMs on inner underwing pylons; four IR R-60 (K-60, AA-8 'Aphid') AAMs on outer underwing pylons, in pairs. R-33s are semi-recessed in fuselage. GSh-6-23M six-barrel Gatling-type 23 mm gun inside fairing on starboard side of lower fuselage, adjacent main landing gear, with 260 linkless rounds.

DIMENSIONS, EXTERNAL:
Wing span	13.465 m (44 ft 2 in)
Wing aspect ratio	2.9
Length overall	22.69 m (74 ft 5¼ in)
Height overall	6.15 m (20 ft 2¼ in)
Tailplane span	8.80 m (28 ft 10½ in)
Wheel track	3.90 m (12 ft 9½ in)
Wheelbase	7.40 m (24 ft 3¼ in)

AREAS:
Wings, gross	61.60 m² (663.1 sq ft)

WEIGHTS AND LOADINGS:
Weight empty	21,820 kg (48,105 lb)
Internal fuel	15,500 kg (34,170 lb)
Max T-O weight:	
with max internal fuel	41,000 kg (90,390 lb)
with max internal fuel and two underwing tanks	46,200 kg (101,850 lb)
Max wing loading	750.0 kg/m² (153.61 lb/sq ft)
Max power loading	152 kg/kN (1.49 lb/lb st)

PERFORMANCE:
Max permitted Mach No. at height	2.83
Max level speed: at 17,500 m (57,400 ft)	1,620 kt (3,000 km/h; 1,865 mph)
at S/L	810 kt (1,500 km/h; 932 mph)
Max cruising speed at height	M2.35
Econ cruising speed	M0.85
Landing speed	141 kt (260 km/h; 162 mph)
Time to 10,000 m (32,810 ft)	3 min
Service ceiling	20,600 m (67,600 ft)
T-O run at max T-O weight	1,200 m (3,940 ft)
Landing run	800 m (2,625 ft)
Radius of action with max internal fuel and four R-33 missiles:	
at M2.35	388 n miles (720 km; 447 miles)
at M0.85	647 n miles (1,200 km; 745 miles)
at M0.85 with two underwing tanks	782 n miles (1,450 km; 901 miles)
at M0.85 with two underwing tanks and one flight refuelling	1,185 n miles (2,200 km; 1,365 miles)
Ferry range, max internal and external fuel, no missiles	1,780 n miles (3,300 km; 2,050 miles)
Max endurance with underwing tanks:	
unrefuelled	3 h 36 min
refuelled in flight	6-7 h
g limit	+5

UPDATED

MiG-35

TYPE: Multirole fighter.

PROGRAMME: MiG-35 designation has been applied to an evolving configuration based on the 9.15 MiG-29M. Work on 9.25 began in 1986-88, adding uprated engines (in 98 kN; 22,000 lb class) and with fuselage stretched by moving engines aft to make space for a new internal fuel tank. Improvements to be added in three phases: MiG-29M1 featured fuselage stretch, giving 20 per cent increase in internal fuel capacity (to 7,000 litres; 1,849 US gallons; 1,540 Imp gallons); MiG-29M2 introduced canard foreplanes and increased wing area, plus enlarged integral tanks; MiG-29M3 added the new einges, new avionics and a further increase in internal fuel tankage (to 8,000 litres; 2,113 US gallons; 1,760 Imp gallons).

Izdelie 9.35 was revealed May 1996 as derivative of *Izdelie* 9.25, with SMT-type increased fuel tankage, inflight refuelling capability, square-cut wingtips and new avionics, but without canards. Provision was to be made for the later incorporation of axisymmetric, two-dimensional thrust vectoring and a Zhuk-F phased-array radar, capable of tracking 24 targets and engaging up to eight. Trainer derivative (with radar) also proposed.

Developed with government funding; first flight due 1997, but failed to take place; to compete against Sukhoi Su-35; almost certainly overtaken by thrust vectoring version of MiG-29SMT and variants thereof – possibly with same new wing, engines, fuselage stretch and radar, but not with -29M-based fuselage structure.

DESIGN FEATURES: Compared with MiG-29ME/MiG-33, the -35 has a new wing of increased root chord and decreased tip chord derived from the MiG-29K; no overwing engine air intake louvres; increased thrust developments of RD-33 engine relocated 0.92 m (3 ft 0¼ in) rearwards to provide space for an additional 1,500 kg (3,307 lb) of fuel, doubling MiG-29 range. By 1996, Zhuk-F had given way to Phazotron RP-35 electronically scanned phased-array radar capable of tracking 24 and engaging four targets simultaneously over a range of 76 n miles (140 km; 87 miles); 10 (increased from eight) weapon stations; and RVV-AE (R-77; AA-12 'Adder') AAMs. Provision in design for foreplanes, though felt to be unnecessary with thrust-vectoring.

FLYING CONTROLS: Fly-by-wire.

POWER PLANT: Initially, two Klimov RD-133 turbofans, each 83.0 kN (18,660 lb st), with thrust-vectoring nozzles and afterburning; later, Klimov RD-333 turbofans, each 98.1 kN (22,050 lb st). The RD-333 is a new engine, but is installationally compatible with the RD-33.

UPDATED

MiG 1.42 (1.44) MFI

TYPE: Multirole fighter.

PROGRAMME: Developed to meet 1983 Soviet I-90 requirement for a fighter for the 1990s (*Istrebitel*-90), under the designation MFI (*mnogofunktsionalnyy frontovoy istrebitel*: multifunctional front-line fighter), as replacement for MiG-29 and Su-27 in Frontal Aviation service. Mikoyan completed preliminary design studies for separate MFI and LFI (heavy and light Frontal fighters) in 1985, and was authorised to proceed in 1986. The OKB received official Air Forces and PVO performance requirements in 1988 and began work on what was planned

Ceramic-coated nozzle petals of the 1.44's AL-41F engines *(Yefim Gordon)* 2000/0062667

MiG—AIRCRAFT: RUSSIAN FEDERATION

MiG 1.44 taxying for a test flight *(Yefim Gordon)*

to become the 1.42 (basic design possibly *Izdelye* 1-4, with selected configuration becoming 1.42) in 1989 under direction of Grigory Sedov, and later (from 1997) Yuri Vorotnikov. The original LFI was first given a lower priority and then suspended in 1988. Final MFI configuration adopted, 1991. One alternative early configuration (possibly 1.41) with upright, closely spaced fins, described in 1997-98 *Jane's*.

MiG 1.42 was finally selected as the Russian Air Forces' fifth-generation fighter after rejection of a competing design from Yakovlev. This was reportedly close to the MiG aircraft in appearance in some respects, though with a kinked (forward/aft/forward) swept wing; F-22-style chined nose; F-22-type canopy; and more sharply-canted trapezoidal fins; also had single engine, with thrust vectoring only in pitch axis. Other configurations featured inward- and outward-canted twin fins; some had foreplanes, but most had a compound delta wing. One of the first OKB drawings showed an aircraft almost identical to 1.44, but lacking foreplanes.

First prototype is aerodynamic/performance/structural stealth proof-of-concept aircraft, possibly one of two flight-standard aircraft built with no mission systems under the designation 1.44; latter originally thought to have been alternative competing configuration. MiG-1.44 was also intended to evaluate airframe strength, and to test integrated flight/power plant control system, and it was envisaged that it might later be fitted with mission avionics and weapons. The OKB also built a dynamic load test airframe and a static test airframe, together with a forward fuselage for ejection seat and other tests. All early airframes jointly built by ANPK workshops and Nizhny Novgorod 'Sokol' plant. First airworthy prototype (Blue 01) completed in early 1991, but first flight delayed by non-availability of flight-cleared engines.

Early reports suggested that 1.42 would be powered by two 181.4 kN (40,785 lb st) Soyuz/Kovchenko R-79M engines derived from the thrust-vectoring power plant of the Yak-141. Some then speculated that the MiG 1.44 would be fitted with two Aviadvigatel D-30F6 afterburning turbofans from MiG-31, probably rated at 93.2 kN (20,943 lb st) dry, 152.0 kN (34,170 lb st) with afterburning, or possibly higher-thrust D-30F6M, as in MiG-31M. Apparently fixed convergent/divergent nozzles. This may have been the original plan, when a first flight in 1994 seemed a realistic expectation. When finally rolled out, the 1.44 was powered by Saturn/Lyulka AL-41F engines.

The Rybinsk-built AL-41 was originally expected to be fitted with rectangular thrust-vectoring nozzles, like the F-15S/MTD, but was probably redesigned with more conventional circular section (but still three-axis thrust-vectoring) nozzles. The inner afterburner nozzle petals are coated with a heat-resistant ceramic material, giving them a characteristic pale cream colour. It has been asserted that 27 AL-41s have been completed, and that the new engine has been tested under a Tupolev Tu-16LL testbed and in one of the engine bays of a converted MiG-25. The same sources claim that the engine has been flown at speeds of up to 1,080 kt (2,000 km/h; 1,243 mph) and at heights up to 20,000 m (65,620 ft).

Aircraft transferred from OKB's Moscow workshops to OKB flight test facility at Zhukovsky in mid-1994. Initial reports that aircraft was 'in 30 tonne class' superseded by description as '35 tonne class fighter', possibly reflecting unplanned weight escalation. Fast taxi trials undertaken late 1994, but first flight delayed by non-delivery of flight control actuators, according to some sources. Others suggest that an attempted take-off was aborted at last moment by project pilot (OKB CTP Roman Taskaev) December 1994, possibly because installed actuators insufficiently powerful for large foreplanes.

May 1995 integration of OKB and MAPO plant led to withdrawal of funding for new programmes (MiG-29M and MiG 1.42) in favour of sales of minimum-change MiG-29 variants. Aircraft then stored at Zhukovsky. OKB attempts to reveal aircraft in static display at Moscow Air Show banned at late stage in both 1995 and 1997, and publicity material prepared but withheld (and destroyed) each time, though aircraft was shown to VIPs, including Premier Victor Chernomyrdin on 24 August 1997. Emergence of the Sukhoi S-37 led to efforts to revive the 1.44 as 'a matter of honour' and employ it as a technology demonstrator for ANPK MiG and for other future fighter programmes. In February 1998, Mikoyan officials predicted a first flight in August 1998, and stated that the Russian Air Forces would provide the Rb15 million (US$2.5 million) needed to return the aircraft to airworthy status. Funding did not materialise, and new first flight date was missed. First photographs released in December 1998; press allowed limited access to '01' at Zhukovsky on 12 January 1999 when it taxied under its own power. Public unveiling of the 1.44 was intended to generate the funding required to install missing flight control surface actuators. First flight date slipped from March to September, but neither was achieved, aircraft eventually taking to the air on 29 February 2000. The aircraft attained an altitude of 20,000 m (65,620 ft) on only its second flight, on 27 April 2000.

Despite its high price, some sources see a possibility that the 1.42/1.44 could be selected to form the basis of a new fighter, and in April 1998, senior air force officers were discussing a competitive evaluation of the S-37 and 1.44. The OKB claims 'foreign interest' in investing in a production programme.

CURRENT VERSIONS: **MiG 1.42:** Planned production version generally similar to 1.44 but to have cranked-delta wing and internal weapons bay, plus RAM coatings and other operational features.

MiG 1.44: Proof-of-concept demonstrator/prototype, initially lacking most operational systems and equipment; simple delta wing and with four side-by-side shallow troughs below the fuselage, possibly aerodynamic representations of planned semi-conformal missile carriage stations.

COSTS: Unit cost of production version estimated at US$70 million (1999)

DESIGN FEATURES: Next-generation fighter; intended counterpart to Eurofighter Typhoon, Dassault Rafale, Lockheed Martin F-22 Raptor and similar combat aircraft. Twin-fin delta canard, with large movable foreplanes and very widely spaced outward-canted twin tailfins. Foreplanes incorporate distinctive 'dogtooth' at root and may have (or have provision for) an articulated trailing-edge. Root fairings which support the canards have an exceptionally sharp leading-edge (presumably for vortex generation), whereas the foreplanes themselves have a more conventional rounded leading-edge profile. There is also a slight step between the wingroot fairing and the canard, giving a pronounced 'double dogtooth', reportedly because the 1.44 is fitted with (larger) foreplanes originally designed for the 1.42. Tailfins are carried on slender oval-section booms extending from trailing-edge of wing, well outboard from fuselage. These continue aft of the fin trailing-edge, culminating in oval-section 'squashed-conical' dielectric fairings. Shallow, long-chord ventral fins are carried below these vestigial booms.

The wing of the production version may incorporate a small change of leading-edge sweep at about one-third span, broadly level with the fin root and the inboard edge of the leading-edge flaps, though the 1.44 appears to have an unbroken leading-edge. The 1.44 wing does not have the bold double-delta planform suggested by intelligence sources, confirmed by spin-test model makers or shown in all drawings of 'future fighter programmes' released by MAPO. Leading-edge sweepback is approximately 48°. Fuselage is of oval section at the nose, with a canopy and windscreen reminiscent of that of the MiG-29, in which the pilot sits quite low. This reflects the importance of speed, acceleration and RCS, and the lower priority accorded to pilot visibility. However, it is alleged that any production aircraft would have Keldysh Research Centre's new 'plasma cloud' stealth system, which is said to reduce RCS by dissipating electromagnetic waves.

The straight-edged engine intake is sharply raked, and is slung below the mid-fuselage. Ram air for avionics bay cooling enters scoops above the air intake lip, passing through heat exchangers and exhausting through vents above the foreplane root, aft of the cockpit. The fuselage is largely flat-bottomed and has a prominent spine, like that of the MiG-31.

FLYING CONTROLS: KSU-142 quadruplex digital fly-by-wire flight control system, possibly derived from BTsK-29 system flown on MiG-29E testbed, which had similar twin nose pitot probes. Claimed 16 control surfaces — presumably including the visible twin rudders, four trailing-edge flaperons/elevons, two inboard trailing-edge flaps, four leading-edge flaps, and two foreplanes. Remaining two perhaps accounted for by two lower intake cowl flaps; some sources suggest that ventral fins have trailing-edge rudders, but detailed examination provides no evidence of this.

STRUCTURE: All-composites wing; high proportion of composites and welded aluminium-lithium alloy in remainder of airframe. Extensive use of radar-absorbent coatings.

LANDING GEAR: Tricycle type. Rearward-retracting nose oleo with twin trailing-link nosewheels. Longitudinally split door on each side of nosewheel bay. Nosewheel tyre size 620×180. Main oleos retract forward into sides of intake trunks. Single mainwheels on trailing oleos. Mainwheel

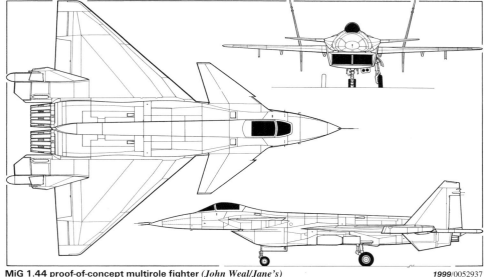

MiG 1.44 proof-of-concept multirole fighter *(John Weal/Jane's)*

tyre size 1,030×350. Mudguards not fitted; brake parachute compartment on 1.44 prototype at rear of dorsal spine.
POWER PLANT: Two Saturn/Lyulka AL-41F engines, each rated at approximately 175 kN (39,350 lb st) with afterburning. Planned derivative engine to be rated at 245 kN (55,077 lb st). Performance figures relate to AL-41F-powered 1.42/1.44.
ACCOMMODATION: Pilot, only, on Zvezda K-36 zero/zero ejection seat (to be replaced by new, variable geometry seat); sharply inclined one-piece, frameless, fixed-curved windscreen and rearward-opening transparent blister canopy. Similar to MiG-29 canopy and windscreen.
AVIONICS: Production aircraft planned to have integrated avionics system based on Phazotron N014 Zhuk-RN radar with active phased-array antenna. Early reports suggested that N014 was related to Zaslon-M used by MiG-31M. ARK-series radio compass with antenna embedded in canopy.
ARMAMENT: Outboard underwing pylons (without launch rails) fitted to prototype. Probable provision for at least one pair of hardpoints further inboard. Aircraft claimed to have internal weapons bay (or bays) for low drag, reduced RCS carriage of unknown number of AAMs, probably folding-fin version of R-77 (AA-12 'Adder') but simple troughs only on prototype. Some sources imply that production aircraft will have tandem bays, each big enough for one folding-fin R-37; others suggest more shallow bays, each capable of carrying single or side-by-side R-77s. Possible provision in production aircraft for 30 mm (probably Gryazev-Shipunov GSh-30-1) cannon, but no evidence of this on 1.44.

DIMENSIONS, EXTERNAL (estimated):
Wing span	17.03 m (55 ft 10½ in)
Length overall	22.83 m (74 ft 10¾ in)
Foreplane span	8.80 m (28 ft 10½ in)
Height overall	5.72 m (18 ft 9¼ in)
Wheel track	3.77 m (12 ft 4½ in)
Wheelbase	6.48 m (21 ft 3 in)

AREAS:
Wings, gross	90.5 m² (974 sq ft)
Foreplanes (each)	5.35 m² (58 sq ft)

WEIGHTS AND LOADINGS:
T-O weight: normal	30,000 kg (66,138 lb)
max	35,000 kg (77,161 lb)
Max wing loading:	
normal T-O weight	309 kg/m² (63 lb/sq ft)
max T-O weight	387 kg/m² (79 lb/sq ft)
Max power loading:	
normal T-O weight	80 kg/kN (0.78 lb/lb st)
max T-O weight	100 kg/kN (0.98 lb/lb st)

PERFORMANCE (estimated):
Max permitted Mach No. at high altitude: clean	2.6
with underwing missiles	2.35
Max sustained Mach No. at max dry power ('supercruise')	1.6-1.8
Max level speed:	
at high altitude	1,350 kt (2,500 km/h; 1,553 mph)
at sea level	810 kt (1,500 km/h; 932 mph)
in supercruise	918 kt (1,700 km/h; 1,056 mph)
Service ceiling	20,000 m (65,620 ft)
Range	2,429 n miles (4,500 km; 2,796 miles)

UPDATED

MiG-AT

TYPE: Advanced jet trainer/light attack jet.
PROGRAMME: Design started late 1980s; selected in May 1992 as one of two finalists (with Yak-130, following rejection of Sukhoi S-54 and Myasishchev contenders) in Russian competition to replace Aero L-29 and L-39 Albatros;

Second MiG-AT, with original air intakes *(Yefim Gordon)*

originally planned (as MiG-ATTA, a designation also applied to abortive joint venture with Promavia) with T tail and R-35 (DV-2) engines, with wing by Daewoo, landing gear by Messier-Bugatti, avionics by Sextant, Thomson-CSF (now Thales) and SFENA. Under October 1992 and subsequent agreements, prototypes and preseries aircraft are powered by an initial five Larzac 04R20 engines supplied by SNECMA of France; production aircraft for domestic market may eventually have Larzacs licence-built by Chernyshev of Moscow, though there is a programme to design a new Russian engine – the 16.7 kN (3,748 lb st) Aviadvigatel (Soyuz/CIAM) RD-1700 – which is stated to be 2.5 times cheaper than the Larzac and which may remedy early shortfalls in performance, ceiling and range. Flight test of MiG-AT with RD-1700 could be undertaken in 2002, if funded. Suggested that MiG-AT has bureau Type number 08, perhaps with original T-tail configuration being 8.1 and later version as 8.2, prototype configuration being 8.21.

Prototype rolled out 18 May 1995; high-speed taxi trials at Zhukovsky August 1995; first 5 minute hop 16 March 1996; first flight 21 March 1996, piloted by Roman Taskaev; three aircraft being used for flight test programme; two static test airframes; production of first series of 16 at MAPO plant well advanced (probably reduced to 12), with seven in final assembly by December 1998. Subsequent progress delayed by funding difficulties; no sign of emergence of production ATs by mid-1999. Russian requirement for 200 to 250 trainers in this category. Request for funds for first 10 aircraft made by Russian Air Forces, late 1996, at which time 10 engines ordered from Turbomeca-SNECMA. Further 20 engines ordered late 1997 for FFr230 million (US$39 million), while Sextant contracted to supply 30 nav/attack systems. Actively promoted in India and South Africa; may have been eliminated from Indian competition due to renewed IAF emphasis on timescale, though BAE Hawk's reported selection may yet be reversed. Beaten by BAE Hawk in South Africa. MiG-AT aircraft now being marketed as one element in AT System, incorporating simulators, computerised classrooms, part-task simulators, and video training packages, all using unified software.

CURRENT VERSIONS (general): **MiG-AT**: Basic trainer; suitable for combat use of unguided weapons against ground and sea targets, and air-to-air missiles in conjunction with helmet-mounted target designation system. Pylons for three weapons/stores in trainer role, seven for combat training. Version for domestic use (ATR) has Russian-built engines and avionics; export version (ATF) has Sextant Topflight avionics (French).
MiG-UTS (MiG-ATR): Planned Russian Air Forces' version, with Russian avionics and RD-1700 engines.
MiG-ATS: Combat trainer, with increased capabilities and underwing hardpoints able to launch guided missiles, using guidance equipment pod. Second flying prototype (823 '83') to this standard and has Russian avionics.
MiG-AS: Single-seat light tactical fighter version with built-in gun and radar for all-weather use of AAMs and ASMs; intended to compete in international market with BAE Hawk 200. Single-seater, aerodynamically similar to original MiG-AT, with rear cockpit covered by metal fairing of similar outline to original canopy.
MiG-AP: Patrol coastguard fishery protection law enforcement and SAR co-ordination aircraft; based on MiG-AS; nose-mounted search radar.

All versions can be modified for deck operation on aircraft carriers at customer's request as **MiG-ATK** (French avionics), **MiG-ATSK** (armed trainer), or **MiG-ASK** (single seater). Span with optional wing folding 6.50 m (21 ft 4 in). Provision for deck arrester hook and catapult launching.

MiG-AT can undertake the following training missions, all with take-off at 50 per cent fuel (plus 150 kg; 331 lb reserves):
Basic flying training: Between 30 and 35 flight manoeuvres, three roller landings and straight-in final landing at up to 27 n miles (50 km; 31 miles) from base, altitude up to 6,000 m (19,680 ft).
Ground attack: Four diving attacks, three advanced manoeuvre attacks and circuit/landing against visual target up to 43 n miles (80 km; 50 miles) from base, transiting at up to 2,500 m (8,200 ft).
Interception/combat manoeuvring: Loiter, directed interception, attack and close combat, day VMC/IMC at up to 140 n miles (260 km; 161 miles) from base, altitudes up to 8,000 m (26,240 ft).

CURRENT VERSIONS (specific): **AT-1/821 '81'**: Prototype; first flight 16 March 1996; first 'official' flight 21 March 1996. Has been modified since its initial flights, briefly gaining in overwing fence level with the leading-edge discontinuity. This since removed, but intakes have been redesigned and now project forward of the wing leading-edge; now have oval cross-section and blunter lips, rather than appearing to be sharp, inverted Us from the front. New intake configuration is understood to have first flown during May 1999; shown at MAKS '99, Moscow, August 1999. Rear cockpit now filled with test instrumentation.
AT-3/823 '83': Second flying prototype shown (unflown) at MAKS '97, Moscow, August 1997; nominally representative of intended UTS in some respects, and of ATS armed trainer configuration in others. Fitted with centreline and underwing hardpoints; Russian avionics and displays. Original intake configuration with overwing fences.
AT-4/824 '84': Third flying prototype; under construction in 1999. Will feature new Sextant avionics suite (full Topflight system, including compatibility with helmet-mounted displays); scheduled to fly during 2000, but apparently did not do so.
COSTS: US$200 million (2000).
DESIGN FEATURES: Intended to have manoeuvrability comparable with front-line combat aircraft, and service life of 15,000 flying hours or 30 years, with not fewer than 25,000 landings, maximum angle of attack of 25°, and sustained 5.4 g turn in 4 km (2.5 mile) radius at M0.7. Able to fly OEI at any stage, including take-off. Reconfigurable FBW FCS to simulate both the handling and limits of other front-line combat types. Onboard simulation of manoeuvring target, meteorological conditions and system failures via HUD, as well as specialised training for all operational modes of individual types of modern combat aircraft of Russian or foreign manufacture.

Conventional low-wing monoplane; wing-root leading-edges swept-forward, with engine air intakes initially overwing to minimise risk of FOD; sweptback vertical tail surfaces; unswept tailplane; tailcone comprises two front-hinged door-type airbrakes.

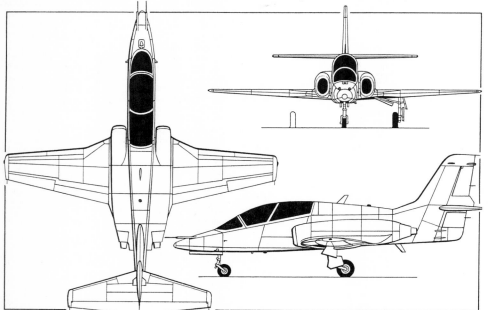

MiG-AT advanced trainer following modification of engine air intakes *(Mike Keep/Jane's)*

MiG—AIRCRAFT: RUSSIAN FEDERATION

Front cockpit of second prototype MiG-AT showing Russian avionics *(Paul Jackson/Jane's)*

Prototype MiG-AT in August 1999, following rebuild with new engine air intakes *(Paul Jackson/Jane's)*

FLYING CONTROLS: Avionika KSU-821 integrated hybrid (digital/analogue) multichannel fly-by-wire flight control system with multiple redundant air data computer. System incorporates autopilot, autothrottle and automatic stall-protection systems. Three-axis controls; three leading-edge slat sections and single-section slotted trailing-edge flaps on each wing. Split two-section rudder above and below tailplane. Hydraulically operated split airbrake beneath lower section of rudder.

STRUCTURE: One-piece, three-panel tapered wing of aluminium alloy, with some honeycomb skin. Three-section fuselage of aluminium alloy, but reportedly with 40 per cent of skin in CFRP/GFRP, including access panels on spine, three-piece engine covers, nose avionics bay cover and parts of wing/nacelle and nacelle/fuselage fairings. CFRP intake ducts and CFRP honeycomb wing trailing-edge control surfaces (elevators and flaps) and fin box, rudders, tailplane and elevators. Fin leading-edge of aluminium alloy and integral with rear fuselage. Titanium alloy rear fuselage forming 'open-box' with sidewalls acting as engine bay fire protection. Split airbrakes and ejection seat mounting tracks also of titanium with intake lips in stainless steel.

LANDING GEAR: Retractable tricycle type; single wheel on each trailing-link unit; wide-track main units retract inward, nosewheel forward; mainwheel tyre size 660×200, pressure 8.80 bars (128 lb/sq in); nosewheel tyre size 500×150, pressure 4.90 bars (71 lb/sq in); high-efficiency brakes; operation practicable from unpaved surfaces of bearing ratio 6 kg/cm² (85.3 lb/sq in).

POWER PLANT: Two Turbomeca-SNECMA Larzac 04-R20 turbofans, each 14.12 kN (3,175 lb st), mounted above wingroots. MRT-931 FADEC controls. Ratings can be reduced by 30 to 40 per cent for primary training. Change to new Chernyshev-built Soyuz RD-1700 originally planned by 2002. One main fuselage fuel tank and one (three-section) tank in each inner wing. Total capacity 2,390 litres (632 US gallons, 527 Imp gallons). Single-point pressure refuelling point in lower rear fuselage adjacent to wing trailing-edge, with five gravity refuelling points (one above the fuselage tank and two above each wing). Tanks pressurised with engine bleed air to ensure high-altitude supply.

ACCOMMODATION: Two crew in tandem, on Zvezda K-93 zero/zero ejection seats operable at up to 485 kt (900 km/h; 559 mph) and M1.5 at altitudes up to 15,000 m (49,215 ft); minimum inverted flight ejection height 50 m (165 ft). Rear seat raised by 400 mm (15.75 in) to improve occupant's forward view. View over the nose −17° from cockpit, −7° from the rear. One-piece birdproof canopy, hinged on starboard side, with 'lock the canopy' warning. Provision for blinds for IFR flight training. Some sources suggest provision of 'built-in ladders', but no evidence for any more than retractable steps.

SYSTEMS: Dual hydraulic systems, driven by independent Messier-Bugatti PR70660-20 engine-driven pumps, actuate control surfaces, landing gear and wheel doors. General system supplied by starboard engine actuates one chamber of each electrohydraulic control actuator and hydraulic actuators of other systems; booster system supplied by port engine actuates other chamber of control actuators. Auxiliary booster system operated by NS-65 emergency hand pump. Electrical system supplies primary 27 V DC and 200/115 V 400 Hz three-phase secondary AC. Two 9 kW Auxilec 8044-31 DC starter/generators with Auxilec WDII-014 control and protection unit; two AC static inverters; 15CTSS-45B lead-zinc batteries.

AVIONICS: Integrated by Russian GoSNIIAS avionics research institute and Sextant Avionique, France. Multifunctional central computer, with all data integrated through MIL-STD-1553B equivalent databus.

Comms: ERA 2000 VHF/UHF transceiver, SG100 IFF, SPU-821 intercom, ALMAZ-UBS voice recorder.

Flight: Automatic control system; UMT 33 air data unit; Totem 3000 laser-gyro INS; NSS 100S-1 GPS. NR 510A 101 VOR/ILS receiver/marker; NC-12B Tacan; EVS 915 video recorder; Tester U3A flight data recorder.

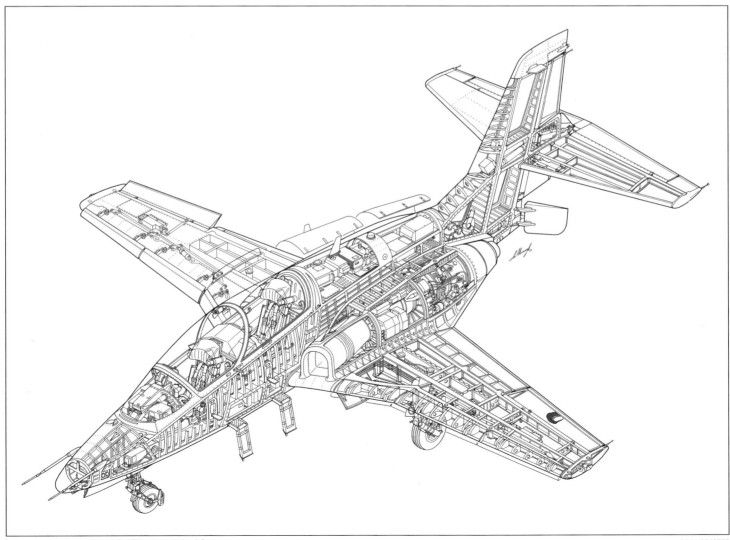

Internal details of the MiG-AT in its original form

Front cockpit of MiG-AT (French avionics)
(Paul Jackson/Jane's) 0024031

Instrumentation: Two MFD 55 multifunctional liquid-crystal colour CRT displays with buttons in each cockpit; helmet-mounted displays; front-cockpit wide-field HUD with input from colour video and TV camera; HSI/ADI.

ARMAMENT (MiG-AS): Seven hardpoints for up to 2,000 kg (4,410 lb) of guided and unguided missiles, guns and bombs, including R-73E (AA-11 'Archer'), R-77 (AA-12 'Adder'), AIM-9L Sidewinder or Magic AAMs; Kh-29TD (AS-14 'Kedge') or Kh-31AE/PE (AS-17 'Krypton') ASMs; UB-16 (16 × 57 mm) or UB-8M (20 × 80 mm) rocket pods; cluster weapons; 100 to 500 kg bombs; UPK-23 twin-barrel 23 mm gun pods; eight Vikhr anti-tank missiles.

DIMENSIONS, EXTERNAL:	
Wing span	10.16 m (33 ft 4 in)
Wing aspect ratio	5.7
Length overall	12.01 m (39 ft 4¾ in)
Height overall	4.42 m (14 ft 6 in)
Wheel track	3.80 m (12 ft 5¾ in)
Wheelbase	4.43 m (14 ft 6½ in)
AREAS:	
Wings, gross	17.67 m² (190.2 sq ft)
WEIGHTS AND LOADINGS:	
Fuel: nominal	850 kg (1,874 lb)
intermediate	1,280 kg (2,822 lb)
max	1,680 kg (3,704 lb)
Normal T-O weight: training	4,610 kg (10,163 lb)
Max T-O weight: training	5,690 kg (12,544 lb)
combat	7,000 kg (15,430 lb)
combat (alternate)*	7,800 kg (17,196 lb)
Max wing loading:	
Normal T-O	260.9 kg/m² (53.43 lb/sq ft)
Max T-O	396.2 kg/m² (81.14 lb/sq ft)
Max power loading:	
Normal T-O	163 kg/kN (1.60 lb/lb st)
Max T-O	248 kg/kN (2.43 lb/lb st)

PERFORMANCE (estimated; A: basic training, B: advanced training):

Max Mach No.: A, B	0.85
Max Mach No.: A, B (alternate)*	0.8
Max level speed: A, B:	
at 2,500 m (8,200 ft)	540 kt (1,000 km/h; 621 mph)
at S/L	460 kt (850 km/h; 528 mph)
T-O speed: A	97 kt (180 km/h; 112 mph)
B	97-119 kt (180-220 km/h; 112-137 mph)
Landing speed:	
A, B	94-125 kt (175-232 km/h; 109-144 mph)
Max rate of climb: at S/L: A	1,680 m (5,510 ft)/min
B	4,140 m (13,580 ft)/min
at 5,000 m (16,400 ft): A	1,320 m (4,330 ft)/min
B	2,400 m (7,875 ft)/min
Service ceiling: A, B	15,500 m (50,860 ft)
A, B (alternate)*	14,000 m (45,940 ft)
T-O run: A	540 m (1,772 ft)
B	310 m (1,020 ft)
Landing run: on concrete: B	570 m (1,870 ft)
on concrete: B (alternate)*	640 m (2,100 ft)
on unpaved runway: B	540 m (1,775 ft)
on unpaved runway: B (alternate)*	600 m (1,970 ft)
Range at M0.5 at 6,000 m (19,680 ft):	
A, B	647 n miles (1,200 km; 745 miles)
Ferry range: A, B	1,400 n miles (2,600 km; 1,615 miles)
A, B (alternate)*	1,079 n miles (2,000 km; 1,242 miles)
g limits: A, B	+8.0/−2.0
Sustained g limit at M0.7 in turn: at S/L: A	+4.5
B	+8
at 5,000 m (16,400 ft): A	+3.4
B	+5
Max side wind for T-O/landing	29 kt (54 km/h; 33 mph)

Figures issued in 1999 by MAPO, although previous data continue to be published officially in parallel

UPDATED

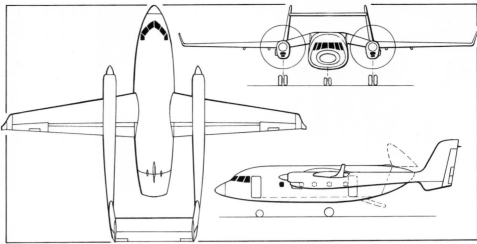

MiG-110 multipurpose transport *(James Goulding/Jane's)* 2000/0062665

MiG-110

TYPE: Twin turboprop transport.

PROGRAMME: Design began February 1992; shown in model form at Moscow Air Show 1993; development received government approval January 1994 and proceeded at low priority until 1997, when completion of MiG-AT design released staff and facilities for more rapid progress; by late 1998, mockup assembled and one-third of construction documentation was complete. On 27 October 1998, ANL Handelsgesellschaft of Austria signed LoI for funding of MiG-110 design and prototype construction; ANL intends to assemble aircraft in Austria from Russian kits. Interest was being maintained in late 2000, involving Austrian purchase of MiG-29 fighters. Main production originally intended by Sokol at Nizhny Novgorod, but Voronin (Moscow) named in February 2000 statement, following injection of funds by Moscow local government; however, preoccupation with Tu-334 may prompt reversion to original plan. Certification planned for 2003; MiG delayed formation of prototype committee from late 2000 to early 2001.

CURRENT VERSIONS: **MiG-110:** Baseline version.
MiG-110A: Austrian-assembled version.
MiG-110M: Western avionics and engines.
MiG-110VT (*Voyenno Transportnaya:* Military Transport): Proposed light tactical transport for Russian Air Forces.
MiG-110PR: Proposed SAR/reconnaissance variant for Russian Air Forces.

CUSTOMERS: Some 200 "ordered" by 20 Russian airlines up to February 2000; estimated world market for 1,900, including 1,200 in CIS.

COSTS: Programme cost estimated as US$100 million in late 2000; unit cost US$6 million to US$7 million.

DESIGN FEATURES: Optimised for ease of rear access to cargo hold and comparatively wide track landing gear. Cabin can be equipped for passenger, combined cargo/passenger or cargo operations, day or night in all weathers.

High wing monoplane; anhedral centre-section between twin booms carrying engines at front and twin fins at rear; horizontal tail surfaces between fin tips; winglets for reduced drag; fuselage pod slung from centre-section, with

Artist's impression of the MiG-110 2000/0062666

Rear loading door of MiG-110 mockup
(Yefim Gordon) 2001/0100416

upward-hinged rear fuselage section and loading ramp. Intended service life 25,000 hours.
FLYING CONTROLS: Conventional and manual. Twin rudders.
LANDING GEAR: Tricycle type, with twin wheels on each unit. Mainwheels retract rearwards into booms; nosewheel retracts forward. Optional floatplane version.
POWER PLANT: Two 1,839 kW (2,466 shp) Klimov TV7-117SV turboprops; six-blade Stupino SV-34 propellers.
ACCOMMODATION: Two crew; seats for up to 48 passengers; or provisions for 15 passengers and 3,500 kg (7,715 lb) of freight, or 5,000 kg (11,023 lb) of freight, including containers (up to four ZAK-1) and vehicles loaded via a rear ramp after tailcone is hinged upwards. Accommodation pressurised. Main passenger door, with integral stairs, port side, front.

DIMENSIONS, EXTERNAL:
Wing span	25.00 m (82 ft 0¼ in)
Length overall	18.90 m (62 ft 1 in)
Height overall	5.385 m (17 ft 8 in)

DIMENSIONS, INTERNAL:
Cargo cabin: Length	7.40 m (24 ft 3¼ in)
Max width	2.20 m (7 ft 2½ in)
Height: min	2.20 m (7 ft 2½ in)
max	2.76 m (9 ft 0¾ in)

WEIGHTS AND LOADINGS:
Max payload: paved runway	5,000 kg (11,023 lb)
unpaved runway	3,500 kg (7,716 lb)
Max T-O weight: paved runway	15,300 kg (33,730 lb)
unpaved runway	12,850 kg (28,329 lb)
Fuel weight: normal	2,000 kg (4,409 lb)
max	3,500 kg (7,716 lb)

PERFORMANCE (estimated):
Max level speed	297 kt (550 km/h 341 mph)
Max cruising speed	270 kt (500 km/h; 310 mph)
Cruising altitude	7,000 m (22,960 ft)
Balanced field length	1,000 m (3,280 ft)
Range, 30 min reserves: with 4,500 kg (9,921 lb) payload	
	907 n miles (1,680 km; 1,043 miles)
with full fuel and 2,660 kg (5,864 lb) payload	
	2,038 n miles (3,775 km; 2,345 miles)

UPDATED

MiG-115

TYPE: Light utility turboprop.
PROGRAMME: Work is believed to be progressing at low priority on a replacement for the venerable and ubiquitous Antonov An-2. Development began in 1994; wind tunnel tests mid-1996. No recent reports, however.
DESIGN FEATURES: Staggered, unbraced biplane with fixed, wide track landing gear and upswept rear fuselage for access to rear loading ramp. Certification planned to AP-23/FAR Pt 23.
Constant chord wings with no dihedral.
FLYING CONTROLS: Conventional and manual. Ailerons and flaps on upper wing; full-span flaps on lower wing. Horn-balanced elevator; trim tab on rudder.
LANDING GEAR: Tricycle type; fixed. Single wheel on each leg.
POWER PLANT: One 1,029 kW (1,380 shp) OMKB TVD-20 turboprop driving a six-blade propeller. Provision for alternative engines, including 969 kW (1,300 shp) P&WC PT6A-67.
ACCOMMODATION: Two (optionally one) crew and 15 passengers or 18 troops. Crew/passenger door port side, front; rear cargo ramp.

DIMENSIONS, EXTERNAL:
Wing span: upper (rear)	15.94 m (52 ft 3½ in)
lower (front)	12.20 m (40 ft 0¼ in)
Length overall	14.66 m (48 ft 1¼ in)
Height overall	6.04 m (19 ft 9¾ in)
Tailplane span	6.40 m (21 ft 0 in)
Wheel track	4.23 m (13 ft 10½ in)
Wheelbase	4.83 m (15 ft 10¼ in)
Propeller diameter	3.00 m (9 ft 10 in)

DIMENSIONS, INTERNAL:
Cabin: Length	5.10 m (16 ft 8¾ in)
Max width	1.96 m (6 ft 5¼ in)
Max height	1.80 m (5 ft 10¾ in)

WEIGHTS AND LOADINGS:
Weight empty	3,332 kg (7,346 lb)
Max payload	1,500 kg (3,306 lb)
Max fuel weight	1,700 kg (3,748 lb)
Max T-O weight	5,700 kg (12,566 lb)
Max power loading	5.54 kg/kW (9.10 lb/shp)

PERFORMANCE (estimated):
Max level speed	221 kt (410 km/h; 255 mph)
Econ cruising speed	184 kt (340 km/h; 211 mph)
Approach speed	62 kt (114 km/h; 71 mph)
Max rate of climb at S/L	540 m (1,771 ft)/min
Service ceiling	7,000 m (22,960 ft)
T-O run	370 m (1,215 ft)
Landing run	245 m (805 ft)
Range: with max payload, 45 min reserves	
	809 n miles (1,500 km; 932 miles)
with max fuel	1,997 n miles (3,700 km; 2,299 miles)
g limits	+3.2/−1.3

UPDATED

MiG-125

There are no known plans to build a prototype of this small business jet. Brief details last appeared in the 2000-01 edition.

UPDATED

MIL

MOSKOVSKY VERTOLETNY ZAVOD IMIENI M L MILYA OAO (Moscow Helicopter Plant named for M L Mil JSC)

Sokolnichyesky val 2, 107113 Moscow
Tel: (+7 095) 264 90 83
Fax: (+7 095) 264 55 71
Telex: 412141 MIL SU
GENERAL MANAGER: Leonid Zapolsky
DESIGNER GENERAL (Acting): Aleksei Samusenko
FIRST DEPUTIES:
 S A Kolupayev
 Aleksei G Samusenko

OKB founded 1947 by Mikhail Leontyevich Mil, who was involved with Soviet gyroplane and helicopter development from 1929 until his death on 31 January 1970, aged 60. His original Mi-1, first flown September 1948 and introduced into service 1951, was first series production helicopter built in former USSR. Twelve basic types of Mil helicopter have been built in 100 subvariants, representing total production of some 30,000, of which over 4,500 exported to 80 countries; 95 per cent of all helicopters in Russian Federation and Associated States (CIS) are Mil designs. Mil helicopters have set 96 world records. The Mil design bureau has its own experimental workshops in Moscow and flight test facilities at Chkalovsky airfield. The company has established a joint venture, Euromil (which see in International section), with Eurocopter and others, for development and production of the Mil Mi-38.

Mil design and production facilities eventually to be integrated into group comprising Mil Moscow, Kazan and Rostov (Rostvertol) plants, a helicopter operating company, financial and insurance interests. Associates will include Ulan-Ude production centre, Arsenyev and Viatka factories. The lighter types of Mil helicopter are marketed by LVM (which see earlier in this section). Currently, however, helicopters of Mil design are available (in near identical form) from as many as four different factories, all of which are energetically competing to market the aircraft with much duplication of effort.

In 1999 Mil joined with Sikorsky for development of a utility (fixed landing gear) version of S-76, designated S-76R. Mil to design composites main rotor blades and undertake assembly in Russia, if demand warrants.

Mil forced into receivership in 1999; new management imposed by Federal Financial Recovery Service returned company to solvency and paid staff wage arrears.

UPDATED

MIL Mi-6 and Mi-22

NATO reporting name: Hook
TYPE: Heavy-lift helicopter.
PROGRAMME: Private venture programme to develop heavy-lift military helicopter for 6,000 kg (13,227 lb) loads began in 1952 under internal designation VM-6 (Vertolet Mil 6 tonnes). Increased payload requirement led to replacement of planned TV-2F engines with TV-2VMs. Development officially authorised 11 June 1954, in competition with Kamov Ka-22 tandem-rotor helicopter. Originally planned

Modified Mil Mi-6VKP command post helicopter *(Paul Jackson/Jane's)*

with fixed wing with high-lift devices and two tractor propellers; subsequently deleted. Prototype, assembled at Zakharkovo and previously used for static and fatigue testing, flew 5 June 1957 as, by far, world's largest helicopter of that time; set many payload and airspeed records; five built for development testing; initial pre-series batch of 30 built by GAZ-23; total of 874 built at GAZ-168 (now Rostvertol) for civil/military use, ending 1981; developments included Mi-10 and Mi-10K flying cranes; Mi-6 dynamic components used in duplicated form on V-12 (Mi-12) of 1967, which remains largest helicopter yet flown. D-25V engines and R-7 gearbox tested first on Mi-6.

All known variants listed for completeness, before the type is grounded. Despite their popularity, versatility and economy, all Mi-6 variants are gradually being withdrawn from use as tail rotors reach the end of their lives. Lend-Lease machine tool (supplied in Second World War) used for manufacture of wooden tail rotor blades is beyond economic repair. Small numbers remained in service in 2000, with military and civil operators.
CURRENT VERSIONS: **Mi-6**: Basic transport.
 Mi-6A: Improved transport for VVS and Aeroflot; manufactured from 1971, incorporating many improvements added to basic aircraft during course of production (from c/n 705301V). Normal T-O weight raised from 39,700 kg (87,523 lb) to 40,500 kg (89,287 lb) and MTOW from 41,700 kg (91,932 lb) to 44,000 kg (97,003 lb). Underslung load raised from 8,000 kg (17,637 lb) to 9,000 kg (19,842 lb); maximum speed from 135 kt (250 km/h; 155 mph) to 164 kt (304 km/h; 189 mph) and cruising speed from 108 kt (200 km/h; 124 mph) to 135 kt (250 km/h; 155 mph).
 Mi-6M: ASW variant, carrying two PLAT or Kondor missiles on each side of the fuselage, in place of the external tanks. Some converted from 1963, but never submitted to acceptance tests; not built in series or deployed; relegated to ASW systems test role.
 Mi-6M: Designation reused to cover radical redesign for carriage of 11,000 to 20,000 kg (24,250 to 44,092 lb) payload. Unbuilt; replaced by Mi-26.
 Mi-6 Burlak: ASW version with dipping sonar or mine countermeasures sled. No acceptance tests, series production or deployment. Relegated to ASW systems test role.
 Mi-6P: Passenger/VIP version with square windows, thermal insulation, extra soundproofing and luxuriously appointed cabin housing 70 to 80 passengers with cloakroom and lavatory. Rear loading doors faired-over. Able to be converted to air ambulance.
 Mi-6PP: ECM version developed in 1962 to prevent enemy Elint-gathering of PVO radar transmissions. Designation reused for unbuilt AWACS-jammer.
 Mi-6PRTV: Prototype strategic missile transporter, carrying warheads or complete 8K11, 8K14, R9 or R-10 missile rounds. Not produced in series.
 Mi-6PS: Specialised factory-built SAR variant used for space programme support, with additional navigation equipment, medical equipment, a new liferaft recovery hoist, and dinghy/survival equipment containers. Used for astronaut and capsule recovery, and decontamination. Existing Mi-6s modified to similar standard as **Mi-6APS**.
 Mi-6PZh: Firefighting version developed in 1967, with internal 12,000 litre (3,170 US gallon; 2,640 Imp gallon) water/retardant tank, pumps, water collection frame and quick-release valve. Could also carry six soft-celled retardant tanks externally. Destroyed during firefighting in South of France, and replaced by Mi-6PZh-2 with nose-mounted water cannon.
 Mi-6RVL: Prototype strategic missile transporter, carrying complete 9K53, 9K73 (R-17V), 9M21 and 8K114 missile rounds. Not produced in series.
 Mi-6S: Casevac version with 41 stretchers.
 Mi-6T ('Hook-A'): Basic airborne forces transport with machine gun in nose glazing. Military use is primarily to haul guns, armour, vehicles, supplies, freight and troops in combat areas, but includes command support roles. *Description applies to this version.* Full description last appeared in 1983-84 *Jane's*.
 Mi-6TZ: Tanker version to refuel aircraft and helicopters on the ground, at forward airstrips. Mi-6TZ conducted in-flight refuelling trials with Mi-4 receivers, but system not adopted. Ground tanker version deployed operationally, however. Two fuel tanks (actually Mi-10 external tanks) of 7,400 litres (1,955 US gallons; 6,279 Imp gallons) capacity with NS-30 pumps. Other

consumables carried in separate containers, **Mi-6TZ-SV** similar.

Mi-6VKP (*vozduzhnyi komandnyi punkt*: airborne command post) ('Hook-B'): Command support helicopter with R-111, R-140, R-155 and R-409 radios; flat-bottom U-shape antenna under tailboom; X configuration blade antennas forward of horizontal stabilisers; large heat exchanger on starboard side of cabin; small cylindrical container aft of starboard rear cabin door. Some seen carrying pole antennas on landing gear struts. Total of 36 converted by 535 Aircraft Repair Depot at Konotop. Performed task while on ground.

Mi-6VR (*Vodolei*: Aquarius): Rotor-systems testbed (Red 88) used in Mi-26 development (with Mi-26 main rotor); subsequently equipped as icing spray rig.

Mi-6VUS ('Hook-C'): Developed command support version with sweptback plate antenna above forward part of tailboom instead of 'Hook- B's' U-shape antenna; small antennas under fuselage; horizontal pole antenna on starboard main landing gear of some aircraft (possibly being carried for erection on the ground). TA-6 APU. Also known as **Izd-50AYa**, **Mi-6AYa** or possibly **Mi-22** in service.

Mi-6AYaSh ('Yakhont Hook-D'): Command support helicopter; flat box-like fairing (reportedly SLAR), forward of external fuel tank on starboard side, and many small antennas. Aircraft numbers 6682604V ('02') and 747306V ('06') originally based at Kaunas, Lithuania, until 1993 and now at Novosibirsk-Severniy.

CUSTOMERS: Operators have included Russian ground forces (100), Aeroflot (training unit at Kremenchug, plus regional divisions at Arkangel'sk, Ashkhabad, Khabarovsk, Krasnoyarsk, Nadym, Nizhnyevartovsk, Noril'sk, Salekhard, Surgut, Syktvkar, Tyumen, Ukhta and Yakutsk), and the armed forces of Algeria (four), Belarus (11, including two Mi-22), Egypt (more than six), Ethiopia (10), Indonesia, Iraq (10), Kazakhstan, Laos (one), Pakistan, Peru (six), Poland, Syria (10+), Ukraine (60), Uzbekistan (27) and Vietnam (10). The Mi-6 remains in use with Russia, Ukraine, Belarus, Kazakhstan, Uzbekistan, Egypt (No. 7 Squadron at Kom Awshim), Laos, Syria and Vietnam, and perhaps in Iraq. Several used in civil role by Tyumen Avia Trans.

DESIGN FEATURES: Five-blade main rotor and four-blade anti-torque tail rotor. Two shoulder wings of CAHI (TsAGI) P-35 aerofoil section, inclined at 14° 15′ (port) and 14° 45′ (starboard), offload rotor by providing some 20 per cent of total lift in cruising flight. Removed when aircraft is operated as flying crane.

FLYING CONTROLS: Mechanical, with dual rods to main rotor; RP-28 dual hydraulic boosters and SDV-5000-0A damper unit. Cables in parts of tail rotor control circuit at rear of tail boom. Four-channel AP-34B Series 2 autopilot. KAU-30B rpm governor unit, and EMT-2M pitch control artificial feel unit.

STRUCTURE: All-metal wing consists of tension-box spar with separate tip, leading- and trailing-edges, with ribs and stringers and a duralumin skin. All-moving tailplane of conventional construction; trailing-edge fabric-covered. Main blade built around single seamless tube spar, with light alloy honeycomb sandwich trailing-edge giving 1,500 flight hour fatigue life. NACA 230 section from root to Rib 18, then TsAG1 high-speed section electrically de-iced. Wooden L63-Kh6BP blades on AV-63B anti-torque tail-rotor, with spray de-icing.

LANDING GEAR: Non-retractable tricycle landing gear with auxiliary tail bumper mounted on Oleo-hydraulic shock absorber. Fully castoring nose gear with twin 720×310 unbraked wheels, fixed shock-absorbed main gear units with single 1328×480 wheels fitted with drum brakes.

POWER PLANT: Two 4,045 kW (5,425 shp) Aviadvigatel/ Soloviev D-25V (TV-2VM) turboshafts, mounted side by side above cabin, with R-6 gearbox forward of main rotor shaft. AI-8 gas-turbine starter unit. Intakes de-iced electrically. Engine access panels fold down hydraulically to double as maintenance platforms. Eleven internal fuel tanks, in five groups, with electrically driven pumps, and pressurised with inert gases for fire-suppression, capacity 6,315 kg (13,922 lb); two external tanks, on each side of cabin, capacity 3,490 kg (7,695 lb); provision for two ferry tanks inside cabin, capacity 3,490 kg (7,695 lb). Oil tanks between inner and outer skins of intake ducts.

ACCOMMODATION: Crew of five: two pilots, navigator, flight engineer and radio operator. Four jettisonable doors and overhead hatch on flight deck. Electrothermal anti-icing system for glazing of flight deck and navigator's compartment. Equipped normally for cargo operation, with easily removable tip-up seats along sidewalls; when these seats are supplemented by additional seats in centre of cabin, 65 to 90 passengers can be carried in Mi- 6A (150 in extreme circumstances; up to 61 in Mi-6), with cargo or baggage in aisles. Normal military seating for 70 combat-equipped troops. As ambulance, 45 stretcher cases and two medical attendants on tip-up seats can be carried; one attendant's station, provided with intercom to flight deck; provision for portable oxygen installations for patients. Cabin floor stressed for loadings of 2,000 kg/m² (410 lb/sq ft); provision for cargo tiedown rings. Rear clamshell doors and ramps operated hydraulically. Standard equipment includes LPG-3 unit, electric winch of 800 kg (1,765 lb) capacity and pulley block system. Central hatch in cabin floor for cargo sling for bulky loads. Maximum underslung load initially restricted to 8,000 kg (17,637 lb). Three jettisonable doors, fore and aft of main landing gear on port side and aft of landing gear on starboard side.

SYSTEMS: Main, back-up and auxiliary hydraulic systems with RP-28 and KAU-30B control units, using AMG-10 frost-resistant fluid at 118 to 152 bars (1,707 to 2,205 lb/sq in) pressure. Main hydraulic system used for flying controls (manual reversion possible in the event of complete hydraulic failure); auxiliary system used for windscreen wipers, gun hatches, pilot seat adjustment and external cargo hook. A pneumatic system (replenished by an AK-50T compressor driven by the port engine) for brakes, compressor release valves, heating ducts and cannon loading. Engine-driven STG-12TM generators supplying 27 V DC power; 12SAM-55 batteries. SGS-90/360 generators supply 360 V 90 kW AC. Five flight deck crew provided with KKO-LS oxygen sets with KP-21 and KP-58 oxygen apparatus and KM-16N masks. Flight engineer and passengers carry portable KP-21 oxygen apparatus. KP-21 also used in ambulance configuration.

AVIONICS: *Comms:* R-802 and R-807 VHF and RSB-70 ('Yadro-1A') HF radio, SPU-7U intercom and MSRP-12 flight data recorder.

Flight: ARK-9 ADF, ARK-UD HF/DF, RU-3 radar altimeter, GMK-1A directional gyro, DAK-DB-5VK astro compass and MRP-56P marker receiver.

Instrumentation: US-35 ASI, VD-10 altimeter and AGK-47VK artificial horizon.

Self-defence: ESPK-46 chaff/flare dispenser.

ARMAMENT: Some military Mi-6s have a nose 12.7 mm machine gun in NUV-1M mounting with 200 to 270 rounds, traversing through ±30° in azimuth and 55° depression.

SOVIET/RUSSIAN Mi-6 OPERATORS 1960-1999

Unit	Base	Variants
36 OVP	Serdovsk	Mi-6A
51 OVP[1]	Alexandria	Mi-6/6A/6PS
65 OVP[2]	Kagan	Mi-6A/ VKP/22
157 OVP		Mi-6/6A/6PS
181 OVP	Dzambul and Afghanistan	Mi-6/6A/PS
239 OVP[3]	Brandis*, Oranienburg*	Mi-6/6A
248 OVP[4]	Minsk	Mi-6VKP/22
280 OVP	Kagan and Afghanistan	Mi-6/6A
320 OVP[5]	Kherson	Mi-6/6A
325 OVP	Yegorlykskaya	Mi-6A
332 OVP	Kluchevoye (Probylovo)	Mi-6A
340 OVP	Kalinov	Mi-6/6A
367 OVP	Serdobsk	Mi-6A
456 OGvSAP	Gavryshevka	Mi-6VKP/22
486 OVP[6]	Ucharal	Mi-6?
696	Torzhok	Mi-6A
111 OAE	Chebenky	Mi-6PS
6 OVE[7]	Dresden-Hellerau*	Mi-6VKP/22
9 OVE[8]	Neuruppin*	Mi-6VKP/22
41 OVE[9]	Finow*	Mi-6VKP/22
248 OVE[3]	Minsk	Mi-22
296 OVE[10]	Mahlwinkel*	Mi-6VKP/22
298 OVE[11]	Hassleben*	Mi-6VKP/22
1169 BRAT[3]	Luninets	Mi-6VKP/22
Unidentified unit	Aral'sk	Mi-6PS
Unidentified squadron	Karaganda	Mi-6PS
Unidentified squadron	Ulan Ude	Mi-6PS
Unidentified squadron	Severnii, Novosibirsk	Mi-6AYSh
Unidentified regiment	Troitsk	Mi-6PS
Unidentified regiment	Sakhalin	Mi-6PS

Notes:

OVE — Otdel'naya Vertolet'naya Eskadrilya (Independent Helicopter Squadron), OVP — Otdel'naya Vertolet'naya Polk (Independent Helicopter Regiement), IIVP — Instruktsiya-Issled ovatelsky Vertolet'nyi Polk (Instruction-Test Helicopter Regiment)

* (Former) East Germany

[1] Regiment participated in Chernobyl operation dropping lead and sand on leaking reactor. Regiment abandoned most of its contaminated aircraft inside 30 km exclusion zone. Became 1 OVB of National Guard of Ukraine

[2] Transferred to Uzbekistan

[3] Regiment moved to Yefremov, where it remains

[4] Transferred to Belarus

[5] Became 2 OVB of National Guard of Ukraine. Grounded 12 June 1998

[6] Transferred to Kazakhstan

[7] Plus many independent Helicopter Squadrons equipped with Mi-6VKP/Mi-22 attached to army HQs. Now at Slokovo

[8] Moved to Berdsk, Novosibirsk

[9] Moved to Tula

[10] Moved to Russia but disbanded

[11] Moved to Kaluga. Plus Ukrainian Army Aviation: 1 OTAP at Boryspil, 2 BAA at Chernobayevka

DIMENSIONS, EXTERNAL:

Main rotor diameter	35.00 m (114 ft 10 in)
Tail rotor diameter	6.30 m (20 ft 8 in)
Length: overall, rotors turning	41.74 m (136 ft 11½ in)
fuselage, excl nose gun and tail rotor	33.18 m (108 ft 10½ in)
Height overall	9.86 m (32 ft 4 in)
Wing span	15.30 m (50 ft 2½ in)
Wheel track	7.50 m (24 ft 7¼ in)
Wheelbase	9.09 m (29 ft 9¾ in)
Rear-loading doors: Height	2.70 m (8 ft 10¼ in)
Width	2.65 m (8 ft 8¼ in)
Passenger doors: Height: front door	1.70 m (5 ft 7 in)
rear doors	1.61 m (5 ft 3½ in)
Width	0.80 m (2 ft 7½ in)
Sill height: front door	1.40 m (4 ft 7 in)
rear doors	1.30 m (4 ft 3¼ in)
Central hatch in floor	1.44 m (4 ft 9 in) × 1.93 m (6 ft 4 in)

DIMENSIONS, INTERNAL:

Cabin: Length	12.00 m (39 ft 4½ in)
Max width	2.65 m (8 ft 8¼ in)
Max height: at front	2.01 m (6 ft 7 in)
at rear	2.50 m (8 ft 2½ in)
Volume	80.0 m³ (2,825 cu ft)

AREAS:

Main rotor disc	962.10 m² (10,356.1 sq ft)
Tail rotor disc	31.17 m² (335.51 sq ft)
Wings	3,500 m² (376.74 sq ft)

WEIGHTS AND LOADINGS (Mi-6A):

Weight empty	27,240 kg (60,055 lb)
Max internal payload	12,000 kg (26,450 lb)
Max slung cargo	9,000 kg (19,842 lb)
Fuel load: internal	6,315 kg (13,922 lb)
with external tanks	9,805 kg (21,617 lb)
Max T-O weight with slung cargo at altitudes below 1,000 m (3,280 ft)	38,400 kg (84,657 lb)
Normal T-O weight	40,500 kg (89,285 lb)
Max T-O weight for VTO	42,500 kg (93,700 lb)
Overload T-O weight	44,000 kg (97,003 lb)
Max disc loading	44.2 kg/m² (9.05 lb/sq ft)

PERFORMANCE (at max T-O weight for VTO):

Max level speed	164 kt (304 km/h; 189 mph)
Max cruising speed	135 kt (250 km/h; 155 mph)
Service ceiling	4,500 m (14,760 ft)
Hovering ceiling IGE	2,250 m (7,380 ft)
Range with 8,000 kg (17,637 lb) payload	334 n miles (620 km; 385 miles)
Range with external tanks and 4,500 kg (9,920 lb) payload	540 n miles (1,000 km; 621 miles)
Max ferry range (tanks in cabin)	674 n miles (1,250 km; 776 miles)

UPDATED

MIL Mi-8 (V-8)

NATO reporting name: Hip

TYPE: Multirole medium helicopter.

PROGRAMME: Development began May 1960, to replace piston-engined Mi-4; first of two prototypes (V-8), with single AI-24V turboshaft and four-blade main rotor, flew 24 June 1961, given NATO reporting name 'Hip-A'; further prototype (V-8A; 'Hip-B'), with two production standard TV2-117 engines and four-blade main rotor, flew 2 August 1962; change to five-blade rotor for production. More than 6,000 Mi-8s, Mi-17s and Mi-171/172s (which see) marketed and delivered from Kazan since 1967 and, since 1970, over 11,000 (about 3,700 Mi-8T and 7,300 Mi-17) from Ulan-Ude (see separate entries for addresses) for civil and military use, including 3,500 exported to 70 countries.

CURRENT VERSIONS: **Mi-8P** ('Hip-C'): Civil passenger helicopter; TV2-117A turboshafts and standard seating for 28 persons in main cabin with large square windows.

Mi-8S ('Hip-C'): Original de luxe ('salon') version of Mi-8; normally 11 passengers, on eight-place inward-facing couch on port side, two chairs and swivelling seat on starboard side, with table; rectangular windows; air-to-ground radiotelephone and removable ventilation fans; compartment for attendant, with buffet and crew wardrobe, forward of cabin; toilet (port) and passenger wardrobe (starboard) to each side of cabin rear entrance; alternative nine-passenger configuration; maximum T-O weight 10,400 kg (22,928 lb); range 205 n miles (380 km; 236 miles) with 30 minute fuel reserve.

Mi-8PS (*Izdelie* 80PS): Military VIP version similar to Mi-8S. Finnish Mi-8PS fitted with undernose weather radar. **Mi-8APS** similar, but with enhanced communications fit and greater luxury; used as Russian Presidential aircraft. Subvariants are -7, -9 and -11 for seven, nine or 11 passengers.

Mi-8T ('Hip-C'): Civil utility transport version with TV2-117A turboshafts and circular cabin windows, marketed by Ulan-Ude plant. Alternative payloads include internal or external freight; 24 passengers on removable folding seats; 26 passengers on conventional seats; 12 stretcher patients; or executive passengers in specially furnished cabin of type described for Mi-8S.

Mil Mi-8PS-11 VIP transport *(Paul Jackson/Jane's)*

Instrument panels on the Mil Mi-14P flight deck *(Paul Jackson/Jane's)*

Mi-8T ('Hip-C'): Similar military transport for Russian Federation and Associated States (CIS) army support forces, carrying 24 fully armed troops. Able to dispense 200 anti-personnel or anti-tank mines in flight, by conveyor belt through rear doors.

Mi-8TP: Military executive version; upgraded comms include R-832 radio with two blade antennas under front fuselage and tailboom, and R-111 with rod antenna lowered under cabin.

Mi-8TV (*vooruzhonnyi*: armed) ('Hip-C' *Izdelie* 80TV): As Mi-8T, but with added twin-rack each side, to carry total of 64 × 57 mm S-5 rockets in four UV-16-57 packs, or bombs, for army assault forces.

Mi-8TB ('Hip-E'): Development of 'Hip-C'; KV-4 flexibly mounted 12.7 mm machine gun, with 700 rounds, in nose; triple stores rack each side, to carry total 192 S-5 rockets in six UV-32-57 packs, plus four 9M17P Skorpion (AT-2 'Swatter') anti-tank missiles (semi-automatic command to line of sight) on rails above racks; about 250 in RFAS ground forces.

Mi-8TBK ('Hip-F'): Export 'Hip-E' originally developed for East Germany; missiles changed to six 9M14M Malyutka (AT-3 'Sagger'; manual command to line of sight).

Mi-8TZ: Tanker variant adapted to deliver fuel to front-line areas.

Mi-8T(K): Dedicated photo-reconnaissance platform with AFA-42/100 or AFA-A87P starboard oblique camera in forward part of cabin, possibly with some onboard processing capability. May be used as fire-correction platform. **Mi-8SKA** believed similar. Quite separate from K version, which has vertical camera.

Mi-8K: Reconnaissance and artillery fire correction version; large window for camera in rear clamshell doors.

Mi-8R: Reconnaissance version.

Mi-8AV: Dedicated minelayer, despatching mines down steep, ladder-like slide projecting from gap between lower corners of clamshell doors.

Mi-8BT: Equipped for minesweeping, towing sled from winch in cabin. Clamshell doors removed for missions.

Mi-8AT ('Hip-C'): Current civil transport version from Ulan-Ude, with TV2-117AG turboshafts. Optional 8A-813 weather radar, DISS-32-90 Doppler and A-723 long-range radio nav. AT designation suffix first used by V-8AT, which was first square-windowed VIP prototype, and of which a small series may have been produced.

Mi-8ATS: Agricultural helicopter with spray hoppers on each side, and with 'wing'-type spraybars.

Mi-8MT: Flying crane version with operator's glazed gondola in place of rear clamshell doors. SSSR-25444 may have been the prototype. Designation reused for Mi-17 (which see).

Mi-8 VIP: Current de luxe version by Kazan Helicopters; three crew and seven to nine passengers; main rotor has vibration damper; hinged airstair door; interior divided into vestibule, passenger cabin, crew compartment, cloakroom and toilet; passenger cabin contains soft seats, sofa, tables and mini-bar; optional water heater, TV and GPS. Maximum T-O weight 12,000 kg (26,455 lb).

Mi-8TM: Upgraded civil transport version of Mi-8T; weather radar and rotor head integrity system.

Mi-8TG: Modified TV2-117TG engines permit operation on both liquefied petroleum gas (LPG) and kerosene. LPG contained in large tanks, on each side of cabin, under low pressure. Engines switch to kerosene for take-off and landing. Reduced harmful exhaust emissions in flight offer anti-pollution benefits. Modification to operate on LPG requires no special equipment and can be effected on in-service Mi-8s at normal maintenance centre. Weights unchanged. Large external tanks, each side of cabin, reduce payload by 100 to 150 kg (220 to 330 lb) over comparable ranges, with little effect on performance. First flight on LPG made 1987. Example (RA-25364) at Moscow Air Show '95 had flown 60 hours of trials using LPG.

Mi-8VZPU (*vozduzhnyi zapastnoi punkt upravlenya*: airborne reserve command post) ('Hip-D'): As 'Hip-C' but rectangular-section battery canisters on outer stores racks; two inverted U-shape antennas above tailboom; no armament.

Mi-9 ('Hip-G'): See separate entry.

Mi-8SMV ('Hip-J'; *Izdelie* 80SMV): ECM version with R-949 jamming system; additional small boxes of equipment on each side of fuselage, fore and aft of main landing gear legs; range of 54 n miles (100 km; 62 miles). Also four containers with total of 32 droppable short-range jammers. Used primarily for border surveillance.

Mi-8PPA ('Hip-K' *Izdelie* 80PP): Active communications jammer and communications intelligence (comint) helicopter, carrying SPS-63, 66 and 68 equipment; rectangular container and array of six cruciform dipole antennas each side of cabin; no Doppler box under tailboom; heat exchangers under front fuselage. About 30 in RFAS service; two with Czech, one with Slovak air forces.

Mi-8AMT, **Mi-8MT** and **Mi-8MTV** are versions of the Mi-17 (which see), with more powerful turboshafts and port-side tail rotor.

All helicopters of Mi-8/Mi-17 series in Russian military service are known as Mi-8s of various subtypes, regardless of engines fitted.

Long-range modification: AEFT (Auxiliary External Fuel Tanks) system by Aeroton adds a further 1,900 litres (502 US gallons; 418 Imp gallons) in two internal tanks, plus another 1,900 litres in four external tanks on the stores pylons of the Mi-8T and Mi-8AT. Operational

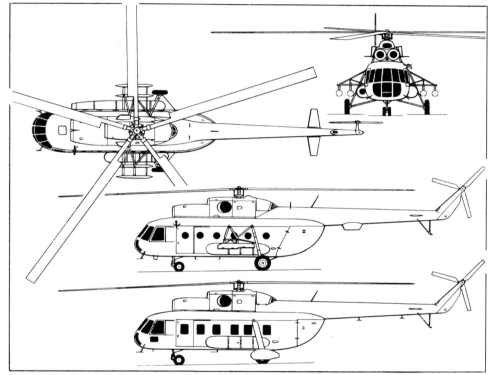
Mil Mi-8TV ('Hip-C') twin-turbine military helicopter, with additional side view (bottom) of original Mi-8S commercial version *(Dennis Punnett/Jane's)*

Winch-equipped Mil Mi-8T 'Hip-C' military transport helicopter *(Paul Jackson/Jane's)*

range with all six auxiliary tanks is 593 n miles (1,100 km; 683 miles); ferry range 863 n miles (1,600 km; 994 miles).

Upgrade: Some surviving Russian Air Forces Mi-8s may receive a limited upgrade, perhaps with improved avionics, instruments, or even composites main rotor blades.

CUSTOMERS: RFAS ground forces (estimated 1,100 Mi-8/17s, including 100 command post and EW); RFAS air forces; more than 50 other air forces, including 200 with Ukrainian Army; civil operators worldwide.

DESIGN FEATURES: Conventional pod and boom configuration; clamshell rear-loading freight doors; five-blade main rotor, inclined forward 4° 30′ from vertical; interchangeable blades of basic NACA 230 section, solidity 0.0777; spar failure warning system; drag and flapping hinges a few inches apart; blades carried on machined spider; pendulum vibration damper; three-blade starboard tail rotor; transmission comprises VR-8A two-stage planetary main reduction gearbox giving main rotor shaft/engine rpm ratio of 0.016:1, intermediate and tail rotor gearboxes, main rotor brake, and drives off main gearbox for tail rotor, fan, AC generator, hydraulic pumps and tachometer generators; tail rotor pylon forms small vertical stabiliser; horizontal stabiliser near end of tailboom. Military versions have smaller, circular cabin windows.

FLYING CONTROLS: Mechanical system, with irreversible hydraulic boosters; main rotor collective pitch control linked to throttles.

STRUCTURE: All-metal; main rotor blades, mainly VD-76 aluminium alloy, each has extruded spar carrying root fitting, 21 honeycomb-filled trailing-edge pockets and blade tip, with titanium abrasion shielding on some of leading-edge; balance tab on each blade; each tail rotor blade made of spar and honeycomb-filled trailing-edge; semi-monocoque fuselage. Composites blades and elastomeric hub components under development.

LANDING GEAR: Non-retractable tricycle type; steerable twin-wheel nose unit, locked in flight; single wheel on each main unit; oleo-pneumatic (gas) shock-absorbers. Mainwheel tyres 865×280; nosewheel tyres 595×185. Pneumatic brakes on mainwheels; pneumatic system can also recharge tyres in the field, using air stored in main landing gear struts. Optional mainwheel fairings.

POWER PLANT: Two 1,250 kW (1,677 shp) Klimov TV2-117AG turboshafts. Main rotor speed governed automatically, with manual override. Original versions have flexible internal fuel tank, capacity 445 litres (117.5 US gallons; 98.0 Imp gallons); two external tanks, each side of cabin, capacity 745 litres (197 US gallons; 164 Imp gallons) in port tank, 680 litres (179.5 US gallons; 149.5 Imp gallons) in starboard tank; total standard fuel capacity 1,870 litres (494 US gallons; 411.5 Imp gallons). Provision for one or two ferry tanks in cabin, raising maximum total capacity to 3,700 litres (977 US gallons; 814 Imp gallons). Current Mi-8T has standard external tank capacity of 2,615 litres (691 US gallons; 575 Imp gallons), comprising 2,170 litres (573 US gallons; 477 Imp gallons) in main tanks and 445 litres (118 US gallons; 98 Imp gallons) in feed tank. Additional 1,830 litres (483 US gallons; 402 Imp gallons) optional auxiliary fuel. Fairing over starboard external tank houses optional cabin air conditioning equipment at front. Engine cowling side panels form maintenance platforms when open, with access via hatch on flight deck. Total oil capacity 60 kg (132 lb).

ACCOMMODATION: Two pilots side by side on flight deck, with provision for flight engineer's station. Military versions can be fitted with external flight deck armour. Windscreen de-icing standard. Basic passenger version furnished with 24 to 26 four-abreast track-mounted tip-up seats at pitch of 72 to 75 cm (28 to 29.5 in), with centre aisle 32 cm (12.5 in) wide; removable bar, wardrobe and baggage compartment. Seats and bulkheads of basic version quickly removable for cargo carrying. Mi-8T and standard military versions have cargo tiedown rings in floor, winch of 150 kg (330 lb) capacity and pulley block system to facilitate loading of heavy freight, an external cargo sling system (capacity 3,000 kg; 6,614 lb), and 24 tip-up seats along sidewalls of cabin. All versions can be converted for air ambulance duties, with accommodation for 12 stretchers and tip-up seat for medical attendant. Large windows on each side of flight deck slide rearward. Sliding, jettisonable main passenger door at front of cabin on port side; electrically operated rescue hoist (capacity 150 kg; 330 lb) can be installed at this doorway. Rear of cabin made up of clamshell freight-loading doors, which are smaller on commercial versions, with downward-hinged passenger airstair door centrally at rear. Hook-on ramps used for vehicle loading.

SYSTEMS: Standard heating system can be replaced by full air conditioning system; heating of main cabin cut out when carrying refrigerated cargoes. Two independent hydraulic systems, each with own pump; operating pressure 44 to 64 bars (640 to 925 lb/sq in). DC electrical supply from two 27 V 18 kW starter/generators and six 28 Ah storage batteries; AC supply for automatically controlled electrothermal de-icing system and some radio equipment supplied by 208/115/36/7.5 V 400 Hz generator, with 36 V three-phase standby system. Engine air intake de-icing standard. Provision for oxygen system for crew and, in ambulance version, for patients. Freon fire extinguishing system in power plant bays and service fuel tank compartments, actuated automatically or manually. Two portable fire extinguishers in cabin.

AVIONICS: *Comms:* R-842 HF transceiver, frequency range 2 to 8 MHz and range up to 540 n miles (1,000 km; 620 miles); R-860 VHF transceiver on 118 to 135.9 MHz effective up to 54 n miles (100 km; 62 miles); SPU-7 intercom, radiotelephone.

Flight: Four-axis autopilot to give yaw, roll and pitch stabilisation under any flight conditions, stabilisation of altitude in level flight or hover, and stabilisation of preset flying speed; Doppler radar box under tailboom.

Instrumentation: For all-weather flying by day and night: two gyro horizons, two airspeed indicators, two main rotor speed indicators, turn indicator, two altimeters, two rate of climb indicators, magnetic compass, astrocompass for Polar flying; ARK-9 automatic radio compass, RV-3 radio altimeter with 'dangerous height' warning.

Self-defence (optional): Infra-red jammer ('Hot Brick') above forward end of tailboom; three ASO-2V flare dispensers above rear cabin window on each side.

ARMAMENT: See individual model descriptions of military versions.

DIMENSIONS, EXTERNAL: See Mi-17 entry

DIMENSIONS, INTERNAL: See Mi-17 entry

AREAS: See Mi-17 entry

WEIGHTS AND LOADINGS:	
Weight empty: civil Mi-8T	7,149 kg (15,760 lb)
military versions (typical)	7,260 kg (16,007 lb)
Max payload: internal	4,000 kg (8,820 lb)
external	3,000 kg (6,614 lb)
Fuel weight, civil Mi-8T:	
standard tanks	2,027 kg (4,469 lb)
with two auxiliary tanks	3,447 kg (7,599 lb)
Normal T-O weight	11,100 kg (24,470 lb)
T-O weight:	
with 28 passengers, each with 15 kg (33 lb) of baggage	11,570 kg (25,508 lb)
with 2,500 kg (5,510 lb) of slung cargo	11,428 kg (25,195 lb)
Max T-O weight for VTO	12,000 kg (26,455 lb)
Max disc loading	33.7 kg/m² (6.90 lb/sq ft)
PERFORMANCE (civil Mi-8T):	
Max level speed at S/L:	
normal AUW	135 kt (250 km/h; 155 mph)
max AUW	124 kt (230 km/h; 142 mph)
with 2,500 kg (5,510 lb) of slung cargo	97 kt (180 km/h; 112 mph)
Max cruising speed:	
normal AUW	121 kt (225 km/h; 140 mph)
max AUW	113 kt (210 km/h; 130 mph)
Service ceiling: normal AUW	4,500 m (14,760 ft)
max AUW	4,000 m (13,120 ft)
Hovering ceiling at normal AUW:	
IGE	1,800 m (5,905 ft)
OGE	850 m (2,785 ft)
Range: with standard fuel, 30 min reserves	307 n miles (570 km; 354 miles)
with auxiliary fuel	531 n miles (985 km; 612 miles)

UPDATED

MIL Mi-9

NATO reporting name: Hip-G

Designation Mi-9 (*Izdelie* 80 IW) applies to command relay platform variant of Mi-8 with Ivolga (oriole) equipment but not used as airborne relay platform; R-111 relay antenna folds upward under fuselage on ground; rearward-inclined 'hockey stick' antennas of R-405MPB datalink project from rear of cabin and from undersurface of tailboom, aft of Doppler radar box; rearward-inclined short whip antenna above forward end of tailboom; two long plate aerials of R-826 HF radio on fuselage undersurface; R-802 VYa UHF mast antenna under forward fuselage.

Operating concept envisages aircraft flying to selected location; landing, shutting down and being linked to two masts — one 11 m (36 ft) for R-111 (deployed 26 m; 85 ft to port side of helicopter), one 16.5 m (54 ft) antenna for R-405M (at 29.5 m; 97 ft) to starboard. Mission crew of six (three controllers, seated aft, at map tables; three radio operators forward), plus pilots and flight engineer. In service with armed forces of Belarus (two), Czech Republic (one), Hungary (one), Kazakhstan, Russia (30), Ukraine (18). Eight in former East Germany now withdrawn. (See also Mil Mi-19.)

Reports suggest that some surviving Mil Mi-9s may have received a limited upgrade, with new instruments and avionics and some improved mission and communications equipment.

UPDATED

MIL Mi-14

NATO reporting name: Haze

TYPE: Amphibious helicopter.

PROGRAMME: Development of Mi-8; initially designated V-8G; prototype SSSR-11051 first flew September 1969, under designation **V-14** and with Mi-8 power plant; changed to TV3-117M engines, with VR-14 gearbox, for production. Built at Ulan-Ude.

CURRENT VERSIONS: **Mi-14PL** ('Haze-A'): Basic ASW version; four crew; large undernose radome; Kalmar ASW suite with OKA-2 retractable, dual-mode, active/passive sonar in starboard rear of planing bottom, forward of two chutes for 18 RGB-NM sonobuoys or signal flares; APM-60 towed magnetic anomaly detection (MAD) bird, streamed by LPG-6-23 electric winch, stowed against rear of fuselage pod (moved to lower position on some aircraft); weapons include one AT-1 ASW or APR-2 torpedo, one 1 kT 'Skat' nuclear depth bomb or eight PLAB-250, PLAB-50-64 or PLAB-MK depth charges or OMAB-25-120 or OMAB-MK day/night markers or Poplavak-1A radar beacons in enclosed bay in bottom of hull; VAS-5M-3 liferaft (in all versions).

Mi-14PL 'Strike': Subvariant with provision for Kh-23 (AS-7 'Kerry') ASMs. Tested from 1983.

Mi-14PLM ('Haze-A'): As Mi-14PL but with updated equipment, including relocated MAD 'bird' and rescue basket. Sometimes **Mi-14 PL-M.**

Mi-14PW: Polish designation of Mi-14PL. Upgrade programme launched in 1995 with modifications to OKA-2 by Air Force Technological Institute and Gdansk Technical University, plus addition of GPS and new radios. Now armed with Western A244S and MU90 torpedoes.

Mi-14BT ('Haze-B'): Mine countermeasures version with three crew (pilot, co-pilot/radar operator and MCM operator) Landysh mission computer and PK-025 datalink; fuselage strake, for hydraulic tubing, and air conditioning pod on starboard side of cabin; no MAD; searchlight and windows in lower part of rear fuselage to observe MCM gear during deployment and retrieval. Marginal performance when towing sled; poor safety record; consequently, most converted for SAR duties.

Mi-14PS ('Haze-C'): Search and rescue version, carrying 10 20-place liferafts; room for 19 survivors in cabin, including two on stretchers; provision for towing many more survivors in liferafts; able to drop floating containers of survival equipment, including food, medicines and clothes; fuselage strake and air conditioning pod as Mi-14BT; double-width sliding door at front of cabin on port side, with retractable rescue hoist able to lift up to three persons in basket; searchlight each side of nose and under tailboom; three crew.

Mi-14PX: Single Polish Mi-14PL (1003) converted for SAR training with all portable ASW equipment removed, extra liferafts and sponson-mounted searchlights.

Mi-14P and Mi-14GP: Conversions of military variant for civil use (*gruzo/passazhirski*: cargo/passenger) offered by Konvers-Avia (subsequently renamed Konvers-Mil; *e-mail:* uuconavi@conavia,msk.ru); 24 seats or 5,000 kg (11,023 lb) payload; seats bolted to floor in Mi-14P, but rapidly removable in Mi-14GP; two 1,434 kW (1,923 shp) TV3-117M turboshafts; maximum T-O weight as all Mi-14s; range 486 to 594 n miles (900 to 1,100 km; 559 to 683 miles). Various nose fairings for weather radar or downward-looking searchlight. Two prototypes; no sales had been notified by early 2000.

Mi-14PZh 'Eliminator': Original OKB Firefighter project, with tank containing 4,600 litres (1,215 US

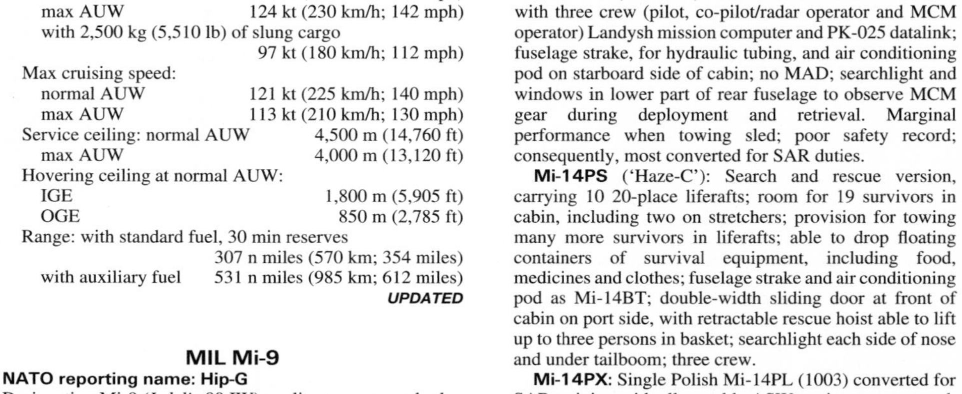

Mil Mi-14PS SAR helicopter of the Russian Navy ***(Yefim Gordon)*** ***2000***/0001135

MIL—AIRCRAFT: RUSSIAN FEDERATION

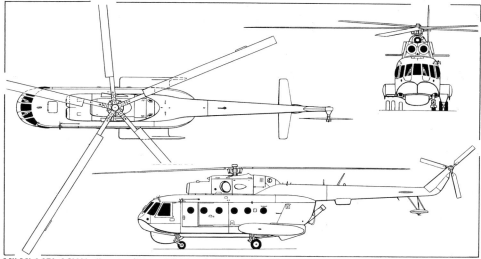

Mil Mi-14PL ASW helicopter ('Haze-A') *(Dennis Punnett/Jane's)*

gallons; 1,012 Imp gallons) of water, plus 400 litres (106 US gallons; 88 Imp gallons) of fire retardant. Water tanks recharged using pump and schnörkel (2 minutes to refill with original electric pumps driven by aircraft DC system; 1 minute using new AC-powered dual hydraulic pumps) or scoops. Latter allowed refill in 12 seconds when taxying at 30 kt (56 km/h; 35 mph), but required removal of undernose radome.

Mi-14 Helitanker: Firefighting conversion by Aerotec GmbH with Isolair firefighting system and new 4,000 litre (1,056 US gallon; 880 Imp gallon) water tank and separate 400 litre (106 US gallon; 88 Imp gallon) retardant tank. System can be installed in 56 man-hours, with no major structural modification.

CUSTOMERS: At least 250 delivered; Russian Naval Aviation has 10 Mi-14PLs and about 55 other versions; Bulgaria (six Mi-14PL, one Mi-14BT), Cuba (14 Mi-14PL), the former East Germany (eight, including Mi-14BT, now withdrawn), North Korea (10 Mi-14PL), Libya (12 Mi-14PL), Poland (nine Mi-14PW, three Mi-14PS, one Mi-14PX), Romania (six Mi-14PL), Syria (20 Mi-14PL), Vietnam (current status unknown), Yugoslavia (four Mi-14PL probably now withdrawn); delivery uncertain, if delivered. Civil aircraft in service in São Tomé and Principe, and elsewhere.

DESIGN FEATURES: Developed from Mi-8; new power plant; modified dynamic components, with tail rotor on port side; VR-14 main gearbox; boat hull giving ability to operate in Sea States 3 to 4, or to plane at up to 32 kt (60 km/h; 37 mph); sponson carrying inflatable flotation bag each side at rear and small float under tailboom; fully retractable landing gear with two forward-retracting single-wheel nose units and two rearward-retracting twin-wheel main units.

POWER PLANT: Two 1,434 kW (1,923 shp) Klimov TV3-117MT turboshafts, with special anti-corrosion finish.

AVIONICS: (Mi-14PL): *Comms:* R-842-M HF transceiver, R-860 VHF transceiver, SBU-7 intercom.
Radar: I-2M or I-2ME Initziativa undernose radar.
Instrumentation: RW3 radio altimeter, ARK-9 and ARK-U2 ADFs, DISS-15 Doppler, Chrom Nikiel IFF, AP34-B autopilot/autohover system and SAU-14 autocontrol system.

AVIONICS (Helitanker): *Comms:* Honeywell KMA 24H Intercom, KT 176A transponder and KY 196A Com 2; R-863 Com 3.
Flight: Honeywell KX 165 nav/com/ILS. Magellan Skynav-5000 or Trimble 2000 GPS; Honeywell KI 202 or KI 525 VOR; Honeywell KEA 129 encoding altimeter, A036 radar altimeter or KAA 1803 altimeter; GMK-1AA compass; AP-34 BS 2 autopilot, and optional DISS-15 Doppler.

DIMENSIONS, EXTERNAL:
Main rotor diameter	21.29 m (69 ft 10¼ in)
Main rotor blade chord	0.52 m (1 ft 8½ in)
Tail rotor diameter	3.91 m (12 ft 9¾ in)
Length: overall, rotors turning	25.30 m (83 ft 0 in)
fuselage: Mi-14PL	18.38 m (60 ft 3½ in)
Mi-14PS	18.78 m (61 ft 7½ in)
Height overall	6.93 m (22 ft 9 in)
Wheelbase	4.13 m (13 ft 6½ in)

AREAS: As for Mi-8

WEIGHTS AND LOADINGS:
Weight empty: Mi-14PL	8,900 kg (19,620 lb)
Max T-O weight	14,000 kg (30,865 lb)
Max disc loading	39.3 kg/m² (8.05 lb/sq ft)

PERFORMANCE:
Max level speed	124 kt (230 km/h; 143 mph)
Max cruising speed	116 kt (215 km/h; 133 mph)
Normal cruising speed	110 kt (205 km/h; 127 mph)
Service ceiling	3,500 m (11,500 ft)
Range with max fuel	612 n miles (1,135 km; 705 miles)
Endurance with max fuel	5 h 56 min

UPDATED

MIL Mi-17 (Mi-8M), Mi-171 and Mi-172
NATO reporting name: Hip-H

TYPE: Multirole medium helicopter.

PROGRAMME: Prototype, known initially as Mi-18, completed 1975 with basic Mi-8 airframe, and power plant and dynamic components of Mi-14. First flew 17 August 1975. Entered service with former Soviet forces in 1977 as Mi-8MT. First displayed at 1981 Paris Air Show as Mi-17 for civil use and export; exports began (to Cuba) 1983; Mi-8MTV followed in 1987; production of improved versions continues at Kazan and Ulan-Ude plants, from where they are being developed and marketed (see these entries for addresses).

CURRENT VERSIONS: **Mi-17** ('Hip-H'): Basic designation of series, with suffix M for military versions. Mi-17P is basic civil version with 28 seats and rectangular cabin windows. All versions in Russian Federation and Associated States' military service retain Mi-8 designations.

Detailed description applies to basic versions, except where indicated.

Mi-8MT ('Hip-H'): Designation of standard Mi-17s in RFAS military service. Twin or triple stores racks, but normal armament is 40 × 80 mm S-8 rockets in two BV-8-20A packs. Afghan experience led to adoption of nose armour, IR jammer, IR suppressors and provision for door-mounted PKT machine gun (rear starboard) and AGS-17 Plamya grenade launcher or NSV 12.7 mm Utyos heavy machine gun (forward port cabin door).

Mi-8AMT: Designation for unarmed version used by RFAS, but also applied to some civil (perhaps ex-military) examples.

Mi-8MT EW variants: More than 30 EW versions of the Mi-8MT serve with RFAS armed forces, under the designations **Mi-8MTSh** (*Shakhta*: mine), **Mi-8MTPSh**, **Mi-8MTU**, **Mi-8MTA**, **Mi-8MTP**, **Mi-8MTPB** (*Bizon*: bison; see below), **Mi-8MTR**, **Mi-8MTI** (*Ikebana*: ikebana; see below), **Mi-8MTPI** and **Mi-8MTTs**.

Mi-8MTPB (or **Mi-17TPB, Mi-17P, Mi-17PP**) ('Hip-H EW'): ECM (radar and communications jammer) and comint helicopter, with three jamming systems in D/F band range over 30° sector and other frequencies over 120°. Operating time 4 hours. Antenna array more advanced than that of Mi-8PPA ('Hip-K'); large 32-element array, resembling vertically segmented panel, aft of main landing gear each side; four-element array to rear on tailboom each side; large radome each side of cabin, below jet nozzle; triangular container in place of rear cabin window each side; six heat exchangers under front fuselage. (Mi-17P designation used also for civil export versions.) Similar versions include **Mi-8MTP**; **Mi-8MTR**; **Mi-8MTSh**; **Mi-8MTPSh**; **Mi-8MTU**; **Mi-8TYa**; **Mi-8MTI** (Mi-17 with small horizontal array on forward part of boom and larger box-like radome on cabin side); **Mi-8MTPI**; **Mi-8MTTs2** and **Mi-8MTTs3** with non-rectangular (teardrop) radome on cabin sides and less regularly shaped arrays on sides of rear cabin.

Mi-8MTV ('Hip-H'): (V=*visotnyi*: high-altitude); 1,633 kW (2,190 shp) TV3-117VM turboshafts for improved 'hot-and-high' operation. Built at Ulan-Ude from 1991; 100 built by 1999. Civil version built at Kazan is **Mi-8MTV-1**; missile-armed, radar-equipped military version with six-hardpoint stub-wing is **Mi-8MTV-2**; export equivalent is **Mi-17-1V**, with optional armament, nose radar, flotation gear and firefighting equipment. Indian Mi-17-1V has enlarged side doors and a rear ramp to facilitate airdrops, as developed at Kazan (which see). Firefighting conversion by Airod in Malaysia adds detachable aluminium boom to starboard side, stowed facing forwards, but trainable through 90°, with hose nozzle at tip. Internal tanks, fillable via own trailing hose and onboard pump in 1½ minutes, carry 2,300 litres (608 US gallons; 506 Imp gallons) of water, which may also be dumped from belly port in 8 seconds. Addition of suffix -**GA** indicates civil version, often with rectangular 'Salon'-type cabin windows.

Mi-8MTO (or **Mi-8N**) ('Hip-H'): (O = *ochki nochnogo videniya*: night vision goggles). Dedicated night attack conversion of Mi-8MT/MTV to meet urgent operational requirement; tested in Chechnya ('Caucasian campaign') with six (some reports suggest four) aircraft equipping BEG (Boyevaya Eperimentalnaya Gruppa: Combat Experimental Group) based at Mozdok. (Unit additionally equipped with four or six upgraded Mi-24s, and also

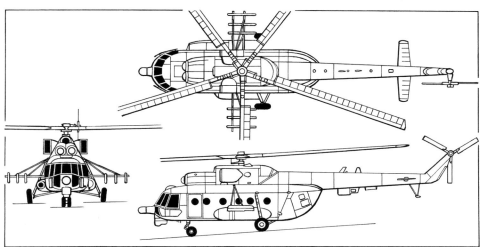

Mil Mi-8AMT(Sh) combat helicopter – the latest Mi-17 variant *(James Goulding/Jane's)*

Mil Mi-8MTV civilianised by Helion Procopter in Germany *(Paul Jackson/Jane's)*

Mi-8AMT(Sh) demonstrator RA-25755 *(Paul Jackson/Jane's)*

parented trial deployment of Ka-50. Originally intended to be equipped with modified Ka-29TB and Ka-50). Upgrade produced by Mil Moscow plant, with collaboration of Russkaya Avionika, NPO Geofizika and Uralsky Optik Mechanichesky Zavod team, and tested at Russian Air Forces flight test centre at Chkalovskaya from April 1999, thereafter undergoing evaluation at Torzhok with the 696th IIVP, equipping BEG with six aircraft from September 1999.

Aircraft has NVG-compatible cockpit and external lights and may also feature NVG-covert, formation-keeping external lights. Crew equipped with Geofizika ONV-1 NVGs, and also use gyrostabilised GOES undernose turret containing AGEMA 1000 or SAGEM-321 FLIR, laser range-finder and LLTV. Incorporates cockpit and navigation system upgrade, with MFI-68 (or, possibly, US-supplied) colour LCD MFDs, A737-00 GPS, full-standard MTO will gain new air data system and Saratov Industrial Automatics OKB PKV-M digital autopilot. May acquire AT-X-16 Vikhr-M ATMs intended for BEG Ka-29TBs, and could operate as laser designation platform for Su-25s and other tactical fast jets. Similar export version offered to India (which has already used its existing Mi-17s in the gunship role in Kashmir, and hopes to upgrade these with an Mi-35-type weapon aiming system, improved avionics and a 'night cockpit') and others as **Mi-17N**.

Mi-17-1VA: Version of Mi-8MTV-1 produced for Ministry of Health of former USSR as flying hospital equipped to highest practicable standards for relatively small helicopter; interior, with equipment developed in Hungary; provision for three stretchers, operating table, extensive surgical and medical equipment, accommodation for doctor/surgeon and three nursing attendants.

Mi-8MTV-3/Mi-172: As Mi-8MTV-2, also from Kazan, but with four-hardpoint stub-wing and other equipment changes and planned for certification to FAR Pt 29 standards; TV3-117VM Srs 2 engines, giving maximum cruising speed of 118 kt (218 km/h; 135 mph) and service ceiling of 6,000 m (19,680 ft); air conditioning and heating systems, main and tail rotor blade de-icing, canopy demisting and heating of engine air intakes standard; options include flotation gear, Doppler, weather radar, DME, GPS, VOR, ILS, transponder and VIP interiors for seven, nine and 11 passengers. Standard seating for up to 26 passengers.

Mi-172 export version first exhibited at 1994 Singapore Air Show; based on 'salon' (rectangular window) cabin. Seven ordered by Mesco, India, 1995; others to Vietnam. Similar Mi-8MTV-5 (Mi-17MD) shipped to Rwanda 1999 with hydraulic ramp and enlarged doors on each side of cabin.

Mi-17PI: As Mi-17P but single D band jamming system able to jam up to eight sources simultaneously over 30° sector.

Interior of Helion Mi-8MTV conversion; door leads to baggage/freight compartment at rear *(Paul Jackson/Jane's)*

Mil Mi-8MTO night attack helicopter conversion *(Paul Jackson/Jane's)*

Mi-17PG: As Mi-17P but with H/I band system for jamming pulse/CW and CW interrupted equipment.

Mi-17AMT: Provisional designation for proposed Ulan-Ude-built variant offered to Malaysia to meet eventual 40-aircraft requirement, with initial order for six CSAR-configured aircraft reportedly already placed. May have rear loading ramp of Mi-8MTV-5. Mi-17 will replace 31 S-61 Nuri helicopters, and two already in use with Malaysian Fire Department. SME Aerospace appointed as local partner responsible for assembly, flight test and local certification.

Mi-8AMT(Sh) 'Terminator': Current counterpart of Mi-8MTV series built at Ulan-Ude. Crew positions protected by armour plating. Armament can include up to eight Igla-V AAMs or 9M114/9M114M Shturm ('Spiral') ASMs, with thimble radome on nose and chin-mounted electro-optics pod. Prototype (RA-255755) exhibited at Farnborough in September 1996. First production helicopter almost complete by late 1998, but delayed by funding shortage. State testing began 10 April 2000.

Mi-171: Export version of Mi-8AMT; first displayed 1989 Paris Air Show. Mostly Western avionics, including Honeywell and Rockwell Collins radios. Also offered as armed **Mi-171(Sh)**. Built at Ulan-Ude.

Mi-17MD (Mi-8 MTV-5): Described under Kazan heading in this section.

Mi-17KF: Described under Kazan heading in this section.

Mi-17LL: Flying testbed (*laboratoriya*: laboratory). One aircraft, RA-70880, employed by the Gromov Flight Research Institute at Zhukovsky on trials including atmospheric sampling for contamination and searching for leakage from oil and gas pipelines. Special equipment includes a sampling probe mounted on a framework extending from the nose and a downward-viewing window on the starboard side.

Mi-18: Originally, was the designation of the prototype Mi-17 (which see). In 1979, two Mi-8MTs (93038 and 93114) were modified by Kazan to meet requirements for the campaign in Afghanistan, under the same designation. These Mi-18s were lengthened by 1.0 m (3 ft 3¼ in) and fitted with an additional sliding door on the starboard side. Payload was 5,000 kg (11,023 lb). Provision for 30 passengers. Later (1984) use of retractable landing gear raised maximum speed to 148 kt (275 km/h; 170 mph). The airframes were used eventually for static training.

Mi-19: Command relay platform; similar to Mi-9.

Mi-17Z-2: Converted from 'Hip-H' in former Czechoslovakia for electronic warfare ECM role; local designation (Z-1 is similarly tasked Antonov An-26). First seen in Czech Air Force service at Dobrany-Line airbase, near Plzen, 1991; each of two examples had a tandem pair of large cylindrical containers mounted each side of cabin; assumed that containers made of dielectric material and contain receivers to locate, analyse and jam hostile electronic emissions; each of two operators' stations in main cabin has large screens, computer-type keyboards and oscilloscope; several blade antennas project from tailboom. Currently in service with Slovak Air Force.

Long-range modification: AEFT (Auxiliary External Fuel Tanks) system by Aeroton adds a further 1,900 litres (502 US gallons; 418 Imp gallons) in two internal tanks, plus 2,850 litres (753 US gallons; 626 Imp gallons) in six tanks on the stores pylons of Mi-8MT, AMT, MTV-1, civil MTV and Mi-17 variants. Operational range with all eight auxiliary tanks is 701 n miles (1,300 km; 807 miles); ferry range 998 n miles (1,850 km; 1,149 miles). Users include Leningrad customs service.

Upgrades: Kazan offers three stages of upgrade for existing Mi-17 variants. The first can include new composites main rotor blades with a 6,000-hour service life, NVG-compatible cockpit and improved avionics; second could include more powerful 1,864 kW (2,500 shp) engines, uprated transmission and Delta-H 'X-type' tail rotor, latter refinements presaging increase in MTOW to 14,000 kg (30,864 lb).

Kronshtadt is offering an avionics and cockpit upgrade, with LCD screens, digital moving map and embedded GPS.

Civil conversion: Helion Procopter Industries GmbH of Bitburg, Germany (joint venture by Mil, Kazan, Ulan-Ude and Wolfgang Wibbelt), demonstrated prototype of ex-military Mi-8MTV conversion with VIP interior at ILA, Berlin, June 2000; then estimating FAA/JAA certification within 18 months; Western avionics optional.

CUSTOMERS: Many operational side by side with Mi-8s in RFAS armed forces; also operated in Afghanistan, Angola, Bangladesh, Bulgaria, Burkina Faso, Cambodia, China,

MIL—AIRCRAFT: RUSSIAN FEDERATION

Colombia, Costa Rica, Croatia, Cuba, Czech Republic, Ecuador, Egypt, Ethiopia, Hungary, India, Indonesia, Laos, Malaysia, North Korea, Nicaragua, Pakistan, Papua New Guinea, Mexico Peru, Poland, Romania, Sierra Leone, Slovak Republic, Sri Lanka, Syria, Turkey, Venezuela, Vietnam, Yugoslavia. More than 810 exported by Aviaexport. Recent announced sales include five Mi-171 to China, 10 Mi-171V to Colombia, 20 Mi-17-1V to Egypt, 40 Mi-17-1V to India, eight Mi-17-1V to Indonesia, five Mi-171s to Iran, 12 Mi-17V to Laos and 19 Mi-17-1V to Turkey. The type was evaluated by the Hellenic Army during mid-2000.

COSTS: Quoted by ITAR-TASS as being from US$4 million to US$5 million; India's 40 Mi-17-1Vs cost a reported US$170 million.

DESIGN FEATURES: Distinguished from basic Mi-8 by port-side tail rotor; shorter engine nacelles, with air intakes extending forward only to mid-point of door on port side at front of cabin; small orifice each side forward of jetpipe; correct rotor speed maintained automatically by system that also synchronises output of the two engines.

POWER PLANT: Two 1,397 kW (1,874 shp) Klimov TV3-117MT turboshafts in basic Mi-17; should one engine stop, output of the other increased automatically to contingency rating of 1,637 kW (2,195 shp), enabling flight to continue. Alternative high-altitude TV3-117VM engine has installed ratings of 1,545 kW (2,071 shp) in emergency; 1,397 kW (1,874 shp) for T-O; 1,250 kW (1,677 shp) normal; and 1,103 kW (1,479 shp) for cruising. Deflectors on engine air intakes prevent ingestion of sand, dust and foreign objects. Fuel as Mi-8T.

ACCOMMODATION: Configuration and payloads generally as Mi-8; but six additional centreline seats optional. Military Mi-17-1V carries up to 30 troops, up to 20 wounded in ambulance role, or weapons on six outrigger pylons; civilian Mi-17 promoted as essentially a cargo-carrying helicopter, with secondary passenger transport role.

SYSTEMS (Mi-17-1V/171): AI-9V APU for pneumatic engine starting; AC electrical supply from two 40 kW three-phase 115/220 V 400 Hz GT40/P-48 V generators.

AVIONICS (Mi-17-1V/171): *Comms:* Baklan-20 and Yadro-1G1 com radio.
Radar: Type 8A-813 weather radar optional.
Flight: Type A-723 long-range nav.
Instrumentation: ARK-15M radio compass, ARK-UD radio compass, DISS-32-90 Doppler; AGK-77 and AGR-74V automatic horizons; BKK-18 attitude monitor; ZPU-24 course selector; A-037 radio altimeter.
Self-defence (optional): ASO-2V chaff/flare dispensers

Mil Mi-8MTV-1 emergency services helicopter, operated for MChS Rossii by TsENTROSPAS *(Yefim Gordon)*   2000/0064902

under tailboom and L-166V IR jammer (NATO 'Hot Brick') at forward end of tailboom.

EQUIPMENT: Options as for Mi-8, plus, on military versions, external cockpit armour, triple-lobe engine nozzle IR suppressors and a VMR-2 fit for air-dropping such stores as mines.

ARMAMENT: Options as for Mi-8, plus 23 mm GSh-23 gun packs. (AAMs and newer ASMs on Mi-8AMT(Sh).)

DIMENSIONS, EXTERNAL:
Main rotor diameter	21.25 m (69 ft 10½ in)
Tail rotor diameter	3.91 m (12 ft 9⅞ in)
Distance between rotor centres: Mi-8	12.65 m (41 ft 6 in)
Mi-17	12.66 m (41 ft 6 in)
Length: overall, rotors turning:	
Mi-8	25.33 m (83 ft 1¼ in)
Mi-17	25.35 m (83 ft 2 in)
fuselage: Mi-8	18.31 m (27 ft 3¼ in)
Mi-17	18.465 m (60 ft 7 in)
nose to tail rotor c/l: Mi-8	18.22 m (59 ft 9¼ in)
Mi-17	18.425 m (60 ft 5½ in)
Mi-172	18.855 m (61 ft 10¼ in)
Mi-17MD	18.99 m (62 ft 3¾ in)
fuselage, excl tail rotor	18.17 m (59 ft 7¼ in)
Width: fuselage, excl fuel tanks	2.50 m (8 ft 2½ in)
over weapon pylons (Mi-17MD)	7.20 m (23 ft 7½ in)
Tailplane span	3.705 m (12 ft 1¾ in)

Height: overall, rotors turning	5.545 m (18 ft 2¼ in)
to top of rotor head: Mi-8	4.73 m (15 ft 6¼ in)
Mi-17	4.745 m (15 ft 6¼ in)
Mi-17MD, Mi-171	4.755 m (15 ft 7¼ in)
Mi-172	4.865 m (15 ft 11½ in)
Wheel track: main	4.51 m (14 ft 9½ in)
nose	0.30 m (11¾ in)
Wheelbase	4.28 m (14 ft 0½ in)
Fwd passenger door: Height	1.41 m (4 ft 7½ in)
Width	0.82 m (2 ft 8¼ in)
Rear passenger door: Height	1.70 m (5 ft 7 in)
Width	0.84 m (2 ft 9 in)
Rear cargo door: Height	1.82 m (5 ft 11½ in)
Width	2.34 m (7 ft 8¼ in)

DIMENSIONS, INTERNAL:
Passenger cabin: Length	6.36 m (20 ft 10¼ in)
Width	2.34 m (7 ft 8¼ in)
Height	1.80 m (5 ft 10¾ in)
Cargo hold (freighter, excl doors):	
Length at floor	5.34 m (17 ft 6¼ in)
Width	2.29 m (7 ft 6¼ in)
Height	1.80 m (5 ft 10¾ in)
Volume	22.5 m³ (795 cu ft)

AREAS:
Main rotor disc	356.16 m² (3,833.7 sq ft)
Tail rotor disc	12.01 m² (129.28 sq ft)

WEIGHTS AND LOADINGS:
Weight empty, equipped: Mi-17	7,100 kg (15,653 lb)
Mi-17-1V/171	7,055 kg (15,555 lb)
Mi-17-1V	7,489 kg (16,510 lb)
Mi-17-1VA	7,586 kg (16,724 lb)
Mi-8AMT(Sh)	8,493 kg (18,724 lb)
Internal fuel weight, Mi-17-1V/171:	
normal	2,027 kg (4,469 lb)
plus one auxiliary tank	2,737 kg (6,034 lb)
plus two auxiliary tanks	3,447 kg (7,600 lb)
Max payload:	
Internal: Mi-17, Mi-17-1V/171, Mi-8AMT(Sh)	4,000 kg (8,820 lb)
external, on sling:	
Mi-17, Mi-17-1V	3,000 kg (6,614 lb)
Mi-171, Mi-8AMT(Sh)	4,000 kg (8,820 lb)
Mi-17-1V	5,000 kg (11,023 lb)
Normal T-O weight:	
Mi-17, Mi-17-1V/171, Mi-8AMT(Sh)	11,100 kg (24,470 lb)
Max T-O weight:	
Mi-17, Mi-17-1V/171, Mi-8AMT(Sh)	13,000 kg (28,660 lb)
Mi-172	11,878 kg (26,186 lb)
Max disc loading:	
Mi-17, Mi-17-1V/171	36.5 kg/m² (7.48 lb/sq ft)

Mil Mi-17-1V firefighting conversion by Airod, Malaysia *(Paul Jackson/Jane's)*   2001/0093707

Mil Mi-171 instrument panels   2000/0064903

Projected Mi-17 instrument panel, with four LCDs, offered by SME of Malaysia *(Paul Jackson/Jane's)*   2000/0064904

PERFORMANCE:
Max level speed:
  Mi-17/172, max AUW   135 kt (250 km/h; 155 mph)
  Mi-17-1V/171, normal AUW
                      141 kt (262 km/h; 163 mph)
  Mi-17-1V/171, max AUW
                      124 kt (230 km/h; 143 mph)
Max cruising speed:
  Mi-17/172, max AUW   129 kt (240 km/h; 149 mph)
  Mi-17-1V, normal AUW 135 kt (250 km/h; 155 mph)
  Mi-8AMT (Sh), normal AUW
                      124 kt (230 km/h; 143 mph)
  Mi-171   113 kt (210 km/h; 130 mph)
  Mi-17-1V/171 and Mi-8AMT(Sh), max AUW
                      116 kt (215 km/h; 134 mph)
Service ceiling:
  Mi-17M, normal AUW   5,600 m (18,380 ft)
  Mi-17M, max AUW   4,400 m (14,440 ft)
  Mi-17-1V, normal AUW   6,000 m (19,680 ft)
  Mi-17-1V, max AUW   4,800 m (15,740 ft)
  Mi-171, normal AUW   5,700 m (18,700 ft)
  Mi-171, max AUW   4,500 m (14,760 ft)
  Mi-172, max AUW   5,650 m (18,535 ft)
Hovering ceiling OGE:
  Mi-17, max AUW   1,760 m (5,775 ft)
  Mi-17M, normal AUW   3,900 m (12,795 ft)
  Mi-17M, max AUW   1,500 m (4,920 ft)
  Mi-17-1V/171, normal AUW 3,980 m (13,055 ft)
  Mi-17-1V/171, max AUW   1,700 m (5,575 ft)
  Mi-172, max AUW   3,300 m (10,825 ft)
Range with max standard fuel, 5% reserves:
  Mi-17, normal AUW 267 n miles (495 km; 307 miles)
  Mi-17, max AUW   251 n miles (465 km; 289 miles)
  Mi-172, max AUW   267 n miles (495 km; 307 miles)
Range at 500 m (1,640 ft), max AUW, 30 min reserves:
  Mi-17M/-1V, standard fuel
              251 n miles (465 km; 289 miles)
  Mi-171 and Mi-8AMT(Sh), standard fuel
              313 n miles (580 km; 360 miles)
  Mi-171, standard fuel plus one auxiliary tank
              440 n miles (815 km; 506 miles)
  Mi-17M, standard fuel plus two auxiliary tanks
              545 n miles (1,010 km; 627 miles)
  Mi-171 and Mi-8AMT(Sh), standard fuel plus two
    auxiliary tanks   675 n miles (1,250 km; 776 miles)
  Mi-17M, standard fuel plus four auxiliary tanks
              809 n miles (1,500 km; 932 miles)

UPDATED

MIL Mi-18
See Mi-17 entry.

MIL Mi-19
See Mi-8MT (Mi-17) and Mi-9.

MIL Mi-22
NATO reporting name: Hook-C
See Mi-6 entry earlier in this section.

MIL Mi-24, Mi-25 and Mi-35
NATO reporting name: Hind
TYPE: Attack helicopter.
PROGRAMME: Development began second half of 1960s, as first fire support helicopter in former USSR, with accommodation for eight armed troops. Complete redesign after construction of 1966 Bell UH-1-sized mockup with skid-type landing gear and side-by-side cockpit. Mil issued with directive to submit new plans in 1967, building three new mockups with five alternative forward fuselage arrangements. All featured replacement of fixed GSh-23 twin-barrelled cannon with faster-firing machine gun in powered turret, and provision for 9M114 Shturm V (AT-6 'Spiral') ATMs; 10.5 tonne aircraft, with two JV3-117A engines, chosen over lighter single-engined alternative. This aircraft (*Izdelie* 240) based on Mi-14 dynamic system, with streamlined new fuselage.

Two V-24 prototypes built by MVZ 329 (the Mil workshops) at Panki; first flight 19 September 1969; 10 preseries Mi-24s followed, five built by MVZ 329, five by Progress (MVZ 116) at Arsenyev. State acceptance trials June 1970 to December 1971. All with TV3-117A engines, not TV2-117 as sometimes reported.

First reported in West and production started 1972; photographs became available 1974, when two units of approximately squadron strength based in East Germany; reconfiguration of front fuselage changed primary role to gunship; new version first observed 1977; used operationally in Afghanistan, Angola, Chad, Chechnya, Iran/Iraq war, Nicaragua and Sri Lanka, when at least one Iranian F-4 Phantom II destroyed by 9M114 (AT-6 'Spiral') anti-tank missile from Mi-24; peak production rate at Progress plant, Arsenyev, was 165 a year but tooling there dismantled, 1989. Late models continue to be available from Rostvertol at Rostov-on-Don, where production continues at low rate for export and for Interior Ministry.

CURRENT VERSIONS: **Mi-24** (*Izdelie* 240 'Hind-B'): Prototypes and pre-series aircraft with simple tapered wing with no

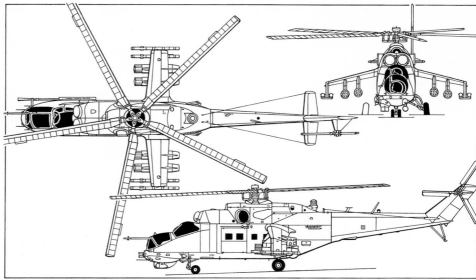

Mil Mi-24P gunship, known to NATO as 'Hind-F' *(Mike Keep/Jane's)*

anhedral and simple underslung BD3-57Kr-V racks. Pilot (offset to port) and WSO (forward) in tandem under heavily glazed cockpit. One modified in 1975 as 'A-10' for successful speed record attempts with wings removed and faired over and with inertia-type dampers on the main rotor head. TV3-117A engines. One later used to test Fenestron tail rotor.

Mi-24A (*Izdelie* 245 'Hind-A'): Initial production version; similar heavily glazed angular cockpit to prototypes and preseries aircraft, but with extended forward fuselage, giving more pointed nose in plan view and with less steeply pitched 'roof' glazing. Single-barrel Afanasyev A-12.7 (TKB-481) 12.7 mm machine gun in NUV-1 flexible mounting in tip of nose underside. Aimed using simple PKV collimator gunsight. Pilot's door (on port side) replaced by large sliding bubble window. WSO still entered through upward-opening side window. Fuselage stretched to accommodate Raduga-F (Rainbow-F) semi-automatic command to line-of-sight missile guidance system, presence of which was indicated by small teardrop fairing in front of nose landing gear. Armed with manually guided 9M17M (AT-2 'Swatter') ATGMs. Anhedral added to stub wings to improve lateral stability and cure high-speed Dutch roll. ATM launch rails relocated from fuselage sides to new endplate pylons at wingtips. Two Mi-24A prototypes produced by grafting new nose on to pre-series Mi-24s; entered production at Arsenyev in 1970.

Mi-24U (*Izdelie* 244 'Hind-C'): Unarmed pilot conversion trainer based on Mi-24A, but lacking nose-mounted gun, wingtip missile launch rails and undernose Raduga antenna. Instructor in former WSO position, with full dual controls and instruments. Built at Arsenyev in small numbers, with others possibly produced by conversion of redundant Mi-24/Mi-24As.

Mi-24F? (*Izdelie* 245M 'Hind-A'): Inadequate tail rotor authority led to replacement of starboard pusher tail rotor by tractor tail rotor on port side from 1972. Seven reinforcing ribs added to port fuselage aft of wing, SRO-2M Khrom ('Odd Rods') IFF antenna relocated from canopy to oil cooler; APU exhaust extended and angled downwards. Total production of Mi-24, Mi-24A, Mi-24U and Mi-24F about 240, ending in 1974.

Mi-24B (*Izdelie* 241 'Hind-A'): Up-gunned and improved model with new 12.7 mm Yakoushev/Borzov YakB (TKB-063 or 9A624) 12.7 mm four-barrel machine gun in USPU-24 powered chin turret, traversable through 120° in azimuth and from +20 to –40° in elevation/depression, and slaved to KPS-53AV sighting system. Manually controlled ATGMs replaced by 9M17P Falanga-P and Falanga-PV with SACLOS guidance. Traversing radio command link antenna moved from centreline to below port side of nose, with gyrostabilised collimated LLTV/FLIR under starboard side in fixed fairing. Passed company trials 1971-72 but overtaken by *Izdelie* 246 ('Hind-D') and abandoned. Full-scale mockup produced from pre-series Mi-24 with undrooped wing, prototype from early Mi-24A with normal anhedral wing. Retained fully retractable landing gear like all previous Mi-24 variants.

Mi-24D (*Izdelie* 246; 'Hind-D'): Interim gunship version combining 'old' weapon system of Mi-24B with new airframe designed for planned Mi-24V due to delays with that aircraft's Shturm-V ATGMs with SPS-24V fire-control system, consisting of KPS-53AV weapons control unit and KS-53 gunsight; design began 1971; two prototypes converted from Mi-24A, with starboard-side tail rotor; entered production at Arsenyev and Rostov plants 1973; about 350 built 1973-77. Basically as late model 'Hind-A' with TV3-117 engines and port-side tail rotor, but entire front fuselage redesigned above floor forward of engine air intakes; separate armoured cockpits for weapon operator and pilot in tandem; flight mechanic optional, in main cabin; transport capability retained; USPU-24 gun system, with range-finding; undernose YakB-12.7 four-barrel 12.7 mm machine gun in turret, slaved to adjacent KPS-53A electro-optical sighting pod, for air-to-air and air-to-surface use; long air data boom with DUAS-V pitch and yaw vanes; Falanga P anti-tank missile system; nosewheel leg extended to increase ground clearance of sensor pods; wing pylons plumbed for 500 litre (132 US gallon; 110 Imp gallon) drop tanks;

Mil Mi-24V 'Hind-E' of the Czech Air Force *(Paul Jackson/Jane's)*

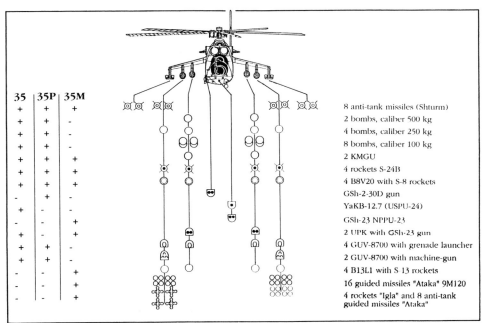

Weapon options for the export Mi-35 variants, as offered by Rostvertol 1999/0051368

nosewheels semi-exposed when retracted; S-13 camera moved from port wingroot to port wingtip/endplate junction. **Mi-24DU** (*Izdelie* 249) dual-control training version has no gun turret. (See also Mi-25.) **Mi-24PTRK** was testbed for Shturm V missile system of Mi-24V.

Detailed description applies to Mi-24D, except where indicated.

Mi-24V (*Izdelie* 242; 'Hind-E'): Up-engined, improved version powered by TV3-117V engines (V = *vysotnii* 'high-altitude') rated at 1,633 kW (2,190 eshp). Airframe as Mi-24D, but modified wingtip launchers and four underwing pylons; empty weight 8,620 kg (19,004 lb); weapons include up to eight 9M114 (AT-6 'Spiral') radio-guided tube-launched anti-tank missiles in pairs in Shturm V (Attack) missile system; fixed enlarged undernose automatic missile guidance pod on port side (antenna inside was articulated), with fixed searchlight to rear; ASP-17V gunsight for pilot; R-60 (K-60; AA-8 'Aphid') air-to-air missiles optional on underwing pylons; pilot's HUD replaces former reflector gunsight. Deliveries to Soviet Air Force began 29 March 1976; about 1,000 built at Arsenyev and Rostov 1976-86. From 1981 usually with PZU filters over engine intakes, and from 1984, with provision for triple-lobe IR filter boxes over downward-pointing exhausts and an L-166V-11E Ispanka (Spaniard) or SOEP-V1A 'Lipa' IR jammer on 'Wendy House' fairing. R-863 VHF, R-828 army radio, SRO-2M IFF replaced by SRO-1P Parol L-006LM Beryoza RHAWS on late aircraft. (See also Mi-35.)

Mi-24V (Mi-24VD; D = *Dorabotanni*: Terminator): High proportion of combat losses in Afghanistan inflicted from the rear hemisphere. Mi-24VD produced in 1985 as testbed for rearward-firing defensive armament. Bulged gondola installed in place of rear avionics bay, accessed via narrow crawlway. Equipped with 12.7 mm NSVT-12.7 Utyos machine gun. Gunner entered turret in flight, legs dangling into the slipstream, encased in built-in rubberised fabric 'trouser' bag. Project abandoned 1986.

Mi-24VM/Mi-24PM/Mi-35M: ('Hind-E'): Mil Design Bureau three-stage, five-block MLU configuration for estimated 200 surviving RFAS Mi-24V/Mi-24P. (Stages 1, 2 and 3 equate to Blocks 2, 3 and 4). Will extend retirement dates from 2004 (Mi-24D) and 2010 (Mi-24V/P). Exact content of RFAS upgrade (and correct designation) confused by appearance of static demonstrators at Paris and MAKS air shows. These had mix of features from different stages of planned RFAS upgrade, and some aspects designed primarily to attract export customers (and later incorporated in Stage 3), including Sextant/Thomson-CSF Nadir 10 navigation system and NOCAS (Night Operation Capable Avionics System). Some sources suggest Mi-24VM designation will apply to Mi-24V or Mi-24VP after upgrade, and Mi-24PM to upgraded Mi-24Ps. Others maintain that the PM suffix applies only to aircraft which retain fixed 30 mm cannon associated with Mi-24P. Three Mi-24VM prototypes flying by late 1999: one with TV3-117VMA-SB3 engines; second for trials of GOES-342 sighting unit; third with new main and tail rotors. Designation **Mi-24VK-1** assigned to night trials prototype with unspecified equipment, flying by late 2000; an **Mi-24VK-2** was due to have joined the programme in December 2000. Actual Mi-24VM upgrade consists of:

Block 1: Life extension to 4,000 flying hours/40 years. May be carried out in parallel with Stage 1/Block 2.

Stage 1: Aircraft brought to Mi-24VM-1/Mi-35M1 standard with Geofizika NVG-compatible cockpit (LCD MFDs, embedded GPS/INS). Glass fibre main rotor blades (with greater thrust) and Mi-28-type Delta-H tail rotor, 1,636 kW (2,194 shp) TV3-117VMA or 1,839 kW (2,466 shp) TV3-117VMA-SB3 engines, extra oil cooler and more powerful KAU-115 control system actuators. Existing machine gun or 30 mm cannon armament replaced by Mi-24VP's NPPU-24 turret with GSh-23l cannon. Outer part of wing, endplate fins and existing missile launchers removed, with APU-8/4U launchers (eight 9M114 ATMs each) on remaining four (BD3-UV) underwing pylons, which have built-in hoists. S-13 attack camera replaced by SSh-45 (monitored in pilot or WSO sight). Strengthened, non-retracting landing gear. (Wing and landing gear, modifications incorporated in prototype, but believed planned as part of second stage). Reduced empty operating weight (8,350 kg/18,409 lb), compared to Mi-24V. Prototype converted from Mi-24VP ('51' of Berkuty formation team) first flew on 8 and 11 February, before official maiden flight on 4 March 1999. Six production conversions to be delivered during 2000.

Stage 2: Aircraft brought to Mi-24VM-2/Mi-35M2 standard with Mi-28N main rotor; GSh-23L replaced by liquid-cooled GSh-23V (*vodyanoye okhlazhdniye*: water cooling), with 450 rounds. Communications system improvements include lightweight tactical datalink and R-999 VHF radio. New or revised ATM guidance compatible with passive IR-homing 9M39 (SA-18 'Grouse') Igla AAM in place of the 9M36 (AT-14 'Gremlin').

Stage 3: Aircraft brought to Mi-24VM/Mi-35M3 standard with new digital PNK-24 avionics suite, closely based on PrNK-28 of Mi-28N. Zenit Tor-24 weapon control system replaces Raduga-F. Shturm V ATM replaced at Ataka V (9M120 AT-12 'Swinger'). L-166V-1E deleted and replaced by Mak-UFM IR- and Otklik laser-MAWS. Beryoza RWR replaced by Pastel RHAWS controlling UV-26 chaff/flare dispensers. Empty operating weight further reduced to 8,200 kg (18,078 lb), with commensurate improvement in hover, speed and range performance.

Block 5: Incorporates NVG-compatible cockpit and FLIR, laser range-finder, helmet-mounted displays and new navigation/display system software to optimise low-level/short duration mission capability. Originally planned to have Sextant/Thomson NOCAS with Chlio FLIR in turret mounted on port side of forward fuselage, TMM-1410 LCD MFD in WSO cockpit and SMD-45H and TMM-1410 in pilot's cockpit, together with VH 100 HUD. French Chlio FLIR replaced by UOMP GOES-342 (OPS-24N) TV/FLIR sighting unit (similar to Shamshit of Ka-50Sh) with Swedish AGEMA FLIR. French MFDs probably replaced by indigenous MFI-10 or MFI-68 152 ×203 mm (6 × 8 in) MFDs. GOES-342 also offered as heart of upgrade proposed to Poland in late 1997.

Arsenal Mi-24V upgrade: Ukrainian upgrade configuration centred around French (Elno) developed Sura HMTDS (helmet-mounted target designation system) for pilot and WSO. NVG-compatible cockpit, new ASP-17VM gunsight computer and INS with embedded GPS. Improved weapon-aiming algorithms. Single aircraft converted as demonstrator by Arsenal, Kiev.

Mi-24VP (*Izdelie* 258): Final basic Army Aviation production version, based on Mi-24V with twin-barrel GSh-23L 23 mm gun in NPPU-24 flexible mount with 450 rounds, in place of four-barrel 12.7 mm gun in nose; photographed 1992; small production series of 25 built at Rostov, entering service in 1989. Production curtailed by ammunition feed problems. One VP flew with Mi-28-type Delta H tail rotor.

'**Mi-24VU**' ('Hind-E'): No dedicated trainer version of Mi-24V produced by OKB or factories for Russian or former Soviet Army. India uses small number of these trainer versions (possibly locally converted) with gun turret removed and faired over, and with dual controls and instruments for instructor in front cockpit. Mi-24VU, Mi-25VU and Mi-35U designations may be unofficial, even in India.

Mi-24P (*Izdelie* 243; 'Hind-F'): Development started 1974; about 620 built 1981-89; first shown in service in 1982 photographs; P designation refers to *pushka*: cannon; as Mi-24V, but nose gun turret replaced by GSh-30K twin-barrel 30 mm gun (with 750 rounds) in semi-cylindrical pack on starboard side of nose; bottom of nose smoothly faired above and forward of sensors. Alternative **Mi-24G** has gun on starboard side.

Mi-24RKhR or **Mi-24R** (*Izdelie* 2462 'Hind-G1'): Dedicated NBC reconnaissance aircraft to replace Mi-8VD. RKhR = *dlya Radiatseeonno-Khimeccheskoi Razvedki* (NBC reconnaissance). Identified at Chernobyl after April 1986 accident at nuclear power station; no undernose electro-optical and RF missile guidance pods, strike camera deleted, but pylons for underwing stores retained; instead of wingtip weapon mounts, has 'clutching hand' excavator mechanisms on lengthened pylons, to obtain six soil samples per sortie, for NBC (nuclear/biological/chemical) warfare analysis; air samples sucked in via pipe on port side, forward of doors exhausting through horizontal slit above; datalink to pass findings to ground; lozenge-shape housing with exhaust pipe of air filtering system under port side of cabin; bubble window on starboard side of main cabin; small rearward-firing marker flag/flare pack on tailskid; crew of four wear NBC

First Mi-24VM upgraded Hind, converted from an Mi-24VP (*Yefim Gordon*) 2001/0099592

Two (MFI-68) screen pilot's cockpit of Mi-24V/Mi-24VM development aircraft 20961
(Paul Jackson/Jane's) 2000/0075929

Ukrainian Mil-Mi-24VP *(Yefim Gordon)* 2001/0099591

'HIND' OPERATORS

Customer	Variant	Delivered/Remaining	Date
Afghanistan	Mi-24A, U, 25, 35	36 plus top-ups/0	1979
Algeria	Mi-24A, 24D	38/c 33-35	1978
Angola	Mi-25, 35, 35P	36+/20-28	1983
Armenia	Mi-24P, K, RKhR	7 P, 3 K, 2 RKR/12	1991
Azerbaijan	Unknown	C 15	1991
Belarus	Mi-24D, V, K, RKhR	Unknown/61	1991
Bosnia	Mi-35	Five reportedly ordered	
Bulgaria	Mi-24D, 24V	38 D, 6 V/35 and 5	1979
Cambodia	Unknown ex-Vietnamese		c 1985
Congo	Unknown	Unknown	1997
Croatia	Mi-24V	15 + 5/15	1993
Cuba	Mi-24D	20/c 15	1982
Czechoslovakia	Mi-24D, DU, V	28 D, 2 DU, 32 V/0	1978
Czech Republic	Mi-24D, DU, V	16 D, 1 DU, 20 V/36[1]	1993
Ethiopia	Mi-24D, Mi-35	40 +?/30	1980
France	Mi-25D, ex-Libyan	3/0	1987
Georgia	Unknown	4/3	1991
East Germany	Mi-24D, Mi-24P	42 D, 12 P/0	1978
Germany	Mi-24D, Mi-24P	39 D, 12 P/0	1990
Hungary	Mi-24D, Mi-24V	30 D, 11 V/29 + 10[2]	1978
India	Mi-25 Akbar, Mi-35	12 D, 20 35/32	1984
Iraq	Mi-25	Unknown/c 30	c 1979
Kazakhstan	Mi-24D, V	42	1991
Kyrgizia	Unknown	50/most stored	1991
Libya	Mi-24A, 25, 35	Unknown/c 65	c 1978
Mongolia	Mi-24V	12/survivors stored	c 1990
Mozambique	Mi-25	15/4	1984
Nicaragua	Mi-25	12-18/0	1983
Nigeria	Mi-35P	6/6	2000
North Korea		Unknown/c 20	1985
Peru	Mi-25	12 + 7[3]/10 + 5	c 1982
Poland	Mi-24D, V	16 + 22D[4], 23 V/41	1978
Russia	Mi-24D, V, P, VP, K, RKhR	c 700*	1972
Rwanda	Unknown	2	1997
Sierra Leone	Mi-24V	3/2 serviceable	
Slovakia	Mi-24D, DU, V	8 D, 1 DU, 10 V/19	1993
Sri Lanka	Mi-24V	13/7	1995
Sudan	Mi-24V ex-Belarus	6/5	1996
Syria	Mi-24D, Mi-25	50/36	1980
Tajikistan	Unknown	c 5	1991
Turkmenistan	Unknown	10	1991
Uganda	Unknown	2 ex-Belarus[5]	
Ukraine	Mi-24D, V, P, K RKhR	278/more than 50 stored	1991
UK	Mi-25	1 ex-Libyan briefly evaluated	c 1989
USA	Mi-24D, Mi-24P	2 ex-German + captured Iraqi[6]	
Uzbekistan	Mi-24D, V	c 45	1991
Vietnam	Mi-24A, 24D, 25	30/?	c 1984
Yemen	Mi-24D/25	15/12?	
Zimbabwe	Mi-35	5/5	1999

Notes: Several Mi-24s were acquired for Papua New Guinea, but none entered service. Massoud guerilla forces in Afghanistan have six operating from Kulob (Tajikistan) (1997). 'Serbian Police' Mi-24s have been reported. Rostvertol notes 'Mi-35' to be in service with 26 air arms by late 2000.
[1] 25 in use, four Ds and eight Vs stored
[2] 16 ex-German Mi-24s (11 24D, five P) acquired for third squadron, but not put into service
[3] ex-Nicaraguan aircraft (possibly only five purchased)
[4] 22 ex-German Mi-24D, 12 of which were in service by late 1997
[5] Returned to CSC in dispute, 1999. Allegedly not overhauled before 1998 delivery
[6] ex-Iraqi aircraft not flown
* some sources

suits; deployed six per helicopter regiment throughout Russian Federation and Associated States (CIS) ground forces. About 152 built 1983-89.

Mi-24RA (*Izdelie* 2462 [or 2463?] 'Hind-G1 Mod'): New series of conversions from Mi-24V. Retained strike camera in wingroot and lacked wingtip excavators; sometimes seen with pod on port station. Crew reduced to three with improved (presumably automated) processing and data transfer. Probably had slightly different and more specialised role – only one known in Group of Soviet Forces in Germany, for example.

Mi-24K (*korrektirovchik*: corrector) (*Izdelie* 201 'Hind-G2'): Dedicated artillery spotter/fire correction aircraft to replace Mi-8TARK. As Mi-24R, but with large A87P or AFA-100 camera in cabin, f8/1,300 mm lens on starboard side; six per helicopter regiment for reconnaissance and artillery fire correction; gun and B-8V-20 rocket pods retained. No target designator pod under nose; upward-hinging cover for IRIS wide-angle IR and optical sensor system. Rita reconnaissance and spotting system with optical target identification, computer and data processor. About 163 built 1983-1989.

Mi-24BMT (*Izdelie* 248): A few modified 1973 for minesweeping.

Mi-24VN (Mi-35O 'Hind-E'): Interim night-attack version based on conversion of Mi-24V in Mi-24VM Stage 1 configuration. Cockpit and external lights compatible with Geo-ONV-1 NVGs and with new RPKB navigation/fire-control system. May also feature GOES-320 gyrostabilised sensor turret containing Sony EVI331 TV and Agema THV1000 FLIR sensors (for navigation/surveillance, not targeting) or a similar GOES-342 turret with a targeting function for the 9M120 (AT-12) ATMs. Some reports suggest that Mi-24VNs were to be used by Experimental Combat Group in Chechnya. A similar upgrade configuration, with A737 GPS and using two MFI-68 cockpit displays for the pilot (one replacing the S-17V gunsight, one functioning as a colour LCD terrain map), and an MFPU console for the gunner, has been prepared for an unnamed customer. Performance generally as for Mi-24V/Mi-24P.

Mi-24PN: Preliminary tests reportedly under way in mid-2000. Presumed to be a 30 mm cannon-armed 'Hind-F' upgraded with Geofizika FLIR, new laser range-finder, mission computer and NVG-compatible cockpit. Possibly equivalent to Mi-24VN.

Mi-24PS (*Patrul'nospasatelny*: patrol/rescue): Transport/law enforcement/SAR variant for Russian Ministry of the Interior. Production or series conversion status unknown. First prototype converted from Mi-24P, retaining 30 mm cannon and wing endplate pylons. Undernose LLTV/FLIR replaced by downward-pointing loudspeaker group, ATGM guidance antenna by FPP-7 searchlight. Nose cut away to allow installation of weather radar and EO turret. LPG-4 winch (120 kg; 264 lb capacity) installed aft of starboard cabin door, grab rails, foot rests and rappel attachment points around sides of doors. Four of six-man squad carried can rappel from the aircraft simultaneously. Satellite communications, secure encrypted voice radios and special police-band radios. Second prototype similar (albeit painted white, with blue cheatlines and Militia titles) but converted from Mi-24V, with USPU-24 turret replaced by FLIR ball. Marketed as Mi-35PS for export.

Mi-24 Ecological Survey Version: Modification by Polyot industrial research organisation, to assess oil pollution on water and seasonal changes of water level. First seen 1991 with large flat sensor 'tongue' projecting from nose in place of gun turret; large rectangular sensor pod on outer starboard underwing pylon; unidentified modification replaces rear cabin window on starboard side.

Mi-25: Export Mi-24D, including those for Afghanistan, Angola, Cuba, India and Peru. Also **Mi-35D**.

Mi-35: Export Mi-24V also known as Mi-25V.

Mi-35D: Upgrade configurations for export. Sub-variants designated Mi-35D, Mi-35D1 and Mi-35D2, using various weapons and systems from the Kamov Ka-50. Proposed by Rostvertol but not supported by Mil OKB. Borrowed equipment to include Shkval fire-control unit, Saturn FLIR and Vikhr ATMs. Probably abandoned and replaced by joint Mil/Rostvertol upgrade configuration with 1,839 kW (2,466 shp) TV3-117VMA-SB3 engines which give 10 per cent greater power output, and increase engine TBO to 3,000 hours, and life to 7,500 hours. This new upgrade configuration could include redesigned main and tail rotors, together with avionics developed by Ramenskoye and Krasnogorsk.

Mi-35P: Export Mi-24P.

Mi-35M: Upgraded night-capable version of Mi-24/35; export counterpart of Mi-24M; designed to meet the latest air mobility requirements of the Russian Army. Features include Mi-28 main and tail rotors and transmission; 1,636 kW (2,194 shp) Klimov TV3-117VMA engines; new avionics; a reduced empty weight resulting from new

MIL—AIRCRAFT: RUSSIAN FEDERATION

Prototype ATE 'Super Hind', featuring IST Dynamics gun turret

Cockpit of Mi-35O demonstrator/mockup 03033, showing MFI-10 LCD, but otherwise unchanged instrumentation *(Paul Jackson/Jane's)*

titanium main rotor head, composites rotor blades, shortened stub-wings and non-retractable landing gear; a 23 mm GSh-23-2 twin-barrel gun in nose turret, with 470 rounds; up to 16 radio-guided 9M114 (AT-6 'Spiral'), or laser-guided 9M-120 anti-tank, 9M-120F blast fragmentation or 9A-220 air-to-air versions of Ataka (AT-12) missile; or a range of armament options including GUV gun/grenade pods; UPK-23-250 gun pods; B-8V-20A and B-13L rocket pods; S-24B rockets; and KMGU pods of anti-armour and anti-personnel mines. Night Operation Capable Avionics System (NOCAS) by Sextant Avionique and Thomson-TTD Optronic integrates Chlio FLIR ball with a TMM-1410 display, providing night vision for target acquisition and identification, missile guidance and gun aiming.

Other equipment includes a VH-100 HUD, NVGs, liquid-crystal MFD, Nadir 10 mission management and navigation system, laser-gyro INS and GPS. The FLIR ball is mounted outboard of the standard missile guidance pod. Ability to carry Igla V air-to-air missiles is optional. Non-flying demonstrator first displayed at 1995 Paris Air Show.

ATE 'Super Hind': Upgrade configuration proposed by South Africa's Advanced Technologies and Engineering. Derived from Denel/Kentron PZL Świdnik W-3WB upgrade. Extended nose in front of cockpit with undernose IST Dynamics turret, Kentron IR/EO sight and Denel Vektor G12 20 mm chain gun (as used by Rooivalk), cheek fairings both sides for ammunition feed, designator, improved displays, new night vision systems and provision for Denel/Kentron Ingwe or Mokopa ATMs. Prototype Mi-24V ZU-BOI delivered for modifications 26 June 1998; rolled out at Grand Central Airport, Midrand, by 15 February 1999. Customer believed to be Algeria.

Tamam Mi-24 HMOSP ('Mission 24'): Israeli upgrade configuration. US$20 million contract placed for upgrade of 25 (probably Indian) Mi-24s based on existing Helicopter Multimission Optronic Stabilised Payload system, with MLM mission computer, TV, FLIR and automatic target tracker, NVG-compatible cockpit integrated with monocular helmet sight, digital moving map, integrated DASS, embedded GPS, and a new mission planning system. FLIR FoV variable from 2° 24′ to 29° 12′ and with built-in automatic tracking. Cannon can be slaved to pilot or WSO's line of sight. Compatible with Rafael Spike ATM and to be fitted with IMI chaff/flare dispensers. Cockpits have new LCD MFDs and can be reorganised to put pilot in front, weapon operator in rear.

CUSTOMERS: Reported 5,200 produced at Arsenyev and Rostov, including 1,000 exported to over 30 countries; about 700 in Russian Army service, most with helicopter attack regiments of Mi-8/17s and Mi-24s; one supplied to Russian army aviation in 1994 and seven in 1995.

DESIGN FEATURES: Typical helicopter gunship configuration, with stepped tandem seating for two crew and heavy weapon load on stub-wings; fuselage unusually wide for role, due to requirement for carrying eight troops; dynamic components and power plant originally as Mi-8, but soon upgraded to Mi-17-type power plant and port-side tail rotor; VR-24 main gearbox. Main rotor blade section NACA 230, thickness/chord ratio 11 to 12 per cent; tail rotor blade section NACA 230M; stub-wing anhedral 12°, incidence 19°; wings contribute approximately 25 per cent of lift in cruising flight; fin offset 3°.

STRUCTURE: Five-blade constant chord main rotor; forged and machined steel head, with conventional flapping, drag and pitch change articulation; each blade has aluminium alloy spar, skin and honeycomb core; spars nitrogen pressurised for crack detection; hydraulic lead/lag dampers; balance tab on each blade; aluminium alloy three-blade tail rotor; main rotor brake; all-metal semi-monocoque fuselage pod and boom; 5 mm hardened steel integral side armour on front fuselage; all-metal shoulder wings with no movable surfaces; swept fin/tail rotor mounting; variable incidence horizontal stabiliser.

LANDING GEAR: Tricycle type; rearward-retracting steerable twin-wheel nose unit; single-wheel main units with oleo-pneumatic shock-absorbers and low-pressure tyres, size 720×320 on mainwheels, 480×200 on nosewheels. Main units retract rearward and inward into aft end of fuselage pod, turning through 90° to stow almost vertically, discwise to longitudinal axis of fuselage, under prominent blister fairings. Tubular tripod skid assembly, with shock-strut, protects tail rotor in tail-down take-off or landing.

POWER PLANT: Two Klimov TV3-117MT turboshafts, each with T-O rating of 1,434 kW (1,923 shp), side by side above cabin, with output shafts driving rearward to main rotor shaft through combining gearbox. There is 5 mm hardened steel armour protection for engines. Main fuel tank in fuselage to rear of cabin, with two bag tanks behind main gearbox and two under floor. Internal fuel capacity 2,130 litres (563 US gallons; 469 Imp gallons), of which 2,050 litres (542 US gallons; 451 Imp gallons) are usable; can be supplemented by two 850 litre (225 US gallon; 187 Imp gallon) auxiliary tanks in cabin (Mi-24D); provision for carrying (instead of auxiliary tank) up to four external tanks, each 500 litres (132 US gallons; 110 Imp gallons), on two inner pylons under each wing. Optional deflectors and separators for foreign objects and dust in air intakes; and IR suppression exhaust mixer boxes over exhaust ducts.

ACCOMMODATION: Pilot (at rear) and weapon operator on armoured seats in tandem cockpits under individual canopies; dual flying controls, with retractable pedals in front cockpit; if required, flight mechanic on jump-seat in cabin, with narrow passage between flight deck and cabin. Front canopy hinged to open sideways to starboard; footstep under starboard side of fuselage for access to pilot's rearward-hinged door; rear seat raised to give pilot unobstructed forward view; anti-fragment shield between cockpits. Main cabin can accommodate eight persons on folding seats, or two stretchers, two seated casualties and a medical attendant; seats often not fitted, allowing larger numbers to be carried; at front of cabin on each side is a door, divided horizontally into two sections hinged to open upward and downward respectively, with integral step on lower portion. Optically flat bulletproof glass windscreen, with wiper, for each crew member.

SYSTEMS: Cockpits and cabin heated and ventilated. Dual electrical system, with three generators, giving 36, 115 and 208 V AC at 400 Hz, and 27 V DC. Retractable landing/taxying light under nose; navigation lights; anti-collision light above tailboom. Stability augmentation system. Electrothermal de-icing system for main and tail rotor blades. AI-9V APU mounted transversely inside fairing aft of rotor head for engine starting and ground services.

AVIONICS: *Comms:* R-860/863 and Karat M24 com; SPU-8 intercom.
Flight: VUAP-1 autopilot, ARK-15M radio compass, ARK-U2 radio compass, RV-5 radio altimeter.
Instrumentation: Blind-flying instrumentation, and ADF navigation system with DISS-15D Doppler-fed mechanical map display. Air data sensor boom forward of top starboard corner of bulletproof windscreen at extreme nose.
Mission: Undernose pods for electro-optics (starboard) and Raduga-F semi-automatic missile guidance (port). Many small antennas and blisters, including SRO-2 Khrom ('Odd Rods') IFF transponder.
Self-defence: SPO-15 Beryoza RWR. IR jammer (L-166V-1E Ispanka microwave pulse lamp: 'Hot Brick') in 'flower pot' container above forward end of tailboom. ASO-2V flare dispensers under tailboom forward of tailskid assembly initially; triple racks (total of 192 flares) on sides of centre-fuselage from 1987, due to combat experience in Afghanistan.

Mockup of proposed Mi-35O (based on grounded airframe 03033) wearing GOES-320 turret *(Paul Jackson/Jane's)*

EQUIPMENT: Gun camera on port wingtip. Colour-coded identification flare system.

ARMAMENT: One remotely controlled YakB-12.7 four-barrel Gatling-type 12.7 mm machine gun, with 1,470 rounds, in VSPU-24 undernose turret with field of fire 60° to each side, 20° up, 60° down; gun slaved to KPS-53AV undernose sighting system with reflector sight in front cockpit; four 9M17P Skorpion (AT-2 'Swatter') anti-tank missiles on 2P32M twin rails under endplate pylons at wingtips; four underwing pylons for UB-32 rocket pods (each 32 S-5 type 57 mm rockets), B-8V-20 pods each containing twenty 80 mm S-8 rockets, B-13L pods each containing five 130 mm S-13 rockets, 240 mm S-24B rockets, UPK-23-250 pods each containing a GSh-23L twin-barrel 23 mm gun, GUV pods each containing either one four-barrel 12.7 mm YakB-12.7 machine gun with 750 rounds and two four-barrel 7.62 mm 9-A-622 machine guns with total 1,100 rounds or an AGS-17 Plamia 30 mm grenade launcher with 300 grenades, up to 1,500 kg (3,300 lb) of conventional bombs, mine dispensers, night flares or other stores. R-60 (AA-8 'Aphid'), R-73 (AA-11 'Archer') and Igla AAMs fitted experimentally. Helicopter can be landed to install reload weapons carried in cabin. PKV reflector gunsight for pilot. Provisions for firing AKMS guns from cabin windows and PK or PKT machine gun from door mounts.

DIMENSIONS, EXTERNAL (A: Mi-24P, B: Mi-35M, C: Mi-24V/-35 alternative official figures):
Main rotor diameter: A 17.30 m (56 ft 9¼ in)
  B 17.20 m (56 ft 5¼ in)
Main rotor blade chord: A 0.58 m (1 ft 10¾ in)
Tail rotor diameter: A 3.91 m (12 ft 10 in)
  B 3.84 m (12 ft 7¼ in)
Wing span: A 6.66 m (21 ft 10¼ in)
  B 5.06 m (16 ft 7¼ in)
Width of fuselage: A, B 1.70 m (5 ft 7 in)
Length:
  excl rotors and gun: A, B 17.51 m (57 ft 5¼ in)
  rotors turning: A 21.35 m (70 ft 0½ in)
  B 21.27 m (69 ft 9½ in)
Height: to main rotor tip, turning:
  A, B 3.97 m (13 ft 0½ in)
  overall: rotors turning: B approx 5.00 m (16 ft 4¾ in)
Span of horizontal stabiliser: A 3.27 m (10 ft 9 in)
Wheel track: A, B 3.03 m (9 ft 11½ in)
Wheelbase: A, B 4.39 m (14 ft 5 in)

DIMENSIONS, INTERNAL:
Main cabin: Length: A 2.825 m (9 ft 3¼ in)
  B 2.60 m (8 ft 6¼ in)
Width: A 1.46 m (4 ft 9½ in)
  B 1.18 m (3 ft 10½ in)
Height: A, B 1.20 m (3 ft 11¼ in)

AREAS:
Main rotor disc: A 235.06 m² (2,530.2 sq ft)
  B 232.35 m² (2,501.0 sq ft)
Tail rotor disc: A 11.99 m² (129.06 sq ft)
  B 11.58 m² (124.66 sq ft)

WEIGHTS AND LOADINGS:
Weight empty: A 8,570 kg (18,894 lb)
  B 8,090 kg (17,835 lb)
Normal internal load: B 1,500 kg (3,307 lb)
Max external stores: A 2,400 kg (5,291 lb)
  B 2,860 kg (6,305 lb)
Max slung cargo: A 2,400 kg (5,291 lb)
Max combat payload: B 2,860 kg (6,305 lb)
Normal T-O weight: A 11,200 kg (24,690 lb)
  B 10,800 kg (23,809 lb)
  C 10,900 kg (24,030 lb)
Max T-O weight: A, B 12,000 kg (26,455 lb)
  attack configuration 11,500 kg (25,353 lb)
  ferry, four external tanks 12,000 kg (26,455 lb)
Max disc loading: A 51.1 kg/m² (10.46 lb/sq ft)
  B 51.7 kg/m² (10.58 lb/sq ft)

PERFORMANCE:
Max level speed: A 172 kt (320 km/h; 198 mph)
  B 167 kt (310 km/h; 193 mph)
  C 168 kt (312 km/h; 194 mph)

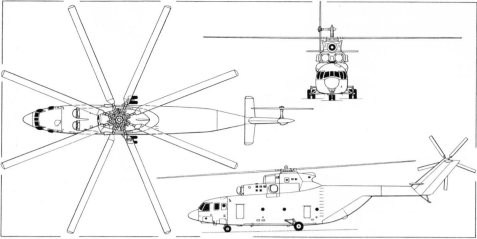
Mil Mi-26 multipurpose heavy-lift helicopter *(Dennis Punnett/Jane's)*

Cruising speed: A 145 kt (270 km/h; 168 mph)
  B 140 kt (260 km/h; 161 mph)
  C 151 kt (280 km/h; 174 mph)
Econ cruising speed: A 117 kt (217 km/h; 135 mph)
Max rate of climb at S/L: A 750 m (2,460 ft)/min
  B 744 m (2,440 ft)/min
Service ceiling: A 4,500 m (14,750 ft)
  B 5,700 m (18,700 ft)
  C 5,750 m (18,860 ft)
Hovering ceiling OGE: A, ISA 1,500 m (4,920 ft)
  C 1,750 m (5,740 ft)
  B, ISA 3,100 m (10,180 ft)
  C 3,000 m (9,840 ft)
  B, ISA + 10°C 2,150 m (7,060 ft)
Combat radius: A, with max military load
  86 n miles (160 km; 99 miles)
  A, with two external fuel tanks
  121 n miles (224 km; 139 miles)
  A, with four external fuel tanks
  155 n miles (288 km; 179 miles)
Range: A, standard internal fuel
  243 n miles (450 km; 279 miles)
  B, standard internal fuel, 5% reserves
  270 n miles (500 km; 450 miles)
  A, B, with four external tanks
  540 n miles (1,000 km; 620 miles)
Max endurance: A 4 h

UPDATED

MIL Mi-26

NATO reporting name: Halo

TYPE: Heavy lift helicopter.

PROGRAMME: Development started early 1970s (initially as Mi-6M); aim was payload capability 1½ to 2 times greater than that of any previous production helicopter; first prototype flew 14 December 1977; one of several prototype or preproduction Mi-26s (SSSR-06141) displayed at 1981 Paris Air Show; in-field evaluation, probably with military development squadron, began early 1982; fully operational 1983; export deliveries started (to India) June 1986; production continues at low rate, with manufacture and marketing by Rostvertol (which see).

CURRENT VERSIONS: **Mi-26** (*Izdelie* 90): Basic military transport.

Detailed description applies to basic Mi-26, except where indicated.

Mi-26A: Modified military Mi-26, tested in 1985, with PNK-90 integrated flight/nav systems for automatic approach and descent to critical decision point, and other tasks. Not adopted.

Mi-26T: Basic civil transport (*Izdelie* 209), generally as military Mi-26. Variants include **Geological Survey** Mi-26 towing seismic gear, with tractive force of 10,000 kg (22,045 lb) or more, at 97 to 108 kt (180 to 200 km/h; 112 to 124 mph) at 55 to 100 m (180 to 330 ft) for up to 3 hours. The mockup of an Mi-26 two-crew flight deck was shown at the 1997 Moscow Air Show.

Mi-26TS (*sertifitsyrovannyi*: certified): Mi-26T (*Izdelie* 219), but prepared for certification and marketed (in West as **Mi-26TC**) from 1996. Preproduction version, with gondola (port, front), positioned a 16,000 kg (35,275 lb) TV tower, 30 m (98 ft) long, in Rostov-on-Don in 1996. One delivered to Samsung Aerospace Industries in South Korea on 13 September 1997.

Mi-26MS: Medical evacuation version of Mi-26T, typically with intensive care section for four casualties and two medics, surgical section for one casualty and three medics, pre-operating section for two casualties and two medics, ambulance section for five stretcher patients, three seated casualties and two attendants; laboratory; and amenities section with lavatory, washing facilities, food storage and recreation unit. Civil version in use by MChS Rossii (Ministry of Emergency Situations). Alternative medical versions available, with modular box-laboratories or fully equipped medical centres that can be inserted into the hold for anything from ambulance to field hospital use. As field ambulance can accommodate up to 60 stretcher patients; or seven patients in intensive care, 32 patients on stretchers and seven attendants; or 47 patients and eight attendants in other configurations, which can include 12 bunks in four tiers forward, or patent Rostvertol box laboratory behind the first row of bunks, with 16 bunks behind.

The box includes an operating table, diagnostic equipment, anaesthetic and breathing equipment and other systems. Another configuration includes a larger theatre box by Heinkel Medizin Systeme and 12 stretchers behind, and the helicopter can be fitted out with an X-ray laboratory or can form the central element of a deployable air-portable field hospital.

Mi-26NEF-M: ASW version with search radar in undernose faired radome, extra cabin heat exchangers and towed MAD housing mounted on ramp.

Mi-26P: Transport for 63 passengers, basically four-abreast in airline-type seating, with centre aisle; lavatory, galley and cloakroom aft of flight deck.

Mi-26PK: Flying crane (*kran*) derivative of Mi-26P with operator's gondola on fuselage side, next to cabin door on port side.

Mi-26PP: Reported ECM version. First noted 1986; current status unknown.

Mi-26TM: Flying crane, with gondola for pilot/sling supervisor under fuselage aft of nosewheels or under rear-loading ramp.

Firefighting (*pozharnyi*) version with internal tanks able to dispense up to 15,000 litres (3,962 US gallons; 3,300 Imp gallons) fire retardant from one or two vents, or 17,260 litres (4,560 US gallons; 3,796 Imp gallons) of water from an underslung VSU-15 bucket, or from two linked EP-8000 containers. Can fill tanks on the ground using pumps with 3,000 litres (793 US gallons; 660 Imp gallons)/min throughput. Prototype RA-06183 operated by Rostvertol. One delivered to Moscow Fire Brigade on 19 August 1999.

Mi-26TZ: Tanker with 14,040 litres (3,710 US gallons; 3,088 Imp gallons) of T2, TS1 or R2 aviation fuel or DL, DZ or DA diesel oil fuel and 1,040 litres (275 US gallons; 228 Imp gallons) lubricants (in 52 jerry cans), dispensed through four 60 m (197 ft) hoses for aircraft, or 10 20 m (66 ft) hoses for ground vehicles. Conversion to/from Mi-26T takes 1 hour 25 minutes for each operation.

Mi-26M: Upgrade under development; all-GFRP main rotor blades of new aerodynamic configuration, new ZMKB Progress D-127 turboshafts (each 10,700 kW; 14,350 shp), and modified integrated flight/nav system with EFIS. Transmission rating unchanged, but full payload capability maintained under 'hot and high' conditions, OEI safety improved, hovering and service

Mil Mi-26 in Ukrainian military markings *(Yefim Gordon)*

ceilings increased, and greater maximum payload (22,000 kg; 48,500 lb) for crane operations.

The 1990 edition of US Department of Defense's *Soviet Military Power* stated ''New variants of 'Halo' are likely in the early 1990s to begin to replace 'Hooks' specialised for command support''. Two prototypes are reported to have been built, with designation **Mi-27**. These have new antennas along lower 'corner' of fuselage, blade and box-type and with long folded masts which are horizontal in flight, vertical when deployed on ground. Prototype wears Aeroflot colour scheme.

CUSTOMERS: Estimated 280 built by 1999; some sources suggest more. Reportedly sold to about 20 countries; operators include Belarus (15), India (10), Kazakhstan, Mexico (two) in 2000, Peru (three), Russian Army (35) Russian Ministry of Emergencies, Mil-Avia and Ukraine (20). Russian Army deliveries included four in 1994 (but none subsequently). Three delivered in 2000 (two to Peru; one to Scorpion of Greece); four scheduled for 2000 (including second for Scorpion).

COSTS: US$10 million to US$12 million (Mi-26TS) (1996).

DESIGN FEATURES: Largest production helicopter; empty weight comparable to that of Mi-6 and, as specified, is approximately 50 per cent of maximum T-O weight; weight saved by in-house design of main gearbox providing multiple torque paths, GFRP tail rotor blades, titanium main and tail rotor heads, main rotor blades of mixed metal and GFRP, use of aluminium-lithium alloys in airframe; conventional pod and boom configuration, but first successful use of eight-blade main rotor, of smaller diameter than Mi-6 rotor; payload and cargo hold size similar to those of Lockheed C-130 Hercules; auxiliary wings not required; rear-loading ramp/doors; main rotor rpm 132; main rotor spindle inclined forwards 4°.

FLYING CONTROLS: Hydraulically powered cyclic and collective pitch controls actuated by small parallel jacks, with redundant autopilot and stability augmentation system inputs. Fly-by-wire system flight tested 1994.

STRUCTURE: Eight-blade constant-chord main rotor; flapping and drag hinges, droop stops and hydraulic drag dampers; no elastomeric bearings or hinges; each blade has one-piece tubular steel spar and 26 GFRP aerofoil shape full-chord pockets, honeycomb filled, with ribs and stiffeners and non-removable titanium leading-edge abrasion strip; blades have moderate twist, taper in thickness toward tip, and are attached to small forged titanium head of unconventional design; each has ground-adjustable trailing-edge tab; five-blade constant-chord tail rotor, starboard side, has GFRP blades, forged titanium head; conventional transmission, with tail rotor shaft inside cabin roof; all-metal riveted semi-monocoque fuselage with clamshell rear doors; flattened tailboom undersurface; engine bay of titanium for fire protection; all-metal tail surfaces; swept vertical stabiliser/tail rotor support profiled to produce sideways lift; ground-adjustable variable incidence horizontal stabiliser.

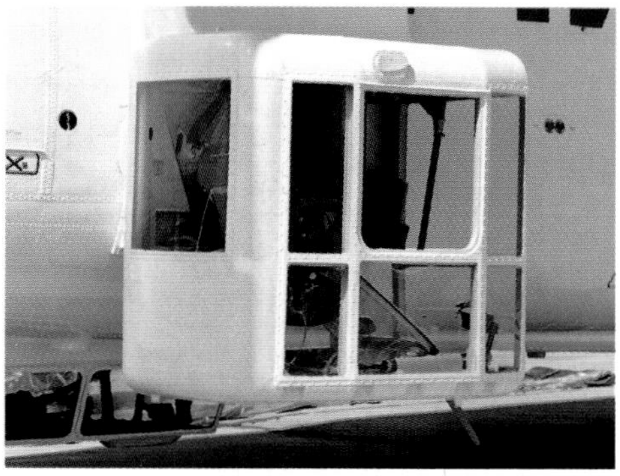

Current style of side-mounted Mi-26TM crane operator's position ***(Paul Jackson/Jane's)*** 0009896

LANDING GEAR: Non-retractable tricycle type; twin wheels on each unit; steerable nosewheels, tyre size 900×300; mainwheel tyres size 1,120×450. Retractable tailskid at end of tailboom to permit unrestricted approach to rear cargo doors. Length of main legs adjusted hydraulically to facilitate loading through rear doors and to permit landing on varying surfaces. Device on main gear indicates take-off weight to flight engineer at lift-off, on panel on shelf to rear of his seat.

POWER PLANT: Two 8,500 kW (11,399 shp) ZMKB Progress D-136 free-turbine turboshafts, side by side above cabin, forward of main rotor driveshaft. Air intakes fitted with particle separators to prevent foreign object ingestion, and have both electrical and bleed air anti-icing systems. Above and behind is central oil cooler intake. VR-26 fan-cooled main transmission, rated at 14,914 kW (20,000 shp), with air intake above rear of engine cowlings. System for synchronising output of engines and maintaining constant rotor rpm; if one engine fails, output of other is increased to maximum power automatically. Independent fuel system for each engine; fuel in eight underfloor rubber tanks, feeding into two header tanks above engines, which permit gravity feed for a period in emergencies; maximum standard internal fuel capacity 12,000 litres (3,170 US gallons; 2,640 Imp gallons); provision for four auxiliary tanks. Mi-26TS normal capacity is 13,020 litres (3,440 US gallons; 2,864 Imp gallons). Two large panels on each side of main rotor mast fairing, aft of engine exhaust outlet, hinge downward as work platforms.

ACCOMMODATION: Crew of four on flight deck: pilot (on port side) and co-pilot side by side, tip-up seat between pilots, and seats for flight engineer (port) and navigator (starboard) to rear. Four-seat passenger compartment aft of flight deck. Loads in hold include two airborne infantry combat vehicles and a standard 20,000 kg (44,090 lb) ISO container; about 20 tip-up seats along each sidewall of hold; maximum military seating for 80 combat-equipped troops; alternative provisions for 60 stretcher patients and four/five attendants. Heated windscreen, with wipers; four large blistered side windows on flight deck; forward pair swing open slightly outward and rearward. Downward-hinged doors, with integral airstairs, at front of hold on port side, and each side of hold aft of main landing gear units. Hold loaded via downward-hinged lower door, with integral folding ramp, and two clamshell upper doors forming rear wall of hold when closed; doors opened and closed hydraulically, with back-up hand pump for emergency use. Two LG-1500 electric hoists on overhead rails, each with capacity of 2,500 kg (5,511 lb), enable loads to be transported along cabin; winch for hauling loads, capacity 500 kg (1,100 lb); roller conveyor in floor and load lashing points throughout hold. Flight deck fully air conditioned.

SYSTEMS: Two main and one emergency hydraulic systems, operating pressure 157 and 206 bars (2,276 and 2,987 lb/sq in). Electrical system three-phase 200/115 V 400 Hz; single-phase 115 V 400 Hz; three-phase, 36 V 400 Hz; single-phase 36 V 400 Hz; DC 27 V. TA-8V 119 kW (160 hp). APU under flight deck, with intake louvres (forming fuselage skin when closed) and exhaust on starboard side, for engine starting and to supply hydraulic, electrical and air conditioning systems on ground. Electrically heated leading-edge of main and tail rotor blades for anti-icing. Only flight deck pressurised.

AVIONICS: All items necessary for day and night operations in all weathers are standard.

Radar: Groza 7A813 weather radar in hinged (to starboard) nosecone.

Flight: Integrated PKV-26-1 flight/nav system and automatic flight control system, Doppler, map display, HSI, and automatic hover system. Optional GPS.

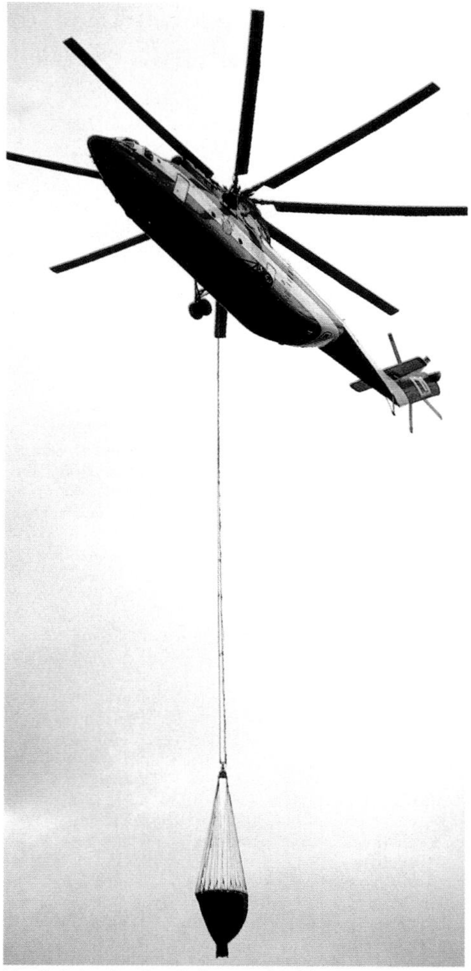

Firefighting bucket below an Mi-26TS ***(Paul Jackson/Jane's)*** ***2000***/0064954

Self-defence: Military versions can have IR jammers and suppressors, IR decoy dispensers and colour-coded identification flare system.

EQUIPMENT: Hatch for load sling in bottom of fuselage, in line with main rotor shaft; sling cable attached to internal winching gear. Closed-circuit TV cameras to observe slung payloads.

ARMAMENT: None.

DIMENSIONS, EXTERNAL:

Main rotor diameter	32.00 m (105 ft 0 in)
Tail rotor diameter	7.61 m (24 ft 11½ in)
Length: overall, rotors turning	40.025 m (131 ft 3¾ in)
nose to turning tail rotor	35.91 m (117 ft 9¾ in)
fuselage, excl tail rotor	33.745 m (110 ft 8½ in)
Height: to top of rotor head	8.145 m (26 ft 8¾ in)
to top of fin	7.45 m (24 ft 5¼ in)
tail rotor turning	11.60 m (38 ft 0¾ in)
Width overall (outsides of mainwheels)	6.15 m (20 ft 2¼ in)
Tailplane span	6.02 m (19 ft 9 in)
Wheel track: c/l shock-absorbers	5.00 m (16 ft 4¾ in)
outer wheels	5.75 m (18 ft 10½ in)
Wheelbase	8.95 m (29 ft 4½ in)

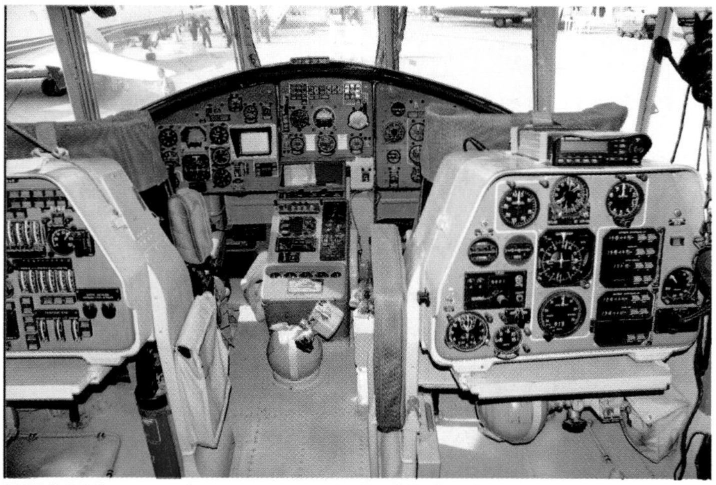

Mil Mi-26 four-crew instrument panels ***(Paul Jackson/Jane's)*** 0015065

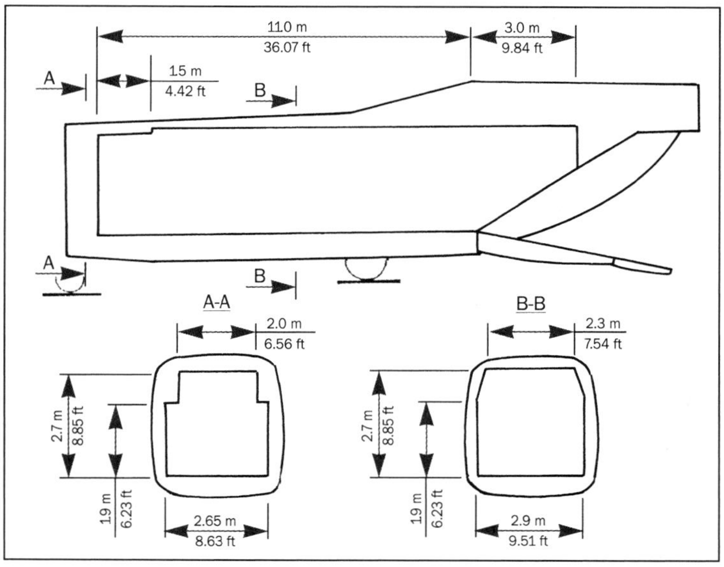

Mil Mi-26 hold dimensions
1999/0051371

Mil Mi-26TS firefighting helicopter *(Paul Jackson/Jane's)* 2000/0064955

Rear (pilot's) cockpit of Mi-28A
(Paul Jackson/Jane's)

DIMENSIONS, INTERNAL:
  Freight hold:
    Length: excl ramp   12.08 m (39 ft 7½ in)
         ramp trailed   15.00 m (49 ft 2½ in)
    Max width   3.23 m (10 ft 7¼ in)
    Height   2.98-3.17 m (9 ft 9¼ in-10 ft 4¾ in)
    Floor area: excl ramp   39.3 m² (423 sq ft)
         ramp trailed   49.2 m² (530 sq ft)
    Volume: excl ramp   121.0 m³ (4,273 cu ft)
         ramp trailed   135.9 m³ (4,799 cu ft)
AREAS:
  Main rotor disc   804.25 m² (8,656.8 sq ft)
  Tail rotor disc   45.48 m² (489.54 sq ft)
WEIGHTS AND LOADINGS:
  Weight empty   28,200 kg (62,170 lb)
  Max payload, internal or external 20,000 kg (44,090 lb)
  Normal T-O weight:
    except Mi-26TS   49,600 kg (109,350 lb)
    Mi-26TS   49,650 kg (109,455 lb)
  Max T-O weight   56,000 kg (123,450 lb)
  Max disc loading: Mi-26TS   69.6 kg/m² (14.26 lb/sq ft)
  Transmission loading at max T-O weight and power:
    Mi-26TS   3.76 kg/kW (6.17 lb/shp)
PERFORMANCE (A: Mi-26, B: Mi-26M, C: Mi-26TS at normal T-O weight):
  Max level speed: A   159 kt (295 km/h; 183 mph)
                   C   146 kt (270 km/h; 168 mph)
  Normal cruising speed: A, C 137 kt (255 km/h; 158 mph)
  Service ceiling: A   4,600 m (15,100 ft)
                   B   5,900 m (19,360 ft)
                   C   4,300 m (14,100 ft)
  Hovering ceiling IGE:
    A, ISA, with 5,100 kg (11,240 lb) payload
                                 1,000 m (3,280 ft)
    B, ISA + 15°C, with 12,300 kg (27,115 lb) payload
                                 1,000 m (3,280 ft)
  Hovering ceiling OGE, ISA: A 1,800 m (5,900 ft)
                             B 2,800 m (9,180 ft)
                             C 1,520 m (4,980 ft)
  Range: A at 2,500 m (8,200 ft) ISA + 15°C, with
    7,700 kg (16,975 lb) payload
                             270 n miles (500 km; 310 miles)
  B at 2,500 m (8,200 ft) ISA + 15°C, with 13,700 kg
    (30,200 lb) payload 270 n miles (500 km; 310 miles)
  A, S/L ISA, with max internal fuel at max T-O weight,
    5% reserves   318 n miles (590 km; 366 miles)
  A, S/L ISA, with four auxiliary tanks
                         1,036 n miles (1,920 km; 1,190 miles)
                                                    UPDATED

MIL Mi-27

Reported designation of command support version of Mi-26 (which see).
                                                    UPDATED

MIL Mi-28

NATO reporting name: Havoc
TYPE: Attack helicopter.
PROGRAMME: Design started 1980 under Marat N Tishchenko; first of two flying Mi-28 prototypes (012) flew 10 November 1982; each prototype different: first and second (022) had upward-pointing exhaust diffusers and fixed undernose fairing for electro-optic equipment; first also had conventional three-blade tail rotor; second replaced this with the definitive 'Delta-H' configuration. The first Mi-28A (032) introduced the definitive downward-pointing exhaust suppressors and flew in January 1988; second Mi-28A prototype (042) demonstrated at Moscow in 1992 and represented the intended production configuration. It had the definitive moving E-O sensor turret undernose, downward-pointing exhaust diffusers and wingtip electronics/chaff dispenser pods; small-scale pre-series production planned, but not yet initiated, by Rostvertol, Rostov-on-Don. Rival Ka-50 officially 'adopted' 5 October 1994, but competition then continued; final decision was due by early 2001.
CURRENT VERSIONS: **Mi-28:** First two prototypes with 1,434 kW (1,923 shp) TV3-117BM engines and VR-28 gearbox. **Mi-28A** (Type 280): Basic version, *as described in detail*. Third and fourth aircraft built.

Mi-28N: Unofficial names: Night Hunter and Night Pirate. Added night/all-weather operating capability. Russian Army funding announced January 1994; demonstrator (014) modified from first Mi-28 prototype (012); first hover 14 November 1995; formal roll-out 16 August 1996; first flight 30 April 1997. Mast-mounted 360° scan millimetre wave Kinzhal V or Arbalet radar (pod soon enlarged in vertical plane); FLIR ball beneath missile-guidance nose radome and above new shuttered turret for optical/laser sensors, including Zenit low-light-level TV. EFIS cockpit. Armament to include 9M114 Shturm (AT-6 'Spiral') ASMs and Igla (SA-16 'Gimlet') AAMs. Uprated TV3-117VK turboshafts, each 1,839 kW (2,466 shp) for T-O; strengthened transmission; new composites rotor with sweptback blade tips added subsequently. Mi-28N introduced uprated VR-29 transmission and IKBO integrated flight/weapon aiming system, with automatic terrain-following and automatic target search, detection, identification and (in formations of Mi-28Ns) allocation. Weights and performance generally unchanged though similarly rated TV3-117VMA-SB3 engine is now undergoing evaluation and will be accompanied by rotors of improved aerodynamic configuration. In 2000, further two Mi-28Ns reportedly under construction at Rostvertol.
Versions projected for naval amphibious assault support and air-to-air missions.
Mi-28NEh: (*Noch, Ehksport:* Night, Export): Version of above offered to South Korea in 2000.
COSTS: Mi-28N development cost US$150 million (2000); unit cost US$12+ million (2000).
DESIGN FEATURES: Conventional gunship configuration, with two crew in stepped cockpits; original three-blade tail rotor superseded by low noise 'scissors' or 'Delta-H' type comprising two independent two-blade rotors set as narrow X (35°/145°) on same shaft with self-lubricating bearings; resulting flapping freedom relieves flight loads; agility enhanced by doubling hinge offset of main rotor blades compared with Mi-24; survivability emphasised; crew compartments protected by titanium and ceramic armour and armoured glass transparencies; single hit will not knock out both engines; vital units and parts are redundant, widely separated and shielded by the less vital; multiple self-sealing fuel tanks in centre-fuselage enclosed in composites second skin, outside metal fuselage skin; no explosion, fire or fuel leakage results if tanks hit by bullet or shell fragment; energy absorbing seats and landing gear protect crew in crash landing at descent rate of 12 m (40 ft)/s; crew doors are rearward-hinged, to open quickly and remain open in emergency; parachutes are mandatory for Russian Federation and Associated States (CIS) military helicopter aircrew; if Mi-28 crew had to parachute, emergency system would jettison doors, blast away stub-wings, and inflate bladder beneath each door sill; as crew jumped, they would bounce off bladders and clear main landing gear; no provision for rotor separation; port-side door, aft of wing, provides access to avionics compartment large enough to permit combat rescue of two or three persons on ground, although it lacks windows, heating and ventilation.

Hand crank, inserted into end of each stub-wing, enables stores of up to 500 kg (1,100 lb) to be winched on to pylons without hoists or ground equipment; current 30 mm gun is identical with that of RFAS army ground vehicles and uses same ammunition; jamming averted by attaching twin ammunition boxes to sides of gun mounting, so that they turn, elevate and depress with gun; main rotor shaft has 5° forward tilt, providing tail rotor clearance; transmission capable of running without oil for 20 to 30 minutes; main rotor rpm 242; with main rotor blades and wings removed, helicopter is air-transportable in An-22 or Il-76 freighter.
FLYING CONTROLS: Hydraulically powered mechanical type; horizontal stabiliser linked to collective; controls for pilot only.
STRUCTURE: Five-blade main rotor; blades have very cambered high-lift section and sweptback tip leading-edge; full-span upswept tab on trailing-edge of each blade; structure comprises numerically controlled, spirally wound glass fibre D-spar, blade pockets of Kevlar-like material with Nomex-like honeycomb core, and titanium erosion strip on leading-edge; each blade has single elastomeric root bearing, mechanical droop stop and hydraulic drag damper; four-blade GFRP tail rotor with elastomeric bearings for flapping; rotor brake lever on starboard side of cockpit; strong and simple machined titanium main rotor head with elastomeric bearings, requiring no lubrication; power output shafts from engines drive main gearbox from each side; tail rotor gearbox, at base of tail pylon, driven by aluminium alloy shaft inside composites duct on top of tailboom; sweptback mid-mounted wings have light alloy primary box structure, leading- and trailing-edges of composites; no wing movable surfaces; provision for countermeasures pod on each wingtip, housing chaff/flare dispensers and sensors, probably RWR; light alloy semi-monocoque fuselage, with titanium armour around cockpits and vulnerable areas; composites access door aft of wing on port side; swept fin has light alloy primary box structure, composites leading- and trailing-edges; cooling air intake at base of fin leading-edge, exhaust at top of trailing-edge; two-position composites horizontal stabiliser.
LANDING GEAR: Non-retractable, tailwheel type; single wheel on each unit; mainwheel tyres size 720×320, pressure 5.40 bars (78 lb/sq in); castoring tailwheel with tyre size 480×200.

Prototype Mil Mi-28N *(Yefim Gordon)* 2001/0099589

POWER PLANT: Two Klimov TV3-117VMA turboshafts, each 1,636 kW (2,194 shp), in pod above each wingroot; three jetpipes inside downward-deflected composites nozzle fairing on each side of third prototype shown in Paris 1989; upward deflecting type also tested. Deflectors for dust and foreign objects forward of air intakes, which are de-iced by engine bleed air. Internal fuel capacity 1,720 litres (454 US gallons; 378 Imp gallons). Provision for four external fuel tanks on underwing pylons.

ACCOMMODATION: Navigator/gunner in front cockpit; pilot behind, on elevated seat; transverse armoured bulkhead between; flat non-glint tinted transparencies of armoured glass; navigator/gunner's door on port side, pilot's door on starboard side.

SYSTEMS: Cockpits air conditioned and pressurised by engine bleed air. Duplicated hydraulic systems, pressure 152 bars (2,200 lb/sq in). 208 V AC electrical system supplied by two generators on accessory section of main gearbox, ensuring continued supply during autorotation. Low-airspeed system standard, giving speed and drift via main rotor blade-tip pitot tubes at −27 to +38 kt (−50 to +70 km/h; −31 to +43 mph) in forward flight, and ±38 kt (±70 km/h; ±43 mph) in sideways flight. Main and tail rotor blades electrically de-iced. Ivchenko AI-9V APU in rear of main pylon structure supplies compressed air for engine starting and to drive small turbine for preflight ground checks.

AVIONICS: *Comms:* UHF/VHF nav/com; small IFF fairing each side of nose and tail.
Instrumentation: Conventional IFR instrumentation, with autostabilisation, autohover, and hover/heading hold lock in attack mode; pilot has HUD and centrally mounted CRT for basic TV; aircraft designed for use with night vision goggles.
Mission: Radio for missile guidance in nose radome. Daylight optical weapons sight and laser range-finder in gyrostabilised and double-glazed nose turret above gun, with which it rotates through ±110°; wiper on outer glass protects inner optically flat panel.
Self-defence: Two fixed IR sensors on initial basic production Mi-28; IR suppressors, radar and laser warning receivers standard; optional countermeasures pod on each wingtip, housing chaff/flare dispensers and sensors, probably RWR. Mi-28N has integrated Vitebsk DASS with Pastel RWR, Mak IR warning system, Platan jammer and UV-26 flare dispensers.

EQUIPMENT: Two slots, one above the other on port side of tailboom, for colour-coded identification flares. Three pairs of rectangular formation-keeping lights in top of tailboom; further pair in top of main rotor pylon fairing.

ARMAMENT: One 2A42 30 mm turret-mounted gun (with 250 rounds in side-mounted boxes) in NPPU-28 mount at nose, able to rotate ±110°, elevate 13° and depress 40°; maximum rate of fire 900 rds/min air-to-air and air-to-ground. (New specially designed gun under development.) Two pylons under each stub-wing, each with capacity of 480 kg (1,058 lb), typically for total of sixteen 9M114 Shturm C (AT-6 'Spiral') radio-guided tube-launched anti-tank missiles and two UB-20 pods of eighty 80 mm S-8 or twenty 122 mm S-13 rockets or two UPK-23-250 gun pods. Alternative ATMs include 9M120/9M121F Vikhr and 9A-2200; up to eight 9M39 Igla-V AAMs in place of ATMs; in minelaying role can carry two KGMU-2 dispensers. Main 2A42 gun fired and guided weapons launched normally only from front cockpit; unguided rockets fired from both cockpits. (When fixed, gun can be fired also from rear cockpit.)

DIMENSIONS, EXTERNAL:
Main rotor diameter	17.20 m (56 ft 5 in)
Main rotor blade chord	0.67 m (2 ft 2½ in)
Tail rotor diameter	3.84 m (12 ft 7¼ in)
Tail rotor blade chord	0.24 m (9½ in)
Length overall, excl rotors, incl gun	17.01 m (55 ft 9¾ in)
Fuselage max width	1.85 m (6 ft 1 in)
Width over stub-wings	4.88 m (16 ft 0¼ in)

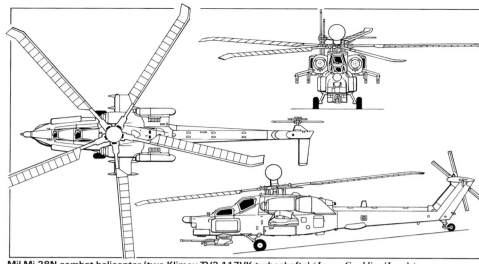

Mil Mi-28N combat helicopter (two Klimov TV3-117VK turboshafts) *(James Goulding/Jane's)*

Height: overall	4.70 m (15 ft 5 in)
to top of rotor head	3.82 m (12 ft 6½ in)
Wheel track	2.29 m (7 ft 6¼ in)
Wheelbase	11.00 m (36 ft 1 in)
AREAS:	
Main rotor disc	232.35 m² (2,501.0 sq ft)
Tail rotor disc	11.58 m² (124.65 sq ft)
WEIGHTS AND LOADINGS:	
Weight empty, equipped: 28	7,900 kg (17,416 lb)
28A	8,095 kg (17,846 lb)
28N	8,590 kg (18,938 lb)
Fuel weight: standard internal	1,337 kg (2,947 lb)
with added tanks	1,782 kg (3,928 lb)
Normal T-O weight: 28	10,200 kg (22,487 lb)
28A	10,400 kg (22,928 lb)
28N	10,700 kg (23,589 lb)
Max T-O weight: 28	11,200 kg (24,691 lb)
28A, 28N	11,500 kg (25,353 lb)
Max disc loading: 28	48.2 kg/m² (9.87 lb/sq ft)
28A, 28N	49.5 kg/m² (10.14 lb/sq ft)
PERFORMANCE:	
Max level speed: 28A	162 kt (300 km/h; 186 mph)
28N	172 kt (320 km/h; 199 mph)
Max cruising speed: 28A	143 kt (265 km/h; 164 mph)
28N	145 kt (270 km/h; 168 mph)
Max rate of climb at S/L	816 m (2,677 ft)/min
Service ceiling: 28A	5,800 m (19,020 ft)
28N (-SB3)	5,700 m (10,700 ft)
Hovering ceiling OGE: 28	3,470 m (11,380 ft)
28A, 28N	3,600 m (11,820 ft)
28N (-SB3)	4,500 m (14,760 ft)
Radius of action, standard fuel, 10 min loiter at target,	
5% reserves	108 n miles (200 km; 124 miles)
Range, max standard fuel, 10% reserves: all	
	234 n miles (435 km; 270 miles)
Ferry range, 5% reserves	593 n miles (1,100 km; 683 miles)
Endurance with max fuel	2 h
g limits	+3/−0.5

UPDATED

MIL Mi-34

NATO reporting name: Hermit

TYPE: Four-seat helicopter.

PROGRAMME: First flight 17 November 1986; two prototypes and structure test airframe completed by mid-1987, when exhibited for first time at Paris Air Show; first helicopter built in former USSR to perform normal loop and roll; series production began at Arsenyev plant in 1993; airframe manufactured by Carpathian Helicopter Production Association; marketed by Mi-Light Helicopters (which see in LVM entry;) Mi-34S/-34C completion was at Moscow plant of LVM, but subsequently reverted to Arsenyev; planned completion of 30 in 1994-95 hampered by lack of funding; five delivered in 1995, one in first half of 1996. State order for resumed manufacture 1996. Marketing initiative, early 1999, planned worldwide establishment of service centres in countries where more than 10 Mi-34s ordered.

Trials in 1999 by civil pilot training school at Omsk showed Mi-34 to be 2.8 times cheaper and more effective to operate than current fleet; school to acquire three Mi-34Cs and obtain up to 10 more in long term. Other schools expected to replace ageing Mi-2 with Mi-34.

CURRENT VERSIONS: **Mi-34S:** Basic version; marketed in Russia as Mi-34; certified by Interstate Aviation Committee Aviation Register (initially at 1,350 kg; 2,976 lb max T-O weight), with helicopter, engine and noise type certificates; meets FAR Pt 27 requirements.

(Note that until 1999, all marketing literature for this version used the hybrid Roman/Cyrillic '**Mi-34C**' to indicate certified status.)

Description applies to Mi-34S (Mi-34C), except where indicated.

Mi-34L: Projected version with 261 kW (350 hp) Textron Lycoming TIO-540J piston engine.

Mi-34P (*patrulnyi:* patrol): Version of Mi-34S, equipped for police duties.

Mi-34A: Originally with 335 kW (450 shp) Rolls-Royce 250-C20R turboshaft; mockup, with luxury interior, exhibited at Moscow Air Show '95. Promotion recommended in 1999, employing 376 kW (504 shp) Turbomeca TM 319 Arrius 2F turboshaft; MTOW 1,450 kg (3,196 lb); usable fuel increased to 340 litres (89.7 US gallons; 74.8 Imp gallons); accommodation for four passengers, plus pilot; 1.0 m³ (35 cu ft) baggage compartment. None yet built.

Instrument panel of the Mil Mi-34S *(Paul Jackson/Jane's)*

Mil Mi-28A combat helicopter second prototype *(Paul Jackson/Jane's)*

RUSSIAN FEDERATION: AIRCRAFT—MIL

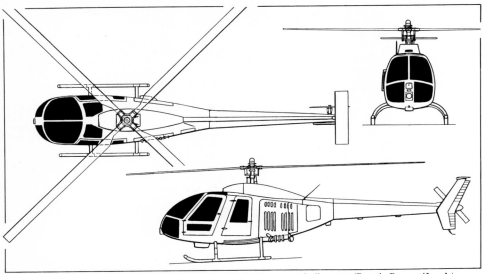

Mil Mi-34 two/four-seat training, competition and light transport helicopter *(Dennis Punnett/Jane's)* 2000/0079310

Mi-34M1/M2: Projected twin-turbine, six-passenger versions; MTOW 2,500 kg (5,511 lb).

Mi-234: Described separately.

CUSTOMERS: Total of 15 sold by May 2000, including 'two or three' built in 1999. First three for Mayor's office, Moscow. Others used by Bashkir Airlines and Mi-Avia for patrol and training.

COSTS: Flyaway (1999): standard, US$316,000; fully equipped, US$350,000. Operating cost US$86.13 per hour (1997).

DESIGN FEATURES: Aerobatic helicopter; intended initially for training and international competition flying; conventional pod and boom configuration; piston engine of same basic type as that in widely used Yakovlev fixed-wing training aircraft and Kamov Ka-26 helicopters; suitable also for light utility, mail delivery, observation and liaison duties, and border patrol; later developments concentrate on light transport role.

Aerobatic capabilities include looping; backwards flight at 70 kt (130 km/h; 81 mph); and rotation about main rotor axis at 120°/s.

FLYING CONTROLS: Manual, with no hydraulic boost.

STRUCTURE: Semi-articulated four-blade main rotor with flapping and cyclic pitch hinges, but natural flexing in lead/lag plane; blades of GFRP with CFRP reinforcement, attached by flexible steel straps to head like that of Boeing MD 500; two-blade tail rotor of similar composites construction, on starboard side; riveted light alloy fuselage; sweptback tailfin with small unswept T tailplane.

LANDING GEAR: Conventional fixed skids on arched support tubes; small tailskid to protect tail rotor.

POWER PLANT: One 239 kW (320 hp) VOKBM M-14V-26V nine-cylinder radial air-cooled engine mounted sideways in centre-fuselage. Fuel capacity 176 litres (46.5 US gallons; 38.7 Imp gallons); system for inverted flight.

ACCOMMODATION: Normally one or two pilots, side by side, in enclosed cabin, with optional dual controls. Rear of cabin contains low bench seat, available for two passengers and offering flat floor for cargo carrying. Forward-hinged door on each side of flight deck and on each side of rear cabin.

SYSTEMS: Primary electric power provided by 27 V 3 kW engine-driven generator; secondary power supplies of 115 V AC, 400 Hz, single-phase and 36 V AC, 400 Hz, three-phase; 27 V 17 Ah battery.

AVIONICS: *Comms:* Briz VHF radio.
Flight: A-037 radio altimeter; ARK-22 radio compass.
Instrumentation: Magnetically slaved compass system incorporating radio magnetic indicator.

EQUIPMENT: Gyro horizon. Version used by Moscow Police has dual controls, two rear seats and loudspeaker under rear of pod.

DIMENSIONS, EXTERNAL:
Main rotor diameter	10.00 m (32 ft 9¾ in)
Main rotor blade chord	0.22 m (8¾ in)
Tail rotor diameter	1.48 m (4 ft 10¼ in)
Tail rotor blade chord	0.16 m (6¼ in)
Length overall, rotors turning	11.415 m (37 ft 5½ in)
Fuselage: Length	8.71 m (28 ft 7 in)
Max width	1.42 m (4 ft 8 in)
Height: overall	2.75 m (9 ft 0¼ in)
to fin-tip	2.45 m (8 ft 0½ in)
Skid track	2.175 m (7 ft 1½ in)
Fuselage ground clearance	0.36 m (1 ft 2¼ in)
Cockpit doors: Height	1.15 m (3 ft 9¼ in)
Max width	0.70 m (2 ft 3½ in)

AREAS:
Main rotor disc	78.70 m² (847.1 sq ft)
Tail rotor disc	1.72 m² (18.52 sq ft)

WEIGHTS AND LOADINGS:
Weight empty	950 kg (2,094 lb)
Fuel weight	128 kg (282 lb)
T-O weight: Aerobatic	1,100 kg (2,425 lb)
Normal	1,280 kg (2,822 lb)
Max	1,450 kg (3,196 lb)
Max disc loading	18.4 kg/m² (3.77 lb/sq ft)
Max power loading	6.08 kg/kW (9.99 lb/hp)

PERFORMANCE (Mi-34S/C at Normal TOW; Mi-34A at MTOW):
Max level speed: Mi-34S/C	113 kt (210 km/h; 130 mph)
Mi-34A	121 kt (225 km/h; 140 mph)
Max cruising speed:	
Mi-34S/C	92 kt (170 km/h; 106 mph)
Mi-34A	113 kt (210 km/h; 130 mph)
Best climbing speed: Mi-34S/C	49 kt (90 km/h; 56 mph)
Service ceiling: Mi-34S/C	4,000 m (13,120 ft)
Mi-34A	5,000 m (16,400 ft)
Hovering ceiling, OGE: Mi-34S/C	900 m (2,960 ft)
Mi-34A	2,750 m (9,020 ft)
Range with max fuel at 500 m (1,640 ft), 5% reserves:	
Mi-34S/C	192 n miles (356 km; 221 miles)
Mi-34A	297 n miles (550 km; 341 miles)
g limits	+3

UPDATED

MIL Mi-35

See Mi-24 entry.

MIL Mi-38

TYPE: Medium transport helicopter.

PROGRAMME: Design begun in 1983; model shown at 1989 Paris Air Show, when aircraft at mockup stage. Modifications in evidence by 1993 included fixed landing gear with wider track and reduced base. Under December 1992 agreement, Eurocopter will integrate flight deck, avionics and passenger systems, and will adapt Mi-38 for international market; Euromil joint stock company (which see in International section) established September 1994 to advance collaboration, adding Kazan production plant (as main manufacturing and final assembly centre); funding for EuroMil granted in October 1994 by European Bank for Reconstruction and Development. Sextant and Pratt & Whitney Canada added as risk-sharing parties for avionics and engine. Funding by Russian Ministry for Defence Industries 1996.

By 1997, Euromil was anticipating first flight in 1999 and start of production two years later, following FAR Pt 29 certification. However, contracts for completion of demonstrator not signed until 18 August 1999, following unilateral decision of Euromil board in December 1998 to launch programme and fly demonstrator at Kazan in 2001. Demonstrator (PT-1) is third airframe, following test articles at Mil Moscow and Kazan. Four prototypes to follow by 2003; deliveries from 2006.

CUSTOMERS: Predicted sales of 200 in CIS, plus 100 exports.

COSTS: Civil export price about US$30 million.

DESIGN FEATURES: Planned as replacement for Mi-8/17 series. Western engines optional. Conventional pod and boom configuration; power plant above cabin; six-blade main rotor with considerable non-linear twist and swept tips; two independent two-blade tail rotors, set as narrow X on same shaft; port-side door at front of cabin; clamshell rear-loading doors and ramp; hatch in cabin floor, under main rotor driveshaft, for tactical/emergency cargo airdrop and for cargo sling attachment; optional windows for survey cameras in place of hatch; sweptback fin/tail rotor mounting; small horizontal stabiliser; for day/night operation over temperature range −60 to +50°C.

FLYING CONTROLS: Fly-by-wire, with manual back-up.

STRUCTURE: Composites main and tail rotors by Kazan; low-profile titanium main rotor head, with elastomeric bearings, built by Stupino; main rotor has hydraulic drag dampers; single lubrication point, at driveshaft; Krasny Oktyabr transmission; fuselage, mainly composites, built by Kazan.

LANDING GEAR: Fixed tricycle type; single wheel on each main unit; twin nosewheels; low-pressure tyres; optional pontoons for emergency use in overwater missions.

POWER PLANT: Those helicopters for CIS customers powered by two Klimov TVA-3000 (TV7-117 derivative) turboshafts, each rated at 1,838 kW (2,465 shp) for T-O; single-engine rating 2,610 kW (3,500 shp) and transmission rated for same power. Demonstrator to have two 2,461 kW (3,300 shp) P&WC PW127 turboshafts, which are also available, in PW127T/S form, as an option for Western customers. Power plant above cabin, to rear of main reduction gear; air intakes and filters in sides of cowling. Bag fuel tanks beneath floor of main cabin; provision for external auxiliary fuel tanks. Liquid petroleum gas fuel planned as alternative to aviation kerosene.

ACCOMMODATION: Crew of two on flight deck, separated from main cabin by compartment for majority of avionics; single-pilot operation possible for cargo missions. Lightweight seats for 30 passengers as alternative to unobstructed hold for 5,000 kg (11,020 lb) freight. Ambulance and air survey versions planned. Provision for hoist over port-side door, remotely controlled hydraulically actuated rear cargo ramp, powered hoist on overhead rails in cabin, and roller conveyor system in cabin floor and ramp.

SYSTEMS: Air conditioning by compressor bleed air, or APU on ground, maintains temperature of not more than 25°C on flight deck in outside temperature of 40°C, and not less than 15°C on flight deck and in main cabin in outside temperature of −50°C. Three independent hydraulic systems; any one able to maintain control of helicopter in emergency. Electrical system has three independent AC generators, two batteries, and transformer/rectifiers for DC supply; electric rotor blade de-icing. Independent fuel system for each engine, with automatic crossfeed; forward part of cowling houses VD-100 APU, hydraulic, air conditioning, electrical and other system components.

AVIONICS: Sextant equipment in export aircraft.
Radar: Weather/nav radar (range 54 n miles; 100 km; 62 miles).
Flight: Preset flight control system allows full autopilot, autohover and automatic landing. Avionics controlled by large central computer, linked also to automatic nav system with Doppler, ILS, satellite nav system, main radar, autostabilisation system and automatic radio compass.

Mil Mi-34SP paramilitary helicopter *(Paul Jackson/Jane's)* 2001/0092554

MIL to MOLNIYA—AIRCRAFT: RUSSIAN FEDERATION

Model of Mi-38 *(Paul Jackson/Jane's)*

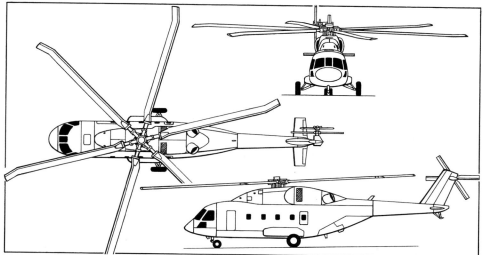

Mil Mi-38 medium transport helicopter *(Mike Keep/Jane's)*

Instrumentation: Six colour CRTs for use in flight and by servicing personnel on ground. Equipment monitoring, failure warning and damage control system. Closed-circuit TV for monitoring cargo loading and slung loads. Options include low-cost electromechanical instrumentation based on that of Mi-8, sensors for weighing and CG positioning of cargo in cabin, and for checking weight of slung loads.

DIMENSIONS, EXTERNAL:
Main rotor diameter	21.10 m (69 ft 2¾ in)
Tail rotor diameter	3.84 m (12 ft 7¼ in)
Length overall, excl rotors	19.95 m (65 ft 5½ in)
Height to top of rotor head	5.20 m (17 ft 0¾ in)
Stabiliser span	4.20 m (13 ft 9½ in)
Wheel track	4.50 m (14 ft 9¼ in)
Wheelbase	5.17 m (16 ft 11½ in)
Forward freight door: Height	1.70 m (5 ft 7 in)
Width	1.50 m (4 ft 11 in)
Floor hatch: Length	1.15 m (3 ft 9¼ in)
Width	0.75 m (2 ft 5½ in)

DIMENSIONS, INTERNAL:
Main cabin: Length to ramp	6.80 m (22 ft 3½ in)
Length, incl fwd part of tailboom	10.70 m (35 ft 1¼ in)
Max width	2.36 m (7 ft 9 in)
Width at floor	2.20 m (7 ft 2½ in)
Height: centre	1.80 m (5 ft 10¾ in)
rear	1.85 m (6 ft 1 in)

AREAS:
Main rotor disc	349.67 m² (3,763.8 sq ft)
Tail rotor disc	11.58 m² (124.65 sq ft)

WEIGHTS AND LOADINGS (provisional):
Max payload: internal	5,000 kg (11,020 lb)
external	6,000 kg (13,225 lb)
Normal T-O weight	14,200 kg (31,305 lb)
Max T-O weight	15,600 kg (34,392 lb)
Max disc loading	44.6 kg/m² (9.14 lb/sq ft)
Transmission loading at max T-O weight and power	5.97 kg/kW (9.82 lb/shp)

PERFORMANCE (estimated):
Max level speed	148 kt (275 km/h; 171 mph)
Cruising speed	135 kt (250 km/h; 155 mph)
Service ceiling	6,500 m (21,325 ft)
Hovering ceiling OGE	2,500 m (8,200 ft)

Range, 30 min reserves:
with 5,000 kg (11,020 lb) payload
175 n miles (325 km; 202 miles)
with 4,500 kg (9,920 lb) payload (30 passengers and baggage) 286 n miles (530 km; 329 miles)
with 3,500 kg (7,715 lb) payload and standard fuel
430 n miles (800 km; 497 miles)
with 1,800 kg (3,965 lb) payload and auxiliary fuel
700 n miles (1,300 km; 808 miles)

UPDATED

MIL Mi-52 SNEGIR
English name: Bullfinch

A description and illustration of this light helicopter project last appeared in the 1998-99 edition. The company unveiled a mockup on 1 June 2000 and announced that it is hoping to resuscitate the venture, with a view to flying a prototype in 2001. Single- (Mi-52-1) and twin-engined (Mi-52-2 with 199 kW; 266 hp VAZ-4265) versions are planned with commercial load capabilities of 350 kg (772 lb) and 400 kg (882 lb), respectively. A military trainer is also envisaged.

UPDATED

MIL Mi-60
TYPE: Three-seat helicopter.
PROGRAMME: Announced July 2000, when mockup under construction at Kazan helicopter factory. Building of prototype to begin in 2001; development cost US$30 million; estimated unit cost US$140,000 to $150,000.
Development by Moscow Aviation Institute; financed by Ministry of Education. Maintenance-free rotor hub.
LANDING GEAR: Skid type.
POWER PLANT: Two 103 kW (138 hp) LOM M332 piston engines.

DIMENSIONS, EXTERNAL:
Main rotor diameter	10.00 m (32 ft 9¾ in)
Main rotor blade chord	0.22 m (8¾ in)

WEIGHTS AND LOADINGS (provisional):
Max T-O weight	1,200 kg (2,645 lb)

PERFORMANCE (estimated):
Max level speed	108 kt (200 km/h; 124 mph)
Max cruising speed	94 kt (175 km/h; 109 mph)
Range	216 n miles (400 km; 248 miles)

NEW ENTRY

MIL Mi-234
TYPE: Four-seat helicopter.
PROGRAMME: Mockup exhibited at Moscow Air Show '92, when known as Mi-34 VAZ (military equivalent Mi-34M). New designation, and confirmation of continued development, released July 2000.
COSTS: US$400,000 (2000).
DESIGN FEATURES: Basically as Mi-34 but completely new rotor head based on carbon fibre starplate, anchoring pitch cases that are integral with main blade section (sleeve over starplate arm); droop stops under blade root; no drag dampers; starplate clamped between upper and lower metal forgings attached to main shaft; pitch-change fittings also metal forgings. Resultant large apparent hinge offset gives rapid control response. Main rotor tip speed 205 m (672 ft)/s.
POWER PLANT: Two VAZ 4265 twin-chamber rotary engines, each 201 kW (270 hp). Internal fuel capacity 245 litres (64.7 US gallons; 53.9 Imp gallons): auxiliary fuel capacity 245 litres (64.7 US gallons; 53.9 Imp gallons). Engines burn Mogas and have power/weight ratio of 0.5 kg (1.1 lb)/hp.
ACCOMMODATION: As Mi-34, plus optional stretcher for EMS duties and optional cargo sling.
SYSTEMS: Cabin heated; de-icing system for main rotor blades and cabin windows.
AVIONICS: *Flight:* Navigation equipment for flying in adverse weather optional.

DIMENSIONS, EXTERNAL:
Main rotor diameter	11.40 m (37 ft 5 in)
Tail rotor diameter	1.50 m (4 ft 11 in)
Fuselage: Length	9.50 m (31 ft 2 in)
Max width	1.45 m (4 ft 9 in)

DIMENSIONS, INTERNAL:
Cabin: Length	2.40 m (7 ft 10½ in)
Width	1.60 m (5 ft 3 in)
Height	1.30 m (4 ft 3¼ in)

AREAS:
Main rotor disc	102.07 m² (1,098.7 sq ft)
Tail rotor disc	1.77 m² (19.02 sq ft)

WEIGHTS AND LOADINGS:
Max payload	550 kg (1,212 lb)
Max T-O weight	1,960 kg (4,320 lb)
Max disc loading	19.2 kg/m² (3.93 lb/sq ft)

PERFORMANCE (estimated):
Max level speed	118 kt (220 km/h; 136 mph)
Nominal cruising speed	97 kt (180 km/h; 112 mph)
Service ceiling	5,000 m (16,400 ft)
Hovering ceiling OGE: ISA	1,500 m (4,920 ft)
ISA + 15°C	800 m (2,625 ft)

Range: with max payload, 30 min reserves
43 n miles (80 km; 50 miles)
with max standard fuel, 340 kg (750 lb) payload 30 min reserves 323 n miles (600 km; 372 miles)
Endurance with max standard fuel, 320 kg (705 lb) payload
5 h to 5 h 30 min

NEW ENTRY

MI-LIGHT HELICOPTERS — see LVM

MOLNIYA

NAUCHNO-PROIZVODSTVENNOYE OBEDINENIE MOLNIYA OAO (Lightning Scientific Production Association JSC)

ulitsa Novoposelkovaya 6, 123459 Moskva
Tel: (+7 095) 492 92 35
Fax: (+7 095) 492 47 23
e-mail: molniya@dol.ru
Web: http://www.buran.ru
MANAGING DIRECTOR: Aleksandr Bashilov
DEPUTY GENERAL DIRECTOR: Valery Verobzhev

CHIEF ENGINEER: Gennady G Kryuchkov
MARKETING DIRECTOR: Prof Mikhail Y Gofin
CHIEF OF MARKETING DEPARTMENT: Dr Vladimir I Fishelovich

To compensate for reduced funding for Buran space shuttle orbiter programme, in which it was much involved, Molniya has diversified its activities to include conveyor equipment (automatic car parking and wheelchair lifts) and aeroplane design. During the early 1990s, it projected a series of civil aircraft of vastly varying size (½ to 450 tonne payloads) all of 'triplane' (canard, wing and tailplane) configuration, as described in the 1997-98 and earlier *Jane's*. The only one so far to fly has been the Molniya-1.

UPDATED

MOLNIYA-1
TYPE: Six-seat/utility transport.
PROGRAMME: Announced at Moscow Air Show '92. Prototype (012001) flew 18 December 1992, was lost subsequently in Moskva River after engine problems; initial production series of 20 begun by Aviacor at Samara; first production aircraft is demonstrator RA-103; certification trials began

RUSSIAN FEDERATION: AIRCRAFT—MOLNIYA to MYASISHCHEV

Molniya-1, following extension of wing tips
(Paul Jackson/Jane's)

Prototype Molniya-1 (VOKBM M-14PM-1 radial engine) *(Paul Jackson/Jane's)*

in 1995; by 1997, RA-103 had an enlarged dorsal scoop for engine cooling air; further development, by mid-1999, resulted in extended wing span and addition of wheel fairings; certification flight testing was due to resume in July 2000. No other production aircraft have been seen, although total of four reported built by 1999.

Cabin, tail surfaces and engine installation used also for an air cushion vehicle marketed by Aviacor.

CURRENT VERSIONS: **Molniya-1:** *As described.*

Molniya-3: Projected version with 261 kW (350 shp) Rolls-Royce 250-B17F turboprop.

CUSTOMERS: Market estimated at 300 to 500 aircraft.

COSTS: US$195,000 (1996), or US$225,000 with Western avionics; US$275,000 with Continental engine and Western avionics. More than US$5 million expended on development by mid-2000.

DESIGN FEATURES: Low-cost light transport of rugged design, with high safety factors. Low-wing configuration, with three lifting surfaces in tandem; foreplanes mid-mounted on nose; wings at rear of fuselage pod, carrying twin tailbooms with tailplane bridging tips of sweptback vertical tail surfaces; engine mounted at rear of fuselage pod. 'Tandem triplane' configuration improves stability, reduces airframe size and weight by 20 per cent for given payload and prevents spin at high angles of attack. Operation practicable from unpaved 500 m (1,640 ft) fields with bearing strength of 4.5 kg/cm² (64 lb/sq in). Designed to permit flying by pilot of average ability after 8 to 16 hours' instruction. Service life 3,000 hours/6,000 sorties.

FLYING CONTROLS: Conventional and manual. Twin full-height rudders, each with ground-adjustable tab; single-piece elevator with two ground-adjustable tabs; ground-adjustable tab on port aileron. Two-segment flaps on each wing, inboard and outboard of tailbooms.

LANDING GEAR: Non-retractable tricycle type; single wheel on each unit; tyre size 500×150-9 on mainwheels, 400×150 on nosewheel; optional fairing over each wheel. Skis and floats optional.

POWER PLANT: One 265 kW (355 hp) VOKBM M-14PM-1 air-cooled radial piston engine, driving Mühlbauer MTV-9 three-blade pusher propeller; or Teledyne Continental TSIO-550-B horizontally opposed engine. Initially unsupercharged but supercharged later.

ACCOMMODATION: Six persons in pairs; seats quickly removable for freight carrying. Alternative configurations include business version with forward-facing seat beside pilot, two armchairs, table, safe, com equipment and small buffet; ambulance for stretcher patient, two attendants and medical equipment; training version with dual controls.

Air conditioning optional. Revised version of the design, shown in mockup form at Moscow Air Show '95, has 'rounded triangular' side windows.

AVIONICS: *Comms:* Radio standard.
Flight: Navaids standard. Autopilot and GPS optional.
Instrumentation: Digital instruments optional.
Mission: Fax optional.

DIMENSIONS, EXTERNAL:
* Wing span	8.50 m (27 ft 10¾ in)
Foreplane span	3.50 m (11 ft 5¾ in)
Length overall	7.86 m (25 ft 9½ in)
Height overall	2.30 m (7 ft 6½ in)
Width, wings removed or folded	3.60 m (11 ft 9¾ in)
Wheel track	3.02 m (9 ft 11 in)
Wheelbase	3.34 m (10 ft 11½ in)

DIMENSIONS, INTERNAL:
Cabin: Length	2.75 m (9 ft 0¼ in)
Max width	1.25 m (4 ft 1¼ in)
Max height	1.24 m (4 ft 0¾ in)

AREAS:
* Wings, gross	11.50 m² (123.8 sq ft)

Before extension to approx 9.3 m (30½ ft) span

WEIGHTS AND LOADINGS:
Max payload	505 kg (1,115 lb)
Max fuel	220 kg (485 lb)
Max T-O weight	1,740 kg (3,835 lb)
Max power loading (M-14 engine)	6.58 kg/kW (10.81 lb/hp)

PERFORMANCE (estimated, with supercharged M-14PM or Continental engine):
Max level speed: M-14AM	172 kt (320 km/h; 199 mph)
Continental	216 kt (400 km/h; 248 mph)
Max cruising speed:	
M-14PM	154 kt (285 km/h; 177 mph)
Continental	175 kt (325 km/h; 202 mph)
Landing speed	71 kt (130 km/h; 81 mph)
T-O run	450 m (1,480 ft)
Landing run	350 m (1,150 ft)
† Range:	
with max payload at 145 kt (270 km/h; 168 mph)	
	270 n miles (500 km; 310 miles)
with max fuel: M-14PM	
	540 n miles (1,000 km; 621 miles)
Continental	648 n miles (1,200 km; 745 miles)

†*Ranges with unsupercharged engine at 121-153 kt (225-285 km/h; 140-177 mph) at 1,500 m (4,920 ft) are close to those listed*

UPDATED

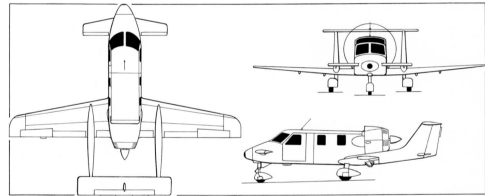

Molniya-1 six-seat light aircraft, as configured in 1999 *(James Goulding/Jane's)*

MYASISHCHEV

EKSPERIMENTALNYI MASHINOSTROITELNYI ZAVOD IMENI V M MYASISHCHEVA (Experimental Engineering Bureau named for V M Myasishchev)

140160 Zhukovsky-5, Moskovskaya oblast
Tel: (+7 095) 556 77 76 and 912 60 41
Fax: (+7 095) 556 52 98 and 728 41 30
e-mail: mdb@mastak.msk.ru
GENERAL DESIGNER: Aleksandr Bruk
CHIEF DESIGNER, M-101: Evgeny Charsky
CHIEF DESIGNER, M-55: Leonid Sokolov
COMMERCIAL DIRECTOR: Aleksandr Gorbunov

The Myasishchev OKB was founded in 1952, and led by Prof Vladimir Mikhailovich Myasishchev until his death on 14 October 1978. In the 1970s and 1980s, the bureau was engaged in development of multipurpose subsonic high-altitude aircraft, but has more recently diversified into civil aviation. In 1981, the bureau was named after Prof Myasishchev.

In 1997, the Myasishchev bureau joined MAPO (now RSK 'MiG') and was reported to have formed a joint venture with General Aircraft USA, each holding a 25 per cent share, the balance having been offered to outside investors. General Aircraft to be responsible for US certification, assembly, engine and avionics installation of the Myasishchev M-101, M-201, M-202 and M-203; site at Williamsburg Airport, Newport News, Virginia, was initially selected, but consideration more recently given to former naval air station

Myasishchev continues to promote the M-55 high-altitude surveillance aircraft (here in formation with an M-101 Gzhel), last described in the 1996-97 edition *(Paul Jackson/Jane's)*

at Cecil Field, Florida. However, in mid-1998, Myasishchev stated that no firm agreement was in place. This was followed, on 18 November 1998, by a preliminary agreement for possible assembly of the M-101 in Malaysia, using Russian fuselages and local/imported avionics and interiors. Nothing further had emerged by the end of 2000, however.

Russian defence ministry approved, in September 1999, plans to complete one of two unfinished Myasishchev M-55 high-altitude surveillance aircraft held at Smolensk production plant, but finance was still awaited in following year.

UPDATED

MYASISHCHEV M-101T GZHEL

TYPE: Light utility turboprop.
PROGRAMME: Derived from Nelli and M-70 projects; first shown in model form at Moscow Air Show '90; developed full-scale mockup exhibited at Moscow Air Show '92 with piston engine and at Moscow Air Show '93 with turboprop. Two static test airframes and three flying prototypes built at Nizhny Novgorod by Sokol; first (RA-15001) first flew 31 March 1995; this and second flying aircraft (RA-15003), shown at Moscow Air Show '95, lack ventral fin. Third prototype (RA-15004) to preproduction standard; first preproduction aircraft (RA-15101) displayed at Paris and Moscow Air Shows 1997. Initial batch of 15 at Sokol plant; five completed by mid-2000; No. 6 is first in full production configuration; No. 7 for reliability tests; No. 8 is first for customer delivery; target production rate 50 per year, but beginning with 20 in 2001.

Russian certification to AP-23 in passenger category was expected late 1998 or early 1999, following delivery of two aircraft to State Civil Aviation Research Institute

MYASISHCHEV—AIRCRAFT: RUSSIAN FEDERATION

Myasishchev M-101T Gzhel multipurpose light aircraft *(Paul Jackson/Jane's)* 2001/0103576

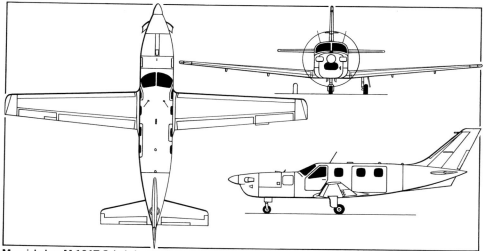

Myasishchev M-101T Gzhel six-seat light aircraft *(Mike Keep/Jane's)* 2000/0079304

(Gos NII GA) on 23 January 1998 for short-range freight services from Moscow/Sheremetyevo by Fenix Air but delayed by lack of funding; latest notified target date for certification was late 2000; meanwhile early production aircraft used on freight services by Sokol plant, 2000, and one reported in use by Central African Airlines, based in Egypt.

Under a 1997 provisional agreement with General Aircraft USA, the M-101 will be re-engined with a Pratt & Whitney Canada PT6A turboprop for marketing in North America; assembly and distribution from Newport News, Virginia, or alternative (see introduction). Similar European operation may be based in Germany.

The aircraft is named for a type of fine porcelain made in the city of Gzhel. Some development funding was provided by Gzhel, other finance coming from Myasishchev, Sokol, Inkombank and a Czech bank. Marketing is by Gzhel-Avia.

CURRENT VERSIONS: **M-101T:** *As described.*

M-101D: Projected variant with diesel engine of Italian origin.

M-101P: Projected version with Textron-Lycoming piston engine.

M-101PW: Projected Westernised version for North American market; 1,178 kW (1,580 shp) P&WC PT6A-64 engine, Hartzell propeller and Honeywell or Becker avionics. Prototype was to have flown before end of 1998, but apparently had not done so by late 2000.

CUSTOMERS: Undisclosed central African customer ordered one in early 2000; minimum estimated market for 300. Reported 100 orders by mid-2000. First Russian customer reported as Gazprom, which requires 10.

COSTS: US$970,000 (2000). By mid-1999, Sokol plant had invested Rb102.4 million in M-101, and in 2000 was planning funding of US$2.5 million to complete certification.

DESIGN FEATURES: Conventional all-metal low-wing monoplane with pressurised cabin; sweptback vertical tail surfaces and ventral fin. Designed in accordance with Russian AP-23 and US FAR Pt 23 airworthiness requirements.

FLYING CONTROLS: Conventional and manual. Trim tabs in port aileron, rudder and each elevator; ground-adjustable tab on starboard aileron.

LANDING GEAR: Hydraulically retractable tricycle type; single wheel on each unit; nosewheel retracts rearward, mainwheels inward into wingroots and fuselage; mainwheels uncovered by doors when retracted; tyre size 500×150-9 on mainwheels, 400×150-5 on nosewheel; levered suspension legs; designed to use paved and unpaved runways.

POWER PLANT: One 559 kW (751 shp) Walter M 601 F turboprop, driving Avia Hamilton Sundstrand V-510 five-blade propeller. Power plant is protected against particle ingestion.

ACCOMMODATION: One or two pilots and four passengers, in pairs in pressurised cabin; max capacity, one pilot and seven passengers. Rear-hinged door to flight deck on port side; large door for passengers and freight loading aft of wing on port side; emergency exit on starboard side above wing; provision for rapid change to cargo/passenger, freight or ambulance configuration.

AVIONICS: Russian or Western equipment for day/night flying under VFR or adverse weather conditions.

DIMENSIONS, EXTERNAL:
Wing span	13.00 m (42 ft 8 in)
Length overall	9.975 m (32 ft 8¾ in)
Height overall	3.72 m (12 ft 2½ in)
Tailplane span	4.32 m (14 ft 2 in)
Wheel track	3.00 m (9 ft 10 in)
Wheelbase	2.825 m (9 ft 3¼ in)
Propeller diameter	2.30 m (7 ft 6½ in)
Passenger/freight door: Height	1.30 m (4 ft 3¼ in)
Width	1.30 m (4 ft 3¼ in)

DIMENSIONS, INTERNAL:
Cabin: Length	4.56 m (14 ft 11½ in)
Max width	1.32 m (4 ft 4 in)
Max height	1.26 m (4 ft 1½ in)

AREAS:
Wings, gross	17.06 m² (183.6 sq ft)

WEIGHTS AND LOADINGS:
Weight empty	2,270 kg (5,004 lb)
Max payload	630 kg (1,389 lb)
Max fuel weight	650 kg (1,433 lb)
T-O weight: normal	2,900 kg (6,393 lb)
max	3,200 kg (7,054 lb)
Max wing loading	187.6 kg/m² (38.41 lb/sq ft)
Max power loading	5.18 kg/kW (8.51 lb/shp)

PERFORMANCE:
Max cruising speed	232 kt (430 km/h; 267 mph)
Stalling speed, flaps down	61 kt (113 km/h; 71 mph)
Cruising altitude	7,600 m (24,940 ft)
T-O run	350 m (1,150 ft)
Landing run with propeller reversal	280 m (920 ft)
Range:	
with max payload	432 n miles (800 km; 497 miles)
with max fuel, 45 min reserves	755 n miles (1,400 km; 869 miles)

UPDATED

MYASISHCHEV M-150

All known details of this turboprop twin last appeared in the 2000-01 edition.

UPDATED

MYASISHCHEV M-201 SOKOL

English name: Falcon

TYPE: Utility turboprop twin.

PROGRAMME: Design study revealed as M-103 Oka at Dubai Air Show, November 1995; twin-engined version of M-101 Gzhel (which see) with stretched fuselage; renamed M-201 Sokol early 1996; design revisions by 1997 included 1.80 m (5 ft 11 in) fuselage stretch, 500 kg (1,102 lb) additional max T-O weight and winglets. It was announced in 1997 that the M-201 would be certified in North America by General Aircraft USA, powered by P&WC PT6A turboprops and with Western avionics; US assembly and distribution from Newport News, Virginia. However, this agreement apparently still requires ratification. No plans for a prototype had been made known by late 2000.

CURRENT VERSIONS: **M-201:** *As described.*

M-201PW: Powered by two P&WC PT6A turboprops; Western avionics.

COSTS: Estimated US$1.8 million, flyaway (1996).

DESIGN FEATURES: Twin-engined derivative of M-101T. Conventional, all-metal, low-wing monoplane intended to meet Russian AP-23 and US FAR Pt 23 requirements for over-water operation. Sweptback vertical tail surfaces with ventral strake. Originally shown with wingtip fuel tanks; these later replaced by winglets.

FLYING CONTROLS: Conventional and manual. Frise ailerons; trim tab in starboard aileron. Horn-balanced elevators with tab in each; tab in horn-balanced rudder.

LANDING GEAR: Retractable tricycle type with twin wheels on each main leg.

POWER PLANT: Two 580 kW (778 shp) Walter M 601F-22 turboprops, driving Hamilton Sundstrand A-510 five-blade propellers.

ACCOMMODATION: Up to 11 persons, including one or two pilots; business configurations for up to eight persons. Upward-opening freight door, port side, rear; crew door port side of flight deck; emergency exit, starboard side, opposite freight door. Cabin pressurised. Versions envisaged for cargo, air ambulance, survey and navaids calibration missions.

DIMENSIONS, EXTERNAL:
Wing span	16.865 m (55 ft 4 in)
Length overall	13.40 m (43 ft 11½ in)
Height overall	4.85 m (15 ft 11 in)
Wheel track (c/l shock-absorbers)	4.50 m (14 ft 9¼ in)
Wheelbase	4.50 m (14 ft 9¼ in)
Propeller diameter	2.30 m (7 ft 6½ in)

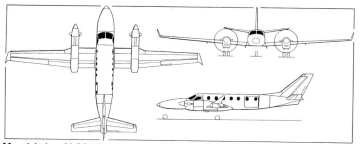

Myasishchev M-201 Sokol 11-seat light transport *(Paul Jackson/Jane's)* 0015081

Model of M-201 Sokol as revealed in 1995 *(Paul Jackson/Jane's)*

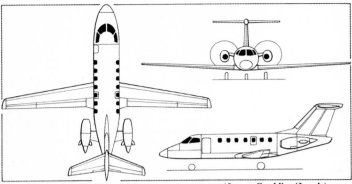

Myasishchev M-202PW multipurpose transport *(James Goulding/Jane's)*

Model of Myasishchev's M-203 Barsuk *(Paul Jackson/Jane's)*

Freight door: Height	1.30 m (4 ft 3¼ in)
Width	1.30 m (4 ft 3¼ in)
WEIGHTS AND LOADINGS:	
Max payload	1,400 kg (3,086 lb)
Max fuel weight	1,490 kg (3,285 lb)
Max T-O weight	5,500 kg (12,125 lb)
Max power loading	4.74 kg/kW (7.79 lb/shp)
PERFORMANCE (estimated):	
Nominal cruising speed	270 kt (500 km/h; 311 mph)
Cruising altitude	7,600 m (24,940 ft)
T-O run	400 m (1,315 ft)
Landing run with propeller reversal	300 m (985 ft)
Range: with max fuel, 45 min reserves	1,717 n miles (3,180 km; 1,976 miles)
with max payload	540 n miles (1,000 km; 621 miles)

UPDATED

MYASISHCHEV M-202PW OLEN
English name: Deer

TYPE: Twin-turboprop airliner.
PROGRAMME: Following the withdrawal of Myasishchev from its partnership with NAL of India on the M-102 Duet/Sāras programme (which see in Indian section) in 1997, details were released of the M-202, which appears to be a slightly enlarged M-102. Subject to formalisation of plans, the aircraft will be promoted in, and assembled for, the North American market by General Aircraft USA of Newport News, Viriginia. By late 2000, this arrangement still had not been confirmed and there were suggestions that the M-102 would be reinstated for Russian production by Sokol at Nizhny Novgorod.
CUSTOMERS: Estimated market for up to 1,000 by 2015.
COSTS: Programme US$40 million; unit cost US$4 million (both 2000).
Structure, controls and other data generally as for NAL Sāras, except the following.
LANDING GEAR: Single nose- and mainwheels.
POWER PLANT: Two 1,062 kW (1,424 shp) P&WC PT6A-67T turboprops.
ACCOMMODATION: Two crew and 19 passengers, plus baggage areas (front and rear of cabin) and lavatory.
DIMENSIONS, EXTERNAL:

Wing span	18.00 m (59 ft 0½ in)
Length overall	15.80 m (51 ft 10 in)
Height overall	5.60 m (18 ft 4½ in)
Propeller diameter	2.30 m (7 ft 6½ in)
WEIGHTS AND LOADINGS:	
Max payload	1,800 kg (3,968 lb)
Max T-O weight	7,400 kg (16,314 lb)
Max power loading	3.49 kg/kW (5.73 lb/shp)
PERFORMANCE (estimated):	
Normal cruising speed	302 kt (560 km/h; 348 mph)
Range: with max fuel	1,727 n miles (3,200 km; 1,988 miles)
with max payload	1,133 n miles (2,100 km; 1,304 miles)

UPDATED

MYASISHCHEV M-203PW BARSUK
English name: Badger

TYPE: Light utility transport.
PROGRAMME: Developed in prototype form (possibly with VOKBM M-14 radial engine) by Myasishchev Design Bureau; stated in mid-1997 to have flown, but no corroboration forthcoming. Production to be jointly financed under an agreement of 1997 with General Aircraft Corporation of the USA; certification prototype, with R-985 engine and several other changes, was due to have begun US trials before end of 1998, but the programme appears to have been delayed, with nothing more heard by end of 2000.
DESIGN FEATURES: Rugged, piston-engined, braced high-wing design for operation from semi-prepared airstrips. Potential duties include passenger and light freight transport, agricultural spraying, paradropping, SAR and glider towing. Meets FAR Pt 23 for VFR operations in all climates.
Wing section CAHI (TsAGI) P4-15M with 12 per cent thickness/chord ratio at root and 15 per cent at tip. Leading-edge sweepback of 1° 47′ on wing; 26° on fin; 7° 58′ on tailplane.
FLYING CONTROLS: Conventional and manual. Aileron deflections +10/−25°; trim tab in starboard aileron; two-position flaps, deflections 20° and 40°. Rudder deflections ±25°; trim tab in rudder. Elevator deflections +15/−20°; tab in port elevator.
LANDING GEAR: Tailwheel type; non-retractable. Cantilever-sprung single mainwheels.
POWER PLANT: One 336 kW (450 hp) Pratt & Whitney R-985 Wasp Junior nine-cylinder air-cooled radial engine driving a Hartzell three-blade propeller.
ACCOMMODATION: Pilot and seven passengers or a combination of freight and passengers; tie-down points in floor. Dual controls for training. Baggage area in rear fuselage. Medevac configuration for one stretcher, two attendants and medical equipment. Crew door each side; cargo/passenger door, port side rear; emergency exit, starboard side rear.
DIMENSIONS, EXTERNAL:

Wing span	16.00 m (52 ft 6 in)
Wing aspect ratio	10.7
Length overall	9.40 m (30 ft 10 in)
Fuselage: Length	8.54 m (28 ft 0¼ in)
Max width	1.52 m (4 ft 11¾ in)
Max height	1.70 m (5 ft 7 in)
Height overall	3.42 m (11 ft 2¾ in)
Tailplane span	6.00 m (19 ft 8¼ in)
Wheel track	3.20 m (10 ft 6 in)
Wheelbase	5.00 m (16 ft 4¾ in)
Propeller diameter	2.59 m (8 ft 6 in)
DIMENSIONS, INTERNAL:	
Cabin: Length	4.46 m (14 ft 7½ in)
Max width	1.30 m (4 ft 3¼ in)
Max height	1.42 m (4 ft 8 in)
AREAS:	
Wings, gross	24.00 m² (258.3 sq ft)
Ailerons (total)	3.12 m² (33.58 sq ft)
Trailing-edge flaps (total)	8.92 m² (96.01 sq ft)
Fin	3.77 m² (40.58 sq ft)
Rudder, incl tab	1.56 m² (16.79 sq ft)
Tailplane	7.50 m² (80.73 sq ft)
Elevators (total, incl tab)	3.00 m² (32.29 sq ft)
WEIGHTS AND LOADINGS:	
Operating weight empty (incl two crew)	1,850 kg (4,079 lb)
Max payload	700 kg (1,543 lb)
Fuel weight: max (reduced payload)	390 kg (860 lb)
with max payload	250 kg (551 lb)
Max T-O and landing weight	2,800 kg (6,172 lb)
Max ramp weight	2,820 kg (6,217 lb)
Max wing loading	116.7 kg/m² (23.90 lb/sq ft)
Max power loading	8.35 kg/kW (13.72 lb/hp)
PERFORMANCE (estimated):	
Max cruising speed	140 kt (260 km/h; 162 mph)
Econ cruising speed at 3,000 m (9,840 ft)	103 kt (190 km/h; 118 mph)
Range, 45 min reserves:	
with max fuel	1,285 n miles (2,380 km; 1,478 miles)
with max payload	723 n miles (1,340 km; 832 miles)

UPDATED

MYASISHCHEV M-500

It was reported during 2000 that this agricultural/forestry and seven-seat transport aircraft is included in the development programme of civil aircraft up to 2015. It will be built at the Sokol factory and at Smolensk beginning 2003. The M-500 will be offered with a 316 kW (424 hp) VOKBM M-14NTK piston engine driving a three-blade MT propeller, or a 331 kW (444 hp) Ufa NPP turbocharged diesel driving a Hartzell propeller. Estimated market for 500 aircraft; unit cost US$140,000 (2000).

UPDATED

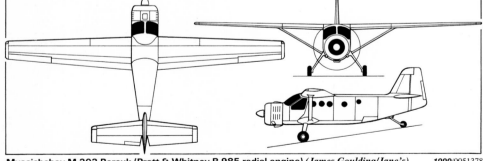

Myasishchev M-203 Barsuk (Pratt & Whitney R-985 radial engine) *(James Goulding/Jane's)*

NAPO
GOSUDARSTVENNOYE UNITARNOYE PREDPRIYATIE NOVOSIBIRSKOYE AVIATSIONNOYE-PROIZVODST-VENNOYE OBEDINENIE IMENI V P CHKALOVA (State Unitary Enterprise Novosibirsk Aircraft Production Association named for V P Chkalov)
ulitsa Polzunova 15, 630051 Novosibirsk
Tel: (+7 3832) 79 80 95 and 77 37 06
Fax: (+7 3832) 77 23 92
e-mail napa@mail.cis.ru
Web: http://www.allsiberia.com/novosibirsk/BUSINESS/Chkalov.html
Telex: 133211 NAPO SU
GENERAL DIRECTOR: Nikolai I Bobritsky
MARKETING DIRECTOR: Valery Sadaev

Founded in 1936 as GAZ 153, NAPO has built about 36,000 aircraft of many types, including I-16, Yak-7, Yak-9, MiG-15, MiG-17, MiG-19, Yak-28, Su-9, Su-15 and, from 1972, Su-24 multirole combat aircraft. The plant has been wholly state-owned, but 49 per cent of shares were due to have been sold by 1999. Its current products include the Antonov An-38 and Sukhoi Su-27IB as well as motor boats. Is promoting Su-24 modernisation programmes for export customers.

UPDATED

PHOENIX — *see Aviacor*

POLYOT

PROIZVODSTVENNOYE OBEDINENIE POLET (Polyot Production Association)
ulitsa Bogdan Khmelnitsky, 226, 644021 Omsk
Tel/Fax: (+7 3812) 57 70 21
GENERAL DIRECTOR: Oleg P Dorofeev

Established in 1941, Polyot is one of the largest aerospace enterprises in Russia, with facilities extending over more than 15 km² (5.8 sq miles) and up to 20,000 employees. During the Second World War it manufactured 3,405 Tupolev Tu-2 bombers and Yakovlev Yak-7 and Yak-9 fighters in 2½ years; post-war products included 758 Ilyushin Il-28 jet bombers and 58 Tupolev Tu-104 jet transports. It remains state-owned. The Antonov An-74 series of STOL transports is now in series production.

Polyot manufactured SS-4, SS-7 and SS-11 strategic missiles, warheads, and first-stage engines for Zenith and Energiya rocket launch vehicles. Other products include many spacecraft and a variety of consumer goods under *konversiya* programmes. Brief details of the projected VK-1 light aircraft appeared in the 1997-98 edition.

In June 1997, Polyot and Baranov Engine Manufacturing Association relaunched the Antonov An-3 programme (An-2 with 1,140 kW; 1,529 shp OMKB TVD-20 turboprop), currently described briefly in Ukrainian section and in detail in *Jane's Aircraft Upgrades*. Certification of factory to build An-3 received by May 2000, but funding unavailable for production line and only conversions of An-2s being considered.

UPDATED

REFLY

KOMPANIYA REFLAI (Refly Company)
a/ya 21, nab Oktyabrskaya 6, 193021 Sankt Peterburg
Tel: (+7 812) 445 08 34
Fax: (+7 812) 968 51 01
Web: http://www.rusavia.ru/refly

Formed in 1997, Refly sells Western aircraft kits (Revolution Mini 500, RotorWay Exec, Skystar Kitfox, American Sportscopter Ultrasport, Aerocomp Comp Air 6) and light aviation supplies on Russian market. Chernov Che-22 (which see in this section) marketed in USA as Refly Pelican.

NEW ENTRY

REFLY DELFIN

Two Delfins were built (first flight 9 January 1991) by an amateur aviation group at Kronstadt, under leadership of Peter Lyavin; by 2000, each had flown some 1,500 hours. Refly plans to begin series production, for sale at US$75,000.

NEW ENTRY

Lyavin Delfin (Dolphin), which Refly plans to manufacture in series

ROSTVERTOL

ROSTOVSKY VERTOLETNYI PROIZVODSTVENNYI KOMPLEKS OAO 'ROSTVERTOL' ('Rostvertol' Rostov Helicopter Production Complex JSC)
ulitsa Novatorov 5, 344038 Rostov-na-Donu
Tel: (+7 8632) 31 74 93
Fax: (+7 8632) 45 05 35
e-mail: rostvertol@rost.ru
DIRECTOR GENERAL: Boris N Slyusar
DEPUTY DIRECTOR GENERAL: Yuri Zaikin
DEPUTY DIRECTOR-GENERAL, MARKETING AND SALES: V Dyatlov
CHIEF ENGINEER: I Semyonov

The company now known as Rostvertol was founded on 1 July 1939, and began by manufacturing wooden propellers. It progressed to aircraft production during the Second World War and to helicopter work in the mid-1950s. Past products at Rostov-on-Don included the UT-2M (1944), Po-2, Yak-14 glider, Il-10M, Il-40, Mi-1, Mi-6, Mi-10 and Mi-10K. Rostvertol is wholly privately owned.

Government decree of 19 February 1996 authorises Rostvertol to export spares and auxiliary equipment for Mil Mi-24 and Mi-25; and to export new Mi-26, Mi-28 and Mi-35 helicopters, as described in the Mil Moscow Helicopter Plant entry. Production in 1999 included one Mi-26 delivered to the Moscow fire brigade for water-bombing and rescue duties. Mi-26 supplied to Mexican Air Force in January 2000 and second ordered in following April. Rostvertol also active in helicopter overhaul and upgrading. However, independent export sales authorisation rescinded 23 July 1999, and Rostvertol exports now made through official Russian agencies. Company has exported 650 aircraft.

UPDATED

RSK MiG

FEDERALNOYE GOSUDARSTVENNOYE UNITARNOYE PREDPRIYATIE, ROSSIYSKAYA SAMOLETOSTROITEL'NAYA KORPORATSIYA 'MiG' (Russian Aircraft-Building Corporation 'MiG')
1-ň Botkinsky proezd, Dom 7, 125040 Moskva
Postal address: a/ya No 1, 103045 Moskva
Tel: (+7/095) 252 88 39
Fax: (+7/095) 250 88 19
CHIEF EXECUTIVE: Sergei Mikheyev
GENERAL DIRECTOR: Nikolai Nikitin
ADVISER TO THE GENERAL DIRECTOR AND MAPO PLANT DIRECTOR: Grígory Nemov
CHAIRMAN, BOARD OF DIRECTORS: Vladimir V Kuzmin
GENERAL DESIGNER: Mikhail Waldenberg

PUBLIC RELATIONS AND MEDIA CENTRE:
Tel: (+7 095) 207 04 76
Fax: (+7 095) 207 07 57
e-mail: migpress@mail.ru

Original Moscow Aircraft Production Organisation (MAPO) formed in January 1996 by a presidential decree to bring together leading Russian military and civil aviation concerns. Integrated their production, intellectual, financial and marketing capabilities, with the common aim of developing, producing and promoting, to the world arms market, advanced aircraft and weapon systems, and to support their after-sales monitoring, maintenance and upgrading. It incorporated several industrial enterprises, each with a wide network of affiliated branches and subsidiary establishments.

As originally organised, the grouping absorbed the MiG design bureau and adopted the Western marketing name MIG MAPO, the first three initials indicating Military-Industrial Group. However, MiG regained its separate identity in July 1997. MAPO (formerly MMZ No. 30 Znamaya Truda) had built 25,000 aircraft of 40 types, from the Nieuport IV of 1913 to the MiG-29. In addition to MiG and Kamov types, current products include Aeroprogress/ROKS-Aero T-101 Grach, Aviatika 890, Interavia I-1L and Ilyushin Il-103 light aircraft. On 5 October 1999, it was nominated as a production centre for the Tupolev Tu-334 airliner and will contribute 50 per cent of certification costs.

Service activities include complete maintenance support and pilot training; is also involved in Chinese Chengdu FC-1 fighter programme, China. It is fifth largest Russian company in terms of export sales volume and first among machine-building complexes. Its aircraft production is centred at the

MiG-29s under construction by MAPO at Lukhovitsy

Moscow plant alongside the Mikoyan complex and in two subsidiaries in the Moscow region.

Following a government decree (No. 14) of 2 February 1999, MAPO was reorganised with members below:

Full members

A I Mikoyan Design Bureau, Moscow (design of new aircraft and upgrades)

P A Voronin Production Centre, Moscow (development and manufacture of prototypes; aircraft upgrades)

Lukhovitsy Machine-Building Plant, Lukhovitsy (series production)

V Ya Klimov Plant, St Petersburg (engines, gearboxes, accessories)

Soyuz Machine-Building Plant, Tushino (engines, gearboxes, accessories)

State Ryazan Instrument Plant, Ryazan (airborne and ground test equipment)

Electroavtomatika Design Bureau, St Petersburg (airborne and ground test equipment)

Associate members

Kamov Company, Lyubertsy (helicopter design and manufacture)

V V Chernyshov Machine-Building Enterprise, Moscow

Krasny Octyabr Machine-Building Enterprise, St Petersburg

Pribor, Kursk

Aviatest, Rostov-on-Don

Instrument-Making Company, Perm

Workforce in late 2000 was 13,000.

From 2000, the organisation has been known as RSK 'MiG'. On 10 November 2000, it formed MiG Light Civil Aviation company to manage assembly of Aeroprogress T-101 Grach and to design and develop new lightplanes.

NEW ENTRY

RSK 'MiG' SL-39

TYPE: Two-seat lightplane.

PROGRAMME: Manufactured at LMZ, Lukhovitsy, since 1994, in parallel with I-1L, although no recent sales have been reported. By 1999, Lukhovitsy had a stockpile of some 30 uncompleted I-1/SL-39 airframes, these also having been used for the Alfa-M A-211 (see 1998-99 *Jane's*). Promotion appears to have been discontinued in favour of the I-1.

CURRENT VERSIONS: **SL-39 VM-1**: With tailwheel landing gear. RA-01460 used as testbed for GosNIPAS RP-3M cropspraying system.

SL-39 VM-2: As VM-1, but tricycle landing gear. None known to have been built.

CUSTOMERS: Six built by mid-1995 (latest data).

DESIGN FEATURES: Generally similar to I-1L (which see under Interavia). Rear fuselage appears to be semi-monocoque.

FLYING CONTROLS: As I-1L, except for controllable trim tab on starboard elevator.

LANDING GEAR: VM-1 same as I-1L, except tailwheel tyre size 200×80. VM-2 has tricycle gear, with single wheel on each unit; steerable nosewheel; cantilever blade-type mainwheel legs.

POWER PLANT: One 103 kW (138 hp) LOM M332 four-cylinder in-line engine; VML-3-08 two-blade propeller. Fuel tanks in wings, capacity 120 litres (31.7 US gallons; 26.4 Imp gallons).

ACCOMMODATION: As I-1L.

DIMENSIONS, EXTERNAL:
Wing span	10.00 m (32 ft 9¾ in)
Length overall	7.20 m (23 ft 7½ in)

WEIGHTS AND LOADINGS (A: VM-1, B: VM-2):
Max payload: A, B	150 kg (330 lb)
Max T-O weight: A	780 kg (1,720 lb)
B	890 kg (1,962 lb)

RSK 'MiG' SL-39 VM-1 two-seat light aircraft
(Paul Jackson/Jane's) 0015023

Max power loading: A	8.17 kg/kW (13.42 lb/hp)
B	8.65 kg/kW (14.22 lb/hp)

PERFORMANCE:
Max level speed: A	108 kt (200 km/h; 124 mph)
B	113 kt (210 km/h; 130 mph)
Nominal cruising speed: A	81 kt (150 km/h; 93 mph)
B	94 kt (175 km/h; 108 mph)
T-O and landing speed: A	52 kt (95 km/h; 59 mph)
B	49 kt (90 km/h; 56 mph)
Service ceiling: A	4,000 m (13,120 ft)
T-O run: A	250 m (820 ft)
Landing run: A	100 m (330 ft)
Range with max fuel: A	323 n miles (600 km; 372 miles)
B	431 n miles (800 km; 497 miles)
g limits	+3.5/−1.5

UPDATED

SAT 'SAVIAT'

SPETSIALNOYE AVIATSIONNOYE TEKHNOLOGY AO (Special Aviation Technologies Co Ltd)

pristan Akademika AN Tupoleva 15, 111250 Moskva
Tel: (+7 095) 265 79 12 and 263 75 14
Fax: (+7 095) 261 87 13
e-mail: ul0409@dialup.podolsk.ru
PRESIDENT: Vladimir S Eger

Trading as Saviat, SAT is developing a series of rhomboid wing light transport aircraft which will be built by VPK MIG.

UPDATED

SAVIAT E-1

No new information has been received concerning this four-seat light aircraft; known details last appeared in the 2000-01 *Jane's*.

UPDATED

SAVIAT E-4

TYPE: Agricultural sprayer/light utility transport.

PROGRAMME: Manufacture of prototype began May 1999; first flight was then due in January 2000, but had not been reported by late 2000.

CUSTOMERS: Orders for eight aircraft had been placed by late 1999.

COSTS: US$60,000 (1999).

DESIGN FEATURES: Intended as small sprayer/utility aircraft with high safety margins, certifiable to JAR-VLA criteria. Specific features of this sesquiplane include newly developed aerofoil; direct side-force control (DSFC) system, which includes unusually large proportion of vertical stabilisation and control areas, for accurate positioning in low-level flight; twin tailbooms; pusher configuration; ballistic parachute; agro-chemical tank in lower wing.

Special aerofoil section; no leading-edge sweep; thickness/chord ratio 8.3 per cent; no dihedral; incidence 4° 30′; no twist.

FLYING CONTROLS: Manual. Conventional ailerons on upper wings; single-piece elevator with trim tab; two-element rudders on each tailfin. Additional fins and rudders between lower wingtips and upper wings, plus inboard ailerons on lower wings, all form DSFC system. Flaps on upper planes only.

STRUCTURE: Fuselage and upper wing centre-section of metal; remainder of composites sandwich.

LANDING GEAR: Tricycle type; fixed. Cantilever mainwheel legs of titanium; nosewheel shock-absorption by helical spring. Single wheel on each unit; all tyres 400×150.

POWER PLANT: One 89 kW (120 hp) Hirth F30-A36C flat-four driving a D&D, fixed-pitch, two-blade propeller. Two fuel tanks in upper wing; normal capacity 100 litres (26.4 US gallons; 22.0 Imp gallons). Provision for chemical hoppers to be used as additional fuel tanks.

ACCOMMODATION: Pilot and provision for two passengers. Single-piece windscreen; forward-hinged door, port side. Passenger seats may be removed for freight-carrying. Cabin heated, ventilated and pressurised to 0.015 bar (0.21 lb/sq in) differential to prevent chemical ingress.

SYSTEMS: Electrical system includes 12 V and 27 V DC generators and 45 Ah battery.

AVIONICS: *Flight:* GPS navigation.

EQUIPMENT: Chemical hoppers in lower wing; spraybars immediately behind, and extending beyond, lower wing. Saviat-designed ballistic parachute recovery system, glide ratio 1.5, in upper wing centre-section.

DIMENSIONS, EXTERNAL:
Wing span (upper)	10.00 m (32 ft 9¾ in)
Wing chord, constant	1.20 m (3 ft 11¼ in)
Wing aspect ratio (upper)	8.3
Length: overall	7.20 m (23 ft 7½ in)
fuselage pod	3.60 m (11 ft 9¾ in)
Fuselage pod: Max width	1.10 m (3 ft 7¼ in)
Max height	1.45 m (4 ft 9 in)
Height overall	2.65 m (8 ft 8¼ in)
Tailplane span	2.00 m (6 ft 6¾ in)
Wheel track	2.10 m (6 ft 10¾ in)
Wheelbase	2.55 m (8 ft 4½ in)
Propeller diameter	1.87 m (6 ft 1½ in)
Propeller ground clearance	0.47 m (1 ft 6½ in)
Passenger door: Max height	1.17 m (3 ft 10 in)
Max width	1.03 m (3 ft 4½ in)
Height to sill	0.40 m (1 ft 3¾ in)

DIMENSIONS, INTERNAL:
Cabin: Length	2.10 m (6 ft 1¾ in)
Max width	1.05 m (3 ft 5¼ in)
Max height	1.22 m (4 ft 0 in)
Floor area	1.35 m² (14.5 sq ft)
Volume: total	2.10 m³ (74 cu ft)
optional freight stowage	0.85 m³ (30.0 cu ft)

AREAS:
Wings, gross	17.20 m² (185.1 sq ft)
Ailerons (total): conventional	1.00 m² (10.76 sq ft)
DSFC	0.43 m² (4.63 sq ft)
Flaps	1.32 m² (14.20 sq ft)
Fins (total)	2.84 m² (30.57 sq ft)
Rudders (total)	0.74 m² (7.97 sq ft)
Tailplane	2.34 m² (25.19 sq ft)
Elevators, incl tab	0.70 m² (7.53 sq ft)

WEIGHTS AND LOADINGS:
Weight empty	420 kg (926 lb)
Max fuel weight	70 kg (154 lb)
Max T-O and landing weight	750 kg (1,653 lb)
Max wing loading	43.6 kg/m² (8.93 lb/sq ft)
Max power loading	8.38 kg/kW (13.78 lb/hp)

PERFORMANCE (estimated):
Never-exceed speed (VNE)	118 kt (220 km/h; 136 mph)
Max level speed at S/L	86 kt (160 km/h; 99 mph)
Max cruising speed at 1,000 m (3,280 ft)	81 kt (150 km/h; 93 mph)
Econ cruising speed at 1,000 m (3,280 ft)	67 kt (125 km/h; 78 mph)
Stalling speed: flaps up	44 kt (80 km/h; 50 mph)
flaps down	39 kt (72 km/h; 45 mph)
Max rate of climb at S/L	264 m (866 ft)/min
Max certified altitude	4,000 m (13,120 ft)
T-O run	120 m (395 ft)
T-O to 15 m (50 ft)	275 m (905 ft)
Landing from 15 m (50 ft)	250 m (820 ft)
Landing run	120 m (395 ft)
Range with max normal fuel	680 n miles (1,259 km; 782 miles)

UPDATED

Model of the Saviat E-4 2000/0079011

SAU

NAUCHNO-PROIZVODSTVENNAYA KORPORATSIYA SAMOLOTY-AMFIBYI UNIVERSALNYYE (Universal Amphibious Aircraft Scientific-Production Corporation)

Construction of a prototype R-50 Robert amphibian is several years behind schedule and may have been abandoned. Details last appeared in the 2000-01 edition.

UPDATED

SAZ

SARATOVSKY AVIATSIONNY ZAVOD ZAO (Saratov Aviation Plant JSC)

ploshchad Ordzhonikidze 1, 410015 Saratov
Tel: (+7 8452) 44 81 01
Fax: (+7 8452) 44 36 07
e-mail: saz@mail.saratov.ru
GENERAL DIRECTOR: Aleksandr V Ermishin
EXECUTIVE DIRECTOR: Nikolay I Dubrovin
DEPUTY GENERAL DIRECTOR, FOREIGN ECONOMIC AFFAIRS AND SALES: Yury N Lyalkov

During the Second World War, the Saratov plant (GAZ 292) manufactured Yakovlev Yak-1 and Yak-3 piston-engined fighters, continuing post-war with Yak-15, Yak-23, Yak-25 and Yak-27 military jet aircraft. In the 1960s and 1970s, production centred on the Yak-40 three-turbofan transport, with Yak-36 and Yak-141 V/STOL prototypes and Yak-38 V/STOL fighters also manufactured in the 1970s and 1980s. Since 1982, the Yak-42 has been in series production, although at a recent rate of only about two per year; it was joined by the Yak-54 trainer/sporting aircraft in 1994, but manufacturing is turning increasingly to agricultural tractors, buses, street cleaning equipment and catamarans.

SAZ also worked on a completely new type of lifting-body jet aircraft known as the EKIP project. It was the first company in the Russian Federation and Associated States (CIS) to receive a production certificate for commercial aircraft in compliance with international standards.

UPDATED

SGAU

SAMARSKY GOSUDARTVENNYI AEROKOSMITSESKY UNIVERSITET (Samara State Aerospace University)

Moskovskoye shosse 34, 443086 Samara
Tel: (+7 8462) 35 18 26
Fax: (+7 8462) 35 18 36

This prominent centre of aerospace learning includes a design bureau, SKB-1, in which students can develop their skills. Although this has concentrated on the small-scale manufacture of lightplanes, it has also designed – under leadership of Albert Elatonsev, its director – an executive transport designated Aist 92 (Stork 92). An accompanying drawing shows the general configuration, the aircraft having a 10 m (33 ft) span and contrarotating propellers driven by a 3,300 kW (4,425 shp) turbine to give a 350 kt (648 km/h; 403 mph) cruising speed with eight persons.

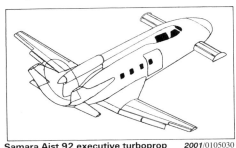

Samara Aist 92 executive turboprop

STUDENTSKOYE KONSTRUCTORSKOYE BURO 1 (Student Design Bureau 1)

Moskovskoye shosse 34, Korpus 3, 443086 Samara
Tel: (+7 8462) 35 72 81
e-mail: skb1@ssau.ru
Web: http://www.aero.ssau.ru/fla/skb1

SKB-1 designs are available for commercial production. Those with a Che- prefix, previously listed under this heading, are now described under the Chernov heading in this section.

Bureau founded 1955; has built Sverchock-1 gyroplane (1971), A-13 motor glider (1974), Shmel lightplane (1977), Strekoza lightplane (1978), Bat glider (1978), A-10 glider (1982) and A-18 Rusich motor glider (1987). Also built Che-15 and Che-25 (see Chernov earlier in this section) and S-202 (last described in 1998-99 *Jane's*).

UPDATED

AEROVOLGA L-06

TYPE: Twin-prop flying-boat.
PROGRAMME: First shown as Gelendzhik hydromarine exhibition in August 2000. No further details revealed. Configuration shown in accompanying illustration; two-step hull.

NEW ENTRY

SKB-1 A-16

TYPE: Single-seat ultralight/kitbuilt.
PROGRAMME: Prototype completed in 1997 and exhibited in unflown state at Moscow Air Show in August of that year. It remained unflown, awaiting an engine, two years later.
COSTS: US$16,000, flyaway.
DESIGN FEATURES: Small, low-cost light aircraft for amateur construction. Stressed to ±11 g.
FLYING CONTROLS: Conventional and manual.
STRUCTURE: Plastics throughout; Plexiglas transparencies.
LANDING GEAR: Tailwheel type; mainwheel brakes.
POWER PLANT: One 70.8 kW (95 hp) Hirth F30 flat-four driving a two-blade wooden propeller.
DIMENSIONS, EXTERNAL:
Wing span 5.50 m (18 ft 0½ in)
Length overall 4.50 m (14 ft 9¼ in)
Height overall 1.45 m (4 ft 9 in)
WEIGHTS AND LOADINGS:
Weight empty 200 kg (441 lb)
Max T-O weight 300 kg (661 lb)
PERFORMANCE (estimated):
Never-exceed speed (V$_{NE}$) 194 kt (360 km/h; 223 mph)
Max level speed 140 kt (260 km/h; 162 mph)
Landing speed 52 kt (95 km/h; 59 mph)

UPDATED

SGAU Favorit two-seat ultralight

SKB-1 FAVORIT

TYPE: Side-by-side ultralight kitbuilt.
PROGRAMME: Designed by V Kurshev for Konversiya Aero; to meet JAR-VLA requirements. Manufacture of prototype began 1998.
DESIGN FEATURES: High-mounted, strut-braced, constant-chord wing; mid-positioned tailplane and sweptback fin.
FLYING CONTROLS: Conventional and manual.
LANDING GEAR: Tricycle type; fixed.
POWER PLANT: One 47.8 kW (64.1 hp) Rotax 582 two-cylinder engine driving a three-blade propeller.
DIMENSIONS, EXTERNAL:
Length overall 6.30 m (20 ft 8 in)
Height overall 2.00 m (6 ft 6¾ in)
AREAS:
Wings, gross 13.78 m² (148.3 sq ft)
WEIGHTS AND LOADINGS:
Weight empty 275 kg (606 lb)
PERFORMANCE (estimated):
Max level speed 84 kt (155 km/h; 96 mph)
Normal cruising speed 65 kt (120 km/h; 75 mph)
Max rate of climb at S/L 270 m (885 ft)/min
Range 189 n miles (350 km; 217 miles)
g limits +4/–2

NEW ENTRY

Aerovolga L-06 flying-boat during its first public showing *(Yefim Gordon)*

The diminutive A-16 on display at Zhukovsky in 1997 *(Paul Jackson/Jane's)*

CKB-1 S-302

TYPE: Two-seat amphibian ultralight kitbuilt.
PROGRAMME: Development of S-202. Design (by D Suslakov) complete by mid-1997; construction had not begun by August 1999.
COSTS: US$22,000 (1997).
DESIGN FEATURES: Shoulder-wing monoplane with strut-braced T tail; single-step flying-boat hull; small fixed float at each wingtip. Airframe life 3,000 hours.
FLYING CONTROLS: Manual, with full-span flaperons. Full-span elevators; large rudder with no aerodynamic balance.
LANDING GEAR: Optional wheels; mainwheel legs, mounted on fuselage sides, turn 90° forward for stowage; mainwheel tyre size 300×125.
POWER PLANT: One 70.8 kW (95 hp) Hirth F30 four-cylinder piston engine, pylon-mounted behind the cockpit, driving a pusher propeller.
ACCOMMODATION: Two persons, side by side.
DIMENSIONS, EXTERNAL:

Wing span	10.80 m (35 ft 5¼ in)
Wing chord, constant	1.30 m (4 ft 3¼ in)
Length overall	6.66 m (21 ft 10¼ in)
Height overall, without wheels	2.21 m (7 ft 3 in)
Tailplane span	2.80 m (9 ft 2¼ in)
Wheel track (optional)	1.52 m (4 ft 11¾ in)
Propeller diameter	1.80 m (5 ft 10¾ in)

WEIGHTS AND LOADINGS:

Weight empty	350 kg (772 lb)
Max T-O weight	600 kg (1,322 lb)
Max power loading	8.47 kg/kW (13.92 lb/hp)

PERFORMANCE (estimated):

Max level speed	92 kt (170 km/h; 106 mph)
Max cruising speed	76 kt (140 km/h; 87 mph)
Landing speed	33 kt (60 km/h; 38 mph)
T-O run: water	200 m (660 ft)
land	100 m (330 ft)
Range: with max fuel	324 n miles (600 km; 372 miles)
with max payload	135 n miles (250 km; 155 miles)
g limits	+4.5/−2.25

UPDATED

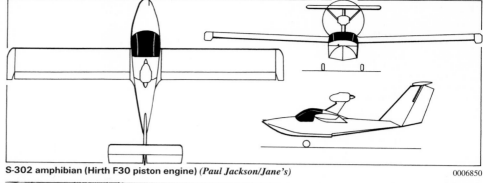

S-302 amphibian (Hirth F30 piston engine) *(Paul Jackson/Jane's)* 0006850

SKB-1 S-400 KAPITAN
English name: Captain

TYPE: Amphibian kitbuilt.
PROGRAMME: Designed by D Suslakov. Prototype flying by 1999.
DESIGN FEATURES: Conforms to FAR Pt 23 and AP-23 Russian criteria. Unswept, constant-chord, mid-mounted wings have downturned tips as stabilising floats. Large, sweptback fin with high, braced tailplane. Engines mounted on outward-canted, faired cabanes above centre-section.
FLYING CONTROLS: Manual. Conventional rudder and elevators, without aerodynamic balances. Full-span flaperons.
LANDING GEAR: Boat hull with removable wheels.
POWER PLANT: Two 70.8 kW (95 hp) Hirth F30-A26 flat-fours driving two-blade, wooden, pusher propellers.
ACCOMMODATION: Four persons, including one or two pilots sharing central instrument panel. Two-piece windscreen; upward-hinged, two-piece canopy each side.
DIMENSIONS, EXTERNAL: No details available
WEIGHTS AND LOADINGS:

Max payload	300 kg (661 lb)
Max T-O weight	1,070 kg (2,359 lb)
Max power loading	7.56 kg/kW (12.42 lb/hp)

PERFORMANCE:

Max level speed	124 kt (230 km/h; 143 mph)
Max cruising speed	97 kt (180 km/h; 112 mph)
Econ cruising speed	65 kt (120 km/h; 75 mph)
Landing speed	41 kt (75 km/h; 47 mph)

Prototype S-400, shortly after completion at Samara 2000/0079306

SKB-1 Krechet prototype at Moscow, August 1999, prior to first flight *(Paul Jackson/Jane's)* 2001/0093700

Max rate of climb at S/L	360 m (1,181 ft)/min
Max range	540 n miles (1,000 km; 621 miles)

UPDATED

SKB-1 KRECHET
English name: Gyrfalcon

TYPE: Side-by-side ultralight kitbuilt.
PROGRAMME: Designed in conjunction with Conversiya Aero. Unflown prototype exhibitied at MAKS, Moscow, August 1999. First flight 23 October 1999. Second aircraft then in planning stage.
DESIGN FEATURES: Simple sport aircraft. Low, constant-chord wing with downturned tips; sweptback fin and mid-mounted tailplane. Windscreen fixed; rear-sliding, single-piece canopy.
FLYING CONTROLS: Manual, by cables and pushrods. Flaperons on approximately 80 per cent of wing trailing-edge. Full-height rudder; no aerodynamic balance on tail surfaces. No inbuilt tabs.
STRUCTURE: Moulded glass fibre throughout, except titanium tube mainwheel legs with glass fibre aerodynamic fairing.
LANDING GEAR: Tailwheel type; fixed. Cable brakes on mainwheels.
POWER PLANT: One 59.6 kW (79.9 hp) Rotax 912 UL flat-four driving a two-blade ground-adjustable pitch propeller with slightly turned-back tips. Fuel capacity 50 litres (13.2 US gallons; 11.0 Imp gallons) in single (starboard) wingroot tank.
DIMENSIONS, EXTERNAL:

Wing span	10.40 m (34 ft 1½ in)
Length overall	6.38 m (20 ft 11¼ in)
Height overall	1.85 m (6 ft 0¾ in)

AREAS:

Wings, gross	13.52 m² (145.5 sq ft)

WEIGHTS AND LOADINGS:

Weight empty	280 kg (617 lb)

PERFORMANCE (estimated):

Max level speed	97 kt (180 km/h; 112 mph)
Cruising speed	81 kt (150 km/h; 93 mph)
Stalling speed	36 kt (65 km/h; 41 mph)
Max rate of climb at S/L	300 m (984 ft)/min
Service ceiling	4,500 m (14,760 ft)
Range	243 n miles (450 km; 279 mph)
g limits	+4/−2

UPDATED

SmAZ

SMOLENSKY AVIATSIONNYI ZAVOD OAO (Smolensk Aircraft Plant PLC)

ulitsa Frunze 74, 214006 Smolensk
Tel: (+7 08122) 293 13
Fax: (+7 0812) 61 97 70
e-mail: smaz@sci.smolensk.ru
GENERAL DIRECTOR: Aleksandr Miroshkin
CHIEF ENGINEER: Yury Litvinov
MARKETING MANAGER: Vladim Merzlyakov

Smolensk Aircraft Plant was established in 1926 as a repair facility. In 1934, Buro Osovikh Konstruktsii, OKB for experimental aircraft, relocated from Moscow to Smolensk GAZ 35, which built the OKB's prototypes. After Second World War, constructed prototypes for various OKBs, Myasishchev M-17/M-55, wings for Yak-40 transports, and more than 700 Yak-18Ts from 1973. Built small number of early production Yak-42s alongside main series at Saratov (SAZ). Government share was 25.5 per cent.

Production of Yak-18T was resumed 1993, five years after entire original fleet had been scrapped; more than 60 built up to 1996, when 15 placed in storage due to financial problems; completion of these resumed in 1998. Built three Yak-112 lightplanes before project transferred entirely to IAPO in Irkutsk. Other work included manufacture of RPVs and cruise missiles; wings for the Yak-42, upgrading of Yak-40s for business use (including additional fuel capacity), and development and production of Technoavia SM-92 Finist seven-seat STOL aircraft. Also participated in Yak-130 programme and overhauled Yak-18/52/55s as well as Antonov An-2s. No evidence of recent production.

UPDATED

SOKOL

SOKOL NIZHEGORODSKY AVIATSTROITELNYI ZAVOD AOOT (Sokol Nizhny Novgorod Aircraft Manufacturing Plant JSC)
ulitsa Chaadaeva 1, 603035 Nizhny Novgorod
Tel: (+7 8312) 46 71 27
Fax: (+7 8312) 22 19 25
e-mail: contact@sokol.nnov.ru
Web: http://www.sokol.nnov.ru
GENERAL DIRECTOR: Vasily H Pankov
CHIEF ENGINEER: Valery Drobyshevski
HEAD OF MARKETING: Aleksandr E Zaitsev

Founded (as GAZ 21) on 1 February 1932, this production plant concentrated on fighter aircraft manufacture until the start of the *konversiya* programme led to retooling for civilian production; registered as a JSC on 22 September 1994. Joined Kaskol group of companies in 2000. From 1949, in collaboration with the Mikoyan OKB, it manufactured 2,148 MiG-15s (1949-51), 2,470 MiG-17s (1951-54), 1,125 MiG-19s (1955-58), 5,278 MiG-21s (1957-85) and 1,186 MiG-25s (1966-85). Russian government holds 38 per cent of shares, which equivalent to 50.5 per cent of voting rights; Rossiysky Credit and Credit Suisse are remaining major (voting) shareholders. Current or recent products include the MiG-29UB, MiG-31 and Yak-130 jet trainer; 90 per cent of Yak-130 tooling in place by mid-2000. It is also responsible for supplying kits for upgrading 125 Indian Air Force MiG-21s to MiG-21-93 standard, with an initial order for 36 (see *Jane's Aircraft Upgrades* for details). Agreement reached in July 2000 on upgrading half of Russian Air Forces' 280 MiG-31s.

Civilian programmes include manufacture of the Myasishchev M-101 Gzhel six-seat turboprop light aircraft, a prototype AeroRIC Dingo amphibious aircraft with air cushion landing gear and Avia Accord twin-engined light multipurpose aircraft; plans also exist for Myasishchev M-202 production. Plans to build MiG-110 twin-turboprop transport apparently not overtaken by announcement in early 2000 of its reallocation to MAPO and discussions on continued participation in programme under way with RSK MiG in mid-2000. Production of the Europace F-15-F Excalibur four-seat touring aircraft was halted at an early stage when a contract for almost 100 was cancelled; some 20 airframes remained at Nizhny Novgorod in June 2000, when 10 sold to undisclosed buyer.

Other products include WIGE craft, air cushion vehicles and hydrofoils. Offers 'familiarisation' flights in MiG-29UB, MiG-21U and L-29 Delfin. Employment (including part-time) was 11,000 in 2000.

UPDATED

STRELA

STRELA PROIZVODSTVENNOYE OBEDINENIE (Strela Production Association)
ulitsa Shevchenko 26, 46005 Orenburg
Tel: (+7 3532) 35 72 09 and 35 61 74
Fax: (+7 3532) 35 54 60

The wholly state-owned Strela plant manufactures UAVs and produces parts of the Kamov Ka-226 helicopter, for which it is also (with KAPP) one of two assembly centres. It has recently produced replica Yak-3/Yak-9s, (which see) for export.

UPDATED

SUKHOI (AVPK SUKHOI)

GOSUDARSTVENNOYE UNITARNOYE PREDPRIYATIE AVIATSIONNYI VOYENNO-PROMYSHLENNYI KOMPLEKS SUKHOI (State Unitary Enterprise, Aviation Military-Industrial Complex Sukhoi)
alleya 604, ulitsa Polikarpova 23A, 125284 Moskva
Tel: (+7 095) 941 76 87
Fax: (+7 095) 945 68 06
GENERAL DIRECTOR: Mikhail A Pogosyan
PRESS OFFICER: Yury Chervakov

AVPK Sukhoi was created, under presidential decree, in August 1996, to bring together the Sukhoi and Beriev design bureaux, the production plants at Irkutsk (IAPO), Komsomolsk-on-Amur (KnAAPO) and Novosibirsk (NAPO), and two banks, Oneximbank and Inkombank. The grouping was formally inaugurated on 30 December 1996. As originally constituted, the group comprises the wholly state-owned KnAAPO and NAPO, plus the partial state shareholdings in Sukhoi OKB (see following entry), IAPO and Taganrog. The replacement of original director Alexei Fyodorov by Pogosyan marked the conclusion of a power struggle between the OKB and the large manufacturing plants, and a move away from total central control of AVPK Sukhoi's disparate elements. Steps are under way which may lead to the merger of Sukhoi, Yakovlev and RSK 'MiG', these including the appointment of Nikolov Nikitin (Pogosyan's deputy) as RSK 'MiG's' General Director. Sukhoi's political acumen continues to be displayed, the company having flown Prime Minister Vladimir Putin in both the Su-25 and Su-27 before he assumed the presidency. Full privatisation remains unlikely, though AVPK Sukhoi is likely to become a Joint Stock Company.

UPDATED

OPYTNYI KONSTRUKTORSKOYE BURO SUKHOGO AOOT (Sukhoi Experimental Design Bureau JSC)
ulitsa Polikarpova 23A, 125284 Moskva
Tel: (+7 095) 945 65 25
Fax: (+7 095) 200 42 43
GENERAL DIRECTOR: Mikhail A Pogosyan
GENERAL DESIGNER: Mikhail Petrovich Simonov
FIRST DEPUTY GENERAL DESIGNER: Aleksandr F Barkovsky
EXECUTIVE DIRECTOR, MANUFACTURE: V V Golovkin
DEPUTY GENERAL DESIGNERS:
V S Konokhov (First Test Base)
Boris V Rakitin (Sport Aviation Projects)
Vladimir M Korchagin (Avionics)
Aleksandr I Blinov (Strength Problems)
O G Kalibabchuk (Aerodynamics)
J I Shenfinkel (Control Systems)

OKB, which is 57 per cent owned by AVPK Sukhoi, named for Pavel Osipovich Sukhoi, who headed it from 1939 until his death in September 1975. It remains one of two primary Russian centres for development of fighter and attack aircraft, and it is notable that 60 per cent of frontline Russian Air Forces aircraft are of Sukhoi design. Bureau has widened its activities to include civilian aircraft, under *konversiya* programme. This offsets the drastic reduction in combat aircraft production and competition in the upgrade and overhaul market. Known deliveries in 2000 included four Su-27s for Kazakhstan and eight for China, and 10 Su-30s for latter.

Sukhoi also active in design of large ground-effect vehicles, including 100/150-passenger A.90.150 Ekranoplan; see *Jane's High-Speed Marine Transportation*; has proposed an 860-seat airliner.

Sukhoi Advanced Technologies
GENERAL DIRECTOR: Boris V Rakitin

This division is responsible for promotion and sales of Sukhoi light aircraft, such as the Su-29, Su-31, Su-38L and Su-49; and in 2000 was planning to provide funds to enable Su-29 and Su-31 production to be relocated to KAPO at Kazan from 2002. Related activities include conversion of Su-26s and Su-29s with SKS-94 extraction systems.

Nauchno-Proizvodstvennoe Kontsern Shturmoviki Sukhogo (Sukhoi Stormovik Scientific-Production Concern)
GENERAL DIRECTOR: Vladimir Babak

Offers of upgrades of Su-25 'Frogfoot' to Russian and export customers. In 2000, Ukraine, Belarus, Georgia and Turkmenistan foreseen as potential purchasers.

Sukhoi Civil Aircraft
GENERAL DIRECTOR: Andrey Il'in

Subdivision formed May 2000; on 5 June 2000 signed agreement with Alliance Aircraft Corporation (which see in US section) for joint development and production of StarLiner regional airliner. However, Sukhoi unilaterally cancelled agreement on 2 November 2000.

UPDATED

SUKHOI Su-24
NATO reporting name: Fencer
TYPE: Attack fighter.
PROGRAMME: Design started 1965 under Yevgeniy S Felsner, Pavel Sukhoi's successor, to replace Il-28 and Yak-28 attack aircraft; T6-1 prototype, first flown 2 July 1967 and now preserved at Monino, had fixed delta-wings with downswept tips, and two pairs of RD-36-35 lift-jets mounted at inclined angle in rear fuselage, to exhaust slightly rearward for improved take-off performance; T6-21G variable geometry prototype, without lift-jets, chosen for production; first flight 17 January 1970; by 1981, delivery rate 60 to 70 a year; 900+ delivered from Komsomolsk; production now terminated. 'Fencer-A/B/C' built at GAZ 153 Novosibirsk; Su-24M versions initially at GAZ 153 (approximately 380), then GAZ 416 Komsomolsk-on-Amur (over 290), although latter plant's literature makes no mention of Su-24. Total production 'about 1,200'.
CURRENT VERSIONS: **Su-24** ('Fencer-A'): Prototype was seventh T6, first flown December 1971. Jetpipes enclosed in slab-sided box with dished underside. Engine air intakes revised, having boundary layer extraction perforations on redesigned splitter plate; variable ramps deactivated from Series 4 (production batch 4) onwards. In service from 1973; with front-line units by 1975. MTOW limited to 36,200 kg (79,807 lb) because of landing gear deficiencies; speed limited to M1.35.
Su-24 ('Fencer-B'): Series 12 introduced extended fin chord below cap, giving kinked profile; heat exchanger inlet above centre fuselage; revised noseprobe with triple side-by-side antenna fairings and dual Filin antennas under forward fuselage. Series 15 (from c/n 1515301) introduced new rear fuselage contours with sides more closely following jetpipes. Braking parachute fairing added below rudder from Series 15, along with finroot ram air intake.
Su-24 ('Fencer-C'): Triangular RWR antennas added to sides of engine air intakes and below fincap from Series 24, terminating at Series 27. Revised noseprobe and redesigned Filin antenna; elint/sigint antennas differed between aircraft within a regiment, with individual aircraft systems tuned to different frequencies. Puma nav/attack system included RWR/elint functions. Empty weight 21,150 kg (46,627 lb).
Su-24M ('Fencer-D'): Second-generation strike/attack version; introduced true terrain-following (rather than terrain-avoidance) capability. New PNS-24M navigation and attack system and with terrain-following radar coupled to autopilot; Kaira laser/TV targeting and weapon guidance system, as in MiG-27, installed in fairing behind nosewheel bay; several other new items of avionics. Noseprobe and undernose antenna replaced by simple probe. Slight reduction of about 85 litres (22.4 US gallons; 18.7 Imp gallons) in internal fuel capacity; increased maximum T-O weight; fuselage lengthened in front of windscreen, partly to accommodate retractable refuelling probe on centreline, replacing Chaika IR sensor.

Eighth prototype (T6-M8) served as first Su-24M, but retained original type of noseprobe and underfuselage probes, plus Kaira-24 sensor; first flew 29 June 1977; other converted 'Fencer-B/Cs' assisted with weapons and systems development, including T6M-21, -22, -23, -24, -27 and -29.

Su-24M ('Fencer-D' Mod): Large overwing fences integral with extended wingroot glove pylons, sometimes incorporating 27-round chaff/flare dispensers (nine-shot launchers); provision for chaff/flare dispensers above rear fuselage, one each side of fin. Compatible with UPAZ-A 'buddy' refuelling pod (on centreline pylon). Abandoned Su-24MM would have had AL-31 engines and further increased maximum T-O weight.

Su-24MK ('Fencer D'): Export version of the above, with downgraded avionics. Some serve with Russian Air Forces (notably at Kubinka with squadron of 237 GvTsPAT) wearing non-standard (export) camouflage, but red star nationality markings.

Su-24 Upgrade Options: Six potential fleet modernisation options have been offered to the Russian Air Forces by Sukhoi and other providers including the Chkalov plant at Novisibirsk and Gefest & T Closed Joint Stock Company, a commercial arm of the VVS's own aircraft repair organisation, ranging from minor armament improvements to the installation of GPS, helmet-mounted sight and R-73 (AA-11 'Archer') self-defence missiles, up to extensive retrofit with avionics and armament developed for the Su-27IB. Novosibirsk received order for initial 10 upgrades on 2 October 2000.

'Su-24 Bis': At Berlin's ILA 2000 exhibition, Sukhoi displayed a model of an advanced upgrade configuration offered to the Russian Air Forces, and a number of export customers. Externally distinguished by 'scabbed-on' armour plate on the sides of the cockpit and rounded conformal fuel tanks on the intake sides, running back below the wingroots. Internally, this upgrade configuration

RUSSIAN FEDERATION: AIRCRAFT—SUKHOI

Cockpit of Sukhoi Su-24 'Fencer-B' *(Yefim Gordon)*

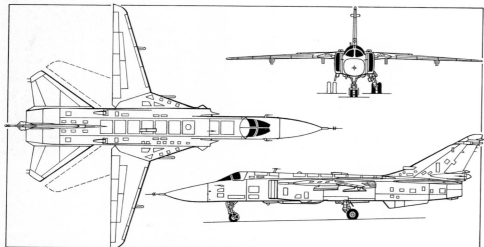

Sukhoi Su-24M ('Fencer-D') variable geometry attack aircraft *(Mike Keep/Jane's)*

includes installation of a new SV-24 computer, CRT radar display, new LCD MFDs in the cockpit, new weapon-aiming algorithms, a digital moving map generator, ILS-31 HUD, and A737 GPS, as well as a new mission planning system, a transfer cartridge for mission data and a new terrain-reference navigation system. Upgraded aircraft gain a 2,400 flying hour, 30 year life extension and compatibility with a range of new weapons, including Kh-31A, Kh-31P and Kh-159MK ASMs, and UAB-500, KAB-500 and KAB-1500 TV-guided bombs. For self-defence, the upgrade includes Shchel helmet-mounted sights for the crew, and compatibility with the R-73 IR-homing AAM.

Su-24MR ('Fencer-E'): Reconnaissance version of Su-24M used by tactical and naval air forces; Kaira and gun omitted; BKR-1 internal equipment includes RDSBO Shtik side-looking airborne multimission radar in nose, Zima IR reconnaissance system under centre-fuselage; Aist-M TV reconnaissance system, AFA AP-402M panoramic camera with 90.5 mm lens in nose and AFA A-100 oblique camera with 100 mm lens in the floor of port intake duct. Can be fitted with Kadr, allowing exposed cassette to be parachuted to the ground, albeit at the cost of some film capacity. A Shpil-2M laser pod can be carried on centreline, with a Tangazh elint pod or Efir-1M radiation detector pod on starboard underwing swivelling pylon and two R-60 (AA-8 'Aphid') AAMs under port wing. Data can be transmitted to ground by datalink. SBU Orbita IO-058R digital computer. Overwing fences on most aircraft; shorter nose radome, with flush dielectric side panels on nose; domed centre-fuselage air intake for heat exchanger; provision for two 3,000 litre (792 US gallon; 660 Imp gallon) underwing auxiliary fuel tanks; flight refuelling capability retained, but weapons control system and attack capability deleted. First of two prototypes (T6MR-26 and -34) flew September 1980; deliveries to Baltic fleet, replacing Yak-16s, began mid-1985. Maximum speed at 200 m (660 ft) limited to 648 kt (1,200 km/h; 746 mph) with full external equipment.

Su-24MP ('Fencer-F'): Electronic warfare/jamming/sigint version to replace Yak-28PP 'Brewer-E'. First of T-6MP-25 and -35 prototypes flew December 1979; dielectric nose panels differ from those of Su-24MR; narrow long-chord antenna fairing under nose; no underside electro-optics; 'hockey stick' communications-jamming antenna at bottom of fuselage under nose section of each engine air intake; centreline SPS-5 Fasol EW pod. Starboard cockpit display panels replaced by EW console. Armament GSh-6-23M gun and four R-60 (AA-8 'Aphid') AAMs. Production reportedly limited to 12 aircraft, though some sources suggest that 13 are in Russian service, with seven more in the Ukraine. Some aircraft had narrow fairing on spine, behind cockpit.

CUSTOMERS: Russian Air Forces and Naval Aviation have 330 for ground attack, 50 for training and 120 for reconnaissance and ECM; others with air forces of Azerbaijan (five Su-24MRs of 11 to 16 left behind by USSR forces plus others acquired clandestinely); Belarus (24 to 42), Kazakhstan (37 or 42), Ukraine (200 to 245, including eight of the 12 Su-24MPs); Uzbekistan (23); Afghanistan (one Su-24 reportedly purchased by Iran for Ahmed Shah Massoud's Northern Alliance; Algeria (10 Su-24MKs and four refurbished RFAS Su-24MRs, with 12 more refurbished ex-Russian Su-24Ms (or 22 according to some sources) following in 2000-2001 (other sources suggest five in service – three delivered in 1999, two in October 2000 – with 20 to follow in 2001, apparently ignoring the original fleet of 14-16 aircraft); Angola (unknown number reportedly sold to UNITA by Belarus in October 1999, with at least one delivered to Catumbela by 22 January 2001); Iraq (24 or 25, survivors now in Iran); Iran (14 plus 18 ex-Iraqi aircraft); Libya (15), Syria (20).

DESIGN FEATURES: Low-level interdictor employing 1960s technology – notably variable geometry. Shoulder wing; wing section SR14S-5.376 on fixed panels, SR14S-9.226 to SR16M-10 on outer panels; 4° 30′ anhedral from roots; incidence 0° at root, –4° at tip; triangular fixed glove box; four-position (16, 35, 45, 69°) pivoted outer panels, each with pivoting stores pylon; slab-sided rectangular section fuselage; integral engine air intake trunks, each with splitter plate and outer lip inclined slightly downward; chord of lower part of tailfin extended forward, giving kinked leading-edge; leading-edge sweepback 59° 30′ on fin, 30° on inset rudder, 50° on horizontal tail surfaces; basic operational task, as designated 'frontal bomber', to deliver wide range of air-to-surface missiles for defence suppression, with some hard target kill potential; specially developed long-range navigation system and electro-optical weapons systems make possible penetration of hostile airspace at night or in adverse weather with great precision, to deliver ordnance within 55 m (180 ft) of target.

FLYING CONTROLS: Automatic flight control system with artificial feel. Full-span leading-edge slats, drooping aileron and two-section double-slotted trailing-edge flaps on each outer wing panel; differential spoilers forward of flaps for roll control at low speeds and use as lift dumpers on landing; airbrake under each side of centre-fuselage; inset rudder; all-moving horizontal tail surfaces operate collectively for pitch control, differentially for roll control, assisted by wing spoilers except when wings fully swept.

Su-24 CURRENT OPERATORS

Unit	Base	Equipment
Russia (Air Forces)		
3rd BAP[1]	Taganrog	Su-24
20th Gv BAP	Kamenka	Su-24
67th BAP	Siverskoye	Su-24
164th GVBAP	Shatalovo	Su-24
207th BAP	Komsomolsk	Su-24
237th GvTsPAT	Kubinka	Su-24, Su-24M, Su-24MR
296th BAP	Marinovka	Su-24
455th BAP	Voronezh	Su-24
559th BAP	Morozovsk	Su-24M
722nd BAP	Smurav'yevo	Su-24
959th BAP	Yeysk	Su-24
968th IISAP	Lipetsk	Su-24 (various)
Unident BAP	Bada	Su-24
Unident BAP	Borzya NW	Su-24
Unident BAP	Nyangi	Su-24
Unident BAP	Verino	Su-24
Unident BAP	Near Komsomolsk	Su-24
1st UBAP	Lebyazhye	Su-24
976th UBAP	Totskoye	Su-24
11th ORAP	Marinovka	Su-24MR
47th ORAP	Shatalovo	Su-24MR
98th ORAP	Monchegorsk	Su-24MR
125th ORAP	Domna	Su-24MR
Russia (Naval Aviation)		
240th GvUAP	Ostrov	Su-24M
4th INShAP	Chernyakhovsk	Su-24M
43rd OBAP	Gvardeskoye (Crimea)	Su-24
846th ORAE	Chkalovsk	Su-24M, Su-24MR
Unident ORAE	Artem	Su-24M, Su-24MR
Algeria		
274th Ftr-Bomber Sqn	Laghouat	Su-24MK
Angola		
Unident unit	Catumbela	Su-24 (unknown sub-variant)
Azerbaijan		
Unident BAP	Kyurdamir or Dallyar	Su-24M
Belarus		
116th BRAB	Ros	Su-24M, Su-24MR
Iran		
72nd Tac Ftr Sqn	Shiraz	Su-24MK
Unident Tac Ftr Sqn	Tehran-Mehrabad	Su-24MK
Detachment (anti-shipping)	Chah Buhar	Su-24MK
Kazakhstan		
149th BAP	Nikolayevka	Su-24, Su-24MR
Libya		
1124 Bomber Sqn	Umm Aitqah	Su-24MK
Syria		
Unident Bomber Sqn	T4	Su-24MK
Ukraine		
6th AVB	Nikolayev	Su-24
7th BAP	Starokonstantinov	Su-24M
44th BAP	Kanatovo	Su-24
69th BAP	Gorodok	Su-24
806th BAP	Luts'k	Su-24
947th BAP	Dubno	Su-24M
48th GvORAP	Kolomiya	Su-24MR, Su-24MP
118th ORAP	Chortkov	Su-24MR, Su-24MP
511th ORAP	Buyalik	Su-24MR
Uzbekistan		
60th BAP	Khanabad	Su-24, Su-24MR

[1] Probably disbanded

Notes: BAP is *Bombardirovch'nyi Aviatsion'nyi Polk* (Bomber Aviation Regiment); ORAP is *Otdel'nyi Razvedivatel'nyi Aviatsion'nyi Polk* (Independent Reconnaissance Aviation Regiment) and ORAE a Squadron); OMShAP is *Otdel'nyi Morskoy Shturmovoy Aviatsion'nyi Polk* (Independent Naval Aviation Attack Regiment). Prefixes are Gv, *Gvardiya* (Guards) and U, *Uchebn'nyi* (Training).

Sukhoi Su-24MR 'Fencer-E' reconnaissance version, showing nose and spine antennas *(Yefim Gordon)*

STRUCTURE: All-metal; semi-monocoque fuselage; two slightly splayed ventral fins.

LANDING GEAR: Hydraulically retractable tricycle type, with twin wheels on each unit; main units retract forward and inward into air intake duct fairings; steerable nose unit retracts rearward. Oleo-pneumatic shock-absorbers. Trailing-link main units; KT-172 mainwheel tyres size 950×300, pressure 12.15 bars (176 lb/sq in); KN-21 nosewheel tyres size 600×200, pressure 9.10 bars (132 lb/sq in); KT-69.430 brakes on mainwheels, with IA-58 anti-skid units; mudguard on nosewheels; two cruciform brake-chutes, each 25 m² (269 sq ft).

POWER PLANT: Two Saturn/Lyulka AL-21F-3A turbojets, each 75.0 kN (16,865 lb st) dry and 109.8 kN (24,690 lb st) with afterburning; fixed engine air intakes. Four internal fuel tanks, capacity 11,883 litres (3,139 US gallons; 2,613 Imp gallons), can be supplemented by two 2,000 litre (528 US gallon; 440 Imp gallon) external tanks under fuselage and two 3,000 litre (792 US gallon; 660 Imp gallon) tanks under wing gloves. Pressure and gravity fuelling. Probe-and-drogue flight refuelling capability, including operation as buddy tanker using UPAZ-A underbelly pod. Oil capacity 24 litres (6.35 US gallons; 5.25 Imp gallons).

ACCOMMODATION: Crew of two (pilot and weapon systems officer) side by side on K-36DM zero/zero ejection seats; cockpit width 1.65 m (5 ft 5 in); jettisonable canopy, hinged to open upward and rearward in two panels, split on centreline.

SYSTEMS: Cockpit air conditioned, with automatic pressure and temperature control. Three independent hydraulic systems. Main and emergency pneumatic systems, each with 18 litre bottle, pressure 152 bars (2,200 lb/sq in). Two electrical generators for 200/115 V AC supply and two for 28.5 V DC supply. Gaseous oxygen system. Hot-air de-icing system.

AVIONICS: *Radar:* Leninets Orion-A pulse Doppler set; two superimposed (over and under) radar scanners in nose, NATO ASCC 'Drop Kick' radar complex with nav/attack and ranging to airborne or ground targets and 'Relief' for terrain avoidance.
Flight: DISS-7 Doppler, MIS-P inertial platform, RSBN-6S Shoran/ILS and ARK-15M radio compass.
Mission: Kaira-24 laser ranger/designator under front fuselage. TsVU IO-058 Orbita 10 computer.
Self-defence: Karpati integrated defensive system with SRO-15 or -155 radar warning receivers on sides of engine air intakes and tailfin; LO-82 Mak-UL missile warning receivers above centre fuselage and below front fuselage; Geran-F active jamming system, all controlled by Neon-F control unit and linked to Automat-F chaff/flare dispensers. Aircraft supplied to Iran have 54 chaff/flares in wing fence dispensers, in addition to standard 24 on sides of rear fuselage.

ARMAMENT (Su-24M): Nine pylons under fuselage, each wingroot glove and outer wings for guided and unguided air-to-surface weapons, including TN-1000 and TN-1200 nuclear weapons, up to four TV- or laser-guided bombs, ASMs such as Kh-23 (AS-7 'Kerry'), Kh-25ML (AS-10 'Karen'), Kh-58 (AS-11 'Kilter'), Kh-25MP (AS-12 'Kegler'), Kh-59 (AS-13 'Kingbolt'), Kh-29 (AS-14 'Kedge') and Kh-31A/P (AS-17 'Krypton'), rockets of 57 to 330 mm calibre, bombs (typically 38 × 100 kg FAB-100), 23 mm gun pods or external fuel tanks; two R-60 (AA-8 'Aphid') AAMs can be carried for self-defence. No internal weapons bay. One GSh-6-23M six-barrel 23 mm Gatling-type gun inside fairing on starboard side of fuselage undersurface; fairing for recording camera on other side. Iranian Su-24MKs also carry indigenous derivatives of US Mk 82, 83 and 84 bombs, Sattar-2 LGBs and C-802K and Fajr Darya anti-ship missiles, and the indigenous Kite area denial mine dispenser.

Model of proposed Su-24bis improved version of Su-24MK, with Kh-31A/P and R-74 missiles and conformal fuel tanks *(Yefim Gordon)*

DIMENSIONS, EXTERNAL:
Wing span: 16° sweep	17.64 m (57 ft 10½ in)
69° sweep	10.365 m (34 ft 0 in)
Wing aspect ratio: 16° sweep	5.6
69° sweep	2.1
Length overall, incl probe: Su-24	22.67 m (74 ft 4½ in)
Su-24M	24.595 m (80 ft 8¼ in)
Height overall: Su-24	5.92 m (19 ft 5 in)
Su-24M	6.19 m (20 ft 3¾ in)
Tailplane span	8.39 m (27 ft 6½ in)
Wheel track	3.31 m (10 ft 10½ in)
Wheelbase	8.51 m (27 ft 11 in)

AREAS:
Wings, gross: 16° sweep	55.17 m² (593.8 sq ft)
69° sweep	51.02 m² (549.2 sq ft)
Leading-edge flaps (total)	3.035 m² (32.68 sq ft)
Trailing-edge flaps (total)	10.21 m² (109.90 sq ft)
Spoilers (total)	2.06 m² (22.17 sq ft)
Fin	9.47 m² (101.94 sq ft)
Rudder	1.37 m² (14.75 sq ft)
Horizontal tail surfaces (total)	13.705 m² (147.55 sq ft)

WEIGHTS AND LOADINGS (Su-24M):
Weight empty, equipped	22,300 kg (49,163 lb)
Max internal fuel	9,764 kg (21,525 lb)
Max external stores	8,100 kg (17,857 lb)
Normal T-O weight	36,000 kg (79,367 lb)
Max T-O weight	39,700 kg (87,523 lb)
Max landing weight	24,500 kg (54,012 lb)
Max wing loading	778.1 kg/m² (159.37 lb/sq ft)
Max power loading	181 kg/kN (1.77 lb/lb st)

PERFORMANCE (Su-24MK):
Max level speed, clean: at height	M1.35
at S/L	M1.08 (712 kt; 1,320 km/h; 820 mph)
Stalling speed, flaps and wheels down	151 kt (280 km/h; 174 mph)
Max rate of climb at S/L	9,000 m (29,525 ft)/min
Service ceiling	17,500 m (57,400 ft)
T-O run	1,300 m (4,265 ft)
T-O to 15 m (50 ft)	1,500 m (4,920 ft)
Landing from 15 m (50 ft)	1,600 m (5,250 ft)
Landing run	950 m (3,120 ft)

Combat radius:
lo-lo-lo over 174 n miles (322 km; 200 miles)
lo-lo-hi with 2,500 kg (5,500 lb) of weapons
515 n miles (950 km; 590 miles)
hi-lo-hi, with 3,000 kg (6,615 lb) of weapons and two external tanks 565 n miles (1,050 km; 650 miles)
g limit +6.5

UPDATED

SUKHOI Su-25 and Su-28
NATO reporting name: Frogfoot

TYPE: Attack fighter.

PROGRAMME: Development began 1968; Novosibirsk-built prototype, known as T8-1, flew 22 February 1975, with two 25.5 kN (5,732 lb st) non-afterburning versions of Tumansky RD-9 turbojet and underbelly twin-barrel depressable GSh-23 23 mm gun in fairing; in eventual developed form, second prototype and rebuilt first aircraft, T8-2D and T8-1D, had more powerful non-afterburning versions of Soyuz/Gavrilov R-13, designated R-95Sh, increased wing span, wingtip avionics/speed brake pods, underwing weapon pylons, and internal AO-17 30 mm gun necessitating offset nosewheel. Also Klen-PS laser range-finder replacing Fone, and new DISS-7 Doppler, new gunsight and new navigation computer; observed by satellite at Zhukovsky flight test centre 1977, given provisional US designation 'Ram-J'; production, with R-95Sh turbojets, began 1978.

T8-1D and preseries T8-3 sent to Afghanistan April 1980, as Rhombus team, for trials and combat testing; followed in mid-1981 by 200th Independent Shturmovik Squadron, Soviet Air Force, flying 12 production Su-25s for co-ordinated low-level close support of Soviet ground forces in mountain terrain, with Mi-24 helicopter gunships;

First production Sukhoi Su-24 'Fencer-B' with redesigned rear fuselage (1515301), currently used by LII research institute of Zhukovsky in a programme to assess the effects of jet engine emissions on the troposphere *(Paul Jackson/Jane's)*

430 RUSSIAN FEDERATION: AIRCRAFT—SUKHOI

Sukhoi Su-25UTG navalised jet trainer *(Paul Jackson/Jane's)* 2000/0062736

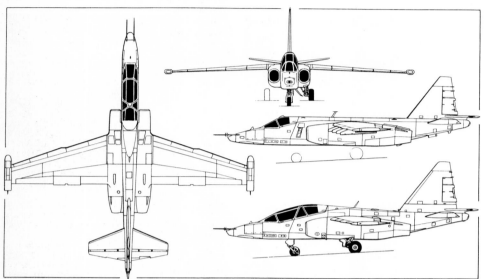
Sukhoi Su-28 (Su-25UT), with added side elevation (centre) of Su-25K *(Dennis Punnett/Jane's)*

fully operational 1984; 200th OShAE and later Su-25 squadrons flew 60,000 sorties during the eight-year war in Afghanistan, losing 23 aircraft and eight pilots killed; all but two of 139 laser-guided ASMs launched in combat achieved direct hits. Attack versions built initially at Factory No 31, Tbilisi, Georgia; production at Tbilisi ended 1989 (approximately 875 built, according to company statement); may subsequently have restarted for small batches, and factory reports sufficient parts "in stock" to assemble up to 75 aircraft; production at Ulan-Ude completed 1991-92; see separate entry on Su-25T.

CURRENT VERSIONS: **Su-25** ('Frogfoot-A'): Single-seat close support aircraft. Cannon muzzle redesigned from ninth production series; aileron control rod fully faired and trim tab deleted from 10th series. Revised airbrakes with double fold on late production aircraft, and by retrofit. Export version **Su-25K** (*kommercheskiy*; commercial).

Cockpit of the Sukhoi Su-25UTG 1999/0022584

Detailed description applies basically to late production Su-25K.

Su-25SM: Russian Air Forces decided to upgrade about 40 per cent of surviving Su-25s (about 80 aircraft) in March 1999. Single-seaters will be modified to Su-25SM standards; two-seaters to Su-25UBM configuration; new variant uses some equipment and systems developed for Su-25TM (Su-39). Su-25SM features Panther fire-control system with Kopyo-25 radar in a rebuilt nose (not pod-mounted, as on Su-25TM), giving compatibility with a wider range of air-to-air and air-to-ground weapons; Klen PS laser range-finder is relocated below fuselage. Revised cockpit with a new HUD and two large colour LCD MFDs, but retains standard analogue flight instruments. Irtysh EW system, with new RWR, MAWS and automatic control of chaff and flares. Structural changes limited to life-extension modifications and provision for nose-mounted radar. Sukhoi Shturmovik Consortium promises improved maintainability and a service life extended to 2,500 flying hours.

Conversion of Russian Su-25s to be undertaken by Air Forces' own Aircraft Repair Works No. 121, at Kubinka; programme management by Sukhoi Stormovik Concern, which agreed in mid-2000 to fund initial two aircraft as means of launching programme. Belarusian aircraft to be upgraded at Baranovichi; Ukrainians also being converted locally. Current timescales call for acceptance tests to be completed during the third quarter of 2000. Czech Su-25s are undergoing a more modest SLEP to allow them to remain productive until replaced by Aero L 159s during 2001-2002.

Su-25UB ('Frogfoot-B'): Tandem two-seat operational conversion and weapons trainer; prototype T8U-1 first flew 10 August 1985; entered production (only at Ulan-Ude) 1987; first photographs early 1989; rear seat raised considerably, giving humpback appearance; separate hinged portion of continuous framed canopy over each cockpit; taller tailfin, increasing overall height to 5.20 m (17 ft 0¾ in); new IFF blade antenna forward of windscreen instead of SRO-2 ('Odd Rods'); weapons pylons and gun retained. Optional periscope over rear cockpit. Export version **Su-25UBK**.

Su-25UBM: Two-seat Su-25s are to be modernised in the same way as the Su-25SM and, in fact, accorded higher priority to allow pilot training for those destined to fly the new single-seat version. First aircraft to be upgraded in the OKB's own Moscow workshops by Sukhoi Stormovik Concern will be a new Su-25UTG which, in April 2000, remained at Ulan-Ude factory, awaiting payment. Cost estimated as US$1 million per aircraft for UBM or US$500,000 for single-seat Su-25SM.

Su-25UT ('Frogfoot-B'): As Su-25UB, but without weapons; intended as L-29/L-39 replacement for air forces and DOSAAF; prototype (converted from T8U-1) first flew 6 August 1985; demonstrated 1989 Paris Air Show as **Su-28**; overall length 15.36 m (50 ft 4¾ in); one or two only. Discontinued.

Su-25UTG (G for *gak*: hook) ('Frogfoot-B'): As Su-25UT, with added arrester hook under tail; first flew September 1988; used initially for deck landing training on dummy flight deck marked on runway at Saki naval airfield, Ukraine; on 1 November 1989 was third aircraft to land for trials on carrier *Admiral of the Fleet Kuznetsov*, after Su-27K and MiG-29K; 10 built in 1989-90; five in Ukrainian use at Saki; one lost; four at Severomorsk, Kola Peninsula, for service individually on *Kuznetsov*. Two on board when ship deployed to Adriatic December 1995. Russian Navy received 11th aircraft in September 1997, while 12th was due in late 1998. These reportedly being given combat role: No. 11 said to be with millimetre-wave SLAR, Pastel-K RWR and TKS coded datalink for reconnaissance and target acquisition; No. 12 reportedly with anti-ship missiles and target acquisition systems also earmarked for Kamov Ka-29 upgrade. Discussions reported in early 1998 on possible further 12 for Russian Navy, but Ulan-Ude foresaw no prospect of further production after 2000.

Su-25UBP: Further aircraft to similar (UTG) standard, produced by conversion of Su-25UBs, to compensate for losses to Ukraine and attrition.

Su-25T/TM (Su-39): See separate entry.

Su-25BM (BM for *buksir misheney*: target towing aircraft) ('Frogfoot-A'). Standard late-series, R-195-engined Su-25 attack aircraft, with added underwing pylons for a winch, and Planyer-M miss-distance detection system and datalink, TL-70 Kometa towed target or PM-6 rocket-propelled targets released for missile training by fighter pilots; prototype (converted T8-9) first flew 1989; 50 built at Tbilisi.

CUSTOMERS: Russian tactical air forces and Naval Aviation have 225, plus 25 for traininng; pre-1985 aircraft now withdrawn. Exports to Angola (12, plus two two-seat), Azerbaijan, Belarus (currently 80), Bulgaria (36 plus four two-seat), Congo (10 ex-Georgian aircraft), Czech Republic (24, plus one two-seat), Georgia (five), Iran (seven), Iraq (45), North Korea (36, plus four two-seat), Peru (10 Su-25 and 8 Su-25UB from Belarus), Slovak Republic (11, plus one two-seat), Turkmenistan (46 currently in storage) and Ukraine (currently 64). Croatia also reported to have received a 'squadron' (never uncrated, according to some reports), but denied by Croatian Air Force. Some sources suggest that Afghanistan received 12, all now out of service. Tbilisi plant quotes total of 875 single-seat Su-25s built (including 50 Su-25BMs), plus Su-25UBs. Combined production stated by Sukhoi to exceed 1,000, of which more than 200 were exported and almost 300 are two-seat.

COSTS: Tbilisi factory quoting US$6 million per aircraft during 2000.

DESIGN FEATURES: Robust structure and systems redundancy increase resistance to small arms fire. Emphasis on survivability led to features accounting for 7.5 per cent of normal T-O weight, including armoured cockpit; pushrods instead of cables to actuate flying control surfaces (duplicated for elevators); damage-resistant main load-bearing members; widely separated engines in stainless steel bays; fuel tanks filled with reticulated foam for protection against explosion.

Russian Sukhoi Su-25 ('Frogfoot-A') armed with a Kh-25 (AS-10 'Karen') ASM *(Yefim Gordon)* 2001/0103529

SUKHOI—AIRCRAFT: RUSSIAN FEDERATION

Su-25 CURRENT OPERATORS

Unit	Base	Equipment
Russia (Air Force)		
16th OShAP	Taganrog	Su-25
18th GvOShAP	Galenki	Su-25
237th GvTsPAT	Kubinka	Su-25
368th OShAP	Budyennovsk	Su-25
461st OShAP	Krasnodar	Su-25
899th OShAP	Buturlinovka	Su-25
968th IISAP	Lipetsk	Su-25
Unident OShAP	Olovyannaya	Su-25
Unident OShAP	Chernigovka	Su-25
Krasnodar Higher Military Aviation School	Krasnodar	Su-25
Russia (Naval Aviation)		
279th KIAP	Severomorsk	Su-25UTG, UBP
Angola		
Unident unit	Saurimo	Su-25, UB
Armenia		
121st ShAE	Kumayari	Su-25
Azerbaijan		
Unident unit	Kyurdamir	Su-25
Belarus		
206th ShAB	Lida	Su-25
Bulgaria		
22nd Shturmova Aviobasa	Bezmer/Yambol	Su-25K, UBK
Czech Republic		
322 Taktica Letka	Namest	Su-25K, UBK
Georgia (Republic)		
Unident unit	Kopitnari	Su-25
Georgia (Abkhazia)		
Unident unit	unknown base	Su-25
Iraq		
Unident unit[1]	unknown base	Su-25
Kazakhstan		
Unident unit	unknown base	Su-25
Korea (North)		
Unident unit	unknown base	Su-25
Peru		
Grupo 11	Vitor	Su-25UB
Slovak Republic		
33/2 Letka	Malacky-Kuchyna	Su-25K, UBK
Turkmenistan		
56 BRS	Kizyl-Arvat	Su-25[2]
Ukraine		
299th OShAP	Saki	Su-25, UTG
452nd OShAP	Chortkov	Su-25
Uzbekistan		
59th APIB	Chirchik	Su-25

Notes: OshAP is *Otdel'nyi Shturmovoy Aviatsion'nyi Polk* (Independent Stormovik Aviation Regiment); KIAP is *Korabel'nyi Istrebitel'nyi Aviatsion'nyi Polk* (Shipborne Fighter Aviation Regiment); APIB is *Aviatsion'nyi Polk Istrebetelei-Bombardirovchikov* (Aviation Fighter-Bomber Regiment); ShAB is *Shturmovoy Aviatsion'nyi Baza* (Stormovik Aviation Base); BRS is *Basis Reserv Samolety* (Base for Reserve Aircraft). Gv prefix indicates *Gvardiya* (Guards).

[1] Probably not in service. Seven survivors fled to Iran in 1991
[2] In storage

Maintenance system packaged into four pods for carriage on underwing pylons; covers onboard systems checks, environmental protection, ground electrical power supply for engine starting and other needs, and pressure refuelling from all likely sources of supply in front-line areas; engines can operate on any fuel likely to be found in combat area, including MT petrol and diesel oil.

Shoulder-mounted wings; approximately 20° sweepback; 2° 30′ anhedral from roots; thickness/chord ratio 10.5 per cent; extended chord leading-edge dogtooth on outer 50 per cent each wing; wingtip pods each split at rear to form airbrakes that project above and below pod when extended; retractable PRF-4M landing light in base of each pod, outboard of small glareshield and aft of dielectric nosecap for ECM; semi-monocoque fuselage, with 24 mm (0.94 in) welded titanium armoured cockpit; pitot on port side of nose, transducer to provide data for fire-control computer on starboard side; conventional tail unit; variable incidence tailplane, with slight dihedral.

FLYING CONTROLS: Conventional and partly assisted. Hydraulically actuated ailerons, with manual back-up; multiple tabs in each aileron; double-slotted two-section wing trailing-edge flaps; full-span leading-edge slats, two segments per wing; manually operated elevators and two-section inset rudder; upper rudder section operated through sensor vanes and transducers on nose probe and automatic electromechanical yaw damping system; tabs in lower rudder segment and each elevator.

STRUCTURE: All-metal (60 per cent duralumin, 13.5 per cent titanium alloys, 19 per cent steel, 2 per cent magnesium); three-spar wings; semi-monocoque slab-sided fuselage.

LANDING GEAR: Hydraulically retractable tricycle type; mainwheels retract to lie horizontally in bottom of engine air intake trunks. Single wheel with low-pressure tyre on each levered suspension unit, with oleo-pneumatic shock-absorber; mudguard on forward-retracting steerable nosewheel, which is offset to port; mainwheel tyres size 840×360, pressure 9.30 bars (135 lb/sq in); nosewheel tyre size 660×200, pressure 7.35 bars (106 lb/sq in); brakes on mainwheels. Twin 25 m² PTK-25 cruciform brake-chutes housed in tailcone.

POWER PLANT: Two Soyuz/Gavrilov R-95Sh turbojets in long nacelles at wingroots, each 40.21 kN (9,039 lb st), superseded in about 50 late production aircraft (from 1989) by R-195s, each 44.18 kN (9,921 lb st); these first trialled by T8-14 and T8-15; 5 mm thick armour firewall between engines; R-195 turbojets have pipe-like fitment (believed not initially cleared for export, thus not fitted to Su-25K) at end of tailcone, from which air is expelled to lower exhaust temperature and so reduces IR signature; non-waisted undersurface to rear cowlings, which have additional small airscoops (as three-view); self-sealing fuel tanks filled with reticulated foam. Nos. 1 and 2 tanks in fuselage between cockpit and wing front spar, and between rear spar and fin leading-edge (latter acting as collector tank) contain 2,386 litres (630 US gallons; 525 Imp gallons); and in wing centre-section, capacity 1,274 litres (337 US gallons; 280 Imp gallons); total capacity 3,660 litres (967 US gallons; 805 Imp gallons); provision for four PTB-1500 external fuel tanks on underwing pylons.

ACCOMMODATION: Single Severin (Zvezda) K-36L ejection seat with zero altitude/54 kt (100 km/h; 62 mph) envelope; sideways-hinged (to starboard) canopy, with small rearview mirror on top; flat bulletproof windscreen. Folding ladder for access to cockpit built into port side of fuselage.

SYSTEMS: 28 V DC electrical system, supplied by two engine-driven generators. Dual independent hydraulic systems, pressure 206 bars (2,987 lb/sq in).

AVIONICS: *Comms:* SRO-1P ('Odd Rods') or (later) SRO-2 IFF transponder, with antennas forward of windscreen and under tail. SO-69 ATC transponder. R-862 V/UHF transceiver, R-855 emergency radio, R-828 air-to-ground radio.
Flight: RSBN Tacan, MRP-56P marker beacon receiver, RV-1S radio altimeter, ARK-15M radio compass SVS-1-72-18 air data system UUAP-72 AoA and acceleration indicator.
Instrumentation: ASP-17 BC-8 weapons sight incorporating AKS-750s camera gun.
Mission: Klyon PS laser range-finder and target designator under flat sloping window in nose; strike camera in top of nosecone.
Self-defence: SPO-15 (L-006LE) Sirena-3 radar warning system antenna above fuselage tailcone; ASO-2V chaff/flare dispensers (total of 256 flares) can be carried above root of tailplane and above rear of engine ducts. Some later aircraft fitted with Gardeniya active jammer.

ARMAMENT: One twin-barrel AO-17A 30 mm gun with rate of fire of 3,000 rds/min in bottom of front fuselage on port side, with 250 rounds (sufficient for a 1 second burst during each of five attacks). Eight large pylons under wings for 4,400 kg (9,700 lb) of air-to-ground weapons (though 1,400 kg; 3,086 lb regarded as normal maximum), including UB-32A rocket pods (each 32 × 57 mm S-5), B-8M1 rocket pods (each 20 × 80 mm S-8), 240 mm S-24 and 330 mm S-25 guided rockets, Kh-23 (AS-7 'Kerry'), Kh-25ML (AS-10 'Karen') and Kh-29L (AS-14 'Kedge') ASMs, laser-guided rocket-boosted 350 kg, 490 kg and 670 kg bombs, conventional bombs, 500 kg incendiary, anti-personnel and other cluster bombs, and SPPU-22 pods each containing a 23 mm GSh-23 gun with twin barrels that can pivot downward for attacking ground targets, and 260 rounds. Some late-production Russian Su-25BMs may have been wired for the carriage of Kh-58U/E (AS-11) ARMs and the associated Vynga datalink pod. Some Soviet aircraft configured to carry RN-61 nuclear weapon or associated IAB-500 'shape' for role training. Two small outboard pylons for R-3S (K-13T; NATO AA-2D 'Atoll') or R-60 (AA-8 'Aphid') self-defence AAMs. SU-25BM carries TL-70 winch and Kometa towed target below port wing, with inert FAB-250 or FAB-500 to starboard.

DIMENSIONS, EXTERNAL (A, Su-25K; B, Su-25UTG):
Wing span: A, B 14.36 m (47 ft 1½ in)

Sukhoi Su-25UBK export two-seat trainer *(Paul Jackson/Jane's)* 2001/0103567

Wing aspect ratio: A, B 6.1
Length overall: A 15.53 m (50 ft 11½ in)
B 15.36 m (50 ft 4¾ in)
Height overall: A 4.80 m (15 ft 9 in)
B 5.20 m (17 ft 0¾ in)
AREAS:
Wings, gross: A, B 33.7 m² (362.75 sq ft)
WEIGHTS AND LOADINGS:
Weight empty: A 9,500 kg (20,950 lb)
T-O weight: normal:
A 14,600 kg (32,187 lb)
B 13,005 kg (28,671 lb)
max: A 17,600 kg (38,800 lb)
B 16,245 kg (35,814 lb)
Max landing weight: A 13,300 kg (29,320 lb)
Max wing loading: A 522.2 kg/m² (106.97 lb/sq ft)
B 482.0 kg/m² (98.73 lb/sq ft)
Max power loading: A 199 kg/kN (1.95 lb/lb st)
B 184 kg/kN (1.80 lb/lb st)
PERFORMANCE:
Max level speed at S/L:
A M0.8 (526 kt; 975 km/h; 606 mph)
B M0.775 (512 kt; 950 km/h; 590 mph)
Max attack speed, airbrakes open:
A 372 kt (690 km/h; 428 mph)
Landing speed (typical): A 108 kt (200 km/h; 124 mph)
Service ceiling: clean: A, B 7,000 m (22,960 ft)
with max weapons: A 5,000 m (16,400 ft)
T-O run: typical: A 600 m (1,970 ft)
A, with max weapon load from unpaved surface
under 1,200 m (3,935 ft)
B, on concrete 750 m (2,460 ft)
B, on carrier ramp 175 m (575 ft)
Landing run: normal: A 600 m (1,970 ft)
A, with brake-chutes 400 m (1,315 ft)
B, arrested 90 m (295 ft)
B, on concrete 1,000 m (3,280 ft)
Range with 4,400 kg (9,700 lb) weapon load and two external tanks:
at S/L: A 405 n miles (750 km; 466 miles)
at height: A 675 n miles (1,250 km; 776 miles)
g limits: with 1,500 kg (3,307 lb) of weapons: A +6.5
with 4,400 kg (9,700 lb) of weapons: A +5.2

UPDATED

SUKHOI Su-25TM (Su-39)

TYPE: Attack fighter.
PROGRAMME: Developed as an ATM-armed derivative of the standard Su-25, with a two-crew concept similar to that used in the Mi-24 'Hind'. Plans to put the WSO in the front cockpit were abandoned at an early stage. Airframe is based on that of the Su-25UB, with which it shares about 85 per cent commonality. Three **Su-25T** development aircraft (Sukhoi T8M; numbered 08, 09 and 10) were converted Su-25UB airframes, with humped rear cockpit faired over and internal space used to house new avionics and extra tonne of fuel; conversions began 1976; first flight by converted first prototype Su-25UB 17 August 1984; construction at GAZ 31 Tbilisi, Georgia, of initial batch of 20 preseries aircraft for air force acceptance testing began 1979; first one flew May 1990 and deliveries to development unit at Akhtubinsk began 1991, although 12 reported in long-term storage at Tbilisi in early 1997, when Georgia reportedly issued a requirement for 50. Passed State Acceptance Tests in 1993. Production transferred to Ulan-Ude (Su-25UB/UTG production plant where two were assembled in 1999, with six more to follow in 2000 and 12 in 2001).

Redesignated **Su-25TM** after provision for 3-CM waveband Kopyo-25 radar, Schkval-M OEPS, Kinzhal MMW radar and Khod IIR pods and other equipment changes; T8M-3 first exhibited at Dubai '91 Air Show as export **Su-25TK**; version (aircraft 20) with Kopyo-25 radar pack, for improved night/bad weather attack capability, shown at Moscow Air Show '95; at that time, one Su-25TM had been completed at Ulan-Ude plant and first two preseries Su-25T aircraft had been updated to same standard. Two further Su-25TMs built at Ulan-Ude in 1997 and 1998, and will be followed by first batch of four production Su-25TMs. Remaining 20 production

RUSSIAN FEDERATION: AIRCRAFT—SUKHOI

Sukhoi Su-25TM (Su-39) all-weather attack aircraft *(Paul Jackson/Jane's)*

Su-25Ts to be similarly upgraded. Known to Sukhoi, but not Russian Air Forces, as **Su-39** 'Strike Shield'; aircraft 21 at Moscow Air Show, 1997, carried only this designation and was also exhibited in 1999, as Su-25TM; RLPK-25 radar, derived from Kopyo-25, fully installed in pack 1996; development continues, although originally scheduled for completion 1997. Licence for production offered to PZL Mielec of Poland, late 1997, as debt resettlement but not taken up.

CUSTOMERS: Russian Air Forces have eight Su-25Ts, delivered to Akhtubinsk test centre (two) and Lipetsk training base (six) in 1990-91; further 12 remain uncompleted at Tbilisi, but all planned to be upgraded at Ulan-Ude to Su-25TM. Four Lipetsk aircraft took part in bombing of Chechnya during 2000. Two of Tbilisi stock delivered to Ethiopia (with two UB trainers) in early 2000. Initial batch of seven Su-25 TMs built at Ulan-Ude; third aircraft was first delivered to Russian Air Forces in early 1998. Initial Russian requirement for 24 Su-25TMs to serve in six regional rapid-deployment groups, each of four Su-25TMs and 12 Su-25s, plus helicopters. Two TMs (Nos. 4 and 5) ordered by Russia in 1999 and due for delivery in 2001; further three planned for funding in 2001. Georgia has requirement for 50 TMs, and reportedly ordered three in 1996 and more in 1998.

COSTS: Rb50 million (US$8.3 million) flyaway (1998).

DESIGN FEATURES: Basically as Su-25 (which see), but embodies lessons learned during war in Afghanistan, particularly survival in intense anti-aircraft defence environment and night/bad weather capability. All sub-variants use increased-thrust R-195 engines. Welded cockpit 'bathtub' of lighter ABVT-20 titanium alloy armour, weight of polyurethane foam fuel tank filling and protective coatings increased. Centre-fuselage fuel pipes and control rods strengthened; gap between fuel tanks and air ducts filled with elastic porous material to avoid spillage into ducts following combat damage; additional armour for fuel system and avionics compartment. Reported to use special radar absorbent paint, previously tested on at least one Su-25. New nav system makes possible flights to and from combat areas under largely automatic control; equipment in widened nose includes TV activated some 5 n miles (10 km; 6 miles) from target; subsequent target tracking to an accuracy of 0.6 m (2 ft), weapon selection and release automatic; wingtip countermeasures pods introduced; gun transferred to underbelly position, on starboard side of farther-offset nosewheel. New, depressible 45 mm cannon originally planned. IR signature reduced by factor of three to four.

Wing leading-edge sweepback 19° 54′, dihedral 2° 30′, incidence 0° 43′.

FLYING CONTROLS: Basically as Su-25. Trim only on rudder; artificial feel in lateral and longitudinal channels. Hydraulically actuated elevator. SAU-8 automatic control system linked to WDNS.

LANDING GEAR: KT-163D mainwheels, with tyre size 840×360, pressure 9.30 bars (135 lb/sq in); KH-27A nosewheel further offset to port, with tyre size 680×260, pressure 7.35 bars (106 lb/sq in); metal and ceramic disc brakes, with anti-skid units. Two PTK-25 cruciform brake-chutes, each 13.0 m² (140 sq ft).

POWER PLANT: Improved Soyuz/Gavrilov R-195Sh turbojets. Ten internal fuel tanks, with new Nos. 3 and 4 tanks in fuselage; total capacity 4,890 litres (1,292 US gallons; 1,076 Imp gallons). Provision for four 800 litre (211 US gallon; 176 Imp gallon) or two 1,150 litre (304 US gallon; 253 Imp gallon) underwing tanks. Single-point pressure fuelling in starboard engine air intake duct; optional gravity fuelling. Additional air intakes on upper surface of engine nacelles to further cool jet exhaust.

ACCOMMODATION: As Su-25 but probably with zero/zero ejection seat.

SYSTEMS: Pressurised cockpit, maximum differential 0.25 bar (3.55 lb/sq in). Hydraulic system using AMG-10 fluid; flow rate 76 litres (20.0 US gallons; 17.7 Imp gallons)/min; pressure 207 bars (3,000 lb/sq in). AC and DC electrical systems. Gaseous oxygen system.

AVIONICS: *Radar:* Kinzhal (dagger) 8 mm MMW radar abandoned and replaced by podded Kopyo-25 pending provision of a new Kinzhal using Russian rather than Ukrainian parts.

Flight: Nav system includes air data system, twin INS, A-312 RSBN radio-nav system combining Tacan and ILS, A-723 long-range Loran/Omega radio-nav, ShO-13A Doppler, RV-21 radar altimeter and GPS.

Instrumentation: HUD; SUV-25T Voskhod nav/attack system on stabilised platform, with two Orbita computers, and Krasnogorsk Schkval I-251 nose-mounted electro-optical system for precision attacks on armour.

Mission: SUV-39 I-251 weapons control system. Larger window in nose for TV, laser range-finder and target designator of Krasnogorsk OMZ I-251 Schkval daylight system, all using same stabilised mirror, ×23 magnification lens and cockpit CRT; Khod (motion) night attack IIR pack, with Merkur LLTV and Prichal laser designator, and Phazotron Kopyo-25 radar pod on centreline pylon. Kopyo-25 detection range 13.5 n miles (25 km; 15.5 miles) for tanks, 40 n miles (75 km; 47 miles) for small ships, 13.5 to 31 n miles (25 to 57 km; 15.5 to 35 miles) air-to-air; able to track eight targets and engage two simultaneously.

Cockpit of Sukhoi Su-25TM *(Paul Jackson/Jane's)*

Self-defence: Irtysh DASS, with Pastel radar warning/emitter location system; UV-26 chaff/flare dispensers (192 cartridges) in top of fuselage tailcone and in large cylindrical housing at base of rudder that also contains L-166SI Sukhogruz 6,000 W Caesium lamp active IR jammer. Optional MSP-410 Omul or Gardeniya underwing active radar jamming pods on outboard underwing pylons.

ARMAMENT: One NNPU-8M twin-barrel 30 mm gun (same AO-17A as Su-25, but in new external VPU-17A mounting) with 200 rounds; 10 external stores attachments; two eight-round underwing clusters of Vikhr M (AT-X-16) tube-launched primary attack missiles able to penetrate 900 mm of reactive armour; other weapons include laser-guided Kh-25ML (AS-10 'Karen') and Kh-29L (AS-14 'Kedge'), TV-guided Kh-29T (AS-14 'Kedge'), anti-ship Kh-35 (AS-20 'Kayak') and anti-radiation Kh-58U (AS-11 'Kilter') or Kh-31P (AS-17 'Krypton') ASMs, KAB-500KR laser-guided bombs, S-25L laser-guided rockets, conventional bombs from 50 to 500 kg each, RBK-250 and RBK-500 cluster bombs, KMGU-2 submunition dispensers, SPPU-687 gun pods (containing a GSh-1-30 with 150 rounds) S-8, S-13 and S-25 rockets, and R-27R/RE (AA-10 'Alamo-A/C'), RVV-AE (R-77; AA-12 'Adder') and R-73 (AA-11 'Archer') AAMs.

DIMENSIONS, EXTERNAL:
Wing span	14.52 m (47 ft 7¾ in)
Wing chord: at root	3.00 m (9 ft 10¼ in)
at tip	1.025 m (3 ft 4¼ in)
Wing aspect ratio	7.0
Length overall	15.35 m (50 ft 4½ in)
Height overall	5.20 m (17 ft 0¾ in)
Tailplane span	4.58 m (15 ft 0½ in)
Wheel track	2.505 m (8 ft 2¾ in)
Wheelbase	3.58 m (11 ft 9 in)

AREAS:
Wings, gross	30.10 m² (324.0 sq ft)
Ailerons (total)	1.51 m² (16.25 sq ft)
Trailing-edge flaps (total)	4.44 m² (47.79 sq ft)
Leading-edge slats (total)	3.16 m² (34.02 sq ft)
Fin	5.28 m² (56.84 sq ft)
Rudder	0.75 m² (8.07 sq ft)
Tailplane	5.61 m² (60.39 sq ft)
Elevators (total)	1.88 m² (20.24 sq ft)

WEIGHTS AND LOADINGS:
Max combat load	5,000 kg (11,023 lb)
Max fuel: internal	3,840 kg (8,465 lb)
external	3,070 kg (6,768 lb)
Max T-O weight	20,500 kg (45,194 lb)
Max landing weight	13,200 kg (29,100 lb)
Max wing loading	681.1 kg/m² (139.49 lb/sq ft)
Max power loading	232 kg/kN (2.28 lb/lb st)

PERFORMANCE:
Never-exceed speed (V_{NE})	M0.82
Max level speed at S/L	M0.77 (512 kt; 950 km/h; 590 mph)
Max cruising speed at 200 m (650 ft)	378 kt (700 km/h; 435 mph)
Econ cruising speed	350 kt (650 km/h; 404 mph)
Landing speed	130 kt (240 km/h; 150 mph)
Max rate of climb at S/L	3,480 m (11,415 ft)/min
Rate of climb at S/L, OEI	1,020 m (3,345 ft)/min
Service ceiling	10,000 m (32,800 ft)
Service ceiling, OEI	9,000 m (29,520 ft)
T-O run	650 m (2,135 ft)
Landing run	750 m (2,465 ft)
Combat radius with 2,000 kg (4,410 lb) weapons:	
at low altitude	216 n miles (400 km; 248 miles)
at height	340 n miles (630 km; 391 miles)
Ferry range	1,214 n miles (2,250 km; 1,398 miles)
g limit	+6.5

UPDATED

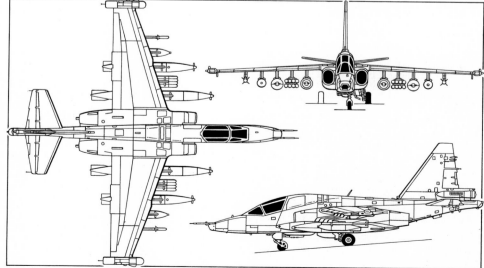

Sukhoi Su-25TM single-seat anti-tank aircraft *(Mike Keep/Jane's)*

SUKHOI Su-27

NATO reporting name: Flanker

TYPE: Air-superiority fighter.

PROGRAMME: Development of long-range heavy fighter to meet PFI (*perspektivnyi frontovoy istrebitel*: advanced frontal fighter) requirement began 1969 under leadership of Pavel Sukhoi; augmented in high-low mix by Mikoyan LPFI (later MiG-29). T10 based on 'integral' layout of unbuilt T4MS. Construction of prototype began in 1974 under Mikhail Simonov's supervision, and it was flown 20 May 1977 by Vladimir Ilyushin. Development undertaken by nine flying **Su-27** 'Flanker-As': four prototypes (T10-1 and T10-2) produced in OKB workshops; T10-3 and T10-4 built by Komsomolsk plant and assembled by OKB); five development (T10-5, -6, -9, -10 and -11) and one static test airframe at Komsomolsk. Individual prototype designations duplicated some competing unbuilt T10 configurations. Prototypes had curved wingtips, rearward-retracting nosewheel and tailfins mounted centrally above engine housings.

Some variation between prototypes; T10-1 originally flew with 'clean' wing; four fences then added, followed by anti-flutter weights on wingtips and fins; inboard fences subsequently removed. T10-3 and T10-4 used AL-31F engines, others had AL-21F-3s. Canted tailfins from T10-3. T10-5, -6, -9, -10, and -11 had eight pylons (two below each wing) and -9 and -11 had long-chord ogival radomes; these five aircraft collectively known as T10-5 subvariant. Development was not easy; original configuration revealed poor controllability with inadequate stability in roll and yaw, and with poor high AoA capability; two pilots lost their lives before major airframe redesign resulted in T-10S production configuration (known internally as T10 junior).

Construction of new model began 1979, with first flight of prototype (T10-7) 20 April 1981, followed by second prototype (T10-12); T10-8 was T10S-0 static test airframe; T10-4 was similar; series production by KnAAPO at Komsomolsk-on-Amur began with T10-15, first flown 2 June 1982; entry into service 22 June 1985 with air defence regiment co-located with Komsomolsk factory airfield; official type acceptance 23 August 1990 (Council of Ministers decree); ground attack role observed in 1991. Following purchase of complete aircraft, China signed licensed production agreement on 6 December 1996, covering further 200. Production continues for export. In late 1999, Sukhoi quoted 567 of Su-27 family in service in eight countries; foresees total of 760 sales by 2005.

CURRENT VERSIONS: **Su-27** ('Flanker-B'): Single-seat land-based production version for air defence force (PVO); full-span leading-edge flaps, trailing-edge flaperons, 5 per cent increase in wing area, straight leading-edges, new aerofoil, square wingtips, carrying anti-flutter weights which

Sukhoi Su-30KI development aircraft 4002 *(Paul Jackson/Jane's)*

doubled as AAM launchers; wider-spaced, uncanted tailfins outboard of engine housings; flatter canopy of reduced cross-section; extended tailcone instead of flat 'beaver-tail'; forward-retracting nosewheel; first flown (T10-7) 20 April 1981. Standard radar tracks 10 targets simultaneously, engages only one. There was early, but probably erroneous, speculation that PVO aircraft were designated Su-27P, with Su-27S designation applied to Frontal Aviation aircraft. All aircraft can carry Sorbtsya-S active ECM jammer pods on wingtips in place of wingtip launch rails. Su-27S designation little used, but differentiates production (Series) from prototype and preproduction. Four initial production aircraft (T10-15, -17, -18 and -22) used for State Acceptance Tests were part of small initial batch featuring horizontally cropped fin caps.

Detailed description applies to the above version, except where indicated.

'Su-27RV' ('Flanker-B'): Six replacement aircraft for Russian Knights aerobatic team feature GPS and Western-compatible communications equipment.

Su-27SK ('Flanker-B'): Export version of basic Su-27, using air-to-ground capabilities not exploited by Soviet/Russian Su-27s and using same weapons options and downgraded avionics. Armament, totalling up to 4,000 kg (8,818 lb), includes 250 kg and 500 kg bombs, B-8M1 packs of 20 × 80 mm rockets, B-13L packs of five 122 mm rockets, S-250FM 250 mm rockets, KMGU-2 cluster bombs, or podded SPPU-22 30 mm gun with optional downward-deflecting barrel for air-to-ground and air-to-air use. Dimensions, weights and performance generally similar to Su-27 but with reinforced landing gear giving increased (33,000 kg; 72,752 lb) MTOW. Chinese aircraft locally designated **J-11**, though this strictly applies only to locally manufactured examples. First Chinese-assembled aircraft flight tested in December 1998. Some later export aircraft may have upgraded radar and 6,200 kg (13,670 lb) weapon load, or even 8,000 kg (17,637 lb) according to some manufacturer's brochures.

Su-27SM: Proposed mid-life update configuration using a derivative of the Zhuk radar, AL-31FM or AL-35F engines and other improvements to avionics and hardpoints. Status unknown in 2000.

Su-27SMK: Single-seat multirole fighter based on the basic Su-27SK, and not the Zhuk-equipped Su-27SM; revealed 1995. Compared with Su-27SK, has 12 instead of 10 hardpoints; 8,000 kg (17,637 lb) weapon load; state-of-the-art nav system, including long-range radio nav, GPS, multichannel comms and latest ECM; larger internal wing fuel tanks and provision for two 2,000 litre (528 US gallon; 440 Imp gallon) underwing tanks; flight refuelling and buddy refuelling capability; retains N001 radar. Most improvements already introduced on other Su-27 series aircraft, enabling virtually full-standard Phase One Su-27SMKs to be offered for immediate delivery. These would be configured for air-to-air roles, with the new wings containing increased internal fuel, provision for underwing tanks, flight refuelling, 12 hardpoints and RVV-AE (R-77; AA-12 'Adder') AAM capability. Su-30KI demonstrator (see below) is regarded as Phase One standard. Phase Two aircraft, with improved avionics and weapon systems for air-to-surface missions, theoretically available later; weapons options to include Kh-29 and Kh-31 ASMs, plus KB-500 LGBs although this seems unlikely without much more extensive upgrade. See accompanying charts for alternative weapon loads.

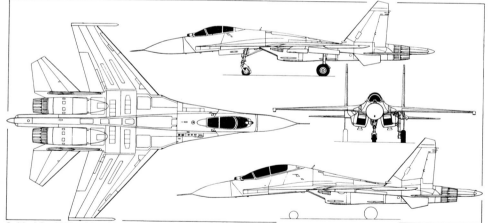

Sukhoi Su-27P, with added side elevation (bottom) of two-seat Su-27B *(Dennis Punnett/Jane's)*

Sukhoi Su-27 'Flanker B' of the Ukrainian Air Force *(Paul Jackson/Jane's)*

Sukhoi Su-27PD '598' of the Gromov Flight Test Centre, Zhukovsky, participated in trials leading to the Su-30KI upgraded version *(Paul Jackson/Jane's)*

Dimensions as Su-27, except wing span over wingtip R-73E missiles 14.95 m (49 ft 0½ in). Weights and performance: see tabulated data.

Su-27PD: Long-endurance test aircraft with retractable AAR probe, offset IRST and recontoured tail 'sting'; sometimes described as single-seat Su-30, originally Su-27P. Probably stripped of weapons system for research at Gromov Flight Research Institute and demonstration flying. Added satellite navigation; no radar or weapon control system; PC 486 with LCD for flight assessment of indication formats. At least one aircraft (3720 '598', last known Russian military Su-27); by early 1998, this was marked 'Su-27 Upgrade' and undertaking development flying for **Su-30KI** (which see). Duties included two flights to North Pole in July and September 1999, testing SRNK satnav system. Another Su-27P (3711 '595' lacked AAR probe.

Su-30KI ('Flanker C'): The Su-30KI prototype (4002, wearing the basic Chinese dark grey colour scheme, but with a disruptive camouflage superimposed on the wings, tailplanes and outer surfaces of the tailfins) first flew on 28 June 1998, and was then sent to Chkalov Flight Test Centre at Akhtubinsk, where it was evaluated and used for launch trials of the RVV-AE (AA-12 'Adder') AAM, and for avionics and refuelling probe verification. The Su-30KI was described as a 'single-seat Su-30', and was tailored to meet an Indonesian requirement, before the Asian economic crisis halted that programme. Extent to which the aircraft incorporated the 'extended endurance' features of the Su-30 (apart from refuelling probe, GPS and Western VOR/DME navigation equipment) remains uncertain, however. Su-30KI also used as basis of Sukhoi's latest proposal to upgrade Russia's in-service Su-27s, replacing Su-27SM. Upgrade programme still confusingly referred to by Sukhoi as the Su-30KI, although Russian Air Forces understood to use neither the Su-30 designation nor the KI suffix.

Su-27UB ('Flanker-C'): Tandem two-seat trainer version of 'Flanker-B' with full combat capability (Sukhoi designation T10U); four prototypes (first of which for static testing) built at Komsomolsk; first flown 7 March 1985 (T10U-1); series manufacture by Irkutsk Aircraft Production Association began 1986; first production UB (T10U-4) flew on 10 September 1986. Instructor in raised rear seat with 6° view forward over the nose; taller fin; length same as 'Flanker-B'; overall height 6.36 m (20 ft 10¼ in), maximum combat load 8,000 kg (17,637 lb). 1,500 kg increase in empty weight, no reduction in internal fuel capacity. Export version is **Su-27UBK**.

'Aircraft 02-01' ('Flanker-C'): Second Su-27UB flying prototype converted as inflight refuelling systems testbed with retractable probe and provision for centreline 'buddy' pod; first flown early 1987. Later used in support of Su-27K and Su-27PU development programmes.

Su-27M: Advanced development of Su-27. Still the official designation for the aircraft known to the OKB as the Su-35. See **Su-35**

Su-27/Su-30 (Su-30KI Minor Modernisation): Proposed upgrade for export Su-27SK, bestowing same multirole standard as Su-30MKK, with inflight refuelling probe, glass cockpit, N-011M radar, GPS, VOR, DME, new mission computer, new navigation system with GPS, expanded EW and provision for new targeting pods. Development undertaken by Su-27PD 3720 '598' and, later, by Su-30KI 4002, first flown 28 June 1998. Phase I of upgrade is known in Russia as Su-30KI, despite the two-seat and Indonesian connotations of that designation. In February 1998, Su-27PD 3720 '598' was displayed at the Singapore Air Show marked 'Su-27 Upgrade' and – notwithstanding the fact that it is a single-seat aircraft – stated to be undertaking development work for a variant to be designated Su-30KI. Latter is phase one of Su-27SMK programme, duties of trials aircraft 4002 including satellite-based navigation; upgraded computers; RVV-AE AAMs; Kh-29T, Kh-31P and Kh-59M ASMs; KAB-500 and KAB-1500 TV-guided bombs; and multifunction cockpit displays.

Su-27/Su-30 Major Modernisation: Proposed advanced upgrade configuration bringing any 'first-generation' 'Flanker' to virtual Su-27M/Su-30MKI standards, with a new MIL-STD-1553B-based avionics system, N011M radar with a phased-array antenna, and with the option of adding thrust vectoring and/or canard foreplanes.

Su-27LL-KS: Su-27 (T10-26; 0702) with axisymmetric afterburner nozzle. Also known as Su-27LLUV(KS) *(Upravlyayayemy Vektor tyagi; Krugloye Soplo*: thrust vector control; axisymmetric nozzle) or Su-27-KSI. Evaluated against two-dimensional nozzle testbed, described below. First flew 21 March 1989.

Su-27UB-PS: Su-27UB (Somsomolsk 0202; '08') modified in 1990 for thrust-vectoring development, with large two-dimensional box nozzle on port tailpipe. Also known as Su-27LL-PS or Su-27LL-UV(PS) *(Upravlyayemy Vektor tyagi; Ploskoye Soplo*: thrust vector control; flat nozzle).

Su-27LMK: CCV *(Lyotno-Modeliruyushchy Kompleks*: flight simulation complex) conversion of production aircraft 2405 with FADEC and side-stick, and fitted with axisymmetric nozzle, tested subsequently with axisymmetric nozzle on starboard tailpipe. Trials at Zhukovsky test centre, under direction of TsIAM and Saturn/Lyulka OKB, began 1990.

Su-27LL-OS: Missile testbed; 1989 conversion of first production Su-27UB 0101 (T10U-4).

Su-27K: Described separately.

Su-27IB: Described separately.

Su-27PU: Two prototypes only; described in 1993-94 *Jane's*. In production as **Su-30** (which see).

P-42: Specially prepared Su-27 (T10-15; first production Su-27; originally flown 2 June 1982); set 31 official world records between 1986 and 1988, including climb to 12,000 m (39,370 ft) in 55.542 seconds, and to 22,250 m (73,000 ft) with 1,000 kg (2,205 lb) payload; some records are in FAI category for STOL aircraft.CUSTOMERS: Approximately 567 in service by December 1999. About 395 in service with Russian Air Forces (or 340 plus 10 with training units by mid-2000, according to some sources); these reports refer to active inventory; evidence suggests that main production batches for former USSR contained 520 Su-27s and 140 Su-27UBs.

China received 26 (24 Su-27SK single-seat, two Su-27UBK two-seat) in 1991-92, followed by 24 more (14 single-seat, 10 two-seat) in 1995-96; original agreements called for licensed manufacture of 200 as J-11, at planned rate of 50 a year, in new purpose-built factory (owned by Shenyang Aircraft Company, which see) from 1998 (when first two shipsets delivered from Komsomolsk; first Chinese-assembled Su-27 rolled out December 1998; only eight delivered by late 2000, following considerable quality control problems, leading to the procurement of 28 extra Russian-built Su-27UBKs, delivered 2000-02, first eight in December 2000. China's licence agreement does not allow exports. The first 50 aircraft are being assembled from Russian-supplied kits of diminishing completeness and the Chinese aircraft will always include at least 30 per cent Komsomolsk content; first 10 kits stated to have been 100 per cent complete but, by 2000, China seeking more in this state to expedite deliveries. Chinese production will change to Su-30 after first 80 aircraft. No licence has been granted for Chinese production of the AL-31 engine. China's 50 Su-27s were initially based at Wuhan, with the 9th Fighter Regiment, 3rd Air Division though 24 subsequently moved to Liancheng, where 17 were damaged (three beyond repair) in an April 1997 hurricane. Two more were written off in accidents. Production will now cease after 80 J-11s, thereafter switching to the Su-30MKK.

Vietnam has first batch of about six (including one Su-27UB, although further two UBs delivered by airfreight on 1 December 1997 and second pair, representing balance

Weapon options for the export Su-27SK

Sukhoi Su-27UB ('Flanker-C') two-seat operational trainer *(Paul Jackson/Jane's)*

of follow-on order, destroyed in An-124 crash six days later, but replaced by 'Su-30s'); second batch of six now delivered, bringing in-service total to seven Su-27SKs and five Su-27UBK/Su-30Ks; 24 more on order. Upgrade expected, since Vietnam requires aircraft to use R-77 AAMs, Kh-29, Kh-31 and Kh-59 ASMs. Kazakhstan is receiving 32 from Russia, most recently six in 1997, four in January 1999 and for during 2000 (with 12 then outstanding), as compensation for return of Tu-95s and for alleged environmental and ecological damage. Belarus has 22 or 23 (though these may have been sold-on, perhaps to Angola); Ukraine 66 to 70 and Uzbekistan 30; others reportedly went to Armenia, Azerbaijan and Georgia; Syria has requested 14 (some reports suggest 17 in service others that only four were in service by mid-2000, two at Minkah AB and two at Damascus). Ethiopia received the first of eight second-hand from Russia in late 1998 and Yemen is negotiating for 'a squadron'. In late 1999, there were reports that Angola had taken delivery of eight Su-27s at Catumbela in August, with the balance of seven expected imminently. Pilots had reportedly trained in Belarus, the presumed source of the aircraft, though technical support came from Ukraine. Japan has been reported to be interested in two aircraft for evaluation/aggressor use, but Sukhoi unwilling to sell less than six. Two Japanese pilots underwent a US$300,000 46-day training programme on the Su-27 in early 1998, however. More recently Malaysia was reportedly 'considering' an Su-27 purchase, while the aircraft was also offered to New Zealand as an alternative to its intended second-hand F-16 buy. Sukhoi offered an Su-30 10-year lease, with targeting pods, PGMs and support, for a quoted cost of NZ$124.8 million (US$68.8 million). On 26 November 1995, the first of two Su-27s was delivered to the USA inside an An-124 for an unknown purpose.

DESIGN FEATURES: Developed to replace Yak-28P, Su-15 and Tu-28P/128 interceptors in APVO, for dual-role ground attack/air combat and to escort Su-24 deep-penetration strike missions; basic requirement was effective engagement of F-15 and F-16 and other future aircraft and cruise missiles; exceptional range on internal fuel made flight refuelling unnecessary until Su-24s received probes when more range was required in the escort role; external fuel tanks still not considered necessary; all-swept blended fuselage/mid-wing configuration, with long curved wing leading-edge root extensions, lift-generating fuselage, twin tailfins and widely spaced engines with wedge intakes; rear-hinged doors in intakes hinge up to prevent ingestion of foreign objects during take-off and landing; integrated fire-control system with pilot's helmet-mounted target designator; exceptional high-Alpha performance; basic wing sweepback 42° on leading-edge, 37° at quarter-chord; no dihedral or incidence.

FLYING CONTROLS: Four-channel analogue SDU-27 fly-by-wire, with no mechanical back-up; artificial feel; relaxed longitudinal stability; no ailerons; full-span leading-edge flaps and plain inboard flaperons controlled manually for take-off and landing, computer-controlled in flight; differential/collective tailerons operate in conjunction with flaperons and rudders for pitch and roll control; flight control system limits g loading to +9 and normally limits angle of attack to 30 to 35°; angle of attack limiter can be overruled manually for certain flight manoeuvres; large door-type airbrake in top of centre-fuselage.

STRUCTURE: All-metal, with extensive use of aluminium-lithium alloys and titanium but no composites; comparatively conventional three-spar wings; basically circular section semi-monocoque fuselage with load-bearing spine, sloping down sharply aft of canopy; cockpit high-set behind drooped nose; large ogival dielectric nosecone; long rectangular steel blast panel forward of gun on starboard side, above wingroot extension; two-spar fins and horizontal tail surfaces; uncanted vertical surfaces on narrow decks outboard of engine housings; fin extensions beneath decks form parallel, widely separated ventral fins.

LANDING GEAR: Hydraulically retractable tricycle type, made by Hydromash, with single wheel on each unit; KT-156D mainwheels turn 90° while retracting forward into wingroots; hydraulically steerable non-braking KN-27 nosewheel, with mudguard, also retracts forward; mainwheel tyres 1,300×350, pressure 12.25 to 15.70 bars (178 to 227 lb/sq in); nosewheel tyre 680×260, pressure 9.30 bars (135 lb/sq in); hydraulic carbon disc brakes with two-signal anti-skid system; electric brake cooling fan in each mainwheel hub; further brake in nosewheel; brake-chute housed in fuselage tailcone.

POWER PLANT: Two Saturn/Lyulka AL-31F turbofans, each 122.6 kN (27,557 lb st) with afterburning. Large spring-loaded auxiliary air intake louvres in bottom of each three-ramp engine duct near primary wedge intake; two rows of small vertical louvres in each sidewall of wedge, and others in top face; fine grille of titanium hinges up from bottom of each duct to shield engine from foreign object ingestion during take-off and landing. Fuel in four integral tanks: three in fuselage and one split between each outer wing. Max internal fuel capacity approximately 11,775 litres (3,110 US gallons; 2,590 Imp gallons); normal operational fuel load 6,600 litres (1,744 US gallons; 1,452 Imp gallons). Higher figure represents internal auxiliary tank for missions in which manoeuvrability not important. No provision for external fuel tanks, except in versions where specifically indicated. Pressure or gravity fuelling. Flight refuelling capability optional; Su-27UB operated as buddy tanker during development of system.

ACCOMMODATION: Pilot only, on Zvezda K-36DM Series 2 zero/zero ejection seat, under large rear-hinged transparent blister canopy, with low sill; 14° view downward over the nose.

SYSTEMS: Automatically regulated cockpit air conditioning. Two independent, duplicated, hydraulic systems, pressure 275 bars (4,000 lb/sq in), for actuation of control surfaces, airbrake, landing gear, wheel brakes, air intake ramps and FOD screens. Electrohydraulic (flight control) parts of system quadruplicated. APU in top of rear fuselage for ground and emergency in-flight power. Pneumatic system pressure 210 bars (3,045 lb/sq in) for back-up landing gear extension avionics bay pressurisation and canopy operation and sealing. Electrical supply 27 V DC, 115 V and 200 400 Hz AC; two type NKBN-25 Ni/Cd batteries. AC supplied by two integral engine-driven generators with three-phase and single-phase converters. Gaseous oxygen for 4 flight hours.

AVIONICS: Systems integrated by NPO Elektroavtomatika.
 Communications: R-800 UHF radio, R-864 HF, intercom and cockpit voice recorder. SO-69 ATC transponder; various IFF fits, according to production batch.
 Flight: PNK-10 flight instrumentation and navigation suite encompasses the usual, traditional flight instruments (the IK-VSP altitude and speed data system), SAU-10 autopilot, Ts-050 computer, ARK-19 or ARK-20 ADF, a radio altimeter and A-317 Uron SHORAN. Aircraft capable of ICAO Cat. I autoland.
 Radar: RLPK-27 radar sighting system with NIIP N-001 Mech (Sword, NATO 'Slot Back') track-while-scan coherent pulse Doppler look-down/shoot-down radar (long-chord twist-cassegrain antenna diameter approximately 1.0 m; 3 ft 4 in) with search range of up to 54 n miles (100 km; 62 miles), tracking range 35 n miles (65 km; 40 miles), in forward hemisphere against MiG-21 size target (ability to track 10 targets and engage two simultaneously in current Su-27SK/UB upgrade configurations) and TsVM-80 computer.
 Instrumentation: Integrated fire-control system enables radar, IRST and laser range-finder to be slaved to pilot's helmet-mounted target designator and displayed on wide-angle HUD; autopilot able to restore aircraft to right-side-up level flight from any attitude when 'panic button' depressed.
 Mission: Duplex SUV-27 weapons targeting complex integrates radar and OEPS-27 electro-optic sighting system with Model 36Sh OLS-27 IR search/track (IRST) sensor, range 27 n miles (50 km; 31 miles), collimated laser range-finder, range 4.3 n miles (8 km; 5 miles), functioning through common optics in transparent housing forward of windscreen and NSTs-27 Schchel-3U helmet-mounted sight. Beryuza tactical/GCI datalink. Provision for reconnaissance pod on centreline pylon.
 Self-defence: SPO-15LM Beryoza 360° radar warning antennas, outboard of each bottom air intake lip and at tail. Gardeniya active ECM jamming system. Three banks of APP-50 chaff/flare dispensers (total 96 cartridges) in bottom of long tailcone extension and top of tailsting. Tailcone widened to provide extra chaff/flare dispensers from batch 18 (mid-1987).

ARMAMENT: One 30 mm Gryazev/Shipunov 9A-407lK GSh-30-1 gun in starboard wingroot extension, with 150 rounds. Up to 10 AAMs in air combat role, on tandem pylons under fuselage between engine ducts, beneath each duct, under each centre-wing and outer-wing, and at each wingtip. Typically, two short-burn semi-active radar homing R-27R (NATO AA-10A 'Alamo-A') in tandem under fuselage; two short-burn IR homing R-27T (AA-10B 'Alamo-B') on centre-wing pylons; and long-burn semi-active radar homing R-27ER (AA-10C 'Alamo-C') or IR R-27ET (AA-10D 'Alamo-D') beneath each engine duct. The four outer pylons carry either R-73A

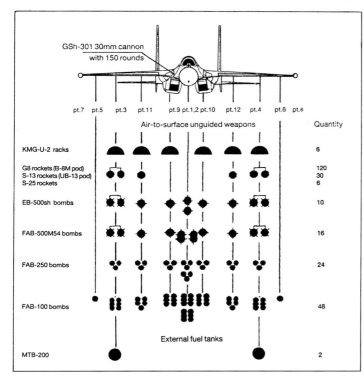

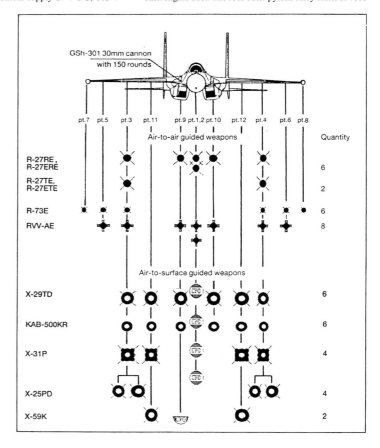

Weapons options for the Su-27SMK multirole fighter: unguided air-to-surface (above) and guided air-to-air and air-to-surface (right). X-29, X-31, X-25 and X-59 have alternative Westernised designations Kh-29, Kh-31, Kh-25 and Kh-59

Avionics upgrades are being evaluated by Su-30KI prototype 4002, formerly destined for Indonesia *(Paul Jackson/Jane's)*

(AA-11 'Archer') or R-60 (AA-8 'Aphid') close-range IR AAMs. R-33 (AA-9 'Amos') AAMs optional in place of AA-10s. Up to eight 500 kg bombs, sixteen 250 kg bombs or four launchers for S-8, S-13 or S-25 rockets.

DIMENSIONS, EXTERNAL (Su-27):
Wing span	14.70 m (48 ft 2¾ in)
Wing aspect ratio	3.5
Length overall, excl nose probe	21.94 m (71 ft 11½ in)
Height overall	5.93 m (19 ft 5½ in)
Fuselage: Max width	1.50 m (4 ft 11 in)
Tailplane span	9.88 m (32 ft 5 in)
Distance between fin tips	4.30 m (14 ft 1¼ in)
Wheel track	4.36 m (14 ft 3¾ in)
Wheelbase	5.88 m (19 ft 3½ in)

AREAS:
Wings, gross	62.00 m² (667.4 sq ft)
Wing leading-edge flaps (total)	4.60 m² (49.51 sq ft)
Flaperons (total)	4.90 m² (52.74 sq ft)
Fins (total)	11.90 m² (128.10 sq ft)
Rudders (total)	3.50 m² (37.67 sq ft)
Horizontal tail surfaces (total)	12.30 m² (132.40 sq ft)

WEIGHTS AND LOADINGS (A: Su-27, B: Su-27UB, C: Su-27SK, D: Su-27SMK):
Operating weight empty: A	16,380 kg (36,110 lb)
B	17,500 kg (38,580 lb)
Max fuel weight: A	9,400 kg (20,723 lb)
Normal T-O weight: A	23,000 kg (50,705 lb)
B	24,140 kg (53,220 lb)
D, with two R-73 and two R-27 missiles	23,700 kg (52,250 lb)
Max T-O weight: A, C, D	33,000 kg (72,750 lb)
B	33,500 kg (73,850 lb)
Max wing loading: A, C, D	532.3 kg/m² (109.01 lb/sq ft)
B	491.1 kg/m² (100.59 lb/sq ft)
Max power loading: A, C, D	135 kg/kN (1.32 lb/lb st)
B	124 kg/kN (1.22 lb/lb st)

PERFORMANCE:
Max level speed: at height:	
A, B	M2.35 (1,350 kt; 2,500 km/h; 1,550 mph)
C	M2.3 (1,319 kt; 2,443 km/h; 1,518 mph)
D	M2.17 (1,241 kt; 2,300 km/h; 1,429 mph)
at S/L:	
A, B	M1.1 (725 kt; 1,345 km/h; 835 mph)
C, D	M1.14 (756 kt; 1,400 km/h; 870 mph)
Stalling speed: A	108 kt (200 km/h; 125 mph)
Acceleration: D:	
from 324 to 594 kt (600 to 1,100 km/h; 373 to 683 mph)	15 s
from 594 to 701 kt (1,100 to 1,300 km/h; 683 to 808 mph)	12 s
Rate of roll: A	approx 270°/s
Service ceiling: A, D	18,000 m (59,060 ft)
B	17,500 m (57,420 ft)
C	18,500 m (60,700 ft)
T-O run: A	450 m (1,475 ft)
B	550 m (1,805 ft)
C, D	650 m (2,135 ft)
Landing run: A, D	620 m (2,035 ft)
B	700 m (2,300 ft)
Combat radius: A, B	810 n miles (1,500 km; 930 miles)
D	840 n miles (1,560 km; 970 miles)
Range with max fuel:	
A, C	1,985 n miles (3,680 km; 2,285 miles)
B	1,620 n miles (3,000 km; 1,865 miles)
Range, D:	
at S/L, internal fuel	745 n miles (1,380 km; 857 miles)
at S/L, with external tanks (dropped when empty)	894 n miles (1,656 km; 1,029 miles)
at height, internal fuel	2,046 n miles (3,790 km; 2,355 miles)
at height, with external tanks (dropped when empty)	2,370 n miles (4,390 km; 2,727 miles)
max, with flight refuelling	2,807 n miles (5,200 km; 3,231 miles)
g limit (operational): A, B, C, D	+9

UPDATED

SUKHOI Su-27IB (Su-32 and Su-34)

TYPE: Attack fighter.

PROGRAMME: Side-by-side two-seat long-range fighter-bomber (*istrebitel bombardiroshchik*) Su-27 variant intended as tactical strike/attack replacement for Su-24 and Su-25; project designation T10V; redesignated Su-34 by Sukhoi ("to stress father-and-son relationship" to Su-24), but Russian Air Forces retained Su-27IB. Su-32FN/Su-32MF assigned to proposed export versions but, in 2000, all Su-34s were redesignated Su-32.

Conceptual design ordered 21 January 1983, production authorised 19 June 1986. Designed under the direction of Rollan Martirosov. Prototype (T10V-1 '42') built in Sukhoi's own workshops and first flown 13 April 1990; first seen in *Tass* photograph showing this aircraft approaching (but not landing on) the carrier *Admiral of the Fleet Kuznetsov*; described as deck landing trainer, but no wing folding or deck arrester hook; although with foreplanes and twin nosewheels like Su-27K. Designation Su-27KU quoted, though dedicated side-by-side carrier trainer then officially known as T10KM-2 or Su-27KM-2 for which the new side-by-side cockpit had been designed. Russian unofficial name 'Platypus'; exhibited to Russian Federation and Associated States (CIS) leaders at Machulishche airfield, Minsk, February 1992, with simulated attack armament on 10 external stores pylons (under each intake duct, on each wingtip, three under each wing); Kh-31A/P (AS-17 'Krypton') ASMs under ducts, R-73A (AA-11 'Archer') AAMs on wingtips; a 500 kg laser-guided bomb inboard, TV/laser-guided Kh-29 (AS-14 'Kedge') ASM on central pylon and RVV-AE (R-77; AA-12 'Adder') AAM outboard under each wing.

Production originally planned for Irkutsk, but eventually located at Novosibirsk. First series production aircraft flew 28 December 1994. Four were in assembly at

Leninets B004 radar, in this instance installed in T10V-4, the Su-32 demonstrator

Novosibirsk by early 1997. Twelve were once scheduled for delivery by 1998; intention was to replace all Su-24s by 2005; reconnaissance and electronic warfare versions reportedly under development. In September 2000, Russian Air Forces C-in-C predicted first Su-32 deliveries in 2004. Side-by-side cockpit to form basis of proposed Su-30-2 long-range interceptor and Su-33UB (Su-27KUB) carrier trainer (which see).

CURRENT VERSIONS (specific): **T10V-1 '42'**: First flying prototype; detailed above. Converted from Su-27UB airframe by Sukhoi OKB workshops, with new nose built at Novosibirsk, and reportedly fitted with Su-33 main landing gear.

T10V-2 '43': Flown 18 December 1993; first aircraft to be built at Novosibirsk; sometimes described as first production Su-34, in that it had Su-35-type four-hardpoint wing panels and larger internal fuel cells, reinforced wing

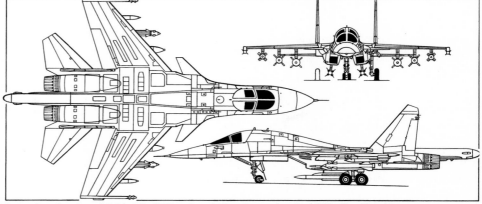

Three-view drawing of Sukhoi Su-27IB twin-turbofan theatre bomber *(Mike Keep/Jane's)*

Sukhoi Su-27IB second prototype, accompanied by potential armament including, in foreground, Yakhont anti-ship missile *(Paul Jackson/Jane's)*

Su-27IB third prototype posing as an Su-32MF, armed with AS-17 ASMs on underfuselage pylons, plus AAM armament of AA-10s, AA-12s and AA-11s (wingtips)

centre-section, new main landing gear and fixed-geometry engine air intakes. Introduced twin mainwheel bogies.

T10V-3: Static test air frame

T10V-4 '44': First flown late 1996. Reported at Leninets radar plant, Pushkino, early 1997; first with full avionics and weapons systems, except EW package. Exhibited at Paris Air Show, June 1997. Also became known as Su-32.

T10V-5 '45': First Su-27IB with full Leninets mission avionics fit. Sometimes described as first series-produced T10V. First flew 28 December 1994.

T10V-6 '46': First flown January 1998.

T10V-7: Reportedly nearing completion at Novosibirsk in August 2000.

CURRENT VERSIONS (general): **Su-27R:** Proposed version to replace Su-24MR and MiG-25RB in tactical reconnaissance roles. BKR (*Bortovoi Kompleks Razvedki:* onboard reconnaissance complex) suite expected to include nose-mounted Pika SLAR and ESM, electro-optical, laser and IRLS reconnaissance equipment.

Su-27IBP: Proposed tactical jammer to replace Yak-28PP and Su-24MP.

'Su-27IB Interceptor': Proposed ultra-long endurance combat air patrol variant. OKB's internal designation not known. Maybe confused reference to Su-33UB-based Su-30K-2 project.

Su-32: Initially export version; applied also to domestic version from 2000.

Su-32FN/MF: Preseries Su-32FN (T10V-4 '45') first flown 28 December 1994, exhibited at 1995 Paris Air Show; then stated to be in production to replace Su-24s of Russian Naval Aviation; programme reportedly suspended early 1997 before a fully equipped true prototype could fly. Su-32MF designation first appeared in 1999 to describe a 'multifunction', export version. Some French sources suggested that the type had received a new ASCC reporting name 'Fallback', by July 2000.

This version designed to attack hostile submarines and surface vessels by day and night in all weathers, although official drawing shows slightly different shape to nose compared with land attack version; was intended for parallel manufacture at Novosibirsk. Probably common to both types are Su-32MF's active artificial intelligence system to support pilot in critical situations; active gust alleviation smooth-flight system to damp turbulence in low-level flight at high speeds; liquid-crystal EFIS with seven CRTs; and Sorbtsya active ECM jamming pods on wingtips. Planned specialised equipment includes Leninets Sea Dragon avionics suite, with 'Sea Snake' coherent maritime search radar, a ventral sonobuoy pod containing 72 buoys of various types, MAD, IIR, IRTV system and laser range-finder.

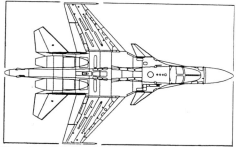

Plan view of Sukhoi Su-32FN showing revised nose contours

Armament of Su-32FN/MF stated to include one 30 mm GSh-301 gun in starboard wingroot extension. Twelve pylons: two on centreline, one under each engine duct, three under each wing and two at wingtips, for high-precision homing and guided ASMs and AAMs and KAB-500 laser-guided bombs, with ranges of 0 to 135 n miles (250 km; 155 miles). ASMs include up to six Kh-25M (AS-10 'Karen'), Kh-29 (AS-14 'Kedge') or Kh-31A/P (AS-17 'Krypton'), three Kh-59M (AS-18 'Kazoo'), two radar-homing Kh-35s (AS-20 'Kayak') or a single Kh-41 Moskit. Recent reports suggest that the Su-32MF will be able to carry two Kh-41s or three AFM-L Alpha ASMs. AAMs include six R-73 (AA-11 'Archer') or eight R-27 (AA-10 'Alamo') or RVV-AE (R-77; AA-12 'Adder'). Bombs range from 34 AB-100s to three KAB-1500s, rockets from 120 S-8s to six S-25s, including laser-guided S-25Ls. Torpedoes, munition or sonobuoy dispensers and mines can be carried.

Su-34: Initially domestic version; discontinued in 2000. *Details generally as for single-/tandem-seat Su-27, except those below.*

DESIGN FEATURES: One third heavier empty weight, with 50 per cent increase in MTOW, 30 per cent increase in internal capacity, 10 per cent increase in mid-section. Completely new and wider front fuselage built as titanium armoured tub, 17 mm ($^{11}/_{16}$ in) thick; armour adds 1,480 kg (3,262 lb); new EFIS cockpit containing two seats side by

Fourth flying Su-32 T10V-5, demonstrated at Farnborough in July 2000 *(Paul Jackson/Jane's)*

side; side-by-side arrangement avoids some duplication of controls and instruments, while promoting better crew co-operation. New avionics suite integrated by Ramenskoye Instrument-making Design Bureau; wing extensions taken forward as chines to blend with dielectric nose housing nav/attack and terrain-following/avoidance radar; deep fairing behind wide humped canopy; small foreplanes; louvres on engine air intake ducts reconfigured; new landing gear; broader chord and thicker tailfins, containing fuel; no ventral fins; and a longer, larger diameter tailcone. This has been raised and now extends as a spine above the rear fuselage to blend into the rear of the cockpit fairing. It houses at its tip a rearward-facing radar to detect aircraft approaching from the rear.

LANDING GEAR: Retractable tricycle type; strengthened twin nosewheel unit with KN-27 wheels, tyre size 680×260, farther forward than on Su-27 and retracting rearward; main units have small tandem KT-206 wheels with tyres size 950×400, carried on links fore and aft of oleo. New down-lock fairings. Twin cruciform brake-chutes repositioned in spine to rear of spine/fairing juncture.

POWER PLANT: Two Saturn/Lyulka AL-31F turbofans; each 74.5 kN (16,755 lb st) dry and 122.6 kN (27,557 lb st) with afterburning. Later, two AL-31FM or AL-35F turbofans, each 125.5 to 137.3 kN (28,220 to 30,865 lb st) with afterburning. Production version to use 175 kN (39,240 lb st) AL-41F with FADEC and TVC, according to some sources. Additional fuel in tailfins and increased capacity No. 1 tank raising total to 12,100 kg (26,676 lb) plus provision for three external tanks totalling 7,200 kg (15,873 lb). Retractable flight refuelling probe beneath port windscreen.

ACCOMMODATION: Two crew side by side on modified K-36DM zero/zero ejection seats with built-in massage function. Access to cockpit via built-in extending ladder to door in nosewheel bay; area protected with 17 mm (⅔ in) thick titanium armour; lavatory and galley with air-stove inside deep fuselage section aft of cockpit.

AVIONICS: *Radar:* Leninets multifunction phased-array radar with high resolution; rearward-facing radar in tailcone.

Instrumentation: Colour CRT, multifunction displays and helmet-mounted sight for pilot and navigator.

Mission: Built-in UOMZ EO IRST sighting system with TV and laser channels, optimised for air-to-ground use. Separate Geofizika padded thermal imaging system planned. New Argon main computer. Sorbtsya active ECM jamming pods under test on Su-27IB prototype 1995.

Self-defence: Internal ECM.

ARMAMENT: One 30 mm GSh-301 gun, as Su-27. Twelve pylons for high-precision self-homing and guided ASMs and KAB-500 laser-guided bombs with ranges of 0 to 135 n miles (250 km; 155 miles); R-73 (AA-11 'Archer') and RVV-AE (R-77; AA-12 'Adder') AAMs. Believed to be the principal platform for Vympel's rearward-firing R-73.

DIMENSIONS, EXTERNAL:
Wing span	14.70 m (48 ft 2¾ in)
Wing aspect ratio	3.5
Foreplane span	6.40 m (21 ft 0 in)
Length (without probe)	23.335 m (76 ft 6¾ in)
Height overall	6.50 m (21 ft 4 in)
Wheel track	4.40 m (14 ft 5¼ in)
Wheelbase	6.60 m (21 ft 7¾ in)

AREAS:
Wings, gross	62.00 m² (667.4 sq ft)

WEIGHTS AND LOADINGS (estimated):
Max external stores	8,000 kg (17,637 lb)
Max T-O weight: normal	39,000 kg (85,980 lb)
max	45,100 kg (99,428 lb)

PERFORMANCE:
Max speed at height	M1.8 (1,025 kt; 1,900 km/h; 1,180 mph)
Max speed at S/L	M1.14 (756 kt; 1,400 km/h; 870 mph)
Service ceiling	19,800 m (65,000 ft)
Combat radius (internal fuel):	
hi-hi-hi	601 n miles (1,113 km; 691 miles)
lo-lo-lo	324 n miles (600 km; 372 miles)
Range with max internal fuel	2,429 n miles (4,500 km; 2,796 miles)

UPDATED

SUKHOI Su-27K

See Sukhoi Su-33.

SUKHOI Su-30 (Su-27PU)

TYPE: Air superiority fighter.

PROGRAMME: Design started 1986; proof-of-concept T10PU first flew 6 June 1987; construction of two prototypes (T10PU-5 '05' and T10PU-6 '06', converted from Su-27UBs) began at Irkutsk in 1987, as Su-27PU; first flown 31 December 1989; prototype flew 7,252 n miles (13,440 km; 8,351 miles) in 15 hours 42 minutes non-stop during round trips Moscow-Novaya Zemlia-Moscow and Moscow-Komsomolsk-Moscow. First preseries Su-27PU flew at Irkutsk on 14 April 1992; intitial two aircraft (27596 and 27597 c/ns 1010101 and 1010102), without military equipment, delivered to 'Test Pilots' aerobatic team at Zhukovsky flight test centre, possibly as Su-27PUDs. Both offered for sale in 1999, although at that time 27597 was flight-testing MFI-68 152 × 203 mm (6 × 8 in) LCDs for Russian 'Flanker' upgrade programmes while 27596 had been used in support of the Su-30MK programme, renumbered '603'.

CURRENT VERSIONS: **Su-30** (Sukhoi T10PU, *Izdelie* 10-4PU): Unofficial OKB designation for basic two-seat long-range interceptor for Russian Air Forces (to which it is still the Su-27PU); deliveries under way by 1996 to 54 Interceptor Air Regiment at Savostleyka advanced training base, though production very limited, and unit relies heavily on Su-27 and Su-27UB. Some sources suggest that only four Su-30s are in frontline use (out of Red 50, 51, 52, 53 and 54 delivered to the PVO). Designed for mission of 10 hours or more with two in-flight refuellings; systems proved for extended duration sorties, including group missions with four Su-27s; only Su-30 would operate radar, enabling it to assign targets to Su-27s by radio datalink; can carry bombs and rockets but not guided air-to-surface weapons. Su-27UB training capability retained. Canards and thrust vectoring to be optional (see Su-30MKI). Export designation **Su-30K** (T10-4PK). Situation confused by tendency to describe all current Irkutsk-built two-seaters as Su-30s, even standard Su-27UB trainers and by emergence of upgrade using MiG-29SMT cockpit, using the same Su-30K designation. This Su-30K was reported to have been tested at Akhtubinsk by June 2000 but may be the same as the Su-30KM described below.

Su-30KI: Single-seat configuration for Indonesia, subsequently offered more widely as an upgrade. Described in main Su-27 entry.

Su-30I: One foreplane-equipped Su-27PU prototype only; served as Su-30MKI prototype (which see); believed offered as upgrade for Su-27 and as naval trainer, but neither taken up.

Su-30K-2: Variant of Su-33UB and described in that entry.

Su-30MKK ('Flanker-C'): Despite its MK-series designation (with the second K standing for Kitaya, or China) the Su-27MKK bears little relation to the canard-equipped MKI; offered to China as upgrade for existing Su-27s. True multirole Su-30M, with an N001VE radar (with expanded air-to-ground capabilities, including mapping); 'glass cockpit' with two 178 × 127 mm (7 × 5 in) MFI-9 colour LCD MFDs in front, plus single MFI-9 and 204 × 152 mm (8 × 6 in) MFI-10 in rear; A737 GPS; expanded EW capability; provision for various new TV- and EO-based targeting pods. Expanded air-to-surface weapon options as described under Su-30KM and provision for R-TI AAMs.

First Su-27PU (T10PU-5) was refurbished and rebuilt to serve as an Su-30MKK development aircraft, first flying in its new guise on 9 March 1999. First KnAAPO-built prototype ('501') flew 19 May 1999 (sometimes reported as 20 February 1999), with second ('502', in basic Chinese camouflage scheme but marked simply as Su-30MK, not MKK) following later in 1999. '501' and '502' representative of planned production configuration, with

Cockpit of upgraded Su-30 '302', showing new LCD at upper right of panel *2000/0075933*

tall, flat-topped Su-35-type tailfins, retractable in-flight refuelling probes and (according to some sources) Su-35-type radomes. Total of 40 on order for the PLA Air Force, the first batch of 10 being delivered from Komsomolsk to Nanjing/Yuxikou on 20 December 2000 although a first aircraft had been handed over in China during August 2000. Chinese production of the 'Flanker' will switch from the J-11 to the Su-30MKK after 80 aircraft. At least 60 Su-30MKKs are currently planned, with at least 40 on firm order.

Su-30KM: The 22nd production Su-30 (10302 '302') first flew in early March 1999 as the testbed (also referred to as Su-30K) for a Russian Air Forces upgrade of UB Su-27s and Su-30s. 'Project 302' adds terrain-mapping and moving target indication to the N001 radar, MFI-55 127 × 127 mm (5 × 5) MFDs, a new MVK computer added to existing SUV-27 weapons control system (permitting new types of AAM and ASM to be carried) and an A737 GPS; joint venture is undertaken by Sukhoi, Irkutsk (IAPO) and Russkaya Avionika and has high commonality with MiG-29SMT upgrade. Prototype tested at Air Forces Research Institute in mid/late 1999. Subsequent modifications to 10302 will involve larger, MFI-68 screens (two for pilot and three for WSO), phased-array radar and equivalent of MIL-STD-1553B databus. Added weapons include Kh-29T short-range, or two Kh-59ME long-range TV-guided ASMs; up to six KAB 500KR bombs; four Kh-31P ARMs; and six Kh-31PA anti-ship missiles. By late 2000, this variant was being referred to as the **Su-30KN** and maritime capabilities were being stressed, including compatibility with Kh-59, Yakhont and Alfa ASMs. 12 Su-30Ks sought by Vietnam believed to be to this standard with Kh-29, Kh-31 and Kh-59 ASMs, KAB-500 PGMs and R-77 AAMs.

Su-30M (Sukhoi T-10PM): Multirole version; described separately.

CUSTOMERS: Russian Air Forces. In addition, the Irkutsk Aircraft Production Association reports that the Su-30 (as well as Su-27UB) has been delivered to China, and will be delivered to Vietnam, though these aircraft are likely to be

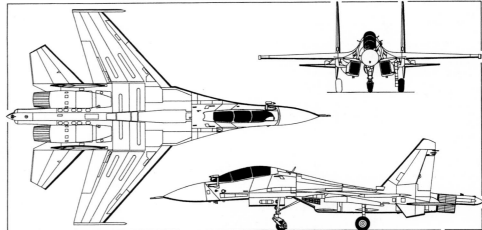

Sukhoi Su-30 two-seat long-range combat aircraft *(Paul Jackson/Jane's)*

Sukhoi Su-30MKK second prototype, wearing Chinese camouflage and Russian insignia *(Photo Link)*
2001/0100801

SUKHOI—AIRCRAFT: RUSSIAN FEDERATION

Russian Air Forces upgrade for the Su-30 is being evaluated in 'Project 302' *(Yefim Gordon)*

standard Su-27UBs. China has requirement for Su-30 but Russia reportedly unwilling to supply new-generation very long-range AAMs requested; China and Vietnam would almost certainly require multirole Su-30M, not Su-30. At least 21 production aircraft built by mid-1997.

COSTS: Chinese order for 40 Su-30MKKs valued at US$1.5 billion (2000).

DESIGN FEATURES: Development of Su-27/27UB, with latter's tandem seating and new avionics, and without Su-35's advanced radar, foreplanes (in basic version), advanced control system and new power plant. Designed for effective engagement of fighters at long distances from base, and to destroy bombers and intercept cruise missiles. Integral configuration similar to Su-27UB, with unstable aerodynamic characteristics. Automatic control system standard.

FLYING CONTROLS: As Su-27UB.
STRUCTURE: As Su-27UB.
LANDING GEAR: As Su-27UB.
POWER PLANT: As Su-27UB, but flight refuelling probe and buddy refuelling capability standard.
ACCOMMODATION: Two crew in tandem in identical cockpits, on K-36 zero/zero ejection seats, with rear seat raised.
SYSTEMS: As Su-27UB, except gaseous oxygen for 10 hours' flight.
AVIONICS: *Radar:* NIIP N001 Myech ('Slot Back') coherent pulse Doppler look-down/shoot-down radar, detection range up to 54 n miles (100 km; 62 miles), tracking range 35 n miles (65 km; 40 miles); ability to track 10 targets and engage two simultaneously offered, but probably not available on current in-service aircraft.
Flight: New navigation system based on GPS, Loran and Omega.
Instrumentation: Integrated fire-control system enables radar, IRST and laser range-finder to be slaved to pilot's helmet-mounted target designator and displayed on wide-angle HUD.
Mission: Provision for fitting foreign-made airborne and weapon systems at customer's request.
Self-defence: SPO-15LM Beryoza 360° radar warning system; chaff/flare dispensers.
ARMAMENT: One 30 mm GSh-301 gun, with 150 rounds; 12 hardpoints for up to six R-27R1E and R-27T1E (AA-10 'Alamo') radar homing and IR long-range AAMs, and six R-73E (AA-11 'Archer') IR close-range AAMs; alternative RVV-AE (R-77; AA-12 'Adder') AAMs; unguided bombs or rockets as Su-27; reconnaissance or EW pods.
DIMENSIONS, EXTERNAL: As Su-27UB
WEIGHTS AND LOADINGS:
Fuel weight: normal	5,090 kg (11,222 lb)
max	9,400 kg (20,723 lb)
Max combat load	8,000 kg (17,635 lb)
Max external stores	8,000 kg (17,635 lb)
Normal T-O weight	24,550 kg (54,123 lb)
Max T-O weight	33,000 kg (72,752 lb)

PERFORMANCE:
Max level speed: at height	M2.35 (1,350 kt; 2,500 km/h; 1,550 mph)
at S/L	M1.14 (756 kt; 1,400 km/h; 870 mph)
Service ceiling	17,500 m (57,420 ft)
T-O run	550 m (1,805 ft)
Landing run	700 m (2,300 ft)
Combat range: with max internal fuel	1,619 n miles (3,000 km; 1,865 miles)
with one flight refuelling	2,805 n miles (5,200 km; 3,230 miles)
g limit	+8.5

UPDATED

SUKHOI Su-30M

TYPE: Air superiority fighter.
PROGRAMME: Design started 1991; demonstration and development work by Su-27UB 1040806 '321' and 04003 '56'; 'Blue 56' first flew in Su-30M configuration on 14 April 1992, but this meant little, since it was not then canard-equipped and did not have full avionics package. Conversion of first true prototype ('06') began 1993; 1010101 '603' (formerly first prototype Su-27PU-5) first demonstrated at Berlin Air Show 1994; thrust vectoring and canards under development as options 1997; canards and AL-37PP engines first flown on '56' 1 July 1997; second prototype (06) flew 23 April 1998. In production for India; China showed interest during 1998 in acquiring 50 (in addition to Su-27s).
CURRENT VERSIONS: **Su-30M:** Basic version, *as described*.
Su-30MK: Irkutsk-built. As Su-30M, for export.
Su-30MK: Designation re-used for advanced two-seater combining two-seat Su-30 concept with avionics, canards and thrust vectoring of Su-37. Known internally (by KnAAPO) as Su-35UB (T10UBM) and Su-37UB. First and second true Su-30MKs, 01 (produced through the retrofit of canards to '56') and 06 (converted from a T10PU-6 in Sukhoi OKB's own workshop) first flown on 1 July 1997 and 23 April 1998, respectively. Equipped with foreplanes and AL-37FP thrust-vectoring engines; demonstrated to Indian officials at Zhukovsky, 15 June 1998 as Su-30MK-1 and Su-30MK-6, respectively. Su-30MK-1 had the new twin-wheel nosegear, which may become a feature of the production MKI, but was lost while displaying at Paris Air Show on 12 June 1999. Additionally, 01 was exhibited at Aero India, December 1998. AL-37FP power plant, as specified for India, extends length by 40 cm (15¾ in) and incurs weight penalty of 110 kg (243 lb), engine life remaining unchanged at 5,000 hours (1,000 hours TBO). Nozzle movement is 15° up or down. Flight control system helps pilot to set power and thrust vector for each engine, according to required manoeuvre. Indian radar choice will be between improved version of Phazotron N010 (Zhuk-27) and NIIP N011M multimode, dual-frequency radar with electronically scanned antenna. Expected to be known as Su-30MKR if produced (at Irkutsk) for Russian Air Forces.

Su-30MKI: Version for India in four configurations, sometimes referred to as Su-30MKI, MKII, MKIII and MKIV. First eight delivered March 1997 to basic Su-30PU (Su-30K or even Su-37UB) standard, with AL-31F engines; eight for 1998 delivery were expected to have French Sextant avionics including VEH 3000 HUDs, high-resolution colour LCD MFDs, a new flight data recorder, a Totem ring laser gyro dual INS with embedded GPS, Israeli EW equipment, a new electro-optic targeting system and rearward-facing radar in tailcone, but batch delayed by Israeli embargo in wake of Indian nuclear test; 12 originally to have been delivered 1999 were to add canards, as on Su-37; final 12 (originally scheduled for delivery in 2000) were to have AL-37FP engines with single-axis thrust-vectoring nozzles inclined outwards 32° from the centreline for improved yaw control, especially in single-engined case. AL-37PP claimed to offer 3-D thrust vectoring, with nozzle actuation via the fuel, and not the hydraulic, system.

Completion and delivery of balance of 32 was repeatedly delayed; decision taken that all these would be completed to final standard before delivery. (In interim, India contracted on 18 December 1998 for 10 standard Su-30Ks which had been cancelled by Indonesia; all had been delivered by October 1999.) First prototype full-specification Su-30MKI flew at Irkutsk on 26 November 2000; under renegotiated contract, IAF to receive six full-specification aircraft by 2002 and balance of 26 in batches beginning 18 to 24 months later; thereafter eight 1997-delivery aircraft will be raised to full standard. Licensed production of 140 Su-30MKIs by HAL was agreed in September 2000, followed by contract signature in Irkutsk on 28 December 2000. Sometimes referred to by KnAAPO as Su-35UB.

CUSTOMERS: Indian Air Force initial US$1.8 billion order for 40 signed 30 November 1996; deliveries from Irkutsk began to No. 24 'Hunting Hawks' Squadron at Pune in March 1997; 10 more ordered September 1998. Option taken up on licensed production of 140 more by HAL, India. On 29 August 1997, Indonesia signed for eight 'single-seat Su-30s' and four two-seat, but this was cancelled on 9 January 1998.

COSTS: Indian aircraft quoted as US$20 million each (flyaway, 1998), with US$8 million extra per aircraft to integrate Indian-specific systems and avionics.

DESIGN FEATURES: Improvement on combat capabilities of Su-30 by compatibility with high-precision guided air-to-surface weapons with standoff launch range up to 65 n miles (120 km; 75 miles), in addition to Su-30's ability to engage two airborne targets simultaneously.

In other details, identical with Su-30 except as follows:

AVIONICS: In addition to standard Su-30 systems, Su-30M has more accurate navigation system, a TV command guidance system, a guidance system for anti-radiation missiles, a larger monochrome TV display system in rear cockpit for ASM guidance, and ability to carry one or two pods, typically for laser designation or ARM guidance in association with Pastel RWR and APK-9 datalink. Western avionics, guidance pods and weapons can be fitted

Sukhoi Su-30PU of India's No. 24 'Hunting Hawks' Squadron

Su-30MKI prototype 01 was the first Indian aircraft with canards *(Yefim Gordon)*

optionally. Sextant Avionique package for Indian aircraft includes VEH3000 or Elop HUD, Totem or Sigma 9SN/MF INS/GPS and liquid-crystal multifunction displays (six 127 × 127 mm; 5 × 5 in MFD 55 and one 152 × 152 mm; 6 × 6 in MFD 66 per aircraft).

ARMAMENT: One 30 mm GSh-301 gun, with 150 rounds; 12 external stations for more than 8,000 kg (17,635 lb) of stores, including AB-500, KAB-500KR and KAB-1500KR bombs; B-8M-1 (20 × 80 mm) and B-13L (5 × 130 mm) rocket packs; 250 mm S-25 rockets; up to six R-27ER (AA-10C 'Alamo-C'), R-27ET (AA-10D 'Alamo-D') or RVV-AE (R-77; AA-12 'Adder') medium-range AAMs; or two R-27ETs and six R-73E (AA-11 'Archer') IR homing close-range AAMs; and a variety of air-to-surface weapons such as four ARMs, six guided bombs or short-range missiles with TV homing, six laser homing short-range missiles, or two long-range missiles with TV command guidance; these include Kh-29L/T (AS-14 'Kedge'), Kh-31A/P (AS-17 'Krypton') and D-9M (probably Kh-59M; AS-18 'Kazoo') with APK-9 pod or single Raduga 3M80E supersonic anti-ship missile.

DIMENSIONS, EXTERNAL: As Su-27 except:
Height overall 6.355 m (20 ft 10¼ in)

WEIGHTS AND LOADINGS:
Weight empty 17,700 kg (39,022 lb)
Normal T-O weight 26,090 kg (57,518 lb)
Max T-O weight 38,000 kg (83,775 lb)

PERFORMANCE:
Max level speed:
 at height M2.35 (1,161 kt; 2,150 km/h; 1,336 mph)
 at S/L M1.14 (729 kt; 1,350 km/h; 839 mph)
Max rate of climb at S/L 13,800 m (45,275 ft)/min
Service ceiling 17,500 m (57,420 ft)
T-O run 550 m (1,805 ft)
Landing run 670 m (2,200 ft)
Combat range:
 internal fuel 1,620 n miles (3,000 km; 1,865 miles)
 with one in-flight refuelling
  2,805 n miles (5,200 km; 3,230 miles)

UPDATED

SUKHOI Su-32FN and MF (Su-27IB)
Described under Su-27IB entry.

SUKHOI Su-33 (Su-27K)
TYPE: Multirole fighter.
PROGRAMME: Development began 1976; based on production Su-27, but embodying folding wings, other features for shipboard operation, and movable foreplanes (last-mentioned first flown on T10-24 in May 1985; the aircraft also used as CCV/aerodynamic research vehicle and for ski-jump take-off trials. Navalised development Su-27 (T10-25), with arrester hook, flew 1984; extensively used for arrested landing trials and ski-ramp take-offs, with T10-3 and T10-24 (first ramp take-off was 22 August 1992 by T10-3; T10-25 followed on 25 September 1985, having made first arrested landing on 1 September 1985). 'Aircraft 02-01' (see Su-27 entry) was fitted with arrester hook and 'Resistor' ACLS and extensively used in support of Su-27K programme.

Full-scale development of definitive Su-27K began in 1984 as T10K, under Konstantin Marbashev. Full-scale (non-flying) T10-20KTM mockup with crude (unpowered) wing and tailplane folding produced by conversion of grounded fifth production T10S. Prototype (T10K-1 '37') flew 17 August 1987; lacked folding wing, double-slotted flaps and strengthened landing gear; retrofitted with folding wing and reflown on 25 August 1988, but written-off on 28 September 1988; second (T10K-2 '39') made initial flight on 22 December 1987; first conventional (non-V/STOL) landing by Soviet aircraft on ship, the *Admiral of the Fleet Kuznetsov* (then *Tbilisi*), 1 November 1989 (26 minutes before MiG-29K); first landing by service pilot followed on 26 September 1991.

Production at Komsomolsk began with static test airframe, and followed by seven pre-series aircraft: T10K-3 (first flight 17 February 1990), T10K-4 '59', T10K-5 '69', T10K-6 '79', T10K-7, T10K-8 (lost 11 July 1991) and T10K-9 '109'. Between 16 and 20, probably 18, delivered of 24-aircraft production batch (plus trials aircraft) and shore-based with 279th KIAP (formerly 100th KIAP) at Severomorsk on Kola Peninsula beginning with initial four in April 1993; five months of intensive flying from *Kuznetsov*, in Barents Sea, in second half 1994; carrier deployed temporarily to Adriatic late 1995, with at least 10 Su-27Ks (including eight of 'red' squadron and T10K-9 operational trials aircraft 'blue 109'). Complement of 24 aircraft in 279th Regiment believed to include at least four preproduction machines, with numbers made up by standard land-based 'Flankers'. Naval variant known as both Su-27K (preferred by Russian Air Forces) and Su-33, until latter was formally adopted on 31 August 1998, coincident with presidential decree of acceptance into service. Reported radar and ESM/ECM problems have reduced aircraft's usefulness while accidents have accounted for several Su-27Ks. Not allocated 'Flanker-D' reporting name, despite some reports to this effect.

CURRENT VERSIONS: **Su-33**: As described.
 Su-33 Upgrade: A 1999 Sukhoi/KnAAPO/Rosvooruzheni brochure described an Su-33 Upgrade configuration, though there is no real evidence that any such programme has been funded. The cockpit is to gain two large colour LCD multifunction displays, with a modernised flight control/navigation system and enhanced air-to-ground capabilities. Upgraded Su-33 is to be compatible with the R-77 (AA-12 'Adder') AAM and with a range of air-to-ground PGMs, including TV-guided Kh-29, Kh-59N, KAB-500Kr and KAB-1500Kr and laser-guided Kh-29L, KAB-500L, and KAB-1500L. Use of the Kh-31P will give an effective SEAD capability.
 Su-28: Early, stillborn AEW variant, with spine-mounted rotodome.
 Su-27KM: Proposed definitive version, with Su-27M FCS, Zhuk radar, full air-to-ground capability and vectoring engine nozzles; probably abandoned.
 Su-27KRT (*Korabelnyi Razvedchik-Tseleukazatel*: shipboard reconnaissance and target acquisition platform): Reportedly flying in prototype form by 1999. Equipped with Leninets MMW radar, encrypted communications and datalink equipment and radio/electronic reconnaissance suite. Production Su-27KRT may be based on Su-27KUB/Su-33UB airframe; configuration of prototype unknown.
 Su-27KPP (*korabelnyi postanovshchik pomekh*: shipboard EW/command post): Reportedly under development.
 Su-27KUB: Described under Su-33UB heading.

DESIGN FEATURES: Airframe generally similar to Su-27, but with folding wings and power-folding tailplane; fins allegedly 'slightly shortened' due to hangar deck constraints; added foreplanes; arrester hook and other features for carrierborne operations; strengthened landing gear with twin nosewheels; long tailcone of land-based versions shortened to prevent tailscrapes during take-off and landing on ship; brake parachute replaced by SPO-15 (L-006) Beryoza RHAWS relocated from the intake sides; IRST with wider angle of view.

FLYING CONTROLS: Basically as Su-27, but with sweptback (52°) foreplanes that operate at all speeds and only collectively, not differentially (7° up/70° down). Standard Su-27 fly-by-wire, automatic control system and central control column. Slow-speed devices comprise drooping ailerons plus two-section Fowler flaps mounted on partly drooping wing trailing-edge.

STRUCTURE: Generally as Su-27, but hydraulically folding outer wings (through 135°) and upward folding horizontal tail surfaces. Riveted and welded structure of aluminium and titanium alloys and steel.

LANDING GEAR: Generally as Su-27, but strengthened and with increased stroke; mainwheel tyres size 1,030×350, twin nosewheel tyres size 620×180; nosewheels steerable through ±60°. Shackles on main oleos for tiedown and for connection to Mustang hangar deck conveyor belt system. Arrester hook under tailcone.

POWER PLANT: Two Saturn AL-31F3 turbofans, each rated at 75.2 kN (16,909 lb st) dry and 125.5 kN (28,220 lb st) with afterburning. Retractable flight refuelling probe beneath windscreen on port side; maximum receipt rate 2,300 litres (608 US gallons; 506 Imp gallons)/min. Provision for UPAZ-1A centreline buddy refuelling pack.

Su-33 demonstrating wing and tailplane folding
(Roman Kondrat'yev via Yefim Gordon)

Sukhoi Su-33 ship-based fighter *(Yefim Gordon)*

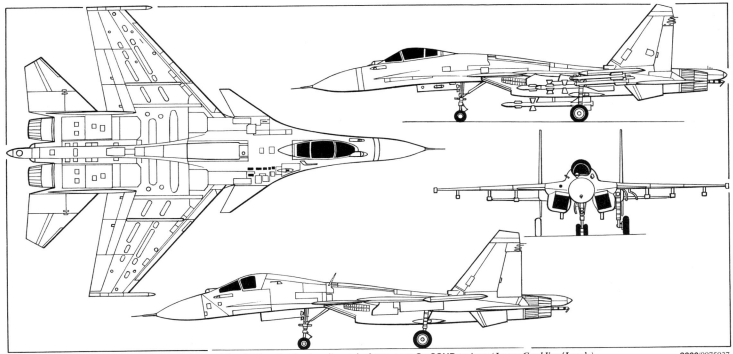

Sukhoi Su-33 single-seat carrierborne fighter, with additional side view (lower) of two-seat Su-33UB trainer *(James Goulding/Jane's)*

ACCOMMODATION: As Su-27.
SYSTEMS: As Su-27.
AVIONICS: *Flight:* Nav systems optimised for use over sea; navalised TKS coded datalink. Resistor K42 ACLS allowing ICAO Cat. II autoland capability. Active radar beacon on nose oleo to increase gear-down RCS.
Self-defence: As Su-27, but with reduced number of chaff/flare dispensers (24 per side).
EQUIPMENT: Lights for optical/laser-based Luna-3 visual approach system.
ARMAMENT: Basically as Su-27; 12 pylons. Normal AAM options expanded to include R27EM for use against sea-skimming targets. Rudimentary air-to-ground capability is equivalent to Su-27SK; is reportedly able to carry Kh-31 (AS-17 'Krypton') ASMs under wing. Exhibited with inert 4,500 kg (9,920 lb) Kh-41 (3M80 Moskit: mosquito) anti-ship missile on centreline; but this considered as impracticable operational load.

DIMENSIONS, EXTERNAL:
Wing span	14.70 m (48 ft 2¾ in)
Wing aspect ratio	3.2
Length: overall, incl noseprobe	21.185 m (69 ft 6 in)
folded	19.20 m (63 ft 0 in)
Width, wings folded	7.40 m (24 ft 3½ in)
Height overall	5.72 m (18 ft 9¼ in)
Tailplane span	9.90 m (32 ft 6 in)
Wheel track	4.44 m (14 ft 6¾ in)
Wheelbase	5.87 m (19 ft 3 in)

AREAS:
Wings, gross	67.84 m² (730.2 sq ft)
Foreplanes (total)	3.00 m² (32.29 sq ft)
Ailerons (total)	2.40 m² (25.83 sq ft)
Trailing-edge flaps (total)	6.60 m² (71.04 sq ft)
Leading-edge flaps (total)	5.40 m² (58.13 sq ft)
Fins (total)	11.60 m² (124.87 sq ft)
Rudders (total)	3.50 m² (37.67 sq ft)
Horizontal tail surfaces (total)	12.30 m² (132.40 sq ft)

WEIGHTS AND LOADINGS:
Max military load	7,045 kg (15,532 lb)
Normal T-O weight	25,000 kg (55,115 lb)
Max carrier T-O weight	30,000 kg (66,135 lb)
Max T-O weight	33,000 kg (72,752 lb)
Max landing weight	24,500 kg (54,013 lb)
Max wing loading	486.4 kg/m² (99.63 lb/sq ft)
Max power loading	131 kg/kN (1.29 lb/lb st)

PERFORMANCE:
Never-exceed speed (V_{NE}) at 11,000 m (36,000 ft)	M2.165 (1,240 kt; 2,300 km/h; 1,430 mph)
Max speed at S/L	M1.06 (702 kt; 1,300 km/h; 807 mph)
Min flying speed	130 kt (240 km/h; 150 mph)
Service ceiling	17,000 m (55,780 ft)
T-O run on carrier: normal	195 m (640 ft)
with 14° ramp	120 m (395 ft)
Range with max internal fuel:	
at S/L	540 n miles (1,000 km; 621 miles)
at altitude	1,620 n miles (3,000 km; 1,865 miles)
g limits	+8/−2

UPDATED

SUKHOI Su-33UB (Su-27KUB)

TYPE: Multirole fighter.
PROGRAMME: Original Su-27IB prototype initially described by Sukhoi design bureau as 'Su-27KU' (*korabelnyi uchebno* or shipborne trainer) and described as carrier trainer, despite lack of arrester hook, folding wings, strengthened landing gear and carrier landing system. Aircraft did make dummy carrier approaches, and may have flown from dummy deck at Saki experimental airbase to evaluate suitability of new cockpit for carrier operation. Naval strike Su-32MF (which see) may once have been intended for carrier operation.

Need for dedicated trainer for Su-33 (which see) became increasingly clear, and development of T10KM-2-based **Su-27KUB** (*korabelnyi uchebno boevoi:* as T10KU shipborne fighter trainer) began in late 1990s. Formally acknowledged 21 October 1998. Probably being developed as a private venture, with no firm commitment from Russian Navy. Likely to become **Su-33UB**, following redesignation of carrierborne single-seat Su-27K as Su-33.

Three prototypes, with noses built by KnAAPO, as T10KU-1, -2, and -3, incorporating lessons from the *Kuznetsov*'s 1996 Mediterranean cruise. Construction of T10KU-1 began in 1998 mating a new nose and new wings and tailplanes to an existing T10K prototype (T10K-4). Powered by AL-31F engines, the aircraft first flew on 29 April 1999 and made first arrested landing on dummy deck at NIUTK ('Nitka') test centure, Saki, 3 September 1999; first take-off from deck ramp followed on 6 September; first landing and take-off from carrier *Kuznetsov* on 6 October 1999. T10KU-2 and -3 reported to be under construction in 1999, perhaps using new-build airframes, but had not been seen by January 2001. T10KU-1 test flown by Indian pilots, September 1999, but Su-27 judged too large for planned carriers.

Any production is likely to be by KnAAPO; baseline variant due to be extrapolated to produce trainer, reconnaissance and AEW versions, the last-mentioned with a phased-array mounted on the spine, between the composites antenna tailfins. Increased thrust, thrust-vectoring AL-31FP, AL-31FM or AL-41F engines mooted for production version.

CURRENT VERSIONS: **Su-30K-2:** Two-seat interceptor version based on Su-33 fuselage under construction at Komsomolsk, late 1999; due to fly late 2000. Temporary designation.
DESIGN FEATURES: Has navalised features of Su-33, including folding wing (with fold further outboard and with a larger fold angle). Some sources suggest that the Su-33UB's tailplanes do not fold, since they reach only as far as the new outboard wing fold. Double slotted flaps, unslotted, 'adaptive' leading-edge, arrester hook, datalink and carrier landing system. However, compared with Su-27IB, forward fuselage is slightly narrower, with seats closer together and has much less pronounced dorsal hump. New 'glass cockpit' with five colour LCD displays (one 53 cm; 21 in diagonally; rest 38 cm; 15 in) with provision for central or sidestick and with helmet-mounted sighting system. Aircraft has OBOGS and OBIGGS and so does not need oxygen or nitrogen bottles. N014 solid-state, phased-array radar, with enhanced air-to-ground and over-water capabilities, planned eventually, but prototypes may initially use NIIR N610-27 (Zhuk 27); circular-section radomes replace flattened 'platypus' nose associated with Su-27IB. Some reports suggest that the Su-33UB's new wing is 12 per cent larger in span (16 m; 52½ ft) and area (70 m²; 750 sq ft) as are the canards and tailplanes. Production Su-27KUB will feature a higher set, square-section, lengthened tail-sting, possibly mounting rear warning radar; tailcone folds (upwards) to reduce stowed length; prototypes have the standard Su-33 tailcone.

UPDATED

SUKHOI Su-34
See Sukhoi Su-27IB entry.

SUKHOI Su-35 (Su-27M)

TYPE: Multirole fighter.
PROGRAMME: Development of Su-27M authorised on 29 December 1993 by Council of Ministers. Experimental version of Su-27 with foreplanes (T10-24) flew May 1985; improved FBW system and refuelling probe tested by T10U-2. Five prototypes produced by conversion of production Su-27s, retaining single nosewheel and standard tailfins: T10M-1 '701' (ex-1602), then lacking radar and weapon control system (successively T10S-70, T10M, Su-27M, Su-35) flew 28 June 1988; T10M-2 flown 18 January 1989; T10M-5, T10M-6 '706' and T10M-7 '707'; used mainly by NII VVS at Akhtubinsk, flown by service pilots.

Production at KnAAPO, Komsomolsk, beginning with static test airframe T10M-4; first flight (T10M-3 '703') 1 April 1992; latter exhibited at 1992 Farnborough Air Show. Of further six ordered (T10M-8 to 13; '708' to '713') '711' and '712' transferred to Su-37 programme

Sukhoi Su-27KUB prototype. New nose results in an obvious change in spine contours *(Paul Jackson/Jane's)*

RUSSIAN FEDERATION: AIRCRAFT—SUKHOI

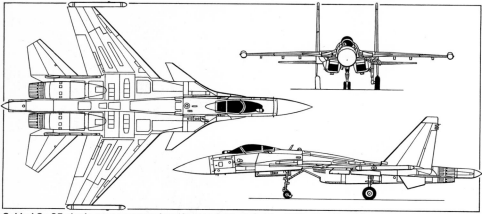

Sukhoi Su-35 single-seat counter-air and ground attack fighter *(Mike Keep/Jane's)*

Late prototype Su-35 No. 709 *(Paul Jackson/Jane's)* 2000/0075885

and '713' cancelled. Large-scale series production originally planned 1996-2005 as interim fighter pending availability of Mikoyan MFI.

Three aircraft (Blue 86, 87 and 88), described by KnAAPO sources as 'production Su-35s', but probably ex-'707', '708' and '709', were delivered to Akhtubinsk from Komsomolsk in 1996 or 1997. It is not known whether they were fitted with thrust-vectoring engines (which could make them more properly Su-37s).

CURRENT VERSIONS: **Sukhoi Su-35**: Baseline single-seater; *as described.*

Sukhoi Su-35UB: Two-seat derivative of the Su-35 with same FCS, canard foreplanes, tall square-topped tailfins, 12-pylon wing, Zhuk main AI radar and N012 rearward-facing, tailcone-mounted radar. Developed as a demonstrator and trainer for the Su-35, the construction of the prototype (Blue 801) may have been prompted by the needs of Sukhoi's campaign to sell the Su-35 to South Korea. The aircraft first flew on 7 August 2000 and was reported to be undergoing trials at Akhtubinsk in October 2000. It is implied that this aircraft has 245 kN (55,115 lb st) AL-31FP (AL-31F) thrust-vectoring engines (which would make it more properly an Su-37UB).

Su35UB has 38,800 kg (83,775 lb) MTOW; 8,000 kg (17,637 lb) combat load; 1,090 kt (2,020 km/h; 1,255 mph) max speed; 1,619 n mile (3,000 km; 1,864 mile) unrefuelled range; wing span 14.70 m (48 ft 2¾ in); length overall 21.94 m (71 ft 11½ in); and height overall 6.355 m (20 ft 10¼ in).

CUSTOMERS: Once scheduled for entry into Russian Air Forces service as Su-27M from 1995 onwards, for effective operation until 2015-2020; programme in suspension and may be superseded by Su-37.

DESIGN FEATURES: Advanced multirole development of Su-27 to counter latest versions of USAF F-15 Eagle and F-16 Fighting Falcon, with better dogfighting characteristics, higher AoA limits, lighter weight and new BVR armament; proposed to include 216 n mile (400 km; 248 mile) range AAM-L (one AAM-L contender was Novator KS- 172). Also planned to have greater autonomy from GCI control. Airframe, power plant, avionics and armament all upgraded; quadruplex digital fly-by-wire controls under development by Avionika, though prototypes retain analogue system; longitudinal static instability; 'tandem triplane' layout, with foreplanes; double-slotted flaperons; taller, square-tip twin tailfins with integral fuel tanks; reprofiled front fuselage for larger-diameter radar antenna; enlarged tailcone for rearward-facing radar; twin-wheel nose landing gear; axisymmetric thrust-vectoring nozzles under development for use on production aircraft (see Su-37).'

STRUCTURE: Higher proportion of carbon fibre and aluminium-lithium alloy in fuselage; composites used for components such as leading-edge flaps, nosewheel door and radomes.

POWER PLANT: Production Su-27M planned to use two Saturn/Lyulka AL-35F (AL-37FM) turbofans; each 125.5 kN (28,218 lb st) with afterburning; prototypes retain standard AL-31F. Increased internal tankage (approximately 1,500 kg; 3,307 lb) through use of welded aluminium-lithium tanks and new tanks in tailfins. Retractable flight refuelling probe on port side of nose.

ACCOMMODATION: Pilot only, on Zvezda K-36MD zero/zero ejection seat, this now angled back 30°.

AVIONICS: *Radar*: Originally planned to incorporate NIIP N011 Zhuk-27 multimode low-altitude terrain-following/avoidance radar, search range up to 54 n miles (100 km; 62 miles) against advancing target, 30 n miles (55 km; 34 miles) against retreating target; able to track 15 targets and engage four to six simultaneously. Phazotron Zhuk-Ph phased-array radar under development as alternatives or for retrofit; search range for fighter-size targets 75 n miles (140 km; 87 miles) with simultaneous tracking of 24 air targets and ripple-fire engagement of six to eight; N012 rearward-facing radar, range approximately 2 n miles (4 km; 2.5 miles), may enable firing of rearward-facing IR homing AAMs.

Flight: Fully automatic flight modes and armament control against ground, maritime and air targets, including automatic low-altitude flight and automatic target designation. RPKB nav system includes laser-gyro INS and Glonass GPS.

Instrumentation: EFIS, with three colour CRTs; HUD.

Mission: New-type IRST moved to starboard; small external TV pod; all combat flight phases computerised. Shown at Farnborough with GEC-Marconi TIALD (thermal imaging airborne laser designator) night/adverse visibility pod fitted for possible future use.

Self-defence: Entirely new integrated EW suite with ECM, including active jammer and wingtip Sorbtsa-S G- or J-band ECM/ESM pods; Pastel RWR; Mak IR-based MAWS.

ARMAMENT: One 30 mm GSh-30 gun in starboard wingroot extension, with 150 rounds. Mountings for up to 14 stores pylons, including R-27 (AA-10 'Alamo-A/B/C/D'), R-40 (AA-6 'Acrid'), R-60 (AA-8 'Aphid'), R-73E (AA-11 'Archer') and RVV-AE (R-77; AA-12 'Adder') AAMs, Kh-25ML (AS-10 'Karen'), Kh-25MP (AS-12 'Kegler'), Kh-29T (AS-14 'Kedge'), Kh-31P (AS-17 'Krypton') and Kh-59 (AS-18 'Kazoo') ASMs, S-25LD laser-guided rockets, S-25IRS IR-guided rockets, GBU-500L and GBU-1500L laser-guided bombs, GBU-500T and GBU-1500T TV-guided bombs, KMGU cluster weapons, KAB-500 bombs and rocket packs. Maximum weapon load 8,200 kg (18,077 lb).

DIMENSIONS, EXTERNAL:
Wing span over ECM pods	15.16 m (49 ft 8¾ in)
Wing aspect ratio	3.5
Length overall	22.185 m (72 ft 9½ in)
Height overall	6.36 m (20 ft 10¼ in)

AREAS:
Wings, gross	62.00 m² (667.4 sq ft)

WEIGHTS AND LOADINGS:
Weight empty	17,000 kg (37,479 lb)
Max fuel	13,400 kg (29,542 lb)
Normal T-O weight	25,670 kg (56,592 lb)
Max T-O weight	34,000 kg (74,957 lb)
Max wing loading	548.4 kg/m² (112.32 lb/sq ft)
Max power loading	124 kg/kN (1.21 lb/lb st)

PERFORMANCE:
Max level speed:
at height	M2.35 (1,350 kt; 2,500 km/h; 1,555 mph)
at S/L	M1.14 (756 kt; 1,400 km/h; 870 mph)
Service ceiling	17,800 m (58,400 ft)
Balanced runway length	1,200 m (3,940 ft)

Range: with max internal fuel
more than 2,160 n miles (4,000 km; 2,485 miles)
with flight refuelling
more than 3,510 n miles (6,500 km; 4,040 miles)
g limit +10

UPDATED

SUKHOI Su-37 (Su-27M)

TYPE: Multirole fighter.

PROGRAMME: Technology demonstrator for proposed production fighter resulting from vectored-thrust programme of the Su-27UB-PS, Su-27UBL, Su-27LMK No. 2405 and the last prototype and pre-series Su-35 (711) which made its first flight, with nozzles fixed, on 2 April 1996 (or 12 April, according to some sources). Designated Su-37 by Sukhoi bureau. By September 1996, when demonstrated at Farnborough Air Show (Western debut), 711 had made 50 flights with hydraulically actuated nozzles able to move ±15° in pitching plane at rate of 30°/s under control of aircraft's flight control system. Probably

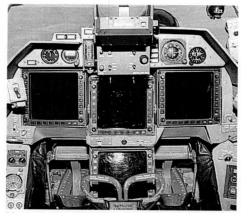

Mockup of instrument panel for advanced versions of the 'Flanker' family 2000/0070019

Prototype Su-35UB two-seat version 2001/0105836

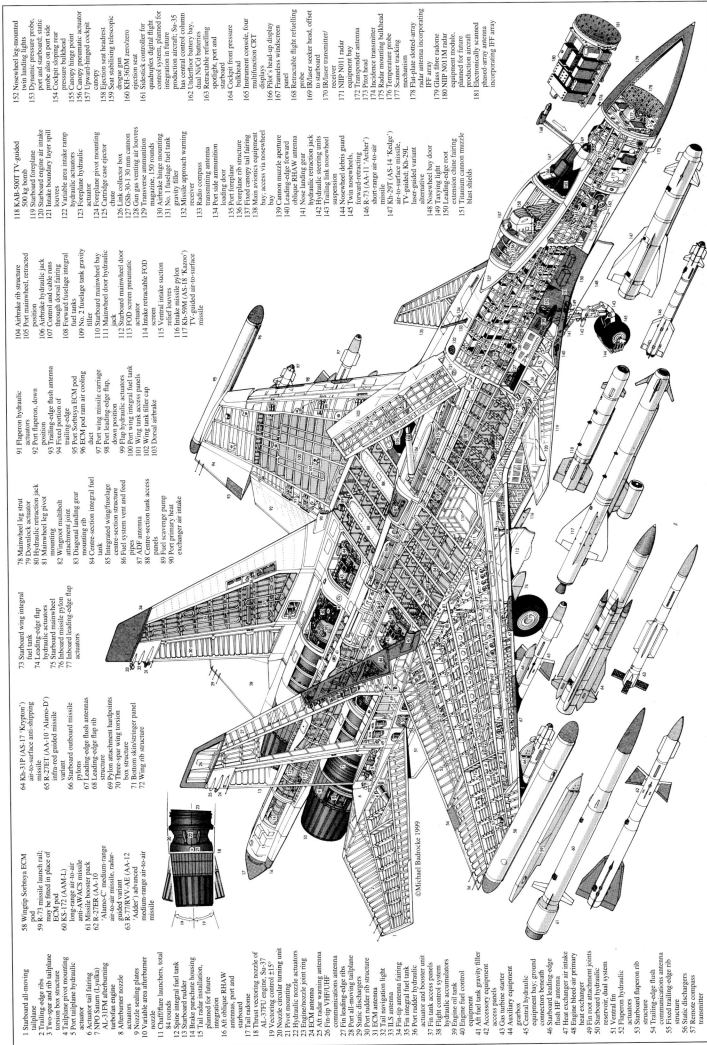

Sukhoi Su-35/Su-37, cutaway drawing key

Instrument panel of the Sukhoi Su-37 *(Yefim Gordon)*

Sukhoi Su-37 thrust-vectoring fighter *(Yefim Gordon)*

toed-out 32° from the centreline, like Su-30 MKI, generating powerful yawing moment when actuated differentially. An emergency pneumatic system returns the nozzles to level flight setting in the event of system failure. Any production Su-37 likely to feature uprated AL-31FU engines (142.2 kN; 31,967 lb st). In 1999, Sukhoi was using designation Su-37MR for proposed production version, apparently reflecting a major avionics upgrade then under way. Three-dimensional vectoring nozzles may not be developed in time for Su-37, although 3-D vectoring AL-37PP engine reportedly at an advanced stage of development by early 1998.

Second Su-37 (T10M-12 '712') reportedly first flew in mid-1998, initially powered by AL-31F; installation of thrust-vectoring AL-37FPs was expected in late 1998; status of this aircraft, if extant, remains unclear.

Production may be at Komsomolsk-on-Amur, although single-seater effectively subordinate to two-seat Su-30MKI – known internally as '**Su-35UB**' despite this designation also being used by a two-seat derivative of the Su-35, described separately.

CURRENT VERSIONS: **Su-37**: Baseline single-seater; *as described*.

Su-37KK: Reported export version for China with Ramenskoye Instrument Design Bureau avionics suite including new computers, Japanese-developed LCD displays, a new digital SMS and integrated W/GPS, and a new UOMZ IRSTS. Ten avionics shipsets had been delivered to KnAPPO by August 2000.

DESIGN FEATURES: Generally as Su-35. Emphasis on super-agility, typified by *kulbit* manoeuvre in which Su-37 pitches up rapidly beyond vertical, through a tight 360° somersault within its own length, and pulls out to resume level flight with minimal height loss.

FLYING CONTROLS: Thrust vectoring is actuated as a fully integrated element of the standard Su-35-type fly-by-wire control system, with manual override. New cockpit controls comprise an articulated sidestick controller and fixed sidebar throttle with thumbswitch actuation.

POWER PLANT: Two Saturn/Lyulka AL-37FU turbofans with thrust-vectoring nozzles; each 83.4 kN (18,739 lb st) dry and 142.2 kN (31,970 lb st) with afterburning.

AVIONICS: *Radar:* NIIP N011M electronically scanned phased-array radar in nose. Ryazan rearward-facing radar in tailcone.
 Flight: Laser-gyro INS and Glonass GPS.
 Instrumentation: Sextant Avionique EFIS, with four liquid-crystal colour MFDs.
ARMAMENT: As for Su-35, plus R-37 and KS-172 (AAM-L) AAMs and supersonic anti-radiation Kh-15P (AS-16 'Kickback') and Kh-65S ASMs.
DIMENSIONS, EXTERNAL: As Su-35
WEIGHTS AND LOADINGS:
 Max T-O weight 25,670-34,000 kg (56,592-74,957 lb)
 Max wing loading 548.4 kg/m² (112.32 lb/sq ft)
 Max power loading 119 kg/kN (1.17 lb/lb st)
PERFORMANCE:
 Speeds as Su-35
 Max rate of climb at S/L 13,800 m (45,275 ft)/min
 Service ceiling 18,800 m (61,680 ft)
 Max range:
  internal fuel, 432 kt (800 km/h; 497 mph) at low
   altitude 750 n miles (1,390 km; 863 miles)
  internal fuel, 513 kt (950 km/h; 590 mph) at high
   altitude 1,781 n miles (3,300 km; 2,050 miles)
  with one in-flight refuelling
   3,509 n miles (6,500 km; 4,039 miles)
 g limit +9

UPDATED

SUKHOI Su-39
See Sukhoi Su-25TM.

SUKHOI S-37 BERKUT
English name: Golden Eagle
TYPE: Multirole fighter.
PROGRAMME: Dedicated 'company-funded' research programme to explore post-stall manoeuvrability and 'supermanoeuvrability' for next-generation fighter; despite original purpose, Sukhoi OKB soon began promoting S-37 as alternative 'heavy' fighter to Mikoyan's MFI and, for this reason, it was designed as large, heavy twin-engined aircraft with some 'stealth' features and with room and/or provision for incorporation of operational and mission equipment and systems. Design bureau and defence ministry sources have recently denied that the S-37 is a prototype of any combat aircraft; daily *Red Star* stated in 1997 that S-37 has zero priority for Russian air force; however, Russian Air Forces' C-in-C quoted in 1998 as foreseeing "a future for the S-37". Funds for programme derived from foreign sales of Su-27 'Flankers'.

First reported as **Su-32** in 1996, then as **S-32**; possibly derived from – or follow-on to – 1982 forward-swept wing (FSW) demonstrator reportedly seen at Saki test airfield and supposedly codenamed SYB-A in West; existence of such an aircraft has been denied by Russian sources, although the Tsybin OKB built several forward-swept gliders and rocket-powered test aircraft in the late 1940s. Reports suggest that the SYB-A was a Sukhoi product, perhaps a converted Su-9. Sukhoi also used a forward-swept wing on the proposed S-86 propeller-driven corporate transport.

Design reportedly began in 1987. S-32 and S-37 originally had no tailplane, instead having shallow, broad, flat surfaces extending back from the wing trailing-edge; these later evolved into very narrow-span tailplanes. Reshuffle of designations continued with use of the Berkut as a 'home' for the S-37 designation already employed by a single-engined light fighter with canard foreplanes and a cranked, cropped-delta wing, probably designed to meet the air forces' LFI requirement. S-37 designation became available when this design project was cancelled.

Prototype (marked '01') first flew at Zhukovsky test airfield, near Moscow, on 25 September 1997; following eighth sortie, on 27 November 1997, was temporarily grounded for increase of horizontal tail span and area, improved avionics integration and better tailoring of engines. Total of at least 31 flights by January 1999 and 50 to August 1999, when reported to be at half-way point of test programme; first phase completed 23 February 2000, after which the aircraft began a second supersonic phase; this included a brief deployment to the Russian Federation MoD Flight Test Centre at Akhtubinsk, where the aircraft exceeded M1.0 for the first time on its 88th flight, in August 2000. A third phase (with an expanded envelope) was due to begin at the end of November 2000, with this planned to include the type's 100th flight. However, in early 2000, testing was expected to continue for further "four to five years". 30° AoA achieved by mid-1999. Public debut at Aviation Day, Tushino, 15 August 1999.

CUSTOMERS: After company testing, S-37 may be passed to LII trials centre at Zhukovsky, where some Russian air force participation is expected. Series production is not planned.

DESIGN FEATURES: Incorporates features and technology applicable to a fifth-generation fighter. Described as 'integrated triplane' with short span, broad chord, tapered foreplanes; forward-swept, slightly tapering wing; and very short span, broad chord horizontal tail surfaces. Wings reported to incorporate provision for powered folding, without which the aircraft would be too large to fit in a standard hardened aircraft shelter. Twin tailfins each canted outwards at approximately 6° FSW improves subsonic agility, high AoA controllability, take-off and landing performance, reduces drag and diminishes frontal

Jet nozzle of Sukhoi Su-37 *(Paul Jackson/Jane's)*

The S-37, pictured taking off for public display at Zhukovsky in August 1999 *(Yefim Gordon)*

SUKHOI—AIRCRAFT: RUSSIAN FEDERATION 445

Sukhoi S-37 Berkut research aircraft

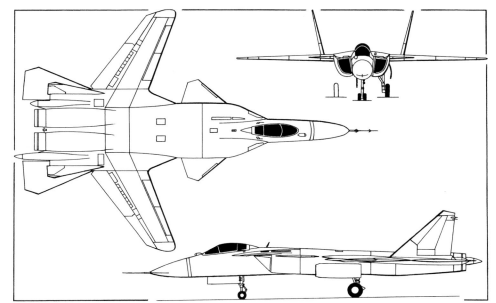

Provisional general arrangement of the Sukhoi S-37 Berkut, following horizontal tail enlargement
(Paul Jackson/Jane's)

radar reflectivity. Other measures, including radar-absorbent materials and engine air intake shape and internal S ducting, taken to reduce frontal radar signature. Large, curved LERX above semicircular engine air intakes resemble those of an earlier Sukhoi light fighter project (a previous carrier of the Su-37 designation). Slightly flattened nosecone enhances supermanoeuvrability. Expected to demonstrate AoA up to 120° at speeds from zero to supersonic. The aircraft reportedly makes extensive use of RAM coatings, these having been developed in the late 1980s under project 'Astra' which used two modified Su-25 preproduction aircraft.

Major components, such as tailfins, canopy, windscreen and landing gear, appear to be standard Su-27 parts. S-37 01 not equipped with radar, weapons or mission systems, but conspicuously has provision for their later fitment: forward- and aft-facing radomes, vented gun bay with cannon port, IRST mockup ahead of windscreen (offset to starboard) and dielectric panels in leading-edges of tailfins, LERX and foreplanes. Of twin, rear-facing radomes, starboard is on extended mounting and protrudes beyond trailing-edge of horizontal tail.

Wing swept forward (all data approximate) by 20° on leading-edge (except 24° sweptback gloves) and 37° on trailing-edge; rounded wingtips. Foreplane leading-edge has 50° sweep and −16° trailing-edge sweep. Horizontal tail swept 70°. Fin leading-edge sweep of 45°.

FLYING CONTROLS: FCS presumed developed from Avionika quadruplex digital FBW system of Su-35 and Su-37, probably with manual override; credited to Petrov. Wing appears to have two-thirds span leading-edge flaps on forward-swept portion, these each being in two parts, of which only the inboard has been noted deployed; trailing-edge has plain flaps inboard and ailerons outboard. Foreplanes may operate differentially and/or in unison. All-moving horizontal tail surfaces; conventional rudder in each fin.

STRUCTURE: Follows Su-27/Su-35 practice, but with wing of 90 per cent composites construction to ensure adequate stiffness.

LANDING GEAR: Generally as Su-33 (Su-27K) naval version of 'Flanker'.

POWER PLANT: Two Aviadvigatel D-30F6M turbofans, each 93.1 kN (20,930 lb st) dry and 153.0 kN (34,392 lb st) with afterburning. Engine choice reportedly dictated by other programmes – all AL-41 engines dedicated to Mikoyan MFI; AL-37FU to Su-30MKI and Su-37 programmes. Potential for replacement with Saturn/Lyulka AL-37FU turbofans with two-dimensional vectoring nozzles; by R-79M engines; or by three-dimensional AL-41Fs rated at 196 kN (44,092 lb st). Last-mentioned reported in early 2000 as candidates for retrofit in 2005. Fixed geometry air intakes; rectangular auxiliary intake door for each engine above mid-fuselage. Unconfirmed reports allege engines have new nozzles which assist cooling of exhaust gases to reduce heat signature.

ACCOMMODATION: Pilot only, on 30° rear-tilted Zvezda K-36DM ejection seat, beneath rear-hinged, Su-27-type transparencies.

AVIONICS: Integrated avionics suite by Technocomplex. May receive new integrated suite by Kronstadt, including Transas Abris navigation and landing system.

Radar: Provision, only, in nose to install multimode, phased-array Phazotron radar of unspecified type, with supplementary, rear-facing radar in starboard tailcone. Planned radar will have range of up to 132 n miles (245 km; 115 miles) and be capable of simultaneously tracking 24 and engaging eight targets.

Instrumentation: Ramenskoye instruments and displays.

Mission: Provision for Pribor weapons control system.

ARMAMENT: Nil. Alleged provision for underwing pylons, wingtip missile launch rails and internal GSh-301 30 mm gun with aperture in upper surface of port LERX. Some sources report provision for conformal weapons carriage to reduce radar signature.

DIMENSIONS, EXTERNAL:
Wing span	16.7 m (55 ft)
Length overall	22.6 m (74 ft)
Height overall	6.4 m (21 ft)

AREAS (approx):
Wings, gross	56 m² (600 sq ft)
Foreplanes (total)	5.7 m² (61 sq ft)
LERX (total)	2.75 m² (30 sq ft)
Fins (each)	7.7 m² (83 sq ft)

WEIGHTS AND LOADINGS:
T-O weight: normal	25,670 kg (56,592 lb)
max	34,000 kg (74,957 lb)

Three-quarter rear aspect of the Sukhoi S-37 *(Yefim Gordon)*

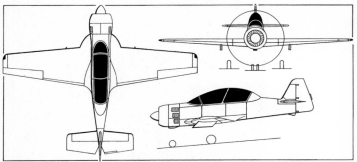

Sukhoi S-49 civil and military primary trainer (Mike Keep/Jane's) 0015117

Model of Sukhoi S-49 shown in its latest configuration (Paul Jackson/Jane's) 0015116

Wing loading:
  normal   approx 458 kg/m² (93.89 lb/sq ft)
  max   approx 607 kg/m² (124.35 lb/sq ft)
Power loading: normal   84 kg/kN (0.82 lb/lb st)
  max   111 kg/kN (1.09 lb/lb st)
PERFORMANCE (estimated):
  Max level speed:
    at S/L   M1.12 (756 kt; 1,400 km/h; 870 mph)
    at altitude   M2.1 (1,188 kt; 2,200 km/h; 1,367 mph)
  Service ceiling   18,000 m (59,060 ft)
  Range   1,781 n miles (3,300 km; 2,050 miles)
  g limit   +9
                                                                 UPDATED

SUKHOI S-49

TYPE: Basic prop trainer.
PROGRAMME: Design of military derivative of Su-26 and Su-29 aerobatic aircraft, to meet trainer requirement, started July 1992, under former designation Su-32; designation changed to Su-39 in 1995, Su-49 in 1996; current promotional material refers to S-49. Construction of prototype began 1994; first flight scheduled second half 1996, but delayed, and now predicted for 2001. Design further refined by mid-1997 to include LERX, raised rear portion of canopy and twin nosewheels. P&W Rus PK6A-25 turboprop reportedly considered as replacement for M-9 (M-14 previously intended) piston engine in at least proportion of production aircraft.
CUSTOMERS: Sukhoi anticipates orders for 1,000 for Russian air force flying schools and ROSTO (formerly DOSAAF); air force version has more complex equipment. Development probably halted, with funding now unlikely in the near term.
COSTS: US$300,000 (baseline) to US700,000 (full avionics) (2000).
DESIGN FEATURES: Based on Su-26 and Su-29 aerobatic aircraft, with objectives of short take-off and landing and high manoeuvrability, plus optional carriage of guided and unguided weapons and suitability for counter-insurgency, patrol and coastal protection missions. Service life 10,000 hours or 30,000 landings.
Wing leading-edge sweepback 5° 30′, wing section NACA 23012, dihedral 1° 30′, incidence 0°.
FLYING CONTROLS: Conventional and manual. Elevators and rudder horn-balanced; pneumatically operated wing trailing-edge slotted area-increasing flaps and ventral airbrake.
STRUCTURE: Two-spar wings; single-spar fin and tailplane. Fuselage longerons and wing spars of CFRP; wing, fuselage and tail unit skin panels of Kevlar-type composites and GFRP.
LANDING GEAR: Pneumatically retractable tricycle type; single wheel on each main unit; twin nosewheels; main units retract inward, nosewheel rearward; oleo-pneumatic shock-absorbers; Rubin mainwheels type KT-214; tyres size 400×150-140, pressure 3.00 to 3.50 bars (43.5 to 50.75 lb/sq in); hydraulic brakes on mainwheels.
POWER PLANT: One 309 kW (414 hp) VOKBM M-9F nine-cylinder radial engine; three-blade MTV-9 propeller. Alternative P&WC Klimov PK6A-25 turboprop. Fuel tank in each wing, capacity 150 litres (39.6 US gallons; 33.0 Imp gallons); header tank in fuselage, capacity 20 litres (5.3 US gallons; 4.4 Imp gallons); total internal fuel capacity 32.0 litres (84.5 US gallons; 70.4 Imp gallons); provision for two 100 litre (26.4 US gallon; 22.0 Imp gallon) underwing tanks; gravity fuelling; oil capacity 20 litres (5.3 US gallons; 4.4 Imp gallons).
ACCOMMODATION: Two seats in tandem; rear (instructor's) seat raised. Zvezda SKS-94 crew extraction system (54 to 215 kt; 100 to 400 km/h; 62 to 248 mph/10 m; 33 ft); crew extracted through canopy on diverging trajectories in emergency, without seats. Canopy hinged to starboard. Baggage space behind cockpits, capacity 0.20 m³ (7.0 cu ft).
SYSTEMS: Cockpit air conditioning and pressurisation standard. Brake hydraulic system pressure 24.50 bars (355 lb/sq in). Pneumatic system pressure 50.60 bars (735 lb/sq in). One 3.5 kW generator for 27 V DC electrical system. Propeller anti-iced and windscreen demisted by hot air.
AVIONICS: To customer's requirements; standard suite described below.
  Comms: Honeywell KX 165 VHF com/nav, KT 46A transponder, KR 87 ADF, GO3 area nav and KLN 90.
  Radar: Provision for external radar pod.
  Flight: Optional automatic approach and satellite nav. Provision for autopilot and even autoland.
  Self-defence: RWR and IR warning system in military versions.
ARMAMENT: Provision for integral gun, bombs, air-to-air and anti-tank missiles in combat versions.
DIMENSIONS, EXTERNAL:

Wing span	8.50 m (27 ft 10¾ in)
Wing chord: at root	1.88 m (6 ft 2 in)
at tip	0.885 m (2 ft 10¾ in)
Wing aspect ratio	5.9
Length overall	7.285 m (23 ft 10¾ in)
Height overall	2.60 m (8 ft 6½ in)
Tailplane span	2.90 m (9 ft 6¼ in)
Wheel track	2.00 m (6 ft 6¾ in)
Wheelbase	1.78 m (5 ft 10 in)
Propeller diameter	2.50 m (8 ft 2½ in)
Propeller ground clearance	0.24 m (9¼ in)

DIMENSIONS, INTERNAL:

Cockpit: Length	2.70 m (8 ft 10¼ in)
Max width	0.60 m (1 ft 11½ in)
Max height	1.10 m (3 ft 7¼ in)
Floor area	2.50 m² (26.9 sq ft)

AREAS:

Wings, gross	13.10 m² (141.0 sq ft)
Ailerons (total)	2.24 m² (24.11 sq ft)
Trailing-edge flaps (total)	3.30 m² (35.52 sq ft)
Fin	0.30 m² (3.23 sq ft)
Rudder	0.90 m² (9.69 sq ft)
Tailplane	0.98 m² (10.55 sq ft)
Elevators (total)	1.56 m² (16.79 sq ft)

WEIGHTS AND LOADINGS:

Weight empty, equipped	850 kg (1,874 lb)
Max fuel: internal	260 kg (573 lb)
external	200 kg (441 lb)
Normal T-O weight	1,300 kg (2,866 lb)
Max T-O and landing weight	1,500 kg (3,307 lb)
Max wing loading	114.5 kg/m² (23.45 lb/sq ft)
Max power loading	4.86 kg/kW (7.99 lb/hp)

PERFORMANCE (estimated, at normal T-O weight):

Never-exceed speed (V$_{NE}$)	286 kt (530 km/h; 329 mph)
Max level speed	243 kt (450 km/h; 280 mph)
Max cruising speed	205 kt (380 km/h; 236 mph)
Landing speed	65 kt (120 km/h; 75 mph)
Stalling speed: flaps up	54 kt (100 km/h; 62 mph)
flaps down	49 kt (90 km/h; 56 mph)
Max rate of climb at S/L	810 m (2,655 ft)/min
Service ceiling	4,000 m (13,120 ft)
T-O run	230 m (755 ft)
Landing run	250 m (820 ft)
Range:	
with max payload	647 n miles (1,200 km; 745 miles)
with max internal fuel	809 n miles (1,500 km; 932 miles)
with max internal and external fuel	1,080 n miles (2,000 km; 1,242 miles)
g limits	+11.0/−8.2

UPDATED

SUKHOI S-54

TYPE: Advanced jet trainer/light attack jet.
PROGRAMME: Began as one of designs by five OKBs to meet official Russian requirement to replace Aero L-29 and L-39 Albatros; programme launched 1990; configuration refined 1992; avionics and systems units being tested on modified Su-25 and Su-27. New version, with airframe enlarged by approximately 25 per cent, revealed at 1996 Farnborough Air Show; *this version described below*. However, by June 1997, Sukhoi had prepared another configuration for the S-54 more closely resembling a scaled-down, single-seat Su-27 equipped with Phazotron Sokol-X radar. This would indicate a change of emphasis to a combat role with secondary advanced training capabilities. No data have been disclosed for the latter, and no further promotion appears to have taken place in the two following years. S-55 described as export version, with raised cockpit indicating a "primary attack specialisation". Described separately. Priority reaffirmed in May 1999.

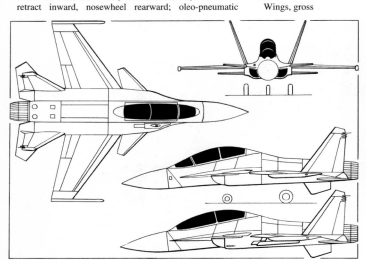

Revised configuration of the Sukhoi S-54 in mid-1997, now resembling a single-seat, scaled-down Su-27 (Paul Jackson/Jane's) 0022126

Two-seat concept for Sukhoi S-54 with additional side view (lower) of S-55 (James Goulding/Jane's)

DESIGN FEATURES: Described by Sukhoi as scaled-down development of Su-27, with speed, operating altitude and manoeuvrability commensurate with combat aircraft; unconventional all-swept mid-wing configuration, with canards, twin outward-canted fins mounted at wing trailing-edges; engine air intakes under wingroots.
FLYING CONTROLS: Fly-by-wire as Su-27, via flaperons, foreplanes, all-moving tailplane and rudders; leading-edge flaps; airbrake; preprogrammable to make aircraft easier or harder to fly, dependent on pupil's ability; optional 'panic button' to return aircraft to straight and level flight from any attitude, and push-button spin recovery; optional playback system to record student's every move in flight.
LANDING GEAR: Retractable tricycle type; single wheel on each unit; mainwheels retract inward, nosewheel forward.
POWER PLANT: One unspecified production engine, possibly Soyuz R-195FS, nominally rated at 62.0 kN (13,940 lb). Reports in 1998 suggested that the Lyulka-designed, two-axis AL-100 thrust vectoring nozzle is earmarked for the fighter version of S-54 or that the aircraft might be re-engined with the AL-41 or R-79.
ACCOMMODATION: Two crew in tandem, on K-36 zero/zero ejection seats, under three-piece blister canopy; rear seat raised.
AVIONICS: Avionics and cockpit interior same as for current and advanced tactical aircraft. Automatic flight control system. Weather radar standard.
ARMAMENT: (optional): Wingtip mounts for two close-range IR homing air-to-air missiles; two hardpoints under each wing for air-to-air and air-to-surface guided weapons.
DIMENSIONS, EXTERNAL:
Wing span	11.23 m (36 ft 10¼ in)
Length overall	15.30 m (50 ft 2¼ in)
Height overall	4.70 m (15 ft 5 in)

WEIGHTS AND LOADINGS: Not available
PERFORMANCE (estimated):
Max level speed:	
at height	M1.55 (890 kt; 1,650 km/h; 1,025 mph)
at S/L	M0.98 (648 kt; 1,200 km/h; 745 mph)
Service ceiling	18,000 m (59,050 ft)
T-O run	360 m (1,185 ft)
Landing run	500 m (1,640 ft)
Range with max fuel:	
at S/L	440 n miles (820 km; 510 miles)
at height	1,620 n miles (3,000 km; 1,864 miles)
g limits	+12/−3

UPDATED

SUKHOI S-55

TYPE: Light fighter.
PROGRAMME: Revealed at Moscow Air Show '95 as export, attack-optimised variant of S-54 (as then envisaged) with ventral strakes, considerably enlarged LERX and two-piece canopy. Radar nose. Prototype under construction by 1996, but had not flown by early 2000. The project has received no recent publicity and may have developed into the S-56.
DESIGN FEATURES: Generally as for S-54.
POWER PLANT: One Saturn/Lyulka AL-37FP or AL-37FU afterburning turbofan with thrust-vectoring nozzle.
ACCOMMODATION: Two crew, on tandem zero/zero ejection seats.
DIMENSIONS, EXTERNAL:
Wing span	9.60 m (31 ft 6 in)
Length overall	13.20 m (43 ft 3¾ in)
Height overall	4.40 m (14 ft 5¼ in)

PERFORMANCE (estimated):
Max level speed:	
at height	890 kt (1,650 km/h; 1,025 mph)
at S/L	648 kt (1,200 km/h; 745 mph)
Range with max fuel:	
at S/L	442 n miles (820 km; 510 miles)
at height	1,619 n miles (3,000 km; 1,864 miles)
g limits	+12/−3

UPDATED

SUKHOI S-56

TYPE: Light fighter.
PROGRAMME: In 1999, India was offered a single-engined, navalised light fighter of this designation for eventual deployment on an aircraft carrier. At that time, no detailed design had been undertaken. The aircraft is officially described as a "carrier-based light multirole fighter for export".

UPDATED

Model of an unarmed S-55, showing ventral strakes and LERX extending forward to radome (*Paul Jackson/Jane's*)

Sukhoi Su-26 of the main production series. Su-26M2 subvariant has retrofitted lighter wings of M series, but additional 40 litres (10.6 US gallons; 8.8 Imp gallons) of internal usable fuel (*Paul Jackson/Jane's*)

SUKHOI LFS

TYPE: Light fighter.
PROGRAMME: Under development by 1998, as confirmed by AVPK Sukhoi general director and former S-37 chief designer, Mikhail Pogosyan. *Legkiy Frontovoy Samolet*: Light Frontline Aircraft. Will derive technology from S-37 programme; options include two RD-133 engines or one AL-31PF, both with thrust vectoring. May even represent continuation of original S-37 programme, possibly now designated S-137 (single-engined) or S-237 (twin-engined).

VERIFIED

SUKHOI Su-26

TYPE: Aerobatic single-seat sportplane.
PROGRAMME: Full details last appeared in the 1996-97 *Jane's*. Production terminated in 1996 after manufacture of 77 aircraft, comprising four Su-26 prototypes (1984; 850 kg empty); eight Su-26Ms (from 1985; 750 kg empty) and 65 production Su-26s (from 1988; 720 kg empty). In 1999, European Agent, Richard Goode Aerobatics, commissioned one additional aircraft, an Su-26M3 (c/n 5401) to be sold for US$215,000, being the last Su-26 to be produced.

UPDATED

SUKHOI Su-29

TYPE: Aerobatic two-seat sportplane.
PROGRAMME: Announced at Moscow Air Show '90; design started 1990; construction of first of three prototypes and two static test airframes began 1991; prototype first flew 1991, first production aircraft May 1992; entered service July 1992. AP-23 type certificate awarded in 1994. Assessed by Russian Air Forces experimental centre at Akhtubinsk in 1996-97, but no orders placed. Sukhoi's first aerobatic lightplane, the Su-26 is mentioned immediately above, but was last described in the 1996-97 *Jane's*; see also Su-31.

Following structural failure of Su-29 wing in 1996; remedial manufacturing practices in place and recertified by 1998. Resumed civilian export production now featured M-14PF engine and new propeller, but M-9F engine introduced in 1999. Retrospective installation of SKS-94 pilot's emergency extraction system offered by Sukhoi from 2000.
CURRENT VERSIONS: **Su-29**: Basic two-seat training/aerobatic aircraft.
Description applies to baseline Su-29.
Su-29AR: Aircraft for Argentina have German propeller, Swiss canopies and US wheel assemblies and avionics, including GPS.
Su-29KS: Development vehicle for Zvezda SKS-94 lightweight crew extraction system. Weight empty, equipped 800 kg (1,764 lb). First exhibited at 1994 Farnborough Air Show. One only.
Su-29M: Production version from 1999 (initial series of 10 under construction); M-9F engine; weights as Su-29; no extraction system.
CUSTOMERS: Total 52 basic Su-29s and one Su-29KS built and sold by 1997; many exported to Pompano Air Center, Florida, USA, for reassembly and delivery worldwide (including eight in 1992; 12 in 1993; seven in 1994 and four in 1995); others to Australia (three), Italy, South Africa (two) and UK. Relaunched production from 1998 (initial batch of 10). Eight ordered by Argentine Air Force, for training, March 1997; delivery between September 1997 and August 1998 to Escuadrilla Cruz del Sur. By mid-1998, Sukhoi had built 164 aerobatic light aircraft of the Su-26/29/31 series. Sukhoi planning in 2000 to relocate production to KAPO at Kazan; previous aircraft built by Sukhoi Advanced Technologies using components manufactured by LMZ, Dubna and others.
COSTS: US$240,000 (1999).
DESIGN FEATURES: Typical aerobatic competition aircraft; mid-wing of specially developed symmetrical section, variable along span, slightly concave in region of ailerons to increase their effectiveness; leading-edge somewhat sharper than usual to improve responsiveness to control surface movement. Two-seat development of Su-26M single-seat aerobatic competition aircraft; wing span and overall length increased; improved aerodynamics and reduced stability margin for enhanced manoeuvrability. Service life 1,250 hours.

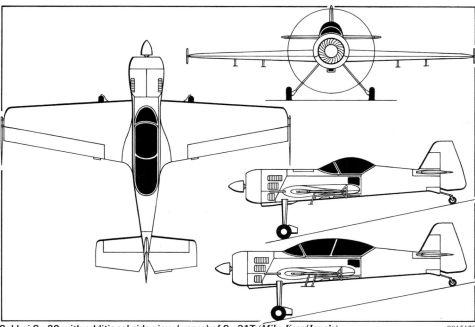

Sukhoi Su-29 with additional side view (upper) of Su-31T (*Mike Keep/Jane's*)

Sukhoi Su-29 training and aerobatic aircraft from 2000 production *(Paul Jackson/Jane's)*

Wing leading-edge sweepback 3° 28′, symmetrical section, thickness/chord ratio 16 per cent at root, 12 per cent at tip, dihedral 0°, incidence 0°.

FLYING CONTROLS: Conventional and manual. Elevators and rudder horn-balanced; elevator trim; two suspended triangular balance tabs under each aileron. Later Su-29s have Su-31-type wing with ailerons extending to wingtips. No flaps.

STRUCTURE: Composites comprise more than 60 per cent of airframe weight; one-piece wing, covered with honeycomb composites panels; foam-filled front box spar with CFRP booms and wound glass fibre webs; channel section rear spar of CFRP; titanium truss ribs; plain ailerons have CFRP box spar, GFRP skin and foam filling; fuselage has basic welded truss structure of VNS-2 high-strength stainless steel tubing; lower nose section of truss removable for wing detachment; quickly removable honeycomb composites skin panels; light alloy engine cowlings; integral fin and tailplane construction same as wings; rudder and elevator construction same as ailerons; titanium exhaust, battery box and firewall; forged magnesium control linkages. Aircraft assembled by Sukhoi from components produced by LMZ, Dubna (DMZ) and NPO Technologia (at Omsk).

LANDING GEAR: Non-retractable tailwheel type; arched cantilever mainwheel legs of titanium alloy; mainwheels size 400×150, with hydraulic disc brakes; steerable tailwheel, on titanium spring, connected to rudder. Optional composites fairings for mainwheels.

POWER PLANT: One 265 kW (355 hp) VOKBM M-14PT or 294 kW (394 hp) M-14PF nine-cylinder radial engine; three-blade MTV-3-8-S/L250-21 or MTV-9-260 propeller. Current production employs VOKBM M-9F (M-14 derivative) rated at 309 kW (414 hp). Steel tube engine mounting. Fuel tank in fuselage forward of front spar; capacity 63 litres (16.6 US gallons; 13.8 Imp gallons); tank in each wing leading-edge; capacity 106.5 litres (28.15 US gallons; 23.4 Imp gallons); total fuel capacity 276 litres (72.9 US gallons; 60.6 Imp gallons); gravity fuelling. Oil capacity 20 litres (5.3 US gallons; 4.4 Imp gallons). Fuel and oil systems adapted for inverted flight; pneumatic engine starting system.

ACCOMMODATION: Pilot only for aerobatic competition, two persons in tandem for training. Canopy normally opens sideways to starboard; but upward and rearward in emergency to jettison. Dual controls standard. Space for 5 kg (11 lb) baggage in rear fuselage.

SYSTEMS: Electrical system 24/28 V, with 3 kW generator, batteries and external supply socket.

AVIONICS: *Comms:* Briz VHF radio; optional Becker or Honeywell com/nav.

EQUIPMENT: Optional provision for smoke generation.

DIMENSIONS, EXTERNAL:
Wing span	8.20 m (26 ft 10¾ in)
Wing chord: at root	1.985 m (6 ft 6¼ in)
at tip	1.04 m (3 ft 4¾ in)
Wing aspect ratio	5.5
Length overall	7.285 m (23 ft 10¾ in)
Height overall	2.885 m (9 ft 5¾ in)
Tailplane span	2.90 m (9 ft 6¼ in)
Wheel track	2.40 m (7 ft 10½ in)
Wheelbase	5.08 m (16 ft 8 in)
Propeller diameter	2.50 m (8 ft 2½ in)
Propeller ground clearance	0.425 m (1 ft 4¾ in)

DIMENSIONS, INTERNAL:
Cockpit: Length	2.60 m (8 ft 6¼ in)
Max width	0.82 m (2 ft 8¼ in)
Max height	1.05 m (3 ft 5¼ in)

AREAS:
Wings, gross	12.20 m² (131.4 sq ft)
Ailerons (total)	2.32 m² (24.97 sq ft)
Fin	0.28 m² (3.01 sq ft)
Rudder	0.90 m² (9.69 sq ft)
Tailplane	0.98 m² (10.55 sq ft)
Elevators (total)	1.56 m² (16.79 sq ft)

WEIGHTS AND LOADINGS (two persons):
Weight: empty	735 kg (1,620 lb)
empty, equipped	780 kg (1,720 lb)
Max fuel	207 kg (456 lb)
Max T-O weight: pilot only	860 kg (1,896 lb)
two persons	1,204 kg (2,654 lb)
Max wing loading	98.7 kg/m² (20.21 lb/sq ft)
Max power loading: M-14PT	4.55 kg/kW (7.48 lb/hp)
M-14PF	4.01 kg/kW (6.73 lb/hp)
M-9F	4.78 kg/kW (6.41 lb/hp)

PERFORMANCE (M-14PT engine):
Never-exceed speed (V_{NE})	242 kt (450 km/h; 279 mph)
Max level speed	175 kt (325 km/h; 202 mph)
Stalling speed	62 kt (115 km/h; 72 mph)
Max rate of climb at S/L	960 m (3,150 ft)/min
Service ceiling	4,000 m (13,120 ft)
Max rate of roll	345°/s
*T-O run	120 m (395 ft)
*Landing run	380 m (1,250 ft)
Range with max fuel	647 n miles (1,200 km; 745 miles)
g limits	+12/–10

*at 914 kg (2,015 lb) AUW

UPDATED

SUKHOI Su-31

TYPE: Aerobatic single-seat sportplane.

PROGRAMME: Design started 1991; prototype construction began 1992, flew June 1992 as Su-29T, demonstrated at 1992 Farnborough Air Show; followed by two more prototypes and two static test airframes; first production aircraft by Sukhoi Advanced Technologies (RA-01405) flown 1994. Su-31M2 introduced in 1999.

CURRENT VERSIONS: **Su-31**: Basic version; non-retractable landing gear. Alternatively known as **Su-31T** (*Turnirnyi:* Competition). See Su-29 drawings for side view.

Su-31M: As Su-31T but with Zvezda SKS-94 pilot's extraction system under modified canopy with deeper frame. Empty weight 760 kg (1,676 lb); normal T-O weight 880 kg (1,940 lb). Prototype RA-01486 converted from Su-31T.

Su-31M2: Available from 2000 (first production batch of five). M-9F engine; lightened structure.

Su-31X: Export version of Su-31T.

Su-31U: As Su-31T but retractable landing gear. None yet built.

CUSTOMERS: Total 25 (including five Su-31Ms) built by late 1998 and exported to Australia, Brazil, Italy, Lithuania, South Africa, Spain, Switzerland, Ukraine, UK and USA. Su-31Ms operated in Italy, Russia and Switzerland. Production batch of five Su-31M2s, 1999-2000, of which at least two to USA.

DESIGN FEATURES: Basically single-seat version of Su-29 with uprated engine; new landing gear; 35° inclination of seat enables pilot to employ repeatedly a g load of +12/–10, giving advantages in controlling aircraft within limited flying area and to perform very complicated manoeuvres; improved field of view; two baggage compartments.

STRUCTURE: More than 70 per cent composites by weight; centre-fuselage is welded truss of high-strength stainless steel tube, with detachable skin panels of honeycomb-filled composite sandwich; rear fuselage is semi-monocoque of composites with honeycomb filler; one-piece two-spar wing with carbon fibre main spar, titanium ribs, covered with honeycomb sandwich skin; tail unit is all-composites; mainwheel legs titanium.

POWER PLANT: One 294 kW (394 hp) VOKBM M-14PF nine-cylinder radial engine in Su-31T/M; 309 kW (414 hp) VOKBM M-9F in Su-31M2; MTV-9 three-blade propeller. Basic fuel capacity 78 litres (20.6 US gallons; 17.2 Imp gallons), in fuselage tank; provision in Russian version for 210 litre (55.5 US gallon; 46.2 Imp gallon) centreline drop tank for ferrying; alternatively, in export version, two tanks in wings, total capacity 200 litres (52.8 US gallons; 44.0 Imp gallons).

Sukhoi Su-31T single-seat aerobatic competition aircraft with separate windscreen and canopy *(Paul Jackson/Jane's)*

Single-piece canopy identifies the extraction system-equipped Sukhoi Su-31M *(Paul Jackson/Jane's)*

SUKHOI—AIRCRAFT: RUSSIAN FEDERATION

ACCOMMODATION: Pilot only; windscreen and separate canopy, opening as Su-29, except single-piece windscreen/canopy and pilot extraction system in Su-31M.

DIMENSIONS, EXTERNAL (Su-31T):
Wing span	7.80 m (25 ft 7 in)
Wing chord: at root	1.99 m (6 ft 6¼ in)
at tip	1.04 m (3 ft 4¾ in)
Wing aspect ratio	5.2
Length overall	6.90 m (22 ft 7¾ in)
Height overall	2.76 m (9 ft 0¾ in)
Tailplane span	2.90 m (9 ft 6¼ in)
Wheel track	2.40 m (7 ft 10½ in)
Wheelbase	4.90 m (16 ft 1 in)
Propeller diameter	2.50 m (8 ft 2½ in)
Propeller ground clearance	0.425 m (1 ft 4¾ in)

AREAS (Su-31T): As Su-29 except:
Wings, gross 11.80 m² (127.0 sq ft)

WEIGHTS AND LOADINGS (Su-31T, except where otherwise indicated):
Weight: empty	670 kg (1,477 lb)
empty, equipped: Su-31T	740 kg (1,631 lb)
Su-31M	750 kg (1,653 lb)
Su-31M2	715 kg (1,576 lb)
Max fuel: internal	53 kg (117 lb)
external	209 kg (461 lb)
T-O weight: normal	780 kg (1,720 lb)
max	968 kg (2,134 lb)
Max wing loading	82.0 kg/m² (16.80 lb/sq ft)
Max power loading: Su-31T/M	3.25 kg/kW (5.34 lb/hp)
Su-31M2	3.14 kg/kW (5.15 lb/hp)

PERFORMANCE (Su-31T):
Never-exceed speed (V_{NE})	243 kt (450 km/h; 280 mph)
Max level speed	178 kt (330 km/h; 205 mph)
Stalling speed	61 kt (113 km/h; 71 mph)
T-O speed	60 kt (110 km/h; 69 mph)
Landing speed	62 kt (115 km/h; 72 mph)
Max rate of climb at S/L	1,440 m (4,725 ft)/min
Service ceiling	4,000 m (13,120 ft)
Max rate of roll	400°/s
T-O run	110 m (360 ft)
Landing run	300 m (985 ft)
Range, internal fuel	156 n miles (290 km; 180 miles)
Ferry range	400-645 n miles (740-1,200 km; 460-745 miles)
g limits	+12/–10

UPDATED

SUKHOI Su-38L

TYPE: Agricultural sprayer.
PROGRAMME: Design started, as Su-38, August 1993; originally based closely on Su-29 sportplane; construction of prototype began January 1994 but curtailed due to financial situation in Russia. Promotion resumed in 1998, under Su-38L designation and with considerable change of detail design, including replacement of VOKBM M-14P radial engine. Announced June 2000 that order signed to build prototype, now known as Su-38L; scheduled for Russian and Western certification in 2002. Manufacture by Razdanmash AOZT of Armenia, subject to acquisition of US$23 million for flight test and tooling; target of five complete (including static test) aircraft by end of 2001; first two under assembly at Sukhoi OKB workshops by September 2000.

CUSTOMERS: Estimated market for 500 in Russia.
COSTS: US$120,000 to US$130,000 (2000).
DESIGN FEATURES: Conventional low-wing monoplane. Constant-chord wings swept slightly forward and with winglets and root nibs; sweptback fin, plus underfin doubling as tailwheel mount. Chemical hopper between engine and cockpit. Service life 3,000 hours/10 years.
FLYING CONTROLS: Conventional and manual. Horn-balanced single-piece elevator and rudder; trim tab in elevator; trailing-edge flaps.
LANDING GEAR: Tailwheel type; fixed.
POWER PLANT: One 154 kW (207 hp) LOM M337 six-cylinder engine; LOM B-546 two-blade variable-pitch propeller. Fuel tank in each wing; total capacity 210 litres (55.5 US gallons; 46.2 Imp gallons).
ACCOMMODATION: Pilot only.
EQUIPMENT: Underwing spraybars; chemicals in 600 litre (159 US gallon; 132 Imp gallon) fuselage hopper.

WEIGHTS AND LOADINGS:
Normal T-O and landing weight 1,200 kg (2,646 lb)

PERFORMANCE (estimated):
Operating speed	97-108 kt (180-200 km/h; 112-124 mph)
Stalling speed	41 kt (76 km/h; 48 mph)
Ferry range:	
with spraybars	486 n miles (900 km; 559 miles)
without spraybars	648 n miles (1,200 km; 745 miles)

SUKHOI S-80

TYPE: Twin-turboprop transport.
PROGRAMME: First, largest and most advanced design by Sukhoi under *konversiya* programme of former Soviet industry; to replace L-410, An-28, Yak-40 and An-24; certification intended to FAR Pt 25, JAR 25 and AP-25. Work began 1989 on medical evacuation aircraft, under order from Ministry of Public Health, by Sukhoi-Europe/Asia joint stock company, with founder members Sukhoi Design Bureau ASIC, KnAAPO (which see), Rybinsk Motor Engineering Design Bureau, Ramenskoye Instrument Engineering Design Bureau, and Instrument Engineering R&D Institute. Included in government's civil

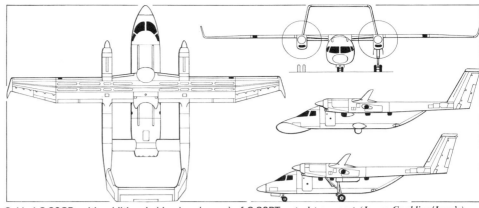

Sukhoi S-80GP, with additional side view (upper) of S-80PT patrol transport *(James Goulding/Jane's)*

aviation plan, but minimal funding by public money. Model displayed 1989 Paris Air Show; funding ended with collapse of USSR; project revived 1992, with priority on more marketable passenger and passenger/cargo variants, with imported engines, propellers and avionics.

Manufacture of flying prototype and test airframe under way at KnAAPO's Komsomolsk-on-Amur plant by 1993; agreement with General Electric to fit CT7-9 turboprops early 1995; design further refined and military versions reintroduced 1996; first flight of prototype (in S-80GP configuration) scheduled second half 1996 but repeatedly postponed, most recently to early 2001 at Zhukovsky test airfield, Moscow; prototype (RA-82911) shown on production line on 15 May 1998 to representatives of Border Guards, MoD, and Magadan, Khabarovsk and Petropavlovsk-Kamchatski airlines; was handed over to Sukhoi for flight test preparation in January 2000; second and third prototypes to follow in late 2001 and 2002; two static airframes (02 and 04) for structural testing, of which first was delivered to SIBNIA at Novosibirsk in early 1998 and second to be completed in third quarter of 2001.

CURRENT VERSIONS: **S-80GP**: Basic cargo/passenger (*gruzo/passazhirski*) version.
Detailed description applies specifically to S-80GP; generally to all versions:
S-80A: Optimised for Arctic operation; otherwise as S-80GP.
S-80GR: Geological survey version.
S-80M: Medical evacuation (10 casualties) version.
S-80P: Passenger (*passazhirski*) version.
S-80PT: Patrol transport, embodying Leninets Strizh (martin) avionics suite, with undernose 360° search radar and rotating electro-optic (FLIR/LLLTV) sensor turret under centre-fuselage; one operator's console in cabin. Able to perform 6 to 9 hour patrol missions over sea and land frontiers up to 189 n miles (350 km; 217 miles) from base; personnel and cargo transportation; and airdropping up to 20 paratroops and/or cargo. Flight crew of two or three. Overall length 17.45 m (57 ft 3 in).
S-80R: Fisheries patrol version.
S-80TD: Troop, paratrooop (total 21), freight and medevac transport, generally similar to civil S-80GP, but with mechanised cargo-handling equipment.
S-80 stretch: Projected version for 32 passengers.

CUSTOMERS: Market for 150 to 200 domestic sales and 360 to 500 exports estimated in 2000. Interest from Delta-K of Yakutia in 2000; export prospects include China, Malaysia, Thailand and Vietnam.

COSTS: Notionally funded by Russian Federation government, but by early 1997, KnAAPO had invested over Rb20 billion in development, compared with Rb1.5 billion received from official sources. Of US$20 million expended by mid-2000, two thirds provided by KnAAPO; thereafter, Sukhoi providing funds for flight test programme, estimated as further US$20 million. Flyaway price US$4 million (2000 estimate).

DESIGN FEATURES: Utility freighter with unobstructed rear access for loading and wide track landing gear. Basically conventional high-wing, podded fuselage, twin-boom, rear-loading configuration, but with short tandem-wing surfaces between each tailboom and rear fuselage; unswept wings of high aspect ratio with no dihedral or anhedral; large-span constant chord inner panels; small sweptback winglets on tapered outer panels; sweptback vertical tail surfaces, toed slightly inward, with bridging horizontal surfaces. Systems, accessories and components of Su-25, Su-27 and Su-35 embodied in S-80. Automatic built-in systems testing.

FLYING CONTROLS: Conventional and manual. Actuation by rods and cables; presumed complex wing trailing-edge surfaces to ensure STOL capability. Electric aileron and rudder trim.

STRUCTURE: No details, but said to employ "advanced and high technology".

LANDING GEAR: Retractable tricycle type; twin wheels on each unit; main units retract into tailbooms; mainwheel tyre size 660×200; nosewheel tyres 500×180; nosewheels steerable ±38°.

Mockup of Sukhoi Su-38L

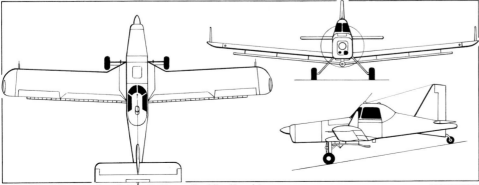

Sukhoi Su-38L agricultural aircraft *(James Goulding/Jane's)*

POWER PLANT: Two 1,305 kW (1,750 shp) General Electric CT7-9B turboprops; production aircraft with engines built locally by GE-Rybinsk Aero Engines; Hamilton Sundstrand 14RF-35 or Dowty feathering and reversible-pitch four-blade propellers. Two fuel tanks, total capacity 2,500 litres (660 US gallons; 550 Imp gallons).

ACCOMMODATION: Pilot only, two crew or pilot and 26th passenger side by side on flight deck; 19 to 25 passengers at 75 cm (29.5 in) seat pitch or 10 stretcher patients in main cabin, or 3,100 kg (6,834 lb) freight or role equipment. All accommodation pressurised. Typical S-80GP carries seven passengers in single seats on port side of cabin and 12 in double seats on starboard side, with baggage space, lavatory, wardrobe or galley at front, as specified by customer. Available in 'Salon' configuration with nine, 12 or 16 passenger seats. Typical freighter has a row of seats behind flight deck and unobstructed main hold for a small vehicle or cargo. Door at centre of cabin on port side; rear-loading ramp.

SYSTEMS: Electrical system developed by Lucas Aerospace and Auxilec. Anti-icing system. APU for autonomous operation at remote, unprepared sites.

AVIONICS: *Comms:* Com/nav, identification and ATC equipment of Russian manufacture. VOR/DME/ILS for ICAO Cat. II operation by Rockwell Collins.
Flight: Elektroavtomatika PNK-80 navigation system. AFCS and autopilot by Rockwell Collins; satellite nav system with Litton computer.
Instrumentation: Rockwell Collins Pro Line 2 five-screen EFIS.

EQUIPMENT: Options include equipment for air photography.

ARMAMENT (S-80PT): Typically 23 mm GSh-23L gun pod pylon-mounted on starboard side of cabin; four underwing pylons for electro-optical ASM, 20-round rocket pack, cluster of eight Vikhr tube-launched missiles and R-60 (AA-8 'Aphid') self-defence AAM.

Mockup of Sukhoi S-80 with rear loading ramp open 2001/0100420

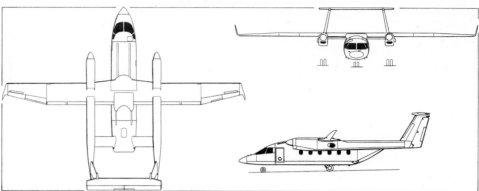

Projected stretched version of Sukhoi S-80 2001/0100419

DIMENSIONS, EXTERNAL:
Wing span	23.18 m (76 ft 0½ in)
Wing chord: at root	2.16 m (7 ft 1 in)
at tip	1.20 m (3 ft 11¼ in)
Wing aspect ratio	12.2
Length overall	16.68 m (54 ft 8¾ in)
Height overall	5.48 m (17 ft 11¾ in)
Tailplane span	4.88 m (16 ft 0¼ in)
Wheel track	5.60 m (18 ft 4½ in)
Wheelbase	6.50 m (21 ft 4 in)
Propeller diameter (each)	2.65 m (8 ft 8½ in)
Propeller ground clearance	1.10 m (3 ft 7¼ in)
Distance between propeller centres	5.60 m (18 ft 4½ in)
Passenger door: Height	1.75 m (5 ft 9 in)
Width	0.90 m (2 ft 11½ in)
Service door: Height	1.34 m (4 ft 4¾ in)
Width	0.64 m (2 ft 1¼ in)

Type III emergency exits (each):
Height	0.905 m (2 ft 11½ in)
Width	0.54 m (1 ft 9¼ in)

DIMENSIONS, INTERNAL:
Cabin: Length	7.80 m (25 ft 7 in)
Max width	2.17 m (7 ft 1½ in)
Max height	1.83 m (6 ft 0 in)
Volume, approx	20.0 m³ (706 cu ft)

AREAS:
Wings, gross	44.00 m² (473.6 sq ft)
Ailerons (total)	3.27 m² (35.20 sq ft)
Trailing-edge flaps (total)	7.05 m² (75.90 sq ft)
Leading-edge slats (total)	4.68 m² (50.38 sq ft)
Fins (total)	6.64 m² (71.47 sq ft)
Rudders (total)	6.22 m² (66.95 sq ft)
Horizontal tail surfaces	5.86 m² (63.08 sq ft)

WEIGHTS AND LOADINGS:
Max payload	3,300 kg (7,275 lb)
Max fuel weight	2,350 kg (5,181 lb)
Max T-O weight	13,500 kg (29,762 lb)
Max wing loading	306.8 kg/m² (62.84 lb/sq ft)
Max power loading	5.18 kg/kW (8.50 lb/shp)

PERFORMANCE (estimated):
Max cruising speed at 6,000 m (19,685 ft)	289 kt (535 km/h; 332 mph)
Balanced field length	850 m (2,790 ft)
T-O run	555 m (1,820 ft)
Landing run with propeller reversal	280 m (920 ft)
Range: with max payload	324 n miles (600 km; 372 miles)
with 1,950 kg (4,299 lb) payload	1,393 n miles (2,580 km; 1,603 miles)

OPERATIONAL NOISE LEVELS: Designed to conform to FAR Pt 36 standards

UPDATED

SUKHOI KR-860

TYPE: High-capacity airliner.

PROGRAMME: On 15 April 1998, Sukhoi was revealed to be working on the design of a nominal 860-seat airliner with the type prefix KR (*Krilatyi Russkii*: Winged Russian). This followed the establishment in Kazan during the previous month of the MAK consortium to co-ordinate this and other civil aviation programmes. Participants include five Moscow-based firms and institutes, five Tatarstani engine and aircraft manufacturers and Moseksimbank. Range of the projected airliner would be 10,000 n miles (18,520 km 11,507 miles); folding wings for access to current airport terminals; variants seat between 700 and 1,000 passengers. Development cost estimated as US$8 billion in 1998, but quoted as US$4.0 billion to US$5.5 billion in early 2000, unit cost then being US$150 million to US$200 million. Market estimated as 300.

UPDATED

Model of the projected Sukhoi KR-860 airliner, shown at Paris in June 1999 *(Paul Jackson/Jane's)* 2000/0054483

TAGANROG (TANTK)

TAGANROGSKY AVIATSIONNYI NAUCHNO-TEKHNICHESKY KOMPLEKS (Taganrog Aviation Scientific-Technical Complex)

Address as for Beriev (which see).

This plant was responsible for production of the Tupolev Tu-142 'Bear' strategic bomber (until 1994) and conversion of Ilyushin Il-76 transports to Beriev A-50 AWACS (until 1992). Current activities centre on Tu-142 overhaul. Conversion of an Il-76 for China (see Beriev A-50I) was terminated in mid-2000; plans for manufacture of Tu-334 airliners were halted following transfer of the programme to RSK MiG by a Presidential decree of 1999; and establishment of a production line for the Beriev Be-32 twin-turboprop transport has yet to show results. The plant is 44 per cent owned by the Russian state, and part by the Tupolev holding company.

UPDATED

TECHNOAVIA

NAUCHNO-KOMMERCHESKY FIRMA TECHNOAVIA (Technoavia Scientific and Commercial Firm)

bulvar Kronshtadsky 7A, 125212 Moskva
Tel: (+7 095) 452 56 03
Fax: (+7 095) 452 56 94
e-mail: techavia@proxel.ru
GENERAL DIRECTOR: Vyatcheslav Kondratiev
CHIEF AERODYNAMICIST: Yuri Kovyzhenko

Technoavia was formed in 1991 and currently produces the SM-92 and SP-55 light aircraft at the Smolensk Aircraft Plant (SmAZ).

UPDATED

TECHNOAVIA SP-55M

TYPE: Aerobatic single-seat sportplane

PROGRAMME: Initial batch of five aircraft under construction in 2000 at Progress plant at Arsenyev; first delivery in January 2001 to Richard Goode Aerobatics in UK; public debut scheduled for Paris Air Show in June 2001.

DESIGN FEATURES: Mid-wing configuration with symmetrical section, no dihedral, anhedral or incidence. Development of Yak-55M, last described in 1998-99 *Jane's*. Modifications include redesigned rear fuselage turtledeck; new fin and tailplane fillets; reprofiled wingtips; redesigned flying surfaces with composites skin; steel engine firewall; new engine cowling; new instrument panel with US instruments and avionics; modified fuel and oil systems to FAR Pt 23 requirements; modified pneumatic system; pilot-controlled oil and carburettor heat doors; new entry steps and assist handle on fuselage and landing

TECHNOAVIA—AIRCRAFT: RUSSIAN FEDERATION

No longer produced (see 1996-97 edition), the rare Technoavia SP-91 continues to evolve in the USA, where four of the five built now reside. N195SF (registered, incorrectly, as an SP-95) has had some 30 cm (12 in) removed from the fintip *(Paul Jackson/Jane's)*   2001/0093702

gear legs; inclined seat for increased g tolerance; new safety harness; baggage compartment and smoke system.

FLYING CONTROLS: Conventional and manual. Ailerons have narrower chord than those of Yak-55M, without horn balances.

STRUCTURE: Metal, single-spar wing with composites-covered ailerons and composites tips. Metal fuselage with steel firewall; composites-covered elevator and rudder.

LANDING GEAR: Non-retractable tailwheel type with spring steel legs on all three units; KT-98 mainwheels with 400×150 tyres; lockable, free-castoring tailwheel.

POWER PLANT: One 265 kW (355 hp) VOKBM M-14P air-cooled, nine-cylinder radial engine driving a three-blade MTV-9 constant-speed propeller. Fuel in wing tanks, total capacity 120 litres (31.7 US gallons; 23.4 Imp gallons); inverted fuel and oil systems.

ACCOMMODATION: One person in inclined seat with recess for backpack parachute; fixed windscreen and one-piece sliding canopy.

SYSTEMS: Electrical system comprises lightweight 10 Ah generator and storage battery, both of US origin

AVIONICS: Avionics fit of US origin.

EQUIPMENT: Optional smoke generation system; or glider hook.

DIMENSIONS, EXTERNAL:
Wing span	8.00 m (26 ft 3 in)
Wing aspect ratio	5.3
Length overall	7.48 m (24 ft 6½ in)
Height overall	2.225 m (7 ft 3½ in)
Tailplane span	3.51 m (11 ft 6¼ in)
Wheel track	2.76 m (9 ft 0½ in)
Propeller diameter	2.50 m (8 ft 2½ in)

AREAS:
Wings, gross	12.17 m² (131.0 sq ft)

WEIGHTS AND LOADINGS:
Weight empty	690 kg (1,521 lb)
Max T-O weight: training	855 kg (1,885 lb)
ferry	955 kg (2,105 lb)
Max wing loading: training	70.3 kg/m² (14.39 lb/sq ft)
ferry	78.5 kg/m² (16.07 lb/sq ft)
Max power loading: training	3.23 kg/kW (5.31 lb/hp)
ferry	3.61 kg/kW (5.93 lb/hp)

PERFORMANCE:
Never-exceed speed (V$_{NE}$)	194 kt (360 km/h; 223 mph)
Max level speed	173 kt (320 km/h; 199 mph)
Stalling speed	57 kt (105 km/h; 66 mph)
Max rate of climb at S/L	1,080 m (3,543 ft)/min
T-O run	170 m (558 ft)

Range with max fuel at econ cruising speed at 1,000 m (3,280 ft), 7% fuel reserves
    415 n miles (770 km; 478 miles)
g limits   +9/−6

NEW ENTRY

TECHNOAVIA SM-92 FINIST

TYPE: Light utility transport.

PROGRAMME: Design (as I-5) by Interavia (which see), to FAR Pt 23 and JAR 23 standards, started July 1992; construction of first of two prototypes (RA-44482) began January 1993; first flight, as SM-92 Finist (name of a magical bird that was transformed into a prince), made 28 December 1993; second Finist (RA-44484) completed round-the-world flight through Europe, Atlantic, Canada, Alaska and Siberia in August 1995 covering 16,200 n miles (30,000 km; 18,640 miles) in 160 flying hours; components production at SmAZ, Smolensk; assembly by VAPO at Voronezh; attempts to interest overseas licencees most recently (late 1998) involved Cuba. Sixth aircraft (RA-44493) completed as **SM-92P** armed version (described separately). PT6 and Walter M 101E turboprop versions under consideration, with possible assembly of former in Canada. AP-23 certification received by late 1998.

CURRENT VERSIONS: Basic aircraft accommodates up to seven persons with baggage. Convertible under field conditions to transport 600 kg (1,323 lb) freight; two stretcher patients and attendant with medical equipment; to drop six trainee parachutists or four firefighters with parachutes and firefighting equipment. Can carry hopper for 600 kg (1,323 lb) agricultural chemicals in cabin, with spraybars, or cameras for forest surveillance, patrolling electric power lines, gas pipeline inspection and similar duties.

CUSTOMERS: Ten ordered, seven built by beginning 1997; production of 16 for Russian Federation border guard (which eventually requires 300) was disrupted by bankruptcy of Smolensk aircraft factory; alternative manufacturer being sought. Regional government of Yakutia ordered 14 in October 1997, but none reported in service by mid-2000. First delivery (aircraft No. 3, RA-44485) to Mike Crymble in UK, 21 January 1995; RA-44487 to Sport-Para Centrum, Antwerp, Belgium, 6 July 1995.

COSTS: US$160,000 to US$200,000 (1998).

DESIGN FEATURES: Rugged light transport; in approximate class of out-of-production DHC-2 Beaver, but less powerful and lower in cost than most used Beavers. Sweptback fin and rudder; small dorsal fin; tailplane mounted on fin, with single bracing strut each side. Wide CG range.

Special wing section by Technoavia and CAHI (TsAGI). No sweep; thickness/chord ratio 15 per cent; dihedral 2°; incidence 3° at root, 1° at tip.

FLYING CONTROLS: Conventional and manual. Ailerons and elevators pushrod-actuated; rudder cable-actuated. Electrically operated, three-position (0, 20, 40°), two-section single-slotted flaps on each wing; horn-balanced rudder and elevators, with fluted skin; trim tab on starboard elevator; ground-adjustable tab on rudder; three-section tab along entire trailing-edge of each aileron, centre section being adjustable on ground.

STRUCTURE: All-aluminium, stressed-skin semi-monocoque construction. Simple, reliable structure, with no expensive or exotic materials; repair possible under field conditions. Airframe life 10,000 hours or 20,000 landings, with 'on condition' extension.

LANDING GEAR: Non-retractable tailwheel type with medium-pressure mainwheel tyres, size 600×180 on KT-317 wheels. Cantilever faired tubular steel main legs; steerable, semi-castoring tailwheel with 255×110 tyre on tubular steel strut. Tyre pressure 2.45 bars (35.5 lb/sq in) on mainwheels; 2.95 bars (43 lb/sq in) on tailwheel. Wheel/skis and amphibious or plain floats optional. Pedal-operated disc brakes, with parking lock. Minimum turning radius 6.6 m (21 ft 8 in).

POWER PLANT: One VOKBM M-14Kh air-cooled nine-cylinder radial engine, rated at 265 kW (355 hp) for take-off and 213 kW (286 hp) maximum continuous, driving a Mühlbauer MTV-3-B-C/L250-21 three-blade variable-pitch propeller. Engine TBO 1,000 hours; total life 3,000 hours. Two fuel tanks in wing leading-edge; usable capacity 380 litres (100.4 US gallons; 83.5 Imp gallons). Oil capacity 30 litres (7.9 US gallons; 6.6 Imp gallons).

ACCOMMODATION: Pilot and six passengers, in pairs; quickly removable seats with folding armrests and back; dual controls standard for pilot training; adjustable rudder pedals. Small baggage container (or medical equipment stowage for ambulance version) on port side at rear of

Technoavia SP-55M aerobatic lightplane   2001/0105833

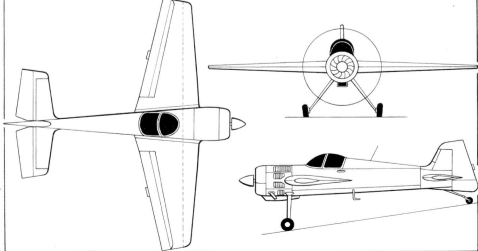

Technoavia SP-55 single-seat sportplane *(James Goulding/Jane's)*   2001/0103522

cabin. Forward-hinged, jettisonable door each side of flight deck; large rearward-sliding passenger/freight door on port side of cabin, openable in flight. Blister windows to flight deck and cabin. Steps on mainwheel legs and cable handholds for access to flight deck; removable tubular steel ladder beneath cabin door.

Convertible in field to transport 600 kg (1,323 lb) of freight; two stretcher patients and two attendants; six trainee parachutists or four smoke-jumpers with parachutes and firefighting equipment. Provision for carrying hopper for 600 kg (1,323 lb) of agricultural chemicals in cabin, or cameras for forest surveillance, patrolling electric power lines, gas pipeline inspection and similar duties. Clearance adequate for underbelly pannier. Cabin heated and ventilated.

SYSTEMS: Pneumatic system for engine starting, pressure 49 bars (710 lb/sq in). Electrical system provides 36/115 V AC power at 400 Hz and 28.5 V DC power, with 20NKBN-25 battery.

AVIONICS: *Comms:* Two Honeywell KY 96A VHF transceivers, KA-134 audio system, KR 87A ADF and KT-76A transponder; Garmin GPS 150.

DIMENSIONS, EXTERNAL:	
Wing span	14.60 m (47 ft 10¾ in)
Wing chord, constant	1.40 m (4 ft 7 in)
Wing aspect ratio	10.4
Length overall	9.30 m (30 ft 6¼ in)
Height: overall	3.08 m (10 ft 1¼ in)
fuselage horizontal	3.94 m (12 ft 11 in)
Tailplane span	5.53 m (18 ft 1¾ in)
Wheel track	3.35 m (11 ft 0 in)
Wheelbase	6.33 m (20 ft 9¼ in)
Propeller diameter	2.50 m (8 ft 2½ in)
Propeller ground clearance (tail up)	0.22 m (8¾ in)
Freight door: Height	1.12 m (3 ft 8 in)
Width	1.35 m (4 ft 5¼ in)
DIMENSIONS, INTERNAL:	
Cabin: Length	3.40 m (11 ft 1¾ in)
Max width	1.27 m (4 ft 2 in)
Max height	1.38 m (4 ft 6¼ in)
Volume	5.2 m³ (184 cu ft)
AREAS:	
Wings, gross	20.50 m² (220.7 sq ft)
Ailerons (total)	2.38 m² (25.62 sq ft)
Flaps (total)	3.28 m² (35.31 sq ft)
Fin	2.13 m² (22.93 sq ft)
Rudder	1.63 m² (17.55 sq ft)
Tailplane	3.73 m² (40.15 sq ft)
Elevators (total)	2.91 m² (31.32 sq ft)
WEIGHTS AND LOADINGS:	
Operating weight empty	1,500 kg (3,307 lb)
Payload: max	600 kg (1,323 lb)
with full fuel	550 kg (1,212 lb)
Max T-O and landing weight	2,350 kg (5,180 lb)
Max wing loading	114.6 kg/m² (23.48 lb/sq ft)
Max power loading	8.87 kg/kW (14.59 lb/hp)
PERFORMANCE:	
Never-exceed speed (VNE)	140 kt (260 km/h; 161 mph)
Max level speed	124 kt (230 km/h; 143 mph)
Max cruising speed	108 kt (200 km/h; 124 mph)
Econ cruising speed	92 kt (170 km/h; 106 mph)
Stalling speed, power off:	
flaps up	63 kt (115 km/h; 72 mph)
T-O flap	57 kt (105 km/h; 66 mph)
full flap	54 kt (100 km/h; 62 mph)
Max rate of climb at S/L	300 m (985 ft)/min
Service ceiling	3,000 m (9,840 ft)
T-O and landing run on grass	250 m (820 ft)
Range, 40 min reserves: with max payload:	
at max level speed	332 n miles (615 km; 382 miles)
at econ cruising speed	594 n miles (1,100 km; 683 miles)
with max fuel and 550 kg (1,212 lb) payload	647 n miles (1,200 km; 745 miles)
Endurance with max fuel	7 h 30 min

UPDATED

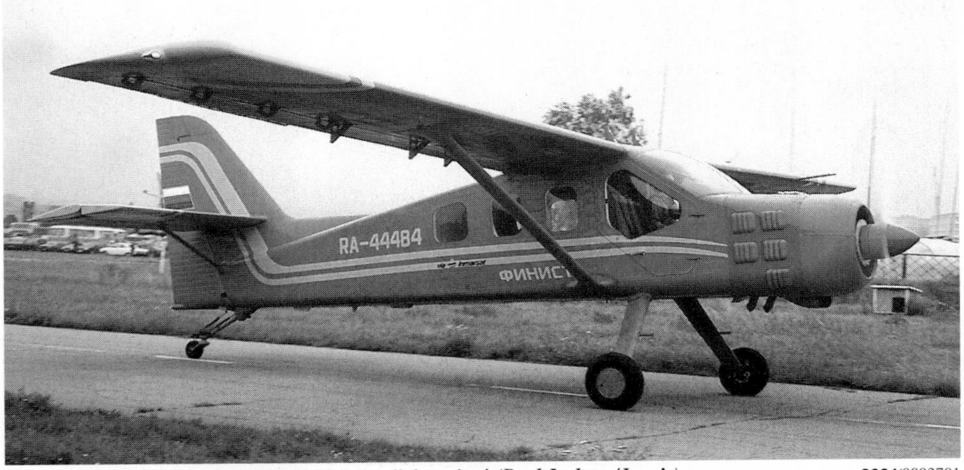

Technoavia SM-92 Finist (VOKBM M-14 radial engine) *(Paul Jackson/Jane's)* **2001**/0093701

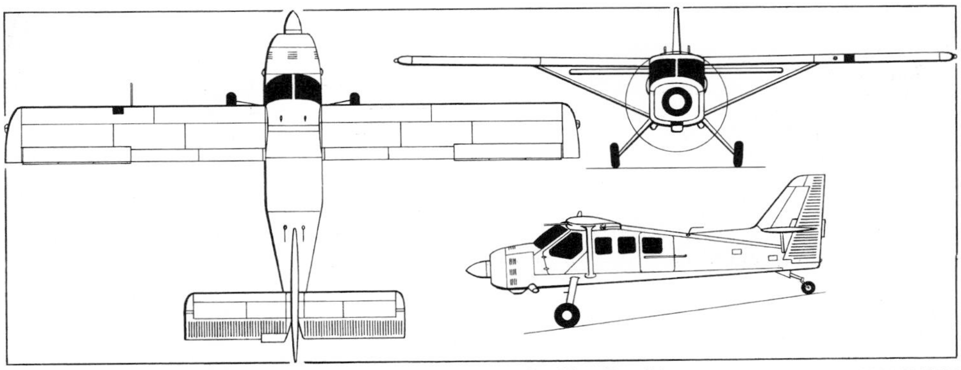

Technoavia SM-92 Finist seven-seat STOL aircraft *(James Goulding/Jane's)* **2000**/0062758

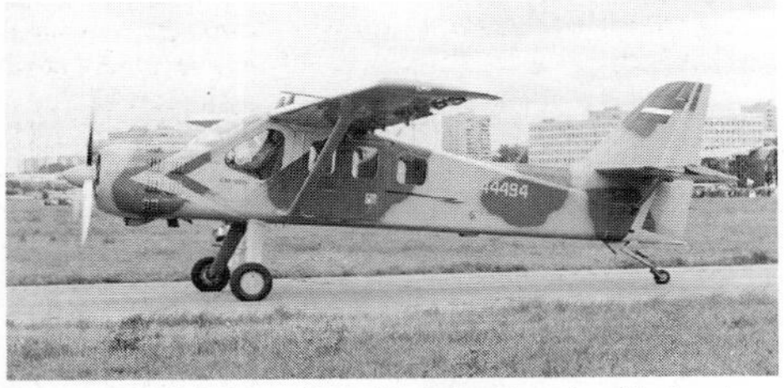

Second SM-92P, ordered by the Russian Federal Border Service *(Paul Jackson/Jane's)* **1999**/0051394

TECHNOAVIA SM-92P

TYPE: Light utility transport.

PROGRAMME: Sixth Finist (RA-44493) completed as SM-92P (*Pogranichni*: Frontier), in brown and green camouflage and with provision for armament; first flight August 1995; flight tests with full armament completed 1995; firing trials completed 1996; RA-44493 displayed at Moscow Air Show '95; second aircraft (RA-44494) complete by 1997, but no further manufacture known.

CUSTOMERS: Initial order for 16 from Federaliya Pogranichiya Sluzhba (Federal Border Service); to be stationed mainly in Baltic region; total orders for up to 300 possible.

DESIGN FEATURES: As standard SM-92, except as follows:

ACCOMMODATION: Crew of three: pilot, navigator and side-door gunner; dual controls, but instruments and PKV gunsight only for pilot in port seat.

ARMAMENT: Quickly removable outrigger pylon on each side for B-8V7 pack of seven 80 mm S-8 rockets and fixed PKT 7.62 mm machine gun with 250 rounds. Two underfuselage hardpoints for 100 kg bombs (or, possibly, auxiliary fuel tanks later). Pintle-mounted 7.62 mm machine gun, firing through open cabin doorway on port side.

DIMENSIONS, EXTERNAL: As SM-92

WEIGHTS AND LOADINGS: As SM-92, except:

Weight empty	1,380 kg (3,042 lb)

PERFORMANCE: As SM-92

UPDATED

TECHNOAVIA SMG-92

TYPE: Light utility turboprop.

PROGRAMME: Developed from SM-92 Finist, with emphasis on sport parachutist dropping; commissioned and financed by G-92 Commerce Ltd, of Hungary. First flight (HA-YDF) 7 November 2000; Hungarian type certificate awarded 14 December 2000; delivery to Wingglider Ltd in UK, January 2001.

Airframe and engine mounting produced by SmAZ at Smolensk; engine installation and cowling, paradrop step and handrails in vicinity of cabin door, new electrical system, instrument panel and engine controls by Aerotech Slovakia AS at Bratislava.

POWER PLANT: One 400 kW (536 shp) Walter M 601D-2 turboshaft, driving a V-508D-2 three-blade, fully feathering, reversible-pitch propeller.

ACCOMMODATION: Pilot and 10 parachutists; alternatively, pilot and up to seven passengers on lightweight seats.

Data generally as for SM-92 Finist, except that below.

DIMENSIONS, EXTERNAL:	
Length overall	9.795 m (32 ft 1¾ in)
WEIGHTS AND LOADINGS:	
Max T-O weight: paradrop	2,700 kg (5,952 lb)
passenger	2,350 kg (5,180 lb)
Max landing weight	2,350 kg (5,180 lb)
PERFORMANCE:	
Never-exceed speed (VNE)	164 kt (305 km/h; 189 mph)
Max operating speed (VMO)	140 kt (260 km/h; 161 mph)
Max level speed at 915 m (3,000 ft)	159 kt (295 km/h; 183 mph)
Stalling speed, flaps down	59 kt (108 km/h; 68 mph)
Time to 4,000 m (13,120 ft)	11 min
Descent from 4,000 m (13,120 ft)	5 min 30 s
T-O to 15 m (50 ft)	457 m (1,500 ft)

NEW ENTRY

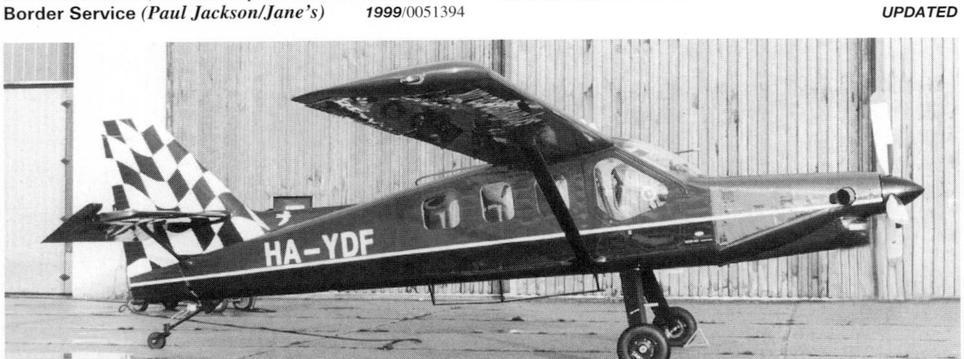

Technoavia SMG-92 sport parachuting turboprop **2001**/0105658

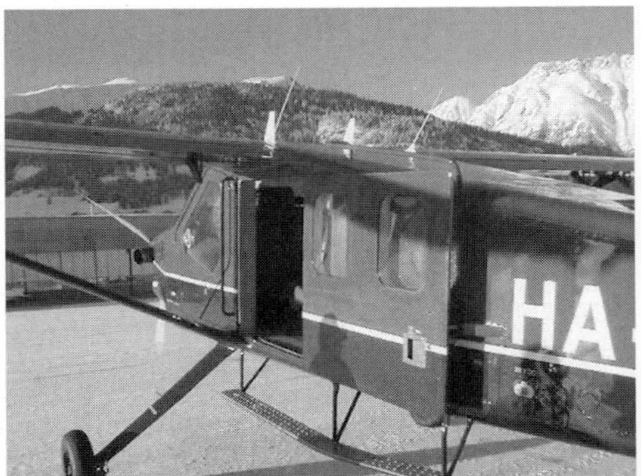

Parachutist door of Technoavia SMG-92 **2001**/0105657

TMZ

TUSHINSKY MASHINOSTROITELNYI ZAVOD AO (Tushino Machine-Building Plant Ltd)
ulitsa Svobod 35, 123362 Moskva
Tel: (+7 095) 493 30 47, 497 48 25 and 493 82 39

TMZ produces the I-3 aerobatic light aircraft, as described under the Interavia design bureau heading.

VERIFIED

TUPOLEV

TUPOLEV OAO (Tupolev JSC)
ulitsa Bakrushina 23, Korpus 1, 11250 Moskva
Tel: (+7 095) 238 34 13
Fax: (+7 095) 238 68 41
e-mail: tu@tupolev.ru
PRESIDENT: Aleksandr Polikov

Tupolev JSC formed 30 June 1999, combining Tupolev design bureau (see below) and Aviastar production plant (which see) at Ulyanovsk, with financial interest from Russian government amounting to 51 per cent, plus 43.6 per cent from Aviastar. JSC also holds intellectual rights to Tupolev designs previously owned by government. Kazan plant will be incorporated later; proposal submitted July 2000.

NEW ENTRY

AVIATSIONNY NAUCHNO-TEKHNISHESKY KOMPLEKS IMENI A N TUPOLEVA OAO (Aviation Scientific-Technical Complex named for A N Tupolev JSC)
Naberezhnaya Akademika Tupoleva 17, 111250 Moskva
Tel: (+7 095) 267 25 33
Fax: (+7 095) 267 27 33
e-mail: centre@tupolev.ru
GENERAL DIRECTOR: Vasili Aleksandrov
CHAIRMAN: Takevos Sourniov
CHIEF DESIGNERS:
  Oleg Alasheev
  Vladimir Andreev
  Valentin Dmitriyev
  Valentin Bliznyuk
  Igor S Kalygine
  Lev A Lakhnovsky
CHIEF ENGINEER: Anatoly V Sakharov
HEAD OF INFORMATION DEPARTMENT: Helen E Koutcherenko

Tupolev Bureau was founded in 1922 and concentrated primarily on large military and civil aircraft until the early 1990s; has designed 300 aircraft, of which 35 placed in production; current effort is 80 per cent civil programmes, although it was suggested in 1998 that Tupolev had begun preliminary design of a new bomber to replace the Tu-160 and, probably, Tu-95/142; this would enter service some time after 2010 and may be based on earlier studies, such as the Tu-202 bomber of the 1980s or the more recent Tu-404 700/850-seat airliner. The Bureau remains heavily committed to the development of a civil supersonic transport aircraft.

ANTK Tupolev has a 5.4 per cent share in Tupolev AO production group.

Tupolev's head office, main design bureau and experimental facility are in Moscow; Tomilino branch and flight research centre at Zhukovsky; and design offices at Samara, Kazan and Voronezh.

UPDATED

TUPOLEV Tu-22M

NATO reporting name: Backfire
TYPE: Medium/maritime reconnaissance bomber.
PROGRAMME: NATO revealed existence of a Soviet variable geometry bomber programme in 1969; development had begun 1962, to meet Soviet Air Force requirement; first of between five and nine Tu-22MO prototypes observed July 1970 on ground near Kazan plant; confirmed subsequently as twin-engined design by Tupolev OKB; first flight 30 August 1969; nine **Tu-22M-1** preproduction models for development testing, weapons trials and evaluation. Tu-22M-1 first flew July 1971; first displayed in West at 1992 Farnborough Air Show. Production at Kazan ended 1992, probably totalling nine Tu-22MO prototypes, nine Tu-22M-1s, 211 Tu-22M-2s and 268 Tu-22M-3s, or 497 in all. Manufacturer's designations prefixed *Izdelie*: Article.
CURRENT VERSIONS: **Tu-22M-2** (*Izdelie* 45-02; 'Backfire-B'): First series production version; entered service with 185th Guards Heavy Bomber Aviation Regiment at Poltava during 1975; operational 1978; differs from original Tu-22M-1 ('Backfire-A') in having increased wing span, and wing trailing-edge pods eliminated except for shallow underwing fairings, no longer protruding beyond trailing-edge; wings and tail surfaces changed to almost supercritical section during production; maximum sweep increased from 60° to 65°; slightly inclined lateral air intakes, with large splitter plates; seen usually with optional flight refuelling nose probe removed and its housing replaced by long fairing. Two Samara/Kuznetsov NK-22 turbofans (each 215.8 kN; 48,500 lb st). Armament normally one Kh-22 (AS-4 'Kitchen') ASM semi-recessed under fuselage; maximum weapon load 21,000 kg (46,300 lb); optional MBDZ-U9-68 external stores racks under engine air intake trunks; two GSh-23 twin-barrel 23 mm guns in UKU-9K-502 tail mounting with barrels side by side horizontally, initially beneath ogival radome, later with drum-shape radome of larger diameter. Two squadrons operational in Afghanistan December 1987 to January 1988.

Tu-22MLL (*Letayouchaya Laboratoriya*: Airborne Laboratory): Converted Tu-M22-3 used by LII at Zhukovsky for supersonic laminar flow research. Aircraft flies with a system which sucks air through multiple perforations in the wing leading-edge; variable geometry allows it to test the system at various sweep angles, and over a large speed range. Trials began in 1994 with wing sweep angle of 20°; future testing planned in two stages: up to 35° in subsonic flight, then 45 to 65° at speeds up to M1.6. Further development will be undertaken by a Tu-154LL.

Tu-22M-2Ye: NK-25-engined aircraft built in prototype form only as cheap, low-risk alternative to Tu-22M-3.

Tu-22M-3 (*Izdelie* 45-03; 'Backfire-C'): Advanced long-range bomber and maritime version; first flown 20 June 1977; entered service with 185th Guards Heavy Bomber Aviation Regiment 1983; deployed with Black Sea Fleet air force 1985. New NK-25 engines of approximately 25 per cent higher rating and strengthened wings, to permit increased weapons load; wedge-type engine air intakes; upturned nosecone; no flight refuelling probe; improved avionics, including INS, active and passive ECM, new radios and electro-optical bombsight; automated flight controls; rack for a Kh-22 (AS-4 'Kitchen') ASM under each fixed-wing centre-section panel, for a maximum three Kh-22s; rotary launcher in weapons bay for six Kh-15P (AS-16 'Kickback') short-range attack missiles, with provision for four more underwing as alternative to Kh-22s; single GSh-23M twin-barrel 23 mm gun with a higher rate of fire, with four barrels superimposed, in aerodynamically improved tail mounting, beneath large drum-shape radome. Guns primarily used for firing chaff/flare cartridges.
Detailed description applies specifically to Tu-22M-3.

Tu-22M-4: Single prototype of a proposed 1990 Tu-22M-3 upgrade. Aircraft (probably the Zhukovsky-based Tu-22M-3 2145345) may have then become the Tu-22M-5 prototype testbed.

Tu-22M-5: Reported designation for Russian Air Forces upgrade of Tu-22M-3, with OKB designation of

Tupolev Tu-22M-3 ('Backfire-C') medium bomber of 240th Guards Training Aviation Regiment, Russian Navy *(Yefim Gordon)*

Navigator's station in Tu-22M-3 *(Yefim Gordon)*

Izdelie 245. Perhaps a less ambitious and less costly alternative to original Tu-22M-4 proposal. A Zhukovsky-based Tu-22M-3 (2145345) may be the prototype for the new upgrade, though this cannot be confirmed. 'Production' conversions are supposedly to be undertaken at 360th Repair Plant at Ryazan, though there is no evidence of any such programme at Ryazan, whose workshops are busy with a Tu-95MS upgrade. Reports that work is 'well advanced' may actually reflect earlier work carried out for Tu-22M-4 upgrade. Funding for a new Tu-22M upgrade would seem unlikely. Some sources suggest that modified aircraft will receive a new attack radar with automatic Terrain Avoidance capabilities as well as a new GPS/INS-based navigation system. The aircraft is intended to be compatible with a range of new weapons, including Raduga Kh-101 CALCMs and nuclear-armed Kh-102s, and possibly also the derived medium range Kh-SD and the supersonic Kh-32 missiles.

Tu-22MP: EW/escort jammer version with MIASS EW system. Included with Tu-22MR (which see) in previous editions of *Jane's*. First prototype believed to have been identified by US satellites or intelligence resources at Kazan in 1986; second and third prototypes built in 1992, but further limited conversion programme using redundant Tu-22M-3 airframes has yet to be undertaken. Aircraft carries a semi-recessed pod in the weapons bay, with bulged dielectric antenna fairings on the sides of the intake ducts and at the root of the dorsal fin fillet. At least one Tu-22MP believed to be undergoing long-term testing at Akhtubinsk.

Tu-22MR: Twelve Tu-22M-3 airframes converted as reconnaissance aircraft for AV-MF (naval aviation) units. Shompol SLAR in belly-mounted 'canoe' fairing and with Tangazh sigint suite, Osen IR reconnaissance equipment and conventional optical cameras. Augmented from 1994 by conversion of Tu-22M-2s to similar standards to Tu-22M-2Rs.

CUSTOMERS: Total of 497 built, including prototypes. Russian Air Forces have 90; Naval aviation 60. Before break-up of USSR, equipped 52nd, 132nd, 184th, 185th, 200th, 260th, 402nd and 840th Regiments at Shaykovka, Tartu (Estonia), Priluki (Ukraine), Poltava (Ukraine), Bobrinsk

Cockpit of Tupolev Tu-22M-3 ('Backfire-C') *(Yefim Gordon)*

454　RUSSIAN FEDERATION: Aircraft—TUPOLEV

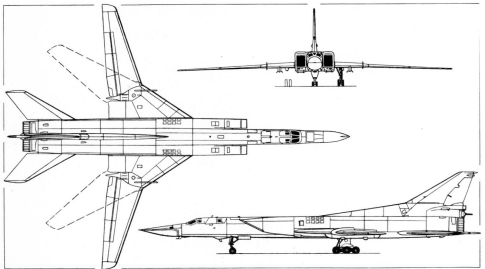

Tupolev Tu-22M-3 ('Backfire-C') bomber and maritime reconnaissance/attack aircraft
(Dennis Punnett/Jane's)

(Belarus), Stryy (Ukraine), Balbasaro (Belarus) and Soltsi, and an unidentified unit at Belaya as well as non-European-based units in the Soviet Far East. The AV-MF had the 5th Regiment at Vesioloye (Ukraine), the 540th at Kulbakino (Ukraine), the 574th at Lakhta, the 924th at Oleni and the 943rd at Oktyabrskoye, and two Pacific Fleet Regiments. Estimated 54 to 70 in Ukraine (16 Tu-22M with 185 Regiment at Poltava, 16 Tu-22M with 184 Regiment at Priluki and 14 Tu-22M-3 with 260 Regiment at Stryy and 24 with 540 Naval Missile Regiment at Kulbakino); first four of 20 former Black Sea Fleet aircraft transferred to Ukraine January 1996, following the disbandment of the 943rd Regiment. Persistent reports suggest that the Tu-22M has been actively marketed to India and Iran, and possibly also to Libya and Syria. India was reported to have discussed leasing Tu-22M-3s during 1999, with some sources suggesting a requirement for only four aircraft. There are suggestions of a purchase option at the end of a three- or four-year lease. Ukraine will reportedly withdraw and scrap its remaining 15 Tu-22Ms by the end of 2001, following a Ukrainian Defence Ministry announcement on 25 May 2000.

Tu-22M CURRENT OPERATORS

Unit	Base	Variant
Russia (Air Forces)		
49th ITBAP	Ryazan	Tu-22M-3
52nd GvTBAP	Shaykovka	Tu-22M-2, M-3
840th TBAP	Sol'tsy	Tu-22M-3
Unident TBAP	Belaya-Tserkov	Tu-22M-3
Unident TBAP	Belaya-Tserkov	Tu-22M-3
Unident TBAP	Ussurisysk	Tu-22M-3
Unident TBAP	Zavitinsk	Tu-22M-3
Russia (Naval Aviation)		
240th GvUAP	Ostrov	Tu-22M[1]
574th MRAP	Lakhta	Tu-22M
924th MRAP	Olenya	Tu-22M
Unident MRAP	Alekseyeva	Tu-22M
Unident MRAP	Alekseyeva	Tu-22M
Unident MRAP	Artem	Tu-22M
Unident MRAP	Artem	Tu-22M
Ukraine		
6th AvB	Nikolaev	Tu-22M-3[1]
184th TBAP	Priluki	Tu-22M-3, Tu-134UBL
185th TBAP	Poltava	Tu-22M-3

[1] Plus other aircraft types; training role
Notes: MRAP is *Morskoy Razvedivatel'nyi Aviatsion'nyi Polk* (Naval Reconnaissance Aviation Division); GvUAP is *Gvardiya Ucheb'nyi Aviatsion'nyi Polk* (Guards Training Aviation Regiment); TBAP is *Tyazhel'nyi Bombardirovch'nyi Aviatsion'nyi Polk* (Heavy Bomber Aviation Regiment); AvB is *Aviatsion'nyi Baza* (Aviation Base)

DESIGN FEATURES: Capable of performing nuclear strike, conventional attack and anti-ship missions; low-level penetration features ensure better survivability than earlier Tupolev bombers; deployment of Kh-15P (AS-16 'Kickback') short-range attack missiles in Tu-22Ms has increased significantly their weapon carrying capability. Low/mid-wing configuration; large-span fixed centre-section and two variable geometry outer wing panels (20, 30, 65° sweepback); no anhedral or dihedral; leading-edge fence towards tip of centre-section each side; basically circular fuselage forward of wings, with ogival dielectric nosecone; centre-fuselage faired into rectangular section air intake trunks, each embodying one fixed and two variable horizontal compression ramps; leading-edge of sidewalls raked at about 65°; 12 auxiliary intake doors in each duct over forward portion of wing; no external area ruling of trunks; all-swept tail surfaces, with large dorsal fin.

FLYING CONTROLS: Automatic high- and low-altitude preprogrammed flight control, with automatic approach. Full-span leading-edge slat, aileron and three-section slotted trailing-edge flaps aft of spoilers/lift dumpers on each outer wing panel; all-moving differential horizontal tail surfaces; inset rudder.

LANDING GEAR: Hydraulically retractable tricycle type; rearward-retracting twin nosewheels; each mainwheel bogie comprises three pairs of wheels in tandem, with varying distances between each pair (narrowest track for front pair, widest track for centre pair); mainwheel tyres size 1,030 × 350 mm, pressure 11.75 bars (170 lb/sq in); nosewheel tyres size 1,000 × 280 mm, pressure 9.32 bars (135 lb/sq in); bogies pivot inward from vestigial fairing under centre-section on each side into bottom of fuselage. Brake-chute housed inside large door under rear fuselage.

POWER PLANT: Two Samara/Kuznetsov NK-25 turbofans, side by side in rear fuselage, each 245.2 kN (55,115 lb st) with afterburning. Integral fuel tanks in centre-fuselage between engine ducts, centre-section carry-through structure, forward portion of each fixed-wing panel, between spars of variable geometry wing panels and in lower portion of fin. APU in dorsal fin. Provision for JATO rockets.

ACCOMMODATION: Pilot and co-pilot side by side, under upward-opening jettisonable gull-wing doors hinged on centreline; navigator and weapon system officer further aft, with similar doors, as indicated by position of windows between flight deck and air intakes. KT-1 ejection seats (200 ft/70 to 162 kt; 60 m/130 to 300 km/h; 200 ft/81 to 186 mph) for all four crew members.

SYSTEMS: Air conditioning system supplies pressurised crew compartment and specialised equipment compartments; air bled from twelfth stage of engine compressors. Three independent hydraulic systems, pressure 207 bars (3,000 lb/sq in), to power flight control system, landing gear actuation, wheelbrakes and wing sweep actuators. Pneumatic system supplied by bottles, pressure 147 bars (2,130 lb/sq in). Three independent electrical systems: 27 V DC system supplied by four GSR-20BK generators (two on each engine), with two storage batteries; 200/115 V three-phase 400 Hz AC system supplied by two GT 60 NZHCH12P generators; 36 V three-phase 400 Hz AC system supplied by two TS 350S04A transformers. Gaseous oxygen system supplied from bottles, pressure 147 bars (2,130 lb/sq in). Electrothermal anti-icing of air intakes, flight deck windows and strike sight window; hot-air anti-icing of inlet guide vanes; radioactive ice detector.

AVIONICS: *Comms:* Secure com.
Radar: Leninets PN/AD missile targeting and nav radar ('Down Beat') inside dielectric nosecone; radar ('Box Tail') for tail turret, above guns.
Flight: Accurate autonomous nav.
Mission: Strike sight fairing with flat glazed front panel under front fuselage, for video camera to provide visual assistance for weapon aiming from high altitude.
Self-defence: IR missile approach warning sensor above fuselage aft of cockpit; eight chaff/flare multiple dispensers in bottom of each engine duct between wingroot and tailplane, another in each tailplane root fairing.

ARMAMENT: Maximum offensive weapon load three Kh-22 (NATO AS-4 'Kitchen') ASMs, one semi-recessed under centre-fuselage, one under fixed centre-section panel of each wing; or 24,000 kg (52,910 lb) of conventional bombs or mines, half carried internally and half on racks under wings and engine air intake trunks. Internal bombs can be replaced by rotary launcher for six Kh-15P (AS-16 'Kickback') short-range attack missiles, with four more underwing as alternative to Kh-22s. Normal weapon load is single Kh-22 or 12,000 kg (26,455 lb) of bombs, Kh-31A/P (AS-17 'Krypton') or Kh-35 (AS-20 'Kayak') ASMs. Typical loads two FAB-3000, eight FAB-1500, 42 FAB-500 or 69 FAB-250 or -100 bombs (figures indicate weight in kg), or eight 1,500 kg or 18 500 kg mines. Upgraded Tu-22M-5 may carry conventional Kh-101 or nuclear Kh-102 long-range subsonic air-launched cruise missiles, medium-range Kh-SDs and supersonic Kh-32s (derived from the Kh-22) in addition to the usual Tu-22M-3 weapons options. All versions may soon receive the upgraded KAB-1500LG, a laser-guided bomb with embedded GPS. Tu-22M-3 and Tu-22M-5 feature one GSh-23M twin-barrel 23 mm gun, with barrels superimposed, in radar-directed tail mounting.

DIMENSIONS, EXTERNAL (Tu-22M-3):
Wing span: fully spread	34.28 m (112 ft 5¼ in)
fully swept	23.30 m (76 ft 5½ in)
Wing aspect ratio: fully spread	6.4
fully swept	3.1
Length overall	42.46 m (139 ft 3¾ in)
Height overall	11.05 m (36 ft 3 in)
Weapons bay: Length	Approx 7.00 m (22 ft 11½ in)
Width	approx 1.80 m (5 ft 10¼ in)

AREAS:
Wings: net	164.00 m² (1,765.3 sq ft)
gross: 20° sweep	183.58 m² (1,976.1 sq ft)
65° sweep	175.80 m² (1,892.4 sq ft)

WEIGHTS AND LOADINGS (Tu-22M-3):
Max weapon load	24,000 kg (52,910 lb)
Fuel load	approx 50,000 kg (110,230 lb)
Max T-O weight	124,000 kg (273,370 lb)
Max T-O weight with JATO	126,400 kg (278,660 lb)
Normal landing weight	78,000 kg (171,955 lb)
Max landing weight	88,000 kg (194,000 lb)
Max wing loading (without JATO):	
20° sweep	675.5 kg/m² (138.34 lb/sq ft)
65° sweep	705.4 kg/m² (144.45 lb/sq ft)
Max power loading (without JATO)	
	253 kg/kN (2.48 lb/lb st)

PERFORMANCE (Tu-22M-3):
Max level speed: at high altitude	
	M1.88 (1,080 kt; 2,000 km/h; 1,242 mph)
at low altitude	M0.86 (567 kt; 1,050 km/h; 652 mph)
Nominal cruising speed at height	
	485 kt (900 km/h; 560 mph)
T-O speed	200 kt (370 km/h; 230 mph)
Normal landing speed	154 kt (285 km/h; 177 mph)
Service ceiling	13,300 m (43,640 ft)
T-O run	2,000-2,100 m (6,560-6,890 ft)
Normal landing run	1,200-1,300 m (3,940-4,265 ft)
Unrefuelled combat radius:	
supersonic, hi-hi-hi, 12,000 kg (26,455 lb) weapons	
	810-1,000 n miles (1,500-1,850 km; 930-1,150 miles)
subsonic, lo-lo-lo, 12,000 kg (26,455 lb) weapons	
	810-900 n miles (1,500-1,665 km; 930-1,035 miles)
subsonic, hi-lo-hi, 12,000 kg (26,455 lb) weapons	
	1,300 n miles (2,410 km; 1,495 miles)
subsonic, hi-hi-hi, max weapons	
	1,188 n miles (2,200 km; 1,365 miles)
g limit	+2.5

UPDATED

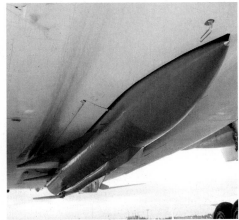

AS-4 'Kitchen' ASM semi-recessed below a Tu-22M-3 'Backfire-C' *(Paul Jackson/Jane's)*
2000/0062699

TUPOLEV Tu-95 and Tu-142
NATO reporting name: **Bear**
TYPE: Strategic/maritime reconnaissance bomber.
PROGRAMME: **Tu-95/1** prototype, with four 8,950 kW (12,000 ehp) Kuznetsov 2TV-2F turboprops, flew 12 November 1952, was destroyed during testing. **Tu-95/2** with 8,950 kW (12,000 ehp) TV-12 turboprops, flew 16 February 1955. Seven **Tu-95**s ('Bear-A') took part in 1955 Aviation Day flypast; operational with strategic attack force 1956; **Tu-95M** ('Bear-A') was production version

with more powerful NK-12 engines to increase speed; experimental **Tu-95K** of 1956 airdropped the MiG-19 SM-20 aircraft equipped to test features of the Kh-20 missile system; production **Tu-95K-20** ('Bear-B') of 1959 was armed with a Kh-20 (AS-3 'Kangaroo') air-to-surface missile; **Tu-95KD** of 1961 was similar to Tu-95K-20, with added flight refuelling noseprobe. Conversions to Tu-95KM standards, with ECM tailcone. 'Bear' series remained in almost continuous, latterly small scale, production for 38 years, ending 1994.

Variants included **Tu-96** of 1956, a high-altitude high-speed bomber development with NK-16 engines, built but not flown; **Tu-116/114D**, with Tu-95 airframe adapted as civil aircraft; **Tu-119**, a Tu-95M converted, but not flown, as testbed for a supposed nuclear engine; and production **Tu-126** ('Moss') AEW&C aircraft (see 1990-91 *Jane's*). All 'Bears' with original Tu-95 airframes, including the **Tu-95RTs** ('Bear-D'), **Tu-95MR** ('Bear-E') and **Tu-95K-22** ('Bear-G') described in the 1996-97 *Jane's*, had been grounded or scrapped by 1997. Current versions listed utilise basic Tu-142 airframe, regardless of their designations.

CURRENT VERSIONS: **Tu-142** (*Izdelie VP*; 'Bear-F'): Anti-submarine version; first of three prototypes (4200-4202) flew July 1968; extensively redesigned compared with Tu-95; leading-edge camber to reduce drag; new, efficient, double-slotted flaps; fuselage forward of wing leading-edge extended; increased rudder chord. Initial production series of 12 briefly fitted with 'rough-field' landing gear with 12-wheel bogies in enlarged inboard engine nacelle 'tails'; subsequently removed. Avionics and mission equipment based on Il-38, with Berkut search radar, RGB-1 and RGB-2 sonobuoys, ANP-3V navigation system, TsVM-263 computer and J-band navigation/weather radar under 'chin'.

Tu-142 ('Bear-F' Mod 1): Weight reduced by 3,630 kg (8,000 lb) with new, strengthened four-wheel landing gear bogies. Four of last five Kuybyshev-built aircraft completed to this standard (from 4231) and earlier aircraft modified to similar configuration. The first one or two Taganrog-built Tu-142s may have been to this standard.

Tu-142M ('Bear-F' Mod 2): The last Kuybyshev Tu-142, completed in 1972, introduced a new cockpit, with a raised roofline. This revised configuration was adopted on the Tu-142s built at Taganrog – designated Tu-142M to differentiate them from Kuybyshev aircraft. Possible increase in fuel capacity. M suffix not used by Russian naval aviation (AV-MF).

Tu-142MK (*Izdelie VPM*; 'Bear-F Mod 3'): Some of the first Tu-142Ms (4243, 4244 and 4264) served as testbeds for modernised ASW system with Korshun-K radar and STS and MMS-106 Ladoga MAD in fin-mounted fairing. This incorporated in production Tu-142MKs from 1978 after protracted development problems. Entered service 1980, known to AV-MF as Tu-142M. NPK-142M navigation system, Strela 142M communications suite, Nerchinsk sonar set. Korshun-equipped aircraft compatible with RGB-15 and RGB-75 detection sonobuoys (usable with explosive sound sources) and with directional RGB-25s (passive) and RGB-55s (active); 11,186 ekW (15,000 eshp) NK-12M turboprops, driving AV-90 contrarotating propellers.

Tu-142MK-E (*Tu-142MK-A* 'Bear-F' Mod 3): Eight built for India. Broadly equivalent to Tu-142MK for AV-MF with some downgraded systems. Efforts reportedly under way to integrate Matra BAe Sea Eagle ASM.

Tu-142M-Z ('Bear-F' Mod 4): Final production ASW platform. Further improved ASW system and improved ECM. NK-12MV engines replaced by more powerful NK-12MPs. TA-12 APU; Korshun-KN-N-STS ASW

Tupolev Tu-142MR VLF communications aircraft (*Yefim Gordon*) 2000/0062655

complex with Korshun radar, Nashaty-Nefrit sonics (compatible with RGB-16 and RGB-26 buoys) and Zarechye sonar system. Able to detect, locate and track submarines. Prototype (converted Tu-142M) first flew 1985; state acceptance trials 1987; fully operational 1993. Final aircraft built 1994.

Data revealed in 1998 indicated that a 'Bear-F' Mod 4 update may be undertaken with the Leninets Sea Dragon sensor suite as a replacement for Korshun. Systems comprise radar, electro-optical (FLIR/LLLTV), sonobuoys, ESM and MAD, managed at three operator consoles; armament augmented by up to eight Kh-35 (AS-20 'Kayak') anti-ship missiles.

Tu-142MP: Single ASW test and trials aircraft.

All versions of Tu-142M have a crew of 10, and can carry eight Kh-35 (AS-20 'Kayak') active radar homing anti-ship missiles in underwing pairs.

Tu-95MS (Tu-95M-55 'Bear-H'): Late production bomber; crew of seven; based on Tu-142 airframe but fuselage shortened; improved NK-12MA engines, with longer TBO, rating unchanged. Initial **Tu-95MS16** ('Bear-H16') version carried six Kh-55 (AS-15A 'Kent') or RKV-500B (AS-15B 'Kent-B') long-range cruise missiles on an MKU-6 internal rotary launcher, two more under each wingroot, and a cluster of three between each pair of engines, for a total of 16. All will be modified to **Tu-95MS6** ('Bear-H6') final production standard, with the pylons for 10 underwing missiles removed to conform with SALT/START Treaty limitations. Built at Kuybyshev; achieved IOC 1984; larger and deeper radome ('Clam Pipe') built into nose; small fin-tip fairing for IR warning receiver; some aircraft have single twin-barrel 23 mm gun instead of usual pair in tail turret. Current plans envisage re-equipment of Tu-95MS with eight conventionally armed Kh-101 opto-electronic/TV homing subsonic ALCMs with range of approximately 1,620 n miles (3,000 km; 1,865 miles) or, alternatively, 14 Kh-SD ASMs with range of less than 324 n miles (600 km; 373 miles); both carry non-nuclear warheads and are intended for high-precision attacks.

Detailed description applies to Tu-95MS, except where indicated.

Tu-95MA: Cruise missile launch platform. Project only.

Tu-142MR ('Bear-J'): Entered service 1980; modified Tu-142M-Z airframe. Soviet counterpart of US Navy E-6A and EC-130Q Tacamo, with VLF communications avionics to maintain on-station/all-ocean link between national command authorities and nuclear missile armed submarines under most operating conditions; large ventral pod for VLF suspended wire antenna, several kilometres long, under centre-fuselage in weapons bay area; undernose fairing as 'Bear-F' Mod 4; fin-tip IR warning pod similar to that on some 'Bear-Hs'; satcom dome aft of flight deck canopy. About 10 operational with Northern and Pacific Fleets.

CUSTOMERS: Russia received a total of 33 'Bear-H6' and 56 'Bear-H16' though only 68 of these were declared under START as being 'deployed', 64 of them as ALCM carriers; that number is believed to have fallen to 60. Between 30 and 40 are expected to be retained beyond 2007. Small numbers of earlier variants remain in use. Naval Aviation has 55 'Bear-F', mostly Mod 3/4 and 24 'Bear-J'; Kazakhstan still has six for training, with sealed bomb bays and other modifications, exempt from START. Further 23 of 25 Tu-95MSs left behind in Uzin, Ukraine, together with two Tu-95M/Ks; these were all to be scrapped with US assistance under the Co-operative Threat Reduction programme, but in October 1999 it was announced that three of the MS variants would be returned to Russia; according to August 2000 report, further seven MSs may be bargained towards cost of natural gas supplied by Russia. Indian Navy (seven 'Bear-F' Mod 3, of eight delivered, with No. 312 INAS). (See table).

Tu-95 CURRENT OPERATORS

Unit	Base	Variant (Quantity)
Russia		
37th Air Army		
49th ITBAP	Ryazan	Tu-95SU (11), Tu-95KU (2)[1]
182nd TBAP	Engels	Tu-95MS-16 (22)
1223rd (?) TBAP	Ukrainka	Tu-95MS-6 (20)
1226th (?) TBAP	Ukrainka	Tu-95MS-6 (7), Tu-95MS-16 (13)
Russia (Naval Aviation)		
240th GvUAP	Ostrov	Tu-142M-Z (4)[1]
76th OPLAP-DD	Kipelovo	Tu-142M-Z, Tu-142MR
Unident OMUAP	Pskov	Tu-142M-Z
Unident OPLAP-DD	Alekseyeva	Tu-142MK (poss also MR, MZ)
Unident OPLAP-DD	Artem	Tu-142MK (poss also MR, MZ)
India		
No. 312 Sqd	Dabolim	Tu-142MK-E (7)
Kazakhstan		
Unident unit	unknown	Tu-95K-22 (6)

[1] And other aircraft types

Note: Ukraine has 21 Tu-95MSs, of which three being transferred to Russia (the first on 5 November 1999) and remainder being scrapped.

GvUAP is *Gvardiya Ucheb'nyi Aviatsion'nyi Polk* (Guards Training Aviation Regiment); ITBAP is *Instruktsiya Tyazhel'nyi Bombardirovch'nyi Aviatsion'nyi Polk* (Heavy Bomber Instruction Aviation Regiment); TBAP is *Tyazhel'nyi Bombardirovch'nyi Aviatsion'nyi Polk* (Heavy Bomber Aviation Regiment); OMUAP is *Otdel'nyi Morskoy Ucheb'nyi Aviatsion'nyi Polk* (Independent Naval Training Aviation Regiment); OPLAP-DD is *Otdel'nyi Protivolodoch'nyi Aviatsion'nyi Polk – Dalneyo Deystviya* (Independent Anti-Submarine Aviation Regiment – Long Range).

DESIGN FEATURES: Unique large, high-performance, four-turboprop combat aircraft; all-swept (30° 30'), high-aspect ratio mid-wing configuration; fuselage same diameter as US Boeing B-29/Soviet Tu-4; main landing gear retracts into wing trailing-edge nacelles; contraprops with high tip speeds; wing anhedral 2° 30'; incidence 1°; sweepback at quarter-chord 35° on inner wings, 33° 30' outer panels.

FLYING CONTROLS: Conventional and hydraulically boosted. Three-segment aileron and two-segment double-slotted area-increasing flap each wing; trim tab in each inboard aileron segment; upper-surface spoiler forward of each inboard aileron; adjustable tailplane incidence (3° up/1° down); trim tab in rudder and each elevator.

STRUCTURE: All-metal; four spars in each inner wing, three outboard; three boundary layer fences above each wing; circular section semi-monocoque fuselage containing three pressurised compartments; tail gunner's compartment not accessible from others.

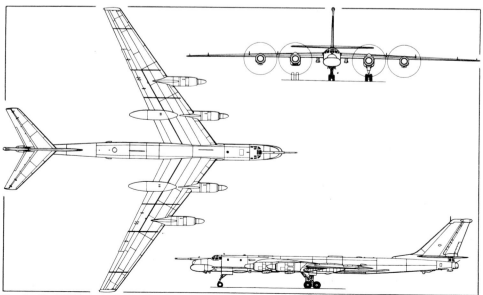

Tupolev Tu-95MS strategic bomber, known to NATO as 'Bear-H' (*Dennis Punnett/Jane's*)

456   RUSSIAN FEDERATION: AIRCRAFT—TUPOLEV

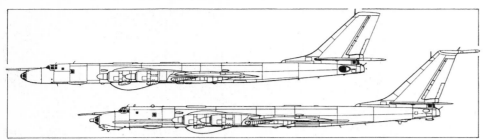

Tu-142 'Bear-F' Mod 1 (top) and Tu-142M-Z 'Bear-F' Mod 4 *(Dennis Punnett/Jane's)*

Tupolev Tu-95MS ('Bear-H') long-range bomber *(Paul Jackson/Jane's)*   2000/0062700

LANDING GEAR: Hydraulically retractable tricycle type; main units consist of four-wheel bogies, tyre size (Tu-95) 1,450×450, pressure 10.35 bars (150 +5/–10 lb/sq in) or (Tu-142) 1,500×500; twin nosewheels, tyre size 1,140×350, pressure 8.27 bars (120 lb/sq in); hydraulic internal expanding brakes; steerable nose unit; all units retract rearward, main units into nacelles built on to wing trailing-edge. No brake-chutes.

POWER PLANT: Four Samara Kuznetsov NK-12MP turboprops, each 11,033 kW (14,795 ehp); eight-blade contrarotating Type AV-60N propellers. Fuel in four integral wing tanks and in two bay tanks in the centre section, with a further tank in the rear fuselage; normal capacity 95,000 litres (25,100 US gallons; 20,900 Imp gallons). Single pressure refuelling point under starboard wingtip. Flight refuelling probe above nose extends forward further 0.5 m (1 ft 7 in) as it enters drogue; fuel flows through duct on starboard side of front fuselage to main tanks; flush light each side of probe in upper part of nose aids night refuelling.

ACCOMMODATION: Crew of seven: pilots side by side on pressurised flight deck; other four crew in forward compartment face rearward; com operator behind pilot, nav/defensive systems operator further aft on port side; flight engineer behind co-pilot, with spare seat for an observer; sixth flight crew member, equivalent to US bombardier/navigator, aft on central seat; gunner in rear turret compartment, with entry via ventral hatch. Remainder of crew enter through hatch in top of nosewheel bay. No ejection seats. Conveyor in flight deck floor carries crew members to hatch in nosewheel bay, with landing gear lowered, in emergency. Astrodome in roof over sixth crew member.

SYSTEMS: Flight crew accommodation pressurised by an ARD-54 regulator. Thermal anti-icing of wing and tailplane leading-edges. Gas-turbine APU in dorsal fin, with exhaust above tailplane leading-edge. KP-24M oxygen unit. DC power provided by eight GSR-18000M engine-driven generators, with Type 12 SAM-55 accumulator batteries. Two secondary AC systems provided by Type PO-4500 and PT-1000 converters and four engine-driven SGO-30U AC generators.

AVIONICS: *Comms:* RSBN short-range nav system 'bow and arrow' antenna under 'Box Tail' radome. Triangular D-band IFF antenna on top of nose refuelling probe.

Radar: Obzor nav/bombing radar ('Clam Pipe') under nose; weather radar above; duct for cooling air from rear of lower radome to rear fuselage on port side; gun fire-control radar ('Box Tail') at base of rudder.

Self-defence: ECM fairing each side of weather radar; small pylon-mounted ECM pod each side at base of tail gunner's compartment. Mak IR warning domes, each with about 200 discrete circular windows, under nose and above wings on top of fuselage; 'Ground Bouncer' ECM jamming system with tooth-shape antennas to port side of undernose IR warning dome and under rear fuselage; RWR each side of front fuselage and fin. Eight three-tube chaff/flare dispensers in two rows aft of landing gear doors on lower side of each wing pod.

EQUIPMENT: LAS-5M liferaft stowage in front of dorsal fin, with two PSN-6A rescue rafts stowed further forward.

ARMAMENT: See notes applicable to individual versions.

DIMENSIONS, EXTERNAL ('Bear-H'):
Wing span	50.04 m (164 ft 2 in)
Wing aspect ratio	8.6
Length overall	49.13 m (161 ft 2¼ in)
Height overall	13.30 m (43 ft 7¾ in)
Fuselage: max diameter	2.90 m (9 ft 6¼ in)
Wheel track	12.53 m (41 ft 1¼ in)
Wheelbase	14.83 m (48 ft 7¾ in)
Propeller diameter	5.60 m (18 ft 4½ in)

AREAS:
Wings, gross	289.90 m² (3,120.5 sq ft)
Ailerons (total)	23.50 m² (252.95 sq ft)
Flaps (total)	45.66 m² (491.48 sq ft)
Tailplane	40.90 m² (440.24 sq ft)
Elevators (total)	13.59 m² (146.28 sq ft)
Vertical tail surfaces (total)	38.53 m² (414.73 sq ft)

WEIGHTS AND LOADINGS (A: 'Bear-F' Mod 4, B: 'Bear-H'):
Weight empty: A	91,800 kg (202,380 lb)
B	94,400 kg (208,115 lb)
Max fuel: A	87,000 kg (191,800 lb)
Max T-O weight: A, B	185,000 kg (407,850 lb)
Max in-flight weight, after refuelling:	
B	187,000 kg (412,258 lb)
Max landing weight: A, B	135,000 kg (297,620 lb)
Max wing loading: B	645.1 kg/m² (132.12 lb/sq ft)
Max power loading: B	4.24 kg/kW (6.96 lb/shp)

PERFORMANCE ('Bear-H'):
Max level speed: at S/L	350 kt (650 km/h; 404 mph)
at 7,620 m (25,000 ft)	M0.83 (499 kt; 925 km/h; 575 mph)
at 11,600 m (38,000 ft)	M0.78 (447 kt; 828 km/h; 515 mph)
Nominal cruising speed	384 kt (711 km/h; 442 mph)
T-O speed	162 kt (300 km/h; 187 mph)
Landing speed	approx 146 kt (270 km/h; 168 mph)
Service ceiling: normal	12,000 m (39,380 ft)
with max weapons	9,100 m (29,860 ft)
T-O run	2,540 m (8,335 ft)

Combat radius with 11,340 kg (25,000 lb) payload:
unrefuelled   3,455 n miles (6,400 km; 3,975 miles)
with one in-flight refuelling
    4,480 n miles (8,300 km; 5,155 miles)

UPDATED

TUPOLEV Tu-160
NATO reporting name: Blackjack
Unofficial name: Belyi Lebed (White Swan)

TYPE: Strategic bomber.

PROGRAMME: Designed as Aircraft 70 under leadership of V I Bliznuk; programme began 1967, but relaunched following issue of more modest specification in 1970; derived from unbuilt Tu-135 bomber and Tu-144 derivatives; some features from rival Myasishchev M-18. First of two prototypes (70-01) observed by intelligence source at Zhukovsky flight test centre 25 November 1981 (photographed from landing airliner; see 1982-83 *Jane's*); first flew 18 or 19 December 1981; first exceeded M1.0 February 1985; second (production-standard) aircraft first flew 7 October 1984. Second production aircraft lost predelivery, March 1987; US Defense Secretary Frank Carlucci invited to inspect 12th aircraft built, at Kubinka airbase, near Moscow, 2 August 1988; deliveries to 184th Guards Heavy Bomber Aviation Regiment, Priluki airbase, Ukraine, began April 1987; equipment of 1096th HBAR at Engels from 16 February 1992, but only six received before production at Kazan airframe plant terminated 1992; of 100 aircraft due to be built, at least 32 (including prototypes) accounted for; unconfirmed reports suggest total of 40 flying, plus three uncompleted at Kazan.

CURRENT VERSIONS: **Tu-160** ('Blackjack'): Strategic bomber.
Tu-160SK: Launch vehicle for proposed Burlak-Diana two-stage space vehicle; described separately.
Tu-160M: Proposed stretched variant carrying two 2,700 n mile (5,000 km; 3,107 mile) range hypersonic Kh-90 missiles.
Tu-160P: Proposed very long-range escort fighter.
Tu-160PP: Proposed escort jammer.
Tu-160R: Proposed strategic reconnaissance platform.

CUSTOMERS: Ukraine government seized 19 at Priluki on achieving independence, pre-empting planned transfer to Engels; purchase of these by Russia was subject to protracted negotiations, and aircraft deteriorated in storage; March 1996 agreement on transfer of 10 best airframes was not implemented; attempts to purchase eight failed in March 1998 and Russia then supposedly abandoned hopes of expanding Tu-160 fleet. However, October 1999 announcement revealed eight to be returned to Russia for refurbishment (together with three Tu-95MSs, for a total of US$285 million) and these were delivered between 5 November 1999 and 21 February 2000. Others will be scrapped with US assistance, although August 2000 report mentioned further three TU-160s which Ukraine could return to Russia in part-payment for natural gas deliveries. One further Ukrainian Tu-160 was flown to Poltava Kondratyuk-Shagray aerospace museum on 5 April 2000.

Russia maintained force of six (declared as ALCM carriers under START) at Engels (where 1096th HBAR was redesignated as part of the 121st Guards HBAR in 1994), plus flying testbed at Zhukovsky; at least four more were derelict at Zhukovsky by 1995. Despite this, 1998 US estimates of Russian combat forces reported total holding of nominally serviceable aircraft as 25, though this would not seem to be possible. Eight more to re-enter service in 2001 after purchase from Ukraine, while plan announced in June 1999 to complete one unfinished Tu-160 at Kazan. This aircraft ('07' *Aleksandr Molodchyi*) was delivered on 5 May 2000. Another, brand new Tu-160 (the first of three more incomplete, unfinished aircraft at Kazan) is now under construction for delivery by the end of 2001 and there are plans to refurbish some (perhaps four) of the grounded aircraft at Zhukhovsky. Tu-160 may re-enter limited production, to meet a stated requirement for 25 operational 'Blackjacks' by 2003, allowing formation of a second regiment. Five of original six at Engels are also named: '01' *Mikhail Gromok*; '02' *Vasily Retsetnikov*; '04' *Ivan Yargin*; '05' *Ilya Muromets* (1) and '06' *Ilya Muromets* (2).

On 3 March 1999, the Russian Commonwealth Aerospace Technology Consortium (RCATC) was authorised by the Ukrainian government to sell three demilitarised Ukrainian Air Force Tu-160s, plus spares, to Platforms International Corporation of the USA, with which it has finalised a strategic partnership. The US$20 million deal includes a 20 per cent interest in Orbital Network Services Corporation, which plans to use the aircraft as re-usable communications satellite launchers in its HAAL-2000 High-Altitude Air Launch programme. The aircraft would probably be modified to Tu-160SK standards and continue to be based at Priluki, maintained and flown by Ukrainian crews, but flown to customer countries for individual space launch missions.

DESIGN FEATURES: Intended for high-altitude standoff role carrying ALCMs and for defence suppression, using short-range attack missiles similar to US Air Force SRAMs, along path of bomber making penetration to attack primary targets with free-fall nuclear bombs or missiles; this implies capability of subsonic cruise/supersonic dash at almost M2 at 18,300 m (60,000 ft) and transonic flight at low altitude. About 20 per cent longer than USAF B-1B, with greater unrefuelled combat radius and much higher maximum speed; low-mounted variable

Trials example of the Tupolev Tu-160 'Blackjack' based at Zhukovsky *(Paul Jackson/Jane's)*   2000/0062701

geometry wings, with very long and sharply swept fixed root panel; small diameter circular fuselage; horizontal tail surfaces mounted high on fin, upper portion of which is pivoted one-piece all-moving surface; large dorsal fin; engines mounted as widely separated pairs in underwing ducts, each with central horizontal V wedge intakes and jetpipes extending well beyond wing centre-section trailing-edge; manually selected outer wing sweepback 20, 35 and 65°; when wings fully swept, inboard portion of each trailing-edge flap hinges upward and extends above wing as large fence; unswept tailfin; sweptback horizontal surfaces, with conical fairing for brake-chute aft of intersection.

FLYING CONTROLS: Quadruplex fly-by-wire with mechanical reversion. Full-span leading-edge flaps, long-span double-slotted trailing-edge flap and inset drooping aileron on each wing; five-section spoilers forward of flaps; all-moving vertical and horizontal one-piece tail surfaces.

STRUCTURE: Slim and shallow fuselage blended with wingroots and shaped for maximum hostile radar signal deflection; 20 per cent titanium, including leading-edges and wing centre-section spar box.

LANDING GEAR: Twin nosewheels retract rearward; main gear comprises two bogies, each with three pairs of wheels; retraction very like that on Tu-154 airliner; as each leg pivots rearward, bogie rotates through 90° around axis of centre pair of wheels, to lie parallel with retracted leg; gear retracts into thickest part of wing, between fuselage and inboard engine on each side; so track relatively small. Nosewheel tyres size 1,080×400; mainwheel tyres size 1,260×425.

POWER PLANT: Four purpose-designed Samara NK-321 turbofans, each 137.3 kN (30,865 lb st) dry, 245 kN (55,115 lb st) with afterburning. In-flight refuelling probe retracts into top of nose. Fuel in centre-section spar box and in outer wings.

ACCOMMODATION: Four crew members in pairs, on individual Zvezda K-36LM zero/zero ejection seats, in pressurised compartment; one window each side of flight deck can be moved inward and rearward for ventilation on ground; flying controls use fighter-type sticks rather than yokes or wheels; crew enter via extending ladder in nosewheel bay. Cooking facilities and toilet.

AVIONICS: Systems utilise around 100 digital processors and eight digital nav computers.
Radar: Obzor (NATO 'Clam Pipe') nav/attack radar in slightly upturned dielectric nosecone with separate Sopka radar providing terrain-following capability.
Flight: K-042K astro-inertial nav with map display.
Instrumentation: Analogue instruments. No HUD or CRTs.
Mission: OPB-15 strike sight fairing with flat glazed front panel, under forward fuselage, for video camera to provide visual assistance for weapon aiming.
Self-defence: Baikal self-protection system, with integrated RHAWS, chaff/flare dispensers in tailcone and active jamming.

ARMAMENT: No guns. Internal stowage for free-fall bombs, mines, short-range attack missiles or ALCMs; two tandem 12.80 m (42 ft) long weapon bays; MKU-6-5U rotary launcher for six (or up to 12) Kh-55MS (AS-15 'Kent') or RKV-500B (AS-15B 'Kent-B') ALCMs or 12 to 24 Kh-15P (AS-16 'Kickback') SRAMs in each bay. Current plans envisage carriage of up to 12 non-nuclear Kh-101 ALCMs, when available.

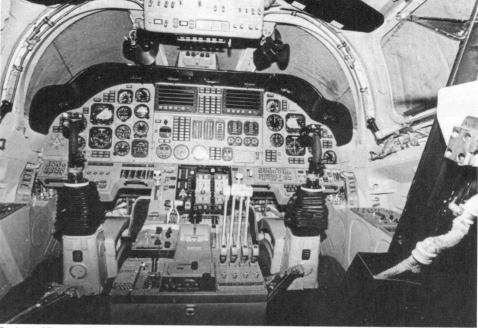
Cockpit of Tupolev Tu-160 supersonic strategic bomber

DIMENSIONS, EXTERNAL:
Wing span: fully spread (20°)	55.70 m (182 ft 9 in)
35° sweep	50.70 m (166 ft 4 in)
fully swept (65°)	35.60 m (116 ft 9¾ in)
Wing aspect ratio: fully spread	8.6
Length overall	54.10 m (177 ft 6 in)
Height overall	13.10 m (43 ft 0 in)
Tailplane span	13.25 m (43 ft 5¾ in)
Wheel track	5.40 m (17 ft 8½ in)
Wheelbase	17.88 m (58 ft 8 in)

DIMENSIONS, INTERNAL:
Weapons bay (each): Volume	43.0 m³ (1,518 cu ft)

AREAS:
Wings, gross: fully swept	360.00 m² (3,875.0 sq ft)
fully spread	approx 400.00 m² (4,305.6 sq ft)
moving areas, fully swept (total)	approx 180.00 m² (1,937.5 sq ft)

WEIGHTS AND LOADINGS:
Weight empty	110,000 kg (242,505 lb)
Weight empty, equipped	117,000 kg (257,940 lb)
Max fuel	171,000 kg (376,990 lb)
Max weapon load	40,000 kg (88,185 lb)
Normal T-O weight	267,600 kg (589,950 lb)
Max T-O weight	275,000 kg (606,260 lb)
Max landing weight	155,000 kg (341,710 lb)
Max power loading	280 kg/kN (2.75 lb/lb st)

PERFORMANCE:
Max level speed at 12,200 m (40,000 ft)	M2.05 (1,200 kt; 2,220 km/h; 1,380 mph)
Cruising speed at 13,700 m (45,000 ft)	M0.9 (518 kt; 960 km/h; 596 mph)
Max rate of climb at S/L	4,200 m (13,780 ft)/min
Service ceiling	15,000 m (49,200 ft)
T-O run at max AUW	2,200 m (7,220 ft)
Landing run at max landing weight	1,600 m (5,250 ft)
Radius of action at M1.5	1,080 n miles (2,000 km; 1,240 miles)
Max unrefuelled range	6,640 n miles (12,300 km; 7,640 miles)
g limit	+2

UPDATED

TUPOLEV Tu-160SK

TYPE: Conversion of Tu-160 bomber as space launch vehicle.
PROGRAMME: Announced at Singapore Air Show '94; proposed by Russian partners MKB-Raduga, OKB MEI and Tupolev, with German company OHB-System. Will operate under control of Ilyushin Il-76SKIP (otherwise known as Beriev and LII '976') radar surveillance aircraft (which see). At the 1995 Paris Air Show, Tu-160 0401 was exhibited statically (the type's Western debut) with a model Burlak rocket below the fuselage. Development of the system continued even after the German government withdrew funding in 1998, and may form the basis of the Ukrainian/PIC HAAL-2000 programme described in the main Tu-160 entry.
COSTS: Payload launch cost US$5,000/kg.
DESIGN FEATURES: Commercialised, demilitarised version of Tu-160 bomber (which see), as carrier component of Burlak aviation space launch complex; Burlak-Diana two-stage rocket, carrying payload, under fuselage on centreline mount.
ACCOMMODATION: Crew of four.
DIMENSIONS, EXTERNAL: As Tu-160 bomber, plus:
Burlak rocket: Length overall	21.00 m (68 ft 10¾ in)
Fin span	5.00 m (16 ft 4¾ in)
Diameter	1.60 m (5 ft 3 in)
Burlak payload module: Length	3.50 m (11 ft 5¾ in)
Diameter	1.40 m (4 ft 7 in)

WEIGHTS AND LOADINGS:
Burlak launch weight	32,000 kg (70,547 lb)
Max Burlak payload:	
200 km (124 mile) equatorial orbit	1,100 kg (2,425 lb)
200 km (124 mile) polar orbit	800 kg (1,763 lb)
1,000 km (620 mile) equatorial orbit	840 kg (1,851 lb)
1,000 km (620 mile) polar orbit	600 kg (1,322 lb)
Max T-O weight	275,000 kg (606,260 lb)

PERFORMANCE:
Max level speed, clean	M2.35
Speed for Burlak launch at 13,500 m (44,300 ft)	M1.7
Nominal cruising speed	M0.77 (459 kt; 850 km/h; 528 mph)
Required runway length	3,500 m (11,500 ft)
Nominal range:	
with Burlak	2,968 n miles (5,500 km; 3,417 miles)
clean	5,936 n miles (11,000 km; 6,836 miles)

VERIFIED

TUPOLEV Tu-54

Development of this agricultural sprayer was suspended for three years because government funding was terminated. However, by September 2000, an undisclosed investor had promised US$1.8 million to complete a prototype.

UPDATED

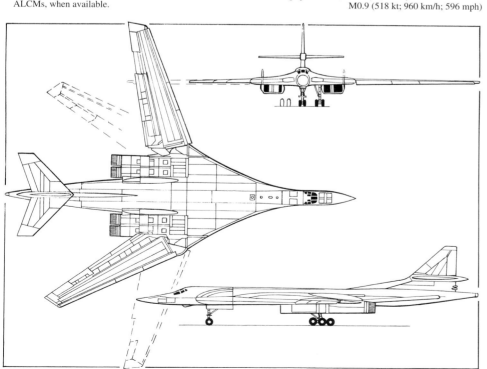
Tupolev Tu-160 strategic bomber *(James Goulding/Jane's)*

TUPOLEV Tu-136

Development of this dual-fuel turboprop twin is not among Tupolev's priorities for the near term, although a model completed trials in the CAHI (TsAGI) wind tunnel in March 2000 and an updated version of the baseline Tu-130 remained under consideration in early 2000. A description last appeared in the 2000-01 *Jane's*.

UPDATED

TUPOLEV Tu-154M

NATO reporting name: Careless

TYPE: Tri-jet airliner.
PROGRAMME: Basic Tu-154 announced second quarter 1966 to replace Tu-104, Il-18 and An-10 on Aeroflot medium/long stages up to 3,240 n miles (6,000 km; 3,725 miles); SSSR-85000, first of six prototype/preproduction models flew 4 October 1968; regular services began 9 February 1972; 606 prototype and production Tu-154s and Tu-154As, Bs and B-2s with uprated turbofans and other refinements delivered, over 500 to Aeroflot (last described 1985-86 *Jane's*); prototype Tu-154M (SSSR-85317), converted from standard Tu-154B-2 (see 1990-91 *Jane's*), first flew February 1982; first two production aircraft delivered to Aeroflot from GAZ 18 Kuybyshev (now known as Samara) December 1984. Aviacor plant to end production in 2002.
CURRENT VERSIONS: **Tu-154M:** Basic airliner with alternative standard configurations for up to 180 passengers; executive version available; with all passenger seats removed can carry light freight.

Detailed description applies to Tu-154M.

Tu-154M-100: Upgraded Zhasmin (Jasmine) avionics system and new Aviacor interior furnishings; 12 ordered by Iranian airlines in 1997; delivered to Iran Air Tours (nine) and Bon Air (three) in 1997-98 from surplus former Aeroflot aircraft, despite which Aviacor was still optimistic in 1998 of up to 10 new-build orders from Iran. Zhasmin includes GPS and collision-avoidance radar, and bestows Cat. II landing capability.

Tu-154-200: Projected version with two Samara NK-93 turbofans.

Tu-154M-LK-1: Two aircraft for head-of-state use.

Tu-154M2: Modernised, twin-engined version; Perm/Soloviev PS-90A turbofans, consuming 62 per cent as much fuel per passenger as Tu-154M. Intended originally to fly 1995, but apparently abandoned. Life 20,000 hours or 15,000 cycles.

Tu-154C: Specialised freight version, announced in third quarter of 1982; offered primarily as Tu-154B conversion; unobstructed main cabin cargo volume 72 m³ (2,542 cu ft); freight door 2.80 m (9 ft 2¼ in) wide and 1.87 m (6 ft 1½ in) high in port side of cabin, forward of wing, with ball mat inside and roller tracks full length of cabin floor; typical load nine standard international pallets 2.24 × 2.74 m (88 × 108 in) plus additional freight in standard underfloor baggage holds, volume 38 m³ (1,341 cu ft); nominal range 1,565 n miles (2,900 km; 1,800 miles) with 20,000 kg (44,100 lb) cargo.

Tu-154 Retrofit: In 1998, Rybinsk Motors was promoting the CFM56 turbofan as a re-engine option.

Tu-154M/OS: Version for Russian 'Open Skies' Treaty observation flights, with side-looking synthetic aperture radar developed under 1995 co-operative agreement between German Ministry of Defence and Defence Ministry of the Russian Federation. Russian designation **Tu-154M-ON** (*okrytoye nebo*: open skies). Sole aircraft is RA-85655, a former Tu-154M-LK-1. Additional duties can include environmental monitoring, ice patrol, mapping and geological survey. Radar developed by Kulon Russian research institute and Dornier GmbH of Germany, but installation plans terminated prior to this being fitted. See *Jane's Aircraft Upgrades* for more details.

Tu-154M Elint: First four of reported 12 ex-airline aircraft entered service by 1999 with Nanjing Military Region, China, following modification by military upgrade plant at Nan Yuan, Beijing.

CUSTOMERS: Total of 318 Tu-154Ms (and 924 of all variants) built or substantially complete by late 1996. Total unchanged by 2000, but some recently supplied from stock, including 1998 deliveries of six: to Azerbaijan government (one), Slovak government (one), SKA Slovak Airlines (three) and Tyumen Avia (one). Sole 1996 delivery was one to the Czech government on 14 December; single aircraft in 1997 was delivered to Tyumen Avia in June. None known in 1999, leaving 15 Tu-154Ms in storage at Samara minus engines and other fitments. These include one for Ukrainian government as VIP transport, ordered 1996, but still awaiting completion in early 2000. Of 15 at Samara, Aviacor will complete four in each year 2000, 2001 and 2002, before abandoning programme. Known Tu-154M civil operators number 73, with some 285 aircraft.

COSTS: US$7 million (2000).

DESIGN FEATURES: Conventional all-swept low-wing configuration; two podded turbofans on sides of rear fuselage, third in extreme rear fuselage with intake at base of fin; nacelle to house retracted main landing gear on trailing-edge of each wing; wing sweep 35° at quarter-chord; anhedral on outer panels; geometric twist along span; circular section fuselage; sweepback at quarter-chord 40° on T tailplane; leading-edge sweep 45° on fin.

FLYING CONTROLS: Conventional and power-operated. Hydraulically actuated ailerons, triple-slotted flaps, four-section spoilers forward of flaps on each wing; electrically actuated slats on outer 80 per cent each wing leading-edge; tab in each aileron; electrically actuated variable incidence tailplane; rudder and elevators hydraulically actuated by irreversible servo controls; tab in each elevator.

STRUCTURE: All-metal; riveted three-spar wings, centre spar extending to just outboard of inner edge of aileron; semi-monocoque fail-safe fuselage; rudder and elevators of honeycomb sandwich construction.

LANDING GEAR: Hydraulically retractable tricycle type; mainwheels 930×305; nosewheels 800×225. Main bogies, each three pairs of wheels in tandem, retract rearward into fairings on wing trailing-edge; rearward-retracting anti-shimmy twin-wheel nose unit, steerable through ±63°; disc brakes and anti-skid units on mainwheels.

POWER PLANT: Three Aviadvigatel D-30KU-154-II turbofans, each 104 kN (23,380 lb st), in pod each side of rear fuselage and inside extreme rear of fuselage; two lateral engines have clamshell thrust reversers. Integral fuel tanks in wings: four tanks in centre-section and two in outer wings; all fuel fed to collector tank in centre-section and thence to engines; single-point refuelling.

ACCOMMODATION: Crew of three; two pilots and flight engineer, with provisions for navigator and five cabin staff. Two passenger cabins, separated by service compartments; alternative configurations for 166 tourist class passengers, at 75 cm (29.5 in) pitch, with hot meal service, or 134 tourist plus separate first class cabin seating 12 persons; mainly six-abreast seating with centre aisle; washable non-flammable materials used for all interior furnishing. Fully enclosed baggage containers. Toilet, galley and wardrobe to customer's requirements. Executive and light cargo configurations available. Passenger doors forward of front cabin and between cabins on port side, with emergency and service doors opposite; all four doors open outward; six emergency exits: two overwing and one immediately forward of engine nacelle each side. Two pressurised baggage holds under floor of cabin, with two inward-opening doors; smaller unpressurised hold under rear of cabin.

SYSTEMS: Air conditioning pressure differential 0.58 bar (8.4 lb/sq in). Three independent hydraulic systems, working pressure 207 bars (3,000 lb/sq in), powered by engine-driven pumps; Nos. 2 and 3 systems each have additional electric back-up pump; systems actuate landing gear retraction and extension, nosewheel steering, and operation of ailerons, rudder, elevators, flaps and spoilers. Three-phase 200/115 V 400 Hz AC electrical system supplied by three 40 kVA alternators; additional 36 V 400 Hz AC and 27 V DC systems and four storage batteries; TA-92 APU in rear fuselage. Hot air anti-icing of wing, fin and tailplane leading-edges, and engine air intakes; wing slats heated electrically. Engine fire extinguishing system in each nacelle; smoke detectors in baggage holds.

AVIONICS: Avionics meet ICAO standards for Cat. II weather minima.

Comms: Dual HF and VHF com and emergency VHF, voice recorder, and transponder.

Radar: Weather radar.

Flight: Automatic flight control system standard until 1995 operates throughout flight except during take-off to 400 m (1,312 ft) and landing from 30 m (100 ft); automatic go-round and automatic speed control provided by autothrottle down to 10 m (33 ft) on landing; triplex INS, Doppler and GPWS. New Zhasmin system available since 1995 offers automatic flight control at all stages of flight including automatic landing to ICAO Cat. II standards; equipment can include TCAS, Orion, Omega and ABSU-154 nav systems.

DIMENSIONS, EXTERNAL:

Wing span	37.55 m (123 ft 2½ in)
Wing aspect ratio	7.0
Length overall	47.92 m (157 ft 2½ in)
Height overall	11.40 m (37 ft 4¾ in)
Diameter of fuselage	3.80 m (12 ft 5½ in)
Tailplane span	13.40 m (43 ft 11½ in)
Wheel track	11.50 m (37 ft 9 in)
Wheelbase	18.92 m (62 ft 1 in)
Passenger doors (each): Height	1.70 m (5 ft 7 in)
Width	0.80 m (2 ft 7½ in)
Height to sill	3.10 m (10 ft 2 in)
Servicing door: Height	1.28 m (4 ft 2½ in)
Width	0.61 m (2 ft 0 in)
Emergency door: Height	1.28 m (4 ft 2½ in)
Width	0.64 m (2 ft 1¼ in)

The sole Tupolev Tu-154M/OS 'Open Skies' monitor *(Paul Jackson/Jane's)*

Tupolev Tu-154M of Varna International Air, Bulgaria *(Paul Jackson/Jane's)*

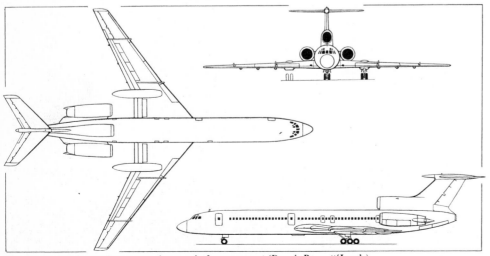

Tupolev Tu-154M medium-range three-turbofan transport *(Dennis Punnett/Jane's)*

Emergency exits (each): Height 0.90 m (2 ft 11½ in)
  Width 0.48 m (1 ft 7 in)
Main baggage hold doors (each):
  Height 1.20 m (3 ft 11¼ in)
  Width 1.35 m (4 ft 5 in)
  Height to sill 1.80 m (5 ft 10¾ in)
Rear (unpressurised) hold: Height 0.90 m (2 ft 11½ in)
  Width 1.10 m (3 ft 7¼ in)
  Height to sill 2.20 m (7 ft 2½ in)
DIMENSIONS, INTERNAL:
Cabin: Width 3.58 m (11 ft 9 in)
  Height 2.02 m (6 ft 7½ in)
  Volume 163.2 m³ (5,763 cu ft)
Baggage hold: Volume: front 21.5 m³ (759 cu ft)
  rear 16.5 m³ (582 cu ft)
Rear underfloor hold: Volume 5.0 m³ (176 cu ft)
AREAS:
Wings, gross 201.45 m² (2,168.4 sq ft)
Horizontal tail surfaces (total) 42.20 m² (454.24 sq ft)
WEIGHTS AND LOADINGS:
Basic operating weight empty 55,300 kg (121,915 lb)
Max payload 18,000 kg (39,680 lb)
Max fuel 39,750 kg (87,633 lb)
Max ramp weight 100,500 kg (221,560 lb)
Max T-O weight 100,000 kg (220,460 lb)
Max landing weight 80,000 kg (176,365 lb)
Max zero-fuel weight 74,000 kg (163,140 lb)
Max wing loading 496.4 kg/m² (101.67 lb/sq ft)
Max power loading 321 kg/kN (3.14 lb/lb st)
PERFORMANCE:
Nominal cruising speed 504 kt (935 km/h; 581 mph)
Max cruising height 11,900 m (39,000 ft)
Service ceiling at 85,000 kg (187,390 lb) AUW
  12,100 m (39,700 ft)
Balanced field length for T-O and landing
  2,500 m (8,200 ft)
Range: with max payload
  1,997 n miles (3,700 km; 2,299 miles)
  with 12,000 kg (26,455 lb) payload
  2,805 n miles (5,200 km; 3,230 miles)
  with max fuel and 5,450 kg (12,015 lb) payload
  3,563 n miles (6,600 km; 4,100 miles)

UPDATED

TUPOLEV Tu-156

TYPE: Cryogenic-fuelled, three-turbofan transport.
PROGRAMME: Details of 1988-89 flight trials of Tu-155, a Tu-154 modified with a Kuznetsov NK-88 turbofan operating on liquid hydrogen and liquefied natural gas fuels, last appeared in 1990-91 *Jane's*. On 23 April 1994, Russian government allocated funding for conversion of three Tu-154s to Tu-156 standard, delivery of 12 NK-89 turbofans and six cryogenic fuel systems; an installation to supply liquefied natural gas (LNG) will be established at Samara. Order reduced to one Tu-156; delivery scheduled for 1998, but had not taken place by June 2000, at which time sources of potential funding were still being examined; requirement is for Rb100 million to 200 million over three years.
CURRENT VERSIONS: **Tu-156S**: Initial kerosene/LNG-powered conversion of Tu-154B, carrying 13,000 kg (28,660 lb) of LNG at rear of cabin and 3,800 kg (8,378 lb) in two underfloor tanks, plus 10,600 kg (23,369 lb) of kerosene in wing tanks; payload 14,500 kg (31,967 lb) or 130 passengers; range 1,457 n miles (2,700 km; 1,677 miles).
Tu-156M: Developed version converted from Tu-154M, carrying 135 passengers; fuel load and performance as Tu-156S.
Tu-156M2: Proposal only; converted from Tu-154M2; 20,000 kg (44,092 lb) of LNG only, carried in two tanks above centre-fuselage allowing passenger load increase to 160; two NK-94 engines; range 2,159 n miles (4,000 km; 2,485 miles).
POWER PLANT: Experimental power plant is 103.0 kN (23,150 lb st) Samara NK-89; tanks for cryogenic fuel mounted in rear of cabin and in forward baggage hold of Tu-156S/M. Aircraft will operate on mixed LNG/kerosene, as Tu-155, kerosene being used for flights out of non-LNG aerodromes and for emergencies, when in-flight switch from LNG can be made in 5 seconds. Tu-156M2 would have only two NK-94 engines, wholly LNG powered.

UPDATED

Model of Tupolev Tu-156M cryogenic aircraft *(Paul Jackson/Jane's)* 0015138

Tupolev Tu-204, executive transport used by Russian government *(Paul Jackson/Jane's)* 2001/0092532

TUPOLEV Tu-204, Tu-214, Tu-224 and Tu-234

TYPE: Twin-jet airliner.
PROGRAMME: Development to replace Tu-154 and Il-62 announced 1983; preliminary details available 1985; programme finalised 1986; first prototype (SSSR-64001), with PS-90AT engines, flown 2 January 1989 by Tupolev chief test pilot A Talalakine; two more prototypes (RA-64003 and 004) followed, plus two (002 and 005) for structural and fatigue testing. Second version, with RB211-535E4-B engines, flew 14 August 1992 (fourth flying prototype, RA-64006) and was demonstrated at Farnborough Air Show in following month.
Production of basic Tu-204 series by Aviastar at Ulyanovsk began 1990, Tu-214 by KAPO at Kazan in 1994; Russian certification of basic Tu-204 received 12 January 1995; used initially only for freight services. First revenue-earning passenger flight, Moscow to Mineralnye Vody, operated by Vnukovo Airlines 23 February 1996. Restrictions on PS-90-engined Tu-204s were eventually removed in mid-1998. Deliveries outside CIS began on 2 November 1998 with handover of -120 and -120C freighter to Air Cairo on seven-year lease.
Avionics upgrade under discussion with NII in mid-2000; goal is six LCD, two-pilot flight deck; estimated cost US$5 million to US$8 million. First application on Tu-204-300.
CURRENT VERSIONS: **Tu-204**: Basic medium-haul airliner for up to 214 passengers or maximum payload of 21,000 kg (46,295 lb); 158.3 kN (35,580 lb st) Aviadvigatel PS-90A turbofans. Marketed 1989.
Tu-204C: Cargo version of basic Tu-204. (Western marketing designation is Tu-204C, but correct Cyrillic designation, used in Russian literature, is **Tu-204S**.) Described separately.
Tu-204-100: Extended-range passenger version; additional fuel in wing centre-section; dimensions, payload and power plant unchanged; maximum T-O weight 103,000 kg (227,070 lb). Marketed 1993.
Tu-204-100C: Freighter, as Tu-204-100. Payload increased at expense of reduced maximum range. Marketed 1994. Bulk freight volume 180.0 m³ (6,357 cu ft), not including fore and aft underfloor holds.
Tu-204-120: As Tu-204-100, but with Rolls-Royce RB211-535-E4 turbofans; maximum T-O weight 103,000 kg (227,070 lb); Russian avionics; prototype (RA-64006) flew 14 August 1992. First production aircraft (RA-64027) flew 7 March 1997; gained limited Russian certification on 16 July 1997 and full certification 12 months later. First five Sirocco 120/120Cs have Russian avionics; later aircraft are Phase II with Honeywell items, including VIA 2000 suite.
Tu-204-120C: As -100C, but RB211 engines; two for Air Cairo (via Sirocco). First aircraft (RA-64028) flew 10 November 1997; certification completed by August 1998. First delivery 2 November 1998. High gross weight version, with 30,000 kg (66,139 lb) payload, was due for roll-out in August 1999 and Russian certification in May 2000.
Tu-204-122: As Tu-204-120, but with Rockwell Collins avionics. None produced.
Tu-204-200: Further increase in payload and T-O weight; small increase in fuel in wing centre-section and adjacent baggage hold; dimensions and power plant as Tu-204-100; strengthened landing gear; marketed 1994. First example of Series 200 was a Tu-214 freighter, first flown in March 1996; first true Tu-204-200 (RA-64036) was nearing completion at Ulyanovsk in late 1998; PS-90 engines. By mid-2000, Tu-214 designation was being used for this aircraft.
Tu-204-200C: As Tu-204-200. Payload increased at expense of reduced maximum range. Marketed 1994.
Tu-204-200C³ (Cargo, Converted, Containerised): Combi version, marketed as **Tu-214**.
Tu-204P: Proposed military patrol (*patrulnyi*) version of Tu-204-200. It was reported in January 1997 that the Russian Navy had approved an anti-submarine version of Tu-204 to replace its Il-38 'May' and shorter-range Be-12 'Mail' patrollers.
Tu-204-220: As Tu-204-200, but 191.7 kN (43,100 lb st) Rolls Royce RB211-535E4 or RB211-535F5 turbofans; Tupolev-funded variant with 185.5 kN (41,700 lb st) Pratt & Whitney PW2240 turbofans; promotion began 1998.
Tu-204-220C: Cargo version of Tu-204-220.
Tu-204-222: As Tu-204-220, but with Rockwell Collins avionics.
Tu-204-230: With 176.5 kN (39,683 lb st) Samara NK-93 propfans. At initial project stage.
Tu-204 Business Jet: Scheduled to fly in 1998, but failed to do so.
Tu-204-300: Upgraded avionics in two-stage programme; prototype to begin 65-sortie certification trials in 2001.
Tu-204-400: Version for 240 passengers; digital avionics; uprated PS-90A2 turbofans for 120,000 kg (264,550 lb) MTOW. Aviastar production.
Tu-204-500: PS-90A2 engines and fuselage stretch to 52.00 m (170 ft 7¼ in); possible wing redesign for higher cruising speeds; private venture outside government's aviation plan. Aviastar production. Designation first reported in early 2000.
Tu-304 and **Tu-306**: Projected 400-seat versions, latter fuelled by LNG. Described in 1999-2000 *Jane's*.
Tu-206 and **Tu-216**: Described separately.
Tu-214C³: Combi version, based on Tu-204-200; also designated -200C³; announced 1995. Typical loads comprise 16 LD3 containers and 30 passengers; or six containers and 130 persons; or 18 LD3s only. Prototype (RA-64501) rolled out at Kazan by KAPO 5 February 1996; first flew 21 March 1996; public debut at Farnborough Air Show, September 1996. Provisional certification awarded 1998. Also in that year, the PW2037 was being assessed as an alternative power plant. By early

Tupolev Tu-204-100C of AirRep 2001/0093379

Tupolev Tu-214C³ combi version of the Tu-204-200 *(Paul Jackson/Jane's)*

2000, Kazan stock included seven partly complete Tu-214s from initial production batch of 18. Flight testing of prototype suspended after some 300 sorties, but resumed in May 2000 with a view to obtaining full certification by end of that year with further series of 85 flights ('Second Class' AP-25 certification awarded March 2000).

Tu-224 (alternatively Tu-204-320) and **Tu-234** (alternatively Tu-204-300): Announced 1994; trunk route versions with shorter fuselage; short-range and mid-range models for 166 passengers, long-range model for 99 to 160 passengers; Tu-224 has RB211-535E4 engines; Tu-234 has 158.3 kN (35,580 lb st) PS-90P engines. Prototype Tu-234 (rebuild of original Tu-204 prototype RA-64001) rolled out on 24 August 1995 at Moscow Air Show, but had not flown by mid-2000. First production aircraft (RA-64026, built by Aviastar at Ulyanovsk) rolled out, minus engines, in mid-1996; fitted with PS-90s in early 2000 and first flew 8 July 2000 (ahead of prototype). **Tu-234C** cargo version planned. First Tu-224 will be RA-64503.

CUSTOMERS: Include GTK Rossiya division of Aeroflot (two), Aeroflot Russian International (four), Vnukovo Airlines (four, three more on order). Russian government authorised guarantee in April 1997 for 20 PS-90-powered Tu-204s for Moscow Aviation International Leasing. The Kato Group of Egypt ordered an initial batch of 13 Tu-204-120s, with 17 options (some of which now taken up), in August 1996, with Honeywell avionics; these to be leased through Sirocco Aerospace to KrasAir (Siberia) (10, subsequently cancelled because of high Russian internal taxes) and Air Cairo (five); by mid-2000, four in service with Air Cairo and two more deliveries due before year's end. Volga-Dnepr ordered two Tu-204C-120 freighters in late 1997 and Armenian Airlines was to have taken two (plus three options) in 1998. KrasAir subsequently ordered two Tu-204-100s from Aviastar with regional government backing; first delivery August 2000; further eight options. Transeuropean (Russia) ordered three, of which first (RA-64018) entered service 15 May 1999. Transaero (Russia) announced intention to order 10 Tu-204-100s from Aviastar plant, 1999; firm commitment awaited, but US$110 million loan being arranged in mid-2000. Sibir (Russia) negotiating in late 1999 for three Tu-214s from Kazan, but leased, then bought, one from Perm engine plant and switched to Ulyanovsk as source of second and third aircraft; Tatarstan Airlines planning for two while Dalavia of Khabarovsk ordered two Tu-214s in June 2000 for delivery in December 2000 and January 2001. AirRep (UK) receiving three Tu-204-100Cs, first of which delivered to Manston on 4 April 2000. Aviastar Asia Corporation of Taipei, Taiwan, plans to order 20 Tu-204-120/220/234 transports annually for sale and lease in Eastern Europe, the Russian Federation and Associated States (CIS), Asia, the Middle East and Africa. Aeroflot Russian Airlines has announced requirement for eight to 10 Tu-204-120s; Aeroflot requires 36 Tu-214s, nine of which will be put into service from 2003. One -100 reportedly ordered by Aviaexpresscruise of Moscow, 2000. Kazan plant plans to build Tu-214 production from one in 2000 to 24 per year from 2006.

In August 2000, of 36 built at Ulyanovsk, allocations were: Tupolev (four), Vnukovo Airlines (three), Aeroflot (two), Rossiya (two), KMV (one), Sibir (one), TransEuropean (one), AirRep (one), Air Cairo (four), KrasAir (one), undelivered (12), converted to Tu-234 (two) and static test airframe (two). KMV (Kaminvodyavia) aircraft owned by Perm engine plant, which took two Tu-204s in lieu of cash payment for PS-90s; one Perm aircraft sold to Sibir in December 1999. Siberia Airlines also planning to acquire three Tu-214s from Kazan line. In late 1999, 15 partly complete Tu-204s were on Ulyanovsk assembly line, although subsequent orders have cleared this backlog.

COSTS: With PS-90 engines and Russian avionics US$27 million (1998); with RB211-535 engines US$38 million (1996).

DESIGN FEATURES: Conventional low/mid-wing configuration, with all surfaces sweptback, and winglets; wing dihedral from roots; semi-monocoque oval section pressurised fuselage; torsion box of fin forms integral fuel tank, used for automatic trimming of CG in flight; design life 45,000 flights, 60,000 flight hours or 20 years.

Wing section supercritical; sweepback 28°; thickness/chord ratio 14 per cent at root, 9 to 10 per cent at tip; negative twist.

FLYING CONTROLS: Triplex digital fly-by-wire, with triplex analogue back-up; conventional 'Y' control yokes selected after evaluation of alternative sidestick on Tu-154 testbed; inset aileron outboard of two-section double-slotted flap on each wing trailing-edge; two-section upper surface airbrake forward of each centre-section flap; five-section spoiler forward of each outer flap; four-section leading-edge slat over full span of each wing; conventional rudder and elevators; no tabs.

STRUCTURE: Approximately 18 per cent of airframe by weight of composites; three-piece two-spar wing, with metal structure, part composites skin; carbon fibre skin on spoilers, airbrakes and flaps; glass fibre wingroot fairings; all-metal fuselage, utilising aluminium-lithium and titanium; nose radome and some access panels of composites; extensive use of composites in tail unit, particularly for leading-edges of fixed surfaces and for rudder and elevators.

LANDING GEAR: Hydraulically retractable tricycle type; electrohydraulically steerable twin-wheel nose unit (±10° via rudder pedals; ±70° by electric steering control) retracts forward; four-wheel bogie main units retract inward into wing/fuselage fairings. Carbon disc brakes, electrically controlled. Tyre size 1,070×390 on mainwheels, 840×290 on nosewheels.

POWER PLANT: Two turbofans (see notes on versions), underwing in composites cowlings. Tu-204-200 series carries fuel in six integral tanks in wings, centre-section and adjacent baggage hold, and in tailfin, total capacity 40,730 litres (10,760 US gallons; 8,960 Imp gallons); torsion box of fin forms integral fuel tank for automatic trimming of CG in flight, as well as for optional standard use.

ACCOMMODATION: Can be operated by pilot and co-pilot, but Aeroflot specified requirement for a flight engineer; provision for fourth seat for instructor or observer. Three basic single-aisle passenger arrangements in Tu-204-200 series: (1) 184 seats, with 30 business class seats six-abreast at pitch of 96 cm (38 in) at front, and 154 tourist class seats six-abreast at pitch of 81 cm (32 in) at rear; (2) 200 six-abreast tourist class seats at pitch of 81 cm (32 in); (3) 212 seats six-abreast at tourist class pitch of 81 cm (32 in). Tu-224/234 series has 166 seats at 81 cm (32 in) pitch in short/mid-range models, 99 seats at 93 cm (36.5 in) or 160 seats at 81 cm (32 in) pitch in long-range model. All configurations have buffet/galley and toilet immediately aft of flight deck; 184-seat and 200-seat configurations have two more toilets in tourist cabin and a further large buffet/galley and service compartment at rear of passenger accommodation; 212-seat configuration has two toilets and galley at rear; overhead stowage for hand baggage. Other layouts optional, with increased galley and toilet provisions.

Passenger doors at front and rear of cabin on port side; service doors opposite. Type I emergency exit doors fore and aft of wing each side; inflatable slide for emergency use at each of eight doors. Tu-204-200 has two underfloor baggage/freight holds for total of eight LD-3-46 international class containers, three in forward hold, five in rear hold (AK-07 containers in Tu-204 and 204-100). Fully automatic container loading system; manual back-up.

SYSTEMS: Triplex fly-by-wire digital control system, with triplex analogue back-up; Barco Display Systems selected in 2000 to supply FMS computers. Three independent hydraulic systems, pressure 207 bars (3,000 lb/sq in). Ailerons, elevators, rudder, spoilers and airbrakes operated by all three systems; flaps, leading-edge slats, brakes and nosewheel steering operated by two systems; landing gear retraction and extension by all three systems. Electrical power supplied by two 200/115 V 400 Hz AC generators and a 27 V DC standby system. Type TA-12-60 APU in tailcone.

AVIONICS: Avionics of Russian design or, optionally, integrated by Sextant Avionique or Honeywell, with standard nav/com equipment by Rockwell Collins.

Comms: VHF and HF radio, intercom.
Radar: Weather radar.
Flight: Triplex automatic flight control, VOR, DME, automatic approach and landing system, for operation to ICAO Cat. IIIa minima; Series 100/200 have RLG INS (Litton LTN-101 or Honeywell HG115OBD02), GPS (Litton LTN-2001) and TCAS II (Honeywell).
Instrumentation: EFIS equipment comprises two colour CRTs for flight and nav information for each pilot, plus two central CRTs for engine and systems data.

DIMENSIONS, EXTERNAL (A: 204, B: 204-100 (and 204C unless otherwise stated), C: 204-120 (and 204-120C unless otherwise stated), D: 204-200 and Tu-214, E: 204-220, F: 234 short-range, G: 234 mid-range, H: 234 long-range, J: 224):

Wing span: all	41.80 m (137 ft 1¾ in)
Wing aspect ratio: all	9.6
Length overall: A, B, C, D, E	46.10 m (151 ft 3 in)
F, G, H, J	40.20 m (131 ft 10¾ in)
Fuselage cross-section:	
all	3.80 × 4.10 m (12 ft 5 1/2 in × 13 ft 5½ in)
Height overall: all	13.90 m (45 ft 7¼ in)
Tailplane span: A, B, C, D, E	15.00 m (49 ft 2½ in)
Wheel track: A, B, C, D, E	7.82 m (25 ft 8 in)
Wheelbase: A, B, C, D, E	17.00 m (55 ft 9¼ in)
Passenger doors (each): Height: all	1.85 m (6 ft 0¾ in)
Width: all	0.84 m (2 ft 9 in)
Service doors (each): Height: all	1.60 m (5 ft 3 in)
Width: all	0.65 m (2 ft ½ in)
Emergency exit doors (each):	
Height: all	1.44 m (4 ft 8¾ in)
Width: all	0.61 m (2 ft 0 in)
Baggage holds: Height to sill: all	2.71 m (8 ft 10¾ in)

DIMENSIONS, INTERNAL:

Cabin, excl flight deck:	
Length: A, B, C, D, E	30.18 m (99 ft 0 in)
Max width: all	3.57 m (11 ft 8½ in)
Max height: all	2.28 m (7 ft 6 in)
Fwd cargo hold: Height: all	1.16 m (3 ft 9¾ in)
Volume: A, B, C	11.0 m³ (388 cu ft)
Rear cargo hold: Height:	
A, B, C, D, E	1.16 m (3 ft 9¾ in)
Volume: A, B, C	15.4 m³ (544 cu ft)

Transeuropean's first Tu-204 entered revenue-earning service on 15 May 1999, flying Moscow-Barcelona *(Paul Jackson/Jane's)*

TUPOLEV—AIRCRAFT: RUSSIAN FEDERATION

AREAS:
Wings, gross: all 182.40 m² (1,963.4 sq ft)

WEIGHTS AND LOADINGS:
Operational weight empty: A	58,300 kg (128,530 lb)
B	58,800 kg (129,630 lb)
C	59,300 kg (130,735 lb)
D, E	59,000 kg (130,070 lb)
Tu-214 freighter	57,000 kg (125,665 lb)
Max payload: A, B, C	21,000 kg (46,296 lb)
204C	27,000 kg (59,524 lb)
204-120C	25,000 kg (55,116 lb)
D, E (weight limited), J	25,200 kg (55,555 lb)
Tu-214 freighter	30,000 kg (66,139 lb)
F, space limited (196 seats)	19,565 kg (43,132 lb)
F, G	18,000 kg (39,682 lb)
H	16,000 kg (35,273 lb)
Payload with max fuel: 204C	13,900 kg (30,644 lb)
204-120C	13,200 kg (29,101 lb)
Max baggage/freight: fwd hold: E	3,625 kg (7,990 lb)
rear hold: E	5,075 kg (11,190 lb)
Max fuel: A	24,000 kg (52,910 lb)
B, C, D, E, J	32,700 kg (72,090 lb)
F	24,300 kg (53,572 lb)
G, H	35,000 kg (77,162 lb)
Max ramp weight: E	111,750 kg (246,360 lb)
Max T-O weight: A	94,600 kg (208,550 lb)
B, C	103,000 kg (227,075 lb)
D, E, J	110,750 kg (244,155 lb)
F	84,800 kg (186,950 lb)
G, H	103,000 kg (227,070 lb)
Max landing weight: B, C	89,500 kg (197,310 lb)
F, G, H	88,000 kg (194,005 lb)
Max zero-fuel weight: E	84,200 kg (185,625 lb)
Max wing loading: A	518.6 kg/m² (106.23 lb/sq ft)
B, C	564.7 kg/m² (115.66 lb/sq ft)
D, E, J	607.2 kg/m² (124.36 lb/sq ft)
F	464.9 kg/m² (95.22 lb/sq ft)
G, H	564.7 kg/m² (115.66 lb/sq ft)
Max power loading: A	299 kg/kN (2.93 lb/lb st)
B	325 kg/kN (3.19 lb/lb st)
C	269 kg/kN (2.63 lb/lb st)
D	350 kg/kN (3.43 lb/lb st)
E	289 kg/kN (2.83 lb/lb st)
F	268 kg/kN (2.63 lb/lb st)
G, H	325 kg/kN (3.19 lb/lb st)

PERFORMANCE:
Nominal cruising speed at 11,100-12,100 m (36,400-39,700 ft):
A, B, C, D, E	437-459 kt (810-850 km/h; 503-528 mph)
F, G, H	448-459 kt (830-850 km/h; 515-528 mph)
Approach speed: F	119 kt (220 km/h; 137 mph)
G, H	122 kt (225 km/h; 140 mph)
T-O run: F	1,450 m (4,760 ft)
G, H	2,050 m (6,725 ft)
Required T-O runway at max T-O weight:	
A	2,030 m (6,660 ft)
B (except 204C), D, E	2,045 m (6,710 ft)
204C	2,160 m (7,090 ft)
C	1,830 m (6,005 ft)
J	2,630 m (8,630 ft)
Required landing runway: F, G, H	2,000 m (6,565 ft)
Range: with max payload:	
A	1,312 n miles (2,430 km; 1,509 miles)
B	2,321 n miles (4,300 km; 2,671 miles)
Tu-204C	2,321 n miles (4,300 km; 2,671 miles)
C	2,213 n miles (4,100 km; 2,547 miles)
Tu-204-120C	1,808 n miles (3,350 km; 2,081 miles)
D	2,591 n miles (4,800 km; 2,982 miles)
Tu-214 passenger	3,374 n miles (6,250 km; 3,883 miles)
E	2,483 n miles (4,600 km; 2,858 miles)
F	1,295 n miles (2,400 km; 1,490 miles)
G	3,585 n miles (6,650 km; 4,130 miles)
H	3,885 n miles (7,200 km; 4,475 miles)
with design payload:	
F	1,835 n miles (3,400 km; 2,110 miles)
G	4,075 n miles (7,550 km; 4,690 miles)
H	4,990 n miles (9,250 km; 5,750 miles)
with max fuel:	
Tu-204C	3,671 n miles (6,800 km; 4,225 miles)
Tu-204-120C	3,542 n miles (6,560 km; 4,076 miles)
J	3,617 n miles (6,700 km; 4,163 miles)

UPDATED

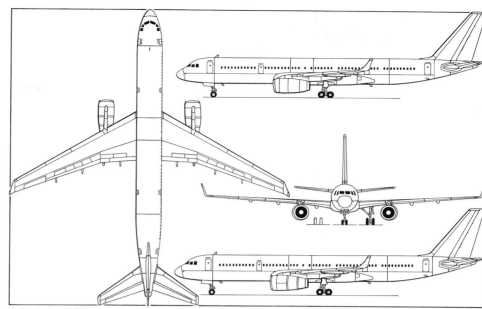

Tupolev Tu-204-200 medium-range transport (two Aviadvigatel PS-90A turbofans), with additional side view (top) of Tu-204-220 *(Mike Keep/Jane's)*

First production Tu-234 awaiting engines at Aviastar, Ulyanovsk *(Yefim Gordon)* 2001/0092555

TUPOLEV/OREL AVIA Tu-204 FREIGHTER

TYPE: Twin-jet freighter.

PROGRAMME: General technical specification for converting Tu-204 passenger aircraft to freighter configuration issued by Tupolev and Orel Avia Airlines May 1993. Offered initially for Tu-204-100 series ordered by Aeroflot, with PS-90A engines, maximum T-O weight of 94,600 kg (208,550 lb), maximum landing weight of 88,200 kg (194,445 lb), a 79,300 kg (174,825 lb) maximum zero-fuel weight, and maximum fuel weight of 24,000 kg (52,910 lb). Applicable to all other Tu-204 versions by means of work listed in Tupolev/ARISC (Aviation Register Interstate Committee) approved supplementary specification. Prototype RA-64010 delivered to Aeroflot Russian International Airlines 7 April 1995, designated Tu-204F.

In 1997, a new-build RB211-powered freighter was offered to Volga-Dnepr Airlines, designated **Tu-204C-120**. Cabin window deletion is optional.

CUSTOMERS: Four (all Tu-204-100 conversions) ordered by Aeroflot, of which only two (RA-64008 and RA-64009) apparently completed; currently used by Vnukovo Airlines and Aeroflot, respectively. Other freighters built from new, as described in previous entry.

DESIGN FEATURES: Main deck cargo door installed on port side forward of wing; redundant passenger doors deactivated; cargo door operating controls, hydraulic, electrical and warning/indication systems installed; cargo door operable in windspeeds up to 29 kt (54 km/h; 33 mph). Courier compartment immediately aft of flight deck, with 9 *g* cargo restraint bulkhead barrier to rear, three forward-facing seats, tip-up seat at front, storage and buffet facilities. Main deck floor reinforced for maximum loading capability of 280 kg/m² (57.3 lb/sq ft); maximum weight of loaded 2.24 × 2.74 m (88 × 108 in) pallet 1,500 kg (3,306 lb). (In later model Tu-204's allowable loading will be 410 kg/m²; 84 lb/sq ft, with maximum weight of 3,100 kg; 6,834 lb for 2.24 × 2.74 m; 88 × 108 in or 2.24 × 3.18 m; 88 × 125 in pallet). New sandwich flooring with non-slip surface, sealed against water ingress. Cargo handling/restraint system installed, typically ball mat and sill protector at door and rollers with side guidance (powered if required). All passenger-related provisions removed. Top-hinged cargo door.

ACCOMMODATION: All existing windows retained, also front toilet.

SYSTEMS: Cabin overhead air conditioning distribution system deleted. Temperature on flight deck and in cabin

Flight deck of Tupolev Tu-204-120C *(Paul Jackson/Jane's)* 0015140

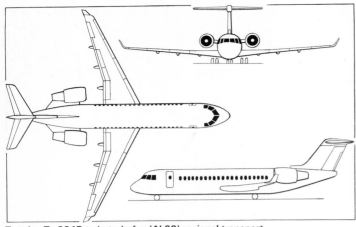
Tupolev Tu-324R twin-turbofan (AI-22) regional transport *(James Goulding/Jane's)* 2000/0079303

Model of the Tu-324A project with CF34 engines *(Paul Jackson/Jane's)* 2001/0092533

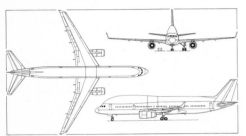
Proposed Tupolev Tu-206 LNG-fuelled version of Tu-204 airliner *(Paul Jackson/Jane's)* 1999/0051414

controllable 16 to 26°C (61 to 79°F); pressurisation unchanged. Flight deck/courier interphone. Fire extinguishers and cabin smoke detection system provided. Cabin lighting compatible with freighter configuration.

DIMENSIONS, EXTERNAL:
Cargo door opening: Length 3.405 m (11 ft 2 in)
Height 2.19 m (7 ft 2¼ in)

WEIGHTS AND LOADINGS:
Weight empty 55,040 kg (121,340 lb)
Operating weight empty 55,840 kg (123,105 lb)
Max payload 23,460 kg (51,720 lb)
Fuel with max payload 15,300 kg (33,730 lb)
Max T-O weight 94,600 kg (208,550 lb)
Max zero-fuel weight 79,300 kg (174,825 lb)
UPDATED

TUPOLEV Tu-206 and Tu-216

TYPE: Twin-jet airliner.
PROGRAMME: Based on Tu-204. **Tu-206** has four tanks, of composites materials, above cabin for 22,500 kg (49,604 lb) of liquid natural gas (LNG); conventional tankage for a further 5,500 kg (12,125 lb) of kerosene as contingency reserve. Aviadvigatel PS-92 turbofans, each of 160.8 kN (36,155 lb st). Range 2,861 n miles (5,300 km; 3,293 miles) with 210 passengers or 1,943 n miles (3,600 km; 2,236 miles) with 25,200 kg (55,556 lb) maximum payload; maximum T-O weight 110,750 kg (244,160 lb).

Tu-216 has cryogenic fuel only; Samara NK-94 turbofans and 21,000 kg (46,297 lb) of liquid natural gas fuel in two tanks above front and centre fuselage; 210 passengers; range 2,321 n miles (4,300 km; 2,671 miles). See also Tu-206. No immediate plans for production, but confirmed in 2000 to be under continued development.

Benefits of LNG include reductions of 150 per cent in NO and 20 per cent in CO emissions; and halving of fuel costs.
UPDATED

TUPOLEV Tu-214
Described under Tu-204 heading.

TUPOLEV Tu-224
Described under Tu-204 heading.

TUPOLEV Tu-234
Described under Tu-204 heading.

TUPOLEV Tu-244

TYPE: Supersonic airliner.
PROGRAMME: Continuing programme based on more than 30 years of Tupolev SST research, development and testing. By 1997, earlier published design had been replaced by an aircraft with increased outboard wing sweep (and consequently reduced span) and reduced thrust, T-O weight and passenger load, while retaining original range.

Having established liaison in the Tu-144LL programme (see 1998-99 *Jane's*), Tupolev advised Boeing on aspects of the USA's SST-2 programme (see International section under Supersonic Airliner Studies) until Boeing abandoned its SST studies. Tupolev continued to promote its design in 2000, although with no prospect of production.
DESIGN FEATURES: Current projected configuration shown in accompanying illustration. Programme directed at estimating SST market capacity; evaluating compatibility with environment in terms of sonic boom, engine noise and emission, and so on; ensuring high fuel/take-off weight ratio (49.5 to 50 per cent in Tu-144); providing high lift/drag ratio at all phases of flight (K9.5 at M2.0 and K15 at M0.9, compared with K8.1 at M2.0 for Tu-144).
POWER PLANT: Four turbofans; each 265 kN (59,525 lb st).
ACCOMMODATION: Nominal seating for 269 passengers.
AVIONICS: For operation to ICAO Cat. IIIa standards.

DIMENSIONS, EXTERNAL:
Wing span 45 m (148 ft)
Wing aspect ratio 2.2
Length overall 88 m (289 ft)
Height overall 15 m (49 ft)

AREAS:
Wings, gross 935 m² (10,000 sq ft)

WEIGHTS AND LOADINGS:
Max payload 25,000 kg (55,100 lb)
Max fuel weight 160,000 kg (352,700 lb)
Max T-O weight 325,000 kg (716,500 lb)
Max wing loading 340 kg/m² (71 lb/sq ft)
Max power loading 307 kg/kN (3.0 lb/lb st)
Required runway length 3,000 m (9,845 ft)

PERFORMANCE (estimated):
Nominal cruising speed at 18,000-20,000 m (59,000-65,600 ft) M2.0
Max range 4,965 n miles (9,200 km; 5,715 miles)
OPERATIONAL NOISE LEVELS: Designed to meet FAR Pt 36 Chapter 3 noise requirements
UPDATED

TUPOLEV Tu-245
Refer to Tu-22M entry.

TUPOLEV Tu-324

TYPE: Regional jet airliner.
PROGRAMME: Announced at 1995 Paris Air Show; proposed manufacture by KAPO at Kazan, Tatarstan, from 2003. Specification revised 1996 and further refined in 1997 when governments of Russia and Tatarstan agreed on 20 August to 50:50 sharing of development costs and latter placed symbolic first order for a presidential transport. Mockup completed by late 1997; technical documentation issued 2000. First of five prototypes to fly in 2002. Is first Russian aircraft programme funded totally by commercial backers and designed entirely with use of computers.
CURRENT VERSIONS: **Tu-324A**: (*Administrativny*: executive) VIP transport, carrying up to 10 passengers.
Tu-324R: Regional airliner; 52 seats in single-class configuration; 10 + 34 in two-class.

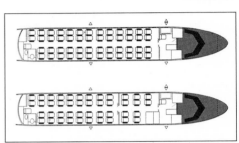
Tupolev Tu-324R configurations include 50 single-class passengers (top) and 10 business-class plus 34 tourist-class passengers (bottom) 0024039

Tu-324 stretch: Long-term project for stretching to 70-seat airliner, possibly with Rolls-Royce Deutschland BR710-48 turbofans.
COSTS: US$30 million (2000).
DESIGN FEATURES: Wings scaled down from Tu-204; illustrations show the aircraft with and without winglets.
LANDING GEAR: Messier-Dowty main contractor.
POWER PLANT: Two General Electric CF34-3B1 turbofans, each 41.0 kN (9,220 lb st), in pods on sides of rear fuselage. Alternatives under consideration include Progress AI-22 and Soyuz TRDD R-126-300.
AVIONICS: Honeywell main supplier.

DIMENSIONS, EXTERNAL:
Wing span 23.20 m (76 ft 1½ in)
Length overall: Tu-324A 23.00 m (75 ft 5½ in)
Tu-324R 25.50 m (83 ft 8 in)
Height overall 7.10 m (23 ft 3½ in)

WEIGHTS AND LOADINGS:
Max payload: Tu-324A 1,800 kg (3,968 lb)
Tu-324R 5,500 kg (12,125 lb)
Max T-O weight: Tu-324A 24,150 kg (53,241 lb)
Tu-324R 23,700 kg (52,249 lb)

PERFORMANCE (estimated):
Max cruising speed 448 kt (830 km/h; 516 mph)
Balanced runway length 1,800 m (5,910 ft)
Range:
Tu-324R: with 52 passengers 1,349 n miles (2,500 km; 1,553 miles)
with 46 passengers 1,619 n miles (3,000 km; 1,864 miles)
Tu-324A 4,022 n miles (7,450 km; 4,629 miles)
UPDATED

TUPOLEV Tu-330 and Tu-338

TYPE: Twin-turbofan freight transport.
PROGRAMME: Announced early 1993 as replacement for Antonov An-12; government support announced April 1994; first 10 to be built at Kazan; first flight scheduled 1997, for service from 1998; neither target achieved. Production at both KAPO and Samara. Stated by commander of Russian Military Transport Aviation to be one of three basic types to be operated by his force in 21st

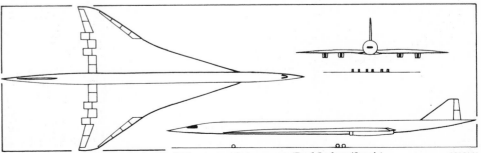
Provisional three-view of projected Tu-244 supersonic transport *(Paul Jackson/Jane's)* 0015141

century. Production drawings released by Tupolev to KAPO at Kazan in late 1997 and construction then begun. By early 2000 the venture was being offered as a potential partnership.

Tu-330 included in federal plans for civil aviation up to 2015, requiring investment of equivalent of US$340 million over seven years. Tupolev expenditure to early 2000 totalled Rb80 million. Kazan plant planning production launch in 2002.

CURRENT VERSIONS: **Tu-330:** *As described.*
Tu-338: Proposed cryogenic-fuelled version with Samara NK-94 engines and 20,000 kg (44,092 lb) of liquid natural gas.
COSTS: US$25 million to 35 million (2000, estimate).
DESIGN FEATURES: See accompanying three-view drawing; said to have 75 per cent commonality with Tu-204. Wing design basically similar to Tu-204; engine pylons as Tu-204; rear-loading ramp.
FLYING CONTROLS: Each wing has four-section leading-edge flaps, three-section trailing-edge flaps, eight spoilers forward of flaps, and aileron. Elevators and rudder each in two sections.
LANDING GEAR: Retractable tricycle type; twin-wheel, rearwards-retracting nose unit; each main unit has three pairs of wheels in tandem. Able to operate from concrete or grass.
POWER PLANT: Two Aviadvigatel PS-90A turbofans. Rolls-Royce and Pratt & Whitney turbofans under consideration for export sales. A version of the standard Tu-330 is projected for LNG transportation, with three tanks in the cabin for 22,800 kg (50,265 lb) of LNG.

DIMENSIONS, EXTERNAL:
Wing span over winglets	43.50 m (142 ft 8½ in)
Wing aspect ratio	10.6
Length overall	42.00 m (137 ft 9½ in)
Height overall	14.00 m (45 ft 11¼ in)
Rear-loading ramp: Length	4.00 m (13 ft 1½ in)
Width	4.00 m (13 ft 1½ in)

DIMENSIONS, INTERNAL:
Cargo hold: Length	19.50 m (63 ft 11½ in)
Width	4.00 m (13 ft 1½ in)
Height: min	3.55 m (11 ft 7¾ in)
max	4.00 m (13 ft 1½ in)
Volume	350.0 m³ (12,360 cu ft)

AREAS:
Wings, gross	177.80 m² (1,913.8 sq ft)

WEIGHTS AND LOADINGS:
Payload: in hold	30,000 kg (66,139 lb)
on ramp	5,000 kg (11,023 lb)
max	35,000 kg (77,162 lb)
Max T-O weight	103,500 kg (228,175 lb)
Max wing loading	582.1 kg/m² (119.23 lb/sq ft)

PERFORMANCE (estimated):
Nominal cruising speed at 11,000 m (36,000 ft)
431-458 kt (800-850 km/h; 497-528 mph)
T-O run 2,200 m (7,220 ft)
Nominal range: with 15,000 kg (33,069 lb) payload
3,779 n miles (7,000 km; 4,349 miles)
with 20,000 kg (44,090 lb) payload
3,020 n miles (5,600 km; 3,480 miles)
with 30,000 kg (66,139 lb) payload
1,620 n miles (3,000 km; 1,865 miles)
UPDATED

TUPOLEV Tu-334, Tu-336 and Tu-354

TYPE: Twin-jet airliner.
PROGRAMME: Launched 1986, with target first flight in 1991 and service entry in 1995; first prototype Tu-334 (RA-94001) eventually rolled out at Zhukovsky during Moscow Air Show, 25 August 1995; then scheduled to fly 1996 with D-436T1 engines, but delayed by funding shortages; flown 8 February 1999 and had achieved eight sorties by time of public debut at Moscow in August 1999 and 43 by July 2000.

Refinancing by Ukrainian government in early 1997 resulted in new target date of May/June 1997 for first Tu-334s built by Aviant at Kiev; this also missed; second and third identical, but non-flying, prototypes under construction at Zhukovsky by 1996; second used for static tests at CAHI (TsAGI) 1996; unidentified test airframe delivered to Aviatest LNK at Riga from Kiev in mid-2000 to simulate 20 year fatigue life; further two prototypes and three production aircraft under construction at Kiev by early 1999; third prototype assembled and awaiting engines by mid-2000, with fourth then yet to be assembled; certification programme due for completion in 2001 after 1,000 flying hours; deliveries begin 2003. Aviacor in Samara is building Tu-334-200 stretched version, four of which nearing completion in early 1997. TANTK at Taganrog was to have been second source for -100, but Russian government decree of 5 October 1999 nominated RSK MiG (which will contribute half of certification costs); first Taganrog airframe transported to Voronin Production Centre, Moscow, arriving 7 April 2000. Completion and final assembly, Khodinka, in second quarter of 2001. Half of all assemblies and components for RSK MiG production being supplied by Aviant.

On 16 June 1997 Rolls-Royce Deutschland agreed loan of BR 710 engines to build prototype of Tu-334-120; first flight of -120 had been expected late 1998, but was delayed by funding shortages. Revised plans for BR 710s to go into -100 prototype, after short test programme with D-436s, were also thwarted by aircraft's continued grounding. R-R support will in future be directed at versions powered by larger BR 715 engines.

During mid-1997, IAIO of Iran showed interest in establishing a kit assembly line at Esfahan for 105 Tu-334s over 20 years. Negotiations were proceeding in 2000, based on eventual fuselage detailed assembly in Iran, with complete wings from Ukraine and empennage from Russia; peak production of 12 per year.

CURRENT VERSIONS: **Tu-334-100:** Basic version, with D-436T1 engines, for 72 mixed class or 102 tourist class passengers. Prototypes are of this version.
Tu-334-100C: Combi version of -100.
Tu-334-100M: Variant of -100; MTOW 47,700 kg (105,160 lb); range 1,684 n miles (3,120 km; 1,938 miles).
Tu-334-100D: Extended-range version, announced at Moscow Air Show '95. Basically similar to Tu-334-100, but with increased wing span and uprated D-436T2 engines of Tu-354; increased fuel and maximum T-O weight; 102 seats.
Tu-334-120: As Tu-334-100, but two 66.7 kN (15,000 lb st) Rolls-Royce Deutschland BR710-48 turbofans. Third flying prototype to be in this configuration.
Tu-334-120D: Rolls-Royce Deutschland BR710-55 engines; increased weight and range.
Tu-334-200: Basically similar to Tu-334-100D and alternatively known as **Tu-354;** D-436T2 or BR 715-55 engines; fuselage lengthened by 3.90 m (12 ft 9½ in) to accommodate up to 126 passengers at 81 cm (32 in) pitch; increased wing span; four-wheel bogies on main landing gear units.
Tu-334-200C: Variant of -200; combi with same payload.

Prototype Tu-334 RA-94001 displaying at Dubai in November 1999 *(Paul Jackson/Jane's)*

Tu-334-220: Further increased MTOW; BR 715-55 engines. Fuselage as Tu-354. Announced 1998, possibly replacing Tu-354.
Tu-334C: Cargo version; at design study stage 1995.
Tu-336: Cryogenic-fuelled version of Tu-334 with 13,000 kg (28,660 lb) of liquid natural gas in two faired tanks extending full length of cabin roof; Ivchenko Progress D-436T2 engines; 102 passengers; range 1,457 n miles (2,700 km; 1,677 miles). Revealed 1998; no production plans announced.
CUSTOMERS: Final assembly at Aviacor, Aviant (Ukraine) and RSK MiG plants. Preliminary contracts signed 1995 for approximately 40 aircraft, for Rossiya Airlines, Bashkiri Airlines, Tyumen Airlines and Tatarstan Airlines; total increased to 160 from 13 airlines by 1997. MoU for 20 Tu-334-120s signed by Aeroflot in June 2000. CIS market estimated as 400 to 500; by 1999, Iranian airlines committed to purchase 50 from local assembly.
COSTS: US$14 million to US$19 million (1999). Break-even at 67th aircraft, assuming US$15 million average price.
DESIGN FEATURES: Replacement for Tu-134, to meet requirements of Russian Federation and Associated States (CIS) airlines. Wings have much in common with those of Tu-204 and the fuselage is shortened version of that of Tu-204, with identical flight deck. Configuration is all-swept low/mid-wing, with rear-mounted engines and T tail; circular-section semi-monocoque fuselage; wings have dihedral from roots. Service life 60,000 hours.

Wings have supercritical section, 24° sweepback, with winglets.
FLYING CONTROLS: Main and standby fly-by-wire; emergency hydraulic and mechanical back-up, except for ailerons. Two-section, single-slotted, trailing-edge flap, two-section airbrakes forward of inner flap and two-section spoilers forward of outer flap on each wing; four-section leading-edge slat over full span each wing; conventional ailerons, elevators and two-section rudder; no tabs. Variable-incidence tailplane.
STRUCTURE: Composites and other lightweight materials make up 20 per cent of structure by weight. Metal alloy wing, fin and horizontal tail, each constructed on two spars, have composites leading-edges and covered by milled panels. Elevators, ailerons, airbrakes, spoilers, flaps, rudder and floor panels of composites honeycomb. Fuselage of metal alloy with riveted skin.
LANDING GEAR: Retractable tricycle type; twin wheels on each unit of Tu-334-100, four-wheel bogies on main units of Tu-334-200; main units retract inward into wing/fuselage fairings; trailing-link mainwheel legs on Tu-334-100; nosewheels retract forwards, forward elements of two-section door each side remaining closed except during cycling. Recent Tupolev illustrations show all versions of Tu-334 with four-wheel main bogies. Tyre size 1,070×390-480R on mainwheels; 680×260-355R on nosewheels. Stout grille behind mainwheels of prototype to prevent ice and slush entering turbofans.
POWER PLANT: Two Ivchenko Progress D-436T1 turbofans, each rated at 73.6 kN (16,535 lb st) on Tu-334-100; Rolls-Royce Deutschland BR710-48 (68.9 kN; 15,500 lb st) on Tu-334-120. Two D-436T2 turbofans, each rated at

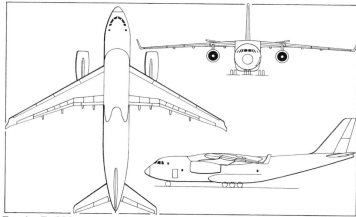

Tupolev Tu-330 twin-turbofan freight transport *(Mike Keep/Jane's)*

Model of the Tu-330 *(Yefim Gordon)*

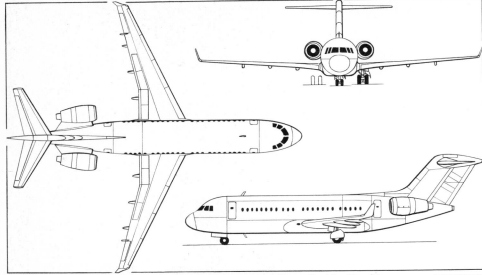

Tupolev Tu-334-100 twin-turbofan medium-range transport *(James Goulding/Jane's)* 2000/0084429

Tupolev Tu-334, with landing gear deployed *(Paul Jackson/Jane's)* 2001/0092534

80.5 kN (18,100 lb st), on Tu-334-100D. Two similar D-436T2 turbofans, or Rolls-Royce Deutschland BR 715-55 turbofans each rated at 88.9 kN (19,995 lb st), on Tu-334-200 and -220. D-436T1 production by consortium of Motor Sich at Zaporozhye (Ukraine), Salyut and UMPO. Fuel in integral wing tanks.

ACCOMMODATION (Tu-334-100 and -120): Crew of two or three on flight deck; provision for fourth seat for instructor or observer. Two basic single-aisle passenger arrangements: (1) 72 seats, with 12 seats four-abreast in first class cabin at front, at pitch of 102 cm (40 in) and with 73 cm (28.75 in) aisle, and 60 tourist class seats six-abreast at 78 cm (30.7 in) pitch and with 47 cm (18.5 in) aisle; (2) 102 seats, all tourist, at 78 cm (30.7 in) pitch. Both configurations have buffet/galley and toilet immediately behind flight deck, a further toilet and service compartment at rear; 72-seater has additional galleys at front and rear; overhead stowage for hand baggage. Other arrangements at customer's option. Passenger doors at front and rear of cabin on port side; service doors opposite. Underfloor baggage/freight holds; doors on starboard side.

ACCOMMODATION (Tu-334-200/Tu-354): Flight deck unchanged. Two basic single-aisle passenger arrangements: (1) 110 seats, with eight seats four-abreast in first class cabin at front, at pitch of 99 cm (39 in) and 102 tourist class seats six-abreast at 81 cm (32 in) pitch; (2) 126 seats, all tourist, at 81 cm (32 in) pitch. Facilities as Tu-334-100/120, plus emergency exit over wing each side.

SYSTEMS: APU in tailcone. Hydraulic system for actuation of flying controls. Air conditioning unit in wing centre-section.

AVIONICS: *Radar:* Nose-mounted weather radar.
Instrumentation: EFIS standard, with six CRTs. Equipment for landings in ICAO Cat. IIIa conditions.

DIMENSIONS, EXTERNAL:
Wing span: 100, 120	29.77 m (97 ft 8 in)
100D, 200	32.61 m (107 ft 0 in)
Wing aspect ratio: 100, 120	10.2
100D, 200	10.6
Length overall: 100, 120	31.26 m (102 ft 6¾ in)
100D	31.80 m (104 ft 4 in)
200	35.16 m (115 ft 4½ in)

Fuselage cross-section:
all	3.80 × 4.10 m (12 ft 5½ in × 13 ft 5½ in)
Height overall: all	9.38 m (30 ft 9¼ in)
Wheelbase: 100, 120	11.75 m (38 ft 6¾ in)
Baggage door, forward: Height	1.375 m (4 ft 6¼ in)
Width	1.165 m (3 ft 9¾ in)

DIMENSIONS, INTERNAL:
Cabin: Length: 100, 120	17.84 m (58 ft 6¼ in)
Height:	
above aisle: all	2.10 m (6 ft 10¾ in)
beneath hand baggage racks: all	1.70 m (5 ft 7 in)
Max width: all	3.505 m (11 ft 6 in)
Volume: 100, 120	118.0 m³ (4,167 cu ft)
Baggage holds:	
Max length: 100, 120, forward	4.10 m (13 ft 5¼ in)
100, 120, rear	3.00 m (9 ft 10 in)
Max height: 100, 120	1.20 m (3 ft 11¼ in)
Volume: 100, forward	11.7 m³ (413 cu ft)
Total volume: 100, 120	16.2 m³ (572 cu ft)

AREAS:
Wings, gross: 100, 120	83.23 m² (895.8 sq ft)
100D, 120D, 200, 220	100.0 m² (1,076.4 sq ft)

WEIGHTS AND LOADINGS:
Weight empty: 100	30,050 kg (66,250 lb)
120	28,350 kg (62,501 lb)
200	34,375 kg (75,785 lb)
Max payload:	
100, 100C, 100D, 120D	9,960 kg (21,958 lb)
120	11,000 kg (24,251 lb)
200, 200C, 220	11,970 kg (26,389 lb)
Max fuel: 100, 120	9,540 kg (21,030 lb)
100D, 200	13,790 kg (30,400 lb)
Max T-O weight: 100, 120, 120C	46,100 kg (101,630 lb)
100D, 120D	54,420 kg (119,975 lb)
200	54,800 kg (120,815 lb)
220: BR715	54,800 kg (120,815 lb)
D-436	55,470 kg (122,290 lb)
Max wing loading:	
100, 120	553.9 kg/m² (113.44 lb/sq ft)
100D, 120D	544.2 kg/m² (111.46 lb/sq ft)
200, 220: BR 715	548.0 kg/m² (112.24 lb/sq ft)
220: D-436	554.7 kg/m² (113.61 lb/sq ft)
Max power loading: 100	313 kg/kN (3.07 lb/lb st)
120	334 kg/kN (3.27 lb/lb st)
100D	338 kg/kN (3.31 lb/lb st)
120D	306 kg/kN (3.00 lb/lb st)
200	340 kg/kN (3.34 lb/lb st)
220: D-436	344 kg/kN (3.38 lb/lb st)
BR 715	308 kg/kN (3.02 lb/lb st)

PERFORMANCE (estimated):
Nominal cruising speed at 10,600-11,100 m (34,775-36,400 ft): 100, 100D, 200
431-442 kt (800-820 km/h; 497-510 mph)
Balanced runway length at 30°C:
100, 100D, 200	2,200 m (7,220 ft)

Range with max payload:
100, 100C	1,080 n miles (2,000 km; 1,242 miles)
100D	2,213 n miles (4,100 km; 2,547 miles)
120	1,685 n miles (3,122 km; 1,939 miles)
120D	2,267 n miles (4,200 km; 2,609 miles)
200, 220	1,187 n miles (2,200 km; 1,367 miles)

OPERATIONAL NOISE LEVELS (ICAO):
T-O	−3 to 4 EPNdB
Approach	−1.5%
Sideline	−3 to 8%

UPDATED

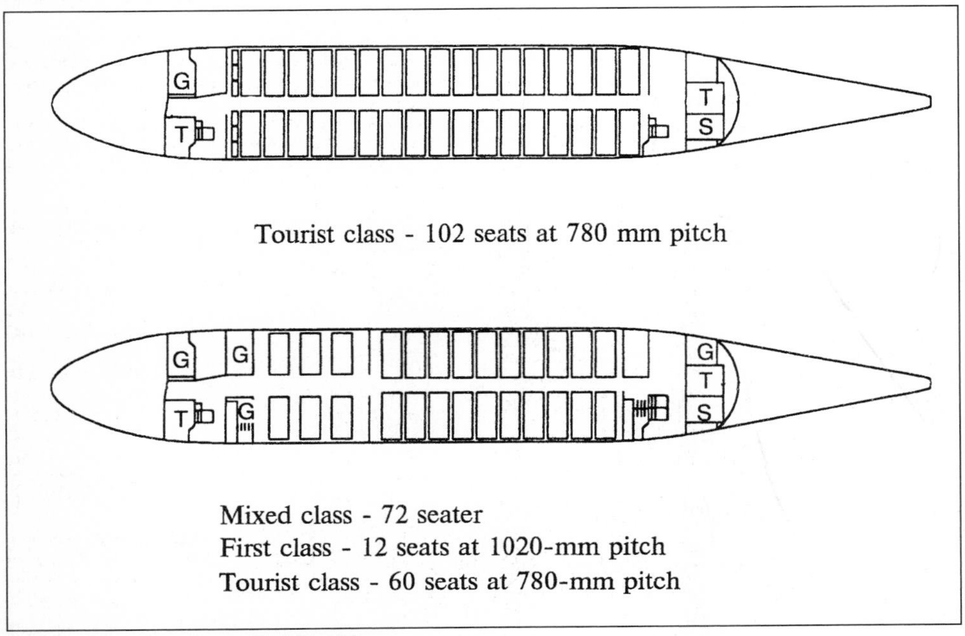

Cabin seating plans for the Tu-334-100
G: galley, S: service compartment, T: lavatory

TUPOLEV Tu-344
TYPE: Supersonic business jet.
PROGRAMME: In 2000, Tupolev showed a model photograph of a projected 12- to 18-seat transport, clearly based on the Tu-22M bomber (which see). Range would be 4,157 n miles (7,700 km; 4,784 miles) cruising at M1.7. There is no indication that this is a currently active project.

UPDATED

TUPOLEV Tu-354
Refer to Tu-334/354 entry.

TUPOLEV Tu-414
TYPE: Regional jet airliner.
PROGRAMME: Announced 1994 and last described and illustrated in 1997-98 edition. No further announcements until confirmed in mid-2000 to be under continued study as 50/70-seat transport with an estimated unit cost between US$11 million and $14 million.

NEW ENTRY

Low-quality image of Tu-344 2001/0103561

UUAP

ULAN-UDENSKY AVIATSIONNYI ZAVOD OAO (Ulan-Ude Aviation Plant JSC)
ulitsa Khorinskaya 1, 670009 Ulan-Ude
Tel: (+7 3012) 25 33 86, 25 74 75 and 25 35 55
Fax: (+7 3012) 25 21 47
e-mail: uuaz@buriatia.ru or uuaz@chat.ru
GENERAL DIRECTOR: Leonid Ya Belykh
TECHNICAL DIRECTOR: Sergey V Solomin
DEPUTY DIRECTOR-GENERAL, SALES AND MARKETING: Ts Ts Galdanov

Founded on 1 January 1939 as GAZ 99, the Ulan-Ude Aviation Plant formed on 26 October 2000, having previously been known as Ulan-Ude Aviation Industrial Association. It manufactures a wide range of items, from helicopters and aeroplanes to spares and domestic equipment. Is only Russian plant currently producing both fixed- and rotary-wing aircraft. Employees in 2000 totalled 5,056. The factory has built An-24 transports, Yak-25RV reconnaissance aircraft and MiG-27 fighter-bombers. Helicopter production began in 1956, progressing through Ka-15, Ka-18 and Ka-25 to the 1970 launch of Mi-8 manufacture; total of 3,700 Mi-8 series built. Current products include modern developments of the Mil Mi-8/Mi-17 series of helicopters (including ten Mi-171s delivered to China in 1999); and the Sukhoi Su-25UBK combat trainer and its related Su-39 attack aircraft. Will build Kamov Ka-62 and Ka-64. Also undertakes overhaul and upgrades, including five refurbished Mi-8AMTs supplied to Iranian Navy in 1999. Government shareholding is 38 per cent. Employees total more than 8,000. Joint venture with SME Aviation of Malaysia, 1999, for promotion of Mi-8/Mi-17 and Su-25.

UPDATED

Mi-8 helicopters under construction at Ulan-Ude

VAPO

VORONEZHSKOYE AVIATSIONNOYE PROIZVODSTVENNOYE OBEDINENIE OAO (Voronezh Aviation Production Association JSC)
ulitsa Tsiolkovskoga 27, Voronezh
Tel: (+7 80732) 44 85 01
Fax: (+7 80732) 49 90 17
Telex: 153244 MARS SU
GENERAL DIRECTOR: Vyatcheslav A Salikov
CHIEF, FOREIGN ECONOMIC RELATIONS: Vyatcheslav K Kudinov

VAPO (Westernised as VASO) is an integral part of the Ilyushin Aviation Complex, responsible for production of the Ilyushin Il-96-300 and Il-96M/N series of transport aircraft. It plans to form a joint holding with Ilyushin, initially by uniting the state shareholdings (30 per cent and 60 per cent, respectively) in each company; this will later be joined by TAPO of Uzbekistan. Russian government holding is 20 per cent. Non-aviation manufacture includes a family of small motor boats.

Plant originated in GAZ 84, established 1932, and later was responsible for Tupolev Tu-144 supersonic airliner and Ilyushin Il-86 wide-bodied airliner.

UPDATED

VITEK

KOMPANIYA VITEK (Vitek Company)
Panfilova D20, Korpus 3, 125080 Moskva
Tel: (+7 095) 158 65 14
Fax: (+7 095) 158 69 52
e-mail: vitek.ltd@relcom.ru
DIRECTOR: Yuri D Bakhenov
CONSTRUCTOR: Eduard B Bakhenov

Vitek is promoting the Ovod which, if demand is sufficient, it is able to produce at the rate of 50 per year.

VERIFIED

VITEK OVOD
English name: Gadfly
TYPE: Two-seat lightplane.
PROGRAMME: Prototype exhibited, complete but unflown, at MAKS '99, Moscow, August 1999.
DESIGN FEATURES: Simple braced high-wing tourer and trainer; deep fuselage and mutually wire-braced empennage with comparatively small fin.
FLYING CONTROLS: Conventional and manual. Actuation by cables. Flight-adjustable trim tab in port elevator; ground-adjustable tab on each aileron. Plain flaps.
STRUCTURE: Metal tube airframe, fabric-covered except for metal engine cowling.
LANDING GEAR: Tricycle type; fixed. Mainwheels size 400×150; nosewheel 300×125. Rubber-in-compression main suspension; wound bungee cord noseleg shock-absorption.
POWER PLANT: One 70.8 kW (95 hp) VAZ 21313 1,700 cc motorcar engine driving a two-blade, fixed-pitch, composites propeller. Fuel capacity 60 litres (15.8 US gallons; 13.2 Imp gallons).
ACCOMMODATION: Two persons, side by side, with dual controls. Upward-hinged door each side. Baggage shelf behind seats.
DIMENSIONS, EXTERNAL:
Wing span	9.80 m (32 ft 1¾ in)
Wing aspect ratio	7.7
Length overall	6.11 m (20 ft 0½ in)

AREAS:
Wings, gross	12.40 m² (133.5 sq ft)

WEIGHTS AND LOADINGS:
Weight empty	433 kg (955 lb)
Baggage capacity	25 kg (55 lb)
Max T-O weight	670 kg (1,477 lb)
Max wing loading	54.0 kg/m² (11.07 lb/sq ft)
Max power loading	9.46 kg/kW (15.55 lb/hp)

PERFORMANCE (estimated):
Max level speed	92 kt (170 km/h; 106 mph)
Normal cruising speed	81 kt (150 km/h; 93 mph)
Max rate of climb at S/L	180 m (591 ft)/min
Service ceiling	3,500 m (11,480 ft)
T-O run	110 m (360 ft)
Landing run	70 m (230 ft)

VERIFIED

Prototype Vitek Ovod, making its debut at Zhukovsky on 20 August 1999 *(Paul Jackson/Jane's)*

YAKOVLEV

YAKOVLEV AVIATSIONNOYE KORPORATSIYA OAO (Yakovlev Aviation Corporation JSC)
Leningradsky prospekt 68, 125315 Moskva
Tel: (+7 095) 157 31 62
Fax: (+7 095) 157 67 47

Yak Aviation Corporation formed in 1992 to engage in aircraft marketing, design, development, production, sales and after-sales support; includes A S Yakovlev Design Bureau, and Saratov (SAZ) and Smolensk (SmAZ) airframe plants. Hyundai-Yak Aerospace Co Ltd created in 1994, as a joint venture with Hyundai Group of South Korea, to develop, market and sell a wide range of small and medium-size aircraft. Collaborative agreement concluded with Aermacchi of Italy on Yak-130, Westernised version of which announced July 2000 as Aermacchi M-346. Yakovlev also provided assistance to Lockheed Martin of USA in connection with the latter's Joint Strike Fighter contender.
UPDATED

OPYTNO-KONSTRUKTORSKOYE BYURO IMENI A S YAKOVLEVA OAO (Experimental Design Bureau named for A S Yakovlev JSC)
Leningradsky prospekt 68, 12315 Moskva
Tel: (+7 095) 157 17 34
Fax: (+7 095) 157 47 26
ACTING CEO: Arkady I Gurtovoy
PRESIDENT AND GENERAL DIRECTOR: Oleg F Demchenko
FIRST VICE-PRESIDENT AND FIRST DEPUTY GENERAL DESIGNER:
  Vladimir G Dmitriev
DEPUTY GENERAL DIRECTOR, FLIGHT TESTS: Andrei A Sinitsin
VICE-PRESIDENTS AND CHIEF DESIGNERS:
  Vladimir G Dmitriev (Yak-46)
  Nicolas N Dolzhenkov (Yak-130)
  Vladimir A Mitkin
  Alexei G Rakhimbaev (Yak-42, Yak-142)
CHIEF DESIGNERS:
  Sergei A Yakovlev (Yak-3, Yak-40TL)
  Dmitry K Drach (Yak-54, Yak-55M, Yak-58, Yak-112)
  Yuri I Yankevich (Yak-18T, Yak-52, UAVs)
CHIEFS OF DEPARTMENTS:
  Anatoly S Ivanov (Marketing and Sales)
  Evgeny M Tarasov (Exterior)
  Yuri V Zasypkin (Information/Press Service)

Yak-130D advanced jet trainer 2001/0067482

More than 200 aircraft designs and variants have been completed by Yakovlev since 1927, of which about 100 have been manufactured in series. About 70,000 aircraft of Yakovlev design have been built.

By mid-2000, Yakovlev working on design of short/medium-range transport conforming to Russian aviation technology plan up to 2015.

UPDATED

YAKOVLEV Yak-130
TYPE: Advanced jet trainer/light attack jet.
PROGRAMME: One of designs by five OKBs to meet official Russian requirement for 200 aircraft to replace Aero L-29 and L-39 Albatros for all aspects of flying training, basic to combat simulation, and combat; known initially as Yak-UTS; developed in partnership with Aermacchi of Italy with definition completed in December 1999 resulting in a common basic configuration, from which all current versions are derived; originally designated Yak/AEM-130 outside Russian Federation and Associated States (CIS) but renamed Aermacchi M-346 (which see) in July 2000; partnership agreement signed 1987, Westernised version being funded 50 per cent by Aermacchi, and 25 per cent each by Yakovlev and Sokol aircraft factory at Nizhny-Novgorod. Aermacchi undertaking 5,000 hours of wind tunnel tests. Aermacchi has rights to establish a second assembly line in Italy and contribute tail section and wing assemblies and FBW control system and avionics to both production lines, all components to be single-sourced. Yakovlev designed, and will manufacture wing, tail surfaces, canopy, and digital FBW FCS and will integrate avionics and weapon systems on non-RFAS aircraft.

Yak-130D development aircraft shown to press 30 November 1994; first flight 25 April 1996 at Zhukovsky (later registered RA-43130). Flew 52 sorties (36 hours) in first year, including six without winglets, pending fitment of differently shaped units. To Italy for flight trials July/August 1997, during which (18 July) the 100th sortie was flown; Low-speed, high-AoA programme (30 flights) completed in Italy by prototype 31 January 1999; AoAs of up to 41° (maximum), 35° in stabilised flight and 28° in landing configuration were flown successfully. Aircraft returned to Russia for maintenance, followed by evaluation at Akhtubinsk and Armavir test centres; again in Italy by 2000. Almost 300 sorties flown by mid-2000.

Ten planned preseries aircraft were to comprise three for Zhukovsky test centre for evaluation originally planned to begin in 1998 and seven for air force evaluation planned for 1999 although development programme transferred to Italy, mid-1998, to escape financial instability in Russia, leading to far-reaching reappraisal. By September 1999 it was expected that funding for four preproduction prototypes would soon be obtained, together with three aircraft for evaluation. Second (which will also be the first series production aircraft) to fly during 2002. First prototype will receive RD-2500 engines in third quarter of 2001.

Navalised version proposed for carrier training; tender now includes Penza (Russia)/CAE Electronics (Canada) simulators and Yak-54 for screening and primary training. Slightly smaller production aircraft to be manufactured at Sokol plant, Nizhny Novgorod from 2002.

CURRENT VERSIONS: **Yak-130D:** Development prototype RA-43130.
  Yak-130: For Russian air force; indigenous equipment; service entry originally scheduled for 2000; delivery now planned by 2004/2005 "at best".
  Yak/AEM-130: Hybrid export version; also known as AY-130; programme frozen. Items listed under Systems and Avionics refer mainly to this version. Viewed as low flyaway cost solution to Western trainer requirements by using Yak's good aerodynamic configuration and by addressing poor man-machine interface, low thrust-to-weight ratio and some other "problem areas".
  Aermacchi M-346: See Italian section
CUSTOMERS: Initial batch of 10 reportedly being produced, with Russian avionics, by Sokol, for operational evaluation by Russian military pilots. US$241 million funding provided by Italian government for development, although IAF has no immediate requirement. Interest expressed by Slovak Republic, which received demonstration in July 1997; Slovakia announced on 5 August 1998 that debts owed by Russia will be repaid in part by delivery of 12 Yak-130s. The Yak-130 remains a competitor in Slovakia's search for a subsonic multirole fighter with the AMX, L 159, Hawk and MiG-AT, but has not been officially selected. Its Slovak engines give it competitive advantage. Sales anticipated of up to 630, of potential market for 2,300, plus 300 for RFAS.
COSTS: US$10 million to US$12 million, depending on standard. US$13 million to US14 million for M-346. US$70 million reportedly already spent by Aermacchi and US$30 million by Yak, with another US$100 million of R&D spending to follow.
DESIGN FEATURES: Designed from outset to conform to Western specifications for structure, flight controls and maintenance (MSG-3 and MIL-STD-470B). All-swept mid-wing monoplane, except for straight wing and tailplane trailing-edges; no dihedral or anhedral; two-piece full-span automatic leading-edge slats originally on prototype, but dogtooth leading-edge discontinuity tested on RA-43130 will be adopted on production version; all-moving low-mounted tailplane, with dogtooth leading-edge; forward-hinged, door-type airbrake on spine, ahead of fin; engines in ducts under wingroots, beneath LERX extending almost to windscreen. Wing leading-edge sweepback 31°. Prototype's winglets reduced in size and moved inboard to wingroot as 'LEX fence'; may be deleted in production version. Strakes ahead of windscreen by 1999 replicate characteristics of planned rounded nose. Series production versions will be 1 tonne lighter than the prototype with a more slender, shallower fuselage.
FLYING CONTROLS: Avionika full-authority, analogue, four-channel fly-by-wire system, with digital 'fourth channel' and 'carefree handling', slaved to laser-gyro platform, developed from that of Yak-141 V/STOL combat aircraft. Demonstrator has two selectable FCS models, one simple, one in MiG-29 class; will be replaced by digital system developed by BAE Italy and Lear Astronics. Prototype inherently stable; production aircraft intended to have 5 per cent longitudinal instability to reproduce handling characteristics of MiG-29/Su-27 series. Handling characteristics can be reprogrammed to simulate other types.
STRUCTURE: Mainly of light alloys, but carbon fibre composites for most control surfaces. Structural design accords to MIL-STD-1530A and MIL-A-8660A. Service life 10,000 hours; extension planned to 15,000 hours.
LANDING GEAR: Retractable tricycle type; single wheel on each unit; mainwheels retract into engine ducts; low-pressure tyres. Nosewheel steering to MIL-STD-8812 Class A. Anti-skid system.
POWER PLANT: Two Povazské Strojàrne ZMK DV-2S (Klimov RD-35) turbofans each 21.58 kN (4,852 lb st). Autonomous engine starting; 20 second negative g fuel supply. Production version of DV-2, designated AI-222, completed design in mid-2000 and to be certified in 2002 and begin deliveries in 2003.

Two fuel tanks in wings; one in centre-fuselage. Provision for two external fuel tanks, each of 590 litres (156 US gallons; 130 Imp gallons) beneath wings, plus 300 litre (79.0 US gallon; 66.0 Imp gallon) tank on centreline. Single-point pressure refuelling; gravity filling optional. Detachable aerial refuelling probe. Prototype has intake FOD doors of MiG-29 type which will be option for production version. Production aircraft may receive new 2,500 kg Tushino-Soyuz RD-2500 engines, based on existing RD-1700. Other engine choices include the Klimov RD-35, Lyulka/Saturn AL-55 and Zaparozhe's AI-222.
ACCOMMODATION: Two crew in tandem under blister canopy, on Zvezda K-36LT3.5 zero/zero ejection seats with built-in computer, side-force rocket engine and extreme attitude performance; export option of Martin-Baker Mk 16 standard on M-346; explosive cord canopy penetration; rear seat raised. Accommodation air

Yakovlev Yak-130D development aircraft in 1999 configuration, with 'LEX fence' on inboard leading-edge and strakes ahead of windscreen *(Paul Jackson/Jane's)* 2000/0075888

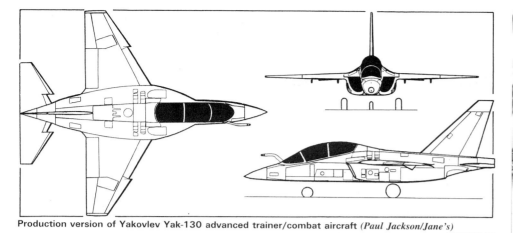

Production version of Yakovlev Yak-130 advanced trainer/combat aircraft *(Paul Jackson/Jane's)*
1999/0051417

conditioned and pressurised up to differential of 0.26 bar (3.8 lb/sq in). Front pilot has maximum view over nose of −16°; rear pilot −6°.

SYSTEMS: Flight control system includes four BAE Systems computers and provides artificial stability, flying qualities in accordance with MIL-STD-1797A, controllability up to at least 35° AoA, carefree handling (g limitation, stall/spin prevention, AoA limitation) and adaptability to various degrees of automation and autopilot modes; includes automatic reversionary modes in the event of damage or failure; and built-in test capability. Honeywell GTCP 36-150 APU, providing normal and emergency electric, hydraulic and pneumatic power. OBOGS. Dual independent hydraulic systems, pressure 207 bars (3,000 lb/sq in); single failure leaves 75 per cent control surface movement; separate accumulators for emergency landing gear extension and brake. Dual Auxilec 20 kVA generators (one per engine); two Auxilec 6 kW transformer/rectifiers; one APU-driven 5 kW 28 V DC generator; system rating 200/115 V AC, 400 Hz. Two Ni/Cd batteries. Anti-icing system for windscreen and engine inlet guide vanes. Fire suppression system for engine and APU.

AVIONICS: Western avionics optional to replace standard Russian equipment. Primary suppliers Leninets (Russia) and GFSA/FIAR (Italy).
Comms: Dual V/UHF transceivers; intercom.
Radar: Optional.
Flight: Nav computer, laser gyro INS with embedded GPS, air data system, short-range radio nav Tacan and VOR/ILS, ADF and radio altimeter; IFF.

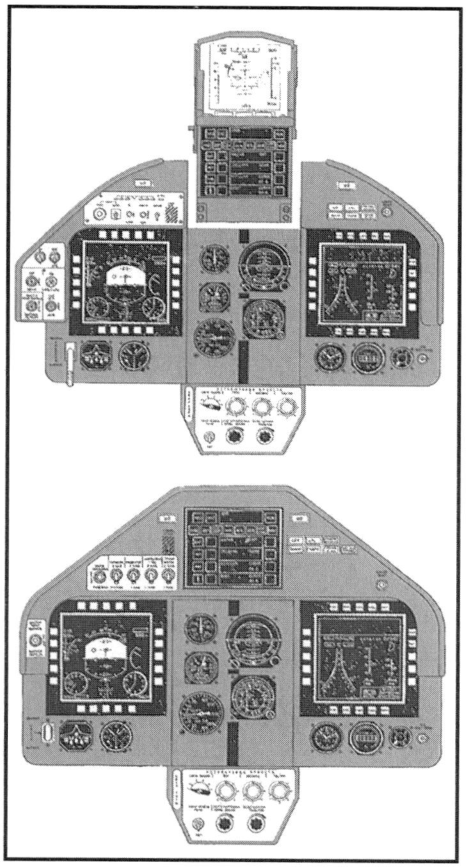

Instrumentation: HUD in front cockpit as part of collimated flight and sighting display in conjunction with pilot's helmet-mounted target designator; two Elektro Avtomatika 130 mm (5.1 in) square colour multifunction liquid-crystal displays in each cockpit (three MFDs in export aircraft) with standby electromechanical flight/nav instruments. Optional helmet-mounted display. NVG compatibility and HOTAS controls.
Mission: Mission data entry system; optical weapon aiming computer and weapons control system; simulator of moving targets, guidance commands, weapons preparation and tactical situations; flight data and crew actions recorder; TV monitor of eyes and hands positions, with video recorder, in front cockpit.
Self-defence: Provision for chaff/flare dispenser, RWR and active electronic countermeasures.

ARMAMENT (optional): Seven (optionally nine, including wingtip AAM stations) hardpoints for up to 3,000 kg (6,614 lb) of rockets, guns, missiles, guided and unguided bombs, including eight 250 kg bombs, four 500 kg laser-guided bombs, submunitions containers, Kh-25ML (AS-10 'Karen') or AGM-65 Maverick ASMs, R-73 (AA-11 'Archer'), AIM-9L Sidewinder or Magic 2 AAMs.

DIMENSIONS, EXTERNAL (production aircraft):
Wing span	9.72 m (31 ft 10¾ in)
Wing aspect ratio	4.0
Length overall	11.49 m (37 ft 8¼ in)
Height overall	4.76 m (15 ft 7½ in)
Wheel track	2.53 m (8 ft 3½ in)
Wheelbase	3.90 m (12 ft 9½ in)

AREAS:
Wings, gross	23.52 m² (253.2 sq ft)

WEIGHTS AND LOADINGS (production aircraft):
Weight empty	4,600 kg (10,141 lb)
Max weapon load	3,000 kg (6,614 lb)
Max internal fuel weight	1,750 kg (3,858 lb)
Max T-O weight: trainer	6,500 kg (14,330 lb)
attack/trainer	9,500 kg (20,943 lb)
Max wing loading: trainer	276.4 kg/m² (56.60 lb/sq ft)
attack/trainer	403.9 kg/m² (82.73 lb/sq ft)
Max power loading: trainer	151 kg/kN (1.48 lb/lb st)
attack/trainer	220 kg/kN (2.16 lb/lb st)

PERFORMANCE (clean, estimated):
Max level speed at 4,570 m (15,000 ft)	560 kt (1,037 km/h; 644 mph)
T-O speed	111 kt (205 km/h; 128 mph)
Landing speed	95 kt (175 km/h; 109 mph)
Stalling speed, 20% fuel	89 kt (165 km/h; 103 mph) CAS
Max rate of climb at S/L	3,048 m (10,000 ft)/min
Time to 9,140 m (30,000 ft)	4 min 0 s
Service ceiling	13,000 m (42,660 ft)
T-O run	340 m (1,115 ft)
Landing run	550 m (1,805 ft)

Range, 10% reserves: max internal fuel
    1,375 n miles (2,546 km; 1,582 miles)
with two external tanks
    1,800 n miles (3,333 km; 2,071 miles)
Radius of action, 5 min combat: interdiction, two Mk 83 1,000 lb bombs, two AAMs, gun pod, two external tanks, hi-lo-hi   440 n miles (815 km; 506 miles)
close air support, two Mk 82 500 lb and two Mk 83 1,000 lb bombs, two AAMs, gun pod, hi-lo-hi, 30 min loiter   120 n miles (222 km; 138 miles)

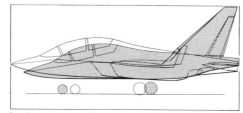

Production Yak-130 (shaded) compared with prototype *(James Goulding/Jane's)*

Front cockpit of the Yak-130 first prototype *(Paul Jackson/Jane's)* 0003355

combat air patrol, two AAMs, gun pod, two external tanks, 2 h loiter at 11,600 m (38,060 ft)
    240 n miles (444 km; 276 miles)
Max rate of roll	200°/s
Sustained g limit at M0.8 at 4,575 m (15,000 ft) with 50% internal fuel	+4.8
g limits	+8/−3

UPDATED

YAKOVLEV Yak-131

TYPE: Light fighter.
PROGRAMME: Derivative of Yak-130 jet trainer (which see) with one or two crew. No known production plans. Optional radar nose and reconnaissance pods. Designations Yak-133 and Yak-135 may apply to subvariants.

UPDATED

YAKOVLEV Yak-3M and Yak-9U-M

TYPE: Single-seat sportplane.
PROGRAMME: Reproduction Second World War fighter. Prototype of original wooden-skinned Yak-3 flew 1943; all-metal version flew 1945; deliveries to Soviet air forces began July 1944, totalled 4,848 (of 36,737 Yakovlev single-engined Second World War fighters); described often as lightest weight and most agile monoplane of 1939-45 period; able to turn 360° in 18.5 seconds. All-metal version now available, with Western power plant and uprated instrumentation; new-build 'prototype' (0470101, later N854DP) displayed 1993 Paris Air Show; 11 built by Strela at Orenburg to meet orders from Gunnell Museum, USA, but at least 14 eventually produced; one later to New Zealand (then Australia in 1999) and one to South Africa.

Production turned in 1996 to Yak-9U of similar appearance. Original Yak-9 entered service in 1942 and 16,769 were built; all-metal Yak-9U-M replicas include at least seven exported to USA by mid-1999, of which five then flying; one to France; distribution by Shadetree Aviation of Carson City, Nevada.

Following data apply to Yak-3M.

DESIGN FEATURES: Precise reproduction in metal of Second World War airframe, except for repositioned carburettor air intake above engine cowling to suit changed engine. Conventional cantilever low-wing configuration, with tapered, round-tipped wings. Mainwheel tyres size 600×180.

POWER PLANT: Reconditioned 925 kW (1,240 hp) Allison V-1710-39 12-cylinder V liquid-cooled piston engine replaces original 925/969 kW (1,240/1,300 hp) VK-105PF-2; three-blade propeller. Fuel capacity 320 litres (84.5 US gallons; 70.4 Imp gallons).

DIMENSIONS, EXTERNAL:
Wing span	9.20 m (30 ft 2¼ in)
Wing aspect ratio	5.7
Length overall	8.49 m (27 ft 10¼ in)
Height overall	2.39 m (7 ft 10 in)

AREAS:
Wings, gross	14.85 m² (159.8 sq ft)

WEIGHTS AND LOADINGS:
Weight empty	1,988 kg (4,383 lb)
Max T-O weight	2,600 kg (5,732 lb)

Front/student's (top) and rear/instructor's instrument panels of proposed Russian preproduction Yak-130. Western versions will have three MFDs

Reproduction Yakovlev Yak-3M fighter with Allison engine *(Paul Jackson/Jane's)* 1999/0051356

Max wing loading	175.1 kg/m² (35.86 lb/sq ft)
Max power loading	2.81 kg/kW (4.62 lb/hp)

PERFORMANCE:
Max level speed: at height 350 kt (648 km/h; 402 mph)
at S/L 307 kt (570 km/h; 354 mph)
Time to turn 360° at 1,000 m (3,280 ft) 19 s
Range 486 n miles (900 km; 559 miles)
UPDATED

YAKOVLEV Yak-18T

TYPE: Four-seat lightplane.
PROGRAMME: Prototype first flew mid-1967 as extensively redesigned cabin version of veteran Yak-18 trainer; over 700 built at Smolensk Aircraft Plant (SmAZ), including trainer, ambulance, communications and light freight versions; production resumed by Smolensk in 1993 to meet new contracts and one converted to Technoavia SM-94 prototype (see 2000-01 and previous editions). Second production run ended in 1996 due to bankruptcy of SmAZ. In 1998, work recommenced on 15 stored, part-built airframes, which being completed with two-piece windscreen, designed for Technoavia SM-94. New aircraft as delivered as either SM-94 or distributed in Western Europe (eight reserved) by Richard Goode Aerobatics of White Waltham, UK ((+44 1963) 36 27 46; fax (+44 1963) 36 37 51; e-mail richard.goode@russianacros.com).
CUSTOMERS: First resumed (1993) deliveries to Luxembourg, Philippines, Russia, Switzerland, Turkey, UAE, UK and USA. Six delivered to Bulgarian Air Force for primary training early 1996.
COSTS: US$95,000 (1999, Goode-sourced).
DESIGN FEATURES: Adaptation of aerobatic trainer. Conventional cantilever low-wing monoplane; three-section two-spar wings; semi-monocoque fuselage of basically square section; strut- and wire-braced tail surfaces. Airframe life 5,000 hours or 20 years.
Wing section Clark YH, thickness/chord ratio 14.5 per cent at root, 9.3 per cent at tip; dihedral 7° 20′ on outer panels only.
FLYING CONTROLS: Conventional and manual. Plain flaps and elevator trim.
STRUCTURE: All-metal construction, except for fabric covering on parts of outer wing panels, ailerons and tail surfaces.
LANDING GEAR: Retractable tricycle type; single wheel on each unit; pneumatic retraction, K-141/T-141 mainwheels inward into centre-section, Type 44-1 nosewheel rearward; Hydromash oleo-nitrogen shock-absorbers; castoring nosewheel is non-steerable; Yaroslavl tyres size 500×150 on mainwheels, 400×150 on nosewheel; tyre pressure (all) 2.95 bars (43 lb/sq in); pneumatic plate-type brakes on mainwheels. Minimum ground turning radius 2.52 m (8 ft 3¼ in).
POWER PLANT: One 265 kW (355 hp) VOKBM M-14P air-cooled radial piston engine; optional 294 kW (394 hp) M-14PF in 1999 version; V-530TA-D35 two-blade variable-pitch metal propeller; wingroot fuel tanks, combined capacity 190 litres (50.2 US gallons; 41.8 Imp gallons); gravity fuelling points on wing upper surface. Optional usable fuel capacities of 321 litres (84.7 US gallons; 70.6 Imp gallons), 360 litres (95.3 US gallons; 79.4 Imp gallons) or 614 litres (162 US gallons; 135 Imp gallons) available in 1999 version. Oil capacity 20 litres (5.3 US gallons; 4.4 Imp gallons). Inverted fuel and oil systems optional.
ACCOMMODATION: Cabin of main production version seats four persons in pairs, with forward-hinged jettisonable door each side; rear bench seat removable for freight carrying; ambulance configuration accommodates pilot, stretcher patient and medical attendant; baggage compartment aft of rear seat. Option to increase seating to six has not been pursued. Cabin heated and ventilated.
SYSTEMS: Pneumatic system for landing gear and flaps, operating pressure 49 bars (711 lb/sq in); 36/115 V AC electrical system at 400 Hz, 28.5 V DC and 20NKBN-25 battery for red panel lighting, nav and landing lights, and fin-tip anti-collision beacon.
AVIONICS: *Comms:* Baklan-5 UHF radio, intercom.

Instrumentation: ARK-15M radio compass, radio altimeter and flight recorder standard.
Customised avionics available in 1999 version.
EQUIPMENT: Optional smoke system for 1999 version.
DIMENSIONS, EXTERNAL:

Wing span	11.16 m (36 ft 7¼ in)
Wing chord: at root	2.00 m (6 ft 6¾ in)
at tip	1.06 m (3 ft 5¾ in)
Wing aspect ratio	6.6
Length overall	8.39 m (27 ft 6½ in)
Height overall	3.40 m (11 ft 1¾ in)
Tailplane span	3.54 m (11 ft 7½ in)
Wheel track	3.12 m (10 ft 2¾ in)
Wheelbase	1.955 m (6 ft 5 in)
Propeller diameter	2.40 m (7 ft 10½ in)
Propeller ground clearance	0.16 m (6¼ in)

DIMENSIONS, INTERNAL:

Cabin: Max width	1.28 m (4 ft 2¼ in)
Max height	1.25 m (4 ft 1¼ in)
Cabin doors (each): Height	0.88 m (2 ft 10¾ in)
Width	1.18 m (3 ft 10½ in)
Baggage door: Height	0.50 m (1 ft 7¾ in)
Width	1.00 m (3 ft 3¼ in)

AREAS:

Wings, gross	18.80 m² (202.4 sq ft)
Ailerons (total)	1.92 m² (20.66 sq ft)
Flaps (total)	1.60 m² (17.22 sq ft)
Fin	0.72 m² (7.73 sq ft)
Rudder	0.98 m² (10.57 sq ft)
Tailplane	1.95 m² (21.00 sq ft)
Elevators (total)	1.235 m² (13.30 sq ft)

WEIGHTS AND LOADINGS (A: instructor and one pupil, B: four persons):

Weight empty	1,217 kg (2,683 lb)
Max payload: A	306 kg (675 lb)
B	338 kg (745 lb)
Max fuel	140 kg (308 lb)
Max T-O weight: A	1,500 kg (3,307 lb)
B	1,650 kg (3,637 lb)
Max landing weight: B	1,650 kg (3,637 lb)
Max wing loading: A	79.8 kg/m² (16.34 lb/sq ft)
B	87.8 kg/m² (17.98 lb/sq ft)
Max power loading: A	5.66 kg/kW (9.32 lb/hp)
B	6.22 kg/kW (10.25 lb/hp)

PERFORMANCE:

Never-exceed speed (V$_{NE}$)	162 kt (300 km/h; 186 mph)
Max level speed: A, B	159 kt (295 km/h; 183 mph)
Max cruising speed: B	135 kt (250 km/h; 155 mph)
Econ cruising speed	113 kt (210 km/h; 130 mph)
Landing speed	68 kt (125 km/h; 78 mph)
Max rate of climb at S/L: B	300 m (985 ft)/min
Service ceiling: A, B	5,520 m (18,120 ft)
T-O run: B	260-280 m (855-920 ft)
Landing run: B	345-365 m (1,135-1,200 ft)
Range with standard max fuel, with reserves:	
A, B	313 n miles (580 km; 360 miles)
g limits: A	+6.4/-3.2

UPDATED

YAKOVLEV Yak-42

NATO reporting name: Clobber
Western marketing name: CityStar
TYPE: Regional jet airliner.
PROGRAMME: Three prototypes ordered initially; first prototype (SSSR-1974) flew 7 March 1975, with 11° wing sweepback and furnished in 100-seat local service form, with carry-on baggage and coat stowage fore and aft of cabin; second prototype (SSSR-1975, later SSSR-42304) had 23° sweepback and more cabin windows, representative of 120-seat version with three more rows of seats and no carry-on baggage areas; third prototype (SSSR-1976, later SSSR-42303) introduced small refinements (described in previous editions of *Jane's*); flight testing proved 23° wing superior; first series of production aircraft, built to replace some Aeroflot Tu-134s, generally similar to SSSR-42303 as exhibited 1977 Paris Air Show; changes for production included substitution of four-wheel main landing gear bogies for twin-wheel units on prototypes; scheduled service on Aeroflot's Moscow-Krasnodar route began late 1980. Substantial redesign during one-year grounding after in-flight failure, June 1982.
Yak-42D introduced from second quarter of 1989 and currently manufactured by Saratov Aviation Plant (SAZ). Yak-42A and other versions have been produced in small numbers or been proposed.
CURRENT VERSIONS: **Yak-42**: Basic standard, *described in detail*. Replaced in production by Yak-42D, to which standard some aircraft were converted.
Yak-42A: Initially an experimental version (RA-42424) displayed at 1993 Paris Air Show; similar to

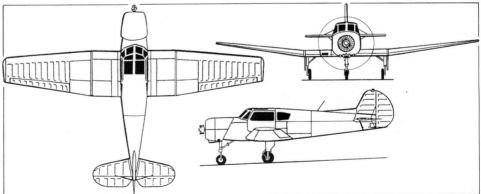

Yakovlev Yak-18T (VOKBM M-14P radial engine) *(Paul Jackson/Jane's)* 0015144

Current production Yak-18T, with two-piece windscreen, as fitted to final 15 aircraft *(Paul Jackson/Jane's)* 2001/0092535

Yak-42D but with Western avionics. Redesignated Yak-142, then Yak-42D-100. Designation now applied to an improved version of the current Yak-42D, due to enter production at Saratov during 1998. Additional fuel capacity in wings; Russian avionics compatible with Cat. II automatic landings and several features of Yak-42D-100 including flight-deployable spoilers and four-position trailing-edge flaps. Maximum T-O weight increased to 57,500 kg (126,765 lb), in which configuration range with 120 passengers is increased by almost 50 per cent to 1,500 n miles (2,778 km; 1,726 miles).

Yak-42B: As Yak-42A but Honeywell nav system and flight deck CRTs. Redesignated Yak-42D-100.

Yak-42D: Second main production variant; increased fuel; range extended to 1,185 n miles (2,200 km; 1,365 miles) with 120 passengers; larger forward passenger door. First production aircraft (SSSR-42359) built July 1988.

Yak-42D-90: Announced early 2000; funded by Russian government. Seating reduced to 90, increasing range to 2,159 n miles (4,000 km; 2,485 miles). Certification was planned for late 2000.

CityStar 100: Yak-42D marketed by Airlines Partners of Switzerland. Announced at Asian Aerospace, Singapore, February 2000, when Yakovlev's executive transport (RA-42423) specially marked. However, production CityStar will have Western avionics and interior, both installed in Switzerland to customer's specifications. Normal seating for 102 at 78 to 81 cm (31 to 32 in) pitch. Details from Via Lugano 18, CH-6982 Agno, Switzerland; telephone (+41 91) 610 26 26; fax (+41 91) 610 26 17; e-mail airlines4@bluewin.ch.

Yak-42D-100: Developed from Yak-42 via experimental Yak-42A/B, RA-42424, displayed at 1993 Paris Air Show; designation later changed to Yak-142, but applied only to this aircraft, which sold to LUKoil in December 1996, following receipt of type certificate. Western avionics. VIP version is available for 39 passengers; maximum payload 3,700 kg (8,157 lb), maximum T-O weight 57,500 kg (126,765 lb), maximum range 2,213 n miles (4,100 km; 2,547 miles) at 405 kt (750 km/h; 466 mph).

Improvements compared with Yak-42 include flight-deployable wing spoilers to increase rate of descent; wide flap setting range for improved T-O performance and rate of climb, even in hot and high conditions; larger port-side cabin door, for improved compatibility with passenger boarding walkways, in addition to rear airstair; improved passenger comfort; state-of-the-art digital avionics; EFIS cockpit displays; redesigned engine air intakes that lower noise levels to meet latest ICAO standards (Ch 3, Appendix 16); and addition of TA-12 APU, able to be started at up to 5,000 m (16,400 ft).

Yak-42DD: Interim designation for variant proposed in response to Russian Federal Aviation Service requirement announced February 1998 for potential Tupolev Tu-134 replacement having seats for 90 passengers and 2,159 n mile (4,000 km; 2,485 mile) range with FAR Pt 25 reserves. Up to 2,300 kg (5,070 lb) of additional fuel capacity; weight saving measures include modern avionics (1,526 kg; 3,365 lb) and 1.60 m (5 ft 3 in) length reduction. Predicted range 3,239 n miles (6,000 km; 3,728 miles), or more if engines fitted with FADEC. Proposal submitted May 1998; no further reports.

Yak-42F: One trials aircraft (SSSR-42644) for Zhukovsky test base.

Yak-42-200: Announced 1997. Projected 6.03 m (19 ft 9½ in) fuselage stretch to increase the all-economy passenger load to 150; maximum T-O weight increased to 65,000 kg (143,300 lb); cruising altitude raised to 10,600 m (34,780 ft).

Yak-42T: Freighter, with 2.50 × 2.00 m (8 ft 2½ in × 6 ft 6¾ in) cargo door in upper deck; maximum freight load 12,000 kg (26,455 lb), including containers in underfloor holds. Design under way by 1996, but no further news by early 2000.

CUSTOMERS: Currently operated by some 42 airlines and executive operators in Russia, Ukraine, Kazakhstan, Turkmenistan, Lithuania, Cuba (four) and China (nine); developments in 1998 included further order from Gazprom and delivery to Government Communications Agency of one specially equipped aircraft. Two deliveries in 1999. Commitments from Donavia (seven) and Bykovo Airlines (15) as yet unfulfilled. Six prototype/preseries aircraft. Production (including first three prototypes) at Saratov, but at least two prototypes/pre-series built at Smolensk, together with some 10 aircraft in 1980-81. Saratov had built 171 (including prototypes) by end of 1999, including about 96 Yak-42Ds; further 15 conversions to Yak-42D. Mid-1990s recipients included Orel Avia, Gazpromavia, Udmurtia and Dnieproavia, several from late production have been given VIP interiors. Total 145 operable Yak-42s in 2000, of which 125 with 33 CIS operators.

DESIGN FEATURES: Basic design objectives were simple construction, reliability in operation, economy, and ability to operate in remote areas with widely differing climatic conditions; design is in accordance with Russian Federation and Associated States (CIS) civil airworthiness standards and US FAR Pt 25; in standard form, Yak-42D conforms with ICAO Ch 2, Appendix 16 noise regulations; after further modifications will meet Ch 3 requirements; aircraft is intended to operate in ambient temperatures from −50 to +50°C; APU for engine starting and services removes need for ground equipment.

Conventional all-swept low-wing configuration; two podded turbofans on sides of rear fuselage, third in extreme rear fuselage with intake at base of fin; basically circular fuselage, blending into oval rear section; T tail.

Wing has no dihedral or anhedral; sweepback 23° at quarter-chord.

FLYING CONTROLS: Conventional and hydraulically actuated; two-section ailerons, each with servo tab on inner section, trim tab on outer section; two-section single-slotted trailing-edge flap; flap settings 20° and 45°, but intermediate 10° and 30° available on Yak-42A and 42D-100; three-section spoiler forward of outer flap, spoilers being deployable in flight on Yak-42A and 42D-100; full-span leading-edge flap on each wing; one-piece variable incidence tailplane (from 1° up to 12° down); trim tab in each elevator, trim tab and spring servo tab in rudder.

STRUCTURE: All-metal; two-spar torsion box wing structure; riveted, bonded and welded semi-monocoque fuselage.

LANDING GEAR: Hydraulically retractable tricycle heavy-duty type, made by Hydromash; four-wheel bogie main units retract inward into flattened fuselage undersurface; twin nosewheels retract forward; hydraulic back-up for extension only; emergency extension by gravity; oleo-nitrogen shock-absorbers; steerable nose unit of levered suspension type; steering angle ±55° for taxying; low-pressure tyres; 930×305 on all 10 wheels; hydraulic disc brakes on mainwheels; nosewheel brakes to stop wheel rotation after take-off. Turning radius 27.5 m (90 ft).

POWER PLANT: Three Ivchenko Progress D-36 three-shaft turbofans, each 63.7 kN (14,330 lb st); centre engine, inside rear fuselage, has S-duct air intake; outboard engines in pod on each side of rear fuselage; no thrust reversers. Integral fuel tanks between spars in wings, capacity 23,330 litres (6,163 US gallons; 5,132 Imp gallons).

ACCOMMODATION: Crew of two side by side, with provision for flight engineer; two or three cabin attendants. Single passenger cabin, with 120 seats six-abreast, at pitch of 75 cm (29.5 in), with centre aisle 45 cm (17.7 in) wide, in high-density configuration. Alternative 104-passenger (96 tourist, eight first class) local service configuration, with carry-on baggage and coat stowage compartments fore and aft of cabin. Main airstair door hinges down from undersurface of rear fuselage; second door forward of cabin on port side, with integral airstairs; service door opposite. Galley and crew coat stowage between flight deck and front vestibule; passenger coat stowage and toilet between vestibule and cabin; second coat stowage and toilet at rear of cabin.

Two underfloor holds for cargo, mail and baggage in nets or standard containers, loaded through door on starboard side, forward of wing; chain-drive system, for handling containers, built into floor; forward hold for six containers, each 2.2 m³ (78 cu ft); rear hold takes two similar containers. Provision for convertible passenger/cargo interior, with enlarged loading door on port side of front fuselage. Two emergency exits, overwing and forward of wing, each side. All passenger and crew accommodation pressurised, air conditioned, and furnished with non-flammable materials.

SYSTEMS: TA-12 APU standard, for engine starting, and for power and air conditioning supply on ground and, if necessary, in flight. TA-6B on CityStar.

AVIONICS (Yak-42D): Flight and navigation equipment for operation by day and night under adverse weather conditions, with landings on concrete or unpaved runways in ICAO Cat. II weather minima down to 30 m (100 ft) visibility at 300 m (985 ft).

Flight: Type SAU-42 automatic flight control system and automatic navigation system standard.

AVIONICS (Yak-42D-100): *Comms:* KHF 950 HF; RTA-44A VHF com.

Radar: RDR-4A weather radar.

Flight: SAU-4201 Cat. II qualified flight control system with autothrottle capability; RIA-35A ILS; RVA-36A VOR; DMA-37A DME; DFA-75A ADF. Honeywell equipment includes KNS 660 flight data control system, KAD 480 air data system. TPA-81A TCAS II, TRA-67A Mode S transponder.

Instrumentation: EFIS-10 system with five colour CRTs and KAH 460 AHR; ALA-52A radio altimeter.

DIMENSIONS, EXTERNAL:
Wing span	34.88 m (114 ft 5¼ in)
Wing aspect ratio	8.1
Length overall	36.215 m (118 ft 9¾ in)
Fuselage max diameter	3.80 m (12 ft 5½ in)
Height overall	9.83 m (32 ft 3 in)
Tailplane span	10.80 m (35 ft 5 in)
Wheel track	5.63 m (18 ft 5¾ in)
Wheelbase	14.96 m (49 ft 1 in)
Passenger door (fwd):	
Height: Yak-42	1.50 m (4 ft 11 in)
Yak-42D	1.70 m (5 ft 7 in)
Width: Yak-42	0.83 m (2 ft 8½ in)
Yak-42D	0.85 m (2 ft 9½ in)
Passenger entrance (rear): Height	1.78 m (5 ft 10 in)
Width	0.81 m (2 ft 7¾ in)
Cargo door (convertible version):	
Height	2.025 m (6 ft 7¾ in)
Width	3.23 m (10 ft 7 in)

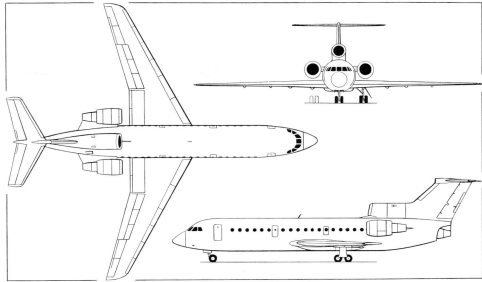

Yakovlev Yak-42D three-turbofan short/medium-range passenger transport *(Dennis Punnett/Jane's)*

Yakovlev Yak-42D used as an executive jet by the Yakovlev company *(Paul Jackson/Jane's)*

Baggage/cargo hold door: Height	1.35 m (4 ft 5 in)
Width	1.145 m (3 ft 9 in)
Height to sill	1.45 m (4 ft 9 in)

DIMENSIONS, INTERNAL:
Cabin: Length	19.89 m (65 ft 3 in)
Max width	3.60 m (11 ft 9¾ in)
Max height	2.08 m (6 ft 9¾ in)
Volume (pressurised)	225.0 m^3 (7,946 cu ft)
Baggage compartment volume (100-seater):	
fwd	19.8 m^3 (700 cu ft)
rear	9.5 m^3 (335 cu ft)

AREAS:
Wings, gross	150.00 m^2 (1,614.6 sq ft)
Vertical tail surfaces (total)	23.29 m^2 (250.7 sq ft)
Horizontal tail surfaces (total)	27.60 m^2 (297.1 sq ft)

WEIGHTS AND LOADINGS:
Weight empty, equipped:	
104 seats	34,500 kg (76,058 lb)
120 seats	34,515 kg (76,092 lb)
Max payload: Yak-42	13,000 kg (28,660 lb)
CityStar	13,500 kg (29,762 lb)
Payload with max fuel: CityStar	5,010 kg (11,045 lb)
Max fuel	18,500 kg (40,785 lb)
Max ramp weight: Yak-42	57,300 kg (126,320 lb)
Yak-42D-100	57,500 kg (126,765 lb)
CityStar	57,800 kg (127,425 lb)
Max T-O weight: Yak-42	57,000 kg (125,660 lb)
Yak-42A, 42D-100, CityStar	57,500 kg (126,765 lb)
Max landing weight: Yak-42	51,000 kg (112,435 lb)
CityStar	50,500 kg (111,335 lb)
Max zero-fuel weight: CityStar	46,800 kg (103,175 lb)
Max wing loading	380.0 kg/m^2 (77.83 lb/sq ft)
Max power loading	298 kg/kN (2.92 lb/lb st)

PERFORMANCE:
Max cruising speed at 7,620 m (25,000 ft)	437 kt (810 km/h; 503 mph)
Long-range cruising speed at 9,600 m (31,500 ft)	373 kt (691 km/h; 429 mph)
Econ cruising speed at 9,100 m (29,850 ft)	399 kt (740 km/h; 460 mph)
T-O speed	119 kt (220 km/h; 137 mph) IAS
Approach speed	114 kt (210 km/h; 131 mph) IAS
Max cruising height	9,600 m (31,500 ft)
T-O run: CityStar	1,540 m (5,055 ft)
T-O to 10.7 m (35 ft) on two engines: CityStar	1,800 m (5,905 ft)
T-O balanced field length	2,200 m (7,220 ft)
Landing from 10.7 m (35 ft) at 49,000 kg (108,025 lb):	
CityStar	1,000 m (3,280 ft)
Range at econ cruising speed, with 3,000 kg (6,614 lb) fuel reserves:	
with max payload:	
Yak-42	745 n miles (1,380 km; 857 miles)
CityStar	917 n miles (1,700 km; 1,056 miles)
with 120 passengers (10,800 kg; 23,810 lb payload)	1,025 n miles (1,900 km; 1,180 miles)
with 104 passengers (9,360 kg; 20,635 lb payload)	1,240 n miles (2,300 km; 1,430 miles)
with max fuel and 42 passengers	2,215 n miles (4,100 km; 2,545 miles)

UPDATED

YAKOVLEV Yak-52

See Aerostar Iak-52 in Romanian section.

YAKOVLEV Yak-54, Yak-56, Yak-57 and Yak-152

TYPE: Aerobatic two-seat sportplane.
PROGRAMME: Yak-54 announced 1992; prototype at 1993 Paris Air Show; first flown 23 December 1993; third prototype at Paris in 1995. In production from 1995 at Saratov Aviation Plant (SAZ), with maximum capacity of more than 100 per year. However, production suspended in 1998; to resume after certification achieved, mid-2001; initial batch of five funded (and to be marketed) by ZAO Gorki Yu-2 company.
CURRENT VERSIONS: **Yak-54:** *As described.*
Yak-56: Primary trainer derivative; also designated **Yak-54M**; redesignated **Yak-152** in late 2000; 294 kW (394 hp) M-14PF, retractable tricycle landing gear and two KS-38 lightweight ejection seats. Prototype under construction in 1999.
Yak-57: Single-seat sportplane; under development by 1999.
CUSTOMERS: Total 48 Yak-54s ordered in January 1997 by Northwest Aerobatic Center, Ephrata, Washington, USA, and Dancing Bear air show team; 10 built up to 1998 suspension of production; five flying in USA, plus one (second-hand) delivered to Australia during 1999. Total of 30 Yak-152s ordered by ROSTO (successor to DOSAAF) for delivery in 2002; eventual Russian market foreseen for between 200 and 800.
COSTS: Yak-54 US$160,000; Yak-152 US$100,000 (2000).
DESIGN FEATURES: Optimised for aerobatics; derived from Yak-55 (1998-99 and earlier *Jane's*). Conventional mid-wing configuration; symmetrical section; no dihedral, anhedral or incidence; almost full-span ailerons, elevators and rudder all horn-balanced; each aileron has large suspended balance tab. Designed on basis of systems and units of Yak-55M.

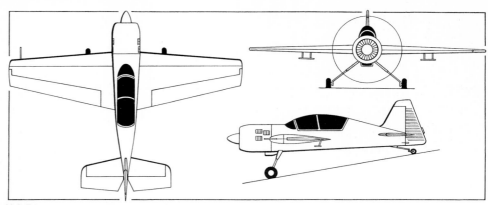

Yak-54 sporting and aerobatic training aircraft *(Mike Keep/Jane's)* 1999/0051406

Yakovlev Yak-54 (VOKBM M-14 radial engine) *(Paul Jackson/Jane's)* 2000/0079358

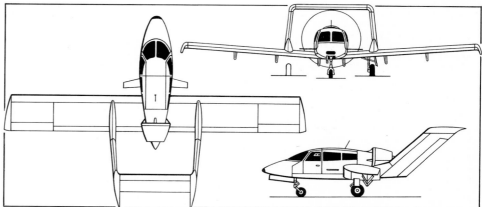

Yakovlev Yak-58 business aircraft (VOKBM M-14PR nine-cylinder piston engine) *(Mike Keep/Jane's)* 0015151

Yakovlev Yak-58 six-seat utility aircraft *(Paul Jackson/Jane's)* 2001/0103521

FLYING CONTROLS: Conventional and manual. Ailerons occupy 90 per cent of wing trailing-edge and have both horn balance and suspended tab; horn-balanced tail surfaces.
STRUCTURE: All-metal; two-spar wings; semi-monocoque fuselage; conventional tail unit.
LANDING GEAR: Non-retractable tailwheel type, with titanium spring cantilever main legs and small wheels, tyre size 400×150; tailwheel tyre size 200×80.
POWER PLANT: One 265 kW (355 hp) VOKBM M-14P nine-cylinder air-cooled radial engine; MTV-9 three-blade variable-pitch propeller.
ACCOMMODATION: Two seats in tandem under continuous transparent canopy, hinged to starboard.

DIMENSIONS, EXTERNAL:
Wing span	8.16 m (26 ft 9¼ in)
Wing aspect ratio	5.2
Length overall	6.91 m (22 ft 8 in)

YAKOVLEV—AIRCRAFT: RUSSIAN FEDERATION

Cockpit of Yak-58 *(Paul Jackson/Jane's)*

Production Yakovlev Yak-112 four-seat multipurpose lightplane from the Irkutsk line *(Paul Jackson/Jane's)*

AREAS:
Wings, gross 12.89 m² (138.75 sq ft)
WEIGHTS AND LOADINGS:
Max T-O weight: one pilot 850 kg (1,874 lb)
  two occupants 990 kg (2,182 lb)
Max wing loading 76.8 kg/m² (15.73 lb/sq ft)
Max power loading 3.74 kg/kW (6.15 lb/hp)
PERFORMANCE:
Never-exceed speed (V_NE) 243 kt (450 km/h; 280 mph)
Stalling speed 60 kt (110 km/h; 69 mph)
Rate of roll 345°/s
Max rate of climb at S/L 900 m (2,950 ft)/min
Ferry range 377 n miles (700 km; 435 miles)
g limits +9/−7

UPDATED

YAKOVLEV Yak-SKh
TYPE: Agricultural sprayer.
PROGRAMME: Announced late 2000, when still at project stage; provisional SKh designation indicates *selskokhozyaistvenny*: agricultural; derived from Yak-54 aerobatic sportplane, retaining M-14 engine. Single-seat version to be standard; second seat optional.

NEW ENTRY

YAKOVLEV Yak-58
TYPE: Six-seat utility transport.
PROGRAMME: Shown in model form at Moscow Air Show '90; full-scale mockup exhibited February 1991; prototype (RA-01003) initially reported to have flown at Tbilisi on 26 December 1993, but date later amended, without explanation, to 17 April 1994; it was lost in accident 27 May 1994; second prototype flew 10 October 1994; two more completed for trials at Zhukovsky test centre, plus two static test airframes, by January 1995; OKB reported receipt of 250 letters of intent to purchase; production by TASA at Tbilisi had begun by 1997 with initial batch of 20 for customers in Georgia, Kazakhstan and Uzbekistan. One aircraft (4L-02010), fitted with Thomson-CSF radio surveillance equipment and operator's console, was exhibited at Paris Air Show in 1997. No deliveries to customers had taken place by late 2000.
DESIGN FEATURES: Constant chord unswept wing with dihedral from roots and cambered tips; fuselage pod mounted above wing, with annular duct at rear to house air-cooled pusher engine; short twin booms carry sweptback and slightly toed-in tailfins and bridging horizontal tail surface. Small foreplane (67.5 cm; 2 ft 2½ in long) fitted to each side of cabin by 1995.
FLYING CONTROLS: Conventional and manual. Twin rudders and electrically actuated trailing-edge flaps.
LANDING GEAR: Pneumatically actuated retractable tricycle type; single wheel and low-pressure tyre on each trailing-link unit, for operation from unprepared strips; mainwheels retract inward, nosewheel forward; mainwheel tyre size 500×50, nosewheel tyre size 400×150.
POWER PLANT: One 265 kW (355 hp) VOKBM M-14PR nine-cylinder air-cooled radial engine enclosed in annular duct; three-blade variable-pitch pusher propeller. Location reduces noise in cabin.
ACCOMMODATION: Pilot and five passengers in pairs in enclosed cabin; large sliding door on starboard side, facilitating freight loading when passenger seats removed, or despatch of parachutists. Forward-hinged door on port side. Planned uses include business, taxi and ambulance transport; surveillance of forests, high-tension cables, oilfields and fisheries; mail and freight operation.
SYSTEMS: No hydraulic system. Two independent pneumatic systems for landing gear actuation and wheel brakes.
AVIONICS: Standard CIS equipment.
DIMENSIONS, EXTERNAL:
Wing span 12.70 m (41 ft 8 in)
Wing aspect ratio 8.1
Length overall 8.55 m (28 ft 0½ in)
Height overall 3.16 m (10 ft 4½ in)
AREAS:
Wings, gross 20.00 m² (215.3 sq ft)
WEIGHTS AND LOADINGS:
Weight empty 1,270 kg (2,800 lb)
Max payload 450 kg (992 lb)
Max fuel 360 kg (793 lb)
Max T-O weight 2,080 kg (4,585 lb)
Max wing loading 104.0 kg/m² (21.30 lb/sq ft)
Max power loading 7.85 kg/kW (12.90 lb/hp)
PERFORMANCE (estimated):
Max level speed 162 kt (300 km/h; 186 mph)
Nominal cruising speed 135 kt (250 km/h; 155 mph)
Landing speed 68 kt (125 km/h; 78 mph)
Service ceiling 4,000 m (13,120 ft)
T-O run 610 m (2,000 ft)
Landing run 600 m (1,970 ft)
Range: with max payload, 45 min reserves
  more than 540 n miles (1,000 km; 620 miles)
with max fuel 971 n miles (1,800 km; 1,118 miles)

UPDATED

YAKOVLEV Yak-112
TYPE: Four-seat lightplane.
PROGRAMME: Winner of 1988 official design competition for two-seat primary trainer for Russian Federation and Associated States (CIS) aero clubs; developed subsequently to current four-seat configuration; model shown at Moscow Air Show '90 and full-scale mockup at 1991 Paris Air Show; first flight of prototype (RA-00001) 20 October 1992; initially powered by a 157 kW (210 hp) Teledyne Continental IO-360ES with two-blade propeller. Two further flight aircraft completed for trials of Textron Lycoming IO-540-A4B5 (first flight 15 December 1995) and Teledyne Continental IO-550-M (23 April 1997). These three manufactured at Smolensk (SmAZ) before Yak-112 activities there were terminated.
  Programme then centred on Irkutsk, where previously parallel manufacture had begun with first flight of IO-360-powered aircraft on19 October 1993. Further three of this version built by mid-1999, but series production will be in higher power range. Trials fleet is one Lycoming and two Continental aircraft. Certification due for completion in first quarter of 2001. Irkutsk plant requires 50 to 70 orders before starting production; will offer programme to other constructors if its workload on other projects becomes prohibitive.
CURRENT VERSIONS: **Yak-112**: Standard version, *as described*. Alternatively Yak-112I, indicating Irkutsk.
  Yak-112P: Proposed floatplane version.
  Yak-112SKh: Cropsprayer version (*selskokhozyaist-venny*: agricultural).
  Yak-112P: Patrol version.
  Six-seat: Enlarged version, under development by 1999.
CUSTOMERS: Seven prototype/pre-series aircraft. Kazakhstan government ordered 10 in mid-1999. Interest expressed by several Russian regional governments. Production aircraft will be manufactured at Irkutsk. Total market estimated at 400 to 500 aircraft.
COSTS: US$120,000 to US$130,000 (2000). Cropsprayer US$30,000 extra.
DESIGN FEATURES: Intended to carry passengers, light cargo and mail, for pilot training, glider towing, ambulance duties, forest, pipeline and cable patrol, fisheries surveillance and agricultural use.
  Conventional all-metal high-wing configuration; constant chord unswept wing, dihedral 1°; cambered wingtips; single bracing strut each side; pod-and-boom fuselage, with heavily glazed cabin offering exceptional all-round view; swept tailfin; composites used extensively.
FLYING CONTROLS: Conventional and manual. Flaps. Single-piece elevator with central tab; ground-adjustable tabs on each aileron.
STRUCTURE: All-metal. Two-spar wing.
LANDING GEAR: Non-retractable tricycle type; single wheel on each unit; cantilever spring mainwheel legs; nosewheel size 310×135; KT-214D mainwheels, size 400×150; floats and skis optional.
POWER PLANT: One 194 kW (260 hp) Textron Lycoming IO-540-A4B5 flat-six driving a Hartzell three-blade, variable-pitch propeller. Alternatively, Teledyne Continental IO-550 of unspecified rating. Proposed later alternatives are Rybinsk DN-200 turbo-diesel and VAZ rotary engines.
ACCOMMODATION: Four persons in pairs in enclosed cabin; dual controls. Provision for stretcher and medical attendant.
AVIONICS: Optional MIKBO-3 flight management system. Honeywell avionics in export version.
EQUIPMENT: Agricultural spray gear optional.
DIMENSIONS, EXTERNAL:
Wing span 10.25 m (33 ft 7½ in)
Wing aspect ratio 6.6
Length overall 6.96 m (22 ft 10 in)
Height overall 2.90 m (9 ft 6 in)
Propeller diameter 2.005 m (6 ft 7 in)
AREAS:
Wings, gross 16.96 m² (182.6 sq ft)
WEIGHTS AND LOADINGS:
Max payload 362 kg (798 lb)
Max T-O weight 1,385 kg (3,053 lb)
Max wing loading 81.7 kg/m² (16.73 lb/sq ft)
Max power loading 7.15 kg/kW (11.74 lb/hp)
PERFORMANCE:
Max level speed 124 kt (230 km/h; 143 mph)
Normal cruising speed 104 kt (193 km/h; 120 mph)
Unstick and landing speed 65 kt (120 km/h; 75 mph)
Max rate of climb at S/L 312 m (1,023 ft)/min
Service ceiling 4,000 m (13,120 ft)
T-O run 260 m (855 ft)
Landing run 340 m (1,115 ft)
Range: with max fuel 523 n miles (970 km; 602 miles)
  with max payload 216 n miles (400 km; 248 miles)

UPDATED

YAKOVLEV SKYLIFTER
TYPE: Space vehicle launch platform.
PROGRAMME: Design study for 545 tonne (1,200,000 lb) MTOW, twin-boom aircraft, employing Antonov An-225 wing and Yak-40 forward fuselage, to be built by Russo-Ukrainian consortium.

VERIFIED

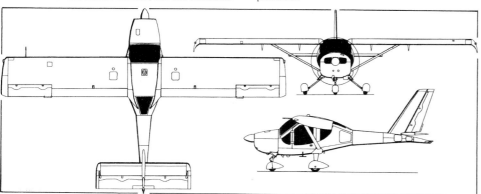

Yakovlev Yak-112 (Teledyne Continental IO-360 engine) *(Paul Jackson/Jane's)*

SINGAPORE

ST Aero

SINGAPORE TECHNOLOGIES AEROSPACE LTD (Division of Singapore Technologies Engineering Ltd)
540 Airport Road, Paya Lebar, Singapore 539938
Tel: (+65) 287 11 11
Fax: (+65) 280 97 13 and 280 82 13
Web: http://www.st.com.sg/STAero
CHAIRMAN AND CEO: C B Lim
DEPUTY CHAIRMAN: S F Boon
PRESIDENT: S K Wee
DEPUTY PRESIDENT AND COO: K K Tay
DEPUTY PRESIDENT AND MILITARY MARKETING PRESIDENT: G Yeo
SENIOR VICE-PRESIDENT, INTERNATIONAL MARKETING: L H Ooi
Tel: (+65) 380 61 83
e-mail: ooilh@st.com.sg
CORPORATE COMMUNICATIONS: Shirley Tan
Tel: (+65) 380 61 76
e-mail: shirley@st.com.sg

Formed early 1982 as government-owned Singapore Aircraft Industries Ltd, controlled by Ministry of Defence Singapore Technology Holding Company Pte Ltd; renamed Singapore Aerospace April 1989; 15,000 m² (161,450 sq ft) new facility at Paya Lebar opened October 1983; workforce 4,000 in late 1998. Currently organised into Commercial Business Group, Military Business Group and Engineering and Development Centre. Provides comprehensive high-echelon commercial and military aircraft maintenance and engineering services, including structural modification and refurbishment, engine and aircraft component repair and overhaul, precision manufacturing, spares and material support. Annual sales S$674 million in 1997 and S$411 million for first half of 1998.

Assembled 30 SIAI-Marchetti S.211s and 17 of 22 Super Pumas for Republic of Singapore Air Force; is partner in EC 120 helicopter programme with Eurocopter and CATIC (see International section); to manufacture 100 shipsets of nosewheel doors worth US$12 million for Boeing 777 (option for another 100); produced passenger doors for Fokker 100. Has recently upgraded F-5Es and F-5Fs of RSAF to F-5S and F-5T standard with new avionics and WDNS (IOC January 1998); now partnering IAI and Elbit of Israel for avionics upgrade of 48 F-5E/Fs of Turkish Air Force; and involved with Lockheed Martin, BAE Systems, FIAR, Rafael, Sextant Avionique and others in promoting 'Falcon ONE' upgrade for F-16A/B Fighting Falcon. Upgrade programme for remaining S.211s in prospect in 1999.

ST Aero is part of the four-member company Singapore Technologies Engineering Ltd. Subsidiaries of ST Aero are:

ST Aerospace Engineering Pte Ltd, 540 Airport Road, Paya Lebar, Singapore 539938
Tel: (+65) 287 11 11
Fax: (+65) 282 32 36
PRESIDENT: Y S Ho

Depot maintenance, major modifications, structural repairs, retrofits and assembly work for a wide range of civil and military aircraft, including helicopters and business jets.

ST Aerospace Engines Pte Ltd, 501 Airport Road, Paya Lebar, Singapore 539931
Tel: (+65) 285 11 11
Fax: (+65) 282 30 10
SENIOR VICE-PRESIDENT/GENERAL MANAGER: K P Chong

Overhaul of civil and military aero-engines.

ST Aerospace Supplies Pte Ltd, 540 Airport Road, Paya Lebar, Singapore 539938
Tel: (+65) 380 62 82
Fax: (+65) 383 47 57
VICE-PRESIDENT/GENERAL MANAGER: Francis Ho

Parts and components for civil and military aircraft.

ST Aerospace Systems Pte Ltd, 505A Airport Road, Paya Lebar, Singapore 539934
Tel: (+65) 287 22 22
Fax: (+65) 284 44 14
VICE-PRESIDENT/GENERAL MANAGER: C K Ang

Specialises in aviation systems reworking.

UPDATED

SLOVAK REPUBLIC

AEROPRO

AEROPRO s.r.o.
Kostolna 42, SK-949 01 Nitra
Tel: (+421 87) 652 63 55
Fax: (+421 87) 652 63 55
e-mail: aeropro@flynet.sk

Light aviation supplies company Aeropro has taken over production of the Eurofox ultralight formerly marketed by Evektor/Aerotechnik of Czech Republic (which see).

NEW ENTRY

AEROPRO EUROFOX

TYPE: Side-by-side ultralight.
PROGRAMME: Adapted from SkyStar (Denney) Kitfox, primarily for sale in Germany by Ikarusflug. Certification to German Bauvorschriften für Ultraleichtflugzeuge (BFU) and Slovakian VLA requirements was achieved in March 1996. Certification of Eurofox Pro followed. The 1999 model features increased cabin width and option of Rotax 912 ULS with provision for glider towing.
Following description refers to version marketed by Ikarusflug.
CURRENT VERSIONS: **Eurofox:** Tailwheel version; two-blade propeller.
Eurofox Pro: Nosewheel version; three-blade propeller.
CUSTOMERS: Total 74 sold by April 1999.
COSTS: Eurofox DM74,600, including tax, basic; Eurofox Pro DM77,950, including tax (both 1999).
DESIGN FEATURES: Extensive redesign of Kitfox (see SkyStar in US section). Main changes are lengthened fuselage to improve longitudinal stability, NACA 4412 modified wing section, Junkers combined flaps/ailerons, more spacious cabin dimensions and revised landing gear.
FLYING CONTROLS: Conventional rudder and elevator, manually actuated; 80 per cent span Junkers-type flaperons, maximum deflection 20°.
STRUCTURE: Folding wings have two alloy tubing spars with diagonal bracing, alloy full and half ribs, glass fibre flaps

Eurofox registered in the Czech Republic *(Paul Jackson/Jane's)*

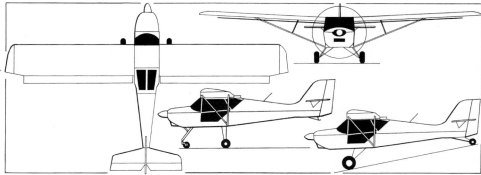

Aeropro Eurofox, a modified version of Kitfox, with additional side view of nosewheel Eurofox Pro *(Paul Jackson/Jane's)*

Slovakian Eurofox Pro nosewheel version *(Paul Jackson/Jane's)*

Revised nose contours of the Rotax 912 ULS-engined Eurofox, announced 1999

and leading-edges and fabric covering. Fuselage of welded steel tubing.
LANDING GEAR: Non-retractable tailwheel type on Eurofox, with hydraulic disc brakes. Eurofox Pro has tricycle gear with smaller mainwheels and fully castoring nosewheel.
POWER PLANT: One 59.6 kW (79.9 hp) Rotax 912 UL flat-four four-stroke driving Kremen SR30 two-blade or SR200 three-blade fixed-pitch propeller via 1:2.273 reduction gear. Alternatively, one 73.5 kW (98.6 hp) Rotax 912 ULS. Fuel capacity 56 litres (14.8 US gallons; 12.3 Imp gallons).
SYSTEMS: Electrical system with 16 Ah 250 W DC generator.
EQUIPMENT: USH 520 or BRS5UL4 parachute recovery system optional.
AVIONICS: Optional avionics by Becker and Honeywell, including GPS, to customer choice.
DIMENSIONS, EXTERNAL (A: Eurofox, B: Eurofox Pro):
Wing span: A, B 9.20 m (30 ft 2¼ in)
Length: overall: A 6.00 m (19 ft 8¼ in)
B 5.76 m (18 ft 10¾ in)
wings folded: A 6.10 m (20 ft 0¼ in)
B 6.20 m (20 ft 4 in)
Height overall: A 1.80 m (5 ft 10¾ in)
B 1.76 m (5 ft 9¼ in)
Width overall, wings folded 2.40 m (7 ft 10½ in)
Propeller diameter, three-blade 1.70 m (5 ft 7 in)
DIMENSIONS, INTERNAL:
Cabin max width: B 1.12 m (3 ft 8 in)

AREAS:
Wings, gross: A, B 11.50 m² (123.8 sq ft)
WEIGHTS AND LOADINGS:
Weight empty, with parachute recovery system:
A 279 kg (615 lb)
B 285 kg (628 lb)
Max T-O weight: A, B 450 kg (992 lb)
PERFORMANCE:
Never-exceed speed (V$_{NE}$) 99 kt (185 km/h; 115 mph)
Cruising speed at 75% power 86 kt (160 km/h; 99 mph)
Stalling speed 35 kt (65 km/h; 41 mph)
Max rate of climb at S/L 270 m (885 ft)/min
Range with max fuel 378 n miles (700 km; 435 miles)
UPDATED

EKOFLUG

EKOFLUG spol s r o
PO Box G-19, SR-043 49 Košice 1
Tel: (+421 95) 644 20 30-1
Fax: (+421 95) 644 20 32

Ekoflug's promotion of the CH 701 ultralight has been assisted since 1998 by an MXP-741 version (OM-AXM) produced in Colombia during 1995 by Agro-Copteros (see 1994-95 *Jane's*) and now employed as a demonstrator, named *Lietajúca Bedňa*: Ugly Duckling.
VERIFIED

EKOFLUG MXP-740
TYPE: Side-by-side ultralight.
PROGRAMME: Modified version of Zenith Zenair STOL CH 701 (which see in US section), with increased wing span and Rotax 912 engine as standard.
DESIGN FEATURES: As CH 701; wing section NACA 650-18 (modified).
POWER PLANT: One 59.6 kW (79.9 hp) Rotax 912 UL flat-four engine; five-blade propeller. Fuel capacity 72 litres (19.0 US gallons; 15.8 Imp gallons) standard, 144 litres (38.0 US gallons; 31.7 Imp gallons) optional.
DIMENSIONS, EXTERNAL:
Wing span 9.00 m (29 ft 6¼ in)
Length overall 6.10 m (20 ft 0¼ in)
Height overall 2.40 m (7 ft 10½ in)

AREAS:
Wings, gross 12.70 m² (136.7 sq ft)
WEIGHTS AND LOADINGS:
Max T-O weight 485 kg (1,069 lb)
PERFORMANCE:
Max level speed 91 kt (170 km/h; 105 mph)
Max cruising speed 86 kt (160 km/h; 99 mph)
Stalling speed: flaps up 31 kt (56 km/h; 35 mph)
flaps down 23 kt (42 km/h; 27 mph)
Max rate of climb at S/L 540 m (1,772 ft)/min
T-O run 15 m (50 ft)
Range: standard fuel 432 n miles (800 km; 497 miles)
max optional fuel 809 n miles (1,500 km; 932 miles)
UPDATED

LOT

LETECKÉ OPRAVOVNE TRENČÍN SP (Trenčín Aircraft Repair Facility)
Legionárska 160, SR-911 04 Trenčín
Tel: (+421 834) 56 51 11 and 56 52 32
Fax: (+421 834) 58 25 26 and 58 18 10

Under a joint co-operation programme with Eurocopter (which see), this Slovakian company is assembling under licence the AS 350 Ecureuil from CKD kits. The first to be completed, an AS 350B2 (OM-SIF), was flown for the first time on 9 March 1998. It was later delivered to the Slovakian electricity authority. Some 80 Ecureuils, civil and military, were planned to be assembled in the country during the first three years of the agreement, but by mid-2000 no evidence was available of subsequent deliveries.

LOT was established in 1949. Main aviation activities include inspection and overhaul of military aircraft and helicopters in Slovakian service, and upgrade of avionics.
UPDATED

SOUTH AFRICA

DENEL

DENEL AVIATION (Division of Denel (Pty) Ltd)
PO Box 11, Kempton Park 1620, Gauteng
Tel: (+27 11) 927 34 14
Fax: (+27 11) 395 15 24
e-mail: clairene@aviation.denel.co.za
Web: http://www.denel.co.za
ACTING DIVISIONAL GENERAL MANAGER: C J Besker
EXECUTIVE MANAGER, MARKETING AND STRATEGIC PLANNING:
Chris Oliver
PROMOTIONS OFFICER, CORPORATE MARKETING:
Claire Nelson-Esch

Atlas Aircraft Corporation (see 1991-92 and earlier *Jane's*) founded 1964 by Bonuskor as private company; delivered first Impala Mk 1 (Aermacchi MB-326) jet trainers to SAAF 1966; manufacturing, design and development facilities for airframes, engines, missiles and avionics; developed Cheetah from Mirage III, Rooivalk from Puma, Oryx (Super Puma) from Puma, V3B and V3C dogfight missiles, and many weapons installations. Incorporated in Armscor Group 1969; restructuring of Armscor on 1 April 1992 created Denel as self-sufficient commercial industrial group, in which Atlas Aviation became military aircraft manufacturing branch of Simera in Denel Aerospace Group.

Denel Group's Simera/Atlas aerospace division renamed Denel Aviation in April 1996 and is now responsible for all Group's military and commercial aviation business. Denel Aviation consists of four business units: Tactical Aircraft Support (for military aircraft); Transport Aircraft Support; Aircraft Manufacturing; and Airmotive (aero-engine and transmission manufacturing). The company is collaborating with DASA of Germany and KAI of South Korea in design of the Mako jet trainer (which see under Germany). An agreement was signed with Aerospatiale Matra in March 1997 for broader co-operation in the areas of aeroplanes, helicopters, tactical missiles, and other business including diversification. MoU in late 1998 provides for co-production of, and export licence for, Agusta A 119 Koala helicopter; July 2000 agreement with Agusta, covering airframe manufacture and systems installation of 30 A109s, ordered by South African Air Force, was expected to be extended shortly thereafter to cover production and sale of Koala as well. Denel Aviation workforce was 2,100 in 1996-97.

The South African government search for a strategic equity partner for a partial privatisation of Denel Aviation ended on 12 October 2000 with the announcement that BAE Systems had been selected. BAE was expected to take a 20 per cent holding in Denel, subject to satisfactory outcome of negotiations due to continue until about March 2001.
UPDATED

DENEL AH-2A ROOIVALK
English name: Red Kestrel
TYPE: Attack helicopter.
PROGRAMME: Considerably modified development of French SA 330 Puma. For background programme, see 1994-95 and earlier editions; see also Current Versions below. XDM and ADM reconfigured during 1995 with original Topaz engines (see 1995-96 *Jane's*) replaced by more powerful Makila 1K2s with digital engine control, making first flight in this form June 1995; more than 1,000 hours flown by XDM, ADM and EDM by December 1998. Airframe changes include modified IR exhaust suppression and intakes on XDM; fuselage skin unchanged on XDM and ADM.

Submitted (partnered by Marshall Aerospace) for British Army requirement in 1993-94, but subsequently eliminated. SAAF order for 12 signed in July 1996. First production Rooivalk (c/n 1001, SAAF serial number 670) left airframe jig 31 July 1997 and made first public flight 17 November 1998; handed over to SAAF on 17 November 1998 and joined No. 16 Squadron on 6 January 1999, formal acceptance being on 7 May 1999. Four delivered by end of 1999; fifth delivered March 2000; trials at Test Flight Development Centre, Bredasdorp; IOC scheduled for late 2000. MoU with Airod of Malaysia provides for co-production in that country in the event of a contract from the Royal Malaysian Air Force.
CURRENT VERSIONS: **XDM** (Experimental Development Model): First prototype (originally designated XH-2); 1,356 kW (1,819 shp) Makila 1A1 engines. Design began late 1984 to meet South African Air Force user requirement specification, leading to public roll-out 15 January and first flight 11 February 1990; ending of war on Namibia/Angola border and subsequent defence cuts removed SAAF need for the aircraft, and programme temporarily halted later in 1990, but restarted as mainly company-funded venture. Served as mechanical systems aircraft.

ADM (Advanced Demonstration Model): Second prototype, with fully integrated avionics and full weapons system; power plant as for XDM; made first flight during second quarter of 1992 and later registered ZU-AHC. Serving as weapons development aircraft including testing

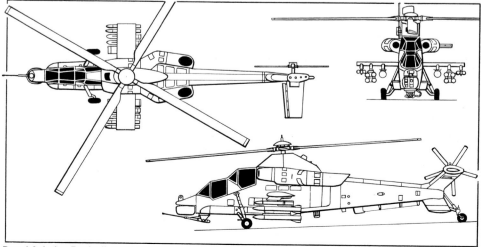
Denel Aviation Rooivalk combat support helicopter *(James Goulding/Jane's)*

SOUTH AFRICA: AIRCRAFT—DENEL

First production AH-2A Rooivalk armed with eight Mokopa ATMs, two 68 mm rocket pods and four Mistral IR air-to-air missiles
1999/0036588

of Denel Mokopa (Black Mamba) long-range (more than 10 km; 6.2 miles) laser-guided anti-tank missile. First round of weapons trials (F2 cannon, HOT 3 missiles and 68 mm rockets) completed 1997. First airborne firing of Mokopa 21 September 1999.

EDM (Engineering Development Model): Third (pre-production) aircraft (672), with 1,376 kW (1,845 shp) Makila 1K2 engines from outset. First flight 17 November 1996. IR signature and weight of intakes and exhaust reduced; MTOW reduced; up to 16 Mokopa anti-tank missiles on stub-wings. Digital avionics include AFCS, Sextant ring laser gyro INS and GPS receiver, single monochrome control and display unit (CDU), two Sextant colour MFDs, health and usage monitoring system (HUMS), Sextant Topowl helmet-mounted sight displays (HMSD) and Kentron tracking system. SFIM nose-mounted sight. First round of avionics qualification completed 1997; second phase started November 1997.

AH-2A: Initial SAAF production version; previously known as CSH-2 but redesignated in 1998.

Description applies to EDM/production AH-2A except where indicated.

Maritime version: Projected version, shown as model at Farnborough Air Show, September 1998, though no commitment yet to build. Would have chin-mounted, 360° maritime search radar in place of undernose cannon; vision-enhancing E-O suite in nose; tailwheel moved forward 2 m (6.6 ft) to facilitate deck operations; flotation gear on forward sponsons and tailboom; manually operated blade folding; and enhanced ECM and communications systems. Shorter stub-wings would support four underwing Penguin or Exocet anti-ship missiles, plus tip-mounted AAMs for self-defence.

CUSTOMERS: South African Air Force, 12 to equip two squadrons; possible longer-term requirement for up to 36. Initial squadron delivery 6 January 1999; deliveries scheduled at four per year until third quarter 2001; No. 16 Squadron reactivated 1 January 1999 at Bloemspruit AB as first squadron. Malaysia may become first export customer (intent to order eight reconfirmed by Malaysian Prime Minister in 1998; possible requirement for up to 30). Algeria, Australia, Singapore and Turkey have also expressed interest (named **RedHawk** for unsuccessful Australian Army bid), and ADM demonstrated to Saudi Arabia in August 1997, but no export orders up to July 2000.

COSTS: SAAF order for 12 priced at R876 million (US$137 million) (1996).

DESIGN FEATURES: Based on reverse engineering of SA 330 Puma dynamics system to Super Puma equivalent standard; engines moved aft to clear stepped tandem cockpits; rear drive taken inboard and forward to modified transmission. Nose-mounted target acquisition turret and chin-mounted gun; non-swept stub-wings for weapon carriage. With Makila 1K2 engines, four-blade fully articulating main rotor rotates at 267 to 290 rpm; five-blade, starboard-mounted tail rotor rotates at 1,290 rpm. Rotor brake fitted. Fixed leading-edge slat on horizontal stabiliser.

Low radar/visual/IR/acoustic signatures; optimised for NOE operation; NVG-compatible 'glass cockpit'; primary missions anti-armour and close air support; integrated digital nav/attack system.

FLYING CONTROLS: Duplex digital AFCS with automatic height hold, hover capture and hover hold; HOCAS (hands on collective and stick) controls.

STRUCTURE: Crash-resistant primary structure, of aluminium alloy, is of I-beam and monocoque construction; rotor blades and secondary structures of composites sandwich. Integral access ladders in fuselage; CFRP engine cowls form work platforms.

LANDING GEAR: Forward-mounted Messier-Dowty main gear with two-stage high-absorption main legs; fully castoring tailwheel at base of lower fin. Mainwheel tyres size 615×225-10. Gear designed to withstand landing impact of up to 6 m (20 ft)/s.

POWER PLANT: Two licence-built Turbomeca Makila 1K2 turboshafts, with digital control, each rated at 1,376 kW (1,845 shp) for T-O, 1,420 kW (1,904 shp) continuous power and 1,573 kW (2,109 shp) for 30 seconds emergency operation. Rear drive turned to drive forward into rear of transmission. Infra-red heat suppressors on exhausts; particle separators on intakes. Main gearbox mounted on vibration isolation system using tuned beam to isolate fuselage from rotor vibrations (said to have one of lowest vibration levels in its class).

Transmission rated at 2,243 kW (3,008 shp) for T-O and 1,826 kW (2,449 shp) maximum continuous from both engines; single-engine transmission ratings are 1,660 kW (2,226 shp) maximum contingency and 1,491 kW (2,000 shp) maximum continuous.

Fuel tankage (self-sealing) in fuselage, under stub-wings (three 618 litre; 163.3 US gallon; 135.9 Imp gallon tanks for total of 1,854 litres; 489.8 US gallons; 407.8 Imp gallons). Pressure refuelling and defuelling. Provision for 750 litre (198 US gallon; 165 Imp gallon) drop tank on each inboard underwing station.

ACCOMMODATION: Pilot (rear) and co-pilot/weapons officer in stepped, tandem cockpits with Martin-Baker armoured, crashworthy seats and armour protection. Access to cockpits via upward-opening gull-wing starboard side window panels. All transparencies flat-plate or single-curvature. Dual flight controls.

SYSTEMS: Environmental control system for cockpit air conditioning. Two independent hydraulic systems, each at 175 bars (2,538 lb/sq in) pressure; at 170 bars (2,465 lb/sq in), flow rate is 12 litres (3.2 US gallons; 2.6 Imp gallons)/min in starboard system and 27 litres (7.1 US gallons; 5.9 Imp gallons)/min in port system.

Electrical power from two 20 kVA alternators providing 200 V three-phase and 115 V single-phase AC at 400 Hz, with two transformer-rectifiers and two 24 V 31 Ah batteries for 28 V DC power. Full or partial electric de-icing/demisting optional.

Crew oxygen system, and fire detection and extinguishing systems, standard.

AVIONICS: All-digital, interfaced to dual-redundant mission computers and MIL-STD-1553B databusses; stores management system conforms to MIL-STD-1760A. Avionics integrator is ATE (Advanced Technologies and Engineering), supported by 15 subcontractors.

Comms: Frequency-agile transceivers: dual V/UHF (30 to 400 MHz) for normal use and single HF (2 to 30 MHz) for NOE flights; intercom/audio system; IFF transponder.

Flight: Duplex four-axis digital AFCS, incorporating (in EDM) dual Sextant Stratus three-axis strapdown ring laser gyro AHRS, Sextant NSS 100-1 eight-channel GPS receiver, Doppler radar velocity sensor, J-band radar altimeter, heading sensor unit (two magnetometers), air data unit and omnidirectional airspeed sensor, all interfaced to two redundant navigation computers. Autopilot has one-touch autohover and attitude hold. Nav/attack system programmed by preloaded cartridge; can hold up to five flight plans (100 waypoints), which can be edited in flight by either crew member.

Instrumentation: Integrated flight management system, with (in EDM) two 160 × 160 mm (6.3 × 6.3 in) Sextant MFD 66 liquid crystal colour MFDs, two Sextant Topowl helmet-mounted sights/displays, and a position management system in each cockpit. Both cockpits have a back-up basic instrument panel for 'get home' capability in the event of computer or power failure. MFDs show flight control, navigation (including moving map on pilot's display), threat warning/EW, fire/weapons control and TDATS imagery, and can copy to each other. Cockpit instruments compatible with NVGs.

Mission: Nose-mounted, gyrostabilised turret contains target detection, acquisition and tracking system (TDATS) incorporating a three fields of view FLIR with automatic guidance and tracking, an LLTV camera and a laser range-finder; associated equipment includes crew members' Sextant helmet sights, missile command link and tracking goniometer. Helmet sights display both flight and weapon data, and can cue both the TDATS turret and the gun turret. Cumulus (South Africa) and Pilkington Optronics (UK) PNVS (Pilot Night Vision Sensor), a two-axis turret with 40 × 30° field of view FLIR, offered as export alternative to Sextant helmet sights.

Self-defence: Optimised ECM can include radar and laser warning receivers, chaff/flare dispensers and RF, IR and laser jammers.

ARMAMENT: Long- or short-barrel version of 20 mm Armscor F2; cleared also for 30 mm weapon. F2 has up to 700 rounds of ammunition, firing rate of 740 rds/min, and slew rate of 90°/s. Gun linked to TDATS and helmet-mounted sight display (HMSD). Three underwing stores stations each side. Two or four M159 19-tube launchers for Forges de Zeebrugge FZ90 70 mm unguided rockets and/or four-round launchers for up to 16 ZT6 Mokopa anti-tank missiles (semi-active laser version) on inner four underwing pylons; two or four Mistral IR air-to-air missiles on outboard pylons.

DIMENSIONS, EXTERNAL:
Main rotor diameter	15.58 m (51 ft 1½ in)
Tail rotor diameter	3.05 m (10 ft 0 in)
Length: fuselage, excl gun, tail rotor turning	16.39 m (53 ft 9¼ in)
overall, rotors turning	18.73 m (61 ft 5½ in)
Wing span (to c/l of AAM pylons)	6.355 m (20 ft 10¼ in)
Fuselage:	
Max width (excl engine fairings)	1.28 m (4 ft 2½ in)
Max depth (belly to top of rotor head)	4.00 m (13 ft 1½ in)
Height: over tail rotor	4.445 m (14 ft 7 in)
to top of rotor head	4.59 m (15 ft 0¾ in)
overall	5.185 m (17 ft 0¼ in)
Wheel track	2.78 m (9 ft 1½ in)
Wheelbase	11.77 m (38 ft 7½ in)

AREAS:
Main rotor disc	190.60 m² (2,052.1 sq ft)
Tail rotor disc	7.27 m² (78.23 sq ft)

WEIGHTS AND LOADINGS:
Weight empty	5,730 kg (12,632 lb)
Max internal fuel weight	1,469 kg (3,238 lb)
External weapons load with full internal fuel	1,563 kg (3,446 lb)
Max external load (with 1,000 kg; 2,205 lb fuel)	2,032 kg (4,480 lb)
Typical mission T-O weight	7,500 kg (16,535 lb)
Max T-O weight	8,750 kg (19,290 lb)
Max disc loading	45.9 kg/m² (9.40 lb/sq ft)
Transmission loading at max T-O weight and power (Makila 1K2)	3.90 kg/kW (6.41 lb/shp)

Rear (pilot's) cockpit of the Rooivalk EDM prototype

PERFORMANCE (at 7,500 kg; 16,535 lb combat weight, except where indicated. A: ISA at S/L, B: ISA + 27°C at 1,525 m; 5,000 ft):

Never-exceed speed (V_{NE}):
A, B	167 kt (309 km/h; 192 mph)
Max cruising speed: A	150 kt (278 km/h; 173 mph)
B	130 kt (241 km/h; 150 mph)
Max sideways speed: A	50 kt (93 km/h; 58 mph)
Max rate of climb at S/L: A	798 m (2,620 ft)/min
B	760 m (2,493 ft)/min
Rate of climb at S/L, OEI: A	344 m (1,128 ft)/min
Service ceiling: A	6,100 m (20,000 ft)
B	5,150 m (16,900 ft)
Hovering ceiling IGE: A	5,850 m (19,200 ft)
B	3,110 m (10,200 ft)
Hovering ceiling OGE: A	5,455 m (17,900 ft)
B	2,410 m (7,900 ft)
Excess hover power margin OGE, S/L anti-tank mission	+39%

Range with max internal fuel, no reserves:
A	380 n miles (704 km; 437 miles)
B	507 n miles (940 km; 584 miles)

Range at max T-O weight with external fuel:
A	680 n miles (1,260 km; 783 miles)
B	720 n miles (1,335 km; 829 miles)

Endurance with max internal fuel, no reserves:
A	3 h 36 min
B	4 h 55 min

Endurance at max T-O weight with external fuel:
A	6 h 52 min
B	7 h 22 min
g limits	+2.6/−0.5

UPDATED

SPAIN

EADS CASA

CONSTRUCCIONES AERONAUTICAS SA
Avenida de Aragón 404, PO Box 193, E-28022 Madrid
Tel: (+34 91) 585 70 00
Fax: (+34 91) 585 76 66/7
e-mail: communications@casa.es
Web: http://www.casa.es
WORKS: Madrid (Barajas), Getafe, Toledo (Illescas), Seville (San Pablo and Tablada) and Cádiz (Puerto Real and Puntales)
CHAIRMAN AND CEO: Alberto Fernández
EXECUTIVE VICE-PRESIDENTS:
 F Fernandez Sainz (Programmes)
 Antonio Fuentes (Operations)
 Pablo de Bergia (Commercial)
 Ramón Madrid (Maintenance Division)
 Pedro Méndez (Space Division)
DIRECTOR, COMMERCIAL COMMUNICATIONS: Miguel Sánchez

CASA, founded in 1923, is owned 99.2852 per cent by the Spanish state holding company Sociedad Estatal de Participaciones Industriales (SEPI), 0.7098 per cent by DASA (Germany) and 0.005 per cent by others. CASA declared a turnover of US$1,305 million in 1999, representing an increase of almost 24 per cent over the preceding year. Its seven factories had a combined workforce of 7,400 in mid-2000.

CASA has a 100 per cent holding in AISA; Compañía Española de Sistemas Aeronáuticas (CESA) formed with 40 per cent holding by Lucas Aerospace; CASA Space Division greatly extended; largely automated composites assembly plant opened at Illescas, near Toledo, January 1991.

Maintenance Division has for nearly 40 years maintained and modernised some 7,900 aircraft and helicopters and 145,000 components for different customers, principally USAF and Spanish Air Force; this Division currently supporting and modernising C-130, P-3, F-4, F/A-18 and Mirage F1.

Largest current programme is 4.2 per cent share in Airbus; makes tailplanes for all Airbus aircraft, plus landing gear doors, wing ribs and skins, leading/trailing-edges and passenger doors for A300/310/320/330/340, many in composites. Second largest programme is 13.67 per cent share in Eurofighter Typhoon with membership of avionics, control systems, flight management and structure teams; builds starboard wing and wing leading-edges of production aircraft; CASA also responsible for integration and software creation for communications system. Own-design programmes include C-212, C-295 and CN-235 (last-named shared with Dirgantara through Airtech); looking for partner for ATX light attack aircraft and trainer.

Designed, developed and manufactured wing of Saab 2000; designed and produced (until mid-1999) tailplanes of Boeing MD-11; makes outer flaps for Boeing 757; produced tail components for Spanish Sikorsky S-70s; member of Airbus Military SAS (designated sole-source assembler of A400M/FLA transport, with 12.4 per cent share also including responsibility for horizontal tail design and manufacture and power plant development and integration); previously participated in US/Italian/Spanish Harrier II Plus programme.

Following approaches by British Aerospace, Aerospatiale, Alenia and DaimlerChrysler Aerospace (DASA), it was announced on 11 June 1999 that SEPI and DaimlerChrysler had signed an MoU to unite CASA and DASA in a new single company. This company, subsequently created as EADS by Aerospatiale Matra and DaimlerChrysler Aerospace (see entry in International section) and joined by CASA as a founder member on 2 December 1999, is the first cross-border merger to take place within the European aerospace industry. Operations as EADS began on 10 July 2000, but CASA continues to trade as such until full integration of assets is achieved.

Measures designed to reinforce CASA's current activities include boosting its participation in Airbus and Eurofighter; consolidating CASA's leadership in European military transport aircraft sector; improving operations in CASA's commercial area by combining capabilities from both companies; enhancing its activities in the space, missile and helicopter business; and utilising joint efforts to generate new business opportunities. Specifically, a military transport aircraft business unit is being created around CASA's military transport aircraft family, based in Spain and responsible for the design, construction and commercialisation of products and services that CASA and EADS Deutschland offer in this sector. Aircraft of this unit include the C-212-400, CN-235-300, C-295 and A400M.

Artist's impression of a possible ATX configuration
1999/0051421

UPDATED

CASA ATX
TYPE: Light fighter/advanced jet trainer.
PROGRAMME: Government-funded prefeasibility and feasibility studies began in 1989 for advanced trainer and light attack aircraft (previously known as AX) to succeed Spanish F-5s from about 2004. Now in conceptual design stage for an aircraft family ranging from a pure advanced trainer to a more capable attack aircraft. Based on a Spanish Air Force requirement for an F-5B trainer replacement, ATX is now CASA candidate for a future European advanced trainer/lead-in fighter trainer. It will have a single 71.2 kN (16,000 lb thrust class) afterburning engine and a basic empty weight of 5,000 to 5,500 kg (11,023 to 12,125 lb). A development partner is being sought.

UPDATED

CASA C-212 SERIES 300 and 400
TYPE: Twin-turboprop transport.
PROGRAMME: Design studies for C-212 began 1964; design accepted by Spanish Air Force 1967; two prototypes (plus one static test) ordered 24 September 1968; first flights 26 March and 23 October 1971. See 1991-92 and previous editions for other early history; Series 100 last described in 1981-82 *Jane's;* for Series 200, see under IPTN heading in Indonesian section of this edition; Series 300 certified to FAR Pts 25, 121 and 135 in December 1987. Former name Aviocar now discontinued.
CURRENT VERSIONS: **100**: Initial version; 158 built (CASA two prototypes plus 128, IPTN 28 from Spanish CKD kits). Customers as listed in 1997-98 *Jane's.*
200: Succeeded Srs 100 from 1980; 151 built by CASA (see 1997-98 *Jane's* for customers). Final examples being produced only in Indonesia (see Dirgantara entry).
300: Differs from Series 200 in having winglets to improve climb performance; enlarged nose for increased payload; and enhanced cruise performance. Available in configurations described below.

300 Airliner: Standard seating for 26 passengers, or 24 if lavatory included.
Detailed description applies to Airliner version except where indicated.
300 Utility: Standard seating for 23 passengers, or 21 if lavatory included, with maximum capacity for 26.
300M: Military troop/cargo/general purpose transport.
400: Initiated mid-1995, but not formally launched until June 1997 (Paris Air Show), following first flight (EC-212) on 4 April that year. EFIS and re-rated TPE331-12JR engines maintaining T-O power under hot/high conditions without APR; other improvements as detailed under Design Features below. Spanish type certificate 30 March 1998. Available in same range of versions as Series 300.
Patrullero: Special missions (anti-submarine and maritime patrol) versions; described separately.

CASA C-212-300/400 PRODUCTION AND ORDERS
(at July 2000)

Country	Operator	Srs 300	Srs 400
Angola	Air Force	4	
Argentina	Coast Guard	5[1]	
Bolivia	Army	1	
Bophuthatswana	Air Force	1[2]	
Cape Verde	TACV	2	
Chile	Air Force	2	
	Army	3	
Colombia	Air Force/Satena	5	
Dominican Republic	Air Force		2[7]
France	Air Force (CEV)	5	
Indonesia	Gatari	4	
Lesotho	Defence Force	3	
Panama	Air Force	3	
Portugal	Air Force	2[3]	
Spain	Min of Agriculture, Fisheries and Food		2[6]
Suriname	Air Force		2[5]
Thailand	Army	2	
	Min of Agriculture	6	
USA	CASA (demonstrator)	1	
	DEA	1	
Venda	Defence Force	1[2]	
Venezuela	Navy		3[4]
Totals		**51**	**9**

Notes:
[1] Includes three Patrulleros
[2] Since absorbed into South African Air Force (No. 8 Squadron)
[3] Patrulleros: No. 401 Squadron
[4] Delivered May 1998
[5] Includes one Patrullero; first delivered December 1998, second (Patrullero) in May 1999
[6] Patrullero. First delivered December 1998; second due June 2001
[7] Delivered on 17 March and 17 May 2000; option on one more

CASA C-212-400M, one of three delivered in 1998 to the Venezuelan Navy 1999/0051419

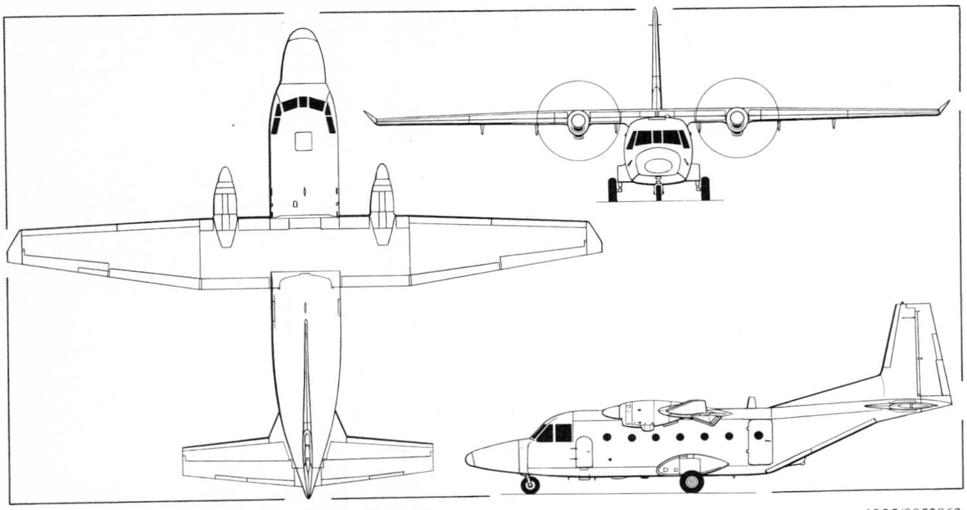

CASA C-212 Series 300 ***(Dennis Punnett/Jane's)*** ***1999***/0052863

CUSTOMERS: See table. Total 460 of all versions ordered by June 1999; all except one Spanish-built aircraft delivered, but total includes some 20 Series 200s still to be completed by Dirgantara of Indonesia (which see). C-212s of all Series have been purchased or leased by 39 military and 51 civil operators worldwide and had accumulated over 2.5 million hours by 1999. Offered to Royal Australian Air Force late 2000 as replacement for DHC-4 Caribou; also reportedly under consideration by South African Air Force as C-47 replacement.

DESIGN FEATURES: Typical high-wing, rear-loading military transport.

Wing section NACA 65_3-218; no dihedral; incidence 2° 30′; swept winglets canted upwards at 45°; meets FAR Pt 36 noise limits.

Series 400 features engine change (see Power Plant); avionics upgrade; underfloor avionics boxes relocated to former nose baggage compartment, facilitating ventral installation of a 360° scan radar and eliminating need for special-shape 'platypus' nose of earlier Patrullero versions, which allows only 270° scan; relocated pressure refuelling point; main cabin refurbished; cargo winch option. Otherwise as for Series 300.

FLYING CONTROLS: Conventional and manual. Trim tab in port aileron; trim and geared tabs in rudder and each elevator; double-slotted flaps.

STRUCTURE: All-metal light alloy fail-safe structure; unpressurised; two-spar tailplane and fin. Wing centre-section, forward and rear passenger doors and dorsal fin manufactured by AISA.

LANDING GEAR: Non-retractable tricycle type, with single mainwheels and single steerable nosewheel. CASA oleo-pneumatic shock-absorbers. Goodyear wheels and tyres, main 11.00-12 (10 ply), nose 24×7.7 (8 ply). Tyre pressure 3.86 bars (56 lb/sq in) on main units, 4.00 bars (58 lb/sq in) on nose unit. Goodyear hydraulic disc brakes on mainwheels. No brake cooling. Anti-skid system optional.

POWER PLANT: *Series 300:* Two Honeywell TPE331-10R-513C turboprops, each flat rated at 671 kW (900 shp) and equipped with automatic power reserve (APR) system providing 690 kW (925 shp) in event of one engine failing during take-off.

Series 400: Fully rated 690 kW (925 shp) TPE331-12JR-701C engines. Dowty Aerospace R-33414-82-F/13 four-blade constant-speed fully feathering reversible-pitch propellers in Series 300 and 400.

Fuel in four integral wing tanks, with total capacity of 2,040 litres (539 US gallons; 449 Imp gallons), of which 2,000 litres (528 US gallons; 440 Imp gallons) usable. Gravity refuelling point above each tank. Single pressure refuelling point under starboard wing leading-edge in Series 300; in starboard mainwheel fairing in Series 400. Additional fuel can be carried in one 1,000 litre or two 750 litre (264 or 198 US gallon; 220 or 165 Imp gallon) optional ferry tanks inside cabin, and/or two 500 litre (132 US gallon; 110 Imp gallon) auxiliary underwing tanks. Oil capacity 4.5 litres (1.2 US gallons; 1.0 Imp gallon) per engine.

ACCOMMODATION: Crew of two on flight deck; cabin attendant in civil version. Series 400 has ergonomically redesigned flight deck; improved cabin soundproofing, sidewall design, lighting and toilet; 400M, foldaway seats along each side of cabin.

For troop transport role, main cabin can be fitted with 25 inward-facing seats along cabin walls, to accommodate 24 paratroops with instructor/jumpmaster; or 25 fully equipped troops. As ambulance, cabin is normally equipped to carry 12 stretcher patients and four medical attendants. As freighter, up to 2,700 kg (5,952 lb) of cargo can be carried in main cabin, including two LD1, LD727/DC-8 or three LD3 containers, or light vehicles. Cargo system, certificated to FAR Pt 25, includes roller loading/unloading system and 9 *g* barrier net. Photographic version equipped with two Leica RC-20/30 vertical cameras and darkroom. Navigation training version has individual desks/consoles for instructor and five pupils, in two rows, with appropriate instrument installations.

Civil passenger transport version has standard seating for up to 26 in mainly three-abreast layout at 72 cm (28.5 in) pitch, with provision for quick change to all-cargo or mixed passenger/cargo interior. Lavatory, galley and 400 kg (882 lb) capacity baggage compartment standard, plus additional 150 kg (330 lb) in nose bay. VIP transport version can be furnished to customer's requirements.

Forward/outward-opening door on port side immediately aft of flight deck; forward/outward-opening passenger door on port side aft of wing; inward-opening emergency exit opposite each door on starboard side. Additional emergency exit in roof of forward main cabin. Two-section underfuselage loading ramp/door aft of main cabin can be opened in flight for discharge of paratroops or cargo, and can be fitted with optional external wheels for door protection during ground manoeuvring. Interior of rear-loading door can be used for additional baggage stowage in civil version. Entire accommodation heated and ventilated; air conditioning optional.

SYSTEMS: Freon cycle or (on special mission versions) engine bleed air air conditioning system optional. Hydraulic system, operating at service pressure of 138 bars (2,000 lb/sq in), provides power via electric pump to actuate mainwheel brakes, flaps, nosewheel steering and rear cargo ramp/door. Hand pump for standby hydraulic power in case of electrical failure or other emergency. Electrical system supplied by two 9 kW starter/generators, three batteries and three static converters. Pneumatic boot and engine bleed air de-icing of wing and tail unit leading-edges; electric de-icing of propellers and windscreens. Oxygen system for crew (including cabin attendant); two portable oxygen cylinders for passenger supply. Engine and cabin fire protection systems.

AVIONICS: *Comms:* Rockwell Collins VHF, ATC transponder, intercom (with Gables control) and PA system standard (new interphone system in Series 400); Rockwell Collins HF, UHF and second transponder, and Fairchild CVR, optional.

Radar: Honeywell weather radar standard. RDR-1500B search radar in ventral radome on Patrullero versions.

Flight: Rockwell Collins VOR/ILS, ADF, DME and radio altimeter, and Honeywell AFCS and directional gyro, standard; second Rockwell Collins ADF, Global Omega nav, Dorne & Margolin marker beacon receiver and Fairchild flight data recorder optional. Flight management system (FMS) in Series 400 incorporates VOR, ADF, DME and GPS nav receiver.

Instrumentation: Blind-flying instrumentation standard. Series 400 has EFIS with four CRTs; IEDS (integrated engine data system) with two colour LCDs.

EQUIPMENT: 1,000 kg (2,205 lb) capacity cargo winch optional.

ARMAMENT (military versions, optional): Two machine gun pods or two rocket launchers, or one launcher and one gun pod, on hardpoints on fuselage sides (capacity 250 kg; 551 lb each).

DIMENSIONS, EXTERNAL:	
Wing span	20.27 m (66 ft 6 in)
Wing chord: at root	2.50 m (8 ft 2½ in)
at tip	1.25 m (4 ft 1¼ in)
Wing aspect ratio	10.0
Length overall	16.15 m (52 ft 11¾ in)
Fuselage max width	2.30 m (7 ft 6½ in)
Height overall	6.60 m (21 ft 7¾ in)
Tailplane span	8.40 m (27 ft 6¾ in)
Wheel track	3.10 m (10 ft 2 in)
Wheelbase	5.46 m (17 ft 11 in)
Propeller diameter	2.79 m (9 ft 2 in)
Propeller ground clearance (min)	1.27 m (4 ft 2 in)
Distance between propeller centres	5.27 m (17 ft 3¾ in)
Passenger door (port, rear): Height	1.58 m (5 ft 2¼ in)
Width	0.70 m (2 ft 3½ in)
Crew and servicing door (port, fwd):	
Max height	1.10 m (3 ft 7¼ in)
Width	0.58 m (1 ft 10¾ in)
Rear-loading door: Max length	3.66 m (12 ft 0 in)
Max width	1.70 m (5 ft 7 in)
Max height	1.80 m (5 ft 10¾ in)
Emergency exit (stbd, fwd): Height	1.10 m (3 ft 7¼ in)
Width	0.58 m (1 ft 10¾ in)
Emergency exit (stbd, rear): Height	0.94 m (3 ft 1 in)
Width	0.55 m (1 ft 9¾ in)
DIMENSIONS, INTERNAL:	
Cabin (excl flight deck and rear-loading door):	
Length: passenger	7.22 m (23 ft 8¼ in)
cargo/military	6.55 m (21 ft 5¾ in)
Max width	2.10 m (6 ft 10¾ in)
Max height	1.80 m (5 ft 10¾ in)
Floor area: passenger	13.5 m² (145 sq ft)
cargo/military	12.2 m² (131 sq ft)
Volume: passenger	23.7 m³ (837 cu ft)
cargo/military	22.0 m³ (777 cu ft)
Cabin: volume incl flight deck and rear-loading door	30.4 m³ (1,074 cu ft)
Baggage compartment volume	3.6 m³ (127 cu ft)
AREAS:	
Wings, gross	41.00 m² (441.3 sq ft)
Ailerons (total, incl tab)	7.50 m² (80.73 sq ft)
Trailing-edge flaps (total)	14.92 m² (160.60 sq ft)
Fin, incl dorsal fin	4.22 m² (45.42 sq ft)
Rudder, incl tab	2.05 m² (22.07 sq ft)
Tailplane	9.01 m² (96.98 sq ft)
Elevators (total, incl tabs)	3.56 m² (38.32 sq ft)
WEIGHTS AND LOADINGS:	
Manufacturer's weight empty:	
300, 400	3,780 kg (8,333 lb)
Weight empty, equipped (cargo):	
300	4,400 kg (9,700 lb)
400	4,550 kg (10,031 lb)

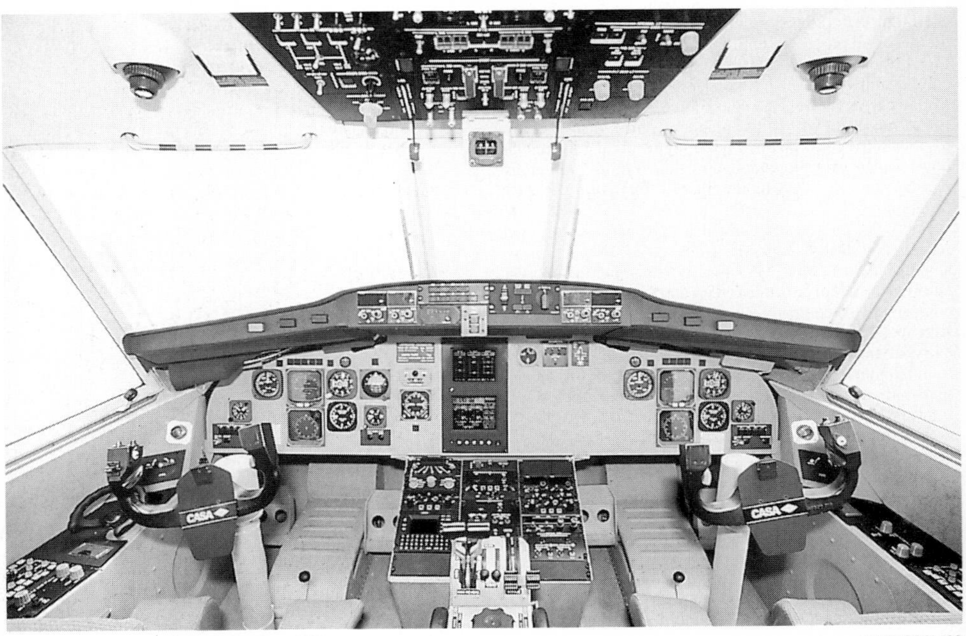

Flight deck of the C-212 Series 400 ***1999***/0051420

Max payload: 300 (cargo)	2,700 kg (5,952 lb)
300M	2,820 kg (6,217 lb)
400	2,950 kg (6,504 lb)
Max fuel (300, 400): standard	1,600 kg (3,527 lb)
with underwing auxiliary tanks	2,400 kg (5,291 lb)
Max T-O weight: 300 standard	7,700 kg (16,975 lb)
300M	8,000 kg (17,637 lb)
400	8,100 kg (17,857 lb)
Max ramp weight: 300 standard	7,750 kg (17,085 lb)
300M, 400	8,150 kg (17,967 lb)
Max landing weight:	
300 standard, 300M	7,450 kg (16,424 lb)
400	8,100 kg (17,857 lb)
Max zero-fuel weight:	
300 standard, 300M	7,100 kg (15,653 lb)
400	7,500 kg (16,535 lb)
Max cabin floor loading	732 kg/m² (150 lb/sq ft)
Max wing loading:	
300 standard	187.8 kg/m² (38.46 lb/sq ft)
300M	195.1 kg/m² (39.96 lb/sq ft)
400	197.6 kg/m² (40.46 lb/sq ft)
Max power loading:	
300 standard	5.74 kg/kW (9.43 lb/shp)
300M	5.96 kg/kW (9.80 lb/shp)
400	6.04 kg/kW (9.92 lb/shp)

PERFORMANCE (Series 300 and 300M at respective max T-O weights): A: passenger version, B: freighter, C: (300):
Max operating speed (V$_{MO}$):
  A, B, C 200 kt (370 km/h; 230 mph)
Max cruising speed at 3,050 m (10,000 ft):
  A, B, C 191 kt (354 km/h; 220 mph)
Econ cruising speed at 3,050 m (10,000 ft):
  A, B, C 162 kt (300 km/h; 186 mph)
Stalling speed in T-O configuration:
  A, B, C 78 kt (145 km/h; 90 mph)
Max rate of climb at S/L: A, B, C 497 m (1,630 ft)/min
Rate of climb at S/L, OEI: A, B, C 95 m (312 ft)/min
Service ceiling: A, B, C 7,925 m (26,000 ft)
Service ceiling, OEI: A, B, C 3,380 m (11,100 ft)
FAR T-O distance: A, B 817 m (2,680 ft)
FAR landing distance: A, B 866 m (2,840 ft)
MIL-7700C T-O to 15 m (50 ft):
  C 610 m (2,000 ft)
MIL-7700C landing from 15 m (50 ft):
  C 462 m (1,520 ft)
MIL-7700C landing run: C 285 m (935 ft)
Required runway length for STOL operation:
  C 384 m (1,260 ft)
Range (IFR reserves):
  A with 25 passengers, at max cruising speed
    237 n miles (440 km; 273 miles)
  B with 1,713 kg (3,776 lb) payload
    773 n miles (1,433 km; 890 miles)
  C with max payload 450 n miles (835 km; 519 miles)
  C with max standard fuel and 2,120 kg (4,674 lb)
    payload 907 n miles (1,682 km; 1,045 miles)
  C with max standard and auxiliary fuel and 1,192 kg
    (2,628 lb) payload
    1,447 n miles (2,680 km; 1,665 miles)

PERFORMANCE (Series 400M):
Max cruising speed at 3,050 m (10,000 ft)
  195 kt (361 km/h; 224 mph)
Econ cruising speed at 3,050 m (10,000 ft)
  163 kt (302 km/h; 188 mph)
Service ceiling 7,620 m (25,000 ft)
Service ceiling, OEI 3,275 m (10,740 ft)
T-O run 402 m (1,320 ft)
T-O to 15 m (50 ft) 589 m (1,935 ft)
Landing from 15 m (50 ft) 520 m (1,710 ft)
Landing run 267 m (875 ft)
Required runway length for STOL operation
  393 m (1,290 ft)
Range:
  with max payload 233 n miles (431 km; 268 miles)
  with max standard fuel and 2,120 kg (4,674 lb)
    payload 824 n miles (1,526 km; 948 miles)
  with max standard and auxiliary fuel and 1,192 kg
    (2,628 lb) payload
    1,295 n miles (2,398 km; 1,490 miles)

UPDATED

CASA C-212 PATRULLERO

TYPE: Maritime surveillance twin-turboprop.
DESIGN FEATURES: Special missions versions of C-212 Series 300/400.
CURRENT VERSIONS: **ASW**: Anti-submarine version; 360° scan radar under fuselage; ESM.
  MP: Maritime patrol version; nose- or belly-mounted search radar in enlarged radome; FLIR; additional antennas. One delivered in December 1998 to Spanish Ministry of Agriculture, Fisheries and Food is first 212-400 with 360° radar and TV/FLIR turret. Other features of this aircraft include sensor operator's console, datalink via Inmarsat, external loudspeakers, searchlight, observation windows, camera, and seating for 10 passengers. Manufacture of a second, for the same customer, was under way in 1999.
CUSTOMERS: See table with main C-212 entry. More than 55 Patrulleros sold to date.
POWER PLANT: As C-212 Series 300 or 400. Two 500 litre (132 US gallon; 110 Imp gallon) underwing auxiliary tanks.

C-212-400MP Patrullero delivered to the Suriname Air Force in 1999

Nose radar (and optional FLIR) detail of the CASA C-212-300MP Patrullero

ACCOMMODATION: Flight crew of two; four systems operators in ASW version; radar operator and two observer stations in MP version. Radar console, weapon controls and intervalometer added to flight deck; nav/com boxes in rack aft of pilot; starboard rack behind pilot contains radar, sonobuoy, MAD and ESM boxes. Three consoles on starboard side of cabin: (1) radar control and display, ESM and intercom; (2) tactical display, MAD recorder and intercom; (3) sonobuoy controls, acoustic controls and displays. Radar repeater console and searchlight controls added to flight deck; nav/com (including UHF/DF and VLF/Omega) and radar in rack behind pilot; observer stations at rear.
AVIONICS (ASW): *Comms*: Dual VHF, single HF and UHF radios; intercom; IFF/SIF.
  Radar: 360° scan underfuselage radar.
  Flight: Autopilot, flight director, UHF/DF, VLF/Omega.
  Mission: MAD, OTPI, tactical and sonobuoy processing systems; ESM.
AVIONICS (MP): *Comms*: As for ASW version.
  Radar: Nose- or belly-mounted search radar with 360° scan.
  Mission: FLIR optional.
EQUIPMENT (MP): Sonobuoy and smoke marker launchers; searchlight.
ARMAMENT (ASW): Can include torpedoes such as Mk 46 and Sting Ray; air-to-surface missiles such as AS15TT and Sea Skua; unguided air-to-surface rockets.

UPDATED

CASA C-295M

TYPE: Twin-turboprop transport.
PROGRAMME: Stretched derivative of Airtech CN-235M (see International section). Initiated November 1996, after market survey, as unilateral development of CASA/IPTN CN-235. Announced at Paris Air Show, June 1997. Prototype (EC-295 *Ciudad de Getafe*), a converted CN-235, flew for first time 28 November 1997 and had accumulated 801 hours in 379 sorties and 555 landings by 17 December 1999. First production C-295 (EC-296 *Ciudad de Sevilla*), designated S-1, made its maiden flight on 22 December 1998; had flown 516 hours in 232 sorties and 379 landings by 17 December 1999.
  Certified by INTA (for military operations) on 30 November 1999; DGAC certification received 3 December 1999 and FAA (FAR Pt 25) certification on 17 December 1999; first deliveries scheduled for late 2000. Clearance for airdrop missions was due in July 2000.
CURRENT VERSIONS: **C-295M**: Military version; *as described*.
CUSTOMERS: Spanish Air Force order for nine announced 30 April 1999 (launch customer) and contract signed January 2000; deliveries to begin in November 2000 and be completed in 2004. Possibility of French Air Force order for up to 20. Market for over 300 foreseen over 10 years. Competed against the CN-235 Srs 300 and Alenia/Lockheed Martin C-27J Spartan for an Australian order of between 10 and 14 aircraft, but requirement suspended in July 2000. Also short-listed by Swiss Air Force, which plans to buy two transport aircraft.
DESIGN FEATURES: Typical rear-loading, high-wing military transport. Conforms to FAR Pt 25 and MIL-STD-7700C. Compared with CN-235, has fuselage lengthened by six frames (three forward and three aft of wing) and pressure differential increased; wing reinforced for higher operating weights and to support up to six optional underwing stores stations; additional 2,380 litres (629 US gallons; 523.5 Imp gallons) of fuel in wings; reinforced landing gear with twin nosewheels to facilitate operation from unprepared airstrips; modernised avionics similar to those of CN-235 Srs 300.
STRUCTURE: Provision for three hardpoints under each wing, capacities 800 kg (1,764 lb) inboard, 500 kg (1,102 lb) centre and 300 kg (661 lb) outboard.
LANDING GEAR: Main gear as CN-235 but reinforced for higher operating weights; twin nosewheels. Designed for sink rate of 3.05 m (10 ft)/s at normal MTOW and 2.74 m (9 ft)/s at overload MTOW; and operation from semi-prepared runways down to CBR-2 category.
POWER PLANT: Two Pratt & Whitney Canada PW127G turboprops, each with T-O rating of 1,972 kW (2,645 shp) normal, 2,177 kW (2,920 shp) with APR. Hamilton Sundstrand HS-568F-5 six-blade composites propellers with autofeathering and synchrophasing. Fuel capacity 7,650 litres (2,021 US gallons; 1,683 Imp gallons). In-flight refuelling probe.
ACCOMMODATION: Crew of two; dual controls standard. Two or three longitudinal rows of foldaway seats in main cabin for 69 troops (high-density, 78 troops); 48 paratroops (each 130 kg; 287 lb with full equipment); as a 12-stretcher intensive care unit or, in medevac configuration, 27 stretchers and four medical personnel. Lavatory, at front on port side, standard. In cargo role, can accommodate up to five 2.24 × 2.74 m (88 × 108 in) pallets, including one on rear ramp/door; or one 88 × 108 in and three 2.24 × 3.18 m (88 × 125 in) pallets; or three Land Rover-type light vehicles. Paratroop door each side at rear of cabin. Crew door aft of flight deck on starboard side, with emergency exit opposite on port side.
SYSTEMS: Pressure differential 0.38 bar (5.52 lb/sq in), giving cabin atmosphere equivalent to 2,395 m (7,850 ft) at altitude of 7,620 m (25,000 ft).

S-1, the first production C-295 *(Paul Jackson/Jane's)*

AVIONICS: All-digital, and compatible with NVGs. Honeywell avionics and Sextant Avionique Topdeck displays standard. Sextant is overall systems integrator.

Comms: Two Sextant U/VHF-AM/FM and one HF radio; Honeywell IFF transponder and solid-state FDR; ELT; air data system. Honeywell solid-state CVR. Alternative com/nav/ident functions optional.

Radar: Honeywell RDR-1400C colour weather radar with search and beacon modes and ground mapping vertical navigation capability.

Flight: Four-dimensional FMS; dual ADU 3000 air data units; dual AHRS; dual multimode receivers; dual GPS receivers; dual multifunction control and display units; Honeywell enhanced GPWS. Two central data processor/interface units and standard digital busses (ARINC 429 and MIL-STD-1553B). Options include Sextant Totem 3000 laser gyro navigation system, enhanced TCAS, Cat. II ILS, MLS and satcom.

Instrumentation: Four 152 × 203 mm (6 × 8 in) Sextant LCDs; one integrated electronic standby instrument; two IFC cabinets with autopilot module; radar altimeter. Video images from E-O sensors such as FLIR or cameras can be displayed on the four LCDs. Full NVG compatibility. Provision for two Sextant HUDs.

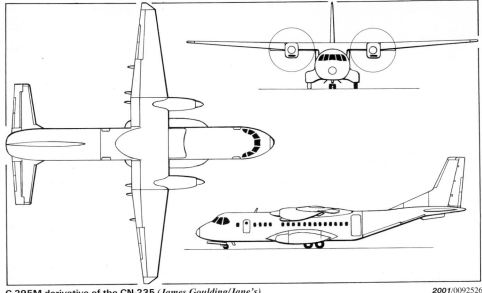

C-295M derivative of the CN-235 *(James Goulding/Jane's)*   2001/0092526

DIMENSIONS, EXTERNAL:
Wing span	25.81 m (84 ft 8¼ in)
Length overall	24.45 m (80 ft 2½ in)
Height overall	8.66 m (28 ft 5 in)
Tailplane span	10.60 m (34 ft 9¼ in)
Wheel track	3.96 m (13 ft 0 in)
Wheelbase	8.435 m (27 ft 8 in)
Propeller diameter	3.89 m (12 ft 9 in)
Distance between propeller centres	7.00 m (22 ft 11½ in)
Paratroop doors (two, each):	
Height	1.75 m (5 ft 9 in)
Width	0.90 m (2 ft 11½ in)
Crew door (fwd, stbd): Height	1.27 m (4 ft 2 in)
Width	0.70 m (2 ft 3½ in)
Emergency exit (fwd, port): Height	0.91 m (3 ft 0 in)
Width	0.50 m (1 ft 7¾ in)

DIMENSIONS, INTERNAL:
Cabin, excl flight deck:	
Length: excl ramp/door	12.69 m (41 ft 7½ in)
incl ramp/door	15.73 m (51 ft 7¼ in)
Max width	2.70 m (8 ft 10¼ in)
Width at floor	2.36 m (7 ft 9 in)
Max height	1.90 m (6 ft 2¾ in)
Volume available for cargo	57.0 m³ (2,013 cu ft)

WEIGHTS AND LOADINGS (A: normal, B: overload):
Max payload: A	7,500 kg (16,535 lb)
B	9,700 kg (21,385 lb)
Max T-O weight: A	21,000 kg (46,297 lb)
B	23,200 kg (51,147 lb)
Max landing weight: A	20,700 kg (45,635 lb)
B	23,200 kg (51,147 lb)
Max zero-fuel weight: A	18,500 kg (40,784 lb)
B	20,700 kg (45,634 lb)
Max power loading: A	5.32 kg/kW (8.75 lb/shp)
B	5.89 kg/kW (9.67 lb/shp)

PERFORMANCE (at normal MTOW except where indicated):
Max cruising speed at optimum altitude	256 kt (474 km/h; 295 mph)
Time to optimum cruising altitude	12 min
Absolute ceiling	9,145 m (30,000 ft)
Service ceiling, OEI	4,125 m (13,540 ft)
T-O run at S/L: A, ISA	843 m (2,765 ft)
A, ISA + 20°C	934 m (3,065 ft)
T-O to 15 m (50 ft) at S/L:	
A, ISA	1,025 m (3,365 ft)
A, ISA + 20°C	1,103 m (3,620 ft)
Landing from 15 m (50 ft)	729 m (2,390 ft)
Landing run	420 m (1,380 ft)
Range, ISA, reserves for 45 min hold at 460 m (1,500 ft):	
with max payload:	
A, B	785 n miles (1,455 km; 904 miles)
with max fuel:	
A with 4,000 kg (8,818 lb) payload	2,683 n miles (4,969 km; 3,087 miles)
B with 6,000 kg (13,228 lb) payload	2,250 n miles (4,167 km; 2,589 miles)
g limits: A	2.53
B	2.25

UPDATED

CASA C-295 with rear ramp lowered in flight   2001/0092506

SRI LANKA

JABIRU

JABIRU ASIA
Koggala

Formation of Jabiru Asia announced mid-1998 as joint venture between Jabiru (see Australian section) and CDE Aviation Company of Sri Lanka (member of Lionair group) to manufacture Jabiru ST3 at new factory in Koggala. Work was due to begin before end of 1998, but start not reported by October 2000. Wings and fuselage to be produced in Sri Lanka; engines, avionics and some equipment to be supplied from parent company. Anticipated eventual production of up to 40 per year.

UPDATED

SWEDEN

SAAB

SAAB AB
SE-581 88 Linköping
Tel: (+46 13) 18 70 00
Fax: (+46 13) 18 18 02
e-mail: infosaab@saab.se
Web: http://www.saab.se
PRESIDENT AND CEO: Bengt Halse
EXECUTIVE VICE-PRESIDENT AND CFO: Göran Sjöblom
GROUP SENIOR VICE-PRESIDENT: Dave Hewitson
CORPORATE INFORMATION: Jan Nygren

Original Svenska Aeroplan AB founded at Trollhättan 1937 to make military aircraft; amalgamated 1939 with Aircraft Division of Svenska Järnvägsverkstäderna rolling stock factory at Linköping; renamed Saab Aktiebolag May 1965; merged with Scania-Vabis 1968 to combine automotive interests; Malmö Flygindustri acquired 1968. Bid for Celsius AB 16 November 1999 and acquisition completed 8 March 2000; enlarged company reorganised in 2000 into five main operating divisions and several independent operations:

Saab Aerospace: Combining military aircraft, future aerospace systems and commercial programmes.

Celsius Aviation Services: Commercial aircraft maintenance and engine and component maintenance. Based at Alexandria, Virginia, USA.

Saab Informatics: Electronic warfare, simulation and training and radar control.

Saab Bofors Dynamics: Missiles, anti-armour and underwater systems.

Technical Support and Services: Including aircraft maintenance.

Saab workforce was 16,037 in mid-2000. Continued subcontract work includes structural floor assemblies and landing gear doors for the Airbus A340-500/600 series. Saab also partner in A3XX programme.

It was announced on 30 April 1998 that Saab's owner, Investor AB, had approved the sale to British Aerospace plc (now BAE Systems) of a 35 per cent voting shareholding. Investor remains Saab's leading owner, with 36 per cent of the votes and 20 per cent of the capital. Purchase rights to the remaining 29 per cent of votes and 45 per cent of capital continue to be offered to Investor shareholders.

More than 4,000 military and commercial aircraft delivered since 1940; has held dealership for MD Helicopters (formerly Schweizer/Hughes, McDonnell Douglas and Boeing) products in Scandinavia and Finland since 1962.

UPDATED

SAAB AEROSPACE

SE-581 88 Linköping
Tel: (+46 13) 18 00 00
Fax: (+46 13) 18 18 02
Web: http://www.gripen.se
HEAD OF BUSINESS AREA: Ake Svensson
DEPUTY HEAD (Commercial): Hans Krüger
MARKETING MANAGER: Kjell Möller
INFORMATION MANAGER: Peter Larsson

Saab-BAE Systems Gripen AB

Saab AB, SE-581 88 Linköping
MANAGING DIRECTOR: Jan Närlinge

Original Industrigruppen JAS AB (JAS Industry Group) formed 1981 to represent (then) Saab-Scania, Ericsson, Volvo Flygmotor and FFV Aerotech in JAS 39 Gripen programme; continues to act as contractor for Försvarets Materielverk (Defence Materiel Administration, FMV) and co-ordinated JAS 39 Gripen programme within Sweden. International marketing of JAS 39 Gripen is now responsibility of Saab and BAE Systems, acting jointly since November 1995; current members of JAS Industry Group are subcontractors.

Workshare on export orders would be Saab 55 per cent/BAE 45 per cent under agreement signed 12 June 1995. BAE involvement also includes manufacture of main landing gear, the first of which was delivered from the Brough factory in April 1996 and incorporated in 49th production Gripen. From May 1999, BAE also responsible for assembly of wing attachment unit and associated component join-up for two-seat aircraft (single-seaters to follow).

UK and Swedish export guarantee departments signed an agreement in August 1996 to finance any third-party sales. A contract of 16 November 1995 provides for Danubian Aircraft Company of Hungary to manufacture Gripen components at Tököl as part of a wider Swedish-Hungarian defence information exchange agreement signed 18 December 1995, which may lead to Hungarian purchase of the JAS 39. First Hungarian parts (tailcone fittings) delivered April 1996. Further Swedish-Hungarian agreement of 22 January 1997 provided for comprehensive industrial offsets in the event of a Gripen purchase. First South African parts – Denel-built assembly set for stores pylon – handed over April 2000. Polish part manufacture (metal subassemblies) began July 2000 at PZL (Polish Aircraft Factory).

Separate operations include Saab Aircraft Leasing AB and Saab Aircraft AB (type certificate holder for Saab 340 and Saab 2000, production of which was terminated in 1999).

Saab Aerospace is also involved in Airbus A380 and A400M programmes.

UPDATED

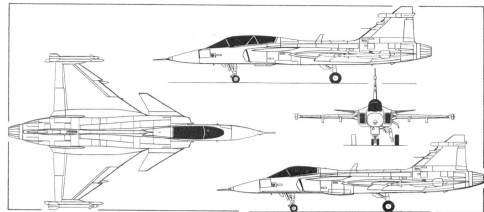

JAS 39A Gripen multirole combat aircraft for the Swedish Air Force, with additional side view (top) of two-seat JAS 39B *(Dennis Punnett/Jane's)*

SAAB JAS 39 GRIPEN

English name: Griffin

TYPE: Single-seat all-weather, all-altitude interceptor, attack and reconnaissance aircraft.

PROGRAMME: Funded definition and development began June 1980; initial proposals submitted 3 June 1981; government approved programme 6 May 1982; initial FMV development contract 30 June 1982 for five prototypes and 30 production aircraft, with options for next 110; overall go-ahead confirmed second quarter 1983; first test runs of RM12 engine January 1985; Gripen HUD first flown in Viggen testbed February 1987; study for two-seat JAS 39B authorised July 1989.

First of five single-seat prototypes (39-1) rolled out 26 April 1987; made first flight 9 December 1988 but lost in landing accident after fly-by-wire problem 2 February 1989; subsequent first flights 4 May 1990 (39-2), 20 December 1990 (39-4), 25 March 1991 (39-3) and 23 October 1991 (39-5); the 2,000th Gripen sortie was flown (by 39-4) on 22 December 1995; modified Viggen (37-51) retired at end of 1991 after assisting with avionics trials (nearly 250 flights); two single-seat fatigue test airframes (39-51 discarded 1993; 39-52 began 16,000 hour programme, August 1993 and achieved 8,000 hours in early 1996). Second production batch (Lot 2: 110 aircraft) approved 3 June 1992; first production Gripen (39101) made first flight 10 September 1992 and joined test programme in lieu of 39-1; flight test programme in 1995-96, included high-AoA (at least 28° achieved) and spin trials by 39-2 and trials of an APU (for Lot 2 production) by 39-4. All development work in the original (Lot 1) contract had been completed by late 1996; total programme was over 1,800 hours in 2,300 sorties by six aircraft. By 1996 had demonstrated M1.08 cruise without reheat. Follow-on trials with mockup aerial refuelling probe conducted by 39-4 on eight sorties between 2 and 17 November 1998 from RAF VC10 K. Mk 4. Captive flight of two KEPD 150 Taurus SOMs on inner wing pylons of 39145 of F7 conducted on 27 August 1998. Live Raytheon AIM-120 AMRAAM firings by 39-5 in April 1998.

First production aircraft for Swedish Air Force (39102) made first flight 4 March 1993 and was handed over to FMV 8 June 1993; flight control software modified following loss of 39102 in crash on 8 August 1993 and installed from December 1994; further software upgrade to new-generation P11 standard (introducing 11 filters to prevent pilot-induced oscillations) first flown 22 March 1995 in trials aircraft and installed in test Gripens from late 1995; in production aircraft built after early 1996 (and retrofitted); modified control stick introduced with production aircraft 39108 (first flight 11 April 1995). R12:3 flight control software under trial by late 1999, bringing Gripen up to original design goals.

PP12 display processor for Lot 2 colour displays first flown August 1995. Thrust-vectoring under consideration for Gripen; first stage is proposed participation in further trials programme of Rockwell/DASA X-31. JAS 39B prototype rolled out 29 September 1995; first flight 29 April 1996.

Initial 30 production JAS 39As delivered 1993-96 (five in 1994; six in 1995, comprising 39108 to 112 and 120; and last of balance – 39130 – on 13 December 1996); deliveries of second lot of 110 began 19 December 1996, and due for completion in 2002; Swedish parliament authorised third batch of 64 on 13 December 1996; contract formally placed on 26 June 1997; deliveries between 2003 and 2007. Saab contracted on 24 November 1999 to supply new EWS 39 defensive aids suite, designed by Ericsson Saab Avionics and with substantial CelsiusTech content. In same month, Gripen made first flight with FADEC, as destined for Lot 3 aircraft.

First unit is F7 Wing at Såtenäs; maintenance training begun May 1994 at Linköping; conversion scheduled to begin 1 October 1995 but postponed to 1996, with pilot training centre at Såtenäs officially opened 9 June 1996; Gripen IOC achieved September 1997, following three-week field exercise by 2/F7 Squadron. Last of F7's previous Viggens withdrawn in October 1998. By mid-2000 SwAF Gripens had flown 16,000 sorties and 12,000 hours.

Gripen being promoted in several fighter competitions, as described in Customers section. On 18 November 1998, it was announced that the aircraft had been selected for purchase by the South African Air Force and this was confirmed by contract signature on 3 December 1999.

CURRENT VERSIONS: **JAS 39A**: Standard single-seater. *Description applies to JAS 39A except where indicated.*

JAS 39B: Two-seater (Gripen SK), with 0.655 m (2 ft 1¾ in) fuselage plug and lengthened cockpit canopy. Primary roles conversion and tactical training, but also combat-capable. Avionics essentially as for JAS 39A and both cockpits identical, except no HUD in rear; instead, HUD image from front seat can be presented on flight data display in rear cockpit. Boosted environmental control system (ventral air intake replaces twin scoops on fuselage sides); inflatable airbag protects rear occupant during pre-ejection canopy fracturing. Reduced fuel; no internal gun.

Prototype entered final assembly 1 September 1994; first flight (39800) 29 April 1996; first production two-seater (39801) completed final assembly on 29 February 1996 and flew on 22 November; production deliveries began with 39802 in mid-1998; two-seat Gripens introduced to training syllabus in late 1999.

Fatigue test specimen (39-71) also built; began a simulated 12,000 hour programme in February 1996.

JAS 39C and D: New features under consideration for Swedish Lot 3 aircraft may be sufficient to warrant revised designations for JAS 39A and B; likely improvements include FADEC, helmet-mounted sight, new (Modular Airborne Computer System) processor for PS-05/A radar, Saab Dynamics IR-OTIS IR search and tracking system and enhanced EW systems. IR-OTIS tested on Saab Viggen in 1997. This and other items could be retrofitted to Lot 2 aircraft. An electronically scanned radar antenna is under development for a potential Gripen MLU in 2010.

Gripen dispersed (road) operations *(Peter Liander/Saab)*

Instrumentation for Lot 3 Gripen, showing enlarged screens

JAS 39X: Potential export version; developed by BAe; likely to include improvements considered for Swedish Lot 3 aircraft as well as colour cockpit displays, integrated EW suite, NATO standard radios, air-to-air refuelling probe (above port air intake trunk), OBOGS, uprated environmental control system and NATO pylons.

CUSTOMERS: Swedish Air Force requirement was originally 280, to equip 16 squadrons, second eight replacing JA and AJS Viggens; this reduced when third batch authorisation (December 1996) covered only four squadrons, thus amending requirement to 204 (including 28 JAS-39B two-seat versions). First 30 ordered with prototypes and full-scale development 30 June 1982; next 110 (of which first – 39131 – flew on 20 August 1996) include 14 JAS 39Bs; black radomes on 39109 to 39127; low-visibility markings from 39128, augmented by light grey (previously medium grey) radomes from 39131. Third lot of 64 (also including 14 JAS 39Bs) for delivery between 2003 and 2007; could be to JAS 39C/D standard with engine and computer upgrades.

Total 89 (including five two-seat) delivered by July 2000 and progressing at 17 per year. First operational squadron was 2/F7, September 1997, followed by 1/F7 12 months later. F10 at Angelholm, followed in 1999-2000, conversion of pilots (by F7) having begun on 11 March 1999, followed by arrival of first two aircraft on 30 September 1999; first squadron (2/F10) operational September 2000, with second following in mid-2001. However, F10 to disband in 2003 and transfer its equipment to F17 at Kallinge. F4 at Froson-Ostersund and F21 at Luleå equipping simultaneously, from 2001 onwards; initial reconnaissance unit (in F21) equips in 2001.

Exports of some 250 anticipated over 20 years from 1996; by then, Saab and BAE engaged in 12 export campaigns; Hungary interested in some 30 aircraft; presentations made to Austria, Brazil, Czech Republic, Chile and Poland in 1996-97. Promotion in Philippines, Slovenia and South Africa, 1997. Final assembly of any Polish Gripens would be by PZL Aircraft Factory, under 1997 agreement.

South Africa selected Gripen on 18 November 1998 for planned purchase of 28; order placed 15 September 1999 for nine two-seat Gripens, with option on further 19 of single-seat version; formal signature 3 December 1999, but first deliveries postponed at that time from 2002 to 2007, with last single-seater following in 2012. However, first aircraft will be built in 2002 for trials in Sweden and South Africa.

COSTS: Planned cost of SEK25.7 billion in 1982, increased to SEK48.5 billion in 1991 by inflation; FMV has reported total cost increase of SEK9.3 billion for period 1982-2001; total budget SEK60.2 billion decided by Swedish Parliament 1993. SEK22.7 billion spent by 1 July 1993, including SEK14.5 billion to IG JAS. Up to SEK300 million approved late 1991 for JAS 39B development. Costs for 300-aircraft programme (subsequently reduced to 204) estimated in 1994 as SEK15 billion for development and SEK48 billion for production. South African purchase of 28 estimated programme cost R10.9 billion (US$1.9 billion) (1998).

DESIGN FEATURES: Intended to replace AJ/SH/SF/JA/AJS versions of Saab Viggen, in that order, and remaining J 35 Drakens; to operate from 800 m (2,625 ft) Swedish V90 road strips; simplified maintenance and quick turnaround with groundcrew comprising one technician and five conscripts.

By 2000, Gripen demonstrating 10 mmh/fh; 7.6 hours MTBF readiness rate; and 10 minute turnround in fighter configuration, 20 minutes in attack role.

Mid-mounted delta wing, with squared tips for missile rails, has three-section leading-edge, of which inboard and outboard sections sweptback at 55° and centre at 52°; foreplanes, independently movable, have leading-edge sweep of approximately 58°. Sweptback fin carries various antennas. Moderately high cockpit. Near-rectangular engine air intakes, each with splitter plate.

FLYING CONTROLS: BAE Systems (Lot 1) or Lockheed Martin SA11 (Lot 2) triplex fly-by-wire system with Moog electrically signalled servo valves on powered control units; Saab Combitech aircraft motion sensors and throttle actuator; mini-stick and HOTAS controls.

Leading-edge with dog-tooth and automatic flaps (one inboard/one outboard of dog-tooth, inner one outboard of canard) on Lucas Aerospace 'geared hinge' rotary actuators; two elevon surfaces at each trailing-edge; individual all-moving foreplanes, which also 'snowplough' for aerodynamic braking after landing; airbrake each side of rear fuselage.

Saab Gripen armed with RBS 15F anti-ship missiles (inboard), Rb75 Maverick and Rb74 Sidewinder (Peter Liander/Saab)

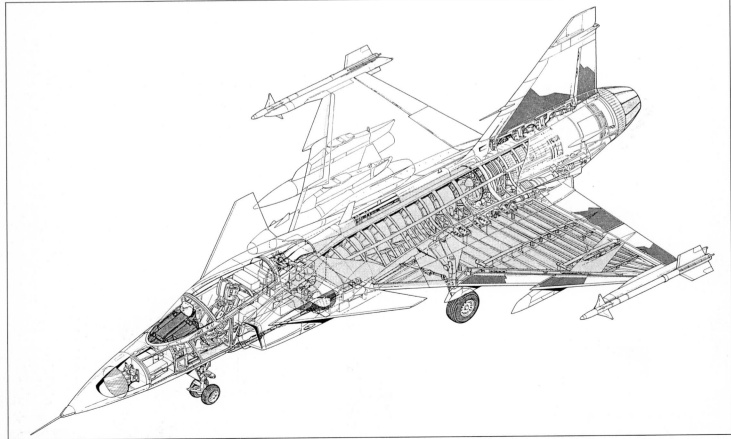

Interior details of the JAS 39A Gripen

STRUCTURE: Airframe is 60 per cent aluminium by weight, 6 per cent titanium and 5 per cent other metals; most of remainder is carbon fibre. First 3½ carbon fibre wing sets produced by BAE; all subsequent carbon fibre parts made by Saab, including wing boxes, foreplanes, fin and all major doors and hatches. BAE produces landing gear and, from 1999, responsible for assembly of centre fuselage for two-seat version, with single-seat to follow.

LANDING GEAR: AP Precision Hydraulics retractable tricycle gear, single mainwheels retracting hydraulically forward into fuselage; steerable twin-wheel nose unit retracts rearward. Goodyear wheels. Carbon disc brakes and ABS anti-skid units. Nosewheel braking. Entire gear designed for high rate of sink. Mainwheel tyres 25.5×8.0-14 (16 ply); nosewheel 14×5.5-6 (8 ply).

POWER PLANT: One General Electric/Volvo Flygmotor RM12 (F404-GE-400) turbofan, rated initially at approximately 54 kN (12,140 lb st) dry and 80.5 kN (18,100 lb st) with afterburning. RM12UP version, in Lot 3 aircraft, incorporates FADEC, improved flame holder and redesigned turbine. Fuel in integral tanks in fuselage and wings. Intertechnique fuel management system; Dowty fuel health monitoring system for emergency (leak/battle damage) conservancy management. Optional FRL telescopic, retractable, hydraulically actuated in-flight refuelling probe mounted in port engine air intake; available on export aircraft.

ACCOMMODATION: Pilot only in JAS 39A, on Martin-Baker Mk 10L zero/zero ejection seat. Hinged canopy (opening sideways to port) and one-piece windscreen by Lucas Aerospace. Two seats in tandem in JAS 39B; command sequence in two-seat aircraft ejects rear occupant first, simultaneously inflating an airbag between the two cockpits to protect the rear pilot from Perspex splinters.

SYSTEMS: Hymatic environmental control system for cockpit air conditioning, pressurisation and avionics cooling. Hughes-Treitler heat exchanger. Two hydraulic systems, with Dowty equipment and Abex pumps. Hamilton Sundstrand main electrical power generating system (40 kVA constant speed, constant frequency at 400 Hz) comprises an integrated drive generator, generator control unit and current transformer assembly. Lucas Aerospace auxiliary and emergency power system, comprising gearbox-mounted turbine, hydraulic pump and 10 kVA AC generator, to provide auxiliary electric and hydraulic power in event of engine or main generator failure. In emergency role, the turbine is driven by engine bleed or APU air; if this is not available, the stored energy mode, using thermal batteries, is selected automatically. Microturbo APU and air turbine starter for engine starting, cooling air and standby electrical power. Optional OBOGS on export aircraft. Lot 3 Gripens have single Ericsson Saab Avionics GECU general electronic control unit, replacing previous three controllers for air, fuel and hydraulic systems.

AVIONICS: *Comms:* CelsiusTech dual VHF/UHF transceivers and IFF in early aircraft; Rohde & Schwarz Series 6000 in third production lot and last 20 Lot 2 aircraft; option for installation in entire fleet; Series 6000 is element of tactical radio system (TARAS), providing secure communications via Data Link 39. Export aircraft to have Avitronics (Grintek) GUS 1000 audio management system.

Radar: Ericsson/BAE PS-05/A multimode pulse Doppler target search and acquisition (lookdown/shootdown) radar (weight 156 kg; 344 lb). For fighter missions, system provides fast target acquisition at long range; search and multitarget track-while-scan; quick scanning and lock-on at short ranges; and automatic fire control for missiles and cannon. In attack and reconnaissance roles, operating functions are search against sea and ground targets; mapping, with normal and high resolution; and navigation.

Flight: Ericsson SDS 80 central computing system (D80 computer, Pascal/D80 high-order language and programming support environment; upgraded D80E computer flown mid-1994 and introduced from 39108); three MIL-STD-1553B databusses, one of which links flight data, navigation, flight control, engine control and main systems; Honeywell laser INS and radar altimeter; Nordmicro air data computer. BAE three-axis strapdown gyromagnetic unit provides standby attitude and heading information. Navigation data fusion in prospect via NINS (new integrated navigation system), under development by 1998; this to be followed by NILS (new integrated landing system) for autonomous Cat. I landing capability.

Saab JAS 39B (nearest) and 39A Gripens *(Katsuhiko Tokunaga)* ***2001***/0092578

Instrumentation: Ericsson EP-17 electronic display system, incorporating Kaiser wide-angle HUD and using advanced diffraction optics to combine symbology and video images; PP1 or PP2 display processors (PP12 in Production Lot 2 makes colour imagery possible, but this facility not required on Swedish Lot 2 aircraft); three Ericsson CRT HDDs, each 120 × 150 mm (4¾ × 6 in), but 152 × 203 mm (6 × 8 in) in Lot 3. Left-hand (flight data) HDD normally replaces all conventional flight instruments; central display shows computer-generated map of area surrounding aircraft with tactical information superimposed; right-hand CRT is a multisensor display showing information on targets acquired by radar, FLIR and weapon sensors. Minimum of conventional analogue instruments for back-up only.

Mission: Zeiss Optronik Litening targeting and navigation pod under consideration in 2000 for carriage under starboard air intake trunk, forward of wing leading-edge, providing heat picture of target on right-hand HDD. IRST under development for possible later installation ahead of windscreen, slightly offset to port. CelsiusTech datalink shares information with up to four aircraft simultaneously.

Vinten Vicon 70 Srs 72C modular reconnaissance pod available for export aircraft.

Self-defence: CelsiusTech RWR in early aircraft replaced by second-generation (Gen 2) version in 2000; Gen 2 also for export. CelsiusTech countermeasures, including chaff/flare and jamming. Ericsson Saab Avionics electronic warfare suite in export version. New EWS 39 suite ordered 1999 to replace existing system in Swedish aircraft is similar to export equipment; includes CelsiusTech BO2D towed radar decoy.

ARMAMENT: Internally mounted 27 mm Mauser BK27 automatic cannon in port side of lower front fuselage and two wingtip-mounted Rb74 (AIM-9L) Sidewinder IR AAMs standard. (No internal gun in JAS 39B.) Six other external hardpoints (two under each wing, one on centreline and one below starboard air intake trunk) for short- and medium-range air-to-air missiles such as Rb74, MICA or Rb99 (AIM-120) AMRAAM; air-to-surface missiles such as Rb75 (Maverick); anti-shipping missiles such as Saab RBS 15F; DWS 39 munitions dispenser; KEPD 150 SOM; conventional or retarded bombs; air-to-surface rockets; or external fuel tanks. New weapons under consideration for integration include IRIS-T, Brimstone and Meteor.

DIMENSIONS, EXTERNAL:

Wing span, incl missile rails	8.40 m (27 ft 6¾ in)
Length, excl pitot tube: JAS 39A	14.10 m (46 ft 3 in)
JAS 39B	14.755 m (48 ft 5 in)
Height overall	4.50 m (14 ft 9 in)
Wheel track	2.40 m (7 ft 10½ in)
Wheelbase: JAS 39A	5.20 m (17 ft 0¾ in)
JAS 39B	5.90 m (19 ft 4¼ in)

WEIGHTS AND LOADINGS:

Operating weight empty: JAS 39A	6,622 kg (14,600 lb)
JAS 39B	8,000 kg (17,637 lb)
Internal fuel weight	2,268 kg (5,000 lb)
T-O weight, clean	approx 8,500 kg (18,740 lb)
Max T-O weight with external stores	approx 13,000 kg (28,660 lb)

PERFORMANCE:

Max level speed	supersonic at all altitudes
T-O and landing strip length	approx 800 m (2,625 ft)
Combat radius	approx 432 n miles (800 km; 497 miles)
g limit	+9

UPDATED

SWITZERLAND

ABS

ABS AIRCRAFT AG

Bühlstrasse 2, CH-8700 Küsnacht/ZH
Tel: (+41 1) 274 20 99
Fax: (+41 1) 274 20 99

VERIFIED

ABS AIRCRAFT ABS RF-9

TYPE: Motor glider.

PROGRAMME: French Fournier RF-9 adapted for production; rights originally obtained by ABS, but programme taken over by Herbert Gomolzig in Germany in 1995; former ABS demonstrator purchased and completed as D-KHGO; first flight 30 June 1995; German production was due to be launched in late 1997, but did not take place; ABS Aircraft reacquired production rights during 1999; first delivery scheduled for early 2000. More powerful versions, suitable for glider-towing, were in development during 1999.

COSTS: €100,000 (1999).

DESIGN FEATURES: Outer wing panels fold inward for hangarage.

FLYING CONTROLS: Conventional and manual. Wing-mounted Schempp-Hirth-type spoilers.

STRUCTURE: All-wood.

LANDING GEAR: Tailwheel layout with inward-retracting wide-track mainwheels; foot-operated wheel brakes.

POWER PLANT: One 59.6 kW (79.9 hp) Rotax 912A-3 water-cooled four-cylinder four-stroke driving a Hoffmann HO-V62 hydraulic constant-speed and feathering propeller.

ACCOMMODATION: Side-by-side seating with adjustable seats and parachute recesses; heating by engine-cooling liquid and three-position fresh air intake; baggage compartment.

DIMENSIONS, EXTERNAL:

Wing span	17.30 m (56 ft 9 in)
Wing aspect ratio	16.6
Width, wings folded	10.00 m (32 ft 9¼ in)

SWITZERLAND: AIRCRAFT—ABS to ACEAIR

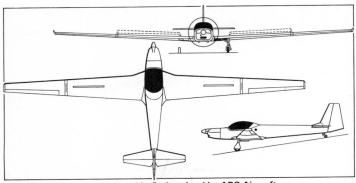

Fournier RF-9, to be produced in Switzerland by ABS Aircraft
(Mike Keep/Jane's)

ABS-built RF-9 formerly used as a demonstrator by Gomolzig
(Paul Jackson/Jane's)

Length overall	8.06 m (26 ft 5¼ in)
Height overall	1.93 m (6 ft 4 in)
AREAS:	
Wings, gross	18.00 m² (193.7 sq ft)
WEIGHTS AND LOADINGS:	
Manufacturer's weight empty	500 kg (1,102 lb)
Max T-O weight	750 kg (1,653 lb)
Max wing loading	41.4 kg/m² (8.47 lb/sq ft)
Max power loading	12.58 kg/kW (20.68 lb/hp)
PERFORMANCE, POWERED:	
Never-exceed speed (V$_{NE}$)	135 kt (250 km/h; 155 mph)
Max level speed	119 kt (220 km/h; 137 mph)
Cruising speed	102 kt (190 km/h; 118 mph)
Stalling speed	38 kt (70 km/h; 44 mph)
Max rate of climb at S/L	216 m (708 ft)/min
Service ceiling	6,500 m (21,320 ft)
T-O to 15 m (50 ft)	400 m (1,312 ft)
Landing from 15 m (50 ft)	300 m (985 ft)
Range, cruising with engine	486 n miles (900 km; 559 miles)
PERFORMANCE, UNPOWERED:	
Best glide ratio	29
Min rate of sink	0.80 m (2.62 ft)/s
NOISE LEVEL:	
To German LSL Chapter 6	55 dBA

UPDATED

ACEAIR

ACEAIR SA
Via Cantonale 35b, CH-6928 Manno
Tel: (+41 91) 605 55 46
Fax: (+41 91) 605 55 19
e-mail: info@aeris.ch
Web: http://www.aeris.ch
CHAIRMAN: Ugo Wyss
GENERAL MANAGER AND CEO: Antonio Latella
PRODUCT MANAGER: Massimo Stoppa

Aceair's unorthodox Aeris 200 kitbuilt sportplane was planned to make its debut at the EAA's Air Venture Oshkosh exhibition in mid-2001.

NEW ENTRY

ACEAIR AERIS 200

TYPE: Tandem-seat kitbuilt sportplane.
PROGRAMME: Concept phase began June 1997; design started 1998; engineering phase, dynamic model flight tests and wind tunnel test activity completed October 1999. Proof-of-concept construction started February 2000; debut planned for EAA Air Venture at Oshkosh, July/August 2001.
DESIGN FEATURES: Three lifting surfaces aerodynamic configuration; single pusher engine. Straight-tapered, low-swept (4° at 25 per cent chord) main wing amidships, with 2° dihedral from roots, features natural laminar flow sections based on NASA NLF(1)-0215F; main wing twist (2° 30′ washout from root chord) and planform are selected for stable low-speed characteristics. Foreplane aerofoil FX-75-141 unswept, 6° anhedral. Lifting surfaces relative setting angles optimised for low drag at normal cruising speed. Standard symmetrical wing sections on T-tail configured 30° swept horizontal surface. Separate fins above and below fuselage (lower portions serve as over-rotation bumper and propeller strike protector); rudder mounted on upper vertical fin.
FLYING CONTROLS: Conventional mechanical via push/pull rods for both roll and T-tail-mounted pitch controls, steel cables for rudder control. Ailerons and elevators have electric trim tab. Electrically actuated single-slotted flaps on main wing, synchronised with plain flaps on foreplanes; all flaps can be partially deflected for T-O; landing flap setting is 40° on main wing, 20° on foreplane (relative to wing chord).
STRUCTURE: All-composites. Main wings detachable for storage and/or transportation; C-section carbon fibre main spars and rear auxiliary spars; 30 per cent chord single-slotted flap and 20 per cent chord aileron in each wing; carbon fibre ribs. Monocoque fuselage. Primary structures built from glass prepreg fabric laid-up on CNC milled all-composites female moulds; all exterior shells of co-cured glass fibre and honeycomb sandwich. Glass fibre flaps and control surfaces. Primary foreplanes and tail assembly of carbon fibre C-spars, composites ribs and glass fibre-based sandwich skins.
LANDING GEAR: Tricycle type. Electrically actuated nose gear retracts rearward into fuselage; fixed main gear units on cantilever self-sprung glass fibre struts with Cleveland heavy-duty disc brakes and wheels. Nosewheel steerable by differential braking. Hydraulic mainwheel brakes and parking brake. Electrically actuated retractable main units optional, retracting forward into fuselage; steel struts with mechanical back-up extension system.
POWER PLANT: One Mid-West AE 110 rotary engine (78.3 kW; 105 hp), driving a three-blade constant-speed MT-Propeller MTV-7-D/157-106 pusher propeller via a carbon fibre driveshaft. Maximum propeller speed 2,500 rpm. Engine is mounted far aft of passenger cabin and has two scoop air inlets under main wingroot. Rotax 912 ULS flat-four engine (73.5 kW; 98.6 hp) optional. Fuel in single 110 litre (29.1 US gallon; 24.2 Imp gallon) integral fuselage tank.
ACCOMMODATION: Pilot and co-pilot or passenger in tandem in individual cockpits; dual sidestick controls. Separate one-piece flush-type canopy hinged on starboard side. Baggage compartment, capacity 20 kg (44 lb), under passenger seat. Both cockpits heated and ventilated. Central carbon fibre roll bar for flip-over protection.
SYSTEMS: Single voltage (14 V DC) electrical system powered by two engine-driven generators to provide power for gear and flap extension/retraction; essential bus powers flaps, trim, landing gear, standby fuel pump, landing light and other flight-essential equipment.
AVIONICS: To customer's requirement. Can be equipped to full IFR standard.
EQUIPMENT: Anti-collision strobe light in each wingtip; one white anti-collision light in tail. Ballistic parachute in central position aft of passenger seat.

DIMENSIONS, EXTERNAL:	
Wing span	8.00 m (26 ft 3 in)
Foreplane span	2.00 m (6 ft 6¾ in)
Wing aspect ratio	11.4
Foreplane aspect ratio	6.7
Length overall	6.41 m (21 ft 0¼ in)
Height over tailplane	2.64 m (8 ft 8 in)
Wheel track	1.44 m (4 ft 8¾ in)
Wheelbase	2.50 m (8 ft 2½ in)
Fuselage ground clearance at mainwheels	0.56 m (1 ft 10 in)
Propeller diameter	1.57 m (5 ft 1¾ in)
Propeller ground clearance	0.565 m (1 ft 10¼ in)
DIMENSIONS, INTERNAL:	
Cockpits: Combined length	2.80 m (9 ft 2¼ in)
Max width	0.68 m (2 ft 2¾ in)
Max height	1.00 m (3 ft 3¼ in)
AREAS:	
Wings, gross	6.20 m² (66.7 sq ft)
Foreplanes, gross	0.60 m² (6.46 sq ft)
Tailplane	1.06 m² (11.41 sq ft)
Fins (total)	1.35 m² (14.53 sq ft)
Ailerons (total, incl tabs)	0.30 m² (3.23 sq ft)
Rudder	0.25 m² (2.69 sq ft)
Elevators (total incl tabs)	0.43 m² (4.63 sq ft)
WEIGHTS AND LOADINGS (with AE 110):	
Weight empty	300 kg (662 lb)
Payload with max fuel	200 kg (441 lb)
Max T-O weight	580 kg (1,278 lb)
Max wing loading	93.5 kg/m² (19.16 lb/sq ft)
Max power loading	7.40 kg/kW (12.17 lb/hp)
PERFORMANCE (estimated with AE 110):	
Never-exceed speed (V$_{NE}$)	200 kt (370 km/h; 230 mph)
Max level speed	180 kt (333 km/h; 207 mph)
Manoeuvring speed (V$_A$)	140 kt (259 km/h; 161 mph)
Max cruising speed at 3,500 m (11,480 ft)	160 kt (296 km/h; 184 mph)
Stalling speed at S/L: flaps up	61 kt (113 km/h; 71 mph)
flaps down (two-seat)	56 kt (104 km/h; 65 mph)
flaps down (solo)	50 kt (93 km/h; 58 mph)
Max rate of climb: at S/L	475 m (1,560 ft)/min
at 3,500 m (11,480 ft)	256 m (840 ft)/min
Service ceiling	4,420 m (14,500 ft)
T-O run at S/L	290 m (950 ft)
T-O to 15 m (50 ft) (JAR 23.59) at S/L	450 m (1,475 ft)
Landing from 15 m (50 ft) (JAR 23.75) at S/L	590 m (1,935 ft)
Landing run at S/L	270 m (885 ft)
Max range at 3,500 m (11,480 ft) at max cruise power, 45 min reserves	877 n miles (1,625 km; 1,009 miles)
Max endurance at 3,500 m (11,480 ft), at max cruise power, 45 min reserves	5 h 45 min
g limits	+6/−3

Artist's impression of Aceair Aeris 200

NEW ENTRY

FFA

FFA FLUGZEUGWERKE ALTENRHEIN AG
CH-9423 Altenrhein
Tel: (+41 71) 858 51 11
Fax: (+41 71) 858 53 30
PRESIDENT: Edwin Naef
PRODUCT SUPPORT AND MARKETING: Bruno Widmer

Originated as Swiss branch of German Dornier company, becoming all Swiss in 1948; privately owned since 1991. Activities include producing spares for Swiss-built Boeing F-18 Hornet and Northrop F-5E/F; parts for Airbus and Pilatus PC-12; overhaul and maintenance for general aviation; and subcontract work for various foreign aircraft manufacturers. Workforce (1998) about 90.

Manufacture of the AS 202 Bravo has been transferred to FFA Bravo (which see).

UPDATED

FFA BRAVO

FFA BRAVO AG
Flughafenstrasse, CH-9423 Altenrhein
Tel: (+41 52) 657 35 83
Fax: (+41 52) 657 24 05
e-mail: b.widmer@ffa-aircraft.com
CHAIRMAN: Carl M Holliger
TECHNICAL DIRECTOR: Bruno Widmer
VICE-PRESIDENT, MARKETING: Rolf Boehm

Company founded in 1999 to manufacture and distribute AS 202 Bravo lightplane, previously managed by FFA Flugzeugwerke Altenrhein AG (which see). Subsidiary, FFA Bravo (Germany) GmbH, will produce the related Bravo 3000.

NEW ENTRY

FFA BRAVO AS 202 BRAVO

TYPE: Aerobatic two-seat lightplane.
PROGRAMME: Originated in joint venture with SIAI-Marchetti. Swiss prototype (HB-HEA) first flew 7 March 1969; Italian prototype (I-SJAI), with 85.8 kW (115 hp) Lycoming, on 7 May 1969; second Swiss prototype (HB-HEC) on 16 June 1969; first production aircraft (HB-HEE) 22 December 1971. Type approval 15 August 1972 for initial AS 202/15 version with 112 kW (150 hp) Lycoming O-320-E2A; FAA certification 16 November 1973; programme assumed entirely by FFA in 1973; marketing via subsidiary company, Repair AG.
Uprated AS 202/18A first flew (HB-HEY) 22 August 1974; certified in Switzerland 12 December 1975; in USA 17 December 1976. Higher-performance AS 202/26A, with Lycoming AEIO-540, flew (conversion of 18A1 HB-HFY) 1978, but not then certified.
AS 202/32TP Turbine Bravo flew (HB-HFJ) 20 July 1992 with 313 kW (420 hp) Rolls-Royce 250-B17D turboprop, three-blade, constant-speed propeller and additional fuel in wings. Plans for 1995 certification were not realised and project discontinued.
Programme transferred to parent company FFT (Flugzeug- und Faserverbund Technologie) at Mengen, Germany, 1 March 1990; FFT placed in receivership 30 September 1992, temporarily ending development of further upgraded version known as Eurotrainer (see following entry). Programmes for 18A and 26A versions reinstated by FFA Bravo AG in 1999.
CURRENT VERSIONS: **AS 202/18A:** Two/three-seat aerobatic version, adopted by several air forces. Baseline (**A1**) version followed by **A2** with higher MTOW and max landing weights, extended canopy and electrical trimming; **A3**, as A2, but with 24 V electronics; and **A4**, as A2, but with special instrumentation for British Aerospace Flying College, by which named **Wren**. UK CAA certified A4 on 10 December 1987 and FAA followed on 4 February 1993.
Description applies generally to AS 202/18A4.
AS 202/26A: Optimised for greater performance, especially in hot-and-high locations. Powered by 194 kW (260 hp) Textron Lycoming AEIO-540-D4B5 flat-six with two-blade, constant-speed, metal propeller. Electrical system as 18A3; optional air conditioning.
CUSTOMERS: Total of 170, including prototypes, built by 1992. Further 15 under production in 2000. See table.
DESIGN FEATURES: Typical low-wing, fixed-gear lightplane configuration; angular flying surfaces; sweptback vertical tail.
Wing section NACA 63_2618 (modified) at centreline, 63_2415 at tip; thickness/chord ratio 17.63 per cent at root,

15 per cent at tip; dihedral 5° 43′ from roots; incidence 3°; quarter-chord sweepback 0° 40′.
FLYING CONTROLS: Conventional and manual. Single-slotted ailerons and mass-balanced rudder; ground-adjustable tab on each aileron, electrically actuated trim tab in starboard elevator; 18A4 also has electrically actuated rudder tab. Single-slotted flaps; fixed incidence tailplane.
STRUCTURE: All-aluminium fail-safe except for glass fibre fairings and engine cowling; single-spar wings and tail with riveted honeycomb laminate skins.
LANDING GEAR: Non-retractable tricycle type, with steerable nosewheel. Rubber cushioned FFA shock-absorber struts. Mainwheel tyres size 6.00-6; nosewheel tyre size 5.00-5. Tyre pressure (all units) 2.41 bars (35 lb/sq in). Independent hydraulically operated disc brake on each mainwheel.
POWER PLANT: One 134 kW (180 hp) Textron Lycoming AEIO-360-B1F flat-four engine, driving a Hartzell HC-C2YK-1BF/F7666A-2 two-blade constant-speed propeller; Hoffmann three-blade propeller optional. Two wing leading-edge rubber fuel tanks with total capacity of 170 litres (44.9 US gallons; 37.4 Imp gallons). Refuelling point above each wing. Starboard tank has additional flexible fuel intake for aerobatics. Christen 801 fully aerobatic oil system, capacity 7.6 litres (2.0 US gallons; 1.6 Imp gallons).
ACCOMMODATION: Seats for two persons side by side in Aerobatic versions, under rearward-sliding jettisonable transparent canopy. Space at rear in Utility versions for third seat or 100 kg (220 lb) of baggage. Dual controls, cabin ventilation and heating standard.
SYSTEMS: Hydraulic system for brake actuation. One 12 V 60 A engine-driven alternator (24 V in A3) and one 25 Ah battery provide electrical power for engine starting, lighting instruments, communications and navigation installations; 28 V electrical system optional.
AVIONICS: To customer's requirements. Provision for VHF radio, VOR, ADF, Nav-O-Matic 200A autopilot, blind-flying instrumentation or other special equipment.

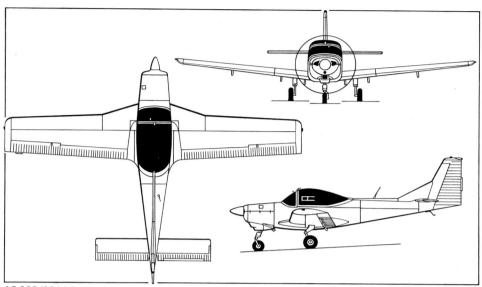

AS 202/26A1 Bravo two/three-seat trainer and sporting aircraft *(Mike Keep/Jane's)* 2001

EQUIPMENT: Clutch and release mechanism for glider towing optional.
DIMENSIONS, EXTERNAL:
Wing span	9.78 m (32 ft 1 in)
Wing aspect ratio	6.90
Length overall	7.50 m (24 ft 7¼ in)
Height overall	2.81 m (9 ft 2¾ in)
Tailplane span	3.67 m (12 ft 0½ in)
Wheel track	2.25 m (7 ft 4½ in)
Wheelbase	1.85 m (6 ft 0¾ in)
Propeller diameter	1.88 m (6 ft 2 in)

AREAS:
Wings, gross	13.86 m² (149.2 sq ft)

WEIGHTS AND LOADINGS:
Weight empty, equipped: 18A4	710 kg (1,565 lb)
26A1	793 kg (1,748 lb)
Max useful load (incl fuel):	
18A4, Aerobatic	177 kg (390 lb)
18A4, Utility	248 kg (546 lb)
26A1	467 kg (1,029 lb)
Max T-O weight:	
18A4, Aerobatic	1,010 kg (2,226 lb)
18A4, Utility	1,080 kg (2,380 lb)
26A1, Aerobatic	1,140 kg (2,513 lb)
26A1, Utility	1,260 kg (2,778 lb)
Max landing weight:	
26A1	1,200 kg (2,645 lb)
Max wing loading:	
18A4, Aerobatic	72.9 kg/m² (14.92 lb/sq ft)
18A4, Utility	77.1 kg/m² (15.96 lb/sq ft)
26A1, Aerobatic	82.25 kg/m² (16.85 lb/sq ft)
26A1, Utility	90.91 kg/m² (18.62 lb/sq ft)
Max power loading:	
18A4, Aerobatic	7.53 kg/kW (12.37 lb/hp)
18A4, Utility	7.84 kg/kW (12.86 lb/hp)
26A1, Aerobatic	5.88 kg/kW (9.67 lb/hp)
26A1, Utility	6.50 kg/kW (10.68 lb/hp)

PERFORMANCE (18A4, at Utility category max T-O weight):
Never-exceed speed (V_{NE})	173 kt (320 km/h; 199 mph)

AS 202 PRODUCTION

Country of registry	Prototype	15	15-1	18A	18A1	18A2	18A3	18A4
Indonesia (Air Force)							40	
Iraq (Air Force)						48		
Italy (SIAI-Marchetti)	1							
Morocco (Air Maroc)				4	2			
(Air Force)						10		
Oman (Air Force)				4				
Switzerland (FFA)	3*							
(various owners)		22	10	3	2	1		
Uganda (Air Force)					8			
UK (BAe Flying College)								11
USA (Japan Airlines)								1
Totals	**4**	**22**	**10**	**11**	**22**	**49**	**40**	**12**

*Two AS 202/15s, one AS 202/32TP

SWITZERLAND: AIRCRAFT—FFA BRAVO to PILATUS

AS 202/18A3 Bravo in service with 101 Squadron, Indonesian Air Force *(Paul Jackson/Jane's)* 2001

Prototype Eurotrainer 2000A now being used as a testbed for the Bravo 3000
2001/0092476

Max level speed at S/L	130 kt (241 km/h; 150 mph)
Max cruising speed (75% power) at 2,440 m (8,000 ft)	122 kt (226 km/h; 141 mph)
Econ cruising speed (55% power) at 3,050 m (10,000 ft)	109 kt (203 km/h; 126 mph)
Stalling speed, engine idling:	
flaps up	62 kt (115 km/h; 71 mph)
flaps down	49 kt (90 km/h; 56 mph)
Max rate of climb at S/L	244 m (800 ft)/min
Service ceiling	5,180 m (17,000 ft)
T-O run at S/L	215 m (705 ft)
T-O to 15 m (50 ft) at S/L	415 m (1,360 ft)
Landing run from 15 m (50 ft)	465 m (1,525 ft)
Landing run	210 m (690 ft)
Range with max fuel, no reserves	615 n miles (1,140 km; 707 miles)
Max endurance	5 h 30 min
g limits	+4.4/−2.2

PERFORMANCE (26A1, at Aerobatic category max T-O weight):

Max permissible diving speed (V_D)	232 kt (429 km/h; 267 mph)
Never-exceed speed (V_{NE})	214 kt (396 km/h; 246 mph)
Max level speed at S/L	146 kt (270 km/h; 168 mph)
Design manoeuvring speed (V_A)	134 kt (248 km/h; 154 mph)
Stalling speed: flaps up	60 kt (112 km/h; 70 mph)
flaps down	51 kt (95 km/h; 59 mph)
Max rate of climb at S/L	426 m (1,398 ft)/min
T-O run at S/L	170 m (558 ft)
T-O to 15 m (50 ft)	310 m (1,017 ft)
Landing from 15 m (50 ft)	460 m (1,510 ft)
Landing run	200 m (657 ft)
g limits	+6/−3

NEW ENTRY

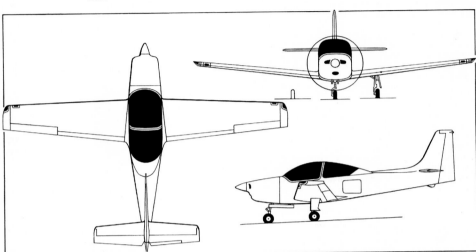

General appearance of the Eurotrainer 2000, on which the Bravo 3000 is based *(Jane's/Mike Keep)*
2001/0092499

FFA BRAVO 3000
TYPE: Four-seat lightplane.
PROGRAMME: Under development by FFA Bravo; over 200 hours of flight testing accumulated by August 2000 using prototype FFT Eurotrainer 2000A, as last described in 1992-93 *Jane's*.

Originally fitted with 200 kW (268 hp) flat-rated Textron Lycoming AEIO-540-L1B5, all-composites prototype D-EJDZ first flew on 29 April 1991; FFT ceased trading on 31 December 1992 and planned purchase of eight by Swissair failed to materialise. D-EJDZ restored to German civil register on 21 December 1999. Bravo 3000 to have 224 kW (300 hp) engine and 200 kt (370 km/h; 230 mph) cruising speed; programme launch scheduled for late October 2000; public debut planned for mid-2001.

NEW ENTRY

PILATUS

PILATUS FLUGZEUGWERKE AG
CH-6371 Stans
Tel: (+41 41) 619 61 11
Fax: (+41 41) 610 61 07
Web: http://www.pilatus-aircraft.com
CHAIRMAN: Peter Küpfer
PRESIDENT AND CEO: O J Schwenk
SENIOR VICE-PRESIDENT: O Bründler (Controlling, Finance and Administration)
VICE-PRESIDENTS:
  Dr O Masefield (R&D)
  J Roche (Marketing and Sales)
  I Gretener (General Aviation)
  M Kälin (Maintenance)
  M Zuberbühler (Logistics)
  H Niederberger (Manufacturing)
  A Waldispühl (Aircraft Assembly)
SALES MANAGER: Karl Scheuber

PC-12 MARKETING:
Pilatus Business Aircraft Ltd
Jefferson County Airport, Terminal Building B-7, 11755 Airport Way, Broomfield, Colorado 80021, USA
Tel: (+1 303) 465 90 99
Fax: (+1 303) 465 91 90
Web: http://www.pc12.com
PRESIDENT AND CEO: Angelo Fiataruolo

Formed December 1939; became part of Oerlikon-Bührle Group; Britten-Norman (Bembridge) Ltd of UK (which see), acquired on 24 January 1979, reverted to UK ownership in 1998, reducing Pilatus workforce to 980. Transairco SA of Geneva was purchased in April 1997. In December 1998, as part of a move to dispose of its aerospace activities, Oerlikon-Bührle (now renamed Unaxis) revealed that it was seeking a buyer for Pilatus. New owners were named in December 2000 as a consortium comprising Jörg Burkart, IHAG Holding, Hoffman-LaRoche pension fund and Hilmar Hilmarson of Iceland. A public share offering is expected in about 2004.

Pilatus PC-6/B2-H4 Turbo Porter of French Army Light Aviation *(Paul Jackson/Jane's)*
2001/0092514

Current products include PC-6 Turbo Porter, PC-7 and PC-7 Mk IIM Turbo Trainer, PC-9M Turbo Trainer, PC-12 and Eagle. Pilatus also overhauls Swiss Air Force aircraft. Modified (Mk II) PC-9 selected to fulfil US Air Force/Navy JPATS trainer requirement as T-6A Texan II. New PC-21 trainer, reportedly based on Texan II, in preliminary design stage.

UPDATED

PILATUS PC-6 TURBO PORTER
US Army designation: UV-20A Chiricahua
TYPE: Light utility turboprop.
PROGRAMME: First flight piston-engined prototype 4 May 1959 (see contemporary *Jane's*); PC-6/A, A1, A2, B, B1, B2 and C2-H2 with various turboprops (see 1974-75 *Jane's*) superseded by PC-6/B2-H4 introduced mid-1985. Production continuing at about eight per year for 1999-2000.
CURRENT VERSIONS: **PC-6/B2-H4:** Gross weight for FAR Pt 23 passenger carrying increased by 600 kg (1,323 lb) over immediately previous B2-H2, giving up to 570 kg (1,257 lb) greater payload for CAR 3 operations; changes include turned-up wingtips, enlarged dorsal fin, uprated mainwheel shock-absorbers, new tailwheel assembly and slight airframe reinforcement; H4 changes can be retrofitted to B2-H2. *Description applies to B2-H4.*
CUSTOMERS: Total of 522 built and at least 534 sold by mid-2000, including 45 of piston-engined versions and 92 of early turboprop versions under licence manufacture by Fairchild in USA; some 290 remain in service in more than 50 countries; two registered in 1997; 10 in 1998, of which two to Slovenian Army on 14 May 1998; and four in 1999. Sales in 2000 include one to Imaginair SA in Switzerland. Current military operators include Angola, Argentina, Austria, Chad, Colombia, Ecuador, France, Indonesia, Iran, Mexico, Myanmar, Peru, Slovenia, South Africa, Switzerland, Thailand, United Arab Emirates and US Army.
DESIGN FEATURES: Rugged, light transport with STOL performance and easily accessible, versatile cabin.

Braced, constant-chord, high wing with single strut each side; unswept fin with large dorsal fillet.

Wing section NACA 64-514 (constant) with span-increasing wingtips; dihedral 1°; incidence 2°.
FLYING CONTROLS: Conventional and manual. Mass-balanced ailerons; horn-balanced elevator and rudder. Servo tabs on

elevator; geared aileron tabs; electrically actuated double-slotted flaps; electrically actuated variable incidence tailplane.

STRUCTURE: All-metal; single-strut wing bracing.

LANDING GEAR: Non-retractable tailwheel type. Oleo shock-absorbers of Pilatus design in all units. Steerable (±25°)/lockable tailwheel. Main tyres size 11.00-12 (8 ply), pressure 1.38 bars (20 lb/sq in), or 7.50-10 (8 ply). Tailwheel with size 5.00-4 (6 ply) tyre, pressure 3.24 bars (47 lb/sq in). Goodyear hydraulic disc brakes. Pilatus wheel/ski gear and (since July 1994) plain or amphibious floats optional.

POWER PLANT: One 507 kW (680 shp) Pratt & Whitney Canada PT6A-27 turboprop (flat rated at 410 kW; 550 shp at S/L), driving a Hartzell HC-B3TN-3D/T-10178 C or CH, or T10173 C or CH, three-blade constant-speed fully feathering reversible-pitch propeller with Beta mode control; four-blade propeller optional for operations in noise-sensitive areas. Standard fuel in integral wing tanks, usable capacity 644 litres (170 US gallons; 142 Imp gallons). Two underwing auxiliary tanks, each of 242.3 litres (64.0 US gallons; 53.3 Imp gallons), available optionally. For self-ferry, up to three 189 litre (50.0 US gallon; 41.6 Imp gallon) fuel tanks can be carried in main cabin. Oil capacity 12.5 litres (3.3 US gallons; 2.75 Imp gallons).

ACCOMMODATION: Cabin has pilot's seat forward on port side, with one passenger seat alongside, and is normally fitted with six quickly removable seats, in pairs, to rear of these for additional passengers. Up to 11 persons, including pilot, can be carried in 2-3-3-3 high-density layout; or up to 10 parachutists; or two stretchers plus three attendants in ambulance configuration. Floor is level, flush with door sill, with seat rails. Forward-opening door beside each front seat. Large rearward-sliding door on each side of main cabin; fixed, external mounting step, starboard side. Optional double door, without central pillar, on port side. Hatch in floor 0.58 × 0.90 m (1 ft 10¾ in × 2 ft 11½ in),

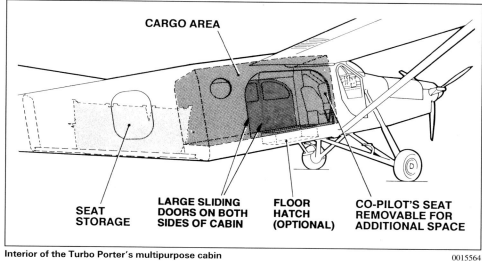

Interior of the Turbo Porter's multipurpose cabin 0015564

openable from inside cabin, for aerial camera or supply dropping. Starboard side door in rear fuselage permits stowage of six passenger seats or accommodation of freight items up to 5.0 m (16 ft 5 in) in length. Walls lined with lightweight soundproofing and heat insulation material. Adjustable heating and ventilation systems. Dual controls optional.

SYSTEMS: Cabin heated by engine bleed air. Scott 8500 oxygen system optional. 200 A 28 V starter/generator and 24 V 36 Ah (optionally 40 Ah) Ni/Cd battery.

AVIONICS: VFR or IFR packages (mainly Honeywell) to customer's requirements.

Instrumentation: Standard instruments comprise ASI, VSI, barometric altimeter, directional gyro and magnetic compass.

EQUIPMENT: Generally to customer's requirements, but can include gear for parajumping role; stretchers for ambulance role; aerial photography and survey gear. Agricultural version no longer produced.

DIMENSIONS, EXTERNAL:
Wing span	15.87 m (52 ft 0¾ in)
Wing chord, constant	1.90 m (6 ft 3 in)
Wing aspect ratio	8.4
Length overall	10.95 m (35 ft 11 in)
Height overall (to top of rudder), tail down	
	3.20 m (10 ft 6 in)
Elevator span	5.12 m (16 ft 9½ in)
Wheel track	3.00 m (9 ft 10 in)
Wheelbase	7.87 m (25 ft 10 in)
Propeller diameter	2.56 m (8 ft 5 in)
Cabin sliding doors (port/starboard, each):	
Max height	1.04 m (3 ft 5 in)
Width	1.58 m (5 ft 2¼ in)
Rear door (starboard): Height	0.80 m (2 ft 7½ in)
Width	0.50 m (1 ft 7¾ in)

DIMENSIONS, INTERNAL:
Cabin, from back of pilot's seat to rear wall:	
Length	2.30 m (7 ft 6½ in)
Max width	1.16 m (3 ft 9½ in)
Max height (at front)	1.28 m (4 ft 2½ in)
Height at rear wall	1.18 m (3 ft 10½ in)
Floor area	2.67 m² (28.6 sq ft)
Volume	3.3 m³ (107 cu ft)

AREAS:
Wings, gross	30.15 m² (324.5 sq ft)
Ailerons (total)	3.83 m² (41.23 sq ft)
Flaps (total)	3.76 m² (40.47 sq ft)
Fin	1.70 m² (18.30 sq ft)
Rudder, incl tabs	0.96 m² (10.33 sq ft)
Tailplane	4.03 m² (43.38 sq ft)
Elevator, incl tab	2.11 m² (22.71 sq ft)

WEIGHTS AND LOADINGS:
Weight empty, equipped	1,270 kg (2,800 lb)
Max fuel weight: internal	508 kg (1,120 lb)
underwing	392 kg (864 lb)
Max payload:	
with reduced internal fuel	1,130 kg (2,491 lb)
with max internal fuel	1,062 kg (2,341 lb)
with max internal and underwing fuel	
	571 kg (1,259 lb)
Max T-O weight, Normal (CAR 3):	
wheels (standard)	2,800 kg (6,173 lb)
skis	2,600 kg (5,732 lb)
Max landing weight: wheels	2,660 kg (5,864 lb)
skis	2,600 kg (5,732 lb)
Max cabin floor loading	488 kg/m² (100 lb/sq ft)
Max wing loading (Normal):	
wheels	92.9 kg/m² (19.03 lb/sq ft)
skis	86.2 kg/m² (17.67 lb/sq ft)
Max power loading (Normal):	
wheels	6.83 kg/kW (11.22 lb/shp)
skis	6.34 kg/kW (10.42 lb/shp)

PERFORMANCE (Normal category):
Never-exceed speed (V_{NE})	
	151 kt (280 km/h; 174 mph) EAS
Max cruising speed at S/L	125 kt (232 km/h; 144 mph)
Econ cruising speed at 3,050 m (10,000 ft)	
	115 kt (213 km/h; 132 mph)
Stalling speed, power off:	
flaps up	58 kt (108 km/h; 67 mph) EAS
flaps down	52 kt (96 km/h; 60 mph) EAS
Max rate of climb at S/L	308 m (1,010 ft)/min
Max operating altitude	7,620 m (25,000 ft)
Service ceiling	6,250 m (20,500 ft)
T-O run at S/L: normal and STOL	197 m (650 ft)
T-O to 15 m (50 ft) at S/L: normal	475 m (1,560 ft)
STOL	440 m (1,445 ft)
Landing from 15 m (50 ft) at S/L at MLW, with propeller reversal	315 m (1,035 ft)
Landing run at S/L at MLW, with propeller reversal	
	127 m (420 ft)
Max range at 115 kt (213 km/h; 132 mph) at 3,050 m (10,000 ft), no reserves:	
with max payload	394 n miles (730 km; 453 miles)
with standard internal fuel	
	500 n miles (926 km; 576 miles)
with standard internal and underwing fuel	
	811 n miles (1,503 km; 933 miles)
g limits	+3.58/−1.43

UPDATED

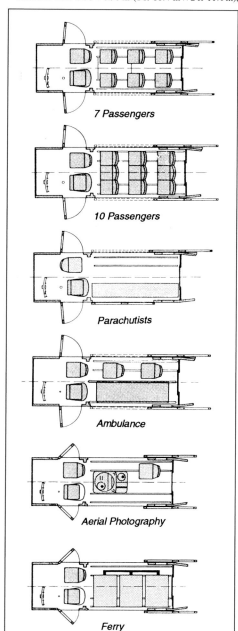

Turbo Porter alternative cabin layouts 2000/0079301

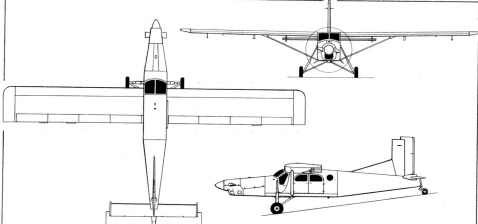

Pilatus PC-6/B2-H4 Turbo Porter with optional port-side double door (Dennis Punnett/Jane's)

PILATUS PC-7 TURBO TRAINER

Swiss Air Force designation: PC-7/CH
Uruguayan Air Force designation: AT-92
TYPE: Basic turboprop trainer.
PROGRAMME: First flight production PC-7, 18 August 1978; Swiss federal civil Aerobatic category certification 5 December 1978; first deliveries December 1978; Swiss Utility category 6 April 1979; French DGAC certification 16 May 1983; US FAR Pt 23 certification 12 August 1983; meets selected group of US military trainer category specifications. Well over 1 million hours flown by in-service PC-7s by end of 1996.
CURRENT VERSIONS: **PC-7:** Initial production version (1978 to 1997); superseded by Mk II M, but sales from stock have continued. Detailed description in 1998-99 and earlier editions of *Jane's*.
 PC-7 Mk II M: Upgraded current version; described separately.
CUSTOMERS: Total of at least 517 produced by August 2000 (451 PC-7 and 66 PC-7 Mk II: see table), including recent deliveries to US civilian owners, comprising one in 1998, one in 1999 and three in 2000. French Patrouille Adecco civil formation display team operated four with smoke generators, two of which to Patrouille Apache in 2000.

PILATUS PC-7 PRODUCTION

Customer	Qty	First aircraft	First delivery
Military:			
Angola	25	R-401	1982
Austria	16	3H-FA	Oct 1983
Bolivia	24	450	Apr 1979
Bophuthatswana	3[1]	T400	Dec 1989
Botswana	7	OD-1	Feb 1990
Brunei	4[2]	ATU-301	1997
Chile	10	N-210	May 1980
France	5	576	1991
Guatemala	12	218	Apr 1980
Iran	35	8-9901	1983
Iraq	52	5000	Jul 1980
Malaysia	46[6]	M33-01	Feb 1983
Mexico	88	501	Jul 1979
Myanmar	17	2300	1979
Netherlands	13	L-01	Jan 1989
South Africa	60[2]	2001	1994
Suriname	3[3]	111	Oct 1986
Switzerland	40	A-902	Aug 1979
UAE (Abu Dhabi)	31	901	Apr 1982
Uruguay	6	301	1992
Subtotal	**497**		
Civil:			
Pilatus	1 (P)	HB-HAO	
	2	HB-HOO	
	1	(stock)	
USA (various)	8	N701RB	
CIPRA	2[4]	F-GDME	
Martini[5]	4	HB-HMA	
Total	**517**		

[1] To South African Air Force; returned to Pilatus and resold to private owners in 1998-99
[2] Mk II
[3] No longer in service
[4] Both to Chad Air Force
[5] 1982-90; to Patrouille Ecco (1991-96); Adecco (1997-98)
[6] Nine Mk II on order; delivery 2001

UPDATED

PILATUS PC-7 Mk II M TURBO TRAINER

SAAF name: Astra
TYPE: Basic turboprop trainer.
PROGRAMME: Launched 1992 and developed as PC-7 Mk II to bid for South African Air Force requirement; two prototypes (HB-HMR and -HMS), first of which made maiden flight 28 September 1992; SAAF contract announced June 1993; first production aircraft started January 1994 and made first flight August 1994. Since 1996, PC-7 Mk IIs and PC-9s (which see) have been built from common fuselages, former being designated PC-7 Mk II M (for 'modular').
CUSTOMERS: South Africa (60); Brunei (four). No known Mk II M deliveries since early 1997, but nine ordered by Royal Malaysian Air Force late 2000 for delivery in 2001.
DESIGN FEATURES: Generally as for PC-9. Reduced-drag airframe based on that of PC-9M; higher-powered PT6A-25C engine with four-blade propeller; Martin-Baker Mk 11A ejection seats standard; optimised flying controls.
FLYING CONTROLS: Generally as for PC-9M, including trim aid device; stall strips and ventral fin added to improve stall characteristics and directional stability.
STRUCTURE: Generally as for PC-9M.
LANDING GEAR: Hydraulically actuated retractable tricycle type, with emergency manual extension. Mainwheels retract inward, nosewheel rearward. Oleo-pneumatic shock-absorber in each unit. Castoring nosewheel, with shimmy dampers. Tyres 6.50-8 (8 ply) all round, pressure 8.62 bars (125 lb/sq in) on mainwheels, 4.83 bars (70 lb/sq in) on nosewheel. Hydraulic disc brakes on mainwheels. Parking brake. Alternatively 6.50-8 main and 6.00-6 tailwheel tubless tyres. Steering ±61° by differential braking; nosewheel steering (±12°) standard. Minimum ground turning radius 11.125 m (36 ft 6 in), based on nosewheel.
POWER PLANT: One 522 kW (700 shp) (flat-rated) Pratt & Whitney Canada PT6A-25C turboprop, driving a Hartzell HC-D4N-2A four-blade constant-speed fully feathering propeller. Two integral wing fuel tanks, each of 253 litres (66.8 US gallons; 55.7 Imp gallons) usable capacity, plus a 12 litre (3.2 US gallon; 2.6 Imp gallon) aerobatic tank in fuselage, giving total internal usable capacity of 518 litres (137 US gallons; 114 Imp gallons).
 Provision for 155 litre (41.0 US gallon; 34.1 Imp gallon) long-range auxiliary tank or 246 litre (65.0 US gallon; 54.1 Imp gallon) ferry tank under each wing. Single overwing gravity fuelling point each side. Oil capacity 16 litres (4.2 US gallons; 3.5 Imp gallons). Capability for 30 seconds of inverted flight.
ACCOMMODATION: Instructor and pupil in tandem on stepped Martin-Baker Mk 11A ejection seats. One-piece canopy opens sideways to starboard. Cockpit unpressurised.
SYSTEMS: Vapour cycle environmental control system. Hydraulic system, pressure 207 bars (3,000 lb/sq in), for landing gear actuation, nosewheel steering and airbrake; system maximum flow rate 18.8 litres (4.97 US gallons; 4.14 Imp gallons)/min. Supplementary system as for PC-9M. Electrical system (28 V DC) powered by 200 A starter/generator and 24 V Ni/Cd battery. OBOGS standard.
AVIONICS (standard): *Comms:* Two VHF com radios, transponder, ELT and audio/intercom to customer's choice. UHF com optional.

Pilatus PC-7 Mk II M Turbo Trainer prototype/demonstrator

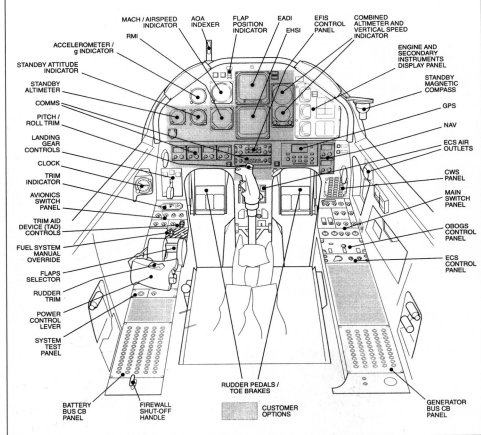

Standard front cockpit of the PC-7 Mk II M and PC-9M; rear cockpit is generally similar

PILATUS—AIRCRAFT: SWITZERLAND   487

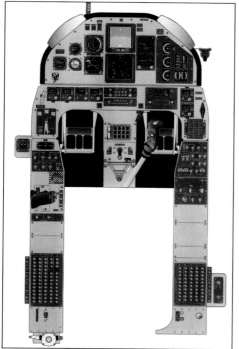
PC-7 Mk II M/PC-9M front cockpit (standard avionics)

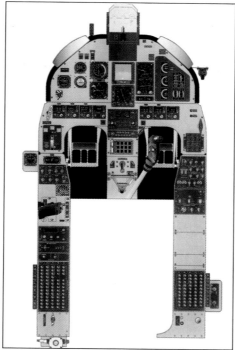
PC-7 Mk II M/PC-9M front cockpit (advanced avionics with HUD)

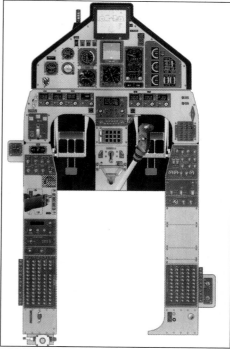
PC-7 Mk II M/PC-9M rear cockpit (advanced avionics and HUD repeater)

Flight: Range of navigation systems including ADF, DME, GPS, dual VHF nav, AHRS, RMI, EADI and EHSI available at customer's option.

Instrumentation: Standard IFR installation includes Mach/airspeed indicators; VSIs; main (encoding) and standby altimeters; standby attitude indicator; standby magnetic compass; combined aileron/rudder/elevator trim indicators; accelerometer indicators; electric clocks; and engine indication system with LCD displays. Advanced IFR with Flight Visions FV-2000 HUD, ADC and radar altimeter optional.

AVIONICS (Astra): See 1998-99 and earlier editions.
EQUIPMENT: As for PC-9M.
DIMENSIONS, EXTERNAL, INTERNAL AND AREAS: As for PC-9M.
WEIGHTS AND LOADINGS (A: Aerobatic, U: Utility category with underwing stores):

Basic weight empty, equipped	1,670 kg (3,682 lb)
Max underwing stores load	1,040 kg (2,293 lb)
Max T-O weight: A	2,250 kg (4,960 lb)
U	2,850 kg (6,283 lb)
Max ramp weight: A	2,260 kg (4,982 lb)
U	2,860 kg (6,305 lb)
Max landing weight: A	2,250 kg (4,960 lb)
U	2,750 kg (6,062 lb)
Max zero-fuel weight: A	1,900 kg (4,189 lb)
Max wing loading: A	138.1 kg/m² (28.29 lb/sq ft)
U	175.0 kg/m² (35.83 lb/sq ft)
Max power loading: A	4.31 kg/kW (7.09 lb/shp)
U	5.46 kg/kW (8.98 lb/shp)

PERFORMANCE (A and U as above):
Max operating Mach number (M_{MO}): A, U	0.60
Max operating speed (V_{MO}): A, U	300 kt (555 km/h; 345 mph) EAS
Max cruising speed:	
A at S/L	242 kt (448 km/h; 278 mph)
A at 3,050 m (10,000 ft)	251 kt (465 km/h; 289 mph)
A at 7,620 m (25,000 ft)	225 kt (417 km/h; 259 mph)
Max manoeuvring speed:	
A	210 kt (389 km/h; 241 mph) EAS
U	200 kt (370 km/h; 230 mph) EAS
Max speed with flaps and/or landing gear down:	
A, U	150 kt (277 km/h; 172 mph) EAS
Stalling speed, engine idling:	
flaps and gear up: A	75 kt (139 km/h; 87 mph) EAS
U	84 kt (156 km/h; 97 mph) EAS
flaps and gear down: A	68 kt (126 km/h; 79 mph) EAS
U	75 kt (139 km/h; 87 mph) EAS
Max rate of climb at S/L: A	866 m (2,840 ft)/min
Time to 4,575 m (15,000 ft)	6 min 55 s
Max operating altitude: A, U	7,620 m (25,000 ft)
Service ceiling: A, U	9,150 m (30,000 ft)
T-O run at S/L: A	260 m (850 ft)
T-O to 15 m (50 ft) at S/L: A	415 m (1,360 ft)
Landing from 15 m (50 ft): A	674 m (2,210 ft)
Landing run: A	336 m (1,100 ft)
Max range at long-range cruising speed, 5% and 20 min fuel reserves:	
A	810 n miles (1,500 km; 932 miles)
Max endurance at 110 kt (204 km/h; 127 mph) IAS, conditions as above: A	4 h 40 min
g limits: A	+7/−3.5
U	+4.5/−2.25

UPDATED

PILATUS PC-9M ADVANCED TURBO TRAINER

Slovene Military Aviation name: Hudournik (Swift)
TYPE: Basic turboprop trainer.
PROGRAMME: Design began May 1982; aerodynamic elements tested on PC-7 1982-83; first flights by two preproduction PC-9s: HB-HPA on 7 May 1984 and HB-HPB on 20 July 1984; aerobatic certification 19 September 1985. PC-9M introduced early 1997.
CURRENT VERSIONS: **PC-9**: Standard production version until 1997. Description in 1997-98 and earlier *Jane's*.
PC-9/A: Australian version of PC-9.
PC-9B: German target towing version of PC-9, operated for Luftwaffe by Condor Flugdienst; increased fuel for a

Hudournik conversion of PC-9M for Slovene military use *(Paul Jackson/Jane's)*

Pilatus PC-9M Advanced Turbo Trainer demonstrator

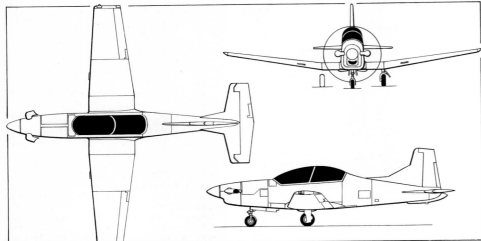

Pilatus PC-7 Mk II M and PC-9M basic/advanced trainer *(James Goulding/Jane's)*

PILATUS PC-9 PRODUCTION
(excluding Raytheon/Beech Mk II)

Customer	Version	Qty	First aircraft	First delivery
Military:				
Angola	PC-9	4	(c/n 115)	1987
Australia	PC-9/A	67[1]	A23-001	Oct 1987
Croatia	PC-9M	17[4]	054	1997
Cyprus	PC-9	2	901	1989
Iraq	PC-9	20	5801	1987
Myanmar	PC-9	10	3601	Apr 1986
Oman	PC-9M	12	426	1999
Saudi Arabia	PC-9	50	2201	Jan 1987
Slovenia	PC-9M	9[5]	L9-61	Nov 1998
Switzerland	PC-9	14[2]	C-401	1987
Thailand	PC-9	36	1/34	1991
US Army	PC-9	3[3]	91-071	1991
Subtotal		**244**		
Civil:				
Pilatus	PC-9	2	HB-HPA	
	PC-9 Mk II	1	HB-HPC	
	PC-9M	1	HB-HPJ	
Condor	PC-9B	10[6]	D-FAMT	Sep 1990
BAE	PC-9	1	ZG969	Jul 1989
Total		**259**		

[1] First 17 (after two from Pilatus) assembled by Hawker de Havilland from Swiss kits; final 48 locally built
[2] First two for evaluation; returned to Pilatus
[3] Transferred to Slovenia in 1995; now L9-51 *et seq*
[4] Plus three second-hand (051 *et seq*)
[5] Modified to Hudournik standard in Israel after delivery
[6] Plus ex-demonstrator, August 2000

3 hour 20 minute mission; two Meggitt RM-24 winches on inboard pylons with targets stowed aft of winch; TAS 06 acoustic scoring system.

PC-9 Mk II: Modified for US Air Force/Navy JPATS trainer programme (T-6A Texan II); described (with export variants) under Raytheon heading in US section.

PC-9M: Upgraded version introduced 1997 and now the standard production model. Enlarged dorsal fin fairing (as on PC-7 Mk II M) improves longitudinal stability and reduces stick forces; modified wingroot fairings improve low-speed characteristics; 'exciter strips' on wing leading-edges improve stall characteristics; new engine/propeller control system separates flight idle and ground idle modes, reducing acceleration time and improving handling.

Detailed description applies to PC-9M.

Hudournik (Swift): Dedicated weapons training version in service with Slovene Military Aviation, modified after delivery by Radom Aviation Systems of Israel. First example to this standard (L9-61) made first flight early May 1999. Mission systems installation includes Litton INS-100G inertial navigation with GPS; HOTAS controls; Flight Visions HUD with up-front control and rear cockpit video repeater; a central flight data recorder; chaff/flare dispenser and control panel; Honeywell EFIS and displays; new weapon selection and armament control unit and displays; MIL-STD-1553B dual databus; and RS-422/ARINC 429 datalink. Maximum external stores load (see Armament paragraph below) increased to 1,250 kg (2,756 lb).

CUSTOMERS: Total of 259 delivered by March 2000; see table. Saudi aircraft prepared for delivery by British Aerospace; first pair from second batch (Nos. 31 and 32) handed over in UK on 4 December 1995.

Slovenian order for nine PC-9M placed December 1997 and completed by December 1998. Omani order placed January 1999; all delivered by March 2000, replacing BAC Strikemasters.

DESIGN FEATURES: Typical turboprop trainer; stepped cockpits; aerobatic; performance and handling ease student's transition to jets. Meets FAR Pt 23 (Amendment 52) and special Swiss federal civil conditions for both Aerobatic and Utility categories; complies with selected parts of US military training specifications.

Conventional, low-wing monoplane; structural commonality with PC-7 Mk II M; parallel chord wing centre-section; tapered outer panels; tailplane LERX.

Wing section PIL15M825 at root, IL12M850 at tip; quarter-chord sweepback 1°; dihedral 7° from centre-section; incidence 1° at root; twist −2°.

FLYING CONTROLS: Conventional and manual. Cable-operated elevator and rudder; ailerons operated by pushrods; mass-balanced, electrically actuated trim tab in each aileron and

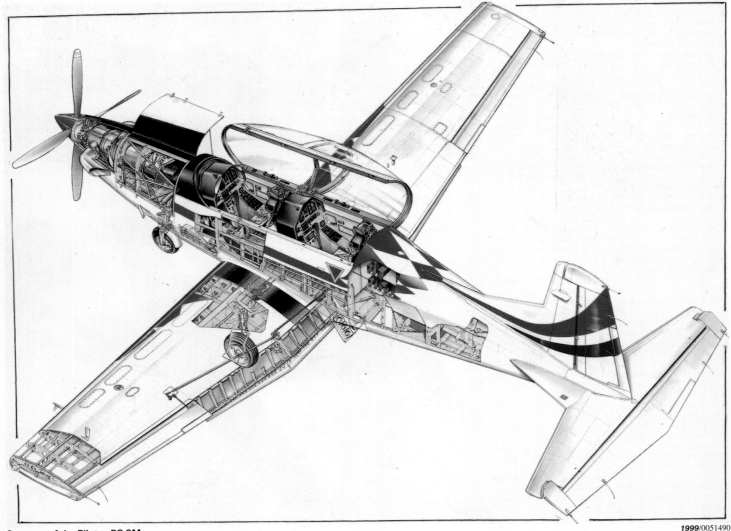

Cutaway of the Pilatus PC-9M

starboard half of elevator; electrically actuated trim/anti-balance tab in rudder controlled from rocker switch on power control lever. Trim aid device (TAD) automatically adjusts rudder trim tab setting in response to changes in engine torque or aircraft airspeed. This counteracts aircraft yaw induced by variations in effect of propeller slipstream, depending upon airspeed and engine power setting. TAD can be selected on or off by switch on trim control panel on left-hand side panel of front cockpit. Electrically operated split flaps; hydraulically operated 'perforated' airbrake under centre-fuselage.

STRUCTURE: All-metal with some GFRP wing/fuselage fairings; one-piece wing with auxiliary spar, ribs and stringer-reinforced skin.

LANDING GEAR: Retractable tricycle type, with hydraulic actuation. Mainwheels retract inward to lie semi-recessed in wing centre-section covered by bulged doors; nosewheel retracts rearward; all units enclosed when retracted. Oleo-pneumatic shock-absorber in each leg. Hydraulically actuated nosewheel steering. Goodrich wheels and tubeless tyres, with Goodrich multipiston hydraulic disc brakes on mainwheels. Main tyres 20×4.4 (8 ply), nose tyre 16×4.4 (8 ply). Low-pressure tyres optional. Parking brake.

POWER PLANT: One 857 kW (1,150 shp) Pratt & Whitney Canada PT6A-62 turboprop, flat rated at 708 kW (950 shp), driving a Hartzell HC-D4N-ZA/09512A four-blade constant-speed fully feathering propeller. Single-lever engine control. Fuel in two integral tanks in wing leading-edges, total usable capacity 518 litres (137 US gallons; 114 Imp gallons). Overwing refuelling point on each side. Fuel system includes 12 litre (3.2 US gallon; 2.6 Imp gallon) aerobatics tank in fuselage, forward of front cockpit, which permits up to 60 seconds of inverted flight. Provision for two 154 or 248 litre (40.7 or 65.5 US gallon; 33.9 or 54.5 Imp gallon) drop tanks on centre underwing attachment points. Total oil capacity 16 litres (4.2 US gallons; 3.5 Imp gallons).

ACCOMMODATION: Two Martin-Baker Mk 11A adjustable ejection seats, each with integrated personal survival pack and fighter-standard pilot equipment. Stepped tandem arrangement with rear seat elevated 15 cm (5.9 in). Seats operable, through canopy, at zero height and speeds down to 60 kt (112 km/h; 70 mph) EAS. Anti-g system optional. One-piece acrylic Perspex windscreen; one-piece framed canopy, incorporating roll-over bar, opens sideways to starboard. Dual controls standard. Cockpit heating, cooling, ventilation and canopy demisting standard. Space for 25 kg (55 lb) of baggage aft of seats, with external access.

SYSTEMS: Vapour cycle air conditioning system and engine bleed air for cockpit heating/ventilation and canopy demisting. Fairey Systems hydraulic system, pressure 207 bars (3,000 lb/sq in), for actuation of landing gear, mainwheel doors, nosewheel steering and airbrake; system maximum flow rate 18.8 litres (4.97 US gallons; 4.14 Imp gallons)/min. Bootstrap oil/oil reservoir, pressurised at 3.45 to 207 bars (50 to 3,000 lb/sq in). A supplementary high-pressure, one-shot nitrogen storage bottle provides power to lower the landing gear in an emergency. This bypasses the main hydraulic systems by feeding directly into shuttle valves in the down side of the main door and landing gear actuators.

Primary electrical system (28 V DC operational, 24 V nominal) powered by 30 V 300 A starter/generator and 24 V 40 Ah Ni/Cd battery; two optional static inverters can supply 115/26 V AC power at 400 Hz. Ground power receptacle. Electric anti-icing of pitot tube, static ports and AoA transmitter standard; electric de-icing of propeller blades optional. Diluter demand oxygen system, capacity 1,350 litres (47.6 cu ft) selected and controlled individually from panel in each cockpit. OBOGS standard.

AVIONICS: Both cockpits fully instrumented to customer specifications, with Logic computer-operated integrated engine and systems data display. Customer specified equipment provides flight environmental, attitude and direction data, and ground-transmitted position determining information.

Comms: Single or dual VHF, UHF and/or HF radios to customer's requirements. Audio integrating system controls audio services from com, nav and interphone systems.

Optional avionics include Honeywell CRT displays (electronic ADI and HSI), Flight Visions FV-2000 HUD, encoding altimeter and ELT.

Instrumentation: See accompanying diagram.

EQUIPMENT: Three hardpoints under each wing: inboard and centre stations each stressed for 250 kg (551 lb), outboard stations for 110 kg (242.5 lb) each. Retractable 250 W landing/taxying light in each main landing gear leg bay. Optional equipment can include smoke generators for aerial displays, target towing kit, electronic jammer for EW training, and blind-flying hood.

ARMAMENT: FN Herstal package for Hudournik/Swift comprises HMP-250 12.7 mm gun pods; LAU-7A seven-round unguided rocket launchers; Alkan Type 65 adaptor and dispensers for IBDU-33 practice bombs; and a laser range-finder pod.

DIMENSIONS, EXTERNAL:
Wing span	10.125 m (33 ft 2½ in)
Wing chord: mean aerodynamic	1.65 m (5 ft 5 in)
mean geometric	1.61 m (5 ft 3½ in)
Wing aspect ratio	6.3
Length overall	10.14 m (33 ft 3¼ in)
Fuselage max width	0.97 m (3 ft 2¼ in)
Height overall	3.26 m (10 ft 8⅓ in)
Tailplane span	3.665 m (12 ft 0¼ in)
Wheel track	2.54 m (8 ft 4 in)
Wheelbase	2.31 m (7 ft 7 in)
Propeller diameter	2.44 m (8 ft 0 in)
Propeller ground clearance	0.18 m (7 in)

DIMENSIONS, INTERNAL:
Baggage compartment volume	0.14 m³ (4.9 cu ft)

AREAS:
Wings, gross	16.29 m² (175.3 sq ft)
Ailerons (total)	1.57 m² (16.90 sq ft)
Trailing-edge flaps (total)	1.77 m² (19.05 sq ft)
Airbrake	0.30 m² (3.23 sq ft)
Fin	0.86 m² (9.26 sq ft)
Rudder, incl tab	0.90 m² (9.69 sq ft)
Tailplane	1.80 m² (19.38 sq ft)
Elevator, incl tab	1.60 m² (17.22 sq ft)

WEIGHTS AND LOADINGS (A: Aerobatic, U: Utility category with underwing stores):
Basic weight empty, equipped	1,725 kg (3,803 lb)
Max underwing stores load	1,040 kg (2,293 lb)
Max T-O weight: A	2,350 kg (5,180 lb)
U	3,200 kg (7,054 lb)
Max ramp weight: A	2,360 kg (5,203 lb)
U	3,210 kg (7,076 lb)
Max landing weight: A	2,350 kg (5,180 lb)
U	3,100 kg (6,834 lb)
Max zero-fuel weight: A	2,000 kg (4,409 lb)
Max wing loading: A	144.3 kg/m² (29.55 lb/sq ft)
U	196.4 kg/m² (40.23 lb/sq ft)
Max power loading: A	3.32 kg/kW (5.45 lb/shp)
U	4.52 kg/kW (7.42 lb/shp)

PERFORMANCE (A and U as above, propeller speed 2,000 rpm):
Max operating Mach number (M_{MO}): A, U	0.65
Max operating speed (V_{MO}):	
A, U	320 kt; (593 km/h; 368 mph) EAS
Max cruising speed:	
A at S/L	271 kt (501 km/h; 311 mph)
A at 3,050 m (10,000 ft)	297 kt (550 km/h; 342 mph)
A at 7,620 m (25,000 ft)	300 kt (556 km/h; 345 mph)
Max manoeuvring speed:	
A	205 kt (380 km/h; 236 mph) EAS
U	200 kt (370 km/h; 230 mph) EAS
Max speed with flaps and/or landing gear down:	
A and U	150 kt (278 km/h; 172 mph) EAS
Stalling speed, engine idling:	
flaps and gear up:	
A	77 kt (143 km/h; 89 mph) EAS
U	90 kt (167 km/h; 104 mph) EAS
flaps and gear down:	
A	69 kt (128 km/h; 80 mph) EAS
U	80 kt (149 km/h; 93 mph) EAS
Max rate of climb at S/L: A	1,247 m (4,090 ft)/min
Time to 4,575 m (15,000 ft): A	4 min 5 s
Max operating altitude: A, U	7,620 m (25,000 ft)
Service ceiling: A, U	11,580 m (38,000 ft)
T-O run at S/L: A	243 m (795 ft)
T-O to 15 m (50 ft) at S/L: A	391 m (1,280 ft)
Landing from 15 m (50 ft) at S/L at MLW:	
A	700 m (2,295 ft)
Landing run at S/L at MLW:	
A (normal braking action)	351 m (1,150 ft)
Max range at 210 kt (389 km/h; 242 mph) at 7,620 m (25,000 ft), 5% and 20 min fuel reserves:	
A	830 n miles (1,537 km; 955 miles)
Max endurance at 110 kt (204 km/h; 127 mph) IAS, conditions as above: A	4 h 30 min
g limits: A	+7/−3.5
U	+4.5/−2.25

UPDATED

PILATUS PC-21

TYPE: Basic turboprop trainer.

PROGRAMME: Launched November 1998; first brief details appeared late 1999. Next-generation (21st century, hence 21 in designation) trainer; variously reported as derivative of T-6A Texan II and as 'significantly new' design; more powerful engine, in development by Pratt & Whitney Canada for initial bench testing in 2000; new propeller. Quoted in October 2000 as promising 'unheralded' performance. Unarmed, but able to provide weapons training by use of simulators. During 2000, a PC-9 development aircraft was flying with a sweptback fin and five-blade propeller, apparently in connection with the PC-21 programme. No other details released at time of going to press.

UPDATED

PILATUS PC-12/45

TYPE: Business turboprop.

PROGRAMME: Announced at NBAA Convention October 1989; first flight P.01 (HB-FOA) 31 May 1991; first flight of P.02 second prototype (HB-FOB) 28 May 1993; Swiss certification to FAR Pt 23 Amendment 42 (covering FAR Pt 135 commercial and Pt 91 general operations) received 30 March 1994, FAA type approval 15 July 1994, and FAR Pt 25 certification for flight into known icing conditions in 1995. Deliveries (N312BC to Carlston Leasing Corporation in USA) began September 1994.

Higher gross weight option (4.5 tonnes, hence PC-12/45) introduced in 1996, this becoming production standard by 1997. FAA FAR Pt 135 approval for commercial single-engined IFR operations announced in third quarter 1997; production increased from three to four per month in August 1997. First airline scheduled service operator (1997) was Kelner Airways of Canada. Total of 76 ordered and 51 delivered in 1998 (including 100th in April); 29 ordered in first half of 1999; 200th sale announced at NBAA Convention in October 1999; production rate increased in 1999 from 48 to 60 per year and targeted to reach 100 or more per year by 2002. Worldwide fleet hours passed 200,000 mark in October 2000. Now certified in 12 countries.

In 1999, Pilatus was designing a new cockpit, featuring a 230 × 250 mm (9 × 10 in) IS&S flat-panel display. Other modifications and improvements, including performance gains from current design studies for PC-21, under development in 2000.

CURRENT VERSIONS: **Standard:** Nine-passenger commuter or passenger/cargo combi.

Detailed description applies to standard PC-12/45 except where indicated.

Pilatus PC-12 operated in the USA (*Paul Jackson/Jane's*)

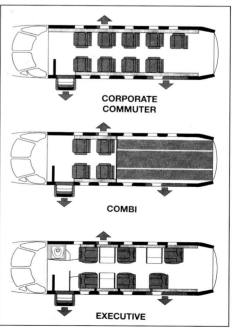

PC-12/45 typical cabin layouts

SWITZERLAND: AIRCRAFT—PILATUS

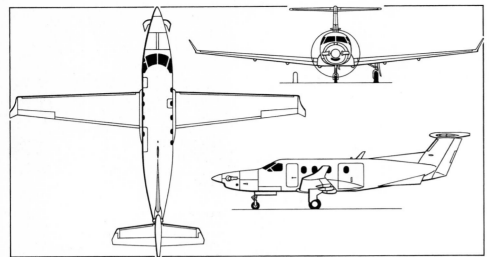

Pilatus PC-12/45 pressurised light utility and business transport *(Mike Keep/Jane's)*

Executive: Six to eight passenger seats.
Eagle: Surveillance and special missions version; described separately.

CUSTOMERS: Delivery of 200th production aircraft effected to Canadian dealer in March 2000; total of 225 delivered by 9 October 2000, at which time order backlog of more than 70. Deliveries in 2000 totalled 70, including 56 for North American market, compared with 55 and 45 respectively in 1999. Some 53 per cent of first 200 aircraft are registered in USA, further 18 per cent in Canada (including nine with Royal Canadian Mounted Police) and 11 per cent in South Africa (including two for Red Cross and one with Air Force). US Drug Enforcement Agency (DEA) received two in transport configuration in December 1999 and October 2000, and predicts further orders. Other customers include those in Argentina (Border Guard), Austria, Bermuda, Brazil, Denmark, France, Germany, Japan, Kenya, Switzerland and Zimbabwe.

Approximately 10 per cent of those sold are in air ambulance configuration. First single-engine turboprop fractional ownership scheme involves PC-12s of Alpha Flying's PlaneSense programme at Nashua, New Hampshire, USA.

COSTS: Basic price US$2.644 million (2000).
DESIGN FEATURES: Able to fly three 200 n mile (370 km; 230 mile) sectors in 6 hour flight; approved for single-pilot VFR/IFR operation into known icing.

Low-wing monoplane with tapered wing and T tail; latter's mounting reduces trim changes with power; CG range 13 to 46 per cent of MAC. Modifications following early flight trials include introduction of winglets, increased wing span, paired elevators instead of single surface, sweptback tailplane/elevator tips, addition of tailplane/fin bullet fairing, and enlarged dorsal fin and ventral strakes.

Wing sections (modified NASA GA(W)-1 series), LS(1)-0417-MOD at root and LS(1)-0313 at tip.
FLYING CONTROLS: Conventional and manual. Actuation by pushrods and cables; servo tab in each aileron; electrically actuated Flettner tab in rudder; short-span mass-balanced ailerons; electrically actuated Fowler flaps cover 67 per cent of wing trailing-edge; electrically actuated (dual redundant) variable incidence tailplane.
STRUCTURE: All-metal basic structure; composites for ventral strakes and dorsal fin (Kevlar/honeycomb sandwich), wingtips (glass fibre), fairings, engine cowling (glass fibre/honeycomb sandwich) and interior trim; titanium firewall; two-spar wing with integral fuel tankage; airframe proved for 20,000 hour life; additional structural tests continuing in late 1997.
LANDING GEAR: Pilatus hydraulically retractable tricycle type, with single wheel on each unit; nosewheel mechanically steerable ±60°. Emergency extension system; toe-operated brakes; parking brake. Suitable for operation from grass strips. Goodrich wheels and low-pressure tyres on all units: size 22×8.50-10 on main gear, 17.5×6.25-6 on nose unit; tyre pressure 4.14 bars (60 lb/sq in) on nose unit, 3.79 bars (55 lb/sq in) on main units. Propeller ground clearance maintained with nose leg compressed and nosewheel tyre flat. Trailing-link main gear retracts inward into wings, nose gear rearward under flight deck. Ground turning radius about wingtip 10.21 m (33 ft 6 in), about main gear 4.50 m (14 ft 9 in).
POWER PLANT: One 1,197 kW (1,605 shp) Pratt & Whitney Canada PT6A-67B turboprop, flat rated to 895 kW (1,200 shp) for T-O and 746 kW (1,000 shp) for climb and cruise. Hartzell HC-E4A-3D/E10477K constant-speed, fully feathering reversible-pitch four-blade aluminium propeller, turning at 1,700 rpm. Two integral fuel tanks in wings, total capacity 1,540 litres (407 US gallons; 339 Imp gallons), of which 1,522 litres (402 US gallons; 335 Imp gallons) are usable. Gravity fuelling point in top of each wing. Oil capacity 11 litres (2.9 US gallons; 2.4 Imp gallons).
ACCOMMODATION: Two-seat flight deck: approved for single pilot, with dual controls; second flight instrument panel optional. Limit of nine passengers under FAR Pt 23, or executive layout for six, both with lavatory. Downward-opening airstair crew/passenger door at front, upward-opening cargo door at rear, both on port side; Type III emergency exit above wing on starboard side.
SYSTEMS: Normalair Garrett engine bleed air ECS, maximum pressure differential 0.4 bar (5.8 lb/sq in). Vickers Systems (Germany) hydraulic system, pressure 207 bars (3,000 lb/sq in), for landing gear actuation. Electrical power system (28 V DC) supplied by 300 A engine-driven starter/generator), 130 A back-up alternator and 24 V 40 Ah Ni/Cd battery. Second Ni/Cd battery optional. Goodrich pneumatic boot de-icing of wing and tailplane leading-edges; Goodrich electric anti-icing of propeller blades; electric heating for windscreen; exhaust air anti-icing of engine air intake. Oxygen system for crew and passengers.
AVIONICS: *Comms:* Honeywell dual KX 155A VHF transceivers, KMA 26 audio control panel and intercom with marker beacon switch and KT 70A transponder, and Narco ELT-910 emergency locator transmitter, standard; Honeywell KHF 950 HF radio, AMS 44 dual audio systems and second KT 70A optional.
Radar: Honeywell RDR 2000 weather radar optional.
Flight: Honeywell KFC 325 autopilot system, KN 63 DME, KR 87 ADF, KNI 582 RMI and KEA 130A encoding altimeter and Litef LCR-92 AHRS standard; Honeywell KLN 90B or KLN 900 GPS optional. Other options include KRA 405 radar altimeter, second LCR-92, WX 1000E Stormscope, TCAS 66A, and Mk VI GPWS with CIC 8800M air data computer.

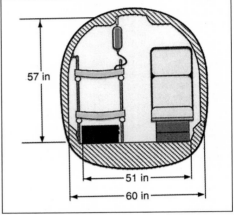

PC-12/45 cabin cross-section in medical evacuation fit

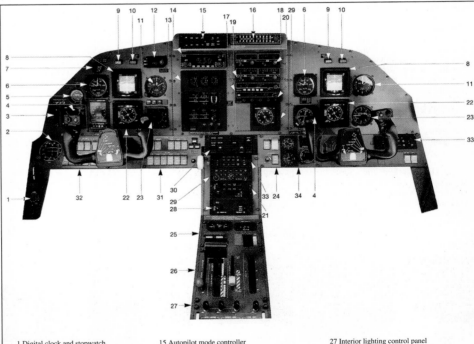

1 Digital clock and stopwatch
2 VOR/LOC glide slope indicator
3 Marker beacon receiver and lights
4 Double-needle RMI
5 Standby attitude indicator
6 Mach and airspeed indicator
7 Flap position indicator
8 Electronic ADI
9 Master caution light
10 Master warning light
11 Encoding altimeter (captain's side only)
12 Altitude and vertical speed preselector
13 Engine instrument system panel
14 GPS unit
15 Autopilot mode controller
16 Audio control console and intercom
17 Digital ADF
18 Com/nav transceiver 1
19 Com/nav transceiver 2
20 Transponder
21 Vertical profile colour weather radar control panel
22 Electronic HSI
23 Vertical speed indicator
24 ECS control panel
25 Triple trim and cabin temperature indicator, and interrupt and emergency switches panel
26 Power and flap control panel
27 Interior lighting control panel
28 HF transceiver
29 Multifunction display with control panel
30 Central advisory and warning system (CAWS) display panel
31 External lights, de-icing and landing gear control panel
32 Engine start, fuel pumps and cockpit cooling switches panel
33 HSI control panel
34 Cabin pressurisation panel

PC-12/45 interior in executive fit

Flight deck of the Pilatus PC-12/45

Current flight deck with optional co-pilot's EFIS
2001/0093858

PC-12/45 standard nine-passenger layout
2001/0093857

Landing from 15 m (50 ft) at MLW	558 m (1,830 ft)
Landing run at MLW	288 m (945 ft)
Max range at 9,150 m (30,000 ft), VFR reserves:	
standard	2,261 n miles (4,187 km; 2,602 miles)
executive	2,210 n miles (4,093 km; 2,543 miles)
combi	2,282 n miles (4,226 km; 2,626 miles)
g limits: flaps up	+3.4/−1.36
flaps down	+2.0

UPDATED

PILATUS PC-12 EAGLE

TYPE: Maritime surveillance turboprop.
PROGRAMME: Announced at Dubai Air Show in November 1995; second PC-12 HB-FOB converted as demonstrator (first flight October 1995). Second demonstrator HB-FOG (c/n 134) followed in September 1996 with airframe modifications and different equipment (see Design Features and Current Versions paragraphs below).
CURRENT VERSIONS: **HB-FOB**: Ventral pannier carrying a variety of electro-optical sensors. A sensor management system (SMS) controls and monitors sensors and displays data on two consoles (one electro-optical, one radar, for comint and elint) inside cabin; includes three COTS processors. Consoles have four displays: colour displays for navigation, situational awareness and ESM/IR/TV; and one 325 mm (14½ in) square LCD XGA high-resolution colour display for synthetic aperture radar (SAR). Sensor operator's map display is 'North up', featuring coastlines, rivers, roads and political boundaries.

Sensor options include: (1) WF-160DS turret with FLIR having $7.2 \times 9.8°$ wide field of view and $2.16 \times 2.95°$

Instrumentation: Honeywell EFS 40 EFIS with 102 mm (4 in) display standard; MFD for EFS 40, and 127 mm (5 in) EFIS on pilot and co-pilot panels, optional.

DIMENSIONS, EXTERNAL:
Wing span	16.23 m (53 ft 3 in)
Wing aspect ratio	10.2
Length overall	14.40 m (47 ft 3 in)
Height overall	4.26 m (13 ft 11¾ in)
Elevator span	5.21 m (17 ft 1 in)
Wheel track	4.52 m (14 ft 10 in)
Wheelbase	3.53 m (11 ft 7 in)
Propeller diameter	2.67 m (8 ft 9 in)
Propeller ground clearance	0.32 m (1 ft 0½ in)
Passenger door: Height	1.35 m (4 ft 5 in)
Width	0.635 m (2 ft 1 in)
Cargo door: Height	1.32 m (4 ft 4 in)
Width	1.35 m (4 ft 5 in)
Emergency exit: Height	0.66 m (2 ft 2 in)
Width	0.48 m (1 ft 7 in)

DIMENSIONS, INTERNAL:
Cabin: Length, excl flight deck	5.16 m (16 ft 11 in)
Max width	1.52 m (5 ft 0 in)
Width at floor	1.295 m (4 ft 3 in)
Max height	1.45 m (4 ft 9 in)
Volume	9.34 m³ (330 cu ft)
Baggage compartment volume	1.13 m³ (40 cu ft)

AREAS:
Wings, gross	25.81 m² (277.8 sq ft)

WEIGHTS AND LOADINGS:
Weight empty: standard	2,600 kg (5,732 lb)
executive	2,835 kg (6,250 lb)
combi	2,536 kg (5,591 lb)
Max usable fuel	1,226 kg (2,703 lb)
Max payload: standard	1,410 kg (3,108 lb)
executive	1,243 kg (2,740 lb)
combi	1,474 kg (3,250 lb)
Payload with max fuel: executive	414 kg (913 lb)
Max ramp weight	4,520 kg (9,965 lb)
Max T-O and landing weight	4,500 kg (9,920 lb)
Max zero-fuel weight	4,100 kg (9,040 lb)

Max wing loading	174.3 kg/m² (35.71 lb/sq ft)
Max power loading	5.03 kg/kW (8.27 lb/shp)

PERFORMANCE:
Max operating Mach number (M_{MO})	0.48
Max operating speed (V_{MO})	240 kt (444 km/h; 276 mph) IAS
Max cruising speed at 7,620 m (25,000 ft)	270 kt (500 km/h; 311 mph)
Stalling speed:	
flaps and gear up	92 kt (171 km/h; 106 mph) IAS
flaps and gear down	65 kt (121 km/h; 75 mph) IAS
Max rate of climb at S/L	512 m (1,680 ft)/min
Max certified altitude	9,150 m (30,000 ft)
Service ceiling	10,670 m (35,000 ft)
T-O run	452 m (1,480 ft)
T-O to 15 m (50 ft)	701 m (2,300 ft)

Eagle special missions version of PC-12, as originally converted from second prototype
2000/0079291

Eagle ventral pannier with covers removed, revealing internal sensors. The WF-160DS turret weighs 43 kg (95 lb) and offers elevations of +50 to −190° for FLIR and +10 to −190° for TV sensors

Operator workstations in the Eagle's cabin

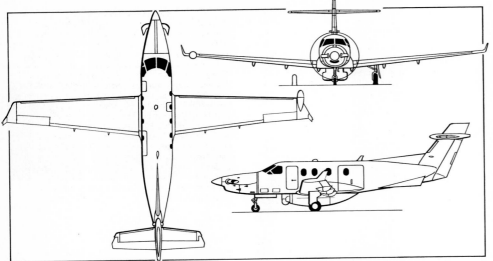

Current configuration of Pilatus PC-12 Eagle special missions aircraft *(James Goulding/Jane's)* 1999/0051495

narrow field, plus daylight camera with 20 to 280 mm zoom lens; (2) SAR installations of several types, either mechanical or electronically scanned (one or two dimensions, from 45 to 200 cm; 18 to 79 in long, giving up to 36 cm; 14¼ in resolution); and (3) RISTA (Reconnaissance, Infra-red Surveillance, Target Acquisition), derived from US Army's Airborne Minefield Detection System and combining FLIR (for battle damage assessment) and IR linescan (covering a 7 n mile; 13 km; 8 mile ground track) with real-time downlink to ground station.

HB-FOG: Sensors include WF-160DS IR/E-O system with tracker and Geotrack cueing; Raytheon Sea Vue SV 1021 radar with ISAR capability 1997 upgrade for SAR; 1998 upgrade for track-while-scan of 100 targets and MTI on SAR; communications digital datalink; 108 n mile (200 km; 124 mile) range video downlink to ground station; and sensor management system.

Eagle is also capable of such other sensors as 2 to 18 GHz (optionally 40 GHz) elint, including automatic jammer; forward- and backward-looking sensor management systems; a passive missile warning system, including automatic chaff/flare dispensers; LOROP; and a bidirectional secure datalink.

CUSTOMERS: None yet announced.

DESIGN FEATURES: FLIR turret at front of ventral pannier; winglets replaced by conventional wingtips on HB-FOB (weather radar in starboard tip pod); undertail strakes enlarged (to maximum depth of 53 cm; 21 in) to maintain stability. HB-FOG fitted 1997 with new 'tiplets' (wingtips with small vertical winglets), newly defined undertail strakes and additional fin area above the elevator, increasing overall height by 48 cm (19 in). Eagle can undertake various reconnaissance and surveillance missions, but retains ability to revert to passenger and other missions at short notice.

DIMENSIONS, EXTERNAL: As standard PC-12/45 except:
  Wing span   16.115 m (52 ft 10½ in)
  Height overall   4.74 m (15 ft 6½ in)
WEIGHTS AND LOADINGS: As standard PC-12/45 except:
  Weight empty   2,688 kg (5,926 lb)
  Max payload   1,402 kg (3,091 lb)
PERFORMANCE:
  Max cruising speed at 6,100 m (20,000 ft)
                     230 kt (426 km/h; 265 mph)
  Stalling speed, flaps and gear down
                     65 kt (121 km/h; 76 mph) IAS
  Max rate of climb at S/L   408 m (1,340 ft)/min
  Max operating altitude   9,145 m (30,000 ft)
  T-O run   450 m (1,475 ft)
  Landing run   280 m (920 ft)
  Max range at 9,145 m (30,000 ft) at 200 kt (370 km/h; 230 mph), 45 min reserves
                     1,633 n miles (3,080 km; 1,913 miles)
                                              UPDATED

SAUER

CHRIS SAUER

It was revealed in late 1999 that Chris Sauer had acquired design and production rights, plus five airframes, of the 1960s **Silvercraft SH-4** two-seat light helicopter, with a view to reinstating it in production in Switzerland. The SH-4, a joint SIAI-Marchetti/Silvercraft design, first flew (prototype I-RAIX) in March 1965, received Italian and FAA certification in 1968, and was powered by a 175 kW (235 hp) Franklin 6A-350 flat-six engine. About 20 (including prototypes) were built before production ended in 1973. It was last described in the Italian section of the 1976-77 *Jane's*.
NEW ENTRY

SF

**SWISS AIRCRAFT AND SYSTEMS ENTERPRISE CORPORATION (SF)
SCHWEIZ UNTERNEHMUNG FÜR FLUGZEUGE UND SYSTEME AG
ENTREPRISE SUISSE D'AERONAUTIQUE ET DE SYSTEMES SA
IMPRESA SVIZZERA D'AERONAUTICA E SISTEMI SA**

CH-6032 Emmen
Tel: (+41 41) 268 41 11
Fax: (+41 41) 260 25 88
Web: http://www.sfaerospace.ch
MANAGING DIRECTOR: Dr Werner Glanzmann
HEAD OF MARKETING:: Michael Perlberger

SF was founded 1996, when the former Swiss Federal Aircraft Factory Emmen merged with the industrial division of the Swiss Air Force Logistic Command and parts of the Federal Ordnance Office. On 1 January 1999, the Swiss Aircraft and Systems Enterprise Corporation was converted from a federal ordnance establishment into a diversified private company within the RUAG Suisse group. It employs 1,600 personnel at several plants.

Main activities are the maintenance, manufacture, modernisation and assembly under licence of military and civil aircraft and guided missiles. R&D activities involve four departments, of which Aerodynamics and Flight Mechanics department has four wind tunnels for speeds up to M4 to M5, and test cells for piston and jet engines with and without afterburners. Structural and Systems Engineering department deals with aircraft and space hardware. Electronics and Missile Systems department deals with all systems aspects of aircraft avionics and missiles. Fourth R&D department is for

One of 32 (from 34 received) F-18 Hornets for Switzerland, built by SF *(Paul Jackson/Jane's)* 2001/0092517

prototype fabrication, flight test, instrumentation and system environmental testing. The Production department can handle mechanical, sheet metal and composites and electronic, electrical, electromechanical and electro-optical subassemblies.

Recent programmes have included structural improvements to Swiss Air Force Tiger F-5E/Fs and the integration of new systems (smoke systems, target towing, missiles); manufacture and delivery of Rafale drop tanks to Dassault Aviation; assembly of 19 BAe Hawk Mk 66s; and integration of AS 532UL Cougars for the Swiss Air Force. On 2 December 1999, it delivered the last (J-5026) of 32 single- and two-seat F-18 Hornets assembled for the Swiss Air Force; it also established engineering support and manufactured leading-edges, rudders and its own low-drag pylons for these aircraft.

Current work includes the serial production of rudders for the Boeing 717, wingtips for the Airbus A319/A320/A321, the tailplane developed for the Pilatus PC-12 and the manufacture and ground support of Ranger UAV and its hydraulic launcher. During 2001 and 2002 it will assemble 10 of the follow-on order for 12 AS 532UL Cougars ordered for the Swiss Air Force in December 1998.

SF is sole manufacturer of Dragon anti-tank missile; TOW missile programme began 1996; Stinger missile programme began 1989.

SF is accredited to maintain GE J85 (Tiger), Adour (Hawk), GE F404 (F/A-18), Makila (Cougar) and PT6A (PC-6, PC-7, PC-9) engines in own workshops. JAR 145 accreditation allows SF to carry out maintenance work at all levels. This includes inspections, repairs, modifications, prototyping and fault elimination. Other specialities include non-destructive testing procedures such as ultrasonic, eddy current and X-ray as well as comprehensive service packages for systems re-engineering.

In collaboration with Contraves, SF co-manufactures all payload fairings for Ariane IV and V space launchers on behalf of the ESA, ESTEC, CNES and private industry.
UPDATED

TAIWAN

AIDC

AEROSPACE INDUSTRIAL DEVELOPMENT CORPORATION
111-6 Lane 68, Fu-Hsing North Road, Taichung 407
Tel: (+886 4) 259 00 01
Fax: (+886 4) 256 22 95
e-mail: hualungchang@ms.aidc.com.tw
Web: http://www.aidc.com.tw

CHAIRMAN: Chuen-Huei Tsai
VICE-PRESIDENT, BUSINESS: Chou Yeuan-Yuan
MANAGER, PUBLIC RELATIONS: Michael H L Chang

Established 1 March 1969; became subsidiary of Chung Shan Institute of Science and Technology (CSIST, which see) under Ministry of National Defense on 1 January 1983 and known as Aero Industry Development Center until June 1996. AIDC produced 118 Bell UH-1H (Model 205) helicopters under licence 1969-76 for Chinese Nationalist Army; built PL-1A prototype (based on Pazmany PL-1) and 55 PL-1B Chien-Shou primary trainers for Republic of China Air Force between 1968 and 1974 (see 1975-76 *Jane's*); built 248 Northrop F-5E Tiger IIs and 36 two-seat F-5Fs under licence between 1974 and 1986 (see 1986-87 edition). Designed and produced T-CH-1 Chung-Hsing turboprop basic trainer (see 1981-82 *Jane's*); developed and produced AT-3 Tsu-Chiang twin-turbofan advanced trainer. New

35,700 m² (384,270 sq ft) assembly facility opened in 1989. Current facilities at Taichung, Kangshan and Sah Lu; total site area 109.3 ha (270 acres); workforce was 3,800 in mid-2000.

From 1 July 1996 was officially transferred under Ministry of Economic Affairs with title as above; has now completed production of IDF Ching-Kuo fighter. AIDC is a participant (with 5 per cent share) in the Sikorsky S-92 Helibus (see US section), responsible for the cockpit structure, and (from 1999) is sole-source supplier of crew and passenger doors for the S-76. Other subcontract work includes wing design of the Ibis Aerospace Ae 270 (see International section); Boeing 717-200 (tail unit; first example delivered mid-December 1997); Dassault Falcon 900 and 2000 (rudders); Eurocopter/CATIC/ST Aero EC 120 Colibri (rear fuselage components); Alenia C-27J Spartan (tail unit); and Bombardier Continental (rear fuselage and tail unit). AIDC currently assembling 13 Bell OH-58Ds for Republic of China Army (first one delivered 16 November 1999; last due in third quarter 2001). It also participates in such programmes as the Lockheed Martin F-16 combat aircraft, and the Honeywell TFE1042 and Rolls-Royce 572K gas-turbine engines. Possibility of participation in Century CA-100 business jet (see US section) being mooted in late 2000.

As a newly government-owned commercial entity, AIDC is dedicated to research and development for Taiwan's aerospace industry (military and commercial aircraft). Full privatisation, originally planned for fourth quarter of 1999, has now been postponed until the end of 2001, following a mid-2000 company restructuring based on three core activities: defence systems and technologies, aerostructures and engines.

UPDATED

AIDC F-CK-1 CHING-KUO

TYPE: Air superiority fighter.
PROGRAMME: Final two aircraft (88-8134 and 88-8135) delivered to RoCAF on 14 January 2000; second wing (1st TFW) declared operational July 2000; detailed description in 2000-01 and earlier editions. An update is under development for existing aircraft and two further versions are proposed to permit production to be restarted.
CURRENT VERSIONS: **F-CK-1A**: Single-seat fighter. Three prototypes; 103 production.
F-CK-1B: Two-seat operational trainer. One prototype; 28 production.
MLU: In 1999 F-CK-1A 1417 was undergoing modification by AIDC as prototype of mid-life update, believed to include GPS antennas ahead of windscreen.
LIFT version: A lead-in fighter trainer version of the Ching-Kuo, based on the two-seater minus its radar, ECM and internal gun, was proposed to the RoCAF in 1995. This engineering modification programme, known as Derivative IDF, in preliminary design stage by late 1999; other changes include simplified avionics and additional 771 kg (1,700 lb) of internal fuel. First flight targeted for mid-2002. AIDC has been seeking an international partner to join a next-generation trainer programme and market this IDF derivative globally.
Strike fighter: Development programme, funded in mid-2000, for version with new avionics and weapons, plus extended range; projected service-entry date 2010. Initial funding reported to be about US$225 million.

UPDATED

CSIST

CHUNG SHAN INSTITUTE OF SCIENCE AND TECHNOLOGY (Aeronautical Systems Research Division)

PO Box 90008-11-15, Fu-Hsing North Road, Taichung 40722
Tel: (+886 4) 252 30 51
Fax: (+886 4) 374 65 35
DEPUTY DIRECTOR, AERONAUTICAL SYSTEMS RESEARCH DIVISION: Henry Chen
DIRECTOR, AIRCRAFT SUBSYSTEMS DEPARTMENT: Dr T C Lee

A part of CSIST since January 1983, the Aeronautical Systems Research Division occupies a 9.3 ha (23 acre) site and had a workforce of approximately 700 in early 2000. It has had a small utility transport, the ARL-1, in the design stage during the past two years.

UPDATED

CSIST ARL-1

TYPE: Twin-turboprop transport.
PROGRAMME: Original small aircraft project (see entry for SAP in 1998-99 *Jane's*) initiated by Aeronautical Research Laboratory of AIDC, but modified 1998 to larger capacity ARL-1 utility transport following domestic market survey. Conceptual design completed mid-1998; wind tunnel tests completed; funded until end of 2000 by Taiwan Ministry of Economic Affairs. Prospective roles include maritime patrol and air ambulance. International partner(s) being sought to enable programme launch.
DESIGN FEATURES: Blended wing/body design (mid-mounted wing, with winglets); sweptback T tail.
POWER PLANT: Two 1,230 kW (1,650 shp) class turboprops; four-blade propellers. Turbofan power plant also being considered.

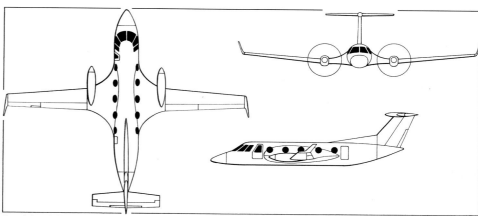

Provisional drawing of the CSIST ARL-1 general purpose transport *(James Goulding/Jane's)*

ACCOMMODATION: Crew of two and up to 19 passengers.
DIMENSIONS, EXTERNAL (provisional):
Wing span	21.03 m (69 ft 0 in)
Wing aspect ratio	10.4
Length overall	17.98 m (59 ft 0 in)
Height overall	4.67 m (15 ft 4 in)

AREAS:
Wings, gross	42.74 m² (460.0 sq ft)

WEIGHTS AND LOADINGS (approx):
Weight empty	4,536 kg (10,000 lb)
Max usable fuel weight	1,814 kg (4,000 lb)
Max payload	1,814 kg (4,000 lb)
Max T-O weight	less than 9,071 kg (20,000 lb)
Max wing loading	212.3 kg/m² (43.48 lb/sq ft)

PERFORMANCE (estimated):
Max level speed at S/L	more than 260 kt (481 km/h; 299 mph)
Max rate of climb at S/L	1,250 m (4,100 ft)/min
Rate of climb at S/L, OEI	488 m (1,600 ft)/min
Max operating altitude	10,980 m (36,000 ft)
T-O distance	less than 549 m (1,800 ft)
Landing distance	less than 488 m (1,600 ft)
Range with max fuel	1,220 n miles (2,259 km; 1,404 miles)
Max endurance	8 h

UPDATED

LASI

LIGHT'S AMERICAN SPORTSCOPTER INC

26 Ta Ho Street, Taichung 407
Tel: (+886 4) 311 80 03
Fax: (+886 4) 311 80 01
e-mail: ultrasport@iname.com
Web: http://www.ultrasport.rotor.com

This company is the Taiwanese production centre for the American Sportscopter **Ultrasport** family of small helicopters, descriptions of which can be found under the ASI heading in the US section. The Taiwan office also handles all marketing except for North and South America. No Taiwanese production had been reported by October 2000.

UPDATED

MODUS

MODUS VERTICRAFT CORPORATION

PO Box 46-155, Taichung Industrial Park, Taichung 407
Tel: (+886 4) 438 35 33
Fax: (+886 4) 328 34 20
Web: http://www.verticraft.com
PRESIDENT: Charles Lin

UPDATED

MODUS MILITARY SCOUT

TYPE: New-concept rotorcraft.
PROGRAMME: Preliminary, tandem-seat design, to evaluate performance potential of VertiJet/Verticraft configuration. Current status not known.
DESIGN FEATURES: See accompanying drawing and main VertiJet/Verticraft entry.
LANDING GEAR: Retractable tricycle type.
POWER PLANT: One Rolls-Royce 250-C30 turboshaft (410 kW; 550 shp for T-O, 336 kW; 450 shp maximum continuous), driving a pusher propeller. Fuel capacity 405 litres (107 US gallons; 89.1 Imp gallons).

Following data are provisional:
DIMENSIONS, EXTERNAL:
Rotor/disc diameter: blades extended	7.92 m (26 ft 0 in)
blades retracted	4.88 m (16 ft 0 in)
Length overall, rotor turning	8.96 m (29 ft 4¾ in)
Height overall	3.23 m (10 ft 7¼ in)

WEIGHTS AND LOADINGS:
Weight empty	816 kg (1,800 lb)
Max T-O weight	1,496 kg (3,300 lb)

PERFORMANCE (estimated):
Max level speed at S/L	250 kt (463 km/h; 287 mph)
Cruising speed	200 kt (370 km/h; 230 mph)
Service ceiling	6,705 m (22,000 ft)
Range at cruising speed	980 n miles (1,815 km; 1,127 miles)

NEW ENTRY

MODUS VERTIJET and VERTICRAFT

TYPE: New-concept rotorcraft.
PROGRAMME: Revealed at HeliExpo 99 air show in February 1999; wind-tunnel tested at Cheng Kung University; plans then were to flight-test quarter-scale remotely controlled prototype by mid-year; prototypes said to be under construction. No further reports of progress by October 2000.
CURRENT VERSIONS: **VertiJet**: Proof-of-concept version; twin turbojets pod-mounted on sides of rear fuselage for forward propulsion.
Verticraft: Forward propulsion by turboshaft instead of side-mounted jet engines; shrouded pusher propeller aft of tailcone.
DESIGN FEATURES: Hybrid design, patented in Taiwan and USA. Instead of conventional rotors, a pair of contrarotating discs is pylon-mounted above fuselage, each with a series of rigid bladelets around its rim. Disc-type rotor is said to generate more lift than a conventional

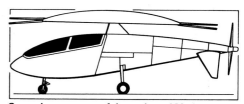

General appearance of the projected Modus military scout design

Model of the Modus Verticraft hybrid VTOL aircraft 1999/0045086

Provisional three-view of the Modus VertiJet (James Goulding/Jane's) 1999/0054206

rotor of same diameter, enabling Verticraft to employ a smaller diameter than traditional rotorcraft and enhancing manoeuvrability. Rotor disc is employed in conjunction with direct propulsion to achieve speeds at which the disc provides sufficient lift to permit disengagement of rotors; bladelets are then retracted into discs, discs are stopped, and because of their close spacing they then act as a single supercritical aerofoil. In this configuration, aircraft is said to be capable of approaching supersonic speeds. Design permits autorotational landing in the event of loss of power.

FLYING CONTROLS: Collective pitch of bladelets is preset; proximity of opposing bladelets to each other minimises separation between rotor discs, enabling dynamic platform to be gimballed and eliminating need for cyclic pitch controls. Vertical movement is accomplished by varying rotational speed of discs; control for hover and horizontal flight by CG transfer and varying angle of attack.

Following data apply to VertiJet POC prototype:
WEIGHTS AND LOADINGS:
  Max payload more than 45.4 kg (100 lb)
PERFORMANCE (estimated):
  Max level speed more than 162 kt (300 km/h; 186 mph)
  Max range more than 810 n miles (1,500 km; 932 miles)

UPDATED

TAC
TAIWAN AEROSPACE CORPORATION
17th Floor, 169 Jen Ai Road, Section 4, Taipei
Tel: (+886 2) 27 71 66 81
Fax: (+886 2) 27 71 67 27
PRESIDENT: Dr George Liu

WORKS: Taichung

Established 27 September 1991 as intended foundation for civil aircraft industry; 29 per cent owned by Taiwan government; start-up capital US$200 million to US$250 million; aim was to develop national manufacturing capability for aircraft, engines, avionics and materials by 2000. Recent work included manufacture of secondary structural components for the AIDC Ching-Kuo fighter.

Joint venture with Swearingen (see Sino-Swearingen Aircraft Corporation entry in US section) as financial partner in SJ30-2 business jet announced late 1994. Also in late 1994, TAC acquired 95 per cent share in Taiwan-based Air Asia overhaul and maintenance company.

VERIFIED

TURKEY

ALP
ALP AVIATION
Ankara Asvalti 8.km (Arcelik Tesisleri Yani), TR-26110 Eskisehir
Tel: (+90 222) 228 03 30
MANAGING DIRECTOR: Tuncer Alpata

Formed May 1999 as joint venture company by Sikorsky Aircraft Turkey Inc and The Alpata Group (50 per cent each) to manufacture high-technology, precision-machined aerospace and defence components and assemblies for global market in modern 6,500 m² (70,000 sq ft) facility. Is already exporting helicopter components to Sikorsky USA under two contracts awarded in 1998 with total potential value of more than US$60 million.

UPDATED

TAI
TURKISH AEROSPACE INDUSTRIES INC
(TUSAS Havacilik ve Uzay Sanayii A S)
PO Box 18, TR-06692 Kavaklidere, Ankara
Tel: (+90 312) 811 18 00
Fax: (+90 312) 811 14 25 and 811 14 08
e-mail: bcelik@tai.com.tr
Web: http://www.tai.com.tr
GENERAL MANAGER: Kaya Ergenç
GENERAL CO-ORDINATOR: Lt Gen (Ret'd) Vural Avar
EXECUTIVE DIRECTOR, PROGRAMMES:
  Brig Gen (Ret'd) Enver Dayanir
MARKETING AND SALES MANAGER: Yilmaz Güldoğan
PUBLIC RELATIONS CO-ORDINATOR: Nursel Köran
PUBLIC RELATIONS SPECIALIST: Bican Çelik

Turkish Aerospace Industries (TAI), the centre of technology in design, development, manufacturing, integration of aerospace systems, modernisation and after-sales support in Turkey, was established on 15 May 1984. TAI is a majority owned Turkish company made up of Turkish (51 per cent) and US (49 per cent) partners. Shareholders are Turkish Aircraft Industries Inc (49 per cent), Turkish Armed Forces Strengthening Foundation (1.9 per cent), Turkish Aeronautical Association (0.1 per cent), Lockheed Martin of Turkey Inc (42 per cent) and General Electric (7 per cent).

TAI's aircraft facilities, located in Ankara, are furnished with high-technology machinery and equipment that provide extensive capabilities ranging from parts manufacturing to aircraft assembly, flight test and delivery from aerodrome and works at Akinci. Quality system meets world standards including NATO AQAP-120, ISO-9001, MIL-Q-9858A and Boeing D1-9000.

With experience in aircraft and aerostructures manufacturing business, TAI is a supplier for Lockheed Martin Tactical Aircraft Systems (LMTAS), Lockheed Martin Vought Systems, Boeing, Airbus, Sikorsky, CASA, Agusta, Eurocopter, Sonaca, Northrop Grumman and others.

TAI's experience includes co-production of F-16 fighters, CN-235 light transports, SF-260 trainers, AS 532 Cougar and MD 902 helicopters, as well as design and development of UAVs, target drones and agricultural aircraft. The first MD 902 fuselage was delivered to MD Helicopters in October 1999. The company is also prime contractor for the Turkish armed forces attack helicopter, Turkish National Police helicopter and Turkish UAV and target drone production programmes. As a full member of the Airbus Military Company, TAI is engaged in the development of the A400M with major European companies.

TAI's core business also includes modernisation, modification and systems integration programmes and after-sales support. Major programmes of this kind are modification of 15 former Turkish Navy S-2E Tracker maritime patrol aircraft into firefighting aircraft; CN-235 modification for Open Skies; and CN-235 modifications for Special Forces, as well as modification of CN-235 platforms for MPA/MSA missions for the Turkish Navy and Coastguard. The 'glass cockpit' retrofit programme for Turkish Army Black Hawk helicopters was completed in November 1999.

UPDATED

TAI (AIRTECH) CN-235M
TYPE: Twin-turboprop transport.
PROGRAMME: Under a contract with CASA of Spain, TAI produced 50 CN-235 transports for the Turkish Air Force between 1991 and 1998, manufacturing 92 per cent of the total airframe, including 20 per cent made of composites. See earlier *Jane's* for further details. Production is continuing with six ASW Persuaders for the Turkish Navy and three maritime patrol Persuaders for the Turkish Coastguard, ordered on 23 September 1998.

UPDATED

TAI (LOCKHEED MARTIN) F-16C/D FIGHTING FALCON
Turkish name: Savasan Sahin (Fighting Falcon)
TYPE: Multirole fighter.
PROGRAMME: TAI was contracted by Lockheed Martin Tactical Aircraft Systems to co-produce 80 Block 50D F-16C/D for the Turkish Air Force from 1996 under the Peace Onyx II programme and to manufacture rear and centre-fuselages and wings for US-built aircraft. Originally ordered as 68 single-seat and 12 two-seat aircraft, these emerged as 60 F-16Cs and 20 F-16Ds, the last of which was delivered on 13 November 1999. ECM (ALQ-178) upgrading and 'Falcon-up' structural modification of 185 Turkish Air Force F-16s had also been completed by the end of that year.

UPDATED

TAI (BELL) AH-1Z KINGCOBRA
TYPE: Attack helicopter.
PROGRAMME: Turkish Army ATAK programme. Bids from Boeing (Apache Longbow) and Eurocopter (Tiger) rejected; AH-1Z (see Bell entry in US section) selected in preference to Agusta A 129 International and Kamov Ka-50-2 (latter reportedly the runner-up) in July 2000; contract negotiations under way in September, with signature expected in early 2001. TAI to be prime contractor, with Bell Helicopter Textron as primary subcontractors.

CUSTOMERS: Turkish Army requirement for 145, to be acquired in batches of 50, 50 and 45; delivery schedule November 2003 to January 2011.
COSTS: Initial 50-aircraft purchase quoted as US$1.5 billion (2000); estimated US$4 billion for full 145.

NEW ENTRY

TAI (EUROCOPTER) COUGAR Mk I
TYPE: Multirole helicopter.
PROGRAMME: Within the framework of the Eurotai consortium, TAI is co-producing 28 of 30 Cougar Mk I helicopters (10 for the Turkish Army and 20 for the Turkish Air Force), under a 1997 contract (Phenix II programme) valued at US$434 million. These comprise 24 AS 532ULs (Army six utility, four SAR; Air Force 14 SAR) and six AS 532ALs (Air Force, combat SAR). The first two helicopters were search and rescue aircraft, manufactured in France by Eurocopter and delivered in April and May 2000; the TAI programme covers fabrication, assembly, systems integration and flight test. First Turkish-assembled Cougar (an AS 532UL) was handed over to the Army on 31 May 2000, at which time 11 others were under construction; deliveries are due to be completed in February 2003.

UPDATED

TAI ZIU
TYPE: Agricultural sprayer and firefighting aircraft.
PROGRAMME: Turkish government preliminary design specification issued 1 April 1997; conceptual and preliminary design phases completed 1 July 1997; programme go-ahead 28 November 1997; construction of prototype began 1998; public debut at IDEF '99 trade show in Ankara, September/October 1999. Partial programme support by Science and Technology Research Council and Ministry of Industry and Technology; also backed by D-8 group of developing countries (Bangladesh, Egypt, Indonesia, Iran, Malaysia, Nigeria, Pakistan and Turkey), which viewed as primary sales targets. Two prototypes, first of which made its maiden flight on 26 June 2000, some six months later than originally planned; second flight two days later; six flights by end of September 2000. Subject to production go-ahead, which still dependent on market research. ZIU will be certificated to JAR 23 (Normal category) for agricultural operation. Two-seat and turboprop versions to be developed later. Name is acronym of *Zirai Ilaçlama Uçağı*: agricultural aircraft.
CUSTOMERS: Potential customers include the other seven Developing Group (D-8) countries: Bangladesh, Egypt, Indonesia, Iran, Malaysia, Nigeria and Pakistan.
DESIGN FEATURES: Low-wing monoplane of typical crop-sprayer configuration, including hopper ahead of cockpit, but untypical in having cantilever wing. 3-D CAD design. Constant-chord wings with 7° dihedral and raked tips; angular tail surfaces. Able to use rough and unprepared airstrips.
FLYING CONTROLS: Conventional and manual. Elevators and rudder horn-balanced; trim tab in port elevator.
STRUCTURE: Wings aluminium alloy stressed skin; tubular metal fuselage main frame; hopper, removable fuselage side panels and wingtips of composites.
LANDING GEAR: Tailwheel type; fixed.
POWER PLANT: One 447 kW (600 hp) Orenda OE-600A turbocharged, liquid-cooled eight-cylinder V-type piston engine; three-blade Hartzell metal propeller. Fuel capacity 400 litres (106 US gallons; 88.0 Imp gallons).
EQUIPMENT: Hopper capacity 1,500 litres (396 US gallons; 330 Imp gallons).
DIMENSIONS, EXTERNAL:

Wing span	15.95 m (52 ft 4 in)
Wing chord, constant	2.34 m (7 ft 8¼ in)
Length overall	10.43 m (34 ft 2¾ in)
Fuselage max width	1.305 m (4 ft 3½ in)
Height overall: tail down	3.03 m (9 ft 11¼ in)
flying attitude	4.01 m (13 ft 1¾ in)
Tailplane span	4.86 m (15 ft 11¼ in)
Wheel track	2.69 m (8 ft 10 in)
Wheelbase	7.075 m (23 ft 2½ in)
Propeller diameter	2.69 m (8 ft 10 in)

AREAS:

Horizontal tail surfaces (total)	5.91 m² (63.61 sq ft)

WEIGHTS AND LOADINGS:

Basic weight empty, equipped	1,675 kg (3,693 lb)
Operating weight empty	1,765 kg (3,891 lb)
Max T-O weight	3,500 kg (7,716 lb)
Zero-fuel weight	3,215 kg (7,088 lb)
Max power loading	7.83 kg/kW (12.86 lb/hp)

PERFORMANCE (estimated):

Max cruising speed at S/L	163 kt (302 km/h; 188 mph)
Stalling speed	57 kt (105 km/h; 66 mph)
T-O run (ISA +16°C)	250 m (820 ft)
Landing run (ISA +16°C)	220 m (720 ft)
Range with max fuel	486 n miles (900 km; 559 miles)
Endurance	4 h

UPDATED

First prototype of the Turkish Aerospace Industries ZIU *2001/0092518*

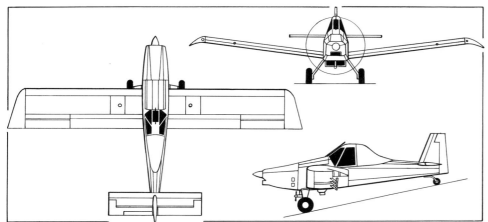

TAI ZIU agricultural and firefighting aircraft (*James Goulding/Jane's*) *2001/0092524*

UKRAINE

AEROPRAKT

AEROPRAKT FIRMA
alya 112, 252148 Kiev
Tel: (+38 044) 457 91 59
Fax: (+38 044) 457 92 93
e-mail: air@prakt.kiev.ua
MANAGER: Oleg V Litovshenko
CHIEF DESIGNER: Yuri V Yakovlev

FAR EAST AGENT:
Revival Technology Pte Ltd
41 Bukit Pasoh Road, Singapore 089855
Tel: (+65) 533 91 15
Fax: (+65) 533 22 06

WEST EUROPEAN AGENT:
Fachschule für Ultraleichtflug GmbH
Hufeisenstrasse 55, D-49401 Damme, Germany
Tel: (+49 54) 91 42 88
Fax: (+49 54) 91 48 01
e-mail: uwerner@os-net.de
Web: http://www.t-online.de/home/FUL.Damme

US AGENT:
Spectrum Aircraft Corporation
PO Box 1381, Sebring, Florida 33872
Tel: (+1 941) 314 97 88
Fax: (+1 941) 314 02 85
e-mail: jhunter@strato.net
Web: http://www.spectrumaircraft.com

Aeroprakt was established in Kiev in 1986, when Yuri Yakovlev, then of the Aeropract organisation of Samara (which see in Russian section) was invited to work at Antonov Design Bureau. Aeroprakt (k for Kiev) is now expanding its world sales network. Production of the range is at the rate of 1.5 per month, with target of 80 per year.

UPDATED

AEROPRAKT-20
US marketing name: Vista
TYPE: Tandem-seat ultralight.
PROGRAMME: Design started September 1990; construction of prototype began November 1990; first flight 5 August 1991; construction of first production aircraft started September 1991, first flown 15 August 1993; named Chervonets; prototype appeared abroad for first time in Hodkovice, Czech Republic, at European Microlight Championship, August 1993, gaining ninth place; fourth production A-20 won third place at World Microlight Championship, Poznan, Poland, August 1994; seventh production A-20 gained second place at EMC '95, Little Rissington, UK.
CURRENT VERSIONS: **A-20:** Basic aircraft, *as described in detail.*
A-20R912: Described separately.
A-20S: Variant under development by 1997, with large multichannel radiometer built into its nose for a wide range of ecological survey work. No known manufacture.

Aeroprakt-20 tandem-seat ultralight (*Paul Jackson/Jane's*) *2001/0100422*

UKRAINE: AIRCRAFT—AEROPRAKT

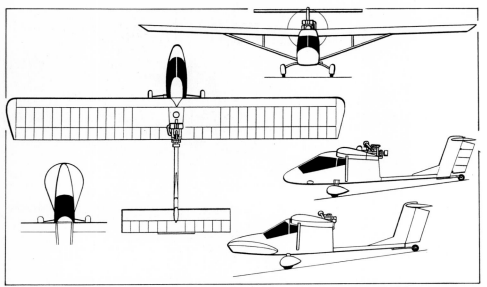

Aeroprakt-20, with additional side and scrap plan views of the radiometer-equipped A-20S *(James Goulding/Jane's)*

A-20SKh: Described separately.
A-20 Vista STOL: With 47.8 kW (64.1 hp) Rotax 582 UL; cruising speed 76 kt (140 km/h; 87 mph).
A-20 Vista SS: Rotax 582 and wing span of 10.20 m (33 ft 5½ in); cruising speed 82 kt (152 km/h; 94 mph).
CUSTOMERS: First customer delivery August 1993; exports to Czech Republic, Germany, Hungary, Jordan, Russia and United Arab Emirates; last-mentioned acquired seven for Umm-al Qaiwain Flying Club. Total of 29 A-20s of all types produced by mid-2000.
COSTS: US$35,000 complete (2000).
DESIGN FEATURES: High-wing, T-tail configuration, optimised for rapid disassembly for transport by trailer. Tail surfaces mounted on aluminium tube which detaches at joint with the nacelle.
Unswept wing; P-IIIa-15 wing section, thickness/chord ratio 15 per cent; chamfered tips; dihedral 1° 30′; incidence at roots 3° 30′, twist 3°.
FLYING CONTROLS: Manual. Single-piece, rod-operated flaperons extend full length of wing trailing-edge; no tabs. Single-piece elevator with protruding, in-flight adjustable tab. Full-height rudder with no tab.
STRUCTURE: Riveted aluminium wing structure; leading-edge D box closed by I-section main spar; stamped wing ribs; bent sheet rear false spar; partially fabric covered. Control surfaces similar to wings. All-composites honeycomb sandwich fuselage pod.
LANDING GEAR: Mainwheels, with speed fairings, on single leaf-spring; steerable tailwheel linked to rudder.
POWER PLANT: One 37.0 kW (49.6 hp) Rotax 503 UL-2V two-stroke piston engine driving a Junkers-Profly (Czech) three-blade ground-adjustable pitch pusher propeller via 3:1 reduction gearbox. Rotax 462, 582, 618 and 912 engines optional. Fuelling point on starboard side of nacelle. Fuel capacity 38 litres (10.0 US gallons; 8.4 Imp gallons).
EQUIPMENT: Ballistic recovery parachute.
DIMENSIONS, EXTERNAL:
Wing span   11.40 m (37 ft 4¾ in)
Length overall   6.67 m (21 ft 10½ in)
Height overall   2.17 m (7 ft 1½ in)
AREAS:
Wings, gross   15.68 m² (168.8 sq ft)
WEIGHTS AND LOADINGS:
Weight empty   220 kg (485 lb)
Max T-O weight   450 kg (992 lb)
PERFORMANCE (Rotax 503 engine):
Max level speed   75 kt (140 km/h; 87 mph)
Max cruising speed   65 kt (120 km/h; 75 mph)
Stalling speed, flaperons down 27 kt (50 km/h; 31 mph)
Max rate of climb at S/L   180 m (590 ft)/min
T-O and landing run   80 m (265 ft)
Max range   216 n miles (400 km; 248 miles)
UPDATED

AEROPRAKT-20SKh

TYPE: Tandem-seat agricultural sprayer ultralight.
PROGRAMME: Development of Aeroprakt-20 ultralight; suffix SKh for *selbskokhozyaistvennyi*: agricultural. Prototype construction began July 1997; first flight September 1997.
CUSTOMERS: One prototype built by early 1999. No subsequent production reported.
Description of A-20 applies generally, except as below.
DESIGN FEATURES: Agricultural version of A-20 with increased power and MTOW; sweptback (1° 30′) wing; and two chemical tanks with spraybars.
POWER PLANT: One 59.6 kW (79.9 hp) Rotax 912 UL flat-four driving a three-blade, ground-adjustable pitch Ivoprop propeller. Fuel as A-20; oil capacity 3 litres (0.8 US gallon; 0.7 Imp gallon).

ACCOMMODATION: Pilot in front seat; rear seat, with dual controls, retained for instructor (in agricultural training role) or for ferrying ground assistant.
EQUIPMENT: Two chemical tanks, each 90 litres (23.8 US gallons; 19.8 Imp gallons), scabbed to fuselage sides; streamlined spraybars with four or six atomisers.
WEIGHTS AND LOADINGS:
Weight empty   320 kg (705 lb)
Max T-O weight   550 kg (1,212 lb)
PERFORMANCE:
Never-exceed speed (V_{NE})   91 kt (170 km/h; 105 mph)
Max level speed   81 kt (150 km/h; 93 mph)
Cruising speed: max   65 kt (120 km/h; 75 mph)
                econ   49 kt (90 km/h; 56 mph)
Stalling speed, power off:
   flaps up   38 kt (70 km/h; 44 mph)
   flaps down   33 kt (60 km/h; 38 mph)
Max rate of climb at S/L   210 m (689 ft)/min
T-O run   110 m (360 ft)
Landing run   100 m (330 ft)
UPDATED

AEROPRAKT-20R912 SKY CRUISER

US marketing name: Vista Cruiser
TYPE: Tandem-seat ultralight.
PROGRAMME: See Aeroprakt-20; designation indicates increased engine power provided by flat-four Rotax 912. Alternatively designated **Aeroprakt-20M**. Construction of prototype Sky Cruiser began February 1997; first flight 11 May 1997. Gained second place in 1st World Air Games, Turkey, September 1997 and second place in 1998 World Microlight Cup, Hungary.
CUSTOMERS: Five aircraft ordered and built by early 1999 (latest data); production continues.
DESIGN FEATURES: Development of A-20 with increased power, shorter wing and tailplane spans, engine cowling, constant-speed propeller, fully balanced control surfaces, smaller wheels and more streamlined wing struts and landing gear. Sweepback 1° 30′.
Description of A-20 applies generally, except as below.
POWER PLANT: One 59.6 kW (79.9 hp) Rotax 912 UL flat-four driving a three-blade Ivoprop constant-speed propeller. Standard fuel as A-20; optional capacity 54 litres (14.3 US gallons; 11.9 Imp gallons); oil capacity 3 litres (0.8 US gallon; 0.7 Imp gallon).
DIMENSIONS, EXTERNAL:
Wing span   9.60 m (31 ft 6 in)
AREAS:
Wings, gross   13.16 m² (141.7 sq ft)
WEIGHTS AND LOADINGS:
Weight empty   260 kg (573 lb)
Max T-O weight   450 kg (992 lb)
PERFORMANCE:
Max level speed   113 kt (210 km/h; 130 mph)
Cruising speed: max   97 kt (180 km/h; 112 mph)
                econ   65 kt (120 km/h; 75 mph)
Stalling speed, power off, flaps down
                             30 kt (55 km/h; 35 mph)
Max rate of climb at S/L   360 m (1,181 ft)/min
T-O and landing run   80 m (265 ft)
Range:
 with normal tankage   297 n miles (550 km; 341 miles)
 with optional tankage 421 n miles (780 km; 484 miles)
UPDATED

AEROPRAKT-22

UK marketing name: Foxbat
US marketing name: Valor
TYPE: Side-by-side ultralight/kitbuilt.
PROGRAMME: Design started February 1990, construction of prototype began September 1994; first flight 21 October 1996; certified to German BFU-95; production began in early 1999.
CURRENT VERSIONS: **Aeroprakt-22 Shark:** Original version, with 59.6 kW (70.9 hp) Rotax 912 UL engine. Data in 1999-2000 and previous *Jane's*.
Following versions have uprated engine.
FUL A22: Marketed from 1999 by FUL of Damme, Germany.
Foxbat: UK version, marketed from 2000 by Small Light Aeroplane Company Ltd at Otherton, Staffordshire (www.foxbat.co.uk). Prototype, G-FBAT (16th airframe) first flown after kit assembly in UK 12 August 2000. Certification under way to BCAR S.
Valor: US version.
Description applies to FUL A22.
CUSTOMERS: Total of eight A-22s built by early 1999, including at least one for Umm-al Qaiwain Flying Club.
COSTS: Foxbat kit £14,375, plus engine £7,300, excluding VAT (2000).

Aeroprakt-22 two-seat ultralight in Foxbat guise *(Paul Jackson/Jane's)*

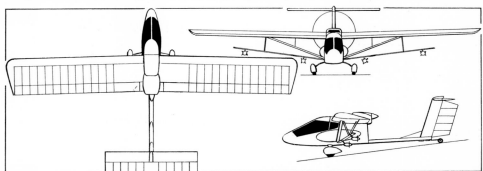

Aeroprakt A-20SKh two-seat agricultural aircraft *(James Goulding/Jane's)*

AEROPRAKT—AIRCRAFT: UKRAINE

DESIGN FEATURES: Constant-chord wings and horizontal tail surfaces. Wings swept forward 2° 30′; P-IIIa-15 wing section, thickness/chord ratio 15 per cent; chamfered tips; dihedral 1° 30′; incidence at root 4°; twist 2° 30′.
FLYING CONTROLS: Manual, by pushrods and cables. Full-span slotted flaperons with trim tab to starboard; single-piece elevator with tab; and sweptback rudder with ground-adjustable tab.
STRUCTURE: Riveted aluminium wing structure; leading-edge D box closed by I-section main spar; stamped wing ribs; bent sheet rear false spar; Ceconite covering on wings (except metal leading-edge), rudder and elevator. Fin and tailplane similar to wings. Aluminium fuselage with profiled sheet and fluted skin, stamped bulkheads, steel and aluminium tubing. Extensive glazing. Glass fibre engine cowling, wheel spats, wing fillets and fin-tip.
LANDING GEAR: Cantilever composites spring mainwheel legs. Hydraulic mainwheel brakes. Steerable nosewheel; small tailwheel protects ventral strake from nose-high landings.
POWER PLANT: One 73.5 kW (98.6 hp) Rotax 912 ULS flat-four driving an Aeroprakt three-blade ground-adjustable pitch propeller or (Foxbat) Newton two-blade. Fuel tank in each wingroot; combined capacity 85 litres (22.5 US gallons; 18.7 Imp gallons). Optional capacity 135 litres (35.6 US gallons; 29.7 Imp gallons).
ACCOMMODATION: Two persons, side by side. Large upward-hinged window/door on each side.
EQUIPMENT: Ballistic parachute.

Note that data vary slightly between German, UK and US variants.

DIMENSIONS, EXTERNAL:
Wing span	10.20 m (33 ft 5½ in)
Length overall	6.30 m (20 ft 8 in)
Height overall	2.30 m (7 ft 6½ in)
Tailplane span	3.00 m (9 ft 10 in)
Propeller diameter	1.68 m (5 ft 6¼ in)

DIMENSIONS, INTERNAL:
Cabin max width	1.20 m (3 ft 11¼ in)

AREAS:
Wings, gross	13.70 m² (147.5 sq ft)

WEIGHTS AND LOADINGS:
Weight empty, equipped	285 kg (628 lb)
Max T-O weight: ultralight	450 kg (992 lb)
design	520 kg (1,146 lb)

PERFORMANCE (Foxbat):
Never-exceed speed (V_{NE})	106 kt (196 km/h; 122 mph)
Max level speed	96 kt (177 km/h; 110 mph)
Cruising speed at 65% power	84 kt (155 km/h; 96 mph)
Stalling speed, flaperons up	33 kt (62 km/h; 38 mph)
Max rate of climb at S/L	305 m (1,000 ft)/min
T-O to 15 m (50 ft)	100 m (330 ft)
Range with max fuel	434 n miles (804 km; 500 miles)
g limits	+4/−2

UPDATED

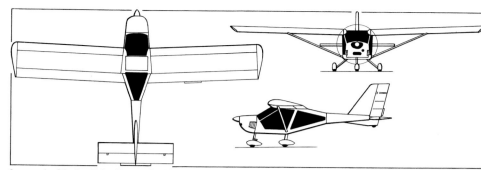
Aeroprakt-22 ultralight *(James Goulding/Jane's)* 2000/0054615

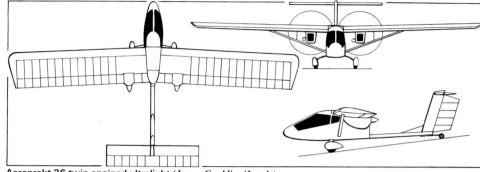
Aeroprakt-26 twin-engined ultralight *(James Goulding/Jane's)* 1999/0051502

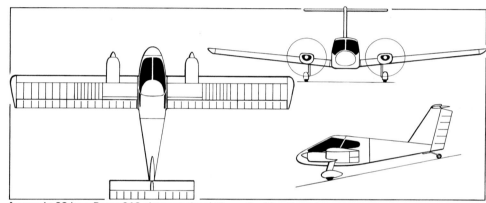
Aeroprakt-28 (two Rotax 912 piston engines) *(James Goulding/Jane's)* 1999/0051503

AEROPRAKT-24

TYPE: Two-seat amphibian ultralight kitbuilt.
PROGRAMME: Design under way by 1996. None known to have been built by 2000, although confirmed to be still under development.
DESIGN FEATURES: Braced parasol monoplane, with light central cabane and single bracing strut and jury strut each side; flying-boat hull with two steps; tubular tailboom carries sweptback vertical surfaces with unswept horizontal surfaces at tip. General concept based on Che-20, amateur-built at Samara, Russia, combined with wing and tail unit of Aeroprakt-20.
FLYING CONTROLS: Manual. Full-span flaperons. Trim tab on single-piece elevator.
LANDING GEAR: Tailwheel type, with retractable, but externally stowed, mainwheels. Stabilising float under each outer wing.
POWER PLANT: One 47.8 kW (64.1 hp) Rotax 582 UL twin-piston engine at forward end of strut projecting from wing centre-section, with fore and aft supports. Fuel capacity 40 litres (10.6 US gallons; 8.8 Imp gallons).
ACCOMMODATION: Two persons, side by side in enclosed cabin.

DIMENSIONS, EXTERNAL:
Wing span	10.60 m (34 ft 9¼ in)
Length overall	approx 6.74 m (22 ft 1½ in)

AREAS:
Wings, gross	14.80 m² (159.3 sq ft)

WEIGHTS AND LOADINGS:
Weight empty	250 kg (551 lb)
Max T-O weight	450 kg (992 lb)

PERFORMANCE (estimated):
Max level speed	70 kt (130 km/h; 81 mph)
Nominal cruising speed	54 kt (100 km/h; 62 mph)
Stalling speed, flaperons down	33 kt (60 km/h; 38 mph)
T-O run: on land	90 m (295 ft)
on water	150 m (495 ft)
Endurance	4 h

UPDATED

AEROPRAKT-26

US marketing name: Vulcan

TYPE: Tandem-seat ultralight twin.
PROGRAMME: Development of A-20 with two engines; commissioned by Gulf Aviation Technologies; prototype construction began March 1996; first flight 18 November 1997.
CUSTOMERS: One prototype built and flown by early 1999. One reported with Umm-al Qaiwain Flying Club, February 1999.

Description of A-20 applies generally, except as below.
DESIGN FEATURES: Introduces twin-engine safety margins to A-20 design; able to take off on one engine. Sweepback increased to 3°; tailfin height and rudder chord both increased.
POWER PLANT: Two 47.8 kW (64.1hp) Rotax 582 UL two-cylinder two-stroke engines, each driving a three-blade, ground-adjustable pitch Ivoprop propeller. Alternatively, two 38.3 kW (51.6 hp) Rotax 462 or 34.0 kW (45.6 hp) Rotax 503 twin-piston engines. Fuel capacity 90 litres (23.8 US gallons; 19.8 Imp gallons) standard; 180 litres (47.5 US gallons; 39.6 Imp gallons) optional.
SYSTEMS: Electrical system with 12 V DC, 14 Ah battery for electric starter.

DIMENSIONS, EXTERNAL:
Wing span	11.40 m (37 ft 4¾ in)
Length: fuselage	5.75 m (18 ft 10½ in)
overall	6.67 m (21 ft 10½ in)
Height overall	1.90 m (6 ft 2¾ in)
Propeller diameter	1.70 m (5 ft 7 in)

AREAS:
Wings, gross	15.68 m² (168.8 sq ft)

WEIGHTS AND LOADINGS:
Weight empty	283 kg (624 lb)
Max T-O weight*	550 kg (1,212 lb)

* or local ultralight limit

PERFORMANCE:
Max level speed	97 kt (180 km/h; 112 mph)
Cruising speed: max	65 kt (120 km/h; 75 mph)
econ	49 kt (90 km/h; 56 mph)
Stalling speed, power off:	
flaps up	36 kt (65 km/h; 41 mph)
flaps down	30 kt (55 km/h; 35 mph)
Max rate of climb at S/L	548 m (1,800 ft)/min
T-O and landing run	100 m (330 ft)
Range with max optional fuel	334 n miles (619 km; 384 miles)

UPDATED

AEROPRAKT-28

TYPE: Four-seat utility twin.
PROGRAMME: Project began in December 1997; first flight was due in early 1999, when eight aircraft were reported to be in production, but had not been reported by early 2000.

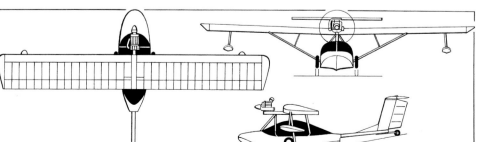
Aeroprakt-24 two-seat light amphibian *(James Goulding/Jane's)* 0015635

UKRAINE: AIRCRAFT—AEROPRAKT to ANTONOV

DESIGN FEATURES: Low-wing, twin-engined configuration with T tail; sole four-seater in current Aeroprakt range.
Unswept, constant-chord wings of P-IIIa-15 section; thickness/chord ratio 15 per cent; chamfered wingtips. Dihedral 5°; incidence 5°; twist 3°.
FLYING CONTROLS: Manual. Conventional ailerons, flaps and rudder; single-piece, mass-balanced elevator.
STRUCTURE: Composites cabin with riveted metal monocoque rear fuselage; tailplane, of composites ribs and skin built on metal spar, is reinforced version of A-20 unit; metal rudder structure with fabric covering.
LANDING GEAR: Tailwheel type; fixed. Speed fairings on each main unit.
POWER PLANT: Two 73.5 kW (98.6 hp) Rotax 912 ULS flat-fours. Fuel capacity 160 litres (42.3 US gallons; 35.2 Imp gallons).
ACCOMMODATION: Up to four persons in side-by-side pairs.
DIMENSIONS, EXTERNAL:
Wing span   12.10 m (39 ft 8½ in)
Wing aspect ratio   8.8
Length overall   7.045 m (23 ft 1¼ in)
Height overall   2.33 m (7 ft 7¾ in)
AREAS:
Wings, gross   16.70 m² (179.8 sq ft)
WEIGHTS AND LOADINGS:
Weight empty   470 kg (1,036 lb)
Max T-O weight   950 kg (2,094 lb)
Max wing loading   56.9 kg/m² (11.65 lb/sq ft)
Max power loading   6.46 kg/kW (10.62 lb/hp)
PERFORMANCE:
Max level speed   119 kt (220 km/h; 137 mph)
Max cruising speed   81 kt (150 km/h; 93 mph)
Stalling speed   41 kt (75 km/h; 47 mph)
Max rate of climb at S/L 300 m (984 ft)/min
T-O and landing run   95 m (315 ft)
Range   809 n miles (1,500 km; 932 miles)
UPDATED

ANTONOV

AVIATSIONNY NAUCHNO-TEKHNICHESKY KOMPLEKS IMENI O K ANTONOVA (Aviation Scientific-Technical Complex named for O K Antonov)
ulitsa Tupoleva 1, 252062 Kiev
Tel: (+38 044) 442 70 98
Fax: (+38 044) 443 00 05
Telex: 131309 OZON
GENERAL DESIGNER AND PRESIDENT, MEDIUM TRANSPORT AIRCRAFT INTERNATIONAL CONSORTIUM: Pyotr Balabuyev
GENERAL DIRECTOR, MEDIUM TRANSPORT AIRCRAFT INTERNATIONAL CONSORTIUM: Leonid Terentyev
CHIEF DESIGNER: Anatoly Vovnyanko
DEPUTY CHIEF DESIGNER: Genrich G Ongirsky
PUBLIC RELATIONS OFFICER: Andrei Sovenko

Antonov OKB was founded in 1946 by Oleg Konstantinovich Antonov, who died 4 April 1984, aged 78. More than 22,000 aircraft of over 100 types and versions of Antonov design have been built; more than 1,500 have been exported, to 42 countries. Other production includes gliders, hang gliders and trikes as well as urban transport vehicles. Antonov received Aviation Register approval on 30 December 1992 to develop civil aircraft. It also operates its own cargo airline with a fleet of eight An-124s, one An-22, three An-12s, two An-32Ps, two An-24s and one each An-26 and An-74. Plans for privatisation of the state-owned Antonov have been in preparation since 1997. Company parents the Medium Transport Aircraft International Consortium created by Russia and Ukraine in February 1996 and formally established on 18 May 1999.

Production plants associated with ANTK Antonov include Aviastar, Aviakor, AAK Progress, KiGAZ, Kharkov Air Carrier, Polyot (Omsk) and TAPO.
UPDATED

Antonov An-12 of Czech Republic Air Force *(Paul Jackson/Jane's)* 2001/0092089

Antonov An-24 of Slovak Air Force *(Paul Jackson/Jane's)* 2001/0092090

ANTONOV An-2
NATO reporting name: Colt
TYPE: Light utility biplane.
PROGRAMME: Prototype flew as SKh-1 on 31 August 1947. An-2s were built at Kiev-Svyetoshino until mid-1960s. Soviet production comprised about 3,480 An-2s and 506 An-2Ms. Production then transferred to PZL Mielec, Poland, from where some 11,650 delivered from 1960 onwards. Antonov estimates over 15,500 built in former USSR, Poland and China, of which over 4,000 (2,315 in Russia, 2000) remain in service with civil operators and with 22 air arms, including 180 with Russian forces; China acquired licence and has built **Yunshuji-5 (Y-5)** versions from 1957 to date at both Nanchang and Shijiazhuang. China is now sole source of new An-2s (see SAMC entry in Chinese section).
UPDATED

ANTONOV An-3
TYPE: Light utility turboprop biplane.
PROGRAMME: Marketing was resumed in 1997 of this TVD-20-engined version of the An-2. Prototype (UR-BWD) demonstrated at MAKS '97, Moscow, August 1997, was Polish-built An-2 converted by Antonov; Polyot-produced prototype (9801) flew 19 February 1998. Original An-3 had flown on 13 May 1980, and passed state testing in 1991; programme was transferred to Polyot in 1993.

First order placed 16 February 1999 by Evenki administrative region of Siberia for four An-3Ts to be delivered from Polyot at Omsk (which see in Russian section); 31 orders by 2000; first production An-3 for Evenki completed April 2000; prospective contracts include 30 for Tyumenaviatrans, six for Russian Federal Border Guard and others for Yakutia. AP-23 certification granted September 2000; cost then quoted as US$450,000 to US$500,000; production target for 2000 was 19. Second conversion line planned at Vinnitsa, Ukraine. An-3SKh agricultural sprayer and An-3P firefighter also offered in addition to passenger An-3TK and utility An-3T. All aircraft are An-2 conversions. Data in *Jane's Aircraft Upgrades*.
UPDATED

ANTONOV An-12
NATO reporting name: Cub
TYPE: Medium transport.
PROGRAMME: Prototype (7900101), built at Irkutsk, flew 16 December 1957, with Kuznetsov NK-4 turboprops, as rear-loading development of An-10 airliner; 1,243 built with AI-20K engines for military and civil use, ending in USSR in 1972; GAZ (Plant) 34 at Tashkent produced 830 between 1961 and 1972; GAZ 40 at Voronezh 258 in 1961-65; and GAZ 90 at Irkutsk 155 in 1957-62.

An-12BP ('Cub') was standard medium-range paratroop and cargo transport of Soviet Military Transport Aviation (VTA) from 1959; replacement with Il-76 began 1974, but about 170 remain in air force service, plus at least 25 An-12BK/PP/PPS 'Cub-A/B/C/D' electronic warfare conversions in air forces and Naval Aviation. More than 100 (of 183 originally exported) are used by nine other air forces; further 190 flown by some 70 freight airlines. Shaanxi Aircraft Company, China, manufactures redesigned **Yunshuji-8 (Y-8)** transport version and derivatives (see SAC entry in Chinese section).
UPDATED

ANTONOV An-24
NATO reporting name: Coke
TYPE: Twin-turboprop transport.
PROGRAMME: An-24 first flew 20 October 1959. Production at Irkutsk (GAZ 91), Ulan Ude (GAZ 99) and Kiev (GAZ 473) ended in 1979, after 1,200 or so had been delivered (mostly from Kiev, which may have produced as many as 985, although Irkutsk contributed 164 An-24Ts). A few remain in Russian Air Forces inventory; nearly 600 are distributed between 125 civil operators. Versions known as **Y7** (which see) continue in production at Xian in China.
UPDATED

ANTONOV An-26
NATO reporting name: Curl
TYPE: Twin-turboprop transport.
PROGRAMME: First exhibited 1969 Paris Air Show; approximately 1,410 built by GAZ 473 at Kiev, 1968-85, before being superseded in production by An-32;

Antonov An-3T demonstrator *(Paul Jackson/Jane's)* 2001/0092111

ANTONOV—AIRCRAFT: UKRAINE

Slovak Air Force An-26 tactical transport *(Paul Jackson/Jane's)*

derivative **Y7H-500** built by Xian Aircraft Company (see XAC in Chinese section). Conversion of military An-26s (and An-30s) into civilian transports by Ukraine's 410 Aviation Repair Depot, in conjunction with Antonov and Aviant, was officially approved in 1999. Details of subvariants and technical specification last appeared in 2000-01 *Jane's*.

UPDATED

ANTONOV An-28

See PZL (Polish Aviation Factory) M-28 in Polish section.

ANTONOV An-32

NATO reporting name: Cline
Indian Air Force name: Sutlej

TYPE: Twin-turboprop, short/medium-range transport.
PROGRAMME: Prototype (SSSR-83966, converted An-26) first flew 9 July 1976; exhibited 1977 Paris Air Show; further two pre-series aircraft; export deliveries to India began July 1984; Indian contract completed before other deliveries; production, originally at 40 a year, largely for Russian Federation and Associated States (CIS) armed forces, from GAZ 473 (now Aviant) at Kiev.
CURRENT VERSIONS: **An-32**: Baseline version. Specialised versions available for fisheries surveillance, agricultural and air ambulance use, last-named complete with operating theatre.

Detailed description refers to An-32.

An-32B: Increase of 500 kg (1,100 lb) in payload.
An-32P (*Protivopozharny*: anti-fire): Firefighting version, marketed as **Firekiller**; quickly removable tank on each side of fuselage; total retardant capacity 8,000 kg (17,635 lb); able to drop 30 smoke-jumpers, special equipment packages and stores at fire sites; provision for MP-26 cartridge packs to induce atmospheric precipitation artificially over fire; targeting assisted by BARPB-7 sight; NK-32P nav/sighting system optional; maximum T-O weight 29,700 kg (65,475 lb); cruising speed 215 kt (400 km/h; 248 mph); operating speed 117 to 124 kt (217 to 230 km/h; 135 to 143 mph) at normal height of 30 m (100 ft) above forest cover; T-O to 15 m (50 ft) 1,800 m (5,905 ft); landing from 15 m (50 ft) 1,250 m (4,100 ft); radius of action 80 n miles (150 km; 93 miles). An-32P has not been built in series.

An-32B-100: First flight October 1999. Upgrade configuration offered by Aviant, with AI-20D Series 5M (MotorSich IA 220) turboprops driving six-blade propellers, enhancing hot-and-high performance, increasing MTOW and allowing 800 kg (1,764 lb) increase in payload to 7,500 kg (16,535 lb) and a range increase of up to 324 n miles (600 km; 372 miles). Option of AI-20M engines, each offering 18 kg (40 lb) per hour fuel burn reduction. Service life increased to 20,000 hours; TBO to 4,000 hours. May form basis of maritime tanker and other subvariants.

An-32V-200: As An-32B-100, but with two 1,500 litre (396 US gallon; 330 Imp gallon) long-range fuel tanks scabbed to fuselage sides, as with An-32P retardant tanks. Range with 3,000 kg (6,613 lb) payload is more than 1,728 n miles (3,200 km; 1,988 miles); further increase of 594 n miles (1,100 km; 684 miles) in ferry range. Westernised version (offered to Greece) has Rolls-Royce AE 2100 engines and six-blade, composites Dowty propellers, and Rockwell Collins avionics for two-crew operation. Demonstrated in 1998/99 to Bolivia and Brazil; interest from Egypt. Unit cost quoted as US$6 million to US$9 million. Exhibited in model form with Stupino SV-34 six-blade propellers.

CUSTOMERS: In quantity production until 1992, by which time 337 (excluding prototypes) delivered; manufacture continues to order; 346th production An-32 delivered to Peruvian Army in July 1997; no later aircraft reported by mid-2000, but at least two were then partly complete. Operators include Russian Air Forces (50); air arms of Afghanistan (six), Angola (three), Bangladesh (three), Cuba (two), Ethiopia (one), India (123, named Sutlej after a Punjabi river), Kazakhstan, Mexico (four, Navy), Peru (18) and Sri Lanka (five); also Peruvian Army (four), Navy (two) and Police (four). Some 38 civil operators around the world have about 80.

DESIGN FEATURES: Development of An-26, with triple-slotted trailing-edge flaps outboard of engines, automatic leading-edge slats, enlarged ventral fins and full-span slotted tailplane; improved landing gear retraction, de-icing and air conditioning, electrical system and engine starting; large increase in power compared with An-26 improves take-off performance, service ceiling and payload under hot-and-high conditions; overwing location of engines reduces possibility of stone or debris ingestion, but requires nacelles of considerable depth to house underwing landing gear; operation possible from unpaved strips at airfields 4,000 to 4,500 m (13,120 to 14,760 ft) above sea level in ambient temperature of ISA + 25°C; APU helps to ensure independence of ground servicing equipment, including onboard engine starting at these altitudes.

FLYING CONTROLS: As An-26 except for high-lift wings (see Design Features).

STRUCTURE: Conventional light alloy; two-spar wing, built in centre, two inner and two detachable outer sections, with skin attached by electric spot welding; bonded/welded semi-monocoque fuselage in front, centre and rear portions, with 'bimetal' (duralumin-titanium) bottom skin for protection during operation from unpaved airfields; blister on each side of fuselage forward of rear ramp carries track to enable ramp to slide forward; large dorsal fin; ventral strake each side of ramp.

LANDING GEAR: Hydraulically retractable tricycle type; twin wheels on each unit. Emergency extension by gravity. All units retract forward. Shock-absorbers of oleo-nitrogen type on main units; nitrogen-pneumatic type on nose unit. Mainwheel tyres size 1,050×400, pressure 5.90 bars (85 lb/sq in). Nosewheel tyres size 700×250, pressure 3.90 bars (57 lb/sq in). Hydraulic disc brakes and anti-skid units on mainwheels.

POWER PLANT: Two ZMKB Progress/Ivchenko AI-20D Series 5 turboprops, each 3,810 kW (5,109 ehp); four-blade constant-speed reversible-pitch propellers; 3,126 kW (4,192 ehp) AI-20M turboprops available optionally. Normal fuel capacity 7,100 litres (1,876 US gallons; 1,562 Imp gallons). Provision for 4,500 litre (1,189 US gallon; 990 Imp gallon) ferry tank in hold.

ACCOMMODATION: Crew of three (pilot, co-pilot and navigator), with provision for flight engineer. Door on starboard side at front of hold, with adjacent ventral emergency exit; emergency exits forward of wing on port side, under wing trailing-edge on starboard side; roof hatch on flight deck. Rear-loading hatch and forward-sliding ramp/door, as An-26, plus winch and hoist, capacity 3,000 kg (6,615 lb), for freight handling. Cargo or vehicles can be airdropped by parachute, including extraction of large loads by drag parachute, with aid of removable roller conveyors and guide rails on floor of hold. Payloads include four freight pallets totalling 6,700 kg (14,770 lb), two 3,000 kg (6,613 lb) or twelve 500 kg (1,102 lb) airdroppable pallets, 50 passengers or 42 parachutists and a jumpmaster on row of tip-up seats along each cabin wall plus two containers totalling 200 kg (441 lb), or 24 stretcher patients in three tiers and one or two medical personnel. Executive/tourist version has removable twin-seats for 28 persons, lounge for four to six persons, compartment for hand baggage and clothes, and galley compartment. A crew rest compartment is optional at front of hold in standard cargo version.

SYSTEMS: Accommodation fully pressurised and air conditioned; cabin pressure differential 0.39 bar (5.7 lb/sq in). Systems basically as An-26 but generally improved. TG-16M APU in rear of starboard landing gear fairing to provide electrical power and for engine starting.

AVIONICS: *Comms:* Two VHF transceivers, HF, intercom.
Radar: Weather/navigation radar.
Flight: Two ADF, radio altimeter, glide path receiver, glide slope receiver, marker beacon receiver. Optional flight director system, astrocompass and autopilot.
Instrumentation: Directional gyro and flight recorder standard.
Self-defence: Provision for chaff/flare dispensers pylon-mounted on each side of lower fuselage below wings (seen on Afghan aircraft).

ARMAMENT: Provision for racks for four bombs or paradrop containers, each up to 500 kg (1,102 lb), two on each side of fuselage below wings (fitted to aircraft for Peruvian Air Force).

DIMENSIONS, EXTERNAL: As for An-26, except:
Wing span 29.20 m (95 ft 9½ in)
Wing aspect ratio 11.7

Antonov An-32B demonstrator UR-48142

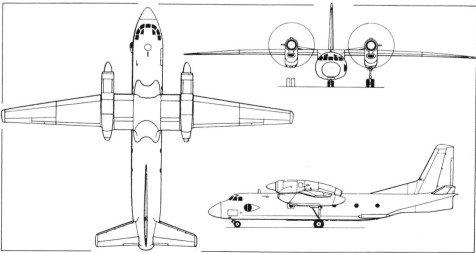

Antonov An-32 transport powered by two ZMKB Progress/Ivchenko AI-20 turboprops
(Dennis Punnett/Jane's)

Length overall	23.68 m (77 ft 8¼ in)
Height overall	8.75 m (28 ft 8½ in)
Fuselage: Max width	2.90 m (9 ft 6 in)
Max depth	2.50 m (8 ft 2½ in)
Tailplane span	10.23 m (33 ft 6¾ in)
Wheel track (c/l shock-struts)	7.90 m (25 ft 11 in)
Wheelbase	7.65 m (25 ft 1¼ in)
Propeller diameter	4.70 m (15 ft 5 in)
Propeller ground clearance	1.55 m (5 ft 1 in)
Crew door (stbd, front): Height	1.40 m (4 ft 7 in)
Width	0.60 m (1 ft 11¾ in)
Height to sill	1.47 m (4 ft 9¾ in)
Cargo ramp: Width: at front	2.40 m (7 ft 10½ in)
at rear	2.00 m (6 ft 6¾ in)
Emergency exit (in floor at front):	
Length	1.02 m (3 ft 4¼ in)
Width	0.70 m (2 ft 3½ in)
Emergency exit (top): Diameter	0.65 m (2 ft 1½ in)
Emergency exits (one each side of hold):	
Height	0.60 m (1 ft 11¾ in)
Width	0.50 m (1 ft 7½ in)

DIMENSIONS, INTERNAL:

Cargo hold: Length: at floor	12.48 m (40 ft 11¼ in)
at roof	15.685 m (51 ft 5½ in)
Width: at floor	2.30 m (7 ft 6½ in)
at roof	1.705 m (5 ft 7¼ in)
max	2.78 m (9 ft 1¼ in)
Height: max	1.84 m (6 ft 0½ in)
min	1.735 m (5 ft 8¼ in)
Area, excl ramp	30.0 m² (322 sq ft)
Volume: excl ramp	60.0 m³ (2,119 cu ft)
incl ramp	66.0 m³ (2,330 cu ft)

AREAS:

Wings, gross	74.98 m² (807.1 sq ft)
Ailerons (total)	6.12 m² (65.88 sq ft)
Flaps (total)	15.00 m² (161.46 sq ft)
Vertical tail surfaces (total, incl dorsal fin)	17.22 m² (185.36 sq ft)
Horizontal tail surfaces (total)	20.30 m² (218.5 sq ft)

WEIGHTS AND LOADINGS (An-32, except where indicated):

Weight empty	16,900 kg (37,258 lb)
Weight empty, equipped	17,405 kg (38,371 lb)
Max droppable payload: An-32	6,700 kg (14,770 lb)
-100, -200	7,500 kg (16,525 lb)
Max fuel: An-32	5,500 kg (12,125 lb)
-100, -200	7,824 kg (17,249 lb)
Max fuel with max payload	2,267 kg (4,998 lb)
Max ramp weight	27,250 kg (60,075 lb)
Max T-O weight: An-32	27,000 kg (59,525 lb)
-100, -200	28,500 kg (62,831 lb)
Max landing weight	25,000 kg (55,115 lb)
Max wing loading: An-32	360.1 kg/m² (73.75 lb/sq ft)
-100, -200	380.1 kg/m² (77.85 lb/sq ft)
Max power loading	3.59 kg/kW (5.90 lb/ehp)

PERFORMANCE (A: ISA, B: ISA + 25°C for An-32, except where indicated):

Max cruising speed:	
A, An-32V-200	286 kt (530 km/h; 329 mph)
B	248 kt (460 km/h; 286 mph)
Econ cruising speed: A	248 kt (460 km/h; 286 mph)
Paratroop drop speed:	
A	108-137 kt (200-255 km/h; 124-158 mph)
Landing speed: A	100 kt (185 km/h; 115 mph)
Optimum cruising height: A, B	8,000 m (26,240 ft)
Time to 6,000 m (19,680 ft): A	11 min
B	14 min
Time to 8,000 m (26,240 ft): A	19 min
B	27 min
Service ceiling: A, -100, -200	9,400 m (30,840 ft)
B	8,300 m (27,230 ft)
Service ceiling, OEI: A	4,800 m (15,750 ft)
B	3,400 m (11,155 ft)
T-O balanced field length: A	1,360 m (4,465 ft)
T-O run on concrete: A	760 m (2,495 ft)
B	880 m (2,888 ft)
-100, -200	950 m (3,120 ft)
T-O to 10.7 m (35 ft): A	1,160 m (3,805 ft)
Landing from 15 m (50 ft): A	875 m (2,870 ft)
Landing run: A	470 m (1,545 ft)
B	505 m (1,657 ft)
-100, -200	600 m (1,970 ft)
Range: transport: with 6,000 kg (13,227 lb) payload:	
An-32	647 n miles (1,200 km; 756 miles)
An-32B-100	971 n miles (1,800 km; 1,118 miles)
An-32V-200	877 n miles (1,625 km; 1,009 miles)
with 6,700 kg (14,700 lb) payload:	
An-32B	486 n miles (900 km; 559 miles)
An-32B-100	755 n miles (1,400 km; 869 miles)
An-32V-200	701 n miles (1,300 km; 807 miles)
with 7,500 kg (16,535 lb) payload:	
An-32B-100, An-32V-200	540 n miles (1,000 km; 621 miles)
ferry: An-32, An-32B-100	1,133 n miles (2,100 km; 1,304 miles)
An-32V-200	1,781 n miles (3,300 km; 2,050 miles)

UPDATED

First production Antonov An-38-100 in the colours of NAPO-Aviatrans *(Paul Jackson/Jane's)* 2001/0092537

ANTONOV An-38

TYPE: Twin-turboprop light transport.

PROGRAMME: Details announced, and model displayed, at 1991 Paris Air Show; initial batch of six built at production factory, NAPO, Novosibirsk, Russia: one prototype (01001; first flight 23 June 1994, with TPE331 engines), four trials aircraft and one (01002) for static testing at Kiev; certification to AP-25 granted 24 April 1997. In December 1995, Antonov and NAPO formed joint venture company, Siberian Antonov Aircraft, to produce, market and provide after-sales service for the An-38. Indian demonstration tour undertaken in July and August 1997, followed by appearance at Aero India in December 1998.

Prototype of An-38-200, with Russian engines, due to have flown at Novosibirsk by late 2000, gaining certification by September 2001.

CURRENT VERSIONS: **An-38-100:** With Honeywell TPE331 engines. First and second (01003; exhibited Moscow 1997) flying aircraft to this standard. Trials of international navigation avionics completed March 2000.

An-38-200: With Omsk MKB 'Mars' TVD-20 engines. Third and fourth prototypes were planned to this standard when engine development complete; however, schedule for flight trials had not been finalised by mid-2000.

An-38K: Convertible version of An-38-100; large upward-hinging side door at rear on port side; able to carry four LD-3 (KMP-500) or five LD-3K containers (= *konteinernyi*); cargo handling equipment removable for conversion to 30-passenger transport.

Versions with RKBM TVD-1500 engines remained under construction in 2000. All versions can be equipped for aerial photography (An-38F: *fotografiya*), survey (An-38GF *geofizichesky*), forest patrol (An-38D: *desantnyi*), VIP transport, ambulance (An-38S: *sanitarnyi*; six stretchers, nine seated, with attendant) and fishery/ice patrol duties (An-38LR: *ledovoi razvedki*).

CUSTOMERS: First three (subsequently increased to eight) An-38-100s ordered for Vostok Airlines and received by mid-1995 for one year of intensive trials before passenger certification. Second firm customer is Chukotavia (10; although initial batch is two); letters of intent from Petropavlovsk-Kamchatsky, Merninsky, Novosibirsky, Ulyanovsky and Nikolaevsk-na-Amur. Planned 1998 deliveries were seven, but by year's end, only known aircraft were Vostok's three, plus unconfirmed report of one with Chukotavia. In 1999, second prototype was being operated by NAPO-Aviatrans, the airline of the NAPO aircraft factory. In 1998, Siberia Airlines was considering purchase of two. Alrosa-Avia of Zhukovsky has ordered five for diamond mining support, of which first in service by early 2000; second followed in July 2000, increasing production deliveries to total of five. Indian Air Force interest expressed in initial six to 10.

COSTS: An-38-100 basic price US$4 million (2000); An-38-200 US$2.6 million to US$3.0 million (2000).

DESIGN FEATURES: Developed from PZL Mielec (Antonov) An-28 (see Polish section) to replace An-24s, Let L 410s and Yak-40s. New high-efficiency engines; lengthened passenger cabin; optional weather radar and automatic flight control system; improved sound and vibration insulation; wheel or ski landing gear; rear cargo door and cargo handling system; able to operate from unpaved runways; operating temperatures from −45 to +45°C, including 'hot and high' conditions. Service life 30,000 hours. Maintenance requirement 4 man-hours/flying hour.

FLYING CONTROLS: Conventional and manual. Single-slotted mass and aerodynamically balanced ailerons (port aileron has trim tab), designed to droop with large, hydraulically actuated, two-segment double-slotted flaps; electrically actuated trim tabs in each elevator have manual back-up; twin rudders each with electrically actuated trim tab; automatic leading edge slats over full span of wing

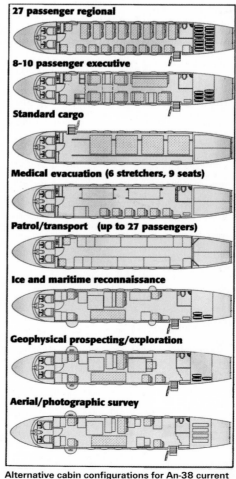

Alternative cabin configurations for An-38 current and projected versions 0052899

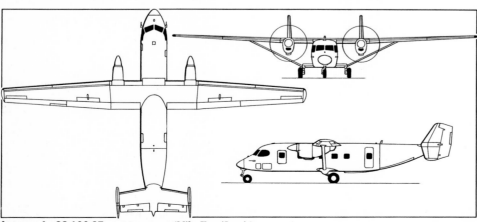

Antonov An-38-100 27-seat transport *(Mike Keep/Jane's)* 1999/0051508

ANTONOV—AIRCRAFT: UKRAINE

Second prototype of the Antonov An-70 pressurised propfan-powered transport *(Paul Jackson/Jane's)*

outboard of engines; slab-type spoiler forward of each aileron and each outer flap segment at 75 per cent chord.

LANDING GEAR: Tricycle type; fixed. Mainwheels 610×320-330; nosewheel 600×320-254.

POWER PLANT: Two Honeywell TPE331-14GR-801E turboprops, each 1,118 kW (1,500 shp), driving Hartzell HC-B5MA five-blade propellers; or two Omsk MKB 'Mars' TVD-20 turboprops, each 1,029 kW (1,380 shp), driving AV-36 quiet reversible-pitch propellers.

ACCOMMODATION: Two crew side by side on flight deck; passenger cabin equipped normally with 26 seats, basically three-abreast, with centre aisle; 27 seats at 75 cm (29½ in) pitch optional; ambulance version for six stretchers, eight seated casualties and medical attendant, executive versions with eight to 10 seats and forest surveillance/paradrop version for 26 smoke-jumpers or trainee paratroops available; seats and baggage compartment can be folded quickly against cabin wall to provide clear space for 2,500 kg (5,510 lb) of freight. Cabin door with airstairs on port side, with service door opposite; emergency exit each side. Optional cargo door under upswept rear fuselage slides forward under cabin for direct loading/unloading of freight.

AVIONICS: Russian, Honeywell and United Instruments or other avionics optional for VFR or IFR flying, including equipment for ICAO Cat. I approach and satellite navigation.

EQUIPMENT: Hand-operated travelling overhead winch in cabin: capacity 500 kg (1,102 lb).

DIMENSIONS, EXTERNAL:
Wing span	22.06 m (72 ft 4½ in)
Length overall	15.67 m (51 ft 5 in)
Height overall	4.60 m (15 ft 1 in)
Span over tailfins	5.14 m (16 ft 10½ in)
Wheel track	3.515 m (11 ft 6½ in)
Wheelbase	6.345 m (20 ft 9¾ in)
Distance between propeller centres	5.58 m (18 ft 3¾ in)
Propeller diameter	2.85 m (9 ft 4 in)

Cargo door: An-38-100:
Length	2.20 m (7 ft 2½ in)
Width: at floor	1.40 m (4 ft 7 in)
at rear	1.00 m (3 ft 3¼ in)

DIMENSIONS, INTERNAL (An-38K):
Cargo hold: Length	8.80 m (28 ft 10½ in)
Max width	1.90 m (6 ft 2¾ in)
Max height	2.14 m (7 ft 0¼ in)
Volume	23.5 m³ (830 cu ft)

WEIGHTS AND LOADINGS (A: An-38-100, K: An-38K):
Weight empty: A	5,300 kg (11,684 lb)
Max payload: A	2,500 kg (5,510 lb)
K	3,200 kg (7,055 lb)
Max fuel: A	2,210 kg (4,872 lb)
Max T-O weight: A	8,800 kg (19,400 lb)

PERFORMANCE (estimated, with TPE331 engines):
Max level speed	219 kt (405 km/h; 252 mph)
Nominal cruising speed: A	205 kt (380 km/h; 236 mph)
Nominal max cruising altitude	4,200 m (13,780 ft)
T-O run: A	350 m (1,150 ft)
K	480 m (1,575 ft)
Landing run: A	270 m (885 ft)
K	440 m (1,445 ft)
Balanced field length: A	900 m (2,955 ft)

Range, 45 min fuel reserves: A, with 27 passengers
324 n miles (600 km; 372 miles)
A, with 17 passengers
782 n miles (1,450 km; 901 miles)
A, max, with nine passengers
890 n miles (1,650 km; 1,025 miles)
K, with max payload 183 n miles (340 km; 211 miles)
K, with max fuel and 1,800 kg (3,968 lb) payload
863 n miles (1,600 km; 994 miles)

UPDATED

ANTONOV An-70

TYPE: Strategic transport.

PROGRAMME: Development began 1975 to replace some An-12s remaining in air force service from 2002-2003; announced by *Izvestia* 20 December 1988; at 1991 Paris Air Show Antonov OKB reported prototype being assembled at Kiev; funding by Russian (80 per cent) and Ukrainian governments under agreement of 24 June 1993; preliminary details released and model displayed at Moscow Aero Engine and Industry Show April 1992; prototype first flight 16 December 1994 (was also delivery flight to Gostomel test airfield); this aircraft lost during fourth sortie following in-flight collision with chase An-72 on 10 February 1995; second prototype (without nose-mounted instrumentation boom) produced by upgrading of static test airframe (for which replacement under construction in 2000); rolled out 24 December 1996; first flight (UR-NTK) 24 April 1997; international debut at Moscow Air Show, August 1997; handed over to Russian Air Forces' test centre at Akhtubinsk, August 1998. Had flown one-third of planned 780-sortie test programme by January 2000. High AoA trials mid-2000. Crash-landed immediately after take-off from Omsk on 27 January 2001 during cold weather trials. Airframe may be used for static exhibition.

Bilateral agreement between Russia and Ukraine revised 18 May 1999, and underlined decision taken to order first 10 (and 50 engines) before planned production at Aviant plant, Kiev, from 1999 and Aviakor plant, Samara, from 2000; each plant building initial batch of five, which to enter service by 2003. Ukrainian order for 65 reported September 2000; at same time, Samara regional government announced that local manufacturers were withdrawing from An-70 project. Russian production specification issued 4 December 1999.

Offered as alternative to Airbus A400M FLA; Germany and Ukraine agreed in December 1997 to explore possible industrial collaboration and German government strongly promoted the An-7X as the basis of the FLA, though this was rejected by other FLA partners and the Luftwaffe. Aircraft evaluated by DaimlerChrysler Aerospace, which determined that it could be modified to meet FLA requirement. Aim of wide co-operation between Russian, Ukrainian and West European aircraft industries (in effect bid to meet European FLA requirement) stated in February 1998 in joint declaration by Russian and Ukrainian Presidents. Certification is planned to FAR Pt 25 and equivalents. Antonov obviated effects of intermittent official funding by investing income from its own An-124 charter operations.

Medium Transport Aircraft International Consortium (MTA; also known as Medium-Size Transport Plane consortium or AirTruck) formed February 1996 to co-ordinate development, certification, sales and after-service, comprising four Ukrainian and six Russian companies: ANTK Antonov (designer and component manufacturer), ZMKB Progress (engine designer), Aviant (airframe manufacturer) and Motor Sich (engine manufacturer) from Ukraine; and Aviacor (airframe manufacturer), Ufa (engine manufacturer), Aviapribor (flight control system manufacturer), Elektroavtomatika (avionics), Leninets (airborne monitoring and diagnostic system) and Aerosila (propfan) from Russia. TAPO (Uzbekistan) builds wings.

Third aircraft to fly will be An-70T commercial variant, construction of which announced mid-1998. In September 2000, Volga-Dnepr airline pledged to fund An-70 production if military contracts not forthcoming. Fourth batch (of seven) D-27 engines, authorised in 2000, includes those for An 70T prototype.

CURRENT VERSIONS: **An-70**: Military STOL transport; proposed production version. Stated to have double the payload of Lockheed Martin C-130J, with similar STOL and rough-field performance, yet only 40 kt (74 km/h; 46 mph) slower than the Boeing C-17A Globemaster III, which carries 15 per cent more payload.
Detailed description applies to baseline An-70.

An-70-100: Proposed military STOL transport, as An-70 but with two-crew cockpit.

An-77: Proposed military STOL transport for export customers; as An-70-100 but with cockpit for two or three crew. Runway length of 1,900 m (6,235 ft) required with 35,000 kg (77,161 lb) payload for 2,051 n mile (3,800 km; 2,361 mile) range.

An-70T: Commercial transport, generally as An-70, with improved runway capability but no requirement for STOL. Two or three crew. To carry 35,000 kg (77,161 lb) payload 2,051 n miles (3,800 km; 2,361 miles) from 1,900 m (6,235 ft) runway, or 20,000 kg (44,092 lb) for 2,915 n miles (5,400 km; 3,555 miles) from 1,300 m (4,265 ft) runway. Stated to have load-carrying capability of Il-76, but runway requirements of An-74. Promoted as Il-76/An-12 replacement. Certification planned to FAR Pt 25 and equivalents. First fuselage delivered to Samara from Kiev November 1999.

An-70T-100: Development of An-70T with two D-27 propfans, two crew and revised landing gear, to carry 30,000 kg (66,138 lb) payload 540 n miles (1,000 km; 621 miles) from 2,500 m (8,205 ft) runway, or 10,000 kg (22,046 lb) for 1,187 n miles (2,200 km; 1,367 miles)

Antonov An-70 transport (four ZMKB Progress D-27 propfans) *(Mike Keep/Jane's)*

Antonov An-70 second prototype, showing trailing-edge double-slotted flaps and slats on the wing and tailplane leading-edges ***(Paul Jackson/Jane's)*** ***2001***/0092093

from 1,300 m (4,265 ft) runway. Promoted as An-12 and An-74 replacement

An-70T-200: Powered by two Kuznetsov NK-93 turbofans.

An-70T-300: With two CFM56-5C4 turbofans.

An-70T-400: With four CFM56-5C4 turbofans.

An-70TK: Convertible cargo/passenger transport for 30,000 kg (66,138 lb) of freight or 150 passengers, with seats in removable modules.

An-7X: Designation for provisional variant offered to meet multinational FLA requirement at 40 per cent of anticipated A400M cost. AirTruck GmbH formed 20 May 1999 by eight supporting German companies (Aerodata, ASL Aircraft Services, Autoflug, BGT, R-R Deutschland, ESG, Liebherr and VDO) to promote An-7X in conjunction with MTA consortium.

Antonov-supplied (on 29 January 1999) data evaluated by DaimlerChrysler Aerospace, which assessed technical risk as minor and performance to be compliant; however, specific fuel consumption targets not met, concern expressed over noise, and power plant felt to need Western FADEC. Further recommendations include wiring insulation change to meet Western standards; landing gear modifications; addition of inflight refuelling and fuel-dumping; completely revised NVG-compatible, two-crew cockpit; permanently installed cargo system; and minor flying control and manufacturing changes.

Assembly of 75 An-7Xs for Luftwaffe would be at Lemwerder, Germany. If rejected by Germany, An-7X programme to continue in readiness for alternative export prospects for non-STOL version.

An-170: Heavy transport derivative carrying 45,000-50,000 kg (99,210-110,230 lb) of cargo.

Adaptation of military An-70 for tanker, AEW, SAR (An-70PS), naval patrol, and of An-70T or An-70T-100 for firefighting, ecological monitoring and ambulance duties being studied.

CUSTOMERS: Requirements originally expressed by Russian Air Forces for up to 500 and by Ukrainian Air Force for 100. This modified by 1999 to 164 for Russia and 65 for Ukraine, with in-service date of 2002. German requirement potentially for 75 aircraft. By mid-1999 potential civil operators had signed documents of intent for some 100 further aircraft. Chinese interest reported in 2000.

COSTS: Military version US$48 million (2000). Civil An-70T to cost 20 to 30 per cent less than military variant.

DESIGN FEATURES: First aircraft to fly powered only by propfans. Slightly larger than projected European FLA transport, much smaller than US Boeing C-17A; conventional high-wing configuration, with wings and tail surfaces slightly sweptback; supercritical wing section; anhedral from roots; loading ramp/doors under upswept rear fuselage with adjustable sill height and built-in cargo handling system; horizontal tail surfaces on rear fuselage; propfans mounted conventionally on wing leading-edge; propeller wash doubles wing lift during take-off and landing. Multiple-section control surfaces provide redundancy in event of battle damage or physical obstruction. Claimed features include independent operational capability at non-equipped airfields for 30 days. Design life 15,000 cycles and 45,000 flying hours in 25 years. Operable 3,500 hours per year, with eight to 10 man-hours of maintenance per flying hour. Cost-effective with only 200 flying hours per month.

FLYING CONTROLS: Prototypes have fly-by-wire system with three digital and six analogue channels; primary controls are quadruplex; back-up by unique fly-by-hydraulics system, in which pilot or autopilot inputs are relayed (via conventional 'mini-wheels') to actuators by commands in hydraulic control channels, unaffected by electromagnetic interference. Production aircraft will have a four-channel, all-digital primary FCS, rather than the hybrid digital/analogue system of the prototypes. Secondary FCS controls two independent flap systems, leading-edge slats and blown flaps. Three-section double-winged rudder. Double-slotted trailing-edge blown Fowler flaps in two sections on each wing; forward element maximum deflection 60°, rear element 80°; intermediate settings (forward element) 5, 10, 15, 20, 25, 30, 35, 40 and 50°. Three-section spoilers forward of each outer flap. Leading-edge has flaps inboard; slats centre and outboard. Two-section, double-hinged elevators forward section maximum deflections +28/–20°, rear section +50/–40°, to enhance low-speed authority; horizontal stabilisers are fitted with automatic leading-edge slats.

STRUCTURE: Approximately 28 per cent of airframe, by weight, made of composites, including complete tail unit, ailerons and flaps. Fuselage stringer/skin joints are spot-welded and hot-bonded, manually. Wings manufactured at Chkalov plant, Tashkent, Uzbekistan.

LANDING GEAR: Twin-wheel nose unit; each main unit has three pairs of wheels in tandem, retracting into large fairing on side of cabin; can operate from unpaved surfaces of bearing ratio 8 kg/cm² (114 lb/sq in). All tyres 1,120×450. Steel-steel brakes. Nosewheel turning angle ±55° for taxying; nosewheel turning radius 16.3 m (53½ ft); wingtip turning radius 29.5 m (97¾ ft); required taxiway width for 180° turn 27.4 m (90 ft).

POWER PLANT: Four ZMKB Progress/Ivchenko D-27 propfans, each 10,290 kW (13,800 shp). Aerosyla Stupino SV-27 contrarotating propellers, each with eight composites blades in front and six at rear. Reversible-pitch blades of scimitar form, with electric anti-icing. Export versions proposed with CFM56-5C4 turbofans.

ACCOMMODATION: Three flight crew (two pilots and flight engineer or 'tactical pilot') plus loadmaster; navigation station on captain's left is optionally operated by fourth member of flight crew; provision for converting cockpit for two-crew operation, with co-pilot operating flight engineer's station; seats in forward fuselage for two cargo attendants; freight loaded via rear ramp using four built-in, powered hoists (each of 3 tonne capacity) reaching out 6.6 m (22 ft) from aircraft. Hoists can be combined for heavier loads. Freight can be carried on PA-5.6 rigid pallets, PA-3, PA-4 and PA-6.8 flexible pallets, in UAK-2.5, UAK-5 and UAK-10 containers; unpackaged freight, wheeled and tracked vehicles, food and perishables can be carried; seats for 300 troops, or 206 stretchers, can be installed using optional, prefabricated (10 section) upper deck or optional, easily removable seven-section upper deck (each segment holding 1.5 tonnes) in cargo hold; vehicles, freight and paratroops can be airdropped; maximum single airdrop item weight 20,000 kg (44,092 lb); crew door at front of cabin on port side; two upper deck doors each side, front and rear; cargo hold pressurised and air conditioned.

SYSTEMS: Aircraft systems automated to simplify operation and decrease probability of crew errors. Electronpribor engine control system; Leninets monitoring and information system.

AVIONICS: Integrated by Aviapribor, Leninets and Elektroavtomatika. Flight data, navigation and radio-navigation systems to ARINC 700 requirements; digital multiplex data interface equivalent to Western MIL-STD-1553B.

Comms: Integrated system by Gorkiski.

Flight: Ring laser INS; SKI-77 HUD; flight management system; designed for operation in adverse weather and for landing in ICAO Cat. II and IIIa conditions. BASK-70 onboard diagnostic system collects data from subsystems, registering and analysing 8,000 in-flight parameters.

Instrumentation: Ten-screen EFIS by Elektroavtomatika comprises six main screens, each 200 × 200 mm (7¾ × 7¾ in), facing pilots and two each at navigation and flight engineer's stations, plus smaller secondary LCD screens and roof-mounted HUDs for pilot and co-pilot on production aircraft.

EQUIPMENT: Four electric hoists in hold.

DIMENSIONS, EXTERNAL:

Wing span	44.06 m (144 ft 6¾ in)
Length overall	40.73 m (133 ft 7½ in)
Fuselage diameter	4.80 m (15 ft 9 in)
Height overall	16.38 m (53 ft 9 in)
Wheel track (bogie centres)	5.21 m (17 ft 1 in)
Wheelbase: front mainwheels	16.65 m (54 ft 7½ in)
centre mainwheels	18.47 m (60 ft 7¼ in)
rear mainwheels	20.43 m (67 ft 0¼ in)
Propeller diameter	4.50 m (14 ft 9 in)
Propeller ground clearance (outer)	3.00 m (9 ft 10 in)
Rear-loading aperture: Height	4.10 m (13 ft 5½ in)
Width	4.00 m (13 ft 1½ in)

DIMENSIONS, INTERNAL:

Cargo hold: Floor length: excl ramp	19.10 m (62 ft 8 in)
incl ramp	22.40 m (73 ft 6 in)
Max width	4.80 m (15 ft 9 in)
Max width at floor	4.00 m (13 ft 1½ in)
Min height	4.00 m (13 ft 1½ in)
Floor area, incl ramp	89.0 m² (958 sq ft)
Volume	425.0 m³ (15,008 cu ft)

WEIGHTS AND LOADINGS:

Weight empty	72,800 kg (160,500 lb)
Normal payload (incl 5,000 kg; 11,025 lb on ramp)	35,000 kg (77,161 lb)
Normal payload from unpaved runway	30,000 kg (66,138 lb)
Payload: max	47,000 kg (103,615 lb)
restricted runway: option	35,000 kg (77,161 lb)
600 m runway option	20,000 kg (44,092 lb)
Max T-O weight	130,000 kg (286,600 lb)
Max power loading	3.16 kg/kW (5.19 lb/shp)

PERFORMANCE (estimated):

Cruising speed: long range	405 kt (750 km/h; 466 mph)
max short range	432 kt (800 km/h; 497 mph)
Nominal cruising height	9,100-11,000 m (29,860-36,080 ft)
Runway length required:	
for normal operation: T-O	1,800 m (5,910 ft)
Landing	2,200 m (7,220 ft)
for STOL operation, T-O from unpaved surface:	
35 tonne load	915 m (2,960 ft)
30 tonne load	600 m (1,970 ft)
Range (runway length A: 1,800 m; 5,905 ft, B: 915 m; 3,005 ft, C: 600 m; 1,970 ft):	
with 47 tonnes: A	728 n miles (1,350 km; 838 miles)
B, C	not an option
with 35 tonnes:	
A	2,051 n miles (3,800 km; 2,361 miles)
B	782 n miles (1,450 km; 901 miles)
C	not an option
with 30 tonnes:	
A	2,699 n miles (5,000 km; 3,106 miles)
B	1,376 n miles (2,550 km; 1,584 miles)
C	378 n miles (700 km; 435 miles)
with 20 tonnes:	
A	3,995 n miles (7,400 km; 4,598 miles)
B	2,618 n miles (4,850 km; 3,013 miles)
C	1,619 n miles (3,000 km; 1,864 miles)
with max fuel: all options	4,319 n miles (8,000 km; 4,971 miles)

UPDATED

ANTONOV An-72 and An-74

NATO reporting name: Coaler

TYPE: Twin-jet transport.

PROGRAMME: First of two prototype An-72s, built at Kiev, flew (SSSR-19744) 31 August 1977; after eight preseries aircraft, manufacture transferred to Kharkov, Ukraine, where first production An-72 flew 22 December 1985; An-74 also produced at Kharkov from December 1989. Russian assembly initially undertaken at Arsenyev; An-74 announced February 1984; An-72P maritime patrol version demonstrated 1992. Production of An-74 also started by Polyot Industrial Association at Omsk, Russia, in 1993; now sole Russian source; first Polyot aircraft (RA-74050) was flown 25 December 1993. Development of An-74-200 and An-74TK-200 by Antonov started 1995; An-74TK certified by Interstate Aviation Committee in August 1995.

CURRENT VERSIONS: **An-72** ('Coaler-C'): Light STOL transport for military use; extended wings, lengthened fuselage and other changes compared with An-72 ('Coaler-A') prototypes; ZMKB Progress/Ivchenko D-36 Series 2A engines; crew of two or three.

An-72P: Maritime patrol version, described separately.

An-72R: One (SSSR-783573), at Akhtubinsk operational research centre, appears to have a large, flat, side-looking airborne radar (SLAR) built into each side of its upswept rear fuselage, possibly for standoff battlefield surveillance. Engineering designation **An-88**. First noted 1995, but possibly by then out of service.

An-72S: Military VIP version with appropriate *'Salon'* interior.

An-72V: Export version; two crew only.

An-72-100: Civilianised An-72, certified 1997; upgraded avionics.

ANTONOV—AIRCRAFT: UKRAINE

An-74: Designation initially applied in 1983 to version optimised for support of Arctic exploration; now refers, generally, to all civilian versions (as under). Wing, tail unit and engine air intake de-icing improved over An-72; advanced navigation aids, including inertial navigation system; provision for wheel/ski landing gear; increased fuel capacity; airframe identical with An-72 except for two blister windows at rear of flight deck and front of cabin on port side; maximum T-O weight increased. Convertible in field for ambulance, firefighting and other duties.

An-74-200: Freight version with D-36 Series 3A engines of unchanged rating; increased payload and maximum T-O weight. Able to carry four YAK-2.5 containers. Crew of four/five.

An-74T-200A ('Coaler-B'): Transport (*transportnyi*) version with longer hold; payload 10,000 kg (22,045 lb); loading winch; roller conveyors in floor; crew of two. Range with maximum payload up to 728 n miles (1,350 km; 838 miles), with 3,500 kg (7,716 lb) payload 2,159 n miles (4,000 km; 2,485 miles).

An-74TK-200 ('Coaler-B'): Convertible transport/passenger aircraft (*transportnyi konvertiruyemnyi*) with twin seats for 52 passengers that fold against cabin walls, and with baggage racks, buffet/galley and toilet. Alternative all-cargo or all-passenger or combi layouts. Typical combi options include 12 passengers plus 6,000 kg (13,228 lb) of freight and 20 plus 4,500 kg (9,921 lb). Built-in loading equipment. Crew of two. Range with 10,000 kg (22,045 lb) payload, 1 hour reserve, 430 n miles (800 km; 497 miles).

An-74TK-200D Salon: Business transport for 10 to 16 passengers, with increased cabin comfort. Equipment includes telephone, fax, video, bar, refrigerator, galley and separate rest area. Optional compartment for car at rear. Power plant and weights as basic An-74. Crew of four/five.

An-74T-100 ('Coaler-B'): As An-74T-200, with navigator station (crew of four).

AN-74TK-100: As An-74TK-200, with navigator and flight engineer stations (crew of four). Russian type certificate issued 4 August 1995.

An-74TK-300: Described separately.

An-74-400: Stretched version; under development by 1998; to be offered in passenger and cargo versions.

An-76: Engineering designation for military An-72Ps.

An-79: US military sources use this designation for a version of An-72 transport.

An-174: Proposed stretched An-74TK-300; described separately.

An-71: AEW version; last described in 1997-98 *Jane's*.

CUSTOMERS: More than 160 An-72/74s built before An-74 production additionally established at Omsk; 20 in Russian Air Forces; four in Peruvian Air Force; 26 in Ukrainian Air Force; others in Kazakhstan and Moldova. Order placed by Iran in 1997 for 12 An-74TKs includes some for Presidential Guard; other recent deliveries to Laotian government, Ukraine Border Guard, MChS Rossii, Vitair and Gazpromavia. By mid-2000, civil operators included five with 12 An-72s and 17 with 37 An-74s. No new civil deliveries reported in 1998; two An-74TK-200s built at Kharkov in 1999. Indian interest being pursued in 2000.

COSTS: Approximately US$12.5 million (An-74).

DESIGN FEATURES: Primary role as STOL replacement for turboprop An-26, with emphasis on freight carrying. High-wing, T-tail configuration, with upswept rear fuselage for freight access. Ejection of exhaust efflux over upper wing surface and down over large multislotted flaps gives considerable increase in lift; high-set engines avoid foreign object ingestion; special ramp/door as An-26; low-pressure tyres and multiwheel landing gear for operation from unprepared strips, ice or snow; sweptback fin and rudder.

Wing leading-edge sweepback 17°; anhedral approximately 10° on outer wings; normal T-O flap setting 25 to 30°, maximum deflection 60°.

FLYING CONTROLS: Conventional and assisted. Power actuated ailerons, with two tabs in port aileron, one starboard; double-hinged rudder, with tab in lower portion of two-section aft panel; during normal flight only lower rear rudder segment is used; both rear segments used in low-speed flight; forward segment is actuated automatically to offset thrust asymmetry; horn-balanced and mechanically actuated, aerodynamically balanced elevators, each with two tabs; hydraulically actuated full-span wing leading-edge flaps outboard of nacelles; trailing-edge flaps double-slotted in exhaust efflux, triple-slotted between nacelles and outer wings; four-section spoilers forward of triple-slotted flaps; two outer sections on each side raised before landing, remainder opened automatically on touchdown by sensors actuated by weight on main landing gear; inverted leading-edge slat on tailplane linked to wing flaps.

STRUCTURE: All-metal; multispar wings mounted above fuselage; wing skin, spoilers and flaps of titanium aft of engine nacelles; circular semi-monocoque fuselage, with rear ramp/door; tapered fairing forward of T tail fin/tailplane junction, blending into ogival rear fairing.

LANDING GEAR: Hydraulically retractable tricycle type, primarily of titanium. Rearward-retracting steerable twin-wheel nose unit. Each main unit comprises two trailing-arm legs in tandem, each with a single wheel, retracting inward through 90° so that wheels lie horizontally in bottom of fairings, outside fuselage pressure cell. Oleo-pneumatic shock-absorber in each unit. Low-pressure tyres, size 720×310 on nosewheels, 1,050×400 on mainwheels. Hydraulic disc brakes. Telescopic strut hinges downward, from rear of each side fairing, to support fuselage during direct loading of hold with ramp/door under fuselage.

POWER PLANT: Two ZMKB Progress/Ivchenko D-36 high-bypass ratio turbofans (Srs 2A in An-74; Srs 3A in An-74-200, T-100/200 and TK-100/200), each 63.74 kN (14,330 lb st). Integral fuel tanks between spars of outer wings. Thrust reversers standard.

ACCOMMODATION: Pilot and co-pilot/navigator side by side on flight deck of basic An-72, plus flight engineer, with provision for fourth person. Heated windows. Two windscreen wipers. Flight deck and cabin pressurised and air conditioned. Main cabin designed primarily for freight, including four YAK-2.5 containers or four PAV-2.5 pallets each weighing 2,500 kg (5,511 lb); An-72 has folding seats along sidewalls and removable central seats for 68 passengers. It can carry 57 parachutists, and has provision for 24 stretcher patients, 12 seated casualties and an attendant in ambulance configuration. An-74 can carry eight mission staff in combi role, with tables and bunks. Bulged observation windows on port side for navigator and hydrologist. Provision for wardrobe and galley. Movable bulkhead between passenger and freight compartments, with provision for 1,500 kg (3,307 lb) of freight in rear compartment. Reinforced, movable bulkhead in combi versions protects passengers from shifting cargo in the event of sudden deceleration.

Downward-hinged and forward-sliding rear ramp/door for loading trucks and tracked vehicles, and for direct loading of hold from trucks. It is openable in flight, enabling freight loads of up to 7,500 kg (16,535 lb), with a maximum of 2,500 kg (5,511 lb) per individual item, to be airdropped by parachute extraction system. In normal freight role, 1,000 kg (2,204 lb) of payload can be placed on ramp. Maximum size of containers up to 1.90 × 2.44 × 1.46 m (6 ft 3 in × 8 ft × 4 ft 9½ in), pallets up to 1.90 × 2.42 × 1.46 m (6 ft 3 in × 7 ft 11 in × 4 ft 9½ in). Main crew and passenger door at front of cabin on port side. Emergency exit and servicing door at rear of cabin on starboard side.

SYSTEMS: Air conditioning system, altitude limit 10,000 m (32,810 ft), with independent temperature control in flight deck and main cabin areas; used to refrigerate main cabin when perishable goods carried. Maximum cabin pressure differential 0.49 bar (7.1 lb/sq in). Hydraulic system for landing gear, flaps and ramp. Electrical system powers auxiliary systems, flight deck equipment, lighting and mobile hoist. Thermal de-icing system for leading-edges of wings and tail unit (including tailplane slat), engine air intakes and cockpit windows. Provision for TA-12 APU in starboard landing gear fairing. This can be used to heat cabin; under cold ambient conditions, servicing personnel can gain access to major electric, hydraulic and air conditioning components without stepping outside.

AVIONICS: *Comms:* HF com, VHF com/nav. 'Odd Rods' IFF standard.

Radar: Navigation/weather radar in nose.

Flight: ADF. Compatible with DME, Tacan, VOR, ILS and SP systems. Doppler-based automatic navigation system, linked to onboard computer, is preprogrammed before take-off on push-button panel to right of map display.

Instrumentation: Failure warning panels above windscreen display red lights for critical failures, yellow lights for non-critical failures, to minimise time spent on monitoring instruments and equipment.

An-74 has enhanced avionics, including INS.

Antonov An-72 ('Coaler-C') twin-turbofan STOL transport in Russian military service, showing deployed thrust-reversers and flaps *(Paul Jackson/Jane's)*

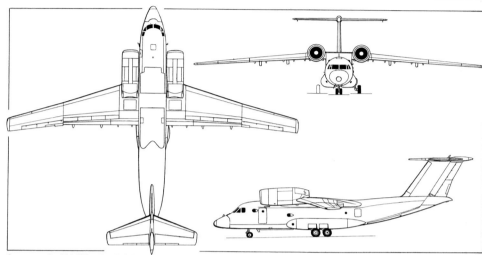
Antonov An-74 ('Coaler-B') STOL transport (two ZMKB Progress/Ivchenko D-36 turbofans) *(Dennis Punnett/Jane's)*

Antonov An-74TK-200 convertible passenger/cargo aircraft *(Paul Jackson/Jane's)*

UKRAINE: AIRCRAFT—ANTONOV

Antonov An-74 employed by Russian civil aid agency MChS Rossii *(Paul Jackson/Jane's)*

EQUIPMENT: Removable mobile winch, capacity 2,500 kg (5,511 lb), assists loading. Cargo straps and nets stowed in lockers on each side of hold when not in use. Provision for roller conveyors in floor.

DIMENSIONS, EXTERNAL:
Wing span	31.89 m (104 ft 7½ in)
Wing aspect ratio	10.3
Length overall	28.07 m (92 ft 1¼ in)
Fuselage: Max diameter	3.10 m (10 ft 2 in)
Height overall	8.65 m (28 ft 4½ in)
Wheel track	4.09 m (13 ft 5 in)
Wheelbase	8.68 m (28 ft 5¾ in)
Min loading clearance beneath rear fuselage	
	2.80 m (9 ft 2¼ in)
Distance between engine centrelines	4.15 m (13 ft 7½ in)
Crew/passenger door: Height	1.65 m (5 ft 5 in)
Width	0.90 m (2 ft 11¼ in)
Rear-loading door: Length	7.10 m (23 ft 3½ in)
Width	2.40 m (7 ft 10½ in)
Height to sill	1.54 m (5 ft 0¾ in)

DIMENSIONS, INTERNAL:
Cabin: Length: excl ramp: An-74	9.50 m (31 ft 2 in)
An-74T	10.50 m (34 ft 5¼ in)
incl ramp: An-74T	14.30 m (46 ft 11 in)
Width: at floor level	2.15 m (7 ft 0½ in)
max	2.50 m (8 ft 2½ in)
Height	2.20 m (7 ft 2½ in)
Floor area: An-74T	22.5 m² (242 sq ft)
Ramp area	8.2 m² (88.3 sq ft)
Volume: An-74T (total)	73.3 m³ (2,589 cu ft)

AREAS:
Wings, gross	98.53 m² (1,060.6 sq ft)

WEIGHTS AND LOADINGS (A: An-72, C: An-74, D: An-74 Salon, E: An-74-200, F: An-74T-200 and An-74TK-200):
Weight empty: A	19,050 kg (42,000 lb)
F	21,820 kg (48,105 lb)
Max fuel: A	12,950 kg (28,550 lb)
C, D, E, F	13,200 kg (29,100 lb)
Max payload: C, D	7,500 kg (16,535 lb)
A, E, F	10,000 kg (22,045 lb)
Max T-O weight: from 1,800 m (5,905 ft) runway:	
A	34,500 kg (76,060 lb)
from 1,500 m (4,920 ft) runway:	
A	33,000 kg (72,750 lb)
from 600-800 m (1,970-2,630 ft) runway:	
A	27,500 kg (60,625 lb)
C, D	34,800 kg (76,720 lb)
E, F	36,500 kg (80,468 lb)
Max landing weight: A	33,000 kg (72,750 lb)
F	36,500 kg (80,468 lb)

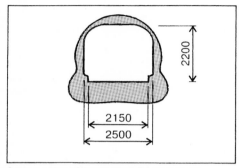

Cross-section of Antonov An-74 cabin/hold, dimensions in milimetres

Max wing loading: A	349.8 kg/m² (71.62 lb/sq ft)
Max power loading: A	271 kg/kN (2.65 lb/lb st)

PERFORMANCE (An-72. A: at T-O weight of 33,000 kg; 72,750 lb, B: at T-O weight of 27,500 kg; 60,625 lb on 1,000 m; 3,280 ft unprepared runway. C, D, E, F, An-74 series as above):
Max level speed at 10,000 m (32,810 ft):	
A	380 kt (705 km/h; 438 mph)
Max level speed at 10,100 m (33,135 ft):	
C, D, E, F	377 kt (700 km/h; 434 mph)
Cruising speed at 10,000 m (32,810 ft):	
A, B	297-324 kt (550-600 km/h; 342-373 mph)
Approach speed: A	97 kt (180 km/h; 112 mph)
Service ceiling: A	10,700 m (35,100 ft)
B	11,800 m (38,720 ft)
Service ceiling, OEI: A	5,100 m (16,740 ft)
B	6,800 m (22,300 ft)
T-O run: A	930 m (3,055 ft)
B	620 m (2,035 ft)
T-O to 10.7 m (35 ft): A	1,170 m (3,840 ft)
B	830 m (2,725 ft)
Landing run: A	465 m (1,525 ft)
B	420 m (1,380 ft)
Max length of runway required:	
C, D	1,200-1,800 m (3,940-5,905 ft)
E, F	1,400-2,150 m (4,595-7,055 ft)
Range, 45 min reserves:	
A with max payload	430 n miles (800 km; 497 miles)
A with 7,500 kg (16,535 lb) payload	
	1,080 n miles (2,000 km; 1,240 miles)
A with max fuel	2,590 n miles (4,800 km; 2,980 miles)
B with 5,000 kg (11,020 lb) payload	
	430 n miles (800 km; 497 miles)
B with max fuel	1,760 n miles (3,250 km; 2,020 miles)
F with 10,000 kg (22,046 lb) payload	
	809 n miles (1,500 km; 932 miles)
F with 5,000 kg (11,023 lb) payload	
	1,943 n miles (3,600 km; 2,236 miles)
Range, 1 hour reserves:	
C, F with 7,500 kg (16,535 lb) payload	
	944 n miles (1,750 km; 1,087 miles)
E with 7,500 kg (16,535 lb) payload	
	1,160 n miles (2,150 km; 1,336 miles)
C, E with 5,000 kg (11,020 lb) payload	
	1,511 n miles (2,800 km; 1,739 miles)
F with 5,000 kg (11,020 lb) payload	
	1,403 n miles (2,600 km; 1,615 miles)
C with max fuel and 800 kg (1,763 lb) payload	
	2,375 n miles (4,400 km; 2,734 miles)
D with max fuel and 16 passengers	
	2,429 n miles (4,500 km; 2,796 miles)
E with max fuel and 2,500 kg (5,511 lb) payload	
	2,294 n miles (4,250 km; 2,640 miles)
F with max fuel and 5,000 kg (11,020 lb) payload	
	2,321 n miles (4,300 km; 2,671 miles)
Endurance: F	6 h 50 min

UPDATED

ANTONOV An-72P

NATO reporting name: **Coaler**

TYPE: Maritime surveillance twin-jet.

PROGRAMME: First seen 1992; displayed at 1992 Farnborough Air Show; in production. Upgraded version, with Israeli avionics and armament, marketed by Israel Aircraft Industries. Engineering designation is **An-76**.

CUSTOMERS: Ukrainian Naval Aviation Force has up to five. No further sales reported by early 2000. Reports of up to 12 in Russian military service cannot be corroborated, although standard An-72 transports operate with both Air Forces and Navy.

COSTS: Israeli upgraded version quoted at US$15 million to US$20 million early 1994.

DESIGN FEATURES: Basically identical to An-72 transport; intended for armed surveillance of coastal areas within 200 n miles (370 km; 230 miles) of shore, day and night in all weathers.

ACCOMMODATION: For maritime missions, flight crew of five, with navigator and radio operator stationed by bulged windows at rear of flight deck on port side and immediately forward of wing leading-edge on starboard side respectively. Provision in main cabin for 40 persons on sidewall and centreline removable seats, including ramp-mounted seats; or 22 fully equipped paratroops; or 16 stretcher patients and medical attendant; or up to 5,000 kg (11,025 lb) of ammunition, equipment or vehicles, with seats stowed.

AVIONICS: Permit automated navigation at all stages of flight, flying the aircraft to a selected point, automatic search, precise determination of co-ordinates, speed and heading of surface ships, air-to-air and air-to-surface communication with aircraft, ships and coastguard to support missions. Digital cockpit avionics in IAI upgraded version. OTV-124 TV scanning system in port main landing gear fairing, plus three stills cameras: one (A-86P) on port side and two (A-86P and night-optimised UA-47) rear/downwards facing, operating from inside hold with cargo door open.

IAI version offered with Elta EL/M-2022A maritime surveillance radar, Elop day/night long-range observation system, and Elisra electronic warfare suite.

EQUIPMENT: Oblique camera on port side opposite radio operator's station; daylight and night mapping cameras in tailcone; SFP-2A flares for night use.

Antonov An-74 flight deck *(Paul Jackson/Jane's)*

Combi interiors of the Antonov An-74TK

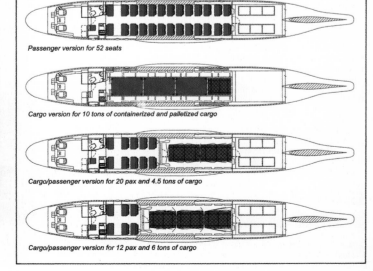

Passenger version for 52 seats

Cargo version for 10 tons of containerized and palletized cargo

Cargo/passenger version for 20 pax and 4.5 tons of cargo

Cargo/passenger version for 12 pax and 6 tons of cargo

ANTONOV—AIRCRAFT: UKRAINE 505

Antonov An-72P STOL maritime patrol aircraft in Ukrainian markings   1999/0052901

ARMAMENT: One 23 mm GSh-23L gun, with 250 rounds, in pod forward of starboard main landing gear fairing; UB-32M rocket pack under each wing; four 100 kg (220 lb) bombs in roof of hold, above rear-loading hatch, with ramp slid forward under cabin to make release practicable. Israeli upgraded version can carry IAI Griffin laser-guided bomb underwing.
Other data generally as for An-72 transport, except:
WEIGHTS AND LOADINGS:
  Mission load   5,000 kg (11,023 lb)
  Max T-O weight   37,500 kg (82,670 lb)
PERFORMANCE:
  Patrol speed at 500-1,000 m (1,640-3,280 ft)
                              162-189 kt (300-350 km/h; 186-217 mph)
  Service ceiling   10,100 m (33,135 ft)
  Field requirement   1,400 m (4,600 ft)
  Max endurance   7 h 0 min-7 hr 18 min
                                                    UPDATED

ANTONOV An-74-300 and An-174
TYPE: Twin-jet transport.
PROGRAMME: First variant, An-74TK-300, announced mid-1998, when prototypes reportedly nearing completion. Model of An-174 shown at Paris, June 1999. Prototype under construction by KhGAPP in 2000 and due to fly before end of that year.
CURRENT VERSIONS: Both are An-74 derivatives in which podded, underslung engines of increased power replace the normal An-72/74 installation.
  An-74T-300: Baseline transport. *As described.*
  An-74TK-300: Combi version. Internal dimensions as for An-74T-100/200.
  An-174: Stretched version with 6 m (19½ ft) fuselage extension provided by plugs fore and aft of wing. Progress D-436T1 engines, each 74.3 kN (916,700 lb st). Studies had begun by 1998. Potential replacement for An-12. Also known as **An-74-400**.
DESIGN FEATURES: Revised engine installation decreases T-O run, despite loss of Coanda effect; new power plants reduce maintenance requirements and fuel consumption (up to 29 per cent); new position simplifies access. Compared to An-74T-200's range with 45 min reserves, An-74T-300 offers additional 372 n miles (690 km; 428 miles) with 10,000 kg (22,046 lb) payload and 518 n miles (960 km; 596 miles) with 5,000 kg (11,023 lb). Cruising speed (at nominal 10,600 to 11,000 m; 34,780 to 36,420 ft) also noticeably improved.
POWER PLANT: Two 63.7 kN (14,330 lb st) Progress D-36-4A turbofans.
PERFORMANCE (estimated):
  Max cruising speed   391 kt (725 km/h; 450 mph)
  Balanced field length   1,900 m (6,235 ft)
                                                    UPDATED

ANTONOV An-124
NATO reporting name: Condor
TYPE: Strategic transport.
PROGRAMME: Bureau design number 305; originally designated An-40. Prototype (SSSR-680125) first flew 26 December 1982; second aircraft registered SSSR-680210 but static test airframe only. First production aircraft (SSSR-82002 *Ruslan*, named after giant hero of Russian folklore immortalised by Pushkin) exhibited 1985 Paris Air Show; lifted payload of 171,219 kg (377,473 lb) to 10,750 m (35,269 ft) on 26 July 1985, exceeding by 53 per cent C-5A Galaxy's record for payload lifted to 2,000 m and setting 20 more records. Entered service January 1986; set closed-circuit distance record 6 to 7 May 1987 by flying 10,880.625 n miles (20,150.921 km; 12,521.201 miles) in 25 hours 30 minutes.
  Deliveries to VTA, to replace An-22, began 1987; in September 1990, during Gulf crisis, an An-124 carried 451 Bangladeshi refugees from Amman to Dacca, after being fitted with chemical toilets, a 570 litre (150 US gallon; 125 Imp gallon) drinking water tank and foam rubber cabin lining in lieu of seats. Carried heaviest single commercial load transported by air: 135.2 tonnes in 1993; and heaviest commercial shipment moved in one flight; 146 tonnes in 1994.
  Service life extension programme begun by Aviastar on first of 17 Antonov Airlines and Volga-Dnepr aircraft in 2000; includes new avionics; upgraded crew rest compartment; and cargo floor and loading equipment strengthening. First, RA-82078 of Volga-Dnepr, redelivered 14 March 2000.
CURRENT VERSIONS: **An-124**: Baseline transport.
Detailed description applies to above version.
  An-124-100: Commercial transport; civil type certificate granted by AviaRegistr of Interstate Aviation Committee of Russian Federation and Associated States (CIS) on 30 December 1992. Civil-operated An-124s are now to this standard. Maximum T-O weight restricted to 392,000 kg (864,200 lb) and maximum payload to 120,000 kg (264,550 lb).
  An-124-100M: As An-124-100, but with Western avionics, including Litton LTN-92 INS, Rockwell Collins GPS, ACARS, weather radar and TCAS-2. Series 3 versions of D-18T engine, offering 6,000 hour overhaul interval and 24,000 hour total life. Crew reduced to four by removal of radio operator and navigator. Prototype (RA-82079) completed at Ulyanovsk late 1995, but not flown until June 2000; delivered Volga-Dnepr 3 August 2000. Version offered to RAF in 1999 as alternative to An-124-210 (which see).
  An-124-102: Flight deck EFIS equipped, with dual sets of CRTs. Crew reduced to three (two pilots and flight engineer).
  An-124-130: Under study in 1996 with General Electric CF6-80 turbofans. Prototype reportedly will be 36th Ulyanovsk aircraft.
  An-124-200: Proposed version with GE CF6-80C2 engines, each 263 kN (592,000 lb st).
  An-124-210: Joint proposal with Air Foyle to meet UK's Short Term Strategic Airlifter (STSA) requirement; 273 kN (60,600 lb st) Rolls-Royce RB211-524H-T engines and Honeywell avionics. Weight empty 184,000 kg (405,650 lb); payload and MTOW as An-124-100. Range (30 min reserves plus 5 per cent) 2,267 n miles (4,200 km; 2,609 miles) with 120,000 kg (264,550 lb) max payload; 3,855 n miles (7,140 km; 4,436 miles) with 80,000 kg (176,375 lb); or 7,424 n miles (13,750 km; 8,543 miles) with max fuel. JAR 25 runway length 2,300 m (7,545 ft). Three flight crew. STSA competition was abandoned in August 1999, reinstated and won by Boeing C-17A.
  An-124FFR: Water-bomber project, able to drop 200 tonnes of fire retardants including 70 tonnes in centre fuel tank. Convertible to freighter.
  An-124 Turboprop: Retrofit with four Aviadvigatel NK-93 propfans considered by Volga-Dnepr in 1997.
  An-124 ORIL: Russian government approval given late 1998 to modify four An-124s to carry the Vozdushny Start booster, capable of placing a 1,630 kg (3,593 lb) satellite into 200 km (124 mile) orbit. ORIL launch system has development cost of US$130 million (2000) and will cost up to US$20 million per launch. Booster is extracted from

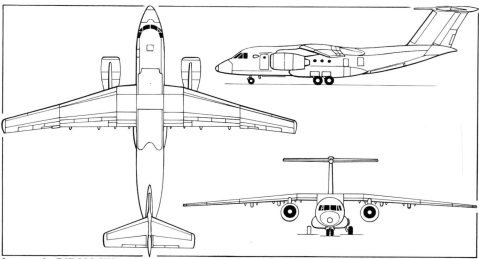
Antonov An-74T-300 STOL transport *(James Goulding/Jane's)*   2000/0079288

Models of Antonov An-74T-300 and An-174 stretched version *(Paul Jackson/Jane's)*   2001/0092559

506 UKRAINE: AIRCRAFT—ANTONOV

Antonov An-124 interior *(Yefim Gordon)*

Antonov An-124 flight deck *(Paul Jackson/Jane's)*

An-124's rear by parachute before rocket ignition. No customers had been reported by mid-2000; IOC by early 2003.

CUSTOMERS: Manufactured at Kiev and (beginning with eighth production aircraft, SSSR-82005, in late 1985) Ulyanovsk; 55 completely or substantially built by late 1995, comprising 19 at Kiev (including prototype, but not test airframe) and 36 at Ulyanovsk; by June 2000, only 18 and 34, respectively, had flown. Production at Kiev temporarily halted by 1991 after 16 aircraft and briefly resumed in 1993-94. Ulyanovsk has continued throughout, producing 33 by 1995; in early 2000, No. 35 was 80 per cent complete (and also targeted for early completion), and No. 36, 60 per cent. No. 35 scheduled for delivery to new operator, Ruslan, in mid-2001.

First international commercial operator was Air Foyle of Luton, UK, which wet leases from Antonov OKB; others available to HeavyLift (UK) from Volga-Dnepr (Russia), for charter operations. Operators in 2000 comprised Antonov Airlines (nine), Polet (six), Atlant (one), Volga-Dnepr (nine), Aeroflot (one, impounded in Netherlands); and Russian Air Forces (21 remaining, most in civil markings, with 566 VTAP at Sescha); all others in storage (one) or written off (four). Polet fleet includes four transferred in 2000 from military use for ORIL programme.

DESIGN FEATURES: World's largest production aircraft; upward-hinged visor-type nose and rear fuselage ramp/door for simultaneous front and rear loading/unloading; titanium floor throughout constant-section main hold, which is lightly pressurised, with a fully pressurised cabin for passengers above; landing gear for operation from unprepared fields, hard packed snow and ice-covered swampland; steerable nosewheels and mainwheels permit turns on 45 m (148 ft) wide runway.

Service life initially 7,500 hours; contract for extension to 12,000 hours signed between Volga-Dnepr and Antonov, July 2000; aircraft delivered from 2000 have 24,000 hour airframe and engine life.

Supercritical wings, with anhedral; sweepback approximately 35° on inboard leading-edge, 32° outboard; all tail surfaces sweptback.

FLYING CONTROLS: All surfaces hydraulically actuated; manual control of automatic control systems, control surface actuators, control system manual linkage and trimming system. Two-section ailerons, three-section single-slotted Fowler flaps (two outer, one inner) and six-section full-span leading-edge flaps on each wing; small slot in outer part of two inner flap sections each side to optimise aerodynamics; 12 spoilers on each wing, forward of trailing-edge flaps (four lateral control spoilers outboard; four glissage spoilers; and four airbrakes inboard); no wing fences, vortex generators or tabs; hydraulic flutter dampers on ailerons; rudder and each elevator in two sections, without tabs but with hydraulic flutter dampers; fixed incidence tailplane; control runs (and other services) channelled along fuselage roof.

STRUCTURE: Basically conventional light alloy, but 5,500 kg (12,125 lb) of composites make up more than 1,500 m² (16,150 sq ft) of surface area, giving weight saving of more than 2,000 kg (4,410 lb); each wing has one-piece root-to-tip upper surface extruded skin panel, strip of carbon fibre skin panels on undersurface forward of control surfaces, and glass fibre tip; front and rear of each flap guide fairing of glass fibre, centre portion of carbon fibre; central frames of semi-monocoque fuselage each comprise four large forgings; fairings over intersection of fuselage double-bubble lobes in line with wing, from rear of flight deck to plane of fin leading-edge, primarily of glass fibre, with central, and lower underwing, portions of carbon fibre; other glass fibre components include tailplane tips, nosecone, tailcone and most bottom skin panels forming blister underfairing between main landing gear legs; carbon fibre components include strips of skin panels forward of each tail control surface, nose and main landing gear doors, some service doors, and clamshell doors aft of rear-loading ramp.

LANDING GEAR: Hydraulically retractable nosewheel type, made by Hydromash, with 24 wheels. Two independent forward-retracting and steerable twin-wheel nose units, side by side. Each main gear comprises five independent inward-retracting twin-wheel units; front two units on each side steerable. Each mainwheel bogie enclosed by separate upper and lower doors when retracted. Nosewheel doors and lower mainwheel doors close when gear extended. All wheel doors of carbon fibre. Main gear bogies retracted individually for repair or wheel change. Mainwheel tyres size 1,270×510. Nosewheel tyres size 1,120×450.

Aircraft can 'kneel', by retracting nosewheels and settling on two extendable 'feet', giving floor of hold a 3.5° slope to assist loading and unloading. Process takes 3 minutes to lower aircraft and 6½ minutes to raise. Rear of cargo hold lowered by compressing main gear oleos. Carbon brakes normally toe-operated, via rudder pedals. For severe braking, pedals depressed by toes and heels. Turning radius (outboard wheels) 19.6 m (64 ft 4 in).

POWER PLANT: Four ZMKB Progress/Ivchenko D-18T turbofans, each 229 kN (51,590 lb st); thrust reversers standard. Optional hush kit certified to ICAO Chapter 3 in mid-1997. Engine cowlings of glass fibre; pylons have carbon fibre skin at rear end. All fuel in 10 integral tanks in wings, total capacity 348,740 litres (92,128 US gallons; 76,714 Imp gallons), not all of which is utilised in civil versions.

ACCOMMODATION: Crew and passenger accommodation on upper deck; freight and/or vehicles on lower deck. Flight crew of six, in pairs, on flight deck, with place for loadmaster in lobby area (10 to 12 cargo handlers and servicing staff carried on commercial flights). Pilot and co-pilot on fully adjustable seats, which rotate for improved access. Two flight engineers, on wall-facing seats on starboard side, have complete control of master fuel cocks, detailed systems instruments, and digital integrated data system with CRT monitor. Behind pilot are navigator and communications specialist, on wall-facing seats. Between flight deck and wing carry-through structure, on port side, are toilets, washing facilities, galley, equipment compartment, and two cabins for up to six relief crew, with table and facing bench seats convertible into bunks.

Aft of wing carry-through is passenger cabin for 88 persons. Hatches in upper deck provide access to wing and tail unit for maintenance when workstands not available. Flight deck and passenger cabin each accessible from cargo hold by hydraulically folding ladder, operated automatically with manual override. Rearward-sliding and jettisonable window each side of flight deck. Primary access to flight deck via airstair door, with ladder extension, forward of wing on port side. Smaller door forward of this and slightly higher. Door from main hold aft of wing on starboard side. Upper deck doors at rear of flight deck on starboard side and at rear of passenger cabin on each side. Emergency exit from upper deck aft of wing on each side.

Hydraulically operated visor-type upward-hinged nose takes 7 minutes to open fully, with simultaneous extension of folding nose loading ramp. When open, nose is steadied by reinforcing arms against wind gusts. No hydraulic, electrical or other system lines broken when nose is open. Radar wiring passes through tube in hinge. Hydraulically operated rear-loading doors take 3 minutes to open, with simultaneous extension of three-part folding ramp. This can be locked in intermediate position for direct loading from truck. Aft of ramp, centre panel of fuselage undersurface hinges upward; clamshell door to each side opens downward.

Completely unobstructed lower deck freight hold has titanium floor, attached 'mobilely' to lower fuselage structure to accommodate changes of temperature, with rollgangs and retractable attachments for cargo tiedowns. Load limits per tiedown fitting: 12,000 kgf (26,455 lbf) on main floor; 5,000 kgf (11,023 lbf) on rear ramp. Narrow catwalk along each sidewall facilitates access to, and mobility past, loaded freight. Payloads include largest CIS main battle tanks, complete missile systems, 12 standard ISO containers, oil well equipment and earth movers; HeavyLift/Volga-Dnepr aircraft previously transported Airbus wings in Europe. No personnel carried normally on

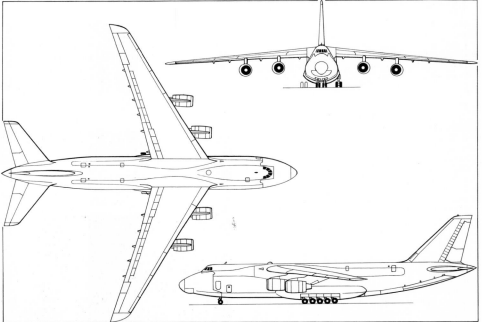

Antonov An-124 (four ZMKB Progress/Ivchenko D-18T turbofans) *(James Goulding/Jane's)*

lower deck in flight, because of low pressurisation, but can accommodate 360 troops (not including 88 on upper deck), plus two lavatories and oxygen bottles. Military aircraft equipped to air-drop up to 16 pallets, each of up to 4,500 kg (9,920 lb); or 268 paratroops in two passes. Medical evacuation capability is 288 stretchers and 28 attendants.

SYSTEMS: Automatic flight control system includes control loading for elevator and ailerons; stability augmentation for elevator and rudder; elevator trim and balance; elevator and rudder gear ratio system; and flight limit condition restriction for elevator. Entire interior of aircraft is pressurised and air conditioned. Maximum pressure differential 0.55 bar (7.8 lb/sq in) on upper deck, 0.25 bar (3.55 lb/sq in) on lower deck. Four independent hydraulic systems. Quadruple redundant fly-by-wire flight control system, with mechanical emergency fifth channel to hydraulic control servos. Special secondary bus electrical system. Landing lights under nose and at front of each main landing gear fairing. APU in rear of each landing gear fairing for engine starting, can be operated in the air or on the ground to open loading doors for airdrop from rear or normal ground loading/unloading, as well as for supplying electrical, hydraulic and air conditioning systems. Bleed air anti-icing of wing leading-edges. Electro-impulse de-icing of fin and tailplane leading-edges.

AVIONICS: *Radar:* Two dielectric areas of nose visor enclose forward-looking weather radar and downward-looking ground-mapping/nav radar.
Flight: Hemispherical dielectric fairing above centre-fuselage for satellite nav receiver; quadruple INS; Loran and Omega.
Instrumentation: Conventional flight deck equipment, including automatic flight control system panel at top of glareshield, weather radar screen and moving map display forward of throttle and thrust reverse levers on centre console. No electronic flight displays. Dual attitude indicator/flight director and HSIs, and vertical tape engine instruments.

EQUIPMENT: Two electric travelling cranes in roof of hold, each with two lifting points, offer total lifting capacity of 20,000 kg (44,092 lb). First trial of 30,000 kg (66,139 lb) system in December 1999; development then launched of 40,000 kg (88,185 lb) lifting system. Two winches each pull a 3,000 kg (6,614 lb) load. Small two-face mirror, of V form, enables pilots to adjust their seating position until their eyes are reflected in the appropriate mirror, which ensures optimum field of view from flight deck.

DIMENSIONS, EXTERNAL:
Wing span	73.30 m (240 ft 5¾ in)
Wing aspect ratio	8.6
Length overall	69.10 m (226 ft 8½ in)
Fuselage max width	7.28 m (23 ft 10½ in)
Height overall	21.08 m (69 ft 2 in)
Wheel track	8.00 m (26 ft 3 in)
Wheelbase (centre row mainwheels)	22.90 m (75 ft 1½ in)
Sill height: front door: normal	up to 2.79 m (9 ft 1¾ in)
kneeling	1.43 m (4 ft 8¼ in)
rear door, normal	up to 2.85 m (9 ft 4¼ in)

DIMENSIONS, INTERNAL:
Cargo hold: Length at floor:	
excl ramps	36.48 m (119 ft 8¼ in)
incl rear ramp	41.54 m (136 ft 3½ in)
Max length	42.68 m (140 ft 0¼ in)
Width: at floor	6.40 m (21 ft 0 in)
max	6.63 m (21 ft 9 in)
at ceiling	4.26 m (13 ft 11¾ in)
Max height	4.40 m (14 ft 5¼ in)
Area, incl ramp	2,650 m² (2,852 sq ft)
Volume, incl ramp	1,160 m³ (40,965 cu ft)
Passenger cabin: Width at floor	3.80 m (12 ft 5½ in)

AREAS:
Wings, gross	628.0 m² (6,760.0 sq ft)

WEIGHTS AND LOADINGS (A: basic An-124, B: An-124-100):
Operating weight empty: A	175,000 kg (385,800 lb)
Max payload: A	150,000 kg (330,700 lb)
B	120,000 kg (264,550 lb)
of which: rear ramp (A and B)	10,000 kg (22,046 lb)
Max fuel weight (An-124-210)	214,000 kg (471,790 lb)
Max T-O weight: A	405,000 kg (892,875 lb)
B	392,000 kg (864,200 lb)
Max ramp weight: B	398,000 kg (877,425 lb)
Max landing weight: B	330,000 kg (727,500 lb)
Max zero-fuel weight: A	325,000 kg (716,500 lb)
Max wing loading: A	644.9 kg/m² (132.09 lb/sq ft)
Max power loading: A	441 kg/kN (4.32 lb/st)

PERFORMANCE:
Max cruising speed: A	467 kt (865 km/h; 537 mph)
Normal cruising speed at 10,000-12,000 m (32,810-39,370 ft):	
A	432-459 kt (800-850 km/h; 497-528 mph)
Air-dropping speed range	127-216 kt (235-400 km/h; 146-249 mph)
Approach speed:	
A	124-140 kt (230-260 km/h; 143-162 mph)
Max certified altitude	12,000 m (39,380 ft)
T-O balanced field length at max T-O weight:	
A	3,000 m (9,840 ft)
T-O run: A	2,520 m (8,270 ft)
Landing run at max landing weight: A	900 m (2,955 ft)
Range: with max payload:	
A	2,430 n miles (4,500 km; 2,795 miles)
B	2,591 n miles (4,800 km; 2,982 miles)
with 80,000 kg (176,375 lb) payload:	
A, B	4,535 n miles (8,400 km; 5,219 miles)
with 40,000 kg (88,184 lb) payload:	
A, B	6,479 n miles (12,000 km; 7,456 miles)
ferry, with max fuel:	
A, B	8,477 n miles (15,700 km; 9,755 miles)

OPERATIONAL NOISE LEVELS:
Stated to meet ICAO requirements

UPDATED

Antonov An-124 heavy freight transport

ANTONOV An-140

TYPE: Twin-turboprop short-range transport.
PROGRAMME: Announced 1993 as An-24 replacement; two prototypes with TV3-117 engines and static test airframe constructed at Kiev; rolled out 6 June 1997; first flight (UR-NTO) at Kiev-Svyatoshino (landing at Gostomel flight test centre, as base for certification trials) 17 September 1997; second airframe (0102) for static tests; second flying prototype (UR-NTP) rolled out 11 December 1997 and flew 26 December. Third (first production), with full systems, due to fly at Kharkov, early 1999, and undertake electromagnetic compatibility and climatic tests but maiden sortie delayed until 11 October 1999; this aircraft flew 41 sorties towards certification.

Certification to FAR/JAR 25/AP-25 and ICAO Part 3 Appendix 16 (FAR Pt 36/AP-36) standards intended; 940 hour certification programme began August 1998 and completed in late 1999; included cold trials at Arkhangelsk (by UR-NTO between 29 March and 1 May 1999, and by UR-NTP in Yakutia for 10 days in January 2000) and hot-and-high trials in Uzbekistan and Kyrgystan (by UR-NTP, concluding 3 September 1999). Following first series of flight testing, tailplanes of prototypes modified to obviate propwash-induced vibration, gaining 6° dihedral and shortened elevator horn balance. An-140's 1,000th hour

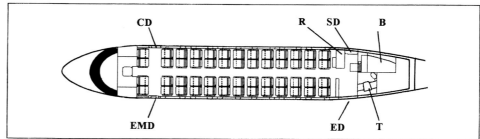
Antonov An-140 cabin seating for 52 passengers *(Paul Jackson/Jane's)*
B: baggage, CD: cargo door, ED: entrance door, EMD: emergency door, R: refreshments, SD: service door, T: lavatory

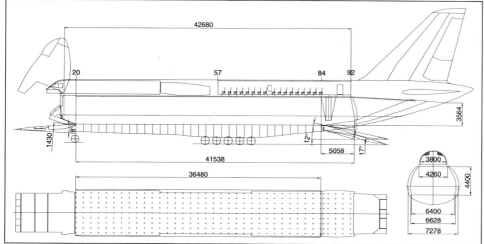
Principal cargo hold dimensions (in millimetres) of the Antonov An-124

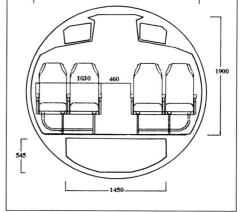
Cross-section of the An-140 cabin; dimensions in millimetres *(Paul Jackson/Jane's)*

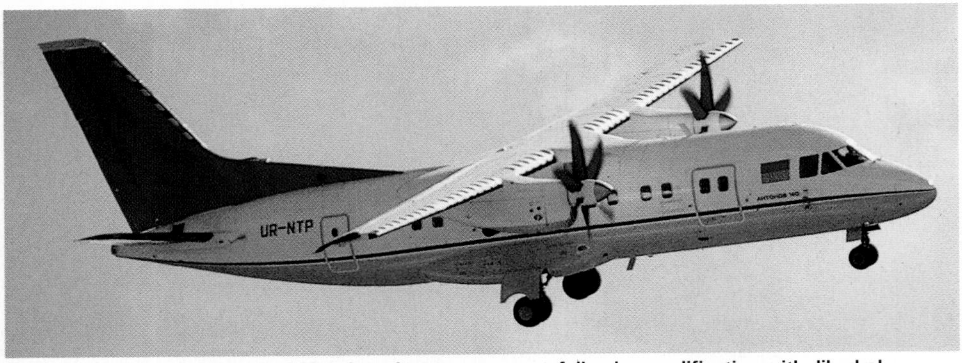

Second prototype Antonov An-140 twin-turboprop transport, following modification with dihedral tailplane and smaller elevator horn balances *(Paul Jackson/Jane's)* ***2001***/0092538

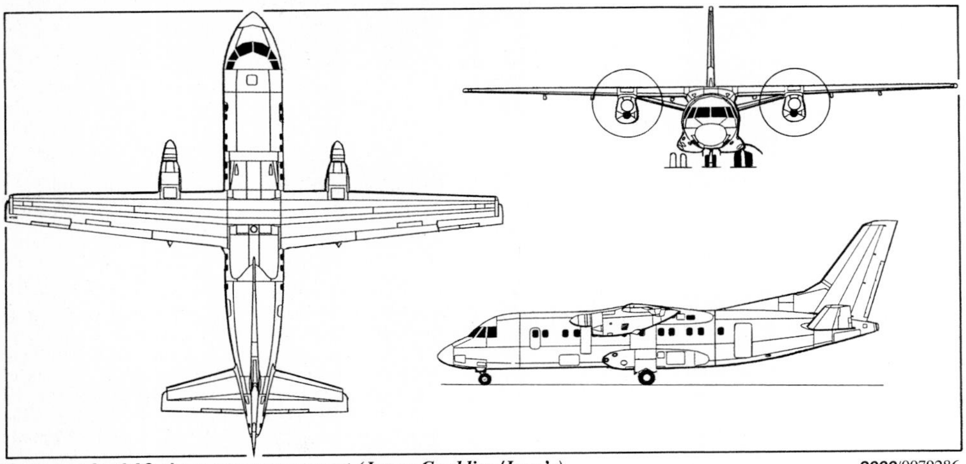

Antonov An-140 short-range transport *(James Goulding/Jane's)* ***2000***/0079286

flown (by UR-NTP) on 12 January 2000. Certification was achieved on 25 April 2000 coincident with that of TV3-117VMA-SBM1 engine and AV-140 propeller; some 1,000 sorties flown by April 2000.

Series production began 1999 at KhGAPP, Kharkov (where wings for prototypes were built) and at Aviacor, Samara, Russian Federation. Aviacor production intended to be 10 in 2000 (first in July) and to reach 30 per year by 2001, but reduced to combined total of eight in 2000-01; five under assembly at Aviacor by May 2000, but none yet ordered and maiden flight of first slipped to late 2000. KhGAPP has capacity for 40 per year, initial batch comprising seven, first of which flying by early 2000 and delivered in May 2000.

Production share agreement of 1998 assigns empennage to Antonov; engine nacelles, wing and associated control surfaces to KhGAPP; landing gear to Aviagregat at Samara and Youzhmash at Dnieprpetrovsk; and fuselage to Aviacor (incorporating Avio Interiors fittings from Italy).

Agreement signed February 1996 for assembly by HESA at rate of 12 a year from 1999, progressing to local parts manufacture, in new plant at Esfahan, Iran, with Ukrainian assistance (see IAIO entry in Iranian section); target completion date for first Iranian aircraft slipped to October 2000. In late 1999, Reims Aviation (which see in French section) was considering promotion of a Westernised version of Kiev-built An-140.

CURRENT VERSIONS: **An-140:** *As described.*

An-140A: Aeroflot version with PW127A engines; for regional services in western Russia.

An-140T: Proposed freighter with large door, port side, rear. Convertible **An-140TK** will be similar.

An-140-100: Proposed 68-seat version; fuselage stretched by 3.80 m (12 ft 5½ in); under development by 1997.

An-142: Under development for 2001 first flight; forward-retracting rear loading ramp similar to An-26.

Iran 140: Licence-built by HESA (which see).

An-140 Military: Patrol, surveillance, photographic and similar variants proposed for military operators.

CUSTOMERS: Air Ukraine letter of intent for up to 40 by 2010 signed 17 September 1997 (first flight); initial four firm conversions made June 1999 (for manufacture at Kharkov), specifying deliveries from 2000; however, initial recipient (2000) from Ukrainian production is Ikar Airlines, with five on order from January 2000 contract. First order for 15 placed by republic of Sakha-Yakutia, Russian Federation, mid-1998, requiring 10 to be delivered to Sakha Airlines from Samara production in 2000-04. Aeroflot letter of intent for 50 PW127-powered An-140As signed mid-1999 (specifying Samara production). Tyumenaviatrans reportedly holds option on 25 An-140s, but this had been relegated to expression of interest (also from Polar Airlines and Mirny) when An-140 toured Siberia in early 2000. Launch order for Iran 140 is 20 for Iran Asseman Airlines; Iran Air also will be operator; Iranian licence initially for 80 aircraft, but interest reported, late 1998, in building further 160. Estimated sales of 645 by 2011, including 430 to Russia and 70 to Ukraine.

COSTS: US$7 million (2000).

DESIGN FEATURES: Light transport, designed to be capable of autonomous operation from airfields with unprepared runways at all altitudes and in all weathers, providing airline-standard comfort. International certification and various engine options to maximise sales prospects. Conventional high-wing monoplane; tapered wing with unswept leading-edge; sweptback fin and tailplane, latter with 6° dihedral; engines mounted underwing. Maintenance target of 6.5 mmh/fh. Service life 50,000 landings/50,000 hours/25 years.

FLYING CONTROLS: Conventional and manual. Control surfaces all horn balanced; ailerons with trim tab in each (two-section tab starboard); elevator with two-section tab in each; rudder with large single tab. Two-section flaps in each wing.

STRUCTURE: Largely of aluminium, with some titanium.

LANDING GEAR: Retractable tricycle type by Pivdennyi; twin Rubin wheels on each unit; nosewheels retract forward, mainwheels into fairings each side of lower fuselage. Mainwheels size 810×320-330; nosewheels 600×220-250. Rubin braking system. Able to operate from gravel or unpaved fields.

POWER PLANT: Two 1,839 kW (2,466 shp) AI-30 Series 1 turboprops (Klimov TV3-117VMA-SBM1 built under licence at Zaporozhye, Ukraine, by Motor-Sich), driving AV-140 propellers; optionally, two 1,864 kW (2,500 shp) Pratt & Whitney Canada PW127A turboprops, driving Hamilton Sundstrand 247F propellers. FED fuel-management system; Star engine control system.

ACCOMMODATION: Flight crew of two, plus cabin attendant; basic seating for 52 passengers, four-abreast with centre aisle, at 75 cm (30 in) pitch, or 48 at 81 cm (32 in) pitch. Main passenger door with airstairs, at rear of cabin on port side, with service door opposite; emergency exit port side at front of cabin; cargo door starboard side, front. Coat stowage, galley and toilet at rear of cabin. Baggage/freight compartment at rear of cabin, plus forward underfloor freight hold, with door on port side. Cargo door on starboard side, forward part of cabin floor reinforced, and detachable equipment provided, enabling 1,900 to 3,650 kg (4,188 to 8,046 lb) of palletised cargo and 36 to 20 passengers to be carried with forward rows of seats removed. Overhead baggage lockers. Accommodation air conditioned and pressurised.

SYSTEMS: Motor-Sich AI-9-3B APU in rear fuselage. FED hydraulic system; Nauka air conditioning. Kommunar anti-icing and air conditioning control systems. Auxilec generators; Eros oxygen system.

AVIONICS: *Comms:* Satori ELT96 ELT. Television Engineering entertainment system.

Radar: Buran A-140 weather radar

Flight: Ukrainian Radio INS; Orizon Navigation SN-3301 GPS; VNIIRA-Navigator VND-94 VOR; Kurs-93M nav computer; other navigation aids by Russian Radio Equipment.

*DIMENSIONS, EXTERNAL:

Wing span	24.505 m (80 ft 4¾ in)
Length overall	22.605 m (74 ft 2 in)
Height overall	8.225 m (26 ft 11¾ in)
Wheel track (c/l shock-absorbers)	3.18 m (10 ft 5¼ in)
Wheelbase	8.01 m (26 ft 3¼ in)
Propeller diameter (AI-30)	3.60 m (11 ft 9¾ in)
Distance between propeller centres	8.20 m (26 ft 10¾ in)
Main cabin door: Height	1.605 m (5 ft 3¼ in)
Width	0.98 m (3 ft 3 in)
Emergency exit: Height	1.20 m (3 ft 11¼ in)
Width	0.515 m (1 ft 8¼ in)
Service door: Height	1.28 m (4 ft 2½ in)
Width	0.63 m (2 ft 0¾ in)
Cargo door: Height	1.34 m (4 ft 4¾ in)
Width	1.00 m (3 ft 3¼ in)
Underfloor freight hold door:	
Height	0.90 m (2 ft 11½ in)
Width	1.02 m (3 ft 4¼ in)

** Data from Antonov design bureau; Kharkov figures mostly different*

DIMENSIONS, INTERNAL:

Cabin: Length:	
excl flight deck, galley/toilet area	10.50 m (34 ft 5¼ in)
incl galley/toilet and baggage	14.30 m (46 ft 11 in)
Max width	2.60 m (8 ft 6¼ in)
Max height	1.90 m (6 ft 2¾ in)
Volume	65.5 m³ (2,313 cu ft)
Aft baggage compartment: Volume	6.0 m³ (212 cu ft)

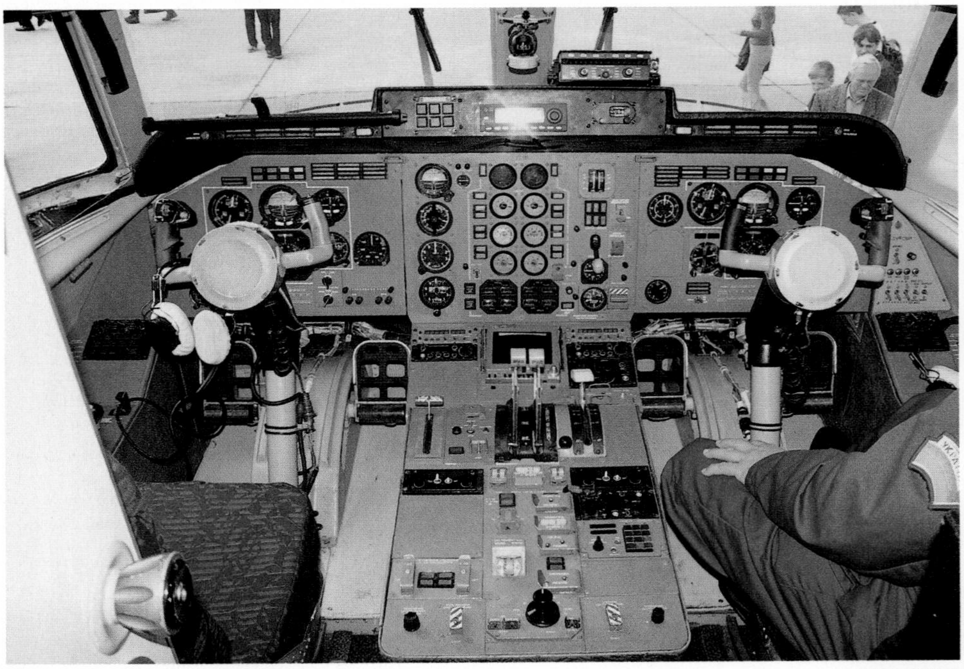

Antonov An-140 flight deck *(Paul Jackson/Jane's)* ***2000***/0084432

Underfloor freight hold: Length	3.98 m (13 ft 0¾ in)
Max width	1.45 m (4 ft 9 in)
Max height	0.545 m (1 ft 9¾ in)
Volume	3.0 m³ (106 cu ft)
Overhead baggage lockers:	
Volume (total)	2.4 m³ (85 cu ft)

WEIGHTS AND LOADINGS:
Baggage capacity	1,840 kg (4,057 lb)
Max payload	6,000 kg (13,227 lb)
Max fuel weight	4,370 kg (9,634 lb)
Max T-O weight	19,150 kg (42,218 lb)
Max landing weight	19,100 kg (42,108 lb)
Max zero-fuel weight	17,800 kg (39,242 lb)
Max power loading	5.21 kg/kW (8.56 lb/shp)

PERFORMANCE:
Max cruising speed at 7,200 m (23,620 ft)
　310 kt (575 km/h; 357 mph)
Econ cruising speed at 7,200 m (23,620 ft)
　280 kt (520 km/h; 323 mph)
Nominal cruising altitude
　7,200-7,500 m (23,620-24,600 ft)
Balanced runway length　1,350 m (4,430 ft)
Design range at 280 kt (520 km/h; 323 mph) at 7,200 m (23,620 ft), no reserves:
　with 6,000 kg (13,227 lb) payload
　　486 n miles (900 km; 559 miles)
　with 52 passengers:
　　PW127　1,349 n miles (2,500 km; 1,553 miles)
　　AI-30　1,133 n miles (2,100 km; 1,304 miles)
　with 46 passengers:
　　PW127　1,646 n miles (3,050 km; 1,895 miles)
　with max fuel and 33 passengers:
　　PW127　2,213 n miles (4,100 km; 2,548 miles)
　　AI-30　1,997 n miles (3,700 km; 2,299 miles)

UPDATED

Prototype Antonov An-140 pictured during a later test sortie

ANTONOV An-225 MRIYA
English name: Dream
NATO reporting name: Cossack
TYPE: Outsize freighter.
PROGRAMME: The world's largest aircraft first flew on 21 December 1988; it was last described in the 1995-96 *Jane's*, the sole prototype having been in storage at Gostomel since May 1994. In 1998, Antonov announced that it was proceeding with assembly of the second prototype, which had previously been abandoned in partly complete condition, and was planning to return the first aircraft to airworthy status; these projects were to cost US$50 million and US$20 million, respectively. By July 2000, Antonov's own freight airline was concentrating on refurbishing the prototype to fly in 2001 (target date June,

Antonov An-225 outsize freighter *(Paul Jackson/Jane's)*

for participation in Paris Air show), estimating a US$20 million programme to replace avionics and six engines removed for use on other aircraft. First batch of replacement items despatched by Aviastar plant at Ulyanovsk, Russia, in December 2000. Volga-Dnepr has requirement for two or three.

Prompted by predictions of a growing market for ultra-large air transports, the Aviastar plant at Ulyanovsk (Russia) has evaluated the prospects for placing the An-225 in series production. Consideration also has been given to replacing the aircraft's 230 kN (51,590 lb st) Progress D-18T engines by 411 kN (91,300 lb st) Rolls-Royce Trent 892s or 436 kN (98,000 lb st) Pratt & Whitney PW4098s.

UPDATED

AVIANT
AVIANT, KIEVSKY GOSUDARSTVENNY AVIATSIONNY ZAVOD (Aviant, Kiev State Aviation Plant)
prospekt Pobedy 100/1, 252062 Kiev
Tel: (+38 044) 442 43 85
Fax: (+38 044) 443 72 45
e-mail: aviant@777.com.ua
Telex: 131265 REPER
GENERAL DIRECTOR: Alexander I Kharlov

The Kiev Aviation Plant has effectively terminated work on the Antonov An-32 twin-turboprop transport and An-124 heavy transport. However, it built two prototypes of the An-70 four-propfan transport and may share production with Aviacor at Samara. The two firms, plus RSK MiG, will also manufacture the Tupolev Tu-334 twin-turbofan transport.

UPDATED

KhGAPP
KHARKOVSKOYE GOSUDARSTVENNOYE AVIATSIONNOYE PROIZVODST-VENNOYE PREDPRIYATIE (Kharkov State Aviation Production Enterprise)
ulitsa Sumskaya 134, 310023 Kharkov
Tel: (+38 0572) 43 08 32
Fax: (+38 0572) 47 80 01
e-mail: itl589@online.kharkov.ua
Telex: 125268 CRAB
GENERAL DIRECTOR: Anatoly Myalitsa

Established in 1926 as GAZ 135, Kharkov Aircraft Production Association has built military, passenger and freight aircraft, including the Su-2, Yak-18, MiG-15, Tu-104 and Tu-134. Export of its products began in 1964. ISO 9002 was gained in 1998.

Current production is centred on the Antonov An-74 in various versions, including the An-74-300, with underslung engines. Manufacture of the twin-turboprop An-140, described in Antonov entry, is now under way.

UPDATED

Antonov An-74 taking shape on the KhGAPP assembly line. Two were delivered in 1999

MOTOR-SICH
OTKRITOYE AKTSIONERNOYE OBSHCHESTVO MOTOR-SICH (Motor-Sich JSC)
ulitsa 8 Marta 15, Zaporozh'e 330068
Tel: (+38 0612) 61 47 77
Fax: (+38 0612) 65 60 07
e-mail: motor@motor.comint.net
Web: http://www.ukrainetrade.com/motorsich
GENERAL DIRECTOR: Vyacheslav Voguslaev

Long-established engine manufacturer Motor-Sich was confirmed in August 2000 as production centre for those Kamov Ka-226 utility helicopters required for use in Ukraine. Details of Ka-226 appear in Russian section.

NEW ENTRY

UNITED KINGDOM

AD

AD AEROSPACE LTD
1 Hilton Square, Pendlebury, Swinton, Manchester M27 4DB
Tel: (+44 161) 727 66 00
Fax: (+44 161) 727 85 67
e-mail: demenquiry@ad-aero.co.uk
Web: http://www.ad-aero.co.uk
CHAIRMAN: Mike Newton
MANAGING DIRECTOR: Mike Horne
MARKETING MANAGER: Vanessa Riding

In 1998, AD Aerospace Ltd (then known as DM Aerospace) acquired the type certificate and design rights to the Thorp T-211 light aircraft, last described in the US section of the 1992-93 edition of *Jane's*. By mid-1999, three US-built T-211s had been imported for company-operated flying clubs at Carlisle, Denham and Liverpool-Hawarden (joint base).

UPDATED

US-built Thorp T-211, used at Denham, UK, by AD Aerospace's Club Thorp *(Paul Jackson/Jane's)*
2001/0092072

AD THORP T-211
TYPE: Side-by-side lightplane/kitbuilt.
PROGRAMME: Originated in T-11 Sky Skooter, designed by John Thorp and first flown 15 August 1946. See 1992-93 and earlier editions for previous development history; production rights passed to Phoenix Aircraft at Mesa, Arizona, in 1992, but that company declared bankrupt in mid-1994. Acquired by AD; market launch (and first sale, to Ken Fowler) of kits at PFA International Air Rally & Exhibition at Cranfield, Bedfordshire, in July 1998. AD markets homebuild kits produced under licence in the USA by Venture Light Aircraft Resources (the previous owner); UK production of complete aircraft planned, but postponed in 1999.
CURRENT VERSIONS: **T-211**: Factory-built version.
T-211A: Continental-engined kitbuilt version, supplied complete, or in eight sub-kits for stage-by-stage assembly. *Data refer to Continental-powered version.*
T-211B: Jabiru-engined kitbuilt. Initial aircraft is first UK-built T-211, G-TZII, which nearing completion in mid-1999.
CUSTOMERS: Five kits sold during PFA International Air Rally in July 1999, increasing total to six. (At least nine built in USA by Tubular Aircraft, one; Aircraft Engineering Associates, one; and Venture.)
COSTS: Kit £15,000, excluding engine and avionics; factory-built £55,000, both excluding VAT (1999).
FLYING CONTROLS: Manual. Plain ailerons; all-moving tailplane with anti-balance tab occupying 90 per cent span; wide-span manually operated three-position trailing-edge flaps.
STRUCTURE: Light alloy fuselage; wing with main spar, false spar and externally stiffened skin; ailerons, flaps, tailplane and rudder have externally stiffened skin.
LANDING GEAR: Non-retractable tricycle type; oleo-pneumatic shock-absorber in each leg; all three wheels size 5.00-5; steerable nosewheel; Cleveland brakes on main units.
POWER PLANT: One 74.6 kW (100 hp) Teledyne Continental O-200A flat-four, driving a Sensenich 69CK two-blade fixed-pitch propeller. Fuel tank aft of cabin, capacity 79.5 litres (21 US gallons; 17.5 Imp gallons); oil capacity 5.7 litres (1.5 US gallons; 1.25 Imp gallons). Jabiru 3300 six-cylinder four-stroke, rated at 89 kW (120 hp), available as an option from 1999.
ACCOMMODATION: Two seats side by side beneath rearward-sliding transparent canopy; fixed windscreen. Baggage compartment behind seats, capacity 8.1 kg (17.8 lb).

DIMENSIONS, EXTERNAL:
Wing span	7.62 m (25 ft 0 in)
Wing aspect ratio	6.0
Length overall	5.50 m (18 ft 0½ in)
Height overall	1.90 m (6 ft 2¾ in)
Propeller diameter	1.70 m (5 ft 7 in)

DIMENSIONS, INTERNAL:
Cabin max width	0.94 m (3 ft 1 in)

AREAS:
Wings, gross	9.72 m² (104.6 sq ft)

WEIGHTS AND LOADINGS:
Weight empty	354 kg (780 lb)
Baggage capacity	18 kg (40 lb)
Max T-O weight	576 kg (1,270 lb)
Max wing loading	59.3 kg/m² (12.14 lb/sq ft)
Max power loading	7.73 kg/kW (12.70 lb/hp)

PERFORMANCE:
Never-exceed speed (V_{NE})	138 kt (255 km/h; 158 mph)
Max level speed	109 kt (202 km/h; 125 mph)
Cruising speed at 75% power	85 kt (157 km/h; 98 mph)
Stalling speed, flaps down, power on	39 kt (73 km/h; 45 mph)
Max rate of climb at S/L	more than 229 m (750 ft)/min
Service ceiling	3,810 m (12,500 ft)
T-O run	137 m (450 ft)
Landing run	151 m (495 ft)
Max range	413 n miles (764 km; 475 miles)

UPDATED

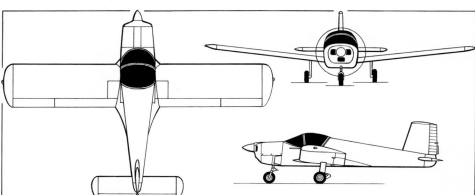

AD Thorp T-211 two-seat kitbuilt *(James Goulding/Jane's)*
2000/0084388

AMF

AMF AVIATION ENTERPRISES LTD
Membury Airfield, Lambourn, Berkshire RG16 7TJ
Tel: (+44 1488) 722 24
Fax: (+44 1488) 722 24
MANAGING DIRECTOR: Angus M Fleming

No evidence has been received of recent production of the Chevvron 2-32 ultralight, last described in the 2000-01 *Jane's*.

UPDATED

AVRO — *see under BAE Systems*

BAE

BAE SYSTEMS plc
HEADQUARTERS: Warwick House, PO Box 87, Farnborough Aerospace Centre, Farnborough, Hampshire GU14 6YU
Tel: (+44 1252) 37 32 32
Fax: (+44 1252) 38 30 00
Web: http://www.baesystems.com
CHAIRMAN: Sir Richard Evans
CEO: John P Weston
CHIEF OPERATING OFFICERS: Mike Turner
  Steve Mogford
MARKETING DIRECTOR: Sir Charles Masefield
ENGINEERING DIRECTOR: Professor Dave Gardner
DIRECTOR, CORPORATE COMMUNICATIONS: Hugh Colver

On 19 January 1999, British Aerospace (BAe) announced that it planned to merge with the Marconi Electronic Systems business, then part of GEC-Marconi. Following regulatory approval, BAE Systems was established on 30 November 1999 as joint-ranking (with EADS) world's third-largest aerospace company, with annual turnover of £12.3 billion, order book of £37.5 billion and (including joint ventures) 115,600 employees in UK (70,000), USA (16,700), Sweden (6,800), Saudi Arabia (5,400), France (4,500), Italy (4,000), Germany (3,500), Australia (3,000) and Canada (1,700).

Principal wholly owned operating companies (aircraft only)

BAE Systems Programmes & Customer Support Group (formerly Military Aircraft and Aerostructures; manufacture of military aircraft, guided weapon systems, ordnance, components and assemblies)
BAE Systems Airbus (Airbus wing design and manufacture)
BAE Systems Aviation Services Ltd (maintenance and modification of Airbus commercial aircraft)
BAE Systems, Customer Support, Flight Training (pilot training in Spain and Australia)
BAE Systems Regional Aircraft (regional airliners)

Subsidiary and associate organisations (aircraft only) (BAE interest percentage)
Airbus Industrie (Airbus airliners; 20 per cent)
Eurofighter Jagdflugzeug GmbH (Eurofighter Typhoon; 33 per cent)
Panavia Aircraft GmbH (Tornado multirole combat aircraft; 42.5 per cent)
Saab AB (35 per cent)
Saab-BAE Systems Gripen AB (Gripen export marketing; 50 per cent)
SEPECAT SA (Jaguar combat aircraft; 50 per cent)

Aircraft design and manufacturing divisions of BAE Systems are as follows:

BAE Systems Programmes & Customer Support Group
Follows this entry
BAE Systems Regional Aircraft
Follows Programmes & Customer Support Group

BAE Systems Airbus
Follows Regional Aircraft

Additionally, BAE has an aviation research establishment with departments assigned to Aerodynamics and Vulnerability, Computational Engineering, Advanced Information Processing, Human Factors, Materials Sciences and Optics and Laser Technology:

BAE Systems Research & Development, PO Box 5, Filton, Bristol, BS34 7QW
Tel: (+44 117) 936 34 00
Fax: (+44 117) 936 67 10

UPDATED

BAE SYSTEMS PROGRAMMES AND CUSTOMER SUPPORT GROUP

GROUP MANAGING DIRECTORS:
Allan Cook (Programmes)
Robin Southwell (Customer Support)
PUBLIC AFFAIRS AND COMMUNICATIONS: Dawn James

Former BAe Defence restructured in 1998 into one manufacturing division and two services divisions. Former Military Aircraft and Aerostructures business reorganised in 2000 under individual programme directors. Aircraft-related directors are given below.
EUROFIGHTER: Allan Cook
FUTURE SYSTEMS AND FOAS: Alison Wood
GRIPEN & SOUTH AFRICA: Nick Franks
HAWK: Jonathan Neale
NIMROD: Tom Nicholson
TORNADO, HARRIER & US MILITARY: Chris Boardman

Aircraft and aerostructures sites reduced from seven to six by closure of Dunsfold Aerodrome. Test flying conducted from Warton and Woodford (latter a detached component of Chadderton).

BROUGH: Brough, East Yorkshire HU15 1EQ
Tel: (+44 1482) 66 71 21
Fax: (+44 1482) 66 66 25

CHADDERTON: Chadderton Works, Greengate, Middleton, Manchester M24 1SA
Tel: (+44 161) 681 20 20
Fax: (+44 161) 955 87 77

FARNBOROUGH: Hertford House, Farnborough Aerospace Centre, Farnborough, Hampshire, GU14 6YU
Tel: (+44 1252) 37 32 32
Fax: (+44 1252) 38 30 00

PRESTWICK: Prestwick International Airport, Ayrshire, KA9 2RW
Tel: (+44 1292) 47 98 88
Fax: (+44 1292) 47 97 03

SAMLESBURY: Samlesbury, Balderstone, Lancashire BB2 7LF
Tel: (+44 1254) 481 23 71
Fax: (+44 1254) 481 36 23

WARTON: Warton Aerodrome, Preston, Lancashire PR4 1AX
Tel: (+44 1772) 63 33 33
Fax: (+44 1772) 63 47 24

Group's main activities include development of Eurofighter Typhoon with Germany, Italy and Spain; support of Panavia Tornado, with EADS Deutschland and Alenia; development and support of Sea Harrier and, with Boeing, AV-8B/GR. Mk 7 Harrier II (see *Jane's Aircraft Upgrades*); design, development, production and support of Hawk and, with Boeing, T-45A Goshawk; and lead contractor for the Nimrod MRA. Mk 4 rebuild programme. Also provides product support for earlier aircraft still in use; offers defence support services and overseas and specialist training facilities, including technical and management courses; and operates the North Sea Range instrumented air combat manoeuvring facility.

Participates in development, component manufacture and export marketing of JAS 39 Gripen multirole fighter under agreement signed with Saab Military Aircraft on 12 June 1995; partner in Saab BAE Systems Gripen AB. Agreement with Dassault of France on 31 October 1995 led to establishment of joint venture military aircraft company EAL (which see in International section). Following elimination of McDonnell Douglas/Northrop Grumman/BAe contender for Joint Strike Fighter (see US Navy entry), BAE announced new teaming with Lockheed Martin on 18 June 1997. Aerostructures business includes components for current family of Boeing 737s.

UPDATED

BAE Hawk Mk 127 test flying before delivery to Australia *(Derek Bower)* 2001/0103519

BAE HAWK 50, 60 and 100 SERIES
RAF designations: Hawk T. Mks 1 and 1A
US Navy designation: T-45A Goshawk
Canadian Forces designation: CT-155
TYPE: Advanced jet trainer/light attack jet.
PROGRAMME: HS P1182 Hawk first flew 21 August 1974; early history in 1989-90 *Jane's*; first-generation Hawk remains available and is marketed with advanced 100 Series and single-seat 200 Series (detailed separately) to meet customers' requirements; Hawk design leadership transferred from Kingston to Brough 1988, and final assembly and flight test from Dunsfold to Warton 1989.

Hawk 50 Series main exports made December 1980 to October 1985; largely supplanted by 60 Series; Hawk 100 enhanced ground attack export model announced mid-1982; first flight of 100 Series aerodynamic prototype (G-HAWK/ZA101 converted as Mk 100 demonstrator) 21 October 1987; trials of wingtip Sidewinder rails started at Warton in April 1990. Warton assembly line officially opened 24 October 1991.

CURRENT VERSIONS: **Hawk T. Mk 1:** Two-seater for RAF flying instruction and weapon training; 23.1 kN (5,200 lb st) Adour 151-01 (-02 in Red Arrows aircraft) non-afterburning turbofan; two dry underwing hardpoints; underbelly 30 mm gun pack; three-position flaps; simple weapon sight in some aircraft of No. 4 FTS. Following basic Tucano stage, future RAF fast-jet pilots undertake 100 hours of advanced flying, weapons and tactical training with No. 4 FTS at Valley. Hawks introduced to navigator training syllabus at No. 6 FTS, Finningley; first delivery 10 September 1992; role to No. 4 FTS in 1995. No. 100 Squadron received 15 Mks 1/1A from September 1991, replacing Canberras in target-towing role; initial seven (subsequently increased to 15) Mk 1s loaned to Royal Navy's Fleet Requirements and Air Direction Unit at Culdrose, first arriving on 6 April 1994. In August 1996, Skyforce Avionics Ltd Skymap II GPS moving map displays were received for Hawks of the Red Arrows and No. 100 Squadron.

Hawk T. Mk 1A: Contract January 1983 to wire 89 Hawks (including Red Arrows) for AIM-9L Sidewinder on each inboard wing pylon and optional activation of previously unused outer wing hardpoints; last conversion redelivered 30 May 1986; 72 (reduced to 50 by 1993 defence cuts) NATO-declared, for point defence and participation in RAF's Mixed Fighter Force, to accompany radar-equipped Tornado ADVs on medium-range air defence sorties.

RAF Hawk re-wing programme began 1989; initial 85 wings completed by BAE in 1993; delivery of second batch of 59 began November 1993 and completed in 1995. Rebuild programme by BAE for 80 RAF Hawks authorised 27 December 1998 to replace centre and rear fuselage, extending service life to 2010; work undertaken at RAF St Athan and BAE Warton; first aircraft, XX348, redelivered April 2000; last due in 2003. Avionics upgrade plans being formulated in Staff Requirement (Air) 449; primary aim is 'glass cockpit'.

Hawk T. Mk 1W: Following re-winging, a small number of RAF Hawk T. Mk 1s gained the ability to carry stores on two underwing pylons, although not the centreline gun pod. The alternative designation **T. Mk 1FTS** is also used for this modification.

Hawk 50 Series: Initial export version with 23.1 kN (5,200 lb st) Adour 851 turbofan; maximum operating weight increased by 30 per cent, disposable load by 70 per cent, range by 30 per cent; revised tailcone shape to improve directional stability at high speed; larger nose equipment bay; four wing pylons, all configured for single or twin store carriage; each pylon cleared for 515 kg (1,135 lb) load; wet inboard pylons for 455 litre (122 US gallon; 100 Imp gallon) fuel tanks; improved cockpit, with angle of attack indication, fully aerobatic twin-gyro AHRS and new weapon control panel; optional braking parachute; suitable for day VMC ground attack and armed reconnaissance with camera/sensor pod.

T-45A/C Goshawk: US Navy version (see Boeing/BAE in International section).

Hawk 60 Series: Development of 50 Series with 25.4 kN (5,700 lb st) Adour 861 turbofan; leading-edge devices and four-position flaps to improve lift capability; low-friction nose leg, strengthened wheels and tyres, and adaptive anti-skid system; 591 litre (156 US gallon; 130 Imp gallon) drop tanks; provision for Sidewinder or Magic AAMs; maximum operating weight increased by further 17 per cent over 50 Series, disposable load by 33 per cent and range by 30 per cent; improved field performance, acceleration, rate of climb and turn rate. Mk 67 is 'long-nosed' version with nosewheel steering, supplied to South Korea. Abu Dhabi upgraded 15 surviving Mk 63s to Mk 63A/B from 1991, incorporating Adour Mk 871 and new combat wing (four pylons and wingtip AAM rails); first two rebuilds at Brough; remainder at Al Dhafra. Surviving Kuwaiti Hawks were refurbished at Dunsfold in 1998-2000.

Description applies to Hawk 60 Series, except where otherwise specified.

Hawk 100 Series: Enhanced ground attack development of 60 Series, announced mid-1982, to exploit Hawk's stores carrying capability; two-seater, with perhaps pilot only on combat missions; 26.0 kN (5,845 lb st) Adour Mk 871 turbofan; new combat wing incorporating fixed leading-edge droop for increased lift and manoeuvrability from M0.3 to M0.7; full-width flap vanes; manually selected combat flaps; detail changes to wing dressing; structural provision for wingtip missile

Hawk T. Mk 1A used by the RAF Centre of Aviation Medicine *(Paul Jackson/Jane's)* 2001/0087981

pylons; MIL-STD-1553B databus; advanced Smiths Industries HUDWAC and new air data sensor package with optional laser ranging and FLIR in extended nose; improved weapons management system allowing preselection in flight and display of weapon status; manual or automatic weapon release; passive radar warning; HOTAS controls; full-colour multipurpose CRT display in each cockpit; provision for ECM pod.

Demonstrator ZA101 (see Programme). Production prototype Mk 102D (ZJ100) flown 29 February 1992. Early orders from Abu Dhabi (placed 1989), Indonesia (signed June 1993), Malaysia (signed 10 December 1990) and Oman (signed 30 July 1990). FLIR, laser ranger and Sky Guardian RWR in Omani aircraft.

Selected by Australia in November 1996 as next-generation lead-in fighter trainer; order signed 24 June 1997 for 33 aircraft (MoU with ASTA March 1991 for licensed production if Hawk 100/200 chosen for RAAF trainer requirement); Australian Hawk Mk 127s have new, advanced instrumentation resembling that of the F-18 Hornet and including an integrated Smiths system of three 127 mm (5 in) square colour screens in each cockpit, HUD, upgraded mission computer, engine life computer and stores management system. First Australian Hawk (A27-01) flew (in UK, temporarily as ZJ632) on 16 December 1999. See BAE Systems entry in Australian section.

BAE, Bombardier and partners supplying Raytheon T-6A Texan II and Hawk Mk 115 as equipment of Canadian-based NATO flying school; contract agreed 17 November 1997 and formally issued on 12 May 1998 for 18, plus eight options, one of latter being taken up, early 1999; further two stated to be required in early 2000, after Singapore joined programme. First aircraft, 155201, arrived in Canada 4 July 2000; officially received 6 July; instructor training began 12 July 2000.

Hawk LIFT: Variant of Series 100 as lead-in fighter trainer (LIFT). Features include 'combat wing', Adour Mk 871 or Mk 900 engine, new three MFD 'glass cockpit' with NVG compatibility, MIL-STD-1553B digital databus, revised mission planning system/data transfer unit, INS/GPS, OBOGS, APU, HUMS, new HUD and HOTAS controls. Provision for FLIR and aerial refuelling. Initial customer is South Africa. Max weapon load 3,084 kg (6,800 lb).

Hawk 200 Series: Single-seat multirole version (described separately).

CUSTOMERS: See table. BAE delivered 20 Hawks in 1996 and 22 in 1997; none in 1998; deliveries resumed in 1999 with 10 to Indonesia before supply of remaining six was suspended by UK government embargo of 11 September 1999; deliveries re-authorised in early 2000, but without certain US-sourced components. Hawk LIFT selected for purchase by South Africa, as announced 18 November 1998; order for 12, plus 12 options, announced 15 September 1999 and signed 3 December 1999; these to be first Hawks with Adour Mk 900 engines. Unconfirmed requirements reportedly include: Abu Dhabi, seven further Hawk Mk 63s; Brunei, six Hawk 100s and four Hawk 200s; India requested quotation in September 1999 for 92 Hawks, later amended to 66, of which eight kits and 42 complete aircraft to be assembled by HAL at Bangalore, and pricing negotiations were continuing in early 2001, following Anglo-Indian political reaffirmation of purchase plans on 12 December 2000; South Korea considering up to 100 Hawks including lead-in trainers to F-16Cs; Kuwait, six attrition replacements; Malaysia, further 12 Mk 203s; Philippines, commitment announced August 1991 for 12 Hawks. Qatari commitment, signed 17 November 1996, for 18 Hawk 100s has not been finalised. In 1998, BAE was anticipating average sales of 48 per year from annual world requirement for 150.

COSTS: South African purchase of 24 Hawk LIFTs estimated at R4.7 billion (1998). Canadian contract for 20 valued at £400 million (1999). Hawk centre/rear fuselage replacement for 80 RAF aircraft valued at £100+ million (2000).

DESIGN FEATURES: Fully aerobatic two-seat advanced jet trainer, adaptable for ground attack and air defence; design capable of other optional roles, with wing improvements on developed Series to enhance combat efficiency. Low wing and mid-mounted, anhedral, sweptback tailplane; air intake on each side of fuselage, forward of wing leading-edge; single non-afterburning engine; elevated rear cockpit to enhance forward view; two strakes below rear fuselage; smurfs (refer Hawk 200 entry) on 100 Series; optional underwing hardpoints; wingtip AAM rails (100 Series).

Wing thickness/chord ratio 10.9 per cent at root, 9 per cent at tip; dihedral 2°; sweepback 26° on leading-edge, 21° 30′ at quarter-chord.

FLYING CONTROLS: Conventional and assisted. Ailerons and one-piece all-moving tailplane actuated hydraulically by tandem actuators; rudder manually actuated, with electric trim tab. Hydraulically actuated double-slotted flaps, outboard 300 mm (12 in) of flap vanes normally deleted; small fence on each wing leading-edge; 100 and 200 Series use special 'combat wing' with full-width flap vanes (refer Hawk 200 entry); large airbrake under rear fuselage, aft of wings. Hydraulic yaw damper on 100 Series rudder.

STRUCTURE: Aluminium alloy; one-piece wing, with machined torsion box of two main spars, auxiliary spar, ribs and skins with integral stringers; most of box forms integral fuel tank; honeycomb-filled ailerons; composites wing fences; frames and stringers fuselage. Wing attached to fuselage by six bolts.

LANDING GEAR: Wide-track, hydraulically retractable tricycle type, with single wheel on each unit. AP Precision Hydraulics oleos and jacks. Main units retract inward into wing, ahead of front spar; castoring (optionally power-steered) nosewheel retracts forward. Dunlop mainwheels, brakes and tyres size 6.50-10 (14 ply) tubeless, pressure 9.86 bars (143 lb/sq in). (Hawk Srs 60 and 100 mainwheel pressure 17.23 bars; 250 lb/sq in at 9,100 kg; 20,061 lb T-O weight.) Nosewheel and tyre size 16×4.4 (8 ply) tubeless, pressure 8.27 bars (120 lb/sq in). Tail bumper fairing under rear fuselage. Anti-skid wheel brakes. Tail braking parachute, diameter 2.64 m (8 ft 8 in), on Mks 52/53 and all 60 and 100 Series aircraft.

POWER PLANT: One Rolls-Royce Turbomeca Adour non-afterburning turbofan, as described under Current Versions. Adour Mk 861A for Switzerland assembled locally by Sulzer Brothers. Adour 900, available for new-build or retrofitted Hawks from 2000, provisionally rated at 28.9 kN (6,500 lb st) and has doubled TBO of 4,000 hours. Engine starting by Microturbo integral gas-turbine starter. Fuel in one fuselage bag tank of 832 litres (220 US gallons; 183 Imp gallons) capacity and integral wing tank of 823 litres (217 US gallons; 181 Imp gallons); total fuel capacity 1,655 litres (437 US gallons; 364 Imp gallons). Pressure refuelling point near front of port engine air intake trunk; gravity point on top of fuselage. Provision for

BAE HAWK CUSTOMERS

Customer	Qty	Mark	First aircraft	Deliveries	Squadrons
Abu Dhabi	16[8]	63	1001	Oct 1984 – May 1985	
	18*	102[1 6 10]	1051	Apr 1993 – Mar 1994	Khalif bin Zayed Air College
	4*	63C	1017	Feb 1995 – Mar 1995	
Australia	33[11]	127	A27-01	Apr 2000 – 2006	76, 79
Canada	19*	115/CT-155	155201	Jul 2000	(CFBs Cold Lake and Moose Jaw)
Dubai	8	61	501	Mar 1983 – Sep 1983	III Shaheen
	1	61	509	Jun 1988	III Shaheen
Finland	50[2]	51	HW301	Dec 1980 – Oct 1985	3/11, 3/21, 3/31, Koulutuslentolaivue,
	7*	51A	HW351	Nov 1993 – Sep 1994	2/Tukilentolaivue
Indonesia	20	53	LL-5301	Sep 1980 – Mar 1984	103
	8*	109	TT-1201	May 1996 – Mar 1997	1, 12
	16*	209	TT-1205	Feb 1996 – Mar 1997	12
	16*	209	TT-0217	Apr 1999 –	1
Kenya	12	52	101	Apr 1980 – May 1982	(Laikipia AB)
Korea, South	20*	67	67-496	Sep 1992 – Aug 1993	216
Kuwait	12	64	140	Nov 1985 – Sep 1986	12
Malaysia	10*	108[1 6 10]	M40-01	Jan 1994 – Sep 1995	3 FTC (15 Sqdn)
	18*	208[6 9 10]	M40-21	Aug 1994 – May 1995	6, 9, 3 FTC (15 Sqdn)
Oman	4*	103[1 6 10]	101	Dec 1993 – Jan 1994	6
	12*	203[9 10]	121	Dec 1994 – May 1995	6
Saudi Arabia	30	65	2110	Aug 1987 – Oct 1988	21, 37
	20	65A	7901	Mar 1997 – Dec 1997	79
South Africa	12	LIFT	–	2005	–
Switzerland	20[3]	66	U-1251	Nov 1989 – Nov 1991	Pilotenschule
UK	176[7]	T. Mk 1	XX154	Nov 1976 – Feb 1982	4 FTS (19 & 208 Sqdns), Red Arrows, 100 Sqdn, FRADU, DERA
USA	189[4]	T-45A/C	162787	Apr 1988 -	VT-7, VT-21, VT-22
Zimbabwe	8	60	600	Jul 1982 – Oct 1982	2
	5*	60A	608	Jun 1992 – Sep 1992	–
Demonstrator	2[5]	60/102D	G-HAWK		–
	3[5]	200/200/200RDA	ZG200		–
Total	**769**				

Notes

* Built at Brough, Hamble and Samlesbury and assembled at Warton; unless stated otherwise, remainder built at Kingston, assembled at Dunsfold (290) and Bitteswell (1)
[1] Laser nose
[2] 46 assembled by Valmet in Finland
[3] 19 assembled by F + W in Switzerland
[4] Production by Boeing in USA (see International section)
[5] One assembled at Warton
[6] Wingtip Sidewinders
[7] 89 converted to T. Mk 1A
[8] 13 converted to 63A; two to 63B
[9] Removable refuelling probe
[10] Radar warning receiver
[11] First 12 assembled at Warton; remainder by BAe Australia at Williamtown, assisted by Hunter Aerospace, Hawker de Havilland and Qantas

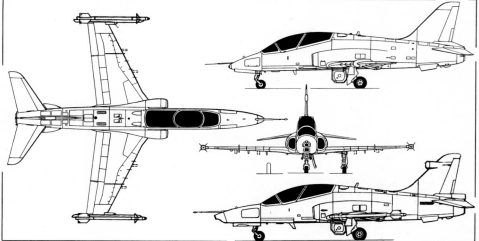

BAE Systems Hawk 100 with wingtip Sidewinders and additional side view (top) of Hawk 60 Series *(Dennis Punnett/Jane's)*

BAE—AIRCRAFT: UK

BAE Systems Hawk LIFT, cutaway drawing

1. Pitot — static probe
2. Optional forward-looking infra-red (FLIR)
3. Nose equipment bay doors
4. In-flight refuelling nozzle
5. In-flight refuelling probe
6. Nose landing gear doors
7. Landing/taxying light (inside of starboard door)
8. Nosewheel leg
9. Lever suspension-type nose landing gear
10. Tacan antenna
11. ILS marker antenna
12. Windscreen
13. Head-up display
14. Martin-Baker Mk 10LH ejection seats
15. Miniature detonating cord
16. HUD monitor/MFD
17. Forward seat frame and pressure bulkhead
18. Air intake duct
19. Navigation light
20. Main landing gear door
21. Main landing gear leg door
22. Inboard stores pylon
23. Outboard stores pylon
24. Wingtip missile launcher
25. Aileron
26. Flap
27. Engine rear access door
28. Airbrake
29. Cockpit rear bulkhead
30. Top longeron
31. Fuel tank walls
32. Fuel tank roof
33. Anti-collision light
34. Single-piece unit canopy
35. Mainplane
36. UHF antenna
37. RR/Turbomeca Adour Mk 871 engine
38. Port bottom longeron
39. Port top longeron
40. Fuselage port strake
41. Tailplane vane
42. IFF antenna
43. Fin
44. Rudder
45. Radar warning receiver (forward coverage)
46. Fin-tip, including VHF aerial
47. VOR/ILS localiser antenna (port); ILS guideline antenna (starboard)
48. Radar warning receiver (aft coverage)
49. Brake parachute housing
50. Tailplane
51. Chaff and flare dispenser

carrying one 455 or 591 litre (120 or 156 US gallon; 100 or 130 Imp gallon) drop tank on each inboard underwing pylon, according to Series.

ACCOMMODATION: Crew of two in tandem under one-piece, fully transparent, acrylic canopy, opening sideways to starboard. Fixed front windscreen able to withstand a 0.9 kg (2 lb) bird at 454 kt (841 km/h; 523 mph). Improved front windscreen fitted retrospectively to RAF Hawks, able to withstand a 1 kg (2.2 lb) bird at 528 kt (978 km/h; 607 mph); this installed on all current export aircraft. Separate internal screen in front of rear cockpit. Rear seat elevated. Martin-Baker Mk 10LH zero/zero rocket-assisted ejection seats, with MDC (miniature detonating cord) system to break canopy before seats eject. MDC can also be operated from outside the cockpit for ground rescue. Dual controls standard. Entire accommodation pressurised, heated and air conditioned.

SYSTEMS: BAE cockpit air conditioning and pressurisation systems, using engine bleed air. Two hydraulic systems; flow rate: System 1, 36.4 litres (9.6 US gallons; 8.0 Imp gallons)/min; System 2, 22.7 litres (6.0 US gallons; 5.0 Imp gallons)/min. Systems pressure 207 bars (3,000 lb/sq in). System 1 for actuation of control jacks, flaps, airbrake, landing gear and anti-skid wheel brakes. Compressed nitrogen accumulators provide emergency power for flaps and landing gear at pressure of 2.75 to 5.50 bars (40 to 80 lb/sq in). System 2 dedicated to powering flying controls. Hydraulic accumulator for emergency operation of wheel brakes. Pop-up Dowty ram air turbine in upper rear fuselage provides emergency hydraulic power for flying controls in event of engine or No. 2 pump failure. No pneumatic system.

DC electrical power from single 12 kW 30 V DC brushless generator, with two 3 kVA 115/26 V 400 Hz three-phase inverters to provide AC power and two batteries for up to 20 minutes of standby power. Gaseous oxygen system for crew; optional Hamilton Sundstrand (HS Marston) OBOGS first installed in Australian Hawk Mk 127s from 2000.

AVIONICS: *Comms:* Mk 1/Srs 50 includes Sylvania UHF and VHF; Cossor 2720 Mk 10A IFF in Finnish aircraft; Srs 60 has Rockwell Collins UHF and VHF, Magnavox UHF and Raytheon 2720 IFF; Srs 100 has Rockwell Collins AN/ARC-182 U/VHF, Magnavox AN/ARC-164 UHF and Raytheon 4720 IFF.

Flight: Mk 1/Srs 50 with Raytheon CAT 7000 Tacan, Cossor ILS having CILS.75/76 localiser/glide slope receiver and marker receiver; Rockwell Collins VOR/ILS and ADF, Rockwell Collins Tacan, Smiths-Newmark 6000-05 AHRS and Smiths radar altimeter in Srs 60; Srs 100 has BAE INS300 inertial platform, Rockwell Collins AN/ARC-118 Tacan, Rockwell Collins VIR-31A VOR/ILS and Smiths 0103-KTX-1 radar altimeter, all integrated via dual redundant MIL-STD-1553B databus. Optional Skyforce Skymap II GPS-driven moving map in some RAF Hawks.

Instrumentation: Smiths-Newmark compass, BAE gyros and inverter, two Honeywell RAI-4 4 in (100 mm) remote altitude indicators and magnetic detector system in Mk 1/Srs 50; Smiths 1500 Series HUDWAC in Srs 100; BAE F.195 weapon sight in approximately 90 RAF aircraft; BAE ISIS 195 sight in Srs 50 and Srs 60, except Saab RGS2 in Finnish Mk 51.

Mission: BAE camera and recorder in F.195-equipped RAF aircraft (requirement announced December 1997 to re-equip 41 with video recorders by late 1999); Vinten camera and recorder in Srs 50 and Srs 60. Srs 100 has Smiths 3000 Series colour MFD, GEC data transfer system and Vinten colour video recording system; plus BAE Type 105H laser range-finder; optional FLIR.

Self-defence (Series 100 only): Racal Prophet RWR in Mk 102s of Abu Dhabi; BAE Sky Guardian in Indonesian, Malaysian and Omani aircraft, and retrofitted to Abu Dhabi's Mk 63s; optional chaff/flare dispenser at base of fin.

Cockpit of BAE Hawk LIFT ***1999***/0051526

ARMAMENT: Underfuselage centreline-mounted 30 mm BAE Aden Mk 4 cannon with 120 rounds (VKT 12.7 mm machine gun beneath Finnish aircraft), and two or four hardpoints under wing, according to Series. Provision for pylon in place of ventral gun pack. In RAF training roles, normal maximum external load is about 680 kg (1,500 lb), but the uprated Hawk 60 and 100 Series are cleared for an external load of 3,000 kg (6,614 lb), or 500 kg (1,102 lb) at 8 *g*. Typical weapon loadings on 60 Series include 30 mm or 12.7 mm centreline gun pod and four packs each containing eighteen 68 mm rockets; centreline reconnaissance pod and four packs each containing twelve 81 mm rockets; five 1,000 lb free-fall or retarded bombs; four launchers each containing four 100 mm rockets; nine 250 lb or 250 kg bombs; thirty-six 80 lb runway denial or tactical attack bombs; five 600 lb cluster bombs; four Sidewinder or two Magic air-to-air missiles; or four CBLS 100/200 carriers, each containing four practice bombs and four rockets. Vinten reconnaissance pod available for centre pylon. Similar options on 100 Series, plus wingtip air-to-air missiles. Mk 102s of Abu Dhabi/UAEAF can carry (but not designate for) two Alenia Marconi PGM-500 ASMs.

DIMENSIONS, EXTERNAL:

Wing span: Mk 1, Srs 50, Srs 60	9.39 m (30 ft 9¾ in)
Srs 100	9.08 m (29 ft 9½ in)
Srs 100 with Sidewinders	9.94 m (32 ft 7⅜ in)
Wing chord: at root	2.65 m (8 ft 8¼ in)
at tip	0.90 m (2 ft 11½ in)
Wing aspect ratio: Mk 1, Srs 50, Srs 60	5.3
Srs 100	4.9
Length: fuselage (incl jetpipe):	
Mk 1, Srs 50, Mks 60-66	10.775 m (35 ft 4¼ in)
Mk 67	11.375 m (37 ft 3¾ in)
Srs 100	11.40 m (37 ft 4¼ in)
fuselage and pitot:	
Mk 1, Srs 50, Mks 60-66	11.455 m (37 ft 7 in)
Mk 67, Srs 100 (non combat)	12.035 m (39 ft 5¾ in)
Srs 100 (incl chaff/flare dispenser)	12.095 m (39 ft 8 in)
overall:	
Mk 1, Srs 50, Mks 60-66	11.845 m (38 ft 10¼ in)
Mk 67, Srs 100	12.425 m (40 ft 9¼ in)
Nose to pitot tip:	
Mk 1, Srs 50, Mks 60-66	0.68 m (2 ft 2¾ in)
Mk 67	0.66 m (2 ft 2 in)
Srs 100	0.635 m (2 ft 1 in)
Tailplane overhang	0.39 m (1 ft 3½ in)

Cockpit of BAE Hawk Mk 100 mockup
(Paul Jackson/Jane's) 0052934

Height overall	3.98 m (13 ft 0¾ in)
Tailplane span	4.39 m (14 ft 4¾ in)
Wheel track	3.47 m (11 ft 5 in)
Wheelbase	4.50 m (14 ft 9 in)

AREAS:

Wings, gross	16.69 m² (179.6 sq ft)
Ailerons (total):	
Mk 1, 50 and 60 Series	1.05 m² (11.30 sq ft)
100 Series	0.97 m² (10.44 sq ft)
Trailing-edge flaps (total)	2.50 m² (26.91 sq ft)
Airbrake	0.53 m² (5.70 sq ft)
Fin: Mk 1, 50 and 60 Series	2.51 m² (27.02 sq ft)
100 Series	2.61 m² (28.10 sq ft)
Rudder, incl tab	0.58 m² (6.24 sq ft)
Tailplane	4.33 m² (46.61 sq ft)

WEIGHTS AND LOADINGS:

Weight empty: 60 Series	4,012 kg (8,845 lb)
100 Series	4,400 kg (9,700 lb)
Max weapon load (60, 100 Series)	3,000 kg (6,614 lb)
Max fuel weight:	
internal (usable)	1,304 kg (2,875 lb)
external (usable)	932 kg (2,055 lb)
Max T-O weight: T. Mk 1	5,700 kg (12,566 lb)
50 Series	7,350 kg (16,200 lb)
60, 100 Series	9,100 kg (20,061 lb)
Max landing weight: T. Mk 1	4,649 kg (10,250 lb)
60 Series	7,650 kg (16,865 lb)
Max wing loading: T. Mk 1	341.5 kg/m² (69.97 lb/sq ft)
50 Series	440.4 kg/m² (90.20 lb/sq ft)
60, 100 Series	545.4 kg/m² (111.70 lb/sq ft)
Max power loading: T. Mk 1	246 kg/kN (2.42 lb/lb st)
50 Series	318 kg/kN (3.12 lb/lb st)
60 Series	359 kg/kN (3.52 lb/lb st)
100 Series	350 kg/kN (3.43 lb/lb st)

PERFORMANCE:

Never-exceed speed (VNE), clean:
at S/L M0.87 (575 kt; 1,065 km/h; 661 mph EAS)
at and above 5,180 m (17,000 ft)
M1.2 (575 kt; 1,065 km/h; 661 mph EAS)

Malaysian BAE Systems Hawk Mk 108 ***2001***/0073322

Max level speed at S/L:
  50 Series   535 kt (990 km/h; 615 mph)
  60 Series   545 kt (1,010 km/h; 627 mph)
  100 Series   540 kt (1,001 km/h; 622 mph)
Max level flight Mach No.   0.88
Stalling speed, flaps down 96 kt (177 km/h; 110 mph)
Max rate of climb at S/L   3,600 m (11,800 ft)/min
Time to 9,150 m (30,000 ft), clean: 60 Series 6 min 54 s
  100 Series   7 min 30 s
Service ceiling: 60 Series   14,020 m (46,000 ft)
  100 Series   13,565 m (44,500 ft)
T-O run (clean): 60 Series   710 m (2,330 ft)
  100 Series   640 m (2,100 ft)
Landing run (10% fuel load): 60 Series 550 m (1,800 ft)
  100 Series   605 m (1,980 ft)
Combat radius, Aden gun pod and two Sidewinder AAMs:
  with four 500 lb bombs and two 130 Imp gallon tanks
                          345 n miles (638 km; 397 miles)
  with four 1,000 lb bombs
                          125 n miles (231 km; 143 miles)
Ferry range:
  60 Series, with two 591 litre (156 US gallon; 130 Imp gallon) drop tanks
                        1,575 n miles (2,917 km; 1,812 miles)
  100 Series, as above
                        1,360 n miles (2,519 km; 1,565 miles)
Endurance, 100 n miles (185 km; 115 miles) from base:
  60 Series   approx 2 h 42 min
  100 Series   approx 2 h 6 min
g limits: full internal fuel, all   +8/−4
  100 series, 60% fuel load:
    1,360 kg (3,000 lb) external load   +8/−4
    2,721 kg (6,000 lb) external load   +6/−3
                                              UPDATED

BAE HAWK 200 SERIES

TYPE: Light fighter.
PROGRAMME: Intention to build demonstrator (ZG200) announced 20 June 1984; first flight 19 May 1986 but lost 2 July 1986 in accident; replaced by first preproduction Hawk 200 (ZH200), first flown 24 April 1987; third demonstrator Series 200RDA (ZJ201) with full avionics and systems, including Lockheed Martin AN/APG-66H radar, flown 13 February 1992. BAe signed a collaborative production agreement with IPTN of Indonesia in June 1991, in anticipation of orders eventually totalling 144 Mks 100/200.

First production Hawk Srs 200 (Oman 121) flew 11 September 1993; first for Malaysia (Mk 208 M40-21) flew 4 April 1994; first Mk 209 for Indonesia followed in early 1996. Hawk 200 cleared to carry AGM-65 Maverick ASM by 1996.

First BAE Systems Hawk for Canada, where it is known as the CT-115

BAE Systems' Hawk Mk 100 demonstrator

CURRENT VERSIONS: Missions can include:
  Airspace denial: Four Sidewinders, gun and two 591 litre (156 US gallon; 130 Imp gallon) drop tanks enabling 2 hour loiter on station at 100 n miles (185 km; 115 miles) from base.
  Close air support: Typically four 1,000 lb bombs precision-delivered up to 115 n miles (213 km; 132 miles) from base in lo-lo-lo-lo mission with gun and wingtip Sidewinder missiles also carried.
  Battlefield interdiction: Typically 907 kg (2,000 lb) load on hi-lo-lo-hi mission over 290 n mile (537 km; 340 mile) radius with gun and wingtip Sidewinder missiles also carried.
  Long-range photo reconnaissance: 490 n mile (908 km; 564 mile) range with two external tanks, pod containing cameras and infra-red linescan and wingtip Sidewinder missiles for self-defence (rapid role change permits follow-on attack by the same aircraft).
  Long-range deployment: 1,365 n mile (2,528 km; 1,571 mile) ferry range using two 591 litre (156 US gallon; 130 Imp gallon) external tanks, unrefuelled and with tanks retained (reserves allow 10 minutes over destination at 150 m; 495 ft).
  Anti-shipping attack: Two rocket pods and two 591 litre (156 US gallon; 130 Imp gallon) external tanks plus wingtip Sidewinder missiles for self-defence, enabling ship attack and return with 10 per cent fuel reserves.
CUSTOMERS: See table. Oman (12 **Mk 203** ordered July 1990), Indonesia (16 **Mk 209** ordered June 1993; second 16 in June 1996), Malaysia (18 **Mk 208** ordered December 1990). Production of second Indonesian batch undertaken in 1998-99; first flight of initial aircraft (temporarily ZJ555) 22 October 1998; other customers' orders completed by March 1995; final six Indonesians built, but embargoed by UK government on 11 September 1999; following lifting of embargo, were awaiting US-sourced components in late 2000. Total orders for, and production of, 200 Series totalled 62 by early 2000, excluding demonstrators. Saudi Arabia signed MoU covering second batch of some 60 Hawks, reportedly including **Mk 205** with APG-66H radar, although none was on firm order by early 2000. All customers also ordered two-seat Hawks.

See Hawk 50, 60 and 100 Series entry for full description; principal differences given below.

Prototype Hawk XX154, now over 26 years old, was replaced in the missile trials support role during 2000 *(Paul Jackson/Jane's)*

DESIGN FEATURES: Hawk 200 virtually identical to current production Hawk two-seater aft of cockpit, giving 80 per cent airframe commonality. Significant changes from two-seat version include taller fin and 'combat wing' with fixed leading-edge droop to enhance lift and manoeuvrability at M0.3 to M0.7, manually selected combat flaps (less than quarter-flap setting) available below 350 kt (649 km/h; 403 mph) IAS to allow sustained 5 g+ at 300 kt (556 km/h; 345 mph) at sea level, full-width flap vanes reinstated and detail modifications to wing dressing; smurfs (strake ahead of each half of tailplane to restore control authority at high angles of attack). Intended to take advantage of new miniaturised, low-cost avionics and intelligent weapons; Hawk 100-type avionics include INS, DPMC, IFF, RWR and optional HOTAS controls; Lockheed Martin AN/APG-66H advanced multimode radar for all-weather target acquisition and navigation fixes; proposed alternative FLIR/laser range-finder nose no longer on offer; intended integral cannon also deleted; all four underwing pylons capable of 907 kg (2,000 lb) load, within maximum 3,493 kg (7,700 lb) external load; wingtip rails make possible four Sidewinders or similar AAMs (inboard pylons not cleared for these missiles).
LANDING GEAR: Mainwheel tyres size H22×6.5-11 (16 ply) tubeless, pressure 16.20 bars (235 lb/sq in). Nosewheel

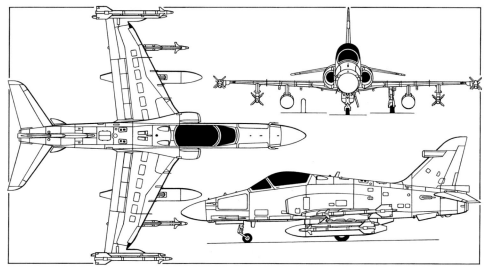

The single-seat BAE Systems Hawk 200 Series with nose-mounted radar *(Mike Keep/Jane's)*

BAE Systems Hawk 200 RDA demonstrator 2000/0080293

tyre size 18×5.5 (8 ply) tubeless, pressure 7.24 bars (105 lb/sq in).

POWER PLANT: One Rolls-Royce Turbomeca Adour Mk 871 non-afterburning turbofan, with uninstalled rating of 26.0 kN (5,845 lb st). Optional removable refuelling probe on starboard side of windscreen.

ACCOMMODATION: Pilot only, on Martin-Baker Mk 10LH zero/zero ejection seat, under starboard-hinged canopy.

SYSTEMS: 25 kVA generator with DC transformer-rectifier. Fairey Hydraulics yaw control system added, comprising rudder actuator and servo control system, incorporating an autostabiliser computer. Lucas Aerospace artificial feel system. 12 kVA APU for engine starting, ground running and emergency power.

AVIONICS: *Comms:* Rockwell Collins AN/ARC-182 U/VHF; Raytheon 4720 IFF; all integrated by MIL-STD-1553 databus.

Radar: Lockheed Martin AN/APG-66H multimode radar.

Flight: Smiths DPMC. Rockwell Collins AN/ARN-153 Tacan and AN/ARN-147 VOR/ILS.

Instrumentation: BAE 127 × 127 mm (5 × 5 in) multifunction display in cockpit; combined com/nav control and display unit allows control of all functions from one panel; HOTAS controls optional.

Self-defence: BAE Sky Guardian 200 RWR in aircraft for Oman. Chaff/flare dispenser (Vinten Vicon 78 Srs 300 or equivalent) at base of fin.

ARMAMENT: None internally. All weapon pylons cleared for 8 *g* manoeuvres with 500 kg (1,102 lb) loads.

DIMENSIONS, EXTERNAL: As Hawk Series 100, except:
Wing span: 200RDA	9.39 m (30 ft 9¾ in)
Length: fuselage: 200	10.95 m (35 ft 11 in)
200 with chaff/flare dispenser	11.01 m (36 ft 1¼ in)
200RDA	10.99 m (36 ft 0¾ in)
200RDA with chaff/flare dispenser	11.05 m (36 ft 3 in)
overall: 200	11.34 m (37 ft 2½ in)
200RDA	11.38 m (37 ft 4 in)
Height overall: 200	4.130 m (13 ft 6¾ in)
200RDA	3.98 m (13 ft 0¾ in)
Wheelbase	3.56 m (11 ft 8 in)

WEIGHTS AND LOADINGS:
Basic weight empty	4,450 kg (9,810 lb)
Max fuel: internal (usable)	1,360 kg (3,000 lb)
external (usable)	932 kg (2,055 lb)
Max weapon load	3,000 kg (6,614 lb)
Max T-O weight	9,100 kg (20,061 lb)
Max wing loading	545.4 kg/m² (111.70 lb/sq ft)
Max power loading	350 kg/kN (3.43 lb/lb st)

PERFORMANCE (no external stores or role equipment unless stated):
Never-exceed speed (V_{NE}):	
at S/L	M0.87 (575 kt; 1,065 km/h; 661 mph EAS)
at and above 5,180 m (17,000 ft)	M1.2 (575 kt; 1,065 km/h; 661 mph EAS)
Max level speed at S/L	540 kt (1,000 km/h; 621 mph)
Econ cruising speed at 12,500 m (41,000 ft)	430 kt (796 km/h; 495 mph)
Stalling speed, flaps down	96 kt (177 km/h; 110 mph) IAS
Max rate of climb at S/L	3,508 m (11,510 ft)/min
Time to 9,150 m (30,000 ft)	7 min 24 s
Service ceiling	13,715 m (45,000 ft)
Runway LCN: flexible pavement	15
rigid pavement	10
T-O run	630 m (2,070 ft)
Landing run	598 m (1,960 ft)
Ferry range (with two drop tanks)	1,365 n miles (2,528 km; 1,570 miles)
g limits	+8/−4

UPDATED

BAE NIMROD MRA. Mk 4

TYPE: Maritime reconnaissance four-jet.

PROGRAMME: Original Hawker Siddeley HS 801 Nimrod first flew 23 May 1967 as adaptation of de Havilland D.H.106 Comet airliner. Total of two prototypes (converted Comet fuselages; now withdrawn), 46 Nimrod MR. Mk 1s and three Nimrod R. Mk 1 elint/sigint aircraft built for the RAF. Of these, 35 converted to MR. Mk 2 from 1979 onwards and 11 became Nimrod AEW. Mk 3s, but failed to enter service and were scrapped. MR. Mk 2 fleet of Nos. 42(R), 120, 201 and 206 Squadrons at Kinloss in 1996 totalled 28 (including three in long-term storage), plus one under conversion to R. Mk 1, four scrapped or used for ground instruction and two lost in accidents. Two original R. Mk 1s remain with No. 51 Squadron at Waddington. These variants were last described in the 1987-88 *Jane's*.

Following RAF issue of Staff Requirement (Air) 420 for a Nimrod replacement, Dassault Atlantique 3, two versions of Lockheed Martin P-3 Orion (new production from manufacturer or refurbished P-3A/Bs from Loral) and upgraded Nimrod offered as potential solutions. UK official announcement on 25 July 1996 nominated last-mentioned proposal under commercial designation (since abandoned) Nimrod 2000. Although based on existing aircraft, the upgrade involves extensive (80 per cent) reconstruction of the airframe, plus incorporation of many new components, including engines, wings, landing gear and general systems, as well as new flight deck and detection systems. Contract MAR21a/100 awarded January 1997. Initial three development batch (DB) aircraft are conversions of stored Mk 2s, on which work began in early 1997; first delivery originally scheduled to RAF in 2001, followed by IOC in April 2003 on delivery of seventh aircraft; by early 1999, these dates had slipped 23 months due, in part, to excessive weight of new wing. New schedule calls for service deliveries to begin in August 2004; IOC in March 2005; and final delivery in December 2008.

Ground testing and verification of engineering plans undertaken at Warton on Nimrod MR. Mk 1 prototype XV147. FR Aviation at Hurn, Bournemouth, selected for airframe relifing and upgrading as subcontractor to BAE, flying 'green' aircraft to Warton for mission avionics installation; first three (DB) aircraft (XV247/PA1, XV234/PA2 and XV242/PA3, for modification in that order) dismantled at RAF Kinloss and flown to Hurn by Antonov An-124 14 to 16 February 1997; redelivery to Warton originally due late 1999, January and February 2000. Fourth conversion subject, XV251/PA4, to Hurn under own power, 2 November 1998.

Programme revised in early November 1999, when FR Aviation contract cancelled and airframe work transferred to BAE at Woodford. PA1 completed at Hurn; PA2, PA3 and PA4 airfreighted to Woodford in An-124 in November/December 1999 for completion. Fifth subject, XV258/PA5, delivered directly ex-RAF to Woodford on 8 November 1999. Wing for initial aircraft delivered from Chadderton to Woodford in September 2000.

Short Brothers and various BAE plants supply airframe components and subassemblies. Ground training systems supplied by Thomson TSL.

Indonesian (second batch) Hawk Mk 209 undertaking carriage trials of AGM-65 Maverick ASMs *(Derek Bower)* 2000/0064960

A prospective new grey colour scheme for the Nimrod MRA. Mk 4 has been applied to this Mk 2 *(Paul Jackson/Jane's)* 2001/0103569

Boeing supplying and integrating TCCS (tactical command and sensor system); 15-month laboratory test began at Warton in November 1998.

BAE offers the Nimrod MRA. Mk 4 for overseas sale, in which eventuality production would be relaunched. An agreement of June 1996 provided for McDonnell Douglas (now Boeing) to promote the Nimrod as a successor to the US Navy's Lockheed Martin P-3 Orion, with production at St Louis, if successful.

CUSTOMERS: Royal Air Force to receive 21. In deference to comprehensiveness of rebuild, aircraft issued with new serial numbers ZJ514 to ZJ534.

COSTS: Ordered as £2 billion programme, of which 75 per cent of work being placed with UK companies. Boeing TCCS mission system avionics contract valued at US$639 million (1996). Programme cost increased to £2.4 billion by March 1999, equivalent to 0.5 per cent above inflation. In mid-2000, BAE agreed to pay £46 million penalty because of 23-month programme slippage.

DESIGN FEATURES: World's first jet-powered, land-based maritime patrol and ASW aircraft on original 1969 service-entry; remains sole four-jet in this role. Optimised for anti-submarine warfare (ASW), anti-surface unit warfare (ASUW), search and rescue (SAR), maritime reconnaissance and provision of aid to civil authorities, including fisheries protection and operations against terrorism, drug smuggling and blockade running. Combines fast transit to operational area with low wing loading (and, hence, manoeuvrability) when on station. Can cruise on two or three engines to extend endurance. No special ground equipment required for off-base deployments.

Mid-mounted wing, with two engines in each root and auxiliary fuel tank in leading-edge at approximately two-thirds span; wingtip ESM pods; two hardpoints under each wing. Dihedral tailplane with auxiliary fins at one-third span; fin has large fillet and electrical equipment/decoy housing pod at tip. MAD tailboom. 'Double bubble' fuselage cross-section, with weapons bay below; latter restructured for Mk 4 and now in two, separated parts. Wing sweepback 20° at quarter-chord.

Nimrod Mk 4 upgrade extends operational lifetime by 25 years. Compared with original version, new power plant improves fuel consumption by over 20 per cent and thrust by 30 per cent. Larger wing and increased thrust restore performance despite 20 per cent increase in maximum take-off weight.

FLYING CONTROLS: Conventional control surfaces actuated hydromechanically. Plain flaps immediately outboard of engines.

STRUCTURE: All-metal, two-spar wing comprises centre-section, two stub-wings and two outer panels. Fuselage is all-metal, semi-monocoque with pressurised upper component and unpressurised pannier including the weapons bay and nose radome. Pannier segments are free to move, so that structural loads are not transmitted to the pressure cell. Cantilever, all-metal tail surfaces surmounted by glass fibre fin-tip pod.

Nimrod MRA. Mk 4 introduces new wing centre box and wing inner panels of increased span designed by BAE Filton, built and assembled at Chadderton. Outer wings, to have been retained under BAE proposal, also new-build, as requested by RAF; designed by Filton, manufactured at Prestwick. Centre fuselage designed at Prestwick and built at Brough; lower (unpressurised) fuselage and front fuselage designed at Farnborough and built at Brough; rear fuselage designed by Dassault (France) and built at Brough; tailplane, elevators, rudder and weapons bay doors designed at Prestwick and Farnborough and manufactured at Prestwick and by FR Aviation. Flap actuation system designed by Dowty. Other new items include wingtip pods, fin fillet, fin-tip pod and (enlarged) finlets. Retained structures, principally the fuselage pressure hull and empennage, are receiving new protective treatment, although all pressure bulkheads and floors are new. Some 80 per cent of the Mk 4 will be newly built.

LANDING GEAR: Retractable tricycle type; all-new Dowty unit in Nimrod MRA. Mk 4. Four-wheel tandem bogie main units; twin nosewheels. Dunlop tyres, wheels and brakes.

POWER PLANT: Four 'marinised', non-afterburning Rolls-Royce BR710 Mk 101 (BR710B3-40) turbofans with FADEC and EICAS, each rated at 68.9 kN (15,500 lb st). Additional 15 to 17 per cent of internal fuel compared with Nimrod MR. Mk 2.

ACCOMMODATION: Two-person flight crew; seven-person tactical team (Tacco 1, Tacco 2, radar, communications, ESM and two acoustics); optional positions for two visual observers and sonobuoy loader. Additional 13 seats for support personnel or replacement crew. Crew areas, front to rear, are flight deck, lavatory, lateral observers' seats with bubble windows, tactical compartment, optional AEO compartment, galley and sonobuoy/general storage area, including crew door.

SYSTEMS: New Normalair-Garrett environmental control system and hydraulic system. Hot air anti-icing system. Portable data storage system for fault analysis and rectification; compatible with Logistics Information Technology System. Honeywell APU; Lucas power-generation system; Normalair-Garrett oxygen system.

AVIONICS: Mission avionics integration by BAE, significant proportion of new equipment being obtained from Airbus and Eurofighter Typhoon programmes. Nimrod MRA. Mk 4 programmed with two million lines of computer software code.

Comms: Digital communications subsystem by Telephonics. Elmer HF, VHF and UHF radios.

Radar: Racal Searchwater 2000MR maritime surveillance radar.

Flight: Smiths Industries navigation and flight management system. Rockwell Collins TCAS II with Thomson-CSF IFF Mode S transponder; Sextant AFCS with autothrottle and vertical hold/select. Two Litton LN-100G laser INS with embedded GPS; ground proximity warning and collision-avoidance. Triplex air data system. Rockwell Collins ADF; BAE ILS/VOR/MLS; Flight Data Co MMS, Smiths NAV/FMS, Thomson-CSF radar altimeter and Rockwell Collins Tacan.

Instrumentation: Sextant seven-screen full-colour LCD EFIS, plus EICAS.

Mission: 'Fourth-generation' sensor suite, including Boeing's tactical command sensor subsystem and Smiths Industries' armament control system conforming to MIL-STD-1760. Tactical display also capable of integrating weather radar, EO/FLIR imagery and defensive aids information. Seven identical, Boeing-supplied operator consoles integrated with Northrop Grumman (EOSDS) Nighthunter FLIR turret under forward fuselage. Retains CAE ASQ-504(V) AIMS MAD and Computing Devices Canada/Ultra Electronics AN/UYS-503 (AQS-970) processor from Nimrod MR. Mk 2, but MAD now fully integrated with tactical system. Ultra Link 11, Rockwell Collins Link 16 (JTIDS), Matra BAe SHF/satcom, Frederick TTY, Raytheon UHF/satcom. Full range of sonobuoys dispensed from two Normalair-Garrett rotary launchers (each holding 10 Size A buoys) in the rear fuselage; two single-barrel launchers for high-level release when fuselage pressurised. Internal racks for 180 buoys (or 360 in emergency), including new SR(SA) 903 active sonobuoy. Active search sonobuoy system (ASSS) to be selected from competing designs by Thomson Marconi Sonar and Ultra Electronics. Elta EL/L-8300UK ESM.

Self-defence: Integrated defensive aids subsystem (DASS) by Lockheed Martin, including a Raytheon AN/ALE-50 towed decoy, Vinten chaff/flare dispensers, Sanders missile approach warning and Lockheed Martin AN/ALR-56M RWR, last mentioned being separate from ESM system. Potential for later addition of DIRCM, laser warning receiver and integrated countermeasures system.

ARMAMENT: Advanced armament control system capable of accepting all current and known future maritime patrol weapons. Four underwing hardpoints, each capable of carrying two side-mounted, self-protection Sidewinder AAMs in addition to an anti-ship missile. Weapons bay, with two pairs of doors, is able to carry up to six lateral rows of ASW weapons, including up to nine torpedoes as well as bombs. Maximum load of Harpoon ASMs is four underwing and one in each of two weapon bays.

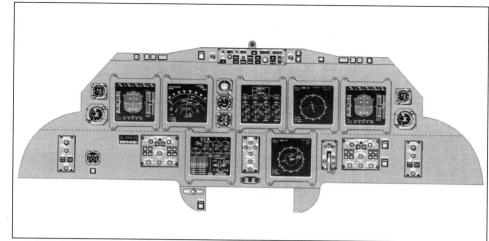

Sextant seven-screen EFIS of the Nimrod MRA. Mk 4

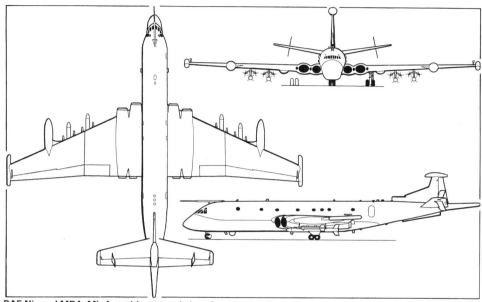

BAE Nimrod MRA. Mk 4 maritime patrol aircraft with reprofiled engine air intakes introduced in 2000 *(Paul Jackson/Jane's)*

DIMENSIONS, EXTERNAL:
Wing span over tip pods	38.71 m (127 ft 0 in)
Length overall (excl refuelling probe)	38.63 m (126 ft 9 in)
Height overall	9.45 m (31 ft 0 in)
Tailplane span	14.51 m (47 ft 7¼ in)

DIMENSIONS, INTERNAL:
Cabin (incl flight deck): Length	26.82 m (88 ft 0 in)
Max width	2.95 m (9 ft 8 in)
Max height	2.08 m (6 ft 10 in)
Volume	124.1 m³ (4,384 cu ft)

AREAS:
Wings, gross	235.80 m² (2,538.0 sq ft)
Ailerons (total)	5.63 m² (60.60 sq ft)
Tailplane	40.41 m² (435.00 sq ft)
Elevators, incl tabs	12.57 m² (135.30 sq ft)

WEIGHTS AND LOADINGS:
Weight empty	46,500 kg (102,515 lb)
Max weapon load	more than 5,443 kg (12,000 lb)
Max fuel weight	50,122 kg (110,500 lb)
Max T-O weight	105,376 kg (232,315 lb)
Max zero-fuel weight	58,287 kg (128,500 lb)
Max wing loading	446.9 kg/m² (91.53 lb/sq ft)
Max power loading	382 kg/kN (3.75 lb/lb st)

PERFORMANCE:
Max operating Mach No. (M_{MO})	0.77
Service ceiling	12,800 m (42,000 ft)
Range with max internal fuel	more than 6,000 n miles (11,112 km; 6,904 miles)
Endurance, unrefuelled	more than 14 h

UPDATED

BAE ADVANCED AIRCRAFT STUDIES
TYPE: High-agility low-observables (HALO) combat aircraft.
PROGRAMME: £100 million spent in 1992-94 on stealth aircraft development; purpose-built research and development facility, including secure hangar, built at Warton plant in 1995. In 1997, BAE urged government go-ahead for stealth demonstrator (to be known as Experimental Aircraft Programme II) which could have flown in 2000 and operated until 2006; production aircraft then to be available in 2013. By mid-1998, however, BAE planning collaboration with Dassault of France, as described under EAL in International section; new joint venture timetable calls for decision in 2003 to build manned demonstrator following four-year technology definition phase begun in January 1999; accordingly, BAE now seeking delay from 2000 to 2008 in force mix decision on Future Offensive Air System (see Royal Air Force entry in this section).
UPDATED

OTHER AIRCRAFT
BAE involvement with **Eurofighter Typhoon** detailed in International section.

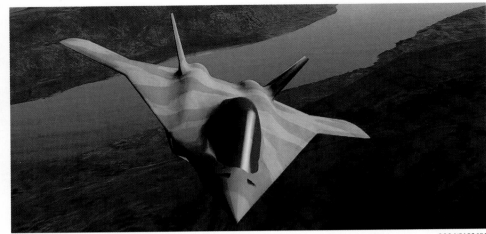
Artist's impression of a possible FOAS configuration

BAE SYSTEMS REGIONAL AIRCRAFT LTD
Woodford Aerodrome, Chester Road, Woodford, Cheshire SK7 1QR
Tel: (+44 161) 439 50 50
Fax: (+44 161) 955 30 08
Web: http://www.bae.regional.co.uk
MANAGING DIRECTOR: Mike O'Callaghan
DIRECTOR, SALES AND ASSET MANAGEMENT: Trevor Hall
SENIOR VICE-PRESIDENT, SALES AND MARKETING: Peter Connolly
VICE-PRESIDENT, MARKETING AND COMMUNICATIONS: Nick Godwin

SALES, MARKETING AND SUPPORT:
BAE Systems Regional Aircraft SA
3 allée Pierre Nadot, BP16, F-31701 Blagnac Cedex, France
Tel: (+33 5) 34 60 70 00
Fax: (+33 5) 34 60 74 90
e-mail: baerasa.marketing@baesystems.com

This division assembles the Avro RJ family of aircraft at Woodford; 185 ha (459 acre) site includes over 93,000 m² (1 million sq ft) of assembly hangars, not including flight test facilities. Also undertakes structural upgrade of Nimrod airframes (which see) as part of Mk 4 modification programme.

Agreement reached in January 1995 (and formally signed on 6 June 1995) for BAe Regional Aircraft to join Franco-Italian ATR (see International Section) in single integrated marketing and support organisation. New venture became operational on 1 January 1996 as Aero International (Regional), but was dissolved in July 1998. Sales, Marketing and Support centre remained in Toulouse, France, and temporarily retained BAe title until 23 February 2000.

Formerly separate Asset Management division became part of Regional Aircraft in 2000; trading of BAe 146 and Avro RJ undertaken from Hatfield, UK; Jetstream turboprop twins from Washington-Chantilly, USA. European logistics centre at Weybridge, UK; customer support facility for Asia at Sydney, Australia; customer training at Prestwick, UK.
UPDATED

Avro RJ100 of Aegean Airlines *(Paul Jackson/Jane's)*

AVRO RJ
Crossair fleet name: Jumbolino
TYPE: Regional jet airliner.
PROGRAMME: Developed from BAe 146 (1993-94 and earlier *Jane's*) with major changes including uprated engines and all-digital avionics; first development (RJ85) aircraft (G-ISEE) flown at Hatfield 23 March 1992 and first RJ100 (G-OIII) on 13 May 1992; formally announced June 1992; first production RJ (RJ85 for Crossair, G-CROS) made first flight at Woodford 27 November 1992, delivered 2 April 1993; CAA and FAA certification completed 1 October 1993 and 10 June 1994, respectively; final assembly of RJs is at Woodford.

Weight savings, together with drag reduction improvements, introduced as standard from 1996, reduce fuel consumption by between 5 and 10 per cent depending on variant and sector length. Cabin pressure differential increased from 1996 to raise cruising altitude by 1,220 m (4,000 ft) to 10,670 m (35,000 ft). The 300th 146/RJ first flew on 1 December 1996 and was delivered to Azzurra Air later the same month. The 100th RJ (N509XJ; the 321st 146/RJ) was delivered to Northwest Airlines on 30 January 1998. By late 2000, 146/RJs had flown over 5.7 million hours and executed some 5.5 million landings. Upgraded RJX variant introduced in 2001.
CURRENT VERSIONS: **RJ70**: Shortest fuselage version, accommodating 70 to 94 passengers; 31.1 kN (7,000 lb st) LF 507 engines. Available on demand.
RJ85: Lengthened version for 85 to 112 passengers; 31.1 kN (7,000 lb st) LF 507 engines.

RJ100: Longest current model, for 100 to 116 passengers (100 at five-abreast seating); engines as for RJ85.

Avro Business Jet: Business, VIP or Corporate Shuttle version of airliner models, launched at 1998 National Business Aviation Association Convention in Las Vegas, Nevada. Public debut at 1999 NBAA Convention in Atlanta, Georgia, based on BAe 146-100 with 27-seat corporate interior.

Avro Business Jet can be based on refurbished BAe 146 or new-build RJ and RJX series. With RJX-85 as basis, has 40,959 kg (90,300 lb) standard or 43,998 kg (97,000 lb) optional MTOW; 38,555 kg (85,000 lb) MLW; and 35,834 kg (79,000 lb) MZFW. Range options with eight passengers are: LR 2,250 n miles (4,167 km; 2,589 miles) or ER 2,850 n miles (5,278 km; 3,279 miles); with 40 passengers: LR 2,090 n miles (3,870 km; 2,405 miles) or ER 2,650 n miles (4,907 km; 3,049 miles). By late 1999, seven BAe 146/RJ variants were in corporate service, in Asia/Pacific region, Middle East, UK and USA.

Avro RJX: Developed in conjunction with risk-sharing partners, engine supplier AlliedSignal Engines (now Honeywell) and pylon/nacelle supplier GKN Westland. 'Conditionally' announced 16 February 1999; formal launch 21 March 2000. Three versions proposed: **RJX-70**, **RJX-85** and **RJX-100**, all powered by four 31.1 kN (7,000 lb st) Honeywell AS977-1A turbofans housed in new nacelles and incorporating an air-start system as part of Integrated Power Plant System (IPPS); noise levels some 18 EPNdB below existing Stage 3 requirements; engine first flew (Boeing 720) January 2000. Engine installation designed for easy retrofit to existing RJ models. Design goals include 227 kg (500 lb) reduction in operating weight empty; 5 per cent increase in climb thrust; 12 to 15 per cent reduction in fuel burn; 20 per cent reduction in direct engine costs; up to 17 per cent increase in range; and common type rating with RJ series.

Preliminary design review completed mid-1999. AS900 series engine first run 30 July 1999. Low-speed wind-tunnel testing of one-eighth scale model completed by

Avro RJ85 of Northwest Airlines *(Norman Pealing)*

Flight deck of Avro RJ series 0052902

October 1999, confirming characteristics of new nacelle system and wing interface. Two RJ airframes, c/ns 376 (G-ORJX) and 378 (G-IRJX), allocated as flight test and certification airframes, as RJX-85 and RJX-100 respectively. In late 2000, first production RJX also earmarked for certification programme. Programme calls for first flight in early 2001; certification in September 2001; and service entry in November or December 2001. Prototype's first engine fitted 24 January 2001.

First firm order was from Druk Air of Bhutan for two, placed 3 April 2000. CityFlyer of UK took option on six, 19 July 2000.

CUSTOMERS: Total 165 RJs, two RJXs and 219 BAe 146s sold by 1 February 2001, see table.

AVRO RJ ORDERS
(at 1 February 2001)

Customer	Variant	Orders	Deliveries
Aegean Airlines	RJ100	6	6
Air Baltic (Latvia)	RJ70	5	2
Air Botnia (SAS)	RJ85	5	0
Air Malta[1]	RJ70	4	4
Azzurra Air (Italy)	RJ70	1	1
	RJ85	3	3
Crossair (Switzerland)	RJ85	4	4
	RJ100	16	16
CityFlyer (UK)	RJ100	16	13
Druk Air	RJX-85	2	0
Lufthansa/Cityline	RJ85	18	18
National Jet Systems (Australia)	RJ70	1	1
Northwest Airlines/ Mesaba	RJ85	36	36
Pelita Air Services (Indonesia)	RJ85	1	1
Sabena	RJ85	14	14
	RJ100	12	12
SAM (Colombia)[2]	RJ100	9	9
Turkish Airlines	RJ70	4	4
	RJ100	10	10
Uzbekistan Airways	RJ85	3	3
Totals		**170**	**157**

[1] Since absorbed by Azzura Air
[2] Fleet transferred to Malmo (Sweden)

COSTS: RJX-70 US$25.4 million, RJX-85 US$27.4 million, RJX-100 US$29.9 million (all estimated, 1999).

DESIGN FEATURES: Low operating noise levels; ability to operate from short or semi-prepared airstrips with minimal ground facilities; four-engine reliability. High-wing layout, with T-tail and four podded engines underwing. Main landing gear in fuselage panniers; petal-type airbrakes. Current version has 'Spaceliner' interior; LF 507 FADEC-controlled engines; digital flight deck with Cat. IIIa all-weather landing capability standard, Cat. IIIb optional.

High lift aerofoil section; thickness/chord ratio 15.3 per cent at root, 12.2 per cent at tip; anhedral 3° at trailing-edge; incidence 3° 6′ at root, 0° at tip; sweepback 15°.

FLYING CONTROLS: Conventional and partly assisted. Manually actuated ailerons and elevators, with trim and servo tabs; powered rudder. Fixed incidence tailplane. Single-section hydraulically actuated Fowler flaps, spanning 78 per cent of trailing-edges, with Dowty actuators; 33° take-off flap setting and steep approach certified for RJ85 and RJ100 for operation into London City Airport; hydraulically operated roll spoiler outboard of three automatically actuated lift dumpers on each wing; no leading-edge lift devices; petal airbrakes form tailcone when closed.

STRUCTURE: All-metal; fail-safe wings with machined skins, integrally machined spars and ribs; fail-safe fuselage with chemically etched skins; strengthened centre-section developed for RJ100; nose free of stringers; remainder of fuselage has top hat stringers bonded to skins above keel area; Z section stringers wet assembled with bonding agent and riveted to skin in keel area; chemically etched tailplane skins bonded to top hat section stringers.

Subcontractors include Denel for rudder and ailerons; GKN Westland for engine pylons, nacelles, fairings and passenger and service doors; Harbin landing gear doors and access panels; and TAPC airbrakes.

LANDING GEAR: Hydraulically retractable tricycle type of Messier Dowty design, with twin Dunlop wheels on each unit. Main units retract inward into fairings on fuselage sides; steerable (±70°) nose unit retracts forward. Oleo-pneumatic shock-absorbers with wheels mounted on trailing axle. Simple telescopic nosewheel strut. Mainwheel tyres size 12.50-16, pressure (RJ70) 8.41 bars (122 lb/sq in). Nosewheel tyres size 7.50-10R, pressure (RJ70) 7.79 bars (113 lb/sq in). Low-pressure tyres optional. Dunlop multidisc carbon brakes operated by duplicated hydraulic systems. Anti-skid units in both primary and secondary brake systems. Minimum ground turning radius about nosewheels: RJ70, 11.53 m (37 ft 10 in); RJ85, 12.55 m (41 ft 2 in); RJ100, 13.97 m (45 ft 10 in).

POWER PLANT: Four Honeywell LF 507 turbofans, each rated at 31.1 kN (7,000 lb st), installed in underwing pylon pods. No reverse thrust. Fuel in two integral wing tanks and integral centre-section tank (latter with vented and drained sealing diaphragm above passenger cabin); combined usable capacity 11,728 litres (3,098 US gallons; 2,580 Imp gallons). Optional auxiliary tanks in wingroot fairings, combined capacity 1,173 litres (310 US gallons; 258 Imp gallons), giving total capacity 12,901 litres (3,408 US gallons; 2,838 Imp gallons). Single-point pressure refuelling, with coupling situated in starboard wing outboard of outer engine.

ACCOMMODATION: Crew of two pilots on flight deck, and two or three cabin staff. Optional observer's seat. RJ70 accommodates 70 passengers five-abreast and up to 94 six-abreast at 74 cm (29 in) pitch; RJ85 accommodates 85

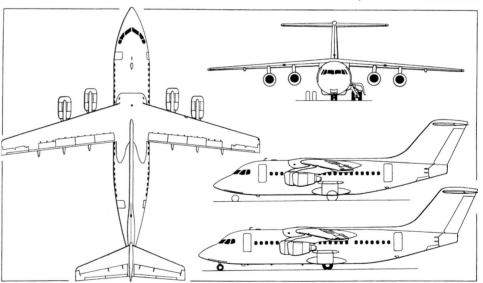
RJ85 with additional side view (upper) of RJ70 *(Dennis Punnett/Jane's)* 1999/0051531

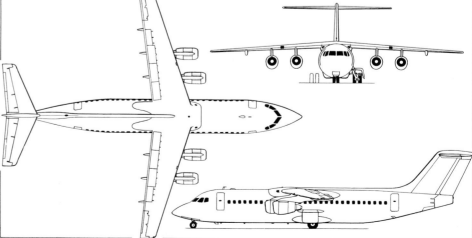
Avro RJ100 regional jet *(Dennis Punnett/Jane's)* 1999/0051533

passengers five-abreast, maximum 112 passengers six-abreast at 74 cm (29 in) pitch; RJ100 standard accommodation for 100 passengers five-abreast at 79 cm (31 in) pitch, and maximum 112 passengers at 81 cm (32 in) pitch. Optional mid-cabin exits and increased capacity air conditioning packs permit increase to maximum 128 passengers at 74 cm (29 in) seat pitch. One outward-opening passenger door forward and one aft on port side of cabin. Built-in airstairs optional. Servicing doors, one forward and one aft on starboard side of cabin. Freight and baggage holds under cabin floor. All accommodation pressurised and air conditioned.

SYSTEMS: Honeywell automatic flight control/flight guidance system. Normalair-Garrett cabin air conditioning and pressurisation system, using engine bleed air. Electropneumatic pressurisation control with discharge valves at front and rear of cabin. Maximum differential 0.51 bar (7.46 lb/sq in), giving 2,440 m (8,000 ft) equivalent altitude at 10,670 m (35,000 ft). Hydraulic system, duplicated for essential services, for landing gear, flaps, rudder, roll and lift spoilers, airbrakes, nosewheel steering, brakes and auxiliary fuel pumps; pressure 207 bars (3,000 lb/sq in).

Electrical system powered by two 40 kVA integrated-drive alternators to feed 115/200 V three-phase 400 Hz primary systems. 28 V DC power supplied by transformer-rectifier in each channel. Hydraulically powered emergency electrical power unit. Hamilton Sundstrand APS 1000 APU for ground and air usable air conditioning and electrical power generation. Kidde Graviner fire suppression system and chemical oxygen system; Intertechnique crew oxygen.

Stall warning and identification system, comprising stick shaker (warning) and stick force (identification) elements, providing soft and hard corrective stick forces at the approach of stall conditions. Hot air de-icing of wing and tailplane leading-edges; windscreen electric anti-icing and demisting and rain repellent coating standard. Pacific Scientific engine fire extinguishing system and Whittaker fire detection system on RJX.

AVIONICS: *Comms:* Dual VHF radios with ARINC 700 interface; Selcal, PA system, CVR. HF radios optional.
Radar: Honeywell RDR4A weather radar.
Flight: Honeywell digital flight guidance system incorporates super fail-passive Cat. IIIb autopilot (Cat. IIIa standard), autothrottle, yaw damper and windshear detection and protection. Rockwell Collins CNLS-910 FMS. Marker beacon receiver, ground proximity warning system; dual DME, radio altimeters, VHF nav and ADF.
Instrumentation: Honeywell four-tube EFIS comprising 127 × 152 mm (5 × 6 in) screens providing primary flight display (PFD) and navigation display (ND) functions with independent processing for pilot and co-pilot; Smiths two-tube EIS provides individual indications for each engine.

DIMENSIONS, EXTERNAL:
Wing span: all versions, excl static dischargers*
    26.34 m (86 ft 5 in)
Wing chord: at root   2.75 m (9 ft 0 in)
  at tip   0.91 m (3 ft 0 in)
Wing aspect ratio   9.0
Length overall: RJ70**   26.16 m (85 ft 10 in)
  RJ85   28.55 m (93 ft 8 in)
  RJ100   30.99 m (101 ft 8¼ in)
Height overall: RJ70, RJ85   8.61 m (28 ft 3 in)
  RJ100   8.59 m (28 ft 2 in)
Fuselage max diameter   3.56 m (11 ft 8 in)
Tailplane span   11.09 m (36 ft 5 in)
Wheel track   4.72 m (15 ft 6 in)
Wheelbase: RJ70   10.09 m (33 ft 1½ in)
  RJ85   11.20 m (36 ft 9 in)
  RJ100   12.52 m (41 ft 1 in)
Passenger doors (port, fwd and rear):
  Height   1.83 m (6 ft 0 in)
  Width   0.85 m (2 ft 9½ in)
  Height to sill: fwd   1.88 m (6 ft 2 in)
    rear   1.98 m (6 ft 6 in)
Servicing doors (stbd, fwd and rear):
  Height   1.47 m (4 ft 10 in)
  Width   0.85 m (2 ft 9½ in)
  Height to sill: fwd   1.88 m (6 ft 2 in)
    rear   1.98 m (6 ft 6 in)
Underfloor freight hold door (stbd, fwd):
  Height   1.09 m (3 ft 7 in)
  Width   1.35 m (4 ft 5 in)
  Height to sill   0.78 m (2 ft 7 in)
Underfloor freight hold door (stbd, rear):
  Height   1.04 m (3 ft 5 in)
  Width   0.91 m (3 ft 0 in)
  Height to sill   0.90 m (2 ft 11½ in)
Freight door (freighter versions):
  Height   1.93 m (6 ft 4 in)
  Width   3.33 m (10 ft 11 in)
  Height to sill   1.93 m (6 ft 4 in)

*Static discharger extends 6.3 cm (2½ in) from each wingtip
**Static dischargers on elevator extend length of all series by 18.4 cm (7¼ in)

DIMENSIONS, INTERNAL:
Cabin (excl flight deck; incl galley and toilets):
  Length: RJ70   15.42 m (50 ft 7 in)
  RJ85   17.81 m (58 ft 5 in)
  RJ100   20.20 m (66 ft 3¼ in)

Latvia's Air Baltic flies the short-fuselage RJ70

Max width   3.43 m (11 ft 3 in)
Max height   2.07 m (6 ft 9½ in)
Baggage/freight holds, underfloor, volume:
  RJ70   13.56 m³ (479 cu ft)
  RJ85   18.25 m³ (644 cu ft)
  RJ100   23.0 m³ (812 cu ft)
AREAS:
  Wings, gross: all versions   77.29 m² (832.0 sq ft)
  Ailerons (total)   3.62 m² (39.00 sq ft)
  Trailing-edge flaps (total)   19.51 m² (210.00 sq ft)
  Spoilers (total)   10.03 m² (108.00 sq ft)
  Fin   15.51 m² (167.00 sq ft)
  Rudder   5.30 m² (57.00 sq ft)
  Tailplane   15.61 m² (168.00 sq ft)
  Elevators, incl tabs (total)   10.03 m² (108.00 sq ft)
WEIGHTS AND LOADINGS (estimated for RJX versions):
  Operating weight empty: RJ70   23,900 kg (52,690 lb)
  RJ85   24,600 kg (54,234 lb)
  RJ100   25,600 kg (56,438 lb)
  RJX-70   23,650 kg (52,139 lb)
  RJX-85   24,500 kg (54,013 lb)
  RJX-100   25,450 kg (56,108 lb)
  Max payload: RJ70   9,893 kg (21,810 lb)
  RJ85   11,234 kg (24,766 lb)
  RJ100   12,275 kg (27,062 lb)
  RJX-70   10,142 kg (22,359 lb)
  RJX-85   11,333 kg (24,985 lb)
  RJX-100   12,425 kg (27,392 lb)
  Max fuel weight: all series:
    standard   9,362 kg (20,640 lb)
    optional   10,298 kg (22,704 lb)
  Max T-O weight: RJ70   43,091 kg (95,000 lb)
  RJ85   43,998 kg (97,000 lb)
  RJ100   46,040 kg (101,500 lb)
  Max ramp weight: RJ70   43,318 kg (95,500 lb)
  RJ85   44,225 kg (97,500 lb)
  RJ100   46,266 kg (102,000 lb)
  Max zero-fuel weight: RJ70   33,793 kg (74,500 lb)
  RJ85   35,834 kg (79,000 lb)
  RJ100   37,875 kg (83,500 lb)
  Max landing weight: RJ70   37,875 kg (83,500 lb)
  RJ85   38,555 kg (85,000 lb)
  RJ100   40,143 kg (88,500 lb)
  Max wing loading: RJ70   557.5 kg/m² (114.19 lb/sq ft)
  RJ85   569.2 kg/m² (116.59 lb/sq ft)
  RJ100   595.7 kg/m² (122.00 lb/sq ft)
  Max power loading: RJ70   346 kg/kN (3.39 lb/lb st)
  RJ85   353 kg/kN (3.46 lb/lb st)
  RJ100   370 kg/kN (3.63 lb/lb st)
PERFORMANCE (estimated for RJX versions):
  Max operating Mach No. (M_{MO}): all versions   0.73
  Max operating speed (V_{MO}):
    RJ70, RJ85   300 kt (555 km/h; 345 mph) IAS
    RJ100   305 kt (565 km/h; 351 mph) IAS
  Cruising speed at 10,670 m (35,000 ft): all versions:
    high speed   412 kt (763 km/h; 474 mph)
    long range   389 kt (720 km/h; 447 mph)
  Stalling speed, 30° flap:
    RJ70   97 kt (179 km/h; 111 mph) EAS
    RJ85   101 kt (187 km/h; 116 mph) EAS
    RJ100   104 kt (192 km/h; 119 mph) EAS
  Stalling speed, 33° flap, at max landing weight:
    RJ70, RJ85   93 kt (172 km/h; 107 mph) EAS
    RJ100   95 kt (176 km/h; 109 mph) EAS
  Max certified altitude   10,670 m (35,000 ft)
  T-O to 10.7 m (35 ft), S/L, ISA: RJ70   1,192 m (3,910 ft)
    RJ85   1,385 m (4,545 ft)
    RJ100   1,535 m (5,035 ft)
  T-O field length for 400 n mile (741 km; 460 mile) sector:
    RJ70   1,092 m (3,580 ft)
    RJ85   1,159 m (3,800 ft)
    RJ100   1,316 m (4,315 ft)
    RJX-70   1,025 m (3,360 ft)
    RJX-85   1,105 m (3,625 ft)
    RJX-100   1,276 m (4,185 ft)
  FAR landing distance from 15 m (50 ft), S/L, ISA, at max landing weight: RJ70   1,180 m (3,871 ft)
    RJ85   1,189 m (3,900 ft)
    RJ100   1,268 m (4,160 ft)
  Range with max fuel, FAR domestic reserves with 100 n mile (185 km; 115 mile) diversion with allowances for engine start, taxi, approach, landing and taxi to gate:
    RJ70   2,107 n miles (3,902 km; 2,424 miles)
    RJ85   2,037 n miles (3,772 km; 2,344 miles)
    RJ100   1,924 n miles (3,563 km; 2,214 miles)
    RJX-70   2,500 n miles (4,630 km; 2,876 miles)
    RJX-85   2,430 n miles (4,500 km; 2,796 miles)
    RJX-100   2,300 n miles (4,259 km; 2,646 miles)
  Range with max payload, allowances as above:
    RJ70   1,377 n miles (2,550 km; 1,584 miles)
    RJ85   1,077 n miles (1,994 km; 1,239 miles)
    RJ100   987 n miles (1,827 km; 1,135 miles)
    RJX-70   1,620 n miles (3,000 km; 1,864 miles)
    RJX-85   1,285 n miles (2,379 km; 1,478 miles)
    RJX-100   1,160 n miles (2,148 km; 1,334 miles)
OPERATIONAL NOISE LEVELS (FAR Pt 36-12, estimated for RJX versions):
  T-O: RJ70   81.9 EPNdB
    RJ85   83.0 EPNdB
    RJ100   84.7 EPNdB
    RJX-70   80.7 EPNdB
    RJX-85   82.9 EPNdB
    RJX-100   85.1 EPNdB
  Approach: RJ70   97.5 EPNdB
    RJ85   97.3 EPNdB
    RJ100   97.6 EPNdB
    RJX-70   95.7 EPNdB
    RJX-85   96.3 EPNdB
    RJX-100   96.6 EPNdB
  Sideline: RJ70   87.2 EPNdB
    RJ85   88.6 EPNdB
    RJ100   88.2 EPNdB
    RJX-70   88.4 EPNdB
    RJX-85   87.6 EPNdB
    RJX-100   86.9 EPNdB

UPDATED

Computer-generated image of the Avro RJX-100

BAE SYSTEMS AIRBUS

New Filton House, Filton, Bristol, BS99 7AR
Tel: (+44 117) 969 38 31
Fax: (+44 117) 936 28 28
MANAGING DIRECTOR: Ray Wilson
HEAD OF PUBLIC AFFAIRS: Howard Berry

ASSOCIATED WORKS:
Chester Road, Broughton, Flintshire CH4 0DR

Tel: (+44 1244) 52 04 44
Fax: (+44 1244) 52 30 00

BAE's ownership interest and industrial responsibilities in Airbus Industrie (which see in International section) are managed by this division. Associated work is undertaken at the Broughton plant, near Chester. Main duties are design of wings and fuel systems for all Airbus airliners; and manufacture and assembly of all Airbus wings, including control surfaces and hydraulic/pneumatic/electrical systems for the A319/A320/A321 family. Also designed, and builds, one fuselage section of A321. The division, which celebrated the completion of its 2,000th set of Airbus wings on 8 February 1999, also participates in Airbus military activities, including A400M and MRTT.

Broughton also builds fuselages and wings for Hawker business jets as subcontractor to Raytheon (see US section).

UPDATED

B-N

B-N GROUP

Bembridge, Isle of Wight PO35 5PR
Tel: (+44 1983) 87 25 11
Fax: (+44 1983) 87 32 46
Web: http://www.britten-norman.com
CHAIRMAN: Alawi Zawawi
BOARD MEMBERS:
  A Munim Zawawi
  Talal Zawawi
CHIEF EXECUTIVE: Paul Bartlett
HEAD OF ENGINEERING: Robert Wilson
HEAD OF SALES AND MARKETING: Ian Wilson
MARKETING SERVICES: Sheila Dewart

Pilatus Aircraft Ltd of Switzerland, which acquired Britten-Norman (Bembridge) Ltd 1979, sold the company on 21 July 1998 to investment group Litchfield Continental Ltd, at which time the original name Britten-Norman Ltd was re-adopted. Company sold almost immediately to Global Spill Management (environmental protection) which reformed as Biofarm (pharmaceutical manufacturer) in Romania. Agreement to purchase Romaero (which see), long-term maker of Islander airframes, made January 1999, but not concluded. Proposed capital injection by UAE-based HSDP, early 2000, also failed to take place.

On 3 April 2000 Britten-Norman was placed in receivership; sale announced 26 April to Alawi Zawawi Enterprises of Oman, and renamed B-N Group Ltd. Order backlog at that time stood at nearly £12 million.

UPDATED

B-N BN2B ISLANDER

TYPE: Light utility twin-prop transport.
PROGRAMME: Prototype (G-ATCT) first flight 13 June 1965 with two 157 kW (210 hp) Rolls-Royce Continental IO-360-B engines and 13.72 m (45 ft) span wings; subsequently re-engined with Textron Lycoming O-540s and flown 17 December 1965; wing span also increased by 1.22 m (4 ft) to initial production standard; production prototype BN2 (G-ATWU) flown 20 August 1966; domestic C of A received 10 August 1967; FAA type certificate 19 December 1967; Romanian manufacture (see Romaero entry) began 1969.
CURRENT VERSIONS: **BN2 Islander:** Initial piston-engined production model (23 built: see earlier *Jane's*).

BN2A Islander: Piston version built from 1 June 1969 until 1989 (see *Jane's Aircraft Upgrades*). Production totalled 890.

BN2B Islander: Standard piston-engined version since 1979; higher maximum landing weight; improved interior design; available with two engine choices and optional wingtip fuel tanks as **BN2B-26** with O-540s and **BN2B-20** with IO-540s (BN2B-27 and -21 no longer available). Features include range of passenger seats and covers, more robust door locks, improved door seals and stainless steel sills, redesigned fresh air system to improve ventilation in hot and humid climates, smaller diameter propellers to decrease cabin noise, and redesigned flight deck and instrument panel. Total of 154 delivered by end December 1999.

B-N, in conjunction with Hartzell Propeller Inc, has been testing a three-blade scimitar-shaped propeller developed as part of the NASA-sponsored AGATE/GAP programme, using a BN-2B operated by the North East Police Air Support Unit. The aim of the test programme is to minimise the aircraft's noise footprint through enhanced propeller aerofoil efficiency and reduced tip speeds.

Detailed description applies to BN2B version.

Series of modification kits available as standard or option for new production aircraft and can be fitted retrospectively to existing aircraft.

BN2B Defender: Described separately.
BN2T Turbine Islander: Described separately.
BN2T-4R Defender: Described in 1997-98 and earlier editions.
BN2T-4S Defender 4000: Described separately.
BN2A Mk III Trislander: Described separately.

CUSTOMERS: By December 1999, deliveries of Islanders and Defenders totalled 1,221; none further in first half of 2000. Recent customers have included the Botswana Defence Force (one attrition replacement), Hawker Pacific (Australia) and Tomen Aerospace (Japan). First to fly post-April 2000 purchase of company was G-BWNG exhibited at Farnborough Air Show, July 2000 having flown on 10 July 2000 and arrived at Bembridge on 19 July. First delivery under new ownership was a BN2B-26 delivered on 10 November 2000 to Aer Arann of Galway, Ireland, followed by another Islander delivered to a Japanese customer on Okinawa on 12 November 2000. See also BN2 status table.

COSTS: BN2B, about £500,000 (1997).

DESIGN FEATURES: Robust STOL light transport, suitable for operation from semi-prepared airstrips. High, unswept, constant-chord wings and tailplane; sweptback fin with small fillet; upswept rear fuselage. Three-blade propellers available on BN2B-20 and BN2B-26 giving quieter noise signature.

NACA 23012 wing section; no dihedral; incidence 2°; no sweepback.

FLYING CONTROLS: Conventional and manual. Actuation by pushrods and cables. Slotted ailerons, with starboard ground-adjustable tab; mass-balanced elevator; trim tabs in rudder and elevator; single-slotted flaps operated electrically; fixed incidence tailplane. Dual controls standard.

STRUCTURE: L72 aluminium-clad aluminium alloys; two-spar wing torsion box in one piece; flared-up wingtips; integral fuel tanks in wingtips optional; four-longeron fuselage of pressed frames and stringers; two-spar tail unit with pressed ribs.

LANDING GEAR: Non-retractable tricycle type, with twin wheels on each main unit and single steerable nosewheel. Cantilever main legs mounted aft of rear spar. All three legs fitted with oleo-pneumatic shock-absorbers. All five wheels and tyres size 16×7-7, supplied by Goodyear. Tyre pressure: main 2.41 bars (35 lb/sq in); nose 2.00 bars (29 lb/sq in). Foot-operated air-cooled Cleveland hydraulic brakes on main units. Parking brake. Wheel/ski gear available optionally. Minimum ground turning radius 9.45 m (31 ft 0 in).

POWER PLANT: Two Textron Lycoming flat-six engines, each driving a Hartzell HC-C2YK-2B or -2C two-blade constant-speed feathering metal propeller; optional three-blade Hartzell HC-C3YR-2UF/FC8468-8R. Propeller synchronisers optional. Standard power plant 194 kW (260 hp) O-540-E4C5, but 224 kW (300 hp) IO-540-K1B5 fitted at customer's option.

Integral fuel tank between spars in each wing, outboard of engine. Total fuel capacity (standard) 518 litres (137 US gallons; 114 Imp gallons). Usable fuel 492 litres (130 US gallons; 108 Imp gallons). With optional fuel tanks in wingtips, total capacity is increased to 814 litres (215 US gallons; 179 Imp gallons). Additional pylon-mounted underwing auxiliary tanks, each of 227 litres (60.0 US gallons; 50.0 Imp gallons) capacity, available optionally. Refuelling point in upper surface of wing above each internal tank. Total oil capacity 22.75 litres (6.0 US gallons; 5.0 Imp gallons).

ACCOMMODATION: Up to 10 persons, including pilot, on side-by-side front seats and four bench seats. No aisle. Seat backs fold forward. Access to all seats via three forward-opening doors, ahead of wing and at rear of cabin on port side and forward of wing on starboard side. Baggage compartment at rear of cabin, with port-side loading door in standard versions. Exit in emergency by removing door windows. Special executive layouts available.

Can be operated as freighter, carrying more than a ton of cargo; in this configuration passenger seats can be stored in rear baggage bay. In ambulance role, up to three stretchers and two attendants can be accommodated. Other layouts possible, including photographic and geophysical survey, parachutist transport or trainer (with accommodation for up to eight parachutists and dispatcher), firefighting, environmental protection and cropspraying.

SYSTEMS: Janaero cabin heater standard. 45,000 BTU Janitrol combustion unit, with circulating fan, provides hot air for distribution at floor level outlets and at windscreen demisting slots. Fresh air, boosted by propeller slipstream, is ducted to each seating position for on-ground ventilation. Electrical DC power, for instruments, lighting and radio, from two engine-driven 24 V 70 A self-rectifying alternators and a controller to main busbar and circuit breaker assembly. Emergency busbar is supplied by a 24 V 25 Ah heavy-duty lead-acid battery in the event of a twin alternator failure. Ground power receptacle provided. Optional electric de-icing of propellers and windscreen, and pneumatic de-icing of wing and tail unit leading-edges. Oxygen system available optionally for all versions.

AVIONICS: *Comms:* Intercom, including second headset, and passenger address system are standard.

Flight: Standard items include autopilot and wide range of VHF and HF communications and navigation equipment.

Instrumentation: IFR standard.

DIMENSIONS, EXTERNAL:
Wing span	14.94 m (49 ft 0 in)
Wing chord, constant	2.03 m (6 ft 8 in)
Wing aspect ratio	7.4
Length overall	10.86 m (35 ft 7¾ in)
Fuselage: Max width	1.21 m (3 ft 11½ in)
Max depth	1.46 m (4 ft 9¾ in)
Height overall	4.18 m (13 ft 8¾ in)
Tailplane span	4.67 m (15 ft 4 in)
Wheel track (c/l of shock-absorbers)	3.61 m (11 ft 10 in)
Wheelbase	3.99 m (13 ft 1¼ in)
Propeller diameter	1.98 m (6 ft 6 in)
Cabin door (front, port): Height	1.10 m (3 ft 7½ in)
Width: top	0.64 m (2 ft 1¼ in)
Height to sill	0.59 m (1 ft 11¼ in)
Cabin door (front, starboard):	
Height	1.10 m (3 ft 7½ in)
Max width	0.86 m (2 ft 10 in)
Height to sill	0.57 m (1 ft 10½ in)
Cabin door (rear, port): Height	1.09 m (3 ft 7 in)
Width: top	0.635 m (2 ft 1 in)
bottom	1.19 m (3 ft 11 in)
Height to sill	0.52 m (1 ft 8½ in)
Baggage door (rear, port): Height	0.69 m (2 ft 3 in)

DIMENSIONS, INTERNAL:
Passenger cabin, aft of pilot's seat:	
Length	3.05 m (10 ft 0 in)
Max width	1.09 m (3 ft 7 in)
Max height	1.27 m (4 ft 2 in)
Floor area	2.97 m² (32.0 sq ft)
Volume	3.7 m³ (130 cu ft)
Baggage space aft of passenger cabin	1.4 m³ (49 cu ft)
Freight capacity:	
aft of pilot's seat, incl rear cabin baggage space	4.7 m³ (166 cu ft)
with four bench seats folded into rear cabin baggage space	3.7 m³ (130 cu ft)

AREAS:
Wings, gross	30.19 m² (325.0 sq ft)
Ailerons (total)	2.38 m² (25.60 sq ft)
Flaps (total)	3.62 m² (39.00 sq ft)
Fin	3.41 m² (36.64 sq ft)

BN2B-20 Islander G-BWNG, still wearing primer, was first new aircraft to fly after company name changed to B-N Group *(Paul Jackson/Jane's)*

UK: AIRCRAFT—B-N

Rudder, incl tab	1.60 m² (17.20 sq ft)
Tailplane	6.78 m² (73.00 sq ft)
Elevator, incl tabs	3.08 m² (33.16 sq ft)

WEIGHTS AND LOADINGS (A: 194 kW; 260 hp engines, B: 224 kW; 300 hp engines):

Weight empty, equipped (without avionics):	
A	1,866 kg (4,114 lb)
B	1,925 kg (4,244 lb)
Max payload: A	929 kg (2,048 lb)
B	870 kg (1,918 lb)
Payload with max fuel: A	692 kg (1,526 lb)
B	633 kg (1,396 lb)
Max fuel weight: standard: A, B	354 kg (780 lb)
with optional tanks in wingtips: A, B	585 kg (1,290 lb)
Max T-O and landing weight: A, B	2,993 kg (6,600 lb)
Max zero-fuel weight (BCAR):	
A, B	2,855 kg (6,300 lb)
Max floor loading, without cargo panels:	
A, B	586 kg/m² (120 lb/sq ft)
Max wing loading: A, B	99.2 kg/m² (20.31 lb/sq ft)
Max power loading: A	7.73 kg/kW (12.69 lb/hp)
B	6.70 kg/kW (11.00 lb/hp)

PERFORMANCE (A and B as above):

Never-exceed speed (VNE):	
A, B	183 kt (339 km/h; 211 mph) IAS
Max level speed at S/L:	
A	148 kt (274 km/h; 170 mph)
B	151 kt (280 km/h; 173 mph)
Max cruising speed at 75% power at 2,135 m (7,000 ft):	
A	139 kt (257 km/h; 160 mph)
B	142 kt (264 km/h; 164 mph)
Cruising speed at 67% power at 2,745 m (9,000 ft):	
A	134 kt (248 km/h; 154 mph)
B	137 kt (254 km/h; 158 mph)
Cruising speed at 59% power at 3,660 m (12,000 ft):	
A	130 kt (241 km/h; 150 mph)
B	132 kt (245 km/h; 152 mph)
Stalling speed:	
flaps up: A, B	50 kt (92 km/h; 57 mph) IAS
flaps down: A, B	40 kt (74 km/h; 46 mph) IAS
Max rate of climb at S/L: A	262 m (860 ft)/min
B	344 m (1,130 ft)/min
Rate of climb at S/L, OEI: A	44 m (145 ft)/min
B	60 m (198 ft)/min
Absolute ceiling: A	4,145 m (13,600 ft)
B	6,005 m (19,700 ft)
Service ceiling: A	3,445 m (11,300 ft)
B	5,240 m (17,200 ft)
Service ceiling, OEI: A	1,525 m (5,000 ft)
B	1,980 m (6,500 ft)
T-O run at S/L, zero wind, hard runway:	
A	278 m (915 ft)
B	215 m (705 ft)
T-O run at 1,525 m (5,000 ft): A	396 m (1,300 ft)
B	372 m (1,225 ft)
T-O to 15 m (50 ft) at S/L, zero wind, hard runway:	
A	371 m (1,220 ft)
B	355 m (1,165 ft)
T-O to 15 m (50 ft) at 1,525 m (5,000 ft):	
A	528 m (1,735 ft)
B	496 m (1,630 ft)
Landing from 15 m (50 ft) at S/L, zero wind, hard runway: A, B	299 m (980 ft)
Landing from 15 m (50 ft) at 1,525 m (5,000 ft):	
A, B	357 m (1,170 ft)
Landing run at 1,525 m (5,000 ft): A, B	171 m (560 ft)
Landing run at S/L, zero wind, hard runway:	
A, B	140 m (460 ft)
Range, block-to-block, plus 10% plus 45 min reserves:	
B, standard fuel: IFR	503 n miles (931 km; 578 miles)
VFR	639 n miles (1,183 km; 735 miles)
B, with optional tanks:	
IFR	896 n miles (1,659 km; 1,031 miles)
VFR	1,075 n miles (1,990 km; 1,237 miles)

UPDATED

B-N BN2T TURBINE ISLANDER

UK Army Air Corps designation: Islander AL. Mk 1
Royal Air Force designation: Islander CC. Mk 2/2A

TYPE: Utility turboprop twin.

PROGRAMME: Turboprop Islander; first flight of prototype (G-BPBN) 2 August 1980 with two Allison (now Rolls-Royce) 250-B17Cs; British CAA certification received end of May 1981; first production aircraft delivered December 1981; FAR Pt 23 US type approval 15 July 1982; full icing clearance to FAR Pt 25 gained 23 July 1984.

CUSTOMERS: Total 65 aircraft delivered by December 1999, mainly of the Defender variant. Recent customers include the Rhine Army Parachute Association.

DESIGN FEATURES: As for piston-engined version (previous entry); turboprops enable use of available low-cost jet fuel; low operating noise level; available for same range of applications as Islander, including military versions (described separately).

Description of BN2B Islander applies also to BN2T, except as follows:

POWER PLANT: Two 298 kW (400 shp) Rolls-Royce 250-B17C turboprops, flat rated at 238.5 kW (320 shp), and each driving a Hartzell three-blade constant-speed fully feathering metal propeller. Usable fuel 814 litres (215 US gallons; 179 Imp gallons). Pylon-mounted underwing tanks, each of 227 litres (60.0 US gallons; 50.0 Imp gallons) capacity, are available optionally for special purposes. Total oil capacity 5.7 litres (1.5 US gallons; 1.25 Imp gallons).

ACCOMMODATION: Generally as for BN2B. In ambulance role can accommodate, in addition to pilot, a single stretcher, one medical attendant and five seated occupants; or two stretchers, one attendant and three passengers; or three stretchers, two attendants and one passenger. Other possible layouts include photographic and geophysical survey; parachutist transport or trainer (with accommodation for up to eight parachutists and a dispatcher); and pest control or other agricultural spraying. Maritime Turbine Islander/Defender versions available for fishery protection, coastguard patrol, pollution survey, search and rescue, and similar applications. In-flight sliding parachute door optional.

AVIONICS: *Comms:* Optional maritime band and VHF transceivers.
Flight: VLF/Omega nav system; radar altimeter.
Self-defence (Military versions): Since at least 1994 Army Air Corps Islander AL. Mk 1s have had provision for AN/ALQ-144 IR jammer under fuselage. Lockheed Martin IRCM suite tested on AAC Islander in early 1998.

EQUIPMENT: According to mission, can include fixed tailboom or towed bird magnetometer, spectrometer, or electromagnetic detection/analysis equipment (geophysical survey); one or two cameras, navigation sights and appropriate avionics (photographic survey); 189 litre (50.0 US gallon; 41.5 Imp gallon) Micronair underwing spraypods complete with pump and rotary atomiser (pest control/agricultural spraying versions); dinghies, survival equipment and special crew accommodation (maritime versions).

DIMENSIONS, EXTERNAL: As for BN2B, except:

Length overall: standard nose	10.86 m (35 ft 7¾ in)
weather radar nose	11.07 m (36 ft 3¾ in)
Propeller diameter	2.03 m (6 ft 8 in)

WEIGHTS AND LOADINGS:

Weight empty, equipped	1,832 kg (4,040 lb)
Payload with max fuel	689 kg (1,519 lb)
Max T-O weight	3,175 kg (7,000 lb)
Max landing weight	3,084 kg (6,800 lb)
Max zero-fuel weight	2,994 kg (6,600 lb)
Max wing loading	105.2 kg/m² (21.54 lb/sq ft)
Max power loading	5.33 kg/kW (8.75 lb/shp)

PERFORMANCE (standard Turbine Islander/Defender):

Max cruising speed:	
at 3,050 m (10,000 ft)	170 kt (315 km/h; 196 mph)
at S/L	154 kt (285 km/h; 177 mph)

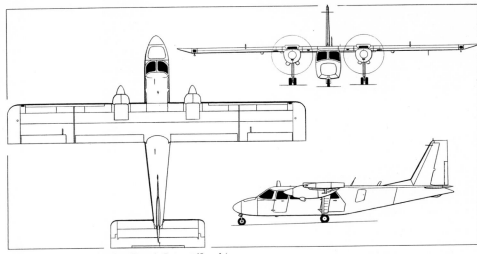

B-N BN2T Turbine Islander *(Dennis Punnett/Jane's)*

Cruising speed at 72% power:	
at 3,050 m (10,000 ft)	150 kt (278 km/h; 173 mph)
at 1,525 m (5,000 ft)	143 kt (265 km/h; 165 mph)
Stalling speed, power off:	
flaps up	52 kt (97 km/h; 60 mph) IAS
flaps down	45 kt (84 km/h; 52 mph) IAS
Max rate of climb at S/L	320 m (1,050 ft)/min
Rate of climb at S/L, OEI	66 m (215 ft)/min
Service ceiling	over 7,010 m (23,000 ft)
Absolute ceiling, OEI	over 3,050 m (10,000 ft)
T-O run	255 m (840 ft)
T-O to 15 m (50 ft)	381 m (1,250 ft)
Landing from 15 m (50 ft)	339 m (1,110 ft)
Landing run	228 m (750 ft)
Range (IFR) with max fuel, reserves for 45 min hold plus 10%	590 n miles (1,093 km; 679 miles)
Range (VFR) with max fuel, no reserves	728 n miles (1,348 km; 838 miles)

BN2 STATUS
(at 1 January 1999)

	Number delivered	Number remaining
BN2	23	3
BN2A	890	596
BN2B	153	147
BN2T	65	64
BN2T-4R	4	4
BN2T-4S	3	3
Trislander	82	52
Totals	**1,220**	**869**

Notes: Number remaining takes into account 11 BN-2s converted to BN2As; eight BN2A-21s converted to BN2B-21s; one BN2B converted to BN2T; and one BN2T converted to BN2T-4R. Six BN2As/BN2Bs operated by the Indian Navy converted to BN2Ts during 1996-98.

UPDATED

B-N DEFENDER

TYPE: See Islander and Turbine Islander.

CUSTOMERS: For piston (BN2B) version, customers include: Indian Navy (six recently converted to turbine power), Belgian Army, Botswana Defence Forces and Jamaican Defence Force.

For turbine (BN2T) version, customers include: Moroccan Ministry of Fisheries, Pakistan Maritime Security Agency, Belgian Gendarmerie, Netherlands

B-N BN2T Islander AL. Mk 1 of the UK Army Air Corps equipped for photographic surveillance (note partly open door with special insert, including camera window) *(Paul Jackson/Jane's)*

Police, Royal Air Force, UK Army Air Corps and Mauritius Coastguard.

DESIGN FEATURES: Similar to those of civil Islander, but with four underwing hardpoints for standard NATO pylons to attach fuel tanks, weapons and other stores; number of additional airframe options, including sliding door on rear port side which can be opened in flight, are also offered. Concept of Defender is to provide a low-cost airframe which can be fitted with best available sensors to meet operational needs of customers.

EQUIPMENT: Optimal installations for British Army Islanders includes photographic surveillance package of door-mounted Zeiss 610 camera, vertical Zeiss Trilens 80 mm and either F126 vertical, RMK vertical or two KS-153 vertical cameras; Vinten F143 panoramic is also available.

UPDATED

B-N BN2T-4S DEFENDER 4000

TYPE: Multisensor surveillance twin-turboprop.

PROGRAMME: Marketing experience with BN2T Defenders revealed that many military and government agencies expressed need for greater payload; announced 1994; prototype (G-SURV) made first flight 17 August 1994; public launch at Farnborough Air Show September 1994; CAA certification achieved 13 November 1995; CAA transport (passenger) certification 4 April 1996.

Britten-Norman and Orenda Recip Inc of Canada announced in September 1999 joint study to re-engine Defender with Orenda OE600 V-8 piston engines, resulting in improved performance and efficiency. Projected specification includes maximum T-O weight 4,241 kg (9,350 lb), maximum cruising speed 176 kt (326 km/h 202 mph) and maximum rate of climb 366 m (1,200 ft)/min.

CUSTOMERS: Seven aircraft registered to manufacturer in April 1996. Irish Ministry of Justice took delivery of one on 14 August 1997; aircraft is operated by the Irish Air Corps at Baldonnel, Dublin, on behalf of the Garda Siochana (National Police), and is equipped with a comprehensive suite of law enforcement communications equipment, thermal imager and observers' removable consoles. Sabah Air of Malaysia took delivery of one in November 1997; primarily equipped for aerial photographic work, with quick-change facility to 12-passenger transport. Police Aviation Services UK took delivery of a single (ex-demonstrator) example in November 1998. Hampshire Police Authority ordered one aircraft in April 1999, for service with its Air Support Unit.

DESIGN FEATURES: As for Islander, but with enlarged wing, based on that of Trislander, although retaining original control surfaces; compared with BN2T, has tailplane and elevator of increased span; modified fin with larger fillet; fuselage stretched by means of 0.76 m (2 ft 6 in) plug forward of wing to seat up to 16; more powerful engines for greater sortie time and payload capacity; fuselage and landing gear strengthened; flight deck windows deepened for enhanced field of view; redesigned nose and tail unit; rear-sliding door with blister window.

POWER PLANT: Two Rolls-Royce 250-B17F turboprops, each flat rated at 298 kW (400 shp) and driving a three-blade propeller. Internal fuel capacity 1,131 litres (299 US gallons; 249 Imp gallons).

ACCOMMODATION: Flight crew of two on airline-type seats; sliding seat rails permit seat and equipment positioning anywhere within fuselage. Space for two or more consoles and operators in tandem along one side of cabin. Up to 16 troops/passengers in tactical transport role. Certified to carry pilot and 11 passengers, subject to local airworthiness requirements.

SYSTEMS: Electrical system includes two 200 A engine-driven generators.

AVIONICS: *Comms:* Full range of open or secure voice com radios from UHF, VHF, HF and VHF-FM.

Radar: Modified nose can accommodate 68.5 cm (27 in), 360° rotating antenna for maritime, simple search or weather radar (BAE Seaspray 2000 in prototype Defender 4000).

Flight: Fully integrated autopilot. GPS nav system, integrated with Omega or INS.

Mission: Sensors can include thermal imagers and/or hand-held or podded video or film cameras. Prototype on debut fitted with Agema FLIR under fuselage on starboard side; alternatives could include BAE pod or FLIR 2000.

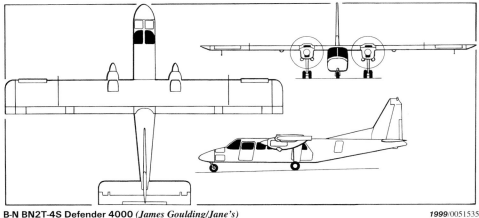

B-N BN2T-4S Defender 4000 *(James Goulding/Jane's)* 1999/0051535

Appropriate radars could include BAE Seaspray, Racal Super Searcher, Thomson-CSF Detexis Ocean Master, Telephonics 143 or Litton 504(V)5.

ARMAMENT: Two hardpoints under each wing, inboard pair stressed for loads of up to 340 kg (750 lb) and outboard pair for up to 159 kg (350 lb) each. Typical weapons on inboard stations can include Sting Ray torpedo or Sea Skua anti-ship missile.

DIMENSIONS, EXTERNAL: As for BN2B, except:
Wing span	16.15 m (53 ft 0 in)
Wing aspect ratio	8.0
Length overall	12.20 m (40 ft 0½ in)
Height overall	4.36 m (14 ft 3½ in)
Tailplane span	5.31 m (17 ft 5 in)
Wheelbase	4.76 m (15 ft 7½ in)

AREAS:
Wings, gross	32.61 m² (351.0 sq ft)

WEIGHTS AND LOADINGS:
Weight empty, equipped	2,223 kg (4,900 lb)
Max usable internal fuel	908 kg (2,002 lb)
Payload with max fuel	724 kg (1,598 lb)
Max T-O and landing weight	3,855 kg (8,500 lb)
Max zero-fuel weight	3,765 kg (8,300 lb)
Max wing loading	109.6 kg/m² (22.45 lb/sq ft)
Max power loading	6.46 kg/kW (10.63 lb/shp)

PERFORMANCE:
Max cruising speed: at S/L	154 kt (285 km/h; 177 mph)
at 3,050 m (10,000 ft)	176 kt (326 km/h; 202 mph)
Transit speed from base to patrol area	160 kt (296 km/h; 184 mph)
Cruising speed at 72% power:	
at 1,525 m (5,000 ft)	149 kt (276 km/h; 171 mph)
at 3,050 m (10,000 ft)	160 kt (296 km/h; 184 mph)
Stalling speed, power off:	
flaps up	53 kt (99 km/h; 61 mph) IAS
flaps down	47 kt (87 km/h; 54 mph) IAS
Max rate of climb at S/L	381 m (1,250 ft)/min
Rate of climb at S/L, OEI	61 m (200 ft)/min
Absolute ceiling	7,620 m (25,000 ft)
Absolute ceiling, OEI	3,660 m (12,000 ft)
T-O run	356 m (1,170 ft)
T-O to 15 m (50 ft)	565 m (1,855 ft)
Landing from 15 m (50 ft)	589 m (1,935 ft)
Landing run	308 m (1,015 ft)
Max range on internal fuel:	
with IFR reserves	861 n miles (1,594 km; 990 miles)
with VFR reserves	1,006 n miles (1,863 km; 1,157 miles)
Max endurance on internal fuel	8 h 30 min

UPDATED

B-N BN2A Mk III TRISLANDER

TYPE: Tri-prop utility transport.

PROGRAMME: Enlarged Islander with additional engine; first flown 1 September 1970; production ended 1979, but some airframes stored uncompleted; one used in reconstruction, 1996 (see 1997-98 edition).

On 29 May 1999, the then Britten-Norman company announced an order for three Trislanders from China Northern Airlines for delivery between September 2000 and January 2001. Programme in question, due to B-N bankruptcy; if reinstated, first aircraft will be assembled at Bembridge from a kit supplied by Romaero; subsequent aircraft will be manufactured in Romania and supplied 'green' to B-N for outfitting and delivery. The new Trislanders will have modern interiors and avionics suites, including GPS, HF, weather radar and de-icing system.

CURRENT VERSIONS: **Trislander:** *As described.*

Integrity: In 1999, Integrity Aircraft of New Zealand proposed a Trislander variant with a Honeywell TPE331-12 turboprop, flat rated at 746 kW (1,000 shp) and driving a five-blade McCauley propeller, mounted in the fin position. Airframe based on Defender 4000, but with 0.76 m (2 ft 6 in) plug immediately behind flight deck. MTOW remains 4,536 kg (10,000 lb), but payload increased by 450 kg (992 lb); passenger capacity of 20. A development and manufacturing licence was obtained from Britten-Norman; aircraft would be built by Romaero.

COSTS: China Northern Airlines contract for three valued at US$4.8 million including crew training and spares (1999).

Differences from twin-engined versions detailed below.

DESIGN FEATURES: As Islander, but third engine mounted on fin, larger wing and stretched fuselage. Flared-up wingtips of B-N design, with raked tips. Fixed-incidence tailplane. NACA 23012 constant wing section. No dihedral. Incidence 2°. No sweepback.

FLYING CONTROLS: Conventional and manual. Slotted ailerons and electrically operated single-slotted, permanently drooped flaps. Ground-adjustable tab in starboard aileron. Fixed-incidence tailplane (with raked tips); trim tab in rudder.

STRUCTURE: Conventional riveted two-spar torsion-box wing structure in one piece using aluminium-clad aluminium alloys. Increases in skin gauges and spar laminates compared with twin-engined versions. Structure is strictly safe-life, but with several fail-safe features and principles. The fuselage is a conventional riveted four-longeron semi-monocoque structure of pressed frames and stringers and metal skin, using L72 aluminium-clad aluminium alloys. Some reinforcement of fuselage aft of wing to support weight of rear engine. Cantilever tail unit using L72

Instrument panel of BN2T-4S Defender 4000 *(Paul Jackson/Jane's)* 0052936

B-N BN2T-4S Defender 4000 of the Irish national police force *(John Dibbs/B-N)* 1999/0015606

524   UK: AIRCRAFT—B-N to CFM

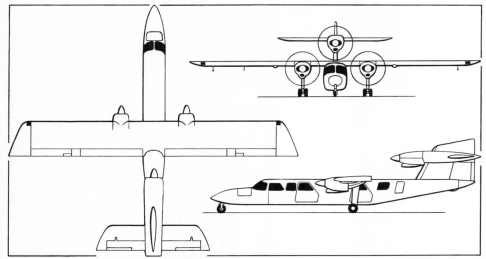

General arrangement of the B-N BN2A Mk III Trislander *(James Goulding/Jane's)*   2000/0064910

aluminium-clad aluminium alloys, with low-aspect ratio main fin which also acts as mount for the third engine.

LANDING GEAR: Non-retractable tricycle type, with twin-wheel main units and single steerable nosewheel. Cantilever main legs mounted aft of rear spar. All five wheels and tyres by Cleveland. Tyre sizes; nose 6.00-6, main 8.00-6; pressure 3.10 bars (45 lb/sq in) on main units, 2.00 bars (29 lb/sq in) on nose unit. Cleveland foot-operated disc brakes on main units. Parking brake. No anti-skid units. Fairings fitted to main gear extension tubes below engine nacelle and above shock-absorber attachment bolts.

POWER PLANT: Three 194 kW (260 hp) Textron Lycoming O-540-E4C5 flat-six engines (two mounted on wings and one on vertical tail), each driving a Hartzell HC-C2YK-2B/C8477-4 two-blade constant-speed fully feathering metal propeller. Automatic feathering device available as an option. Fuel in two integral tanks between front and rear wing spars, outboard of engine nacelles, total capacity 746 litres (197 US gallons; 164 Imp gallons). Overwing refuelling point above each tank. Oil capacity 34 litres (9.0 US gallons; 7.5 Imp gallons).

ACCOMMODATION: Up to 18 persons, including pilot, in pairs on bench seats at 79 cm (31 in) pitch. Access to all seats provided by five broad-hinged rearward-opening car-type doors, two on port side and three on starboard side. Baggage compartment at rear of cabin, with external baggage door on port side. Exit in emergency by removing window panels in front four passenger doors. Heating, ventilation and sound insulation standard. Ambulance or VIP interior layouts at customer's option. Dual controls optional.

SYSTEMS: One Janaero cabin heater fitted as standard. DC electrical system includes two 24 V 70 A self-rectifying alternators, supplying the instruments, lighting and radio, and a 24 V 17 Ah battery. No hydraulic or pneumatic systems, except for self-contained hydraulic brakes. Goodrich pneumatic de-icing boots optional for wings and tailplane.

DIMENSIONS, EXTERNAL:
Wing span	16.15 m (53 ft 0 in)
Wing chord (constant)	2.03 m (6 ft 8 in)
Wing aspect ratio	7.95
Length overall	15.01 m (49 ft 3 in)
Fuselage: Max width	1.21 m (3 ft 11½ in)
Max depth	1.46 m (4 ft 9¾ in)
Height overall	4.29 m (14 ft 1 in)
Tailplane span	6.43 m (21 ft 1½ in)
Tailplane chord (constant)	1.45 m (4 ft 9 in)
Wheel track (c/l of shock-absorbers)	3.60 m (11 ft 10 in)
Wheelbase	7.12 m (23 ft 4¼ in)
Propeller diameter	2.03 m (6 ft 8 in)
Propeller ground clearance	0.69 m (2 ft 3 in)
Distance between propeller centres (wing engines)	3.61 m (11 ft 10 in)
Passenger doors (stbd, fwd and centre):	
Height	1.10 m (3 ft 7½ in)
Max width	0.89 m (2 ft 11 in)
Height to sill	0.57 m (1 ft 10½ in)
Passenger doors (port, fwd and rear):	
Height	1.09 m (3 ft 7 in)
Max width	1.21 m (3 ft 11⅞ in)
Height to sill	0.57 m (1 ft 10½ in)
Passenger door (stbd, rear): Height	1.09 m (3 ft 7 in)
Width	0.75 m (2 ft 5½ in)
Baggage compartment door (rear, port):	
Height	0.66 m (2 ft 2 in)
Width	0.44 m (1 ft 5⅓ in)
Nose baggage compartment door (port, optional):	
Width	0.79 m (2 ft 7 in)

DIMENSIONS, INTERNAL:
Cabin: Length excl flight deck but incl rear baggage compartment	8.24 m (27 ft 0½ in)
Max width	1.09 m (3 ft 7 in)
Max height	1.27 m (4 ft 2 in)
Floor area	7.85 m² (84.45 sq ft)
Volume	9.54 m³ (337 cu ft)
Rear baggage compartment volume	0.71 m³ (25.0 cu ft)
Nose baggage compartment volume (optional)	0.62 m³ (22.0 cu ft)

AREAS:
Wings, gross	31.31 m² (337.0 sq ft)
Ailerons (total)	2.38 m² (25.60 sq ft)
Trailing-edge flaps (total)	3.62 m² (39.00 sq ft)
Fin	5.83 m² (62.70 sq ft)
Rudder, incl tab	1.13 m² (12.20 sq ft)
Tailplane	8.36 m² (90.00 sq ft)
Elevators	2.42 m² (26.00 sq ft)

WEIGHTS AND LOADINGS:
Weight empty, equipped	2,650 kg (5,842 lb)
Max T-O and landing weight	4,536 kg (10,000 lb)
Max wing loading	144.8 kg/m² (29.67 lb/sq ft)
Max power loading	7.80 kg/kW (12.82 lb/hp)

PERFORMANCE:
Max level speed at S/L	156 kt (290 km/h; 180 mph)
Cruising speed at 75% power at 1,980 m (6,500 ft)	144 kt (267 km/h; 166 mph)
Cruising speed at 67% power at 2,470 m (9,000 ft)	138 kt (256 km/h; 159 mph)
Cruising speed at 50% power at 3,960 m (13,000 ft)	130 kt (241 km/h; 150 mph)
Max rate of climb at S/L	298 m (980 ft)/min
Rate of climb at S/L, OEI	86 m (283 ft)/min
Absolute ceiling	4,450 m (14,600 ft)
Service ceiling	4,010 m (13,150 ft)
Service ceiling, one engine out	2,105 m (6,900 ft)
T-O run at S/L, hard runway	393 m (1,290 ft)
T-O to 15 m (50 ft) at S/L, hard runway	594 m (1,950 ft)
Landing from 15 m (50 ft) at S/L, hard runway	440 m (1,445 ft)
Landing run at S/L, hard runway	259 m (850 ft)
Max still-air range at 59% cruising power	868 n miles (1,610 km; 1,000 miles)

UPDATED

B-N BN2A Mk III Trislander of Aurigny Airways   2000/0064909

CFM

CFM AIRCRAFT
Unit 2D, Eastlands Industrial Estate, Leiston, Suffolk IP16 4LL
Tel: (+44 1728) 83 23 53
Fax: (+44 1728) 83 29 44
e-mail: cfmaircraft@compuserve.com
Web: http://www.cfmaircraft.co.uk
MANAGING DIRECTOR: David Moore
DESIGN ENGINEER: John Wighton
GENERAL MANAGER: Anthony Preston

Cook Flying Machines (trading as CFM MetalFax Ltd) went into receivership during 1996; assets were purchased from the liquidator by a small group of investors and the company was relaunched as CFM Aircraft, which continues to manufacture and market the Shadow series of microlights and light aircraft in kitbuilt and production versions as described below. By end of 1998, 330 CFM ultralights were flying in 40 countries. In 1999, CFM secured the world's largest ultralight order when the Indian Air Force contracted for 24 Streak Shadows.

Workforce was 10 in 1999, with annual production capacity of 12 to 18 assembled aircraft, and a similar number of kits, at the company's 465 m² (5,000 sq ft) factory at Leiston. Flight testing is undertaken at Parham (Framlingham). Airborne Innovations (which see in the US section) is licensed to produce the Shadow and Starstreak.

UPDATED

CFM SHADOW SERIES D

TYPE: Tandem-seat ultralight/kitbuilt.
PROGRAMME: Shadow first flew as prototype in 1983; in production since 1984; type approval to BCAR CAP 482 Section S gained May 1985.
CURRENT VERSIONS: **Series C:** Rotax 503 engine. **Series CD:** has dual controls. Suitable for cropspraying with 64 litre (16.8 US gallon; 14.0 Imp gallon) chemical tank and multisensor surveillance for photography (Hasselblad survey camera), video recording, linescan or thermal imagery, or closed-circuit TV microwave transmission to a command vehicle/station. Total of 112 delivered. Now superseded by Series D.
Series D: Cockpit width increased by 10 per cent; electric starter and elevator trim as standard. Dual controls in **Series DD**; all aircraft to DD standard, unless otherwise requested by customer. Certified to BCAR CAP 482 Section S in 1998. *As described.*
CUSTOMERS: Total of 16 Series D delivered by December 1998, including one fitted with hand controls for a disabled pilot.
COSTS: Assembled: £23,995, including VAT (1999). Kit: £14,995, including VAT (1999).
DESIGN FEATURES: Design stated to have no 'defined stall' and to be unspinnable. Pod-and-boom layout; slightly tapered wings; three vertical stabilisers; sweptback tailplane with large rudder beneath; downturned wingtips. Options include agricultural spraygear and other specialised equipment. Quoted kit assembly time is 500 hours.
FLYING CONTROLS: Conventional and manual. Three-position flaps: 0, 15 and 30°.
STRUCTURE: Wings of aluminium alloy and wood, with foam/glass fibre ribs, plywood covering on forward section and polyester fabric aft. Fuselage pod of Fibrelam; aluminium tube tailboom.
LANDING GEAR: Non-retractable tricycle type with brakes. Options include floats.
POWER PLANT: One 47.8 kW (64.1 hp) Rotax 582 UL driving a three-blade pusher propeller; four-blade propeller optional. Standard fuel capacity 35 litres (9.2 US gallons; 7.7 Imp gallons). Options include a 73 litre (19.2 US gallon; 16.0 Imp gallon) auxiliary fuel tank.
EQUIPMENT: ASI, VSI, altimeter, tachometer, fuel and temperature gauges, slip ball indicator, electric elevator trim and front cockpit brakes standard. Alloy wheels, wheel spats, rear cockpit brakes, seat cushions, electric starter, dual magneto switches, oil injection, intercom and GPS, optional. Modified cockpit for disabled pilots comprises combined throttle/rudder hand control on left

CFM Shadow DD tandem-seat ultralight *(Paul Jackson/Jane's)* 2001/0092073

CFM Streak two-seat ultralight, owned by D and J S Grint *(Paul Jackson/Jane's)* 2001/0099163

Star Streak SA-II cockpit *(Paul Jackson/Jane's)* 1999/0051537

console, sidestick controller for pitch/rudder on right console, each with hand-operated brake lever.

DIMENSIONS, EXTERNAL:
Wing span	10.03 m (32 ft 11 in)
Length overall	6.40 m (21 ft 0 in)
Height overall	1.75 m (5 ft 9 in)

AREAS:
Wings, gross	15.42 m² (166.0 sq ft)

WEIGHTS AND LOADINGS:
Weight empty	190 kg (419 lb)
Max T-O weight	386 kg (851 lb)

PERFORMANCE (speeds IAS):
Max level speed	88 kt (163 km/h; 101 mph)
Cruising speed at 75% power	76 kt (141 km/h; 87 mph)
Stalling speed, flaps down	30 kt (56 km/h; 35 mph)
Max rate of climb at S/L	253 m (820 ft)/min
Range with standard fuel	180 n miles (333 km; 207 miles)

UPDATED

CFM STREAK

TYPE: Tandem-seat ultralight/kitbuilt.

PROGRAMME: Construction of prototype started 1987 and this first flew June 1988; first flight of production aircraft (also known as Streak Shadow) October 1988. Also built under licence in South Africa, with Jabiru as engine option.

CURRENT VERSIONS: **Streak 582**: Rotax 582 UL engine, *as described*.

Streak 618: Rotax 618 UL engine of 55.0 kW (73.8 hp) with four-blade propeller.

Streak 912: Rotax 912 UL four-cylinder four-stroke engine of 59.6 kW (79.9 hp). Max level speed 112 kt (207 km/h; 129 mph).

CUSTOMERS: Total 95 built by end of 1998; at least 12 in South Africa. Recent customers include the Indian Air Force, which ordered 24 in September 1999 for delivery over a 15-month period.

COSTS: Assembled: £23,995, including VAT, 1999. Kit: £15,195 including VAT, 1999. Not available in complete form in the UK.

DESIGN FEATURES: Derivative version of Shadow (which see), designated an ultralight in Europe and Experimental homebuilt in USA; has new wing design, light airframe weight and more powerful engine; no 'defined stall' and is spin-resistant.

FLYING CONTROLS: See Structure. Electric trim.

STRUCTURE: Similar to Shadow but with foam/glass fibre wings (CFM aerofoil section), and control surfaces with aluminium alloy ribs and polyester fabric covering (ailerons, flaps, rudder and elevators).

POWER PLANT: One 47.8 kW (64.1 hp) Rotax 582. Standard fuel capacity 70 litres (18.5 US gallons; 15.4 Imp gallons).

DIMENSIONS, EXTERNAL:
Wing span	8.53 m (28 ft 0 in)
Length overall	6.40 m (21 ft 0 in)
Height overall	1.75 m (5 ft 9 in)

AREAS:
Wings, gross	13.01 m² (140.0 sq ft)

WEIGHTS AND LOADINGS:
Weight empty	205 kg (452 lb)
Max T-O weight	408 kg (900 lb)

PERFORMANCE (Streak 618; all speeds IAS):
Max level speed	98 kt (181 km/h; 113 mph)
Cruising speed at 75% power	85 kt (157 km/h; 98 mph)
Stalling speed	33 kt (62 km/h; 38 mph)
Max rate of climb at S/L	292 m (960 ft)/min
T-O to 15 m (50 ft)	122 m (400 ft)
Landing from 15 m (50 ft)	183 m (600 ft)
Range	425 n miles (787 km; 489 miles)

UPDATED

CFM SA-II STAR STREAK

TYPE: Tandem-seat ultralight/kitbuilt.

PROGRAMME: Development of Streak with reduced wing chord, wider cockpit and uprated engine. Prototype flew in 1992; first deliveries February 1994. Total of seven delivered by end of 1998.

COSTS: Kit: £16,195, inclusive of engine, propeller and VAT (1998). Available for export in complete form at £23,495 (1998).

POWER PLANT: One 55.0 kW (73.8 hp) Rotax 618 UL driving an Arplast four-blade composites pusher propeller. Standard fuel capacity 70 litres (18.5 US gallons; 15.4 Imp gallons).

DIMENSIONS, EXTERNAL:
Wing span	8.53 m (28 ft 0 in)
Length overall	6.25 m (20 ft 6 in)
Height overall	1.75 m (5 ft 9 in)

AREAS:
Wings, gross	11.15 m² (120.0 sq ft)

WEIGHTS AND LOADINGS:
Weight empty	205 kg (452 lb)
Max T-O weight	408 kg (900 lb)

PERFORMANCE (all speeds IAS):
Max level speed at S/L	102 kt (87 km/h; 54 mph)
Cruising speed at 75% power	87 kt (161 km/h; 100 mph)
Stalling speed, flaps down	38 kt (71 km/h; 44 mph)
Max rate of climb at S/L	320 m (1,050 ft)/min
T-O run to 15 m (50 ft)	122 m (400 ft)
Landing from 15 m (50 ft)	183 m (600 ft)
Range with max fuel	435 n miles (805 km; 500 miles)

UPDATED

CMC

CHICHESTER-MILES CONSULTANTS LTD

HEAD OFFICE: West House, Ayot St Lawrence, Welwyn, Hertfordshire AL6 9BT
Tel: (+44 1438) 82 03 41
Fax: (+44 1438) 82 00 30

RESEARCH AND DEVELOPMENT CENTRE: 4 The Woodford Centre, Lysander Way, Old Sarum, Salisbury, Wiltshire SP4 6BU
Tel: (+44 1722) 32 87 77
Fax: (+44 1722) 33 58 88
CHAIRMAN AND CEO: Ian Chichester-Miles
CHIEF ENGINEER: I A Townsend

Ian Chichester-Miles, formerly Chief Research Engineer of BAe Hatfield, established Chichester-Miles Consultants to develop the Leopard high-performance light business jet. Two prototypes have now flown.

UPDATED

CMC LEOPARD

TYPE: Light business jet.

PROGRAMME: Design began January 1981; mockup completed early 1982; detail design and construction of prototype by Designability Ltd of Dilton Marsh, Wiltshire, began July 1982 under CMC contract; Noel Penny Turbines NPT 301-3A turbojets, each of nominal 1.33 kN (300 lb st); first flight of unpressurised prototype (001/G-BKRL) 12 December 1988 at RAE Bedford; by December 1991 had made 50 flights investigating basic handling qualities at speeds up to 200 kt (371 km/h; 230 mph) IAS; new tailplane incorporating AS&T liquid anti-icing system on leading-edge subsequently installed on prototype before resumption of flight testing aimed at expanding airspeed, altitude and CG envelopes. Testing halted due to Noel Penny engine company going out of business; change of engine for second aircraft.

Design of second (preproduction) aircraft, 002/G-BRNM, began April 1989; displayed statically at Farnborough Air Show, September 1996; first flight 9 April 1997 followed by low-speed (up to 261 kt; 483 km/h; 300 mph) performance phase, which was completed in September 1997. US debut at EAA AirVenture, Oshkosh, July 1998. Flight envelope was due to be expanded up to 434 kt (805 km/h; 500 mph) and 10,670 m (35,000 ft) in 1999.

Second aircraft features strengthened structure, new landing gear, pressurised cabin, de-icing, reprofiled nose for EFIS avionics, substitution of oleo-pneumatic main landing gear legs for current rubber-in-compression units, and powered by Williams FJX-1 turbofans. Definitive FJX-2 engines scheduled to be installed in planned third (first production standard) aircraft in 2000. Type certification and sales launch planned for 2002-2003.

CURRENT VERSIONS: **Leopard**: *As described*.

Leopard Jet 3: Proposed basic production version; US$1 million (1999).

Leopard Jet 4: Proposed full-performance production version; US$1.35 million (1999).

Leopard Jet 6: Proposed production version with oval side windows and two-piece, wraparound windscreen.

Leopard T-Jet: Proposed military trainer. Side-by-side two-seat cockpit with windscreen and single-piece canopy; strengthened wing for increased g limits and provision for drop-tanks and other underwing stores.

DESIGN FEATURES: All-composites airframe; sweptback supercritical wings; sweptback tail unit; twin low-cost turbofans; pressurised cabin; AS&T liquid anti-icing and decontamination system on wing and tailplane leading-edges of production aircraft (see Programme); warm air de-icing of engine intake leading-edges. First prototype had lower-powered engines; lacked full pressurisation/air conditioning system, anti-icing, advanced avionics and

526   UK: AIRCRAFT—CMC to EUROPA

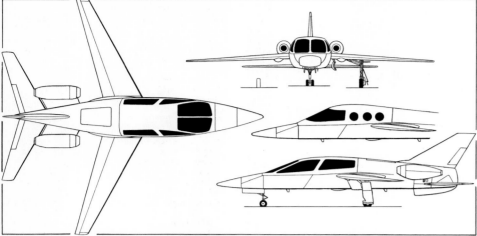

CMC Leopard preproduction version (two Williams FJX-1 turbofans), with scrap view of proposed Jet 6 version *(Dennis Punnett/Jane's)* 2000/0064911

instrumentation of second aircraft and planned production model.

ARA designed wing section and 3-D profiles combining laminar flow and supercritical technology; thickness/chord ratio 14 per cent at root, 11 per cent at tip; wing sweepback at quarter-chord 25°.

FLYING CONTROLS: Manual. All-moving fin; two independent tailplane sections operated collectively for pitch control and differentially for roll control; no ailerons. Full-span electrically actuated trailing-edge plain flaps, with ±45° deflections for high drag landing and airbraking/lift dumping; no spoilers.

STRUCTURE: Two-spar wings, primarily of carbon fibre; carbon fibre flaps; fuselage built in three sections as unpressurised nose, housing avionics and nosewheel gear; pressurised cabin (production aircraft); and unpressurised rear, housing baggage bay, with fuel tanks below and equipment bays to rear; fuselage primarily GFRP with some carbon fibre reinforcement (fore and aft bulkheads, engine and tailplane axle frames moulded in); pressure cabin section divided approximately along aircraft horizontal datum, with upper section formed by electrically actuated upward-opening canopy hinged at windscreen leading-edge; bonded-in acrylic side windows carry pressurisation tension; nose opens for access to avionics; light alloy engine nacelles, with stainless steel firewalls; composites fin and tailplane; fin sternpost projects to bottom of rear fuselage; low-set tailplane in two independent sections, each mounted on steel axle projecting from side of rear fuselage; carbon fibre tabs.

LANDING GEAR: Electrically retractable tricycle type, main units retracting inward into wingroot wells, nosewheels forward. Gravity extension assisted by bias springs and aerodynamic drag. Long-stroke shock-absorber in each unit on 001, using synthetic elastomers in compression. CMC-designed oleo-pneumatic main landing gear installed in 002 and for production aircraft. Main units, each with single Cleveland wheel, size 5.00-5, have tyres size 11×4, pressure 4.82 bars (70 lb/sq in) on prototype, 11.56 bars (170 lb/sq in) on production aircraft. Unpowered steerable twin-wheel nose unit has wheels size 4.00-3 and tyres size 8.5×2.75, pressure 2.75 bars (40 lb/sq in) on prototype, 3.8 bars (55 lb/sq in) on production aircraft. Hydraulic disc brakes. Parking brake.

POWER PLANT: Two Williams FJX-1 turbofans, each of 3.11 kN (700 lb st), in second aircraft. Production aircraft will have two Williams FJX-2 turbofans, although this power plant (known as EJ-22 in production form) exclusively assigned to Eclipse 500 (US section) for unspecified period. Each engine in nacelle, mounted on crossbeam located in rear fuselage. Fuel tanks in fuselage, below baggage bay. First prototype has total fuel capacity of 341 litres (90.1 US gallons; 75.0 Imp gallons). Production aircraft will have maximum capacity of 673 litres (178 US gallons; 148 Imp gallons). Refuelling point on upper surface of fuselage.

ACCOMMODATION: Cabin seats four, in two pairs, on semi-reclining (35°) seats beneath upward-opening canopy. Options include dual controls, and accommodation for pilot, stretcher and attendant in medevac role. Unpressurised baggage bay aft of cabin, capacity 63 kg (140 lb), with external door in upper surface of fuselage.

SYSTEMS: (production aircraft): Air conditioning and pressurisation (maximum differential 0.66 bar: 9.6 lb/sq in) by engine bleed air. Electrical system powered by dual engine-driven 3 kVA starter/generators. Hydraulic system for brakes only. Fluid anti-icing for wings, tailplane and windscreen, hot air anti-icing for engine intakes.

AVIONICS: (production aircraft): Honeywell avionics mounted in nose bay.
Radar: Weather radar.
Instrumentation: Two CRTs in pilot's instrument panel. Electromechanical standby flight instruments.

DIMENSIONS, EXTERNAL (A: first prototype; B: production aircraft):
Wing span: A	7.16 m (23 ft 6 in)
B	7.62 m (25 ft 0 in)
Wing chord: at root: A, B	1.14 m (3 ft 9 in)
at tip: A	0.36 m (1 ft 2 in)
B	0.30 m (1 ft 0 in)
Wing aspect ratio: A	8.8
B	9.7
Length overall: A	7.54 m (24 ft 9 in)
B	7.85 m (25 ft 9 in)
Height: overall: A	2.06 m (6 ft 9 in)
to canopy sill: A	0.76 m (2 ft 6 in)
Tailplane span: A	3.91 m (12 ft 10 in)
Wheel track: A	3.45 m (11 ft 4 in)
Wheelbase: A	3.20 m (10 ft 6 in)

DIMENSIONS, INTERNAL:
Cabin: Length: A	2.74 m (9 ft 0 in)
Max width: A	1.14 m (3 ft 9 in)
Max height: A	0.94 m (3 ft 1 in)
B	1.02 m (3 ft 4 in)
Baggage bay volume: A	0.40 m³ (14.0 cu ft)
B	0.48 m³ (17.0 cu ft)

AREAS:
Wings, gross: A	5.85 m² (62.9 sq ft)
B	5.97 m² (64.3 sq ft)
Trailing-edge flaps (total): A	1.24 m² (13.30 sq ft)
Fin: A	0.90 m² (9.72 sq ft)
B	1.06 m² (11.43 sq ft)
Tailplane (incl tabs)	2.14 m² (23.00 sq ft)

WEIGHTS AND LOADINGS (A: first prototype, B: production aircraft, estimated):
Weight empty, equipped: A	862 kg (1,900 lb)
B	998 kg (2,200 lb)
Max fuel weight: A	367 kg (810 lb)
B	544 kg (1,200 lb)
Max T-O weight: A	1,156 kg (2,550 lb)
B	1,814 kg (4,000 lb)
Max zero-fuel weight: A	1,043 kg (2,300 lb)
B	1,361 kg (3,000 lb)
Max landing weight: A	1,156 kg (2,550 lb)
B	1,701 kg (3,750 lb)
Max wing loading: A	197.9 kg/m² (40.54 lb/sq ft)
B	303.7 kg/m² (62.21 lb/sq ft)
Max power loading: A	434 kg/kN (4.25 lb/lb st)
B	240 kg/kN (2.35 lb/lb st)

PERFORMANCE (production aircraft, estimated, ISA):
Never-exceed speed (V_{NE})	M0.81 (300 kt; 556 km/h; 345 mph EAS)
Max level speed at 9,450 m (31,000 ft)	469 kt (869 km/h; 540 mph)
Max and econ cruising speed at 13,715 m (45,000 ft)	434 kt (804 km/h; 500 mph)
Stalling speed, full flap, at AUW of 1,497 kg (3,300 lb)	84 kt (156 km/h; 97 mph)
Max rate of climb at S/L	1,830 m (6,000 ft)/min
Rate of climb at S/L, OEI	631 m (2,070 ft)/min
Service ceiling	15,545 m (51,000 ft)
Service ceiling, OEI	9,150 m (30,000 ft)
T-O to 15 m (50 ft)	727 m (2,385 ft)
T-O balanced field length	838 m (2,750 ft)
Landing factored field length	854 m (2,800 ft)
Landing from 15 m (50 ft) at AUW of 1,497 kg (3,300 lb)	732 m (2,400 ft)
Range: max payload, with reserves	1,500 n miles (2,778 km; 1,726 miles)
max fuel and reduced payload	1,915 n miles (3,547 km; 2,204 miles)

UPDATED

Preproduction CMC Leopard business jet *(Paul Jackson/Jane's)* 2001/0092134

EUROPA

EUROPA AIRCRAFT COMPANY LTD
Unit 2A, Dove Way, Kirby Mills Industrial Estate, Kirkbymoorside, North Yorkshire YO62 6NR
Tel: (+44 1751) 43 17 73
Fax: (+44 1751) 43 17 06
e-mail: enquiries@europa-aircraft.com
Web: http://www.europa-aircraft.com
CHAIRMAN: Graham Walker
GENERAL MANAGER: Keith Wilson

US OFFICE:
Europa Aircraft Inc
3925 Aero Place, Lakeland, Florida 33811
Tel: (+1 863) 647 53 55
Fax: (+1 863) 646 28 77
e-mail: europa@gate.net
Web: http://www.europa-aircraft.com

Company founder Ivan Shaw previously built three Rutan canards before designing Europa for European environment, including operating from grass strips. Don Dykins, formerly British Aerospace Chief Aerodynamicist, defined aerodynamics; Barry Mellers (Chief Designer of Slingsby Aviation Ltd) made structural calculations.

UPDATED

EUROPA AVIATION EUROPA
TYPE: Side-by-side lightplane/motor glider kitbuilt.
PROGRAMME: Design started January 1990; prototype (G-YURO) made first flight 12 September 1992. PFA certification achieved May 1993; produced mainly in kit form under PFA auspices, but two assembled by Europa Aviation in 1994 and further three followed by 1996; first customer-built aircraft (G-OPJK) flown 14 October 1995.
CURRENT VERSIONS: **Europa:** Standard aircraft; has options of several types of engine, and monowheel, tailwheel or tricycle landing gear.
Europa XS: Improved version introduced June 1997. Features include higher aspect ratio premoulded wings; larger ailerons; enlarged cabin providing additional legroom and seat width; Rotax 912 ULS or turbocharged Rotax 914 engine in **Turbo XS**, with new cowlings giving improved cooling, lower drag, increased propeller clearance and better field of view from cockpit; non-steerable tailwheel permitting full rudder deflection during take-off and landing; and supplementary tank holding 35 litres (9.1 US gallons; 7.6 Imp gallons) of usable fuel. Maximum take-off weight increased by 32 kg (70 lb), making it possible to fit a child's seat in the baggage area; maximum speed in excess of 174 kt (322 km/h; 200 mph); maximum range more than 869 n miles (1,609 km; 1,000 miles). Prototype (G-EUXS) first flown 21 March 1997; public debut at EAA Convention at Oshkosh July 1997. Trial installation undertaken in 1999 on G-WWWG with 89 kW (120 hp) Wilksch WAM 120 three-cylinder Airmotive diesel.
Europa Motor Glider: Alternative long-span glider wings, interchangeable in quoted 5 minutes with standard wings, and featuring trailing-edge airbrakes, maximum extension 80°, inboard of the ailerons, but no flaps; public debut, installed on demonstrator G-ODTI, at PFA International Rally at Cranfield, July 1997; first flight in this configuration on 20 November 1998; production-standard wing, which made public debut at PFA

Europa Turbo XS monowheel (top) and XS, tri-gear *(Keith Wilson/Europa)*

International Rally, July 2000, has winglets and extended span of 14.40 m (47 ft 3 in).

CUSTOMERS: More than 700 kits sold by mid-2000 to customers in 32 countries, of which nearly 200 then flown.

COSTS: Airframe kits (XS monowheel) £16,380, (XS tri-gear) £16,820, (XS motor glider) £18,850; engine packages (Rotax 912) £9,160, (Rotax 912 ULS) £9,840, (Rotax 914) £11,920 (all 1999).

DESIGN FEATURES: Objectives included low-cost, economical cruise at IAS up to 120 kt (222 km/h; 138 mph) over 500 n miles (926 km; 575 miles), grass field capability, and ability to rig and de-rig quickly for storage in a trailer similar to a shorter version of glider trailer. Rig/de-rig by six pip-pins, two for tail, two for each wing. Quoted kit building time is 700 hours. Designed to JAR-VLA, stressed for maximum in-flight normal g of 4.3; proof factor of 2 used in design instead of the more usual 1.5, because of extensive use of composites materials in primary structure.

Low/mid-wing layout; moderately tapered wings and tailplane; large, sweptback fin and short fuselage.

Aerofoil is laminar flow Dykins design with 12 per cent thickness/chord ratio and 1° 30′ washout at tips (2° 30′ washout on XS). Modified Dykins section (15 per cent t/c ratio) on motor glider wings. Wing dihedral 3° (2° 30′ on motor glider); no tail dihedral.

Speed Kit introduced in 1999 features outrigger mechanism and wheel fairings and flap hinge fairings for monowheel versions; and wheel, landing gear leg and flap hinge fairings for tricycle version resulting in 9 to 10 kt (17 to 18 km/h; 10 to 11 mph) speed increase for Europa XS.

FLYING CONTROLS: Manual. Conventional ailerons and rudder; all-moving tailplane for pitch control with tab geared for balance, under pilot control for trim. Ground-adjustable tabs in ailerons and rudder. Two control columns, one centrally mounted at each seat; two pairs of rudder pedals. Central console between seats has throttle, combined flap and landing gear levers. Pitch trim switch next to throttle. Slotted flaps with settings of 0° and 25°, electrically actuated on tricycle landing gear version.

STRUCTURE: General construction of GFRP.

LANDING GEAR: Monowheel version has large, single, semi-retractable mainwheel and steerable tailwheel; outriggers at about half-span mounted on nylon stalks which retract with flaps. Mainwheel uses standard 6 in hub, as in many light aircraft; and an 8.00-6 Tundra tyre. Tricycle landing gear version has fixed, spring steel legs, castoring nosewheel, and toe brakes on port side rudder pedals; mainwheel tyre size 5.00-5, nosewheel 4.00-5. Tailwheel landing gear kit offered by Aero Developments of Kemble, UK.

POWER PLANT: One 59.6 kW (79.9 hp) Rotax 912 UL flat-four engine, directly driving a three-blade fixed-pitch Warp Drive propeller; propeller pitch is adjustable on ground to match operating environment (fine pitch for good take-off distances and rate of climb but higher noise and lower cruising speeds; reverse for coarse pitch); alternatively, one 73.5 kW (98.6 hp) Rotax 912 ULS or one 84.6 kW (113.4 hp) turbocharged Rotax 914, driving a three-blade variable-pitch NSI propeller. Ivoprop Magnum and Hoffmann constant-speed propellers were scheduled for trial installation during 1997. Alternative Subaru/NSI EA81 installation in 73 kW (98 hp) and 88 kW (118 hp) form first flown (G-NDOL) 18 November 1995. MWAE rotary (73 kW; 98 hp), BMW 1100RS (67.1 kW; 90 hp) and 89 kW (120 hp) Wilksch WAM 120 engine installations under development, the first-mentioned installed in UK prototype G-YURO and UK- and US-built tricycle gear demonstrators (G-KITS and N496TG); BMW engine began ground running trials in the third quarter of 1996. Jabiru 3300 offered from 1998.

Normal fuel capacity 68.2 litres (18.0 US gallons; 15.0 Imp gallons); optional 104.6 litres (27.6 US gallons; 23.0 Imp gallons).

ACCOMMODATION: Enclosed cabin seating two side by side under individual upward-opening canopies, hinged on centreline. Baggage compartment at rear of cabin.

SYSTEMS: Hydraulics: mainwheel brake. Electrical: 12 V 30 Ah battery; alternator fit appropriate to engine.

AVIONICS: Customer choice.

DIMENSIONS, EXTERNAL:
Wing span: standard	7.92 m (26 ft 0 in)
XS	8.28 m (27 ft 2 in)
motor glider	13.00 m (42 ft 8 in)
Width, wings removed	1.17 m (3 ft 10 in)
Wing aspect ratio: standard	7.1
XS	7.2
motor glider	13.5
Length overall	5.84 m (19 ft 2 in)
Height overall: monowheel	1.32 m (4 ft 4 in)
tricycle	2.13 m (7 ft 0 in)
Tailplane span	2.44 m (8 ft 0 in)
Wheel track (tricycle)	1.83 m (6 ft 0 in)
Wheelbase (tricycle)	1.47 m (4 ft 9¾ in)
Propeller diameter	1.57 m (5 ft 2 in)
Propeller ground clearance:	
standard	0.36 m (1 ft 2 in)
XS	0.39 m (1 ft 3½ in)

DIMENSIONS, INTERNAL:
Cabin: Length: standard	1.42 m (4 ft 8 in)
XS	1.83 m (6 ft 0 in)
Max width	1.12 m (3 ft 8 in)
Max height	0.97 m (3 ft 2 in)
Baggage volume: XS	0.64 m³ (22.5 cu ft)

AREAS:
Wings, gross: standard	8.83 m² (95.0 sq ft)
XS	9.48 m² (102.0 sq ft)
motor glider	12.54 m² (135.0 sq ft)
Ailerons (total): standard	0.57 m² (6.10 sq ft)
XS	0.62 m² (6.65 sq ft)
Trailing-edge flaps (total)	1.37 m² (14.70 sq ft)
Fin	1.03 m² (11.10 sq ft)
Rudder	0.48 m² (5.20 sq ft)

WEIGHTS AND LOADINGS (standard aircraft unless otherwise stated):
Weight empty, equipped:	
XS monowheel	340 kg (750 lb)
XS tricycle	354 kg (780 lb)
motor glider	358 kg (790 lb)
Baggage capacity	36 kg (80 lb)
Max T-O and landing weight: XS	621 kg (1,370 lb)
motor glider	635 kg (1,400 lb)
Payload with full fuel	197 kg (435 lb)
Max wing loading: XS	65.6 kg/m² (13.43 lb/sq ft)
motor glider	50.6 kg/m² (10.37 lb/sq ft)
Max power loading: XS	8.34 kg/kW (13.70 lb/hp)

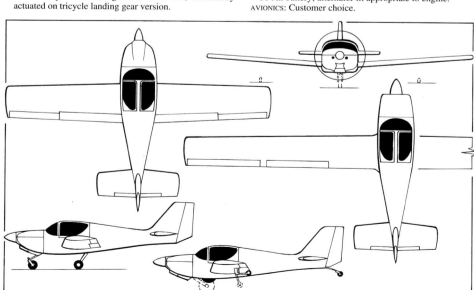

Standard (monowheel) version of Europa, with additional side elevation of tricycle option and partial plan view of Europa Motor Glider *(James Goulding/Jane's)*

Europa Motor Glider *(Keith Wilson/Europa)*

528　UK: AIRCRAFT—**EUROPA** to **INTORA**

PERFORMANCE, POWERED (XS and motor glider with Rotax 912 ULS engine, unless otherwise stated):
Max cruising speed at 75% power:
　Rotax 912 ULS at 2,440 m (8,000 ft):
　　monowheel　　　140 kt (259 km/h; 161 mph)
　　tricycle　　　　135 kt (250 km/h; 155 mph)
　　motor glider　　130 kt (241 km/h; 150 mph)
　Rotax 914 at 3,050 m (10,000 ft):
　　monowheel　　　174 kt (322 km/h; 200 mph)
　　tricycle　　　　166 kt (307 km/h; 191 mph)
Stalling speed, power off:
　flaps up　　　　　50 kt (92 km/h; 57 mph)
　flaps down　　　　45 kt (83 km/h; 51 mph)

Rate of climb at S/L:
　Rotax 912 ULS　　　305 m (1,000 ft)/min
　motor glider　　　　335 m (1,100 ft)/min
　Rotax 914　　　　　396 m (1,300 ft)/min
T-O run: Rotax 912 ULS　180 m (590 ft)
　motor glider　　　　183 m (600 ft)
　Rotax 914　　　　　153 m (500 ft)
Landing run: all　　　183 m (600 ft)
Range at econ cruising speed:
　standard fuel　　732 n miles (1,355 km; 842 miles)
　optional fuel　1,093 n miles (2,024 km; 1,257 miles)
g limits (all): design　　　　+3.8/−1.9
　ultimate　　　　　　　　　+8.55/−4.27

PERFORMANCE, UNPOWERED (motor glider):
Best glide ratio　　　　　　　　　　27
Min rate of sink at 47 kt (87 km/h; 52 mph)
　　　　　　　　　　　　0.90 m (2.94 ft)/s
UPDATED

EUROPA PRODUCTION LIGHT AIRCRAFT
Refer to Liberty entry in this section.
UPDATED

FARNBOROUGH
FARNBOROUGH-AIRCRAFT.COM
PO Box 19, Farnborough, Hampshire GU14 6WJ
Tel: (+44 1252) 51 81 28
Fax: (+44 1252) 51 81 24
e-mail: Information@Farnborough-aircraft.com
Web: http://www.Farnborough-Aircraft.com
CHIEF EXECUTIVE: Richard Noble
DESIGN TEAM LEADER: Ralph Hooper

Orders are being taken for the company's first product, the F1.
UPDATED

FARNBOROUGH F1
TYPE: Business turboprop.
PROGRAMME: Launched 14 October 1999, at which time wind tunnel testing had been completed; first flight scheduled for fourth quarter of 2002; FAR/JAR 23 certification in 2003; production in 2004.
CUSTOMERS: Aimed at air taxi operators providing on-demand, Internet-booked service from a network of small airfields. Launch orders for one each from separate British customers announced in September and October 2000.
COSTS: Development US$38.5 million; unit cost, equipped US$1.9 million; estimated operating cost: USA operation, US$408 per hour, European operation £371 per hour (all 2000).
DESIGN FEATURES: Designed, via Internet communications, using computational fluid dynamics; design goals include ability to operate quietly from small airstrips carrying five passengers at up to 330 kt (611 km/h; 380 mph) over a range of 1,000 n miles (1,852 km; 1,150 miles).
Wing aerofoil proprietary HLLF29140 high-lift, laminar flow section.
FLYING CONTROLS: Conventional and manual; 30 per cent chord Fowler flaps occupy approximately two-thirds span.
STRUCTURE: Primarily composites; fuselage of carbon fibre skins with foam honeycomb core, moulded in two vertically split halves; carbon fibre/epoxy wing with aluminium alloy landing gear pick-ups, wing/fuselage fittings, flap tracks and aileron attachments.
LANDING GEAR: Retractable tricycle type; trailing-link main units retract inwards, nosewheel rearwards.
POWER PLANT: One Pratt & Whitney PT6A-60A turboprop, flat rated at 634 kW (850 shp).
ACCOMMODATION: Pilot and five passengers in pressurised cabin, maximum pressure differential 0.55 bar (8.0 lb/sq in) maintaining sea level cabin altitude to 5,790 m (19,000 ft). Four cabin windows on each side; door on port side aft of wing; overwing emergency exit on starboard side. Interior layout, designed by Coventry University, provides alternative configurations for business/executive, commuter, medevac and cargo applications.
AVIONICS: 'Glass cockpit' compatible with existing and planned ATC and GPS-based en route navigation and airfield approach systems. Weather radar in wingtip pod.

DIMENSIONS, EXTERNAL:
Wing span　　　　　　12.41 m (40 ft 8½ in)
Wing aspect ratio　　　　　　　　　　8.4
Length overall　　　　　10.70 m (35 ft 1¼ in)
Height overall　　　　　　3.47 m (11 ft 4¾ in)

Artist's impression of Farnborough F1 turboprop business aircraft *(Michael Turner/Farnborough)*
2000/0085738

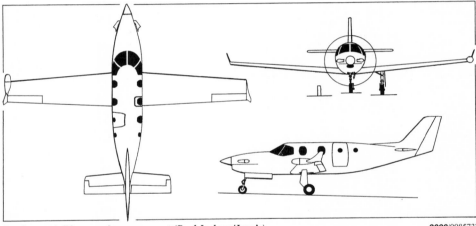

Farnborough F1 general arrangement *(Paul Jackson/Jane's)*
2000/0085737

DIMENSIONS, INTERNAL:
Cabin:
　Length　　　　　　　3.44 m (11 ft 3½ in)
　Max width　　　　　　1.40 m (4 ft 7¼ in)
　Max height　　　　　　1.34 m (4 ft 4¾ in)
AREAS:
Wings, gross　　　　　18.26 m² (196.6 sq ft)
WEIGHTS AND LOADINGS:
Operating weight empty　　1,805 kg (3,980 lb)
Max fuel weight　　　　　　763 kg (1,682 lb)
Max payload　　　　　　　567 kg (1,250 lb)
Payload with max fuel　　　107 kg (236 lb)
Max T-O weight　　　　　2,675 kg (5,898 lb)
Max landing weight　　　　2,541 kg (5,603 lb)
Max ramp weight　　　　　2,688 kg (5,927 lb)
Max zero-fuel weight　　　2,372 kg (5,230 lb)
Max wing loading　　　146.5 kg/m² (30.00 lb/sq ft)
Max power loading　　　4.22 kg/kW (6.94 lb/shp)

PERFORMANCE (estimated):
Max operating speed　　295 kt (546 km/h; 339 mph)
High cruising speed at 9,145 m (30,000 ft)
　　　　　　　　　　329 kt (609 km/h; 379 mph)
Long-range cruising speed at 9,145 m (30,000 ft)
　　　　　　　　　　242 kt (448 km/h; 278 mph)
Stalling speed, flaps and landing gear down
　　　　　　　　　　59 kt (110 km/h; 68 mph)
Time to 7,620 m (25,000 ft)　　　　　8 min
Service ceiling　　　　　　9,145 m (30,000 ft)
T-O run　　　　　　　　　466 m (1,530 ft)
Range: with max fuel
　　　　　　　1,645 n miles (3,046 km; 1,893 miles)
　with 363 kg (800 lb) payload, NBAA IFR reserves, 100 n mile (185 km; 115 mile) alternate
　　　　　　　　975 n miles (1,805 km; 1,122 miles)
Ferry range　1,675 n miles (3,102 km; 1,927 miles)
UPDATED

FLS
FLS AEROSPACE LTD
The Optica and Sprint light aircraft remain out of production, as last detailed in the 2000-01 *Jane's*.
UPDATED

INTORA
INTORA-FIREBIRD plc
PO Box 8000, Aviation Way, London-Southend Airport, Essex SS2 6DE
Tel: (+44 1702) 56 18 18
Fax: (+44 1702) 56 18 19
e-mail: bluesky@spectraweb.ch
DIRECTOR: Brian Nalborough

There has been no reported production of the Firebird ultralight helicopter, last described in the 2000-01 edition.
UPDATED

LIBERTY

LIBERTY AIRCRAFT plc

Kings Avenue, Hamble-le-Rice, Southampton, Hampshire SO31 4NF
Tel: (+44 2380) 74 49 27
Fax: (+44 2380) 43 17 06
Web: http://www.libertyaircraft.com
PRESIDENT AND CEO: Anthony Tiarks
VICE-PRESIDENT: Paul Bartlett
CHIEF DESIGNER: Ivan Shaw
MARKETING DIRECTOR: Keith Wilson

US SALES OFFICE:
Liberty Aerospace Inc
3925 Aero Place, Lakeland, Florida 33811
Tel: (+1 863) 709 05 55
Fax: (+1 863) 709 05 56
Web: http://www/libertyaerospace.com

MANUFACTURING FACILITY:
Scaled Technology Works, One Creative Place, Montrose, Colorado 81401
Tel: (+1 970) 252 30 27
Fax: (+1 863) 249 14 10

NEW ENTRY

Mockup of Liberty XL-2 two-seat light trainer 2001/0103517

LIBERTY XL-2

TYPE: Two-seat lightplane.
PROGRAMME: Design, by the team that created the Europa kitbuilt (which see in this section), began in 1997. Announced 26 May 2000; mockup displayed at NBAA Convention in New Orleans and AOPA-USA Convention in Long Beach during October 2000, when Scaled Technology Works announced as manufacturing partner; FAR Pt 23 certification anticipated in fourth quarter 2001, followed shortly thereafter by JAR-VLA certification and first customer deliveries. Target production rate up to 400 aircraft per year.
CUSTOMERS: Launch customer Civil Flying School at Moorabbin Airport, Melbourne, Australia, ordered two in July 2000; Bill Crokaris of Melbourne has ordered one. First 50 aircraft (for Founders Club members, at special price of US$85,000) all sold by October 2000.
COSTS: US$97,500 (2000).
DESIGN FEATURES: Based on Europa, but with modifications to optimise airframe for mass production; cabin is 10 cm (4 in) wider than that of Europa. Design goals included low price, efficient high-speed cruise with STOL performance, economy and ease of maintenance, advanced structural materials offering strength and durability, excellent handling with positive stability, and ability to store at home on a purpose-designed transporter.
FLYING CONTROLS: Conventional and manual, via pushrods. Dual controls and adjustable rudder pedals standard. All-moving tailplane with electrically actuated 1.3:1 geared anti-balance/trim tab; ailerons with differential action, maximum deflection +24/–20°; slotted flaps occupying 70 per cent span and 30 per cent chord, maximum deflection 27°; rudder maximum deflection ±30°.
STRUCTURE: Welded steel tube forward fuselage and centre-section to carry engine, landing gear and wing attachment loads, with modular carbon fibre skin; flying surfaces comprise riveted subassembly with bonded aluminium skin; single-spar wings have optional folding facility, quick connect attachment to centre-section and self-connecting flap and aileron controls; de-riggable tailplane optional. Scaled Technology Works will undertake manufacturing, final assembly and flight testing at Montrose, Colorado.
LANDING GEAR: Non-retractable tricycle type; aluminium main legs; tubular steel nose leg; steering via castoring nosewheel and differential brakes.
POWER PLANT: One 73.5 kW (98.6 hp) Rotax 912 ULS flat-four, driving a purpose-designed two-blade, fixed-pitch carbon fibre Dowty propeller. Fuel capacity 87 litres (23.0 US gallons; 19.15 Imp gallons).
ACCOMMODATION: Two persons, side by side. Upward-hinged door on port side.

DIMENSIONS, EXTERNAL:
Wing span 8.23 m (27 ft 0 in)
Wing aspect ratio 6.5
Length overall 6.10 m (20 ft 0 in)
Height overall 2.24 m (7 ft 4 in)
Propeller diameter 1.60 m (5 ft 3 in)
DIMENSIONS, INTERNAL:
Cabin max width 1.22 m (4 ft 0 in)
Baggage volume 0.68 m³ (24.0 cu ft)
AREAS:
Wings, gross 10.41 m² (112.0 sq ft)
WEIGHTS AND LOADINGS:
Weight empty 363 kg (800 lb)
Baggage capacity 45 kg (100 lb)
Max T-O weight 635 kg (1,400 lb)
Max wing loading 61.0 kg/m² (12.50 lb/sq ft)
Max power loading 8.64 kg/kW (14.20 lb/hp)
PERFORMANCE (estimated):
Cruising speed at 75% power at 2,440 m (8,000 ft)
 120 kt (222 km/h; 138 mph)
T-O and landing speed
 45 to 50 kt (84 to 93 km/h; 52 to 58 mph)
Max rate of climb at S/L 229 m (750 ft)/min
T-O and landing run 229 m (750 ft)
Range with max fuel 500 n miles (926 km; 575 miles)

NEW ENTRY

RAF

ROYAL AIR FORCE

Main Building, Ministry of Defence, Whitehall, London SW1A 2HB
Tel: (+44 020) 72 18 90 00

The RAF selected Boeing's C-17A Globemaster III as a short-term strategic airlifter in 2000. Other programmes in the formative stage are detailed below.

UPDATED

FUTURE OFFENSIVE AIR SYSTEM (FOAS)

TYPE: Attack fighter.
PROGRAMME: Staff Target (Air) 425 formulated as replacement for the RAF's Panavia Tornado GR. Mk 4; prefeasibility studies, 1993-95, under title of Future Offensive Aircraft (FOA). Programme launched on 16 December 1996 as FOAS, when £35 million assigned for allocation in 1997 to feasibility studies of broadest possible range of options, including variants of Eurofighter Typhoon, Lockheed Martin F-22 and JSF; new manned combat aircraft; uninhabited air vehicles, both combatant and non-combatant; and standoff missiles launched from transport aircraft. Development will be in close parallel with the Royal Navy's FCBA (which see).
Technologies being considered include fly-by-light, stealth, virtual reality cockpits and integrated modular avionics. Study contracts awarded to several contractors, including the Defence Evaluation and Research Agency; data sources include an Anglo-French technology demonstration programme permitting computer modelling of weapons systems. Studies into a crewed FOAS, with particular reference to LO technologies, have been undertaken by BAE at Warton (which see). BAE and partners received two-thirds of study funds, other contracts going to Logica and GKN Aerosystems.
Some 100 companies were working on FOAS feasibility by mid-1999. A third-phase study was launched that year; concept phase was then due to end in March 2001 with 'initial gate' decision to proceed to assessment phase. Force mix between different systems (if more than one solution chosen, as appeared likely in 1999) will be decided in time for a firm requirement to be issued in 2008. Assessment phase and risk-reduction studies are due to begin in 2001. By late 1999, French companies were working on joint studies with UK, while Germany was showing interest in joining.
IOC, originally due in 2015, and since moved to October 2017, may be rescheduled to 2020 following 1997 decision not to include Harrier GR. Mk 7 replacement in FOAS. If only a crewed aircraft is selected, the RAF requirement would be for approximately 200.

UPDATED

FUTURE STRATEGIC TANKER AIRCRAFT (FSTA)

TYPE: Tanker-transport.
PROGRAMME: Planned replacement for 35 VC10 and Tristar tankers under Staff Requirement (Air) 447 by privately funded fleet operated under contract. Two consortia offering Boeing 767 conversion; one the Airbus A330. Invitations to negotiate (ITN) issued 21 December 2000; responses were due by 3 July 2001; contract award 2003; service entry 2007 to 2009.

UPDATED

SUPPORT, AMPHIBIOUS AND BATTLEFIELD ROTORCRAFT (SABR)

TYPE: Medium transport helicopter.
PROGRAMME: On 27 September 1999, the Royal Navy's FASH requirement for replacement of 37 Sea King (Commando) support helicopters was merged with the RAF's Future Support Rotorcraft to produce SABR. In concept phase during 2000; no Staff Target then assigned; 'initial gate' due June 2001 for service entry in 2009.
The RAF requirement covers 40 existing Pumas and 25 SAR-configured Sea Kings, although it is possible that the latter need will be satisfied by a different type of rotorcraft from the main purchase, probably owned and maintained by a civilian contractor. Contenders for SABR could include the Bell Boeing V-22 Osprey tiltrotor (promoted by BAE Systems), Boeing CH-47 Chinook, EHI EH 101, Eurocopter Cougar, NHI NH 90, Sikorsky S-92 Helibus and Sikorsky CH-53E Super Stallion. Funding allocation is £5 billion over 30-year life cycle.

UPDATED

RN

ROYAL NAVY

Main Building, Ministry of Defence, Whitehall, London SW1A 2HB
Tel: (+44 020) 72 18 90 00

Fleet Air Arm requirements include two types of equipment for a new class of 40,000 tonne aircraft carrier. Transport helicopter support will be provided by the SABR programme, undertaken jointly with the RAF (which see).

UPDATED

FUTURE CARRIER-BORNE AIRCRAFT (FCBA)

TYPE: Multirole fighter.
PROGRAMME: User Requirement Document 6464 for Future Carrier-Borne Aircraft (FCBA) replacement for Sea Harrier FA. Mk 2 presumes purchase of some 60 Joint Strike Fighters (JSF) of type eventually selected by US armed forces (see US Navy entry and related entries under Boeing and Lockheed Martin). RN will require STOVL variant to be procured for US Marine Corps. First deliveries in 2010; in-service date is 2012, to coincide with first of two new carriers.
US-UK MoU of December 1995 covers UK participation in JSF; BAe currently teamed with Lockheed

Martin. UK participation in JSF EMD phase confirmed 18 January 2001 with £1,900 million commitment securing 8 per cent share.

In 1997, RAF showed interest in replacing its Harrier GR. Mk 7s by FCBA; URD 6464 now officially includes RAF Harriers, for replacement from 2015.

Throughout 1999, the MoD continued the assessment of potential alternatives to JSF; these officially included a further Harrier development, navalised Eurofighter Typhoon, Rafale M and F/A-18E Super Hornet. Programme cost estimated as £7 billion (2000) for 150 aircraft.

UPDATED

FUTURE ORGANIC AIRBORNE EARLY WARNING (FOAEW)

TYPE: AEW helicopter.
PROGRAMME: Studies into Future Organic Airborne Early Warning aircraft to replace the Sea King AEW. Mks 2/7 were awarded in February 1998 to Racal Radar Defence Systems and GKN Westland Helicopters; submission by 31 August 1998; funding provided by FCBA programme (which see in adjacent entry). Requirements include 6,100 m (20,000 ft) operational altitude, favouring compound helicopter or tilt-rotor. By mid-1999, requirement had expanded to include airborne command, control and communications. In concept phase during 2000; deadline for expressions of interest having been 7 April 2000; Staff Target (Sea) 6849; Staff Requirement due to be issued in 2003, with prime contract award following in late 2006. In service date 2012.

UPDATED

A compound lift EH 101 Merlin is one potential answer to FOAEW *1999/0014652*

SLINGSBY

SLINGSBY AVIATION LIMITED

Ings Lane, Kirkbymoorside, York YO62 6EZ
Tel: (+44 1751) 43 24 74
Fax: (+44 1751) 43 11 73
e-mail: SAL1@slingsby.co.uk
Web: http://www.slingsby.co.uk
Telex: 57911 SELG
MANAGING DIRECTOR: Jeff Bevan
CHIEF DESIGNER: David Goddard
CONTRACTS DIRECTOR: Simon Cooper
CUSTOMER SERVICE MANAGER: Malcolm Drinkell

Formerly a subsidiary of ML Holdings, Slingsby was sold to Cobham plc in 1997. Company specialises in application of composites materials; was previously manufacturer of sailplanes but now concentrates on development and low-rate production of T67 Firefly series of aerobatic training aircraft.

Other activities include design and manufacture of air cushion vehicles in composites materials; and design, development and manufacture of high-performance composites structures for marine and aerospace industries. Supplier to Lockheed Martin of C-130J cockpit interior trim panels in sculptured composites.

Work for UK MoD includes technical support for RAF Air Cadet gliders and Firefly aircraft.

UPDATED

SLINGSBY T67 FIREFLY
US Air Force designation: T-3A Firefly

TYPE: Primary prop trainer/sportplane.
PROGRAMME: Current composites constructed Firefly developed from wooden Slingsby T67A (licence-built version of French Fournier RF6B — see 1982-83 *Jane's*); T67B gained CAA certification 18 September 1984; T67C was CAA certified 15 December 1987. Subsequent versions were T67M, T67M200 and T67M260.

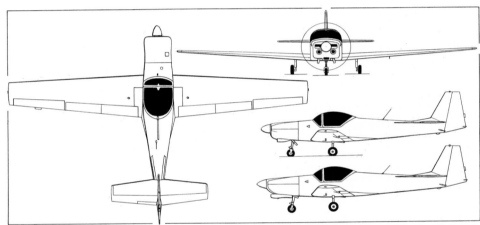

Slingsby T67M Mk II Firefly (Textron Lycoming AEIO-320-D1B engine) with additional side view (bottom) of T67M260 *(Dennis Punnett/Jane's)*

Last-mentioned selected by USAF to meet Enhanced Flight Screener (EFS) requirement. Slingsby prime contractor for both acquisition contract and seven-year contractor logistic support (CLS) contract. Northrop Grumman subcontractor for final assembly at Hondo, Texas, and for operation of CLS activities at Hondo and USAF Academy at Colorado Springs.

CURRENT VERSIONS: **T67C:** Basic version. See 1997-98 and previous editions for earlier C1, C2 and C3 subvariants. One 119 kW (160 hp) Textron Lycoming O-320-D2A flat-four engine, driving a Sensenich M74DM6-O-64 two-blade fixed-pitch metal propeller. Fuel tank in each wing leading-edge, combined capacity 159 litres (42 US gallons; 35 Imp gallons). Nine T67Cs purchased by Netherlands government Civil Aviation Flying School for KLM and Royal Netherlands Navy pilot training; 12 T67Cs for Canadian Department of National Defence for military primary flying training; plus other T67C variants for UK schools.

T67M: Military variant. First flight of T67M Firefly 160 (G-BKAM) 5 December 1982; CAA certification 20 September 1983; designation changed to T67M Mk II as wing fuel tanks and two-piece canopy introduced. Powered by 119 kW (160 hp) Textron Lycoming AEIO-320-D1B flat-four engine, driving a Hoffmann HO-V72 two-blade constant-speed composites propeller.

Sold to Netherlands for military grading; Japan and UK for airline training; and Switzerland for aerobatic and

Slingsby T67M260 Firefly used for training UK military pilots *(Paul Jackson/Jane's)* *2000/0093704*

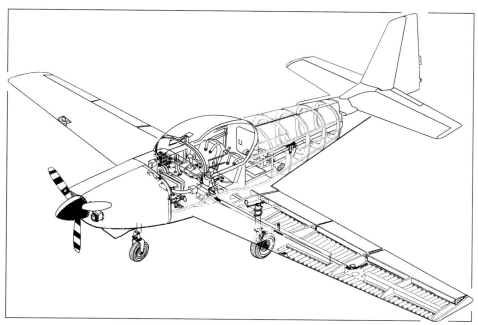
Structural details of the Slingsby T67M260 Firefly

general flying training. Used by RAF, RN and British Army for elementary flying training; Joint Elementary FTS formed at Topcliffe, July 1993, with first of eventual 18 (including some second-hand) T67M Mk IIs operated under contract by Hunting Aircraft Ltd (now Hunting Contract Services); transferred to Barkston Heath in April 1995.

T67M200: Development of T67M; 149 kW (200 hp) Textron Lycoming AEIO-360-A1E engine; Hoffmann HO-V123 three-blade variable-pitch composites propeller; wing fuel capacity as for T67C. First flight 16 May 1985; CAA certification 13 October 1985. First customer Turkish Aviation Institute, Ankara (16 delivered from 1985); others ordered by Dutch operator King Air (three T67M200s, plus one T67M Mk II) as screening trainers for prospective RNethAF pilots; Royal Hong Kong Auxiliary Air Force (now Government Flying Service) (four); and Norwegian government's Flying Academy (six).

T67M260: Development of T67M; higher-powered Lycoming engine; electric elevator trim and electric flaps optional; cabin air conditioning system optional; higher maximum T-O and aerobatic weights, to allow 227 kg (500 lb) for two pilots and equipment plus full fuel load.

Prototype (G-BLUX) flown May 1991, evaluated at Wright-Patterson AFB; hot and high trials at USAF Academy in mid-1991. Preproduction aircraft (G-EFSM) flown September 1992. First flight of first USAF aircraft (92-0625/N7020D) 4 July 1993. T-3A type certificate awarded by CAA and FAA December 1993. Official USAF acceptance 25 February 1994; student pilot training began March 1994. Also acquired for UK military pilot training.

CUSTOMERS: Total of 267 civil/military T67s of all subtypes (including 10 T67As) delivered to customers in 13 countries by late 2000; main production ended March 1997 with 262nd; one built in 1998, two in 1999, and two registered in UK in 2000.

US Air Force acquired 113 T-3As in three lots (38, 42 and 33) to replace Cessna T-41. USAF pilot conversion completed September 1993. Deliveries began January 1994 and were completed in November 1995. First operating unit (with 56 aircraft) was 1st Flight Screening Squadron of 12th Flying Training Wing based at Hondo, Texas; second and final operating unit (with 57 aircraft) is 557th FTS of USAF Academy at Colorado Springs; deliveries to this unit from August 1994; operational January 1995. All T-3As wore dual military/civilian identities. Remaining fleet of 110 grounded in July 1997 for fuel system modifications, and remained thus in late 2000, despite re-assessment proving aircraft safe. At that time, USAF announced that it was contracting out future flight screening to civilian operators and was considering options for the T-3A fleet, including disposal.

Hunting Contract Services (see T67M, above) took delivery of 25 T67M260s between June 1996 and March 1997 for extension to UK tri-service training programme, followed by a further two in September 1999. One each delivered to Belize Defence Force and a UK civilian customer in 1996.

DESIGN FEATURES: Conventional low-wing monoplane; design translated from wood to GFRP. Tapered wings and mid-mounted tailplane; sweptback fin.

Wing section NACA 23015 at root, 23013 at tip; dihedral 3° 30′; incidence 3°.

FLYING CONTROLS: Conventional and manual. Mass-balanced Frise ailerons, without tabs; mass-balanced elevators with manually operated port trim tab (electric trim optional); trailing-edge fixed hinge flaps; spin strakes forward of tailplane roots.

STRUCTURE: GFRP; single-spar wings with double skin (corrugated inner skin bonded to plain outer skin) and ribs in heavy load positions; frame and top-hat stringer fuselage; stainless steel firewall between cockpit and engine; fixed incidence tailplane of similar construction to wings (built-in VOR antenna); fin incorporates VHF antenna.

LANDING GEAR: Non-retractable tricycle type. Oleo-pneumatic shock-absorber in each unit. Steerable nosewheel. Mainwheel tyres size 6.00-6, pressure 1.38 bars (20 lb/sq in). Nosewheel tyre size 5.00-5, pressure 2.55 bars (37 lb/sq in). Hydraulic disc brakes. Parking brake.

Data follow for T67M260/T-3A; refer to 1999-2000 and earlier Jane's for other variants.

POWER PLANT: One 194 kW (260 hp) Textron Lycoming AEIO-540-D4A5 flat-six engine, driving a Hoffmann HO-V123K-KV/180DT three-blade constant-speed composites propeller. Fuel 159 litres (42.0 US gallons; 35.0 Imp gallons) in wing leading-edges; refuelling point in upper wing surface.

ACCOMMODATION: Two seats side by side, fixed windscreen and rearward-hinged upward-opening rear-canopy section. Optional raised canopy and towered seats to accommodate crew wearing military-style helmets. Dual controls standard. Adjustable rudder pedals. Cockpit heated and ventilated. Baggage space aft of seats.

SYSTEMS: Hydraulic system for brakes only. Vacuum system for blind-flying instrumentation. Electrical power supplied by 28 V 70 A engine-driven alternator and 24 V 15 Ah battery.

AVIONICS: Optional avionics, available to customer requirements, include equipment by Honeywell, up to full IFR standard.

Instrumentation: Standard avionics include artificial horizon and directional gyro, with vacuum system and vacuum gauge, electric turn and slip indicator, rate of climb indicator, recording tachometer, stall warning system, clock, outside air temperature gauge, accelerometer.

EQUIPMENT: Includes tiedown rings and towbar; cabin fire extinguisher, crash axe, heated pitot; instrument, landing, navigation and strobe lights. Optional equipment includes external power socket, and wingtip-mounted smoke system.

DIMENSIONS, EXTERNAL:
Wing span	10.59 m (34 ft 9 in)
Wing chord: at root	1.53 m (5 ft 0¼ in)
at tip	0.83 m (2 ft 8¾ in)
Wing aspect ratio	8.9
Length overall	7.57 m (24 ft 10 in)
Height overall	2.36 m (7 ft 9 in)
Tailplane span	3.40 m (11 ft 1¾ in)
Wheel track	2.44 m (8 ft 0 in)
Wheelbase	1.50 m (4 ft 11 in)
Propeller diameter	1.80 m (5ft 10¾ in)

DIMENSIONS, INTERNAL:
Cockpit: Length	2.05 m (6 ft 8¾ in)
Max width	1.08 m (3 ft 6½ in)
Max height	1.08 m (3 ft 6½ in)

AREAS:
Wings, gross	12.63 m² (136.0 sq ft)
Ailerons (total)	1.24 m² (13.35 sq ft)
Trailing-edge flaps (total)	1.74 m² (18.73 sq ft)
Fin	0.80 m² (8.61 sq ft)
Rudder	0.82 m² (8.80 sq ft)
Tailplane	1.65 m² (17.76 sq ft)
Elevators (incl tab)	0.99 m² (10.66 sq ft)

WEIGHTS AND LOADINGS (with air conditioning installed):
Weight empty	794 kg (1,750 lb)
Max fuel weight	114 kg (252 lb)
Max T-O and landing weight:	
Utility/Aerobatic	1,157 kg (2,550 lb)
Max wing loading	91.5 kg/m² (18.75 lb/sq ft)
Max power loading	5.97 kg/kW (9.81 lb/hp)

PERFORMANCE:
Never-exceed speed (V_NE)	195 kt (361 km/h; 224 mph)
Max level speed at S/L	152 kt (281 km/h; 175 mph)
Max cruising speed at 75% power at 2,590 m (8,500 ft)	140 kt (259 km/h; 161 mph)
Stalling speed, power off, flaps down	54 kt (100 km/h; 63 mph)
Max rate of climb at S/L	480 m (1,380 ft)/min
T-O run	334 m (1,095 ft)
T-O to 15 m (50 ft)	689 m (2,265 ft)
Landing from 15 m (50 ft)	984 m (3,228 ft)
Landing run	401 m (1,315 ft)
Range with max fuel at 65% power at 2,440 m (8,000 ft), allowances for T-O, climb and 30 min reserves:	407 n miles (753 km; 468 miles)
Endurance at best econ setting at 2,440 m (8,000 ft), allowances as above	5 h 40 min
g limits	+6/−3

UPDATED

SPEEDTWIN

SPEEDTWIN DEVELOPMENTS LTD

DEVELOPMENT FACILITY:
Upper Cae Garw Farm, Trelleck, Monmouth, Gwent NP5 4PJ
Tel: (+44 1600) 86 90 46
Fax: (+44 1600) 86 91 27
MANAGING DIRECTOR: Malcolm Ducker

SALES OFFICE:
77 Culverden Down, Tunbridge Wells, Kent TN4 9SL
Tel: (+44 1892) 53 73 40
Fax: (+44 1892) 57 07 29
e-mail: speedtwin@ontel.net.uk
Web: http://www.speedtwin.co.uk

This company was formed to market the Speedtwin twin-engined aerobatic light sporting aircraft designed by the late Peter Phillips, a former demonstration pilot with Britten-Norman and test pilot with the Norman Aeroplane Company. Speedtwin was bought by the present managing director in March 2000. A design contract was signed in May 2000 with RCS Aviation to complete a full stress analysis and to obtain design approval for the Mk II.

UPDATED

SPEEDTWIN E2E SPEEDTWIN

TYPE: Two-seat kitbuilt twin.

PROGRAMME: Design started 1981; E2E designation stands for 'engineered to excel'; prototype (G-GPST) first flew 30 September 1991; construction requirements arranged to meet the '51 per cent rule' for homebuilt aircraft. Speedtwin Developments Ltd will be seeking a partner for series production, while retaining design and development authority.

CURRENT VERSIONS: **Mark I:** Prototype; Continental engines.
Mark II: Second prototype; LOM engines; oleo-pneumatic main landing gear; nearing completion in 2001.
Mark III: Proposed version with retractable landing gear and aerodiesel engines operating on Avtur.

CUSTOMERS: Deposits for kits will be accepted following receipt of UK CAA approval for Mk II.

COSTS: Quick-build kit approx US$65,000, including substantially complete wing, fuselage, canopy, empennage and landing gear.

DESIGN FEATURES: Designed to meet FAR/JAR 23 standards for twin-engine flying and aerobatics. Low-wing, twin-engined monoplane with (initially) fixed landing gear. Patented variable servo system for ailerons; moderately tapered wing; V_MCA of Mk I stated to be below the stall.

Wing section NACA 23012; washout to NACA 2412 on outer panels. Tailplane NACA 0011.

FLYING CONTROLS: Conventional and manual, assisted by patented RAP system of two servo spades beneath cockpit, adjustable to vary required aileron input. Horn-balanced rudder and elevators. Elevator tab for pitch trim. Ground-adjustable tabs on ailerons and rudder. Electrically powered plain flaps; deflections 0 to 40°.

STRUCTURE: Mk II generally of aluminium monocoque, flush riveted and butt jointed.

LANDING GEAR: Tailwheel type; fixed. Mk I has DHC-1 Chipmunk main gear; Mk II has oleo-pneumatic. Speed fairings on mainwheels.
POWER PLANT: *Mk I:* Two 74.6 kW (100 hp) Teledyne Continental O-200, each driving a two-blade, fixed-pitch wooden propeller.
Mk II: Two 119 kW (160 hp) LOM M332 four-cylinder, supercharged, in-line piston engines driving P-tip propellers. Alternative engines in range 75 to 82 kW (100 to 180 hp).
Fuel in two wing tanks, total 303 litres (80.0 US gallons; 66.6 Imp gallons). Fuel filler ports in upper surface of each wing.
ACCOMMODATION: Two in tandem, beneath single-piece canopy. Baggage area behind rear seat. Dual controls.
SYSTEMS: Electrical system has two alternators and two 12 V batteries, providing full redundancy.
AVIONICS: Customer specified.
DIMENSIONS, EXTERNAL:
Wing span 7.92 m (26 ft 0 in)
Wing aspect ratio 5.6
Length overall 6.96 m (22 ft 10 in)
Width, wing outer panels removed for stowage
 4.88 m (16 ft 0 in)
Height overall 2.08 m (6 ft 10 in)
DIMENSIONS, INTERNAL:
Cabin min width 0.63 m (2 ft 1 in)
AREAS:
Wings, gross 11.38 m² (122.5 sq ft)
WEIGHTS AND LOADINGS:
Weight empty: Mk I 640 kg (1,410 lb)
 Mk II 680 kg (1,500 lb)
Baggage capacity 45 kg (100 lb)
Max weight for aerobatics 907 kg (2,000 lb)
Max T-O weight 1,111 kg (2,450 lb)
Max wing loading at aerobatic weight
 79.7 kg/m² (16.33 lb/sq ft)
Max power loading at aerobatic weight:
 Mk I 6.09 kg/kW (10.00 lb/hp)
 Mk II 3.80 kg/kW (6.25 lb/hp)
PERFORMANCE (Mk II estimated):
Never-exceed speed (V$_{NE}$) 214 kt (397 km/h; 247 mph)
Max level speed at S/L:
 Mk I 150 kt (278 km/h; 173 mph)
 Mk II 187 kt (346 km/h; 215 mph)

Econ cruising speed at 75% power:
 Mk I 139 kt (257 km/h; 160 mph)
 Mk II 174 kt (322 km/h; 200 mph)
Stalling speed, power off:
 flaps up 59 kt (110 km/h; 68 mph)
 flaps down 53 kt (97 km/h; 60 mph)
Landing crosswind limit 25 kt (46 km/h; 28 mph)
Max rate of climb at S/L, aerobatic weight:
 Mk I 365 m (1,200 ft)/min
 Mk II 929 m (3,050 ft)/min
Service ceiling, OEI, aerobatic weight:
 Mk I 915 m (3,000 ft)
 Mk II 3,505 m (11,500 ft)
T-O run, aerobatic weight: Mk I 200 m (660 ft)
 Mk II 150 m (495 ft)
Landing run, aerobatic weight 156 m (510 ft)
Range with max internal fuel at 61% power, Mk I
 1,129 n miles (2,092 km; 1,300 miles)
g limits, aerobatic weight, Mk II +6/−3
UPDATED

Speedtwin Mk I aerobatic sporting aircraft *(Paul Harrison)* 2000/0079282

WARRIOR

WARRIOR (AERO-MARINE) LTD
IRC House, The Square, Pennington, Lymington, Hampshire SO41 8GN
Tel: (+44 1590) 67 66 22
Fax: (+44 1590) 67 55 99
e-mail: jlabouchere@centaurseaplane.com
Web: http://www.centaurseaplane.com
MANAGING DIRECTOR: James Labouchere
TECHNICAL DIRECTOR: Leon Eversfield

NORTH AMERICAN OFFICE:
 Warrior (Aero-Marine) Inc
 23 Oceanwood Drive, Scarborough, Maine 04074, USA
 Tel: (+1 207) 885 99 20
 e-mail: dverril@centaurseaplane.com
 SECRETARY AND VICE-PRESIDENT: David Verrill

UK SALES OFFICE:
 Bob Crowe Aircraft Sales Ltd, Cranfield Airport, Bedford, MK43 0JR
 Tel: (+44 1234) 75 04 42
 Fax: (+44 1234) 75 19 44
 e-mail: aircraftsales@centaurseaplane.com

In April 2000, Warrior announced first orders for the Centaur amphibian and revealed plans to transfer assembly to North America, where an office was opened in January 2000.
UPDATED

WARRIOR CENTAUR
TYPE: Six-seat amphibian.
PROGRAMME: First design studies 1992; Mk 1 hull design, and first flight by one-fifth scale model, in 1993; Mk 2 hull configuration developed and trialled 1995; dynamically refined (and still current) one-fifth model first flown 1996; cockpit mockup completed 1997; initial funding by Royal Aeronautical Society Handley Page award. Revealed at Farnborough Air Show, September 1998, when development and planning continuing. Agreement in principle for Chichester-Miles Consultants Ltd (see CMC entry earlier in this section) to be the prototyping facility; £15 million funding for construction of a full-scale prototype being sought from 1999. Certification will be to JAR/FAR 23. North American assembly is probable.
CUSTOMERS: Initial 10 production aircraft reserved for demonstration; Nos. 11 and 12 sold by March 2000 to undisclosed London businessman and Active Funds of Delaware, USA, for use by its Avlamar transportation business. Production target is 68 in first full year; 172 in second.

Scale proof-of-concept model of the Warrior Centaur 2001/0103516

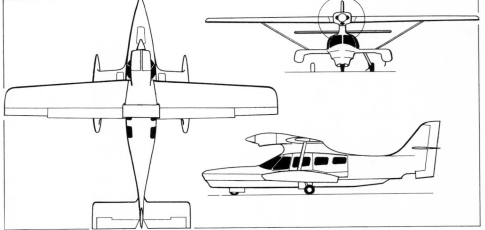
General arrangement of the Warrior Centaur amphibian *(Paul Jackson/Jane's)* 2000/0084503

COSTS: Expected to be between US$500,000 and US$575,000 at 2000 prices and rates. Seat-mile cost estimated at 35 per cent less than comparable amphibians.

Following description applies to projected full-size aircraft:
DESIGN FEATURES: Objectives of design included low hydrodynamic drag and structural weight, use of composites materials, folding wing, comfortable operation in coastal wave conditions, competitive seat-mile costs and payload-range performance, safety, saline impermeability and access to all areas frequented by boats.

Slender boat hull without planing step improves rough-water ride and increases aerodynamic efficiency; parasol-mounted high-lift wing with centrally installed engine nacelle; stub-wing/sponsons with end-mounted floats, strut-braced to main wing; conventional tail surfaces. Main panels of wing fold back to within beamwidth of sponsons to permit docking in standard 12 m (40 ft) yacht berths.

FLYING CONTROLS: Conventional and manual. Electrohydraulically actuated, slotted trailing-edge flaps.
STRUCTURE: Composites construction (carbon, glass and Kevlar fibres). Top surfaces of stub-wings laminated with energy absorbent tread; door sills reinforced with stainless steel.
LANDING GEAR: Flying-boat or amphibian. Electrohydraulically retractable tricycle type in amphibian, nosewheel retracting rearward, mainwheels outward into sponson tip-floats. Stern-mounted 7 kW (9 hp) water-jet for surface manoeuvring.
POWER PLANT: One 224 kW (300 hp) Textron Lycoming IO-540 flat-six engine; three-blade, constant speed variable-pitch propeller. Alternatives could include 231 kW (310 hp) Textron Lycoming TIO-540 or similar engines in 209 to 231 kW (280 to 310 hp) range. See under Weights and Loadings for fuel details.
ACCOMMODATION: Pilot and six passengers in flying-boat; five passengers in amphibian. Baggage compartment behind seats. Three doors on port side; one on starboard.
SYSTEMS: Electrically driven hydraulic actuation of landing gear and flaps. Electrical systems powered by 28 V DC alternator and 24 V lead-acid batteries. Air conditioning optional.
AVIONICS: Honeywell Silver Crown avionics and VFR instrumentation as standard; options include IFR avionics, marine HF radio, two-axis autopilot and bow-mounted weather radar.

DIMENSIONS, EXTERNAL:
Wing span 12.87 m (42 ft 2½ in)
Width over sponsons 4.64 m (15 ft 2½ in)
Length overall 11.15 m (36 ft 7 in)
Height overall, on land 3.51 m (11 ft 6 in)
DIMENSIONS, INTERNAL:
Cabin, excl cockpit: Length 3.81 m (12 ft 6 in)
Max width 1.37 m (4 ft 6 in)
Max height 1.22 m (4 ft 0 in)
Volume 4.1 m³ (144 cu ft)
WEIGHTS AND LOADINGS:
Weight empty, equipped 1,100 kg (2,426 lb)
Operating weight empty 1,196 kg (2,637 lb)
Baggage capacity 90 kg (200 lb)
Fuel weight: max 354 kg (780 lb)
with max payload 68 kg (150 lb)
Payload: max: amphibian 550 kg (1,213 lb)
flying-boat 622 kg (1,372 lb)
with max fuel 264 kg (583 lb)
Max T-O weight 1,814 kg (4,000 lb)
Max zero-fuel weight 1,746 kg (3,850 lb)
Max power loading 8.11 kg/kW (13.33 lb/hp)
PERFORMANCE (amphibian with IO-540, estimated):
Cruising speed:
at max cruise power at 1,525 m (5,000 ft)
127 kt (235 km/h; 146 mph)
at recommended cruise power at S/L
119 kt (220 km/h; 137 mph)
Unstick speed 40 kt (74 km/h; 46 mph)
Stalling speed, power off:
flaps up 49 kt (91 km/h; 57 mph)
flaps down 44 kt (82 km/h; 51 mph)
Stalling speed, power on:
flaps down 39 kt (73 km/h; 45 mph)
Time to 3,050 m (10,000 ft) at max cruise power
13 min 24 s
T-O run: on water 173 m (569 ft)
T-O to 15 m (50 ft): on land 287 m (940 ft)
on water 363 m (1,192 ft)
Landing from 15 m (50 ft) at 1,700 kg (3,748 lb):
on land 348 m (1,140 ft)
Range at max cruising speed, VFR reserves: with max payload 150 n miles (278 km; 172 miles)
with 419 kg (923 lb) payload*
600 n miles (1,111 km; 690 miles)
with 281 kg (620 lb) payload*
1,200 n miles (2,222 km; 1,380 miles)
* *For flying-boat, add 72 kg (159 lb) to payload*

UPDATED

WESTLAND

GKN WESTLAND HELICOPTERS LIMITED

Lysander Road, Yeovil, Somerset BA20 2YB
Tel: (+44 1935) 47 52 22
Fax: (+44 1935) 70 21 31
CEO: Richard Case
PUBLIC RELATIONS DIRECTOR: Christopher Loney

Westland Aircraft Ltd (later expanded to Westland Group plc) formed July 1935, taking over aircraft branch of Petters Ltd (known previously as Westland Aircraft Works) that had designed/built aircraft since 1915; entered helicopter industry having acquired licence to build US Sikorsky S-51 as Dragonfly 1947.

GKN shareholding in Westland Group progressively increased until overall control achieved on 18 April 1994.

Merger of GKN's helicopter interests and those of Italy's Finmeccanica finalised on 26 July 2000 with announcement of **AgustaWestland** (see International section).

GKN Westland Helicopters Ltd included GKN Westland Industrial Products Ltd at Weston-super-Mare, Somerset, and has 50 per cent interest in Aerosystems International Ltd at Yeovil (aerospace software; partnership with BAE Systems) and EH Industries Ltd. Last-mentioned was formed as partnership with Agusta of Italy for development and manufacture of EH 101 helicopter (see EHI in International section). Employment in helicopter and transmissions business was 4,800 in mid-2000.

Under June 1989 agreement with former McDonnell Douglas, Westland obtained co-production rights for Boeing AH-64 Apache; this selected for British Army Air Corps, July 1995; Westland is prime contractor to UK MoD. Production now under way in newly built 5,200 m² (56,000 sq ft) assembly hall at Yeovil, opened 14 January 1999.

Aviation Training International (ATI) joint venture company formed with Boeing on 3 August 1998 to provide air, ground and maintenance crews for British Army Apaches under £650 million, 30 year contract; began operations 29 July 1999 at Sherborne HQ; is opening branches at Middle Wallop, Dishforth and Wattisham flying bases, plus engineering school at Arborfield.

Pending full production of EH 101 and Apache, Westland continued low-rate production of existing products; delivered four helicopters in 1993; none in 1994; three in 1995; and 11 in 1997, including the final Sea King. Production in 1999 comprised 10 Lynx and 13 EH 101s.

Support of Sea King, Puma and Gazelle undertaken at Weston-super-Mare by Westland Industrial Products Ltd.

UPDATED

GKN WESTLAND WAH-64D APACHE LONGBOW

TYPE: Day/night twin-engined attack helicopter.
PROGRAMME: Westland selected, 13 July 1995, to build McDonnell Douglas (now Boeing) AH-64D Apache for Army Air Corps (AAC) and Royal Marines. Total 67 ordered, although original requirement was for 91; contract H12b/400 finalised 25 March 1996; £2,500 million programme involves some 240 UK companies (including 60 direct subcontractors) that will receive over 50 per cent of UK Apache work. Value increased to some £3,100 million by 1999, in part through addition of Marconi Avionics (now part of BAE Systems) to provide HIDAS self-protection. Apache description in US section; UK version similar, including Longbow radar, but powered by Rolls Royce Turbomeca RTM 322 turboshafts. Other differences from US Army version given below.

Engine integration trials conducted on leased sixth preproduction AH-64D (85-25408) at Boeing's Mesa plant; first engine run 6 April 1998; first flight 29 May 1998. Eight WAH-64Ds built and flown at Mesa; ZJ166/N9219G first flew 25 September 1998 and handed-over to GKN Westland 28 September; initial trials and pilot training conducted in US, where four WAH-64s were retained by January 2000; further four supplied, complete, to Westland; remainder following in kit form from 30 August 1999 (see Current Versions; first delivery to AAC 15 March 2000; IOC with nine aircraft in December 2000; initial weapons clearance planned for July 2001; HIDAS clearance December 2001; cannon and CRV-7 clearance August 2002, allowing full deployment; shipboard clearance November 2003; final delivery in December 2003; full environmental clearance in November 2004.

Shorts contracted in September 1996 to supply engine nacelles, stub-wings and horizontal stabiliser; heat exchangers manufactured by IMI Marston under Hughes-Treitler licence. UK and Netherlands agreed on 5 September 1996 on joint support of their Apache fleets. UK contract also includes 68 Longbow radars, 980 Hellfire missiles and 204 launchers.

CURRENT VERSIONS (specific): **WAH1/ZJ166** (N9219G): US built; first flight at Mesa 25 September 1998; handed over to Westland 28 September 1998 for RTM 322 engine trials in US.
WAH2/ZJ167 (N3266B): US built; pilot training in US. Delivered mid-2000.
WAH3/ZJ168 (N3123T): US built; airfreighted to Yeovilton 28 May 1999; thence Yeovil; first flight of WAH-64 in UK 26 August 1999; DERA Boscombe Down. Delivered June 2000.
WAH4/Z1169 (N3114H): US built; pilot training in US. Delivered mid-2000.
WAH5/ZJ170 (N3065U): US built; HIADS trials in US.
WAH6/ZJ171 (N3266T): US built; transferred to Yeovil 16 December 1999; instrumented aircraft; first army 'delivery' 15 March 2000, but not allocated to DERA at Boscombe Down until April 2000.
WAH7/ZJ172: First UK kit; arrived Yeovil 30 August 1999. First flight 18 July 2000; delivered 31 July 2000.
WAH8/ZJ173 (N3267A): US built; transferred to Yeovil 30 December 1999. General trials and pilot training.
WAH9/ZJ174: Second UK kit; arrived Yeovil 16 December 1999.
WAH10/ZJ175 (N3218V): US built; arrived Yeovil 6 April 2000. Delivered June 2000.
WAH11/ZJ176: Third UK kit.
CUSTOMERS: British Army. Service allocation comprises 19 for operational evaluation, training (of which eight at Middle Wallop) and attrition replacement, plus 16 for each of three Army Air Corps regiments: No. 9 at Dishforth (656 and 657 Squadrons from third quarter of 2001) and Nos. 3 (654 and 669 Squadrons from mid/late 2002) and 4 (662 and 663 Squadrons from early/mid-2003) at Wattisham. Units have additional responsibility for supporting Royal Marines with 'navalised' WAH-64s and integrated (Army/RM) crews.
DESIGN FEATURES: Manual blade folding (windspeed limit 45 kt; 83 km/h; 52 mph).
FLYING CONTROLS: Manual back-up to FBW system.
POWER PLANT: Two 1,566 kW (2,100 shp) RRTI RTM 322-01/12 turboshafts.
SYSTEMS: Rotor blade de-icing system.
AVIONICS: *Comms:* Mode S IFF.
Instrumentation: Honeywell IHADSS helmet sight to be upgraded from about 2004.
Mission: Video recorder.
Self-defence: BAE Systems HIDAS helicopter integrated defensive aids system, including Sky Guardian 2000 RWR, Type 1223 LWR, Vinten 455 chaff/flare dispenser and Sanders AN/AAR-57(V) common missile warning system (CMWS). Standard US fit of AN/APR-48 RFI also included.
ARMAMENT: Additional provision for Bristol Aerospace CRV-7 70 mm rocket pods and Shorts Starstreak AAM.

UPDATED

US-built WAH-64D Apache exercising in UK

GKN WESTLAND LYNX

TYPE: Twin-engined multipurpose helicopter.
PROGRAMME: Developed within Anglo-French helicopter agreement confirmed 2 April 1968; Westland given design leadership; first flight of first of 13 prototypes (XW835) 21 March 1971; first flight of fourth prototype (XW838) 9 March 1972, featuring production-type monobloc rotor head; first flights of British Army Lynx prototype (XX153) 12 April 1972, French Navy prototype (XX904) 6 July 1973, production Lynx (RN HAS. Mk 2 XZ229) 20 February 1976; first RN operational unit (No. 702 Squadron) formed on completion of intensive flight trials December 1977; AH. Mk 5 first flew (ZE375) 23 February 1985; other development details and records in 1975-76 and subsequent *Jane's*. Production shared 70 per cent Westland, 30 per cent Aerospatiale; for details of G-LYNX's 1986 world helicopter absolute speed record, and Lynx AH. Mk 7 XZ170's 1989 agility trials, see 1990-91 *Jane's*.

Second phase of development was Super Lynx (naval) and Battlefield Lynx, to which standards later UK military Lynx (Mks 8 and 9) were built; also exported. Battlefield Lynx mockup displayed at 1988 Farnborough Air Show (converted demonstrator G-LYNX), featuring wheeled landing gear, exhaust diffusers and provision for anti-helicopter missiles each side of fuselage; first flight of wheeled prototype (converted trials AH. Mk 7 XZ170) 29 November 1989; first flight of South Korean Super Lynx (90-0701, temporarily ZH219) 16 November 1989 (also first Lynx with Seaspray Mk 3 radar); first flight of Portuguese Super Lynx (9201, temporarily ZH580, ex-RN ZF559) 27 March 1992.

In September 1996, GKN Westland formally launched versions of Lynx powered by CTS800 turboshafts, with or without a six-screen EFIS. Current production versions were simultaneously redesignated, those now on offer being Srs 100, 200 or 300. Fuselages built from 1999 have potential for MTOW extension to 5,443 kg (12,000 lb) and beyond. First Super Lynx 300 (ZT800) flew 27 January 1999.

CURRENT VERSIONS: **Lynx AH. Mk 1**: British Army general purpose and utility version; 113 built and 107 converted to Mk 7 (see below). Described in 1986-87 and earlier *Jane's*.

Lynx HAS. Mk 3: Second Royal Navy version for advanced shipborne anti-submarine and other duties; similar to Mk 2, with BAE Seaspray search and tracking radar in modified nose; capable of anti-submarine classification and strike, air-to-surface-vessel search and strike, SAR, reconnaissance, troop transport fire support, communications and fleet liaison, and vertrep; can carry Sea Skua; Mk 2's Gem 2 engines replaced by two 835 kW (1,120 shp) Gem 41-1 engines; 23 delivered March 1982 to April 1985; eight more in **HAS. Mk 3S** configuration (first flight, ZF557, 12 October 1987) delivered November 1987 to November 1988; this version has two BAE AD3400 UHF radios with secure speech facility; additionally, ZD560 built in approximately Mk 7 configuration, delivered to Empire Test Pilots' School April 1988; further 53 obtained through modification of all existing HAS. Mk 2s by 1989. **Lynx HAS. Mk 3ICE** is Mk 3 lacking Sea Skua capability, and with survey camera pod for general duties aboard Antarctic survey vessel, HMS *Endurance*; four converted, of which three to **Mk 3SICE** with AD3400 radios.

Main RN operators, Nos. 815 and 829 Squadrons, combined on 26 March 1993 as No. 815 Squadron, which moved from Portland to Yeovilton on 12 February 1999. Training unit, 702 Squadron, to Yeovilton on 15 January 1999. Those used by Armilla Patrol in Persian Gulf modified to **HAS. Mk 3GM** (Gulf Mod), with better cooling and provision for Yellow Veil jamming pods and Sandpiper FLIR, or **HAS. Mk 3S/GM**, also with Mk 3S

GKN Westland Lynx HMA. Mk 8, with Sea Owl thermal imager and chin radome *(Paul Jackson/Jane's)*

modifications (to which standard all 3GMs converted). Augmenting new-build Mk 3Ss, 36 modified by RN Aircraft Yard (Naval Aircraft Repair Organisation from September 1996) at Fleetlands from April 1989; Mk 3S was Phase 1 of Mk 8 conversion programme, involving AD3400 secure speech radios (blade aerial beneath midpoint of tailboom) and upgraded ESM; programme continues, including Mk 3S/GM. Phase 2 was Lynx **HAS. Mk 3CTS**, adding RAMS 4000 central tactical system; prototype (XZ236 ex-Mk 3) flew 25 January 1989; further six for RN trials (one ex-Mk 3; five ex-Mk 3S); deliveries to Operational Flight Trials Unit, Portland, from April 1989; unit became No. 700L Squadron 6 July 1990 with three Lynx; remaining three deployed to destroyers and frigates at sea from 3 December 1990 (HMS *Newcastle*); CTS service clearance granted August 1991; Mk 3CTS has flotation bag each side of nose. RN Lynx status mid-2000: one Mk 3, 15 Mk 3S, 15 Mk 3SGM, four Mk 3SICE/ICE and 40 Mk 8. (Excludes two Mk 3S sold to Portugal as Mk 95s in 1993 and three Mk 3s sold to Pakistan in August 1994.)

Lynx AH. Mk 7: Uprated British Army version, meeting GSR 3947; with improved systems, reversed-direction tail rotor with improved composites blades to reduce noise and enhance extended period hover at high weights; redesigned tailcone; 4,876 kg (10,750 lb) AUW; 13 ordered, including eight from Mk 5 contract, but two were cancelled; first flight (a conversion, ZE376) 7 November 1985; first new-build Mk 7 (ZE377) flew 13 June 1986; 11th delivered July 1987. At least one (ZE381) fitted with 'Chancellor' FLIR ball on port side of cabin for surveillance in internal security role. Seven converted to Mk 9.

RN workshops at Fleetlands converted Mk 1s to Mk 7s; first (XZ641) redelivered 30 March 1988; box-type exhaust diffusers added from early 1989; last conversion mid-1994; total 107 conversions from Mk 1 and one each from Mks 3 and 5; one to Mk 9; 94 extant by mid-2000. Interim version was Lynx **AH. Mk 1GT** with uprated engines and rotors, but lacking Mk 7's improved electronic systems; first conversion (XZ195) 1991. Marconi (later

BAE Systems) AWARE-3 radar warning receiver selected 1989 for retrofit, designated ARI 23491 Rewarder; Mk 1 XZ668 to Westland for trial installation 22 November 1991. (Sky Guardian Mk 13 installed in some Lynx AH. Mk 7s for Gulf War, 1990-91; later uprated to Mk 15.) BERP (extended tip chord) blades retrofitted to Mk 7 from 1993. Upgrade proposal, involving 1,193 kW (1,600 shp) Gem 62 engines, was not pursued.

Subject to results of study contract awarded to GKN Westland on 1 December 1998, up to 78 Army Air Corps Mk 7s and all 24 Mk 9s will be modified to a common equipment standard (wheeled landing gear, MIL-STD-1553B databus; avionics management system; GPS/ INS/Doppler navigation; improved communications; civil navigation equipment; improved defensive aids); Mk 7 will replace Gazelles and be designated LUH (light utility helicopter); Mk 9 role to be unchanged; service life extended to 2018. Associated options include new fuselages and CTS800 engines.

Super Lynx: Upgraded export naval Lynx, approximately equivalent to Lynx HMA. Mk 8 (see below).

Battlefield Lynx: Upgraded export army Lynx; approximately equivalent to Lynx AH. Mk 9 (see below).

Lynx HMA. Mk 8: For RN, designated HAS. Mk 8 until late 1995; equivalent to export **Super Lynx**; passive identification system; engines uprated to Gem 42 Srs 200 (686 kW; 920 shp); original 5,125 kg (11,300 lb) maximum T-O weight later increased to 5,330 kg (11,750 lb); improved (reversed-direction) tail rotor control; BERP composites main rotor blades; Racal RAMS 4000 central tactical system (CTS eases crew's workload by centrally processing sensor data and presents mission information on multifunction CRT display; 15 systems ordered 1987, 106 more in September 1989); original Seaspray Mk 1 radar repositioned in new chin radome; BAE Sea Owl thermal imager (×5 or ×30 magnifying system on gimballed mount, with elevation +20 to -30° and azimuth +120 to -120°; ordered October 1989) in former radar position; MIR-2 ESM updated; three Mk 3s used in development programme as tactical system (XZ236), dummy Sea Owl/chin radome (ZD267) and avionics (ZD266) testbeds; see Lynx Mk 3 for Phases 1 and 2 of Lynx Mk 8 programme.

Definitive Mk 8 (Phase 3) conversions begun 1992 with addition of Sea Owl, further radar and navigation upgrades, (including Racal RNS252 'Super TANS' with associated TNL8000 GPS), composites BERP main rotor blades and reversed-direction tail rotor (internal MAD not adopted for economy reasons).

Conversion planned of 44 Mks 3/3S/3CTS to Mk 8; contract awarded to Westland, May 1992 for first seven conversions; initial delivery (XZ732) in July 1994 to Operational Evaluation Unit within 815 Squadron at Portland; all received by mid-1995. Others for conversion by RN at Fleetlands; first Westland-produced conversion kit for RN delivered in early 1995; initial contract covers 18 kits; first conversions in hand at NARO Fleetlands by 1995; conversions to be completed by 2003. Mk 8 issued to 702 Squadron (OCU) from February 1996; began sea deployments on HMS *Montrose* late 1995 and serving aboard 10 frigates by late 1997.

Lynx AH. Mk 9: UK Army Air Corps LBH (light battlefield helicopter); equivalent of export **Battlefield Lynx**; tricycle wheel landing gear; maximum T-O weight 5,125 kg (11,300 lb); advanced technology composites main rotor blades; exhaust diffusers; no TOW capability; first flight of prototype (converted company demonstrator XZ170) 29 November 1989; 16 new aircraft (beginning ZG884, flown 20 July 1990) ordered for delivery from

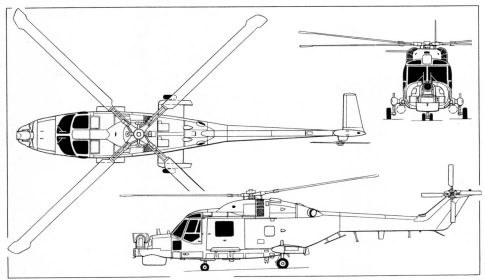

GKN Westland Lynx HMA. Mk 8, based on the Super Lynx advanced export version *(Dennis Punnett/Jane's)*

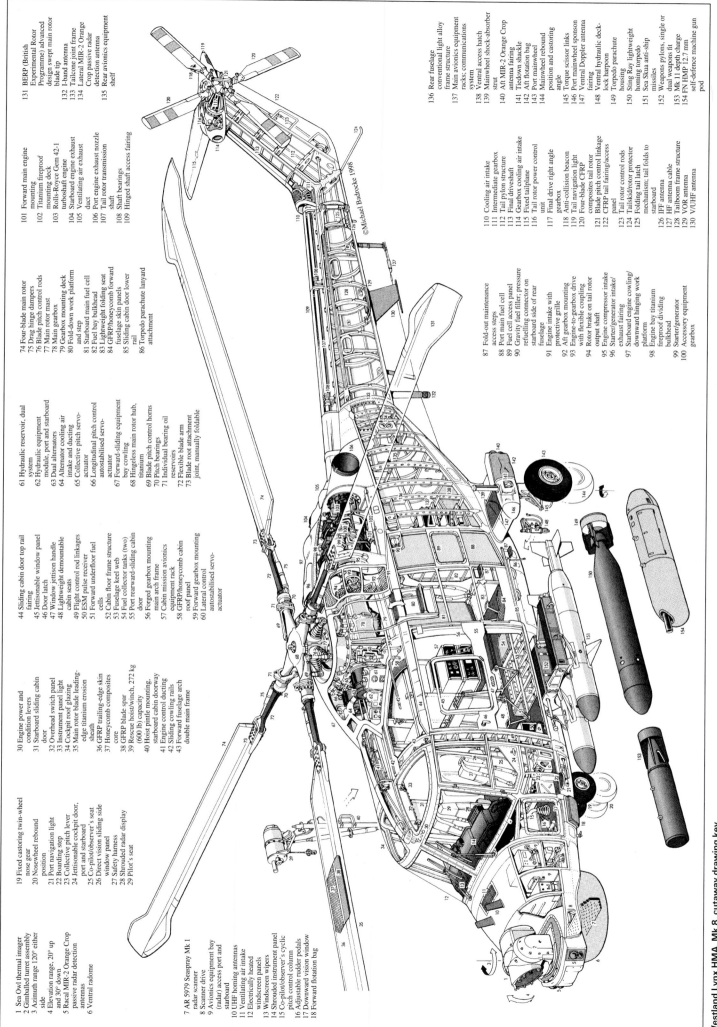

Westland Lynx HMA. Mk 8, cutaway drawing key

536 UK: AIRCRAFT—WESTLAND

1991, plus nine Mk 7 conversions (main contract for eight awarded November 1991); for support of 16 Air Assault Brigade; five outfitted as advanced command posts (with two secure radios and Tacan), remainder for tactical transport role.

Deliveries to A&AEE Boscombe Down from 22 May 1991 (ZG884); to No. 672 Squadron from 19 December 1991 (ZG889); No. 664 Squadron from June 1992; final aircraft (ZG923) flown 30 June 1992; first 'production' conversion (ZF538) to Westland for modifications 3 February 1992; all Mk 9s reassigned to 653 and 659 Squadrons in late 1993; all extant in early 2000.

Super Lynx Series 100: Introduced September 1996; export version of HMA. Mk 8; conventional cockpit instrumentation, Rolls-Royce Gem 42-1 turboshafts and 360° radar. Applies retrospectively to aircraft sold to Brazil (Mk 21A), Portugal (Mk 95) and South Korea (Mk 99) as well as to seven Mk 88As sold to Germany, also in September 1996.

Super Lynx Series 200: Conventional cockpit, but electronic power system displays and two LHTEC CTS800-4N turboshafts with FADEC. Original Lynx demonstrator G-LYNX fitted with two 1,007 kW (1,350 shp) T800 turboshafts as **Battlefield Lynx 800** private venture (LHTEC funding power plants and gearboxes, Westland provided airframe for full flight demonstration programme); first flight 25 September 1991; programme terminated early 1992 after 17 hours. CTS800 is derived from the T800 (as in the Boeing Sikorsky RAH-66 Comanche) and offers an additional 30 per cent of power to operators using the Lynx in hot climates. Version for production Lynx is CTS800-4N.

Super Lynx Series 300: Prototype first flown (ZT800) 27 January 1999, initially with Gem 42 engines. Six-screen EFIS cockpit; dual redundant MIL-STD-1553B and ARINC 429 databusses; new navigation system and AHRS; revised communications suite; CTS800 engines. Ordered by Malaysia; second customer expected to be South Africa, which announced intention to purchase on 18 November 1998, but later deferred expected contract for four to unspecified date; South African Srs 300s will be designated Mk 64 and equipped with Telephonics AN/APS-143 radar.

Lynx ACH: Advanced Compound Helicopter. Technology demonstrator project, publicly announced 22 May 1998 and due to have begun in 1999, partly funded by UK MoD. However, military funding withdrawn, prompting search for alternative source. Target is 50 per cent speed increase by means of wings attached at cabin roof level and variable area exhaust nozzles; additional thrust derived from RRTI RTM322 turboshafts in place of Gems (uprated gearbox taken from W30-200); BERP rotor blades; flaps on wing trailing-edges, with pitching moment neutralised by all-moving tailplane; and trimming rudder to reduce tail rotor loads.

Performance will include maximum level speed of 250 kt (463 km/h; 288 mph); ceiling of 6,100 m (20,000 ft); 20 per cent additional payload/range; 50 per cent more propulsive efficiency; and between 25 and 50 per cent improvement in lift/drag ratio.

Other versions and operators where orders completed, see 1990-91 *Jane's*.

CUSTOMERS: See table. Contracts for later versions include Super Lynx ordered by South Korea 1988 (12 **Mk 99** with Racal Avionics Doppler 71/TANS N nav system, Seaspray Mk 3 360° radar, AN/AQS-18 dipping sonar and Sea Skua), handed over between 26 July 1990 and May 1991 for 'Sumner' and 'Gearing' class destroyers; further 13 delivered 1999-2000 as Mk 99As to order confirmed in June 1997. Mk 99A has composites tailplane; first aircraft flown 3 July 1999 and departed Yeovil for surface delivery 1 September 1999. Portugal ordered five (first two ex-Royal Navy modified airframes) Super Lynx **Mk 95** 1990 (plus three options) with Racal RNS252 GPS-aided INS and Doppler 91 navigation systems and some US equipment including AN/AQS-18 sonar and Honeywell RDR 1500 radar; first two handed over 29 July 1993 for 'Vasco da Gama' class (MEKO 200) frigates; final two delivered 16 November 1993. Brazil ordered nine **Mk 21As** (plus five conversions from Mk 21). First for upgrade, N3027 delivered to Yeovil on 1 February 1995 and reflown as Mk 21A (N4010) on 22 December 1995; last two redelivered in late April 1998. First new-build helicopter (N4001) flew 12 June 1996; last delivered in August 1997. Mk 21 avionics include 360° Seaspray 3000 radar, RNS252 INS and Doppler 71 (but no FLIR or CTS); armament includes Sea Skua missiles. Mk 21A introduced 5,330 kg (11,750 lb) MTOW.

German Navy ordered seven Super Lynx **Mk 88As** in September 1996 and simultaneously took an option (confirmed on 25 June 1998) on upgrading (with new-build airframes) of its existing 17; new deliveries began with roll-out on 14 July 1999 (post first flight) of initial aircraft; upgrades due for delivery in 2001-2003, all except first undertaken by Eurocopter at Donauwörth; Mk 88A is a Srs 100 aircraft with 360° Seaspray Mk 3000 radar, FLIR turret (or nose fairing when not fitted), Rockwell Collins GPS and Racal Doppler 91 and RNS252; Sea Skua ASM armament. Eight Danish aircraft to be upgraded to **Mk 90B** with new airframes under contract announced 20 January 1998; completion due in 2004; all except first

LYNX PRODUCTION

Variant	Customer	Qty	First aircraft	First flights	Operators
AH. Mk 1	Army Air Corps	113*	XZ170	11 Feb 1977-24 Jan 1984	Note A
HAS. Mk 2	Fleet Air Arm	60	XZ227	20 Feb 1976-26 May 1981	Converted to Mk 3
HAS. Mk 2(FN)	French Navy	26	260	4 May 1977-5 Sep 1979	31F, 34F, ERCE
HAS. Mk 3	Fleet Air Arm	31	ZD249	4 Jan 1982-21 Oct 1988	815, 702 Sqdns
HAS. Mk 4(FN)	French Navy	14	801	1 Apr 1982-26 Aug 1983	31F, 34F, ERCE
Mk 5	MoD(PE)	3*	ZD285	21 Nov 1984-23 Feb 1985	DERA
AH. Mk 6	Royal Marines			None built	-
AH. Mk 7	Army Air Corps	11*	ZE376	23 Apr 1985-5 Jun 1987	Note A
HMA. Mk 8	Fleet Air Arm	–		Conversions	815, 702 Sqdns
AH. Mk 9	Army Air Corps	16*†	ZG884	20 Jul 1990-21 Jun 1992	653, 659 Sqdns
Mk 21	Brazilian Navy	9	N3020	30 Sep 1977-14 Apr 1978	Five upgraded to 21A
Mk 21A	Brazilian Navy	9†	N4001	12 Jun 1996-24 Apr 1997	1° EHEAA
Mk 22	Egyptian Navy			None built	
Mk 23	Argentine Navy	2	0734	17 May 1978-23 Jun 1978	Withdrawn from service
Mk 24	Iraqi Army			None built	–
Mk 25[1]	Netherlands Navy	6	260	23 Aug 1976-16 Sep 1977	7/860 Sqdns
Mk 26	Iraqi Army (armed)			None built	–
Mk 27[2]	Netherlands Navy	10	266	6 Oct 1978-12 Nov 1979	7/860 Sqdns
Mk 28	Qatari Police	3*	QP-31	2 Dec 1977-12 Apr 1978	Withdrawn from service[7]
Mk 64	South African Navy[10]				
Mk 80[12]	Danish Navy	8[11]	S-134	3 Feb 1980-15 Sep 1981	See Mk 90
Mk 81[3]	Netherlands Navy	8	276	9 Jul 1980-24 Mar 1981	7/860 Sqdns
Mk 82	Egyptian Army			None built	–
Mk 83	Saudi Army			None built	–
Mk 84	Qatari Army			None built	–
Mk 85	UAE Army			None built	–
Mk 86	Norwegian Coast Guard	6	207	23 Jan 1981-11 Sep 1981	Skv 337
Mk 87	Argentine Navy			Embargoed	–
Mk 88	German Navy	19[9]	8301	26 May 1981-10 Dec 1988	3/MFG 3
Mk 88A	German Navy	7†	8320	30 Apr 1999 – Mar 2000	3/MFG 3
Mk 89	Nigerian Navy	3	01-F89	29 Sep 1983-14 Mar 1984	101 Sqdn
Mk 90[12]	Danish Navy	1[4]	S-256	19 Apr 1988[6]	Søværnets Flyvetjeneste
Mk 95	Portuguese Navy	3†[5]	9203	9 Jul 1993-1993	EHM[8]
Mk 99	South Korean Navy	12†	90-0701	16 Nov 1989-14 May 1991	627 Sqdn
Mk 99A	South Korean Navy	13†	99-0721	3 Jul 1999-Aug 2000	
Srs 300	Malaysian Navy	6			499 Sqdn
Subtotal		399			
Lynx 3		1	ZE477	14 Jun 1984	
Demonstrators		3	G-LYNX	18 May 1979-27 Jan 1999	
Prototypes		13	XW835	21 Mar 1971-5 Mar 1975	
Total		416			

Notes:
Note A: Army Air Corps 651, 652, 654, 655, 656, 657, 659, 661, 662, 663, 664, 665, 667, 669 and 671 Squadrons; 847 Squadron (Royal Marine Commando)
[1] Netherlands designation UH-14A; all to SH-14D
[2] Netherlands designation SH-14B; all to SH-14D
[3] Netherlands designation SH-14C; all to SH-14D
[4] Plus one conversion from demonstrator; these two to be upgraded to Super Lynx Mk 90B with new airframes
[5] Plus two conversions from Mk 3
[6] Completion and first flight at Vaerløse, Denmark
[7] Sold to Royal Navy for spares recovery and use as ground instructional airframes
[8] Esquadrilha de Helicopteros de Marinha
[9] Surviving 17 to be upgraded to Super Lynx with new airframes in 2001-2003
[10] Intends to order four, but programme delay announced in 1999
[11] Surviving six to be converted to Super Lynx Mk 90A with new airframes
[12] Converted during 1990s to Mk 80A/90A with Gem 42-1 engines
* Army version
† Super/Battlefield Lynx

GKN Westland Lynx AH. Mk 7 (nearest) and Mk 9 operated by the Army Air Corps *(Paul Jackson/Jane's)*
2001/0092540

German Navy Super Lynx Mk 88A, wearing optional Sea Owl turret *(Paul Jackson/Jane's)*

undertaken locally. Super Lynx also offered to Australia and New Zealand (both unsuccessfully) and Malaysia, last-mentioned announcing order for six on 7 September 1999. By August 2000, production totalled 393 production aircraft, 14 prototypes and three demonstrators, or 410 in all; backlog then six complete and 23 rebuilt aircraft.

COSTS: £3.5 million (1991) to upgrade eight Mk 7s to Mk 9s. £20 million (1992) to convert seven Mk 3s to Mk 8s; £150 million for nine new Lynx and five upgrades, 1993 (Brazil). £100 million for seven Mk 88As (Germany), 1996. Proposed four Srs 300s for South Africa estimated to cost total of £80 million (1998).

DESIGN FEATURES: Compact design suited to hunter-killer ASW and missile-armed anti-ship naval roles from frigates or larger ships (superseding ship-guided helicopters), armed/unarmed land roles with cabin large enough for squad, or other tasks; manually folding tail pylon on most (but not all) naval versions; single four-blade semi-rigid main rotor (foldable), each blade attached to main rotor hub by titanium root plates and flexible arm; rotor drives taken from front of engines into main gearbox mounted above cabin ahead of engines; in flight, accessory gears (at front of main gearbox) driven by one of two through shafts from first stage reduction gears; four-blade tail rotor, drive taken from main ring gear; single large window in each main cabin sliding door; provision for internally mounted armament, and for exterior universal flange mounting each side for other weapons/stores. Super Lynx has all-weather day/night capability; extended payload/range; advanced technology swept-tip (BERP) composites main rotor blades offering improved speed and aerodynamic efficiency and reduced vibration; and reversed direction tail rotor for improved control.

FLYING CONTROLS: Rotor head controls actuated by three identical tandem servojacks and powered by two independent hydraulic systems; control system incorporates simple stability augmentation system; each engine embodies independent control system providing full-authority rotor speed governing, pilot control being limited to selection of desired rotor speed range; in event of one engine failure, system restores power up to single-engine maximum contingency rating; main rotor can provide negative thrust to increase stability on deck after touchdown on naval versions; hydraulically operated rotor brake mounted on main gearbox; sweptback fin/tail rotor pylon, with starboard half-tailplane.

STRUCTURE: Conventional semi-monocoque pod and boom, mainly light alloy; glass fibre access panels, doors, fairings, pylon leading/trailing-edges, and bullet fairing over tail rotor gearbox; composites main rotor blades; main rotor hub and inboard flexible arm portions built as complete unit, as titanium monobloc forging; tail rotor blades have light alloy spar, stainless steel leading-edge sheath and rear section as for main blades.

LANDING GEAR (general purpose military version): Non-retractable tubular skid type. Provision for a pair of adjustable ground handling wheels on each skid. Flotation gear optional. Battlefield Lynx and AH. Mk 9 equivalent have non-retractable tricycle gear with twin nosewheels.

LANDING GEAR (naval versions): Non-retractable oleo-pneumatic tricycle type. Single-wheel main units, carried on sponsons, fixed at 27° toe-out for deck landing; can be manually turned into line and locked fore and aft for movement of aircraft into and out of ship's hangar. Twin-wheel nose unit steered hydraulically through 90° by the pilot to facilitate independent take-off into wind. Sprag brakes (wheel locks) fitted to each wheel prevent rotation on landing or inadvertent deck roll. These locks disengaged hydraulically and re-engage automatically in event of hydraulic failure. Maximum vertical descent 1.83 m (6 ft)/s; with lateral drift 0.91 m (3 ft)/s for deck landing. Flotation gear, and hydraulically actuated harpoon deck lock securing system, optional.

POWER PLANT: Current option of two Rolls-Royce Gem 42-1 turboshafts, each rated at 835 kW (1,120 shp) or two LHTEC CTS800-4Ns, each of 995 kW (1,334 shp). Transmission rating 1,372 kW (1,840 shp). Exhaust diffusers for IR suppression optional on Battlefield Lynx.

Original installation was two Rolls-Royce Gem 2 turboshafts, each with maximum contingency rating of 671 kW (900 shp) in Lynx AH. 1, HAS. 2 and early export variants. Later versions had Gem 41-1, 41-2, or 42-1 engines, all with maximum contingency rating of 835 kW (1,120 shp); same transmission rating. Engines of British and French Lynx in service converted to Mk 42 standard during regular overhauls from 1987 onwards. Danish, Netherlands and Norwegian Lynx similarly retrofitted.

Fuel in five internal tanks; usable capacity 957 litres (253 US gallons; 210 Imp gallons) when gravity-refuelled; 985 litres (260 US gallons; 217 Imp gallons) when pressure-refuelled. For ferrying, two tanks each of 441 litres (116 US gallons; 97.0 Imp gallons) in cabin, replacing bench tank. Maximum usable fuel 1,867 litres (493 US gallons; 411 Imp gallons). Engine oil tank capacity 6.8 litres (1.8 US gallons; 1.5 Imp gallons). Main rotor gearbox oil capacity 28 litres (7.4 US gallons; 6.2 Imp gallons).

ACCOMMODATION: Pilot and co-pilot or observer on side-by-side seats. Dual controls optional. Individual forward-hinged cockpit door and large rearward-sliding cabin door on each side; cockpit doors jettisonable; windows of cabin doors also jettisonable. Cockpit accessible from cabin area. Maximum high-density layout (military version) for one pilot and 10 armed troops or paratroops, on lightweight bench seats in soundproofed cabin. Alternative VIP layouts for four to seven passengers, with additional cabin soundproofing. Seats can be removed quickly to permit carriage of up to 907 kg (2,000 lb) of freight internally. Tiedown rings provided. In casualty evacuation role, with a crew of two, Lynx can accommodate up to six Alphin stretchers and a medical attendant. Both basic versions have secondary capability for search and rescue (up to nine survivors) and other roles.

SYSTEMS: Two independent hydraulic systems, pressure 141 bars (2,050 lb/sq in). Third hydraulic system provided in naval version when sonar equipment installed. No pneumatic system. 28 V DC electrical power supplied by two 6 kW engine-driven starter/generators and an alternator. External power sockets. 24 V 23 Ah (optionally 40 Ah) Ni/Cd battery fitted for essential services and emergency engine starting. 200 V three-phase AC power available at 400 Hz from two 15 kVA transmission-driven alternators. Cabin heating and ventilation system. Optional supplementary cockpit heating system. Electric anti-icing and demisting of windscreen, and electrically operated windscreen wipers, standard; windscreen washing system.

AVIONICS (general): Avionics common to all roles (general purpose and naval versions).

Comms: Rockwell Collins VOR/ILS; DME; Rockwell Collins AN/ARN-118 Tacan; I-band transponder (naval version only); BAE PTR446, Rockwell Collins APX-72, DaimlerChrysler STR 700/375 or Italtel APX-77 IFF.

Flight: BAE duplex three-axis automatic stabilisation equipment; BAE GM9 Gyrosyn compass system; Racal tactical air navigation system (TANS); Racal 71 Doppler, E2C standby compass. BAE Mk 34 AFCS. Additional units fitted in naval version, when sonar is installed, to provide automatic transition to hover and automatic Doppler hold in hover.

AVIONICS (Army): *Flight:* Latest versions have Racal Doppler 91 and RNS 252 navigation; Honeywell/Smiths AN/APN-198 radar altimeter; Rockwell Collins 206A ADF; Rockwell Collins VIR 31A VOR/ILS.

Mission: British Army Lynx equipped with TOW missiles have roof-mounted Hughes sight manufactured under licence by British Aerospace. Roof sight upgraded with night vision capability in far IR waveband; first test firing of TOW with added Marconi (now BAE) thermal imager took place in October 1988. (Sextant 250 sight offered on export aircraft for fixed armament.) Optional equipment, according to role, can include lightweight sighting system with alternative target magnification, vertical and/or oblique cameras, flares for night operation, low-light level TV, IR linescan, searchlight, and specialised communications equipment. Some have IR formation flying lights and provision for crew's NVGs. For surveillance, some AAC Lynx carry Chancellor Helitele in external (port) ball housing, complete with datalink.

Self-defence: Sanders AN/ALQ-144 IR jammer installed beneath tailboom of some British Army Lynx from 1987; later augmented by exhaust diffusers. Requirement for RWR satisfied by 1989 selection of BAE AWARE-3 (ARI 23491) system; BAE Sky Guardian Mk 13 (later Mk 15) on some aircraft from 1990.

AVIONICS (Navy): *Comms:* RN helicopters have two BAE AD3400 VHF/UHF transceivers, Dowty D403M standby UHF radio, Rockwell Collins 718U-5 HF transceiver, BAE PTR446 D-band transponder and BAE ARI 5983 I-band transponder. Racal contracted in early 2000 to supply Saturn secure comms equipment, including Rockwell Collins AN/ARC-210 radios, RA800L control systems and AMS 2000 display units.

Radar: Super Lynx has Seaspray Mk 3000 or Honeywell RDR 1500 360° scan radar in chin fairing. (UK Mk 8 has original Seaspray Mk 1 upgraded to Mk 3000 under 1994 contract and repackaged below fuselage, leaving space for BAE Sea Owl thermal imaging equipment above nose.) BAE ARI 5979 Seaspray Mk 1 lightweight search and tracking radar in earlier versions.

GKN Westland Lynx HAS. Mk 3SGM continues to serve the Royal Navy in significant numbers, pending completion of the Mk 8 programme *(Paul Jackson/Jane's)*

Flight: GPS on Royal Navy and Netherlands Lynx from 1997.

Mission: Optional Honeywell AN/AQS-18 or Thomson HS-312 sonars. Detection of submarines by dipping sonar or magnetic anomaly detector. Dipping sonar operated by hydraulically powered winch and cable hover mode facilities within the AFCS. Racal MIR-2 Orange Crop passive radar detection system in RN Lynx; similar Racal Kestrel retrofitted to Danish Mk 90. (CAE Electronics AN/ASQ-504(V) internal MAD ordered for RN Lynx in 1990 but not taken up.) SFIM AF 530 or APX-334 stabilised sight in French naval Lynx. Optional Marconi Sandpiper FLIR on RN Lynx; FLIR Systems 2000HP installed in Netherlands SH-14D from 1996; FLIR Systems Safire optional for Danish Lynx. Vinten Vipa 1 reconnaissance pod; or Agiflite reconnaissance camera system.

Self-defence: BAE M-130 chaff/flare dispensers and Rodale AN/ALQ-167(V) D- to J-band anti-ship missile jamming pods installed on RN Lynx patrolling Persian Gulf, 1987. Two Lockheed Martin Challenger IR jammers above cockpit of RN Lynx during 1991 Gulf War. BAE AWARE-3 RWR in Netherlands SH-14Ds from 1996.

EQUIPMENT: All versions equipped as standard with navigation, cabin and cockpit lights; adjustable landing light under nose; and anti-collision beacon. For search and rescue, with three crew, both versions can have a waterproof floor and a 272 kg (600 lb) capacity clip-on hydraulic or electric hoist on starboard side of cabin; cable length 30 m (98 ft).

ARMAMENT: For armed escort, anti-tank or air-to-surface strike missions, army version can be equipped with two 20 mm cannon mounted externally so as to permit fitment of pintle-mounted 7.62 mm machine gun inside cabin. External pylon can be fitted on each side of cabin for variety of stores, including two Minigun or other self-contained gun pods; two rocket pods; or up to eight HOT, Hellfire, TOW, or similar air-to-surface missiles. Additional six or eight reload missiles carried in cabin. For ASW role, armament includes two Mk 44, Mk 46, A244S or Sting Ray homing torpedoes, one each on an external pylon on each side of fuselage, and six marine markers; or two Mk 11 depth charges. Alternatively, up to four Sea Skua semi-active homing missiles; on French Navy Lynx, four AS.12 or similar wire-guided missiles. Self-protection FN HMP 0.50 in machine gun pod optional on RN Lynx.

Super Lynx as standard naval Lynx, including four Sea Skua or two Penguin or Marte anti-ship missiles. Battlefield Lynx may carry two Giat 20 mm cannon pods; two FN Herstal pods with two 7.62 mm machine guns each; or two M.159C pods containing nineteen 2.75 in rockets each.

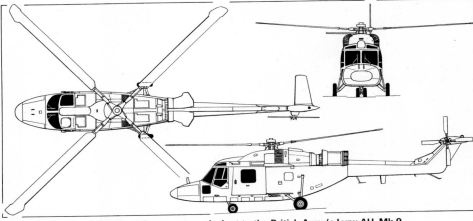

GKN Westland Battlefield Lynx for export, equivalent to the British Army's Lynx AH. Mk 9 *(Dennis Punnett/Jane's)*

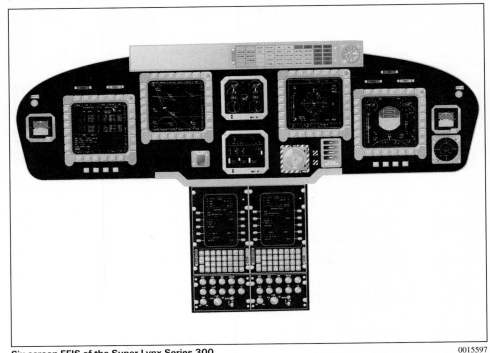

Six-screen EFIS of the Super Lynx Series 300   0015597

DIMENSIONS, EXTERNAL (A: early military version, B: Battlefield Lynx, N: early naval version, S: Super Lynx):

Main rotor diameter	12.80 m (42 ft 0 in)
Tail rotor diameter: A, N	2.21 m (7 ft 3 in)
B, S	2.36 m (7 ft 9 in)
Distance between rotor centres	7.66 m (25 ft 1½ in)
Length overall:	
both rotors turning: A, N	15.165 m (49 ft 9 in)
B, S	15.24 m (50 ft 0 in)
main rotor blades and tail folded:	
N	10.62 m (34 ft 10 in)
S	10.85 m (35 ft 7¼ in)
B, rotors folded	13.24 m (43 ft 5¼ in)
Width overall, main rotor blades folded:	
A	3.75 m (12 ft 3¾ in)
B	3.02 m (9 ft 11 in)
N, S	2.94 m (9 ft 7¾ in)
Height overall: both rotors stopped:	
A	3.505 m (11 ft 6 in)
N	3.48 m (11 ft 5 in)
tail rotor turning:	
B	3.73 m (12 ft 3 in)
S	3.67 m (12 ft 0½ in)
main rotor blades and tail folded:	
N	3.20 m (10 ft 6 in)
S	3.25 m (10 ft 8 in)
Tailplane half-span: A, N	1.78 m (5 ft 10 in)
B, S	1.32 m (4 ft 4 in)
Skid track: A	2.03 m (6 ft 8 in)
Wheel track: N	2.78 m (9 ft 1½ in)
B, S	2.80 m (9 ft 2¼ in)
Wheelbase: N	2.94 m (9 ft 7¾ in)
B, S	3.02 m (9 ft 11 in)

DIMENSIONS, INTERNAL:

Cabin, from back of pilots' seats:	
Min length	2.055 m (6 ft 9 in)
Max width	1.78 m (5 ft 10 in)
Max height	1.42 m (4 ft 8 in)
Floor area	3.72 m² (40.0 sq ft)
Volume	5.2 m³ (184 cu ft)
Cabin doorway: Width	1.37 m (4 ft 6 in)
Height	1.19 m (3 ft 11 in)

AREAS:

Main rotor disc	128.71 m² (1,385.4 sq ft)
Tail rotor disc: A, N	3.84 m² (41.28 sq ft)
B, S	4.37 m² (47.04 sq ft)

WEIGHTS AND LOADINGS (A, B, N and S as above):

Manufacturer's empty weight: A	2,578 kg (5,683 lb)
N	2,740 kg (6,040 lb)
Manufacturer's basic weight: A	2,658 kg (5,860 lb)
B	3,178 kg (7,006 lb)
N	3,030 kg (6,680 lb)
S	3,291 kg (7,255 lb)
Operating weight empty (including crew and appropriate armament):	
B, anti-tank (eight TOW)	3,949 kg (8,707 lb)
B, reconnaissance	3,444 kg (7,592 lb)
B, transport (unladen)	3,496 kg (7,707 lb)
S, ASW (two torpedoes)	4,618 kg (10,181 lb)
S, ASV (four Sea Skuas)	4,373 kg (9,641 lb)
S, surveillance and targeting	3,708 kg (8,174 lb)
S, search and rescue	3,778 kg (8,329 lb)
Max underslung load: B, S	1,361 kg (3,000 lb)
Max T-O weight: A	4,535 kg (10,000 lb)
N (Gem Mk 41)	4,763 kg (10,500 lb)
N (Gem Mk 42)	4,876 kg (10,750 lb)
B, S (early)	5,125 kg (11,300 lb)
B with external load	5,307 kg (11,700 lb)
S (1995+ build)	5,330 kg (11,750 lb)
Max disc loading: A	35.2 kg/m² (7.22 lb/sq ft)
N (Gem Mk 41)	37.0 kg/m² (7.58 lb/sq ft)
N (Gem Mk 42)	37.9 kg/m² (7.76 lb/sq ft)
B, S (early)	39.8 kg/m² (8.16 lb/sq ft)
S (1995+ build)	41.4 kg/m² (8.48 lb/sq ft)
Transmission loading at max T-O weight and power:	
A	3.31 kg/kW (5.43 lb/shp)
N (Gem Mk 41)	3.47 kg/kW (5.71 lb/shp)
N (Gem Mk 42)	3.55 kg/kW (5.84 lb/shp)
B, S	3.74 kg/kW (6.14 lb/shp)
S (1995+ build)	3.89 kg/kW (6.39 lb/shp)

PERFORMANCE (at normal max T-O weight at S/L, ISA, except where indicated, Gem 41/42 engines, A, B, N and S as above):

Never-exceed speed (VNE), Mk 9:	
clean	156 kt (289 km/h; 180 mph)
IR exhaust diffusers fitted	145 kt (269 km/h; 167 mph)
Max continuous cruising speed:	
A	140 kt (259 km/h; 161 mph)
N	125 kt (232 km/h; 144 mph)
A (ISA + 20°C)	130 kt (241 km/h; 150 mph)
N (ISA + 20°C)	114 kt (211 km/h; 131 mph)
S	138 kt (256 km/h; 159 mph)
Speed for max endurance: A, N (ISA and ISA + 20°C)	
	70 kt (130 km/h; 81 mph)
Max forward rate of climb: A	756 m (2,480 ft)/min
N	661 m (2,170 ft)/min
A (ISA + 20°C)	536 m (1,760 ft)/min
N (ISA + 20°C)	469 m (1,540 ft)/min
Max vertical rate of climb: A	472 m (1,550 ft)/min
N	351 m (1,150 ft)/min
A (ISA + 20°C)	390 m (1,280 ft)/min
N (ISA + 20°C)	244 m (800 ft)/min
Hovering ceiling OGE: A	3,230 m (10,600 ft)
N	2,575 m (8,440 ft)
Typical range, with reserves:	
A, troop transport	292 n miles (540 km; 336 miles)
B, tactical transport	370 n miles (685 km; 426 miles)
Radius of action:	
B, anti-tank, 2 h on station with four TOWs	
	25 n miles (46 km; 29 miles)
S, anti-submarine, 2 h on station, dipping sonar and one torpedo	20 n miles (37 km; 23 miles)
S, point attack with four Sea Skuas	
	125 n miles (232 km; 143 miles)
S, surveillance, 3 h 50 min on station	
	75 n miles (139 km; 86 miles)
Max range: A	340 n miles (630 km; 392 miles)
N	320 n miles (593 km; 368 miles)
Max ferry range with auxiliary cabin tanks:	
A	724 n miles (1,342 km; 834 miles)
N	565 n miles (1,046 km; 650 miles)
Max endurance: A	2 h 57 min
N (ISA + 20°C)	2 h 50 min

UPDATED

UNITED STATES OF AMERICA

AASI

ADVANCED AERODYNAMICS AND STRUCTURES INC

3205 Lakewood Boulevard, Long Beach, California 90808
Tel: (+1 562) 938 86 18
Fax: (+1 562) 938 86 20
e-mail: AASI@AasiAircraft.com
Web: http://www.AasiAircraft.com
CHAIRMAN, PRESIDENT AND CEO: Dr Carl L Chen
EXECUTIVE VICE-PRESIDENT: Gene Comfort
VICE-PRESIDENT, MANUFACTURING: Auts Ruff
VICE-PRESIDENT AND CHIEF FINANCIAL OFFICER: David Turner
VICE-PRESIDENT, ENGINEERING: Mike Lai

AASI succeeded Aerodynamics and Structures Inc (ASI), formed by former airline pilot Darius Sharifzadeh to develop Jetcruzer. Construction of new 18,580 m² (200,000 sq ft) factory started in November 1997 for production of aircraft at a rate of 10 per month; factory completed November 1998; in mid-1999 it was sold for US$9.8 million and immediately leased back, the balance of funds being used to finance continued development of the Jetcruzer 500. Company also designs UAVs.

Development of the Stratocruzer 1250 continues; this project was last described in the 1997-98 *Jane's*.

UPDATED

Model of Stratocruzer 1250-ER (Williams FJ44 turbofans), which is AASI's next project *(Paul Jackson/Jane's)*

AASI JETCRUZER 500

TYPE: Business turboprop.
PROGRAMME: Design began March 1983; aerodynamic design undertaken in UK 1984-85 by 'Sandy' Burns; layout prepared by Ladislao Pazmany; structural design by David Kent of Light Transport Design in UK; wind tunnel tests by University of San Diego. Prototype construction started June 1988; first exhibited at NBAA show October 1988 with 313 kW (420 shp) Rolls-Royce 250-C20S engine; first flight of Jetcruzer 450 forerunner 11 January 1989 (N5369M); preproduction prototype (N102JC) made first flight April 1991; first flight in production form (N450JC) 13 September 1992; certification to FAR Pt 23 on 14 June 1994; intending single-engine FAA Pt 135 public transport IFR certification. Final assembly of aircraft for customers in the Asia-Pacific region may be undertaken in Taiwan.

Prototype Jetcruzer 500 (N102JC modified from 450) made first flight 22 August 1997, second prototype (N200JC, modified from 450 N450JC) on 7 November 1997. First public appearance at NBAA Las Vegas 19 October 1998. Third prototype (N136JC) new-build scheduled to have flown in early 2000; FAA pressurisation test certificate awarded June 2000 to second prototype. US certification to FAA Pt 135 planned for June 2001.

CURRENT VERSIONS: **Jetcruzer 500:** Pressurised model, *as described below*. Corporate, executive, air ambulance and freight models available.
Jetcruzer ML-1: Unmanned military model.
Jetcruzer ML-2: Piloted military model.

CUSTOMERS: Total of 188 firm orders by October 2000, plus a letter of intent for 30 aircraft and 20 options from China Eastern Aviation Educational Training. Initial production rate to be 130 per year.

COSTS: US$1,495,000 (2001). Development cost of Jetcruzer 450 programme up to certification was US$25 million; further US$32 million to certify 500.

DESIGN FEATURES: Canard layout; spin-resistant design; high speed; short field T-O/landing performance; all CAD; rear-mounted main wing; tip-mounted fins; unswept foreplane. Compared to prototype, production Jetcruzer 500P has 25 cm (10 in) fuselage stretch, addition of fuel-carrying strakes at wingroots with 34° sweepback, 43 cm (17 in) nose extension to allow forward retraction of nosewheel, and vertical tail surfaces canted 5° inboard to improve aileron control.

Mainplane section NACA 2412 with 20° sweepback at quarter-chord and 4° dihedral; foreplane NASA LS 0417 (Mod).

FLYING CONTROLS: Manual. Actuated by pushrods. Ailerons in main wing; elevators on foreplane; rudders in tip-mounted fins.

STRUCTURE: Aluminium alloy wings, foreplane and vertical tail surfaces; monocoque fuselage of graphite composite/Nomex honeycomb sandwich with embedded aluminium mesh.

LANDING GEAR: Retractable tricycle oleo-pneumatic type with steerable nosewheel. Mainwheels retract inwards; nosewheel forwards.

POWER PLANT: One 634 kW (850 shp) Pratt & Whitney Canada PT6A-66A turboprop, driving a Hartzell five-blade pusher propeller. Fuel capacity 946 litres (250 US gallons; 208 Imp gallons), of which 931 litres (246 US gallons; 205 Imp gallons) are usable.

ACCOMMODATION: Pilot plus five passengers in three pairs of side-by-side leather seats certified crash-resistant to 21 g. Pilot's access is through door in port side of forward fuselage; main cabin entry via starboard side airstep door at mid-fuselage point. Four cabin windows on each side. Dual controls. Baggage compartments in nose bay and, in cabin, behind rear seating.

SYSTEMS: Cabin pressurisation 0.41 bar (6.0 lb/sq in) gives equivalent of 2,500 m (8,200 ft) altitude when flying at 9,140 m (30,000 ft). 24 V DC battery, 300 A starter/generator. R134A refrigerant air conditioning, de-icing, anti-icing and supplemental oxygen standard.

AVIONICS: *Comms:* Honeywell dual KX 165 VHF com, KT 70 Mode S transponder and KMA 24H audio panel.
Radar: Four-colour weather radar and/or 3M Stormscope optional.
Flight: Honeywell dual KX 165 VHF nav, dual ILS, KI 208 VOR with glide slope, KR 21 marker beacon receiver, KR 87A ADF, KI 482 ADI, ED 461 HSI, KLN GPS, RMI and KEA 130A encoding altimeter; fully coupled KFC 325 three-axis autopilot.
Instrumentation: Honeywell Silver Crown IFR package standard; EFIS with 102 mm (4 in) displays optional.

DIMENSIONS, EXTERNAL:
Wing span	12.85 m (42 ft 2 in)
Wing aspect ratio	9.2
Foreplane span	5.77 m (18 ft 11 in)
Length overall	9.19 m (30 ft 2 in)
Fuselage: Max depth	1.46 m (4 ft 9½ in)
Height: to top of fuselage	2.13 m (7 ft 0 in)
overall	3.20 m (10 ft 6 in)
Wheel track	3.05 m (10 ft 0 in)
Wheelbase	3.91 m (12 ft 10 in)
Propeller diameter	2.16 m (7 ft 1 in)
Ground clearance: fuselage	0.70 m (2 ft 3½ in)
ventral fin	0.305 m (1 ft 0 in)
propeller	0.58 m (1 ft 11 in)
Cabin doors (each): Height	1.09 m (3 ft 7 in)
Width	0.965 m (3 ft 2 in)

DIMENSIONS, INTERNAL:
Cabin, incl flight deck: Length	4.62 m (15 ft 2 in)
Max width	1.24 m (4 ft 1 in)
Max height	1.27 m (4 ft 2 in)
Volume	7.00 m³ (247.1 cu ft)
Baggage hold, total volume	1.56 m³ (55.2 cu ft)

AREAS:
Wings, gross	17.95 m² (193.2 sq ft)
Ailerons (total)	0.81 m² (8.75 sq ft)
Foreplanes (total)	3.71 m² (39.90 sq ft)
Elevators (total, incl tabs)	0.42 m² (4.53 sq ft)
Fins (total)	0.81 m² (8.75 sq ft)
Rudders (total)	0.33 m² (3.60 sq ft)

WEIGHTS AND LOADINGS:
Weight empty	1,760 kg (3,880 lb)
Max fuel weight	726 kg (1,600 lb)
Max T-O weight	2,812 kg (6,200 lb)
Max wing/foreplane loading	129.9 kg/m² (26.60 lb/sq ft)
Max power loading	4.44 kg/kW (7.29 lb/shp)

PERFORMANCE (provisional):
Max operating speed (V_{MO})	312 kt (579 km/h; 360 mph)
Econ cruising speed	280 kt (518 km/h; 322 mph)
Stalling speed	65 kt (121 km/h; 75 mph)
Max rate of climb at S/L at MTOW	975 m (3,200 ft)/min
Max certified altitude	9,140 m (30,000 ft)
T-O run	579 m (1,900 ft)
T-O to 15 m (50 ft)	707 m (2,320 ft)
Landing from 15 m (50 ft)	534 m (1,750 ft)
Landing run with propeller reversal	376 m (1,231 ft)
Range with max fuel at econ cruising speed, no reserves	1,280 n miles (2,372 km; 1,474 miles)
Endurance	6 h

UPDATED

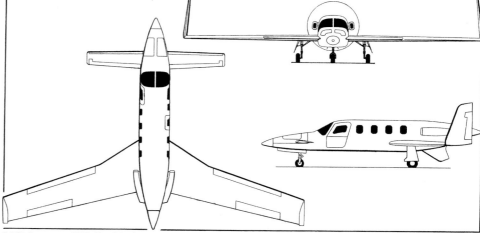

Second prototype AASI Jetcruzer 500 six-seat business aircraft *(Paul Jackson/Jane's)*

AASI Jetcruzer 500 *(Paul Jackson/Jane's)*

ADAM

ADAM AIRCRAFT INDUSTRIES

6560 South Greenwood Plaza, Suite 220, Englewood, Colorado 80111
Tel: (+1 303) 406 59 00
Fax: (+1 303) 406 59 50
e-mail: info@vansaircraft.com
Web: http://www.adamaircraft.net
CEO: George F 'Rick' Adam
COO: Cecil Miller
PRESIDENT: John Knudsen
INFORMATION CONTACT: Mary Hammack

Adam Aircraft Industries formed in 1998 and is developing the M-309 centreline-thrust twin-engine business aircraft, which it will manufacture in a new 2,975 m² (32,000 sq ft) facility at Denver Centennial Airport, Colorado. In mid-2000 the workforce numbered 12 but will rise when production begins.

NEW ENTRY

ADAM M-309

TYPE: Business twinprop.
PROGRAMME: Designed by Burt Rutan, designation M-309 denoting his 309th aircraft design; development started September 1999; proof-of-concept aircraft (N309A), manufactured by Scaled Composites Inc, first flown at Mojave, California 21 March 2000; formally rolled-out 5 April 2000; total of 80 flight hours completed by the time of public debut at EAA AirVenture 2000 at Oshkosh in July 2000; prototype scheduled to fly in third quarter of 2001, followed by a second, production conforming, prototype, leading to FAA FAR Pt 23 IFR certification, including flight into known icing and first customer deliveries expected in 2003.
  Future growth versions are planned.
CUSTOMERS: Orders for five aircraft received by October 2000; delivery positions between six and 25 then on option.
COSTS: Unit cost less than US$1 million (2000).
DESIGN FEATURES: Twin-boom configuration with swept fins and high-set tailplane; unswept low-wings have dihedral on outboard panels. Tractor/pusher power plant simplifies handling with single engine failure.
FLYING CONTROLS: Conventional and manual.
STRUCTURE: Single-cure graphite carbon composite with no secondary bonds or fasteners. Proof-of-concept aircraft has three-spar main wing spar, but production aircraft will have a single-piece spar, with wing mounted lower than on POC aircraft to minimise spar carry-through in cabin.
LANDING GEAR: Retractable tricycle type, with single wheel on each unit; trailing link suspension on main units.
POWER PLANT: Two 261 kW (350 hp) Teledyne Continental TSIO-550 flat-six piston engines, with FADEC, mounted in centreline thrust configuration and driving three-blade Hartzell propellers. Future versions may be powered by diesel and turboprop engines. Fuel capacity 946 litres (250 US gallons; 208 Imp gallons).
ACCOMMODATION: Pilot and five passengers in pressurised cabin on three pairs of seats. Door on port side, above wing leading edge.
SYSTEMS: Pressurisation system maintains a 2,440 m (8,000 ft) cabin environment at 7,620 m (25,000 ft).
AVIONICS: Meggitt Avionics MAGIC flat-panel display standard. Nexrad weather radar mounted in forward end of one tailboom.

DIMENSIONS, EXTERNAL:
Wing span	12.80 m (42 ft 0 in)
Wing aspect ratio	10.9
Length overall	9.69 m (31 ft 9½ in)
Height overall	2.68 m (8 ft 9½ in)
Wheel track	3.57 m (11 ft 8½ in)
Wheelbase	3.08 m (10 ft 1¼ in)
Propeller diameter	1.93 m (6 ft 4 in)

DIMENSIONS, INTERNAL:
Cabin:
Length	4.15 m (13 ft 7¼ in)
Max width	1.37 m (4 ft 6 in)
Max height	1.29 m (4 ft 3 in)

Proof-of-concept prototype Adam M-309
2001/0099067

AREAS:
Wings, gross	15.05 m² (162.0 sq ft)

WEIGHTS AND LOADINGS:
Weight empty	1,533 kg (3,380 lb)
Max fuel	499 kg (1,100 lb)
Max T-O weight	2,449 kg (5,400 lb)
Max wing loading	162.7 kg/m² (33.33 lb/sq ft)
Max power loading	4.70 kg/kW (7.71 lb/hp)

PERFORMANCE:
Max level speed at 6,100 m (20,000 ft)
  250 kt (463 km/h; 288 mph)
Cruising speed at 75% power at 6,100 m (20,000 ft)
  220 kt (407 km/h; 253 mph)
Econ cruising speed at 60% power at 6,100 m (20,000 ft)
  190 kt (352 km/h; 219 mph)
Stalling speed:
flaps up	75 kt (139 km/h; 87 mph)
flaps down	70 kt (130 km/h; 81 mph)
Max rate of climb at S/L:	518 m (1,700 ft)/min
Rate of climb at S/L, OEI	113 m (370 ft)/min
Service ceiling	7,620 m (25,000 ft)

Range with max fuel, economy power setting, IFR reserves 1,500 n miles (2,778 km; 1,726 miles)

NEW ENTRY

AEROCOMP

AEROCOMP

2335 Newfound Harbor Drive, Merritt Island, Florida 32952
Tel: (+1 407) 453 66 41
Fax: (+1 407) 543 88 47
e-mail: info@AerocompInc.com
Web: http://www.AerocompInc.com
OWNERS:
  Ron Lueck
  Stephen Young

Formed in 1993 and previously known for its production of floats, Aerocomp markets the Comp Monster kitbuilt, together with six-, seven-, eight- and 10-seat versions, known respectively as the Comp Air 6, 7, 8 and 10. Aerocomp acquired production rights for the Merlin GT and E-Z Flyer from Merlin Aircraft Inc, and has passed the latter to Blue Yonder Aviation of Canada (which see).

UPDATED

AEROCOMP COMP MONSTER

Renamed Comp Air 4.

UPDATED

AEROCOMP COMP AIR 3

TYPE: Three-seat kitbuilt.
PROGRAMME: Announced 1998; smaller version of Comp Air 4.
CUSTOMERS: At least one flying by early 2000.
COSTS: Kit US$25,595 (2000).
Data generally as for Comp Air 4, except as noted.
STRUCTURE: Wings removable for storage. Tapered wings, optional.
POWER PLANT: One 119 kW (160 hp) Textron Lycoming O-320 in prototype; engines in range 112 to 134 kW (150 to 180 hp) recommended. Fuel capacity: standard 170 litres (45.0 US gallons; 37.5 Imp gallons).

ACCOMMODATION: Pilot and passenger side by side; third occupant, or cargo, in back of cockpit.

DIMENSIONS, EXTERNAL:
Wing span	10.52 m (34 ft 6 in)
Wing aspect ratio	6.6
Length overall	7.32 m (24 ft 0 in)
Max width of fuselage	0.91 m (3 ft 0 in)
Height overall	2.46 m (8 ft 1 in)

DIMENSIONS, INTERNAL:
Cabin max width	1.09 m (3 ft 7 in)

AREAS:
Wings, gross	16.35 m² (176.0 sq ft)

WEIGHTS AND LOADINGS:
Weight empty	590 kg (1,300 lb)
Max T-O weight	1,111 kg (2,450 lb)
Max wing loading	68.0 kg/m² (13.92 lb/sq ft)

PERFORMANCE (119 kW; 160 hp Textron Lycoming engine):
Max operating speed	152 kt (281 km/h; 175 mph)
Normal cruising speed	126 kt (233 km/h; 145 mph)
Stalling speed	40 kt (73 km/h; 45 mph)
Max rate of climb at S/L	335 m (1,100 ft)/min
Service ceiling	4,724 m (15,500 ft)
T-O run	107 m (350 ft)
Landing run	183 m (600 ft)
Range	725 n miles (1,342 km; 834 miles)

NEW ENTRY

AEROCOMP COMP AIR 4

TYPE: Four-seat kitbuilt.
PROGRAMME: First flew 3 April 1995. Public debut, then named Comp Monster, at Sun 'n' Fun 1995, when powered by 82 kW (110 hp) Hirth F 30 four-cylinder two-stroke engine.
CURRENT VERSIONS: **150G:** Powered by 112 kW (150 hp) Textron Lycoming O-320.
  180G: 134 kW (180 hp) Textron Lycoming O-360; *as described.*
  180SF: Float-equipped version of 180G.
  Trainer: 'Two plus two' seat trainer version.
CUSTOMERS: Total of 14 flying by end 1998 (latest information).
COSTS: US$47,995 with 150 hp, US$46,995 with 180 hp Lycoming; US$26,995 basic kit without engine, propeller and instruments; Trainer kit US$22,900. (All 2000.)
DESIGN FEATURES: Easy-to-build composites aircraft; quoted build time 350 to 400 hours. High, braced wing; unswept, but with optional tapered trailing-edge.
  Modified Clark Y aerofoil wings with turned-down tips.
FLYING CONTROLS: Conventional and manual. Flaps.
STRUCTURE: Composites construction with carbon and Kevlar reinforcement. Single-strut braced wings; braced tailplane. Optional tapered wing. Fuselage width upgrades available up to maximum of 1.21 m (3 ft 11½ in).
LANDING GEAR: Spring steel; available with nosewheel, tailwheel, floats or amphibious gear, last-mentioned using Matco wheels and hydraulic brakes with 5.00-5 in tyres. Non-retracting 0.15 m (6 in) Matco tailwheel acts as water rudder. Tailwheel version uses 6.00-6 in Matco wheels, tyres and brakes with a 0.20 m (8 in) spring-legged tailwheel.
POWER PLANT: One 134 kW (180 hp) Textron Lycoming O-360-A1A flat-four engine, driving a Sensenich 76 × 57 three-blade metal propeller. Options exist for engines ranging from 82 to 186 kW (110 to 250 hp); prototype employed 82 kW (110 hp) Hirth 95. Fuel capacity 197 litres (52.0 US gallons; 43.3 Imp gallons).
ACCOMMODATION: Pilot and three passengers in two side-by-side pairs. Glass-panelled door on each side of cabin for improved downwards visibility.

DIMENSIONS, EXTERNAL:
Wing span	11.46 m (37 ft 7 in)
Wing chord (constant version)	1.88 m (6 ft 2 in)
Length overall	7.92 m (26 ft 0 in)
Height overall	2.44 m (8 ft 0 in)
Tailplane span	3.05 m (10 ft 0 in)
Doors (each): Height	0.76 m (2 ft 6 in)
Width	0.91 m (3 ft 0 in)

DIMENSIONS, INTERNAL:
Cabin: Length	2.74 m (9 ft 0 in)

Aerocomp Comp Air 4 trainer, with nosewheel option 0015571

Aerocomp Comp Air 4 four-seat kitbuilt 0015572

AEROCOMP—AIRCRAFT: USA

Max width	1.07 m (3 ft 6 in)
Max height	1.30 m (4 ft 3 in)

AREAS:
Wings, gross	19.70 m² (212.0 sq ft)

WEIGHTS AND LOADINGS (G: 180G landplane, SF: 180SF floatplane):
Weight empty: G	630 kg (1,390 lb)
SF	753 kg (1,660 lb)
Baggage capacity	81.6 kg (180 lb)
Max T-O weight, both	1,292 kg (2,850 lb)
Max wing loading, both	65.6 kg/m² (13.44 lb/sq ft)
Max power loading, both	9.64 kg/kW (15.83 lb/hp)

PERFORMANCE:
Max operating speed: G	129 kt (239 km/h; 149 mph)
SF	110 kt (204 km/h; 127 mph)

Normal cruising speed at 70% power:
G	115 kt (212 km/h; 130 mph)
SF	100 kt (185 km/h; 115 mph)
T-O speed	36 kt (68 km/h; 42 mph)
Stalling speed: G	34 kt (63 km/h; 39 mph)
SF	35 kt (65 km/h; 40 mph)
Max rate of climb at S/L: G	442 m (1,450 ft)/min
SF	366 m (1,200 ft)/min
Service ceiling	4,880 m (16,000 ft)
T-O run: G	91 m (300 ft)
SF	221 m (725 ft)
Landing run	213 m (700 ft)
Range	660 n miles (1,222 km; 759 miles)

NEW ENTRY

Prototype Aerocomp Comp Air 6 six-seat composites kitplane *(Paul Jackson/Jane's)* 2001/0089484

AEROCOMP COMP AIR 6

TYPE: Six-seat kitbuilt.
PROGRAMME: Development of Comp Air 4; first flew January 1996.
CURRENT VERSIONS: **CA6G**: Normal landing gear, *as described.*
CA6SF: Float-equipped version.
CA6AF: Amphibian.
CA6TW: Tapered wing.
CA6TWHG: Tapered wing, high gross (1,451 kg; 3,200 lb MTOW).
CA6SHG: Super high gross (1,632 kg; 3,600 lb MTOW), wide-bodied version.
CUSTOMERS: At least 15 flying by early 2000.
COSTS: Kit US$29,995, without engine (2000).
DESIGN FEATURES: As Comp Air 4. Quoted build time 350 hours.
Data for CA6G generally as Comp Air 4, except those below.
STRUCTURE: Wings removable for storage. Optional wide-body fuselage and tapered wing upgrades available.
LANDING GEAR: Choice of tricycle or tailwheel configuration; optional floats. Mainwheels 6.00-6 in.
POWER PLANT: Prototype had 164 kW (220 hp) Franklin engine; design compatible with engines in the 164 to 224 kW (220 to 300 hp) class. Automotive engine conversions are also possible. Fuel capacity 310 litres (82.0 US gallons; 68.3 Imp gallons).

DIMENSIONS, EXTERNAL:
Wing span	10.52 m (34 ft 6 in)
Wing aspect ratio	5.6
Length overall	7.47 m (24 ft 6 in)
Height overall	2.44 m (8 ft 0 in)

DIMENSIONS, INTERNAL:
Cabin: Length	3.30 m (10 ft 10 in)
Max width	1.08 m (3 ft 6½ in)
Max height	1.30 m (4 ft 3 in)

AREAS:
Wings, gross	19.70 m² (212.0 sq ft)

WEIGHTS AND LOADINGS:
Weight empty	676 kg (1,490 lb)
Max T-O weight	1,293 kg (2,850 lb)
Max wing loading	65.6 kg/m² (13.44 lb/sq ft)
Max power loading	7.88 kg/kW (12.95 lb/hp)

PERFORMANCE (164 kW; 220 hp Franklin engine):
Max operating speed	145 kt (268 km/h; 167 mph)
Normal cruising speed	133 kt (246 km/h; 153 mph)
Stalling speed	35 kt (65 km/h; 40 mph)
Max rate of climb at S/L	366 m (1,200 ft)/min
Service ceiling	5,480 m (18,000 ft)
T-O run	107 m (350 ft)
Landing run	168 m (550 ft)
Range	695 n miles (1,287 km; 800 miles)

UPDATED

Turboprop-powered Comp Air 7 *(Paul Jackson/Jane's)* 2001/0089485

AEROCOMP COMP AIR 7

TYPE: Utility kitbuilt.
PROGRAMME: Development of Comp Air 6. First shown 1998.
CURRENT VERSIONS: **Comp Air 7**: Piston-engine version.
Comp Air 7T: Turbine version.
CUSTOMERS: At least two of each version flying by early 2000.
COSTS: Comp Air 7 US$39,995; Comp Air 7T US$49,995; both minus engine (2000).
Data generally as for Comp Air 6, except those below.
DESIGN FEATURES: Suited for bush operations. Quoted build time 700 hours.
FLYING CONTROLS: Horn-balanced tail surfaces. Ground-adjustable tab on each aileron and on rudder. Trim tab in starboard elevator.
LANDING GEAR: Tailwheel type; fixed. Speed fairings optional. Mainwheels 8.00-6.
POWER PLANT: *Comp Air 7:* One Textron Lycoming TIO-540 rated between 194 and 261 kW (260 and 350 hp), driving a two-blade metal propeller. Fuel capacity 333 litres (88.0 US gallons; 73.3 Imp gallons); optional extra tank increases capacity to 455 litres (120 US gallons; 100 Imp gallons).
Comp Air 7T: One 490 kW (657 shp) Walter M 601D turboprop driving an three-blade, constant-speed, feathering propeller. Fuel capacity 568 litres (150 US gallons; 125 Imp gallons).
ACCOMMODATION: Pilot and up to six passengers in enclosed cabin. Third (rear passengers') door, starboard side.

DIMENSIONS, EXTERNAL (Comp Air 7 and Comp Air 7T):
Wing span: CA 7, CA 7T	10.67 m (35 ft 0 in)
Wing aspect ratio: CA 7, CA 7T	6.9
Length overall: CA 7	8.08 m (26 ft 6 in)
CA 7T	8.99 m (29 ft 6 in)
Height overall: CA 7, CA 7T	2.44 m (8 ft 0 in)

DIMENSIONS, INTERNAL:
Cabin max width: CA 7 standard	1.08 m (3 ft 6½ in)
CA 7 optional; CA 7T standard	1.17 m (3 ft 10 in)
CA 7 optional	1.21 m (3 ft 11½ in)

AREAS:
Wings, gross: CA 7, CA 7T	16.54 m² (178.0 sq ft)

WEIGHTS AND LOADINGS:
Weight empty: CA 7	953 kg (2,100 lb)
CA 7T	1,157 kg (2,550 lb)
Max T-O weight: CA 7	1,678 kg (3,700 lb)
CA 7T: standard	1,710 kg (3,770 lb)
high gross upgrade	2,177 kg (4,800 lb)
Max landing weight:	
CA 7T high gross upgrade	2,087 kg (4,600 lb)
Max wing loading:	
CA 7	101.5 kg/m² (20.79 lb/sq ft)
CA 7T: standard	103.4 kg/m² (21.18 lb/sq ft)
high gross upgrade	131.7 kg/m² (26.97 lb/sq ft)
Max power loading:	
CA 7T: standard	3.49 kg/kW (5.74 lb/shp)
high gross upgrade	4.45 kg/kW (7.31 lb/shp)

PERFORMANCE:
Never-exceed speed (V_NE):
CA 7	191 kt (354 km/h; 220 mph)
CA 7T	223 kt (413 km/h; 257 mph) IAS

Max operating speed:
CA 7	178 kt (330 km/h; 205 mph)

Max cruising speed:
CA 7T	239 kt (443 km/h; 275 mph) TAS
Stalling speed: CA 7	46 kt (86 km/h; 53 mph)
CA 7T	48 kt (89 km/h; 55 mph)

Max rate of climb at S/L:
CA 7	457 m (1,500 ft)/min
CA 7T	1,219 m (4,000 ft)/min
Service ceiling: CA 7	7,620 m (25,000 ft)
T-O run: CA 7	145 m (475 ft)
CA 7T	122 m (400 ft)
Landing run: CA 7	244 m (800 ft)

Range:
CA 7	1,100 n miles (2,037 km; 1,265 miles)
g limits: CA 7T	+6/−4

UPDATED

AEROCOMP COMP AIR 8

TYPE: Utility kitbuilt.
PROGRAMME: Prototype first flew mid-June 1999 and publicly displayed at Oshkosh in July 1999.
CUSTOMERS: Two flying by end 1999.
COSTS: Kit US$59,995 (2000), excluding engine.
DESIGN FEATURES: Generally similar to Comp Air 7.
FLYING CONTROLS: Mass-balanced controls for high-speed handling.

Aerocomp Comp Air 8 (high gross weight) owned by Stephen Darrow of Merritt Island, Florida *(Paul Jackson/Jane's)* 2001/0089486

542   USA: AIRCRAFT—AEROCOMP

Aerocomp Comp Air 10 floatplane *(Paul Jackson/Jane's)*

STRUCTURE: Generally as Comp Air 7, but with strengthened carbon fibre fuselage and tail unit. Tapered wings standard. Fuselage width upgrades available.
LANDING GEAR: Tailwheel type; fixed.
POWER PLANT: One 490 kW (657 shp) Walter M 601D turboprop driving an Avia three-blade, feathering constant-speed propeller. Fuel capacity 681 litres (180 US gallons; 150 Imp gallons).
ACCOMMODATION: Pilot and five adults in three pairs of bucket seats, plus rear bench seat for two children.
DIMENSIONS, EXTERNAL::
  Wing span   10.97 m (36 ft 0 in)
  Wing aspect ratio   5.5
  Length overall   9.60 m (31 ft 6 in)
  Height overall   2.46 m (8 ft 1 in)
DIMENSIONS, INTERNAL:
  Cabin max width: standard 1.17 m (3 ft 10 in)
    wide option   1.21 m (3 ft 11½ in)
AREAS:
  Wings, gross   22.02 m² (237.0 sq ft)
WEIGHTS AND LOADINGS:
  Weight empty   1,134 kg (2,500 lb)
  Max T-O weight: standard 2,177 kg (4,800 lb)
    high gross option   2,540 kg (5,600 lb)
  Max wing loading: standard 98.9 kg/m² (20.25 lb/sq ft)
    high gross option   115.4 kg/m² (23.63 lb/sq ft)
  Max power loading: standard 4.45 kg/kW (7.31 lb/shp)
    high gross option   5.19 kg/kW (8.52 lb/shp)
PERFORMANCE:
  Never-exceed speed (V$_{NE}$) 197 kt (365 km/h; 227 mph)
  Max cruising speed at 6,400 m (21,000 ft)
                           217 kt (402 km/h; 250 mph) IAS
  Normal cruising speed   182 kt (338 km/h; 210 mph)
  Stalling speed, power off, flaps down
                           42 kt (78 km/h; 48 mph)
  Max rate of climb at S/L 610 m (2,000 ft)/min
  Service ceiling   8,230 m (27,000 ft)
  T-O run   122 m (400 ft)
  Landing run   183 m (600 ft)
  Range   990 n miles (1,833 km; 1,139 miles)
                           NEW ENTRY

AEROCOMP COMP AIR 10
TYPE: Utility kitbuilt.
PROGRAMME: Announced 1997; based on Comp Air 6.
CURRENT VERSIONS: **CA10**: Standard version, *as described*.
  CA10XL: Stretched version; length 9.88 m (32 ft 5 in); maximum T-O weight 2,722 kg (6,000 lb).
COSTS: US$69,995 turbopowered kit version (2000). CA-10XL US$5,000 extra.
Data generally as for Comp Air 6, except that below.
DESIGN FEATURES: Twin outward-canted fins in addition to conventional horizontal tail surfaces.
FLYING CONTROLS: Conventional and manual. Twin rudders operating in unison. Balance tab in starboard elevator.
STRUCTURE: Extensive use of composites throughout. Single-strut braced wings. Quoted build time 600 hours.
LANDING GEAR: Fixed tricycle type; sprung steel main legs. Mainwheel size 8.00-6 in; nosewheel 6.00-6 in. Floats optional.
POWER PLANT: One 213 kW (285 hp) Teledyne Continental O-520 piston engine driving a Hartzell three-blade constant-speed propeller; turboprop version has 490 kW (657 shp) Walter M 601D driving an Avia three-blade, constant-speed propeller. Standard fuel capacity of piston version 363 litres (96.0 US gallons; 78.9 Imp gallons); optional capacity 568 litres (150 US gallons; 125 Imp gallons); fuel capacity of turboprop version 455 litres (120 US gallons; 100 Imp gallons). Oil capacity 5.7 litres (1.5 US gallons; 1.25 Imp gallons).

ACCOMMODATION: Pilot and up to nine passengers; passenger door, forward, each side; horizontally split, two-piece freight door, starboard, rear side; extra passenger door on port rear side. Optional luggage pannier.
DIMENSIONS, EXTERNAL:
  Wing span   11.43 m (37 ft 6 in)
  Length overall   9.40 m (30 ft 10 in)
  Height overall   2.74 m (9 ft 0 in)
  Wheel track   2.79 m (9 ft 2 in)
DIMENSIONS, INTERNAL:
  Cabin: Length: CA10   3.30 m (10 ft 10 in)
    CA10XL   3.96 m (13 ft 0 in)
  Max width   1.54 m (5 ft 0½ in)
  Max height   1.35 m (4 ft 5 in)
AREAS:
  Wings, gross   22.56 m² (243.0 sq ft)
WEIGHTS AND LOADINGS (A: 213 kW; 285 hp piston engine, B: 490 kW; 657 shp turboprop):
  Weight empty: A   1,225 kg (2,700 lb)
    B   1,247 kg (2,750 lb)
  Baggage capacity: A, B   136 kg (300 lb)
  Max T-O and landing weight: A 2,358 kg (5,200 lb)
    B   2,449 kg (5,400 lb)
  Max wing loading: A   104.5 kg/m² (21.40 lb/sq ft)
    B   108.5 kg/m² (22.22 lb/sq ft)
  Max power loading: A   11.11 kg/kW (18.25 lb/hp)
    B   5.00 kg/kW (8.22 lb/shp)
PERFORMANCE (A, B, as above):
  Max level speed: A   139 kt (257 km/h; 160 mph)
    B   186 kt (346 km/h; 215 mph)
  Max cruising speed at 75% power:
    A   130 kt (241 km/h; 150 mph)
    B   167 kt (309 km/h; 192 mph)
  Normal cruising speed at 65% power:
    A   124 kt (230 km/h; 143 mph)
    B   152 kt (282 km/h; 175 mph)
  Stalling speed: flaps up, A, B 51 kt (95 km/h; 59 mph)
    flaps down: A   46 kt (86 km/h; 53 mph)
  Max rate of climb at S/L: A 305 m (1,000 ft)/min
    B   610 m (2,000 ft)/min
  Service ceiling: A, B   6,400 m (21,000 ft)
  T-O run: A   244 m (800 ft)
    B   91 m (300 ft)
  Landing run: A, B   213 m (700 ft)

Range with max fuel at 65% power:
  A   1,162 n miles (2,152 km; 1,337 miles)
  B   795 n miles (1,472 km; 915 miles)
Endurance at 65% power, 60 min reserves: B 3 h 42 min
                           UPDATED

AEROCOMP MERLIN GT
TYPE: Side-by-side kitbuilt.
PROGRAMME: Developed by Macair Aircraft in Canada; later Merlin Aircraft.
CURRENT VERSIONS: Available as floatplane, trainer, agricultural sprayer.
CUSTOMERS: At least 250 flying by mid-2000.
COSTS: Complete kits: Standard (with Rotax 582) US$24,995; Rotax 912 powered version US$32,995; Rotax 912S powered version US$33,995 (2000).
DESIGN FEATURES: Strut-braced high wing; braced tailplane. No aerodynamic balances or tabs on empennage.
FLYING CONTROLS: Manual. Full span Junkers flaperons; conventional rudder and elevator. Actuation by pushrods and cables.
STRUCTURE: Fuselage of welded 4130 chromoly steel tubing, with wooden floor and glass fibre engine cowling, otherwise fabric-covered. Metal wings with fabric covering.
LANDING GEAR: Non-retractable mainwheels and tailwheel. Metal tube mainwheel legs with bungee-bound shock-absorption; hydraulic brakes; steerable tailwheel. Full Lotus or Superfloats floats optional.
POWER PLANT: One 47.8 kW (64.1 hp) Rotax 582 engine, driving a two- or three-blade propeller; or a choice of 59.6 kW (79.9 hp) Rotax 912, 73.5 kW (98.6 hp) Rotax 912 ULS, 55.0 kW (73.8 hp) Rotax 618, 74.6 kW (100 hp) Canadian Automotive (CAM) 100, 59.7 kW (80 hp) Jabiru 2000 or 82.0 kW (110 hp) Formula Power Subaru EA81 engines. Fuel capacity 61 litres (16.0 US gallons; 13.3 Imp gallons).
SYSTEMS: 12 V electrical system, with battery.
ACCOMMODATION: Two persons, side by side with baggage stowage behind seats. Upward-opening door each side. Dual controls.
DIMENSIONS, EXTERNAL:
  Wing span   9.75 m (32 ft 0 in)
  Wing chord, constant   1.52 m (5 ft 0 in)
  Wing aspect ratio   6.0
  Length overall   6.10 m (20 ft 0 in)
  Height overall   1.98 m (6 ft 6 in)
  Tailplane span   2.15 m (7 ft 0½ in)
  Wheel track   2.21 m (7 ft 3 in)
  Wheelbase   6.10 m (20 ft 0 in)
DIMENSIONS, INTERNAL:
  Cabin max width   1.04 m (3 ft 5 in)
AREAS:
  Wings, gross   14.86 m² (160.0 sq ft)
WEIGHTS AND LOADINGS (Rotax 912):
  Weight empty   261 kg (575 lb)
  Max baggage capacity   45 kg (100 lb)
  Max T-O weight: landplane 590 kg (1,300 lb)
    floatplane   635 kg (1,400 lb)
  Max wing loading: landplane 39.7 kg/m² (8.13 lb/sq ft)
    floatplane   42.7 kg/m² (8.75 lb/sq ft)
  Max power loading: landplane 9.89 kg/kW (16.25 lb/hp)
    floatplane   10.65 kg/kW (17.5 lb/hp)
PERFORMANCE (Rotax 912):
  Never exceed speed (V$_{NE}$) 104 kt (193 km/h; 120 mph)
  Cruising speed   81 kt (150 km/h; 93 mph)
  Stalling speed   34 kt (63 km/h; 38 mph)
  Max rate of climb at S/L 455 m (1,500 ft)/min
  T-O distance   30 m (100 ft)
  Landing distance   61 m (200 ft)
  Max range   350 n miles (648 km; 402 miles)
                           UPDATED

Aerocomp Merlin GT, assembled in Malaysia by Langkawi Recreation Club and powered by a Rotax 912 ULS *(Paul Jackson/Jane's)*

AEROLITES

AEROLITES INC
12104 David Road, Welsh, Louisiana 70591
Tel/Fax: (+1 318) 734 38 65
e-mail: aerolites@centurytel.net
Web: http://www.aerolites.com
Telex: 6503079915 (WUI)

NEW ENTRY

AEROLITES AEROMASTER

TYPE: Single-seat agricultural sprayer ultralight kitbuilt.
CUSTOMERS: Total of 12 flying by mid-2000.
COSTS: Kit US$23,600 (2000).
DESIGN FEATURES: Strut-braced aluminium wing has ladder construction with internal diagonal bracing tested to +6/–4 g; aluminium ribs slot into pockets within fabric covering.
FLYING CONTROLS: Conventional and manual.
STRUCTURE: Fuselage of welded 4130 chromoly with fabric covering.
LANDING GEAR: Tailwheel type; fixed.
POWER PLANT: One 47.8 kW (64.1 hp) Rotax 582 two-cylinder engine driving Warp Drive carbon fibre three-blade propeller through C type 3:1 reduction gear. Fuel capacity 38 litres (10.0 US gallons; 8.3 Imp gallons).
EQUIPMENT: 12-nozzle SprayMiser system fitted for agricultural operations; belly tank capacity 114 litres (30.0 US gallons; 25.0 Imp gallons). Optional ballistic parachute.
DIMENSIONS, EXTERNAL:
  Wing span   8.74 m (28 ft 8 in)
  Length overall   5.64 m (18 ft 6 in)
  Height overall   2.11 m (6 ft 11 in)
AREAS:
  Wings, gross   13.54 m² (145.7 sq ft)
WEIGHTS AND LOADINGS::
  Weight empty   193 kg (425 lb)
  Max T-O weight   453 kg (1,000 lb)
PERFORMANCE (with SprayMiser system fitted):
  Never-exceed speed (V$_{NE}$)   95 kt (177 km/h; 110 mph)
  Normal cruising speed   52 kt (97 km/h; 60 mph)
  Spray speed   56 kt (105 km/h; 65 mph)
  Stalling speed   28 kt (52 km/h; 32 mph)
  Max rate of climb at S/L   244 m (800 ft)/min
  T-O run   153 m (500 ft)
  Landing run   46 m (150 ft)
  Range with max fuel   150 n miles (277 km; 172 miles)
  Swath width   9-30 m (30-100 ft)

NEW ENTRY

AEROLITES AEROSKIFF

TYPE: Two-seat amphibian ultralight kitbuilt.
CUSTOMERS: Three flying by mid-2000.
COSTS: Kit US$23,775 (2000).
DESIGN FEATURES: Pusher layout. Wings detach for storage.
FLYING CONTROLS: Conventional and manual. High-lift flaps.
STRUCTURE: Glass fibre fuselage and fabric-covered, strut-braced wings.
LANDING GEAR: Tailwheel type; retracts for water landings. Drum brakes. Floats mounted on tubular outriggers at wing mid-point.
POWER PLANT: One 47.8 kW (64.1 hp) Rotax 582 water-cooled engine is standard; optionally, 55.0 kW (73.8 hp) Rotax 618. Fuel capacity 45 litres (12.0 US gallons; 10.0 Imp gallons).
SYSTEMS: Optional bilge pump and electric starter.
DIMENSIONS, EXTERNAL:
  Wing span   9.04 m (29 ft 8 in)
  Length overall   6.81 m (22 ft 4 in)
  Height overall   2.13 m (7 ft 0 in)
AREAS:
  Wings, gross   14.49 m² (156.0 sq ft)
WEIGHTS AND LOADINGS::
  Weight empty   256 kg (565 lb)
  Max T-O weight   510 kg (1,125 lb)
PERFORMANCE (Rotax 582):
  Never-exceed speed (V$_{NE}$)   82 kt (152 km/h; 95 mph)
  Normal cruising speed   56 kt (105 km/h; 65 mph)
  Stalling speed   33 kt (62 km/h; 38 mph)
  Max rate of climb at S/L   183 m (600 ft)/min
  T-O run: water   168 m (550 ft)
        land   122 m (400 ft)
  Landing run: land   92 m (300 ft)
  Range   190 n miles (351 km; 218 miles)

NEW ENTRY

AEROLITES BEARCAT

TYPE: Single-seat ultralight kitbuilt.
CUSTOMERS: Total of 15 completed by mid-2000.
COSTS: Kit US$13,695, including engine and propeller (2000).
DESIGN FEATURES: Replica Corben Baby Ace. Wings are strut-braced and similar to AeroMaster (which see). Quoted build time 60 to 90 hours.
FLYING CONTROLS: Conventional and manual.
STRUCTURE: Fabric-covered 4130 chromoly fuselage and aluminium extended-wing spar.
LANDING GEAR: Tailwheel type; fixed. Bungee shock absorption on main legs; optional hydraulic disc brakes.
POWER PLANT: One 31.0 kW (41.6 hp) Rotax 447 in-line two-cylinder piston engine, driving a two-blade propeller. Optionally, 37.0 kW (49.6 hp) Rotax 503 UL-2V. Fuel capacity 19 litres (5.0 US gallons; 4.2 Imp gallons).
DIMENSIONS, EXTERNAL:
  Wing span   9.14 m (30 ft 0 in)
  Length overall   5.33 m (17 ft 6 in)
  Height overall   1.96 m (6 ft 5 in)
AREAS:
  Wings, gross   13.93 m² (150.0 sq ft)
WEIGHTS AND LOADINGS:
  Weight empty   125 kg (275 lb)
  Max T-O weight   318 kg (700 lb)
PERFORMANCE:
  Never-exceed speed (V$_{NE}$)   95 kt (177 km/h; 110 mph)
  Max operating speed   61 kt (113 km/h; 70 mph)
  Normal cruising speed   48 kt (89 km/h; 55 mph)
  Stalling speed   24 kt (44 km/h; 27 mph)
  Max rate of climb at S/L   366 m (1,200 ft)/min
  T-O and landing run   46 m (150 ft)
  Range with max fuel   120 n miles (222 km; 138 miles)

NEW ENTRY

Aerolites Aeromaster agricultural ultralight *(Geoffrey P Jones)*   2001/0092107

AEROSTAR

AEROSTAR AIRCRAFT CORPORATION LLC
10555 Airport Drive, Coeur d'Alene Airport, Hayden Lake, Idaho 83835-9742
Tel: (+1 208) 762 03 38 or (+1 800) 442 42 42
Fax: (+1 208) 762 83 49
e-mail: info@aerostarjet.com
Web: www.aerostarjet.com
PRESIDENT: Steve Speer
VICE-PRESIDENT: Jim Christy
SENIOR PROJECT ENGINEER: William V Leeds

Aerostar Aircraft Corporation was formed by former Ted Smith employees and acquired the rights to the Aerostar series of pressurised piston-engined twins from Piper in 1991. It is planning to extend its 2,975 m² (32,000 sq ft) factory to 10,400 m² (112,000 sq ft) to accommodate production of a jet version.

A new company, Aerostar Jet LLC, was planned for launch in 2001 for the development of the FJ-100; in February 2001 it announced it was seeking US$8.4 million as the first stage towards certification.

UPDATED

AEROSTAR FJ-100

TYPE: Business jet.
PROGRAMME: Jet derivative of Aerostar 600/700 series, of which over 1,000 were built (see *Jane's* 1985-86); will be certified as a new-build design, but first flight-test example will be modified from an existing piston-engined aircraft. First flight originally scheduled for August 2001, with deliveries due to begin July 2002; has been revised (due to engine unavailability) to a first flight in March 2002 and deliveries by June 2003.
CURRENT VERSIONS: Family planned, ranging from four to eight seats.
CUSTOMERS: Total 25 deposits received by October 2000. First year (2003) production set at 17, rising to 34 in 2004 and 50 the following year; production expected to reach 70 per year.
COSTS: Development costs estimated at US$40 million. Unit price US$1.95 million (2001).
DESIGN FEATURES: Improved, jet-powered version of original piston-engined Aerostar; FAR Pt 23 compliant.
FLYING CONTROLS: Conventional and manual. Single-slotted Fowler flaps; maximum deflection 45°. Electrically operated trim tabs on port aileron, rudder and both elevators.
STRUCTURE: Metal throughout; semi-monocoque fuselage with flush-riveted skins; two-spar wing mounted mid-fuselage with 2° dihedral; contains integral fuel tanks. Mid-mounted tailplane; glass fibre dorsal fillet.
LANDING GEAR: Retractable, hydraulically actuated tricycle type; oleo-pneumatic suspension. Electrically actuated nosewheel steering. Dual calliper disc brakes on main wheels.
POWER PLANT: Two 5.35 kN (1,200 lb st) Williams FJ33-1 turbofans on elastomeric engine mounts on pylons on each side of rear fuselage. Fuel in integral wing tanks, plus bladder-type fuselage tank between rear of cabin and baggage compartment; all three tanks crossfeed for balanced fuel load.
ACCOMMODATION: Pilot and up to seven passengers in pairs of seats; standard layout is four club seats in cabin. Door on port side of fuselage behind cockpit. Baggage compartment to rear of cabin; optional lavatory on starboard side at rear of cabin; optional fold-out table.
SYSTEMS: 28 V DC battery for engine starting and as

Artist's impression of FJ-100 (two Williams FJ33-1 turbofans)   2000/0084552

USA: AIRCRAFT—AEROSTAR to AIR COMMAND

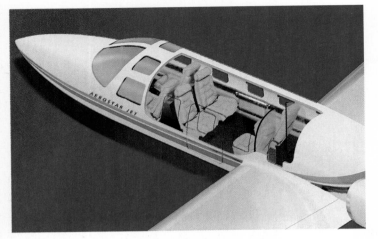

Cabin layout of Aerostar FJ-100

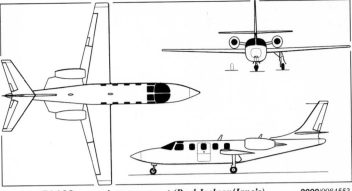

Aerostar FJ-100 general arrangement (Paul Jackson/Jane's)

emergency/back-up source; DC external power point on aft fuselage underside. Cabin pressurised to 0.52 bar (7.6 lb/sq in) at 3,050 m (10,000 ft) and cooled using Freon 134 vapour-cycle system. De-icing system to be fitted to engine inlets, surface leading edges and windscreen; exact type to be determined.

AVIONICS: Cockpit modelled on original Aerostar; instrument panel canted to enhance readability by solo pilot.
Comms: Dual VHF nav/com; Dual Mode S transponders; audio control panel.
Radar: Colour weather radar.
Flight: GPS; DME; three-axis autopilot.

DIMENSIONS, EXTERNAL:
Wing span	11.18 m (36 ft 8 in)
Wing aspect ratio	7.6
Length overall	11.76 m (38 ft 7 in)
Height overall	4.42 m (14 ft 6 in)
Tailplane span	4.37 m (14 ft 4 in)
Wheel track	4.37 m (14 ft 4 in)
Wheelbase	3.10 m (10 ft 2 in)

DIMENSIONS, INTERNAL:
Cabin: Length	4.23 m (13 ft 10½ in)
Max width	1.16 m (3 ft 9½ in)
Max height	1.21 m (3 ft 11½ in)
Baggage compartment volume	0.68 m³ (24.0 cu ft)
Freight hold volume	0.85 m³ (30.0 cu ft)

AREAS:
Wings, gross	16.54 m² (178.0 sq ft)

WEIGHTS AND LOADINGS:
Weight empty	1,905 kg (4,200 lb)
Max fuel weight	1,131 kg (2,494 lb)
Max T-O and landing weight	3,311 kg (7,300 lb)
Max ramp weight	3,334 kg (7,350 lb)
Max zero-fuel weight	2,676 kg (5,900 lb)
Max wing loading	200.2 kg/m² (41.01 lb/sq ft)
Max power loading	310 kg/kN (3.04 lb/lb st)

PERFORMANCE:
Max operating speed: S/L to 8,840 m (29,000 ft)	260 kt (482 km/h; 299 mph)
above 8,840 m (29,000 ft)	M0.71
Manoeuvring speed	167 kt (309 km/h; 192 mph)
Max cruising speed	415 kt (769 km/h; 478 mph)
Econ cruising speed	377 kt (698 km/h; 434 mph)
Stalling speed, landing configuration	78 kt (145 km/h; 90 mph)
Max rate of climb at S/L	975 m (3,200 ft)/min
Rate of climb at S/L, OEI	274 m (900 ft)/min
Service ceiling	12,495 m (41,000 ft)
T-O run	579 m (1,900 ft)
Landing run	488 m (1,600 ft)
Max range: VFR	1,520 n miles (2,815 km; 1,749 miles)
IFR with 100 n mile (185 km; 87 mile) alternate	1,550 n miles (2,870 km; 1,783 miles)
g limits	+3.6/–1.6

UPDATED

AIR & SPACE

AIR & SPACE AMERICA INC
4460 Shemwell Lane, Paducah, Kentucky 42003
Tel: (+1 502) 898 24 03
Fax: (+1 502) 898 86 91
e-mail: fac@farringtonacft.com
Web: http://www.farringtonacft.com
PRESIDENT: James A Gleason
VICE-PRESIDENT SALES: John Potter
VICE-PRESIDENT ENGINEERING: Tom Davey

Manufacturing and marketing subsidiary of Farrington Aircraft Corporation, also providing flight training for autogyro and fixed-wing pilots. Additionally offers the Air & Space 20A Heliplane, a development of the Model 18 last described in the 1997-98 *Jane's*. Farrington company has 420 m² (4,500 sq ft) facility for kit production. Company founder Don Farrington was killed in April 2000 in an autogyro crash.

UPDATED

AIR & SPACE TWINSTARR
TYPE: Two-seat autogyro kitbuilt.
PROGRAMME: Prototype first flown 1990, but kit production did not begin until 1994. Conforms to FAA 51 per cent kit ruling.
CUSTOMERS: Total of 25 sold, of which nine flying, by mid-2000.
COSTS: Basic kit US$10,995 less engine (January 2001).
DESIGN FEATURES: Open-cockpit autogyro provided in kit form; quoted build time 200 hours. Choice of engine and rotor blades is left to constructor (factory recommended items listed below). Optional hydraulic prerotator for rotors. Twin-fin tail unit.
STRUCTURE: Welded 4130 steel tube with composites nacelle. Two-blade rotors of between 8.23 m (27 ft 0 in) and 9.14 m (30 ft 0 in) diameter can be fitted. Rotor construction of aluminium, with foam core and composites skin.
LANDING GEAR: Tricycle type using Goodyear wheels, all size 5.00×5. Main legs are sprung steel. Cleveland brakes are optional. Castor under engine prevents tipping when unoccupied.
POWER PLANT: Choice of engines ranging from 48 kW (65 hp) to 149 kW (200 hp) can be mounted on four-strut pylon.

Air & Space Twinstarr autogyro

Prototype has one 112 kW (150 hp) Textron Lycoming O-320 engine with Sensenich fixed-pitch two-blade wooden pusher propeller. Fuel tank of 75.7 litres (20.0 US gallons; 16.7 Imp gallons) located under rear seat.
Description below refers to large rotor diameter version.

DIMENSIONS, EXTERNAL:
Rotor diameter	9.14 m (30 ft 0 in)
Length, without rotor	3.96 m (13 ft 0 in)
Height: overall	2.59 m (8 ft 6 in)
for storage	1.73 m (5 ft 8 in)
Wheel track	2.16 m (7 ft 1 in)
Wheelbase	2.16 m (7 ft 1 in)
Propeller diameter	1.73 m (5 ft 8 in)

AREAS:
Rotor disc	65.70 m² (707.1 sq ft)

WEIGHTS AND LOADINGS:
Weight empty	318 kg (700 lb)
Max T-O weight	544 kg (1,200 lb)
Max disc loading	8.3 kg/m² (1.70 lb/sq ft)

PERFORMANCE:
Never-exceed speed (VNE)	95 kt (177 km/h; 110 mph)
Max level speed	78 kt (145 km/h; 90 mph)
Cruising speed at 75% power at 2,135 m (7,000 ft)	56 kt (105 km/h; 65 mph)
Min level flight speed	26 kt (49 km/h; 30 mph)
Max rate of climb at S/L	305 m (1,000 ft)/min
Service ceiling	3,660 m (12,000 ft)
T-O run	61 m (200 ft)
T-O to 15 m (50 ft)	152 m (500 ft)
Landing from 15 m (50 ft)	15 m (50 ft)
Range with reserves	113 n miles (209 km; 130 miles)
Endurance with 30 min reserves	2 h 30 min

UPDATED

AIRBORNE INNOVATIONS

AIRBORNE INNOVATIONS LLC
Airborne Innovations acquired the assets and licences of the former Laron Aviation Technologies, which was US distributor for CFM Shadow and Streak Shadow (see UK section). No recent reports have been received from this company, but the Laron Tundra is now produced by Joplin Light Aircraft (which see).

UPDATED

AIR COMMAND

AIR COMMAND INTERNATIONAL INC
PO Box 1177, Caddo Mills Municipal Airport Building B, Caddo Mills, Texas 75135
Tel: (+1 903) 527 33 35
Fax: (+1 903) 527 38 05
e-mail: aircmd@aircommand.com
Web: http://www.aircommand.com

Air Command currently produces kits of autogyros. For details of earlier products, including the currently available 582, see the 1992-93 edition of *Jane's All the World's Aircraft*.

UPDATED

AIR COMMAND COMMANDER 147A
TYPE: Two-seat autogyro kitbuilt.
PROGRAMME: Prototype (N147GY) introduced at Sun 'n' Fun 1995 and awarded Best New Rotorcraft Design at Oshkosh 1995.
CUSTOMERS: One flying by end 1998 (latest data supplied). No further registrations up to February 2001.
COSTS: Kit US$30,000, including engine (2001).
DESIGN FEATURES: Available in both side-by-side and tandem cockpit configuration; *description applies to former, unless otherwise stated.* Quoted build time 200 hours.

STRUCTURE: Aluminium tubing throughout; fuselage enclosure of glass fibre. Skyweels main rotor.
LANDING GEAR: Fixed; nosewheel; hydraulic brakes on mainwheels.
POWER PLANT: One 119 kW (160 hp) Mazda Rotary 13B engine producing 324 kg (715 lb) thrust, driving a Warp Drive five-blade ground-adjustable pitch propeller via 2.5:1 colinear planetary gear reduction drive. Fuel capacity 114 litres (30.0 US gallons; 25.0 Imp gallons).
ACCOMMODATION: Pilot and passenger in fully enclosed cabin. Cabin doors optional.
SYSTEMS: Cabin heating optional.
DIMENSIONS, EXTERNAL:
Rotor diameter	9.45 m (31 ft 0 in)
Fuselage length	4.27 m (14 ft 0 in)
Fuselage width	2.11 m (6 ft 11 in)
Height to top of rotor head	2.69 m (8 ft 10 in)
Propeller diameter	1.73 m (5 ft 8 in)

AREAS:
Rotor disc	70.12 m² (754.8 sq ft)

WEIGHTS AND LOADINGS:
Weight empty	318 kg (700 lb)
Max payload	261 kg (575 lb)
Max T-O weight	680 kg (1,500 lb)
Max disc loading	9.7 kg/m² (1.99 lb/sq ft)
Max power loading	5.71 kg/kW (9.38 lb/hp)

PERFORMANCE (two occupants):
Max operating speed	104 kt (193 km/h; 120 mph)
Cruising speed	65 kt (121 km/h; 75 mph)
Min flying speed	27 kt (49 km/h; 30 mph)
Max rate of climb at S/L	183 m (600 ft)/min
Service ceiling	3,050 m (10,000 ft)
T-O run	107-213 m (350-700 ft)
Landing run	0-6 m (0-20 ft)
Range with max fuel	150 n miles (278 km; 172 miles)

UPDATED

Air Command Commander 147A in side-by-side form *(Geoffrey P Jones)* 1999/0054215

AIR COMMAND COMMANDER 618
TYPE: Single-/two-seat autogyro kitbuilt.
PROGRAMME: Announced late 1998.
COSTS: US$14,255 for single-seat; tandem US$17,935; side-by-side US$16,755 (2001).
DESIGN FEATURES: Open cockpit. Available as single-seat and also with two-seat tandem and side-by-side seating.
Description applies to side-by-side model, unless otherwise stated.
STRUCTURE: Fuselage of 6061-T6 anodised tubular steel. Glass fibre enclosure.
LANDING GEAR: Fixed tricycle type with speed fairings. Brakes on mainwheels.
POWER PLANT: One 55.0 kW (73.8 hp) Rotax 618 two-cylinder, liquid-cooled piston engine driving a three-blade Warp Drive propeller. Electric starter. Fuel capacity 18.9 litres (5.0 US gallons; 4.2 Imp gallons) in single-seater; 37.9 litres (10.0 US gallons; 8.3 Imp gallons) in two-seater models; all contained in seat tanks.

DIMENSIONS, EXTERNAL:
Rotor diameter: side-by-side	7.62 m (25 ft 0 in)
tandem	8.23 m (27 ft 0 in)
Fuselage length: side-by-side	3.25 m (10 ft 8 in)
tandem	4.09 m (13 ft 5 in)
Height to top of rotor head:	
side-by-side	2.13 m (7 ft 0 in)
tandem	2.69 m (8 ft 10 in)
Width overall	1.70 m (5 ft 7 in)

AREAS:
Rotor disc	45.60 m² (490.9 sq ft)

WEIGHTS AND LOADINGS:
Weight empty: side-by-side	172 kg (380 lb)
tandem	181 kg (400 lb)
Max T-O weight: side-by-side	444 kg (980 lb)
tandem	523 kg (1,155 lb)

PERFORMANCE (two occupants):
Never-exceed speed (V_NE):	
side-by-side	82 kt (152 km/h; 95 mph)
tandem	91 kt (169 km/h; 105 mph)
Cruising speed	56-65 kt (105-121 km/h; 65-75 mph)
Min flying speed	18 kt (33 km/h; 20 mph)
T-O run	11-46 m (35-150 ft)
Landing run	0-6 m (0-20 ft)

UPDATED

AIRCRAFT DESIGNS

AIRCRAFT DESIGNS INC
5 Harris Court, Building 5, Monterey, California 93940
Tel: (+1 831) 649 62 12
Fax: (+1 831) 649 57 38
e-mail: aircraft@mbay.net
Web (1): http://www.aircraftdesigns.com
Web (2): http://www.superstallion.com
PRESIDENT: Martin Hollmann

This company produces kits for the Hollmann-designed Stallion; sells plans for the Hollmann HA-2M Sportster and Hollmann Bumble Bee (see 1992-93 *Jane's*, Private Aircraft section); and is a major contributor to the Lancair series of kitbuilt aircraft (which see), as well as others including Seawind, KIS, Prowler, Thunder Mustang, Kitfox and Condor.

UPDATED

AIRCRAFT DESIGNS SUPER STALLION
TYPE: Six-seat kitbuilt.
PROGRAMME: Construction of prototype began in 1990; first flew July 1994. Designed to FAR Pt 23.
CUSTOMERS: One aircraft flying and 34 kits sold by early 1999.
COSTS: US$68,400 (2000) excluding engine, instruments and avionics.
DESIGN FEATURES: High-performance tourer for amateur construction. Extensive use of composites. Employs Lancair IV wings and landing gear. Wing easily removed for storage and transportation. Quoted build time 1,500 hours.
Laminar flow, cantilever wing with 2° washout at tip. Wing section NACA 64-212 at tip, Jacosky RXM5-217 at root.
FLYING CONTROLS: Conventional and manual. Fowler flaps, width 2.84 m (9 ft 4 in). Aileron width 1.78 m (5 ft 10 in).
STRUCTURE: Prebuilt centre-fuselage area of welded 4130 steel tubing; remainder of airframe primarily of graphite/Nomex honeycomb core/epoxy.
LANDING GEAR: Hydraulically retractable tricycle type; power is provided by an electric motor. Cleveland 6.00-6 mainwheels and tyres.
POWER PLANT: Choice of either 224 kW (300 hp) Teledyne Continental IO-550-G or 261 kW (350 hp) turbocharged Teledyne Continental TSIO-550-B six-cylinder engine driving a McCauley or Hartzell three-blade constant-speed propeller. Fuel capacity 681 litres (180 US gallons; 150 Imp gallons) in two wing tanks. Other engine options include 268 kW (360 hp) Textron Lycoming TIO-540-AE2A and 560 kW (751 shp) Walter M 601E.
ACCOMMODATION: Two pilots and up to four passengers in three pairs of seats. Two rear pairs can be swivelled to provide club seating arrangement. Alternatively, centre row of seats can be quickly removed for carriage of cargo. Large door on port side. Removable 1.88 m (6 ft 2 in) × 0.91 m (3 ft 0 in) panel on starboard side for loading.
AVIONICS: To customer's specification.

DIMENSIONS, EXTERNAL:
Wing span	10.67 m (35 ft 0 in)
Wing chord: at root	1.52 m (5 ft 0 in)
at tip	0.91 m (3 ft 0 in)
Wing aspect ratio	8.8
Length overall	7.62 m (25 ft 0 in)
Max width of fuselage	1.27 m (4 ft 2 in)
Height overall	2.90 m (9 ft 6 in)
Tailplane span	4.17 m (13 ft 8 in)
Tailplane chord: at root	0.91 m (3 ft 0 in)
at tip	0.56 m (1 ft 10 in)

DIMENSIONS, INTERNAL:
Cabin max width	1.24 m (4 ft 1 in)

AREAS:
Wings, gross	13.00 m² (140.0 sq ft)
Vertical tail surfaces (total)	1.78 m² (19.20 sq ft)
Tailplane	3.07 m² (33.00 sq ft)

WEIGHTS AND LOADINGS:
Weight empty	998 kg (2,200 lb)
Max payload	499 kg (1,100 lb)
Max T-O weight	1,724 kg (3,800 lb)
Max wing loading	132.5 kg/m² (27.14 lb/sq ft)
Max power loading: 300 hp	7.71 kg/kW (12.67 lb/hp)
350 hp	6.61 kg/kW (10.86 lb/hp)

PERFORMANCE:
Max operating speed (V_MO):	
300 hp	220 kt (407 km/h; 253 mph)
350 hp	266 kt (493 km/h; 306 mph)
Max cruising speed at 2,500 rpm:	
300 hp	200 kt (370 km/h; 230 mph)
350 hp	256 kt (474 km/h; 295 mph)
Stalling speed, power off, flaps down	62 kt (115 km/h; 71 mph)
Max rate of climb at S/L: 300 hp	610 m (2,000 ft)/min
350 hp	792 m (2,600 ft)/min
Service ceiling	9,750 m (32,000 ft)
T-O run: flaps 10°	549 m (1,800 ft)
Landing run	213 m (700 ft)
Range at cruising speed, 22% reserves	2,300 n miles (4,259 km; 2,646 miles)

UPDATED

Prototype Aircraft Designs Super Stallion *(Paul Jackson/Jane's)* 2001/0089489

AIR TRACTOR

AIR TRACTOR INC
PO Box 485, Municipal Airport, Olney, Texas 76374
Tel: (+1 940) 564 56 16
Fax: (+1 940) 564 23 48
e-mail: airmail@airtractor.com
Web: http://www.airtractor.com
PRESIDENT: Leland Snow
VICE-PRESIDENT FINANCE: David Ickert

Air Tractor agricultural aircraft based on 46-year experience of Leland Snow, who produced Snow S-2 series, which later became Rockwell S-2R (see earlier *Jane's*); AT-300/301/302 (587 built) no longer in production; 1,800th aircraft delivered in April 2000. Eight versions available, powered by various P&W PT6A and R-1340 engines. Company now has 170 employees and total of 13,660 m² (147,000 sq ft) of manufacturing space at three plants.

UPDATED

Air Tractor AT-401B

AIR TRACTOR AT-401B AIR TRACTOR

TYPE: Agricultural sprayer.
PROGRAMME: AT-401 developed 1986 from AT-301 (see *Jane's Aircraft Upgrades*), with increased wing span and larger hopper. AT-401A version with Polish PZL-3S radial engine abandoned, with just one aircraft produced (see 1992-93 *Jane's*). AT-401B version replaced the 401 and has Hoerner wingtips and increased wing span.
CURRENT VERSIONS: **AT-401B:** Standard version.
  Increased T-O weight: Optional version with landing gear of AT-402A and 517 kg (1,140 lb) of additional disposable load.
CUSTOMERS: By early 2000, 198 AT-401s, one AT-401A and 53 AT-401Bs delivered to Argentina, Australia, Brazil, Canada, Colombia, Mexico, Spain and USA. Production of AT-400/400A totalled 72 and 14, respectively.
COSTS: Standard AT-401B US$240,900; with customer-supplied engine US$204,900 (2001).
DESIGN FEATURES: Purpose-designed sprayer with cantilever low wing and hopper at CG. Constant-chord, unswept wings and tailplane; latter braced; moderately sweptback fin. High-mounted cockpit with protective reinforcement.
  Wing aerofoil NACA 4415; dihedral 3° 30′; incidence 2°.
FLYING CONTROLS: Conventional and manual. Balance tabs on ailerons, elevators and rudders; ailerons droop 10° when electrically operated Fowler flaps deflected to their maximum of 26°.
STRUCTURE: Two-spar wing structure of 2024-T3 light alloy, with alloy steel lower spar cap; bonded doubler inside wing leading-edge to resist impact damage; glass fibre wingroot fairings and skin overlaps sealed against chemical ingress; wing ribs and skins zinc chromated before assembly; flaps and ailerons of light alloy. Fuselage of 4130N steel tube, oven stress relieved and oiled internally, with skin panels of 2024-T3 light alloy attached by Camloc fasteners for quick removal; rear fuselage lightly pressurised to prevent chemical ingress; cantilever fin and strut-braced tailplane of light alloy, metal-skinned and sealed against chemical ingress.
LANDING GEAR: Tailwheel type; non-retractable. Cantilever heavy-duty E-4340 spring steel main gear, thickness 28.6 mm (1.125 in); flat spring suspension for castoring and lockable tailwheel. Cleveland mainwheels with tyre size 8.50-10 (8 ply), pressure 2.83 bars (41 lb/sq in). Tailwheel tyre size 5.00-5. Cleveland four-piston brakes with heavy-duty discs. Optional AT-402A-type landing gear.
POWER PLANT: One remanufactured 447 kW (600 hp) Pratt & Whitney R-1340 air-cooled radial engine with speed ring cowling, driving a Hamilton Sundstrand 12D40/6101A-12 two-blade propeller; optional propellers include a Pacific Propeller 22D40/AG200-2 Hydromatic two-blade constant-speed metal and Hydromatic 23D40 three-blade propeller. Air Tractor has designed and is producing new FAA-approved replacement crankshaft for R-1340; other new replacement parts available include main and thrust bearings, and blower (impeller) bearings. Over 300 replacement crankshafts delivered by early 1997. 477 litre (126 US gallon; 105 Imp gallon) fuel tanks. Oil capacity 30 litres (8.0 US gallons; 6.7 Imp gallons). Orenda OE600 V-8 liquid-cooled piston engine of 447 kW (600 hp) tested in December 1999; development towards certification is continuing, with aim of deliveries during 2001.
ACCOMMODATION: Single seat with nylon mesh cover in enclosed cabin which is sealed to prevent chemical ingress. Downward-hinged window/door on each side. 'Line of sight' instrument layout, with swing-down lower instrument panel for ease of access for instrument maintenance. Baggage compartment in bottom of fuselage, aft of cabin, with door on port side. Cabin ventilation by 0.10 m (4 in) diameter airscoop.
SYSTEMS: 24 V electrical system, supplied by 35 A engine-driven alternator; 60 A alternator optional.
AVIONICS: Optional avionics include Honeywell KX 155 nav/com and ACK E-01 emergency locator transmitter.
EQUIPMENT: Agricultural dispersal system comprises a 1,514 litre (400 US gallon; 333 Imp gallon) Derakane vinylester resin/glass fibre hopper mounted in forward fuselage with hopper window and instrument panel-mounted hopper quantity gauge; 0.97 m (3 ft 2 in) wide Transland gatebox; Transland 5 cm (2 in) bottom loading valve; Agrinautics 6.4 cm (2½ in) spraypump with Transland on/off valve and five-blade variable-pitch plastics fan, and 38-nozzle stainless steel spray system with streamlined booms; swath width 24.39 m (80 ft 0 in). Ground start receptacle and hopper rinse system are standard.
  Optional equipment includes night flying package comprising strobe and navigation lights, night working lights, retractable 600 W landing light in port wingtip; and ferry fuel system. Alternative agricultural equipment includes Transland 22358 extra high volume spreader, Transland 54401 NorCal Swathmaster, and 40 extra spray nozzles for high-volume spraying. Three-piece safety plate glass centre windshield and washer system optional, as is a new crew seat.

DIMENSIONS, EXTERNAL:
Wing span 15.57 m (51 ft 1¼ in)
Wing chord, constant 1.83 m (6 ft 0 in)
Wing aspect ratio 8.5
Length overall 8.23 m (27 ft 0 in)
Height overall 2.59 m (8 ft 6 in)
Propeller diameter: standard 2.77 m (9 ft 1 in)
  optional 2.59 m (8 ft 6 in)
AREAS:
Wings, gross 28.43 m² (306.0 sq ft)
Ailerons (total) 3.55 m² (38.20 sq ft)
Trailing-edge flaps (total) 3.75 m² (40.40 sq ft)
Fin 0.90 m² (9.70 sq ft)
Rudder 1.30 m² (14.00 sq ft)
Tailplane 2.42 m² (26.00 sq ft)
Elevators, incl tabs 2.36 m² (25.40 sq ft)
WEIGHTS AND LOADINGS (S: standard, IGW: increased gross weight):
Weight empty, spray equipped 1,925 kg (4,244 lb)
Max T-O weight: S 3,565 kg (7,860 lb)
  IGW 4,082 kg (9,000 lb)
Max landing weight: S 2,721 kg (6,000 lb)
  IGW 3,175 kg (7,000 lb)
Max wing loading: S 125.4 kg/m² (25.69 lb/sq ft)
  IGW 143.6 kg/m² (29.41 lb/sq ft)
Max power loading: S 7.97 kg/kW (13.10 lb/hp)
  IGW 9.13 kg/kW (15.00 lb/hp)
PERFORMANCE (at standard max T-O weight, except where indicated):
Max cruising speed at S/L, hopper empty
  135 kt (251 km/h; 156 mph)
Cruising speed at 1,220 m (4,000 ft)
  124 kt (230 km/h; 143 mph)
Typical working speed
  104-122 kt (193-225 km/h; 120-140 mph)
Stalling speed at 2,721 kg (6,000 lb):
  flaps up 64 kt (118 km/h; 73 mph)
  flaps down 53 kt (99 km/h; 61 mph)
Stalling speed as usually landed 47 kt (87 km/h; 54 mph)
Max rate of climb at S/L:
  at max landing weight 335 m (1,100 ft)/min
  at max T-O weight 158 m (520 ft)/min
T-O run 402 m (1,320 ft)
Range at econ cruising speed at 2,440 m (8,000 ft), no reserves 547 n miles (1,014 km; 630 miles)

UPDATED

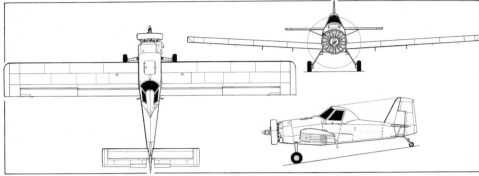

Air Tractor AT-401B (Pratt & Whitney R-1340 radial engine) *(Paul Jackson/Jane's)*

AIR TRACTOR AT-402 TURBO AIR TRACTOR

TYPE: Agricultural sprayer.
PROGRAMME: Follow-on from AT-400 of 1980, AT-402 first flight August 1988; certified November 1988; first delivery late 1988. Current model 402B has Hoerner wingtips and increased span.
CURRENT VERSIONS: **AT-402A:** Introduced in mid-1997, supplementing AT-402B, powered by a 410 kW (550 shp) P&WC PT6A-11AG. Aimed at first-time turbine buyer and priced accordingly, complete with spray dispersal equipment, lights and air conditioning.
  AT-402B: Combines fuselage, tail surfaces and landing gear of AT-400 with turboprop engine and wing of AT-401B.
Description refers to AT-402B, and is generally as for AT-401, except that below.
CUSTOMERS: Total of 96 AT-402s, 68 AT-402As and 28 AT-402Bs delivered by early 2000.

Air Tractor AT-402B Turbo Air Tractor *(Paul Jackson/Jane's)*

AIR TRACTOR—AIRCRAFT: USA

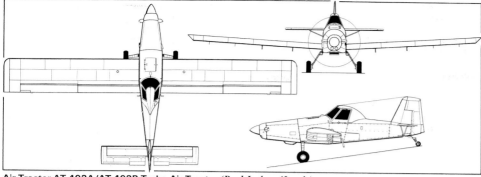

Air Tractor AT-402A/AT-402B Turbo Air Tractor (Paul Jackson/Jane's)

COSTS: Standard AT-402A with PT6A-11AG US$399,500; standard AT-402B US$531,500; or US$257,900 with customer-supplied PT6A-15AG, -27, -28 or -34AG engine (2000).

DESIGN FEATURES: Broadly as AT-401. All versions have steel alloy lower wing spar caps for long fatigue life and reinforced leading-edges to prevent bird strike damage. Size 29×11.0-10 high-flotation tyres and wheels as standard.

POWER PLANT: One 507 kW (680 shp) P&WC PT6A-15AG, -27 or -28, either new or customer-furnished, driving a Hartzell HCB3TN-3D/T1028N+4 three-blade constant-speed reversible-pitch propeller. Standard fuel capacity 644 litres (170 US gallons; 142 Imp gallons); optional fuel tankage 818 litres (216 US gallons; 180 Imp gallons) or 886 litres (234 US gallons; 195 Imp gallons).

SYSTEMS: 250 A starter/generator and two 24 V 21 Ah batteries.

EQUIPMENT: Hopper and gatebox size as for AT-401; optional equipment includes Transland extra high-volume dispersal system; engine-driven air conditioning system.

DIMENSIONS, EXTERNAL:
Length overall	9.32 m (30 ft 7 in)
Height overall	2.90 m (9 ft 6 in)
Tailplane span	5.23 m (17 ft 2 in)
Wheel track	2.62 m (8 ft 7 in)

WEIGHTS AND LOADINGS:
Weight empty, spray equipped	1,783 kg (3,930 lb)
Certified gross weight (FAR Pt 23)	3,175 kg (7,000 lb)
Typical operating weight (CAM 8):	
AT-402A	3,901 kg (8,600 lb)
AT-402B	4,159 kg (9,170 lb)
Max wing loading	137.2 kg/m² (28.10 lb/sq ft)
Max power loading	7.70 kg/kW (12.65 lb/shp)

PERFORMANCE:
Max level speed at S/L:	
clean	174 kt (322 km/h; 200 mph)
with dispersal equipment	160 kt (298 km/h; 185 mph)
Cruising speed 283 kW (380 shp) at 2,440 m (8,000 ft)	142 kt (264 km/h; 164 mph)
Typical working speed	113-126 kt (209-233 km/h; 130-145 mph)
Stalling speed at 2,721 kg (6,000 lb) AUW:	
flaps up	64 kt (118 km/h; 73 mph)
flaps down	53 kt (99 km/h; 61 mph)
Stalling speed at 2,041 kg (4,500 lb) typical landing weight	46 kt (86 km/h; 53 mph)
Max rate of climb at S/L, dispersal equipment installed:	
AUW of 2,721 kg (6,000 lb)	495 m (1,625 ft)/min
AUW of 3,565 kg (7,860 lb)	320 m (1,050 ft)/min
T-O run at AUW of 3,565 kg (7,860 lb)	247 m (810 ft)
Landing run at 2,041 kg (4,500 lb)	122 m (400 ft)
Range at econ cruising speed at 2,440 m (8,000 ft), no reserves	573 n miles (1,062 km; 660 miles)

UPDATED

AIR TRACTOR AT-502 TURBO

TYPE: Agricultural sprayer.

PROGRAMME: Developed as stretched AT-401. First flight of AT-502 April 1987; certified 23 June 1987; 38 sold in 1997.

CURRENT VERSIONS: **AT-502**: Original version. *Description refers to AT-502, and is generally as for AT-401, except that below.*

AT-502A: Similar to AT-502B but with 820 kW (1,100 shp) PT6A-45R; slow turning (1,425 rpm) five-blade Hartzell (HCB5MP-3C/M10876AS) propeller; enlarged vertical tail surfaces. For operation in mountainous terrain or short strips. Prototype first flight February 1992; certified April 1992; 25 delivered by end 1999.

AT-502B: Stronger wing than AT-502, with 0.61 m (2 ft) longer span and Hoerner wingtips, which are stated to increase width of spray pattern by 0.91 m (3 ft).

AT-503A: Dual control, tandem-seat trainer; PT6A-34AG engine. Prototype designated AT-503. In low-rate production.

CUSTOMERS: 515 AT-500 series delivered by October 1999, including out-of-production AT-501 (nine), AT-502 (208), AT-503 (one), plus AT-503A (three).

COSTS: Standard AT-502B US$547,500; or US$272,500 with customer-supplied PT6A-15AG, -27, -28, -34AG (2000).

DESIGN FEATURES: See AT-401. 1,892 litre (500 US gallon; 416 Imp gallon) hopper handles low-density nitrogen-based fertilisers such as urea; safety glass centre windscreen with wiper. Alloy steel lower spar cap and bonded doubler on inside of wing leading-edge for increased resistance to impact damage, and glass fibre wingroot fairings.

LANDING GEAR: Non-retractable tailwheel type. Heavy-duty E-4340 spring steel main gear, thickness 37.2 mm (1.31 in); flat spring for castoring and lockable tailwheel. Cleveland mainwheels, tyre size 29×11.0-10, pressure 3.45 bars (50 lb/sq in); tailwheel tyre size 5.00-5; Cleveland six-piston brakes with heavy-duty discs.

POWER PLANT: One 507 kW (680 shp) Pratt & Whitney Canada PT6A-15AG, PT6A-27 or PT6A-28, or 559 kW (750 shp) PT6A-34 or PT6A-34AG turboprop, driving a Hartzell HCB3TN-3D/T10282+4 three-blade metal propeller. Standard fuel capacity 644 litres (170 US gallons; 142 Imp gallons). Optional capacities 818 litres (216 US gallons; 180 Imp gallons) and 886 litres (234 US gallons; 195 Imp gallons). AT-502A has 818 litre (216 US gallon; 180 Imp gallon) tanks as standard.

ACCOMMODATION: One or two persons (see Current Versions). Has quickly detachable instrument panel and removable fuselage skin panels for ease of maintenance.

SYSTEMS: Two 24 V 42 Ah batteries and 250 A starter/generator.

AVIONICS: *Comms:* Optional avionics include Honeywell KX 155 nav/com and KR 87 ADF, KY 196 com radio, KT 76A transponder and ACK E-01 emergency locator transmitter.

EQUIPMENT: Agricultural dispersal system comprises a 1,892 litre (500 US gallon; 416 Imp gallon) hopper mounted in forward fuselage with hopper window and instrument panel-mounted hopper quantity gauge; 0.97 m (3 ft 2 in) wide Transland gatebox; 6.4 cm (2½ in) bottom loading valve; Agrinautics 6.4 cm (2½ in) spraypump with Transland on/off valve and five-blade variable-pitch plastics fan and 38-nozzle stainless steel spray system with streamlined booms. Optional dispersal equipment includes 7.6 cm (3 in) spray system with 119 spray nozzles, and automatic flagman; spray swath 25.91 m (85 ft 0 in).

Standard equipment includes safety glass centre windscreen panel, ground start receptacle, three-colour polyurethane paint finish, strobe and navigation lights; windscreen washer and wiper; and twin nose-mounted landing/taxi lights. Optional equipment includes engine-driven air conditioning system, night flying package, comprising night working lights, retractable 600 W landing light in port wingtip; fuel flowmeter, fuel totaliser and ferry fuel system. Alternative agricultural equipment includes Transland 22356 extra high-volume spreader, Transland 54401 NorCal Swathmaster, 40 extra spray nozzles for high-volume spraying, and eight- or 10-unit Micronair Mini Atomiser unit; hopper rinse tank is standard. New crew seat is also optional.

DIMENSIONS, EXTERNAL:
Wing span: AT-502A/502B	15.85 m (52 ft 0 in)
AT-502	15.24 m (50 ft 0 in)
Wing chord, constant	1.83 m (6 ft 0 in)
Wing aspect ratio	8.7
Length overall	10.11 m (33 ft 2 in)
Height overall	3.12 m (10 ft 3 in)
Tailplane span	5.23 m (17 ft 2 in)
Wheel track	3.12 m (10 ft 3 in)
Wheelbase	6.64 m (21 ft 9½ in)
Propeller diameter: AT-502B	2.69 m (8 ft 10 in)

AREAS:
Wings, gross: AT-502A/502B	28.99 m² (312.0 sq ft)
AT-502	27.87 m² (300.0 sq ft)
Ailerons (total)	3.53 m² (38.0 sq ft)
Trailing-edge flaps (total)	3.75 m² (40.4 sq ft)
Fin	0.90 m² (9.7 sq ft)
Rudder	1.30 m² (14.0 sq ft)
Tailplane	2.41 m² (26.0 sq ft)
Elevators (total, incl tab)	2.44 m² (26.3 sq ft)

WEIGHTS AND LOADINGS:
Weight empty, spray equipped	1,949 kg (4,297 lb)
Max T-O weight	4,399 kg (9,700 lb)
Max landing weight	3,629 kg (8,000 lb)
Max wing loading:	
AT-502A/502B	151.8 kg/m² (31.09 lb/sq ft)
Max power loading: 502A	5.37 kg/kW (8.82 lb/shp)
502B (PT6A-34AG)	7.87 kg/kW (12.93 lb/shp)
502B (PT6A-15AG)	8.68 kg/kW (14.26 lb/shp)

PERFORMANCE (AT-502B with spray equipment installed):
Never-exceed speed (VNE) and max level speed at S/L, hopper empty	156 kt (290 km/h; 180 mph)
Cruising speed at 2,440 m (8,000 ft), 283 kW (380 shp)	136 kt (253 km/h; 157 mph)
Typical working speed	104-126 kt (193-233 km/h; 120-145 mph)
Stalling speed at 3,629 kg (8,000 lb):	
flaps up	72 kt (132 km/h; 82 mph)
flaps down	59 kt (110 km/h; 68 mph)
Stalling speed at 1,978 kg (4,360 lb) typical landing weight	46 kt (86 km/h; 53 mph)
Max rate of climb at S/L, AUW of 3,629 kg (8,000 lb):	
with PT6A-15AG	311 m (1,020 ft)/min
with PT6A-34AG	360 m (1,180 ft)/min
Max rate of climb at S/L, AUW of 4,309 kg (9,500 lb):	
with PT6A-15AG	232 m (760 ft)/min
with PT6A-34AG	282 m (925 ft)/min
T-O run at AUW of 3,629 kg (8,000 lb):	
with PT6A-15AG	236 m (775 ft)
with PT6A-34AG	222 m (730 ft)
T-O run at AUW of 4,309 kg (9,500 lb):	
with PT6A-15AG	356 m (1,170 ft)
with PT6A-34AG	302 m (990 ft)
Range with max fuel	538 n miles (998 km; 620 miles)

UPDATED

Air Tractor AT-502B spraying a field

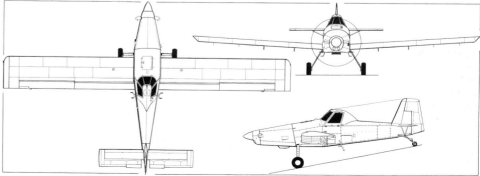

Air Tractor AT-502B agricultural aircraft (Paul Jackson/Jane's)

AIR TRACTOR AT-602

TYPE: Agricultural sprayer.
PROGRAMME: Prototype first flew 1 December 1995. Certification completed 6 June 1996 and deliveries began the following month. Aircraft fits into Air Tractor range between AT-502B and AT-802A, being AT-502 with increased wing and tail spans, plus taller fin.
CUSTOMERS: Total of 92 sold by early July 2000.
COSTS: Standard AT-602 with PT6A-60AG US$777,500; with customer-supplied engine US$334,500 (2000).
Description generally as for AT-401, except that following.
POWER PLANT: Choice of new 783 kW (1,050 shp) Pratt & Whitney Canada PT6A-60AG turboprop, or used PT6A-45R of 783 kW (1,050 shp) or PT6A-65AG of 966 kW (1,295 shp), driving a Hartzell five-blade constant-speed reversing propeller. Fuel capacity 818 litres (216 US gallons; 180 Imp gallons); optional fuel capacity 1,105 litres (292 US gallons; 243 Imp gallons).
SYSTEMS: Three batteries and 250 A starter/generator.
EQUIPMENT: Glass fibre hopper, capacity 2,385 litres (630 US gallons; 525 Imp gallons); Transland 0.96 m (3 ft 2 in) wide gatebox; five-blade ground-adjustable spraypump fan; pump shut-off valve. Optional crew seat.

DIMENSIONS, EXTERNAL:
Wing span	17.07 m (56 ft 0 in)
Wing chord, constant	1.83 m (6 ft 0 in)
Wing aspect ratio	9.3
Length overall	10.41 m (34 ft 2 in)
Height overall	3.38 m (11 ft 1 in)
Tailplane span	5.66 m (18 ft 7 in)
Wheel track	3.07 m (10 ft 1 in)
Wheelbase	6.71 m (22 ft 0 in)

AREAS:
Wings, gross	31.22 m^2 (336.0 sq ft)

WEIGHTS AND LOADINGS:
Weight empty, equipped	2,540 kg (5,600 lb)
Max T-O weight	5,670 kg (12,500 lb)
Max landing weight	5,443 kg (12,000 lb)
Max wing loading	181.6 kg/m^2 (37.20 lb/sq ft)
Max power loading	7.25 kg/kW (11.90 lb/shp)

PERFORMANCE:
Never-exceed speed (V$_{NE}$)	189 kt (350 km/h; 217 mph)
Working speed	130 kt (241 km/h; 150 mph)
Stalling speed: flaps up	86 kt (160 km/h; 99 mph)
flaps down	75 kt (139 km/h; 87 mph)

UPDATED

AT-602, second-largest of the Air Tractor range   0015583

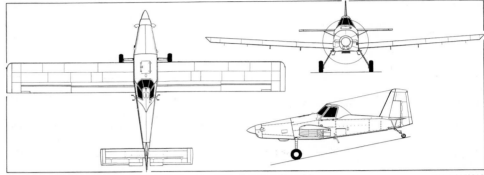

Air Tractor AT-602 (P&WC PT6A-60AG engine) *(Paul Jackson/Jane's)*   2000/0079015

AIR TRACTOR AT-802

TYPE: Agricultural sprayer.
PROGRAMME: Design started July 1989; optional configuration as firefighter; first flight of prototype (N802LS) 30 October 1990; second aircraft flew November 1991, with PT6A-45R and configured as agricultural model with spraybooms, pump and Transland gatebox. Production deliveries started second quarter 1993.
CURRENT VERSIONS: **AT-802**: Tandem two-seater powered by P&WC PT6A-45R; certified for gross weight of 6,804 kg (15,000 lb) 27 April 1993. PT6A-65AG and -67AG versions certified April 1993 at maximum T-O weight of 7,257 kg (16,000 lb).
AT-802A: Single-seat version; third production aircraft in this configuration; can be powered by refurbished PT6A-65AG or -67AG engine. First flight 6 July 1992; FAA certification gained 17 December 1992; certified for gross weight of 6,804 kg (15,000 lb) with PT6A-45R on 27 April 1993. PT6A-65AG versions certified March 1993 at maximum T-O weight of 7,257 kg (16,000 lb).
AT-802F: Two-seat firefighting version; PT6A-67AG engine. FAA certification at 5,670 kg (12,500 lb) maximum T-O weight 17 December 1992; certified at 7,257 kg (16,000 lb) on 27 April 1993, giving useful load of 3,987 kg (8,790 lb).
Data apply to AT-802F version, except where indicated.
CUSTOMERS: 100 delivered by July 2000. Deliveries include 21 to Australia, 26 to Europe and five to Canada plus two to Saudi Arabia for oil slick eradication. Total of 45 engaged in firefighting duties in six countries during 2000.
COSTS: Standard AT-802A US$916,500 with new PT6A-65AG; with customer-supplied PT6A-45R, -65AG or -67R US$419,900. Standard AT-802AF US$1,141,900 with PT6A-67AG; or US$563,500 with customer-supplied PT6A-67AG. Two-seat version with dual controls US$31,500 extra (2000).

DESIGN FEATURES: Generally as for AT-401. Largest aircraft built by company to date; full dual controls for training; also designed for firefighting; programmable logic computer with cockpit control panel and digital display enables pilot to select coverage level and opens hydraulically operated 'bomb bay' drop doors to prescribed width, closing them when selected amount of retardant released. Drop doors adjust automatically for changing head pressure and aircraft acceleration to provide a constant flow rate and even ground coverage.
Wing aerofoil section NACA 4415; dihedral 3° 30′; incidence 2°.
FLYING CONTROLS: Manually operated ailerons, elevators and rudder with balance tabs; electrically operated Fowler trailing-edge flaps deflect to maximum 30°.
STRUCTURE: Two-spar wing of 2024-T3 light alloy with alloy steel upper and lower spar caps and bonded doubler on inside of leading-edge for impact damage resistance; ribs and skins zinc chromated before assembly; glass fibre wingroot fairing and skin overlaps sealed against chemical ingress; flaps and ailerons of light alloy. Fuselage of welded 4130N steel tube, oven stress relieved and oiled internally, with skin panels of 2024-T3 light alloy attached by Camloc fasteners for quick removal; rear fuselage lightly pressurised to prevent chemical ingress. Cantilever fin and strut-braced tailplane of light alloy, metal skinned and sealed against chemical ingress.
LANDING GEAR: Non-retractable tailwheel type. Cantilever heavy-duty E-4340 spring steel main legs, thickness 44.5 mm (1.75 in); flat spring suspension for castoring and lockable tailwheel. Cleveland mainwheels with tyre size 11.00-12 (10 ply), pressure 4.14 bars (60 lb/sq in). Tailwheel tyre size 6.00-6. Cleveland eight-piston brakes with heavy-duty discs.
POWER PLANT: One Pratt & Whitney Canada PT6A-65AG or -67AG turboprop, rated at 966 kW (1,295 shp) for -65AG and 1,007 kW (1,350 shp) for -67AG, both driving a Hartzell five-blade feathering and reversible-pitch constant-speed metal propeller. Fuel in two integral wing tanks, total usable capacity 961 litres (254 US gallons; 211 Imp gallons); optional tanks increase capacity to 1,438 litres (380 US gallons; 317 Imp gallons). Engine air is filtered through two large pleated paper industrial truck filters.
ACCOMMODATION: One or two seats in enclosed cabin, which is sealed to prevent chemical ingress and protected with overturn structure. Four downward-hinged doors, two on each side. Windscreen is safety-plate auto glass, with washer and wiper. Air conditioning system standard.
SYSTEMS: Hydraulic system, pressure 207 bars (3,000 lb/sq in).
AVIONICS: Advanced nav/com, including GPS.
EQUIPMENT: Two removable Derakane vinylester hoppers forward of cockpit and 227 litre (60 US gallon; 50 Imp gallon) gate tank in ventral bulge, for agricultural chemical, fire retardant or water; total capacity 3,104 litres (820 US gallons; 683 Imp gallons).

DIMENSIONS, EXTERNAL:
Wing span	18.06 m (59 ft 3 in)
Wing chord, constant	2.07 m (6 ft 9½ in)
Wing aspect ratio	8.8
Length overall	10.95 m (35 ft 11 in)
Height overall	3.89 m (12 ft 9 in)
Tailplane span	6.03 m (19 ft 9½ in)
Wheel track	3.10 m (10 ft 2 in)
Wheelbase	7.25 m (23 ft 9½ in)
Propeller diameter (-65AG)	2.92 m (9 ft 7 in)

AREAS:
Wings, gross	37.25 m^2 (401.0 sq ft)
Ailerons (total)	4.61 m^2 (49.60 sq ft)
Trailing-edge flaps (total)	5.54 m^2 (59.60 sq ft)
Fin	1.24 m^2 (13.40 sq ft)
Rudder	1.57 m^2 (16.90 sq ft)

Air Tractor AT-802 releasing fire-retardant *(Paul Jackson/Jane's)*   2000/0062931

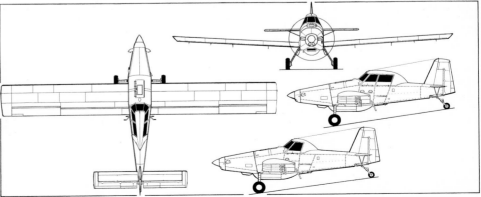

Air Tractor AT-802 two-seat agricultural aircraft with extra side view (lower) of AT-802A single-seater *(Paul Jackson/Jane's)*   2001/0100424

Tailplane	3.44 m² (37.00 sq ft)
Elevators (total, incl tab)	3.00 m² (32.30 sq ft)

WEIGHTS AND LOADINGS (AT-802A):
Weight empty, equipped: sprayer	2,951 kg (6,505 lb)
firefighter	3,270 kg (7,210 lb)
Max T-O and landing weight	7,257 kg (16,000 lb)
Max wing loading	194.8 kg/m² (39.90 lb/sq ft)
Max power loading (-67AG)	7.21 kg/kW (11.85 lb/shp)

PERFORMANCE:
Max level speed at S/L	182 kt (338 km/h; 210 mph)
Max cruising speed at 2,440 m (8,000 ft)	192 kt (356 km/h; 221 mph)
Stalling speed, power off, flaps down, at max landing weight	79 kt (147 km/h; 91 mph)
Max rate of climb at S/L	259 m (850 ft)/min
Service ceiling	3,965 m (13,000 ft)
T-O run	610 m (2,000 ft)
Range with max fuel	696 n miles (1,289 km; 800 miles)

UPDATED

Air Tractor AT-802 cropsprayer *(Paul Jackson/Jane's)*

AKROTECH

AKROTECH AVIATION INC

53774 Airport Road, Scappoose, Oregon 97056
Tel: (+1 503) 543 79 60
Fax: (+1 503) 543 79 64
e-mail: info@akrotech.com
PRESIDENT: Richard Giles
DIRECTOR OF MARKETING: Rich Ebers

AkroTech Aviation marketed the Giles G-200, G-202 and G-300 composites construction competition aerobatic monoplanes. The French company CAP (which see) builds and markets the G-202 as CAP 222.

In June 2000, Akrotech laid off its workforce and suspended operations, while it explored ways of restructuring.

UPDATED

AKROTECH GILES G-200

TYPE: Aerobatic single-seat sportplane.
PROGRAMME: First customer-built aircraft first flew 26 May 1996. Prototype G-200 (N5296E) made public debut at Sun 'n' Fun 1994. One completed in 2000. No known new production since 1998.
CUSTOMERS: First customer delivery May 1995. One completed in 2000. No known new production since 1998. More than 30 kits ordered by late 1998, 13 completed at this time.
COSTS: US$58,985 for airframe kit (2000).
DESIGN FEATURES: Optimised for agility and speed. Designed to produce highest possible performance from four-cylinder engine while complying with FAR Pt 21. Available as fast-build kit (quoted build time 2,000 hours). Rate of roll in excess of 400°/s.
FLYING CONTROLS: Conventional and manual. Elevator deflection ±30°; rudder deflection ±30°; aerodynamically and mass-balanced full-span ailerons, deflection ±20°. Wing leading-edge sweepback 8° 30′.
STRUCTURE: Carbon fibre construction throughout with exception of engine mount, landing gear and control hardware.
LANDING GEAR: Tailwheel configuration; fixed. Leaf-spring aluminium mainwheel legs; mainwheel spats. Cleveland 5.00-5 mainwheels and brakes; lockable 10 cm (4 in) Haigh tailwheel.
POWER PLANT: Choice of Textron Lycoming engines from 112 kW (150 hp) to 172 kW (230 hp), driving a two-blade fixed-pitch or constant-speed propeller; standard fitting is 149 kW (200 hp) Textron Lycoming AEIO-360 with MT two-blade constant-speed propeller. Fuselage fuel tank, capacity 72 litres (19.0 US gallons; 15.8 Imp gallons), of which 68 litres (18.0 US gallons; 15.0 Imp gallons) are usable; optional extra 114 litres (30.0 US gallons; 25.0 Imp gallons) in two wing tanks. Oil capacity 7.6 litres (2.0 US gallons; 1.66 Imp gallons).
ACCOMMODATION: Pilot only, in 45° angled seat, under one-piece side-hinged acrylic bubble canopy. Pilot parameters from 1.55 m (5 ft 1 in) to 1.93 m (6 ft 4 in) standard; other heights by optional customisation.
SYSTEMS: 14 V DC electrical system.
AVIONICS: VFR equipped.

DIMENSIONS, EXTERNAL:
Wing span	6.10 m (20 ft 0 in)
Wing chord: at root	1.45 m (4 ft 9 in)
at tip	0.84 m (2 ft 9 in)
Wing aspect ratio	5.3
Length overall	5.49 m (18 ft 0 in)
Height overall	1.63 m (5 ft 4 in)
Wheel track	1.52 m (5 ft 0 in)
Wheelbase	3.84 m (12 ft 7 in)
Propeller diameter	1.93 m (6 ft 4 in)

DIMENSIONS, INTERNAL:
Cockpit max width	0.58 m (1 ft 11 in)

AREAS:
Wings, gross	6.97 m² (75.0 sq ft)

WEIGHTS AND LOADINGS (AEIO-360 engine):
Weight empty	340 kg (750 lb)
Max T-O weight	589 kg (1,300 lb)
Max wing loading	84.6 kg/m² (17.33 lb/sq ft)
Max power loading	3.96 kg/kW (6.50 lb/hp)

PERFORMANCE (AEIO-360 engine):
Never-exceed speed (V_{NE})	220 kt (407 km/h; 253 mph)
Max cruising speed	185 kt (343 km/h; 213 mph)
Cruising speed at 75% power	180 kt (333 km/h; 207 mph)
Stalling speed at max T-O weight	57 kt (106 km/h; 66 mph)
Max rate of climb at S/L	1,067 m (3,500 ft)/min
T-O run	183 m (600 ft)
T-O to 15 m (50 ft)	244 m (800 ft)
Landing from 15 m (50 ft)	457 m (1,500 ft)
Range, no reserves	950 n miles (1,759 km; 1,093 miles)
g limits	±10

UPDATED

AKROTECH GILES G-202

TYPE: Aerobatic two-seat sportplane.
PROGRAMME: Developed from single-seat G-200; prototype (N50AL) displayed at Oshkosh 1995 and made first flight on 22 December 1995. First customer delivery March 1996. Approved by FAA under 51 per cent rule for homebuilts.
CUSTOMERS: At least 60 kits delivered by 2000, of which some 35 then registered (includes French assembly).
COSTS: Kit US$63,529 (2000).
Description and data generally as for G-200; differences detailed below.
POWER PLANT: Three-blade MT propeller and extra 151 litres (40.0 US gallons; 33.3 Imp gallons) of fuel in two wing tanks.
FLYING CONTROLS: Aileron deflection ±22°.
ACCOMMODATION: Pilot and passenger in tandem.

DIMENSIONS, EXTERNAL:
Wing span	6.71 m (22 ft 0 in)
Wing chord: at root	1.52 m (5 ft 0 in)
at tip	0.91 m (3 ft 0 in)
Wing aspect ratio	5.4
Length overall	6.10 m (20 ft 0 in)
Height overall	1.70 m (5 ft 7 in)
Wheelbase	4.22 m (13 ft 10 in)

DIMENSIONS, INTERNAL:
Cabin max width	0.71 m (2 ft 4 in)

AREAS:
Wings, gross	8.36 m² (90.0 sq ft)

WEIGHTS AND LOADINGS:
Weight empty	431 kg (950 lb)
Max T-O weight	725 kg (1,600 lb)
Max wing loading	86.8 kg/m² (17.78 lb/sq ft)
Max power loading	4.87 kg/kW (8.00 lb/hp)

PERFORMANCE:
Stalling speed	51 kt (95 km/h; 59 mph)
Max rate of climb at S/L	1,067 m (3,500 ft)/min

UPDATED

AkroTech Aviation Giles G-202 *(Paul Jackson/Jane's)*

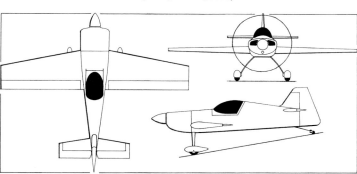

AkroTech Aviation Giles G-200 *(Paul Jackson/Jane's)*

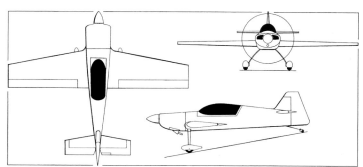

AkroTech Aviation Giles G-202 *(Paul Jackson/Jane's)*

AKROTECH GILES G-300

TYPE: Aerobatic single-seat sportplane.
PROGRAMME: Prototype (N300NW) first flew August 1997.
CUSTOMERS: One aircraft only.
COSTS: Price fixed at US$218,000 for first three orders (1999).
DESIGN FEATURES: Generally as for G-200, but with stretched fuselage and additional power. Wing leading-edge sweepback 8.3°.
FLYING CONTROLS: Conventional and manual. Elevator deflection ±30°; rudder deflection ±27°; aileron deflection ±20°.
STRUCTURE: Carbon fibre construction throughout with exception of engine mount, landing gear and control actuators.
LANDING GEAR: Tailwheel configuration with spring steel main gear. Speed fairings on mainwheels; steerable 12.7 cm (5 in) tailwheel. Mainwheels 5.00-5.
POWER PLANT: One Textron Lycoming AEIO-540, modified by Ly-Con to produce 246 kW (330 hp), driving a Hartzell three-blade constant-speed propeller. Fuel capacity 68 litres (18.0 US gallons; 15.0 Imp gallons) in fuselage tank; optional wing tanks together add 151 litres (40.0 US gallons; 33.3 Imp gallons).
ACCOMMODATION: Pilot only, under one-piece side-hinged acrylic bubble canopy. Maximum pilot height 1.88 m (6 ft 2 in).
SYSTEMS: 14 V DC electrical system.
AVIONICS: Basic VFR equipped.

Akrotech Giles G-300 *(James Goulding/Jane's)* 2001/0100425

DIMENSIONS, EXTERNAL:
Wing span	6.71 m (22 ft 0 in)
Wing chord: at root	1.55 m (5 ft 1 in)
at tip	0.94 m (3 ft 1 in)
Wing aspect ratio	5.4
Length overall	6.25 m (20 ft 6 in)
Height overall	1.70 m (5 ft 7 in)
Wheel track	1.68 m (5 ft 6 in)
Wheelbase	4.32 m (14 ft 2 in)

DIMENSIONS, INTERNAL:
Cabin max width	0.58 m (1 ft 11 in)

AREAS:
Wings, gross	8.36 m² (90.0 sq ft)

WEIGHTS AND LOADINGS:
Weight empty	499 kg (1,100 lb)
Max T-O weight	635 kg (1,400 lb)
Max wing loading	75.9 kg/m² (15.56 lb/sq ft)
Max power loading	2.58 kg/kW (4.24 lb/hp)

PERFORMANCE:
Never-exceed speed (V_{NE})	220 kt (407 km/h; 253 mph)
Max operating speed	215 kt (398 km/h; 247 mph)
Stalling speed	52 kt (97 km/h; 60 mph)
Max rate of climb at S/L	1,463 m (4,800 ft)/min

UPDATED

Akrotech Giles G-300 participating at World Aerobatic Championships, France, August 2000 *(Paul Jackson/Jane's)* 2000/0100500

AKROTECH GILES G-750

TYPE: Undisclosed.
PROGRAMME: A prototype (N17HE) of a new Pratt & Whitney PT6A-25C turboprop-powered Giles high-performance light aircraft was built in 1998 and during 1999 set four Class C-1 Group 2 world records. However, it was badly damaged in a crash at Salinas, Kansas on 3 October 1999. No data had been released by the manufacturer by end 2000.

UPDATED

ALLIANCE

ALLIANCE AIRCRAFT CORPORATION
100 Main Street, Courtyard Entrance, Suite 222, PO Box 9000, Dover, New Hampshire 03821-9000
Tel: (+1 603) 334 80 00
Fax: (+1 603) 334 81 01
Web: http://www.allianceaircraftcorp.com
PRESIDENT AND CEO: Earl Robinson
VICE-PRESIDENT, CORPORATE DEVELOPMENT: Tim Ryan
CHIEF FINANCIAL OFFICER: Scott Gaul
CHIEF DESIGNER: Guido Pessotti
VICE-PRESIDENT, MARKETING: Michael E Cardellichio

Negotiations were under way in 2000 to construct a US$20 million, 37,160 m² (400,000 sq ft) factory at Pease International Tradeport, New Hampshire (the former Pease AFB), where Alliance will assemble (from overseas-sourced components) the first of a family of jet airliners. The company is looking to hire up to 2,000 employees over the next two years and has discussed provision of hardware and engineering systems by Sukhoi of Russia. Provisional agreement signed with newly-formed Sukhoi Civil Aircraft Company on 5 June 2000; according to Russian reports, Sukhoi will "design and produce the aircraft".

UPDATED

ALLIANCE REGIONAL JET
Renamed StarLiner.

UPDATED

ALLIANCE STARLINER

TYPE: Regional jet airliner.
PROGRAMME: Announced 20 January 2000; launched, and StarLiner name revealed, at ILA, Berlin, 6 June 2000; 'fast track' development schedule; preliminary design review due late 2000; critical design review to be completed April 2001, with first flight of 90-seater in February 2002 followed two weeks later by 70-seater. Certification fleet of three StarLiner 200s and three 300s; FAA/JAA certification target March 2003. Assuming go-ahead received, a third family member could be in service by mid-2004. Risk/revenue-sharing partners are being sought for the wing, fuselage and empennage.
Family of airliners ranging from 55 to 110 seats envisaged; first two members will be **70-seat** and **90-seat** versions, offered simultaneously, later followed by **110-seat** and **55-seat** 'stretch' and 'shrink' models.
CURRENT VERSIONS: **StarLiner 100/150:** Later development; available late 2004/early 2005; 50-seat 'shrunk' version with three-abreast seating; length 26.39 m (86 ft 7 in), span 21.62 m (70 ft 11 in), wing area 50.0 m² (538.0 sq ft); cabin maximum height 1.88 m (6 ft 2 in). Further reductions planned with 44 seats (length 24.76 m; 81 ft 3 in), 35 seats (22.33 m; 73 ft 3 in) and, as StarLiner 100/130, 30 seats (20.70 m; 67 ft 11 in). Honeywell AS900 turbofan as baseline; ratings between 26.7 and 35.6 kN (6,000 and 8,000 lb st).
StarLiner 200: Initial 70-seat version.
StarLiner 300: Initial 90-seat version.
StarLiner 400: Later 110-seat development. Wing area 93 m² (1,000 sq ft); engines in 78 kN (17,535 lb st) class.
CUSTOMERS: Letters of intent from undisclosed potential customers for up to 30 aircraft held at beginning 2000.
COSTS: Development costs for StarLiner 200 and 300 estimated as US$660 million. Unit cost, 70-seater US$18.9 million and 90-seater US$24.3 million (2000). Further US$320 million to develop StarLiner 100.
DESIGN FEATURES: Low wing with two podded engines; typical regional jet airliner configuration. Wing sweep 27° 18' at 25 per cent chord. Targets include 17-minute turnround time; common crew rating for all variants; 1 hour engine change; and 1 mmh/fh.
FLYING CONTROLS: Not finalised.
LANDING GEAR: Retractable tricycle type. Trailing-link main gear.
POWER PLANT: Two engines in 60 kN (13,500 lb st) class required for StarLiner 200, and 71 kN (16,000 lb st) for StarLiner 300, mounted on underwing pylons. To be selected; contenders include Rolls-Royce Deutschland BR700 (preferred) and General Electric CF34-8.
ACCOMMODATION: Both Starliner 200 and 300 have 51 cm (20 in) aisle offset to accommodate five-abreast seating at

Model of projected Alliance StarLiner *(Paul Jackson/Jane's)* 2001/0089488

ALLIANCE to AMERICAN CHAMPION—AIRCRAFT: USA

81 cm (32 in) pitch. Lavatory and galley front and rear. Cabin doors at front and rear. Underfloor baggage holds.
SYSTEMS: Not finalised.
AVIONICS: Five display screens. Contents include Honeywell, Rockwell Collins and Sextant.
All data are provisional.
DIMENSIONS, EXTERNAL (StarLiner 200 and 300):
Wing span	28.00 m (91 ft 10 in)
Wing aspect ratio	9.2
Length overall: SL 200	26.2 m (86 ft)
SL 300	31.57 m (103 ft 7 in)
Height overall	10.03 m (32 ft 11 in)
Tailplane span	10.87 m (35 ft 8 in)
Wheel track	5.54 m (18 ft 2 in)
Wheelbase	11.79 m (38 ft 8 in)

DIMENSIONS, INTERNAL:
Cabin: Length: SL 200	18.3 m (60 ft)
SL 300	21.3 m (70 ft)
Max width	3.45 m (11 ft 4 in)
Max height	2.13 m (7 ft 0 in)
Baggage hold max height	1.04 m (3 ft 5 in)

AREAS:
Wings, gross: SL 200, SL 300	85 m² (915 sq ft)

WEIGHTS AND LOADINGS:
Weight empty, equipped: SL 200	20,276 kg (44,700 lb)
SL 300	23,269 kg (51,300 lb)
Max payload: SL 200	8,165 kg (18,000 lb)
SL 300	9,979 kg (22,000 lb)
Max fuel weight: SL 200, SL 300	10,886 kg (24,000 lb)
Max T-O weight: SL 200	35,607 kg (78,500 lb)
SL 300	41,957 kg (92,500 lb)
Max landing weight: SL 200	34,019 kg (75,000 lb)
SL 300	40,369 kg (89,000 lb)
Max wing loading: SL 200	422 kg/m² (86 lb/sq ft)
SL 300	492 kg/m² (101 lb/sq ft)
Max power loading:	
SL 200, SL 300	295 kg/kN (2.9 lb/lb st)

PERFORMANCE:
Max cruising Mach No	0.84
Initial cruising altitude	11,280 m (37,000 ft)
Max certified altitude	12,495 m (41,000 ft)
Max cruising speed at 7,925 m (26,000 ft):	
CAS	350 kt (648 km/h; 402 mph)
TAS	more than 500 kt (926 km/h; 575 mph)
T-O run: SL 200	1,372 m (4,500 ft)
SL 300	1,524 m (5,000 ft)
Landing run: SL 200	1,067 m (3,500 ft)
SL 300	1,220 m (4,000 ft)
Range	2,000 n miles (3,704 km; 2,301 miles)

UPDATED

AMD

AIRCRAFT MANUFACTURING & DEVELOPMENT

415 Airport Road, Heart of Georgia Regional Airport, PO Box 639, Eastman, Georgia 31023
Tel: (+1 912) 374 27 59
Fax: (+1 912) 374 27 93
e-mail: info@newplane.com
Web: http://www.NewPlane.com
PRESIDENT: Mathieu Heintz
SALES MANAGER: Lisa Lewis

AMD was established in 1999 to produce and market the Zenair Zenith CH 2000 and CH2T. Its purpose-built, 2,600 m² (28,000 sq ft) factory became operational in December 1999; the first US-produced CH 2000 was completed in January 2000.

VERIFIED

Fourth US-built AMD Zenith CH 2000 *(Paul Jackson/Jane's)*

AMD ZENITH CH 2000

TYPE: Two-seat lightplane.
PROGRAMME: Prototype C-FQCU first flown 26 June 1993; Canadian and US certification of prototypes 26 and 31 July 1994, respectively; first two production models (C-FRSK and C-FRSV) completed April 1994. First delivery September 1994; US certification 25 July 1995. A test example fitted with 104.5 kW (140 hp) Lycoming O-320 was produced in 1995. Certified for IFR operations and spins on 2 October 1996. JAA certification received March 2000.
CURRENT VERSIONS: **CH 2000**: Standard production model, *as described*.
CH2T: Basic model intended for flight training market; avionics are not included.
CUSTOMERS: Total of 43 delivered by time production moved from Canada to US in June 1999. First US-built aircraft (N17KA, one of 11 CH2Ts for KeyFlite Academy at Utica, New York, and Nashua, New Hampshire) received C of A on 14 January 2000 and delivered following day. Production expected to reach 200 annually by 2001; target for 2000 is 100.
COSTS: CH 2000 US$94,900; CH2T US$69,900 (2000).
DESIGN FEATURES: Side-by-side two-seat and smaller-span derivative of Tri-Z CH 300. Designed to conform to FAR Pt 23 (JAR-VLA equivalent level of safety). Conventional, low-wing configuration, with mid-mounted tailplane and sweptback fin.
Wing section LS (1) 0417 (mod).
FLYING CONTROLS: Manual. All-moving tailplane with anti-balance tabs; all-moving fin/rudder with horn balance. Electrically actuated split flaps.
STRUCTURE: Aluminium alloy, with stressed skins.
LANDING GEAR: Non-retractable tricycle type with steerable nosewheel. Shock-cord absorption on nosewheel. Cleveland 5.00-5 in wheels with hydraulic disc brakes on mainwheels. Optional wheel fairings.
POWER PLANT: One 86.5 kW (116 hp) Textron Lycoming O-235-N2C, driving metal Sensenich 72-CK-0-48 propeller. Fuel capacity 106 litres (28.0 US gallons; 23.3 Imp gallons) standard, 129 litres (34.0 US gallons; 28.3 Imp gallons) optional.
SYSTEMS: 12 V 60 A, heavy-duty battery.
ACCOMMODATION: Two persons, side by side, with dual controls; upward-hinged door each side; fixed windscreen; rear side windows. Baggage shelf behind seats.
AVIONICS: *Comms:* Honeywell KX 155 nav/com, KT 76A transponder and PS 1000II intercom, KMA 24 audio panel, KR 87 ADF with KI 227 indicator.
Flight: KN 62A DME and KI 209 indicator.

DIMENSIONS, EXTERNAL:
Wing span	8.79 m (28 ft 10 in)
Wing aspect ratio	5.4
Length overall	7.01 m (23 ft 0 in)
Height overall	2.08 m (6 ft 10 in)
Propeller diameter	1.83 m (6 ft 0 in)

DIMENSIONS, INTERNAL:
Cabin max width	1.17 m (3 ft 10 in)

AREAS:
Wings, gross	12.73 m² (137.0 sq ft)

WEIGHTS AND LOADINGS:
Weight empty	476 kg (1,050 lb)
Max T-O weight	728 kg (1,606 lb)
Max wing loading	57.2 kg/m² (11.72 lb/sq ft)
Max power loading	8.43 kg/kW (13.84 lb/hp)

PERFORMANCE:
Never-exceed speed (V$_{NE}$)	147 kt (273 km/h; 170 mph)
Econ cruising speed	104 kt (193 km/h; 120 mph)
Stalling speed	44 kt (82 km/h; 51 mph)
Max rate of climb at S/L	250 m (820 ft)/min
Service ceiling	3,660 m (12,000 ft)
T-O to 15 m (50 ft)	468 m (1,535 ft)
Landing run	183 m (600 ft)
Landing from 15 m (50 ft)	518 m (1,700 ft)
Range with max payload:	
standard fuel	434 n miles (804 km; 500 miles)
optional fuel	851 n miles (1,577 km; 980 miles)
g limits, Utility category	+4.4/−2.2

UPDATED

AMERICAN CHAMPION

AMERICAN CHAMPION AIRCRAFT CORPORATION

PO Box 37, 32032 Washington Avenue, Highway D, Rochester, Wisconsin 53167
Tel: (+1 414) 534 63 15
Fax: (+1 414) 534 23 95
e-mail: acac@execpc.com
Web: http://www.amerchampionaircraft.com
PRESIDENT AND CEO: Jerry K Mehlhaff
GENERAL MANAGER: Dale Gauger

ACAC offers new-build Citabria (now called Aurora, Adventure and Explorer), Super Decathlon and Scout, all designed by Aeronca. Model 7 (based on Second World War L-3 Grasshopper liaison/observation machine) certified 18 October 1945. Series later marketed by Champion Aircraft Corporation (1954); then, from 1970, Bellanca Aircraft Corporation (see 1979-80 *Jane's*) and then, in 1982, Champion Aircraft Company (see 1985-86 *Jane's*). Assets transferred to Jerry Mehlhaff's American Champion Aircraft Corporation, which restarted production in 1990. By early 1999, various manufacturers had built 14,750 of the Model 7 and Model 8 families. American Champion delivered a total of 75 aircraft in 1998 and nearly 100 in 1999. In addition, ACAC offers its new metal spar wing for retrofit on all models, including the original 7AC Champion. A new model, the 7ACA, was displayed at Oshkosh in July 2000; the company is investigating ways of making the aircraft financially viable for production.

UPDATED

AMERICAN CHAMPION 7ACA CHAMP

TYPE: Two-seat lightplane.
PROGRAMME: Prototype (N82AC) displayed at Oshkosh in July 2000, shortly after first flight; powered by 59.7 kW (80 hp) Jabiru engine; original 44.7 kW (60 hp) Franklin-engined Bellanca 7ACA was underpowered. Will remain experimental while company explores demand. Provisional maximum take-off weight is 533 kg (1,220 lb) and maximum level speed 100 kt (185 km/h; 115 mph); cost estimated in region of US$60,000 (2000 prices), or US$45,000 for quick-build kit, if offered.

NEW ENTRY

AMERICAN CHAMPION 7ECA CITABRIA AURORA

TYPE: Two-seat lightplane.
PROGRAMME: Original Aeronca Model 7 and successors built in 16 subvariants, with various letter suffixes. Currently the lowest powered of the American Champion range, the 7ECA was introduced in the mid-1950s; 1,353 had been built up to 1984; was re-introduced into range during 1995 as the cheapest aircraft. Current production aircraft has improved ventilation and heating; redesigned instrument panel and quick-jettison door; and dual controls and brakes. Total of 29 built by June 2000.
COSTS: Standard aircraft US$66,900 (2000).
DESIGN FEATURES: High-wing cabin monoplane; wings braced by V-struts and two secondary struts each side; wire-braced fin and tailplane; constant-chord wings; sweptback fin with (current versions) squared tip.
Wing section NACA 4412; dihedral 1°.
FLYING CONTROLS: Conventional and manual. Horn-balanced elevators and rudder. Flight-adjustable trim tab in port elevator. No flaps.
STRUCTURE: Stainless steel exhaust system. Fuselage and empennage are welded chromoly steel tube with Dacron covering; two-spar wing has aluminium spars and ribs, Dacron covering, GFRP tips and steel tube V struts.
LANDING GEAR: Tailwheel type; non-retractable. Cantilever spring steel main gear; mainwheels 6.00-6 (4 ply); tailwheel 5.00, steerable. Hydraulic, toe-operated disc and parking brakes.
POWER PLANT: One 88 kW (118 hp) Textron Lycoming O-235-K2C flat-four piston engine driving a Sensenich 72 CKS8-0-52 fixed-pitch propeller. Fuel in two wing tanks, total capacity 136 litres (36.0 US gallons; 30.0 Imp gallons) of which 132 litres (35.0 US gallons; 29.1 Imp gallons) usable; overwing refuelling point for each tank. Oil capacity 5.7 litres (1.5 US gallons; 1.3 Imp gallons).
ACCOMMODATION: Pilot and passenger in tandem; five-point safety harness; quick jettison door on starboard side; pilot's port window opens. Space for baggage behind seats. Dual controls. Options include split door for photographic work.

USA: AIRCRAFT—AMERICAN CHAMPION

SYSTEMS: Electric starter; 60 A alternator; 12 V gel-cell battery; voltage regulator with protector. Optional lighting system.
AVIONICS: To customer's specification. Optional packages available from Honeywell and Narco.
EQUIPMENT: Oil pressure gauge, oil temperature gauge, tachometer, stall warning and cabin heating standard. Navigation, landing, cabin and wingtip strobe lighting optional.

DIMENSIONS, EXTERNAL:
Wing span	10.21 m (33 ft 6 in)
Wing aspect ratio	6.8
Length overall	6.73 m (22 ft 1 in)
Height overall	2.35 m (7 ft 8½ in)
Propeller diameter	1.83 m (6 ft 0 in)

DIMENSIONS, INTERNAL:
Cabin: Length	2.69 m (8 ft 10 in)
Max width	0.76 m (2 ft 6 in)
Max height	1.19 m (3 ft 11 in)

AREAS:
Wings, gross	15.33 m² (165.0 sq ft)

WEIGHTS AND LOADINGS:
Weight empty, equipped	508 kg (1,120 lb)
Max baggage weight	45 kg (100 lb)
Max T-O weight	748 kg (1,650 lb)
Max wing loading	48.8 kg/m² (10.00 lb/sq ft)
Max power loading	8.51 kg/kW (13.98 lb/hp)

PERFORMANCE:
Never-exceed speed (V_NE)	140 kt (260 km/h; 162 mph)
Max level speed at S/L	104 kt (193 km/h; 120 mph)
Cruising speed at 75% power	100 kt (185 km/h; 115 mph)
Stalling speed	44 kt (81 km/h; 50 mph)
Max rate of climb at S/L	226 m (740 ft)/min
Service ceiling	3,810 m (12,500 ft)
T-O run	138 m (450 ft)
T-O to 15 m (50 ft)	272 m (890 ft)
Landing from 15 m (50 ft)	233 m (765 ft)
Landing run	122 m (400 ft)
Range with max fuel, allowance for start, taxi, S/L T-O, climb and descent, no reserves:	
at 85% power	556 n miles (1,030 km; 640 miles)
at 55% power	617 n miles (1,142 km; 710 miles)
g limits	+5/–2

UPDATED

AMERICAN CHAMPION 7GCAA CITABRIA ADVENTURE

TYPE: Two-seat lightplane.
PROGRAMME: American Champion 7GCAA Adventure launched 1997, following on from agricultural 7GCA, built up to 1980.
CUSTOMERS: First series totalled 396 built; 48 built by American Champion up to June 2000.
COSTS: US$75,900 (2000).
DESIGN FEATURES: As for Aurora.
FLYING CONTROLS: As for Aurora.
STRUCTURE: As for Aurora.
LANDING GEAR: As for Aurora.
POWER PLANT: One 119 kW (160 hp) Textron Lycoming O-320-B2B flat-four engine driving a two-blade fixed-pitch Sensenich 74DM6S8-1-56 propeller. Fuel and oil capacities as for Aurora.
ACCOMMODATION: As for Aurora.
SYSTEMS: As for Aurora.
AVIONICS: As for Aurora.
EQUIPMENT: As for Aurora.
DIMENSIONS, EXTERNAL: As for Aurora, except:
Propeller diameter	1.85 m (6 ft 1 in)

DIMENSIONS, INTERNAL: As for Aurora
AREAS: As for Aurora

WEIGHTS AND LOADINGS: As for Aurora, except:
Weight empty	544 kg (1,200 lb)
Max power loading	6.28 kg/kW (10.31 lb/hp)

PERFORMANCE:
Never-exceed speed (V_NE)	140 kt (260 km/h; 162 mph)
Max level speed	121 kt (225 km/h; 140 mph)
Max cruising speed at 75% power	117 kt (217 km/h; 135 mph)
Stalling speed, power off	44 kt (81 km/h; 50 mph)
Max rate of climb at S/L	387 m (1,270 ft)/min
Service ceiling	5,180 m (17,000 ft)
T-O run	108 m (355 ft)
T-O to 15 m (50 ft)	186 m (610 ft)
Landing from 15 m (50 ft)	230 m (755 ft)
Landing run	122 m (400 ft)
Range with max fuel:	
at 75% power, no reserves	482 n miles (893 km; 555 miles)
at 55% power	595 n miles (1,102 km; 685 miles)
g limits	+5/–2

UPDATED

AMERICAN CHAMPION 7GCBC CITABRIA EXPLORER

TYPE: Two-seat lightplane.
PROGRAMME: Formerly Citabria 150S (7GCBC). Certified late 1993.
CUSTOMERS: Total of 101 built since 1994, including eight in 1997 and 19 in 1998. Overall production of 7GCBC totalled 1,298 by June 2000.
COSTS: Standard aircraft US$78,900 (2000).
DESIGN FEATURES: As for Aurora. Specific 7GCBC features include optional wheel fairings. NACA 4412 aerofoil section. Dihedral 2°; incidence 1°.
FLYING CONTROLS: As for Aurora, plus flaps.
STRUCTURE: Aluminium alloy (front) and steel tube (aft) bracing struts.
LANDING GEAR: As for Aurora.
POWER PLANT: As for Adventure. Fuel as for Aurora. Oil capacity 7.5 litres (2.0 US gallons; 1.7 Imp gallons).
ACCOMMODATION: As for Aurora.
SYSTEMS: As for Aurora.
AVIONICS: To customer's specification.
EQUIPMENT: As for Aurora.

DIMENSIONS, EXTERNAL:
Wing span	10.49 m (34 ft 5 in)
Wing chord, constant	1.52 m (5 ft 0 in)
Wing aspect ratio	6.9
Length overall	6.74 m (22 ft 1 in)
Height overall	2.35 m (7 ft 8½ in)
Wheel track	1.93 m (6 ft 4 in)
Wheelbase	4.90 m (16 ft 1 in)
Propeller diameter	1.85 m (6 ft 1 in)

DIMENSIONS, INTERNAL: As for Aurora

AREAS:
Wings, gross	15.97 m² (171.9 sq ft)
Ailerons (total)	1.53 m² (16.50 sq ft)
Fin	0.65 m² (7.02 sq ft)
Rudder	0.63 m² (6.83 sq ft)
Tailplane	1.14 m² (12.25 sq ft)
Elevators (total, incl tab)	1.35 m² (14.58 sq ft)

WEIGHTS AND LOADINGS:
Weight empty, equipped	567 kg (1,250 lb)
Baggage capacity	45 kg (100 lb)
Max T-O and landing weight	816 kg (1,800 lb)
Max wing loading	51.1 kg/m² (10.47 lb/sq ft)
Max power loading	6.85 kg/kW (11.25 lb/hp)

PERFORMANCE:
Never-exceed speed (V_NE)	140 kt (261 km/h; 162 mph)
Max level speed at S/L	115 kt (212 km/h; 132 mph)
Cruising speed:	
at 75% power, at optimum height	111 kt (206 km/h; 128 mph)
at 65% power	107 kt (198 km/h; 123 mph)
Stalling speed, flaps up	45 kt (82 km/h; 51 mph)
flaps down	40 kt (74 km/h; 46 mph)
Max rate of climb at S/L	344 m (1,130 ft)/min
Service ceiling	4,725 m (15,500 ft)
T-O run	126 m (412 ft)
T-O to 15 m (50 ft)	200 m (656 ft)
Landing from 15 m (50 ft)	226 m (740 ft)
Landing run	110 m (360 ft)
Range with max fuel, allowance for start, taxi, S/L T-O, climb and descent, no reserves:	
at 75% power	431 n miles (799 km; 496 miles)
at 55% power	521 n miles (966 km; 600 miles)
g limits	+5/–2

UPDATED

AMERICAN CHAMPION 8GCBC SCOUT

TYPE: Two-seat lightplane.
PROGRAMME: Model 8 series of Champion developed by Bellanca; first was 8GCBC version, which received type approval in 1974.
CURRENT VERSIONS: **Scout**: Fixed-pitch propeller.
Super Scout: Constant-speed propeller.
CUSTOMERS: Total 420 built by June 2000.
COSTS: Scout: US$98,900; Scout CS: US$100,900 (2000).
DESIGN FEATURES: Upgraded version of Bellanca/Champion Scout, with new lighter and stronger metal spar wing with 300 per cent less deflection than previous wooden wings; circuit breakers; modern avionics; revised interior; and

American Champion 7GCAA Citabria Adventure *(Paul Jackson/Jane's)*

American Champion 7GCBC Citabria two-seat tourer *(Paul Jackson/Jane's)*

high-gloss weather-resistant exterior finish. Can operate from short fields while towing glider or banner; low speed assists pipeline, border patrol, forestry and wildlife management roles.
Wing section NACA 4412; dihedral 1°; incidence 1°.
FLYING CONTROLS: As for Explorer. Four-position trailing-edge flaps droop 7, 16, 21 and 27°.
STRUCTURE: As for Explorer.
LANDING GEAR: As for Aurora; main tyres 8.50-6 (4/6 ply); tail 5.00-5 (4 ply). Optional floats or skis.
POWER PLANT: One 134 kW (180 hp) Textron Lycoming O-360-C1G flat-four engine, driving either a McCauley 1A200HFA80 fixed-pitch (in Scout) or Hartzell HC-C2YR-1BF/F7666A constant-speed (Super Scout) propeller; three-blade 1.83 m (6 ft 0 in) Hartzell HC-C3YR-1RF/F7282 constant-speed propeller optional for Super Scout. Standard fuel capacity as for Aurora; optional additional tank increased total to 265 litres (70.0 US gallons; 58.3 Imp gallons). Oil capacity 7.5 litres (2.0 US gallons; 1.7 Imp gallons).
ACCOMMODATION: As for Aurora. Heated and ventilated.
SYSTEMS: As for Aurora.
AVIONICS: As for Aurora.
EQUIPMENT: Options include cropduster package, long-range fuel tank and glider tow assembly.
DIMENSIONS, EXTERNAL (A: fixed-pitch propeller, B: constant-speed propeller):
Wing span 11.02 m (36 ft 2 in)
Wing chord, constant 1.52 m (5 ft 0 in)
Wing aspect ratio 7.3
Length overall: A 6.93 m (22 ft 9 in)
B 7.01 m (23 ft 0 in)
Height overall 2.96 m (9 ft 8½ in)
Tailplane span 3.10 m (10 ft 2¼ in)
Propeller diameter: A 1.93 m (6 ft 4 in)
B 2.03 m (6 ft 8 in)
DIMENSIONS, INTERNAL: As for Aurora
AREAS:
Wings, gross 16.70 m² (180.0 sq ft)
WEIGHTS AND LOADINGS (A and B as above):
Weight empty: A 597 kg (1,315 lb)
B 635 kg (1,400 lb)
Baggage capacity 45 kg (100 lb)
Max T-O weight: A and B, Normal 975 kg (2,150 lb)
A, Restricted 1,179 kg (2,600 lb)
Max wing loading:
A and B, Normal 58.3 kg/m² (11.94 lb/sq ft)
A, Restricted 70.5 kg/m² (14.44 lb/sq ft)
Max power loading:
A and B, Normal 7.28 kg/kW (11.94 lb/hp)
A, Restricted 8.80 kg/kW (14.44 lb/hp)
PERFORMANCE (A and B as above):
Never-exceed speed (V$_{NE}$) 140 kt (260 km/h; 162 mph)
Max level speed at S/L:
A, B 121 kt (225 km/h; 140 mph)
Cruising speed at 75% power:
A 106 kt (196 km/h; 122 mph)
B 113 kt (209 km/h; 130 mph)

Stalling speed, flaps up 47 kt (87 km/h; 54 mph)
flaps down: A, B 44 kt (81 km/h; 50 mph)
Max rate of climb at S/L: A 328 m (1,075 ft)/min
B 433 m (1,420 ft)/min
Service ceiling 4,420 m (14,500 ft)
T-O run: A, B 149 m (490 ft)
T-O to 15 m (50 ft) 312 m (1,025 ft)
Landing from 15 m (50 ft): A, B 376 m (1,235 ft)
Landing run: A, B 128 m (420 ft)
Range with standard tankage, no reserves:
A, at 75% power 333 n miles (618 km; 384 miles)
A, at 75% power, optional fuel
680 n miles (1,260 km; 783 miles)
A, at 55% power, optional fuel
779 n miles (1,444 km; 897 miles)
B, at 75% power 360 n miles (668 km; 415 miles)
B, at 75% power, optional fuel
732 n miles (1,355 km; 842 miles)
B, at 55% power, optional fuel
838 n miles (1,551 km; 964 miles)
UPDATED

AMERICAN CHAMPION 8KCAB SUPER DECATHLON

TYPE: Two-seat lightplane.
PROGRAMME: Original Decathlon was powered by a 112 kW (150 hp) Lycoming AEIO-320-E2B; Super introduced by American Champion; first flight (N38AC) July 1990. The Fixed Pitch Decathlon (Sensenich propeller) is no longer marketed.
CUSTOMERS: Total of 225 built 1990 to June 2000 (868 overall).
COSTS: Standard price US$103,900 (2000).

American Champion 8KCAB Super Decathlon *(Paul Jackson/Jane's)*

DESIGN FEATURES: Generally as for Aurora. Cleared for limited inverted flight; constant-speed propeller. Wing section NACA 1412 (modified); dihedral 1°; incidence 1° 30′.
FLYING CONTROLS: As for Aurora.
STRUCTURE: As for Aurora. Optional metal wing spar.
LANDING GEAR: Tailwheel type; non-retractable. Cantilever spring steel main legs; mainwheel tyres size 17×6-6, pressure 1.66 bars (24 lb/sq in); steerable tailwheel, tyre size, 8.30×2.50-2.80, pressure 2.07 bars (30 lb/sq in); optional main tyres 6.00-6 or 8.00-6 (4 ply), tail tyre 5.00-5 (4 ply). Cleveland disc brakes; optional wheel fairings.
POWER PLANT: One 134 kW (180 hp) Textron Lycoming AEIO-360-H1B flat-four engine, driving a Hartzell HC-C2YR-4CF/FC7666A-2 constant-speed propeller. Fuel capacity 151 litres (40.0 US gallons; 33.3 Imp gallons), of which 148 litres (39.0 US gallons; 32.5 Imp gallons) are usable. Oil capacity 7.5 litres (2.0 US gallons; 1.7 Imp gallons).
ACCOMMODATION: As for Aurora.
SYSTEMS: As for Aurora, plus electric fuel boost pump.
DIMENSIONS, EXTERNAL:
Wing span 9.75 m (32 ft 0 in)
Wing chord, constant 1.63 m (5 ft 4 in)
Wing aspect ratio 6.1
Length overall 6.98 m (22 ft 10¾ in)
Height overall 2.36 m (7 ft 9 in)
Tailplane span 3.10 m (10 ft 2¼ in)
Propeller diameter 1.88 m (6 ft 2 in)
DIMENSIONS, INTERNAL: As for Aurora
AREAS:
Wings, gross 15.71 m² (169.1 sq ft)
WEIGHTS AND LOADINGS:
Weight empty 608 kg (1,340 lb)
Baggage capacity 45 kg (100 lb)
Max T-O weight 816 kg (1,800 lb)
Max wing loading 52.0 kg/m² (10.64 lb/sq ft)
Max power loading 6.09 kg/kW (10.00 lb/hp)
PERFORMANCE:
Never-exceed speed 173 kt (321 km/h; 200 mph)
Max level speed at 2,135 m (7,000 ft)
135 kt (249 km/h; 155 mph)
Cruising speed:
at 75% power 128 kt (237 km/h; 147 mph)
at 55% power 111 kt (206 km/h; 128 mph)
Stalling speed 46 kt (86 km/h; 53 mph)
Max rate of climb at S/L 390 m (1,280 ft)/min
Service ceiling 4,815 m (15,800 ft)
T-O run 151 m (495 ft)
T-O to 15 m (50 ft) 276 m (904 ft)
Landing from 15 m (50 ft) 321 m (1,051 ft)
Landing run 130 m (425 ft)
Range, no reserves:
at 75% power 509 n miles (944 km; 587 miles)
at 55% power 542 n miles (1,006 km; 625 miles)
g limits +6/−5
UPDATED

American Champion 8GCBC Super Scout with three-blade propeller *(Paul Jackson/Jane's)*

AMERICAN HOMEBUILTS'

AMERICAN HOMEBUILTS' INC
10419 Vander Karr Road, Hebron, Illinois 60034
Tel: (+1 815) 648 46 17
Fax: (+1 815) 648 16 07
VERIFIED

AMERICAN HOMEBUILTS' JOHN DOE

TYPE: Tandem-seat kitbuilt.
PROGRAMME: Proof-of-concept aircraft first flown 1994; preproduction examples first flown early 1997; kit became available from August 1997 onwards.
CUSTOMERS: Two flying by mid-2000.
COSTS: Kit US$19,500 without coverings (2000).

DESIGN FEATURES: Designed for operation from 'bush' airstrips; many repairs can be carried out 'off base'. Ribs replaceable without removing wing. Quoted build time 400 hours.
FLYING CONTROLS: Ailerons and flaps are interchangeable, as are elevators and rudders. Patented flap/aileron mechanism allows ailerons to droop progressively (8, 15 and 22°) as three-position flaps (15, 25 and 40°) are lowered. Slats and stall fences also fitted to reduce stall speeds further.
STRUCTURE: Fabric-covered welded tube fuselage and wing; nose area metal panelled. High wing, with endplates, supported by two streamline steel V-struts.
LANDING GEAR: Tailwheel type; fixed. Bungee cord suspension.

POWER PLANT: One 93 kW (125 hp) Teledyne Continental IO-240 flat-four, driving an IVO three-blade propeller; alternatively an 89.5 kW (120 hp) LOM 132A can be fitted; design will accept other engines in the 48.5 to 119 kW (65 to 160 hp) range. Fuel capacity 98 litres (26.0 US gallons; 21.7 Imp gallons).
ACCOMMODATION: Pilot and passenger in tandem; can be flown solo from rear seat.
DIMENSIONS, EXTERNAL:
Wing span 9.32 m (30 ft 7 in)
Wing chord, constant 1.30 m (4 ft 3 in)
Wing aspect ratio 7.2
Length overall 6.64 m (21 ft 9½ in)
Height overall 1.98 m (6 ft 6 in)

554   USA: AIRCRAFT—AMERICAN HOMEBUILTS' to ARCHEDYNE

DIMENSIONS, INTERNAL:
Cabin: Length	2.18 m (7 ft 2 in)
Max width	0.69 m (2 ft 3 in)
Max height	1.19 m (3 ft 11 in)
Cargo area volume	0.51 m³ (18.0 cu ft)

AREAS:
Wings, gross	12.13 m² (130.6 sq ft)

WEIGHTS AND LOADINGS:
Weight empty	431 kg (950 lb)
Max T-O weight	680 kg (1,500 lb)
Max wing loading	56.1 kg/m² (11.49 lb/sq ft)
Max power loading	7.30 kg/kW (12.00 lb/hp)

PERFORMANCE:
Never-exceed speed (V$_{NE}$)	139 kt (257 km/h; 160 mph)
Max operating speed	122 kt (225 km/h; 140 mph)
Normal cruising speed	102 kt (188 km/h; 117 mph)
Stalling speed, flaps down	27 kt (49 km/h; 30 mph)
Max rate of climb at S/L	335 m (1,100 ft)/min
Service ceiling	3,960 m (13,000 ft)
T-O run	30 m (100 ft)
Landing run	38 m (125 ft)
Range with max fuel	434 n miles (804 km; 500 miles)
g limits	+4.4/−3

UPDATED

Preproduction example of American Homebuilts' John Doe *(Geoffrey P Jones)*
1999/0054214

ARCHEDYNE

ARCHEDYNE AEROSPACE
PO Box 568, Cape Canaveral, Florida 32920-0568
Tel: (+1 407) 631 50 67
Fax: (+1 407) 631 82 66
Web: http://www.archedyne.com
PRESIDENT AND CEO: G Leonard Gioia
COO: Thomas Marsh
BUSINESS DEVELOPMENT MANAGER: Denis Bonneaux

Archedyne, previously known as Amjet, planned to merge with Lake Aircraft in 1999 in order to launch production of its NauticAir 450 amphibian. No agreement was reached and the merger was abandoned. In October 2000, Archedyne disclosed it was in negotiations to purchase another aerospace competitor.

UPDATED

ARCHEDYNE NAUTICAIR 450

TYPE: Business jet amphibian.
PROGRAMME: Project originally a six-seater known as Amjet 400 and powered by single TFE731-20 turbofan; renamed NauticAir 450 early 1999 when redesigned with twin engines and increased capacity. Subject to acquisition of a new source of funding, first flight is now planned for late 2003, followed by certification to FAR Pt 23 or Pt 25 standards; customer deliveries to start in 2005.
CUSTOMERS: 'Serious' inquiries for 100 aircraft. Production expected to reach 20 per year.
COSTS: US$5 million (2001).
DESIGN FEATURES: Low-mounted, sweptback, tapered gull wings, roots of which form sponsons. Engines pylon-mounted on fuselage sides, aft of trailing-edge; sharply swept T tail. Tunnel hull design for improved water stability without need of deep hull or underwing floats.
Aerofoil section NACA 66A212.

Artist's impression of the Archedyne NauticAir 450
2000/0084558

FLYING CONTROLS: Conventional. Electromechanically operated, wide chord Fowler flaps.
STRUCTURE: Aluminium alloy and composites construction to reduce corrosion.
LANDING GEAR: Tricycle type; retractable. Hydraulic operating system with mechanical uplocks to prevent deployment during flight. Sponsons have retractable steps for water operation.
POWER PLANT: Two Williams FJ44-2 turbofans, each nominally 10.23 kN (2,300 lb st).

ACCOMMODATION: Pilot plus up to eight passengers in pressurised cabin which includes lavatory and galley. Door in port forward fuselage behind cabin. Escape hatches on each side of fuselage.
SYSTEMS: De-icing standard. 360° bow and stern thrusters for use during docking manoeuvres, powered by APU. Cabin pressurisation to sea level while flying at 12,190 m (40,000 ft).
All data provisional.

DIMENSIONS, EXTERNAL:
Wing span	14.48 m (47 ft 6 in)
Wing aspect ratio	9.8
Length overall	16.00 m (52 ft 6 in)
Height overall	4.52 m (14 ft 10 in)
Tailplane span	4.34 m (14 ft 3 in)
Wheelbase	6.17 m (20 ft 3 in)

DIMENSIONS, INTERNAL:
Cabin: Length	6.32 m (20 ft 9 in)
Max width	2.08 m (6 ft 10 in)
Max height	1.52 m (5 ft 0 in)

AREAS:
Wings, gross	21.37 m² (230.0 sq ft)

WEIGHTS AND LOADINGS:
Weight empty	3,003 kg (6,620 lb)
Max payload	739 kg (1,630 lb)
Max fuel weight	980 kg (2,160 lb)
Max T-O weight	5,216 kg (11,500 lb)
Max wing loading	244.1 kg/m² (50.00 lb/sq ft)
Max power loading	255 kg/kN (2.50 lb/lb st)

PERFORMANCE:
Max cruising speed	448 kt (830 km/h; 516 mph)
Stalling speed: flaps up	90 kt (167 km/h; 104 mph)
flaps down	70 kt (130 km/h; 81 mph)
Max rate of climb at S/L	1,524 m (5,000 ft)/min
T-O run: on land	640 m (2,100 ft)
on water	914 m (3,000 ft)
Range with 45 min reserves	1,800 n miles (3,333 km; 2,071 miles)

UPDATED

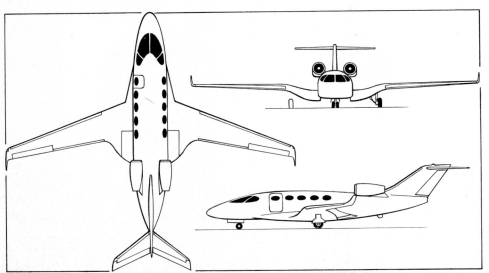

Archedyne NauticAir 450 jet amphibian, showing original 17.70 m (58 ft 1 in) wing span
(James Goulding/Jane's)
2000/0084557

ASI

AMERICAN SPORTSCOPTER INTERNATIONAL INC

Suite 228, 11712 Jefferson Avenue, Newport News, Virginia 23606
Tel: (+1 757) 872 87 78
Fax: (+1 757) 872 87 71
e-mail: ASII@visi.com
Web: http://www.ultrasport.rotor.com

FAR EAST OFFICE:
Light's American Sportscopter Inc, 26 Ta Ho Street, Taichung 407, Taiwan
Tel: (+886 4) 311 80 03
Fax: (+886 4) 311 80 01
e-mail: ultrasport@iname.com
Web: http://www.ultrasport.rotor.com

ASI was founded in 1990. North and South American markets are covered by US office; rest of world by Taiwanese office. Helicopters are also assembled by Ultra Helicopters (NZ) Ltd in New Zealand. An unmanned surveillance prototype named Vigilante 496 (first flight February 1998) has also been developed (see *Jane's Unmanned Aerial Vehicles and Targets*).

UPDATED

Two-seat Ultrasport 496 trainer *(Paul Jackson/Jane's)*

ASI ULTRASPORT 254 and 331

TYPE: Sport helicopter kitbuilt.
PROGRAMME: Prototype (254 version) first flight 24 July 1993 and publicly displayed at Oshkosh that year. Two prototypes built and tested. Prototype 331 (N331UV) first flew December 1993.
CURRENT VERSIONS: Designations reflect empty weight in pounds.
    Ultrasport 254: Single-seat, ultralight to FAR Pt 103.
    Ultrasport 331: Experimental category FAR Pt 21.191(g); single-seat 'growth' model. Meets FAA 51 per cent amateur-built kit rules.
CUSTOMERS: Total of 17 single-seaters flying by end 1999 (latest data available).
COSTS: 254: US$31,900; 331: US$33,900 (2000).
DESIGN FEATURES: Design objective of 254 was basic weight not to exceed 115 kg (254 lb) in order to comply with FAR Pt 103. Two-blade composites construction main rotor with tip weights for momentum conservation in event of engine failure, 8° linear twist and infinite life; shielded two-blade tail rotor; tailplane with fins at tip; tail rotor drive carried in narrow streamlined tailboom. Centrifugal sprag clutch for starting engages rotors at 2,000 engine rpm and automatically disengages in the event of engine failure. Quoted build time 80 hours. Broad (2.44 m; 8 ft 0 in) skid track helps to prevent rollovers.
FLYING CONTROLS: Conventional collective, cyclic and yaw pedals. Floor-mounted cyclic option available since 1998; early models had top-mounted stick.
STRUCTURE: Generally of epoxy resin, graphite fabric and Nomex honeycomb; aluminium tailboom.
LANDING GEAR: Aluminium skids stressed for landings at up to 2.5 g; floats (weight 18 kg; 40 lb) optional.
POWER PLANT: *254:* One 41 kW (55 hp) Hirth 2703 dual-carburettor two-stroke engine with pull starter and 12:1 planetary transmission. Normal fuel capacity 19 litres (5.0 US gallons; 4.1 Imp gallons).
    331: One 48.5 kW (65 hp) Hirth 2706 dual-carburettor two-stroke engine with electric starter (also option for 254); 12:1 planetary transmission. Fuel capacity 38 litres (10 US gallons; 8.3 Imp gallons).
ACCOMMODATION: Single seat partially enclosed (254 version); enclosed (331 version).
SYSTEMS: Electrical: 12 V battery; 14 V alternator fit appropriate to engine.
AVIONICS: Customer choice.
DIMENSIONS, EXTERNAL:
    Main rotor diameter   6.40 m (21 ft 0 in)
    Main rotor blade chord   0.17 m (6¾ in)
    Tail rotor diameter   0.76 m (2 ft 6 in)
    Tail rotor blade chord   0.05 m (2 in)
    Fuselage: Length, main rotor folded   5.84 m (19 ft 2 in)
    Height overall   2.39 m (7 ft 10 in)
    Skid track   2.44 m (8 ft 0 in)
DIMENSIONS, INTERNAL:
    Cabin: Length   1.32 m (4 ft 4 in)
        Max width   0.76 m (2 ft 6 in)
        Max height   1.50 m (4 ft 11 in)
AREAS:
    Main rotor disc   32.2 m² (346.4 sq ft)
WEIGHTS AND LOADINGS:
    Weight empty: 254   114 kg (252 lb)
        331   150 kg (330 lb)
    Max T-O and landing weight: 254   238 kg (525 lb)
        331   294 kg (650 lb)
PERFORMANCE:
    Max level speed: 254   55 kt (101 km/h; 63 mph)
        331   90 kt (167 km/h; 104 mph)
    Cruising speed: 254, 331   55 kt (101 km/h; 63 mph)
    Max rate of climb at S/L   305 m (1,000 ft)/min
    Service ceiling   3,660 m (12,000 ft)
    Hovering ceiling: IGE   3,290 m (10,800 ft)
        OGE   2,135 m (7,000 ft)
    Range: normal fuel, 55 kt:
        254   65 n miles (120 km; 75 miles)
        331   130 n miles (241 km; 150 miles)

UPDATED

ASI ULTRASPORT 496

TYPE: Light helicopter trainer.
Details as for 254 and 331, except as follows:
PROGRAMME: Developed from the 331, the Ultrasport 496 (N496AS) first flew in July 1995. Deliveries began in April 1997; second production (third overall) aircraft to New Zealand as demonstrator.
CURRENT VERSIONS: **Ultrasport 496:** *As described.*
    Sportscopter 600: New Zealand assembly; initial four registered in July 1999.
    Vigilante 500: Remotely piloted version developed by Science Applications International Corporation (SAIC). Prototype (N496UV) evaluated as optionally piloted vehicle (OPV) by US Navy during first and second quarters of 1998. First Vigilante 500, with reduced-size fuselage module, under construction in 1999. Planned shipboard version is **Vigilante 600**, equipped for high-resolution optical or radar surveillance.
CUSTOMERS: By end of 1997, 32 had been delivered, 20 of which were flying by end 1999.
COSTS: US$49,900 (2000).
DESIGN FEATURES: Generally as for 254/331. Quick-build kit quoted build time 80 hours; dual controls standard.
STRUCTURE: Infinite-life composites rotor blades and fuselage. Shaft-driven tail rotor. Vertically mounted direct-drive Helical Spur Gears 11:1 two-stage main transmission.
LANDING GEAR: Aluminium skids; track as for 254/331. Floats optional.
POWER PLANT: One 85.8 kW (115 hp) Hirth F30 quad-carburettor engine with electric starter. Fuel capacity 61 litres (16.0 US gallons; 13.3 Imp gallons).
EQUIPMENT: Ballistic parachute under consideration.
DIMENSIONS, EXTERNAL:
    Main rotor diameter   7.01 m (23 ft 0 in)
    Main rotor blade chord   0.17 m (6¾ in)
    Length, blades folded   6.02 m (19 ft 9 in)
    Tail rotor blade chord   0.05 m (2 in)
    Height   2.39 m (7 ft 10 in)
DIMENSIONS, INTERNAL:
    Cabin: Length   1.35 m (4 ft 5 in)
        Max width   1.22 m (4 ft 0 in)
        Max height   1.50 m (4 ft 11 in)
AREAS:
    Main rotor disc   38.60 m² (415.5 sq ft)
WEIGHTS AND LOADINGS:
    Weight empty   245 kg (540 lb)
    Max T-O and landing weight   512 kg (1,130 lb)
PERFORMANCE (at max T-O weight, ISA):
    Never-exceed speed (V_{NE})   90 kt (167 km/h; 104 mph)
    Max cruising speed   60 kt (111 km/h; 69 mph)
    Max rate of climb at S/L   305 m (1,000 ft)/min
    Hovering ceiling: IGE   3,290 m (10,800 ft)
        OGE   2,135 m (7,000 ft)
    Range   130 n miles (240 km; 149 miles)
    Endurance   2 h 30 min

UPDATED

ASI Ultrasport 331 single-seat ultralight helicopter with doors removed and optional floor-mounted cyclic (thereby resembling the 254)

ASI Ultrasport 254

AUC

AMERICAN UTILICRAFT CORPORATION
Suite B, 300 Petty Road NE, Lawrenceville, Georgia 30043
Tel: (+1 678) 376 08 98
Fax: (+1 678) 376 90 93
e-mail: info@utilicraft.com
Web: http://www.utilicraft.com
PRESIDENT AND CEO: John DuPont
EXECUTIVE VICE-PRESIDENT AND SENIOR VICE-PRESIDENT, MARKETING: James Carey

Company formed August 1990 to design and produce Freight Feeder cargo transport. Production will be undertaken at Gwinnett County, Atlanta, Georgia.

UPDATED

AUC FF-1080-200 FREIGHT FEEDER
TYPE: Twin-turboprop freighter.
PROGRAMME: Design started 1990; patents filed 1991; original capacity for four LD3 containers increased to six in 1997. Two flying prototypes planned; formal programme launch March 1998. Mockup completed end 1999. Construction of preproduction prototype was to begin in January 2001; FAA Pt 25 certification process starting mid-2001. First flight scheduled before end 2001, with certification completed 2003.
CURRENT VERSIONS: **FF-1080-100**: Short fuselage version; four LD3 containers; PW121 engines. Not intended for construction at present; weights and loadings in 1999-2000 *Jane's*.
FF-1080-200: Standard version; *as described*.
FF-1080-500: Enlarged version; BR 715 engines.
CUSTOMERS: AUC currently in discussion with various US and international customers for fleet purchases. First production batch will comprise 48 aircraft. Company estimates market for 5,000 such aircraft and aims to capture 10 per cent.
COSTS: Development and certification to FAR Pt 25 estimated at US$75 million (1998).
DESIGN FEATURES: High-mounted wings, tapered outboard of overwing engine nacelles; box-section fuselage with ventral pannier; upswept rear fuselage with rear-loading door; angular tail surfaces with large dorsal fin fairing. Onboard freight management system.
FLYING CONTROLS: Conventional and manual. Upper surface blowing (USB) using engine bleed air.
STRUCTURE: All-aluminium construction. Interest in component manufacture shown by companies in South Korea, Taiwan and USA. Fuselage subassemblies constructed by Metalcraft Technologies Inc of Utah. Other risk-sharing partners include Goodrich, UPS Aviation Technologies, AAR Cargo Systems, Auxilec Inc, Shaw Aero, HS Dynamic Controls, Securaplane Technologies, Lord Mounts, Hi-Temp Insulation Inc and General Electrodynamics Corp. Wing subassemblies by Aerostructures. Detailed engineering undertaken by Aircraft Design Services International.
LANDING GEAR: Non-retractable tricycle type; twin nosewheels and tandem pairs of mainwheels.
POWER PLANT: *FF-1080-200:* Two 2,051 kW (2,750 shp) Pratt & Whitney Canada PW127F turboprops, each driving a Hamilton Sundstrand 568F six-blade propeller. FADEC fitted as standard. Fuel capacity 6,853 litres (1,810 US gallons; 1,507 Imp gallons), optional fuel 10,601 litres (2,800 US gallons; 2,332 Imp gallons).
FF-1080-100: PW121 engines each rated at 1,567 kW (2,100 shp).
FF-1080-500: Turboprop version of Rolls-Royce Deutschland BR 715 turbofan, (each 3,424 kW; 4,591 shp).
ACCOMMODATION: Flight crew of two on IPECO seats. Main cargo hold accommodates up to six LD3 containers. Cargo roller floor from AAR Cargo Systems. Large cargo double door on port side, forward of wing; rear-loading door; crew airstair door on port side of flight deck. Flight deck pressurised; main cargo bay unpressurised. Additional capacity in cargo pannier and nose compartment.
SYSTEMS: Securaplane smoke detection system; HS Dynamic Controls anti-icing and de-icing systems; electrical system integration by Auxilec. Shaw Aero Devices fuel management system.
AVIONICS: Flight and engine displays and autopilot by Meggitt Avionics of UK; GPS by UPS Aviation Technologies.
DIMENSIONS, EXTERNAL:
Wing span 27.10 m (88 ft 11 in)
Length overall 23.80 m (78 ft 1 in)
Height overall 9.04 m (29 ft 8 in)
Side cargo door: Height 2.13 m (7 ft 0 in)
Width 2.59 m (8 ft 6 in)
Rear cargo door: Height 2.13 m (7 ft 0 in)
Width 1.68 m (5 ft 6 in)
DIMENSIONS, INTERNAL:
Cargo bay: Length 13.31 m (43 ft 8 in)
Max width 1.93 m (6 ft 4 in)
Max height 2.13 m (7 ft 0 in)
Volume: cargo bay 54.8 m³ (1,936 cu ft)
pannier 5.61 m³ (198 cu ft)
nose compartment 5.47 m³ (193 cu ft)
WEIGHTS AND LOADINGS (estimated):
Weight empty: 200 8,117 kg (17,894 lb)
500 14,515 kg (32,000 lb)
Payload: for 400 n mile (740 km; 460 mile) range:
200 6,619 kg (14,592 lb)
500 9,979 kg (22,000 lb)
for max range: 200 4,019 kg (8,860 lb)
500 7,348 kg (16,200 lb)
Max T-O weight: 200 17,236 kg (38,000 lb)
Max landing weight: 200 16,783 kg (37,000 lb)
Max zero-fuel weight: 200 12,247 kg (27,000 lb)
Max power loading: 200 4.21 kg/kW (6.91 lb/shp)
PERFORMANCE (estimated):
Cruising speed: 200 250 kt (463 km/h; 287 mph)
500 270 kt (500 km/h; 311 mph)
Stalling speed, power off, 200:
T-O flaps 85 kt (158 km/h; 98 mph)
flaps down 73 kt (136 km/h; 84 mph)
Minimum single-engine control speed (V_{MC}):
200 82 kt (152 km/h; 95 mph)
Max rate of climb at S/L: 200 671 m (2,200 ft)/min
Rate of climb at S/L, OEI: 200 213 m (700 ft)/min
Service ceiling: 200 7,620 m (25,000 ft)
Service ceiling, OEI: 200 4,572 m (15,000 ft)
T-O run: 200 558 m (1,830 ft)
T-O to 15 m (50 ft): 200 856 m (2,810 ft)
Landing from 15 m (50 ft): 200 634 m (2,080 ft)
Landing run: 200 396 m (1,300 ft)
Range: with max fuel:
200 1,500 n miles (2,778 km; 1,726 miles)
500 2,000 n miles (3,704 km; 2,301 miles)
with max payload:
200 500 n miles (926 km; 575 miles)

UPDATED

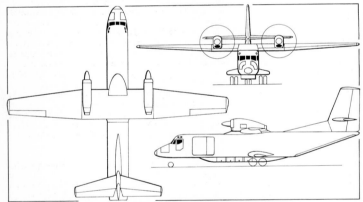
General appearance of the AUC FF-1080-200 Freight Feeder *(James Goulding/Jane's)* 0015552

Model of the mid-range Freight Feeder 200 *(Paul Jackson/Jane's)* 2001/0100501

AVIABELLANCA

AVIABELLANCA AIRCRAFT CORPORATION
PMB 47, 2315 B Forest Drive, Annapolis, Maryland 21401
Tel: (+1 410) 266 55 18
Fax: (+1 410) 266 86 97
e-mail: avbellanca@aol.com
Web: http://www.aviabellanca.com
PRESIDENT: August T Bellanca

The original Bellanca Aircraft Corporation was established in 1927 by Giuseppe Mario Bellanca and produced low-wing cabin monoplanes such as the Cruisair and Cruisemaster both before and after the Second World War. Bellanca left the original firm to set up Bellanca Development Company in 1954, which was renamed Bellanca Aircraft Engineering in 1963 and AviaBellanca in 1983. AviaBellanca's designer is August Bellanca, son of Giuseppe. New, all-composites August Bellanca-designed Skyrocket II introduced 1975 and set five NAA/FAI class speed records; improved Skyrocket III made public debut at Oshkosh 1997. At the beginning of 1997, 3,842 Bellanca-designed aircraft were registered in the USA.

UPDATED

AVIABELLANCA 19-25 SKYROCKET III
TYPE: Six-seat kitbuilt.
PROGRAMME: Design started 1969; prototype first flight (N14666) 1974; set five FAI international closed circuit speed records.

AviaBellanca 19-25 Skyrocket II prototype *(Paul Jackson/Jane's)* 2001/0099629

CURRENT VERSIONS: **Skyrocket II**: Original prototype. Currently used as flight test vehicle (re-registered N771AB).
Skyrocket III: Projected factory-manufactured version; available with or without pressurisation.
Description refers to Skyrocket III.
CUSTOMERS: By early 2000, one prototype was flying and 20 more were on order. Preproduction static test and flight test aircraft were under construction. An MoU was signed in July 1999 with Grupo EB-RIM of Spain to produce composite parts for the Skyrocket.
DESIGN FEATURES: Designed for rapid, uncomplicated assembly. Intended to form the foundation for an entire family of aircraft from single-engined general aviation to twin-engined turboprop business aircraft. Aerodynamically clean, low-wing design with laminar flow wing section.
FLYING CONTROLS: Conventional and manual. Elevator with electric trim on entire stabiliser; actuation by cable and bellcrank; horn-balanced rudder and elevators; no trim tabs. Plain flaps.
STRUCTURE: All-composites. Skyrocket II wings comprise upper and lower covers and forward and rear spars; last-

mentioned are spliced at centreline; all of sandwich construction, with glass fibre skins and aluminium honeycomb, plus glass fibre spar cap. Ailerons, flaps and empennage similar, including left and right fin cover shells. Fuselage in left and right halves, with flat sandwich bulkheads and floor. Skyrocket III will have graphite fibre over Nomex core.

LANDING GEAR: Tricycle hydraulically retractable type; mainwheels retract inboard; nosewheel aft. Oleo-pneumatic shock-absorption; hydraulic brakes.

POWER PLANT: One 324 kW (435 hp) Teledyne Continental GTSIO-520F flat-six, driving a Hartzell three-blade constant-speed propeller. Later option of 261 kW (350 hp) Teledyne Continental TSIO-550-B flat-six and 540 kW (724 shp) Walter M 601D turboprop. Fuel capacity 666 litres (176 US gallons; 147 Imp gallons) in two wing tanks, of which 644 litres (170 US gallons; 142 Imp gallons) are usable. Gravity-fed refuelling points in wingtips.

ACCOMMODATION: Pilot and five passengers in three rows of side-by-side seats; door starboard side, above wing.

SYSTEMS: Air conditioning and pressurisation available on production version and on de luxe kitbuilt examples.

AVIONICS: To customer's specification; optional advanced IFR fit.

DIMENSIONS, EXTERNAL:
Wing span	10.67 m (35 ft 0 in)
Wing aspect ratio	6.7
Length	11.28 m (37 ft 0 in)
Height overall	2.74 m (9 ft 0 in)

DIMENSIONS, INTERNAL:
Cabin: Length	3.94 m (12 ft 11 in)
Max width	1.14 m (3 ft 9 in)
Max height	1.22 m (4 ft 0 in)

AREAS:
Wings, gross	16.96 m² (182.6 sq ft)

WEIGHTS AND LOADINGS:
Weight empty	1,129 kg (2,490 lb)
Baggage capacity	91 kg (200 lb)
Max T-O weight	1,905 kg (4,200 lb)
Max wing loading	112.3 kg/m² (23.00 lb/sq ft)
Max power loading	5.88 kg/kW (9.66 lb/hp)

PERFORMANCE:
Max operating speed (V$_{MO}$)	295 kt (547 km/h; 340 mph)
Max cruising speed at 75% power at 7,620 m (25,000 ft)	284 kt (526 km/h; 327 mph)
Normal cruising speed at 65% power at 7,925 m (26,000 ft)	273 kt (505 km/h; 314 mph)
Stalling speed, flaps and landing gear down	59 kt (110 km/h; 68 mph)
Max rate of climb at S/L	634 m (2,080 ft)/min
Service ceiling	9,140 m (30,000 ft)
T-O run	207 m (680 ft)
T-O to 15 m (50 ft)	344 m (1,130 ft)
Landing from 15 m (50 ft)	546 m (1,790 ft)
Landing run	282 m (925 ft)
Range at 65% power, 45 min reserves:	
with max fuel	1,998 n miles (3,701 km; 2,300 miles)
with max payload	820 n miles (1,519 km; 944 miles)

UPDATED

AVIAT

AVIAT AIRCRAFT INC
672 South Washington, PO Box 1240, Afton, Wyoming 83110
Tel: (+1 307) 885 31 51
Fax: (+1 307) 885 96 74
e-mail: aviat@aviataircraft.com
Web: http://www.aviataircraft.com
CHAIRMAN AND PRESIDENT: Stuart Horn
VICE-PRESIDENT, MARKETING: Bob James
PUBLIC AFFAIRS: Lynn Thomas

The Pitts Aerobatics company was acquired by Christen Industries in 1983 along with manufacturing and marketing rights for the Pitts Special aerobatic aircraft. In turn, Aviat Aircraft Inc acquired Christen Industries in 1991 and now owns production and type certificates for the Christen range. Aviat itself was bought by Stuart Horn in December 1995. The company has acquired manufacturing rights to the 1930s/1940s Globe Swift and Monocoupe 110. Aviat delivered a total of 83 aircraft in 1998, 85 in 1999 and was aiming to produce 90 Huskies in 2000.

UPDATED

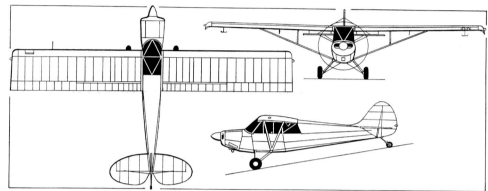

Aviat Husky A-1 (Textron Lycoming O-360 flat-four) *(James Goulding/Jane's)*

AVIAT HUSKY A-1

TYPE: Two-seat lightplane.

PROGRAMME: First flight (N6070H) 1986; FAA certification under FAR Pt 23 1987, including Edo 2000 floats and skis, glider/banner towing hook approved.

CURRENT VERSIONS: **A-1** Original version, now superseded; 400 built.
A-1A Standard version from c/n 1401.
Description refers to A-1A unless otherwise stated.
A-1B Mission-specific version for government agencies. Has higher maximum take-off weight of 907 kg (2,000 lb). 55 built by December 1999; 100 built by mid-2000.

CUSTOMERS: Total 400 A-1s and 68 A-1As delivered by December 1999, including two to US Department of Interior and four to US Department of Agriculture; also operated by US police agencies and in Kenya for wildlife protection patrols. Production for 1999 totalled 85 (44 A-1A and 41 A-1B).

COSTS: A-1A US$112,964; A-1B US$125,354 (2000).

DESIGN FEATURES: Conventional high-wing monoplane; constant-chord wing with V-strut bracing and auxiliary struts; wire-braced empennage with elliptical surfaces. Wing has modified Clark Y US 35B section; optional drooped Plane Booster wingtips.

FLYING CONTROLS: Conventional and manual. Symmetrical section ailerons with spade-type mass balance; trim tabs in elevators; slotted high-lift flaps. Fixed tailplane; trim by adjustable bungee.

STRUCTURE: Tubular welded 4130 steel fuselage. Wing has two aluminium spars, metal ribs and metal leading-edge, Dacron covering overall. Twin bracing struts each side of wings and wire- and strut-braced tail unit. Light alloy slotted flaps and ailerons, with Dacron covering. Fuselage and tail have chrome molybdenum steel tube frames, covered in Dacron except for metal skin to rear fuselage. Seven-coat corrosion protection.

LANDING GEAR: Non-retractable tailwheel type. Two faired side Vs and half-axles hinged to bottom of fuselage, with internal (under front seat) bungee cord shock-absorption. Cleveland mainwheels, tyres size 8.00-6 as standard; 6.00-6 or 8.50-6 tyres and 0.79 m (2 ft 7 in) Tundra tyres optional. Cleveland mainwheel hydraulic disc brakes. Steerable leaf-spring tailwheel. Wheel-replacement or wheel-retract skis and floats optional.

POWER PLANT: One 134 kW (180 hp) Textron Lycoming O-360-A1P flat-four engine, driving a Hartzell HC-C2YK-1BF two-blade constant-speed metal propeller. Fuel in two metal tanks, one in each wing, total capacity 208 litres (52.0 US gallons; 45.75 Imp gallons), of which 189 litres (50.0 US gallons; 41.6 Imp gallons) are usable. Fuel filler point in upper surface of each wing, near root.

ACCOMMODATION: Enclosed cabin seating two in tandem, with dual controls. Five-point safety harness. Downward-hinged door on starboard side, with upward-hinged window door. Skylight window in roof. A-1B has luggage door on starboard side as standard; optional on A-1A. Rear compartment for extra 13.6 kg (30 lb) baggage standard from 2000; certified retrofitting kits available.

SYSTEMS: 12V electrical system includes lights and 60 A alternator.

AVIONICS: VFR standard; Loran C transponder, GPS, nav/com and IFR instrumentation optional.

EQUIPMENT: Optional equipment includes Straight Edo, Baumann and Wipline 2100 floats, Aero wheel replacement skis, Aero wheel skis and Flairdyne hydraulic wheel-retracting skis.

DIMENSIONS, EXTERNAL:
Wing span	10.82 m (35 ft 6 in)
Wing aspect ratio	6.9
Length overall	6.88 m (22 ft 7 in)
Height overall	2.01 m (6 ft 7 in)
Propeller diameter	1.93 m (6 ft 4 in)

DIMENSIONS, INTERNAL:
Baggage hold volume	0.28 m³ (10 cu ft)

AREAS:
Wings, gross	17.00 m² (183.0 sq ft)
Ailerons (total)	1.43 m² (15.40 sq ft)
Trailing-edge flaps (total)	2.09 m² (22.50 sq ft)
Fin	0.43 m² (4.66 sq ft)
Rudder	0.62 m² (6.76 sq ft)
Tailplane	1.48 m² (15.90 sq ft)
Elevators, incl tabs	1.31 m² (14.10 sq ft)

WEIGHTS AND LOADINGS:
Weight empty	540 kg (1,190 lb)
Max baggage weight	36 kg (80 lb)
Max T-O weight:	
landplane	857 kg (1,890 lb)
floatplane: A-1A	939 kg (2,070 lb)
A-1B	998 kg (2,200 lb)
Max wing loading:	
landplane	50.4 kg/m² (10.33 lb/sq ft)
floatplane: A-1A	55.2 kg/m² (11.31 lb/sq ft)
A-1B	58.7 kg/m² (12.02 lb/sq ft)
Max power loading:	
landplane	6.39 kg/kW (10.50 lb/hp)
floatplane: A-1A	7.00 kg/kW (11.50 lb/hp)
A-1B	7.44 kg/kW (12.22 lb/hp)

PERFORMANCE (landplane):
Never-exceed speed (V$_{NE}$)	132 kt (245 km/h; 152 mph)
Max level speed	126 kt (233 km/h; 145 mph)
Cruising speed: at 75% power at 1,220 m (4,000 ft)	122 kt (225 km/h; 140 mph)
at 55% power	113 kt (209 km/h; 130 mph)
Landing speed: A-1A	42 kt (77 km/h; 48 mph)
A-1B	51 kt (94 km/h; 58 mph)
Max rate of climb at S/L	457 m (1,500 ft)/min
Service ceiling	6,100 m (20,000 ft)

Aviat Husky A-1B on Wipline 2100 floats *(Paul Jackson/Jane's)*

558 USA: AIRCRAFT—AVIAT

T-O run: flaps down	61 m (200 ft)
T-O to 15 m (50 ft): flaps down	122 m (400 ft)
Landing from 15 m (50 ft): flaps down	244 m (800 ft)
Landing run, full flap	107 m (350 ft)
Range with max fuel at 75% power, no reserves	
	608 n miles (1,126 km; 700 miles)

UPDATED

AVIAT PITTS S-1-11B SUPER STINKER

This single-seat aerobatic biplane was last described in the 1998-99 edition; plans and components are available and Aviat has announced that it will build a limited number of S-1-11Bs to special order. A radial-engined version is built by Kimball as the Model 12 (which see).

VERIFIED

AVIAT PITTS S-2B

TYPE: Aerobatic two-seat biplane.
PROGRAMME: Successor to S-2A. Prototype completed September 1982; certified in FAR Pt 23 Aerobatic category 1983; won first place Advanced Category of 1982 US Nationals with two occupants.
CUSTOMERS: Total of 357 factory-built examples produced by mid-1998. These include limited edition version marketed in 1997; seven produced with special black and gold paint scheme, gold tinted canopy and leather upholstery, plus additional instruments and three-blade Hartzell propeller option. Three produced in 1998; none in 1999.
COSTS: US$137,000 (1997).
DESIGN FEATURES: Compact, positive-stagger, single-bay biplane; constant-chord wings, upper of which sweptback; wire-braced, elliptical tail surfaces. Optimised for aerobatic performance; roll rate 240°/s.
Wing sections NACA 6400 series on upper wing, 00 series on lower wings. Both wings have 1° 30′ incidence and lower wing has 3° dihedral.
FLYING CONTROLS: Conventional and manual. Ailerons on upper and lower wings with aerodynamic spade-type balances on lower ailerons; trim tab on each elevator; no flaps.
STRUCTURE: Fuselage 4130 steel tube with wooden stringers, aluminium top decking and side panels; remainder Dacron covered. Steel tube, Dacron-covered fixed tail surfaces; Dacron-covered control surfaces.
LANDING GEAR: Non-retractable tailwheel type. Rubber cord shock-absorption. Steerable tailwheel. Streamline fairings on mainwheels.
POWER PLANT: One 194 kW (260 hp) Textron Lycoming AEIO-540-D4A5 flat-six engine, driving a Hartzell HC-C2YR-4CF/FC 8477A-4 two-blade constant-speed metal propeller; MT and Hartzell three-blade propellers are optional. Fuel tank in fuselage, immediately aft of firewall, capacity 110 litres (29.0 US gallons; 24.1 Imp gallons), of which 106 litres (28.0 US gallons; 23.3 Imp gallons) are usable. Refuelling point on fuselage upper surface forward of windscreen. Oil capacity 11.4 litres (3.0 US gallons; 2.5 Imp gallons). Inverted fuel and oil systems standard.
ACCOMMODATION: Two seats in tandem cockpits, with dual controls. Sideways-opening one-piece canopy covers both cockpits. Space for 9.1 kg (20 lb) baggage aft of rear seat when flown in non-aerobatic category.
SYSTEMS: Electrical system powered by 12 V 40 A alternator and non-spill 12 V battery.

DIMENSIONS, EXTERNAL:
Wing span: upper	6.10 m (20 ft 0 in)
lower	5.79 m (19 ft 0 in)
Wing chord, constant, both	1.02 m (3 ft 4 in)
Wing aspect ratio	3.2
Length overall	5.71 m (18 ft 9 in)
Height overall	2.02 m (6 ft 7½ in)

AREAS:
Wings, gross	11.61 m² (125.0 sq ft)

WEIGHTS AND LOADINGS:
Weight empty	521 kg (1,150 lb)
Max T-O weight: aerobatic	737 kg (1,625 lb)
Max wing loading	63.5 kg/m² (13.00 lb/sq ft)
Max power loading	3.80 kg/kW (6.25 lb/hp)

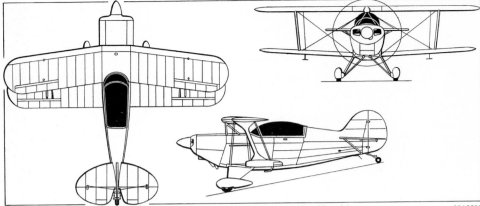

Aviat Pitts S-2B (Textron Lycoming AEIO-540 flat-six) *(James Goulding/Jane's)* 0015593

Aviat Pitts S-2C aerobatic biplane *(Paul Jackson/Jane's)* 2001/0099631

PERFORMANCE:
Never-exceed speed (VNE)	184 kt (340 km/h; 212 mph)
Max level speed at S/L	161 kt (299 km/h; 186 mph)
Max cruising speed	151 kt (280 km/h; 174 mph)
Stalling speed	53 kt (99 km/h; 61 mph)
Max rate of climb at S/L	823 m (2,700 ft)/min
Service ceiling	6,400 m (21,000 ft)
Range with max fuel at 55% power, 30 min reserves	
	277 n miles (513 km; 319 miles)
g limits	+6/−3

UPDATED

AVIAT PITTS S-2C

TYPE: Aerobatic two-seat biplane.
PROGRAMME: Developed from Pitts S-2 series of two-seat biplanes using wing design from Curtis Pitts. Uses S-2B basic fuselage structure.
CUSTOMERS: 35 built by 31 December 1999, including 16 in that year.
COSTS: Basic price US$179,550 (2000).
DESIGN FEATURES: Generally as for S-2B, except symmetrical ailerons; squared wingtips; metal fuselage undersides removable for easier maintenance. Stretch-formed compound-curved engine cowling. Rate of roll is greater than 300°/s.
FLYING CONTROLS: Conventional and manual. Rudder and elevators aerodynamically balanced
STRUCTURE: Generally as S-2B, with extra strengthening where required.
LANDING GEAR: Conventional tailwheel type, with bungee suspension in protective legs. Mainwheels 5.00-5; tyre pressure 2.34 bars (34 lb/sq in).
POWER PLANT: One 194 kW (260 hp) Textron Lycoming AEIO-540 flat-six driving a Hartzell three-blade constant-speed composites propeller. Fuel capacity 110 litres (29.0 US gallons; 24.0 Imp gallons) of which 106 litres (28.0 US gallons; 23.3 Imp gallons) are usable; for aerobatic flight, the 19 litre (5.0 US gallon; 4.2 Imp gallon) wing tank is not used. Oil capacity 11.4 litres (3.0 US gallons; 2.5 Imp gallons).
ACCOMMODATION: Pilot and passenger in tandem under one-piece rearward-sliding canopy; single-piece windscreen.

DIMENSIONS, EXTERNAL:
Wing span	6.10 m (20 ft 0 in)
Length overall	5.41 m (17 ft 9 in)
Height overall	1.96 m (6 ft 5 in)
Wheel track	1.54 m (5 ft 0¾ in)
Wheelbase	4.14 m (13 ft 7 in)
Propeller diameter	1.98 m (6 ft 6 in)

DIMENSIONS, INTERNAL:
Cabin: Length	2.11 m (6 ft 11 in)
Max width	0.71 m (2 ft 4 in)

AREAS:
Wings, gross	11.85 m² (127.5 sq ft)

WEIGHTS AND LOADINGS:
Weight empty	522 kg (1,150 lb)
Baggage capacity	9 kg (20 lb)
Max T-O and landing weight	771 kg (1,700 lb)
Max wing loading	65.1 kg/m² (13.33 lb/sq ft)
Max power loading	3.98 kg/kW (6.54 lb/hp)

PERFORMANCE:
Never-exceed speed (VNE)	185 kt (342 km/h; 212 mph)
Max level speed	169 kt (313 km/h; 194 mph)
Manoeuvring speed	134 kt (248 km/h; 154 mph) IAS
Cruising speed at 55% power	
	147 kt (272 km/h; 169 mph)
Stalling speed, power off, flaps up	
	55 kt (102 km/h; 63 mph)
Max rate of climb at S/L	884 m (2,900 ft)/min
T-O run	169 m (555 ft)
T-O to 15 m (50 ft)	262 m (860 ft)
Landing from 15 m (50 ft)	647 m (2,124 ft)
Landing run	402 m (1,320 ft)
Max range at 55% power, 30 min reserves	
	300 n miles (555 km; 345 miles)
Endurance, 30 min reserves	1 h 36 min
g limits	+6/−5

UPDATED

AVIAT PITTS S-2S

This S-2B variant remains available to order. Details last appeared in the 2000-01 edition.

UPDATED

AVIAT (CHRISTEN) EAGLE II

TYPE: Aerobatic two-seat biplane kitbuilt.
PROGRAMME: First flown in February 1977.
CUSTOMERS: More than 350 completed and flown by mid-2000.
COSTS: Kit US$74,000 (2000) less engine and propeller.
DESIGN FEATURES: Generally as for S-2B. Available in 24 individual subassemblies; quoted build time 1,800 hours.

Christen Eagle II in appropriate markings *(Paul Jackson/Jane's)* 2000/0084593

AVIAT to AVIATION DEVELOPMENT—AIRCRAFT: USA

Sixth production Globe GC-1B Swift, built August 1946, is representative of Aviat's Millennium Swift *(Paul Jackson/Jane's)* 2001/0099632

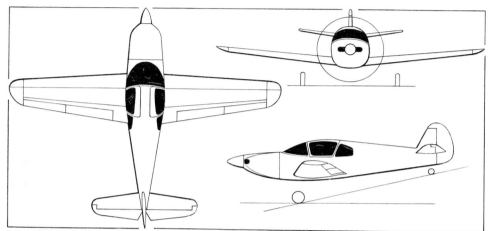

Aviat Millennium Swift (Continental IO-360 flat-six) *(James Goulding/Jane's)* 2001/0100868

FLYING CONTROLS: Conventional and manual. Ailerons on both sets of wings.
STRUCTURE: Fabric-covered wings with wooden spars and ribs plus metal leading- and trailing-edges. Welded 4130 steel tube fuselage, fabric covered aft of cockpit; metal panelled before.
LANDING GEAR: Tailwheel type; fixed.
POWER PLANT: One 149 kW (200 hp) Textron Lycoming AEIO-360-A1D driving a Hartzell constant-speed two-blade metal propeller. Fuel capacity 94.6 litres (25.0 US gallons; 20.8 Imp gallons), of which 90.8 litres (24.0 US gallons; 20.0 Imp gallons) are usable.
ACCOMMODATION: Pilot and passenger in tandem; flown solo from rear seat.
SYSTEMS: Inverted oil system and manual fuel pump system as standard.
DIMENSIONS, EXTERNAL:
Wing span	6.07 m (19 ft 11 in)
Length overall	5.64 m (18 ft 6 in)
Height overall	1.98 m (6 ft 6 in)

AREAS:
Wings, gross	11.61 m² (125.0 sq ft)

WEIGHTS AND LOADINGS:
Weight empty	476 kg (1,050 lb)
Max T-O and landing weight	725 kg (1,600 lb)
Max wing loading	62.5 kg/m² (12.80 lb/sq ft)
Max power loading	4.87 kg/kW (8.00 lb/hp)

PERFORMANCE:
Never-exceed speed (VNE)	182 kt (338 kt; 210 mph)
Max operating speed	159 kt (296 km/h; 184 mph)
Cruising speed	143 kt (266 km/h; 165 mph)
Stalling speed	51 kt (90 km/h; 58 mph)
Max rate of climb at S/L	645 m (2,120 ft)/min
Service ceiling	5,180 m (17,000 ft)
T-O run	244 m (800 ft)
Landing from 15 m (50 ft)	480 m (1,575 ft)
Range with 30 min reserves	380 n miles (703 km; 437 miles)
g limits	+6/−3

UPDATED

Prototype Aviat 110 Special *(Paul Jackson/Jane's)* 2000/0084596

AVIAT MILLENNIUM SWIFT
TYPE: Two-seat lightplane.
PROGRAMME: Globe Aircraft Corporation GC-1 Swift first flew in 1941, but did not enter production until 1946, with type certificate (A-766) issued 7 May. Production by Globe and Temco totalled 408 GC-1As (63.4 kW; 85 hp Continental C-85) and 1,094 GC-1Bs (93.2 kW; 125 hp Continental C-125) up to 1951; type certificate then to Universal Aircraft Industries; passed to Swift Association; planned return to production by LoPresti-Piper as SwiftFury in early 1990s later shelved and relaunched in late 1990s as LoPresti Fury, similar visually, but larger than the Swift.
Considerable redesign of GC-1B by Aviat before January 1999 relaunch as Millennium Swift; features new engine, wing, landing gear and instrumentation; parts count reduced from some 5,000 to about 3,000. First flight planned for mid-1999 and certification to FAR Pt 23 before year end, but neither achieved due to (now resolved) legal dispute with LoPresti (which see).
CUSTOMERS: Planned production of 50 per year.
COSTS: US$180,000 to US$200,000 (1999).
DESIGN FEATURES: Conventional low-wing monoplane. Wing aerofoil NLF(1) 0414F/M1 at root and NLF(1) 0414F/M1S at tip; washout −1.06°; no twist.
FLYING CONTROLS: Conventional and manual. Horn-balanced tail surfaces; flaps.
STRUCTURE: Aluminium monocoque.
LANDING GEAR: Tailwheel type. Mainwheels retract inwards.
POWER PLANT: One 157 kW (210 hp) Teledyne Continental IO-360-ES flat-six; 239 kW (320 hp) Textron Lycoming AEIO-580-L1B5 optional. Fuel capacity 295 litres (78.0 US gallons; 65.0 Imp gallons) in two wing tanks.
ACCOMMODATION: Side-by-side seating for two. Gull-wing cabin doors on each side. 45 kg (100 lb) baggage capacity behind seats.

DIMENSIONS, EXTERNAL:
Wing span	8.03 m (26 ft 4 in)
Wing aspect ratio	6.8
Length overall	6.81 m (22 ft 4 in)
Max diameter of fuselage	1.12 m (3 ft 8 in)
Tailplane span	3.12 m (10 ft 2¾ in)
Propeller diameter	1.93 m (6 ft 4 in)

AREAS:
Wings, gross	9.49 m² (102.2 sq ft)
Ailerons (total)	0.70 m² (7.57 sq ft)
Trailing edge flaps (total)	1.04 m² (11.20 sq ft)
Fin	0.29 m² (3.17 sq ft)
Rudder, incl tab	0.55 m² (5.94 sq ft)
Tailplane	1.87 m² (20.09 sq ft)
Elevators, incl tab	0.70 m² (7.52 sq ft)

WEIGHTS AND LOADINGS:
Weight empty	708 kg (1,561 lb)
Max T-O weight	1,122 kg (2,474 lb)
Max wing loading	118.2 kg/m² (24.21 lb/sq ft)
Max power loading	7.17 kg/kW (11.78 lb/hp)

PERFORMANCE:
Never-exceed speed (VNE)	275 kt (509 km/h; 316 mph)
Normal cruising speed	200 kt (370 km/h; 230 mph)
Stalling speed, flaps down	59 kt (110 km/h; 68 mph)
Endurance	6 h 30 min
g limits	+6/−3

UPDATED

AVIAT 110 SPECIAL
TYPE: Two-seat lightplane.
PROGRAMME: Revised version of Monocoupe 110, certified 16 June 1930, as design of Mono Aircraft. Prototype (N110XZ) first flown 25 July 1999; publicly displayed at Oshkosh end July 1999. Certification to FAR Pt 23 planned for 2000 with deliveries soon after, but programme suspended in late 2000.

UPDATED

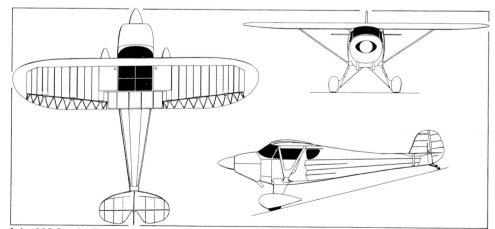

Aviat 110 Special (Textron Lycoming IO-360 flat four) *(Paul Jackson/Jane's)* 2000/0084542

AVIATION DEVELOPMENT

AVIATION DEVELOPMENT INTERNATIONAL LTD
110 Homestead Avenue, Hillsboro, Ohio 45133
Tel: (+1 937) 393 26 42
Fax: (+1 740) 335 89 96

NEW ENTRY

AVIATION DEVELOPMENT ALASKAN BUSHMASTER
TYPE: Four-seat kitbuilt.
CUSTOMERS: Total 12 flying by early 2000.
COSTS: Kit US$22,500 (2000).
DESIGN FEATURES: Piper Cub family development. High wing braced with two struts; wire-braced tailplane and fin.
FLYING CONTROLS: Conventional and manual. Horn-balanced elevators and rudder; flight-adjustable tabs in both elevators.
STRUCTURE: Conventional metal tube and fabric construction.
LANDING GEAR: Tailwheel type; fixed. Bungee cord suspension. Mainwheels 8.50-6 in. Options include floats, amphibious floats, skis and tundra tyres.

POWER PLANT: Prototype had 149 kW (200 hp) Textron Lycoming engine driving a two-blade propeller; recommended engine power between 112 and 224 kW (150 and 300 hp), including engines from Continental and LOM. Standard fuel capacity in wings, 167 litres (44.0 US gallons; 36.6 Imp gallons), with options up to 500 litres (132 US gallons; 110 Imp gallons).
ACCOMMODATION: Pilot and three passengers in two pairs of seats. Dual controls. Door each side, plus freight door port, rear.
DIMENSIONS, EXTERNAL:
Wing span 12.19 m (40 ft 0 in)
Wing aspect ratio 7.6
Length overall 7.62 m (25 ft 0 in)
Height overall 2.13 m (7 ft 0 in)
AREAS:
Wings, gross 19.51 m² (210.0 sq ft)
WEIGHTS AND LOADINGS:
Weight empty 635 kg (1,400 lb)
Max T-O weight 1,360 kg (3,000 lb)
Max wing loading 69.75 kg/m² (14.29 lb/sq ft)
PERFORMANCE (149 kW; 200 hp Textron Lycoming engine):
Never-exceed speed (VNE) 130 kt (241 km/h; 150 mph)
Normal cruising speed 109 kt (201 km/h; 125 mph)
Stalling speed, power off, flaps up
 39 kt (71 km/h; 44 mph)
Max rate of climb at S/L 366 m (1,200 ft)/min
Service ceiling 5,485 m (18,000 ft)
T-O run 244 m (800 ft)
Landing run 107 m (350 ft)
Range 1,200 n miles (2,222 km; 1,381 miles)

Aviation Development Alaskan Bushmaster, showing LOM engine *(Paul Jackson/Jane's)* 2001/0089491

NEW ENTRY

AVID

AVID AIRCRAFT INC
5057 Hwy 287 North, Ennis, Montana 59714
Tel: (+1 406) 682 56 15
Fax: (+1 406) 682 55 54
e-mail: avidair@avidair.com
Web: http://www.avidair.com
PRESIDENT: Jim Tomash
GENERAL MANAGER: Robert Stone

Avid Aircraft was formed in 1983 and currently markets the Avid Flyer Mark IV, Catalina (to special order only), Bandit, Magnum and Champion. Avid Flyer also assembled in Europe by Yalo of Czech Republic (which see). In May 1999, the company was sold to a stock holder who intends to restart production in Ennis, Montana.

UPDATED

AVID FLYER MARK IV
TYPE: Side-by-side ultralight kitbuilt.
PROGRAMME: First flown 1983; available as single kit, or six separate kits to spread cost of purchase. Strongly influenced design of Indaer-Peru Chuspi light aircraft (see under Peru in 1992-93 *Jane's*), and the SkyStar series of kitbuilt aircraft. Avid **Bandit** identical, except for 454 kg (1,000 lb) maximum T-O weight.
CUSTOMERS: More than 1,600 Avid Flyer kits delivered.
COSTS: Kit: US$12,940 (1999).
DESIGN FEATURES: Strut-braced, high-wing ultralight of traditional configuration and construction. Two forms available (interchangeable), as original **High Gross STOL** with unique near full-span auxiliary aerofoil flaperons, and shorter span **Aerobatic Speedwing** using new wing section; cruising and stalling speeds raised with Aerobatic Speedwing fitted. Wings fold for storage. Quoted build time 650 hours.
FLYING CONTROLS: Manual. Junkers flaperons (see Design Features), elevators with adjustable trim tab, and rudder.
STRUCTURE: Aluminium wing spars and plywood ribs, covered with heat-shrunk Dacron. Welded steel tube fuselage, rudder, tailplane and elevators, Dacron covered except for fuselage nose which has premoulded GFRP cowlings. Fin integral with fuselage.
LANDING GEAR: Non-retractable tricycle or tailwheel landing gear, with Tundra tyres and brakes. Wheel rim size 12.7 cm (5 in). Steerable tailwheel. Mainwheels have bungee cord shock absorption. Optional Aqua 1500 floats, skis and wheel/skis.
POWER PLANT: One 47.8 kW (64.1 hp) Rotax 582 two-stroke engine, driving a two- or three-blade propeller; fixed-pitch wooden propeller for STOL, three-blade ground-adjustable for Speedwing. Fuel capacity 53 litres (14.0 US gallons; 11.7 Imp gallons) for High Gross STOL and 68 litres (18.0 US gallons; 15.0 Imp gallons) for Aerobatic Speedwing. Similar capacity fuel tanks may be added in port wing.
DIMENSIONS, EXTERNAL (A: STOL, B: Aerobatic Speedwing):
Wing span: A 9.11 m (29 ft 10½ in)
 B 7.32 m (24 ft 0 in)
Length overall 5.46 m (17 ft 11 in)
Height overall: A 1.80 m (5 ft 11 in)
 B: 2.11 m (6 ft 11 in)
DIMENSIONS, INTERNAL:
Cabin max width 1.02 m (3 ft 4 in)
AREAS (A and B as above):
Wings, gross: A 11.38 m² (122.5 sq ft)
 B 9.04 m² (97.3 sq ft)
WEIGHTS AND LOADINGS (A and B as above):
Weight empty: A 200-231 kg (440-510 lb)
 B 231 kg (510 lb)

Avid Aircraft Avid Flyer Speedwing Mark IV in nosewheel configuration *(Paul Jackson/Jane's)* 2001/0092076

Baggage capacity 16 kg (35 lb)
Max T-O weight: A, B 522 kg (1,150 lb)
PERFORMANCE (A and B as above):
Max level speed at 1,525 m (5,000 ft):
 A, B 117 kt (217 km/h; 135 mph)
Max cruising speed: A 83 kt (153 km/h; 95 mph)
 B 104 kt (193 km/h; 120 mph)
Stalling speed:
 A, flaps down, engine idling 31 kt (58 km/h; 36 mph)
 B 40 kt (74 km/h; 46 mph)
Max rate of climb at S/L: A 305 m (1,000 ft)/min
 B 259 m (850 ft)/min
Service ceiling: A, B 3,180 m (12,500 ft)
T-O run: A 43 m (140 ft)
 B 92 m (300 ft)
Landing run: A 61 m (200 ft)
 B 183 m (600 ft)
Range, no reserves: A 340 n miles (629 km; 391 miles)
 B 566 n miles (1,048 km; 651 miles)

UPDATED

AVID MAGNUM
TYPE: Side-by-side kitbuilt.
CUSTOMERS: About 200 under construction or completed.
COSTS: Kit US$22,500 without engine (2000).
DESIGN FEATURES: High-wing braced monoplane, similar to Avid Flyer, but in certified category, using a Textron Lycoming engine. Recent upgrades include larger rudder and new external baggage door. Quoted build time 750 hours.
FLYING CONTROLS: Similar to Avid Flyer.
STRUCTURE: Similar to Avid Flyer, with Poly Fiber covering.
LANDING GEAR: Available as tricycle or tailwheel, with Cleveland wheels and brakes; wheel fairings. Spring aluminium landing gear standard. Optional floats.
POWER PLANT: One Textron Lycoming engine in 80.5 to 134 kW (108 to 180 hp) range, including O-235, O-320 and O-360. Tricycle version also accepts 119 kW (160 hp) Subaru EJ22 engine. Fuel capacity 106 litres (28.0 US gallons; 23.3 Imp gallons). Optional 45 litre (12.0 US gallon; 10.0 Imp gallon) wingtip tanks available.
ACCOMMODATION: Two seats side by side, with dual controls, plus optional jump seat for small adult or two children in 0.81 m³ (28.5 cu ft) baggage area.
DIMENSIONS, EXTERNAL:
Wing span 9.75 m (32 ft 0 in)
Wing chord, constant 1.30 m (4 ft 3 in)
Wing aspect ratio 7.8

Tailwheel version of Avid Flyer, sporting rectangular rudder and wheel speed fairings *(Paul Jackson/Jane's)* 2000/0062932

Length overall 6.40 m (21 ft 0 in)
Height overall 1.86 m (6 ft 1¼ in)
DIMENSIONS, INTERNAL:
Cabin max width 1.12 m (3 ft 8 in)
AREAS:
Wings, gross 13.03 m² (140.3 sq ft)
WEIGHTS AND LOADINGS:
Weight empty 431 kg (950 lb)
Baggage capacity 68 kg (150 lb)
Max T-O weight 794 kg (1,750 lb)
Max wing loading 60.9 kg/m² (12.47 lb/sq ft)
Power loading (119 kW; 160 hp)
 6.66 kg/kW (10.94 lb/hp)
PERFORMANCE (119 kW; 160 hp O-320 engine):
Never-exceed speed (VNE) 134 kt (249 km/h; 155 mph)
Cruising speed 113 kt (209 km/h; 130 mph)
Stalling speed 35 kt (65 km/h; 40 mph)
Max rate of climb at S/L 549 m (1,800 ft)/min
Service ceiling more than 5,340 m (17,500 ft)
T-O run approx 46 m (150 ft)
T-O to 15 m (50 ft) 99 m (325 ft)
Landing run 61 m (200 ft)
Range 450 n miles (833 km; 517 miles)

UPDATED

AVID CHAMPION
TYPE: Single-seat kitbuilt.
CUSTOMERS: Six flying by late 2000.
COSTS: Kit without engine US$9,450.
DESIGN FEATURES: Generally as for Flyer. Conforms to FAR Pt 103. Wings fold for storage. Quoted build time 200 hours.
FLYING CONTROLS: As for Flyer.

STRUCTURE: Fuselage 4130 chromoly steel tube covered with fabric; wing has aluminium spar and wooden ribs. Glass fibre cowling.
LANDING GEAR: Tailwheel type; fixed. Mainwheels 4.00-6; cable brakes.
POWER PLANT: One Rotax engine rated between 20.9 and 37.3 kW (28 and 50 hp) and driving two-blade composites propeller. Fuel capacity 19 litres (5.0 US gallons; 4.2 Imp gallons).

DIMENSIONS, EXTERNAL:
Wing span	8.17 m (26 ft 9½ in)
Length: overall	5.36 m (17 ft 7 in)
wings folded	5.71 m (18 ft 9 in)
Height overall	1.83 m (6 ft 0 in)

AREAS:
Wings, gross	10.64 m² (114.5 sq ft)

WEIGHTS AND LOADINGS:
Weight empty	115 kg (254 lb)
Max T-O weight	269 kg (594 lb)

PERFORMANCE (37.3 kW; 50 hp engine):
Max level speed	56 kt (105 km/h; 65 mph)
Cruising speed	55 kt (101 km/h; 63 mph)
Stalling speed	23 kt (42 km/h; 26 mph)
Max rate of climb at S/L	213 m (700 ft)/min
T-O run	23 m (75 ft)
Landing run	30 m (100 ft)
Range	90 n miles (166 km; 103 miles)

UPDATED

Avid Champion single-seat kitbuilt *(Paul Jackson/Jane's)* 2001/0092077

AVTEKAIR

AVTEKAIR INC

555 Airport Way, Suite A, Camarillo Airport, Camarillo, California 93010
Tel: (+1 805) 482 27 00
Fax: (+1 805) 987 00 68
e-mail: AvtekAir@aol.com
PRESIDENT: Quinten E Ward
SENIOR VICE-PRESIDENT, ENGINEERING: Niels Andersen
VICE-PRESIDENT, MARKETING: Robert D Honeycutt

Company founded 1982 as Avtek to develop Avtek 400, last described in 1995-96 *Jane's*. Planned preproduction aircraft failed to achieve 1995 first flight. However, in 1998 Avtek secured finance to certify a stretched, commuter version, the Avtek 419 Express and in 1999 changed its name to AvtekAir, renaming its product AvtekAir 9000T.

VERIFIED

AVTEKAIR 9000T

TYPE: Utility turboprop twin.
PROGRAMME: Design of Avtek 400 six/10-seat all-composites business aircraft started March 1981 and programme launched June 1982; construction of proof-of-concept aircraft N400AV (see 1985-86 *Jane's*) began January 1983; flew 17 September 1984; N400AV then fitted with P&WC PT6A-135M engines (PT6A-35s with counter-rotating gearboxes from PT6A-66); extensive changes made Spring 1985, including fuselage stretch 20 cm (8 in) forward and 76 cm (2 ft 6 in) aft of front pressure bulkhead, widened cabin, new outer wing and enlarged fuel tanks in forward-swept root extensions, foreplane with greater span and reduced chord ventral strakes (known as delta fins), relocated main landing gear legs and specially developed P&WC PT6A-3s mounted closer to wings. Preproduction second prototype abandoned and late 1996 FAA certification failed to take place.
Programme relaunched in 1998, with priority to stretched 419 version for commuter airlines; renamed AvtekAir 9000T in 1999.
CUSTOMERS: 62 ordered by August 2000.
COSTS: Standard aircraft US$2,795,000; fully equipped IFR (2001).
DESIGN FEATURES: Twin-engine pusher configuration with foreplane. Initial design by Al W Mooney, founder of Mooney Aircraft; refined by Niels Anderson, Ford Johnston and Irvin Culver; computer analysis of configuration and wind tunnel testing by NASA; materials research by Dow Chemical and Dr Leo Windecker, Dow Chemical basic patents on Windecker Eagle (first all-composites aircraft to receive civil certification) licensed to Avtek.
Avtek 12 aerofoil section; anhedral 2° 30′ from roots; sweepback 50° inboard, 15° 30′ outboard; foreplane dihedral 1°; fuselage pressurised.
FLYING CONTROLS: Manual. Actuation by pushrods and cranks throughout; mass-balanced elevators on foreplane; mass-balanced rudder without trim tab; two-section ailerons, with inboard sections also electrically actuated as pitch-axis trim surfaces; no flaps.
STRUCTURE: All composites structure: 90 per cent graphite/carbon fibre, smaller quantitites of R-glass, S-glass, aluminium and nickel fibres; wire mesh incorporated to protect against lightning.
LANDING GEAR: Hydraulically retractable tricycle type, main units retracting inward and nosewheel forward. Emergency extension system. Oleo-pneumatic shock-absorber in each unit. Single wheel on each unit. Cleveland hydraulic disc brakes.
POWER PLANT: Two 582 kW (780 shp) Pratt & Whitney Canada PT6A-135 turboprops, one mounted within nacelle above each wing. Hartzell four-blade constant-speed fully feathering reversible-pitch pusher propellers.
ACCOMMODATION: Up to nine passengers. Unpressurised baggage compartment in nose with external door on port side. Accommodation pressurised, air conditioned, heated and ventilated.
SYSTEMS: AiResearch bleed air pressurisation system and air cycle air conditioning system. Electrically driven hydraulic pump provides pressure of 138 bars (2,000 lb/sq in) for landing gear actuation. Anti-icing of windscreen by electrical system, of propellers by engine efflux, electrically heated pitot. Choice of wing and foreplane pneumatic, alcohol, electric or engine bleed de-icing. Dual anti-icing inlets for each engine.

DIMENSIONS, EXTERNAL:
Wing span	10.67 m (35 ft 0 in)
Wing aspect ratio	8.5
Foreplane span	6.92 m (22 ft 8½ in)
Length overall	16.62 m (54 ft 5 in)
Passenger door (port): Height	1.17 m (3 ft 10 in)
Width	0.76 m (2 ft 6 in)
Height to sill	0.76 m (2 ft 6 in)
Baggage door (port, nose): Height	0.51 m (1 ft 8 in)
Width	0.56 m (1 ft 10 in)
Height to sill	0.79 m (2 ft 7 in)
Emergency exit (stbd): Height	0.51 m (1 ft 8 in)
Width	0.67 m (2 ft 2½ in)

DIMENSIONS, INTERNAL:
Cabin: Max width	1.40 m (4 ft 7 in)
Max height	1.37 m (4 ft 6 in)

AREAS:
Wings, gross	13.40 m² (144.2 sq ft)
Foreplane, gross	4.52 m² (48.7 sq ft)
Elevators (total)	0.92 m² (9.9 sq ft)
Ailerons (total)	0.60 m² (6.5 sq ft)
Fin	1.16 m² (12.5 sq ft)
Rudder	0.90 m² (9.7 sq ft)

WEIGHTS AND LOADINGS:
Max T-O weight	5,670 kg (12,500 lb)
Max wing loading	423.2 kg/m² (88.69 lb/sq ft)
Max power loading	4.88 kg/kW (8.01 lb/shp)

PERFORMANCE (estimated):
Cruising speed	364 kt (674 km/h; 419 mph)
T-O run	463 m (1,520 ft)
Range with 4 passengers	1,911 n miles (3,540 km; 2,200 miles)

UPDATED

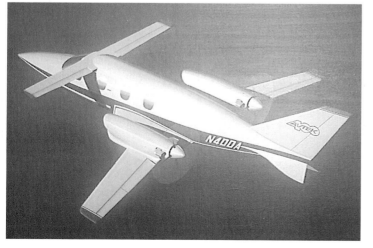

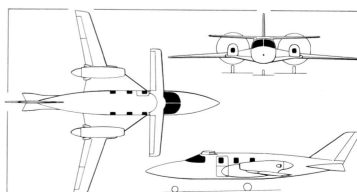

Avtek 400A six/10-seat twin-turboprop aircraft which is to be lengthened to become the AvtekAir 9000T *(Dennis Punnett/Jane's)* 1999

Artist's impression of Avtek 400A, precursor of the 9000T 1999

AYRES

AYRES CORPORATION
PO Box 3090, 1 Ayres Way, Albany, Georgia 31707-3090
Tel: (+1 912) 883 14 40
Fax: (+1 912) 439 97 90
e-mail: ayres-corp.com
Web: http://www.ayrescorp.com
PRESIDENT AND CEO: Fred P Ayres
VICE-PRESIDENT, MANUFACTURING: Milton R Humphries
VICE-PRESIDENTS, PRODUCT SUPPORT: Marvin H Wilson
  Don Murphy
VICE-PRESIDENT, SALES: Daniel Lewis
THRUSH SALES: Terry Humphries

Ayres Corporation bought manufacturing and world marketing rights to Thrush Commander-600 and -800 from Rockwell International General Aviation Division in November 1977. Acquired 93 per cent shareholding in Let Kunovice of Czech Republic (which see) in 1998, but stopped providing finance in July 2000 after Czech bank cancelled a debt repayment freeze; Let declared bankrupt in October 2000.

Ayres filed for Chapter 11 bankruptcy protection in late November 2000 and is seeking further investment of US$80 million; work restarted on the Loadmaster programme after a financing agreement with GATX, as debtor-in-possession, was concluded.

UPDATED

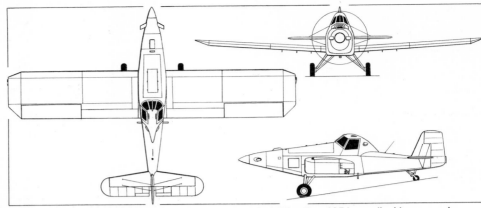

Ayres Turbo-Thrush S2R-T34 with optional 1,930 litre (510 US gallon; 425 Imp gallon) hopper and no longer available wing span option of 13.54 m (44 ft 5 in) *(Dennis Punnett/Jane's)*

Ayres Turbo-Thrush S2R-T34 agricultural aircraft   1999/0051558

AYRES TURBO-THRUSH S2R
TYPE: Agricultural sprayer.
PROGRAMME: Leland Snow designed S-1 prototype, first flown with radial engine on 17 August 1953; entered production as S-2; over 1,800 built by Snow and Rockwell. Ayres introduced turboprop version, having previously specialised in turbine conversions.
CURRENT VERSIONS: **S2R-T11:** 373 kW (500 shp) PT6A-11AG turboprop; standard 1,514 litre (400 US gallon; 333 Imp gallon) chemical hopper. Available to special order only.
  S2R-T15: 507 kW (680 shp) P&WC PT6A-15AG turboprop; standard or optional 1,930 litre (510 US gallon; 425 Imp gallon) hopper.
  S2R-T34: 559 kW (750 shp) P&WC PT6A-34AG turboprop; standard or optional hoppers.
  S2R-T45: Powered by P&WC PT6R-45AG.
  S2R-T65 NEDS: Narcotics Eradication Delivery System; 1,026 kW (1,376 shp) P&WC PT6A-65AG turboprop and 2.82 m (9 ft 3 in) five-blade propeller; 19 delivered to US State Department (see Customers).
  S2R-G1: Powered by 496 kW (665 shp) TPE-331-1 turboprop; 1,514 litre (400 US gallon; 333 Imp gallon) hopper.
  S2R-G6: Introduced 1992; 559 kW (750 shp) TPE331-6 turboprop; standard or optional hopper; two 435 litre (115 US gallon; 95.75 Imp gallon) fuel tanks, with 863 litres (228 US gallons; 190 Imp gallons) usable; optional hopper capacity 1,930 litres (510 US gallons; 425 Imp gallons); pilot, with dual-cockpit option.
  S2R-G10: First flown November 1992; CAM 8 certification; 701 kW (940 shp) TPE331-10 turboprop and four-blade propeller.
  V-1A Vigilante: Surveillance and close air support version; TPE331-14-GR turboprop; first flight (N3100A) 15 April 1989. Still carries hopper; has five-blade propeller for quiet operation; four hardpoints beneath wings; can be fitted with range of equipment including loudspeaker systems for policing duties. Fuel capacity 1,136 litres (300 US gallons; 250 Imp gallons) plus optional external tank containing 1,514 litres (400 US gallons; 333 Imp gallons).
CUSTOMERS: US State Department ordered 19 S2R-T65 NEDS during 1983-88 for use by International Narcotics Matters Bureau. Vigilante used in Africa for game and poacher surveillance. By late 2000, 15 G1s, 54 G6s, 65 G10s, 39 T15s, 269 T34s and 15 T45s had been produced and over 2,500 of all versions delivered to 80 countries.
COSTS: About US$600,000, depending on engine and customer fit.
DESIGN FEATURES: Conventional cropsprayer configuration, with hopper ahead of cockpit; cantilever low wing of constant chord; braced empennage with sweptback leading-edges; robust landing gear.
  Advantages over piston-engined versions include much improved take-off and climb, 454 kg (1,000 lb) higher

Ayres Turbo-Thrush S2R with optional dual cockpit   1999/0051557

payload because of lower engine weight, operation on aviation turbine fuel or diesel, 3,500 hour TBO, quieter operation and ability to feather propeller without stopping engine while refuelling and reloading.
FLYING CONTROLS: Conventional and manual. Plain ailerons; servo tab in each elevator; electrically actuated flaps.
STRUCTURE: Two-spar light alloy wing with 4130 chrome molybdenum steel spar caps; welded chrome molybdenum steel tube fuselage structure covered with quickly removable light alloy skin panels; underfuselage skin of stainless steel; all-metal tail surfaces and strut-braced tailplane; metal ailerons and flaps. Extended wing, originally optional, is now standard, replacing earlier 13.54 m (44 ft 5 in), 30.34 m² (326.6 sq ft) version.
LANDING GEAR: Tailwheel type; fixed. Main units have rubber-in-compression shock-absorption and 8.50-10 (10 ply) tyres. Hydraulically operated Cleveland dual calliper disc brakes. Parking brakes. Wire cutters on main gear. Steerable, spring steel locking tailwheel, size 5.00-5.
POWER PLANT: One turboprop (see under Current Versions for engine and fuel details). All but NEDS, G10 and Vigilante have Hartzell three-blade, constant-speed, feathering and reversing propellers; usable fuel capacity 515 litres (136 US gallons; 113 Imp gallons) in 400 gallon hopper models and 863 litres (228 US gallons; 190 Imp gallons) in 510 gallon hopper models.
ACCOMMODATION: Single adjustable mesh seat in 'safety pod' sealed cockpit enclosure, with steel tube overturn structure. Tandem seating optional, with forward-facing second seat. Dual controls optional with forward-facing rear seat, for pilot training. Adjustable rudder pedals. Downward-hinged door on each side. Tempered safety glass windscreen. Cockpit wire cutter. Dual inertia reel safety harness with optional second seat. Baggage compartment standard on single-seat aircraft.
SYSTEMS: Electrical system powered by a 24 V 50 A alternator. Dual lightweight 24 V 35 Ah batteries.
AVIONICS: To customer's requirements.
EQUIPMENT: GFRP hopper forward of cockpit can hold 1,514 litres (400 US gallons; 333 Imp gallons) of liquid or 1,487 kg (3,280 lb) of dry chemical. Optional 1,931 litre (510 US gallon; 425 Imp gallon) hopper can be installed. Hopper has a 0.33 m² (3.56 sq ft) lid, openable by two handles, and cockpit viewing window. Standard equipment includes Universal spray system with external 50 mm (2 in) stainless steel plumbing, 50 mm pump with wooden fan, Transland gate, 50 mm valve, quick-disconnect pump mount and strainer. Streamlined spraybooms with outlets for 68 nozzles. Micro-adjust valve control (spray) and calibrator (dry). A 51 mm (2 in) side-loading system is installed on the port side. Stainless steel rudder cables. Navigation lights, instrument lights and two strobe lights. Windscreen washer/wiper.
  Optional equipment includes a rear cockpit to accommodate forward-facing seat for passenger, or flying instructor if optional dual controls installed; space can be used alternatively for cargo. Other optional items are a Transland high-volume spreader; agitator installation; 10-unit AU5000 Micronair installation in lieu of standard booms and nozzles; Transland gatebox with stiffener casting; quick-disconnect flange and kit; night working lights including landing light and wingtip turn lights; cockpit fire extinguisher; and water-bomber configuration.

DIMENSIONS, EXTERNAL:
Wing span	14.48 m (47 ft 6 in)
Wing aspect ratio	6.4
Length overall	10.06 m (33 ft 0 in)
Height overall: except Vigilante	2.79 m (9 ft 2 in)
Vigilante	2.90 m (9 ft 6 in)
Wheel track	2.74 m (9 ft 0 in)

DIMENSIONS, INTERNAL:
Hopper volume	1.9 m³ (68 cu ft)

AREAS:
Wings, gross	32.52 m² (350.0 sq ft)

WEIGHTS AND LOADINGS: (A: standard hopper and PT6A, B: optional hopper and PT6A, C: with TPE331-6, D: with TPE331-10, E: Vigilante):
Weight empty: A	1,633 kg (3,600 lb)
B	1,769 kg (3,900 lb)
C	1,950 kg (4,300 lb)
D	2,041 kg (4,500 lb)
E	2,223 kg (4,900 lb)
Max T-O weight (CAR 3): A, B	2,721 kg (6,000 lb)
E	4,763 kg (10,500 lb)
Typical operating weight (CAM 8):	
A	3,719 kg (8,200 lb)
B	3,856 kg (8,500 lb)
C	4,400 kg (9,700 lb)
D	4,082 kg (9,000 lb)
CAM 8 wing loading:	
A	114.4 kg/m² (23.43 lb/sq ft)
B	118.6 kg/m² (24.29 lb/sq ft)
C	135.3 kg/m² (27.71 lb/sq ft)
D	125.5 kg/m² (25.71 lb/sq ft)
Max wing loading: E	146.5 kg/m² (30.00 lb/sq ft)
Max power loading: A	7.30 kg/kW (12.00 lb/shp)
B	5.37 kg/kW (8.82 lb/shp)
E	6.80 kg/kW (11.17 lb/shp)

PERFORMANCE (A and B: with PT6A-34AG engine, C: with TPE331-6, D: with TPE331-10, E: Vigilante):
Never-exceed speed (V_{NE}):	
D	138 kt (256 km/h; 159 mph)
E	191 kt (354 km/h; 220 mph)
Max level speed with spray equipment:	
A, B, D	138 kt (256 km/h; 159 mph)
Max level speed: E	163 kt (302 km/h; 188 mph)
Cruising speed:	
A, B and E at 50% power	130 kt (241 km/h; 150 mph)
C at 55% power	130 kt (241 km/h; 150 mph)
Working speed at 30-50% power	
	78-130 kt (145-241 km/h; 90-150 mph)

AYRES—AIRCRAFT: USA

Stalling speed: flaps up	61 kt (113 km/h; 70 mph)
flaps down	57 kt (106 km/h; 66 mph)
Stalling speed at normal landing weight:	
flaps up	51 kt (95 km/h; 59 mph)
flaps down	50 kt (92 km/h; 57 mph)
Max rate of climb at S/L: A, B, C	530 m (1,740 ft)/min
D	762 m (2,500 ft)/min
E	1,067 m (3,500 ft)/min
Service ceiling: A, B, C, E	7,620 m (25,000 ft)
D	3,660 m (12,000 ft)
T-O run: A, B, D	183 m (600 ft)
C at 4,400 kg (9,700 lb)	366 m (1,200 ft)
Landing from 15 m (50 ft)	366 m (1,200 ft)
Landing run: A, B, C	183 m (600 ft)
D	244 m (800 ft)
Landing run with propeller reversal:	
A, B, C	122 m (400 ft)
Range: D	500 n miles (926 km; 575 miles)
Ferry range at 45% power:	
A, B, C	669 n miles (1,239 km; 770 miles)
E	900 n miles (1,667 km; 1,036 miles)

UPDATED

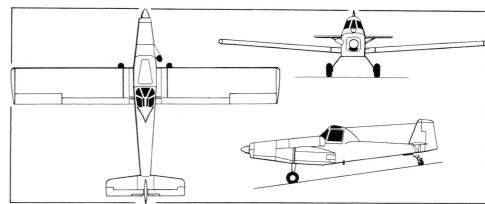

TPE331-powered version of Ayres 660 Turbo-Thrush *(Paul Jackson/Jane's)* 1999/0051560

AYRES 660 TURBO-THRUSH

TYPE: Agricultural sprayer.
PROGRAMME: Prototype produced in 1997 to verify new hopper and landing gear designs, but was otherwise as S2R. Certification achieved 14 March 2000 with first delivery same day.
CUSTOMERS: Nine delivered by October 2000.
DESIGN FEATURES: Based on S2R (which see) but with enlarged hopper, made possible by change to cantilever main landing gear. Enlarged wing centre-section extends span.

Data as for Turbo-Thrush S2R, except as below.

FLYING CONTROLS: Aileron droop system, plus enlarged flaps.
LANDING GEAR: Tailwheel type; non-retractable. Cantilever mainwheel legs of spring steel, with heavy-duty 29 in tyres (31 in tyres optional) and Cleveland dual calliper hydraulic disc brakes. Wire cutters on main undercarriage legs. Tailwheel 6.00×6.
POWER PLANT: One 917 kW (1,230 shp) Pratt & Whitney Canada PT6A-65AG turboprop. Optionally, one 783 kW (1,050 shp) P&WC PT6A-60 turboprop. Hartzell five-blade, constant-speed feathering and reversing propeller. Fuel capacity 863 litres (228 US gallons; 190 Imp gallons) usable; optional capacity 1,136 litres (300 US gallons; 250 Imp gallons) usable.
ACCOMMODATION: Optional air conditioning.
EQUIPMENT: GFRP hopper can hold 2,500 litres (660 US gallons; 550 Imp gallons) of liquid. Side loader size 76 mm (3 in) on port side. Standard equipment includes Universal spray system with external 76 mm (3 in) stainless steel plumbing, five-blade Weath-Aerofan, 76 mm Transland pump, 68-nozzle spraybom extending over 70 per cent of span. Optional equipment as per Turbo-Thrush S2R.

DIMENSIONS, EXTERNAL:
Wing span	16.46 m (54 ft 0 in)
Wing aspect ratio	7.3
Wing chord, constant	2.29 m (7 ft 6 in)
Length overall	10.97 m (36 ft 0 in)
Height overall	3.73 m (12 ft 3 in)
Tailplane: span	5.18 m (17 ft 0 in)
chord: at root	1.40 m (4 ft 7 in)
at tip	1.09 m (3 ft 7 in)
Wheel track	2.90 m (9 ft 6 in)
Wheelbase	6.74 m (22 ft 1¼ in)

DIMENSIONS, INTERNAL:
Hopper volume	2.5 m³ (88 cu ft)

AREAS:
Wings, gross	37.16 m² (400.0 sq ft)

WEIGHTS AND LOADINGS (PT6A-65AG):
Weight empty	2,880 kg (6,350 lb)
Max T-O weight	6,418 kg (14,150 lb)
Max landing weight	5,670 kg (12,500 lb)
Max wing loading	172.7 kg/m² (35.38 lb/sq ft)
Max power loading	7.00 kg/kW (11.50 lb/shp)

PERFORMANCE (PT6A-65AG):
Never-exceed speed (V_{NE})	191 kt (354 km/h; 220 mph)
Cruising speed at 55% power	152 kt (282 km/h; 175 mph)
Working speed	86-152 kt (185-282 km/h; 100-175 mph)
Stalling speed, power off, flaps down	50 kt (92 km/h; 57 mph)
Max rate of climb at S/L	381 m (1,250 ft)/min
T-O run	457 m (1,500 ft)
Landing run	183 m (600 ft)
Ferry range at 50% power	521 n miles (966 km; 600 miles)

UPDATED

AYRES LM200 LOADMASTER

TYPE: Twin-turboprop freighter.
PROGRAMME: Design officially launched 1996, following considerable detail redesign (see 1996-97 edition); Federal Express announced as launch customer November 1996; contract placed in 1997 for LHTEC to deliver 100 power plants from May 1999 onwards; three prototypes to be built at Albany. Production line started at Dothan, Alabama; Let was expected to build wings and tail unit subassemblies in Czech Republic, and was due to deliver first sets in December 1998; however Let's bankruptcy has delayed programme; was also to have built fuselages and provided second production line. Ayres has suggested this may move to Poland or Romania.

First flight due early 1999 but later rescheduled to November 2000; first CTP800-4T engine delivered to Ayres July 2000; certification will be for single pilot operation, and FAR/JAR 23.

Work halted during 2000 after financial problems with Let and Ayres itself; these were partially resolved in November 2000 when the first flight was rescheduled for the end of 2001 with first deliveries during first half of 2003.

CURRENT VERSIONS: **LM200**: Civil version; *as described*. Applications include container freighter and passenger version carrying 19 passengers at ordinary seating pitch or 34 in high density. Firefighting, reconnaissance and long-range patrol variants also under development.

Searchmaster: Proposed remote earth sensing aircraft; two operators' consoles in cargo hold.

LM250: Military freighter; promotion discontinued; last described fully in 1999-2000 *Jane's*..

CUSTOMERS: Ayres foresees sales of 600 aircraft by 2010. Federal Express signed letter of intent for 50, plus 200 options, in November 1996, and converted 25 options to firm orders in May 1999. By mid-1998, additional orders received from Corporate Air of Montana (three, plus 20 options); Duijvestijn Aviation, Netherlands (five, plus five options which were later converted to firm orders); Orsmond Aviation, South Africa (four); and WestAir of Fresno, California (two), and Kelner Air Centers (10). Orders stood at 104 in late 1999.

COSTS: Basic price US$4 million to US$5 million (2000).

DESIGN FEATURES: Meets requirements of mixed freight operators seeking aircraft sized between Cessna Caravan and Fokker F27/50. Single-propeller configuration, but circumvents several countries' prohibitions of single-engine commercial night flights by employing two linked

Prototype Ayres 660 Turbo-Thrush 1999/0051559

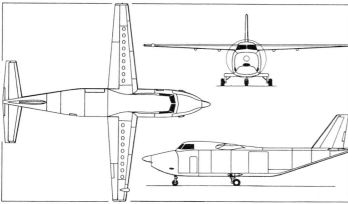

Ayres LM200 Loadmaster *(Paul Jackson/Jane's)* 2001/0100427

Model of Ayres LM200 Loadmaster in alternative floatplane configuration 2000/0084549

turboshafts. Shoulder-wing design, with constant chord centre-section and tapered outer panels with outward-canted endplates. No cabin windows on freight version.

LM 200 wing section; thickness/chord ratio 20 per cent at root, 13 per cent at tip; taper ratio 0.4; dihedral 1° 30′. Rudder, cockpit fittings, control wheels and engine inlets from Let L-410. Tailplane section NACA 0014 at root and NACA 0012 at tip.

FLYING CONTROLS: Conventional and manual. Actuation by pushrods and cables. Electromechanical single-slotted, three-position flaps. Cable-operated trim tabs in starboard aileron, rudder and both elevators.

STRUCTURE: All metal, semi-monocoque fuselage with aluminium alloy skins, having clad aluminium frames and stringers. Two-spar, fully cantilevered, tapered wing with integral fuel tank; spars at 18 and 63 per cent section. Aluminium alloy leading-edge skins; composites wingtips. One-piece horizontal stabiliser constructed on two full-span spars. Special protection against corrosive cargo. All-metal flying controls.

LANDING GEAR: Fixed tricycle type, with sponson-mounted trailing-link main gear having oleo-pneumatic shock-absorbers and differential three-rotor 'tri-metallic' brakes from Aircraft Braking Systems, which also provides wheel assemblies. Nose leg has oleo-pneumatic shock-absorber; torque link disconnects for towing. Parking brake; nosewheel steerable ±75°. Mainwheel tyres 28×9.00-12, pressure 6.76 bars (98 lb/sq in); nosewheel 18×5.50-12, pressure 6.27 bars (91 lb/sq in). Ski, float and retractable landing gear are under consideration.

POWER PLANT: One 2,013 kW (2,700 shp) LHTEC CTP800-4T 'twin-barrel propulsion system' of two in-parallel coupled engines, each with FADEC, driving a six-blade Hamilton Sundstrand 568F-11 constant-speed propeller through a GKN Westland combining gearbox. A 'hot-and-high' version using T801 engine is being considered. Two fuel tanks, one in each wing, between the spars, combined capacity 3,407 litres (900 US gallons; 749 Imp gallons). Single pressure refuelling/defuelling point in starboard sponson; optional gravity defuelling.

ACCOMMODATION: Pilot and optional co-pilot and observer on flight deck. Four LD3 or Demi freight containers, plus general freight. Four cargo areas: Main, comprising constant section of fuselage; Upper, behind flight deck; Forward, below flight deck; and Aft, in tailcone. Main area contains four seat tracks for total of 34 passengers, or 19 passengers and bulk cargo, or 28 paratroops. Cabin windows and two emergency exits optional. Crew and passenger door forward of wing, port side, gives access to forward cargo area, flight deck (via ladder) and main area. Main freight door, port side, aft of wing; aft baggage door, starboard side. Internal door on passenger version (optional for freight version) between forward cargo area and main hold. Flight deck side windows are hinged at top for ground ventilation and emergency exit. Accommodation ventilated and heated; air conditioning optional.

SYSTEMS: Electrical system 28 V DC, powered by two engine-driven 400 A starter/generators; two 24 V 43 Ah lead/acid batteries. Optional 24 V battery for cargo hold lighting. Ground power inlet, port side sponson, 28 V DC. Hydraulic system, 138 bars (2,000 lb/sq in), powers mainwheel brakes and nosewheel steering; electric pump. Fire detection and suppression system in each engine compartment. Environmental system of cargo version comprises outside air ventilation for cooling, and either engine bleed air or electric elements for heat; optional vapour cycle air conditioning. Passenger version has engine bleed air heat and vapour cycle cooling with automatic temperature controls. Pneumatic rubber de-icing boots on wing and tail leading-edges; electric heating of propeller blades, engine air inlets and windscreens for flight into known icing.

AVIONICS: Honeywell SPZ-5000 as core system.
Comms: Primus II system including two RCZ-851 com units, RM-850 radio management unit, CD-850 Clearance Delivery Head and AV-850A audio panel.
Radar: Primus 440 weather radar antenna on starboard wing leading-edge.
Flight: IC-500 autopilot, Goodrich AGH-3000 integrated standby instrument system, AA-300 radio altimeter, AM-250 barometric altimeter, AH-800 AHRS and dual Trimble 2101 GPS. Optional TCAS 2000.
Instrumentation: Two ED-600 127 × 127 mm (5 × 5 in) CRT displays; Goodrich GH-300 electronic standby instrument system.

EQUIPMENT: Ball-mat cargo handling floor; cargo restraints around floor.

DIMENSIONS, EXTERNAL:
Wing span	19.51 m (64 ft 0 in)
Wing chord: at root	3.12 m (10 ft 3 in)
at tip	1.24 m (4 ft 1 in)
Wing aspect ratio	8.9
Length overall	19.61 m (64 ft 4 in)
Height overall	7.00 m (22 ft 11½ in)
Tailplane span	8.13 m (26 ft 8 in)
Wheel track	3.96 m (13 ft 0 in)
Wheelbase	5.84 m (19 ft 2 in)
Propeller diameter	3.96 m (13 ft 0 in)
Propeller ground clearance	0.62 m (2 ft 0½ in)
Freight door: Height	2.11 m (6 ft 11 in)
Width	1.88 m (6 ft 2 in)
Aft baggage door: Height	1.07 m (3 ft 6 in)
Width	0.76 m (2 ft 6 in)

DIMENSIONS, INTERNAL:
Main cargo area: Length	7.01 m (23 ft 0 in)
Width at floor	2.28 m (7 ft 5¾ in)
Max width	2.54 m (8 ft 4 in)
Max height	2.40 m (7 ft 10½ in)
Floor area	15.9 m² (171 sq ft)
Volume	38.2 m³ (1,350 cu ft)
Upper cargo area: Volume	2.4 m³ (85 cu ft)
Forward cargo area: Volume	8.0 m³ (284 cu ft)
Aft cargo area: Floor area	4.1 m² (43.7 sq ft)
Volume	8.1 m³ (285 cu ft)

AREAS:
Wings, gross	42.55 m² (458.0 sq ft)
Ailerons (total)	4.33 m² (46.60 sq ft)
Trailing-edge flaps (total)	8.53 m² (91.80 sq ft)
Fin	4.65 m² (50.00 sq ft)
Rudder, incl tab	2.45 m² (26.40 sq ft)
Tailplane	11.61 m² (125.0 sq ft)
Elevators (total, incl tabs)	2.36 m² (25.40 sq ft)

WEIGHTS AND LOADINGS:
Weight empty, equipped	4,082 kg (9,000 lb)
Max payload	3,992 kg (8,800 lb)
Max fuel weight	2,735 kg (6,030 lb)
Max T-O and landing weight	8,618 kg (19,000 lb)
Max ramp weight	8,704 kg (19,190 lb)
Max wing loading	202.5 kg/m² (41.48 lb/sq ft)
Max power loading	4.28 kg/kW (7.03 lb/shp)

PERFORMANCE (estimated):
Never-exceed speed (V_{NE})	231 kt (427 km/h; 265 mph)
Max cruising speed	205 kt (380 km/h; 236 mph)
Normal cruising speed	180 kt (333 km/h; 207 mph)
Econ cruising speed at 3,660 m (12,000 ft)	160 kt (296 km/h; 184 mph)
Max rate of climb at S/L	567 m (1,861 ft)/min
Max certified altitude	7,620 m (25,000 ft)
T-O to 15 m (50 ft)	465 m (1,525 ft)
Landing from 15 m (50 ft)	532 m (1,745 ft)
Range: with max payload	280 n miles (518 km; 322 miles)
with 1,814 kg (4,000 lb) payload	1,590 n miles (2,944 km; 1,829 miles)
ferry	1,630 n miles (3,018 km; 1,875 miles)
g limits	+2.9/−1.1

UPDATED

BARR

BARR AIRCRAFT
900 Airport Road, Montoursville, Pennsylvania 17754
Tel: (+1 570) 368 36 55
Fax: (+1 570) 368 20 95
e-mail: six@barraircraft.com
Web: http://www.barraircraft.com
MANAGING DIRECTOR: James L Barr

Barr Aircraft holds type certificates for the Piper PA-12, PA-14, PA-20 and PA-22 series of aircraft and is presently developing the BarrSix.

UPDATED

BARR BARRSIX
TYPE: Six-seat kitbuilt.
PROGRAMME: Prototype was due to fly late 1999, but delayed by hangar construction; new completion target was late 2000 but this has slipped into 2001.
COSTS: US$69,900 (2000).
DESIGN FEATURES: Designed to meet FAR Pt 23 and 51 per cent kitplane rules; extensive use of composites results in lightweight design. High wing with single bracing strut each side.
Dihedral 1° 44′; incidence 1° 30′ at wing root; −1° 30′ at tip.
FLYING CONTROLS: Conventional and manual. Frise-type ailerons; NACA single-slotted three-position (10, 20, 40°) flaps; horn-balanced rudder and elevators.
STRUCTURE: Primary structure is 7781 pre-preg 36-38 per cent resin with Nomex core and graphite rods.
LANDING GEAR: Fixed tricycle type; speed fairings optional. Skis, floats and tailwheel layout optional.
POWER PLANT: One 298 kW (400 hp) Textron Lycoming IO-720-A1B flat-eight driving a Hartzell HC-E3YR-1RF/F8468A-6R constant-speed propeller. Standard fuel capacity 341 litres (90.0 US gallons; 74.9 Imp gallons), optionally increased to 526 litres (140 US gallons; 116 Imp gallons) with auxiliary tank in place of rear cabin baggage area.
ACCOMMODATION: Pilot and up to five passengers; club seating for four in rear cabin. Passenger/cargo double door on starboard side; optional door in port side for pilot. Dual controls.
AVIONICS: Onboard computer.

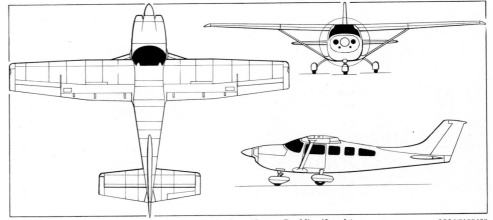

Barr BarrSix six-seat tourer with optional pilot's door *(James Goulding/Jane's)*

Prototype BarrSix under construction

BARR to BEL-AIRE—AIRCRAFT: USA

Floatplane BarrSix, showing underfin of this version and starboard freight door   2001/0100429

DIMENSIONS, EXTERNAL:	
Wing span	10.92 m (35 ft 10 in)
Wing chord: at root	1.63 m (5 ft 4 in)
at tip	1.07 m (3 ft 6 in)
Wing aspect ratio	7.4
Length overall	9.12 m (29 ft 11 in)
Height overall	2.18 m (7 ft 2 in)
Tailplane span	3.96 m (13 ft 0 in)
Propeller diameter	2.11 m (6 ft 11 in)
Propeller ground clearance	0.25 m (9¾ in)
Passenger door: Height	1.02 m (3 ft 4 in)
Width	0.94 m (3 ft 1 in)
Cargo double doors: Height	0.98 m (3 ft 2½ in)
Width	1.12 m (3 ft 8 in)
DIMENSIONS, INTERNAL:	
Cabin: Length	3.89 m (12 ft 9 in)
Max width	1.35 m (4 ft 5 in)
Max height	1.24 m (4 ft 1 in)
AREAS:	
Wings, gross	16.17 m² (174.0 sq ft)
Ailerons (total)	1.61 m² (17.32 sq ft)
Trailing-edge flaps (total)	2.69 m² (29.00 sq ft)
WEIGHTS AND LOADINGS:	
Weight empty: One seat	1,086 kg (2,395 lb)
Fully IFR equipped	1,134 kg (2,501 lb)
Baggage capacity	91 kg (200 lb)
Max T-O weight	2,041 kg (4,500 lb)
Max wing loading	124.8 kg/m² (25.86 lb/sq ft)
Max power loading	6.85 kg/kW (11.25 lb/hp)
PERFORMANCE:	
Never-exceed speed (V_{NE})	238 kg (441 km/h; 274 mph)
Max level speed	206 kt (381 km/h; 237 mph)
Max cruising speed	196 kt (362 km/h; 225 mph)
Normal cruising speed	180 kt (333 km/h; 207 mph)
Stalling speed, power off:	
flaps up	62 kt (115 km/h; 71 mph)
flaps down	54 kt (100 km/h; 62 mph)
Max rate of climb at S/L	274 m (900 ft)/min
Service ceiling	7,010 m (23,000 ft)
T-O run	274 m (900 ft)
T-O to 15 m (50 ft)	549 m (1,800 ft)
Landing from 15 m (50 ft)	427 m (1,400 ft)
Landing run	229 m (750 ft)
Range with max fuel	1,251 n miles (2,317 km; 1,440 miles)
Endurance	6 h 24 min
g limits	+3.86/−1.96

UPDATED

BEDE

BEDE AIRCRAFT CORPORATION
1021 N Jefferson Street, Medina, Ohio 44256
Tel: (+1 330) 725 13 82
Fax: (+1 330) 764 79 59
PRESIDENT: James R Bede

Bede Aircraft Corporation specialises in design of high-performance jet aircraft for private ownership; civilian version of a previous product, BD-10, was transferred to Vortex Aircraft Company as the Phoenix. In June 2000, Bede announced it was working on the BD-17 Nugget, an all-metal low-wing single-seater.

UPDATED

BEDE BD-12
TYPE: Single-seat sportplane kitbuilt.
PROGRAMME: Prototype (N112BD) revealed to public at EAA Fly-In, Oshkosh, 1995. Uprated **BD-14** planned.
CURRENT VERSIONS: **BD-12A**, **BD-12B** and **BD-12C**: Three versions differing by engine power; see Power Plant below.
COSTS: BD-12A US$21,900; BD-12B US$24,900; BD-12C US$33,450 (1998). Engine, propeller and instrumentation extra.
DESIGN FEATURES: Similar layout to earlier BD-5, but enlarged some 60 per cent. Underfuselage airscoop and supercharger. Upward-curving wingtips.
FLYING CONTROLS: Normal control surfaces; split flaps.
STRUCTURE: Honeycomb construction using aluminium and composites.
LANDING GEAR: Tricycle type; retractable.
POWER PLANT: Prototype BD-12C uses one 112 kW (150 hp) Textron Lycoming O-320-E2D driving two-blade Q-tip pusher propeller. Fuel capacity 91 litres (24 US gallons; 20 Imp gallons). Optional engines range from BD-12A's 60 kW (80 hp) through BD-12B's 75 kW (100 hp). BD-14 will feature engines from 112 to 261 kW (150 to 350 hp).
ACCOMMODATION: Pilot in enclosed cockpit under two-piece canopy/windscreen, of which rear portion opens forward.

DIMENSIONS, EXTERNAL:	
Wing span	6.82 m (22 ft 4½ in)
Length overall	6.55 m (21 ft 6 in)
Height overall	2.38 m (7 ft 9½ in)
Propeller diameter	1.45 m (4 ft 9 in)
DIMENSIONS, INTERNAL:	
Cockpit: Length	1.73 m (5 ft 8 in)
Max width	1.17 m (3 ft 10 in)
Max height	1.07 m (3 ft 6 in)
AREAS:	
Wings, gross	8.60 m² (92.6 sq ft)
WEIGHTS AND LOADINGS:	
Weight empty	308 kg (679 lb)
Max T-O weight	592 kg (1,305 lb)
Max wing loading	68.8 kg/m² (14.09 lb/sq ft)
Max power loading (BD-12C)	5.30 kg/kW (8.70 lb/hp)
PERFORMANCE:	
Max cruising speed	187 kt (346 km/h; 215 mph)
Econ cruising speed	174 kt (322 km/h; 200 mph)
Stalling speed: flaps up	54 kt (100 km/h; 63 mph)
flaps down	47 kt (87 km/h; 54 mph)
Max rate of climb at S/L	567 m (1,860 ft)/min
Service ceiling	6,400 m (21,000 ft)
T-O run	213 m (700 ft)
Landing run	244 m (800 ft)
Range with max fuel and VFR reserves	816 n miles (1,511 km; 939 miles)

UPDATED

BEDE BD-17 NUGGET
TYPE: Single-seat sportplane kitbuilt.
PROGRAMME: Announced June 2000; prototype due to be completed by end of that year.
COSTS: Estimated around US$10,000 including engine (2000).
DESIGN FEATURES: Intended to be extremely easy to build, with approximately 110 parts. Constant-chord wings. Fin fillet.
FLYING CONTROLS: Conventional and manual.
STRUCTURE: All-metal. Tubular spar and honeycomb fuselage covered with 5 mm (0.20 in) aluminium sheet; control surfaces of urethane foam.
LANDING GEAR: Fixed; choice of tricycle or tailwheel layout. Speed fairings optional.
POWER PLANT: Power plants in the range 29.8 to 52.2 kW (40 to 70 hp) being considered. Fuel capacity 76 litres (20.0 US gallons; 16.7 Imp gallons).
ACCOMMODATION: Pilot only, in open cockpit; canopy optional.
All data provisional.

DIMENSIONS, EXTERNAL:	
Wing span	6.55 m (21 ft 6 in)
Wing chord, constant	0.76 m (2 ft 6 in)
Wing aspect ratio	8.6
Length overall	5.33 m (17 ft 6 in)
Height overall	2.36 m (7 ft 9 in)
AREAS:	
Wings, gross	4.97 m² (53.5 sq ft)
WEIGHTS AND LOADINGS:	
Weight empty	193 kg (425 lb)
Max T-O weight	340 kg (750 lb)
PERFORMANCE:	
Never-exceed speed (V_{NE})	160 kt (296 km/h; 184 mph)

NEW ENTRY

Prototype Bede BD-12 high-performance single-seat monoplane *(Howard Levy)*

BEECH — see Raytheon Aircraft Company

BEL-AIRE

BEL-AIRE AVIATION
3891 Maxison Drive, Lyons, New York 14489
Tel: (+1 315) 923 77 45 and 946 50 58
PRESIDENT: Gerald Belcher

Having acquired and reverse-engineered a 1926 Travel Air Model B, Mr Belcher launched production of kits at the 2000 Sun 'n' Fun convention.

VERIFIED

BEL-AIRE 2000
TYPE: Three-seat biplane kitbuilt.
PROGRAMME: Original Travel Air Model B designed by Walter Beech, Clyde Cessna and Lloyd Stearman, powered by Curtiss OX-5 engine; 1,550 built between 1925 and 1930; widely used for stunt/film flying.
First production Bel-Aire 2000 built on remanufactured jigs, flew (N221BA) July 1999; achieved 60 hours by April 2000.
DESIGN FEATURES: Single-bay, positive-stagger biplane; N-type interplane struts and cabane. Period construction, apart from engine and other machined items. Quoted build time 1,100 hours.
FLYING CONTROLS: Conventional and manual; pushrods replace original cables. Horn-balanced rudder. Interconnecting strut between aileron actuator in lower wing and ailerons on upper wing only. No flaps.
STRUCTURE: Wood and fabric wings; steel tube fuselage with wood floor and turtle back attached by bolts. Steel struts and wires; composites engine cowling and wheel fairings. Mutually braced empennage.
LANDING GEAR: Tailwheel type. Main legs of tube steel, fabric covered, with bungee cord suspension and motorcar wheels and 850×15 tyres. Tailwheel 4.10/3.50-6. Mainwheel brakes.
POWER PLANT: One 246 kW (330 hp) Chevrolet 6,276 cc (383 cu in) V-8 motorcar engine with 2:1 reduction gearbox and external, underslung radiator, driving a Bel-Aire Aviation two-blade, wooden propeller. Fuel tank ahead of cockpits, capacity 133 litres (35.0 US gallons; 29.2 Imp gallons).
ACCOMMODATION: Three, in two separate cockpits with individual windscreens.

DIMENSIONS, EXTERNAL:	
Wing span	10.36 m (34 ft 0 in)
Length overall	7.47 m (24 ft 6 in)
Height overall	3.04 m (10 ft 0 in)
Propeller diameter	2.44 m (8 ft 0 in)

WEIGHTS AND LOADINGS:
Weight empty	971 kg (2,140 lb)
Baggage capacity	18 kg (40 lb)
Max T-O weight	1,320 kg (2,910 lb)
Max power loading	5.37 kg/kW (8.82 lb/hp)

PERFORMANCE:
Max level speed	83 kt (153 km/h; 95 mph)
Stalling speed	48 kt (89 km/h; 55 mph)

UPDATED

Prototype Bel-Aire 2000 *(Paul Jackson/Jane's)*
2001/0099633

BELL

BELL HELICOPTER TEXTRON INC
(Subsidiary of Textron Inc)
PO Box 482, Fort Worth, Texas 76101
Tel: (+1 817) 280 20 11
Fax: (+1 817) 280 23 21
Web: http://www.bhti.com
CHAIRMAN AND CEO: Terry D Stinson
PRESIDENT: John R Murphey
EXECUTIVE VICE-PRESIDENT AND COO: P D Shabay
SENIOR VICE-PRESIDENT BUSINESS DEVELOPMENT AND STRATEGIC
  PLANNING: Dr Leo Mackay
SENIOR VICE-PRESIDENT, ACQUISITIONS AND INTERNATIONAL
  BUSINESS DEVELOPMENT: Fred N Hubbard
VICE-PRESIDENT, RESEARCH AND ENGINEERING: Troy M Gaffey
DIRECTOR, PUBLIC AFFAIRS AND ADVERTISING: Carl L Harris

During 1970-81, Bell Helicopter Textron was unincorporated division of Textron Inc; became wholly owned subsidiary of Textron Inc from 3 January 1982. Bell Helicopter Canada (see Canada) formed at Montreal/Mirabel under contract with Canadian government October 1983; transfer to Mirabel, completed January 1987, of Bell 206B JetRanger and 206L LongRanger production. Production of Bell 212/412 transferred mid-1988 and early 1989 respectively; Bell 230, 430, 427 and 407 programmes also undertaken in Canada. Global workforce exceeds 7,500.

More than 34,000 Bell helicopters manufactured worldwide, including over 9,500 commercial models.

Bell and Boeing collaborate in design and manufacture of V-22 Osprey tiltrotor aircraft, as described in the following entry. New 41,800 m² (450,000 sq ft) factory at Amarillo International Airport, Texas, completed in 1999 as a Tiltrotor Assembly Centre (TAC) for the V-22 and the commercial BA609; latter tiltrotor, previously also a joint venture with Boeing, became a Bell programme on 1 March 1998 and is now (with the AB139) the core of a joint venture effort undertaken in conjunction with Agusta of Italy; an agreement establishing the Bell/Agusta Aerospace Company (see International section) was signed in November 1998.

Bell helicopters built in USA detailed here. Those currently built in Canada listed under Canada; other models built under licence by IPTN in Indonesia (currently suspended) and Agusta in Italy (which see); KAI (which see) is co-producing Bell 427 as SB 427 in Republic of Korea; Bell Helicopter Asia (Pte) Ltd is wholly owned Singapore-based company for marketing and support in Southeast Asia.

MoU signed in mid-1996 in connection with planned acquisition by Bell of majority share of Romanian IAR Brasov aircraft manufacturing company. Formal takeover was to occur after signature of contract in late 1996 covering procurement of 96 AH-1RO Dracula attack helicopters by armed forces of Romania. However, subsequent protracted negotiations failed to result in agreement being reached with the Romanian government and in the fourth quarter of 1999, Bell revealed that it was abandoning plans for the IAR Brasov takeover and AH-1RO production in Romania. At same time, Bell revealed possibility of closer co-operation with PZL Swidnik of Poland, which may ultimately include creation of a SuperCobra production line; no recent news has been received concerning this.

UPDATED

BELL 209 and SUPERCOBRA
US Navy/Marine Corps designations: AH-1W and AH-1Z
TYPE: Attack helicopter.
PROGRAMME: Prototype Bell 209, derived from single-engined UH-1, first flew as tandem-seat combat aircraft on 7 September 1965. Built for US armed forces and export and under licence in Japan, as described in previous editions of *Jane's*. Universally known as HueyCobra.

First twin-engined Cobra was AH-1J SeaCobra, delivered from mid-1970; AH-1T Improved SeaCobra followed from 1977. All surviving US Marine Corps AH-1J SeaCobras withdrawn and 44 (including one ground-based trainer) AH-1T Improved SeaCobras converted to AH-1W to augment new production.

Rear (pilot's) cockpit of the AH-1W SuperCobra
(Paul Jackson/Jane's)

CURRENT VERSIONS: **AH-1W SuperCobra:** Bell flew AH-1T powered by two GE T700-GE-700; first flight of improved AH-1T+, including GE T700-GE-401 engines, 16 November 1983. Production for USMC has now ended, but continues for export; USMC also received two composites maintenance trainers. Missions of AH-1W include anti-armour, escort, multiple-weapon fire support, including air-to-air with Sidewinder, armed reconnaissance, search and target acquisition.

AH-1W Upgrades: Following abandonment of the proposed Integrated Weapon System (IWS) project in July 1995 and the Marine Observation and Attack Aircraft programme which was intended to provide a replacement for both the AH-1W SuperCobra and the UH-1N Iroquois, the US Marine Corps has opted for a two-stage upgrade of the AH-1W. Under present plans, this will be completed by FY11, allowing it to be retained in the active inventory beyond 2020. Phase I is the Night Targeting System (NTS), under which USMC AH-1Ws are currently being fitted with the Israeli Tamam laser NTS for dual TOW/Hellfire day, night and adverse weather capability. NTS integrates FLIR laser designation and range-finding, plus automatic boresighting, and includes an onboard video cassette recorder in the gunner's cockpit.

Conversion of a prototype (162533) was authorised in December 1991, with an initial batch of 25 sets being built by Tamam for delivery from January 1993; joint production with Kollsman was approved in May 1994 and a total of 250 sets is required by the USMC, with a further 12 ordered for Turkey and 53 for Taiwan in 1994. Deliveries of modified aircraft to operational units of the USMC began in June 1994.

A further improvement programme, involving installation of an Embedded Global Positioning System/Inertial Navigation System (EGI), is being undertaken. Two prototype conversions (162532 and 163936) were delivered to test units for trials in November 1995 and March 1996, with EGI installed on new-build aircraft from Lot IX onwards, as well as older AH-1Ws as a retrofit programme.

Phase II will involve installation of the Bell 680 four-blade rotor, offering a 70 per cent reduction in vibration; formerly designated **AH-1W (4BW)**, but now known as **AH-1Z**. Initial trials of the four-blade rotor system were undertaken with AH-1W 161022; bench testing of the new drive system began in second quarter of 1999 and was completed successfully in first quarter of 2000. Bell also demonstrated 30-minute run dry capability of new intermediate and tail rotor gearboxes in March 2000. The definitive aircraft will be fitted with a new four-blade, all-composites, hingeless/bearingless rotor system; four-blade composites tail rotor; a new transmission rated at 1,957 kW (2,625 shp) and new wing assemblies with the ability to carry twice the number of anti-armour missiles, 379 litres (100 US gallons; 83.0 Imp gallons) more fuel and additionally permitting concurrent carriage of two air-to-air self-defence missiles.

Lockheed Martin selected to develop and manufacture advanced target sighting system (TSS), with work on US$8 million, 54 month, engineering development and integration programme beginning in July 1998. TSS will feature imaging technology by Wescam of Canada and Lockheed Martin's Sniper third-generation FLIR, as well as colour TV camera, laser ranger, spot-tracker and designator. This equipment will also be integrated with an advanced BAE Systems helmet-mounted display and sighting system.

Also provided in the second phase will be 'glass cockpits'; Litton Industries has been selected as prime contractor for this aspect of the upgrade programme. Digital transfer of information on tactical situation,

Bell AH-1W SuperCobras of Turkish Army
2001/0014922

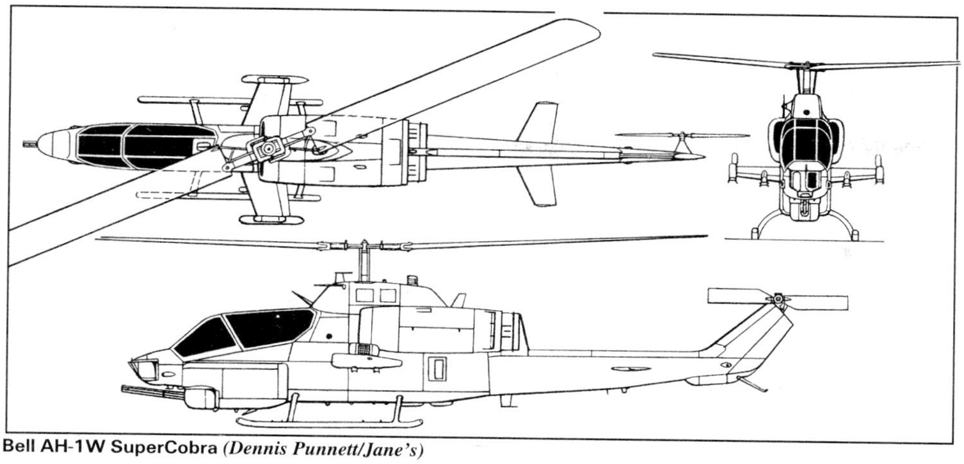

Bell AH-1W SuperCobra *(Dennis Punnett/Jane's)*

weaponry and flight data will enable crew interchangeability and allow AH-1Z to be flown from either front or rear seat. Major subcontractors include Rockwell Collins, which will supply active matrix liquid crystal displays (AMLCDs); Smiths Industries (fire-control system); Meggitt Avionics (standby air data and inertial sensing devices); and BAE (air data computers and advanced helmets). Other elements of the upgrade include new stores management system, onboard systems monitoring, mission data loader, HOTCC (hands on throttle, collective and cyclic) controls, airborne target handover system and a new EW suite including AN/APR-39A (XE2) RWR, AN/AAR-47 MAWS and AN/AVR-2 LWS. Maximum gross weight will rise to 8,391 kg (18,500 lb).

Cost of implementing Phase II in its entirety will be of the order of US$1.99 billion for 180 AH-1Zs, of which US$472 million will go on research and development. A US$310 million cost-plus-fixed-fee contract was awarded to Bell in November 1996, for design, development, fabrication, installation, test and delivery of three engineering development AH-1W SuperCobra Upgrade Aircraft, with work to be undertaken at Forth Worth, Texas. Assembly of first AH-1Z begun at Hurst, Texas, in April 1999, by which time 85 per cent of drawings had been released, with design work due for completion by end of 1999. Initial AH-1Z completed final assembly in second quarter of 2000 and moved to Bell Flight Research Center at Fort Worth for functional testing and installation of instrumentation in readiness for first flight, which is due to take place in October 2000; second AH-1Z to follow by end 2000, with flight test and development expected to be completed by September 2003. Programme will include flight test and evaluation at Patuxent River, Maryland; weapons testing at Yuma Proving Ground, Arizona; and other trials at China Lake, California. Testing of full-scale AH-1Z fatigue and static test article at Arlington, Texas, was due to begin in April 2000 and include verification of predicted 10,000-hour service life. Finalisation of the cockpit upgrade design occurred in FY99, with aircraft remanufacture starting in FY03, when six AH-1Ws are to be modernised to AH-1Z configuration at an estimated cost of US$84 million; deliveries should begin with five in 2005. Planned expenditure for FY04 is US$135 million for 12 conversions, with 24 AH-1Ws to be upgraded annually from FY07 onwards, before the programme terminates with 19 aircraft in FY13.

AH-1RO Dracula: Derivative of AH-1W for Romania, which intended to purchase initial batch of 96. Project abandoned by Bell in fourth quarter of 1999.

AH-1Z King Cobra: Proposed for Turkey, which plans to acquire 145 attack helicopters at cost of US$4 billion; bids for initial batch of 50 (including two prototypes) submitted by end 1997. Announcement of winning contender was expected at start of 1999 but decision deferred to mid-2000, following delays in flight evaluations of competing types. AH-1Z selected, with announcement made at Farnborough 2000 in late July, when revealed that initial batch of 50 to be purchased at approximate US$1.5 billion cost. Licensed production to be undertaken in Turkey by TAI at Ankara; current plan stipulates follow-on batches of 50 and 45 helicopters, with deliveries to begin in mid-2003 and continue until 2011.

MH-1W: In April 1998, Bell revealed it had conceived a reconnaissance, armed escort and fire support 'multimission' version of the SuperCobra under this designation. Evolved in response to a perceived need for armed helicopters to undertake anti-drug and counter-insurgency support operations, marketing efforts were then principally aimed at Latin American countries, with presentations having been given to Argentina, Brazil, Chile, Colombia and Venezuela. Configuration of the MH-1W includes a nose-mounted sighting system, with a steerable FLIR sensor, laser range-finder, video recorder and automatic target tracker. Proposed weaponry is comparable to that of the standard AH-1W and includes a 20 mm cannon as well as up to four 70 mm rocket pods, anti-armour missiles and air-to-air missiles. US government authorisation to begin demonstrations to potential customers was received by Bell in first quarter of 2000, with Chile and Colombia emerging as likely purchasers.

CUSTOMERS: US Marine Corps (see under Current Versions); total of 179 AH-1W, including 10 diverted to Turkey.

Deliveries to USMC began on 27 March 1986 to Camp Pendleton, California, for HMLA-169, -267, -367 and -369, plus HMT-303 for training; further aircraft issued to USMC Reserve, beginning with HMA-775 (now HMLA-775) at Camp Pendleton from June 1992; followed by HMA-773 (now HMLA-773) at Atlanta, Georgia, and HMLA-767 (now HMLA-775 Detachment A) at New Orleans, Louisiana. Procurement augmented by remanufacture of 43 AH-1Ts to AH-1W for HMLA-167 and -269 at New River, North Carolina; last completed in 1993; 100th new/converted AH-1W delivered 8 August 1991. Last of 169 new-build aircraft for USMC delivered in fourth quarter of 1998.

Turkish Land Forces received five AH-1Ws in 1990 and five in 1993; all diverted from USMC contracts. Taiwan signed letter of offer and acceptance February 1992, for 42 over five years (plus one ground trainer); deliveries began 1993 for training with USMC; first aircraft 501 (ex-164913); in service with 1st Attack Helicopter Squadron at Lung Tan and 2nd AHS at Shinsur. Further 21 AH-1Ws subject of Taiwanese re-order announced in mid-1997; contract for first nine awarded in October 1997.

Potential customers include Japan, which is now seeking a replacement for the AH-1S HueyCobra; and South Korea, which is considering the AH-1W to meet its AHX attack helicopter requirement. AH-1Z was originally a contender for Australian Project Air 87 combat helicopter programme, but was eliminated from short list in March 1999, only to be reinstated in early 2000, when competition reopened to all bidders following a programme review; Australia requires 25 to 30 helicopters and anticipates IOC of first squadron to occur in 2007.

COSTS: US$10.7 million (1996) projected unit cost. Total value of 1997 follow-on order for 21 AH-1Ws by Taiwan is US$479 million. Unit cost of new-build AH-1Z estimated at US$14 million (Turkish proposal, 1998). Total cost of upgrade programme (including UH-1Y) anticipated to be in excess of US$7 billion.

DESIGN FEATURES: Essentially first-generation attack helicopter, with slightly stepped tandem seating and stub-wings for armament. Two-blade main rotor, similar to that of Bell 214, with strengthened rotor head incorporating Lord Kinematics Lastoflex elastomeric and Teflon-faced bearings. Blade aerofoil Wortmann FX-083 (modified); normal 311 rpm. Tail rotor also similar to that of Bell 214 with greater diameter and blade chord; normal 1,460 rpm. Rotor brake standard. Stub-wings have NACA 0030 section at root; NACA 0024 at tip; incidence 14°; sweepback 14.7°. AH-1Z will incorporate new four-blade rotor system and transmission (see current versions section for details).

STRUCTURE: Main rotor blades have aluminium spar and aluminium-faced honeycomb aft of spar; tail rotor has aluminium honeycomb with stainless steel skin and leading-edge. Airframe conventional all-metal semi-monocoque.

LANDING GEAR: Non-retractable tubular skid type on AH-1W; AH-1Z will incorporate new, lighter, design with rectangular cross tubes. Ground handling wheels optional.

POWER PLANT: Two General Electric T700-GE-401 turboshafts, each rated at 1,285 kW (1,723 shp). Transmission rating 1,515 kW (2,032 shp) for take-off; 1,286 kW (1,725 shp) continuous. Fuel (JP5) contained in two interconnected self-sealing rubber fuel cells in fuselage, with protection from damage by 0.50 in ballistic ammunition, total usable capacity 1,128 litres (298 US gallons; 248 Imp gallons). Gravity refuelling point in forward fuselage, pressure refuelling point in rear fuselage. Provision for carriage on underwing stores stations of two or four external fuel tanks each of 291 litres (77.0 US gallons; 64.1 Imp gallons) capacity; or two 378 litre (100 US gallon; 83.3 Imp gallon) tanks; or two 100 and two 77 US gallon tanks; large tanks on outboard pylons only. Oil capacity 19 litres (5.0 US gallons; 4.2 Imp gallons).

AH-1W PROCUREMENT

Lot	Qty	First Aircraft	Delivered
USMC New-Build			
I	22	162532	Mar 1986 to Jun 1987
II	22[1]	162554	Sep 1987 to Nov 1988
III	34[2]	163921	Jul 1990 to Nov 1991
IV	5	164572	Mar 1992 to Jun 1992
V	14	164586	Jun 1992 to Feb 1993
VI	20[2]	165037	Oct 1993 to Sep 1994
VII	22	165271	Aug 1994 to Sep 1995
VIII	17	165317	Sep 1995 to Apr 1996
IX	12	165358	1996 to 1997
X	11	165392	1997 to 1998
Total	**179**		
Taiwan New-Build			
I	8[3]	164913/501	Mar 1993 to Jul 1993
II	9	164921/509	Nov 1993 to Jul 1994
III	9	164930/518	May 1995 to Aug 1995
IV	8	165334/527	Apr 1996 to Aug 1996
V	8	165370/535	1996 to 1997
VI et seq	21	165545/543	
Total	**63**		
USMC AH-1T to AH-1W Retrofit			
Prototypes	2		Jan 1989
I	12		Jun 1989 to Dec 1989
II	9		Dec 1989 to May 1990
III	5[4]		Mar 1991 to May 1991
IV	7		Nov 1991 to Mar 1992
V	8		Jul 1993 to Oct 1993
Total	**43**		

[1] One non-flying training airframe not included in Lot II total
[2] Includes five transferred to Turkey
[3] One non-flying training airframe not included in Lot I total
[4] One non-flying training airframe (159229) not included in Lot III total

ACCOMMODATION: Crew of two in tandem, with co-pilot/gunner in front seat and pilot at rear in AH-1W; crew stations will be interchangeable in AH-1Z. Cockpit is heated, ventilated and air conditioned. Dual controls; lighting compatible with night vision goggles, and armour protection standard. Forward crew door on port side and rear crew door on starboard side, both upward-opening. Inflatable body and head restraint system by Simula of Phoenix, Arizona, nearing end of development in mid-1995; retrofit provisions installed in 1996 production, with system incorporated in 1997 production.

SYSTEMS: Three independent hydraulic systems, pressure 207 bars (3,000 lb/sq in), for flight controls and other services. Electrical system comprises two 28 V 400 A DC generators, two 24 V 34.5 Ah batteries and three inverters: main 115 V AC, 1 kVA, single-phase at 400 Hz, standby 115 V AC, 750 VA, three-phase at 400 Hz and a dedicated 115 V AC 365 VA single-phase for AIM-9 missile system. AiResearch environmental control unit.

AVIONICS: *Comms:* Two AN/ARC-210(V) radios, KY-58 TSEC secure voice set; AN/APX-100(V) IFF.

Flight: AN/ASN-75 compass set, AN/ARN-89B ADF, AN/ARN-118 Tacan on AH-1W and AN/ARN-153(V-4) Tacan on AH-1Z, AN/APN-154(V) radar beacon set and Teledyne AN/APN-217 Doppler-based navigation system being replaced by Embedded Global Positioning System/Inertial Navigation System (EGI), which adds Honeywell CN-1689(V) EGI and Rockwell Collins AN/ARN-153(V) Tacan. Installation began with Lot IX production in October 1996 and will also involve retrofit of earlier aircraft. AN/APN-194 radar altimeter. Rockwell Collins CDU-800 control/display unit and dual Rockwell Collins ICU-800 processors.

Instrumentation: Kaiser HUD compatible with PNVS-5 and Elbit ANVIS-7 night vision goggles. Helmet-mounted sighting/aiming device. Flat panel colour LCD displays to be installed on AH-1Z.

Mission: Tamam/Kollsman Night Targeting System (NTSF-65) comprising FLIR, laser range-finder/designator, TV camera, day/night video tracker and full in-flight boresighting; being retrofitted (within M-65 sighting system for TOW missiles) from 1993 (see Design Features); first redelivery in June 1994; alternative Boeing Electronic Systems NightHawk system also offered for export SuperCobras, following 1992-93 flight testing.

Self-defence: AN/APR-39(V) pulse radar signal detecting set, AN/APR-44(V) CW radar warning system, and AN/ALQ-144(V) IR countermeasures set. Dual AN/ALE-39 chaff system with one MX-7721 dispenser mounted on top of each stub-wing. Improved countermeasures suite in USMC AH-1Ws will replace AN/APR-39 and AN/APR-44 by AN/APR-39(XE2) radar warning and adds AN/AVR-2 laser warner and AN/AAR-47 plume detecting set.

ARMAMENT: Electrically operated General Electric undernose A/A49E-7(V4) turret housing an M197 three-barrel 20 mm gun. A 750-round ammunition container is located in the fuselage directly aft of the turret; firing rate is 675 rds/min; a 16-round burst limiter is incorporated in the firing switch. Either crew member can fire the gun, which can be slaved to a helmet-mounted sight/aiming device. Gun can be tracked 110° to each side, 18° upward, and 50° downward, but barrel length of 1.52 m (5 ft 0 in) makes it imperative that the M197 is centralised before wing stores are fired. Underwing attachments for up to four LAU-61A (19-tube), LAU-68A, LAU-68A/A, LAU-68B/A or LAU-69A (seven-tube) 2.75 in Hydra 70 rocket launcher pods; two CBU-55B fuel-air explosive weapons; four SUU-44/A flare dispensers; two M118 grenade dispensers; Mk 45 parachute flares; or two GPU-2A or SUU-11A/A Minigun pods.

Provision for carrying totals of up to eight TOW missiles, eight AGM-114 Hellfire missiles, two AIM-9L Sidewinder or AGM-122A Sidearm missiles, on outboard underwing stores stations. Canadian Marconi TOW/Hellfire control system enables AH-1W to fire both TOW and Hellfire missiles on same mission. AH-1Z expected to include FIM-92 Stinger AAM for self-defence.

Following data applicable to AH-1W except where stated.

DIMENSIONS, EXTERNAL:
Main rotor diameter	14.63 m (48 ft 0 in)
Main rotor blade chord	0.84 m (2 ft 9 in)
Tail rotor diameter	2.97 m (9 ft 9 in)
Tail rotor blade chord	0.305 m (1 ft 0 in)
Distance between rotor centres	8.89 m (29 ft 2 in)
Wing span	3.28 m (10 ft 9 in)
Wing aspect ratio	3.7
Length: overall, rotors turning	17.68 m (58 ft 0 in)
fuselage	13.87 m (45 ft 6 in)
Width overall	3.28 m (10 ft 9 in)
Height: to top of rotor head	4.11 m (13 ft 6 in)
overall	4.44 m (14 ft 7 in)
Ground clearance, main rotor turning	2.74 m (9 ft 0 in)
Elevator span	2.11 m (6 ft 11 in)
Width over skids	2.24 m (7 ft 4 in)

AREAS:
Main rotor blades (each)	6.13 m² (66.0 sq ft)
Tail rotor blades (each)	0.45 m² (4.835 sq ft)
Main rotor disc	168.11 m² (1,809.6 sq ft)
Tail rotor disc	6.94 m² (74.70 sq ft)
Vertical fin	2.01 m² (21.70 sq ft)
Horizontal tail surfaces	1.41 m² (15.20 sq ft)

WEIGHTS AND LOADINGS:
Weight empty: AH-1W	4,953 kg (10,920 lb)
AH-1Z	5,579 kg (12,300 lb)
Mission fuel load (usable): AH-1W	946 kg (2,086 lb)
AH-1Z	1,256 kg (2,768 lb)
Max useful load (fuel and disposable ordnance):	
AH-1W	1,736 kg (3,828 lb)
AH-1Z	2,812 kg (6,200 lb)
Max T-O and landing weight:	
AH-1W	6,690 kg (14,750 lb)
AH-1Z	8,391 kg (18,500 lb)
Max disc loading: AH-1W	39.8 kg/m² (8.15 lb/sq ft)
AH-1Z	49.9 kg/m² (10.22 lb/sq ft)
Transmission loading at max T-O weight and power:	
AH-1W	4.42 kg/kW (7.26 lb/shp)
AH-1Z	5.54 kg/kW (9.10 lb/shp)

PERFORMANCE:
Never-exceed speed (V_{NE}):	
AH-1W	190 kt (352 km/h; 219 mph)
AH-1Z	222 kt (411 km/h; 255 mph)
Max level speed at S/L	152 kt (282 km/h; 175 mph)
Max cruising speed: AH-1W	150 kt (278 km/h; 173 mph)
AH-1Z	157 kt (291 km/h; 181 mph)
Rate of climb at S/L, OEI	244 m (800 ft)/min
Service ceiling	more than 4,270 m (14,000 ft)
Service ceiling, OEI	more than 3,660 m (12,000 ft)
Hovering ceiling: IGE	4,495 m (14,740 ft)
OGE	915 m (3,000 ft)
Range at S/L with standard fuel, no reserves:	
AH-1W	280 n miles (518 km; 322 miles)
AH-1Z	370 n miles (685 km; 425 miles)
Max endurance with standard fuel: AH-1W	2 h 48 min
AH-1Z	3 h 30 min
g limits: AH-1W	+2.5/–0.5
AH-1Z	+3.5/–0.5

UPDATED

BELL 406 (AHIP)

US Army designations: OH-58D Kiowa and OH-58D(I) Kiowa Warrior

TYPE: Armed observation helicopter.

PROGRAMME: Bell won US Army Helicopter Improvement Program (AHIP) 21 September 1981; first flight of OH-58D 6 October 1983; deliveries started December 1985; first based in Europe June 1987. Production is terminating in FY00.

CURRENT VERSIONS: **Prime Chance:** Fifteen special armed OH-58Ds (86-8908 to -8922) modified from September 1987 under Operation Prime Chance for use against Iranian high-speed boats in Gulf; delivery started after 98 days, in December 1987; firing clearance for Stinger, Hellfire, 0.50 in gun and seven-tube rocket pods completed in seven days. Further conversion (85-24716) for development trials. All transferred to 4-2nd Cavalry at Fort Bragg, North Carolina.

OH-58D(I) Kiowa Warrior: Armed version, introduced to production with effect from 202nd conversion and to which 188 surviving OH-58Ds have been modified to give US Army total fleet of 411; integrated weapons pylons, uprated transmission and engine, lateral CG limits increased, raised gross weight, EMV protection of avionics bays, localised strengthening, RWR, IR jammer, video recorder, SINCGARS radios, laser warning receiver and tilted vertical fin; armament same as Prime Chance; integrated avionics and lightened structure. See illustration in 1995-96 edition for stealth kit modification (details of kit in 1992-93 *Jane's*); converted aircraft (of which 89-0090 was first) reportedly since reverted to standard, but low-observable OH-58Ds of 1-17th Cavalry at Fort Bragg, North Carolina, operating with undisclosed modifications by 1993; changes include coated rotor blades, new windscreen material and modified rotor sail, rotor cuffs and hubs for substantial reduction in frontal radar signature.

Further modifications for Warrior, currently under consideration, include improved SINCGARS (secure speech) radios; improved data modem; improved master processor and mast-mounted sight processor; video cross-link; INS/GPS and digital map display. Simula Safety Systems awarded US$5.4 million contract in May 1998 for engineering work to apply existing cockpit airbag system and/or COTS airbag technologies to OH-58D; contract

Artist's impression of upgraded AH-1Z SuperCobra

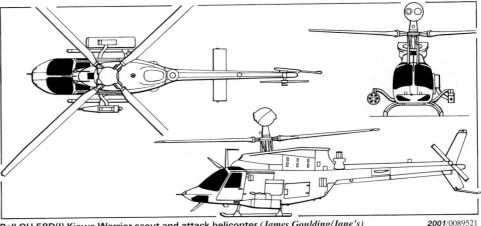
Bell OH-58D(I) Kiowa Warrior scout and attack helicopter *(James Goulding/Jane's)*

Bell OH-58D(I) Kiowa Warrior with current shape of infra-red suppressing engine cowling

option for subsequent delivery of up to 500 airbag ship sets for installation on OH-58D.

Safety Enhancement Program (SEP) launched in 1998, when Bell began work on initial batch of 28 OH-58D(I) Kiowa Warriors, with first upgraded aircraft delivered on 22 December 1998. Programme involves installation of improved Rolls-Royce 250-C30R/3 engine, crashworthy energy-attenuating seats, with cockpit airbag system to follow; US Army currently plans to modify total of 310 OH-58D(I)s at rate of up to 48 per year, during 2000 to 2006.

MultiPurpose Light Helicopter (MPLH): Further modification of Kiowa Warrior; features include squatting landing gear, quick-folding rotor blades, horizontal stabiliser and tilting fin to allow helicopter to be transported in cargo aircraft and flown to cover 10 minutes after unloading from C-130. Later additions include cargo hook for up to 907 kg (2,000 lb) slung load and fittings for external carriage of six outward-facing troop seats or two stretchers. All Kiowa Warriors have MPLH capability.

OH-58X Light Utility Variant: Contender for anticipated US Army requirement; fourth development OH-58D (69-16322) modified in 1992 with partial stealth features (including chisel nose); concept subsequently discarded.

CUSTOMERS: US Army initial plan to modify 592 OH-58A/C to OH-58D reduced to 477; again reduced to 207, but Congressionally mandated re-orders increased total to 424 (excluding five prototypes) by FY98, when procurement terminated with final batch of 13. See table below for OH-58D acquisition.

OH-58D PROCUREMENT

FY	Lot	Qty	First aircraft
83	1	16	83-24129
85	2	44	85-24690
86	3	39	86-8901
87	4	36	87-0725
88	5	36	88-0285
89	6	36	89-0082
90	7	36	90-0346
91	8	36	91-0536
92	9	36	92-0571
93	10	36	93-0935
94	11	15	94-0050
95	12	16	95-0072
96	13	16	96-0113
97	14	13	97-1321
98	15	13	
Total		424†	

† Plus five prototypes

First 117 conversions at Amarillo; from 87-0743 at Fort Worth; produced as Kiowa Warrior from 202nd aircraft (89-0112), May 1991; initially to 'C' and 'D' Troops of 4-17 Aviation, Fort Bragg. Retrofit of surviving examples of initial 201 OH-58Ds to OH-58D(I) begun with award of first contract to Bell in January 1992; subsequent contracts increased total to 188 helicopters in six lots (see table), with retrofit programme completed in second quarter of FY99.

KIOWA WARRIOR RETROFIT

FY	Lot	Qty	First aircraft
92	1	28	92-0518
93	2	38	93-0971
94	3	38	94-0149
95	4	38	95-0002
96	5	33	96-0009
97	6	13	97-0124
Total		188	

Taiwan ordered 12 OH-58Ds plus maintenance trainer in February 1992 and reserved 14 options, which subsequently converted to firm order; deliveries from July 1993 (s/n 601). Additional 13 requested by Taiwan in third quarter of 1997, with deliveries from November 1999. Operated by 1st Reconnaissance Helicopter Squadron at Lung Tan and 2nd RHS at Shinsur. Taiwanese helicopters are new-build; US Army aircraft are ex-OH-58As and OH-58Cs with new serial numbers and receive a third serial number on further upgrading to Warrior.

COSTS: US$9.42 million (1990) programme unit cost. Flyaway US$4.9 million (Kiowa) or US$6.7 million (Kiowa Warrior). Estimated cost of newest batch of 13 for Taiwan is US$172 million, including basic airframe plus mast-mounted sight, engine, Hellfire launchers, Hydra 70 rockets and launchers, ammunition, spares and associated support. Retrofit to Warrior configuration US$1.34 million (1992-93 average).

DESIGN FEATURES: Four-blade Bell soft-in-plane rotor with carbon composites yoke, elastomeric bearings and composites blades. Main rotor rpm 395; tail rotor rpm 2,381. Boeing mast-mounted sight containing TV and IR optics and laser designator/ranger; Honeywell integrated control of mission functions, navigation, communications, systems and maintenance functions based on large electronic primary displays for pilot and observer/gunner; hands-on cyclic and collective controls for all combat functions; automatic target hand-off system in some OH-58Ds operates air-to-air as well as air-to-ground using digital frequency hopping; system indicates location and armament state of other helicopters; some OH-58Ds have real-time video downlink capable of relaying to US Army Guardrail aircraft, to headquarters 22 n miles (40 km; 25 miles) away or, via satellite, to remote locations.

FLYING CONTROLS: Fully powered controls, including tail rotor, with four-way trim and trim release; stability and control augmentation system (SCAS) using AHRS gyro signals; automatic bob-up and return to hover mode; Doppler blind hover guidance mode; co-pilot/observer's cyclic stick can be disconnected from controls and locked centrally.

STRUCTURE: Basic OH-58 structure reinforced; armament cross-tube fixed above rear cabin floor; avionics occupy rear cabin area, baggage area and nose compartment.

LANDING GEAR: Light alloy tubular skids bolted to extruded cross-tubes.

POWER PLANT: One Rolls-Royce 250-C30R/1 (T703-AD-700) turboshaft (C30R/3 with upgraded hot section to eliminate power drops and improve performance in high-altitude/high-temperature regions in late production Kiowa Warrior and for retrofit to earlier aircraft), with an intermediate power rating of 485 kW (650 shp) at S/L ISA. Transmission rating: Kiowa Warrior 410 kW (550 shp) continuous. Engine incorporates FADEC and is currently the subject of a three-phase modification programme which provides a 19.5 per cent improvement at high ambient temperature; this permits OGE hover at 1,220 m (4,000 ft) at 35°C (95°F) with four Hellfire ATMs. One self-sealing crash-resistant fuel cell, capacity 424 litres (112 US gallons; 93.0 Imp gallons), located aft of cabin area. Refuelling point on starboard side of fuselage. Oil capacity 5.7 litres (1.5 US gallons; 1.2 Imp gallons).

ACCOMMODATION: Pilot and co-pilot/observer seated side by side. Door on each side of fuselage. Accommodation is heated and ventilated. OH-58D Safety Enhancement Program includes installation of crashworthy seats and cockpit airbag systems from FY99.

SYSTEMS: Single hydraulic system with three-axis SCAS, pressure 69 bars (1,000 lb/sq in), for main and tail rotor controls. Maximum flow rate 11.4 litres (3.0 US gallons; 2.5 Imp gallons)/min. Open-type reservoir. Primary electrical power provided by 10 kVA 400 Hz three-phase 120/208 V AC alternator with 200 A 28 V DC transformer-rectifier unit for secondary DC power. Back-up power provided by 500 VA 400 Hz single-phase 115 V AC solid-state inverter and 200 A 28 V DC starter/generator.

AVIONICS: *Comms:* Five com transceivers, datalink and secure voice equipment. Phase 1 additions, introduced on production line in 1991 in preparation for Kiowa Warrior, include AN/ARC-201 SINCGARS secure voice/data radio and Have Quick II radio.

Flight: Plessey (PESC) AN/ASN-157 Doppler strapdown INS. Honeywell embedded GPS/INS unit installed from mid-point of Lot 12 and on all subsequent production and retrofit Kiowa Warriors.

Instrumentation: Equipped for day/night VFR. Multifunction displays for vertical and horizontal situation indication, mast-mounted sight day/night viewing and communications control, with selection via control column handgrip switches.

Mission: Boeing mast-mounted sight houses ×12 magnification TV camera, autofocusing IR thermal imaging sensor and laser range-finder/designator, with automatic target tracking and in-flight automatic boresighting; turret may be trained 190° port and 190° starboard in azimuth; ±30° in elevation. Elop Group received US$3 million contract in 1997 to cover development of new laser range-finder/designator; based on equipment developed for the RAH-66 Comanche, this replacing existing unit from 1999 onwards. Night vision goggles; Litton AHRS; airborne target handoff subsystem (ATHS); and Honeywell embedded tactical information control system. Phase 1 additions include doubled computer capacity to 88 kbits, added weapons selection/aiming and multitarget acquisition/track displays, video recorder, data transfer system, ANVIS display and symbology system and EMV hardening.

Self-defence: AN/APR-39(V)1 or -39A(V)1 RWR. Phase I adds AN/ALQ-144 IR jammer, second RWR (AN/APR-44(V)3) and AN/AVR-2 laser detection system.

EQUIPMENT: M-43 NBC mask in Phase 1 and Warrior aircraft.

ARMAMENT: Four FIM-92A Stinger air-to-air or AGM-114B Hellfire air-to-surface missiles, or two seven-round M260 2.75 in Hydra 70 rocket pods, or one XM296 pod for 0.50 in machine gun, mounted on outriggers on cabin sides (port side only for gun).

DIMENSIONS, EXTERNAL:
Main rotor diameter	10.67 m (35 ft 0 in)
Main rotor blade chord (mean)	0.24 m (9½ in)
Tail rotor diameter	1.65 m (5 ft 5 in)
Length: overall, rotors turning	12.58 m (41 ft 2½ in)
fuselage (pitot to skid)	10.44 m (34 ft 3 in)
Width: rotors folded, clean	1.97 m (6 ft 5½ in)
armament fitted, pylons folded for air transport (MPLH)	2.39 m (7 ft 10 in)
Height: overall	3.93 m (12 ft 10½ in)
squatting and folded for air transport (MPLH)	2.73 m (8 ft 11½ in)
Horizontal stabiliser span	2.29 m (7 ft 6 in)
Skid track	1.97 m (6 ft 5½ in)
Cabin doors (port and stbd, each):	
Height	1.04 m (3 ft 5 in)
Width	0.91 m (3 ft 0 in)
Height to sill	0.66 m (2 ft 2 in)

AREAS:
Main rotor blades (each)	1.38 m² (14.83 sq ft)
Tail rotor blades (each)	0.13 m² (1.43 sq ft)
Main rotor disc	89.37 m² (962.0 sq ft)
Tail rotor disc	2.14 m² (23.04 sq ft)
Horizontal stabiliser	1.11 m² (11.92 sq ft)
Fin	0.87 m² (9.33 sq ft)

WEIGHTS AND LOADINGS (armed OH-58D(I) Kiowa Warrior):
Weight empty	1,492 kg (3,289 lb)
Max useful load	998 kg (2,200 lb)
Max fuel weight	341 kg (752 lb)
Mission weight	2,359 kg (5,200 lb)
Max T-O and landing weight	2,495 kg (5,500 lb)
Max disc loading	27.9 kg/m² (5.72 lb/sq ft)
Transmission loading at max T-O weight and power	6.09 kg/kW (10.00 lb/shp)

PERFORMANCE (Kiowa Warrior at mission weight):
Never-exceed speed (V_{NE})	130 kt (241 km/h; 149 mph)
Max level speed at 1,220 m (4,000 ft)	125 kt (232 km/h; 144 mph)
Max cruising speed	114 kt (211 km/h; 131 mph)

Artist's impression of proposed Bell Quad TiltRotor

USA: AIRCRAFT—BELL to BELL BOEING

Econ cruising speed at 1,220 m (4,000 ft)
　　　　　　　　　110 kt (204 km/h; 127 mph)
Max rate of climb: at S/L, ISA　　469 m (1,540 ft)/min
　at 1,220 m (4,000 ft), 35°C (95°F)
　　　　　　　　　over 366 m (1,200 ft)/min
Vertical rate of climb: at S/L, ISA　488 m (1,600 ft)/min
Service ceiling　　　　　　4,575 m (15,000 ft)
Hovering ceiling: IGE, ISA　　3,050 m (10,000 ft)
　OGE, ISA　　　　　　　2,105 m (6,900 ft)
　OGE, 35°C (95°F)*　　　　1,220 m (4,000 ft)
Range　　　　　268 n miles (496 km; 308 miles)
Endurance　　　　　　　　　　3 h 5 min
at 2,176 kg (4,798 lb)

UPDATED

BELL QUAD TILTROTOR

Bell announced in early 1999 that it was studying a proposed Quad TiltRotor (QTR) to meet Future Transport Rotorcraft (FTR) requirements. As projected, the aircraft would feature a fuselage approximately the size of that of a Lockheed Martin C-130-30, mated to two sets of wings, engines and tiltrotors from the Bell Boeing V-22 Osprey, the rear units mounted on stub wings to extend span and ensure adequate fuselage clearance. Rear tiltrotors could fold in cruising flight, with their engines providing supplemental thrust. The Quad TiltRotor would be able to accommodate up to 90 passengers, or an AH-64, AH-1Z, RAH-66, UH-1Y or UH-60 helicopter, or three HMMWVs, or up to eight 463L pallets.

Provisional specifications include VTOL maximum T-O weight 45,360 kg (100,000 lb); STOL maximum T-O weight 63,505 kg (140,000 lb); and a payload up to 18,144 kg (40,000 lb). The Quad TiltRotor was in the conceptual design phase in mid-2000, with water tunnel testing of a 1/48th scale model to visualise complex airflow patterns around the tandem wings and four tiltrotors completed. Bell was then seeking NASA and DARPA funding for construction of wind tunnel models. The first of two prototypes could be flown by 2005, with production deliveries starting in 2010. Potential customers include the US Marine Corps (to replace Sikorsky CH-53E helicopters and KC-130 Hercules) and USAF (MH-53J combat SAR/ special forces helicopters).

UPDATED

OTHER AIRCRAFT

Bell **TH-67 Creek** and **civilian helicopters** appear under Bell Helicopter Textron Canada in that country's section. **Bell/Agusta 609** appears in the International section.

UPDATED

BELL BOEING

BELL HELICOPTER TEXTRON and THE BOEING COMPANY

Bell Boeing V-22 Program Office, PO Box 70, Patuxent River, Maryland 20670-0070
Tel: (+1 301) 757 66 34
Web: http://www.boeing.com/rotorcraft/military/v22 or http://www.bellhelicopter.textron.com/products/tiltRotor/v22
BELL BOEING PROGRAMME DIRECTOR: Stuart D Dodge
V-22 PROGRAMME MANAGERS:
　Colonel Nolan Schmidt, USMC (Naval Air Systems Command)
　Colonel Dick Wohls, USAF (Deputy for CV-22)
　Mike Tkach (Bell Boeing)
COMMUNICATIONS MANAGERS:
　Gidge Dady (US Navy Public Affairs)
　Carl Harris (Bell Helicopter Textron)
　Douglas C Kinneard (Boeing)

UPDATED

BELL BOEING V-22 OSPREY

US Air Force designation: CV-22
US Navy designation: HV-22
US Marine Corps designation: MV-22
Manufacturer's model: 901
TYPE: Multimission tiltrotor.
PROGRAMME: Based on Bell/NASA XV-15 tiltrotor; initiated as US Department of Defense Joint Services Advanced Vertical Lift Aircraft (JVX), run by US Army, FY82; programme transferred to US Navy January 1983; 24 month US Navy preliminary design contract 26 April 1983; aircraft named V-22 Osprey January 1985; seven year full-scale development (FSD) began 2 May 1986 with order for six prototypes (Nos. 1, 3 and 6 by Bell at Arlington, Texas; Nos. 2, 4 and 5 by Boeing at Wilmington, Delaware) plus static test airframes.

Prototype (163911) first flew 19 March 1989; joined by four further aircraft by June 1991; sixth not flown; all now retired (last flight 27 March 1997 by 163913); details of individual airframes and early development history appear in 1997-98 and previous editions of *Jane's*.

An EMD Bell Boeing V-22 demonstrates carriage of XM777 lightweight howitzer　　　*2001*/0073043

Osprey passed critical design review 13 December 1994; simultaneous defence review authorised V-22 production for both Marines and special forces, but latter version subsequently delayed, with decision to proceed with EMD phase not reached until January 1997. In meantime, contract for five low-rate initial production (LRIP) aircraft awarded June 1996.

All four EMD Ospreys flown in 1997-98. EMD Ospreys – see Current Versions (specific) – have significant changes from earlier aircraft, including substantial reduction in empty weight to approximately 14,800 kg (32,628 lb); aluminium cockpit cage, replacing titanium, but with smaller windows to preserve structural strength; upgraded flight controls; enhanced engine and drive system; improved tail unit construction (built by Aerostructures) including fibre placement aft fuselage; redesigned rotor system; absence of fin tuning weights; improved wing constructional techniques; redesigned wiring; and pyrotechnic escape hatches.

Manufacture of first LRIP MV-22B (165433) began 7 May 1997; splicing of three major fuselage sections took place in Philadelphia on 25 February 1998, with completed fuselage airlifted by C-17 to Arlington on 8 September 1998 for final assembly. Second LRIP aircraft also completed at Arlington, whereupon assembly transferred to new factory at Amarillo, Texas.

First flight of first LRIP MV-22B on 30 April 1999, followed by official roll-out and handover to US Marine Corps on 14 May, with delivery to New River later in May; subsequently to Patuxent River for flight testing. Next major milestone was operational evaluation; seven-month opeval began 2 November 1999 with first two LRIP MV-22Bs, which joined by two more by January 2000. These four aircraft accumulated 820 flight hours in 350 sorties by end of opeval in July 2000; launch of full-rate production in FY01 dependent upon successful conclusion of opeval, but approval still awaited at beginning of October 2000 and may be deferred until 2001.

Opeval included trials in USS *Essex* with four MV-22Bs in first quarter 2000 and one aircraft temporarily sent to Kirtland AFB, New Mexico, in March 2000 for trials with USAF 58th Special Operations Wing, before programme abruptly halted on 8 April 2000 when type grounded following fatal crash of fourth LRIP MV-22B (165436) at Avra Valley Airport, Arizona. Subsequent investigation established most likely cause as 'power settling', a condition in which it becomes difficult to stop descent because of recirculating air from rotor downwash. Clearance to return to flight given on 25 May 2000, with Opeval resuming on 5 June. Final phase included trials at China Lake, California, and New River, North Carolina, before being concluded in late July 2000.

By mid-September 2000, V-22 total flight time was 2,900 hours, during which aircraft had demonstrated speed of 342 kt (633 km/h; 394 mph), 7,620 m (25,000 ft) height, 27,442 kg (60,5000 lb) MTOW and 3.9 *g* load factor.

CURRENT VERSIONS (specific): **No. 7/164939**: Assembly began 15 February 1995; fuselage sections mated at Boeing's Philadelphia plant on 1 August 1995; followed two months later by wing and nacelle mating at Fort Worth. Fuselage to Fort Worth for fitting of wings and engines. Ground vibration testing completed in late July 1996; tests of hydraulic lines concluded soon after, with integrated functional testing commencing in August. First flight in helicopter mode 5 February 1997 (at Fort Worth); first transition to conventional flight 6 March 1997; aircraft to V-22 Integrated Test Team at Patuxent River on 15 March 1997; ensuing objectives (structural, load and vibration tests) completed by mid-1998. Prepared at Arlington for trials of CV-22 version's auxiliary fuel tanks and terrain-following/terrain-avoidance radar; modification work began 1 July 1999; flight status regained on 28 February 2000 and by 6 July had completed 427 hours and 214 sorties.

No. 8/164940: First flight 15 August 1997; to Patuxent River on 13 September 1997 and allocated to propulsion and systems testing and envelope expansion, including high altitude trials and external loads demonstrations in 1998. During latter, it set new unofficial world record in August 1998 by carrying 4,536 kg (10,000 lb) external load at 220 kt (407 km/h; 253 mph). By 6 July 2000, had completed 489 hours of flight testing in 276 sorties.

No. 9/164941: First flight 17 July 1997; to Patuxent River 30 October 1997; allocated to validation of FLIR, navigation and other mission systems, as well as government technical evaluation. Had achieved total of 318 flight hours in 140 sorties, when arrived at Arlington on 7 June 1999 to be remanufactured as prototype for CV-22 special missions version, at cost saving of US$50 million; work scheduled for completion in mid-2000, with aircraft due to have joined 418th Flight Test Squadron at Edwards AFB, California, on 18 September 2000 for 18-month trials programme.

No. 10/164942: Wing and fuselage mating accomplished at Fort Worth in September 1996. To Patuxent River 15 February 1998, following first flight on

V-22 Osprey production version flight deck　　　*2001*/0093859

15 January; underwent modification in mid-1998, before participating (with No. 9) in operational evaluation during September and October 1998; performed sea trials in USS *Saipan* in January-February and USS *Tortuga* and *Saipan* in August-September 1999. By 6 July 2000, had completed 450 hours of flight testing in 198 sorties.

CURRENT VERSIONS (general): **MV-22B:** Basic US Marine Corps transport; original requirement for 552 (now 360), to replace CH-46 Sea Knight and CH-53 Sea Stallion. First delivery (to Patuxent River via New River) in late May 1999 with three more handed over to test unit HMX-1 by January 2000; IOC to follow in August 2001. First unit to be VMMT-204 at New River, North Carolina, which officially redesignated (from HMT-204) on 10 June 1999; received four opeval aircraft from HMX-1 at end of July 2000 and expected to have become operational with 12 MV-22Bs in January 2000; will function as fleet replacement squadron and also train USAF pilots; first deployable squadron expected to be VMM-264, also at New River. USMC eventually plans to have 18 regular and four Reserve MV-22 squadrons (each with 12 aircraft) by 2013, plus training unit already mentioned.

HV-22B: US Navy combat search and rescue (CSAR), special warfare and fleet logistics model. Requirement for 48 (originally 50); deliveries from FY10.

CV-22B: US Air Force long-range special missions aircraft to replace MH-53J helicopter and augment MC-130 (Hercules) with Air Force Special Operations Command (AFSOC). Original requirement for 80 reduced to 55, then 50; should carry 12 troops or 1,306 kg (2,880 lb) internal cargo over 520 n mile (964 km; 599 mile) radius at 250 kt (463 km/h; 288 mph), with ability to hover OGE at 1,220 m (4,000 ft) at 35°C (95°F). EMD go-ahead authorised in January 1997, with award of US$490 million contract; critical design review completed in mid-December 1998. Flight testing will use remanufactured first and third EMD Ospreys at Edwards AFB, California, and is scheduled for completion in 2002. First five due for delivery in 2003, to 58th Special Operations Wing at Kirtland AFB, New Mexico; IOC of first AFSOC squadron at Hurlburt Field, Florida in September 2004. Procurement scheduled for FY01 to FY07, with final delivery in 2009.

Initial eight Lot 5/6 CV-22Bs to 'baseline' Block 0 configuration, but will have provisions for Block 10 upgrades, including dual digital map display capability, AN/AVR-2A laser detector, directional IR countermeasures (DIRCM), second AN/ALE-47 dispenser unit and hoist control from cockpit; subsequent lots will incorporate Block 10 improvements, although DIRCM lasers will not be installed until Lot 7. Block 10 perceived as first phase of three-stage upgrade programme for CV-22B, with testing of new features to be accomplished by third EMD Osprey in 2003.

V-22 also now under consideration for USAF combat search and rescue (CSAR) mission as potential replacement for Sikorsky HH-60G; response to USAF request for information submitted on 3 February 2000, with analysis of alternatives due to be completed by March 2001 and procurement following from 2004. V-22 in competition with five helicopter types (HH-60, H-53, S-92, NH90 and EH 101).

V-22 PROCUREMENT

FY	Lot	MV-22	CV-22
97	1 (LRIP)	5	
98	2 (LRIP)	7	
99	3 (LRIP)	7	
00	4 (LRIP)	11	
01	5	16	4
Total		46	4

Note: Overall procurement programme comprises 360 MV-22, 50 CV-22 and 48 HV-22

US Army: Original requirement for 231 V-22s, based on USMC transport, withdrawn. Documented requirement remains for V-22 in medevac, special operations and combat assault support roles.

CUSTOMERS: See above. Initial increment of production funding in FY96 budget, when US$48 million requested. First five production aircraft funded in FY97; further seven in FY98, seven in FY99 and 11 in FY00, to complete LRIP phase. Marketing in Europe has begun, with first objective being to determine interest of foreign governments with similar requirement. Bell Boeing has also proposed V-22 as suitable for UK's joint RAF/Navy Support Amphibious Battlefield Rotorcraft requirement, for which total of over 40 will be needed. In-service date currently expected to be 2008. V-22 could also satisfy Royal Navy's Future Organic Airborne Early Warning (FOAEW) requirement, although Northrop Grumman Hawkeye also a candidate.

COSTS: Estimated (1991) to complete full-scale development, US$2,750 million. Unit cost US$32.3 million (November 1997); FY97 budget included US$1.385 billion for five MV-22s; FY98 budget included US$661 million for seven MV-22s; FY99 budget included US$664.1 million for seven MV-22s. Separate US$490 million contract modification awarded by USN in

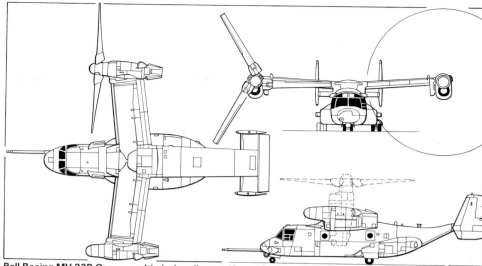

Bell Boeing MV-22B Osprey multimission tiltrotor *(James Goulding/Jane's)*

Bell Boeing V-22 Osprey in aeroplane mode *(Kevin Flynn/Boeing)*

December 1996 for EMD of CV-22 special operations version; acquisition costs of 50 CV-22s estimated at US$3.72 billion based on FY01-08 procurement plan. Unit cost of MV-22B reported as US$57 million (2000).

DESIGN FEATURES: Unconventional design, with proprotors and engines mounted at tips of wings. Fuselage optimised for transport, featuring upswept rear, with loading ramp and twin fins of moderate sweepback. High-mounted, constant-chord wings with slight forward sweep; unswept tailplane; prominent landing gear sponsons.

During vertical take-off, wing begins to produce lift and ailerons, elevators and rudders become effective at between 40 and 80 kt (74 and 148 km/h; 46 and 92 mph). At this point, rotary-wing controls are gradually phased out by the flight control system. At approximately 100 to 120 kt (185 to 222 km/h; 115 to 138 mph), wing is fully effective and cyclic pitch control of proprotors is locked out.

In conversion from aeroplane flight to hover, fuselage and wing are free to remain in level attitude, eliminating tendency for wing to stall as speed decreases. Rotor lift fully compensates for decrease in wing lift. Because of great variability between aircraft and nacelle attitude, conversion corridor (range of permissible airspeeds for each angle of nacelle tilt) is very wide (about 100 kt; 185 km/h; 115 mph).

Engines are connected by shaft through wing. Under dual engine operations, shaft transmits very little power, but if one engine is lost, half remaining power is transferred to opposite proprotor. In event of double engine failure, can maintain proprotor rpm while descending without power. Pilot has options of making wingborne or rotorborne descent.

Engines, transmission and proprotors tilt through 97° 30′ between forward flight and steepest approach gradient or tail-down hover; cross-shaft keeps both proprotors turning after engine loss. Three-blade, contrarotating proprotors have special high-twist tapered format blades with elastomeric bearings and powered folding mechanisms; separate swashplates produce respectively yaw and fore-and-aft translation in hover and sideways flight in level altitude; tip speed 202 m (662 ft)/s.

Wing-fold sequence from helicopter mode involves power-folding of blades parallel to wing leading-edge, tilting engine nacelles down to horizontal and rotating entire wing/engine/proprotor group clockwise on stainless steel carousel to lie over fuselage; entire procedure for stowage takes about 90 seconds and MV-22 occupies same amount of deck space as Sikorsky CH-53E.

FLYING CONTROLS: Three-lane fly-by-wire (Moog actuators) with automatic stabilisation, full autopilot and formation-flying modes. Conventional aeroplane stick, rudder pedals and throttle automatically function as in helicopter cyclic stick, yaw pedals and collective control. Automatic control of configuration change during transition and of transfer of control from aerodynamic surfaces to rotor-blade pitch changing; flaperons and ailerons droop during hover to reduce download on wing. Pilot also controls nacelle angle. Failure of one nacelle actuator automatically results in both nacelles reverting to helicopter mode for a vertical or run-on emergency landing.

Rotors have separate cyclic control swashplates for sideways flight and fore-and-aft control (symmetrical for forward and rearward flight and differential for yaw) in hover. Lateral attitude controlled in hover by differential rotor thrust. Integrated electronic cockpit with six electronic display screens; helicopter-style control columns rather than aileron wheels, but left-handed power levers move forward for full power in opposite sense to helicopter collective lever. Using nacelle control provides additional method of manoeuvring, completely independent of fuselage attitude or cyclic pitch. Minimum time to accomplish a full conversion from hover to aeroplane flight mode is 12 seconds.

STRUCTURE: Approximately 43 per cent of airframe is composites; main composites are Hercules IM-6 graphite/epoxy in wing and AS4 in fuselage and tail; proprotor blades of graphite/glass fibre; nacelle cowlings and pylon supports are of GFRP. Cabin has composites floor panels and aluminium window frames. Crew seats of boron carbide/polyethylene laminate.

Different design approach adopted for EMD and subsequent production aircraft, whereby integrated product teams (IPTs) were tasked with using best available material to reduce cost and weight and simultaneously improve quality. In consequence, fuselage is now a hybrid structure, with mainly aluminium frames and composite skins. Benefits of IPT process are a 1,200 kg (2,645 lb) reduction in weight and a reduction in cost of over 22 per cent, with part count falling by 36 per cent and fastener count by 34 per cent.

High-strength wing, torsion box made up from one-piece upper and lower skins with moulded ribs and bonded stringers; two-segment graphite single-slotted flaperons with titanium fittings; three-segment detachable leading-edge of aluminium alloy with Nomex honeycomb core. Wing locking and unlocking with Lucas Aerospace actuators; fuselage sponsons contain landing gear, air conditioning unit and fuel; tail unit of Hercules AS4 graphite/epoxy, built by Bell from first EMD aircraft onwards. Bell also contributes wing, nacelles, proprotor, ramp, overwing fairing and transmission systems and integrates engines. Boeing responsible for fuselage, landing gear, electric and hydraulic systems and integrates avionics.

LANDING GEAR: Dowty hydraulically actuated retractable tricycle type, with twin wheels and oleo-pneumatic shock-absorbers on each unit, Menasco Canada steerable nose unit. Main gear accommodated in sponsons on lower sides of centre-fuselage. Dowty Toronto two-stage shock-absorption in main gear is designed for landing impacts of up to 3.66 m (12 ft)/s normal, 4.48 m (14.7 ft)/s maximum, and has been drop tested to 7.32 m (24 ft)/s. Nosewheels retract rearward, mainwheels forward (was rearward on FSD aircraft only). Manual and nitrogen pressurised standby systems for emergency extension. Parker Bertea wheels and multidisc hydraulic carbon brakes. Main tyres 8.50-10 (12 ply) tubeless; nose tyres 18×5.7-8 (14 ply) tubeless.

Landing approach by Bell Boeing MV-22 Osprey

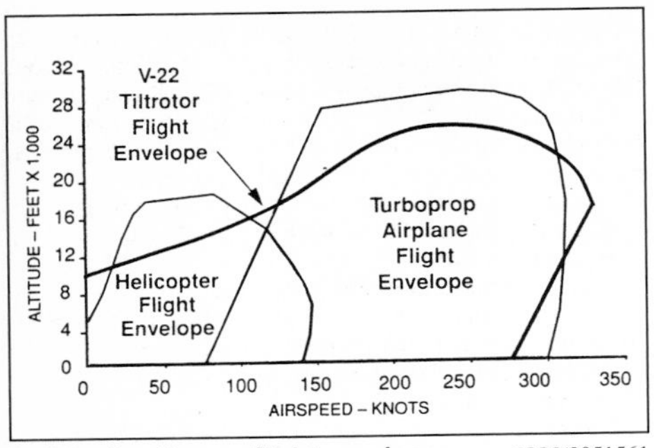

Bell Boeing Osprey flight envelope ***1999*/0051561**

POWER PLANT: Two Rolls-Royce T406-AD-400 (501-M80C) turboshafts, each with T-O and intermediate rating of 4,586 kW (6,150 shp) and maximum continuous rating of 4,392 kW (5,890 shp), installed in Bell-built hydraulically actuated tilting nacelles at wingtips and driving a three-blade proprotor incorporating substantial proportion of graphite/epoxy and glass fibre materials.

Engine output transferred to each rotor via proprotor gearbox which also drives tilt-axis gearbox, each of which absorbs 315 kW (423 shp) for electric and hydraulic pressure generation. Tilt-axis gearboxes connected by segmented shaft, driving (in fuselage) single mid-wing gearbox and APU. Transmission T-O rating 3,408 kW (4,570 shp) in MV-22; 3,706 kW (4,970 shp) in USN and USAF versions. OEI rating 4,415 kW (5,920 shp); rotor rpm remains constant. With total engine failure, aircraft generators can maintain essential systems on 67 per cent of rotor rpm. Transmission has 30 minute run dry capability.

Each nacelle has a Honeywell IR emission suppressor at rear. Air particle separator and Lucas inlet/spinner ice protection system for each engine. Lucas Aerospace FADEC for each engine, with analogue electronic back-up control. Pratt & Whitney originally named as second production source for engines, starting with production Lot 5.

Fuel capacity varies according to role; basic internal usable fuel (JP-5) in four crash-resistant, self-sealing (nitrogen pressurised) ILC Dover flexible fuel tanks: one 1,809 litre (478 US gallon; 398 Imp gallon) forward cell in each sponson and a 333 litre (88.0 US gallon; 73.3 Imp gallon) feed tank in each outer wing; basic fuel (MV-22) 4,285 litres (1,132 US gallons; 943 Imp gallons). Additional fuel for long range contained in 1,196 litre (316 US gallon; 263 Imp gallon) cell in rear of starboard sponson (MV-22 option 5,481 litres; 1,448 US gallons; 1,206 Imp gallons) and four 280 litre (74 US gallon; 61.6 Imp gallon) auxiliary cells in each wing leading-edge (CV-22 baseline 7,722 litres; 2,040 US gallons; 1,699 Imp gallons). Self deployment aided by up to four 2,328 litre (615 US gallon; 512 Imp gallon) auxiliary tanks in cabin. (Not all versions have all tanks.) Pressure refuelling point in starboard sponson leading-edge; gravity point in upper surface of each wing. Simmonds fuel management system. In-flight refuelling probe in lower starboard side of forward fuselage; standard item on CV-22 and available in kit form for MV-22. Fixed probe installed initially, but will be replaced by retractable type with effect from Lot 6; new probe likely to be provided by Flight Refuelling; will give wider field of fire for proposed gun turret and facilitate deck handling at sea.

ACCOMMODATION: Normal crew complement of pilot (in starboard seat), co-pilot and crew chief in USMC variant. USAF CV-22 will have third seat for flight engineer. Flight crew accommodated on Simula Inc crashworthy armoured seats capable of withstanding strikes from 0.30 in armour-piercing ammunition, 30 *g* forward and 14.5 *g* vertical decelerations. Flight deck has overhead and knee-level side transparencies in addition to large windscreen and main side windows, plus an overhead rearview mirror.

Main cabin can accommodate up to 24 combat-equipped troops, on inward-facing crashworthy foldaway seats, plus two gunners; up to 12 litters plus medical attendants; or a 9,070 kg (20,000 lb) cargo load with energy absorbing tiedowns. Cargo handling provisions include a 907 kg (2,000 lb) capacity cargo winch and pulley system and removable roller rails. Main cabin door at front on starboard side, top portion of which opens upward and inward, lower portion (with built-in steps) downward and outward. Full-width rear-loading ramp/door in underside of rear fuselage, operated by Parker Bertea hydraulic actuators. Emergency exit windows on port side; escape hatch in fuselage roof aft of wing.

SYSTEMS: Environmental control system, utilising engine bleed air; control unit in rear of port main landing gear sponson. Three hydraulic systems (two independent main systems and one standby), all at operating pressure of 345 bars (5,000 lb/sq in), with Parker Bertea reservoirs.

Electrical power supplied by two Leland 40 kVA constant frequency AC generators, two 50/80 kVA variable frequency DC generators (one driven by APU), rectifiers, and a 24 Ah lead-acid battery. Latter provides 20 minutes emergency flight power.

GE Aerospace triple redundant digital fly-by-wire flight control system, incorporating triple primary FCS (PFCS) and triple automatic FCS (AFCS) processors, and triple flight control computers (FCC) each linked to a MIL-STD-1553B databus; two PFCSs and one AFCS are fail-operational. FBW system signals hydraulic actuation of flaperons, elevator and rudders, controls aircraft transition between helicopter and aeroplane modes, and can be programmed for automatic management of airspeed, nacelle tilting and angle of attack. FCCs provide interfaces for swashplate, conversion actuator, flaperon, elevator, rudder and engine nacelle primary actuators, flight deck central drive, force feel, and nosewheel steering. Dual 1750A processors for PFCS and single 1750A for AFCS incorporated in each FCC. Non-redundant standby analogue computer (in development aircraft only) provides control of aircraft, including FADEC and pylon actuation, in the event of FBW system failure.

Hamilton Sundstrand Turbomach T-62T-46-2 224 kW (300 shp) APU, in rear portion of wing centre-section, provides power for mid-wing gearbox which, in turn, drives two electrical generators and an air compressor. Anti-icing of windscreens and engine air intakes; de-icing of proprotors and spinners. Clifton Precision combined oxygen (OBOGS) and nitrogen (OBIGGS) generating systems for cabin and fuel tank pressurisation respectively. Systron Donner pneumatic fire protection systems for engines, APU and wing dry bays.

AVIONICS: *Comms:* VHF/AM-FM, HF/SSB and (USAF only) UHF secure voice com; IFF. CV-22 will have four DCS2000 radios, combining AN/ARC-210 transceiver and KY-58 communications security encoder. Crash position indicator also to be installed on CV-22.

Radar: (USAF and USN only): Raytheon AN/APQ-174D terrain-following, terrain-avoidance multifunction radar in offset (to port) nose thimble for USN; advanced version, designated AN/APQ-186, with lower-altitude TF capability, in USAF aircraft.

Flight: AN/ARN-153(V) Tacan, AN/ARN-147 VOR/ILS, AHRS, AN/APN-194(V) radar altimeter, OA-8697/ARC UHF/VHF automatic direction-finder, lightweight inertial navigation system (LWINS), miniature airborne GPS receiver (MAGR) and digital map displays; Jet Inc ADI-350W standby attitude indicator; L-3 Communications data acquisition and storage system. Two Control Data AN/AYK-14 mission computers, with Boeing/IBM software. Low probability of intercept/detection radar altimeter may be added to CV-22 version at later date.

Instrumentation: Elbit/BAE Systems North America AN/AVS-7 pilots' night vision and integrated display system. Four full-colour. multifunction displays (MFDs) in cockpit provide pilots with primary flight symbology to control and navigate aircraft, plus video imagery such as FLIR and digital map data; these originally relied on CRTs, but replaced by EFW Inc flat panel active matrix liquid crystal displays (AMLCDs) on US Marine Corps MV-22Bs starting in FY99, followed by USN HV-22s. AMLCDs to be installed on USAF CV-22B from outset. Lighting compatible with night vision goggles. Additional AMLCD control display unit and engine indication and crew alerting system (CDU/EICAS) in centre of console.

Mission: Raytheon AN/AAQ-27 FLIR incorporating laser range-finder in undernose fairing, with wide and narrow fields of view. Additional equipment expected to satisfy unique mission requirements of special operations forces and Navy combat search and rescue.

Self-defence: Honeywell AN/AAR-47 missile warning system; AN/APR-39A radar warning system; IR warning system; BAE Systems North America AN/ALE-47 countermeasures dispenser system (CMDS). CV-22 for USAF to utilise ITT Avionics AN/ALQ-211 integrated RF countermeasures suite, incorporating radar warning receiver, ESM radar location and jammers and will also have AN/AVR-2A laser detector system and a second, forward-firing, AN/ALE-47 dispenser system. Dedicated electronic warfare display (DEWD) unit also under development for CV-22 by Meggitt Avionics.

EQUIPMENT: Provision for internally stowed rescue hoist over forward (starboard) cabin door on CV-22. USMC and USAF Ospreys have fast rope/rope ladders to facilitate insertion/extraction operations.

ARMAMENT: Provision stipulated in December 1994 for nose cannon; budget constraints resulted in this being deleted, but consideration again given in late 1998 to installing nose-mounted turreted defensive gun on MV-22 and this was evaluated in 2000, with General Dynamics Armament Systems 12.7 mm weapon selected in September 2000. Based on GAU-19/A, it consists of three-barrel weapon system weighing 209 kg (460 lb) and capable of firing 1,200 rounds per minute; magazine containing 750 rounds to be positioned under the floor and can be replenished in flight.

US Air Force also considering self-defence gun for CV-22. USMC plans to request funding in 2001, followed by flight trials in 2004, with weapon to be installed on new-build MV-22Bs and retrofitted to earlier aircraft.

DIMENSIONS, EXTERNAL:
Rotor diameter, each 11.58 m (38 ft 0 in)
Rotor blade chord: at root 0.90 m (2 ft 11½ in)
at tip 0.56 m (1 ft 10 in)
Wing span: excl nacelles 14.02 m (46 ft 0 in)
incl nacelles 15.52 m (50 ft 11 in)
Width: rotors turning 25.40 m (83 ft 4 in)
stowed 5.61 m (18 ft 5 in)
Wing chord, constant 2.54 m (8 ft 4 in)
Wing aspect ratio, excl nacelles 5.5
Distance between proprotor centres 14.25 m (46 ft 9 in)
Length: fuselage, excl probe 17.47 m (57 ft 4 in)
overall, wings stowed/blades folded 19.08 m (62 ft 7 in)
Height: over tailfins 5.38 m (17 ft 7¾ in)
wings stowed/blades folded 5.51 m (18 ft 1 in)
overall, nacelles vertical 6.63 m (21 ft 9 in)
Tail span, over fins 5.61 m (18 ft 5 in)
Wheel track (c/l of outer mainwheels) 4.62 m (15 ft 2 in)
Wheelbase 6.59 m (21 ft 7½ in)
Nacelle ground clearance, nacelles vertical 1.31 m (4 ft 3¾ in)
Proprotor ground clearance, nacelles vertical 6.35 m (20 ft 10 in)
Dorsal escape hatch: Length 1.02 m (3 ft 4 in)
Width 0.74 m (2 ft 5 in)

DIMENSIONS, INTERNAL:
Cabin: Length 7.37 m (24 ft 2 in)
Max width 1.80 m (5 ft 11 in)
Max height 1.83 m (6 ft 0 in)
Usable volume 24.3 m³ (858 cu ft)

AREAS:
Rotor discs, each 105.36 m² (1,134.1 sq ft)
Rotor blades: each 4.05 m² (43.58 sq ft)
total 24.30 m² (261.5 sq ft)
Wing, total incl flaperons and fuselage centre-section 35.49 m² (382.0 sq ft)
Flaperons, total 8.25 m² (88.80 sq ft)
Fins, total 21.63 m² (232.80 sq ft)
Rudders, total 3.27 m² (35.20 sq ft)
Tailplane 8.22 m² (88.50 sq ft)
Elevators, total 4.79 m² (51.54 sq ft)

WEIGHTS AND LOADINGS:
Weight empty 15,032 kg (33,140 lb)
Max fuel weight: MV-22 baseline 3,493 kg (7,700 lb)
MV-22 option 4,468 kg (9,850 lb)
CV-22 baseline 6,282 kg (13,850 lb)
CV-22 with cabin tanks 11,970 kg (26,390 lb)
Max internal payload (cargo) 9,072 kg (20,000 lb)
Cargo hook capacity: single 4,536 kg (10,000 lb)
two hooks (combined weight) 6,804 kg (15,000 lb)
Rescue hoist capacity 272 kg (600 lb)
Normal mission T-O weight: VTO 21,545 kg (47,500 lb)
STO 24,947 kg (55,000 lb)
Max VTO weight 23,981 kg (52,870 lb)
Max STO weight for self-ferry 27,442 kg (60,500 lb)
Max floor loading (cargo) 1,464 kg/m² (300 lb/sq ft)
Max disc loading: VTO 113.8 kg/m² (23.31 lb/sq ft)
STO 130.2 kg/m² (26.67 lb/sq ft)
Transmission loading at max T-O weight and power 3.55 kg/kW (5.84 lb/shp)

PERFORMANCE:
Max level speed at S/L 275 kt (509 km/h; 316 mph)
Max cruising speed: at S/L, helicopter mode 100 kt (185 km/h; 115 mph)
Max forward speed with max slung load 214 kt (396 km/h; 246 mph)
Max rate of climb at S/L: vertical 332 m (1,090 ft)/min
inclined 707 m (2,320 ft)/min
Service ceiling 7,925 m (26,000 ft)
Service ceiling, OEI 3,441 m (11,300 ft)
Hovering ceiling OGE 4,331 m (14,200 ft)
T-O run at normal mission STO weight less than 152 m (500 ft)
Range:
amphibious assault 515 n miles (953 km; 592 miles)
VTO with 4,536 kg (10,000 lb) payload 350+ n miles (648+ km; 403+ miles)
VTO with 2,721 kg (6,000 lb) payload 700+ n miles (1,296+ km; 806+ miles)
STO with 4,536 kg (10,000 lb) payload 950+ n miles (1,759+ km; 1,093+ miles)
STO at 27,442 kg (60,500 lb) self-ferry gross weight, no payload 2,100 n miles (3,892 km; 2,418 miles)
g limits +4/−1

UPDATED

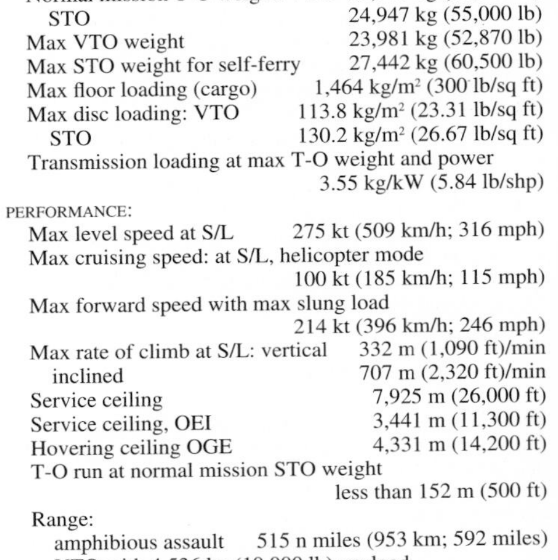

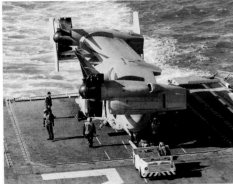

Bell Boeing V-22 Osprey folded for stowage
***2001*/0093861**

BOEING

THE BOEING COMPANY

7755 East Marginal Way South, Seattle, Washington 98108
Tel: (+1 206) 655 21 21
Fax: (+1 206) 655 11 77
Web: http://www.boeing.com
CHAIRMAN AND CEO: Philip M Condit
PRESIDENT AND COO: Harry C Stonecipher
CHIEF FINANCIAL OFFICER: Michael M Sears
VICE-PRESIDENT, COMMUNICATIONS AND INVESTOR RELATIONS: Larry Bishop

Company founded July 1916. Currently organised into four major business segments as detailed below. Net earnings in 1999 were US$2,309 million; revenues US$58 billion. Unchallenged as leading US aerospace firm, but has recently experienced difficulties with airliner production and is expected to reduce workforce by up to 10 per cent and consolidate laboratory and test facilities as well as fabrication centres during 2000-01. Deliveries of 563 commercial aircraft in 1998 increased to 620 in 1999, but fell to 489 in 2000.

Boeing and McDonnell Douglas merger under the Boeing name was completed on 4 August 1997, elevating Boeing to status of largest aerospace company in the world, with about 238,000 employees (reduced to about 180,000 by end of 2000) and business backlog exceeding US$100 billion. Further expansion of global interests occurred in 1997, with Boeing entering into alliances with AVIC and Taikoo Aircraft Engineering Co of China; with CSA Czech Airlines to establish joint venture company and acquire stake in Aero Vodochody; and with Instytut Lotnictwa of Poland for potential future collaboration on advanced technologies. In 1998, similar agreements and arrangements concluded with Israel Military Industries concerning military and civil programmes and with Hellenic Aerospace Industry to promote F-15H as new fighter for Greek Air Force, though latter failed to secure order. In January 1999, Boeing reached agreement with Netherlands-owned company MD Helicopters (which see in US section) on previously announced disposal of light civil helicopter range.

Operating components of The Boeing Company include:

Military Aircraft and Missile Systems Group
Follows this entry
Boeing Commercial Airplanes Group
Follows Military Aircraft and Missile Systems Group
Boeing Space and Communications Group
Follows Boeing Commercial Airplanes Group
Shared Services Group
Follows Boeing Space and Communications Group

UPDATED

MILITARY AIRCRAFT AND MISSILE SYSTEMS GROUP

PO Box 516, St Louis, Missouri 63166-0516
Tel: (+1 314) 232 28 00
Fax: (+1 314) 234 82 96
PRESIDENT: Gerald E Daniels
MEDIA CONTACTS:
GENERAL MANAGER: Doug Kennett, *Tel:* (+1 314) 234 35 00
MEDIA RELATIONS: Jo Anne Davis, *Tel:* (+1 314) 233 89 57
MEDIA/INTERNATIONAL PROGRAMMES: Mary Ann Brett, *Tel:* (+1 314) 234 71 11
INTERNATIONAL PROGRAMMES: Jim Schlueter, *Tel:* (+1 314) 234 21 49
DIRECTOR OF COMMUNICATIONS, SOUTHERN CALIFORNIA: Richard L Fuller, *Tel:* (+1 562) 496 51 95

Created as part of the major reorganisation that accompanied the merger between Boeing and McDonnell Douglas, Military Aircraft and Missile Systems Group has overall responsibility for military business activities. Management of aircraft and helicopter programmes as detailed below reflects the military customer base, but several subordinate organisations exist to oversee other key areas. Foremost among these is **Weapons Programs**, with its major assembly centre at St Charles, Missouri; this is responsible for production of tactical missile systems (including the AGM-84 Harpoon and AGM-84E SLAM) as well as continued support of strategic missile systems (notably the AGM-129 Advanced Cruise Missile and BGM-109 Tomahawk).

Other elements include **Aerospace Support** (service and logistics support, plus modifications) and the **Aerospace Operations Division** (support for Space Shuttle and other NASA/USAF projects).

Group earnings in 1999 were US$1,193 million; revenues US$12.2 billion.

Major elements of Military Aircraft and Missile Systems Group are as follows:

Phantom Works

PRESIDENT: David O Swain

Tasked with improving Boeing's competitive position through use of innovative technologies, improved processes and creation of new products, the Phantom Works was originally established by McDonnell Douglas at St Louis and currently has its headquarters at Seattle, with facilities in Arizona, California, Missouri, Pennsylvania and Washington. Technologies and expertise being expanded in support of company-wide activities, including commercial transport aircraft, and this division has personnel at every major Boeing facility.

Joint Strike Fighter Program

MANAGER: Frank Statkus (Seattle)
MEDIA CONTACT: Randy Harrison, *Tel:* (+1 206) 655 86 55

Operating as separate entity and reporting directly to Military Aircraft and Missile Systems President, this reflects importance of Joint Strike Fighter project to US DoD and future of Boeing.

USAF Fighter and Bomber Programs

MANAGER: Mike Marks (St Louis)
MEDIA CONTACT: Todd Blecher, *Tel:* (+1 314) 233 02 06

Responsible for continued production at St Louis of F-15 Eagle for USAF. Boeing participation in F-22 programme.

US Navy and Marine Corps Programs

MANAGER: Pat Finneran (St Louis)
MEDIA CONTACT: Ellen LeMond-Holman, *Tel:* (+1 314) 232 64 96

Responsible for production at St Louis of F/A-18 Super Hornet for US Navy, as well as AV-8B Harrier II Plus rebuild, T-45 Goshawk and Boeing participation in V-22 Osprey.

USAF Airlift and Tanker Programs

VICE-PRESIDENT: David Spong (Long Beach)
MANAGER, MD-17 PROGRAM: David Bowman
MEDIA CONTACT: Larry Whitley, *Tel:* (+1 562) 496 51 97

Production of C-17 Globemaster III strategic airlift aircraft for USAF continues at Long Beach. Also responsible for development of civil BC-17X equivalent and of other derivatives such as proposed tanker version. Marketing of military tanker versions of DC-10 and Boeing 767.

US Army Programs

SITE MANAGER, BOEING PHILADELPHIA: Charles Vehlow
MEDIA CONTACT, PHILADELPHIA: Madelyn Bush, *Tel:* (+1 610) 591 28 64
SITE MANAGER, BOEING MESA: Marty Stieglitz
MEDIA CONTACT, MESA: Hal Klopper, *Tel:* (+1 480) 891 55 19

Philadelphia factory produces and upgrades CH-47 Chinook for US Army and export customers and is also partnered by Bell on V-22 Osprey and by Sikorsky on RAH-66 Comanche projects. Mesa factory responsible for building and remanufacturing AH-64D Apache Longbow for US Army and export.

Aerospace Support

MANAGER: Jim Restelli (St Louis)
MEDIA CONTACT: Paul Guse, *Tel:* (+1 314) 232 15 20

Responsible for Boeing Aerospace Support Centers at San Antonio, Texas, which perform maintenance and modification work on KC-10, C-17, KC-135, DC-10 and other types, and at Cecil, Florida, which is responsible for modification work on F/A-18 Hornet. Also involved in T-38 avionics upgrade programme, Aero L 159 advanced avionics programme, and flight and maintenance training systems.

UPDATED

BOEING DRAGONFLY CANARD ROTOR/WING

The Canard Rotor/Wing (CR/W) began as a McDonnell Douglas Helicopters project for a VTOL reconnaissance and surveillance UAV in 1992; concept based on NASA-funded studies into high-speed rotorcraft and earlier (Hughes) rotor/wing studies. Design of the CR/W undertaken by Phantom Works personnel at Mesa, Arizona, under a March 1998 DARPA contract. US Marine Corps is reported to have expressed interest in a manned version armed with missiles and other weaponry for use in the escort role.

As currently envisaged, the CR/W will possess VTOL attributes of the helicopter and the capability to fly like a conventional fixed-wing aircraft, with the addition of foreplanes being a key factor in bestowing this potential. Power will be provided by a single turbofan engine. In helicopter mode, exhaust gases will pass through channels in the two-blade rotor before exiting through vents in the blade tips. In aeroplane mode, as speed increases, exhaust efflux progressively transfers from driving the rotor to a nozzle at the rear, thus providing forward thrust.

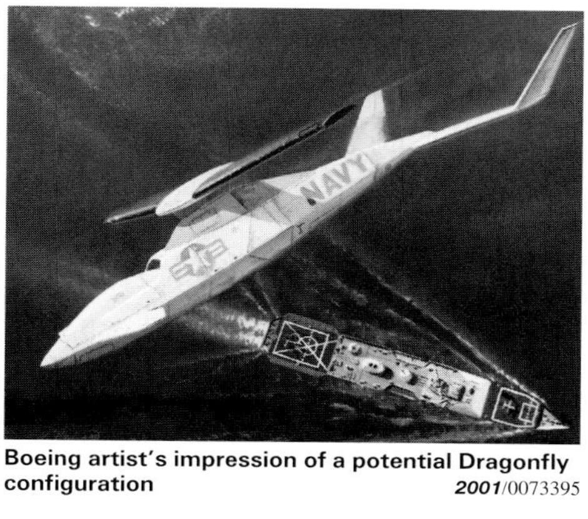

Boeing artist's impression of a potential Dragonfly configuration ***2001***/0073395

With no tail rotor, directional control in helicopter mode will rely on what Boeing calls 'reaction drive', in which exhaust gases are diverted to nozzles on both sides of the tail; for aeroplane mode, conventional control surfaces will be provided, although consideration is being given to employing reaction jets for directional control in both modes of flight.

Following vertical take-off as a helicopter, the CR/W will accelerate to about 122 kt (225 km/h; 140 mph) before making the transition to aeroplane. This will be accomplished with the assistance of flap surfaces on the foreplanes and the aft-mounted wings. These will increase the amount of lift that is generated by the wings and reduce rotor loadings, whereupon the rotor will be slowed to a stop before being locked in position across the fuselage to act as an extra lifting surface. The flaps will then be retracted, with the three lifting surfaces sharing the load. For landing, the process will be reversed, enabling the CR/W to operate safely from confined areas, such as a carrier deck or helicopter platform. In conventional aeroplane mode, maximum speed is predicted to exceed 375 kt (695 km/h; 432 mph).

Initial research to include studies into ways and means of using exhaust gases to perform multiple functions, with US$37 million allocated jointly by Boeing and DARPA to pay for a three-year research effort that will involve manufacture and flight trials of two technology demonstrator aircraft in a programme known as Dragonfly. Boeing's Mesa facility is leading the project and will have responsibility for assembly; further support is provided by the sites in Philadelphia and St Louis.

Initial technology demonstrator to be a subscale, unmanned aerial vehicle (UAV), which will fly in 2001. Likely to measure 5.38 m (17 ft 8 in) in length and weigh about 590 kg (1,300 lb), this will be powered by a 3.11 kN (700 lb st) Williams F112 turbofan previously used on the X-36 project, and have a 3.65 m (12 ft) diameter rotor. Control systems will be based on those of the X-36, other spin-offs from this project possibly including flight software and hardware in order to reduce development costs. Any subsequent production CR/W is likely to use advanced composites materials and could include manned and unmanned versions; Boeing studies include a 1,100 kg (2,425 lb) maritime UAV and a 10,000 to 11,000 kg (22,046 to 24,250 lb) manned aircraft that could function as an armed escort fighter in support of the MV-22B Osprey.

UPDATED

BOEING X-32 JOINT STRIKE FIGHTER

TYPE: Multirole fighter.

PROGRAMME: Boeing among five contenders in ARPA (US Advanced Research Projects Agency) contest, late 1992; initially rejected, but Boeing continued private funding of own CALF (Common Affordable Lightweight Fighter) design, hoping to re-enter competition in 1994; technology-sharing agreement with ARPA, May 1993; one-tenth scale model testing begun, July 1993; early Boeing partners comprised Pratt & Whitney for single F119 derivative engine and exhaust nozzle and Rolls-Royce for direct lift system and attitude control system. Subsequently, on 30 September 1998, Boeing announced establishment of Joint Strike Fighter Industry Team (also known as 'One Team'); other key partners of original group and their responsibilities comprised Flight Refuelling Ltd (fuel system), Marconi (now BAE Systems; vehicle management system and cockpit displays including HUD), Messier-Dowty Ltd (main and nose landing gear) and Raytheon (select mission systems). Team members added since then include Goodrich (fuel system); Dowty Aerospace (flight control actuators); EDO (weapons bay swing-arm system); Fokker (airframe structural details and wire bundles); Hamilton Sundstrand (engine subsystems); Hexcel (composites raw materials); BAE (fuel system); Martin-Baker (ejector seat); Sanders (EW systems) and TRW (CNI system).

ARPA realigned STOVL Strike Fighter (SSF) programme in January 1994 on Congressional instructions,

USA: AIRCRAFT—BOEING

Boeing X-32A on its maiden flight, 18 September 2000

allowing Boeing's direct-lift design to compete against remote fans; Boeing awarded US$32 million 26 month research agreement on 25 March 1994 to perform operational testing of a STOVL/CTOL CALF using direct-lift technology. Boeing also secured a further US$32 million concept definition contract in January 1995 for Joint Advanced Strike Technology (JAST) project, following Congress-directed merger of CALF with separate JAST programme, retaining latter title. Dassault of France was subcontractor to Boeing, tasked with study of multiservice common airframe concepts.

JAST subsequently renamed JSF (Joint Strike Fighter) (which see under US Navy heading for further details) in late 1995. Numerous subscale wind-tunnel tests completed in 1994; tests of jet effects undertaken on models in NASA and Rolls-Royce facilities, January 1995; further trials at NASA Langley and NASA Ames into 1996. Testing began in June 1995 at Tulalip (near Seattle) of 94 per cent scale YF119-powered model of Boeing design, with results verifying design predictions and trials with small-scale models.

On 16 November 1996, Boeing selected as one of two contenders to progress to next stage of JSF programme (Lockheed Martin being the other) and secured a US$661.8 million contract for the 51 month Concept Demonstration Program (CDP). This involves construction and flight testing of two X-32 aircraft as well as ground test demonstrations of systems and subsystems for the Preferred Weapon System Concept (PWSC). First aircraft is USAF X-32A version, later to be converted to USN configuration; second is STOVL X-32B version. Boeing announced on 10 September 1997 that Initial Design Review successfully passed; fabrication of mid-fuselage components began in fourth quarter of 1997, with assembly of modules beginning at St Louis on 8 July 1998 when first parts for X-32A forebody module were loaded. Assembly of mid-fuselage began at Palmdale on 21 August 1998. X-32B forebody assembly began on 23 September 1998. Final assembly of X-32A started at Palmdale on 1 April 1999, with X-32B following in mid-1999. Structural assembly of X-32A completed in July 1999; proof load testing completed in early October 1999; with formal unveiling of X-32A and X-32B at Palmdale on 14 December.

Maiden flight of X-32A **first prototype** then expected by April 2000, but initial phase of engine runs not completed until late in month, with low- and medium-speed taxi tests following in mid-May, while high-speed taxi tests were not concluded until 15 September. These cleared way for 20 minute first flight from Palmdale to Edwards AFB on 18 September, marking start of five month test programme. Second flight took place on 23 September and lasted 50 minutes; further eight sorties made by 13 October, when X-32A had logged over seven hours; subsequently used for simulated carrier landing tests between 15 November and 2 December. At end of CV testing, aircraft had completed 33 flights. CTOL testing began immediately thereafter, and early achievement, on 49th sortie, 21 December 2000, was aerial refuelling from KC-10A, followed two days later by speed of Mach 1. Flight with loaded weapons bay demonstrated 26 January 2001.

Trials of first prototype ended on 3 February 2001, after 66 sorties and 50.4 hours flown by six pilots.

Second prototype X-32B originally scheduled to fly for first time in early 2000, but also delayed, with revised target of first quarter of 2001, following initial engine runs in September 2000 and completion of structural mode interaction testing in early December. Taxying trials began early January 2001.

Boeing 737 Avionics Flying Laboratory (AFL) first flew 26 March 1999 and delivered to Seattle for equipment installation; fitted with development Raytheon synthetic aperture radar and other sensors in modified nose and also has representative cockpit in cabin; began seven-month test programme on 5 December 1999 and successfully demonstrated avionics multisensor fusion by May 2000, when it had completed 24 of planned 50 mission test programme. In early June, same aircraft used for weapon system capability demonstration during live-fire exercise at White Sands Missile Range by gathering, processing and disseminating targeting data that allowed an F-15E to deliver a GPS-aided JDAM. Validation of radar, antenna and stealth attributes began April 2000, using full-scale Supportable Electromagnetics Test Aircraft (SETA) model, pole-mounted on test range in Seattle.

Decision also reached in 1997 that St Louis will be responsible for design and manufacture of X-32 forebody module; mid- and aft-body modules, plus wing, to be produced at Seattle. If Boeing candidate emerges as eventual winner in October 2001, final assembly line to be located at St Louis, although in early 2000, US Department of Defense was considering award of second production line to whichever contender lost JSF contract; this proposal rejected by both JSF contenders and has apparently been dropped. During CDP, Boeing will also fully define the multirole, multiservice JSF PWSC before its low technical risk entry into EMD in late 2001.

In February 1999, Boeing announced revised configuration for X-32 production version. First two demonstrators reflect Configuration 372, with modified delta planform and twin fins; Configuration 373, stated to be within 10 per cent of the presumed eventual production shape, adopts smaller, swept wing with horizontal tail surfaces in addition to two outward-canted fins. Considerations in the change included weight-saving and improved pitch control for aircraft carrier operations; additionally, engine air inlet lip is swept back, instead of forward, and cockpit canopy shape modified. Features such as leading-edge sweep, wing span and propulsion systems remain unchanged. Further redesign revealed in fourth quarter of 1999, with Configuration 374 incorporating backward swept, straight trailing-edge from root to tip, plus new intakes adjacent to cockpit canopy to provide cooling air for environmental control system; latter replaced inlets within main engine air inlet duct. This configuration within 2 per cent of definitive JSF production shape.

Final configuration (375) still to be determined, but will feature weight-saving measures in order to comply with 1,814 kg (4,000 lb) payload 'bring-back' requirement stipulated for STOVL variant.

CURRENT VERSIONS: **X-32A:** CTOL derivative to demonstrate characteristics of USAF/USN aircraft. First flight 18 September 2000, from Palmdale to Edwards AFB, California.

X-32B: STOVL derivative to demonstrate characteristics of USMC and UK Royal Navy aircraft.

DESIGN FEATURES: Compact, modular combat aircraft; empty weight approximately 9,070 kg (20,000 lb); VTOL penalty only 5 per cent of empty weight. One-piece blended wing with twin vertical tails having inset rudders, and all-moving horizontal tail surfaces; flaperons on inboard two-thirds of wing trailing-edge. Wing and tailplane leading-edge sweepback approximately 55°; trailing-edges 20° and −20°, respectively. Fins each canted outwards at 28°; leading-edge sweepback 42°. Wing may incorporate Boeing-developed cellular design to reduce vulnerability to ground fire and SAMs. Single jet engine; STOVL version has two directional, ventral exhaust nozzles at CG for vertical or short take-offs and transition to wingborne flight; large chin inlet translates to permit increased area for STOVL operation; high position of wing minimises propulsion-induced lift loss commonly experienced with STOVL. Weapons bay suspension and release equipment by EDO Corporation.

Variants of production aircraft available for conventional operation from land (USAF); catapult-and-hook aircraft carrier operation (USN); and full STOVL (Marine Corps and Royal Navy); Boeing estimates that the three basic variants will have 90 per cent commonality of parts. Consideration being given to covering one of the demonstration aircraft with non-hydroscopic appliqué finish developed by 3M and Boeing; this material is lighter than conventional paint and weight remains constant throughout aircraft life.

LANDING GEAR: Messier-Dowty is prime contractor for landing gear to be used on production aircraft. STOVL and CTOL versions to have single wheels on main and nose landing gear; CV version for US Navy to have twin-wheel nose gear including catapult launch bar and holdback.

POWER PLANT: One Pratt & Whitney JSF119-614 turbofan engine, derived from F119 as used by F-22 Raptor, installed on X-32 prototypes; -614C for CTOL version, while STOVL -614S incorporates Rolls-Royce lift module and spool duct. Fuel system will include retractable AAR probe by Flight Refuelling Ltd of UK.

ACCOMMODATION: Pilot only, on derivative of Martin-Baker Mk 16 zero-zero ejector seat, in production aircraft.

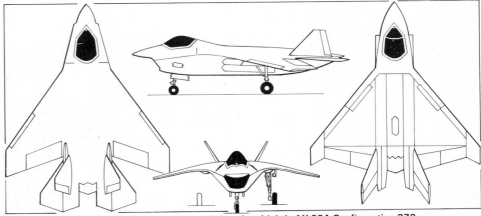

Boeing JSF Configuration 374, with additional plan view (right) of X-32A Configuration 372 *(James Goulding/Jane's)*

Boeing X-32A on landing approach
2001/0073504

ARMAMENT: Weapons payload 6,350 to 8,165 kg (14,000 to 18,000 lb) accommodated internally and externally; internal bays in fuselage sides for missiles (including AIM-120 AMRAAM) and bombs; also four wing stations able to carry in excess of 4,536 kg (10,000 lb) of bombs or missiles, with two being plumbed for auxiliary fuel tanks. Advanced version of Mauser BK 27 cannon, subject of teaming arrangement concluded by Boeing, Mauser and Primex Technologies on 29 July 1998, selected in May 1999; 27 mm weapon to be built at Mesa, Arizona, by Boeing under licence from Mauser, with Primex Technologies responsible for production of ammunition, including new high-explosive, dual-purpose cartridge.

DIMENSIONS, EXTERNAL (approximate):
Wing span: CTOL variant	11.0 m (36 ft)
STOVL variant	9.1 m (30 ft)
Length overall: CTOL variant	14.3 m (47 ft)
STOVL variant	14.0 m (46 ft)
Height overall	4.0 m (13 ft)
Tailplane span	6.4 m (21 ft)

AREAS (approximate):
Wings, gross	50 m² (540 sq ft)

UPDATED

BOEING F-15C/D EAGLE
Israel Defence Force names: F-15A/B Baz (Eagle); F-15C/D Akef (Buzzard)

US production of interceptor F-15A/B/C/D Eagles by McDonnell Douglas ended in 1992, and F-15C/D manufacture (in F-15J/DJ form) completed in Japan by Mitsubishi (which see) in December 1999. Refer also to production tables with F-15E entry which follows. Although out of production in USA, F-15C/D is still premier air superiority asset, and some of the 400 aircraft in the USAF inventory (as well as some of the 100-plus F-15A/Bs with the Air National Guard) are being retrofitted with the AN/APG-63(V)1 radar. Flight trials of this upgraded system began on 18 July 1997, following award of a US$175 million engineering and manufacturing development (EMD) contract to McDonnell Douglas in October 1994. Subsequently, in September 1997, a further US$23.05 million contract was awarded for low-rate initial production (LRIP) of the AN/APG-63(V)1; this followed in October 1999 by US$79.7 million contract to launch full-rate production for USAF. At present, Raytheon anticipates delivering approximately 162 radar systems by 2005, but could produce further units for installation on F-15s operated by Israel, Japan and Saudi Arabia. In addition, a new advanced display core processor is under development and may also be selected for the F-15. Variant of radar designated AN/APG-63(V)2 incorporating active electronically scanned array (AESA), installed on 18 aircraft of 3rd Wing of Elmendorf AFB, Alaska, from late 2000; these are first fighters to enter service with AESA and provide USAF with operational experience of system and technology in readiness for deployment of F-22A Raptor.

Multifunctional Information Distribution System Fighter Data Link (MIDS FDL) Common Configuration Terminal now being retrofitted to F-15C/D; similar equipment under consideration for F-15E. Data Link Solutions awarded contract in 1997 for production, test, certification and delivery of six MIDS FDL units; option for 46 more units taken up in September 1998.

Boeing has proposed so-called **F-15R** (renewed) version for Eastern European air arms such as Czech Republic, Hungary and Poland; these would be refurbished F-15As and F-15Bs taken from storage and could be delivered with at least 6,500 hour life. Refurbishment proposals include installation of new radar to customer specification, re-engining with F100-PW-220E and updating avionics to F-15C MISP (multistage improvement programme) standard.

In 2000, Boeing working on design of **F-15C+** reduced-cost fighter, employing lower-priced materials and techniques; study shows Air National Guard could save between US$6.49 billion and US$8.89 billion over 20 years if 115 F-15C+s purchased new instead of similar number of old F-15Cs handed down by USAF and upgraded. Proposed next step is US$286 million programme to design, build and flight test two F-15C+s.

UPDATED

BOEING F-15E EAGLE
Israel Defence Force name: Ra'am (Thunder)

TYPE: Multirole fighter.

PROGRAMME: F-15 initially employed as air superiority fighter, having first flown on 27 July 1972. Demonstration of industry-funded Strike Eagle prototype (71-0291) modified from F-15B, including accurate blind weapons delivery, completed at Edwards AFB and Eglin AFB during 1982; product improvements for the F-15E were tested on four Eagles, among which were the Strike Eagle prototype, an F-15C and an F-15D, between November 1982 and April 1983, including first take-off at 34,019 kg (75,000 lb), 3,175 kg (7,000 lb) more than F-15C with conformal tanks; new weight included conformal tanks, three other external tanks and eight 500 lb Mk 82 bombs; 16 different stores configurations tested, including 2,000 lb Mk 84 bombs, and BDU-38 and CBU-58 weapons delivered visually and by radar.

Full programme go-ahead announced 24 February 1984; first flight of first production F-15E (86-0183) 11 December 1986; first delivery to Luke AFB, Arizona, 12 April 1988; first delivery 29 December 1988 to 4th Wing at Seymour Johnson AFB, North Carolina. Small number of F-15Es used for trials with 3246th Test Wing at Eglin AFB, Florida, and 6510th TW (412th TW from October 1992) at Edwards AFB, California.

Trials include 87-0180 with GE F110-GE-129 engines in place of F100s; US$5 million allocated in 1995 to allow USAF to complete test project, with USAF undertaking field evaluation of two re-engined F-15Es at Nellis AFB, Nevada. These began flying at Nellis on 11 April 1997; test programme ended in January 1999 after 1,915 hours and demonstrated 99.8 per cent reliability. GE also funding upgraded F110-GE-129EFE enhanced fighter engine, which will offer either a 50 per cent increase in engine life or significant additional thrust in 151 to 160 kN (33,950 to 35,970 lb st) region; no plans to adopt any variant of F110 at present, but could be retrofitted or offered as alternative to F100 on future aircraft. P&W F100-PW-229 first flown in F-15E of 6510th TW on 2 May 1990.

CURRENT VERSIONS: **F-15E**: Basic version, *as detailed*.

F-15F: Proposed single-seat version, optimised for air combat; not built.

F-15H: Proposed export version, lacking specialised air-to-ground capability; supplanted by F-15S. Designation subsequently allocated to proposed version for Greece, which eventually selected F-16 Fighting Falcon.

F-15I Ra'am: Israeli export version of F-15E; selected November 1993; confirmed 27 January 1994; 21 ordered 12 May 1994; option on four more converted to firm order in November 1995. First flight of initial (unpainted) aircraft on 12 September 1997; this was subject of formal roll-out and handover ceremony at St Louis on 6 November 1997, with first pair of aircraft to be delivered leaving St Louis on 16 January 1998 and arriving in Israel three days later; total of 16 delivered during 1998, with final nine aircraft following during 1999. Operating unit No. 69 Squadron. Tactical electronic warfare system deleted; being replaced by Israeli-built SPS-2100 integrated system including active jamming, radar and missile warning, and dispenser subsystems.

Otherwise identical to USAF F-15E, with F100-PW-229 engines, LANTIRN pods, full capability AN/APG-70 radar, Kaiser holographic HUD, Litton ring laser INS and VHSIC central computer. Associated equipment includes four Sanders mission planning subsystems and one Sanders common mapping production system (CMPS) to assist ground planning, briefing and debriefing activities at total cost of US$6.2 million. F-15I includes significant number of co-produced components, such as airframe and wing subassemblies, heat exchangers and weapons pylons.

F-15K: Proposed version to satisfy South Korea's F-X fighter requirement; preliminary bids submitted in June 2000, with final selection in second half of 2001. Initial batch of 40 aircraft needed to replace F-4D/E Phantom, with IOC anticipated in 2004-05 and possibility of follow-on purchase of 40 more. F-15K incorporates AN/APG-63(V)1 radar, expanded weapons capability (including JDAM and AGM-130) and improved environmental control system, plus infra-red search and track system, VSI Joint Helmet-Mounted Cueing System and revised cockpit with seven 152 × 152 mm (6 × 6 in) active matrix LCDs. In competition with Dassault Rafale, Eurofighter Typhoon and Sukhoi Su-35. In October 2000, South Korean personnel undertook evaluation project involving three F-15Es provided on lease to Boeing by 3rd Wing at Elmendorf AFB, Alaska.

F-15L: Proposed less costly version offered to Israel by Boeing in unsuccessful, last-minute, attempt to secure order for new combat aircraft; F-16I eventually selected.

F-15S: Saudi Arabian export version of F-15E, lacking some air-to-air and air-to-ground capabilities; Saudi Arabian request for 72 aircraft approved by US government in December 1992; initially designated **F-15XP**; first funds assigned by US government on 23 December 1992; contract signature by Saudi government May 1993; planned delivery rate halved, early 1994, to one per month, beginning 1995. First F-15S flown 19 June 1995; official roll-out and handover 12 September 1995. Initial aircraft briefly retained in USA for trials; first two examples delivered to Saudi Arabia in November 1995

F-15E Eagle launching a training version of AGM-130 powered standoff bomb
2000/0103982

USA: AIRCRAFT—BOEING

Boeing F-15I Ra'am of the Israel Defence Force transiting the UK on delivery *(Paul Jackson/Jane's)*

were second and third built; total of 49 received by end of 1998 and 70 by end of 1999, with final two following in late July 2000. In April 2000, it was reported that Saudi Arabia was considering the purchase of up to 24 more aircraft, with Saudi military delegation expected to visit USA in June 2000 to discuss this and other matters.

Saudi versions comprise 24 optimised for air-to-air missions and 48 optimised for air-to-ground; AN/APG-70 radar. Despite earlier planned restrictions, manufacturer states that aircraft delivered with full F-15E capability, plus conformal fuel tanks and associated tangential stores attachments. Armament includes AGM-65D/G Maverick, AIM-9M and AIM-9S Sidewinder missiles, CBU-87 submunitions dispenser and GBU-10/12 bombs. Saudi programme includes about 154 Pratt & Whitney F100-PW-229 engines.

F-15SE: Proposed single-seat version of F-15E for USAF, for which pricing data supplied to Department of Defense as part of Quadrennial Defense Review. Not proceeded with.

F-15U: Version conceived to satisfy United Arab Emirates requirement for 20 to 80 long-range interdictor aircraft, in which it was competing against Lockheed Martin F-16 (selected), Dassault Rafale, Eurofighter Typhoon and Sukhoi Su-30MK. **F-15U Plus** proposal anticipated extended range, with additional 2,570 kg (5,665 lb) of fuel in thicker clipped-delta, 50° leading-edge sweep wing; more stores stations and internally situated IR navigation and targeting sensor suite in lieu of LANTIRN. Typical ordnance loads would have comprised nine 2,000 lb Mk 84 bombs or seven laser-guided GBU-24s.

F-15MANX: Stealthy, tail-less proposal based on technology developed for ACTIVE thrust-vectoring programme (which see in NASA entry).

CUSTOMERS: USAF funding for originally planned 392 reduced to 200; however, further nine funded in FY91 and FY92, comprising three Gulf War loss replacements and six with proceeds of sale to Saudi Arabia of 24 surplus F-15C/Ds. Additional six examples in FY96 budget, despite not being requested by USAF. Further six aircraft purchased in FY97 and five in FY98, raising total USAF buy to 226, with another five in FY00 and five in FY01, these final 10 for delivery between May 2002 and mid-2004. Saudi Arabia 72 (F-15S); Israel 25 (F-15I). Potential future customer is South Korea, which has still to select a type to satisfy its F-X requirement (see F-15K).

USAF F-15E PROCUREMENT

FY	Batch	Qty	First aircraft
86	Lot 1	8	86-0183
87	Lot 2	42	87-0169
88	Lot 3	42	88-1667
89	Lot 4	36	89-0471
90	Lot 5	36	90-0227
91	Lot 6	36	91-0300
91	Lot 7	6	91-0600
92	Lot 8	3	92-0364
96	Lot 9	6	96-0200
97	Lot 10	6	97-0217
98	Lot 11	5	98-0131
00	Lot 12	5	
01	Lot 13	5	
Total		**236**	

Initial USAF combat-capable unit, 4th Wing, declared operational October 1989, currently with 333, 334, 335 and 336 Fighter Squadrons; and others with 57th Wing USAF Weapons School at Nellis AFB, Nevada; 90th FS of 3rd Wing at Elmendorf, Alaska, received first F-15E on 29 May 1991; 391st FS, sole F-15E squadron in multitype 366th Wing at Mountain Home AFB, Idaho, received first aircraft (reallocated from early production) 6 November 1991; 492nd and 494th FS of 48th FW at Lakenheath, UK, received first aircraft on 21 February 1992 (46th arrival at Lakenheath, 16 June 1993, was 200th F-15E, 91-0335). Deliveries of Lot 7 began to 48th FW on 13 April 1994 and 209th USAF F-15E (92-0366) to 57th Wing on 11 July 1994. First aircraft from Lot 9 flown on 1 April 1999 and delivered to 57th Wing on 7 June; remaining five F-15Es from this batch all assigned to 48th Fighter Wing by November 1999; subsequent deliveries to 48th FW comprised all Lot 10 and Lot 11 aircraft, with final pair arriving on 26 August 2000.

Initial contract placed with McDonnell Douglas by US government on 18 December 1992 for 72 Saudi aircraft; project name, Peace Sun IX.

COSTS: US$35 million, flyaway; US$2,000 million for 21 F-15Is (1993), Israel; US$288.5 million appropriation for five F-15Es in FY00.

DESIGN FEATURES: Conceived as air superiority fighter (F-15A to F-15D), but further developed for strike/attack mission, with provision for increased weight of ordnance. Typical 1970s fighter configuration, with twin fins positioned to receive vortex flow off wing and maintain directional stability at high angles of attack. Straight two-dimensional external compression engine air inlet each side of fuselage. High-mounted cockpit for all-round vision

Mission includes approach and attack at night and in all weathers; main systems of F-15E include new high-resolution, synthetic aperture Raytheon AN/APG-70 radar, wide field of view FLIR, Lockheed Martin LANTIRN navigation (AN/AAQ-13) and targeting (AN/AAQ-14) pods beneath starboard and port air intakes respectively; air-to-air capacity with AIM-7 Sparrow, AIM-9 Sidewinder and AIM-120 AMRAAM retained; rear cockpit has four multipurpose CRT displays for radar, weapon selection, and monitoring enemy tracking systems; front cockpit modifications include redesigned up-front controls, wide field of view HUD, colour CRT multifunction displays for navigation, weapon delivery, moving map, precision radar mapping and terrain-following. Engines have digital electronic control, engine trimming and monitoring; fuel tanks are foam-filled; more powerful generators; better environmental control.

NACA 64A aerofoil section with conical camber on leading-edge; sweepback 38° 42′ at quarter-chord; thickness/chord ratio 6.6 per cent at root, 3 per cent at tip; anhedral 1°; incidence 0°.

FLYING CONTROLS: Powered. Plain ailerons and all-moving tailplane with dog-tooth extensions, both powered by National Water Lift hydraulic actuators; rudders have Ronson Hydraulic Units actuators; no spoilers or trim tabs; Moog boost and pitch compensator for control column; plain flaps; upward-opening airbrake panel in upper fuselage between fins and cockpit. Digital triple-redundant BAE Systems flight control system capable of automatic coupled terrain-following.

STRUCTURE: Wing based on torque box with integrally machined skins and ribs of light alloy and titanium; aluminium honeycomb wingtips, flaps and ailerons; airbrake panel of titanium, aluminium honeycomb and graphite/epoxy composites skin. F-15E version includes 60 per cent of earlier F-15 structure redesigned to allow 9 *g* and 16,000 hours fatigue life; superplastic forming/diffusion bonding used for upper rear fuselage, rear fuselage keel, main landing gear doors, and some fuselage fairings, plus engine bay structure.

New wing design that would provide 33 per cent range increase and give double the number of weapons stations of existing F-15E revealed in 1994. This could be incorporated in future production aircraft or installed on existing F-15s as retrofit programme, but neither now seems likely.

LANDING GEAR: Hydraulically retractable tricycle type, with single wheel on each unit. All units retract forward. Cleveland nose and main units, each incorporating an oleo-pneumatic shock-absorber. Honeywell wheels and Michelin AIR X radial tyres on all units. Nosewheel tyre size 22×7.75-9 or 22×7.75R9 (26 ply) tubeless; mainwheel tyres size 36×11-18 or 36×11R18 (30 ply) tubeless; tyre pressure 21.03 bars (305 lb/sq in) on all units. Honeywell five-rotor carbon disc brakes.

POWER PLANT: Initially, two Pratt & Whitney F100-PW-220 turbofans, each rated for take-off at 104.3 kN (23,450 lb st), installed, with afterburning. USAF aircraft 135 onwards (90-0233), built from August 1991, have 129.4 kN (29,100 lb st) Pratt & Whitney F100-PW-229s, which also ordered for Saudi F-15S and Israeli F-15I. Air inlet controllers by Hamilton Sundstrand. Air inlet actuators by National Water Lift.

Internal fuel in foam-filled structural wing tanks and six Goodyear fuselage tanks, total capacity 7,643 litres (2,019 US gallons; 1,681 Imp gallons). Simmonds fuel gauge system. Optional conformal fuel tanks (CFT) attached to side of engine air intakes, beneath wing, each containing 2,737 litres (723 US gallons; 602 Imp gallons). Provision for up to three additional 2,309 litre (610 US gallon; 508 Imp gallon) external fuel tanks. Maximum total internal and external fuel capacity 20,044 litres (5,295 US gallons; 4,409 Imp gallons).

ACCOMMODATION: Two crew, pilot and weapon systems officer, in tandem on Boeing ACES II zero/zero ejection seats. Single-piece, upward-hinged, bird-resistant canopy.

SYSTEMS: Lucas Aerospace generating system for electrical power, with Hamilton Sundstrand 60/75/90 kVA constant-speed drive units. Litton molecular sieve oxygen generating system (MSOGS) introduced in 1991 to replace liquid oxygen system. Honeywell air conditioning system. Three independent hydraulic systems (each 207 bars; 3,000 lb/sq in) powered by Abex engine-driven pumps; modular hydraulic packages by Hydraulic Research and Manufacturing Company. AlliedSignal APU for engine starting, and for provision of limited electrical or hydraulic power on the ground independently of main engines.

AVIONICS: *Comms:* Raytheon AN/ARC-164 UHF transceiver and UHF auxiliary transceiver with cryptographic capability; Honeywell AN/APX-101 IFF transponder; BAE Systems AN/APX-76 IFF interrogator with Litton reply evaluator.

Radar: Raytheon AN/APG-70 I-band pulse Doppler radar provides air-to-air capability equal to F-15C, plus high-resolution synthetic aperture mode for air-to-ground.

Flight: Triple redundant BAE Systems digital flight control system with automatic terrain-following standard. IBM CP-1075C very high-speed integrated circuit (VHSIC) central computer introduced in 1992 (replacing CP-1075). Honeywell AN/ASK-6 air data computer, Honeywell AN/ASN-108 AHRS, Honeywell CN-1655A/ASN ring laser gyro INS providing basic navigation data and serving as primary attitude reference system, Rockwell Collins AN/ARN-118 Tacan, Rockwell Collins HSI presenting aircraft navigation information on a symbolic pictorial display, Rockwell Collins AN/ARN-112 ILS receiver, Rockwell Collins ADF receiver, Dorne & Margolin glide slope localiser antenna and Teledyne Avionics angle of attack sensors. Rockwell Collins Miniaturised Airborne GPS Receiver installed from 1995. Latest aircraft to join USAF (FY96 and subsequent production) feature new embedded GPS/INS.

In July 1998, Boeing began flight test of commercial computing technology in F-15E as part of Phantom Works Bold Stroke project, whereby commercial technology is being used to provide non-proprietary computer systems for installation on military aircraft. F-15E installation embodied advanced display core processor (ADCP) using PowerPC hardware to replace existing central computer and multipurpose display processor; initial trial demonstrated same capabilities as current military hardware and was due to lead to flight trials in late 1999 of preproduction ADCP with ability to perform additional F-15E operational flight programme (OFP) functions.

Instrumentation: FLIR imagery displayed on Kaiser IR-2394/A wide field of view HUD; Honeywell vertical situation display set using CRT to present radar, electro-optical identification and attitude director indicator formats to pilot under all light conditions; moving map display by Honeywell RP-341/A remote map reader. Honeywell digital map system intended to replace remote map reader from 1996. F-15I has Elbit helmet-mounted sight and other cockpit displays.

Mission: Lockheed Martin LANTIRN externally mounted sensor package comprising AN/AAQ-13 navigation pod and AN/AAQ-14 targeting pod. Advanced targeting pod system in prospect for Air Combat Command and could be fielded on F-15E by about 2005.

Self-defence: Northrop Grumman Enhanced AN/ALQ-135(V) internal countermeasures set provides automatic jamming of enemy radar signals; Lockheed Martin AN/ALR-56C RWR, Raytheon AN/ALQ-128 EW warning set, BAE Systems AN/ALE-45 chaff dispenser. Unique SPS-2100 EW system developed by Elisra and Rokar for Israeli F-15I. Sanders and ITT Avionics AN/ALQ-214 Integrated Defensive Electronic CounterMeasures (IDECM) under development and will be installed on F-15E; IDECM package incorporates Sanders AN/ALE-55 fibre optic towed decoy (FOTD).

ARMAMENT: 20 mm M61A1 six-barrel gun in starboard wing-root, with 512 rounds. General Electric lead computing gyro. Provision on underwing (one per wing) and centreline pylons for air-to-air and air-to-ground weapons and external fuel tanks. Wing pylons use standard rail and launchers for AIM-9 Sidewinder (Israeli F-15I also compatible with Rafael Python 4) and AIM-120 AMRAAM air-to-air missiles; AIM-7 Sparrow and AIM-120 AMRAAM can be carried on ejection launchers on the fuselage or on tangential stores carriers on CFTs. Maximum aircraft load (with or without CFTs) is four each AIM-7 and AIM-9, or up to eight AIM-120. Single or triple rail launchers for AGM-65 Maverick air-to-ground missiles can be fitted to wing stations only.

Tangential carriage on CFTs provides for up to six bomb racks on each tank, with provision for multiple ejector racks on wing and centreline stations. Edo BRU-46/A and BRU-47/A adaptors throughout, plus two LAU-106A/As each side of lower fuselage. F-15E can carry a wide variety and quantity of guided and unguided air-to-ground weapons, including Mk 20 Rockeye (26), Mk 82 (26), Mk 84 (seven), BSU-49 (26), BSU-50 (seven), GBU-10 (seven), GBU-12 (15), GBU-15 (two), GBU-24 (five), CBU-52 (25), CBU-58 (25), CBU-71 (25), CBU-87 (25) or CBU-89 (25) bombs; SUU-20 training weapons (three); A/A-37 U-3 tow target (one); and B57 and B61 series nuclear weapons (five). An AN/AXQ-14 datalink pod is used in conjunction with the GBU-15; LANTIRN pod illumination is used to designate targets for laser-guided bombs; AGM-130 powered standoff bomb integrated in 1993; AGM-88 HARM capability in 1996. Integration of JDAM, JSOW and WCMD weapons currently under way. AN/AWG-27 armament control system.

Pneumatic weapon ejection system under development in early 2000; makes use of compressed air for weapons separation from aircraft.

DIMENSIONS, EXTERNAL:
Wing span	13.05 m (42 ft 9¾ in)
Wing aspect ratio	3.0
Length overall	19.43 m (63 ft 9 in)
Height overall	5.63 m (18 ft 5½ in)
Tailplane span	8.61 m (28 ft 3 in)
Wheel track	2.75 m (9 ft 0¼ in)
Wheelbase	5.42 m (17 ft 9½ in)

AREAS:
Wings, gross	56.49 m² (608.0 sq ft)
Ailerons (total)	2.46 m² (26.48 sq ft)
Flaps (total)	3.33 m² (35.84 sq ft)
Fins (total)	9.78 m² (105.28 sq ft)
Rudders (total)	1.85 m² (19.94 sq ft)
Tailplanes (total)	10.34 m² (111.36 sq ft)

WEIGHTS AND LOADINGS (F100-PW-220 engines):
Operating weight empty (no fuel, ammunition, pylons or external stores)	14,515 kg (32,000 lb)
Max weapon load	11,113 kg (24,500 lb)
Max fuel weight: internal (JP4)	5,952 kg (13,123 lb)
CFTs (two, total)	4,265 kg (9,402 lb)
external tanks (three, total)	5,396 kg (11,895 lb)
max internal and external	15,613 kg (34,420 lb)
Max T-O weight	36,741 kg (81,000 lb)
Max landing weight: unrestricted	20,094 kg (44,300 lb)
at reduced sink rates	36,741 kg (81,000 lb)
Max zero-fuel weight	28,440 kg (62,700 lb)
Max wing loading	650.5 kg/m² (133.22 lb/sq ft)
Max power loading	176 kg/kN (1.73 lb/lb st)

PERFORMANCE:
Max level speed at height	M2.5
Max combat radius	685 n miles (1,270 km; 790 miles)
Max range	2,400 n miles (4,445 km; 2,762 miles)

UPDATED

US F-15 PRODUCTION

	F-15A	F-15B	F-15C	F-15D	F-15DJ	F-15E	F-15I	F-15J	F-15S	Total
Israel	19	2	18	13			25			77
Japan					12			10[1]		22
Saudi Arabia			55	19					72	146
USA	365[2]	59[3]	409	61		236[4]				1,120
Total	384	61	482	93	12	236	25	10	72	1,375

Note: Mitsubishi of Japan also produced 155 F-15Js and 36 F-15DJs to complete total JASDF procurement of 165 and 48, respectively

[1] Including eight kits to Mitsubishi
[2] Including 10 YF-15s
[3] Including two YF-15s, of which one converted to F-15E prototype
[4] Total includes five being purchased in FY00 and five in FY01

BOEING F/A-18 HORNET

US Navy/Marine Corps designations: F/A-18A, B, C, D
Royal Australian Air Force designations: AF-18A and ATF-18A
Canadian Forces designations: CF-188A and CF-188B
Spanish Air Force designations: C.15 and CE.15

TYPE: Multirole fighter.

PROGRAMME: US Navy study of VFAX low-cost, lightweight multimission fighter accepted 1974; VFAX study terminated August 1974 and replaced by derivative of either General Dynamics YF-16 or Northrop YF-17 lightweight fighter prototypes; McDonnell Douglas proposed F-17 derivative with Northrop as associate; resultant Navy Air Combat Fighter called Hornet accepted in two versions, F-18 fighter and A-18 attack aircraft; single F/A-18 selected to fill both roles; McDonnell Douglas prime contractor and Northrop principal subcontractor for all versions agreed 1985; programme passed into Boeing hands in 1997. First Hornet flight (160775) 18 November 1978; 11 development aircraft flying by March 1980; delivery of F/A-18A/B (TF-18A designation dropped) to US Navy and Marines began May 1980 and completed 1987; millionth flying hour achieved 10 April 1990; two millionth on 17 September 1993.

Enhancements to Raytheon AN/APG-65 radar funded (US$65.7 million) May 1990; new signal and data processors, upgraded receiver/exciter. Resultant AN/APG-73 installed in production F/A-18s from 1994 (first two F/A-18Cs with this radar delivered to VFA-146 and -147 at NAS Lemoore, California, on 25 and 26 May 1994).

Last F/A-18D, for US Marine Corps, delivered to VMFA(AW)-121 on 25 August 2000, ending production of first-generation Hornet derivatives.

CURRENT VERSIONS: **F/A-18A and F/A-18B:** Single- and two-seat versions respectively. Total of 371 F/A-18As and 39 F/A-18Bs (plus 11 prototypes, including two tandem-seat trainers) for USN and USMC as escort fighters to replace F-4s and as attack aircraft replacing A-7s under FY79 to FY85 contracts; two-seater, originally designated TF-18A, has internal fuel capacity reduced by 6 per cent; first training squadron (VFA-125) formed at NAS Lemoore, California, November 1980; in service 7 January 1983 with Marine Fighter/Attack Squadron VMFA-314 at MCAS El Toro, California; first Navy squadron, VFA-113 of Pacific Fleet, October 1983; first Atlantic Fleet squadrons formed NAS Cecil Field, Florida, 1 February 1985; same month VFA-113 and VFA-25 embarked in USS *Constellation*.

At least 28 Navy Reserve aircraft upgraded to F/A-18A+ in Engineering Change Proposal 560 (ECP 560) programme during 1999, bringing them to a standard closely approximating that of the F/A-18C. Similar, but more elaborate, programme launched by US Marine Corps in FY00 budget, which featured request for funds to initiate Engineering Change Proposal 583 (ECP 583); project to begin with 24 F/A-18As and expected to total 76, in two active and four reserve squadrons. Work entails installation of AN/APG-73 radar and Rockwell Collins AN/ARC-210(V) UHF/VHF AM/FM jam-resistant radio suite, including Have Quick II, SINCGARS and MIDS Link 16 communications. Additional avionics changes concern provision of BAE Systems AN/APX-111(V) combined IFF interrogater/transponder, Smiths Industries AN/AYQ-9 stores management system, night vision display system, upgraded General Dynamics XN-8 mission computer and digital display indicator; upgraded aircraft able to employ newer sensors, such as Lockheed Martin AN/AAS-38B FLIR pod, and more capable weaponry, such as AIM-120 AMRAAM and JDAM.

ECP 583 also to be adopted by Canada at cost of more than C$900 million and may eventually be implemented by others that have yet to address issue of avionics upgrading.

F/A-18C and F/A-18D: Single- and two-seat versions respectively. Purchased from FY86 onwards; F/A-18Cs and F/A-18Ds bought under FY86 and FY87 procurements are baseline non-night attack models; overall total of 627 F/A-18C/Ds (including Night Attack – see below) built to complete first-generation Hornet production for US Navy/Marine Corps, though this total includes replacement for one F/A-18C destroyed in pre-delivery accident and eight FY97 aircraft originally ordered for Thailand (four each of F/A-18C and F/A-18D) which were completed as F/A-18D for USMC. First flight of production F/A-18C (163427) 3 September 1987.

Main description applies to F/A-18C except where indicated.

Modifications include provision for up to 10 AIM-120 AMRAAM missiles (two on fuselage and two on each inboard and outboard pylon); up to four imaging IR Maverick missiles (one on each wing pylon); provision for reconnaissance version; upgraded stores management set with 128 kbits memory, Intel 8086 processor, MIL-STD-1553B armament multiplex bus with MIL-STD-1760 weapons interface capability; flight incident recorder and monitoring set (FIRAMS) with integrated fuel/engine indicator, data storage set for recording maintenance and flight incidents data, signal data processor interfacing with fuel system to provide overall system control, enhanced built-in test capability and automatic CG adjustment as fuel is consumed; maintenance status panel isolating faults to card level; and new faster XN-6 mission computer with twice memory of previous XN-5; upgraded to XN-8 from FY91. Small rectangular fence retrofitted to US Navy aircraft above LEX strake just ahead of wing leading-edge broadens LEX vortices, reduces fatigue and improves directional control at angles of attack higher than 45°. AIM-120A AMRAAM cleared for use by Hornet from September 1993, initially on aircraft of VFA-22, VFA-94 and VMFA-314 in Arabian Gulf.

F/A-18C/D Night Attack: First flight of prototype 6 May 1988; one Night Attack F/A-18C (163985) and one D (163986) delivered to Naval Air Test Center, Patuxent River, on 1 and 14 November 1989 respectively; all F/A-18Cs and Ds delivered subsequently (commencing

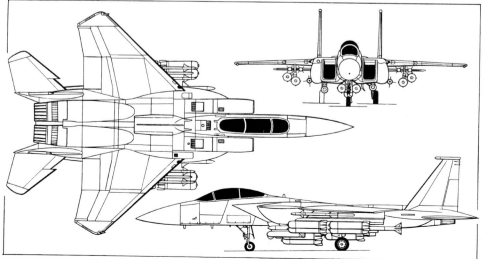

Boeing F-15E Eagle equipped for high ordnance payload air-to-ground mission *(Dennis Punnett/Jane's)* 1987

HORNET PRODUCTION

FY	Block	US Navy/Marines F/A-18A/C	US Navy/Marines F/A-18B/D	Canada CF-18A	Canada CF-18B	Australia AF-18A	Australia AF-18B	Spain EF-18A	Spain EF-18B	Kuwait KAF-18C	Kuwait KAF-18D	Finland F-18C	Finland F-18D	Switzerland F-18C	Switzerland F-18D	Malaysia F-18D	Cum Total
76	1	3															3
	2	3	1														7
	3	3	1														11
79	4	7	2														20
80	5	3	4														27
	6	8	1														36
	7	9															45
81	8	11	4		4												64
	9	17	4	1	5												91
	10	22	2	5	3												123
82	11	17	4	7													151
	12	23	1	8	2												185
	13	19		6	1												211
83	14	18	3	7	1	3	7										250
	15	29	2	6	2	4											293
	16	32		7	1	4											337
84	17	24	3	7	2	7		2									382
	18	25	4	7	1		5	1	2								427
	19	27	1	7	1	3	2	4									472
85	20	24	3	6	2	6		3	1								517
	21	26	1	8		5		5	3								565
	22	30		8		4	2	9									618
86	23	23[1]	8[2]	8		4	2	9									672
	24	21	8		9	4		6									720
	25	17	7		6	5		2									757
87	26	25	3			4		5									794
	27	26	2			3		5									830
	28	25	3			1		7									866
88	29	20[3]	10[4]					2									898
	30	16	10					4									928
	31	18	10					2									958
89	32	17	7														982
	33	23	7														1,012
	34	24	6														1,042
90	35	22								5	3						1,072
	36	11	10							3	3						1,099
	37	13	10							2	2						1,126
91	38	12	4							8							1,150
	39	12	4							8							1,174
	40	12	4							6							1,196
92	41	9	9														1,214
	42	12	3														1,229
	43	13	2														1,244
93	44	7	8														1,259
	45	14	4														1,277
	46	3										4					1,284
94	47	12										6	3				1,305
	48	12										2		8	6		1,333
	49	12										3		18	2		1,368
95	50	24										11				8	1,411
96	51	10	8									17					1,446
97	52	1[8]	14[7]									18					1,479
Subtotals		846[5]	202[6]	98	40	57	18	60	12	32	8	57	7	26	8	8	
Totals		1,048		138		75		72		40		64		34		8	1,479

Notes:
[1] Begins F/A-18C production
[2] Begins F/A-18D production
[3] Begins F/A-18C Night Attack production
[4] Begins F/A-18D Night Attack production
[5] Nine prototypes, 371 As, 137 Cs and 329 C Nights
[6] Two prototypes, 39 Bs, 31 Ds and 130 D Nights
[7] Total includes eight aircraft originally ordered by Thailand but delivered to US Marine Corps
[8] Replacement for aircraft destroyed on pre-delivery test flight

FY88 procurement) have all-weather night attack avionics. Raytheon AN/APG-73 radar (first F/A-18 flight test 15 April 1992) standard from May 1994; initially on aircraft of VFA-146 and VFA-147. Deliveries of night-capable versions to fleet units began 1990. Marine Corps has replaced six squadrons of Grumman A-6Es, McDonnell Douglas OA-4s and RF-4Bs in attack, reconnaissance and forward air controller roles, with 96 Hornets, of which first 48 authorised in 1990; remainder in F/A-18D(RC) configuration (see below). Navy squadrons unchanged, with two-seaters used only by specialist training units. 1,000th Hornet was F/A-18D 164237, which was delivered to VMFA(AW)-242 on 22 April 1991. USN/USMC operating squadron status in accompanying table.

Night Attack system includes BAE Systems Cat's Eyes pilot's night vision goggles, Raytheon AN/AAR-50 thermal imaging navigation set (TINS) presenting forward view in Kaiser AN/AVQ-28 raster HUD, colour multifunction displays and Smiths colour digital moving map; external sensor pods comprise Lockheed Martin AN/AAS-38B NITE Hawk targeting FLIR and TINS; NITE Hawk added laser target designator/ranger subsystem from January 1993, initially for squadrons VFA-146 and -147, operating in Arabian Gulf; USMC version of F/A-18D has mission-capable rear cockpit with no control column, but two sidestick weapons controllers and two Kaiser 12.7 cm (5 in) colour MFDs in addition to Smiths Srs 2100 colour map display; can be converted to dual control, with stick and throttles, for pilot training.

F/A-18D(RC): Simple reconnaissance version, launched 1982 and first flown 1984, included a twin-sensor package replacing gun in nose; was to be fitted with Martin Marietta Advanced Tactical Airborne Reconnaissance System (ATARS) centreline pod; first sensor-capable aircraft delivered to El Toro, February 1992; original ATARS programme suspended June 1993. As replacement, 31 F/A-18Ds of Marine Corps to receive partial ATARS fit, comprising a pallet with an IR linescanner, low- and medium-altitude electro-optical sensors, long-range optical sensor; modified version of Raytheon AN/APG-73 radar capable of producing high-resolution strip-maps and a Lockheed Martin Aeronautics pod-mounted digital datalink. Testing of ATARS under way at China Lake and Patuxent River since 1992.

US Navy authorisation to proceed was received in December 1996, with US$50 million prime contractor contract awarded to McDonnell Douglas (now Boeing) in early 1997 for the first four Lockheed Martin Fairchild Systems AN/ASD-10(V) ATARS units comprising the initial phase of low-rate production; second phase followed, with four more units ordered in FY99 at total cost of US$23 million and five in FY00 for US$35.3 million. Full-scale production now expected in early 2000s. Delivery of the first pods took place in 1998 for operational testing, followed by deployment with elements of the Marine Corps in mid-1999. Transfer of imagery to ground exploitation system by datalink in real time successfully demonstrated in mid-2000. Production anticipated of 31 ATARS units and 24 datalink pods.

F/A-18E and F/A-18F Super Hornet: Described separately.

AF-18A and ATF-18A: Royal Australian Air Force versions; decision to purchase 75 announced 20 October 1981; deliveries started 17 May 1985; first flight of ATF-18A assembled by AeroSpace Technologies of Australia (ASTA), 26 February 1985; first flight of Australian-manufactured aircraft (ATF-18A, A21-104) 3 June 1985; last of 57 single-seat and 18 two-seat Hornets delivered 16 May 1990; Hornet replaced Dassault Mirage IIIO; units are No. 2 OCU, Williamtown, No. 3 Squadron (formed August 1986) and No. 77 Squadron at same base, and No. 75 Squadron, Tindal. Weapons include AIM-7M, AIM-9M, AGM-88 HARM, AGM-84 Harpoon, GBU-10/ GBU-12/Paveway II LGBs, Mk 82 bombs and 70 mm rockets. From 1990, remaining 74 aircraft being fitted with F/A-18C/D type avionics and (from 1991) provision for Lockheed Martin AN/AAS-38B NITE Hawk IR tracking and laser designating pod.

Hornet Upgrade programme (HUG; Project Air 5376) will bring 71 aircraft to F/A-18C/D configuration in two phases at total cost of about US$700 million to US$725 million. First is new mission computer (General Dynamics XN-8+ selected) and com/nav suites (including Rockwell Collins AN/ARC-210 UHF/VHF), plus installation of Litton embedded GPS/INS units and new IFF. Two Raytheon AN/APG-73 radars delivered in July 1997 for evaluation, but BAE Systems also offered version of Blue Vixen pulse Doppler radar; AN/APG-73 selected for second phase, which also to involve new EW suite featuring AN/ALR-67(V)3 RWR and either AN/ALQ-165 ASPJ or AN/ALQ-214(V)4 RF countermeasures system as well as Joint Helmet-Mounted Cueing System (JHMCS) and Multifunction Information Distribution System/Low Volume Terminal (MIDS/LVT), plus spares, support and test equipment. New medium- and short-range AAMs also to feature in upgrade, with BAE ASRAAM selected in February 1998.

Two aircraft modified in mid-2000 as part of phase one for test and evaluation; first of these delivered to RAAF in early September for trials, with main upgrade effort to begin immediately thereafter. Phase two improvements to start in late 2001 or early 2002 and due for completion by 2005; subsequent developments may include structural upgrade, post-2006, as phase three.

CF-18A and B: Canada's purchase of 138 Hornets (finalised as 98 CF-18As and 40 two-seat CF-18Bs, known respectively as **CF-188A** and **CF-188B**) announced 10 April 1980; first flight of CF-18 29 July 1982; deliveries between 25 October 1982 and September 1988; CF units were No. 410 OCU and Nos. 416 and 441 Squadrons at CFB Cold Lake, Alberta, 425 and 433 at Bagotville, Quebec, and 439 and 421 Squadrons of No. 4 Fighter Wing/No. 1 Air Division at Baden Sollingen, Germany; last aircraft left Europe 26 January 1993. Currently active with Nos. 425 and 433 Squadrons at Bagotville and Nos. 410 (OCU), 416 and 441 at Cold Lake. Differences from US Navy F/A-18A/B include ILS, in-flight identification spotlight in port side of fuselage, and provision for LAU-5003 19-tube pods for CRV-7 70 mm (2.75 in) high-velocity submunition rockets; other weapons are AIM-7M and AIM-9L air-to-air missiles, 500 lb Mk 82 bombs and Hunting BL755 CBUs. Pilot has comprehensive cold weather land survival kit. Upgrade known as CF-18 Incremental Modernization Project to extend service life by 17 to 20 years begun in 1999; total cost expected to be around C$1.8 billion, with Engineering Change Proposal 583 (ECP 583) being most significant element. Work to be undertaken as part of ECP 583 includes installation of AN/APG-73 radar, new XN8 mission computer and software, Have Quick II jam-resistant radios and some structural work, including overhaul of fuselage bulkheads following discovery of cracks in June 1998. Funds also to be sought for new IFF, JTIDS, new RWR and electronic jammers, stores management upgrade and helmet-mounted sight. Canada contemplating sale of some aircraft to help fund the upgrade of about 80 CF-18s.

EF-18A and B: Spanish versions; purchase of 60 single-seat Hornets and 12 two-seaters, known respectively as **C.15** and **CE.15**, under *Futuro Avion de Combate y Ataque* programme announced 30 May 1983; financial restrictions reduced number from 84 and deliveries then stretched from 36, 24 and 12 during 1986 to 1988 to 11, 26, 15, 12 and eight during 1986 to 1990; maintenance performed in Spain by CASA, which also works on USN Hornets with 6th Fleet in Mediterranean; first flight 4 December 1985; deliveries began 10 July 1986; all 12 trainers delivered by early 1987; armament includes GBU-10/16 LGBs, AGM-65G Maverick ASM, AIM-7F/M and AIM-9L/M air-to-air missiles, AGM-84C/D Harpoon, AGM-88 HARM and free-fall bombs; AIM-120 AMRAAM ordered 1990 and delivered late 1995. First 36 aircraft have Sanders AN/ALQ-126B deception jammers ordered in 1987; final 36 received Northrop Grumman AN/ALQ-162(V) systems. All had AN/ALR-67 RWR, though this subsequently replaced by AN/ALR-67B(V)2.

Units currently equipped are Ala de Caza 15 (15th Fighter Wing) formed at Zaragoza December 1985 and operational December 1987, with Escuadrones 151, 152, and from October 1995, 153 as OCU; Ala de Caza 12 at Torrejon (Escuadrones 121 and 122) completed re-equipping in July 1990; Grupo 21 at Moron (Escuadron 211) re-equipped with ex-USN aircraft in mid-1990s and will be joined by Escuadron 212. Agreement in 1992 to upgrade Spanish aircraft to **F-18A+/B+** standard, close to F/A-18C/D; of 71 available, 46 were converted by McDonnell Douglas between September 1992 and March 1994; final 25 by CASA by 1995. Engineering Change Proposal 287 includes later mission and armament computers, databusses, data-storage set, new wiring, pylon modifications and software. New capabilities include AN/AAS-38B NITE Hawk targeting FLIR pods, although this to be replaced by Rafael Litening system, with total of 24 pods to be acquired at cost of US$38 million under deal agreed in mid-2000; Rafael LOROP (long-range oblique photography) system also selected by Spain in third quarter 2000 for use by EF-18.

Spain decided in January 1995 to obtain 24 former USN F/A-18A Hornets, with six more on option. First six were handed over in December 1995 and have late standard F404 engines plus other Spanish-specified modifications. Second batch of six followed in 1996, with one in 1997 and five in 1999; remaining six all delivered by early 2000.

Hornet upgrades: McDonnell Douglas proposed a series of optional upgrades in early 1996. These include upgrading to 9 *g* from current 7.5 *g* manoeuvre limit; 2,271 litre (600 US gallon; 500 Imp gallon) external tanks (land-based operations only); F/A-18E fuel tanks of polyurethane, offering up to 160 kg (353 lb) additional internal capacity; six additional chaff/flare dispensers; and cockpit upgrades using some F/A-18E technology, such as LCD upfront display, colour tactical situation display, helmet-mounted cueing system and advanced targeting FLIR; some of these features are incorporated in US Navy F/A-18A+ (see current versions). Northrop Grumman has also received first order for structural assemblies for Center Barrel Replacement Plus (CBR+) programme, whereby major centre/aft fuselage assembly will be replaced to extend service life by up to eight years; first CBR+ installation scheduled to occur in final quarter of 2000 and as many as 200 F/A-18C/D aircraft may be involved over 10 year period.

CUSTOMERS: See table of production.

USN proposed procurement of 1,156 production Hornets: first-generation models total 1,048 including prototypes, with follow-on F/A-18E/F versions set to raise procurement to more than 1,500 for USN/MC. Additional 431 of first-generation models completed for export customers by time production ceased.

In addition to US Navy/Marine Corps, Australia, Canada and Spain (see Current Versions), **Switzerland** selected 26 F/A-18Cs and eight F/A-18Ds powered by GE F404-GE-402 engines and with AN/APG-73 radar, AN/

US HORNET OPERATING SQUADRONS

Unit	Base	Version	Remarks
Regular Navy			
VFA-15	Oceana	C Night	
VFA-22	Lemoore	C Night	
VFA-25	Lemoore	C Night	
VFA-27	Atsugi, Japan	C Night	
VFA-34	Oceana	C Night	
VFA-37	Oceana	C Night	
VFA-81	Oceana	C	
VFA-82	Beaufort	C Night	
VFA-83	Oceana	C Night	
VFA-86	Beaufort	C	
VFA-87	Oceana	C Night	
VFA-94	Lemoore	C Night	
VFA-97	Lemoore	A	
VFA-105	Oceana	C Night	
VFA-106	Oceana	A, B, C, C Night, D, D Night	Training
VFA-113	Lemoore	C Night	
VFA-125	Lemoore	A, B, C, C Night, D, D Night	Training
VFA-131	Oceana	C Night	
VFA-136	Oceana	C Night	
VFA-137	Lemoore	C Night	
VFA-146	Lemoore	C	
VFA-147	Lemoore	C	
VFA-151	Lemoore	C Night	
VFA-192	Atsugi, Japan	C Night	
VFA-195	Atsugi, Japan	C Night	
VX-9	Point Mugu/China Lake	A, C, C Night, D, D Night	Trials unit
NFDS 'Blue Angels'	Pensacola	A, B	Display team
NSAWC	Fallon	A, B	Combat training
Marine Corps			
VMFAT-101	Miramar	A, B, C, C Night, D, D Night	Training
VMFA-115	Beaufort	A	
VMFA(AW)-121	Miramar	D Night	
VMFA-122	Beaufort	A	
VMFA-212	Iwakuni	C	
VMFA(AW)-224	Beaufort	D Night	
VMFA(AW)-225	Miramar	D Night	
VMFA-232	Miramar	C	
VMFA(AW)-242	Miramar	D Night	
VMFA-251	Beaufort	C Night	
VMFA-312	Beaufort	C Night	
VMFA-314	Miramar	C Night	
VMFA-323	Miramar	C Night	
VMFA(AW)-332	Beaufort	D Night	
VMFA(AW)-533	Beaufort	D Night	
Naval Reserve			
VFC-12	Oceana	A, B	Aggressor
VFA-201	Fort Worth	A	
VFA-203	Atlanta	A, B	
VFA-204	New Orleans	A	
Marine Corps Reserve			
VMFA-112	Fort Worth	A	
VMFA-134	Miramar	A	
VMFA-142	Atlanta	A	
VMFA-321	Washington	A	

Disbanded units are VFA-132 on 1 June 1992; VFA-303 and VFA-305 both October 1994; VMFA-333 and VMFA-531 both 31 March 1992; VMFA-235 in April 1996; VMFA-451 in 1997; VAQ-34 on 1 October 1993; VX-4 and VX-5 both September 1994 (replaced by VX-9); VF-45 in April 1996; VF-127 in March 1996. VFC-13 has transitioned to F-5E/F Tiger II. VFA-115 began transition from F/A-18C to F/A-18E in third quarter of 2000. All F/A-18A operators, except VFA-115 and VMFA-235 which last flew F/A-18C and VX-4 and VX-5 which used several versions.

Malaysian Boeing F/A-18D Hornet, operated by No. 18 Squadron *(Paul Jackson/Jane's)*

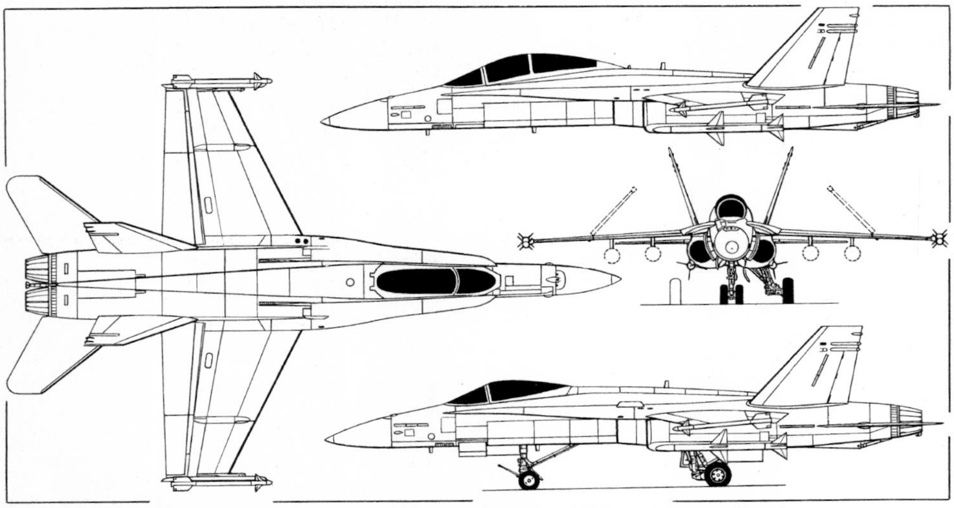

Boeing F/A-18C Hornet with additional side view (top) of F/A-18D ***(Dennis Punnett/Jane's)***

AAS-38B NITE Hawk targeting FLIR, AIM-120 AMRAAMs and night vision systems in October 1988, as its Neue Jagdflugzeug to replace front line F-5Es; purchase ratified by parliament 12 June 1992; confirmed by referendum 6 June 1993; maiden flight of first aircraft (F/A-18D J-5231) on 20 January 1996, followed by formal roll-out ceremony on 25 January 1996. First F/A-18C (J-5001) first flown 8 April 1996, with both aircraft being retained in USA for weapons system verification at China Lake, California, and Patuxent River, Maryland; Phase One testing was completed at Patuxent River in early October 1996; and included launches of AIM-9 Sidewinder AAMs; second stage of assessment at China Lake, beginning in October 1996; J-5231 delivered to Switzerland on 16 December 1996. Remaining 32 assembled at Emmen by SF (which see), with the first flight by a Swiss-assembled aircraft (F/A-18D J-5232) being made on 3 October 1996; deliveries to Swiss Air Force began 23 January 1997 and completed in fourth quarter of 1999. Operating squadrons are Fliegerstaffel 11 at Dubendorf, FlSt 17 at Payerne and FlSt 18 at Sion. Swiss Hornets first of the type rated for 9 *g*. Switzerland has requirement for additional batch of about 12 two-seat aircraft to perform reconnaissance/attack missions; however, these not required until about 2005 and will therefore have to be acquired second-hand and refurbished before entry into service.

Kuwaiti contract signed September 1988 for 32 F/A-18Cs and eight F/A-18Ds (also referred to as **KAF-18C** and **KAF-18D** respectively) together with AGM-65G Maverick, AGM-84 Harpoon, AIM-7F Sparrow and AIM-9L Sidewinder; first flight 19 September 1991; first three delivered to No. 25 Squadron 25 January 1992; final delivery to No. 9 Squadron 23 August 1993; F404-GE-402 power plants.

Finland selected seven F-18Ds (built by McDonnell Douglas) and 57 F-18Cs (for assembly from kits by Patria Finavitec); announcement 6 May 1992; letter of offer signed 5 June 1992; -402 engines and AN/APG-73 radar; initial procurement of four aircraft in 1993; first US-built aircraft (F-18D HN-461) flown 21 April 1995, before formal roll-out on 7 June 1995; initial four F-18Ds delivered to Tampere-Pirkkala on 7 November 1995. Patria Finavitec began assembly of first F-18C (HN-401) in September 1995; initial flight made from Halli, Finland, on 14 May 1996 with delivery following on 28 June; final aircraft delivered on 8 August 2000.

Malaysia confirmed order for eight USMC-standard F-18Ds (AN/APG-73 radar and -402 engines) 29 June 1993; FMS contract placed with manufacturer 7 April 1994; first aircraft (M45-01) made initial flight on 1 February 1997 and was formally handed over on 19 March 1997; this aircraft and three others delivered to Malaysia on 27 May 1997, with remaining four following on 28 August 1997; all for use by 18 Squadron.

Thailand confirmed order in February 1996; contract for four F-18Cs and four F-18Ds signed 30 May 1996. Deliveries originally to commence in October 1999, but economic crisis forced reappraisal and Thailand requested US Navy to find new buyer. US government assumed control of programme on 1 May 1998 and aircraft produced as F/A-18Ds for US Marine Corps.

Other potential customers include Austria, Brazil, Chile, Czech Republic, Hungary, the Philippines and Poland, all of which have conducted evaluations of the Hornet; surplus USN/MC aircraft available for resale, but Boeing seeking early release of Super Hornet for export.

COSTS: US$25 million, flyaway unit cost, 1991. US$55,632 million (1991) US programme, 1,167 aircraft. FY97 batch of six F/A-18Ds valued at US$159.4 million. Swiss purchase valued at approximately US$2.3 billion, including aircraft, spares, AIM-120 missiles, technical support, training and a weapons tactics trainer.

DESIGN FEATURES: Developed version of an original light fighter design. Sharp-edged, cambered leading-edge extensions (LEX), slots at fuselage junction and outward-canted twin fins are designed to produce high agility and docile performance at angles of attack over 50°; wings have 20° sweepback at quarter-chord and fold up 90° at inboard end of ailerons, even on land-based F/A-18s; landing gear designed for unflared landings on runways as well as on carriers.

FLYING CONTROLS: Full digital fly-by-wire controls using ailerons and tailerons for lateral control, plus flaps in flaperon form at low airspeeds; leading- and trailing-edge flaps scheduled automatically for high manoeuvrability, fast cruise and slow approach speed; both rudders turned in at take-off and landing to provide extra nose-up trim effort; fly-by-wire returns towards 1 *g* flight if pilot releases controls; lateral and then directional control progressively washed out as angle of attack reaches extreme values; height, heading and airspeed holds provided in fly-by-wire system; US Navy aircraft can land automatically using carrier-based guidance system; airbrake panel located on top of fuselage, between fins. Bertea hydraulic actuators for trailing-edge flaps; Hydraulic Research actuators for ailerons; National Water Lift actuators for tailerons.

STRUCTURE: Multispar wing mainly of light alloy, with graphite/epoxy inter-spar skin panels and trailing-edge flaps; tail surfaces mainly graphite/epoxy skins over aluminium honeycomb core; graphite/epoxy fuselage panels and doors; titanium engine firewall. Northrop Grumman responsible for rear and centre fuselages; assembly and test at St Louis factory; CASA produces horizontal tail surfaces, flaps, leading-edge extensions, speedbrakes, rudders and rear side panels for all F/A-18s.

LANDING GEAR: Messier-Dowty retractable tricycle type, with twin-wheel nose and single-wheel main units. Nose unit retracts forward, mainwheels rearward, turning 90° to stow horizontally inside the lower surface of the engine air ducts. Bendix wheels and brakes. Nosewheel tyres size 22×6.6-10 (20 ply) tubeless, pressure 24.13 bars (350 lb/sq in) for carrier operations, 10.34 bars (150 lb/sq in) for land operations. Mainwheel tyres size 30×11.5-14.5 (24/26 ply) tubeless, pressure 24.13 bars (350 lb/sq in) for carrier operations, 13.79 bars (200 lb/sq in) for land operations. Ozone nosewheel steering unit. Nose unit towbar for catapult launch. Arrester hook, for carrier landings, under rear fuselage.

POWER PLANT: Two General Electric F404-GE-400 low-bypass turbofans initially, each producing approximately 71.2 kN (16,000 lb st) with afterburning. F404-GE-402 EPE (Enhanced Performance Engine) standard from early 1992; rated at approximately 78.3 kN (17,600 lb st). Self-sealing fuel tanks and fuel lines; foam in wing tanks and fuselage voids. Internal fuel capacity (JP5) approximately 6,061 litres (1,600 US gallons; 1,333 Imp gallons). Provision for up to three 1,250 litre (330 US gallon; 275 Imp gallon) external tanks. Canadian aircraft carry three 1,818 litre (480 US gallon; 400 Imp gallon) tanks. Flight refuelling probe retracts into upper starboard side of nose. Simmonds fuel gauging system. Fixed ramp air intakes. Fit checks of larger 2,271 litre (600 US gallon; 500 Imp gallon) external tanks had been conducted at Patuxent River by 1996, and these may be offered as an option to overseas air arms that require longer-range capability.

ACCOMMODATION: Pilot only, on Martin-Baker SJU-5/6 zero/zero ejection seat, in pressurised, heated and air conditioned cockpit. Upward-opening canopy, with separate windscreen, on all versions. Two pilots in F/A-18B and USN F/A-18D; pilot and Naval Flight Officer in USMC F/A-18D.

SYSTEMS: Two completely separate hydraulic systems, each at 207 bars (3,000 lb/sq in). Maximum flow rate 212 litres (56 US gallons; 46.6 Imp gallons)/min. Bootstrap-type reservoir, pressure 5.86 bars (85 lb/sq in). Honeywell air conditioning system. General Electric electrical power system. Honeywell GTC36-200 APU for engine starting and ground pneumatic, electric and hydraulic power. Oxygen system. Fire detection and extinguishing systems.

AVIONICS: *Comms:* AN/ARC-182 UHF/VHF, AN/ARC-210 SECOS 610 UHF/VHF, Conrac communications system control; AN/APX-100 IFF; BAE Systems AN/APX-111 combined interrogator transponder (CIT) for Kuwait; AN/APX-111 evaluated on F/A-18D by Naval Air Warfare Center at China Lake commencing mid-1995 and was incorporated into production aircraft from mid-1997, with up to 500 existing Hornets to be retrofitted for Navy and Marine Corps.

Radar: Raytheon AN/APG-65 multimode digital air-to-air and air-to-ground tracking radar, with air-to-air modes which include velocity search (VS), range while search (RWS), track-while-scan (TWS – track 10 targets and display eight to pilot) and raid assessment mode (RAM). Improved Raytheon AN/APG-73 replaced AN/APG-65 in F/A-18C/D for USN, USMC, Finland, Malaysia and Switzerland from May 1994. AN/APG-73 key feature of ECP 583 upgrade for US Marine Corps F/A-18A and will be retrofitted to Australian and Canadian aircraft as part of upgrade packages.

Flight: Automatic carrier landing system (ACLS) for all-weather carrier operations; Rockwell Collins AN/ARN-118 Tacan, DF-301E UHF/DF, Telephonics AN/ARA-63 receiver/decoder, JET ID-1791/A flight director indicator, Honeywell HSI; General Electric quadruple-redundant fly-by-wire flight control system, with direct electrical back-up to all surfaces and direct mechanical back-up to tailerons; Litton AN/ASN-130A inertial navigation system (plus GPS from FY93), being replaced by Litton AN/ASN-139 ring laser system (including retrofits); Honeywell digital data recorder and maintenance recording system; flight incident recording and monitoring system (FIRAMS), Smiths standby altimeter, Kearflex standby airspeed indicator, standby vertical speed indicator, cockpit pressure altimeter. Night Attack F/A-18 has Raytheon AN/AAR-50 thermal imaging navigation set (TINS). Litton embedded GPS/inertial (EGI) system being installed in all Navy and Marine Corps and Australian aircraft as retrofit. Finnish aircraft have Telephonics Corporation tactical instrument landing system, permitting operation from road landing strips in bad weather. New Raytheon joint precision approach and landing system (JPALS) under development, with initial sea trials on F/A-18A due to have taken place in November 2000 on USS *Enterprise*. If trials successful and funds available, USN anticipates introducing JPALS to fleet service in about 2005-06.

Instrumentation: Smiths Industries multipurpose colour map display, two Kaiser monochrome MFDs (colour on Night Attack Hornets), central BAE Systems-Honeywell CRT, Kaiser AN/AVQ-28 HUD, BAE Systems FID 2035 horizontal situation display, plus provision for BAE Systems Cat's Eyes NVGs.

Mission: Two Control Data Corporation AN/AYK-14 digital computers, BAE Systems Type 117 laser designator, Harris AN/ASW-25 radio datalink and Lockheed Martin AN/AAS-38B NITE Hawk targeting FLIR. Raytheon selected by Boeing in November 1997 to develop advanced targeting forward-looking infra-red (ATFLIR) sensor system for F/A-18. Conceived as a private venture, the third-generation FLIR was one of five candidates for source selection in 1997 and is also to be used by the Super Hornet; EMD contract award in March 1998 for production and delivery of 10 development units to Boeing for qualification and flight testing beginning in mid-1999. Options for the first two years of low-rate initial production also part of this contract (24 in first year; 33 in second), with IOC planned for mid-2002.

Lockheed Martin Aeronutronics precision direction-finding targeting avionics system (TAS), which permits the AGM-88 HARM to be launched in its most effective 'range-known' mode, has been under development and was successfully demonstrated at China Lake in late 1996. Consideration given in late 1997 to integrating Raytheon AN/ASQ-213 HARM targeting system on US Navy F/A-18C/D aircraft. DRS Technologies WRR-818 cockpit video recording system on US Navy F/A-18C/D. BAE Systems/Rockwell Collins multifunctional information distribution system (MIDS) datalink being installed on F/A-18C/D as upgrade programme.

Self-defence: Raytheon AN/ALR-50 RWR, Raytheon AN/ALR-67(V)2 RHWR, Goodyear AN/ALE-39 chaff dispenser (AN/ALE-47 from FY93 including exports to Finland, Malaysia and Switzerland), Sanders AN/ALQ-126B deception jammers (additionally, Northrop Grumman AN/ALQ-162(V) CW jammers in Canadian, Kuwaiti and later Spanish aircraft). Decision in 1998 to upgrade some AN/ALR-67(V)2s to (V)3 standard. Northrop Grumman AN/ALQ-165 ASPJ jammer selected by Finland and Switzerland. AN/ALQ-165 first installed as temporary measure on 12 US Marine Corps F/A-18Ds for operations over Bosnia in 1995, utilising systems placed in storage when US Navy cancelled ASPJ programme in 1992; two US Navy F/A-18C squadrons also subsequently equipped for carrier-based operation and service recommended to Senate in early 1997 that all F/A-18Cs should be given either ASPJ or limited version of the AN/ALQ-214 integrated defensive electronic countermeasures (IDECM) system installed on the F/A-18E/F Super Hornet.

In September 1997, US General Accounting Office recommended adoption of AN/ALE-50 towed radar decoy, but US Navy reluctant to surrender weapon/fuel capacity and has some concerns about technical problems; in consequence, Navy currently favours ASPJ for F/A-18C/D and is seeking to place new production contracts. AN/ALQ-126B for Malaysia.

ARMAMENT: Nine external weapon stations, comprising two wingtip stations for AIM-9 Sidewinder air-to-air missiles; two outboard wing stations for an assortment of air-to-air or air-to-ground weapons, including AIM-7 Sparrows, AIM-9 Sidewinders, AIM-120 AMRAAMs (launch trials by VX-4 in 1992; cleared for squadron use mid-1993), AGM-84 Harpoon, AGM-65F Maverick and Boeing Standoff Land Attack Missile (SLAM); two inboard wing stations for external fuel tanks, air-to-ground weapons or IMI ADM-141A TALD tactical air-launched decoys; two nacelle fuselage stations for Sparrows or Lockheed Martin AN/ASQ-173 laser spot tracker/strike camera (LST/SCAM) or AN/AAS-38B and AN/AAR-50 sensor pods (see Avionics); and a centreline fuselage station for external fuel or weapons. Air-to-ground weapons include GBU-10 and -12 laser-guided bombs, Mk 82 and Mk 84 general purpose bombs, and CBU-59 cluster bombs. Trials with SLAM ER (SLAM Expanded Response) were under way in 1997-98 on the F/A-18 and it was planned to upgrade existing SLAMs to this configuration for operational introduction in 1999. Australia selected Matra BAe Dynamics ASRAAM short-range AAM for AF-18 in February 1998. An M61A1 20 mm six-barrel gun, with 570 rounds, is mounted in the nose and has a Boeing director gunsight, with a conventional sight as back-up.

DIMENSIONS, EXTERNAL:
Wing span	11.43 m (37 ft 6 in)
Wing span over missiles	12.31 m (40 ft 4¾ in)
Wing chord: at root	4.04 m (13 ft 3 in)
at tip	1.68 m (5 ft 6 in)
Wing aspect ratio	3.5
Width, wings folded	8.38 m (27 ft 6 in)
Length overall	17.07 m (56 ft 0 in)
Height overall	4.66 m (15 ft 3½ in)
Tailplane span	6.58 m (21 ft 7¼ in)
Distance between fin tips	3.60 m (11 ft 9½ in)
Wheel track	3.11 m (10 ft 2½ in)
Wheelbase	5.42 m (17 ft 9½ in)

AREAS:
Wings, gross	37.16 m² (400.0 sq ft)
Ailerons (total)	2.27 m² (24.40 sq ft)
Leading-edge flaps (total)	4.50 m² (48.40 sq ft)
Trailing-edge flaps (total)	5.75 m² (61.90 sq ft)
Fins (total)	9.68 m² (104.20 sq ft)
Rudders (total)	1.45 m² (15.60 sq ft)
Tailerons (total)	8.18 m² (88.10 sq ft)

WEIGHTS AND LOADINGS:
Weight empty	10,810 kg (23,832 lb)
Max fuel weight: internal (JP5)	4,926 kg (10,860 lb)
external: F/A-18 (JP5)	3,053 kg (6,732 lb)
CF-18 (JP4)	4,246 kg (9,360 lb)
Max external stores load	7,031 kg (15,500 lb)
T-O weight: fighter mission	16,651 kg (36,710 lb)
attack mission	23,541 kg (51,900 lb)
max	approx 25,401 kg (56,000 lb)
Max wing loading (attack mission)	600.8 kg/m² (123.06 lb/sq ft)
Max power loading (attack mission)	157 kg/kN (1.54 lb/lb st)

PERFORMANCE:
Max level speed	more than M1.8
Max speed, intermediate power	more than M1.0
Approach speed	134 kt (248 km/h; 154 mph)
Acceleration from 460 kt (850 km/h; 530 mph) to 920 kt (1,705 km/h; 1,060 mph) at 10,670 m (35,000 ft)	under 2 min
Combat ceiling	approx 15,240 m (50,000 ft)
T-O run	less than 427 m (1,400 ft)
Min wind over deck: launching	35 kt (65 km/h; 40 mph)
recovery	19 kt (35 km/h; 22 mph)
Combat radius, interdiction, hi-lo-lo-hi	290 n miles (537 km; 340 miles)
Combat endurance, CAP 150 n miles (278 km; 173 miles) from aircraft carrier	1 h 45 min
Ferry range, unrefuelled	more than 1,800 n miles (3,333 km; 2,071 miles)

UPDATED

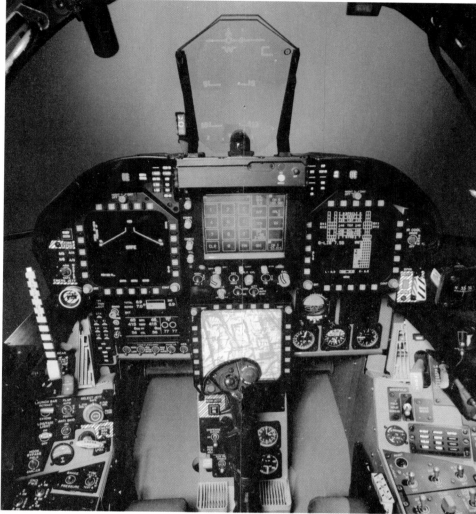
Cockpit of the Boeing F/A-18 Super Hornet

BOEING F/A-18E/F SUPER HORNET

TYPE: Multirole fighter.

PROGRAMME: Proposed 1991 as replacement for cancelled GD/MDC A-12 and follow-on for early F/A-18As and other USN/MC tactical aircraft as they phase out; based on earlier versions of Hornet; development funding approved by Congress for FY92; US$4.88 billion engineering and manufacturing development contract awarded June 1992, covering seven flight test aircraft (five Es; two Fs) and three ground test articles, plus associated 7½-year test programme; US$754 million award in 1992 to GE for F414 engine development.

Critical design review (CDR) undertaken 13 to 17 June 1994 at St Louis by team of independent government evaluators; successfully negotiated, with F/A-18E/F satisfying or surpassing all timescale, cost, technical, reliability and maintainability requirements.

Principal subcontractor Northrop Grumman launched assembly of first aircraft on 24 May 1994 with start of work on centre/aft fuselage section at Hawthorne, California. First forward fuselage section followed suit on new assembly line at St Louis, Missouri, 23 September 1994; completed 12 January 1995, except for wiring; mating with centre/aft section from Hawthorne effected 8 May 1995. Roll-out of prototype (E1/165164) on 18 September 1995; first flight 29 November 1995.

Prototype delivered to Naval Air Warfare Center at Patuxent River, Maryland, on 14 February 1996 for start of three year flight test programme involving seven aircraft; initial flight from Patuxent River made by E1 on 4 March 1996. Second F/A-18E prototype (E2/165165) made first flight on 26 December 1995 and first F/A-18F (F1/165166) on 1 April 1996. Supersonic speed exceeded for first time on 12 April 1996 when E1 achieved M1.1. Carrier suitability trials began mid-1996 and F1 completed three successful catapult launches at Patuxent River on 6 August that year.

By 1 February 1997, when the last development aircraft was delivered to Patuxent River, test fleet had flown 390 sorties, for a total of 631 flight hours. Subsequent milestones were passed on 13 May 1997 (1,000th flight hour); 9 September 1997 (1,500th flight hour); 24 September 1997 (1,000th sortie); 11 December 1997 (2,000th flight hour and 1,300th sortie); 1 July 1998 (3,000th hour) and 12 January 1999 (4,000th hour). By 5 December 1998, the seven development aircraft had completed 2,615 sorties and accumulated 3,909.6 flying hours, extending envelope to M1.75, +7.5/−3 g and also demonstrating controlled flight at +180/−180° angles of attack. First missile launch (an AIM-9 Sidewinder) accomplished by aircraft F2 on 5 April 1997; other weapons expended by mid-May 1997 included AIM-7 Sparrow, AIM-120 AMRAAM, SLAM/SLAM-ER, AGM-84 Harpoon, Mk 82 and Mk 83 bombs plus Rockeye CBUs. Successful deployment of AN/ALE-50 towed radar decoy also accomplished by mid-May 1997.

Flight test programme briefly halted in October 1997 for engine inspection after discovery of cracks in stator vanes.

Canadian Armed Forces CF-18B (nearest) and CF-18A *(Lindsay Peacock)*

582   USA: AIRCRAFT—BOEING

Two-seat F/A-18Fs are included in the strength of training squadron VFA-122

More intractable problem concerned wing-drop, with uncommanded departures from controlled flight evident as early as the seventh sortie in March 1996. Boeing and US Navy considered three solutions to eradicate this in January 1998, including fitting stall strips on upper surface; adding a chord-wise fence just inboard of the hinge fairing; and switching to a porous hinge fairing with slots that allow air to flow in both directions through wing fold fairing. Last-mentioned option selected in early 1998; subsequent flight testing confirmed efficacy of solution.

EMD phase of test programme completed at end of April 1999, by which time the fleet of seven aircraft had accumulated 4,673 flight hours in 3,172 sorties; in the process, over 15,000 test points completed and 29 weapons configurations cleared for flight.

Static testing began in August 1995 with airframe ST50; shock loading assessment from February 1996 with DT50; fatigue testing from 30 June 1997 with FT50, which completed first lifetime (6,000 hours) one month ahead of schedule, on 27 August 1998. ST50 transferred to Lakehurst, New Jersey, for series of six emergency barricade engagements; first successfully completed on 3 September 1997, but ST50 damaged during third test on 23 September when restraint cable failed, allowing aircraft to overturn before coming to rest in a wood. ST50 subsequently repaired and returned to test duty, for live-fire testing at China Lake, California; these trials included firing large armour-piercing incendiary projectile through inlet duct into aft fuel tank, with resultant hole showing little evidence of leakage.

Approval for low-rate initial production (LRIP) of 62 aircraft in three lots given on 26 March 1997; first lot composed of eight F/A-18Es and four F/A-18Fs. Assembly of first LRIP F/A-18E (165533) started at Northrop Grumman in May 1997 and at Boeing in September 1997; final assembly began at St Louis on 19 June 1998, with mating of centre/aft and forward fuselage sections; first flight 6 November 1998, six weeks ahead of schedule; this aircraft officially accepted by US Navy on 18 December 1998 and flown to Patuxent River to join flight test programme before attachment to VX-9 squadron at China Lake, California, for operational evaluation. Latter programme began 27 May 1999 and involved total of 1,233 flight hours in 866 sorties by seven LRIP aircraft (three F/A-18Es and four F/A-18Fs) in six-month period, including testing of all mission capabilities in varying climates as well as operations at sea aboard USS *John C Stennis* and participation in 'Red Flag' exercise at Nellis AFB, Nevada. Results of operational evaluation announced 15 February 2000, with report stating aircraft to be "operationally effective and operationally suitable" and recommending a go-ahead for fleet introduction with the US Navy.

First USN squadron is VFA-122, at Lemoore, California, as specialist FRS (Fleet Replacement Squadron), with responsibility for training pilots and ground crew; VFA-122 received first seven aircraft (of eventual 34) 17 November 1999. First operational squadron to be VFA-115, which began transition from F/A-18C to F/A-18E in third quarter 2000; currently expected to make first deployment in USS *Abraham Lincoln* from June 2002. Next two squadrons to be VF-14 and VF-41, with transition from F-14 Tomcat scheduled for early 2001; VF-14 will equip with F/A-18E, while VF-41 to receive F/A-18F, but both will be assigned to Carrier Air Wing 11 (CVW-11) in USS *Nimitz*. VFA-97 also earmarked to get F/A-18E by FY04, with the F/A-18F to replace F-14 with VF-102 (FY03), VF-154 (FY07), VF-2 (FY08) and VF-32 (FY09). Second FRS scheduled to establish at Oceana, Virginia, in second half of 2003.

CURRENT VERSIONS (general): **F/A-18E**: Single-seat.

F/A-18F: Two-seat.

F/A-18 C²W: Private venture development of F/A-18F as two-seat electronic warfare aircraft; announced 7 August 1995; Following merger, Boeing is prime contractor, with Northrop Grumman as principal subcontractor with special responsibility for integration of electronic warfare suite; potential replacement for Grumman EA-6B Prowler; minimal structural changes; wideband receiver pods replace wingtip Sidewinder AAMs; other pods and antennas on weapon pylons; satcom receiver behind cockpit. Studies continue, with go-ahead decision not expected before FY02; in 1997, US Navy formulated requirements for new EW aircraft to replace Prowler from 2007 onwards, although Navy currently considering retention of EA-6B until at least 2015. In the meantime, Boeing continued with simulator work to ensure good crew/vehicle interface on EW-dedicated variant and has also been investigating feasibility of using EA-6B equipment such as AN/ALQ-99 jamming system pods on Super Hornet to augment Prowler aircraft. Future of this proposal, unofficially known as **F/A-18G Growler**, unlikely to be resolved until two-year Analysis of Alternatives study is completed in late 2001; this tasked with establishing joint service needs in EW mission, with various options available, including Growler/Super Hornet, UCAVs and larger aircraft of Boeing 757/767 size.

CURRENT VERSIONS (specific): Integrated test team (including five pilots from US Navy and five from industry) was established at Naval Air Warfare Center at Patuxent River in early 1996 for operation from the Hazelrigg Flight Test Facility. Duties of five F/A-18E and two F/A-18F engineering and manufacturing development aircraft during three year programme described in 2000-01 and earlier editions of *Jane's*.

CUSTOMERS: US Navy. Seven prototypes; approval for 12 LRIP aircraft in FY97 (Lot 1), with manufacture rising incrementally to peak rate of 48 per annum during FY98 to FY02. See table for details of three LRIP batches and first multiyear procurement (MYP) batch; contract for latter signed on 15 June 2000. Broad requirement originally identified for over 1,000 by FY15, but Quadrennial Defense Review reduced this to minimum of 548 and maximum of 785. Revised total excludes proposed EW version. Boeing also seeking export sales, following approval by US government; Super Hornet will be offered to Australia, which seeking new multirole fighter under Project Air 6000; Chile also a potential customer, while other possible export orders could originate from Hungary, Malaysia and Poland.

SUPER HORNET PROCUREMENT

FY	Batch	Total	Remarks
97	LRIP 1	12	Eight F/A-18E and four F/A-18F (Block 52)
98	LRIP 2	20	Eight F/A-18E and 12 F/A-18F (Block 53)
99	LRIP 3	30	14 F/A-18E and 16 F/A-18F
00	MYP1	36	
01		42	
02		48	
03		48	
04		48	
Total		284	

COSTS: Development estimated US$4.8 billion (1992); US$1,089 million in FY92 budget; US$943 million in FY93; approximately US$1,500 million in FY94 and US$1,348 million requested for FY95. US$2,600 million in provisional FY97 budget for first 12 aircraft; with US$2,100 million requested for FY98 procurement of 20 aircraft. US GAO estimated flyaway unit cost as being of the order of US$43.6 million in 1996, based on original planned procurement of 1,000. Most recent figures, for first MYP batch, put total cost at US$8.9 billion for 222 aircraft. Unit cost quoted as US$48 million in mid-2000, when Boeing pursuing measures to drive cost down to around US$40 million.

DESIGN FEATURES: Generally as for first-generation Hornet (which see). Stretched versions of F/A-18C/D; gross landing weight increased by 4,536 kg (10,000 lb); 0.86 m (2 ft 10 in) fuselage plug; wings photometrically increased in size to provide 9.29 m² (100.0 sq ft) extra area and 1.31 m (4 ft 3½ in) span increase; control surfaces disproportionately enlarged and dogtooth added to leading-edge for increased aileron authority. Wings 2.5 cm (1 in) deeper at root; larger horizontal tail surfaces; LEX size substantially increased in early 1993 (from total 5.8 m²; 62.4 sq ft to 7.0 m²; 75.3 sq ft, compared with 5.2 m²; 56.0 sq ft on F/A-18C/D), ensuring full manoeuvre capability at beyond 40° AoA; also incorporates spoilers on upper surface of LEX as speedbrake and to increase nose-down control authority. Nevertheless, has 42 per cent fewer parts than immediate predecessor.

Additional 1,637 kg (3,600 lb) of internal and 1,406 kg (3,100 lb) of external fuel; 40 per cent extra range; further two (making 11) weapon hardpoints (stations 2 and 10, inboard of wingtips, for AAMs and ASMs of up to 520 kg; 1,146 lb); 'bring-back' weapons load increased to 4,082 kg (9,000 lb); additional survivability measures; air intakes redesigned and slewed to increase mass flow to more powerful F414-GE-400 engines and also changed to 'caret' shape to reduce radar signature. Incorporates several other 'affordable' stealth features to reduce radar cross-section, including saw-toothed doors and panels, realigned joints and edges and angled antennas.

FLYING CONTROLS: As F/A-18C/D, except that horizontal stabilisers automatically assume neutral position if one is

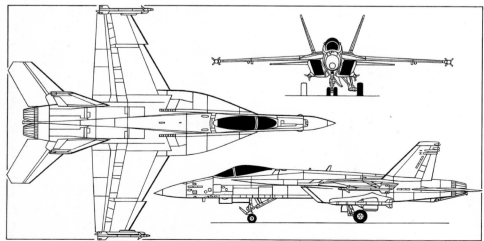

Boeing F/A-18E Super Hornet *(Mike Keep/Jane's)*

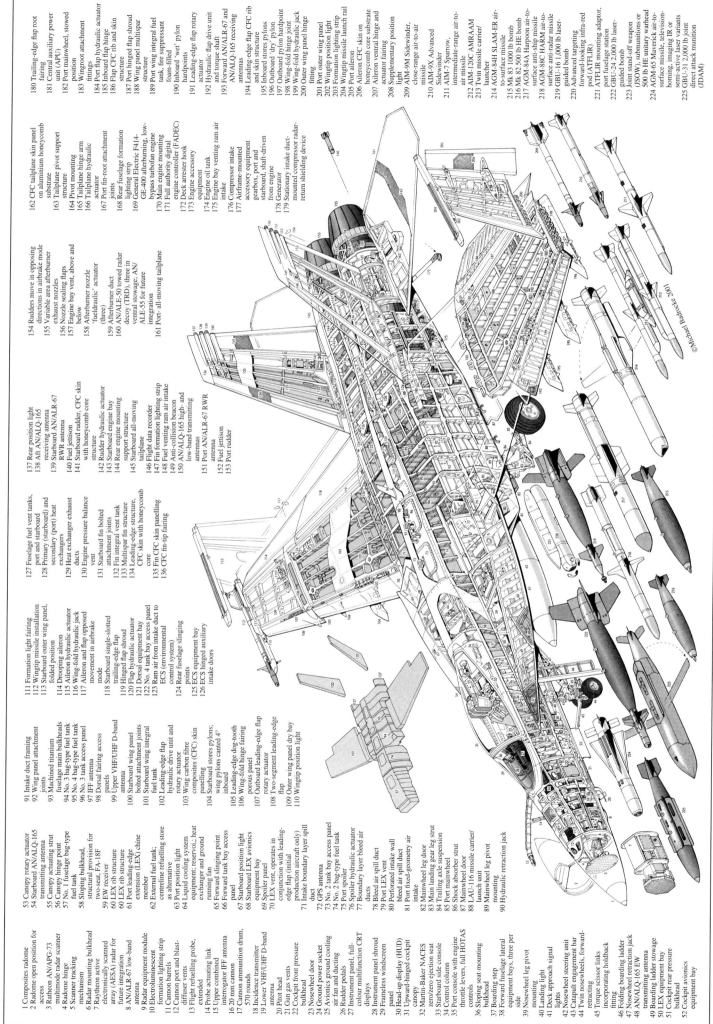

Boeing F/A-18E Super Hornet, cutaway drawing key

1 Composites radome
2 Radome open position for access
3 Raytheon AN/APG-73 multimode radar scanner
4 Radome hinge
5 Scanner tracking mechanism
6 Radar mounting bulkhead
7 Raytheon active electronically scanned array (AESA) radar for future integration
8 AN/ALR-67 low-band antenna
9 Radar equipment module
10 Electroluminescent formation lighting strip
11 Cannon barrels
12 Cannon port and blast-diffuser vents
13 Flight refuelling probe, extended
14 Probe actuating link
15 Upper combined interrogator IFF antenna
16 20 mm cannon
17 Cannon ammunition drum, 570 rounds
18 Incidence transmitter
19 Lower VHF/UHF D-band antenna
20 Pitot head
21 Gun gas vents
22 Cockpit front pressure bulkhead
23 Nosewheel door
24 Ground power socket
25 Avionics ground cooling air fan and ducting
26 Rudder pedals
27 Port spoiler
28 Instrument panel, full-colour multifunction CRT displays
29 Frameless windscreen
30 Head-up display (HUD)
31 Upward-hinged cockpit canopy
32 Martin-Baker NACES zero/zero ejection seat
33 Starboard side console
34 Control column
35 Port console with engine throttle levers, full HOTAS controls
36 Sloping seat mounting bulkhead
37 Boarding step
38 Forward fuselage lateral equipment bays, three per side
39 Nosewheel leg pivot mounting
40 Landing light
41 Deck approach signal lights
42 Nosewheel steering unit
43 LAU-116 missile carrier/launch unit
44 Twin nosewheels, forward-retracting
45 Torque scissor links
46 Folding boarding ladder
47 Nosewheel retraction jack
48 AN/ALQ-165 EW transmitting antenna
49 Boarding ladder stowage
50 LEX equipment bay
51 Cockpit rear pressure bulkhead
52 Cockpit avionics equipment bay
53 Canopy rotary actuator
54 Starboard AN/ALQ-165 transmitting antenna
55 Canopy actuating strut
56 Canopy hinge point
57 No. 1 fuselage bag-type fuel tank
58 Sloping bulkhead, structural provision for two-seat, FA-18F
59 EW receiver
60 LEX rib structure
60 LEX rib structure
61 Port leading-edge extension (LEX) chine member
62 External fuel tank; centreline refuelling store as alternative
63 Port position light
64 Liquid cooling system
65 Forward slinging point
66 Forward tank bay access panel
67 Starboard position light
68 Starboard LEX avionics equipment bay
69 Spoiler panel
70 LEX vent, operates in conjunction with leading-edge flap (initial production aircraft only)
71 Intake boundary layer spill duct
72 GPS antenna
73 No. 2 tank bay access panel
74 No. 2 bag-type fuel tank
75 Port spoiler
76 Spoiler hydraulic actuator
77 Boundary layer bleed air ducts
78 Bleed air spill duct
79 Port LEX vent
80 Perforated intake wall bleed air spill duct
81 Port fixed-geometry air intake
82 Mainwheel leg door
83 Main landing gear leg strut
84 Trailing axle suspension
85 Shock absorber strut
86 Shock absorber strut
87 Mainwheel door
88 LAU-116 missile carrier/launch unit
89 Mainwheel leg pivot mounting
90 Hydraulic retraction jack
91 Intake duct framing
92 Wing panel attachment joints
93 Machined titanium fuselage main bulkheads
94 No. 3 bag-type fuel tank
95 No. 4 bag-type fuel tank
96 No. 3 tank access panel
97 IFF antenna
98 Dorsal fairing access panels
99 Upper VHF/UHF D-band antenna
100 Starboard wing panel bolted attachment joints
101 Starboard wing integral fuel tank
102 Leading-edge flap hydraulic drive unit and rotary actuator
103 Wing carbon fibre composites (CFC) skin panelling
104 Starboard stores pylons; wing pylons canted 4° inboard
105 Leading-edge dog-tooth
106 Wing-fold hinge fairing porous panel
107 Outboard leading-edge flap rotary actuator
108 Two-segment leading-edge flap
109 Outer wing panel dry bay
110 Wingtip position light
111 Formation light fairing
112 Wingtip missile installation
113 Starboard outer wing panel, folded position
114 Drooping aileron
115 Aileron hydraulic actuator
116 Wing-fold hydraulic jack
117 Aileron and flap opposed movement in airbrake mode
118 Starboard single-slotted trailing-edge flap
119 Hinged flap shroud
120 Flap hydraulic actuator
121 Dorsal equipment bay
122 No. 4 tank bay access panel
123 Ram air from intake duct to ECS (environmental control system)
124 Rear fuselage slinging points
125 ECS equipment bay
126 ECS auxiliary intake doors
127 Fuselage fuel vent tanks, port and starboard
128 Primary (starboard) and secondary (port) heat exchangers
129 Heat exchanger exhaust ducts
130 Engine pressure balance vent
131 Starboard rudder, CFC skin with honeycomb core structure
131 Starboard fin bolted attachment joints
132 Fin integral vent tank
133 Multispar fin structure
134 Leading-edge structure, CFC skin with honeycomb core
135 Fin CFC skin panelling
136 CFC fin-tip fairing
137 Rear position light
138 Aft AN/ALQ-165 receiving antenna
139 Starboard AN/ALR-67 RWR antenna
140 Fuel jettison
141 Starboard rudder, CFC skin with honeycomb core structure
142 Rudder hydraulic actuator
143 Starboard engine bay
144 Rear engine mounting support structure
145 Starboard all-moving tailplane
146 Flight data recorder
147 Fin formation lighting strip
148 Fuel venting ram air intake
149 Anti-collision beacon
150 AN/ALQ-165 high- and low-band transmitting antennas
151 Port AN/ALR-67 RWR antenna
152 Fuel jettison
153 Port rudder
154 Rudders move in opposing directions in airbrake mode
155 Variable area afterburner exhaust nozzles
156 Nozzle sealing flaps
157 Engine bay vent, above and below
158 Afterburner nozzle 'fueldraulic' actuator (three)
159 Afterburner duct
160 AN/ALE-50 towed radar decoy (TRD); three in ventral stowage; AN/ALE-55 for future integration
161 Port- all-moving tailplane
162 CFC tailplane skin panel on aluminium honeycomb substrate
163 Tailplane pivot support structure
164 Pivot mounting
165 Tailplane hinge arm
166 Tailplane hydraulic actuator
167 Port fin-root attachment joints
168 Rear fuselage formation lighting strip
169 General Electric F414-GE-400 afterburning, low-bypass turbofan engine
170 Main engine mounting
171 Full authority digital engine controller (FADEC)
172 Deck arrester hook
173 Engine accessory equipment
174 Engine oil tank
175 Engine bay venting ram air intake
176 Compressor intake
177 Airframe-mounted accessory equipment gearbox, port and starboard, shaft-driven from engine
178 Generator
179 Stationary intake duct-mounted compressor radar-return shielding device
180 Trailing-edge flap root fairing
181 Central auxiliary power unit (APU)
182 Port mainwheel, stowed position
183 Wingroot attachment fittings
184 Port flap hydraulic actuator
185 Inboard flap hinge
186 Flap CFC rib and skin structure
187 Port hinged flap shroud
188 Wing panel multispar structure
189 Port wing integral fuel tank, fire suppressant foam-filled
190 Inboard 'wet' pylon hardpoints
191 Leading-edge flap rotary actuator
192 Hydraulic flap drive unit and torque shaft
193 Forward AN/ALR-67 and AN/ALQ-165 receiving antennas
194 Leading-edge flap CFC rib and skin structure
195 Outboard stores pylons
196 Outboard 'dry' pylon
197 Outboard pylon hardpoint
198 Wing-fold hinge rib
199 Wing-fold hydraulic jack
200 Outer wing panel hinge fitting
201 Port outer wing panel
202 Wingtip position light
203 Formation lighting strip
204 Wingtip missile launch rail
205 Port aileron
206 Aileron CFC skin on honeycomb core substrate
207 Aileron ventral hinge and actuator fairing
208 Supplementary position light
209 AIM-9M Sidewinder, close-range air-to-air missile
210 AIM-9X Advanced Sidewinder
211 AIM-7 Sparrow, intermediate-range air-to-air missile
212 AIM-120C AMRAAM
213 Twin missile carrier/launcher
214 AGM-84H SLAM-ER air-to-surface missile
215 Mk 83 1000 lb bomb
216 Mk 82 500 lb HE bomb
217 AGM-84A Harpoon air-to-surface anti-ship missile
218 AGM-88C HARM air-to-surface anti-radar missile
219 GBU-16 1,000 lb laser-guided bomb
220 Advanced targeting forward-looking infra-red pod (ATFLIR)
221 ATFLIR mounting adaptor, port fuselage station
222 GBU-24 2,000 lb laser-guided bomb
223 Joint stand-off weapon (JSOW), submunitions or 500 lb HE unitary warhead
224 AGM-65 Maverick air-to-surface missile, television-homing, imaging IR or semi-active laser variants
225 GBU-31 2,000 lb joint direct attack munition (JDAM)

USA: AIRCRAFT—BOEING

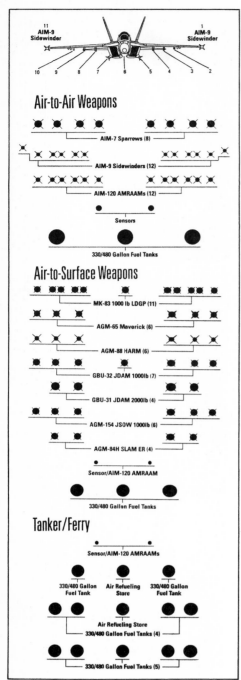

Sample of Super Hornet weapon options

F/A-18F Super Hornet *(Paul Jackson/Jane's)*

damaged, pitch control being then passed to other surfaces. Upper fuselage airbrake deleted on F/A-18E/F; replaced by rudder and flap deflection for aerodynamic braking on landing. No mechanical back-up to FBW system.

STRUCTURE: As F/A-18C/D.

LANDING GEAR: As F/A-18C/D.

POWER PLANT: Two General Electric F414-GE-400 turbofans, each rated at approximately 97.9 kN (22,000 lb st) with afterburning. Internal fuel capacity (JP-5 fuel) 8,063 litres (2,130 US gallons; 1,774 Imp gallons). Wing tanks protected from combat damage by low-density foam. Provision for five 1,250 litre (330 US gallon; 275 Imp gallon) or Lincoln Composites 1,817 litre (480 US gallon; 400 Imp gallon) external tanks, giving maximum fuel capacity of 17,148 litres (4,530 US gallons; 3,772 Imp gallons) and ability to operate in air refuelling role. Normal operational fit anticipated as three external tanks of either size.

ACCOMMODATION: As F/A-18C/D.

SYSTEMS: High commonality with F/A-18C/D, but more powerful actuators to accommodate enlarged control surfaces with increased deflections. Leland Electrosystems power generating system; provides 60 per cent more electrical power than in F/A-18C. Hamilton Sundstrand air conditioning; Vickers hydraulic pumps.

AVIONICS: Over 90 per cent commonality with F/A-18C, but differences include following:

Comms: Rockwell Collins AN/ARC-210 secure UHF/VHF radio to be incorporated in 1999. Multifunction information distribution system (MIDS), Rockwell Collins digital communication system and Hazeltine combined interrogator transponder all to be installed as part of upgrade package in 2001, before first operational deployment.

Radar: Raytheon AN/APG-73 multimode, digital air-to-air and air-to-ground radar as standard. Raytheon active electronically scanned array (AESA) radar under development for use on Super Hornet from 2006, initially with company funding, though US Navy is expected to sign EMD contract in early 2001. Current plans anticipate procurement of at least 258 AESA radars, following trials in Boeing 737 avionics testbed from early 2002; low-rate initial production likely to involve 42 units in FY03-FY05, with full-rate production starting in FY06.

Flight: Litton embedded GPS to be incorporated in 1999. Smiths/Harris tactical aircraft moving map capability (TAMMAC) to be provided in 2001. DRS Technologies deployable flight incident recorder set (DFIRS).

Instrumentation: Cockpit upgraded with 76 × 127 mm (3 × 5 in) touch-panel LCD upfront display and 159 mm (6¼ in) square colour LCD tactical situation display; retains two 127 mm (5 in) square monochrome displays and will have monochrome programmable LCD in place of existing F/A-18C engine/fuel display. Planar Advance/dpiX awarded joint development contract for Eagle-6 multifunction AMLCD in first half of 1998.

Mission: Raytheon developing advanced targeting forward-looking infra-red (ATFLIR) targeting and navigation system, with award of EMD contract in March 1998 and IOC to follow in mid-2002. Lockheed Martin Super Hornet Advanced Reconnaissance Pod (SHARP) system under development. DRS Technologies WRR-818 cockpit video recording system.

Advanced mission computers based on commercial hardware and software and colour flat panel LCD displays are in development for incorporation by 2004, with DY4 Systems awarded US$1 million order to upgrade existing system which relies on Control Data Corporation AN/AYK-14 digital computers; upgrade will involve replacement of AN/AYK-14 by DMV-179 single board computers and PMC-642 fibre channel network interface module. Kaiser/Elbit joint helmet-mounted cueing system also to be installed by 2003 to target the AIM-9X high off-boresight missile.

Self-defence: Management by Sanders integrated defensive electronic countermeasures suite (IDECM); interfaces with Raytheon AN/ALR-67(V)3 radar warning receiver; Sanders AN/ALE-55 fibre optic towed radar decoy (with triple dispenser stowed between jetpipes); and four BAE Systems AN/ALE-47 chaff/flare dispensers. However, delay and cost growth of IDECM means that first three operational squadrons will have aircraft fitted with less sophisticated Raytheon AN/ALE-50 towed decoy in conjunction with either AN/ALQ-165 ASPJ (first two squadrons) or AN/ALQ-214 radio frequency countermeasures system (third squadron). Definitive AN/ALQ-214 and AN/ALE-55 combination to be deployed with effect from fourth squadron, by the end of 2003, while earlier aircraft will be updated.

ARMAMENT: See diagram. Full range of USN offensive and defensive ordnance. At least 29 weapons combinations cleared before service entry. Lightweight internal cannon in place of M61A1 as used by earlier Hornet; is also to be compatible with forthcoming AIM-9X missile.

DIMENSIONS, EXTERNAL (approx):
Wing span over missiles	13.62 m (44 ft 8½ in)
Wing aspect ratio	4.0
Width, wings folded	9.94 m (32 ft 7¼ in)
Length overall	18.38 m (60 ft 3½ in)
Height overall	4.88 m (16 ft 0 in)

AREAS:
Wings, gross	46.45 m² (500.0 sq ft)

WEIGHTS AND LOADINGS:
Weight empty	13,864 kg (30,564 lb)
Max fuel weight: internal (JP-5)	6,559 kg (14,460 lb)
external (JP-5)	7,430 kg (16,380 lb)
Max external stores load: T-O	8,028 kg (17,700 lb)
landing	4,082 kg (9,000 lb)
T-O weight, attack mission	29,937 kg (66,000 lb)
Max wing loading	644.5 kg/m² (132.00 lb/sq ft)
Max power loading	153 kg/kN (1.50 lb/lb st)

PERFORMANCE (estimated):
Max level speed at altitude	more than M1.8
Approach speed	125 kt (232 km/h; 144 mph)
Combat ceiling	15,240 m (50,000 ft)
Min wind over deck:	
launching	30 kt (56 km/h; 34.5 mph)
recovery	15 kt (28 km/h; 17.5 mph)

Combat radius: interdiction with four 1,000 lb bombs, two Sidewinders and three 1,817 litre (480 US gallon; 400 Imp gallon) external tanks, navigation FLIR and targeting FLIR, hi-hi-hi
665 n miles (1,231 km; 765 miles)

Combat endurance: maritime air superiority, six AAMs, three 1,818 litre (480 US gallon; 400 Imp gallon) external tanks, 150 n miles (278 km; 173 miles) from aircraft carrier 2 h 15 min

g limit +7.5

UPDATED

BOEING C-17A GLOBEMASTER III

TYPE: Strategic transport.

PROGRAMME: US Air Force selected McDonnell Douglas to develop C-X cargo aircraft 28 August 1981; full-scale development called off January 1982 and replaced on 26 July 1982 by slow-paced preliminary development order; development and three prototypes (one flying) ordered 31 December 1985; fabrication of first C-17A (T1/87-0025) began 2 November 1987; first production C-17A contract 20 January 1988; assembly started at Long Beach 24 August 1988; assembly of prototype completed 21 December 1990. Programme transferred from Douglas Aircraft Company to McDonnell Douglas Aerospace in 1992; to McDonnell Douglas Military Transport Aircraft in 1996 and to Boeing following merger of August 1997.

First flight (87-0025) 15 September 1991 – also delivery to Edwards AFB; first flight of initial production aircraft (P1/88-0265) 18 May 1992; first delivery to operational unit 14 June 1993 (for details of test milestones see 1998-99 *Jane's*). First overseas service flight (P11/92-3291) to Mildenhall, UK, 25 May 1994.

C-17 was named Globemaster III on 5 February 1993; peak production target 15 per year; assembly in new 102,200 m² (1.1 million sq ft) facility at Long Beach, California, dedicated 13 August 1987. Feasibility study in 1991 for hose-drogue tanker/transport combi; promotion of this variant began in 1996.

Development flight testing completed 15 December 1994, by which time 16 production aircraft delivered and 22nd record set. Initial AMC Squadron (17th AS) received its 12th C-17A 22 December 1994; achieved IOC 17 January 1995 with acceptance of 13th (nominal spare) aircraft. Second squadron (14th AS) received its first aircraft 17 February 1995. Air Education and Training Command's 97th AMW at Altus AFB, Oklahoma, subsequently took delivery of eight aircraft between March 1996 and November 1997, all of which had previously been assigned to the 437th AW. Deliveries to 7th AS of 62nd AW at McChord AFB, Washington, began at end of July 1999. C-17 fleet passed 150,000 flying hours in 1999, excluding more than 1,000 hours accumulated by the test-dedicated prototype aircraft.

Further procurement of C-17 for USAF a possibility, in light of statement in late 1997 by commander of US Transportation Command that an additional 15 aircraft (of which funding for 14 has been identified in the post-FY00 period) are required to support special operations; this in direct contradiction to earlier US General Accounting Office report advocating an end to deliveries to USAF after 100th C-17 at a saving of about US$7 billion on grounds

BOEING—AIRCRAFT: USA

Desert airstrip landing by a Boeing C-17A Globemaster III of 437th Airlift Wing   2000/0064919

that this would be sufficient to perform missions which entail use of short and austere airfields.
CURRENT VERSIONS: **C-17A**: Basic standard.
Detailed description applies to C-17A.
    C-17B: Not assigned. Had been used unofficially (with C-17ER) to identify aircraft with extended-range fuel tanks (Lot 12 and upwards).
    KC-17: Private venture tanker/transport project, offered as replacement for USAF KC-135R/T Stratotanker. Additional fuel in wing centre-section and in palletised cargo hold tank; interchangeable cargo hold upper door fitted with 'flying boom' and/or hose/drum unit refuelling equipment; optional underwing hose/drum pods at 75 per cent span using hardpoints fitted as standard to C-17A. Palletised refuelling operators' station in cargo hold receives imagery from closed-circuit TV pod scabbed to hold upper door. Total normal usable fuel capacity 102,294 litres (27,024 US gallons; 22,502 Imp gallons). Optional 37,853 litre (10,000 US gallon; 8,327 Imp gallon) centre-section tank in KC-17; further 25,362 litres (6,700 US gallons; 5,579 Imp gallons) in optional palletised cargo hold tank of KC-17. Total usable capacity (optional) 165,509 litres (43,724 US gallons; 36,408 Imp gallons).
    BC-17X: Projected civil cargo version, known until 2000 as MD-17, described separately.
CUSTOMERS: US Air Force; original requirement 210; cut to 120 by 1991; capped at 40 in January 1994 for two year probationary period, during which contractor to achieve performance, cost and delivery targets; decision taken on 3 November 1995 to acquire the balance of 80 C-17s. Multiyear procurement contract signed 31 May 1996 for production through 2004; first aircraft from this contract (97-0041) delivered 10 August 1998; last due on 30 November 2004.
    Basing plans were revealed soon after the decision to procure 120 in all, when the USAF announced that the 437th AW at Charleston AFB, South Carolina, and the 62nd AW at McChord AFB, Washington, will each receive 48 Globemaster IIIs, with these respectively being supported by Air Force Reserve Command personnel of the 315th AW and 446th AW. A further eight C-17As are assigned to the 97th Air Mobility Wing at Altus AFB, Oklahoma, for training duties. Finally, six will be assigned from 2003 to the Air National Guard's 172nd AW at Jackson, Mississippi. The remaining 10 aircraft are to be distributed among those units as reserves to cover scheduled depot level maintenance.
    The first C-17A to be handed over to Air Mobility Command was the final aircraft from Lot 2 (P6/89-1192), which made its first flight on 8 May 1993 and was ferried to Charleston AFB for the 437th AW on 14 June 1993. Subsequent deliveries to Air Mobility Command comprise four Lot 3 aircraft between 26 August 1993 and 8 February 1994; four Lot 4 aircraft between 8 April and 20 August 1994; six Lot 5 aircraft between 28 September 1994 and 19 June 1995; six Lot 6 aircraft between 31 July 1995 and May 1996; six Lot 7 aircraft between August 1996 and June 1997; eight Lot 8 aircraft between August 1997 and May 1998; eight Lot 9 aircraft between August 1998 and February 1999; nine Lot 10 aircraft between April and December 1999, including the 50th C-17A, which was accepted by the USAF on 21 May 1999; and 13 Lot 11 aircraft during 2000. By end of 2000, 70 C-17As had been delivered.
    In early 1997, UK Ministry of Defence reportedly considering acquisition of a small number of C-17s for service with RAF in strategic airlift role; no formal requirement then existed, but cost data were received from the USA. Invitation to submit bids for supply on lease basis of up to four 'C-17 equivalent' Short Term Strategic Airlift aircraft issued at end September 1998, but subsequent criticism by other potential bidders that requirement was weighted in favour of C-17 prompted UK MoD to waive some requirements. Submissions made in early 1999, but competition then suspended in August 1999, when none of competing bids proved acceptable. Restarted, late 1999; announced that C-17 selected on 16 May 2000, with total of four aircraft to be delivered to No. 99 Squadron at Brize Norton during 2001. These are identical to USAF C-17s and taken from Lot 12 production batch; assembly of first officially begun on 28 June 2000, although lease agreement with Chase Manhattan Bank not signed until 12 January 2001. Each aircraft valued at £197 million.
    C-17 also a contender for separate RAF programme to replace C-130K Hercules; formal RFP issued 31 July 1998; Boeing/BAE proposal reportedly includes possibility of re-engining with version of Rolls-Royce RB211 as part of bid to supply up to 25 aircraft. Competing aircraft are Antonov An-7X (An-70 variant) and Airbus A400M. Canada also has requirement for new strategic transport and is believed to be examining several options, including lease or outright purchase of C-17.
COSTS: Multiyear procurement contract for 80 aircraft signed in May 1996 worth US$14.2 billion; separate US$1.6 billion contract followed in December 1996 for 320 F117 engines. Unit price proposed in November 1995 was US$190 million, but further cost-cutting initiatives are expected to result in average unit price of US$175 million (1996 dollars); recent figures (early 2000) refer to unit cost of US$199 million). Boeing proposal for 60 additional C-17s at average unit flyaway cost of US$149 million (FY99 dollars) examined by USAF and Congress in 1999-2000 but eventually rejected by USAF on grounds of overall cost being too high. However, further procurement not ruled out to satisfy special forces requirement for 15 aircraft from FY04.
DESIGN FEATURES: Typical T-tail, upswept rear fuselage, high-wing, podded-engine transport, deriving much detail, including externally blown flap system, from McDonnell Douglas YC-15 medium STOL transport prototypes (see 1979-80 *Jane's*), which involves extended flaps being in exhaust flow from engines during take-off and landing. Combines load-carrying capacity of Lockheed C-5 with STOL performance of Lockheed Martin C-130; required to operate routinely from 915 m (3,000 ft) long and 27.45 m (90 ft) wide runways, complete 180° three-point turn in 25 m (82 ft) and back-taxi up 1 in 50 gradient when fully loaded using thrust reversers. Structure designed to survive battle damage and protect crew; essential line-replaceable units (LRU) to be replaceable in flight; rear-loading ramp.
    Supercritical wing with 25° sweepback; 2.72 m (8.9 ft) high NASA winglets, angled at 15° from vertical and with 30° sweep.
FLYING CONTROLS: First military transport with all-digital FBW control system and two-crew cockpit with central stick controllers; outboard ailerons and four spoilers per wing; four elevator sections; two-surface rudder split into upper and lower segments; full-span leading-edge slats; two-slot, fixed-vane, simple hinged flaps over about two-thirds of trailing-edge; small strakes under tail. Quadruple-redundant Lockheed Martin digital fly-by-wire flight control system, with mechanical back-up.
STRUCTURE: Major subassemblies produced in factory at Macon, Georgia; some 50 subcontractors, of which 21 for

C-17 PROCUREMENT

FY	Lot	Qty	Cum	First aircraft	Delivery
-	Proto	1	(1)	87-0025	15 Sep 1991
88	1	2	2	88-0265	18 May 1992
89	2	4	6	89-1189	7 Sep 1992
90	3	4	10	90-0532	26 Aug 1993
91	-	-	-	-	-
92	4	4	14	92-3291	8 Apr 1994
93	5	6	20	93-0599	29 Sep 1994
94	6	6	26	94-0065	31 Jul 1995
95	7	6	32	95-0102	5 Nov 1996
96	8	8	40	96-0001	29 Aug 1997
97	9	8	48	97-0041	10 Aug 1998
98	10	9	57	98-0049	22 April 1999
99	11	13	70	99-0058	17 March 2000
00	12	15	85	00-0171	Feb/Mar 2001
	12	4 (RAF)	89	00-0201/ZZ171	May 2001
01[1]	13	15	104	-	Aug 2002
02[1]	14	15	119	-	Aug 2003
03[1]	15	5	124	-	Aug 2004
Total		**124 + 1**			

Notes:
[1] Quantities to be purchased in FY01-03 may be altered to 12, 15 and eight respectively

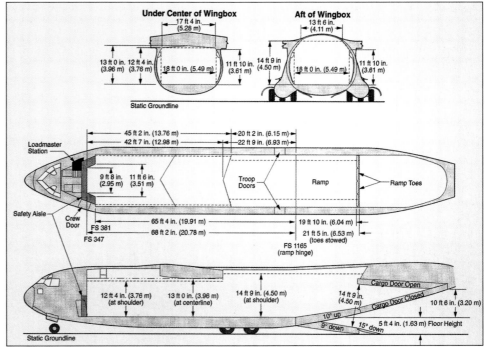

C-17A cargo area dimensions   0015910

586 USA: AIRCRAFT—BOEING

airframe; subcontractors include Northrop Grumman (composites ailerons, rudder, elevators, vertical and horizontal stabilisers, engine nacelles and thrust reversers), Reynolds Metals Company (wing skins), Contour Aerospace (wing spars and stringers), Kaman Aerospace (wing ribs and bulkheads), Hitco Technologies (tailcone), Heath Tecna (wing-to-fuselage fillet), Aerostructures Hamble (composites flap hinge fairings and trailing-edge panels) and Northwest Composites (main landing gear pod panels). Raytheon was original supplier of composites winglets and landing gear doors, but replaced by Marion Composites with effect from 41st aircraft; Marion Composites also fabricates nose and tail radomes for last 80 aircraft of USAF order. C-17A structure is 69.3 per cent aluminium; 12.3 per cent steel alloys; 10.3 per cent titanium and 8.1 per cent composites. Wings of P1-P10 underwent local strengthening as consequence of load test of 1 October 1992; first rework to P8/90-0533 at McDonnell Douglas, Tulsa, between 3 January and 9 April 1994.

Proposal, early 1994, to design all-composites horizontal tail surfaces for weight-saving resulted in McDonnell Douglas (later Boeing) securing a US$40.7 million contract to build new unit; revised tail manufactured by Northrop Grumman using AS-4 carbon fibre for spars and skins and machined 7075 aluminium for ribs, resulting in assembly said to be 50 per cent cheaper and 20 per cent lighter than existing tail. Incorporates 2,000 fewer components and 42,000 fewer fasteners. Following testing on prototype T1, completed April 1999, it was incorporated on production aircraft beginning with P51.

Landing gear pod entirely redesigned in 1995, drastically reducing number of parts and fasteners and simplifying process of attachment to aircraft; introduced on first Lot 8 aircraft (P33/96-0001) and expected to save US$45 million over remainder of production programme for USAF. New, lighter and less costly engine nacelle, saving 113 kg (250 lb) per unit, flight tested at Edwards AFB between December 1997 and February 1998; first production aircraft with new nacelles (P41) delivered to USAF on 10 August 1998.

LANDING GEAR: Hydraulically retractable tricycle type, with free-fall emergency extension; designed for sink rate of 3.81 m (12 ft 6 in)/s and suitable for operation from paved runways or unpaved strips. Mainwheel units, each consisting of two legs in tandem with three wheels on each leg, rotate 90° to retract into fairings on lower fuselage sides; tyre size 50×21.0-20 (30 ply) tubeless; pressure 9.52 bars (138 lb/sq in). Menasco (Goodrich from 41st aircraft) twin-wheel nose leg retracts forwards; tyre size 40×16-14 (26 ply) tubeless; pressure 10.69 bars (155 lb/sq in). Honeywell wheels and carbon brakes. Minimum ground turning radius at outside mainwheels 17.37 m (57 ft 0 in); minimum taxiway width for three-point turn 27.43 m (90 ft 0 in); wingtip/tailplane clearance 74.24 m (237 ft 0 in).

POWER PLANT: Four Pratt & Whitney F117-PW-100 (PW2040) turbofans, with maximum flat rating of 179.9 kN (40,440 lb st), pylon-mounted in individual underwing pods and each fitted with a directed-flow thrust reverser deployable both in flight and on the ground. With effect from the 20th production aircraft, an improved version of the F117 was adopted, this embodying single-crystal turbine blade technology, a supercharged compressor and enhanced thermal barrier coatings; benefits include a 20 per cent reduction in cost of maintenance, increased reliability and slightly better sfc. Provision for in-flight refuelling. Two outboard wing fuel tanks of 21,210 litres (5,603 US gallons; 4,665 Imp gallons) each; two inboard wing fuel tanks of 30,056 litres (7,940 US gallons; 6,611 Imp gallons) each; total capacity 102,532 litres (27,086 US gallons; 22,554 Imp gallons).

Flight deck of the C-17A Globemaster III 0022134

BAE Systems fuel pumps. Upgrade, included in the 71st (first Lot 12) and subsequent aircraft and for retrofit to earlier C-17s, involves adoption of the extended-range fuel tank containment system (ERFCS); this will convert current wing dry bay into additional fuel tank containing approximately 36,339 litres (9,600 US gallons; 7,994 Imp gallons) of fuel, thus increasing range with typical 40,823 kg (90,000 lb) payload by 600 n miles (1,111 km; 690 miles).

ACCOMMODATION: Normal flight crew of pilot and co-pilot side by side and two observer positions on flight deck, plus loadmaster station at forward end of main floor; access to flight deck via downward-opening airstair door on port side of lower forward fuselage. Bunks for crew immediately aft of flight deck area; crew comfort station at forward end of cargo hold.

Main cargo hold able to accommodate US Army wheeled and tracked vehicles up to M1 main battle tank, including 5 ton expandable vans in two rows, or up to three AH-64 Apache attack helicopters, with loading via hydraulically actuated rear-loading ramp which forms underside of rear fuselage when retracted. Aircraft fitted with 27 stowable tip-up seats along each sidewall and another 48 seats carried on board which can be erected along the centreline; optionally up to 36 litters for medical evacuation mission or up to 100 passengers on 10-passenger pallets in addition to 54 sidewall seats. Air delivery system capability for nine 463L pallets plus two on ramp in single row; or logistics handling system for 18 463L pallets in double row. Airdrop capability includes single platforms of up to 27,215 kg (60,000 lb), multiple platforms of up to 49,895 kg (110,000 lb), or up to 102 paratroops; aircraft originally configured for single-row airdrop, but dual-row capability operationally certified in 1998 and is standard feature of 51st and subsequent C-17s, as well as being retrofitted to earlier aircraft. Equipped for low-altitude parachute extraction system (LAPES) drops.

Cargo handling system includes rails for airdrops and rails/rollers for normal cargo handling. Each row of rails/rollers can be converted quickly by a single loadmaster from one configuration to the other. Total of 295 cargo tiedown rings, each stressed for 11,340 kg (25,000 lb), all over cargo floor forming grid averaging 74 cm (29 in) square. Three quick-erecting litter stanchions, each supporting four (USAF, three) litters, permanently carried. Main access to cargo hold is via rear-loading ramp, which is itself stressed for 18,145 kg (40,000 lb) of cargo in flight. Underfuselage door aft of ramp moves upward inside fuselage to facilitate loading and unloading. Paratroop door at rear on each side; three overhead FEDS (flotation equipment deployment system) escape hatches and liferaft.

SYSTEMS: Include Honeywell computer-controlled integrated environmental control system and cabin pressure control system; 2,440 m (8,000 ft) equivalent cabin pressure up to 11,280 m (37,000 ft); quad-redundant flight control and four independent 276 bar (4,000 lb/sq in) hydraulic systems; independent fuel feed systems; electrical system; Honeywell GTCP331-250G APU (at front of starboard landing gear pod), provides auxiliary power for environmental control system, engine starting, and on-ground electronics requirements; onboard inert gas generating system (OBIGGS) for the explosion protection system, pressurised by engine bleed air at 4.14 bars (60 lb/sq in) to produce NEA (nitrogen-enriched air) and governed by a Parker-Gull system controller; fire suppression system; smoke detection systems. All phases of cargo operation and configuration change capable of being handled by one loadmaster.

Electrical system includes single 90 kVA generator per engine and an APU, providing 115/200 V, three-phase, 400 Hz AC power; four 200 A transformer-rectifiers providing 28 V DC; single-phase, 1,000 VA inverter for ground refuelling and emergency AC; and two 40 Ah Ni/Cd batteries for APU starting and emergency DC. Aeromedical equipment provided with 60 Hz power.

AVIONICS: *Comms:* Telephonics Corporation IRMS integrated radio management system; Rockwell Collins AN/ARC-210 secure radio (from P50, replacing Raytheon AN/ARC-187 plus Honeywell KY 58 encoder); satcom; VHF-AM/FM; HF; secure voice and jam-resistant UHF/VHF-FM intercom; IFF/SIF; en route army/marine UHF LOS and satcom hook-up; cockpit voice recorder; crash

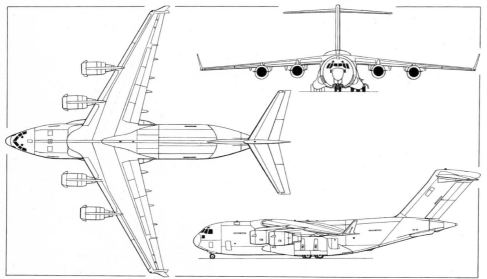

Boeing C-17A Globemaster III long-range heavy cargo transport *(Dennis Punnett/Jane's)*

Cargo hold of the Boeing C-17A, looking aft, with full complement of 102 paratroops
2000/0064920

Manoeuvrability demonstration by a lightly laden Globemaster III
(*Paul Jackson/Jane's*)
2001/0103565

position indicator. Automatic communications processor and multiband radio in new-build aircraft starting with P50 and retrofitted to earlier C-17As. In December 1998, Rockwell Collins selected by Boeing to provide SAT-2000 Aero-1 satcom systems and CMU-900 communication management units for future global air traffic management (GATM) environment; initial order for support and preproduction hardware, with options to install equipment on all 120 aircraft. Production line GATM introduction at P71, involving TCAS, Mode S IFF, ADS-A (automatic dependent surveillance) and Inmarsat Aero I satcom, plus LCD displays replacing all head-down CRTs. GATM augmented from P101 by HF datalink and RNP 4 (required navigation performance) measures. Under development in 1999 were two SOLL II (special operations low-level) roll-on/roll-off communications suites for use on specific missions; and carry-on radio suite for use on special forces flights, beginning in 2002.

Radar: Honeywell AN/APS-133(V) weather/mapping radar.

Flight: Three Delco Electronics mission computers with MDC software and electronic control system on P1 to P40; beginning with P41, two Lockheed Martin core integrated processors (CIPs) replaced Delco Electronics equipment and retrofitted to earlier aircraft by September 1998; additional memory added from P50 onwards. Hamilton Sundstrand aircraft and propulsion data management computer; Honeywell dual air data computers; Litton warning and caution system; master warning system provides aural and voice alerts plus visual alerts on glareshields; General Dynamics automatic test equipment, and support equipment data acquisition and control system; GPS and four Honeywell inertial reference units, although new, embedded GPS INS introduced from P50; VOR/DME; Tacan; ILS/marker beacon; UHF-DF; ground proximity warning system; radar altimeter; flight data recorder; flight plan entry manually or via laptop computer. BAE Systems terrain awareness warning system (TAWS) to be installed in conjunction with Lockheed Martin Control Systems video integrated processor (VIP), currently under development, from P86 onwards.

Instrumentation: Advanced digital avionics and four Honeywell full-colour cathode ray tube (CRT) displays; new 152 × 152 mm (6 × 6 in) AMLCD colour multifunction displays (MFDs) currently under development by Honeywell for installation in 71st and subsequent production aircraft; two BAE Systems full flight regime foldable head-up displays.

Mission: Litton integrated mission and communications keyboards (MCKs) and displays (MCDs); Sierra Technologies AN/APN-243A(V) station-keeping equipment (SKE 2000) fitted as standard commencing with P33, retrofitted to earlier aircraft and upgraded from P86 (first Lot 13) onwards to increase capacity from 17 to 100 aircraft. Loadmaster's laptop computer from P41 onwards; this replaced from P58 by Dolch aircrew data device.

Self-defence: Development of defensive electronic systems completed in 1994. ATK, Lockheed Martin or Honeywell AN/AAR-47 missile approach warning system and associated BAE Systems AN/ALE-47 automatic flare dispenser installed on five aircraft; USAF to adopt this equipment across the fleet. USAF seeking to counter IR homing missiles with laser jamming system and has plans to deploy these on C-17 from FY03 onward.

EQUIPMENT: Removable crew armour can be fitted around flight deck and loadmaster's area.

DIMENSIONS, EXTERNAL:
Wing span: wings only	50.29 m (165 ft 0 in)
at winglet tips	51.74 m (169 ft 9 in)
Wing aspect ratio	7.2
Length: overall	53.04 m (174 ft 0 in)
fuselage	48.49 m (159 ft 1¼ in)
Height: to flight deck roof	7.34 m (24 ft 1 in)
to winglet tips	6.93 m (22 ft 9 in)
overall	16.79 m (55 ft 1 in)
Fuselage diameter	6.85 m (22 ft 6 in)
Tailplane span	19.81 m (65 ft 0 in)
Wheel track	10.26 m (33 ft 8 in)
Wheelbase: to front mainwheel	17.60 m (57 ft 8¾ in)
to rear mainwheel	20.05 m (65 ft 9½ in)
Height to sill	1.63 m (5 ft 4 in)
Ground clearance under engine pods:	
inboard	2.71 m (8 ft 10½ in)
outboard	2.35 m (7 ft 8½ in)
Distance between fuselage centreline and engine	
centreline: inboard	7.44 m (24 ft 5 in)
outboard	13.94 m (45 ft 9 in)
Height of winglets	2.72 m (8 ft 11 in)

DIMENSIONS, INTERNAL:
Cargo compartment: Length, incl 6.05 m (19 ft 10 in)	
rear-loading ramp	26.82 m (88 ft 0 in)
Loadable width	5.49 m (18 ft 0 in)
Max height: under wing	3.96 m (13 ft 0 in)
aft of wing	4.50 m (14 ft 9 in)
Volume	591.8 m³ (20,900 cu ft)

AREAS:
Wings, gross	353.03 m² (3,800.0 sq ft)
Winglets (total)	3.33 m² (35.85 sq ft)
Ailerons (total)	11.83 m² (127.34 sq ft)
Tailplane	78.50 m² (845.00 sq ft)

WEIGHTS AND LOADINGS:
Operating weight empty	125,645 kg (277,000 lb)
Max payload (2.5 g load factor)	76,655 kg (169,000 lb)
Max weight on rear-loading ramp	18,143 kg (40,000 lb)
Max usable fuel weight	82,125 kg (181,054 lb)
Max T-O weight	265,350 kg (585,000 lb)
Max ramp weight	265,800 kg (586,000 lb)
Max wing loading	751.6 kg/m² (153.95 lb/sq ft)
Max power loading	366 kg/kN (3.59 lb/lb st)

PERFORMANCE:
Normal cruising speed at 8,535 m (28,000 ft) M0.74-0.77
Max cruising speed at low altitude
    350 kt (648 km/h; 403 mph) CAS
Airdrop speed: at S/L
    115-250 kt (213-463 km/h; 132-288 mph) CAS
at 7,620 m (25,000 ft)
    130-250 kt (241-463 km/h; 150-288 mph) CAS
Approach speed with max payload
    115 kt (213 km/h; 132 mph) CAS
Service ceiling   13,715 m (45,000 ft)
Runway LCN (paved surface)   better than 49
T-O field length at MTOW   2,360 m (7,740 ft)
Landing field length with 72,575 kg (160,000 lb)
payload, using thrust reversal   915 m (3,000 ft)
Range with payloads indicated, with no in-flight refuelling:
    18,144 kg (40,000 lb) payload
        4,400 n miles (8,148 km; 5,063 miles)
    72,575 kg (160,000 lb), T-O in 2,286 m (7,500 ft),
    land in 915 m (3,000 ft), load factor of 2.25 g
        2,400 n miles (4,444 km; 2,761 miles)
    self-ferry (zero payload), T-O in 1,128 m (3,700 ft),
    land in 701 m (2,300 ft), load factor of 2.5 g
        4,700 n miles (8,704 km; 5,408 miles)

UPDATED

BOEING 737 AEW&C

TYPE: Airborne early warning and control system.

PROGRAMME: Adaptation of Boeing Business Jet (BBJ), which combines 737-700 fuselage with strengthened wing and landing gear of 737-800. Additional features include extra fuel tanks in former baggage hold. Proposed for Australian Project Air 5077 Wedgetail by Boeing, Northrop

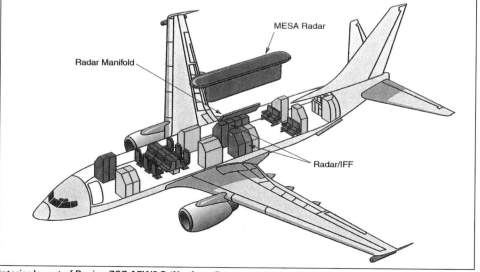

Interior layout of Boeing 737 AEW&C (*Northrop Grumman*)
2001/0079023

588 USA: AIRCRAFT—BOEING

Grumman Electronic Sensor Systems Division (ESSD) and BAE Australia. Competed against proposals from Lockheed Martin and Raytheon. Secured initial design activity contract worth about US$6 million in December 1997, paving way for submission of full tenders in early 1999. Selection of Boeing submission officially announced on 21 July 1999. RAAF initially planned to acquire seven aircraft for US$1.32 billion; cost concerns resulted in reduction to six plus one option in May 2000, but when contract signed on 19 December, this had further reduced to four and three options Delivery of first two scheduled for 2006, with operating unit to be No. 2 Squadron. Main base at Williamtown, New South Wales, with two aircraft permanently deployed to forward operating location at Tindal, Northern Territories.

ESSD L-band multirole electronically scanned array (MESA) radar mounted above rear fuselage ('top hat' configuration) providing 360° coverage from stationary antenna 10.7 m (35 ft) long and 2.35 m (7 ft 9 in) high. Operating modes will include acute long-range or broad short-range scanning and track-while-scan. In late 1998, demonstrator BBJ temporarily fitted with full-scale replica of MESA radar, six operator consoles and equipment cabinets for inspection by Australian defence department officials; mockup also featured in-flight refuelling probe above cockpit and EW/ECM sensors. Definitive aircraft will, however, feature 10 operator consoles initially, with potential to add further two. Aerodynamic effects of MESA offset by two large strakes below rear fuselage. Separate competition to be organised for electronic warfare self-protection (EWSP) system; this will be controlled by the ALR-2001 computer. Other mission equipment to include Link 11, JTIDS and satcom; flight deck tactical displays, three HF and eight VHF/UHF radios. Patrol endurance of 8 hours at 300 n miles (555 km; 345 miles) from base can be extended by airborne refuelling, with modification to include installation of flying boom receptacle and a fixed probe. Maximum T-O weight 77,110 kg (170,000 lb); service ceiling in excess of 12,190 m (40,000 ft).

South Korea reportedly interested in AEW configured 737 as less costly solution to E-X requirement in lieu of original proposal of Boeing 767, which now considered too expensive, although latter type apparently still in contention; selection was expected in 1999, but has been deferred. After studying proposals for AEW aircraft involving Airbus A310/Phalcon and Boeing 737/MESA combinations, Turkey announced selection of latter in early December 2000 and revealed intent to obtain total of six aircraft (with option on a seventh) at cost of US$1.5 billion; these will include some indigenous equipment. Delivery of first aircraft due in 2005, following contract signature, which expected in early 2001.

UPDATED

BOEING 737 MMMA

TYPE: Maritime surveillance twinjet.
PROGRAMME: Design study for 737 MMMA (multimission maritime aircraft) revealed 18 April 2000 as potential replacement for Lockheed Martin P-3 Orion. Based on C-40A (see Boeing 737 entry under Boeing Commercial Airplane Group), combining 737-800 wing and 737-700 fuselage, latter with internal weapons bay and former with hardpoints for air-to-surface missiles. Full range of maritime patrol equipment, including enlarged nose accommodating search radar, up to seven operators' consoles and rotary sonobuoy launcher, plus additional fuel in aft baggage hold.

Maximum T-O weight 77,565 kg (171,000 lb); mission radius 2,000 n miles (3,704 km; 2,301 miles). US Navy awarded US$493,000 concept exploration study contract to Boeing in July 2000; one of four contracts awarded to industry, results of five month studies will be used by Center for Naval Analysis in selection of preferred concept.

NEW ENTRY

BOEING 767 MILITARY VERSIONS

TYPE: Tanker-transport; airborne ground surveillance system.
PROGRAMME: Variants of Boeing 767 twin-turbofan airliner (which see under Boeing Commercial Airplane Group for technical description).
CURRENT VERSIONS: **767 AWACS**: Described separately.

KC-767: Tanker-transport version announced by Boeing, February 1995, in anticipation of Japanese order; Boeing in discussions with Kawasaki by mid-1996, concerning co-operative venture to offer KC-767 to Japan ASDF. Kawasaki involvement expected to take form of post-production work (including fitment of boom refuelling gear, extra tanks and associated plumbing) on 767-300 derivative. JASDF intended to request funding from FY99, but economic downturn and reductions in defence budget caused delay. At beginning of 1999, Japan Defence Agency (JDA) still considering purchase of four KC-767 or similar aircraft at estimated cost of US$710 million, with head of JDA, Hosei Norota, confirming on 21 July 1999 the intention to seek funding in FY00. However, request overruled by political authority. Service reportedly has requirement for six to 12 aircraft in longer term.

Artist's impression of Boeing 737 AEW&C variant in Turkish insignia   2000/0073806

Version of 767 with three hose drum units offered to UK Royal Air Force which has FSTA requirement for a new tanker aircraft to replace VC-10 and Tristar. Total of 20 to 30 aircraft needed, with six consortia invited in September 1999 to tender submissions for a public-private finance initiative to satisfy requirement. Combinations had reduced this to four by late 1999, of which three offering 767 solution (two wholly, one partly), probably using ex-British Airways 767-300s.

Tanker-transport proposals most recently (1999) based on 767-200ER (see Boeing Commercial Airplanes Group entry); fuel dispensed through 'flying boom' and two underwing pods; boom remotely controlled from cabin, assisted by CCTV; first aircraft available 40 months from go-ahead. Fuel capacity 91,380 litres (24,140 US gallons; 20,101 Imp gallons), including 31,059 litres (8,205 US gallons; 6,832 Imp gallons) in supplementary underfloor tanks. As freighter (side cargo door and reinforced floor), can carry up to 18 463L pallets on main deck or 216 passengers, with additional cargo capacity on lower deck dependent upon auxiliary tank configuration. Version offered to RAF, with HDU in place of 'flying boom', has reduced underfloor fuel of 29,942 litres (7,910 US gallons; 6,586 Imp gallons). Formal invitations to tender issued by UK MoD on 21 December 2000.

767 Joint STARS: Boeing 767 proposed as alternative airborne platform to Boeing 707 for export version of this surveillance system (see Northrop Grumman E-8 in this section). No recent news received.
CUSTOMERS: Anticipated sales of 95, including about 20 AWACS variants.
Data follow for Tanker-Transport.
WEIGHTS AND LOADINGS (estimated):

Operational weight, empty	84,260 kg (185,760 lb)
Max fuel weight: wing	71,173 kg (156,910 lb)
underfloor	24,190 kg (53,330 lb)
Max T-O weight	179,170 kg (395,000 lb)
Max ramp weight	179,625 kg (396,000 lb)
Max zero-fuel weight	117,934 kg (260,000 lb)

UPDATED

BOEING 767 AWACS
Japan ASDF designation: E-767
TYPE: Airborne early warning and control system.
PROGRAMME: Boeing announced, December 1991, definition studies for modified 767-200ER airliner (which see) with Northrop Grumman AN/APY-2 radar; replacement platform for Boeing E-3 Sentry (707), which then out of production; project sustained by Japanese interest; engine selection October 1992. Japanese government purchase decision, December 1992; parliamentary approval given early 1993; total requirement for four; first two ordered November 1993; second two in October 1994.

First flight of clean airframe (N767JA; 557th Boeing 767) 10 October 1994 at Seattle; to Wichita, November 1994, for modifications to accommodate mission equipment; second aircraft flew 18 July 1995 at Seattle and was ferried to Wichita on 18 August 1995. Initial 767 AWACS (now JASDF 64-3501) returned to Seattle in June 1996 for installation of rotodome. First flight with the rotodome took place on 9 August 1996; test programme completed in January 1997, after which Boeing installed prime mission equipment including computers, radar, sensors, displays and communications gear. Systems integration and mission suite evaluation, using second aircraft, occupied latter half of 1997, clearing the way for handover on 11 March 1998 of the first two aircraft to Japan, where they arrived on 23 March 1998; second pair delivered 5 January 1999.

At various times, Italy, South Korea, Saudi Arabia and Turkey have expressed interest in the 767 AWACS, but no firm orders have been placed. South Korea expected to finalise selection in 1999, but decision postponed, while Turkey now to obtain less costly Boeing 737 derivative. Israel also contemplating acquisition of Boeing 767 with different radar system based on Elta Phalcon, although this reportedly subject of objections from USAF, allegedly because it would give Israeli air force an independent capability.
CUSTOMERS: Japan ASDF (Model 767-27C), four (two funded in 1993 budget; two in 1994); based at Hamamatsu with

Artist's impression of Boeing KC-767 Tanker-Transport refuelling an F-15 Eagle   2000/0064922

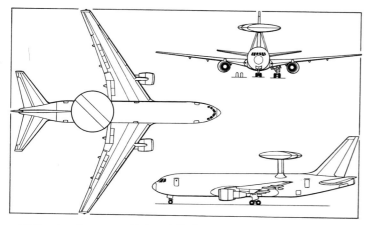

The first JASDF Boeing 767 AWACS during its initial flight with the rotodome installed
1999/0014111

AWACS version of Boeing 767 airliner *(James Goulding/Jane's)*

601 Squadron. Japan reported in latter half of 1998 to be considering purchase of four additional aircraft with which to establish second full AWACS 'orbit', but no follow-on order yet placed.

COSTS: Japan (1993) US$840 million for two airframes, systems, installation and testing; further US$773 million for second pair (1994); unit cost US$403 million.

DESIGN FEATURES: See Boeing 767 airliner entry. Additionally, substantial structural modifications (replaced two frames by bulkheads; deepened two frames; reinforced four floor beams) to accommodate rotodome, computer equipment and operators' consoles; ventral fins may be added on rear fuselage. Boeing 767 airframe offers twice floor space and three times internal volume of 707/E-3.

POWER PLANT: Two 273.6 kN (61,500 lb st) General Electric CF6-80C2B6FA turbofans, modified for additional electrical generation. In-flight refuelling receptacle not fitted to Japan ASDF aircraft, but provisions made for AAR should customers specify it.

ACCOMMODATION: Two-man flight crew; up to 19 mission crew, though latter number can vary for tactical and defence missions. Full crew complement is two pilots, mission director, tactical director, fighter allocator, two weapon controllers, surveillance controller, link manager, seven surveillance operators, communications operator, radar technician, communications technician and computer display technician. Aft of flight deck, from front to rear of fuselage, are communications, data processing and other equipment bays; multipurpose consoles; communications, navigation and identification equipment; and crew rest area, galley and storage space.

SYSTEMS: Two 150 kVA generators on each engine replace single 90 kVA units; APU has single 90 kVA generator; total 690 kVA. Typical additional systems in the E-3 Sentry include a liquid cooling system providing protection for the radar transmitter. An air cycle pack system, a draw-through system, and two closed loop ram-cooled environmental control systems to ensure a suitable environment for crew and avionics equipment. Distribution centre for mission equipment power and remote avionics in lower forward cargo compartment. Rear cargo compartment houses radar transmitter. External sockets allow intake of power when aircraft is on ground.

AVIONICS: Based on those of E-3 Sentry, a description of which follows.

Comms: HF, VHF and UHF channels through which information can be transmitted or received in clear or secure mode, in voice or digital form.

Radar: Weather radar in nose. Main system is Northrop Grumman AN/APY-2; elliptical cross-section rotodome of 9.14 m (30 ft) diameter and 1.83 m (6 ft) maximum depth, mounted above fuselage, comprises four essential elements: a turntable, strut-mounted above rear fuselage, supporting rotary joint assembly to which are attached sliprings for electrical and waveguide continuity between rotodome and fuselage; strengthened centre-section supporting AN/APY-2 surveillance radar and IFF/TADIL-C antennas, radomes, auxiliary equipment for radar operation and environmental control of the rotodome interior; liquid cooling of the radar antennas; and two radomes of multilayer glass fibre sandwich material, one for surveillance radar and one for IFF/TADIL-C array. In operation, rotodome is hydraulically driven at 6 rpm, but during non-operational flights it is rotated at only ¼ rpm, to keep bearings lubricated.

Radar operates in E/F-band and can function as both a pulse and/or a pulse Doppler radar for detection of aircraft targets. A similar pulse radar mode with additional pulse compression and sea clutter adaptive processing is used to detect maritime/ship traffic. Radar is operable in six modes: PDNES (pulse Doppler non-elevation scan), when range is paramount to elevation data; PDES (pulse Doppler elevation scan), providing elevation data with some loss of range; BTH (beyond the horizon), giving long-range detection with no elevation data; Maritime, for detection of surface vessels in various sea states; Interleaved, combining available modes for all-altitude longer-range aircraft detection, or for both aircraft and ship detection; and Passive, which tracks enemy ECM sources without transmission-induced vulnerability. Radar antennas, spanning about 7.32 m (24 ft), and 1.52 m (5 ft) deep, scan mechanically in azimuth, and electronically from ground level up into the stratosphere.

Flight: Navigation by two modified Litton LN-100G INS/GPS platforms, providing position and velocity data.

Mission: Heart of the data processing capability is a Lockheed Martin CC-2E computer with a main storage capacity of 3,145,725 words. Data display and control are provided by Hazeltine high-resolution colour situation display consoles (SDC) and auxiliary display units (ADU). Identification is based on a Telephonics AN/APX-103 interrogator set with airborne IFF interrogator with AIMS Mk X SIF air traffic control and Mk XII military identification friend or foe (IFF) in a single integrated system. Simultaneous Mk X and Mk XII multitarget and multimode operations allow the operator to obtain instantaneously the range, azimuth and elevation, code identification, and IFF status, of all targets within radar range. Options include AN/AYR-1 ESM, IRCM, satcom, JTIDS and Have Quick secure radios, with latter installed on JASDF E-767s (four per aircraft) as part of upgrade completed by mid-2000.

DIMENSIONS, EXTERNAL:
Wing span	47.57 m (156 ft 1 in)
Length overall	48.51 m (159 ft 2 in)
Height overall	15.85 m (52 ft 0 in)
Radome: Diameter	9.14 m (30 ft 0 in)
Max thickness	1.83 m (6 ft 0 in)

WEIGHTS AND LOADINGS:
Max T-O weight	174,635 kg (385,000 lb)

PERFORMANCE (estimated):
Max cruising speed
more than 434 kt (805 km/h; 500 mph)
Service ceiling 10,360-12,220 m (34,000-40,100 ft)
Range, unrefuelled
4,500-5,000 n miles (8,334-9,260 km; 5,178-5,754 miles)
Endurance:
at 1,000 n mile (1,852 km; 1,151 mile) radius   8 h
at 300 n mile (556 km; 345 mile) radius   13 h

UPDATED

BOEING AL-1A

TYPE: Missile defence system.

PROGRAMME: Prototype YAL-1A (USAF serial number 00-0001) ordered on 12 November 1996; purchase of 747-400F airframe from Boeing confirmed on 30 January 1998. Formal authority to proceed received on 26 June 1998, after TRW successfully demonstrated laser firing and missile tracking. First metal cut on YAL-1A on 10 August 1999; rolled out at Everett on 12 December 1999; first flight 6 January 2000; delivered to Boeing Wichita on 21 January 2000 for outfitting with strengthened floor and modifications to take nose-mounted laser turret; subsequent critical design review completed in late April. Installation of six-module COIL laser to be undertaken at Edwards AFB, California, from March 2002, followed by trials against various missiles, with PDRR phase culminating with demonstration against a ballistic missile fired from Vandenberg AFB, California, in September 2003; EMD phase expected to begin in 2004, with IOC of first three aircraft set for late 2007. However, funding cuts could result in delay of up to two years in each of these milestones being achieved.

CUSTOMERS: US Air Force (seven required).

COSTS: Programme cost (1997, estimated) US$5 billion for one concept prototype and one EMD (both eventually to be fully upgraded) and five production aircraft. Initial programme definition and risk reduction (PDRR) contract of November 1996 valued at US$1.3 billion.

DESIGN FEATURES: Based on airframe of Boeing 747-400F (which see under Boeing Commercial Airplanes Group). Equipped with TRW multihundred kW chemical oxygen iodine long-range laser and Lockheed Martin electron-bombarded charge-coupled device (EBCCD) camera system and fire control. Intended to destroy theatre ballistic missiles during their boost phase, but with additional capability against low-flying cruise missiles; other

Prototype YAL-1A pictured before leaving for Wichita, where internal equipment is being installed
2001/0067821

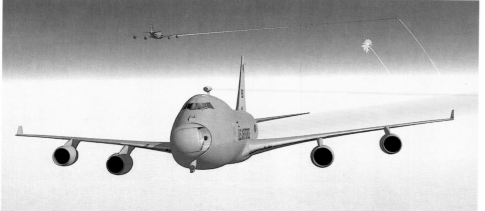

Artist's impression of Boeing AL-1As destroying missiles by laser fire
2000/0064923

USA: AIRCRAFT—BOEING

Artist's impression of the Boeing Advanced Theater Transport, employing original sweptback wing
2000/0036944

potential applications began emerging in 1998, these including protection of high-value aircraft such as AWACS and Joint STARS by destroying SAMs and AAMs; imaging and reconnaissance with aid of optical telescope; SEAD, by combining intelligence data to engage hostile missile sites and radar control systems; and command and control through search/detection of infrared signatures to aid cueing of other weapons. Laser able to fire 30 times per mission; titanium to be used in certain areas instead of aluminium to protect against heat damage to undersides of AL-1A caused by laser exhaust gases.

UPDATED

BOEING ATT/SSTOL 'SUPER FROG'

TYPE: Medium transport.
PROGRAMME: Design study revealed by McDonnell Douglas (now Boeing) in September 1996 for a potential replacement for the Lockheed Martin C-130 Hercules. Original study envisaged aircraft with four 8,950 kW (12,000 shp) turboprops fixed to wing, which tilted 15° for take-off and 45° for landing; potential in-service date of 2020. Those studies provided basis for No-Tail Advanced Theater Transport (NOTAIL ATT) Tilt-wing Super Short Takeoff and Landing (SSTOL) or 'Super Frog' as Douglas Products division project also known. This also borrowed from BWB studies, as described under UHCA/VLCT entry in International section. Tail-less design reduces surface area by 33 per cent in comparison to conventional aircraft of similar capability.

Concept revealed in September 1998, with 39 m (128 ft) span 'Super Frog' designed to carry up to four times the payload of the C-130J Hercules and use runways as short as 183 m (600 ft). Baseline requirement is delivery of 27,215 kg (60,000 lb) load on to 229 m (750 ft) of rough airstrip, 1,220 m (4,000 ft) AMSL in 35°C (95°F) ambient temperature. Signature of co-operative research and development agreement (CRADA) between Boeing's Phantom Works and USAF Research Laboratory at Wright-Patterson AFB, Ohio, in first quarter of 1999 paved way for development and demonstration of 'enabling technologies' that could lead to production-configured aircraft.

Fuselage interior width is 6.4 m (21 ft) and will be able to accommodate various loads up to a maximum of about 36,285 kg (80,000 lb), including one AH-64 Apache or two RAH-66 Comanche helicopters; or up to 11 cargo pallets; or 40 fully equipped soldiers. Wing features widely separated podded turboprop engines with contrarotating propellers and originally pivoted about lateral axis near rear spar, with tilt used to increase lift during take-off and landing. Directional control of tail-less design uses similar system to B-2 bomber, which has split flaps. By early 2000, however, design had progressed to forward-swept (7 to 9°) wing, pivoted at leading-edge and tilting downwards at trailing-edge.

Initial wind tunnel testing of subscale model completed by end of 1998, followed by further trials with a 7 per cent subscale model during 1999; these included tethered and untethered tests which began at Gray Butte, California, on 5 June. Boeing intends to conduct flight trials with larger models in 2001 and in early 2000 was seeking involvement of US research agency such as NASA or DARPA to assist with funding of larger manned or unmanned concept demonstrator. If project goes ahead, Boeing optimistic that production aircraft could be operational by 2013 to 2015, with 'Super Frog' a potential candidate for USAF MC-X special operations aircraft requirement.

UPDATED

BOEING 114 and 414

US Army designations: **CH-47 and MH-47 Chinook**
Royal Air Force designations: **Chinook HC. Mk 2, HC. Mk 2A and HC. Mk 3**
Spanish Army designation: **HT.17**

TYPE: Medium-lift helicopter.
PROGRAMME: Design of all-weather medium transport helicopter for US Army began 1956; first flight of YCH-47A 21 September 1961; for details of CH-47A and CH-47B see 1974-75 *Jane's*. Performance increased in CH-47C by uprated transmissions and 2,796 kW (3,750 shp) T55-L-11A; internal fuel capacity increased; first flight 14 October 1967; delivered to US Army from 1968; full details in 1980-81 *Jane's*.

CURRENT VERSIONS: **CH-47D**: US Army contract to modify one each of CH-47A, B and C to prototype Ds placed 1976; first flight 11 May 1979; first production contract October 1980; first flight 26 February 1982; first delivery 31 March 1982; initial operational capability (IOC) achieved 28 February 1984 with 101st Airborne Division; first multiyear production contract for 240 CH-47Ds awarded 8 April 1985; second multiyear production contract for 144 CH-47Ds (including 11 MH-47Es) awarded 13 January 1989, bringing total CH-47D (and MH-47E) ordered to 472; two Gulf War attrition replacements authorised August 1992 (these new-build); seven ex-Australian rebuilds funded June 1993 for delivery January to November 1995. Additional new-build CH-47D ordered for US Army Reserve in June 1999; at that time CH-47D fleet totalled 433, of which 302 scheduled for CH-47F conversion.

CH-47D update included strip down to bare airframe, repair and refurbish, fit Honeywell T55-L-712 turboshafts, uprated transmissions with integral lubrication and cooling, composite rotor blades, new flight deck compatible with night vision goggles (NVG), new redundant electrical system, modular hydraulic system, advanced automatic flight control system, improved avionics and survivability equipment, Solar T62-T-2B APU operating hydraulic and electrical systems through accessory gear drive, single-point pressure refuelling, and triple external cargo hooks. Principal external change is large, rectangular air intake in leading-edge of rear sail. Composites account for 10 to 15 per cent of structure. About 300 suppliers involved.

At maximum gross weight of 22,680 kg (50,000 lb), CH-47D has more than double useful load of CH-47A. Sample loads include M198 towed 155 mm howitzer, 32 rounds of ammunition and 11-man crew, making internal/external load of 9,980 kg (22,000 lb); D5 caterpillar bulldozer weighing 11,225 kg (24,750 lb) on centre cargo hook; US Army Milvan supply containers carried at up to 130 kt (256 km/h; 159 mph); up to seven 1,893 litre (500 US gallon, 416 Imp gallon), 1,587 kg (3,500 lb) rubber fuel blivets carried on three hooks.

Test programme began late 1995 of Chinook with vibration-reducing dynamically tuned fuselage, as well as the first two production T55-GA-714A engines; this joint Boeing/US Army project offered benefits for the existing CH-47D and forthcoming CH-47F (which see). CH-47D 86-1678 was modified into the vibration reduction test aircraft (VRTA) with structural improvements that will be adopted for the CH-47F. T55-GA-714A engines with FADEC and magneto-restrictive torquemeter (IMT) also installed and are to be retrofitted to CH-47D fleet. After initial testing by Boeing, the VRTA moved to Fort Rucker, Alabama, for a 66 hour flight test programme that showed a reduction of 30 per cent in vibration. Additional testing for air transportability modifications, flight control modification, new rotor hubs and an engine barrier filter undertaken by this joint Boeing/US Army testbed which later known as 'Bearcat 3'; at end of trials programme, it was decided to modify it to MH-47E configuration for US Special Forces.

MH-47D Special Operations Aircraft: Element of 160th Special Operations Aviation Regiment (at Hunter AAF, Georgia) equipped with 11 CH-47Ds modified to **MH-47D SOA** standard with refuelling probes (first refuelling July 1988), thermal imagers, Honeywell RDR-1300 weather radar, Rockwell Collins 'glass cockpits', improved communications and navigation

US Army CH-47D Chinook instrument panel *(Paul Jackson/Jane's)* 0015907

EFIS cockpit of the Royal Netherlands Air Force CH-47D
0022137

equipment, and two pintle-mounted 7.62 mm six-barrel miniguns. Navigator/commander's station also fitted.

GCH-47D: Limited number of Chinooks grounded for maintenance training at Fort Eustis, Virginia.

US ARMY CH-47D PROCUREMENT

FY	Qty	First aircraft	
81	9	81-23381	
82	19	82-23762	
83	24	83-24102	
84	36*	84-24152	
85	48	85-24322	
86	48	86-1635	
87	48	87-0069	MYP1
88	48†	88-0062	
89	48	89-0130	
90	48[1]	90-0180	
91	48[2]	91-0230	MYP2
92	48[3]	92-0280	
	2[4]	92-0367	
93	7[5]	93-0928	
98	1		
Total	**482**		

* One crashed on test
† One to YMH-47E

Notes: All except one prototype issued with new serial numbers on remanufacture to CH-47D
MYP1 First multiyear procurement contract, 8 April 1985
MYP2 Second MYP, 13 January 1989
[1] This batch includes one aircraft subsequently further modified to MH-47E configuration under separate contract
[2] Total includes six which were updated to CH-47D standard and then further modified to MH-47E under separate contract
[3] This batch includes two of original three prototypes (remaining prototype used as maintenance trainer at Fort Eustis, Virginia) and 11 from Italian production; total includes four which were updated to CH-47D standard and then further modified to MH-47E under separate contract
[4] New-build Gulf War attrition replacements, authorised 2 August 1992
[5] Former Australian Air Force CH-47Cs; contract 2 June 1993

MH-47E: Special Forces variant; planned procurement 51, all originally to be deducted from total 472 CH-47D conversions, but only 25 initially received, including 14 not covered by multiyear production contract, but excluding one currently under conversion after use as VRTA testbed; prototype development contract 2 December 1987; long-lead items for next 11 helicopters authorised 14 July 1989; firm order for 11, plus option on next 14, awarded 30 June 1991; Lot 2 (14 helicopters) confirmed 23 June 1992. Prototype (88-0267) flew 1 June 1990; delivered 10 May 1991; initial production aircraft (90-0414) flown 1992; first 11 (of 24 intended) originally due to be delivered from November 1992 to 2 Battalion of 160th Special Operations Aviation Regiment at Fort Campbell, Kentucky. Following mission software problems, deliveries began January 1994 with 91-0498 to Fort Campbell; last of original 26 (including prototype) received at Fort Campbell April 1995. The 27th conversion (86-1678) is expected to enter service by early 2003.

Mission profile 5½ hour covert deep penetration over 300 n mile (560 km; 345 mile) radius in adverse weather, day or night, all terrain with 90 per cent success probability. Requirements include self-deployment to Europe in stages of up to 1,200 n miles (2,222 km; 1,381 miles), 44-troop capacity, powerful defensive weapons and ECM. Equipment includes IBM-Honeywell integrated avionics with four-screen NVG-compatible EFIS; dual MIL-STD-1553B digital databusses; AN/ASN-145 AHRS; jamming-resistant radios, Rockwell Collins CP1516-ASQ automatic target handoff system; inertial AN/ASN-137 Doppler, Rockwell Collins AN/ASN-149(V)2 GPS receiver and terrain-referenced positioning navigation systems; Rockwell Collins ADF-149; laser (Raytheon AN/AVR-2), radar (Lockheed Martin AN/APR-39A) and missile (Honeywell AN/AAR-47) warning systems; ITT AN/ALQ-136(V) pulse jammer and Northrop Grumman AN/ALQ-162 CW jammer; BAE Systems M-130 chaff/flare dispensers; Raytheon AN/APQ-174A radar with modes for terrain-following down to 30 m (100 ft), terrain-avoidance, air-to-ground ranging and ground-mapping; Raytheon AN/AAQ-16 FLIR in chin turret; digital moving map display; Elbit ANVIS-7 night vision goggles; uprated T55-L-714 turboshafts with FADEC; increased fuel capacity; additional troop seating (44 maximum); OBOGS; rotor brake; 272 kg (600 lb) rescue hoist with 61 m (200 ft) usable cable; two six-barrel miniguns in forward port and starboard positions (also on MH-47D); provisions for Stinger AAMs using FLIR for sighting. This system largely common with equivalent Sikorsky MH-60K (which see).

MH-47E has nose of Commercial Chinook to allow for weather radar, if needed; forward landing gear moved 1.02 m (3 ft 4 in) forward to allow for all-composites

Boeing CH-47D military transport helicopter with additional side view (lower) of MH-47E special forces' variant (James Goulding/Jane's)

external fuel pods (also from Commercial Chinook) that double fuel capacity; Brooks & Perkins internal cargo handling system. See also Chinook HC. Mk 3, below.

US ARMY MH-47E PROCUREMENT

FY	Qty	First aircraft
90	1	90-0414
91	6	91-0496
92	4	92-0400
	14	92-0464
01	1	(86-1678)
Total	**26**	

Notes: First three batches included in MYP2 (see CH-47D Procurement)
Excludes prototype (88-0267)

Chinook HC. Mk 2/2A: RAF version; Mk 1 designation CH47-352; all survivors of original 41 HC. Mk 1s upgraded to HC. Mk 1B (see 1989-90 and earlier Jane's); UK MoD authorised Boeing to update 33 (later reduced to 32) Mk 1Bs to Mk 2, equivalent to CH-47D, October 1989; changes include new automatic flight control system, updated modular hydraulics, T55-L-712F power plants with FADEC, stronger transmission, improved Solar 71 kW (95 shp) T62-T-2B APU, airframe reinforcements, low IR paint scheme, fuel system and standardisation of defensive aids package (IR jammers, chaff/flare dispensers, missile approach warning and machine gun mountings). Smiths Industries Health and Usage Monitoring System (HUMS) being installed on fleet-wide basis at cost of about £100 million during 1997 to 2000. Requirement exists for FLIR.

Conversion continued from 1991 to July 1995. Chinook HC. Mk 1B ZA718 began flight testing Chandler Evans/Hawker Siddeley dual-channel FADEC system for Mk 2 in October 1989. Same helicopter to Boeing, March 1991; rolled out as first Mk 2 19 January 1993; arrived RAF Odiham 20 May 1993; C(A) clearance November 1993. Initial deliveries pooled at Odiham by Nos 7 and 27 (Reserve) Squadrons – latter for training; first delivery to No 18 Squadron at Laarbruch, Germany, 1 February 1994; to No 78 Squadron, Falkland Islands, February 1994. Final Mk 1 withdrawn from service, May 1994, at which time 11 Mk 2s received. Further three new-build Mk 2s ordered 1993 (as Lot 2), for delivery from late 1995; decision to order further 14 Mks 2A/3 for delivery 1997-2000 at cost of US$365 million announced March 1995; these comprise Lots 3 and 4, respectively; Mk 2A has dynamically tuned fuselage (refer above). Latest purchase raised total RAF procurement to 58 CH-47C/D/Es. First Lot 3 HC. Mk 2A (ZH891) handed over in USA 6 December 1997; shipped to UK and arrived Boscombe Down for clearance trials 18 December 1997. Rest of Lot 3 delivered by end 1998. Retrofits announced in 1996 comprise Smiths HUMS and Racal RA 800 secure communications control system. Current squadron dispositions are Nos. 7, 18 and 27 at Odiham; and No. 78 in Falkland Islands.

Chinook HC. Mk 3: Eight of 14 additional RAF Chinooks announced March 1995 assigned to Special Forces; build standard as CH-47SD, but with MH-47E's large fuel panniers, weather radar and refuelling probes. First flight of initial aircraft (ZH897, as N2045G) in mid-October 1998. Following flight test, all transferred to temporary store at Shreveport, Louisiana, for six to eight months, pending fitment of avionics (including Sky Guardian RWR) which form subject of separate contract; delivery was expected to begin in February 2000, but UK formally refused to accept initial aircraft, citing software problems as cause. RAF service entry with Joint Helicopter Command not now expected before 2002.

HT.17 Chinook: Spanish Army version.

Boeing 234: Commercial version, currently out of production.

Boeing 414: Export military version, described in 1985-86 Jane's. Superseded by CH-47D International Chinook and CH-47SD versions (see below).

CH-47D International Chinook: Boeing 414-100 first sold to Japan; Japan Defence Agency ordered two for JGSDF and one for JASDF in 1984; first flight (N7425H) January 1986 and, with second machine, delivered to Kawasaki April 1986 for fitting out; co-production arrangement (see under Japan: Kawasaki **CH-47J**). International Chinook was available in four versions with combinations of standard or long-range (MH-47E-type) fuel tanks and T55-L-712 SSB or T55-L-714.

CH-47SD 'Super D': Latest export variant, first flown on 25 August 1999; embodies some improvements first installed on the MH-47E for US special operations forces, including increased max T-O weight. Honeywell T55-GA-714A turboshaft with Coltec Industries FADEC chosen as standard power plant; single-point pressure refuelling and jettison capability on both sides of aircraft, with fuel contained in two ballistic and crash-resistant tanks; total usable capacity is 7,828 litres (2,068 US gallons; 1,722 Imp gallons); CH-47SD also has Smiths digital fuel quantity gauging system in place of Ragen analogue system.

Simplified structure offers benefits in maintainability and reliability. Specific changes include machined frames instead of standard frames.

CH-47SD also incorporates modernised NVG-compatible cockpit with avionics control management system (ACMS), utilising proven military and commercial off-the-shelf equipment on single console to reduce pilot workload and with provisions for growth. ACMS organised into three separate functional groups, specifically air vehicle group (including engines, drive system, fuel, electrics, flight controls and warning indicators); mission group (including communications, navigation and survivability equipment such as INS/GPS, weather radar and digital map display); and pilotage group (including EFIS). Radar nose (but not necessarily radar) fitted as standard.

Avionics suite is comparable to that of baseline CH-47D, but features two embedded INS/GPS units as well as AN/ARN-147 VOR/ILS and AN/ARN-149 ADF, plus space and power provisions for Tacan. AN/ARC-210 frequency-hopping radios to be installed on CH-47SD for Taiwan.

Initial order for four from Singapore received shortly before March 1998 announcement of formal SD launch, with first delivery then due in fourth quarter of 1999;

Royal Air Force Chinook HC. Mk 2

Singapore has revealed intent to obtain additional four. Second customer is Taiwan which, in January 2000, announced purchase of nine CH-47SDs for use by Army; formal letter of agreement for initial batch of three for delivery in 2001 signed 29 April 1999.

CH-47F Improved Cargo Helicopter (ICH): Boeing programme for improved Chinook configuration for US Army involving the development and production of a new version that will remain operational and cost-effective until 2033, when last examples replaced by a new cargo helicopter to be developed between 2015 and 2020 under the current Army Aviation Modernization Plan. Programme will coincide with the existing Chinook beginning to reach planned life cycle limits in 2002. Benefits envisaged include greater airframe and systems reliability arising from lower levels of vibration; reduced operating and support costs; reduced pilot workload; and more efficient cargo handling. Weights unchanged, although US Army aircraft, like CH-47D, will have peacetime MTOW of 22,680 kg (50,000 lb). Key goal of upgrade is ability to airlift 7,257 kg (16,000 lb) load over 50 n miles (92 km; 56 miles) under 35°C (95°F) and 1,220 m (4,000 ft) altitude daytime conditions.

Studies indicated that further remanufacture, rather than new production, offered best degree of affordability, with a key factor being fleet-wide fitment of the T55-GA-714A, an improved variant of the T55-L-714 engine, as installed in the MH-47E and the International Chinook. Provision of Coltec Industries FADEC will result in lower specific fuel consumption and a modest reduction (about 3 per cent) in fuel burn. The T55-GA-714A results from a Honeywell (at Greer, South Carolina) kit upgrade programme begun December 1997 and affecting approximately 1,150 existing T55-L-712 engines in the Army inventory; upgrading brings power ratings up to those of the T55-L-714 and adds marinisation features.

Increased versatility also derived from replacement of the existing cargo handling system by integral flip-over roller panels in the cargo deck; Boeing contracted to install this equipment in one MH-47D Chinook for demonstration purposes during 1997.

Cockpit modernisation by Rockwell Collins will include adding a MIL-STD-1553B databus, four 127 mm (5 in) square MFD 255 EFIS LCDs, two 152 × 203 mm (6 × 8 in) MFD 268 displays, two CDU 900 controller/processors, DR-200 data loader, moving digital map display, digital communications and electronic instrument displays. This allows for updated communications and navigation, enabling the Chinook to meet US Army Force XXI Battlefield requirements. Rockwell Collins selected by Boeing in April 1998 to provide integrated avionics suite, which will feature open architecture to facilitate future growth and updating; CH-47F will be first US Army helicopter to incorporate improved data modem (IDM), allowing it to link with digital battlefield and automatically send and receive information concerning targeting and aircraft position.

Army award of a development contract was completed in late 1997, with the objective of delivering the first upgraded Chinook in 2003. Two Chinooks (83-24107 and 83-24115) are being refurbished for the engineering and manufacturing development (EMD) phase, for which contract worth US$76 million was awarded to Boeing in May 1998. Both EMD aircraft arrived for conversion on 5 January 1999 and first is expected to be rolled out in May 2001 and enter flight test in June; second should complete upgrade in July 2001 and fly several weeks later, with EMD phase planned to end in December 2002. Thereafter, low-rate initial production (LRIP) expected to begin with initial batch of 11 aircraft, followed by second batch of 17; subsequent procurement of up to 26 per year until 2015 will give a total fleet of 300; delivery of first LRIP CH-47F due in third quarter of FY03, with first unit fully equipped by fourth quarter of FY04.

CUSTOMERS: US Army has received five YCH-47A, 349 CH-47A, 108 CH-47B, 270 CH-47C and three CH-47D as new-build aircraft from US production plus 11 CH-47C from Italian production; further 479 CH-47D/MH-47E obtained from conversion programme involving CH-47A/B/C models as detailed in table elsewhere. CH-47D, MH-47D and MH-47E inventory was 466 in September 1999.

Exports include five CH-47Cs to Argentina (three Air Force; two Army); Australia 12 CH-47Cs with crashworthy fuel system (four refurbished as CH-47D and redelivered from May 1995 to C Squadron of Army's 5 Aviation Regiment at Townsville; seven sold to US Army and converted to CH-47D for Hawaii National Guard; contract for further two signed 19 June 1998, with delivery to Townsville for further modification on 10 March 2000); Canada nine CH-47Cs designated **CH-147**, delivered from September 1974 (details in 1985-86 *Jane's*), but withdrawn from use and seven sold to Netherlands in July 1993; Egypt finalised contract for four CH-47Ds on 20 August 1998, with delivery scheduled for third quarter of 1999, but actually effected in mid-2000; Greece ordered seven CH-47Ds in third quarter of 1998 for delivery in 2001; Japan (two CH-47D International Chinooks, plus five kits to initiate licensed production, with further 54 co-produced by Kawasaki); Netherlands, six ordered in December 1993 for delivery in 1998-99 (first, D-101, on 29 May 1998), plus seven ex-Canadian CH-147s upgraded to CH-47D (including nose radar) by Boeing and redelivered from December 1995 for service with No. 298 Squadron at Soesterberg; Spanish Army 19, designated **HT.17**, of which 10 CH-47Cs (one lost), three Boeing 414-176s and six CH-47D International Chinooks delivered to 5th Helicopter Transport Battalion (Bheltra-V) at Colmenar Viejo, Madrid (last six have Honeywell RDR-1400 weather radar), nine survivors of original 10 upgraded by Boeing in USA to CH-47D between August 1990 and mid-1993, first redelivery to Spain on 27 September 1991, with three Boeing 414s also upgraded in 1997; South Korea, total 30 International Chinooks, comprising 24 for Army (deliveries from 1988); and six for Air Force (ordered early 1990; delivered by February 1992); Singapore, six ordered April 1994 for delivery in 1996-97, with further six SDs ordered in 1998 (training operations conducted in 1996-97 at Grand Prairie, Texas, before transfer to Singapore base); Taiwan (three, notionally civil Boeing 234MLR; finalised contract for nine CH-47SDs in January 2000, with deliveries to begin in 2001); Thailand, six for Army (three ordered August 1988, three early 1990, deliveries 1990 to February 1992; all are International Chinook); UK (58, see entries for Chinook HC. Mk 2/2A/3); 16 civilian, of which few left operating in original oilfield support role; two for trials; 46 in kits, comprising 40 for assembly in Italy and six for Japan.

Agusta sold licence-built CH-47Cs to Egypt (15; six survivors under consideration in 1998 for CH-47D upgrade), Greece (10, of which nine converted to CH-47D by Boeing between March 1992 and 1995; first redelivery in October 1993), Iran (68), Italy (37, including 10 CH-47C Plus, to which standard 23 earlier helicopters converted), Libya (20), Morocco (nine) and Pennsylvania US Army National Guard (11). Further six Italian helicopters to Civil Protection Agency. Kawasaki (which see) has firm orders for 61.

Total Chinook orders, including civil, 1,164, excluding additional order for unknown quantity of CH-47SD version from unspecified customer. Six CH-47Ds, ordered by China January 1989, embargoed by US government.

CHINOOK PRODUCTION

Customer	Boeing	Boeing kits	Agusta	Kawasaki
Civilian	10*		6*	
Argentina	5			
Australia	14			
Canada	9			
Egypt	4		15	
Greece	7		10	
Iran		38	30	
Italy		2	35	
Japan	2	5		54†
Libya			20	
Morocco			9	
Netherlands	6			
Singapore	12			
South Korea	30			
Spain	19			
Taiwan	12‡			
Thailand	6			
UK (RAF)	58†			
US Army	735		11	
(Total 1,164)	**929**	**45**	**136**	**54**
Rebuilds to CH-47D				
Australia	4			
Greece	9			
Italy			23	
Netherlands	7			
Spain	12			
UK (RAF)	32			
US Army	479			
(Total 566)	**543**		**23**	

* Civilian standard
† Deliveries in progress
‡ Includes three to notional civilian standard and nine CH-47SD

COSTS: First US Army MYP (1985) US$1,200 million for 240 upgrades from CH-47C; second MYP (1989) US$773 million for 144 upgrades; US$67 million (1993) for 11 upgrades (four Australian Army; seven US Army); approximately US$23 million for two new-build CH-47Ds (1992). Estimated cost of four CH-47Ds requested by Egypt in 1997 is US$149 million including spares, training and support; Australia, US$60 million for two (1998).

MH-47E: US$81.8 million (placed 1987) for development of MH-47E and conversion of one prototype; US$422 million (1989-91) for 11 MH-47Es and option on further 14. US$25.6 million allocated for conversion of single CH-47D to MH-47E (2000). £140 million upgrade for RAF aircraft (32 to Mk 2 standard, 1993-95).

US Army estimates cost of CH-47F programme, including R&D, long-lead production and upgrade of 300 Chinooks, as approximately US$3.3 billion (1999).

DESIGN FEATURES: Two three-blade intermeshing contrarotating tandem rotors; front rotor turns anti-clockwise, viewed from above; rotor transmissions driven by connecting shafts from engine gearbox, which is driven by rear-mounted engines. Classic rotor heads with flapping and drag hinges; manually foldable blades, using Boeing VR7 and VR8 aerofoils with cambered leading-edges; blades can survive hits from 23 mm HEI and API rounds; rotor brake optional.

Development of new, low-maintenance elastomeric rotor hub begun at end August 1999, following signature of contract with US Army; development, testing and installation to be completed over four-year period, with new assembly to be installed on all US Army regular and reserve force Chinooks. Also under consideration for CH-47F and will be incorporated in CH-47SD; UK expected to participate in development programme, with view to adoption. Will have longer fatigue life (4,500 hours), 75 per cent fewer parts and offer 70 per cent reduction in requirement for special maintenance tools; will also retain same rotor flight dynamics and be fully interchangeable with existing hub.

Constant cross-section cabin with side door at front; rear-loading ramp that can be opened in flight; underfloor section sealed to give flotation after water landing; access to flight deck from cabin; main cargo hook mounting covered by removable floor panel so that load can be observed in flight.

FLYING CONTROLS: Differential fore and aft cyclic for pitch attitude control; differential lateral cyclic pitch (from rudder pedals) for directional control; automatic control to keep fuselage aligned with line of flight. Dual hydraulic rotor pitch-change actuators: secondary hydraulic actuators in control linkage behind flight deck for autopilot/autostabiliser input; autopilot provides stabilisation, attitude hold and outer-loop holds.

STRUCTURE: Blades based on D-shaped glass fibre spar, fairing assembly of Nomex honeycomb core and crossply glass fibre skin.

LANDING GEAR: Non-retractable quadricycle type, with twin wheels on each front unit and single wheels on each rear unit. Oleo-pneumatic shock-absorbers in all units. Rear units fully castoring; power steering on starboard rear unit. All wheels are size 24×7.7, with tyres size 8.50-10, pressure 6.07 bars (88 lb/sq in). Single-disc hydraulic brakes on all six wheels. Provision for fitting detachable wheel/skis.

POWER PLANT: Two Honeywell T55-L-712 turboshafts, pod-mounted on sides of the rear pylon, each with a standard power rating of 2,237 kW (3,000 shp) and maximum rating of 2,796 kW (3,750 shp). Honeywell T55-L-712 SSB engine has standard power rating of 2,339 kW (3,137 shp) and maximum of 3,217 kW (4,314 shp); transmission capacity (CH-47D/SD and MH-47E) 5,617 kW (7,533 shp) on two engines and 3,430 kW (4,600 shp) OEI; rotor rpm 225.

Self-sealing pressure refuelled crashworthy fuel tanks in external fairings on sides of fuselage. Total usable fuel capacity 3,914 litres (1,034 US gallons; 861 Imp gallons) in CH-47D. Provision for up to three additional long-range tanks in cargo area, each of 3,028 litres (800 US gallons; 666 Imp gallons); maximum fuel capacity (fixed and auxiliary) 12,998 litres (3,434 US gallons; 2,859 Imp gallons). Oil capacity 14 litres (3.7 US gallons; 3.1 Imp gallons).

From January 1991, more than 100 CH-47Ds fitted with engine air particle separator (also available for RAF variant). Standard in MH-47E and International Chinook are two Honeywell T55-L-714 turboshafts, each with a standard power rating of 3,108 kW (4,168 shp) continuous and emergency rating of 3,629 kW (4,867 shp). CH-47SD (and Netherlands CH-47Ds) have T55-GA-714A turboshafts, with maximum continuous rating of 3,039 kW (4,075 shp). FADEC installed on late production CH-47Ds and CH-47SD. Normal fuel (MH-47E, International Chinook and CH-47SD) 7,828 litres (2,068 US gallons; 1,722 Imp gallons) but MH-47E can also operate with three long-range tanks in cargo area, each containing 3,028 litres (800 US gallons; 666 Imp gallons), raising total fuel capacity to 16,913 litres (4,468 US gallons; 3,720 Imp gallons). MH-47 SOA and MH-47E have 8.97 m (29 ft 5 in) refuelling probe on starboard side of forward fuselage, which extends 5.41 m (17 ft 9 in) forward of the nose. Refuelling probe also an option for International Chinook.

ACCOMMODATION: Two pilots on flight deck, with dual controls. Lighting compatible with pilots' NVGs (Nite-Op in RAF variant). Jump seat for crew chief or combat commander. Jettisonable door on each side of flight deck. Depending on seating arrangement, 33 to 55 troops can be accommodated in main cabin, or 24 litters plus two attendants, or (see under Current Versions) vehicles and freight. Rear-loading ramp can be left completely or partially open, to permit transport of extra-long cargo and in-flight parachute or free-drop delivery of cargo and equipment.

Main cabin door, at front on starboard side, comprises upper hinged section which can be opened in flight, and lower section with integral steps. Upper section has a panel with window that is jettisonable. Triple external cargo hook system, with US Army aircraft having centre hook rated to carry maximum load of 11,793 kg (26,000 lb) and the forward and rear hooks 7,711 kg (17,000 lb) each, or 10,433 kg (23,000 lb) in unison, while International Chinook and CH-47SD ratings are 12,701 kg (28,000 lb) for centre hook and 9,072 kg (20,000 lb) each for forward and rear hooks, or 11,340 kg (25,000 lb) in unison. Options are available for a power-down ramp and water dam to permit ramp operation on water, internal ferry fuel tanks, external rescue hoist, and windscreen washers.

SYSTEMS: Hydraulic system contains a utility system, a No. 1 flight control system and a No. 2 flight control system. Each includes a separate, variable delivery pump and a reservoir cooler module. Utility system contains a pressure control module and a return control module. Both flight control systems contain a power control module, incorporating pressure and return in one module for each system. Each flight control system can be driven by the utility system for ground checkout through a power transfer unit without intermixing of hydraulic fluids. All hydraulic systems have a pressurised reservoir to prevent pump cavitation and are serviced by a common filter.

The CH-47D has a significant reduction in the number of hydraulic lines and fittings; approximately 200 tubes and hoses eliminated, thereby obviating approximately 700 leak points. Majority of all lines now swaged together for permanent joining; Rosan fluid adaptors for hardpoint connections. All systems capable of being monitored, both in flight and on the ground, for servicing prechecks and fault isolation. Subsystems designed for pressurisation on demand; power transfer units now used for flight control ground system checkout. Electrical system includes two 40 kVA oil-cooled alternators driven by transmission drive system. Solar T62-T-2B APU drives a 20 kVA generator and hydraulic motor pump, providing electrical and hydraulic power for main engine start and system operation on the ground.

AVIONICS: Baseline CH-47D. Specific MH-47E avionics listed under that heading. Avionics for RAF HC. Mk 1 listed in 1985-86 and earlier editions. Netherlands aircraft have Honeywell advanced cockpit management system (ACMS) and EFIS, latter adopted by Boeing as baseline for subsequent sales.

Comms: Honeywell AN/ARC-199 HF com radio, Rockwell Collins AN/ARC-186 VHF/FM-AM, Magnavox AN/ARC-164 UHF/AM com; C-6533 intercom (Netherlands aircraft have Telephonics Corp STARCOM); Honeywell AN/APX-100 IFF.

Flight: Honeywell AN/APN-209 radar altimeter; AN/ARN-89B ADF; BAE Systems AN/ASN-128 Doppler; AN/ARN-123 VOR/glide slope/marker beacon receiver; and AN/ASN-43 gyromagnetic compass. AFCS maintains helicopter stability, eliminating the need for constant small correction inputs by the pilot to maintain desired attitude. The AFCS is a redundant system using two identical control units and two sets of stabilisation actuators. RAF Chinooks have Racal RNS252 Super TANS INS including GPS.

Instrumentation: Flight instruments are standard for IFR, and include an AN/AQU-6A horizontal situation indicator.

Mission: Chelton 19-400 satellite communications antenna on some RAF helicopters.

Self-defence: Provisions for Lockheed Martin AN/APR-39A RWR, Sanders AN/ALQ-156 missile warning equipment and BAE Systems M-130 chaff/flare dispensers. RAF Chinooks have BAE Systems M-206/M-1 chaff/flare dispensers, BAE Systems ARI 18228 Sky Guardian RWR and (from 1990) Lockheed Martin AN/ALQ-157 IR jammers and Honeywell AN/AAR-47 missile approach warning equipment.

EQUIPMENT: Hydraulically powered winch for rescue and cargo handling, rearview mirror, plus integral work stands and step for maintenance.

ARMAMENT: Provision for two machine guns or miniguns in crew door (starboard) and forward hold window (port).

DIMENSIONS, EXTERNAL:
Rotor diameter (each) 18.29 m (60 ft 0 in)
Rotor blade chord (each) 0.81 m (2 ft 8 in)
Distance between rotor centres: CH-47SD
11.85 m (38 ft 10¾ in)
Length: overall, rotors turning: MH-47E and CH-47SD
30.14 m (98 ft 10¾ in)
fuselage: MH-47E and CH-47SD 15.87 m (52 ft 1 in)
MH-47D/E, incl probe 21.08 m (69 ft 2 in)
Width, rotors removed: MH-47E 4.78 m (15 ft 8 in)

Royal Netherlands Air Force (former Canadian) Boeing CH-47D Chinook *(Paul Jackson/Jane's)*

USA: AIRCRAFT—BOEING

Height to top of rear rotor head:
 CH-47SD   5.70 m (18 ft 8½ in)
 MH-47E   5.59 m (18 ft 4 in)
Ground clearance, rotors turning:
 rear approach   5.77 m (18 ft 11 in)
Ground clearance, static: rear approach:
 MH-47E and CH-47SD   4.90 m (16 ft 0¾ in)
 forward approach: CH-47SD   2.29 m (7 ft 6 in)
Wheelbase: MH-47E and CH-47SD   7.87 m (25 ft 10 in)
Passenger door (fwd, stbd): Height   1.68 m (5 ft 6 in)
 Width   0.91 m (3 ft 0 in)
 Height to sill   1.09 m (3 ft 7 in)
Rear-loading ramp entrance: Height   1.98 m (6 ft 6 in)
 Width   2.31 m (7 ft 7 in)
 Height to sill   0.79 m (2 ft 7 in)
DIMENSIONS, INTERNAL (MH-47E and CH-47SD):
 Cabin, excl flight deck: Length   9.19 m (30 ft 2 in)
  Width: mean   2.29 m (7 ft 6 in)
  at floor   2.51 m (8 ft 3 in)
  Height   1.98 m (6 ft 6 in)
 Floor area   21.0 m² (226 sq ft)
 Usable volume   41.7 m³ (1,474 cu ft)
AREAS:
 Rotor blades (each)   7.43 m² (80.0 sq ft)
 Rotor discs (total): MH-47E and CH-47SD
    525.3 m² (5,655 sq ft)
WEIGHTS AND LOADINGS:
 Weight empty: MH-47E   12,210 kg (26,918 lb)
  CH-47SD   11,550 kg (25,463 lb)
 Useful load: MH-47E   12,284 kg (27,082 lb)
  CH-47SD   12,944 kg (28,537 lb)
 Max underslung load: CH-47SD   12,700 kg (28,000 lb)
 Max fuel weight: MH-47E, CH-47SD
    6,815 kg (15,025 lb)
 Max T-O weight: MH-47E, CH-47SD
    24,494 kg (54,000 lb)
 Transmission loading at max T-O weight and power:
  MH-47E, CH-47SD   4.36 kg/kW (7.17 lb/shp)
PERFORMANCE (at 22,680 kg; 50,000 lb):
 Max level speed: MH-47E   154 kt (285 km/h; 177 mph)
  CH-47SD   155 kt (287 km/h; 178 mph)
 Max cruising speed at S/L:
  MH-47E, CH-47SD   140 kt (259 km/h; 161 mph)
 Max rate of climb: MH-47E   561 m (1,840 ft)/min
  CH-47SD   563 m (1,846 ft)/min
 Service ceiling: MH-47E   3,090 m (10,140 ft)
  CH-47SD   3,385 m (11,100 ft)
 Hovering ceiling: IGE: MH-47E   2,985 m (9,800 ft)
  CH-47SD   2,835 m (9,300 ft)
 IGE, ISA + 20°C (68°F): MH-47E   2,410 m (7,900 ft)
  CH-47SD   2,180 m (7,160 ft)
 OGE: MH-47E, CH-47SD   1,675 m (5,500 ft)
 OGE, ISA + 20°C (68°F):
  MH-47E, CH-47SD   1,005 m (3,300 ft)
 Radius of action, MH-47E, deploy special forces team (1,814 kg; 4,000 lb) at 1,220 m (4,000 ft), 35°C (95°F) ambient temperature   505 n miles (935 km; 581 miles)
 Range, MH-47E, self-deployment at 24,494 kg (54,000 lb) T-O weight
    1,260 n miles (2,333 km; 1,449 miles)
 Range, CH-47SD, with 12,558 kg (27,686 lb) payload
    651 n miles (1,207 km; 750 miles)
UPDATED

BOEING AH-64 APACHE
US Army designations: AH-64A and D
Israel Defence Force name: Pethen (Cobra)

TYPE: Attack helicopter.
PROGRAMME: Original Hughes Model 77 entered for US Army advanced attack helicopter (AAH) competition; first flights of two development prototype YAH-64s 30 September and 22 November 1975; details of programme in 1984-85 and earlier *Jane's*; selected by US Army December 1976; named Apache late 1981.

Deliveries started 26 January 1984; 900th delivered in October 1995, at which time US Army had ordered 821 AH-64As (excluding prototypes) with export contracts totalling 104; latter total increased to 221 (116 AH-64A; 105 AH-64D) by June 1999. Last of 821 AH-64As delivered to US Army on 30 April 1996; aircraft concerned was production vehicle 915, and manufacture of AH-64A variant terminated with completion of 937th example (for Egypt), in November 1996; details of initial service use in 1999-2000 and earlier *Jane's*. Delivery of 1,000th Apache (including AH-64D rebuilds) effected on 30 March 1999.

Boeing awarded four year, US$15.9 million, contract on 27 May 1998 to design, manufacture and flight test new centre fuselage section incorporating advanced composites materials; Phantom Works responsible for design leadership, with support from Boeing facilities at Long Beach, Philadelphia and St Louis. If successful, new section from rear of aft cockpit to just behind engines will be simpler to manufacture as well as lighter and more durable than existing all-metal structure.

In separate development, on 5 October 1998, a modified AH-64D Apache Longbow prototype made initial flight with Rotorcraft Pilot's Associate (RPA) advanced cockpit management system. Also developed by Phantom Works over 60 month period, under terms of US$80 million advanced technology demonstration contract, this featured updated controls and displays, including Boeing-developed four-axis, full authority, advanced digital flight control system. Flight and mission data presented to pilots on three large multipurpose colour displays; RPA also features advanced data fusion and an advanced pilotage system, as well as the ability to recognise and respond to verbal commands. Company flight trials continued throughout remainder of 1998, with test aircraft then visiting US Army's Yuma Proving Grounds for demonstration flights in January-February 1999.

CURRENT VERSIONS: **AH-64A**: Produced for US Army and export; 937 built. First 603 had two 1,265 kW (1,696 shp) T700-GE-701 engines. Total of 530 US Army examples to be upgraded to AH-64D. Retrofit from 1993 with SINCGARS secure radios and GPS; first installed in Apaches of 5-501 Aviation Regiment on deployment to Camp Eagle, South Korea.

AH-64B: Cancelled in 1992. Was planned near-term upgrade of 254 AH-64As with improvements derived from operating experience in 1991 Gulf War, including GPS SINCGARS radios, target handover capability, better navigation, and improved reliability including new rotor blades.

AH-64C: Previous designation for upgrade of AH-64As to near AH-64D standard, apart from omission of Longbow radar and retention of -701 engines; provisions for optional fitment of both; Army requested draft proposal, August 1991; funding for two prototype conversions awarded in September 1992. Designation abandoned late 1993.

AH-64D Apache Longbow: Current improvement programme based on Lockheed Martin and Northrop Grumman joint venture development of mast-mounted AN/APG-78 Longbow millimetre-wave radar and Hellfire missile with RF seeker; Northrop Grumman has lead on Longbow with Lockheed Martin taking principal role for Hellfire. Programme also includes more powerful engines, larger generators, MIL-STD-1553B databus allied to dual 1750A processors, and a vapour cycle cooling system for avionics; early user tests completed April 1990.

Detailed description applies to AH-64D.

Full-scale development programme, lasting 51 months, authorised by Defense Acquisition Board August 1990, but airframe work extended in December 1990 to 70 months to coincide with missile development; supporting modifications being incorporated progressively; first flight of AH-64A (82-23356) with dummy Longbow radome 11 March 1991; first (89-0192) of six AH-64D prototypes flown 15 April 1992; second (89-0228) flew 13 November 1992; fitted with radar in mid-1993 and flown 20 August 1993; No.3 (90-0324) flown 30 June 1993; No.4 (90-0423) on 4 October 1993; No.5 (85-25477; formerly AH-64C No.1) 19 January 1994 (first Apache with new Hamilton Sundstrand lightweight flight management computer); No.6 (85-25408) flown 4 March 1994; last two mentioned lack radar. Following redesignation of AH-64C in late 1993, original plan was to convert 748 to AH-64D, although only 227 (original AH-64D total) to carry Longbow radar; subsequent review, revealed at start of 1999, proposed reducing total number of conversions to 530 and increasing purchase of Longbow radars to 500; these to equip regular Army units, with remaining 218 AH-64As of Army National Guard and Army Reserve undergoing service life extension programme. Most recent news indicates that US Army may eventually receive 600 AH-64Ds.

Under original programme, AH-64D to equip 26 battalions, although this may be reduced to 16 of regular Army; three companies, each with eight helicopters, per

Boeing AH-64A (top) and AH-64D Apache attack helicopters

Co-pilot/gunner's cockpit on the AH-64D Apache Longbow 0022142

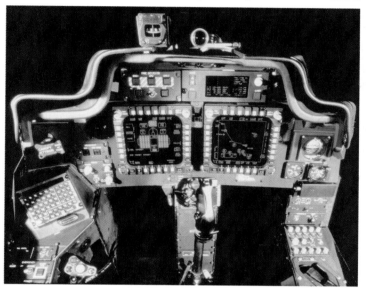
Pilot's cockpit on the AH-64D Apache Longbow 0022141

Experimental Rotorcraft Pilot's Associate cockpit, first flown in an AH-64D in October 1998 under a US$80 million programme by Boeing Phantom Works for US Army, intended to improve efficiency and situational awareness *(US Army)* 2000/0048663

battalion. Longbow can track flying targets and see through rain, fog and smoke that defeat FLIR and TV. RF Hellfire, which delivered to US Army from November 1996, can operate at shorter ranges; it can lock on before launch or launch on co-ordinates and lock on in flight; Longbow scans through 360° for aerial targets or scans over 270° in 90° sectors for ground targets; mast-mounted rotating antenna weighs 113 kg (250 lb). Longbow radar transmitter subject of redesign in late 1997 to overcome poor performance of some electrical components in low temperatures; eliminate lengthy and costly manual integration necessary to achieve required output; and avoid shortages of critical components which suppliers reported in 1995 that they would no longer provide. New transmitter meets or surpasses original specification and is fully interchangeable with original unit.

Further modifications include 'manprint' cockpit with large displays, air-to-air missiles, digital autostabiliser, integrated GPS/Doppler/INS/air data/laser/radar altimeter navigation system, digital communications, faster target handoff system, and enhanced fault detection with data transfer and recording. Cockpit displays initially monochrome, but these replaced by colour displays with effect from 27th production conversion; first flight of AH-64D with colour displays on 12 September 1997. AH-64D No.1 made first Hellfire launch on 21 May 1993; first RF Hellfire launch 4 June 1994; first demonstration of digital air-to-ground data communications with Symetrics Industries improved data modem, 8 December 1993.

First preproduction AH-64D conversion completed remanufacture in September 1995; aircraft made successful first flight at Mesa on 29 September 1995.

Advanced acquisition phase contract for remanufacture programme, worth US$279.6 million and covering 18 helicopters (later increased to 24), awarded on 14 December 1995; predated by arrival of first two AH-64As (85-25387 and 85-25394) at Mesa on 27 November 1995 for stripping to basic fuselage in readiness for start of remanufacture in early 1996. First fuselage moved to final assembly area on 15 August 1996; first flight (85-25387 with new identity 96-5001) 17 March 1997; formal roll-out at Mesa on 21 March.

Multiyear contract, worth US$1.87 billion, covering 232 AH-64Ds (retrospectively including advanced acquisition aircraft) over five year period signed 16 August 1996; further 269 conversions to be acquired in second multiyear contract covering FY01 to FY05, which signed 29 September 2000, at total cost of US$2.3 billion. US Army currently expects to complete AH-64D acquisition with final batch of 99, to give total buy of 600 for operational deployment. Initial contract also included 227 Longbow radars, 13,311 Hellfire missiles and 3,296 launchers. AH-64D deliveries to US Army began 31 March 1997 and total of 16 handed over by end November 1997. Delivery of 24th and last Lot 1 aircraft accomplished 4 March 1998, with US Army having accepted total of 48 by end of October 1998, when production rate was three per month; 100th remanufactured AH-64D delivered to US Army in early December 1999 and, by start of October 2000, almost 150 had been accepted by US Army, with production running at six per month.

Initial AH-64D battalion (1-227 AvRgt) at Fort Hood, Texas fully equipped by end July 1998 and attained combat-ready status on 19 November 1998, after eight month training programme at company and battalion level which included four live fire exercises and more than 2,500 flight hours. Second unit is 2-101 AvRgt at Fort Campbell, Kentucky, which certified as combat-ready on 28 October 1999; third is 1-3 AvRgt at Hunter AAF, Fort Stewart, Georgia, in first quarter of 2001.

GAH-64A: At least 17 AH-64As grounded for technical instruction.

JAH-64A: Seven AH-64As for special testing, of which one reverted to standard.

WAH-64D: British Army version with Longbow radar and two Rolls-Royce/Turbomeca RTM 322 turboshafts; see Westland entry in the UK section.

CUSTOMERS: US Army 827 AH-64A (including six prototypes), of which last delivered 30 April 1996; see Programme and Current Versions for details; operational fleet in November 1999 totalled 743. Confirmed export orders and firm commitments totalled 221 at end of 1999. Israel ordered 18 in March 1990; first two delivered 12 September 1990 to 113 Squadron, powered by T700-GE-701s; further 24 second-hand (including 18 US Army Europe AH-64As delivered in September 1993), for 113 and 127 Squadrons; first two units based at Ramon; deliveries to 190 Squadron at Ramat David, 1995. After rejecting purchase of new-build AH-64D Apache Longbow on cost grounds, Israel considered upgrade of existing AH-64A to AH-64D; however, subsequent change of plan involved abandoning upgrade proposal in favour of acquisition of new-build AH-64D and initial batch of eight requested from USA in September 2000, along with 10 AN/APG-78 Longbow radars.

Deliveries began in April 1993 to Saudi Arabia (12) and in 1994 to Egypt (24). Further orders from Greece (20) and United Arab Emirates (20), both in December 1991; six handed over to UAE on 30 October 1993; 14 followed in 1994. Greek deliveries from February 1995 for training in USA; initial six Apaches ferried to New Orleans, Louisiana, for shipment to Greece on 9 June 1995. Ten more ordered for UAE in June 1994; 12 more to Egypt approved early 1995, with Egypt revealing intent in July 2000 to upgrade 35 AH-64A to AH-64D; signature of contract expected in early 2001, with deliveries to begin in July 2003. Additional four AH-64A requested by Greece in second quarter of 1999 at estimated cost of US$111 million, including spares, technical support and HADSS.

Version of AH-64D Apache Longbow offered to British Army by consortium of McDonnell Douglas, Westland, Lockheed Martin, Northrop Grumman and Shorts; UK announced order for 67 on 13 July 1995. Netherlands signed contract on 24 May 1995 for 30 Apaches for Nos. 301 and 302 Squadrons at Gilze-Rijen; radar not required at outset, but formal request made in November 1999 and this will now be installed; US Army provided 12 AH-64As for use by No. 301 Squadron on lease basis for period 1996-99; all delivered 13 November 1996. First AH-64Ds accepted by Netherlands at Mesa on 15 May 1998; No. 302 Squadron first unit to be equipped, with initial deliveries to Gilze-Rijen in mid-May 2000; No. 301 Squadron due to become operational in 2003. Latest customer is Singapore, which selected AH-64D on 14 June 1999; total of eight to be acquired, with AGM-114K Hellfire 2 laser-guided missile, but lacking Longbow radar. Kuwait was authorised in late 1997 to receive 16 AH-64Ds under FMS programme; letter of acceptance due for signature in last quarter of 1998, but debate continued over inclusion of Longbow radar and deal still to be finalised at end of 2000; proposed package includes 384 Hellfire missiles (310 combat rounds, 50 dummies and 24 training rounds), plus spares, training, government and contractor support. AH-64D also candidate for sales to Australia, whose Project Air 87 calls for procurement of 25 to 30 attack helicopters to replace Bell 206B and UH-1H in Army service; and South Korea, which initially needs 18 combat helicopters, but could eventually acquire total of 36 to 48.

596   USA: AIRCRAFT—BOEING

Japan, Norway, Spain and Sweden have also expressed interest in AH-64D.

APACHE REMANUFACTURE PROGRAMME

Year	Qty	First aircraft
Prototypes	6	89-0192
Preproduction	2	
FY96	24	96-5001
FY97	24	97-5025
FY98	44	98-5049
FY99	66	99-5093
FY00	74	00-5159
FY01-05	269	
Total	509	

APACHE PROCUREMENT

	Qty	Remarks
US Army AH-64A		
FY73	3	Prototypes
FY79	3	Preproduction
FY82	11	
FY83	48	
FY84	112	
FY85	138	
FY86	116	
FY87	101	
FY88	77	
FY89	54	
FY90	132	
FY91	12	
FY92	10	
FY94	10	
Subtotal	827[1]	

Exports		First aircraft and date	
Israel	18[2]	801	Sep 1990
Saudi Arabia	12[3]	90-0291	Apr 1993
Egypt (Lot 1)	24[3]	3701	Feb 1994
Egypt (Lot 2)	12[3]	3725	1996
Greece	20[3]	E-1001	Feb 1995
UAE (Lot 1)	20[3]	050	Oct 1993
UAE (Lot 2)	10[3]	070	1996
Netherlands	30[4]	Q-01/98-0101	May 1998
UK	67[5]	ZJ166	Sep 1998
Singapore	8[6]		
Subtotal	221		
Total	1,048		

Notes:
[1] 501 programmed to be remanufactured as AH-64D, with further 99 in prospect, for total of 600
[2] AH-64A; further 24 obtained from US Army, eight AH-64D requested in September 2000, not included in total
[3] AH-64A
[4] Non-radar AH-64D; 12 US Army AH-64A operated during 1996-99 on lease pending delivery of AH-64D
[5] WAH-64D
[6] AH-64D

APACHE PRODUCTION

Lot	FY	Qty US	Qty Export	First aircraft
-	73	3[1]		73-22247
-	79	3[2]		79-23257
1	82	11		82-23355
2	83	48		83-23787
3	84	112		84-24200
4	85	138		85-25351
5	86	116		86-8940
6	87	101		87-0407
7	88	67		88-0197
		10		88-0275
8	89	54	18	89-0192
9	90	60	6	90-0280
10	90	72	36	90-0415
11	90/92	22	34	91-0112
12	94	10[3]	10	94-0328
13	95		12	95-0108
-	98 et seq		105[4]	98-0101
Subtotals		827	221	
Total		1,048		

Notes: [1] Prototypes; 73-22247 was static test airframe
[2] Preproduction
[3] Partial replacement for 24 transferred to Israel, 1993
[4] AH-64D/WAH-64D; all other aircraft are AH-64A version

AH-64A version of Boeing Apache in service with Israel Defence Force/Air Force *(Simon Watson)*
2001/0103525

COSTS: New US$18 million. Longbow radar costs US$4 million (1996) without supporting modifications compared to US$14 million flyaway cost of AH-64A. Longbow R&D contract (for four prototypes) US$194.6 million. Programme cost (807 aircraft at 1991 values) US$1,169 million. Egyptian Lot 2 request costed at US$318 million for 12 Apaches plus four spare Hellfire launchers, thirty-four 70 mm (2.75 in) rocket launchers, six spare T700 engines, one spare TADS/PNVS system and miscellaneous spares. Second multiyear procurement contract valued at US$2.3 billion, including training devices, spares, logistics and support services.

DESIGN FEATURES: Modern, tandem-seat, armoured and damage-resistant combat helicopter; is required to continue flying for 30 minutes after being hit by 12.7 mm bullets coming from anywhere in the lower hemisphere plus 20°; also survives 23 mm hits in many parts; target acquisition and designation sight (TADS) and pilot night vision sensor (PNVS) sensors mounted in nose; low-airspeed sensor above main rotor hub; avionics in lateral containers; chin-mounted Chain Gun fed from ammunition bay in centre-fuselage; four weapon pylons on stub-wings (six when air-to-air capability is installed); engines widely separated, with integral particle separators and built-in exhaust cooling fittings; four-blade main rotor with lifting aerofoil blade section and swept tips; blades can be folded or easily removed; tail rotor consists of two teetering two-blade units crossed at 55° to reduce noise; airframe meets full crash-survival specifications. Two AH-64s will fit in C-141, six in C-5 and three in C-17A.

Main transmission, by Litton Precision Gear Division, can operate for 1 hour without oil; tail rotor drive, by Aircraft Gear Corporation, has grease-lubricated gearboxes with Bendix driveshafts and couplings; gearboxes and shafts can operate for 1 hour after ballistic damage; main rotor shaft runs within airframe-mounted sleeve, relieving transmission of flight loads and allowing removal of transmission without disturbing rotor; AH-64A has flown aerobatic manoeuvres and is capable of flying at 0.5 g.

FLYING CONTROLS: Fully powered controls with stabilisation and automatic flight control system; automatic hover hold; tailplane incidence automatically adjusted by Hamilton Sundstrand control to streamline with downwash during hover and to hold best fuselage attitude during climb, cruise, descent and transition.

STRUCTURE: Main rotor blades (by Tool Research and Engineering Corporation, Composite Structures Divisions) tolerant to 23 mm cannon shells, have five U-sections forming spars and skins bonded with structural glass fibre tubes, laminated stainless steel skin and composites rear section; blades attached to hub by stack of laminated steel straps with elastomeric bearings. Northrop Grumman produces all fuselages, wings, tail, engine cowlings, canopies and avionics containers.

LANDING GEAR: Menasco trailing arm type, with single mainwheels and fully castoring, self-centring and lockable tailwheel. Mainwheel tyres size 8.50-10 (10 ply) tubeless, tailwheel tyre size 5.00-4 (14 ply) tubeless. Hydraulic brakes on main units. Main gear is non-retractable, but legs fold rearward to reduce overall height for storage and transportation. Energy-absorbing main and tail gears are designed for normal descent rates of up to 3.05 m (10 ft)/s and heavy landings at up to 12.8 m (42 ft)/s. Take-offs and landings can be made at structural design gross weight on terrain slopes of up to 12° (head-on) and 10° (side-on).

POWER PLANT: Two General Electric T700-GE-701C turboshafts, each rated at 1,409 kW (1,890 shp) for 10 minutes, 1,342 kW (1,800 shp) for 30 minutes, 1,238 kW (1,660 shp) maximum continuous and 1,447 kW (1,940 shp) 2½ minutes OEI. Engines mounted one on each side of fuselage, above wings, with key components armour-protected. Upper cowlings let down to serve as maintenance platforms.

AH-64 modernisation may eventually lead to installation of new 2,237 kW (3,000 shp) engine; US Army effort to prepare requirement for Common Engine Program (CEP) likely to result in adoption on H-60 series first in about 2007, but Army expects to draw up operational requirements document for AH-64 during 2001 or 2002. In near term, Boeing proposing adoption of new five-blade main rotor system allied to new drive system, offering improved performance, greater reliability and simplified maintenance.

Two crash-resistant fuel cells in fuselage, combined capacity 1,421 litres (375 US gallons; 312 Imp gallons). Modifications ordered September 1993 for carriage of four 871 litre (230 US gallon; 192 Imp gallon) Brunswick Corporation external tanks on 437 Apaches. Total internal and external fuel 4,910 litres (1,295 US gallons; 1,078 Imp gallons). New crashworthy, ballistically self-sealing internal auxiliary fuel tank entered evaluation phase in fourth quarter of 1997; tank holds 492 litres (130 US gallons; 108 Imp gallons) and is interchangeable with ammunition storage magazine, enabling all four weapons pylons to carry ordnance on long-range missions. Testing of preproduction system undertaken in 1998 in addition to formal test and qualification programme by US Army; total of 48 units ordered from Robertson Aviation in 1999; smaller 379 litre (100 US gallon; 83.3 Imp gallon) internal tank to be utilised by AH-64D, this permitting carriage of up to 300 rounds of 30 mm ammunition for Chain Gun. 'Black Hole' IR suppression system protects aircraft from heat-seeking missiles: this eliminates an engine bay cooling fan, by operating from engine exhaust gas through ejector nozzles to lower the gas plume and metal temperatures.

ACCOMMODATION: Crew of two in tandem: co-pilot/gunner (CPG) in front, pilot behind on 48 cm (19 in) elevated seat. Crew seats, by Simula Inc, are of lightweight Kevlar. Northrop Grumman canopy, with PPG transparencies and transparent acrylic blast barrier between cockpits, is designed to provide optimum field of view. Crew stations are protected by Ceradyne Inc lightweight boron armour shields in cockpit floor and sides, and between cockpits, offering protection against 12.7 mm armour-piercing rounds. Sierracin electric heating of windscreen. Seats and structure designed to give crew a 95 per cent chance of surviving ground impacts of up to 12.8 m (42 ft)/s. Simula of Phoenix, Arizona, under contract by US Army to develop cockpit-airbag system; development phase was due to be completed in early 1997, but was cancelled in favour of modifications to gunner's sighting equipment to lessen risk of injury.

SYSTEMS: Honeywell totally integrated pneumatic system includes a shaft-driven compressor, air turbine starters, pneumatic valves, temperature control unit and environmental control unit. Fairchild Controls improved environmental control system comprises a distributed vapour-cycle cooling and heating unit, with two redundant systems incorporating dual-speed compressors, digital databus controllers and multiple heat exchangers, fans and control valves. Parker dual-hydraulic systems, operating at 207 bars (3,000 lb/sq in), with actuators ballistically tolerant to 12.7 mm direct hits. Redundant flight control system for both rotors. In the event of a flying control system failure, the system activates Honeywell secondary fly-by-wire control. Honeywell electrical power system, with two 45 kVA fully redundant engine-driven AC generators, two 300 A transformer-rectifiers, and URDC standby DC battery. Honeywell GTP 36-155(BH) 93 kW (125 shp) APU for engine starting and maintenance checking. DASA (TST) electric blade de-icing. Smiths Industries integrated electrical power management system (IEPMS) installed on AH-64D is currently being upgraded to incorporate multichannel remote interface unit (RIU) that will replace core electronics and wiring associated with conventional electrical control systems; improved IEPMS to become available in 2001 and will be installed on approximately 300 Apaches for US Army.

AVIONICS: *Comms:* AN/ARC-164 UHF, AN/ARC-222 SINCGARS secure UHF/VHF; AN/ARC-220 UHF to be retrofitted; KY-28/58/TSEC crypto secure voice, C-8157 secure voice control; AN/APX-100 IFF unit with KIT-1A secure encoding; C-10414 Tempest intercom.

Radar: Optional Lockheed Martin/Northrop Grumman AN/APG-78 Longbow mast-mounted 360° radar, presenting up to 256 targets on tactical situation display; detects air targets in air-to-ground mode; air-to-air mode for flying targets only.

Flight: BAE North America AN/ASN-157 lightweight Doppler navigation system, Litton LR-80 (AN/ASN-143) strapdown AHRS, AN/ARN-89B ADF, GPS, Honeywell digital automatic stabilisation equipment (DASE), Astronautics Corporation HSI, Pacer Systems omnidirectional, low-airspeed air data system, remote

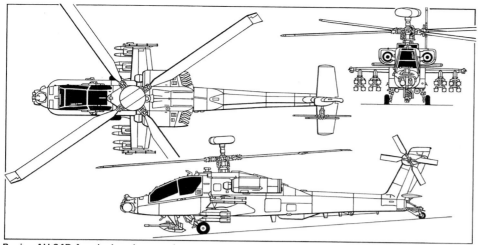
Boeing AH-64D Apache Longbow tandem-seat advanced attack helicopter *(Mike Keep/Jane's)*

magnetic indicator, BITE fault detection and location. Doppler system, with AHRS, permits nap-of-the-earth navigation and provides data for storing target locations. BAE Systems air data system, comprising two omnidirectional airspeed and direction sensors (AADSs) mounted on engine cowlings and a high integration air data computer (HIADC) installed in avionics bay.

Instrumentation: Honeywell all-raster symbology generator processes TV data from IR and other sensors, superimposes symbology, and distributes the combination to CRT and helmet-mounted displays; Honeywell AN/APN-209 radar altimeter video display unit. 'Manprint' (manpower integration) instrumentation including Litton Canada upfront display and two Honeywell 152 × 152 mm (6 × 6 in) monochrome CRT displays in each cockpit of early aircraft; with effect from 27th AH-64D, Honeywell flat-panel, colour, active matrix LCD multipurpose displays (MPDs) installed in both cockpits, as well as in aircraft for UK and Netherlands. AH-64A's 1,200 cockpit switches reduced to approximately 200 on AH-64D.

Mission: Lockheed Martin target acquisition and designation sight and AN/AAQ-11 pilot's night vision sensor (TADS/PNVS) comprises two independently functioning, fully integrated systems mounted on nose.

TADS consists of a rotating turret (±120° in azimuth, +30/−60° in elevation) housing sensor subsystems, optical relay tube (being replaced under 1998 contract by Planar Advance/dpiX flat panel display) in the CPG's cockpit, three electronic units in the avionics bay, and cockpit-mounted controls and displays; used principally for target search, detection and laser designation, with CPG as primary operator (can also provide back-up night vision to pilot in event of PNVS failure). Once acquired by TADS, targets can be tracked manually or automatically for autonomous attack with gun, rockets or Hellfire missiles. TADS daylight sensor consists of TV camera with narrow (0° 50') and wide angle (4° 0') fields of view; direct view optics (4° narrow and 18° wide angle); laser spot tracker; and International Laser Systems laser range-finder/designator. New switchable eyesafe laser range-finder designator (SELRD) currently being developed by Kollsman Inc under US$2.8 million contract awarded at start of 2000; to be installed in existing TADS turret on AH-64H and AH-64D and will have 80 per cent commonality with Kiowa Warrior SELRD. Night sensor, in starboard half of turret, incorporates FLIR sight with narrow, medium and wide angle (3° 6', 10° 6' and 50°) fields of view.

PNVS consists of FLIR sensor (30 × 40° field of view) in rotating turret (±90° in azimuth, +20/−45° in elevation) mounted above TADS; electronic unit in the avionics bay, and pilot's display and controls; provides pilot with thermal imaging for nap-of-the-earth flight to, from and within battle area at night or in adverse daytime weather, at altitudes low enough to avoid detection. Second-generation FLIR sensor eventually to be installed on AH-64D, but selection still awaited in late 2000. Lockheed Martin Arrowhead and Raytheon FIREsight systems under development to satisfy this requirement. PNVS imagery displayed on monocle in front of one of pilot's eyes; flight information including airspeed, altitude and heading is superimposed on this imagery to simplify piloting. Monocle is part of Honeywell integrated helmet and display sighting system (HADSS) worn by both crew members. Symetrics Industries improved data modem for transmission of target data (and eventually real-time imagery) between helicopters, tactical jet, Joint STARS airborne command posts, HQs and ground units at 16,000 bits/s, plus radio frequency interferometer beneath radome for identification of hostile transmitters.

Self-defence: Aircraft survivability equipment (ASE) consists of Litton AN/APR-39 passive RWR, Sanders AN/ALQ-144 IR jammer, Raytheon AN/AVR-2 laser warning receiver, ITT AN/ALQ-136 radar jammer and chaff dispensers and Lockheed Martin AN/APR-48A radar frequency interferometer. Sanders AN/ALQ-212 Advanced Threat Infra-Red Countermeasures (ATIRCM) system and ITT AN/ALQ-211 suite of integrated RF countermeasures (SIRFC) system currently under development. ATIRCM combines next-generation directable IRCM system with Sanders AN/AAR-57 Common Missile Warning System (CMWS); SIRFC at EMD stage, with contractor tests on Apache Longbow undertaken in latter half of 1999, followed by operational test and evaluation from early 2000; production decision due to be taken in mid-2000. Elisra began flight test of passive airborne warning system in second half of 1999 and Israel plans to install this on its AH-64 fleet. Elisra-supplied EW equipment to be installed on IDF/AF AH-64D, which may also be fitted with LWS-20V-2 laser warning system.

EQUIPMENT: Avpro of UK cleared Exint transport pod for use with AH-64 at start of 2000, but certification to carry personnel still required. In special forces insertion role, Apache can carry maximum of four pods, each able to accommodate 226 kg (500 lb) payload.

ARMAMENT: Boeing M230 Chain Gun 30 mm automatic cannon, located between the mainwheel legs in an underfuselage mounting with Smiths Industries electronic controls. Normal rate of fire is 625 rds/min of HE or HEDP (high-explosive dual-purpose) ammunition, which is interoperable with NATO Aden/DEFA 30 mm ammunition. Maximum ammunition load is 1,200 rounds. New 'Sideloader' system demonstrated June 1994 and now installed in starboard forward avionics bay; cuts normal loading time of 30 minutes by up to half and reduces number of personnel required from three to one. Gun mounting is designed to collapse into fuselage between pilots in the event of a crash landing.

New electric turret under development by Boeing, which received two year, US$5 million contract in first half of 1999; objective is to achieve accuracy of 0.5 mrads compared with current 3.0 mrads. Gun, mount and feed system to be retained in conjunction with redesigned mechanical system featuring electric rather than hydraulic drive as well as digital control; result should be at least 10 per cent lighter and require one instead of two electrical boxes. HR Textron responsible for controls, with Boeing providing the rest. Prototype delivery scheduled for September 2001. Four underwing hardpoints, with Aircraft Hydro-Forming pylons and ejector units, on which can be carried up to 16 AGM-114 Hellfire/Hellfire 2 anti-tank missiles or up to seventy-six 2.75 in FFAR (folding fin aerial rockets) in their launchers or a combination of Hellfires and FFAR. Planned modification adds two extra hardpoints for four Stinger, four Mistral or two Sidewinder (including Sidearm anti-radiation variant) missiles; Shorts Starstreak high-velocity AAM system completed initial 17 month test programme on Apache in February 1997 and being promoted for use by US and UK, with funding allocated for follow-on two year test programme; second round of live fire tests completed October/November 1998. Starstreak to participate in competitive evaluation against Stinger, despite problems with debris from missile container and pressure wave caused by first-stage rocket motor; trials of both weapons began in late 1999 and continued into 2000; Hellfire remote electronics by Rockwell Collins; Honeywell aerial rocket control system; multiplex (MUX) system units by Honeywell. Co-pilot/gunner (CPG) has primary responsibility for firing gun and missiles, but pilot can override his controls to fire gun or launch missiles.

DIMENSIONS, EXTERNAL:

Main rotor diameter	14.63 m (48 ft 0 in)
Main rotor blade chord	0.53 m (1 ft 9 in)
Tail rotor diameter	2.79 m (9 ft 2 in)
Length overall: tail rotor turning	15.54 m (51 ft 0 in)
both rotors turning	17.76 m (58 ft 3¼ in)
Wing span: clean	5.23 m (17 ft 2 in)
over empty weapon racks	5.82 m (19 ft 1 in)
Height: over tailfin	3.55 m (11 ft 7½ in)
over tail rotor	4.30 m (14 ft 1¼ in)
to top of rotor head	3.84 m (12 ft 7 in)
overall (top of air data sensor)	4.66 m (15 ft 3½ in)
overall, Longbow radar	4.95 m (16 ft 3 in)
Main rotor ground clearance (turning)	3.59 m (11 ft 9¼ in)
Distance between c/l of pylons:	
inboard pair	3.20 m (10 ft 6 in)
outboard pair	4.72 m (15 ft 6 in)
Tailplane span	3.40 m (11 ft 2 in)
Wheel track	2.03 m (6 ft 8 in)
Wheelbase	10.59 m (34 ft 9 in)

AREAS:

Main rotor disc	168.11 m² (1,809.5 sq ft)
Tail rotor disc	6.13 m² (66.0 sq ft)

WEIGHTS AND LOADINGS:

Weight empty:	
without Longbow	approx 5,165 kg (11,387 lb)
with Longbow	5,352 kg (11,800 lb)
Max fuel weight: internal	1,108 kg (2,442 lb)
external (four Brunswick tanks)	2,712 kg (5,980 lb)
Primary mission gross weight	7,480 kg (16,491 lb)
Design mission gross weight	8,006 kg (17,650 lb)
Max T-O weight: -701 engines	9,525 kg (21,000 lb)
-701C engines, ferry mission, full fuel	10,432 kg (23,000 lb)
Max disc loading	60.1 kg/m² (12.31 lb/sq ft)

PERFORMANCE (A: with -701 engines, without Longbow at 6,552 kg; 14,445 lb AUW, L: Apache Longbow at 7,530 kg; 16,601 lb with -701C engines):

Never-exceed speed (V_{NE})	197 kt (365 km/h; 227 mph)
Max level and max cruising speed:	
A	158 kt (293 km/h; 182 mph)
L	143 kt (265 km/h; 165 mph)
Max rate of climb at S/L: L	736 m (2,415 ft)/min
Max vertical rate of climb at S/L:	
A	762 m (2,500 ft)/min
L	450 m (1,475 ft)/min
Service ceiling: A	6,400 m (21,000 ft)
L	5,915 m (19,400 ft)
Service ceiling, OEI: A	3,290 m (10,800 ft)
Hovering ceiling:	
IGE: A	4,570 m (15,000 ft)
L	4,170 m (13,690 ft)
OGE: A	3,505 m (11,500 ft)
L	2,890 m (9,480 ft)
Max range, internal fuel: 30 min reserves:	
A	260 n miles (482 km; 300 miles)
L	220 n miles (407 km; 253 miles)
no reserves: L	257 n miles (476 km; 295 miles)
Ferry range, max internal and external fuel, still air, 45 min reserves	1,024 n miles (1,899 km; 1,180 miles)
Endurance at 1,220 m (4,000 ft) at 35°C	1 h 50 min
Max endurance, L: internal fuel	2 h 44 min
internal and external fuel	8 h 0 min
g limits at low altitude and airspeeds up to 164 kt (304 km/h; 189 mph)	+3.5/−0.5

WEIGHTS FOR TYPICAL MISSION PERFORMANCE (all without Longbow; A: anti-armour at 1,220 m/4,000 ft and 35°C, four Hellfire and 320 rounds of 30 mm ammunition; B: as A, but with 1,200 rounds; C: as A, but with six Hellfire and 540 rounds; D: anti-armour at 610 m/2,000 ft and 21°C, 16 Hellfire and 1,200 rounds; E: air cover at 1,220 m/4,000 ft and 35°C, four Hellfire and 1,200 rounds; F: as E but at 610 m/2,000 ft and 21°C, four Hellfire, 19 rockets, 1,200 rounds; G: escort at 1,220 m/4,000 ft and 35°C, 19 rockets and 1,200 rounds; H: escort at 610 m/2,000 ft and 21°C, 38 rockets and 1,200 rounds):

Mission fuel: A	727 kg (1,602 lb)
G	741 kg (1,633 lb)
E	745 kg (1,643 lb)
C	902 kg (1,989 lb)
B	1,029 kg (2,269 lb)
D	1,063 kg (2,344 lb)
H	1,077 kg (2,374 lb)
F	1,086 kg (2,394 lb)
Mission gross weight: A	6,552 kg (14,445 lb)
E	6,874 kg (15,154 lb)
G	6,932 kg (15,282 lb)
B, C	7,158 kg (15,780 lb)
D	7,728 kg (17,038 lb)
F	7,813 kg (17,225 lb)
H	7,867 kg (17,343 lb)

TYPICAL MISSION PERFORMANCE (A-H as above):

Cruising speed at intermediate rated power:	
C	147 kt (272 km/h; 169 mph)
D	148 kt (274 km/h; 170 mph)
F	150 kt (278 km/h; 173 mph)
B	151 kt (280 km/h; 174 mph)
E, H	153 kt (283 km/h; 176 mph)
A	154 kt (285 km/h; 177 mph)
G	155 kt (287 km/h; 178 mph)
Max vertical rate of climb at intermediate rated power:	
B, C	137 m (450 ft)/min
H	238 m (780 ft)/min
F, G	262 m (860 ft)/min
E	293 m (960 ft)/min
D	301 m (990 ft)/min
A	448 m (1,470 ft)/min
Mission endurance (no reserves): A, E, G	1 h 50 min
C	1 h 47 min
D, F, H	2 h 30 min
B	2 h 40 min

UPDATED

USA: AIRCRAFT—BOEING

BOEING COMMERCIAL AIRPLANE GROUP (BCAG)
1901 Oaksdale Avenue Southwest, Renton, Washington 98055
Tel: (+1 206) 655 98 00
Fax: (+1 206) 655 97 00
Web: http://www.boeing.com
PRESIDENT: Alan R Mulally
EXECUTIVE VICE-PRESIDENTS:
  AIRPLANE PROGRAMS: James M Jamieson
  COMMERCIAL AVIATION SERVICES: Thomas E Schick
VICE-PRESIDENT, COMMUNICATIONS: Donna Mikov

Single-aisle aircraft are 717, 737, 757 and MD-11; twin-aisle are 747, 767 and 777.

Group revenue for 1999 was US$38.4 billion, an increase of 4.1 per cent over the previous year. Employment stood at 93,000 at the end of 1999, compared to 118,000 at the end of 1998.

Divisions of the Boeing Commercial Airplane Group comprise the following:

737/757 Programs
North 8th and Park Avenue, Renton, Washington 98055
Tel: (+1 206) 237 21 21
Fax: (+1 206) 237 13 79
VICE-PRESIDENT AND GENERAL MANAGER: Gary R Scott
Products described below in numerical order.

747/767 Programs
3003 West Casino Road, Everett, Washington 98203
Tel: (+1 206) 294 40 88
Fax: (+1 206) 342 17 56
VICE-PRESIDENT AND GENERAL MANAGER: Edward J Renouard
Products described below in numerical order.

BOEING AIRLINER FIVE YEAR DELIVERIES

Year	717	737	747	757	767	777	MD-11	MD-80	MD-90	Total
1996		76	26	42	43	32	15	12	24	**270**
1997		135	39	46	42	59	12	16	26	**375**
1998		281	53	54	47	74	12	8	34	**563**
1999	12	320	47	67	44	83	8	26	13	**620**
2000	32	281	25	45	44	55	4		3	**489**

BOEING COMMERCIAL AIRPLANE GROUP ORDERS AND DELIVERIES (at 1 January 2001)

	Orders Total	Deliveries Total	Orders in 2000	Deliveries in 2000
707/720	1,010	1,010	0	0
Total	**1,010**	**1,010**	**0**	**0**
717-200	151	44	21	32
Total	**151**	**44**	**21**	**32**
727	1,831	1,831	0	0
Total	**1,831**	**1,831**	**0**	**0**
737-100	30	30	0	0
737-200	1,114	1,114	0	0
737-300	1,113	1,113	0	0
737-400	486	486	0	2
737-500	389	389	0	0
737-600	86	38	0	6
737-700	776	259	274	75
737-700BBJ	56	42	10	10
737-700/C-40A	5	3	1	3
737-800	772	383	106	185
737-900	46	0	0	0
Total	**4,873**	**3,857**	**391**	**281**
747-100	250	250	0	0
747-200	393	393	0	0
747-300	81	81	0	0
747-400	438	407	8	9
747-400D	19	19	0	0
747-400ER	6	0	6	0
747-400F	88	51	13	15
747-400M	63	60	0	1
Total	**1,338**	**1,261**	**27**	**25**
757-200	917	852	31	37
757-200M			0	0
757-200PF	80	80	0	0
757-300	29	15	12	8
Total	**1,027**	**948**	**43**	**45**
767-200	128	128	0	0
767-200ER	112	104	1	3
767-300	104	103	0	0
767-300ER	471	432	4	24
767-300F	38	34	4	1
767-400ER	48	16	3	16
Total	**901**	**817**	**12**	**44**
777-200	89	78	1	9
777-200ER	378	203	66	42
777-200LR	3	0	3	0
777-300	47	35	1	4
777-300ER	46	0	46	0
Total	**563**	**316**	**117**	**55**
Grand Totals	**11,694**	**10,084**	**611**	**482**

Note: Table does not include one each of 727-100, 747-100, 757-200, 767-200 and 777-200 owned by Boeing. Annual order total is net (after deduction of cancellations)

MCDONNELL DOUGLAS (BOEING) AIRLINER ORDERS AND DELIVERIES
(at 1 January 2001)

	Orders	Deliveries
DC-8-10	26	26
DC-8-20	36	36
DC-8-30	57	57
DC-8-40	32	32
DC-8-50	89	89
DC-8-50C	39	39
DC-8-50F	15	15
DC-8-61	78	78
DC-8-61C	10	10
DC-8-62	51	51
DC-8-62C	10	10
DC-8-62F	6	6
DC-8-63	47	47
DC-8-63C	53	53
DC-8-63F	7	7
Total	**556**	**556**
DC-9-10	113	113
DC-9-10C	24	24
DC-9-20	10	10
DC-9-30	585	585
DC-9-30C	30	30
DC-9-30F	6	6
DC-9-40	71	71
DC-9-50	96	96
C-9A	21	21
C-9B	17	17
VC-9C	3	3
Total	**976**	**976**
MD-81	132	132
MD-82	562	562
MD-83	264	264
MD-87	75	75
MD-88	158	158
Total	**1,191**	**1,191**
MD-90-30	114	114
MD-90-30T	0	0
Total	**114**	**114**
DC10-10	123	123
DC-10-10C	8	8
DC-10-15	7	7
DC-10-30	166	166
DC-10-30CF	26	26
DC-10-30ER	3	3
DC-10-30F	11	11
DC-10-40	42	42
KC-10A	60	60
Total	**446**	**446**
MD-11	136	136
MD-11C	5	5
MD-11F	59	57[1]
Total	**200**	**198**
Grand totals	**3,483**	**3,481**

[1] Final two delivered to Lufthansa Cargo on 25 January and 22 February 2001.

Note: Activity during 2000 comprised deliveries of four MD-11s and three MD-90s. There were no new orders.

777 Program
3303 West Casino Road, Everett, Washington 98204
Tel: (+1 206) 294 40 88
VICE-PRESIDENT AND GENERAL MANAGER: Ron Ostrowski
Described below.

Douglas Products Division
3855 Lakewood Boulevard, Long Beach, California 90846
Tel: (+1 562) 982 26 03
Described separately following Boeing 777.

Customer Services Division
2925 South 112th Street, Seattle, Washington 98168
Tel: (+1 206) 544 83 01
Fax: (+1 206) 544 95 50
VICE-PRESIDENT AND GENERAL MANAGER: Brad Cvetovich
Provides training, field support and spare parts for Boeing customers.

Engineering Division
North 8th and Park Avenue, Renton, Washington 98055
Tel: (+1 425) 234 35 35
VICE PRESIDENT, ENGINEERING: Walt Gillette
Manufacturing for other divisions.

Fabrication Division
1002 15th Street SW, Auburn, Washington 98002
Tel: (+1 206) 931 33 30
Fax: (+1 206) 931 95 00
VICE-PRESIDENT AND GENERAL MANAGER: Howard Fitz
Manufacturing for other divisions.

Materiel Division
20818 44th Avenue W, Lynnwood, Washington 98036
Tel: (+1 206) 655 21 21
VICE-PRESIDENT AND GENERAL MANAGER: Russell Bunio
Responsible for purchasing, quality and vendor supplies.

Propulsion Systems Division
7600 212th Avenue SW, Kent, Washington 98032
Tel: (+1 206) 393 80 00
VICE-PRESIDENT AND GENERAL MANAGER: Richard H Pearson

Wichita Division
PO Box 7730, M/S K12-12, Wichita, Kansas 67277-7730
Tel: (+1 316) 526 31 53
VICE-PRESIDENT AND GENERAL MANAGER: Jeffrey L Turner
Produces engine strut, nacelle and fuselage components and is principal site for aircraft modifications.

Jointly-owned company:

Boeing Business Jets
Described following Boeing Commercial Airplane Group entry.

BOEING 737
US Navy designation: C-40A
TYPE: Twin-jet airliner.
PROGRAMME: Original Boeing 737 first flew 9 April 1967; -100 and -200 with Pratt & Whitney JT8D engines; -300 entered service with CFM56 engines, November 1984, followed by -400 and -500; 3,132nd and last of this generation delivered February 2000. These versions, including military T-43 and Surveiller, described in *Jane's Aircraft Upgrades*.

Current 'Next-Generation' of family (initially 737X) originated in 1991, when Boeing asked more than 30 airlines to help define improved series; company board authorised offer for sale June 1993; Southwest Airlines ordered 63 737-700s (32 converted from options for 737-300s) plus 63 new options (all, and more, taken up) 18 November 1993; roll-out (737-700) 8 December 1996; first flight (N737X) 9 February 1997; certification 7 November 1997, immediately followed by first deliveries. CFM56-7B power plant first flew on Boeing 747 testbed on 16 January 1996. Flight test programme involved 10 aircraft: four 737-700s, three 737-800s and three 737-600s. FAA approval for 180-minute ETOPS granted in September 1999. By January 2000, B737s of all subtypes had flown over 100 million hours.

CURRENT VERSIONS: **737-600**: Smallest of current 737 family. Known as 737-500X until officially launched 15 March 1995; 110 two-class passengers; final assembly of prototype began at Renton 29 August 1997; roll-out December 1997; first flight 22 January 1998; FAA certification 18 August 1998; first delivery (SE-DNM to launch customer SAS) 18 September 1998. Total of 104 ordered and 32 delivered by 31 March 2000.

737-700: First to be ordered and manufactured; mid-size version of family, equivalent to previous ('Classic') 737-300, seating 126 passengers in two-class layout. First aircraft (N737X) rolled out 2 (officially 8) December 1996; first flight 9 February 1997, followed by second aircraft 27 February that year; second aircraft attained maximum certified altitude of 12,500 m (41,000 ft) for the first time on 19 March 1997; FAA certification 7 November 1997, with first delivery (fourth built, N700GS to Southwest Airlines) on 17 December 1997. Total of 600 ordered (plus 49 BBJs) and 195 delivered (plus 34 BBJs) by 31 March 2000.

737-700IGW: Formerly 737-700X; increased gross weight version, under study in 1997; based on Boeing Business Jet airframe; three versions under consideration: passenger, equipped with three underfloor auxiliary fuel tanks extending range beyond 3,496 n miles (6,475 km; 4,023 miles); quick-change -700C version with cargo door aft of port forward and cabin door and cargo handling system ordered as **C-40A** to meet US Navy requirement for C-9B Skytrain II replacement; and all-cargo version. First C-40A flew 17 April 2000.

737-800: Known as 737-400X Stretch until launched 5 September 1994; seats 162 two-class passengers; roll-out (N737BX) 30 June 1997; first flight 31 July 1997; certified by FAA 13 March and JAA 9 April 1998; first delivery (D-AHFC to launch customer Hapag-Lloyd) 22 April 1998. Total of 655 ordered and 224 delivered by 31 March 2000.

737-900: Formerly 737-900X; launched 10 November 1997 with an order for 10, plus 10 options, from Alaska Airlines; largest 737 variant to date, with (compared to -800) stretch by means of 1.57 m (5 ft 2 in) forward plug and 1.07 m (3 ft 6 in) aft plug and strengthened fuselage; seating for 177 two-class passengers; deliveries from early 2001. Total of 45 ordered by 31 March 2000. One prototype/certification aircraft only; rolled out 23 July 2000; first flight (N737X) 3 August 2000, beginning (supported by 20 hours of ground testing by second -900), leading to first delivery (Alaska Airlines) in April 2001.

Boeing Business Jet (BBJ): Corporate versions, described separately.

Special missions: Military versions described under Boeing Aircraft and Missile Systems Group heading.
CUSTOMERS: See table.
COSTS: Approximately US$50 million (737-800; 2000).
DESIGN FEATURES: Conventional, medium-size airliner with podded engines and sweptback wing and tail surfaces. Dihedral 6° at root; sweepback 25° at quarter-chord. Greater range and speed than previous 737s, with less noise and fewer emissions; wing area increased by some 25 per cent by means of 0.43 m (1 ft 5 in) increase in wing chord and about 4.83 m (15 ft 10 in) increase in wing span; new high-lift systems; larger tail surfaces; increased tankage gives US transcontinental range; new aircraft can use same runways, taxiways, ramps and gates as preceding variants; new variant of CFM56 turbofan derated from nominal thrust to suit smaller versions of the family. Noise on ground reduced by approximately 12 dB by new diffuser duct and cooling vent silencer on APU, new ECS fan and duct and new electrical/electronics cooling fan.

FLYING CONTROLS: Conventional and powered. All surfaces actuated by two independent hydraulic systems with manual reversion for ailerons and elevator; elevator servo tabs unlock on manual reversion; rudder has standby hydraulic actuator and system. Three outboard-powered overwing spoiler panels on each wing assist lateral control and also act as airbrakes. Variable incidence tailplane has two electric motors and manual standby.

Leading-edge Krueger flaps inboard and four sections of slats outboard of engines; two airbrake/lift dumper panels on each wing, inboard and outboard of engines; continuous-span, double-slotted trailing-edge flaps inboard and outboard of engines.

FAA Cat. II landing minima system standard using SP-300 dual digital integrated flight director/autopilot; Cat. IIIa capability optional.

STRUCTURE: Aluminium alloy dual-path fail-safe two-spar wing structure with corrosion-resistant 7055-T77 upper skin. Aluminium alloy two-spar tailplane. Graphite composites ailerons, elevators and rudder. Aluminium honeycomb spoiler/airbrake panels and trailing-edges of slats and flaps. Fuselage structure fail-safe aluminium. Elevators, rudder and ailerons contain graphite/Kevlar; other, unstressed, components in GFRP and CFRP include nosecone, wing/fuselage fairing, fin fillet, fintip and flap actuator fairings. Rears of engine nacelles are of graphite/Kevlar/glass fibre.

LANDING GEAR: Hydraulically retractable tricycle type, with Boeing oleo-pneumatic shock-absorbers; inward-retracting main units have no doors, wheels forming wheel well seal; nose unit retracts forward; free-fall emergency extension. Twin nosewheels have tyres size 27×7.75. Main units have heavy-duty twin wheels, H40×14.5-19 heavy-duty tyres, and Honeywell or Goodrich heavy-duty wheel brakes as standard. Mainwheel tyre pressure 13.45 to 14.00 bars (195 to 203 lb/sq in). Nosewheel tyre pressure 11.45 to 11.85 bars (166 to 172 lb/sq in).

POWER PLANT: *737-600*: Two CFM International CFM56-7B18 turbofans, each rated at 86.7 kN (19,500 lb st) standard, or two CFM56-7B22s, each rated at 101 kN (22,700 lb st) in high gross weight version.

737-700: Two CFM56-7B20s, each rated at 91.6 kN (20,600 lb st) standard, or two CFM56-7B24s, each rated at 101 kN (22,700 lb st) in high gross weight version.

737-800: Two CFM56-7B24s, each rated at 107.6 kN (24,200 lb st) standard, or two CFM56-7B27s, each rated at 121.4 kN (27,300 lb st) in high gross weight version.

737-900: Two CFM56-7B26s, each rated at 117 kN (26,300 lb st) standard, or two CFM56-7B27s, each rated at 121.4 kN (27,300 lb st) in high gross weight version.

BOEING—AIRCRAFT: USA

BOEING 737 NEXT-GENERATION ORDER BOOK
(at 30 June 2000)

Customer	Variant	Qty
Air Algerie	737-600	5
	737-800	7
Air Berlin	737-800	20
Air China	737-800	11
Air Pacific	737-700	1
	737-800	2
Alaska Airlines	737-700	18
	737-900	11
American Airlines	737-800	105
American Trans Air	737-800	20
Ansett Worldwide	737-700	4
Aramco Services Co	737-700	5
Bavaria	737-700	6
Bouillion Aviation	737-700	28
	737-800	2
Braathens	737-700	8
Britannia Airways	737-800	2
China Airlines	737-800	15
China Southwest	737-800	3
China Xinjiang	737-700	3
China Yunnan	737-700	4
CIT Group	737-700	10
Continental	737-700	36
	737-800	44
	737-900	15
Copa Airlines	737-700	8
Delta Air Lines	737-800	132
Eastwind Airlines	737-700	2
EasyJet	737-700	32
El Al	737-700	2
	737-800	3
GATX	737-800	21
GECAS	737-600	7
	737-700	42
	737-800	31
Germania	737-700	12
Hainan Airlines	737-800	3
Hapag-Lloyd	737-800	24
ILFC	737-600	49
	737-700	113
	737-800	42
Itochu AirLease	737-800	10
Jet Airways	737-800	16
Kenya Airways	737-700	2
KLM	737-800	13
	737-900	4
Korean Air	737-800	6
	737-900	16
LAPA	737-700	6
Lauda Air	737-600	2
	737-700	4
	737-800	2
LOT	737-800	2
Maersk	737-700	10
Midway Airlines	737-700	15
Olympic Airways	737-800	8
Pegasus Airlines	737-800	1
Pembroke Capital Ltd	737-600	1
	737-700	8
Royal Air Maroc	737-700	5
	737-800	4
Ryanair	737-800	25
SAS	737-600	33
	737-700	6
	737-800	15
Shanghai Airlines	737-700	2
Shenzhen Airlines	737-700	2
South African Airways	737-800	5
Southwest	737-700	236
Sumitomo Trust	737-800	2
Sunrock	737-700	1
	737-800	4
Taiwan Air Force	737-800	1
TAROM	737-700	4
	737-800	4
Tombo Aviation	737-700	10
	737-800	5
Transavia	737-800	12
TunisAir	737-600	7
Turkish Airlines	737-800	26
US Navy	737-700	5
Varig Airlines	737-700	4
	737-800	10
Wuhan Airlines	737-800	2
Xiamen Airlines	737-700	6
Various unidentified	737-700BBJ	50†
	737-700	33
	737-800	49
Totals	737-600	104
	737-700	743
	737-800	709
	737-900	46
Grand Total		**1,602**

† Completed as BBJ, which see in Boeing Business Jet entry

USA: AIRCRAFT—BOEING

Boeing 737-800 of Delta Air Lines *(Paul Jackson/Jane's)*

Fuel capacity (all) 26,025 litres (6,875 US gallons; 5,725 Imp gallons).

ACCOMMODATION: *All:* Crew of two side by side on flight deck. One plug-type door at each corner of cabin, with passenger doors on port side and service doors on starboard side. Airstair for forward cabin door optional. Overwing emergency exit on each side. One or two galleys and one modular vacuum lavatory forward and one or two galleys and lavatories aft; all lavatories interconnected to single waste collection tank at port side rear. Lightweight interior, of crushed core materials, has movable class divider, overnight seating-pitch flexibility and modular passenger service unit (PSU) including fold-down video screen in underside of baggage bin. Centreline stowage bins optional for emergency equipment and crew baggage.

Two underfloor baggage holds, forward and aft of wing. Rear hold has provision for post-delivery installation of telescopic baggage conveyor system (when additional fuel tanks not fitted). One baggage door in starboard side of each hold.

737-600: Alternative cabin layouts seat from 110 to 132 passengers. Typical arrangements offer eight first class seats four-abreast at 91 cm (36 in) pitch and 102 tourist class seats six-abreast at 81 cm (32 in) pitch in mixed class; and 132 all-tourist class at 76 cm (30 in) pitch. Total overhead baggage capacity of 6.1 m³ (216 cu ft), equivalent to 0.045 m³ (1.6 cu ft) per passenger.

737-700: Alternative cabin layouts seat from 126 to 149 passengers. Typical arrangements offer eight first class seats four-abreast at 91 cm (36 in) pitch and 118 tourist class seats six-abreast at 81 cm (32 in) pitch in mixed class; and 149 all-tourist class at 76 cm (30 in) pitch. Total overhead baggage capacity of 7.0 m³ (248 cu ft), equivalent to 0.05 m³ (1.8 cu ft) per passenger. C-40A options comprise 121 passengers, all-cargo (eight pallets) and combinations of 70 passengers and three pallets.

737-800: Alternative cabin layouts seat from 162 to 189 passengers. Typical arrangements offer 12 first class seats four-abreast at 91 cm (36 in) pitch and 150 tourist class seats six-abreast at 81 cm (32 in) pitch in mixed class; and 189 all-tourist class at 76 cm (30 in) pitch. Total overhead baggage capacity of 9.3 m³ (328 cu ft), equivalent to 0.05 m³ (1.7 cu ft) per passenger.

737-900: Alternative cabin layouts seat from 177 to 189 passengers. Typical arrangements offer 12 first class seats four-abreast at 91 cm (36 in) pitch and 165 tourist class seats six-abreast at 81 cm (32 in) pitch in mixed class; and 189 all-tourist class at 81 cm (32 in) pitch.

SYSTEMS: Honeywell 131-9(B) APU with air start capability to maximum certified altitude and 90 kVA electrical load capability to 11,278 m (37,000 ft). Three-wheel air cycle environmental control system with optional ozone converter and digital cabin pressure controls.

AVIONICS: *Flight:* Optional GPS, satcom and dual FMS (single standard) integrated with GPS.
Instrumentation: Honeywell Air Transport Systems Common Display System (CDS) with six-screen flat-panel liquid crystal display (LCD) technology and programmable software, enables operators to emulate previous 737 electronic flight instrument system (EFIS) and 747-400/777 primary flight display-navigation display (PFD-ND) flight deck formats. Optional HUD.

Boeing 737-700 twin-turbofan airliner *(Paul Jackson/Jane's)*

DIMENSIONS, EXTERNAL:
Wing span: all versions	34.31 m (112 ft 7 in)
Wing aspect ratio	9.4
Length overall: 600	31.24 m (102 ft 6 in)
700	33.63 m (110 ft 4 in)
800	39.47 m (129 ft 6 in)
900	42.11 m (138 ft 2 in)
Height overall: 600	12.57 m (41 ft 3 in)
700, 800, 900	12.55 m (41 ft 2 in)
Tailplane span: all	14.35 m (47 ft 1 in)
Wheel track (c/l shock-struts): all	5.71 m (18 ft 9 in)
Main passenger door (port, fwd), all:	
Height	1.83 m (6 ft 0 in)
Width	0.86 m (2 ft 10 in)
Height to sill	2.64 m (8 ft 8 in)
Passenger door (port, rear):	
Height: all	1.83 m (6 ft 0 in)
Width: all	0.76 m (2 ft 6 in)
Height to sill: 600, 700	3.00 m (9 ft 10 in)
800	2.97 m (9 ft 9 in)
Emergency exits (overwing, port and stbd, each), all:	
Height	0.96 m (3 ft 2 in)
Width	0.51 m (1 ft 8 in)
Galley service door (stbd, fwd), all:	
Height	1.65 m (5 ft 5 in)
Width	0.76 m (2 ft 6 in)
Height to sill	2.64 m (8 ft 8 in)
Service door (stbd, rear):	
Height: all	1.65 m (5 ft 5 in)
Width: all	0.76 m (2 ft 6 in)
Height to sill: 600, 700	3.00 m (9 ft 10 in)
800	2.97 m (9 ft 9 in)
Baggage hold door (stbd, fwd), all:	
Height	0.89 m (2 ft 11 in)
Width	1.22 m (4 ft 0 in)
Baggage hold door (stbd, rear), all:	
Height	0.84 m (2 ft 9 in)
Width	1.22 m (4 ft 0 in)

DIMENSIONS, INTERNAL::
Cabin, aft of flight deck to rear pressure bulkhead:	
Length: 600	21.79 m (71 ft 6 in)
700	24.18 m (79 ft 4 in)
800	30.02 m (98 ft 6 in)
Max height: all	2.13 m (7 ft 0 in)
Floor area: 600	67.3 m² (725 sq ft)
700	75.1 m² (808 sq ft)
800	94.0 m² (1,012 sq ft)
Baggage hold:	
Max width, all: at roof	3.15 m (10 ft 4 in)
at floor	1.22 m (4 ft 0 in)
Min height, all	1.13 m (3 ft 8½ in)
Volume:	
600: front	7.0 m³ (248 cu ft)
rear	13.4 m³ (472 cu ft)
700: front	10.9 m³ (386 cu ft)
rear	16.4 m³ (580 cu ft)
800: front	19.0 m³ (672 cu ft)
rear	25.0 m³ (883 cu ft)
900: front	23.3 m³ (823 cu ft)
rear	28.7 m³ (1,012 cu ft)

AREAS:
Wings, gross	125.00 m² (1,345.5 sq ft)
Vertical tail surfaces (total)	26.40 m² (284.2 sq ft)
Horizontal tail surfaces (total)	32.80 m² (353.1 sq ft)

WEIGHTS AND LOADINGS: (600: A: CFM56-7B18 engines, B: CFM56-7B22s; 700: A: CFM56-7B20s, B: CFM56-7B24s; 800: A: CFM56-7B24s, B: CFM56-7B27s; 900: A: CFM56-7B24s, B: CFM56-7B26s, B: CFM56-7B27s):

Operating weight empty:
600, 110 passengers: A, B	37,104 kg (81,800 lb)
700, 126 passengers: A, B	38,147 kg (84,100 lb)
800, 162 passengers: A, B	41,145 kg (90,710 lb)
900, 177 passengers: A, B	42,493 kg (93,680 lb)

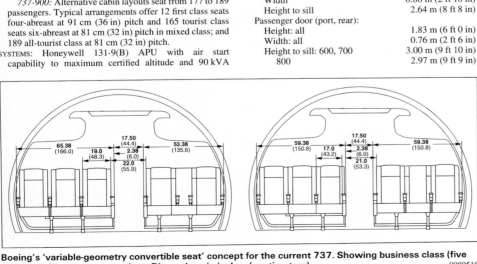

Boeing's 'variable-geometry convertible seat' concept for the current 737. Showing business class (five abreast; left) and economy class. Dimensions in inches (centimetres)

Max T-O weight:
- 600: A   56,245 kg (124,000 lb)
- B   65,090 kg (143,500 lb)
- 700: A   60,330 kg (133,000 lb)
- B   70,080 kg (154,500 lb)
- 800: A   70,535 kg (155,500 lb)
- B   79,015 kg (174,200 lb)
- 900: A   74,840 kg (164,000 lb)
- B   79,015 kg (174,200 lb)

Max ramp weight:
- 600: A   56,470 kg (124,500 lb)
- B   65,315 kg (144,000 lb)
- 700: A   60,555 kg (133,500 lb)
- B   70,305 kg (155,000 lb)
- 800: A   70,760 kg (156,000 lb)
- B   79,245 kg (174,700 lb)
- 900: A   74,615 kg (164,500 lb)
- B   79,245 kg (174,700 lb)

Max landing weight:
- 600 A, B   54,655 kg (120,500 lb)
- 700: A   58,060 kg (128,000 lb)
- B   58,605 kg (129,200 lb)
- 800: A   65,315 kg (144,000 lb)
- B   66,360 kg (146,300 lb)
- 900: A, B   66,360 kg (146,300 lb)

Max zero-fuel weight:
- 600: A, B   51,480 kg (113,500 lb)
- 700: A   54,655 kg (120,500 lb)
- B   55,200 kg (121,700 lb)
- 800: A   61,690 kg (136,000 lb)
- B   62,730 kg (138,300 lb)
- 900: A, B   62,730 kg (138,300 lb)

Max wing loading:
- 600: A   450.0 kg/m² (92.16 lb/sq ft)
- B   520.7 kg/m² (106.65 lb/sq ft)
- 700: A   482.6 kg/m² (98.85 lb/sq ft)
- B   560.6 kg/m² (114.83 lb/sq ft)
- 800: A   564.3 kg/m² (115.57 lb/sq ft)
- B   632.1 kg/m² (129.47 lb/sq ft)
- 900: A   595.1 kg/m² (121.89 lb/sq ft)
- B   632.1 kg/m² (129.47 lb/sq ft)

Max power loading:
- 600: A   324 kg/kN (3.18 lb/lb st)
- B   322 kg/kN (3.16 lb/lb st)
- 700: A   329 kg/kN (3.23 lb/lb st)
- B   347 kg/kN (3.40 lb/lb st)
- 800: A   327 kg/kN (3.21 lb/lb st)
- B   325 kg/kN (3.19 lb/lb st)
- 900: A   318 kg/kN (3.12 lb/lb st)
- B   325 kg/kN (3.19 lb/lb st)

PERFORMANCE (A, B as above):
- Max operating Mach No. (M_{MO}): all   0.82
- Cruising speed: all   M0.785
- Approach speed:
  - 600: A, B   125 kt (232 km/h; 144 mph)
  - 700: A   130 kt (241 km/h; 150 mph)
  - B   131 kt (243 km/h; 151 mph)
  - 800: A   141 kt (261 km/h; 162 mph)
  - B   142 kt (263 km/h; 163 mph)
  - 900: A, B   145 kt (268 km/h; 167 mph)
- Max certified altitude: all   12,500 m (41,000 ft)
- Initial cruising altitude, ISA +10°C:
  - 600: A   12,500 m (41,000 ft)
  - B   12,220 m (40,100 ft)
  - 700: A   12,500 m (41,000 ft)
  - B   10,710 m (35,140 ft)
  - 800: A   11,675 m (38,300 ft)
  - B   10,955 m (35,940 ft)
  - 900: A   11,310 m (37,100 ft)
  - B   10,975 m (36,000 ft)
- T-O field length, S/L, 30°C:
  - 600: A   1,616 m (5,300 ft)
  - B   1,796 m (5,890 ft)
  - 700: A   1,744 m (5,720 ft)
  - B   1,921 m (6,300 ft)
  - 800: A   2,100 m (6,890 ft)
  - B   2,308 m (7,570 ft)
  - 900: A   2,622 m (8,600 ft)
  - B   2,500 m (8,200 ft)
- Landing field length at max landing weight:
  - 600: A, B*   1,342 m (4,400 ft)
  - 700: A   1,418 m (4,650 ft)
  - B*   1,433 m (4,700 ft)
  - 800: A   1,634 m (5,360 ft)
  - B   1,659 m (5,440 ft)
  - 900: A, B   1,704 m (5,590 ft)
- Design range:
  - 600 with 110 passengers:
    - A   1,340 n miles (2,481 km; 1,542 miles)
    - B   3,050 n miles (5,648 km; 3,509 miles)
  - 700 with 126 passengers:
    - A   1,540 n miles (2,852 km; 1,772 miles)
    - B   3,260 n miles (6,037 km; 3,751 miles)
  - 800 with 162 passengers:
    - A   1,990 n miles (3,685 km; 2,290 miles)
    - B   2,940 n miles (5,444 km; 3,383 miles)
  - 900 with 177 passengers:
    - A   2,060 n miles (3,815 km; 2,370 miles)
    - B   2,745 n miles (5,083 km; 3,158 miles)

* Category D Honeywell brakes

UPDATED

Flight deck of Boeing 737-700

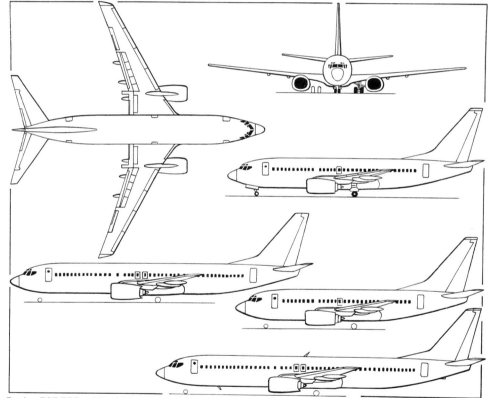
Boeing 737-700, with additional side views of stretched -800, shorter -600 and further stretched -900 (bottom) *(James Goulding/Jane's)*

Prototype Boeing 737-900

BOEING 747-400

TYPE: Wide-bodied airliner.
PROGRAMME (original): Announced 13 April 1966 (first ever wide-body jet airliner), with Pan American order for 25; official programme launch 25 July 1966; first flight 9 February 1969; FAA certification 30 December 1969; first delivery (to Pan Am) 12 December 1969; first route service New York-London flown 21 January 1970. 747-400 announced October 1985. In May 1990, Boeing decided to market only the -400; last -200 (a -200F Freighter for Nippon Cargo Air Lines) delivered 19 November 1991.

For all variants before 747-400, see *Jane's Aircraft Upgrades*. Production variants totalled 724 (205 -100, 45 SP, 393 -200 and 81 -300). Nineteen Pan American 747s modified as passenger/cargo **C-19As** by Boeing Military Airplanes for Civil Reserve Air Fleet (see 1990-91 edition). Boeing board approved launch of 747-400IGW in December 1997. By 21 January 2000 (30 years after first commercial service) 1,238 Boeing 747s (including 500 -400s) had been delivered, of which 1,100 remained in service; the worldwide fleet of 747s (all models) had logged more than 50 million hours in 12 million flights, carrying 2.2 billion passengers, by September 1998.

PROGRAMME (current): Series 400 announced October 1985 as 747 development with extended capacity and range; design go-ahead July 1985; roll-out 26 January 1988; first flight 29 April 1988; certified with P&W PW4056 on 10 January 1989; certified with GE CF6-80C2B1F on 8 May 1989; R-R RB211-524G on 8 June 1989; R-R RB211-524H on 11 May 1990. Since May 1990, -400 is the only 747 marketed. 1,100th 747 rolled out 16 December 1996 and delivered to ILFC/Virgin Atlantic in January 1997. Boeing scheduled to build 12 747s in 2000, compared with 49 in 1998 and 47 in 1999.

CURRENT VERSIONS: **747-400**: Basic passenger version; standard and three optional gross weights (see below).

Detailed description applies to -400, except where indicated.

747-400M Combi: Passenger/freight version; certified 1 September 1989; maximum 266 three-class passengers with freight, 413 without; port-side rear freight door; main deck limit is seven pallets at 27,215 kg (60,000 lb); underfloor and fuel capacities as for passenger 747; 49 delivered by 31 December 1996. For all gross weights, maximum landing weight 285,763 kg (630,000 lb) and maximum zero-fuel weight 256,280 kg (565,000 lb). All three engine options available.

747-400F: All-freight version. See separate entry.

747-400 Domestic: Special high-density two-class 568-passenger version; certified 10 October 1991; ordered by Japan Air Lines (six), All Nippon (six) and Japan Air System (one). Maximum T-O weight 272,155 kg (600,000 lb) but can be certified to 394,625 kg (870,000 lb). Structurally reinforced; no winglets; lower engine thrust; five more upper deck windows; revised avionics software and cabin pressure schedule; brake cooling fans; five pallets, 14 LD-1 containers and bulk cargo under floor; GE or P&W engines.

747-400 Performance Improvement Package (PIP): Announced April 1993, and first stage implemented in July 1993. Included gross weight increase of 2,268 kg (5,000 lb). Second stage, implemented in December 1993, included longer-chord dorsal fin made of CFRP, and wing spoilers held down more tightly to reduce profile drag and leakage. These improvements were immediately applied to production aircraft and are retrofittable; PIP flight tested in leased United Airlines 747-400 May 1993.

747-400IGW: Offered from December 1997 in response to Qantas requirement; not formally launched; replaced by 747-400X. One or two additional fuel tanks in hold, each of 11,600 litres (3,064 US gallons; 2,552 Imp gallons); maximum fuel capacity 240,000 litres (63,403 US gallons; 52,794 Imp gallons). Range 7,500 n miles (13,890 km; 8,630 miles) with one additional tank; 7,700 n miles (14,260 km; 8,861 miles) with two. Structural strengthening around centrebody, wing/fuselage joint, flaps and landing gear for 413,140 kg (910,825 lb) MTOW.

747-400X: Developed version in passenger and Combi variants: physical size unchanged (except for 2.5 cm; 1 in height reduction); will be first of 21st Century 747s. Based on 747-400 airframe, but with revised (B777-style) flight

Boeing 747-400 flight deck

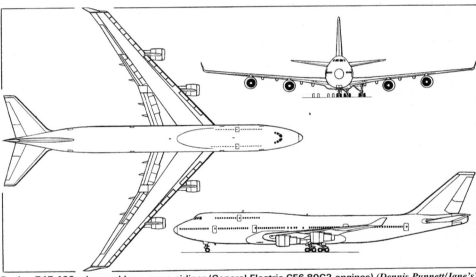

Boeing 747-400 advanced long-range airliner (General Electric CF6-80C2 engines) *(Dennis Punnett/Jane's)*

deck, crew rest/passenger sleeping area in upper aft fuselage, and 747-400F's thicker gauge outboard wing. MD-11-type trailing-edge wedges (of which flight tests began in October 1998), strengthened fuselage sections and landing gear and modifications to cargo and fuel systems to permit installation of additional fuselage tank forward of centre wing tank, with second additional tank of same capacity optional; fuel load thus increased to 228,273 litres (60,305 US gallons; 50,214 Imp gallons) basic, or 240,537 litres (63,545 US gallons; 52,912 Imp gallons) optional. Power plant options comprise 272 kN (61,100 lb st) General Electric CF6-80C2B5F, 282 kN (63,300 lb st) Pratt & Whitney PW4062 and 265 kN (59,500 lb st) Rolls-Royce RB211-524H6-T-19. Maximum T-O weight 412,770 kg (910,000 lb). In typical three-class configuration the 747-400X will carry 416 passengers and in Combi version 266 passengers and six or seven main deck cargo pallets. High-density seating for 568. Range 7,690 n miles (14,241 km; 8,849 miles) at M 0.855. Assuming launch in mid-2000, deliveries could begin from third quarter 2002.

747X: Replaces more ambitious -500/-600 projects abandoned in late 1990s. However, first version to be marketed will be 747X Stretch, described in following subsection. Based on 747-400X airframe, but with strengthened wing of 69.77 m (228 ft 11 in) span (derived from root inserts), and area of 632 m² (6,800 sq ft); fuselage length increased slightly to 73.48 m (241 ft 1 in); height increased to 21.44 m (70 ft 4 in); extended tailplane tips based on 747SP and increasing span to 25.03 m (82 ft 1½ in); mean wheelbase 26.99 m (88 ft 6½ in); additional fuel in enlarged wing torsion box, for total of 274,710 litres (72,573 US gallons; 60,430 Imp gallons); MD-11-type trailing-edge wedges (of which flight tests began in October 1998), providing 6 per cent lift/drag improvement; double (replacing triple) slotted flaps; and aileron drooping at T-O. Power plant expected to be Alliance GP7167, or Rolls-Royce Trent, each rated at 302 kN (68,000 lb st). Maximum T-O weight 473,100 kg (1,043,000 lb). In typical three-class configuration, the 747X will carry 430/442 passengers. Cargo volume 150.2 m³ (5,304.0 cu ft). Range 8,975 n miles (16,621 km; 10,328 miles) with 430 passengers, or 8,810 n miles (16,316 km; 10,138 miles) with 442 passengers. Cruising speed M0.86.

747X Stretch: Extended version of 747X with fuselage length increased by 7.06 m (23 ft 2 in) by means of plugs at fore and aft of wing box; rudder of increased

Boeing 747-400 of KLM Royal Dutch Airlines *(Paul Jackson/Jane's)*

chord, increasing overall length to 80.55 m (264 ft 3¼ in), but reducing height to 19.86 m (65 ft 2 in); new main leading gear. In typical three-class configuration, the 747X Stretch will carry 504 to 522 passengers, increasing to 660 for Japanese domestic market. Cargo volume 191.1 m³ (6,749 cu ft). Maximum take-off weight 473,095 kg (1,043,000 lb); fuel and engines as for 747X; mean wheelbase 31.56 m (103 ft 6½ in); range 7,800 n miles (14,445 km; 8,976 miles) with 504 passengers, 7,600 n miles (14,075 km; 8,745 miles) with 522 passengers. Cruising speed M0.86. Projected seat-mile costs lower than those of 747-400. Interest reported from Asian carriers in 1999/2000. Available 2004-05. Freighter version under consideration by 2000.

CUSTOMERS: See table. Launch customer Northwest Orient Airlines ordered 10 -400s with PW4000s and 420-passenger interior October 1985; first delivery 26 January 1989. Total 1,338 of all 747 variants ordered (and 1,261 delivered) by 1 January 2001.

COSTS: July 1996 Air China contract for three aircraft valued at US$510 million.

DESIGN FEATURES: Wide-bodied extrapolation of Boeing intercontinental jet configuration of low wing and four podded engines, optimised for greater passenger numbers and increased efficiency. Twin-deck forward fuselage; four mainwheel bogies for weight distribution.

According to engine type, fuel burn per seat over 3,000 n mile (5,556 km; 3,452 mile) sector varies between 135.8 kg (299.3 lb) and 138.5 kg (305.4 lb).

Sweepback at quarter-chord 37° 30′; thickness/chord ratio 13.44 per cent inboard, 7.8 per cent at mid-span, 8 per cent outboard; dihedral at rest 7°; incidence 2°; winglets, canted 22° outward and swept 60°, increase range by 3 per cent; upper deck extended rearward by 7.11 m (23 ft 4 in).

FLYING CONTROLS: Conventional and powered.

Elevators: Four elevator sections mechanically linked with breakable shear devices; each elevator has dual hydraulic-powered control units; control feel and three individual autopilot input servos mounted on central elevator quadrant; all surfaces have position transmitters; feel computer-operated by pitot pressure and tailplane angle.

Rudder: Upper rudder surface operated by three hydraulic actuators served by two hydraulic systems, lower surface by two actuators fed by remaining two hydraulic systems; no balance weights; each rudder has separate yaw damper module; left and right digital air data computers provide signals for controlling rudder ratio changer on each rudder surface according to air data and tailplane angle; combined feel actuator, rudder centring and trim actuator in rear servo area; mechanical cable linkage between rudder pedals and aft actuator area; rudder trim control switches on centre console. Maximum rudder deflection ±30°.

Tailplane: Tailplane angle set by hydraulic motor-driven shaft and ball screw with primary and secondary hydraulic brakes; flight control unit and air data computer signals sent to tailplane through dual stabiliser, trim and rudder ratio modules, which automatically apply Mach trim, and by dual-stabiliser control modules; tailplane trim limits computed according to flap positions.

Lateral control: Pilot and co-pilot aileron linkage can be physically separated if necessary; all four ailerons operate at low speeds; outboard ailerons are locked out at cruising speed; the inboard spoiler panel on each wing used on ground only; remainder have variable ratio response and spoiler mixer units; there are trim, centring and feel units.

Leading-edge and trailing-edge devices: Three-section Krueger flaps inboard of engines; variable camber slats between (five-section) and outboard (six-section) of engines lie flat when retracted and adopt camber curvature when extended. Two flap assemblies on each wing, one inboard of engines and the other between engines; three sections, fore flap, mid-flap and aft flap, move rearwards as single flat panel up to 5° deflection; thereafter, three sections separate progressively to form three slots, and

BOEING 747-400 ORDER BOOK
(at 1 January 2001)

Customer	Variant	First order	Engine	Qty
Air Canada	400M	20 Jan 89	PW4056	3
Air China	400	31 May 90	PW4056	6
	400M	16 May 86	PW4056	8
Air France	400	16 Dec 87	CF6-80	7
	400M	16 Dec 87	CF6-80	5
Air India	400	14 Aug 91	PW4056	6
Air Namibia	400M	21 Apr 99	CF6-80	1
Air New Zealand	400	30 Jul 84	RB211-524	3
	400	1 Mar 91	CF6-80	1
All Nippon Airways	400	21 Oct 86	CF6-80	12
	400D	21 Jan 86	CF6-80	11
Amiri Flight	400	30 Nov 99	CF6-80	1
Asiana Airlines	400	12 Jun 89	CF6-80	2
	400F	3 Sep 90	CF6-80	6
	400M	12 Jun 89	CF6-80	6
Atlas Air	400F	9 Jun 97	CF6-80	16
British Airways	400	15 Aug 86	RB211-524	57
Canadian Airlines Intnl	400	28 Jul 88	CF6-80	4
Cargolux Airlines	400F	6 Dec 90	CF6-80	5
	400F	13 Jun 95	RB211-524	7
Cathay Pacific Airways	400	3 Jun 86	RB211-524	17
	400F	28 Feb 90	RB211-524	5
China Airlines	400	21 Jul 87	PW4056	13
	400F	11 Aug 99	PW4056	13
El Al	400	11 Dec 90	PW4056	4
EVA Air	400	6 Oct 89	CF6-80	7
	400F	6 May 99	CF6-80	3
	400M	6 Oct 89	CF6-80	8
Garuda Indonesia	400	15 Nov 90	CF6-80	2
GE Capital	400	22 Dec 95	CF6-80	1
	400F	15 Dec 99	CF6-80	5
ILFC	400	16 May 88	CF6-80	10
	400	16 May 88	RB211-524	2
Japan Airlines	400	21 Sep 87	CF6-80	37
	400D	30 Jun 88	CF6-80	8
Japan Air SDF	400	23 Dec 87	CF6-80	2
KLM	400	9 Apr 86	CF6-80	5
	400M	9 Apr 86	CF6-80	19
Korean Air Lines	400	29 Aug 86	PW4056	28
	400F	11 Jun 90	PW4056	6
	400F	31 Dec 99	RB211	1
	400F	28 Nov 00	n/k	2
	400M	14 Apr 88	PW4056	1
Kuwait Airways	400M	12 Apr 92	CF6-80	1
Lufthansa	400	21 May 86	CF6-80	24
	400M	21 May 86	CF6-80	7
Malaysia Airlines	400	19 Oct 88	CF6-80	2
	400	12 Jan 89	PW4056	19
	400M	30 Oct 87	CF6-80	2
Mandarin Airlines	400	15 Sep 94	PW4056	1
Northwest Airlines	400	22 Oct 85	PW4056	14
Philippine Airlines	400	29 Oct 92	CF6-80	7
	400M	18 Jan 96	CF6-80	1
Qantas Airways	400	2 Mar 87	RB211-524	21
	400ER	19 Dec 00	n/k	6
Saudia	400	18 Jun 95	CF6-80	5
Singapore Airlines	400	27 Mar 86	PW4056	42
	400F	16 Jan 90	PW4056	11
South African Airways	400	6 May 89	RB211-524	6
	400	30 Dec 98	CF6-80	2
Thai Airways Intnl	400	16 Jun 87	CF6-80	14
United Airlines	400	7 Nov 85	PW4056	44
US Air Force (AL-1A)	400F	30 Jan 98	CF6-80	1
UTA	400	3 Jul 86	CF6-80	1
	400M	3 Jul 86	CF6-80	1
Virgin Atlantic Airways	400	20 Dec 96	CF6-80	9
Undisclosed	400	31 Jul 00	n/k	1
	400F	28 Nov 00	n/k	6
Total				**614**

Business Class seating on the 747's upper deck

Boeing 747-400 economy class accommodation

USA: AIRCRAFT—BOEING

Members of the next-generation Boeing 747 family: 747-400X (top), 747X and 747X Stretch

camber angles relative to each other increase progressively.

Automatic flight control system: Combines autopilot, flight director and automatic tailplane trim and sends commands through triple independent flight control computers; system automates all flight phases except take-off; dual digital air data computers; pilots' primary flight and navigation displays are large-size cathode-ray tubes; two engine indicating and crew alerting screens, one on main deck, one on console; three multifunction control and display panels control flight management system, navigation and communications; flight control computers (autopilot) and inertial reference units are triplicated; new features include full-time autothrottle and dual-thrust management system included in flight management computer; integrated radio control panels and automatic start and shutdown of APU.

STRUCTURE: Wing and tail surfaces are aluminium alloy dual-path fail-safe structures; advanced aluminium alloys in wing torsion box save 2,721 kg (6,000 lb); advanced aluminium honeycomb spoiler panels; CFRP winglets and main deck floor panels; advanced graphite/phenolic and Kevlar/graphite in cabin fittings and engine nacelles; frame/stringer/stressed skin fuselage with some bonding. Improved corrosion protection and further coverage with compound introduced from 1993.

LANDING GEAR: Twin-wheel nose unit retracts forward; main gear consists of four four-wheel bogies; two, mounted side by side under fuselage at wing trailing-edge, retract forward; two, mounted under wings, retract inward; nosewheel steerable up to 70° left or right from tillers; full rudder pedal travel gives up to 7° for use at high speed; two centre main legs steer up to 13° when nosewheels are steered more than 20° and speed is less than 20 kt (37 km/h; 23 mph); carbon disc brakes on all mainwheels, with individually controlled digital anti-skid units; one of three brake pressure supplies automatically selected; main and nose tyres H49×19.0-20 or -22 (32 ply). Minimum ground turning radius, with body gear steering, is 48.46 m (159 ft 0 in) at wingtip and 27.73 m (91 ft 0 in) at nosewheels.

POWER PLANT: Four turbofans: 252 kN (56,750 lb st) Pratt & Whitney PW4056; 258 kN (57,900 lb st) General Electric CF6-80C2B1F or 276 kN (62,100 lb st) CF6-80C2B5F; or 258 kN (58,000 lb st) Rolls-Royce RB211-524G or 270 kN (60,600 lb st) RB211-524H.

Further optional engines (subject to certification) are 267 kN (60,000 lb st) PW4060, 276 kN (62,000 lb st) PW4062 and 274 kN (61,500 lb st) CF6-80C2B1F1.

Fuel in four main tanks in wings can feed to any engine; in addition there are a centre-wing tank and reserve tanks in outer wing; optional tailplane tank; vent and surge tanks in outer wings and starboard tailplane; jettison pumps in inner main tanks; APU fed from port inner tank; automatic refuelling through two receptacles under each wing leading-edge between engines; automatic condensate scavenging and flame arresters in vent outlets.

Basic fuel capacity 204,355 litres (53,985 US gallons; 44,952 Imp gallons) with P&W and R-R engines; 203,523 litres (53,765 US gallons; 44,769 Imp gallons) with GE engines. At alternative higher T-O weights above 394,625 kg (870,000 lb) use of 12,492 litre (3,300 US gallon; 2,748 Imp gallon) tailplane/centre section tank is mandatory; fuel capacity including tailplane tank is 216,846 litres (57,285 US gallons; 47,700 Imp gallons) with P&W and R-R engines and 216,013 litres (57,065 US gallons; 47,516 Imp gallons) with GE engines.

ACCOMMODATION: Two-crew flight deck, with seats for two observers; two-bunk crew rest cabin accessible from flight deck. Optional (but currently available on 90 per cent of B747-400 fleet) overhead cabin crew rest compartments above rear of main deck cabin (four bunks, four seats; eight bunks, two seats; two bunks, two seats, five sleeper seats). Typical 416-seat, three-class, long-range configuration accommodates 40 business class on upper deck; 23 first class in front cabin, 38 business class in middle cabin and 315 economy class in rear cabin on main deck. Maximum upper deck capacity 69 economy class. First class seating six abreast with two 86 cm (34 in) aisles, each twin-seat unit 1.45 m (4 ft 9 in) wide. Business passengers four abreast with 72 cm (28½ in) aisle and 1.37 m (4 ft 6 in) wide seat pairs on upper deck or two-three-two on lower deck with two 63 cm (24¾ in) aisles and 2.08 m (6 ft 10 in) triple seat. Economy seating three-four-three, with 49.5 cm (19½ in) aisles, two 1.51 m (4 ft 11½ in) triple seats and 2.07 m (6 ft 9½ in) quad seat.

Centre overhead stowage bins 0.16 m³ (5.7 cu ft) volume per 1.02 m (40 in) long bin; outboard bins 0.45 m³ (15.9 cu ft) volume per 1.52 m (60 in) long bin; 0.083 m³ (2.95 cu ft) bin volume per passenger (three-class). Two modular upper deck lavatories, 14 on main deck, relocatable (six upper deck optional locations; 33 on lower deck) and vacuum-drained into four waste tanks with combined volume of 1,136 litres (300 US gallons; 250 Imp gallons). Single-point drainage. Basic galley configuration, one on upper deck, seven centreline and two sidewall on main deck; lavatories and galleys can be quickly relocated if required fittings are installed; advanced integrated audio/video/announcement system.

Underfloor freight: forward compartment, five 2.44 m (96 in) × 3.18 m (125 in) pallets (totalling 58.8 m³, 2,075 cu ft) or 16 LD-1 containers (totalling 78.4 m³; 2,768 cu ft); aft compartment, 14 LD-1 containers (totalling 68.6 m³; 2,422 cu ft) or four pallets (totalling 47.0 m³; 1,660 cu ft); and bulk storage behind aft compartment 23.6 m³ (835 cu ft). Max capacity (16 LD-1s forward, 14 LD-1s aft and bulk storage) 170.6 m³ (6,025 cu ft).

SYSTEMS: Each engine drives a hydraulic pump feeding an independent system; services are connected to supplies in such a way that loss of one supply cannot disable one system; two hydraulic systems also have air-driven pumps to maintain pressure and two have electric pumps; one electric pump can be run to provide braking when the aircraft is being towed on the ground; all four hydraulic reservoirs can be filled from a single location in the port main landing gear bay.

Hot air bled from the low-pressure and high-pressure compressors of all four engines is precooled by fan exit air and fed via a manifold to the cabin pressurisation and air conditioning system and to provide de-icing of wing leading-edge and engine nose cowling and to pressurise hydraulic tanks. Three conditioning packs in wing/fuselage fairing provide cabin air; five cabin zones, each with digital temperature control.

Each engine drives an integrated drive generator supplying 90 kVA power to respective AC busses; three generators are a dispatch item, but one will supply essential loads; APU drives two further generators; automatic start-up, load transfers and load shedding reduce crew workload; power systems may be isolated from each other for triple-channel Cat. III autoland.

Completely self-contained 1,081 kW (1,450 shp) P&WC PW901A APU, mounted clear of all flight-critical structure and flight controls in the extreme tail, drives two 90 kVA generators that can supply electrical power for

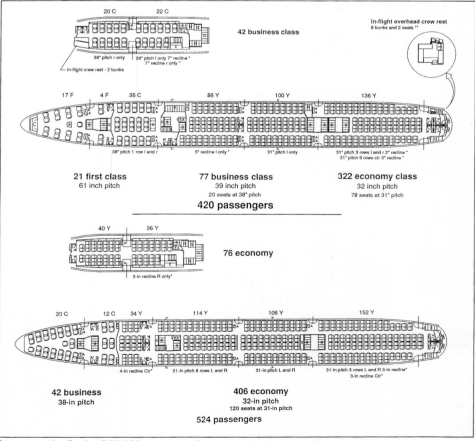

Representative Boeing 747-400 interior configurations for dual- or multiple-class travel

Boeing 747X Stretch, with additional side elevation (upper) of 747X *(James Goulding/Jane's)*

whole aircraft; also supplies compressed air to operate pneumatic components; can run at up to 6,100 m (20,000 ft) and supply compressed air below 4,575 m (15,000 ft). Capabilities include maintenance of 24°C (75°F) ground cabin temperature in 38°C (100°F) ambient conditions.

Forward underfloor cargo compartment heated to 5°C by hot air exhausted from flight deck cooling equipment and avionics in main equipment centre, boosted as necessary by two electrical heaters; rear underfloor hold heated to minimum 5°C or 18°C (selected by crew) by engine bleed.

Overheat detection and automatic extinguishing provided in all toilets; APU automatically shut down and fire extinguisher bottles initiated on detection of fire; each engine has three dual fire detectors in series and a fourth detector for overheating. Underfloor freight compartments and upper deck hold of Combi have smoke detectors and extinguisher systems; wheel wells have overheat detectors.

AVIONICS: Boeing launched development of new Flight Management Computer software in January 1993 to match existing aircraft to international Future Air Navigation System (FANS-1) during 1995. Standard avionics fit as follows:

Comms: Dual VHF and HF transceivers with Selcal; dual transponders; flight intercom with air-to-ground facility, connectable also to satcom system; cabin entertainment and passenger address and service units.

Radar: Colour weather radar transmitting in I- and G-bands.

Flight: Dual VOR; triple ILS receivers with single marker beacon receiver; dual ADF; dual DME; all nav radios automatically tuned by flight management computer system (FMCS). Automatic flight control system (AFCS) integrates autopilot, flight director and automatic stabiliser trim functions; dual digital air data computers with dual selectable pressure sensors, angle of attack sensors and total air temperature probes; FMCS allows crew to preselect flight plan using standard air traffic control language; FMCS incorporates database, updated every 28 days, which includes data on waypoints, airports, standard instrument departures (SIDs), standard terminal arrival routes (STARs), airline routes and information on specific geographic areas; triple ring laser gyro inertial reference units provide navigation input on EFIS, flight management displays or radio magnetic indicators; other systems include ground proximity warning, triple low-range radio altimeters and TCAS.

Central maintenance computer monitors over 75 electrical and electromechanical systems, performs tests and centralises maintenance data; failures are indicated in EICAS displays and stored for future reference for in-flight use or line or hangar maintenance. Satcom datalink allows ground crews to interrogate system for additional information while aircraft in flight.

Instrumentation: Electronic flight instrument system (EFIS) comprising six (left/right inboard/outboard and central upper/lower) 20.3 × 20.3 cm (8 × 8 in) integrated display units (IDU), two each for primary flight display (PFD), navigation display (ND) and engine indicating and crew alerting (EICAS) functions; all IDUs receive data from all three EFIS/EICAS interface units (EIU), updated via software data loader; PFD and EICAS primary formats automatically switch to inboard and lower IDUs respectively, with facility for manual selection of formats on different IDUs as required. B747-400 has 181 switches, 171 lights and 13 gauges, compared with 284, 555 and 132 of earlier variants; total 365 is below average 450 for typical two-crew jet transport.

DIMENSIONS, EXTERNAL:
Wing span	64.44 m (211 ft 5 in)
Wing span, fully fuelled	64.92 m (213 ft 0 in)
Wing aspect ratio	7.7
Length: overall	70.67 m (231 ft 10¼ in)
fuselage	68.63 m (225 ft 2 in)
Height overall	19.41 m (63 ft 8 in)
Tailplane span	22.17 m (72 ft 9 in)
Wheel track	11.00 m (36 ft 1 in)
Wheelbase: mean	25.60 m (84 ft 0 in)
to forward main bogie	24.07 m (78 ft 11½ in)
to rear main bogie	27.14 m (89 ft 0½ in)
Passenger doors (10, each):	
Height	1.93 m (6 ft 4 in)
Width	1.07 m (3 ft 6 in)
Height to sill	approx 4.88 m (16 ft 0 in)
Baggage door (front hold):	
Height	1.68 m (5 ft 6 in)
Width	2.64 m (8 ft 8 in)
Height to sill	approx 2.64 m (8 ft 4 in)
Baggage door (forward door, rear hold):	
Height	1.68 m (5 ft 6 in)
Width	2.64 m (8 ft 8 in)
Height to sill	approx 2.69 m (8 ft 10 in)
Bulk loading door (rear door, rear hold):	
Height	1.19 m (3 ft 11 in)
Width	1.12 m (3 ft 8 in)
Height to sill	approx 2.90 m (9ft 6 in)
Freighter cargo door (port):	
Height	3.05 m (10 ft 0 in)
Width	3.40 m (11 ft 2 in)
Height to sill	4.87 m (16 ft 0 in)

DIMENSIONS, INTERNAL:
Passenger cabin volume	885.9 m³ (31,285 cu ft)

AREAS:
Wings, gross	541.16 m² (5,825.0 sq ft)
Ailerons (total)	20.90 m² (225.00 sq ft)
Trailing-edge flaps (total)	78.69 m² (847.00 sq ft)
Leading-edge flaps (total)	43.85 m² (472.00 sq ft)
Inboard spoilers (total)	12.78 m² (137.60 sq ft)
Outboard spoilers (total)	15.46 m² (166.40 sq ft)
Fin	77.11 m² (830.00 sq ft)

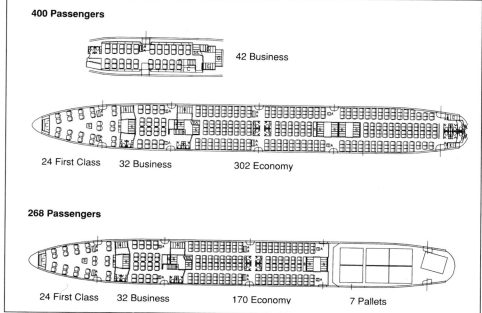

Boeing 747 Combi interiors, showing all-passenger and passenger/freight arrangements

606   USA: AIRCRAFT—BOEING

Rudder	21.37 m² (230.00 sq ft)
Tailplane	136.57 m² (1,470.00 sq ft)
Elevators (total, incl tabs)	30.38 m² (327.00 sq ft)

WEIGHTS AND LOADINGS (GB: CF6-80C2-B1F, GM: CF6-80C2-B5F, PB: PW4056, PM: PW4062, RB: RB211-524G, RM: 211-524M engines; all 416 passengers, five cargo pallets and 14 containers):

Operating weight: empty:
GB	180,395 kg (397,700 lb)
GM	180,880 kg (398,775 lb)
PB	180,620 kg (398,200 lb)
PM	181,110 kg (399,275 lb)
RB	181,390 kg (399,900 lb)
RM	181,880 kg (400,980 lb)

Baggage/freight capacity, all:
forward compartment	26,490 kg (58,400 lb)
aft compartment	22,938 kg (50,570 lb)
bulk compartment	6,749 kg (14,880 lb)

Max fuel weight:
GB	162,580 kg (358,425 lb)
GM	172,560 kg (380,425 lb)
PB, RB	163,250 kg (359,900 lb)
PM, RM	173,225 kg (381,900 lb)

Max T-O weight:
GB, PB, RB	362,875 kg (800,000 lb)
GM, PM, RM	396,895 kg (875,000 lb)

Max ramp weight:
GB, PB, RB	364,235 kg (803,000 lb)
GM, PM, RM	397,800 kg (877,000 lb)

Max landing weight:
GB, PB, RB	260,360 kg (574,000 lb)
GM, PM, RM	295,745 kg (652,000 lb)

Max zero-fuel weight:
GB, PB, RB	242,670 kg (535,000 lb)
GM, PM, RM	251,745 kg (555,000 lb)

Max wing loading:
GB, PB, RB	670.5 kg/m² (137.34 lb/sq ft)
GM, PM, RM	733.4 kg/m² (150.21 lb/sq ft)

Max power loading:
GB, RB	352 kg/kN (3.45 lb/lb st)
GM, PB	359 kg/kN (3.52 lb/lb st)
PM	360 kg/kN (3.53 lb/lb st)
RM	368 kg/kN (3.61 lb/lb st)

PERFORMANCE (as above; landing at MLW):
Cruising Mach No. 0.85
Approach speed: GB, PB, RB
146 kt (270 km/h; 168 mph)
GM, PM, RM 157 kt (291 km/h; 181 mph)

Initial cruising altitude:
GB, PB, RB	10,575 m (34,700 ft)
GM, PM, RM	10,000 m (32,800 ft)

T-O field length, 30°C (86°F):
GB	2,805 m (9,200 ft)
GM	3,018 m (9,900 ft)
PB	2,820 m (9,250 ft)
PM	2,990 m (9,800 ft)
RB	2,835 m (9,300 ft)
RM	3,185 m (10,450 ft)

Landing field length:
GB, PB, RB	1,905 m (6,250 ft)
GM, PM, RM	2,180 m (7,150 ft)

Design range:
GB	6,210 n miles (11,500 km; 7,146 miles)
GM*	7,280 n miles (13,482 km; 8,377 miles)
PB	6,220 n miles (11,519 km; 7,157 miles)
PM*	7,335 n miles (13,584 km; 8,441 miles)
RB	6,080 n miles (11,260 km; 6,996 miles)
RM*	7,210 n miles (13,352 km; 8,297 miles)

*fuel volume limited

UPDATED

BOEING 747-400F
USAF designation: AL-1A
TYPE: All-freight version of 747-400.
PROGRAMME: First flight (N6005C) 4 May 1993; FAA certification October 1993; JAR certification followed; first delivery (Cargolux) 17 November 1993. During 2000, Boeing began design of a freighter conversion of passenger 747-400s as an eventual partner to the first 747-300F conversion, begun in that year. However, this would differ from new-production-400F in several respects.
CURRENT VERSIONS: **747-400F**: As described.
AL-1A: Anti-missile defence aircraft; details under Boeing Military Aircraft and Missile Systems heading.

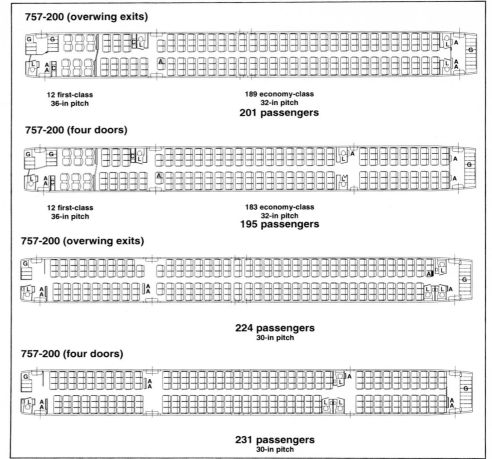

Typical seating arrangements of the Boeing 757-200 in two- and single-class configuration
A: attendant's seat, G: galley, L: lavatory

Delta Air Lines Boeing 757-200 *(Paul Jackson/Jane's)*

CUSTOMERS: See table with main 747-400 entry. First 747-400F delivered to Cargolux 17 November 1993; total of 34 delivered by October 1999. Production of 16 envisaged in 2000. Recent customers include China Airlines, which ordered 13 on 11 August 1999, for delivery between 2000 and 2007 (first on 6 July 2000); EVA Air, three for delivery from 2000 (first on 20 July); and Cathay Pacific Airways, which ordered two in October 1999 (first delivered on 12 September 2000), subsequently increasing its order to five.
COSTS: China Airlines order for 13 aircraft valued at US$2.5 billion (1999).
DESIGN FEATURES: 747-200F fuselage (short upper deck) with additional changes combined with stronger and larger 747-400 wing; strengthened floor of short upper deck, as offered for -200F, also integrated into 747-400F; further developed freight handling system; total cargo volume 777.8 m³ (27,467 cu ft), of which 604.5 m³ (21,347 cu ft) on main deck, 158.6 m³ (5,600 cu ft) in lower hold and 48.3 m³ (520 cu ft) available for bulk cargo. Compared to -200F, empty weight saving of 2,000 kg (4,409 lb) has raised maximum revenue freight load to about 113,000 kg (249,125 lb), at which range is 4,400 n miles (8,149 km; 5,063 miles); fuel consumption more than 15 per cent lower than 747-200F. Same gross weights as passenger 747-400; maximum landing weight at optional T-O weight, 302,090 kg (666,000 lb); maximum zero-fuel weight, 276,690 kg (610,000 lb), can be increased on condition T-O weight is decreased.
ACCOMMODATION: Two-pilot crew, as 747-400. Upward-opening nose cargo door and optional port-side rear cargo door; underfloor cargo doors fore and aft of wing and bulk cargo door aft of rear underfloor door; two crew doors to port. Capacity for 30 pallets on main deck and 32 LD-1 containers plus bulk cargo under floor.

UPDATED

BOEING 757-200
USAF designation: VC-32A
TYPE: Twin-jet airliner.
PROGRAMME: Announced early 1978; has 707/727/737 fuselage cross-section and two large turbofans; Eastern Air Lines and British Airways ordered 21 firm and 24 optioned and 19 + 18 respectively 13 August 1978; first flight (N757A) 19 February 1982 powered by 166.4 kN (37,400 lb st) Rolls-Royce RB535Cs and designated 757-200; first Boeing airliner launched with foreign engine.
FAA certification 21 December 1982; CAA certification 14 January 1983; revenue services began 1 January 1983 (EAL) and 9 February 1983 (BA). First flight of 757 powered by P&W PW2037s, 14 March 1984; certified October 1984 and delivered to Delta; first 757 with RB535E4s delivered to EAL 10 October 1984; first extended-range model delivered to Royal Brunei Airlines

Boeing 747-400F freighter of Singapore Airlines *(Paul Jackson/Jane's)*

May 1986; 757 with RB535E4 engines approved FAA ETOPS December 1986 (extended to 180 minutes July 1990); 757 with PW2037/2040 ETOPS approved April 1990 (180 minutes for PW2037 April 1992); Boeing windshear guidance and detection system approved by FAA January 1987. Certified for operation in the Russian Federation and Associated States (CIS) September 1993.

CURRENT VERSIONS: **757-200**: Initial production passenger airliner; extended range available.

Main description applies to -200 version, except where indicated.

757-200PF Package Freighter: Developed for United Parcel Service. Large freight door forward, single crew door and no windows; up to 15 standard 2.24 × 3.18 m (88 × 125 in) cargo pallets on main deck; same higher operating weights as Freighters. UPS ordered 20 on 31 December 1985; deliveries began 17 September 1987.

757-200M Combi: Boeing's mixed cargo/passenger configuration with windows; upward-opening cargo door to port (forward) 3.40 × 2.18 m (134 × 86 in); carries up to three 2.24 × 2.74 m (88 × 108 in) cargo containers and 150 passengers; one delivered to Royal Nepal Airlines on 15 September 1988; no others by mid-2000.

757-200 Freighter: Developed by Pemco Aeroplex in 1992 as conversions of existing 757s; all-freight, combi and quick-change versions available; same weights as Boeing 757-200PF Package Freighter (see data below); choice of more powerful engines; large freight door forward on port side.

757SF: Under an agreement announced 5 October 1999, Boeing Airplane Services will purchase 757-200s from British Airways and other operators and modify them to 757SF Special Freighter configuration for lease to DHL Worldwide Express. The modification, which is also available to other customers, provides 226.5 m³ (8,000 cu ft) of cargo space with payload of 27,215 kg (60,000 lb) and range of over 2,000 n miles (3,704 km; 2,301 miles). Eventual DHL fleet will be 44, of which first two are 757-200PFs bought from Ansett on 15 November 1999.

757-200 'Catfish': Boeing's own 757-200 prototype (N757A) fitted with radar nose in Lockheed Martin F-22A profile and representative F-22A swept wing section above flight deck containing conformal radar antennas for advanced radar trials; first flight in this configuration 11 March 1999. See also Lockheed Martin entry.

757-200X: Projected extended-range version under study in 1999; would combine fuselage of 757-200 with strengthened wing structure of 757-300; two auxiliary fuel tanks in aft cargo hold, combined capacity 3,785 litres (1,000 US gallons; 833 Imp gallons); maximum take-off weight 123,375 kg (272,000 lb); range 5,000 n miles (9,260 km; 5,753 miles).

757-300: Stretched version. *Described separately.*

VC-32A: Boeing 757-2G4. Four, with PW2040 engines, ordered 8 August 1996 as replacements for VC-137s of USAF's 89th Airlift Wing at Andrews AFB, Maryland. First aircraft (98-0001) flew 11 February 1998 and was delivered to 89th AW on 19 June. Further three followed on 23 June, 20 November and 25 November 1998. Post-production modifications, performed at Boeing's Wichita facility and completed on first aircraft on 2 April 1999, include installation of auxiliary fuel tanks, capacity 6,984 litres (1,845 US gallons; 1,536 Imp gallons) in forward and aft cargo holds, increasing range to 5,000 n miles (9,260 km; 5,753 miles); self-deploying forward airstair; crew ladder; satcom upgrade; and 378 litre (100 US gallon; 83.0 Imp gallon) potable water tank.

CUSTOMERS: See table Totals of 1,025 ordered and 940 delivered by 1 October 2000.

Production totalled 54 in 1998 and 67 in 1999 before scheduled fall to 48 in 2000.

COSTS: Price of four 757s ordered by US Air Force in August 1996 reported as US$365 million. Supplementary order for one -200 from Turkmenistan Airlines valued at US$67 million (2000).

DESIGN FEATURES: Low-wing, single-aisle airliner, with two podded turbofans below wings. Common design philosophy – including flight deck – with Boeing 767. Design goals also included multimarket performance, fuel efficiency, low noise and emissions, employment of advanced, digital avionics and 'dark' flight deck.

All flying surfaces sweptback and tapered; Boeing aerofoils; wing sweepback at quarter-chord 25°; dihedral 5°; incidence 3° 12′.

FLYING CONTROLS: Conventional and hydraulically powered. All-speed outboard ailerons assisted by five flight spoilers on each wing also acting variously as airbrakes and ground spoilers; one additional ground spoiler inboard on each wing; elevators and rudder; double-slotted trailing-edge flaps; full-span leading-edge slats, five sections each wing; variable incidence tailplane.

STRUCTURE: Aluminium alloy two-spar fail-safe wing box; centre-section continuous through fuselage; ailerons, flaps and spoilers extensively of honeycomb, graphite composites and laminates; tailplane has full-span light alloy torque boxes; fin has three-spar, dual-cell light alloy torque box; elevators and rudder have graphite/epoxy honeycomb skins supported by honeycomb and laminated spar and rib assemblies; CFRP wing/fuselage and flap track fairings. All landing gear doors of CFRP/Kevlar.

BOEING 757 ORDER BOOK
(as at 1 October 2000)

Customer	Variant	Engine	Qty
Air 2000	757-200	RB211	1
Air Europe	757-200	RB211	14
Air Holland	757-200	RB211	3
Airtours International	757-200	RB211	1
America West	757-200	RB211	4
American Airlines	757-200	RB211	123
American Trans Air	757-200	RB211	13
	757-300		10
Ansett Worldwide	757-200	RB211	19
	757-200	PW2037	2
	757-200PF	RB211	4
Aries Intl Leasing	757-200	RB211	2
Arkia Israeli Airlines	757-300	RB211	2
Azerbaijan Airlines	757-200	RB211	2
Britannia Airways	757-200	RB211	11
British Airways	757-200	RB211	50
CAAC	757-200	RB211	3
China Southern	757-200	RB211	16
China Southwest	757-200	RB211	12
China Xinjian Airlines	757-200	RB211	6
Condor Flugdienst	757-200	PW2040	17
	757-300	RB211	13
Continental Airlines	757-200	RB211	41
Delta Air Lines	757-200	PW2037	110
	757-200	PW2040	6
Eastern Air Lines	757-200	RB211	25
El Al	757-200	RB211	7
Ethiopian Airlines	757-200	PW2040	4
	757-200PF	PW2040	1
Far Eastern Air Transport	757-200	PW2037	7
GATX Capital Corp	757-200	RB211	3
GE Capital	757-200	RB211	4
GPA Group Ltd	757-200	RB211	12
Iberia Airlines	757-200	RB211	24
Icelandair	757-200	RB211	7
	757-300	RB211	2
ILFC	757-200	RB211	43
	757-200	PW2037	15
	757-200	PW2040	25
JMC Airline	757-300		2
Kawasaki Leasing	757-200	RB211	2
LTU/LTE	757-200	RB211	12
Mexican Government	757-200	RB211	1
Mid East Jet	757-200	PW2040	1
Monarch Airlines	757-200	RB211	7
Northwest Airlines	757-200	PW2037	73
Republic Airlines	757-200	RB211	6
Rolls-Royce Aircraft Mgmt	757-200	RB211	4
Royal Air Maroc	757-200	PW2037	2
Royal Brunei Airlines	757-200	RB211	3
Royal Nepal Airlines	757-200	RB211	1
	757-200M	RB211	1
Shanghai Airlines	757-200	PW2037	8
Singapore Airlines	757-200	PW2037	4
Starflite Corp	757-200	RB211	1
Sterling European	757-200	RB211	3
Sunrock Aircraft	757-200	RB211	2
Transavia Airlines	757-200	RB211	3
Trans World Airlines	757-200	PW2037	14
Turkmenistan Airlines	757-200	RB211	3
United Airlines	757-200	PW2037	47
	757-200	PW2040	51
UPS	757-200PF	RB211	40
	757-200PF	PW2040	35
US Airways	757-200	RB211	23
USAF	757-200	PW2040	4
Uzbekistan Airways	757-200	PW2037	3
Xiamen Airlines	757-200	RB211	5
Total			**1,025**

Subcontractors include Hawker de Havilland (wing in-spar ribs), Shorts (inboard flaps), CASA (outboard flaps), various Boeing divisions (leading-edge slats, main cabin sections, fixed leading-edges and flight deck), Northrop Grumman (fin and tailplane, extreme rear fuselage, overwing spoiler panels), Heath Tecna (wing/fuselage and flap track fairings), Schweizer (wingtips), Rohr Industries (engine support struts), IAI (dorsal fin) and Fleet Industries (APU access doors).

LANDING GEAR: Retractable tricycle type, with main and nose units manufactured by Menasco. Each main unit carries a four-wheel bogie, fitted with Dunlop or Goodrich wheels, carbon brakes and tyres. Twin-wheel nose unit, also with Dunlop or Goodrich tyres. Nose tyres H31×13.0-12 (20

Boeing 757-200 in Finnair colours (*Paul Jackson/Jane's*)

USA: AIRCRAFT—BOEING

Boeing 757 flight deck

Windowless Boeing 757-200PF freighter *(Paul Jackson/Jane's)*

ply); main tyres either H42×14.5-19 (24/26 ply) or, for higher weight options H42×16.0-19 (24 ply). Minimum ground turning radius 21.64 m (71 ft) at nosewheels, 29.87 m (98 ft) at wingtip.

POWER PLANT: Two 162.8 kN (36,600 lb st) Pratt & Whitney PW2037, 178.4 kN (40,100 lb st) PW2040, 178.8 kN (40,200 lb st) Rolls-Royce RB211-535E4, 189.5 kN (42,600 lb st) PW2043 or 193.5 kN (43,500 lb st) RB211-535E4-B turbofans, mounted in underwing pods. Standard fuel capacity 42,684 litres (11,276 US gallons; 9,389 Imp gallons), optional 43,489 litres (11,489 US gallons; 9,566 Imp gallons).

ACCOMMODATION: Crew of two on flight deck, with provision for an observer; common crew qualification with Boeing 767. Five to seven cabin attendants. Standard interior arrangements for 201 (12 first class/189 economy) or 195 (12 first class/183 economy) mixed-class passengers, or 224 or 231 all-economy passengers. First class seats are four-abreast, at 91 cm (36 in) pitch; business class five-abreast; economy seat pitch is 81 cm (32 in) in mixed class or 76 cm (30 in) in all-economy, mainly six-abreast. New cabin interior includes twin-door overhead luggage bins of 203 cm (80 in) width, replacing single-door, 152 cm (60 in) type; optional ceiling-mounted stowage compartments and video screens; aesthetic improvements; and (typically, five) vacuum lavatories, as in 757-300. Choice of two cabin door configurations, with either three or four passenger doors each side; two overwing emergency exits each side on three-door version. Up to nine galleys at forward, mid-cabin and aft locations on starboard side; nine lavatory position options in typical three-door configuration, or up to 12 in four-door configuration. Movable class dividers. Cargo hold doors forward and aft on starboard side.

SYSTEMS: Honeywell ECS; General Electric engine thrust management system; Honeywell-Vickers engine-driven hydraulic pumps; four Abex electric hydraulic pumps. Hydraulic system maximum flow rate 140 litres (37.0 US gallons; 30.8 Imp gallons)/min at T-O power on engine-driven pumps; 25.4 to 34.8 litres (6.7 to 9.2 US gallons; 5.6 to 7.7 Imp gallons)/min on electric motor pumps;

42.8 litres (11.3 US gallons; 9.4 Imp gallons)/min on ram air turbine. Independent reservoirs, pressurised by air from pneumatic system, maximum pressure 207 bars (3,000 lb/sq in) on primary pumps. Hamilton Sundstrand electrical power generating system and ram air turbine; and Honeywell GTCP331-200 APU. Wing thermally anti-iced.

AVIONICS: *Flight:* Honeywell inertial reference system (IRS) (first commercial application of laser gyros); IRS provides position, velocity and attitude information to flight deck displays, and the flight management computer system (FMCS) and digital air data computer (DADC) supplied by Honeywell; FMCS provides automatic en route and terminal navigation capability, and also computes and commands both lateral and vertical flight profiles for optimum fuel efficiency, maximised by electronic linkage of the FMCS with automatic flight control and thrust management systems; CAT. IIIb instrument landing capability; Boeing windshear detection and guidance system is optional. Future Air Navigation System (FANS) FMS.

Instrumentation: Rockwell Collins EFIS-700 six-tube display with engine indication and crew alerting system (EICAS), EADI and EHSI functions; Rockwell Collins FCS-700 autopilot flight director system (AFDS).

DIMENSIONS, EXTERNAL:	
Wing span	38.05 m (124 ft 10 in)
Wing chord: at root	8.20 m (26 ft 11 in)
at tip	1.73 m (5 ft 8 in)
Wing aspect ratio	7.8
Length: overall	47.32 m (155 ft 3 in)
fuselage	46.96 m (154 ft 10 in)
Height overall	13.56 m (44 ft 6 in)
Tailplane span	15.21 m (49 ft 11 in)
Wheel track	7.32 m (24 ft 0 in)
Wheelbase	18.29 m (60 ft 0 in)
Passenger doors (two, fwd, port):	
Height	1.83 m (6 ft 0 in)
Width	0.84 m (2 ft 9 in)
Passenger door (rear, port): Height	1.83 m (6 ft 0 in)
Width	0.76 m (2 ft 6 in)
Service door (fwd, stbd): Height	1.65 m (5 ft 5 in)
Width	0.76 m (2 ft 6 in)
Service door (stbd, opposite second passenger door):	
Height	1.83 m (6 ft 0 in)
Width	0.84 m (2 ft 9 in)
Service door (rear, stbd): Height	1.83 m (6 ft 0 in)
Width	0.76 m (2 ft 6 in)
Emergency exits (four, overwing):	
Height	0.97 m (3 ft 2 in)
Width	0.51 m (1 ft 8 in)
Emergency exits, optional (two, aft of wings):	
Height	1.32 m (4 ft 4 in)
Width	0.61 m (2 ft 0 in)
DIMENSIONS, INTERNAL:	
Cabin, aft of flight deck to rear pressure bulkhead:	
Length	36.09 m (118 ft 5 in)
Max width	3.53 m (11 ft 7 in)
Height: max	2.13 m (7 ft 0 in)
to ceiling video screen	1.85 m (6 ft 1 in)
Floor area	116.0 m² (1,249 sq ft)
Passenger section volume	230.5 m³ (8,140 cu ft)
Underfloor cargo hold volume (bulk loading):	
fwd	19.8 m³ (699 cu ft)
rear	27.5 m³ (971 cu ft)
AREAS:	
Wings, gross	185.25 m² (1,994.0 sq ft)
Ailerons (total)	4.46 m² (48.00 sq ft)
Trailing-edge flaps (total)	30.38 m² (327.00 sq ft)
Leading-edge slats (total)	18.39 m² (198.00 sq ft)
Flight spoilers (total)	10.96 m² (118.00 sq ft)
Ground spoilers (total)	12.82 m² (138.00 sq ft)
Fin	34.37 m² (370.00 sq ft)
Rudder	11.61 m² (125.00 sq ft)
Tailplane	50.35 m² (542.00 sq ft)
Elevators (total)	12.54 m² (135.00 sq ft)

WEIGHTS AND LOADINGS (with 201 passengers. A: PW2037 engines, B: PW2040s, C: RB211-535E4s, D: RB211-535E4-Bs):

Operating weight empty: A	58,325 kg (128,580 lb)
B	58,390 kg (128,730 lb)
C	58,550 kg (129,080 lb)
D	58,620 kg (129,230 lb)
Baggage capacity, underfloor:	
fwd hold	4,672 kg (10,300 lb)
aft hold	7,394 kg (16,300 lb)
Max T-O weight: A, C	99,790 kg (220,000 lb)
B, D	115,665 kg (255,000 lb)
Max landing weight: A, C	89,815 kg (198,000 lb)
B, D	95,255 kg (210,000 lb)
Max ramp weight: A, C	100,245 kg (221,000 lb)
B, D	116,120 kg (256, 000 lb)
Max zero-fuel weight: A, C	83,460 kg (184,000 lb)
B	84,370 kg (186,000 lb)
D	85,275 kg (188,000 lb)
Max wing loading: A, C	538.7 kg/m² (110.33 lb/sq ft)
B, D	624.4 kg/m² (127.88 lb/sq ft)
Max power loading: A	306 kg/kN (3.01 lb/lb st)
B	324 kg/kN (3.18 lb/lb st)
C	279 kg/kN (2.74 lb/lb st)
D	299 kg/kN (2.93 lb/lb st)

PERFORMANCE (with 201 passengers; at max basic T-O weight except where indicated):

Max operating Mach No. (M_{MO}): A, B, C, D	0.86
Cruising speed: A, B, C, D	M0.80
Approach speed at S/L, flaps down, max landing weight:	
A, C	132 kt (245 km/h; 152 mph) EAS
B, D	137 kt (254 km/h; 158 mph)
Initial cruising height: A	11,675 m (38,300 ft)
B	10,790 m (35,400 ft)

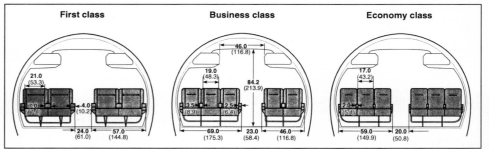

Boeing 757 cabin cross-sections. Dimensions in inches (centimetres)

C	11,795 m (38,700 ft)
D	10,880 m (35,700 ft)

Runway LCN at ramp weight of 100,244 kg (221,000 lb), optimum tyre pressure and subgrade C flexible pavement: H40×14.5-19.0 tyres 36

T-O field length (S/L, 29°C):
A	1,814 m (5,950 ft)
B	2,378 m (7,800 ft)
C	1,677 m (5,500 ft)
D	2,104 m (6,900 ft)

Landing field length at max landing weight:
A	1,463 m (4,800 ft)
B	1,555 m (5,100 ft)
C	1,418 m (4,650 ft)
D	1,494 m (4,900 ft)

Range with 201 passengers:
A	2,570 n miles (4,759 km; 2,957 miles)
B	3,930 n miles (7,278 km; 4,522 miles)*
C	2,375 n miles (4,398 km; 2,733 miles)
D	3,695 n miles (6,843 km; 4,252 miles)*

OPERATIONAL NOISE LEVELS (FAR Pt 36 Stage 3):
T-O, at max basic T-O weight, cutback power:
A	82.2 EPNdB
B	86.2 EPNdB
C (estimated)	84.7 EPNdB

Approach at max landing weight, 30° flap:
A	95.0 EPNdB
B, C	97.7 EPNdB

Sideline:
A	93.3 EPNdB
B	94.0 EPNdB
C (estimated)	94.6 EPNdB

Fuel volume limited

UPDATED

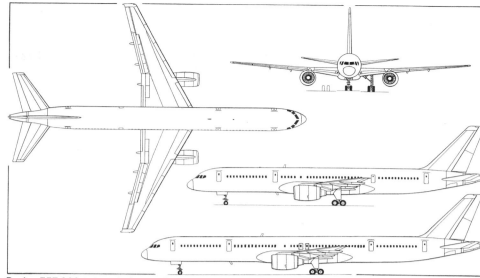
Boeing 757-200 twin-turbofan medium-range transport, with additional side view of stretched -300 (Dennis Punnett/Jane's)

BOEING 757-300

TYPE: Medium-range twin-turbofan airliner.

PROGRAMME: Stretched 757-200. Launched 2 September 1996; major assembly began 9 August 1997; prototype (N757X; 804th B757; earmarked for Condor Flugdienst as D-ABOA) rolled out at Renton on 19 May 1998 (officially 31 May); first flight (transmitted live on the Internet) 2 August 1998. 912-hour, 356-sortie flight test programme undertaken by three aircraft. N757X/NU701, employed on initial airworthiness and basic controllability tests; D-ABOB/NU721, first flown 4 September 1998, for ground effect, autoland and fuel consumption tests; and D-ABOC/NU722, first flown 2 October 1998, initially for HERF testing, before installation of standard interior for a four-day period of simulated airline operations with launch customer Condor Flugdienst. FAA certification with 180-minute ETOPS approval 27 January 1999, followed by first delivery (D-ABOE) to Condor on 10 March 1999 and first revenue-earning service 19 March 1999. Achieved 8.7 hours daily utilisation and 99.64 per cent reliability rate in first 12 months of operation.

CUSTOMERS: See table accompanying Boeing 757-200 entry. Launch customer Condor Flugdienst ordered 12 firm, with 12 options, on 2 September 1996; Icelandair ordered two, with Rolls-Royce engines, in June 1997, for delivery in second quarter 2001 and second quarter 2002. Orders totalled 17 by 1 January 1999.

COSTS: Condor launch order for 12 aircraft valued at US$875 million (1996). Seat-mile operating costs estimated at 10 per cent lower than those of 757-200.

Description generally as for 757-200, except that below.

DESIGN FEATURES: Fuselage stretched by 7.11 m (23 ft 4 in) – 4.06 m (13 ft 4 in) ahead of the wing and 3.05 m (10 ft 0 in) aft – to provide 20 per cent more passenger accommodation and nearly 50 per cent more cargo volume than 757-200; typical accommodation 240 passengers in two-class configuration or up to 289 in single class with pitch of between 71 and 74 cm (28 and 29 in). Other features include strengthened wings, engine pylons, high-lift devices and landing gear; strengthened overwing exit body section; new wheels, 26-ply tyres and brakes; nosewheel spray deflector; retractable tailskid, similar to those of 767-300 and 777-300, with 'body contact' indicator on flight deck; redesigned cabin interior based on 'Next-Generation' 737s, with upgraded environmental control system, larger precooler, more powerful fans, and vacuum lavatories with single-point drainage. Flight deck and operating systems generally as for 757-200, but with Pegasus FMS (including software options for GPS, FANS and satcom) and enhanced EICAS with improved BITE functions.

POWER PLANT: As for 757-200. Total fuel capacity 43,402 litres (11,466 US gallons; 9,547 Imp gallons) for all options.

ACCOMMODATION: Typical cabin arrangements provide for 243 passengers (12 first class, 231 economy) in mixed class configuration, four-abreast in first class, at 91 cm (36 in) seat pitch, six-abreast in economy at 81 cm (32 in) seat pitch; or 279 passengers in inclusive tour configuration, six-abreast at 71/74 cm (28/29 in) seat pitch.

Economy seating aboard Boeing 757

DIMENSIONS, EXTERNAL: As for 757-200 except:
Length: overall	54.43 m (178 ft 7 in)
fuselage	54.08 m (177 ft 5 in)
Wheelbase	22.35 m (73 ft 4 in)

Passenger doors (three each side, each):
Height	1.83 m (6 ft 0 in)
Width	0.84 m (2 ft 9 in)

Service doors (one each side, each):
Height	1.32 m (4 ft 4 in)
Width	0.61 m (2 ft 0 in)

Cargo door (fwd, port): Height	1.08 m (3 ft 6½ in)
Width	1.40 m (4 ft 7 in)
Cargo door (aft, port): Height	1.12 m (3 ft 8 in)
Width	1.40 m (4 ft 7 in)

DIMENSIONS, INTERNAL:
Cabin (rear of flight deck to rear pressure bulkhead):
Length	43.21 m (141 ft 9 in)

Underfloor cargo volume (bulk loading):
fwd	30.3 m³ (1,071 cu ft)
rear	36.8 m³ (1,299 cu ft)

WEIGHTS AND LOADINGS (with 243 passengers: A: PW2040 engines, B: PW2043s, C: RB211-535E4s, D: RB211-535E4-Bs):

Operating weight empty:
A, B	63,655 kg (140,330 lb)
C, D	63,855 kg (140,780 lb)

Baggage capacity, underfloor:
fwd hold	7,303 kg (16,100 lb)
aft hold	8,845 kg (19,500 lb)

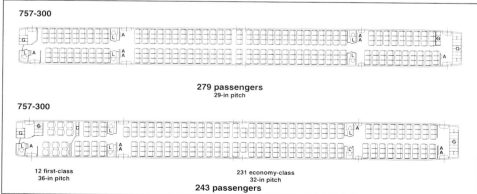
Typical Boeing 757-300 seating options
A: attendant's seat, G: galley, L: lavatory

USA: AIRCRAFT—BOEING

Boeing 757-300 of launch customer, Condor *(Paul Jackson/Jane's)*

Max T-O weight: A, C		108,860 kg (240,000 lb)
B, D		123,605 kg (272, 500 lb)
Max landing weight: A, B, C, D		101,605 kg (224,000 lb)
Max ramp weight: A, C		109,315 kg (241,000 lb)
B, D		124,055 kg (273,500 lb)
Max zero-fuel weight: A, B, C, D		95,255 kg (210,000 lb)
Max wing loading: A, C		587.7 kg/m² (120.36 lb/sq ft)
B, D		667.2 kg/m² (136.66 lb/sq ft)
Max power loading: A		305 kg/kN (2.99 lb/lb st)
B		326 kg/kN (3.20 lb/lb st)
C		304 kg/kN (2.99 lb/lb st)
D		319 kg/kN (3.13 lb/lb st)

PERFORMANCE (243 passengers):
Max operating Mach No. (M_{MO}): A, B, C, D   0.86
Cruising speed: A, B, C, D   M0.80
Approach speed, S/L, flaps down, at max landing weight:
A, B, C, D   143 kt (265 km/h; 165 mph)
Initial cruising height: A   11,005 m (36,100 ft)
B   10,400 m (34,120 ft)
C   11,200 m (36,740 ft)
D   10,420 m (34,180 ft)
T-O field length, S/L, 86°F: A   2,082 m (6,830 ft)
B   2,625 m (8,610 ft)
C   2,119 m (6,950 ft)
D   2,607 m (8,550 ft)
Landing field length at max landing weight:
A, B   1,750 m (5,740 ft)
C, D   1,723 m (5,650 ft)
Range: A   2,235 n miles (4,139 km; 2,572 miles)
B   3,465 n miles (6,417 km; 3,987 miles)*
C   2,055 n miles (3,805 km; 2,364 miles)
D   3,240 n miles (6,000 km ; 3,728 miles)*

* Fuel volume limited

UPDATED

BOEING 763

TYPE: Wide-bodied airliner.
PROGRAMME: Under study in 1998-99; target service entry in 2006. No further announcements by mid-2000.
CURRENT VERSIONS: Several different configurations have been examined, of which the following are known.
763-241: Evaluated in 1998; resembled stretched Boeing C-17A; span 80 m (262 ft); length 76 m (250 ft); height 21 m (70 ft). Replaced by more conventional ('four-engined 777') 763-246.
763-246C: Standard version; four engines from currently available 777 options in 297 kN (68,800 lb st) class, derated to 261 kN (58,700 lb st) for take-off, pod mounted below wings; fuel capacity 378,541 litres (100,000 US gallons; 83,267 Imp gallons); accommodation for 454 passengers in single-deck, three-class, three-aisle layout, 12 or 13 abreast in economy cabin; optional 58 personal space units/sleeper berths on second deck above first class and business class cabins; below-deck area could accommodate berths and 'personal areas', two lower deck galleys with service cart lifts, and cargo hold capable of accommodating LD-3 containers three abreast.
763-264CER: Extended-range version; strengthened fuselage, wing, tail unit and landing gear; 331 kN (74,500 lb st) engines; accommodation for 450 passengers in three-class layout, or up to 524 passengers in two-class configuration; expanded provisioning, water and waste systems.
763-246CS: Stretched version, with two plugs of ten frames each fore and aft of wing box, adding about 9.88 m (32 ft 5 in) to overall length; engines as for 763-246CER; strengthened fuselage with two additional Type A doors aft and provision for an optional seventh pair forward; retractable 777-type tailskid; ground manoeuvring camera; accommodation for 547 passengers in three-class layout and optional overhead sleeper berths for 82 passengers.
COSTS: Target operating costs up to 10 per cent lower than those of Airbus A3XX-100.
DESIGN FEATURES: Aims include high systems commonality with 747-400 and 777 to reduce costs and development time; take-off, climb, cruise and approach performance and turnaround time equal to or better than those of 747-400; pavement loading lower than that of 777 or MD-11; and all doors, sills and related equipment compatible with existing airport loading bridges and ground service equipment. Design features include laminar flow engine nacelles, riblets and programmable flaps, resulting in two per cent reduction in drag compared to 777.
Wing sweepback 35° 30′, dihedral 6°.
FLYING CONTROLS: Fly-by-wire, with wing load alleviation system.
STRUCTURE: Chemically milled fuselage panels; carbon composites floor beams; composites wing spar box 20 per cent lighter than alloy structure; co-cured graphite tail unit similar in configuration to that of 777.
LANDING GEAR: Retractable tricycle type; three fuselage- and two wing-mounted units have four wheels each, nosewheel two wheels; wing and centre-body units all retract into underfuselage area, wing units inwards, fuselage units forwards; aft unit steerable, ±26°.

Data below are provisional.

DIMENSIONS, EXTERNAL (A: 763-246C, B: 763-246CER, C: 763-246CS):
Wing span: A, B, C   79.86 m (262 ft 0 in)
Wing aspect ratio: A, B, C   8.9
Length overall: A, B   73.46 m (241 ft 0 in)
C   83.34 m (273 ft 5 in)
Height overall: A, B, C   23.14 m (75 ft 11 in)
Tailplane span: A, B, C   30.81 m (101 ft 1 in)
Wheel track (c/l shock struts):
A, B, C   13.38 m (43 ft 11 in)
DIMENSIONS, INTERNAL (A, B, C as above):
Cabin: Length: A, B   about 54.86 m (180 ft 0 in)
Max width: all   7.77 m (25 ft 6 in)
AREAS (A, B, C as above):
Wings, gross: A, B, C   715.35 m² (7,700.0 sq ft)
WEIGHTS AND LOADINGS (A, B, C as above):
Max T-O weight: A   470,825 kg (1,038,000 lb)
B   532,060 kg (1,173,000 lb)
Max landing weight: A   356,520 kg (786,000 lb)
B   369,675 kg (815,000 lb)
C   410,045 kg (904,000 lb)
Max zero-fuel weight: A   331,575 kg (731,000 lb)
B   340,195 kg (750,000 lb)
C   385,100 kg (849,000 lb)
PERFORMANCE (A, B, C as above):
Design range:
A, C   7,750 n miles (14,353 km; 8,918 miles)
B   more than 9,000 n miles (16,668 km; 10,357 miles)

UPDATED

BOEING 767

TYPE: Wide-bodied airliner.
PROGRAMME: Launched on receipt of United Air Lines order for 30 on 14 July 1978; construction of basic 220-passenger 767-200 began 6 July 1979; first flight (N767BA) 26 September 1981 with P&W JT9D turbofans; first flight fifth aircraft with GE CF6-80A 19 February 1982; 767 with JT9D-7R4D certified 30 July 1982; with CF6-80A 30 September 1982.
First delivery with JT9D (United Air Lines) 19 August 1982 (initial service 8 September); first delivery with CF6 (Delta) 25 October 1982. ETOPS approval for 767-200 with JT9D-7R4 or CF6-80A or -80A2 granted January 1987; ETOPS approval for 767-200 and -300 with PW4000 obtained April 1990; 180-minute ETOPS approval with PW4000 engines obtained August 1993. Joint 757/767 crew rating approved 22 July 1983. Boeing windshear detection and guidance system FAA approved for 767-200 and -300 February 1987.
Boeing 767-400 first flew 9 October 1999; FAA certification and 180-minute ETOPS granted 20 July 2000; JAA certification 24 July 2000; FAA common type rating with 767-200/300 and 757-200/300 issued 21 August 2000.
CURRENT VERSIONS: **767-200**: Basic model; no longer available. Medium-range variant (MTOW 136,080 kg; 300,000 lb) has reduced fuel; higher gross weight variant (142,880 kg; 315,000 lb) certified June 1983. Weight and performance data in 2000-01 and previous *Jane's*.
767-200ER: Extended-range version; announced January 1983; first flight 6 March 1984; basic -200ER with centre-section tankage and gross weight increased initially to 156,490 kg (345,000 lb) first delivered to El Al 26 March 1984; 175,540 kg (387,000 lb) certified 1988; current gross weights are 159,211 kg (351,000 lb), and 179,169 kg (395,000 lb).
767-300: Stretched 269-passenger version, with 3.07 m (10 ft 1 in) plug forward of wing and 3.35 m (11 ft) plug aft, and same gross weight as 767-200; strengthened landing gear and thicker metal in parts of fuselage and underwing skin; same flight deck and systems as other 767s; same engine options as 767-200ER; first ordered (by Japan Airlines) 29 September 1983. First flight with JT9D-7R4D engines 30 January 1986; certified with JT9D-7R4D and CF6-80A2 22 September 1986. First delivery (Japan Airlines) 25 September 1986. British Airways ordered 11 in August 1987, later increased to total 25, with Rolls-Royce RB211-524H engines; delivered from 8 February 1990. No longer available; weight and performance data in 2000-01 and previous *Jane's*.
767-300ER: Extended-range, higher gross weight version; development began January 1985; optional gross weights 172,365 kg (380,000 lb) and, from 1992, 186,880 kg (412,000 lb); further increased centre-section tankage. Engine choice CF6-80C2, PW4000, RB211-524H; structural reinforcement; certified late 1987. Launch customer American Airlines (15), delivered from 19 February 1988. New interior introduced late 2000; based on Boeing 777; first recipient Lauda Air.
767-300ERX: Further range extension, under study from 1998; addition of tailplane fuel tank, capacity 7,571 litres (2,000 US gallons; 1,665 Imp gallons) would increase range to 6,695 n miles (12,400 km; 7,705 miles).
767-300 General Market Freighter: See separate entry.

Delta Air Lines 767-200 *(Paul Jackson/Jane's)*

BOEING—AIRCRAFT: USA   611

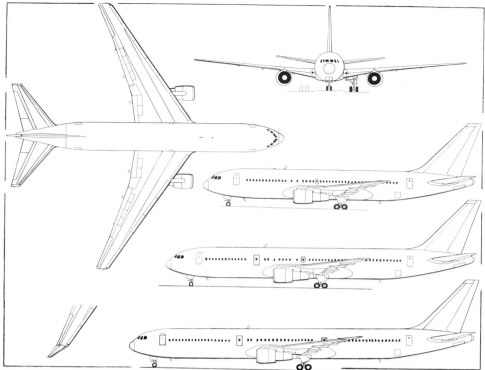

Boeing 767-200 wide-bodied airliner, with additional side view of stretched -300 and further stretched -400, including last mentioned's revised wingtips *(Dennis Punnett/Jane's)*   2000/0085775

767-400ER: Stretched version with 10 to 15 per cent increase in passenger accommodation, seating 245 passengers in three-class configuration and 304 in two-class. Features include strengthened wing with thicker ribs, spars and skin; updated flight deck based on Boeing 777; fuselage lengthened 6.43 m (21 ft 1 in) by means of plugs forward (3.36 m; 11 ft 0¼ in) and aft (3.07 m; 10 ft 0¾ in) of centre-section; stringerless window belt with elliptical 777-type cabin windows; wing span increased by 4.42 m (14 ft 6 in) with highly sweptback (9° 50′) wingtips of composites construction which reduce take-off distance, increase climb rate and improve fuel consumption; redesigned interior; cargo volume 129.7 m³ (4,580 cu ft); new landing gear with 46 cm (18 in) longer main legs, Boeing 777 brakes and 127 cm (50 in) tyres and revised, hydraulically actuated tail skid; 120 kVA AC generators and more powerful Honeywell 331-400 APU also of 120 kVA. Engine choice 276 kN (62,100 lb st) CF6-80C2B7F1 or 282 kN (63,300 lb st) CF6-80C2B8, with PW4000 series as option; fuel capacity as currently offered on 767-300ER. Maximum T-O weight increase to 204,115 kg (450,000 lb) together with aerodynamic improvements to provide maximum range of approximately 5,600 n miles (10,371 km; 6,444 miles), enabling 767-400ER to operate most existing 767-300ER routes.

Offered from January 1997; launch customer Delta Air Lines announced intention to order 21 on 20 March 1997; confirmed 28 April 1997; Continental ordered 26 on 10 October 1997. Assembly of first aircraft began at Everett on 9 February 1999; roll-out 26 August 1999; first flight (N76400 No. 1) 9 October 1999. Four aircraft took part in test programme, comprising 1,150 flight hours and 1,200 ground testing hours: prototype used primarily to test and certify basic handling qualities; N76401 served as aerodynamics and avionics certification article; N87402, with full cabin interior used for systems development and certification; N47403 (first flown June 2000) for cabin entertainment and related evaluation. World tour (by N76400 No. 2) in July-August 2000. First delivery (N828MH), to launch customer Delta Airlines, 11 August 2000; deliveries to Continental Airlines began 30 August 2000 with refurbished second prototype (N76401/N66051). Also on 30 August 2000, Delta received first 767-400ER with Rockwell Collins Large Format Display System, comprising six 203 × 203 mm (8 × 8 in) LCDs.

767-400ER Shrink: Under study in 1999 as alternative to 767-300ERX; would feature tailplane fuel tank, capacity 4,536 kg (10,000 lb); 302 kN (68,000 lb st) class engines providing a three to five per cent increase in cruising speed; and range up to 6,479 n miles (12,000 km; 7,456 miles).

767-400ERX: Extended-range version; launched 13 September 2000; first deliveries planned in second quarter of 2004; range with 245 passengers increased to 6,150 n miles (11,389 km; 7,077 miles).

767 AWACS: See Boeing 767 AWACS earlier in Boeing Company entry. Also under consideration are a tanker version for boom and hose-reel refuelling systems and a carrier for Joint STARS radar.

767 AST: Boeing contracted by US Army on 11 October 1994 to supply company-owned 767 as Airborne Sensor Testbed for long wavelength infra-red surveillance system; operating contract was extended for 12 months in September 1998 at cost of US$4.1 million.

767 SF: Special Freighter conversion of 767-200 airliner; available from 2000; payload 39,010 kg (86,000 lb); freight door as 767-300F; strengthened floor, main landing gear and forward fuselage.

CUSTOMERS: See table for orders. Original prototype became 767 Airborne Surveillance Testbed (formerly AOA) for US Army (see 1991-92 *Jane's*). One reconfigured by E-Systems as medevac aircraft for Civil Reserve Air Fleet.

DESIGN FEATURES: Low-wing, wide-bodied airliner with twin, podded turbofans underwing. Boeing aerofoils; quarter-chord sweepback 31° 30′; thickness/chord ratio 15.1 per cent at root, 10.3 per cent at tip; dihedral 6°; incidence 4° 15′.

FLYING CONTROLS: Conventional and hydraulically powered. Inboard, all-speed (between inner and outer flaps) and outboard low-speed ailerons supplemented by flight spoilers (four-section outboard; two-section inboard) also acting as airbrakes and lift dumpers; single-slotted, linkage-supported outboard trailing-edge flaps, double-slotted inboard; track-mounted leading-edge slats; variable incidence tailplane driven by hydraulic srewjack; two-piece elevators each side; no trim tabs; roll and yaw trim through spring feel system; triple digital flight control computers and EFIS; Boeing windshear detection and guidance system optional. Control surface deflections: outboard ailerons +30/−15°, inboard ailerons ±20°, inboard flaps 61° (first element 36°), outboard flaps 36°, spoilers +60°, elevators +28/−20°, rudder ±26°; tailplane incidence +2/−12°.

STRUCTURE: Fail-safe structure. Conventional aluminium structure augmented by graphite ailerons, spoilers, elevators, rudder and floor panels; advanced aluminium alloy keel beam chords and wing skins; composites engine cowlings, wing/fuselage fairing and rear wing panels; CFRP landing gear doors; and aramid flaps and engine pylon fairings.

Subcontractors include Boeing Military Aircraft (wing fixed leading-edges); Northrop Grumman (wing centre-section and adjacent lower fuselage section; fuselage bulkheads); Vought Aircraft (horizontal tail); Canadair

Latest version of Boeing 767, the -400ER   2001/0093711

Boeing 767-300ER owned by GECAS, registered in Ireland and on lease to Aeroflot *(Paul Jackson/Jane's)*   2001/0093666

www.janes.com   Jane's All the World's Aircraft 2001-2002

612 USA: AIRCRAFT—BOEING

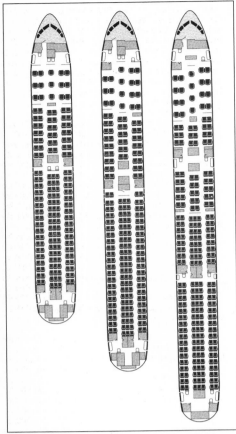

Typical seating plans for Boeing 767-200 (181 passengers), 767-300 (218) and 767-400 (245)
2001/0093604

BOEING 767 ORDER BOOK
(at 1 November 2000)

Customer	Variant	Engine	Qty
Aeromaritime	200ER	PW	2
	300ER	PW	1
Air Algerie	300ER	GE	3
Air Canada	200	PW	10
	200ER	PW	9
	300ER	PW	6
Air China	200ER	PW	3
	300	PW	4
Air France	300ER	PW	3
Air Mauritius	200ER	GE	2
Air New Zealand	200ER	GE	3
	300ER	GE	5
Airtours International	300ER	GE	3
Air Zimbabwe	200ER	PW	2
All Nippon Airways	200	GE	25
	300	GE	34
	300ER	GE	8
American Airlines	200	GE	13
	200ER	GE	17
	300ER	GE	49
Amiri Flight	300ER	GE	1
Ansett Australia	200	GE	5
Ansett Worldwide	200ER	GE	4
	300ER	PW	12
	300ER	GE	12
Asiana Airlines	300	GE	9
	300ER	GE	2
	300F	GE	1
Avianca	200ER	PW	2
Braathens	200	PW	2
Britannia Airways	200	GE	11
	300ER	GE	7
British Airways	300ER	RR	28
CAAC	200ER	PW	3
Canadian Airlines Int'l	300ER	GE	14
China Airlines	200	PW	2
China Yunnan	300ER	RR	3
Condor Flugdienst	300ER	PW	11
Continental Airlines	200ER	GE	10
	400ER	GE	24
Delta Airlines	200	GE	15
	300	GE	24
	300	PW	4
	300ER	PW	31
	300ER	GE	21
	400ER	GE, PW	21
Egyptair	200ER	PW	3
	300ER	PW	2
El Al Israel Airlines	200	PW	2
	200ER	PW	2
Ethiopian Airlines	200ER	PW	3
Eva Airways	200	GE	4
	300ER	GE	4
Flightlease	300ER	PW	2
GATX Capital Corp	200ER	GE	1

Customer	Variant	Engine	Qty
GECAS	300ER	GE	34
GPA Ltd	200ER	PW	1
	300ER	PW	13
Gulf Air	300ER	GE	18
ILFC	200	GE	1
	200ER	GE	4
	300ER	GE	34
	300ER	PW	20
Itochu Air Lease Corp	300ER	GE	3
Itochu Corp	200ER	GE	4
Japan Airlines	200	PW	3
	300	PW	13
	300	GE	9
Kenya Airways	400ER		3
Kuwait Airlines	200ER	PW	3
LAM Mozambique	200ER	PW	1
LAN Chile	300ER	GE	3
	300F	GE	3
Lauda Air	300ER	PW	7
LOT Polish Airlines	200ER	GE	2
	300ER	GE	3
LTU	300ER	PW	5
Malev Hungarian Airlines	200ER	GE	2
Martinair Holland	300ER	PW	6
Mid East Jet	200ER	GE	1
Pacific Western	200	PW	2
Piedmont	200ER	GE	6
Polaris Aircraft	300ER	GE	2
Qantas Airways	200ER	GE, PW	7
	300ER	GE, PW	22
Royal Brunei Airways	300ER	GE	2
SAS	200ER	PW	2
	300ER	PW	14
Shanghai Airlines	300	PW	4
Singapore Lease	300ER	GE	3
TACA Int'l Airlines	200	GE	1
	300ER	GE	2
Transbrasil	200	GE	3
TWA	200	PW	10
unidentified	300ER	PW	2
	300ER	RR	4
	300ER		1
	300F		2
United Airlines	200	PW	19
	300ER	PW	37
United Parcel Service	300F	GE	30
US Airways	200ER	GE	6
Uzbekistan	300ER	PW	2
Varig	200ER	GE	6
	300ER	GE	10
Total			**899[1]**

[1] Of which 805 delivered by 1 November 2000

(rear fuselage); Alenia (wing control surfaces, flaps and leading-edge slats, wingtips, elevators, fin and rudder, nose radome); Fuji (wing/body fairings and main landing gear doors); Kawasaki (forward and centre fuselage; exit hatches; wing in-spar ribs); Mitsubishi (rear fuselage body panels and rear fuselage doors).

LANDING GEAR: Hydraulically retractable tricycle type; Menasco twin-wheel nose unit retracts forward; Cleveland Pneumatic main gear, with two four-wheel bogies, retracts inward; oleo-pneumatic shock-absorbers; Honeywell wheels and brakes; mainwheel tyres of current production versions H46×18.0-20 (26/28 ply for -200/300; 32 ply for -200ER/300ER); nosewheel tyres size H37×14.0-15 (22/24 ply) for all; steel disc brakes on all mainwheels; electronically controlled anti-skid units. Nosewheel steerable ±16°; ±65° for towing.

POWER PLANT: Two high-bypass turbofans in pods, pylon-mounted on the wing leading-edges.
  General Electric options: 225 kN (50,600 lb st) CF6-80C2B2F, 251 kN (56,500 lb st) CF6-80C2B4F, 268 kN (60,200 lb st) CF6-80C2B6F and 276 kN (62,100 lb st) CF6-80C2B7F.
  Pratt & Whitney options: 233 kN (52,300 lb st) PW4052, 254 kN (57,100 lb st) PW4056, 268 kN (60,200 lb st) PW4060 and 282 kN (63,300 lb st) PW4062. P&W JT9D-7R4D of 213.5 kN (48,000 lb st) no longer offered.
  Rolls-Royce options: 251 kN (56,400 lb st) RB211-524G4-T and 265 kN (59,500 lb st) RB211-524H2-T.
  Fuel in one integral tank in each wing, and in centre tank, with total capacity of 63,216 litres (16,700 US gallons; 13,905 Imp gallons) in 200/300; 767-200ER and -300ER have additional 28,163 litres (7,440 US gallons; 6,195 Imp gallons) in second centre-section tank, raising total capacity to 91,379 litres (24,140 US gallons; 20,100 Imp gallons). Refuelling point in port outer wing.

ACCOMMODATION: Operating crew of two on flight deck; observer's seat and optional second observer's seat. Basic accommodation in -200 models for 224 passengers, made up of 18 first class passengers forward in six-abreast seating at 96.5 cm (38 in) pitch, and 206 tourist class in seven-abreast seating at 81 cm (32 in) pitch. Window or aisle seats comprise 86 per cent of total. Type A inward-opening plug doors provided at both front and rear of cabin on each side of fuselage, with options of Type A, I or III emergency exits at various mid-cabin locations on each side. Total of five lavatories installed, two centrally in main cabin, two aft in main cabin, and one forward in first class section. Galleys situated at forward and aft ends of cabin. Alternative single-class layouts provide for 255 tourist passengers seven-abreast (two-three-two) at 81 cm (32 in) pitch (one overwing exit each side) and maximum (requiring two additional overwing emergency exits) 290, mainly eight-abreast (two-four-two), at 76 cm (30 in) pitch. Three-class layout for 181 passengers: 15 first class (two-one-two) at 152 cm (60 in) pitch; 40 business class (two-two-two) at 91 cm (36 in); and 126 tourist class (two-three-two) at 81 cm (32 in).

Basic accommodation in -300 models for 269 passengers, made up of 24 first class passengers forward in six-abreast seating at 96.5 cm (38 in) pitch, 245 tourist

Boeing 767 first class sleeper
2001/0093605

Boeing 767 inclusive tour
2001/0093606

Jane's All the World's Aircraft 2001-2002 www.janes.com

Boeing 767-300 flight deck

class in seven-abreast at 78.7 cm (31 in) pitch, six lavatories and five galleys. Alternatives include 286 in two-three-two seating at 81 cm (32 in) pitch and 218 in three-class layout comprising 18 first, 46 business and 154 tourist class passengers arranged as in -200. Maximum seating capacity in -300 models is 350 passengers at 71 cm (28 in), six lavatories and four galleys; capacities from 291 upwards require standard -300 door configuration (each side) of Type A front and rear and two Type Is overwing to be replaced by two Type As, plus third Type A ahead of wing and Type I adjacent to trailing-edge.

Underfloor cargo holds (forward and rear, combined) of -200 versions can accommodate, typically, up to 22 LD2 or 11 LD1 containers; 767-300 underfloor cargo holds can accommodate 30 LD2 or 15 LD1 containers. Starboard side forward and rear cargo doors of equal size on 767-200 and 767-300, but larger forward door standard on 767-200ER and 767-300ER and optional on 767-200 and 767-300. Bulk cargo door at rear on port side. Overhead stowage for carry-on baggage is 0.08 m³ (3.0 cu ft) per passenger. Cabin air conditioned, cargo holds heated.

SYSTEMS: Honeywell dual air cycle air conditioning system. Pressure differential 0.59 bar (8.6 lb/sq in). Electrical supply from two engine-driven 90 kVA three-phase 400 Hz constant frequency AC generators, 115/200 V output. 90 kVA generator mounted on APU for ground operation or for emergency use in flight. Three hydraulic systems at 207 bars (3,000 lb/sq in), for flight control and utility functions, supplied from engine-driven pumps and a Honeywell bleed air-powered hydraulic pump or from APU. Maximum generating capacity of port and starboard systems is 163 litres (43 US gallons; 35.8 Imp gallons)/min; centre system 185.5 litres (49 US gallons; 40.8 Imp gallons)/min, at 196.5 bars (2,850 lb/sq in). Reservoirs pressurised by engine bleed air via pressure regulation module. Reservoir relief valve pressure nominally 4.48 bars (65 lb/sq in). Additional hydraulic motor-driven generator, to provide essential functions for extended-range operations, standard on 767-200ER and 767-300ER and optional on 767-200 and 767-300. Nitrogen chlorate oxygen generators in passenger cabin, plus gaseous oxygen for flight crew. APU in tailcone to provide ground and in-flight electrical power and pressurisation. Anti-icing for outboard wing leading-edges (none on tail surfaces), engine air inlets, air data sensors and windscreen.

AVIONICS: *Radar:* Honeywell RDR-4A colour weather radar in aircraft for All-Nippon, Britannia and Transbrasil.

Flight: Standard ARINC 700 series equipment, including Honeywell VOR/ILS/marker beacon receivers, ADF, DME, RMI-743 radio magnetic indicator and radio altimeter. Honeywell IRS, FMCS and DADC, as described in Boeing 757 entry; dual digital flight management systems, and triple flight control computers, including FCS-700 flight control system; certified for Cat. IIIb landings; options include Boeing's windshear protection and guidance system.

Instrumentation: Honeywell EFIS-700 electronic flight instrument system.

DIMENSIONS, EXTERNAL:
Wing span: except 400ER	47.57 m (156 ft 1 in)
400ER	51.99 m (170 ft 7 in)
Wing chord: at root	8.57 m (28 ft 1¼ in)
at tip	2.29 m (7 ft 6 in)
Wing aspect ratio: except 400ER	8.0
400ER	9.3
Length: overall: 200/200ER	48.51 m (159 ft 2 in)
300/300ER	54.94 m (180 ft 3 in)
400ER	61.37 m (201 ft 4 in)
fuselage: 200/200ER	47.24 m (155 ft 0 in)
300/300ER	53.67 m (176 ft 1 in)
Fuselage: Max width	5.03 m (16 ft 6 in)
Height overall	15.85 m (52 ft 0 in)
Tailplane span	18.62 m (61 ft 1 in)
Wheel track	9.30 m (30 ft 6 in)
Wheelbase: 200/200ER	19.69 m (64 ft 7 in)
300/300ER	22.76 m (74 ft 8 in)
Passenger doors (two, fwd and rear, port):	
Height	1.88 m (6 ft 2 in)
Width	1.07 m (3 ft 6 in)
Galley service door (two, fwd and rear, stbd):	
Height	1.88 m (6 ft 2 in)
Width	1.07 m (3 ft 6 in)
Emergency exits (two, each): Height	0.97 m (3 ft 2 in)
Width	0.51 m (1 ft 8 in)
Cargo door (rear, stbd; fwd, stbd 200/300):	
Height	1.75 m (5 ft 9 in)
Width	1.78 m (5 ft 10 in)
Cargo door (fwd, stbd, 200ER/300ER):	
Height	1.75 m (5 ft 9 in)
Width	3.40 m (11 ft 2 in)
Bulk cargo door (port,rear): Height	1.22 m (4 ft 0 in)
Width	0.97 m (3 ft 2 in)

DIMENSIONS, INTERNAL:
Cabin, excl flight deck:	
Length: 200/200ER	33.93 m (111 ft 4 in)
300/300ER	40.36 m (132 ft 5 in)
Max width	4.72 m (15 ft 6 in)
Max height	2.87 m (9 ft 5 in)
Floor area: 200/200ER	154.9 m² (1,667 sq ft)
300/300ER	184.0 m² (1,981 sq ft)
Volume: 200/200ER	428.2 m³ (15,121 cu ft)
300/300ER	483.9 m³ (17,088 cu ft)
Volume, flight deck	13.5 m³ (478 cu ft)
Baggage holds (containerised), volume:	
200/200ER: fwd	40.8 m³ (1,440 cu ft)
rear	34.0 m³ (1,200 cu ft)
300/300ER: fwd	54.4 m³ (1,920 cu ft)
rear	47.6 m³ (1,680 cu ft)
Bulk cargo hold volume:	
all versions	12.2 m³ (430 cu ft)
Combined baggage hold/bulk cargo hold volume:	
200/200ER	87.0 m³ (3,070 cu ft)
300/300ER	114.1 m³ (4,030 cu ft)
Total cargo volume:	
200/200ER	111.3 m³ (3,930 cu ft)
300/300ER	147.0 m³ (5,190 cu ft)

AREAS:
Wings, gross: except 400ER	283.3 m² (3,050.0 sq ft)
400ER	290.70 m² (3,129.0 sq ft)
Ailerons (total)	11.58 m² (124.60 sq ft)
Trailing-edge flaps (total)	36.88 m² (397.00 sq ft)
Leading-edge slats (total)	28.30 m² (304.60 sq ft)
Spoilers (total)	15.83 m² (170.40 sq ft)
Fin	30.19 m² (325.00 sq ft)
Rudder	15.95 m² (171.70 sq ft)
Tailplane	59.88 m² (644.50 sq ft)
Elevators (total)	17.81 m² (191.70 sq ft)

WEIGHTS AND LOADINGS (2GB: 767-200ER basic with CF6-80C2B2F engines, 2GM: 767-200ER max optional T-O weight and CF6-80C2B7Fs, 2PB: 767-200ER basic PW4052, 2PM: 767-200ER max PW4062, 3GB: 767-300ER basic CF6-80C2B4F, 3GM: 767-300ER max CF6-80C2B7F, 3PB: 767-300ER basic PW4056, 3PM: 767-300ER max PW4062, 3RB: 767-300ER basic RB211-524G4, 3RM: 767-300ER max RB211-524H; all -200 with 181 or 224 passengers, all -300ER with 218 or 269 passengers as indicated, where different):

Operating weight empty:	
2GB (181), 2GM (181)	84,685 kg (186,700 lb)
2GB (224), 2GM (224)	83,960 kg (185,100 lb)
2PB (181), 2PM (181)	84,730 kg (186,800 lb)
2PB (224), 2PM (224)	83,915 kg (185,000 lb)
3GB (218), 3GM (218)	90,535 kg (199,600 lb)
3GB (269), 3GM (269)	89,855 kg (198,100 lb)
3PB (218), 3PM (218)	90,585 kg (199,700 lb)
3PB (269), 3PM (269)	89,900 kg (198,200 lb)
3RB (218), 3RM (218)	91,580 kg (201,900 lb)
3RB (269), 3RM (269)	90,900 kg (200,400 lb)
Baggage capacity, underfloor:	
200ER: fwd: standard	9,798 kg (21,600 lb)
alternate	15,309 kg (33,750 lb)
rear: standard	8,165 kg (18,000 lb)
alternate	12,247 kg (27,000 lb)
300ER: fwd: standard	13,063 kg (28,800 lb)
alternate	20,412 kg (45,000 lb)
rear: standard	11,431 kg (25,200 lb)
alternate	17,574 kg (38,745 lb)
Bulk hold capacity (all versions)	2,926 kg (6,450 lb)
Max fuel weight:	
200, 300	51,130 kg (112,725 lb)
200ER, 300ER	73,635 kg (162,340 lb)
Max T-O weight: 2GB, 2PB	156,490 kg (345,000 lb)
2GM, 2PM	179,170 kg (395,000 lb)
3GB, 3PB, 3RB	172,365 kg (380,000 lb)
3GM, 3PM, 3RM	186,880 kg (412,000 lb)
Max ramp weight: 2GB, 2PB	156,945 kg (346,000 lb)
2GM, 2PM	179,625 kg (396,000 lb)
3GB, 3PB, 3RB	172,820 kg (381,000 lb)
3GM, 3PM, 3RM	187,335 kg (413,000 lb)
Max landing weight:	
2GB, 2PB	126,100 kg (278,000 lb)
2GM, 2PM	136,080 kg (300,000 lb)
3GB, 3GM, 3PB, 3PM, 3RB, 3RM	145,150 kg (320,000 lb)
Max zero-fuel weight:	
2GB, 2PB	114,755 kg (253,000 lb)
2GM, 2PM	117,935 kg (260,000 lb)
3GB, 3GM, 3PB, 3PM, 3RB, 3RM	133,810 kg (295,000 lb)
Max wing loading:	
2GB, 2PB	552.3 kg/m² (113.11 lb/sq ft)
2GM, 2PM	632.3 kg/m² (129.51 lb/sq ft)
3GB, 3PB, 3RB	608.3 kg/m² (124.59 lb/sq ft)
3GM, 3PM, 3RM	659.5 kg/m² (135.08 lb/sq ft)
Max power loading:	
2GB	348 kg/kN (3.41 lb/lb st)
2GM	324 kg/kN (3.18 lb/lb st)
2PB	336 kg/kN (3.30 lb/lb st)
2PM	318 kg/kN (3.12 lb/lb st)
3GB	343 kg/kN (3.36 lb/lb st)
3GM	338 kg/kN (3.32 lb/lb st)
3PB	339 kg/kN (3.33 lb/lb st)
3PM	332 kg/kN (3.25 lb/lb st)
3RB	344 kg/kN (3.37 lb/lb st)
3RM	353 kg/kN (3.46 lb/lb st)

PERFORMANCE:
Normal cruising speed, all versions	M0.80
Approach speed at MLW:	
2GB, 2PB	137 kt (254 km/h; 158 mph)
2GM, 2PM	139 kt (258 km/h; 160 mph)
3GB, 3GM, 3PB, 3PM	145 kt (269 km/h; 167 mph)
3RB, 3RM	148 kt (275 km/h; 171 mph)
Initial cruising altitude at max T-O weight:	
2GB	11,550 m (37,900 ft)
2GM, 2PM	10,670 m (35,000 ft)
2PB	11,310 m (37,100 ft)
3GB	10,700 m (35,100 ft)
3GM, 3PM	10,180 m (33,400 ft)
3PB	10,730 m (35,200 ft)
3RB	10,730 m (35,200 ft)
3RM	10,210 m (33,500 ft)
Service ceiling, OEI: 2GB	5,090 m (16,700 ft)
2GM	4,205 m (13,800 ft)
2PB	5,395 m (17,700 ft)
2PM	4,845 m (15,900 ft)
3GB	4,510 m (14,800 ft)
3GM	3,930 m (12,900 ft)
3PB	4,785 m (15,700 ft)
3PM	4,085 m (13,400 ft)
3RB	3,900 m (12,800 ft)
3RM	3,505 m (11,500 ft)
T-O field length at S/L, 86°F: 2GB	2,301 m (7,550 ft)
2GM, 3PB	2,485 m (8,150 ft)
2PB	2,180 m (7,150 ft)
2PM	2,439 m (8,000 ft)
3GB	2,530 m (8,300 ft)
3GM	2,713 m (8,900 ft)
3PM	2,652 m (8,700 ft)

614   USA: AIRCRAFT—BOEING

3RB	2,500 m (8,200 ft)
3RM	2,896 m (9,500 ft)
Landing field length at MLW: 2GB	1,524 m (5,000 ft)
2GM	1,555 m (5,100 ft)
2PB	1,509 m (4,950 ft)
2PM	1,540 m (5,050 ft)
3GB, 3GM, 3PB, 3PM	1,677 m (5,500 ft)
3RB, 3RM	1,707 m (5,600 ft)
3RM	2,896 m (9,500 ft)
Design range:	
2GB (181)	5,125 n miles (9,491 km; 5,897 miles)
2GB (224)	4,830 n miles (8,945 km; 5,558 miles)
2GM (181)	6,655 n miles (12,325 km; 7,658 miles)
2GM (224)	6,545 n miles (12,121 km; 7,531 miles)
2PB (181)	5,030 n miles (9,315 km; 5,788 miles)
2PB (224)	4,740 n miles (8,778 km; 5,454 miles)
2PM (181)	6,555 n miles (12,139 km; 7,543 miles)
2PM (224)	6,450 n miles (11,945 km; 7,422 miles)
3GB (218)	5,230 n miles (9,686 km; 6,018 miles)
3GB (269)	4,890 n miles (9,056 km; 5,627 miles)
3GM (218)	6,150 n miles (11,389 km; 7,077 miles)
3GM (269)	5,990 n miles (11,093 km; 6,893 miles)
3PB (218)	5,155 n miles (9,547 km; 5,932 miles)
3PB (269)	4,820 n miles (8,926 km; 5,546 miles)
3PM (218)	6,075 n miles (11,250 km; 6,991 miles)
3PM (269)	5,915 n miles (10,954 km; 6,806 miles)
3RB (218)	4,880 n miles (9,037 km; 5,615 miles)
3RB (269)	4,555 n miles (8,435 km; 5,241 miles)
3RM (218)	5,850 n miles (10,834 km; 6,467 miles)
3RM (269)	5,620 n miles (10,408 km; 6,467 miles)

UPDATED

BOEING 767-300 GENERAL MARKET FREIGHTER

TYPE: Twin-jet freighter.

PROGRAMME: First 767 specialised package freighter launched January 1993 by United Parcel Service order; mockup completed early 1994; first flight (N301UP) 20 June 1995. The 767-300F for general operation was ordered by Asiana in November 1993 and differs from UPS version in having mechanical freight handling on main and lower decks, air conditioning for animals and perishables on main and forward lower decks and more elaborate crew facilities.

CUSTOMERS: UPS ordered 30 parcel freighters 15 January 1993; first delivery 12 October 1995; last on 9 September 1999; further 30 options not yet taken up. Asiana ordered two 767-300Fs in November 1993; contract reduced to one, which was delivered 23 August 1996. One to LAN Chile on 23 September 1998; second in December 1999; third ordered same month. Total 34.

DESIGN FEATURES: Modifications include reinforced landing gear and internal wing structure; main deck floor strengthened to take 24 containers, each 2.24 × 3.18 m (88 × 125 in); no passenger windows; 2.67 × 3.40 m (8 ft 9 in × 11 ft 1¾ in) freight door forward to port; pilot-type rating and extensive component commonality with 757 Freighter.

POWER PLANT: Two General Electric CF6-80C2B6F/B7F, Pratt & Whitney PW4060/4062 or Rolls-Royce RB211-524/H turbofans; 267 kN (60,000 lb st) each.

DIMENSIONS, INTERNAL:
Cabin: Length	39.80 m (130 ft 7 in)
Main deck container capacity	339.5 m³ (11,990 cu ft)
Lower deck cargo capacity	92.9 m³ (3,282 cu ft)

WEIGHTS AND LOADINGS:
Max T-O weight: standard	185,065 kg (408,000 lb)
optional	186,880 kg (412,000 lb)
Max ramp weight	187,335 kg (413,000 lb)
Max zero-fuel weight	140,160 kg (309,000 lb)
Max landing weight	147,871 kg (326,000 lb)
Max wing loading	659.5 kg/m² (135.08 lb/sq ft)
Max power loading	350 kg/kW (3.43 lb/lb st)

PERFORMANCE:
Range: with 40,823 kg (90,000 lb) payload
 4,000 n miles (7,408 km; 4,603 miles)
with 50,800 kg (112,000 lb) payload
 3,000 n miles (5,556 km; 3,452 miles)

UPDATED

BOEING 777

TYPE: Wide-bodied airliner.

PROGRAMME: Formerly known as 767-X, initial variant now 777-200; Boeing board authorised firm offer for sale 8 December 1989; launch order by United Airlines 15 October 1990 (see Customers); Boeing launched production programme of initial market and Increased Gross Weight 777s and formed 777 Division 29 October 1990; configuration frozen March 1991; Boeing signed final agreement with Mitsubishi, Kawasaki and Fuji, making them risk-sharing programme partners for about 20 per cent of the 777 structure, on 21 May 1991; roll-out occurred 9 April 1994.

First flight (line no. WA001/N7771 with PW4084 engines), 12 June 1994, WA002 15 July, WA003 2 August, WA004 28 October, WA005 11 November; WA006/G-ZZZA (first with GE90, for British Airways) 2 February 1995, same day as FAA granted engine approval; PW-engined aircraft accumulated some 3,235 hours in 2,340 cycles in test programme, leading to joint FAA/JAA certification on 19 April 1995; FAA awarded 180-minute ETOPS approval 30 May 1995; first delivery, to United Airlines, on 15 May 1995; service entry with United on 7 June 1995 with inaugural revenue flight (by N777NA) from London to Washington, DC. Second GE90 aircraft for British Airways (WA010/G-ZZZB) joined WA006 in 1,750 hour, 1,260 cycle test programme; certification with GE engine 9 November 1995, with deliveries to BA commencing two days later.

Flight testing of Rolls-Royce Trent 800 on Boeing 747-100 testbed began late March 1995, with first flight of Trent-powered 777 (Boeing test aircraft) 26 May; certification and first delivery (to Thai International) on 3 April 1996 (formal hand-over 31 March), with ETOPS clearance following in October. The US National Aeronautic Association awarded Boeing and the 777 its 1995 Robert J. Collier Trophy for aeronautical achievement. On 2 April 1997, a Trent-powered 777 landed at Boeing Field having captured the eastbound world circumnavigation record in 41 hours 59 minutes with a single stop at Kuala Lumpur, Malaysia, its arrival at the latter also having secured the non-stop great circle distance record of 10,266.9 n miles (19,014.3 km; 11,814.9 miles).

Two new derivatives launched 29 February 2000 as 777-200LR and 777-300ER, both as extended-range versions of current production types. Anticipated market of 500 for these derivatives, 45 per cent of which with Asian operators.

CURRENT VERSIONS: **777-100X**: Under study in 2000 to meet Singapore Airlines requirements for shortened, 250-seat version.

777-200: Initial production version. Maximum T-O weight 247,210 kg (545,000 lb); up to 440 passengers at maximum density.

777-200ER: Formerly -200-IGW (Increased Gross Weight). Maximum T-O weight initially 286,895 kg (632,500 lb); increased to 293,930 kg (648,000 lb) in March 1998 and 297,555 kg (656,000 lb) in January 1999; additional 53,828 litres (14,220 US gallons; 11,840 Imp gallons) of fuel in centre-section tank; same passenger capacity as basic aircraft; configuration frozen January 1994, including strengthened wing, fuselage, empennage, landing gear and engine pylons. First flew (G-VIIA, one of two for British Airways) 7 October 1996 (GE engines); FAA/JAA ETOPS certification 5 February 1997; delivery 6 February. First Trent-powered -200ER (A6-EMI, for Emirates) flew 21 November 1996. January 1999 MTOW increase resulted from fitting 777-300 main landing gear and restricting CG travel to 4 per cent, for increased taxying weight. First flight with GE90-94B engines 12 June 2000; FAA certification 14 November 2000, followed immediately by first delivery (to Air France).

First 777 for private use is a -200ER (N777AS) first flown 3 November 1998 and delivered to Raytheon Systems at Waco, Texas, for outfitting; customer delivery to Middle East Jet due late 2000.

777-200LR: Ultra-long-range version (previously 777-200X) powered by two General Electric GE90-110B engines, each rated at 489 kN (110,000 lb/st); launched 29 February 2000. Maximum T-O weight 341,100 kg (752,000 lb); fuel capacity 195,283 litres (51,590 US gallons; 42,958 Imp gallons), with addition of tanks in rear cargo hold; range 8,860 n miles (16,408 km; 10,195 miles) with 301 passengers. Height 18.62 m (61 ft 1 in); see 777-300LR for further features. Firm launch order from EVA Airlines, 27 June 2000, for three. Project milestones trail 777-300ER by four to six months.

777-300: Initially known as 777 Stretch; revealed at Paris Air Show, 14 June 1995; launched by Boeing board 26 June 1995 when 36 commitments held (see 1999-2000 and earlier *Jane's*); configuration frozen 7 April 1997; roll-out 8 September 1997; first flight (Rolls-Royce engines) 16 October 1997 (N5014K); first with P&W engines (HL-7534 of Korean Air) 4 February 1998; FAA certification achieved 4 May 1998 with 180-minute ETOPS approval; first delivery (B-HNH, 136th 777) to Cathay Pacific 22 May 1998.

Compared with first-generation 747s, 777-300 carries same number of passengers, but at two-thirds of fuel cost and with 40 per cent less maintenance. Features strengthened airframe, inboard wing and landing gear, ground-manoeuvring cameras on horizontal tail surfaces and wing/fuselage fairing; tailskid; Type A emergency door over each wing; and fuselage stretched by 19 frames, 10.13 m (33 ft 3 in) longer (5.33 m; 17 ft 6 in ahead of wing and 4.80 m; 15 ft 9 in aft of wing) than that of 777-200 to increase passenger capacity to 550 in single-class high-density configuration.

777-300ER: Stretched equivalent of 777-200R (previously 777-300X); launched 29 February 2000. GE90-115B engines each rated at 512 kN (115,000 lb st); height 18.57 m (60 ft 11 in). Maximum T-O weight as for 777-200LR; fuel capacity 181,283 litres (47,890 US gallons; 39,877 Imp gallons); range 7,200 n miles (13,334 km; 8,285 miles) with 359 passengers. Roll-out October 2002; first flight December 2002; certification August 2003. Launch customer Japan Airlines ordered eight on 31 March 2000 for delivery from September 2003; EVA Airlines followed with four on 27 June 2000. Both -200LR and -300ER have strengthened horizontal and vertical tail surfaces, strengthened wings with new tips which increase span to 64.80 m (212 ft 7 in) and strengthened landing gear, including semi-levered main gear on -300ER. Both will be certified for 207-minute ETOPS.

777-300ER Stretch: Under study in 1999 as possible replacement for 747-400 with Asian carrier on routes to Europe; fuselage stretch of about 7.9 m (26 ft) to accommodate an additional 60 passengers, but airframe otherwise unchanged; range as for 777-300ER. Also referred to as **777-400X**.

CUSTOMERS: Total 502 ordered and 303 delivered by 1 November 2000. Latter comprises 13, 32, 59, 74 and 83 in 1995-99, plus 42 in first 10 months of 2000.

COSTS: Estimated development cost US$4 billion (1990); aircraft cost: 777-200 US$128 million to US$146 million; 777-200ER US$135 million to US$153 million; 777-300 US$151 million to US$170 million. 777-300 estimated to burn 33 per cent less fuel and have 40 per cent lower maintenance cost than 747-100/200, resulting in direct operating cost savings of around 30 to 35 per cent.

DESIGN FEATURES: Objectives of design included replacement of McDonnell Douglas DC-10 Srs 10 and Lockheed

Boeing 777 two-man flight deck has five flat panel displays in horizontal row on main panel and two FMS control and display panels and a multifunction display on the centre console

TriStar in regional market, as well as DC-10 Srs 30 and Boeing 747SP in intercontinental service; also is replacement for early 747s. All features required for 180-minute ETOPS incorporated and tested in basic aircraft design. Cylindrical fuselage wider than 767 to allow twin aisle seating for from six- to 10-abreast; toilets and overhead baggage bins designed to allow rapid change of layout.

Wing, of 31° 30′ sweepback at quarter-chord, incorporates new technology to allow minimum M0.83 cruise in combination with high thickness for economical structure and large internal volume, long span for improved take-off and payload/range and large area for high cruise altitude and low approach speed; no winglets. Outer 6.48 m (21 ft 3 in) of each wing can, as an option, be folded to vertical to reduce gate width requirement at airports.

FLYING CONTROLS: Fly-by-wire. Hydraulically actuated, with trim tab in rudder. Six-segment slats in each wing leading-edge. Single-slotted flaps mid-wing, double-slotted flaps inboard; flaperon between inboard and outboard flaps. Five-segment spoilers ahead of single-slotted flaps; two-segment spoilers ahead of double-slotted flaps.

Boeing's first airliner fly-by-wire system; fully powered control surface actuators (31 by Teijin Seiki America) electrically signalled from full FBW system; this signals slats, flaps, spoilers and control feel unit as well as inboard flaperons, outboard ailerons, elevators and rudder; system provides flight envelope protection as well as stabilisation and autopilot inputs, but the normal control columns and rudder pedals in cockpit are back-driven by the system to give the pilots direct appreciation of the activity of the automatic system.

In normal mode, flight guidance commands are generated by Rockwell Collins triple redundant digital autopilot/flight directors and the control laws and envelope protection commands are shaped by the Marconi Avionics triple digital primary flight computers; each of the three primary flight computers contains three 32-bit microprocessors (a Motorola 68040, an Intel 80486 and an AMD 29050), all three programmed in Ada to perform all FBW functions; with power supply and ARINC 629 modules, each microprocessor module constitutes a lane and the three lanes constitute a channel; each lane is compared with the others in its channel; the system not only has high fault tolerance, but allows deferred maintenance, by which failures can be carried over until the next scheduled maintenance.

Commands to the powered control units are produced by three BAE Systems and Teijin Seiki actuator control electronics units, which have a fourth analogue channel directly signalled from the sticks and pedals in the cockpit; normal operating mode is for the aircraft to be flown through autopilots, primary flight computers and actuator control electronics, which simultaneously back-drive the sticks and pedals in the cockpit; first degraded (secondary) mode is used if inertial units and standby attitude sensors all become disabled and the pilots take manual control through the digital primary flight computers; second degraded (direct) mode bypasses the main FBW system with the direct analogue link between cockpit and actuator control electronics; ultimate standby is mechanical control of tailplane incidence for the pitch axis and two wing spoiler panels for lateral control. Some powered control units produced by Parker-Bertea and Moog; tailplane trim module and hydraulic brake by Raytheon Systems.

Pitch axis control law is C* U, effectively tending to make the aircraft hold an airspeed and to respond in pitch attitude to a departure from that airspeed; trim changes due to configuration changes are suppressed; the system returns the bank angle to 35° if that angle is exceeded by the pilots and the controls then released; the system prevents exceeding the limiting airspeed and stalling; asymmetric thrust is automatically countered; the variable feel system adjusts control forces to warn of approach to flight envelope limits in manual flight; the FBW system is linked to the ARINC 629 dual triplex digital databusses (see also aircraft information management system under Avionics heading).

STRUCTURE: Composites of carbon and toughened resin used in skins of tailplane and fin torsion boxes and cabin floor beams; CFRP used for rudder, elevators, ailerons, flaps, engine nacelles and landing gear doors; hybrid composites in wingroot fairing; GFRP in fixed-wing leading-edge, tailplane and fin fore and aft panels, wing aft panels, engine pylon fairings and radome. Toughened materials have high damage resistance and allow simple low-temperature bolted repairs. Metal structure includes thick skins without need for tear straps; no bonding; single-piece fuselage frames; fuselage skin of advanced 2000-series aluminium alloy; wing, empennage and engine nacelle leading edges of 2000-series; wing top skin and stringers made in advanced 7055 aluminium alloy with greater compression strength; tailcone in standard 7000-series; 10 per cent of structure weight is composites.

Fully digital product definition with all parts created by Dassault/IBM CATIA CAD/CAM and communicated to manufacturing and publications; structure and systems integration, tube and cable run design completed before design release; 238 design/build teams have ensured that design, fabrication and test have proceeded concurrently

BOEING 777 ORDER BOOK
(at 1 November 2000)

Customer	Variant	Engine	First Order	First Delivery	Qty
Air China	777-200	PW4000	24 Mar 97	26 Oct 98	10
Air France	777-200	GE90	10 Nov 96	27 Mar 98	14
All Nippon Airways	777-200	PW4000	19 Sep 90	4 Oct 95	17
	777-200ER	PW4000	19 Dec 90	6 Oct 99	4
	777-300	PW4000	12 Sep 95	30 Jun 98	14
American Airlines	777-200ER	Trent 800	21 Nov 96	21 Jan 99	45
Asiana Airlines	777-200ER	PW4000	20 Dec 96	–	5
	777-300	PW4000	20 Dec 96	–	1
British Airways	777-200	GE90	21 Aug 91	11 Nov 95	5
	777-200ER	GE90	21 Aug 91	6 Feb 97	24
	777-200ER	Trent 800	1 Aug 98	7 Jun 00	16
Cathay Pacific Airways	777-200	Trent 800	6 May 92	9 May 96	5
	777-300	Trent 800	6 May 92	22 May 98	7
China Southern Airlines	777-200	GE90	17 Dec 92	28 Dec 95	4
	777-200ER	GE90	17 Dec 92	28 Feb 97	2
Continental Airlines	777-200ER	GE90	12 May 93	28 Sep 98	14
Delta Air Lines	777-200ER	Trent 800	13 Nov 97	23 Mar 99	13
Egyptair	777-200ER	PW4090	23 Aug 95	23 May 97	3
El Al Israel Airlines	777-200ER	Trent 800	14 Dec 99	–	3
Emirates	777-200	Trent 800	4 Jun 92	5 Jun 96	3
	777-200ER	Trent 800	4 Jun 92	11 Apr 97	6
EVA Airways	777-200LR	GE90	27 Jun 00	–	3
	777-300ER	GE90	27 Jun 00	–	4
Garuda Indonesian	777-200ER	GE90	25 Jun 96	–	6
GECAS	777-200ER	GE90	26 Sep 00	–	5
	777-300	GE90	26 Sep 00	–	10
ILFC	777-200ER	GE90	15 Dec 92	8 Jan 98	18
	777-200ER	PW4000	15 Dec 92	16 Jul 98	8
	777-200ER	Trent 800	6 Mar 96	–	6
	777-300	GE90	6 Mar 96	–	1
	777-300	Trent 800	2 Sep 97	–	1
Japan Air System	777-200	PW4000	29 Jun 93	3 Dec 86	7
Japan Airlines	777-200	PW4000	24 Jan 92	15 Feb 96	10
	777-300	PW4000	22 Dec 95	28 Jul 98	5
	777-300ER	GE90	31 Mar 00	–	8
Korean Air Lines	777-200ER	PW4000	16 Dec 93	21 Mar 97	6
	777-300	PW4000	16 Dec 93	12 Aug 99	6
Kuwait Airways	777-200	GE90	10 Jul 96	30 Mar 98	2
Lauda Air	777-200ER	GE90	13 Dec 91	24 Sep 97	4
Malaysia Airlines	777-200ER	Trent 800	8 Jan 96	23 Apr 97	15
Mid East Jet	777-200ER	GE90	1 Oct 97	24 Nov 98	1
Saudia	777-200ER	GE90	18 Jun 95	26 Dec 97	23
Saudi Oger	777-200ER	GE90	30 Nov 98	22 Oct 99	1
Singapore Aircraft Leasing	777-200ER	Trent 800	22 Dec 95	–	4
	777-300	Trent 800	22 Dec 95	12 Nov 99	2
Singapore Airlines	777-200ER	Trent 800	22 Dec 95	5 May 97	33
	777-300	Trent 800	22 Dec 95	10 Dec 98	7
Thai Airways International	777-200	Trent 800	20 Jun 91	31 Mar 96	8
	777-300	Trent 800	22 Dec 95	23 Dec 98	6
unidentified	777-200ER	*	–	–	7
	777-200ER	Trent 800	19 Oct 00	–	3
	777-200ER	–	31 Oct 00	–	2
United Airlines	777-200	PW4000	15 Oct 90	15 May 95	22
	777-200ER	PW4000	15 Oct 90	7 Mar 97	39
Varig	777-200ER	GE90	8 Sep 98	–	4
Total					**502**

* Not announced/undecided

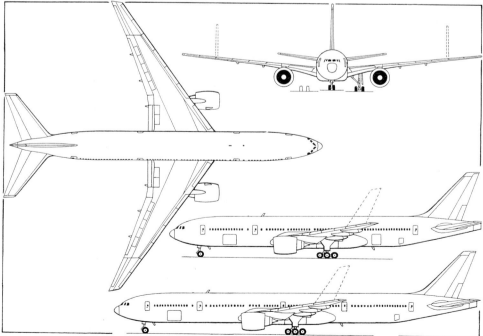

Boeing 777-200 twin-turbofan high-capacity airliner, with additional side view of -300 stretched version; optional folding wing shown by broken lines *(Mike Keep/Jane's)*

USA: AIRCRAFT—BOEING

Prototype Boeing 777-300

for structure and systems. Whole aircraft defined in computer system; no mockup built.

Centre and rear fuselage barrel sections, tailcone, doors, wingroot fairing and landing gear doors made in Japan. Wing and tail leading-edges and moving wing parts, landing gear, floor beams, nose landing gear doors, wingtips, dorsal fin and nose radome made by Northrop Grumman, Kaman, Alenia (Italy), Embraer (Brazil), Short Brothers (UK), Singapore Aerospace Manufacturing, HDH and ASTA (Australia), Korean Air and other subcontractors. Boeing manufactures flight deck and forward cabin, basic wing and tail structures and engine nacelles; assembles and tests completed aircraft.

LANDING GEAR: Retractable tricycle type (Menasco/Messier-Bugatti joint design for main gear); two main legs carrying six-wheel bogies with steering rear axles automatically engaged by nose gear steering angle; six-wheel bogies avoid need for third leg in fuselage and simplify braking system; twin-wheel steerable nose gear; mainwheel tyres H49×19.0-22 or 50×20.0R22 (32 ply); nosewheel tyres 42×17.0R18 (26 ply). Honeywell Carbenix 4000 mainwheel brakes arranged so that initial toe-pedal pressure used during taxying applies brakes to alternate sets of three wheels to save brake wear; full toe-pedal pressure applies all six brakes together.

POWER PLANT: Two turbofans.

777-200: 343 kN (77,000 lb st) General Electric GE90-77B, 331 kN (74,500 lb st) Pratt & Whitney PW4074 or 343 kN (77,200 lb st) PW4077, or 327 kN (73,400 lb st) Rolls-Royce Trent 875 or 338 kN (76,000 lb st) Trent 877.

777-200ER: 377 kN (84,700 lb st) GE90-85B, 400 kN (90,000 lb st) GE90-90B, 417 kN (93,700 lb st) GE90-94B, 376 kN (84,600 lb st) PW4084, 401 kN (90,200 lb st) PW4090, 436 kN (98,000 lb st) PW4098, 372 kN (83,600 lb st) Trent 884, 400 kN (90,000 Trent 892) or 415 kN (93,400 kN) Trent 895.

777-300: PW4090, PW4098 or Trent 892, as above.

All fuel of baseline version contained in integral tanks in wing torsion box, with reserve tank, surge tank and fuel vent and jettison pipes all inboard of wing fold; combined capacity of main, centre and reserve tanks in 777-200 is 117,348 litres (31,000 US gallons; 25,813 Imp gallons); 777-200ER and 777-300 fuel capacity increased by centre-section tank of 53,829 litres (14,220 US gallons; 11,841 Imp gallons) to maximum of 171,176 litres (45,220 US gallons; 37,653 Imp gallons).

ACCOMMODATION: Two-pilot crew; cabin cross-section, which is between that of 747 and 767, chosen to allow widest selection of twin-aisle class and seating layouts ranging from six- to 10-abreast; galleys and lavatories can be located at a selection of fixed points in front and rear cabins or freely positioned within large footprints in which they can be moved in 2.5 cm (1 in) increments and attached to prepositioned mounting, plumbing and electric fittings; overhead bins open downward and provide each passenger with 0.08 m³ (3 cu ft) volume; bins can be removed without disturbing ceiling panels, ducts or support structure; advanced cabin management system simplifies cabin management and includes digital sound system of hi-fi quality.

Typical configurations for 200/200ER include 261 passengers in three-class layout: 12 first class (two-two-two-abreast), 79 business class (two-three-two) and 170 economy class (three-three-three); 367 in two classes: 14 business (two-three-two) and 353 economy (three-three-three); 375 in two classes: 30 first class (two-two-two at 97 cm; 38 in pitch) and 345 economy (two-five-two at 79 cm; 31 in or 81 cm; 32 in pitch); 400 in two classes: 30 first class, as immediately previously and 370 economy (three-four-three, pitches as previously); and 440 passengers in single class (three-four-three).

For 777-300, options include 368 in three classes: 30 first (two-two-two at 152 cm; 60 in), 84 business (two-three-two at 97 cm; 38 in) and 254 economy (two-five-two at 79 cm; 31 in or 81 cm; 32 in); 386 in first (30), business (77) and economy (279) classes, as previously, except last-mentioned at three-four-three, with only four seats at minimum pitch; 451 passengers in two classes: 40 first, as above, and 411 economy (two-five-two, at usual pitches); 479, in first (44) and economy (435) classes, the former with 97 cm; 38 in pitch and latter three-four-three at usual pitches; and ultimate 550 single-class passengers, mostly two-four-two. Underfloor cargo compartments have mechanical handling system and can accommodate all LD formats and 88 in or 96 in width pallets; up to 32 LD-3 containers plus 17.0 m³ (600 cu ft) bulk cargo can be loaded in 777-200, or 44 LD-3s and some bulk cargo in 777-300.

Flight crew rest module on flight deck, port, contains two bunks. Optional underfloor crew rest module with same footprint as 96 in pallet contains six bunks, and stowage space and requires only electrical connection and hatch in passenger cabin floor, level with wing trailing-edge.

SYSTEMS: Honeywell air drive unit, using bleed air from engines, APU or ground supply, drives central hydraulic system; cabin air supply and pressure control by Honeywell; Hamilton Sundstrand variable-speed, constant frequency AC electrical power generating system, with two 120 kVA integrated drive generators, one APU-driven generator and Honeywell ram air turbine system. Honeywell GTCP331-500 APU; Hamilton Sundstrand air conditioning; Smiths Industries ultrasonic fuel quantity gauging system and electrical load management system; optional wingtip folding by Raytheon Montek Division and Frisby Airborne Hydraulics.

AVIONICS: *Radar:* Honeywell weather radar standard.

Flight: Main navigation system is Honeywell air data and inertial reference (ADIRS) containing the Hexad skewed axis arrangement of six ring laser gyros; standby system is the secondary attitude and air data reference unit (SAARU) containing interferometric fibre optic gyros (using light transmitted in two directions along fibre optic paths), which produces a secondary flight director attitude display, airspeed and altimeter; both are linked to the ARINC 629 digital databus (777 being first aircraft thus equipped); Honeywell TCAS; Honeywell/BAE Systems Canada global navigation satellite sensor with 12-channel receiver; Honeywell/Racal multichannel satcom system optional.

Dual Honeywell aircraft information management system (AIMS) contains the processing equipment required to collect, format and distribute onboard avionic information, including the flight management system (FMS), engine thrust control, digital communications management, operation of flight deck displays and monitoring of aircraft condition; both pilots and ground engineers can assess the condition of all onboard avionics systems.

Instrumentation: Based on five-screen EFIS using Honeywell 203 mm (18 in) ARINC D-size colour liquid crystal flat panel displays (two primary flight displays, two navigation displays and EICAS display); three multipurpose control and colour display units on centre console provide interface with integrated aircraft information management system, which handles flight management, thrust control and communications control as well as all systems information.

EQUIPMENT: Boeing 777-300 has Ground Maneuver Camera System with TV cameras in leading-edges of both horizontal stabilisers and underside of fuselage.

DIMENSIONS, EXTERNAL:
Wing span	60.93 m (199 ft 11 in)
Wing span with tips folded	47.32 m (155 ft 3 in)
Wing aspect ratio	8.7
Length overall: 200	63.73 m (209 ft 1 in)
300	73.86 m (242 ft 4 in)
Fuselage: Length	62.74 m (205 ft 10 in)
Max diameter	6.20 m (20 ft 4 in)
Max height: overall: 200	18.51 m (60 ft 9 in)
300	18.49 m (60 ft 8 in)
to tip of folded wing	14.22 m (46 ft 8 in)
Tailplane span	21.52 m (70 ft 7½ in)

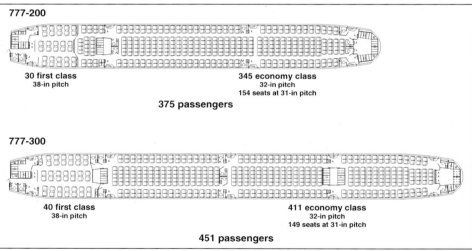

Boeing 777 medium density seating

Boeing 777-200ER of British Airways *(Paul Jackson/Jane's)*

Jane's All the World's Aircraft 2001-2002

Boeing 777 first class accommodation / Option for Boeing 777 economy class

Wheel track	10.97 m (36 ft 0 in)	
Wheelbase	25.88 m (84 ft 11 in)	
Passenger doors (four port, four stbd, each):		
Height	1.88 m (6 ft 2 in)	
Width	1.07 m (3 ft 6 in)	
Max height to sill	5.51 m (18 ft 1 in)	
Forward cargo door, stbd: Height	1.70 m (5 ft 7 in)	
Width	2.69 m (8 ft 10 in)	
Max height to sill	3.05 m (10 ft 0 in)	
Rear cargo door, stbd, standard:		
Height	1.70 m (5 ft 7 in)	
Width	1.78 m (5 ft 10 in)	
Max height to sill	3.40 m (11 ft 2 in)	
Rear cargo door, stbd, optional width	2.69 m (8 ft 10 in)	
Bulk cargo door, stbd: Height	0.91 m (3 ft 0 in)	
Width	1.14 m (3 ft 9 in)	
Max height to sill	3.48 m (11 ft 5 in)	

DIMENSIONS, INTERNAL:
Cabin: Length 49.10 m (161 ft 1 in)
  Max width 5.87 m (19 ft 3 in)
  Floor area: 200, 200ER 279.1 m² (3,004 sq ft)
Max underfloor cargo hold volume:
  200, 200ER: forward 80.5 m³ (2,844 cu ft)
    aft 62.6 m³ (2,212 cu ft)
    bulk 17.0 m³ (600 cu ft)
    total 160.2 m³ (5,656 cu ft)
  300: forward 107.4 m³ (3,792 cu ft)
    aft 89.5 m³ (3,160 cu ft)
    bulk 17.0 m³ (600 cu ft)
    total 213.8 m³ (7,552 cu ft)

AREAS:
  Wings, projected 427.8 m² (4,605.0 sq ft)
  Ailerons (total) 7.11 m² (76.50 sq ft)
  Trailing-edge flaps (total) 67.13 m² (722.60 sq ft)
  Slats (total) 36.84 m² (396.50 sq ft)
  Inboard spoilers (total) 8.67 m² (93.30 sq ft)
  Outboard spoilers (total) 14.34 m² (154.40 sq ft)
  Flaperons 6.69 m² (72.00 sq ft)
  Horizontal tail surfaces, projected 101.26 m² (1,090.0 sq ft)
  Vertical tail surfaces, projected 53.23 m² (573.00 sq ft)
  Elevators, incl tabs (total) 25.48 m² (274.30 sq ft)
  Rudder, incl tab 18.16 m² (195.50 sq ft)

WEIGHTS AND LOADINGS (777-200 with 320 passengers, 24/61/235, 2GB: GE90-77B engines at basic MTOW, 2GM: GE90-77B maximum, 2PB: PW4074 basic, 2PM: PW4077 maximum, 2RB: Trent 875 basic, 2RM: Trent 877 maximum; 777-200ER, as above, 2ERGB: GE90-85B basic, 2ERGM: GE90-94 maximum, 2ERPB: PW4084 basic, 2ERPM: PW4090 maximum, 2ERRB: Trent 884 basic, 2ERRM: Trent 895 maximum; 777-300 with 386 passengers, 30/77/279, 3PB: PW4090 basic, 3PM: PW4098 maximum, 3RB: Trent 892 basic, 3RM: Trent 892 maximum):

Operating weight empty:
  2GB 140,615 kg (310,000 lb)
  2GM 141,115 kg (311,100 lb)
  2PB 139,210 kg (306,900 lb)
  2PM 139,345 kg (307,200 lb)
  2RB 137,350 kg (302,800 lb)
  2RM 137,485 kg (303,100 lb)
  2ERGB 143,925 kg (317,300 lb)
  2ERGM 144,105 kg (317,700 lb)
  2ERPB 142,155 kg (313,400 lb)
  2ERPM 142,925 kg (315,100 lb)
  2ERRB 140,295 kg (309,300 lb)
  2ERRM 140,480 kg (309,700 lb)
  3PB 158,170 kg (348,700 lb)
  3PM 158,620 kg (349,700 lb)
  3RB, 3RM 155,675 kg (343,200 lb)
Max fuel weight: 200 94,210 kg (207,700 lb)
  200ER/300 135,845 kg (299,490 lb)
Max T-O weight:
  2GB, 2PB, 2RB 229,575 kg (506,000 lb)
  2GM, 2PM, 2RM 247,205 kg (545,000 lb)
  2ERGB, 2ERPB, 2ERRB, 3PB, 3RB 263,080 kg (580,000 lb)
  2ERGM, 2ERPM, 2ERRM 297,555 kg (656,000 lb)
  3PM, 3RM 299,370 kg (660,000 lb)
Max ramp weight allowance:
  200, 200ERB, 300 907 kg (2,000 lb)
  200ERM 454 kg (1,000 lb)
Max landing weight:
  200 201,845 kg (445,000 lb)
  200ER 208,650 kg (460,000 lb)
  300 237,680 kg (524,000 lb)
Max zero-fuel weight:
  200 190,510 kg (420,000 lb)
  200ER 195,045 kg (430,000 lb)
  300 224,525 kg (495,000 lb)
Max wing loading:
  2GB, 3PB, 2RB 536.5 kg/m² (109.88 lb/sq ft)
  2GM, 3PM, 2RM 577.8 kg/m² (118.35 lb/sq ft)
  2ERGB, 2ERPB, 2ERRB, 3PB, 3RB 614.9 kg/m² (125.95 lb/sq ft)
  2ERGM, 2ERPM, 2ERRM 695.5 kg/m² (142.45 lb/sq ft)
  3PM, 3RM 699.8 kg/m² (143.32 lb/sq ft)
Max power loading:
  2GB 335 kg/kN (3.29 lb/lb st)
  2GM 361 kg/kN (3.54 lb/lb st)
  2PB 346 kg/kN (3.40 lb/lb st)
  2PM 360 kg/kN (3.53 lb/lb st)
  2RB 351 kg/kN (3.45 lb/lb st)
  2RM 366 kg/kN (3.59 lb/lb st)
  2ERGB 349 kg/kN (3.42 lb/lb st)
  2ERGM 357 kg/kN (3.50 lb/lb st)
  2ERPB 350 kg/kN (3.43 lb/lb st)
  2ERPM 371 kg/kN (3.64 lb/lb st)
  2ERRB 354 kg/kN (3.47 lb/lb st)
  2ERRM 358 kg/kN (3.51 lb/lb st)
  3PB, 3RB 328 kg/kN (3.22 lb/lb st)
  3PM 343 kg/kN (3.37 lb/lb st)
  3RM 374 kg/kN (3.67 lb/lb st)

PERFORMANCE:
Cruising speed: all M0.84
Approach speed: 200 136 kt (252 km/h; 157 mph)
  200ER 138 kt (256 km/h; 159 mph)
  300 149 kt (276 km/h; 171 mph)
Initial cruising altitude (ISA + 10°C):
  2GB 11,980 m (39,300 ft)
  2GM, 2PB 11,550 m (37,900 ft)
  2PM, 2ERGB 11,155 m (36,600 ft)
  2RB 11,645 m (38,200 ft)
  2RM 11,370 m (37,300 ft)
  2ERGM 10,605 m (34,800 ft)
  2ERPB 10,820 m (35,500 ft)
  2ERPM 10,270 m (33,700 ft)
  2ERRB 11,005 m (36,100 ft)
  2ERRM 10,455 m (34,300 ft)
  3PB 10,975 m (36,000 ft)
  3PM 10,485 m (34,400 ft)
  3RB 11,245 m (36,900 ft)
  3RM 10,395 m (34,100 ft)
Service ceiling, OEI (ISA +10°C):
  2GB 5,515 m (18,100 ft)
  2GM 4,755 m (15,600 ft)
  2PB 4,940 m (16,200 ft)
  2PM 4,905 m (16,100 ft)
  2RB 4,815 m (15,800 ft)
  2RM 5,365 m (17,600 ft)
  2ERGB 3,995 m (13,100 ft)
  2ERGM 3,720 m (12,200 ft)
  2ERPB 4,235 m (13,900 ft)
  2ERPM 3,660 m (12,000 ft)
  2ERRB 4,785 m (15,700 ft)
  3RRM 3,749 m (12,300 ft)
  3PB 4,480 m (14,700 ft)
  3PM 3,535 m (11,600 ft)
  3RB 4,725 m (15,500 ft)
  3RM 3,415 m (11,200 ft)
T-O field length (30° C):
  3GB 2,073 m (6,800 ft)
  3GM 2,530 m (8,300 ft)
  2PB, 2RB 2,164 m (7,100 ft)
  2PM, 2RM 2,576 m (8,450 ft)
  2ERGB 2,515 m (8,250 ft)
  2ERGM 3,018 m (9,900 ft)
  2ERPB 2,591 m (8,500 ft)
  2ERPM 3,582 m (11,750 ft)
  2ERRB 2,561 m (8,400 ft)
  2ERRM 3,368 m (11,050 ft)
  3PB 2,759 m (9,050 ft)
  3PM 3,414 m (11,200 ft)
  3RB 2,652 m (8,700 ft)
  3RM 3,704 m (12,150 ft)
Landing field length:
  2GB, 2GM 1,570 m (5,150 ft)
  2PB, 2PM, 2RB, 2RM 1,555 m (5,100 ft)
  2ERGB, 2ERGM 1,616 m (5,300 ft)
  2ERPB, 2ERPM, 2ERRB, 2ERRM 1,601 m (5,250 ft)
  300 1,844 m (6,050 ft)
Design range: 2GB 3,770 n miles (6,982 km; 4,338 miles)
  2GM 4,920 n miles (9,111 km; 5,661 miles)
  2PB 3,860 n miles (7,148 km; 4,442 miles)
  2PM 4,990 n miles (9,241 km; 5,742 miles)
  2RB 3,995 n miles (7,398 km; 4,597 miles)
  2RM 5,065 n miles (9,380 km; 5,828 miles)
  2ERGB 5,645 n miles (10,454 km; 6,496 miles)
  2ERGM 7,625 n miles (14,121 km; 8,774 miles)
  2ERPB 5,695 n miles (10,547 km; 6,553 miles)
  2ERPM 7,505 n miles (13,899 km; 8,636 miles)
  2ERRB 5,765 n miles (10,676 km; 6,634 miles)
  2ERRM 7,595 n miles (14,065 km; 8,740 miles)
  3PB 3,750 n miles (6,945 km; 4,315 miles)
  3PM 5,515 n miles (10,213 km; 6,346 miles)
  3RB 3,915 n miles (7,250 km; 4,505 miles)
  3RM 5,820 n miles (10,778 km; 6,697 miles)

UPDATED

777-200

30 first class — 38-in pitch
370 economy class — 32-in pitch; 90 seats at 31-in pitch
400 passengers

777-300

44 first class — 38-in pitch
435 economy class — 32-in pitch; 98 seats at 31-in pitch
479 passengers

Typical Boeing 777 high-density seating options

BOEING 20XX

TYPE: Jet airliner.
PROGRAMME: Revealed in January 2001 to be a broad-ranging study for potential 757/767 replacement in 180- to 300-seat class, with cruising speed as high as M0.95 and range options from 1,000 n miles (1,852 km; 1,150 miles) to 6,000 n miles (11,112 km; 6,904 miles).

NEW ENTRY

DOUGLAS PRODUCTS DIVISION (Division of Boeing Commercial Airplane Group)

The Douglas Aircraft Company became Douglas Products Division when Boeing absorbed McDonnell Douglas in August 1997. After reviewing the partial overlap of its enlarged product line, Boeing announced in November 1997 that it would close the MD-80 and MD-90 production lines by end 1999, by which time all existing orders would have been completed. However, in mid-1999, MD-90 line closure was postponed to end 2000. Final MD-80 series, an MD-83, was delivered on 21 December 1999. MD-11 termination revised from early 2000 to 2001; last aircraft delivered 22 February 2001. Former MD-95 now marketed as the Boeing 717.

UPDATED

BOEING MD-11

TYPE: Wide-bodied airliner.
PROGRAMME: Follow-on to DC-10; revealed at Paris Air Show 1985; British Caledonian ordered nine 3 December 1986; official programme launch 30 December 1986; five aircraft in flight test programme (four with GE engines, one with P&W); first flight (N111MD) 10 January 1990 powered by CF6s; first flight of third prototype powered by P&W PW4460s, 26 April 1990; certified 8 November 1990; first delivery to Finnair 29 November 1990, entering service 20 December. 100th MD-11 delivered 30 June 1993. Certification with R-R Trent 650 discontinued. Fuselage production line moved from San Diego to Long Beach during early 1996. Production terminated at 200th aircraft (D-ALCN) which delivered to Lufthansa Cargo on 22 February 2001.

CURRENT VERSIONS: **MD-11:** Standard passenger version for 298 passengers in three-class layout; maximum range 7,000 n miles (12,964 km; 8,055 miles) with maximum optional T-O weight.

Detailed description applies to improved MD-11, MD-11F, MD-11 Combi and MD-11 convertible Freighter.

MD-11 Performance Improvement Programme (PIP): Continuous improvement programme aimed at weight and drag reduction and extended range under way since 1990, resulting in recovery and extension of MD-11's design range. First delivery November 1990 with initial choice of gross weights between 273,289 kg (602,500 lb) and 276,691 kg (610,000 lb); range shortfall 440 n miles (815 km; 506 miles) with GE engines and 710 n miles (1,315 km; 817 miles) with P&W engines.

Successive drag reduction, weight-saving and fuel consumption and engine installation improvements (see 1997-98 and earlier *Jane's* for details), have included the following: airframe weight reduced progressively by 1,020 kg (2,250 lb); maximum T-O weight increased to 280,320 kg (618,000 lb) in January 1991 and further optional increase to 283,720 kg (625,500 lb) in July 1993; cumulative range improvement of four-stage drag reduction programmes, extending range to more than 6,911 n miles (12,800 km; 7,953 miles) with 29,000 kg (63,934 lb) payload; from April 1992, one or two 7,472 litre (1,974 US gallon; 1,644 Imp gallon) auxiliary fuel tanks which can be mounted in rear of forward underfloor cargo compartment, displacing two or four LD3 cargo containers (tanks can be removed through normal cargo door); T-O distance reduced by up to 137 m (450 ft) to accommodate increased T-O weight by deflecting inboard and outboard ailerons with flaps at take-off; internal improvements in the CF6-80C2 engine, saving some 1.5 per cent fuel consumption, equivalent to about 1,360 kg (3,000 lb) of payload; P&W three-phase engine and intake improvement sequence, introduced in June 1992 and November 1993, which together gave 2.7 per cent improvement; and P&W certified optional thrust increase to 276 kN (62,000 lb st).

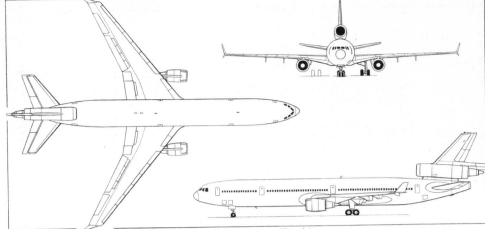

Boeing MD-11 medium/long-range transport *(Dennis Punnett/Jane's)*

Total range increase for these improvements is 600 n miles (1,111 km; 690 miles) for GE-powered aircraft and 690 n miles (1,278 km; 794 miles) for P&W-powered aircraft.

MD-11 Combi: Mixed cargo/passenger version for four to 10 cargo pallets and 168 to 240 passengers; ranges from 5,180 n miles (9,593 km; 5,961 miles) to 6,860 n miles (12,705 km; 7,894 miles). Main deck cargo door at rear on port side. Certified April 1992 to latest FAA Class C smoke and fire containment requirements.

MD-11CF: Convertible freighter; launched August 1991 with four aircraft order from Martinair-Holland. Main deck cargo door at front on port side.

MD-11F: All-freight version; no cabin windows, except observation windows. Freighter conversions are offered by various companies, see *Jane's Aircraft Upgrades* for details.

MD-11ER: Extended-range version launched February 1994; maximum T-O weight increased to 285,989 kg (630,500 lb) and fuel capacity increased by up to 11,583 litres (3,060 US gallons; 2,548 Imp gallons) in removable auxiliary tank in lower cargo compartment; offers either 480 n miles (889 km; 552 miles) greater range or 2,721 kg (6,000 lb) more payload; offered in passenger, Combi, convertible or all-freight versions; drag reduction and 680 kg (1,500 lb) weight saving will reduce fuel burn by 1.5 per cent; intended to provide lower costs on very long routes with lower passenger traffic; reportedly, MD-11ER costs 26 per cent less to operate than Boeing 747-400; MD-11ER extends MD-11 range with 298 passengers from 7,000 n miles (12,964 km; 8,055 miles) to 7,240 n miles (13,408 km; 8,331 miles), which is said to be very slightly greater than Boeing 747-400 with 421 passengers. Launch customers World Airways (two, on lease, with PW4462 engines; delivered on 11 March 1996), and Garuda (three, with CF6-80C2s, converted from MD-11 order); first delivery 19 December 1996.

CUSTOMERS: Deliveries totalled three in 1990, 31 in 1991, 42 in 1992, 36 in 1993, 17 in 1994, 18 in 1995, 15 in 1996, 12 in 1997, 12 in 1998, eight in 1999, four in 2000 and two in 2001. See table.

BOEING MD-11 ORDER BOOK

Customer	Variant	First order date	Qty
Alitalia	MD-11	Apr 1987	3
	MD-11C	Apr 1987	5
American Airlines	MD-11	Aug 1989	19
China Airlines	MD-11	Sep 1988	4
China Eastern	MD-11	Dec 1988	5
	MD-11F	Dec 1988	1
City Bird	MD-11	Oct 1996	3
Delta Air Lines	MD-11	Sep 1988	15
Eva Air	MD-11	Dec 1989	2
	MD-11F	Dec 1989	9
Federal Express	MD-11F	Jun 1987	22
Finnair	MD-11	Mar 1987	4
Garuda Indonesia	MD-11	Dec 1988	9
GATX Capital Corporation	MD-11	Oct 1990	1
Guinness Peat	MD-11	Dec 1986	2
ILFC	MD-11	Dec 1988	6
Japan Airlines	MD-11	Apr 1990	10
KLM – Royal Dutch Airlines	MD-11	Mar 1990	10
Korean Air Lines	MD-11	Dec 1986	5
LTU	MD-11	Dec 1988	4
Lufthansa	MD-11F	Feb 1997	14
Martinair Holland	MD-11F	Sep 1991	6
Mitsui & Co	MD-11	Dec 1986	5
Saudi Government	MD-11	Jun 1993	2
Saudia	MD-11F	Oct 1995	4
Swissair	MD-11	Mar 1987	16
Thai Airways Int'l	MD-11	Jun 1988	4
Varig	MD-11	Dec 1988	4
VASP	MD-11	Apr 1995	4
World Airways	MD-11	Jan 1996	2
Total			**200**

DESIGN FEATURES: Compared with DC-10, MD-11 has winglets above and below each wingtip; tailplane has advanced cambered aerofoil, modified trailing-edge camber, reduced sweepback and 7,571 litre (2,000 US gallon; 1,665 Imp gallon) fuel trim tank; extended tailcone of low-drag chisel profile; two-crew all-digital flight deck; restyled interior; choice of GE CF6-80C2D1F and P&W PW4460 engines. Wing has Douglas aerofoil section; sweepback at quarter-chord 35°; dihedral 6°; incidence at root 5° 51′; tailplane sweepback 33°.

FLYING CONTROLS: Conventional and powered. Ailerons have Parker Hannifin actuators; electrohydraulically actuated variable incidence tailplane with slotted elevators in two sections each side powered by Parker Hannifin and Teijin Seiki actuators; inboard all-speed ailerons and outboard low-speed ailerons droop with flaps on take-off; dual-section rudder split into vertical segments; near full-span leading-edge slats; double-slotted trailing-edge flaps with offset external hinges; five spoilers in groups of four and one on each wing. Cat. IIIb automatic landing with ground roll control (certified April 1991) standard.

MD-11 two-crew flight deck with six large displays and two flight management system panels on console. The third FMS panel at rear is for ground crew use when testing aircraft avionics

Boeing MD-11CF leased to Malaysia Airlines *(Paul Jackson/Jane's)* ***2001***/0103515

STRUCTURE: Composites used in virtually all control surfaces, engine inlets and cowlings, and wing/fuselage fillets; wing has two-spar structural box with chordwise ribs and skins with spanwise stiffeners; upper winglet of ribs, spars and stiffened aluminium alloy skin with carbon fibre trailing-edge; lower winglet carbon fibre; inboard ailerons have metal structure with composites skin; outboard ailerons all-composites; inboard flaps composites-skinned metal; outboard flaps all-composites; spoilers aluminium honeycomb and composites skin; tailplane has CFRP trailing-edge; elevators CFRP.

Rear engine inlet duct and fan cowl doors, and nose cowl outer barrels on wing-mounted engines, are of composites construction. Inner surfaces of engine nacelles are acoustically treated.

Suppliers include Alenia (fin, rudder, fuselage panels, winglets), AP Precision Hydraulics (centreline and nose landing gear), Bendix (mainwheels and carbon brakes), CASA (horizontal tail surfaces), Embraer (outboard flap sections), Fischer GmbH (composites flap hinge fairings), Pneumo Abex Corporation (main landing gear), Rohr Industries (engine pylons), Honeywell (advanced flight deck and avionics), and GKN Westland Aerospace (flap vane and inlet duct extension rings).

LANDING GEAR: Hydraulically retractable tricycle type, with additional twin-wheel main unit mounted on the fuselage centreline; heaviest proposed variants might have four-wheel centreline bogie; nosewheel and centreline units retract forward, main units inward into fuselage. Twin-wheel steerable nose unit (±70°). Main gear has four-wheel bogies. Oleo-pneumatic shock-absorbers in all units. Loral nosewheels and Goodyear tyres size 40×15.5-16 (28 ply), pressure 13.44 bars (195 lb/sq in). Main and centreline units have Bendix wheels and Goodyear tyres size H54×21.0-24 (36 ply), pressure 13.79 bars (200 lb/sq in). Bendix carbon brakes with air convection cooling; Loral anti-skid system. Minimum ground turning radius about nosewheels 26.67 m (87 ft 6 in); about wingtip 35.90 m (117 ft 9½ in). Optional Taxi Brake Select system increases brake life by up to 40 per cent by cycling alternate brake pads during low-speed taxying.

POWER PLANT: Three Pratt & Whitney PW4460 turbofans, each rated at 267 kN (60,000 lb st), or three PW4462 turbofans, each rated at 276 kN (62,000 lb st), or three General Electric CF6-80C2D1F turbofans, each rated at 274 kN (61,500 lb st); two engines mounted on underwing pylons, the third above the rear fuselage aft of the fin torsion box. Refuelling point in leading-edge of each wing. Standard MD-11 fuel capacity 146,174 litres (38,615 US gallons; 32,154 Imp gallons); one or two 7,472 litre (1,974 US gallon; 1,643 Imp gallon) tanks can be added in cargo hold.

ACCOMMODATION: Crew of two, plus two observer seats. Standard class seating for 250, two-class for 356 and all-economy for up to 410; Combi carries 214 passengers plus six pallets. Crew door and three passenger doors each side, all eight of which open sliding inward and upward. Two rear doors are deactivated in Combi configuration. Two freight holds in lower deck, forward and aft of wing, and one bulk cargo compartment in rear fuselage. Forward freight hold is heated and ventilated; rear freight hold heated only. MD-11 Combi has a lower deck cargo door in centre compartment on starboard side of fuselage for loading of pallets, an upward-opening main deck cargo door on port side at rear of cabin. MD-11F/CF have port side forward main deck cargo door.

SYSTEMS: Air conditioning system includes three Honeywell air-bearing air cycle units with two automatic digital pressure controllers and electromanual back-up. Cabin maximum pressure differential 0.59 bar (8.6 lb/sq in). Three independent hydraulic systems for operation of flight controls and braking, with motor/pump interconnects to allow one system to power another. Electrical system comprises three 400 Hz, 100/120 kVA integrated drive generators, one per engine; one 90 kVA generator in APU; 50 Ah battery; four transformer-rectifiers to convert AC power to DC; and 25 kVA drop-out air-driven emergency generator. Pneumatic system, maximum controlled pressure 3.17 bars (46 lb/sq in) at 230°C, supplies air conditioning, engine bleed air anti-icing for wing (outer slats) and tailplane leading-edges, galley vent jet pump, and cargo compartment floor heating. EROS plumbed gaseous oxygen system for crew; chemical oxygen generators with automatically deploying masks for passengers. Portable oxygen cylinders for attendants and first aid. De-icing for windscreens, angle of attack sensors, TAT probe and static port plate. Honeywell TSCP700-4E APU.

AVIONICS: *Flight:* Avionics integrator, Honeywell, responsible for flight guidance/flight deck system consisting of 44 line-replaceable units. These include aircraft system controllers (ASC) that perform flight engineer control and monitoring functions, providing automated hydraulic, electrical, environmental and fuel systems; laser inertial reference system (IRS) for navigation; digital air data computer (DADC). Flight control computer includes auto-throttle and longitudinal stability augmentation; windshear detection and guidance.

Instrumentation: Six-tube EFIS and systems displays; 'dark cockpit' philosophy with lights only showing to indicate abnormal states; no need to look on overhead panels to check systems status; hydraulic, electrical, environmental and fuel systems segregated and each system configured in normal and abnormal conditions by a pair of computers.

DIMENSIONS, EXTERNAL:

Wing span	51.64 m (169 ft 5 in)
Wing chord: at root	10.71 m (35 ft 1¾ in)
at tip	2.73 m (8 ft 11½ in)
Wing aspect ratio	7.9
Length overall: with PW4460	61.16 m (200 ft 8 in)
with CF6-80	61.62 m (202 ft 2 in)
Fuselage: Length	58.65 m (192 ft 5 in)
Max diameter	6.02 m (19 ft 9 in)
Height overall	17.58 m (57 ft 8 in)
Tailplane span	18.03 m (59 ft 2 in)
Wheel track	10.56 m (34 ft 8 in)
Wheelbase	24.61 m (80 ft 9 in)
Crew doors (two, each): Height	1.93 m (6 ft 4 in)
Width	0.81 m (2 ft 8 in)
Passenger doors:	
Height: front pair	1.93 m (6 ft 4 in)
rear six	1.93 m (6 ft 4 in)
Width: front pair	0.81 m (2 ft 8 in)
rear six	1.07 m (3 ft 6 in)
* Lower deck forward freight door:	
Height	1.68 m (5 ft 6 in)
Width	2.64 m (8 ft 8 in)
Lower deck centre freight door (standard):	
Height	1.68 m (5 ft 6 in)
Width	1.78 m (5 ft 10 in)
Lower deck bulk cargo door: Height	0.91 m (3 ft 0 in)
Width	0.76 m (2 ft 6 in)
Combi main deck cargo door (port, rear):	
Height	2.59 m (8 ft 6 in)
Width	4.06 m (13 ft 4 in)
CF and F main deck cargo door (port, forward):	
Height	2.59 m (8 ft 6 in)
Width	3.56 m (11 ft 8 in)

** Centre freight door of Combi also this size, available as an option on other models*

DIMENSIONS, INTERNAL:

Cabin:	
Length, flight deck door to rear bulkhead	46.51 m (152 ft 7¼ in)
Max width	5.71 m (18 ft 9 in)
Max height	2.41 m (7 ft 11 in)
Floor area, incl galleys and toilets	244.7 m² (2,634 sq ft)
Volume, incl galleys and toilets	599.3 m³ (21,165 cu ft)
Lower deck freight holds, volume	194.0 m³ (6,850 cu ft)
MD-11 freight volume total	633.7 m³ (22,380 cu ft)

AREAS:

Wings, gross	338.91 m² (3,648.0 sq ft)
Winglets (total)	7.42 m² (80.00 sq ft)
Vertical tail surfaces (total)	56.21 m² (605.00 sq ft)
Horizontal tail surfaces (total)	85.47 m² (920.00 sq ft)

WEIGHTS AND LOADINGS:

Operating weight empty: -11*	130,165 kg (286,965 lb)
-11F	113,920 kg (251,150 lb)
-11 Combi	131,035 kg (288,885 lb)
-11CF passenger	131,525 kg (289,965 lb)
-11CF freight	115,380 kg (254,372 lb)
Weight-limited payload:	
-11	51,272 kg (113,035 lb)
-11F	90,787 kg (200,151 lb)
-11 Combi	64,009 kg (141,115 lb)
-11CF passenger	73,180 kg (161,335 lb)
-11CF freight	89,552 kg (197,428 lb)
Max T-O weight: standard**	273,314 kg (602,555 lb)
optional	285,990 kg (630,500 lb)
Max landing weight: -11	195,040 kg (430,000 lb)
-11F, -11CF, standard	213,870 kg (471,500 lb)
-11F, -11CF, optional	218,178 kg (481,000 lb)
-11 Combi	207,745 kg (458,000 lb)
Max zero-fuel weight: -11	181,435 kg (400,000 lb)
-11F/-11CF	204,700 kg (451,300 lb)
-11 Combi	195,040 kg (430,000 lb)
Max wing loading:	
standard	806.5 kg/m² (165.17 lb/sq ft)
optional	843.8 kg/m² (172.83 lb/sq ft)

** Empty weights with P&W engines about 317 kg (700 lb) lower than with GE engines*
*** All versions*

PERFORMANCE:

Max operating Mach No. (MMO): all	0.945
Max level speed at 9,450 m (31,000 ft):	
all	M0.87 (511 kt; 945 km/h; 588 mph)
FAA T-O field length, PW4462 engines, MTOW, S/L, 30°C:	
all	3,115 m (10,220 ft)
FAA landing field length, MLW, S/L:	
-11	2,118 m (6,950 ft)
-11F, -11CF	2,323 m (7,620 ft)
-11 Combi	2,234 m (7,330 ft)
Design range, FAA international reserves:	
-11, 298 passengers, three-class	6,821 n miles (12,633 km; 7,850 miles)
-11F, with weight-limited payload:	3,867 n miles (7,161 km; 4,450 miles)
-11 Combi, 183 passengers, three-class, plus six main deck freight pallets	6,717 n miles (12,440 km; 7,730 miles)
-11CF, 298 passengers	6,621 n miles (12,263 km; 7,620 miles)
-11CF, weight-limited freight	3,945 n miles (7,306 km; 4,540 miles)
Design range, as above, but with 11,356 litre (3,000 US gallon; 2,498 Imp gallon) auxiliary fuel tank:	
-11	7,213 n miles (13,358 km; 8,300 miles)
-11 Combi, 183 passengers, space-limited payload	5,735 n miles (10,622 km; 6,600 miles)

UPDATED

BOEING MD-90

The 114th and last MD-90, a Series 30 (HZ-AP4), was handed over to Saudi Arabian Airlines on 23 October 2000 and delivered to Jeddah between 23 and 25 October. An illustrated description of the type last appeared in the 2000-2001 *Jane's*.

UPDATED

BOEING 717-200

TYPE: Twin-jet airliner.

PROGRAMME: Announced at Paris Air Show 1991 as MD-95; potential airline customers briefed and manufacturing partners announced in Berlin, November 1994; modification of former Eastern Airlines DC-9-30 into development prototype began late 1994; design acquired by Boeing in August 1997 and renamed Boeing 717 in January 1998. Three flight test aircraft (T1, T2 and T3); first nose section (built by MDC) delivered to Huntington Beach on 11 December 1996 for T1, which began final assembly in May 1997 and was rolled out on 10 June 1998. BR 715 engine certified 1 September 1998.

First flight of prototype (N717XA; T1) 2 September 1998; second prototype (N717XB; T2) first flew 26 October 1998; third (N717XC; T3) on 16 December 1998, at which time first two had flown 361 hours in 193 sorties. First production aircraft (N717XD; P1) rolled out 23 January 1999; JAA/FAA certification awarded 1 September 1999 at end of five aircraft, 2,000 hour, 1,900 sortie programme; first aircraft (N942AT; third production) delivered to AirTran on 23 September 1999, after demonstration flights; entered service on Atlanta-Washington, DC, route on 12 October 1999. Updated FMS certified 20 October 2000, adding GPS, fuel prediction, vertical guidance and automatic calculation of V-speeds. CIS certification awarded 20 February 2001.

CURRENT VERSIONS: **Boeing 717-200**: Initial production version, *to which following description applies*. Available in basic (**BGW**) and high (**HGW**) gross weight versions.

Boeing 717-100: Proposed 86-seat version, formerly MD-95-20; studies will be four frames (1.9 m; 6 ft 3 in) shorter. Renamed **-100X**; wind tunnel tests began in early 2000; revised mid-2000 to eight-frame (3.86 m; 12 ft 8 in) shrink. Launch decision was deferred in December 2000 and again thereafter to an undisclosed date.

Boeing 717-100X Lite: Proposed 75-seat version, powered by Rolls-Royce Deutschland BR 710 turbofans; later abandoned.

Boeing 717-300: Proposed 130-seat version, formerly MD-95-50; studies suggest will be five frames (2.38 m; 7 ft 10 in) longer. Currently in abeyance.

CUSTOMERS: Manufacturer foresees a market for around 1,500 aircraft in this class over the next 20 years at a rate of 10 a month on two parallel production lines. Launched 19 October 1995 with order for 50, plus 50 options, from

Boeing 717 in service with first Australian customer, Impulse Airlines *(Paul Jackson/Jane's)*

ValuJet Airlines (later renamed AirTran). Second customer was Bavaria International Leasing Company (five), followed by TWA (50, plus 50 options), Pembroke Capital (10, later increased to 25) and Hawaiian (13, plus seven options). Boeing delivered 12 in 1999 and 32 in 2000. By 1 March 2001, orders stood at 154, of which 49 had been delivered.

COSTS: AirTran order for 50 aircraft worth US$1 billion. January 2001 estimate US$35 million to US$39.5 million per aircraft.

DESIGN FEATURES: All major elements of airframe based on DC-9/MD-80 series; systems and avionics are blend of low cost and advanced technology. Conventional low-wing, rear-engine, T-tail configuration.

Wing from DC-9-34, but with 1° 34′ additional incidence; sweep 24° 30′ at quarter-chord; thickness/chord ratio 11.6.

FLYING CONTROLS: Conventional and partly assisted. Elevator and ailerons are manually actuated via cables; rudder powered hydraulically with manual reversion; fly-by-wire trimming of rudder, two-section spoilers and elevator; three-section, double-slotted flaps; full-span, two-position, five-section leading-edge slats.

STRUCTURE: All-metal, two-spar wing with riveted spanwise stringers; glass fibre trailing-edges on wings, ailerons, flaps, elevators and rudder. Variations from MD-80/MD-90 include thicker skins on tail surfaces; MD-87 fuselage lengthened forward of wing by three frames 1.45 m (4 ft 9 in) and fin tip by 250 mm (10 in); wing/fuselage fillet extended forwards by three frames, using composites structure. Composites also for fuselage tailcone, fintip, elevator and aileron tabs, radome and wing trailing-edge panels; otherwise of 2024T3 aluminium alloy.

Partners are: Alenia (fuselage sections), Korean Air Lines Aerospace Division (nose structure and main passenger door/entry area), Hyundai Space and Aircraft Co (wings, in conjunction with Boeing Canada, which built initial sets of wings for flight test aircraft and early production units), Rolls-Royce Deutschland (power plant), Goodrich (engine nacelles), ShinMaywa Industries Ltd (horizontal tail surfaces and engine pylons), Fischer Advanced Composite Components GmbH (cabin furnishings), Andalucia Aerospacial (slats, landing gear components, aft pressure bulkhead), Israel Aircraft Industries SHL Servo Systems (landing gear), Honeywell (environmental control system, wheels and brakes, (flight guidance and avionics systems), Parker-Hannifin Corp (hydraulic and control systems), AIDC (empennage), Labinal (electric assemblies), Hamilton Sundstrand (electrical power generating system), and (in partnership with Hamilton Sundstrand) Auxiliary Power International Corporation (APU). Final assembly undertaken in Long Beach.

LANDING GEAR: Hydraulically retractable tricycle with steerable nosewheels; twin wheels on all legs. All-steel brakes; anti-skid units. Main tyres H41×15.0-19 (24 ply); nose tyres 26×6.6 (12 ply).

POWER PLANT: Two Rolls-Royce Deutschland BR 715 A1-30 turbofans, each 82.3 kN (18,500 lb st) at T-O at 30°C ambient, for 717-200 BGW; BR 715 C1-30 of 93.4 kN (21,000 lb st) for 717-200 HGW; optional 89.9 kN (20,000 lb st) available; switching between three thrust ratings does not involve engine hardware changes. Integrated drive generators system. BFG single-pivot door-type reversers for ground use only. Engine pylon based on MD-80, but thinner and without powered flap, although with extra frame for additional strength. Standard (BGW) fuel capacity 13,904 litres (3,673 US gallons; 3,058 Imp gallons) in three tanks in wingroots and fuselage centre section; HGW 16,667 litres (4,403 US gallons; 3,666 Imp gallons) with two underfloor (one in each cargo hold) auxiliary fuel tanks. Fuel recirculation system prevents wing upper surface ice accumulation and aids cooling.

ACCOMMODATION: Crew of two on advanced flight deck optimised for reduced parts count and high reliability. Cabin cross-section as for MD-80. Typical two-class seating for 106 passengers: eight first-class in four-abreast seating with 0.91 m (36 in) pitch and 98 standard class with 0.81 m (32 in) pitch in five-abreast arrangement in new modern cabin, alternatively 117 single-class in 0.79/0.81 m (31/32 in) pitch. Seating designed by Avio Interiors and complies with 16 g impact regulations. Cabin designed with inputs from 500 airline executives, flight attendants and passengers; interior, manufactured by Fischer Advanced Composite Components of Austria, features wider and deeper overhead baggage bins, full-grip handrail throughout length of cabin, and optional video monitors that drop down from passenger service units on both sides of cabin at every third seat row. Two lavatory units at rear of cabin, plus optional unit at front for first-class. Galley units positioned at front of cabin. Integral cabin door, with optional airstairs, behind flight deck on port side; emergency exit on opposite side, plus two above each wing. Underfloor baggage and cargo hold; latter reduced in HGW version by auxiliary fuel tanks. Conveyor system and movable bulkheads facilitate cargo loading.

SYSTEMS: Honeywell dual air cycle air conditioning and pressurisation system with digital cabin air controllers, utilising engine bleed air, maximum differential 0.54 bar (7.77 lb/sq in). Three-wheel air cycle machine and modified water separator in rear fuselage. Two separate 207 bar (3,000 lb/sq in) hydraulic systems for operation of spoilers, flaps, slats, rudder, landing gear, nosewheel steering, brakes, thrust reversers and ventral stairway. Maximum flow rate 30.3 litres (8.0 US gallons; 6.7 Imp gallons)/min. Airless bootstrap-type reservoirs, output

BOEING 717-200 ORDERS AND DELIVERIES
(at 1 March 2001)

Customer	First order date	Qty	First delivery date	Qty
Aebal	16 May 2000	3	22 Jun 2000	3
AirTran	19 Oct 1995	50	23 Sep 1999	18
Bavaria IL	4 May 1998	5	29 Dec 1999	5
Hawaiian Airlines	1 Mar 2000	13		1
Impulse Airlines	29 Dec 2000	3	29 Dec 2000	3
Pembroke Capital	31 Dec 1998	27	17 Aug 2000	4
TWA	31 Dec 1998	50	18 Feb 2000	15
Turkmenistan Airlines	25 Jul 2000	3		
Totals		**154**		**49**

Note: Quantities are cumulative

Boeing 717 flight deck

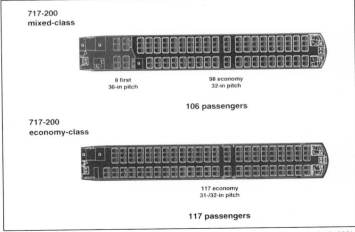

Seating options of Boeing 717

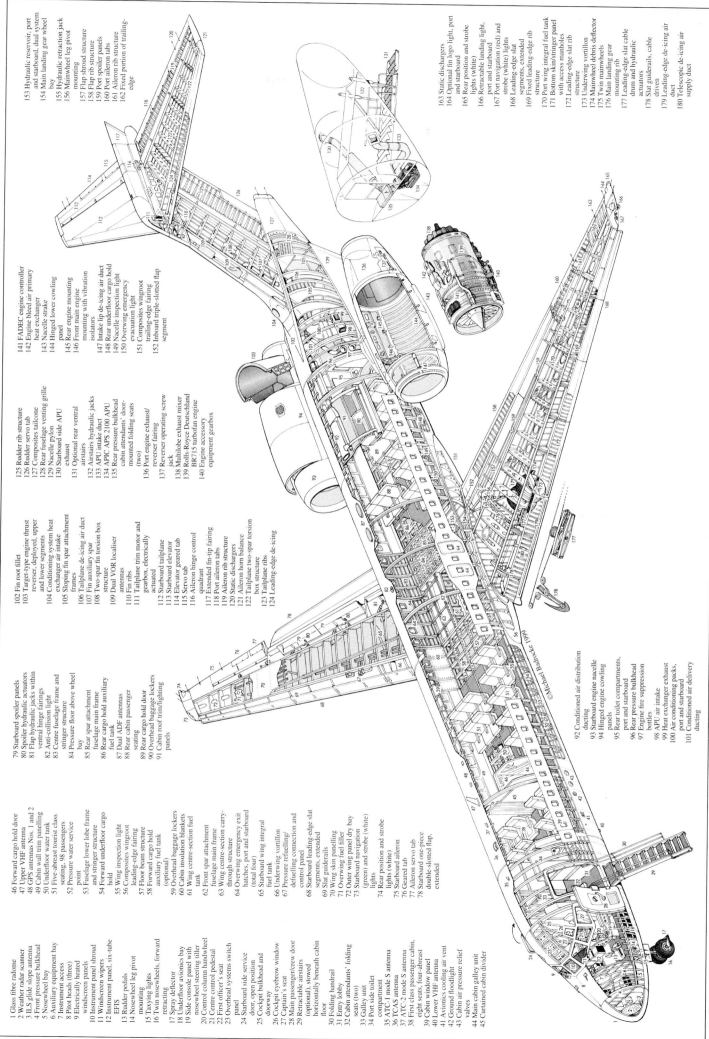

Boeing 717-200, cutaway drawing

USA: AIRCRAFT—BOEING

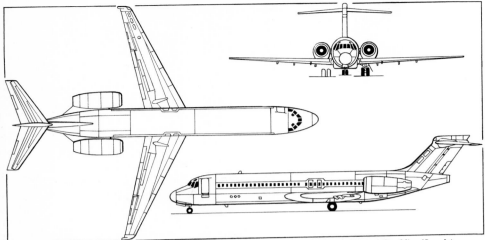

Boeing 717-200 airliner (two Rolls-Royce Deutschland BR 715 turbofans) *(James Goulding/Jane's)*

pressure 2.07 bars (30 lb/sq in). Pneumatic system, for air conditioning/pressurisation, engine starting and ice protection, utilises engine bleed air and/or APU. Electrical system includes two 35/40 kVA integrated drive generators, plus 60 kVA APU generator. Oxygen system of diluter demand type for crew on flight deck; continuous flow chemical canister type with automatic mask presentation for passengers. Anti-icing of wing, engine inlets and tailplane by engine bleed air. Electric windscreen de-icing. Thermal anti-icing of leading-edges. APIC APS 2100 APU.

AVIONICS: Honeywell Versatile Integrated Avionics VIA 2000 computer as core avionics management system; full Cat. IIIa capability will be upgraded later to IIIb with addition of radalt, ILS receiver and inertial reference unit, SFE ARINC 700 avionics.

Flight: Honeywell flight management system (FMS), inertial reference system (IRS), digital flight guidance system (DFGS), digital air data computer and windshear detection system. To be certified for Cat. IIIb automatic landings. Flight computer system upgraded and certified 20 October 2000; changes include improved GPS.

Instrumentation: Six-tube EFIS with 203 × 203 mm (8 × 8 in) LCD screens providing navigation, flight management and systems data. Flight deck features include MD-11 AFCS with glareshield-mounted controls enabling crew to fly the aircraft automatically with only push-button and thumbwheel inputs. Simplified integrated flightcrew warning and alerting panel (IFWAP) overhead control panel with four LCDs replacing 13 gauges, meters and switch panels. Central fault display system for reduced maintenance time.

DIMENSIONS, EXTERNAL:
Wing span	28.45 m (93 ft 3 in)
Wing aspect ratio	8.7
Length: overall	37.80 m (124 ft 0 in)
fuselage	34.34 m (112 ft 8 in)
Fuselage max width	3.34 m (10 ft 11½ in)
Height overall	8.86 m (29 ft 1 in)
Tailplane span	11.18 m (36 ft 8 in)
Wheel track	4.88 m (16 ft 0 in)
Wheelbase	17.60 m (57 ft 8¾ in)
Baggage door: forward:	
Height	1.27 m (4 ft 2 in)
Width	1.34 m (4 ft 4¾ in)
rear:	
Height	1.27 m (4 ft 2 in)
Width	0.91 m (3 ft 0 in)

DIMENSIONS, INTERNAL (BGW: standard, HGW: with extended-range tanks):
Cabin: Max width	3.12 m (10 ft 3 in)
Height at aisle	2.03 m (6 ft 8 in)
Underfloor baggage/freight hold: BGW	
front: Length	11.00 m (36 ft 1 in)
Volume	18.3 m³ (646 cu ft)
rear: Length	5.21 m (17 ft 1 in)
Volume	8.2 m³ (289 cu ft)
HGW: front: Volume	20.7 m³ (730 cu ft)
rear: Volume	5.7 m³ (203 cu ft)

AREAS:
Wings, gross	92.97 m² (1,000.7 sq ft)

WEIGHTS AND LOADINGS (BGW: standard, HGW: with extended-range tanks):
Operating weight empty: BGW	30,447 kg (67,124 lb)
HGW	30,901 kg (68,124 lb)
Space-limited payload: BGW	12,220 kg (26,940 lb)
HGW	11,059 kg (24,380 lb)
Max fuel weight: BGW	10,913 kg (24,059 lb)
HGW	13,381 kg (29,500 lb)
Max T-O weight: BGW	49,895 kg (110,000 lb)
HGW	54,885 kg (121,000 lb)
Max ramp weight: BGW	50,349 kg (111,000 lb)
HGW	55,340 kg (122,000 lb)
Max landing weight: BGW	45,359 kg (100,000 lb)
HGW	49,895 kg (110,000 lb)
Max zero-fuel weight: BGW	43,545 kg (96,000 lb)
HGW	45,586 kg (100,500 lb)
Max wing loading: BGW	536.7 kg/m² (109.92 lb/sq ft)
HGW	590.4 kg/m² (120.92 lb/sq ft)
Max power loading: BGW	303 kg/kN (2.97 lb/lb st)
HGW	294 kg/kN (2.88 lb/lb st)

PERFORMANCE:
Max level speed: BGW, HGW	438 kt (811 km/h; 504 mph) (M0.76)
Approach speed	134 kt (248 km/h; 154 mph)
Initial cruise altitude, MTOW, ISA +10°C:	
BGW	10,211 m (33,500 ft)
HGW	9,815 m (32,200 ft)
Service ceiling	11,278 m (37,000 ft)
T-O field length at MTOW, S/L, ISA +10°C:	
BGW	1,905 m (6,250 ft)
HGW	1,789 m (5,870 ft)
Landing field length at MLW: BGW	1,402 m (4,600 ft)
HGW	1,484 m (4,870 ft)

Design range, domestic reserves, 106 passengers and baggage:
BGW	1,430 n miles (2,648 km; 1,645 miles)
HGW	2,060 n miles (3,815 km; 2,370 miles)

OPERATIONAL NOISE LEVELS: (ICAO Annex 16 Ch 3):
T-O with cutback	−7.0 dB
Approach	−6.4 dB
Sideline	−7.1 dB

UPDATED

BOEING MD-17
Redesignated BC-17X (which see).

UPDATED

BOEING BC-17X GLOBEMASTER

TYPE: Four-jet freighter.

PROGRAMME: Model displayed at Farnborough Air Show, September 1996. Civil version of C-17A, intially designated MD-17, lacking specific military equipment, such as refuelling receptacle, OBIGGS, ECM and paratroop doors, but otherwise identical. Standard version able to carry a 78,607 kg (173,300 lb) payload over 2,500 n miles (4,630 km; 2,877 miles); optional fuel tank in wing centre-section contains an additional 37,853 litres (10,000 US gallons; 8,327 Imp gallons). Cargo capacity eighteen 2.24 × 2.74 m (88 × 108 in) pallets, including four on ramp. Formal launch awaits initial customer(s), although as few as between five and 10 commitments would be sufficient to initiate the programme; Boeing looking at fractional ownership scheme to encourage customers; civil certification would rely largely on military proving trials already accomplished; first delivery two years after launch. In third quarter of 1997, Boeing received Organizational Designated Airworthiness Representative delegation from FAA as part of progress towards type certification and announced it was building two 'white tail' aircraft which would be taken over by USAF if unsold, but these were not produced. Cost quoted as US$170 million at 2001 prices. Decision of launch was due in July 2001; if the project proceeds, the first aircraft will be completed in early 2004 and ready for delivery in late 2004; maximum production rate of four per year envisaged; Boeing and USAF estimate market for 10 aircraft over period of 10 years; guaranteed DoD business being offered with aircraft: from 50 per cent of utilisation in first year, gradually reducing to 13.7 per cent by 2015.

NEW ENTRY

BOEING BLENDED WING/BODY STUDIES
Refer to International section entry for Ultra-High Capacity Airliner/Very Large Commercial Transport.

OTHER AIRCRAFT
See International section for description of **T-45 Goshawk** programme undertaken jointly with BAE Systems. Details of **V-22 Osprey** will be found under Bell Boeing entry in this section; **RAH-66 Comanche** under Boeing Sikorsky.

UPDATED

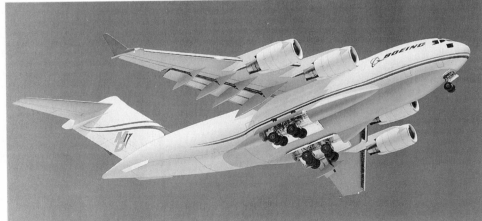

Artist's impression of the proposed MD-17, now renamed BC-17X

SPACE AND COMMUNICATIONS GROUP
2201 Seal Beach Boulevard, Seal Beach, California 90740-1515
Tel: (+1 562) 797 56 30
Fax: (+1 562) 797 52 81
PRESIDENT: James F Albaugh
PRESIDENT, INFORMATION AND COMMUNICATIONS SYSTEMS: James W Evatt

Space and Communications Group is responsible for Boeing's prime and subcontract roles on various Department of Defense and NASA space programmes. Noteworthy among these are the International Space Station, Shuttle Orbiter and Delta II, III and IV rockets.

Corporate reorganisation in 1998 resulted in former independent Information and Communications Systems being absorbed into expanded Space and Communications Group, adding responsibility for a variety of product areas, including integration of information and surveillance systems such as on E-3 Sentry and 767 AWACS and airborne laser on forthcoming AL-1A. Also responsible for production and support of communications and information management systems; commercial communications systems like the Teledesic programme; advanced integrated defence systems; satellite systems involving GPS; electronic products such as commercial avionics and cabin management systems, development of phased-array antennas and military electronics. In addition, it conducts special projects and investigations into advanced system concepts.

UPDATED

SHARED SERVICES GROUP

2810 160th Avenue SE, Bellevue, Washington 98008
Tel: (+1 425) 865 51 66
Fax: (+1 425) 865 29 58
PRESIDENT: James F Palmer
PUBLIC RELATIONS DIRECTOR: David R Suffia

Shared Services Group provides information management services and computing resources to Boeing operating divisions and government customers on a worldwide basis.

UPDATED

BOEING BUSINESS JETS

BOEING BUSINESS JETS

PO Box 3707, Seattle, Washington 98124-2207
Tel: (+1 206) 655 98 00
Fax: (+1 206) 655 97 00
e-mail: business.jets@boeing.com
PRESIDENT: Borge Boeskov
VICE-PRESIDENT, SALES: Lee Monson
COMMUNICATIONS MANAGER: William J Cogswell
DIRECTOR, PUBLIC RELATIONS AND ADVERTISING: Fred L Kelley
PARTICIPATING COMPANIES:
  Boeing Company: see this entry
  General Electric Company: see *Jane's Aero Engines*

In July 1996 The Boeing Company and The General Electric Company announced formation of a joint venture, Boeing Business Jets, to develop and market corporate versions of the Next-Generation 737, deliveries of which began in 1999. By 1999, corporate adaptations of other Boeing airliners were receiving preliminary consideration.

Company headquarters is at the corporate air services facility of Boeing Field's Flight Center.

UPDATED

BOEING BUSINESS JET (BBJ)

TYPE: Large business jet.
PROGRAMME: Launched July 1996. Aircraft are assembled at Boeing Commercial Airplane Group's Renton facility and supplied to Boeing Business Jets, which hands them over in 'green' condition to customers and delivers them to PATS Inc at Georgetown, Delaware, for installation of long-range fuel tanks before the aircraft are delivered to the customer's chosen completion centre for interior outfitting and painting; designated completion centres are KC Aviation and Associated Air Center in Dallas, Texas; The Jet Center in Van Nuys, California; Raytheon in Waco, Texas; Lufthansa Technik in Hamburg, Germany; and Jet Aviation in Basle, Switzerland, but other completion centres may be used at the discretion of the customer. After completion, Boeing ferries the aircraft to the customer's base and carries out crew training.

First BBJ (101st N-G 737, N737BZ) rolled out 26 July 1998; first flight 4 September 1998; FAA and JAA certification achieved 29 October 1998; supplementary type certificate for long-range tanks awarded 20 May 1999, following demonstration non-stop flight of 6,252.5 n miles (11,580 km; 7,200.4 miles) in 13 hours 57 minutes 42 seconds.

CUSTOMERS: Total of 71 firm orders by October 2000, at which time 46 'green' airframes had been delivered and 17 completed aircraft were in customer service where they had accumulated some 4,500 flight hours. Launch customer General Electric ordered two in July 1996, of which first (N366G) flew 23 October 1993 and delivered 23 November 1998 for outfitting. First delivery of a completed aircraft to Dubai Air Wing Royal Flight 4 September 1999. Production rate 24 per year.

Other announced customers include Atlas Air, which took delivery of one aircraft in October 1999; Aramco of Saudi Arabia, which ordered five (including two passenger/freighters) for delivery from September 2000; Atlas Air (one, delivered after completion in October 1999); PrivatAir of Switzerland (three, with one option, in September 1997 which were delivered in June 1999, November 1999 and January 2000); Malaysia Airlines System (one, delivered November 1999, for use by company chairman Tajudin Ramli and Malaysian government VIPs), and Executive Jet Inc, which has ordered nine BBJs, plus six options, for its NetJets fractional ownership scheme, with first delivery of 'green' aircraft in mid-1999. The BBJ has been proposed for USAF's Commander-in-Chief (CINC) support aircraft requirement.

COSTS: US$35.875 million 'green', estimated US$42 million to 47 million typically equipped (1999). Direct operating cost estimated at US$1,350 per hour based on operation within the USA and utilisation of 900 hours per year.

DESIGN FEATURES: Combines fuselage of 737-700, strengthened in aft section, with centre-section, wing and landing gear of 737-800. Aviation Partners Inc winglets standard, affording 5 to 7 per cent reduction in cruise drag, resulting in four to five per cent increase in range; winglets evaluated in mid-1998 by 737-800 (N737BX), first flown on BBJ prototype (N737BZ) on 20 February 1999 and received FAA approval on 6 September 2000.

POWER PLANT: Two CFM International CFM56-7 turbofans, each rated at 117.4 kN (26,400 lb st). Standard N-G 737 fuel of 26,025 litres (6,875 US gallons; 5,725 Imp gallons) contained in wing, plus between three and nine belly tanks; maximum combined capacity 40,484 litres (10,695 US gallons; 8,905 Imp gallons).

ACCOMMODATION: To customer's choice; operating weights based on allowance of 5,624 kg (12,400 lb). Typical configuration includes forward lounge and private suite with double bed; mid-section conference room; 12 first class sleeper seats at 152 cm (60 in) pitch in two rows with centre aisle, and galley, lavatory and service area at rear, with crew rest area, galley and lavatory aft of flight deck. Alternative arrangements provide for exercise room/gymnasium, office, 24 first-class sleeper seats or high-density seating for up to 63 passengers, three abreast in two rows.

AVIONICS: Rockwell Collins Series 90 as core system.
  Comms: Triple VHF comm with 8.33 kHz channel spacing; dual HF comm; L-3 Communications 120-minute CVR and Coltech Selcal.
  Flight: Dual Rockwell Collins multimode GPS/ILS/VOR/DME receivers; dual ADF; TCAS II; predictive windshear; dual Smiths Industries flight management computers; dual Honeywell ADIRU; Honeywell EGPWS; L-3 Communications FDR; Teledyne airborne navigation data recorder, digital flight data acquisition unit and quick-access recorder. Optional avionics include two additional navigation computers and electronic standby horizon.
  Instrumentation: Honeywell flat-panel LCD displays.
  Mission: Optional Teledyne Telelink.

DIMENSIONS, EXTERNAL AND AREAS: As for 737-700 except:
  Wing span, incl winglets　　35.79 m (117 ft 5 in)
DIMENSIONS, INTERNAL:
  Cabin: Length　　24.13 m (79 ft 2 in)
    Height　　2.16 m (7 ft 1 in)
    Width　　3.53 m (11 ft 7 in)
    Floor area　　75.0 m² (807 sq ft)
    Volume　　148.7 m³ (5,250 cu ft)
WEIGHTS AND LOADINGS:
  Operating weight empty, typically equipped
    　　42,896 kg (94,570 lb)

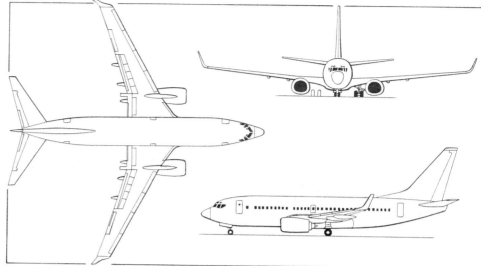

Boeing Business Jet corporate transport *(James Goulding/Jane's)*　　2001/0100433

Typical BBJ 2 floor plan　　2000/0084730

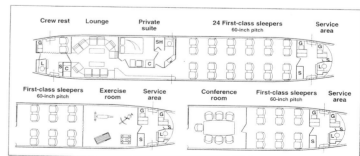

Three typical interior configurations for Boeing Business Jet. Forward half is identical for all
G: galley, L: lavatory, S: storage, SH: shower

Typical cabin interior of BBJ
2001/0100432

USA: AIRCRAFT—BOEING BUSINESS JETS to BOEING SIKORSKY

Artist's impression of Boeing Business Jet 2   2000/0084736

Interior completion allowance	5,625 kg (12,400 lb)
Max fuel weight (incl supplementary tanks)	32,745 kg (72,190 lb)
Max T-O weight	77,565 kg (171,000 lb)
Max ramp weight	77,790 kg (171,500 lb)
Max landing weight	60,780 kg (134,000 lb)
Max zero-fuel weight	57,155 kg (126,000 lb)
Max wing loading	620.5 kg/m² (127.09 lb/sq ft)
Max power loading	330 kg/kN (3.24 lb/lb st)

PERFORMANCE:
Max operating Mach No. (M_{MO}) 0.82
Cruising speed: normal M0.80
long range M0.79
Approach speed 133 kt (246 km/h; 153 mph)
Max rate of climb 980 m (3,215 ft)/min
Initial cruising altitude 11,278 m (37,000 ft)
Max certified altitude 12,500 m (41,000 ft)
Service ceiling, OEI 7,070 m (23,200 ft)
T-O field length, S/L:
 fuel for range of 4,000 n miles (7,408 km; 4,603 miles)
  1,384 m (4,540 ft)
 fuel for range of 5,000 n miles (9,260 km; 5,754 miles)
  1,475 m (4,840 ft)
 fuel for range of 6,000 n miles (11,112 km; 6,905 miles)
  1,720 m (5,645 ft)
Landing run at typical landing weight 777 m (2,550 ft)
Range (nine belly tanks):
 with 8 passengers
  6,200 n miles (11,482 km; 7,134 miles)
 with 25 passengers
  5,980 n miles (11,075 km; 6,882 miles)
 with 50 passengers
  5,510 n miles (10,204 km; 6,341 miles)
OPERATIONAL NOISE LEVELS (FAR Pt 36 Stage 3):
T-O 86.5 EPNdB
Approach 96.2 EPNdB
Sideline 94.8 EPNdB
UPDATED

BOEING BUSINESS JET 2 (BBJ 2)

TYPE: Large business jet.
PROGRAMME: Launched 11 October 1999; roll-out and first 'green' delivery scheduled for early 2001; forecast production rate eight per year.
CUSTOMERS: Total of six firm orders received by October 2000.
COSTS: US$43 million, estimated US$49 million to US$55 million, typically equipped (1999).
DESIGN FEATURES: Based on 737-800 airframe (which see), affording 25 per cent more cabin volume and twice the cargo volume of the BBJ; Aviation Partners Inc winglets standard.
POWER PLANT: Standard 737 'Next-Generation' fuel of 26,025 litres (6,875 US gallons; 5,725 Imp gallons) contained in wing, plus between three and seven belly tanks; maximum combined capacity 39,531 litres (10,443 US gallons; 8,696 Imp gallons).
ACCOMMODATION: Up to 78 passengers, with executive lounge and private suite. Operating weights based on completion allowance of 7,031 kg (15,500 lb).
DIMENSIONS, EXTERNAL: As for BBJ, except:
Length overall 39.47 m (129 ft 6 in)
DIMENSIONS, INTERNAL:
Cabin: Length 29.97 m (98 ft 4 in)
 Floor area 93.27 m² (1,004 sq ft)
 Max cargo volume 34.7 m³ (1,224 cu ft)
WEIGHTS AND LOADINGS:
Operating weight empty, typically equipped
 45,729 kg (100,815 lb)
Max fuel weight (incl supplementary tanks)
 31,974 kg (70,490 lb)
Max T-O weight 79,015 kg (174,200 lb)
Max ramp weight 79,245 kg (174,700 lb)
Max landing weight 66,360 kg (146,300 lb)
Max zero-fuel weight 62,730 kg (138,300 lb)
PERFORMANCE:
Max cruising speed M0.82
Long-range cruising speed M0.79
Max rate of climb at S/L 948 m (3,110 ft)/min
Initial cruising altitude 11,505 m (37,750 ft)
Max certified altitude 12,500 m (41,000 ft)
Service ceiling, OEI 6,090 m (22,600 ft)
T-O field length, S/L:
 fuel for range of 4,000 n miles (7,408 km; 4,603 miles)
  1,619 m (5,310 ft)
 fuel for range of 5,000 n miles (9,260 km; 5,753 miles)
  1,872 m (6,140 ft)
 fuel for range of 5,735 n miles (10,621 km; 6,599 miles)
  2,134 m (7,000 ft)
Landing run at typical landing weight 834 m (2,735 ft)
Range (seven belly tanks):
 with 8 passengers
  5,735 n miles (10,621 km; 6,599 miles)
 with 25 passengers
  5,465 n miles (10,121 km; 6,289 miles)
 with 50 passengers
  4,935 n miles (9,139 km; 5,679 miles)
OPERATIONAL NOISE LEVELS (FAR Pt 36 Stage 3):
T-O 87.5 EPNdB
Approach 96 EPNdB
Sideline 95 EPNdB
UPDATED

BOEING BUSINESS JET 3 (BBJ 3)

At the NBAA Convention in Atlanta, Georgia, in October 1999, Boeing Business Jets president Borge Boeskov predicted a further addition to the BBJ range. Tentatively designated BBJ 3, this aircraft would not be based on the Boeing 737 airframe, but might combine the fuselage of the Boeing 757-200 with the wing of the 757-300, or could be based on the 767-200. At NBAA 2000, Boeing predicted a launch of the BBJ 3 in 2001.
UPDATED

Boeing Business Jet prototype, retrofitted with winglets   2001/0100431

BOEING SIKORSKY

BOEING COMPANY and SIKORSKY AIRCRAFT

Boeing Sikorsky Comanche Joint Program Office, 5030 Bradford Drive NW, Building 48-99, Suite 100, Mailstop JE-01, Huntsville, Alabama 35805
Tel: (+1 256) 217 00 00
Fax: (+1 256) 217 00 50
Web: http://www.rah66comanche.com
RAH-66 PROGRAMME DIRECTOR: Charles Allen
VICE-PRESIDENT, RAH-66 PROGRAMME: Robert L Moore
ARMY PROGRAMME MANAGER: Colonel Robert Birmingham
PUBLIC AFFAIRS OFFICERS:
 John R Satterfield (Boeing)
 William S Tuttle (Sikorsky)

Boeing and Sikorsky began collaboration on what later became the RAH-66 in June 1985 and were awarded a contract for demonstration/validation programme in 1991.
UPDATED

BOEING SIKORSKY RAH-66 COMANCHE

TYPE: Attack helicopter.
PROGRAMME: Light Helicopter Experimental (LHX) design concepts requested by US Army 1981; numerous changes of programme, given later in this account; original plan for 5,000 to replace UH-1, AH-1, OH-58 and OH-6; reduced in 1987 to 2,096 scout/attack only, replacing 3,000 existing helicopters; to 1,292 in 1990 (with further 389 possible) and then to 1,096 in 1999, although subsequently raised to 1,213 (including eight for US Army operational evaluation) by mid-2000. LHX request for proposals issued 21 June 1988; 23 month demonstration/validation contracts issued to Boeing Sikorsky First Team and Bell/McDonnell Douglas Super Team. Boeing Sikorsky selected 5 April 1991; to build four (reduced to three in 1992, then two in December 1994) YRAH-66 demonstration/validation prototypes in 78 month programme, plus static test article (STA) and propulsion system testbed (PSTB).

LHTEC T800 engine specified October 1988. LHX designation changed to LH early 1990, then US Army designation RAH-66 Comanche in April 1991. Original timetable was: 39 month FSD phase with two additional prototypes, starting August 1995; first low-rate production contract due October 1996; initial 72 helicopters to be built by 2002; first full-rate production contract due November 1998; balance of 1,220 built by 2010. IOC in December 1998. Longbow radar to be installed from production Lot 4 onwards.

FY93 budget delayed development and production phases indefinitely, anticipating Longbow integration from first aircraft; development and production phases reinstated January 1993; further three prototypes then scheduled to be built for engineering and manufacturing development (EMD) phase, in FY98-03; first production procurement in FY99; low-rate production of first 24 in FY01, followed by 48 in FY02, 96 in FY03, then 120 per year; IOC in January 2003.

Prototype critical design review, completed in December 1993, authorised production of three YRAH-66 prototypes (first item for which manufactured in September 1993). At same time, however, further R&D economies under study; December 1994 decision reduced dem/val phase to two prototypes (lacking Longbow/Hellfire capability) and recommended deferment of production; FY94 funding postponed maiden flight target date from August 1995 to November 1995.

In early 1995, US Army restructured programme to reinstate production phase; this successful (see 1998-99 *Jane's* for details) but further restructuring occurred in mid-1998, when 'preproduction prototype' (PPP) programme adopted in order to speed development and align more closely with US Army digitisation plans.

Subsequently, in second quarter of 1999, programme again revised and accelerated, with approximate value of EMD phase costed at US$3.6 billion; start of EMD also brought forward to 2000, subject to satisfactory negotiation of Defense Acquisition Board programme review in March 2000.

Internal arrangement of the Boeing Sikorsky RAH-66 Comanche

Prototype construction began 29 November 1993 with forward fuselage at Sikorsky, Stratford; Boeing built aft fuselage in Philadelphia. STA airframe delivered to Stratford 1994, at which time PSTB under construction there; front and rear sections of prototype joined at Stratford 25 January 1995; completed helicopter (94-0327) rolled out 25 May 1995. Following transfer to Sikorsky's Development Flight Test Center in West Palm Beach, Florida, during June 1995, first flight accomplished on 4 January 1996, when airborne for 36 minutes.

Testing of main rotor system included 100 hour endurance demonstration, which also undertaken by tail rotor system, using whirl stand facilities at Stratford, Connecticut. PSTB trials also commenced in 1995 at West Palm Beach, with 100 per cent torque from both engines achieved during first 10 hours of running. PSTB subsequently suffered failure of left input bevel gear, which disintegrated and punched hole in main gearbox housing during 110 per cent power test; resonance was blamed for the failure, which further delayed second flight of prototype.

Imposition of power limit to avoid possibility of encountering resonance permitted resumption of flight testing on 24 August 1996 when Comanche prototype was airborne for 54 minutes. By mid-October, over seven flight hours had been completed, during which forward speeds of over 100 kt (185 km/h; 115 mph) were achieved as envelope was steadily expanded. By mid-December 1996, prototype had recorded forward speed of 146 kt (270 km/h; 168 mph) and accomplished various flight manoeuvres using AFCS. Testing in first half of 1997 evaluated flight controls and handling qualities; original transmission retained for initial 45 hours of flight testing, with prototype grounded during mid-1997 for installation of revised gearbox. Flight trials resumed in August 1997, with next stage of trials programme including demonstration of 170 kt (315 km/h; 196 mph) forward speed and 45 kt (83 km/h; 52 mph) sideways and rearward speed. By late 1997, prototype had accumulated 62 flight hours in 56 sorties, with a further 3 hours to be completed before start of planned maintenance in late December.

Total flight time had increased to 105 hours by July 1998, when first aircraft temporarily grounded for inspection and refit; by then, it had flown at bank angles of 60°, achieved 204 kt (378 km/h; 235 mph) in a dive, 175 kt (324 km/h; 201 mph) in level flight, 75 kt (139 km/h; 86 mph) sideways and 70 kt (130 km/h; 81 mph) rearward speed. Flight trials resumed on 24 October 1998 to assess modifications before installation of new software for flight controls, the mission equipment package and the engine control unit. By 1999, was testing new main rotor pylon and other modifications, including revised horizontal tail with endplates, to reduce empennage buffeting. Subsequent testing intended to expand flight regime and include operation at higher load factors and evaluation of new main rotor pylon shape as well as development of weapon door-open envelope. By beginning of May 1999, first prototype had logged 144 flight hours in 130 sorties; in mid-July 2000, totals had risen to 233.3 and 197 respectively.

Aft fuselage section of second YRAH-66 prototype, but third airframe (95-0001), delivered by Boeing to Sikorsky's Stratford assembly plant in early December 1996 for mating with forward fuselage; completed helicopter exhibited at Army Aviation Association's annual meeting in April 1998 and then delivered to West Palm Beach. Made international debut when displayed statically at Farnborough Air Show in September 1998; flew for first time on 30 March 1999. Completed initial test schedule in April 1999, recording 4.9 hours in five sorties before temporary lay-up, due to funding constraints; also used for vertical rate of climb demonstration later in year and will, in 2001, be tasked with testing integrated mission equipment package (MEP), including digital avionics, communications, navigation and target acquisition systems. By mid-July 2000, had logged 22.1 flight hours in 25 sorties.

EMD phase of programme officially began 1 June 2000, following RAH-66 meeting (on 4 April 2000) seven key Defense Acquisition Board Milestone 2 criteria, including a 107 m (350 ft)/min vertical climb rate, a specified detection range for the FLIR sensors, a radar cross-section specification, ballistic vulnerability and tolerance specifications and tower-testing of the selected FCR. EMD expected to take six years and will include production of five RAH-66s specifically for EMD testing, followed by further eight for operational test and evaluation by the US Army; initial EMD aircraft due to fly in April 2004; first operational test aircraft (actually the seventh RAH-66) will follow in August 2004, with all 13 having flown by April 2005. In meantime, first YRAH-66 will continue test work and is planned to fly 340 hours by mid-FY03, when it will be grounded for training; second YRAH-66 expected to assume increasing burden of test duty and is to fly 630 hours by mid-FY06, notable milestones including first flight with night vision pilotage system in March 2002 and first with electro-optical target acquisition and designation system (EOTADS) in December 2002.

First PPP aircraft (third RAH-66) to complete 495 hours of flight testing of structural loads; second (fourth RAH-66) will complete 520 hours of propulsion testing; third (fifth RAH-66) to undergo 460 hours of armament testing; and fourth (sixth RAH-66) will accumulate 454 hours of full-system testing. Of remaining nine PPP aircraft, four (Nos. 7-10) will go to US Army for limited user testing and one (No. 12) will be assigned to live firing and tested to destruction; final four (Nos. 11, 13-15) will also go to US Army for IOT&E. Some PPP aircraft also to participate in US Army Force XXI digitisation exercise in 2004.

Revised plan facilitates accelerated development of MEP and fire-control radar, with latter now earmarked for installation on IOC aircraft in 2006 rather than being introduced on Lot 6 production in 2010. FCR unit to be installed on one-third of eventual force; this will be a version of the Northrop Grumman/Lockheed Martin

First two YRAH-66 Comanches during flight testing from West Palm Beach (*US Army*)

USA: AIRCRAFT—BOEING SIKORSKY

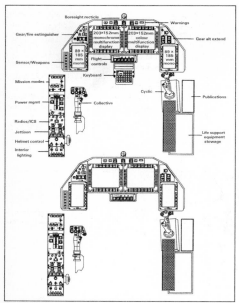

Boeing Sikorsky RAH-66 Comanche front and rear cockpit layout

Longbow radar with a 560 mm (22 in) diameter electronically scanned antenna.

Following EMD machines, low-rate initial production (LRIP) expected to be approved by October 2004; Lot 1 currently planned to involve 14 aircraft with Lots 2 and 3 adding 26 and 44 more respectively and paving way for full rate production from FY08.

First -801 growth version of T800 turboshaft began bench runs in March 1994; -801 preliminary design review completed May 1993; critical design review March 1995.

CUSTOMERS: US Army, two prototypes, five EMD and eight initial operational test and evaluation (IOT&E) aircraft for test duties and operational trials. Planned procurement of 1,205 production aircraft between FY05 and FY23; see accompanying table for details.

RAH-66 COMANCHE PRODUCTION

FY	Lot	Total
94	- (Prototype)	1
95	- (Prototype)	1
-	- (EMD)	5
-	- (IOT&E)	8
05	1 (LRIP 1)	14
06	2 (LRIP 2)	26
07	3 (LRIP 3)	44
08	4	60
09	5	60
10	6	72
11	7	72
12	8	72
13	9	72
14	10	72
15	11	72
16	12	72
17	13	72
18	14	72
19	15	72
20	16	72
21	17	72
22	18	72
23	19	65
Total		1,220

Notes: US Army production total of 1,213 excludes both prototypes and five aircraft for EMD, but includes eight IOT & E aircraft

COSTS: US$34,000 million programme, including US$1,960 million dem/val and US$900 million FSD but reduced to US$2,240 million dem/val/FSD between 1993 and 1997 by cancellation of three of six planned prototypes; US$8.9 million flyaway unit cost (1988 values), increased to US$13 million by early 1993. By 1993 (in 1994 dollars, estimated), procurement unit cost US$21 million, programme unit cost US$27 million. Appropriations for R&D in FY97-99 comprised US$338.6 million, US$282.0 million and US$367.8 million, with US$467 million in FY00 and US$614 million requested for FY01. EMD phase currently projected to cost US$3.15 billion, with initial contract worth US$147.5 million awarded to Boeing Sikorsky on 1 June 2000.

DESIGN FEATURES: First combat helicopter designed from outset to have 'stealth' features and target acquisition radar. Embodies low-observables (LO) attributes and stated to have radar cross-section (RCS) lower than that of Hellfire missile; frontal RCS reportedly 360 times smaller than AH-64, 250 times smaller than OH-58D and 32 times smaller than OH-58D with mast-mounted sight. Also has quarter of AH-64D's IR emissions and is six times quieter, head-on.

RAH-66 is lighter, but only slightly smaller, than AH-64 Apache; specified empty weight of 3,402 kg (7,500 lb) increased to 3,522 kg (7,765 lb) by early 1992, as result of Army add-ons, including allowance for Longbow radar; mission equipment package has maximum commonality with F-22A ATF technology. Design has faceted appearance for radar reflection; downward-angled engine exhausts; T tail; eight-blade fan-in-fin shrouded tail rotor; and five-blade all-composites bearingless main rotor system, with latter increased in diameter by 0.3 m (1 ft 0 in) and gaining noise-reducing anhedral tips on forthcoming EMD aircraft. RAH-66 also features internal weapon stowage.

Split torque transmission, obviating need for planetary gearing. Upper part of T tail folds down for air transportation. Detachable stub-wings for additional weapon carriage and/or auxiliary fuel tanks (EFAMS: external fuel and armament management system). Radar, infra-red, acoustic and visual signature requirements set to defeat threats postulated by US Army. Eight deployable inside Lockheed C-5 Galaxy with only removal of main rotor; ready for flight 20 minutes after transport lands. Combat turnround time 13 minutes.

Subcontractors to Boeing Sikorsky include Boeing Military Aircraft and Missile Systems Group (flight control computer), Hamilton Sundstrand (environmental control system, air vehicle interface controller, air data system, electrical power generation, control and distribution system), Harris Corporation (3-D digital map, controls and displays, super high-speed fibre optic databusses), Hughes-Link (operator training systems), Kaiser Electronics (helmet integrated display sight system), BAE Systems (flight control computer, controller grips), Litton Guidance and Control Systems (inertial navigation), Lockheed Martin Armament Systems (turreted gun system and ammunition feed system), Lockheed Martin Electronics and Missiles (target acquisition/designation system and night vision pilotage system), Moog (flight control actuators), Northrop Grumman (signal and data processors, aided target detection/classification systems), Smiths Industries (crash survivable memory unit), TRW Military Electronics and Avionics (communication, navigation and identification system and survivability systems) and Williams (subsystem power unit).

FLYING CONTROLS: Dual triplex fly-by-wire, with sidestick cyclic pitch controllers and normal collective levers. Main rotor blades removable without disconnecting control system.

STRUCTURE: Largely composites airframe and rotor system. Fuselage built around composites internal box-beam; non-load-bearing skin panels, more than half of which can be hinged or removed for access to interior (for example, weapons bay doors can double as maintenance work platforms). Eight-blade Fantail rear rotor operable with 12.7 mm calibre bullet hits; or for 30 minutes with one blade missing. Main rotor blades and tail section by Boeing, forward fuselage and final assembly by Sikorsky.

LANDING GEAR: Tailwheel type, retractable, with single wheel on each unit. Main units can 'kneel' for air transportability.

POWER PLANT: Two LHTEC T800-LHT-801 turboshafts, each rated at 1,165 kW (1,563 shp). Transmission rating 1,639 kW (2,198 shp). Internal fuel capacity 1,142 litres (301.6 US gallons; 251.1 Imp gallons). Two external tanks totalling 3,407 litres (900 US gallons; 749.4 Imp gallons) for self-deployment; total fuel capacity 4,548 litres (1,201.6 US gallons; 1,000.5 Imp gallons). Additional fuel to be contained in two 424 litre (112 US gallon; 93.3 Imp gallon) tanks in side weapon bays, which in preliminary development in mid-1999. Main rotor tip speed (100 per cent N_r) 221 m (725 ft)/s; 355 rpm; normal operation from 95 per cent (quiet mode) to 107 per cent (load factor enhancement).

ACCOMMODATION: Pilot (in front) and WSO in identical stepped cockpits, pressurised for chemical/biological warfare protection. Crew seats resist 11.6 m (38 ft)/s vertical crash landing.

AVIONICS: Maximum commonality required with USAF Lockheed Martin F-22 Raptor programme.

Comms: Package known as Comanche Integrated Communications, Navigation and Identification (ICNIA) systems comprises dual anti-jam VHF-FM and UHF-AM Have Quick tactical communications, VHF-AM, anti-jam HF-SSB; IFF.

Flight: GPS and radar altimeter. Litton AHRS, comprising two fibre optic LN-210C and one LN-100C gyro platforms.

Instrumentation: Lockheed Martin electro-optical sensor system (EOSS) and Kaiser Electronics helmet-mounted display; NVG-compatible lighting; integrated cockpit, second-generation FLIR targeting and night vision pilotage systems, and digital map display. Two 152 × 203 mm (6 × 8 in) multifunction flat-screen LCDs in each cockpit (left is monochrome for FLIR/TV, right is colour for moving map, tactical situation and night operations), plus two 89 × 185 mm (3½ × 7¼ in) multipurpose display (MPD) flat-screen monochrome LCDs for fuel, armament and communications information in each cockpit. Microvision awarded contract by US Army to provide in 1998 a prototype helmet-mounted display incorporating virtual retinal-display technology as potential alternative to conventional helmet-mounted displays. Three redundant databusses: one low-speed (MIL-STD-1553B), one high-speed and one very high-speed (fibre optic-based) for signal data distribution.

Mission: Airborne target handover system. Miniaturised version of Longbow radar in one-third of fleet although all to have carriage provision. Laser range-finder/designator. Nose-mounted sight with IR and TV camera.

Self-defence: Laser warning, chemical warning and radar warning receivers; RF and IR jammers.

EQUIPMENT: Side armour for cockpit fitted as standard; optional armour kit available for floor of cockpit.

ARMAMENT: General Dynamics stowable XM-301 three-barrel 20 mm cannon in Giat undernose turret, with up to 500 rounds (320 rounds normal for primary mission) and either 750 or 1,500 rounds per minute firing rates. Aiming coverage of gun is +15 to −45° in elevation and ±120° in azimuth. Integrated retractable aircraft munitions system (IRAMS) features side-opening weapons bay door in each side of fuselage, on each of which can be mounted up to three Hellfire or six Stinger missiles or other weapons. Four more Hellfires and eight Stingers can be deployed from multiple carriers under tip of each optional stub-wing, or auxiliary fuel tank for self-deployment. Maximum of 56 Hydra 70 2.75 in FFARs. All weapons can be fired, and targets designated, from push-buttons on collective and sidestick controllers.

DIMENSIONS, EXTERNAL:
Main rotor diameter: YRAH-66		11.90 m (39 ft 0½ in)
RAH-66		12.19 m (40 ft 0 in)
Fantail diameter		1.37 m (4 ft 6 in)
Fantail blade chord		0.17 m (6¾ in)

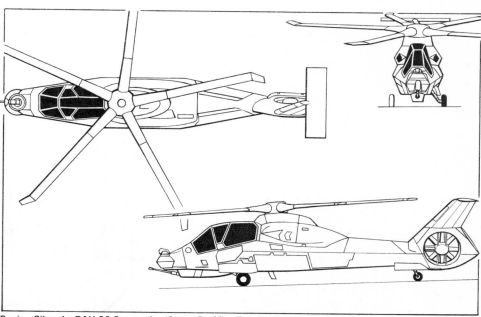

Boeing Sikorsky RAH-66 Comanche *(James Goulding/Jane's)*

Length: overall, rotor turning	14.28 m (46 ft 10¼ in)	Max internal fuel weight	870 kg (1,918 lb)	Cruising speed	160 kt (296 km/h; 184 mph)
fuselage (excl gun barrel)	13.20 m (43 ft 3¾ in)	T-O weight: primary mission*	5,799 kg (12,784 lb)	Vertical rate of climb	273 m (895 ft)/min
Fuselage: Max width	2.04 m (6 ft 8¼ in)	max alternative	5,850 kg (12,897 lb)	Masking	1.6 s
Width over mainwheels	2.31 m (7 ft 7 in)	max (self-deployment)	7,896 kg (17,408 lb)	180° hover turn to target	4.7 s
Height over tailplane	3.37 m (11 ft 0¾ in)	Max disc loading: YRAH-66	71.0 kg/m² (14.54 lb/sq ft)	Snap turn to target at 80 kt (148 km/h; 92 mph)	4.5 s
Tailplane span	2.82 m (9 ft 3 in)	Transmission loading at max T-O weight and power		Operational radius, internal fuel	
AREAS:			4.82 kg/kW (7.92 lb/shp)		150 n miles (278 km; 173 miles)
Main rotor disc: YRAH-66	111.22 m² (1,197.1 sq ft)	*with two crew, full internal fuel, 320 rds gun ammunition,		Ferry range with external tanks	
Fantail disc	1.48 m² (15.90 sq ft)	radar, four Hellfires and two Stingers			1,260 n miles (2,334 km; 1,450 miles)
WEIGHTS AND LOADINGS (estimated):		PERFORMANCE (at 1,220 m; 4,000 ft and 35°C; 95°F,		Endurance (standard fuel)	2 h 30 min
Weight empty	4,218 kg (9,300 lb)	estimated):		g limits	+3.5/–1
Max useful load	2,296 kg (5,062 lb)	Max level (dash) speed	175 kt (324 km/h; 201 mph)		**UPDATED**

BRANTLY

BRANTLY INTERNATIONAL INC

12399 Airport Drive, Wilbarger County Airport, Vernon, Texas 76384
Tel: (+1 940) 552 54 51
Fax: (+1 940) 552 27 03
e-mail: sales@brantly.com
Web: http://www.brantly.com
PRESIDENT AND CEO: Henry Yao
VICE-PRESIDENT: Cy A Russum
VICE-PRESIDENT, MARKETING AND SALES: George Stokes

In 1994, Brantly International obtained the type and production certificates for the Brantly B-2B and 305 helicopters from Japanese-American businessman James T Kimura's Brantly Helicopter Industries, which had acquired them in May 1989. In 2000, Brantly employed 38 in its 2,790 m² (30,000 sq ft) facility.

UPDATED

BRANTLY B-2B

TYPE: Two-seat helicopter.
PROGRAMME: Developed from coaxial twin-rotor B-1 by Newby O Brantly. First flight (B-2) 21 February 1953; FAA certification 27 April 1959. Total of 194 B-2s and 18 B-2As (with additional headroom) produced between 1960 and 1963. Improved Model B-2B with metal rotor blades and fuel-injected Lycoming IVO-360-A1A engine certified in July 1963; total of 165 built between 1963 and 1967 and a further one (as H-2) in 1975 by Brantly-Hynes Helicopter. Brantly Helicopter Industries took over manufacturing and marketing rights and production facilities in 1989. First new-build B-2B (N25411 c/n 2001) flew 12 April 1991. Production continues under Brantly International, which received FAA production certificate on 19 July 1996.
CUSTOMERS: Total of 15 of current series registered by January 2001. Notified deliveries were two in 1998, none in 1999 and seven (to China) in 2000. Seven remained registered in USA at January 2001, and one in Australia. (Over 150 survive from earlier production.)
COSTS: US$150,000 basic equipped (2000). Direct operating cost US$80 per hour (2000).
DESIGN FEATURES: Simple design, with blown main transparency and constant-taper fuselage. Double-articulated three-blade main rotor with flapping hinges close to hub and pitch-change/flap/lag hinges at 40 per cent blade span; symmetrical, rigid, inboard blade section with 29 per cent thickness/chord ratio, outboard section NACA 0012; outer blades quickly removable for compact storage; rotor brake standard; two-blade tail rotor mounted on starboard side, with guard. Transmission through automatic centrifugal clutch and planetary reduction gear. Bevel gear take-off from main transmission, with flexible coupling to tail rotor drive-shaft. Main rotor/engine rpm ratio 1:6.158; tail rotor ratio 1:1.
FLYING CONTROLS: Conventional and manual; small fixed tailplanes on port and starboard sides of tailcone.
STRUCTURE: Fuselage has welded steel tube centre-section with alloy stressed skin tailcone. Inboard rotor blades have stainless steel leading-edge spar; outboard blades have extruded aluminium spar; polyurethane core with bonded aluminium envelope riveted to spar. All-metal tail rotor blades.
LANDING GEAR: Fixed skid type with oleo-pneumatic shock-absorbers; small retractable ground handling wheels, size 10×3.5, pressure 4.12 bars (60 lb/sq in); fixed tailskid. Optional inflatable pontoons attach to standard skids for over-water operation.
POWER PLANT: One 134 kW (180 hp) Textron Lycoming IVO-360-A1A flat-four air-cooled piston engine, mounted vertically. Fuel contained in two interconnected bladder tanks behind cabin, total capacity 117.3 litres (31.0 US gallons; 25.8 Imp gallons) of which 116 litres (30.6 US gallons; 25.5 Imp gallons) are usable. Oil capacity 5.7 litres (1.5 US gallons; 1.25 Imp gallons).
ACCOMMODATION: Two, side by side in enclosed cabin; forward hinged door on each side. Dual controls and cabin heater standard. Ground accessible baggage compartment, maximum capacity 22.7 kg (50 lb) in forward end of tailcone.
SYSTEMS: 60A alternator.
AVIONICS: To customer choice; GPS is standard.
EQUIPMENT: Twin landing lights in nose standard.

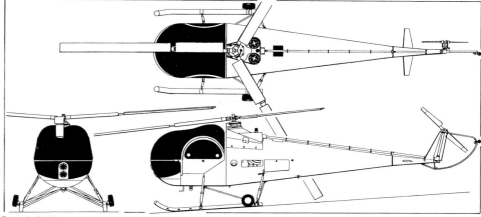

Brantly B-2B two-seat light helicopter *(Paul Jackson/Jane's)* 2001/0105040

DIMENSIONS, EXTERNAL:	
Main rotor diameter	7.24 m (23 ft 9 in)
Main rotor blade chord: inboard	0.22 m (8¾ in)
outboard	0.20 m (8 in)
Tail rotor diameter	1.30 m (4 ft 3 in)
Length: fuselage	6.43 m (21 ft 1 in)
overall, rotors turning	8.53 m (28 ft 0 in)
Height overall	2.11 m (6 ft 11 in)
Skid track	2.08 m (6 ft 10 in)
Passenger doors (each): Height	0.79 m (2 ft 7 in)
Width	0.86 m (2 ft 9¾ in)
Baggage door: Height	0.25 m (9¾ in)
Width	0.55 m (1 ft 9¾ in)
DIMENSIONS, INTERNAL:	
Cabin: Length	1.83 m (6 ft 0 in)
Max width	1.19 m (3 ft 11 in)
Max height	0.99 m (3 ft 3 in)
Floor area	2.60 m² (28.0 sq ft)
Volume	2.78 m³ (98.0 cu ft)
Baggage compartment volume	0.17 m³ (6.0 cu ft)
AREAS:	
Main rotor blades (each)	0.69 m² (7.42 sq ft)
Main rotor disc	41.16 m² (443.0 sq ft)
Tail rotor blades (each)	0.05 m² (0.50 sq ft)
Tail rotor disc	1.32 m² (14.19 sq ft)

Mid-rotor hinge on Brantly B-2B *(Paul Jackson /Jane's)* 2001/0092655

WEIGHTS AND LOADINGS:	
Weight empty: with skids	476 kg (1,050 lb)
with pontoons	494 kg (1,090 lb)
Max fuel	82 kg (180 lb)
Max T-O weight	757 kg (1,670 lb)
Max baggage weight	22.7 kg (50 lb)
Max disc loading	18.4 kg/m² (3.77 lb/sq ft)
Max power loading	5.65 kg/kW (9.28 lb/hp)
PERFORMANCE:	
Never-exceed speed (VNE)	87 kt (161 km/h; 100 mph)
Max cruising speed at 75% power	78 kt (144 km/h; 90 mph)
Max rate of climb at S/L	427 m (1,400 ft)/min
Service ceiling	1,981 m (6,500 ft)
Hovering ceiling IGE	1,074 m (3,525 ft)
Range with max fuel, without reserves	191 n miles (353 km; 219 miles)
with reserves	176 n miles (326 km; 202 miles)

UPDATED

Brantly B-2B from the current production series *(Paul Jackson/Jane's)* 2001/0089497

Rotor hub of Brantly B-2B *(Paul Jackson/Jane's)* 2001/0089498

CAMERON

CAMERON & SONS AIRCRAFT
1605 W 130th Street, Unit 6, Gardena, California 90249
Tel: (+1 310) 769 16 90
Fax: (+1 310) 640 84 90
CEO: Murdo Cameron

Cameron Aircraft's P-51 Mustang derivative is the company's first product. The kits are produced by Exclusive Aviation, under which heading this entry previously appeared.

NEW ENTRY

CAMERON P-51 G
TYPE: Turboprop sportplane/kitbuilt.
PROGRAMME: Full-size, turbine-powered representation of North American P-51 Mustang. Design began 1988; publicly announced end 1997; first flight 1998 and displayed at Oshkosh in July 1998 as Grand 51; later renamed P-51G.
CUSTOMERS: Prototype and two production aircraft under construction by early 1998.
COSTS: Development costs US$4.5 million; kit price US$245,000; completed airframes US$600,000 (2000).
DESIGN FEATURES: Wing span matches racing P-51s; option to extend to 'normal' 11.38 m (37 ft 4 in) available; aircraft kit comprises 12 major components. Quoted build time 1,500 to 2,000 hours.
STRUCTURE: Airframe and flight control surfaces of carbon fibre epoxy; 85 per cent is composites. One-piece 'wet' wing.
FLYING CONTROLS: Conventional and manual. No horn balances, no flaps. Electric trim tab on starboard elevator.
LANDING GEAR: Tailwheel configuration; mainwheels retract inward; tailwheel rearward. Bridgestone 24×7.7 mainwheel tyres; 12.5 in tailwheel type. Cleveland wheels and brakes.
POWER PLANT: One 1,081 kW (1,450 shp) Textron Lycoming T53-L-701A turboprop, driving a Hamilton Sundstrand three-blade, fully feathering and reversible propeller. Fuel capacity 1,703 litres (450 US gallons; 375 Imp gallons) in prototype; kitbuilts have 946 litres (250 US gallons; 208 Imp gallons).
ACCOMMODATION: Pilot and passenger in tandem under one-piece hinged canopy. Dual controls. Impact Dynamics leather seating. Dummy belly scoop provides baggage compartment with starboard side access door.
AVIONICS: To customer's specification; IFR is standard.

DIMENSIONS, EXTERNAL:
Wing span 9.80 m (32 ft 2 in)
Wing chord at tip 1.52 m (5 ft 0 in)
Wing aspect ratio 4.7
Length overall 9.98 m (32 ft 9 in)
Height overall 3.89 m (12 ft 9 in)
Propeller diameter 3.05 m (10 ft 0 in)
Baggage door width 0.91 m (3 ft 0 in)
DIMENSIONS, INTERNAL:
Cockpit max width 0.76 m (2 ft 6 in)
AREAS:
Wings, gross 20.44 m² (220.0 sq ft)
WEIGHTS AND LOADINGS:
Weight empty 1,860 kg (4,100 lb)
Max T-O weight 3,628 kg (8,000 lb)
Max wing loading 177.5 kg/m² (36.36 lb/sq ft)
Max power loading 3.36 kg/kW (5.52 lb/shp)
PERFORMANCE (provisional):
Max operating speed 391 kt (724 km/h; 450 mph)
Normal cruising speed 313 kt (579 km/h; 360 mph)
Stalling speed 79 kt (145 km/h; 90 mph)
Max rate of climb at S/L 1,371 m (4,500 ft)/min
Service ceiling 7,620 m (25,000 ft)
T-O and landing run 366 m (1,200 ft)
Range 1,130 n miles (2,092 km; 1,300 miles)
g limits (prototype) ±11

UPDATED

Cameron P-51G (Textron Lycoming T53-L-701A) 0016207

CARLSON

CARLSON AIRCRAFT INC
PO Box 88, East Palestine, Ohio 44413-0088
Tel: (+1 330) 426 39 34
Fax: (+1 330) 426 11 44
e-mail: mlc@sky-tek.com
Web: http://www.sky-tek.com

In addition to the Skycycle and newly added Criquet, Carlson also markets the Sparrow series of ultralight aircraft (see 1992-93 *Jane's*) and a replacement wing for Baby Great Lakes and Piper Cub and Vagabond lightplanes. An ultralight version of the Cub was under construction in 2000.

Founder Ernest W Carlson was killed in the crash of the prototype Criquet in May 2000.

UPDATED

CARLSON SKYCYCLE
TYPE: Single-seat kitbuilt.
PROGRAMME: Based on Piper PA-8 Skycycle of 1945. Production of original aircraft totalled two prototypes. Second Piper prototype (NX47Y) restored as demonstrator in 1995 by Neal Carlson, but crashed on 8 September 2000. Production tooling completed by April 2000.
CUSTOMERS: Options held on 800 kits by late 1998.
COSTS: Kit US$13,900 (2000).
DESIGN FEATURES: Experimental category. Detachable wings for easy storage. (PA-8 fuselage was based on surplus F4U Corsair drop-tank.)
FLYING CONTROLS: Conventional and manual. Large ailerons; no aerodynamic balance or tabs.
STRUCTURE: Aluminium wing with GFRP cockpit section, steel reinforced; 2024-T3 aluminium tailcone; steel empennage, fabric covered.
LANDING GEAR: Tailwheel type; fixed. Mainwheels 5.00-5. Hydraulic brakes; steerable tailwheel.
POWER PLANT: One 48.5 kW (65 hp) Textron Lycoming O-145-B2 engine driving a Sensenich 70/36 two-blade propeller. Fuel capacity 38 litres (10.0 US gallons; 8.3 Imp gallons); optional increase to 76 litres (20.0 US gallons; 16.7 Imp gallons).

DIMENSIONS, EXTERNAL:
Wing span 6.10 m (20 ft 0 in)
Length overall 4.80 m (15 ft 9 in)
Height overall 1.63 m (5 ft 4¼ in)
AREAS:
Wings, gross 8.83 m² (95.0 sq ft)
WEIGHTS AND LOADINGS:
Weight empty 227 kg (500 lb)
Max T-O weight 362 kg (800 lb)
PERFORMANCE:
Never-exceed speed (VNE) 120 kt (223 km/h; 139 mph)
Normal cruising speed 87 kt (161 km/h; 100 mph)
Stalling speed 47 kt (87 km/h; 54 mph)
Max rate of climb at S/L 267 m (875 ft)/min
T-O run 107 m (350 ft)
Landing run 183 m (600 ft)
Range 250 n miles (463 km; 287 miles)

UPDATED

CARLSON CA6 CRIQUET
TYPE: Tandem-seat kitbuilt.
PROGRAMME: Prototype N22CA under trial in 1999; lost in accident 24 May 2000. Construction of second aircraft was continuing in November 2000. Programme in abeyance, pending accident investigation.
DESIGN FEATURES: Three-quarter scale replica of Morane-Saulnier MS.500 Criquet, itself a copy of Fieseler Fi 156 Storch; uses Carlson wing, however.
POWER PLANT: One 103 kW (138 hp) LOM M332A four-cylinder piston engine driving a two-blade propeller. Fuel capacity 152 litres (40.0 US gallons; 33.3 Imp gallons). Optional 119 kW (160 hp) fuel-injected and supercharged M332B for higher gross weight version (998 kg; 2,200 lb) with 197 litre (52.0 US gallon; 43.3 Imp gallon) tankage.

NEW ENTRY

Piper PA-8 Skycycle, employed as a demonstrator for the Carlson version *(Geoffrey P Jones)* 2001/0054043

CARTER

CARTERCOPTERS LLC
5720 Seymour Highway, Wichita Falls, Texas 76310
Tel: (+1 940) 691 08 19
Fax: (+1 940) 691 59 77
e-mail: carter@wf.net
Web: http://www.cartercopters.com
CEO: Jay Carter

Company founder Jay Carter previously worked on design of the Bell XV-15 tiltrotor. CarterCopters has produced a flying prototype of what it anticipates will be a family of convertiplanes, although its ultimate purpose is to market the design, not establish a manufacturing programme itself.

UPDATED

CARTER CARTERCOPTER CC1
TYPE: Convertiplane.
PROGRAMME: With some US$670,000 funding from NASA, Carter has built a prototype of the CarterCopter, a five-seat pressurised convertiplane powered by a modified, twin-turbocharged V-6 NASCAR racing car engine. The

Prototype CarterCopter CC1 2001/0105828

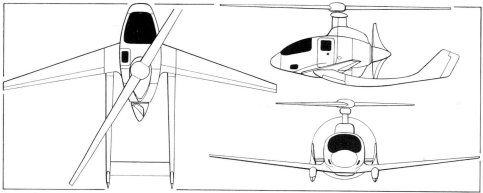

General arrangement of the Carter CarterCopter *(James Goulding/Jane's)* 2000/0087603

Projected CCH-T military transport, based on the CarterCopter principle 2001/0100517

aircraft, which is of twin-boom configuration with high-aspect ratio sweptback wings, has a pylon-mounted two-blade rotor which is powered only for take-off and landing, free-rotating in forward flight, when all power is transmitted to a pusher propeller. Maximum T-O weight is 1,451 kg (3,200 lb). Carter predicts a maximum cruising speed of 348 kt (643 km/h; 400 mph) at 15,240 m (50,000 ft).

First flight (N121CC) September 1998; development delayed following damage to prototype during aborted rolling take-off on 16 December 1999; flight testing resumed in late October 2000. Attempts are planned on rotary-wing world records. Production versions expected to sell for US$250,000 to US$300,000 (1997). A two-seat kitbuilt version of the CarterCopter may also be developed. Commercial/corporate and military versions with maximum take-off weight of 68,039 kg (150,000 lb), powered by turbofan or turboprop engines and carrying up to 50 passengers, are envisaged, designated CCH-T (CarterCopter Heliplane Transport). A smaller, UAV version was submitted for US Navy consideration in early 2000.

UPDATED

CENTURY

CENTURY AEROSPACE CORPORATION

3250 University Boulevard SE, Access Road B, Albuquerque, New Mexico 87106
Tel: (+1 505) 246 82 00
Fax: (+1 505) 246 83 00
e-mail: info@centuryaero.com
Web: http://www.centuryaero.com
PRESIDENT AND CEO: Bill Northrup
VICE-PRESIDENT AND GENERAL MANAGER: Roy Johnston
VICE-PRESIDENT, ENGINEERING: Dale Ruhmel
SALES DIRECTOR: Kevin Whited
MARKETING MANAGER: J Sheldon 'Torch' Lewis

UPDATED

CENTURY CA-100 CENTURY JET 100

TYPE: Light business jet.
PROGRAMME: Formerly known as Paragon Spirit; design started 1993; engineering work undertaken by North American Aeronautical Consultants (NAAC) and team from The Ohio State University; wind tunnel testing completed and detailed design, then as single-turbofan, started January 1996; relaunched as twin-turbofan at NBAA Convention in Las Vegas, Nevada, in October 1998. Funding of some US$60 million being sought in October 2000, at which time negotiations were under way with the Committee for Aviation and Space Technology Development of Taiwan leading to possible investment and establishment of partnership with Taiwanese companies; first flight anticipated 20 months after final signature of partnership agreements, with FAR Pt 23 certification and first customer deliveries in a further 18 months. Two flying and two static test prototypes planned.
CUSTOMERS: Total of 51 orders and deposits by October 2000.
COSTS: Programme US$40 million to US$50 million. Unit cost US$2.7 million (2000). Estimated direct operating cost US$335 per hour (2000).
DESIGN FEATURES: Design goals include low direct operating costs, combined with high-speed cruise and short field capability. New laminar flow aerofoil for wing and horizontal and vertical tail surfaces; T tail with sweptback fin.

Wing section CAC CW(1)-55615 at root, CAC CW(1)-45613 at tip; leading-edge sweepback 5° at 40 per cent chord; dihedral 5°; incidence 1° 30′; twist 2° 24′.
FLYING CONTROLS: Conventional and manual. Cable actuation for elevators, ailerons, spoilerons, rudder and airbrakes; geared trim tabs in elevators; bungee trim for ailerons and rudder; electrically actuated single-slotted Fowler flaps, 28 per cent chord, with asymmetry detection system; hydraulically operated wing-mounted airbrakes/lift dumpers.
STRUCTURE: Main wing structure consists of single carry-through primary spar of aluminium 2024 and 7045Al at 30 per cent chord, with secondary spars at 6 per cent and 65 per cent chord; wing skins of aluminium with aluminium

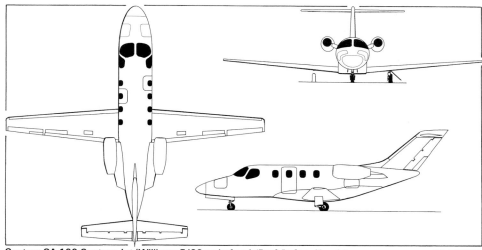

Century CA-100 Century Jet (Williams FJ33 turbofans) *(Paul Jackson/Jane's)* 2001/0100435

ribs. Fuselage of carbon fibre honeycomb sandwich construction with internal frames and bulkheads. Composites materials are high-temperature autoclave-cured carbon fibre pre-impregnated composites with honeycomb cores. Fuselage designed and built by risk-sharing partner SGL Carbon Composites/Hitco of California.
LANDING GEAR: Retractable, with single wheel on each unit; built by AllTool. Trailing-link main legs, oleo strut nose leg. Hydraulic actuation, with free-fall emergency back-up. Main units retract inwards, steerable nosewheel forwards; gear door on nose unit only. Mainwheel diameter 20.3 cm (8 in), tyre diameter 0.44 m (1 ft 5½ in), pressure 6.61 bars (96 lb/sq in); nosewheel diameter 12.7 cm (5 in), tyre diameter 0.36 m (1 ft 2¼ in), pressure 3.45 bars (50 lb/sq in). Nosewheel maximum steering angle ±30°. Turning circle about nosewheel is 5.76 m (18 ft 10¾ in). Disc brakes.
POWER PLANT: Two Williams FJ33-1 turbofans, each rated at 5.34 kN (1,200 lb st), pod-mounted on fuselage aft of wing. Fuel contained in two fuselage tanks, with usable capacity of 113 litres (30.0 US gallons; 25.0 Imp gallons) and two wing tanks, each of 568 litres (150 US gallons; 125 Imp gallons) usable capacity. Total of 1,272 litres (336 US gallons; 280 Imp gallons), of which 1,249 litres (330 US gallons; 275 Imp gallons) are usable. Single-point gravity fuelling on fuselage port side aft of wing.
ACCOMMODATION: Crew of one. Standard seating for pilot and five passengers: one in co-pilot's seat and four in club arrangement in cabin; optional private lavatory replaces aft cabin seat on starboard side. Cabin is pressurised, heated and air conditioned. Flight-accessible baggage area at rear of cabin; externally accessed, unpressurised, heated baggage areas in nose and aft fuselage, last-mentioned with cargo nets. Four cabin windows on each side. Windscreen has electric anti-misting and de-icing. Forward-hinged crew/passenger door with integral airstair on port side forward of wing; emergency exit directly opposite on starboard side.
SYSTEMS: Vapour-cycle pressurisation system, maximum differential 0.63 bar (9.1 lb/sq in), supplied by high-pressure bleed air. Two-pump electrohydraulic system for landing gear operation, maximum flow rate 7.6 litres (2.0 US gallons; 1.7 Imp gallons)/min, pressure 207 bars (3,000 lb/sq in). Master cylinder-operated hydraulic braking system, operating pressure 51.7 bars (750 lb/

Artist's impression of Century CA-100 twin-turbofan business aircraft 2001/0100434

sq in). Electrical system comprises 28 V 44 Ah DC battery and two 200 A generators. Pneumatic de-icer boots on horizontal tail leading-edges; bleed air anti-icing for wing leading-edges and engine inlets.

AVIONICS: Honeywell avionics suite.
 Comms: Dual KX 165 nav/com, dual KT 70 Mode S transponders, KMA 24H audio panel.
 Radar: Honeywell RDR-2000 colour weather radar.
 Flight: KFC integrated AFCS; KR 21 marker beacon receiver, KR 87 ADF, KN 63 DME, KLN 90B GPS.
 Instrumentation: Meggitt Avionics LCD PFDs and engine instrumentation displays.

DIMENSIONS, EXTERNAL:
Wing span	12.01 m (39 ft 4¾ in)
Wing chord: at root	1.73 m (5 ft 8 in)
at tip	0.66 m (2 ft 2 in)
Wing aspect ratio	9.5
Length overall	11.65 m (38 ft 2½ in)
Fuselage: Length	10.36 m (34 ft 0 in)
Diameter (constant)	1.55 m (5 ft 1 in)
Max width	1.55 m (5 ft 1 in)
Height overall	3.78 m (12 ft 5 in)
Tailplane span	4.94 m (16 ft 2½ in)
Wheel track	3.57 m (11 ft 8½ in)
Wheelbase	4.27 m (14 ft 0 in)
Crew/passenger door: Height	1.28 m (4 ft 2¼ in)
Width	0.66 m (2 ft 2 in)
Height to sill	0.79 m (2 ft 7¼ in)
Emergency exit: Height	0.56 m (1 ft 10 in)
Width	0.61 m (2 ft 0 in)
Height to sill	1.45 m (4 ft 9 in)
Nose baggage bay door (port):	
Max height	0.46 m (1 ft 6 in)
Width	0.61 m (2 ft 0 in)
Height to sill	1.05 m (3 ft 5½ in)
Systems bay access door (starboard, rear):	
Max height	0.46 m (1 ft 6 in)
Width	0.71 m (2 ft 4 in)
Height to sill	1.04 m (3 ft 5 in)
Aft baggage compartment door (port):	
Max height	0.46 m (1 ft 6 in)
Width	0.71 m (2 ft 4 in)
Height to sill	1.08 m (3 ft 6½ in)

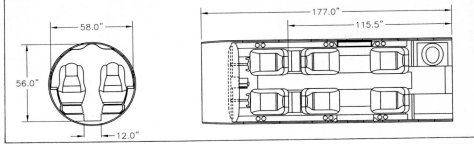

Seating arrangements in the Century Jet business aircraft; dimensions in inches 2001/0100436

DIMENSIONS, INTERNAL:
Cabin (incl flight deck): Length	4.50 m (14 ft 9 in)
Max width	1.47 m (4 ft 10 in)
Max height	1.42 m (4 ft 8 in)
Floor area	4.37 m² (47.0 sq ft)
Volume	9.06 m³ (320 cu ft)
Baggage compartment volume:	
Cabin	0.28 m³ (10.0 cu ft)
Aft	0.85 m³ (30.0 cu ft)
Nose	0.08 m³ (3.0 cu ft)

AREAS:
Wings, gross	15.19 m² (163.5 sq ft)
Ailerons (total)	0.78 m² (8.40 sq ft)
Spoilers (total)	0.14 m² (1.50 sq ft)
Trailing-edge flaps (total)	2.60 m² (28.00 sq ft)
Fin	2.62 m² (28.15 sq ft)
Rudder, incl tab	0.65 m² (7.00 sq ft)
Tailplane	3.76 m² (40.50 sq ft)
Elevators (total incl tabs)	1.41 m² (15.20 sq ft)

WEIGHTS AND LOADINGS (estimated):
Weight, empty, equipped	1,851 kg (4,080 lb)
Max payload	667 kg (1,470 lb)
Max fuel weight	1,011 kg (2,228 lb)
Max T-O weight	3,175 kg (7,000 lb)
Max ramp weight	3,198 kg (7,050 lb)
Max landing weight	3,016 kg (6,650 lb)
Max zero-fuel weight	2,608 kg (5,750 lb)
Max wing loading	209.0 kg/m² (42.81 lb/sq ft)
Max power loading	297 kg/kN (2.92 lb/lb st)

PERFORMANCE (estimated):
Never-exceed speed (V_{NE}): S/L to 4,725 m (15,500 ft)	345 kt (638 km/h; 397 mph) CAS
above 4,725 m (15,500 ft)	M0.77
Max operating speed (V_{MO}): S/L to 8,230 m (27,000 ft)	275 kt (509 km/h; 316 mph) CAS
above 8,230 m (27,000 ft)	M0.71
Max cruising speed at 10,670 m (35,000 ft)	370 kt (685 km/h; 426 mph)
Econ cruising speed at 13,715 m (45,000 ft)	310 kt (574 km/h; 357 mph)
Stalling speed, flaps down	73 kt (136 km/h; 84 mph)
Max rate of climb at S/L	945 m (3,100 ft)/min
Rate of climb at S/L, OEI	244 m (800 ft)/min
Max certified altitude	13,715 m (45,000 ft)
Service ceiling, OEI	6,705 m (22,000 ft)
T-O balanced field length	884 m (2,900 ft)
Landing balanced field length	732 m (2,400 ft)

Range with max fuel, 336 kg (740 lb) payload, econ cruising speed at 12,500 m (41,000 ft), 45 min reserves: 1,500 n miles (2,778 km; 1,726 miles)
Range with max payload, conditions as above: 1,000 n miles (1,852 km; 1,150 miles)

UPDATED

CESSNA

CESSNA AIRCRAFT COMPANY
(Subsidiary of Textron Inc)
PO Box 7706, Wichita, Kansas 67277-7706
Tel: (+1 316) 941 60 00
Fax: (+1 316) 941 78 12
Web: http://www.cessna.textron.com
CHAIRMAN: Russell W Meyer Jr
CEO AND VICE-CHAIRMAN: Gary W Hay
PRESIDENT AND COO: Charles B Johnson
SENIOR VICE-PRESIDENT, PRODUCT ENGINEERING: Milt Sills
SENIOR VICE-PRESIDENT, SALES AND MARKETING: Roger Whyte
SENIOR VICE-PRESIDENT, FINANCE AND CHIEF FINANCIAL OFFICER:
 Michael Shonka
VICE-PRESIDENT, MARKETING: Phil Michel
VICE-PRESIDENT, INTERNATIONAL SALES: Mark Paolucci
VICE-PRESIDENT, CORPORATE COMMUNICATIONS:
 Marilyn Richwine

SINGLE-ENGINE PISTON AIRCRAFT PRODUCTION PLANT:
One Cessna Boulevard, PO Box 1996, Independence, Kansas 67301
Tel: (+1 316) 332 00 00
Fax: (+1 316) 332 03 88
GENERAL MANAGER: Pat Boyarski

Founded by late Clyde V Cessna 1911; incorporated 7 September 1927; acquired by General Dynamics as wholly owned subsidiary 1985; acquisition by Textron announced February 1992.

Suspended manufacture of some piston-engined light aircraft types in mid-1980s; plans for resumed production of Cessna 172 Skyhawk, 182 Skylane, 206 Stationair and T206 Turbo Stationair announced 1994; 46,450 m² (500,000 sq ft) new factory at Independence Municipal Airport, Montgomery County, Kansas, opened 3 July 1996 to house final assembly, painting, flight test, engineering, finance, marketing and human resource operations for single-engined light aircraft range. First new model 172 rolled out 6 November 1996.

Owned subsidiaries include Cessna Finance Corporation in Wichita and Cessna-Columbus Division in Columbus, Georgia. Sold 49 per cent interest in Reims Aviation of France to Compagnie Française Chaufour Investissement (CFCI) February 1989; Reims continues manufacturing Cessna F406 Caravan II (which see in French section).

Total 182,798 aircraft produced by end of December 2000. Total of 1,256 delivered in 2000, comprising 912 single-engined piston aircraft, 92 Caravans and 252 Citations. 3,000th piston single from resumed production, a 182T Skylane, delivered in early February 2000; 3,000th Citation executive jet, a Citation X, delivered 19 November 1999. Workforce 12,000 in 1999.

UPDATED

CESSNA 172 SKYHAWK

TYPE: Four-seat lightplane.
PROGRAMME: Development of previous models described in 1985-86 and earlier editions of *Jane's*; total of 35,643 commercial aircraft in the earlier Model 172/Skyhawk series built between 1955 (first flight of prototype, 12 June) and 1985, when production was suspended. Prototype 'Restart 172' (N6786R), modified from 1978 Model 172N with IO-360 engine, made its first flight at Wichita 19 April 1995; certification flight testing began 10 January 1996; three preproduction 172Rs (N172KS, N172SE and N172NU) assembled at East Pawnee, Wichita, plant under Single-Engine Pilot Program, of which the first flew on 16 April 1996; FAA FAR Pt 23 certification achieved 21 June 1996 after 145 flight test hours; first production aircraft (c/n 80004; N172FN) rolled out at Independence 6 November 1996 and delivered 18 January 1997. First export delivery began on 30 January 1997 when second Independence-built aircraft (c/n 80005; VH-NMN) left US for air ferry to Australia.

CURRENT VERSIONS: **172R Skyhawk**: Standard version. Compared with pre-1986 (172Q) versions, features fuel-injected engine; specially developed McCauley propeller optimised for reduced rpm operation; new Honeywell avionics; metal instrument panel with backlit, non-glare instruments; all-electric engine gauges, dual vacuum system, digital clock, EGT gauge, hour meter and centrally mounted annunciator panel; stainless steel control cables; epoxy corrosion-proofing; cabin improvements as detailed below; steps for visual checking of fuel tanks, which have quick-reference quantity tabs, and consolidation of all primary electrical components into a single junction box on the firewall for simplified servicing.

New standard features announced at the NBAA Convention at New Orleans on 9 October 2000, and introduced on 172R and 172S from 2001 model year, include redesigned wingtips, restyled and reclocked main landing gear fairings, improved nose gear fairing, low-drag wing strut fairings, streamlined refuelling and entrance steps, and improved interiors featuring sculptured composite side panels with integrated cupholders and cushioned armrests, removable overhead panels for simplified maintenance, Rosen sun visors, lightweight door handles and a 12 V electrical port for use with a portable GPS, CD player or laptop computer. New optional avionics include a 12.7 cm (5 in) Honeywell KMD 550 LCD colour MFD with Stormscope that interfaces with the previously available KLN 94 colour IFR GPS.

Detailed description applies to 172R.

172S Skyhawk SP: Special performance version, announced 19 March 1998; features Textron Lycoming IO-360-L2A engine rated at 134 kW (180 hp) driving a higher-performance McCauley fixed-pitch, 1.93 m (6 ft 4 in) diameter propeller; avionics and equipment as for baseline version except leather seats standard. Standard empty weight 745 kg (1,642 lb); maximum T-O and landing weight 1,157 kg (2,550 lb); maximum level speed at S/L 126 kt (233 km/h; 145 mph); cruising speed, 75 per cent power at 2,590 m (8,500 ft) 124 kt (230 km/h; 143 mph); maximum rate of climb at S/L 222 m (730 ft)/

Cessna 172R Skyhawk used for training by Embry-Riddle Aeronautical University *(Paul Jackson/Jane's)*
2001/0100480

Cabin interior of Cessna 172S Skyhawk SP *1999/0054028*

Instrument panel of Cessna 172R Skyhawk *1999/0054027*

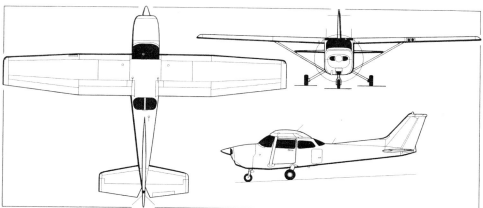
Cessna 172R Skyhawk *(Paul Jackson/Jane's)* 0015674

min; T-O run 293 m (960 ft); T-O to 15 m (50 ft) 497 m (1,630 ft); landing from 15 m (50 ft) 407 m (1,335 ft); landing run 175 m (575 ft); range with maximum fuel: at 75 per cent power at 2,590 m (8,500 ft), 518 n miles (959 km; 596 miles); at 45 per cent power at 3,050 m (10,000 ft), 638 n miles (1,181 km; 734 miles). First delivery 31 July 1998. Further improvements introduced in October 2000, as noted above.

Millennium Edition: Special configuration limited edition Skyhawk SP, announced in February 2000 and available from March 2000. Cessna announced on 9 April 2000 that limited edition would be extended, due to demand. Features redesigned cabin interior with leather seats; floor mats embossed with Cessna Millennium logo; new cup- and chart-holders; Rosen sun visors; polished spinner and cowling fasteners; Goodyear Flight Custom II tyres; avionics package plus Honeywell KLN 94 colour IFR GPS with moving map display; new exterior paint scheme; pilot's accessory collection.

CUSTOMERS: Total of 180 172Rs and 279 172SPs delivered in 1999 and 150 and 340 respectively in 2000. The 1,000th 172R/SP was delivered in October 1999 to North Florida Medical. Recent customers include Embry-Riddle Aeronautical University (initial batch of 73 172Rs out of 300 piston-engined Cessnas of all models to be ordered over a 12 year period), TAM Training (10), Western Michigan University (38), US AFB Flight Training Centre (22 delivered from 30 March 1998), Civil Air Patrol (42 172Rs/SPs ordered by August 2000, Daniel Webster College of Nashua, New Hampshire (20 172Rs delivered by August 2000, Central Missouri State University (16 172Rs delivered by March 2000), Kansas State University (15 172Rs delivered), and Singapore Flying College at Jandakot, Perth, Australia (six 172Rs ordered in February 2000 in addition to five delivered in early 1998).

COSTS: 172R Skyhawk US$139,900 standard equipped (2000); Skyhawk SP US$146,900 (2000).

DESIGN FEATURES: Classic, all-metal, braced high-wing, touring lightplane. Tapered, mid-mounted tailplane and wing outer panels; sweptback fin.
NACA 2412 wing section. Dihedral 1° 44′. Incidence 1° 30′ at root, −1° 30′ at tip; fin sweepback 35° at quarter-chord.

FLYING CONTROLS: Conventional and manual. Actuation by stainless steel cables; modified Frise all-metal ailerons; electrically actuated single-slotted Para Lift flaps; trim tab on elevator.

STRUCTURE: Conventional light alloy, with composites for some non-structural components such as nose-bowl, wing-, tailplane- and fin-caps, and fairings. Epoxy corrosion proofing and polyurethane exterior paint standard.

LANDING GEAR: Fixed tricycle type with Cessna Land-O-Matic cantilever tapered steel tube main legs and oleo-pneumatic nose leg; steerable nosewheel, tyre size 5.00×5, mainwheel tyres size 6.00×6; hydraulic brakes. Wheel fairings optional, but included free when Nav I and II packages are purchased.

POWER PLANT: One 119 kW (160 hp) Textron Lycoming IO-360-L2A flat-four piston engine driving a two-blade, fixed-pitch McCauley metal propeller. Fuel contained in two integral wing tanks, combined capacity 212 litres (56.0 US gallons; 46.6 Imp gallons) of which 201 litres (53.0 US gallons; 44.1 Imp gallons) are usable. Refuelling points on upper surface of each wing. Oil capacity 7.6 litres (2.0 US gallons; 1.7 Imp gallons).

ACCOMMODATION: Four persons in two pairs on contoured, energy-absorbing seats dynamically tested to 26 g, those for pilot and front seat passenger being vertically adjustable, reclining and mounted on enlarged seat rails with dual locking pins; dual controls standard; rear bench seat with reclining back; door each side of cabin, hinged forward; inertia reel harnesses for all occupants; tinted windows with hinged side windows; ambient noise reduction soundproofing; composites headliner, and double-pin latching system for cabin doors. Baggage compartment to rear of cabin, with access door on port side. Cabin is heated and ventilated.

SYSTEMS: Electrical system comprising 28 V 60 A alternator and 24 V 12.75 Ah battery.

AVIONICS: *Comms:* Honeywell KX 155A nav/com; KMA 28 audio panel/marker beacon receiver and four-position voice-activated intercom; KT 76C Mode C transponder all standard.
Flight: KI 208 VOR/LOC indicator standard. Optional NAV I package comprises KLN 89 GPS-VFR, second KX 155A with glide slope, KI 209 VOR/LOC/GS indicator and KR 87 ADF; optional NAV II package comprises KLN 89B GPS-IFR, MD 41-228 GPS-NAV selector/annunciator, second KX 155A with glide slope, KI 209A VOR/LOC/GS indicator with GPS interface, KR 87 ADF and KAP-140 single-axis autopilot. Honeywell KMD 550 MFD optional.

DIMENSIONS, EXTERNAL:
Wing span	11.00 m (36 ft 1 in)
Wing chord: at root	1.63 m (5 ft 4 in)
at tip	1.12 m (3 ft 8½ in)
Wing aspect ratio	7.5
Length overall	8.28 m (27 ft 2 in)
Height overall	2.72 m (8 ft 11 in)
Wheel track	2.53 m (8 ft 3½ in)
Wheelbase	1.63 m (5 ft 4 in)
Propeller diameter	1.90 m (6 ft 3 in)
Cabin doors: Height	1.02 m (3 ft 4½ in)
Max width	0.94 m (3 ft 1 in)
Baggage door: Height	0.56 m (1 ft 10 in)
Max width	0.39 m (1 ft 3¼ in)

DIMENSIONS, INTERNAL:
Cabin (firewall to baggage area):	
Length	3.61 m (11 ft 10 in)
Max width	1.00 m (3 ft 3½ in)
Max height	1.22 m (4 ft 0 in)

AREAS:
Wings, gross	16.17 m² (174.0 sq ft)
Ailerons (total)	1.70 m² (18.3 sq ft)
Trailing-edge flaps (total)	1.98 m² (21.26 sq ft)
Fin	1.04 m² (11.24 sq ft)
Rudder, incl tab	0.69 m² (7.43 sq ft)
Tailplane	2.00 m² (21.56 sq ft)
Elevators, incl tab	1.35 m² (14.53 sq ft)

WEIGHTS AND LOADINGS:
Weight empty, standard	743 kg (1,639 lb)
Baggage capacity	54 kg (120 lb)
Max T-O and landing weight:	
Normal	1,111 kg (2,450 lb)
Utility	952 kg (2,100 lb)
Max ramp weight: Normal	1,114 kg (2,457 lb)
Utility	956 kg (2,107 lb)
Max wing loading	68.7 kg/m² (14.08 lb/sq ft)
Max power loading	9.32 kg/kW (15.31 lb/hp)

PERFORMANCE:
Max level speed at S/L	123 kt (227 km/h; 141 mph)
Cruising speed at 80% power at 2,440 m (8,000 ft)	122 kt (226 km/h; 140 mph)

Latest standard of Cessna 172S Skyhawk SP exhibited at NBAA Convention, October 2000 *(Paul Jackson/ Jane's)*
2001/0100481

Stalling speed, power off:
flaps up	51 kt (95 km/h; 59 mph)
flaps down	47 kt (87 km/h; 54 mph)
Max rate of climb at S/L	219 m (720 ft)/min
Service ceiling	4,115 m (13,500 ft)
T-O run	288 m (945 ft)
T-O to 15 m (50 ft)	514 m (1,685 ft)
Landing from 15 m (50 ft)	395 m (1,295 ft)
Landing run	168 m (550 ft)

Range with max fuel, 45 min reserves:
at 80% power at 2,440 m (8,000 ft)
580 n miles (1,074 km; 667 miles)
at 60% power at 3,050 m (10,000 ft)
687 n miles (1,272 km; 790 miles)
UPDATED

CESSNA 182S SKYLANE

TYPE: Four-seat lightplane.
PROGRAMME: Development of previous models described in 1985-86 and earlier editions of *Jane's*; total of 19,773 earlier Model 182/Skylanes were built between 1956 and 1985, when production was suspended. Manufacture resumed when three preproduction 182Ss assembled at East Pawnee, Wichita, plant under Single-Engine Pilot Program, of which the first (N182NU) flew on 16 July 1996; FAA FAR Pt 23 certification achieved 3 October 1996 after 164 flight test hours; assembly of first new production aircraft (N182FN) began at Independence 25 September 1996, with delivery 24 April 1997.
CURRENT VERSIONS: **182S Skylane:** Standard version until 2001 model year. New features introduced on 182S include a fuel-injected engine; Honeywell avionics; metal instrument panel with backlit, non-glare instruments; all-electric engine gauges, dual vacuum system, digital clock, EGT gauge, hour meter and centrally mounted annunciator panel; stainless steel control cables; epoxy corrosion-proofing; cabin improvements as detailed below; steps for visual checking of fuel tanks, which have quick-reference quantity tabs; consolidation of all primary electrical components into a single junction box on the firewall for simplified servicing; static wicks; GPU receptacle and fire extinguisher.

182T: Replaced 182S as standard version from 2001 model year; announced at the NBAA Convention at New Orleans on 9 October 2000; features redesigned wingtips, restyled and reclocked main landing gear fairings, improved nose gear fairing, low-drag wing strut fairings, streamlined refuelling and entrance steps, and improved interiors featuring sculptured composite side panels with integrated cupholders and cushioned armrests, removable overhead panels for simplified maintenance, Rosen sun visors, lightweight door handles, and a 12 V electrical port for use with a portable GPS, CD player or laptop computer. New optional avionics include a 12.7 cm (5 in) Honeywell KMD 550 LCD colour MFD with Stormscope that interfaces with the previously available KLN 94 colour IFR GPS. Standard empty weight 860 kg (1,897 lb); maximum level spped at S/L 149 kt (276 km/h; 171 mph); cruising speed at 80 per cent power at 1,829 m (6,000 ft), 144 kt (267 km/h; 166 mph); range with maximum fuel, 45 min reserves at 75 per cent power at 1,830 m (6,000 ft), 845 n miles (1,565 km; 972 miles).

T182T Turbo Skylane: Turbocharged model. Described separately.

Millennium Edition: Special configuration limited edition of 182S, announced in February 2000 and available from April 2000. Features redesigned cabin interior with leather seats; floor mats embossed with Cessna Millennium logo; new cup- and chart-holders; Rosen sun visors; polished spinner and cowling fasteners; Goodyear Flight Custom II tyres; Honeywell KLN 94 colour IFR GPS with moving map display; new exterior paint scheme; and pilot's accessory collection.
CUSTOMERS: Recent customers include the Mexican Air Force, which had taken delivery of 76 Skylanes by June 2000. Total of 248 delivered in 1999, and 267 in 2000.
COSTS: US$227,900 standard equipped (2000).
DESIGN FEATURES: Generally as for Cessna 172.
Wing section NACA 2412 (modified); incidence 0° 47' at root, −2° 50' at tip; dihedral 1° 44'.
FLYING CONTROLS: Conventional and manual. Actuation by stainless steel control cables; modified Frise ailerons; electrically actuated single-slotted Para Lift flaps; trim tabs on elevator and rudder.
STRUCTURE: Conventional light alloy, with composites for some non-structural components such as nose-bowl, wing-, tailplane- and fin-caps, and fairings. Epoxy corrosion proofing and polyurethane exterior paint standard.
LANDING GEAR: Fixed tricycle type with Cessna Land-O-Matic cantilever tapered steel tube main legs and oleo-pneumatic nose leg; steerable nosewheel, tyre size 5.00×5, mainwheel tyre size 6.00×6; hydraulic brakes. Wheel fairings optional.
POWER PLANT: One 172 kW (230 hp) Textron Lycoming IO-540-AB1A5 flat-six piston engine driving a three-blade, constant-speed McCauley metal propeller. Fuel contained in two integral wing tanks, combined capacity 348 litres (92.0 US gallons; 76.6 Imp gallons) of which 333 litres (88.0 US gallons; 73.3 Imp gallons) are usable.

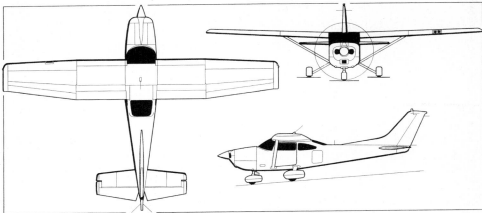

Cessna 182S Skylane *(Paul Jackson/Jane's)* 0015672

Cessna 182S Skylane Millennium Edition *(Paul Jackson/Jane's)* 2001/0100482

Refuelling points on upper surface of each wing. Oil capacity 8.5 litres (2.25 US gallons; 1.9 Imp gallons).
ACCOMMODATION: Four persons in two pairs on contoured, energy-absorbing seats, dynamically tested to 26 g, those for pilot and front seat passenger being vertically adjustable, reclining and mounted on enlarged seat rails with dual locking pins; dual controls standard; rear bench seat with reclining back; inertia reel harnesses for all occupants; tinted windows with hinged side windows; ambient noise reduction soundproofing; composites headliner, and double-pin latching system for cabin doors. Baggage compartment to rear of cabin, with access door on port side. Cabin is heated and ventilated.
SYSTEMS: Electrical system comprising 28 V 60 A alternator and 24 V 12.75 Ah battery.
AVIONICS: *Comms:* Dual Honeywell KX 155A nav/com, one with glide slope; KMA 26 audio panel/marker beacon receiver and four-position voice-activated intercom; KT 76C Mode C transponder all standard.
Flight: KLN 89 GPS-VFR and KAP 140 single-axis autopilot standard. KI 209A VOR indicator with GPS interface; KI 208 VOR/LOC indicator and MD 41-230 GPS-NAV selector all standard. Optional NAV package comprises KLN 89B GPS-IFR (exchange for GPS-VFR), exchange MD 41-228 GPS-NAV selector/annunciator, Honeywell KCS 55A, KR 87 ADF and KAP-140 two-axis autopilot with electric trim and altitude preselect (exchange for single-axis unit).

DIMENSIONS, EXTERNAL:
Wing span	10.97 m (36 ft 0 in)
Wing chord: at root	1.63 m (5 ft 4 in)
at tip	1.09 m (3 ft 7 in)
Wing aspect ratio	7.4
Length overall	8.84 m (29 ft 0 in)
Height overall	2.84 m (9 ft 4 in)
Wheel track	2.74 m (9 ft 0 in)
Wheelbase	1.69 m (5 ft 6½ in)
Propeller diameter	2.01 m (6 ft 7 in)
Cabin doors: Height	1.04 m (3 ft 5 in)
Max width	0.93 m (3 ft 0½ in)
Baggage door: Height	0.56 m (1 ft 10 in)
Width	0.40 m (1 ft 3¾ in)

DIMENSIONS, INTERNAL:
Cabin (firewall to baggage compartment):
Length	3.40 m (11 ft 2 in)
Max width	1.07 m (3 ft 6 in)
Max height	1.23 m (4 ft 0½ in)

AREAS:
Wings, gross	16.30 m² (175.50 sq ft)
Ailerons (total)	1.70 m² (18.3 sq ft)
Trailing-edge flaps (total)	1.97 m² (21.20 sq ft)
Fin	1.08 m² (11.62 sq ft)
Rudder	0.65 m² (6.95 sq ft)
Tailplane	2.13 m² (22.96 sq ft)
Elevators	1.54 m² (16.61 sq ft)

WEIGHTS AND LOADINGS:
Weight empty	875 kg (1,928 lb)
Baggage capacity	91 kg (200 lb)
Max T-O weight	1,406 kg (3,100 lb)
Max ramp weight	1,411 kg (3,110 lb)
Max landing weight	1,338 kg (2,950 lb)
Max wing loading	86.2 kg/m² (17.66 lb/sq ft)
Max power loading	8.20 kg/kW (13.48 lb/hp)

PERFORMANCE:
Max level speed at S/L	145 kt (268 km/h; 166 mph)

Cruising speed at 80% power at 1,830 m (6,000 ft)
140 kt (259 km/h; 161 mph)

Stalling speed, power off:
flaps up	54 kt (100 km/h; 63 mph)
flaps down	49 kt (91 km/h; 57 mph)
Max rate of climb at S/L	282 m (924 ft)/min
Service ceiling	5,515 m (18,100 ft)
T-O run	242 m (795 ft)
T-O to 15 m (50 ft)	462 m (1,514 ft)
Landing from 15 m (50 ft)	412 m (1,350 ft)
Landing run	180 m (590 ft)

Range with max fuel, 45 min reserves:
at 75% power at 1,981 m (6,500 ft)
820 n miles (1,518 km; 943 miles)
at 55% power at 3,050 m (10,000 ft)
968 n miles (1,792 km; 1,114 miles)
UPDATED

CESSNA T182T TURBO SKYLANE

TYPE: Four-seat lightplane.
PROGRAMME: Announced at NBAA Convention at New Orleans on 9 October 2000.
Description for Cessna 182S and 182T Skylanes applies also to the T182T Turbo Skylane, except as follows.
CUSTOMERS: Turbo Skylane N72US, delivered in early February 2001 to Lincoln Park Aviation of New Jersey, was the 3,000th single-engined piston aircraft delivered by Cessna since production resumed.
POWER PLANT: One 175.2 kW (235 hp) Textron Lycoming TIO-540-AK1A turbocharged flat-six engine, driving a three-blade, constant-speed McCauley metal propeller.

WEIGHTS AND LOADINGS:
Standard weight empty	929 kg (2,048 lb)
Max ramp weight	1,412 kg (3,112 lb)
Max power loading	8.03 kg/kN (13.19 lb/hp)

PERFORMANCE:
Max level speed at 6,100 m (20,000 ft)
170 kt (315 km/h; 196 mph)
Cruising speed at 88% power at 3,660 m (12,000 ft)
160 kt (296 km/h; 184 mph)
Max rate of climb at S/L	323 m (1,060 ft)/min
Max operating altitude	6,100 m (20,000 ft)
T-O run	241 m (790 ft)
T-O to 15 m (50 ft)	450 m (1,475 ft)

Computer-generated image of Cessna T182T Turbo Skylane

Instrument panel of Cessna T182T Turbo Skylane with optional multifunction display

Range with max fuel, 45 min reserves:
at 75% power at 3,050 m (10,000 ft)
700 n miles (1,296 km; 805 miles)

NEW ENTRY

CESSNA 206H STATIONAIR and T206H TURBO STATIONAIR

TYPE: Six-seat utility transport.
PROGRAMME: Development of previous models described in 1984-85 and earlier editions of *Jane's*; beginning in 1964, total of 7,556 earlier Model 206/Skywagons/Super Skylanes/Stationairs/Turbo Stationairs had been built when production was suspended in 1985. New model T206H prototype (N732CP) first flown from Wichita 28 August 1996; new model 206H prototype first flown 6 November 1996; assembly of first Pilot Program T206H began 3 December 1996.

FAA certification of 206H achieved 9 September 1998, followed by approval for T206H on 1 October 1998. First delivery, of 10 206Hs to Uruguayan Air Force, began 8 December 1998.

CURRENT VERSIONS: **206H Stationair** and **T206H Turbo Stationair (Stationair TC):** Standard versions. New features introduced on 206H and T206H comprise fuel-injected engines; new Honeywell avionics; metal instrument panel with backlit, non-glare instruments plus back-up cold fluorescent light system; all-electric engine gauges; stainless steel control cables; epoxy corrosion-proofing; cabin improvements as detailed below; strobe lights; handles and steps to check fuel levels; static wicks; GPU receptacle; and fire extinguisher.

New standard features announced at the NBAA Convention at New Orleans on 9 October 2000, and introduced on 206H and T206H from 2001 model year, include redesigned wingtips, restyled and reclocked main landing gear fairings, improved nose gear fairing, low-drag wing strut fairings, streamlined refuelling and entrance steps, and improved interiors featuring sculptured composite side panels with integrated cupholders and cushioned armrests, removable overhead panels for simplified maintenance, Rosen sun visors, lightweight door handles, and a 12 V electrical port for use with a portable GPS, CD player or laptop computer. New optional avionics include a 12.7 cm (5 in) Honeywell KMD 550 LCD colour MFD with Stormscope that interfaces with the previously available KLN 94 colour IFR GPS.

Description applies to the above version.

Millennium Edition: Special configuration limited edition, announced in February 2000 and available from April 2000. Features redesigned cabin interior with leather seats; floor mats embossed with Cessna Millennium logo; new cup- and chart-holders; Rosen sun visors; polished spinner and cowling fasteners; Goodyear Flight Custom II tyres; Honeywell KLN 94 IFR GPS; new exterior paint scheme; and a pilot's accessory collection.

Cessna T206H turbocharged version of Stationair *(Paul Jackson/Jane's)*

CUSTOMERS: Total of 116 Stationairs and 213 Turbo Stationairs delivered by 31 December 2000.
COSTS: Stationair US$311,000, Turbo Stationair US$347,700, both standard equipped (2000).
DESIGN FEATURES: Generally as for Cessna 172.
Wing section NACA 2412 (modified). Dihedral 2° 14'; incidence 1° 30' at root, −1° 30' at tip.
FLYING CONTROLS: As Cessna 182.
LANDING GEAR: As Cessna 182. Wipline 3450 floatplane conversion certified April 2000.
POWER PLANT: Stationair has one 224 kW (300 hp) Textron Lycoming IO-540-AC1A flat-six piston engine; Turbo Stationair has one 231 kW (310 hp) Textron Lycoming TIO-540-AJ1A turbocharged flat-six piston engine; each drives a three-blade, constant-speed McCauley metal propeller. Total fuel capacity 348 litres (92.0 US gallons; 76.6 Imp gallons), of which 333 litres (88.0 US gallons; 73.3 Imp gallons) are usable. Oil capacity 10.4 litres (2.75 US gallons; 2.3 Imp gallons).
ACCOMMODATION: Forward-hinged door, port side; double cargo doors starboard side, each opening away from centre. Six persons in three pairs on contoured, energy-absorbing seats, dynamically tested to 26 g, those for pilot and front seat passenger being vertically adjustable, reclining and mounted on enlarged seat rails with dual locking pins; other seats have reclining backs; inertia reel harnesses for all occupants; tinted windows; hinged front side windows; composites headliner; multilevel ventilation system; ambient noise reduction soundproofing and double-pin latching system on pilot's door, all standard. Utility interior optional.
AVIONICS: Standard Honeywell package.

Comms: Dual nav/com, audio panel with marker beacon receiver, digital transponder with encoder and six-position voice-activated intercom.
Flight: Glide slope receiver, VFR-GPS and single-axis autopilot. Optional NAV package includes IFR-GPS (exchange), ADF and dual-axis autopilot (exchange).
EQUIPMENT: Optional equipment includes electric propeller anti-icing, 0.45 m³ (16 cu ft) cargo pod, floatplane provisions kit and oversized rough terrain tyres.

DIMENSIONS, EXTERNAL:
Wing span	10.97 m (36 ft 0 in)
Wing chord: at root	1.63 m (5 ft 4 in)
at tip	1.13 m (3 ft 8½ in)
Wing aspect ratio	7.4
Length overall	8.61 m (28 ft 3 in)
Height overall	2.83 m (9 ft 3½ in)
Wheel track	2.46 m (8 ft 1 in)
Wheelbase	1.76 m (5 ft 9¼ in)
Pilot's door (port): Max height	1.04 m (3 ft 5 in)
Max width	0.94 m (3 ft 1 in)
Cargo double door (stbd):	
Max height	1.00 m (3 ft 3¼ in)
Max width	1.09 m (3 ft 7 in)
Height to sill	0.64 m (2 ft 1 in)

DIMENSIONS, INTERNAL:
Cabin: Length	3.68 m (12 ft 1 in)
Max width	1.12 m (3 ft 8 in)
Max height	1.26 m (4 ft 1½ in)
Volume available for payload	2.87 m³ (101.2 cu ft)

AREAS:
Wings, gross	16.30 m² (175.5 sq ft)
Ailerons (total)	1.61 m² (17.35 sq ft)
Trailing-edge flaps (total)	2.68 m² (28.85 sq ft)
Fin	1.08 m² (11.62 sq ft)
Rudder, incl tab	0.65 m² (6.95 sq ft)
Tailplane	2.31 m² (24.84 sq ft)
Elevators, incl tab	1.86 m² (20.08 sq ft)

WEIGHTS AND LOADINGS:
Weight empty, standard:	
206H	1,002 kg (2,210 lb)
T206H	1,031 kg (2,274 lb)
Max baggage	82 kg (180 lb)
Max fuel: 206H, T206H	239 kg (528 lb)
Max T-O and landing weight:	
206H, T206H	1,632 kg (3,600 lb)
Max ramp weight: 206H	1,639 kg (3,614 lb)
T206H	1,641 kg (3,617 lb)
Max wing loading:	
206H, T206H	100.2 kg/m² (20.51 lb/sq ft)
Max power loading: 206H	7.30 kg/kW (12.00 lb/hp)
T206H	7.07 kg/kW (11.61 lb/hp)

PERFORMANCE:
Max level speed: 206H at S/L	151 kt (280 km/h; 174 mph)

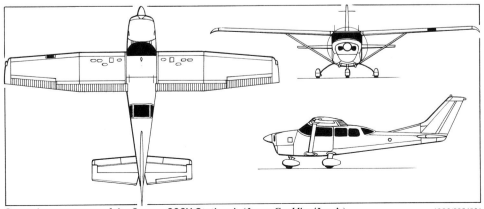

General arrangement of the Cessna 206H Stationair *(James Goulding/Jane's)*

634 USA: AIRCRAFT—CESSNA

T206H at 5,182 m (17,000 ft)
178 kt (330 km/h; 205 mph)
Cruising speed at 75% power:
206H at 1,890 m (6,200 ft) 142 kt (263 km/h; 163 mph)
T206H at 6,100 m (20,000 ft)
164 kt (304 km; 189 mph)
Stalling speed:
206H, T206H: flaps up   62 kt (115 km/h; 72 mph)
flaps down   54 kt (100 km/h; 63 mph)
Max rate of climb at S/L: 206H   301 m (988 ft)/min
T206H   320 m (1,050 ft)/min
Service ceiling: 206H   4,785 m (15,700 ft)
T206H   8,230 m (27,000 ft)
T-O run: 206H, T206H   277 m (910 ft)
T-O to 15 m (50 ft): 206H   567 m (1,860 ft)
T206H   530 m (1,740 ft)
Landing from 15 m (50 ft)   425 m (1,395 ft)
Landing run: 206H, T206H   224 m (735 ft)
Range with max fuel, 45 min reserves:
206H: at 75% power at 1,890 m (6,200 ft)
605 n miles (1,120 km; 696 miles)
max range at 1,981 m (6,500 ft)
730 n miles (1,352 km; 840 miles)
T206H: at 75% power at 3,050 m (10,000 ft)
541 n miles (1,002 km; 622 miles)
at 75% power at 6,100 m (20,000 ft)
568 n miles (1,052 km; 653 miles)
max range at 6,100 m (20,000 ft)
692 n miles (1,281 km; 796 miles)

UPDATED

CESSNA 208 CARAVAN U-27A
Brazilian Air Force designation: C-98
TYPE: Light utility turboprop.
PROGRAMME: First flight of engineering prototype (N208LP) 9 December 1982; first production Caravan rolled out August 1984; FAA certification October 1984; full production started 1985; wheeled float version certified March 1986. First single-engined aircraft to achieve FAA certification for ILS in Cat. II conditions (Federal Express aircraft equipped); approval for IFR cargo operations 1989 made France and Ireland first European countries to allow single-engined public transport day/night IFR operation; also since approved in Canada, Denmark and Sweden. Russian certification of 208 and 208B achieved 4 September 1998.
The 1,000th Cessna 208, a Grand Caravan, delivered to Tropic Air of Belize during 1998 NBAA Convention at Las Vegas, Nevada. 100th Caravan equipped with Wipaire 8000 floats delivered 16 May 2000. Total fleet operating time exceeded 4.5 million hours by October 2000, at which time Caravans were in service in 67 countries.
CURRENT VERSIONS: **208:** Basic utility model for passengers or cargo. Engine flat rated at 447 kW (600 shp) to 3,800 m (12,500 ft). Commissioned by Federal Express Corporation as **208A Cargomaster** freighter with special features including T-O weight 3,629 kg (8,000 lb), Honeywell avionics, no cabin windows or starboard rear door, more cargo tiedowns, additional cargo net, underfuselage cargo pannier of composites materials, 15.2 cm (6 in) vertical extension of fin/rudder, jetpipe deflected to carry exhaust clear of pannier.
208-675: Combines airframe of 208 with fully rated engine of 208B. Announced at NBAA Convention in Dallas, Texas, September 1997; FAA certification achieved in April 1998 with first delivery (N900RG; c/n 00277) 15 April to Riversville Aviation Company of New York in amphibious configuration; replaces 208.
Detailed description applies to 208-675, except where indicated.
208B: Stretched version, lacking cabin windows, and with ventral cargo pod as standard, developed at request of Federal Express. Commissioned by FedEx as **Super Cargomaster**; first flight (N9767F) 3 March 1986; certified October 1986; first delivery to FedEx 31 October 1986. Features include 1.22 m (4 ft) fuselage plug aft of wing, payload of 1,587 kg (3,500 lb) and 12.7 m³ (450 cu ft) of cargo volume; 503 kW (675 shp) P&WC PT6A-114A from 1991.
Grand Caravan: Announced at NBAA 1990; passenger version of 208B, with cabin windows, accommodating up to 14 in quick-change interior.
U-27A: Military utility/special mission derivative of 208 Caravan I; announced 1985; US DoD designation assigned for potential FMS contract purposes only (sponsoring service, US Army); roles include cargo, logistic support, paratroop or supply dropping, medevac, electronic surveillance, forward air control, passenger/troop transport, C³I, maritime patrol, SAR, psychological warfare, radio relay/RPV control, military base support, range safety patrol, reconnaissance and fire patrol. Fittings can include six underwing and one centreline hardpoints, observation windows and bubble windows for downward view, centreline reconnaissance pod, 2.8 m³ (84 cu ft) cargo pannier from 208 and two-part electrically actuated upward- and downward-rolling shutter door with slipstream deflector.
Soloy Pathfinder 21: Dual-engine conversion; described in *Jane's Aircraft Upgrades*.
CUSTOMERS: Federal Express Corporation received 40 208As and 260 208Bs. Recent customers include Ben Air Inc of Stauning, Denmark, which will take delivery of five Grand Caravans between 2000 and 2004 for onward sale to Scandinavian customers; and Shandong Airlines of the People's Republic of China, which ordered one Grand Caravan and two Caravan 675s on amphibious floats for delivery in first quarter 2001, plus 37 options. Other customers include Royal Canadian Mounted Police (first amphibious version), Brazilian Air Force (eight), Chilean Army (three delivered from 7 February 1998 onwards), Liberian Army (one) and Malaysian Police (six delivered in 1994).
Total of 1,209 Caravans (all versions) delivered by 31 December 2000, including 102 in 1998; 87 in 1999 and 92 in 2000.
COSTS: US$1,265,000 (208-675); US$1,285,000 (Grand Caravan); US$1,335,000 (Super Cargomaster) (2001).
DESIGN FEATURES: Launched as first all-new single-engined turboprop general aviation aircraft; intention was to replace de Havilland Canada Beavers and Otters, Cessna 180s, 185s and 206s in worldwide utility role. Main qualities advertised are high speed with heavy load,

Cessna Grand Caravan   2001/0100439

compatibility with unprepared strips, economy and reliability with minimum maintenance; can also carry weather radar, air conditioning and oxygen systems; optional packs for firefighting, photography, spraying, ambulance/hearse, border patrol, parachuting and supply dropping, surveillance and government utility missions; optional wheel or float landing gear.
Braced, high-wing design; tapered wings and tailplane; sweptback fin with dorsal fillet; auxiliary fins on floatplane.
Wing aerofoil NACA 23017.424 at root, 23012 at tip; dihedral 3° from root; incidence 2° 37' at root, −0° 36' at tip.
FLYING CONTROLS: Conventional and manual. Lateral control by small ailerons and slot-lip spoilers ahead of outer section of flaps; aileron trim standard; all tail control surfaces horn balanced; fixed tailplane with upper surface vortex generators ahead of elevator; elevator trim tabs; electrically actuated single-slotted flaps occupy more than 70 per cent of trailing-edge and deflect to maximum 30°.
STRUCTURE: All-metal. Fail-safe two-spar wing; conventional fuselage.
LANDING GEAR: Non-retractable tricycle type, with single wheel on each unit. Tubular spring cantilever main units; oil-damped steerable nosewheel. Mainwheel tyres size 6.50-10 (8 ply); nosewheel 6.50-8 (8 ply). Oversize tyres, optional, mainwheels 8.50-10, nosewheel 22×8.00-8, and extended nosewheel fork, optional. Hydraulically actuated single-disc brake on each mainwheel. Certified with Wipline 8000 floats (with or without retractable land wheels).
POWER PLANT: One 503 kW (675 shp) P&WC PT6A-114A turboprop. McCauley three-blade constant-speed reversible-pitch and feathering metal propeller. Integral fuel tanks in wings, total capacity 1,268 litres (335 US gallons; 279 Imp gallons), of which 1,257 litres (332 US gallons; 276.5 Imp gallons) are usable.
ACCOMMODATION: Pilot and up to nine passengers or 1,360 kg (3,000 lb) of cargo. Maximum seating capacity with FAR Pt 23 waiver is 14. Cabin has a flat floor with cargo track attachments for a combination of two- and three-abreast seating, with an aisle between seats. Forward-hinged door for pilot, with direct vision window, on each side of forward fuselage. Airstair door for passengers at rear of cabin on starboard side. Cabin is heated and ventilated. Optional air conditioning. Two-section horizontally split cargo door at rear of cabin on port side, flush with floor at bottom and with square corners. Upper portion hinges upward, lower portion forward 180°. Optional electrically operated, flight openable tambour roll-up door with air flow deflecting spoiler. In a cargo role, cabin will accommodate, typically, two D-size cargo containers or up to ten 208 litre (55 US gallon; 45.8 Imp gallon) drums.
SYSTEMS: Electrical system is powered by 28 V 200 A starter/generator and 24 V 45 Ah lead-acid battery (24 V 40 Ah Ni/Cd battery optional). Standby electrical system, with 75 A alternator, optional. Hydraulic system for brakes only. Oxygen system, capacity 3,315 litres (117 cu ft), optional. Vacuum system standard. Cabin air conditioning system optional from c/n 208-00030 onwards. De-icing system, comprising electric propeller de-icing boots, pneumatic wing, wing strut and tail surface boots, electrically heated windscreen panel, heated pitot/static probe, ice detector light and standby electrical system, all optional.
AVIONICS: Standard Honeywell Silver Crown package.
  Comms: Single VHF transceiver standard.
  Radar: Honeywell RDR-2000 colour weather radar optional; housed in pod on starboard wing leading-edge.
  Flight: Nav receiver, ADF and transponder standard.
  Instrumentation: Sensitive altimeter, electric clock, magnetic compass, attitude and directional gyros, true airspeed indicator, turn and bank indicator, vertical speed

Cessna Caravan 675 floatplane *(Paul Jackson/Jane's)*   2001/0100484

Jane's All the World's Aircraft 2001-2002   www.janes.com

Cessna 208B Caravan instrument panel *(Paul Jackson/Jane's)*

indicator, ammeter/voltmeter, fuel flow indicator, ITT indicator, oil pressure and temperature indicator.

EQUIPMENT: Standard equipment includes windscreen defrost, ground service plug receptacle, variable intensity instrument post lighting, map light, overhead courtesy lights (three) and overhead floodlights (pilot and co-pilot), approach plate holder, cargo tiedowns, internal corrosion proofing, vinyl floor covering, emergency locator beacon, partial plumbing for oxygen system, pilot's and co-pilot's adjustable fore/aft/vertical/reclining seats with armrest and five-point restraint harness, tinted windows, control surface bonding straps, heated pitot and stall warning systems, retractable crew steps (port side), rudder gust lock, tiedowns and towbar.

Optional equipment includes passenger seats, stowable folding utility seats, digital clock, fuel totaliser, turn co-ordinator, flight hour recorder, fire extinguisher, dual controls, co-pilot flight instruments, floatplane kit (from c/n 208-00030 onwards), hoisting rings (for floatplane), inboard fuel filling provisions (included in floatplane kit), ice detection light, courtesy lights on wing underside, passenger reading lights, flashing beacon, retractable crew step for starboard side, oversized tyres, electric trim system, oil quick drain valve and fan-driven ventilation system.

DIMENSIONS, EXTERNAL (L: landplane, A: amphibian):
Wing span	15.88 m (52 ft 1 in)
Wing chord: at root	1.98 m (6 ft 6 in)
at tip	1.22 m (4 ft 0 in)
Wing aspect ratio	9.7
Length overall: L	11.46 m (37 ft 7 in)
Height overall: L	4.52 m (14 ft 10 in)
A (on land)	4.98 m (16 ft 4 in)
Tailplane span	6.25 m (20 ft 6 in)
Wheel track: L	3.56 m (11 ft 8 in)
A	3.25 m (10 ft 8 in)
Wheelbase: L	3.54 m (11 ft 7½ in)
A	4.44 m (14 ft 7 in)
Propeller diameter	2.69 m (8 ft 10 in)
Airstair door: Height	1.27 m (4 ft 2 in)
Width	0.61 m (2 ft 0 in)
Cargo door: Height	1.27 m (4 ft 2 in)
Width	1.24 m (4 ft 1 in)

DIMENSIONS, INTERNAL (208):
Cabin: Length, excl baggage area	4.57 m (15 ft 0 in)
Max width	1.57 m (5 ft 2 in)
Max height	1.30 m (4 ft 3 in)
Volume	9.7 m³ (341 cu ft)
Cargo pannier: Length	5.58 m (18 ft 3½ in)
Width	1.28 m (4 ft 2½ in)
Depth	0.50 m (1 ft 7½ in)
Volume	2.4 m³ (83 cu ft)

AREAS:
Wings, gross	25.96 m² (279.4 sq ft)
Vertical tail surfaces (total, incl dorsal fin)	3.57 m² (38.41 sq ft)
Horizontal tail surfaces (total)	6.51 m² (70.04 sq ft)

WEIGHTS AND LOADINGS (civil 208-675; L: landplane, F: floatplane, A: amphibian):
Weight empty: L	1,780 kg (3,925 lb)
F	2,063 kg (4,548 lb)
A	2,220 kg (4,895 lb)
Baggage capacity	147 kg (325 lb)
Cargo pannier capacity	372 kg (820 lb)
Max fuel	1,009 kg (2,224 lb)
Max ramp weight	3,645 kg (8,035 lb)
Max T-O weight	3,629 kg (8,000 lb)
Max landing weight	3,538 kg (7,800 lb)
Max wing loading	139.8 kg/m² (28.63 lb/sq ft)
Max power loading	7.21 kg/kW (11.85 lb/shp)

WEIGHTS AND LOADINGS (U-27A; L and A as above): As civil 208 except:
Weight empty, standard: L	1,780 kg (3,925 lb)
A	2,220 kg (4,895 lb)

PERFORMANCE (civil 208-675; L, F and A as above):
Max operating speed (V_{MO})	175 kt (325 km/h; 202 mph) IAS
Max cruising speed at 3,050 m (10,000 ft):	
L	186 kt (344 km/h; 214 mph)
A	163 kt (302 km/h; 188 mph)
Stalling speed, power off:	
L: flaps up	75 kt (139 km/h; 87 mph) CAS
flaps down	61 kt (113 km/h; 71 mph) CAS
F, A, landing configuration	59 kt (109 km/h; 68 mph) CAS
Max rate of climb at S/L: L	320 m (1,050 ft)/min
F, A	251 m (823 ft)/min
Max certified altitude: F, A	6,100 m (20,000 ft)
L	7,620 m (25,000 ft)
T-O run: L	354 m (1,160 ft)
T-O run, water: F, A	585 m (1,919 ft)
T-O to 15 m (50 ft): L	626 m (2,053 ft)
A, F, from water	919 m (3,015 ft)
Landing from 15 m (50 ft) at S/L, without propeller reversal: L	505 m (1,655 ft)
A	590 m (1,935 ft)
Landing run at S/L, without propeller reversal:	
L	227 m (745 ft)
A	319 m (1,045 ft)

Range with max fuel, at max cruise power, allowances for start, taxi and reserves stated:
L at 3,050 m (10,000 ft), 45 min	932 n miles (1,726 km; 1,072 miles)
L at 6,100 m (20,000 ft), 45 min	1,220 n miles (2,259 km; 1,404 miles)
F, A at 3,050 m (10,000 ft), 30 min	990 n miles (1,833 km; 1,139 miles)

Range with max fuel at max range power, allowances as above: L at 3,050 m (10,000 ft)
1,085 n miles (2,009 km; 1,248 miles)
L at 6,100 m (20,000 ft)
1,295 n miles (2,398 km; 1,490 miles)
g limits +3.8/−1.52

PERFORMANCE (208B; A: standard; B: with cargo pannier):
Max cruising speed:	
at 3,050 m (10,000 ft): A	184 kt (341 km/h; 212 mph)
B	175 kt (324 km/h; 201 mph)
at 6,100 m (20,000 ft): A	174 kt (322 km/h; 200 mph)
B	164 kt (304 km/h; 189 mph)
Stalling speed, power off:	
flaps up: A, B	75 kt (139 km/h; 87 mph)
flaps down: A, B	61 kt (113 km/h; 71 mph)
Max rate of climb at S/L: A	297 m (975 ft)/min
B	282 m (925 ft)/min
Service ceiling: A, B	7,620 m (25,000 ft)
T-O run: A	416 m (1,365 ft)
B	428 m (1,405 ft)
T-O to 15 m (50 ft): A	738 m (2,420 ft)
B	762 m (2,500 ft)
Landing from 15 m (50 ft), without propeller reversal:	
A	547 m (1,795 ft)
B	530 m (1,740 ft)
Landing run, without propeller reversal:	
A	290 m (950 ft)
B	279 m (915 ft)

Range with max fuel, max cruise power, allowances for start, taxi, climb and descent, plus 45 min reserves:
at 3,050 m (10,000 ft):	
A	907 n miles (1,679 km; 1,043 miles)
B	862 n miles (1,596 km; 992 miles)
at 5,490 m (18,000 ft), conditions as above:	
A	1,109 n miles (2,053 km; 1,276 miles)
B	1,044 n miles (1,933 km; 1,201 miles)

Range with max fuel, max range power, allowances for start, taxi, climb and descent, plus 45 min reserves:
at 3,050 m (10,000 ft):	
A	1,026 n miles (1,900 km; 1,180 miles)
B	963 n miles (1,783 km; 1,108 miles)
at 5,490 m (18,000 ft), conditions as above:	
A	1,163 n miles (2,153 km; 1,338 miles)
B	1,076 n miles (1,992 km; 1,238 miles)

UPDATED

REIMS-CESSNA F406/CARAVAN II

See Reims Aviation in French section.

CESSNA 525 CITATION CJ1

TYPE: Light business jet.

PROGRAMME: Original CitationJet announced at NBAA Convention 1989 as replacement for Citation 500 and I (production of which stopped 1985); first flight of FJ44 turbofans in Citation 500 April 1990; first flight of CitationJet (N525CJ) 29 April 1991; first flight of second (preproduction) prototype 20 November 1991; FAA certification for single-pilot operation received 16 October 1992; first customer delivery 30 March 1993. Russian certification achieved 4 September 1998. RVSM Group approval granted by FAA in January 2000.

Replacement Citation CJ1 announced at NBAA Convention 18 October 1998; first aircraft (N31CJ) completed late 1999; FAA certification 16 February 2000; first customer delivery 31 March 2000 to the Commercial Envelope Company of Deer Park, New York.

CUSTOMERS: Total 359 CitationJets, plus three prototypes, delivered up to early 2000, when manufacture turned to CJ1. Total of 56 CJ1s delivered by 31 December 2000, when order backlog accounted for planned production until first quarter 2002. Recent customers include Avemex

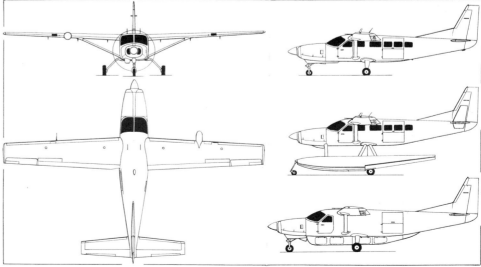

Cessna Caravan 675 with side views, top to bottom, of 208 Caravan 675, 208 wheeled floatplane and 208B Super Cargomaster *(Dennis Punnett/Jane's)*

SA of Toluca, Mexico (one); Taxi Marilia (TAM) of Brazil (three, including c/n 0500); Atlas Air Service GmbH of Germany (four), and Cessna's and TAG Aviation USA's fractional ownership operation, CitationShares, which has received one.

COSTS: US$3.716 million (2000).

DESIGN FEATURES: Small business jet of conventional appearance; wings attached below fuselage; T tail; tapered wings and tailplane; podded engines mounted clear of upper rear fuselage. Natural Laminar Flow (NLF) aerofoil section. CJ1 is generally identical to CitationJet; increased maximum T-O, ramp and landing weights to provide improved range/payload; Pro Line 21 avionics.

FLYING CONTROLS: Conventional and manual. Ailerons and elevators horn balanced; trim tab on port aileron; trim tabs on rudder and both elevators; single-slotted flaps.

STRUCTURE: All-metal. Three-spar wing.

LANDING GEAR: Hydraulically retractable tricycle type, with single wheel on each unit. Trailing-link main units retract inward into wing; nose gear retracts forward. Main tyres 22×7.75-10 (8 ply) tubeless; nose 18×4.4 (6 ply) tubeless.

POWER PLANT: Two 8.45 kN (1,900 lb st) Williams FJ44-1A turbofans. Fuel: see under Weights and Loadings.

ACCOMMODATION: Crew of two on flight deck. Main cabin with standard seating for five passengers, one on sideways-facing seat at front of cabin, with four in club arrangement; tray tables which stow in elbow rails, and individual reading lights and ventilation ducts, standard.

SYSTEMS: Pressurisation system, maximum differential 0.58 bar (8.5 lb/sq in).

AVIONICS: Honeywell Pro Line 21 as core system.
Comms: Honeywell CNI-5000 radios and radio altimeter.
Radar: Rockwell Collins RTA 800 colour weather radar.
Flight: Honeywell KLN-900 FMS. Dual VHF nav receivers, ADF, DME, RMI, Honeywell KLN-90B GPS.
Instrumentation: Two-tube EFIS with 203 × 253 mm (8 × 10 in) adaptive flight displays (third optional), LCD attitude director indicator (ADI) and horizontal situation indicator (HSI) for co-pilot.

DIMENSIONS, EXTERNAL:
Wing span	14.26 m (46 ft 9½ in)
Wing aspect ratio	9.1
Length overall	12.98 m (42 ft 7 in)
Height overall	4.20 m (13 ft 9¼ in)
Tailplane span	5.73 m (18 ft 9½ in)
Wheel track	3.96 m (13 ft 0 in)
Wheelbase	4.67 m (15 ft 4 in)
Crew/passenger door: Height	1.29 m (4 ft 2¾ in)
Width	0.60 m (1 ft 11½ in)

DIMENSIONS, INTERNAL:
Cabin: Length between pressure bulkheads	4.85 m (15 ft 11 in)
Max width	1.47 m (4 ft 10 in)
Max height	1.45 m (4 ft 9 in)
Baggage compartment volume:	
nose	0.69 m³ (24.4 cu ft)
cabin	0.11 m³ (4.0 cu ft)
tailcone	0.86 m³ (30.2 cu ft)
total	1.66 m³ (58.6 cu ft)

AREAS:
Wings, gross	22.30 m² (240.0 sq ft)
Vertical tail surfaces (total, incl tab)	4.35 m² (46.8 sq ft)
Horizontal tail surfaces (total, incl tabs)	5.64 m² (60.7 sq ft)

WEIGHTS AND LOADINGS:
Weight empty, typically equipped	3,016 kg (6,650 lb)
Max usable fuel weight	1,460 kg (3,220 lb)
Payload with max fuel, single pilot	306 kg (675 lb)
Max T-O weight	4,808 kg (10,600 lb)
Max landing weight	4,445 kg (9,800 lb)
Max ramp weight	4,853 kg (10,700 lb)

Cessna CitationJet six/seven-seat light business jet *(Paul Jackson/Jane's)* 2000/0075901

Flight deck of Cessna CitationJet 0015668

Max wing loading	215.6 kg/m² (44.17 lb/sq ft)
Max power loading	284 kg/kN (2.79 lb/lb st)

PERFORMANCE:
Max operating speed (V_{MO}):	
S/L to 9,300 m (30,500 ft)	263 kt (487 km/h; 302 mph) IAS
above 9,300 m (30,500 ft)	M0.70
Max cruising speed at 3,992 kg (8,800 lb) AUW at 10,670 m (35,000 ft)	380 kt (704 km/h; 437 mph)
Stalling speed, landing configuration at max landing weight	82 kt (152 km/h; 94 mph) CAS
Max rate of climb at S/L	985 m (3,230 ft)/min
Rate of climb at S/L, OEI	259 m (850 ft)/min
Max certified altitude	12,500 m (41,000 ft)
Service ceiling, OEI	6,462 m (21,200 ft)
Range with single pilot, three passengers, 45 min reserves	1,475 n miles (2,731 km; 1,697 miles)

UPDATED

CESSNA 525A CITATION CJ2

TYPE: Light business jet.

PROGRAMME: Design began 1 May 1998; construction of prototype started August 1998; announced at NBAA Convention at Las Vegas, Nevada, 18 October 1998; first flight of prototype (N2CJ, a rebuilt CitationJet) 27 April 1999, followed by first preproduction aircraft (N525AZ) on 15 October and second (N765CT) in mid-December 1999; total of 1,008 flight test hours accumulated by three aircraft by 9 April 2000; first production aircraft c/n 525A-0003/(N132CJ) rolled out 14 January 2000; FAA FAR Pt 23 certification achieved 21 June 2000; public debut at EAA AirVenture Oshkosh in July 2000; first customer delivery November 2000 to King Pharmaceuticals of Tennessee (second production aircraft).

CUSTOMERS: Total of 76 firm orders held at time of launch; by September 2000 order backlog accounted for planned production until fourth quarter of 2003; total of eight delivered by 31 December 2000. Recent customers include Avemex SA of Toluca, Mexico (one); Taxi Aereo Marilia (TAM) of Brazil (five); and Atlas Air Service GmbH of Germany (three).

COSTS: US$4.529 million (2000).

DESIGN FEATURES: Development of CitationJet/CJ1 (which see) with fuselage stretched by 1.30 m (4 ft 3 in) in cabin and tailcone areas (extra two cabin windows), wing span increased by 0.84 m (2 ft 9 in), swept tailplane of greater span, Williams FJ44-2C turbofans and Rockwell Collins Pro Line 21 avionics. Design goals included improvements in cabin room and comfort, speed and range over CitationJet. Common type rating with CJ1.

Description for the Citation CJ1 applies also to the Citation CJ2, except as follows:

FLYING CONTROLS: Spoilers, 60° flap deflection and engine nacelle thrust attenuators provide lift dump function during landing roll.

STRUCTURE: Primarily metal, with composites in non-critical areas such as fairings, wing and tailplane tips and exhaust nozzles.

POWER PLANT: Two Williams FJ44-2C turbofans, each flat rated at 10.68 kN (2,400 lb st) at 22°C (72°F). Integral fuel tank in each wing, total capacity 2,245 litres (593 US gallons; 494 Imp gallons); overwing gravity refuelling.

ACCOMMODATION: One or two crew on flight deck; six passengers in standard centre club layout with refreshment centre on starboard side; customised interiors to customer's choice. Flight accessible baggage area at rear of cabin; other baggage compartments in nose and tailcone. Cabin is pressurised, heated and air conditioned. Six cabin windows per side. Main door on port side forward of wing; emergency exit starboard aft.

SYSTEMS: Pressurisation system, maximum pressure differential 0.61 bar (8.9 lb/sq in). Vapour cycle air conditioning system. Electrical system supplied by battery and two engine-driven starter/generators. Oxygen system, capacity 623 litres (22.0 cu ft) standard, 1,417 litres (50.0 cu ft) optional.

AVIONICS: Rockwell Collins Pro Line 21 suite as core system.
Comms: Honeywell CNI-5000 panel-mounted radios.
Radar: Rockwell Collins RTA-800 solid-state colour weather radar.
Flight: Honeywell KLN 900 GPS FMS, VOR, DME, ADF, Rockwell Collins digital autopilot, digital air data computer and AHRS standard.
Instrumentation: Two-tube EFIS with 203 × 254 mm (8 × 10 in) PFD and MFD active matrix LCDs.

DIMENSIONS, EXTERNAL:
Wing span	15.09 m (49 ft 6 in)
Wing aspect ratio	9.3
Length overall	14.30 m (46 ft 11 in)
Height overall	4.24 m (13 ft 10¾ in)
Tailplane span	6.34 m (20 ft 9½ in)
Wheel track	4.88 m (16 ft 0 in)
Wheelbase	5.59 m (18 ft 4 in)
Passenger door: Height	1.29 m (4 ft 2¾ in)
Width	0.60 m (1 ft 11½ in)

DIMENSIONS, INTERNAL:
Cabin: Length: overall	5.77 m (18 ft 11 in)
excl cockpit	4.19 m (13 ft 9 in)
Max width	1.47 m (4 ft 10 in)
Max height	1.45 m (4 ft 9 in)
Baggage volume: nose	0.58 m³ (20.4 cu ft)
cabin	0.11 m³ (4.0 cu ft)
tailcone	1.42 m³ (50.0 cu ft)
total	2.11 m³ (74.4 cu ft)

AREAS:
Wings, gross	24.53 m² (264.0 sq ft)
Vertical tail surfaces	4.35 m² (46.8 sq ft)
Horizontal tail surfaces	6.50 m² (70.0 sq ft)

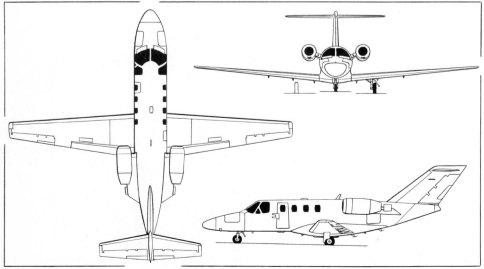
Cessna 525 Citation CJ1 (two Williams FJ44 turbofans) *(Dennis Punnett/Jane's)*

Prototype Cessna 525A Citation CJ2 with flight test probe on nose

WEIGHTS AND LOADINGS:
Weight empty, typically equipped	3,402 kg (7,500 lb)
Max fuel	1,797 kg (3,961 lb)
Payload with max fuel	363 kg (800 lb)
Baggage capacity: nose	181 kg (400 lb)
cabin	45 kg (100 lb)
tailcone	272 kg (600 lb)
total	499 kg (1,100 lb)
Max T-O weight	5,613 kg (12,375 lb)
Max landing weight	5,216 kg (11,500 lb)
Max ramp weight	5,670 kg (12,500 lb)
Max zero-fuel weight	4,218 kg (9,300 lb)
Max wing loading	228.9 kg/m² (46.88 lb/sq ft)
Max power loading	263 kg/kN (2.58 lb/lb st)

PERFORMANCE (estimated):
Max operating Mach No. (M_{MO})	0.72
Max level speed (V_{MO}):	
S/L to 2,440 m (8,000 ft)	260 kt (481 km/h; 299 mph)
S/L to 8,930 m (29,300 ft)	275 kt (509 km/h; 316 mph)
above 8,930 m (29,300 ft)	M0.72
Max cruising speed at 10,060 m (33,000 ft)	410 kt (759 km/h; 472 mph)
Econ cruising speed at 12,500 m (41,000 ft)	332 kt (615 km/h; 382 mph)
Stalling speed in landing configuration at MLW	86 kt (160 km/h; 99 mph)
Max rate of climb at S/L	1,180 m (3,870 ft)/min
Rate of climb at S/L, OEI	360 m (1,180 ft)/min
Time to: 11,280 m (37,000 ft)	17 min
13,105 m (43,000 ft)	36 min
Max certified altitude	13,715 m (45,000 ft)
FAR Pt 25 T-O balanced field length	1,042 m (3,420 ft)
Landing field length at max landing weight	908 m (2,980 ft)
Range, 45 min reserves	1,724 n miles (3,192 km; 1,984 miles)

UPDATED

CESSNA 550 CITATION BRAVO

TYPE: Business jet.
PROGRAMME: Announced at Farnborough Air Show September 1994; replaced Citation II; prototype (N550BB, c/n 0734) first flight 19 April 1995; initial production aircraft (N801BB, c/n 0801) flew mid-1996; FAA certification in January 1997; first delivery, to Firebond Corporation of Minden, Louisiana, on 25 February 1997. Russian certification achieved 4 September 1998.

Cessna Citation Bravo twin-turbofan business jet *(Paul Jackson/Jane's)*

CUSTOMERS: First 18 months' production had been sold by time of initial delivery. Earlier production in this series accounted for 621 Model 550 Citation IIs, 97 Model 551 Citation II SPs and 15 Model 552 T-47A Citations, or 733 in all. Total of 152 Bravos delivered by 31 December 2000, comprising 28 in 1997, 34 in 1998, 36 in 1999 and 54 in 2000. Recent customers include Avemex SA of Toluca, Mexico (two, for delivery from May 2002); Taxi Aereo Marilia (TAM) of Brazil (three, for delivery from April 2002); Flying Partners CV of Antwerp, Belgium, which has ordered two for its shared ownership programme, for delivery from January 2002; and Cessna's and TAG Aviation USA's fractional ownership operation, CitationShares, which placed an initial order for six, for delivery before the end of 2000.
COSTS: US$5.184 million, typically equipped (2000).
DESIGN FEATURES: Small/mid-size business jet. Based on Citation II airframe; tapered wing with sweptback inboard leading-edge; tapered, mid-set tailplane; sweptback fin; podded engines mounted clear of upper rear fuselage; wide track landing gear.
  Wing aerofoil NACA 23014 (modified) at centreline, NACA 23012 at wing station 247.95; dihedral 4°; tailplane dihedral 9°.
FLYING CONTROLS: Conventional and manual. Trim tab on port aileron is manually operated; manual rudder trim; electric elevator trim tab with manual standby; electrically actuated single-slotted flaps; hydraulically actuated airbrake.
STRUCTURE: Two primary, one auxiliary metal wing spars; three fuselage attachment points; conventional ribs and stringers. All-metal pressurised fuselage with fail-safe design providing multiple load paths.
LANDING GEAR: Hydraulically retractable tricycle type with single wheel on each unit. Trailing-link main units retract inward into the wing, nose gear forward. Free-fall and pneumatic emergency extension systems. Mainwheel tyres H22.0×8.25-10 (14 ply); steerable nosewheel with tyre size 18×4.4DD (10 ply); all tubeless. Hydraulic brakes. Parking brake and pneumatic emergency brake system. Anti-skid system optional.
POWER PLANT: Two 12.84 kN (2,887 lb st) Pratt & Whitney Canada PW530A turbofans; Nordam target-type thrust reversers standard. Integral fuel tanks in wings, with usable capacity of 2,725 litres (720 US gallons; 600 Imp gallons).
ACCOMMODATION: Crew of two on separate flight deck, on fully adjustable seats, with seat belts and inertia reel shoulder harness. Sun visors standard. Fully carpeted main cabin equipped with seats for seven to 10 passengers (seven standard, on pedestal-mounted seats). Main baggage areas in nose and tailcone; flight accessible baggage area at rear of cabin. Refreshment centre standard.

Standard six-seat cabin of Cessna 525A Citation CJ2

Flight deck of Cessna Citation CJ2

USA: AIRCRAFT—CESSNA

Flight deck of Cessna Citation Bravo　　0015667

Standard cabin interior of Cessna Citation Bravo, facing rear　　0015666

Cabin is pressurised, heated and air conditioned. Individual reading lights and air inlets for each passenger. Dropout constant-flow oxygen system for emergency use. Plug-type door with integral airstair at front on port side and one emergency exit on starboard side. Doors on each side of nose baggage compartment. Tinted windows, each with curtains. Pilot's storm window, birdproof windscreen with de-fog system, anti-icing, standby alcohol anti-icing and bleed air rain removal system.

SYSTEMS: Pressurisation system supplied with engine bleed air, maximum pressure differential 0.61 bar (8.9 lb/sq in), maintaining a sea level cabin altitude to 6,720 m (22,040 ft), or a 2,440 m (8,000 ft) cabin altitude to 13,110 m (43,000 ft). Hydraulic system, pressure 103.5 bars (1,500 lb/sq in), with two pumps to operate landing gear and speed brakes. Separate hydraulic system for wheel brakes. Electrical system supplied by two 28 V 400 A DC starter/generators, with two 350 VA inverters and 24 V 40 Ah Ni/Cd battery. Oxygen system of 1,814 litre (64 cu ft) capacity includes two crew demand masks and five dropout constant-flow masks for passengers. Engine fire detection and extinguishing systems. Wing leading-edges electrically de-iced ahead of engines; pneumatic de-icing boots on outer wings.

AVIONICS: Honeywell Primus 1000 with dual digital flight directors and IC-600 integrated avionics computer as core system.
　Comms: Dual Honeywell CN15000 transceivers; dual Mode S transponders; Lockheed Martin cockpit voice recorder.
　Radar: Honeywell P-650 colour weather radar.
　Flight: Dual Honeywell CN15000 receivers, DME, ADF and radio altimeter; Global GNS-XLS flight management system with GPS and Vnav.
　Instrumentation: EFIS with two 178 × 203 mm (7 × 8 in) primary flight display (PFD) screens combining attitude and HSI formats with additional air data; additional multifunction display (MFD) screen of same size with map/plan/checklist capability.

DIMENSIONS, EXTERNAL:
Wing span	15.90 m (52 ft 2 in)
Wing aspect ratio	8.4
Length overall	14.39 m (47 ft 2½ in)
Height overall	4.57 m (15 ft 0 in)
Tailplane span	5.79 m (19 ft 0 in)
Wheel track	4.06 m (13 ft 4 in)
Wheelbase	5.64 m (18 ft 6 in)
Crew/passenger door: Height	1.29 m (4 ft 2¾ in)
Width	0.60 m (1 ft 11½ in)

DIMENSIONS, INTERNAL:
Cabin: Length (between pressure bulkheads)
　　　　　　　　　　　　6.37 m (20 ft 11 in)
Max width	1.48 m (4 ft 10¼ in)
Max height	1.43 m (4 ft 8¼ in)
Baggage volume: nose	0.50 m³ (17.6 cu ft)
cabin	0.78 m³ (27.7 cu ft)
tailcone	0.80 m³ (28.2 cu ft)

AREAS:
Wings, gross	30.00 m² (322.9 sq ft)
Vertical tail surfaces (total)	4.73 m² (50.9 sq ft)
Horizontal tail surfaces (total, incl tab)	6.48 m² (69.8 sq ft)

WEIGHTS AND LOADINGS:
Weight empty, typically equipped	4,060 kg (8,950 lb)
Max fuel weight (usable)	2,204 kg (4,860 lb)
Max T-O weight	6,713 kg (14,800 lb)
Max ramp weight	6,804 kg (15,000 lb)
Max landing weight	6,123 kg (13,500 lb)
Max zero-fuel weight	5,126 kg (11,300 lb)
Max wing loading	223.8 kg/m² (45.83 lb/sq ft)
Max power loading	262 kg/kN (2.56 lb/lb st)

PERFORMANCE:
Max operating speed (V_{MO}):
　S/L to 2,440 m (8,000 ft)　260 kt (481 km/h; 299 mph)
　2,440 m (8,000 ft) to 8,500 m (27,880 ft)
　　　　　　　　　　275 kt (509 km/h; 316 mph)
　at 8,500 m (27,880 ft) and above　　M0.70
Max cruising speed at 9,450 m (31,000 ft) at average cruise weight of 5,670 kg (12,500 lb)
　　　　　　　　　　402 kt (745 km/h; 463 mph)
Stalling speed in landing configuration at max landing weight　　86 kt (160 km/h; 99 mph) CAS
Max rate of climb at S/L	974 m (3,195 ft)/min
Rate of climb at S/L, OEI	345 m (1,133 ft)/min
Max certified altitude	13,715 m (45,000 ft)
Service ceiling, OEI	8,458 m (27,750 ft)
T-O balanced field length (FAR Pt 25)	1,098 m (3,600 ft)

FAR Pt 25 landing field length at max landing weight
　　　　　　　　　　970 m (3,180 ft)
Range with four passengers, allowances for T-O, climb, cruise, descent and 45 min reserves
　　　　　　　　　2,000 n miles (3,704 km; 2,301 miles)
UPDATED

CESSNA 560 CITATION ULTRA
US Army designation: UC-35A
US Marine Corps designation: UC-35C
USAF designation: OT-47B

TYPE: Business jet.
PROGRAMME: Improved version of Citation V with digital autopilot, EFIS and increased payload and performance. First flight of Citation V engineering prototype (N560CC) August 1987; announced at NBAA Convention 1987; first flight of preproduction prototype early 1988; FAA certification 9 December 1988; first delivery April 1989; customers included Spanish Air Force with two SLAR-equipped aircraft, locally designated **TR.20**; total 260 built before mid-1994 change to Citation Ultra, which announced September 1993; FAA certification June 1994; Russian certification achieved 4 September 1998. Replaced by Citation Encore (which see, below) from c/n 0539 in March 2000. Described in 2000-01 *Jane's*.
UPDATED

CESSNA 560 CITATION ENCORE
US Army designation: UC-35B
US Marine Corps designation: UC-35D

TYPE: Business jet.
PROGRAMME: Prototype, based on rebuilt Citation Ultra, first flown (N560VU) 9 July 1998; announced at NBAA Convention at Las Vegas, Nevada 18 October 1998; first production aircraft (c/n 560-0539) rolled out in early March 2000. FAA certification achieved 26 April 2000; first delivery (N539CE/N5108G) on 29 September 2000 to J R Simplot Company of Boise, Idaho.
CURRENT VERSIONS: **Citation Encore:** Civilian business jet; *as described*.
　UC-35B: In January 1996, US Army selected Ultra for its C-XX medium-range transport aircraft programme, for which 35 aircraft are required over a five year period; equipped with 0.90 × 1.15 m (2 ft 11½ in × 3 ft 9 in) upward-opening clamshell cargo door on port forward fuselage; first order was for two UC-35As, of which one (95-0123) was delivered in last quarter of 1996; first two serve with 207 AvnCo at Heidelberg, Germany. Subsequent orders for five in each of FY96, FY97, FY98 and FY99; last two aircraft of final batch are first UC-35Bs.
CUSTOMERS: Deliveries of Ultra began July 1994; total of 258 delivered, comprising 24 in 1994, 56 in 1995, 52 in 1996, 47 in 1997, 41 in 1998, 32 in 1999 and six in 2000. Announced customers for Encore include J R Simplot Co of Boise, Idaho, which has ordered two; Taxi Aereo Marilia (TAM) of Brazil (one); Flying Partner CV of Antwerp, Belgium (one).
COSTS: US$7.150 million (2000).
DESIGN FEATURES: Stretched version of Citation S/II for full eight-seat cabin and with fully enclosed toilet/vanity area; seventh cabin window each side; two baggage compartments outside main cabin. Rear-engined executive jet with low-mounted wing of tapered planform with root gloves; tapered horizontal tail with sweptback fin and fillet.
　Encore generally as Citation Ultra and Citation Bravo (which see). Compared to Ultra, has increased wing span; bleed air anti-icing for wing leading-edges; boundary layer energisers and stall fences to improve stall characteristics; trailing-link main landing gear with narrower track; increased fuel capacity; Pratt & Whitney Canada PW535 turbofans offering 10 per cent increase in thrust and 15 per cent improvement in specific fuel consumption; reduced fuel capacity; fuel heaters obviating the need for additives; digital pressurisation system; improved braking system; single level access electrical junction box; and redesigned interior with increased headroom, new passenger service units, pin-mounted seats for easy removal/replacement and increased serviceability.
Description for Citation Ultra applies also to Citation Encore except as follows:
FLYING CONTROLS: Conventional and manual. Trim tab on port aileron manually actuated; manual rudder trim; electric elevator trim tab with manual standby; hydraulically actuated Fowler flaps; hydraulically actuated airbrake.
STRUCTURE: Two primary, one auxiliary metal wing spars; four fuselage attachment points, conventional ribs and stringers. All-metal pressurised fuselage with fail-safe design providing multiple load paths.

US Army UC-35A military communications version of Cessna Citation Ultra *(Paul Jackson/Jane's)*　　2001/0100486

First production Cessna 560 Citation Encore in flight test markings *2001/0100441*

LANDING GEAR: Hydraulically retractable trailing-link tricycle type with single wheel on each unit. Main units retract inward into the wing, nose gear forward. Free-fall and pneumatic emergency extension systems. Goodyear mainwheels with tubeless tyres size 22×8.0-10 (12 ply), pressure 6.90 bars (100 lb/sq in). Steerable nosewheel (±20°) with Goodyear wheel and tyre size 18×4.4DD (10 ply), pressure 8.27 bars (120 lb/sq in). Goodyear hydraulic brakes. Parking brake and pneumatic emergency brake system. Anti-skid system optional. Minimum ground turning radius about nosewheel 8.38 m (27 ft 6 in).

POWER PLANT: Two Pratt & Whitney Canada PW535A turbofans, each rated at 15.12 kN (3,400 lb st) at 27°C (80°F). Usable fuel capacity approximately 2,953 litres (780 US gallons; 649 Imp gallons).

ACCOMMODATION: Standard seating for seven passengers in three forward-facing and four in club arrangement, or eight passengers in double-club arrangement, on swivelling and fore/aft/inboard-tracking pedestal seats; refreshment centre in forward cabin area; lavatory/vanity centre with sliding doors to rear, metallic plating on cabin fittings, veneer overlay on armrests, optional pleated window shades and cabin divider mirror standard; space in aft section of cabin for 272 kg (600 lb) of baggage, in addition to baggage compartments in nose and rear fuselage.

SYSTEMS: Pressurisation system supplied with engine bleed air, maximum pressure differential 0.61 bar (8.8 lb/sq in), maintaining a sea level cabin altitude to 6,720 m (22,040 ft), or a 2,440 m (8,000 ft) cabin altitude to 12,495 m (41,000 ft). Hydraulic system, pressure 103.5 bars (1,500 lb/sq in), with two pumps to operate landing gear and speed brakes. Separate hydraulic system for wheel brakes. Electrical system supplied by two 28 V 300 A DC starter/generators, with two 350 VA inverters and 24 V 40 Ah Ni/Cd battery. Oxygen system of 0.62 m³ (22 cu ft) capacity includes two crew demand masks and five dropout constant flow masks for passengers. High-capacity oxygen system optional. Engine fire detection and extinguishing system. Bleed air anti-icing system for wing leading-edges and engine inlets; pneumatic de-icing boots on tailplane leading-edges.

AVIONICS: Standard avionics package based on Honeywell Primus 1000 digital flight control system with integrated avionics computer.

Comms: Dual Honeywell Primus II transceivers, dual TDR-94 transponders, dual altitude reporting systems and Lockheed Martin A200S cockpit voice recorder standard.

Radar: Honeywell Primus 660 colour weather radar standard.

Flight: Dual Honeywell Primus II, single Honeywell ADF, ALT-55 radio altimeter, coupled vertical navigation system, Global GNS-XL with GPS, expanded keyboard and colour CDU display standard. Universal Avionics UNS-1Csp FMS optional.

Instrumentation: Three-tube EFIS with 203 × 178 mm (8 × 7 in) CRTs comprising pilot's and co-pilot's primary flight displays (PFDs) and centrally mounted multifunction display (MFD); PFDs integrate functions of five flight instruments and several sources of navigation data, and provide trend data for airspeed, altitude and rate of climb.

DIMENSIONS, EXTERNAL:
Wing span	16.48 m (54 ft 1 in)
Wing aspect ratio	7.2
Length overall	14.90 m (48 ft 10¾ in)
Height overall	4.57 m (15 ft 0 in)
Tailplane span	6.55 m (21 ft 6 in)
Wheelbase	6.06 m (19 ft 10¾ in)
Wheel track	4.04 m (13 ft 3 in)

DIMENSIONS, INTERNAL:
Cabin:
Length:	
between pressure bulkheads	6.89 m (22 ft 7¼ in)
excl cockpit	5.28 m (17 ft 4 in)
Max width	1.49 m (4 ft 10¾ in)
Max height	1.40 m (4 ft 7 in)

AREAS:
Wings, gross	29.94 m² (322.3 sq ft)
Vertical tail surfaces (total, incl tab)	4.73 m² (50.90 sq ft)
Horizontal tail surfaces (total)	7.88 m² (84.80 sq ft)

WEIGHTS AND LOADINGS:
Weight empty, typically equipped	4,593 kg (10,125 lb)
Max fuel	2,468 kg (5,440 lb)
Max T-O weight	7,543 kg (16,630 lb)
Max ramp weight	7,634 kg (16,830 lb)
Max landing weight	6,895 kg (15,200 lb)
Max zero-fuel weight	5,715 kg (12,600 lb)
Max wing loading	251.9 kg/m² (51.60 lb/sq ft)
Max power loading	249 kg/kN (2.45 lb/lb st)

PERFORMANCE:
Max cruising speed at 10,670 m (35,000 ft)
   429 kt (795 km/h; 494 mph)
Stalling speed, flaps down   83 kt (154 km/h; 96 mph)
Max rate of climb at S/L   1,414 m (4,640 ft)/min
Time to climb:
  to 10,670 m (35,000 ft)   12 min
  to 13,715 m (45,000 ft)   28 min
Rate of climb at S/L, OEI   439 m (1,440 ft)/min
T-O balanced field length (FAR Pt 25) 1,064 m (3,490 ft)
Landing from 15 m (50 ft) at max landing weight
   844 m (2,770 ft)
Range with max fuel, 45 min reserves
   2,000 n miles (3,704 km; 2,301 miles)

UPDATED

CESSNA 560XL CITATION EXCEL

TYPE: Business jet.

PROGRAMME: Announced at National Business Aircraft Association Convention in New Orleans, October 1994. Construction of prototype (N560XL) began February 1995; first flight 29 February 1996; public debut at NBAA convention at Orlando, Florida, November 1996, by which time prototype and preproduction aircraft (N561XL) had completed 350 hours of flight testing in 400 sorties; first production aircraft rolled out 21 November 1997; FAA certification achieved 22 April 1998; first delivery 2 July 1998 to Swift Transportation Inc of Phoenix, Arizona. First export, September 1998, to Automobilvertriebs AG, Austria; 100th production Excel rolled out 17 April 2000.

CUSTOMERS: Over 200 orders received by mid-1998. Total market expected to exceed 1,000. Total of 15 delivered in 1998, 39 in 1999, and 79 in 2000. Recent customers include Chauffair of Farnborough, UK, which ordered seven Excels in November 1999 for delivery in 2001 to its ChauffAir Share business jet leasing operation, Avemex SA of Toluca, Mexico (three) and Taxi Aereo Marilia (TAM) of Brazil (one).

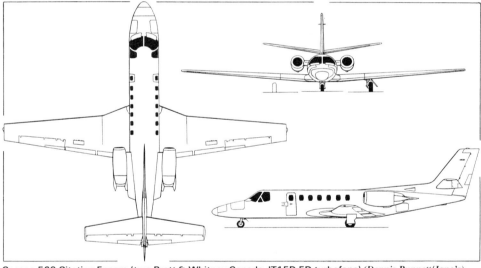

Cessna 560 Citation Encore (two Pratt & Whitney Canada JT15D-5D turbofans) *(Dennis Punnett/Jane's)*

Cabin interior of Cessna 560 Citation Encore *1999/0054037*

Flight deck of Cessna 560 Citation Encore *(Paul Jackson/Jane's)* *2000/0075949*

USA: AIRCRAFT—CESSNA

Early production Cessna Citation Excel *(Paul Jackson/Jane's)* 2001/0100487

COSTS: US$8.795 million (2000).
DESIGN FEATURES: Combines systems and wing and tail surfaces of Citation Ultra (Encore) with shortened version of Citation X's fuselage, providing 10-seat cabin with stand-up headroom.
FLYING CONTROLS: As Citation Encore.
LANDING GEAR: Hydraulically retractable tricycle type with single wheel on each unit; trailing-link suspension on main legs. Mainwheel tyre size 22×8.0-10 (12 ply) tubeless, pressure 10.48 bars (152 lb/sq in); mechanically steerable nosewheel (±20°) with chined tyre, size 18×4.4DD (10 ply) tubeless, pressure 9.65 bars (140 lb/sq in). Hydraulic multiple disc carbon brakes with anti-skid and pneumatic emergency system.
POWER PLANT: Two 16.92 kN (3,800 lb st) Pratt & Whitney Canada PW545A turbofans. Nordam thrust reversers standard. Integral fuel tank in each wing, total capacity 3,804 litres (1,005 US gallons; 837 Imp gallons) of which 3,668 litres (969 US gallons; 807 Imp gallons) usable; single-point pressure refuelling.
ACCOMMODATION: Choice of four standard seating configurations for up to 10 passengers in various layouts; seats recline, swivel and track forward and aft and laterally; forward refreshment centre and cupboard; aft lavatory and centreline cupboard. Baggage compartment in rear fuselage with external access door incorporating integral step.
SYSTEMS: Honeywell RE-100(XL) APU optional from c/n 5021 (mid-1999) onwards. Pressurisation system maximum differential 0.64 bar (9.3 lb/sq in), maintaining a sea level cabin altitude to 7,700 m (25,200 ft) or a 2,070 m (6,800 ft) cabin altitude to 13,715 m (45,000 ft). Hydraulic system, pressure 103.5 bars (1,500 lb/sq in), with two engine-driven pumps to operate landing gear, flaps, horizontal stabiliser, speed brakes and thrust reversers; separate hydraulic system for wheel brakes and anti-skid. Electrical system supplied by two 28 V 300 A DC starter/generators, with 24 V 44 Ah Ni/Cd battery. Vapour cycle air conditioning system. Oxygen system, capacity 1,417 litres (50 cu ft) with pressure demand masks for crew and dropout constant-flow masks for passengers. Engine inlets and wing leading-edges supplied by engine bleed air for anti-icing; tailplane has de-icer boots; fin unprotected; electrically heated windscreen and cockpit side windows with PPG SurfaceSeal coating for rain dispersal, with electric blower assistance.
AVIONICS: Standard Honeywell Primus 1000 integrated digital avionics suite with IC-600 avionics computer as core system. Rockwell Collins system optional.
Comms: Dual Honeywell TR-850 transceivers, dual XS-852B Mode S transponders, Artex 110-4 ELT, Lockheed Martin A200S cockpit voice recorder, airborne telephone system.
Radar: Honeywell Primus 880 colour weather radar.
Flight: Dual Honeywell NV-850, dual DM-850 DME, single DF-850 ADF, long-range navigation management system incorporating GPS, and Rockwell Collins ALT-55 radio altimeter.
Instrumentation: Three-tube EFIS with 178 × 203 mm (7 × 8 in) CRT screens comprising dual primary flight displays (PFDs) showing attitude/heading and all air data information, and single multifunction display (MFD) for map/plan, weather and checklist data.

DIMENSIONS, EXTERNAL:
Wing span	16.98 m (55 ft 8½ in)
Wing aspect ratio	8.4
Length: overall	15.79 m (51 ft 9½ in)
Height overall	5.24 m (17 ft 2½ in)
Tailplane span	6.55 m (21 ft 6 in)
Wheelbase	6.67 m (21 ft 10¾ in)
Wheel track	4.54 m (14 ft 10¾ in)

DIMENSIONS, INTERNAL:
Cabin: Length:
between pressure bulkheads	7.01 m (23 ft 0 in)
excl cockpit	5.69 m (18 ft 8 in)
Max width	1.70 m (5 ft 7 in)
Max height	1.73 m (5 ft 8 in)
Baggage capacity (aft)	2.4 m³ (83 cu ft)

AREAS:
Wings, gross	34.35 m² (369.7 sq ft)
Vertical tail surfaces (total, incl tab)	4.73 m² (50.9 sq ft)
Horizontal tail surfaces (total, incl tab)	7.88 m² (84.8 sq ft)

WEIGHTS AND LOADINGS:
Weight empty, typically equipped	5,579 kg (12,300 lb)
Max fuel weight (usable)	3,080 kg (6,790 lb)
Max T-O weight	9,071 kg (20,000 lb)
Max ramp weight	9,163 kg (20,200 lb)
Max landing weight	8,482 kg (18,700 lb)
Max zero-fuel weight	6,804 kg (15,000 lb)
Max wing loading	264.1 kg/m² (54.10 lb/sq ft)
Max power loading	269 kg/kN (2.64 lb/lb st)

PERFORMANCE:
Max operating speed (V$_{MO}$):
S/L to 2,440 m (8,000 ft) 260 kt (481 km/h; 299 mph)
2,440 m (8,000 ft) to 8,082 m (26,515 ft)
305 kt (564 km/h; 351 mph) CAS
above 8,082 m (26,515 ft) M0.75
Max cruising speed at 10,670 m (35,000 ft)
429 kt (795 km/h; 494 mph)
Stalling speed in landing configuration, at max landing weight 90 kt (167 km/h; 104 mph)
Max rate of climb at S/L 1,155 m (3,790 ft)/min
Rate of climb at S/L, OEI 259 m (850 ft)/min
Max certified altitude 13,715 m (45,000 ft)
Service ceiling, OEI 8,717 m (28,600 ft)
T-O balanced field length (FAR Pt 25) 1,095 m (3,590 ft)
FAR Pt 25 landing field length at max weight
969 m (3,180 ft)
Range with max fuel, 45 min reserves
2,080 n miles (3,852 km; 2,393 miles)

UPDATED

CESSNA 650 CITATION VII

TYPE: Business jet.
PROGRAMME: Citation VII derives from Model 650 Citation III, which was Cessna's initial mid-size business jet when first flown (N650CC) 30 May 1979; this replaced in 1991 by Citation VI; more powerful Citation VII announced

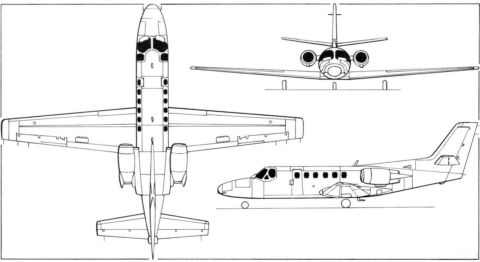

Cessna Citation Excel (PW545 turbofans) *(James Goulding/Jane's)* 2001/0100442

Eight-seat cabin interior of Cessna Citation Excel 1999/0054039

Cockpit of Cessna Citation Excel showing Honeywell Primus 1000 avionics with three-tube EFIS 1999/0054038

Eight-seat double club cabin of Cessna 680 Citation Sovereign (1999/0054041)

Cessna 750 Citation X flight deck (0015654)

1990; engineering prototype first flew February 1991; FAR Pt 25 certification January 1992; Russian certification achieved 4 September 1998. First production aircraft (N701CD) as demonstrator; first customer delivery (N1AP) to golfer Arnold Palmer April 1992. Production complete; full description last appeared in 2000-01 edition.

CUSTOMERS: Total deliveries 115 by September 2000, including 14 in 1994, 14 in 1995, 19 in 1996, eight in 1997, 11 in 1998, 14 in 1999, and 12 in 2000. Customers include Executive Jet Aviation (EJA), which ordered 20 for delivery from 1997 for its NetJets fractional ownership operation in USA and Europe. Predecessors totalled 202 Citation IIIs and 38 VIs. Citation VII production terminated in September 2000 at s/n 7119, delivered to United Foods; total one prototype and 119 production aircraft.

COSTS: Basic, without FMS or interior, US$9.395 million (2000).

UPDATED

CESSNA 680 CITATION SOVEREIGN

TYPE: Business jet.

PROGRAMME: Design started mid-1998; announced at NBAA Convention at Las Vegas, Nevada, 18 October 1998; critical design review completed in late 1999; structural testing of fatigue test fuselage began in late 1999; official launch 3 January 2000; construction of cyclic fatigue test airframe started October 2000; manufacture of prototype scheduled to start in 2001; certification expected late 2003, followed by first customer delivery in early 2004.

CUSTOMERS: Launch customer Swift Air of Phoenix, Arizona, ordered six on day of launch announcement; Executive Jet Aviation ordered 50, with 50 options, on 20 October for its NetJets fractional ownership scheme. Other announced customers include Atlas Air Service GmbH of Germany, which has ordered one.

COSTS: US$12.695 million, basic equipped (2000).

DESIGN FEATURES: Design goals included large cabin, good short-field performance and US coast-to-coast range. Low wing with sweptback leading-edge; mid-mounted tailplane with leading-edge sweepback; podded engines on rear fuselage shoulders.

New wing design; sweepback 16° 18′ at leading-edge, dihedral 3°. Mid-mounted tailplane, sweepback 17° 36′ at leading-edge, dihedral 5°.

FLYING CONTROLS: Conventional. Four hydraulically actuated spoiler panels per wing; two augment ailerons for roll; three outermost act as speed brakes; all four serve as lift dumpers when aircraft is on the ground. Trim tabs in ailerons and rudder; trimmable tailplane; yaw damper.

STRUCTURE: Primarily metal. Fokker Aerostructures of the Netherlands selected in 2000 to manufacture the tail surfaces of production aircraft.

LANDING GEAR: Hydraulically retractable tricycle type with twin wheels on trailing-link main units, single nosewheel; main units retract inboard, nosewheel forwards. Carbon brakes; anti-skid standard.

POWER PLANT: Two Pratt & Whitney Canada PW306C turbofans with FADEC, each developing 25.3 kN (5,686 lb st) flat rated to ISA +15°C; target-type thrust reversers. Integral fuel tanks in wings; single-point pressure fuelling.

ACCOMMODATION: Crew of two on flight deck and up to 11 passengers in cabin; standard accommodation for eight passengers in double club arrangement with forward galley and aft lavatory. Baggage compartment in tailcone with external access. Cabin is pressurised, heated and air conditioned. Airstair door at front on port side; two emergency exits aft, above wings.

SYSTEMS: Pressurisation system, maximum differential 0.64 bar (9.3 lb/sq in). Hydraulic system, pressure 207 bars (3,000 lb/sq in). Electrical system, supplied by two 300 A starter/generators and two AC alternators. Oxygen system standard. Wing and tailplane leading-edges anti-iced by engine bleed air; electrically anti-iced windscreens.

AVIONICS: Honeywell Primus Epic as core system.

Radar: Honeywell Primus 880 colour weather radar.

Flight: VOR, ILS, ADF, GPS, FMS and three-axis autopilot standard.

Instrumentation: Four-tube EFIS with 203 × 254 mm (8 × 10 in) active matrix quartz PFD, MFD and EICAS displays.

DIMENSIONS, EXTERNAL:
Wing span	19.24 m (63 ft 1½ in)
Wing aspect ratio	7.7
Length overall	19.35 m (63 ft 6 in)
Height overall	6.07 m (19 ft 11 in)
Tailplane span	8.38 m (27 ft 6 in)
Wheel track	3.11 m (10 ft 2½ in)
Wheelbase	8.51 m (27 ft 11 in)

DIMENSIONS, INTERNAL:
Cabin: Length	7.32 m (24 ft 0 in)
Max width	1.70 m (5 ft 7 in)
Max height	1.73 m (5 ft 8 in)
Volume	19.25 m³ (680 cu ft)
Baggage compartment volume	2.83 m³ (100 cu ft)

AREAS:
Wings, gross	47.94 m² (516.0 sq ft)
Vertical tail surfaces	8.85 m² (95.3 sq ft)
Horizontal tail surfaces	12.87 m² (138.5 sq ft)

WEIGHTS AND LOADINGS:
Payload with max fuel	725 kg (1,600 lb)

PERFORMANCE (estimated):
Max cruising speed at 11,890 m (39,000 ft)
    444 kt (822 km/h; 511 mph)
Econ cruising speed at 12,500 m (41,000 ft)
    430 kt (796 km/h; 495 mph)
Time to 13,105 m (43,000 ft)   26 min
Max certified altitude   14,326 m (47,000 ft)
T-O balanced field length (FAR Pt 25) 1,219 m (4,000 ft)
Landing from 15 m (50 ft)   975 m (3,200 ft)
Range: NBAA IFR, with 8 passengers
    2,500 n miles (4,630 km; 2,877 miles)
VFR with max fuel, allowances for T-O, climb, cruise, descent and 45 min reserves
    2,820 n miles (5,222 km; 3,245 miles)

UPDATED

CESSNA 750 CITATION X

TYPE: Business jet.

PROGRAMME: Announced at NBAA Convention in New Orleans in October 1990; engine flew on Citation VII testbed (N650) 21 August 1992; first flight (N750CX) 21 December 1993; two preproduction aircraft to aid integration of production systems; first of these (N751CX) flown 27 September 1994, second (N752CX) flown 11 January 1995; FAA certification 3 June 1996 after flight test programme totalling more than 3,000 hours; JAA certification achieved during 1999. First customer delivery July 1996. Citation X design team awarded the National Aeronautic Association's Robert J. Collier Trophy in February 1997. Cessna delivered its 3,000th Citation, a Citation X, on 19 November 1999.

In October 2000 Cessna announced improvements to the Citation X aimed at boosting range/payload performance and enabling the aircraft to operate from shorter runways. Improvements, to be incorporated on all aircraft delivered after 1 January 2002, beginning with c/n 0173 include uprated 30.01 kN (6,764 lb st) Rolls-Royce AE 3007C-1 turbofans; 181 kg (400 lb) increase in maximum take-off weight to 16,374 kg (36,100 lb), enabling a typically equipped aircraft to carry seven passengers with maximum fuel; and take-off balanced field length at MTOW of 1,585 m (5,200 ft). Several currently optional items of avionics will become standard, including Honeywell TCAS II and EGPWS, CVR and provisions for an FDR, second HF transceiver and satcom.

CUSTOMERS: First delivery (0003/N1AP) to Arnold Palmer July 1996; 100th Citation X delivered 23 December 1999 to Townsend Engineering of Des Moines, Iowa; total 142

Computer-generated image of Cessna 680 Citation Sovereign (1999/0054040)

by 31 December 2000, comprising seven, 28, 30, 36 in 1996-99 and 37 in 2000; most to US operators, but others exported to Canada, Finland, Germany, Mexico, South Africa and UK. Executive Jet Aviation (EJA) ordered 31 for delivery beginning in 1997 and extending beyond 2000, for its NetJets fractional ownership operation. Other early recipients included General Motors (five), Honeywell (two) and Williams Companies (three).
COSTS: US$17.372 million (2000).
DESIGN FEATURES: Optimised for high maximum operating Mach number; US transcontinental and transatlantic range. Design generally as for Citation VII (which see), but with greater angles of sweepback on all flying surfaces and of increased size and weight.
Wing sweepback 37°; dihedral 3°.
FLYING CONTROLS: Dual hydraulically powered controls with manual reversion. One-piece all-moving tailplane; two-piece rudder, lower portion hydraulically powered, upper portion electrically powered; speed brakes/spoilers with manual back-up. Five spoiler panels per wing, operating in combination as aileron augmentors, airbrakes and lift dumpers.
STRUCTURE: Thick wing skins, milled from solid; all control surfaces, spoilers, speedbrakes and flaps are of composites construction.
LANDING GEAR: Trailing-link main units, each with twin wheels; powered anti-skid carbon brakes; hydraulically steerable nose unit with twin wheels. Main tyres 26×6.6R14 (14 ply) tubeless; nose tyres 16×4.4D (6 or 10 ply) tubeless.
POWER PLANT: Two Rolls-Royce AE 3007C turbofans, each rated at 28.66 kN (6,442 lb st) for take-off, pod-mounted on sides of rear fuselage; FADEC. Hydraulically operated target-type thrust reversers standard. Fuel contained in three separate tanks, one in each outer wing and one in centre-section/forward fairing; two independent fuel supply systems; fuel is fed from centre tank to wing tanks; single point refuelling.
ACCOMMODATION: Crew of two on separate flight deck, and up to 12 passengers; interior custom designed; cabin is pressurised, heated and air conditioned; heated and pressurised baggage compartment in rear fuselage with external door. Windscreen electrically heated and demisted.
SYSTEMS: Pressurisation system, maximum pressure differential 0.64 bar (9.3 lb/sq in), maintains 2,440 m (8,000 ft) cabin altitude at 15,545 m (51,000 ft). Dual isolated hydraulic systems, pressure 207 bars (3,000 lb/sq in), maintained by pressure-compensated pumps. Electrical system is powered by two engine-driven 400 A DC generators, with two 24 V 44 Ah Ni/Cd batteries; wiring designed to minimise susceptibility of critical systems to HIRF interference. Wing and tail leading-edges and engine inlets heated by engine bleed air for ice protection; wing cuffs, pitot/static system, AoA system and windscreen electrically heated.
AVIONICS: Honeywell Primus 2000 autopilot/flight director as core system.
Flight: Dual flight management systems (FMS), dual attitude and heading reference systems, and Honeywell GPS standard.
Instrumentation: Five-tube EFIS with 178 × 203 mm (7 × 8 in) primary flight display (PFD) and multifunction display (MFD) for pilot and co-pilot; similarly sized engine instrument and crew alerting system (EICAS) display in centre.

Cessna 750 Citation X business jet with thrust reversers deployed *(Paul Jackson/Jane's)* 2000/0075905

DIMENSIONS, EXTERNAL:
Wing span	19.38 m (63 ft 7 in)
Wing aspect ratio	7.8
Length overall	22.05 m (72 ft 4 in)
Height overall	5.84 m (19 ft 2 in)
Tailplane span	7.95 m (26 ft 1 in)
Wheel track	3.23 m (10 ft 7 in)
Wheelbase	8.74 m (28 ft 8 in)

DIMENSIONS, INTERNAL:
Cabin (front to mid pressure bulkhead):
Length: incl flight deck	8.89 m (29 ft 2 in)
excl flight deck	7.16 m (23 ft 6 in)
Max width	1.70 m (5 ft 7 in)
Max height	1.73 m (5 ft 8 in)
Baggage compartment volume (aft, including ski compartment)	2.32 m³ (82 cu ft)

AREAS:
Wings, gross	48.96 m² (527.0 sq ft)
Vertical tail surfaces (total)	10.31 m² (111.0 sq ft)
Horizontal tail surfaces (total)	11.15 m² (120.0 sq ft)

WEIGHTS AND LOADINGS:
Weight empty, typically equipped	9,798 kg (21,600 lb)
Max fuel weight	5,913 kg (13,035 lb)
Max T-O weight	16,193 kg (35,700 lb)
Max ramp weight	16,329 kg (36,000 lb)
Max landing weight	14,424 kg (31,800 lb)
Max zero-fuel weight	11,068 kg (24,400 lb)
Max wing loading	330.7 kg/m² (67.74 lb/sq ft)
Max power loading	283 kg/kN (2.77 lb/lb st)

PERFORMANCE:
Max operating Mach No. (M$_{MO}$)	0.92
Max operating speed (V$_{MO}$):	
S/L to 2,440 m (8,000 ft)	270 kt (500 km/h; 310 mph)
2,440 m (8,000 ft) to 9,340 m (30,650 ft)	350 kt (648 km/h; 403 mph)
Max cruising speed, mid-cruise weight at 11,275 m (37,000 ft)	M0.91
Max rate of climb at S/L	1,134 m (3,720 ft)/min
Max certified altitude	15,545 m (51,000 ft)
T-O balanced field length (FAR Pt 25)	1,615 m (5,300 ft)
FAR Pt 25 landing field length	1,039 m (3,410 ft)
Range with 45 min reserves	3,430 n miles (6,352 km; 3,947 miles)

OPERATIONAL NOISE LEVELS (FAR Pt 36 Amendment 20):
T-O	72.3 EPNdB
Sideline	83.0 EPNdB
Approach	90.2 EPNdB

UPDATED

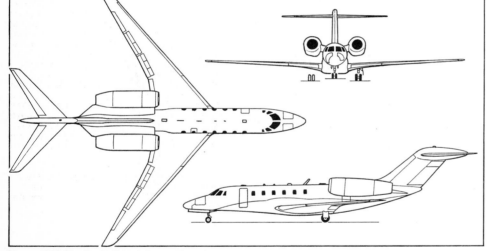

Cessna 750 Citation X *(Dennis Punnett/Jane's)* 2000/0075950

CIRRUS

CIRRUS DESIGN CORPORATION
4515 Taylor Circle, Duluth International Airport, Duluth, Minnesota 55811
Tel: (+1 218) 727 27 37
Fax: (+1 218) 727 21 48
e-mail: cmaddy@cirrusdesign.com
Web: http://www.cirrusdesign.com
PRESIDENT: Alan Klapmeier
EXECUTIVE VICE-PRESIDENT: Dale Klapmeier
VICE-PRESIDENT, MANUFACTURING: Cecil Miller
VICE-PRESIDENT, ENGINEERING: Patrick Waddick
VICE-PRESIDENT, RESEARCH AND TECHNOLOGY: Dean Vogel
VICE-PRESIDENT, SALES AND MARKETING: Thomas Shea
DIRECTOR, COMMUNICATIONS: Chris Maddy

Founded 1984 by Klapmeier brothers. Previously engaged in production of kits, Cirrus is concentrating on fully certified factory-built aircraft; first such product was ST-50 (see under Israviation in Israeli section of 1998-99 and earlier editions). Support for the VK-30 (see 1996-97 *Jane's*) continues.
Current products are SR20 and SR22. Cirrus purchased 20 per cent of parachute systems company BRS in September 1999. Workforce was 608 in September 2000, based at Duluth and Hibbing, Minnesota, and Grand Forks, North Dakota, though 20 per cent were laid off during early 2001.

UPDATED

CIRRUS DESIGN SR20 and SR22
TYPE: Four-seat lightplane.
PROGRAMME: Development began 1990; mockup revealed at Oshkosh 1994; first flight (N200SR) 31 March 1995; second prototype (N202CD) flown November 1995; FAR Pt 23 certification aircraft (N203FT), designated C-1, made first flight 28 January 1998 after completion of wing redesign to lower stall speed and improve lateral control. By end 1997, the two prototypes had accumulated 1,500 hours of test flying. A second production-standard aircraft (N204CD; C-2) joined the flight test programme on 3 June 1998, committed to trials of fuel and electrical systems and avionics. Recovery parachute trials involved eight deployments, including three for FAA. Late 1998 start of deliveries delayed by decision of avionics supplier, Trimble, to withdraw from general aviation. FAR Pt 23 certification Amdt 47 received 23 October 1998.
Initial production aircraft first flew (N115CD) 22 March 1999, but lost on following day. First delivery (c/n 1005, N415WM) 20 July 1999, at which time 325 on order. Production certificate to enable Cirrus to carry out its own inspections awarded 12 June 2000.
CURRENT VERSIONS: **SR20:** *As described.*
SR22: SR20 airframe with modified wing, 231 kW (310 hp) Teledyne Continental IO-550 engine and all-electric instrumentation (no vacuum system); certification target was December 2000, with deliveries beginning shortly afterwards; prototype (N140CD) displayed at AOPA convention in October 2000; FAA certification awarded 30 November 2000.
CUSTOMERS: Orders for 644 received by September 2000, of which 18 per cent from Europe; 111 delivered by February 2001, at which time backlog was 639. Production reached six per month in March 2000 and two per week by September; target was five per week by end 2000. Total of 95 delivered in 2000. Launch of SR22 saw approximately 20 per cent of SR20 customers upgrade to new model; first SR 22 delivered 6 February 2001.
COSTS: SR20 US$188,300, SR22 US$276,600 (2000).
DESIGN FEATURES: Low-wing, composites construction monoplane with upturned wingtips; mid-set tailplane with horn-balanced elevators; horn-balanced rudder; fixed tab on starboard aileron. First certified aircraft with ballistic parachute as standard equipment (Cirrus Airplane Parachute System or CAPS), fitted just aft of baggage compartment. Wing dihedral 4° 30′.
STRUCTURE: Composites monocoque; one-piece main spar attached to fuselage on two locations under front seats; spar carry-through attached to fuselage side-walls. Wing upper and lower surfaces bonded to spars and ribs, forming torsion box. All flying control surfaces are flush-riveted aluminium. Fin integral with fuselage; one-piece tailplane. Cabin incorporates composites roll-cage; acrylic windscreen and windows.
FLYING CONTROLS: Conventional and manual. Actuation by pushrods, cables and bellcranks. Electric trim tabs for roll

and pitch activated through springed centring devices; yaw trim is factory-set. Three-position flaps.
- LANDING GEAR: Fixed, tricycle; composites cantilever main legs; single wheels and speed fairings throughout. Main tyres 15×6.00-6, nosewheel tyre 5.00-5. Hydraulic calliper disc brakes on mainwheels. Castoring nosewheel.
- POWER PLANT: *SR20*: One 149 kW (200 hp) Teledyne Continental IO-360-ES flat-six engine driving a two-blade Hartzell BHC-J2YF-1BF/F7694 propeller (three-blade 1.88 m; 6 ft 2 in PHC-J3YF-1MF/F77392-1 propeller optional). Single fuel tank in each wing, each of 113.5 litres (30.0 US gallons; 25.0 Imp gallons) capacity, of which 212 litres (56.0 US gallons; 46.7 Imp gallons) usable. Oil capacity 7.6 litres (2.0 US gallons; 1.7 Imp gallons).

SR22: One 231 kW (310 hp) Teledyne Continental IO-550-N flat-six engine driving a three-blade Hartzell constant-speed propeller. Single fuel tank in each wing, total usable capacity 303 litres (80.0 US gallons; 66.6 Imp gallons). Oil capacity as SR20.
- ACCOMMODATION: Four 26 *g* energy attenuating seats in pairs; two forward-hinged passenger doors. Dual controls. Baggage compartment door aft of cabin on port side of fuselage. Interior reflects modern motorcar design.
- SYSTEMS: Single-axis Meggitt System FortyX autopilot; dual axis Meggitt System FiftyX and Fifty FiveX optional; 24 V DC power system with 28 V 75 A alternator and 24 V 10 Ah battery. 12 V power port.
- AVIONICS: Garmin integrated package.

Comms: GNS 430 colour GPS/com/nav IFR approach-certified GPS plus back-up, GNS 250XL GPS/com, GMA 340 audio panel and GTX 327 transponder. Options include GNS 420 colour GPS/com and/or additional GNS 430 (back-up IFR GPS).

Flight: GPS; Meggitt System FortyX autopilot. Optional Goodrich Stormscope WX-500. Options include Meggitt System FiftyX or Fifty FiveX

Instrumentation: ARNAV ICDS 2000 265 mm (10.4 in) MFD with moving map, flight plan, checklist and other functions.
- EQUIPMENT: Integral ballistic recovery parachute (Cirrus Airplane Parachute System); vertical descent rate 7.3 m (24 ft)/s at S/L. Navigation, landing and strobe lights standard.

DIMENSIONS, EXTERNAL:

Wing span: SR20	10.82 m (35 ft 6 in)
SR22	11.73 m (38 ft 6 in)
Wing aspect ratio: SR20	9.1
SR22	10.2
Length overall: SR20, SR22	7.92 m (26 ft 0 in)
Height overall: SR20, SR22	2.80 m (9 ft 2¼ in)
Tailplane span: SR20, SR22	3.93 m (12 ft 10¾ in)
Wheel track: SR20	3.38 m (11 ft 1 in)
SR22	3.23 m (10 ft 7¼ in)
Wheelbase: SR20	2.29 m (7 ft 6 in)
SR22	2.20 m (7 ft 2½ in)
Propeller diameter (two blade): SR20	1.93 m (6 ft 4 in)
SR22	1.98 m (6 ft 6 in)
Cabin doors (each):	
Height, SR20, SR22	0.94 m (3 ft 1 in)
Width SR20, SR22	0.86 m (2 ft 10 in)

DIMENSIONS, INTERNAL (SR20, SR22):

Cabin: Length	3.30 m (10 ft 10 in)
Max width	1.25 m (4 ft 1¼ in)
Max height	1.27 m (4 ft 2 in)

Cirrus Design SR20 four-seat lightplane *(Paul Jackson/Jane's)*

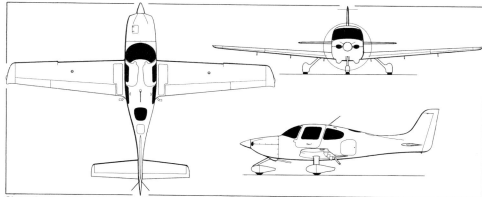
Cirrus Design SR20 (Teledyne Continental IO-360-ES) *(Paul Jackson/Jane's)*

AREAS:

Wings, gross: SR20	12.56 m² (135.2 sq ft)
SR22	13.46 m² (144.9 sq ft)
Horizontal tail: SR20, SR22	3.48 m² (37.50 sq ft)

WEIGHTS AND LOADINGS:

Weight empty: SR20	885 kg (1,950 lb)
SR22	1,021 kg (2,250 lb)
Max fuel weight: SR20	163 kg (360 lb)
SR22	218 kg (480 lb)
Max T-O weight: SR20	1,315 kg (2,900 lb)
SR22	1,542 kg (3,400 lb)
Max wing loading: SR20	104.7 kg/m² (21.45 lb/sq ft)
SR22	114.6 kg/m² (23.46 lb/sq ft)
Max power loading: SR20	8.82 kg/kW (14.50 lb/hp)
SR22	6.68 kg/kW (10.97 lb/hp)

PERFORMANCE:

Never-exceed speed (V_{NE}):	
SR20	200 kt (370 km/h; 230 mph)
SR22	204 kt (377 km/h; 234 mph)
Cruising speed at 75% power:	
SR20	160 kt (296 km/h; 184 mph)
SR22	180 kt (333 km/h; 207 mph)
Manoeuvring speed:	
SR20	135 kt (250 km/h; 155 mph) IAS
SR22	142 kt (263 km/h; 163 mph) IAS
Stalling speed: flaps up:	
SR20	64 kt (119 km/h; 74 mph) IAS
SR22	70 kt (130 km/h; 81 mph) IAS
flaps down: SR20	54 kt (100 km/h; 63 mph) IAS
SR22	59 kt (110 km/h; 68 mph) IAS
Max rate of climb at S/L: SR20	280 m (920 ft)/min
SR22	427 m (1,400 ft)/min
Service ceiling: SR20, SR22	5,335 m (17,500 ft)
T-O run: SR20	409 m (1,341 ft)
SR22	336 m (1,100 ft)
T-O to 15 m (50 ft): SR20	597 m (1,958 ft)
SR22	488 m (1,600 ft)
Landing from 15 m (50 ft): SR20	622 m (2,040 ft)
SR22	701 m (2,300 ft)
Landing run: SR20	309 m (1,014 ft)
SR22	311 m (1,020 ft)
Range with 45 min reserves:	
SR20	800 n miles (1,481 km; 920 miles)
SR22	1,000 n miles (1,852 km; 1,150 miles)
g limits: SR20, SR22	+3.8/−1.9

UPDATED

CLASSIC

CLASSIC FIGHTER INDUSTRIES INCORPORATED

Snohomish County Airport, Washington
Tel: (+1 425) 290 78 78
Web: http://www.stormbirds.com/classic
PROJECT MANAGER: Bob Hammer

Founded by the late Stephen L Snyder, Classic launched a programme in 1993 to build five exact flying replicas of a Second World War German jet fighter, the Messerschmitt Me 262. Work was originally contracted to the Texas Airplane Factory, but halted during 1997 because of a legal dispute. The production line was relocated to Paine Field (Snohomish County Airport) in early 1999.

NEW ENTRY

CLASSIC MESSERSCHMITT Me 262
TYPE: Two-seat jet sportplane.
PROGRAMME: Original prototype first flew 25 March 1942; entered operational service June 1944; flying terminated May 1945, except for limited number assigned to Allied technical evaluation or flown by Czechoslovak Air Force.

Classic programme based on reverse-engineering of WkNr 110639, former US Me 262B-1a evaluation aircraft, refurbished by Classic from 1993 onwards and redelivered to US Navy at Willow Grove, Pennsylvania, on 8 September 2000. Completion of first aircraft rescheduled from late 2000, due to company restructuring.

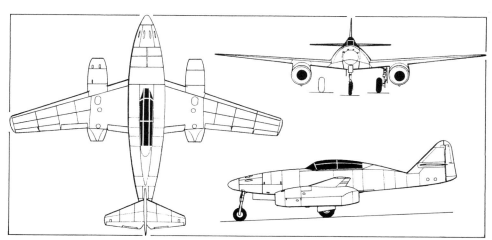
Classic Messerschmitt Me 262B-1c two-seat trainer version of Second World War jet fighter *(Paul Jackson/Jane's)*

CURRENT VERSIONS: **Me 262A-1c**: Single-seat version; one (501245) under construction by Classic; 60 per cent complete in mid-2000, but suspended, pending customer interest.

Me 262B-1c: Two-seat trainer; two (501241/N262AZ and 501243) under construction; first airframe 90 per cent complete for private owner in US; second 60 per cent complete, but suspended.

'**Me 262A/B-1c**': Further two aircraft (501242 and 501244); can be converted to either of above variants; second machine earmarked for private owner in Germany and was 70 per cent complete in 2000.

Current production aircraft assigned 'c' type suffix by Messerschmitt Foundation to indicate employment of CJ610 engine; maker's serial numbers follow on from German 1945 production.

Descriptions of Me 262 appear in 1945-46 Jane's and in various historical references. Differences from original are noted below.

DESIGN FEATURES: Modifications minimal, except where required to enhance safety. Replacement of Jumo 004 engine involves smaller, lighter and more powerful CJ610 being located in rear of original-sized nacelles; interior air duct and false Jumo engine exterior shape below access panels restore weight to original for aerodynamic and balance purposes. Landing gear mountings strengthened to obviate original weakness; modern disc brakes within mainwheel hubs; nosewheel brake deleted.

POWER PLANT: Two 12.7 kN (2,850 lb st) General Electric CJ610 turbojets; soft throttle stop at 8.0 kN (1,800 lb st) to emulate Jumo engine max output; full estimated installed thrust rating of up to 11.1 kN (2,500 lb st) employed for initiating (only) take-off roll and may be re-engaged above 226 kt (418 km/h; 260 mph) asymmetric safety speed.

PERFORMANCE:
Never-exceed speed (VNE)* 434 kt (804 km/h; 500 mph)
Max level speed (estimated) 539 kt (998 km/h; 620 mph)
Range with max fuel
    more than 869 n miles (1,609 km; 1,000 miles)
* *voluntary limitation*

NEW ENTRY

COMMANDER

COMMANDER AIRCRAFT COMPANY
Wiley Post Airport, 7200 North-West 63rd Street, Bethany, Oklahoma 73008
Tel: (+1 405) 495 80 80
Fax: (+1 405) 495 83 83
Web: http://www.commanderair.com
CEO: Wirt Walker
PRESIDENT AND COO: Dean N Thomas
VICE-PRESIDENT, MARKETING: Jay McFarland

Company acquired manufacturing, marketing and support rights for Rockwell Commander 112 and 114 from Gulfstream Aerospace Corporation in 1988; spares and support services for existing aircraft and manufacturing based in Oklahoma; 120 employees in 1999. Commander 114 remains in production, including TC version introduced in 1995. Dealerships/representatives in Abu Dhabi, Australia, Germany, Netherlands, Switzerland, UK and USA (nine). Deliveries in 1995-99 comprised 25, 15, 14, 13 and 13. In March 2000 Commander announced the introduction of the 115, an updated and refined replacement for the 114.

UPDATED

COMMANDER 114B
Replaced from 2000 by Commander 115.

UPDATED

Commander 115 four-seat touring aircraft *2001/0092108*

COMMANDER 115
TYPE: Four-seat lightplane.
PROGRAMME: Original Rockwell 112 first flew (N112AC) 4 December 1970; 149 kW (200 hp) Lycoming IO-360; production deliveries began August 1972 and ended in February 1978; more powerful Rockwell 114 subsequently added to range; all production terminated in September 1979. Following formation of Commander Aircraft, 114B recertified 5 May 1992. Commander 114 series was withdrawn in early 2000 in favour of the 115; see 2000-01 edition for details. New version uses type certificate for 114 and continues to be registered as such.
CURRENT VERSIONS: **Commander 115:** Basic version, announced in early 2000. Improvements over 114B include increased fuel capacity; wing-mounted landing lights; taxying lights on landing gear; flush fuel and panel access covers; and redesigned electrical system, instrumentation and cabin environment.
Decription applies to Commander 115.
  Commander 115TC: Replaces turbocharged 114TC in product line. Powered by 201 kW (270 hp) Textron Lycoming TIO-540-AG1A turbocharged six-cylinder engine driving McCauley B3D32C419/82NHA-5 three-blade constant-speed propeller. Total fuel capacity 341 litres (90.0 US gallons; 74.9 Imp gallons) in two integral wing tanks, of which 333 litres (88.0 US gallons; 73.3 Imp gallons) are usable. Oil capacity 9.46 litres (2.5 US gallons; 2.1 Imp gallons).
CUSTOMERS: Rockwell built 801 Model 112s, 429 Model 114/114As and two fixed-gear Model 111s. Commander Aircraft had added at least 126 Model 114Bs and 28 Model 114TCs (including eight 114Bs and five 114TCs in 1999) before change to Model 115. Production plan for 2000 was 20 to 25.
COSTS: 115 US$425,000; 115TC US$472,000 (2000).
DESIGN FEATURES: Rockwell 111 (fixed gear; two prototypes only) and 112 conceived as competitors to Piper Cherokee and Raytheon (Beech) Bonanza.
  Low wing with root glove, unswept leading-edge and sweptforward trailing-edge; sweptback fin with mid-mounted, tapered tailplane and small ventral strake.
  Wing section NACA 63-415 modified; dihedral 7°; incidence 2° at root; sweepforward 2° 30′ at quarter-chord.
FLYING CONTROLS: Conventional and manual. Electrically actuated single-slotted flaps; horn-balanced tail control surfaces; anti-servo tabs in both elevators.
LANDING GEAR: Retractable tricycle type; mainwheels retract inward; nosewheel aft. Nosewheel (steerable) tyre size 5.00-5 (6 ply); mainwheel tyres size 6.00-6 (6 ply); dual brakes and oleo-pneumatic shock-absorbers.
POWER PLANT: One 194 kW (260 hp) Textron Lycoming IO-540-T4B5 flat-six engine, driving a McCauley B3D32C419/82NHA-5 three-blade constant-speed metal propeller. Fuel in two integral wing tanks, total capacity 341 litres (90.0 US gallons; 74.9 Imp gallons), of which 333 litres (88.0 US gallons; 73.3 Imp gallons) are usable. Refuelling points in upper surface of each wing. Maximum oil capacity 7.6 litres (2.0 US gallons; 1.7 Imp gallons).
ACCOMMODATION: Four persons on two individual seats (forward, with dual controls) and bench seat. Forward-hinged door each side. Externally accessed baggage compartment, port side, aft of wing. Provision for heating and ventilation.
SYSTEMS: 28 V DC electrical; 85A alternator. Hydraulic system with electric pump. Optional air conditioning in 114TC. TKS fluid anti-icing system for wings and tail surfaces.
AVIONICS: *Comms:* Standard includes PS Engineering PMA 7000MS audio and intercom panel and KT 76A transponder; Honeywell Silver Crown suite. Aircell phone.
  Flight: KX 155 digital nav and Terra AT-3000 altitude encoder. Garmin GNS 430 GPS.

DIMENSIONS, EXTERNAL:
Wing span	9.98 m (32 ft 9 in)
Wing chord: at c/l	1.78 m (5 ft 10¼ in)
at tip	0.89 m (2 ft 11 in)
mean aerodynamic	1.40 m (4 ft 7½ in)
Wing aspect ratio	7.1
Length overall	7.59 m (24 ft 11 in)
Height overall	2.57 m (8 ft 5 in)
Tailplane span	4.11 m (13 ft 6 in)
Wheel track	3.34 m (10 ft 11 in)
Wheelbase	2.11 m (6 ft 11 in)
Propeller diameter	1.96 m (6 ft 5 in)
Propeller ground clearance	0.19 m (7½ in)
Passenger doors (each): Height	0.97 m (3 ft 2 in)
Width	0.71 m (2 ft 4 in)
Baggage door: Height	0.51 m (1 ft 8 in)
Width	0.71 m (2 ft 4 in)

DIMENSIONS, INTERNAL:
Cabin: Length	1.91 m (6 ft 3 in)
Max width	1.19 m (3 ft 11 in)
Max height	1.24 m (4 ft 1 in)
Volume	2.8 m³ (100 cu ft)
Baggage compartment volume	0.62 m³ (22.0 cu ft)

AREAS:
Wings, gross	14.12 m² (152.0 sq ft)
Ailerons (total)	1.02 m² (11.0 sq ft)
Trailing-edge flaps (total)	1.67 m² (18.00 sq ft)
Fin	1.58 m² (17.00 sq ft)

WEIGHTS AND LOADINGS:
Weight empty: 115	953 kg (2,102 lb)
115TC	976 kg (2,152 lb)
Baggage capacity	91 kg (200 lb)
Max T-O weight: 115	1,474 kg (3,250 lb)
115TC	1,499 kg (3,305 lb)
Max ramp weight	1,479 kg (3,260 lb)
Max wing loading: 115	104.4 kg/m² (21.38 lb/sq ft)
115TC	106.2 kg/m² (21.74 lb/sq ft)
Max power loading: 115	7.61 kg/kW (12.50 lb/hp)
115TC	7.45 kg/kW (12.24 lb/hp)

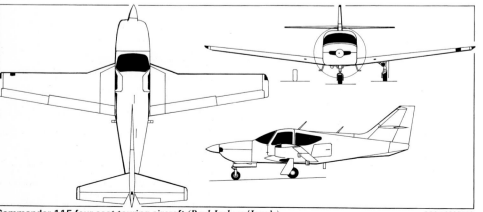

Commander 115 four-seat touring aircraft *(Paul Jackson/Jane's)* *2001/0092112*

PERFORMANCE:
Max level speed: 115 164 kt (304 km/h; 189 mph)
115TC 197 kt (364 km/h; 226 mph)
Cruising speed at 75% power:
115 160 kt (296 km/h; 184 mph)
115TC 177 kt (328 km/h; 204 mph)
Econ cruising speed at 65% power:
115 155 kt (287 km/h; 178 mph)
115TC 169 kt (313 km/h; 194 mph)
Long-range cruising speed at 55% power:
115 149 kt (276 km/h; 171 mph)
Stalling speed:
flaps and wheels up: 115 60 kt (112 km/h; 69 mph)
115TC 66 kt (123 km/h; 76 mph)
flaps and wheels down:
115 54 kt (100 km/h; 63 mph)
115TC 59 kt (110 km/h; 68 mph)
Max rate of climb at S/L: 115 326 m (1,070 ft)/min
115TC 320 m (1,050 ft)/min
Service ceiling: 115 5,120 m (16,800 ft)
115TC 7,620 m (25,000 ft)
T-O run: 115 349 m (1,145 ft)
115TC 429 m (1,410 ft)
T-O to 15 m (50 ft): 115 605 m (1,985 ft)
115TC 678 m (2,225 ft)
Landing from 15 m (50 ft): 115 366 m (1,200 ft)
115TC 400 m (1,312 ft)
Landing run: 115 220 m (720 ft)
115TC 224 m (734 ft)
Range: at 75% power:
115 854 n miles (1,581 km; 982 miles)
115TC 670 n miles (1,240 km; 771 miles)
at 65% power:
115 921 n miles (1,705 km; 1,060 miles)
115TC 780 n miles (1,444 km; 897 miles)
at 55% power:
115 1,007 n miles (1,865 km; 1,159 miles)
115TC 870 n miles (1,611 km; 1,001 miles)

NEW ENTRY

CULP'S

CULP'S SPECIALTIES
PO Box 7542, Shreveport, Louisiana 71137
Tel/Fax: (+1 318) 222 08 50
e-mail: culpspecial@yahoo.com
Web: http://www.culpsspecialties.com
PRESIDENT: Steve Culp

In addition to producing the Special biplane, Culp's Specialties restores and rebuilds classic aircraft. In 1998, Culp's began building six Monocoupes, and in 2000 announced a Sopwith Pup replica.

UPDATED

CULP'S SPECIAL
TYPE: Aerobatic, three-seat biplane/kitbuilt.
PROGRAMME: Design started September 1993; work on first prototype began in following November and first flight achieved (N367LS) on 15 April 1996. Made public debut at Sun 'n' Fun 1996; initially had 670 kg (1,478 lb) empty weight and 930 kg (2,050 lb) MTOW. Manufacture of first production aircraft began in August 1994. FAA Experimental certification granted April 1996 and aircraft embarked on intensive demonstration programme.
CUSTOMERS: By May 2000, four aircraft were flying and seven were under construction. 40 sets of plans and six kits sold.
COSTS: Factory-built US$145,000; kit US$44,000; plans US$315 (2000).
DESIGN FEATURES: Fully aerobatic, single-bay biplane with 1930s styling, but offering modern performance. Loosely based on the Steen Skybolt. Cabane-mounted upper wing with single interplane strut and wire bracing; wire-braced empennage. Quoted build time 2,500 hours.
FLYING CONTROLS: Conventional and manual. Actuation by pushrods; electric rudder trim and manual elevator trim. Aileron design influenced by Curtis Pitts. Horn-balanced elevator and rudder.
STRUCTURE: Fuselage is 4130 steel tubing covered with spruce; metal-skinned forward of cockpit, with fabric to rear. Wings are of fabric-covered wooden construction.
LANDING GEAR: Tailwheel type; fixed. Braced main landing gear legs are of 4130 steel tubing and use bungee shock cords. Goodyear tyres: mainwheels 10.00-6, pressure 1.86 bars (27 lb/sq in); tailwheel 8.00-4, pressure 3.10 bars (45 lb/sq in). Cleveland brakes. Minimum ground turning radius 7.92 m (26 ft 0 in).
POWER PLANT: One 265 kW (355 hp) VOKBM M-14PT radial engine driving a constant-speed two-blade wooden propeller. Main fuel tank in fuselage contains 121 litres (32.0 US gallons; 26.6 Imp gallons); wing tank contains 38 litres (10.0 US gallons; 8.3 Imp gallons) and ferry tank a further 151 litres (40.0 US gallons; 33.3 Imp gallons); total usable fuel 299 litres (79.0 US gallons; 65.8 Imp gallons). Oil capacity 15 litres (4.0 US gallons; 3.3 Imp gallons).
ACCOMMODATION: Pilot and either one or two passengers, depending on model, in open cockpit. Baggage compartment behind headrest.
SYSTEMS: 10 A electrical system for radio and lights; pneumatic engine starting system.
AVIONICS: *Comms:* Honeywell radio, transponder and GPS.
DIMENSIONS, EXTERNAL:
Wing span: upper 7.32 m (24 ft 0 in)
lower 7.01 m (23 ft 0 in)
Wing chord, constant 1.07 m (3 ft 6 in)
Length overall 6.40 m (21 ft 0 in)
Max diameter of fuselage 1.05 m (3 ft 5½ in)
Height overall 2.44 m (8 ft 0 in)
Tailplane span 2.29 m (7 ft 6 in)
Wheel track 1.98 m (6 ft 6 in)
Wheelbase 5.18 m (17 ft 0 in)
Propeller diameter 2.39 m (7 ft 10 in)
Propeller ground clearance 0.91 m (3 ft 0 in)
DIMENSIONS, INTERNAL:
Cabin: Length 2.44 m (8 ft 0 in)
Max width 0.81 m (2 ft 8 in)
Max height 0.81 m (2 ft 8 in)
Floor area 0.56 m² (6.0 sq ft)
Baggage hold volume 0.11 m³ (4.0 cu ft)
AREAS:
Wings, gross 14.96 m² (161.0 sq ft)
Ailerons (total) 2.60 m² (28.00 sq ft)
Rudder, incl tab 1.02 m² (11.00 sq ft)
Tailplane 2.97 m² (32.00 sq ft)
Elevators, incl tab 1.49 m² (16.00 sq ft)
WEIGHTS AND LOADINGS:
Weight empty 635 kg (1,400 lb)
Max fuel weight 218 kg (480 lb)
Max T-O and landing weight 1,043 kg (2,300 lb)
Max zero-fuel weight 848 kg (1,870 lb)
Max wing loading 69.7 kg/m² (14.29 lb/sq ft)
Max power loading 3.94 kg/kW (6.48 lb/hp)
PERFORMANCE:
Never-exceed speed (V_{NE}) at S/L
  208 kt (386 km/h; 240 mph)
Max operating speed (V_{MO}) 191 kt (354 km/h; 220 mph)
Econ cruising speed at 2,590 m (8,500 ft)
  130 kt (241 km/h; 150 mph)
Stalling speed, power off 63 kt (116 km/h; 72 mph)
Max rate of climb at S/L 1,371 m (4,500 ft)/min
Service ceiling 3,660 m (12,000 ft)
T-O run 91 m (300 ft)
T-O to 15 m (50 ft) 122 m (400 ft)
Landing from 15 m (50 ft) 213 m (700 ft)
Landing run 244 m (800 ft)
Range: with max fuel at econ cruising speed
  955 n miles (1,770 km; 1,100 miles)
with max payload 487 n miles (902 km; 561 miles)
g limits ±10

UPDATED

Prototype Culp's Special *(Paul Jackson/Jane's)* 2001/0092078

CULP'S MONOCULP
TYPE: Four-seat kitbuilt.
PROGRAMME: Announced at Sun 'n' Fun, April 2000, when prototype under construction.
CUSTOMERS: Two kits and three sets of plans sold by May 2000.
COSTS: Kit price US$58,000 with wings completed (2000).
DESIGN FEATURES: Scaled-up (125 per cent) version of 1930s Monocoupe 90A.
FLYING CONTROLS: Conventional and manual.
STRUCTURE: Strut-braced high wing; wire-braced tailplane.

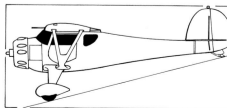
Culp's MonoCulp *(Paul Jackson/Jane's)* 2001/0092144

LANDING GEAR: Tailwheel type; fixed. Speed fairings on mainwheels.
POWER PLANT: One 294 kW (394 hp) VOKBM M-14PF nine-cylinder radial engine. Fuel capacity 318 litres (84.0 US gallons; 70.0 Imp gallons).
ACCOMMODATION: Pilot and three passengers in two pairs of seats; cabin door on port side.
DIMENSIONS, EXTERNAL:
Wing span 9.75 m (32 ft 0 in)
Length overall 7.47 m (24 ft 6 in)
Height overall 2.54 m (8 ft 4 in)
DIMENSIONS, INTERNAL:
Cabin: max width 1.22 m (4 ft 0 in)
AREAS:
Wings, gross 17.84 m² (192.0 sq ft)
WEIGHTS AND LOADINGS:
Weight empty 771 kg (1,700 lb)
Max T-O weight 1,315 kg (2,900 lb)
Max wing loading 73.7 kg/m² (15.10 lb/sq ft)
Max power loading 4.48 kg/kW (7.36 lb/hp)
PERFORMANCE:
Never-exceed speed (V_{NE}) 234 kt (434 km/h; 270 mph)
Normal cruising speed 156 kt (290 km/h; 180 mph)
Stalling speed, power off 54 kt (100 km/h; 62 mph)
Max rate of climb at S/L 914 m (3,000 ft)/min
Service ceiling 3,660 m (12,000 ft)
T-O run 122 m (400 ft)
Landing run 274 m (900 ft)
Range with max fuel 827 kt (1,532 km; 952 miles)
Endurance 4 h 0 min

UPDATED

CULP'S SOPWITH PUP
TYPE: Tandem-seat biplane kitbuilt.
PROGRAMME: Announced at Sun 'n' Fun, April 2000, prototype then under construction.
CUSTOMERS: Orders for 12 kits and 18 sets of plans received by May 2000.
COSTS: Plans US$200; kit US$40,000 including engine (2000).
DESIGN FEATURES: Based on First World War Sopwith Pup fighter, but using all-new materials. Quoted build time 2,500 hours.
STRUCTURE: Wood, fabric and tube construction. Biplane wings with interplane struts.
LANDING GEAR: Tailwheel type; fixed.
POWER PLANT: One VOKBM M-14P nine-cylinder radial derated to 179 kW (240 hp); engines in range 164 to 268 kW (220 to 360 hp) can be fitted. Fuel capacity 189 litres (50.0 US gallons; 41.7 Imp gallons).
ACCOMMODATION: Pilot and passenger in tandem in open cockpit.

DIMENSIONS, EXTERNAL:
Wing span 8.08 m (26 ft 6 in)
Length overall 5.49 m (18 ft 0 in)
Height overall 2.74 m (9 ft 0 in)
DIMENSIONS, INTERNAL:
Cabin max width 0.71 m (2 ft 4 in)
AREAS:
Wings, gross 24.62 m² (265.0 sq ft)

WEIGHTS AND LOADINGS:
Weight empty 590 kg (1,300 lb)
Max T-O weight 1,088 kg (2,400 lb)
Max wing loading 44.2 kg/m² (9.06 lb/sq ft)
Max power loading 6.09 kg/kW (10.00 lb/hp)
PERFORMANCE:
Never-exceed speed (V$_{NE}$) 191 kt (354 km/h; 220 mph)
Normal cruising speed 130 kt (241 km/h; 150 mph)

Stalling speed 33 kt (62 km/h; 38 mph)
Max rate of climb at S/L 1,372 m (4,500 ft)/min
Service ceiling 3,660 m (12,000 ft)
T-O run 76 m (250 ft)
Landing run 152 m (500 ft)
Range with max fuel 521 n miles (965 km; 600 miles)

NEW ENTRY

DERRINGER

DERRINGER AIRCRAFT COMPANY LLC
1246 Sabovich Street, Mojave, California 93501
Tel: (+1 661) 824 22 22
Fax: (+1 661) 824 22 02
e-mail: derringer@derringeraircraft.com
Web: http://www.derringeraircraft.com
MARKETING MANAGER: Roger Lewis

SALES DIVISION:
558 S Broad Street, Mobile, Alabama 36603
Tel: (+1 888) 909 82 22
Fax: (+1 334) 433 33 64

Derringer expects to employ around 100 once full production rate is achieved.

UPDATED

DERRINGER NEW DERRINGER

TYPE: Four-seat utility twin.
PROGRAMME: Original two-seat Derringer designed by John Thorp in 1958 as T-17 twin-engined version of T-11 Sky Skooter (see AM Aerospace, UK section); first flight 1 May 1962; type certificate 20 December 1966; marketed by Wing Aircraft; after two attempts at series production (the last failing in 1982), 12 had been built; for further details see 1972-73 and 1982-83 *Jane's*.
Prototype New Derringer rolled out at Mojave August 1998 and certification was expected by end of 1998. Production aircraft were due for delivery from January 1999; but this timetable has slipped. Derringer expects to build about 50 per year.
CURRENT VERSIONS: **New Derringer GT-320**: Baseline two-seat version using Textron Lycoming IO-320 engine.
New Derringer T-320: Training version of GT-320, offered with optional EFIS cockpit.
New Derringer GT-400: Standard four-seat model, *as described*.
COSTS: New Derringer GT-400 US$349,000 VFR, US$400,000 IFR (1999); New Derringer T-320 US$320,000 VFR (1999).
DESIGN FEATURES: High-speed business aircraft, employing (for 1960s) advanced methods of construction to ensure high strength and surface smoothness. Wings constant-chord, but with leading-edge root extension. Constant-chord tailplane and sweptback fin.
Wings of NACA 65$_2$-415 section; dihedral 6°; incidence 1°.
FLYING CONTROLS: Conventional and manual. Electrically operated differential aileron control. Horn-balanced rudder and elevators. Flaps.
STRUCTURE: Aluminium semi-monocoque fuselage. Single-spar wing; two-spar control surfaces. Glass fibre nose-cone, wing tips and engine nacelles.
LANDING GEAR: Retractable tricycle type; steerable nosewheel retracts forwards; mainwheels forwards into nacelles. Hydraulic disc brakes.
POWER PLANT: Two 149 kW (200 hp) Textron Lycoming IO-360-C1A engines, flat rated to 119 kW (160 hp), driving Hartzell two-blade constant-speed fully feathering propellers. Fuel in two outboard wing leading-edge tanks; combined capacity 333 litres (88.0 US gallons; 73.3 Imp gallons); filler caps at wingtips. Oil capacity 8.5 litres (2.25 US gallons; 1.9 Imp gallons (per engine)).
ACCOMMODATION: Pilot and three passengers in 2+2 configuration; trainer version allows for pilot and student in front seats with second student or baggage area in rear. Upward-opening canopy; single-piece windscreen. Taxi-latch for on-ground cabin ventilation.
SYSTEMS: 12 V 35 A battery, 60 A engine-driven alternator. MIL-W-5086 wiring standard.
AVIONICS: *Comms:* Standard VFR comprises Honeywell suite of KA 134 audio control panel, KY 96 transceiver and KT 76A transponder. Options include KX 165 transceiver.
Flight: Optional KLN 89B GPS, KR 87 ADF and KAP 150 autopilot.

DIMENSIONS, EXTERNAL:
Wing span 8.89 m (29 ft 2 in)
Wing chord, constant 1.27 m (4 ft 2 in)
Wing aspect ratio 7.0
Length overall 7.01 m (23 ft 0 in)
Height overall 2.44 m (8 ft 0 in)
Tailplane span 3.30 m (10 ft 10 in)
Tailplane chord, constant 0.85 m (2 ft 9½ in)
Wheel track 3.30 m (10 ft 10 in)
Wheelbase 1.64 m (5 ft 4½ in)
Propeller diameter 1.68 m (5 ft 6 in)
Propeller ground clearance 0.23 m (9 in)
DIMENSIONS, INTERNAL:
Cabin: Max width 1.14 m (3 ft 9 in)
Max height 1.22 m (4 ft 0 in)
Baggage volume (two-seat) 0.62 m³ (22.0 cu ft)

AREAS:
Wings, gross 11.24 m² (121.0 sq ft)
Ailerons (total) 0.74 m² (8.00 sq ft)
Flaps (total) 1.11 m² (12.00 sq ft)
Fin 1.08 m² (11.65 sq ft)
Rudder 0.48 m² (5.18 sq ft)
Horizontal tail surfaces 2.59 m² (27.86 sq ft)
WEIGHTS AND LOADINGS:
Weight empty 903 kg (1,990 lb)
Baggage capacity 113 kg (250 lb)
Max T-O weight 1,383 kg (3,050 lb)
Max wing loading 123.1 kg/m² (25.21 lb/sq ft)
Max power loading 46.54 kg/kW (9.53 lb/hp)
PERFORMANCE:
Max operating speed 216 kt (400 km/h; 248 mph)
Normal cruising speed 203 kt (376 km/h; 234 mph)
Stalling speed, power off:
landing gear and flaps up 68 kt (126 km/h; 79 mph)
landing gear and flaps down 60 kt (112 km/h; 69 mph)
Max rate of climb at S/L 457 m (1,500 ft)/min
Rate of climb at S/L, OEI 122 m (400 ft)/min
Service ceiling 7,620 m (25,000 ft)
Service ceiling, OEI 2,745 m (9,000 ft)
T-O run 229 m (750 ft)
T-O to 15 m (50 ft) 366 m (1,200 ft)
Landing from 15 m (50 ft) 579 m (1,900 ft)
Landing run 244 m (800 ft)
Range at 3,050 m (10,000 ft), max fuel, 45 min reserves
750 n miles (1,389 km; 863 miles)
Endurance 3 h 45 min

UPDATED

Derringer light twin, now returned to production

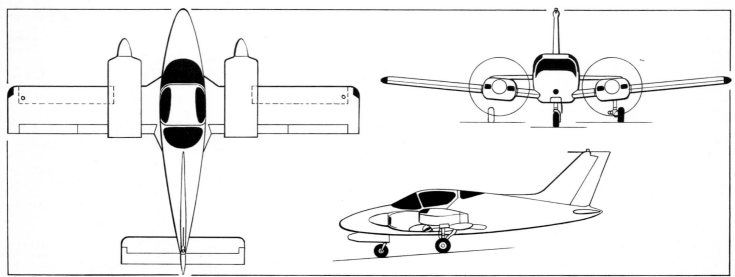

General arrangement of the Derringer New Derringer (*James Goulding/Jane's*)

DFL

DFL HOLDINGS INC
Team Tango Division, 3001 Northeast 20th Way, Gainesville, Florida 32609
Tel: (+1 800) 340 40 35 or (+1 352) 373 41 07
Fax: (+1 352) 373 53 07
e-mail: ttango@fdn.com
Web: http://www.teamtango.com

Current projects of DFL are the two-seat Tango-2 and a four-seat development, the Foxtrot-4; the latter is due to fly during 2001.

UPDATED

DFL TANGO-2
TYPE: Side-by-side kitbuilt.
PROGRAMME: Developed from the Aero Mirage TC-2 of 1983. Prototype flew November 1996. First two production aircraft displayed at Oshkosh in July 1999; third flew March 2000.
CUSTOMERS: Total 11 kits sold by April 2000.

COSTS: Basic kit US$18,995, plus US$5,990 assisted assembly, US$36,500 O-360 engine kit and US$11,995 VFR avionics kit (2001).
DESIGN FEATURES: Conventional design optimised for amateur assembly; low-mounted, constant-chord wings and mid-positioned tailplane. Steeply raked windscreen. Quoted build time 570 hours for quick-build kit version; company has building facility to enable purchasers to complete in 30 to 45 days.
FLYING CONTROLS: Conventional and manual. Roll rate 120°/s. Flaps deflect 10/20/35°.
STRUCTURE: Generally of composites. Two-spar wing has modified NACA64415 aerofoil and 3° dihedral.
LANDING GEAR: Tricycle type; fixed. Cantilever mainwheel legs are fibre glass; nose leg is 4130 steel tube; speed fairings on all wheels. Main units have Cleveland wheels and brakes with Goodyear 5.00-5 tyres; nosewheel is steerable and has 4.50-5 Lamb tyre.
POWER PLANT: One 134 kW (180 hp) Textron Lycoming O-360 flat-four driving a two-blade Hartzell HC2YR7666A constant-speed propeller. Fuel capacity 152 litres (40.0 US gallons; 33.3 Imp gallons) in two wing tanks; two optional extra 38 litre (10.0 US gallon; 8.3 Imp gallon) tanks can be fitted. Optionally, 149 kW (200 hp) IO-360 or 194 kW (260 hp) IO-540.
ACCOMMODATION: Two, side by side. Upward-opening, centreline hinged door each side; fixed windscreen. Baggage stowage behind seats.
AVIONICS: To customer's specification. Standard VFR option includes Honeywell KLX 135 GPS/com, KT 76A transponder and Sigtronics intercom; IFR version adds KX 155 nav/com and KI 209 VOR/LOC.

DIMENSIONS, EXTERNAL:
Wing span	7.62 m (25 ft 0 in)
Wing aspect ratio	8.3
Length overall	6.31 m (20 ft 8½ in)
Height overall	2.21 m (7 ft 3 in)
Wheel track	2.54 m (8 ft 4 in)
Propeller diameter	1.93 m (6 ft 4 in)

DIMENSIONS, INTERNAL:
Cockpit max width	1.12 m (3 ft 8 in)
Baggage compartment volume	0.34 m³ (12.0 cu ft)

AREAS:
Wings, gross	6.97 m² (75.0 sq ft)

WEIGHTS AND LOADINGS:
Weight empty	488 kg (1,075 lb)
Baggage capacity	45 kg (100 lb)
Max T-O weight	839 kg (1,850 lb)
Max wing loading	120.4 kg/m² (24.67 lb/sq ft)
Max power loading	6.26 kg/kW (10.28 lb/hp)

AREAS:
Wings, gross	6.97 m² (75.0 sq ft)

PERFORMANCE:
Never-exceed speed (V_{NE})	212 kt (394 km/h; 245 mph)
Max level speed	191 kt (354 km/h; 220 mph)
Max cruising speed	174 kt (322 km/h; 200 mph)
Stalling speed	55 kt (102 km/h; 63 mph) IAS
Max rate of climb at S/L	549 m (1,800 ft)/min
Service ceiling	7,315 m (24,000 ft)
T-O run	183 m (600 ft)
Landing run	244 m (800 ft)
Max range	869 n miles (1,609 km; 1,000 miles)
g limits	+6/−4

UPDATED

DFL Tango-2 kitbuilt tourer *(Paul Jackson/Jane's)*
2001/0103550

DREAMWINGS

DREAMWINGS LLC
2550 North 7th Street, Lawrence Municipal Airport, Lawrence, Kansas 66044
Tel: (+1 785) 842 65 26
Fax: (+1 785) 842 65 27
e-mail: dreamwings@dreamwings.com
Web: http://www.dreamwings.com
PRESIDENT: John Hunter
VICE-PRESIDENT: Leslie Hunter

DreamWings occupies a 1,115 m² (12,000 sq ft) plant with the capacity to build 20 kitplanes per month. Sales in 1999 totalled nearly 100, representing turnover of US$2.5 million; the company was forecasting sales of US$7 million for 2000-01. The Valkyrie and forthcoming Rhapsody are hailed as first of the next generation of ultralights.

UPDATED

DREAMWINGS VALKYRIE
TYPE: Tandem-seat kitbuilt; tandem-seat ultralight kitbuilt.
PROGRAMME: Prototype (N6406V) first flew 1999 and had achieved 15 hours by time of debut at Sun 'n' Fun in April 2000.
COSTS: US$17,000 to US$18,000 excluding engine (2000).
DESIGN FEATURES: Aerodynamically efficient design to FAR Pt 23 requirements; FAR Pt 103 legal. Mid-wing configuration with twin tailbooms and canard. Quoted build time 100 to 200 hours. Dismantled within 10 minutes for transport or storage.
Wing and canard section NACA 63615a; horizontal tail NACA 63012; vertical tail NACA 63006. Mainplane dihedral 3°; incidence 1° 30′; washout 0°; taper ratio 1.72:1.
FLYING CONTROLS: Manual. Differential ailerons; single-piece, horn-balanced elevator; twin rudders. Slotted flaps. Flight adjustable pitch trim tab on canards. Actuation by torque tubes and pushrods.
STRUCTURE: Generally of carbon fibre.
LANDING GEAR: Tricycle type; fixed. Cantilever main legs of metal with rubber-in-compression suspension and 5.00-5 tyres. Nosewheel 11×4.00, steerable. Mainwheel hydraulic brakes.
POWER PLANT: One pusher engine. Rating 20.9 to 26.1 kW (28 to 35 hp) with 19 or 38 litres (5.0 or 10.0 US gallons; 4.2 or 8.3 Imp gallons) of fuel to FAR Pt 103. Alternatively, engines up to 93 kW (125 hp) and 133 litres (35.0 US gallons; 29.2 Imp gallons) of fuel in one fuselage and two wing tanks. Options include 34.0 kW (45.6 hp) Rotax 503 UL-2V and 59.7 kW (80 hp) Jabiru 2200. Prototype has three-blade, ground-adjustable pitch propeller.
ACCOMMODATION: Two, beneath single-piece, rear-hinged canopy activated by gas struts. Baggage compartments fore and aft of cockpit; rear seat removable for additional stowage space. Sidestick control.

Prototype DreamWings Valkyrie ultralight *(Paul Jackson/Jane's)*
2000/0084600

DIMENSIONS, EXTERNAL:
Wing span	9.60 m (31 ft 6 in)
Wing aspect ratio	10.2
Canard span	1.98 m (6 ft 6 in)
Length overall	7.14 m (23 ft 5¼ in)
Fuselage max width	0.72 m (2 ft 4½ in)
Height overall	2.11 m (6 ft 11 in)
Tailplane span	3.87 m (12 ft 8½ in)
Propeller max diameter	1.83 m (6 ft 0 in)

AREAS:
Wings, gross	9.01 m² (97.0 sq ft)
Canards, gross	0.79 m² (8.50 sq ft)
Elevator	1.64 m² (17.66 sq ft)
Vertical tails, gross	1.86 m² (20.06 sq ft)

WEIGHTS AND LOADINGS (A: 20.8 kW; 28 hp engine, B: 93 kW; 125 hp engine):
Weight empty: A	115 kg (254 lb)
B	168 kg (370 lb)
Max T-O weight: A	453 kg (1,000 lb)
B	567 kg (1,250 lb)

PERFORMANCE, POWERED (A, B as above):
Never-exceed speed (V_{NE}): A, B	
	152 kt (281 km/h; 175 mph)
Cruising speed at 70% power:	
A	65 kt (121 km/h; 75 mph)
B	122 kt (225 km/h; 140 mph)
Stalling speed: A	25 kt (46 km/h; 28 mph)
B	28 kt (52 km/h; 32 mph)
T-O run: A	957 m (3,140 ft)
B	76 m (250 ft)
Max rate of climb at S/L: A	51 m (166 ft)/min
B	755 m (2,477 ft)/min
Range: A	1,042 n miles (1,931 km; 1,200 miles)
g limits: A	+6/−4
B	+9/−6

PERFORMANCE, UNPOWERED:
Best glide ratio	12

UPDATED

DREAMWINGS RHAPSODY
TYPE: Single-seat kitbuilt; single-seat ultralight kitbuilt; single-seat kitbuilt twin; two-seat kitbuilt twin.
PROGRAMME: Project; details published at Sun 'n' Fun, April 2000.
CURRENT VERSIONS: **Rhapsody Mark 1:** Single 20.9 kW (28 hp) engine with two- or three-blade pusher propeller, meeting FAR Pt 103 with single seat and 19 or 38 litres (5.0 or 10.0 US gallons; 4.2 or 8.3 Imp gallons) of fuel. Alternative Experimental category lightplane with engine of up to 59.7 kW (80 hp) and 133 litres (35.0 US gallons; 29.2 Imp gallons) of fuel.
Rhapsody Mark 2: Two 29.8 kW (40 hp) engines with three-blade pusher propellers mounted on strengthened wing and 38 litre (10.0 US gallon; 8.3 Imp gallon) fuel capacity as ultralight two-seat trainer. Alternative Experimental, single-seat amateur-built with engines of up

648 USA: AIRCRAFT—DREAMWINGS to ECLIPSE

to 59.7 kW (80 hp) and 133 litre (35.0 US gallon; 29.2 Imp gallon) tankage.
COSTS: Mark 1 US$11,000 to US$12,000; Mark 2 US$14,000; both less engine(s) (2000).
DESIGN FEATURES: High, unbraced wing; empennage, with mid-mounted tailplane, carried on small diameter boom.
Wing section NACA 63615a; dihedral 2°; incidence 1°; washout 0°; taper ratio 1.6. Horizontal tail section NACA 63009; vertical tail section NACA 63009.
FLYING CONTROLS: Conventional and manual. Horn-balanced elevators and slotted flaps. Electric trim.
STRUCTURE: Carbon fibre.
LANDING GEAR: Generally as for Valkyrie.
POWER PLANT: See Current Versions. Options as for Valkyrie.
DIMENSIONS, EXTERNAL:
Wing span: Mark 1	9.14 m (30 ft 0 in)
Mark 2	10.06 m (33 ft 0 in)
Length overall	7.05 m (23 ft 1½ in)
Height overall	1.90 m (6 ft 3 in)
Propeller max diameter	1.57 m (5 ft 2 in)

AREAS:
Wings, gross: Mark 1	9.85 m² (106.0 sq ft)
Mark 2	10.22 m² (110.0 sq ft)

WEIGHTS AND LOADINGS:
Weight empty: Mark 1	136 kg (300 lb)
Mark 2	215 kg (475 lb)
Max T-O weight: Mark 1	567 kg (1,250 lb)
Mark 2	680 kg (11,500 lb)

PERFORMANCE, POWERED (Mark 1 with 59.7 kW; 80 hp, Mark 2 with two 59.7 kW; 80 hp engines):
Never-exceed speed (V_{NE}): both	152 kt (281 km/h; 175 mph)
Cruising speed: Mark 1	86 kt (159 km/h; 99 mph)
Stalling speed: Mark 1	28 kt (52 km/h; 32 mph)
Mark 2	31 kt (57 km/h; 35 mph)
Max rate of climb at S/L: Mark 1	451 m (1,480 ft)/min
Mark 2	975 m (3,200 ft)/min
T-O run: Mark 1	53 m (173 ft)
Mark 2	31 m (100 ft)
Range: Mark 1	695 n miles (1,287 km; 800 miles)
Mark 2	617 n miles (1,142 km; 710 miles)
g limits	+9/−6

PERFORMANCE, UNPOWERED:
Best glide ratio: Mark 1	12:1
Mark 2	11:1

UPDATED

DreamWings Rhapsody Mark 1 (top) and Mark 2 2000/0084544

ECLIPSE

ECLIPSE AVIATION CORPORATION

2503 Clark Carr Loop SE, Albuquerque, New Mexico 87106
Tel: (+1 505) 245 75 55
Fax: (+1 505) 245 78 88
e-mail: info@eclipseaviation.com
Web: http://www.eclipseaviation.com
CHAIRMAN: Harold A Poling
PRESIDENT AND CEO: Vern Raburn
VICE-PRESIDENT, ENGINEERING: Oliver Masefield
VICE-PRESIDENT, SALES: Chris Finnoff
VICE-PRESIDENT, BUSINESS AFFAIRS: Jack Harrington
CFO: Peter C Reed

Eclipse Aviation formed May 1998 to market the Eclipse 500, which supersedes the Williams V-Jet II (1999-2000 *Jane's*). Eclipse currently has exclusivity rights to the EJ-22 engine and, by April 2000, had initial funding of US$60 million in place; further US$65 million secured by 4 December 2000. The company has subcontracted Williams to design, build, develop and certify the Eclipse 500 and to establish a plant for its manufacture in series; 140 Williams employees are currently working on the project.

UPDATED

ECLIPSE ECLIPSE 500
TYPE: Light business jet.
PROGRAMME: Announced early 2000; 22 per cent scale model wind tunnel testing undertaken February to April 2000; preliminary design review September 2000; full-size mockup displayed at EAA AirVenture 2000 and NBAA 2000. First flight due June 2002, with certification to FAR Pt 23 plus take-off and landing to Pt 25 standard in June 2003; applications submitted to FAA 29 September 2000; deliveries will start August 2003.
CUSTOMERS: Estimated production rate of 110 per year, rising to 200 per year by 2005; 160 deposits taken by July 2000.
COSTS: Unit cost US$837,500 (2001). Development programme US$300+ million (estimated, 2000).
DESIGN FEATURES: Designed to fulfil goals of NASA's Small Air Transportation System programme; extensive use of computerised welding and riveting to reduce costs. Unswept, low-mounted wing and T tail. Friction stir welding to be used for first time on high-volume production, obviating need for large numbers of rivets and reducing weight of finished article.
STRUCTURE: Principally of aluminium; sharply tapered rear fuselage means distance between engine centres will be 1.04 m (3 ft 5 in).
LANDING GEAR: Retractable tricycle type.
POWER PLANT: Two rear-mounted 3.43 kN (770 lb st) Williams EJ-22 turbofan engines (a development of the FJX-2). Engine bypass ratio 4.6. Optional wingtip fuel tanks on extended-range model, available from early 2004.
ACCOMMODATION: Pilot and five passengers; cabin laid out in club format. Polarised windows throughout; door on port side of cabin behind flight deck. Optional lavatory.

Eclipse 500 (Williams EJ-22 turbofans) *(James Goulding/Jane's)* 2001/0103510

SYSTEMS: Cabin pressurised to 2,440 m (8,000 ft) at 12,500 m (41,000 ft) altitude. De-icing system for leading-edges and engine intakes. Three-axis autopilot.
AVIONICS: All-glass, three-screen, fully IFR cockpit provided by Avidyne and BAE Systems, including dual VHF com and nav, Mode S transponder, dual GPS, dual AHRS, three-axis autopilot and colour weather radar.

All data are provisional.

DIMENSIONS, EXTERNAL:
Wing span	10.97 m (36 ft 0 in)
Length overall	10.08 m (33 ft 1 in)
Height overall	3.35 m (11 ft 0 in)

DIMENSIONS, INTERNAL:
Cabin: Length	3.61 m (11 ft 10 in)
Max width	1.47 m (4 ft 10 in)
Max height	1.30 m (4 ft 3 in)
Volume	5.4 m³ (191 cu ft)
Baggage compartment volume	1.20 m³ (42.5 cu ft)

WEIGHTS AND LOADINGS:
Weight empty: standard	1,225 kg (2,700 lb)
with extended-range tip tanks	1,256 kg (2,770 lb)
Max fuel weight: standard	603 kg (1,330 lb)
Max T-O weight	2,131 kg (4,700 lb)
Max ramp weight	2,143 kg (4,725 lb)
Max landing weight	2,086 kg (4,600 lb)
Max power loading	311 kg/kN (3.05 lb/lb st)

PERFORMANCE:
Max operating speed (M_{MO}/V_{MO})	M0.68 (285 kt; 528 km/h; 328 mph)
Normal cruising speed	355 kt (657 km/h; 409 mph)
Stalling speed	62 kt (115 km/h; 72 mph)
Max rate of climb at S/L	869 m (2,850 ft)/min
Rate of climb at S/L, OEI	259 m (850 ft)/min
Service ceiling	12,495 m (41,000 ft)
Service ceiling, OEI	8,535 m (28,000 ft)
T-O run	732 m (2,400 ft)
Landing run	625 m (2,050 ft)
Range: with max fuel and tip tanks	1,825 n miles (3,380 km; 2,100 miles)
without tip tanks	1,600 n miles (2,963 km; 1,841 miles)
with four occupants, NBAA IFR alternate	1,300 n miles (2,407 km; 1,496 miles)

UPDATED

Eclipse 500 instrument panel 2001/0103511

Mockup of projected Eclipse 500 business jet 2001/0103512

ENSTROM

THE ENSTROM HELICOPTER CORPORATION

PO Box 490, 2209 North 22nd Street, Twin County Airport, Menominee, Michigan 49858-0490
Tel: (+1 906) 863 12 00
Fax: (+1 906) 863 68 21
e-mail: enstrom@cybrzn.com
Web: http://www.enstromhelicopter.com
PRESIDENT AND CEO: Robert M Tuttle
VICE-PRESIDENT, ENGINEERING: Robert L Jenny
VICE-PRESIDENT, MARKETING AND SERVICE: Robert J Cleland
DIRECTOR, MANUFACTURING: James Stutson

Enstrom Helicopter Corporation was established in 1959 to develop and produce light helicopters, based on Rudy Enstrom's initial design work. Extensive development and certification efforts led to start of production of the F28 helicopter in April 1965; development of turbine-powered model was initiated in 1988, culminating in certification of the TH28 in September 1992 and the 480, to FAR Pt 27 standards, in December 1994. Enstrom is currently introducing the 480B, which features increased gross weight, take-off power and continuous power. Certification was planned for end 2000.

Initially publicly owned, the company has passed through a period as a subsidiary of a large corporation and is now privately owned. By the beginning of 2000, Enstrom had produced over 1,000 helicopters of which over 700 remain in operation; deliveries in 1999 comprised five F28/280s and three 480s. The fleet has amassed over three million flying hours. Negotiations continue with the Wuhan Helicopter Company regarding possible licensed production in China; see China section.

UPDATED

ENSTROM F28 and 280

TYPE: Three-seat helicopter.
PROGRAMME: Basic F28A and 280 described in 1978-79 *Jane's*; replaced by turbocharged F28C and 280C, certified by FAA 8 December 1975 and last described in 1984-85 *Jane's*; production of these models ceased November 1981; succeeded by F28F and 280F Shark, described in 1985-86 *Jane's*, and 280FX; current models detailed here.
CURRENT VERSIONS: **F28F Falcon:** Basic model certified to FAR Pt 6 on 31 December 1980. The two major changes incorporated into the F28F and 280F/280FX over the earlier F28C/280C series were an increase in power from 153 kW (205 shp) to 168 kW (225 shp) and the addition of a throttle correlator to reduce pilot workload. Most recent changes to the F28F and 280FX are the redesigned main gearbox with a heavy wall main rotor shaft (standard equipment on all new aircraft and retrofittable to all existing F models); optional lightweight exhaust silencer, which reduces noise in the hover by 40 per cent and gives a 30 per cent reduction when flying at 152 m (500 ft) (can also be retrofitted to F28F, 280F and 280FX); and a lightweight starter motor.
F28F-P Sentinel: Dedicated police patrol version; first delivery October 1986; can be fitted with a PA and siren system, Spectrolab SX-5 or Carter searchlight, specialised police equipment, FLIR and datalink. F28F-P has same specification and performance as F28F.
280FX Shark: Certified to FAR Pt 6 on 14 January 1985. Improvements over previous 280 series (see also F28F entry) include new seats with lumbar support and energy-absorbing NASA foam, new tailplane with endplate fins, tail rotor guard, covered tail rotor drive shaft, redesigned air inlet system, and completely faired landing gear. Optional internal auxiliary fuel tank extends range to 339 n miles (627 km; 390 miles).
CUSTOMERS: Private owners and fleet operations. Chilean Army operates 15 280FXs for primary and instrument training; Peruvian Army has 10 F28Fs for flight training; Colombian Air Force operates 12 F28Fs for primary and instrument training. Numerous US police departments operate F28F-P for patrol and surveillance missions.

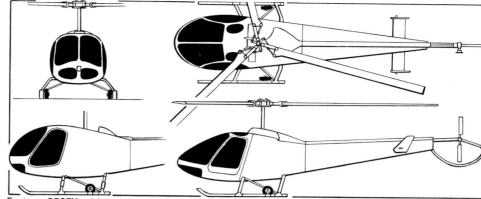

Enstrom 280FX, with scrap side view of F28F (*James Goulding/Jane's*)

Over 960 piston-engined helicopters produced by early 2000. Current production models are F28F (122 built by end 1998) and 280FX (85 built by end 1999). Production of both models continues at low rate.
COSTS: Basic 2000 price US$267,900 for F28F Falcon; US$279,900 for 280FX Shark.
DESIGN FEATURES: Conventional light helicopter with skid landing gear and tubular metal tail rotor protector; horizontal stabiliser with fins at tips. High inertia, three-blade fully articulated rotor head with blades attached by retention pin and drag link; control rods pass inside tubular rotor shaft to swashplate inside fuselage; no rotor brake; blade section NACA 0013.5; blades do not fold; two-blade teetering tail rotor. Goodyear 30-groove belt drive from horizontally mounted engine to transmission.
FLYING CONTROLS: Conventional and manual. Trim system absorbs feedback from rotor and repositions stick datum as required by pilot.
STRUCTURE: Bonded light alloy blades. Fuselage has glass fibre and light alloy cabin section, steel tube centre-section frame, and stressed skin aluminium tailboom.
LANDING GEAR: Skids carried on Enstrom oleo-pneumatic shock-absorbers. Air Cruiser inflatable floats available optionally.
POWER PLANT: One 168 kW (225 hp) Textron Lycoming HIO-360-F1AD flat-four engine with Rotomaster 3BT5EE10J2 turbocharger. Two fuel tanks, each of 79.5 litres (21.0 US gallons; 17.5 Imp gallons). Total standard fuel capacity 159 litres (42.0 US gallons; 35.0 Imp gallons), of which 151 litres (40.0 US gallons; 33.3 Imp gallons) are usable. Auxiliary tank, capacity 49 litres (13.0 US gallons; 10.8 Imp gallons), can be installed in the baggage compartment. Oil capacity 9.5 litres (2.5 US gallons; 2.1 Imp gallons).
ACCOMMODATION: Pilot and two passengers, side by side on bench seat; centre place removable. Removable door on each side of cabin. Baggage space aft of engine compartment, with external door. Cabin heated and ventilated.
SYSTEMS: Electrical power on F28F provided by 12 V 70 A engine-driven alternator; 24 V 70 A system optional on F28F, standard on 280FX. No hydraulic system.
AVIONICS: Variety of fits from Honeywell and other avionics suppliers.
Instrumentation: Standard equipment includes airspeed indicator, sensitive altimeter, compass, outside air temperature gauge, turn and bank indicator, rotor/engine tachometer, manifold pressure/fuel flow gauge, EGT gauge, oil pressure gauge, gearbox and oil temperature gauge, ammeter, cylinder head temperature gauge, and fuel quantity gauge. Eight-light annunciator panel consisting of low rotor rpm, chip detectors (main and tail rotor transmissions), overboost, clutch not fully engaged, low fuel pressure, starter and low-voltage warning lights.
EQUIPMENT: Shoulder harnesses for three seats. Night lighting is optional for F28F and standard on 280FX. Night lighting includes instrument lighting with dimmer control, position light on each horizontal stabiliser tip, anti-collision light and nose-mounted landing light. Optional equipment for both F28F and 280FX includes fixed float kit, wet or dry agricultural spray kit and cargo hook for utility missions. Wide instrument panel available for IFR training.

DIMENSIONS, EXTERNAL (F28 and 280, except where specified):
Main rotor diameter	9.75 m (32 ft 0 in)
Tail rotor diameter	1.42 m (4 ft 8 in)
Distance between rotor centres	5.56 m (18 ft 3 in)
Main rotor blade chord	0.24 m (9½ in)
Tail rotor blade chord	0.11 m (4½ in)
Length: fuselage: F28	8.56 m (28 ft 1 in)
280	8.75 m (28 ft 8½ in)
overall, rotors stationary	8.92 m (29 ft 3 in)
Fuselage max width: F28	1.55 m (5 ft 1 in)
280	1.52 m (5 ft 0 in)
Height: to top of cabin: F28	1.85 m (6 ft 1 in)
280	1.83 m (6 ft 0 in)
to top of rotor head	2.74 m (9 ft 0 in)
Tailplane span	1.60 m (5 ft 3 in)
Skid track	2.21 m (7 ft 3 in)
Cabin doors (each): Height	1.04 m (3 ft 5 in)
Width	0.84 m (2 ft 9 in)
Height to sill	0.64 m (2 ft 1 in)
Baggage door: Height	0.55 m (1 ft 9½ in)
Width	0.39 m (1 ft 3½ in)
Height to sill	0.86 m (2 ft 10 in)

DIMENSIONS, INTERNAL:
Cabin max width: F28F	1.55 m (5 ft 1 in)
280FX	1.50 m (4 ft 11 in)
Baggage compartment volume	0.18 m³ (6.3 cu ft)

AREAS:
Main rotor disc	74.72 m² (804.25 sq ft)
Tail rotor disc	1.66 m² (17.88 sq ft)

WEIGHTS AND LOADINGS (F28F Normal category):
Weight empty, equipped: F28F	712 kg (1,570 lb)
280FX	719 kg (1,585 lb)
Baggage capacity: tailboom	49 kg (108 lb)
Max T-O weight: F28F, 280FX	1,179 kg (2,600 lb)
Max disc loading: F28F, 280FX	15.8 kg/m² (3.23 lb/sq ft)

PERFORMANCE (both versions at AUW of 1,066 kg; 2,350 lb, except where indicated):
Never-exceed speed (V_{NE}):	
F28F	97 kt (180 km/h; 112 mph)
280FX	102 kt (189 km/h; 117 mph)
Max level speed, S/L to 915 m (3,000 ft):	
F28F	97 kt (180 km/h; 112 mph) IAS
280FX	102 kt (189 km/h; 117 mph) IAS
Econ cruising speed:	
F28F	89 kt (165 km/h; 102 mph)
280FX	93 kt (172 km/h; 107 mph)
Max rate of climb at S/L:	
at AUW of 1,066 kg (2,350 lb)	442 m (1,450 ft)/min
at AUW of 1,179 kg (2,600 lb)	350 m (1,150 ft)/min
Max certified altitude	3,660 m (12,000 ft)
Hovering ceiling:	
IGE: at 1,066 kg (2,350 lb)	4,020 m (13,200 ft)
at 1,179 kg (2,600 lb)	2,345 m (7,700 ft)
OGE: at 1,066 kg (2,350 lb)	2,650 m (8,700 ft)
at 1,179 kg (2,600 lb)	1,707 m (5,600 ft)
Max range, standard fuel, no reserves:	
F28F	229 n miles (423 km; 263 miles)
280FX	260 n miles (483 km; 300 miles)
Max endurance	3 h 30 min

UPDATED

ENSTROM 480 and TH-28

TYPE: Light utility helicopter.
PROGRAMME: Proof-of-concept 280FX, powered by Allison 250-C20W turbine engine, flown December 1988; first flight of wide-cabin 480/TH-28 prototype (N8631E) in October 1989; second 480/TH-28 (production prototype) flew December 1991; TH-28 FAA certified September 1992 (three-seat), 480 FAA certified in June 1993 and recertified to FAR Pt 27 in December 1994.
CURRENT VERSIONS: **480:** Five-seat (two plus three) configuration with staggered seating for forward visibility from all seats. Easily reconfigured to three seats for

Enstrom 280FX Shark (Textron Lycoming HIO-360-F1AD engine) (*Paul Jackson/Jane's*)

training or executive transport; or pilot's seat only for transporting light cargo. Instrument panel centrally located above avionics console.

480B: Certification was due by end 2000; has increased gross weight and incorporates cyclic control and airframe vibration damping, providing significant improvement in comfort. Demonstrator, N480EN (43rd production 480) exhibited at NBAA Convention, New Orleans, October 2000.

TH-28: Specially configured as a training/light patrol helicopter; unique features include three crashworthy seats, crashworthy fuel system, and a large instrument panel which will accommodate dual instrumentation for either VFR or IFR training. Seating configuration allows training of two students simultaneously.

CUSTOMERS: Production began in 1994 with initial deliveries to Europe. By late 2000, 43 480s and six TH-28s had been built. Now certified and operating in 13 countries: Belgium, Brazil, Canada, China, France, Germany, Japan, South Africa, Sweden, Switzerland, Thailand, UK and USA. In addition, 480s are also operating in Burkino Faso and Russian Federation.

COSTS: Basic price US$580,000 (2000) for 480B; TH-28 price on application.

Description generally as for F28/280, except that below (for 480B).

DESIGN FEATURES: High-inertia, three-blade, fully articulated main rotor system with upgraded and sturdier main rotor gearbox than piston models; larger tail rotor; unboosted flight controls and high skid gear are standard equipment. Rotor speed 300 rpm nominal; 334 rpm maximum permissible; anti-nodal vibrating cantilever beam to damp vibration at low speeds. Extensive crashworthiness and safety features incorporated into basic design. The 480 cabin layout can be quickly reconfigured to seat one, three or five persons.

First Enstrom 480B improved version of five-seat turbine-powered light helicopter *(Paul Jackson/Jane's)*
2001/0100512

POWER PLANT: One 313 kW (420 shp) Rolls-Royce 250-C20W turboshaft engine derated to 227 kW (305 shp) for take-off and 207 kW (277 shp) for continuous operation. Engine air inlet particle separator is standard. Fuel capacity 340 litres (90.0 US gallons; 74.9 Imp gallons) in two interconnected tanks.

ACCOMMODATION: See Current Versions. Options include air conditioning (two or three evaporator system). Baggage compartment in tailboom.

EQUIPMENT: Latest additions to optional equipment include emergency pop-out floats, air conditionig and cargo hook rated for external loads up to 454 kg (1,000 lb); 480 can also be fitted with police patrol equipment including special police radios, searchlight, siren/PA system, FLIR and data recording facility.

DIMENSIONS, EXTERNAL:
Main rotor diameter	9.75 m (32 ft 0 in)
Tail rotor diameter	1.54 m (5 ft 0½ in)
Distance between rotor centres	5.64 m (18 ft 6 in)
Length of fuselage	9.09 m (29 ft 10 in)
Fuselage max width	1.79 m (5 ft 10½ in)
Height: to top of cabin	2.11 m (6 ft 11¼ in)
to top of rotor head	3.00 m (9 ft 10 in)
Skid track	2.50 m (8 ft 2½ in)

DIMENSIONS, INTERNAL:
Cabin max width	1.80 m (5 ft 10¾ in)

AREAS:
Main rotor disc	74.72 m² (804.25 sq ft)
Tail rotor disc	1.85 m² (19.96 sq ft)

WEIGHTS AND LOADINGS:
Weight empty	769 kg (1,695 lb)
Baggage capacity: cabin	23 kg (50 lb)
tailboom	70 kg (155 lb)
Max T-O weight	1,360 kg (3,000 lb)
Max disc loading	18.2 kg/m² (3.73 lb/sq ft)

PERFORMANCE:
Never-exceed speed (V$_{NE}$) 125 kt (231 km/h; 144 mph)
Cruising speed: at AUW of 1,360 kg (3,000 lb)
  109 kt (202 km/h; 125 mph)
 at AUW of 1,134 kg (2,500 lb)
  115 kt (213 km/h; 132 mph)
Max rate of climb at S/L:
 at AUW of 1,360 kg (3,000 lb) 360 m (1,180 ft)/min
 at AUW of 1,134 kg (2,500 lb) 457 m (1,500 ft)/min
Service ceiling 3,965 m (13,000 ft)
Hovering ceiling:
 IGE:
  at AUW of 1,360 kg (3,000 lb) 3,350 m (11,000 ft)
  at AUW of 1,134 kg (2,500 lb) 4,755 m (15,600 ft)
 OGE:
  at AUW of 1,360 kg (3,000 lb) 1,370 m (4,500 ft)
  at AUW of 1,134 kg (2,500 lb) 4,265 m (14,000 ft)
Max range (no reserves) at 914 m (3,000 ft):
 at 1,360 kg (3,000 lb) 350 n miles (648 km; 402 miles)
 at 1,134 kg (2,500 lb) 435 n miles (806 km; 501 miles)
Endurance (no reserves):
 at 1,360 kg (3,000 lb) 4 h 24 min
 at 1,134 kg (2,500 lb) 4 h 36 min

UPDATED

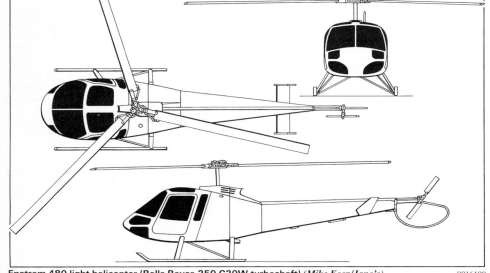

Enstrom 480 light helicopter (Rolls-Royce 250-C20W turboshaft) *(Mike Keep/Jane's)* 0016199

EXCLUSIVE

EXCLUSIVE AVIATION

Details of the Exclusive Grand 51, now renamed P-51G, appear under the Cameron heading in this section.

UPDATED

EXPLORER

EXPLORER AIRCRAFT INC

357 Madison Street, Denver, Colorado 80206
Tel: (+1 303) 388 06 00
Fax: (+1 303) 388 10 05
e-mail: info@exploreaircraft.com
Web: http://www.exploreaircraft.com
CEO: Graham Swannell
PRESIDENT: Donald G Joseph
DIRECTOR, OPERATIONS, RESEARCH AND DEVELOPMENT:
 Geoffrey P Danes

Explorer Aircraft Inc was established in 1998 to develop and market the Australian-designed Explorer range of single-engined utility aircraft.

NEW ENTRY

EXPLORER EXPLORER

TYPE: Light utility turboprop.
PROGRAMME: Designed by Aeronautical Engineers Australia (AEA), beginning in 1993, with assistance from US aerodynamicist John Roncz; company renamed Explorer Aircraft Corporation. Proof-of-concept (POC) prototype (VH-OKA), designated Explorer 350R, made first flight 23 January 1998 and had flown more than 70 hours up to award of Developmental C of A on 27 April 1998. Prototype (re-registered VH-ONA in May 1998) and given Experimental C of A on 27 April 1999; IFR certification 6 May 1999; departed for USA 20 May 1999; remained based at Denver following relocation of parent company from Australia.

Re-engined with 447 kW (600 shp) Pratt & Whitney Canada PT6-135B turboprop as prototype Explorer 500T, in which form it first flew on 9 June 2000; simultaneously fitted with six-screen EFIS; public debut at EAA AirVenture Oshkosh in August 2000; flown 50 hours by October 2000; will later be re-engined with Orenda OE600A V-8 as prototype 500R; manufacture on North American production line expected to begin in 2002; FAA certification and first customer deliveries of launch model 500T in 2004, with deliveries of 500R following six months later and 750T in 2006.

CURRENT VERSIONS: **350R:** Proof-of-concept prototype, powered by one 261 kW (350 hp) Teledyne Continental TSIO-55-E3B piston engine. No production decision taken by mid-2000. Described in 2000-01 edition of *Jane's*.

500T: Initial production version employing airframe similar to that of 350R but with 447 kW (600 shp) Pratt & Whitney Canada PT6-135B turboprop. *As described.*

500R: Airframe, payload and performance as for 500T, but with 447 kW (600 hp) Orenda OE600 liquid-cooled V-8 piston engine.

750T: Stretched version with plugs fore and aft of wing; up to 16 passengers; PT6-60A engine flat rated to 597 kW (800 shp).

COSTS: 500T US$1.035 million; 500R US$820,000; 750T US$1.16 million (2000).

DESIGN FEATURES: High-mounted, strut-braced wing with Roncz advanced aerofoil profile and constant chord, thick centre-section; outer panels (approximately 40 per cent of each half-span) sharply tapered; tall, sweptback fin and rudder. Constant cross-section cabin. 'Productionisation' to be accelerated by use of state-of-the-art computer modelling. Wing thickness/chord ratio 19.6; washout 2°.

FLYING CONTROLS: Conventional and manual. Fowler flaps with three settings for take-off, approach and landing, plus

EXPLORER AIRCRAFT to EXPLORER AVIATION—AIRCRAFT: USA

Instrument panel of Explorer 500T
(Paul Jackson /Jane's) 2001/0100506

Explorer Aircraft multirole utility transport in its latest guise as prototype 500T *(Paul Jackson/Jane's)*
2001/0100505

neutral; three-axis trim via cable/screw-jack actuated tabs in starboard elevator, starboard aileron and rudder.
STRUCTURE: Carbon fibre fuselage shell over metal frame; remainder of primary structure is of high-grade aluminium alloy, with composites flaps, elevators, ailerons and rudder.
LANDING GEAR: Fully retractable tricycle type, with single wheel on each unit. Mainwheel tyre size 22×7.75-10, nosewheel 6.00-6. Steerable nose oleo unit retracts rearward; glass fibre mainwheel legs retract by first extending downward and then crossing over and upward so that port wheel is housed in streamline fairing on starboard side, and vice versa. Mainwheels do not intrude into cabin. Hydraulic brakes.
POWER PLANT: *500T*: One Pratt & Whitney Canada PT6-135B turboprop, flat rated at 447 kW (600 shp), driving a four-blade, constant-speed, fully reversing Hartzell D9511FK-2 propeller. Fuel tanks in wing centre-section, total capacity 1,049 litres (277 US gallons; 231 Imp gallons).
500R: One 447 kW (600 hp) Orenda V-8 OE600A piston engine driving a three-blade constant-speed propeller. Fuel capacity 795 litres (210 US gallons; 175 Imp gallons).
750T: One Pratt & Whitney Canada PT6-60A turboprop, flat rated at 597 kW (800 shp), driving a four-blade constant-speed Hartzell reversing propeller. Fuel capacity 1,514 litres (400 US gallons; 333 Imp gallons).
All versions have refuelling port on top of each wing.
ACCOMMODATION: *500T, 500R*: 11 seats, including pilot(s); *750T*: 17 seats, including pilot(s). Forward-opening crew door on each side of forward fuselage. Large cargo door at rear of cabin on port side. Cabin is extensively glazed with five (500T, 500R) or seven (750T) large, square windows on each side. Fuselage hardpoints for tiedowns and cargo nets. Air conditioning standard.
SYSTEMS: 28 V DC electrical system. Engine-driven hydraulic pump. De-icing system, comprising leading-edge boots and heated propeller.
AVIONICS: *Comms*: Dual Garmin GNS 430 GPS/com/nav; GTX 327 transponder; GMA 340 audio panel.
Flight: Garmin GI 106A VOR/LOC/GS; Meggitt System 55 programmer/computer; remote annunciator; altitude selector/alerter; 6446 flux gate; Meggitt autopilot.
Instrumentation: Dual Meggitt Avionics 84-132-1 EDU; 84-133-1 PFD; 84-134-1 ND.
DIMENSIONS, EXTERNAL (A: 500T, B: 500R, C: 750T):

Wing span: A, B	14.43 m (47 ft 4 in)
C	17.68 m (58 ft 0 in)
Wing aspect ratio: A, B	11.3
Length overall: A, B	9.68 m (31 ft 9 in)
C	12.34 m (40 ft 6 in)
Height overall: A, B, C	4.72 m (15 ft 6 in)
Passenger door: A, B: Height	1.24 m (4 ft 1 in)
Width	1.27 m (4 ft 2 in)
Height to sill	0.96 m (3 ft 2 in)
Baggage door: A, B: Height	1.24 m (4 ft 1 in)
Width	1.27 m (4 ft 2 in)
Height to sill	0.96 m (3 ft 2 in)

DIMENSIONS, INTERNAL:

Cabin: Length: A, B	3.35 m (11 ft 0 in)
C	5.11 m (16 ft 9 in)
Max width: A, B, C	1.55 m (5 ft 1 in)
Max height: A, B, C	1.35 m (4 ft 5 in)
Volume: A, B	7.1 m³ (250 cu ft)
C	11.3 m³ (400 cu ft)
Cargo pod volume: A, B	1.42 m³ (50 cu ft)
C	2.265 m³ (80 cu ft)

AREAS:

Wings, gross: A, B	18.36 m² (197.6 sq ft)
Horizontal tail surfaces (total): A, B	5.29 m² (56.90 sq ft)
Vertical tail surfaces (total): A, B	2.87 m² (30.90 sq ft)

WEIGHTS AND LOADINGS:

Weight empty: A, B	1,724 kg (3,800 lb)
C	2,268 kg (5,000 lb)
Max T-O weight: A, B	2,812 kg (6,200 lb)
C	4,082 kg (9,000 lb)
Max wing loading: A, B	153.2 kg/m² (31.38 lb/sq ft)
Max power loading: A	6.29 kg/kW (10.33 lb/lb shp)
B	6.29 kg/kW (10.33 lb/lb shp)
C	6.85 kg/kW (11.25 lb/lb shp)

PERFORMANCE (estimated):

Cruising speed: A	200 kt (370 km/h; 230 mph)
Cruising speed: A, B	175 kt (324 km/h; 201 mph)
C	190 kt (352 km/h; 219 mph)
Stalling speed, flaps down: A, B, C	61 kt (113 km/h; 71 mph)
Service ceiling: A, B, C	7,620 m (25,000 ft)
Max rate of climb at S/L: A, B	305 m (1,000 ft)/min
C	366 m (1,200 ft)/min
T-O run: A, B, C	366 m (1,200 ft)
Max range: A, B	950 n miles (1,759 km; 1,093 miles)
C	900 n miles (1,666 km; 1,035 miles)

UPDATED

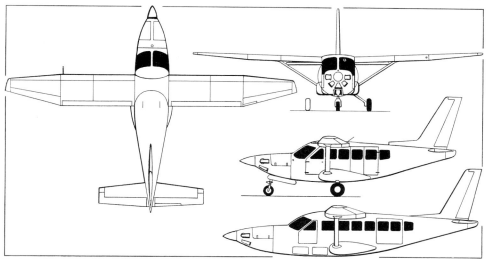

Explorer Aircraft Explorer 500T, with additional side view (lower) of stretched Explorer 750T
(James Goulding/Jane's) 2001/0105701

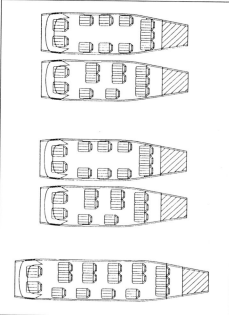

Alternative seating arrangement for, Explorer 500R (upper pair), 500T (centre pair) and stretched 750T
2000/0085627

EXPLORER

EXPLORER AVIATION

Idaho County Airport, Grangeville, Idaho 83530
Tel: (+1 208) 983 24 23
Fax: (+1 208) 983 37 67

EUROPEAN AGENT:
Hubert Ferte
Ferme de Montrodoise, F-10150 Montsuzain, France
Tel: (+33 3) 25 37 50 28
Fax: (+33 3) 25 37 50 27
e-mail: explorer@camasnet.com

Explorer Aviation is co-located with Private Explorer (which see).

UPDATED

EXPLORER AVIATION ELLIPSE

TYPE: Four-seat kitbuilt.
PROGRAMME: First flight in early 1997; three flying by end 1998.
COSTS: US$35,000 (1999).
DESIGN FEATURES: Elliptical wing shape for higher performance; similarly shaped tail surfaces; wings strut-braced and fold for storage; tailplane strut-braced. Quoted build time 1,000 hours.
FLYING CONTROLS: Conventional and manual. Horn-balanced rudder and elevator. Flaps.
STRUCTURE: Fuselage fabric-covered 4130 steel tube frame; wings of wooden construction and wood-covered. Control surfaces fabric-covered.
LANDING GEAR: Tailwheel type; fixed. Strut-braced mainwheel legs; steerable tailwheel. Pontoon and ski fitting points provided.
POWER PLANT: One 112 kW (150 hp) Textron Lycoming O-320 flat-four driving a two-blade propeller; alternative engines from 93 to 149 kW (125 to 200 hp) can be fitted. Fuel capacity 155 litres (41.0 US gallons; 34.1 Imp gallons).
ACCOMMODATION: Pilot and up to three passengers in enclosed cabin; door each side. Rear seats fold forward to give access to baggage area.
DIMENSIONS, EXTERNAL:

Wing span	11.16 m (36 ft 7½ in)
Wing aspect ratio	10.5

USA: AIRCRAFT—EXPLORER AVIATION to EXPRESS

Width, wings folded	2.55 m (8 ft 4⅜ in)
Length: overall	6.10 m (20 ft 0 in)
wings folded	7.92 m (26 ft 0 in)
Height overall	1.93 m (6 ft 4 in)

DIMENSIONS, INTERNAL:
Cabin max width	1.12 m (3 ft 8 in)

AREAS:
Wings, gross	11.89 m² (128.0 sq ft)

WEIGHTS AND LOADINGS:
Weight empty	534 kg (1,177 lb)
Max T-O weight	997 kg (2,200 lb)
Max wing loading	83.9 kg/m² (17.19 lb/sq ft)
Max power loading	8.93 kg/kW (14.67 lb/hp)

PERFORMANCE, POWERED:
Max operating speed	142 kt (263 km/h; 164 mph)
Cruising speed at 75% power	130 kt (241 km/h; 150 mph)
Stalling speed	48 kt (89 km/h; 55 mph)
Max rate of climb at S/L	290 m (950 ft)/min
Service ceiling	4,630 m (15,200 ft)
T-O and landing run	335 m (1,100 ft)
Range with 45 min reserves	650 n miles (1,203 km; 748 miles)

PERFORMANCE, UNPOWERED:
Glide ratio	14

UPDATED

Explorer Aviation Ellipse (N8069X) *(Geoffrey P Jones)*

EXPRESS

EXPRESS AIRCRAFT COMPANY

PO Box 236, Olympia, Washington 98507
Tel: (+1 360) 352 05 06
Fax: (+1 360) 352 05 53
Web: http://www.express-aircraft.com
OWNERS: Larry Olson and Paul Fagerstrom

Assets of former Wheeler Technology Inc (see 1992-93 *Jane's*), which became bankrupt in late 1990, were acquired by Express Design Inc; EDI resumed production and customer support for former Wheeler Express kitbuilt aircraft. Factory floor area 1,950 m² (21,000 sq ft). EDI sold in late 1996 and renamed Express Aircraft Company. Also holds rights to Series 90, based on the Express, of which seven were flying by early 1999. A development of the Express, the **Auriga**, was shown at Sun 'n' Fun in April 1999.

UPDATED

Express Express 2000 kit aircraft *(Paul Jackson/Jane's)*

EXPRESS EXPRESS

TYPE: Four/six-seat kitbuilt.
PROGRAMME: Designed by Wheeler Technology Inc as high-speed cross-country kitplane. Three prototypes built from kits of premoulded parts; first flight 28 July 1987. First, larger, production line Express demonstrator made first flight May 1990, powered by Teledyne Continental engine. Delivered kits incomplete at time of Wheeler bankruptcy; deficiencies made good by EDI, and at least six aircraft completed by 1994; further two registered in 1995 and 12 in 1996. Complete kits available from Express incorporate some modifications, which can also be retrofitted to existing aircraft.
CURRENT VERSIONS: **Express FT:** Four-seat, original version; no longer available. Had 157 kW (210 hp) Teledyne Continental IO-360-ES1.
Express CT: Replaced FT.
Express 2000: Originally S-90; became 2000 during 1999; four-seats; higher-powered version.
Express 2000RG: Available from 2000; retractable landing gear.
Loadmaster 3200: Six seats; 194 kW (260 hp) engine; optional ventral pannier; no longer available.
CUSTOMERS: Over 300 kits sold, of which around 35 had been completed by January 1999. Quoted build time 1,500 hours.

COSTS: Airframe kit US$47,500; engine, propeller, upholstery, paint and instruments not included (2000).
DESIGN FEATURES: Conforms to FAR Pt 23. Streamlined low-wing monoplane; tapered wing with downturned tips; sweptback fin and tailplane.
Wing section NASA NFL-1 0215-F (laminar flow); dihedral 5°.
FLYING CONTROLS: Conventional and manual. Horn-balanced rudder and elevators; trim tab in starboard elevator. Flaps.
STRUCTURE: Constructed of composites sandwich material, comprising polyurethane foam core, glass fibre, unidirectional glass fibre tape and vinylester resin. CT has cruciform tail; 2000 has larger, conventional tail unit.
LANDING GEAR: Non-retractable tricycle type standard; speed fairings. Differential Cleveland brakes; castoring nosewheel. Retractable gear 2000RG available from mid-2000.
POWER PLANT: *CT:* One 186 kW (250 hp) Textron Lycoming IO-540-C4B5 recommended.
2000: 224 kW (300 hp) Textron Lycoming IO-540-K1G5, or 231 kW (310 hp) Teledyne Continental IO-550-N1B both driving Hartzell three-blade constant-speed propeller. Optional engine is 261 kW (350 hp) Teledyne Continental TSIO-550-E, giving cruise speed of 225 kt (417 km/h; 259 mph). Fuel capacity of CT and 2000 is 310 litres (82.0 US gallons; 68.3 Imp gallons). Oil capacity 6.62 litres (1.75 US gallons; 1.46 Imp gallons).

ACCOMMODATION: Unusual seating arrangement of one forward- and one aft-facing seat in rear, behind two side-by-side front seats with dual controls. Two additional seats in enlarged versions. Upward-hinged door each side; baggage door on port side.

DIMENSIONS, EXTERNAL:
Wing span	9.60 m (31 ft 6 in)
Wing chord: at root	1.52 m (5 ft 0 in)
at tip	0.95 m (3 ft 1½ in)
Wing aspect ratio	7.6
Length overall: CT	7.92 m (26 ft 0 in)
2000	7.62 m (25 ft 0 in)
Height overall	2.13 m (7 ft 0 in)
Tailplane span	3.05 m (10 ft 0 in)
Wheel track	3.43 m (11 ft 3 in)
Wheelbase	1.97 m (6 ft 5½ in)

DIMENSIONS, INTERNAL:
Cabin max width	1.17 m (3 ft 10 in)

AREAS:
Wings, gross	12.08 m² (130.0 sq ft)
Rudder	0.40 m² (4.33 sq ft)

WEIGHTS AND LOADINGS:
Weight empty	839 kg (1,850 lb)
Baggage capacity, with four 77 kg (170 lb) occupants	29 kg (64 lb)
Max T-O weight	1,451 kg (3,200 lb)
Max wing loading	120.2 kg/m² (24.62 lb/sq ft)
Max power loading (IO-540)	7.79 kg/kW (12.80 lb/hp)

PERFORMANCE (CT with IO-540, 2000 with IO-550, where different):
Never-exceed speed (V_{NE})	230 kt (426 km/h; 264 mph)
Cruising speed at 75% power at 2,290 m (7,500 ft):	
CT	185 kt (343 km/h; 213 mph)
2000	195 kt (361 km/h; 224 mph)
Stalling speed: flaps up	58 kt (108 km/h; 67 mph)
flaps down	55 kt (102 km/h; 64 mph)
Demonstrated crosswind speed for T-O and landing	29 kt (54 km/h; 33 mph)
Max rate of climb at S/L	366 m (1,200 ft)/min
Service ceiling	6,100 m (20,000 ft)
T-O run	305 m (1,000 ft)
Landing run	244 m (800 ft)
Range: at 55% power, standard fuel, no reserves:	
CT	1,110 n miles (2,056 km; 1,227 miles)
2000	1,170 n miles (2,166 km; 1,346 miles)
with auxiliary fuel tank	1,688 n miles (3,125 km; 1,942 miles)
g limits	+4.4/−2.2
	+8.8/−4.4 ultimate

UPDATED

Tailwheel version of Express 90, built by John Kee *(Paul Jackson/Jane's)*

FAIRCHILD DORNIER

FAIRCHILD DORNIER CORPORATION

In June 1996, Fairchild Aerospace and DaimlerChrysler Aerospace of Germany concluded an agreement for Fairchild to acquire 80 per cent of Dornier Luftfahrt GmbH and create a joint venture company, which was renamed Fairchild Dornier Aerospace on 13 April 2000. See main entry under Germany. The final Merlin 23 turboprop twin was delivered in early 2001.

UPDATED

FISHER

FISHER FLYING PRODUCTS

PO Box 468, Industrial Park, Edgeley, North Dakota 58433
Tel: (+1 701) 493 22 86
Fax: (+1 701) 493 25 39
e-mail: ffpjac@daktel.com
Web: http://www.fisherflying.com
PRESIDENT: Darlene Jackson-Hanson

Fisher Flying Products markets a range of homebuilt light aircraft. For information on the FP-202 Koala, Super Koala, FP-404 and Classic, see the Private Aircraft section of *Jane's* 1992-93. Also produces FP-303, 505 and 606 microlights. In 1998, Fisher took over Fisher Aero of Portsmouth, Ohio, and so added the Celebrity, Horizon 1 and Horizon 2 (see *Jane's* 1992-93) to the product line.

UPDATED

FISHER FLYING PRODUCTS AVENGER

TYPE: Single-seat kitbuilt.
PROGRAMME: First flown 1994.
CURRENT VERSIONS: **Avenger:** Baseline version.
   Avenger V: Powered by one 37.3 kW (50 hp) Volkswagen four-cylinder engine.
CUSTOMERS: Total of 150 sold by mid-2000.
COSTS: Avenger US$3,950 less engine (2000); Avenger V US$4,250 less engine (2000).
DESIGN FEATURES: Low-wing monoplane; wings braced by V-struts to mainwheel hubs; mutually braced tail surfaces. Quoted build time 300 hours.
FLYING CONTROLS: Conventional and manual.
STRUCTURE: Wood and fabric.
LANDING GEAR: Tailwheel type; fixed.
POWER PLANT: One 37.0 kW (49.6 hp) two-cylinder Rotax 503 UL-2V; other engines from 20.9 to 48.5 kW (28 to 65 hp) are optional. Fuel capacity 19 litres (5.0 US gallons; 4.2 Imp gallons).
DIMENSIONS, EXTERNAL:
  Wing span   8.23 m (27 ft 0 in)
  Length overall   4.95 m (16 ft 3 in)
  Height overall   1.52 m (5 ft 0 in)
AREAS:
  Wings, gross   11.24 m² (121.0 sq ft)
WEIGHTS AND LOADINGS:
  Weight empty: Avenger   113 kg (250 lb)
    Avenger V   159 kg (350 lb)
  Max T-O weight: Avenger   272 kg (600 lb)
    Avenger V   294 kg (650 lb)
PERFORMANCE:
  Max level speed: Avenger   82 kt (152 km/h; 95 mph)
    Avenger V   86 kt (160 km/h; 100 mph)
  Normal cruising speed   52 kt (97 km/h; 60 mph)
  Stalling speed: Avenger   23 kt (42 km/h; 26 mph)
    Avenger V   28 kt (52 km/h; 32 mph)
  Max rate of climb at S/L: Avenger   274 m (900 ft)/min
    Avenger V   259 m (850 ft)/min
  T-O run: Avenger   30 m (100 ft)
    Avenger V   61 m (200 ft)
  Landing run: Avenger   61 m (200 ft)
    Avenger V   76 m (250 ft)
  Range: Avenger   150 n miles (277 km; 172 miles)
    Avenger V   250 n miles (463 km; 287 miles)

UPDATED

Replica Tiger Moth produced by Fisher Flying Products

FISHER FLYING PRODUCTS DAKOTA HAWK

TYPE: Side-by-side ultralight kitbuilt.
PROGRAMME: Introduced to Fisher range in 1992.
CUSTOMERS: Total of 50 kits sold by mid-2000. Ten flying by end 1999.
COSTS: Airframe kit US$9,900; quick-build airframe kit US$11,900; both excluding engine (2000).
DESIGN FEATURES: Folding wings with V-struts for easy storage; I-beam wing spar. Braced tailplane. Quoted build time 300 to 400 hours.
FLYING CONTROLS: Conventional and manual.
STRUCTURE: Precut, shaped and slotted wooden pieces; glass fibre cowling.
LANDING GEAR: Tailwheel type; fixed. Matco brakes. Steerable tailwheel; shock-absorbers and Matco speed fairings on main legs.
POWER PLANT: One 59.6 kW (79.9 hp) Rotax 912 UL four-stroke engine driving a two-blade propeller. Optional alternatives include Teledyne Continental engines between 48.4 and 63 kW (65 and 85 hp) or 74.6 kW (100 hp) Subaru. Fuel capacity 45.4 litres (12.0 US gallons; 10.0 Imp gallons).
DIMENSIONS, EXTERNAL:
  Wing span   8.69 m (28 ft 6 in)
  Length overall   6.02 m (19 ft 9 in)
  Height overall   1.85 m (6 ft 1 in)
AREAS:
  Wings, gross   11.89 m² (128.0 sq ft)
WEIGHTS AND LOADINGS:
  Weight empty   272 kg (600 lb)
  Max T-O weight   522 kg (1,150 lb)
PERFORMANCE:
  Max cruising speed   87 kt (161 km/h; 100 mph)
  Normal cruising speed   78 kt (145 km/h; 90 mph)
  Stalling speed   31 kt (57 km/h; 35 mph)
  Max rate of climb at S/L   366 m (1,200 ft)/min
  T-O run   107 m (350 ft)
  Landing run   122 m (400 ft)
  Range with max fuel   217 n miles (402 km; 250 miles)

UPDATED

FISHER FLYING PRODUCTS YOUNGSTER and YOUNGSTER V

TYPE: Single-seat biplane ultralight kitbuilt.
PROGRAMME: First flown 1994.
CURRENT VERSIONS: **Youngster:** As described.
   Youngster V: Uprated version with 48.5 kW (65 hp) Volkswagen motor car engine.
CUSTOMERS: Total of 10 flying by end 1999.
COSTS: Youngster kit US$4,950 less engine. Youngster V US$5,200 less engine (2000).
DESIGN FEATURES: Open cockpit, strut-braced biplane with detachable cockpit transparency for winter operations. Constant-chord wings. Wire-braced low-mounted tailplane. Stainless steel control wires. Additional central strut between nose and upper wing forward of cockpit. Aerofoil type 2315. Quoted build time 400 hours.
FLYING CONTROLS: Conventional and manual.
STRUCTURE: Wood and fabric.
LANDING GEAR: Tailwheel type; fixed. Bungee cord shock absorption.
POWER PLANT: One 37.3 kW (50 hp) Great Plains four-cylinder engine; Rotax engines up to 48.5 kW (65 hp) are optional. Fuel capacity 22.7 litres (6.0 US gallons; 5.0 Imp gallons).
DIMENSIONS, EXTERNAL:
  Wing span   5.49 m (18 ft 0 in)
  Length overall   4.72 m (15 ft 6 in)
  Height overall   1.85 m (6 ft 1 in)
AREAS:
  Wings, gross   11.71 m² (126.0 sq ft)
WEIGHTS AND LOADINGS:
  Weight empty   181 kg (400 lb)
  Max T-O weight   294 kg (650 lb)
PERFORMANCE:
  Max level speed   95 kt (177 km/h; 110 mph)
  Normal cruising speed   74 kt (137 km/h; 85 mph)
  Stalling speed   30 kt (55 km/h; 34 mph)
  Max rate of climb at S/L   244 m (800 ft)/min
  T-O run   61 m (200 ft)
  Landing run   76 m (250 ft)
  Range   225 n miles (416 km; 259 miles)

UPDATED

FISHER FLYING PRODUCTS R-80 and RS-80 TIGER MOTH

TYPE: Tandem-seat ultralight biplane kitbuilt.
PROGRAMME: Scale (80 per cent) version of de Havilland D.H.82A Tiger Moth. Launched third quarter 1994.
CURRENT VERSIONS: **R-80 Tiger Moth:** As described.
   RS-80 Tiger Moth: Introduced 1999; has steel tube fuselage and 89.5 kW (120 hp) LOM M 132 engine.
CUSTOMERS: Total of 45 kits sold by mid-2000, of which five were then flying. Four under construction in Zimbabwe for use on aerial safaris.
COSTS: R.80 standard airframe kit US$10,500; quick-build airframe kit US$13,500; both excluding engine (2000); RS-80 standard airframe kit US$16,500; quick-build airframe kit US$18,500; both excluding engine (2000).
DESIGN FEATURES: Can be built using the contents of a basic tool kit. Quoted build time 550 hours.
   Single-bay, staggered biplane with N-type interplane struts and slight sweepback to constant-chord wings.
FLYING CONTROLS: Conventional and manual, with horn-balanced rudder.
STRUCTURE: Wood construction using epoxy adhesives on preformed components.
LANDING GEAR: Tailwheel type; fixed. Tubular steel mainwheel legs; Matco wheels with hydraulic brakes.

Fisher Dakota Hawk two-seat kitbuilt

POWER PLANT: Prototype has one 70.8 kW (95 hp) Mid West AE 100 rotary, but any engine between 55.9 and 74.6 kW (75 and 100 hp) can be fitted, including Rotax 618 and 912 and Subaru EA81-100. One 72 litre (19.0 US gallon; 15.8 Imp gallon) centre-section fuel tank.

DIMENSIONS, EXTERNAL:
Wing span	7.01 m (23 ft 0 in)
Length overall	5.79 m (19 ft 0 in)
Height overall	2.24 m (7 ft 4 in)

DIMENSIONS, INTERNAL:
Cockpit max width	0.66 m (2 ft 2 in)

AREAS:
Wings, gross	15.79 m² (170.0 sq ft)

WEIGHTS AND LOADINGS:
Weight empty: R-80	254 kg (560 lb)
RS-80	363 kg (800 lb)
Max T-O weight: R-80	522 kg (1,150 lb)
RS-80	612 kg (1,350 lb)

PERFORMANCE:
Max level speed: R-80	95 kt (177 km/h; 110 mph)
RS-80	104 kt (193 km/h; 120 mph)
Cruising speed: R-80	69 kt (129 km/h; 80 mph)
RS-80	78 kt (145 km/h; 90 mph)
Stalling speed: R-80	31 kt (57 km/h; 35 mph)
RS-80	35 kt (65 km/h; 40 mph)
Max rate of climb: both	244 m (800 ft)/min
T-O run: R-80	91 m (300 ft)
RS-80	76 m (250 ft)
Landing run: R-80	122 m (400 ft)
RS-80	91 m (300 ft)
Range with max fuel: both	250 n miles (463 km; 287 miles)

UPDATED

FLIGHTSTAR

FLIGHTSTAR SPORTSPLANES
PO Box 760, Ellington, Connecticut 06029
Tel: (+1 860) 875 87 85
Fax: (+1 860) 875 54 99
e-mail: fstar@mail2.nai.net
Web: http://www.flyflightstar.com
PRESIDENT: Thomas A Peghiny
TECHNICAL DIRECTOR: Mark Lamontagne

Pioneer International formed as a division of Pioneer Parachute Company in the mid-1980s to produce the initial Flightstar; design was then sold to Pampa's Bull of Argentina as the Aviastar I (see 1992-93 *Jane's*). Flightstar formed in the 1990s initially to import the design, and has since developed it into the current range of ultralights. The initial Flightstar Classic is no longer available.

NEW ENTRY

FLIGHTSTAR IISL and IISC

TYPE: Side-by-side ultralight kitbuilt.
PROGRAMME: Originally designed by Pioneer International; construction of prototype Flightstar started 1992; first flight 1993. Certified to German BFU-95 standards 1997. Flightstar IISL introduced in 1995.
CURRENT VERSIONS: **IISL:** Standard version, *as described*.
IISC: Introduced 1999; cabin version with doors for cold-weather operations.
CUSTOMERS: Over 1,000 of original Flightstar II built. Current version orders total 350, of which 330 were flying by end 2000.
COSTS: IISL US$11,995; IISC US$12,895, both without engines (2001).
DESIGN FEATURES: Wings fold for storage and transportation. Optional BRS parachute recovery system. Quoted build time 100 hours for IISL; 130 hours for IISC.
Flightstar FS-I-2 aerofoil section; dihedral 2° 48′; incidence 4°, twist 2°.
FLYING CONTROLS: Conventional and manual. Push-pull cables; optional flaps on IISL.
STRUCTURE: Aluminium tube spar, strut-braced wing with Dacron covering. Composites-covered chromoly 4130 steel fuselage cage with stainless steel brackets; engine attached to heavy-duty aluminium tube running length of aircraft. Strut-braced, low-mounted tailplane.
LANDING GEAR: Tricycle type; fixed; steerable nosewheel. Azusa wheels and optional drum brakes. Shock cord absorption. Optional Full Lotus floats.
POWER PLANT: One 37.0 kW (49.6 hp) Rotax 503 UL-2V driving three-blade Powerfin propeller; optional engines include 47.8 kW (64.1 hp) Rotax 582 UL and 44.7 kW (60.0 hp) four-stroke HKS 700E. Fuel capacity 37.9 litres (10.0 US gallons; 8.3 Imp gallons), of which 36.0 litres (9.5 US gallons; 7.9 Imp gallons) usable.
ACCOMMODATION: Optional Fastback fabric rear fairing for IISL.

DIMENSIONS, EXTERNAL:
Wing span	9.75 m (32 ft 0 in)
Length overall	5.79 m (19 ft 0 in)
Height overall	2.39 m (7 ft 10 in)

AREAS:
Wings, gross	14.59 m² (157.0 sq ft)

WEIGHTS AND LOADINGS:
Weight empty	204 kg (450 lb)
Max T-O weight	450 kg (992 lb)

PERFORMANCE:
Never-exceed speed (V$_{NE}$)	83 kt (154 km/h; 96 mph)
Max operating speed	74 kt; (137 km/h; 85 mph)
Normal cruising speed	56 kt; (105 km/h; 65 mph)
Stalling speed	32 kt (58 km/h; 36 mph)
Max rate of climb at S/L	213 m (700 ft)/min
T-O run	76 m (250 ft)
Landing run	91 m (300 ft)
Range with max fuel	156 n miles (289 km; 180 miles)

NEW ENTRY

FLIGHTSTAR SPYDER and FORMULA

TYPE: Single-seat ultralight kitbuilt.
PROGRAMME: Derived from Flightstar II. Can conform to FAR Pt 103.
CURRENT VERSIONS: **Spyder:** Standard version, *as described*.
Formula: High-performance upgraded model with cockpit doors for cold weather flying.

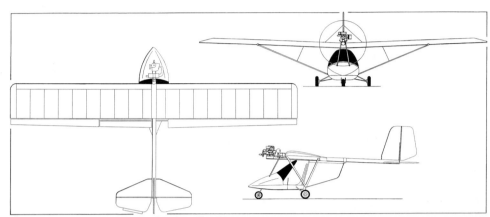
Open cockpit Flightstar IISL *(Paul Jackson/Jane's)* 2001/0105029

Two-seat Flightstar IISC version with cabin doors 2001/0077412

Flightstar Spyder single-seat ultralight 2001/0077411

CUSTOMERS: Total of 102 kits sold, and 93 flying, by end 2000.
COSTS: Spyder US$9,995; Formula US$10,895, both without engines (2001).
DESIGN FEATURES: Flightstar FS-I-2 aerofoil. Wings fold for storage and transportation. Optional BRS parachute recovery system.

FLYING CONTROLS: As Flightstar IISL.
STRUCTURE: As Flightstar IISL.
LANDING GEAR: As Flightstar IISL.
POWER PLANT: One 31.0 kW (41.6 hp) Rotax 447 UL-2V driving Tennessee two-blade or Ivo three-blade propeller. Options include 37.0 kW (49.6 hp) Rotax 503 UL-2V.

Standard fuel capacity of Spyder 18.9 litres (5.0 US gallons; 4.2 Imp gallons); optionally 37.9 litres (10.0 US gallons; 8.3 Imp gallons); Formula fuel capacity 37.9 litres (10.0 US gallons; 8.3 Imp gallons).

DIMENSIONS, EXTERNAL:
Wing span	9.14 m (30 ft 0 in)
Length overall	5.03 m (16 ft 6 in)
Height overall	2.39 m (7 ft 10 in)

AREAS:
Wings, gross	13.38 m² (144.0 sq ft)

WEIGHTS AND LOADINGS:
Weight empty: Spyder	127 kg (280 lb)
Formula	181 kg (400 lb)
Max T-O weight: Spyder	294 kg (650 lb)
Formula	431 kg (950 lb)

PERFORMANCE:
Never-exceed speed (V_{NE})	83 kt (154 km/h; 96 mph)
Normal cruising speed: Spyder	48 kt (89 km/h; 55 mph)
Formula	56 kt (105 km/h; 65 mph)
Stalling speed: both	28 kt (52 km/h; 32 mph)
Max rate of climb at S/L: both	201 m (660 ft)/min
T-O run: both	62 m (205 ft)
Landing run: both	61 m (200 ft)
Range with standard fuel:	
Spyder	86 n miles (161 km; 100 miles)
Formula	191 n miles (354 km; 220 miles)

NEW ENTRY

FLYING K

FLYING K ENTERPRISES, INC
3403 Arthur Street, Caldwell, Idaho 83605
Tel: (+1 208) 455 75 29
Fax: (+1 208) 455 76 78
e-mail: flyingk@micron.net
Web: http://www.skyraider.com

Flying K Enterprises was established by former SkyStar employee, the late Kenny Schrader, and is run by his widow, Kyna Schrader.

UPDATED

FLYING K SKY RAIDER and SKY RAIDER II
TYPE: Single-seat ultralight kitbuilt/tandem-seat ultralight kitbuilt.
PROGRAMME: First flown in 1996.
CURRENT VERSIONS: **Sky Raider**: Single-seat model, available as certified aeroplane or under FAR Pt 103 ultralight regulations.
Sky Raider II: Tandem-seat version; introduced in 1999.
CUSTOMERS: Total of 60 flying by mid-2000.
COSTS: Sky Raider kit US$8,500; Sky Raider II kit US$9,950 (2000).
DESIGN FEATURES: High wing with two tubular main, and two secondary bracing struts to each wing. Mutually wire-braced fin and tailplane. Wings foldable for storage without disconnection of control runs. Quoted build time 300 hours for Sky Raider and 350 hours for Sky Raider II.
FLYING CONTROLS: Conventional and manual. Frise ailerons, cable actuated. Slotted flaps.
STRUCTURE: Fabric-covered steel tube. Optional wingtip extensions to Sky Raider add 55 cm (21½ in) to span; choice of rounded or square fin on Sky Raider. Heavier wing spar on Sky Raider II.
LANDING GEAR: Tailwheel type: fixed. Mainwheels 8.00-6. Optional floats and skis.
POWER PLANT: Sky Raider has one 20.1 kW (27 hp) Rotax 277 two-cylinder two-stroke driving a two-blade propeller for Pt 103; 29.8 kW (40 hp) Rotax 477 for normal operations. Optionally, engines up to 37.3 kW (50 hp) can be fitted. Fuel capacity 18.9 litres (5.0 US gallons; 4.2 Imp gallons). Sky Raider II has one 44.7 kW (60 hp) HKS engine driving a GSC ground-adjustable propeller. Fuel capacity 37.9 litres (10.0 US gallons; 8.3 Imp gallons).
ACCOMMODATION: *Sky Raider:* Pilot only, in open-sided cabin.
Sky Raider II: Two seats in tandem within fully enclosed cabin.

DIMENSIONS, EXTERNAL (A: Sky Raider, B: Skyraider II):
Wing span: A	7.99 m (26 ft 2½ in)
B	8.53 m (28 ft 0 in)
Length overall: A, B	5.18 m (17 ft 0 in)
Height overall: A	1.57 m (5 ft 2 in)
B	1.73 m (5 ft 8 in)

AREAS:
Wings, gross, A, B:	
without extensions	9.10 m² (98.0 sq ft)
wth extensions	9.94 m² (107.0 sq ft)

WEIGHTS AND LOADINGS:
Weight empty: A with Rotax 447	127 kg (279 lb)
B with HKS	184 kg (405 lb)
Max T-O weight: A	249 kg (550 lb)
B	431 kg (950 lb)

PERFORMANCE:
Normal cruising speed: A	55 kt (101 km/h; 63 mph)
B	65 kt (121 km/h; 75 mph)
Stalling speed, flaps down: A	20 kt (37 km/h; 23 mph)
B	26 kt (49 km/h; 30 mph)
Max rate of climb at S/L: A	244 m (800 ft)/min
B	396 m (1,300 ft)/min
T-O and landing run: A, B	23 m (75 ft)
Range: A	120 n miles (222 km; 138 miles)
B	173 n miles (321 km; 200 miles)

UPDATED

FLYING K SKY ROCKET II
Renamed Sky Raider II.

UPDATED

Flying K Sky Raider (*Paul Jackson/Jane's*)   2001/0099634

FREEBIRD

FREEBIRD SPORT AIRCRAFT
Integrated with ProSport Aviation (which see).

UPDATED

GALAXY

GALAXY AEROSPACE CORPORATION
One Galaxy Way, Alliance Airport, Fort Worth, Texas 76177
Tel: (+1 817) 837 33 33
Fax: (+1 817) 837 38 62
e-mail: jmiller@galaxyaero.com
PRESIDENT AND CEO: Brian Barents
EXECUTIVE VICE-PRESIDENT, SALES AND MARKETING: Roger Sperry
VICE-PRESIDENT, GOVERNMENT AND INTERNATIONAL OPERATIONS: Barry L Harris
VICE-PRESIDENT, CUSTOMER SERVICE: Mike Wuebbling
VICE-PRESIDENT, CORPORATE COMMUNICATIONS: Jeff Miller

Galaxy Aerospace is the type certificate holder for both the Galaxy and Astra SPX. These are manufactured in Israel and flown 'green' to Fort Worth for outfitting at the company's 13,935 m² (150,000 sq ft) facility at Alliance Airport, opened in 1998. In July 2000, Galaxy Aerospace was negotiating an agreement with FFA Altenrhein of Switzerland for the latter to become its European parts distribution centre.

UPDATED

Galaxy Aerospace's Fort Worth service centre maintains Westwind business jets (foreground) as well as outfitting the later Astra SPX and Galaxy
2001/0099586

GREAT PLAINS

GREAT PLAINS AIRCRAFT SUPPLY CO INC
PO Box 545, Boystown, Nebraska 68010
Tel: (+1 402) 493 65 07
Fax: (+1 402) 333 77 50
e-mail: gpasc@earthlink.net
Web: http://www.greatplainsas.com
MARKETING DIRECTOR: Steve Bennett

Company formed in 1982, specialising in supply of hardware for homebuilt aircraft. Hardware division sold in 1984; company concentrated on Volkswagen engines. Took over supply of Sonerai plans, kits and parts in 1987; original Sonerai designed by John Monnett. Annual sales average 30 kits and 110 sets of plans.

VERIFIED

Great Plains Sonerai II two-seat homebuilt *(W J Bushell)*

GREAT PLAINS SONERAI II SERIES

TYPE: Tandem-seat sportplane kitbuilt.
PROGRAMME: Prototype of tandem two-seat Sonerai II first flown July 1973.
CURRENT VERSIONS: **Sonerai I:** Single-seat version; described separately.
  Sonerai II: Standard mid-wing model.
Description applies to Sonerai II, except where indicated.
  Sonerai II-L: Low-wing instead of mid-wing configuration and 3° dihedral; first flight June 1980.
  Sonerai II-LT: Similar to Sonerai II-L but has 2,200 cc Volkswagen engine standard, tricycle landing gear and larger front cockpit for taller pilot; retrofit kit was produced to allow existing Sonerais to be fitted with tricycle gear; first flight January 1983.
  Sonerai II-LS: Low-wing stretched version (6.20 m; 20 ft 4 in).
  Sonerai II-LTS: Stretched version of Sonerai II-LT; Volkswagen 2,200 cc engine; first flight June 1984. Further details in 1997-98 and earlier *Jane's*.
CUSTOMERS: Company estimates around 2,300 Sonerais of all versions completed so far, including at least 500 Sonerai IIs.
COSTS: Airframe kit: US$6,000. Plans: US$99.95 (1999).
DESIGN FEATURES: Mid- or low-wing monoplane with closely cowled engine and narrow, streamlined cockpit canopy. Many components, complete kits for fuselage, tail and wings, and materials, available to amateur constructors; quoted build time 850 hours.
  Wing section NACA 64A212.
FLYING CONTROLS: Conventional and manual. Flaps. No tabs.
STRUCTURE: Wings of aluminium alloy; full-span aluminium alloy ailerons. Welded steel tube fuselage and tail unit, fabric covered except for glass fibre engine cowling.
LANDING GEAR: Non-retractable; optional tailwheel or nosewheel type. Disc brakes.
POWER PLANT: One 44.7, 56 or 61 kW (60, 75 or 82 hp) Great Plains converted Volkswagen 1,600, 1,834 or 2,200 cc motorcar engine; 2,200 cc version is standard for Sonerai II-LT. Fuel capacity 38 litres (10.0 US gallons; 8.3 Imp gallons). Sonerai II can accommodate optional auxiliary tank of 23 to 38 litres (6.0 to 10.0 US gallons; 5.0 to 8.3 Imp gallons) in place of passenger.

DIMENSIONS, EXTERNAL:
  Wing span   5.69 m (18 ft 8 in)
  Length overall   5.74 m (18 ft 10 in)
  Height overall   1.65 m (5 ft 5 in)
AREAS:
  Wings, gross   7.80 m² (84.0 sq ft)
WEIGHTS AND LOADINGS:
  Weight empty   236 kg (520 lb)
  Max T-O weight   431 kg (950 lb)
PERFORMANCE (Sonerai II/II-L with 1,700 cc engine):
  Max level speed   152 kt (282 km/h; 175 mph)
  Econ cruising speed at S/L   122 kt (226 km/h; 140 mph)
  Stalling speed   39 kt (73 km/h; 45 mph)
  Max rate of climb at S/L   152 m (500 ft)/min
  T-O run   274 m (900 ft)
  Landing run   274 m (900 ft)
  Range, 45 min reserves   213 n miles (394 km; 245 miles)

UPDATED

GREAT PLAINS SONERAI I

TYPE: Single-seat sportplane kitbuilt.
PROGRAMME: First flown July 1971.
Description generally as for two-seat Sonerai II.
COSTS: US$5,500 kit (2000).
STRUCTURE: Folding wings.
POWER PLANT: Fuel capacity 41.6 litres (11.0 US gallons; 9.2 Imp gallons).

DIMENSIONS, EXTERNAL:
  Wing span   5.08 m (16 ft 8 in)
  Length overall   5.08 m (16 ft 8 in)
  Height overall   1.52 m (5 ft 0 in)
AREAS:
  Wings, gross   6.97 m² (75.0 sq ft)
WEIGHTS AND LOADINGS:
  Weight empty   200 kg (440 lb)
  Max T-O weight   340 kg (750 lb)
PERFORMANCE:
  Max level speed   148 kt (274 km/h; 170 mph)
  Cruising speed at 75% power at S/L
                                    130 kt (241 km/h; 150 mph)
  Stalling speed   39 kt (73 km/h; 45 mph)
  Max rate of climb at S/L   305 m (1,000 ft)/min
  T-O run   183 m (600 ft)
  Landing run   274 m (900 ft)
  Range, 45 min reserves   260 n miles (482 km; 300 miles)

UPDATED

GRIFFON

GRIFFON AEROSPACE INC
301C Nick Fitchard Road, Huntsville, Alabama 35806
Tel: (+1 256) 859 38 80
Fax: (+1 256) 859 34 97
e-mail: info@griffon-aerospace.com
Web: http://www.griffon-aerospace.com
PRESIDENT: Larry French

Griffon's first product, the Lionheart, is stated to be the first kitplane designed (by Larry French) entirely on 3-D CAD.

UPDATED

GRIFFON LIONHEART

TYPE: Utility biplane kitbuilt.
PROGRAMME: Representation of 1930s Beech 17 'Staggerwing'. Design began 1993; prototype (N985L) first flown 1997 and publicly displayed at Sun 'n' Fun 1998.
CUSTOMERS: Eight kits sold by April 2000; three flying by end 1999, first being N985CC (c/n 3), built by Chuck Cianchette in 1999 and second, N223TC (c/n 2), built by Tom Constantino.
COSTS: US$99,900, excluding engine (2000); 'express build' kits available at extra US$18,000.

DESIGN FEATURES: Classic, equal-span negative stagger biplane. While retaining superficial Beech 17 appearance, is entirely new, computer-aided design. Changes from original include additional cabin glazing, addition of elevator horn balance and enlargement of rudder horn, unbraced wings lacking elliptical tips, and replacement of full span upper ailerons and lower flaps by separate flaps and ailerons on each wing. Aerobatics permissible below 1,905 kg (4,200 lb).
  Aerofoil modified NACA 64-215 at root, tapering to 64-212 at tip. Laminar flow maintained over 45 per cent of chord. Upper wing dihedral 0° 54'; lower wing dihedral 0° 24'. Washout 1°.
FLYING CONTROLS: Conventional and manual. Ailerons and elevator actuated by pushrods; cable-operated rudder. Flaps and ailerons on all wings. Flight-adjustable tab in port elevator.
STRUCTURE: Extensive use of composites; mainly Nomex honeycomb core with carbon skins and 7781 glass fibre. Cantilever wings; fuselage contains four longerons and underfloor keel beam.
LANDING GEAR: Tailwheel type; all three units electrohydraulically retractable. Mainwheels 6.50-8. Pressurised gas damping on main legs; gas-charged air/oil cylinder on tail units.

Prototype Griffon Lionheart accompanied by its progenitor, the Beech 17

Lionheart instrument panel

GRIFFON to GROEN—AIRCRAFT: USA

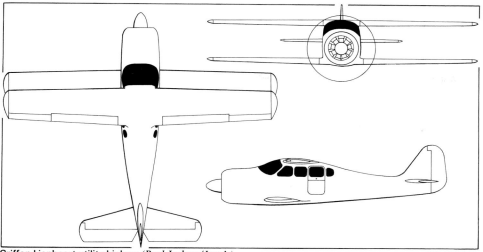

Griffon Lionheart utility biplane *(Paul Jackson/Jane's)* 2001/0099164

POWER PLANT: One 336 kW (450 hp) Pratt & Whitney R-985 Wasp Junior nine-cylinder radial engine (as in Beech D17S), driving a constant-speed two-blade Hamilton Sundstrand or three-blade Hartzell R10152-5.5 propeller. Alternative engines in the range 261 to 447 kW (350 to 600 hp). Fuel capacity 704 litres (186 US gallons; 155 Imp gallons) in two upper wing tanks, each holding 208 litres (55.0 US gallons; 45.8 Imp gallons) and two lower wing tanks each of 144 litres (38.0 US gallons; 31.7 Imp gallons); total usable fuel 681 litres (180 US gallons; 150 Imp gallons). Oil capacity 28.4 litres (7.5 US gallons; 6.25 Imp gallons).

ACCOMMODATION: Pilot and up to six passengers in enclosed cabin in either forward-facing or club seating. Dual controls. Two-piece, horizontally split door behind wing on port side.

SYSTEMS: Vision Microsystems VM1000 engine monitoring system.

DIMENSIONS, EXTERNAL:
Wing span	9.45 m (31 ft 0 in)
Wing chord, mean	1.19 m (3 ft 10¾ in)
Wing aspect ratio (effective)	5.2
Negative stagger	0.63 m (2 ft 1 in)
Length overall	8.31 m (27 ft 3 in)
Height overall	2.54 m (8 ft 4 in)
Propeller diameter	2.44 m (8 ft 0 in)

DIMENSIONS, INTERNAL:
Cabin: Length	3.91 m (12 ft 10 in)
Max width	1.30 m (4 ft 3 in)
Max height	1.32 m (4 ft 4 in)

AREAS:
Wings, gross (each)	10.96 m² (118.0 sq ft)

WEIGHTS AND LOADINGS:
Weight empty	1,412 kg (3,114 lb)
Max T-O weight	2,358 kg (5,200 lb)
Max wing loading	107.6 kg/m² (22.03 lb/sq ft)
Max power loading	7.03 kg/kW (11.56 lb/hp)

PERFORMANCE:
Never-exceed speed (V_{NE})	278 kt (514 km/h; 320 mph)
Max level speed	200 kt (370 km/h; 230 mph)
Max cruising speed	182 kt (338 km/h; 210 mph)
Normal cruising speed at 60% power at 2,895 m (9,500 ft)	170 kt (315 km/h; 196 mph)
Stalling speed, power off:	
flaps up	61 kt (113 km/h; 71 mph)
flaps down	56 kt (104 km/h; 65 mph)
Max rate of climb at S/L	533 m (1,750 ft)/min
Service ceiling	6,095 m (20,000 ft)
T-O run	320 m (1,050 ft)
T-O to 15 m (50 ft)	488 m (1,600 ft)
Landing from 15 m (50 ft)	640 m (2,100 ft)
Landing run	396 m (1,300 ft)
Range with max fuel at 75% power, 45 min reserves	1,450 n miles (2,685 km; 1,668 miles)
Endurance	7 h 12 min
g limits	+6/−3

UPDATED

GROEN

GROEN BROTHERS AVIATION INC
2640 West California Avenue, Suite A, Salt Lake City, Utah 84104
Tel: (+1 801) 973 01 77
Fax: (+1 801) 973 40 27
e-mail: mktg@gbagyros.com
Web: http://www.gbagyros.com
PRESIDENT AND CEO: David Groen
CHAIRMAN: Jay Groen
VICE-PRESIDENT AND COO: James P Mayfield III
DIRECTOR OF MARKETING: Scott Muir

MARKETING:
18000 Pacific Highway South, Suite 408, Seattle, Washington 98188
Tel: (+1 206) 241 81 16
Fax: (+1 206) 241 81 17
MARKETING MANAGER: Cyndi Upthegrove

SUBSIDIARY COMPANY:
Nanjing Groen Aviation Industrial Ltd
140 Guangzhou Road, Suite 22D, Nanjing, Jiangsu 210024, People's Republic of China
Tel: (+86 25) 360 50 73
Fax: (+86 25) 360 49 53

Groen Brothers has been working on its Hawk Gyroplane family since 1986. Prototypes leading up to the Hawk 4, currently in the process of FAA type certification, include the Hawk I and H2X, as described in previous editions of *Jane's*. The Hawk 4 Gyroplane will be the first in a series of VTOL-capable aircraft. Development and FAA certification of the Hawk 6 (six place) and Hawk 8 (eight place) will follow the Hawk 4, and preliminary designs of larger aircraft have been completed. The Hawk 6G uses the fuselage of a Cessna 337 as the basis of its fuselage. The company has a flight test facility at Buckeye, Arizona where, on 12 July 2000, the prototype Jet Hawk 4T made its initial flight.

UPDATED

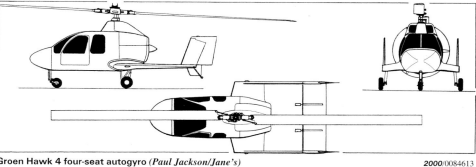

Groen Hawk 4 four-seat autogyro *(Paul Jackson/Jane's)* 2000/0084613

GROEN HAWK 4 and JET HAWK 4T
TYPE: Autogyro.

PROGRAMME: Design started April 1996 and prototype two-seat H2X first flew 4 February 1997; later converted to three-seat Hawk III standard. First deliveries had been due June 1998, but design changes to H2X and later Hawk III in October 1998 resulted in Hawk 4. Initial aircraft (N402GB) first flew 29 September 1999 and made first vertical take-off on 9 December 1999; had flown 120 hours in 200 sorties by early April 2000. FAA certification (450 hour programme by two aircraft in six months) expected by second quarter of 2001.

CURRENT VERSIONS: Applications include airborne law enforcement, electronic news gathering, aerial surveillance, utility/passenger transport and aerial application.

Hawk 4: Piston-engined version. *Description applies to this version.*

Jet Hawk 4T: Powered by 313 kW (420 shp) Rolls-Royce 250-B17C turboprop; first flew 12 July 2000; intended for certification late 2001. Two further prototypes under construction.

CUSTOMERS: Contract for licensed assembly signed with the Shanghai Energy and Chemicals Corporation in July 1996 to cover 200 autogyros. Order reconfirmed in August 1998, plus options on further 300. By September 2000, deposits on an additional 148 aircraft had been taken, via 10 dealerships.

COSTS: Around US$295,000 (2000) for Hawk 4 and US$749,000 for Jet Hawk 4T.

DESIGN FEATURES: Twin tailbooms supported by stub-wings which also house main landing gear; large twin rudders; fixed horizontal tail surface. Two-blade, semi-rigid aluminium teetering rotor with swashplate. Rotor speed 270 rpm.

FLYING CONTROLS: Patented collective pitch-controlled rotor head allows smooth vertical take-off (zero ground roll) and enhanced flight performance. Rotor brake is standard. Fixed horizontal stabiliser; yaw control by two rudders in tailfins, plus two all-moving rudders on horizontal stabiliser. Actuation by pushrods.

STRUCTURE: Steel mast and engine mounts; stressed skin aluminium semi-monocoque fuselage, tail unit, hub structure and propeller; composites nose, engine cowling and wingtips; acrylic windscreen and doors; glass fibre nosecone and engine cowling.

LANDING GEAR: Initially fixed tricycle type; with electrohydraulically operated retractable gear to be offered later. Mainwheel tyres 6.00×6; nosewheel 5.00×5; Cleveland hydraulic brakes. Twin safety wheels at rear of tailbooms.

Prototype Groen Hawk 4 on a test flight with cabin door removed 2001/0100445

USA: AIRCRAFT—GROEN to GULFSTREAM AEROSPACE

POWER PLANT: Production prototype for Hawk series was powered by a 134 kW (180 hp) Textron Lycoming O-360-A4M flat-four. Hawk III had one 335 kW (450 hp) Geschwinder V-8 aluminium liquid-cooled engine, derated to 261 kW (350 hp) at 2,500 rpm, driving a Hartzell three-blade constant-speed propeller.

Hawk 4 has air-cooled, six-cylinder Teledyne Continental TSIO-550 rated at 261 kW (350 hp) at 2,700 rpm; prototype had four-blade MTV propeller but production models will have Hartzell three-blade constant-speed propeller. Engine provides power to rotor for prerotation to provide for short and vertical take-off capability; power to rotor system never engaged during flight. Hawk 4T version planned with Rolls-Royce turbine engine.

Fuel capacity 284 litres (75.0 US gallons; 62.5 Imp gallons) in single tank at rear of fuselage; refuelling point at top of fuselage. Oil capacity 11.4 litres (3.0 US gallons; 2.5 Imp gallons).
ACCOMMODATION: Pilot and up to three passengers in enclosed cabin in two pairs of seats; rear seats fold to provide baggage space.
SYSTEMS: Electrical system 28 V DC.
AVIONICS: *Comms:* Honeywell KLX 135A GPS/COM, KT 76A transponder with Mode C, PM 3000 intercom.
Flight: United Instruments suite.
DIMENSIONS, EXTERNAL:
Rotor diameter	12.80 m (42 ft 0 in)
Rotor blade chord	0.34 m (1 ft 1½ in)
Propeller diameter	1.93 m (6 ft 4 in)
Length of fuselage: Hawk 4	6.71 m (22 ft 0 in)
Jet Hawk 4T	7.31 m (24 ft 0 in)
Height to top of rotor head	3.35 m (11 ft 0 in)
Wheel track	2.72 m (8 ft 11 in)
Wheelbase	2.47 m (8 ft 1¼ in)
Passenger doors: Width	0.94 m (3 ft 1 in)
Height to sill	0.89 m (2 ft 11 in)

DIMENSIONS, INTERNAL:
Cabin: Length	1.91 m (6 ft 3 in)
Max width	1.37 m (4 ft 6 in)

AREAS:
Main rotor disc	128.71 m² (1,385.4 sq ft)
Vertical tails	1.88 m² (20.2 sq ft)
Horizontal stabiliser	1.49 m² (16.0 sq ft)

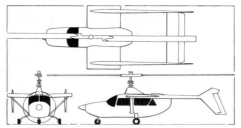

Groen Hawk 6G, based on Cessna 337 airframe *(Paul Jackson/Jane's)*

WEIGHTS AND LOADINGS:
Weight empty	835 kg (1,840 lb)
Max T-O weight	1,270 kg (2,800 lb)
Max disc loading	9.87 kg/m² (2.02 lb/sq ft)
Max power loading	4.87 kg/kW (8.00 lb/hp)

PERFORMANCE:
Never-exceed speed (V$_{NE}$)	113 kt (209 km/h; 130 mph)
Cruising speed at 75% power	100 kt (185 km/h; 115 mph)
Max rate of climb at S/L	457 m (1,500 ft)/min
Service ceiling	4,875 m (16,000 ft)
T-O to 15 m (50 ft)	46 m (150 ft)
Landing from 15 m (50 ft)	46 m (150 ft)
Range with max fuel at 75% power	521 n miles (965 km; 420 miles)

UPDATED

GROEN HAWK 6 and HAWK 8
TYPE: Autogyro.
PROGRAMME: Projected developments of Hawk 4 (see previous entry); will benefit from Hawk 4 certification. Preliminary design work has begun on both.
CURRENT VERSIONS: **Hawk 6:** Six-seat version.
Hawk 6G: Announced late 2000 with an expected first flight (prototype N9112A) in January 2001. Based around Cessna 337 fuselage with single 335 kW (420 shp) Rolls-Royce 250-B17F2 turboprop in nose, inverted Cessna 337 fins and rudders, shortened wing with spoilers and two-blade rotor system from Hawk 4. Estimated cruising speed 113 to 130 kt (209 to 241 km/h; 130 to 150 mph); useful load 907 kg (2,000 lb).
Hawk 8: Eight-seat version.
DESIGN FEATURES: Basically similar to Hawk 4, but Hawk 8 will have four rotor blades.
FLYING CONTROLS: Broadly similar to Hawk 4.
STRUCTURE: As Hawk 4.
POWER PLANT: Each powered by a 559 kW (750 shp) turboprop engine. Fuel capacity will be 416 litres (110 US gallons; 91.6 Imp gallons) in Hawk 6 and 606 litres (160 US gallons; 133 Imp gallons) in Hawk 8.
AVIONICS: *Instrumentation:* Full IFR.
DIMENSIONS, EXTERNAL:
Main rotor diameter: Hawk 6	14.63 m (48 ft 0 in)
Hawk 8	12.80 m (42 ft 0 in)

AREAS:
Main rotor disc: Hawk 6	168.11 m² (1,809.6 sq ft)
Hawk 8	128.71 m² (1,385.4 sq ft)

WEIGHTS AND LOADINGS (estimated):
Weight empty: Hawk 6	964 kg (2,125 lb)
Hawk 8	1,406 kg (3,100 lb)
Max T-O weight: Hawk 6	1,678 kg (3,700 lb)
Hawk 8	2,495 kg (5,500 lb)
Max disc loading: Hawk 6	9.98 kg/m² (2.04 lb/sq ft)
Hawk 8	19.38 kg/m² (3.97 lb/sq ft)
Max power loading: Hawk 6	3.00 kg/kW (4.93 lb/shp)
Hawk 8	4.46 kg/kW (7.33 lb/shp)

PERFORMANCE (estimated):
Never-exceed speed (V$_{NE}$)	147 kt (273 km/h; 170 mph)
Cruising speed	122 kt (225 km/h; 140 mph)
Service ceiling	4,875 m (16,000 ft)
Range:	
Hawk 6 at 50% power	360 n miles (667 km; 415 miles)
Hawk 8 at 75% power	347 n miles (643 km; 400 miles)
Endurance: Hawk 6	2 h 58 min
Hawk 8	2 h 53 min

UPDATED

GROSSO

GROSSO AIRCRAFT INC
229 Yarrow Hill Drive, Cottage Grove, Wisconsin 53527
Tel: (+1 608) 345 04 06
PRESIDENT: Ron Grosso

Grosso Aircraft was set up to market the Easy Eagle ultralight.

NEW ENTRY

GROSSO EASY EAGLE
TYPE: Single-/two-seat biplane ultralight kitbuilt.
PROGRAMME: Construction began 1996; Easy Eagle first displayed at Oshkosh 1998. Two-seat Easy Eagle II introduced in 1999.
CURRENT VERSIONS: **Easy Eagle:** Single-seat version.
Easy Eagle II: Two-seat version.
CUSTOMERS: Two Easy Eagles and one Easy Eagle II flying by end 2000.
COSTS: Kit US$9,900 (Easy Eagle) or US$13,900 (Easy Eagle II) (2000).
DESIGN FEATURES: Classic open-cockpit, single-bay biplane with steel tube cabane and flying struts and steel landing wires. Modified Clark Y aerofoil. No aerofoil on tailplane. Quoted build time 300 hours.
FLYING CONTROLS: Conventional and manual. Ailerons on lower wings only.
STRUCTURE: Fabric-covered 4130 steel tube fuselage and two-spar wooden wings with aluminium leading-edge.
LANDING GEAR: Tailwheel configuration; one-piece sprung aluminium mainwheel legs. Azusa wheels and brakes.
POWER PLANT: *Easy Eagle:* One 1,914 cc Volkswagen four-cylinder air-cooled motorcar engine. Fuel capacity 45.4 litres (12.0 US gallons; 10.0 Imp gallons).
Easy Eagle II: One 83.5 kW (112 hp) Textron Lycoming O-235 flat-four. Fuel capacity 60.6 litres (16.0 US gallons; 13.3 Imp gallons).
DIMENSIONS, EXTERNAL:
Wing span: Easy Eagle	5.64 m (18 ft 6 in)
Easy Eagle II	6.71 m (22 ft 0 in)

Grosso Easy Eagle I *(Geoffrey P Jones)*

Length overall: Easy Eagle	4.42 m (14 ft 6 in)
Easy Eagle II	4.72 m (15 ft 6 in)
Height overall: Easy Eagle	1.78 m (5 ft 10 in)
Easy Eagle II	1.98 m (6 ft 6 in)

AREAS:
Wings, gross: Easy Eagle	9.57 m² (103.0 sq ft)
Easy Eagle II	13.38 m² (144.0 sq ft)

WEIGHTS AND LOADINGS:
Weight empty: Easy Eagle	206 kg (454 lb)
Easy Eagle II	318 kg (700 lb)
Max T-O weight: Easy Eagle	362 kg (800 lb)
Easy Eagle II	521 kg (1,150 lb)

PERFORMANCE:
Max operating speed:	
Easy Eagle	91 kt (169 km/h; 105 mph)
Easy Eagle II	95 kt (177 km/h; 110 mph)
Normal cruising speed: both	87 kt (161 km/h; 100 mph)
Stalling speed: both	40 kt (73 km/h; 45 mph)
Max rate of climb at S/L: Easy Eagle	213 m (700 ft)/min
Easy Eagle II	274 m (900 ft)/min
T-O run: Easy Eagle	137 m (450 ft)
Easy Eagle II	152 m (500 ft)
Landing run: Easy Eagle	122 m (400 ft)
Easy Eagle II	137 m (450 ft)
Range with max fuel:	
Easy Eagle	260 n miles (482 km; 300 miles)
Easy Eagle II	173 n miles (321 km; 200 miles)

NEW ENTRY

GULFSTREAM AEROSPACE

GULFSTREAM AEROSPACE CORPORATION
500 Gulfstream Road, Savannah, Georgia 31408
Tel: (+1 912) 965 30 00
Fax: (+1 912) 965 37 75/41 71
Web: http://www.gulfstream.com

CHAIRMAN: Theodore (Ted) J Forstmann
VICE-CHAIRMAN: Bryan T Moss
PRESIDENT AND COO: William (Bill) W Boisture Jr
EXECUTIVE VICE-PRESIDENT AND CFO: Chris A Davis
SENIOR VICE-PRESIDENT, SALES AND MARKETING: Joseph K Walker
SENIOR VICE-PRESIDENT, OPERATIONS: Joe Lombardo
SENIOR VICE-PRESIDENT, AIRCRAFT SERVICES: Larry R Flynn
SENIOR VICE-PRESIDENT, PROGRAMS: Pres Henne
VICE-PRESIDENT, CORPORATE COMMUNICATIONS AND INVESTOR RELATIONS: Tricia Bergeron
DIRECTOR OF CORPORATE COMMUNICATIONS: Keith Mordoff
COMMUNICATIONS SPECIALIST: Kelly Holland

Originally an offshoot of Grumman Corporation; passed through several owners before purchase by General

GULFSTREAM AEROSPACE—AIRCRAFT: USA

Dynamics for US$4.8 billion in May 1999; deal completed 31 July 1999.

Corporate HQ and production centre at Savannah, Georgia, has 4,430 employees. Subassembly and support locations are Oklahoma City, Oklahoma (700) and Mexicali, Mexico (600). Completions and service undertaken at Long Beach, California (865), Dallas, Texas (705), Appleton, Wisconsin (400) and Brunswick, Georgia (215). Service and refurbishment centre at Westfield, Massachusetts (150); a 4,985 m (53,650 sq ft) refurbishment support facility was added at Savannah in April 2000, at which time the total direct workforce was 8,200.

The 1,000th Gulfstream, a Gulfstream V, was delivered to the company's Long Beach completion centre in September 1997. During 1998, Gulfstream and Lockheed Martin's Skunk Works were revealed to be undertaking preliminary studies into the potential of a supersonic business jet.

Gulfstream delivered 51 aircraft (22 Gulfstream IV-SPs and 29 Vs) in 1997 and 61 (32 Gulfstream IV-SPs and 29 Vs) in 1998. Gulfstream simultaneously handed over the 400th Gulfstream IV and 100th Gulfstream V at Savannah, Georgia, on 25 April 2000. Sales for 2000 were 82, comprising 27 new-generation V-SPs, 35 IV-SPs and 20 Vs; deliveries for same period were 71. Backlog in January 2001 was worth US$3.5 billion.

UPDATED

GULFSTREAM AEROSPACE GULFSTREAM IV-SP, IV-MPA and IV-B

TYPE: Long-range business jet.

PROGRAMME: Design of Gulfstream IV started March 1983; manufacture of four production prototypes (one for static testing) began 1985; first aircraft (N404GA) rolled out 11 September 1985; first flight 19 September 1985; first flight of second prototype 11 June 1986 and third prototype August 1986; FAA certification 22 April 1987 after 1,412 hours flight testing. Westbound round-the-world flight from Le Bourget Airport, Paris, on 12 June 1987, covering 19,887.9 n miles (36,832.44 km; 22,886.6 miles), took 45 hours 25 minutes at average speed of 437.86 kt (811.44 km/h; 504.2 mph) and set 22 world records; eastbound round-the-world flight in N400GA from Houston, Texas, on 26 and 27 February 1988 covered 20,028.68 n miles (37,093.1 km; 23,048.6 miles) in 36 hours 8 minutes 34 seconds at average speed of 554.15 kt (1,026.29 km/h; 637.71 mph), setting 11 records.

In March 1993, Gulfstream IV-SP N485GA set new world speed and distance records in class, at 503.57 kt (933.21 km/h; 579.87 mph) and 5,139 n miles (9,524 km; 5,918 miles) respectively, on routine business flight from Tokyo, Japan, to Albuquerque, USA. Russian Federation and Associated States (CIS) certification achieved in June 1996. FAA approval for RVSM operation of Gulfstream IV series granted 11 August 1997. Gulfstream IV-SP holds 74 flight records; by April 2000 fleet had accumulated 1.2 million flying hours.

CURRENT VERSIONS: **Gulfstream IV:** Built until 1992. Described in earlier editions.

Gulfstream IV-SP: Improved (Special Performance), higher-weight version announced at NBAA convention, Houston, in October 1991. Prototype (N476GA, converted from standard) first flown 24 June 1992; designation applies to all new IVs sold after 6 September 1992; maximum payload increased by 1,134 kg (2,500 lb) and maximum landing weight increased by 3,402 kg (7,500 lb), with no increase in guaranteed manufacturer's bare weight empty. Payload/range envelope extended; expanded capability Honeywell SPZ-8400 flight guidance and control system.

Detailed description applies to Gulfstream IV-SP, except where otherwise indicated.

Gulfstream IV-MPA: Multipurpose Aircraft, announced September 1994; derived from US Navy C-20G Operational Support Aircraft (which see, below) to provide commercial operators with quick-change interior for up to 26 passengers in high-density shuttle layout, low-density executive configuration, 2,177 kg (4,800 lb) cargo capacity, or combination; large cargo door and larger/additional emergency exits standard.

Gulfstream IV-B: Improved, longer-range version, announced September 1994 as subject of development study by Gulfstream Aerospace and Texas Aerostructures; would feature increased wing span incorporating winglet design from Gulfstream V and additional fuel capacity, providing 400 n mile (741 km; 460 mile) increase in range; project indefinitely postponed, June 1995.

SRA-4, C-20, S 102, Tp 102, U-4: Special missions aircraft, described separately.

CUSTOMERS: Total more than 400 Gulfstream IV and IV-SPs delivered by January 2001 including 22 in 1997, 32 in 1998, 39 in 1999 and 37 in 2000. Major customers include Executive Jet International which placed order in January 1995 for seven Gulfstream IV-SPs for delivery in 1995 and 1996, plus a further 11 in early 1997 for delivery in 1997 (one), 1999 (five) and 2000 (five); letter of intent signed in November 1997 for up to 12 additional aircraft over a six year period and confirmed in mid-1998 as 14 for delivery by 2004. Executive Jet's aircraft used in Gulfstream Shares fractional ownership programme. Two delivered to Egyptian Air Force in November and December 1998; one to Gabonese government in November 1998. In all 50 Gulfstream IVs are operated by 19 governments and military units. Recent order is repeat for two IV-SPs for Egyptian Air Force. On short list for Indian Air Force Boeing 737 replacement.

COSTS: US$24.652 million 'green' (1996).

DESIGN FEATURES: Rear-engined, T-tail configuration, with all flying surfaces sweptback; wing mounted below cabin. Differences from Gulfstream III (1987-88 and earlier *Jane's*) include aerodynamically redesigned wing, with winglets, contributing to lower cruise drag; wing also structurally redesigned with 30 per cent fewer parts, 395 kg (870 lb) lighter and carrying 544 kg (1,200 lb) more fuel; increased tailplane span; fuselage 1.37 m (4 ft 6 in) longer, with sixth window each side; Rolls-Royce Tay turbofans; flight deck with electronic displays; digital avionics; and fully integrated flight management and autothrottle systems.

Advanced sonic rooftop aerofoil; sweepback at quarter-chord 27° 40'; thickness/chord ratio 10 per cent at wing station 50, 8.6 per cent at station 414; dihedral 3°; incidence 3° 30' at root, −2° at tip; NASA (Whitcomb) winglets.

Gulfstream Aerospace Gulfstream IV-SP (Special Performance) with Matador missile jammer in tailcone
2001/0100446

FLYING CONTROLS: Conventional. Hydraulically powered flying controls with manual reversion; trim tab in port aileron and both elevators; two spoilers on each wing act differentially to assist aileron and, with third spoiler each side, act collectively as airbrakes and lift dumpers; single-slotted Fowler flaps; four vortillons and a single 'tripper' strip under leading-edge of each wing ensure inboard part of wing stalls before outboard section; variable incidence tailplane.

STRUCTURE: Light alloy airframe except for carbon composites ailerons, spoilers, rudder and elevators, some tailplane parts, some cabin floor structure, and parts of flight deck; winglets of aluminium honeycomb. Wing manufactured by Textron Aerostructures.

LANDING GEAR: Retractable tricycle type with twin wheels on each unit. Main units retract inward, steerable nose unit forward. Mainwheel tyres size 34×9.25-16 (18 ply) tubeless; pressure 12.07 bars (175 lb/sq in). Nosewheel tyres size 21×7.25-10 (10 ply) tubeless, pressure 7.93 bars (115 lb/sq in); maximum steering angle ±82°. Dunlop air-cooled carbon brakes; Aircraft Braking Systems anti-skid units and digital electronic brake-by-wire system. Dowty electronic steer-by-wire system. Turning circle about wingtip 14.43 m (47 ft 4 in); about nosewheel 12.04 m (39 ft 6 in).

POWER PLANT: Two Rolls-Royce Tay Mk 611-8 turbofans, each flat rated at 61.6 kN (13,850 lb st) to ISA +15°C. Target-type thrust reversers. Fuel in two integral wing tanks, with total capacity of 16,542 litres (4,370 US gallons; 3,639 Imp gallons). Single pressure fuelling point in leading-edge of starboard wing. In 1999 Gulfstream considered re-engining of IV-SP as part of product improvement policy, possibilities including Rolls-Royce Deutschland BR710, and General Electric CF34; in May 2000 it placed on order for US$1.4 billion with Rolls-Royce for improved Tays.

ACCOMMODATION: Crew of two plus cabin attendant. Standard seating for up to 19 passengers (typically 12 to 14 in corporate configuration) in pressurised and air conditioned cabin. 'Quick Change' cargo/passenger version, certified for up to 26 passengers, announced 22 December 1993. Galley, lavatory and large baggage compartment, capacity 907 kg (2,000 lb), at rear of cabin. Integral airstair door at front of cabin on port side. Baggage compartment door on port side. Electrically heated wraparound windscreen. Six cabin windows, including two overwing emergency exits, on each side.

SYSTEMS: Cabin pressurisation system maximum differential 0.65 bar (9.45 lb/sq in) maintains 1,980 m (6,500 ft) cabin altitude at 13,715 m (45,000 ft); dual air conditioning systems. Two independent hydraulic systems, each 207 bars (3,000 lb/sq in). Maximum flow rate 83.3 litres (22 US gallons; 18.3 Imp gallons)/min. Two bootstrap-type hydraulic reservoirs, pressurised to 4.14 bars (60 lb/sq in). Honeywell GTCP36-100 APU in tail compartment, flight rated to 12,500 m (41,000 ft) since s/n 1156. Electrical system includes two 36 kVA alternators with two solid-state 30 kVA converters to provide 23 kVA 115/200 V 400 Hz AC power and 250 A of regulated 28 V DC power; two 24 V 40 Ah Ni/Cd storage batteries and external power socket. Wing leading-edges and engine inlets anti-iced.

AVIONICS: *Comms:* Dual VHF/HF transceivers, transponders and cockpit audio systems; cockpit voice recorder; Calquest CD-400 satellite communications equipment optional.

Radar: Digital colour weather radar.

Nav: Dual VOR/LOC/GS with marker beacon receivers; dual DME; dual ADF; dual radio altimeters; optional MLS, GPS and VLF Omega. Optional Northstar Technologies CT-1000 flight deck organiser.

Flight: Honeywell SPZ-8400 digital AFCS; Honeywell SPZ-8000 flight management system (FMS); dual fail-operational flight guidance systems including autothrottles; dual air data systems; dual flight guidance and performance computers; dual laser IRS; AHRS; VNAV; flight data recorder. System integration is accomplished through a Honeywell avionics standard

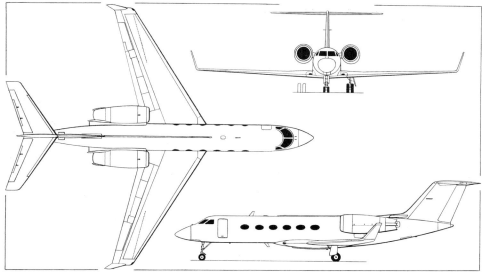

Gulfstream Aerospace Gulfstream IV twin-turbofan business transport *(Dennis Punnett/Jane's)*

USA: AIRCRAFT—GULFSTREAM AEROSPACE

Cabin interior of Gulfstream IV-SP business jet

Interior of Japan Air Self-Defence Force U-4 showing medical evacuation installation

communications bus (ASCB). Optional TCAS and Honeywell EGPWS.

Instrumentation: Six 203 × 203 mm (8 × 8 in) colour CRT EFIS screens, two each for primary flight display (PFD), navigation display (ND) and engine instrument and crew alerting system (EICAS). Honeywell/BAE HUD 2020 head-up display received FAA approval for Cat. II operations in early 1997, with first installation in Gulfstream IV-SP completed in May 1997. EVS system used on Gulfstream V will be available in near future.

Self-defence: Optional Sanders AN/ALQ-204 Matador IRCM available at cost of US$3.5 million.

DIMENSIONS, EXTERNAL:
Wing span over winglets	23.72 m (77 ft 10 in)
Wing chord: at root (fuselage c/l)	5.94 m (19 ft 5¾ in)
at tip	1.85 m (6 ft 0¾ in)
Wing aspect ratio	6.4
Length overall	26.92 m (88 ft 4 in)
Fuselage: Length	24.03 m (78 ft 10 in)
Max diameter	2.39 m (7 ft 10 in)
Height overall	7.44 m (24 ft 5 in)
Tailplane span	9.75 m (32 ft 0 in)
Wheel track	4.17 m (13 ft 8 in)
Wheelbase	11.61 m (38 ft 1¼ in)
Passenger door (fwd, port): Height	1.57 m (5 ft 2 in)
Width	0.91 m (3 ft 0 in)
Baggage door (rear): Height	0.90 m (2 ft 11¾ in)
Width	0.72 m (2 ft 4½ in)

DIMENSIONS, INTERNAL:
Cabin:
Length, incl galley, toilet and baggage compartment	13.74 m (45 ft 1 in)
Max width	2.24 m (7 ft 4 in)
Max height	1.88 m (6 ft 2 in)
Floor area	22.9 m² (247 sq ft)
Volume	43.2 m³ (1,525 cu ft)
Flight deck volume	3.5 m³ (124 cu ft)
Rear baggage compartment volume	4.8 m³ (169 cu ft)

AREAS:
Wings, gross	88.29 m² (950.4 sq ft)
Ailerons (total, incl tab)	2.68 m² (28.86 sq ft)
Trailing-edge flaps (total)	11.97 m² (128.84 sq ft)
Spoilers (total)	7.46 m² (80.27 sq ft)
Winglets (total)	2.38 m² (25.60 sq ft)
Fin	10.92 m² (117.53 sq ft)
Rudder, incl tab	4.16 m² (44.75 sq ft)
Horizontal tail surfaces (total)	18.83 m² (202.67 sq ft)
Elevators (total, incl tabs)	5.22 m² (56.22 sq ft)

WEIGHTS AND LOADINGS:
Manufacturer's weight empty	16,102 kg (35,500 lb)
Allowance for outfitting	3,175 kg (7,000 lb)
Typical operating weight empty	19,278 kg (42,500 lb)
Max payload	2,948 kg (6,500 lb)
Payload with max fuel	1,361 kg (3,000 lb)
Max usable fuel	13,381 kg (29,500 lb)
Max T-O weight	33,838 kg (74,600 lb)
Max ramp weight	34,019 kg (75,000 lb)
Max landing weight	29,937 kg (66,000 lb)
Max zero-fuel weight	22,226 kg (49,000 lb)
Max wing loading	383.2 kg/m² (78.49 lb/sq ft)
Max power loading	275 kg/kN (2.69 lb/lb st)

PERFORMANCE:
Max operating speed (V_{MO}/M_{MO})
  340 kt (629 km/h; 391 mph) CAS or M0.88
Max cruising speed at 9,450 m (31,000 ft)
  505 kt (936 km/h; 582 mph) or M0.85
Normal cruising speed at 13,715 m (45,000 ft)
  M0.80 (459 kt; 850 km/h; 528 mph)
Approach speed at max landing weight
  149 kt (276 km/h; 172 mph)
Stalling speed at max landing weight:
  wheels and flaps up 130 kt (241 km/h; 150 mph)
  wheels and flaps down 115 kt (213 km/h; 133 mph)
Max rate of climb at S/L 1,256 m (4,122 ft)/min
Rate of climb at S/L, OEI 314 m (1,030 ft)/min
Initial cruising altitude 12,500 m (41,000 ft)
Max certified altitude 13,715 m (45,000 ft)
Runway PCN 25
FAA balanced T-O field length at S/L 1,662 m (5,450 ft)
Landing run 973 m (3,190 ft)
Range:
  with max payload, normal cruising speed and NBAA IFR reserves 3,338 n miles (6,182 km; 3,841 miles)
  with max fuel, eight passengers, at M0.80 and with NBAA IFR reserves
    4,220 n miles (7,815 km; 4,856 miles)
OPERATIONAL NOISE LEVELS (FAR Pt 36):
T-O	77.5 EPNdB
Approach	92.0 EPNdB
Sideline	86.6 EPNdB

UPDATED

GULFSTREAM AEROSPACE SRA-4

Swedish Air Force designations: S 102B Korpen and Tp 102
US military designations: C-20F/G/H
JASDF designation: U-4

TYPE: Multirole twin-jet.

PROGRAMME: Adaptations of Gulfstream IV. Development aircraft (N413GA) for electronic warfare support version, integrated by Electrospace Systems Inc (later CTAS), exhibited at Farnborough Air Show 1988.

CURRENT VERSIONS: **Electronic warfare support:** Development aircraft had forward underfuselage pod for jamming antennas; cabin contains operators' consoles and microwave generator and amplifier rack, modulation generator rack, radio racks, chaff supply and cutters to simulate EW from adversary aircraft and missiles; aircraft could be used to test and evaluate weapon systems and to develop electronic warfare tactics.

Electronic surveillance/reconnaissance: Possible sensors include side-looking synthetic aperture radar in belly-mounted pod under forward fuselage, long-range oblique photographic camera (LOROP), ESM, VHF/UHF/HF communications for C^3, chaff dispensers, IR countermeasures in tailcone, SAR equipment and accommodation for operators for each system. Typical mission profile with 1,950 kg (4,300 lb) payload allows 10½ hours on station at loiter altitudes between 10,670 and 15,550 m (35,000 and 51,000 ft).

Maritime patrol: Equipment includes high-definition surface search radar, forward-looking IR detection system (IRDS), electronic support measures (ESM), flare/marker launch tubes, nav/com and ESM consoles, positions for up to eight observers/console operators, stowage and deployment for survival equipment, and crew rest area. SRA-4 with 1,950 to 4,173 kg (4,300 to 9,200 lb) mission payload including 272 kg (600 lb) expendable stores can operate at 600 n mile (1,112 km; 690 mile) radius for 4 to 6 hours; outbound flight at 12,500 m (41,000 ft) and 454 kt (841 km/h; 523 mph); search at 3,050 m (10,000 ft), but spend one-third of time at 61 m (200 ft); return flight at 13,715 m (45,000 ft).

ASW: Equipment includes nose radar able to detect periscope and snorkels, FLIR, sonobuoy launchers, acoustic processor, magnetic anomaly detector (MAD) in tail, ESM, torpedo stowage in weapon bay under forward fuselage, and anti-shipping missile carried on each underwing hardpoint. Mission profile with six crew and 2,503 kg (5,518 lb) payload can stay over 4¼ hours in hi-lo loitering and manoeuvring at 1,000 n mile (1,852 km; 1,151 mile) radius. Mission profile for **anti-shipping** with two missiles allows 1,350 n mile (2,500 km; 1,553 mile) outbound flight at high altitude; descent to 61 m (200 ft) for 100 n mile (185 km; 115 mile) attack run at 350 kt (649 km/h; 403 mph); launch missile 50 n miles (93 km; 57 miles) from target; return to base at 13,715 m (45,000 ft).

Medical evacuation: Accommodation for 15 stretchers and attendants.

Priority cargo transport: Cargo door (see below) plus floor-mounted cargo roller system.

C-20F: US Army administrative transport (one); delivered to Priority Air Transport Flight Detachment, Andrews AFB, Maryland, 1991.

C-20G: Operational Support Aircraft (OSA) for US Navy (four) and USMC (one). Convertible interior for passengers and cargo (26 passengers; large cargo door, larger/additional emergency exits). First two for USN delivered 4 February and 7 March 1994 and equip VR-48 at Naval Air Facility at Washington, DC; second pair with VR-51 at Kanehoe Bay, Hawaii. USMC aircraft destroyed by hurricane on 2 February 1998.

C-20H: USAF aircraft; one delivered March 1994 and based at Andrews AFB, Maryland with 99th Airlift Squadron, to which second aircraft also delivered late 1995.

U-4: Selected late 1994 by Japan Air Self-Defence Force for U-X multimission aircraft requirement; funds for two VIP/transport/training/support aircraft, valued at US$72 million, included in FY95 budget; third and fourth U-4s ordered in FY96 and FY97 and fifth in FY98; total order will be for nine aircraft, but none bought in 1999-2000; configuration includes medevac capability, high-density seating for up to 18 passengers plus crew, and cargo door. (Last-mentioned installed by Marshall Aerospace in UK.) First delivery (75-3251) November 1996; in service January 1997, fifth delivered March 2000.

Gulfstream IV-SP of Saudi Ministry of Defence and Aviation used in Medevac role

GULFSTREAM AEROSPACE—AIRCRAFT: USA

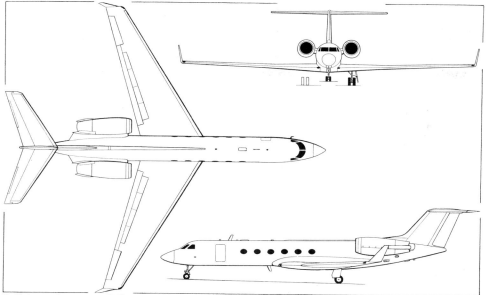

Gulfstream Aerospace Gulfstream V long-range business transport (*Dennis Punnett/Jane's*)

CUSTOMERS: US services, as detailed above. Swedish Defence Materiel Administration (FMV) concluded contract 29 June 1992 for three Gulfstream IVs; one as **Tp 102** transport (delivered 24 October 1992 to F16 Wing at Uppsala); remaining two modified as **S 102B Korpen** for electronic intelligence gathering; operated by Swedish Air Force; delivered August 1995, for service entry following installation and trials of TRW sigint system 1997-98, replacing Caravelles. Japan Civil Aviation Bureau SRA airways flight inspection aircraft delivered December 1993; Turkish Air Force acquired one in 1993; Saudi Arabia acquired one (based on Gulfstream IV-SP) for Medevac duties in early 2000. Other sales to undisclosed customers; total of 50 aircraft in use with 19 operators.

DESIGN FEATURES: Cabin arranged for rapid role changes; upward-opening cargo door 1.60 m (5 ft 3 in) high by 2.11 m (6 ft 11 in) wide can be fitted to starboard ahead of wing to allow for bulky cargo, mission equipment or stretchers.

UPDATED

GULFSTREAM AEROSPACE GULFSTREAM V and V-SP

US military designation: C-37

TYPE: Long-range business jet.

PROGRAMME: Study announced at NBAA convention, Houston, in October 1991. Go-ahead commitment and engine selection (BR710) announced at Farnborough Air Show in September 1992. Risk-sharing agreement with wing designers/manufacturers (Vought and ShinMaywa) announced at Paris Air Show in June 1993, and with tail (and later floor panel) manufacturer (Fokker) at NBAA convention in September 1993.

Prototype (N501GV) rolled out 22 September 1995; first flight 28 November 1995; second aircraft, c/n 502 (N502GV) was structural test article before completion as company demonstrator; c/n 503 (N503GV), first flown 10 March 1996, used for systems testing; c/n 504 (N504GV), first flown May 1996, for engine, flight loads and environmental trials and JAA certification testing; c/n 505 (N505GV), first flown August 1996, outfitted in standard production configuration for operational testing and HIRF evaluation. Public debut (N502GV) at NBAA convention at Orlando, Florida, in November 1996.

Provisional FAA certification achieved 16 December 1996 after more than 1,100 hours of flight testing in 550 sorties. Full FAA type certification granted on 11 April 1997, and FAA production certificate awarded in June. First fully completed aircraft for a customer (c/n 507) delivered to Walter Annenberg, former US Ambassador to UK, on 1 July 1997.

By 17 November 1997 company demonstrators had achieved 39 world records in time to climb and maximum altitude with payload, and city pair categories, including the first ever non-stop flights by a business jet between Los Angeles and London, London and Hong Kong, Tokyo and New York, and Washington, DC and Dubai; last-named 6,330 n mile (11,723 km; 7,284 mile) journey completed on 13 November 1997 in 12 hours 40 minutes 48 seconds with four crew and seven passengers. On 19 February 1998, N502GV became first business jet to fly non-stop from New York to Hawaii. By October 2000, the Gulfstream V held 65 world and national records. RVSM approval January 2000. Fleet dispatch reliability 99% in July 2000.

CURRENT VERSIONS: **Gulfstream V:** *As described.*

Gulfstream V-SP: Announced on eve of NBAA Convention, 8 October 2000; larger cabin, enhanced performance and better range by reason of aerodynamic refinement of existing V. First flight due end 2001, with certification under new certificate by end 2002 and deliveries in 2003. Range 6,750 n miles (12,500 km; 7,765 miles); operating weight (empty) 21,908 kg (48,300 lb), maximum T-O weight 41,277 kg (91,000 lb), maximum payload 3,220 kg (7,100 lb). Flight deck uses Honeywell Primus Epic suite with four 360 mm (14 in) LCDs, including HUD and EVS. First customer is Executive Jets, which ordered 20 for delivery up to 2008, at cost of US$800 million; unit cost with most popular options US$24.9 million.

Special Missions: Lockheed Martin, teamed with (then) GEC-Marconi Defence, Logica, Marshall Aerospace, MSI, Racal and CAE Electronics Montreal, chose the Gulfstream V as its platform for the UK's Airborne Stand-Off Radar (ASTOR) requirement, involving a Racal dual-mode, electronically scanned surveillance radar in a ventral fairing, with satcom antennas above and below the fuselage for real-time data transfer to ground stations, and provision for in-flight refuelling.

Additionally, when an upgraded version of the Joint STARS system was re-admitted to the ASTOR competition in January 1998, the Gulfstream V was selected by Northrop Grumman as platform in preference to Boeing E-8 (707). Neither application was successful. Currently competing for NATO Ground Surveillance System programme.

C-37A: Military version; two (plus four options, one of which converted to firm order in 1998, another in April 1999 and a third subsequently, with fourth due for conversion in 2001) ordered by USAF as part of VCX requirement to replace Boeing VC-137s of 89th Airlift Wing, Andrews AFB, Maryland. Announced 5 May 1997. First (97-0400) delivered 14 October 1998; second in January 1999 and third (for C-in-C USAF) 21 February 2000. US Army's Priority Air Transport Squadron operates the remaining in-service example. Five C-37As leased by US Air Force for delivery between July 2001 and September 2003. In December 2000 US Coast Guard ordered a C-37A for delivery during mid-2002.

EC-37A: Proposed airborne combat support aircraft; Gulfstream announced at RIAT 2000, Cottesmore, UK, that it was looking for funding during late 2000 and partner to develop systems. Aimed at two roles: standoff electronic jammer and intelligence gathering, using interchangeable underwing and underfuselage pods and in-flight workstation reconfiguration.

CUSTOMERS: Break-even on non-recurring research and development costs expected at or before 65th aircraft. Three produced in 1996, 29 in 1997, 25 in 1998, 31 in 1999 and 34 in 2000; of these (including prototypes) the 10th was delivered to BMW on 7 January 1998. Executive Jet International took options on two aircraft in January 1995, for delivery to its Gulfstream Shares fleet in 1998-99. Recent customers include Kuwait Airways, (three, first aircraft delivered November 1999); Brunei government (one); Time Warner (two); Chrysler Corporation (two) and Executive Jet International (10, for delivery by 2004, plus 12 options); Saudi Arabian Ministry of Defence and Aviation (one in medevac role, handed over 15 May 2000). By end 1999 total sales exceeded 150 and by October 2000, 170. 100th Gulfstream V was delivered April 2000; by October 2000 over 90 were in service.

COSTS: US$29.5 million (fixed price) for first 24 aircraft, then US$30.5 million (fixed price) up to 39th aircraft; typically equipped price after outfitting, US$35 million; aircraft sold on Internet during December 1999 in deal valued at US$40 million. Cost of EVS system estimated at US$1 million.

Contract for two C-37As valued at US$68.9 million (1997).

DESIGN FEATURES: Gulfstream IV fuselage re-engineered to increase length by 2.13 m (7 ft 0 in); larger wing of same basic shape and interior structure, but 10 per cent more efficient than Gulfstream IV's; larger vertical and horizontal tail surfaces; flight deck volume increased by moving bulkhead 0.30 m (1 ft) aft to provide more space for pilots and to accommodate full-size jump seat; cockpit layout and instrumentation generally similar to Gulfstream IV-SP, but redesigned to incorporate human engineering changes in system control functions; airstair door moved aft by 1.52 m (5 ft); avionics bay relocated. Computational

Flight deck of Gulfstream V

Cabin interior of Gulfstream V in typical executive layout

fluid dynamics and CATIA design system used extensively in development.

FLYING CONTROLS: As Gulfstream IV.
LANDING GEAR: As Gulfstream IV. Turning circle about wingtip 17.07 m (56 ft 0 in); about nosewheel 14.15 m (46 ft 5 in).
POWER PLANT: Two 65.6 kN (14,750 lb st) Rolls-Royce Deutschland BR710-48 turbofans with FADEC. Fuel capacity 23,417 litres (6,186 US gallons; 5,151 Imp gallons) in integral wing tanks, of which 22,993 litres 6,074 US gallons; 5,058 Imp gallons) are usable.
ACCOMMODATION: Crew of two/three plus cabin attendant. Standard seating for 15 to 19 passengers in pressurised and air conditioned cabin. Rear windows, each side, are emergency exits. Customised interiors according to customer requirements.
SYSTEMS: Digitally controlled automatic cabin pressurisation system; maximum pressure differential 0.703 bar (10.2 lb/sq in). Hamilton Sundstrand electrical power generating system profile integrated with the flight management system (FMS), and will maintain equivalent of 1,830 m (6,000 ft). Honeywell RE220 APU, designed specifically for Gulfstream V, provides engine-starting capability up to 13,110 m (43,000 ft), 40 kVA of electrical power for ground and flight use up to 13,715 m (45,000 ft), and ground air conditioning, with almost twice the cooling air flow rate of the Gulfstream IV's APU.
AVIONICS: Honeywell SPZ-8500 as core system.
Flight: Three independent IRS integrated into FMS; GPS. Honeywell enhanced ground proximity warning system (EGPWS); TCAS; turbulence-detecting Doppler radar; Hamilton Sundstrand maintenance data acquisition unit; optional Northstar Technologies CT-1000 flight deck organiser system.
Instrumentation: Honeywell SPZ-8500 digital AFCS/FMS with six 20.3 × 20.3 cm (8 × 8 in) colour LCD EFIS displays with EICAS; Honeywell/BAE Model 2020 HUD. Kollsman All Weather Window IR sensor flight testing began September 1999 to provide enhanced vision system (EVS) in conjunction with Honeywell 2020 HUD, facilitating operations in Cat. III weather on Cat. I runway, with decision height of 30 m (100 ft) and RVR of 220 m (722 ft). IR sensor flight is mounted beneath radome. Certification was due early 2001.
Mission: IBM satellite-based international communications system, including airborne voice, data, networking, fax and teleconferencing facilities, provides Gulfstream V with 'office in the sky' capability; optional Sanders AN/ALQ-204 Matador IRCM system..

DIMENSIONS, EXTERNAL:
Wing span: basic	27.69 m (90 ft 10 in)
over winglets	28.50 m (93 ft 6 in)
Length overall	29.39 m (96 ft 5 in)
Height overall	7.87 m (25 ft 10 in)
Tailplane span	10.72 m (35 ft 2 in)
Wheel track (c/l shock-absorbers)	4.37 m (14 ft 4 in)
Wheelbase	13.72 m (45 ft 0 in)

DIMENSIONS, INTERNAL:
Cabin: Length, aft of flight deck	15.57 m (50 ft 1 in)
Max width	2.24 m (7 ft 4 in)
Max height	1.88 m (6 ft 2 in)
Volume	47.3 m³ (1,669 cu ft)
Baggage compartment volume	6.4 m³ (226 cu ft)

AREAS:
Wings, gross	105.63 m² (1,137.0 sq ft)

WEIGHTS AND LOADINGS:
Weight empty, 'green'	17,917 kg (39,500 lb)
Operating weight empty	21,772 kg (48,000 lb)
Allowance for outfitting	3,856 kg (8,500 lb)
Baggage capacity	1,134 kg (2,500 lb)
Max payload	2,948 kg (6,500 lb)
Max fuel	18,733 kg (41,300 lb)
Payload with max fuel	726 kg (1,600 lb)
Max T-O weight	41,050 kg (90,500 lb)
Max ramp weight	41,231 kg (90,900 lb)
Max landing weight	34,155 kg (75,300 lb)
Max zero-fuel weight	24,720 kg (54,500 lb)
Max wing loading	382.2 kg/m² (78.28 lb/sq ft)
Max power loading	308 kg/kN (3.02 lb/lb st)

PERFORMANCE (at max T-O weight, except where indicated):
Max operating Mach No. (V$_{MO}$)	M0.885
Cruising speed: max	499 kt (924 km/h; 574 mph)
normal	488 kt (904 km/h; 562 mph)
long-range	459 kt (850 km/h; 528 mph)
Max rate of climb at S/L	1,276 m (4,188 ft)/min
Initial cruising altitude	12,500 m (41,000 ft)
Max certified altitude	15,545 m (51,000 ft)
T-O balanced field length	1,862 m (6,110 ft)
Landing distance, S/L, max landing weight	841 m (2,760 ft)
Max range, eight passengers and four crew, M0.80, NBAA IFR reserves	6,500 n miles (12,038 km; 7,480 miles)

UPDATED

Gulfstream V long-range twin-turbofan business aircraft 2001/0100518

GULFSTREAM SBJ
TYPE: Supersonic business jet.
PROGRAMME: Gulfstream Aerospace and Lockheed Martin (which see) revealed at Farnborough on 7 September 1998 that they were jointly conducting an 18 to 24 month feasibility study into an SBJ. As currently envisaged, the aircraft, similar in size to the Gulfstream V, would feature a stand-up-headroom cabin accommodating eight passengers, and would cruise at M1.6 to M2.0 over a range of more than 4,000 n miles (7,408 km; 4,603 miles). Engines in the class of P&W F119 and GE F120 could be considered. Key design goals are the ability to operate out of existing business aviation airfields; take-off noise compatible with anticipated future emissions regulations; fuel-efficient operation at subsonic speeds; and an initial cruising altitude above that of subsonic traffic.

The feasibility phase of the project, completed in mid-2000, is to be followed by a further two years of wind-tunnel testing and project definition before a launch decision is taken; studies carried out by the company confirm that there is a market, but sonic boom suppression

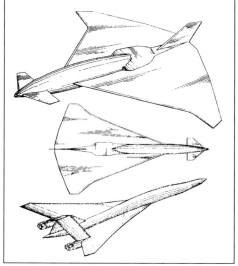
Potential configurations for a Gulfstream SBJ, as revealed by patent applications 2001/0100448

and propulsion present technological challenges. Gulfstream and Lockheed Martin, through its Advanced Development Projects division, in defining the SBJ, will have access to data from NASA's (now discontinued) High-Speed Research (HSR) programme for a future supersonic airliner. The partners lobbied the US government for support via the Defense Advanced Research Projects Agency which received US$15 million for its quiet supersonic aircraft technology (QSAT) budget. A model of a twin-engined aircraft, displayed at the time of the project's announcement at the Farnborough Air Show in September 1998, was only 'notional', according to the two companies, and does not necessarily reflect their current thinking on configuration; at 1999 NBAA Convention the companies agreed the model was no longer accurate; in 2000 patents were filed with different layouts including tailplane mounted at tail fintip and drooping to attach to engine nacelles; further patents included one with large delta wing shape plus nose canard; in October Gulfstream displayed, at NBAA, the Quiet Supersonic Jet (QSJ) which has swept wing and wingtip-mounted swept tail.

If launched, the partners are estimating a first flight in 2005, but it will be at least 2010 before a supersonic Gulfstream enters service. Estimated unit cost is in the region of US$70 million to US$80 million (1998), based on 200 produced. The proposed aircraft is not related to the abortive Gulfstream-Sukhoi SSBJ project from which Gulfstream withdrew in 1992. Executive Jet Aviation, which operates the NetJets fractional ownership programme, is reportedly interested in the concept of supersonic business jets.

UPDATED

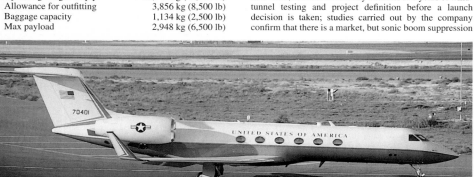
Second Gulfstream C-37A for the US Air Force *(Paul Jackson/Jane's)* 2000/0084575

HENDERSON

HENDERSON AERO SPECIALTIES INC
7481 Canterbury Road, Felton, Delaware 19943
Tel: (+1 302) 284 81 07
Fax: (+1 302) 284 81 84
e-mail: Litbear@aol.com
CEO: David Henderson

UPDATED

HENDERSON LITTLE BEAR
TYPE: Two-seat lightplane.
PROGRAMME: Replica of 1937 Piper J3 Cub. Prototype construction started 1993; first flight 1 July 1993. Also has been built in Thailand.
CUSTOMERS: 13 ordered and six built by mid-1997 (latest information supplied).
COSTS: US$24,500 (1999).
FLYING CONTROLS: Conventional and manual.
STRUCTURE: Fabric-covered fuselage of 4130 steel. Choice of metal or wooden spar wings; metal ribs. All-steel tail surfaces.
LANDING GEAR: Fixed, tailwheel type. Wheels 8.00-4; Cleveland or original Piper brakes.

POWER PLANT: One Teledyne Continental engine in the 48.5 to 74.6 kW (65 to 100 hp) range, driving a two-blade fixed-pitch propeller. Fuel capacity 45 litres (12.0 US gallons; 10.0 Imp gallons), with optional extra 61 litre (16.0 US gallon; 13.3 Imp gallon) tank. Oil capacity 5.7 litres (1.5 US gallons; 1.25 Imp gallons).
ACCOMMODATION: Pilot and passenger in tandem in enclosed cabin.
DIMENSIONS, EXTERNAL:
Wing span 10.72 m (35 ft 2 in)
Wing aspect ratio 6.9
Length overall 6.78 m (22 ft 3 in)
Height overall 2.04 m (6 ft 8½ in)

AREAS:
Wings, gross 16.58 m² (178.5 sq ft)
WEIGHTS AND LOADINGS:
Weight empty 318 kg (700 lb)
Max T-O weight 589 kg (1,300 lb)
Max wing loading 35.6 kg/m² (7.28 lb/sq ft)
Max power loading (100 hp engine)
 7.90 kg/kW (13.00 lb/hp)
PERFORMANCE:
Never-exceed speed (V_{NE}) 108 kt (201 km/h; 125 mph)
Max level speed 74 kt (137 km/h; 85 mph)
Normal cruising speed 65 kt (121 km/h; 75 mph)
Econ cruising speed 61 kt (113 km/h; 70 mph)
Stalling speed, power off 33 kt (62 km/h; 38 mph)
Max rate of climb at S/L:
 65 hp engine 244 m (800 ft)/min
 100 hp engine 305 m (1,000 ft)/min
Service ceiling 3,500 m (11,500 ft)
T-O run 122 m (400 ft)
T-O to 15 m (50 ft) 223 m (730 ft)
Landing run 91 m (300 ft)
Range with max fuel 340 n miles (629 km; 391 miles)

UPDATED

HILLBERG

HILLBERG HELICOPTERS
PO Box 8974, Fountain Valley, California 92728-8974
Tel: (+1 909) 279 56 78
PRESIDENT: Don Hillberg

Company founded 1990 to provide maintenance and modification services to helicopter operators, and moved into kit and experimental fields to help builders.
Hillberg has also developed the four-seat 1-04, which awaits funding for production. Produces a retrofit kit for the RotorWay Exec (which see), upgrading it to turbine power with a Solar T62-T32 of 112 kW (150 shp) and reducing empty weight to 340 kg (750 lb).

VERIFIED

HILLBERG EH 1-01 ROTORMOUSE and EH 1-02 TANDEMMOUSE

TYPE: Single-seat helicopter kitbuilt/two-seat helicopter kitbuilt.
PROGRAMME: Design and construction of prototype began in August 1990; first flew (N10TE) August 1992. Construction of first production aircraft began in September 1996. Type is currently certified as experimental by FAA.
CURRENT VERSIONS: **EH 1-01 RotorMouse**: Single-seat version.
Description applies to EH 1-01.
EH 1-02 TandemMouse: Tandem two-seat version; under development; was due to fly 1999 but this date appears to have slipped. Powered by 112 kW (150 shp) Solar T62-T32 turbine engine. Provisional figures appear below.
CombatMouse: Proposed military version.
CUSTOMERS: By end 1998, 27 had been ordered, of which at least three had been built.
COSTS: Kit: EH-1-01 US$86,000 (2000) ex-works US$100,000.
DESIGN FEATURES: Forward fuselage resembles Bell AH-1 HueyCobra. Stub-wing behind cockpit for carriage of external loads. Main rotor speed 510 rpm; tail rotor 3,250 rpm.
FLYING CONTROLS: Auto-governed hydromechanical flying controls. Optional rotor brake.
STRUCTURE: Aluminium alloy (2024-T3) monocoque fuselage, bulkheads and keel; stainless steel engine deck and firewall. Rotor has Gyrodyne QH-50 modified blades of NACA 1200 aerofoil section with 8° twist and block tip caps. Modified Robinson R22 hub and transmission. Conventional tail rotor using aluminium-skinned blades.

Hillberg EH 1-01 RotorMouse (one Honeywell 36-55-C)

LANDING GEAR: Fixed steel cross tube skid type; floats optional.
POWER PLANT: One 108 kW (145 shp) Honeywell 36-55-C turboshaft engine driving a non-folding, two-blade, semi-rigid wood-laminated main rotor. Engines in range 108-186 kW (145-250 hp) are suitable. Single 125 litre (32.9 US gallon; 27.4 Imp gallon) crash-resistant bladder fuel tank around transmission; refuelling point behind main mast. Optional 76 litre (20 US gallon; 16.7 Imp gallon) fuel tank on stub-wing. EH 1-02 has 170 litre (45 US gallon; 37.5 Imp gallon) fuel capacity.
ACCOMMODATION: Single pilot in enclosed cockpit. Entry through upward-hinged port and starboard side windows. Baggage compartment behind seat.
SYSTEMS: 24 V 60 A electrical system; Electrocraft alternator.
EQUIPMENT: Optional cargo hook.
DIMENSIONS, EXTERNAL:
Main rotor diameter: EH 1-01 6.10 m (20 ft 0 in)
 EH 1-02 7.32 m (24 ft 0 in)
Tail rotor diameter 0.91 m (3 ft 0 in)
Length: overall, rotors turning: EH 1-01
 7.32 m (24 ft 0 in)
 fuselage: EH 1-01 6.35 m (20 ft 10 in)
 EH1-02 8.53 m (28 ft 0 in)
Height overall 2.26 m (7 ft 5 in)
DIMENSIONS, INTERNAL:
Cabin max width: EH 1-01 0.61 m (2 ft 0 in)
 EH 1-02 0.89 m (2 ft 11 in)
AREAS:
Main rotor disc 32.18 m² (346.4 sq ft)

WEIGHTS AND LOADINGS:
Weight empty: EH 1-01 288 kg (635 lb)
 EH 1-02 386 kg (850 lb)
Max T-O and landing weight:
 EH 1-01 589 kg (1,300 lb)
 EH 1-02 816 kg (1,800 lb)
PERFORMANCE:
Max level speed: EH 1-01 139 kt (257 km/h; 160 mph)
 EH 1-02 148 kt (274 mph; 170 mph)
Cruising speed: both 113 kt (209 km/h; 130 mph)
Max rate of climb at S/L:
 EH 1-01 1,433 m (4,700 ft)/min
Vertical rate of climb at S/L:
 EH 1-01 823 m (2,700 ft)/min
 EH 1-02 457 m (1,500 ft)/min
Service ceiling: EH 1-02 4,115 m (13,500 ft)
Range, 20 min reserves, EH 1-01:
 with max payload 260 n miles (482 km; 300 miles)
 with max internal and external fuel
 695 n miles (1,287 km; 800 miles)
EH 1-02: with max payload
 347 n miles (643 km; 400 miles)

UPDATED

HILLBERG 1-04 BABY HUEY

TYPE: Four-seat helicopter kitbuilt.
PROGRAMME: Proposed development of RotorMouse/TandemMouse with four seats and external cabin styling resembling Bell UH-1 'Huey'. MTOW 1,134 kg (2,500 lb).

VERIFIED

JAMMAERO/STARCORP

JAMMAERO/STARCORP
Greenbackville, Virginia
INFORMATION: Geoffrey L Bland

PO Box 236, Horntown, Virginia 23395
Tel: (+1 757) 824 05 14

All known details of the company's Rainbow Chaser last appeared in the 2000-01 *Jane's*.

UPDATED

JOPLIN

JOPLIN LIGHT AIRCRAFT
523 North Schifferdecker, Joplin, Missouri 64801
Tel: (+1 417) 623 29 50
Fax: (+1 417) 623 76 06
e-mail: jla@compnmore.com
Web: http://www.compnmore/jla
SALES MANAGER: John Lukey

Joplin Light Aircraft manufactures the Tundra and ½ TUN previously produced by Laron Aviation Technologies; it also markets the Suzi-Air engine.

NEW ENTRY

JOPLIN TUNDRA

TYPE: Tandem-seat ultralight kitbuilt.
PROGRAMME: Originally produced by Laron.

Tundra kitbuilt, previously produced by Laron

2001

CUSTOMERS: 30 sold and 25 flying by mid-2000.
COSTS: US$9,500 (2001).
DESIGN FEATURES: High-wing monoplane with high-life wing and pusher engine; cockpit enclosure optional. Quoted build time 250 hours.
FLYING CONTROLS: Manual. Full-span Junkers ailerons.
STRUCTURE: Welded metal tube fuselage.
LANDING GEAR: Fixed, wide-track tricycle type; ski and float options. Steerable nosewheel.
POWER PLANT: One 47.8 kW (64.1 hp) Rotax 582 two-cylinder two-stroke engine; options include 34.0 kW (45.6 hp) Rotax 503 and 47.7 kW (64 hp) Suzi Air. Fuel capacity 38 litres (10.0 gallons; 8.3 Imp gallons).
DIMENSIONS, EXTERNAL:
Wing span	9.75 m (32 ft 0 in)
Length overall	6.40 m (21 ft 0 in)
Height overall	1.98 m (6 ft 6 in)

AREAS:
Wings, gross	15.79 m² (170.0 sq ft)

WEIGHTS AND LOADINGS:
Weight empty	181 kg (400 lb)
Max T-O weight	408 kg (900 lb)

PERFORMANCE:
Max level speed	104 kt (193 km/h; 120 mph)
Normal cruising speed	65 kt (121 km/h; 75 mph)
Stalling speed	31 kt (57 km/h; 35 mph)
Max rate of climb at S/L	366 m (1,200 ft)/min
T-O run	30 m (100 ft)
Landing run	61 m (200 ft)
Range at max level speed	202 n miles (374 km; 232 miles)

NEW ENTRY

JOPLIN ½ TUN

TYPE: Single-seat ultralight kitbuilt.
CUSTOMERS: Three flying by mid-2000.
COSTS: US$5,995 (2001).
DESIGN FEATURES: As Tundra, but meets FAR Pt 103 weight rules. Quoted build time 90 hours.
FLYING CONTROLS: As Tundra.
STRUCTURE: As Tundra.
LANDING GEAR: Tailwheel type.

POWER PLANT: One 29.5 kW (39.6 hp) two-cylinder, liquid cooled Rotax 447; options include 2 SI 460 F35 and Suzi Air. Fuel capacity 18.9 litres (5.0 US gallons; 4.2 Imp gallons).
DIMENSIONS, EXTERNAL:
Wing span	7.92 m (26 ft 0 in)
Length overall	6.40 m (21 ft 0 in)
Height overall	1.68 m (5 ft 6 in)

AREAS:
Wings, gross	10.87 m² (117.0 sq ft)

WEIGHTS AND LOADINGS:
Weight empty	113 kg (250 lb)
Max T-O weight	249 kg (550 lb)

PERFORMANCE:
Max level speed	54 kt (101 km/h; 63 mph)
Normal cruising speed	48 kt (89 km/h; 55 mph)
Stalling speed	22 kt (41 km/h; 25 mph)
Max rate of climb at S/L	305 m (1,000 ft)/min
T-O run	46 m (150 ft)
Landing run	31 m (100 ft)

NEW ENTRY

KAMAN

KAMAN AEROSPACE CORPORATION
(Subsidiary of Kaman Corporation)

Old Windsor Road, PO Box No. 2, Bloomfield, Connecticut 06002
Tel: (+1 860) 243 71 00 or 242 44 61
Fax: (+1 860) 243 75 14
Web: http://www.kaman.com
CHAIRMAN: Charles Kaman
PRESIDENT AND CEO, KAMAN CORPORATION: Paul R Kuhn
PRESIDENT, KAMAN AEROSPACE: Walter R Kozlow
VICE-PRESIDENT, ENGINEERING: Michael Bowes
VICE-PRESIDENT, MILITARY HELICOPTERS:
    M Terry Higginbotham
DIRECTOR, BUSINESS DEVELOPMENT: William D Brown
VICE-PRESIDENT, PUBLIC RELATIONS: J Kenneth Nasshan
PUBLIC RELATIONS MANAGER: David M Long

Founded 1945 by Charles H Kaman. Developed servo-flap control of helicopter main rotor, initially in contrarotating two-blade main rotors, and still used in SH-2G Super Seasprite four-blade main rotor and on the K-MAX. R&D programmes sponsored by US Army, Air Force, Navy and NASA include advanced design of helicopter rotor systems, blades and rotor control concepts, component fatigue life determination and structural dynamic analysis and testing. Kaman has undertaken helicopter drone programmes since 1953; is continuing advanced research in rotary-wing unmanned aerial vehicles (UAVs).

Kaman is major subcontractor on many aircraft and space programmes, including design, tooling and fabrication of components in metal, metal honeycomb, bonded and composites construction, using techniques such as filament winding, braiding and RTM. Participates in programmes including Northrop Grumman E-2C, Bell Boeing V-22, Boeing 737, 747, 757, 767, 777 wing trailing-edge, KC-135 and C-17, Bell Boeing Sikorsky RAH-66, Sikorsky UH-60 and SH-60, and NASA Space Shuttle Orbiter main fuel tank. It also designs and produces fuzes for a range of weaponry.

Kaman designed and, since 1977, has been producing all-composites rotor blades for Bell AH-1 Cobras for US and foreign forces. Appointed in April 2000 to build fuselages for MD Helicopters' single-engine line (MDH 500, 520, 530 and 600) under 10-year contract; a further contract to build MDH Explorer rotor blade systems was awarded in July 2000.

Kaman Aerospace International Corporation was established in 1995 as the international affiliate of Kaman Aerospace with offices in Canberra, Australia; Kuala Lumpur, Malaysia; Cairo, Egypt; and Bloomfield, Connecticut.

Net sales of Kaman's Aerospace business were US$382.7 million in 1998, US$371.8 million in 1999 and US$381.9 million in 2000. The company employs 1,500 people.

UPDATED

KAMAN (K894) SH-2G SUPER SEASPRITE

TYPE: Naval combat helicopter.
CURRENT VERSIONS: **SH-2G Super Seasprite**: Available as either SH-2F remanufactured airframe or as new-build helicopter. For full specification and detailed description see *Jane's All the World's Aircraft 1994-95* and the current edition of *Jane's Aircraft Upgrades*.
CUSTOMERS: Most recent customers include Egypt, for which a US$150 million contract was awarded on 22 February 1995; first of 10 **SH-2G(E)**s, remanufactured from SH-2Fs, was delivered 21 October 1997; deliveries were completed November 1998. Royal Australian Navy has ordered 11 **SH-2G(A)**s remanufactured from SH-2Fs for delivery from early 2001 in a US$600 million contract for use on its eight new 'ANZAC' class frigates; the first (N351KA, ex-163210) made its first flight 6 January 2000 and was due for delivery in March 2001, but it and next two will be to an interim standard, due to systems development delays; fully operational versions will be delivered in 2003.

In 1997 Royal New Zealand Navy ordered four new-build **SH-2G(NZ)**s (later increased to five) for delivery from 2000 as replacements for Westland Wasp, also for use on 'ANZAC' class frigates as well as older 'Leander' class, equipped with Raytheon AGM-65 Maverick missiles; in the interim it is using four SH-2F Seasprites for training. Initial RNZAF SH-2G(NZ) (N352KA) made first flight 2 August 2000. Kaman negotiating throughout 1999 and 2000 with Thai Navy for maritime helicopter programme; it is offering ten SH-2Fs upgraded to SH-2G standard. Also proposing SH-2 to Malaysia for anti-submarine helicopter requirement; to Canada as CH-124 Sea King replacement; and to Singapore for new 'La Fayette' class frigates. Four SH-2Gs delivered to Poland in 2000-01 are second-hand stock from US Navy.

Following details refer specifically to SH-2G(NZ):
STRUCTURE: Composites main rotor blades.
LANDING GEAR: Emergency flotation system.
ACCOMMODATION: Crew of two: pilot and co-pilot/TACCO; no instructor's seat. Three-person bench seat and two single seats in cabin, plus one stretcher.
SYSTEMS: Hydraulic system rated at 207 bars (3,000 lb/sq in); maximum flow rate 30.3 litres (8.0 US gallons; 6.7 Imp gallons)/min. Electrical system powered by two 30 kVA AC generators, two 100 A DC converters and an Ni/Cd battery. Honeywell GTCP 36-150(BH) APU.
AVIONICS: *Comms:* Rockwell Collins AN/ARC-210 V/UHF (AM/FM) and HF-9000D radios; ITT KY 100 for HF (embedded in ARC-210 for V/UHF); Honeywell AN/APX-100 IFF with KIT 1C encoder; Telephonics Starcom intercom. Three-frequency (121.5/243.0/406 MHz) ELT.
Radar: Telephonics AN/APS-143(V)3 surveillance radar.
Flight: Rockwell Collins AN/ARN-147(V) VOR/ILS and marker beacon receiver, DME-42 and DF-301E VHF/UHF ADF; Trimble Tasman GPS/INS; Litton AN/APN-217(V)6 Doppler velocity sensor and AN/ASN-150 tactical data system; Honeywell AN/APN-171B(V) radar altimeter; Rosemount 542CBAY8 air data computer; AN/APQ-107 RAWS; AN/ASN-50 AHRS.
Instrumentation: Standby instruments and magnetic compass for pilot and co-pilot.
Mission: FLIR Systems AN/AAQ-22 SAFIRE FLIR; Panasonic AG-750 video cassette recorder.
Self-defence: Litton LR-100 ESM suite; BAE North America AN/ALE-47 chaff/flare dispenser.
EQUIPMENT: Two external launchers each for four Mk 25 marine markers; provision for LUUIIB flares on outboard stores stations. Fairey Hydraulics harpoon and Scott three-wire deck securing and handling system. NVG-compatible internal and external lighting. Pilot's and TACCO's personal survival packs. Hoist and cargo hook as for standard SH-2G.
ARMAMENT: Two AGM-65D Maverick ASMs; one Mk 46 torpedo; Mk 11 depth charge; provision for door-mounted machine gun.

UPDATED

KAMAN K-1200 K-MAX

TYPE: Light lift helicopter.
PROGRAMME: First flight (N3182T) 23 December 1991; first public showing 22 March 1992; first flights of second prototype (N131KA) 18 September 1993 and first production aircraft (N132KA) 12 January 1994; third prototype is static and drop test aircraft to prove 20 year life at 1,000 hours per year and with 30 return logging sorties per hour. Certification to FAR Pts 27 and 133 achieved 30 August 1994 after 800 hour/32 month programme. Production rate, six per year. N133KA of Scott Paper (now Kimberly-Clark) achieved 1,000 hours in eight months, 22 June 1995, as first K-MAX to reach this total. Canadian certification awarded 23 November 1994. Uses include logging, firefighting, agricultural spraying, constructing and surveying.

A US$690,000 contract to demonstrate Vertrep (Vertical Replenishment) to US Navy awarded August 1995: two month demonstration period saw two K-MAXs (c/n 0010 and 0013) lift 453,592 kg (1,000,000 lb) with follow-on assessment in Guam during May 1996 during which 142 hours were flown and 2,449,400 kg (5,400,000 lb) was lifted. Two helicopters then began US$5.7 million six-month deployment in Arabian Gulf on 3 June 1996 aboard USS *Niagara Falls*. During late August 1998, the prototype K-MAX was involved in trials for Magic Lantern mine detection systems. FAA Pt 27 IFR certification received 14 May 1999 as part of bid for US Navy Vertrep contract. On 14 July 1999 USMC placed contract worth US$4.2 million for remote piloting package and another worth US$2.7 million to May 2000: for further details see *Jane's Unmanned Aerial Vehicles and Targets*.

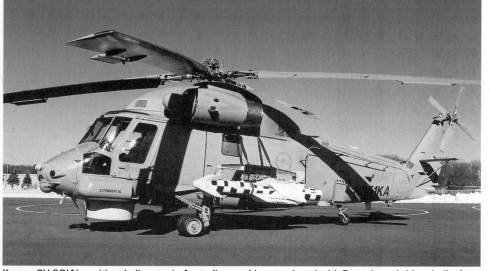

Kaman SH-2G(A) maritime helicopter in Australian markings equipped with Penguin anti-ship missile, for which Kaman manufactures the fuze

2001/0103508

Kaman received certification for an external seat in June 1999; up to two may be fitted; first customer announced October 1999.

CUSTOMERS: Kaman has moved from its original helicopter lease programme to a sales programme. Current operators include Mountain West (two), Woody Contracting, Kachina Aviation (two), Superior Helicopter, Rainier Helicopter Logging, Central Copters and Grizzly Mountain Aviation in the USA; Midwest Helicopters (two) and Cariboo Chilcotin in Canada; Helog of Switzerland, HeliAir Zagel of Germany, Japan Royal Helicopters (two) of Japan, Rotex Helicopters AG of Liechtenstein and Wucher Helikopter of Austria. The Royal Malaysian Air Force deployed a K-MAX (c/n 0013) for firefighting during April and May 1998. Peruvian National Police indicated interest in purchasing five aircraft in early 2000 and in December 2000 the US State Department ordered these five at a contract cost of US$21 million for anti-drug operations in Peru. Four were sold in the US, Europe and Taiwan (for Petroleum Helicopters Inc) during 2000. Orders for 36 had been received by 1 January 2001; four had been written off in non-fatal accidents and a further two substantially damaged by July 2000.

COSTS: US$4.2 million (2001).

DESIGN FEATURES: Kaman intermeshing, contrarotating rotors ensure all engine power produces lift; rotor disc loading is very low and airspeed is limited to reduce rotor stresses; all internal spaces painted white and provided with lights to ease night-time servicing; freight compartment beneath transmission casing allows carriage of special tools or parts; no hydraulics; transmission and engine fluid lines located on opposite sides of helicopter to avoid servicing errors; fuselage is narrow to give good downward view; panels in domed side windows can be opened to give direct vision or doors can be removed altogether. Electronic engine instruments and hook load measuring sensor record engine cycles and any overloads, but also give the pilot an immediate record of operations for billing purposes, avoiding paperwork for pilot. Inter-blade friction damper stops can be removed so that blades swing sideways to align for parking in narrow spaces.

Normal rotor rpm range between 250 and 270 unloaded, 260 and 270 with external load, giving maximum blade tip speed of 200 m (658 ft)/min; translational lift is attained at 12 kt (22 km/h; 14 mph); rpm reduced to 200 for autorotation and airspeed of 50 kt (92 km/h; 57 mph) then gives a power-off descent rate of 366 to 427 m (1,200 to 1,400 ft)/min.

FLYING CONTROLS: Blade angle of attack controlled by trailing-edge flaps and light control linkage, avoiding need for hydraulic power; an electric actuator in each flap control run is operated by pilot to track the blades on ground or in flight.

In normal powered flight, turns at or near the hover are effected by applying differential torque to the rotors by means of differential collective pitch commanded from the foot pedals; a small fore-and-aft cyclic pitch change also occurs; at low powers near autorotation, this would produce directional control reversal, because differential collective pitch change would cause drag on the unwanted side; so a non-linear cam between collective lever and rotor blade controls phases out differential collective from a 'dwell zone' at about 25 per cent collective demand and replaces it progressively with differential cyclic.

Intermeshing rotors cause pronounced pitch attitude change in response to collective pitch change; the K-MAX tailplane is connected to collective to alleviate this, to reduce blade stresses and produce touchdown and lift-off in level attitude; fixed fins on tailplane are in rotor downwash and rear fin is outside rotor disc; rudder is connected to foot pedals to help balance turns.

STRUCTURE: Light alloy airframe; GFRP and CFRP rotor blades and flaps. Tail assembly weighs 36.3 kg (80 lb) and can be removed quickly by two people.

LANDING GEAR: Fixed, tricycle; nosewheel is out of pilot's field of view; impact sustaining suspension with transverse mounting tube for mainwheels; rubber-in-compression suspension for mainwheels; oleo for nosewheel; bear paw plate round each wheel for operation from soft ground and snow; nosewheel swivels and locks; mainwheels have individual foot-powered brakes and parking brake.

POWER PLANT: One 1,119 kW (1,500 shp) Honeywell T53-17A-1 turboshaft (civil equivalent of military T53-L-703), with particle separator; flat rated at 1,007 kW (1,350 shp) for take-off up to approximately 8,875 m (29,120 ft). Transmission designed for 1,119 kW (1,500 shp), but operated at 1,007 kW (1,350 hp) with a 10,000 hour life with 1,800 hour overhaul intervals, raised to 2,500 hours in September 1998. Fuel capacity 865 litres (228.5 US gallons; 190.0 Imp gallons) located at aircraft CG. Hot refuelling capable; dual electric fuel pumps.

ACCOMMODATION: Pilot only, in Simula crash impact-absorbing seat with five-point harness; external seat for one passenger can be attached rapidly to either side, immediately ahead of the mainwheel legs. Pilot's seat and rudder pedals adjustable; heater and windscreen demister; doors removable for operation in hot weather; curved windscreen in production version. Tool/cargo compartment 0.74 m³ (26 cu ft) fitted with 2,268 kg (5,000 lb) stress tiedown rings.

SYSTEMS: DC electrical system with starter/generator; no hydraulics.

EQUIPMENT: Pilot-controlled swivelling landing light; standard configuration is for lifting slung loads, but fittings provided for 2,650 litre (700 US gallon; 583 Imp gallon) Bambi firefighting bucket, Loadcell and long-line hook gear; kits planned for patrol, firefighting tank and snorkel system, agricultural applications and similar missions. Optional rear-view mirror, port side of nose.

DIMENSIONS, EXTERNAL:
Rotor diameter, each	14.73 m (48 ft 4 in)
Length overall: rotors turning	15.85 m (52 ft 0 in)
fuselage	12.73 m (41 ft 9 in)
Height, centre of hubs	4.14 m (13 ft 7 in)
Width, rotors turning	15.67 m (51 ft 5 in)
Wheel track	3.68 m (12 ft 1 in)
Wheelbase	4.11 m (13 ft 6 in)

AREAS:
Rotor discs (total)	340.9 m² (3,669.0 sq ft)

WEIGHTS AND LOADINGS:
Operating weight empty	2,334 kg (5,145 lb)
Max hook capacity	2,721 kg (6,000 lb)
Max payload, ISA +15°C: at S/L	2,721 kg (6,000 lb)
at 1,525 m (5,000 ft)	2,568 kg (5,663 lb)
at 3,050 m (10,000 ft)	2,341 kg (5,163 lb)
at 3,810 m (12,100 ft)	2,273 kg (5,013 lb)
at 4,920 m (15,000 ft)	1,956 kg (4,313 lb)
Max fuel weight	699 kg (1,541 lb)
Max T-O weight:	
without jettisonable load	2,948 kg (6,500 lb)
with jettisonable load	5,443 kg (12,000 lb)
Max disc loading	17.20 kg/m² (3.52 lb/sq ft)
Max transmission power loading (T-O power)	5.83 kg/kW (9.57 lb/shp)

PERFORMANCE:
Never-exceed speed (V_{NE}):	
clean	100 kt (185 km/h; 115 mph)
with external load	80 kt (148 km/h; 92 mph)
Max rate of climb at S/L, normal flat-rated torque	762 m (2,500 ft)/min
Service ceiling	4,572 m (15,000 ft)
Hovering ceiling at AUW of 2,721 kg (6,000 lb) ISA:	
IGE	8,020 m (26,300 ft)
OGE	8,875 m (29,120 ft)
Range with max fuel	300 n miles (556 km; 345 miles)

OPERATIONAL NOISE LEVELS (FAR Pt 36), limit 87 dB(A):
Flyover	Average of 82 dB(A)

UPDATED

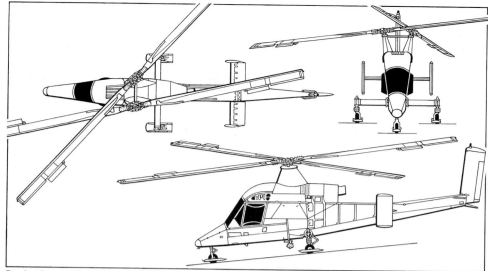

Production version of Kaman K-MAX intermeshing rotor helicopter *(Mike Keep/Jane's)*

Kaman K-MAX used for firefighting trials in Malaysia *(Paul Jackson/Jane's)*   2001/0103509

KESTREL

KESTREL AIRCRAFT COMPANY

Max Westheimer Airport, PO Box 720960, Norman, Oklahoma 73070
Tel: (+1 405) 573 00 90
Fax: (+1 405) 329 88 44

PRESIDENT AND CEO: Donald L Stroud
VICE-PRESIDENT, MARKETING: Charles Sallaway
RESOURCES DIRECTOR: Michael M Humphreys

Formed in 1991 by former Cessna Sales Manager Donald Stroud. The company's first product is the K-250, previously known as the KL-1. In mid-1998, Kestrel announced a two year delay in its plans, caused by the requirement to secure additional development funds. Nothing further had been heard by mid-2000, however. Full description last appeared in the 1999-2000 *Jane's*.

UPDATED

KIMBALL

JIM KIMBALL ENTERPRISES INC
PO Box 849, 5354 Cemetary Road, Zellwood, Florida 32798-0849
Tel: (+1 407) 889 34 51
Fax: (+1 407) 889 71 68
e-mail: jwkimball@aol.com
Web: http://www.jimkimballenterprises.com
PRESIDENT: Jim Kimball
VICE-PRESIDENT: Kevin Kimball

Kimball Enterprises offers kits of the Pitts Model 12 Super Stinker after purchasing design rights from Mid-America Aircraft. Plans are marketed through Model 12 Inc at same address.

UPDATED

KIMBALL (PITTS) MODEL 12
TYPE: Aerobatic, two-seat biplane/kitbuilt.
PROGRAMME: Designed by Curtis Pitts and originally named Macho Stinker and, later, Pitts Monster. Design started in 1994; construction of prototype began 1996; first flight March 1996; first delivery April 1997.
CUSTOMERS: Total of 45 under construction and four flying by November 1999. At least one sold to South Africa.
COSTS: Kit US$45,000; complete aircraft US$180,000 (2000).
DESIGN FEATURES: Traditional Pitts design, optimised for aerobatics; contours dictated by M-14P radial engine. Wire- and strut-braced wings of unequal span; upper wing mounted on cabane; single interplane strut each side; wire-braced tailplane and fin.
FLYING CONTROLS: Conventional and manual. Rod-connected ailerons on upper and lower wings. Flight-adjustable trim tab in each elevator.
STRUCTURE: 4130 steel tube fuselage; wooden spars and ribs; fabric covering.
LANDING GEAR: Tailwheel type; non-retractable. Optional mainwheel fairings. Aluminium sprung main legs; mainwheels 6.00-6; Cleveland disc brakes.
POWER PLANT: One 265 kW (355 hp) VOKBM M-14P nine-cylinder radial engine driving an MT three-blade constant-speed propeller. Fuel capacity 204 litres (54.0 US gallons; 45.0 Imp gallons) in 68 litre (18.0 US gallon; 15.0 Imp gallon) wing tank and 136 litre (36.0 US gallon; 30.0 Imp gallon) fuselage tank. Oil capacity 15 litres (4.0 US gallons; 3.3 Imp gallons).
ACCOMMODATION: Pilot and passenger in enclosed cockpit under one-piece Perspex canopy hinged on right side. Fixed windscreen.
SYSTEMS: 14 V or 28 V battery.

Kimball (Pitts) Model 12 aerobatic biplane *(Paul Jackson/Jane's)*   2001/0092079

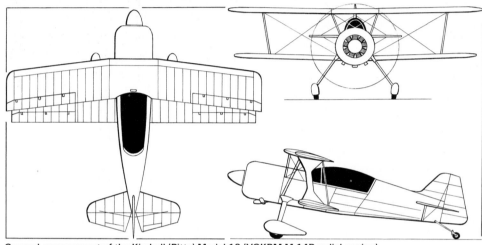
General arrangement of the Kimball (Pitts) Model 12 (VOKBM M-14P radial engine) *(James Goulding/Jane's)*   2000/0079018

DIMENSIONS, EXTERNAL:
Wing span: upper	6.71 m (22 ft 0 in)
lower	6.40 m (21 ft 0 in)
Wing chord at tip	1.07 m (3 ft 6 in)
Length overall	5.99 m (19 ft 8 in)
Max width of fuselage	1.07 m (3 ft 6 in)
Height overall	2.95 m (9 ft 8 in)
Tailplane span	2.44 m (8 ft 0 in)
Wheel track	2.29 m (7 ft 6 in)
Wheelbase	4.42 m (14 ft 6 in)
Propeller diameter	2.49 m (8 ft 2 in)
Propeller ground clearance	0.36 m (1 ft 2 in)

AREAS:
Wings, gross	13.94 m² (150.1 sq ft)
Ailerons (total)	1.88 m² (20.21 sq ft)
Fin	0.34 m² (3.69 sq ft)
Rudder, incl tab	0.63 m² (6.82 sq ft)
Elevators, incl tab	0.68 m² (7.30 sq ft)

WEIGHTS AND LOADINGS:
Weight empty	658 kg (1,450 lb)
Baggage capacity	18 kg (40 lb)
Max T-O weight	1,020 kg (2,250 lb)
Max wing loading	73.2 kg/m² (14.99 lb/sq ft)
Max power loading	3.86 kg/kW (6.34 lb/hp)

PERFORMANCE:
Never-exceed and max operating speed (V_{NE}, V_{MO})	207 kt (384 km/h; 239 mph)
Max cruising speed	178 kt (330 km/h; 205 mph)
Econ cruising speed	148 kt (274 km/h; 170 mph)
Stalling speed, power off	55 kt (102 km/h; 63 mph)
Roll rate	300°/s
Max rate of climb at S/L	more than 914 m (3,000 ft)/min
T-O run	76 m (250 ft)
T-O to 15 m (50 ft)	91 m (300 ft)
Landing from 15 m (50 ft)	274 m (900 ft)
Range with max fuel	456 n miles (845 km; 525 miles)

UPDATED

KINETIC

KINETIC AVIATION
7481 Northland Avenue, San Ramon, California 94583
Tel: (+1 925) 373 93 96
e-mail: mtgoatacft@aol.com
Web: http://www.members.aol.com/mtgoatacft/Kinetic Aviation
PRESIDENT: Bill Montagne

Kinetic Aviation has selected Wasilla in Alaska as the production site for the Mountain Goat and its proposed Big Horn development.

NEW ENTRY

KINETIC MOUNTAIN GOAT
TYPE: Two-seat lightplane.
PROGRAMME: Prototype currently flying; certification awaits raising of finance.
CUSTOMERS: Orders being taken. Company intends construction of 40 to 50 aircraft in first full year of production.
COSTS: US$120,000, equipped (2000).
DESIGN FEATURES: Similar in appearance to Piper PA-18 Super Cub. Heavy-duty construction for inhospitable terrain.
FLYING CONTROLS: Conventional and manual. Actuation bellcrank and pushrod. Flaps and flaperons.
STRUCTURE: Fuselage of fabric-covered steel tube. All-metal high wing has twin bracing struts. Strut-braced tailplane. Optional metal panels on underside.
LANDING GEAR: Tailwheel type; fixed. Bungee cord suspension. Mainwheels have Goodyear 26×10.5-6 tyres and toe-operated single piston brakes; dual-piston brakes optional. Scott 3200 tailwheel with heavy-duty tail spring. Float mountings standard; skis optional.
POWER PLANT: One 134 kW (180 hp) Textron Lycoming IO-360-B2E flat-four driving a Kinetic Aviation two-blade, ground-adjustable pitch propeller. Fuel capacity 265 litres (70.0 US gallons; 58.3 Imp gallons) of which 246 litres (65.0 US gallons; 54.1 Imp gallons) usable.
ACCOMMODATION: Two seats in tandem. One-piece full Plexiglas door on each side of fuselage. Dual controls. Four-point seat belts. Baggage space behind rear seat.
SYSTEMS: Skytec lightweight starter; 40 A alternator; electric auxiliary fuel pump.
AVIONICS: *Comms:* Garmin GPS/COM 250 XL, Mode C transponder. VISIN Microsystems panel optional.
EQUIPMENT: Wingtip strobe lighting; landing light and anti-collision beacon.

DIMENSIONS, EXTERNAL:
Wing span	10.82 m (35 ft 6 in)
Wing aspect ratio	6.7
Length overall	7.42 m (24 ft 4 in)
Height overall	2.06 m (6 ft 9 in)
Propeller diameter	2.03 m (6 ft 8 in)

DIMENSIONS, INTERNAL:
Baggage hold volume	0.685 m³ (24.2 cu ft)

AREAS:
Wings, gross	17.47 m² (188.0 sq ft)

WEIGHTS AND LOADINGS:
Weight empty	556 kg (1,225 lb)
Baggage capacity	159 kg (350 lb)
Max T-O weight	1,122 kg (2,475 lb)
Max wing loading	64.3 kg/m² (13.16 lb/sq ft)
Max power loading	8.37 kg/kW (13.75 lb/hp)

PERFORMANCE:
Max cruising speed at 75% power	138 kt (256 km/h; 159 mph)
Econ cruising speed at 55% power	111 kt (206 km/h; 128 mph)
Stalling speed, power off:	
flaps up	44 kt (81 km/h; 50 mph)
flaps down	28 kt (52 km/h; 32 mph)
Max rate of climb at S/L	366 m (1,200 ft)/min
T-O run	92 m (300 ft)
Landing run	84 m (275 ft)
Range with max fuel, 45 min reserves	659 n miles (1,287 km; 800 miles)

NEW ENTRY

KINETIC BIG HORN
The proposed Big Horn will be a six-seat aircraft powered by a 261 kW (350 hp) engine and with a useful load of 794 kg (1,750 lb). Development costs are estimated at US$2.5 million.

NEW ENTRY

KNR

KNR AIRCRAFT CO INC
6831 Oxford Street, St Louis Park, Minnesota 55426-4412
Tel: (+1 877) 304 19 68 and (+1 612) 920 98 11
Fax: (+1 612) 924 11 11
Web: http://www.aeropony.com

KNR Aircraft is marketing the G10 Aeropony, an updated version of the 1939 Bücker Bü 181 Bestmann, manufactured by the Kader factory at Nasser City, Egypt.

UPDATED

KNR BESTMANN and AEROPONY
TYPE: Side-by-side kitbuilt.
PROGRAMME: Bücker Bü 181 Bestmann (nautical: most capable sailor on deck) entered production in 1939; licence-built in Sweden, Czechoslovakia (as Zlin Z6, Z-281 and Z-381) and (from 1950) Egypt (Heliopolis Gomhouria); various power plants, including 78.3 kW (105 hp) Walter Minor in Egypt. Return to production as S-10 Aeropony announced by Shadin Co Inc of Minneapolis in 1994 (see 1995-96 edition) but not achieved. KNR adopted programme (initially as G10) and launched two new-build versions, to be fabricated in Egypt, at EAA AirVenture '99 at Oshkosh in July 1999.
CURRENT VERSIONS: **Bestmann Mark 6:** 'WW2 warbird replica' version with Bü 181-style multipane 'greenhouse' canopy and Continental O-300A engine.
Description applies to Bestmann, unless otherwise stated.
Aeropony Mark 8: Two-piece windscreen and two-piece canopy. Generally as for Bestmann except empty weight 570 kg (1,256 lb), MTOW 848 kg (1,870 lb), stalling speed, flaps up 42 kt (78 km/h; 49 mph).
COSTS: Bestmann kit, less engine and avionics, US$35,000 (2000).
DESIGN FEATURES: Conceived as advanced (for 1930s) monoplane trainer. Low-mounted, sharply tapered wings; generally similar to Bü 181D; modern Honeywell avionics. Dihedral 5° 30′. Quoted build time 1,000 hours.
FLYING CONTROLS: Conventional and manual. Pushrod-and-cable actuated; split flaps; horn-balanced rudder; mass-balanced rudder and elevators; trim tabs on both ailerons and rudder (ground-adjustable) and port elevator (flight-adjustable).
STRUCTURE: Forward and centre fuselage of welded Chromoloy tubing with fabric and aluminium skinning with large inspection hatch in fuselage bottom; rear of wooden monocoque with fabric covering; wooden wings with ply skinning to leading-edge and madapolan fabric covering (optionally plywood) to rear of spar; aluminium flaps; wooden tail unit with fabric-covered control surfaces. Aluminium engine cowling.
LANDING GEAR: Tailwheel type; fixed. Coil-spring suspension on main units; steerable tailwheel. Hydraulic brakes; parking brake. Mainwheels 6.00-6; tailwheel 4.00-3½.
POWER PLANT: One 108 kW (145 hp) Teledyne Continental O-300A flat-six, driving a two-blade wooden or aluminium propeller optimised for either cruise or aerobatics. Fuel contained in single tank at rear of seats, capacity 123 litres (32.5 US gallons; 27.1 Imp gallons).
ACCOMMODATION: Two persons side by side under upward-opening canopy with Plixidur T transparencies. Dual controls standard.
SYSTEMS: 12 V DC electrical power supplied by 60 A engine-driven alternator and 35 Ah battery; external power receptacle standard.
AVIONICS: Honeywell nav/com/GPS suite (KLX 135A-12) optional.
EQUIPMENT: Landing and taxying lights in port wing leading edge.

Egyptian-built demonstrator for KNR Bestmann

KNR Bestmann instrument panel
(Paul Jackson/Jane's)

DIMENSIONS, EXTERNAL:
Wing span	10.59 m (34 ft 9 in)
Wing chord: at root	1.80 m (5 ft 10¾ in)
at tip	0.60 m (2 ft 0 in)
Wing aspect ratio	8.3
Length overall	7.90 m (25 ft 11 in)
Height overall	2.06 m (6 ft 9 in)
Wheel track	1.83 m (6 ft 0 in)
Propeller diameter	1.98 m (6 ft 6 in)

DIMENSIONS, INTERNAL:
Cockpit max width	1.24 m (4 ft 1 in)

AREAS:
Wings, gross	13.50 m² (145.3 sq ft)
Ailerons (total)	1.65 m² (17.75 sq ft)
Trailing edge flaps (total)	1.28 m² (13.77 sq ft)
Fin	0.65 m² (6.99 sq ft)
Rudder	1.00 m² (10.76 sq ft)
Tailplane	1.37 m² (14.74 sq ft)
Elevators (total)	1.25 m² (13.45 sq ft)

WEIGHTS AND LOADINGS:
Weight empty	602 kg (1,327 lb)
Max T-O weight	850 kg (1,874 lb)
Max wing loading	62.97 kg/m² (12.90 lb/sq ft)
Max power loading	7.87 kg/kW (12.92 lb/hp)

PERFORMANCE:
Never-exceed speed (VNE)	185 kt (342 km/h; 212 mph)
Manoeuvring speed	125 kt (232 km/h; 144 mph)
Max level speed	117 kt (217 km/h; 135 mph)
Cruising speed	105 kt (194 km/h; 121 mph)
Econ cruising speed	95 kt (176 km/h; 109 mph)
Stalling speed: flaps up	42 kt (78 km/h; 49 mph)
flaps down	36 kt (67 km/h; 42 mph)
Max rate of climb at S/L	210 m (689 ft)/min
Service ceiling	4,000 m (13,120 ft)
T-O run	200 m (660 ft)
T-O to 15 m (50 ft)	320 m (1,050 ft)
Landing run	120 m (395 ft)
Range with max fuel at econ cruising speed	385 n miles (713 km; 443 miles)
Endurance	4 h 5 min
g limits	+6/−3

UPDATED

Aeropony Mk 8, with later style of canopy

KOLB

NEW KOLB AIRCRAFT COMPANY INC
8375 Russell Dyche Highway, London, Kentucky 40741
Tel: (+1 606) 862 96 92
Fax: (+1 606) 862 96 22
e-mail: customersupport@tnkolbaircraft.com
Web: http://www.tnkolbaircraft.com
PRESIDENT AND CEO: Norm Labhart

The New Kolb Aircraft Company purchased design rights to Kolb series of light aircraft in 1999. For details of the FireStar, see 1992-93 *Jane's*; the Laser has recently been relaunched; TwinStar has been renamed Kolb Mark III. Kolbra and King Kolbra launched at EAA convention, Oshkosh, July 2000. Total sales of all products exceed 1,500.

UPDATED

KOLB FIREFLY
TYPE: Single-seat ultralight.
PROGRAMME: Introduced at Sun 'n' Fun 1995.
CUSTOMERS: At least 75 flying by end 1999.
COSTS: US$9,388 (2000).
DESIGN FEATURES: High, constant-chord wing and boom-mounted empennage, based on FireStar; meets FAR Pt 103.7. Folding wings and tail for easy storage. Quoted build time 400 hours.
Modified NACA 4412 series aerofoil.
FLYING CONTROLS: Full-span flaperons.
STRUCTURE: Welded 4130 chromoly steel fuselage cage; 127 mm (5 in) rear fuselage tube and wing spar. Glass fibre nosecone; aluminium tube, fabric-covered wings.
LANDING GEAR: Tailwheel type; brakes optional.
POWER PLANT: One 29.5 kW (39.6 hp) Rotax 447 UL two-cylinder piston engine with Rotax 2.58:1 reduction unit, driving IVO two-blade wooden propeller (three-blade IVO propeller optional). Fuel capacity 18.9 litres (5.0 US gallons; 4.2 Imp gallons).
ACCOMMODATION: Optional full enclosure cockpit.

DIMENSIONS, EXTERNAL:
Wing span	6.71 m (22 ft 0 in)
Width, wings folded	1.68 m (5 ft 6 in)
Length overall	5.94 m (19 ft 6 in)
Height overall	1.75 m (5 ft 9 in)

AREAS:
Wings, gross	10.87 m² (117.0 sq ft)

WEIGHTS AND LOADINGS:
Weight empty	115 kg (253 lb)
Max T-O weight	226 kg (500 lb)

PERFORMANCE:
Max operating speed	55 kt (101 km/h; 63 mph)
Stalling speed	24 kt (44 km/h; 27 mph) CAS
Max rate of climb at S/L	305 m (1,000 ft)/min
T-O and landing run	46 m (150 ft)
g limits	+4/−2

UPDATED

KOLB KOLBRA
TYPE: Tandem-seat ultralight kitbuilt.
PROGRAMME: Announced July 2000 and publicly displayed at Oshkosh that month. Kits available by end 2000.
CURRENT VERSIONS: **Kolbra Trainer:** Ultralight trainer under FAR Pt 103.
King Kolbra: Heavier version with Jabiru engine; in FAA Experimental category.
COSTS: Kolbra Trainer kit US$16,288; King Kolbra kit US$20,540, both including engine (2000).
DESIGN FEATURES: Based on Kolb Mark III series (formerly Twinstar; see 1992-93 edition) and optimised for training.

USA: AIRCRAFT—KOLB to LANCAIR

Kolb FireFly (29.8 kW; 40 hp Rotax 447 engine)　　1999/0045104

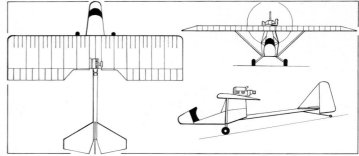

Kolb FireFly single-seat ultralight (*James Goulding/Jane's*)　　1999/0045105

FLYING CONTROLS: Manual. Full-span flaperons.
STRUCTURE: Welded 4130 chromoly steel fuselage cage; 152 mm (6 in) diameter rear fuselage tube and wing spar. Glass fibre nosecone; aluminium tube wings, fabric-covered.
LANDING GEAR: Tailwheel type; fixed. Chromoly construction for high strength. High flotation tyres. Optional hydraulic disc brakes.
POWER PLANT: *Kolbra Trainer:* One 47.8 kW (64.1 hp) Rotax 528 UL two-cylinder piston engine driving a two- or three-blade IVO or Warp Drive pusher propeller.
　King Kolbra: One 59.7 kW (80 hp) Jabiru 2200 flat-four driving a two- or three-blade IVO or Warp Drive pusher propeller. Options include 59.6 kW (79.9 hp) Rotax 912 UL.
　Fuel capacity (both) 37.9 litres (10.0 US gallons; 8.3 Imp gallons).
SYSTEMS: Optional electric starter.
EQUIPMENT: Optional BRS ballistic parachute.
DIMENSIONS, EXTERNAL:
　Wing span　　8.84 m (29 ft 0 in)
　Length overall　　7.32 m (24 ft 0 in)
　Height overall, three-blade propeller　　2.18 m (7 ft 2 in)
AREAS:
　Wings, gross　　14.49 m² (156.0 sq ft)
WEIGHTS AND LOADINGS (A: Kolbra Trainer with Rotax 582, B: King Kolbra with Jabiru):
　Weight empty: A　　225 kg (496 lb)
　B　　249 kg (550 lb)
　Max T-O weight: A, B　　454 kg (1,000 lb)
PERFORMANCE:
　Never-exceed speed (V_{NE}): A, B
　　　　95 kt (177 km/h; 110 mph)
　Normal cruising speed: A　　65 kt (121 km/h; 75 mph)
　B　　74 kt (137 km/h; 85 mph)
　Stalling speed, flaps down: A　　35 kt (65 km/h; 40 mph)
　B　　40 kt (73 km/h; 45 mph)
　Max rate of climb at S/L: A　　229 m (750 ft)/min
　T-O run: A　　76 m (250 ft)
　B　　61 m (200 ft)
NEW ENTRY

KOLB LASER
TYPE: Side-by-side ultralight.
PROGRAMME: Initial design began in 1989 and prototype shown at 'n' Fun 1991, but not put into production due to need for further design development and work on other products. Marketing began during 1999.
COSTS: US$20,126 with Rotax 582; US$25,446 with Rotax 912.
DESIGN FEATURES: Wings fold upwards for easy storage. Quoted build time 400 hours; quick-build options available.
FLYING CONTROLS: Conventional and manual. Mechanically actuated flaps deflect to 45°. Horn-balanced rudder.
STRUCTURE: 4130 chromoly steel tubing throughout, covered with Stits Poly-Fiber fabric. Wing centre-section has 152 mm (6 in) 4130 I-beam spar; outer panels span 2.20 m (7 ft 2½ in) and have 127 mm (5 in) 6061-T6 aluminium tube spars; entire wing has wraparound aluminium leading-edge.
　Wing section NACA 23012.
LANDING GEAR: Tailwheel type; fixed. Main legs 32 mm (1¼ in) 7075-T6 aluminium, with Matco aluminium wheels, 6.00-6 tyres and hydraulic brakes. 127 mm (5 in) steerable tailwheel.
POWER PLANT: Prototype has 47.8 kW (64.1 hp) Rotax 582, driving two-blade Prince wooden propeller. Optional engines include 59.6 kW (79.9 hp) Rotax 912 with two- or three-blade Warp Drive propeller. Standard fuel capacity 37.9 litres (10.0 US gallons; 8.3 Imp gallons) in single tank behind seats; optional capacity 56.8 litres (15.0 US gallons; 12.5 Imp gallons).
EQUIPMENT: BRS parachute system standard.
DIMENSIONS, EXTERNAL:
　Wing span　　6.71 m (22 ft 0 in)
　Wing chord　　1.37 m (4 ft 6 in)
　Width, wings folded　　2.44 m (8 ft 0 in)
　Length overall　　5.94 m (19 ft 6 in)
　Height overall　　1.47 m (4 ft 10 in)
DIMENSIONS, INTERNAL:
　Cabin max width　　0.99 m (3 ft 3 in)
AREAS:
　Wings, gross　　9.20 m² (99.0 sq ft)
WEIGHTS AND LOADINGS:
　Weight empty　　215 kg (475 lb)
　Baggage capacity　　6.8 kg (15 lb)
　Max T-O weight　　408 kg (900 lb)
PERFORMANCE (Rotax 582 engine):
　Never-exceed speed (V_{NE})　　108 kt (201 km/h; 125 mph)
　Max operating speed　　100 kt (185 km/h; 115 mph)
　Normal cruising speed at 75% power
　　　　91 kt (169 km/h; 105 mph)
　Stalling speed, flaps down　　37 kt (68 km/h; 42 mph)

　Max rate of climb at S/L　　305 m (1,000 ft)/min
　T-O and landing run　　91 m (300 ft)
　g limits　　+4.4/−4
NEW ENTRY

KOLB SLING SHOT
TYPE: Tandem-seat ultralight.
PROGRAMME: Added to range in 1996.
CUSTOMERS: Eight flying by early 2000.
COSTS: US$13,782 with Rotax 582 engine; US$19,834 with Rotax 912 engine (2000).
DESIGN FEATURES: Generally as for FireFly. Quoted build time 400 hours.
FLYING CONTROLS: Full-span flaperons.
STRUCTURE: Welded 4130 chromoly steel fuselage cage; 127 mm (5 in) rear fuselage tube and wing spar. Glass fibre nosecone and steel-reinforced, aluminium tube, fabric-covered wings.
LANDING GEAR: Tailwheel type; fixed; hydraulic disc brakes optional.
POWER PLANT: One 47.8 kW (64.1 hp) Rotax 582 using 2.58:1 or 3.47:1 reduction unit driving IVO two- or three-blade wooden propeller (metal propeller optional); further options are 37.0 kW (49.6 hp) Rotax 503 with 2.58:1 reduction unit; and 59.6 kW (79.9 hp) Rotax 912. Fuel capacity 37.0 litres (10.0 US gallons; 8.3 Imp gallons). Electric starter optional.
EQUIPMENT: Optional ballistic parachute and strobe lighting.
DIMENSIONS, EXTERNAL:
　Wing span　　6.71 m (22 ft 0 in)
　Width, wings folded　　1.91 m (6 ft 3 in)
　Length overall　　5.79 m (19 ft 0 in)
　Height overall　　2.08 m (6 ft 10 in)
AREAS:
　Wings, gross　　10.22 m² (110.0 sq ft)
WEIGHTS AND LOADINGS:
　Weight empty　　156 kg (345 lb)
　Max T-O weight　　385 kg (850 lb)
PERFORMANCE (Rotax 582 engine):
　Max operating speed　　87 kt (161 km/h; 100 mph)
　Normal cruising speed　　76 kt (140 km/h; 87 mph)
　Stalling speed, power off　　36 kt (66 km/h; 41 mph)
　Max rate of climb at S/L　　396 m (1,300 ft)/min
　T-O run　　53 m (175 ft)
　Landing run　　61 m (200 ft)
　g limits　　+4/−2
UPDATED

Kolb Sling Shot (*Paul Jackson/Jane's*)　　2001/0089501

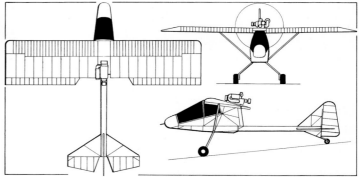

Kolb Sling Shot two-seat ultralight (*James Goulding/Jane's*)　　1999/0045107

LANCAIR
LANCAIR GROUP INC
22550 Nelson Road, Bend, Oregon 97701
Tel: (+1 541) 318 11 44
Fax: (+1 541) 318 11 77
e-mail: lancair@lancair.com
Web: http://www.lancair.com
PRESIDENT, LANCAIR GROUP: Lance Neibauer

More than 1,525 Lancairs (including out-of-production Model 235) sold to customers in 40 countries; around 800 are currently flying. Augmenting the manufacture of kits, the company has moved into the certified aircraft market with the four-seat Lancair Columbia 300, which was unveiled at Oshkosh in August 1997 and is being assembled at a new facility at Bend Municipal Airport, Redmond, Oregon. Development of the Tigress has been abandoned and the prototype donated to the EAA Museum; for details see *Jane's* 1999-2000.

In July 1999 Lancair announced the Legacy 2000, superseding the early 320/360 series, which last appeared in the 1999-2000 *Jane's;* deliveries began mid-2000.
UPDATED

LANCAIR LANCAIR 320 MARK II
No longer available, although completions continue. Last described in 2000-01 *Jane's*.
UPDATED

LANCAIR LEGACY 2000
TYPE: Side-by-side performance kitbuilt.
PROGRAMME: Prototype (N199L) first flown 16 July 1999, publicly displayed at Oshkosh 28 July 1999.
CUSTOMERS: Ten sold during Oshkosh debut; 16 by end of 1999 and 70 by April 2000.

Lancair IV, holder of four US cross-country records *(Paul Jackson/Jane's)* 2001/0089502

COSTS: US$32,900 in glass fibre; $37,900 in carbon fibre (2000).
DESIGN FEATURES: Improved version of Lancair series of two-seat kitbuilts, offering increased cabin area, new wing and better performance; only part common to Lancair 360 is horizontal stabiliser. Quoted build time under 1,000 hours.
Streamlined design. Sweptforward wing trailing-edge, tapered tailplane and sweptback fin; low wing and mid-mounted horizontal tail.
Wing section NLF(1)-0215F; incidence 1°; dihedral 3°; twist 1° 30′; thickness/chord ratio 15 per cent.
FLYING CONTROLS: Conventional and manual. Fowler flaps. Three-axis electric trim system optional.
STRUCTURE: Glass fibre, with carbon fibre for load-carrying components; option to build all-carbon airframe.
LANDING GEAR: Tricycle type. Mainwheels retract inwards; nosewheel rearwards. Cleveland 5.00-5 wheels and brakes.
POWER PLANT: One 149 kW (200 hp) Textron Lycoming IO-360-C1D6 flat-four driving a two-blade Hartzell or three-blade MTV-12B constant-speed propeller. Optional engines include 119 kW (160 hp) Lycoming IO-320-D1B; and 134 kW (180 hp) Lycoming IO-360-B1F and IO-360-M1A. Design will accept engines up to 224 kW (300 hp). Standard fuel capacity 189 litres (50.0 US gallons; 41.7 Imp gallons) in two wing tanks; optional long-range capacity 250 litres (66.0 US gallons; 55.0 Imp gallons) in tanks forward of main spar.
ACCOMMODATION: Pilot and passenger side by side beneath forward-hinging, one-piece canopy.

DIMENSIONS, EXTERNAL:
Wing span 7.77 m (25 ft 6 in)
Wing aspect ratio 7.9
Length overall 6.71 m (22 ft 0 in)
Height overall 2.54 m (8 ft 4 in)
DIMENSIONS, INTERNAL:
Cabin: Length 1.60 m (5 ft 3 in)
Max width 1.10 m (3 ft 7½ in)
Max height 1.13 m (3 ft 8½ in)
AREAS:
Wings, gross 7.66 m² (82.5 sq ft)
WEIGHTS AND LOADINGS (IO-360-C1D6 engine):
Weight empty 612 kg (1,350 lb)
Baggage capacity 41 kg (90 lb)
Max T-O weight: Normal category 997 kg (2,200 lb)
Utility category 861 kg (1,900 lb)
Max wing loading:
Normal category 130.2 kg/m² (26.67 lb/sq ft)
Utility category 112.4 kg/m² (23.03 lb/sq ft)
Max power loading:
Normal category 6.70 kg/kW (11.00 lb/hp)
Utility category 5.78 kg/kW (9.50 lb/hp)
PERFORMANCE (estimated; IO-360-C1D6 engine):
Max cruising speed at 5,485 m (18,000 ft)
 251 kt (465 km/h; 289 mph)
Normal cruising speed at 3,660 m (12,000 ft)
 217 kt (402 km/h; 250 mph)
Stalling speed, flaps down 53 kt (97 km/h; 60 mph)
Max rate of climb at S/L 594 m (1,950 ft)/min
Service ceiling 7,315 m (24,000 ft)
T-O run 153 m (500 ft)
Landing run 275 m (900 ft)
Range with max fuel, no reserves
 1,303 n miles (2,414 km; 1,500 miles)
g limits: Normal +3.8/-1.7
 Utility +4.4/-2.2

UPDATED

LANCAIR LANCAIR IV and IV-P

TYPE: Four-seat performance kitbuilt.
PROGRAMME: Kit deliveries began 1990. On 20 February 1991, prototype set NAA world speed record between San Francisco and Denver of 314.7 kt (583.2 km/h; 362.4 mph). First kit completed July 1991.
CURRENT VERSIONS: **Lancair IV:** Standard version.
Description mainly applies to Lancair IV.
Lancair IV-P: Pressurised version; first flight 1 November 1993. Provides 0.34 bar (5.0 lb/sq in) differential; equipment from Dukes Research of California.
CUSTOMERS: Total of 500 kits sold; 110 flying by late 1998 (latest information supplied).
COSTS: Standard US$52,400; New Fast-Build US$73,700. Pressurised version: Standard US$76,500; New Fast-Build US$97,800. Wingmate US$3,000 extra; Firewall Fast-Build options and builder workshop available.
DESIGN FEATURES: Conventional seating for four persons in airframe of new design, but following Lancair formula for high cruising speeds. Optional winglets. Quoted Fast-Build construction time 2,000 hours.
FLYING CONTROLS: Conventional and manual. Fowler flaps. Sidestick controllers for both pilots. Trim tabs in port aileron and port elevator. Optional speed brakes.
STRUCTURE: Carbon fibre/epoxy airframe, with Nomex honeycomb cores.
LANDING GEAR: As Lancair 320. Mainwheels 15 × 6.00-6; nosewheel is 5.00-5. Cleveland dual-piston caliper brakes.
POWER PLANT: One 261 kW (350 hp) Teledyne Continental TSIO-550-E1B, -E2B or -E3B twin-turbocharged flat-six engine driving a three- or four-bladed MT or Hartzell

Lancair IV cockpit, showing comprehensive avionics fit and sidestick controllers *(Paul Jackson/Jane's)* 0016190

constant-speed propeller. Fuel in integral wing tanks; total standard usable capacity 310 litres (82.0 US gallons; 68.3 Imp gallons); long-range tanks, total capacity 341 litres (90.0 US gallons; 74.9 Imp gallons), of which 333 litres (88.0 US gallons; 73.3 Imp gallons) are usable. Oil capacity 11.4 litres (3.0 US gallons; 2.5 Imp gallons). Two versions of PT-6A tested during 1999.
SYSTEMS: Electro-Explosive Separation System electromagnetic de-icing system was tested during 1999.

DIMENSIONS, EXTERNAL:
Wing span: plain tips 9.19 m (30 ft 2 in)
 winglets 9.93 m (32 ft 7 in)
Wing chord: at root 1.18 m (3 ft 10½ in)
 at tip 0.77 m (2 ft 6¼ in)
Wing aspect ratio 9.3
Width, wings removed 2.41 m (7 ft 11 in)
Length overall 7.62 m (25 ft 0 in)
Height overall 2.44 m (8 ft 0 in)
Tailplane span 3.35 m (11 ft 0 in)
Wheel track 2.13 m (7 ft 0 in)
Wheelbase 1.88 m (6 ft 2 in)
Propeller diameter 1.93 m (6 ft 4 in)
DIMENSIONS, INTERNAL:
Cabin: Length 3.20 m (10 ft 6 in)
 Max width 1.17 m (3 ft 10 in)
 Max height 1.22 m (4 ft 0 in)
AREAS:
Wings, gross: plain tips 9.10 m² (98.0 sq ft)
 winglets 10.03 m² (108.0 sq ft)
Winglets, total 1.30 m² (14.00 sq ft)
Flaps, total 1.12 m² (12.07 sq ft)
WEIGHTS AND LOADINGS:
Weight empty: IV 907 kg (2,000 lb)
 IV-P 998 kg (2,200 lb)
Baggage capacity 68 kg (150 lb)
Max T-O weight 1,451 kg (3,200 lb)
Max landing weight 1,360 kg (3,000 lb)
Max wing loading, plain tips 159.4 kg/m² (32.65 lb/sq ft)
Max power loading 5.56 kg/kW (9.14 lb/hp)
PERFORMANCE:
Never-exceed speed (V_{NE}) 274 kt (507 km/h; 315 mph)
Cruising speed at 75% power at 7,315 m (24,000 ft)
 291 kt (539 km/h; 335 mph)
Stalling speed 66 kt (121 km/h; 75 mph)
Max rate of climb at S/L 792 m (2,600 ft)/min
Service ceiling 8,840 m (29,000 ft)
T-O run 457 m (1,500 ft)
T-O to 15 m (50 ft) 549 m (1,800 ft)
Landing run 518 m (1,700 ft)
Range, no reserves 1,260 n miles (2,334 km; 1,450 miles)
Endurance 4 h 48 min
g limits +4.4/-2.3 (allowable), +8/-4 ultimate

UPDATED

LANCAIR LANCAIR ES and SUPER ES

TYPE: Four-seat performance kitbuilt.
PROGRAMME: Prototype unveiled at Oshkosh in 1992; flight testing completed 1995. Certification of Fast-Build kit granted early 1996 under simplified version of FAR Pt 23 governing light aircraft.
CUSTOMERS: Total of 127 kits sold by late 1999, of which 22 were then flying. Production continues.
COSTS: Standard kit US$39,900; Fast-Build US$53,800 (1998).
DESIGN FEATURES: Generally as for earlier Lancairs; hardtop cabin with paired side-by-side seats. Sidestick controls. Quoted Fast-Build kit construction time 1,800 hours.
FLYING CONTROLS: Generally as for Lancair IV.
STRUCTURE: All-composites.
LANDING GEAR: Non-retractable tricycle type. Main tyres 6.00-15 × 6; nose 5.00-5. Nosewheel steerable through 90°.
POWER PLANT: *ES:* One 149 kW (200 hp) Teledyne Continental IO-360-ES flat-six engine; two-blade McCauley constant-speed propeller.
Super ES: One 209 kW (280 hp) Teledyne Continental IO-550-G or IO-550-N with three-blade Hartzell constant-

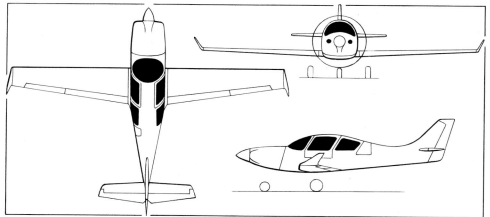

Lancair IV four-seat light aircraft *(James Goulding/Jane's)* 1999/0051771

USA: AIRCRAFT—LANCAIR

Lancair Super ES fixed-gear four-seater *(Paul Jackson/Jane's)* 2000/0085830

speed propeller. Standard fuel capacity 288 litres (76.0 US gallons; 63.3 Imp gallons), of which 280 litres (74.0 US gallons; 61.6 Imp gallons) are usable, in two wing tanks. Long-range fuel capacity 546 litres (100 US gallons; 120 Imp gallons) usable. Oil capacity 7.6 litres (2.0 US gallons; 1.7 Imp gallons).

ACCOMMODATION: Pilot and three passengers in two pairs of seats.

DIMENSIONS, EXTERNAL:
Wing span	10.82 m (35 ft 6 in)
Wing chord: at root	1.50 m (4 ft 11 in)
at tip	0.91 m (3 ft 0 in)
Wing aspect ratio	9.0
Width, wings removed	2.41 m (7 ft 11 in)
Length overall	7.62 m (25 ft 0 in)
Height overall	2.44 m (8 ft 0 in)
Tailplane span	3.96 m (13 ft 0 in)
Wheel track	2.36 m (7 ft 9 in)
Wheelbase	2.13 m (7 ft 0 in)
Propeller diameter	1.93 m (6 ft 4 in)

DIMENSIONS, INTERNAL:
Cabin: as Lancair IV	
Baggage hold volume	1.0 m³ (35 cu ft)

AREAS:
Wings, gross	13.01 m² (140.0 sq ft)

WEIGHTS AND LOADINGS:
Weight empty	771 kg (1,700 lb)
Max T-O weight	1,361 kg (3,000 lb)
Max wing loading	104.6 kg/m² (21.43 lb/sq ft)
Max power loading: ES	9.13 kg/kW (15.00 lb/hp)
Super ES	6.52 kg/kW (10.71 lb/hp)

PERFORMANCE:
Never-exceed speed (V_{NE})	240 kt (444 km/h; 276 mph)
Cruising speed at 75% power at 3,050 m (10,000 ft):	
ES	174 kt (332 km/h; 200 mph)
Super ES	196 kt (362 km/h; 225 mph)
Stalling speed	50 kt (92 km/h; 57 mph)
Rate of climb at S/L: ES:	
pilot only	457 m (1,500 ft)/min
at MTOW	381 m (1,250 ft)/min
Service ceiling	5,490 m (18,000 ft)
T-O run	183 m (600 ft)
Landing run	244 m (800 ft)
Range, max fuel, no reserves:	
ES	1,220 n miles (2,259 km; 1,403 miles)
Super ES	1,175 n miles (2,176 km; 1,352 miles)
Endurance: ES	7 h
Super ES	6 h
g limits	+9.0/−4.5

UPDATED

LANCAIR COLUMBIA 300

TYPE: Four-seat lightplane.

PROGRAMME: Announced 1996 as LC-40; aerodynamic prototype flown July 1996; first flight of certification prototype (N140LC), early 1997. Second prototype first flew 14 April 1997; public debut at Oshkosh August 1997; modified for FAA testing to confirm spin-resistance. Certification completed 18 September 1998.
First production aircraft (N424CH) delivered 24 February 2000.

CURRENT VERSIONS: **Tourer:** *As described.*
Trainer: Lower-powered version with 116 kW (155 hp) engine, planned for flight training market, rivalling Cessna 172 and Cirrus SR20.
Columbia IV: Proposed version with pressurisation and retractable landing gear.
Columbia Turbo 400: Turbocharged Columbia 300 with Continental TSIO-550E and (when available) FADEC. Announced 9 April 2000; prototype to be completed mid-2000; certification in 2001 and deliveries from mid-2002. Provisional empty weight 1,066 kg (2,350 lb), max T-O weight 1,588 kg (3,500 lb) and service ceiling 7,315 m (24,000 ft). Other data as Columbia 300.

CUSTOMERS: Deposits for 295 aircraft taken by end 1998. May be offered in fractional ownership schemes. Target production of 170 in second full year.

COSTS: US$285,500 to US$299,700 (1999). Columbia Turbo 400 estimated at US$365,000.

DESIGN FEATURES: Low-wing monoplane of typical Lancair appearance and design.

FLYING CONTROLS: Generally as for Lancair IV.

STRUCTURE: Extensive composites construction throughout. One-piece dual-spar wing.

LANDING GEAR: Fixed tubular steel tricycle type. Mainwheels 6.00-6, nosewheel 5.00-5. Cleveland brakes. Speed fairings on all wheels. Retractable version planned. Minimum ground turning radius 4.72 m (15 ft 6 in); nosewheel steerable ±60°.

POWER PLANT: One 224 kW (300 hp) Teledyne Continental IO-550-N2B flat-six engine, driving a three-blade constant-speed Hartzell propeller. Total fuel capacity 379 litres (100 US gallons; 83.3 Imp gallons) in two integral wing tanks, of which 371 litres (98.0 US gallons; 81.6 Imp gallons) are usable. Oil capacity 7.6 litres (2.0 US gallons; 1.7 Imp gallons).

ACCOMMODATION: Pilot and three passengers in two pairs of seats in enclosed cabin. Double gull-wing doors hinge upwards. Emergency exit through large baggage door behind rear passenger seats.

SYSTEMS: Single 12 V battery; 14 V 60 A alternator; air conditioning. FADEC optional from 2001.

AVIONICS: *Comms:* Apollo SL30 nav/com, SL 70 transponder.
Flight: Apollo nav indicator, S-Tec 30 or 55 autopilot, Apollo GX60 GPS and BFG Stormscope WX-950 lightning sensor. Options include Honeywell KR 87 ADF, KI 227 remote indicator, KCS 55A HSI and KI 256 flight director.
Instrumentation: AvroTec 265 mm (10½ in) flat panel display.

EQUIPMENT: Fire extinguisher standard.

DIMENSIONS, EXTERNAL:
Wing span	11.00 m (36 ft 1 in)
Wing chord: at root	1.40 m (4 ft 7 in)
at tip	1.02 m (3 ft 4 in)
Wing aspect ratio	9.2
Length overall	7.67 m (25 ft 2 in)
Fuselage max width	1.28 m (4 ft 2½ in)
Height overall	2.74 m (9 ft 0 in)
Tailplane span	4.17 m (13 ft 8 in)
Wheel track	2.24 m (7 ft 4 in)
Wheelbase	2.04 m (6 ft 8¼ in)
Propeller diameter	1.96 m (6 ft 5 in)

DIMENSIONS, INTERNAL:
Cabin: Length	3.54 m (11 ft 7½ in)
Max width	1.24 m (4 ft 1 in)
Max height	1.30 m (4 ft 3 in)

AREAS:
Wings, gross	13.12 m² (141.2 sq ft)

WEIGHTS AND LOADINGS:
Weight empty	998 kg (2,200 lb)
Baggage capacity	54 kg (120 lb)
Max fuel weight	272 kg (600 lb)
Max T-O weight	1,542 kg (3,400 lb)
Max landing weight	1,465 kg (3,230 lb)
Max wing loading	117.6 kg/m² (24.08 lb/sq ft)
Max power loading	6.90 kg/kW (11.33 lb/hp)

PERFORMANCE:
Never-exceed speed (V_{NE})	235 kt (435 km/h; 270 mph)
Normal cruising speed at 75% power	
	191 kt (354 km/h; 220 mph)
Stalling speed: flaps up	71 kt (132 km/h; 82 mph)
flaps down	57 kt (106 km/h; 66 mph)
Max rate of climb at S/L	408 m (1,340 ft)/min
Service ceiling	5,480 m (18,000 ft)
T-O run	213 m (700 ft)
T-O to 15 m (50 ft)	381 m (1,250 ft)
Landing from 15 m (50 ft)	716 m (2,350 ft)
Landing run	472 m (1,550 ft)
Range with max fuel	
	1,100 n miles (2,037 km; 1,265 miles)
g limits	+4.4/−1.76

UPDATED

First Lancair Columbia 300 to be delivered *(Paul Jackson/Jane's)* 2001/0089503

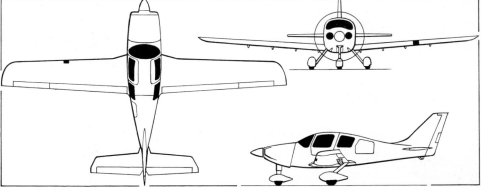

Lancair Columbia 300 (Teledyne Continental IO-550 flat-six) *(James Goulding/Jane's)* 2001/0089517

LEARJET

BOMBARDIER AEROSPACE LEARJET
(Subsidiary of Bombardier Inc)
One Learjet Way, PO Box 7707, Wichita, Kansas 67277
Tel: (+1 316) 946 20 00
Fax: (+1 316) 946 22 20
Telex: 417441
VICE-PRESIDENT AND GENERAL MANAGER: Jim Ziegler
VICE-PRESIDENT, OPERATIONS: Ghislain Bourque
VICE-PRESIDENT, FINANCE: Chris Chrashaw
VICE-PRESIDENT, ENGINEERING: Keith Miller
DIRECTOR, PUBLIC RELATIONS: Dave Franson

Company originally founded 1960 by Bill Lear Snr as Swiss American Aviation Corporation (SAAC); transferred to Kansas 1962 and renamed Lear Jet Corporation; prototype Lear Jet 23 (N801L) first flew 7 October 1963; Gates Rubber Company bought about 60 per cent share in 1967 and renamed it Gates Learjet Corporation, moving much of manufacturing to Tucson, Arizona; 64.8 per cent acquired by Integrated Acquisition Inc September 1987 and renamed Learjet Corporation; all manufacturing moved from Tucson to Wichita during 1988, leaving customer service completion and modification centre in Tucson.

Acquisition of Learjet by Canada's Bombardier announced April 1990 and concluded 22 June 1990 for US$75 million; name changed to Learjet Inc; Bombardier assumed responsibility for Learjet's line of credit; now part of Bombardier Inc (which see under Canada). Learjet 31A, 45 and 60 assembled in Wichita and flown to Tucson for completion and delivery.

Over 2,100 Learjets built, including 105 LJ23s, 259 LJ24s, 368 LJ25s, five LJ28s, four LJ29s, 184 LJ31s, 673 LJ35s, 62 LJ36s, 50 LJ45s, 147 LJ55s and 162 LJ60s. 2,000th Learjet, Model 45 N158PH, delivered to Parker Hannifin Corporation in August 1999.

Learjet bought manufacturing and marketing rights and tooling of Aeronca thrust reversers, for application to Learjet and other aircraft, March 1989.

Total of 36 deliveries in 1994; 43 in 1995; 34 in 1996; 45 in 1997; 61 in 1998; 99 in 1999, and 134 in 2000.

UPDATED

LEARJET 31A

TYPE: Business jet.
PROGRAMME: Learjet 31 introduced September 1987 following first flight of aerodynamic prototype (modified Learjet 35A) on 11 May 1987; first production aircraft (N311DF) used as systems testbed; FAA certification 12 August 1988. Learjet 31A and 31A/ER announced October 1990 to replace Learjet 31. Certified in 21 countries, including Argentina, Australia, Bermuda, Brazil, Canada, Czech Republic, Denmark, Germany, Grand Cayman, Guatemala, Indonesia, Italy, Japan, Luxembourg, Mexico, Namibia, Pakistan, Philippines, South Africa, Switzerland and USA. Total fleet time 400,000 hours by 1 October 2000.
CURRENT VERSIONS: **Learjet 31**: See 1990-91 *Jane's*.
Learjet 31A: Current version.
Description applies to Learjet 31A, except where indicated.
Learjet 31A/ER: Optional extended-range version with 2,627 litres (694 US gallons; 578 Imp gallons) of fuel.
CUSTOMERS: Total of 206 built by December 2000, including 36 of original Learjet 31; 14 Learjet 31As delivered in 1994, 19 in 1995, 13 in 1996, 21 in 1997, 22 in 1998, 24 in 1999, and 28 in 2000. Twenty-one were in the Bombardier FlexJets fleet in early 1999. 200th Learjet 31A delivered 3 October 2000 to Falcon Air Services of Phoenix, Arizona.
COSTS: US$6.3 million, equipped (1999).
DESIGN FEATURES: Small, rear-engined business jet; low wing with leading-edge sweepback; T tail and two large strakes below rear fuselage.

Learjet 31 (1990-91 *Jane's*) combined fuselage/cabin and power plant of Learjet 35A/36A with wing of Learjet 55; delta fins added to eliminate Dutch roll, stabilise aircraft at high airspeeds, induce docile stall and reduce approach speeds and field lengths; stick pusher/puller and dual yaw dampers no longer required; stick shaker and single yaw damper retained for comfort.

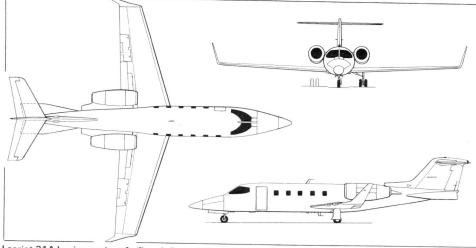

Learjet 31A business aircraft *(Dennis Punnett/Jane's)*

Additional features of Learjet 31A include cruise Mach number up to 13,105 m (43,000 ft) increased 4 per cent to 0.81 and V_{MO} increased from 300 kt (556 km/h; 345 mph) to 325 kt (602 km/h; 374 mph) IAS. Increases mainly benefit descent from high altitudes. Learjet 31A also features integrated digital avionics package.

Improvements announced at the NBAA Convention in Atlanta in October 1999, and incorporated from c/n 31A-194, include: increased maximum T-O weight of 7,711 kg (17,000 lb) and maximum landing weight of 7,257 kg (16,000 lb), latter available for retrofit on earlier aircraft; revised winglet design, based on that of Learjet 60, for enhanced high-speed/high-altitude performance and handling; thrust reversers standard; Universal UNS-1C FMS standard; lighter and more reliable AHRS; dual R134a air conditioning systems to provide increased capacity and redundancy, with one system dedicated to cockpit; and digital engine control for increased reliability and provision of trend monitoring.

FLYING CONTROLS: Conventional and manual. Ailerons have brush seals and geared tabs; electrically actuated trim tab in port aileron. Electrically actuated tailplane incidence control has separate motors for pilot and co-pilot and single-fault survival protection; aircraft can be manually controlled following tailplane runaway and landed with reduced flap. Rudder has electric trim tab; automatic electric rudder assist servo operates automatically if rudder pedal loads exceed 22.6 kg (50 lb). Full-chord fences bracket the ailerons. Single spoiler panel in each wing used as airbrake and lift dumper. Hydraulically actuated flaps extend to 40°. Optional drag parachute mounted on inside of baggage hatch under tail.

STRUCTURE: Semi-monocoque fuselage with eight-spar internal structure; multispar wing with machined skins; winglets have full-depth honeycomb core bonded to skin.

LANDING GEAR: Retractable tricycle gear; main legs retract inward, nose leg forward; twin-wheel main units with anti-skid disc brakes; nosewheel has full-time digital steer-by-wire. Tyres 17.5×5.75-8 (12 ply) main; 18×4.4 (10 ply) nose. Ground turning radius about nosewheel 11.91 m (39 ft 1 in).

POWER PLANT: Two 15.56 kN (3,500 lb st) Honeywell TFE731-2-3B turbofans with N_1 digital electronic engine controller giving engine trend monitoring, automatic retention of power settings above 4,575 m (15,000 ft) and special idling control for descent from 15,545 m (51,000 ft). Engine synchroniser fitted. Dee Howard 4000 thrust reverser system standard. One integral fuel tank in each wing standard or ER tank in fuselage (see under Weights and Loadings for quantities); fuselage fuel transferred by gravity or pump; single-point pressure refuelling standard.

ACCOMMODATION: Cabin furnishings include a three-seat divan, four Erda 10-way adjustable individual seats in club seating arrangement, side-facing seat with toilet, two folding tables, baggage compartment, overhead panels with reading lights, indirect lighting, air vents and oxygen masks. Revised interior offering more headroom and three cabin configurations announced October 1994 and introduced on 100th aircraft May 1995. Aft lavatory option introduced in 1996, providing an additional 0.28 m³ (10.0 cu ft) of baggage space. Optional external baggage locker increases total baggage volume to 1.76 m³ (62.0 cu ft).

SYSTEMS: Hydraulic system operates flaps, landing gear, airbrakes, wheelbrakes and thrust reversers; system pressure 69 to 120.6 bars (1,000 to 1,750 lb/sq in); pneumatic standby for gear extension and wheelbrakes. Normal cabin pressure differential 0.64 bar (9.4 lb/sq in) with automatic flood engine bleed if cabin altitude exceeds 2,820 m (9,250 ft); pop-out emergency oxygen for passengers and masks for crew. Electrical system based on two starter/generators, two lead-acid batteries and two inverters; both busses can run from one engine; electrics operate tailplane incidence, rudder assister and nosewheel steering. Anti-icing by bleed air for wing, engine intakes and windscreen; tailplane electrically heated; fin not protected. Alcohol spray for radome to stop shed ice entering engines; controls prevent internal ice and condensation during long descents.

AVIONICS: Honeywell integrated digital avionics package with five-tube EFIS 50, Universal UNS-1M flight management system (FMS), dual Series III nav/comms, and dual KFC 3100 autopilots/flight directors.

EQUIPMENT: Throttle-mounted landing gear warning mute and go-around switches, nacelle heat annunciator, engine synchroniser and synchroscope, recognition light, wing ice light, emergency press override switches, transponder ident switch in pilot's control wheel, flap preselect, crew lifejackets, cockpit dome lights, cockpit speakers, crew oxygen masks and fire extinguisher are standard.

DIMENSIONS, EXTERNAL:
Wing span	13.36 m (43 ft 10 in)
Wing aspect ratio	7.2
Length overall	14.83 m (48 ft 8 in)
Max diameter of fuselage	1.63 m (5 ft 4 in)
Height overall	3.75 m (12 ft 3½ in)
Tailplane span	4.48 m (14 ft 8½ in)

DIMENSIONS, INTERNAL:
Cabin: Length:
incl flight deck: 31A	6.63 m (21 ft 9 in)
31A/ER	6.27 m (20 ft 7 in)
excl flight deck: 31A	5.21 m (17 ft 1 in)
31A/ER	4.85 m (15 ft 11 in)
Max width: 31A, 31A/ER	1.50 m (4 ft 11 in)
Max height: 31A, 31A/ER	1.32 m (4 ft 4 in)
Floor area, excl flight deck	3.60 m² (38.75 sq ft)
Volume, cockpit divider to rear pressure bulkhead	7.7 m³ (271 cu ft)
Baggage compartment volume: 31A	1.1 m³ (40 cu ft)
31A/ER	0.76 m³ (26.7 cu ft)
Optional baggage locker	0.34 m³ (12.0 cu ft)

AREAS:
Wings, gross	24.57 m² (264.5 sq ft)
Horizontal tail surfaces (total)	5.02 m² (54.0 sq ft)
Vertical tail surfaces (total)	3.57 m² (38.4 sq ft)

WEIGHTS AND LOADINGS:
Weight empty	4,651 kg (10,253 lb)
Basic operating weight empty	5,053 kg (11,140 lb)
Max payload	1,070 kg (2,360 lb)
Payload with max fuel:	674 kg (1,486 lb)
Max fuel weight:	
wing tanks: 31A	1,272 kg (2,804 lb)
31A/ER	1,282 kg (2,826 lb)
fuselage tank: 31A	599 kg (1,320 lb)
31A/ER	829 kg (1,827 lb)
total: 31A	1,871 kg (4,124 lb)
31A/ER	2,111 kg (4,653 lb)
Max T-O weight: 31A (standard)	7,711 kg (17,000 lb)
31A (optional)	8,028 kg (17,700 lb)

Learjet 31A (two Honeywell TFE731-2 turbofans)

672   USA: AIRCRAFT—LEARJET

Max ramp weight: 31A (standard)	7,802 kg (17,200 lb)
31A (optional)	8,119 kg (17,900 lb)
Max landing weight	7,257 kg (16,000 lb)
Max zero-fuel weight	6,124 kg (13,500 lb)
Payload with max fuel: 31A (standard)	896 kg (1,976 lb)
31A (optional)	1,089 kg (2,400 lb)
Max wing loading:	
31A (standard)	313.8 kg/m² (64.27 lb/sq ft)
31A (optional)	326.7 kg/m² (66.92 lb/sq ft)
Max power loading:	
31A (standard)	248 kg/kN (2.43 lb/lb st)
31A (optional)	258 kg/kN (2.53 lb/lb st)

PERFORMANCE (A: standard max T-O weight, B: optional max T-O weight):
Max operating Mach No. (M_{MO}):
8,380-13,105 m (27,500-43,000 ft)	M0.81
13,105-14,020 m (43,000-46,000 ft)	M0.81-0.80
14,020-14,325 m (46,000-47,000 ft)	M0.80-0.79
above 14,325 m (47,000 ft)	0.79
Max cruising speed	481 kt (891 km/h; 554 mph)

Typical cruising speeds at mid-cruise weight:
at 12,495 m (41,000 ft)	463 kt (857 km/h; 533 mph)
at 13,105 m (43,000 ft)	458 kt (848 km/h; 527 mph)
at 13,715 m (45,000 ft)	449 kt (832 km/h; 517 mph)
at 14,325 m (47,000 ft)	440 kt (815 km/h; 506 mph)
Max speed with landing gear extended	260 kt (481 km/h; 299 mph)
Tyre limiting speed	183 kt (339 km/h; 210 mph)
Approach speed	122 kt (226 km/h; 140 mph) IAS
Stalling speed, flaps and landing gear down	93 kt (173 km/h; 107 mph) IAS
Max rate of climb at S/L: A	1,554 m (5,100 ft)/min
B	1,490 m (4,890 ft)/min
Rate of climb at S/L, OEI: A	494 m (1,620 ft)/min
B	462 m (1,515 ft)/min
Max certified altitude	15,545 m (51,000 ft)
Service ceiling, OEI: A	8,015 m (26,300 ft)
B	7,620 m (25,000 ft)
T-O field length: A	1,000 m (3,280 ft)
B	1,158 m (3,800 ft)
FAR Pt 91 landing distance: A, B	844 m (2,770 ft)

Range with two crew and four passengers:
VFR: 31A	1,562 n miles (2,892 km; 1,797 miles)
31A/ER	1,782 n miles (3,300 km; 2,050 miles)
IFR reserves:	
31A	1,460 n miles (2,703 km; 1,680 miles)
31A/ER	1,690 n miles (3,129 km; 1,944 miles)

Learjet 31A instrument panel   0016192

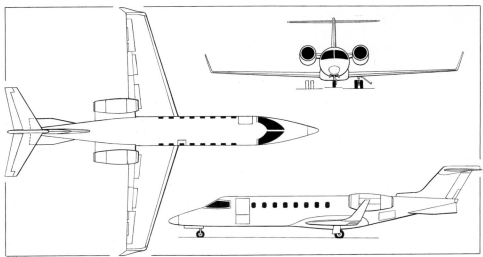

Learjet 45 business jet (two Honeywell TFE731-20 turbofans) *(Dennis Punnett/Jane's)*

OPERATIONAL NOISE LEVELS (FAR Pt 36):
T-O	81.9 EPNdB
Approach	92.8 EPNdB
Sideline	86.9 EPNdB

UPDATED

LEARJET 45

TYPE: Business jet.
PROGRAMME: Design started September 1992; unveiled at NBAA Convention 20 September 1992. Other members of Bombardier group are involved; Learjet is responsible for project co-ordination, final assembly, testing and certification.

Engine testing began January 1995 with TFE731-20 installed in one nacelle of Learjet 31A testbed. Wing and fuselage of first production aircraft (N45XL) mated at Wichita 4 November 1994; first flight 7 October 1995; second prototype (N452LJ) first flown 6 April 1996, assigned to flutter testing; third aircraft (N453LJ), first flown 24 April 1996, assigned to avionics testing; fourth aircraft assigned to HIRF and lightning-strike testing, engine testing and fuel system operation; fifth aircraft fitted with production interior and assigned to function and reliability testing, including interior noise measurement.

Initial FAA certification granted 22 September 1997 followed by full approval in May 1998; JAA certification achieved 3 July 1998; FAA RVSM approval granted 25 July 2000. First customer delivery (N903HC) 28 July 1998, to Hytrol Conveyor of Jonesboro, Arkansas. First delivery to Europe (N459LJ) 8 September 1998 for Eifel Holdings Ltd of Jersey. More than 50 delivered by February 2000, including 2,000th Learjet, N158PH, to Parker Hannifin Corporation in August 1999. 100th Learjet 45 delivered in October 2000, initially as Bombardier company demonstrator, but destined for a Philippine operator. Total fleet time stood at 25,000 hours by 1 October 2000.

CUSTOMERS: Over 155 ordered by May 1998 including 40, with 10 options, by JetSolutions for its FlexJets fractional ownership scheme. More than 100 in service by end of 2000, including 71 delivered in 2000.
COSTS: US$8.2 million (1999).
DESIGN FEATURES: Generally as Model 31; Learjet 45 designed to combine docile handling characteristics of 31/31A and 60 with exceptional fuel efficiency and good overall performance, and offer increased maintainability and reliability; new larger fuselage, wing and tail unit; increased head and shoulder room; wing carry-through spar recessed beneath floor; latest technology systems. Wing designed with NASA; winglets; sweepback 13° 25′ at 25 per cent chord.

Performance enhancement package announced at the Paris Air Show in June 1999 includes maximum T-O weight increased by 136 kg (300 lb); reductions in T-O speeds; improved nosewheel steering and removal of 40 kt (74 km/h; 46 mph) steering system limitation; improved brake-by-wire effectiveness; reconfiguration of flap selector module to permit use of 8° flap setting, instead of 20°, for approach climb in the event of a go-around; improved climb performance with bleed air anti-icing systems operating; and updated Honeywell avionics software.
FLYING CONTROLS: Conventional and manual. Two spoiler/lift dumper panels in each wing. Horn-balanced elevators. Trim tab in rudder; two in each aileron. Flaps.

Further improvements introduced in late 2000 include restyled seats providing greater freedom of movement in the cabin, a 10 to 12 dB reduction in cabin noise levels and improvements in the cabin air distribution system.
STRUCTURE: Unigraphics, CATIA and Computervision digital design systems adopted by Learjet, de Havilland of Canada and Shorts for engineering design. Short Brothers of UK manufactures the fuselage and de Havilland Canada the wings.
LANDING GEAR: Retractable tricycle type; semi-articulated trailing-link main legs retract inward, nose leg forward;

Learjet 45, twin-turbofan business jet   2001/0100450

Learjet 45 flight deck with Honeywell Primus 1000 integrated avionics system

Learjet 45 cabin

twin-wheel main units, size 22×5.75-12 (10 ply) with brake-by-wire anti-skid carbon multidisc brakes; nosewheel has dual-chine tyre, size 18×4.4 (10 ply), with steer-by-wire.

POWER PLANT: Two Honeywell TFE731-20 turbofans, each flat rated at 15.57 kN (3,500 lb st); Dee Howard target-type thrust reversers; digital electronic engine control. Total fuel capacity 3,392 litres (896 US gallons; 746 Imp gallons).

ACCOMMODATION: Two crew and up to nine passengers; eight-passenger cabin typically with PMP fully adjustable swivelling and reclining seats in double club arrangement, galley/refreshment centre and storage cabinet closet at front of cabin; lavatory, doubling as optional ninth seat, at rear; cabin pressurised, maximum differential 0.65 bar (9.43 lb/sq in); clamshell door with integral steps on port side at front of cabin, upper part serves as emergency exit; eight cabin windows per side, one forward of starboard wing leading-edge serving as emergency exit; externally accessible heated and lined baggage compartment, capacity 230 kg (507 lb), in aft fuselage accessed via door on port side beneath engine nacelle.

SYSTEMS: Honeywell air conditioning and pressurisation system, maximum differential 0.65 bar (9.4 lb/sq in), with dual independent digital control system and pneumatic redundancy; dual-zone automatic temperature control. Gaseous oxygen system, pressure 127.6 bars (1,850 lb/sq in). Main and auxiliary back-up hydraulic systems, pressure 207 bars (3,000 lb/sq in). Dual independent anti-icing and de-icing systems comprising bleed air anti-icing on wing and tailplane leading-edges and engine inlets, electric de-icing on pitot-static probes and electric anti-icing and de-fogging on windscreen. Honeywell RE 100 APU.

AVIONICS: Honeywell Primus 1000 integrated avionics system.
Comms: Dual Primus II nav/ident radios.
Radar: Primus 650 weather radar standard.
Flight: Dual Primus II nav radios. Primus 1000 digital autopilot/flight director standard; Honeywell traffic-alert and collision-avoidance (TCAS II) optional.
Instrumentation: Primus 1000 with EICAS; dual PFDs and MFDs; flight and navigation information displayed on four 203 × 178 mm (8 × 7 in) EFIS screeens; heart of system is IC-600 integrated avionics computer, which combines EFIS and EICAS processor.

DIMENSIONS, EXTERNAL:
Wing span	14.58 m (47 ft 10 in)
Wing aspect ratio	7.3
Length overall	17.81 m (58 ft 5 in)
Max diameter of fuselage	1.75 m (5 ft 9 in)
Height overall	4.37 m (14 ft 4 in)
Wheel track	2.85 m (9 ft 4¼ in)
Wheelbase	7.92 m (26 ft 0 in)

DIMENSIONS, INTERNAL:
Cabin:
Length: incl flight deck	7.54 m (24 ft 9 in)
excl flight deck	6.02 m (19 ft 9 in)
Max width	1.55 m (5 ft 1 in)
Max height	1.50 m (4 ft 11 in)
Floor area, excl flight deck	6.11 m² (65.8 sq ft)
Volume, excl flight deck	11.6 m³ (410 cu ft)
Baggage compartment volume	1.4 m³ (50 cu ft)

AREAS:
Wings, basic	28.95 m² (311.6 sq ft)

WEIGHTS AND LOADINGS:
Weight empty	5,693 kg (12,550 lb)
Basic operating weight empty	6,146 kg (13,550 lb)
Max payload	1,111 kg (2,450 lb)
Payload with max fuel	516 kg (1,138 lb)
Max fuel weight	2,750 kg (6,062 lb)
Max T-O weight	9,298 kg (20,500 lb)
Max ramp weight	9,412 kg (20,750 lb)
Max landing weight	8,709 kg (19,200 lb)
Max zero-fuel weight	7,257 kg (16,000 lb)
Max wing loading	321.2 kg/m² (65.79 lb/sq ft)
Max power loading	299 kg/kN (2.93 lb/lb st)

PERFORMANCE:
Max operating speed (V_{MO})	330 kt (611 km/h; 379 mph)
Max operating Mach No. (M_{MO})	0.81
High cruising speed	468 kt (867 km/h; 539 mph)
Normal cruising speed	457 kt (846 km/h; 526 mph)
Long-range cruising speed	425 kt (787 km/h; 489 mph)

Time to height, at T-O weight of 9,072 kg (20,000 lb), from S/L:
to 12,500 m (41,000 ft)	23 min
to 13,105 m (43 000 ft)	27 min
to 13,715 m (45,000 ft)	37 min
Max certified altitude	15,545 m (51,000 ft)
T-O field length	1,326 m (4,350 ft)
FAR Pt 91 landing distance	811 m (2,660 ft)

Range with two crew and four passengers:
VFR	2,380 n miles (4,407 km; 2,738 miles)
NBAA IFR reserves	2,120 n miles (3,926 km; 2,439 miles)

OPERATIONAL NOISE LEVELS:
T-O	75.1 EPNdB
Sideline	93.4 EPNdB
Approach	85.2 EPNdB

UPDATED

LEARJET 60

TYPE: Business jet.

PROGRAMME: Announced 3 October 1990 as Learjet 55C successor; first flight of proof-of-concept aircraft with one PW305 turbofan 18 October 1990; flight testing resumed 13 June 1991 with two PW305s and stretched fuselage (more than 300 hours flown by May 1992); first production aircraft first flight (N601LJ) 15 June 1992; certification awarded 15 January 1993; deliveries started immediately. Certified in Argentina, Austria, Bermuda, Brazil, Canada, China, Denmark, Germany, Grand Cayman, Malaysia, Mexico, the Philippines, South Africa, Switzerland, Turkey, the United Arab Emirates and USA. Production increased from two to three per month in 2000.

CUSTOMERS: Total of 22 delivered in 1994, 24 in 1995, 23 in 1996, 24 in 1997, 32 in 1998, 32 in 1999 and 35 in 2000. Recent customers include Publix Super Markets Inc, Canadian National Railway Company and Krispy Kreme Doughnut Corporation.

COSTS: US$11.4 million (1999).

DESIGN FEATURES: Largest Learjet; otherwise generally as for earlier versions.

FLYING CONTROLS: As Learjet 31A. Spoilers can be partially extended to adjust descent rates.

LANDING GEAR: Retractable tricycle type, hydraulically actuated and electrically controlled; single wheel on nose unit and twin wheels on main units; nosewheel retracts forward, mainwheels inward; steerable nosewheel (±60°). Mainwheel tyre size 17.5×5.75-8, pressure 14.76 bars (214 lb/sq in); nosewheel tyre size 18×4.4 (10 ply), pressure 7.24 bars (105 lb/sq in).

POWER PLANT: Two Pratt & Whitney Canada PW305A turbofans with FADEC, each flat rated at 20.46 kN (4,600 lb st) at up to 27°C (80°F). Fuel details under Weights and Loadings below. Single-point pressure refuelling system standard; gravity-feed fuel filler ports in each wing.

ACCOMMODATION: Two crew and six to nine passengers; gross pressure cabin volume 15.57 m³ (550 cu ft); compared with 55C, main cabin is 0.71 m (2 ft 4 in) longer and rear baggage hold section 0.38 m (1 ft 3 in) longer; full-across aft lavatory has flat floor, large mirror, coat closet and external servicing; total 1.67 m³ (59 cu ft) baggage capacity divided between an externally accessible hold (larger than that of Learjet 55C) and internal pressurised, heated compartment that is accessible in flight; galley cabinet has warming oven, cold liquid dispensers, ice compartment and storage for dinnerware; entertainment centre; 10-way adjusting seating is standard. New interior introduced in 1998 (from c/n 115), featuring redesigned passenger service unit, wider headliner, revised beverage cabinet, optional television monitor and wider choice of upholstery fabrics and leather trim.

SYSTEMS: Environmental control system uses engine bleed air for air conditioning and pressurisation, maximum differential 0.65 bars (9.4 lb/sq in), maintaining sea level cabin altitude to 7,835 m (25,700 ft) and 2,440 m (8,000 ft) cabin altitude to 15,545 m (51,000 ft). Hydraulic system, pressure 103.4 bars (1,500 lb/sq in), provided by two engine-driven, variable volume pumps, with electrically driven auxiliary pump. Electrical system comprises two 30 V DC 400 A engine-driven starter-generators and two 24 V DC batteries. Oxygen system, capacity 2.18 m³ (77.0 cu ft), with demand-type masks for crew and drop-down constant-flow masks for passengers.

Learjet 60 (Pratt & Whitney PW305 turbofans)

Bleed air anti-icing on wing leading-edges, engine inlets, engine spinners and inlet guide vanes, electric anti-icing on windscreen. Hamilton Sundstrand T-20G-10C3A APU.

AVIONICS: Standard fully integrated all-digital Rockwell Collins Pro Line 4.

Radar: Rockwell Collins WXR840 colour weather radar.

Flight: Four-tube EFIS, dual digital air data computers, dual navigation and communications radios, dual automatic AHRS, Rockwell Collins AMS-850 avionics management system, advanced autopilot and long-range navaid as standard; circuit breaker and controls panels redistributed, as in Learjet 31A.

DIMENSIONS, EXTERNAL:
Wing span	13.34 m (43 ft 9 in)
Wing chord: at root	2.74 m (9 ft 0 in)
at tip	1.12 m (3 ft 8 in)
Wing aspect ratio	7.2
Length: overall	17.88 m (58 ft 8 in)
fuselage	17.02 m (55 ft 10 in)
Fuselage max diameter	1.96 m (6 ft 5 in)
Height overall	4.47 m (14 ft 8 in)
Tailplane span	4.47 m (14 ft 8 in)
Wheel track	2.51 m (8 ft 3 in)
Wheelbase	7.72 m (25 ft 4 in)
Cabin door: Width	0.64 m (2 ft 1 in)
Height to sill	0.69 m (2 ft 3 in)

DIMENSIONS, INTERNAL:
Cabin: Length: incl flight deck	7.04 m (23 ft 1 in)
excl flight deck	5.38 m (17 ft 8 in)
Max width	1.80 m (5 ft 11 in)
Max height	1.73 m (5 ft 8 in)
Floor area, excl flight deck	6.00 m² (64.6 sq ft)
Volume, excl flight deck	12.8 m³ (453 cu ft)

AREAS:
Wings, gross	24.57 m² (264.5 sq ft)
Horizontal tail surfaces (total)	5.02 m² (54.00 sq ft)
Vertical tail surfaces (total)	4.67 m² (50.29 sq ft)

WEIGHTS AND LOADINGS:
Weight empty	6,282 kg (13,850 lb)
Basic operating weight empty	6,641 kg (14,640 lb)
Max payload	1,070 kg (2,360 lb)
Payload with max fuel	544 kg (1,200 lb)
Max fuel weight: total	3,588 kg (7,910 lb)
usable	3,062 kg (6,750 lb)
Max T-O weight	10,659 kg (23,500 lb)
Max ramp weight	10,773 kg (23,750 lb)
Max landing weight	8,845 kg (19,500 lb)
Max wing loading	433.8 kg/m² (88.85 lb/sq ft)
Max power loading	260 kg/kN (2.55 lb/lb st)

PERFORMANCE:
Max operating speed (V_{MO}):	
S/L–2,440 m (8,000 ft)	300 kt (555 km/h; 345 mph) IAS
2,440-6,100 m (8,000-20,000 ft)	340 kt (629 km/h; 391 mph) IAS
6,100-7,010 m (20,000-23,000 ft)	340-330 kt (630-611 km/h; 391-378 mph) IAS
7,010-8,155 m (23,000-26,750 ft)	330 kt (611 km/h; 378 mph) IAS
Max operating Mach No. (M_{MO}):	
8,155-11,280 m (26,750-37,000 ft)	0.81
11,280-13,105 m (37,000-43,000 ft)	0.81-0.78
above 13,105 m (43,000 ft)	0.78
Cruising speed: high	464 kt (859 km/h; 534 mph)
normal	453 kt (839 km/h; 521 mph)
Long-range cruising speed	422 kt (782 km/h; 486 mph)
Stalling speed, flaps and landing gear down	106 kt (197 km/h; 122 mph) CAS
Approach speed	139 kt (257 km/h; 160 mph) IAS
Max rate of climb at S/L	1,371 m (4,500 ft)/min
Rate of climb at S/L, OEI	378 m (1,240 ft)/min
Max certified altitude	15,545 m (51,000 ft)
Service ceiling, OEI	7,195 m (23,600 ft)
T-O field length	1,661 m (5,450 ft)
FAR Pt 91 landing distance	1,043 m (3,420 ft)
Range with two crew and four passengers:	
VFR	2,735 n miles (5,065 km; 3,147 miles)
NBAA IFR	2,499 n miles (4,628 km; 2,875 miles)

OPERATIONAL NOISE LEVELS:
T-O	78.8 EPNdB
Sideline	83.2 EPNdB
Approach	87.7 EPNdB

UPDATED

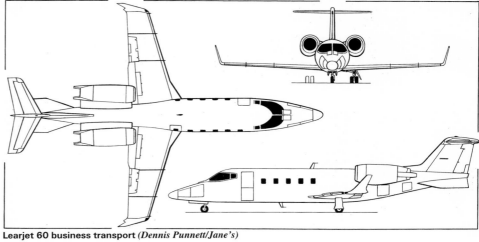
Learjet 60 business transport *(Dennis Punnett/Jane's)*

Learjet 60 flight deck

Learjet 60 cabin

LEZA-LOCKWOOD

LEZA-LOCKWOOD CORPORATION

1 Leza Drive, Sebring, Florida 33870
Tel: (+1 941) 655 42 42
Fax: (+1 941) 655 03 10
e-mail: aircam@ct.net
Web: http://www.lezalockwood.com
CEO: Antonio Leza
CHIEF DESIGNER: Phil Lockwood

Leza-Lockwood has a 3,900 m² (42,000 sq ft) factory at Sebring Airport and also produces the Super Drifter ultralight.

NEW ENTRY

LEZA-LOCKWOOD AIR CAM

TYPE: Two-seat kitbuilt twin.
PROGRAMME: Initially designed as camera platform for National Geographic magazine and first flew 1994. Redesigned as a kit and made public debut at Sun 'n' Fun 1996. First float-equipped version made initial flight 9 June 1999.

Leza-Lockwood Air Cam *(Paul Jackson/Jane's)*

CUSTOMERS: More than 100 sold by March 2000.
COSTS: US$49,975 including Rotax 582 engines; US$68,500 with Rotax 912 (2000).
DESIGN FEATURES: Configured for slow flight. Cabane-mounted, strut-braced, constant-chord wings and mid-mounted tailplane wire-braced to tall, narrow-chord fin.
FLYING CONTROLS: Conventional and manual. Electric flaps.
STRUCTURE: All-metal monocoque fuselage with glass fibre cockpit enclosure. Wings of 6061-T6 aluminium tubing with aluminium spars and ribs, and fibric covering. Tailfin has glass fibre fillet.
LANDING GEAR: Tailwheel type; fixed. Spring steel mainwheel legs; hydraulic disc brakes and 6.00-6 tyres. Maule steerable tailwheel. Full Lotus floats optional.
POWER PLANT: Two 47.8 kW (64.1 hp) Rotax 582s mounted above wings, driving three-blade pusher propellers. Optionally two 73.5 kW (98.6 hp) Rotax 912 ULSs or two 85.8 kW (115 hp) turbocharged Rotax 914s can be fitted. Fuel in two wing-mounted aluminium tanks, total capacity 106 litres (28.0 US gallons; 23.3 Imp gallons).

ACCOMMODATION: Pilot and passenger/observer in tandem seats within open cockpit. Dual controls. Open baggage compartment behind second seat.
AVIONICS: To customer's specification.
EQUIPMENT: Fire extinguisher standard.
DIMENSIONS, EXTERNAL:
Wing span 11.07 m (36 ft 4 in)
Wing aspect ratio 6.5
Length overall 8.23 m (27 ft 0 in)
Height overall 2.54 m (8 ft 4 in)
Tailplane span 3.96 m (13 ft 0 in)
Wheel track 2.59 m (8 ft 6 in)
Min distance between propeller blade tips 0.15 m (6 in)
DIMENSIONS, INTERNAL:
Baggage compartment: Length 1.07 m (3 ft 6 in)
Max width 0.63 m (2 ft 0¾ in)
AREAS:
Wings, gross 18.95 m² (204.0 sq ft)
WEIGHTS AND LOADINGS:
Weight empty 433 kg (955 lb)
Max T-O weight 680 kg (1,500 lb)
Max wing loading 35.90 kg/m² (7.35 lb/sq ft)
Max power loading (Rotax 582) 7.13 kg/kW (11.72 lb/hp)
PERFORMANCE (Rotax 582):
Never-exceed speed (V_{NE}) 95 kt (177 km/h; 110 mph)
Max operating speed 87 kt (161 km/h; 100 mph)
Normal cruising speed 43-70 kt (80-129 km/h; 50-80 mph)
Stalling speed, power off, flaps down 33 kt (60 km/h; 37 mph)
Max rate of climb at S/L 488 m (1,600 ft)/min
Rate of climb at S/L, OEI 61 m (200 ft)/min
Service ceiling 5,485 m (18,000 ft)
T-O run less than 61 m (200 ft)
Landing run 61 m (200 ft)
Range at 57 kt (106 km/h; 66 mph) 212 n miles (392 km; 244 miles)
Endurance 4 h 0 min

NEW ENTRY

LIBERTY

LIBERTY AERONAUTICAL
42650 State Road, 64 East Myakka City, Florida 34251
Tel: (+1 941) 322 88 86
e-mail: rberstfly@aol.com
PRESIDENT: Richard C Berstling

MARKETING:
Bellaire Monoplane Company, 144 Pinetree Drive, Lake Placid, Florida 33852
Tel: (+1 863) 699 68 38

UPDATED

LIBERTY BELLAIRE
TYPE: Side-by-side ultralight kitbuilt.
PROGRAMME: Launched as family of ultralights with high commonality of components.
CURRENT VERSIONS: **Bellaire ST:** Single-seat high-wing version; 35.8 kW (48 hp) Rotax 503 engine.
Bellaire AL: Single-seat mid-wing version.
Bellaire SE: Two-seat high-wing version.
Following description applies to Bellaire SE.
DESIGN FEATURES: Traditional high-wing design with V-strut bracing.
FLYING CONTROLS: Conventional and manual.
STRUCTURE: Welded 4130 chromoly steel fuselage and tail surfaces; wooden wings; fabric covering.
LANDING GEAR: Tailwheel type; fixed; Cleveland wheels and brakes; speed fairings on mainwheels.
POWER PLANT: One 59.6 kW (79.9 hp) Rotax 912 UL flat-four piston engine. Fuel capacity 53 litres (14.0 US gallons; 11.7 Imp gallons).
DIMENSIONS, EXTERNAL:
Wing span 9.14 m (30 ft 0 in)
Length overall 6.40 m (21 ft 0 in)
Height overall 1.75 m (5 ft 9 in)
WEIGHTS AND LOADINGS:
Weight empty 245 kg (540 lb)
Max T-O weight 476 kg (1,050 lb)
PERFORMANCE:
Max level speed 104 kt (193 km/h; 120 mph)
Cruising speed 83 kt (153 km/h; 95 mph)
Stalling speed 31 kt (57 km/h; 35 mph)
Max rate of climb at S/L 290 m (950 ft)/min
T-O run 53 m (175 ft)
Landing run 76 m (250 ft)
Range with max fuel 295 n miles (547 km; 340 miles)

UPDATED

Liberty Bellaire ultralight kitplane *(Geoffrey P Jones)* 1999/0051772

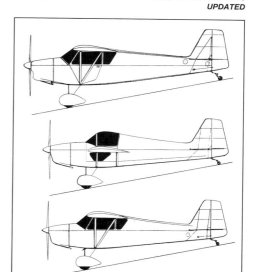

Profiles of the Liberty Bellaire ST (top), AL and SE 1999/0051773

LITTLE WING

LITTLE WING AUTOGYROS INC
746 Highway 89 North, Mayflower, Arkansas 72106
Tel: (+1 501) 470 74 44
Fax: (+1 501) 470 74 87
e-mail: rotopup@aol.com
PRESIDENT: Ronald Herron

The company is developing a series of autogyros based on the Piper Cub Special fuselage.

VERIFIED

LITTLE WING ROTO-PUP
TYPE: Single-/two-seat autogyro kitbuilt.
PROGRAMME: Initial design work began in 1980 and construction of LW-1 prototype (N45LW), using a Piper PA-11 fuselage, started December 1990; this made its first flight 21 October 1994. LW-2, scaled-down version of LW-1, first flew April 1995; LW-3, controlled by full universal tilting of rotor head, first flew (N46LW) 22 September 1996. Certified by FAA in Experimental category.
CURRENT VERSIONS: **Roto-Pup LW-2:** Controlled by elevator, rudder and laterally tilting rotor head.
Roto-Pup LW-2+2: Tandem two-seat version. One flying by end 1999.
Roto-Pup LW-3: Controlled by rudder and universally tilting rotor head. Single seat.
Description applies to LW-3, unless otherwise stated.
Roto-Pup Ultralight: First flew April 1995.
CUSTOMERS: At present, four prototypes are undertaking flight test programme, which will be followed by first customer sales. Nine sets of plans also sold.

COSTS: LW-3 kit approximately US$10,000; LW-2+2 US$15,000; Ultralight US$8,995 including engine, rotor system and instruments (1999).
DESIGN FEATURES: Fuselage based on Piper PA-11 Cub Special, giving classic autogyro appearance at low development cost. Two-blade Bensen 603 rotor; each blade has extruded aluminium spar with riveted and bonded skin, NACA 2300 aerofoil and no twist. Rotor speed 390 rpm, blade tip speed 140 m (459 ft)/s. Pitch link to airframe is isolated by elastomeric dampers, determining rotor axis inclination relative to airframe. Rotor brake standard on LW-3, optional on LW-2.
FLYING CONTROLS: Floor-mounted control stick; cable and pulley actuated elevator and rudder; pushrod for lateral tilting of rotor axis. Ground-adjustable horizontal stabiliser for airframe pitch; tip shields at ends of horizontal stabilisers.
STRUCTURE: Full chromoly 4130 steel tube welded fuselage and rotor pylon; fuselage covered with Dacron fabric and polyurethane finish. Hub structure of 2024-T3 aluminium bar with bolt attachments, pivoted at rotor attachment point for ground adjustment of blade pitch.
LANDING GEAR: Non-retractable type with tailwheel. Mainwheels Hegar 6.50-6 with 1.03 bars (15 lb/sq in) tyre pressure. Matco 3.00-2 steerable tailwheel of solid rubber. Internal expanding go-kart style brakes by Leaf. Floats optional.
POWER PLANT: *LW-3:* One 52.2 kW (70 hp) TEC converted Volkswagen four-stroke engine with dual ignition; options include 2si 52.2 kW (70 hp) two-stroke water-cooled engine and various Rotax, Subaru and Hirth engines. Fuel capacity 38 litres (10.0 US gallons; 8.3 Imp gallons) in seat tank. Refuelling position on starboard side of cabin. Oil capacity 2.4 litres (5.0 US pints; 4.2 Imp pints).

LW-2+2: One 82.0 kW (110 hp) Hirth F30; optional power plants in the range 67.1 to 112 kW (90 to 150 hp) can be fitted. Fuel capacity 45 litres (12.0 US gallons; 10.0 Imp gallons).
ACCOMMODATION: One or two occupants, according to version, in enclosed cockpit; forward-opening door on starboard side. Baggage compartment behind single seat. Lexan polycarbonate windscreen and side windows. Tandem two-seat model under development.
SYSTEMS: 1,000 A 12 V battery for ignition back-up and rotor pre-spin.
DIMENSIONS, EXTERNAL:
Rotor diameter: LW-2/3 7.01 m (23 ft 0 in)
LW-2+2 8.53 m (28 ft 0 in)
Rotor blade chord 0.18 m (7 in)
Fuselage length 5.49 m (18 ft 0 in)
Height to top of rotor head 2.59 m (8 ft 6 in)
Tail unit span 2.13 m (7 ft 0 in)
Wheel track 2.13 m (7 ft 0 in)
Wheelbase 3.96 m (13 ft 0 in)
Propeller diameter: VW engine 1.57 m (5 ft 2 in)
2si engine 1.73 m (5 ft 8 in)
DIMENSIONS, INTERNAL:
Cabin: Length 1.40 m (4 ft 7 in)
Max width 0.57 m (1 ft 10½ in)
Max height 1.02 m (3 ft 4 in)
AREAS:
Rotor blades (each) 0.54 m² (5.83 sq ft)
Rotor disc: LW-3 38.61 m² (415.6 sq ft)
LW-2+2 57.21 m² (615.8 sq ft)
Dorsal fin 0.42 m² (4.52 sq ft)
Auxiliary tip plates (each) 0.065 m² (0.70 sq ft)
Rudder 0.28 m² (3.03 sq ft)
Tailplane 0.92 m² (9.94 sq ft)
Elevators (total) 0.89 m² (9.58 sq ft)

676 USA: AIRCRAFT—LITTLE WING to LOCKHEED MARTIN

Little Wing LW-1 prototype 0016181

Little Wing LW-3 Roto-Pup 0016180

WEIGHTS AND LOADINGS:
Weight empty: LW-2 160 kg (352 lb)
    LW-2+2 215 kg (475 lb)
    LW-3 205 kg (452 lb)
Baggage capacity 11 kg (25 lb)
Max fuel weight 24 kg (54 lb)
Max T-O weight: LW-2 340 kg (750 lb)
    LW-2+2 476 kg (1,050 lb)
Max disc loading: LW-2+2 8.33 kg/m² (1.71 lb/sq ft)
PERFORMANCE:
Never-exceed speed (V$_{NE}$) 86 kt (160 km/h; 100 mph)

Max operating speed at S/L:
    LW-3 70 kt (129 km/h; 80 mph)
    LW-2+2 78 kt (145 km/h; 90 mph)
Econ cruising speed at 305 m (1,000 ft):
    LW-3 52 kt (97 km/h; 60 mph)
    LW-2+2 65 kt (121 km/h; 75 mph)
Touchdown speed for power-off landing
    9-13 kt (16-24 km/h; 10-15 mph)
Max rate of climb at S/L: LW-3 305 m (1,000 ft)/min
    LW-2+2 183 m (600 ft)/min
Service ceiling 3,050 m (10,000 ft)

T-O run: LW-3 61 m (200 ft)
    LW-2+2 92 m (300 ft)
Landing run: LW-3 15 m (50 ft)
    LW-2+2 3-6 m (10-20 ft)
Range: LW-3 86 n miles (160 km; 100 miles)
    LW-2+2 130 n miles (241 km; 150 miles)
UPDATED

LOCKHEED MARTIN

LOCKHEED MARTIN CORPORATION
6801 Rockledge Drive, Bethesda, Maryland 20817
Tel: (+1 301) 897 60 00
Fax: (+1 301) 897 60 28
Web: http://www.lmco.com
CHAIRMAN AND CEO: Vance D Coffman
PRESIDENT AND COO: Robert J Stevens
CHIEF FINANCIAL OFFICER: Robert J Stevens
EXECUTIVE VICE-PRESIDENT, AERONAUTICS COMPANY:
  Dain M Hancock
EXECUTIVE VICE-PRESIDENT, SPACE SYSTEMS:
  Albert E Smith
EXECUTIVE VICE-PRESIDENT, SYSTEMS INTEGRATION:
  Robert B Coutts
EXECUTIVE VICE-PRESIDENT, TECHNOLOGY SERVICES:
  Michael F Camardo
VICE-PRESIDENT, CORPORATE COMMUNICATIONS: Dennis Boxx

Former Lockheed Aircraft Corporation renamed Lockheed Corporation in September 1977. Merger with Martin Marietta announced 30 August 1994 and completed 15 March 1995. Further expansion resulted from the acquisition of Loral Corporation's defence electronics and systems integration businesses in April 1996 for approximately US$9.1 billion. Workforce total of about 149,000 at end of 1999, with sales of approximately US$25.5 billion and net earnings of US$382 million in 1999. Activities include design and production of aircraft, electronics, satellites, space systems, missiles, ocean systems, information systems, and systems for strategic defence and for command, control, communications and intelligence. Is also largest US DoE contractor.

Company reported mounting losses in some sectors during 1999 and predicted reduced earnings for 1999 and 2000; efforts to improve financial standing have included divestiture of some non-core businesses during 2000, with most recent instance concerning sale of Aerospace Electronics Systems business to BAE Systems North America for US$1.67 billion.

Following major strategic and organisational review, revised corporate structure came into being in early 2000; under this, four major business groups of Lockheed Martin are:
Aeronautics Company (see next entry)
Space Systems, comprising:
  Astronautics Operations
  Michoud Operations
  Missiles & Space Operations
    Communications and Power Center
    Advanced Technology Center
    Commercial Satellite Center
  Lockheed Martin International Launch Services (joint venture with Khrunichev Energia International)
  Space Imaging Inc (joint venture with Raytheon)
  United Space Alliance (joint venture with Boeing)
Systems Integration, comprising:
  Lockheed Martin Air Traffic Management
  Lockheed Martin Electronics Platform Integration
    Lockheed Martin Electronic Systems – UK
    Lockheed Martin Electronics Platform Integration – Owego, NY
  Lockheed Martin Canada
  Lockheed Martin Postal Systems
  Lockheed Martin Information Systems
  Lockheed Martin Management & Data Systems
  Lockheed Martin Management & Data Systems – Southwest Region
  Lockheed Martin Management & Data Systems – Western Region
  Lockheed Martin Missiles and Fire Control – Orlando
  Lockheed Martin Missiles and Fire Control – Dallas
  Lockheed Martin Mission Systems
  Naval Electronics & Surveillance Systems
    Lockheed Martin Naval Electronics & Surveillance Systems – Manassas
    Lockheed Martin Naval Electronics & Surveillance Systems – Moorestown
    Lockheed Martin Naval Electronics & Surveillance Systems – Syracuse
  Perry Technologies
  Lockheed Martin Tactical Defense Systems – Akron
  Lockheed Martin Tactical Defense Systems – Eagan
Technology Services, comprising:
  Lockheed Martin Aircraft & Logistics Centers
  Lockheed Martin Energy Research Corporation
  Lockheed Martin Energy Systems Inc
  Lockheed Martin Technology Services
    KAPL Inc
  Lockheed Martin Information Support Services
  Lockheed Martin Space Operations
  Lockheed Martin Support & Training Services
  Lockheed Martin Technical Operations Inc
Sandia Corporation
Technology Ventures Corporation
UPDATED

LOCKHEED MARTIN AERONAUTICS COMPANY (LM Aero)
1 Lockheed Boulevard, Fort Worth, Texas 76108
Tel: (+1 817) 777 20 00
PRESIDENT AND COO: Dain M Hancock
EXECUTIVE VICE-PRESIDENT FOR CUSTOMER REQUIREMENTS:
  Tom Burbage
EXECUTIVE VICE-PRESIDENT FOR PROGRAMS: Bob Elrod
EXECUTIVE VICE-PRESIDENT FOR FINANCE: John McCarthy
EXECUTIVE VICE-PRESIDENT FOR OPERATIONS: Ralph Heath

LM Aero includes the following operating units:
Marietta Operations
  Next entry
Palmdale Operations
  Follows Marietta Operations
Fort Worth Operations
  Follows Palmdale Operations

Before merger that culminated in creation of Lockheed Martin Aeronautics Sector (subsequently renamed Lockheed Martin Aeronautical Systems Division Area and, since early 2000, Lockheed Martin Aeronautics Company), Lockheed aircraft manufacturing activity consolidated at Marietta, Georgia in 1991, with move of P-3 Orion assembly line from Palmdale; Orion now out of production, but established work on C-130 programme, plus F-22 development, continues at Marietta under management of Marietta Operations, which is also co-operating with Alenia of Italy on C-27J tactical airlift project (see Alenia entry in Italian section). Teaming arrangement agreed with Northrop Grumman to collaborate in development and marketing of AEW aircraft; also with Samsung Aerospace to develop the KTX-2 advanced trainer/light combat aircraft (see entry in Korea, South section); future co-operation with France has been suggested on projects such as the Airbus Industrie A3XX and a new tanker aircraft to replace USAF KC-135s; approach made to Airbus Military Company in 1997, suggesting transatlantic airlifter programme in event that A400M is cancelled.

LM Aero grouping has workforce of about 30,000.
VERIFIED

MARIETTA OPERATIONS
86 South Cobb Drive, Marietta, Georgia 30063-0264
Tel: (+1 770) 494 44 11
Fax: (+1 770) 494 76 56
Web: http://www.lmasc.com
MANAGER, MEDIA RELATIONS: Sam Grizzle

In April 1991, Lockheed won competition to produce F-22 with General Dynamics (now Lockheed Martin Tactical Aircraft Systems) and (then) Boeing Military Airplanes. Marietta Operations (formerly Lockheed Martin Aeronautical Systems – LMAS) also involved in studies into advanced mobility aircraft for 21st century strategic transport; was subcontractor to Boeing on NASA High-Speed Civil Transport (HSCT) programme; working with NASA on advanced subsonic technology transport programme; and engaged in other, classified projects.

Further long-term activities at Marietta include production of C-130J Hercules and F-22 Raptor aircraft. LMAS signed joint venture agreement with Alenia of Italy in September 1996 concerning the development and marketing of the C-27J tactical transport (see entry in Italian section) that incorporates systems developed for the C-130J Hercules.

Marietta Operations working to develop standard avionics suite for future versions (upgrades) of P-3 Orion and AEW aircraft; open architecture design for maximum flexibility and system growth; elements include AN/APS-145 radar, GPS and communications/navigation system. Also interest from several countries on possible new-build aircraft.

In 1999, secured contract worth up to US$450 million for C-5 Avionics Modernization Program (AMP), which will entail integration of advanced flight control and com/nav system as well as installation of new instrument displays on fleet-wide basis; C-5 Reliability Enhancement and Re-engining Program (RERP) contract worth US$2.6 billion will be awarded by USAF in January 2001; General Electric CF6-80C2 turbofan selected in preference to bids from Rolls-Royce and Pratt & Whitney.

Aeronautical Systems Support, with headquarters in Smyrna, Georgia, is subordinate element, with responsibility for after sales support, including provision of spare parts and technical back-up for variety of aircraft types.

Marietta workforce approximately 8,500 at end of 1999.

UPDATED

LOCKHEED MARTIN (645) F-22 RAPTOR

TYPE: Air superiority fighter.

PROGRAMME: US Air Force Advanced Tactical Fighter (ATF) requirement called for 750 McDonnell Douglas F-15 Eagle replacements incorporating low observables technology and supercruise (supersonic cruise without afterburning); parallel assessment of two new power plants; request for information issued 1981; concept definition studies awarded September 1983 to Boeing, General Dynamics, Grumman, McDonnell Douglas, Northrop and Rockwell; requests for proposals issued September 1985; submissions received by 28 July 1986; USAF selection announced 31 October 1986 of demonstration/validation phase contractors: Lockheed YF-22 and Northrop YF-23 (see 1991-92 *Jane's*); each produced two prototypes and ground-based avionics testbed; first flights of all four prototypes 1990. Competing engine demonstration/validation programmes launched September 1983; ground testing began 1986-87; flight-capable Pratt & Whitney YF119s and General Electric YF120s ordered early 1988; all four aircraft/engine combinations flown.

Decision of 11 October 1989 extended evaluation phase by six months; draft request for engineering and manufacturing development (EMD) proposals issued April 1990; first artists' impressions released May 1990; Lockheed teamed with General Dynamics (Fort Worth) and Boeing Military Airplanes to produce two YF-22 prototypes, which first flew on 29 September and 30 October 1990. Details of flight testing, configuration and dimensions appeared in the 1996-97 and previous editions. Second aircraft placed in USAF Museum, Wright Patterson AFB, 31 March 1998.

Acquisition of former General Dynamics gave Lockheed Martin control of 67.5 per cent of programme; this involves 1,150 suppliers in 44 US states. Final engineering and manufacturing development (EMD) requests issued for both weapon system and engine 1 November 1990; proposals submitted 2 January 1991; F-22 and F119 power plant announced by USAF as winning combination, 23 April 1991; EMD contract given 2 August 1991 for 11 (since reduced to nine) flying prototypes (including two tandem-seat F-22Bs which will now be completed as F-22A single-seat aircraft), plus one static and one fatigue test airframes; design underwent several detail refinements through early 1990s, immediately previous layout being Configuration 644.

Combat roles reassessment of May 1993 added air-to-ground attack with precision-guided munitions (PGMs) to F-22's roles. Under US$6.5 million contract addition on 25 May 1993, main weapon bay and avionics to be adapted for delivery of AIM-9X missile and 454 kg (1,000 lb) GBU-32 Joint Direct Attack Munition (JDAM); two JDAMs will replace two AIM-120 AMRAAMs in main weapon bay; required provision for two AGM-137A Tri-Service Standoff Attack Missiles (TSSAM) on underwing pylons later cancelled, as was TSSAM programme. Decision of April 1995 permitted 3 per cent weight increase as consequence of Critical Design Review completed two months previously; new limit is 14,365 kg (31,670 lb) including uncertainty allowance of 227 kg (500 lb).

Target first flight date delayed some 12 months, until May 1997, due to three consecutive fiscal year budget cuts; preliminary design review, covering all aspects of the design, completed 30 April 1993; critical design review undertaken February 1995; preproduction verification (PPV) batch of four aircraft was scheduled to be ordered 1997 but these deleted from overall programme at start of that year; long lead items for first production batch (two aircraft), subsequently reclassified as Production Representative Test Vehicles (PRTV), ordered in 1998, with full funding in FY99. Minor design changes for production F-22 announced July 1991. Suggested name of SuperStar rejected in 1991 and F-22 remained unnamed until occasion of roll-out in April 1997 when it was announced that the name Raptor had been chosen.

Fabrication of first component for first EMD aircraft (91-4001, c/n 4001) began 8 December 1993 at Boeing's facility in Kent, Washington; assembly of forward fuselage launched at Marietta on 2 November 1995 with start of work on nose landing gear well; assembly work also begun at Fort Worth in mid-1995 with mating of three assemblies that comprise the mid-fuselage of first EMD aircraft taking place in early 1996, followed by road transfer of entire section to Marietta in September 1996 for start of final assembly. Delivery by Boeing to Lockheed Martin at Marietta of subassemblies, including wing and aft fuselage, for first EMD aircraft took place on schedule in third quarter of 1996. Pratt & Whitney also delivered first F119 flight test power plant in September. Prototype rolled out 9 April 1997; planned May 1997 first flight not achieved due to several delays attributed to fuel leaks and other hardware-related anomalies. First flight accomplished 7 September 1997, with aircraft airborne for 58 minutes during which initial handling evaluation undertaken before landing gear was retracted and further handling assessment performed at speeds of up to 250 kt (463 km/h; 288 mph); F-22 also completed simulated powered approach at medium altitude before landing at Marietta. Second sortie of 35 minutes took place on 14 September 1997, after which aircraft underwent minor structural modifications and was then placed in structural test fixture for load ground tests and strain gauge calibration.

Low-rate initial production (LRIP) decision originally dependent upon accumulating 183 hours of flight testing; this milestone passed on 23 November 1998 and cleared way for release of US$195.5 million in late December 1998 for advance procurement (long lead items) for six Lot 1 LRIP aircraft, which subsequently reclassified as PRTV 2 aircraft after wrangle over funding in mid-1999. Earlier, in December 1998, Lockheed Martin received contract worth US$503 million for two PRTVs and associated programme support. First production delivery expected in first quarter of 2002; high-rate production decision due August 2003 with IOC by first unit of 32 aircraft scheduled for December 2005.

Wind tunnel testing occupied 19,195 hours up to YF-22 stage and a further 16,930 hours up to mid-1995, when Configuration 645 was finalised; six major categories were investigated (aerodynamic loads and weapons bay acoustics, inlet and engine compatibility, mission/manoeuvre performance, inlet icing, stability and control flying qualities, weapons and stores separation), with last-named comprising majority of 900 hours that remained to be done in 1995-97.

Avionics trials from November 1997 in a Boeing 757 (N757A) Flying Test Bed (FTB) with AN/APG-77 radar in F-22-type nosecone; block 1 software, permitting basic radar operation including simultaneous search-and-track modes, delivered to Boeing, April 1998 and subsequently tested on 757; block 2 software delivered on 7 December 1998, with block 3S beginning flight testing on 24 April 2000; this is early version of block 3.0 software, which

First two EMD F-22As during trials at Edwards AFB; 4001 (upper) was retired from flying in 2000

Second prototype Lockheed Martin F-22A EMD aircraft undertaking high AoA trials with weapon bay doors open

678   USA: AIRCRAFT—LOCKHEED MARTIN

Maiden Flight of 4003, the third Lockheed Martin F-22A Raptor

delivered to Seattle on 11 August 2000 in readiness for airborne trials programme that began in mid-September. Communications/navigation/identification (CNI) system and EW suite being tower-tested on full-scale model of forward fuselage at Fort Worth, Texas, during 1998-99; subsequently tested on Boeing 757 FTB, which is fitted with three common integrated processors (CIPs) for 1,400 hour software flight test programme. Further modification of Boeing 757 to mount representative F-22 wing section above forward fuselage was completed in late 1998, and flight testing of conformal antennas began on 11 March 1999; by December 2000, 757 FTB had accumulated 641.9 flight test hours in 126 sorties. Radar testing has also been accomplished using T-39 Sabreliner as target aircraft and has involved a Lockheed T-33 for calibrated airborne trials since December 1998.

Milestones in 1999 included delivery of first AN/ALR-94 EW system on 15 February by Sanders to the Avionics Integration Laboratory in Seattle; 100th sortie in May; successful completion of design load limit testing of article 3999 on 25 September; and compliance with all five major 1999 flight test objectives (including flight at altitude of 50,000 ft; opening of side and main weapon bay doors in flight; supercruise demonstration; and flutter envelope expansion) by 24 September. Future of programme more secure in near-term, after House of Representatives' July 1999 vote to withhold almost US$2 billion of funds for six Lot 1 aircraft; eventual compromise resulted in funding cut of US$500 million in FY00, with initial six LRIP aircraft being redesignated as Production Representative Test Vehicles (PRTV). Clearance to begin low-rate initial production now dependent on successful completion of first flight of F-22 with Block 3.0 software, but opponents in Congress are expected to attempt to impose financial cuts in FY01 and FY02.

Flight test programme interrupted at least twice during 2000, with most serious occurrence arising in May after discovery of hairline cracks in canopies of first two aircraft; grounding order lifted on 5 June, when second EMD F-22 resumed testing with some restrictions while awaiting replacement unit. Major event of year was expected to be Pentagon Defense Acquisition Board review to culminate in award of contract for initial batch of LRIP aircraft; this was scheduled for 21 December, but slipped to 3 January and was further delayed by poor weather that prevented remaining three out of 11 critical test objectives being achieved; funding release dependent upon compliance with several 'exit criteria', including first flight of F-22 with block 3.0 software, first AMRAAM launch and first flights of aircraft 4004, 4005 and 4006. By early November 2000, only six of 15 objectives achieved, prompting changes to some test plans. First AMRAAM launch (of 60 planned) achieved 24 October 2000; static tests completed 28 December 2000.

CURRENT VERSIONS (specific): Test programme began with first flight of initial EMD aircraft; extends over five years, involving total of nine EMD aircraft, plus two non-flying test articles. Airframe 3999 for static loads testing built between second and third flying F-22s, and airframe 4000 for fatigue testing built between third and fourth flying F-22s. Final assembly of 3999 began in July 1998; following completion in January 1999, it was transferred to the structural test facility at Marietta and began load testing in March 1999. Static testing was expected to have been completed in November 2000, with fatigue testing concluding in February 2001.

Clear division of test assignments will result in first three aircraft (4001-03) being allocated to airframe structure evaluations, with remaining six concentrating on avionics test taskings. Use of separate instrumentation configurations for airframe and avionics-dedicated test articles provides back-up for almost every F-22 aircraft and offers potential to switch missions between test fleet if necessary. Tasks allocated to EMD aircraft (numbered 4001 onwards) are:

4001/91-4001: 'Spirit of America'. Structural assembly of forward fuselage section completed at Marietta by July 1996, with installation of wiring, tubing, instrumentation and avionics racks having begun. Mid-fuselage section received from Fort Worth in September 1996, with aft section airlifted from Boeing in Seattle on 18 October, after which mating occurred. Wing assemblies completed by Boeing and shipped to Marietta on 9 November, with mating to fuselage following two days later.

Rolled out on 9 April 1997 at Lockheed Martin's Marietta, Georgia, facility and made first flight on 7 September 1997. Following completion of structural ground tests, disassembled and airlifted to Edwards AFB on 5 February 1998; used for evaluation of flying qualities, flutter and loads characteristics. First flight from Edwards made by 411th Test Squadron, 412th Test Wing on 17 May 1998; first aerial refuelling (KC-135 tanker) 30 July 1998; exceeded M1.0 for first time on 10 October 1998 and by 30 October 2000, had recorded 173 flights (367.6 hours) during which it expanded the flutter and handling qualities envelope and also demonstrated supercruise at M1.5 without use of afterburner. Flown to Wright-Patterson AFB, Ohio, on 2 November 2000 and formally retired from flight test duty; 175 flights and 372.7 hours; stripped of useful components and used for live-fire testing, involving exploding shells and missile fragments, during 2001.

Horizontal stabilisers replaced in January 1999, following discovery of 'disbonding' between core and skin materials arising from mismatch in thermal coefficients of expansion; new stabilisers (also fitted to aircraft 4002 in March 1999) use materials that are more thermally compatible, with longer-term solution concerning development of third stabiliser design that will be fitted to remainder of the EMD fleet as well as static and fatigue test articles. Shortly before being retired from test flying, same aircraft used to investigate unpredicted tail buffet and vibration; this involved study of airflow patterns impinging on fin surfaces, although manufacturer confident that structural changes unlikely to be required, as first two EMD aircraft completed to less robust configuration. Rudimentary avionics were installed, with Block 0 software; 89 k avionics source lines of code (SLOC) from total 272 k air vehicle SLOC; basic functions only, including Tacan. Software package for first operational flight programme (OFP) electronically transmitted via secure link from Fort Worth to Marietta in early February 1997; initial release included 275,000 lines of software code; additional 1.45 million lines to be developed during EMD process for F-22 avionics systems.

4002/91-4002: 'Old Reliable'. Rolled out 10 February 1998; maiden flight on 29 June 1998 at Marietta; flew to Edwards AFB on 26 August and by end October 1998 had made 27 flights (66.1 hours), expanding flutter and handling qualities envelope, including 26° AoA. By 1 December 2000, had completed 179 flights (including nine from Marietta and ferry flight to Edwards) and accumulated 421.0 hours. Used to launch first AIM-9M Sidewinder on 25 July 2000, followed by first AIM-120C AMRAAM on 24 October and had previously been employed for fit checks and captive-carry trials with AGM-88 HARM in April 1999. Tasks include performance assessment (propulsion; high AoA) plus stores separation and jettison as well as some electronic warfare and IR signature evaluations.

4003/91-4003: First Block 2 aircraft, with internal structure fully representative of production version; rolled out at Marietta on 25 May 1999; first engine runs in October 1999; taxi trials completed at beginning of March 2000, with first flight following on 6 March and delivery to

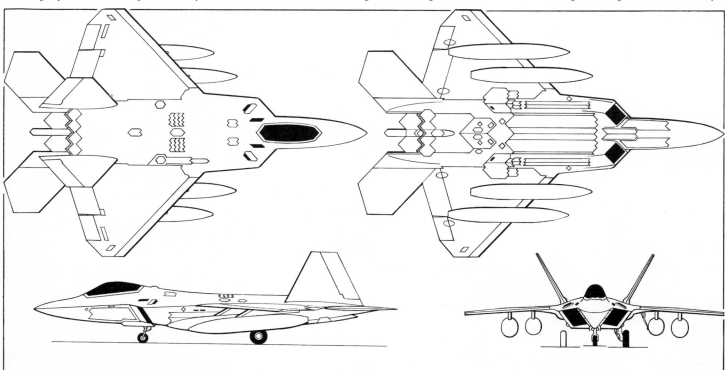

Production (Configuration 645) Lockheed Martin F-22A in ferry configuration (*James Goulding/Jane's*)

Jane's All the World's Aircraft 2001-2002

LOCKHEED MARTIN—AIRCRAFT: USA

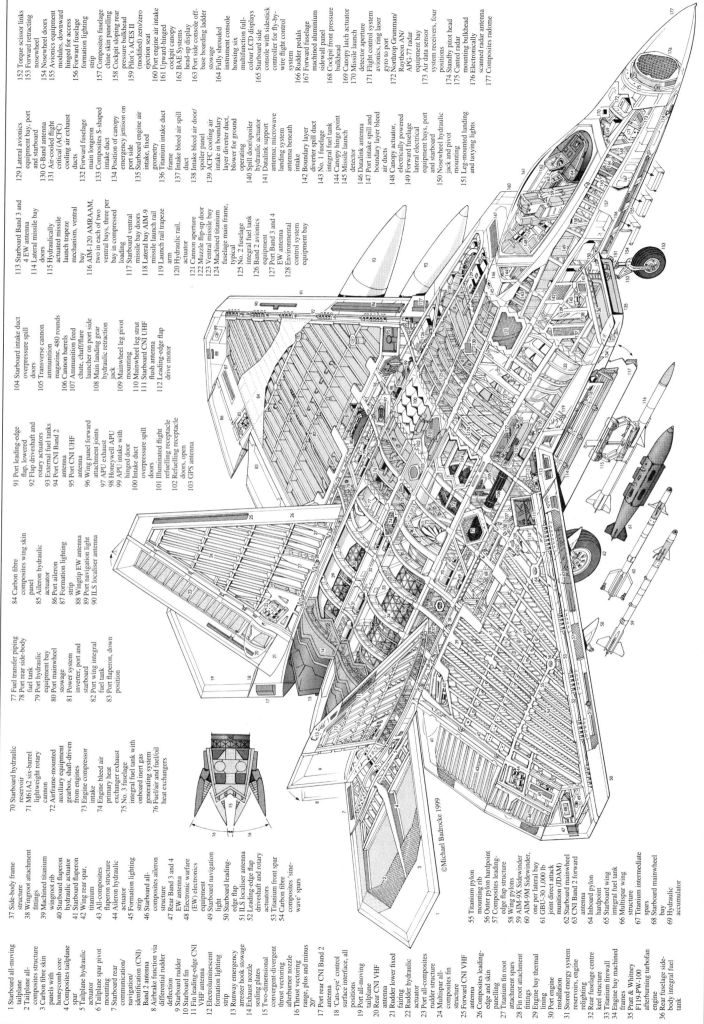

Lockheed Martin F-22 Raptor, cutaway drawing key

1. Starboard all-moving tailplane
2. Tailplane all-composites structure
3. Carbon fibre skin panels with honeycomb core
4. Composites tailplane spar
5. Tailplane hydraulic actuator
6. Tailplane spar pivot mounting
7. Starboard rear communication/navigation (CNI) Band 2 antenna
8. Airbrake function via differential rudder deflection
9. Starboard rudder
10. Starboard fin
11. Fin leading-edge CNI VHF antenna
12. Electroluminescent formation lighting strip
13. Runway emergency arrester hook stowage
14. Exhaust nozzle sealing plates
15. Two-dimensional convergent-divergent thrust vectoring afterburner nozzle
16. Thrust vectoring range, plus and minus 20°
17. Port rear CNI Band 2 antenna
18. 'Cats-eye' control surface interface, all positions
19. Port all-moving tailplane
20. Rear CNI VHF antenna
21. Rudder lower fixed fairing
22. Rudder hydraulic actuator
23. Port all-composites rudder structure
24. Multispar all-composites fin structure
25. Forward CNI Band 2 antenna
26. Composites leading-edge flap panelling
27. Titanium fin root attachment spars
28. Fin root attachment fittings
29. Engine bay thermal lining
30. Port engine installation
31. Stored energy system reservoirs, engine relighting
32. Rear fuselage centre keel structure
33. Titanium firewall
34. Engine bay machined frames
35. Pratt & Whitney F119-PW-100 afterburning turbofan engine
36. Rear fuselage side-body integral fuel tank
37. Side-body frame structure
38. Wingroot attachment fittings
39. Machined titanium wingroot rib
40. Starboard flaperon hydraulic actuator
41. Starboard flaperon
42. Wing rear spar, titanium
43. All-composites flaperon structure
44. Aileron hydraulic actuator
45. Formation lighting strip
46. Starboard all-composites aileron structure
47. Rear Band 3 and 4 EW antenna
48. Electronic warfare (EW) electronics equipment
49. Starboard navigation light
50. Starboard leading-edge flap
51. ILS localiser antenna
52. Leading-edge flap drivers drive and rotary actuators
53. Titanium front spar
54. Carbon fibre composites 'sine-wave' spars
55. Titanium pylon mounting rib
56. Outer pylon hardpoint
57. Composites leading-edge flap structure
58. Wing pylons
59. AIM-9X Sidewinder, one per lateral bay
60. AIM-9M Sidewinder, one per lateral bay
61. GBU-30 1,000 lb joint direct attack munition (JDAM)
62. Starboard mainwheel
63. CNI Band 2 forward antenna
64. Inboard pylon hardpoint
65. Starboard wing integral fuel tank
66. Multispar wing structure
67. Titanium intermediate spars
68. Starboard mainwheel bay
69. Hydraulic accumulator
70. Starboard hydraulic reservoir
71. M61A2 six-barrel lightweight rotary cannon
72. Airframe-mounted auxiliary equipment gearbox, shaft-driven from engines
73. Engine compressor intake
74. Engine bleed air primary heat exchanger exhaust
75. No. 3 fuselage integral fuel tank with onboard inert gas generating system
76. Fuel/air and fuel/oil heat exchangers
77. Fuel transfer piping
78. Port rear side-body fuel tank
79. Port hydraulic equipment bay
80. Port mainwheel stowage
81. Power system inverter, port and starboard
82. Port wing integral fuel tank
83. Port flaperon, down position
84. Carbon fibre composites wing skin panel
85. Aileron hydraulic actuator
86. Port aileron
87. Formation lighting strip
88. Wingtip EW antenna
89. Port navigation light
90. ILS localiser antenna
91. Port leading-edge flap, lowered
92. Flap driveshaft and rotary actuators
93. External fuel tanks
94. Port CNI Band 2 antenna
95. Port CNI UHF antenna
96. Wing panel forward attachment joints
97. APU exhaust
98. Honeywell APU
99. APU intake with hinged door
100. Intake duct overpressure spill doors, open
101. Illuminated flight refuelling receptacle
102. Refuelling receptacle doors, open
103. GPS antenna
104. Starboard intake duct overpressure spill doors
105. Transverse cannon ammunition magazine, 480 rounds
106. Cannon barrels
107. Ammunition feed chute, chaff/flare launcher on port side
108. Main landing gear hydraulic retraction jack
109. Mainwheel leg pivot mounting
110. Mainwheel leg strut
111. Starboard CNI UHF flush antenna
112. Leading-edge flap drive motor
113. Starboard Band 3 and 4 EW antenna
114. Lateral missile bay doors
115. Hydraulically actuated missile launch trapeze mechanism, one per bay
116. AIM-120 AMRAAM, two in each of two ventral bays, three per bay in compressed loading
117. Starboard ventral missile bay doors
118. Lateral bay AIM-9 missile launch rail
119. Launch rail trapeze arm
120. Hydraulic rail, actuator
121. Cannon aperture
122. Muzzle flip-up door
123. Ventral missile bay
124. Machined titanium fuselage main frame, typical
125. No. 2 fuselage integral fuel tank
126. Band 2 avionics equipment
127. Port Band 3 and 4 EW antenna
128. Environmental control system equipment bay
129. Lateral avionics equipment bay, port and starboard
130. G-Band antenna
131. Air-cooled flight critical (ACFC) cooling air exhaust ducts
132. Forward fuselage main longeron
133. Composites S-shaped intake duct
134. Position of canopy emergency jettison on port side
135. Starboard engine air intake, fixed geometry
136. Titanium intake duct
137. Intake bleed air spill
138. Intake bleed air door/spoiler panel
139. ACFC cooling air intake in boundary layer diverter duct, blower for ground operating
140. Spill door/spoiler hydraulic actuator
141. Datalink support antenna; microwave landing system antenna beneath intake
142. Boundary layer diverter spill duct
143. No. 1 fuselage integral fuel tank
144. Canopy hinge point
145. Missile launch detector
146. Datalink antenna
147. Port intake spill and boundary layer bleed air doors
148. Canopy actuator, electrically powered
149. Forward fuselage lateral electrical equipment bays, port and starboard
150. Nosewheel hydraulic jack and pivot mounting
151. Leg-mounted landing and taxying lights
152. Torque scissor links
153. Forward retracting nosewheel
154. Nosewheel doors
155. Avionics equipment modules, downward hinged for access
156. Forward fuselage formation lighting strip
157. Composites fuselage chine skin panelling
158. Cockpit sloping rear pressure bulkhead
159. Pilot's ACES II (modified) zero/zero ejection seat
160. BAE Systems head-up display
161. Upward-hinged cockpit canopy
162. BAE Systems head-up display
163. Port side console off-base boarding ladder stowage
164. Fully shrouded instrument console housing six multifunction full-colour LCD displays
165. Starboard side console with sidestick controller for fly-by-wire flight control system
166. Rudder pedals
167. Forward fuselage machined aluminium sidewall panel
168. Cockpit front pressure bulkhead
169. Canopy latch actuator
170. Missile launch detector aperture
171. Flight control system avionics, ring laser gyro to port
172. Northrop Grumman/Raytheon AN/APG-77 radar equipment bay
173. Air data sensor system receivers, four positions
174. Canted radar
175. Standby pitot head
176. Electronically scanned radar antenna
177. Composites radome

Edwards AFB (fourth flight) on 15 March; regained flight status on 19 September 2000, being previously engaged on planned ground testing and upgrade programme. By 1 December 2000, had completed 15 flights and accumulated 28.7 hours. Earmarked for envelope expansion, replacing 4001, including loads testing, crosswind landings, validation of arrester hook and weapons bay environment work. Mid-fuselage section used for fit checks of Sidewinder and AMRAAM weapons at Fort Worth in July 1998 shortly before delivery to Marietta for final assembly. This aircraft expected to undertake initial AMRAAM launch trials at Edwards AFB in mid-2000. Will also test M61A2 cannon and JDAM integration.

4004/91-4004: First EMD aircraft with Hughes CIP software, including AN/APG-77 radar and ILS. Allocated to avionics development, but also to be used for low observables evaluation including radar cross-section and IR signature assessments; and comms/nav/ident (CNI) testing. Mid-fuselage section delivered from Fort Worth to Marietta on 28 December 1998. Block 1.1 avionics installed at Marietta, with electrical power applied for first time on 31 August 1999; Block 1.2 avionics then incorporated to support taxi trials and initial flight testing at Dobbins Air Reserve Base, from where first flight took place on 15 November 2000.

4005/91-4005: Radar, CNI and armament development tasks. Assigned to initial testing of Block 3.0 software by F-22. To be primary fire-control evaluation aircraft, with first flight made on 5 January 2001.

4006/91-4006: Avionics development tasks. Primarily for integrated avionics testing (first Block 3 software), RCS testing and, eventually, for systems effectiveness/military utility evaluation. Expected to fly for first time in December 2000, but had not done so by end of January 2001.

4007/91-4007: Was to have been initial two-seat F-22B but will be completed as single-seat F-22A and assigned to integrated avionics testing.

4008/91-4008: Allocated to avionics development tasks and also destined to validate F-22 observability specification data. Subsequently to be used for IOT&E between mid-2002 and early 2003, with aircraft 4009 and two PRTV aircraft (4010 and 4011).

4009/91-4009: Was to have been the second F-22B but will be completed as F-22A. Avionics development and observability trials.

4010 and **4011:** First PRTV aircraft; initial example due for delivery for Force Development Evaluation (FDE) activity at Nellis AFB, Nevada, by March 2002.

Trials by 4001 to 4009 will occupy 4,337 hours in 2,409 sorties. Of these totals, 2,110 hours and 1,200 sorties dedicated to airframe and systems testing, with balance allocated to mission avionics testing.

CURRENT VERSIONS (general): **F-22A:** Single-seat production version for USAF. Planned in-service date is December 2005.

Block 10 F-22: Multirole version which currently in planning stage and development of which may eventually be funded. If go-ahead is given, will retain existing air-to-air capabilities and also be capable of attacking airfields and anti-aircraft defences, using new standoff weapons which could include new long-range version of AGM-88 HARM. Could also be employed to direct UCAVs against heavily defended targets.

F-22B: Projected two-seat version for USAF; development terminated 10 July 1996 to reduce overall programme costs. Near-term economies of US$150 million expected, with longer-term production savings anticipated to be around US$350 million. Programme restructuring means that both EMD F-22Bs will be completed as F-22As.

NATF: Projected US Navy variant to replace Grumman F-14 Tomcat; development abandoned.

CUSTOMERS: US Air Force: two YF-22 demonstrators; planned 11 EMD aircraft reduced to nine plus one static and one fatigue test airframes in January 1993; original 648 production aircraft programme reduced to 442 in January 1994; latter originally to be funded from 1997 (long lead), beginning with four preproduction verification (PPV) aircraft in FY98, followed by series production of 438 (but PPV aircraft cancelled in early 1997); Quadrennial Defense Review (QDR) report in May 1997 resulted in planned procurement falling to 339, but production still expected to continue until 2013, although further reduction in total number to be acquired for USAF not yet ruled out on grounds of cost. Contracts for two PRTV 1 aircraft and long lead items for batch of six PRTV 2 aircraft signed December 1998. Decision to authorise LRIP expected in December 2000 but delayed until early 2001 and still awaited at end of January. Operational deployment plans still to be revealed, but pilot training to be accomplished at Tyndall AFB, Florida, by 325th Fighter Wing from 2003.

Second EMD F-22A over the wastes of Arizona 2000/0048744

Pentagon approval for export sales to selected friendly nations is expected at an early stage in order to reduce overall cost..

F-22 PLANNED PROCUREMENT

FY	Lot	F-22A
91	EMD	9
99	PRTV 1	2
00	PRTV 2	6
01	1	10
02	2	16
03	3	24
04	4	36
05	5	36
06	6	36
07	7	36
08	8	36
09	9	36
10	10	36
11	11	29
Total		**348**

Note: Lots 1-3 are low-rate initial production; lots 4 *et seq* are full-rate production.

COSTS: US$818 million contracts to both ATF teams, October 1986, for 54-month studies; each airframe team investing own funds (Lockheed/Boeing/GD team investment totalled US$675 million in addition to DoD funding); each engine contractor, about US$50 million; total US$3,800 million spent by USAF on both ATFs up to April 1991; programme cost for 648 aircraft was US$13,000 million for development (1991 base year) and US$52,500 million for production (1994 then-year); programme acquisition cost US$162 million (1994); flyaway cost US$61.2 million at 1991 prices. EMD contract, 2 August 1991, comprised US$9,550 million for 11 (subsequently nine) airframes, plus US$1,375 million to P&W for 33 (later amended to 27) engines. FY94 US Congressional appropriation of US$2,100 million was US$163 million below expectations, resulting in slippage of critical design review and first flight. Similarly, FY95 appropriation of US$2,300 million was US$110 million below expected figure, leading to further delay in maiden flight; and on 9 December 1994, Defense Secretary William Perry announced 10 per cent cut (approximately US$210 million) in FY96 budget for F-22 programme, necessitating a third restructuring of the F-22 programme. December 1996 announcement by USAF revealed that additional US$1.5 billion would be allocated for development phase and that total procurement cost of 438 aircraft by FY10 would be US$87 billion. Flyaway cost per aircraft stated to be US$92.6 million at same time.

Most recent cost figures (April 1999), in then-year dollars, adjusted for inflation, value total programme at US$62.6 billion, comprising US$22.7 billion for RDT & E (including Dem/Val and EMD), US$39.7 billion for production of 339 aircraft and US$0.2 billion for F-22 related construction projects at military bases. At that time, unit flyaway cost of one production aircraft, including engines but excluding fuel and consumables, quoted as US$84.7 million; latter figure rises to US$184.0 million if RDT&E expenditures taken into account.

On 30 December 1999, Lockheed Martin awarded contract for approximately US$1.3 billion, covering procurement of six PRTV 2 aircraft, augmenting earlier

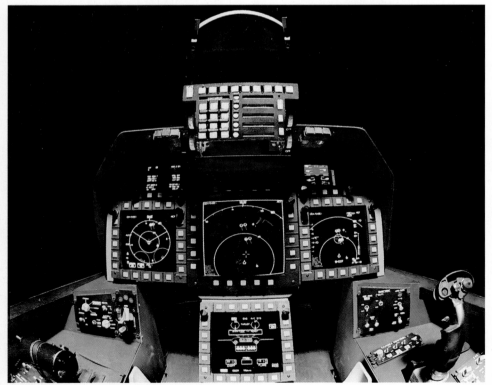

F-22 cockpit concept demonstrator at Marietta 0105079

Antenna of the F-22's Northrop Grumman AN/APG-77 radar

US$195.5 million contract for long-lead items; at same time, Pratt & Whitney received separate US$180 million award for 12 F119 engines. Additional appropriation of US$277.1 million allocated to Lockheed Martin for long-lead items for Lot 1 aircraft.

DESIGN FEATURES: Low-observables configuration and construction; stealth/agility trade-off decided by design team. Antennas located in leading- or trailing-edges of wings and fins, or flush with surfaces, to minimise radar signature. Target thrust/weight ratio 1.4 (achieved ratio 1.2 at T-O weight); greatly improved reliability and maintainability for high sortie-generation rates, including under 20 minutes combat turnround time; enhanced survivability through 'first-look, first-shot, first-kill' capability; short T-O and landing distances; supersonic cruise and manoeuvring (supercruise) in region of M1.5 without afterburning; internal weapons storage and generous internal fuel; conformal sensors.

Highly integrated avionics for single-pilot operation and rapid reaction. Radar, RWR and comms/ident managed by single system presenting relevant data only, and with emissions controlled (passive to fully active) in stages, according to tactical situation. CIP handles all avionics functions, including self-protection and radio, and automatically reconfigures to compensate for faults and failures. Two CIPs, with space for third, linked by 400 Mbits/s fibre optic network (see Avionics).

Wing and horizontal tail leading-edge sweep 42°; trailing-edge 17° forward, increased to 42° outboard of ailerons; all-moving five-edged horizontal tail. Vertical tail surfaces canted outwards at 28°; leading- and trailing-edge sweep 22.9°; biconvex aerofoil. Wing taper ratio 0.169; leading-edge anhedral 3.25°; root twist 0.5°; tip twist −3.1°; thickness/chord ratio 5.92 per cent at root, 4.29 per cent at tip; custom-designed aerofoil. Horizontal tails have no dihedral or twist.

Sidewinder AAMs stored internally in sides of intake ducts, with AMRAAMs, Sidewinders or GBU-32 JDAM 1000 precision-guided munitions in ventral weapons bay. Diamond-shaped cheek air intakes with highly contoured air ducts; single-axis thrust vectoring included on F119, but most specified performance achievable without.

Production aircraft expected to be coated with new, Boeing-developed, stealthy paint intended to enhance low-visibility attributes; this first applied to second EMD F-22 in March 2000.

FLYING CONTROLS: Triplex, digital, fly-by-wire system with GEC sidestick control, using line-replaceable electronic modules to enhance maintainability; thrust vectoring utilised to augment aerodynamic pitch control power and provide firm control even at low speeds and high angles of attack. Technology and control concepts demonstrated throughout the flight envelope during prototype air vehicle test programme, including flight at AoA greater than 60°; wind tunnel testing with models of production aircraft successfully attained AoAs greater than 85°.

Ailerons and flaperons occupy almost entire wing trailing-edge; full-span leading-edge flaps; conventional rudders in vertical tail surfaces; slab taileron surfaces; airbrake not included (differential rudder and wing trailing-edge surfaces for speed control). Control surface authorities: leading-edge flaps 0° up/35° down (2°/37° overtravel); trailing-edge flaperons 20° up/35° down; ailerons ±25°; horizontal tail leading-edges 30° up/25° down; rudders ±30°; speedbrake (rudder) 30° out.

STRUCTURE: Lockheed Martin at Marietta constructs forward fuselage, including cockpit (with avionics architecture, displays, controls, air data system), apertures, edges, tail assembly, landing gear and environmental control system and undertakes final assembly. Lockheed Martin's Fort Worth plant builds the centre-fuselage; Palmdale, the radome. Boeing responsible for wings, fuselage aft sections, power plant installation, auxiliary power generation system, radar, arresting gear system and avionics integration laboratory.

Total airframe weight comprises approximately 36 per cent titanium 64, three per cent titanium 62222, 24 per cent thermoset composites (both epoxy resin and bismaleimide), one per cent thermoplastic composites, 16 per cent aluminium, 6 per cent steel and 14 per cent other materials. Forward fuselage substructure is aluminium and composites; centre-fuselage includes titanium, aluminium and composites; both forward and centre-fuselage skins primarily graphite bismaleimide. Four main mid-fuselage section bulkheads are single-piece, closed-die, titanium forgings. Rear fuselage approximately 67 per cent titanium, 22 per cent aluminium and 11 per cent composites. Engine bay doors titanium honeycomb, produced by liquid-interface diffusion bonding. Tailbooms assembled by electron beam welding. Titanium vent screens (to reduce radar reflectivity) contain thousands of precisely shaped holes in special alignment, each cut by abrasive water-jet. Thermoplastics used in areas requiring high tolerance to damage, examples including doors for landing gear and weapon bay.

Wing skins are monolithic graphite bismaleimide. Main (front) wing spars are machined titanium forgings; intermediate spars are mix of resin transfer moulded (RTM) sine-wave composites and titanium for strength to meet vulnerability requirements in wing fuel tank; rear spars composites and titanium. Wingroot and control surface actuator attachment fittings are titanium HIP castings. Horizontal stabiliser incorporates 'tow placed' composites pivot shaft in addition to aluminium honeycomb core and graphite bismaleimide skins; vertical stabilisers use solid graphite bismaleimide skins over graphite epoxy RTM spars. Wing control surfaces are combination of co-cured composites skins/substructure and non-metallic honeycomb core construction.

LANDING GEAR: Menasco retractable tricycle type, stressed for no-flare landings of up to 3.05 m (10 ft)/s. Honeywell wheels, brakes and anti-skid system. Nosewheel tyre Goodyear or Michelin 23.5×7.5-10 (22 ply) tubeless; mainwheel tyres Goodyear or Michelin 37×11.50-18 (30 ply) tubeless. Kaiser airfield arrester hook in enclosed fairing between engines.

POWER PLANT: Two 156 kN (35,000 lb st) class Pratt & Whitney F119-PW-100 advanced technology reheated turbofans, each fitted with a two-dimensional, convergent/divergent thrust vectoring (±20° in vertical plane) exhaust nozzle for enhanced performance and manoeuvrability.

Fuel in eight tanks located in forward fuselage, mid-fuselage, wings and tailbooms; provision for later addition of fuel in saddle and fin tanks. Additionally, up to four external fuel tanks, each 2,271 litres (600 US gallons; 500 Imp gallons), on underwing hardpoints. Dorsal Xar Industries aerial refuelling receptacle covered by doors, except when required. Fuel grade JP-8.

ACCOMMODATION: Pilot only, on zero/zero modified Boeing ACES II ejection seat and wearing tactical life support system with improved g-suits, pressure breathing and arm restraint. Pilot's view over nose is −15°. Canopy manufactured by Sierracin Sylmar Corp as single-piece unit, hinged at rear.

SYSTEMS: Twin 276 bar (4,000 lb/sq in) hydraulic systems with four pumps (each 273 litres; 72.0 US gallons; 60.0 Imp gallons per minute), but single Parker Bertea actuator on each control surface to save weight and cost. Curtiss Wright actuators on leading-edge flaps. Two 65 kW engine-driven generators. Honeywell G250 335 kW (450 hp) APU driving a Hamilton Sundstrand 27 kW generator and 100 litre (26.5 US gallon; 22.0 Imp gallon)/min pump. Smiths 270 V DC electrical distribution system. Honeywell environmental control system for life support system and flight avionics (open loop air cycle), mission avionics (closed loop vapour cycle) and fuel cooling (thermal management). OBIGGS and Normalair-Garrett OBOGS.

AVIONICS: Final integration, as well as integration of entire suite with non-avionics systems, undertaken at F-22 Avionics Integration Laboratory, Seattle, Washington; airborne integration supported by Boeing 757 flying testbed and the Air Vehicle Integration Facility (including coherent RF stimulation) at Marietta.

Comms: TRW AN/ASQ-220 communication/navigation/identification (CNI) system includes Mk 12 IFF and UHF/VHF. CNI system uses modules contained in two integrated CNI racks, CIP assets and integrated display/control panels. Inter/Intra-Flight Data Link (IFDL) allows all F-22s in a flight to share target and system data without voice communication. Supplier team, responsible for electronics and software, also includes Rockwell Collins, ITT and GEC.

Radar: Northrop Grumman/Raytheon AN/APG-77 multimode radar incorporates active electronically scanned array (AESA), capable of interleaving air-to-air search and multitarget track functions. Also has weather mapping mode and provisions for air-to-ground modes and side arrays. Radar reported to be capable of detecting 1 m^2 (10.76 sq ft) target at range of approximately 109 n miles (201 km; 125 miles).

Flight: Vehicle management system (VMS) combines flight and propulsion controls; integrated vehicle subsystem control (IVSC) operates utilities via digital databus. Total of 18 Raytheon 1750A common processor modules used for VMS, IVSC and stores management system. F-22 is first fighter with triplex digital flight control computers and no electrical or mechanical back-up; control reconfiguration modes provide safe flying after actuator or hydraulic failures. VMS controls 14 surfaces (horizontal tail, ailerons, flaperons, rudders, leading-edge flaps and inlet bleed and bypass doors). No AoA limitation, but overstressing made impossible by restrictions to roll rate and load factor, according to fuel state, stores carriage and flight condition. Lear Astronics VMS integrated with Rosemount low-observable air data system, including two AoA probes and four sideslip plates on nose. CNI system includes GPS, Tacan and ILS. Twin Litton LN-100F INS. Throttle and stick contain 20 controls with 63 functions.

Software Block 0 for initial flight tests. Block 1.1 installed on aircraft 4004 in 1999, is primarily for radar, but includes more than half of the avionics suite's source lines of code. Block 2, installed in the 757 FTB in October 1999,

Boeing 757 Flying Test Bed fitted with F-22 nose and (above flight deck) wing for avionics trials

begins sensor fusion, including radio frequency co-ordination and some electronic warfare functions. Block 3S brings CNI and ECCM; Block 3.0 provides full sensor fusion and weapon delivery function; Block 3.1 adds GBU-32 JDAM, JTIDS receive and GPS and will be available to IOC aircraft. Block 4 will incorporate helmet cueing, AIM-9X and JTIDS send. Work is also under way on a Block 5 upgrade that should provide enhanced air-ground capability from around 2006.

Instrumentation: Fused situational awareness information is displayed to pilot via four Lockheed Martin colour liquid crystal multifunction displays (MFD); MFD bezel buttons provide pilot format control. Centre screen measures 203 × 203 mm (8 × 8 in) and typically will function as situation display; right and left screens measure 152 × 152 mm (6 × 6 in) and function as attack and defensive displays respectively; fourth screen (directly below situation display) can be used to provide global side-view depiction of tactical situation and may also present other data such as fuel status, engine parameters, stores data, BIT reports and electronic checklists. Additionally, 76 × 102 mm (3 × 4 in) upfront display screens are each side of the integrated control panel, immediately below the BAE Systems HUD. Illumination is fully NVG-compatible.

Mission: Two Hughes common integrated processors (CIP); CIP also contains mission software that uses tailorable mission planning data for sensor emitter management and multisensor fusion; mission-specific information delivered to system through Fairchild data transfer equipment/mass memory (DTE/MM) system that contains mass storage for default data and air vehicle operational flight programme; stores management system. General purpose processing capacity of CIP is rated at more than 700 million instructions per second (Mips) with growth to 2,000 Mips; signal processing capacity greater than 20 billion operations per second (Bops) with expansion capability to 50 Bops; CIP contains more than 300 Mbytes of memory with growth potential to 650 Mbytes. Of 132 slots available in CIPs 1 and 2, 41 are initially vacant and thus available for growth. Intra-flight datalink automatically shares tactical information between two or more F-22s. CNI system includes JTIDS (receive-only terminal). Lockheed Martin airborne video recorder. Lockheed Martin stores management system. Airframe contains provisions for IRST and side-mounted phased-array radar.

Self-defence: Sanders AN/ALR-94 electronic warfare (RF warning and countermeasures and missile launch detection functions) subsystem. AN/ALE-52 flare dispenser.

ARMAMENT: Internal long-barrel General Dynamics M61A2 20 mm cannon with hinged muzzle cover and 480-round magazine capacity (production F-22). Three internal bays (see Design Features) for AIM-9M Sidewinder or next-generation AIM-9X (one in each side bay on Hughes LAU-141/A trapeze-type launcher) and six AIM-120C AMRAAM AAMs and/or 460 kg (1,015 lb) GBU-32 JDAM PGMs on Edo LAU-142/A hydraulic ejection launchers in main weapons bay. Four underwing stores stations at 317 mm (125 in) and 442 mm (174 in) from centreline of fuselage capable of carrying 2,268 kg (5,000 lb) each. Other weaponry envisaged for use by the F-22 includes the BLU-109 Penetrator, the wind-corrected munitions dispenser (WCMD), AGM-88 HARM, GBU-22 Paveway 3 guidance unit (with 500 lb bomb) and the low-cost autonomous attack system (LOCAAS) submunitions dispenser package.

DIMENSIONS, EXTERNAL:
Wing span	13.56 m (44 ft 6 in)
Wing chord: at root (theoretical)	9.85 m (32 ft 3½ in)
at tip (reference)	1.66 m (5 ft 5½ in)
at tip (actual)	1.14 m (3 ft 9 in)
Wing aspect ratio	2.4
Length overall	18.92 m (62 ft 1 in)
Height overall	5.02 m (16 ft 5 in)
Tail span: horizontal surfaces	8.84 m (29 ft 0 in)
vertical surfaces	5.97 m (19 ft 7 in)
Wheelbase	6.04 m (19 ft 9¾ in)
Weapon bay ground clearance	0.94 m (3 ft 1 in)

AREAS:
Wings, gross	78.0 m² (840.0 sq ft)
Leading-edge flaps (total)	4.76 m² (51.20 sq ft)
Flaperons (total)	5.10 m² (55.00 sq ft)
Ailerons (total)	1.98 m² (21.40 sq ft)
Vertical tails (total)	16.54 m² (178.00 sq ft)
Rudders/speedbrakes (total)	5.09 m² (54.80 sq ft)
Stabilators (total)	12.63 m² (136.00 sq ft)

WEIGHTS AND LOADINGS (estimated):
Weight empty (target)	14,365 kg (31,670 lb)
Max T-O weight	almost 27,216 kg (60,000 lb)
Max wing loading	348.7 kg/m² (71.43 lb/sq ft)
Max power loading	87 kg/kN (0.86 lb/lb st)

PERFORMANCE (YF-22, demonstrated):
Max level speed: supercruise	M1.58
with afterburning	M1.7 at 9,150 m (30,000 ft)
Ceiling	15,240 m (50,000 ft)
g limit	+7.9

PERFORMANCE (F-22A, design target, estimated):
Max level speed at S/L	800 kt (1,482 km/h; 921 mph)
g limit	+9

UPDATED

F-22 cockpit mockup inside the Boeing 757 FTB

LOCKHEED MARTIN 382U/V HERCULES
US Air Force designation: C-, EC-, WC-130J
US Marine Corps designation: KC-130J
RAF designations: Hercules C. Mk 4 (C-130J-30); Hercules C. Mk 5 (C-130J)

TYPE: Medium transport/multirole.

PROGRAMME: US Air Force specification issued 1951, leading to first-generation Hercules (Allison T56 turboprops); first production contract for C-130A to Lockheed September 1952; first flight 23 August 1954; two YC-130 prototypes, 231 C-130As, 230 C-130Bs, 491 C-130Es, 1,089 C-130Hs and 113 L-100s manufactured before introduction of C-130J and commercial L-100J equivalent (details in earlier *Jane's* and *Jane's Aircraft Upgrades*). Official total of 2,156 (including prototypes) delivered by January 1998, when final C-130H handed over.

Privately funded development and flight test programme for next-generation version began in 1991 as Hercules II, but now known as C-130J Hercules, or L-100J in equivalent civilian form. Early development detailed in 2000-01 and previous editions of *Jane's*.

Initial British delivery accomplished on 24 August 1998; aircraft involved was ZH865, which arrived at Boscombe Down on 26 August for start of clearance trials by UK Defence Evaluation Research Agency (DERA). Further 11 flown to UK by end 1998 for temporary storage and final preparation for delivery to RAF by Marshall Aerospace at Cambridge; of these, one (ZH871) delivered to Boscombe Down on 30 November to participate in DERA trials programme. First aircraft ferried to RAF transport force at Lyneham was ZH878 on 21 November 1999. By end 1999, total of 22 had arrived in UK, with number delivered to RAF by mid-April 2000 having risen to seven. Final RAF aircraft flown from Marietta to Cambridge in late May 2000.

First 'operational' mission accomplished by Lockheed Martin test crew in late November 1998, when C-130J completed three sorties from Marietta and airlifted 37,650 kg (83,000 lb) of hurricane relief supplies to Tegucigalpa, Honduras. Subsequently, 1999 witnessed start of deliveries to US Air Force Reserve Command and Air National Guard, as well as to operating units in Australia and the UK.

CURRENT VERSIONS: **C-130J:** Baseline model. Dimensionally similar to preceding C-130H, but incorporating new equipment and features as subsequently described. Subject of initial order for two from USAF in FY94, with subsequent contract for two in FY96; these earmarked for trials, while further eight funded in FY97 and FY98 assigned to Air National Guard's 135th Airlift Squadron at Martin State Airport, Baltimore, Maryland, which fully equipped by mid-July 2000 and then expecting to become operational in second quarter 2001; initial aircraft delivered to Air Force Reserve Command's 403rd Wing at Keesler AFB, Mississippi, on 31 March 1999 for training.

Description applies mainly to baseline C-130J except where indicated.

C-130J-30: Stretched version of current production C-130; fuselage lengthened by 4.57 m (15 ft 0 in), offering increases in capability of between 31 and 50 per cent, dependent upon mission and configuration (see accompanying diagram). Orders received from Australia, Italy, UK and USA. Last named could ultimately purchase as many as 150 to replace C-130E version with regular USAF units.

EC-130J: 'Commando Solo' psychological warfare version; first two funded in FY98 budget, with first (99-1933) handed over on 17 October 1999; after flight testing, it moved to Palmdale, California, in July 2000 for fitting out and is not expected to enter operational service with the 193rd Special Operations Squadron, Air National Guard, at Harrisburg IAP, Pennsylvania, until mid-2001. Additional procurement comprises third example in FY99 budget, fourth in FY00 and fifth in FY01.

HC-130J: Replacement for earlier US Coast Guard HC-130s; funding for initial six contained in FY01 budget.

KC-130J: Tanker/transport version for US Marine Corps, which has requirement for about 50. Fitted with two Flight Refuelling Mk32B-901E hose-and-drogue wing-

Lockheed Martin C-130J-30 Hercules C. Mk 4 accompanied by a Fighting Falcon

LOCKHEED MARTIN — AIRCRAFT: USA

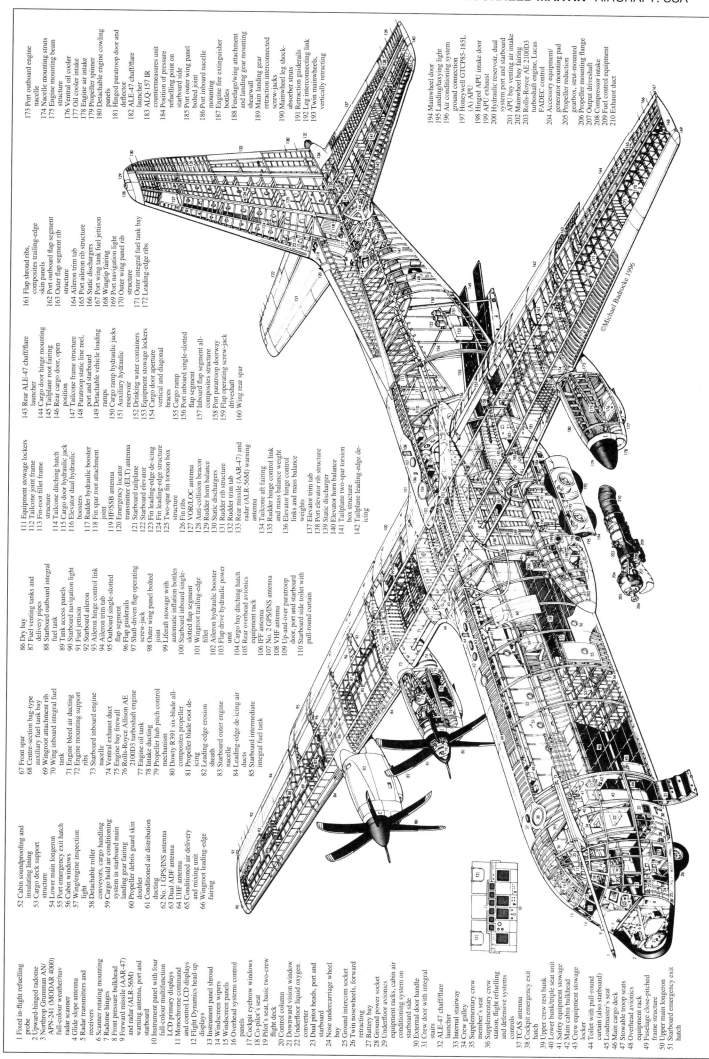

C-130J-30 Hercules C. Mk 4, cutaway drawing key

1 Fixed in-flight refuelling probe
2 Upward-hinged radome
3 Northrop Grumman AN/APN-241 (MODAR 4000) full-colour weather/nav radar scanner
4 Glide slope antenna
5 Radar transmitters and receivers
6 Scanner rotating mounting
7 Radome hinges
8 Front pressure bulkhead
9 Forward missile (AAR-47) and radar (ALR-56M) warning antenna, port and starboard
10 Instrument panel with four full-colour multifunction LCD primary displays
11 Monochrome command and control LCD displays
12 Flight Dynamics head-up displays
13 Instrument panel shroud
14 Windscreen wipers
15 Windscreen panels
16 Overhead systems control panels
17 Cockpit eyebrow windows
18 Co-pilot's seat
19 Pilot's seat, basic two-crew flight deck
20 Control column
21 Downward vision window
22 Up-and-over paratroop door, port and starboard
23 Dual pilot heads, port and starboard
24 Nose undercarriage wheel bay
25 Ground intercom socket
26 Twin nosewheels, forward retracting
27 Battery bay
28 Ground power socket
29 Underfloor avionics equipment racks, cabin air conditioning system on starboard side
30 External door handle
31 Crew door with integral stairs
32 ALE-47 chaff/flare launcher
33 Internal stairway
34 Crew galley
35 Supplementary crew member's seat
36 Supplementary crew station, flight refuelling and defensive systems controls
37 TCAS antenna
38 Cockpit emergency exit hatch
39 Upper crew rest bunk
40 Lower bunk/triple seat unit
41 Safety equipment stowage
42 Main cabin bulkhead
43 Crew equipment stowage locker
44 Toilet with pull-round curtain (also starboard)
45 Loadmaster's seat
46 Main cargo deck
47 Stowable troop seats
48 Overhead avionics equipment rack
49 Fuselage close-pitched frame structure
50 Upper main longeron
51 Starboard emergency exit hatch
52 Cabin soundproofing and insulating lining
53 Cargo deck support structure
54 Lower main longeron
55 Port emergency exit hatch
56 Cabin windows
57 Wing/engine inspection light
58 Detachable roller conveyors, cargo handling
59 Cargo hold air conditioning system in starboard main landing gear fairing
60 Propeller debris guard skin doubler
61 Conditioned air distribution ducting
62 No. 1 GPS/INS antenna
63 Dual ADF antenna
64 UHF antenna
65 Conditioned air delivery and mixing unit
66 Wingroot leading-edge fairing
67 Front spar
68 Centre-section bag-type auxiliary fuel tank bay
69 Wingroot attachment rib
70 Wing inboard integral fuel tank
71 Engine bleed air ducting
72 Engine mounting support ribs
73 Starboard inboard engine nacelle
74 Ventral exhaust duct
75 Engine bay firewall
76 Rolls-Royce Allison AE 2100D3 turboshaft engine
77 Engine oil tank
78 Intake ducting
79 Propeller hub pitch control mechanism
80 Dowty R391 six-blade all-composites propeller
81 Propeller blade root de-icing
82 Leading-edge erosion sheath
83 Starboard outer engine nacelle
84 Leading-edge de-icing air ducts
85 Starboard intermediate integral fuel tank
86 Dry bay
87 Fuel venting tanks and delivery pipes
88 Starboard outboard integral fuel tank
89 Tank access panels
90 Starboard navigation light
91 Fuel jettison
92 Starboard aileron
93 Aileron hinge control link
94 Aileron trim tab
95 Outboard single-slotted flap segment
96 Flap guiderails
97 Shaft-driven flap operating screw-jack
98 Outer wing panel bolted joint
99 Liferaft stowage with automatic inflation bottles
100 Starboard inboard single-slotted flap segment
101 Wingroot trailing-edge fillet
102 Aileron hydraulic booster
103 Flap drive hydraulic power unit
104 Cargo bay ditching hatch
105 Rear overhead avionics equipment rack
106 IFF antenna
107 No. 2 GPS/INS antenna
108 VHF antenna
109 Up-and-over paratroop door, port and starboard
110 Starboard side toilet with pull-round curtain
111 Equipment stowage lockers
112 Tailcone joint frame
113 Fin-root fillet frame structure
114 Tailcone ditching hatch
115 Cargo door hydraulic jack
116 Elevator dual hydraulic boosters
117 Rudder hydraulic booster
118 Fin spar root attachment joint
119 HF/SSB antenna
120 Emergency locator transmitter (ELT) antenna
121 Starboard tailplane
122 Starboard elevator
123 Fin leading-edge de-icing
124 Fin leading-edge structure
125 Two-spar fin torsion box structure
126 Fin ribs
127 VOR/LOC antenna
128 Anti-collision beacon
129 Rudder horn balance
130 Rudder rib structure
131 Rudder trim tab
132 Rudder
133 Rear missile (AAR-47) and radar (ALR-56M) warning antenna
134 Tailcone aft fairing
135 Rudder hinge control link and mass balance weight
136 Elevator hinge control links and mass balance weights
137 Elevator trim tab
138 Port elevator rib structure
139 Static discharger
140 Elevator horn balance
141 Tailplane two-spar torsion box structure
142 Tailplane leading-edge de-icing
143 Rear ALE-47 chaff/flare launcher
144 Cargo door hinge mounting
145 Tailplane root fairing
146 Rear cargo door, open position
147 Tailcone frame structure
148 Paratroop static line reel, port and starboard
149 Detachable vehicle loading ramps
150 Cargo ramp hydraulic jacks
151 Auxiliary hydraulic reservoir
152 Drinking water containers
153 Equipment stowage lockers
154 Cargo door aperture vertical and diagonal braces
155 Cargo ramp
156 Port inboard single-slotted flap segment
157 Inboard flap segment all-composites structure
158 Port paratroop doorway
159 Flap operating screw-jack drivshaft
160 Wing rear spar
161 Flap shroud ribs, composites trailing-edge skin panels
162 Port outboard flap segment
163 Outer flap segment rib structure
164 Aileron trim tab
165 Port aileron rib structure
166 Static dischargers
167 Port wing tank fuel jettison
168 Wingtip fairing
169 Port navigation light
170 Outer wing panel rib structure
171 Outer integral fuel tank bay
172 Leading-edge ribs
173 Port outboard engine nacelle
174 Nacelle mounting struts
175 Engine mounting beam structure
176 Ventral oil cooler
177 Oil cooler intake
178 Engine air intake
179 Propeller spinner
180 Detachable engine cowling panels
181 Hinged paratroop door and deflector
182 ALE-47 chaff/flare launcher
183 ALQ-157 IR countermeasures unit
184 Position of pressure refuelling point on starboard side
185 Port outer wing panel bolted joint
186 Port inboard nacelle mounting
187 Engine fire extinguisher bottles
188 Fuselage/wing attachment and landing gear mounting shearwall
189 Main landing gear retraction interconnected screw-jacks
190 Mainwheel leg shock-absorber struts
191 Retraction guiderails
192 Leg interconnecting link
193 Twin mainwheels, vertically retracting
194 Mainwheel door
195 Landing/taxying light
196 Air conditioning system ground connection
197 Honeywell GTCP85-185L (A) APU
198 Hinged APU intake door
199 APU exhaust
200 Hydraulic reservoir, dual system port and starboard
201 APU bay venting air intake
202 Mainwheel bay fairing
203 Rolls-Royce AE 2100D3 turboshaft engine, Lucas FADEC control
204 Accessory equipment/generator mounting pad
205 Propeller reduction gearbox, strut-mounted
206 Propeller mounting flange
207 Output driveshaft
208 Compressor intake
209 Fuel control equipment
210 Exhaust duct

High-performance take-off by a USAF Lockheed Martin C-130J Hercules *(Paul Jackson/Jane's)* 2000/0075908

mounted refuelling pods; three aircraft funded in FY97 budget, two in FY98, two in FY99, one in FY00 and three in FY01. First contract for five USMC conversions (to be accomplished by end of 2000) announced 21 July 1998. Final assembly of first KC-130J (165735) began 22 March 1999, with first flight on 9 June 2000; total of three aircraft assigned to test programme at Patuxent River in latter half of 2000; first drogue engagement accomplished by Navy F/A-18 Hornet on 30 August 2000. On completion of trials, all three will be assigned to USMC tanker/transport squadron VMGR-252 at MCAS Cherry Point, North Carolina, which due to receive first example, for maintenance training, in late 2000. Design studies of aircraft fitted with telescopic refuelling boom under rear fuselage/ramp have been discontinued. Pod refuelling rate (each) 1,136 litres (300 US gallons; 250 Imp gallons) per minute; total offload capability is 32,005 litres (8,455 US gallons; 7,040 Imp gallons). Provisions for installation of refuelling probe incorporated in basic aircraft. Additional fuel in underwing and cargo hold tanks; see Power Plant.

WC-130J: Weather reconnaissance version to be equipped with aerial reconnaissance weather officer console, dropsonde system operator console and dropsonde launch tube; initial batch of four included in FY96 budget, with three more in FY97, two in FY98 and one in FY99; for 53rd WRS, Air Force Reserve Command at Keesler AFB, Mississippi. US$46.9 million contract for modification of first six, with option for further four, signed 18 September 1998; deliveries began on 30 September 1999; formal acceptance 12 October 1999.

C-130J-30 AEW&C: In September 1996, Lockheed Martin and Northrop Grumman announced a teaming agreement with the objective of capturing additional business in international AEW&C market and plan to promote jointly a C-130J variant with an AEW system. Further work suspended following Australian decision in 1999 to purchase Boeing/Northrop Grumman proposal based on Boeing 737; further details in 2000-01 *Jane's*.

Lockheed Martin also studied floatplane version of the C-130J, with a removable catamaran hull fixed to the undersides of a conventional transport aircraft; conversion from landplane to floatplane and vice versa would require several hours. Mission applications could include covert deployment of special forces, SAR and firefighting, with ramp-mounted deployable water scoops employed in last-mentioned role. Programme dormant.

CUSTOMERS: See tables. First customer was RAF, which ordered 25 in December 1994; of these, 15 are stretched C-130J-30, designated C. Mk 4, with final 10 as standard C-130Js, designated Hercules C. Mk 5. Delivery of first example to the trials unit at Boscombe Down was due to take place in November 1996, but was delayed until 26 August 1998. First service recipient is J Conversion Flight of No. 57 (Reserve) Squadron, followed by No. 24 and then No. 30 Squadrons at RAF Lyneham. Deliveries to operating base at Lyneham began on 21 November 1999, when C. Mk 4 ZH878 arrived for duty as temporary ground procedures trainer; two days later, on 23 November, C. Mk 4 ZH875 was formally handed over in official ceremony at Lyneham. Remainder of original order now due for delivery by mid-2001. Five additional aircraft are subject to option and total value of contract is just over £1 billion.

Second order covered two C-130Js for evaluation by the USAF and was finalised on 13 October 1995, one week before the first was rolled out at Marietta. Further two funded in FY96, eight in FY97/98 and three requested in FY99 for Air Force Reserve Command and Air National Guard units; first ANG squadron is 135th AS at Martin State Airport, Baltimore. USAF orders will begin in 2001 with contract for two C-130J-30 aircraft, with FY01 procurement including initial batch of six C-130Js for long-range SAR duties with US Coast Guard. First USAF -30 rolled out 25 January 2001.

One further order received before the end of 1995 was from Australia, for 12 C-130J-30s to replace an identical number of C-130Es of No. 37 Squadron at Richmond, at total cost of US$660 million. Order was placed on 21 December 1995 and includes options for an additional 24 aircraft, plus eight options for New Zealand, which is guaranteed pricing based on Australia's larger order. First Australian C-130J (A97-440/N130JQ, c/n 5440) flew on 15 February 1997; delivery to RAAF began on 7 September 1999, when A97-464 arrived at Richmond; last of initial batch handed over on 1 June 2000.

Recent firm orders, placed in January 2000 and March 2000, were for batches of two additional aircraft from Italy, which had previously contracted for 18 C-130Js; first aircraft formally rolled out at Marietta on 11 July 2000. Italian military certification also awarded in July 2000, with first delivery (actually second aircraft, MM 62176, c/n 5497) to 46° Aerobrigata at Pisa departing USA on 16 August; formal acceptance 21 September 2000. All will be tanker-capable and configured to receive fuel, although only six likely to be operated as such at any one time. Newest orders specified C-130J-30, which also subject of option for up to two further aircraft. At same time as placing latest firm order, Italy also revised overall procurement plan and will acquire 12 C-130J and 10 C-130J-30 versions.

Other potential customers reported to be in discussion with Lockheed Martin include Denmark, Israel, Kuwait and Portugal. Danish decision was due in first half of 2000, but was delayed until end of year, when order placed for three C-130J-30s, with option on fourth. Israel considering purchase of 12, but early order appears unlikely. Norway has also expressed interest in obtaining six as replacements for its current fleet of six C-130Hs. Extensive world promotional tour undertaken by C-130J between February and June 1998, visiting 32 countries and completing 130 demonstration flights.

MILITARY/GOVERNMENT C-130J SALES
(To January 2001, excluding USA)

Country	J	J-30	Total
Australia		12	12
Denmark		3	3
Italy	12	10	22
UK	10	15	25
Total	**22**	**40**	**62**

Note: Options not included

COSTS: US$55 million programme unit cost (Australia) (1995). Italian order of November 1997 valued at US$1.2

US C-130J PROCUREMENT
(to 2 January 2001)

FY	C-130J	EC-130J	KC-130J	WC-130J	Total
94	2				2
96	2			4	6
97	4		3	3	10
98	4	2	2	2	10
99	3[1]	1	2	1	7
00		1	1		2
01	8[2]	1	3		12
Total	**23**	**5**	**11**	**10**	**49**

Note: FY01 aircraft requested but not yet on contract
[1] Includes one C-130J-30 for USAF
[2] Comprises six HC-130J for US Coast Guard and two C-130J-30 for USAF

Hercules Model	Cargo Length	463L Pallets	Medical Litters	CDS Bundles	Combat Troops	Para-Troops
C-130J	40.4 ft	5	74	16	92	64
C-130J-30	55.4 ft	7	97	24	128	92
Increase	37%	40%	31%	50%	39%	44%

C-130J Hercules accommodation; standard and (illustrated) stretched versions compared 0105080

Flight deck of the C-130J, showing HUDs and EFIS 2000/0075954

billion. Purchase of two USAF C-130J-30 aircraft in FY01 budget costed at total of US$149 million.

DESIGN FEATURES: Archetypal tactical transport: upswept rear fuselage for ramp access; high wing for propeller ground clearance despite low floor height; latter provided by pannier-mounted landing gear, which obviates long mainwheel legs stowed in wings. Can deliver loads and parachutists over open ramp and parachutists through side doors; cargo hold pressurised.

Significant changes introduced with C-130J, which optimised for economical operation, justifying customers' substitution for earlier C-130s on 30 year lifetime savings alone. Entirely revised flight deck reduces LRUs by half and wire assemblies by 53 per cent, with wire terminations cut by 81 per cent; has four MFDs, plus HUD for both pilots; lighting compatible with NVGs. Most systems have digital interfaces with the main mission computer to include unmodified mechanical systems like the hydraulics, which are largely unaltered from those of C-130H; provision for integrated self-defence suite (RWR, MAW, chaff/flare dispensers and IR jammers). Propulsion system provides 29 per cent more take-off thrust and is 15 per cent more efficient; fuel efficiencies obviate requirement for external tanks on most types of mission; propeller has 50 per cent fewer parts and weighs 15 per cent less. Manpower requirements of typical 16-aircraft squadron cut by 38 per cent compared with earlier versions of C-130, as result of reduced flight crew and 50 per cent better maintainability. Comprehensive computerised maintenance system employs a hand-held data module as interface between aircraft's BITE and operating base's central technical records.

Wing section NACA 64A318 at root and NACA 64A412 at tip; dihedral 2° 30′; incidence 3° at root, 0° at tip.

FLYING CONTROLS: All flying controls integrated with digital autopilot/flight director and comprise control surfaces boosted by dual hydraulic units; trim tabs on ailerons, both elevators and rudder; elevator tabs have AC main supply and DC standby; Lockheed-Fowler composites trailing-edge flaps.

STRUCTURE: All-metal two-spar wing with integrally stiffened taper-machined skin panels up to 14.63 m (48 ft 0 in) long. Incorporates carbon fibre composites materials for flaps and 32 graphite-epoxy trailing-edge panels.

LANDING GEAR: Hydraulically retractable tricycle type. Each main unit has two wheels in tandem, retracting into fairing built on to fuselage side. Nose unit has twin wheels and is steerable ±60°. Mainwheels 20.00-20 (26 ply) tubeless; nose 12.50-16 (12 ply) tubed or tubeless. Oleo shock-absorbers. Minimum ground turning radius: C-130J, 11.28 m (37 ft) about nosewheel and 25.91 m (85 ft) about wingtip; C-130J-30 14.33 m (47 ft) about nosewheel and 27.43 m (90 ft) about wingtip.

POWER PLANT: Four Rolls-Royce AE 2100D3 turboprops, flat rated to 3,424 kW (4,591 shp) (manufacturer's rating 3,458 kW; 4,637 shp at ISA + 25°C), fitted with Dowty Aerospace R391 six-blade composites propellers and Lucas Aerospace FADEC. Automatic thrust control system (ATCs) and autofeather systems, plus engine monitoring system (EMS) which is incorporated into aircraft's integrated diagnostic system (IDS).

Total internal fuel capacity of 25,552 litres (6,750 US gallons; 5,621 Imp gallons) without foam and 24,363 litres (6,436 US gallons; 5,359 Imp gallons) with foam. Provisions only for two optional underwing pylon tanks, each with capacity of 5,220 litres (1,379 US gallons; 1,148 Imp gallons) without foam and 4,883 litres (1,290 US gallons; 1,074 Imp gallons) with foam. Total fuel capacity 35,992 litres (9,508 US gallons; 7,917 Imp gallons) without foam and 34,129 litres (9,016 US gallons; 7,507 Imp gallons) with foam. Tanker versions can carry cargo hold tank with additional 13,578 litres (3,587 US gallons; 2,987 Imp gallons). Single pressure refuelling point and overwing gravity fuelling. In-flight refuelling probe fitted as standard on port side of RAF aircraft; optional for all others, with Italian aircraft currently unique in being configured as receiver/tankers.

ACCOMMODATION: Crew of two on flight deck, comprising pilot and co-pilot, with provisions for optional third workstation. Ergonomic problems suffered by short pilots necessitated a number of alterations to cockpit, including redesign of seat and seat track, HUD and main control yoke, with throttle quadrant also modified. Two crew bunks, with lower incorporating three additional seats and harnesses for relief crew/flight deck passengers. Galley. Separate loadmaster's station in cargo hold, including folding desk, is standard equipment on RAF aircraft; optional for all others. Flight deck and main cabin pressurised and air conditioned.

Standard complements for C-130J are as follows: 92 troops, 64 paratroopers, 74 litter patients plus two attendants, 54 passengers on palletised airline seating. Corresponding data for C-130J-30 are 128 troops, 92 paratroopers, 97 litter patients plus four attendants and 79 passengers on palletised airline seating. Airdrop loads comparable to C-130H/H-30 and include light armoured vehicles. Light and medium towed artillery pieces, wheeled and tracked vehicles and 463L palletised loads (five in C-130J and seven in C-130J-30, plus one on ramp in each model) are transportable. Hydraulically operated (with dual actuators) main loading door and ramp at rear of hold; can be opened in flight at up to 250 kt (463 km/h; 288 mph). Crew door on port forward fuselage side. Paratroop door on each side aft of landing gear fairing. Two emergency exit doors standard. Optional cargo handling system ordered by the USAF includes flush-mounted winch; in-ramp towplate for airdrop operations; low-profile rails and electric locks; flip-over rollers with covers; container delivery system centre vertical restraint rails. Capable of automatic preprogrammed cargo drops.

SYSTEMS: Lucas generator. Environmental control system in starboard undercarriage fairing is similar to that of the C-130H, incorporating dual air cycle machines, but with 30 per cent greater cooling capacity and a digital electronic control system. Honeywell GTCP85-185L(A) auxiliary power unit in port undercarriage fairing furnishes ground electrical power and bleed air for environmental control system. MIL-STD-1553B digital databus architecture. Integrated diagnostic system (IDS) incorporating fault detection/isolation subsystem with BIT (built-in test) facility. Goodrich pneumatic fin anti-icing system.

AVIONICS: *Comms:* Honeywell com/nav/ident management system, with Intel 80960 processor. AN/ARC-222 VHF, HF/UHF radio, intercom and IFF, with provisions for satcom system. All communication radios have secure features.

Radar: Northrop Grumman AN/APN-241 low-power colour radar incorporates Doppler beam-sharpening ground mapping mode, air-to-air skin paint mode and protective windshear mode as well as conventional colour weather radar.

Flight: HG-9550 radar altimeter, AN/ARN-153(V) Tacan, digital autopilot/flight director (DA/FD), dual Honeywell laser INS with embedded GPS receivers, Doppler velocity sensor, VOR, ILS, marker beacon receiver, UHF/VHF DF, ADF, E-TCAS, ground collision avoidance system (GCAS), global digital map display units and provision for microwave landing system.

Instrumentation: 'Dark cockpit' concept. Flight Dynamics HUD as certified primary flight instrument at pilot and co-pilot positions, four 152 × 203 mm (6 × 8 in) Avionics Display Corporation active matrix liquid-crystal display (AMLCD) colour multipurpose display systems (CMDSs) which are NVG-compatible for flight instrumentation, navigation and engine information and five Avionics Display Corporation 58 × 76 mm (2.3 × 3 in) monochrome AMLCDs for digital selector panels.

Mission: Sierra Technologies AN/APN-243(V) station-keeping equipment. Provision for secure voice communication system.

Self-defence: Provisions for Lockheed Martin AN/AAR-47 missile warning system, Sanders AN/ALQ-157 IR countermeasures system, BAE AN/ALE-47 chaff/flare dispensing systems and AN/ALR-56M radar warning receiver or AN/ALR-69 enhanced radar warning system. RAF aircraft originally ordered with US-supplied systems, but RWR and countermeasures dispensing systems now subject to bidding from UK defence sector, with formal contest beginning early 1998.

EQUIPMENT: USMC and Italian KC-130Js have Flight Refuelling Mk 32B-901E hose pods underwing.

Weather reconnaissance Lockheed Martin WC-130J *2001*/0100452

DIMENSIONS, EXTERNAL:

Wing span	40.41 m (132 ft 7 in)
Wing aspect ratio	10.1
Length overall: C-130J	29.79 m (97 ft 9 in)
C-130J-30	34.37 m (112 ft 9 in)
Height overall: C-130J	11.84 m (38 ft 10 in)
C-130J-30	11.81 m (38 ft 9 in)
Tailplane span	16.05 m (52 ft 8 in)
Wheel track	4.34 m (14 ft 3 in)
Propeller diameter	4.11 m (13 ft 6 in)
Main cargo door (rear of cabin):	
Height	2.77 m (9 ft 1 in)
Width	3.12 m (10 ft 3 in)
Height to sill	1.03 m (3 ft 5 in)
Paratroop doors (each): Height	1.83 m (6 ft 0 in)
Width	0.91 m (3 ft 0 in)
Height to sill	1.03 m (3 ft 5 in)
Emergency exits (each): Height	1.22 m (4 ft 0 in)
Width	0.71 m (2 ft 4 in)

DIMENSIONS, INTERNAL:

Cabin, excl flight deck:	
Length excl ramp: C-130J	12.19 m (40 ft 0 in)
C-130J-30	16.76 m (55 ft 0 in)
Length incl ramp: C-130J	15.32 m (50 ft 3 in)
C-130J-30	19.89 m (65 ft 3 in)
Max width	3.12 m (10 ft 3 in)
Max height	2.74 m (9 ft 0 in)
Total usable volume: C-130J	128.9 m^3 (4,551 cu ft)
C-130J-30	170.5 m^3 (6,022 cu ft)

AREAS:

Wings, gross	162.12 m^2 (1,745.0 sq ft)
Ailerons (total)	10.22 m^2 (110.00 sq ft)
Trailing-edge flaps (total)	31.77 m^2 (342.00 sq ft)
Fin	20.90 m^2 (225.00 sq ft)
Rudder, incl tab	6.97 m^2 (75.00 sq ft)
Tailplane	35.40 m^2 (381.00 sq ft)
Elevators, incl tabs	14.40 m^2 (155.00 sq ft)

WEIGHTS AND LOADINGS (internal fuel only, except where specified):

Operating weight empty:	
C-130J	34,274 kg (75,562 lb)
C-130J-30	35,966 kg (79,291 lb)
Max fuel weight: internal	20,819 kg (45,900 lb)
external (optional)	8,506 kg (18,754 lb)
Max payload, 2.5 *g*: C-130J	18,955 kg (41,790 lb)
C-130J-30	17,264 kg (38,061 lb)
Max normal T-O weight	70,305 kg (155,000 lb)
Max overload T-O weight:	
C-130J, C. Mk 4	79,380 kg (175,000 lb)
Max normal landing weight	58,965 kg (130,000 lb)
Max overload landing weight	70,305 kg (155,000 lb)
Max zero-fuel weight, 2.5 *g*	53,230 kg (117,350 lb)
Max wing loading (normal)	433.7 kg/m^2 (88.83 lb/sq ft)
Max power loading (normal)	5.14 kg/kW (8.44 lb/shp)

PERFORMANCE (C-130J except where indicated):

Never-exceed speed (VNE) at 3,050 m (10,000 ft)	
C. Mk 4	378 kt (700 km/h; 435 mph)
Max cruising speed:	
C-130J	348 kt (645 km/h; 400 mph)
C. Mk 4 at 7,620 m (25,000 ft)	356 kt (659 km/h; 410 mph)
Econ cruising speed	339 kt (628 km/h; 390 mph)
Stalling speed	100 kt (185 km/h; 115 mph)
Max rate of climb at S/L	640 m (2,100 ft)/min
Time to 6,100 m (20,000 ft)	14 min
Cruising altitude	8,535 m (28,000 ft)
Service ceiling at 66,680 kg (147,000 lb) AUW	9,315 m (30,560 ft)
Service ceiling, OEI, at 66,680 kg (147,000 lb) AUW	6,955 m (22,820 ft)
T-O run	930 m (3,050 ft)
T-O to 15 m (50 ft)	1,433 m (4,700 ft)
T-O run using max effort procedures	549 m (1,800 ft)
Landing from 15 m (50 ft) at 58,967 kg (130,000 lb) AUW	777 m (2,550 ft)
Landing run at 58,967 kg (130,000 lb) AUW	427 m (1,400 ft)
Runway LCN: asphalt	37
concrete	42
Range with 18,144 kg (40,000 lb) payload and Mil-C-5011A reserves	2,835 n miles (5,250 km; 3,262 miles)

UPDATED

LOCKHEED MARTIN ADVANCED MOBILITY AIRCRAFT (AMA)

TYPE: Medium transport/multirole.

PROGRAMME: Studies under way since early 1995; initially known as 'World Airlifter' and described in 1998-99 *Jane's* as New Strategic Aircraft (NSA). Lockheed Martin's primary objective is to develop a replacement for Boeing KC-135 Stratotanker in-flight refuelling aircraft, Lockheed C-141 StarLifter strategic transport and tanker/transport types such as the Lockheed L-1011 TriStar and McDonnell Douglas KC-10 Extender, although it is also intended to offer commercial freighter versions as well as special mission aircraft configured for AEW and battlefield surveillance tasks.

Lockheed Martin is proposing to develop the AMA as a private venture and is seeking two or three risk-sharing

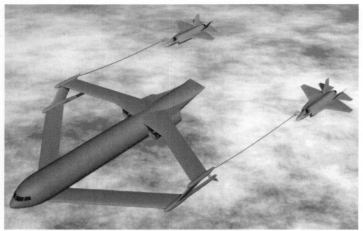
Computer-generated image of Lockheed Martin box-wing tanker/transport design simultaneously conducting hose-and-drogue refuelling 2000/0062394

Artist's impression of a possible Lockheed Martin ATT 2000/0062395

Strut-braced wing design for potential Lockheed Martin 325-seat transport 2000/0075955

Lockheed Martin 325-seat, box wing airliner design 2000/0075956

international partners to form a consortium; discussions have taken place with a number of potential partners, which are understood to include Aerospatiale Matra, DaimlerChrysler Aerospace and BAE Systems.

At present, the company is looking to develop a twin-engined design with high bypass turbofans in the 267 to 311 kN (60,000 to 70,000 lb st) class, an M0.85 cruise speed, with 30 per cent greater fuel offload than the KC-135R/T Stratotanker. In the cargo role, AMA will carry a payload of 45,360 to 54,430 kg (100,000 to 120,000 lb) over 4,000 n miles (7,408 km; 4,603 miles).

Over 40 advanced aircraft designs examined, leading to further study of four basic concepts, all of which feature modular design using common basic structure and systems in order to reduce initial manufacturing costs and enable airframe upgrades during service life. Modular systems and avionics bus architecture will easily accommodate mission-orientated equipment for specific roles. Configurations studied include a conventional high-wing aircraft; a blended wing/body aircraft; a box-wing aircraft with two refuelling booms and two hose-and-drogue assemblies (see accompanying illustration); and a global transport with an unrefuelled range of 12,000 n miles (22,220 km; 13,810 miles).

Design effort directed to the box-wing aircraft concept during 1997-2000, by virtue of aerodynamic and structural efficiency, combined with greatly reduced aircraft size. As now envisaged, it will have two flight deck crew, plus advanced refuelling and loadmaster workstations; incorporate roll-on/roll-off cargo handling capability and be compatible with 20 and 40 ft ISO containers; and embody fly-by-light/power-by-wire flight control systems. Testing of a radio-controlled scale model was begun on 7 March 1997, exhibiting excellent flight characteristics and meeting, or surpassing, test objectives during its 18-sortie programme.

Current planning anticipates AMA development effort to reach a peak during 2004-13, with resultant production aircraft ready for delivery to USAF and other customers from 2013; world market estimated at 700 to 1,000 by 2030.

UPDATED

LOCKHEED MARTIN ADVANCED THEATER TRANSPORT

TYPE: Medium transport/multirole.
PROGRAMME: Artist's impression of private venture studies published in 1999; envisaged as complement to C-130J Hercules. No recent news received.
CURRENT VERSIONS: **'C-X'**: Baseline transport.
'MC-X': Special forces' support aircraft.
'AC-X': Gunship.
'EC-X': Battlefield management aircraft.
'RC-X': Reconnaissance version.
'P-X': Maritime patrol aircraft.

UPDATED

LOCKHEED MARTIN ADVANCED TRANSPORT STUDIES

TYPE: New-concept airliner.
PROGRAMME: In 1998, Lockheed Martin began two NASA-funded studies into advanced concepts for use in future passenger and cargo aircraft. Lockheed Martin concentrated on box wing and strut-braced wing transports with ability to carry 325 passengers over range of 7,500 n miles (13,890 km; 8,630 miles). These revolutionary design concepts are expected to result in reductions of more than 20 per cent in fuel burn, weight and operating cost when compared with existing transport aircraft.

Lockheed Martin also developed concept for a box wing transport able to carry 600 passengers or comparable cargo payload over the same range. So-called 'Super Freighter' concept provides 50 per cent more payload and lower operating costs than Boeing 747, while having similar wing span. Cargo volume 991 m³ (35,000 cu ft); cargo capacity 158,750 kg (350,000 lb).

Studies entered their second phase in 2000, with potential of flying demonstrators of box wing and/or strut-braced wing aircraft in 2003 as part of NASA's Revolutionary Concepts programme.

Following data are approximate:
DIMENSIONS, EXTERNAL:
  Wing span: Box wing: 325 seat   56.4 m (185 ft)
    600 seat   68.6 m (225 ft)
  Strut-braced wing: 325 seat   66.8 m (219 ft)
  Length: Box wing: 325 seat   56.7 m (186 ft)
    600 seat   85.3 m (280 ft)
  Strut-braced wing: 325 seat   73.5 m (241 ft)
WEIGHTS AND LOADINGS:
  Operating weight empty:
    Box wing: 325 seat   127,400 kg (280,900 lb)
    600 seat   200,575 kg (442,200 lb)
    Strut-braced wing: 325 seat   122,525 kg (270,100 lb)
  Max T-O weight:
    Box wing: 325 seat   243,100 kg (535,950 lb)
    600 seat   416,625 kg (918,500 lb)
    Strut-braced wing: 325 seat   244,975 kg (540,100 lb)
PERFORMANCE:
  Design range, with reserves: all
    7,500 n miles (13,890 km; 8,630 miles)

UPDATED

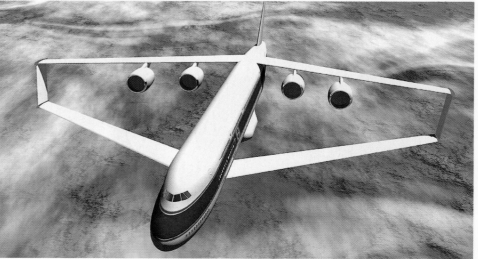
Four-engine, box wing 'Super Freighter' concept 2000/0084013

LOCKHEED MARTIN—AIRCRAFT: USA

PALMDALE OPERATIONS
1011 Lockheed Way, Palmdale, California 93599-3740
Tel: (+1 805) 572 41 55
Fax: (+1 805) 572 41 63
SITE MANAGER: Bob Elrod
DIRECTOR, COMMUNICATIONS: Ronald C Lindeke

Nickname, Skunk Works became official title when former Lockheed Advanced Development Company renamed in 1995, but formally renamed Palmdale Operations in January 2000. Functions autonomously at Palmdale; has specialised in 'black' or covert development programmes, including U-2/TR-1, SR-71 and F-117A and retains responsibility for support of U-2 and F-117 aircraft. Participant in Joint Strike Fighter (JSF) project and responsible for the production of both X-35 prototypes, even though Fort Worth Operations (which see) has programme leadership within Lockheed group. Selected in July 1996 to build the X-33 VentureStar technology demonstrator for future Reusable Launch Vehicle (RLV), as described under NASA later in this section.

Organisation expanded in latter half of 1995 through assumption of responsibility for Lockheed Martin Aircraft Services at Ontario, California.

Former separate division has designed, fabricated and installed major aircraft structural modifications and integrated complete avionic systems for US military, foreign governments and commercial customers.

UPDATED

LOCKHEED MARTIN AEROCRAFT
TYPE: New-concept airliner.
PROGRAMME: Revealed February 1999. Original 'pure' airship concept abandoned in favour of hybrid (and thus heavier than air) design. Expected to require combined civil/military development funding in event of a go-ahead. Projected in both rigid and non-rigid forms. In August 1999, NASA announced plans for US$10 million funding for a 7.8 per cent scale AeroCraft piloted testbed to fly in 2001 at its Dryden base. Studies continued in 2000.

DESIGN FEATURES: Hybrid design, essentially an airship but deriving only half its lift from helium gas, remainder being provided by vertically orientated turboshaft engines driving proprotors. Envelope broadly of aerofoil shape in side elevation, with twin fins and rudders; elliptical cross-section, approximately half of overall length at its widest point. Multiple-bogie landing gear. Flight deck inside envelope nose. Roll-on/roll-off cargo loading. Envisaged as possible future strategic airlifter with transatlantic range.
POWER PLANT: Four tiltrotors, mounted on stub-wings.
Following data all provisional:
DIMENSIONS, EXTERNAL:
   Length overall   255 m (835 ft)
WEIGHTS AND LOADINGS:
   Payload   453,600 kg (1 million lb)
   Max T-O weight   1,400,000 kg (3.1 million lb)
PERFORMANCE:
   Typical cruising speed   125 kt (231 km/h; 144 mph)
UPDATED

FORT WORTH OPERATIONS
1 Lockheed Boulevard, Fort Worth, Texas 76108
Tel: (+1 817) 777 20 00
Fax: (+1 817) 763 47 97
Web: http://www.lmtas.com
MANAGER, F-16 PROGRAMMES: William B Anderson
EXECUTIVE VICE-PRESIDENT, JSF PROGRAMME:
   C T 'Tom' Burbage

General Dynamics' Fort Worth Division sold to Lockheed; became Lockheed Fort Worth Company on 1 March 1993; renamed LMTAS following merger between Lockheed and Martin Marietta in 1995 and further renamed as Fort Worth Operations in January 2000.

Activities include production of F-16 Fighting Falcon; previously shared development of F-22A ATF (with Lockheed and Boeing). Also principal US subcontractor for Japanese Mitsubishi F-2 fighter and anticipates over US$2,000 million of production work related to manufacture in Japan; provides technical support for Taiwan's AIDC Ching-Kuo combat aircraft programme. Has programme leadership of Lockheed Martin team for JSF programme (see below and US Navy entry).

VERIFIED

LOCKHEED MARTIN (220) X-35 AND (235) JOINT STRIKE FIGHTER
TYPE: Multirole fighter.
PROGRAMME: US Department of Defense designation X-32 originally allocated for technology demonstrator phase of Advanced Research Projects Agency (ARPA) programme. CALF project initially launched as STOVL Strike Fighter (SSF; envisaged as X-32B) to provide US Navy/Marine Corps with replacement for F/A-18 Hornet and AV-8B Harrier II; later expanded to include USAF F-16 replacement candidate (X-32A), in which vertical lift system eliminated in favour of additional fuel to produce longer-range, conventional take-off aircraft.

Lockheed Martin one of three teams originally in contention for X-32 contract; others headed by Boeing and McDonnell Douglas; Lockheed Martin partnered by Allison, Pratt & Whitney and Rolls-Royce engine manufacturers. Congressional pressure in late 1994 resulted in CALF project merging with Joint Advanced Strike Technology (JAST) programme before being renamed as Joint Strike Fighter (JSF) in late 1995 (see US Navy entry). Design competition for JSF based on elimination of one proposal, leaving two to progress to concept demonstration (CD) phase involving comparative evaluation and eventual selection of one for engineering and manufacturing development (EMD) and eventual service with USAF, US Navy/Marine Corps and Royal Navy.

Lockheed Martin produced 91 per cent scale powered model of JAST demonstrator for wind-tunnel tests and in June 1994 revealed agreement with Yakovlev of Russia to purchase data on cancelled Yak-141 programme, which employed similar articulating rear nozzle as part of its propulsion system. Model JAST, with F100 engine, began trials at Pratt & Whitney's West Palm Beach, Florida, facility in February 1995; subsequently to outdoor hover test rig at NASA Ames and then installed in 24 × 36 m wind tunnel at Mountain View, California, for series of powered hover and transition tests which ran from December 1995 to 5 March 1996; total of 196 hours accumulated, representative of approximately 2,400 take-offs and landings with the vertical lift system. By early 1996, Lockheed Martin's contender had been reconfigured to eliminate canards.

Final Lockheed Martin JSF CD phase proposal submitted to US DoD on 13 June 1996; company rewarded on 16 November 1996 with US$718.8 million contract covering design, development, construction and flight testing of two full-scale demonstrator aircraft, designated X-35, while original X-32 designation allocated to Boeing contender (which see). Contract required one initially to be flown as CTOL version (X-35A) to demonstrate land-based USAF model, before being reconfigured to serve as STOVL version (X-35B) for US Marine Corps, Royal Air Force and Royal Navy; other to carrier-capable US Navy model (X-35C). Although Fort Worth Operations is team leader, both X-35 aircraft built at Palmdale, California, using rapid prototyping techniques; subsequent production, if Lockheed Martin is successful in this competition, will probably be concentrated at Fort Worth, Texas.

Design of X-35 frozen as **Configuration 220** 13 June 1997 after Initial Design Review, at which time 11,000 hours of model testing had been accumulated. Development team joined by Northrop Grumman on 8 May 1997 and BAe (now BAE Systems) on 18 June 1997.

Release of engineering drawings in early September 1997 heralded start of parts production at Palmdale for both demonstrator aircraft; by end of 1997, Lockheed Martin had completed about 70 per cent of required tooling and had conducted second interim programme review. X-35 final design review completed in September 1998 and coincided with roll-out of full-scale mockup at Fort Worth. Assembly of first aircraft began in April 1998, with main wing carry-through bulkhead installed in early August 1998, by which time manufacturing of large composites skins for upper and lower wing surfaces also complete; aircraft moved from assembly testing to factory floor on 18 September 1999, in readiness for installation of flight control surfaces and landing gear as well as systems checks.

Flight control system software tested on NF-16D VISTA in 1998 as part of integrated subsystem technology demonstration. Further trials using AFTI/F-16 in 1999-2000 entailed testing of all-electric flight control system and modular electric power system planned for JSF. JSF avionics also tested on Northrop Grumman's BAe One-Eleven Co-operative Avionics Test Bed (CATB), which was fitted with sensors, processors and software in 1999; trials begun of Northrop Grumman radar and distributed infra-red sensor system, Kaiser helmet-mounted display and Lockheed Martin core processor in first quarter 2000.

More than 100 hours of flight time accumulated by early September 2000 using CATB, when avionics development and integration process completed; successful demonstrations included automatic target cueing (ATC), whereby sensors acquire targets rapidly and automatically; electro-optical targeting system (EOTS); electronic warfare suite and electronically scanned radar. Subsequent testing included co-operative engagement between CATB and Northrop Grumman E-8C Joint STARS, demonstrating all-weather precision targeting and combat identification techniques for fixed and moving targets.

Design of JSF continued to be refined after selection of Configuration 220. By 1998, third version of **Configuration 230 (233)** had reduced area of USAF/USMC JSF wings by seven per cent, but increased USN

Lockheed Martin X-35 prototype

USA: AIRCRAFT—LOCKHEED MARTIN

X-35A conducting its maiden flight on 24 October 2000

variant's wing area by 11 per cent. Further redesign occurred in 1999, culminating in September with Configuration 235, which has enlarged wing to satisfy 9 g stress requirement; further main change involved redesign of lift-fan nozzle from D-shape extendible box to venetian blind-type box, offering dual benefits of simpler design and reduced weight.

Redesign process continued into 2000, with Configuration 235 adding further refinements aimed at reducing weight and increasing payload bringback capability, which reportedly 227 kg (500 lb) below specified figure of 2,268 kg (5,000 lb), although Lockheed Martin claimed to have comfortably exceeded that; resultant structural modification entailed weight reduction in several areas such as weapon bay and landing gear door assemblies. Configuration 235 finally submitted as Preferred Weapons System Concept design in mid-2000.

Significant programme events in latter part of 1998 and 1999 included first test run of basic Pratt & Whitney JSF119 engine (designated FX661) at West Palm Beach, Florida, facility on 11 June 1998; over 330 hours of developmental testing performed by end August 1999, validating complete X-35 flight envelope. Final altitude flight qualification testing undertaken with another engine (designated FX663) in fourth quarter of 1999. National Aerospace Laboratory low-speed wind tunnel in the Netherlands used for testing of scale models of JSF starting in August 1998; initial trials of lift fan system's clutch, fan and gearbox rigs at Indianapolis, Indiana, in May and June 1998; and installation of engine inlet duct in assembly tool at Palmdale at beginning of July 1998. Subsequently, also in July, over 50 hours of preliminary engine testing were completed at West Palm Beach, including vibration surveys, fan and core running-in, operating performance calibration and engine control stability assessment; next key phase of engine test programme involved altitude testing at Arnold Engine Development Center, Tennessee. Testing of first STOVL engine (designated FX662) undertaken initially at West Palm Beach; first run with lift fan engaged accomplished on 10 November 1998, with operation to 100 per cent speed following on 22 November; first series of tests included stress surveys and performance calibration and occupied about six weeks. Validation of STOVL propulsion system achieved in late August 1999, with high-power clutch engagement of shaft-driven lift fan, simulating seamless conversion from conventional wing-borne flight configuration to jet-borne flight configuration for STOVL approach ending with vertical landing.

Following this, on 9 December 1999, flight engine YF001 successfully installed on the X-35A demonstrator at Palmdale, with integration tests, including plumbing connections, data communications and electrical checks, beginning immediately thereafter.

In UK, DERA vectored-thrust advanced aircraft control (VAAC) Harrier T. Mk 4 completed 20 hour, 36 sortie, flight test programme in November 1998, during which a sidestick control column was evaluated by civilian and military pilots. Two-phase programme began with initial calibration to validate control laws and stick characteristics; subsequent evaluation included pattern work, approach and transition to hover and precision and aggressive hover tasks, resulting in confirmation that side-stick provides satisfactory control of STOVL aircraft at low speeds.

Lockheed Martin also reached preliminary agreement with partners over work-sharing arrangements to be implemented if X-35 selected for large-scale production; 1998 plan anticipates Lockheed Martin having responsibility for cockpit and wing design, with fuselage and tail sections allocated to Northrop Grumman and BAE Systems, respectively. Wider international involvement strategy also drawn up by Lockheed Martin in closing stages of 1998.

Radar signature testing of full-scale pole model began in late 1999 at Helendale, California; this work will include measurement of radar cross-section, assessment of antenna performance and demonstration of the robustness of supportable low-observable materials.

Significant events in 2000 included X-35A flight readiness review in March; at that time, maiden flight expected to take place in June or July. Subsequently, in April, testing associated with development and flight qualification of the JSF119-PW-611 engine for the X-35A and X-35C was completed after 193 hours of operating time. Final assembly and painting of X-35A accomplished in late May and was followed by lengthy series of ground tests, including systems checkout, engine running at full military power and with full afterburner augmentation, and low- and medium-speed taxi tests. These extended into October and culminated in a successful first flight from Palmdale to Edwards; 22 minute sortie by X-35A Article 301 occurred on 24 October and included landing gear retraction, as well as preliminary assessment of handling characteristics and verification of test data transfer systems; by 6 November, further four flights had been made, including first by USAF pilot, and envelope had been expanded to 390 kt (722 km/h; 449 mph); first aerial refuelling on 8 November; maiden supersonic flight 21 November, when 25 hours had been flown in 25 sorties. On completion of initial envelope expansion, X-35A was planned to return to Palmdale for conversion to X-35B; this expected to begin at about the end of November 2000, at which time the X-35C (Article 300) was being prepared to make its first flight, which took place on 16 December 2000 from Palmdale to Edwards AFB. Previously, X-35C used for structural loads testing programme, which was completed in June 2000.

CURRENT VERSIONS: Three variants of basic design, optimised for the mission requirements of different armed services. As currently envisaged, company designation for operational JSF is Configuration 235.

US Air Force: Land-based CTOL. To be evaluated by **X-35A.**

US Marine Corps, Royal Air Force and **Royal Navy:** STOVL. Engine-driven lifting fan behind cockpit replaces some fuel. Wing folding originally specified on Royal Navy version only, but requirement since dropped. To be evaluated by **X-35B.**

US Navy: Carrier-based CTOL. Wing, fin and elevator areas increased by chord extension; ailerons in addition to flaperons on wing; enlarged control surfaces and modified control system; strengthened landing gear with catapult launch bar on nose leg; arrester hook; and folding wing. To be evaluated by **X-35C.**

CUSTOMERS: Two X-35 demonstrators; potential for some 3,000 JSFs for USA and UK.

DESIGN FEATURES: Trapezoidal mid-wing configuration, optimised for low observability from frontal aspect. Twin tailfins; internal weapon bays. Wing and tailplane leading-edges swept back approximately 35°; trailing-edges swept forward approximately 15°; fins swept back approximately 35° and canted outward at tips by approximately 25°. Twin 'divertless' fibre-placed graphite/epoxy composites engine air intakes with no moving parts produced by ATK. All-electric flight control system.

STOVL version employs a lifting fan behind the cockpit, driven by a shaft from the single engine; inlet and outlet are covered by doors, except when in use. The resultant cold air cushion prevents hot air from being reingested when on or near the ground.

FLYING CONTROLS: All-electric flight control system for movement of primary flight control surfaces (flaps, tailerons and rudder) incorporating Parker Hannifin electric actuators. Moog leading-edge flap drive system. Flight control computer originally to be supplied by Honeywell, but replaced by advanced Lockheed Martin unit in third quarter of 1998 to eliminate anticipated throughput problems arising from growth in flight control software.

LANDING GEAR: Retractable tricycle type; mainwheels retract inwards; nosewheel(s) forward. Single wheels on each unit, except twin nosewheels and catapult towbar on naval (220C) variant, which also has reinforced gear for deck landings. Tyres by Goodyear will feature embedded transponder including integrated circuit and capacitive pressure sensor to facilitate monitoring of pressure and condition.

POWER PLANT: One Pratt & Whitney JSF119-PW-611 (F119 derivative) turbofan. Axisymmetric (thrust-vectoring) exhaust nozzle on USAF and US Navy -611C versions. Rolls-Royce three-bearing swivel-duct nozzle on -611S version to deflect thrust downwards for STOVL, plus a Rolls-Royce engine-driven fan behind cockpit and a bleed air reaction control valve in each wingroot to provide stability at low speeds. Total vertical lift of 164.6 kN (37,000 lb st) comprises 66.7 kN (15,000 lb st) from main nozzle, 80.1 kN (18,000 lb st) from fan and combined 17.8 kN (4,000 lb st) from reaction control valves. USAF version has in-flight refuelling receptacle on spine; US Navy specifies retractable probe on starboard side.

ACCOMMODATION: Pilot only; canopy by Sierracin; STOVL versions have canopy of reduced length. Martin-Baker Mk 16E lightweight ejection seat in X-35s; production version expected to have new seat resulting from Joint Ejection Seat Program (JESP).

SYSTEMS: Hamilton Sundstrand 80 kW engine-driven switched-reluctance starter/generator providing two independent channels of 270 V DC electrical power; electric distribution units, power panels, power distribution centres, batteries and battery charger equipment provided by Smiths Industries; Honeywell thermal- and energy-management module (T/EMM) combining functions of auxiliary and emergency power units and environmental control system; 270 V DC emergency battery. Weapons bay door drive system by TRW Aeronautical Systems.

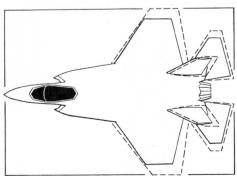

Plan view of Lockheed Martin JSF USAF/USMC version and broken outline of USN version *(Paul Jackson/Jane's)*

LOCKHEED MARTIN—AIRCRAFT: USA

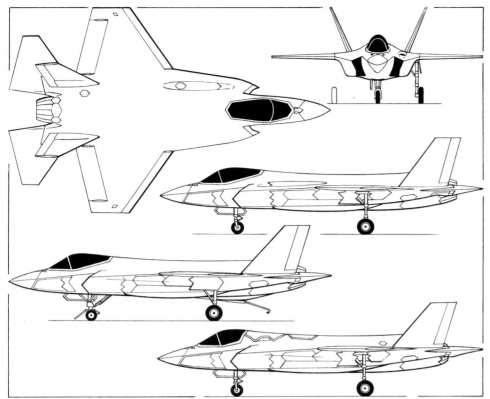

USAF (Configuration 220A) version of X-35 JSF, with additional side views of US Navy (220C; centre) and Marine Corps/Royal Navy (220B; bottom) versions *(James Goulding/Jane's)* 0003306

AVIONICS: *Radar:* Northrop Grumman MIRFS/MFA (multifunction integrated RF system/multifunction nose array) active electronically scanned array combining radar, electronic warfare and communications functions.

Flight: Lockheed Martin Tactical Defense Systems Integrated Core Processor (ICP) incorporating open system architecture.

Instrumentation: Kaiser 200 × 500 mm (8 × 20 in) single flat panel MFD. Meggitt secondary flight display system. Kaiser/Vision Systems International selected in third quarter of 1999 to provide advanced integrated helmet-mounted display system.

Mission: Northrop Grumman conformal array multifunction imaging IR sensor system under development for JSF; distributed architecture IR system (DAIRS) functions include air-to-air search-and-track, target cueing and missile warning, and air-to-ground surface-target tracking; uses six conformal compact lightweight imaging IR sensors around airframe, with combined data providing all-aspect multifunction imaging to pilot via wide-angle helmet-mounted display, overlaid with target and threat data information.

Self-defence: EW capability incorporated into MIRFS/MFA. Sanders is prime contractor for EW equipment and systems integration; Litton Amecom to supply low-cost ESM equipment.

ARMAMENT: Internal cannon specified by USAF and under consideration by US Navy; Boeing/Mauser BK 27 (27 mm) weapon selected in second quarter of 1999; US Marine Corps and UK variants to have 'missionised' cannon in low-observables pod. Internal weapon bays; preferred USAF weapons fit is two AIM-120 AMRAAMs and two 1,000 lb JDAM bombs; US Navy specifies 2,000 lb JDAM (which optional on all variants).

Data are provisional and apply to Configuration 220/X-35 and Configuration 230/production JSF.

DIMENSIONS, EXTERNAL:
Wing span: 220	10.05 m (33 ft 0 in)
230A, 230B	10.70 m (35 ft 1¼ in)
230C	13.10 m (42 ft 11¾ in)
Wing aspect ratio: 220A, 220B	2.4
230A, 230B	2.7
220C	2.0
230C	3.0
Width, wings folded:	
220C	9.14 m (30 ft 0 in)
Length overall: 220, 230	15.47 m (50 ft 9 in)

WEIGHTS AND LOADINGS:
Empty weight:
220A	11,033 kg (24,324 lb)
230A	12,010 kg (26,477 lb)
220C	12,117 kg (26,713 lb)
230C	13,536 kg (29,842 lb)

AREAS:
Wings, gross: 220A, 220B	41.81 m² (450.0 sq ft)
230A, 230B	42.7 m² (460 sq ft)
220C	50.17 m² (540.0 sq ft)
230C	57.6 m² (620 sq ft)

UPDATED

LOCKHEED MARTIN F-16 FIGHTING FALCON

Israel Defence Force names: F-16A/B Netz (Falcon), F-16C Barak (Lightning) and F-16D Brakeet (Thunderbolt)
Turkish Air Force name: Savaşan Şahin (Fighting Falcon)

TYPE: Multirole fighter.

PROGRAMME: Emerged from YF-16 of US Air Force Lightweight Fighter prototype programme 1972 (details under General Dynamics in 1977-78 and 1978-79 *Jane's*); first flight of prototype YF-16 (72-01567) 2 February 1974; first flight of second prototype (72-01568) 9 May 1974; selected for full-scale development 13 January 1975; day fighter requirement extended to add air-to-ground capability with radar and all-weather navigation; production of six single-seat F-16As and two two-seat F-16Bs began July 1975; first flight of full-scale development aircraft 8 December 1976; first flight of F-16B 8 August 1977. F-16 achieved 5 millionth flying hour late in 1993; 7 millionth flying hour passed in early 1997; 3,900th aircraft delivered 26 March 1999, coincidentally 3,035th from Fort Worth factory and 2,202nd for USAF. 4,000th aircraft delivered (to Egyptian Air Force) on 28 April 2000.

USAF still procuring small quantities for attrition reserve and plans to obtain an additional 30 to replace F-16A/Bs in the Air National Guard. Latest contract, signed 26 December 2000, is for four aircraft acquired with FY01 funds and delivered between October 2002 and January 2003.

Lifting of US embargo on sales of major weapons systems to Latin America offers potential of export to Argentina, Brazil and Chile; and supply of additional F-16s to Venezuela, which is only current operator in this region. Elsewhere, sales announced in 1999-2000 to Egypt, Greece, Israel, Singapore, South Korea and the United Arab Emirates are sufficient to assure production in medium-term. Total of 234 F-16s ordered in 2000, increasing sales to 4,285 as of 1 January 2001.

CURRENT VERSIONS: **F-16A:** First production version for air-to-air and air-to-ground missions; production for USAF completed March 1985, but still available for other customers; powered since late 1988 (Block 15OCU) by P&W F100-PW-220 turbofan; Westinghouse AN/APG-66 range and angle track radar; first flight of first aircraft (78-0001) 7 August 1978; entered service with 388th TFW at Hill AFB, Utah, 6 January 1979; combat ready October 1980, when named Fighting Falcon; most now serving ANG and USAF test community; power plants upgraded to F100-PW-220E, between 1991 and 1996. Also produced in Europe. Built in Blocks 01, 05, 10 and 15, of which Blocks 01 and 05 retrofitted to Block 10 standard 1982-84; Block 15 retrofitted to OCU standard from late 1987. First **'GF-16A'** (unofficial designation) ground trainers relegated to instructional use at 82nd Training Wing, Sheppard AFB, by 1993.

Taiwanese F-16A/Bs, delivery of which began July 1996 for pilot training in USA and to Taiwan in April 1997, are to Block 20 standard; avionics configuration similar to MLU, including modular mission computer, AN/ALE-47 chaff/flare dispenser, colour displays and AN/APX-111 IFF; Block 20 combat potential exceeds that of early F-16Cs.

Operational Capabilities Upgrade (OCU): USAF programme to equip F-16A/B for next-generation BVR air-to-air and air-to-surface weapons; radar and software updated, fire-control and stores management computers improved, data transfer unit fitted, combined radar/barometric altimeter fitted, and provision for AN/ALQ-131 jamming pods. Ring laser INS planned for 1990s. FMS exports since 1988 to Block 15OCU standard with F-16C features including ring laser INS, wide-angle conventional optics HUD, F100-PW-220 power plant, certification for 17,009 kg (37,500 lb) maximum take-off weight and AIM-9P-4 Sidewinder AAM capability.

Mid-Life Update (MLU): Development authorised 3 May 1991 (signature of final partner); US government contract to GD 15 June 1991; originally planned to be applied to 533 aircraft of USAF (130), Belgium (110), Denmark (63), Netherlands (172) and Norway (58) from 1996 in co-development/co-production programme. USAF withdrew 1992, but ordered 223 modular computer retrofit kits from MLU to equip Block 50/52 aircraft. European share re-negotiated on 28 January 1993; letters of offer and acceptance finalised 30 June 1993. Lockheed awarded contract for 301 MLU kits for European air forces (including one USAF prototype) on 17 August 1993; additional kits ordered in 1998 and 1999, raising total to 360, comprising 92 for Belgium, 70 for Denmark, 139 for Netherlands, 58 for Norway and one USAF prototype. Kit deliveries to air forces began in October 1996, with completion of original MLU project set for 2000. First production MLU aircraft flew in mid-1997. Unofficial designations **F-16AM** and **F-16BM** for upgraded aircraft adopted by Netherlands and also thought to be in use by Belgium, Denmark and Norway.

Cockpit similar to F-16C/D Block 50 with wide-angle HUD, night vision goggle compatibility, modular mission computer replacing existing three core computers, Fairchild digital terrain module (incorporating Terprom algorithms developed by BAe), AN/APG-66(V)2A fire-control radar, GPS, improved data modem and two Honeywell 10 cm (4 in) square LCD colour displays. Inlet hardpoints and wiring for FLIR pods will be added to Block 10 aircraft of Netherlands and Norway. Marconi Hazeltine AN/APX-113 IFF interrogator/transponder also selected as standard, with provisions for microwave landing system (MLS), helmet-mounted display and reconnaissance pod.

Five aircraft for prototype conversion to GD in September 1992, comprising Belgian F-16A FA-93, Danish F-16B ET-204, Netherlands F-16B J-650, Norwegian F-16A 299 and USAF/ANG F-16A 80-0584. First flight (by last-mentioned aircraft) 28 April 1995; formal debut 11 May 1995. Three prototypes (Belgian, Netherlands and Norwegian aircraft) were due to begin trials at Leeuwarden, Netherlands, in early May 1996, but

Lockheed Martin X-35C US Navy version

US Air Force Lockheed Martin F-16D Fighting Falcon *(Paul Jackson/Jane's)*

start of programme was delayed until late July because of necessity to rectify below-specification wire harnesses; additional four 'lead the fleet' aircraft converted in Europe (one at each of the four depots) were unaffected by this problem and began to join the trials programme on schedule in August 1996. Additional testing (Danish and USAF aircraft) undertaken at Edwards AFB, California, and was completed in November 1997. Further test on 12 December 1997 involved successful launch of AIM-120 AMRAAM at Eglin AFB, Florida, from Danish F-16B, with captive missile used to simulate second launch; live weapon destroyed drone target and captive weapon also adjudged successful following analysis of test data. Flight testing completed in late 1997. IOC (No 322 Squadron, Royal Netherlands Air Force) declared 13 June 1998. All four countries currently modifying aircraft in respective depots. Software update to M2 standard due to occur in 2000 and M3 in 2003.

Portugal became first outside country to purchase MLU in 1998, with order for 20 kits as part of package involving acquisition of 25 former USAF aircraft (21 F-16As and four F-16Bs). Total of 16 F-16As and all four F-16Bs to be upgraded, with Portugal planning to establish in-country modification line; remaining five F-16As to be broken down and used as spares source.

F-16(ADF): Modification of 279 (actually 272 because of preconversion attrition) Block 15 F-16A/Bs as USAF air defence fighters to replace F-4s and F-106s with 11 Air National Guard squadrons; ordered October 1986. Modifications include upgrade of AN/APG-66 radar to improve small target detection and provide continuous-wave illumination, provision of AMRAAM datalink, improved ECCM, AlliedSignal AN/ARC-200 HF/SSB radio (F-16A only), Teledyne/E-Systems Mk XII advanced IFF, provision for Navstar GPS Group A, low-altitude warning, voice message unit, night identification light (port forward fuselage of F-16A only), and ability to carry and guide two AIM-7 Sparrow missiles. First successful guided launch of AIM-7 over Point Mugu range, California, February 1989; F-16(ADF) can carry up to six AIM-120 AMRAAM or AIM-9 Sidewinder or combinations of all three missiles; retains internal M61 20 mm gun. GD converted one prototype, then produced modification kits for installation by USAF Ogden Air Logistics Center, Utah, in conjunction with upgrade to OCU avionics standard; first Ogden aircraft, F-16B 81-0817, completed October 1988. Development completed at Edwards AFB during 1990; operational test and evaluation with 57th Fighter Weapons Wing at Nellis AFB, Nevada; first F-16(ADF), 81-0801, delivered to 114th Fighter Training Squadron at Kingsley Field, Oregon, 1 February 1989; 194th Fighter Interceptor Squadron, California ANG, Fresno, achieved IOC in 1989, following receipt of first aircraft (F-16B 82-1048) on 13 April 1989; first AIM-7 launch by ANG (159th FS) June 1991. Programme completed early 1992; included approximately 30 F-16Bs. Majority of F-16(ADF) aircraft now in storage.

F-16/UCAV: This LMTAS proposal, based on feasibility studies into adapting the F-16A for the Uninhabited Combat Aerial Vehicle (UCAV) role, emerged in late 1996. To date, minimal customer interest expressed and LMTAS is not actively marketing this concept, for which more details can be found in 2000-01 *Jane's*. However, in November 1997, LMTAS revealed arrangement with Tracor Flight Systems for mutual support on adapting F-16 as potential candidate for USAF Next-Generation Aerial Target and UCAV programmes.

F-16B: Standard tandem two-seat version of F-16A; fully operational both cockpits; fuselage length unaltered; reduced fuel.

'GF-16B': Unofficial designation allocated to non-flying aircraft in use for instructional tasks at Sheppard AFB by 1993.

F-16C/D: Single-seat and two-seat USAF Multinational Staged Improvement Program (**MSIP**) aircraft respectively, implemented February 1980. MSIP expands growth capability to allow for ground attack and BVR missiles, and all-weather, night and day missions; **Stage I** applied to Block 15 F-16A/Bs delivered from November 1981 included wiring and structural changes to accommodate new systems; **Stage II** applied to Block 25 F-16C/Ds from July 1984 includes core avionics, cockpit and airframe changes. Block 25 aircraft originally delivered with F100-PW-200 engine, but all surviving examples now have F100-PW-220E following retrofit, as well as other improvements as detailed under Stage III of MSIP. **Stage III** includes installation of systems as they become available, beginning 1988 and extending to Block 50/52, including selected retrofits back to Block 25. Changes include Northrop Grumman AN/APG-68 multimode radar with improved range, resolution, more operating modes and better ECCM than AN/APG-66; advanced cockpit with multifunction displays and upfront controls, BAE Systems wide-angle HUD, Fairchild mission data transfer equipment and radar altimeter; expanded base of fin giving space for proposed later fitment of AN/ALQ-165 Airborne Self-Protection Jamming system (since cancelled, though now being installed in Korean aircraft); increased electrical power and cooling capacity; structural provision for increased take-off weight and manoeuvring limits; and smart weapons such as AIM-120A AMRAAM and AGM-65D IR Maverick.

Common engine bay introduced at **Block 30/32** (deliveries from July 1986) to allow fitting of either P&W F100-PW-220 (Block 32) or GE F110-GE-100 (Block 30) Alternate Fighter Engine. Other changes include computer memory expansion and seal-bonded fuselage fuel tanks. First USAF wing to use F-16C/Ds with F110 engines was 86th TFW at Ramstein AB, Germany, from October 1986. Additions in 1987 included voice message unit, doubled chaff/flare capacity, repositioning of RWR antennas to provide better coverage in forward hemisphere, Shrike anti-radiation missiles (from August 1997), crash survivable flight data recorder and modular common inlet duct allowing full thrust from F110 at low airspeeds.

Software upgraded for full Level IV multitarget compatibility with AMRAAM early 1988. Industry-sponsored development of radar missile capability for several European air forces resulted in firing of AIM-7F and AIM-7M missiles from F-16C in May 1988; capability introduced mid-1991; missiles guided using pulse Doppler illumination while tracking targets in a high PRF mode of the AN/APG-68 radar.

Block 40/42 Night Falcon (deliveries from December 1988) upgrades include AN/APG-68(V) radar allowing 100 hour operation before maintenance, full compatibility with Lockheed Martin low-altitude navigation and targeting infra-red for night (LANTIRN) pods, four-channel digital flight control system, expanded capacity core computers, diffractive optics HUD, enhanced envelope gunsight, GPS, improved leading-edge flap drive system, improved cockpit ergonomics, high gross weight landing gear, structural strengthening, increased performance battery and provision for improved EW equipment, including advanced interference blanker. LANTIRN targeting pod gives day/night standoff target identification, automatic target handoff for multiple launch of Mavericks, autonomous laser-guided bomb delivery and precision air-to-ground laser ranging. LANTIRN navigation pod provides real-world IR view through HUD for night flight plus automatic/manual terrain following with dedicated radar sensor. Combat Edge pressure breathing system installed 1991 for higher pilot g tolerance.

A total of 39 Block 40 F-16C/Ds of the 31st FW was involved in a quick response capability (QRC) modification effort, known as 'Sure Strike', to install Improved Data Modem (IDM) equipment. Work was undertaken by a joint USAF/LMTAS team at Aviano AB, Italy, and was completed in December 1995, these being the first Block 40 aircraft to receive the IDM which is standard equipment on current production Block 50 F-16s. In mid-1998, a demonstration programme was conducted to adapt existing IDM with Lockheed Martin kit to provide 'Gold Strike' system capable of two-way transmission of digitised video imagery of targets and thus enhance pilot's situational awareness. Separate retrofit programme involves installation of AN/ALR-56M radar warning receiver and AN/ALE-47 chaff/flare dispensers on USAF Block 40/42 F-16C and F-16D aircraft.

First Block 40 F-16C/Ds issued in late 1990 to 363rd FW (Shaw AFB, South Carolina); first LANTIRN pods to 36th FS/51st FW at Osan, South Korea, in 1992. USAF Block 40/42 F-16Cs and F-16Ds unofficially designated **F-16CG** and **F-16DG**, respectively. Following retirement of F-111F, Block 40/42 F-16Cs and F-16Ds with LANTIRN comprise more than 50 per cent of USAF night/precision strike force.

Block 50/52 (deliveries began with F-16C 90-0801 in October 1991 for operational testing) upgrades include F110-GE-129 and F100-PW-229 increased performance engines (IPE), AN/APG-68(V5) radar with advanced programmable signal processor employing VHSIC technology, Have Quick IIA UHF radio and AN/ALR-56M advanced RWR. Changes initiated at **Block 50D/52D** in 1993 include full integration of HARM anti-radiation missiles via HARM aircraft launcher interface computer (ALIC), improved data modem (IDM), upgraded programmable display generator with growth potential for colour and map, expanded data transfer cartridge, ring laser INS (Honeywell and Litton units both used) and improved VHF/FM antenna. AN/ALE-47 advanced chaff/

Delivery of Bahrain's second batch of F-16 Block 40s began on 22 June 2000

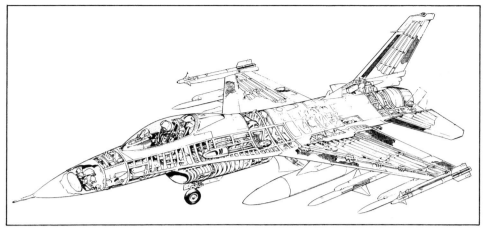

Cutaway drawing of Lockheed Martin F-16

flare dispenser fitted to all FMS Block 50 aircraft delivered after mid-1996 and incorporated as standard on USAF aircraft with effect from FY97 purchase; will also be retrofitted to earlier USAF machines.

First Block 50D/52D (91-0360) delivered to USAF on 7 May 1993; optimised for defence suppression missions, having software for horizontal situation display on existing two MFDs and provision for one of 112 HARM (AGM-88 High-Speed Anti-Radiation Missile) targeting systems ordered by USAF; sensor in pod on starboard side of engine inlet; AN/ASQ-213 HARM targeting system (HTS) has capability similar to F-4G 'Wild Weasel' which it replaced in SEAD role. USAF has programme to raise HARM targeting system inventory to 150 pods and incorporated an upgrade in FY99 that features software and hardware improvements enabling more targets to be tracked with enhanced ambiguity resolution and speedier reaction time.

Deliveries of Block 50/52 began to 4th FS of 388th FW at Hill AFB, Utah, from October 1992; others to 52nd FW at Spangdahlem, Germany, replacing Block 30 aircraft from (first delivery) 20 February 1993. Block 50D/52D aircraft initially to 309th FS (now 79th FS) of 363rd (now 20th) FW at Shaw AFB, South Carolina; followed by 23rd FS/52nd FW at Spangdahlem, Germany, from 14 January 1994, then squadrons at Mountain Home AFB, Idaho, and Misawa, Japan. Production for USAF was due to terminate with FY94 batch, but six additional F-16Cs funded in each of FY96 and FY97, plus three in FY98, one in FY99, 10 in FY00 and four requested for FY01; FY96 aircraft to Block 50 standard. FY97 and subsequent aircraft to improved configuration, incorporating advanced mission computer by Raytheon (replacing three core avionics processors) that was developed for F-16A/B MLU programme, Honeywell colour liquid crystal multifunction displays (replacing monochrome CRT MFDs), Honeywell colour programmable display generator, Teac colour airborne video tape recorder, colour cockpit TV sensor and Litton onboard oxygen generating system (OBOGS). USAF Block 50/52 F-16Cs and F-16Ds unofficially designated **F-16CJ** and **F-16DJ**, respectively.

First Samsung-assembled F-16C Block 52D rolled out in South Korea on 7 November 1995. First flight of initial TAI-built F-16C Block 50D in late May 1996, with delivery to Turkish Air Force following on 29 July; last IAI-produced F-16 delivered October 1999. First Block 50D delivered to Greece 29 January 1997. First Block 52D delivered to Singapore 30 January 1998. First production lease (PL) Block 52D aircraft for Singapore was delivered 28 May 1998; this was under terms of commercial contract, with delivery achieved in less than 24 months from placing of order.

Block 52+: Improved version for Greece, which revealed intention to buy as many as 58 at estimated cost of US$2 billion in May 1999 and subsequently placed firm order for 50. Aircraft will feature upgraded AN/APG-68(V)XM radar with greater detection range, plus an improved modular mission computer, two 102 mm (4 in) colour cockpit displays, helmet-mounted cueing system and a digital terrain system. EW equipment to be supplied indigenously, following decision to cancel international competition in third quarter 2000. Conformal 2,271 litre (600 US gallon; 500 Imp gallon) fuel tanks will be fitted to single- and two-seat versions and Greece is also contemplating acquisition of an off-boresight missile system. Engine will be F100-PW-229, with deliveries expected to begin in late 2002. Greece reportedly also considering retrofit programme, whereby its existing F-16 fleet could receive some or all of the Block 52+ modifications.

A 'generic' export version known as the **Viper 2000** is also being offered by Lockheed Martin. Intended to reduce costs by providing popular features in a common core configuration, this will include a modular mission computer, colour displays, conformal tanks and the Northrop Grumman AN/APG-68(V)X radar.

'GF-16C': Unofficial designation allocated to non-flying aircraft in use at Sheppard AFB for instructional purposes from 1993.

USAF F-16C/D Retrofit Programmes: All USAF F-16C/D aircraft are undergoing structural upgrade in a programme called 'Falcon UP'. This intended to ensure an 8,000 hour service life without depot inspection; late Block 50/52s receiving this during production, with earlier aircraft retrofitted concurrently with other depot-level upgrades. Follow-up effort called 'Falcon Star', in preparation in mid-1999, is expected to involve Block 40/42 and earlier aircraft delivered during 1984-91 and, possibly, the 100 or so remaining ANG F-16A/Bs. 'Falcon Star' will be a structural enhancement to sustain operational life; current USAF planning anticipates the modification programme to take six years to complete at a cost of between US$250 million and US$500 million. No start date yet determined, but work could begin in FY02.

ANG/AFRC Block 25/30/32 F-16C/Ds in process of receiving following upgrades: situation awareness datalink (SADL), improved airborne videotape recorder, colour camera, initial NVG-compatible cockpit lighting, LANTIRN targeting pod capability, AN/ALQ-213 countermeasures system and provisions for GPS/laser INS. Future improvements include expanded central computer, Rafael Litening II FLIR targeting pod for ANG/AFRC aircraft, joint helmet-mounted cueing system, AIM-9X missile, follow-on NVIS capability, PIDS-3 pylon upgrade for smart weapon compatibility, ACES II ejection seat improvements, enhanced main battery and software upgrades.

Block 40/42 F-16C/Ds are currently being upgraded with AN/ALR-56M RWR, AN/ALE-47 advanced chaff/flare dispenser and an improved battery system. Near-term upgrades include NVG compatibility improved data modem, digital terrain system, AN/ALE-50 towed decoy, common missile warning system (CMWS) and smart weapons compatibility (including the JDAM, JSOW and WCMD). Block 50/52 F-16C/Ds already have (or will receive) these capabilities.

USAF has also launched an upgrade effort known as common configuration implementation program (CCIP), which is intended to provide common hardware and software capability to almost 700 Block 40/42/50/52 aircraft. Modification is scheduled to begin in late 2003 and improvements that will be incorporated include modular mission computer core avionics suite, colour multifunction displays, Link 16 (JTIDS) datalink, joint helmet-mounted cueing system, AIM-9X, Joint Air-to-Surface Standoff Missile (JASSM), and various pilot/vehicle interface improvements. Block 50/52 additionally will receive AN/APX-113 advanced IFF system and a FLIR targeting system for delivering laser-guided bombs.

Block 60: On 12 May 1998, the United Arab Emirates announced selection of the Block 60 F-16, although signature of contract delayed by UAE decision to reopen competition for EW system; further announcement, on 5 March 2000, revealed that contractual agreement had been signed by UAE and Lockheed Martin. In third quarter of 2000, Israel reported to be considering acquisition of Block 60 version; USAF also contemplating it as interim fighter until JSF becomes available in substantial numbers from 2010. After an extensive development programme, 80 **Desert Falcon** aircraft will be delivered between 2004 and 2007. Major differences from the Block 50/52 F-16 include active electronically scanned array (AESA) radar, internal FLIR targeting system, advanced internal electronic countermeasures system, enhanced performance General Electric F110-GE-132 engine in the 144.6 kN (32,500 lb st) class, conformal tanks carrying an additional 1,893 litres (500 US gallons; 416 Imp gallons) of fuel, an advanced cockpit and Thomson-CSF secure radio and datalink equipment. These systems could be available for future F-16 versions.

A 'generic' export version of the Block 60 known as the **Viper 2100** is offered by Lockheed Martin, including Northrop Grumman active array agile beam radar and internal FLIR targeting system.

NF-16D: Variable stability in-flight simulator test aircraft (**VISTA**) modified from Block 30 F-16D (86-0048) ordered December 1988 to replace NT-33A testbed. Features include Calspan variable stability flight control system, fully programmable cockpit controls and displays,

Late standard F-16 cockpit, as installed in MLU aircraft

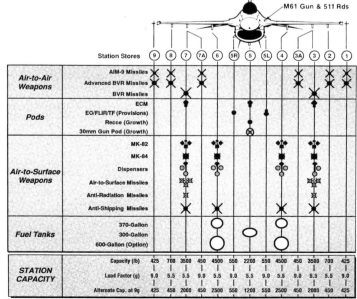

Lockheed Martin F-16C Fighting Falcon weapon options

692   USA: AIRCRAFT—LOCKHEED MARTIN

additional computer suite, permanent flight test data recording system, variable feel centrestick and sidestick in front cockpit, with latter also available for use in F-16 mode; safety pilot in rear cockpit. Internal gun, RWR and chaff/flare equipment removed, providing space for Phase II and III growth including additional computer, reprogrammable display generator and customer hardware allowance. Dorsal avionics compartment in bulged spine. First flight 9 April 1992; delivery due July 1992 but aircraft stored after five flights because of funding shortage; fitted with axisymmetric thrust vectoring engine nozzle (AVEN) for 90 flight, 120 hour MATV (Multi-Axis Thrust Vectoring) trials programme beginning 2 July 1993; thrust vectoring first used on 30 July; all except first few sorties flown at Edwards AFB; demonstrated transient 180° AoA and sustained 80°; nose-pointing authority demonstrated at 90° AoA at zero airspeed; MATV test programme completed 15 March 1994 after 95 sorties and 135.7 flight hours; NF-16D subsequently returned to VISTA configuration and used to replicate certain F-22 flight characteristics in series of trials undertaken by Calspan from Niagara Falls, New York, in 1996; has also performed similar trials in support of India's Light Combat Aircraft (LCA) project.

USAF intended to install F100-PW-229 engine in 1996; however, funding constraints imposed in late 1996 resulted in delay but re-engining accomplished before NF-16D returned to flight status on 29 May 1997, thereafter being handed back to Calspan on 13 June 1997. Incorporation of a pitch-yaw balanced beam nozzle has been proposed; this will provide axisymmetric thrust capability for use in conjunction with high AoA and variable stability flight control testing. NF-16D supported Boeing and Lockheed Martin Joint Strike Fighter designs during 1998.

F-16I: Unofficial designation of two-seat version (to F-16D-52+ standard) for Israel, featuring Pratt & Whitney F100-PW-229 engine, conformal fuel tanks and SAR. Final configuration still to be determined, but will incorporate Northrop Grumman AN/APG-68(V)XM radar and significant amount of Israeli avionics, including EW suite, cockpit displays and helmet-mounted sight; EW equipment by Elisra will probably include radar and missile lock warning systems plus self-protection jammers, while Elbit Systems to provide central mission computer, advanced display processor, DASH IV display and sight helmet system and stores management systems. Aircraft also to utilise Rafael Litening II targeting pod. Total of 50 to be purchased in US$2.5 billion deal, with delivery to begin in 2003; further 60 aircraft subject to option, but these may be Block 60 standard. Approval for order given by Israeli prime minister Ehud Barak on 16 July 1999, with formal letter of agreement signed on 10 September.

F-16N: US Navy supersonic adversary aircraft (SAA) modified from F-16C/D Block 30; selected January 1985; deliveries of 26 aircraft started 1987 and completed 1988; features include AN/APG-66 instead of AN/APG-68 radar, F110-GE-100 engine, deletion of M61 gun, AN/ALR-69 RWR, titanium in lower wing fittings instead of aluminium and cold working of lower wing skin holes to resist greater frequency of high *g*; wingtips fitted only for AIM-9 practice missiles and ACMI AIS pods, but normal tanks and stores on other stations. Four of 26 were two-seat **TF-16N**. F/TF-16Ns served with 'Top Gun' Fighter Weapons School (eight) and with VF-126 (six) at NAS Miramar, California, VF-45 (six) at NAS Key West, Florida, and VF-43 (six) at NAS Oceana, Virginia. Inactivations and reductions in size of Navy adversary training force resulted in entire fleet being retired during 1994-95.

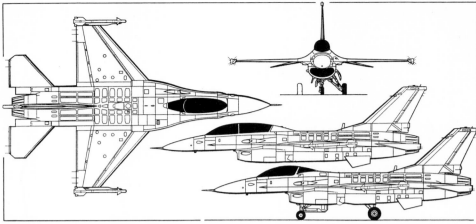

F-16C (GE F110 turbofan) with extra side view (top) of two-seat F-16D (P&W F100 turbofan) *(Paul Jackson/Jane's)* 0016166

FS-X and TFS-X: F-16 derivatives selected by Japan Defence Agency for its FS-X (now F-2) requirement 19 October 1987; details under Mitsubishi in Japanese section.

F-16 Recce: Four existing European reconnaissance pods, including that for German Navy Tornado IDS, demonstrated in flight on F-16 fighters with minimum changes; RNethAF reconnaissance **F-16A(R)** operational since 1983 with Orpheus pod, but decision taken to purchase a new airborne reconnaissance system for use by aircraft following MLU; Elop of Israel and Lockheed Martin Fairchild Systems and Recon/Optical of USA competing to supply sensors, management system, recorder, ground exploitation stations and systems integration. The RNethAF originally intended to order 24 pod systems for service with No. 306 Squadron from 1999, but failure of offers to satisfy specification requirements resulted in decision to restage competition with revised objective to acquire six low-level and 12 medium-level pod systems, plus three ground stations instead of four. In June 1999, eight companies were competing for contract, these comprising three aforementioned plus Honeywell, Northrop Grumman, Stork Aerospace, Terma Industries Grenaa and Vinten; however, in first quarter 2000, RNethAF delayed programme for at least two years. Four medium-altitude reconnaissance systems (MARS) pods were obtained in 1996 as interim equipment. Using Per Udsen modular reconnaissance pod (MRP), MARS contains Recon/Optical KS-87B wet film cameras. Belgium purchased eight MRP pods in 1996, deliveries of low-level version, with film cameras previously used on Mirage 5BR, began mid-1998; electro-optical (E-O) sensor version due in 1999. Some Danish F-16s given reconnaissance tasking from January 1994, using six locally designed Per Udsen MRPs with interchangeable wet film and E-O sensors; Denmark also purchasing six Per Udsen pods configured with E-O and IR sensors.

Following trials with a prototype system in mid-1995, an LMTAS-designed reconnaissance pod for the US Air National Guard was flown for the first time on 29 September 1995 and subsequently tested on an F-16C (86-227) of the 149th Fighter Squadron, Virginia ANG. Four pods were built and then deployed with the 149th FS to Aviano AB, Italy, in May 1996 for operational validation in support of NATO missions over Bosnia. This successful trial culminated in decision to procure a total of 20 podded systems for service with F-16C Block 30 aircraft of five ANG squadrons, each of which will have four pods and one ground exploitation system. The core of the programme, known as the Theater Airborne Reconnaissance System (TARS), is the Per Udsen MRP with USAF-supplied KS-87 cameras incorporating E-O video back instead of wet film; the Lockheed Martin Fairchild Systems medium-altitude E-O (MAEO) camera; an Ampex DCRsi-240 digital data recorder and a TERMA Elektronik cockpit control device. The TARS pods were delivered in 1999 and certified for operational use by the F-16 in first half of 2000.

Elta EL/M-2060P pod also cleared for use by F-16 in 1999. System is contained in standard 300 US gallon drop tank and comprises autonomous, all-weather, day and night high-resolution reconnaissance synthetic aperture radar sensor with ability to transmit imagery to ground station via bi-directional datalink.

Most F-16s delivered since the early 1990s have provisions for a reconnaissance pod. Taiwan is seeking reconnaissance pods for use by about 10 F-16s; proposals have been received from Lockheed Martin and Recon/Optical, with decision expected in late 2000 or early 2001.

AFTI/F-16: Modified FSD F-16A (75-0750) used for US Air Force Systems Command Advanced Fighter Technology Integration (AFTI); first flight 10 July 1982; previous achievements detailed in 1986-87, 1987-88 and 1991-92 *Jane's*; currently with Block 15 horizontal tail surfaces, Block 25 wing and Block 40 avionics. Trials programmes include automatic target designation and attack (1988), night navigation and map displays (1988-89), digital datalink and two-aircraft operations (1989), autonomous attack (1989-91) and night attack (1989-92); tested automatic ground collision avoidance system (Auto GCAS) and pilot-activated, low-level pilot disorientation recovery system 1991; LANTIRN pod and Falcon Night FLIR trials, 1992. AFTI/F-16 has also undertaken live firing of AGM-88 HARM as part of demonstration of advanced technology applicable to suppression of enemy air defence (SEAD) mission. Total 700 plus sorties by early 1997, when test effort concentrating on Auto GCAS.

AFTI/F-16 tasked with support of Joint Strike Fighter integrated subsystems technology (J/IST) programme throughout 1998-2000; aircraft featured electric

Late production F-16C Block 50 aircraft in special markings for the commander, 52nd Fighter Wing at Spangdahlem, Germany carries an HTS pod adjacent to the air inlet *(Lindsay Peacock)* 2001/0100453

USAF F-16C wearing stick-on electro-optical turrets and conformal tanks to show Desert Falcon configuration *(Paul Jackson/Jane's)*

2001/0100494

hydrostatic actuators for all five primary flight control surfaces; ground testing of 'power-by-wire' system started in 1999, with flight trials beginning on 24 October 2000; exceeded M1.0 9 November 2000. System incorporates 270 V DC engine-driven Hamilton Sundstrand generator and Parker Aerospace actuators; same manufacturers expected to provide equipment to production JSF in event of Lockheed Martin contender being selected. AFTI/F-16 flown to USAF Museum for retirement 9 January 2001, after 756 sorties and 1,446 hours.

Also in 1998, an Edwards AFB test fleet F-16D Block 52 (83-1176) was additionally employed on Auto GCAS development.

F-16XL: Two F-16XL prototypes, in flyable storage since 1985, leased from General Dynamics by NASA; first flight of single-seat N848NA/75-0749, 9 March 1989; NASA modified this aircraft at Dryden with wing glove having laser-perforated skin to smooth air flow over cranked arrow wing in supersonic flight, reducing drag and turbulence and saving fuel. Two-seat N849NA/75-0747, with GE F110-GE-129 engine, similarly converted early 1992. Second aircraft tested laminar flow panel installed above port wing as part of NASA's High-Speed Research Program. Project concluded in late 1996 and aircraft placed in storage. First F-16XL upgraded with digital flight control system and returned to flight status in December 1997. F-16XL described in 1985-86 *Jane's*.

F-16B-2: Second FSD F-16B (75-0752) converted to private-venture testbed of close air support and night navigation and attack systems; equipment includes F-16C/D HUD, helmet sight or Marconi Cat's Eyes NVGs, Falcon Eye head-steered FLIR or LANTIRN nav/attack pods, digital terrain system (Terprom), and automatic target handoff system. Alternative nav/attack FLIR pods comprise GEC-Marconi Atlantic and Lockheed Martin Pathfinder (LANTIRN derivative). NVG-compatible cockpit lighting. In 1994, this aircraft demonstrated a helmet-mounted cueing system and Raytheon's Box Office II high off-boresight missile in a live-fire programme known as 'Look and Shoot'. It has now been retired and replaced by an F-16B (81-0813) modified with Block 40/42 avionics and an F-16D Block 50 (90-0848).

F-16ES: Enhanced Strategic two-seat, long-range interdictor F-16 proposal; now defunct (see 1998-99 *Jane's* for details), but provided basis for Israeli F-16I.

F-16U: Proposed two-seat version unsuccessfully offered to United Arab Emirates. See 1995-96 *Jane's* for more details.

F-16 'Falcon 2000': Private venture design of early 1990s, similar to F-16U, which Lockheed Martin aimed at USAF as a follow-on to the F-16 before introduction of JSF. Also known as the **F-16X**; now defunct (see 1998-99 *Jane's* for details).

F-16 Testbeds: Augmenting aircraft that are permanently configured for research tasks (see entries elsewhere for NF-16D, AFTI/F-16 and F-16XL), a number of F-16s has been adapted to evaluate new equipment, systems and engine-related modifications. Trials undertaken during 1996 included ground testing of the F100-PW-200 and F110-GE-129 engines in conjunction with a low-observable axisymmetric nozzle (LOAN); both demonstrations were associated with the Joint Strike Fighter project, although General Electric is also proposing installation of LOAN on some versions of the F-15 and F-16. In addition, during early 1997, the third production F-16C (83-1120) completed a series of flights from Fort Worth with an advanced propulsion system inlet which features a radically different, lower intake lip that extends forward and a bulged, contoured upper surface. Frontal radar cross-section is believed to be reduced by this revised layout, which is said by LMTAS to be applicable to future tactical aircraft projects. Auto GCAS testing previously undertaken by AFTI/F-16 was reassigned to an F-16D of the USAF; Auto GCAS utilises digital terrain system coupled to the digital control system and provides both a manual pilot warning and automatic recovery to prevent ground collision during all types of manoeuvre. Testing successfully completed in November 1998.

Israel also considering Avionics Capabilities Enhancement (ACE) upgrade for older F-16s and planned to fly F-16B testbed aircraft in fourth quarter of 2000. IAI is leading the project which integrates systems produced by Elbit, Elta, Elisra, IMI, Rafael, Elop, Rokar, BVR and RADA. Technology demonstration programme will feature upgraded cockpit including three 127 × 178 mm (5 × 7 in) multifunction LCDs, a wide-angle HUD, upfront control panels, digital moving map display and embedded global positioning/inertial navigation systems, plus integration with helmet-mounted display/sight. ACE upgrade and structural work likely to be offered as complete package to F-16 operators, but Israeli Air Force reportedly having second thoughts about fleetwide upgrade following decision to purchase F-16I.

CUSTOMERS: See tables. Total 4,285 production aircraft ordered or requested by beginning of November 2000, including planned USAF procurement of 2,230 and 28 embargoed Pakistan Air Force examples that are to be distributed equally between the USAF and US Navy, following collapse of planned lease by New Zealand.

COSTS: Approximately US$24 million, flyaway, USAF version (1999). UAE purchase of 80 aircraft and associated equipment valued at US$7.9 billion.

DESIGN FEATURES: Conceived as 'lo' complement to Boeing F-15 Eagle in hi-lo fighter mix; optimised for high agility in air combat. Cropped delta wings blended with fuselage, with highly swept vortex control strakes along fuselage forebody and joining wings to increase lift and improve directional stability at high angles of attack; wing section NACA 64A-204; leading-edge sweepback 40°; relaxed stability (rearward CG) to increase manoeuvrability; deep wing-roots increase rigidity, save 113 kg (250 lb) structure weight and increase fuel volume; fixed geometry engine intake; pilot's ejection seat inclined 30° rearwards; single-piece birdproof forward canopy section; two ventral fins below wing trailing-edge.

Baseline F-16 airframe life planned as 8,000 hours with average usage of 55.5 per cent in air combat training, 20 per cent ground attack and 24.5 per cent general flying; structural strengthening programme for pre-Block 50 aircraft was required during 1990s.

FLYING CONTROLS: Four-channel digital fly-by-wire (analogue in earlier variants); pitch/lateral control by pivoting monobloc tailerons and wing-mounted flaperons; maximum rate of flaperon movement 52°/s; automatic wing leading-edge manoeuvring flaps programmed for Mach number and angle of attack; flaperons and tailerons interchangeable left and right; sidestick control column with force feel replacing almost all stick movement.

STRUCTURE: Wing, mainly of light alloy, has 11 spars, five ribs and single-piece upper and lower skins; attached to fuselage by machined aluminium fittings; leading-edge flaps are one-piece bonded aluminium honeycomb and driven by rotary actuators; fin is multispar, multirib with graphite epoxy skins; brake parachute or ECM housed in fairing aft of fin root; tailerons have graphite epoxy laminate skins, attached to corrugated aluminium pivot shaft and removable full-depth aluminium honeycomb leading-edge; ventral fins have aluminium honeycomb and skins; split speedbrakes in fuselage extensions inboard of tailerons open to 60°. Nose radome by Brunswick Corporation.

LANDING GEAR: Menasco hydraulically retractable type, nose unit retracting rearward and main units forward into fuselage. Nosewheel is located aft of intake to reduce the risk of foreign objects being thrown into the engine during ground operation, and rotates 90° during retraction to lie horizontally under engine air intake duct. Oleo-pneumatic struts in all units. Aircraft Braking Systems mainwheels and brakes; Goodyear or Goodrich tubeless mainwheel tyres, size 27.75×8.75-14.5 (or 27.75×8.75R14.5) (24 ply), pressure 14.48 to 15.17 bars (210 to 220 lb/sq in) at T-O weights less than 13,608 kg (30,000 lb). Steerable nosewheel with Goodyear, Goodrich or Dunlop tubeless tyre, size 18×5.7-8 (18 ply), pressure 20.68 to 21.37 bars (300 to 310 lb/sq in) at T-O weights less than 13,608 kg (30,000 lb). All but two main unit components interchangeable. Brake-by-wire system on main gear, with Aircraft Braking Systems anti-skid units. Runway arresting hook under rear fuselage; Irvin 7.01 m (23 ft 0 in) diameter braking parachute fitted in Greek, Indonesian, Netherlands (retrofit completed December 1992), Norwegian, Taiwanese, Turkish (Block 30/40 only) and Venezuelan F-16s. Belgian (Block 15 only), Israeli (F-16C/D models only) and Singaporean (F-16B/Ds only) aircraft have braking parachute compartment configured for electronic equipment. Landing/taxying lights on nose landing gear door.

POWER PLANT: One 131.6 kN (29,588 lb st) General Electric F110-GE-129, or one 129.4 kN (29,100 lb st) Pratt & Whitney F100-PW-229 afterburning turbofan as alternative standard. These Increased Performance Engines (IPE) installed from late 1991 in Block 50 and Block 52 aircraft. Pratt & Whitney has proposed F100-PW-229A version, which features a new fan module among other radical improvements that will raise airflow by more than 10 per cent, lower turbine temperatures by almost 50°C (122°F) and permit inspection intervals to rise from 4,300 cycles to 6,000. New version offers potential to increase maximum augmented thrust rating to about 142 kN (31,860 lb st), although this would require larger inlet on F-16. General Electric also engaged in improvement efforts, using company funding to begin development of F110-GE-129 EFE (Enhanced Fighter Engine) in October 1997; flight qualification currently scheduled for December 1999, with EFE initially to be rated at up to 151.0 kN (33,945 lb st) and with further growth potential to 160.0 kN (35,970 lb st); alternatively, improved thrust levels can be sacrificed for up to a 50 per cent increase in TBO and servicing intervals. Production derivative known as F110-GE-132 and rated at 144.6 kN (32,500 lb st) to be installed in F-16 Block 60 aircraft for UAE. Immediately prior standard was 128.9 kN (28,984 lb st) F110-GE-100 or 105.7 kN (23,770 lb st) F100-PW-220 in Blocks 40/42.

Of 1,446 F-16Cs and F-16Ds ordered by USAF, 556 with F100 and 890 with F110. IPE variants have half share each in FY92 procurement of 48 F-16s for USAF, following eight reliability trial installations including six Block 30 aircraft which flew 2,400 hours between December 1990 and September 1992. F100s of ANG and AFRes F-16A/Bs upgraded to -220E standard from late 1991. Fixed geometry intake, with boundary layer splitter plate, beneath fuselage. Apart from first few, F110-powered aircraft have intake widened by 30 cm (1 ft 0 in) from 368th F-16C (86-0262); Israeli second-batch F-16D-30s have power plants locally modified by Bet-Shemesh Engines to F110-GE-110A with provision for up to 50 per cent emergency thrust at low level. USAF AVEN (axisymmetric vectoring engine nozzle) trials, 1992, involved F100-IPE-94 and F110; possible application to F-16.

Standard fuel contained in wing and five seal-bonded fuselage cells which function as two tanks; 3,986 litres (1,053 US gallons; 876 Imp gallons) in single-seat aircraft; 3,297 litres (871 US gallons; 726 Imp gallons) in two-seat aircraft. Halon inerting system. In-flight refuelling receptacle in top of centre-fuselage, aft of cockpit. Auxiliary fuel can be carried in drop tanks: one 1,136 litre (300 US gallon; 250 Imp gallon) under fuselage; 1,402 litre (370 US gallon; 308 Imp gallon) under each wing. Optional Israel Military Industries 2,271 litre (600 US gallon; 500 Imp gallon) underwing tanks initially adopted only by Israel, but have since been selected by one or two other operators including USAF, which expected to award contract for 140 tanks by end of 1999; also adopted for F-16 Block 60 version.

ACCOMMODATION: Pilot only in F-16C, in pressurised and air conditioned cockpit. Boeing (formerly McDonnell Douglas) ACES II zero/zero ejection seat. Bubble canopy made of polycarbonate advanced plastics material. Inside of USAF F-16C/D canopy (and most Belgian, Danish, Netherlands and Norwegian F-16A/Bs) coated with gold film to dissipate radar energy. In conjunction with radar-absorbing materials in air intake, this reduces frontal radar signature by 40 per cent. Windscreen and forward canopy are an integral unit without a forward bow frame, and are separated from the aft canopy by a simple support structure which serves also as the breakpoint where the forward section pivots upward and aft to give access to the cockpit. A redundant safety lock feature prevents canopy loss. Windscreen/canopy design provides 360° all-round view, 195° fore and aft, 40° down over the side, and 15° down over the nose.

To enable the pilot to sustain high *g* forces, and for pilot comfort, the seat is inclined 30° aft and the heel line is raised. In normal operation the canopy is pivoted upward and aft by electrical power; the pilot is also able to unlatch the canopy manually and open it with a back-up handcrank. Emergency jettison is provided by explosive unlatching devices and two rockets. A limited displacement, force-sensing control stick is provided on the right-hand console, with a suitable armrest, to provide precise control inputs during combat manoeuvres.

USA: AIRCRAFT—LOCKHEED MARTIN

F-16 TABLES

Table 1 provides a rapid reference to customers and quantities; more detailed information on production block numbers appears in Tables 2 and 3.

TABLE 1: F-16 CUSTOMERS

Operator	Total	Single-seat	Qty	Two-seat	Qty	Power plant	First aircraft A/C	First aircraft B/D	First delivery	Squadrons (or base)
Bahrain	22	F-16C-40	18	F-16D-40	4	F110-GE-100	101	150	March 1990	1, 2
Belgium	160[1]	F-16A-10	55	F-16B-10	12	F100-PW-200	FA01	FB01	January 1979	1, 2, 23, 31, 349, 350
		F-16A-15	41	F-16B-15	8	F100-PW-200	FA56	FB13	October 1982	
		F-16A-15OCU	40	F-16B-15OCU	4	F100-PW-220	FA97	FB21	January 1988	
Denmark	70	F-16A-10	30[1]	F-16B-10	8[1]	F100-PW-200	E-174	ET-204	January 1980	723, 727, 730
		F-16A-15	16[1]	F-16B-15	4[1]	F100-PW-200	E-596	ET-613	May 1982	723, 727, 730
		F-16A-15OCU	8[3]	F-16B-15OCU	4[3]	F100-PW-220	E-004	ET-197	December 1987	726
Egypt	220	F-16A-15	34	F-16B-15	8[2]	F100-PW-200	9301	9201	March 1982	72, 74
		F-16C-32	34	F-16D-32	6	F100-PW-220	9501	9401	August 1986	68, 70
		F-16C-40	34	F-16D-40	7	F110-GE-100	9901	9801	October 1991	60, 64
		F-16C-40	1	F-16D-40	5	F110-GE-100	9935	9808	1994	
		F-16C-40	34[7]	F-16D-40	12[7]	F110-GE-100	9951	9851	March 1994	75, 77
		F-16C-40	21			F110-GE-100B	(96-0086)	—	May 1999	71, 73
		F-16C-40	12	F-16D-40	12	F110-GE-100B	(99-0105)	(99-0117)	January 2001	
Greece	130	F-16CG-30	34	F-16DG-30	6	F110-GE-100	110	144	November 1988	330, 346
		F-16CG-50	32	F-16DG-50	8	F110-GE-129	045	077	January 1997	341, 347
		F-16CG-52+	34	F-16DG-52+	16	F100-PW-229			Late 2002	
Indonesia	12	F-16A-15OCU	8	F-16B-15OCU	4	F100-PW-220	S-1605	S-1601	December 1989	3
Israel	260	F-16A-10	67	F-16B-10	8	F100-PW-200	100	001	January 1980	140, 147, 253
		F-16C-30	51	F-16D-30	24	F110-GE-100A	301	020	December 1986	101, 105, 109, 110, 117
		F-16C-40	30	F-16D-40	30	F110-GE-100A	502	601	July 1991	101, 105, 109, 110, 117
				F-16D-52+ (F-16I)	50	F110-PW-229			2003	
Korea, South	180	F-16C-32	30	F-16D-32	10	F100-PW-220	85-574	84-370	March 1986	161, 162, 155
		F-16C-52	95[8]	F-16D-52	45[8]	F100-PW-229	92-000	92-028	December 1994	
Lockheed Martin	12	F-16C-52	4	F-16D-52	8	F100-PW-229	96-5033	96-5025	June 1998	Singapore (lease in USA)
Netherlands	213[3]	F-16A-10	46	F-16B-10	13	F100-PW-200	J-212	J-259	June 1979	306, 311, 312, 313, 315, 322, 323
		F-16A-15	84	F-16B-15	18	F100-PW-200	J-258	J-649	May 1982	
		F-16A-15OCU	47	F-16B-15OCU	5	F100-PW-200	J-141	J-065	1988	
Norway	74	F-16A-10	28[3]	F-16B-10	7[3]	F100-PW-200	272	301	January 1980	332, 338
		F-16A-15	32[3]	F-16B-15	5[3]	F100-PW-200	300	690	June 1982	331, 334
				F-16B-15OCU	2	F100-PW-220	—	711	July 1989	331
Pakistan	68	F-16A-15	28	F-16B-15	12	F100-PW-200	82-701	82-601	January 1983	9, 11, 14
		F-16A-15OCU	13	F-16B-15OCU	15	F100-PW-220	91-729	91-613	—	See note 10
Portugal	20	F-16A-15OCU	17	F-16B-15OCU	3	F100-PW-220E	15101	15118	February 1994	201
Singapore	58	F-16A-15OCU	4	F-16B-15OCU	4	F100-PW-220	880	884	February 1988	140
		F-16C-52	18	F-16D-52	12	F100-PW-229	608	624	January 1998	
		F-16C-52	20[11]			F110-PW-229			Late 2003	
Taiwan	150	F-16A-20	120	F-16B-20	30	F100-PW-220	6601	6801	July 1996	12, 14, 17, 21, 22, 23, 26, 27
Thailand	36	F-16A-15OCU	14	F-16B-15OCU	4	F100-PW-220	10305	10301	June 1988	103
		F-16A-15OCU	12	F-16B-15OCU	6	F100-PW-220	07020	07032	September 1995	403
Turkey	240[4]	F-16C-30	34	F-16D-30	9	F110-GE-100	86-0066	86-0191	May 1987	142, Oncel Flight
		F-16C-40	102	F-16D-40	15	F110-GE-100	88-0033	88-0014	July 1990	141, 161, 162, 191, 192, 181, 182
		F-16C-50	60	F-16D-50	20	F110-GE-129	93-0657	93-0691	July 1996	141, 151, 152
UAE	80	F-16C-60	55	F-16D-60	25	F110-GE-132			2004	
USAF	2,230[9]	F-16A-10	255	F-16B-10	74	F100-PW-200	78-0001	78-0077	August 1978	See note A
		F-16A-15	409[5]	F-16B-15	46[6]	F100-PW-200	80-0541	80-0635	September 1981	See note A
		F-16C-25	209	F-16D-25	35	F100-PW-200	83-1118	83-1174	July 1984	See note B
		F-16C-30/32	360/56	F-16D-30/32	48/5	both (100/220)	85-1398	85-1509	July 1986	See note B
		F-16C-40/42	234/150	F-16D-40/42	31/47	both (100/220)	87-0350	87-0391	December 1988	See note B
		F-16C-50/52	189/42	F-16D-50/52	28/12	both (129/229)	90-0801	90-0834	October 1991	See note B
US Navy	26	F-16N-30	22	TF-16N-30	4	F110-GE-100	163268	163278	June 1987	Stored
Venezuela	24	F-16A-15	18	F-16B-15	6	F100-PW-200	1041	1715	September 1983	161, 162
Totals	**4,285**		**3,441**		**844**					

Notes:
[1] Built by Sabca (Belgium); 222nd and last Sabca F-16 (BAF FA-136) delivered 22 October 1991
[2] One built by Fokker (Netherlands)
[3] Built by Fokker; 300th and last Fokker F-16 (RNethAF J-021) delivered 27 February 1992
[4] Two F-16Cs and six F-16Ds built by GD; remainder by TAI
[5] Two built by Fokker
[6] Four built by Sabca
[7] TAI production
[8] 12 built by GD, 36 CKD kits and 92 produced locally by Samsung Aerospace
[9] Table excludes full-scale development (FSD) aircraft (six F-16A and two F-16B)
[10] Embargoed. Pakistan offered compromise batch of 38 aircraft. Offer rejected and 28 completed aircraft of follow-on order for 71 placed in storage at Davis-Monthan AFB, Arizona. To be distributed between US Navy (10 F-16A and four F-16B) for aggressor training and US Air Force (three F-16A and 11 F-16B) for test support duties
[11] Includes unspecified number of F-16D-52 aircraft

Note A: Currently operated by US Air Force Flight Test Center/412th Test Wing, Edwards AFB, California; US Air Force Air Armament Center/46th Test Wing, Eglin AFB, Florida (39, 40 FITSs); 56th FW, Luke AFB, Arizona (21 FS/Taiwan AF training squadron); 148, 178, 179, 184, 186 and 195 FSs, Air National Guard.

Note B: Currently operated by US Air Force Flight Test Center/412th TW, Edwards AFB, California; US Air Force Air Armament Center/46th Test Wing, Eglin AFB, Florida (39, 40 FITSs); 8th FW, Kunsan AB, South Korea (35, 80 FSs); 20th FW, Shaw AFB, South Carolina (55, 77, 78, 79 FSs); 27th FW, Cannon AFB, New Mexico (522, 523, 524 FSs and 428 FS/Singapore AF training squadron); 31st FW, Aviano AB, Italy (510, 555 FSs); 35th FW, Misawa AB, Japan (13, 14 FSs); 51st FW, Osan AB, South Korea (36 FS); 52nd FW, Spangdahlem AB, Germany (22, 23 FSs); 53rd W, Eglin AFB, Florida (422 Test and Evaluation Squadron, at Nellis AFB, Nevada); 56th FW, Luke AFB, Arizona (61, 62, 63, 308, 309, 310, 425 FSs); 57th W, Nellis AFB, Nevada (F-16 Fighter Weapons School, 414 Combat Training Sqdn, Thunderbirds Air Demonstration Sqdn); 347th W, Moody AFB, Georgia (68, 69 FSs); 354th FW, Eielson AFB, Alaska (18 FS); 366th W, Mountain Home AFB, Idaho (389 FS); 388th FW, Hill AFB, Utah (4, 34, 421 FSs); and also 93, 301, 302, 457 and 466 FSs, Air Force Reserve Command; 107, 111, 112, 113, 119, 120, 121, 124, 125, 134, 138, 149, 152, 157, 160, 162, 163, 170, 174, 175, 176, 182, 184, 188 and 194 FSs, Air National Guard.

Blocks 1 and 5 retrofitted to Block 10 standard 1982-84; Block 15 retrofitted to Block 15OCU (avionics standard only) from 1987. New-build F-16As are Block 15OCU from November 1987.

Export programme codenames are: Bahrain – Peace Crown I-II; Egypt – Peace Vector I, II, III, IIIA and IV; Greece – Peace Xenia I-II; Indonesia – Peace Bima-Sena; Israel – Peace Marble I-III; Jordan – Peace Falcon; South Korea – Peace Bridge I-II; Pakistan – Peace Gate I-IV; Portugal – Peace Atlantis I-II; Singapore – Peace Carvin I-IV; Taiwan – Peace Fenghuang; Thailand – Peace Naresuan I-II; Turkey – Peace Onyx I-II (and CAE-Link simulator, Peace Onyx III); and Venezuela – Peace Delta.

Recent orders: Egypt ordered 12 F-16Cs and 12 F-16Ds in 1999. Contracts agreed in 2000 with Greece (34 F-16C, 16 F-16D); Israel (50 F-16I); South Korea (15 F-16C, 5 F-16D); Singapore (20 F-16C/D) and the United Arab Emirates (55 F-16C, 25 F-16D). USAF resumed procurement with batch of six aircraft in FY96 plus six in FY97, three in FY98, one in FY99, 10 in FY00 and four in FY01. Up to 20 more may be purchased by 2010.

Transfers: Israel received 36 F-16As and 14 F-16Bs from surplus USAF stocks August 1994 to March 1995 for 140, 144 and 253 Squadrons. Denmark received three ex-USAF F-16As in July 1994, plus three F-16As and one F-16B in 1997 and plans to obtain another three in the post-2000 period. **Jordan** took delivery of an initial batch of ex-USAF ADF aircraft (12 F-16As and four F-16Bs) on a lease basis in December 1997/April 1998 and may eventually receive 36 to 48 examples; Portugal to receive 25 second-hand aircraft (21 F-16As and four F-16Bs, of which five F-16As to be broken-down for spares); **Thailand** request for 16 surplus USAF F-16A/B approved by US Congress at beginning of 2000; Italy appears likely to acquire 30 F-16A and four F-16B on five-year lease arrangement; and Austria, Croatia, Czech Republic, Hungary, Oman, Philippines, Poland, Romania, Slovenia and Venezuela may obtain F-16s on purchase/lease from this source.

LOCKHEED MARTIN—AIRCRAFT: USA

TABLE 2: F-16A/B PRODUCTION

Block	USAF A	USAF B	Belgium A	Belgium B	Denmark A	Denmark B	Egypt A	Egypt B	Indonesia A	Indonesia B	Israel A	Israel B	Netherlands A	Netherlands B	Norway A	Norway B	Pakistan A	Pakistan B	Portugal A	Portugal B	Singapore A	Singapore B	Taiwan A	Taiwan B	Thailand A	Thailand B	Venezuela A	Venezuela B
01	21	22	17	6	3	2							12	6	3	2												
05	89	27	8	4	12	3					18	8	14	2	10	2												
10	26	4	5	–	6	1					10	–	5	1	5	1												
10A	28	5	7	1	3	1					10	–	7	1	4	–												
10B	29	6	6	1	2	1					4	–	4	1	6	1												
10C	27	4	6	–	4	–					16	–	4	2	–	1												
10D	35	6	6	–							9	–																
Subtotal	**255**	**74**	**55**	**12**	**30**	**8**					**67**	**8**	**46**	**13**	**28**	**7**												
15	31	2			2	1							4	–	3	1												
15A	25	1							3	4																		
15B	26	1	1	1	5	1	2	1					5	2	4	1												
15C	26	1					2	1																				
15D	25	2	4	4	4	–	3	–					6	1	4	1	–	2										
15E	25	2					3	–							2	2												
15F	25	2	5	3	3	1	3	–					6	2	4	1												
15G	24	2					4	–																				
15H	24	2	6	–	2	–	4	–					6	2	4	1												
15J	25	3					2	–																				
15K	20	5	8	–			3	–					4	2	8	–									1	3		
15L	21	4					3	–																	2	–		
15M	21	4	8	–			2	–					4	1	5	–	2	2										
15N	21	3					–	1									1	2										
15P	18	4	5	–			–	1					5	–			3	–										
15Q	22	6			–	1											5	–										
15R	19	2	4	–									10	2			4	–										
15S	11	–															5	–										
15T													7	2			2	–									3	3
15U													10	2			4	1									8	–
15V																	–	3									4	–
15W													9	2														
15X													8	–														
Subtotal	**664**	**120**	**96**	**20**	**46**	**12**	**34**	**8**			**67**	**8**	**130**	**31**	**60**	**12**	**28**	**12**									**18**	**6**
15Y			7	–	–	4							7	1														
15Z																												
15AA			7	1	4	–							7	1							–	1	2	2	1	3		
15AB																							2	1	7	1		
15AC			7	1	4	–							7	1														
15AD																			–	2								
15AE			6	2					–	2			7	2														
15AF									2	2																		
15AG			8	–					6	–			8	–														
15AH																												
15AJ			5	–									8	–											6	–		
15AK																												
15AL													3	–														
15AM																	6[1]	5[1]										
15AN																												
15AP																												
15AQ																	5[1]	1[1]										
15AR																	2[1]	3[1]										
15AS																	–	3[1]										
15AT																			3	3								
15AU																	–	3[1]										
15AV																			10	–								
15AW																			4	–								
15AX																									–	4		
15AY																									4	–		
15AZ																									8	2		
Subtotals	**664**	**120**	**136**	**24**	**54**	**16**	**34**	**8**	**8**	**4**	**67**	**8**	**177**	**36**	**60**	**14**	**41**	**27**	**17**	**3**	**4**	**4**	–	–	**26**	**10**	**18**	**6**
20																							**120**	**30**				
Totals	**664**	**120**	**136**	**24**	**54**	**16**	**34**	**8**	**8**	**4**	**67**	**8**	**177**	**36**	**60**	**14**	**41**	**27**	**17**	**3**	**4**	**4**	**120**	**30**	**26**	**10**	**18**	**6**
Grand total													**1,736**															

Notes: Table excludes FSD aircraft (six F-16A and two F-16B); OCU standard from Block 15Y
[1] Embargoed aircraft; currently in storage at Davis-Monthan AFB, Arizona; to be operated by USAF and US Navy in test support and aggressor roles respectively

The F-16D has two cockpits in tandem, equipped with all controls, displays, instruments, avionics and life support systems required to perform both training and combat missions. The layout of the F-16D second station is similar to the F-16C, and is fully systems-operational. A single-enclosure polycarbonate transparency, made in two pieces and spliced aft of the forward seat with a metal bow frame and lateral support member, provides outstanding view from both cockpits.

F-16Ds supplied to Israel and Singapore are configured with weapon system operator station in rear cockpit, plus a large dorsal equipment compartment extending from the rear of the canopy to the leading edge of the fin; compartment houses avionics unique to each operator, plus additional chaff/flare dispensers and an in-flight refuelling receptacle.

SYSTEMS: Regenerative 12 kW environmental control system, with digital electronic control, uses engine bleed air for pressurisation and cooling of crew station and avionics compartments. Two separate and independent hydraulic systems supply power for operation of the primary flight control surfaces and the utility functions. System pressure (each) 207 bars (3,000 lb/sq in), rated at 161 litres (42.5 US gallons; 35.4 Imp gallons)/min. Bootstrap-type reservoirs, rated at 5.79 bars (84 lb/sq in).

Electrical system powered by engine-driven 60 kVA main generator and 10 kVA standby generator (including ground annunciator panel for total electrical system fault reporting), with Sundstrand constant speed drive and powered by a Sundstrand accessory drive gearbox. 17 Ah battery. Four dedicated, sealed cell batteries provide transient electrical power protection for the fly-by-wire flight control system.

An onboard Sundstrand/Solar jet fuel starter is provided for engine self-start capability. Simmonds fuel measuring system. AlliedSignal emergency power unit automatically drives a 5 kVA emergency generator and emergency pump to provide uninterrupted electrical and hydraulic power for control in the event of the engine or primary power systems becoming inoperative.

AVIONICS: *Comms:* Magnavox AN/ARC-164 UHF transceiver (AN/URC-126 Have Quick IIA in Block 50/52); provision for Magnavox KY-58 secure voice system; Rockwell Collins AN/ARC-186 VHF AM/FM transceiver, ARC-190 HF radio, government-furnished AN/AIC-18/25 intercom and SCI advanced interference blanker, Teledyne Electronics AN/APX-101 IFF transponder with government-furnished IFF control, government-furnished National Security Agency KIT-1A/TSEC cryptographic equipment. F-16C/D Block 52 aircraft of Singapore and South Korea have Litton AN/APX-109+ advanced interrogator/transponder. AN/APX-113 advanced interrogator/transponder installed in F-16A/B MLU retrofit as well as Greek Block 50, Taiwanese Block 20 and Turkish Block 50 aircraft and is standard equipment on USAF aircraft procured in FY00.

Radar: Northrop Grumman AN/APG-68(V) pulse Doppler range and angle track radar, with planar array in nose. Provides air-to-air modes for range-while-search, uplook search, velocity search with ranging, air combat,

USA: AIRCRAFT—**LOCKHEED MARTIN**

TABLE 3: F-16C/D PRODUCTION

Block	USAF F-16C	USAF F-16D	USN F-16N	USN TF-16N	Bahrain C	Bahrain D	Egypt C	Egypt D	Greece C	Greece D	Israel C	Israel D	Korea C	Korea D	Singapore C	Singapore D	Turkey C	Turkey D	UAE C	UAE D	LM C	LM D
25	7	4																				
25A	16	3																				
25B	25	5																				
25C	35	5																				
25D	40	4																				
25E	47	4																				
25F	39	10																				
Subtotal	**209**	**35**																				
30	16	2									4	–	2	6								
32							1	4														
30A	40	2									15	1	4	–								
32A							13	2									17	6				
30B	47	3	4	–							15	–	4	–								
32B							20	–														
30C	35	2	8	–							16	–	4	–								
32C	20	3																				
30D	42	4	8	–							1	4	4	–								
32D	13	–																				
30E	55	5	2	4							–	10	4	–			17	3				
32E																						
30F	51	6									–	9	4	–								
32F	2	1																				
30H	34	12							2	4			4	–								
32H	11	1																				
30J	25	8							12	2												
32J	10	–																				
30K	15	3							16	–												
30L									4	–												
32Q													–	4								
30VISTA	–	1																				
Subtotal	**625**	**88**	**22**	**4**			**34**	**6**	**34**	**6**	**51**	**24**	**30**	**10**			**34**	**9**				
40	2	–															17	3				
40A	6	3																				
42A	5	3																				
40B	22	–																				
42B	7	13																				
40C	38	3																				
42C	17	2																				
40D	38	2			2	4											17	3				
42D	17	3																				
40E	37	–			6	–																
42E	18	5																17	3				
40F	33	4																				
42F	15	8																				
40G	31	5					2	–														
42G	21	3																				
40H	14	5					–	4					6	4								
42H	27	6																				
40J	7	5					9	2					8	8			16	3				
42J	21	4																				
40K	6	4					15	1					8	8								
42K	2	–																				
40L							8	–					8	10			17	3				
40M							1	1														
40N							–	8														
40P																	18	–				
40Q							3	3														
40R							12	5														
40S							19															
40T							5															
40U							8															
40V							8															
40W					10																	
40							12	12														
Subtotal	**1,009**	**166**	**22**	**4**	**18**	**4**	**136**	**42**					**81**	**54**			**136**	**24**				
50	4	–																				
50A	7	7																				
52A	1	1																				
50B	24	8																				
50C	21	4																				
50D	133	9							32	8			95	45	38	12	60	20			4	8
52D	41	11																				
52+									34	16		50										
Subtotal	**1,240**	**206**	**22**	**4**	**18**	**4**	**136**	**42**	**100**	**30**	**81**	**104**	**125**	**55**	**38**	**12**	**196**	**44**	**4**	**8**		
60																			**55**	**25**		
Totals	**1,240**	**206**	**22**	**4**	**18**	**4**	**136**	**42**	**100**	**30**	**81**	**104**	**125**	**55**	**38**	**12**	**196**	**44**	**55**	**25**	**4**	**8**
Grand total											**2,549**											

track-while-scan (10 targets), raid cluster resolution, single target track and pulse Doppler track to provide target illumination for AIM-7 missiles, plus air-to-surface modes for ground-mapping, Doppler beam-sharpening, ground moving target, sea target, fixed target track, target freeze after pop-up, beacon, and air-to-ground ranging. Improved AN/APG-68(V)XM radar derivative will be installed on new aircraft from 2002 and Northrop Grumman plans to offer an upgrade kit enabling existing radars to be brought to latest standard, which has synthetic aperture radar (SAR) mapping and terrain following (TF) modes, plus interleaving of all modes; if internal FLIR targeting system is selected, this could share processor with ABR. Elta EL/2032 multimode pulse Doppler radar could be installed in future aircraft for Israel and may also be adopted as part of an upgrade package for existing Israeli F-16s.

Flight: Litton LN-39 standard inertial navigation system (ring laser Litton LN-93 or Honeywell H-423 in Block

Per Udsen MRP on the centreline of a Danish F-16B *(Lindsay Peacock)*

50/52 and current FMS F-16A/B: LN-93 for Egypt, Indonesia, Israel, South Korea, Pakistan, Portugal and Taiwan, plus Netherlands retrofit and Greek second batch); Rockwell Collins AN/ARN-108 ILS, Rockwell Collins AN/ARN-118 Tacan, Rockwell Collins GPS, Honeywell central air data computer, Elbit Fort Worth enhanced stores management computer, Gould AN/APN-232 radar altimeter. Fairchild digital terrain system (incorporating BAe Terprom algorithms) to be installed in all new USAF F-16s, USAF Reserve F-16C/Ds and European aircraft destined for MLU. Danish aircraft to be equipped with Thomson-CSF TLS 2010 multimode receiver airborne radio navigation system from early 2001. Optional equipment includes Rockwell Collins VIR-130 VOR/ILS.

Instrumentation: Marconi wide-angle holographic electronic HUD with raster video capability (for LANTIRN) and integrated keyboard; data entry/cockpit interface and dedicated fault display by Litton Canada and Elbit Fort Worth; Astronautics cockpit/TV set. Total of 43 aircraft (F-16C/D Block 40) of 31st FW at Aviano AB, Italy, modified between September 1997 and March 1998 as part of quick response capability (QRC) programme to provide night vision information system (NVIS) compatible lighting and facilitate use of NVGs; it is eventually planned to fit NVIS lighting on all USAF F-16Cs and F-16Ds.

Mission: Honeywell multifunction displays. Lockheed Martin LANTIRN package comprises AN/AAQ-13 (navigation) and AN/AAQ-14 (targeting) pods. Turkish aircraft (150+ modified by 1996) to share 60 LANTIRN pod systems; LANTIRN also purchased by Belgium, Greece, South Korea and Singapore. Sharpshooter pod (down-rated export version of AAQ-14 LANTIRN targeting system) acquired by Bahrain and Israel, but latter also required indigenous Rafael Litening IR targeting and navigation pod as replacement. Netherlands selected Marconi Atlantic navigation pod, with enhancements such as laser spot tracker and electronic boresight capability; also procuring 10 Lockheed Martin LANTIRN targeting pods with added TV capability; delivery of latter to begin in early 2000. Pakistan F-16s carry Thomson-CSF Atlis laser designator pods. Total of 168 Litening II navigation/targeting pods ordered from Rafael and Northrop Grumman to equip Block 25/30/32/40/42 F-16C/Ds of the Air National Guard (136) and Air Force Reserve Command (32); programme called Precision Attack Targeting System, with first AFRC unit (457th FS at Fort Worth, Texas) receiving initial batch of four pods in February 2000; this and three more AFRC squadrons expected to be fully equipped by fourth quarter of 2000. ANG also accepted first pods during 2000; goal is to allocate eight pods to each 15-aircraft squadron.

Raytheon AN/ASQ-213 HARM Targeting System (HTS) pod introduced on Block 50D/52D aircraft and subsequently retrofitted to entire Block 50/52 fleet. Entered service 1994 and currently deployed by USAF units in USA, Japan and Germany. Some Netherlands F-16As equipped to carry Delft Orpheus reconnaissance pods; Danish F-16As with Per Udsen recce pods. Historical details in 1986-87 and earlier editions. New modular mission computer and colour displays for retrofit as part of MLU and new production in Block 20 F-16A/Bs and Block 50/52 F-16C/Ds.

Self-defence: Dalmo Victor AN/ALR-69 radar warning system replaced in USAF Block 50/52 by Lockheed Martin AN/ALR-56M advanced RWR, which also ordered for USAF Block 40/42 retrofit and (first export) Korean Block 52s. Korean aircraft being retrofitted with the ITT Avionics/Northrop Grumman AN/ALQ-165 Airborne Self-Protection Jammer (ASPJ) from mid-2000. Provision for Northrop Grumman AN/ALQ-131 or Raytheon AN/ALQ-184 jamming pods. AN/ALQ-184 supplied to Bahrain, Egypt, Netherlands, Pakistan and Thailand. Taiwan receiving more than 100 Raytheon AN/ALQ-184 (first export order and first foreign use). Israeli Air Force F-16s extensively modified with locally designed and manufactured equipment, as well as optional US equipment to tailor them to the IAF defence role. This includes Elisra SPS 3000 self-protection jamming equipment in enlarged spines of F-16D-30s and Elta EL/L-8240 ECM in third batch of F-16C/Ds, replacing Lockheed Martin AN/ALQ-178(V)1 Rapport ECM in Israeli F-16As.

Belgian F-16s have Thomson-CSF Carapace passive ECM system in fin-root housing on 100 aircraft (with some reserve systems) from April 1995 (to be used in conjunction with active AN/ALQ-131 jamming pods to be obtained from surplus US stocks). Lockheed Martin AN/ALQ-178(V)3 Rapport III integral self-protection system in Turkish F-16C/Ds; total of 79 Block 50 aircraft to be updated with Thomson-CSF/Aselsan Fast 16 EW system following signature of contract worth US$190 million on 19 September 2000. In March 1993, Greece ordered integrated Litton ASPIS self-defence system, comprising AN/ALR-93 RWR, Tracor chaff/flare dispensers and (initially on only 35 aircraft) Raytheon AN/ALQ-187 I-DIAS jammer.

Denmark, Netherlands and Norway procuring Per Udsen PIDS wing weapon pylon that houses additional chaff/flare dispensers.

Denmark is integrating the Northrop Grumman AN/ALQ-162 jammer in one wing pylon in programme called ECIPS. Tracor AN/ALE-40(V)-4 chaff/flare dispensers (AN/ALE-47 in FY97 Block 50, FMS Block 20/50 since mid-1996 and for retrofit to Block 40/42 and 50/52 of USAF). Raytheon AN/ALE-50(V)2 towed decoy installed in AMRAAM missile pylons and adopted by USAF for all Block 40/42/50/52 aircraft and was to be widely fielded in 1999; USAF plan to buy total of 961. Norway, Netherlands and Denmark all intend to install passive missile approach warning system in Per Udsen PIDS+/ECIPS+ wing weapon pylon; concept demonstration of Northrop Grumman AN/AAR-54(V) undertaken in 1997-98; decision on system to be selected had been expected in second quarter 1999, with common unit likely to be adopted and fielded by 2003.

ARMAMENT: General Dynamics M61A1 20 mm multibarrel cannon in the port side wing/body fairing, equipped with a General Dynamics ammunition handling system and an enhanced envelope gunsight (part of the head-up display system) and 511 rounds of ammunition. There is a mounting for an air-to-air missile at each wingtip, one underfuselage centreline hardpoint, and six underwing hardpoints for additional stores. For manoeuvring flight at 5.5 g the underfuselage station is stressed for a load of up to 1,000 kg (2,200 lb), the two inboard underwing stations for 2,041 kg (4,500 lb) each, the two centre underwing stations for 1,587 kg (3,500 lb) each, the two outboard underwing stations for 318 kg (700 lb) each, and the two wingtip stations for 193 kg (425 lb) each. For manoeuvring flight at 9 g the underfuselage station is stressed for a load of up to 544 kg (1,200 lb), the two inboard underwing stations for 1,134 kg (2,500 lb) each, the two centre underwing stations for 907 kg (2,000 lb) each, the two outboard underwing stations for 204 kg (450 lb) each, and the two wingtip stations for 193 kg (425 lb) each. There are mounting provisions on each side of the inlet shoulder for the specific carriage of sensor pods (electro-optical, FLIR and so on); each of these stations is stressed for 408 kg (900 lb) at 5.5 g, and 250 kg (550 lb) at 9 g.

Typical stores loads can include two wingtip-mounted AIM-9L/M/P Sidewinders, with up to four more on the outer underwing stations; Rafael Python 3 on Israeli F-16s from early 1991 and Python 4 from mid-1997; centreline GPU-5/A 30 mm cannon; drop tanks on the inboard underwing and underfuselage stations; HARM targeting system pod along the starboard side of the nacelle; and bombs, air-to-surface missiles or flare pods on the four inner underwing stations. Stores can be launched from Aircraft Hydro-Forming MAU-12C/A bomb ejector racks, Hughes LAU-88 launchers, Orgen triple or multiple ejector racks and Lucas Aerospace Flight Structures Twin Store Carrier (TSC). New BRU-57 bomb racks to be fitted to Block 50/52 aircraft of USAF in fourth quarter of 2001; to be installed at mid-span hardpoint, each BRU-57 will be able to carry two smart munitions such as JSOW, JDAM and WCMD. Non-jettisonable centreline GPU-5/A 30 mm gun pods on dedicated USAF ground-attack F-16As.

Weapons launched successfully from F-16s, in addition to AIM-9 Sidewinder and AIM-120A AMRAAM, include radar-guided AIM-7 Sparrow and Sky Flash air-to-air missiles, AIM-132 ASRAAM and Magic 2 IR homing air-to-air missiles, AGM-65A/B/D/G Maverick air-to-surface missiles, AGM-88 HARM and AGM-45 Shrike anti-radiation missiles, AGM-84 Harpoon anti-ship missiles (clearance trials 1993-94), and, in Royal Norwegian Air Force service, the Penguin Mk 3 anti-ship missile. LGBs include GBU-10, GBU-12, GBU-22, GBU-24 and GBU-27; F-16 can also deliver GBU-15 glide bomb, which used in conjunction with datalink pod. Israeli IMI STAR-1 anti-radiation weapon has also begun carriage trials on F-16D, although full-scale development is dependent upon receipt of a firm order. CMS Defense Systems Autonomous Free-flight Dispenser System (AFDS) was tested at Eglin AFB, Florida, during 1992-93 and can be loaded with a variety of submunitions, including cratering bombs, shaped charge bomblets, anti-tank mines, area denial submunitions and general purpose bomblets.

Newest capability, introduced on Block 50/52 aircraft of USAF, incorporates 50T5 software upgrade, allowing F-16 to carry and deliver latest family of precision munitions; release to service occurred in mid-2000. New weapons comprise GBU-31 Joint Direct Attack Munition (JDAM), AGM-154 Joint StandOff Weapon (JSOW) and

The 4,000th Fighting Falcon, delivered to Egypt in 2000

CBU-103, CBU-104 and CBU-105 wind-corrected munitions dispensers (WCMDs). First operational unit with JSOW and WCMD is 20th FW at Shaw AFB, South Carolina.

DIMENSIONS, EXTERNAL (F-16C, D):
Wing span: over missile launchers	9.45 m (31 ft 0 in)
over missiles	10.00 m (32 ft 9¾ in)
Wing aspect ratio	3.2
Length overall	15.03 m (49 ft 4 in)
Height overall	5.09 m (16 ft 8½ in)
Tailplane span	5.58 m (18 ft 3¾ in)
Wheel track	2.36 m (7 ft 9 in)
Wheelbase	4.00 m (13 ft 1½ in)

AREAS (F-16C, D):
Wings, gross	27.87 m² (300.0 sq ft)
Flaperons (total)	2.91 m² (31.32 sq ft)
Leading-edge flaps (total)	3.41 m² (36.72 sq ft)
Fin, incl dorsal fin	4.00 m² (43.10 sq ft)
Rudder	1.08 m² (11.65 sq ft)
Horizontal tail surfaces (total)	5.92 m² (63.70 sq ft)

WEIGHTS AND LOADINGS:
Weight empty:
F-16C: F100-PW-229	8,433 kg (18,591 lb)
F110-GE-129	8,581 kg (18,917 lb)
F-16D: F100-PW-229	8,645 kg (19,059 lb)
F110-GE-129	8,809 kg (19,421 lb)

Max internal fuel (JP-8): F-16C   3,249 kg (7,162 lb)
F-16C Block 60	3,084 kg (6,800 lb)
F-16D	2,687 kg (5,924 lb)
F-16D Block 60	2,521 kg (5,558 lb)

Max external fuel (JP-8), both:
normal	3,208 kg (7,072 lb)
optional	4,627 kg (10,200 lb)
F-16C/D Block 60	5,987 kg (13,200 lb)

Max external load (full internal fuel):
F-16C: F100-PW-229	7,226 kg (15,930 lb)
F110-GE-129	7,072 kg (15,591 lb)

Typical combat weight (two AAMs, 50% fuel):
F-16C: F100-PW-229	10,659 kg (23,498 lb)
F110-GE-129	10,812 kg (23,837 lb)

Max T-O weight: with two AAMs, no tanks:
F-16C: F100-PW-229	12,138 kg (26,760 lb)
F110-GE-129	12,292 kg (27,099 lb)

with full external load:
F-16C/D Block 50/52	19,187 kg (42,300 lb)
F-16C/D Block 60	22,679 kg (50,000 lb)

Wing loading: at 12,927 kg (28,500 lb) AUW
463.8 kg/m² (95.00 lb/sq ft)
at 19,187 kg (42,300 lb) AUW
688.4 kg/m² (141.00 lb/sq ft)
Thrust/weight ratio (clean)   1.1 to 1

PERFORMANCE:
Max level speed at 12,200 m (40,000 ft)   above M2.0
Service ceiling   more than 15,240 m (50,000 ft)
Radius of action:
F-16C Block 50, two 907 kg (2,000 lb) bombs, two Sidewinders, 3,940 litres (1,040 US gallons; 867 Imp gallons) external fuel, tanks retained, hi-lo-lo-hi
676 n miles (1,252 km; 780 miles)
F-16C Block 50, armament as above, 5,678 litres (1,500 US gallons; 1,249 Imp gallons) external fuel, tanks retained, hi-lo-lo-hi
802 n miles (1,485 km; 923 miles)
F-16C Block 50, two BVR missiles, two Sidewinders, 3,940 litres (1,040 US gallons; 867 Imp gallons) external fuel, tanks dropped when empty, combat air patrol mission   866 n miles (1,604 km; 997 miles)
Ferry range:
F-16C Block 50, with 3,940 litres (1,040 US gallons; 867 Imp gallons) external fuel
1,961 n miles (3,632 km; 2,257 miles)
F-16C Block 50, with 5,678 litres (1,500 US gallons; 1,249 Imp gallons) external fuel
2,276 n miles (4,215 km; 2,619 miles)
Symmetrical *g* limit with full internal fuel   +9

UPDATED

LOPRESTI

LOPRESTI INC

2620 North Airport Drive, Vero Beach, Florida 32960
Tel: (+1 561) 562 47 57 or (+1 800) 859 47 57
Fax: (+1 561) 563 04 46
e-mail: Roy@LoPrestiFury.com
Web: http://www.LoPrestiFury.com
PRESIDENT: LeRoy P LoPresti

Original LoPresti Piper Aircraft Engineering Co (see 1990-91 *Jane's*) formed as Piper subsidiary to produce re-engineered version of GC-1B Swift as SwiftFury, but operations suspended December 1990 when Piper abandoned programme; design team retained in LoPresti Speed Merchants (performance improvement kits for Piper small singles and twins); LoPresti Flight Concepts formed October 1991 to acquire rights to SwiftFury and carry through to certification and production; project considerably delayed by legal disputes over name and design rights; these largely resolved in March 1999, when aircraft renamed as below.

VERIFIED

LOPRESTI LP-1 FURY

TYPE: Aerobatic two-seat lightplane.
PROGRAMME: Modernised, high-performance development of 1940s Globe/Temco GC-1B Swift (1951-52 *Jane's*), with airframe improvements and Textron Lycoming IO-360 instead of 93 kW (125 hp) Teledyne Continental; first flight (converted 1946 Swift, N217LP) 23 March 1989. Following delays caused by Piper bankruptcy and legal disputes, Fury rights wholly acquired by LoPresti in March 1999 and project relaunched. Funding for certification being raised in 1999-2000; target date of January 2001 for launch of two-year FAR Pt 23 certification programme; delivery of factory-built new aircraft thereafter.
CURRENT VERSIONS: **Fury:** Basic version, *to which detailed description applies.*
SwiftThunder: Tricycle landing gear version; developed by original company as contender in USAF flight screener competition (won by Slingsby Firefly). Project terminated.
SwiftFire: Turboprop version (313 kW; 420 shp Rolls-Royce 250-B17C), preceding SwiftFury; first flight (N345LP) 19 July 1988; details in 1990-91 *Jane's*. Terminated.
CUSTOMERS: Total of 569 SwiftFuries ordered before Piper abandonment. Deposits for Fury accepted from 1999.
COSTS: US$175,000 (Fury).
DESIGN FEATURES: Low wing, 10.2 cm (4 in) farther forward than on original Globe design, with large root fillets; dihedral wings and tailplane. LoPresti improvements include 40 per cent reduction in drag, HOTAS controls, low-friction control linkages via pushrods, new fuel system, new structural members, elimination of ballast, revised cockpit and instrumentation, sequenced main landing gear doors, dual braking system and ram boost air induction system for increased engine output. Design now retains only original wing; no parts commonality with Swift.
Wing sections NACA 23015 (root) and 23009 (tip); dihedral 6°; incidence 2°. Tailplane dihedral 8°.
FLYING CONTROLS: Conventional and manual. Horn-balanced rudder and elevator. Electrically operated trim tab in port elevator, anti-servo tab in starboard elevator; ground adjustable tab on rudder; aileron travel +12/−25°; electrically operated single-slotted flaps deflect to maximum of 40°; optional electrically operated speedbrake in upper surface of each wing. Stall strip on each inboard wing leading-edge; fence at inboard end of each aileron.
STRUCTURE: Conventional aluminium alloy stressed skin construction, butt-jointed and flush riveted.
LANDING GEAR: Tailwheel type; retractable. Main units fully enclosed when retracted, tailwheel remaining partially exposed. Wheel doors close when gear fully extended. Prototype's tailwheel non-steerable, with solid tyre; production aircraft to have steerable tailwheel with pneumatic tyre. Dual redundant braking system.
POWER PLANT: One 149 kW (200 hp) Textron Lycoming IO-360-A1B6 flat-four engine, driving a McCauley two-blade constant-speed metal propeller. Integral fuel tank in each wing, combined usable capacity 227 litres (60.0 US gallons; 50.0 Imp gallons).
ACCOMMODATION: Two seats side by side, with dual control sticks embodying HOTAS technology. Wraparound windscreen; one-piece rearward sliding canopy. Space for baggage aft of seats.

Prototype LoPresti Fury *(Paul Jackson/Jane's)*   2000/0084603

AVIONICS: Analogue/digital instrumentation for VFR and IFR flight.

DIMENSIONS, EXTERNAL:
Wing span	8.93 m (29 ft 3½ in)
Wing aspect ratio	6.2
Length overall	6.86 m (22 ft 6 in)
Height overall	1.80 m (5 ft 10¾ in)

AREAS:
Wings, gross	12.88 m² (138.6 sq ft)

WEIGHTS AND LOADINGS:
Weight empty	658 kg (1,450 lb)
Baggage capacity	90 kg (200 lb)
Max T-O weight	1,043 kg (2,300 lb)
Max wing loading	81.02 kg/m² (16.59 lb/sq ft)
Max power loading	7.00 kg/kW (11.50 lb/hp)

PERFORMANCE (estimated):
Never-exceed speed	235 kt (435 km/h; 270 mph)
Max level speed at S/L	189 kt (349 km/h; 217 mph)
Max cruising speed at 75% power	182 kt (338 km/h; 210 mph)
Max flying speed, wheels down	117 kt (217 km/h; 135 mph)
Stalling speed, wheels and flaps down, engine idling	53 kt (99 km/h; 62 mph)
Max rate of climb at S/L	411 m (1,350 ft)/min
Service ceiling	6,400 m (21,000 ft)
Range, 45 min reserves	869 n miles (1,609 km; 1,000 miles)
Max rate of roll	180°/s
g limits	+6/−3 (aerobatic)
	+10/−4 (ultimate)

UPDATED

LUSCOMBE

LUSCOMBE AIRCRAFT CORPORATION

6340 South Sandhill 5, Las Vegas, Nevada 89120
Tel: (+1 702) 434 67 22
Fax: (+1 702) 434 10 04
e-mail: sales@luscombeaircraft.com
Web: http://www.luscombeaircraft.com
WORKS: 5333 North Main Street, Altus, Oklahoma 73521
FOUNDER AND CEO: Bill McKown
PRESIDENT: Billy McCoy
SENIOR VICE-PRESIDENT: Charles Gibson Jr

Original Luscombe company formed in 1934, specialising in all-metal private light aircraft; ownership passed to Temco in 1949, Silvaire Aircraft in 1955, and to Luscombe Aircraft Corporation on termination of aircraft production in 1960. Type certificate of Model 11 transferred to Classic Air of Lansing, Michigan; relaunch of modified version undertaken from 1998 in newly built 11,150 m² (120,000 sq ft) factory at Altus. In late 1998, company announced it was looking for production partners in Australia, Brazil and Netherlands. US production could reach 500 per year, subject to demand. Production has been delayed, but deliveries expected during 2001.

A second design of the original Luscombe company, the Model 8, is marketed in the US by Renaissance (which see).

UPDATED

LUSCOMBE 11E SPARTAN

TYPE: Four-seat lightplane.
PROGRAMME: Model 11A Sedan first flew (NX74202) 11 September 1946; total of 198 built with tailwheel and 123 kW (165 hp) Continental E-165 engine; production ended in 1949; planned 1970 relaunch as Alpha Aviation Alpha IID failed to materialise. Considerably redesigned Spartan being produced at Altus, Oklahoma, following

conversion of Sedan N1674B as proof-of-concept followed by construction of preproduction prototype for certification. Structural test fuselage completed December 1997; production of new aircraft due to have started April 2000, but this has slipped. Certification for sales and marketing purposes awarded November 1998, with full certification expected by end 2000.

CURRENT VERSIONS: **Spartan 185**: Basic version; *as described.*

Spartan 210: Proposed higher-powered version using same airframe; provisional data below, where appropriate; certification targeted for first quarter 2001.

CUSTOMERS: By late 1998, orders believed to exceed 300.

COSTS: US$144,500 basic (2000).

DESIGN FEATURES: High-wing braced monoplane; original 1946 design modified for nosewheel. Constant-chord inner wing, with sweptforward outer trailing-edge; tapered tailplane; slightly sweptback fin with ventral fillet.

FLYING CONTROLS: Conventional and manual. Stainless steel control cables. All control surfaces aluminium skinned. Three-position flaps; maximum deflection 35°.

STRUCTURE: Steel tube semi-monocoque corrosion-proofed fuselage and strut-braced two-spar wings, all covered with aluminium skins. Dorsal tail fillet.

LANDING GEAR: Alloy spring leaf main legs with 6.00-6 tyres, pressure 2.4 bars (35 lb/sq in). Air/oil shock-absorbing strut nose gear with 5.00-5 tyre, pressure 1.7 bars (25 lb/sq in), steerable ±10°. Hydraulic disc brakes and parking brake on mainwheels; turning radius approximately 8.23 m (27 ft 0 in).

POWER PLANT: *Spartan 185:* One 138 kW (185 hp) Teledyne Continental IO-360-ES4 six-cylinder fuel-injected piston engine driving a Sensenich 76EC8S10 two-blade fixed-pitch metal propeller. Fuel capacity 152 litres (40.0 US gallons; 33.3 Imp gallons) in two bladder wing tanks, of which 148 litres (39.0 US gallons; 32.5 Imp gallons) are usable. Oil capacity 7.5 litres (2.0 US gallons; 1.6 Imp gallons). Proof-of-concept aircraft has three-blade propeller, which is optional on production airframes.

Spartan 210: Same engine, uprated to 157 kW (210 hp), with optional turbocharger; same fuel capacity, but a further 76 litre (20.0 US gallon; 16.7 Imp gallon) tank is optional.

ACCOMMODATION: Pilot and three passengers in side-by-side pairs; door each side; front seats fully adjustable; dual controls. Inertia reel harnesses for all seats. Cabin windows are tinted, as are two vision ports in cabin roof.

SYSTEMS: 28 V 70 A alternator; 24 V 12.75 Ah battery; dual vacuum pumps.

Luscombe 11E Spartan preproduction aircraft used for certification

AVIONICS: II Morrow/Meggitt and Honeywell options. Representative (but not exclusive) equipment listed below.
Comms: Honeywell KX155A nav/com; KI 209A VOR/receiver; KT 76C transponder with ACK-30 Mode C encoder; KI 227 ADF; KN 64 DME; emergency locator transmitter; PS Engineering 6000C audio panel.
Flight: Altimeter, compass, ASI, OAT, VSI, turn co-ordinator; GX60 GPS with moving map display.
Instrumentation: Ammeter, vacuum gauge, fuel capacity, tachometer, EGT, oil pressure.

EQUIPMENT: Strobe position lights.

Data apply to both versions, except when otherwise specified.

DIMENSIONS, EXTERNAL:
Wing span	11.73 m (38 ft 6 in)
Wing aspect ratio	8.9
Length overall	7.24 m (23 ft 9 in)
Height overall	2.69 m (8 ft 10 in)
Tailplane span	3.58 m (11 ft 9 in)
Wheel track	2.24 m (7 ft 4 in)
Wheelbase	2.24 m (7 ft 4 in)
Propeller diameter	1.93 m (6 ft 4 in)
Passenger door: Max height	1.19 m (3 ft 11 in)
Max width	0.81 m (2 ft 8 in)

DIMENSIONS, INTERNAL:
Cabin: Length	2.70 m (8 ft 10¼ in)
Max width	1.16 m (3 ft 9½ in)
Max height	1.30 m (4 ft 3¼ in)

AREAS:
Wings, gross	15.51 m² (167.0 sq ft)

WEIGHTS AND LOADINGS:
Weight empty: 185	612 kg (1,350 lb)
210	658 kg (1,450 lb)
Baggage capacity	45 kg (100 lb)
Max T-O and landing weight	1,034 kg (2,280 lb)
Max ramp weight	1,037 kg (2,286 lb)
Max wing loading	66.7 kg/m² (13.65 lb/sq ft)
Max power loading	7.50 kg/kW (12.32 lb/hp)

PERFORMANCE:
Never-exceed speed (V_{NE})	157 kt (290 km/h; 180 mph)
Max level speed: 185	130 kt (241 km/h; 150 mph)
210	139 kt (257 km/h; 160 mph)
Normal cruising speed: 185	113 kt (209 km/h; 130 mph)
T-O speed	60 kt (111 km/h; 69 mph)
Stalling speed, power off:	
flaps up	47 kt (87 km/h; 54 mph)
flaps down	43 kt (79 km/h; 49 mph)
Max rate of climb at S/L:	290 m (950 ft)/min
Service ceiling	5,486 m (18,000 ft)
T-O run: 185	229 m (750 ft)
210	168 m (550 ft)
T-O to 15 m (50 ft): 185	381 m (1,250 ft)
210	282 m (925 ft)
Landing from 15 m (50 ft)	275 m (900 ft)
Landing run	153 m (500 ft)
Range with max fuel:	
185	460 n miles (853 km; 530 miles)
210	809 n miles (1,499 km; 932 miles)
Endurance	3 h 30 min
g limits	+3.8/−1.52
Noise level	72.4 dB(A)

UPDATED

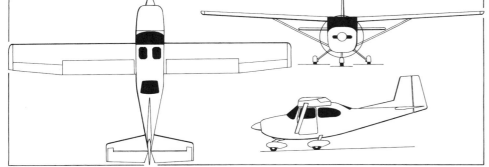

General arrangement of the Luscombe 11E Spartan (*Paul Jackson/Jane's*)

MAULE

MAULE AIR INC

2099 GA Highway 133 South, Lake Maule, Moultrie, Georgia 31768
Tel: (+1 912) 985 20 45
Fax: (+1 912) 985 96 28
e-mail: bdmaule@alltel.net
Web: http://members.surfsouth.com/~mauleair
CHAIRMAN: June D Maule
VICE-PRESIDENT: David Maule
ENGINEER: Don Richie
SALES MANAGER: Brent Maule

Original Maule Aircraft Corporation formed by the late Belford D Maule to manufacture M-4, a four-seat extrapolation of Piper Cub; transferred to Moultrie, Georgia, 1968; production ceased 1975; Maule Air Inc formed 1984 to produce uprated M-5 Lunar Rocket and M-7 Super Rocket; Lunar Rocket discontinued, but variants listed below currently available. Delivered 63 aircraft in 1998, and 69 in 1999 and 44 in the first nine months of 2000, including one M-6-235.

UPDATED

MAULE M-7 SERIES

TYPE: Five-seat lightplane/turboprop.

PROGRAMME: Introduced 1984 as M-7-235 Super Rocket (prototype N5656A; based on M-6-235 but with extended cabin and additional windows). Produced with short (M-5), mid-length (M-6) or long (A Model) wing span and engine/landing gear options as detailed below; short option deleted on 1999 versions, while piston-engined and turboprop/nosewheel aircraft have only M-6 wing from that year, the turboprop/tailwheel variants employing A Model wings. Designation prefixes are T = tricycle and X = short span.

CURRENT VERSIONS: **MX-7-160 Sportplane**: Four-seater, 119 kW (160 hp) Textron Lycoming O-320-B2D engine, Sensenich two-blade fixed-pitch metal propeller, and four-position flaps. One delivered in 1999.

MXT-7-160 Comet: As MX-7-160, but tricycle landing gear; two seats standard, four seats optional. Previously known as Maule Trainer.

MX-7-160C: As M-7-160, but spring aluminium landing gear. Available from 2000.

MX-7-180A Sportplane: As MX-7-160, but 134 kW (180 hp) Textron Lycoming O-360-C4F engine.

MXT-7-180A Comet: As MX-7-180A, but tricycle landing gear and four seats standard. Introduced and first production aircraft (N1002N) flown 1996; optimised for flight training schools. Total of 18 delivered in 1999, and five in first nine months of 2000.

MX-7-180B Star Rocket: 134 kW (180 hp) Textron Lycoming O-360-C1F engine, Hartzell two-blade constant-speed metal propeller and five-position flaps.

MX-7-180C Millenium: As MX-7-180B, but with spring aluminium main landing gear. Two delivered in 1999, and two in first nine months of 2000.

MXT-7-180 Star Rocket: As MX-7-180B, but with tricycle landing gear. Previously known as Star Craft. One delivered in 1999, and six in first nine months of 2000.

MX-7 Rocket: 153 kW (205 hp) PZL-Franklin 6A-350-C1R engine, McCauley two-blade propeller. Prototype was being test flown by second quarter of 1999. Engine may be offered as an option on Textron Lycoming O-360-powered models.

M-7-235B Super Rocket: Five-seater; choice of 175 kW (235 hp) carburetted Textron Lycoming O-540-J1A5, low-compression Mogas approved O-540-B4B5 or fuel-injected IO-540-W1A5 engines; McCauley constant-speed propeller, five-position flaps; fuselage raised 7.6 cm (3 in) at trailing-edge of wing and baggage area moved aft 12.7 cm (5 in) to accommodate fifth seat; recommended for high gross weight short-field operation, and for floatplane operation. Eight delivered in 1999, and five in first nine months of 2000.

M-7-235C Orion: As for M-7-235B but with spring aluminium main landing gear. Sixteen delivered in 1999, and 16 in first nine months of 2000.

MT-7-235 Super Rocket: As M-7-235B, but tricycle landing gear, four-position flaps and IO-540-W1A5 engine only. Four delivered in 1999, and four in first nine months of 2000. Civil Air Patrol approved purchase of 15 in glider towing configuration mid-2000.

M-7-260: As M-7-235B, but 194 kW (260 hp) Textron Lycoming IO-540-V4A5 engine; McCauley two-blade constant-speed propeller standard, Hartzell and MT propellers optional. Seven delivered in 1999.

USA: AIRCRAFT—MAULE

M-7-260C: As M-7-235C, but 194 kW (260 hp) Textron Lycoming IO-540-V4A5 engine; McCauley two-blade constant-speed propeller standard, Hartzell and MT propellers optional. Eight delivered in 1999, and four in first nine months of 2000.

MT-7-260: As MT-7-235, but 194 kW (260 hp) Textron Lycoming IO-540-V4A5 engine; McCauley two-blade constant-speed propeller standard, Hartzell and MT propellers optional. Three delivered in 1999, and one in first nine months of 2000.

M-7-420AC: M-7 fuselage with long-span wings, 313 kW (420 shp) Rolls-Royce 250-B17C turboprop, spring aluminium tailwheel landing gear. One delivered in 1999.

MT-7-420: As M-7-420AC, but tricycle gear.

CUSTOMERS: Total of 840 produced by third quarter of 2000, plus 908 of earlier M-5 and M-6 series.

COSTS: MX-7-160 US$99,069; MX-7-180A US$104,499; MXT-7-160 US$108,570; MXT-7-180A US$113,999; MX-7-180B US$116,435; MX-7-180C US$121,435; MXT-7-180 US$126,888; M-7-235B (O-540 engine) US$129,868; M-7-235B (IO-540 engine) US$137,568; M-7-235C (O-540 engine) US$135,706; M-7-235C (IO-540 engine) US$143,406; MT-7-235 US$149,278; M-7-260 US$147,568; M-7-260C US$153,988; MT-7-260 US$159,278; M-7-420AC US$450,000. All 1998; standard equipment.

DESIGN FEATURES: Rugged, STOL utility aircraft. High, constant chord, wing braced by V-struts; mid-mounted tailplane; large sweptback fin.

USA 35B (modified) wing section; dihedral 1°; incidence 0° 30'; cambered wingtips standard.

FLYING CONTROLS: Conventional and manual. Ailerons linked to rudder servo tab to reduce adverse yaw; trim tab in port elevator; rudder trim by spring to starboard rudder pedal; flap deflection 40° down for slow flight (further setting of 48° down on all except -420), 24° down, 0 and 7° up for improved cruise performance; underfin on floatplane and amphibious versions.

STRUCTURE: All-metal two-spar wing with dual struts and glass fibre tips; fuselage frame of welded 4130 steel tube with Ceconite covering aft of cabin and metal doors and skin forward of cabin; glass fibre engine cowling.

LANDING GEAR: Non-retractable tailwheel or nosewheel type. Maule oleo-pneumatic shock-absorbers in narrow track main units on MX-7-160, MX-7-180A, MX-7-180B, M-7-235B and M-7-260 models; wide chord main units standard. Maule steerable tailwheel. Cleveland mainwheels with Goodyear or McCreary tyres size 7.00-6, pressure 1.79 bars (26 lb/sq in). Tailwheel tyre size 8×3.5-4, pressure 1.03 to 1.38 bars (15 to 20 lb/sq in). Cleveland hydraulic disc brakes. Parking brake. Oversize tyres, size 20×8.5-6 (pressure 1.24 bars; 18 lb/sq in), and fairings aft of mainwheels optional

Provisions for fitting optional Aqua 2400 Baumann 2790, Edo 2400B or Edo 797-2500 amphibious or Wipline 2350 or Wipline 3000 amphibious floats (also available on some tricycle models). Float option available for MX-7-160, MX-7-180A/B/C, M-7-235B/C, M-7-260 and M-7-420AC. Ski option available for M-7-235B.

POWER PLANT: One flat-four or flat-six engine as described under Current Versions, driving a Sensenich two-blade fixed-pitch or Hartzell two-blade constant-speed propeller (three-blade McCauley propeller optional on 175 kW; 235 hp models; or Rolls-Royce 250-B17C turboprop with three-blade, Hartzell constant-speed propeller in -420 models. Piston versions have two fuel tanks in wings with total usable capacity of 163 litres (43.0 US gallons; 35.8 Imp gallons). Auxiliary fuel tanks in outer wings (standard on all 1999 models), to provide total capacity of 276 litres (73.0 US gallons; 60.8 Imp gallons). Turboprop versions have total fuel capacity of 322 litres (85.0 US gallons; 70.8 Imp gallons). Refuelling points on wing upper surface.

ACCOMMODATION: Four or five seats according to model, as described under Current Versions; individual, adjustable front seats; rear bench seat. Dual controls standard. Baggage compartment, capacity 113 kg (250 lb), aft of seats; cargo capacity with passenger seats removed 349 kg (770 lb). One front-hinged door on port side; three doors on starboard side, forward and centre doors hinged at front edge, rear baggage door hinged at rear edge to form double cargo door providing an opening 1.30 m (4 ft 3 in) wide to facilitate loading of bulky cargo; aircraft may be flown with doors removed. Accommodation heated and ventilated.

SYSTEMS: Hydraulic system for brakes only; electrical system powered by 60 A engine-driven alternator; 12 V battery (24 V battery on turboprops).

AVIONICS: *Comms:* Single Honeywell KX 125-01 nav/com or KLX-135A GPS/com, KT 76A-00 transponder, Narco AR-850 remote altitude encoder, PS Engineering PM1000 intercom, ELT, broadband antenna, omni antenna, avionics master switch, cabin overhead speaker, microphone and microphone/headset jacks front and rear, all standard. Options on most models include KX-155-42, KX-165-21, KI-206, KI VOR indicator, KT-76C transponder, Garmin GNC 300 GPS and PS Engineering PMA-6000 audio panel.

Flight: Option of KI 206-04 or KI-209-01 VOR/LOC/GS indicators, KN 62A-01 or KN 34-00 DME, KR 87-16 digital ADF with KI 1227 indicator and KA 33-00 avionics cooling blower.

Instrumentation: Standard (MX-7, MXT-7 and M-7 models, but others generally similar) includes full gyro panel/vacuum system, carburettor air temperature gauge, acoustic stall warner, vertical speed indicator, altimeter, compass tachometer, electric turn co-ordinator, electric fuel gauge, fuel pressure gauge, cylinder head temperature, manifold pressure and oil temperature/pressure gauges, OAT gauge, ammeter, clock, and stall warning light.

EQUIPMENT: (MX-7, MXT-7, and M-7; others similar): Includes instrument and dome lights, auxiliary cabin heater, auxiliary fuel pump, auxiliary power plug, heated pitot tube, cabin soundproofing, cloth velour upholstery, cabin steps, cargo tiedowns, landing light in port wing, navigation lights, wingtip strobe lights, tinted windscreen, pilot's swing-out window, windscreen defroster, airframe powder coating, wing corrosion proofing and standard external paint scheme in Insignia White base colour with two-tone trim. Optional equipment includes dual-calliper brakes, dynamically balanced propeller, McCauley three-blade constant-speed metal propeller (175 kW; 235 hp models only), Hartzell three-blade constant-speed propeller (turboprop models), wing corrosion proofing, Alcor EGT gauge (one-, four- or six-cylinder), Aviall hour meter, jump seat in baggage compartment, cabin door pockets, Plexiglas door, observation window, skylight, co-pilot's swing-out window, shoulder harnesses, landing light in starboard wing, fuselage grab handles, fire extinguisher, float reinforcement, lift rings and fairing, Schweizer glider tow/release kit and aircraft towbar.

MAULE M-7 OPTIONS

Version	Power Plant							Gear		
	B2D	C4F	C1F	B4B5	W1A5	V4A5	B17	TW/S	TW/A	TR/A
MX-7-160	•							•		
MXT-7-160	•									•
MX-7-160C	•							•		
MX-7-180A		•						•		
MX-7-180B			•					•		
MX-7-180C		•						•		
MXT-7-180			•							•
MXT-7-180A		•								•
M-7-235B				•				•		
M-7-235C				•				•		
MT-7-235					•					•
M-7-260						•		•		
M-7-260C						•		•		
MT-7-260						•				•
M-7-420AC							•		•	
MT-7-420							•			•

Notes: Power plants are Lycoming O-320-B2D, O-360-C4F, O-360-C1F, O-540-B4B5, IO-540-W1A5, IO-540-V4A5 (or earlier equivalents) and Rolls-Royce (Allison) 250-B17-C.
Landing gears are tailwheel/oleo main gear, tailwheel/aluminium and tricycle/aluminium.
'X' designations indicate shorter wing span.

DIMENSIONS, EXTERNAL:
Wing span: mid-length (M-6)	10.01 m (32 ft 10 in)
long (A Model)	10.31 m (33 ft 10 in)
Wing chord, constant: all	1.60 m (5 ft 3 in)
Wing aspect ratio: M-6	6.5
A Model	6.7
Length overall:	
piston-engined versions	7.16 m (23 ft 6 in)
turboprop versions	7.32 m (24 ft 0 in)
Height overall: tailwheel versions	1.93 m (6 ft 4 in)
tricycle versions	2.54 m (8 ft 4 in)
floatplane	3.05 m (10 ft 0 in)
amphibian	3.20 m (10 ft 6 in)
Wheel track:	
MX-7-160, MX-7-180A, MX-7-180B, M-7-235B, M-7-260	1.83 m (6 ft 0 in)
other versions	2.39 m (7 ft 10 in)
Propeller diameter:	
MX-7-160, MXT-7-160	1.88 m (6 ft 2 in)
MX-7-180A, MX-7-180C, MXT-7-180A, MX-7-180B, MXT-7-180	1.93 m (6 ft 4 in)
M-7-235, MT-7-235, MX-7-235	2.06 m (6 ft 9 in)
M-7-235, MT-7-235, MX-7-235 three-blade:	
option 1	2.03 m (6 ft 8 in)
option 2	1.98 m (6 ft 6 in)
M-7-420, MT-7-420	2.03 m (6 ft 8 in)

DIMENSIONS, INTERNAL:
Cabin max width	1.07 m (3 ft 6 in)

AREAS:
Wings, gross: M-6	15.38 m² (165.6 sq ft)
A Model	15.72 m² (169.2 sq ft)

WEIGHTS AND LOADINGS:
Weight empty: MX-7-160	603 kg (1,330 lb)
MXT-7-160	649 kg (1,430 lb)
MX-7-180A	612 kg (1,350 lb)
MXT-7-180A	658 kg (1,450 lb)
MX-7-180B	633 kg (1,395 lb)
MXT-7-180	640 kg (1,410 lb)
MX-7-180C landplane	680 kg (1,500 lb)
M-7-235B	694 kg (1,530 lb)
M-7-235C	726 kg (1,600 lb)
M-7-235B floatplane	826 kg (1,821 lb)
MX-7-420	626 kg (1,380 lb)
Max T-O weight:	
MX-7-160, MXT-7-160	998 kg (2,200 lb)
MX-7-180A, MXT-7-180A	1,089 kg (2,400 lb)
all other landplanes	1,134 kg (2,500 lb)
M-7-235B floatplane	1,247 kg (2,750 lb)
Max wing loading:	
MX-7-180B, MX-7-180C, M-7-235B, M-7-235C	73.7 kg/m² (15.10 lb/sq ft)
M-7-235B floatplane	81.1 kg/m² (16.61 lb/sq ft)
Max power loading:	
MX-7-160, MXT-7-160	8.37 kg/kW (13.75 lb/hp)
MX-7-180A, MXT-7-180A	8.12 kg/kW (13.33 lb/hp)
MX-7-180B, MXT-7-180	8.45 kg/kW (13.89 lb/hp)
M-7-235B, MX-7-235	6.47 kg/kW (10.64 lb/hp)
M-7-235B floatplane	7.12 kg/kW (11.70 lb/hp)
MX-7-420	3.62 kg/kW (5.95 lb/shp)

PERFORMANCE (at max T-O weight, ISA):
Max level speed:	
M-7-235B floatplane	130 kt (241 km/h; 150 mph)
MX-7-420	174 kt (322 km/h; 200 mph)
Cruising speed at 75% power at optimum altitude, TAS:	
MXT-7-160	113 kt (209 km/h; 130 mph)
MX-7-160, MXT-7-180A	117 kt (217 km/h; 135 mph)
MXT-7-180, MX-7-180A	122 kt (225 km/h; 140 mph)
MX-7-180, MX-7-180C	126 kt (233 km/h; 145 mph)
M-7-235B, M-7-235C	139 kt (257 km/h; 160 mph)

Maule MX-7-180B Star Rocket four/five-seat light aircraft *(Paul Jackson/Jane's)*

MAULE M-7 PRODUCTION

Version	Annual Deliveries																	Totals
	1983	1984	1985	1986	1987	1988	1989	1990	1991	1992	1993	1994	1995	1996	1997	1998	1999	
MX-7-160											15	16	9	3			1	44
MXT-7-160											1	1	1			5		8
MX-7-180			20	12	8	17	6	7	12	9	4	1						96
MX-7-180A											9	17	22	5	7	1		61
MX-7-180B												1	4	8		4		17
MX-7-180C													1	2	6	2		11
MXT-7-180							3	21	7	16	4	4	18	3	4	1		81
MXT-7-180A										1	1	3	6	22	24	18		75
M-7-235	2	17	13	11	16	15	13	9	13	3	13	7						132
M-7-235B												5	17	11	7	6	8	54
M-7-235C													7	11	5	16		39
MT-7-235										2	6	9	6	4	2	6	4	39
MX-7-235		2	23	25	18	14	5	3	16	9	2		1					118
M-7-260																1	7	8
M-7-260C															1	8		9
MT-7-260																	3	3
M-7-420																1		1
MMT-7-420AC																		0
MX-7-420								2	1		1							4
Yearly totals	**2**	**19**	**56**	**48**	**42**	**46**	**24**	**24**	**63**	**30**	**67**	**63**	**67**	**63**	**54**	**63**	**69**	**800**

M-7-235 floatplane 125 kt (232 km/h; 144 mph)
MX-7-420: at 50% power 156 kt (290 km/h; 180 mph)
at 75% power 169 kt (314 km/h; 195 mph)
Stalling speed, flaps down, power off:
 MX-7-160, MXT-7-160, MX-7-180A, MX-7-180C,
  MXT-7-180A, MX-7-180B, MXT-7-180, MX-7-
  235, M-7-235C 35 kt (64 km/h; 40 mph)
 M-7-235B 31 kt (57 km/h; 35 mph)
 M-7-235B floatplane 47 kt (87 km/h; 54 mph)
 MX-7-420 44 kt (81 km/h; 50 mph)
Max rate of climb at S/L:
 MX-7-160, MXT-7-160 251 m (825 ft)/min
 MX-7-180A, MXT-7-180A 280 m (920 ft)/min
 MX-7-180B, MX-7-180C, MXT-7-180
   365 m (1,200 ft)/min
 M-7-235B, M-7-235C, MT-7-235
   609 m (2,000 ft)/min
 M-7-235B floatplane 411 m (1,350 ft)/min
 MX-7-420 1,432 m (4,700 ft)/min
Service ceiling:
 MX-7-160, MXT-7-160 3,965 m (13,000 ft)
 MX-7-180A, MX-7-180C, MXT-7-180A, MX-7-
  180B, MXT-7-180 4,575 m (15,000 ft)
 M-7-235B, MT-7-235, M-7-235C, MX-7-420
   6,100 m (20,000 ft)
 M-7-235B floatplane 5,180 m (17,000 ft)
T-O run (solo, half fuel):
 MX-7-160, MXT-7-160 183 m (600 ft)
 MX-7-180C, MX-7-180B 92 m (300 ft)
 MXT-7-180, MX-7-420 61 m (200 ft)
 MX-7-180A, MXT-7-180A 168 m (550 ft)
 M-7-235C 77 m (250 ft)
 M-7-235B, M-7-235B floatplane 229 m (750 ft)
T-O to 15 m (50 ft) at max T-O weight:
 MX-7-160, MXT-7-160 360 m (1,180 ft)
 MX-7-180A, MXT-7-180A 350 m (1,150 ft)
 MX-7-180B, MX-7-180C, MXT-7-180, M-7-235B,
  M-7-235C 183 m (600 ft)
 M-7-235B floatplane 381 m (1,250 ft)
Landing from 15 m (50 ft):
 all landplane models 152 m (500 ft)
 M-7-235B floatplane 305 m (1,000 ft)
 Landing run: MX-7-420 91 m (300 ft)
Range with main tank fuel only, optimum altitude,
 30 min reserves:
 MX-7-160, MXT-7-160
   469 n miles (869 km; 540 miles)
 MX-7-180A, MXT-7-180A
   434 n miles (804 km; 500 miles)
 MXT-7-180, all tanks
   825 n miles (1,528 km; 950 miles)

Maule M-7-235C Orion *(Paul Jackson/Jane's)* 2001/0099622

Range with max fuel, no reserves:
 MX-7-180C 951 n miles (1,762 km; 1,095 miles)
 M-7-235B:
  O-540 engine 792 n miles (1,467 km; 912 miles)
  IO-540 engine 829 n miles (1,537 km; 955 miles)
 M-7-235C:
  O-540 engine 792 n miles (1,467 km; 912 miles)
  IO-540 engine 864 n miles (1,601 km; 995 miles)
 M-7-235B floatplane:
  O-540 engine 708 n miles (1,311 km; 815 miles)
  IO-540 engine 772 n miles (1,430 km; 889 miles)
 MX-7-420:
  at 75% power 551 n miles (1,022 km; 635 miles)
  at 50% power 782 n miles (1,448 km; 900 miles)

UPDATED

MAVERICK

MAVERICK AIR INC
Fremont County Airport, 100 Twinjet Way, Penrose, Colorado 81240
Tel: (+1 719) 784 02 55
e-mail: MavAir@aol.com
Web: http://www.twinjet.com
PRESIDENT: Robert Bornhofen
CHIEF FINANCIAL OFFICER: Michael Mock

The company moved into new premises in May 1998 and immediately began manufacture of the first Twinjet. Kits are now available to builders.

UPDATED

MAVERICK TJ-1500 TWINJET 1500
TYPE: Light business jet/kitbuilt.
PROGRAMME: Launched at EAA Convention at Oshkosh, Wisconsin, July 1997; then known as Twinjet 1200; construction of proof-of-concept prototype, undertaken by Composites Unlimited of Scappoose, Oregon, started September 1999; first flight (N750TJ) 4 August 1999; public debut at EAA AirVenture, Oshkosh, August 2000.
CUSTOMERS: Orders for 14 received by October 1999.
COSTS: Kit, less engines, US$219,000; engines (two), US$110,000; projected operating cost 50 cents per mile (2000).
DESIGN FEATURES: Designed using advanced CAD and three-dimensional modelling techniques; design goals include comfort and performance of a traditional business jet in a light piston twin configuration and good short-field performance; unswept, mid-mounted, constant-chord wing with optional tip tanks; T tail; engines pod-mounted on rear fuselage shoulders.
FLYING CONTROLS: Conventional and manual. Flaps, approximately two-thirds span, standard; speedbrakes in upper wing optional; single control column between front seats.
STRUCTURE: Composites, employing prepreg glass fibre with carbon fibre reinforcement in high-stress areas. Supplied as kit, with fast-build options; quoted build time less than 2,000 hours.
LANDING GEAR: Retractable tricycle type with single wheel on each unit; main units retract inwards; nosewheel, forwards; Cleveland brakes.

POWER PLANT: Two reconditioned General Electric T58 turbojets, converted from turboshafts used in Bell UH-1 helicopters, each derated to 3.34 kN (750 lb st). Fuel in two wing tanks and one fuselage tank, combined capacity 1,022 litres (270 US gallons; 225 Imp gallons); gravity refuelling point on each wing; tip tanks, each of 95 litres (25 US gallons; 20.8 Imp gallons) optional.
ACCOMMODATION: Four or five persons in pressurised cabin; single upward-opening gullwing door on port side; baggage compartments in nose and to rear of cabin.
SYSTEMS: Pressurisation system, maximum differential 0.39 bar (5.6 lb/sq in) maintains 3,050 m (10,000 ft) cabin altitude to 9,150 m (30,000 ft). Electrical system supplied by two 24 V batteries and two alternators. Airframe de-icing and oil-heated engine inlets standard.
AVIONICS: Sierra Flight Systems EFIS-2000 as core system; Garmin 430 GPS.

DIMENSIONS, EXTERNAL:
Wing span: standard	10.21 m (33 ft 6 in)
with tip tanks	10.52 m (34 ft 6 in)
Wing aspect ratio	7.5
Length overall	8.69 m (28 ft 6 in)
Height overall	2.74 m (9 ft 0 in)

DIMENSIONS, INTERNAL:
Cabin: Length	2.54 m (8 ft 4 in)
Max width	1.35 m (4 ft 5 in)
Max height	1.22 m (4 ft 0 in)
Baggage volume	0.57 m³ (20.0 cu ft)

AREAS:
Wings, gross	13.94 m² (150.0 sq ft)

WEIGHTS AND LOADINGS:
Weight empty	1,134 kg (2,500 lb)
Max T-O weight	2,268 kg (5,000 lb)
Max landing weight	2,086 kg (4,600 lb)
Max wing loading	162.7 kg/m² (33.33 lb/sq ft)
Max power loading	340 kg/kN (3.33 lb/lb st)

PERFORMANCE (estimated):
Max level speed	417 kt (772 km/h; 480 mph)
Econ cruising speed at max certified altitude	348 kt (644 km/h; 400 mph)
Cruising speed, OEI	200 kt (370 km/h; 230 mph)
Stalling speed, flaps down	79 kt (145 km/h; 90 mph)
Max rate of climb at S/L	1,067 m (3,500 ft)/min
Rate of climb at S/L, OEI	305 m (1,000 ft)/min
Max certified altitude	9,150 m (30,000 ft)
Service ceiling, OEI	6,095 m (20,000 ft)
T-O run	671 m (2,200 ft)
Landing run	607 m (1,990 ft)
Range: with max internal fuel	1,000 n miles (1,852 km; 1,150 miles)
with tip tanks	1,300 n miles (2,407 km; 1,496 miles)

UPDATED

Prototype Maverick Twinjet 1500 *(Don Mickey/Maverick Air)*

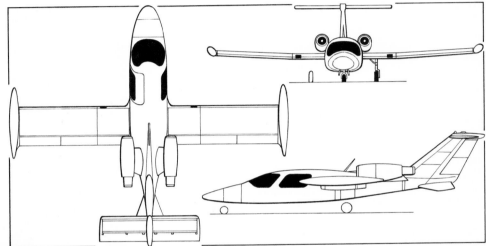

General arrangement of Maverick Twinjet 1500 *(James Goulding/Jane's)*

MD

MD HELICOPTERS INC

5000 East McDowell Road, Mesa, Arizona 85215-9797
Tel: (+1 480) 891 80 14
Fax: (+1 480) 891 80 18
e-mail: helicoptersales@mdhelicopters.com
Web: http://www.mdhelicopters.com
CHAIRMAN AND CEO: Henk Schaeken
COO: Albert Halder
VICE-PRESIDENT OF SALES AND MARKETING: Colin Whicher

McDonnell Douglas line of light helicopters, derived from the Hughes company's products, was acquired by Boeing as part of its purchase of the former company in August 1997. Having no part in the Boeing business strategy, all except the AH-64 Apache were offered for sale, but interest by Bell in the 500/600 series floundered early in 1998 when it became apparent that the transfer would be prevented by anti-trust legislation. Boeing then sought international interest and in January 1999 it was announced that Netherlands-based holding company MD Helicopters had been successful in its bid.

MD now owns all production jigs and tooling for the MD 500, 530F, 520N, 600N and Explorer, as well as a licence to employ the NOTAR system in future helicopters (the technology remaining in Boeing ownership) and in August 2000 acquired the former Boeing facility at Falcon Field, Mesa, Arizona, and began a 6,975 m² (75,000 sq ft) expansion to include a spares warehouse, completion and delivery centre and customer training facility.

TAI (which see in Turkish section) to build Explorer fuselages at Akinci, from where they will be shipped to St Truiden, Belgium, for assembly by Heli Fly. MD Helicopters and Heli Fly are both owned by the RDM group. In April 2000, Kaman (which see) was contracted to build fuselages for MD's single-engine product range at its Moosup and Jacksonville plants, deliveries beginning June 2000, and in July 2000 was contracted also to supply rotor blades for the MD Explorer. Composite Solutions of Auburn, Washington, supplies tailbooms for the NOTAR range of helicopters.

MD's production targets are 64 aircraft in 2000, 90 in 2001 and 100 per year from 2002. Employment stood at 300 in August 2000.

MD 500E five/seven-seat utility and executive helicopter *(Paul Jackson/Jane's)*

In 1999 MD Helicopters received orders for 50 aircraft and delivered 33 (five MD 500Es, six MD 530Fs, five MD 520Ns, six MD 600Ns, and 11 Explorers).

UPDATED

MD 500 and MD 530

TYPE: Light utility helicopter.
PROGRAMME: Derived from Hughes OH-6A/civil Model 500 first flown in 1963; see McDonnell Douglas Helicopters entries in earlier *Jane's* for MD 500D and previous versions. First flight MD 500E (N5294A) 28 January 1982; first flight MD 530F, 22 October 1982.
CURRENT VERSIONS: **MD 500E:** Replaced MD 500D in production 1982; deliveries started December 1982; Rolls-Royce 250-C20R became optional replacement for standard 250-C20B in late 1988; window area of forward canopy increased in 1991 model. MD 500E introduced many cabin improvements including more space for front and rear seat occupants, lower bulkhead between front and rear seats, T tail and optional four-blade Quiet Knight tail rotor.

MD 530F Lifter: Powered by Rolls-Royce 250-C30; transmission rating increased from 280 kW (375 shp) to 317 kW (425 shp) from 11 July 1985; diameter of main rotor increased by 0.3 m (1 ft 0 in) and of tail rotor by 5 cm (2 in); cargo hook kit for 907 kg (2,000 lb) external load available; certified 29 July 1983; first delivery 20 January 1984.

Detailed description applies to MD 500E and 530F, except where indicated.

MD 500/530 Defender: Described separately.
CUSTOMERS: Some 4,845 OH-6/MD 500/530 series (excluding licensed manufacture) produced by late 1999. Recent customers for the MD 500E include the San Diego County, California, Sheriff's Department (one); Columbus, Ohio, Police Department (one); and an

unnamed US operator (two), all delivered in the first half of 2000. The San Diego County Sheriff's Department also took delivery of an MD 530F. Five MD 500s and six MD 530s delivered in 1999.

COSTS: Typical MD 500E US$835,000 (1998); MD 530F US$998,000 (1999). Direct operating cost: MD 500E US$189, MD 530F US$211 per hour (both 1999).

DESIGN FEATURES: Fully articulated five-blade main rotor with blades retained by stack of laminated steel straps; blade aerofoil section NACA 015; blades can be folded after removing retention pins; two-blade tail rotor with optional X-pattern four-blade Quiet Knight tail rotor to reduce external noise; optional high-skid landing gear to protect tail rotor in rough country; protective skid on base of lower fin; narrow chord fin with high-set tailplane and endplate fins introduced with MD 500D. Main rotor rpm (500E/530F) 492/477 normal; main rotor tip speed 207 to 208 m (680 to 684 ft)/s; tail rotor rpm, 2,933/2,848.

FLYING CONTROLS: Plain mechanical without hydraulic boost.

STRUCTURE: Airframe based on two A frames from rotor head to landing gear legs, enclosing rear-seat occupants; front-seat occupants protected within straight line joining rotor hub and forward tips of landing skids; engine mounted inclined in rear of fuselage pod, with access through clamshell doors; main rotor blades have extruded aluminium spar hot-bonded to wraparound aluminium skin; tail rotor blades have swaged tubular spar and metal skin. Thicker fuselage skins, to reduce surface rippling, introduced during 2001.

LANDING GEAR: Tubular skids carried on oleo-pneumatic shock-absorbers. Utility floats, snow skis and emergency inflatable floats optional.

POWER PLANT: MD 500E powered by 313 kW (420 shp) Rolls-Royce 250-C20B or 335.5 kW (450 shp) 250-C20R turboshaft, derated in both cases to 280 kW (375 shp) for T-O; maximum continuous rating 261 kW (350 shp). MD 530F has 485 kW (650 shp) Rolls-Royce 250-C30 turboshaft, derated to 317 kW (425 shp) up to 50 kt (92 km/h; 57 mph) and 280 kW (375 shp) above 50 kt (92 km/h; 57 mph) and for maximum continuous power (MCP).

MCP transmission rating 261 kW (350 shp); improved, heavy-duty transmission, rating 447 kW (600 shp), derated to 280 kW (375 shp) optional on production aircraft or for retrofit from June 1995. Two interconnected bladder fuel tanks with combined usable capacity of 242 litres (64.0 US gallons; 53.3 Imp gallons). Self-sealing fuel tank optional. Refuelling point on starboard side of fuselage. Auxiliary fuel system, with 79.5 litre (21.0 US gallon; 17.5 Imp gallon) internal tank, available optionally. Oil capacity 5.7 litres (1.5 US gallons; 1.2 Imp gallons). Greater capacity internal fuel tanks also available.

ACCOMMODATION: Forward bench seat for pilot and two passengers, with two or four passengers, or two litter patients and one medical attendant, in rear portion of cabin. Pilot sits on left instead of normal right-hand seating. Low-back front seats and individual rear seats, with fabric or leather upholstery, optional. Baggage space, capacity 0.31 m³ (11 cu ft), under and behind rear seat in five-seat form. Clear space for 1.2 m³ (42 cu ft) of cargo or baggage with only three front seats in place. Two doors on each side. Interior soundproofing optional.

SYSTEMS: Aero Engineering Corporation air conditioning system or Fargo pod-mounted air conditioner optional.

AVIONICS (MD 500E): Optional avionics listed below.

Comms: Dual Honeywell KY 195 or Rockwell Collins VHF-251 transceivers; Honeywell KT 76 or Rockwell Collins TDR-950 transponder; intercom system, headsets, microphones; and optional public address system.

Radar: Optional installation.

Flight: Dual Honeywell KX 175 or Rockwell Collins VHF-251/231 nav receivers, latter with IND-350 nav indicator; Honeywell KR 85 or Rockwell Collins ADF-650 ADF.

Instrumentation: Basic VFR instruments and night flying lighting, attitude and directional gyros and rate of climb indicator.

Mission: Optional, FLIR and 30 Mcd Spectrolab SX-16 Nightsun searchlight.

EQUIPMENT: Optional equipment includes shatterproof glass, heating/demisting system, nylon mesh seats, dual controls, cargo hook, cargo racks, underfuselage cargo pod, heated pitot tube, extended landing gear, blade storage rack, litter kit, emergency inflatable floats and inflated utility floats.

DIMENSIONS, EXTERNAL:

Main rotor diameter: 500E	8.05 m (26 ft 5 in)
530F	8.33 m (27 ft 4 in)
Main rotor blade chord	0.171 m (6¾ in)
Tail rotor diameter: 500E	1.37 m (4 ft 6 in)
530F	1.42 m (4 ft 8 in)
Distance between rotor centres:	
500E	4.67 m (15 ft 4 in)
530F	4.88 m (16 ft 0 in)
Length overall, rotors turning:	
500E	8.61 m (28 ft 3 in)
530F	9.94 m (32 ft 7¼ in)
Length of fuselage	7.49 m (24 ft 7 in)
Height to top of rotor head:	
standard skids	2.67 m (8 ft 9 in)
extended skids	2.97 m (9 ft 8¾ in)
Tailplane span	1.65 m (5 ft 5 in)

MD 530MG Defender equipped with TOW missile tubes and nose-mounted Hughes M65 sight 0022145

Skid track (standard)	1.91 m (6 ft 3 in)
Cabin doors (each): Height	1.13 m (3 ft 8½ in)
Max width	0.76 m (2 ft 6 in)
Height to sill: 500E	0.79 m (2 ft 7 in)
530F	0.76 m (2 ft 6 in)
Cargo compartment doors (each):	
Height	1.12 m (3 ft 8¼ in)
Width	0.88 m (2 ft 10½ in)
Height to sill: 500E	0.71 m (2 ft 4 in)
530F	0.66 m (2 ft 2 in)

DIMENSIONS, INTERNAL:

Cabin: Length	2.44 m (8 ft 0 in)
Max width	1.31 m (4 ft 3½ in)
Max height	1.52 m (5 ft 0 in)

AREAS:

Main rotor blades (each): 500E	0.62 m² (6.67 sq ft)
530F	0.65 m² (6.96 sq ft)
Tail rotor blades (each): 500E	0.063 m² (0.675 sq ft)
530F	0.066 m² (0.711 sq ft)
Main rotor disc: 500E	50.89 m² (547.8 sq ft)
530F	54.58 m² (587.5 sq ft)
Tail rotor disc: 500E	1.53 m² (16.47 sq ft)
530F	1.65 m² (17.72 sq ft)
Fin	0.56 m² (6.05 sq ft)
Tailplane	0.76 m² (8.18 sq ft)

WEIGHTS AND LOADINGS:

Weight empty: 500E	672 kg (1,481 lb)
530F	722 kg (1,591 lb)
Max normal T-O weight: 500E	1,361 kg (3,000 lb)
530F	1,406 kg (3,100 lb)
Max overload T-O weight:	
500E, 530F	1,610 kg (3,550 lb)
Max T-O weight, external load:	
530F	1,701 kg (3,750 lb)
Max hook capacity: 530F	907 kg (2,000 lb)
Disc loading at max normal T-O weight:	
500E	26.8 kg/m² (5.48 lb/sq ft)
530F	25.8 kg/m² (5.28 lb/sq ft)
Transmission loading at max T-O weight and power:	
500E standard	5.22 kg/kW (8.57 lb/shp)
500E optional	4.87 kg/kW (8.00 lb/shp)
530F standard	5.39 kg/kW (8.86 lb/shp)
530F optional	5.03 kg/kW (8.27 lb/shp)

PERFORMANCE (at max normal T-O weight, except where indicated):

Never-exceed speed (VNE) at S/L:	
500E, 530F	152 kt (282 km/h; 175 mph)
Max cruising speed at S/L:	
500E	134 kt (248 km/h; 154 mph)
530F	133 kt (247 km/h; 154 mph)
Max cruising speed at 1,525 m (5,000 ft):	
500E	132 kt (245 km/h; 152 mph)
530F	135 kt (249 km/h; 155 mph)
Econ cruising speed at S/L:	
500E	129 kt (239 km/h; 149 mph)
530F	131 kt (243 km/h; 151 mph)
Econ cruising speed at 1,525 m (5,000 ft):	
500E, 530F	123 kt (228 km/h; 142 mph)
Max rate of climb at S/L, ISA:	
500E	536 m (1,760 ft)/min
530F	631 m (2,070 ft)/min
Max rate of climb at S/L, ISA+20°C:	
530F	628 m (2,061 ft)/min
Vertical rate of climb at S/L: 500E	248 m (813 ft)/min
530F	446 m (1,462 ft)/min
Service ceiling: 500E	4,575 m (15,000 ft)
530F	5,700 m (18,700 ft)
Hovering ceiling IGE: ISA: 500E	2,590 m (8,500 ft)
530F	4,875 m (16,000 ft)
ISA+20°C: 500E	1,830 m (6,000 ft)
530F	4,358 m (14,300 ft)
Hovering ceiling OGE: ISA: 500E	1,830 m (6,000 ft)
530F	4,389 m (14,400 ft)
ISA+20°C: 500E	975 m (3,200 ft)
530F	3,535 m (11,600 ft)
Range, 2 min warm-up, standard fuel, no reserves:	
500E at S/L	233 n miles (431 km; 268 miles)
530F at S/L	200 n miles (371 km; 231 miles)
500E at 1,525 m (5,000 ft)	264 n miles (488 km; 303 miles)
530F at 1,525 m (5,000 ft)	232 n miles (429 km; 267 miles)
Endurance, 530F at S/L	2 h 0 min

UPDATED

MD 500 and MD 530 DEFENDER

US Army designations: AH-6, EH-6, MH-6

TYPE: Armed observation helicopter.

PROGRAMME: Earlier 500MD Scout Defender, TOW Defender, 500MD/ASW Defender and 500MD Defender II described in the McDonnell Douglas Helicopters' entry in 1987-88 and earlier *Jane's*. Except for TOW Defender, military Defenders have same airframe as current civil 500/530; versions available detailed below.

CURRENT VERSIONS: **MD 500MG Defender:** As 530MG, but with 313 kW (420 shp) Rolls-Royce 250-C20B and MD 500E rotor system.

TOW Defender: Retains round nose of original MD 500; M65 TOW sight in nose; carries four TOW missiles; available with Rolls-Royce 250-C20B, 250-C20R or 250-C30.

Paramilitary MG Defender: Introduced July 1985 as low-cost helicopter suitable for police, border patrol, rescue, narcotics control and internal security use; available in either 500E or 530F configurations.

MD 530MG Defender: Based on MD 530F Lifter; first flight of prototype/demonstrator (N530MG) 4 May 1984; designed mainly for point attack and anti-armour, but also suitable for scout, day and night surveillance, utility, cargo lift and light attack. Integrated crew station with multifunction display allows hands on lever and stick (HOLAS) control of weapon delivery, communications and flight control; HOLAS based on Racal RAMS 3000, designed for all-weather and NOE flight and connected to MIL-STD-1553B digital databus linking the processor interface unit (PIU), control and display unit (CDU) and data transfer unit (DTU); CDU used for flight planning, navigation, frequency selection and subsystem management and has its own monochrome display and keyboard; multifunction display is high-definition monochrome tube with symbolic and alphanumeric capability; data input to DTU using ground loader unit inserted in cockpit receptacle.

Other equipment includes Astronics Corporation autopilot; Decca Doppler navigator with Racal Doppler velocity sensor; BAE FIN 1110 AHRS; twin Rockwell Collins VHF/UHF-AM/FM radios; Honeywell HF radio, ADF/VOR, radar altimeter and transponder; Telephonics intercom; SFENA attitude indicator. Options include mast-mounted Hughes TOW sight, FLIR, RWR, IFF, GPWS and laser ranger.

Weapons qualified or tested include TOW 2, FN Herstal pods containing 7.62 mm or 0.50 in machine guns, and 2.75 in rockets in seven- or 12-tube launchers; stores attached by standard 14 in NATO racks. Future armament will include General Dynamics air-to-air Stinger and 7.62 mm McDonnell Douglas (now Boeing) Chain Gun. Chaff and flare dispensers with automatic chaff discharge available. Both cyclic sticks have triggers for gun or rocket firing; co-pilot/gunner's visual image display has two handgrips for TOW/FLIR operation.

Nightfox: Introduced 1986 for low-cost night surveillance and military operations; equipment includes FLIR Systems Series 2000 thermal imager and night vision goggles, with same weapons as 530MG; available in both 500MG and 530MG forms.

Several versions of **H-6** have been used by the US Army's 160th Special Operations Aviation Regiment at Fort Campbell, Kentucky, as described in the 1998-99 and earlier editions.

CUSTOMERS: Operated by US Army (EH-6E, MH-6E, AH-6F, MH-6H, AH-6J and MH-6J) and also by Colombia (six 500MG and three 530MG delivered from 1986) and Philippines (22 520MGs from October 1990, plus six in 1992 and five ordered in 1993). Earlier Model 500s in military service with Argentina (500D/M), Colombia (500D/E), Iraq (500D), Israel (500MD), Kenya (500D/M/MD/ME), North Korea (500D/E) and South Korea (500MD); OH-6 and civilian standard Model 500/530 in other air arms. See also Korean Air Lines entry in 1994-95 *Jane's* for co-developed 520MK Black Tiger (non-NOTAR version of 500/530, despite designation) and Kawasaki entry in Japanese section for OH-6D/DA.

AVIONICS: See Current Versions.

ARMAMENT: See Current Versions.

DIMENSIONS, EXTERNAL: As for 500E/530F except:

Length of fuselage: 500MD/TOW	7.62 m (25 ft 0 in)
530MG	7.29 m (23 ft 11 in)
Height to top of rotor head:	
500MD/TOW	2.64 m (8 ft 8 in)
530MG	2.62 m (8 ft 7 in)
530MG with MMS	3.41 m (11 ft 2½ in)
Height over tail (endplate fins):	
500MD/TOW	2.71 m (8 ft 10½ in)
530MG	2.59 m (8 ft 6 in)
Width over skids: 500MD/TOW	1.93 m (6 ft 4 in)
530MG	1.96 m (6 ft 5 in)
Width over TOW pods: 500MD/TOW, 530MG	3.23 m (10 ft 7¼ in)

USA: AIRCRAFT—MD

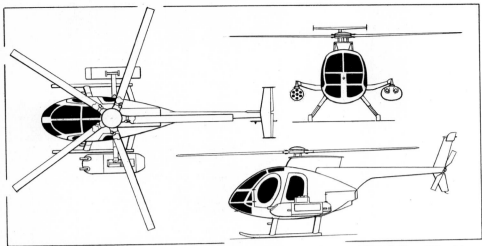

MD 530MG Defender, with seven-tube rocket launcher and FN Herstal gun pod *(Dennis Punnett/Jane's)*

Tailskid ground clearance:
- 500MD/TOW 0.64 m (2 ft 1¼ in)
- 530MG 0.61 m (2 ft 0 in)

WEIGHTS AND LOADINGS:
Weight empty, equipped:
- 500MD/TOW 849 kg (1,871 lb)
- 530MG 898 kg (1,979 lb)

Max T-O weight:
- 500MD/TOW, normal 1,361 kg (3,000 lb)
- 500MD/TOW, max overload 1,610 kg (3,550 lb)
- 530MG, normal 1,406 kg (3,100 lb)
- 530MG, max overload 1,701 kg (3,750 lb)

Max disc loading:
- 500MD/TOW 31.6 kg/m² (6.48 lb/sq ft)
- 530MG 31.2 kg/m² (6.38 lb/sq ft)

PERFORMANCE (at max normal T-O weight, except where indicated):
Never-exceed speed (V_{NE}) at S/L:
- 500MD/TOW, 530MG 130 kt (241 km/h; 150 mph)

Max cruising speed at S/L:
- 500MD/TOW, 530MG 121 kt (224 km/h; 139 mph)

Max cruising speed at 1,525 m (5,000 ft):
- 500MD/TOW 120 kt (222 km/h; 138 mph)
- 530MG 123 kt (228 km/h; 142 mph)

Max rate of climb at S/L, ISA:
- 500MD/TOW 520 m (1,705 ft)/min
- 530MG 626 m (2,055 ft)/min

Vertical rate of climb at S/L:
- 500MD/TOW 248 m (813 ft)/min
- 530MG 445 m (1,660 ft)/min

Service ceiling: 500MD/TOW 4,635 m (15,210 ft)
- 530MG over 4,875 m (16,000 ft)

Hovering ceiling IGE:
- ISA: 500MD/TOW 2,590 m (8,500 ft)
- 530MG 4,360 m (14,300 ft)
- ISA+20°C: 500MD/TOW 1,830 m (6,000 ft)
- 530MG 3,660 m (12,000 ft)
- ISA+30°C: 500MD/TOW 1,340 m (4,400 ft)
- 530MG, 35°C 2,680 m (8,800 ft)

Hovering ceiling OGE:
- ISA: 500MD/TOW 1,830 m (6,000 ft)
- 530MG 3,660 m (12,000 ft)
- ISA+20°C: 500MD/TOW 975 m (3,200 ft)
- 530MG 2,970 m (9,750 ft)
- ISA+30°C: 500MD/TOW 732 m (2,400 ft)
- 530MG, 35°C 2,120 m (6,950 ft)

Range, 2 min warm-up, standard fuel, no reserves:
- 500MD/TOW at S/L 203 n miles (376 km; 233 miles)
- 530MG at S/L 176 n miles (326 km; 202 miles)
- 500MD/TOW at 1,525 m (5,000 ft) 227 n miles (420 km; 261 miles)
- 530MG at 1,525 m (5,000 ft) 200 n miles (370 km; 230 miles)

Endurance with standard fuel, 2 min warm-up, no reserves:
- 5000MD/TOW at S/L 2 h 23 min
- 530MG at S/L 1 h 56 min
- 500MD/TOW at 1,525 m (5,000 ft) 2 h 35 min
- 530MG at 1,525 m (5,000 ft) 2 h 7 min

UPDATED

MD 520N

TYPE: Light utility helicopter.
PROGRAMME: First flight OH-6A NOTAR (no tail rotor) testbed 17 December 1981; programme details in 1982-83 *Jane's*; extensive modifications during 1985 with second blowing slot, new fan, 250-C20B engine and MD 500E nose; flight testing resumed 12 March 1986 and completed in June; retired to US Army Aviation Museum, Fort Rucker, Alabama, October 1990. Commercial MD 520N and uprated (485 kW; 650 shp Rolls Royce 250-C30) MD 530N NOTAR helicopters announced February 1988 and officially launched January 1989; first flight of MD 530N (N530NT) 29 December 1989, but this variant not pursued; first flight 520N (N520NT) 1 May 1990 and first production 520N 28 June 1991; 520N certified 13 September 1991; first production 520N (N521FB) delivered to Phoenix Police Department 31 October 1991. MD 520N set new Paris to London speed record in September 1992, at 1 hour 22 minutes 29 seconds. Now certified in 21 countries.

CURRENT VERSIONS: **MD 520N**: NOTAR version of MD 500, offering more power, higher operating altitude and greater maximum T-O weight than MD 500E.
Description applies to 520N.
MD 600N: Stretched version, described separately.
MD 520N Defender: Military variant, being developed.

CUSTOMERS: Production by mid-2000 totalled 92. Law enforcement agencies flying MD 520Ns include Phoenix, Arizona; Burbank, Glendale, Huntington Beach, Los Angeles, Ontario and San Jose, California; Hernando County, Florida; Orange County, Florida; Prince George County, Clinton, Maryland, which took delivery of two on 4 October 2000; Hamilton County, Ohio, which has seven 520Ns; El Salvador, whose Policia Nacional Civil took delivery of two in 1996; San Juan, Puerto Rico; Honolulu, Hawaii; and Calgary, Alberta, Canada. Other operators include the Tata Group of Mumbai, India, Weetabix Ltd, UK, and Belgian Gendarmerie (two).

COSTS: US$900,000 (1999). Direct operating cost US$192 per hour (1999).

DESIGN FEATURES: NOTAR system provides anti-torque and steering control without an external tail rotor, thus eliminating the danger of tail strikes; air emerging through Coanda slots and steering louvres is cool and at low velocity. Believed to be currently the quietest turbine helicopter, based on FAA certification test noise figures. Main rotor rpm 477; main rotor tip speed 208 m (684 ft)/s; NOTAR system fan rpm 5,388. Emergency floats among options. Redesigned diffuser in NOTAR tailboom and revised fan rigging introduced from early 2000 and available for retrofit; combined with uprated Rolls-Royce 250-C20R+ engine (see below); these improvements increase the aircraft's payload capability.

FLYING CONTROLS: Unboosted mechanical, as in earlier versions. Rotor downwash over tailboom deflected to port by two Coanda-type slots fed with low-pressure air from engine-driven variable-pitch fan in root of tailboom; this counters normal rotor torque; some fan air is also vented at tail through variable-aperture louvres controlled by pilot's foot pedals, giving steering control in hover and forward flight. Port moving fin on tailplane connected to foot pedals, primarily to increase directional control during autorotation and allow touchdown at under 20 kt (37 km/h; 23 mph); starboard fin operated independently by yaw damper.

STRUCTURE: Same as for MD 500E/530F, except graphite composites tailboom; metal tailplane and fins; new high-efficiency fan with composites blades fitted in production aircraft. NOTAR system components now have twice the lifespan of conventional tail rotor system assemblies. During 1993, NOTAR system components warranty increased from two to three years. Thicker fuselage skins, to reduce surface rippling, will be introduced during 2001.

POWER PLANT: One Rolls-Royce 250-C20R turboshaft, derated to 317 kW (425 shp) for T-O and 280 kW (375 shp) maximum continuous. Improved, heavy-duty transmission, rating 447 kW (600 shp), derated to 317 kW (425 shp) for T-O, 280 kW (375 hp) maximum continuous, on production aircraft from June 1995. Rolls-Royce 250-C20R+ engine, providing improved hot weather performance and 3 to 5 per cent more power, introduced from early 2000. Fuel capacity 235 litres (62.0 US gallons; 51.6 Imp gallons).

SYSTEMS: Electrical system includes 87 Ah starter-generator. Aero Aire Inc air conditioning system optional.

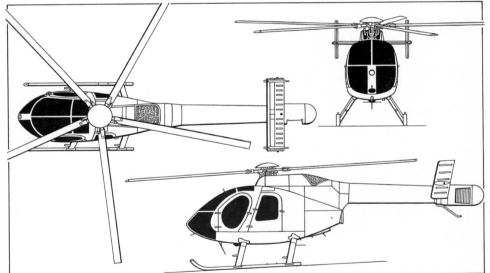

MD 520N five-seat NOTAR helicopter *(Mike Keep/Jane's)*

MD 520N operated in Belgium by Helifly *(Paul Jackson/Jane's)*

EQUIPMENT: Phoenix Police Department pioneered use of MD 520Ns for firefighting, using 341 litre (90.0 US gallon; 75.0 Imp gallon) 'Bambi' buckets to drop 151,416 litres (40,000 US gallons; 33,307 Imp gallons) of water on fires in remote desert and mountain areas.

DIMENSIONS, EXTERNAL:
Rotor diameter	8.33 m (27 ft 4 in)
Length: overall, rotor turning	9.78 m (32 ft 1¼ in)
fuselage	7.77 m (25 ft 6 in)
Height to top of rotor head:	
standard skids	2.74 m (9 ft 0 in)
extended skids	3.01 m (9 ft 10¾ in)
Height to top of fins	2.83 m (9 ft 3½ in)
Tailplane span	2.01 m (6 ft 7¼ in)
Skid track	1.92 m (6 ft 3½ in)

AREAS:
Rotor disc	54.51 m² (586.8 sq ft)

WEIGHTS AND LOADINGS:
Weight empty: standard	719 kg (1,586 lb)
Max fuel weight	183 kg (404 lb)
Max hook capacity	1,004 kg (2,214 lb)
Max T-O weight: normal	1,519 kg (3,350 lb)
with external load	1,746 kg (3,850 lb)
Max normal disc loading	27.9 kg/m² (5.71 lb/sq ft)
Transmission loading at max T-O weight and power	4.80 kg/kW (7.88 lb/shp)

PERFORMANCE (at normal max T-O weight, ISA, except where indicated):
Never-exceed speed (V_NE)	152 kt (281 km/h; 175 mph)
Max cruising speed at S/L	135 kt (249 km/h; 155 mph)
Max rate of climb at S/L: ISA	564 m (1,850 ft)/min
ISA + 20°C	480 m (1,575 ft)/min
Service ceiling	4,320 m (14,175 ft)
Hovering ceiling IGE	3,414 m (11,200 ft)
Hovering ceiling OGE: ISA	1,830 m (6,000 ft)
ISA +20°C	1,292 m (4,240 ft)
Range at S/L	229 n miles (424 km; 263 miles)
Endurance at S/L	2 h 24 min

OPERATIONAL NOISE LEVELS:
T-O	85.4 EPNdB
Approach	87.9 EPNdB
Flyover	80.2 EPNdB

UPDATED

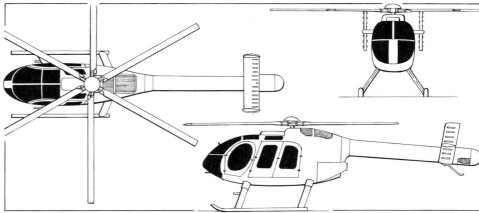

MD 600N eight-seat NOTAR helicopter *(James Goulding/Jane's)*

MD 600N

TYPE: Light helicopter.

PROGRAMME: Stretched version of MD 520N. Announced as 'concept', 8 November 1994; prototype, then known as MD 630N (N630N, converted from MD 530F demonstrator), first flight 22 November 1994, 138 days after project approval; public debut at Heli-Expo in Las Vegas January 1995; production go-head 28 March 1995, at which time designation changed to MD 600N; prototype first flown with production standard engine and rotor system 6 November 1995; production prototype (N600RN) first flown 15 December 1995, and became certification test vehicle leading to FAR Pt 27 certification, but was destroyed in ground fire 28 May 1996, following emergency landing after rotor/tailboom strike during abrupt control reversal tests. This resulted in changes to tailboom/rotor clearance; third prototype (N605AS) flown 9 August 1996; further accidents to N630N on 4 and 21 November 1996 and on 18 January 1997, all during autorotational descents, culminated in total loss and delayed certification and first delivery, originally scheduled for 18 December 1996, to 15 May and 6 June 1997, respectively.

In July 1998, Boeing completed a year-long envelope expansion programme for the MD 600N leading to FAA approval for operation at a density altitude of 2,135 m (7,000 ft) at a T-O weight of 1,746 kg (3,850 lb) and at a density altitude of 1,220 m (4,000 ft) at a T-O weight of 1,860 kg (4,100 lb). Other performance enhancements approved by the FAA included provision for doors-off operation at speeds up to 115 kt (213 km/h; 132 mph), operation at temperatures −40°C/52°C (−40°F/126°F), lifting up to 968 kg (2,134 lb) on the external cargo hook, making slope landings up to 10° in any direction, operation with emergency floats and for installation of a movable landing light and additional wire strike protection on the fuselage. The MD 600N also completed HIRF trials at NAS Patuxent River, Maryland. Yaw-stability augmentation system (Y-SAS) under development during 2000, aimed at reducing pilot workload during extended flights and in turbulent conditions; Y-SAS was expected to be available from March 2001, following FAA certification, and will be field-installable.

CUSTOMERS: Launch customer AirStar Helicopter of Arizona (two, of which first delivered 6 June 1997); Saab Helikopter AB of Sweden and Rotair Limited of Hong Kong ordered one each in June 1995; other customers include Guangdong General Aviation Company (GGAC) of the People's Republic of China, which took delivery of one MD 600N in November 2000, during Airshow China 2000 in Zhuhai, Los Angeles County Sheriff's Department Aero Bureau (three), Orange County, California, Sheriff's Department, Presta Services of France (one), Turkish National Police, which ordered 10 in December 2000 for delivery during 2001, UND (University of North Dakota) Aerospace (one), West Virginia State Police (one) and the US Border Patrol (45, of which 11 delivered by end of 1998, when procurement halted pending evaluation of UAVs for border patrol role). Deliveries totalled 15 in 1997, 21 in 1998, six in 1999 and five in the first eight months of 2000 (by which time 60 had been registered, including prototypes and forward construction); total fleet time by mid-December 2000 was 32,000 hours recorded by the 52 aircraft then in service.

COSTS: US$1.2 million (1999). Direct operating cost US$230.83 per hour (1999).

DESIGN FEATURES: Stretched MD 520N airframe (less than 1 per cent new parts) by means of 0.76 m (2 ft 6 in) plug aft of cockpit/cabin bulkhead and 0.71 m (2 ft 4 in) plug in tailboom, combined with more powerful engine, uprated transmission and six-blade main rotor. Cabin has flat floor to assist cargo handling, and will feature quick-change interior configurations to suit multiple-use operators. Intended for civil, utility, offshore, executive transport, medevac, aerial news gathering, touring, law enforcement and other noise-sensitive operations; also adaptable for armed scout, utility and other military missions.

Description generally as for MD 520N except as follows:

POWER PLANT: One 603 kW (808 shp) Rolls-Royce 250-C47M turboshaft, derated to 447 kW (600 shp) for T-O and 395 kW (530 shp) maximum continuous, with FADEC. Transmission manufactured from WE43A magnesium alloy for lower weight, greater strength and enhanced corrosion resistance, rating 447 kW (600 shp). Fuel contained in two crashworthy bladder tanks in lower fuselage, total capacity 435 litres (115 US gallons; 95.8 Imp gallons).

ACCOMMODATION: Standard seating for eight, including pilot, in 3+3+2 configuration with centre bench; alternative six-seat club and coach layouts; utility with main cabin seats removed; EMS with accommodation for pilot, one stretcher case and two medical attendants; or electronic news gathering, with accommodation for pilot and passenger in cockpit and operator with cabin and monitor in cabin. All facilitated by quick-release seat mechanisms. Two removable centre-opening doors on each side of cabin; door to cockpit on each side. Bubble 'comfort' windows and sliding windows for cabin, custom soundproofing and Integrated Flight Systems Inc air conditioning optional.

SYSTEMS: Electrical system comprises 28 V 200 Ah starter-generator and 28 V 17 Ah Ni/Cd battery. 24 V auxiliary power receptacle inside starboard cockpit door standard.

AVIONICS: To customer's choice, including VHF/UHF/FM/AM transceivers, GPS, transponder, ELT, intercom, stereo tape system and PA system.

EQUIPMENT: Optional equipment includes dual controls, particle separator, heated pitot tube, rotor brake, cargo hook, wire strike kit and emergency floats.

DIMENSIONS, EXTERNAL:
Rotor diameter	8.38 m (27 ft 6 in)
Rotor ground clearance	2.65 m (8 ft 8½ in)
Length: overall, rotor turning	10.79 m (35 ft 4¾ in)
fuselage	8.99 m (29 ft 6 in)
Fuselage width at cabin	1.40 m (4 ft 7 in)
Height to top of rotor head:	
standard skids	2.65 m (8 ft 8½ in)
extended skids	2.74 m (9 ft 0 in)
Height to top of fins:	
standard skids	2.68 m (8 ft 9½ in)
extended skids	2.83 m (9 ft 3½ in)
Fuselage ground clearance, extended skids	0.58 m (1 ft 10¾ in)
Skid track	2.47 m (8 ft 1¼ in)
Tailplane span, over fins	2.01 m (6 ft 7¼ in)

DIMENSIONS, INTERNAL:
Cabin, excl cockpit:	
Length	1.83 m (6 ft 0 in)
Width	1.22 m (4 ft 0 in)
Height	1.22 m (4 ft 0 in)

AREAS:
Rotor disc	55.18 m² (594.0 sq ft)

WEIGHTS AND LOADINGS (estimated):
Weight empty, standard	952 kg (2,100 lb)
Useful load: internal	907 kg (2,000 lb)
external	1,179 kg (2,600 lb)
Max hook capacity	1,361 kg (3,000 lb)
Max T-O weight: internal load	1,859 kg (4,100 lb)
slung load	2,131 kg (4,700 lb)
Max disc loading:	
internal load	33.7 kg/m² (6.90 lb/sq ft)
slung load	38.6 kg/m² (7.91 lb/sq ft)
Transmission loading at max T-O weight and power:	
internal load	4.16 kg/kW (6.83 lb/shp)
slung load	4.77 kg/kW (7.83 lb/shp)

PERFORMANCE:
Never-exceed speed (V_NE)	135 kt (250 km/h; 155 mph)
Max cruising speed, S/L to 1,525 m (5,000 ft), ISA	134 kt (248 km/h; 154 mph)
Max rate of climb at S/L, ISA	411 m (1,350 ft)/min
Max operating altitude	6,100 m (20,000 ft)
Hovering ceiling, ISA: IGE	3,383 m (11,100 ft)
OGE	1,829 m (6,000 ft)
Max range:	
at S/L	342 n miles (633 km; 393 miles)
at 1,525 m (5,000 ft), ISA	382 n miles (707 km; 439 miles)
Endurance: at S/L	3 h 36 min
at 1,525 m (5,000 ft)	3 h 54 min

UPDATED

MD 600N operated by Fuchs Helicopter of Switzerland *(Paul Jackson/Jane's)*

MD EXPLORER
US Coast Guard designation: MH-90 Enforcer

TYPE: Light utility helicopter.

PROGRAMME: Initially known as MDX, then MD 900 (proposed MD 901 with Turbomeca engines was not pursued); McDonnell Douglas design; announced February 1988; launched January 1989; Hawker de Havilland of Australia designed and manufactures airframe; Canadian Marconi tested initial version of integrated instrumentation display system (IIDS) early 1992; Kawasaki completed 50 hour test of transmission early 1992. Other partners include Aim Aviation (interior), IAI (cowling and seats) and Lucas Aerospace (actuators). Ten prototypes and trials aircraft, of which seven (Nos. 1, 3-7 and 9) for static tests; first flight (No.2/N900MD) 18 December 1992, followed by No.8/N900MH 17 September 1993 and No.10/N9208V 16 December 1993; first production/demonstrator Explorer (No.11/N92011) flown 3 August 1994. FAA certification 2 December 1994; first delivery 16 December 1994; JAA certification July 1996; FAA certification for single-pilot IFR operation achieved January 1997.

FAA certification of uprated PW207E engine achieved in July 2000, providing 11 per cent more power for take-off and 610 m (2,000 ft) increase in hovering capability OEI in hot and high conditions; first delivery of PW207E-engined Explorer to Police Aviation Services, UK, 27 September 2000.

CURRENT VERSIONS: **MD Explorer:** Initial civilian utility version, as described.

Details apply to civilian version except where indicated.

MD Enhanced Explorer: Improved version, announced September 1996; originally MD 902. Main features include Pratt & Whitney Canada PW206E engines with increased OEI ratings; transmission approved for dry running for 30 minutes at 50 per cent power; improved engine air inlets, NOTAR inlet design and engine fire suppression system, and more powerful stabiliser control system, resulting in 7 per cent increase in range, 4 per cent increase in endurance and 113 kg (250 lb) increase in payload over Explorer. First flight (N9224U; 41st Explorer) 5 September 1997. FAA certification to Category A performance standards (including continued take-off with one failed engine) and single-pilot IFR operation achieved 11 February 1998; JAA certification for Category A performance achieved July 1998. Retrofit kits to convert Explorers to Category A standard. First Enhanced Explorer delivery in May 1998 to Tomen Aerospace of Japan. PW206E replaced by PW207E from 2000.

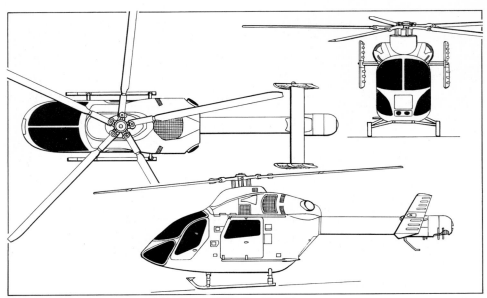

MD Explorer eight-seat commercial helicopter *(Mike Keep/Jane's)*

MH-90 Enforcer: Beginning March 1999, under a programme code-named Operation New Frontier, the US Coast Guard used two leased MD 900 Explorers for shipboard anti-drug smuggling operations. Armed with a pintle-mounted M240 7.62 mm minigun at the door station. In September 1999 the MD900s were exchanged for two leased MD 902 Enhanced Explorers. The USCG has a requirement for eight to 12 helicopters in the Explorer class. Six delivered to Mexican Navy at Acapulco (two each respectively in May and December 1999 and April 2000) for anti-drug operations, equipped with 0.50 in General Dynamics GAU-19/A Gatling guns, M2 gun pod and 70 mm rocket pods; further four in process of delivery. Weapons qualification trials were completed at Fort Bliss, Texas in November 2000.

Combat Explorer: Displayed at Paris Air Show, June 1995; demonstrator N9015P (No.15), an MD 900 variant. Can be configured for utility, medevac or combat missions; armament and mission equipment may include seven- or 19-tube 70 mm rocket pods, 0.50 calibre machine gun pods, chin-mounted FLIR night pilotage system and roof-mounted NightHawk surveillance and targeting systems. Combat weight 3,130 kg (6,900 lb); two P&WC PW206A engines. No customers announced by January 2000, but N9015P became one of initial two MH-90s (with third prototype, N9208V.

CUSTOMERS: Total of 102 ordered, of which 62 delivered, by May 2000. Market estimated at 800 to 1,000 in first decade; first delivery 16 December 1994 to Petroleum Helicopters Inc (PHI) which ordered five; second delivery (N901CF) December 1994 to Rocky Mountain Helicopters for EMS duties with affiliate Care Flight unit of Regional Emergency Medical Services Authority (REMSA) in Reno, Nevada. Total of two delivered in 1994, 12 in 1995, 15 in 1996, one in 1997, four in 1998 and 11 in 1999; initial (MD 900) series comprised 40 aircraft, including three flying prototypes; PW207E engine from 64th production (67th overall) aircraft. Other disclosed customers include Aero Asahi of Japan (15, of which the first, JA6757, was delivered in July 1995 and seven in service by October

MD Explorer light utility helicopter *(Paul Jackson/Jane's)*

Dual-pilot instrument panels of MD Explorer

2000), Allegheny General Hospital, Pittsburgh, Pennsylvania (four, delivered mid-1996), Belgian government (two delivered in 1996, in law enforcement configuration for Rijkswacht/Gendarmerie), German State Police of Lower Saxony operates three, based at Hanover, and State Police of Baden-Württemberg at Stuttgart has ordered three, with two options, for delivery in 2001. IBCOL Group of Germany (one, delivered in early 1996, operated by Air Lloyd of Bonn), Idaho Helicopters Inc/Life Flight (one, for EMS service at St Alphonsus Regional Medical Center, Boise, Idaho), Japan Digital Laboratory (one), Luxemburg Air Rescue (one, in EMS configuration), Police Aviation Services (PAS) UK (10, of which initial delivery was made in June 1998; by March 2000, PAS-operated Explorers were in service with the police forces of Dorset, Sussex, West Midlands and Wiltshire, and with the Kent Air Ambulance Service, with deliveries to the Greater Manchester Police and West Yorkshire Police scheduled for December 2000 and early 2001 respectively); Tata Group of India (one, delivered in fourth quarter 2000); Televisa of Mexico (one); Japanese distributor Tomen (two); UND (University of North Dakota) Aerospace (one); and Virgin HEMS (London) Ltd (one, delivered in October 2000).

COSTS: US$2.6 million (1999); direct operating cost US$368.85 (1999) per hour.

DESIGN FEATURES: NOTAR anti-torque system; all-composites five-blade rotor of tapered thickness with parabolic swept tip with bearingless flexbeam retention and pitch case; tuned fixed rotor mast and mounting truss for vibration reduction; replaceable rotor tips; maximum rotor speed 392 rpm; modified A-frame construction from rotor mounting to landing skids protects passenger cabin; energy-absorbing seats absorb 20 g vertically and 16 g fore and aft; onboard health monitoring, exceedance recording and blade track/balance.

FLYING CONTROLS: NOTAR tailboom (see details under MD 520N); mechanical engine control from collective pitch lever is back-up for electronic FADEC. Automatic stabilisation and autopilot available for IFR operation.

STRUCTURE: Cockpit, cabin and tail largely carbon fibre; top fairings Kevlar composites; no magnesium; lightning strike protection embedded in composites skin. Transmission overhaul life 5,000 hours; glass fibre blades have titanium leading-edge abrasion strip and are attached to bearingless hub by carbon fibre encased glass fibre flexbeams; rotor blades and hub on condition.

LANDING GEAR: Fixed skids with replaceable abrasion pads; emergency floats optional.

POWER PLANT: Two Pratt & Whitney Canada PW206E turboshafts with FADEC, each rated at 463 kW (621 shp) for 5 minutes for T-O, 489 kW (656 shp) for 2½ minutes OEI and 410 kW (550 shp) maximum continuous. Transmission rating 820 kW (1,100 shp) for T-O, 746 kW (1,000 shp) maximum continuous, 507 kW (680 shp) for 2½ minutes OEI and 462 kW (620 shp) maximum continuous OEI.

Fuel contained in single tank under passenger cabin, usable capacity 602 litres (159 US gallons; 132 Imp gallons); optional 818 litres (216 US gallons; 180 Imp gallons); single-point refuelling; self-sealing fuel lines.

ACCOMMODATION: Two pilots or pilot/passenger in front on energy-absorbing adjustable crew seats with five-point shoulder harnesses/seat belts; six passengers in club-type energy-absorbing seating with three-point restraints; rear baggage compartment accessible through rear door; cabin can accept long loads reaching from flight deck to rear door; hinged, jettisonable door to cockpit on each side; sliding door to cabin on each side.

SYSTEMS: Hydraulic system, operating pressure 34.475 bars (500 lb/sq in).

AVIONICS: Full IFR capability for single- or two-pilot operation.

Comms: Two headsets standard. Honeywell Silver Crown VHF transceiver, audio control panel, ELT, cockpit voice recorder and Wulfsberg Flexcomm II optional.

Radar: Honeywell RDR 2000 vertical profile colour weather radar optional with IFR package.

Flight: Optional equipment includes Honeywell Silver Crown VOR/ILS, HSI, ADF, DME, marker beacon receiver, radar altimeter, Loran C and KLN 90B GPS. Coupled three-axis autopilot optional.

Instrumentation: Single- or two-pilot instrument panels incorporate Canadian Marconi integrated instrumentation display system (IIDS) with high-resolution sunlight-readable LCD screen displaying engine and system information including engine condition trend monitoring, exceedance recording, caution annunciators, onboard track and balance of rotor and fan, weight on cargo hook, outside air temperature, digital clock, running time meter and RS-232 download modem interface for personal computer. Other standard instrumentation includes airspeed indicator, encoding altimeter, vertical speed indicator, turn and slip indicator, wet compass and clock. EFIS 40 electronic flight information system and parallel IIDS monitor for long-line hook operations from left seat optional.

Mission: Law enforcement panel with space for FLIR screen available. Avionics and equipment options for law enforcement conversions are described in *Jane's Aircraft Upgrades.*

EQUIPMENT: Standard equipment includes magnetic chip detectors on engines, tiedown fittings, flush-mounted cargo tiedowns, rotor blade tiedowns, right side passenger step, utility beige colour carpet, trim, wall and ceiling panels, soundproofing, tinted windows, map case, recessed hover and approach light, wander and white dome lights in cockpit, white dome light in cabin, utility light in baggage compartment, single 28 V DC power outlet each in cockpit and cabin, single colour external paint with two-colour accent stripes, FOD covers, pitot tube cover and cockpit fire extinguisher.

Optional equipment includes dual controls, heated pitot head, rotor brake, pilot-activated engine fire extinguisher, engine air particle separator, maintenance hand pump for hydraulics, external cargo hook with 1,361 kg (3,000 lb) capacity 272 kg (600 lb) personnel hoist, wire strike kit, emergency floats for skids, retractable landing light, port side cabin step, landing gear and rotor fairing canopy cover, heater/defogger, vapour-cycle air conditioner, upgraded soundproofing, passenger service unit with air gaspers and reading lights, window reveal panels, matching close-out panel for aft baggage area, upgraded passenger seats, smoke detector in baggage compartment, jack-point fittings and ground handling wheels.

DIMENSIONS, EXTERNAL:
Rotor diameter	10.31 m (33 ft 10 in)
Length: overall, rotor turning	11.83 m (38 ft 10 in)
fuselage	9.85 m (32 ft 4 in)
Fuselage width at cabin	1.63 m (5 ft 4 in)
Height: to top of rotor head	3.66 m (12 ft 0 in)
to top of fins	2.79 m (9 ft 2 in)
Min fuselage ground clearance	0.38 m (1 ft 3 in)
Tailplane span	2.84 m (9 ft 4 in)
Skid track	2.24 m (7 ft 4 in)
Cabin door width	1.27 m (4 ft 2 in)

DIMENSIONS, INTERNAL:
Cabin: Length overall, incl baggage compartment	3.93 m (12 ft 10¾ in)
Length, passenger compartment only	1.91 m (6 ft 3 in)
Max height	1.24 m (4 ft 1 in)
Max width	1.45 m (4 ft 9 in)
Volume	4.9 m³ (173 cu ft)
Baggage (if closed off)	1.39 m³ (49 cu ft)

AREAS:
Rotor disc	83.52 m² (899.0 sq ft)

WEIGHTS AND LOADINGS:
Weight empty, standard configuration	1,531 kg (3,375 lb)
Standard fuel load	489 kg (1,078 lb)
Max internal payload	1,292 kg (2,848 lb)
Max slung load	1,361 kg (3,000 lb)
Max T-O weight: internal load	2,835 kg (6,250 lb)
external load	3,129 kg (6,900 lb)
Max disc loading:	
internal load	33.9 kg/m² (6.95 lb/sq ft)
external load	37.5 kg/m² (7.68 lb/sq ft)
Transmission loading at max T-O weight and power:	
internal load	3.80 kg/kW (6.25 lb/shp)
external load	3.82 kg/kW (6.27 lb/shp)

PERFORMANCE (at max internal load T-O weight, ISA, except where indicated):
Never-exceed speed (V$_{NE}$) at S/L, ISA	140 kt (259 km/h; 161 mph)
Max cruising speed at S/L, 38°C (100°F)	134 kt (248 km/h; 154 mph)
Max rate of climb at S/L	579 m (1,900 ft)/min
Vertical rate of climb at S/L	411 m (1,350 ft)/min
Rate of climb at S/L, OEI	305 m (1,000 ft)/min
Service ceiling: both engines	5,335 m (17,500 ft)
OEI	3,200 m (10,500 ft)
Hovering ceiling:	
IGE, ISA	3,353 m (11,000 ft)
IGE, ISA + 20°C	2,042 m (6,700 ft)
OGE, ISA	2,743 m (9,000 ft)
OGE, ISA + 20°C	1,433 m (4,700 ft)
Hovering ceiling, OEI:	
IGE, ISA at 87% max T-O weight	1,220 m (4,000 ft)
Max range: at S/L	257 n miles (476 km; 295 miles)
at 1,525 m (5,000 ft), ISA	293 n miles (542 km; 337 miles)
Max endurance: at S/L	2 h 54 min
at 1,525 m (5,000 ft), ISA	3 h 12 min

OPERATIONAL NOISE LEVELS:
T-O	84.1 EPNdB
Approach	88.9 EPNdB
Flyover	83.1 EPNdB

UPDATED

MD Explorer of Police Aviation Services/Dorset & Wiltshire constabulary in the UK

MICCO

MICCO AIRCRAFT COMPANY
3100 Airman's Drive, Fort Pierce, Florida 34946
Tel: (+1 561) 465 99 96 or (+1 800) 647 95 35
Fax: (+1 561) 465 99 97
e-mail: micco_aircraft@msn.com
Web: http://www.miccoair.com
PRESIDENT: F DeWitt Beckett

Micco is owned by the Seminole tribe of native Americans and is descended from the Meyers company, formed in 1936, which produced the OTW (102 built), MAC-125/145 (22 built), Meyers 200 (46 built) and Meyers 400 Interceptor turboprop; after purchase by Rockwell in July 1965, the Meyers 200 became the Aero Commander 200 (88 built), until production ceased in 1967. Reformed in 1994, it was initially known as New Meyers until named Micco (Seminole for chief) in 1997. The Micco plant has 3,720 m² (40,000 sq ft) of factory space and 465 m² (5,000 sq ft) of engineering and administrative offices and employed 73 people in 2000. Production is now under way of the SP20 and its derivative, the SP26.

UPDATED

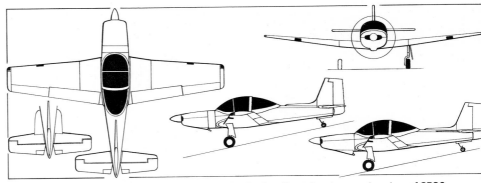

Micco SP20 two-seat light aircraft, with additional side view (lower) and scrap plan view of SP26 *(James Goulding/Jane's)* 2000/0084907

MICCO SP20 and SP26

TYPE: Aerobatic two-seat sportplane.
PROGRAMME: Based on original 1947 Meyers MAC-145A, but with increased power. Redesign started March 1994, using converted MAC-145 (N520SP); construction started June 1995 and prototype (N720SP) first flown 17 December 1997. First production aircraft (N820SP) construction started January 1999. Certification to FAR Pt 23 issued 15 January 2000, with deliveries soon after. Second production aircraft shown at Sun 'n' Fun, Lakeland, Florida, April 2000, before delivery to TE Aero, Texas; third production aircraft to Clipper Aviation of Oneonta, New York, June 2000. Aircraft registered using MAC-145A type certificate.

CURRENT VERSIONS: **SP20:** Baseline version.
SP26: Uprated version; engineering designation MAC-145B. Six-cylinder engine in longer cowling, but firewall moved rearwards by identical distance (30 cm; 12 in), resulting in no overall change of fuselage length. Prominent strakes ahead of tailplane. Prototype converted from N820SP; debut at Sun 'n' Fun, April 2000. Certification granted in October 2000 for 113 kg (250 lb) extension of MTOW over SP20.

CUSTOMERS: Over 50 on order by mid-2000.
COSTS: VFR equipped US$150,000; IFR equipped US$165,000 (2000).
DESIGN FEATURES: Low-wing, high-performance, metal sport aircraft; low development cost achieved by employment of existing design. Major changes from MAC-145A include replacement of the 108 kW (145 hp) Continental engine; substitution of a Meyers 200 fin and rudder; wings and flaps from the Meyers 400 Interceptor; and provision of a sliding canopy in place of the fixed canopy and side doors.
FLYING CONTROLS: Conventional and manual. Horn-balanced cable-operated rudder with ground-adjustable trim tab; horn-balanced push/pull tube-controlled elevators with flight-adjustable tabs in each; electrically operated Fowler flaps.
STRUCTURE: Steel tube centre section and all-aluminium wings.
LANDING GEAR: Tailwheel type; main units retract inward into wings, tailwheel rearward; mainwheel doors cover legs only when retracted, wheel wells remaining exposed. Electrohydraulic Parker Hannifin hydraulic brakes and leaf spring shock-absorbers. 6.00-6 tyres on mainwheels; 10 cm (4 in) solid tailwheel. Ground turning radius about wingtip 6.07 m (19 ft 11 in).
POWER PLANT: *SP20:* One 149 kW (200 hp) Textron Lycoming IO-360-C1E6 flat-four driving a three-blade, constant-speed McCauley QZP (B3D36C424/74SA-O) propeller.

Prototype Micco SP26, converted from first production SP20 *(Paul Jackson/Jane's)* 2000/0084751

SP26: One 194 kW (260 hp) Textron Lycoming IO-540-T4B5 flat-six driving a three-blade, constant-speed Hartzell Compact propeller.

Fuel (both versions) in integral wing tanks, combined capacity 273 litres (72.0 US gallons; 60.0 Imp gallons), of which 257 litres (68.0 US gallons; 56.6 Imp gallons) are usable. Three drain points. Oil capacity 7.6 litres (2.0 US gallons; 1.67 Imp gallons in SP20, 9.5 litres (2.5 US gallons; 2.1 Imp gallons) in SP26.
ACCOMMODATION: Two, side by side, under rearward-sliding canopy. Baggage area behind seats.
SYSTEMS: 28 V 70 A alternator and battery; landing lights.
AVIONICS: VFR or IFR standards available, recommended Trimble Terra suite; S-Tec 30 three-axis autopilot and Apollo MX20 multifunction display optional.
EQUIPMENT: Landing/taxying light in each wing leading-edge.
Data apply to both versions, except where stated.
DIMENSIONS, EXTERNAL:
Wing span	9.25 m (30 ft 4 in)
Wing chord: at root	2.29 m (7 ft 6 in)
at tip	1.02 m (3 ft 4 in)
Wing aspect ratio	5.9
Length overall	7.34 m (24 ft 1 in)
Max diameter of fuselage	1.22 m (4 ft 0 in)
Height overall	1.98 m (6 ft 6 in)
Tailplane span	3.76 m (12 ft 4 in)
Wheel track	3.15 m (10 ft 4 in)
Wheelbase	4.02 m (13 ft 2¼ in)
Propeller diameter: SP20	1.88 m (6 ft 2 in)
Propeller ground clearance: SP20	0.76 m (2 ft 6 in)

DIMENSIONS, INTERNAL:
Cabin: Length	1.65 m (5 ft 5 in)
Max width	1.09 m (3 ft 7 in)
Max height	0.99 m (3 ft 3 in)
Baggage volume	0.40 m³ (14.0 cu ft)

AREAS:
Wings, gross	14.55 m² (156.6 sq ft)

WEIGHTS AND LOADINGS:
Weight empty: SP20	839 kg (1,850 lb)
SP26	862 kg (1,900 lb)
Baggage capacity	45 kg (100 lb)
Max aerobatic weight: SP26	1,202 kg (2,650 lb)
Max T-O weight: SP20	1,179 kg (2,600 lb)
SP26*	1,292 kg (2,850 lb)
Max ramp weight: SP20	1,193 kg (2,630 lb)
Max wing loading: SP20	81.1 kg/m² (16.60 lb/sq ft)
SP26*	88.9 kg/m² (18.20 lb/sq ft)
Max power loading: SP20	7.91 kg/kW (13.00 lb/hp)
SP26*	6.67 kg/kW (10.96 lb/hp)

Limited to SP20 weights, pending certification

PERFORMANCE:
Never-exceed speed (V_{NE})	202 kt (374 km/h; 232 mph)
Max level speed: SP20	140 kt (259 km/h; 161 mph)
SP26	165 kt (306 km/h; 190 mph)
Max cruising speed at 75% power:	
SP20	130 kt (241 km/h; 150 mph)
SP26	155 kt (287 km/h; 178 mph)
Normal cruising speed at 65% power:	
SP20	128 kt (237 km/h; 147 mph)
SP26	148 kt (274 km/h; 170 mph)
Econ cruising speed at 55% power:	
SP20	120 kt (222 km/h; 138 mph)
SP26	135 kt (250 km/h; 155 mph)
Stalling speed, power off: flaps and landing gear up	55 kt (102 km/h; 64 mph)
flaps and landing gear down	49 kt (91 km/h; 57 mph)

Micco SP20 production prototype *(Paul Jackson/Jane's)* 2000/0084750

Cockpit of production SP20 *(Paul Jackson/Jane's)* 2000/0084752

Max rate of climb at S/L: SP20	290 m (950 ft)/min
SP26	458 m (1,500 ft)/min
Service ceiling: SP20	3,660 m (12,000 ft)
SP26	5,485 m (18,000 ft)
T-O run: SP20	460 m (1,510 ft)
SP26	244 m (800 ft)
T-O to 15 m (50 ft): SP20, SP26	618 m (2,025 ft)
Landing from 15 m (50 ft): SP20, SP26	610 m (2,000 ft)
Landing run: SP20, SP26	213 m (700 ft)
Range with 30 min reserves: at max level speed:	
SP20, SP26	750 n miles (1,389 km; 863 miles)
at max cruising speed:	
SP20	850 n miles (1,574 km; 978 miles)
SP26	1,040 n miles (1,926 km; 1,196 miles)
at econ cruising speed:	
SP20	1,004 n miles (1,859 km; 1,155 miles)
SP26	1,150 n miles (2,129 km; 1,323 miles)

UPDATED

MOONEY

MOONEY AIRCRAFT CORPORATION
PO Box 72, Louis Schreiner Field, Kerrville, Texas 78028
Tel: (+1 830) 896 60 00
Fax: (+1 830) 896 81 80
e-mail: sales@mooney.com
Web: http://www.mooney.com
CHAIRMAN AND CEO: Paul S Dopp
PRESIDENT AND SECRETARY: Christian E Dopp
VICE-PRESIDENT, SALES AND MARKETING: William C Kolloff

Original Mooney company formed in Wichita, Kansas, 1948; produced single-seat M-18 Mite until 1952 (total 282); later history recorded in 1987-88 *Jane's*. Alexandre Couvelaire, President of Euralair/Avialair Paris, France, and Michel Seydoux, President of MSC, jointly acquired Mooney in 1985; Mooney and Aerospatiale (Socata) announced joint development of TBM 700 June 1987 (see under Socata in French section), but Mooney withdrew in 1991. By early 2000, Mooney had produced 10,220 aircraft, including 9,840 of M20 family. New Mooney Encore announced October 1996; Eagle in 1998. Company decided in 1999 not to proceed with a turboprop, but is considering reintroducing a pressurised aircraft with 240 kt (444 km/h; 276 mph) cruising speed. Decision in 2000 could result in initial deliveries from 2003. Mooney delivered 97 aircraft in 1999, compared with 93 in 1998, 86 in 1997, 73 in 1996 and 84 in 1995. Personnel totalled 416 in 1999, including those engaged on Lockheed Martin F-16 and C-130, Boeing, Air Tractor, Aviat Pitts Special and Commander subcontracts.

UPDATED

MOONEY M20J ALLEGRO
TYPE: Four-seat lightplane.
PROGRAMME: First flight of original Mooney 201 version of M20J in June 1976; certified September 1976; improved 201 SE (Special Edition) followed in 1987 by modified 205; Allegro version introduced (as MSE) August 1990. Production of 2,137 of original version with 7.52 m (24 ft 8 in) fuselage was completed in October 1998; longer fuselage version was due to be launched in late 1998, but did not materialise. Details of the Allegro appear in the 2000-2001 edition.

UPDATED

MOONEY M20K ENCORE
TYPE: Four-seat lightplane.
PROGRAMME: Prototype M20K, a turbocharged M20J, first flew October 1976 and received FAA certification on 16 November 1978. Produced until mid-1980s, variously as Mooney 231, 231SE and 252TSE. Improved version of the M20K announced October 1996; prototype (N20XK) first flew 3 March 1997 and received supplementary type certificate in April; deliveries began mid-1997 with first production aircraft N20MK. Mooney announced plans to batch-produce the M20K in mid-1998 to rationalise its product line but had not restarted production by mid-2000.
CUSTOMERS: Total 35 delivered by January 1999, including 18 in 1998; no more in 1999. (Further 1,119 built in earlier series.)
COSTS: US$329,950 standard IFR equipped; US$369,900 typically equipped (1998).
LANDING GEAR: Retractable tricycle; main units retract inward, nosewheel rearward.
Description generally as for M20M Bravo, except the following:
POWER PLANT: One 164 kW (220 hp) Teledyne Continental TSIO-360-SB turbo-intercooled flat-four driving a two-blade constant-speed McCauley propeller.
ACCOMMODATION: Pilot and three passengers in two pairs of seats. Door on starboard side. Baggage door on port side aft of cabin.
SYSTEMS: Electrical system includes 10 Ah 24 V battery, dual 70 A 28 V alternators, voltage regulator with high and low warning light, and protective circuit breakers.
DIMENSIONS, EXTERNAL:
Length overall	7.75 m (25 ft 5 in)

WEIGHTS AND LOADINGS:
Weight empty	966 kg (2,130 lb)
Max T-O and landing weight	1,419 kg (3,130 lb)
Max wing loading	87.3 kg/m² (17.89 lb/sq ft)
Max power loading	8.66 kg/kW (14.23 lb/hp)

PERFORMANCE:
Never-exceed speed (V_{NE})	196 kt (363 km/h; 225 mph)
Max operating speed	174 kt (322 km/h; 200 mph)
Max cruising speed at 6,700 m (22,000 ft)	213 kt (394 km/h; 245 mph)
Normal cruising speed at 75% power at 6,700 m (22,000 ft)	198 kt (367 km/h; 228 mph)
Stalling speed, power off, flaps down	60 kt (112 km/h; 70 mph)
Max rate of climb at S/L	396 m (1,300 ft)/min
Service ceiling	7,620 m (25,000 ft)
Range with max fuel	1,100 n miles (2,037 km; 1,265 miles)

UPDATED

Mooney M20K Encore *(Paul Jackson/Jane's)*

MOONEY M20M BRAVO
TYPE: Four-seat lightplane.
PROGRAMME: Announced 2 February 1989 as TLS (Turbo Lycoming Sabre, with 201 kW; 270 hp TIO-540-AF1A); certified 1989; Bravo introduced in 1996.
CUSTOMERS: Total of 301 TLS/Bravos delivered by March 2000; production in 1999 totalled 25.
COSTS: Standard IFR equipped US$481,950 (2000).
DESIGN FEATURES: High-efficiency touring aircraft, originally designed by Mooney brothers. Low wing and mid-mounted tailplane; flying surfaces of characteristic Mooney 'reversed' configuration with unswept leading-edges and sharply forward-swept trailing-edges.
Wing section NACA $63_2 182\text{-}215$ at root, $64_1 181\text{-}412$ at tip; dihedral 5° 30′; incidence 2° 30′ at root, 1° at tip; wing swept forward 2° 29′ at quarter chord.
FLYING CONTROLS: Conventional and manual. Sealed gap, differential ailerons; fin and tailplane integral so that both tilt, varying tailplane incidence for trimming; no trim tabs; electrically actuated single-slotted flaps; electric rudder trim; speed brakes.
STRUCTURE: Single-spar wing with auxiliary spar out to mid-position of flaps; wing and tail surfaces covered with stretch-formed wraparound skins. Steel tube cabin section covered with light alloy skin; semi-monocoque rear fuselage with extruded stringers and sheet metal frames.
LANDING GEAR: Electrically retractable levered suspension tricycle type with airspeed safety switch bypass. Nosewheel retracts rearward, main units inward into wings. Rubber disc shock-absorbers in main units. Cleveland mainwheels, size 6.00-6, and steerable nosewheel, size 5.00-5. Tyre pressure, mainwheels 2.07 bars (30 lb/sq in), nosewheel 3.38 bars (49 lb/sq in). Cleveland hydraulic single-disc (double optional) brakes on mainwheels. Parking brake.
POWER PLANT: One 201 kW (270 hp) turbocharged Textron Lycoming TIO-540-AF1B flat-six engine, driving a McCauley three-blade metal propeller. Two integral fuel tanks in inboard wing leading-edges, with a combined capacity of 363 litres (96.0 US gallons; 79.9 Imp gallons), of which 337 litres (89.0 US gallons; 74.1 Imp gallons) usable. Two-piece nose cowling is of glass fibre/graphite composites construction.
ACCOMMODATION: Cabin accommodates four persons in pairs on individual vertically adjusting seats with reclining back, armrests, lumbar support and headrests; side armrests removable in rear seats. Front seats have inertia reel shoulder harnesses. Dual controls standard. Overhead ventilation system. Cabin heating and cooling system, with adjustable outlets and illuminated control. One-piece wraparound windscreen. Tinted Plexiglas windows. Rear seats removable for freight stowage. Rear seats fold forward for carrying cargo. Single door on starboard side. Compartment for 54 kg (120 lb) baggage behind cabin, with access from cabin or through door on starboard side.
SYSTEMS: Dual 70 Ah 28 V alternators, dual 24 V 10 Ah batteries, voltage regulator and warning lights, together with protective circuit breakers. Oxygen system, capacity 3,259 litres (115 cu ft), with masks and overhead outlets, standard. Windscreen defrosting system standard. TKS known ice protection system, air conditioning and Shadin fuel flow system.
AVIONICS: *Comms:* Basic avionics comprise PS Engineering PMA 7000 audio panel, two Honeywell KX 155A nav/coms, KI 203 VOR/LOC, KT 76C transponder and Trimble blind encoder.
Flight: Honeywell KN 64 DME, KFC 225 two-axis autopilot, KCS 55A slaved HSI and KLN 89B GPS; Classic/Plus group upgrade adds EFIS upgrade Goodrich Stormscope WX-950 adverse weather warning system. Additional options are available.
EQUIPMENT: Standard equipment includes attitude indicator, IFR directional gyro, two electric fuel quantity gauges, electric OAT gauge, CHT and EGT gauges, speed brakes, push-pull tube-actuated flight control system, annunciator panel; navigation lights, wing-mounted dual landing/taxying lights, internally lit instruments with rheostat; sound-damping composites interior, vertically adjusting front seats, wing jackpoints and external tiedowns, fuel tank quick drains, auxiliary power plug, heated pitot tube, epoxy polyimide and conversion anti-corrosion treatment, external polyurethane paint finish, dual static ports with drains, and alternate static air source.
DIMENSIONS, EXTERNAL:
Wing span	11.00 m (36 ft 1 in)
Wing chord, mean	1.50 m (4 ft 11¼ in)
Wing aspect ratio	7.4
Length overall	8.15 m (26 ft 9 in)
Height overall	2.54 m (8 ft 4 in)
Tailplane span	3.58 m (11 ft 9 in)
Wheel track	2.79 m (9 ft 2 in)
Wheelbase	2.02 m (6 ft 7½ in)
Propeller diameter	1.91 m (6 ft 3 in)
Baggage door: Width	0.53 m (1 ft 9 in)
Height	0.43 m (1 ft 5 in)

DIMENSIONS, INTERNAL:
Cabin: Length	3.20 m (10 ft 6 in)
Max width	1.10 m (3 ft 7½ in)

Mooney Bravo four-seat turbocharged light aircraft *(Paul Jackson/Jane's)*

Max height	1.13 m (3 ft 8½ in)
Floor area	3.53 m² (38.0 sq ft)
Volume	3.9 m³ (137 cu ft)
Baggage compartment volume	0.64 m³ (22.6 cu ft)
AREAS:	
Wings, gross	16.26 m² (175.0 sq ft)
Ailerons (total)	1.06 m² (11.40 sq ft)
Trailing-edge flaps (total)	1.66 m² (17.90 sq ft)
Fin	0.73 m² (7.92 sq ft)
Rudder	0.58 m² (6.23 sq ft)
Tailplane	1.99 m² (21.45 sq ft)
Elevators (total)	1.11 m² (12.05 sq ft)
WEIGHTS AND LOADINGS:	
Weight empty	1,029 kg (2,268 lb)
Max T-O weight	1,528 kg (3,368 lb)
Max landing weight	1,452 kg (3,200 lb)
Max wing loading	94.0 kg/m² (19.25 lb/sq ft)
Max power loading	7.59 kg/kW (12.47 lb/hp)
PERFORMANCE:	
Max cruising speed:	
at 3,960 m (13,000 ft)	195 kt (361 km/h; 224 mph)
at 7,620 m (25,000 ft)	220 kt (407 km/h; 253 mph)
Stalling speed:	
flaps and wheels up	66 kt (122 km/h; 76 mph)
flaps and wheels down	59 kt (109 km/h; 68 mph)
Max rate of climb at S/L	344 m (1,130 ft)/min
Max certified altitude	7,620 m (25,000 ft)
Range with reserves, at 3,960 m (13,000 ft)	
	1,070 n miles (1,982 km; 1,231 miles)
Endurance	6 h 42 min

UPDATED

Mooney Ovation 2 *(Paul Jackson/Jane's)*   2001/0089506

Mooney M20R Ovation *(James Goulding/Jane's)*   2001/0092656

MOONEY M20R OVATION 2 and M20S EAGLE

TYPE: Four-seat lightplane.
PROGRAMME: Prototype (N20XR) rolled out April 1994; first flight May 1994; FAA certification June 1994 (Ovation) and February 1999 (Eagle).
CURRENT VERSIONS: **Mooney M20R Ovation**: Baseline version, superseded by Ovation 2 in 1999.
Description applies to M20R Ovation 2 unless otherwise stated.
Mooney M20S Eagle: Entry-level version, using basic M20R fuselage and power plant, derated to 182 kW (244 hp), two-blade McCauley propeller and lower-cost avionics (PS Engineering PMA 7000 audio panel, single Honeywell KX 155A nav/com, KT 76C transponder, KLN 89B GPS, S-Tec 30 two-axis autopilot, KCS 55A HSI). Cruising speed 180 kt (333 km/h; 207 mph); maximum climb rate 350 m (1,150 ft)/min; range 1,200 n miles (2,222 km; 1,380 miles) at 3,048 m (10,000 ft); service ceiling 5,640 m (18,500 ft); max T-O weight 1,451 kg (3,200 lb). Price US$343,950 (2000). 47 delivered by March 2000.

CUSTOMERS: Most sold in USA, but others exported to Brazil, France, Germany, Israel, Italy, Paraguay, South Africa, Switzerland, Thailand and UK; orders include three for Western Michigan University. The 100th Ovation (N100DX) was delivered on 16 December 1996; by March 2000, 239 had been delivered, including 24 Ovations and 10 Ovation 2s in 1999.
First Eagle certified and delivered in February 1999; 38 delivered by 1 January 2000.
COSTS: US$418,950 typically equipped (2000).
DESIGN FEATURES: Combines airframe of M20M Bravo (which see) with normally aspirated flat-six engine. Completely restyled instrument panel, seats and cabin interior with sandwich-core rigid trim/soundproofing panels and leather and wool fabric upholstery.

Description generally as for M20M Bravo except:
POWER PLANT: M20R Ovation has one 224 kW (300 hp) Teledyne Continental IO-550-G5B flat-six engine, derated to 209 kW (280 hp), driving a McCauley three-blade constant-speed metal propeller; Ovation 2 has redesigned propeller giving slight performance improvement. Usable fuel 337 litres (89.0 US gallons; 74.1 Imp gallons).
AVIONICS: As Mooney M20M Bravo; additional option package includes EFIS upgrade to Classic/Plus Group.
DIMENSIONS, EXTERNAL:
Propeller diameter	1.85 m (6 ft 1 in)

WEIGHTS AND LOADINGS:
Weight empty	1,009 kg (2,225 lb)
Max usable fuel	242 kg (534 lb)
Max T-O weight	1,528 kg (3,368 lb)
Max landing weight	1,451 kg (3,200 lb)
Max wing loading	94.0 kg/m² (19.25 lb/sq ft)
Max power loading	7.32 kg/kW (12.03 lb/hp)

PERFORMANCE:
Max cruising speed	192 kt (356 km/h; 221 mph)
Stalling speed:	
flaps and wheels up	66 kt (123 km/h; 76 mph)
flaps and wheels down	59 kt (110 km/h; 68 mph)
Max rate of climb at S/L	381 m (1,250 ft)/min
Service ceiling	6,100 m (20,000 ft)
Max range, economy cruise, with reserves, at 2,745 m	
(9,000 ft)	1,280 n miles (2,370 km; 1,473 miles)

UPDATED

MOONEY M20T PREDATOR

Anticipated return to production has not taken place. A brief description last appeared in the 2000-01 *Jane's*.

UPDATED

Early production Mooney M20S Eagle *(Paul Jackson/Jane's)*   2001/0089507

MORROW

MORROW AIRCRAFT CORPORATION
3280 25th Street NE, PO Box 12665, Salem, Oregon 97389
Tel: (+1 503) 365 02 00
Fax: (+1 503) 365 03 00
e-mail: comments@morrowaircraft.com
Web: http://www.morrowaircraft.com
PRESIDENT: Ray Morrow
VICE-PRESIDENT, PRODUCT MANAGEMENT: Dale Johnson
VICE-PRESIDENT, MARKETING AND SALES: Dan Waldron

Morrow Aircraft Corporation formed 1996 to develop general aviation aircraft using advanced composites materials and state-of-the-art technology. Ray Morrow is co-founder of American Blimp Corporation (which see) and established II Morrow avionics company (now UPS Aviation Technology). The company's first product is a development of the Rutan Boomerang, which it also intends to use for its own Sky Taxi programme.

NEW ENTRY

MORROW MB-300 BOOMERANG

TYPE: Business twin-prop.
PROGRAMME: Development of Rutan 202-11 Boomerang (see 1994-95 *Jane's* for details); Morrow example is 15 per cent longer than original with approximately 20 per cent greater payload. In October 1998, Scaled Composites contracted to build one prototype and two preproduction examples for certification; prototype one-third complete by early 2001. Due to enter service during 2003.
CUSTOMERS: Company estimates market for up to 4,500 aircraft for its own SkyTaxi programme by 2011.
DESIGN FEATURES: Unconventional layout with twin booms of unequal size, each with engine at front; starboard boom contains passenger cabin and flight deck; port boom offers extra storage space. Booms joined at wingroot by sweptforward wing. Tailplane at rear mounted between booms with greater span on starboard side; upswept wingtip and fintip extensions of original Boomerang replaced with conventional tips.
FLYING CONTROLS: Conventional and manual.

STRUCTURE: Extensive use of composites throughout; airframe components manufactured by Scaled Technology Works.
LANDING GEAR: Retractable tricycle type. Mainwheels under each boom, nosewheel at front of starboard boom.
POWER PLANT: Two 261 kW (350 hp) Teledyne Continental TSIO-550F engines flat-rated to 242 kW (325 hp) for take-off. FADEC.
ACCOMMODATION: Two pilots plus four passengers in club seating in main cabin. Lavatory behind flight deck. Small baggage compartment at rear of passenger cabin, with larger baggage compartment in separate pod. Cabin door on starboard side behind wing.
SYSTEMS: Pressurisation system differential 0.48 bar (7.0 lb/sq in). Cabin pressurised to 1,370 m (4,500 ft) cabin altitude at 7,620 m (25,000 ft). Full icing protection.
AVIONICS: Full IFR fit.
DIMENSIONS, EXTERNAL:
Wing span	13.00 m (42 ft 8 in)

Length overall	10.39 m (34 ft 1 in)
Height overall	3.23 m (10 ft 7 in)
DIMENSIONS, INTERNAL:	
Cabin (excl flight deck): Length	4.19 m (13 ft 9 in)
Port boom baggage hold: Length	3.05 m (10 ft 0 in)
Max width	0.76 m (2 ft 6 in)
WEIGHTS AND LOADINGS:	
Weight empty	1,746 kg (3,850 lb)
Max payload	1,089 kg (2,400 lb)
Max fuel weight	653 kg (1,440 lb)
Max T-O weight	2,835 kg (6,250 lb)
Max power loading	5.85 kg/kW (9.61 lb/hp)
PERFORMANCE:	
Max cruising speed	300 kt (556 km/h; 345 mph)
Normal cruising speed	284 kt (526 km/h; 327 mph)
Econ cruising speed	246 kt (456 km/h; 283 mph)
Stalling speed	85 kt (158 km/h; 98 mph)
Max rate of climb at S/L	480 m (1,575 ft)/min
Rate of climb at S/L, OEI	140 m (460 ft)/min
Service ceiling, OEI	more than 6,100 m (20,000 ft)
T-O run	610 m (2,000 ft)
T-O to 15 m (50 ft)	914 m (3,000 ft)
Landing from 15 m (50 ft)	1,097 m (3,600 ft)
Landing run	762 m (2,500 ft)
Range: with max fuel	2,200 n miles (4,074 km; 2,531 miles)
with max payload	1,680 n miles (3,111 km; 1,933 miles)

NEW ENTRY

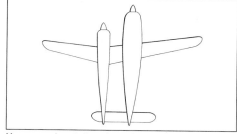

Unconventional plan view of the Morrow Boomerang *2001*/0105042

NASA

NATIONAL AERONAUTICS AND SPACE ADMINISTRATION
(Office of Aeronautics)
600 Independence Avenue SW, Washington, DC 20546
Tel: (+1 202) 453 26 93
Fax: (+1 202) 426 42 56
Web: http://www.nasa.gov

Telex: 89530 NASA WSH
ADMINISTRATOR: Daniel S Goldin
CHIEF ENGINEER: Daniel R Mulville
ASSOCIATE ADMINISTRATOR, AEROSPACE TECHNOLOGY: Spence M Armstrong
CHIEF FINANCIAL OFFICER: Arnold G Holz

NASA's FY01 budget is US$14.2 billion, of which only US$1.19 billion related to aerospace technology. For FY02, total of US$14.6 billion will be appropriated. Personnel totalled approximately 17,800 in 2000. Prime NASA concerns are R&D, rather than operationally driven. In addition to selected programmes described below, NASA funds are also allocated to the Lockheed Martin AeroCraft and Boeing Blended Wing/Body programmes described elsewhere in this edition.

UPDATED

NASA SPACE PROGRAMMES
Although properly belonging in *Jane's Space Directory*, several NASA space ventures involve craft which, albeit briefly, employ wings or lifting bodies for atmospheric flight. Accordingly, they are described below with commensurate brevity.

REUSABLE LAUNCH VEHICLE (X-33)
Joint NASA/industry project to develop single-stage-to-orbit, half-scale technology demonstrator reusable launch vehicle (RLV), paving way for unmanned orbital RLV to enter service in 2005 and with capability to carry 18,000 kg (39,683 lb) payload. Industry briefed on funding by NASA at Marshall Space Flight Center, Alabama, on 19 October 1994. NASA funding for the X-33 project capped at US$942 million, of which US$115 million still to be spent in early 2000, by which time Lockheed Martin and its subcontractors had invested US$356 million in project. Intent is for any follow-on RLV to be privately financed, with government commitment limited to guaranteeing to buy agreed number of launches. Decision on whether to go ahead with full-scale **Venture-Star** vehicle to be taken after X-33 flight test programme.

Three teams selected by NASA in April 1995 for 15 month concept definition and design phase ending in June 1996. Phase 1 expenditure by NASA totalled US$24 million, split equally between these competing teams. Selection of Lockheed Martin contender as winning submission announced 3 July 1996; lifting-body design for vertical take-off and conventional landing.

Subsequent programme elements are Phase 2 entailing design, manufacture and flight trials of small-scale version of successful design; and Phase 3 production of operational full-scale RLV. Lockheed Martin formed partnership arrangement with Rocketdyne and Rohr; design leadership vested in Lockheed Martin at Palmdale, California; Sverdrup provides launch facilities. Preliminary design and initial testing of Boeing Rocketdyne XRS-2200 linear-aerospike engine was completed in mid-1997; test firings of 916 kN (206,000 lb st) engine began at Stennis Space Center, Mississippi, in third quarter of 1999 and have demonstrated 125 second run at 100 per cent power as well as thrust vector control. Full-scale VentureStar to have two engines, generating total of 1,807 kN (406,200 lb st) at sea level. Testing of thermal protection system materials undertaken by NASA F-15 during series of six flights at speeds of up to M1.4 in 1998.

Second of two preliminary design reviews (PDRs) completed on 18 December 1996, clearing way for detailed design of ground support system and launch facility (GSSLF) at Edwards AFB; one month earlier, NASA and industry team members endorsed baseline configuration for detailed vehicle design and approved long-lead procurement. Selection of preferred landing sites announced in October 1997, these being Michael Army Air Field, Utah, and Malmstrom AFB, Montana. Construction of launch facility completed at Edwards in last quarter of 1998.

Between January and August 1997, a series of critical design reviews (CDRs) was scheduled; however, technical problems concerning excess weight and aerodynamic instability caused delay and postponement of final CDR to October 1997. Design then frozen, clearing way for start of fabrication and assembly at Palmdale; vehicle originally due to be rolled out in May 1999 and go to Edwards for ground testing in July, but delay occurred and roll-out postponed. Rocketdyne engine to be installed at Edwards. VentureStar design evolved further in 1999, culminating in decision to employ an external payload bay. First flight test of X-33 initially expected in December 1999, but slipped to July 2000 and may not now occur until late 2002 as consequence of failure in November 1999 of composites liquid hydrogen tank during pressure and loading test at Marshall Space Flight Center. Decision then taken to build and install aluminium propellant tanks as replacement. Test programme planned to include 15 sorties in a six month period, during which speeds of up to M13.5 and altitudes of about 80 km (50 miles) will be attained. First five flights will end in recovery at Michael Army Air Field, with sixth and seventh terminating at Malmstrom AFB; at that point, programme review will be undertaken to determine if additional flights are needed. These could be with a modified vehicle, provisionally designated X-33B.

DIMENSIONS, EXTERNAL:
Wing span: X-33	23.5 m (77 ft)
RLV	39 m (127 ft)
Length overall: X-33	21 m (69 ft)
RLV	39 m (127 ft)
DIMENSIONS, INTERNAL:	
Payload bay: Length: X-33	3 m (10 ft)
RLV	14 m (45 ft)
Width: X-33	1.5 m (5 ft)
RLV	5 m (15 ft)
WEIGHTS AND LOADINGS (estimated):	
Weight empty: X-33	34,020 kg (75,000 lb)
Max T-O weight: X-33	129,275 kg (285,000 lb)
RLV	1,188,410 kg (2,620,000 lb)
Payload: RLV	at least 25,400 kg (56,000 lb)

UPDATED

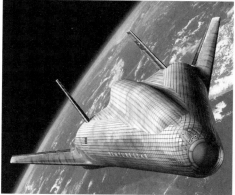

Manufactured image of the Lockheed Martin VentureStar, intended follow-on to X-33
2000/0075960

REUSABLE LAUNCH VEHICLE (X-34)
Begun as joint NASA/industry project for small RLV with capability to deliver 544 to 1,134 kg (1,200 to 2,500 lb) payload into low Earth orbit. Co-operative agreement notice issued 19 October 1994; at least 16 companies developed concepts for X-34. NASA had allocated US$70 million

Research vehicles and carriers based at Dryden include (from left) the veteran Boeing NB-52A launch aircraft (due for replacement by a B-52H), Boeing X-37, Boeing X-40A behind MicroCraft X-43 mockup, Orbital Sciences X-34 and Lockheed L-1011 TriStar carrier for X-34 *(Tony Landis/NASA)* *2001*/0100542

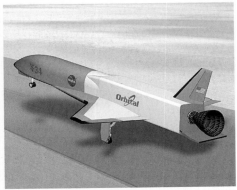

Artist's impression of NASA/OSC X-34

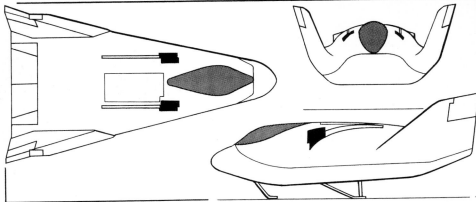

General arrangement of the X-38 Crew Return Vehicle *(Paul Jackson/Jane's)*

budget for development and testing of system and, in March 1995, selected Orbital Sciences Corporation/Rockwell consortium to take part in joint programme, with industry contributing further US$100 million. Studies in 1995 examined alternative approaches of 35,000 kg (77,162 lb) X-34A carried by specially modified Lockheed L-1011 TriStar; and larger, 50,000 kg (110,230 lb) X-34A fitted on top of Boeing 747. Project targets included suborbital test flight in late 1997; first orbital launch in 1998; and two test missions in 1998 and 1999. By 1996, however, future of project in doubt, following withdrawal of Orbital Sciences Corporation/Rockwell on grounds that X-34 was not commercially viable. Consequently, NASA elected to use US$62 million of unspent funds to build and fly a technology demonstrator of its own.

In June 1996, NASA chose Orbital Sciences to build a smaller, suborbital X-34 test vehicle, to be air-launched from L-1011 TriStar. Initial contract, worth US$50 million, was awarded on 28 August 1996, for design, development and testing, with NASA using residual funds from original programme; further US$10 million allocated for in-house support of X-34 development. Revised X-34 design frozen in May 1997, at which time only one test vehicle to be built, but in January 1999 NASA revealed intention to obtain second example; since then, a third vehicle has been added as a back-up and this will incorporate wing, control surfaces and landing gear taken from the first X-34.

X-34 will be about 17.8 m (58 ft 4 in) long, with a wing span of 8.4 m (27 ft 8 in) and a gross weight of approximately 21,320 kg (47,000 lb). Main engine is a government-furnished 267 kN (60,000 lb) thrust version of Fastrac (now MS-1) liquid oxygen/kerosene engine originally developed by the Marshall Space Flight Center. Assembly of first X-34 under way in 1998, with wing being mated with full-scale X-34 test article fuselage in July. Delivery of first X-34 (A-1) to NASA accomplished in early 1999, with formal roll-out ceremony at Dryden, California, at end April 1999. Wing fuselage mate of second X-34 (A-2) accomplished in late 1999, with vehicle then shipped to Holloman AFB, New Mexico, for installation of engine and subsequent testing of propulsion system. A-2 will be first X-34 to undertake powered flight and is due to complete series of eight sorties during which speed will gradually increase from M2.2 to about M5; third X-34 (A-3) will extend envelope to M8 and altitude of 76,200 m (250,000 ft). Contract terms require Orbital to accomplish two suborbital flights; ultimate goal is to develop vehicle with at least M8 and 76,200 m (250,000 ft) capabilities, plus capacity to complete 25 flights per year.

Basic programme objective is flight demonstration of key reusable launch vehicle operations and technologies directed at RLV goals of low-cost access to space and competitiveness in commercial space launch operations. More specific objectives include integration of new technologies and demonstration of these throughout flight envelope; ability to undertake autonomous flight operations and demonstrate safe abort; ability to undertake powered flight to at least 76,200 m (250,000 ft); incorporation of some composites structures; use of composites propellant tanks and cryo insulation; use of advanced low-cost avionics (including GPS/INS) plus low-cost flight software development tools; and employment as low Mach number testbed for items such as pulse detonation wave and dual-expansion engines. Conventional aircraft-type elevons, rudder, body flaps and speed brake used for control and trim during unpowered flight, with reaction control during high-altitude coast phase and thrust vector control for pitch and yaw when flying under main engine propulsion.

Test programme to be undertaken in two phases. First involves design and construction and also includes two envelope expansion flights restricted to maximum speed of M3.8; second involves total of 25 flights throughout full envelope and in variety of weather and environmental conditions during 12 month period. Typical flight will begin with air launch from TriStar, followed by engine start and acceleration to planned Mach number and altitude; subsequent coast phase will take X-34 as far as 500 n miles (926 km; 575 miles) downrange for autonomous approach and landing. Initial powered testing to be undertaken from Dryden, but some later flights may be made from the Kennedy Space Center in Florida in order to test vehicle capability in adverse weather conditions.

First captive-carry flight by Orbital Sciences TriStar undertaken 29 June 1999, using X-34 A-1, which to make further captive flights and also be used for ground tow tests at Dryden following upgrade to A-1A configuration. A-1 to be used for unpowered flights in New Mexico; series of five flights planned for 2000, with vehicle to be released from TriStar at altitude of about 10,670 m (35,000 ft) and fly for several minutes; first powered flight by A-2 was scheduled for third quarter of 2000, but delayed by several months as consequence of additional testing of vehicle and engine, and may not have occurred before end of 2000.

UPDATED

EXPERIMENTAL CREW RETURN VEHICLE (X-38)

Design of the experimental X-38 Crew Return Vehicle (formerly known as the 'X-35' and X-CRV) was begun in early 1995 as an in-house technology demonstration programme by NASA personnel at the Johnson Space Center, Houston, Texas, with initial flight testing of the parafoil concept being undertaken with models at the Yuma Proving Grounds, Arizona. In early 1996, Scaled Composites of Mojave, California, received a contract to build three full-scale atmospheric test airframes, making extensive use of commercial-off-the-shelf components.

The primary mission envisaged for the definitive CRV is to act as a 'lifeboat' or 'rescue pod' for space station crews, and the X-38 is intended to demonstrate capability and validity of the CRV for the crew return mission. After studying various alternatives, NASA selected a lifting-body design that incorporates a 511 m² (5,500 sq ft) Pioneer Aerospace parafoil to slow the vehicle during the final stages of descent for a fully automated landing. Length of the X-38 is approximately 8.7 m (28 ft 6 in), while it has a span of about 4.4 m (14 ft 6 in) and weighs around 7,260 kg (16,000 lb). The proposed operational version is expected to be about 20 per cent bigger to accommodate up to six astronauts. Four movable surfaces are provided, comprising body flaps for control in pitch and roll and rudders for yaw.

Delivery of the first X-38 (Vehicle 131) took place in September 1996 and it went initially to the Johnson Space Center for installation of avionics, computer systems and other hardware. On 4 June 1997, V131 was airlifted by C-17 to Edwards AFB for flight trials with NASA's Dryden Flight Research Facility. The first captive-carry flight by NASA's NB-52B was accomplished on 30 July 1997, with a second flight following on 2 August, after which operations ceased in order to solve electrical problems. Further captive-carry tests took place in the latter half of 1997, but the initial drop test was delayed until 12 March 1998, from an original target of August 1997, following problems with pyrotechnic system initiators and parafoil durability. V131 made its second (and last) flight on 6 February 1999, before being returned to the Johnson Space Center for installation of systems and instrumentation in readiness for further flight research duty at Dryden as Vehicle 131R. In this configuration, it will also use larger 697 m² (7,500 sq ft) parafoil, which completed successful first flight at Yuma Proving Ground, Arizona, on 19 January 2000, with 8,165 kg (18,000 lb) pallet to simulate weight of Vehicle 131R. Further parafoil tests conducted with 11,340 kg (25,000 lb) payload that approximates to weight of Vehicle 201. V131R returned to Dryden on 11 July 2000, following modification; first captive-carry flight by NB-52B took place on 4 August, with first free flight then expected to occur in October.

The second X-38 (Vehicle 132) to be completed by Scaled Composites was handed over to the Johnson Space Center in December 1996 and was delivered to Dryden in September 1998. Fitted with the full lifting-body flight control system, this made its first sortie on 5 March 1999 and is capable of autonomous flight before parafoil deployment. Drop tests with this X-38 progressively increased in altitude to a maximum of about 13,715 m (45,000 ft), thus allowing longer flight time for control and manoeuvrability assessment; fifth and final flight completed successfully on 30 March 2000, with V132 dropped from NB-52B at altitude of 11,890 m (39,000 ft).

The third and last X-38 (Vehicle 201) will embody additional space-qualified systems and approximate closely in size to eventual CRV. Aerojet GenCorp of Sacramento, California, secured a US$16.4 million contract in August 1998 to produce a deorbit propulsion stage that will be fired after the X-38 has been deployed in orbit from a Space Shuttle. The propulsion unit will initiate the X-38's descent from orbit and will then be jettisoned to burn up, while the X-38 re-enters the atmosphere and lands automatically. First orbital return test flight expected in late 2001; definitive CRV was scheduled to become operational on the International Space Station in December 2003, although future of programme now in doubt, following NASA decision to proceed with five-year, US$4.5 billion Space Launch Initiative project that is intended to culminate in a privately developed next-generation launch vehicle. Final decision on whether to proceed with CRV not now expected to be taken until about 2002.

UPDATED

X-38 Vehicle 131R arrives at Dryden by Aero Spacelines Guppy on 11 July 2000 *(Tony Landis/NASA)*

NASA—AIRCRAFT: USA 713

NASA HYPERSONIC FLIGHT PROGRAMMES

Since the days of NACA and the Bell X-1, NASA has been involved in experiments to increase the speed of atmospheric and stratospheric flight. Latest projects include those described below. In addition, the agency announced in August 1999 that it is to provide US$9.6 million for live, small-scale tests of a pulse detonation engine attached to a Lockheed SR-71 in 2002 as part of the research effort allocated to the Revolutionary Concepts programme.

UPDATED

Artist's impression of the X-43 Hyper-X scramjet-powered aircraft 2000/0067147

'WAVERIDER' PROJECTS

Initial NASA testing of 'Waverider' hypersonic concept undertaken by NASA Langley before the closure of full-scale wind-tunnel facilities there in September 1995. At that time, 'Waverider' studies envisaged aircraft powered by air-breathing engines and capable of speed in range M4 to M6; applications then being considered by NASA/USAF included cruise missile carrier/launch platform, high-altitude reconnaissance, long-range strike and transport.

Joint NASA/USAF project now aimed at producing 'Waverider' aircraft capable of speeds in excess of M10.

First is the **Low Observable Flight Test Experiment (LoFLYTE)** developed by Accurate Automation Corporation of Chattanooga, Tennessee; this is 1.57 m (5 ft 2 in) span, 2.54 m (8 ft 4 in) long, 32 kg (70 lb) subscale prototype of M5.5 design and is powered by 0.16 kN (35 lb st) micro gas-turbine engine produced by SWB Turbines of Appleton, Wisconsin, which is expected to allow speeds of up to 250 kt (463 km/h; 287 mph) to be achieved. Test objectives include assessment of low-speed handling qualities and evaluation of a neural network fly-by-wire control system that will 'teach' the 'Waverider' to remain stable in flight.

Six test models of remotely piloted LoFLYTE vehicle are to be built; first sustained minor damage during course of 34 second maiden flight at Mojave, California, on 16 December 1996, when control problems necessitated premature termination of flight and subsequent wheels-up landing, but at least 10 further sorties flown from Edwards AFB, California during 1997. Majority relied on conventional computer for control of engine and moving surfaces, with first use of neural network control system occurring in December 1997. By then, a second LoFLYTE had been completed and work on a third, with an M10 shape, was 80 per cent complete. Later versions of LoFLYTE will be larger, with research team intending to use 7 m (23 ft) model for tests at high-subsonic speeds; this will lack vertical tail surfaces of initial LoFLYTE models.

Second venture is NASA's **X-43A Hyper-X**, which has objective of allowing testing of subscale 'Waverider' powered by hydrogen-fuelled GASL Inc scramjet engine at speeds of M7 to M10. Forebody is shaped to generate shock waves which compress air entering scramjet intake; hydrogen-based fuel combusts spontaneously on contact with oxygen contained in the compressed air.

Request for proposals was issued on 13 October 1996, with contract worth US$33.4 million over 55 month period following in March 1997 to MicroCraft Inc of Tullahoma, Tennessee, although total project cost, including flight testing, has since risen to around US$170 million. Phantom Works was unsuccessful contender, having invested its own funds in the development of a subscale 'Waverider' known as 'Phantom X', which was due to fly for the first time in early 1997; MicroCraft heads a consortium that includes Accurate Automation, Boeing North American and GASL, and is to produce three (originally to have been four) test vehicles, each approximately 3.7 m (12 ft) long, with a span of about 1.5 m (5 ft) and gross weight of 1,270 kg (2,800 lb). Preliminary design review completed in June 1997, with fabrication commencing in third quarter of 1997. First flight vehicle delivered to NASA's Dryden Flight Research Center in October 1999 and has since completed various tests, including leak-proofing, structural/mechanical interference checks and verification of CG and inertia in all three axes.

Three flights are scheduled, with first originally due to take place in May 2000, after captive-carry flight on NASA's NB-52B launch platform; however, development delays resulted in postponement to September 2000. The initial flight is planned to reach M7, followed by second flight to M7 in late 2000 or early 2001 and final flight to M10; latter originally scheduled for September 2001, but date has still to be determined. After drop launch at altitude of 6,100 m (20,000 ft), an Orbital Sciences' Pegasus booster rocket will accelerate test vehicle to specified speed at approximately 30,500 m (100,000 ft), whereupon X-43 will separate to fly under own power. Flight path is entirely over water and it is not planned to recover test vehicles from Pacific Ocean.

UPDATED

LoFLYTE is the first in the 'Waverider' series of hypersonic projects

NASA THRUST-VECTORING CONTROL PROGRAMME

A new series of X-craft was proposed in 1999 to continue from earlier studies by NASA and the aerospace industry.

X-44A

Possible new design for research aircraft, building on experience gained with F-15B assigned to ACTIVE project; feasibility study under way in fourth quarter of 1999, involving NASA, Lockheed Martin, Pratt & Whitney and the USAF. Notional X-44A design resembles F-22A Raptor, but without horizontal and vertical tail surfaces and also lacking moving control surfaces on wings; this would rely entirely on thrust-vectoring for aerodynamic control.

VERIFIED

A potential X-44, based on the Lockheed Martin F-22
2000/0067007

NASA AIRCRAFT TRIALS PROGRAMMES

Of some 100 aircraft operated by NASA, most are standard types assigned to specific research programmes. The fleet comprises about 35 dedicated to research and 65 to programme support; most carry civilian registrations between N400NA and N999NA, assigned according to base. Main operating bases are:

Dryden Flight Research Center, Edwards AFB, California (N800-899NA)

Langley Research Center, Langley AFB, Hampton, Virginia (N500-599NA)

Glenn Research Center, Cleveland-Hopkins IAP, Ohio (N600-699NA)

Other significant aircraft operations bases are:

Ames Research Center, Moffett Federal Airfield, California (N700-799A) (Helicopters operated by US Army personnel in support of NASA test and research projects; will also be base for Boeing 747SP Stratospheric Observatory from 2002.)

Johnson Space Center, Ellington ANGB, Houston, Texas (N900-999NA)

Wallops Flight Facility, Wallops Island, Virginia (N400-499NA)

Some recent and forthcoming developments are summarised hereunder:

Boeing F-15B: Original F-15B prototype (NASA837/71-0290) modified for ACTIVE (Advanced Control Technology for Integrated Vehicles) project involving thrust vectoring nozzles developed by Pratt & Whitney. Ground trials completed in November 1995; flight evaluation began February 1996 and then expected to last into 1997, accumulating approximately 100 flying hours in 60 sorties. Second phase of trials began 27 March 1996 following completion of initial exploration of pitch and yaw vectoring at subsonic speeds. Engine thrust vectoring at supersonic speed accomplished for first time on 24 April 1996, when aircraft flying at M1.2 at 9,150 m (30,000 ft); further testing included thrust vectoring at M1.95 at 13,715 m (45,000 ft) on 31 October and 1 November 1996. Phase III of ACTIVE programme accomplished during 1998, with pitch yaw balance beam thrust vectoring nozzles integrated with aircraft's FBW flight control system to achieve 'inter-loop' thrust vectoring control. Total of 14 sorties undertaken using reversionary FCS mode in event of problems or envelope limits being exceeded; integrated system reported to have performed 'flawlessly' during trials that included advanced manoeuvring and tracking in vectoring mode only. Further trials, in non-reversionary mode, undertaken in 1999, including take-offs and landings using vectoring only as well as flight to unlimited angles of attack and speeds of M2. New nozzles can turn up to 20° in any direction, permitting thrust

The NASA test fleet is not restricted to high-performance aircraft, as this Raytheon King Air 200 demonstrates *(Lori Losey/NASA)* 2001/0100543

USA: AIRCRAFT—NASA to NORTHROP GRUMMAN

F-15B NASA836 carries a laminar flow experiment on its centreline pylon *(Jim Ross/NASA)*

control in pitch and yaw. ACTIVE programme is a joint venture involving NASA, the USAF's Wright Laboratory, Boeing (formerly McDonnell Douglas) and Pratt & Whitney.

ACTIVE F-15B also used in early 1999 to evaluate 'neural network'-based controller as opening part of intelligent flight control system (IFCS) programme; total of 15 flights accomplished during March and April 1999.

A second F-15B (NASA836/74-0141), known as the Aerodynamic Flight Facility and equipped with a flight test fixture (airborne test-bench) on the centreline pylon, flew seven sorties between 26 February and 1 April 1996 to test materials for the X-33's thermal protection system; used during 1999 for investigation of supersonic natural laminar flow of advanced wing design, carrying 0.91 by 1.22 m (3 by 4 ft) test article on centreline pylon during four test flights at speeds approaching M2 and altitudes of up to 13,715 m (45,000 ft). This aircraft recently involved in experiments into gust monitoring and aeroelasticity, with 0.45 m (1 ft 6 in) long, flexible, test wing mounted on the flight test fixture.

Boeing F/A-18: Includes F/A-18A High Angle of Attack Research Vehicle (HARV) (NASA840/161251) with thrust vectoring system; first flight 16 January 1991 at Dryden; first flight with vectoring 15 July 1991; reached 79° AoA in sustained flight, 1992; initial series of Phase 2 trials completed in January 1993. Aircraft then further modified by January 1994, initially with inlet-mounted pressure sensor and new programmable Ada software for quadruplex digital FCS; addition of vortex-flow nose strakes, late 1994; strakes 1.22 m (4 ft 0 in) long and 0.15 m (6 in) wide, hinged on lower edges and rotated outwards in flight to control nose vortices. After second series of Phase 2 testing between January and June 1994, HARV was fitted with strakes and commenced Phase 3 trials, involving approximately 65 flights. These began in April 1995 and concluded with final sortie on 29 May 1996, by which time HARV had undertaken total of 383 research flights over nine-year period; final test objectives included evaluation of nose strakes for yaw control at high AoA.

NASA840 currently being refurbished to serve as testbed for Active Aeroelastic Wing Program (AAWP); trials started in 1999 and will examine use of 'flexible wings' to research airflow over the wing and assess how this affects controllability.

Two NASA F/A-18s to be used for demonstration of autonomous close formation flight in 2001.

Boeing 747SP: Engineering work under way to build Stratospheric Observatory For Infra-red Astronomy (SOFIA) with 2.69 m (8 ft 10 in) telescope housed in unpressurised cavity in aft fuselage and looking to port; partner is German Ministry of Science (BMFT); cruise altitude 12,500 to 13,700 m (41,000 to 45,000 ft); to fly up to 160 sorties (8 hours each) per year for 20 years. US$484 million, 10 year contract secured in December 1996 by a team led by Universities Space Research Association (USRA) and including Raytheon Systems and United Airlines to design, assemble, test and operate SOFIA. Former United Airlines Boeing 747SP (N145UA) chosen from several in storage at Las Vegas, Nevada, and ferried to San Francisco in February 1997 for attention by United before being handed over to NASA; following acceptance, 747SP delivered to Waco, Texas, in May 1997 for modification and fitting out by Raytheon. Conversion now under way, with 747SP due to fly in modified form during 2001 and enter operational service with Ames Research Center in 2002.

Lockheed SR-71: Two SR-71As (NASA832/64-17971 and NASA844/64-17980) and one SR-71B (NASA831/64-17956) acquired in 1991 in support of high-speed research and support to X-30 NASP; one SR-71A (NASA832) returned to USAF operational inventory. SR-71 will be used to carry Pulse Detonation Engine for series of live-fire tests at M3 in 2002 and is also earmarked to undertake tests of two-dimensional, mixed compression engine inlet for potential future supersonic transport aircraft with objective of reducing sonic boom to level that will permit flight over land.

McDonnell Douglas DC-8-72: Joined Airborne Science Program in April 1998 at Dryden, following outfitting with Jet Propulsion Lab (Pasadena, California) advanced synthetic aperture radar and other sensor systems including imaging spectroradiometer and thermal emission and reflection radiometer. Used for convection and moisture experiments, Earth science studies and also supporting NASA investigation into ozone depletion.

UPDATED

Earth resources research is conducted by NASA using two Lockheed ER-2s *(Tony Landis/NASA)*

NORTHROP GRUMMAN

NORTHROP GRUMMAN CORPORATION

1840 Century Park East, Los Angeles, California 90067
Tel: (+1 310) 553 62 62
Fax: (+1 310) 201 30 23 and (+1 310) 553 20 76
Web: http://www.northgrum.com
CHAIRMAN, PRESIDENT AND CEO: Kent Kresa
VICE-PRESIDENT, CORPORATE COMMUNICATIONS:
  Rosanne O'Brien

Northrop company formed 1939 to produce military aircraft; activities extended to missiles, target drones, electronics, space technology, communications, support services and commercial products. Grumman Aircraft Engineering Corporation incorporated 6 December 1929; became major supplier of carrierborne aircraft to US Navy.

Acquisition of Grumman by Northrop completed 1 May 1994 and new corporation formed 18 May. Northrop Grumman subsequently completed acquisition of Vought Aircraft Company, with US$130 million purchase of Carlyle Group's 51 per cent interest, in August 1994. Further expansion announced on 3 January 1996, when Northrop Grumman revealed agreement to acquire the defence and electronic systems businesses of the Westinghouse Electric Corporation for US$3 billion. Finalisation of this purchase in March 1996 resulted in significant increase in products and technologies offered by Northrop Grumman. Merger with Logicon Inc followed in 1997, subsequent acquisitions being the Inter-National Research Institute in mid-1998 and the Information Systems Division of California Microwave Inc in April 1999. A major recent addition, finalised in July 1999, was that of Teledyne Ryan Aeronautical at cost of approximately US$140 million. More recently, on 12 June 2000, Northrop Grumman announced plans for disposal of commercial aerostructures business to The Carlyle Group for sum in excess of US$1 billion; this completed July 2000 under resurrected name of Vought. Following sale, workforce totalled about 36,000 at end of July 2000. Net income for 1999 was US$483 million, with sales worth US$9.0 billion recorded.

Collaboration with Lockheed Martin announced in September 1996 concerning development and marketing of airborne early warning and control system aircraft. Further co-operation occurred in May 1997, when Northrop Grumman joined Lockheed Martin Joint Strike Fighter team, having previously been associated with unsuccessful McDonnell Douglas/British Aerospace JSF submission. Closer collaboration with DASA on surveillance and reconnaissance systems to result from MoU signed on 19 April 2000.

Northrop Grumman now organised into three major business units as follows:
Integrated Systems Sector
  Follows this entry.
Electronic Sensors and Systems Sector
  Follows Integrated Systems Sector.
Logicon Inc
  Based in Herndon, Virginia, this wholly owned subsidiary provides technical engineering, project management and support resource services for information technology systems, plus technical and professional services in operations and maintenance.

UPDATED

INTEGRATED SYSTEMS SECTOR

9314 West Jefferson Boulevard, Dallas, Texas 75211-9300
Postal Address: Box 655907, Dallas, Texas 75265-5907
Tel: (+1 972) 946 20 11
Fax: (+1 972) 946 57 61
CORPORATE VICE-PRESIDENT AND PRESIDENT, INTEGRATED SYSTEMS SECTOR: Ralph D Crosby Jr
MANAGER, COMMUNICATIONS: Georgia Calaway

Having recently divested itself of the aerostructures business, the Integrated Systems Sector now functions as prime contractor for the B-2A Spirit stealth bomber and the E-8C Joint STARS airborne targeting and battlefield management system, while continuing production of the E-2C Hawkeye for the US Navy and export customers. Serves as principal subcontractor for the F/A-18 (see Boeing) and is teamed with Lockheed Martin on the JSF programme. Other work includes modification and support of the EA-6B Prowler, F-5 Tiger, F-14 Tomcat and Fairchild A-10 Thunderbolt II. This business sector is also engaged in production and support of UAVs for reconnaissance, surveillance and deception as well as aerial target systems.

Operating elements of the Integrated Systems Sector are now organised into business areas, with responsibilities as detailed below; last mentioned includes unmanned version of Schweizer 330 helicopter (which see).

Air Combat Systems

One Hornet Way, El Segundo, California 90245-2804
Tel: (+1 310) 331 36 16
CORPORATE VICE-PRESIDENT AND GENERAL MANAGER:
  William H Lawler
MANAGER, COMMUNICATIONS: James F Hart

B-2 Spirit, F/A-18 Hornet/Super Hornet, Joint Strike Fighter, F-5 Tiger/T-38 Talon and UAV systems. Has sites at Palmdale and San Diego, California; New Town, North Dakota; Whiteman AFB, Missouri; Tinker AFB, Oklahoma; and Hill AFB, Oklahoma, Utah.

Airborne Early Warning and Electronic Warfare Systems

South Oyster Bay Road, Bethpage, New York 11714
Tel: (+1 516) 575 51 19
MANAGER, COMMUNICATIONS: John Vosilla

Production of E-2C Hawkeye, plus modification, repair and support of other types, including EA-6B Prowler, A-10 Thunderbolt II, C-2 Greyhound, F-5 Tiger, F-14 Tomcat and S-2T Turbo-Tracker. Has sites at St Augustine and Cecil Field, Florida; Point Mugu, California, plus field support services at many other locations.

Airborne Ground Surveillance and Battle Management Systems

2000 NASA Boulevard, Melbourne, Florida 32902
Tel: (+1 516) 575 51 19
MANAGER, COMMUNICATIONS: John Vosilla

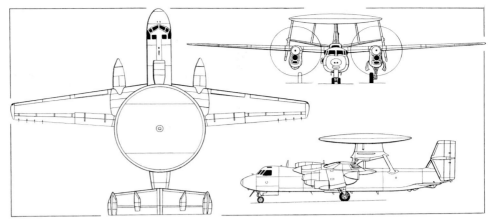

Production of E-8C Joint STARS, plus associated advanced surveillance and battle management systems. Has additional site at Lake Charles, Louisiana.

UPDATED

NORTHROP GRUMMAN E-2C HAWKEYE
Israel Defence Force name: Daya (Kite)

TYPE: Airborne early warning and control system.

PROGRAMME: First flight of first of three prototypes 21 October 1960; total 59 production E-2As, of which 51 updated to E-2B by end 1971 apart from two TE-2A trainers and two converted to E-2C prototypes (see earlier *Jane's*); first flight of E-2C prototype 20 January 1971; production started mid-1971; first flight production aircraft 23 September 1972; 211 of all E-2C versions ordered, of which 184 had been delivered by mid-2000. AN/APS-145 radar replaces AN/APS-120, AN/APS-125, AN/APS-138 and AN/APS-139 in new-built E-2Cs.

Evaluation of Lockheed Martin AN/APS-145 began 1986; radar tracks more than 2,000 targets and operates at longer ranges, has improved jamming resistance and sharper, fully automated/optimised overland detection. AN/APS-145 radar and new main operator displays, IFF, mission computer processor, JTIDS tactical software and upgraded engines form core of Update Development Program (Groups I and II), which fully developed and delivered in new production aircraft from October 1991. Second production Group I aircraft (163535 – 123rd E-2C; 102nd USN E-2C) set 20 (broke 14 and established six new) records for time-to-height, altitude and 100 km closed circuit, 17 to 19 December 1991. Group II aircraft expected to be phased out in 2002-2003, when improved Hawkeye 2000 joins fleet.

CURRENT VERSIONS: **E-2C:** Current service and production version (*as detailed*).

AN/APS-139 and Allison T56-427 engines form **Group I** update; first operational aircraft (163538) delivered to VAW-112 on 8 August 1989; 18 built; AN/APS-139 can detect cruise missiles at ranges exceeding 100 n miles (185 km; 115 miles); also monitors maritime traffic; radar coverage extended by AN/ALR-73 passive detection system (PDS), detecting electronic emitters at twice radar detection range; more detailed description in 1979-80 *Jane's*; AN/APS-145 in **Group II** aircraft from December 1991; other enhancements give Group II 96 per cent expansion in radar volume, 400 per cent extra target tracking capability, 40 per cent more radar and identification range and 960 per cent increase in numbers of targets displayed. Group II added JTIDS in 1993-94; also has GPS.

TE-2C: Training model, based on E-2C; two conversions originally undertaken, of which one (158639) was assigned to JTIDS development with Northrop Grumman. At least three further conversions done later for training purposes with VAW-120 at Norfolk, Virginia.

E-2T: Originally reported to be conversion of E-2B for Taiwan, but new-build aircraft actually supplied; AN/APS-138 radar and electronic warfare upgrades. Delivery of initial four aircraft began in 1995, with further two to be obtained following agreement in July 1999; these will be to Hawkeye 2000E standard, with AN/APS-145 radar.

Hawkeye 2000: In December 1994, company received US$155 million contract to redefine E-2C as the Hawkeye 2000. Key element is mission computer upgrade (MCU), with new equipment based on Raytheon's Model 940, itself a modification of Digital Equipment Corporation 2100 Model A500MP processing system, which is same family of hardware used by E-8C Joint STARS.

Initial trials of upgraded mission computer installed on second Group II aircraft (164109) began with first flight on 24 January 1997 and were completed in mid-1997, at which time authorisation given for low-rate initial production of new mission computer. However, early flight trials revealed software problems that will delay production of new mission computer until at least March 2003 – about a year later than planned. More ambitious technical and operational evaluations undertaken with five modified aircraft in 1999-2000. All are Group II aircraft fitted with MCU and ACIS (see below) elements of proposed Hawkeye 2000; first two delivered to Patuxent River for initial evaluation by May 1999, with remainder due to follow by July. At least four to Point Mugu from August 1999, joining VAW-117 for operational evaluation from October, with latter phase expected to include deployed duty aboard a carrier for full battlegroup operations. In meantime, another E-2C serving as a testbed for satcom, vapour-cycle cooling upgrade and Navy's USG-3 co-operative engagement capability (CEC) package, following first flight in April 1998.

New mission computer is less than half the weight of current L-304, one-third of volume, and offers 15 times the processing power; other improvements for Hawkeye 2000 include government-furnished advanced control indicator set (ACIS), satellite-based voice and data communications capability, a new Honeywell vapour-cycle cooling system, air-to-air refuelling capability (if required) and inclusion of equipment and systems that form part of Navy CEC package. MCU and ACIS make use of commercial off-the-shelf technology incorporating open architecture.

In April 1999, contract awarded for 24 Hawkeye 2000s for US Navy (21) as five-year procurement package, Taiwan (two) and France (one). Retrofit planned with eight-blade propellers.

Hawkeye 2005: Also referred to as '**Improved Hawkeye**' and '**Advanced Hawkeye**', this is currently in proposal stage and is unlikely to be operationally available until post-2007 if project goes ahead; US Navy seeking to begin engineering and manufacturing development (EMD) in FY03. As now envisaged, will incorporate improved radar, surveillance infra-red search and track (SIRST) system, modular communications equipment, multisensor integration and tactical (glass) cockpit, with latter enabling co-pilot to augment system operators by performing some tactical functions. New systems will offer improved detection capability against theatre air and missile threats, including overland cruise missiles. US Navy expected to be prime customer if Hawkeye 2005 comes to fruition, but other potential customer is UK Royal Navy, which has emerging requirement for new AEW platform to operate from future aircraft carrier.

CUSTOMERS: US Navy orders for E-2C originally totalled 139 by FY92, of which all delivered by March 1994. Further procurement authorised December 1994, when Northrop Grumman awarded US$122.5 million contract for start-up of new assembly line at St Augustine, Florida. Initial orders for seven Group II aircraft (four with FY95 funds, three with FY96 funds); first of these new aircraft (165293) was rolled out on 24 February 1997, flew on 22 March, and was delivered to VAW-120 at Norfolk, Virginia, after brief period of company and Navy testing. Further contract awarded in December 1996 to cover advance acquisition costs of four Group II aircraft with FY97 funds; four more ordered before April 1999 award of multiyear contract for 24 Hawkeye 2000s for US Navy, France and Taiwan. Delivery of first new-build Hawkeye 2000 scheduled for October 2001, following handover of single upgraded aircraft that is being converted from existing E-2C under US$17.8 million contract awarded in April 2000; four more new production aircraft will follow in 2002, then five per year in 2003-05, with last in February 2006. Sole example for France to be handed over in 2003 and French Navy proposes upgrading current two E-2Cs to similar standard. Current US Navy programme status in adjacent table:

Variant	Qty	First aircraft
E-2C 'Group 0'	100	158638
E-2C Group I	18[1]	163029
E-2C Group II	36	164108
Hawkeye 2000	21	165648
Total	**175**	

[1] Two aircraft later converted to TE-2C and 16 upgraded to Group II standard

Major retrofit programme for USN and FMS aircraft was planned, including upgrade of all 18 Group I and minimum of 36 older ('Group 0') aircraft to Group II standard from FY95. Instead, 1994 defence review established new E-2Cs more economical than retrofit; however, 16 Group I E-2Cs have been upgraded to Group II at St Augustine, Florida; first returned to service on 21 December 1995. First operational squadron with upgraded Group II E-2C was VAW-123 which accepted its fourth and final aircraft on 29 April 1996; VAW-121 followed suit later in year. Total cost of upgrade programme originally to be US$135 million; however, an amendment

Northrop Grumman E-2C Hawkeye (Allison T56 turboprop engines) (Dennis Punnett/Jane's)

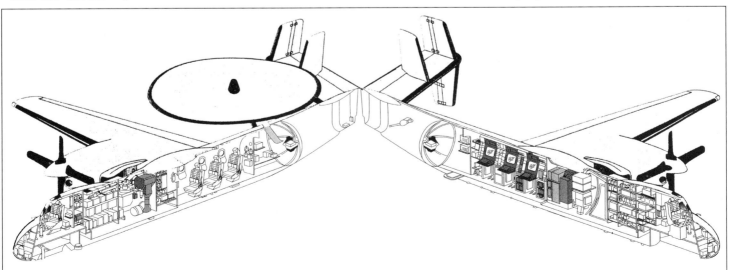

Interior layout of E-2C Hawkeye, showing three combat staff positions directly beneath the rotodome

USA: AIRCRAFT—NORTHROP GRUMMAN

E-2C EXPORTS

Customer	Qty	Group	First aircraft	Delivery	Unit (or base)
Egypt[4]	6	0	162791	1987-88 (5)	(Cairo West)
		0	164626	1993 (1)	
France[5]	3	II	1(165455)	1998	4 Flottille
Israel[1]	4	0	160771	1978	192 Sqdn
Japan[2]	13	0	34-3451	1982 (4)	601 Sqdn
		0	54-3455	1984 (4)	
		0	34-3459	1992-93 (3)	
		0	44-3462	1993 (2)	
Singapore	4	0	162793	1987	111 Sqdn
Taiwan[3]	6		2501	1995	78 Sqdn
Total	**36**				

Notes:
[1] Israeli aircraft no longer in service; stored pending disposal
[2] Japanese aircraft being upgraded to Group II standard, with some elements of Hawkeye 2000 technology also to be incorporated
[3] Taiwanese aircraft known as E-2Ts; total includes two on order as E-2T Hawkeye 2000E
[4] Five Egyptian aircraft to be upgraded to Hawkeye 2000 standard
[5] Third French aircraft due for delivery in 2003 and will be to Hawkeye 2000 standard; original two E-2Cs to be upgraded to similar configuration

to contract in third quarter of 1997 added US$8.5 million to cover conversion of an additional aircraft by May 1999.

E-2C entered service with VAW-123 at NAS Norfolk, Virginia, November 1973 and went to sea on board USS *Saratoga* late 1974; E-2C issued to 19 other squadrons, including three of Naval Reserve; current squadrons are VAW-112, 113, 116 and 117 at Point Mugu, California; VAW-115 at Atsugi, Japan; VAW-120, 121, 123 to 126 with VAW-78 of Reserves at Norfolk, Virginia, plus VAW-77 of Reserves at Atlanta, Georgia, for anti-drug smuggling surveillance duty. VAW-120 is training unit. Miramar was base of first Group II squadron, VAW-113, June 1992. VAW-113 first operational evaluation cruise in USS *Carl Vinson*, 1993. New Build Group II aircraft also issued to VAW-110 (disbanded September 1994), 112, 116 and 117. Final two of original 139 E-2Cs delivered to USN in March 1994; first from production at St Augustine joined training squadron in first half of 1997.

See table for exports. Singaporean aircraft with AN/APS-138 radar; Taiwan received E-2T with AN/APS-138 radar. Israeli aircraft in storage by 1996 and reportedly for sale in mid-2000. France signed Letter of Offer and Acceptance (LOA) in June 1995 for two aircraft; first flight of first Aéronavale aircraft on 12 March 1998, with formal roll-out ceremony at St Augustine, Florida, on 28 April 1998. Both aircraft initially retained in USA for crew training, with first E-2C (actually No 2/165456) delivered to Lann-Bihoué, France, on 14 December 1998; second in April 1999. Operating unit, 4 Flottille, formed at Lann-Bihoué January 1999.

No further export orders received in 1995-98, but French Navy given authorisation in 1998 to purchase third aircraft, which will be to latest Hawkeye 2000 configuration. E-2C offered to Gulf Collaboration Council as solution to regional requirement for AEW platform, but no firm order forthcoming. Italy a potential customer in longer term, with requirement for at least two AEW aircraft by 2007; Turkey now in preliminary stages of acquiring AWACS system, but has apparently eliminated the Hawkeye 2000 as a contender to satisfy requirement for four aircraft. Egypt and Singapore said to be considering acquisition of surplus US Navy aircraft from storage facility in mid-2000.

COSTS: Total cost of French procurement of two aircraft approximately US$562 million including logistics support. US$122.5 million awarded December 1994 for production start-up for seven new E-2Cs. Upgrading of 13 Japanese aircraft to Group II standard expected to cost approximately US$400 million, with cost of Egyptian upgrade of five aircraft estimated at US$210 million. Procurement of 21 new Hawkeye 2000s for US Navy will cost US$1.47 billion.

DESIGN FEATURES: E-2C can cover naval task force in all weathers flying at 9,150 m (30,000 ft) and can detect and assess approaching aircraft at in excess of 300 n miles (556 km; 345 miles); AN/APS-145 has total radiation aperture control antenna (TRAC-A) to reduce sidelobes to offset jamming; radar sweeps 6 million cu mile envelope and simultaneously monitors surface ships; long-range, automatic target track initiation and high-speed processing enable each E-2C to track more than 2,000 targets simultaneously and automatically, and control more than 40 intercepts; Randtron Systems AN/APA-171 antenna housed in 7.32 m (24 ft) diameter radome, rotating at 5 to 6 rpm above rear fuselage; antenna arrays in rotodome provide radar sum and difference signals and IFF.

High-mounted tapered wing; moderately sweptback tailplane with pronounced dihedral; twin fins and twin auxiliary fins, all with sweptback leading-edges and at right angles to tailplane. Rotating, circular radar housing mounted on cabane above rear fuselage. Nose-tow catapult attachment, arrester hook and tail bumper; parts of tail made of composites to reduce radar reflection; wings fold hydraulically on skewed hinges to lie parallel to fuselage.

Wing incidence 4° at root, 1° at tip.

FLYING CONTROLS: Conventional and fully powered, with artificial feel; tailplane has 11° dihedral; four fins and three double-hinged rudders; long-span ailerons droop automatically when hydraulically operated Fowler flaps are extended; autopilot provides autostabilisation or full flight control. Empennage (including horizontal stabilisers, elevators, rudders, vertical fins and tabs) produced by Potez Aéronautique of France, following award of contract in mid-1997 for initial batch of five assemblies; first completed assembly delivered in second quarter of 1999.

STRUCTURE: Wing centre-section has three beams, ribs and machined skins; hinged leading-edge provides access to flying and engine controls. Fuselage conventional light metal. Composites used in parts of tail.

LANDING GEAR: Hydraulically retractable tricycle type. Pneumatic emergency extension. Steerable nosewheel unit retracts rearward. Mainwheels retract forward and rotate to lie flat in bottom of nacelles. Twin wheels on nose unit only. Oleo-pneumatic shock-absorbers. Mainwheel tyres size 36×11 (24 ply) tubeless, pressure 17.93 bars (260 lb/sq in) on ship, 14.48 bars (210 lb/sq in) ashore. Nosewheel tyres 20×5.5 (12/14 ply) tubeless. Hydraulic brakes. Hydraulically operated retractable tailskid. A-frame arrester hook under tail.

POWER PLANT: Two 3,803 kW (5,100 ehp) Allison T56-A-427 turboprops, driving Hamilton Sundstrand Type 54460-1 four-blade fully feathering reversible-pitch constant-speed propellers. These have foam-filled blades which have a steel spar and glass fibre shell. T56-A-427 engines provide 15 per cent improvement in efficiency, compared with -425 installed before 1989. Israeli Hawkeyes fitted locally with fixed in-flight refuelling probes; first aircraft (941) seen 1993. Hamilton Sundstrand NP2000 eight-blade, all-composites propeller being flight tested on E-2C during latter part of 2000, following successful ground running on T56 and completion of critical design review in September 1998; US Navy has selected NP2000 for E-2C and C-2A Greyhound and placed US$44.5 million contract for 187 propellers, with option on further 54. Service entry currently expected in late 2001 and new propellers will be installed on all Hawkeye 2000s as well as retrofitted to surviving E-2C Group II aircraft by 2004.

ACCOMMODATION: Normal crew of five, consisting of pilot and co-pilot on flight deck, plus ATDS Combat Information Center (CIC) staff of combat information centre officer, air control officer and radar operator. Downward-hinged door, with built-in steps, on port side of centre-fuselage and three overhead escape hatches.

SYSTEMS: Pneumatic boot de-icing on wings, tailplane and fins. Spinners and blades incorporate electric anti-icing.

AVIONICS: *Comms:* AN/AIC-14A intercom.
Radar: Lockheed Martin AN/APS-145 advanced radar processing system (ARPS) with fully automatic overland/overwater detection capability, Randtron AN/APA-171 rotodome (radar and IFF antennas).
Flight: Lockheed Martin AN/ASN-92 CAINS carrier aircraft inertial navigation system, GPS, AN/ASN-50 heading and attitude reference system, Rockwell Collins AN/ARA-50 UHF ADF, AN/ASW-25B ACLS, BAE Systems standard central air data computer, ASM-400 in-flight performance monitor, Honeywell AN/APN-171(V) radar altimeter.
Mission: BAE Systems/Hazeltine AN/APA-172 control indicator group with Lockheed Martin enhanced (colour) main display units, Litton OL-77/ASQ computer programmer (L-304) with Lockheed Martin enhanced high-speed processor, BAE Systems/Hazeltine OL-483/AP airborne interrogator system, Litton AN/ALR-73 passive detection system, AN/ARC-158 UHF datalink, AN/ARQ-34 HF datalink and JTIDS Class 2 HP terminal. Barco Display Systems to supply graphics controllers with radar display capability and colour flat panel displays for Hawkeye 2000 upgrade. Hawkeye 2000 has Lockheed Martin AN/ALQ-217 ESM.

DIMENSIONS, EXTERNAL:
Wing span	24.56 m (80 ft 7 in)
Wing chord: at root	3.96 m (13 ft 0 in)
at tip	1.32 m (4 ft 4 in)
Wing aspect ratio	9.3
Width, wings folded	8.94 m (29 ft 4 in)
Length overall	17.60 m (57 ft 8¾ in)
Height overall	5.58 m (18 ft 3¾ in)
Diameter of rotodome	7.32 m (24 ft 0 in)
Tailplane span	7.99 m (26 ft 2½ in)
Wheel track	5.93 m (19 ft 5¾ in)
Wheelbase	7.06 m (23 ft 2 in)
Propeller diameter	4.11 m (13 ft 6 in)

AREAS:
Wings, gross	65.03 m² (700.0 sq ft)
Ailerons (total)	5.76 m² (62.00 sq ft)
Trailing-edge flaps (total)	11.03 m² (118.75 sq ft)
Fins, incl rudders and tabs:	
outboard (total)	10.25 m² (110.36 sq ft)
inboard (total)	4.76 m² (51.26 sq ft)
Tailplane	11.62 m² (125.07 sq ft)
Elevators (total)	3.72 m² (40.06 sq ft)

WEIGHTS AND LOADINGS:
Weight empty	18,363 kg (40,484 lb)
Max fuel (internal, usable)	5,624 kg (12,400 lb)
Max T-O weight	24,687 kg (54,426 lb)
Max wing loading	379.6 kg/m² (77.75 lb/sq ft)
Max power loading	3.25 kg/kW (5.34 lb/hp)

PERFORMANCE:
Max level speed	338 kt (626 km/h; 389 mph)
Max cruising speed	325 kt (602 km/h; 374 mph)
Cruising speed (ferry)	259 kt (480 km/h; 298 mph)
Approach speed	103 kt (191 km/h; 119 mph)
Stalling speed (landing configuration)	75 kt (138 km/h; 86 mph)
Service ceiling	11,275 m (37,000 ft)
Min T-O run	564 m (1,850 ft)
T-O to 15 m (50 ft)	793 m (2,600 ft)
Min landing run	439 m (1,440 ft)
Ferry range	1,541 n miles (2,854 km; 1,773 miles)
Time on station, 175 n miles (320 km; 200 miles) from base	4 h 24 min
Endurance with max fuel	6 h 15 min

UPDATED

NORTHROP GRUMMAN E-8 JOINT STARS

TYPE: Airborne ground surveillance system.
PROGRAMME: Full-scale development contract for Joint Surveillance Target Attack Radar System (Joint STARS) programme awarded to Grumman (later Northrop Grumman), 27 September 1985; two Boeing 707-328Cs used as EC-18C testbeds, later redesignated E-8A; first aircraft for modification reached Boeing Military Airplanes, Wichita, January 1986; delivered to Grumman Melbourne Systems Division, Florida, 31 July 1987;

Hawkeye 2000 demonstrator

second aircraft to Wichita June 1986 and then to Grumman in late 1988.

First flight in full Joint STARS configuration (86-0416/N770JS) 22 December 1988; first flight of second aircraft (86-0417/N8411) 31 August 1989; first instantaneous transmission to ground station August 1989; interoperability studies with France, Italy and UK 1989-90; European operational field demonstration (OFD) February and March 1990 (N770JS) and September 1990 (86-0417). IOC originally planned for September 1997 with five aircraft, but emergency interim capability achieved for Gulf War against Iraq. Both E-8As deployed to Saudi Arabia in December 1990 along with six ground station modules; assigned to 4411th Joint STARS Squadron (Provisional) and accomplished first operational mission on 14 January 1991. Temporary USAF unit completed 49 combat sorties/534.6 hours with 100 per cent mission effectiveness rate by end of conflict.

Low-rate initial production advanced procurement contract signed by USAF 24 April 1992; progress review, May 1993, authorised first six production aircraft at two per year; approval to proceed with full production given by US DoD on 26 September 1996, despite criticisms voiced by USAF evaluation team which identified problems with radar masking and clutter, communications equipment, software deficiencies and airframe structural defects. Airframe modifications done at company's Lake Charles, Louisiana, plant following Boeing's withdrawal from programme; avionics installation at Melbourne; first E-8C (90-0175/N526SJ) completed December 1993 and made first flight 25 March 1994.

NATO considered integrated Joint STARS force; planning group established December 1992; programme office established November 1994; short European visit by second E-8A in October 1994 for demonstration to Belgium, France, Germany and UK; requirement for up to 12 Joint STARS to be based at Geilenkirchen, alongside E-3A Sentry AWACS unit; may use Airbus A321 or similar aircraft as avionics platform, with current concept likely to employ improved radar now under development for USAF E-8C.

Further interest expressed by Japan (possibly four aircraft), South Korea (four) and Saudi Arabia (up to six). Grumman teamed with BAe in 1993 to offer Joint STARS to UK for SR(L/A) 925 (ASTOR) requirement. UK received two sets of data from Northrop Grumman but E-8 airframe not selected as finalist. Joint STARS system re-admitted to ASTOR in January 1998, using more advanced radar and Gulfstream V as platform, but was again unsuccessful.

MultiStage Improvement Program (MSIP) formulated by 1994 and first phase funded; involves TADIL-J datalink with dedicated operator's position for monitoring of air threats, followed by 'End Game' radar- and IR-countermeasures; Phase 2 options include satellite communications, improved data modem and automatic target recognition.

Northrop Grumman awarded US$132 million (two contracts) in June 1997 for computer replacement programme taking advantage of commercial off-the-shelf (COTS) technology and intended to reduce costs; will result in integration of new, more powerful central computers that will be significantly faster than those originally installed. It is also intended to replace programmable signal processors and substitute high-capacity switch and fibre-optic cable for existing copper-wired workstation network. This expected to produce savings of about US$19.3 million per aircraft and will involve existing five central computers (three operating and two 'hot' spares) giving way to just two (one operating and one 'hot' spare).

Major radar upgrade programme begun in December 1998, with award of US$14.5 million contract to Northrop Grumman for pre-Engineering and Manufacturing Development (EMD) phase of radar technology insertion program (RTIP) that will replace existing AN/APY-3 with new 2-D electronically scanned array radar, offering upgraded signal processing and operation and control performance. Research and development effort likely to be divided equally between Northrop Grumman (prime contractor) and Raytheon (subcontractor), with former taking overall responsibility for next-generation radar and latter working on RTIP antenna development. New radar expected to offer a five-fold increase in resolution in MTI mode, a 10-fold acuity increase in SAR mode, a faster scan rate and additional operating modes to improve ground clutter elimination and resistance to jamming. Also likely to feature smaller antenna with resolution improved from current 3.66 m (12 ft) to 0.30 to 0.91 m (1 to 3 ft).

RTIP just one of a number of upgrades and improvements planned for implementation by 2006 that will take E-8C from initial **Block 0** (essentially, test and evaluation) configuration to Block 50 configuration with full battle management capabilities, as described under Current Versions.

CURRENT VERSIONS: **E-8A**: Two development aircraft, as above. Fitted with consoles for 10 operators; Pratt & Whitney JT3D-3B engines. Returned to Grumman after participation in Gulf War for completion of performance testing in 100 sorties and several thousand hours of ground testing. Both accepted by USAF, December 1993, for trials

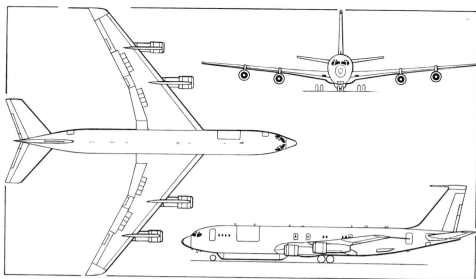

Northrop Grumman E-8C Joint STARS surveillance aircraft (*James Goulding/Jane's*)

Northrop Grumman E-8C Joint STARS

at Edwards AFB. To be eventually upgraded as E-8Cs, but first currently serving as **TE-8A** crew trainer with 93rd Wing.

E-8B: Originally proposed production version, based on new-built airframe; F108 turbofan engines; 15 operator consoles. One prototype YE-8B, 88-0322, flown 12 June 1990 in 'green' state; avionics not installed; overtaken by decision to use remanufactured Boeing 707 airframes; delivered to USAF 3 October 1991 for storage at Davis-Monthan AFB, Arizona; bartered with Omega Air, 1993, for five used 707s.

E-8C: Production version; 18 operator consoles (17 operations, one navigation/self-defence). Initial two aircraft to Lake Charles for conversion, May-June 1992. First E-8C is permanent testbed, not for delivery to USAF. First operational aircraft (P1/92-3289) commenced flight testing on 17 August 1995 and was delivered to USAF on 22 March 1996, making operational debut with mission from Frankfurt over former Yugoslavia on 15 November 1996.

Following description applies to E-8C, except where otherwise indicated.

E-8C Block 10: Baseline USAF aircraft, featuring interoperability with other US and allied aircraft; specific enhancements include TADIL-J secure communications, a digital autopilot and Y2K computer compliance. First example was P5/94-0284, in service 1999; first four Block 0 aircraft to be upgraded to same standard, which also applicable to P6-P10.

E-8C Block 20: Modified to accept COTS hardware, with computer replacement programme tied to networked communications; introduces US Army improved data modem, basic SINCGARS radio and radar advanced signal processing. Planned for 2000, but first aircraft to this configuration reportedly P11, which due for delivery in third quarter of 2001.

E-8C Block 30: Baseline attack control platform with UHF satcom, broadcast intelligence, TADIL-J attack support upgrade (ASU) and phase one of global air traffic management (GATM). Planned for 2002.

E-8 JOINT STARS PLANNED PROCUREMENT

FY Advance Acquisition	FY Full Funding	Lot	Qty	Type	Serial Numbers
85[1]		FSD	2	E-8A	86-0416, 86-0417
90[2]		FSD	1	E-8C	90-0175
92	93	1	2	E-8C	92-3289, 92-3290
93	94	2	2	E-8C	93-0597, 93-1097
94	95	3	2	E-8C	94-0284, 94-0285
95	96	4	2	E-8C	95-0121, 95-0122
96	97	5	2	E-8C	96-0042, 96-0043
97	98	6	1	E-8C	97-0066
98	99	7	2	E-8C	
99	00	8	1	E-8C	
00	01	9	1	E-8C	
01		10	1	E-8C	
Total			19[3]		

Notes: [1] Year of Full-Scale Development (FSD) contract award
[2] Year of Follow-On FSD contract award; aircraft serves as permanent testbed
[3] Includes two E-8A development aircraft, one E-8C permanent testbed and 16 production E-8Cs; does not include originally planned E-8B version, sole example of which was disposed of before avionics installation. Both E-8As expected to be upgraded to E-8C standard and assigned to operational fleet

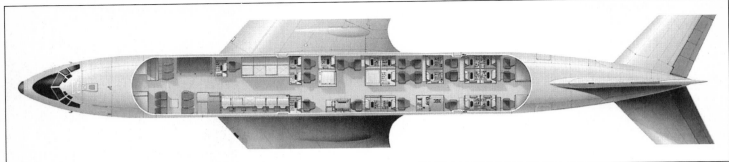

Inboard view of E-8C Joint STARS, showing operators' positions

E-8C Block 40: Baseline information dominance platform featuring enhanced and inverse SAR, improved MTI and EHF/SHF satcom. Planned for 2004.

E-8C Block 50: Baseline battle management platform with RTIP, elint capability, combat identification, helicopter detection/tracking, maritime detection, automatic target recognition and Kinematic Auto Tracker capability. Planned for 2006.

CUSTOMERS: US Air Force and US Army (aircraft operated by USAF). See table for details of planned USAF procurement; total 20 originally expected, including two upgraded E-8As and one E-8C permanent testbed, but proposed fleet reduced to 13 by 1997 Quadrennial Defense Review. Group of Senate and House members subsequently advocated restoration of full 19-aircraft programme by US DoD; this apparently not favoured, but recent budget requests have increased fleet to 15 (with long-lead funding for a 16th requested in FY01). Senior USAF personnel reportedly intend to seek extra funding and aircraft until original planned fleet of 19 is reached.

Operating unit is 93rd Air Control Wing which activated at Robins AFB, Georgia, in January 1996. IOC achieved with delivery of third aircraft (93-0597) on 18 December 1997; fourth (93-1097) delivered on 18 August 1998; fifth (94-0284) accepted by USAF at Melbourne, Florida on 13 August 1999, but retained for installation and testing of system upgrades for connectivity and newly modified engines. On completion of this work, it joined the 93rd ACW by October, with sixth E-8C (actually seventh production aircraft) handed over to USAF on 6 December 1999 and seventh (actually ninth production aircraft) following on 7 March 2000. Most recent delivery was eighth E-8C (10th production aircraft), which joined 93rd ACW on 27 July 2000.

COSTS: E-8A development costs US$1,006.2 million; E-8C development costs US$868.7 million (FY91). Procurement of Lot 1 to Lot 4 (eight aircraft) totalled US$2,141 million; advance acquisition costs (long lead) for two aircraft in FY96 US$104.6 million. Unit cost of eighth aircraft quoted as US$240 million (August 2000).

DESIGN FEATURES: Boeing 707-320C airliner converted with 7.32 m (24 ft) antenna covered by a 12.19 m (40 ft) 'canoe' radome and fairing under forward fuselage, for phased-array SLAR.

FLYING CONTROLS, STRUCTURE, LANDING GEAR: As, or similar to, commercial Boeing 707, last described in 1980-81 *Jane's* and detailed in the current *Jane's Aircraft Upgrades*.

POWER PLANT: Four 80.1 kN (18,000 lb st) Pratt & Whitney TF33-P-102B turbofans initially installed; now being upgraded to 85.4 kN (19,200 lb st) TF33-P-102C standard following award of US$10.5 million contract in December 1998 to United Technologies for 42 modification kits. See Weights and Loadings for fuel. Consideration being given to re-engining entire fleet to reduce need for aerial refuelling, increase time on station and raise operating ceiling to 12,800 m (42,000 ft) to increase radar coverage significantly; funding improbable before FY02. Engines likely to be considered include CFM56, Rolls-Royce Deutschland BR700 and Pratt & Whitney JT8D-200.

ACCOMMODATION: Standard mission crew of 21, comprising pilot, co-pilot, flight engineer, navigator/self-defence suite operator and 17 Air Force/Army operators; on longer missions replacement flight deck crew and system operators can be carried up to maximum of 34.

SYSTEMS: As Boeing 707; additional electrical generating capacity. World Auxiliary Power Company APU.

AVIONICS: *Comms*: Telephonics multiple intercom net control system; Raytheon UHF communications system.

Radar: Northrop Grumman (Norden Systems) AN/APY-3 multimode side-looking phased-array I-band radar, scanned electronically in azimuth and steered mechanically in elevation from either side of aircraft to provide 120° field of view. Synthetic aperture (SAR) mode used to detect stationary objects, such as parked tanks. Can interleaf Doppler mode to detect moving targets in moving target indicator/wide-area search (MTI/WAS), which is primary operating mode. Coverage is approximately 50,000 km² (19,305 sq miles) per minute, cruising at 9,150 to 12,200 m (30,000 to 40,000 ft), with ability to detect targets at range of 50 km (31 miles) to 250 km (155 miles) from aircraft.

Flight: Litton INS, Rockwell Collins flight management system.

Mission: Five Raytheon Model 920/866 supermini computers per aircraft, Computing Devices International programmable signal processors (three), Interstate Electronics graphic displays, 18 Raytheon Model 920 workstations, Orbit International workstation keyboards, Miltope message printers (to be replaced by Data Metrics printers on fifth and subsequent aircraft), Cubic Defense Systems surveillance and control datalink, JTIDS for TADIL-J generation and processing. Satellite communications link, Magnavox encrypted UHF radios (12), two encrypted HF radios, three encrypted VHF radios with single channel ground and airborne radio system (SINCGARS) provision. Radar data can be transmitted instantaneously to ground stations; or attacks by aircraft and ground forces directed via JTIDS datalink.

Self-defence: Unspecified, but includes chaff/flare dispensers and probably also incorporates threat-sensing equipment.

ARMAMENT: Nil.

DIMENSIONS, EXTERNAL (abbreviated):
Wing span	44.42 m (145 ft 9 in)
Length overall	46.61 m (152 ft 11 in)
Height overall	12.95 m (42 ft 6 in)

WEIGHTS AND LOADINGS:
Weight empty	77,564 kg (171,000 lb)
Max fuel weight	70,307 kg (155,000 lb)
Max T-O weight	150,139 kg (331,000 lb)

PERFORMANCE:
Max operating Mach No. (M_{MO})	0.84
Service ceiling	12,800 m (42,000 ft)
Endurance: internal fuel	11 h
with one in-flight refuelling	20 h

UPDATED

Close up of E-8C radome *(Paul Jackson/Jane's)*
2001/0093509

NORTHROP GRUMMAN GREYHOUND 21

No recent news has been received of this project, last described and illustrated in the 2000-01 edition.

UPDATED

ELECTRONIC SENSORS AND SYSTEMS SECTOR

1580-A West Nursery Road, Linthicum, Baltimore, Maryland 21090
Tel: (+1 410) 993 24 63
Fax: (+1 410) 993 23 94
CORPORATE VICE-PRESIDENT AND PRESIDENT, ELECTRONIC SENSORS AND SYSTEMS: James G Roche
COMMUNICATIONS OFFICER: Jack Martin Jr

Electronic Sensors and Systems Sector (ESSS) responsibilities include the development and production of radar and electronic systems for installation in the F-16 Fighting Falcon, F-22 Raptor, B-1B Lancer, AH-64D Apache Longbow, C-130 Hercules, E-3B/C Sentry and Boeing 737 AWACS and E-8C Joint STARS. Teamed with Rafael of Israel for sale and production of Litening II target-designation and navigation pod.

ESSS development, integration and manufacturing expertise also embraces military airborne, space and undersea radar, surveillance and reconnaissance systems, air defence systems, tactical communications equipment, anti-submarine warfare sensors and systems, submersibles, mine countermeasures equipment, marine systems and shipboard instrumentation. Radar systems are produced for civil air traffic control agencies in the USA, Europe, Africa, the Middle and Far East, Asia and Latin America.

California Microwave Systems was acquired in April 1999, retaining its original name.

UPDATED

PAM

PERFORMANCE AVIATION MANUFACTURING GROUP

PO Box 80, Williamsburg, Virginia 23187
Tel: (+1 757) 229 03 67
Fax: (+1 757) 229 73 80
e-mail: pamgroup@erols.com
CEO: Clement Makowski
PRESIDENT: Robert Pegg

UPDATED

PAM 100B INDIVIDUAL LIFTING VEHICLE

TYPE: Lifting vehicle.

PROGRAMME: Project initiated October 1989; construction of prototype started January 1993; first flight (N6172N) June 1994, powered by single engine; rebuilt with two Hirth engines and reflown May 1998; FAA certification April 1999. Total 50 hours (tethered and free) flown between 1994 and mid-1999; over 90 hours by April 2000.

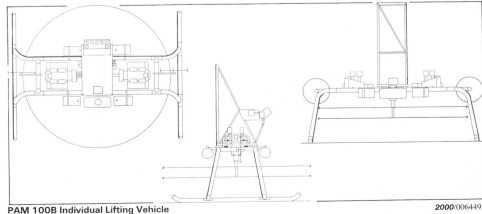

PAM 100B Individual Lifting Vehicle
2000/0064492

CURRENT VERSIONS: **PAM 100B:** *As described.*
 UAV: Unmanned military version projected, 1999.
COSTS: Kit approximately US$38,000 (1999).
DESIGN FEATURES: Two co-axial counter-rotating two-blade rotors mounted below main open-frame structure. Rotor section NACA 0012, constant; no twist; end caps. Rotor speed 1,200 rpm. Tip speed 172 m (565 ft)/s. Two small propellers at ends of cross-tubes provide directional control.
FLYING CONTROLS: Kinesthetic control for pitch and roll functions; manual throttle on engine for vertical control.
STRUCTURE: Tubular main structure. Rotors of extruded 6063T6 aluminium; aluminium flexure rotor hub.
LANDING GEAR: Fixed skid type.
POWER PLANT: Two 78.3 kW (105 hp) Hirth F30A flat-four, two-stroke engines with two-spur 1:2.64 reduction gear on each engine; over-running and centrifugal clutches on transmission. Fuel in two tanks, total capacity 26.5 litres (7.0 US gallons; 5.8 Imp gallons).
ACCOMMODATION: One person, standing within protective tubular framework.
DIMENSIONS, EXTERNAL:
Rotor diameter (each)	2.79 m (9 ft 2 in)
Rotor blade chord	0.20 m (0 ft 8 in)
Height overall	2.59 m (8 ft 6 in)
Skid track	3.05 m (10 ft 0 in)

AREAS:
Rotor blades (each)	0.20 m² (2.11 sq ft)
Rotor discs (total)	12.26 m² (132.0 sq ft)

WEIGHTS AND LOADINGS:
Weight empty	272 kg (600 lb)
Max fuel	20 kg (45 lb)
Max T-O weight	431 kg (950 lb)
Max disc loading	35.1 kg/m² (7.20 lb/sq ft)
Max power loading	2.75 kg/kW (4.52 lb/hp)

PERFORMANCE:
Max level speed	52 kt (97 km/h; 60 mph)
Max cruising speed	39 kt (72 km/h; 45 mph)
Service ceiling	1,830 m (6,000 ft)
Range with max fuel	21 n miles (40 km; 25 miles)

UPDATED

PAM 100B Individual Lifting Vehicle in hovering flight

PAPA 51

PAPA 51 LTD

This company was formed to market a scale replica of the classic P-51 Mustang fighter. It ceased trading on 31 November 2000, whereupon 25 kitbuilders (six of them with almost complete aircraft) formed Thunder Builder Group LLC and acquired company's intellectual rights and production moulds with a view to sustaining their own projects and resuming manufacture.

UPDATED

PAPA 51 THUNDER MUSTANG

TYPE: Single-seat sportplane kitbuilt.
PROGRAMME: Design started 1993; first flight of prototype (N151TM) 16 November 1996; prototype destroyed in crash 30 May 1998. First customer aircraft (N7TR) owned by Tommy Rose of Hickory, Mississippi, first flown from Nampa factory, where assembly took place, 11 January 1999. In September 1999 this aircraft qualified in the Sport Class of the US National Championship Air Races at Reno, Nevada, at a speed of 269 kt (499 km/h; 310 mph).
CUSTOMERS: Total of 30 aircraft under construction by May 1999, including at least two in South Africa.
COSTS: Complete kit, including engine, but not including instruments, avionics, upholstery or paint, US$285,000 (2000). Production limited to maximum of 300 kits.
DESIGN FEATURES: Three-quarters scale replica of Second World War North American P-51D Mustang fighter; faithful to the original in outline. Low-wing, mid-tailplane configuration with sweptback fin and fillet, plus prominent belly air intake for radiator.

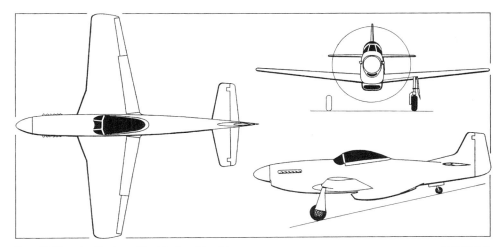

Papa 51 Thunder Mustang *(Paul Jackson/Jane's)*

Wing section NACA 65-series laminar flow with thickness/chord ratio 15 per cent at root, 12 per cent at tip, 1 per cent camber and 1.25 per cent leading-edge droop.
FLYING CONTROLS: Conventional and manual. Electrically actuated flaps; trim tab in rudder.
STRUCTURE: Primary structure of carbon/graphite composites, supplied in kit of over 200 ready-formed components; assembled with two-part epoxy adhesive; quoted build time approximately 1,000 hours.
LANDING GEAR: Tailwheel type; main units retract inwards into wings, tailwheel rearwards. Mainwheel tyre size 6.50-8; tailwheel tyre size 4.10-4.
POWER PLANT: One 477 kW (640 hp) Ryan Falconer V-12 with 2.8:1 reduction gearbox, driving a four-blade

Second Thunder Mustang scale kitbuilt replica of the P-51D

constant-speed wood/composites MT propeller. Supercharged 895 kW (1,200 hp) version of Falconer engine under development during 1999 for planned attempts on several class world records. Fuel contained in integral wing tanks, combined capacity 386 litres (102 US gallons; 84.9 Imp gallons). External wing tanks, each of 114 litres (30.0 US gallons; 25.0 Imp gallons) capacity, optional.
ACCOMMODATION: Two in tandem under rearward-sliding bubble canopy with fixed three-piece windscreen. Baggage compartment at rear of cockpit. Custom upholstery optional.
SYSTEMS: Optional oxygen system.
AVIONICS: Instrument and avionics packages and autopilot optional.
EQUIPMENT: Optional gas-operated dummy cannon.

DIMENSIONS, EXTERNAL:
Wing span	7.24 m (23 ft 9 in)
Wing aspect ratio	5.4
Length overall	7.37 m (24 ft 2¼ in)
Height overall	2.77 m (9 ft 1¼ in)
Tailplane span	2.70 m (8 ft 10¼ in)
Propeller diameter	2.44 m (8 ft 0 in)

AREAS:
Wings, gross	9.66 m² (104.0 sq ft)

WEIGHTS AND LOADINGS:
Weight empty	998 kg (2,200 lb)
Baggage capacity	23 kg (50 lb)
Payload with max fuel	181 kg (400 lb)
Max T-O weight	1,451 kg (3,200 lb)
Max wing loading	150.2 kg/m² (30.77 lb/sq ft)
Max power loading	3.04 kg/kW (5.00 lb/hp)

PERFORMANCE (provisional):
Never-exceed speed (V_{NE})	439 kt (813 km/h; 505 mph)
Max level speed at S/L	326 kt (604 km/h; 375 mph)
Cruising speed at 75% power	300 kt (556 km/h; 345 mph)
Manoeuvring speed	222 kt (411 km/h; 255 mph)
Stalling speed, power off:	
flaps and landing gear up	77 kt (143 km/h; 89 mph)
flaps and landing gear down	68 kt (126 km/h; 79 mph)
Max rate of climb at S/L	1,585 m (5,200 ft)/min
Service ceiling	7,620 m (25,000 ft)
Range at econ cruising speed	approx 1,300 n miles (2,407 km; 1,496 miles)
g limits: at 1,179 kg (2,600 lb)	+9/–6
at 1,451 kg (3,200 lb)	+7.3/–4.9

UPDATED

PERFORMANCE

PERFORMANCE AIRCRAFT
12901 West 151 Street, Suite C, Olathe, Kansas 66062
Tel: (+1 913) 780 91 40
Fax: (+1 913) 782 10 92
Web: http://www.performanceaircraft.com
PRESIDENT: Jeff Ackland

VERIFIED

PERFORMANCE LEGEND and TURBINE LEGEND
TYPE: Tandem-seat sportplane kitbuilt/turboprop sportplane kitbuilt.
PROGRAMME: Prototype (N620L) first flown in 1996 with Chevrolet V-8 engine; subsequently converted to Turbine Legend with Walter M 601E turboprop; public debut in turboprop form at EAA Sun 'n' Fun at Lakeland, Florida, in April 1999.
CURRENT VERSIONS: **Legend:** Piston-engined version, powered by one 429 kW (575 hp) liquid-cooled Chevrolet V-8 driving a three-blade Hartzell propeller via a Geschwender 2:1 reduction gearbox. Wing span 8.22 m (27 ft 0 in).
Turbine Legend: Turboprop version, *as described.*
COSTS: Standard kit, less engine US$94,450; fast-build kit, less engine US$115,700 (both 1999). Engine (new) US$104,000 (1999).
DESIGN FEATURES: Highly streamlined, low-wing monoplane with sweptback tail surfaces, steeply raked windscreen and, in Chevrolet-engined version, low-drag (Mustang-type) ventral radiator air intake. Tapered wings and mid-mounted tailplane. Quoted build time: 2,500 hours, standard; 1,900 hours fast-build.
FLYING CONTROLS: Conventional and mechanical; horn-balanced elevators and rudder; three-axis electric trim; electrically actuated slotted flaps, maximum deflection 38°.
STRUCTURE: Mostly CFRP.
LANDING GEAR: Retractable tricycle type; mainwheels retract inwards, nosewheel rearwards; steering by differential braking; dual brakes standard.
POWER PLANT: One Walter M 601E turboprop, derated to 490 kW (657 shp), driving an Avia V 508E three-blade,

Prototype Performance Aircraft Legend with Chevrolet V-8 piston engine *(Geoffrey P Jones)* 2000/0079277

constant-speed, reversible-pitch propeller. Standard fuel capacity 379 litres (100 US gallons; 83.3 Imp gallons); optional tip tanks, combined capacity 95 litres (25.0 US gallons; 20.8 Imp gallons).
ACCOMMODATION: Two persons in tandem under rear-hinged, upward-opening canopy with fixed single-piece windscreen. Dual controls standard.

DIMENSIONS, EXTERNAL:
Wing span	8.69 m (28 ft 6 in)
Wing aspect ratio	8.0
Length overall	7.84 m (25 ft 8½ in)
Propeller diameter	2.13 m (7 ft 0 in)

DIMENSIONS, INTERNAL:
Cabin max width	0.75 m (2 ft 5½ in)

AREAS:
Wings, gross	9.38 m² (101.0 sq ft)

WEIGHTS AND LOADINGS:
Weight empty	885 kg (1,950 lb)
Baggage capacity	59 kg (130 lb)
Max T-O weight	1,496 kg (3,300 lb)
Max wing loading	159.5 kg/m² (32.67 lb/sq ft)
Max power loading	3.06 kg/kW (5.02 lb/shp)

PERFORMANCE:
Never-exceed speed (V_{NE})	391 kt (724 km/h; 450 mph)
Max level speed	309 kt (573 km/h; 356 mph)
Max cruising speed at 7,620 m (25,000 ft)	340 kt (630 km/h; 392 mph)
Econ cruising speed at 7,620 m (25,000 ft)	288 kt (533 km/h; 331 mph)
Manoeuvring speed	261 kt (482 km/h; 300 mph)
Stalling speed, landing configuration	58 kt (107 km/h; 66 mph)
Max rate of climb at S/L	1,981 m (6,500 ft)/min
Range with auxiliary fuel:	
at max cruising speed	887 n miles (1,643 km; 1,021 miles)
at econ cruising speed	1,207 n miles (2,236 km; 1,390 miles)
g limits	+6/–4

UPDATED

PIASECKI

PIASECKI AIRCRAFT CORPORATION
Second Street West, Essington, Pennsylvania 19029-0360
Tel: (+1 610) 521 57 00
Fax: (+1 610) 521 59 35
e-mail: piac@concentric.net
PRESIDENT: Frank N Piasecki
VICE-PRESIDENTS:
 Frederick W Piasecki (Technology)
 John W Piasecki (Contracts)
DIRECTOR, BUSINESS DEVELOPMENT: Joseph P Cosgrove

Formed in 1955 by Frank Piasecki, former Chairman of the Board and President of Piasecki Helicopter Corporation; latter now part of Boeing Aircraft and Missile Systems Group.
Piasecki assisted PZL Swidnik of Poland (which see) in FAA certification of the W-3A helicopter, which was awarded in May 1993. Piasecki has exclusive sales agreement for W-3A in the Americas and Pacific Rim countries. W-3A Sokół complies with FAR Pt 29, is certified for full IFR operations and has US instrumentation.

UPDATED

PIASECKI VECTORED THRUST DUCTED PROPELLER (VTDP)
TYPE: Experimental compound helicopter.
PROGRAMME: Began with US Army contract to develop a compound helicopter incorporating the Piasecki VTDP concept for the AH-64 Apache and AH-1W SuperCobra. Programme objectives were met or exceeded by both the AH-64 VTCAD (Vectored Thrust Combat Agility Demonstrator) and AH-1W VTCAD configurations, resulting in increased maximum level flight speed; 50 per cent improvement in longitudinal acceleration and deceleration capability in level flight; 50 per cent decrease in turn and pull-up radii at speeds in excess of 95 kt (176 km/h; 109 mph); and handling qualities at least as good as those of the baseline AH-64A and AH-1W. In addition, tactical simulations confirmed superiority of the VTCADs over the standard Apache and AH-1W SuperCobra.
A separate US Navy contract, awarded in April 1997 and valued at US$16.1 million, involved investigation into application of VTDP technology to the AH-1W(4BW) four-blade rotor configuration. The Navy contract, recently completed, included ground testing of the full-scale VTDP and additional flight controls simulation and testing of the 4BW/VTDP configuration.
Piasecki then proposed flight demonstration of this technology to the Navy on an AH-1W(4BW), but instead was awarded a four-year US$26.1 million contract on 28 September 2000 for integration, testing and flight demonstration of VTDP on a modified Sikorsky YSH-60F. This flight test programme, to begin in 2003, will be conducted jointly by Piasecki and the US Navy. The VTDP concept is being investigated by the DoD as an affordable means of upgrading the capabilities extending the service life of existing single main rotor helicopters such as the UH-1, UH/SH-60 and AH-64, until the follow-on Joint Replacement Aircraft is fielded some time after 2025. Most recently, the US Air Force selected the H-60/VTDP concept as one of a number of alternatives being considered as an upgrade or replacement, to be fielded as early as 2007, for its ageing HH-60G combat search and rescue helicopters.
DESIGN FEATURES: The VTDP comprises a ducted propeller with integral vanes and spherical sectors to vector propeller thrust. It provides lateral thrust for anti-torque and lateral control, in lieu of a tail rotor, as well as forward thrust for auxiliary propulsion. Combined with a lifting wing, it provides for increasing speed to more than 200 kt (370 km/h; 230 mph); greater manoeuvrability; reduced vibration and fatigue loads; and consequent lowering of maintenance costs.

UPDATED

Model of AH-1W(4BW) VTCAD with PiAC vectored thrust ducted propeller and lifting wings

PIPER

THE NEW PIPER AIRCRAFT INC
2926 Piper Drive, Vero Beach, Florida 32960
Tel: (+1 561) 567 43 61
Fax: (+1 561) 778 21 44
Web: http://www.newpiper.com
PRESIDENT AND CEO: Charles (Chuck) M Suma
VICE-PRESIDENT, PRODUCT SERVICES: Werner Hartlieb
MARKETING AND SALES DIRECTOR: Larry Bardon

In July 1995, after Piper had stabilised its financial position during four years of Chapter 11 protection, the US Bankruptcy Court approved a new reorganisation plan under which Piper's assets were bought for US$95 million by Newco Pac Inc, a new company jointly owned by Philadelphia-based investment firm Dimeling, Schreiber and Park, Teledyne Continental Motors (which was Piper's largest creditor), and the remaining creditors. Under this ownership, The New Piper Aircraft Inc was established. Earlier history of Piper last appeared in the 1995-96 *Jane's*.

During 1999, Piper delivered 329 aircraft, 395 in 2000 and expects to deliver 530 in 2001. Workforce 1,400 by October 2000, rising to 1,500 in 2001. Gross revenue forecast at US$200 million in 2000, rising to US$300 million in 2001.

UPDATED

Piper PA-28-161 Warrior *(Paul Jackson/Jane's)*   2001/0100488

PIPER PA-28-161 WARRIOR III

TYPE: Four-seat lightplane.
PROGRAMME: As replacement for Cherokee 140 series, Piper redesigned airframe with several refinements, including fuselage stretch and new wing; first flight of prototype PA-28-151 17 October 1972; FAA certification 9 August 1973; PA-28-161 Warrior II first flown 27 August 1976; two/four-seat Cadet trainer version introduced April 1988, but no longer in production; Warrior III introduced late 1994.
CUSTOMERS: 18 delivered in 1995, five in 1996, seven in 1997, 20 in 1998, 25 in 1999 and 43 in 2000. Piper has built some 30,000 of PA-28 series since prototype Cherokee flew on 10 January 1960.
COSTS: Standard equipped price US$152,500 (2000).
DESIGN FEATURES: Classic tourer and trainer. Low, moderately tapered wing (with root glove) and low-set tailplane; sweptback fin.
NACA 65_2-415 wing section on inboard panels, Mod No. 5 of NACA TN 2228 on leading-edge of outboard panels; dihedral 7°, incidence 2° at root, −1° at tip; sweepback at quarter-chord 5°.
FLYING CONTROLS: Manual. Conventional ailerons and rudder, plus all-moving tailplane with combined anti-servo and trim tab. Four-position manually operated flaps.
STRUCTURE: Conventional light alloy, with semi-monocoque fuselage, single-spar wings and ribbed light alloy skins on fin and tailplane; glass fibre nose cowl and wing/tailplane/fin-tips.
LANDING GEAR: Non-retractable tricycle type. Steerable nosewheel. Piper oleo-pneumatic shock-absorbers; single wheel on each unit. Cleveland wheels with tyres size 6.00-6 (4 ply) on main units, pressure 1.65 bars (24 lb/sq in). Cleveland nosewheel and tyre size 5.00-5 (4 ply), pressure 2.06 bars (30 lb/sq in). Cleveland disc brakes. Parking brake. Glass fibre speed fairings standard.
POWER PLANT: One 119 kW (160 hp) Textron Lycoming O-320-D3G flat-four engine, driving a Sensenich 74DM6-0-60 two-blade fixed-pitch metal propeller. Fuel in two wing tanks, with total capacity of 189 litres (50.0 US gallons; 41.6 Imp gallons), of which 181.5 litres (48.0 US gallons; 40.0 Imp gallons) are usable. Refuelling point on upper surface of each wing. Oil capacity 7.5 litres (2.0 US gallons; 1.7 Imp gallons).
ACCOMMODATION: Four persons in pairs in enclosed cabin. Individual horizontally adjustable front seats with seat belts and shoulder harnesses; bench-type rear seat with seat belts and shoulder harnesses. Dual controls standard. Large door on starboard side. Baggage compartment at rear of cabin, with external baggage door on starboard side. Heating, ventilation and windscreen defrosting standard.
SYSTEMS: Hydraulic system for brakes only. Electrical system powered by 28 V 60 A engine-driven alternator. 24 V 10 Ah battery standard.
AVIONICS: Standard and optional Advanced Training Group packages by Garmin/Meggitt, as described.
  Comms: Standard: Garmin GNS-430 com/nav/IFR GPS; GMA-340 audio panel with marker beacon lights and four-position intercom; GTX-320 transponder; ELT; avionics master switch; cabin speaker; Telex 100T microphone and Airman 760 headset. Advanced Training Group: as above but with dual GNS-430.
  Flight: Standard: GI-106A VOR/LOC/GS/GPS indicator; Narco AR-850 altitude reporter. Advanced Training Group: as above, plus dual GI-106A and RCR 650A ADF with IND 650A indicator.
  Instrumentation: Piper true airspeed indicator, magnetic compass, sensitive altimeter, gyro horizon, directional gyro, rate of turn indicator, rate of climb indicator, OAT gauge, digital ammeter, electric clock, annunciator panel with push-to-test, recording tachometer, three-way oil temperature/pressure and fuel-pressure gauge, dual fuel quantity gauges and vacuum gauge.

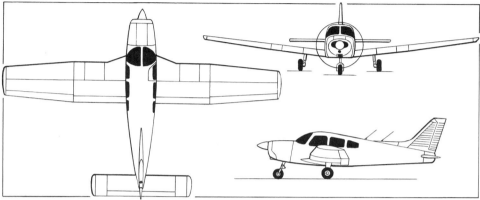

Piper PA-28-161 Warrior III (Lycoming O-320 flat-four) *(James Goulding/Jane's)*   2001/0100458

EQUIPMENT: Standard equipment includes engine hour recorder, alternate static source, heated pitot head, colour co-ordinated control wheels, internally lit rocker switches, external power receptacle, electrical engine primer system, instrument panel lighting package, cabin dome light, navigation lights, landing/taxying light, wingtip Comet strobe lights, avionics dimming switch, pilot's and co-pilot's vertically adjustable seats in fabric and vinyl with magazine storage pockets on backs, rear bench seat, four floor-mounted cabin fresh-air vents, vinyl cabin side panels, wall-to-wall carpet and headliner, crew armrests, pilot's storm window, two sun visors, two map pockets, 'Quietised' soundproofing, carpeted baggage compartment with security straps, halon fire extinguisher, tiedown points, jack pads, static discharge wicks, and Du Pont Imron polyurethane exterior paint in base colour with two contrasting trim stripes. Meggitt electric trim, carburettor ice detector, second altimeter, two-tone base colour exterior paint and foreign certification gross weight kit optional.

DIMENSIONS, EXTERNAL:
Wing span	10.67 m (35 ft 0 in)
Wing chord: at root, excl glove	1.60 m (5 ft 3 in)
at tip	1.07 m (3 ft 6¼ in)
Wing aspect ratio	7.2
Length overall	7.25 m (23 ft 9½ in)
Height overall	2.22 m (7 ft 3½ in)
Tailplane span	3.96 m (12 ft 11¾ in)
Wheel track	3.05 m (10 ft 0 in)
Wheelbase	2.03 m (6 ft 8 in)
Propeller diameter	1.88 m (6 ft 2 in)
Propeller ground clearance	0.21 m (8¼ in)
Cabin door: Height	0.89 m (2 ft 11 in)
Width	0.91 m (3 ft 0 in)
Baggage door: Height	0.51 m (1 ft 8 in)
Max width	0.56 m (1 ft 10 in)
Height to sill	0.71 m (2 ft 4 in)

DIMENSIONS, INTERNAL:
Cabin (instrument panel to rear bulkhead):	
Length	2.49 m (8 ft 2 in)
Max width	1.05 m (3 ft 5¼ in)
Max height	1.14 m (3 ft 8¾ in)
Floor area	2.28 m² (24.5 sq ft)
Volume (incl baggage)	3.0 m³ (106 cu ft)
Baggage compartment volume	0.68 m³ (24.0 cu ft)

AREAS:
Wings, gross	15.79 m² (170.0 sq ft)
Ailerons (total)	1.23 m² (13.20 sq ft)
Trailing-edge flaps (total)	1.36 m² (14.60 sq ft)
Fin	0.69 m² (7.40 sq ft)
Rudder	0.38 m² (4.10 sq ft)
Tailplane, incl tab	2.76 m² (29.70 sq ft)

WEIGHTS AND LOADINGS:
Weight empty, equipped	695 kg (1,533 lb)
Baggage capacity	91 kg (200 lb)
Max T-O weight	1,106 kg (2,440 lb)
Max wing loading	70.1 kg/m² (14.35 lb/sq ft)
Max power loading	9.28 kg/kW (15.25 lb/hp)

PERFORMANCE:
Max level speed	117 kt (216 km/h; 134 mph)
Normal cruising speed at 75% power	
	115 kt (213 km/h; 132 mph)
Stalling speed: flaps up	50 kt (93 km/h; 58 mph)
flaps down	44 kt (82 km/h; 51 mph)
Max rate of climb at S/L	196 m (644 ft)/min
Service ceiling	3,355 m (11,000 ft)
T-O to 15 m (50 ft)	494 m (1,620 ft)
Landing from 15 m (50 ft)	354 m (1,160 ft)
Range, 45 min reserves:	
at 75% power	426 n miles (789 km; 490 miles)
at 55% power at 3,050 m (10,000 ft)	
	513 n miles (950 km; 590 miles)

UPDATED

PIPER PA-28-181 ARCHER III

TYPE: Four-seat lightplane.
PROGRAMME: Introduced (as Cherokee Challenger 180) 9 October 1972 as successor to Cherokee 180; Archer 180, featuring minor changes, introduced 1974; PA-28-181 Archer II launched 1976, and in 1978 introduced the tapered wings of Warrior II; Archer III from 1994, with axisymmetric engine inlets, redesigned windscreen and cabin side windows, and interior restyling and improvements.
CUSTOMERS: 37 Archer IIIs delivered in 1995, and 45 each in 1996 and 1997; 91 in 1998; 100 in 1999 and 102 in 2000. Recent customers include Sabena Airlines, which took delivery of five during 1999 for its training centre at Scottsdale, Arizona, and Westwind Aviation Academy of Phoenix, Arizona, which ordered 15 for delivery commencing July 1999.

Description of the Warrior III applies also to Archer III except as follows:

COSTS: Standard, equipped US$179,900 (2000).
LANDING GEAR: Tricycle type. Tyres size 6.00-6 (4 ply), on all three wheels. Mainwheel tyre pressure 1.65 bars (24 lb/sq in), nosewheel 1.24 bars (18 lb/sq in). Cleveland high-capacity disc brakes. Parking brake. Wheel speed fairings standard.

Instrument panel of 1999 Piper PA-28-181 Archer III   1999/0051814

USA: AIRCRAFT—PIPER

Piper PA-28-181 Archer III *(Paul Jackson/Jane's)*   2001/0100489

POWER PLANT: One 134 kW (180 hp) Textron Lycoming O-360-A4M flat-four engine, driving a Sensenich 76EM8S14-O-62 two-blade fixed-pitch metal propeller. Fuel in two tanks in wing leading-edges, with total capacity of 189 litres (50.0 US gallons; 41.6 Imp gallons), of which 181.5 litres (48.0 US gallons; 40.0 Imp gallons) are usable. Oil capacity 7.5 litres (2.0 US gallons; 1.7 Imp gallons).

ACCOMMODATION: Four persons in pairs in enclosed cabin. Individual adjustable front seats, with dual controls; individual rear seats. Large door on starboard side. Baggage compartment at rear of cabin; door on starboard side. Rear seats removable to provide 1.3 m³ (44 cu ft) cargo space. Accommodation heated and ventilated. Windscreen defrosting.

SYSTEMS: 28 V 70 A alternator and 24 V 10 Ah battery.

AVIONICS: Standard Garmin International and Meggitt package, Optional Premium Select package adds second GNS-430 and GI-106A VOR/LOC/GS/GPS indicator, Meggitt System 55 dual-axis autopilot with automatic electric trim, DG with heading bug and trim indicator and Meggitt ST-360 altitude preselect.

Comms: Garmin GMA-340 audio panel with marker beacon receiver and stereo intercom; GTX-320 solid-state transponder; Telex 100T microphone and Airman 760 headset.

Flight: Integrated VOR/LOC/GS/COM and GPS.

Instrumentation: Standard GNS-430 high-resolution colour LCD MFD with GI-106A VOR/LOC/GS/GPS indicator.

EQUIPMENT: Standard equipment generally as for Warrior III except: polished propeller spinner, metal instrument panel, illuminated side-mounted OAT gauge, overhead switch panel, EGT gauge, flush locking fuel caps, overhead vent fan system, restyled windscreen and window lines with tinted transparencies, full chemical corrosion protection, Aztec Silver interior decor, DuPont Imron 6000 clearcoat/basecoat and exterior paint, wool carpeting and Hobnail side panels, passenger armrests and headrests. Additional options include Piper Aire air conditioning, carburettor ice detector, stainless steel cowling fasteners; Meggitt electric trim low-noise exhaust system and leather seats in choice of three colours.

DIMENSIONS, EXTERNAL: As for Warrior III except:
Length overall	7.32 m (24 ft 0 in)
Tailplane span	3.92 m (12 ft 10½ in)
Wheelbase	2.00 m (6 ft 7 in)
Propeller diameter	1.93 m (6 ft 4 in)

DIMENSIONS, INTERNAL: As for Warrior III except:
Cabin: Width	1.06 m (3 ft 5¾ in)
Height	1.14 m (3 ft 9 in)
Baggage compartment volume, incl hatshelf	0.74 m³ (26.0 cu ft)

AREAS: As for Warrior III

WEIGHTS AND LOADINGS:
Weight empty, equipped	766 kg (1,689 lb)
Baggage capacity	90 kg (200 lb)
Max T-O weight	1,156 kg (2,550 lb)
Max ramp weight	1,160 kg (2,558 lb)
Max wing loading	73.2 kg/m² (15.00 lb/sq ft)
Max power loading	8.62 kg/kW (14.17 lb/hp)

PERFORMANCE:
Max level speed	133 kt (246 km/h; 153 mph)
Normal cruising speed at 75% power at 2,410 m (7,900 ft)	128 kt (237 km/h; 147 mph)
Stalling speed, flaps down	45 kt (84 km/h; 52 mph)
Max rate of climb at S/L	203 m (667 ft)/min
Service ceiling	4,300 m (14,100 ft)
T-O run	346 m (1,135 ft)
T-O to 15 m (50 ft)	490 m (1,608 ft)
Landing from 15 m (50 ft)	427 m (1,400 ft)
Landing run	280 m (920 ft)
Range at S/L, allowances for taxi, T-O, climb and descent and 45 min reserves, at 1,830 m (6,000 ft):	
at 75% power	444 n miles (822 km; 511 miles)
at 65% power	487 n miles (901 km; 560 miles)
at 55% power	522 n miles (966 km; 600 miles)

UPDATED

PIPER PA-28R-201 ARROW

TYPE: Four-seat lightplane.

PROGRAMME: Derived from Cherokee Archer II, but with retractable landing gear, more powerful engine and untapered wings of 1975 PA-28-180 Archer; replaced original retractable gear Cherokee, the PA-28R Arrow 180, which had first flown 1 February 1967; PA-28R-201 Arrow III with increased span, tapered wings first flown 16 September 1975; first production aircraft 7 January 1977; new Arrow IV and Turbo Arrow IV models introduced 1979 with all-moving T tails; these subsequently discontinued and earlier low-tail Arrow III model restored to production.

CUSTOMERS: Four delivered during 1995, seven in 1996, two in 1997, two in 1998, six in 1999 and 18 in 2000.

Description of Archer III applies also to Arrow except as follows:

COSTS: Standard, equipped US$228,700 (2000).

LANDING GEAR: Tricycle type, retracted hydraulically with an electrically operated pump supplying the hydraulic pressure. Main units retract inward into wings, nose unit rearward. All units fitted with oleo-pneumatic shock-absorbers. Mainwheels and tyres size 6.00-6 (6 ply), pressure 2.07 bars (30 lb/sq in). Nosewheel and tyre size 5.00-5 (4 ply), pressure 1.86 bars (27 lb/sq in). High-capacity dual hydraulic disc brakes and parking brake.

POWER PLANT: One 149 kW (200 hp) Textron Lycoming IO-360-C1C6 flat-four engine, driving a McCauley two-blade constant-speed metal propeller. Fuel tanks in wing leading-edges with total capacity of 291 litres (77.0 US gallons; 64.1 Imp gallons), of which 273 litres (72.0 US gallons; 60.0 Imp gallons) are usable. Oil capacity 7.5 litres (2.0 US gallons; 1.7 Imp gallons).

SYSTEMS: Generally as for Archer III and Warrior III except for electrohydraulic system for landing gear actuation.

AVIONICS: Standard and optional Advanced Training Group (IFR) packages by Honeywell, as described.

Comms: Standard: Honeywell KX 155 nav/com with audio amplifier and broadband and VOR antennas; KT 76A transponder; ELT; avionics master switch; cabin speaker; Telex 100T microphone; Telex headset; Telex ProCom four-position intercom; microphone and microphone and headset jacks. Advanced Training Group: As above but with second KX 155 nav/com and PS Engineering PM6000M audio selector panel.

Flight: Standard: KI 203 VOR/LOC indicator; Narco AR-850 altitude reporter. Advanced Training Group: As above, plus KI 204-02 VOR/LOC/GS indicator; KLN 89B GPS, KR 87 digital ADF and KI 227-00 and KCS 55A slaved compass system.

DIMENSIONS, EXTERNAL:
Wing span	10.80 m (35 ft 5 in)
Wing chord: at root, excl glove	1.60 m (5 ft 3 in)
at tip	1.07 m (3 ft 6¼ in)
Wing aspect ratio	7.4
Length overall	7.52 m (24 ft 8¼ in)
Height overall	2.39 m (7 ft 10¼ in)
Tailplane span	3.92 m (12 ft 10½ in)
Wheel track	3.19 m (10 ft 5½ in)
Wheelbase	2.39 m (7 ft 10¼ in)

AREAS:
Wings, gross	15.79 m² (170.0 sq ft)

WEIGHTS AND LOADINGS:
Weight empty, equipped	812 kg (1,790 lb)
Max T-O weight	1,247 kg (2,750 lb)
Max wing loading	79.0 kg/m² (16.18 lb/sq ft)
Max power loading	8.37 kg/kW (13.75 lb/hp)

PERFORMANCE:
Max level speed	145 kt (268 km/h; 166 mph)
Normal cruising speed	137 kt (253 km/h; 157 mph)
Stalling speed: flaps up	60 kt (112 km/h; 69 mph)
flaps down	55 kt (102 km/h; 64 mph)
Max rate of climb at S/L	253 m (831 ft)/min
Service ceiling	4,935 m (16,200 ft)
T-O to 15 m (50 ft)	488 m (1,600 ft)
Landing from 15 m (50 ft)	463 m (1,520 ft)
Cruising range at 55% power, at 2,743 m (9,000 ft), 45 min reserves	880 n miles (1,630 km; 1,013 miles)

UPDATED

PIPER PA-32R-301 SARATOGA II HP

TYPE: Six-seat utility transport.

PROGRAMME: Saratoga family of fixed and retractable landing gear light aircraft announced 17 December 1979 to replace earlier PA-32 Cherokee SIX 300 and T tail Lance series; fixed-gear models now out of production; Saratoga II HP introduced in 1993, featuring axisymmetric engine inlets, aerodynamic clean-up and interior improvements and restyling.

CURRENT VERSIONS: **Saratoga II HP:** Standard version, *as described.*

Saratoga II TC: Turbocharged version with Textron Lycoming TIO-540 engine; described separately.

CUSTOMERS: Total of 38 HPs delivered in 1997, 28 in 1998, 30 in 1999 and 28 in 2000. Recent customers include Sabena Airlines, which took delivery of three in 1999 for its training centre at Scottsdale, Arizona.

Piper PA-28R-201 Arrow *(Paul Jackson/Jane's)*   2000/0075911

Piper PA-32R-301 Saratoga II HP, clearly showing 'semi-tapered' wing *(Paul Jackson/Jane's)*   2000/0075912

Jane's All the World's Aircraft 2001-2002   www.janes.com

Piper has built over 7,200 of PA-32 family since prototype first flew on 6 December 1963.

COSTS: Standard, equipped US$409,800 (2000).

DESIGN FEATURES: Larger counterpart to PA-28 series, featuring stretched fuselage. Current versions have 'semi-tapered' wing, with root glove, unswept inboard panel and tapered outer panel with small fence inboard of tip.

Wing section NACA 66_2-415, dihedral 7°, thickness/chord ratio 15°, twist 2°, incidence 2°, washout 0°.

FLYING CONTROLS: Conventional. Ailerons, rudder and all-moving tailplane with combined anti-servo and trim tab. Rudder trim; four-position electrically actuated flaps with preselect.

STRUCTURE: Conventional light alloy, with semi-monocoque fuselage, single-spar wings and ribbed light alloy skins on fin and tailplane; glass fibre nose cowl and wing/tailplane/fin-tips.

LANDING GEAR: Hydraulically retractable tricycle type with single wheel on each unit. Main units retract inward, nosewheel aft. Emergency free-fall extension system. Piper oleo-pneumatic shock-absorbers. Steerable nosewheel. Mainwheels and tyres size 6.00-6 (8 ply), pressure 2.62 bars (38 lb/sq in). Nosewheel and tyre size 5.00-5 (6 ply), pressure 2.41 bars (35 lb/sq in).

POWER PLANT: One 224 kW (300 hp) Textron Lycoming IO-540-K1G5 flat-six engine, driving a Hartzell three-blade constant-speed metal propeller. Two fuel tanks in each wing with combined capacity of 405 litres (107 US gallons; 89.1 Imp gallons), of which 386 litres (102 US gallons; 84.9 Imp gallons) are usable. Refuelling points on wing upper surface. Oil capacity 11.5 litres (3.0 US gallons; 2.5 Imp gallons).

ACCOMMODATION: Enclosed cabin seating six people, rear four in club arrangement; dual controls standard. Two forward-hinged doors, one on starboard side forward, overwing, and one on port side at rear end of cabin. Space for 45 kg (100 lb) baggage at rear of cabin, with external, lockable baggage/utility door on port side. Additional baggage space, capacity 45 kg (100 lb), between engine fireproof bulkhead and instrument panel, with external door on starboard side. Pilot's storm window. Accommodation heated and ventilated. Piper Aire air conditioning system optional. Windscreen defroster standard.

SYSTEMS: Electrically driven hydraulic pump for landing gear actuation. Electrical system includes 28 V 90 A engine-driven alternator and 24 V 10 Ah battery. Standby vacuum system standard. Optional oxygen system and Piper Aire air conditioning.

AVIONICS: Standard Garmin International and Meggitt package.

Comms: Garmin GMA-340 audio panel with marker beacon receiver and stereo intercom; GTX-320 solid-state transponder.

Flight: Dual Garmin GNS-430 integrated VOR/LOC/GS/COM and GPS, GI-106A VOR/LOC/GD/GPS indicator, Meggitt ST-180 HSI, DME 450 and Meggitt 55 two-axis autopilot with automatic electric trim, DG with heading bug and trim indicator, ST-361 ADI with flight director features.

Instrumentation: Dual GNS-430 high-resolution colour LCD MFD; Piper true airspeed indicator, illuminated magnetic compass, sensitive altimeter (with second optional) ADI, HSI, rate of turn indicator, rate of climb indicator, OAT gauge, digital ammeter, annunciator panel with push-to-test, recording tachometer, manifold pressure/fuel flow gauge, oil pressure gauge, oil temperature gauge, dual fuel quantity gauges, fuel quantity sight gauges, CHT gauge, EGT gauge and vacuum gauge. Co-pilot's 76 mm (3 in) instrument panel with electric attitude indicator optional.

EQUIPMENT: Standard equipment includes engine hour recorder, alternate static source, electric clock, internally lit rocker switches, heated pitot head, resettable circuit breakers in CB panel, standby electric vacuum pump, electric emergency fuel pump, external power receptacle, instrument panel lighting package, avionics dimming, map lights, navigation lights, landing/taxying light, wingtip strobe lights, pulsating recognition lights, pilot's and co-pilot's vertically adjustable all-leather seats, pilot's vent window, sun visors, four all-leather passenger seats with headrests, shoulder harnesses, seat belts, and quick-release facility, fold-down armrests (fifth and sixth seats), refreshment console, executive writing table, Hobnail fabric side panels, super 'Quietised' soundproofing, utility door for rear baggage access and cargo loading, static discharge wicks, and Du Pont Imron 6000 clearcoat/basecoat exterior paint. Optional equipment includes entertainment/executive console; provision for AM/FM radio and CD player, multimedia entertainment system and laptop computer workstation; and metallic paint.

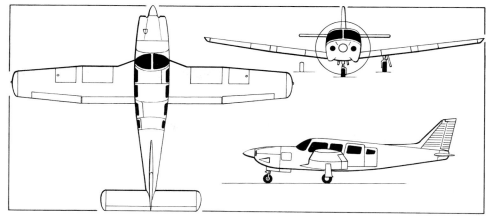

General arrangement of the Piper PA-32R-301 Saratoga II HP *(James Goulding/Jane's)*

DIMENSIONS, EXTERNAL:
Wing span	11.02 m (36 ft 2 in)
Wing aspect ratio	7.3
Length overall	8.50 m (27 ft 10½ in)
Height overall	2.59 m (8 ft 6 in)
Tailplane span	3.94 m (12 ft 11 in)
Wheel track	3.38 m (11 ft 1 in)
Wheelbase	2.41 m (7 ft 11 in)
Cabin door (fwd, stbd): Height	0.89 m (2 ft 11 in)
Width	0.91 m (3 ft 0 in)
Cabin door (rear, port): Height	0.72 m (2 ft 4½ in)
Width	0.71 m (2 ft 4 in)
Baggage door (fwd): Height	0.41 m (1 ft 4 in)
Width	0.56 m (1 ft 10 in)
Baggage/utility door (fwd): Height	0.52 m (1 ft 8½ in)
Width	0.66 m (2 ft 2 in)

DIMENSIONS, INTERNAL:
Cabin: Length (instrument panel to rear bulkhead)	3.16 m (10 ft 4¼ in)
Max width	1.24 m (4 ft 0¾ in)
Max height	1.07 m (3 ft 6 in)
Volume (incl rear baggage area)	5.5 m³ (195 cu ft)
Baggage compartment volume:	
forward	0.20 m³ (7.0 cu ft)
rear	0.49 m³ (17.3 cu ft)

AREAS:
Wings, gross	16.56 m² (178.3 sq ft)
Ailerons (total)	0.98 m² (10.60 sq ft)
Trailing-edge flaps (total)	1.36 m² (14.60 sq ft)
Fin	0.70 m² (7.50 sq ft)
Rudder, incl tab	0.40 m² (4.30 sq ft)
Horizontal tail surfaces (total)	2.94 m² (31.60 sq ft)

WEIGHTS AND LOADINGS:
Weight empty, equipped	1,087 kg (2,396 lb)
Max T-O weight	1,633 kg (3,600 lb)
Max ramp weight	1,639 kg (3,615 lb)
Max wing loading	98.6 kg/m² (20.19 lb/sq ft)
Max power loading	7.30 kg/kW (12.00 lb/hp)

PERFORMANCE:
Max level speed	175 kt (324 km/h; 201 mph)
Normal cruising speed	166 kt (307 km/h; 191 mph)
Stalling speed: flaps up	65 kt (121 km/h; 75 mph) CAS
flaps down	63 kt (117 km/h; 73 mph)
Max rate of climb at S/L	340 m (1,116 ft)/min
Service ceiling	4,750 m (15,580 ft)
T-O run	366 m (1,200 ft)
T-O to 15 m (50 ft)	540 m (1,770 ft)
Landing from 15 m (50 ft)	464 m (1,520 ft)
Landing run	195 m (640 ft)
Range, long-range cruise power, allowances for start, taxi, T-O, climb and descent, and 45 min reserves	859 n miles (1,590 km; 988 miles)

UPDATED

PIPER PA-32R-301T SARATOGA II TC

TYPE: Six-seat utility transport.

PROGRAMME: Development of Saratoga II HP. Rolled out 15 July 1997; FAA certification 21 July 1997; first customer deliveries August 1997; total of 26 delivered in 1997, 48 in 1998, 46 in 1999 and 70 in 2000.

COSTS: Standard, equipped US$439,600 (2000).

The description for the Saratoga II HP applies also to the Saratoga II TC except:

DESIGN FEATURES: New interior and relocated battery and ground power receptacle.

POWER PLANT: One 224 kW (300 hp) Textron Lycoming TIO-540-AH1A turbocharged flat-six engine, with automatic wastegate, driving a Hartzell three-blade constant-speed metal propeller. Fuel as for Saratoga II HP.

WEIGHTS AND LOADINGS:
Weight empty, equipped	1,118 kg (2,465 lb)

PERFORMANCE:
Max level speed	187 kt (346 km/h; 215 mph)
Cruising speed:	
at 2,440 m (8,000 ft)	172 kt (318 km/h; 198 mph)
at 3,050 m (10,000 ft)	175 kt (324 km/h; 201 mph)
at 4,575 m (15,000 ft)	185 kt (343 km/h; 213 mph)
Max certified altitude	6,100 m (20,000 ft)
T-O run	473 m (1,550 ft)
T-O to 15 m (50 ft)	756 m (2,480 ft)
Landing from 15 m (50 ft)	519 m (1,700 ft)
Landing run	269 m (880 ft)
Range, 45 min reserves and normal allowances, normal (N) or long-range (LR) cruise power:	
at 2,440 m (8,000 ft):	
N	842 n miles (1,559 km; 969 miles)
LR	950 n miles (1,759 km; 1,093 miles)

Piper Saratoga II TC instrument panel *(Piper/Scott Wohrman)*

Piper Seneca V instrument panel *(Paul Jackson/Jane's)*

USA: AIRCRAFT—PIPER

Piper PA-32R-301T Saratoga II TC (Paul Jackson/Jane's)

at 3,050 m (10,000 ft):
- N   844 n miles (1,563 km; 971 miles)
- LR   948 n miles (1,756 km; 1,091 miles)

at 4,575 m (15,000 ft):
- N   844 n miles (1,563 km; 971 miles)
- LR   945 n miles (1,750 km; 1,087 miles)

UPDATED

PIPER PA-34-220T SENECA V

TYPE: Six-seat utility twin-prop.

PROGRAMME: Original PA-34 Seneca announced 23 September 1971, having been preceded by prototypes of fixed-gear PA-34-180 Twin Six (flown 25 April 1967) and retractable gear version in following year; redesignated Seneca II from 1975; improved Seneca III with more powerful counter-rotating (C/R) engines introduced 15 February 1981; Seneca IV with axisymmetric engine inlets, aerodynamic refinements and interior improvements and restyling introduced in 1994. Seneca V with L/TSIO-360-RB engines unveiled 16 January 1997 (having been secretly certified in the previous month); engines, equipped with intercoolers and density control system (automatic wastegate), maintain S/L rated power to 5,945 m (19,500 ft). Other new features include redesigned instrument panel with digital display monitoring panel (DDMP), overhead switch/dome light/speaker panel, entertainment/executive console with extendable work table and in-flight phone/fax option replacing the second row seat on the starboard side. Senecas are also built in Brazil (which see), and until recently in Poland.

CUSTOMERS: Total of 28 Seneca IVs delivered in 1995, and 18 in 1996; 39 Seneca Vs delivered in 1997, 55 each in 1998 and 1999 and 42 in 2000. Widely used as twin-conversion trainer; recent customers include Sabena Airlines, which took delivery of two in 1999 for its training centre at Scottsdale, Arizona.

Production of the PA-34 series in the USA totals some 4,700; the prototype first flew on 30 August 1968.

COSTS: Standard equipped price US$19,900 (2000).

DESIGN FEATURES: Twin-engined version of Cherokee SIX/Saratoga. Low-mounted, constant-chord wings and tailplane; leading-edge gloves inboard of engines; constant-chord tailplane.

Wing section NACA 65₂-415; dihedral 7°; thickness/chord ratio 15 per cent; washout 0° 41′; twist 2° 41′.

FLYING CONTROLS: Manual. Frise ailerons, rudder and one-piece all-moving tailplane with combined anti-balance and trim tab; anti-servo tab in rudder; wide-span electrically operated slotted flaps.

STRUCTURE: Conventional light alloy, with semi-monocoque fuselage; single-spar wings; glass fibre wingtips.

LANDING GEAR: Hydraulically retractable tricycle type. Main units retract inward, nose unit forward. Oleo-pneumatic shock-absorbers. Steerable nosewheel. Emergency free-fall extension system. Mainwheels and tyres size 6.00-6 (8 ply), pressure 3.79 bars (55 lb/sq in); nosewheel and tyre size 6.00-6 (6 ply), pressure 2.76 bars (40 lb/sq in). Nosewheel safety mirror. High-capacity disc brakes. Parking brake.

POWER PLANT: One 164 kW (220 hp) Teledyne Continental TSIO-360-RB and one 164 kW (220 hp) LTSIO-360-RB flat-six turbocharged counter-rotating engine, each driving a Hartzell two-blade constant-speed fully feathering metal propeller; McCauley three-blade propellers optional, but mandatory when de-icing equipment installed. Propeller synchrophasers standard. Fuel in four tanks in wings, with a total capacity of 485 litres (128 US gallons; 107 Imp gallons), of which 462 litres (122 US gallons; 102 Imp gallons) are usable. Oil capacity 7.5 litres (2.0 US gallons; 1.7 Imp gallons). Glass fibre engine cowlings.

ACCOMMODATION: Enclosed cabin, seating five people on individual seats with 0.25 m (10 in) centre aisle; sixth seat (second row, starboard side) optional, replacing entertainment/executive console. Dual controls standard. Pilot's storm window. Two forward-hinged doors, one on starboard side at front, the other on port side at rear. Large utility door adjacent rear cabin door provides an extra-wide opening for loading bulky items. Passenger seats easily removable to provide different seating/baggage/cargo combinations. Space for 39 kg (85 lb) baggage at rear of cabin, and for 45 kg (100 lb) in nose compartment with external door on port side. Cabin heated and ventilated. Windscreen defrosters standard.

SYSTEMS: Electrohydraulic system for landing gear actuation. Electrical system powered by two 28 V 85 A alternators; 24 V 19 Ah battery. Optional pneumatic de-icing boots on wing, tailplane and fin leading-edges. Oxygen system optional.

AVIONICS: *Comms:* Dual Garmin GNS-430 com/nav/GS/GPS; GMA-340 audio panel with marker beacon receiver and intercom; GTX-320 transponder; Telex 100T microphone and Airman 760 headset.

Radar: Provision in nose and on instrument panel for weather radar installation.

Flight: Dual Garmin GNS-430; Meggitt ST-361 ADI; ST-180 slaved HSI; dual GI-106A VOR/LOC/GS/GPS indicators; AR-850 altitude encoder; Meggitt System 55 dual-axis FCS.

Instrumentation: Piper true airspeed indicator, magnetic compass, sensitive altimeter, ADI, HSI, rate of turn indicator, rate of climb indicator, illuminated OAT gauge, digital ammeter, annunciator panel with push-to-test, recording tachometer, dual manifold pressure gauges, three-way oil pressure, temperature and CHT gauge, dual fuel quantity gauges and vacuum gauge with warning indicator. Co-pilot's 75 mm (3 in) instrument and United 5035-P40 encoding altimeter panel optional.

EQUIPMENT: Standard equipment as listed for Saratoga II HP, plus emergency landing gear extension system, cabin dome light, pilot's storm window, sun visors with power setting table and checklist, chemical corrosion protection, flush fuel caps and nose gear safety mirror. Piper Aire air conditioning system, built-in oxygen system and foreign certification gross weight kit optional.

DIMENSIONS, EXTERNAL:
Wing span	11.85 m (38 ft 10¾ in)
Wing chord, constant	1.60 m (5 ft 3 in)
Wing aspect ratio	7.3
Length overall	8.72 m (28 ft 7½ in)
Height overall	3.02 m (9 ft 10¾ in)
Tailplane span	4.14 m (13 ft 6¾ in)
Wheel track	3.38 m (11 ft 1 in)
Wheelbase	2.13 m (7 ft 0 in)
Propeller diameter	1.93 m (6 ft 4 in)
Distance between propeller centres	3.80 m (12 ft 5½ in)
Cabin door (stbd, fwd): Height	0.89 m (2 ft 11 in)
Width	0.91 m (3 ft 0 in)
Cabin door (port, rear): Height	0.72 m (2 ft 4½ in)
Width	0.71 m (2 ft 4 in)
Baggage door (stbd, rear): Height	0.52 m (1 ft 8½ in)
Width	0.66 m (2 ft 2 in)
Baggage door (port, fwd): Height	0.53 m (1 ft 9 in)
Width	0.61 m (2 ft 0 in)

DIMENSIONS, INTERNAL:
Cabin (incl flight deck): Length	3.15 m (10 ft 4¼ in)
Max width	1.24 m (4 ft 0¾ in)
Max height	1.07 m (3 ft 6 in)
Volume	5.5 m³ (195 cu ft)
Forward baggage compartment	0.43 m³ (15.3 cu ft)
Rear baggage compartment	0.49 m³ (17.3 cu ft)

AREAS:
Wings, gross	19.39 m² (208.7 sq ft)
Ailerons, incl tab (total)	1.17 m² (12.60 sq ft)
Trailing-edge flaps (total)	1.94 m² (20.84 sq ft)
Fin	1.14 m² (12.32 sq ft)
Rudder, incl tab	0.71 m² (7.62 sq ft)
Horizontal tail surfaces (total)	3,60 m² (38.74 sq ft)

WEIGHTS AND LOADINGS:
Weight empty, equipped	1,548 kg (3,413 lb)
Max T-O weight	2,154 kg (4,750 lb)
Max ramp weight	2,165 kg (4,773 lb)
Max zero-fuel weight	2,031 kg (4,479 lb)
Max landing weight	2,047 kg (4,513 lb)
Max wing loading	111.1 kg/m² (22.76 lb/sq ft)
Max power loading	6.57 kg/kW (10.80 lb/hp)

PERFORMANCE:
Max level speed	205 kt (379 km/h; 236 mph)
Cruising speed: at 3,050 m (10,000 ft):	
at max cruise power	182 kt (337 km/h; 209 mph)
at normal cruise power	176 kt (326 km/h; 202 mph)
at 5,640 m (18,500 ft):	
at max cruise power	197 kt (365 km/h; 227 mph)
at normal cruise power	190 kt (352 km/h; 219 mph)
Stalling speed, flaps down	61 kt (113 km/h; 71 mph)
Max rate of climb at S/L	446 m (1,462 ft)/min
Rate of climb at S/L, OEI	76 m (250 ft)/min
Max certified altitude	7,620 m (25,000 ft)
Service ceiling, OEI	5,030 m (16,500 ft)
T-O run	349 m (1,145 ft)
T-O to 15 m (50 ft)	521 m (1,710 ft)
Landing from 15 m (50 ft)	665 m (2,180 ft)
Landing run	427 m (1,400 ft)

Range at long-range cruise power, incl allowance for taxi, T-O, climb and 45 min reserves:
- at 3,050 m (10,000 ft)   812 n miles (1,503 km; 934 miles)
- at 4,575 m (15,000 ft)   826 n miles (1,529 km; 950 miles)
- at 5,640 m (18,500 ft)   819 n miles (1,516 km; 942 miles)

UPDATED

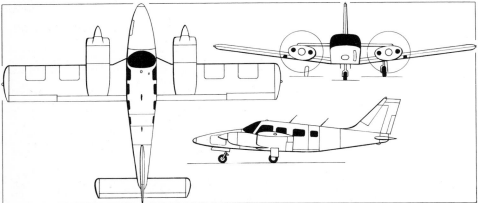

Piper PA-34-220T Seneca V six-seat twin (James Goulding/Jane's)

Piper PA-34-220T Seneca V (Paul Jackson/Jane's)

PIPER—AIRCRAFT: USA 725

PIPER PA-44-180 SEMINOLE

TYPE: Four-seat utility twin.
PROGRAMME: Prototype first flown May 1976; production version announced 21 February 1978; two versions, Seminole and Turbo Seminole, produced until 1982; normally aspirated Seminole restored to production in 1988, suspended in 1990 and restored again in 1995.
CUSTOMERS: Four delivered in 1995, eight in 1996, six in 1997, four in 1998, six in 1999 and 11 in 2000.
  Manufacture of all PA-44 variants reached 512 in early 2000. Many used in twin-conversion and IFR training. Recent customers include Westwind Aviation Academy of Phoenix, Arizona, which took delivery of two during 1999.
COSTS: Standard, equipped US$356,600 (2000).
DESIGN FEATURES: Twin-engined derivation of PA-28 Arrow (which see). Constant-chord low wing and T tail; leading-edge gloves inboard of engines; sweptback fin with small fillet.
  Wing section NACA 65_2-415; thickness/chord ratio 15 per cent; dihedral 7°; incidence 2°; twist 3°; washout −1°.
FLYING CONTROLS: Manual. Ailerons, rudder and all-moving tailplane with full-span anti-servo tab; rudder tab; four-position manually operated flaps.
STRUCTURE: Conventional light alloy, with semi-monocoque fuselage; single-spar wings.
LANDING GEAR: Hydraulically retractable tricycle type. Free-fall emergency extension system. Piper oleo-pneumatic shock-absorbers. Mainwheels and tyres size 6.00-6 (8 ply), with tubes. Steerable nosewheel with tyre size 5.00-5 (6 ply), with tube. Dual toe-operated high-capacity disc brakes. Heavy-duty brakes and tyres optional.
POWER PLANT: Two 134 kW (180 hp) Textron Lycoming flat-four counter-rotating engines (one O-360-A1H6 and one LO-360-A1H6), each driving a Hartzell two-blade constant-speed fully feathering metal propeller. One bladder-type fuel tank in each engine nacelle, with total capacity of 416 litres (110 US gallons; 91.6 Imp gallons), of which 409 litres (108 US gallons; 89.9 Imp gallons) are usable. Refuelling point on upper surface of each nacelle. Oil capacity 11.5 litres (3.0 US gallons; 2.5 Imp gallons).
ACCOMMODATION: Cabin seats four in two pairs of individual seats. Dual controls standard. Emergency exit on port side. Pilot's storm window. Baggage compartment at rear of cabin, capacity 91 kg (200 lb). Accommodation heated and ventilated. Windscreen defrosters.
SYSTEMS: Electrohydraulic system for landing gear actuation and brakes. Electrical system includes two engine-driven 14 V 60 A alternators and 12 V 35 Ah battery. Janitrol combustion heater of 45,000 BTU capacity. Dual vacuum systems standard.
AVIONICS: Standard and optional Advanced Training Group (IFR) packages by Garmin, as described.
  Comms: Standard: Garmin GNS-430 com/nav/IFR GPS; GTX-320 transponder; GMA-340 audio panel with marker beacon receiver and four-position intercom; Telex 100T noise-cancelling microphone and Airman 760 headset.
  Flight: Standard: Garmin GNS-430 com/nav/IFR GPS; Narco AR-850 altitude reporter. Advanced IFR training group: as above, plus second GNS 430; Meggitt ST-180 slaved compass system; and RCR-650A ADF with IND-650A indicator. Optional equipment includes ST-361 ADI; ST-360 altitude select/alerter; DME-450; Meggitt System 55 dual-axis autopilot; and Meggitt electric trim (standard with autopilot).
  Instrumentation: Garmin GI-106A VOR/LOC/GPS indicator. Advanced IFR training group adds second GI-106A.
EQUIPMENT: Metal instrument panel, engine hour recorders, alternate static source, heated pitot head, instrument panel lights and overhead blue lighting, avionics dimming, dome light, navigation lights, landing/taxying light, wingtip strobe lights, pilot's and co-pilot's vertically adjustable seats in fabric and vinyl, with optional lumbar support, two reclining rear passenger seats, vinyl cabin side panels, crew armrests, tinted windscreen and windows, pilot's storm window, 'Quietised' soundproofing, tiedown points, jack pads, nose gear safety mirror, external power receptacle, static discharge wicks, and Du Pont Imron polyurethane exterior paint in white base colour with two contrasting trim stripes. Metallic paint optional.

DIMENSIONS, EXTERNAL:
Wing span	11.75 m (38 ft 6½ in)
Wing aspect ratio	8.1
Length overall	8.41 m (27 ft 7¼ in)
Height overall	2.59 m (8 ft 6 in)
Tailplane span	3.05 m (10 ft 0 in)
Wheel track	3.21 m (10 ft 6½ in)
Wheelbase	2.56 m (8 ft 4¾ in)
Propeller diameter	1.88 m (6 ft 2 in)
Distance between propeller centres	3.86 m (12 ft 7¼ in)
Cabin door (stbd): Height	0.89 m (2 ft 11 in)
Width	0.91 m (3 ft 0 in)
Baggage door: Height	0.51 m (1 ft 8 in)
Width	0.56 m (1 ft 10 in)

DIMENSIONS, INTERNAL:
Cabin (instrument panel to rear bulkhead):	
Length	2.46 m (8 ft 1 in)
Max width	1.05 m (3 ft 5½ in)
Max height	1.25 m (4 ft 1 in)
Volume	3.0 m³ (106 cu ft)
Baggage compartment volume	0.74 m³ (26.0 cu ft)

AREAS:
Wings, gross	17.08 m² (183.8 sq ft)
Ailerons (total)	1.13 m² (12.12 sq ft)
Trailing-edge flaps (total)	1.36 m² (14.60 sq ft)
Fin	1.26 m² (13.56 sq ft)
Rudder, incl tab	0.76 m² (8.21 sq ft)
Horizontal tail surfaces (total)	2.27 m² (24.40 sq ft)

WEIGHTS AND LOADINGS:
Weight empty, equipped	1,179 kg (2,600 lb)
Max T-O weight	1,723 kg (3,800 lb)
Max wing loading	100.9 kg/m² (20.67 lb/sq ft)
Max power loading	6.42 kg/kW (10.55 lb/hp)

PERFORMANCE:
Max level speed	168 kt (311 km/h; 193 mph)
Cruising speed:	
at 75% power	162 kt (300 km/h; 186 mph)
at 65% power	157 kt (291 km/h; 181 mph)
Stalling speed: flaps up	57 kt (106 km/h; 66 mph) IAS
flaps down	55 kt (102 km/h; 63 mph) IAS
Max rate of climb at S/L	408 m (1,340 ft)/min
Rate of climb at S/L, OEI	65 m (212 ft)/min
Service ceiling	4,575 m (15,000 ft)
Service ceiling, OEI	1,155 m (3,800 ft)
T-O to 15 m (50 ft)	671 m (2,200 ft)
Landing from 15 m (50 ft)	454 m (1,490 ft)
Cruising range at 55% power, 45 min reserves	
	770 n miles (1,426 km; 886 miles)

UPDATED

Piper PA-44-180 Seminole used for training by Embry Riddle Aeronautical University *(Paul Jackson/Jane's)*
2001/0100492

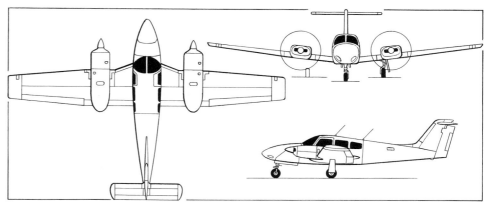
Piper PA-44-180 Seminole four-seat light twin *(James Goulding/Jane's)*
2000/0100460

Instrument panel of Piper Seminole
(Paul Jackson/Jane's) 2000/0075915

PIPER PA-46-350P MALIBU MIRAGE

TYPE: Business prop.
PROGRAMME: Prototype Malibu first flew 30 November 1979; FAA certification of original PA-46-310P Malibu (TSIO-520 engine) received September 1983; production deliveries began December 1983; 404 built before replaced by PA-46-350P Malibu Mirage October 1988. Production, temporarily suspended in 2000 to enable the company to concentrate on launching the turboprop Malibu Meridian; resumption then planned for July 2001.
CURRENT VERSIONS: **Malibu Mirage:** *As described.* Improvements introduced on 1995 model include pilot's heated glass windscreen, inflatable lumbar support on

Piper PA-46-350P Malibu Mirage pressurised, single piston-engined aircraft *(Paul Jackson/Jane's)*
2001/0100493

Jane's All the World's Aircraft 2001-2002

pilot/co-pilot's seats, colour co-ordinated control wheels and restyled interior trim, cabinetry and seats. Strengthened wing structure of Malibu Meridian (which see) introduced as standard from 1999, affording 18 kg (40 lb) increase in maximum take-off weight.

Malibu Meridian: Described separately.

CUSTOMERS: Total of 40 delivered in 1995, 57 in 1996, 54 in 1997, 55 in 1998, 61 in 1999 and 63 in 2000.

COSTS: Standard, equipped US$869,800 (2000).

DESIGN FEATURES: High-speed, long-range Piper single, with streamlined appearance and moderately high-aspect ratio wing, which is low mounted and tapered; mid-tailplane, also tapered; sweptback fin; wide track landing gear.

Wing section NASA 23016 at root, 23009 at tip; dihedral 4° 30'; thickness/chord ratio 16 per cent; twist 2° 57'; incidence 3° 38'.

FLYING CONTROLS: Conventional and manual. Horn-balanced elevators and rudder; stainless steel control cables. Electronic trim tab in elevator; electrically operated trailing-edge flaps. Precise Flight speed brakes standard.

STRUCTURE: Cantilever high-aspect ratio all-metal wings; light alloy fuselage, fail-safe construction in pressurised area; light alloy tail surfaces.

LANDING GEAR: Hydraulically retractable tricycle type with single wheel on each unit; main units retract inward into wingroots, nosewheel rearward, rotating 90° to lie flat under baggage compartment. Mainwheel size 6.00-6 (8 ply), pressure 3.80 bars (55 lb/sq in); nosewheel 5.00-6 (6 ply) pressure 3.45 bars (50 lb/sq in). Toe-operated brakes.

POWER PLANT: One 261 kW (350 hp) Textron Lycoming TIO-540-AE2A turbocharged and intercooled flat-six engine, driving a Hartzell three-blade constant-speed propeller with polished spinner. Composites (Kevlar) propeller replaced metal from 1998. Fuel system capacity 462 litres (122 US gallons; 101.6 Imp gallons), of which 454 litres (120 US gallons; 100 Imp gallons) are usable. Oil capacity 11.5 litres (3.0 US gallons; 2.5 Imp gallons).

ACCOMMODATION: Pilot and five passengers in pressurised, heated and air conditioned cabin; dual controls standard; front two occupants have vertical, fore-and-aft adjusting and reclining leather seats with inflatable lumbar supports, stowaway armrests, inertia reel shoulder harnesses and map-holders. Leather reclining passenger seats in club arrangement with stowaway armrests and inertia reel shoulder harnesses; unpressurised baggage compartment in nose, and pressurised space at rear of cabin. Door with integral steps on port side aft of wing.

SYSTEMS: Pressurisation, maximum differential 0.38 bar (5.5 lb/sq in), to provide a cabin altitude of 2,400 m (7,900 ft) to a height of 7,620 m (25,000 ft). Hydraulic system pressure 107 bars (1,550 lb/sq in). Dual engine-driven vacuum pumps standard. Split bus electrical system has two 28 V/70 A alternators; 24 V 10 Ah battery; full icing protection standard. Optional electrically heated pilot's windscreen and fuel management system.

AVIONICS: Standard Garmin New Generation IFR avionics package, as detailed.
 Comms: Dual GNS-430 com/nav/IFR GPS with MFDs; GTX-320 transponder (second optional); GMA-340 audio panel with marker beacon receiver and intercom; Telex 100T noise-cancelling microphone and Airman 760 headset.
 Radar: RDR-2000 vertical profile colour weather radar in wing-mounted pod. Goodrich WS-1000+ Stormscope optional.
 Flight: Garmin GNS-430 com/nav/IFR GPS; ST-361 ADI; ST-180 HSI with slaved compass system; ST-360 altitude select/alerter; United 5035P-P40 altitude encoder; Meggitt System 55 three-axis autopilot with electric trim, turn indicator and yaw damper; co-pilot's longitudinal electric trim button. Options include single or dual RCR-650A ADF with IND-650A indicators; DME 450; Honeywell KI 229 RMI (exchange for IND-650A): and KRA-10A radar altimeter.
 Instrumentation: Dual MFDs with GNS-430 installation; dual GI-106A VOR/LOC/GS/GPS indicators.

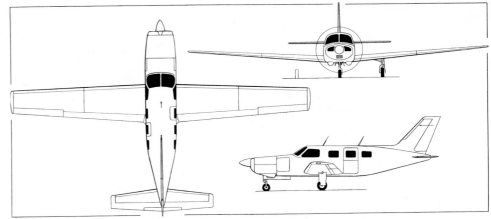

Piper PA-46-350P Malibu Mirage (Textron Lycoming TIO-540-AE2A) *(Dennis Punnett/Jane's)*

EQUIPMENT: Standard equipment includes heated lift detector; stall warning computer and horn; digital ammeter/voltmeter; six-channel CHT monitoring system with selectable cylinder readout; gyro air filter; alternate static source; heated pitot head; nosewheel light; wingtip taxying light; navigation lights, wingtip strobe lights; cockpit dome lights; solid-state dimming landing gear position and instrument panel lights; seven cabin overhead lights; PPG pilot's heated windscreen; windscreen defrosters; pilot's opening storm window; pilot's relief tube; supplemental electric heater; emergency oxygen system; leather sidewalls; stowaway executive writing table; forward refreshment/entertainment centres; super 'Quietised' soundproofing with inner passenger windows; halon fire extinguisher; Truax static discharge wicks; chemical corrosion protection; and Du Pont Imron polyurethane exterior paint in single- or two-tone base colour with graphics striping in choice of three trim colours. Optional equipment includes 'Infinity' paint scheme and stainless steel cowling fasteners.

DIMENSIONS, EXTERNAL:
Wing span	13.11 m (43 ft 0 in)
Wing aspect ratio	10.6
Length overall	8.81 m (28 ft 10¾ in)
Height overall	3.44 m (11 ft 3½ in)
Tailplane span	4.42 m (14 ft 6 in)
Wheel track	3.75 m (12 ft 3½ in)
Wheelbase	2.44 m (8 ft 0 in)
Propeller diameter	2.03 m (6 ft 8 in)
Passenger door (port, rear): Height	1.17 m (3 ft 10 in)
Width	0.61 m (2 ft 0 in)
Baggage door (port, nose): Height	0.58 m (1 ft 11 in)
Width	0.48 m (1 ft 7 in)

DIMENSIONS, INTERNAL:
Cabin (instrument panel to rear pressure bulkhead):
Length	3.76 m (12 ft 4 in)
Max width	1.26 m (4 ft 1½ in)
Max height	1.19 m (3 ft 11 in)
Baggage compartment volume:	
nose	0.37 m³ (13.0 cu ft)
rear cabin	0.57 m³ (20.0 cu ft)

AREAS:
Wings, gross	16.26 m² (175.0 sq ft)

WEIGHTS AND LOADINGS:
Weight empty, equipped	1,416 kg (3,121 lb)
Baggage capacity: nose	45 kg (100 lb)
rear cabin	45 kg (100 lb)
Max T-O weight	1,968 kg (4,340 lb)
Max ramp weight	1,976 kg (4,358 lb)
Max landing weight	1,870 kg (4,123 lb)
Max zero-fuel weight	1,870 kg (4,123 lb)
Max wing loading	121.1 kg/m² (24.80 lb/sq ft)
Max power loading	7.55 kg/kW (12.40 lb/hp)

PERFORMANCE:
Max level speed at mid-cruise weight
  220 kt (407 km/h; 253 mph)
Cruising speed at optimum altitude, mid-cruise weight, high-speed cruise power 213 kt (394 km/h; 245 mph)
Stalling speed, flaps and wheels down
  58 kt (108 km/h; 67 mph)
Max rate of climb at S/L 372 m (1,220 ft)/min
Max certified altitude 7,620 m (25,000 ft)
T-O run 332 m (1,090 ft)
T-O to 15 m (50 ft) 637 m (2,090 ft)
Landing from 15 m (50 ft) 597 m (1,960 ft)
Landing run 311 m (1,020 ft)
Range with max fuel, allowances for start, T-O, climb and descent, plus 45 min reserves, at optimum altitude, normal cruise power
  1,055 n miles (1,953 km; 1,214 miles)

UPDATED

PIPER PA-46-500TP MALIBU MERIDIAN

TYPE: Business turboprop.

PROGRAMME: Launched at 1997 National Business Aviation Association Convention in Dallas, Texas, where a full-scale fuselage mockup was displayed; prototype (N400PT, converted from second Malibu Mirage) rolled out 13 August 1998; first flight 21 August 1998. Static test airframe and three further certification flight test aircraft (N403MM, first flown in July 1999, N402MM first flown 27 August 1999 and N401MM first flown in September 1999) built on production tooling. N401MM was dedicated to stability and autopilot testing and ice shape testing; N402MM, first with production standard interior and exterior paint finish, conducted performance and avionics testing and certification for flight into known icing, as well as serving as marketing demonstrator, N403MM was responsible for systems and power plant testing, high-speed flight testing and flutter testing. Public debut (N402MM) at the NBAA Convention in Atlanta, Georgia, in October 1999. First flight of production aircraft (c/n 003/N375RD) 30 June 2000. FAA certification achieved 27 September 2000. Production targets are 35 in 2000, rising to more than 100 per year in 2001 and 2002.

CUSTOMERS: Total of 239 orders held by October 2000, of which 18 delivered by 31 December 2000. First delivery (N375RD) to Richard Dumais of Richardson, Texas, in November 2000.

COSTS: US$1.5 million typically equipped (2000).

Description generally as for Malibu Mirage, except that below.

DESIGN FEATURES: Strengthened wing, incorporating wingroot leading-edge gloves to increase area and reduce stalling speed; strengthened tail surfaces; and 37 per cent increase in area of horizontal stabiliser.

FLYING CONTROLS: Flight-adjustable trim tab in rudder. Ground-adjustable tab on each flap.

POWER PLANT: One Pratt & Whitney Canada PT6A-42A turboprop, thermodynamic rating 901 kW (1,209 shp), flat rated at 373 kW (500 shp) for take-off and 298 kW (400 shp) maximum continuous, driving a Hartzell four-blade constant-speed reversible propeller. Usable fuel capacity 644 litres (170 US gallons; 141.5 Imp gallons).

Third prototype turboprop Piper Malibu Meridian *(Carl Miller/Piper)*

Piper Malibu Meridian EFIS instrument panel

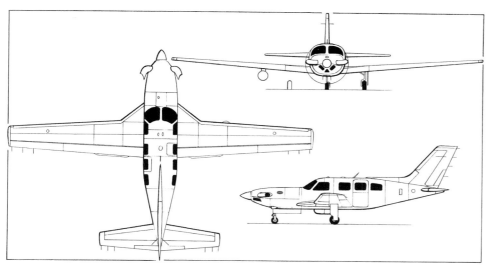

Piper Malibu Meridian (Pratt & Whitney Canada PT6A turboprop) *(James Goulding/Jane's)* 2000/0075961

Piper Malibu Meridian cabin with optional entertainment centre 2001/0100470

ACCOMMODATION: Pilot and five passengers in standard club layout; passenger capacity reduced to four with optional entertainment centre comprising beverage cooler, storage cabinets, AM/FM/CD stereo and provision for VCR with flat-panel monitor. Fully automatic bleed-air conditioning and temperature control systems.

SYSTEMS: Upgraded electrical system, with 200 A starter/generator and 130 A standby alternator.

AVIONICS: Standard Garmin International and Meggitt package with dual Garmin GNS 530 nav/com/GPS as core system.
Comms: Garmin transceiver with 8.33 kHz spacing; GMA-340 audio panel; GTX-327 transponder.
Radar: Honeywell RDR 2000 colour weather radar.
Flight: Meggitt System 550 autopilot/flight director, fully coupled to GNS 530; air data, attitude, heading and reference system (ADAHRS) provides digital readout of pitch and roll attitudes, roll and yaw rates, altitude, rate of altitude change, airspeed and heading.
Instrumentation: Meggitt Avionics engine instrument display system (EIDS) standard. Optional dual MAGIC EFDS fit comprising pilot's and co-pilot's colour flat panel LCD primary flight display and navigation display driven by dual air data attitude heading reference systems (ADAHRS).

DIMENSIONS, EXTERNAL: As for Malibu Mirage except:
Length overall 9.02 m (29 ft 7¼ in)
Height overall 3.45 m (11 ft 4 in)
Propeller diameter 2.095 m (6 ft 10½ in)
DIMENSIONS, INTERNAL: As for Malibu Mirage
AREAS:
Wings, gross 17.00 m² (183.0 sq ft)
WEIGHTS AND LOADINGS:
Weight empty, equipped 1,471 kg (3,243 lb)
Max T-O weight 2,200 kg (4,850 lb)
Max ramp weight 2,219 kg (4,892 lb)
Max wing loading 129.4 kg/m² (26.50 lb/sq ft)
Max power loading 7.38 kg/kW (12.13 lb/shp)

PERFORMANCE (estimated):
Max cruising speed at 9,150 m (30,000 ft) at mid-cruise weight 262 kt (485 km/h; 302 mph)
Max rate of climb at S/L 531 m (1,741 ft)/min
Max certified altitude 9,150 m (30,000 ft)
T-O run 467 m (1,530 ft)
T-O to 15 m (50 ft) 724 m (2,375 ft)
Landing from 15 m (50 ft) 595 m (1,950 ft)
Landing run 311 m (1,020 ft)
Max range at 9,150 m (30,000 ft), max cruise power, mid-cruise weight, 45 min reserves
1,070 n miles (1,981 km; 1,231 miles)
Endurance 4 h 22 min
UPDATED

PIPER NEW PROJECTS

At the NBAA Convention at New Orleans in October 2000, New Piper Aircraft CEO Chuck Suma announced that the company is "actively developing new platforms at the lighter end of our product line" that will "embrace advanced propulsion technologies, aerodynamic refinements and cutting edge avionics capabilities". Among these are expected to be FADEC-equipped piston engines across the company's range and a new model to fill the market gap between the Seneca and Malibu Mirage and Meridian. New Piper would also consider a joint venture to develop a new turboprop aircraft bigger than the Meridian, Suma said.

NEW ENTRY

PRECISION TECH

PRECISION TECH AIRCRAFT INC
155E Hangar 33, Highway 61 SE, Cartersville, Georgia 30120
Tel: (+1 770) 607 40 09
Fax: (+1 770) 386 63 55
e-mail: info@fergy.net
Web: http://www.fergy.net

Precision Tech took over manufacture and distribution rights for the Ferguson F-II from its designer during 1999.
NEW ENTRY

PRECISION TECH FERGUSON F-IIB FERGY
TYPE: Side-by-side ultralight kitbuilt.
PROGRAMME: Designed by Bill Ferguson; awarded Best New Design at Sun 'n' Fun 91.
CURRENT VERSIONS: **Ferguson F-II**: Initial version, no longer produced.
F-IIB Fergy: Current production model, *as described*.
CUSTOMERS: At least 25 sold by early 2000.
COSTS: Kit US$10,900 (2000).
DESIGN FEATURES: Pod-and-boom configuration; high wing and low tailplane. Wings and tail fold for transportation. Quoted build time 250 hours.
FLYING CONTROLS: Conventional and manual. Dual controls. Full-span flaperons.
STRUCTURE: Fuselage of 4130 chromoly steel tubing with glass fibre cowling; 6063-T6 aluminium tube tailboom and wing spars; wing ribs and tail 6061-T6 aluminium tubing with chromoly steel brackets. Fabric covering; aluminium wing struts.
LANDING GEAR: Tailwheel type; fixed. Brakes on mainwheels; optional speed fairings.
POWER PLANT: Pusher engine driving two-blade (optional three-blade) propeller mounted above wing; engines in the range 33.6 to 48.5 kW (45 to 65 hp) recommended. Fuel capacity 47.3 litres (12.5 US gallons; 10.4 Imp gallons).

Precision Tech F-IIB Fergy kitplane *(Paul Jackson/Jane's)* 2001/0089508

DIMENSIONS, EXTERNAL:
Wing span 8.99 m (29 ft 6 in)
Length overall 6.71 m (22 ft 0 in)
Height overall 1.73 m (5 ft 8 in)
AREAS:
Wings, gross 13.01 m² (140.0 sq ft)
WEIGHTS AND LOADINGS:
Weight empty 181 kg (400 lb)
Max T-O weight 453 kg (1,000 lb)

PERFORMANCE (48.5 kW; 65 hp engine):
Never-exceed speed (V_{NE}): 86 kt (161 km/h; 100 mph)
Normal cruising speed range
 43-70 kt (80-129 km/h; 50-80 mph)
Stalling speed, power off:
 flaperons up 30 kt (55 km/h; 34 mph)
 flaperons down 25 kt (45 km/h; 28 mph)
Max rate of climb at S/L 366 m (1,200 ft)/min
T-O run 53 m (175 ft)
NEW ENTRY

PRIVATE EXPLORER

PRIVATE EXPLORER INC
120 E South 6th, PO Box 94, Grangeville, Idaho 83530
Tel: (+1 208) 839 27 21
Fax: (+1 208) 839 25 50
e-mail: PvExplorer@aol.com
DIRECTORS: Hubert DeChevigny
Dean Wilson

First product of this partnership was the twin-engined DeChevigny/Wilson Explorer, last described in the 1994-95 *Jane's*. Design is by Dean Wilson, originator of the Avid Flyer and other light aircraft.
The Private Explorer company has been formed to market a single-engine aircraft of similar concept to its larger predecessor. See also Explorer Aviation Ellipse earlier in this section; the two companies have a joint address.
VERIFIED

PRIVATE EXPLORER
TYPE: Light utility transport.
PROGRAMME: Prototype built in production jigs; first flew (N7065H) January 1998; public debut at Sun 'n' Fun, April 1998.
COSTS: Complete airframe kit US$96,820 (1998).
DESIGN FEATURES: Rugged construction with wide, deep fuselage to provide living accommodation. High wing with A-frame main bracing struts and N-shape auxiliary struts

USA: AIRCRAFT—PRIVATE EXPLORER to PROSPORT

each side; tailplane braced to fin by A-struts, each side. Kit quoted build time 1,000 hours.

FLYING CONTROLS: Conventional and manual. Horn-balanced tail surfaces; electric elevator and rudder trim.

STRUCTURE: Welded 4130 steel tube fuselage, wooden (mahogany and spruce) wing with Ceconite covering.

LANDING GEAR: Fixed, tricycle type; 8.50×10 tyres throughout; provision for addition of skis; inbuilt fittings for twin floats. Main gear attached to two heat-treated steel 'wishbone' springs each side, giving additional strength on rough terrain. Cleveland brakes.

POWER PLANT: Prototype has one 175 kW (235 hp) Textron Lycoming O-540 flat-six, driving a two-blade, fixed-pitch propeller. Optionally, 224 kW (300 hp) IO-540 with constant-speed propeller. Fuel in wing tanks, total 318 litres (84.0 US gallons; 70.0 Imp gallons).

ACCOMMODATION: Optimised for two occupants; dual controls standard; further two seats optional; internal fittings can include 203 × 177 cm (80 × 54 in) bed. Door forward of wing, each side, forward opening; cabin door, port side, rear of wing, forward opening.

DIMENSIONS, EXTERNAL:
Wing span 14.40 m (47 ft 3 in)
Wing chord, constant 1.92 m (6 ft 3¾ in)
Wing aspect ratio 7.7
Length overall 9.25 m (30 ft 4 in)
Height overall 3.38 m (11 ft 1 in)
Propeller diameter 2.54 m (8 ft 4 in)
Propeller ground clearance 0.41 m (1 ft 4 in)

DIMENSIONS, INTERNAL:
Cabin (excl cockpit): Length 3.35 m (11 ft 0 in)
Max width 2.01 m (6 ft 7 in)
Max height 1.93 m (6 ft 4 in)
Bed: Length 1.98 m (6 ft 6 in)
Width 1.37 m (4 ft 6 in)
Cockpit: Length 1.22 m (4 ft 0 in)
Max width 1.68 m (5 ft 6 in)
Max height 1.40 m (4 ft 7 in)
Volume: cabin 11.6 m^3 (408 cu ft)
cockpit 4.3 m^3 (152 cu ft)

AREAS:
Wings, gross 26.96 m^2 (290.2 sq ft)

WEIGHTS AND LOADINGS:
Weight empty: O-540 973 kg (2,145 lb)
IO-540 1,049 kg (2,313 lb)
Max T-O weight 1,860 kg (4,102 lb)
Max wing loading 69.0 kg/m^2 (14.14 lb/sq ft)
Max power loading (O-540):
A 10.62 kg/kW (17.45 lb/hp)
B 9.52 kg/kW (15.64 lb/hp)

PERFORMANCE (A: O-540 engine, B: IO-540 engine):
Never-exceed speed (V_{NE}): A, B
146 kt (270 km/h; 168 mph)
Max level speed: A 110 kt (204 km/h; 127 mph)
B 117 kt (217 km/h; 135 mph)
Cruising speed: A 97 kt (180 km/h; 112 mph)
B 109 kt (201 km/h; 125 mph)
Stalling speed: A, B 42 kt (78 km/h; 48 mph)
Max rate of climb at S/L: A 253 m (829 ft)/min
B 355 m (1,166 ft)/min
Service ceiling: A 4,420 m (14,500 ft)
B 5,790 m (19,000 ft)
T-O run: A 255 m (837 ft)
B 174 m (570 ft)
Range with max fuel:
A 681 n miles (1,261 km; 784 miles)
B 575 n miles (1,065 km; 662 miles)
Endurance: A 7 h 0 min
B 5 h 18 min
g limits: A, B +3.8/−2.25
UPDATED

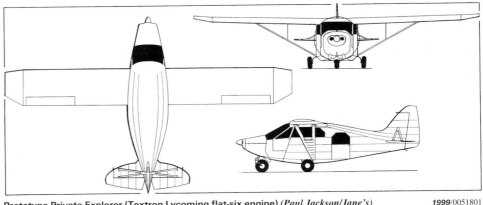

Prototype Private Explorer (Textron Lycoming flat-six engine) *(Paul Jackson/Jane's)*

Prototype Private Explorer 'flying motor-home'

PROSPORT

PROSPORT AVIATION
PO Box 3429, Tega Cay, South Carolina 29715
Tel: (+1 803) 802 01 98
Fax: (+1 803) 802 01 96
e-mail: eagle@sportlite103.com
Web: www.sportlite103.com

Prosport also markets the Sportlite 103 ultralight.
NEW ENTRY

PROSPORT FREEBIRD II

TYPE: Side-by-side kitbuilt.
CUSTOMERS: Over 50 under construction and 33 flying by mid-2000.
COSTS: US$9,995 (2000) excluding engine.
DESIGN FEATURES: Wings removable for storage.
FLYING CONTROLS: Conventional and manual; flaps. Large horn balance on rudder.
STRUCTURE: Fabric-covered steel tubing and composites nose skin. Strut-braced wings and tailplane. Wing-to-tailplane bracing struts augment slim rear fuselage.
LANDING GEAR: Tricycle type; fixed. Semi-recessed nosewheel.
POWER PLANT: One 37.0 kW (49.6 hp) Rotax 503 UL-2V two-cylinder engine driving a wooden fixed-pitch pusher propeller. Fuel capacity 37.8 litres (10.0 US gallons; 8.3 Imp gallons).
ACCOMMODATION: Optional enclosed cockpit with provision for heater.
SYSTEMS: Optional electric starter.

DIMENSIONS, EXTERNAL:
Wing span 8.84 m (29 ft 0 in)
Length overall 5.33 m (17 ft 6 in)
Height overall 1.37 m (4 ft 6 in)

AREAS:
Wings, gross 12.26 m^2 (132.0 sq ft)

WEIGHTS AND LOADINGS:
Weight empty 175 kg (385 lb)
Max T-O weight 394 kg (870 lb)

PERFORMANCE (enclosed cockpit):
Max level speed 70 kt (129 km/h; 80 mph)
Normal cruising speed at 75% power
61 kt (113 km/h; 70 mph)
Stalling speed, flaps down 28 kt (52 km/h; 32 mph)
Max rate of climb at S/L 244 m (800 ft)/min
T-O run 38 m (125 ft)
Landing run 46 m (150 ft)
Range at 75% power 173 n miles (321 km; 200 miles)
g limits +6/−3
UPDATED

Prosport Freebird II *(Geoffrey P Jones)*

PULSAR

PULSAR AIRCRAFT CORPORATION
4233 North Santa Anita Avenue, Suite 5, El Monte Airport, El Monte, California 91731
Tel: (+1 626) 443 10 19
Fax: (+1 626) 443 13 11
e-mail: info@pulsaraircraft.com
Web: http://www.pulsaraircraft.com
PRESIDENT: Solly Melyon

Pulsar Aircraft Corporation acquired rights to Pulsar series of light aircraft from Skystar in July 1999 and announced the Super Pulsar 100, based on the Pulsar III, at that time. It has also obtained rights to the former Tri-R KIS and KIS Cruiser, which were due for release in late 2000.

UPDATED

Tailwheel version of earlier Pulsar, now superseded by Super Pulsar *(Paul Jackson/Jane's)* 2001/0103552

PULSAR SUPER PULSAR 100
TYPE: Side-by-side kitbuilt.
PROGRAMME: Designed by Mark Brown. Prototype first flew 3 April 1988 as two-seat version of Aero Designs Star-Lite. Originally with 47.8 kW (64.1 hp) Rotax 582 engine. Until suspended in 1999, in favour of Super Pulsar, parallel main production versions were Pulsar II, with 59.6 kW (79.9 hp) Rotax 912 UL and Pulsar III with 84.6 kW (113.4 hp) turbocharged Rotax 914, giving improved performance. Larger canopy and redesigned cowling with chin intake; 100 per cent mass-balanced controls. Nose- and tailwheel versions of both were available. Programme acquired by SkyStar in late 1996 and by Pulsar in 1999.
CURRENT VERSIONS: **Super Pulsar 100:** Refined version of Pulsar III with repositioned tail unit for improved stability and higher, wider cabin. Prototype (N601SP) first flew August 2000 but damaged 20 August 2000. Tailwheel version under consideration. Quoted build time 1,000 hours.

Description applies to Super Pulsar 100.

Wega: Version of Pulsar II-914 Turbo, certified to JAR-VLA with 600 kg (1,323 lb) maximum T-O weight, marketed by HK Aircraft Technology in Germany, which see.
CUSTOMERS: Approximately 200 of all Pulsar variants flying by late 1999, with a further 300 then under construction.
COSTS: Basic Super Pulsar kit (no engine) US$21,950 (2001).
DESIGN FEATURES: Highly streamlined low-wing configuration, with sweptback fin. Designed for ease of home construction with premoulded components where possible. Wings removable in about 15 minutes for mounting on trailer.
FLYING CONTROLS: Conventional and manual; 60 per cent mass-balanced surfaces operated by rods and cables.
STRUCTURE: GFRP with carbon fibre and foam cores. Wing skins of composites.
LANDING GEAR: Tricycle type; fixed. Hydraulic brakes.
POWER PLANT: One 59.6 kW (79.9 hp) Rotax 912 UL, 84.6 kW (113.4 hp) Rotax 914 UL, 59.7 kW (80 hp) Jabiru 2200 or 89 kW (120 hp) Jabiru 3300 piston engine. Fuel capacity 106 litres (28.0 US gallons; 23.3 Imp gallons).
DIMENSIONS, EXTERNAL:
Wing span　　　　　　　　　　　7.62 m (25 ft 0 in)
Wing aspect ratio　　　　　　　　7.8
Length overall　　　　　　　　　6.10 m (20 ft 0 in)
Height overall　　　　　　　　　1.80 m (5 ft 11 in)
DIMENSIONS, INTERNAL:
Cabin max width　　　　　　　　1.12 m (3 ft 8 in)
AREAS:
Wings, gross　　　　　　　　　　7.43 m² (80.0 sq ft)
WEIGHTS AND LOADINGS (Rotax 914):
Weight empty, equipped　　　　329 kg (725 lb)
Baggage capacity　　　　　　　27 kg (60 lb)
Max T-O weight　　　　　　　589 kg (1,300 lb)
Max wing loading　　　　　　79.3 kg/m² (16.25 lb/sq ft)
Max power loading　　　　　6.98 kg/kW (11.46 lb/hp)
PERFORMANCE (Rotax 914):
Never-exceed speed (V$_{NE}$)　180 kt (334 km/h; 208 mph)
Max cruising speed at 75% power
　　　　　　　　　　　　　174 kt (322 km/h; 200 mph)

Prototype Super Pulsar, pictured while still under construction *(Paul Jackson/Jane's)* 2001/0103551

Econ cruising speed　　　　156 kt (290 km/h; 180 mph)
Stalling speed, power off　　48 kt (89 km/h; 55 mph)
Max rate of climb at S/L　　610 m (2,000 ft)/min
Service ceiling　　　　　　5,330 m (17,500 ft)
T-O run　　　　　　　　　152 m (500 ft)
Landing run　　　　　　　244 m (800 ft)
Range　　　　　738 n miles (1,367 km; 850 miles)
g limits　　　　　　　　　　+4.4/−2.5

UPDATED

PULSAR SPORT 150
TYPE: Side-by-side kitbuilt.
PROGRAMME: First flown 1991 and initially known as Tri-R KIS; name derived from 'keep it simple'. Programme sold to Pulsar in 1999; new owner modifying design with revised canopy and empennage.
CURRENT VERSIONS: **TR-1:** Tricycle landing gear version, *as described*.
TD: Tailwheel version, introduced 1992; powered by 88.0 kW (118 hp) Textron Lycoming O-235-C1B; seven flying by end of 1996.
CUSTOMERS: At least 31 flying by end 1996 (latest data supplied).
COSTS: Basic kit US$18,500, without engine, avionics, propeller, spinner, upholstery, battery, instruments and paint (1998).
DESIGN FEATURES: Conventional low-wing monoplane of composites construction. Upturned wingtips and sharply tapered fin. Quoted build time 1,000 hours.
FLYING CONTROLS: Conventional and manual. Horn-balanced rudder and elevators, latter with trim tab.
STRUCTURE: Constructed of high-temperature epoxy preimpregnated GFRP/CFRP premoulded components, with either Divinycell or honeycomb core. All metal components prewelded or premachined. Aerofoil NACA 63215 slightly modified.
LANDING GEAR: Non-retractable, with single wheels and cantilever main legs; optional speed fairings. Tricycle or tailwheel versions available. Matco wheels and brakes; McCreary 5.00-5 tyres. Tricycle version has steerable nosewheel.
POWER PLANT: One piston engine driving a two- or three-blade propeller; alternatives include 59.7 kW (80 hp) Limbach L 2000; 74.6 kW (100 hp) Fire Wall Forward CAM 100 four-cylinder converted Honda motorcar engine; 88 kW (118 hp) Textron Lycoming O-235-C1B; and 93 kW (125 hp) Teledyne Continental IO-240. Fuel capacity 76 litres (20.0 US gallons; 16.7 Imp gallons); optionally 129 litres (34.0 US gallons; 28.3 Imp gallons) for Lycoming- and Continental-powered versions.
ACCOMMODATION: Two, side by side. Upward-hinged door each side.
DIMENSIONS, EXTERNAL:
Wing span　　　　　　　　　7.01 m (23 ft 0 in)
Wing chord　　　　　　　　　1.17 m (3 ft 10 in)
Wing aspect ratio　　　　　　6.0
Length overall　　　　　　　6.71 m (22 ft 0 in)
Height overall: TR-1　　　　1.83 m (6 ft 0 in)
　　　　　　　TD　　　　　1.89 m (6 ft 2½ in)
Propeller diameter　　　　　1.42 m (4 ft 8 in)
DIMENSIONS, INTERNAL:
Cabin: Length　　　　　　　1.65 m (5 ft 5 in)
　　　Max width　　　　　　1.07 m (3 ft 6 in)
　　　Max height　　　　　　1.02 m (3 ft 4 in)
AREAS:
Wings, gross　　　　　　　　8.18 m² (88.0 sq ft)
WEIGHTS AND LOADINGS (L: Limbach engine, C: Continental engine):
Weight empty: L　　　　　　308 kg (680 lb)
　　　　　　　C　　　　　　367 kg (810 lb)
Max T-O weight: L　　　　　544 kg (1,200 lb)
　　　　　　　　C　　　　　658 kg (1,450 lb)
Max wing loading: L　　　　66.6 kg/m² (13.64 lb/sq ft)
　　　　　　　　C　　　　80.4 kg/m² (16.48 lb/sq ft)
Max power loading: L　　　　9.13 kg/kW (15.00 lb/hp)
　　　　　　　　　C　　　　7.06 kg/kW (11.60 lb/hp)
PERFORMANCE:
Never-exceed speed (V$_{NE}$)　191 kt (354 km/h; 220 mph)
Max level speed: L　　　　135 kt (249 km/h; 155 mph)
　　　　　　　　C　　　　165 kt (306 km/h; 190 mph)
Cruising speed at 75% power:
　L　　　　　　　　　　117 kt (217 km/h; 135 mph)
　C　　　　　　　　　　148 kt (274 km/h; 170 mph)
Stalling speed: L, C　　　46 kt (86 km/h; 53 mph)
Max rate of climb at S/L: L　198 m (650 ft)/min
　　　　　　　　　　　C　　366 m (1,200 ft)/min
Service ceiling: C　　　　5,180 m (17,000 ft)
T-O run: L　　　　　　　305 m (1,000 ft)
　　　　C　　　　　　　229 m (750 ft)
Landing run: C　　　　　396 m (1,300 ft)
Range at cruising speed, standard tankage:
　L　　　　　　521 n miles (965 km; 600 miles)
　C　　　　　　651 n miles (1,205 km; 749 miles)
g limits　　　　　　　　　　+4.4/−2.2

UPDATED

PULSAR CRUISER and SUPER CRUISER
TYPE: Four-seat kitbuilt.
PROGRAMME: First flown 1994, when known as Tri-R Cruiser TR-4. Design purchased by Pulsar in 1999.
CURRENT VERSIONS: **Cruiser:** Baseline model; being replaced by Super Cruiser.
Super Cruiser: Higher-powered version with minor redesign to dorsal fin to give larger area. First aircraft to this standard (N98WG) flew 8 March 1999.

Tri-R KIS TD (now known as Pulsar Sport 150) two-seat kitplane *(Paul Jackson/Jane's)* 2000/0103553

USA: AIRCRAFT—PULSAR to QUIKKIT

Pulsar Cruiser four-seat tourer *(Paul Jackson/Jane's)* 2000/0085834

CUSTOMERS: Three flying by end 1996 (latest data supplied).
DESIGN FEATURES: Larger version of the KIS TR-1. Quoted build time 1,500 hours.
Description generally as for TR-1 (see following entry) except that below.
POWER PLANT: Choice of piston engines between 119 and 149 kW (160 and 200 hp). Usable fuel capacity 189 litres (50.0 US gallons; 41.6 Imp gallons). Super Cruiser has 157 kW (210 hp) six-cylinder Teledyne Continental IO-360.

DIMENSIONS, EXTERNAL:
Wing span	8.83 m (29 ft 0 in)
Length overall	7.77 m (25 ft 6 in)
Height overall	2.29 m (7 ft 6 in)

DIMENSIONS, INTERNAL:
Cabin: Length	1.98 m (6 ft 6 in)
Max width	1.12 m (3 ft 8 in)
Max height	1.17 m (3 ft 10 in)

AREAS:
Wings, gross	1.25 m² (135.0 sq ft)

WEIGHTS AND LOADINGS:
Weight empty: Cruiser	544 kg (1,200 lb)
Super Cruiser	590 kg (1,300 lb)
Baggage capacity	29 kg (65 lb)
Max T-O weight: Cruiser	1,088 kg (2,400 lb)
Super Cruiser	1,134 kg (2,500 lb)
Max wing loading: Cruiser	86.8 kg/m² (17.78 lb/sq ft)
Super Cruiser	90.4 kg/m² (18.52 lb/sq ft)
Max power loading (119 kW; 160 hp):	
Cruiser	9.13 kg/kW (15.00 lb/hp)
Super Cruiser	7.25 kg/kW (11.90 lb/sq ft)

PERFORMANCE:
Never-exceed speed (V$_{NE}$)	186 kt (346 km/h; 215 mph)
Max level speed: Cruiser	165 kt (306 km/h; 190 mph)
Super Cruiser	174 kt (322 km/h; 200 mph)
Cruising speed: Cruiser	152 kt (282 km/h; 175 mph)
Super Cruiser	156 kt (290 km/h; 180 mph)
Stalling speed	48 kt (89 km/h; 55 mph)
Max rate of climb at S/L: Cruiser	305 m (1,000 ft)/min
Super Cruiser	457 m (1,500 ft)/min
Service ceiling	5,180 m (17,000 ft)
T-O run: Cruiser	305 m (1,000 ft)
Super Cruiser	274 m (900 ft)
T-O to 15 m (50 ft): Cruiser	366 m (1,200 ft)
Super Cruiser	335 m (1,100 ft)
Landing run: Cruiser	395 m (1,300 ft)
Super Cruiser	411 m (1,350 ft)
Range, with reserves:	
Cruiser	803 n miles (1,487 km; 924 miles)
Super Cruiser	695 n miles (1,287 km; 799 miles)

UPDATED

QUIKKIT

RAINBOW FLYERS INC, QUIKKIT DIVISION
9002 Summer Glen, Dallas, Texas 75243-7445
Tel/Fax: (+1 214) 349 04 62
e-mail: quikkit@glassgoose.com
Web: http://www.glassgoose.com
PRESIDENT: Thomas W Scott

WORKS:
Lakeview Airport, Lake Dallas, Texas 75065

UPDATED

QUIKKIT GLASS GOOSE
TYPE: Two-seat amphibian kitbuilt.
PROGRAMME: Extensively redesigned Aero Composite Sea Hawker (see 1992-93 *Jane's*), obviating deficiencies of that discontinued aircraft. Modifications initially applied to Tom Scott's Sea Hawker and later marketed for incorporation by other Sea Hawker operators, although no longer available.
    Prototype new-build Glass Goose flew 1996, but lost to flaperon flutter. Second prototype, in definitive configuration, first flew (N96GG) 20 March 1999. Production undertaken on Sea Hawker tools and jigs, acquired by Quikkit in 1993.
CUSTOMERS: Total 35 sold, of which two flying, by late 2000. At least six others extant by 2000, reflecting Sea Hawker conversions.
COSTS: US$27,000, excluding engine, avionics and equipment.
DESIGN FEATURES: Pusher, cantilever biplane, with single-step hull, sponsons and underwing pods for retractable main landing gear. Dished upper rear fuselage; mid-mounted tailplane. Wing positive stagger of 20 cm (8 in); upswept lower wingtips; downswept upper wingtips.
    Modifications to Sea Hawker design include extended (91 cm; 3 ft 0 in) upper wing span; sponsons of aerofoil section (replacing wingtip floats); and extended pylon chord, with associated fillets, vortex generators and augmentation slots in reshaped cowling to eliminate propeller vibration.

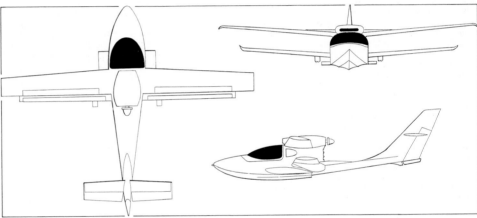

Quikkit Glass Goose two-seat amphibian *(Paul Jackson/Jane's)* 2000/0084616

FLYING CONTROLS: Conventional and manual, with drooping ailerons on all wings, max deflection +10/−40°. Actuation by pushrods and cables.
STRUCTURE: Glass fibre, with Kevlar reinforcement. Forward wing spars of glass fibre beam type, with carbon caps, at 27 per cent chord; glass fibre C-type rear spar at 85 per cent chord; glass fibre/foam core sandwich ribs. Intersecting fuselage bulkheads for additional strength. Quoted build time 1,000 hours.
LANDING GEAR: Retractable tricycle type for land operation. Two, side-by-side Cheng Shin 11×4.00-5 tyres on each main unit, underside of landing gear stowage pod also being forward-hinged leg/door. Mainwheel brakes.
    Rearward retracting, stainless steel, fully castoring nose leg, with 4.00-5 tyre, extends against spring-loaded single door. Electrohydraulic actuation. Ventral rudder, cable operated, for water manoeuvring.
POWER PLANT: One 119 kW (160 hp) Textron Lycoming O-320-B2B flat-four driving a Warp Drive, four-blade, ground-adjustable pitch propeller. Four equal-size, integral fuel tanks in wings (upper pair main; lower pair for long range), total capacity 265 litres (70.0 US gallons; 58.3 Imp gallons). No crossfeed. Optional provision for total of 455 litres (120 US gallons; 100 Imp gallons) involves fuselage tank.
ACCOMMODATION: Two, side by side, with baggage stowage areas behind seats and between lower wings.

DIMENSIONS, EXTERNAL:
Wing span: upper	8.23 m (27 ft 0 in)
lower	7.62 m (25 ft 0 in)
Wing chord: at root	0.84 m (2 ft 9 in)
at tip	0.65 m (2 ft 1½ in)
Length overall	5.94 m (19 ft 6 in)
Height overall	2.29 m (7 ft 6 in)
Tailplane span	2.44 m (8 ft 0 in)

DIMENSIONS, INTERNAL:
Cockpit max width	1.07 m (3 ft 6 in)
Baggage volume: cockpit	0.24 m³ (8.5 cu ft)
between wings	0.45 m³ (16.0 cu ft)

AREAS:
Wings, gross	12.59 m² (135.5 sq ft)
Sponsons (lifting)	1.44 m² (15.50 sq ft)
Flaperons (total)	1.86 m² (20.0 sq ft)

WEIGHTS AND LOADINGS:
Weight empty	474 kg (1,045 lb)
Max T-O weight	816 kg (1,800 lb)
Max wing loading	58.2 kg/m² (11.92 lb/sq ft)
Max power loading	6.85 kg/kW (11.25 lb/hp)

PERFORMANCE:
Max level speed	139 kt (257 km/h; 160 mph)
Max cruising speed at 75% power	122 kt (225 km/h; 140 mph)
Stalling speed: flaps up	43 kt (79 km/h; 49 mph)
flaps down	37 kt (68 km/h; 42 mph)
Max rate of climb at S/L	366 m (1,200 ft)/min
Service ceiling	3,960 m (13,000 ft)
T-O run	244 m (800 ft)
Landing run	274 m (900 ft)
Max range, 30 min reserves	955 n miles (1,770 km; 1,100 miles)

UPDATED

Second prototype Quikkit Glass Goose *(Paul Jackson/Jane's)* 2001/0103554

RANS

RANS INC
4600 Highway 183 Alternate, Hays, Kansas 67601
Tel: (+1 785) 625 63 46
Fax: (+1 785) 625 27 95
e-mail: rans@media-net.net
Web: http://www.rans.com
PRESIDENT AND CEO: Randy J Schlitter
VICE-PRESIDENT: Paula Schlitter

Details of S-4 Coyote, S-6 Coyote II, S-9 Chaos, S-10 Sakota, S-11 Pursuit and S-14 Airaile can be found in the 1992-93 *Jane's*. Rans delivered its 3,000th kitplane during 1998 and in 1999 achieved FAA '51 per cent rule' certification of the S-7, S-12 and S-16 kits.

UPDATED

RANS S-7 COURIER

TYPE: Tandem-seat kitbuilt.
PROGRAMME: Prototype first flew November 1985; in production from 1986.
CURRENT VERSIONS: **S-7**: Kitbuilt version, *as described*.
S-7C: Production version; changes include additional fuselage side stringers, smaller ailerons, larger flaps, taller (15 cm; 6 in) fin and rudder, extended dorsal strake, control surface gap seals, improved low-friction control cable runs, tab for pitch trim, tapered spring steel main landing gear legs with improved fairings, dual calliper brakes, larger baggage compartment and corrosion-proofing as standard. First flight (N11632) 20 December 1996.
CUSTOMERS: Total of 296 kits sold, of which 266 completed, by December 2000.
COSTS: S-7 kit, with engine: Rotax 912UL, US$26,495; Rotax 912ULS, US$27,995.
DESIGN FEATURES: High-wing, mid-tailplane configuration with sweptback fin, constant-chord wings and tapered tailplane. Wings braced by A-struts; empennage mutually wire-braced. Wings fold rearwards and tailplanes upwards for stowage; usual rigging/de-rigging time 30 to 45 minutes. Quoted build time for kit version is 500 to 700 hours.
NACA 2412 wing section, thickness/chord ratio 14 per cent; dihedral 1°; twist 25°.
FLYING CONTROLS: Conventional and manual. Cable-operated rudder and ailerons, the last-named spade-assisted, and torque tube-operated elevators; trim tab on starboard elevator; cable-operated four-position flaps (0, 8, 16 and 26°).
STRUCTURE: Wings have two tubular 6061-T6 alloy spars, preformed aluminium tube ribs riveted and clipped to spars and sheet alloy leading-edges; fuselage and tail surfaces of welded 4130 steel; glass fibre forward decking and engine cowling. Airframe is fabric-covered.
LANDING GEAR: Non-retractable tailwheel type with tubular spring steel main legs; steerable, full-swivelling tailwheel; mainwheel size 16×5, maximum pressure 2.41 bars (35 lb/sq in); tailwheel with solid tyre, size 8×2; Matco/Cleveland hydraulic brakes. Optional speed fairings on main legs increase cruising speed by 2 to 4 kt (5 to 8 km/h; 3 to 5 mph). Tundra tyres, skis, floats and Stoddart-Hamilton Aerocet amphibious floats optional.
POWER PLANT: One 59.6 kW (79.9 hp) Rotax 912 UL or one 73.5 kW (98.6 hp) Rotax 912 ULS four-cylinder piston engine, each with 2.27:1 reduction gear, driving a two-blade, ground-adjustable Warp Drive composites propeller. S-7C has Rotax 912 ULS. Fuel contained in two wing tanks, combined capacity 68 litres (18.0 US gallons; 15.0 Imp gallons), of which 58 litres (15.2 US gallons; 12.7 Imp gallons) are usable. Oil capacity 2.9 litres (0.75 US gallon; 0.6 Imp gallon).
ACCOMMODATION: Pilot and passenger in tandem on fore- and aft-adjustable seats; lap and shoulder harness standard; optional deluxe seats. Dual controls standard. Cabin is heated and ventilated. Upward-hinged door on each side is flight-openable with uplocks; quick-release hinge pins optional. Baggage compartment at rear of cabin.
AVIONICS: To customer's choice; optional GPS, VHF com and intercom on S-7C.
DIMENSIONS, EXTERNAL: (all versions, except: A: S-7 with Rotax 912 UL, B: S-7 with Rotax 912 ULS, C: S-7C):

Wing span	8.92 m (29 ft 3 in)
Wing chord, constant	1.52 m (5 ft 0 in)
Wing aspect ratio	5.8
Length overall: A, B	6.96 m (22 ft 10 in)
C	7.09 m (23 ft 3 in)
Length overall, wings folded: A, B	7.11 m (23 ft 4 in)
Height overall: A, B	1.90 m (6 ft 3 in)
C	2.01 m (6 ft 7 in)
Tailplane span	2.44 m (8 ft 0 in)
Wheel track	1.93 m (6 ft 4 in)
Wheelbase	5.18 m (17 ft 0 in)
Propeller diameter:	
A, B	1.73 m to 1.83 m (5 ft 8 in to 6 ft 0 in)
C	1.83 m (6 ft 0 in)
Cabin door: Max height	0.97 m (3 ft 2 in)
Max width	1.52 m (5 ft 0 in)
Height to sill	0.81 m (2 ft 8 in)

DIMENSIONS, INTERNAL (all versions):
Cabin: Length 1.93 m (6 ft 4 in)

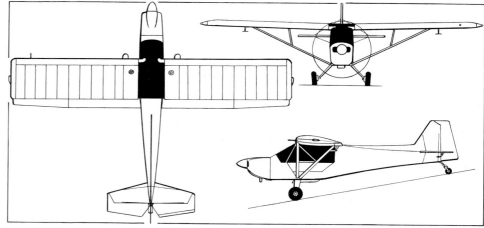

Rans S-7C Courier two-seat factory-built light aircraft *(James Goulding/Jane's)* 2001/0089516

Rans S-7C Courier development aircraft 2000/0064495

Max width	0.76 m (2 ft 6 in)
Max height	1.07 m (3 ft 6 in)
Floor area	1.11 m² (12.00 sq ft)
Volume	1.4 m³ (48 cu ft)
Baggage compartment:	
Volume	0.28 m³ (10.0 cu ft)
AREAS (A, B and C as above):	
Wings, gross: all	13.67 m² (147.1 sq ft)
Ailerons (total): A, B	1.11 m² (12.00 sq ft)
Trailing-edge flaps (total): A, B	1.11 m² (12.00 sq ft)
Fin: A, B	0.74 m² (8.00 sq ft)
Rudder: A, B	0.56 m² (6.00 sq ft)
Tailplane: A, B	1.11 m² (12.00 sq ft)
Elevators, incl tabs (total): A, B	0.93 m² (10.00 sq ft)
WEIGHTS AND LOADINGS (A, B and C as above):	
Weight empty: A, B	306 kg (675 lb)
C	318 kg (700 lb)
Max T-O weight: all	544 kg (1,200 lb)
Max wing loading: all	39.8 kg/m² (8.16 lb/sq ft)
Max wing loading: A	9.14 kg/kW (15.02 lb/hp)
B, C	7.41 kg/kW (12.17 lb/hp)
PERFORMANCE (A, B and C as above):	
Never-exceed speed (V_{NE})	113 kt (209 km/h; 130 mph)
Cruising speed: A	96 kt (177 km/h; 110 mph)
B	103 kt (190 km/h; 118 mph)
C	104 kt (193 km/h; 120 mph)
Stalling speed:	
flaps up	40 kt (74 km/h; 46 mph)
flaps down	36 kt (66 km/h; 41 mph)
Max rate of climb at S/L: A	244 m (800 ft)/min
B	335 m (1,100 ft)/min
C	305 m (1,000 ft)/min
Service ceiling: A	4,270 m (14,000 ft)
B, C	4,420 m (14,500 ft)
T-O run: A, C	72 m (235 ft)
B	69 m (225 ft)
Landing run	91 m (300 ft)
Range: A	420 n miles (779 km; 484 miles)
B	410 n miles (759 km; 472 miles)
C	417 n miles (772 km; 480 miles)
Endurance: A	4 h 24 min
B, C	4 h 0 min
g limits	+4/−2

UPDATED

RANS S-12XL AIRAILE and S-12S SUPER AIRAILE

TYPE: Side-by-side ultralight kitbuilt.
PROGRAMME: Improved, production versions of S-12 Airaile kitbuilt, last described in 1992-93 *Jane's*; R&D prototype of S-12XL (N8045X) made its debut at Sun 'n' Fun in Lakeland, Florida, April 1995.
CURRENT VERSIONS: **S-12XL Airaile**: Standard version. *Description applies to S-12XL, except where indicated.*
S-12S Super Airaile: Improved version, first flown (N80887) early 1999. 59.6 kW (79.9 hp) Rotax 912 or 73.5 kW (98.6 hp) Rotax 912 ULS engines, with thrust line lowered 25 mm (1 in) and driving ground-

Rans S-12S Super Airaile 2000/0064494

adjustable, composites Warp Drive propeller; new wing with improved aerofoil section, identical in rib construction to that of S-7C Courier; stamped aluminium ribs in tail surfaces; reduced tailplane incidence; cabin height raised 50 mm (2 in), with more sharply raked windscreen incorporating a new flow fence; increased range of seat tilt to improve headroom and comfort; hydraulic brakes standard; pre-sewn Dacron skins replaced by traditional doped fabric secured by pop rivets.

Performance improvements include 9 kt (16 km/h; 10 mph) increase in cruising speed. Marketed in parallel with S-12XL.

CUSTOMERS: Total of 50 ordered at time of launch; total of 858 S-12s (all models) ordered by December 1998 (latest data supplied), of which 772 then delivered.

COSTS: S-12XL kits, with engine: Rotax 503SC, US$14,500; Rotax 503DC, US$14,625; Rotax 582, US$16,230; Rotax 912, US$23,200; Rotax 912S, US$24,700.
S-12S kits, with engine: Rotax 912, US$28,250; Rotax 912S, US$29,750 (all 2000).

DESIGN FEATURES: Pod and boom fuselage; V-strut-braced wings; wire-braced tail surfaces; wings fold rearwards and tailplanes upwards for storage. Quoted build time 175 to 225 hours; or 325 to 500 hours with full enclosure cockpit. (Extra 200 hours for S-12S.)

FLYING CONTROLS: Conventional and manual. Cable-operated; ground-adjustable flaps primarily used for setting best trim, but can also be lowered for landing; tailplane incidence is ground-adjustable. Dual controls standard.

STRUCTURE: Welded steel tube cockpit cage; aluminium alloy tubular tailboom; wings have two tubular alloy spars and preformed ribs; ailerons, flaps and tail surfaces of alloy tube with stamped alloy or moulded ABS ribs; fabric covering.

LANDING GEAR: Fixed tricycle type with small tailwheel/bumper; tubular spring steel main legs, spring-loaded telescoping nose leg; pedal-operated hydraulic brakes on mainwheels; steerable nosewheel. Optional spats, tundra tyres and floats.

POWER PLANT: One air-cooled 34.0 kW (45.6 hp) Rotax 503 or liquid-cooled 47.8 kW (64.1 hp) Rotax 582 two-cylinder engine with 2.58:1 reduction gear, or 59.6 kW (79.9 hp) Rotax 912 or 73.5 kW (98.6 hp) Rotax 912 ULS flat-fours with 2.27:1 reduction gear, driving a two-blade wooden propeller; three-blade Warp Drive propeller optional; electric start optional on 503 and 582, standard on 912. Version with 48.5 kW (65 hp) Hirth 2706 two-cylinder two-stroke is available in Germany. Fuel in single wing tank, maximum capacity 34 litres (9.0 US gallons; 7.5 Imp gallons); dual tanks optional, (standard in S-12S) combined capacity 68 litres (18.0 US gallons; 15.0 Imp gallons).

ACCOMMODATION: Pilot and passenger side by side in open cockpit; transparent mini pod fairing standard; optional partial- and full-enclosure cockpits provide increased leg, shoulder and head room and lower noise levels and, combined with flow fences attached to fuselage pod and boom, raise cruising speed.

DIMENSIONS, EXTERNAL:
Wing span	9.60 m (31 ft 6 in)
Length overall	6.25 m (20 ft 6 in)
Height overall	2.36 m (7 ft 9 in)
Propeller diameter:	
S-12XL	1.73 to 1.83 m (5 ft 8 in to 6 ft 0 in)
S-12S	1.73 m (5 ft 8 in)

AREAS:
Wings, gross	14.12 m² (152.0 sq ft)

WEIGHTS AND LOADINGS (A: S-12XL with Rotax 912 UL, B: S-12S with Rotax 912 ULS):
Weight empty: A	261 kg (575 lb)
B	295 kg (650 lb)
Max T-O weight: A	499 kg (1,100 lb)
B	521 kg (1,150 lb)

PERFORMANCE (A and B as above):
Cruising speed: A	74 kt (137 km/h; 85 mph)
B*	87 kt (161 km/h; 100 mph)
Stalling speed, flaps down: A	31 kt (57 km/h; 35 mph)
Max rate of climb at S/L: A	274 m (900 ft)/min
B	366 m (1,200 ft)/min
T-O run: A	87 m (285 ft)
B	81 m (265 ft)
Landing run: A, B	61 m (200 ft)
Range: A, B	325 n miles (601 km; 374 miles)
Endurance: A	4 h 24 min
B	3 h 54 min

* With cockpit and wheel fairings

UPDATED

RANS S-15 PURSUIT II

Promotion of this side-by-side ultralight kitbuilt was discontinued in 2000. Prototype only was built, as described in 2000-01 edition.

UPDATED

RANS S-16 SHEKARI

TYPE: Side-by-side kitbuilt.
PROGRAMME: Prototype (N8072U), powered by 59.6 kW (79.9 hp) Rotax 912 UL engine, first flew 25 March 1997 in tricycle landing gear configuration, and won its class in the 1997 Sun 'n' Fun Race at Lakeland, Florida, averaging 113 kt (209 km/h; 130 mph) over a 52 n mile (97 km; 60 mile) course; re-engined with Teledyne Continental IO-240B and flown April 1998; subsequently converted to tailwheel configuration, although this configuration not marketed. Further prototype, N8073U, (plus one static test airframe) completed for trial installation of 119.3 kW (160 hp) Textron Lycoming O-320. FAA approval under homebuilt rules achieved 16 December 1998; first kit delivered to customer two days later; production capacity 50 kits per year.

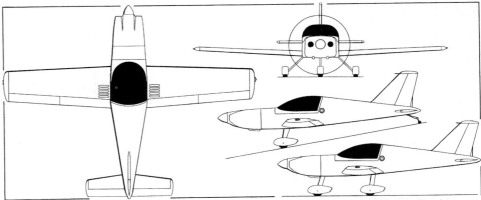

Rans S-16 Shekari, with additional side view of optional nosewheel version *(Paul Jackson/Jane's)*

Rans S-16 Shekari with nosewheel landing gear *(Paul Jackson/Jane's)*

CUSTOMERS: Total 22 sold and four flying by December 2000.
COSTS: Kit US$25,000, not including engine (2000).
DESIGN FEATURES: Conventional configuration low-wing monoplane with tailwheel or nosewheel landing gear. Quoted build time 500 to 1,000 hours.
FLYING CONTROLS: Conventional and manual. Pushrod-operated elevators, cable-operated rudder; half-span, five-position flaps. Triangular trim tab at base of rudder.
STRUCTURE: Composites fuselage, made up of four moulded shells, with powder-coated welded chromoly steel cockpit cage; aluminium wings, elevators and rudder with composites tips.
LANDING GEAR: Non-retractable nosewheel configuration; nosewheel steerable; hydraulic toe brakes standard; leg and wheel speed fairings on all units.
POWER PLANT: One 93.2 kW (125 hp) Teledyne Continental IO-240B flat-four or one 119.3 kW (160 hp) Textron Lycoming O-320 flat-four, driving a three-blade ground-adjustable Warp Drive wooden propeller. Fuel in two fuselage tanks, total capacity 121 litres (32.0 US gallons; 26.6 Imp gallons); oil capacity 4.7 litres (1.25 US gallons; 1.04 Imp gallons).
ACCOMMODATION: Two persons, side by side under one-piece forward-hinged canopy supported by gas struts; dual controls, central throttle, shoulder harnesses, lap belts and roll-over protection standard.

DIMENSIONS, EXTERNAL:
Wing span	7.32 m (24 ft 0 in)
Wing chord, mean	1.12 m (3 ft 8 in)
Wing aspect ratio	6.6
Length overall	5.89 m (19 ft 4 in)
Height overall	2.24 m (7 ft 4 in)
Tailplane span	2.54 m (8 ft 4 in)
Propeller diameter	2.08 m (6 ft 10 in)

DIMENSIONS, INTERNAL:
Cabin max width	1.02 m (3 ft 4 in)
Baggage volume	0.20 m³ (7.0 cu ft)

AREAS:
Wings, gross	8.08 m² (87.0 sq ft)

WEIGHTS AND LOADINGS (IO-240 engine):
Weight empty	422 kg (930 lb)
Baggage capacity	23 kg (50 lb)
Max T-O weight: Normal	657 kg (1,450 lb)
Aerobatic	590 kg (1,300 lb)
Max wing loading	81.4 kg/m² (16.67 lb/sq ft)
Max power loading	7.06 kg/kW (11.60 lb/hp)

PERFORMANCE (IO-240 engine):
Never-exceed speed (V_{NE})	191 kt (354 km/h; 220 mph)
Max cruising speed	139 kt (257 km/h; 160 mph)
Stalling speed: flaps up	54 kt (100 km/h; 62 mph)
flaps down	51 kt (94 km/h; 58 mph)
Max rate of climb at S/L	305 m (1,000 ft)/min
Service ceiling	4,420 m (14,500 ft)
T-O run	152 m (500 ft)
Landing run	160 m (525 ft)
Range	741 n miles (1,372 km; 853 miles)
Endurance	5 h 18 min
g limits: Normal	+4/−4
Aerobatic	+6/−6

UPDATED

RANS S-17 STINGER

TYPE: Single-seat ultralight/kitbuilt.
PROGRAMME: First flight April 1996; development continuing towards FAA FAR Pt 103 certification. Had undergone fundamental redesign by 2000, changing from tractor to pusher propeller and relocating tailboom to low position, complete with tailwheel.
CURRENT VERSIONS: **S-17**: *As described.*
S-18: Two-seat; described separately.

Rans S-17 Stinger *(Paul Jackson/Jane's)*

COSTS: Kit, with engine: Rotax 447, US$10,500; Rotax 503 SC, US$11,120; Rotax 503 DC, US$11,270 (all 2001). Flyaway US$3,000 to US$4,000 extra.
DESIGN FEATURES: Pod and boom fuselage; V-strut-braced wings; quoted building time 100 hours.
FLYING CONTROLS: Conventional and manual.
STRUCTURE: Light alloy fuselage, pre-assembled with all controls in situ; wing has tubular spar. Fuselage is 12.7 cm (5 in) aluminium tube with 4130 welded steel cockpit cage; optional composites nosecone. Dacron skin; spring steel landing gear.
LANDING GEAR: Fixed tailwheel type; tapered spring steel main legs; hand-lever operated brakes on mainwheels.
POWER PLANT: Prototype has one 34.0 kW (45.6 hp) Rotax 503 two-cylinder piston engine driving a three-blade IVO propeller. Alternatives include 31.0 kW (41.6 hp) Rotax 447 UL-2V. Fuel capacity 19 litres (5.0 US gallons; 4.2 Imp gallons) standard; 34 litres (9.0 US gallons; 7.5 Imp gallons) optional.
ACCOMMODATION: One person in open cockpit with Lexan windscreen/nosecone; four-point Hooker harness.
DIMENSIONS, EXTERNAL:
Wing span 8.99 m (29 ft 6 in)
Length overall 5.30 m (17 ft 4¾ in)
Height overall 2.13 m (7 ft 0 in)
AREAS:
Wings, gross 11.80 m² (127.0 sq ft)

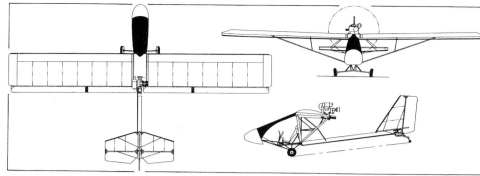

Rans S-17 general arrangement 2001/0089513

WEIGHTS AND LOADINGS (Rotax 447):
Weight empty 114 kg (252 lb)
Max T-O weight 239 kg (527 lb)
PERFORMANCE (Rotax 447):
Never-exceed speed (V_{NE}) 82 kt (152 km/h; 95 mph)
Cruising speed 48 kt (89 km/h; 55 mph)
Stalling speed, power off, flaps up 25 kt (45 km/h; 28 mph)
Max rate of climb at S/L 274 m (900 ft)/min
T-O and landing run 30 m (100 ft)
Range with standard fuel 61 n miles (114 km; 71 miles)
UPDATED

RANS S-18 STINGER II

TYPE: Tandem-seat ultralight kitbuilt.
PROGRAMME: Developed from single-seat S-17 Stinger. First flight of prototype S-18 (N24052) 9 September 2000; production from December 2000; first kit deliveries scheduled for March 2001.
COSTS: Kit, less engine and exhaust: Rotax 503DC installation: US$13,640; Rotax 582 installation US$14,300; Rotax 912 installation US$16,340 (all 2000). Engine US$2,860 to US$8,685 extra, according to type.
DESIGN FEATURES: Pod-and-boom fuselage; V-strut braced wings are identical to those of S-12XL Airaile and fold rearwards for transport or storage; wire-braced tail surfaces. Quoted build time under 200 hours.
FLYING CONTROLS: Conventional and manual. Flaps.
STRUCTURE: Welded 4130 steel tube fuselage cage and 15 cm (6 in) light alloy tailboom, pre-assembled and painted; wing has tubular spar; all parts prefabricated.
LANDING GEAR: Tailwheel type; fixed. One-piece tapered aluminium main legs; manually operated brakes on mainwheels.
POWER PLANT: Prototype has one 37.0 kW (49.6 hp) Rotax 503 DC two-cylinder piston engine with 1:2.58 reduction gearing, driving a three-blade Ivo pusher propeller; alternatives are 47.8 kW (64.1 hp) Rotax 582 and 73.5 kW (98.6 hp) Rotax 912. Fuel contained in single tank, capacity 34 litres (9.0 US gallons, 7.5 Imp gallons); second tank of similar capacity optional.
ACCOMMODATION: Two persons in tandem in open cockpit with Lexan windscreen/nosecone; full windscreen optional; four-point Hooker harness.
DIMENSIONS, EXTERNAL:
Wing span 9.45 m (31 ft 0 in)
Width, wings folded 1.75 m (5 ft 9 in)
Length overall 6.71 m (22 ft 0 in)
Height overall 2.06 m (6 ft 9 in)
Propeller diameter 1.68 m (5 ft 6 in)
AREAS:
Wings, gross 14.12 m² (152.0 sq ft)
WEIGHTS AND LOADINGS:
Weight empty 217 kg (478 lb)
Max T-O weight 417 kg (920 lb)
PERFORMANCE (Rotax 503 DC):
Never-exceed speed (V_{NE}) 78 kt (144 km/h; 90 mph)
Cruising speed 52 kt (97 km/h; 60 mph)
Stalling speed, power off 32 kt (58 km/h; 36 mph)
Max rate of climb at S/L 152 m (500 ft)/min
T-O run 101 m (330 ft)
Landing run 61 m (200 ft)
Endurance with max fuel, with reserves 3 h 0 in
NEW ENTRY

Rans S-18 Stinger II ultralight kitbuilt 2001/0097141

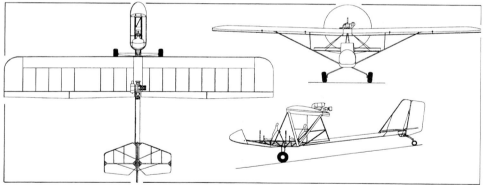

Two-seat Rans S-18 Stinger II (Paul Jackson/Jane's) 2001/0105043

RAYTHEON

RAYTHEON AIRCRAFT COMPANY
(Subsidiary of Raytheon Company)
10511 East Central, Wichita, Kansas 67201-0085
Tel: (+1 316) 676 71 11
Fax: (+1 316) 676 82 86
Web: http://www.raytheon.com
BRANCH DIVISIONS: Salina, Kansas and Little Rock, Arkansas.
CHAIRMAN, PRESIDENT AND COO: Hansel E Tookes II
VICE-PRESIDENTS:
Richard Danforth (Aircraft Business)
Thomas J Sarama (Engineering)
Paul Schumacher (Operations)
Doug Mahin (Strategic Planning and New Business Management)
C Don Cary (Customer Relations)
Tariq 'Ted' Farid (Worldwide Jet Sales)
Bradley A Hatt (Worldwide Beech Sales)
Tom Bertels (Marketing)
David H Riemer (Government Business)
Jim Siebauer (Chief Information Officer)
Mike Scheidt (Airline Program)
Jim Link (Global Sales)
Peter Herr (International Sales, Europe, Africa, Middle East and Far East)
Tamara Christen (International Special Mission Sales)
Stephen Lynch (International Special Mission Hawker Deliveries)
DIRECTOR, CORPORATE AFFAIRS: James M Gregory

EUROPEAN SUPPORT OF HAWKERS:
Raytheon Aircraft Services – Chester
Hawarden Airport, Broughton, Clwyd CH4 0BA, UK
Tel: (+44 1244) 52 04 44
Fax: (+44 1244) 52 30 04
MANAGING DIRECTOR, UK OPERATIONS: Ian Atkinson

BEECH
Aircraft marketed under the Beech name are described in the first part of this entry.

HAWKER
Products marketed under this name follow Beech aircraft.

Raytheon Aircraft Company (RAC) formed 15 September 1994, combining Raytheon Company subsidiaries Beech Aircraft Corporation and Raytheon Corporate Jets Inc. Beech Aircraft Corporation founded 1932 by Walter and Olive Ann Beech; became wholly owned subsidiary of Raytheon 8 February 1980. Raytheon acquired British Aerospace's Corporate Jets division 6 August 1993; founded Raytheon Corporate Jets, headquartered in Little Rock, Arkansas, with responsibility for design, development, production, marketing and support of Hawker family of corporate jets. RAC builds civil and military aircraft, missile targets, and components for aircraft and missiles; Salina division supplies all wings, non-metallic interior components, ventral fins, nosecones and tailcones used in Wichita piston-engined and turboprop Beech production and builds major subassemblies for the Beechjet.

Raytheon company purchased BAe Corporate Jets for US$372 million, adding the Hawker 800 and 1000 types (the

USA: AIRCRAFT—RAYTHEON

BEECH/HAWKER FIVE YEAR DELIVERIES

Type	1996	1997	1998	1999	2000	Total
T-6A	–	–	–	1	49	50
Bonanza	105	99	93	96	103	496
Baron	44	35	42	49	50	220
King Air	114[1]	117[2]	124[3]	128[4]	151[5]	634
Beech 1900	69	42	45	24	54	234
Starship	8	–	–	–	–	8
Beechjet (civil)	29	43	43	48	51	214
Jayhawk	35	15	–	–	–	50
Hawker 800	26	33	48	55	67	229
Hawker 1000	3	2	–	–	–	5
Annual totals	433	386	395	401	525	2,140

Notes:
[1] 38 King Air C90Bs; 49 King Air 200Bs, including 16 RC-12K/C-12R; and 27 King Air 350s
[2] 41 King Air C90Bs; 46 King Air B200s, including one RC-12K/C-12R; and 30 King Air 350s
[3] 37 King Air C90Bs, 45 King Air B200s and 42 King Air 350s
[4] 41 King Air C90Bs; 44 King Air B200s and 43 King Air 350s
[5] 46 King Air C90Bs; 59 King Air B200s and 46 King Air 350s

1000 since discontinued) to the company's aircraft products offered by what was then its Beech aircraft unit. The Hawker variants were developed from the BAe 125 series, of which some 850 of all versions had been sold to 44 countries by the time of Raytheon's purchase; final assembly of these models transferred from UK to new 41,800 m² (450,000 sq ft) plant at Wichita beginning last quarter of 1995 and finalised by April 1997, when last UK-built Hawker was completed; 17,745 m² (191,000 sq ft) Training Systems Division plant added for Texan II assembly line. Little Rock continues to provide custom interiors, avionics, sales and customer support services for Hawker models and aircraft painting services for Hawkers. Beech and Hawker product names are retained.

Wholly owned subsidiaries include Raytheon Aerospace of Madison, Mississippi (worldwide logistic support for USAF T-1As and Beech MQM-107 targets, and of US Navy T-34C and T-44 trainers in the USA); Raytheon Aircraft Credit Corporation Inc (business aircraft retail financing and leasing); Travel Air Insurance Company Ltd (aircraft liability insurance); Raytheon Aircraft Services network of fixed-base operations; and Raytheon Travel Air, formed in 1997, which offers fractional ownership in King Air 200Bs, Beechjet 400As and Hawker 800XPs.

In January 2001 RAC had 18,000 employees worldwide, of whom 8,800 were in Kansas, and occupied 380,900 m² (4,100,000 sq ft) of plant area at its two major facilities in Wichita.

Total production by RAC more than 53,400 by 4 January 2001.

UPDATED

BEECH T-6A TEXAN II
Canadian Forces name: CT-156 Harvard II
TYPE: Basic turboprop trainer.
PROGRAMME: US version of Pilatus PC-9 (which see in Swiss section). For participation in USAF/USN Joint Primary Aircraft Training System (JPATS) competition, Beech and Pilatus reached agreement on joint approach in August 1990; Beech received two standard PC-9s from Pilatus, one of which (N26BA) converted as engineering development prototype and completed more than 260 hours' flight testing to reduce programme risk and develop engineering design baseline before return to Pilatus. Followed by two Beech-built production prototypes (Beech designation PD373; later, Beech 3000), first flights December 1992 (N8284M/PT-2) and July 1993 (N209BA/PT-3); PT-2 used to complete flight test programme and evaluate systems performance; PT-3 incorporated several improvements and was principal aircraft for USAF/USN flight evaluation; one prototype and four production aircraft completed more than 1,400 hours of flight testing to achieve FAA certification in August 1999. Promotional name for the purposes of competition was **Beech Mk II**.

US FORCES' T-6A TEXAN II PROCUREMENT

FY	Lot	Qty	First Aircraft	Order	Delivery
95	1	1	95-3000	Jun 95	1999
96	2	3	95-3001	Jul 96	1999-2000
96	3	6	95-3004	Sep 96	2000
97	4	15	96-3010	Apr 97	2000-01
98	5	22		Feb 98	
99	6	22		May 99	
00	7	40		Jun 00	2002-03
Total		109			

Note: Serial number prefixes do not indicate year of order, as is usual in USAF. Early aircraft numbered 95-3000 to 3009, 96-3010 to 3012 and 97-3013 onwards

Selection as JPATS winner announced 22 June 1995; total requirement for 711 aircraft (372 USAF, 339 US Navy) by 2014; all to be built in USA; contract valued at US$4.7 billion awarded February 1996; allocated designation T-6 Texan (in advance of T-4 and T-5) to honour North American AT-6 Texan of Second World War era; Canadian name reflects British Commonwealth name for AT-6.

First metal cut February 1997; roll-out of manufacturing development aircraft (N23262/95-3000/PT-4) 29 June 1998; first flight 15 July 1998; second production aircraft flew 2 September 1998; first delivery (95-3003/PT-7) to Randolph AFB, June 1999, for technical evaluation; further two were due to have followed in third quarter of 1999 for six month multiservice operational test and evaluation programme (by first two Lot-3 aircraft), but were delayed by engineering fault with PT6 engine; First handover was 95-3004 on 1 March 2000; equipment of first squadron of 12th FTW, the 559th FTS, began on 23 May 2000.

Initial production rate 12 aircraft per year, rising to maximum of 43 by 2004; early aircraft to 12th FTW at Randolph AFB for instructor training; IOC with USAF at Laughlin AFB, Texas, 2001; IOC with Navy 2003; all USAF units equipped by 2011. FlightSafety selected on 21 April 1997 to provide associated ground training system at cost over 24 years of US$500 million.

CURRENT VERSIONS: **T-6A Texan II:** USAF/USN version, as described.
T-6A-1/CT-156 Harvard II: Version for NATO Flying Training in Canada (NFTC) programme; generally similar to Texan II, but with blind-flying hood, dual VOR and ADF, and back-up VHF comm in place of UHF.
Beech/Pilatus PC-9 Mk II: Designation of Greek aircraft; first 25 are in USAF/US Navy configuration, remainder in Greek-specified New Trainer Aircraft (NTA) configuration.

CUSTOMERS: USAF initial operator is 12th Flying Training Wing at Randolph AFB, Texas; second wing will be new unit at Moody AFB, Georgia, in 2001; subsequent deliveries to 47th FTW, Laughlin AFB, Texas; 71st FTW, Vance AFB, Oklahoma and 80th FTW, Sheppard AFB, Texas. Navy deliveries to Whiting Field, Florida, for TW-5 begin November 2002; further deliveries to TW-4 at Corpus Christi, Texas, and TW-6 at Pensacola, Florida.

Chilean Air Force signed a letter of intent in late 1996 for future purchase of 16 to 25 Beech Mk IIs. Bombardier Services of Canada ordered 24 T-6A-1s in December 1997 for its NFTC programme for delivery to No. 2 FTS at Moose Jaw between April and December 2000; first delivery (156103) 29 February 2000; NTFC inaugurated 6 July 2000. On 9 October 1998, Greek Air Force announced selection of T-6 and ordered 45, to be delivered between 2000 and 2003; option held on further five. First delivery (001) 17 July 2000 to 361 Squadron at Kalamata.

Raytheon holds marketing rights to PC-9 Mk II in all countries except Switzerland.

COSTS: Programme US$7 billion, of which US$362 million contracted by March 1998 for first 47 aircraft, comprising one manufacturing development aircraft and 46 production aircraft in five lots. Flyaway cost US$5.41 million over 740 aircraft (1998); USAF share of procurement is US$1,750

Beech T-6A Texan II delivered to the USAF   2000/0075962

Instrument panel of Beech T-6A Texan II

million (1996). Contracts for first 109 aircraft amounted to US$639 million.

DESIGN FEATURES: Approximately 90 per cent redesign including strengthened fuselage; aerodynamic changes to rudder and elevator; improved birdproofing; stronger landing gear; more powerful engine with linear power response; trim aid device; increased fuel capacity and pressure refuelling; ergonomically improved pressurised cockpits; numerous maintainability changes to decrease DOC including engine/systems diagnostics available through a central computer download connection; and new digital avionics. Certified to FAR Pt 23 in Aerobatic category.

FLYING CONTROLS: Conventional and manual. Selectable trim aid device (TAD) by Aeromach Labs automatically trims rudder in conjunction with throttle position to minimise effect of torque. Stall strip on starboard wing. Control deflections include rudder ±24°; elevators 18° up/16° down; ailerons 20° up/11° down; and flaps 23° for take-off, 50° for landing. Airbrake maximum deflection of 70°.

STRUCTURE: Generally all-metal; durability and damage-tolerant (DADT) structure includes wing and tailplane leading-edges reinforced for bird resistance; 18,000 hour service life. Avionics bay behind rear cockpit; forward-hinged door each side. Airframe built entirely in USA.

LANDING GEAR: As for PC-9, but Goodrich wheels and brakes. Capable of withstanding sink rates up to 4 m (13 ft)/s. Mainwheels 20×4.4 (14 ply (tubeless); nosewheel 16×4.4 (8 ply) tubeless.

POWER PLANT: One 1,274 kW (1,708 shp) Pratt & Whitney Canada PT6A-68 turboprop flat rated at 820 kW (1,100 shp), driving a four-blade, Hartzell propeller at a constant 2,000 rpm. Raytheon/P&WC power management unit provides jet-type linear throttle response. Integral 298 litre (78.5 US gallon; 65.5 Imp gallon) tank in each wing leading-edge, plus 26.5 litre (7.0 US gallon; 5.8 Imp gallon) collector tank below front cockpit; total fuel capacity 621 litres (164 US gallons; 137 Imp gallons). Provision for underwing auxiliary fuel tanks. Pressure refuelling/defuelling point in lower fuselage, adjacent to port wingroot; gravity refuelling points at three quarters' span in each wing upper surface.

ACCOMMODATION: Two pilots in stepped cockpits on Martin-Baker Mk 16LA zero/zero ejection seats. Three-piece, Pilkington Aerospace acrylic starboard-hinged canopy; integral windscreen 19 mm (¾ in) thick for resistance to strike by 1.8 kg (4 lb) bird at 270 kt (500 km/h; 311 mph); other panels 9 mm (⅓ in) thick. ATK canopy fracturing initiation unit and Northrop Grumman fracturing system. Accommodation pressurised and air conditioned; warm air canopy defrosting/demisting. Baggage compartment in rear fuselage, behind avionics bay; upward-hinged door port side.

SYSTEMS: Generally as PC-9. Dowty Aerospace hydraulics; Enviro Systems air conditioning unit; Litton OBOGS. Cockpit pressurisation system has 0.24 bar (3.5 lb/sq in) max differential. Battery 28 V. Goodrich emergency power system.

AVIONICS: Honeywell primary contractor.

Comms: VHF/UHF transceiver, plus back-up VHF with separate antenna; intercom and interphone. MST 67A Mode S transponder.

Flight: Dual VNS-41 navigation systems (including VN-411B receivers) for VOR, localiser, glide slope and marker beacon functions through common antenna. DFS-43A ADF (DF-431 receiver and AT-434 loop/sense antenna); KLN 90 GPS; and Rockwell Collins 252-channel DM-441B DME. Litef fibre optic gyro for AHRS. Goodrich collision warning system. Flight data recorder.

Instrumentation: Two independent EFIS 50 127 mm (5 in) square active matrix LCDs, plus control panel, in each cockpit; attitude director indicator portion of EFIS provides primary attitude display, turn rate, mode selection annunciation, localiser/glide slope deviation as appropriate; horizontal situation indicator for primary heading display, primary navigation display, course select indication, navigation source annunciation, DME, localiser and glide slope deviation, remote map presentation and selected heading reference marker. Additional three 76 mm (3 in) square MFDs per cockpit provide engine and auxiliary instrument information, including fuel state and pressurisation. Standby instrumentation comprises airspeed, attitude, turn-and-slip and magnetic compass. Korry Electronics warning panels. Provision for HUD in export aircraft.

ARMAMENT: Provision for three hardpoints under each wing.

DIMENSIONS, EXTERNAL: As for PC-9M

AREAS:
Wings, gross	16.29 m² (175.3 sq ft)

WEIGHTS AND LOADINGS:
Weight empty	2,087 kg (4,600 lb)
Baggage capacity	54 kg (120 lb)
Max fuel weight	499 kg (1,100 lb)
Max T-O weight	2,857 kg (6,300 lb)
Max wing loading	175.5 kg/m² (35.94 lb/sq ft)
Max power loading	3.49 kg/kW (5.73 lb/shp)

PERFORMANCE:
Never-exceed speed (VNE)	350 kt (648 km/h; 402 mph)
Max level speed: at altitude	310 kt (574 km/h; 357 mph)
at low level	270 kt (500 km/h; 311 mph) IAS
Max cruising speed at 2,285 m (7,500 ft)	230 kt (426 km/h; 265 mph)
Approach speed	100 kt (185 km/h; 116 mph) IAS
Stalling speed, power off:	
flaps up	82 kt (152 km/h; 95 mph)
flaps down	74 kt (137 km/h; 86 mph)
Max rate of climb at S/L	1,372 m (4,500 ft)/min
Service ceiling	10,670 m (35,000 ft)
T-O run	541 m (1,775 ft)
T-O to 15 m (50 ft)	610 m (2,000 ft)
Landing from 15 m (50 ft)	732 m (2,400 ft)
Landing run	580 m (1,900 ft)
Range at altitude	850 n miles (1,574 km; 978 miles)
Endurance at max cruising speed	3 h
g limits	+7/−3.5

UPDATED

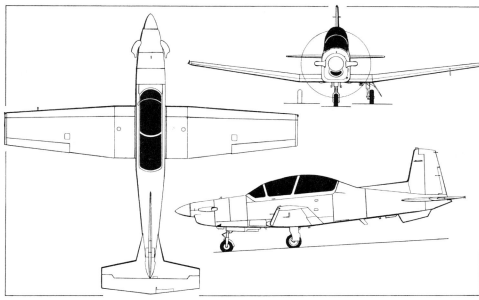

Beech T-6A Texan II primary trainer *(Paul Jackson/Jane's)*

Beech CT-156 Harvard II in Canadian markings *(CAF)*

Third prototype Raytheon Premier I light business jet

RAYTHEON PREMIER I

TYPE: Light business jet.

PROGRAMME: Design started early 1994 as PD374 (later PD390) and approved early 1995; originated in former Beech design offices, but is first aircraft to carry only the Raytheon name; brief details of 'new light business jet' revealed June 1995; launched at National Business Aircraft Association Convention in Las Vegas 26 September 1995 with full-scale fuselage/cabin mockup; wind-tunnel tests of one-eighth model conducted early 1996 at Boeing, Boeing V/STOL, NASA-Lewis and Wichita State University facilities; to compete with Cessna CitationJet.

First forward fuselage completed in February 1997 and mated to aft fuselage in April 1998; roll-out (N390RA, c/n RB-1) 19 August 1998; first flight 22 December 1998. Second aircraft (N704T) first flown 4 June 1999, followed by third (N390TC), first with complete interior, on 17

USA: AIRCRAFT—RAYTHEON (Beech)

Raytheon Premier I cabin mockup

September 1999; public debut (N390TC) at National Business Aviation Association Convention at Atlanta, Georgia, October 1999; more than 720 flight test hours accumulated by 23 December 1999, at which time eight production aircraft were in final assembly; static testing of wing to 150 per cent of design load completed on 17 December 1999; four aircraft undertaking 1,400 hour flight test programme culminating in FAA FAR Pt 23 certification and first delivery in second quarter of 2001. By November 2000, 15 Premier Is were in final assembly, with the fuselage of the 34th aircraft being manufactured; target production rate 60 per year.

CUSTOMERS: More than 260 orders received by November 2000, representing a backlog until 2005. Customers include Raytheon Travel Air, the fractional ownership subsidiary of Raytheon Aircraft, which has ordered 71 for delivery beginning 2001; the Jordan Grand Prix racing team, which has ordered one for delivery in mid-2002; and Aviation Leasing Group (ALG Transportation Inc) of London, which ordered three in August 2000, two of which will be used by the Civil Aviation Training Centre (CATC) in Thailand for training student pilots for Thai Airways International and other Pacific Rim carriers.

COSTS: US$4.8 million (2000). Estimated direct operating cost US$592 per hour.

DESIGN FEATURES: Conventional small business jet, developed with assistance of CATIA programmes. Rear-mounted engines, T tail and wing mounted below fuselage for additional cabin space. Sweepback 20° at 25 per cent chord; 2° 30′ dihedral; tailplane sweepback 25° at 25 per cent chord.

FLYING CONTROLS: Conventional and manual. Activation via pushrods and cables. Pitch trim via electrically actuated, variable incidence tailplane and mechanically driven geared tab on each elevator; electrically actuated trim tab on port aileron, ground-adjustable trim tab on starboard aileron; electrically actuated rudder trim tab. Electrically signalled, hydraulically powered, three-segment spoilers on upper surface of each wing augment aileron roll control; outboard and middle panels provide roll, airbrake and post-landing lift-dump functions; inboard panel provides lift-dump function only. 75 per cent span, four-segment, electrically controlled Fowler flaps, deflections 0, 10, 20 and 30°. Rudder boost, for asymmetric thrust and yaw damper, standard.

STRUCTURE: Fuselage of graphite/epoxy laminate and honeycomb composites, formed by Cincinnati Milacron Viper automatic fibre-placement machines over aluminium mandrel, placing fibres at speeds up to 46 m (150 ft) per minute, enabling entire fuselage to be completed in one week; elimination of all internal frames, and skin thickness of 20 mm (0.78 in), increase cabin volume by 13 per cent and afford a weight saving of some 20 per cent over conventional alloy construction. Wing of aluminium alloy with six-spar wing box, manufactured using high-speed equipment capable of machining more than 93 m² (1,000 sq ft) of material per minute, and automatic riveting machines; with exception of three small bays along trailing-edge, entire wing is used for fuel storage. Ailerons and flaps of graphite/epoxy composites; fin has aluminium alloy spars and ribs with graphite/epoxy honeycomb skin; tailplane has one-piece, composites forward-and-rear spar with alloy centre rib, composites mid- and tip ribs and Nomex composites skin.

LANDING GEAR: Hydraulically actuated, retractable tricycle type with free-fall emergency extension system; single wheel on each unit. Mainwheels retract inwards; nosewheel forwards. Steerable nosewheel, maximum deflection ±45°. Hydraulic disc brakes with electric anti-skid system.

POWER PLANT: Two pod-mounted Williams FJ44-2A turbofans, each rated at 10.23 kN (2,300 lb st). Fuel contained in integral wing tanks, total usable capacity 2,059 litres (544 US gallons; 453 Imp gallons), with gravity filling point on each wing. Single-point pressure refuelling/defuelling optional.

ACCOMMODATION: Crew of one or two, with dual controls standard; six passengers in cabin, comprising four in standard club seating arrangement with tracking, swivelling and reclining capability and stowable writing tables and two on fixed forward-facing seats to rear;

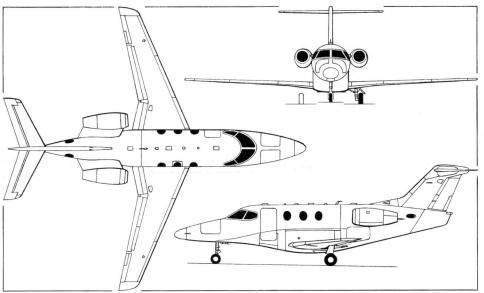

Raytheon Premier I (two Williams FJ44 turbofans) (James Goulding/Jane's)

lavatory at rear, doubling as flight-accessible baggage compartment, maximum capacity 68 kg (150 lb). Refreshment/hang-up baggage cabinet on forward starboard side of cabin. Airstair door on port side to rear of flight deck; single plug-type emergency exit on starboard side. Three cabin windows on each side. Accommodation is air conditioned and pressurised. Externally accessible main baggage compartment to rear of cabin, with upward-opening door on port side, can accommodate large items such as skis; heating optional; forward baggage compartment in nose on port side with swing-up door.

SYSTEMS: Pressurisation system, maximum differential 0.58 bar (8.4 lb/sq in), maintains 2,440 m (8,000 ft) cabin altitude to 12,500 m (41,000 ft). Vapour cycle, ozone-safe R134a air conditioning system. Hydraulic system, maximum pressure 207 bars (3,000 lb/sq in), for landing gear, brakes, anti-skid and spoilers. Electrical system comprises two 28 V 300 A engine-driven starter/generators, 24 V 40 Ah lead/acid main battery, 24 V 5 Ah standby battery and 28 V external power receptacle; system is configured so that load-shedding is primarily automatic in the event of failure of any or all main electrical power sources. Oxygen system, capacity 1,134 litres (40 cu ft) standard, 2,182 litres (77 cu ft) optional, with diluter-demand masks for crew and continuous flow masks for passengers. Engine bleed-air anti-icing for wing leading-edges and nacelle inlets; electromagnetic expulsion de-icing (EMEDI) for tailplane leading-edges, automatically activated by nose-mounted, heated, ice detectors; electrically heated windscreens (with silicone coating for rain dispersal), pitot tubes and AoA probes.

AVIONICS: Rockwell Collins Pro Line 21 EFIS avionics suite as core system.

Comms: Dual Rockwell Collins VHF-422A transceivers, TDR-94 Mode S transponders and DB Model 438 audio systems; single CTL-23 nav/com tuning unit; four-speaker cabin paging unit.

Radar: Rockwell Collins RTA-800 colour weather radar.

Flight: Dual Rockwell Collins AHC-3000 AHRS, ADC-3000 air data computers, CDU-3000 control/display units and VIR-432 nav receivers; IAPS-3000 lightweight, integrated avionics processing system; FGC-3000 flight guidance system, FMS-3000 flight management system with database, ADF-462, DME-442, GPS-4000, ALT-4000 radio altimeter and MDC-3000 maintenance diagnostic computer.

Instrumentation: Rockwell Collins AFD-3010 integrated EFIS comprising two 254 × 203 mm (10 × 8 in) active matrix LCD adaptive flight displays providing PFD and MFD functions, with CRT HSI and back-up electromechanical rate/sensor/attitude instrument and ASI on right side; third LCD optional, but mandatory for RVSM compliance.

DIMENSIONS, EXTERNAL:
Wing span	13.56 m (44 ft 6 in)
Wing aspect ratio	8.0
Length overall	14.02 m (46 ft 0 in)
Height overall	4.67 m (15 ft 4 in)
Tailplane span	4.90 m (16 ft 1 in)
Wheel track	2.79 m (9 ft 2 in)
Wheelbase	5.36 m (17 ft 7 in)
Crew/passenger door: Height	1.27 m (4 ft 2 in)
Width	0.64 m (2 ft 1½ in)

DIMENSIONS, INTERNAL:
Cabin:
Length: between pressure bulkheads	5.69 m (18 ft 8 in)
excl flight deck	4.11 m (13 ft 6 in)
Max width	1.68 m (5 ft 6 in)
Max height	1.65 m (5 ft 5 in)
Volume	9.4 m³ (331 cu ft)

External baggage compartment volume:
main	1.4 m³ (50 cu ft)
forward	0.28 m³ (10.0 cu ft)

AREAS:
Wings, gross	22.95 m² (247.0 sq ft)
Horizontal tail surfaces (total)	4.65 m² (50.00 sq ft)

WEIGHTS AND LOADINGS:
Basic operating weight	3,627 kg (7,996 lb)
Baggage capacity: main	181 kg (400 lb)
forward	68 kg (150 lb)
internal	27 kg (60 lb)
Max fuel weight	1,655 kg (3,648 lb)
Max T-O weight	5,670 kg (12,500 lb)
Max ramp weight	5,710 kg (12,590 lb)
Max landing weight	5,386 kg (11,875 lb)
Max zero-fuel weight	4,536 kg (10,000 lb)
Max wing loading	247.1 kg/m² (50.61 lb/sq ft)
Max power loading	277 kg/kN (2.72 lb/lb st)

PERFORMANCE:
Max operating speed:	
S/L to 8,990 m (29,500 ft)	320 kt (593 km/h; 368 mph)
above 8,990 m (29,500 ft)	M0.83
Max cruising speed at 10,060 m (33,000 ft)	461 kt (854 km/h; 530 mph)
Max operating altitude	12,500 m (41,000 ft)
T-O run	915 m (3,000 ft)
Landing run	747 m (2,450 ft)
Range with single pilot, four passengers, NBAA IFR reserves	1,500 n miles (2,778 km; 1,726 miles)
g limits	+3.2/−1.28

UPDATED

BEECH BONANZA A36

TYPE: Six-seat utility transport.

PROGRAMME: Descended from Model 35 of 1945 (10,403 built in V tail configuration), but now has little more than the name in common; Model 36 introduced in 1968; developed from Bonanza E33A; current A36 introduced 3 October 1983, succeeding model powered by 212.5 kW (285 hp) Continental IO-520-BB; certified in FAA Utility category.

CURRENT VERSIONS: **Bonanza A36:** Standard version. Features introduced for 1999 model year comprise Honeywell Silver Crown Plus avionics package, including KFC 225 AFCS, dual KX 155A nav/coms and KMA 26 audio control system with six-point intercom; Teledyne Continental Motors Raytheon Special Edition engines with internal engine balancing, balanced fuel injectors, airflow enhancements, valve train improvements, fine wire spark plugs and custom features; new Hartzell aluminium propeller; more robust door locks; and silicon rubber door seals.

Description applies to A36, except where otherwise stated.

Bonanza A36AT (Airline Trainer): Certified 1991 for use in Europe; modifications to propeller, engine and exhaust systems to reduce external noise; flyover sound level 71.8 dBA; Teledyne Continental IO-550-B, limited to 216 kW (290 hp) at 2,550 rpm, driving Hartzell three-blade constant-speed propeller; special exhaust silencers; additional engine cooling louvres; four seats.

Bonanza A36 Jaguar Special Edition: Announced 13 June 1999 in joint marketing venture with Jaguar Cars; features interior mirroring that of Jaguar motorcar, with extra padding on seat armrests, cashmere carpet and seat headrest covers embossed with Jaguar's leaping cat logo; figured cherry wood cabinetry, hand-stitched leather-covered control yokes and exterior paint in Jaguar green and gold house colours with Jaguar logos; 13 were scheduled for production in 1999.

Bonanza B36TC: Described separately.

CUSTOMERS: Total approximately 3,350 Bonanza 36/A36s delivered by September 2000, including 89 in 1995, 83

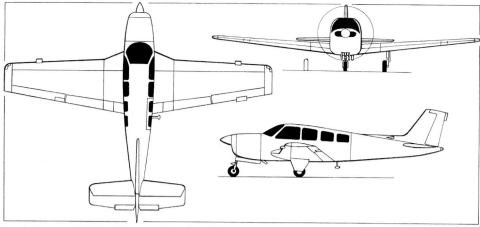

Beech Bonanza A36 four/six-seat utility aircraft *(James Goulding/Jane's)*

in 1996, 85 in 1997, 71 in 1998, 77 in 1999 and 85 in 2000.

DESIGN FEATURES: Conventional low-wing cabin monoplane. Tapered wings with leading-edge gloves; tapered tailplane and sweptback fin.

Beech modified NACA 23016.5 wing section at root, modified 23012 at tip; dihedral 6°; incidence 4° at root and 1° at tip.

FLYING CONTROLS: Conventional and manual. Electrically actuated trim tab in each elevator; ground-adjustable tabs in ailerons and rudder; single-slotted, three-position flaps.

STRUCTURE: Light alloy, with two-spar wing torsion box and stressed-skin tail surfaces. Conventional construction.

LANDING GEAR: Electrically retractable tricycle type, with steerable nosewheel. Mainwheels retract inward into wings, nosewheel rearward. Beech oleo-pneumatic shock-absorbers in all units. Cleveland mainwheels, size 6.00-6, and tyres, size 7.00-6 (6 ply), pressure 2.28 to 2.76 bars (33 to 40 lb/sq in). Cleveland nosewheel and tyre, size 5.00-5 (4 ply), pressure 2.76 bars (40 lb/sq in). Cleveland ring disc hydraulic brakes. Parking brake. Magic Hand landing gear system optional.

POWER PLANT: One 224 kW (300 hp) Teledyne Continental IO-550-B Raytheon Special Edition flat-six engine, driving a Hartzell three-blade constant-speed metal propeller. The engine is equipped with an altitude-compensating fuel pump which automatically makes the fuel/air mixture leaner and richer during climb and descent respectively. Usable fuel capacity 280 litres (74.0 US gallons; 61.6 Imp gallons).

ACCOMMODATION: Enclosed cabin seating four to six persons on individual seats. Pilot's seat is vertically adjustable. Dual controls standard. Two rear removable seats and two folding seats permit rapid conversion to utility configuration. Optional club seating with rear-facing third and fourth seats, executive writing desk, refreshment cabinet, headrests for third and fourth seats, reading lights and fresh air outlets for fifth and sixth seats. Double doors of bonded aluminium honeycomb construction on starboard side facilitate loading of cargo. As an air ambulance, one stretcher can be accommodated with ample room for a medical attendant and/or other passengers. Extra windows provide improved view for passengers.

SYSTEMS: Optional 12,000 BTU refrigeration-type air conditioning system comprises evaporator located beneath pilot's seat, condenser on lower fuselage, and engine-mounted compressor. Air outlets on centre console, with two-speed blower. Electrical system supplied by 28 V 100 A alternator, 24 V 15.5 Ah battery; optional standby generator. Hydraulic system for brakes only. Standby vacuum pump standard. Pneumatic system for instrument gyros optional. Oxygen system and electric propeller de-icing optional.

AVIONICS: Honeywell Silver Crown Plus package.
Comms: Dual KX 155A nav/comms; KT 70 Mode S transponder; microphone; headset and cabin speaker.
Flight: Dual KX 155A nav/glide slope receiver/converter with KI 209 VOR/ILS indicator; KI 208 VOR/LOC indicator; KR 87 ADF with 227-00 indicator; KN 63 DME with KDI 572 indicator; DME hold and nav 1/nav 2 switching; KEA 130A encoding altimeter; KMA 26 audio control/marker beacon receiver; and KFC 225 AFCS. KLN 90B GPS and Goodrich WX-1000 Stormscope optional.

EQUIPMENT: Standard equipment includes LCD digital chronometer, EGT and OAT gauges, rate of climb indicator, turn co-ordinator, 3 in horizon and directional gyros, four fore- and aft-adjustable and reclining seats, armrests, headrests, single diagonal strap shoulder harness with inertia reel for all occupants, pilot's storm window, ultraviolet-proof windscreen and windows, sun visors, large cargo door, emergency locator transmitter, stall warning device, alternate static source, heated pitot, rotating beacon, three-light strobe system, carpeted floor, super soundproofing, control wheel map lights, entrance door courtesy light, internally lit instruments, coat hooks, glove compartment, in-flight storage pockets, approach plate holder, utility shelf, cabin dome light, reading lights, instrument post lights, control wheel map light, electroluminescent subpanel lighting, landing light, taxying light, full-flow oil filter, three-colour polyurethane exterior paint, external power socket, static wicks and towbar.

Optional equipment includes dual controls, leather seats, co-pilot's wheel brakes, air conditioning, fifth passenger seat, fresh air vent blower and ground com switch.

DIMENSIONS, EXTERNAL:
Wing span	10.21 m (33 ft 6 in)
Wing chord: at root	2.13 m (7 ft 0 in)
at tip	1.07 m (3 ft 6 in)
Wing aspect ratio	6.2
Length overall	8.38 m (27 ft 6 in)
Height overall	2.62 m (8 ft 7 in)
Tailplane span	3.71 m (12 ft 2 in)
Wheel track	2.92 m (9 ft 7 in)
Wheelbase	2.39 m (7 ft 10¼ in)
Propeller diameter: A36	2.03 m (6 ft 8 in)
A36AT	1.93 m (6 ft 4 in)
Forward cockpit door: Height	0.91 m (3 ft 0 in)
Width	0.94 m (3 ft 1 in)
Baggage compartment door: Height	0.61 m (2 ft 0 in)
Width	0.99 m (3 ft 3 in)
Rear passenger/cargo door: Height	0.89 m (2 ft 11 in)
Width	1.14 m (3 ft 9 in)

DIMENSIONS, INTERNAL:
Cabin (aft of firewall, incl extended baggage compartment): Length	3.84 m (12 ft 7 in)
Max width	1.07 m (3 ft 6 in)
Max height	1.27 m (4 ft 2 in)
Volume	3.9 m³ (136 cu ft)

AREAS:
Wings, gross	16.80 m² (181.0 sq ft)
Ailerons (total)	1.06 m² (11.40 sq ft)
Trailing-edge flaps (total)	1.98 m² (21.30 sq ft)
Fin	0.93 m² (10.00 sq ft)
Rudder, incl tab	0.52 m² (5.60 sq ft)
Tailplane	1.75 m² (18.82 sq ft)
Elevators, incl tabs	1.67 m² (18.00 sq ft)

WEIGHTS AND LOADINGS:
Weight empty, standard:	
A36, A36AT	1,052 kg (2,320 lb)
Baggage capacity	181 kg (400 lb)
Max T-O weight: A36	1,655 kg (3,650 lb)
A36AT	1,633 kg (3,600 lb)
Max ramp weight: A36	1,661 kg (3,663 lb)
A36AT	1,639 kg (3,613 lb)
Max wing loading: A36	98.5 kg/m² (20.17 lb/sq ft)
A36AT	97.1 kg/m² (19.89 lb/sq ft)
Max power loading: A36	7.40 kg/kW (12.17 lb/hp)
A36AT	7.59 kg/kW (12.46 lb/hp)

PERFORMANCE:
Max level speed (min weight)	184 kt (340 km/h; 212 mph)
Max cruising speed (mid-cruise weight):	
2,500 rpm at 1,830 m (6,000 ft)	176 kt (326 km/h; 202 mph)
2,300 rpm at 2,440 m (8,000 ft)	167 kt (309 km/h; 192 mph)
2,100 rpm at 1,830 m (6,000 ft)	160 kt (296 km/h; 184 mph)
2,100 rpm at 3,050 m (10,000 ft)	153 kt (283 km/h; 176 mph)
Stalling speed, power off:	
flaps up	68 kt (126 km/h; 78 mph) IAS
30° flap	59 kt (109 km/h; 68 mph) IAS
Max rate of climb at S/L	368 m (1,208 ft)/min
Service ceiling	5,640 m (18,500 ft)
T-O run: flaps up: A36	360 m (1,185 ft)
A36AT	450 m (1,475 ft)
12° flap: A36	296 m (970 ft)
A36AT	402 m (1,320 ft)
T-O to 15 m (50 ft): flaps up: A36	640 m (2,100 ft)
A36AT	662 m (2,170 ft)
12° flap: A36	583 m (1,915 ft)
A36AT	618 m (2,025 ft)
Landing from 15 m (50 ft)	442 m (1,450 ft)
Landing run	280 m (920 ft)
Range with max usable fuel, with allowances for engine start, taxi, T-O, climb and 45 min reserves at econ cruise power: 2,500 rpm at 3,660 m (12,000 ft)	875 n miles (1,621 km; 1,008 miles)
2,300 rpm at 3,660 m (12,000 ft)	903 n miles (1,672 km; 1,039 miles)
2,100 rpm at 1,830 m (6,000 ft)	914 n miles (1,694 km; 1,052 miles)

UPDATED

BEECH TURBO BONANZA B36TC

TYPE: Six-seat utility transport.

PROGRAMME: Derived from A36 (see previous entry). Certified as A36TC 7 December 1978; 271 A36TCs delivered; improved B36TC introduced 1982.

CURRENT VERSIONS: **Bonanza B36TC:** Standard version, *as described.*

Bonanza B36TC Jaguar Special Edition: Announced 13 June 1999 in joint marketing venture with Jaguar Cars; features interior mirroring that of Jaguar motorcars, with extra padding on seat armrests, cashmere carpet and seat headrest covers embossed with Jaguar's leaping cat logo; figured cherry wood cabinetry, hand-stitched leather-covered control yokes, and exterior paint in Jaguar green and gold house colours with Jaguar logos; three were scheduled for production in 1999.

CUSTOMERS: Total of some 670 A/B36TCs delivered by September 2000, including 14 per year in 1993-97, 22 in 1998, 20 in 1999, and 18 in 2000.

DESIGN FEATURES: Compared with A36TC, B36TC has greater span; wing section NACA 23010.5 at tip; 0° incidence at tip; greater fuel capacity.

Data below summarise differences from A36TC:

POWER PLANT: One 223.7 kW (300 hp) Teledyne Continental TSIO-520-UB Raytheon Special Edition turbocharged flat-six engine, driving a three-blade constant-speed Hartzell metal propeller. Fixed engine cowl flaps. Two fuel tanks in each wing leading-edge, with total usable capacity of 386 litres (102 US gallons; 84.9 Imp gallons). Refuelling points above tanks. Oil capacity 11.5 litres (3.0 US gallons; 2.5 Imp gallons).

SYSTEMS: Air conditioning optional.

EQUIPMENT: EGT gauge not available. Turbine inlet temperature gauge is standard.

DIMENSIONS, EXTERNAL:
Wing span	11.53 m (37 ft 10 in)
Wing chord at tip	0.91 m (3 ft 0 in)

Beech Bonanza A36

USA: AIRCRAFT—RAYTHEON (Beech)

Wing aspect ratio	7.6
Propeller diameter	1.98 m (6 ft 6 in)

AREAS:
Wings, gross	17.47 m² (188.1 sq ft)

WEIGHTS AND LOADINGS:
Weight empty, standard	1,116 kg (2,460 lb)
Max T-O and landing weight	1,746 kg (3,850 lb)
Max ramp weight	1,753 kg (3,866 lb)
Max wing loading	99.9 kg/m² (20.47 lb/sq ft)
Max power loading	7.81 kg/kW (12.83 lb/hp)

PERFORMANCE (speeds at mid-cruise weight):
Max level speed at 6,700 m (22,000 ft)	213 kt (394 km/h; 245 mph)
Cruising speed at 7,620 m (25,000 ft):	
at 79% power	200 kt (370 km/h; 230 mph)
at 75% power	195 kt (361 km/h; 224 mph)
at 69% power	188 kt (348 km/h; 216 mph)
at 56% power	173 kt (320 km/h; 199 mph)
Stalling speed, power off:	
flaps up	65 kt (120 km/h; 75 mph) IAS
30° flap	57 kt (106 km/h; 66 mph) IAS
Max rate of climb at S/L	321 m (1,053 ft)/min
Service ceiling	over 7,620 m (25,000 ft)
T-O run, 15° flap	311 m (1,020 ft)
T-O to 15 m (50 ft), 15° flap	649 m (2,130 ft)
Landing run	298 m (980 ft)

Range at 7,620 m (25,000 ft) with max fuel, allowances for engine start, taxi, T-O, cruise climb, descent, and 45 min reserves at 50% power:
at 79% power	956 n miles (1,770 km; 1,100 miles)
at 75% power	984 n miles (1,822 km; 1,132 miles)
at 69% power	1,022 n miles (1,893 km; 1,176 miles)
at 56% power at 6,100 m (20,000 ft)	1,087 n miles (2,013 km; 1,250 miles)

UPDATED

BEECH BARON 58

TYPE: Six-seat utility twin.
PROGRAMME: Prototype Beech 55 Baron flew 29 February 1960; total 3,728 built, not including 94 Model 56 Turbo Barons. Model 58 derived from E55 with 25 cm (10 in) fuselage stretch, plus internal rearrangement to lengthen cabin by 76 cm (30 in); certified in FAA Normal category 19 November 1969; marketing began in following month. Beech 58P Pressurised Baron and 58TC turbocharged version no longer produced.
CURRENT VERIONS: **Baron 58:** Standard version, *as described*. Features introduced for 1999 model year comprise Honeywell Silver Crown Plus avionics package including KFC 225 AFCS, dual KX 155A nav/coms and KMA 26 audio control system with six-point intercom; Teledyne Continental Motors Raytheon Special Edition engines with internal engine balancing, balanced fuel injectors, airflow enhancements, valve train improvements, fine wire spark plugs and custom features; new Hartzell aluminium propellers; more robust door locks; and silicon rubber door seals.

Baron 58 Jaguar Special Edition: Announced 13 June 1999 in joint marketing venture with Jaguar Cars; features interior mirroring that of Jaguar motorcars, with extra padding to seat armrests, cashmere carpet and seat headrest covers embossed with Jaguar's leaping cat logo; figured cherry wood cabinetry, hand-stitched leather-covered control yokes and exterior paint in Jaguar green and gold house colours with Jaguar logos; nine were scheduled for production in 1999, of which first (N158JC) delivered July 1999 to Dr Wendell Row of Eureka, California.

CUSTOMERS: Operators include several commercial aviation flying schools. Total 2,610 Baron 58s delivered by September 2000, including 29 in 1995, 44 in 1996, 35 in 1997, 42 in 1998, 47 in 1999, and 50 in 2000. (Incorporates 497 of 58P and 151 of 58TC versions.)

DESIGN FEATURES: Conventional low-wing monoplane, ultimately derived from Beech Bonanza. Tapered wing with leading-edge gloves inboard of engines; sweptback fin.
Wing section NACA 23015.5 at root, 23010.5 at tip; dihedral 6°; incidence 4° at root and 0° at tip.

FLYING CONTROLS: Conventional and manual. Manually operated trim tabs in elevators, rudder and port aileron; electrically operated single-slotted flaps.

STRUCTURE: Light alloy with two-spar wing box; elevators have smooth magnesium alloy skins.

LANDING GEAR: Electrically retractable tricycle type. Main units retract inward into wings, nosewheel aft. Beech oleo-pneumatic shock-absorbers in all units. Steerable nosewheel with shimmy damper. Cleveland wheels, with mainwheel tyres size 6.50-8 (8 ply) tubed, pressure 3.59 to 3.96 bars (52 to 56 lb/sq in) or 19.5×6.75-8 (10 ply) tubeless. Nosewheel tyre size 5.00-5 (6 ply) tubed, pressure 3.79 to 4.14 bars (55 to 60 lb/sq in) or 15×6.0-6 (4 ply) tubed. Cleveland ring disc hydraulic brakes. Heavy-duty brakes optional. Parking brake.

POWER PLANT: Two 224 kW (300 hp) Teledyne Continental IO-550-C Raytheon Special Edition flat-six engines, each driving a Hartzell three-blade constant-speed fully feathering metal propeller. The standard fuel system has a usable capacity of 628 litres (166 US gallons; 138 Imp gallons). Optional 'wet wingtip' installation also available, increasing usable capacity to 734 litres (194 US gallons; 161.5 Imp gallons).

ACCOMMODATION: Standard version has four individual seats in pairs in enclosed, soundproofed, heated and ventilated cabin, with door on starboard side. Single diagonal strap shoulder harness with inertia reel standard on all seats. Pilot's vertically adjusting seat is standard. Co-pilot's vertically adjusting seat, folding fifth and sixth seats, or club seating comprising folding fifth and sixth seats and aft-facing third and fourth seats, are optional. Adjustable rudder pedals (retractable on starboard side). Executive desk available as option with club seating. Baggage compartment in nose. Double passenger/cargo doors on starboard side of cabin provide access to space for baggage or cargo behind the third and fourth seats. Pilot's storm window. Openable windows adjacent the third and fourth seats are used for ground ventilation and as emergency exits. Windscreen defrosting standard.

SYSTEMS: Cabin heated by Janitrol 50,000 BTU heater, which serves also for windscreen defrosting. Oxygen system of 1,389 litres (49 cu ft) or 1,814 litres (64 cu ft) capacity optional. Electrical system includes two 28 V 100 A engine-driven alternators with alternator failure lights and two 12 V 25 Ah batteries. Two 100 A alternators optional. Hydraulic system for brakes only. Pneumatic pressure system for air-driven instruments, and optional wing and tail unit de-icing system. Oxygen system, cabin air conditioning and windscreen electric anti-icing optional.

AVIONICS: Honeywell Silver Crown package standard.
Comms: KX 155-09 760-channel com transceiver with audio amplifier, microphone, headset and cabin speaker; KT 70 Mode S transponder. Optional avionics by Honeywell or Rockwell Collins to customer requirements, including full IFR packages. Emergency locator transmitter.
Radar: RDR 2000 VP colour weather radar.
Flight: KX 155A 200-channel nav receiver with KI 208 VOR/LOC converter/indicator, KR 87 ADF with KI 227-00 indicator, combined loop/sense antenna, nav and com antennas. KFC 225 autopilot, yaw damper and Goodrich Stormscope optional. KLN 90B GPS, certified for IFR operation, optional.
Instrumentation: Blind-flying instruments, outside air temperature gauge, sensitive altimeter, turn co-ordinator, EGT and CHT gauges and synchroscope. Optional true airspeed indicator, engine and flight hour recorders and instantaneous vertical speed indicator.

EQUIPMENT: Standard equipment includes ultraviolet-proof windscreen and cabin windows, super soundproofing, heated pitot head, instrument panel floodlights, navigation and position lights, steerable taxying light, dual landing lights, heated fuel vents, heated fuel and stall warning vanes and external power socket.
Options include alternate static source, internally illuminated instruments, strobe lights, electric windscreen anti-icing, wing ice detection light, static wicks, leather seat trim, cabin club seating, cockpit relief tube, cabin fire extinguisher and ventilation blower.

Standard instrument panel of Beech Bonanza and Baron *2000/0075966*

DIMENSIONS, EXTERNAL:
Wing span	11.53 m (37 ft 10 in)
Wing chord: at root	2.13 m (7 ft 0 in)
at tip	0.90 m (2 ft 11½ in)
Wing aspect ratio	7.2
Length overall	9.09 m (29 ft 10 in)
Height overall	2.97 m (9 ft 9 in)
Tailplane span	4.85 m (15 ft 11 in)
Wheel track	2.92 m (9 ft 7 in)
Wheelbase	2.72 m (8 ft 11 in)
Propeller diameter	1.96 m (6 ft 5 in)
Rear passenger/cargo doors:	
Max height	0.89 m (2 ft 11 in)
Width	1.14 m (3 ft 9 in)
Baggage door (fwd): Height	0.56 m (1 ft 10 in)
Width	0.64 m (2 ft 1 in)

DIMENSIONS, INTERNAL:
Cabin (incl rear baggage area):	
Length	3.84 m (12 ft 7 in)
Max width	1.07 m (3 ft 6 in)
Max height	1.27 m (4 ft 2 in)
Floor area	3.72 m² (40.0 sq ft)
Volume	3.9 m³ (136 cu ft)
Baggage compartment: nose	0.51 m³ (18.0 cu ft)

AREAS:
Wings, gross	18.51 m² (199.2 sq ft)
Ailerons (total)	1.06 m² (11.40 sq ft)

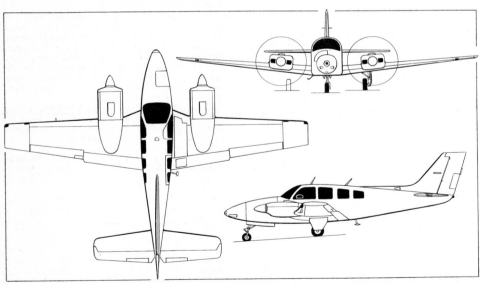

Beech Baron 58 (two Teledyne Continental IO-550-C piston engines) *(Dennis Punnett/Jane's)*

Beech Baron 58, owned by 123 Holdings of Dover, Delaware *(Paul Jackson/Jane's)* *2001/0103580*

Trailing-edge flaps (total)	1.98 m² (21.30 sq ft)
Fin	1.46 m² (15.67 sq ft)
Rudder, incl tab	0.81 m² (8.75 sq ft)
Tailplane	4.95 m² (53.30 sq ft)
Elevators, incl tabs	1.84 m² (19.80 sq ft)

WEIGHTS AND LOADINGS:
Weight empty	1,633 kg (3,600 lb)
Baggage capacity: cabin	181 kg (400 lb)
nose	136 kg (300 lb)
Max T-O weight	2,495 kg (5,500 lb)
Max ramp weight	2,506 kg (5,524 lb)
Max landing weight	2,449 kg (5,400 lb)
Max wing loading	134.8 kg/m² (27.61 lb/sq ft)
Max power loading	5.58 kg/kW (9.17 lb/hp)

PERFORMANCE (cruising speeds at average cruise weight):
Max level speed at S/L	208 kt (386 km/h; 239 mph)
Max cruising speed, 2,500 rpm at 1,525 m (5,000 ft)	203 kt (376 km/h; 234 mph)
Cruising speed, 2,500 rpm at 2,440 m (8,000 ft)	192 kt (356 km/h; 221 mph)
Econ cruising speed, 2,100 rpm at 3,660 m (12,000 ft)	162 kt (300 km/h; 186 mph)
Stalling speed, power off:	
flaps up	84 kt (156 km/h; 97 mph) IAS
flaps down	75 kt (139 km/h; 86 mph) IAS
Max rate of climb at S/L	529 m (1,735 ft)/min
Rate of climb at S/L, OEI	119 m (390 ft)/min
Service ceiling	6,305 m (20,680 ft)
Service ceiling, OEI	2,220 m (7,280 ft)
T-O run	427 m (1,400 ft)
T-O to 15 m (50 ft)	701 m (2,300 ft)
Landing from 15 m (50 ft)	747 m (2,450 ft)
Landing run	434 m (1,425 ft)

Range with 628 litres (166 US gallons; 138 Imp gallons) usable fuel, power/altitude settings as above, with allowances for engine start, taxi, T-O, climb and 45 min reserves at econ cruise power:
at max cruising speed	860 n miles (1,593 km; 990 miles)
at cruising speed	974 n miles (1,804 km; 1,121 miles)
at econ cruising speed	1,313 n miles (2,432 km; 1,511 miles)

UPDATED

BEECH KING AIR C90B and C90SE

TYPE: Business twin-turboprop.
PROGRAMME: King Air 90 first flew 21 November 1963. C90B announced at NBAA Convention, October 1991; superseded King Air 90, A90, B90, C90, C90-1, C90A; introduced four-blade McCauley propellers, special interior soundproofing, updated and redesigned interior, updated cockpit features and interior noise and vibration levels substantially reduced. A model delivered on 24 June 1996 to Jeld-Wen of Klamath Falls, Oregon, was the 5,000th King Air of all versions to be produced.
CURRENT VERSIONS: **C90B**: Standard version from 1991.
Detailed description applies to C90B.
C90B Jaguar Special Edition: Announced 7 January 1998; features interior mirroring that of Jaguar cars, with ivory Connolly hide seats and sidewalls, green carpets incorporating the motorcar manufacturer's leaping cat logo, walnut and boxwood cabinetry and hand-stitched leather-covered control wheels; exterior paint trim in Jaguar green and gold house colours with Jaguar logos; Jaguar has FAA registrations for Special Edition production run reserved with '8XJ' suffixes (first was N18XJ of Jewett Scott Truck Line). At least eight produced in 1998, including one exported to Colombia.
King Air 200 and 350: Described in following entries.
CUSTOMERS: Total 1,583 commercial and 226 military King Air 90/A90/B90/C90/C90-1/C90A/C90B/C90SE delivered by 1 January 2000, including 40 C90Bs and (now discontinued) C90SEs in 1995, 38 C90Bs in 1996, 41 C90Bs in 1997, 37 C90Bs in 1998, 41 C90Bs in 1999, and 46 C90Bs in 2000; the 1,500th, delivered January 1998, was first Jaguar SE; 1,600th to Larry V Plummer at Camarillo, California, September 2000.
COSTS: C90SE US$1.747 million (1995).
DESIGN FEATURES: Conventional low-wing, twin-engined monoplane. Wing section NACA 23014.1 (modified) at root, 23016.22 (modified) at outer end of centre-section, 23012 at tip; dihedral 7°; incidence 4° 48′ at root, 0° at tip; tailplane 7° dihedral.
FLYING CONTROLS: Conventional and manual. Trim tabs on port aileron, in both elevators and rudders; single-slotted aluminium flaps.
STRUCTURE: Generally light alloy; magnesium ailerons. Internal corrosion-proofing.
LANDING GEAR: Hydraulically retractable tricycle type. Nosewheel retracts rearward, mainwheels forward into engine nacelles. Mainwheels protrude slightly beneath engine nacelles when retracted, for safety in a wheels-up emergency landing. Fully castoring steerable nosewheel with shimmy damper. Beech oleo-pneumatic shock-absorbers. Goodrich mainwheels with tyres size 8.50-10 (8 ply) tubeless, pressure 3.79 bars (55 lb/sq in). Goodrich nosewheel with tyre size 6.50-10 (6 ply) tubeless, pressure 3.59 bars (52 lb/sq in). Goodrich heat-sink and air-cooled multidisc hydraulic brakes. Parking brake. Minimum ground turning radius 10.82 m (35 ft 6 in).

Beech King Air C90B

POWER PLANT: Two 410 kW (550 shp) Pratt & Whitney Canada PT6A-21 turboprops, each driving a McCauley four-blade constant-speed fully feathering propeller. Propeller auto ignition system, environmental fuel drain collection system, magnetic chip detector, automatic propeller feathering and propeller synchrophaser standard. Fuel in two tanks in engine nacelles, each with usable capacity of 231 litres (61.0 US gallons; 50.8 Imp gallons), and auxiliary bladder tanks in outer wings, each with capacity of 496 litres (131 US gallons; 109 Imp gallons). Total usable fuel capacity 1,454 litres (384 US gallons; 320 Imp gallons). Refuelling points in top of each engine nacelle and in wing leading-edge outboard of each nacelle. Oil capacity 13.2 litres (3.5 US gallons; 2.9 Imp gallons) per engine.
ACCOMMODATION: Two four-way adjustable seats side by side in cockpit with dual controls standard. Normally, four reclining seats in main cabin, in two pairs facing each other fore and aft. Standard furnishings include cabin forward partition, with fore and aft partition and coat rack, hinged nose baggage compartment door, seat belts and inertia reel shoulder harnesses for all seats. C90B has standard accommodation for six passengers, four in club arrangement, one on sideways-facing seat adjacent door, and one on lavatory/seat in baggage area. C90SE has standard accommodation for five passengers, as above except no sideways-facing seat. Baggage racks at rear of cabin on starboard side, with lavatory on port side. Door on port side aft of wing, with built-in airstairs. Emergency exit on starboard side of cabin. Entire accommodation pressurised, heated and air conditioned. Electrically heated windscreen, windscreen defroster and windscreen wipers standard.
SYSTEMS: Pressurisation by dual engine bleed air system with pressure differential of 0.34 bar (5.0 lb/sq in). Cabin heated by 45,000 BTU dual engine bleed air system and auxiliary electrical heating system. Hydraulic system for landing gear actuation. Electrical system includes two 28 V 250 A starter/generators, 24 V 34 Ah air-cooled Ni/Cd battery with failure detector. Oxygen system, 623 litres (22 cu ft) or 1,814 litres (64 cu ft) capacity, optional. Vacuum system for flight instruments. Automatic pneumatic de-icing of wing/fin/tailplane leading-edges standard. Engine and propeller anti-icing systems standard. Engine fire detection and extinguishing system optional.
AVIONICS: Standard Rockwell Collins Pro Line II package with EFIS-84 in sectional instrument panel.
Comms: Dual VHF-22A transceivers with CTL-22 controls; dual TDR-64 transponders; dual DB Systems Model 415 audio systems; dual Lockheed Martin Fairchild CVR A1005; dual Flite-Tronics PC-250 inverters; edgelight radio panel; ELT; avionics master switch; ground clearance switch on com 1; control wheel push-to-talk switches; dual hand microphones and cockpit speakers; dual Telex headsets.
Radar: WXR-270 colour weather radar.
Flight: Dual VIR-32 VOR/LOC/GLS/MKR receivers with CTL-32 controls; ADF-60A with CTL-62 control; DME-42 with IND-42 indicator; RMI-30; dual MCS-65 compass systems. Goodrich WX-1000+ Stormscope optional. APS-65 autopilot/flight director with EADI-84 and EHSI.
Instrumentation: EFIS-84; ALT-50A radio altimeter; pilot's encoding altimeter; dual 2 in electric turn and bank indicators; co-pilot's 75 mm (3 in) horizon indicator; co-pilot's HSI. Dual blind-flying instrumentation with dual instantaneous VSIs; standby magnetic compass; digital OAT gauge; LCD digital chronometer clock; vacuum gauge; de-icing pressure gauge; cabin rate of climb indicator; cabin altitude and differential pressure indicator; flight hour recorder; automatic solid-state warning and annunciator panel. Primary and secondary instrument lighting systems.
EQUIPMENT: Standard equipment includes fresh air outlets; oxygen outlets with overhead-mounted diluter demand masks with microphones; removable low-profile lavatory with shoulder harness and lap belt; two cabin tables; cabin fire extinguisher; wing ice lights; dual landing lights; nosewheel taxying light; flush position lights; dual rotating beacons; rheostat-controlled white cockpit lighting; wingtip recognition lights; wingtip and tail strobe lights; and vertical tail illumination lights.
Optional equipment includes electric flushing lavatory; engine fire detection system; oxygen bottle; cabinet with three drawers and stereo tape deck storage; and pilot-to-cabin paging with four stereo speakers.

DIMENSIONS, EXTERNAL:
Wing span	15.32 m (50 ft 3 in)
Wing chord: at root	2.15 m (7 ft 0½ in)
at tip	1.07 m (3 ft 6 in)
Wing aspect ratio	8.6
Length overall	10.82 m (35 ft 6 in)
Height overall	4.34 m (14 ft 3 in)
Tailplane span	5.26 m (17 ft 3 in)
Wheel track	3.89 m (12 ft 9 in)
Wheelbase	3.73 m (12 ft 3 in)
Propeller diameter	2.29 m (7 ft 6 in)
Propeller ground clearance	0.34 m (1 ft 1½ in)
Passenger door: Height	1.30 m (4 ft 3½ in)
Width	0.69 m (2 ft 3 in)
Height to sill	1.22 m (4 ft 0 in)

DIMENSIONS, INTERNAL:
Pressurised area: Length	5.43 m (17 ft 10 in)
Cabin: Length	3.94 m (12 ft 11 in)
Max width	1.37 m (4 ft 6 in)
Max height	1.45 m (4 ft 9 in)
Floor area	6.50 m² (70.0 sq ft)
Volume	4.9 m³ (174 cu ft)
Baggage compartment volume, rear	1.5 m³ (54 cu ft)

AREAS:
Wings, gross	27.31 m² (293.9 sq ft)
Ailerons (total)	1.29 m² (13.90 sq ft)
Trailing-edge flaps (total)	2.72 m² (29.30 sq ft)
Fin	2.20 m² (23.67 sq ft)
Rudder, incl tab	1.30 m² (14.00 sq ft)
Tailplane	4.39 m² (47.25 sq ft)
Elevators, incl tabs (total)	1.66 m² (17.87 sq ft)

WEIGHTS AND LOADINGS:
Weight empty	3,040 kg (6,702 lb)
Max T-O weight	4,581 kg (10,100 lb)
Max ramp weight	4,608 kg (10,160 lb)

740 USA: AIRCRAFT—RAYTHEON (Beech)

Max landing weight	4,354 kg (9,600 lb)
Max wing loading	167.8 kg/m² (34.36 lb/sq ft)
Max power loading	5.59 kg/kW (9.18 lb/shp)

PERFORMANCE:
Max cruising speed at AUW of 3,855 kg (8,500 lb):
at 4,880 m (16,000 ft)	247 kt (457 km/h; 284 mph)
at 6,400 m (21,000 ft)	243 kt (450 km/h; 280 mph)
at 7,315 m (24,000 ft)	239 kt (442 km/h; 275 mph)
Approach speed	101 kt (187 km/h; 116 mph)

Stalling speed, power off:
wheels and flaps up	88 kt (163 km/h; 101 mph) IAS
flaps down	78 kt (144 km/h; 90 mph) IAS
Max rate of climb at S/L	610 m (2,003 ft)/min
Rate of climb at S/L, OEI	151 m (494 ft)/min
Service ceiling	8,810 m (28,900 ft)
Service ceiling, OEI	3,990 m (13,100 ft)
T-O run	620 m (2,035 ft)
T-O to 15 m (50 ft)	826 m (2,710 ft)
Accelerate/stop distance	1,113 m (3,650 ft)
Landing from 15 m (50 ft) at max landing weight, with propeller reversal	698 m (2,290 ft)
Landing run at max landing weight, with propeller reversal	384 m (1,260 ft)

Range with max fuel at max cruising speed, incl allowance for starting, taxi, take-off, climb, descent and 45 min reserves at max range power, ISA, at:
4,875 m (16,000 ft)	911 n miles (1,687 km; 1,048 miles)
6,400 m (21,000 ft)	1,039 n miles (1,924 km; 1,195 miles)
7,315 m (24,000 ft)	1,121 n miles (2,076 km; 1,290 miles)

Max range at econ cruising power, allowances as above:
at 4,875 m (16,000 ft)	1,118 n miles (2,070 km; 1,286 miles)
at 6,400 m (21,000 ft)	1,229 n miles (2,276 km; 1,414 miles)
at 7,315 m (24,000 ft)	1,282 n miles (2,374 km; 1,475 miles)

UPDATED

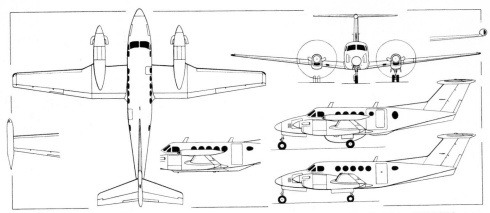

Beech King Air B200 twin-turboprop transport, with additional side view of Maritime Patrol B200T (centre right); scrap views of wingtip tanks and centre-fuselage of photo survey aircraft for IGN *(Dennis Punnett/Jane's)*

BEECH KING AIR B200
Swedish Air Force designation: Tp 101
TYPE: Utility turboprop twin.
PROGRAMME: Design of Super King Air 200 began October 1970; 'Super' prefix deleted from all 200, 300 and 350 series King Airs in 1996; first flight (c/n BB1) 27 October 1972; certified to FAR Pt 23 plus icing requirements of FAR Pt 25, 14 December 1973; design of B200 (prototype c/n BB343) began March 1980; production started May 1980; FAA certification 13 February 1981; on sale March 1981.
CURRENT VERSIONS: **King Air B200:** Baseline version.
Detailed description applies to B200.
King Air B200C: As B200 but with 1.32 × 1.32 m (4 ft 4 in × 4 ft 4 in) cargo door. Two, identified as military C-12R/AP, ordered by US Army on behalf of Greece, January 2000; special missions fit includes cameras; delivery due before July 2001.
King Air B200T: Standard provision for removable tip tanks, adding total 401 litres (106 US gallons; 88.25 Imp gallons), making total 2,460 litres (650 US gallons; 541 Imp gallons). Span without tip tanks 16.92 m (55 ft 6 in). Total 38 built up to 1993. Four ordered by US Army Missile Command in October 1998 for delivery by June 2000; however, these appear to be part of batch of five for Israel Defence Force, also completed in June 2000 and increasing production total to 43.
King Air B200CT: Combines tip tanks and cargo door as standard. Four built by 1983; none subsequently.
Maritime patrol B200T: Described in 1998-99 and previous editions.
Beech 200 HISAR: Radar surveillance platform. Launched 1997; based on Beech 200T airframe; demonstrator (N4277E) undertook 16-country tour in 1997-98. First international sale to Traffic 2000 of Germany, November 1997, for North Sea environmental monitoring. Ventral radome for Hughes Integrated Synthetic Aperture Radar (HISAR) suitable for border surveillance, remote sensing, pollution monitoring, EEZ patrol and agricultural monitoring. Equipment operator's console in cabin.
C/RC/UC-12: Military versions; described separately.
King Air 300LW: Described in 1996-97 and earlier editions of *Jane's*.
King Air 350: Described separately.
CUSTOMERS: Deliveries by 1 January 2000 totalled 1,789 for commercial and private orders, plus 397 military versions (described separately) to US armed forces and foreign customers. Total of 23 King Air B200s delivered in 1994, 28 in 1995, 33 in 1996, 45 in 1997, 45 in 1998, 55 in 1999, and 59 in 2000.
French Institut Géographique National has three B200Ts fitted with twin Wild RC-10 Superaviogon camera installations and Doppler navigation; maximum endurance 10 hours 20 minutes; high flotation landing gear; special French certification for maximum T-O weight 6,350 kg (14,000 lb) and maximum landing weight 6,123 kg (13,500 lb). Egyptian government acquired one King Air 200 in 1978 for water, uranium and other natural resources exploration over Sinai and Egyptian deserts as follow-up to satellite surveys; fitted with remote sensing gear, specialised avionics and special cameras. Navaid checking versions supplied to Taiwan government (one) and Malaysian government (two). Other special missions aircraft delivered to Taiwan Ministry of Interior May 1979; Royal Hong Kong Auxiliary Air Force (later Government Air Service) (two) 1986 and 1987; four King Air 200s operated by Swedish Air Force since 1988 as **Tp 101**.
COSTS: B200SE US$2.995 million (1996); B200 US$3.699 million (1996); B200T US$4.18 million US Army programme unit cost (1999).
DESIGN FEATURES: Generally as for earlier versions. Wing aerofoil NACA 23018 to 23016.5 over inner wing, 23012 at tip; dihedral 6°; incidence 3° 48′ at root, −1° 7′ at tip; swept vertical and horizontal tail.
FLYING CONTROLS: Conventional and manual. Trim tabs in port aileron and both elevators; anti-servo tab in rudder; single-slotted trailing-edge flaps; fixed tailplane.
STRUCTURE: Two-spar light alloy wing; safe-life semi-monocoque fuselage.
LANDING GEAR: Hydraulically retractable tricycle type, with twin wheels on each main unit. Single wheel on steerable nose unit, with shimmy damper. Main units retract forward, nosewheel rearward. Beech oleo-pneumatic shock-absorbers. Goodrich mainwheels and tyres size 18×5.5 (10 ply) tubeless, pressure 7.25 bars (105 lb/sq in). Oversize and/or 10 ply mainwheel tyres optional. Goodrich nosewheel size 6.50×10 (8 ply) tubeless, with tyre size 22×6.75-10, pressure 3.93 bars (57 lb/sq in). Goodrich hydraulic multiple-disc brakes. Parking brake.
POWER PLANT: Two 634 kW (850 shp) Pratt & Whitney Canada PT6A-42 turboprops, each driving a Hartzell four-blade constant-speed reversible-pitch metal propeller with autofeathering and synchrophasing. Bladder fuel cells in each wing, with main system capacity of 1,461 litres (386 US gallons; 321.5 Imp gallons) and auxiliary system capacity of 598 litres (158 US gallons; 131.5 Imp gallons). Total usable fuel capacity 2,059 litres (544 US gallons; 453 Imp gallons). Two refuelling points in upper surface of each wing. Wingtip tanks optional, providing an additional 401 litres (106 US gallons; 88.3 Imp gallons) and raising maximum usable capacity to 2,460 litres (650 US gallons; 541 Imp gallons). Oil capacity 29.5 litres (7.8 US gallons; 6.5 Imp gallons).
ACCOMMODATION: Pilot only, or crew of two side by side, on flight deck, with full dual controls and instruments as standard. Seven cabin seats standard, each equipped with seat belts and inertia reel shoulder harness; optional seats in baggage compartment raise passenger capacity to nine. Partition with sliding door between cabin and flight deck, and partition at rear of cabin. Door at rear of cabin on port side, with integral airstair. Large cargo door optional. Inward-opening emergency exit on starboard side over wing. Lavatory and stowage for baggage in rear fuselage. Maintenance access door in rear fuselage; radio compartment access doors in nose. Cabin is air conditioned and pressurised, with electric heat panels to warm cabin before engine starting.
SYSTEMS: Cabin pressurisation by engine bleed air, with a maximum differential of 0.44 bar (6.5 lb/sq in). Cabin air conditioner of 32,000 BTU capacity. Auxiliary electric cabin heating. Oxygen system for flight deck, and 623 litre (22 cu ft) oxygen system for cabin, with automatic drop-down face masks; 2,182 litre (77 cu ft) system optional. Dual vacuum system for instruments. Hydraulic system for landing gear retraction and extension, pressurised to 171 to 191 bars (2,475 to 2,775 lb/sq in). Separate hydraulic system for brakes. Electrical system has two 250 A 28 V starter/generators and a 24 V 34 Ah air-cooled Ni/Cd battery with failure detector. AC power provided by dual 250 VA inverters. Engine fire detection system standard; engine fire extinguishing system optional. Pneumatic de-icing of wings and tailplane standard. Anti-icing of engine air intakes by hot air from engine exhaust, electrothermal anti-icing for propellers.
AVIONICS: Standard Rockwell Collins Pro Line II package.
  Comms: Cockpit-to-cabin paging standard. Honeywell KHF 950 transceiver, Fairchild A-100A cockpit voice recorder standard, Wulfsberg Flitefone optional.
  Radar: Rockwell Collins WXR-270 colour weather radar standard, WXR-840 or WXR-850 turbulence detecting radar optional.
  Flight: Universal UNS-1D and UNS-1K navigation management systems with GPS optional.
  Instrumentation: Pilot's ALT-80A encoding altimeter; dual maximum allowable airspeed indicators, and flight director standard. Options include Rockwell Collins three-tube EFIS-85B with MFD.
EQUIPMENT: Standard/optional equipment generally as for King Air C90B except fluorescent cabin lighting, one-place couch with storage drawers, flushing toilet (B200) or chemical toilet (B200C), cabin electric heating, cockpit/cabin partition with sliding doors, and airstair door with hydraulic snubber and courtesy light, standard. FAR Pt 135 operational configuration includes cockpit fire extinguisher and 2.2 m³ (77 cu ft) oxygen bottle with

Beech King Air B200 seven-passenger pressurised transport

cockpit oxygen pressure indicator as standard. A range of optional cabin seating and cabinetry configurations is available, including quick-removable fold-up seats.

DIMENSIONS, EXTERNAL:
Wing span	16.61 m (54 ft 6 in)
Wing chord: at root	2.18 m (7 ft 1¾ in)
at tip	0.90 m (2 ft 11½ in)
Wing aspect ratio	9.8
Length overall	13.36 m (43 ft 10 in)
Height overall	4.52 m (14 ft 10 in)
Tailplane span	5.61 m (18 ft 5 in)
Wheel track	5.23 m (17 ft 2 in)
Wheelbase	4.56 m (14 ft 11½ in)
Propeller diameter	2.39 m (7 ft 10 in)
Propeller ground clearance	0.43 m (1 ft 4¾ in)
Distance between propeller centres	5.23 m (17 ft 2 in)
Passenger door: Height	1.31 m (4 ft 3½ in)
Width	0.68 m (2 ft 2¾ in)
Height to sill	1.17 m (3 ft 10 in)
Cargo door (optional): Height	1.32 m (4 ft 4 in)
Width	1.24 m (4 ft 1 in)

Nose avionics service doors (port and stbd):
Max height	0.57 m (1 ft 10½ in)
Width	0.63 m (2 ft 1 in)
Height to sill	1.37 m (4 ft 6 in)
Emergency exit (stbd): Height	0.66 m (2 ft 2 in)
Width	0.50 m (1 ft 7¾ in)

DIMENSIONS, INTERNAL:
Cabin (from forward to rear pressure bulkhead):
Length	6.71 m (22 ft 0 in)
Max width	1.37 m (4 ft 6 in)
Max height	1.45 m (4 ft 9 in)
Floor area	7.80 m² (84 sq ft)
Volume	11.1 m³ (392 cu ft)

Baggage hold, rear of cabin:
Volume	1.5 m³ (54 cu ft)

AREAS:
Wings, gross	28.15 m² (303.0 sq ft)
Ailerons (total)	1.67 m² (18.00 sq ft)
Trailing-edge flaps (total)	4.17 m² (44.90 sq ft)
Fin	3.46 m² (37.20 sq ft)
Rudder, incl tab	1.40 m² (15.10 sq ft)
Tailplane	4.52 m² (48.70 sq ft)
Elevators, incl tabs (total)	1.79 m² (19.30 sq ft)

WEIGHTS AND LOADINGS:
Weight empty	3,716 kg (8,192 lb)
Baggage capacity	249 kg (550 lb)
Max fuel	1,653 kg (3,645 lb)
Max T-O and landing weight	5,670 kg (12,500 lb)
Max ramp weight	5,710 kg (12,590 lb)
Max zero-fuel weight	4,990 kg (11,000 lb)
Max wing loading	201.4 kg/m² (41.25 lb/sq ft)
Max power loading	4.48 kg/kW (7.35 lb/shp)

PERFORMANCE:
Never-exceed speed (V_{NE})	259 kt (480 km/h; 298 mph) IAS
Max operating Mach No.	0.52
Max level speed at 7,620 m (25,000 ft), average cruise weight	292 kt (541 km/h; 336 mph)
Max cruising speed at 8,230 m (27,000 ft), average cruise weight	287 kt (531 km/h; 330 mph)
Econ cruising speed at 8,230 m (27,000 ft), average cruise weight, normal cruise power	222 kt (411 km/h; 255 mph)
Stalling speed: flaps up	99 kt (183 km/h; 114 mph) IAS
flaps down	75 kt (139 km/h; 86 mph) IAS
Max rate of climb at S/L	747 m (2,450 ft)/min
Rate of climb at S/L, OEI	226 m (740 ft)/min
Service ceiling	over 10,670 m (35,000 ft)
Service ceiling, OEI	6,675 m (21,900 ft)
T-O run, 40% flap	567 m (1,860 ft)
T-O to 15 m (50 ft), 40% flap	786 m (2,580 ft)

Landing from 15 m (50 ft):
without propeller reversal	867 m (2,845 ft)
with propeller reversal	632 m (2,075 ft)
Landing run	536 m (1,760 ft)

Range with max fuel, allowances for start, taxi, climb, descent, and 45 min reserves at max range power, ISA:
max cruise power:
at 5,485 m (18,000 ft)	1,157 n miles (2,142 km; 1,331 miles)
at 8,230 m (27,000 ft)	1,477 n miles (2,735 km; 1,699 miles)
at 9,450 m (31,000 ft)	1,611 n miles (2,983 km; 1,854 miles)
at 10,670 m (35,000 ft)	1,850 n miles (3,426 km; 2,129 miles)

econ cruise power:
at 5,485 m (18,000 ft)	1,469 n miles (2,720 km; 1,690 miles)
at 8,230 m (27,000 ft)	1,773 n miles (3,283 km; 2,040 miles)
at 9,450 m (31,000 ft)	1,859 n miles (3,442 km; 2,139 miles)

UPDATED

BEECH KING AIR 200/A200/B200 (MILITARY VERSIONS)

US basic military designation: C-12

TYPE: Twin-turboprop light transport/multisensor surveillance twin-turboprop.

Beech RC-12P Guardrail Common Sensor (System 2) *(US Army)* 2001/0034481

PROGRAMME: US Army procured first three King Air 200s designated RU-21Js in 1971; ordered 60 military passenger-carrying King Air A200s designated C-12A beginning FY73; worldwide deployment began July 1975. Total of 380 ordered by US armed forces; final airframe delivery in 1997, but outfitting of battlefield reconnaissance versions was continuing in 2000. Further eight exports.

Total of 126 Army C-12s modified in the field with Raisbeck Engineering's Ram Air Recovery Systems (RARS) and Dual Aft Body Strakes (DABS) between October 1997 and April 1998.

CURRENT VERSIONS: **C-12A**: Initial A200 version, powered by 559 kW (750 shp) P&WC PT6A-38 turboprops with Hartzell three-blade fully feathering reversible-pitch propellers; auxiliary tanks. Total 91 delivered (US Army 60; US Air Force 30; Greek Air Force one); entered service July 1975; USAF aircraft mainly with embassy support flights; details in 1980-81 *Jane's*. See C-12C and C-12E for C-12A re-engining.

UC-12B: US Navy/Marine Corps version (Model A200C) with 634 kW (850 shp) PT6A-41 turboprops, cargo door, high flotation landing gear. US Navy (49), US Marines (17), delivered by May 1982 for various base communications flights. Of these, 14 converted to **TC-12B** (unofficial designation) crew trainers for VT-31 (Squadron) of Training Wing 4 at Corpus Christi, Texas.

C-12C: As C-12A, but with PT6A-41 turboprops. Deliveries (US Army 14) complete; US Army C-12A fleet re-engined as C-12Cs; five civilianised 1989 for covert operations by US Customs Service, total having increased to 13 by 1998; additional 20 sold in 1996, including 11 to US police forces; further 10 sales in 2000.

C-12D: Model A200CT. As US Army C-12C but cargo door, high flotation landing gear and provision for tip tanks. US Army (24), US Air Force (six for embassy support flights); additional 21 built but converted to RC-12 before delivery to US Army (16) and Israel (five in early 1985), as described below. Wing span (over tip tanks) 16.92 m (55 ft 6 in).

RC-12D Improved Guardrail V: Model A200CT. US Army special mission version; carries AN/USD-9 Improved Guardrail remote-controlled communications intercept and direction-finding system with direct reporting to tactical commanders at corps level and below; aircraft survivability equipment (ASE) system, Carousel IV-E INS and Tacan system, radio datalink, AN/ARW-83(V)5 airborne relay with antennas above and below wings, wingtip elint/comint pods; associated equipment includes AN/TSQ-105(V)4 integrated processing facility, AN/ARM-63(V)4 AGE flightline van and AN/TSC-87 tactical commander's terminal. System prime contractor ESL (now TRW). US Army had 13 RC-12D Improved Guardrail Vs converted from C-12Ds, with deliveries starting in mid-1983; one to HQ Forces Command at Fort McPherson, Georgia, remainder to 1st Military Intelligence Battalion, Wiesbaden, Germany; German-based aircraft reassigned late 1991 and currently with 3rd, 15th and 304th MIBs at Camp Humphreys (South Korea), Fort Hood (Texas) and Fort Huachuca (Arizona), although one converted to JC-12D testbed and one civilianised in 1999. Five new-build aircraft to Israel for 191 Squadron at Sde Dov, designated **FWC-12D**. Wing span (over tip pods) 17.63 m (57 ft 10 in).

C-12E: Designation of 29 US Air Force C-12As retrofitted with PT6A-42 turboprops; two crew plus nine passengers. Assigned to various embassies; further modified to C-12C standard.

C-12F: Operational support aircraft (OSA), similar to Model B200C with PT6A-42 engines; payload choices include two crew and eight passengers, more than 1,043 kg (2,300 lb) freight, two litter patients plus attendants; cargo door standard. First delivery May 1984. US Air Force purchased 40 after initial five year lease, but many transferred to US Army in 1995; US Army National Guard (20 ordered FY85-87) and delivered as 12 **C-12F-1**s and eight **C-12F-2**s; Air National Guard (six ordered FY84). Ex-USAF/ANG aircraft currently in US Army service are designated **C-12F-3**. Planned upgrade of 23 F-3s to Global Air Traffic Management standard authorised June 2000; involves Rockwell Collins WXR-850 weather radar, TCAS, FMS-3000, ADC-3000 and GPS-4000A.

UC-12F: US Navy equivalent of USAF C-12F with PT6A-42 turboprops. US Navy received first of 12 in 1986; two modified with surface search radar and operator's console to **RC-12F** Range Surveillance Aircraft (RANSAC) for Pacific Missile Range Facilities, Barking Sands, Hawaii.

RC-12G: US Army special mission aircraft; based on A200CT; similar to RC-12D but maximum T-O weight increased to 6,804 kg (15,000 lb). Provides real-time intelligence support to field commanders; two crew. Mission equipment contractor Sanders Associates. Three delivered in 1985, after conversion from C-12D; served with 138 MI Company at Orlando, Florida; currently with 304th MIB (one), converted to **JRC-12G** testbed (one) and withdrawn from use (one).

RC-12H Guardrail Common Sensor (System 3 Minus): US Army special mission aircraft, similar to

US MILITARY C-12s

Model	Type	USAF	Army	USN	First Aircraft
A200	C-12A	30			73-1205
			60		73-22250
A200C	UC-12B			66	161185
A200	C-12C		14		78-23126
A200CT	C-12D	6			83-0494
			24		78-23140
A200CT	RC-12D		13		78-23141
-	C-12E	(29[1])			-
B200	C-12F	46[2]			84-0143
			20		85-1261
B200C	UC-12F			10	163553
B200C	RC-12F			2	163563
A200CT	RC-12G		3		80-23372
A200CT	RC-12H		6		83-24313
A200CT	RC-12K		9		85-0147
200	C-12L		3[3]		71-21058
B200C	UC-12M			10	163836
B200C	RC-12M			2	163846
A200CT	RC-12N		15		88-0325
A200CT	RC-12P		9		91-0518
A200CT	RC-12Q		3		93-0697[4]
B200C	C-12R		29		92-3327
Totals		**82**	**208**	**90**	

Notes: [1] Conversions
[2] At least 40 to Army in 1995
[3] Previously RU-21J
[4] Other two are 93-0699 and 93-0701
List excludes C-12J, which is Beech 1900 variant, and exports to Greece (one C-12A) and Israel (seven RC-12)

742 USA: AIRCRAFT—RAYTHEON (Beech)

RC-12D but maximum T-O weight increased to 6,804 kg (15,000 lb); system contractor ESL Inc. Six delivered in 1988 to 3rd MIB, Camp Humphreys, Pyongtaek, South Korea and current in 2000.

C-12J: Variant of Beech 1900 (see later in this section).

RC-12K Guardrail Common Sensor (System 4): Similar to RC-12H except 820 kW (1,100 shp) PT6A-67 turboprops and maximum T-O weight 7,257 kg (16,000 lb). US Army ordered nine in October 1985, of which eight replaced RC-12Ds in 1st MIB, May 1991, one subsequently being lost in accident; ninth retained by Raytheon for RC-12N conversion; further two (4X-FSF and 'FSG) delivered to Israel in May-June 1991.

C-12L: Three RU-21Js (71-21058 to 21060) stripped of Guardrail equipment in 1979 for transport duties but not redesignated until mid-1980s. One civilianised in 1996; other two in storage.

UC-12M: US Navy designation of C-12F but with unique cockpit instruments, lighting and voice communications. Twelve delivered from 1987 to equip base flights; two conversions with surface search radar and operator's console to **RC-12M** RANSAC; ordered 1988 for Pacific Missile Test Center, Point Mugu; now based at Roosevelt Roads, Puerto Rico.

RC-12N Guardrail Common Sensor (System 1): Generally similar to RC-12K, but with 7,348 kg (16,200 lb) maximum T-O weight and equipped with dual EFIS and aircraft survivability equipment/avionics control system (ASE/ACS). ASE suite includes AN/APR-39 radar warning receiver, AN/APR-44 radar warning system, AN/ALQ-136, AN/ALQ-156 and AN/ALQ-162 countermeasure sets chaff/flare and M130 dispensers. Avionics suite includes AN/ARC-186 or AN/ARC-201 VHF-FM radio, AN/ARC-164 Have Quick II UHF-AM radio; AN/APX-100 IFF transponder; three KY-58 and one KIT-1A secure communications systems; Carousel IV INS; AN/ASN-149 GPS receiver. Prototype (85-0149) converted from RC-12K; 15 converted by E-Systems delivered 1992-93 to 224th MIB at Hunter AAF, Georgia and 304th MIB at Libby AAF, Fort Huachuca, Arizona. One lost in accident.

RC-12P Guardrail Common Sensor (System 2): RC-12P has identical avionics and power plant to RC-12N, but different mission equipment (including datalink capability and CHALS-X: communication high-accuracy location system-exploitable), fibre optic cabling, smaller, lighter wing pods and increased T-O weight of 7,484 kg (16,500 lb). Nine aircraft delivered to ESL (now TRW) at Moffett Federal Airfield in 'green' condition late 1994 and 1995; to 15th MIB, Fort Hood, Texas, February 2000.

RC-12Q Direct Air Satellite Relay: Three RC-12Ps modified by Raytheon and TRW. As RC-12P, but with satcom antenna on spine for out-of-theatre transmission of collected data. Operating procedure is one RC-12Q and two RC-12Ps flying with predetermined separation for immediate triangulation of intercepted signals. To TRW in 1996 for outfitting; prominent dorsal radome for satcom antenna; to 15th MIB, February 2000.

C-12R: Off-the-shelf Model B200C for US Army Reserve and National Guard. Initial 15 ordered FY94; option for further 14 exercised 1995; latter for delivery between July 1996 and August 1997. First deliveries in late 1994 to A Company 2-228 Aviation Battalion at Willow Grove, Pennsylvania, followed by B company at McCoy AAF, Wisconsin and D Company of 1-228 at Columbia, South Carolina, and State Area Commands of Alabama, Arizona, Mississippi, Montana, New Mexico and Washington.

C-12R/AP: Two camera-equipped special mission aircraft for Greek government ordered via US Army on 31 January 2000, for delivery by mid-2001.

C-12S: See King Air 350.

CUSTOMERS: See table and under Current Versions above.
UPDATED

BEECH KING AIR 350
US Army designation: C-12S
JGSDF designation: LR-2
TYPE: Utility turboprop twin.
PROGRAMME: Replaced King Air 300 (1991-92 *Jane's*); first flight (N120SK) September 1988; introduced at NBAA Convention 1989; certified to FAR Pt 23 (commuter category); first delivery 6 March 1990; Russian certification to AP 23 in November 1995; FAA approval for operation from unprepared runways granted during 1997.
CURRENT VERSIONS: **King Air 350:** Baseline version.
Detailed description applies to King Air 350.
King Air 350C: Has 132 × 132 cm (52 × 52 in) freight door with built-in airstair passenger door.
King Air 350 Special Missions: Versions available for aerial photography and airways and ground-based navaid checking. Single hardpoint under each wing for external stores; optional bubble windows and belly camera bay. Maritime patrol version has 7,257 kg (16,000 lb) maximum weight and a 1,292 kg (2,850 lb) mission payload, including ventral radar and FLIR; survival kits can be dispensed through an optional dropping hatch.
Australian Army received KingAir 350 VH-HPJ in 1998 as follow-on (to Douglas C-47 Dakota) test platform for Ingara SAR/MTI radar in ventral pannier; installation by Hawker Pacific.

Beech King Air 350 eight/11-passenger turboprop business aircraft 2001/0103503

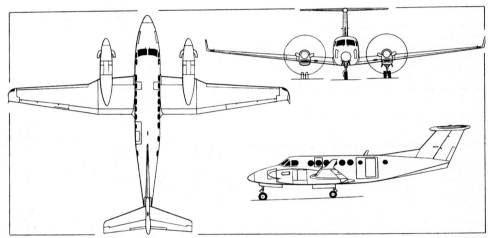
Beech King Air 350 (*Dennis Punnett/Jane's*) 2000/0085295

RC-350 Guardian: Elint version, converted from 350 prototype 1991 by Beech Aircraft Corporation; mission avionics include Raytheon AN/ALQ-142 ESM, Watkins-Johnson 9195C communications interceptor, Honeywell laser INS, GPS receiver and Cubic secure digital datalink; can loiter on station at 10,670 m (35,000 ft) for more than 6 hours; can locate/monitor radar emitters in 20 MHz to 18 GHz range, and intercept communications within 20 to 1,400 MHz bandwidths. Wingtip pods house AN/ALQ-142 antennas; underfuselage bulge contains antenna for comint system.

LR-2: Japan Ground Self-Defence Force funded two in FY97 for liaison and reconnaissance (undisclosed sensor in ventral radome). Total requirement for 20, of which third funded in 1999 and fourth in FY00. First delivery (23051) 22 January 1999; initial operator is the HQ Flight of 1st Helicopter Brigade at Kisarazu.

C-12S: US Army version with quick-change cargo capability and seating for up to 15 passengers. By late 2000, no aircraft of this type had been identified in service.

CUSTOMERS: Total 289 King Air 350s and 350Cs delivered by 1 January 2000; first 350C delivery in 1990 to Rossing Uranium, Namibia; 15 King Air 350 deliveries 1995, 27 in 1996, 30 in 1997, 42 in 1998, 45 in 1999, and 46 in 2000. Recent orders include three, with three options, for the US Drug Enforcement Administration, for delivery during 1999.

DESIGN FEATURES: Compared with King Air 300, fuselage stretched 0.86 m (2 ft 10 in) by plugs 0.37 m (1 ft 2½ in) forward of main spar and 0.49 m (1 ft 7½ in) aft; wing span increased by 0.46 m (1 ft 6 in) with NASA winglets 0.61 m (2 ft 0 in) high; two additional cabin windows each side. Can depart with full payload and full tanks. Raisbeck dual aft body strakes (DABS), standard on production aircraft from c/n FL-312 (N3165M) in first quarter 2001, reduce drag, improve handling and stability and relax or eliminate restrictions on operations with inoperative yaw damper.

FLYING CONTROLS: Automatic cable tensioner in aileron circuit and larger elevator bobweight; larger rudder anti-servo tab; ailerons and rudder cleaned up.
STRUCTURE: As for B200.
LANDING GEAR: As for B200.
POWER PLANT: Two 783 kW (1,050 shp) Pratt & Whitney Canada PT6A-60A turboprops, each driving a Hartzell four-blade constant-speed fully feathering reversible-pitch metal propeller. Bladder cells and integral tanks in each wing, with usable capacity of 1,438 litres (380 US gallons; 316.5 Imp gallons); auxiliary tanks inboard of engine nacelles, capacity 601 litres (159 US gallons; 132.5 Imp gallons). Total fuel capacity 2,040 litres (539 US gallons; 449 Imp gallons). No provision for wingtip tanks. Oil capacity 30.2 litres (8.0 US gallons; 6.7 Imp gallons).
ACCOMMODATION: Double club seating for eight passengers; optionally two more seats in rear of cabin and one passenger on side-facing toilet seat making maximum 11 passengers; certified for maximum 17 occupants including crew. Ultra Electronics UltraQuiet active noise control system installed as standard from 1998.
SYSTEMS: As for B200, except for automatic bleed air-type heating and 22,000 BTU cooling system with high-capacity ventilation system; 2,182 litre (77 cu ft) oxygen system standard; hydraulic landing gear retraction and extension; two 300 A 28 V starter/generators with triple bus electrical distribution system. Ultra Electronics Ltd UltraQuiet active noise control system introduced as standard from 1998, comprising 12 loudspeakers, 24 microphones and a high-speed digital processor which cancel propeller noise and reduce in-flight cabin sound level to less than 80 dB(A).
AVIONICS: Generally as for B200.
EQUIPMENT: Generally as for B200.
DIMENSIONS, EXTERNAL: As for B200 except:
 Wing span over winglets   17.65 m (57 ft 11 in)
 Wing aspect ratio   10.8

Jane's All the World's Aircraft 2001-2002   www.janes.com

Australian Army King Air 350 carrying Ingara SAR/MTI radar   2001/0105830

Length overall	14.22 m (46 ft 8 in)
Height overall	4.37 m (14 ft 4 in)
Propeller diameter	2.67 m (8 ft 9 in)
Propeller ground clearance	0.29 m (11½ in)
Emergency exit (each side of cabin, above wing):	
Height	0.66 m (2 ft 2 in)
Width	0.50 m (1 ft 7¾ in)
DIMENSIONS, INTERNAL:	
Cabin (excl cockpit): Length	5.84 m (19 ft 2 in)
Max width	1.37 m (4 ft 6 in)
Height	1.45 m (4 ft 9 in)
Baggage hold volume	1.5 m³ (54 cu ft)
AREAS:	
Wings, gross	28.80 m² (310.0 sq ft)
WEIGHTS AND LOADINGS:	
Weight empty	4,132 kg (9,110 lb)
Max fuel weight	1,638 kg (3,611 lb)
Max T-O and landing weight	6,804 kg (15,000 lb)
Max ramp weight	6,849 kg (15,100 lb)
Max zero-fuel weight	5,670 kg (12,500 lb)
Max wing loading	236.2 kg/m² (48.39 lb/sq ft)
Max power loading	4.35 kg/kW (7.14 lb/shp)
PERFORMANCE:	
Max level speed	315 kt (584 km/h; 363 mph)
Max cruising speed, AUW of 5,896 kg (13,000 lb) at:	
7,315 m (24,000 ft)	311 kt (576 km/h; 358 mph)
10,670 m (35,000 ft)	290 kt (537 km/h; 334 mph)
Cruising speed, normal cruising power, AUW of 5,896 kg (13,000 lb) at:	
7,315 m (24,000 ft)	301 kt (558 km/h; 347 mph)
10,670 m (35,000 ft)	281 kt (521 km/h; 324 mph)
Cruising speed, max range power, AUW of 5,896 kg (13,000 lb) at:	
5,485 m (18,000 ft)	210 kt (389 km/h; 242 mph)
10,670 m (35,000 ft)	240 kt (445 km/h; 276 mph)
Stalling speed at max landing weight, flaps and wheels down	81 kt (150 km/h; 94 mph)
Max rate of climb at S/L	832 m (2,731 ft)/min
Rate of climb at S/L, OEI, AUW of 6,350 kg (14,000 lb)	236 m (775 ft)/min
Service ceiling	above 10,670 m (35,000 ft)
Service ceiling, OEI	6,555 m (21,500 ft)
T-O balanced field length	1,006 m (3,300 ft)
Landing from 15 m (50 ft)	821 m (2,695 ft)
Landing run	441 m (1,450 ft)

Range with 2,040 litres (539 US gallons; 449 Imp gallons) usable fuel, allowances for start, T-O, climb and descent plus 45 min reserves:

max cruising power at: 5,485 m (18,000 ft)
    1,032 n miles (1,911 km; 1,187 miles)
7,315 m (24,000 ft)
    1,211 n miles (2,242 km; 1,393 miles)
8,535 m (28,000 ft)
    1,358 n miles (2,515 km; 1,562 miles)
10,670 m (35,000 ft)
    1,666 n miles (3,085 km; 1,917 miles)
normal cruising power, allowances as above:
5,485 m (18,000 ft)
    1,054 n miles (1,952 km; 1,212 miles)
7,315 m (24,000 ft)
    1,265 n miles (2,342 km; 1,455 miles)
8,535 m (28,000 ft)
    1,424 n miles (2,637 km; 1,638 miles)
10,670 m (35,000 ft)
    1,713 n miles (3,172 km; 1,971 miles)
max range power, allowances as above:
5,485 m (18,000 ft)
    1,385 n miles (2,565 km; 1,593 miles)
8,535 m (28,000 ft)
    1,694 n miles (3,137 km; 1,949 miles)
10,670 m (35,000 ft)
    1,813 n miles (3,357 km; 2,086 miles)
Range, NBAA VFR, four passengers, 30 min reserves
    1,983 n miles (3,672 km; 2,282 miles)

UPDATED

BEECH 1900D
US Army designation: C-12J
TYPE: Twin-turboprop airliner.
PROGRAMME: Original Beech 1900 first flew on 3 September 1982; three prototypes followed by 74 1900Cs and 174 wet-wing 1900C-1s by late 1991. Current 1900D announced at US Regional Airlines Association meeting 1989; development of 1900C (1991-92 *Jane's*); prototype (converted from 1900C-1 N5584B) first flight 1 March 1990; certification to FAR Pt 23 Amendment 34 received March 1991; full certification with supplements received, and deliveries (to Mesa Airlines) began, November 1991; contract signed February 1997 with Xian Aircraft Company of People's Republic of China for supply to Raytheon of 800 metal-bonded subassemblies for 1900D, to be delivered between 1997 and 2001; 500th Model 1900 delivered in March 1997; 400th Model 1900D (N44640) rolled out 21 March 2000. Beech also offers special mission versions for signals intelligence, maritime patrol and similar duties.

CURRENT VERSIONS: **1900D**: As described.
1900D Executive: Features custom-designed executive interior ranging from twin double-club to corporate shuttle configuration; refreshment bar, entertainment system and flight phones optional.
C-12J: US Air Force ordered six Beech 1900C-1s (see 1991-92 *Jane's*) in 1986, of which four currently assigned to 3rd, 46th and 51st (two) Wings for support, and two with US Army (HQ Europe and 78th Aviation Battalion in Japan). In March 1997, Army received a further Beech 1900D (96-0112) for Chemical and Biological Defense Command at Aberdeen Proving Ground. C-12 designation more properly belongs to King Air.

CUSTOMERS: Total 65 delivered in 1995, 69 in 1996, 42 in 1997, 45 in 1998, 24 in 1999, and 54 in 2000.

COSTS: US$ 895 million (1998). Six ordered in 2000 by the Algerian Air Force in sigint configuration, equipped with Northrop Grumman Systems radar and FLIR.

DESIGN FEATURES: Flat floor with stand-up headroom; cabin volume increased by 28.5 per cent compared to 1900C; winglets add better hot-and-high performance; tailplane and fin swept; each tailplane carries small fin (tailet) on underside near tip; auxiliary horizontal fixed tail surface (stabilon) each side of rear fuselage improve centre of gravity range; twin ventral strakes improve directional stability and turbulence penetration; small horizontal vortex generator on fuselage ahead of wingroots.

Wing aerofoil NACA 23018 (modified) at root, 23012 (modified) at tip; dihedral 6°; incidence 3° 29′ at root, −1° 4′ at tip; no sweepback at quarter-chord.

FLYING CONTROLS: Conventional and manual. Automatic cable tensioner in aileron circuit; trim tabs in elevators, rudder and port aileron; primary and secondary controls routed separately to improve protection from possible engine-failure damage; single-slotted trailing-edge flaps in two sections on each wing.

STRUCTURE: Generally of light alloy. Wing has continuous main spar with fail-safe structure riveted and bonded; fuselage pressurised and mainly bonded.

LANDING GEAR: Hydraulically retractable tricycle type; main units retract forward and nose unit rearward; Beech oleo-pneumatic shock-absorber in each unit. Twin Goodyear wheels on each main unit, size 6.50×10, with Goodyear tyres size 22×6.75-10 (10 ply) tubeless, pressure 6.69 bars (97 lb/sq in); Goodyear steerable nosewheel size 6.50×8, with Goodyear tyres size 19×6.75-8 (10 ply) tubeless, pressure 4.14 bars (60 lb/sq in). Multiple-disc hydraulic brakes. Optional Beech Hydro-Aire anti-skid units, power steering and brake de-icing. Ground turning radius based on wingtip clearance 12.55 m (41 ft 2 in).

POWER PLANT: Two Pratt & Whitney Canada PT6A-67D turboprops, each flat rated at 954 kW (1,279 shp) and driving a Hartzell four-blade constant-speed fully feathering reversible-pitch composites propeller. Wet wing fuel storage with a total capacity of 2,528 litres (668 US gallons; 556 Imp gallons), of which 2,519 litres (665 US gallons; 554 Imp gallons) usable. Refuelling point in each wing leading-edge, inboard of engine nacelle. Oil capacity (total) 29.5 litres (7.8 US gallons; 6.5 Imp gallons).

ACCOMMODATION: Crew of one (FAR Pt 91) or two (FAR Pt 135) on flight deck, with standard accommodation in cabin of commuter version for 19 passengers in single airline-standard seats on each side of centre aisle. Forward carry-on baggage lockers, underseat baggage stowage, rear

Beech 1900D operating feeder service for US Air *(Paul Jackson/Jane's)*   2001/0103581

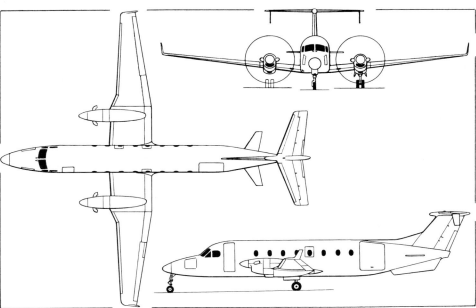

Beech 1900D regional transport *(Dennis Punnett/Jane's)*

744 USA: AIRCRAFT—RAYTHEON (Beech)

baggage compartment. Forward door, incorporating airstairs, on port side. Upward-hinged rear cargo door, also on port side. Three emergency exits over wing (two starboard, one port). Accommodation air conditioned, heated, ventilated and pressurised. Ultra Electronics UltraQuiet active noise cancellation system optional. Executive and corporate shuttle options seat between 10 and 18 passengers with options for forward and rear compartments, combination lavatory/passenger seat and two beverage bars at cabin compartment division. Club, double club and triple club seating optional. Customised interiors to customer's choice.

SYSTEMS: Bleed air cabin heating and pressurisation, maximum differential 0.35 bar (5.1 lb/sq in). Air cycle and vapour cycle air conditioning. Hydraulic system, pressure 207 bars (3,000 lb/sq in), for landing gear actuation. Electrical system includes two 300 A engine-driven starter/generators and one 34 Ah Ni/Cd battery. Constant flow oxygen system of 4,420 litre (156 cu ft) capacity standard. Engine inlet screen anti-ice protection, exhaust heated engine inlet lips, fuel vent heating, electric propeller and windscreen de-icing systems standard. Brake de-icing optional. Pneumatic de-icing boots on wings, tailplane, tailets and stabilons.

AVIONICS: *Comms:* Rockwell Collins Pro Line II digital technology radios; cabin briefer; cockpit voice recorder.
Flight: Dual flight directors; GPWS; provision for TCAS1. GPS optional.
Instrumentation: Rockwell Collins EFIS-84 four-tube EFIS. Primary display consists of multicolour CRT panels, remote display processor unit and system control units; CRT displays provide conventional electronic attitude director indicator (EADI) and electronic horizontal situation indicator (EHSI) functions.

DIMENSIONS, EXTERNAL:
Wing span over winglets	17.67 m (57 ft 11¾ in)
Wing chord: at root	2.18 m (7 ft 1¾ in)
at tip	0.91 m (3 ft 0 in)
Wing aspect ratio	10.9
Length overall	17.63 m (57 ft 10 in)
Height overall	4.72 m (15 ft 6 in)
Tailplane span	5.63 m (18 ft 5¾ in)
Wheel track	5.23 m (17 ft 2 in)
Wheelbase	7.25 m (23 ft 9½ in)
Propeller diameter	2.78 m (9 ft 1½ in)
Propeller ground clearance	0.35 m (1 ft 1¾ in)
Distance between propeller centres	5.23 m (17 ft 2 in)
Passenger door: Height	1.63 m (5 ft 4¼ in)
Width	0.64 m (2 ft 1¼ in)
Cargo door: Height	1.45 m (4 ft 9 in)
Width	1.32 m (4 ft 4 in)
Emergency exits (each): Height	0.80 m (2 ft 7½ in)
Width	0.51 m (1 ft 8 in)

DIMENSIONS, INTERNAL:
Cabin (incl flight deck and rear baggage compartment):
Length	12.03 m (39 ft 5½ in)
Max width	1.37 m (4 ft 6 in)
Max height	1.80 m (5 ft 10¾ in)
Floor area	15.3 m² (165 sq ft)
Pressurised volume	26.0 m³ (918 cu ft)
Volume of passenger cabin	18.1 m³ (640 cu ft)
Baggage volume, cabin: forward	0.34 m³ (12.0 cu ft)
underseat	0.91 m³ (32.0 cu ft)
rear	5.0 m³ (175 cu ft)

AREAS:
Wings, gross	28.80 m² (310.0 sq ft)
Ailerons (total)	1.67 m² (18.00 sq ft)
Trailing-edge flaps (total)	4.17 m² (44.90 sq ft)
Fin	4.86 m² (52.30 sq ft)
Rudder (incl tab)	1.40 m² (15.10 sq ft)
Tailets (total)	0.63 m² (6.80 sq ft)
Tailplane	6.32 m² (68.00 sq ft)
Elevator (incl tab)	1.79 m² (19.30 sq ft)
Stabilons (total)	1.44 m² (15.50 sq ft)

WEIGHTS AND LOADINGS:
Weight empty (typical)	4,831 kg (10,650 lb)
Max fuel (usable)	2,022 kg (4,458 lb)
Max baggage	939 kg (2,070 lb)
Max ramp weight	7,738 kg (17,060 lb)
Max T-O weight	7,688 kg (16,950 lb)
Max landing weight	7,530 kg (16,600 lb)
Max zero-fuel weight	6,804 kg (15,000 lb)
Max wing loading	266.9 kg/m² (54.68 lb/sq ft)
Max power loading	4.03 kg/kW (6.63 lb/shp)

PERFORMANCE:
Max cruising speed at AUW of 6,804 kg (15,000 lb):
at 2,440 m (8,000 ft)	272 kt (504 km/h; 313 mph)
at 4,875 m (16,000 ft)	283 kt (524 km/h; 326 mph)
at 7,620 m (25,000 ft)	274 kt (507 km/h; 315 mph)

Unstick speed, T-O flap setting
    105 kt (195 km/h; 121 mph) IAS
Approach speed at max landing weight
    117 kt (217 km/h; 135 mph)
Stalling speed at max T-O weight:
  wheels and flaps up   101 kt (187 km/h; 116 mph)
  wheels down, T-O flap setting
    90 kt (167 km/h; 104 mph)
Stalling speed at max landing weight, wheels and flaps
  down   84 kt (156 km/h; 97 mph)
Max rate of climb at S/L   800 m (2,625 ft)/min
Rate of climb at S/L, OEI   192 m (630 ft)/min

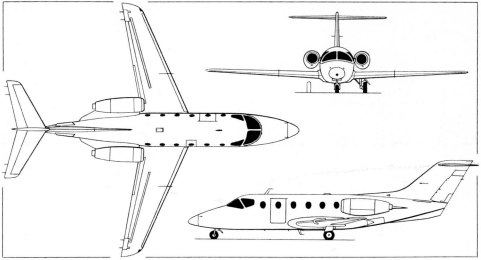
Beechjet 400A (two P&WC JT15D-5 turbofans) *(Dennis Punnett/Jane's)*

Service ceiling	10,058 m (33,000 ft)
Service ceiling, OEI	5,181 m (17,000 ft)
T-O field length, T-O flap setting	1,140 m (3,740 ft)

Landing from 15 m (50 ft) at max landing weight
    844 m (2,770 ft)
Range with 10 passengers, at long-range cruise power, with allowances for starting, taxi, T-O, climb and descent, with reserves (45 min hold at 1,525 m; 5,000 ft)   1,476 n miles (2,733 km; 1,698 miles)

UPDATED

BEECH BEECHJET 400A
US Air Force designation: T-1A Jayhawk
JASDF designation: T-400

TYPE: Business jet.
PROGRAMME: Conceived as Mitsubishi MU-300 Diamond; first flight 29 August 1978; two prototypes; FAR Pt 25 certification awarded 6 November 1981; production aircraft fabricated in Japan and assembled at San Angelo, Texas; deliveries totalled 63 Diamond Is (JT15D-4 engines), 27 Diamond IAs (JT15D-4D) and one Diamond II (JT15D-5).

Beech acquired rights to Diamond II from Mitsubishi Heavy Industries and Mitsubishi Aircraft International, December 1985; made improvements to aircraft and renamed it Beechjet 400. First Beech-assembled Beechjet rolled out 19 May 1986; initial 64 used Japanese components. During 1989, Beech moved entire manufacturing operation to Wichita. Announced new Beechjet 400A November 1989, featuring certification to 13,715 m (45,000 ft), larger and more comfortable cabin, all Collins avionics with digital EFIS; customer deliveries began November 1990.

CURRENT VERSIONS: **Beechjet 400:** Initial production version (64 built; see earlier *Jane's*); superseded by 400A.
Beechjet 400A: Announced at 1989 NBAA show; production 400A first flight 22 September 1989; FAA certification received 20 June 1990; deliveries began November 1990. Also certified by July 1993 in Australia, Canada, France, Germany, Italy and UK; Brazilian and Pakistani type approval April 1994; Civil Aviation Authority of China certification achieved in second quarter of 1999.
Description applies to 400A, except where indicated.
Beechjet T-1A Jayhawk: US Air Force selected McDonnell Douglas, Beech and Quintron to supply Tanker Transport Training System (TTTS) on 21 February 1990, including requirement for 180 Beechjet 400Ts, valued at US$755 million and designated T-1A Jayhawk; represents missionised version of 400A, sharing many components and characteristics with commercial counterpart; differences include cabin-mounted avionics, increased air conditioning capability, greater fuel capacity with single-point refuelling, strengthened windscreen and leading-edges for low-level birdstrike protection. First production aircraft (90-0400) delivered 17 January 1992; deliveries at approximately three per month; final delivery 23 July 1997. By then total fleet time exceeded 182,000 flying hours, with 90 per cent operational availability, and more than 680 pilots had been trained on the Jayhawk.

IOC for USAF Jayhawks January 1993, for Air Education and Training Command Specialised Undergraduate Pilot Training (SUPT) programme at Reese AFB (52nd FTS/64th FTW) where establishment of 41 received by October 1993; Reese closed in 1997. Second recipient was 99th FTS/12th FTW at Randolph AFB, Texas, where 16 delivered for instructor training in 1993; third unit was 86th FTS/47th FTW at Laughlin AFB, Texas, from late 1993 with training courses beginning May 1994; fourth was 71st FTW at Vance AFB, Oklahoma (first aircraft December 1994); fifth was 14th FTW at Columbus AFB, Mississippi (early 1996). T-1A used for training crews for KC-10, KC-135, C-5 and C-17, with total fleet experience of more than 376,000 hours and more than 733,000 landings by October 1999.

In February 1997 Raytheon Aircraft and its subsidiary Raytheon Aerospace were awarded a contract, valued at US$6.2 million, to retrofit 62 Jayhawks with GPS; two further options to retrofit the entire fleet would bring the total value of the GPS upgrade to about US$25.3 million.
Beechjet 400T: JASDF T-400 version, featuring thrust reversers, long-range inertial navigation and direction-finding systems; interior changes. Meets TC-X trainer requirement; three, three, two and one ordered in 1992-95, plus one in 1998; first (41-5051) delivered 31 January 1994; 10th in 2000.
CUSTOMERS: The 500th Beechjet/Jayhawk (N500TH; 246th 400A) was delivered to Global Financial Services Group of Anderson, South Carolina, on 12 October 1999, during the NBAA Convention at Atlanta, Georgia. Total 30 delivered in 1995, 29 in 1996, 43 in both 1997 and 1998, 48 in 1999, and 51 in 2000. Recent customers include Hainan Airlines of China, which took delivery of one aircraft in February 1999. US Air Force 180 T-1As ordered, of which delivery completed 23 July 1997. JASDF had received 10 by December 1999 and ordered further two in 2000; operated by 41 Hikotai at Miho. By October 2000, 303 civil 400As had been registered, in addition to 64 Model 400s, 180 Jayhawks, 10 400Ts and 93 Diamonds, or 650 in all.

BEECHJET T-1A JAYHAWK PROCUREMENT

FY	Lot	Qty	First Aircraft	Delivery
89	–	1	89-0284	1991
90	1	14	90-0400	1992
91	2	28	91-0075	1992-93
92	3	34	92-0330	1993-94
93	4	36	93-0621	1994-95
94	5	35	94-0114	1995-96
95	6	32	95-0040	1996-97
Total		**180**		

Beechjet 400A twin-turbofan business aircraft *(Paul Jackson/Jane's)*

COSTS: Jayhawk programme cost US$1.3 billion; Beech contracts for 180 aircraft, US$755 million. Civil 400A quoted at US$5.4 million in early 1995.

DESIGN FEATURES: Typical low-wing, T tail, rear-engined small business jet, with sweptback wings and empennage, plus small underfin. Compared to Diamond, Beechjet has increased payload and certified ceiling, greater cabin volume achieved by moving rear-fuselage fuel tank forward under floor (balanced by moving lavatory to rear of cabin), improved soundproofing, and emergency door moved one window forward to facilitate forward club seating.

Wing has computer-designed three-dimensional Mitsubishi MAC510 aerofoil; thickness/chord ratio 13.2 per cent at root, 11.3 per cent at tip; dihedral 2° 30′; incidence 3° at root, −3° 30′ at tip; sweepback 20° at quarter-chord.

T-1A Jayhawk features include student pilot in left seat, instructor on right and pupil/observer behind instructor; more bird-resistant windscreen and leading-edges; fewer cabin windows; strengthened wing carry-through structure and engine attachment points to meet low-level flight stresses; rails for four passenger seats in cabin for personnel transport; avionics relocated from nose to rack in cabin to facilitate nose installation of air conditioning; emergency door moved forward to position opposite main cabin door to allow straight-through egress; improved brakes; additional fuel tank; single-point pressure refuelling; Rockwell Collins five-tube EFIS; digital autopilot; weather radar; central diagnostic and maintenance system; Tacan with air-to-air capability.

FLYING CONTROLS: Conventional and manual. Variable incidence tailplane and elevators for pitch axis; lateral control by small ailerons and almost full semi-span, narrow chord spoilers used also as airbrakes and lift dumpers; rudder with trim tab; narrow chord Fowler-type flaps, double-slotted inboard and single-slotted outboard, occupy most of trailing-edges and are hydraulically actuated; mid-span leading-edge fences on wing; small horizontal strakes on fuselage at base of fin; small ventral fin.

STRUCTURE: Wings include integrally machined metal upper and lower skins joined to two box spars forming integral fuel tank; tailplane and fin similar. Wing, fuselage and tail unit certified fail-safe for unlimited life (with periodic inspections and maintenance).

LANDING GEAR: Retractable tricycle type, with single wheel and oleo-pneumatic shock-absorber on each unit. Hydraulic actuation, controlled electrically. Emergency free-fall extension. Main tyres 24×7.7 (16 ply) tubeless; nose tyre 18×4.4 (10 ply) tubeless. Nosewheel, which is steerable by rudder pedals, retracts forward; mainwheels retract inward into fuselage. Goodyear wheels and tyres; Aircraft Braking Systems brakes.

POWER PLANT: Two Pratt & Whitney Canada JT15D-5 turbofans, each rated at 13.19 kN (2,965 lb st) for take-off. Nordam thrust reversers optional on 400A, but not fitted to T-1A. Total usable fuel capacity: 400A 2,775 litres (733 US gallons; 610 Imp gallons); 400T 2,998 litres (792 US gallons; 656 Imp gallons). One refuelling point in top of each wing, and one in rear fuselage for fuselage tank, capacity 1,158 litres (306 US gallons; 255 Imp gallons). (T-1A, single-point refuelling.) Oil capacity 7.7 litres (2.0 US gallons; 1.7 Imp gallons).

ACCOMMODATION: Crew of two on flight deck of 400A on vertically and horizontally adjustable reclining seats with five-point safety harnesses; T-1A has seats for trainee pilot, co-pilot/instructor and observer. Improved interior introduced 1996, featuring redesigned trim panels, enhanced acoustic panels and vibration-damping engine mounts.

Standard 'centre club' layout of 400A seats eight passengers in pressurised cabin. Of these, seven are on tracking, 360° swivelling, reclining seats: four in facing pairs, two forward-facing and one aft-facing; each with integral headrest, armrest and shoulder harness. Fold-out writing table between each pair of seats. Private flushing lavatory at rear with sliding doors and optional illuminated vanity unit and hot water supply. With seat belts, this compartment can serve as eighth passenger seat.

Interior options include substitution of carry-on baggage compartment for one of the forward centre seats, and hot and cold service refreshment centre with integral stereo entertainment system. Independent temperature control for flight deck and cabin heating systems standard. In-flight telephone optional. Tailcone baggage compartment with external access. Optional four passenger seats in main cabin of T-1A. The 400T has an aft club arrangement with swivel chairs.

SYSTEMS: Pressurisation system, with normal differential of 0.63 bar (9.1 lb/sq in). Back-up pressurisation system, using engine bleed air, for use in emergency. Hydraulic system, pressure 103.5 bars (1,500 lb/sq in), for actuation of flaps, landing gear and other services. Each variable volume output engine-driven pump has a maximum flow rate of 14.76 litres (3.9 US gallons; 3.25 Imp gallons)/min, and one pump can actuate all hydraulic systems. Reservoirs, capacity 4.16 litres (1.1 US gallons; 0.9 Imp gallon), pressurised by filtered engine bleed air at 1.03 bars (15 lb/sq in). All systems are, wherever possible, of modular conception: for example, entire hydraulic installation can be removed as a single unit. Stick shaker as back-up stall warning device.

AVIONICS: *Flight:* GPS retrofitted to some T-1As.

Instrumentation: Standard avionics include pilot's integrated Rockwell Collins Pro Line 4 EFIS featuring three-tube (optional four-tube) colour CRT primary flight display (PFD) and multifunction display (MFD) units mounted side by side, and control/display unit. PFD displays airspeed, altitude, vertical speed, flight director, attitude and horizontal situation information, while MFD displays navigation, radar, map, checklist and fault annunciation information. Smaller, single or dual CRTs mounted on central console function as independent navigation sensor displays or back-up displays for main CRTs. EFIS installation features strapdown attitude/heading referencing system, electronic map navigation display, airspeed trend information and V-speeds on Mach airspeed display, and solid-state Doppler turbulence detection radar.

DIMENSIONS, EXTERNAL:	
Wing span	13.25 m (43 ft 6 in)
Wing aspect ratio	7.8
Length overall	14.75 m (48 ft 5 in)
Fuselage: Length	13.15 m (43 ft 2 in)
Max width	1.68 m (5 ft 6 in)
Max depth	1.85 m (6 ft 1 in)
Height overall	4.24 m (13 ft 11 in)
Tailplane span	5.00 m (16 ft 5 in)
Wheel track	2.84 m (9 ft 4 in)
Wheelbase	5.86 m (19 ft 3 in)
Crew/passenger door: Height	1.27 m (4 ft 2 in)
Width	0.71 m (2 ft 4 in)
DIMENSIONS, INTERNAL:	
Cabin: Length: incl flight deck	6.32 m (20 ft 9 in)
excl flight deck	4.72 m (15 ft 6 in)
Max width	1.50 m (4 ft 11 in)
Max height	1.45 m (4 ft 9 in)
Volume: incl flight deck	11.3 m³ (400 cu ft)
excl flight deck	8.6 m³ (305 cu ft)
Baggage compartment volume:	
tailcone	0.75 m³ (26.4 cu ft)
cabin, optional	0.34 m³ (12.0 cu ft)
AREAS:	
Wings, net	22.43 m² (241.4 sq ft)
Trailing-edge flaps (total)	4.22 m² (45.40 sq ft)
Spoilers (total)	0.57 m² (6.20 sq ft)
Fin, incl dorsal fin	5.91 m² (63.60 sq ft)
Rudder, incl yaw damper	0.99 m² (10.70 sq ft)
Tailplane	5.25 m² (56.50 sq ft)
Elevators, incl tab	1.55 m² (16.70 sq ft)
WEIGHTS AND LOADINGS:	
Basic operating weight, incl crew, avionics and interior fittings	4,921 kg (10,850 lb)
Baggage capacity, tailcone	204 kg (450 lb)
Max fuel weight	2,228 kg (4,912 lb)
Max T-O weight	7,303 kg (16,100 lb)
Max ramp weight	7,393 kg (16,300 lb)
Max landing weight	7,121 kg (15,700 lb)
Max zero-fuel weight	5,896 kg (13,000 lb)
Max power loading	284 kg/kN (2.78 lb/lb st)
PERFORMANCE:	
Max limiting Mach No.	0.78
Max level speed at 8,230 m (27,000 ft)	468 kt (867 km/h; 539 mph)
Typical cruising speed at 12,500 m (41,000 ft)	450 kt (834 km/h; 518 mph)
Long-range cruising speed at 12,500 m (41,000 ft)	392 kt (726 km/h; 451 mph)
Stalling speed, flaps down, idling power	93 kt (173 km/h; 107 mph) CAS
Max rate of climb at S/L	1,149 m (3,770 ft)/min
Rate of climb at S/L, OEI	306 m (1,005 ft)/min
Service ceiling	13,240 m (43,450 ft)
Service ceiling, OEI	6,279 m (20,600 ft)
FAA (FAR Pt 25) T-O to 10.7 m (35 ft) at S/L	1,271 m (4,169 ft)
FAA landing distance from 15 m (50 ft) at S/L, max landing weight	902 m (2,960 ft)
Range with max fuel, incl 45 min reserves:	
430 kt (796 km/h; 495 mph) cruise	1,574 n miles (2,915 km; 1,811 miles)
403 kt (746 km/h; 464 mph) cruise	1,673 n miles (3,098 km; 1,925 miles)

UPDATED

HAWKER 800XP

TYPE: Business jet.

PROGRAMME: Derived from the de Havilland/Hawker Siddeley/British Aerospace 125, which was built in the UK from 1962 onwards, progressing through Srs 1 to 3 and 400 to 700, as described under various headings in earlier *Jane's*. Prototype Srs 800 first flew (G-BKTF) 26 May 1983; type certificate gained 4 May 1984 and Public Transport Category C of A on 30 May 1984; FAA certification 7 June 1984; Russian certification May 1993; Canadian certification (800XP) awarded in third quarter of 1997. Adopted Hawker nomenclature when programme purchased by Raytheon in 1993.

Final assembly of Hawker 800XP gradually transferred to Wichita; first US-assembled aircraft flew on 5 November 1996, being N297XP, the 297th Series 800; second (N1105Z; No. 301) followed on 24 November 1996; transition complete with flight of last UK-assembled aircraft, No. 337, 29 April 1997; FAA production certificate awarded to Raytheon in May 1997; UK is providing components sets for Wichita for at least five years thereafter. One thousandth 125/Hawker series aircraft, an 800XP, was delivered as the 'Millennium Hawker' to Gainey Corporation of Grand Rapids, Michigan (N984GC), October 1998.

CURRENT VERSIONS: **800:** Original version superseded by 800XP in late 1995; 275 built, of which last delivered December 1995. Detailed description can be found in 1995-96 and earlier editions of *Jane's*.

800XP (Extended Performance): Announced March 1995, when prototype (G-BVYW, modified from 800) completed; this and preproduction 800XP used in development programme, culminating in CAA and FAA certification in July 1995; first delivery (to Green Tree Financial of St Paul, Minnesota) October 1995 after public debut at National Business Aviation Association Convention in Las Vegas during previous month. First seven delivered in 1995, followed by 26, 33, 48 and 55 in 1996-99.

Detailed description applies to this version.

800FI, SM, RA and SIG: Special missions versions; described separately.

CUSTOMERS: By December 2000 more than 500 Series 800/800XPs built. Largest contract for 125/Hawker placed in May 1997 when Executive Jet Inc ordered 20 800XPs, followed in September 1998 by order for a further 20, plus 16 options for NetJets fractional ownership scheme; deliveries between 1997 and 2004. Recent customers include National Air Service (NAS) of Jeddah, Saudi Arabia, which ordered 14 in November 1999 for its NetJets Middle East fractional ownership programme, five of these being delivered in 2000, including the first (HZ-KSRA) at the Farnborough International Air Show on 24 July 2000, and three per year thereafter until 2003; and Hainan Airlines of China, which took delivery of one aircraft on 19 July 1999.

COSTS: 800XP: US$10.845 million (1998).

DESIGN FEATURES: Classic small business jet; sweptback wing mounted below cabin floor; podded engines on rear fuselage sides; and high tailplane.

Improvements of baseline Series 800, compared with earlier 700 variant, include curved windscreen, sequenced nosewheel doors, extended fin leading-edge, larger ventral fuel tank, and increased wing span which reduces induced drag, enhances aerodynamic efficiency and carries extra fuel; outboard 3.05 m (10 ft) of each wing redesigned.

In XP version, TFE731-5BR-1H turbofans boost performance, including 14 kt (26 km/h; 16 mph) increase in cruising speed at 11,800 kg (26,015 lb) at 11,890 m (39,000 ft); 225 kg (496 lb) payload increase with eight passengers; 15 to 23 per cent reduction in time to cruising

Hawker 800XP owned by Bank of America *(Paul Jackson/Jane's)* **2001**/0103583

USA: AIRCRAFT—RAYTHEON (Hawker)

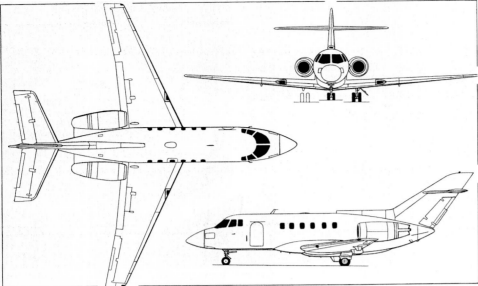

Hawker 800XP (two Honeywell TFE731-5BR-1H turbofans) *(Dennis Punnett/Jane's)*

altitude, to reach 11,280 m (37,000 ft) in 23 minutes at maximum take-off weight in ISA + 10°C conditions; and enhanced take-off performance. Other improvements include installation of vortillons in place of wing fences, permitting lower V-speeds and reducing drag; enhanced TKS de-icing system with increased fluid capacity; improved high-energy brakes; restyled cabin interior to maximise use of available volume; and improved environmental control system; and redesigned interior, with increased headroom by relocating oxygen dropout units to sidewall panels, and 12.2 cm (4.8 in) extra width at shoulder level by sculpturing sidewall panels around fuselage frame. Interior further improved in 1998 – see Accommodation paragraph.

Wing thickness/chord ratio 14 per cent at root, 8.35 per cent at tip; dihedral 2°; incidence 2° 5′ 42″ at root, −3° 5′ 49″ at tip; sweepback 20° at quarter-chord; small fairings on tailplane undersurface eliminate turbulence around elevator hinge cutouts.

FLYING CONTROLS: Conventional and manual. Each control surface with geared tab; port aileron tab trimmed manually via screwjack. Hydraulically actuated four-position double-slotted flaps; mechanically operated hydraulic cutout prevents asymmetric flap operation; upper and lower airbrakes, with interconnected controls to prevent asymmetric operation, form part of flap shrouds and provide lift dumping. Fixed incidence tailplane.

STRUCTURE: All-metal. One-piece wings, dished to pass under fuselage and attached by four vertical links, side link and drag spigot; two-spar fail-safe wings, with partial centre spar of approximately two-thirds span, to form integral fuel tankage; single-piece skins on each upper and lower wing semi-spans; detachable leading-edges; fail-safe fuselage structure of mainly circular cross-section, incorporating Redux bonding.

LANDING GEAR: Retractable tricycle type, with twin wheels on each unit. Hydraulic retraction; nosewheels forward, mainwheels inward into wings. Oleo-pneumatic shock-absorbers. Fully castoring nose unit, steerable ±45°. Dunlop mainwheels size 23×7-12 (12 ply) tubeless tyres. Dunlop nosewheels size 18×4.25-10 (6 ply) tubeless tyres. Dunlop triple-disc hydraulic brakes with Maxaret anti-skid units on all mainwheels. Minimum ground turning radius about nosewheel 9.14 m (30 ft 0 in).

POWER PLANT: Two 20.73 kN (4,660 lb st) Honeywell TFE731-5BR-1H turbofans, mounted on sides of rear fuselage in pods designed and manufactured by Northrop Grumman. Thrust reversers developed by Dee Howard fitted as standard. Integral fuel tanks in wings, with combined capacity of 4,818 litres (1,273 US gallons; 1,060 Imp gallons). Rear underfuselage tank of 882 litres (233 US gallons; 194 Imp gallons) capacity, giving total capacity of 5,700 litres (1,506 US gallons; 1,254 Imp gallons). Single pressure refuelling point at rear of ventral tank. Overwing refuelling point near each wingtip.

ACCOMMODATION: Flight deck crew of two. Dual controls standard. Seat for third crew member. Executive layout has forward baggage compartment, forward galley comprising automatic coffee maker, microwave oven, and miscellaneous storage. Seats swivel through 360°. Seating for eight passengers, with club four seating at the front of the cabin, three-place settee on the right side rear cabin and single seat opposite. Airliner style lavatory at rear with external servicing as standard. Maximum seating for 14. Interior options include differing seating layouts; microwave oven; entertainment system including CD player and video LCD screen. New interior introduced at National Business Aviation Convention at Las Vegas, Nevada, in October 1998 features oval internal window frames, additional sidewall lighting and restyled side panels and work tables.

SYSTEMS: Honeywell air conditioning and pressurisation system. Maximum cabin differential 0.59 bar (8.55 lb/sq in). Oxygen system standard, with dropout masks for passengers. Hydraulic system, pressure 186 to 207 bars (2,700 to 3,000 lb/sq in), for operation of landing gear, mainwheel doors, flaps, spoilers, nosewheel steering, mainwheel brakes and anti-skid units. Two accumulators, pressurised by engine bleed air, one for main system pressure, other providing emergency hydraulic power for wheel brakes in case of main system failure. Independent auxiliary system for lowering landing gear and flaps in event of main system failure.

DC electrical system utilises two 30 V 12 kW engine-driven starter/generators and two 24 V 23 Ah Ni/Cd batteries. A 24 V 4 Ah battery provides separate power for standby instruments. AC electrical system includes two 1.25 kVA static inverters, providing 115 V 400 Hz single-phase supplies, one 250 VA standby static inverter for avionics, and two engine-driven 208 V 7.4 kVA frequency-wild alternators for windscreen anti-icing. Ground power receptacle on starboard side at rear of fuselage for 28 V external DC supply. Honeywell 36-150 W APU. TKS liquid system de-icing/anti-icing on leading-edges of wings and tailplane. Engine ice protection system supplied by engine bleed air. Kidde-Graviner triple FD Firewire fire warning system and two BCF engine fire extinguishers. Stall warning and stick pusher system fitted.

AVIONICS: Standard Honeywell SPZ 8000; Rockwell Collins EFIS-86 fit as option.

Comms: Dual RCZ-851-E integrated communication system with 8.33 kHz channel spacing; Motorola N1335B Selcal; Fairchild A110A CVR.

Radar: Primus 880 weather radar.

Flight: Honeywell NZ2000 FMS; dual Omega; L-3 Com TCAS 2000; dual GPS optional.

Instrumentation: Dual EDZ-818 five-tube EFIS with centrally mounted multifunction display for flight plans and automated check lists; electronic standby instrument system (ESIS) combining attitude, altitude and ASI in one active matrix LCD; dual ADZ-810 air data system; dual AHZ-600 AHRS, AA-300 radio altimeter; Safeflight AoA system.

DIMENSIONS, EXTERNAL:
Wing span	15.66 m (51 ft 4½ in)
Wing chord, mean	2.29 m (7 ft 6¼ in)
Wing aspect ratio	7.1
Length overall	15.60 m (51 ft 2 in)
Height overall	5.36 m (17 ft 7 in)
Fuselage: Max diameter	1.93 m (6 ft 4 in)
Tailplane span	6.10 m (20 ft 0 in)
Wheel track (c/l of shock-absorbers)	2.79 m (9 ft 2 in)
Wheelbase	6.41 m (21 ft 0½ in)
Passenger door (fwd, port): Height	1.30 m (4 ft 3 in)
Width	0.69 m (2 ft 3 in)
Height to sill	1.07 m (3 ft 6 in)
Emergency exit (overwing, stbd):	
Height	0.91 m (3 ft 0 in)
Width	0.51 m (1 ft 8 in)

DIMENSIONS, INTERNAL:
Cabin (excl flight deck): Length	6.50 m (21 ft 4 in)
Max width	1.83 m (6 ft 0 in)
Max height	1.75 m (5 ft 9 in)
Floor area	5.11 m² (55.0 sq ft)
Volume	17.1 m³ (604 cu ft)
Baggage compartments: forward	0.93 m³ (33.0 cu ft)
rear	0.74 m³ (26.0 cu ft)
pannier (optional)	0.79 m³ (28.0 cu ft)

AREAS:
Wings, gross	34.75 m² (374.0 sq ft)
Ailerons (total)	2.05 m² (22.10 sq ft)
Airbrakes: upper (total)	0.74 m² (8.00 sq ft)
lower (total)	0.46 m² (5.00 sq ft)
Trailing-edge flaps (total)	4.83 m² (52.00 sq ft)
Fin (excl dorsal fin)	6.43 m² (69.20 sq ft)
Rudder	1.32 m² (14.20 sq ft)
Horizontal tail surfaces (total)	9.29 m² (100.00 sq ft)

WEIGHTS AND LOADINGS:
Basic weight empty	7,380 kg (16,270 lb)
Typical operating weight empty	7,303 kg (16,100 lb)
Max payload	989 kg (2,180 lb)
Max ramp weight	12,755 kg (28,120 lb)
Max T-O weight	12,701 kg (28,000 lb)
Max zero-fuel weight	8,369 kg (18,450 lb)
Max landing weight	10,591 kg (23,350 lb)
Max wing loading	365.5 kg/m² (74.86 lb/sq ft)
Max power loading	306 kg/kN (3.00 lb/lb st)

PERFORMANCE:
Max limiting Mach No.	0.87
Max level speed and max cruising speed at 8,840 m (29,000 ft)	456 kt (845 km/h; 525 mph)
Econ cruising speed at 11,900-13,100 m (39,000-43,000 ft)	400 kt (741 km/h; 461 mph)
Stalling speed in landing configuration at typical landing weight	92 kt (170 km/h; 106 mph)
Max rate of climb at S/L	945 m (3,100 ft)/min
Time to 10,670 m (35,000 ft)	19 min
Max certified altitude	13,100 m (43,000 ft)
T-O balanced field length at max T-O weight	1,640 m (5,380 ft)
Landing from 15 m (50 ft) at typical landing weight (six passengers and baggage)	1,372 m (4,500 ft)
Range: with max payload	2,280 n miles (4,222 km; 2,623 miles)
with max fuel, NBAA VFR reserves	2,955 n miles (5,472 km; 3,400 miles)

UPDATED

HAWKER 800FI

US Air Force designation: C-29A
JASDF designation: U-125

TYPE: Multirole twin-jet.

CURRENT VERSIONS: **C-29A:** Development of Hawker 800 for US Air Force, fitted with fully automatic flight inspection system (FIS). First of six C-29As for USAF delivered on 24 April 1990, equipped with LTV (Sierra Technologies Inc) inspection system for combat flight inspection and navigation (C-FIN) mission, replacing CT-39A and C-140A calibration fleet with 1866th FCS at Scott AFB, Illinois. In September 1991 control of the six C-29As transferred to FAA headquarters in Oklahoma City. They are currently deployed on a worldwide basis in support of FAA flight inspection operations.

U-125: Selected by Japan Air Self-Defence Force (JASDF) for its FC-X flight inspection requirement. First of three U-125s handed over 18 December 1992; delivery of second U-125 to JASDF took place 16 December 1993; third delivered 22 September 1994. Operated by Hiko Tenkentai (Flight Check Group) at Iruma. See also KAC under Japan.

EU-93: Earlier versions of BAe 125 serve the Brazilian Air Force as calibrators under this designation; first of four new 800XPs (EU-93A 6050) was handed over in May 2000.

DESIGN FEATURES: Version of Hawker 800 with Rockwell Collins EFIS-85 B2 avionics suite, and integrated inertial/GPS system for precision navigation in addition to dual VOR/DME/ADF. Maximum ramp weight up from 12,483 kg (27,520 lb) of standard Hawker 800 to 12,746 kg (28,100 lb); maximum take-off weight up from 12,428 kg (27,400 lb) to 12,701 kg (28,000 lb).

UPDATED

HAWKER 800SM

JASDF designation: U-125A

TYPE: Multirole twin-jet.

CURRENT VERSIONS: **U-125A:** Selected by JASDF for its HS-X search and rescue requirement (see KAC entry under Japan). JASDF requirement is for total of 27 aircraft for delivery beginning early 1995 through to 2004. First three ordered in FY92, one each in FY93 and 94; two in FY95; three in FY96; four in FY97; three in FY98; two in FY99; and two in FY00. First aircraft (52-3001) flew 19 July 1994; 350 hours of development flying in UK; first delivery (52-3003) 11 December 1994 to Japan for completion by KAC; handover to JASDF March 1995; first Wichita-assembled U-125A (seventh built) delivered to KAC in November 1997. KAC had received 16 by December 2000.

DESIGN FEATURES: Hawker 800 fitted with two large observation windows in cabin, a 360° scan surveillance radar and retractable, steerable FLIR sensor. Capable of dropping marker flares and liferafts. Rear ventral fuselage fuel tank replaced by avionics pannier. Maximum zero-fuel weight increased from 8,165 kg (18,000 lb) of Hawker 800FI to 8,369 kg (18,450 lb).

UPDATED

HAWKER 800RA and 800SIG

TYPE: Multirole twin-jet.

CURRENT VERSIONS: **800RA:** Surveillance version of Hawker 800XP, with synthetic aperture radar, defensive aids subsystem and military communications equipment; capable of high altitude (up to 13,100 m; 43,000 ft) and long endurance (6½ hours). Maximum take-off weight 12,701 kg (28,000 lb), maximum zero-fuel weight 8,369 kg (18,450 lb).

800SIG: Signals intelligence version of Hawker 800XP to collect communications intelligence and electronic intelligence for airborne analysis or datalink to ground station for near-realtime assessment or, with recording systems for onboard storage, for post-flight analysis; estimated performance as 800RA.

CUSTOMERS: Four of each version ordered by South Korea on 24 July 1996; total value US$461 million. Sigint version known as **Paekdu**; US project name Peace Pioneer. In early 1998, first three aircraft were at Lockheed Martin Tactical Defense Systems, Goodyear, Arizona, undergoing equipment installation. Contract suspended in mid-1998, but reactivated in mid-1999; four sigint aircraft delivered February 2000; first of surveillance aircraft delivered November 2000.

UPDATED

HAWKER HORIZON

TYPE: Business jet.

PROGRAMME: Design started in 1993; initially thought to have had preliminary designation PD1000Y, but now known to have been PD376; briefly identified as Horizon 1000; announced at NBAA Convention, Orlando, Florida, 19 November 1996; full-size cabin mockup exhibited at NBAA convention, Dallas, Texas, September 1997. First production wing delivered to Wichita from Fuji Heavy Industries in December 1998; first three fuselage sections mated in October 2000 and mated to wing on 16 January 2001. Honeywell is risk-sharing avionics integrator. FAA certification and first customer deliveries scheduled for 2002; production target 36 per year.

CUSTOMERS: More than 150 ordered by July 2000, including 50, plus 50 options, for Executive Jet Inc's NetJets fractional ownership programme, announced at the Paris Air Show in June 1999; 27 for Raytheon Travel Air, and one for the Jordan Grand Prix racing team, for delivery in 2004.

COSTS: US$16.3 million (2000).

DESIGN FEATURES: All-new design sharing only slight family resemblance with the BAe 125-derived Hawker 800 and (discontinued) 1000. An 'advisory council' of business jet operators assisted RAC in defining the aircraft, a major requirement being a flat-floor stand-up cabin. Conventional swept-wing, T tail design with small 'overfin' housing antenna for optional satellite communications system. Wing of supercritical aerofoil section designed using computational fluid dynamics (CFD) techniques; CFD also used to design area-ruled aft fuselage to minimise engine nacelle drag and to reprofile Hawker 1000-based nose section.

Wing sweepback 28° 24' at 25 per cent chord; dihedral 4°; tailplane sweepback 33° 30' at 25 per cent chord.

FLYING CONTROLS: Ailerons and elevators manually operated via pushrods and cables. Pitch trim via electrically acutated, variable incidence tailplane and mechanically driven geared tab on each elevator; electrically powered geared trim tab on port aileron, geared tab on starboard aileron; fly-by-wire hydraulically powered rudder with boost for asymmetric thrust. Electrically signalled, hydraulically powered, three-segment spoilers on upper surface of each wing augment aileron roll control; outboard and middle panels provide roll and speed brake functions, with maximum deflection 35°; lift dump function provided by all spoilers at 60° deflection. Four-segment, electrically controlled and powered double-slotted flaps, deflections 0, 20, 30 and 45°. Dual controls standard.

STRUCTURE: Fuselage of graphite/epoxy laminate and honeycomb composites; 25 mm (1 in) thick shell is formed by Cincinnati Milacron Viper automatic fibre-placement machines over aluminium mandrel in three sections: nose,

Computer-generated image of the Hawker Horizon, which will enter service in 2001

centrebody and tailcone, including dorsal fairing and engine pylons mated using aluminium splice plates in a Nova-Tech Engineering Inc fuselage automated splice tool which automatically seals the joints and installs some 1,800 hi-shear fasteners. Wing, designed using CFD, has a supercritical airfoil and is manufactured by Fuji Heavy Industries at Utsunomiya, Japan, as a complete unit including all systems, integral fuel tanks and leading-edge bleed-air anti-icing, and shipped to Wichita for final assembly. Horizontal stabiliser is of light alloy, two-spar construction with graphite/epoxy composites sandwich skins; vertical stabiliser has three alloy spars and graphite/epoxy composites sandwich skins.

LANDING GEAR: Retractable tricycle type by Messier-Dowty; landing gear electrically signalled and hydraulically actuated, with free-fall emergency extension system; twin wheels on each unit. Trailing-link suspension on main units, which retract inwards; nosewheel forwards. Steer-by-wire nosewheel, maximum deflection 70°, with disconnect for towing. Hydraulic carbon disc brakes with digital brake-by-wire and electric anti-skid systems.

POWER PLANT: Two pod-mounted Pratt & Whitney Canada PW308A turbofans with FADEC, each flat-rated at 28.9 kN (6,500 lb st) at ISA + 20°C. Nordam target-type thrust reversers. Fuel in two integral wing tanks, maximum capacity 7,912 litres (2,090 US gallons; 1,740 Imp gallons). Single point fuelling/defuelling; gravity filler ports in top of each wing.

ACCOMMODATION: Crew of two, plus eight to 12 passengers. Standard accommodation for eight passengers in double

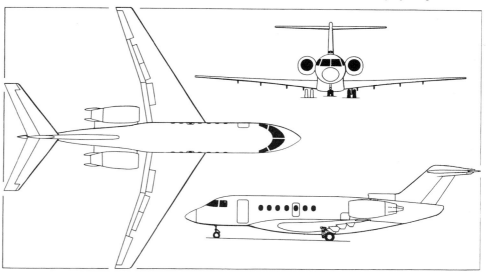
Hawker Horizon business aircraft (two P&WC PW308A turbofans) *(Paul Jackson/Jane's)*

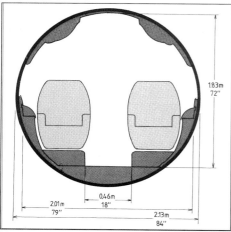
Hawker Horizon cabin cross-section *(Paul Jackson/Jane's)*

Fuselage mating of prototype Hawker Horizon

Hawker Horizon cabin mockup

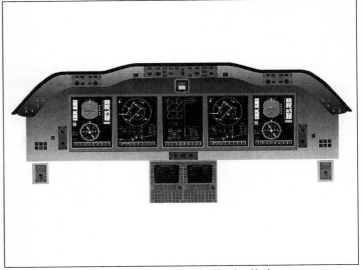

Five-screen instrument panel designed for the Hawker Horizon

'club four' arrangement on reclining and side tracking seats, each pair having stowable writing table. Two closets and a galley immediately aft of flight deck; lavatory and flight accessible baggage compartment at rear of cabin. Accommodation is air conditioned and pressurised. Door on port side aft of flight deck; single plug-type emergency exit on each side over wing; external baggage compartment door on port side aft of wing. Four-panel windscreen; seven cabin windows per side.

SYSTEMS: Digitally controlled pressurisation system, maximum pressure differential 0.66 bars (9.64 lb/sq in), maintains an effective cabin altitude of 1,830 m (6,000 ft) to 13,715 m (45,000 ft). Engine bleed air anti-icing for wing and tailplane leading-edges and engine inlet lips; electrically heated pitot and static masts, AoA and TAT probes; windscreen anti-icing and demisting and cabin window demisting via electrically conductive transparent film embedded between panels. Automatically controlled electrical power generation and distribution system (EPGDS) comprises two high-speed, variable frequency, engine-driven AC generators, one APU-driven AC generator and two 43 Ah lead acid batteries. AC/DC external power receptacles standard. Two independent hydraulic systems, operating pressure 206.85 bars (3,000 lb/sq in), for normal and emergency landing gear operation, braking, thrust reversers, spoilers, rudder, nosewheel steering and emergency electrical generation. Four bottle oxygen system, total capacity 4,080 litres (144 cu ft), with quick-donning diluter-demand masks for crew and auto-deploy constant flow masks in overhead boxes for passengers. Tailcone-mounted Honeywell AE-36-150(HH) APU approved for in-flight operation from sea level to 10,670 m (35,000 ft) and for main engine starting from sea level to 7,925 m (26,000 ft).

AVIONICS: Honeywell Primus Epic lightweight modular avionics system based on Virtual Backplane Network architecture, which combines the cabinet-based modular capabilities of Honeywell's 777 AIMS system with the aircraft-wide network capabilities of its Primus 2000 system and built-in maintenance recording with portable access terminal. The system will have 'point and click' capability via human Cursor Control Devices (CCDs) which include touchpad, joystick, light pen, trackball and on-screen 'soft keys'. Voice Command will be a future option for some Epic functions.

The Horizon's avionics system will provide functions which Raytheon states have never previously been offered in a mid-size business jet, including vertical navigation, full-authority autothrottle and moving map displays. All avionics boxes will be installed in a cabinet behind the co-pilot's seat, in an environmentally controlled, pressurised area with easy access for maintenance.

Comms: Dual VHF comms; single Honeywell KHF-950 HF with Selcal; dual Mode S Diversity transponders; dual audiophone/interphone/PA systems; airborne telephone.

Radar: Primus 880 colour weather radar.

Flight: Primus Epic AFCS and FMS with integrated performance computer; dual VHF nav VOR/LOC/GS/Markers; dual DME; dual INS; dual GPS; EGPWS; TCAS II; solid-state FDR and CVR.

Instrumentation: Five 203 × 254 mm (8 × 10 in) colour, active, flat panel LCD screens comprising two PFDs, two MFDs and one EICAS, plus two smaller multifunction control and display units (MCDUs).

DIMENSIONS, EXTERNAL:
Wing span	18.82 m (61 ft 9 in)
Mean aerodynamic chord	2.92 m (9 ft 7 in)
Wing aspect ratio	7.2
Length overall	21.08 m (69 ft 2 in)
Height overall	5.97 m (19 ft 7 in)
Tailplane span	7.90 m (25 ft 11 in)
Wheel track	2.79 m (9 ft 2 in)
Wheelbase	8.53 m (28 ft 0 in)
Passenger door: Height	1.58 m (5 ft 6 in)
Width	0.76 m (2 ft 6 in)

DIMENSIONS, INTERNAL:
Cabin:
Length, excl flight deck	7.62 m (25 ft 0 in)
Max width	1.97 m (6 ft 5½ in)
Width at floor	1.27 m (4 ft 2 in)
Max height	1.83 m (6 ft 0 in)
Volume	9.37 m³ (331 cu ft)
Baggage compartment volume	2.83 m³ (100 cu ft)

AREAS:
Wings, gross	49.3 m² (531.0 sq ft)
Horizontal tail surfaces	13.01 m² (140.00 sq ft)
Vertical tail surfaces	10.26 m² (110.40 sq ft)

WEIGHTS AND LOADINGS:
Basic operating weight (incl crew)	9,494 kg (20,930 lb)
Max fuel weight	6,350 kg (14,000 lb)
Payload with max fuel	544 kg (1,200 lb)
Max zero fuel weight	11,113 kg (24,500 lb)
Max T-O weight	16,329 kg (36,000 lb)
Max ramp weight	16,420 kg (36,200 lb)
Max wing loading	331.0 kg/m² (67.80 lb/sq ft)
Max power loading	282 kg/kN (2.77 lb/lb st)

PERFORMANCE (estimated):
Max operating speed	M0.84
Max certified altitude	13,715 m (45,000 ft)
Cruising speed: S/L to 2,440 m (8,000 ft)	280 kt (519 km/h; 322 mph)
2,440 m (8,000 ft) to 8,057 m (26,435 ft)	350 kt (648 km/h; 403 mph)
T-O run	1,600 m (5,250 ft)
Landing run	713 m (2,340 ft)
Max range	3,400 n miles (6,297 km; 3,912 miles)
Range at cruising speed of M0.82, NBAA IFR reserves	3,100 n miles (5,741 km; 3,567 miles)

UPDATED

HAWKER 450

TYPE: Business jet.

PROGRAMME: Announced at NBAA Convention in New Orleans, 9 October 2000; Hawker 450 is interim designation pending announcement of name; full-scale development began in late 2000; service entry predicted for 2006.

CUSTOMERS: Launch customer Raytheon Travel Air ordered 50, with 25 options, on 10 October 2000; total of 105 orders held by mid-October 2000.

COSTS: US$7.8 million to US$9 million (2000).

DESIGN FEATURES: Conventional swept-wing, T tail design.

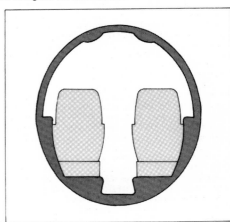

Hawker 450 cabin cross-section
(Paul Jackson/Jane's)

STRUCTURE: Composites fuselage formed over aluminium mandrel by automatic fibre-placement machines. Aluminium wing.

LANDING GEAR: Retractable tricycle type.

POWER PLANT: Two pod-mounted Honeywell TFE731-40 turbofans, each flat-rated at 18.9 kN (4,250 lb st) at ISA +16°C.

ACCOMMODATION: Crew of two, standard accommodation for eight passengers in double 'club four' or 'centre club' arrangement; baggage area and lavatory at rear of cabin. Door on port side to rear of flight deck; six cabin windows on each side.

AVIONICS: Honeywell Primus Epic suite as core system.
Flight: EICAS, TCAS II and EGPWS standard.
Instrumentation: EFIS panel with four 203 × 254 mm (8 × 10 in) active matrix LCD screens for PFD and MFD functions.

DIMENSIONS, INTERNAL:
Cabin (excl flight deck): Length	5.89 m (19 ft 4 in)
Max width	1.83 m (6 ft 0 in)
Max height	1.80 m (5 ft 11 in)

PERFORMANCE (estimated):
Max speed	Mach 0.80+
Range	more than 2,000 n miles (3,704 km; 2,302 miles)

NEW ENTRY

Computer-generated image of Hawker 450 mid-size business jet

RENAISSANCE

RENAISSANCE AIRCRAFT LLC
PO Box 696, Eastman, Georgia 31023
Tel: (+1 912) 374 21 22
Fax: (+1 912) 374 21 24
Web: http://www.renaissanceaircraft.com

Renaissance Aircraft has plans to relaunch production of the classic Luscombe 8F light aircraft, supported by the worldwide service centres of the Zenair company. However, by late 2000, these appeared to have been delayed.
UPDATED

RENAISSANCE 8F
TYPE: Two-seat lightplane.
PROGRAMME: Luscombe Airplane Corp Model 8 Silvaire first flew 18 December 1938; progressed through several subvariants, built by Luscombe, Temco Engineering and Silvaire Aircraft until 1960; total of some 5,867 produced. In 1999, Renaissance refurbished an original Luscombe 8F (N999RA) as prototype for relaunched production. First deliveries were due late 2000, but apparently not effected.
COSTS: US$71,200, basic VFR; US$78,700 Mode C VFR; US$90,200 IFR (all 2000).
DESIGN FEATURES: High wing, all-metal touring lightplane of late 1930s technology, but incorporating modifications and some modern features. Changes include considerable increase in power from original 67.1 kW (90 hp) Continental C-90; 60 per cent more baggage area or optional seating for two children; additional fuel; and new cockpit interior.
FLYING CONTROLS: Conventional and manual. Flight-adjustable tab in port elevator; ground-adjustable tabs on both ailerons. No horn balances. Plain flaps; deflections 0, 10, 20 and 30°.
STRUCTURE: Alclad aluminium monocoque fuselage; oval section duralumin bulkhead stampings and riveted Alclad skin. Wings constructed on two I-type spars of extruded aluminium and ribs of riveted T-section extrusions with Alclad skin. Ailerons covered with beaded Alclad sheet riveted to a single duralumin spar. Wings attached to upper sides of cabin and braced by single streamlined strut each side. Wing struts and landing gear attached to aluminium forgings riveted to forward section of metal seat bottom on each side. Duralumin/Alclad tailplane bolted to fuselage. Steel tube main landing gear legs hinged at fuselage sides and with single oleo-spring unit within fuselage.
LANDING GEAR: Tailwheel type; fixed. Mainwheel tyres 6.00-6; hydraulic brakes; optional speed fairings. Steerable tailwheel, size 2.30/2.50-4. Optional skis, floats or tundra tyres.
POWER PLANT: One 112 kW (150 hp) Textron Lycoming O-320 flat-four driving a Sensenich 74-62, two-blade,

Prototype Renaissance 8F *(Paul Jackson/Jane's)*

fixed-pitch aluminium propeller. Fuel tank in each wing, total 117 litres (30.0 US gallons; 25.0 Imp gallons).
ACCOMMODATION: Two, on individual seats, with dual controls. Forward-hinged door each side. Baggage compartment behind seats may be replaced by two child seats.
AVIONICS: To customer's specification.
 Comms: ELT standard. Optional Mode C VFR package includes Honeywell KX 155 nav/com with KI 208 VOR, KT 76A transponder and PS 1000 II intercom. IFR option additionally includes KMA 24 audio panel and second KX 155 with KI 209A VOR/LOC.
 Flight: Optional Honeywell KMD 150 GPS and moving map. IFR option includes KLN 89B GPS as interim, pending KLN 94.
EQUIPMENT: Landing light in starboard wing leading edge.
DIMENSIONS, EXTERNAL:
 Wing span 6.71 m (35 ft 0 in)
 Wing aspect ratio 8.8
 Length overall 6.70 m (22 ft 0 in)
 Height overall 2.13 m (7 ft 0 in)
DIMENSIONS, INTERNAL:
 Cabin max width 1.01 m (3 ft 4 in)
AREAS:
 Wings, gross 13.01 m² (140.0 sq ft)
WEIGHTS AND LOADINGS:
 Weight empty 449 kg (990 lb)
 Max T-O weight: interim 635 kg (1,400 lb)
  planned 816 kg (1,800 lb)
 Max wing loading: interim 48.8 kg/m² (10.00 lb/sq ft)
  planned 62.8 kg/m² (12.86 lb/sq ft)
 Max power loading: interim 5.68 kg/kW (9.33 lb/hp)
  planned 7.30 kg/kW (12.00 lb/hp)
PERFORMANCE:
 Max level speed 130 kt (241 km/h; 150 mph)
 Max cruising speed at 75% power:
  at S/L 122 kg (225 km/h; 140 mph)
  at 2,440 m (8,000 ft) 126 kt (233 km/h; 145 mph)
 Econ cruising speed at 65% power
  119 kt (220 km/h; 137 mph)
 Stalling speed: flaps up 41 kt (76 km/h; 47 mph)
  flaps down 38 kt (70 km/h; 43 mph)
 Max rate of climb at S/L 442 m (1,450 ft)/min
 Service ceiling 6,400 m (21,000 ft)
 g limits +4.6/−2.2
UPDATED

RENAISSANCE

RENAISSANCE COMPOSITES INC
3025A Airport Avenue, Santa Monica, California 90405
Tel: (+1 310) 391 19 43
Fax: (+1 310) 391 86 45
e-mail: berkut@primenet.com
Web: http://www.berkut.com
CHIEF DESIGNER: Dave Ronneberg
VICE-PRESIDENT, OPERATIONS: Diane Moser

Experimental Aviation (which see in 1997-98 edition) filed for Chapter 7 bankruptcy in November 1996, but was re-established as Renaissance Composites in mid-1997.
UPDATED

RENAISSANCE BERKUT
TYPE: Tandem-seat sportplane kitbuilt.
PROGRAMME: Named after Asian eagle; first flight 11 July 1991.
CURRENT VERSIONS: **Berkut 360:** 134 kW (180 hp) Textron Lycoming IO-360 flat-four engine.
 Berkut 540: 210 kW (270 hp) Textron Lycoming IO-540 flat-six engine.
 Berkut Four-Seat: Four-seat version. Proof-of-concept prototype under construction during 2000; features include 201 kW (270 hp) Textron Lycoming IO-540 engine, moulded wings, extensive use of carbon fibre in airframe and baggage compartment in nosecone.
CUSTOMERS: Total of 47 kits sold by early 1995. No further production notified. Seven reported to be flying by mid-1999. One completed in 1998 (first flight 26 May) in Olympic Aviation facility in Manila, Philippines; donated to nation by Cannard Aviation as part of that year's centennial celebrations. First European Berkut completed in UK, early 1999, by Glen Waters.
COSTS: Kit: US$34,490 (2000).
DESIGN FEATURES: Rear-mounted wings with 23° of sweep on outer panels; winglets. Straight foreplane. Resembles Rutan Long-EZ, of which Dave Ronneberg built seven examples, but features larger cabin and wing strakes with upper camber lift. Intended first for sport and recreational flying; seen to have business applications

G-REDX, the first European-built Berkut *(Paul Jackson/Jane's)*

as well as military. Quoted build time 1,500 to 2,000 hours.
FLYING CONTROLS: Manual. Ailerons and rudders on main wing and winglets, elevators on canard; electrically actuated airbrake under fuselage.
STRUCTURE: Carbon fibre main spar; solid foam core in first 50 wing sets; separate carbon fibre ribs and skins thereafter; canard has solid core with no ribs. GFRP, Styrofoam, PVC foam and Kevlar also used.
LANDING GEAR: Electrohydraulically retractable tricycle type (nosewheel partially exposed when retracted); outward-retracting main units. Matco calliper brakes and wheels; mainwheels 11×4.00; nosewheel 2.80/2.50-4.
POWER PLANT: Textron Lycoming IO-360 or IO-540 piston engine, as described under Current Versions, each driving a Light Speed Engineering two-blade composites pusher propeller. Usable fuel capacity 219 litres (58.0 US gallons; 48.3 Imp gallons).
ACCOMMODATION: Two seats in tandem; 0.31 m³ (11.0 cu ft) baggage space in wing strakes and nose.
SYSTEMS: Hydraulic system (96.5 bars; 1,400 lb/sq in) for landing gear; 12 V electrical system with 30 Ah battery.

750　USA: AIRCRAFT—RENAISSANCE COMPOSITES to RFW

DIMENSIONS, EXTERNAL:
  Wing span   8.13 m (26 ft 8 in)
  Length overall   5.64 m (18 ft 6 in)
  Height overall   2.29 m (7 ft 6 in)
  Propeller diameter   1.70 m (5 ft 7 in)
AREAS:
  Wings, gross   10.22 m² (110.0 sq ft)
WEIGHTS AND LOADINGS (A: Berkut 360, N: Berkut 540):
  Weight empty: A   469 kg (1,035 lb)
  B   522 kg (1,150 lb)
  Max T-O weight: A, B   907 kg (2,000 lb)
  Max wing loading: A, B   88.8 kg/m² (18.18 lb/sq ft)
  Max power loading: A   6.76 kg/kW (11.11 lb/hp)
  B   4.51 kg/kW (7.41 lb/hp)
PERFORMANCE (A, B, as above):
  Never-exceed speed (V_{NE}):
    A, B   304 kt (563 km/h; 350 mph)
  Max level speed: A   215 kt (398 km/h; 247 mph)
  B   230 kt (426 km/h; 265 mph)
  Cruising speed at 2,440 m (8000 ft): at 75% power:
    A   208 kt (385 km/h; 239 mph)
    B   222 kt (410 km/h; 255 mph)
  at econ power: A, B   187 kt (346 km/h; 215 mph)
  Manoeuvring speed   200 kt (370 km/h; 230 mph)
  Min controllable speed   57 kt (105 km/h; 65 mph)
  Max demonstrated crosswind limit
                           25 kt (46 km/h; 29 mph)
  Max rate of climb at S/L: A   610 m (2,000 ft)/min
    B   915 m (3,000 ft)/min
  Service ceiling: A, B   8,992 m (29,500 ft)
  T-O and landing run: A, B   335 m (1,100 ft)
  Range with 30 min reserves:
    at 75% power:
      A   1,091 n miles (2,021 km; 1,256 miles)
      B   740 n miles (1,371 km; 852 miles)
    at econ power: A, B
                     1,313 n miles (2,433 km; 1,512 miles)
UPDATED

REVOLUTION

REVOLUTION HELICOPTER CORPORATION INC

The company ceased trading on 25 October 1999. Mini 500 and Voyager 500 helicopters last described and illustrated in the 2000-01 edition.

Support and spares for existing 500 owners are available from Stitt Industries Inc, 217 Virginia Road, Ex Spgs, MO 64024; Tel: (+1 816) 637 32 83; e-mail: rickay@aol.com; Web: http://www.sittind.com.

UPDATED

RFW

REFLEX FIBERGLASS WORKS INC

538 Rentz Road, PO Box 497, Walterboro, South Carolina 29488-0494
Tel: (+1 843) 538 66 82
Fax: (+1 843) 538 66 83
e-mail: pdantzler@reflexfiber.com
Web: http://www.reflexfiber.com
PRESIDENT: Howell C Jones Jr

RFW took over production of the Lightning Bug and White Lightning kitbuilt sport aircraft in early 1996 and is housed in a 1,400 m² (15,000 sq ft) purpose-built factory.

UPDATED

RFW LIGHTNING BUG

TYPE: Single-seat sportplane kitbuilt.
CUSTOMERS: At least nine completed, of which five current in January 2001; two confirmed lost in accidents. Most recent completions, in 1999, by Roger Pearce of Easley, South Carolina, and John Gibbs of Monterey, California.
COSTS: US$25,000 (2001) for kit which includes engine and propeller. US$1,000 deposit on ordering.
DESIGN FEATURES: Compact, mid-wing monoplane; conforms to FAR Pt 23 Utility category at 430 kg (950 lb); Aerobatic category at 340 kg (750 lb). Quoted build time 400 hours.
FLYING CONTROLS: Conventional and manual. Elevator tab for pitch trim; ground-adjustable tabs on ailerons and rudder.
STRUCTURE: General construction of composites, stainless steel and aluminium 2024-T3 or 7075-T6. Moulded parts use E glass, epoxy resin, PVC foam and Baltek mat.
LANDING GEAR: Fixed mainwheels with spats; option of (retractable) nosewheel, or tailwheel. Same wheels and tyres as Long-EZ.
POWER PLANT: One 74.6 kW (100 hp) AMW 808 FIAGD three-cylinder liquid-cooled two-stroke engine, with fuel injection and dual ignition, driving fixed-pitch propeller through reduction gear; variable-pitch propeller under investigation. Fuel capacity 87 litres (23.0 US gallons, 19.2 Imp gallons); grades 100 LL or Mogas.
ACCOMMODATION: Single seat accommodates pilot with maximum weight of 127 kg (280 lb) and maximum height 1.98 m (6 ft 6 in).
SYSTEMS: Electrical system includes 12 V 30 Ah battery.
AVIONICS: Customer specified.
DIMENSIONS, EXTERNAL:
  Wing span   5.44 m (17 ft 10 in)
  Length overall   5.31 m (17 ft 5 in)
  Fuselage max width   0.63 m (2 ft 1 in)
  Height overall   1.55 m (5 ft 1 in)
AREAS:
  Wings, gross   3.72 m² (40.0 sq ft)
WEIGHTS AND LOADINGS:
  Weight empty   215 kg (475 lb)
  Max T-O weight: Aerobatic   340 kg (750 lb)
    Utility   430 kg (950 lb)
PERFORMANCE:
  Max level speed   217 kt (402 km/h; 250 mph)
  Econ cruising speed at 75% power
                           196 kt (362 km/h; 225 mph)
  Stalling speed, power off:
    flaps up   74 kt (137 km/h; 85 mph)
    flaps down   54 kt (104 km/h; 62 mph)
  Max rate of climb at S/L at 139 kt (257 km/h; 160 mph)
                           366 m (1,200 ft)/min
  T-O run   244 m (800 ft)
  Landing run   305 m (1,000 ft)
  Range with max fuel at 75% power
                           764 n miles (1,416 km; 880 miles)

UPDATED

Lightning Bug high-performance monoplane (AMW 808 two-stroke engine)

RFW WLAC-1 WHITE LIGHTNING

TYPE: Four-seat kitbuilt.
PROGRAMME: Previously produced by the White Lightning Aircraft Corporation. Prototype first flew 8 March 1986. In same year it established several world speed records in Classes C1b and C1c.
CUSTOMERS: 32 kits sold, 15 aircraft flying by mid-1999. By January 2001, 10 remained on US civil register; most recent completions in 1998 by Leonard Moore of Hurst, Texas.
COSTS: Kit: US$45,000 (2001) including 40-hour builders' course. Many optional subkits, including electrical for US$1,110.
DESIGN FEATURES: Highly streamlined mid-wing monoplane. Options include electrical, various engine installation, light, strobe light and interior subkits, plus other items. Quoted build time 1,500 hours.
  Wing section NACA 66_2-215.
FLYING CONTROLS: Conventional and manual. Horn-balanced elevators. Ground-adjustable tab on starboard aileron and rudder, trim tab on port elevator.
STRUCTURE: Wings have graphite tubular 'wet' main spar, and glass fibre/epoxy front and rear spars and ribs, all precast in the lower glass fibre/epoxy skin. Glass fibre/epoxy fuselage moulded in upper and lower halves. Spars and ribs of fin and tailplane precast into one skin of each.
LANDING GEAR: Retractable tricycle type, with brakes. Main and nosewheels retract rearwards into fuselage. Single wheel on each unit.
POWER PLANT: One 156.6 kW (210 hp) Teledyne Continental IO-360-CB flat-six engine. Fuel capacity 265 litres (70.0

RFW WLAC-1 White Lightning four-seat composites kit aircraft *(Geoffrey P Jones)*

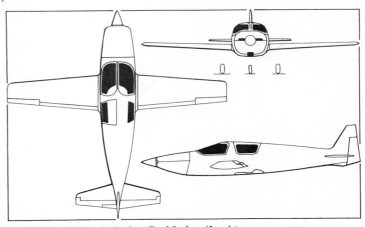
RFW WLAC-1 White Lightning *(Paul Jackson/Jane's)*

US gallons; 58.0 Imp gallons) of which 257 litres (68.0 US gallons; 56.6 Imp gallons) are usable.
ACCOMMODATION: Four persons in two side-by-side pairs. Door port side, rear.
DIMENSIONS, EXTERNAL:

Wing span	8.43 m (27 ft 8 in)
Wing aspect ratio	8.6
Length overall	7.11 m (23 ft 4 in)
Height overall	2.18 m (7 ft 2 in)
Tailplane span	2.74 m (9 ft 0 in)
Propeller diameter	1.85 m (6 ft 1 in)

AREAS:

Wings, gross	8.29 m² (89.2 sq ft)
Tailplane	1.81 m² (19.50 sq ft)

WEIGHTS AND LOADINGS:

Weight empty	635 kg (1,400 lb)
Max baggage capacity	154 kg (340 lb)
Max T-O weight	1,224 kg (2,700 lb)
Max wing loading	147.8 kg/m² (30.27 lb/sq ft)
Max power loading	7.83 kg/kW (12.86 lb/hp)

PERFORMANCE:

Max level speed at S/L	243 kt (450 km/h; 280 mph)
Max cruising speed at 2,440 m (8,000 ft)	230 kt (426 km/h; 265 mph)
Econ cruising speed at 3,200 m (10,500 ft)	217 kt (402 km/h; 250 mph)
Stalling speed: flaps up	80 kt (148 km/h; 92 mph)
flaps down	53 kt (99 km/h; 61 mph)
Max rate of climb at S/L	610 m (2,000 ft)/min
Service ceiling	6,100 m (20,000 ft)
T-O and landing run, full flap	400 m (1,300 ft)

Range with max fuel:

at 75% power	1,423 n miles (2,636 km; 1,638 miles)
at 55% power	1,738 n miles (3,218 km; 2,000 miles)
Endurance	6 h 10 min
g limits	+4.4/−2 utility

UPDATED

ROBINSON

ROBINSON HELICOPTER COMPANY
2901 Airport Drive, Torrance, California 90505
Tel: (+1 310) 539 05 08
Fax: (+1 310) 539 51 98
Web: http://www.robinsonheli.com
PRESIDENT: Franklin D Robinson
VICE-PRESIDENT, MARKETING: Wayne Walden
MARKETING DIRECTOR: Tim Goetz
PUBLIC RELATIONS: Barbara K Robinson

Robinson expanded in 1994 with a new 24,620 m² (265,000 sq ft) plant for R44 production and received ISO 9001 certification in November 1997. The company produced 278 helicopters in 1999, increasing its overall total to 3,712 and on 25 September 2000 delivered its 4,000th helicopter. Workforce totals 800, producing 10 helicopters (seven R44s and three R22s) per week. Deliveries for 2000 totalled 126 R22s and 264 R44s.

UPDATED

ROBINSON R22 BETA II

TYPE: Two-seat helicopter.
PROGRAMME: Design began 1973; first flight 28 August 1975; first flight of second R22 early 1977; FAA certification 16 March 1979; UK certification June 1981; deliveries began October 1979; early versions were R22 (O-320-A2B engine; 199 built) and R22HP (O-320-B2C; 151 built), both with 621 kg (1,300 lb) max T-O weight. R22 Alpha (O-320-B2C and 621 kg; 1,370 lb MTOW) certified October 1983 (further 151 built); R22 Beta certified 5 August 1985. Current Beta II introduced in 1995.
CURRENT VERSIONS: **R22 Beta II**: More powerful Textron Lycoming O-360 engine provides better high-level hover performance and allows take-off power to be sustained up to 2,285 m (7,500 ft). Previously optional, tinted windscreen and door windows fitted as standard. Production began at c/n 2571 in 1995.
Main description applies to Beta II except where indicated.
R22 Mariner II: Float-equipped Beta II; introduced in 1995, replacing original Mariner introduced 1985. Total of 219 R22 Mariners of both versions produced by October 1999.
R22 Police: Special communications and other equipment including removable port-side controls, 70 A alternator, searchlight, loudspeaker, siren and transponder.

Robinson R22 Beta two-seat helicopter *(Paul Jackson/Jane's)*

R22 IFR: Equipped with flight instruments and radio to allow training for helicopter IFR flying; includes Honeywell KCS 55A HSI, KR 87 ADF, KX 165 nav/com, KT 76A transponder; KR 22 marker, plus optional KN 63 DME and KLN 89B GPS (replacing KR 87). Three delivered in 1995, four in 1997, six in 1998 and six in 1999.
External load R22: Hook kit certified for 181 kg (400 lb) produced by Classic Helicopter Corporation, Boeing Field, Seattle, Washington; weighs 2.3 kg (5 lb); used for slung load training.
R22 Agricultural: Equipped with Apollo Helicopter Services DTM-3 spray system; FAA approved December 1991. Low-drag belly tank contains 151 litres (40.0 US gallons; 33.3 Imp gallons); boom length 7.31 m (24 ft 0 in); tank frame attached to landing gear mounting points with four bolts and wing nuts; installation of entire system requires no tools and can be completed by one person in approximately 5 minutes.
CUSTOMERS: Total production of 3,167 by 1 January 2001. 3,000th handed over 15 October 1999; deliveries in 1999 totalled 128 and 126 in 2000.
COSTS: US$148,000 for basic R22 Beta II (1999). Direct operating cost US$24.40 per hour; total cost (500 hours annually) US$68.52 per hour (1998). Mariner II basic cost US$157,000; R22 IFR US$173,060 (both 1999).
DESIGN FEATURES: Simple, pod-and-boom light helicopter; horizontal stabiliser, starboard side only; vertical stabiliser above and below boom; offset to starboard; tall rotor mast. Horizontally mounted piston engine drives transmission through multiple V belts and sprag-type overrunning clutch; main and tail gearboxes use spiral bevel gears; maintenance-free flexible couplings of proprietary manufacture used in both main and tail rotor drives. Two-blade semi-articulated main rotor, with tri-hinged underslung rotor head to reduce blade flexing, rotor vibration and control force feedback, and an elastic teeter hinge stop to prevent blade-boom contact when starting or stopping rotor in high winds; blade section NACA 63-015 (modified); two-blade tail rotor on port side; rotor brake standard.
FLYING CONTROLS: Manual. Cyclic stick mounted between pilots with handgrips on swing arm for comfortable access from either seat.
STRUCTURE: All-metal bonded blades with stainless steel spar and leading-edge, light alloy skin and light alloy honeycomb filling; cabin section of steel tube with metal and plastics skinning; full monocoque tailboom.
LANDING GEAR: Welded steel tube and light alloy skid landing gear, with energy-absorbing crosstubes. Twin float/skid gear on Mariner with additional tailplane surface on lower tip of fin.
POWER PLANT: One Textron Lycoming O-360 flat-four engine, derated to 97.5 kW (131 hp), mounted in the lower rear section of the main fuselage, with cooling fan. Light alloy main fuel tank in upper rear section of the fuselage on port side, usable capacity 72.5 litres (19.2 US gallons; 16.0 Imp gallons). Optional auxiliary fuel tank, capacity 39.75 litres (10.5 US gallons; 8.7 Imp gallons). Oil capacity 5.7 litres (1.5 US gallons; 1.25 Imp gallons).
ACCOMMODATION: Two seats side by side in enclosed cabin, with inertia reel shoulder harness. Curved two-panel, tinted windscreen. Removable door, with tinted window, on each side. Police version has observation doors with bubble windows, which are also available as options on

Instrument panel of Robinson R22

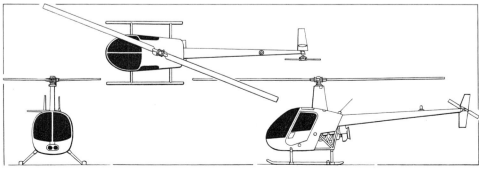

Robinson R22 Beta two-seat helicopter *(James Goulding/Jane's)*

USA: AIRCRAFT—ROBINSON

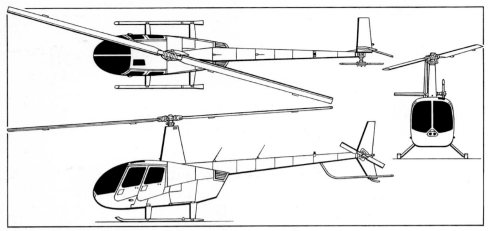

Robinson R44 light helicopter *(Mike Keep/Jane's)*

other models. Baggage space beneath each seat. Cabin heated and ventilated.
SYSTEMS: Electrical system, powered by 12 V DC alternator, includes navigation, panel and map lights, dual landing lights, anti-collision light and battery. Second battery optional.
AVIONICS: *Comms:* Honeywell KY197A VHF com radio. Optional KX 155 nav/com digital display radio with KI 208 indicator (replacing KY 197A) KT 76A transponder and remote altitude encoder.
Flight: Optional KR 22 marker beacon receiver, KN 53 nav receiver, Garmin 150XL/160XL or Honeywell KLN 89 GPS and KR 87 ADF. IFR trainer avionics listed under Current Versions above.
Instrumentation: AIM 305-1 AL DVF artificial horizon, Honeywell KEA 129 encoding altimeter, Astronautics DC turn indicator, rate of climb indicator, sensitive altimeter, quartz clock, hour meter, low rotor rpm warning horn, temperature and chip warning lights for main gearbox and chip warning light for tail gearbox.
EQUIPMENT: Landing and anti-collision lights.

DIMENSIONS, EXTERNAL:
Main rotor diameter	7.67 m (25 ft 2 in)
Tail rotor diameter	1.07 m (3 ft 6 in)
Main rotor blade chord	0.18 m (7¼ in)
Distance between rotor centres	4.39 m (14 ft 5 in)
Length overall, rotors turning	8.76 m (28 ft 9 in)
Fuselage: Length	6.30 m (20 ft 8 in)
Max width	1.12 m (3 ft 8 in)
Height: overall	2.72 m (8 ft 11 in)
to top of cabin	1.75 m (5 ft 9 in)
Skid track	1.93 m (6 ft 4 in)

DIMENSIONS, INTERNAL:
Cabin max width	1.12 m (3 ft 8 in)

AREAS:
Main rotor blades (each)	0.70 m² (7.55 sq ft)
Tail rotor blades (each)	0.04 m² (0.40 sq ft)
Main rotor disc	46.21 m² (497.4 sq ft)
Tail rotor disc	0.89 m² (9.63 sq ft)
Fin	0.21 m² (2.28 sq ft)
Stabiliser	0.14 m² (1.53 sq ft)

WEIGHTS AND LOADINGS:
Weight empty (without auxiliary fuel tank):	
Beta II	388 kg (855 lb)
Mariner II	405 kg (892 lb)
Fuel weight: standard	52 kg (115 lb)
auxiliary	28.6 kg (63 lb)
Max T-O and landing weight	621 kg (1,370 lb)
Max zero-fuel weight	569 kg (1,255 lb)
Max disc loading	13.4 kg/m² (2.75 lb/sq ft)
Max power loading at T-O	6.37 kg/kW (10.46 lb/hp)

PERFORMANCE (Beta II):
Never-exceed speed (V_NE):	
without sling load	102 kt (190 km/h; 118 mph)
with sling load	75 kt (139 km/h; 86 mph)
Max level speed	97 kt (180 km/h; 112 mph)
Cruising speed at 70% power at 2,440 m (8,000 ft)	
	96 kt (177 km/h; 110 mph)
Econ cruising speed	82 kt (153 km/h; 95 mph)
Max rate of climb at S/L more than 305 m (1,000 ft)/min	
Rate of climb at 3,050 m (10,000 ft)	
	more than 183 m (600 ft)/min
Service ceiling	4,265 m (14,000 ft)
Hovering ceiling IGE	2,865 m (9,400 ft)
Range with max payload, no reserves:	
normal fuel	
	more than 173 n miles (321 km; 200 miles)
max fuel	more than 260 n miles (482 km; 300 miles)
Endurance at 65% power, auxiliary fuel, no reserves	
	3 h 20 min

UPDATED

ROBINSON R44

TYPE: Four-seat helicopter.
PROGRAMME: Development began 1986; first flight (N44RH) 31 March 1990; first flight of third R44 March 1992; marketing began 22 March 1992; certification completed 10 December 1992. In August 1997 an R44 became first piston helicopter flown around the world.
CURRENT VERSIONS: **R44 Astro** and **Raven**: Versions available for passenger transport, water bombing, law enforcement and logging support. Raven, introduced in April 2000, has hydraulic flight control system as standard, plus elastomeric tail rotor bearings and adjustable pedals for pilot.
R44 Clipper: Float-equipped version; certified 17 July 1996. Standard landing gear set extra US$6,000. Optional avionics include ICOM IC-M59 marine radio, IC-271 2 m radio and Garmin GPS MAP 130. 63 delivered by October 1999, five of them with pop-out floats, certified in June 1999.
R44 Police: Specially modified for law enforcement, with choice of IR sensors and television camera systems mounted in gyrostabilised nose turret; video monitor; Spectrolab SX-5E searchlight; bubble door windows; empty weight, equipped, approx 715 kg (1,570 lb). Certified July 1997; first customer El Monte Police Agency of Los Angeles.
R44 Newscopter: Electronic news gathering (ENG) version intended for media companies and fitted with FLIR Systems UltraMedia RS gyrostabilised video-camera and microwave transmitter; empty weight, equipped, 725 kg (1,598 lb). Deliveries began January 1998; 20 sold by November 1999.
R44 IFR: Equipped for IFR helicopter training; generally as for R22 IFR.
CUSTOMERS: Total of 868 delivered by 25 September 2000. Deliveries in 1999 totalled 150 and in 2000, 264. 1,000th delivery was expected February 2001.
COSTS: Standard price US$294,000 (2001). Clipper US$310,000 with utility floats, US$316,000 with pop-out floats (2001). Direct operating cost US$40.70 per hour; total (500 hours annually) US$120.93 per hour (2001). R44 IFR US$299,570; Newscopter, typically US$495,000 (both 1999).
DESIGN FEATURES: New design, incorporating general configuration (but larger cabin) and some proven concepts of R22; designed to requirements of FAR Pt 27; features for comfort and safety include electronic throttle governor to reduce pilot workload by controlling rotor and engine rpm during normal operations, rotor brake, advanced warning devices, automatic clutch engagement to simplify and reduce start-up procedure and reduce chance of overspeed, T-bar cyclic control, and crashworthy features including energy-absorbing landing gear and lap/shoulder strap restraints designed for high forward *g* loads. High reliability and low maintenance; patented rotor design with tri-hinge (see R22); low noise levels.
FLYING CONTROLS: Conventional, with Robinson central cyclic stick; rpm governor; rotor brake standard; left-hand collective lever and pedals removable. Customised hydraulic flight control system certified July 1999 and offered from September 1999 as option at US$13,000 (2000).
LANDING GEAR: Fixed skids or, on R44 Clipper, twin utility floats or pop-out helium floats which inflate in 2 seconds.
POWER PLANT: One 194 kW (260 hp) Textron Lycoming O-540 flat-six engine, derated to 165 kW (225 hp) at T-O, 153 kW (205 hp) continuous. Standard fuel capacity 116 litres (30.6 US gallons; 25.5 Imp gallons) in tank on port side; optional extra 69 litres (18.3 US gallons; 15.2 Imp gallons) in starboard tank.
ACCOMMODATION: Four persons seated 2 + 2. Concealed baggage compartments. Dual controls. Cabin heated and ventilated. Tinted windscreen and door windows.
AVIONICS: Selection of optional avionics and instruments.
Comms: Honeywell KY 197A VHF radio; Inter-VOX AA-80 intercom for all four occupants. Optional KT 76A or 76C transponder with remote altitude encoder.
Flight: Optional KX 155 nav/com with KI 203 indicator (replacing KY 197A), Garmin 150XL/250XL or KLN 89 GPS and KR 87 ADF.
EQUIPMENT: Standard equipment includes auxiliary fuel system and heater and rotor brake. Options include belly hardpoint.

DIMENSIONS, EXTERNAL:
Main rotor diameter	10.06 m (33 ft 0 in)
Tail rotor diameter	1.47 m (4 ft 10 in)
Length overall, rotors turning	11.76 m (38 ft 7 in)
Fuselage: Length	9.07 m (29 ft 9 in)
Max width	1.28 m (4 ft 2½ in)
Height: overall	3.28 m (10 ft 9 in)
to top of cabin	1.83 m (6 ft 0 in)
Skid track	2.18 m (7 ft 2 in)

DIMENSIONS, INTERNAL:
Cabin max width	1.19 m (3 ft 11 in)

AREAS:
Main rotor disc	79.46 m² (855.3 sq ft)
Tail rotor disc	1.70 m² (18.35 sq ft)

WEIGHTS AND LOADINGS:
Weight empty: standard	654 kg (1,442 lb)
Clipper: with utility floats	676 kg (1,490 lb)
with pop-out floats	684 kg (1,509 lb)
Fuel weight: standard	83 kg (184 lb)
auxiliary	50 kg (110 lb)
Max T-O and landing weight	1,088 kg (2,400 lb)
Max disc loading	13.7 kg/m² (2.81 lb/sq ft)
Max power loading at T-O	6.49 kg/kW (10.67 lb/hp)

PERFORMANCE:
Cruising speed at 75% power	
	113 kt (209 km/h; 130 mph)
Max rate of climb at S/L	over 305 m (1,000 ft)/min
Service ceiling	4,270 m (14,000 ft)
Hovering ceiling:	
IGE at 1,088 kg (2,400 lb)	1,950 m (6,400 ft)
OGE at 998 kg (2,200 lb)	1,555 m (5,100 ft)
Max range, no reserves	
	approx 347 n miles (643 km; 400 miles)

UPDATED

Robinson R44 Raven, latest version of this four-seat helicopter *(Paul Jackson/Jane's)*

ROTORWAY

ROTORWAY INTERNATIONAL

4140 West Mercury Way, Stellar Air Park, Chandler, Arizona 85226
Tel: (+1 480) 961 10 01
Fax: (+1 480) 961 15 14
e-mail: rotorway@primenet.com
Web: http://www.rotorway.com
PRESIDENT: Elbert Wolter
VICE-PRESIDENT, PRODUCTION: Rusty Mueller
VICE-PRESIDENT, SALES: Brent Marshall
MARKETING DIRECTOR: Susie Bell

Former RotorWay Aircraft Inc purchased and re-established on 1 June 1990 by the late John Netherwood; assets later acquired by company's employees, who are current owners. Original product was Exec 90, the former Exec with some 23 modifications, which was certified in UK in 1990 and in Poland in 1993. In August 1994, Exec 162F introduced and replaced the Exec 90; 162F improvements can be retrofitted to Exec 90.

RotorWay currently manufactures Exec 162F helicopter kits for final assembly by amateur builders and operators; over 500 had been sold by June 2000 to customers in 50 countries. The Exec is powered by the company's own engine, which has been in production for more than 15 years and has recently been upgraded to improve performance at altitude.

Some foreign distributors also sell helicopters; approximately 50 per cent of sales are outside the USA. Kruz company of Sofrina, Moscow Region, established assembly facility in 2000. RotorWay offers flight orientation and maintenance training, and customer service programme; 3,440 m² (37,000 sq ft) facility near Phoenix, Arizona, incorporates all departments, including manufacturing, sales and flight school.

UPDATED

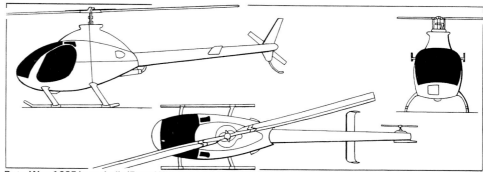

RotorWay 162F homebuilt (RotorWay International RI 162 engine) *(Paul Jackson/Jane's)* 1999/0051838

ROTORWAY INTERNATIONAL EXEC 162F

TYPE: Two-seat helicopter kitbuilt.
PROGRAMME: First flown in 1980. Kits, which lack only avionics and paint, are currently marketed in USA and 49 other countries; 50 per cent of production is shipped abroad.
CUSTOMERS: Including earlier Exec and Exec 90 models, over 650 kits sold and 500 flying; 350th 162F delivered November 1998; 400th 162F delivered June 1999; 500th delivered to Mexican Navy June 2000. Russian version (50 on order) designated **162E**; first example (FLARF-02133) completed and flown in June 2000. Uses include police surveillance, forestry observation, powerline inspection, ranch work, cropspraying and flight training. One purchased in 2000 by Mexican Navy for pilot training.
COSTS: Kit: US$64,350 complete (2001). ACIS US$4,000 extra. Also available in four batches of parts to spread cost.
DESIGN FEATURES: Asymmetrical aerofoil section two-blade main rotor. All-metal aluminium alloy blades attached to aluminium alloy teetering rotor hub by retention straps. Teetering tail rotor, with two blades each comprising steel spar and aluminium alloy skin. Elastomeric bearing rotor hub system with dual push/pull cable-controlled swashplate for cyclic pitch control. Quoted build time 300 to 400 hours. Company offers on-line diagnostic testing by modem.
STRUCTURE: Blades as detailed under Design Features. Basic 4130 steel tube airframe structure, with wraparound glass fibre fuselage/cabin enclosure. Aluminium alloy monocoque tailboom.
LANDING GEAR: Twin-skid type. Floats optional.
POWER PLANT: One 113 kW (152 hp) RotorWay International RI 162F 2.66 litre (162 cu in) liquid-cooled engine with FADEC. Optional Altitude Compensation Induction System (ACIS) supercharger supplies pressurised air to engine at altitude, obviating need for turbocharging and prolonging engine life. Standard fuel capacity 64.4 litres (17.0 US gallons; 14.2 Imp gallons).
ACCOMMODATION: Two persons side by side; optional external Helipac baggage pod fits between skids, capacity 0.21 m³ (7.4 cu ft).
AVIONICS: Extra navigational equipment standard on helicopters for Russian market.
EQUIPMENT: Optional Helipac cargo hold. Optional cropspraying kit, including two 42 litre (11.0 US gallon; 9.2 Imp gallon) tanks.

DIMENSIONS, EXTERNAL:
Main rotor diameter	7.62 m (25 ft 0 in)
Tail rotor diameter	1.28 m (4 ft 2½ in)
Length of fuselage	6.71 m (22 ft 0 in)
Length overall, rotors turning	8.99 m (29 ft 6 in)
Height to top of main rotor	2.44 m (8 ft 0 in)
Skid track	1.65 m (5 ft 5 in)

DIMENSIONS, INTERNAL:
Cabin max width	1.12 m (3 ft 8 in)

AREAS:
Main rotor disc	45.60 m² (490.9 sq ft)
Tail rotor disc	1.17 m² (12.57 sq ft)

WEIGHTS AND LOADINGS:
Weight empty	442 kg (975 lb)
Crew weight	193 kg (425 lb)
Max T-O weight	680 kg (1,500 lb)
Max disc loading	14.9 kg/m² (3.06 lb/sq ft)
Max power loading at T-O	6.01 kg/kW (9.87 lb/hp)

PERFORMANCE:
Never-exceed (VNE) and max level speed	100 kt (185 km/h; 115 mph)
Normal cruising speed	82 kt (153 km/h; 95 mph)
Max rate of climb at S/L	305 m (1,000 ft)/min
Service ceiling	3,050 m (10,000 ft)
Hovering ceiling, with two persons:	
IGE	2,135 m (7,000 ft)
OGE	1,525 m (5,000 ft)
Range with max fuel at optimum cruising power	156 n miles (289 km; 180 miles)
Endurance with max fuel at optimum cruising power	2 h

UPDATED

Rotorway Exec 162F kitbuilt helicopter *(Paul Jackson/Jane's)* 2001/0103555

SADLER

SADLER AIRCRAFT CORPORATION

8225 East Montebello Avenue, Scottsdale, Arizona 85250
Tel: (+1 602) 994 46 31
Fax: (+1 602) 481 05 74
PRESIDENT AND CEO: William G Sadler

Original American Microflight Vampire won the Grand Champion Design Award at Oshkosh in 1982; small batch of SV-1 and SV-2 variants produced in Australia by Skywise Ultraflight Pte Ltd. A-22 Piranha military version currently then promoted; plans for production in Turkey as the TG-X1 (see TAI entry in Turkish section of 1998-99 edition) were abandoned in 1998; the A-22 was last described in *Jane's 2000-01*. No further information received.

UPDATED

SAFIRE

SAFIRE AIRCRAFT COMPANY

400 Clematis Street, Suite 207, West Palm Beach, Florida 33401
Tel: (+1 561) 650 08 30 and (+1 800) 862 86 80
Fax: (+1 561) 659 63 19
e-mail: info@safireaircraft.com
Web: http://www.safireaircraft.com
PRESIDENT: Dimitri Margaritoff
EXECUTIVE VICE-PRESIDENT: David Humphries
DIRECTOR, SALES AND MARKETING: Bruce Hamilton
MANUFACTURING ENGINEERING MANAGER: Joe DiVito

Safire, formed in September 1998, is developing the S-26 six-seat business jet. Later projects will be the two-seat, private-owner S-12 and four-seat S-14. The company is currently evaluating manufacturing facilities.

UPDATED

SAFIRE S-26

TYPE: Light business jet.
PROGRAMME: Preliminary details released early 1999. Second-stage funding, completed 7 July 1999, will take project to first flight in mid-2002 and FAR Pt 23 certification for single pilot operation in May 2003.
CUSTOMERS: More than 700 delivery positions reserved by 1 January 2001.
COSTS: Estimated flyaway cost US$800,000 (2001).

DESIGN FEATURES: Low-wing monoplane with rear-mounted engines and sweptback vertical tail.
STRUCTURE: Composites construction.
POWER PLANT: Original choice was two Williams FJX-2 turbofans, each approximately 3.11 kN (700 lb st); this was changed to two Agilis TF-800 turbofans each of 4.00 kN (900 lb st). Fuel capacity 897 litres (237 US gallons; 197 Imp gallons).
ACCOMMODATION: Six persons, including pilot. Private lavatory in rear of cabin. Cabin pressurised to 2,440 m (8,000 ft) at certified ceiling. Dual controls.
AVIONICS: Full EFIS with three 203 × 254 mm (8 × 10 in) flat panel displays; compatible with Synthetic Vision and Highway-in-the-Sky (HITS) programmes.

Following data are provisional:

DIMENSIONS, EXTERNAL:
Wing span	10.91 m (35 ft 9½ in)
Length overall	10.39 m (34 ft 1 in)
Height overall	4.24 m (13 ft 11 in)

DIMENSIONS, INTERNAL:
Cabin: Length	3.96 m (13 ft 0 in)
Max width	1.45 m (4 ft 9 in)
Max height	1.35 m (4 ft 5 in)

WEIGHTS AND LOADINGS:
Weight empty	1,465 kg (3,230 lb)
Payload: max	635 kg (1,400 lb)
with max fuel	217 kg (478 lb)
Max T-O weight	2,326 kg (5,130 lb)
Max power loading	290 kg/kN (2.85 lb/lb st)

PERFORMANCE (estimated):
Cruising speed	330 kt (611 km/h; 380 mph)
Stalling speed, flaps down	69 kt (128 km/h; 80 mph)
Max rate of climb at S/L	884 m (2,900 ft)/min
Rate of climb at S/L, OEI	183 m (600 ft)/min
Service ceiling	11,280 m (37,000 ft)
T-O to 15 m (50 ft)	762 m (2,600 ft)
Landing from 15 m (50 ft)	762 m (2,500 ft)
Range: with max fuel	1,550 n miles (2,870 km; 1,783 miles)
with 454 kg (1,000 lb) payload	610 n miles (1,129 km; 702 miles)

UPDATED

Artist's impression of Safire S-26 (two Agilis TF-800 turbofans) 2001/0105829

SAFIRE S-12
There are no immediate plans to produce this aircraft, brief details of which last appeared in the 2000-01 edition.
UPDATED

SAFIRE S-14
There are no immediate plans to build this four-seat jet.
UPDATED

SCALED

SCALED COMPOSITES
1624 Flight Line, Mojave, California 93501-1663
Tel: (+1 661) 824 45 41
Fax: (+1 661) 824 41 74
e-mail: info@scaled.com
Web: http://www.scaled.com
PRESIDENT: Burt (Elbert L) Rutan
VICE-PRESIDENT AND GENERAL MANAGER: Michael W Melvill

Scaled Composites founded 1982; bought by Beech Aircraft Corporation (now Raytheon) June 1985; sold back to Burt Rutan November 1988 and integrated in joint venture with Wyman-Gordon Company of North Grafton, Massachusetts; is now owned by private investors following buy-out by Burt Rutan and 10 other investors; continues to provide R&D facilities to individuals and companies; several projects developed for Beech retained by Scaled.

Past projects summarised in 1994-95 *Jane's* and detailed in earlier editions; company also active in UAV field (refer *Jane's Unmanned Aerial Vehicles and Targets*). Current programmes include proprietary products.

For details of the Visionaire Vantage business jet, see that company's entry; Scaled produced the prototype and undertook early test flights. Similarly, Scaled was involved in the Adam M-309 (which see), which it designed, built and flight tested.

The company's V-Jet II testbed for the Williams FJX-2 turbofan, described briefly in the 1999-2000 and earlier editions, has been replaced by the Eclipse Eclipse 500 (which see).

Burt Rutan's private venture **202-11 Boomerang** light transport, described in the 1994-95 *Jane's* and first flown (N24BT) on 17 June 1996, will enter production in a modified form with Morrow Aircraft of Salem, Oregon (which see); Scaled is building a prototype and two certification test aircraft.

Company currently has 120 employees and annual revenue of around US$15 million.

UPDATED

SCALED 281 PROTEUS
TYPE: High-altitude platform.
PROGRAMME: Announced 24 October 1997. First flight (N281PR) 26 July 1998. In 1998, Angel Technologies of St Louis teamed with Raytheon to develop air and ground elements for HALO fixed and mobile radio, data and video services, and signed agreement with Scaled owners Wyman-Gordon for 100 Proteus-type production aircraft. By late 2000, this project was understood to have lapsed.

During 1999, participated in NASA's Environmental Research Aircraft and Sensor Technology (ERAST) programme, carrying (from 2 February onwards) ventral pod containing HyperSpectral Sciences Inc Airborne Real-Time Imaging System camera. NASA's Dryden facility also assisted with development of station-keeping autopilot and satcom-based uplink/downlink data system.

On 25 and 27 October 2000, the Proteus set three world altitude records – peak altitude 19,137 m (62,786 ft), sustained altitude in horizontal flight 18,873 m (61,919 ft) and peak altitude of 17,032 m (55,878 ft) with 1,000 kg (2,205 lb) payload.

In January 2001 aircraft was sold to Northrop Grumman as a sensor and mission platform.
CURRENT VERSIONS: **Telecommunications:** *As described.*
Atmospheric Research: Single pilot operation; two scientists in cabin; able to loiter for 18 hours at 500 n mile (926 km; 575 mile) radius or 6 hours at 3,000 n mile

Scaled Proteus, overflying Mojave Desert (*Patrick Wright/NASA*) 2001/0104604

(5,556 km; 3,452 mile) radius. Service ceiling 19,810 m (65,000 ft) at 3,175 kg (7,000 lb) weight.

Reconnaissance/surveillance: Single pilot plus one crew member; carries integrated aircraft/payload thermal management system. Max T-O weight 7,167 kg (15,800 lb) including payload of 2,268 kg (5,000 lb). Service ceiling 19,510 m (64,000 ft) at 2,438 kg (8,000 lb); loiter time 22 hours at 500 n mile (926 km; 575 mile) radius or 12 hours at 2,000 n mile (3,704 km; 2,301 mile) radius. Fuel weight up to 4,082 kg (9,000 lb) can be carried using fuselage tank. Can also be operated as a UAV.

Micro Satellite Launcher: Capable of carrying micro satellites up to 29 kg (65 lb) using two-stage rocket, including reusable booster. Rocket would be launched at 10,973 m (36,000 ft) and speed of 180 kt (333 km/h; 207 mph) up to 70°γ. Satellite launch cost US$600,000. Crew of pilot plus launch officer. Gross weight 6,350 kg (14,000 lb); launch weight 2,948 kg (6,500 lb). Port aft wing and canard wingtips installed for asymmetric loads.

DESIGN FEATURES: Optimised for loiter at 18,000 to 20,000 m (59,050 to 65,620 ft) carrying underslung antenna. Highly unconventional configuration, incorporating large foreplane; twin, unbraced tailbooms carrying vertical fins; and twin, podded turbofans mounted on rear fuselage shoulders. Anhedral on outboard wing panels. Provision for extended wing- and foreplane tips. Fuselage centre section interchangeable depending on mission.

FLYING CONTROLS: Manual ailerons in foreplanes with servo tab on port aileron; flaps in centre-section of each wing, outboard of booms; twin rudders with trim tab to port.

STRUCTURE: Composites throughout. Participating companies include Hickok Inc (engine instruments), Puroflow Inc (fuel filters), Concorde Battery Corp (batteries), Hamilton Sundstrand (engine and throttle controls), Whittaker Corp, Dodson International and Arens Controls.

LANDING GEAR: Retractable tricycle type. Twin steerable nosewheels (size 600-6) retract rearwards; single mainwheels (6.50-10) retract rearwards into tailbooms. Hydraulic disc brakes on mainwheels. Wide wheelbase improves crosswind take-off capability.

POWER PLANT: Two 10.2 kN (2,293 lb st) Williams FJ44-2 turbofans.

ACCOMMODATION: Pilot and relief pilot in cabin pressurised by engine bleed air system to 0.69 bar (10.0 lb/sq in) differential, providing 2,440 m (8,000 ft) equivalent altitude at cruising height. Crew door starboard side.

SYSTEMS: Electrical system 28 V. 30 kW electrical power available for payload. Primary coolant loop and power conditioning unit housed in fuselage.

EQUIPMENT: Ventral relay antenna in 4.57 m (15 ft 0 in) suspended ventral pod, stabilised mechanically in pitch and roll, and liquid cooled, provides 120 km (75 mile) diameter communications footprint; flying 9 to 15 km (6 to 9 mile) circle keeps aircraft on station.

DIMENSIONS, EXTERNAL:
Wing span: normal	23.65 m (77 ft 7 in)
extended	27.99 m (91 ft 10 in)
Foreplane span: normal	16.66 m (54 ft 8 in)
extended	19.71 m (64 ft 8 in)
Length overall	17.17 m (56 ft 4 in)
Height overall	5.38 m (17 ft 8 in)

DIMENSIONS, INTERNAL:
Cabin: Length	2.74 m (9 ft 0 in)
Max diameter	1.55 m (5 ft 1 in)

AREAS:
Wings, gross, normal	27.92 m² (300.5 sq ft)
Foreplanes, gross, normal	16.60 m² (178.7 sq ft)

WEIGHTS AND LOADINGS:
Weight empty	2,658 kg (5,860 lb)
Mission payload	1,043 kg (2,300 lb)
Max fuel weight	2,676 kg (5,900 lb)
Max T-O weight: FAR Pt 23	5,670 kg (12,500 lb)
military	7,166 kg (15,800 lb)
Max landing weight	4,900 kg (11,000 lb)
Max wing/foreplane loading, normal span, military	161.0 kg/m² (32.97 lb/sq ft)
Max power loading, military	351 kg/kN (3.45 lb/lb st)

PERFORMANCE (estimated, FAR Pt 23, relay role):
Max level speed	272 kt (504 km/h; 313 mph)
Econ cruising speed:	
at 6,100 m (20,000 ft)	190 kt (352 km/h; 219 mph)
at 12,190 m (40,000 ft)	280 kt (519 km/h; 322 mph)
Unstick speed	71 kt (130 km/h; 81 mph)
Max rate of climb at S/L	1,036 m (3,400 ft)/min
Cruising altitude, mid-mission	18,590 m (61,000 ft)
T-O run: at 3,629 kg (8,000 lb)	223 m (730 ft)
at FAR Pt 23 MTOW	433 m (1,420 ft)
Endurance on station, 1,000 n miles (1,852 km); 1,150 miles) from base	14 h
g limits	+3.2/−1.8

UPDATED

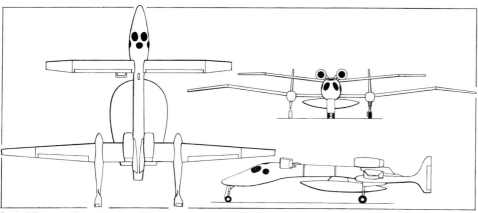

Scaled Proteus high-altitude radio relay aircraft *(James Goulding/Jane's)*

SCHWEIZER

SCHWEIZER AIRCRAFT CORPORATION
PO Box 147, Elmira, New York 14902
Tel: (+1 607) 739 38 21
Fax: (+1 607) 796 24 88
e-mail: schweizer@sacusa.com
Web: http://sacusa.com
PRESIDENT: Paul H Schweizer
EXECUTIVE VICE-PRESIDENTS:
  Leslie E Schweizer
  W Stuart Schweizer
VICE-PRESIDENT: Michael D Oakley
MARKETING DIRECTORS: Rocky G Peters
  David K Savage
  Barbara J Tweedt

Established 1939 by Ernest Schweizer (1912-2000) to produce sailplanes; from mid-1957 to 1995 made Grumman (later Gulfstream American) Ag-Cat, initially under subcontract, but as Schweizer-owned product from January 1981; manufacturing rights sold to Ag-Cat Corporation of Malden, Missouri, but production terminated soon thereafter.

Schweizer acquired rights for sole US manufacture of Hughes 300 light helicopter 13 July 1983; continues to support earlier Hughes 300s; first Elmira-built 300C completed June 1984; Schweizer purchased US rights for whole 300C programme from McDonnell Douglas Helicopter Company (formerly Hughes Helicopters) 21 November 1986. Deliveries totalled 56 helicopters in 1996, 44 in 1997, 41 in 1998 and 43 in 1999; by September 2000, Schweizer had delivered over 800 helicopters.

Schweizer also manufactures the Model 300CB and 330SP helicopters and the 2-37 and 2-38 family of surveillance aircraft. An unmanned version of the 300CB, known as the RoboCopter 300, has been developed in collaboration with Kawada Industries Inc of Japan and is described in *Jane's Unmanned Aerial Vehicles and Targets*.

Subcontracts include aircraft structural components and assemblies for Northrop Grumman, Sikorsky and Boeing. The company employed 430 in 1999 in its 17,650 m² (190,000 sq ft) facility.

UPDATED

SCHWEIZER SA 2-37A
US military designation: RG-8A

TYPE: Multisensor surveillance lightplane.

PROGRAMME: First flight 1985; prototype fitted with Hughes AN/AAQ-16 and other manufacturers' thermal imaging systems. Full description in 1991-92 and earlier *Jane's*.

CUSTOMERS: Total 14 built by mid-2000, including demonstrator. Three for US Army, of which one lost in accident; remaining two transferred in 1987 to US Coast Guard at Opa Locka for anti-narcotics operations; one converted to SA 2-38A (see next entry).

Fifth, sixth and seventh 2-37As built for Central Intelligence Agency late 1989/early 1990; one aircraft each to Colombian (since returned) and Mexican air forces; further five built for undisclosed customer(s) between 1997 and 1999, at least one for Colombia. See table of production.

Schweizer RG-8A of the Colombian Air Force

CIA aircraft employed as airborne communications relay platforms for long-range reconnaissance UAVs; believed used over former Yugoslavia in 1994, supporting General Atomics Gnat 750 vehicles; also fitted with surface surveillance equipment to identify sites for NATO air strikes in same operational theatre. Participated in hostage rescue at Japanese embassy in Peru, 1997. Colombian and Mexican aircraft believed supplied by US government for drugs interdiction.

USA: AIRCRAFT—SCHWEIZER

SCHWEIZER SA 2-37A PRODUCTION

c/n	Reg/serial	Date	Operator
1	N3623C	1985	Schweizer
	85-0047	1986	US Army; to
	8101	1987	USCG; to SA 2-38A
2	N3623F	1985	Schweizer
	85-0048	1986	US Army; lost
3	N9237A		Schweizer; cancelled Apr 94
4	86-0404	1987	US Army; to
	8102	1987	USCG; lost 1996
5	N7508U	Jul 89	CIA; to
	N701AN	Dec 96	CIA
6	N7508W	Nov 89	CIA; to
	N87260	Feb 97	CIA
7	N7508Y	Jan 90	CIA; to
	N4122L	Dec 96	CIA
8	N7508Z	Feb 91	Schweizer; to
	FAC 2046		Colombian AF; to
	N12139	Mar 97	CIA
9	N7510U	May 94	Schweizer; to
	OES-2252	Jul 94	Mexican AF
010	N6147U	Oct 97	LAIRD International
011	N61474	Nov 97	Crashed 19 Feb 98
012	N61499	Oct 98	Schweizer; to
	FAC 5751		Colombian AF
014	N20086	May 99	Schweizer
015	N6150U	Jun 00	Schweizer

Notes: CIA aircraft registered to Vantage Leasing Inc of Dover, Delaware.
Nos. 1 and 2 remain currently registered (2001) as N3623C and N3623F.
No. 010 owner at Marietta, Georgia
No. 012 remains currently registered.
No. 013: This serial number not allocated by Schweizer company.

DESIGN FEATURES: Modification of Schweizer SGM 2-37 motor glider, but with slightly greater wing span, drooped leading-edge and leading-edge fences on outer wing panels to improve stall, much more powerful engine with large exhaust silencers on fuselage sides, three-blade quiet propeller, fuselage modified to accept bulged canopy and larger engine, streamlined fairings and hydraulic parking brake on mainwheels; more than treble standard fuel capacity, but optional extra tank also available; removable underfuselage skin and hatches give access to 1.84 m³ (65 cu ft) payload bay behind cockpit, in which pallets holding LLLTV, FLIR or camera payloads can be quickly removed and installed; other engines and larger payloads available for surveillance, basic and advanced training, operator training, glider and banner towing, and priority cargo delivery. Certified to FAR Pt 23 for day and night IFR; inaudible when overflying at 'quiet mode' speed at about 610 m (2,000 ft) using 38.8 kW (52 hp) from its 175 kW (235 hp) Textron Lycoming IO-540 engine.

Wing section Wortmann FX-61-163 at root and FX-60-126 (modified) at tip; outer wing panels and horizontal tail can be removed for transport.

FLYING CONTROLS: Manual. All-moving tailplane with anti-balance tab; externally mass-balanced ailerons; rudder.
LANDING GEAR: Tailwheel type; fixed. Cleveland disc brakes.
POWER PLANT: Initially one 175 kW (235 hp) Textron Lycoming IO-540-W3A5D flat-six engine, driving a McCauley three-blade constant-speed propeller. Reportedly re-engined in service with 186 kW (250 hp) turbocharged TIO-540. Standard fuel capacity of 196.8 litres (52.0 US gallons; 43.3 Imp gallons), increasable optionally to 253.6 litres (67.0 US gallons; 55.8 Imp gallons). Shadin fuel flow system.
ACCOMMODATION: Seats for two persons side by side under two-piece upward-opening canopy, hinged on centreline. Dual controls, seat belts and inertia reel harnesses standard. Compartment aft of seats enlarged to accommodate pallet containing up to 340 kg (750 lb) of sensors or other equipment.

DIMENSIONS, EXTERNAL:
Wing span	18.745 m (61 ft 6 in)
Wing aspect ratio	19
Fuselage length	8.46 m (27 ft 9 in)
Height overall (tail down)	2.36 m (7 ft 9 in)
Wheel track	2.79 m (9 ft 2 in)
Wheelbase	5.99 m (19 ft 8 in)
Propeller diameter	2.18 m (7 ft 2 in)

AREAS:
Wings, gross	18.52 m² (199.4 sq ft)

WEIGHTS AND LOADINGS:
Weight empty	918 kg (2,025 lb)
Max mission payload	340 kg (750 lb)
Max T-O weight	1,587 kg (3,500 lb)
Max wing loading	85.65 kg/m² (17.55 lb/sq ft)
Max power loading	9.06 kg/kW (14.89 lb/hp)

PERFORMANCE (IO-540):
Max permissible diving speed (V_D)	176 kt (326 km/h; 202 mph) CAS
Cruising speed at 1,525 m (5,000 ft):	
75% power	138 kt (256 km/h; 159 mph)
65% power	129 kt (239 km/h; 148 mph)
Approach speed	117 kt (217 km/h; 135 mph) CAS

Schweizer RU-38A Twin Condor surveillance aircraft
2001/0077028

Quiet mode speed	70-80 kt (130-148 km/h; 80-92 mph)
Optimum climbing speed	77 kt (142 km/h; 88 mph) CAS
Stalling speed:	
airbrakes open	71 kt (132 km/h; 82 mph)
airbrakes closed	67 kt (124 km/h; 77 mph)
Max rate of climb at S/L	292 m (960 ft)/min
Service ceiling: IO-540	5,490 m (18,000 ft)
TIO-540	7,315 m (24,000 ft)
T-O run (S/L, ISA): hard surface	387 m (1,270 ft)
grass	533 m (1,750 ft)
T-O to 15 m (50 ft) (S/L, ISA):	
hard surface	612 m (2,010 ft)
grass	759 m (2,490 ft)
Landing from 15 m (50 ft) (S/L, ISA):	
hard surface	680 m (2,230 ft)
grass	732 m (2,400 ft)
Best glide ratio	20
g limits	+6.6/−3.3

UPDATED

SCHWEIZER SA 2-38A
US military designation: RU-38A Twin Condor
TYPE: Multisensor surveillance lightplane.
PROGRAMME: Development started 1993; substantial redesign of single-engined SA 2-37A/RG-8A; first US Coast Guard 2-37A returned to Schweizer 24 January 1994 for conversion beginning in April; first flight (N61428) 31 May 1995; second aircraft was due to fly in mid-1996, but was destroyed in a crash earlier the same year while still an RG-8; third was to be first 'new' build, but with wing taken from stock; all three for USCG. Type publicly revealed 20 July 1995. First delivery due to US Coast Guard in late 1996; however, programme suspended following RG-8 crash and not resumed until December 1996 when contract amended to two aircraft with engines uprated with turbochargers. Modifications introduced in 1997 comprised a redesigned tailplane and improved inlet ducts for the rear engine. Second (new-build) aircraft registered N61449 in July 1997. Flight testing (aircraft reserialled 8103) for USCG acceptance began at Edwards AFB on 10 July 1998 by 445th Flight Test Squadron, USAF; programme was 100 sorties in three/four months; however, test programme was terminated in the following year. Both aircraft remain registered to Schweizer.

CURRENT VERSIONS: **Two-seat** version, *as described*, for USCG; **three-seat** version reportedly under design for an export customer; turbocharged version also under development, to increase operating altitude to 9,150 m (30,000 ft).

CUSTOMERS: US Coast Guard (two); second US customer, and one overseas, also reported. USCG aircraft expected to be deployed mainly over Caribbean and Gulf of Mexico from the USCG base at Miami, Florida. By late 1999, only two known to have been built.

COSTS: US$450,000 (1993) for initial redesign; US$3.5 million (1994) USAF contract to convert existing two aircraft; approximately US$1 million each (excluding sensors) for any additional procurement. Renegotiated contract (December 1996) covers two aircraft for US$5.3 million.

DESIGN FEATURES: Main objectives were to increase night patrol capability and reduce engine coking problems compared with SA 2-37A; additional engine also increases safety factor for overwater operation, though normal mode is single-engine cruise with second engine shut down.

Utilises wings and cockpit forward section of SA 2-37A; wing section Wortmann FX-61-163 at root and FX-60-126 (modified) at tip; principal design changes are adoption of pod and twin tailboom configuration with twin engines in push-pull layout; cabin slightly widened.

LANDING GEAR: Fixed tricycle type; single wheel on each unit. Main units Cleveland 6.50-10 wheel assemblies with Cleveland 30-95A brakes; nosewheel is Cleveland 6.00-6.

POWER PLANT: Prototype originally with two heavily muffled Teledyne Continental GIO-550A flat-six engines, each rated at 261 kW (350 hp) at 3,400 rpm, with 3:2 reduction gear to limit propeller speed to 2,267 rpm. One engine in nose and one in rear of fuselage pod, respectively driving a

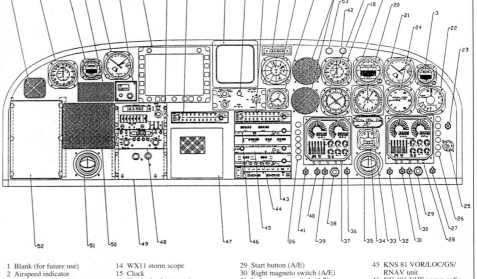

1 Blank (for future use)	14 WX11 storm scope	29 Start button (A/E)	45 KNS 81 VOR/LOC/GS/
2 Airspeed indicator	15 Clock	30 Right magneto switch (A/E)	RNAV unit
3 Attitude gyro	16 Blank (for future use)	31 Left magneto switch (A/E)	46 KY 196 VHF comm radio
4 Video recorder control	17 KI 229 RMI	32 Fuel pump switch (A/E)	47 Tracor 7880 Omega/VHF/
5 Altimeter	18 Airspeed indicator	33 KDI 572 DME indicator	GPS navigation unit
6 KFS 594 HF control	19 KI 256 attitude gyro and	34 Turn and bank indicator	48 ARC-182 VHF/UHF radio
7 FLIR screen	autopilot flight director	35 Air vent	49 KY 58 encryption device
8 Wolfsburg RT9600 HF radio	20 KI 525A HSI	36 Alternator switch (F/E)	50 Blank (for future use)
9 KMA 24H-65 audio selector	21 Altimeter	37 Engine cluster (F/E)	51 Air vent
10 Radar screen	22 KI 250 radar altimeter	38 Start button (F/E)	52 FLIR control panel
11 Radar control	23 Radar altimeter switch	39 Right magneto switch (F/E)	53 Master caution light
12 KMA 24H-65 audio selector	24 7030 VSI	40 Left magneto switch (F/E)	
13 KR 21 marker beacon	25 Suction gauge	41 Fuel pump switch (F/E)	A/E denotes aft engine
	26 Warning lights	42 Low fuel warning	F/E denotes front engine
	27 Alternator switch (A/E)	43 KC 192 autopilot control	
	28 Engine cluster (A/E)	44 KR 87 ADF receiver	

Instrument panel of the Schweizer SA 2-38A

SCHWEIZER—AIRCRAFT: USA

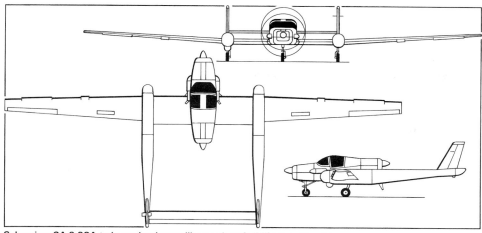
Schweizer SA 2-38A twin-engined surveillance aircraft *(Mike Keep/Jane's)* 2000/0064927

tractor and pusher constant-speed, fully feathering three-blade propeller. Turbocharged engines in service version. Usable fuel capacity 375 litres (99.0 US gallons; 82.5 Imp gallons).

ACCOMMODATION: Pilot and sensor operator, side by side.
SYSTEMS: Three independent 28 V power sources.
AVIONICS: *Comms:* Rockwell Collins AN/ARC-182 VHF/UHF and Honeywell HF 990 radios, KY 196 VHF, KMA 24H-65 audio selector (two) and KFS 594 HF control; Wolfsburg RT9600 marine band radio.
Radar: Honeywell AN/APN-215(V) colour weather radar with search and mapping modes in nose of port tailboom.
Flight: Honeywell KI 229 RMI, KI 256 flight director, KI 525A HSI, KI 250 radar altimeter, KDI 572 DME, KC 192 autopilot, KR 87 ADF and KNS 81 VOR/LOC/GS/RNAV. BAE Systems 7880 Omega/GPS.
Mission: Surveillance radar, as above. FLIR/LLLTV and dual recorder in nose of starboard boom; Honeywell KY 58 and KY 75 communications encryption devices. Fully integrated sensor suite under development.

DIMENSIONS, EXTERNAL:
Wing span	19.51 m (64 ft 0 in)
Wing aspect ratio	18.2
Length overall	9.19 m (30 ft 2 in)
Height overall	2.90 m (9 ft 6 in)
Wheelbase	5.79 m (19 ft 0 in)

AREAS:
Wings, gross	20.98 m² (225.9 sq ft)
Fins (total)	2.26 m² (24.30 sq ft)
Tailplane	3.26 m² (35.10 sq ft)

WEIGHTS AND LOADINGS:
Weight empty	1,524 kg (3,360 lb)
Max payload	408 kg (900 lb)
Max fuel weight	272 kg (600 lb)
Max T-O weight	2,404 kg (5,300 lb)
Max wing loading	114.6 kg/m² (23.46 lb/sq ft)
Max power loading	4.61 kg/kW (7.57 lb/hp)

PERFORMANCE (GIO-550 engine):
Never-exceed speed (V$_{NE}$)	165 kt (305 km/h; 189 mph)
Cruising speed at 1,525 m (5,000 ft):	
at 75% power	136 kt (252 km/h; 157 mph)
at 63% power	127 kt (235 km/h; 146 mph)
Mission speed	90 kt (167 km/h; 104 mph)
Best climbing speed	87 kt (161 km/h; 100 mph)
Stalling speed	75 kt (139 km/h; 87 mph)
Max rate of climb at S/L	670 m (2,200 ft)/min
Rate of climb at S/L, nose engine only	350 m (1,150 ft)/min
Service ceiling	7,315 m (24,000 ft)
T-O run	295 m (960 ft)
T-O to 15 m (50 ft)	430 m (1,400 ft)
Landing from 15 m (50 ft)	755 m (2,475 ft)
Landing run	410 m (1,350 ft)
Endurance with no reserves: typical mission	6 h
max	10 h

UPDATED

SCHWEIZER 300C
Engineering designation: Model 269
TYPE: Three-seat helicopter.
PROGRAMME: Hughes 269 first flew 2 October 1956; production deliveries began October 1961; marketed later as Model 300. Developed 300C first flew August 1969; first flight Hughes production model December 1969; FAA certification May 1970 (basic Hughes 300 described in 1976-77 *Jane's*). Production of 300C transferred from Hughes Helicopters to Schweizer July 1983; first flight Schweizer 300C (1,166th 300C) June 1984; Schweizer bought entire programme November 1986. Deliveries of 300CB began in August 1995 with initial three of 10 ordered by launch customer Helicopter Adventures, of Concord, California. By late 1999 nearly 3,500 Hughes/Schweizer 300s of all models had been produced.
CURRENT VERSIONS: **300C:** Standard civil version.
Main description applies to 300C, unless otherwise indicated.

300C Sky Knight: Special police version; options include safety mesh seats with inertia reel shoulder harnesses, public address/siren system, searchlight, integrated communications, infra-red sensor, heavy-duty 28 V 100 A electrical system, cabin heater, night lights with strobe beacons, cabin utility light, fire extinguisher, first aid kit and map case.
TH-300C: Military training version.
300CB: 'Basic' version (otherwise designated 269C-1) for training role; 134 kW (180 hp) Textron Lycoming HO-360-C1A engine, 132 litre (35 US gallon; 29.1 Imp gallon) fuel tank and 2,000 hour TBO; stated to be cheaper to operate than 300C. First flight (N6002X) 28 May 1993; type certificate awarded August 1995. Changes from standard model include one-piece upper pulley hub; improved air filtration system; lightweight exhaust; and improved oil filter. 100,000 hours flown by August 1999, at which time five had been lost in accidents without fatalities.
RoboCopter: Unmanned version developed by Kawada Industries in Japan for cropspraying, pipeline patrol, aerial crane and reconnaissance; first flight October 1996; may be offered as new production by Schweizer, which teamed with Northrop Grumman to offer helicopter for US Navy VTOL UAV requirement. Further details in *Jane's Unmanned Aerial Vehicles and Targets*.

CUSTOMERS: Hughes produced 2,775 of all versions, including TH-55A Osage for US Army, before transfer to Schweizer. Royal Thai Army received 58 TH-300C helicopters between March 1986 and mid-1989; 500th Schweizer-manufactured 300C delivered at Heli Expo 1994. Recent Sky Knight customers include police departments of Baltimore, Maryland; Lakewood, California; and Kansas City, Missouri. By September 2000, Schweizer had built some 645 300Cs and 114 300CBs.
COSTS: 300C: US$223,500 (1998); 300CB US$189,500 (1998).
DESIGN FEATURES: Simple, pod-and-boom configuration, with exposed power plant and fuel tanks. Aerofoil-section stabiliser, starboard side, rear, at approx 45° dihedral; triangular underfin. Fully articulated three-blade main rotor, turning at 471 rpm; fully interchangeable blades; blade section NACA 0015; elastomeric dampers; two-blade teetering tail rotor; limited blade folding; no rotor brake; multiple V-belt and pulley reduction gear/drive system between horizontally mounted engine and transmission, with electrically controlled belt-tensioning system instead of clutch; braced tubular tailboom; separate dihedral tailplane and fin.
FLYING CONTROLS: Manual. Electric cyclic trim.
STRUCTURE: Main rotor blades bonded with constant-section extruded aluminium spar, wraparound skin and trailing-edge; tail rotor blades have steel tube spar and glass fibre skin; steel tube cabin section with light alloy and stainless steel skin and Plexiglas transparencies.
LANDING GEAR: Skids carried on oleo-pneumatic shock-absorbers. Replaceable heavy-duty skid shoes. Two ground handling wheels with 0.25 m (10 in) balloon tyres,

Schweizer 300C light utility helicopter 2000/0077025

Schweizer 300CB training helicopter *(Paul Jackson/Jane's)* 2001/0103556

USA: AIRCRAFT—SCHWEIZER

pressure 4.14 to 5.17 bars (60 to 75 lb/sq in). Available optionally on floats made of polyurethane-coated nylon fabric, 4.70 m (15 ft 5 in) long and with a total installed weight of 27.2 kg (60 lb). Heavy-duty skid plates optional.

POWER PLANT: One 168 kW (225 hp) Textron Lycoming HIO-360-D1A flat-four engine, derated to 142 kW (190 hp), mounted horizontally aft of seats. Normal fuel capacity of 300C is 123 litres (32.5 US gallons; 27.1 Imp gallons) of which 114 litres (30.0 US gallons; 25.0 Imp gallons) are usable; usable capacity of 300CB is 133 litres (35.0 US gallons; 29.2 Imp gallons); auxiliary tank of 133 litres (35.0 US gallons; 29.2 Imp gallons) can be fitted to 300C and tank of 114 litres (30.0 US gallons; 25.0 Imp gallons) to 300CB. Oil capacity 9.5 litres (2.5 US gallons; 2.1 Imp gallons).

ACCOMMODATION: Three persons side by side on sculptured and cushioned bench seat, with shoulder harness, in Plexiglas enclosed cabin. Accommodation ventilated. Carpet and tinted canopy standard. Forward-hinged, removable door on each side. Dual controls optional in 300C, standard in 300CB; captain sits on left in 300C and on right in 300CB. Exhaust muff heating kit available.

SYSTEMS: Standard electrical system includes 24 V 70 A alternator, 24 V battery, starter and external power socket.

AVIONICS: *Comms:* Avionics include Honeywell KY 96A com transceiver and headsets, KR 86 ADF and GTX 320 transponders and ACK A-30 blind encoder.

EQUIPMENT: Standard for all models are fire extinguisher, first aid kit and engine hour meter. Model 300C has external power plug and fuel pressure indicator; Model 300CB has night lights with strobes. Optional equipment includes stretcher kits, cargo racks with combined capacity of 91 kg (200 lb), external load sling of 408 kg (900 lb) capacity, Simplex Model 5200 agricultural spray or dry powder dispersal kits, instrument training package, throttle governor, start-up overspeed control unit, night flying kit, single or dual exhaust mufflers, door lock and dual oil coolers.

DIMENSIONS, EXTERNAL:
Main rotor diameter	8.18 m (26 ft 10 in)
Main rotor blade chord	0.17 m (6¾ in)
Tail rotor diameter	1.30 m (4 ft 3 in)
Distance between rotor centres	4.66 m (15 ft 3½ in)
Length: fuselage	6.77 m (22 ft 2½ in)
overall, rotors turning	9.40 m (30 ft 10 in)
Height: to top of rotor head	2.66 m (8 ft 8¾ in)
to top of cabin	2.19 m (7 ft 2 in)
Width: rotor partially folded	2.44 m (8 ft 0 in)
cabin	1.30 m (4 ft 3 in)
Skid track	1.99 m (6 ft 6½ in)
Length of skids	2.51 m (8 ft 3 in)
Passenger doors (each): Height	1.09 m (3 ft 7 in)
Width	0.97 m (3 ft 2 in)
Height to sill	0.91 m (3 ft 0 in)

DIMENSIONS, INTERNAL:
Cabin: Length	1.50 m (4 ft 11 in)
Max width	1.45 m (4 ft 9 in)

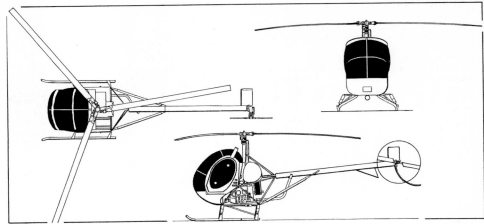

General arrangement of the Schweizer 300 *(Paul Jackson/Jane's)*

AREAS:
Main rotor blades (each)	0.70 m² (7.55 sq ft)
Tail rotor blades (each)	0.08 m² (0.86 sq ft)
Main rotor disc	52.54 m² (565.5 sq ft)
Tail rotor disc	1.32 m² (14.20 sq ft)
Fin	0.23 m² (2.50 sq ft)
Horizontal stabiliser	0.246 m² (2.65 sq ft)

WEIGHTS AND LOADINGS:
Weight empty: 300C	499 kg (1,100 lb)
300CB	494 kg (1,088 lb)
Baggage capacity	45 kg (100 lb)
Max T-O weight:	
Normal category: 300C	930 kg (2,050 lb)
300CB	794 kg (1,750 lb)
external load	975 kg (2,150 lb)
Max disc loading:	
Normal category: 300C	17.7 kg/m² (3.63 lb/sq ft)
300CB	15.1 kg/m² (3.09 lb/sq ft)
external load	18.6 kg/m² (3.80 lb/sq ft)
Max power loading at T-O:	
Normal category: 300C	6.57 kg/kW (10.79 lb/hp)
300CB	5.61 kg/kW (9.21 lb/hp)
external load	6.89 kg/kW (11.32 lb/hp)

PERFORMANCE (at max Normal T-O weight, ISA):
Never-exceed speed (V$_{NE}$) at S/L:	
300C	95 kt (176 km/h; 109 mph) IAS
300CB	94 kt (174 km/h; 108 mph) IAS
Max cruising speed: 300C	86 kt (159 km/h; 99 mph)
300CB	80 kt (148 km/h; 92 mph)
Speed for max range, at 1,220 m (4,000 ft)	67 kt (124 km/h; 77 mph)
Max rate of climb at S/L: 300C	229 m (750 ft)/min
300CB	381 m (1,250 ft)/min
Hovering ceiling: IGE: 300C	3,290 m (10,800 ft)
300CB	2,135 m (7,000 ft)
OGE: 300C	2,620 m (8,600 ft)
300CB	1,465 m (4,800 ft)
Range at 1,220 m (4,000 ft), 2 min warm-up, max normal fuel, no reserves	208 n miles (386 km; 240 miles)
Max endurance at S/L: 300C	3 h 48 min
300CB	3 h 18 min

UPDATED

SCHWEIZER 330SP

TYPE: Four-seat helicopter.

PROGRAMME: Announced 1987; first flight in public (N330TT; converted from a 300C) 14 June 1988; FAA certification September 1992. Deliveries started mid-1993. Three sets of controls available for student training. Model 330SP is marketing designation; type is 269D.

CURRENT VERSIONS: **330:** Original version delivered from 1993.

330SP: Announced May 1997, incorporates larger main rotor hub, increased blade chord and raised landing gear. Modifications can be retrofitted to existing 330s.

333: Announced early 2000 as an upgraded version of the 330SP; lead customer was San Antonio Police Department (two aircraft). FAA certification awarded 28 September 2000, at which time first three already delivered. 330SP can be upgraded to 333 standard. Differences include new cambered aerofoil main rotor blades and increased transmission rating.

Details apply to 333 unless otherwise stated.

CUSTOMERS: Include Kawada Industries, Saab Helikopter, AER of Bladensburg, Maryland, Heli-fly of Belgium, Heli-Holland, Hindustan Aeronautics of India, and West Palm Beach Police Department and San Antonio Police Department (two, including the first 330SP); 330SP/333 also selected as airframe basis for Northrop Grumman Model 379 Fire Scout VTOL tactical UAV under development for US Navy (see *Jane's Unmanned Aerial Vehicles and Targets*) for which a US$93.7 million contract was awarded on 10 February 2000. Three aircraft delivered in 1993; five in 1994, plus one prototype and one demonstrator; five in 1996, six in 1997, three in 1998 and five in 1999, for a total of 29 by 1 January 2000. Production continues at low rate, but no known registrations in 2000.

COSTS: US$535,000 for 330SP and US$604,750 for 333 (2001).

DESIGN FEATURES: Developed from Model 300, with turbine power, enclosed engine and tapered tailboom. Civil roles include law enforcement, scout/observation, aerial photography, light utility, agricultural spraying and personal transport; uses turbine fuel rather than scarcer Avgas; extremely low flat rating of engine for good hot-and-high performance; streamlined external envelope and large tailplane with endplate fins added during development; rotor rpm 471.

FLYING CONTROLS: Similar to 300C; see Design Features.

STRUCTURE: Generally similar to 300C.

POWER PLANT: *330/330SP:* One 313.2 kW (420 shp) Rolls-Royce 250-C20W turboshaft derated to 175 kW (235 shp). Transmission rating 175 kW (235 hp) for T-O; 164 kW (220 hp) maximum continuous. Maximum fuel capacity 276 litres (73.0 US gallons; 60.8 Imp gallons).

333: One 313.2 kW (420 shp) Rolls-Royce 250-C20W turboshaft derated to 188 kW (252 shp). Transmission rating 188 kW (252 shp) for T-O; 173 kW (232 shp) maximum continuous. Fuel capacity as 330SP.

ACCOMMODATION: Three persons side by side (central occupant slightly forward), with up to three sets of controls; four point inertia reel seat belts. Fourth occupant in 'high density' configuration (central seat replaced by double seat). Cabin ventilated.

SYSTEMS: Engine inlet anti-ice system; 24 V lead-acid battery; 150 A starter/generator; external APU plug; circuit breaker protection.

AVIONICS: *Instrumentation:* Tachometer, engine torquemeter, engine oil temperature and pressure indicators; electric

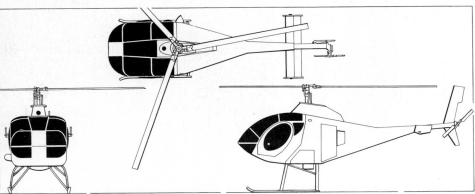

Schweizer 330SP three/four-seat light helicopter *(Mike Keep/Jane's)*

Schweizer 330SP turbine-powered helicopter

digital OAT, turbine outlet temperature indicator; annunciator warning panel.
 Mission: Law enforcement options include FLIR, video recorder, Spectrolab SX-5 Starburst searchlight, Pronet and Litton computer system.

DIMENSIONS, EXTERNAL:
Main rotor diameter	8.38 m (27 ft 6 in)
Tail rotor diameter	1.30 m (4 ft 3 in)
Length overall, rotors turning	9.54 m (31 ft 3½ in)
Fuselage: Length	6.82 m (22 ft 4½ in)
Max width	1.72 m (5 ft 7¾ in)
Height overall	3.35 m (11 ft 0 in)
Width overall	2.13 m (7 ft 0 in)
Tailplane span	2.04 m (6 ft 8½ in)
Skid track	1.92 m (6 ft 3½ in)

DIMENSIONS, INTERNAL:
Cabin: Width at seat	1.72 m (5 ft 7¾ in)
Height at seat	1.35 m (4 ft 5¼ in)

Fire Scout differs considerably from its progenitor *(Northrop Grumman)* 2001/0073052

AREAS:
Main rotor disc area	55.20 m² (594.2 sq ft)
Tail rotor disc area	1.32 m² (14.20 sq ft)

WEIGHTS AND LOADINGS:
Weight empty	549 kg (1,210 lb)
Max T-O weight	1,156 kg (2,550 lb)
Max disc loading	20.95 kg/m² (4.29 lb/sq ft)
Max power loading	6.16 kg/kW (10.12 lb/shp)
Transmission loading at max T-O weight and power	
	6.16 kg/kW (10.12 lb/shp)

PERFORMANCE:
Never-exceed speed (V$_{NE}$)	110 kt (203 km/h; 126 mph)
Normal cruising speed	101 kt (187 km/h; 116 mph)
Econ cruising speed	94 kt (174 km/h; 108 mph)
Max rate of climb at S/L	420 m (1,380 ft)/min
Hovering ceiling: IGE	2,500 m (8,200 ft)
OGE	1,158 m (3,800 ft)
Max range at 1,220 m (4,000 ft), no reserves	
	300 n miles (555 km; 345 miles)
Max endurance, no reserves	3 h 54 min

UPDATED

SEASTAR

SEASTAR AIRCRAFT INC
PO Box 2031, Thomas Stafford Airport, Weatherford, Oklahoma 73096
Tel/Fax: (+1 580) 772 21 40
e-mail: info@seastarplane.com
Web: http://www.seastarplane.com
DESIGNER: Craig Easter

Seastar has undertaken a fundamental redesign of the Seawind amphibian which it now offers complete or as a kit. The next product will be the Adventurer turbo-powered version.
UPDATED

SEASTAR SEASTAR
TYPE: Six-seat amphibian kitbuilt.
PROGRAMME: Production prototype Seawind first flew 23 August 1982. Progressed through several versions, last of which described in 1990-91 *Jane's* under Seawind International Inc of Canada.
 Computer-aided redesign of Seawind produced several refinements, including changed aerofoil, single-piece Fowler flaps, gull-wing cabin doors and stainless steel landing gear. Adapted for optional pressurisation and alternative diesel and turboprop engines. Prototype (N601SS) exhibited, partly complete, at Sun 'n' Fun, April 2000; first flown 12 November 2000.
COSTS: Basic kit, less engine and avionics, US$90,000, pressurised US$99,000. Factory-built, M 601 turboprop but without avionics, US$590,000 (2000).

DESIGN FEATURES: Unconventional configuration; shoulder wing with stabilising floats at tips; long-chord fin mounts engine and tailplane at mid position, shielding propeller from water spray.
FLYING CONTROLS: Conventional and manual.
STRUCTURE: Carbon fibre throughout; stainless steel landing gear.
POWER PLANT: *Piston version:* One 261 kW (350 hp) Textron Lycoming TIO-540 flat-six driving a four-blade, reversible pitch MT propeller, with fuel capacity of 474 litres (125 US gallons; 104 Imp gallons).
 Turbine version: One 490 kW (657 shp) derated Walter M 601D turboprop driving a five-blade, reversible pitch MT propeller, with fuel capacity of 758 litres (200 US gallons; 167 Imp gallons).
ACCOMMODATION: Two, four or six persons, according to version; large, upward-opening door each side.
ARMAMENT: Military version has provision for total of four underwing pylons.

DIMENSIONS, EXTERNAL:
Wing span	11.58 m (38 ft 0 in)
Length overall	8.29 m (27 ft 2½ in)
Height overall	3.35 m (11 ft 0 in)
Baggage door: width	0.84 m (2 ft 9 in)
height	0.81 m (2 ft 8 in)

DIMENSIONS, INTERNAL:
Cabin: Length	3.25 m (10 ft 8 in)
Width: at front	1.35 m (4 ft 5 in)
at rear	1.44 m (4 ft 8 in)
Baggage compartment: Length	1.83 m (6 ft 0 in)
Volume	0.76 m³ (27.0 cu ft)

AREAS:
Wings, gross	15.89 m² (171.0 sq ft)

WEIGHT AND LOADINGS:
Weight empty	1,089 kg (2,400 lb)
Max T-O weight: civil	1,814 kg (4,000 lb)
military turboprop	1,886 kg (4,600 lb)
Max wing loading: civil	114.2 kg/m² (23.39 lb/sq ft)
military turboprop	131.3 kg/m² (26.90 lb/sq ft)
Max power loading: Lycoming	6.96 kg/kW (11.43 lb/hp)
Walter: military	4.26 kg/kW (7.00 lb/shp)
civil	3.71 kg/kW (6.09 lb/shp)

PERFORMANCE:
Max level speed: Lycoming	183 kt (339 km/h; 211 mph)
Walter	228 kt (422 km/h; 262 mph)
Max cruising speed: Lycoming at 75% power:	
at 3,050 m (10,000 ft)	177 kt (328 km/h; 204 mph)
at 6,100 m (20,000 ft)	188 kt (348 km/h; 216 mph)
Walter at 90% power:	
at 3,050 m (10,000 ft)	233 kt (432 km/h; 268 mph)
at 6,100 m (20,000 ft)	244 kt (452 km/h; 281 mph)
Stalling speed	52 kt (97 km/h; 60 mph)
Service ceiling	7,010 m (23,000 ft)
Max rate of climb at S/L: Lycoming	427 m (1,400 ft)/min
Walter	549 m (1,800 ft)/min
T-O run on land: Lycoming	244 m (800 ft)
Walter	198 m (650 ft)
Max range, no reserves:	
Lycoming	1,215 n miles (2,251 km; 1,399 miles)
Walter	1,203 n miles (2,228 km; 1,385 miles)

UPDATED

Seastar Seastar multipurpose amphibian during taxying trials 2001/0106860/0106859

SEQUOIA

SEQUOIA AIRCRAFT CORPORATION
2000 Tomlynn Street, PO Box 6861, Richmond, Virginia 23230
Tel: (+1 804) 353 17 13
Fax: (+1 804) 359 26 18
e-mail: seqair@aol.com
Web: http://www.SeqAir.com
PRESIDENT: Alfred P Scott

Sequoia has marketed the F.8L Falco since 1979. By late 2000, 65 had been completed and flown.
UPDATED

SEQUOIA FALCO F.8L
TYPE: Side-by-side lightplane/kitbuilt.
PROGRAMME: Sequoia Aircraft markets plans and kits to build improved version of Falco F.8L high-performance monoplane, designed in Italy by Ing Stelio Frati of General Avia (which see) and first flown on 15 June 1955. Italian production totalled 77; kit version first flown 1974.
CUSTOMERS: 65 flown by late 2000.
COSTS: Kit: US$77,000 (2000). Plans: US$400. Extra costs to complete estimated by manufacturer as US$13,000.

Sequoia F.8L, built in Australia by Ian Ferguson 2001/0103499

DESIGN FEATURES: Streamlined low-wing, retractable-gear monoplane. Quoted build time 2,000 hours.

NACA $64_2212.5$ wing section at root, NACA 64_2210 at tip. Wing dihedral 5°, washout 3°. Tailplane section NACA 65009.

FLYING CONTROLS: Conventional and manual.

STRUCTURE: Entire airframe has plywood-covered wood structure, with overall fabric covering and glass fibre wing fillets. Optional metal control surfaces.

LANDING GEAR: Retractable tricycle type. Steerable nosewheel. Cleveland single calliper disc brakes. 5.00-5 mainwheels and 4.00-5 nosewheel.

POWER PLANT: One 119 kW (160 hp) Textron Lycoming IO-320-B1A is standard for kitbuilt aircraft, driving Hartzell HCC-ZYL-1BF/F7663-4 two-blade constant-speed metal propeller. Optional engines for which Sequoia offers installation kits are the 112 kW (150 hp) IO-320-A1A and 134 kW (180 hp) IO-360-B1E. Fuel capacity 151 litres (40.0 US gallons; 33.3 Imp gallons). Optional 11.4 litre (3.0 US gallon; 2.5 Imp gallon) header tank to permit inverted flight.

ACCOMMODATION: Two seats, side by side; child's seat or baggage area to rear. Dual controls.

DIMENSIONS, EXTERNAL:	
Wing span	8.00 m (26 ft 3 in)
Wing chord: at root	1.65 m (5 ft 5 in)
at tip	0.83 m (2 ft 8½ in)
Wing aspect ratio	6.4
Length overall	6.63 m (21 ft 9 in)
Height overall	2.29 m (7 ft 6 in)
Tailplane span	2.71 m (8 ft 10½ in)
Wheel track	2.08 m (6 ft 10 in)
Wheelbase	1.50 m (4 ft 11 in)
Propeller diameter	1.83 m (6 ft 0 in)
DIMENSIONS, INTERNAL:	
Cabin max width	1.07 m (3 ft 6 in)
AREAS:	
Wings, gross	10.00 m^2 (107.6 sq ft)
Rudder	0.48 m^2 (5.20 sq ft)
Tailplane	2.17 m^2 (23.40 sq ft)
Elevator	0.83 m^2 (9.00 sq ft)
WEIGHTS AND LOADINGS (119 kW; 160 hp engine):	
Weight empty	550 kg (1,212 lb)
Payload with max fuel	194 kg (428 lb)
Baggage capacity	41 kg (90 lb)
Max aerobatic weight	748 kg (1,650 lb)
Max T-O weight	852 kg (1,880 lb)
Max wing loading	85.3 kg/m^2 (17.47 lb/sq ft)
Max power loading	7.15 kg/kW (11.75 lb/hp)
PERFORMANCE (119 kW; 160 hp engine):	
Never-exceed speed (VNE)	208 kt (386 km/h; 240 mph)
Max level speed at S/L	184 kt (341 km/h; 212 mph)
Cruising speed at 75% power	165 kt (306 km/h; 190 mph)
Stalling speed: clean	66 kt (121 km/h; 75 mph)
flaps and wheels down	54 kt (100 km/h; 62 mph)
Max rate of climb at S/L	347 m (1,140 ft)/min
Service ceiling	5,790 m (19,000 ft)
T-O run	174 m (570 ft)
T-O to 15 m (50 ft)	351 m (1,150 ft)
Landing from 15 m (50 ft)	351 m (1,150 ft)
Landing run	229 m (750 ft)
Range at econ cruising speed	869 n miles (1,609 km; 1,000 miles)
Endurance	4 h 24 min
g limits	+6/−3

UPDATED

SIKORSKY

SIKORSKY AIRCRAFT (Subsidiary of United Technologies Corporation)

6900 Main Street, PO Box 9729, Stratford, Connecticut 06497-9129
Tel: (+1 203) 386 40 00
Fax: (+1 203) 386 73 00
Web: http://www.sikorsky.com
Telex: 96 4372

OTHER WORKS: Troy, Alabama; South Avenue, Bridgeport, Connecticut; Shelton, Connecticut; West Haven, Connecticut; Development Flight Test Center, West Palm Beach, Florida

PRESIDENT AND CEO: Dean C Borgman
SENIOR VICE-PRESIDENT:
Mark J Evans (Production Operations)
VICE-PRESIDENT, PROGRAMS: Kenneth J Kelly
VICE-PRESIDENT, CIVIL PROGRAMS: Tommy H Thomason
VICE-PRESIDENT, DEVELOPMENT ENGINEERING AND ADVANCED PROGRAMS: Dr Kenneth M Rosen
VICE-PRESIDENT, PRODUCTION ENGINEERING: Donald P Gover
VICE-PRESIDENT, RESEARCH AND ENGINEERING: Paul Martin
VICE-PRESIDENT, FINANCE AND CFO: Jay Haberland
VICE-PRESIDENT, GOVERNMENT BUSINESS DEVELOPMENT:
Rene I Beauchamp
COMMUNICATIONS MANAGER: Ed Steadham
MANAGER OF PUBLIC RELATIONS: William S Tuttle

Founded as Sikorsky Aero Engineering Corporation by late Igor I Sikorsky; has been division of United Technologies Corporation since 1929, but established as a subsidiary with effect 1 January 1995; began helicopter production in 1940s. Workforce around 7,000 at end of 2000.

Headquarters and main plant at Stratford, Connecticut; also has manufacturing facilities elsewhere in Connecticut; other smaller facilities in Alabama and Florida; total area of buildings owned/leased in 1998 by Sikorsky over 325,150 m^2 (3,500,000 sq ft); 1999 revenue totalled US$1.4 billion. Main current programmes include UH-60 Black Hawk and derivatives, CH-53E Super Stallion, S-76 series and, in co-operation with international partners, development of S-92 Helibus. Sikorsky delivered total of 105 helicopters in 1996, 101 in 1997, 87 in 1998 and 65 in 1999. Sikorsky and Boeing Helicopters won US Army RAH-66 Comanche light helicopter demonstration/validation order on 5 April 1991 (see under Boeing Sikorsky in this section).

Sikorsky licensees include Agusta of Italy, Eurocopter of France and Germany, Korean Air Lines of South Korea, Mitsubishi of Japan, Pratt & Whitney Canada Ltd and UK element of AgustaWestland. Sikorsky and Embraer of Brazil signed agreement in mid-1983 to transfer technology covering design and manufacture of composites components. Sikorsky and CASA of Spain signed MoU in June 1984 covering long-term helicopter industrial co-operation programme; CASA builds tail rotor pylon, tailcone and stabiliser components for H-60 and S-70, with first CASA S-70 components delivered to Sikorsky January 1986. Most recent overseas venture, in collaboration with the Alpata Group of Turkey, concerns creation of Alp Aviation, which will manufacture high technology, precision-machined aerospace and defence components in 6,500 m^2 (70,000 sq ft) facility in Eskisehir; formal announcement of this made at Paris Air Show in June 1999.

In July 1998, Sikorsky announced intent to purchase Helicopter Support Inc, so as to offer enhanced support to helicopter operators on worldwide basis; signature of contract followed in October 1998, by which time Sikorsky had also secured US$150 million deal with US Navy covering contractor maintenance of jet trainer aircraft as well as HH-1N Iroquois and UH-3H Sea King helicopters. Overall responsibility for latter allocated to Sikorsky Support Services Inc, with the work undertaken at Meridian, Mississippi; Pensacola, Florida; and Corpus Christi, Texas.

UPDATED

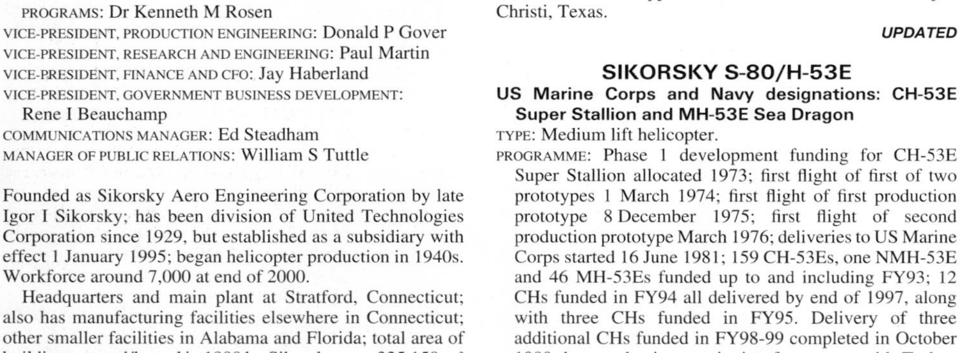

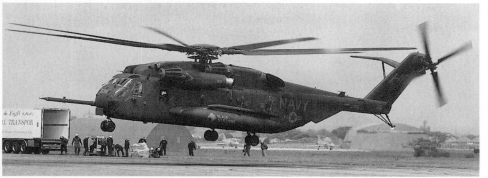

Sikorsky CH-53E Super Stallion heavy-lift helicopter of HC-4 (squadron) US Navy ***(Paul Jackson/Jane's)***
2000/0064510

SIKORSKY S-80/H-53E

US Marine Corps and Navy designations: CH-53E Super Stallion and MH-53E Sea Dragon

TYPE: Medium lift helicopter.

PROGRAMME: Phase 1 development funding for CH-53E Super Stallion allocated 1973; first flight of first of two prototypes 1 March 1974; first flight of first production prototype 8 December 1975; first flight of second production prototype March 1976; deliveries to US Marine Corps started 16 June 1981; 159 CH-53Es, one NMH-53E and 46 MH-53Es funded up to and including FY93; 12 CHs funded in FY94 all delivered by end of 1997, along with three CHs funded in FY95. Delivery of three additional CHs funded in FY98-99 completed in October 1999, but production continuing for export, with Turkey having ordered eight for delivery in 2002.

US Marine Corps reportedly contemplating major CH-53E upgrade in case proposed Joint Transport Rotorcraft programme is delayed; extent of upgrade work still undefined but could include adoption of 'glass cockpit' similar to that which will be incorporated in S-80E version for Turkey, as well as re-engining with Rolls-Royce T406 turboshaft engine as used by MV-22B Osprey.

CURRENT VERSIONS: **CH-53E Super Stallion:** Used by US Marine Corps for amphibious assault, carrying heavy equipment and armament, and recovering disabled aircraft; also used by US Navy for vertical onboard delivery and recovery of damaged aircraft from aircraft carriers; went into operation in Mediterranean area mid-1983 with HC-4 at Sigonella, Italy (since re-equipped with MH-53E). Current Marine Corps squadrons are HMH-361, HMH-462, HMH-465 and HMH-466 of Pacific Fleet and HMH-461 and HMH-464 of Atlantic Fleet; also serves with Marine Corps Reserve squadrons HMH-769 and HMH-772, plus training squadron HMT-302 and experimental squadron HMX-1.

Detailed description applies to CH-53E, but applicable also to MH-53E except where indicated.

Planned improvements for CH-53E include composites tail rotor blades, Omega navigation system, ground proximity warning system, flight crew night vision system (Elbit ANVIS 7 NVG/HUD system already fitted), improved internal cargo handling system and external lifting device, missile alerting system, chaff/flare dispensers, nitrogen fuel inerting system, and facility for refilling hydraulic system inside cargo compartment.

Full-scale development of helicopter night vision system (HNVS) for CH-53E began June 1986, in co-operation with Northrop Electro-Mechanical Division; HNVS includes Lockheed Martin pilot night vision system (PNVS) and Honeywell integrated helmet and display sighting system (IHADSS) from Bell AH-1S surrogate trainer; HNVS will allow low-level operations at night and in adverse weather; HNVS ground testing began 1988; operational evaluation began August 1989. Smaller scale HNVS capability authorised 1993, with contract to Engineering and Economics Research Systems of Seabrook, Maryland, for installation of Hughes (now Raytheon Systems) AN/AAQ-16B FLIR, Teledyne Ryan Electronics AN/APN-217 Doppler and Rockwell Collins GPS 3A; total 24 upgrades initially authorised; subsequent contracts include US$7.65 million for procurement and installation of 50 HNVS in FY96 and procurement of further 49 with FY97 funding, with work scheduled for completion by May 1999.

In 1997, Sikorsky conducting studies with objective of reducing IR signature by means of adopting engine-suppression system, in order to enhance survivability; other measures that may also be adopted by USN/MC include provision of threat warning sensors, IR jammers and chaff/flare dispensers.

USMC has also evaluated new multiple cargo hook concept that will permit several drops per sortie; device developed by Skyhook Technologies and uses tetrahedral frame attached to usual single-hook pendant. Each of the frame's three base corners features individual load-bearing hook, but system can be modified for maximum of six hooks. Test device delivered for evaluation from Patuxent River in October 1997; testing completed in 1998.

Trials with helicopter health monitoring system began in second half of 1999 on CH-53E at Patuxent River, Maryland. Installation comprises Goodrich integrated mechanical diagnostics health and usage monitoring system (HUMS) that is expected to cut operational and support costs and also improve safety and readiness rates. On completion of operational evaluation, a retrofit programme is expected to include entire CH-53E and MH-53E fleet.

CH-53X Future Stallion: Sikorsky proposal for combined service life extension/product improvement programme that may be implemented in FY03, when early production CH-53Es begin to reach end of their service lives. At present, Sikorsky expects to begin limited production in FY09 and has plans for 15 aircraft to be remanufactured annually; likely improvements include new engines, main rotor blades, elastomeric rotor head and electric blade-folding as well as a modernised cockpit based on that of the MV-22 Osprey or UH-1Y Iroquois. In addition to the Marine Corps, the improved version has been suggested as a candidate for the US Army, which has a requirement for an advanced heavy lift helicopter.

MH-53E Sea Dragon: Airborne mine countermeasures (AMCM) helicopter able to tow, through water, a hydrofoil sledge carrying mechanical, acoustic and magnetic

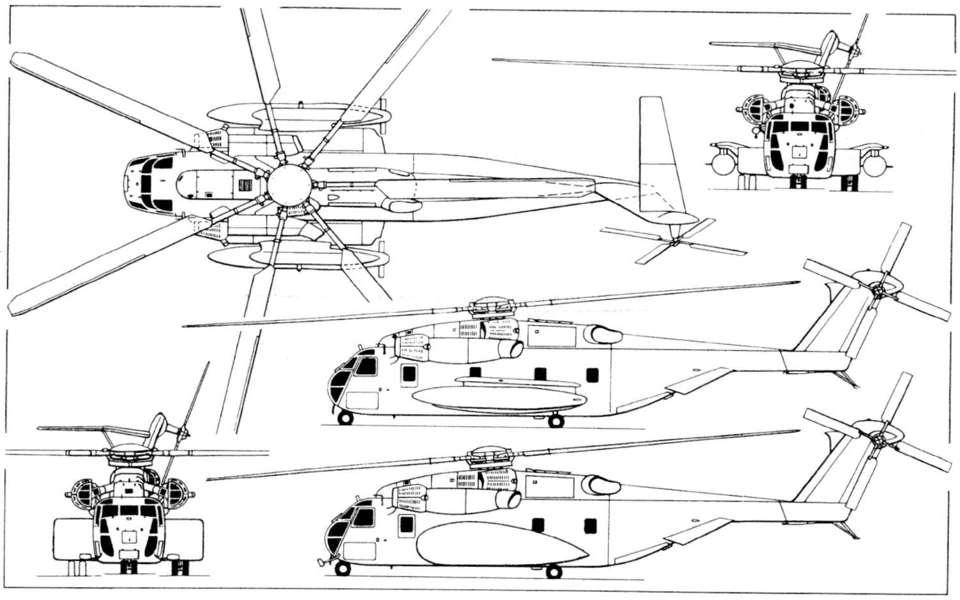

Sikorsky CH-53E Super Stallion, with additional lower side view and lower front view of MH-53E Sea Dragon ***(Dennis Punnett/Jane's)***

sensors; early history in 1982-83 *Jane's*; nearly 3,785 litres (1,000 US gallons; 833 Imp gallons) extra fuel carried in enlarged sponsons made of composites; improved hydraulic and electrical systems; minefield, navigation and automatic flight control system with automatic towing and approach and departure from hover modes. First flight of first preproduction MH-53E, 1 September 1983; first delivery to US Navy 26 June 1986; in operational service with HM-14 at Norfolk, Virginia, 1 April 1987; another MH-53E delivered to HM-12 (since disestablished) in 1987 was 100th H-53E; 32 delivered to USN by December 1991, including some with HM-15 then at Alameda, California, for Pacific Fleet, but since moved to Corpus Christi, Texas; first carrier deployment by HM-15 on board USS *Tripoli*, 9 December 1989. HM-18 at Norfolk, Virginia, and HM-19 (USN Reserve) at Alameda both re-equipped from RH-53Ds in 1993 but subsequently integrated with regular force units HM-14 and HM-15 respectively; transport squadron HC-4 at Sigonella, Sicily, re-equipped with MH-53Es (from CH- variants) in 1995; also partly equips Marines training squadron HMT-302 at New River, North Carolina. Total requirement was 56; number acquired believed to be 47 (including one NMH-53E development aircraft).

Delivery effected 1994 of MH-53E retrofitted with upgraded avionics package by EER Systems, comprising two 152 mm (6 in) horizontal situation display colour screens, Fairchild mission data loader and Rockwell Collins GPS 3A; USN developmental testing began June 1994; upgrade of entire MH-53E fleet planned but may be reduced to around 30 as result of defence budget trimming. One MH-53E to West Palm Beach, Florida, for installation of T64-GE-419 engines, late 1993. Trials during 1994 verified performance gains, including recovery and flyaway capability in event of engine failure during hover; retrofit under way of entire MH-53E fleet, with Sikorsky delivering first eight engine kits in late 1996.

Survivability measures mentioned for CH-53E may also be considered for installation on MH-53E.

S-80E: Export version of CH-53E, as ordered by Turkey in June 2000. Will be similar to standard CH-53E of US Navy/Marine Corps, but is to incorporate revised Rockwell Collins avionics based on commercial ProLine 21 system and including four 152 × 254 mm (6 × 10 in) liquid crystal displays, dual CDU-900 flight management systems, embedded GPS and a data loading facility.

S-80M: Export MH-53E supplied to Japan; see 1997-98 edition.

CUSTOMERS: US Navy and Marine Corps (see Programme and Current Versions). Japan acquired 11 S-80M for JMSDF as MH-53EJ. In June 2000, Turkey signed contract for eight S-80Es and may eventually obtain as many as 20 for service with Land Forces Command. Approximately 765 of Sikorsky S-65/S-80/H-53 series built by end of 1999.

COSTS: US$24.36 million (1992) projected average unit cost. Total cost of Turkish order for eight aircraft put at about US$350 million.

DESIGN FEATURES: Fully articulated seven-blade main rotor; blade twist 14°; hydraulic-powered blade folding for main rotor; tail pylon folds hydraulically to starboard; four-blade tail rotor on pylon canted 20° to port to derive some lift from tail rotor and extend CG range; cranked, strut braced tailplane; rotor brake standard; fuselage stressed for 20 *g* vertical and 10 *g* lateral crash loads.

FLYING CONTROLS: Fully powered, with autostabilisation and autopilot. See also Current Versions and Avionics.

STRUCTURE: Fuselage has watertight primary structure of light alloy, steel and titanium; glass fibre/epoxy cockpit section; extensive use of Kevlar in transmission fairing and engine cowlings; main rotor blades have titanium spar, Nomex honeycomb core and glass fibre/epoxy composites skin; titanium and steel rotor head; Sikorsky blade inspection method (BIM) sensors detect blade spar cracks occurring in service; tail rotor of aluminium; pylon and tailplane of Kevlar composites.

LANDING GEAR: Retractable tricycle type, with twin wheels on each unit. Main units retract into rear of sponsons on each side of fuselage. Fully castoring nosewheels.

POWER PLANT: Three General Electric T64-GE-416 or -416A turboshafts, each with a maximum rating of 3,266 kW (4,380 shp) for 10 minutes, intermediate rating of 3,091 kW (4,145 shp) for 30 minutes and maximum continuous power rating of 2,756 kW (3,696 shp). Retrofit of MH-53E now under way with 3,542 kW (4,750 shp) T64-GE-419, which has contingency rating of 3,729 kW (5,000 shp) for up to two minutes. Transmission rated at 10,067 kW (13,500 shp) for take-off.

Self-sealing bladder fuel cell in forward part of each sponson, each with capacity of 1,192 litres (315 US gallons; 262 Imp gallons). Additional two-cell unit, with capacity of 1,465 litres (387 US gallons; 322 Imp gallons), brings total standard internal capacity to 3,849 litres (1,017 US gallons; 847 Imp gallons). (Total internal capacity of MH-53E is 12,113 litres; 3,200 US gallons; 2,664 Imp gallons.)

Optional drop tank outboard of each sponson of CH-53E, total capacity 4,921 litres (1,300 US gallons; 1,082 Imp gallons). (MH-53E can carry seven internal range extension tanks, total capacity 7,949 litres; 2,100 US gallons; 1,748 Imp gallons.) Forward extendable probe for in-flight refuelling. Alternatively, aircraft can refuel by hoisting hose from surface vessel while hovering.

ACCOMMODATION: Crew of three. Main cabin of CH-53E will accommodate up to 55 troops on folding canvas seats along walls and in centre of cabin. New crashworthy seats being delivered by Martin Baker under US$20 million contract awarded in July 1999; up to 210 shipsets to be supplied, with each comprising 31 seats, at rate of 52 helicopters per year; entire fleet to be outfitted by FY03. Door on forward starboard side of main cabin. Hydraulically operated rear-loading ramp. Typical freight loads include seven standard 1.02 × 1.22 m (3 ft 4 in × 4 ft) pallets. Single-point central hook for slung cargo, capacity 16,330 kg (36,000 lb); improved external lift system that will permit maximum of three separate underslung loads to be carried and delivered to different locations will be incorporated from FY01.

SYSTEMS: Hydraulic system, with four pumps, for collective, cyclic pitch/roll, yaw and feel augmentation flight control servo mechanisms; engine starters; landing gear actuation; cargo winches; loading ramp; and blade and tail pylon folding. System pressure 207 bars (3,000 lb/sq in), except for engine starter system which is rated at 276 bars (4,000 lb/sq in). (Separate hydraulic system in MH-53E to power AMCM equipment.) Electrical system includes three 115 V 400 Hz 40 to 60 kVA AC alternators, and two 28 V 200 A transformer-rectifiers for DC power. Solar APU.

AVIONICS: *Comms:* Rockwell Collins AN/ARC-210 tactical radios currently being installed, with completion expected in FY00.

Flight: Hamilton Sundstrand automatic flight control system, using two digital onboard computers and a four-axis autopilot. Retrofit test flown late 1993, comprising four Canadian Marconi CM A-2082 152 mm (6 in) square colour displays, tied with GPS, Doppler and AHRS; installation by Teledyne Ryan. S-80E for Turkey features revised 'glass cockpit' based on commercial Rockwell Collins ProLine 21 system.

Instrumentation: Currently being retrofitted with compatible lighting for AN/AVS-6 NVGs, with completion scheduled for FY00; will also feature NVG-compatible HUD by FY04.

Self-defence: BAE Systems AN/AAR-47 missile warning system installation currently under way, with completion expected in FY00.

EQUIPMENT: MH-53E equipment includes Northrop Grumman AN/AQS-14 towed sonar, AN/AQS-17 mine neutralisation device, AN/ALQ-141 electronic sweep, and Edo AN/ALQ-166 towed hydrofoil sled for detonating magnetic mines.

DIMENSIONS, EXTERNAL (CH-53E and MH-53E):
Main rotor diameter 24.08 m (79 ft 0 in)
Main rotor blade chord 0.76 m (2 ft 6 in)
Tail rotor diameter 6.10 m (20 ft 0 in)
Length overall: rotors turning 30.19 m (99 ft 0½ in)
rotor and tail pylon folded 18.44 m (60 ft 6 in)
Fuselage: Length 22.35 m (73 ft 4 in)
Width 2.69 m (8 ft 10 in)
Width overall, rotor and tail pylon folded:
CH-53E 8.66 m (28 ft 5 in)
MH-53E 8.41 m (27 ft 7 in)
Height: to top of main rotor head 5.32 m (17 ft 5½ in)
tail rotor turning 8.97 m (29 ft 5 in)
rotor and tail pylon folded 5.66 m (18 ft 7 in)
Wheel track (c/l of shock-struts) 3.96 m (13 ft 0 in)
Wheelbase 8.31 m (27 ft 3 in)

DIMENSIONS, INTERNAL (CH-53E and MH-53E):
Cabin: Length (rear ramp/door hinge to fwd bulkhead) 9.14 m (30 ft 0 in)
Max width 2.29 m (7 ft 6 in)
Max height 1.98 m (6 ft 6 in)

AREAS (CH-53E and MH-53E):
Main rotor disc 455.38 m² (4,901.7 sq ft)
Tail rotor disc 29.19 m² (314.2 sq ft)

WEIGHTS AND LOADINGS:
Weight empty: CH-53E 15,072 kg (33,228 lb)
MH-53E 16,482 kg (36,336 lb)
Internal payload (100 n mile; 185 km; 115 mile radius):
CH-53E 13,607 kg (30,000 lb)
External payload (50 n mile; 92.5 km; 57.5 mile radius):
CH-53E 14,515 kg (32,000 lb)
Max external payload: CH-53E 16,330 kg (36,000 lb)
Useful load, influence sweep mission:
MH-53E 11,793 kg (26,000 lb)
Max T-O weight (CH-53E and MH-53E):
internal payload 31,640 kg (69,750 lb)
external payload 33,340 kg (73,500 lb)
Max disc loading:
internal payload 69.5 kg/m² (14.23 lb/sq ft)
external payload 73.2 kg/m² (14.99 lb/sq ft)
Transmission loading at max T-O weight and power:
internal payload 3.14 kg/kW (5.17 lb/shp)
external payload 3.31 kg/kW (5.44 lb/shp)

PERFORMANCE (CH-53E and MH-53E, ISA, at T-O weight of 25,400 kg; 56,000 lb):
Max level speed at S/L 170 kt (315 km/h; 196 mph)
Cruising speed at S/L 150 kt (278 km/h; 173 mph)
Max rate of climb at S/L, 11,340 kg (25,000 lb) payload 762 m (2,500 ft)/min
Service ceiling at max continuous power 5,640 m (18,500 ft)
Hovering ceiling at max power: IGE 3,515 m (11,540 ft)
OGE 2,895 m (9,500 ft)
Self-ferry range, unrefuelled, at optimum cruise condition for best range:
CH-53E 1,120 n miles (2,074 km; 1,289 miles)

UPDATED

SIKORSKY S-70A

US Army designations: UH-60A, UH-60C, UH-60L, UH-60M, UH-60Q, UH-60X Black Hawk, AH-60L, EH-60A, MH-60A, MH-60K and MH-60L

US Air Force designations: UH-60A, HH-60G, MH-60G Pave Hawk

US Navy designation: CH-60S

US Marine Corps designation: VH-60N

Israel Defence Force name: Yanshuf (Owl)

Japanese Self-Defence Forces designations: UH-60J and UH-60JA

TYPE: Multirole medium helicopter.

PROGRAMME: UH-60A declared winner of US Army Utility Tactical Transport Aircraft System (UTTAS) competition against Boeing Vertol YUH-61A 23 December 1976; first flight of first of three YUH-60A competitive prototypes 17 October 1974; early development history recorded in 1982-83 and earlier *Jane's*; 2,000th H-60 delivered May 1994. For retrofitted improvements, see UH-60L below.

AgustaWestland in UK, Korean Air Lines of South Korea and Mitsubishi of Japan have licences to build the S-70 series.

CURRENT VERSIONS: **UH-60A Black Hawk:** Initial production version, designed to carry crew of three and 11 troops; also can be used without modification for medevac, reconnaissance, command and control, and troop supply; cargo hook capacity 4,082 kg (9,000 lb); one UH-60A can be carried in C-130, two in C-141 and six in C-5.

Medevac kits delivered from 1981; missile qualification completed June 1987, with day and night firing of Hellfire in various flight conditions; airborne target handover

762   USA: AIRCRAFT—SIKORSKY

system (ATHS) qualified; cockpit lighting suitable for night vision goggles fitted to production UH-60s since November 1985 and retrofitted to those built earlier; US Army began testing Alliant Techsystems Volcano mine dispensing system July 1987; modular Volcano container is disposable and dispenses 960 Gator anti-tank and anti-personnel mines; deployment of 2,940 kg (6,482 lb) system began in FY95; usage monitor to measure certain rotor loads installed in 30 UH-60As; wire strike protection added to UH-60s and EH-60s during 1987; accident data recorders also fitted. Total 1,049 built for US Army (including 66 conversions to EH-60A) before production change to UH-60L in 1989; more than 600 UH-60Ls since produced for US Army, with procurement continuing.

Detailed description applies to UH-60A/L except where indicated.

Enhanced Black Hawk: Incorporates active and passive self-defence systems, retrofitted by Corpus Christi Army Depot, Texas, to new-build UH-60A/Ls; first 15 delivered to US Army in South Korea November 1989. Equipment includes BAE Systems AN/ARN-148 Omega navigation receiver, Motorola AN/LST-5B satellite UHF communications transceiver, Honeywell AN/ARC-199 HF-SSB, and AEL AN/APR-44(V)3 specific threat RWR complementing existing AN/APR-39 general threat RWR; M134 Minigun can be fitted on each of two pintle mounts, replacing M60 machine gun.

JUH-60A: At least seven used temporarily for trials.

GUH-60A: At least 20 grounded airframes for technical training.

HH-60D Night Hawk: One prototype (82-23718) completed for abandoned USAF combat rescue variant; subsequently became an HH-60G.

EH-60A: (Designation EH-60C reserved, but not adopted by US Army.) Prototype YEH-60A (79-23301) ordered in October 1980 to carry 816 kg (800 lb) Quick Fix IIB battlefield ECM detection and jamming system. TRW Electronic Systems Laboratories was prime contractor for AN/ALQ-151(V)2 ECM kit, with installation by Tracor Aerospace; four dipole antennas on fuselage and deployable whip antenna; hover IR suppressor system (HIRSS) standard. YEH-60A first flight 24 September 1981; order for Tracor Aerospace to modify 40 UH-60As to EH-60A standard under US$51 million contract placed October 1984; first delivery July 1987 as part of US Army Special Electronics Mission Aircraft (SEMA) programme; 66 funded by FY87 excluding prototype; programme completed 1989. Under current Army planning, EH-60A to be phased out of use in 2005.

Intercepts/locates AM, FM, CW and SSB radio emissions from upper HF to mid-VHF ranges over bandwidths of 8, 30 or 50 kHz; jams VHF communications. Protective systems of UH-60A/L (M-130 chaff/flare and AN/ALQ-144 IR jammer) augmented by Sanders AN/ALQ-156(V) missile approach warning system, ITT AN/ALQ-136(V) pulsed transmitter, Northrop Grumman AN/ALQ-162(V) CW transmitter and Litton AN/APR-39(V) RWR.

Contract for conversion of 32 EH-60A Quick Fix IIBs to Advanced Quick Fix EH-60L configuration was due to be issued in 1995 for 1997 deployment, but deferred in favour of low-rate initial production (LRIP) upgrading of seven (three planned in FY96 and four in FY97); latter batch subsequently reduced to one, with all four systems due to be assigned to test duties; current status of programme uncertain, but may have been postponed indefinitely.

AN/ALQ-151(V)3 Advanced Quick Fix mission system tasked with providing ESM capability to forward ground units at division level; at least one EH-60A (85-24468) has been adapted to serve as prototype for **EH-60L**. As fielded for Task Force XXI trials (1997), the EH-60L utilised the following equipment: Sensor subsystem comprised Sanders TACJAM-A ESM for detection, direction-finding, identification and tracking of communications signals in HF, UHF, VHF and SHF frequency bands; Lockheed Martin Federal Systems communications high-accuracy locating system – exploitable (CHALS-X) for direction-finding in HF, UHF and SHF frequency bands; and signal location subsystem (SILO) for direction-finding in VHF frequency band. Navigation and timing subsystem featured INS, GPS and control display unit in cockpit. Mission control and interface subsystem comprised control, navigation, workstation and graphics processors, mass storage unit, keyboards with trackball and 483 × 483 mm (19 × 19 in) colour monitors. Communications subsystem comprised modified PRC-118 wideband control datalink, two AN/ARC-201A SINCGARS radios, AN/ARC-164 tasking and reporting datalink, intercom and AN/UYH-15 digital temporary storage recorder/reproducer set. Antenna subsystem. Airborne survivability subsystem comprised AN/ALQ-144(V)1 IR jammer, AN/ALQ-156(V)2 missile approach warning system (MAWS), AN/ALQ-162(V)2 CW transmitter and AN/APR-39(V)2 RWR.

Improvements to airframe include installation of UH-60L engines and gearbox for increase in maximum weight from 7,845 kg (17,295 lb) to 10,206 kg (22,500 lb).

MH-60A: About 30 modified for Army 160th Special Operations Aviation Regiment (SOAR); fitted with Raytheon Systems AN/AAQ-16 FLIR, BAE Systems AN/ARN-148 Omega/VLF navigation, M-130 chaff/flare dispensers, AN/ALQ-144 IR jammer, night vision equipment, multifunction displays, auxiliary fuel tanks and door-mounted Minigun; fitted with -701C engines; interim equipment, pending MH-60K, but replaced by MH-60L in late 1990 and passed to 1-245 Aviation Regiment, Oklahoma ArNG.

Sikorsky Firehawk firefighting helicopter   2000/0064929

UH-60C: Designation for command and control version; now under development at the US Naval Research Laboratory on behalf of the US Army, with operational trials scheduled for first quarter of 2002; if successful, could lead to eventual conversion of more than 200 UH-60As. Rockwell Collins AN/ASC-15B/C consoles have been fitted to more than 50 US Army aircraft as **UH-60A(C)**. Army Aviation Command and Control System (A2C2S) prototypes tested at Fort Hood, Texas, as part of Force XXI Developmental Test Exercise. System architecture incorporates enhanced communications interface terminal (ECIT) with five operator workstations for staff of commander, operations officer, artillery fire support officer, intelligence officer, forward air controller, staff aviation officer and communications specialist; all stations interfaced on wideband fibre optic bus and system also includes 760 mm (30 in) colour, flat panel display. UH-60C will be able to communicate digitally with tactical aircraft, other helicopters, armoured fighting vehicles and troops; this capability successfully demonstrated in September 1994 during field exercise with a development AH-64D Apache. Communications suite will include Have Quick II, SINCGARS/SIP and JTIDS; other capabilities also to be incorporated, including FLIR, NVG compatibility, digital map display, mission planner facility, improved ECCM and storage space for ground power generators and antennas.

HH/MH-60G Pave Hawk: Replaced US Air Force HH-60D Night Hawk rescue helicopters, which were not funded (see 1987-88 *Jane's*); converted from UH-60A/L, including 10 originally delivered to 55th Aerospace Rescue and Recovery Squadron (later Special Operations Squadron) at Eglin AFB, Florida, in 1982-83, initially remaining as UH-60As; all progressively fitted by Sikorsky Support Services at Troy, Alabama, with aerial refuelling probe, 443 litre (117 US gallon; 97.5 Imp gallon) internal auxiliary fuel tank and fuel management panel; then to Pensacola NAD for mission avionics and modified

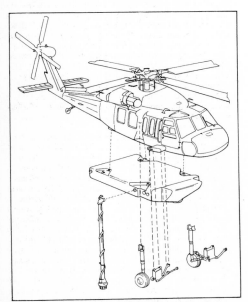
Installation of Firehawk water equipment takes some 4 hours   2000/0064930

Prototype UCN/IEW (or Block 152) version of Sikorsky HH-60G   2001/0093620

instrument panel; some retrofitted with replacement internal tank of 700 litres (185 US gallons; 154 Imp gallons); -701C engines fitted to 10 special operations examples and later production aircraft (FY89 onwards); current retrofit programme, begun in November 1999, entails installation of -701C engine on 11 HH-60Gs of California and New York ANG by April 2001, with remaining 22 aircraft likely to follow suit, if funding can be found.

Further procurement began with batch of nine in FY87, followed by purchases of 16, 18, 22, 15 and 13 in FY88-92; eight more funded in FY97 and delivered in 1998. All designated MH-60G until 1 January 1992, when 82 in combat rescue role redesignated HH-60G, balance of 16 remaining as MH-60G for special operations units; by fourth quarter of 1998, only nine still in MH-60G configuration and at start of 2000, all 102 in inventory were using HH-60G designation. All have rescue hoist, Doppler/INS, electronic map display, Tacan, Honeywell AN/APN-239 lightweight weather/ground-mapping radar, secure HF, and satcom; MH-60G had ESSS (see Armament paragraph) for weapons and additional fuel carrying capability, plus door-mounted 0.50 in machine guns and Raytheon AN/AAQ-16 Pave Low III FLIR as fitted in MH-53; however, this version no longer assigned to USAF Special Operations Command, following inactivation of 55th Special Operations Squadron.

Upgraded version of HH-60G, known as **Block 152**, made debut at Stratford, Connecticut, on 29 April 1999, when first of 49 planned aircraft rolled out in Upgraded Communication, Navigation/Integrated Electronic Warfare (UCN/IEW) configuration; new features include enhanced com/nav system and EW suite integrated into MIL-STD-1553 databus to reduce workload of crew. Contractor trials undertaken in May and June 1999, after which modified HH-60G (possibly 92-26460) delivered to Nellis AFB, Nevada, for operational test and evaluation with 422nd TES. This six-month programme expected to pave way for contract covering retrofit of 48 more by 2007. Retrofit also includes installation of revised, externally mounted, armament system with 0.50 in machine guns, expendable chaff/flare defensive countermeasures and repositioned nose radar.

MH-60K: US Army special operations aircraft (SOA); prototype (89-26194) ordered in January 1988; first flight 10 August 1990. US Army funded two batches of 11 with options for another 38, which not taken up; first production aircraft (91-26368) completed, February 1992; trials at Patuxent River and Edwards AFB before intended first deliveries in June 1992 to 160th Special Operations Aviation Group (part of 160 SOA Regiment). Deliveries delayed by software problems with special operations equipment; first 10 accepted in 1992 in non-operational state; remaining 12 initially stored, then delivered with new software installed, October to December 1993, to permit start of training by 160 SOA Group, February 1994.

Features include provision for additional 3,141 litres (829 US gallons; 691 Imp gallons) of internal and external fuel (see Power Plant), plus flight refuelling capability, integrated avionics system with electronic displays, Raytheon AN/AAQ-16 FLIR and AN/APQ-174B terrain-following, ground-mapping and air-to-ground ranging radar, T700-GE-701C engines and uprated transmission, external hoist, wire-strike protection, rotor brake, tiedown points, folding tailplane, AFCS similar to that of SH-60B, strengthened pintle mounts for 0.50 in machine guns, provision for Stinger missiles, missile warning receiver, pulse radio frequency jammer, CW radio jammer, laser detector, chaff/flare dispensers, and IR jammer.

SH-60B Seahawk: US Navy ASW/ASST helicopter, described separately.

SH-60F Seahawk: US Navy carrierborne inner-zone ASW helicopter to replace SH-3D Sea King. See Seahawk entry.

Standard US Army UH-60L in medical evacuation markings *(Paul Jackson/Jane's)*

HH-60H and HH-60J Jayhawk: Search and rescue/special warfare helicopters; see Seahawk entry.

UH-60J: Designation of Japanese-built S-70A-12 for Air and Maritime Self-Defence Forces; procurement details under Mitsubishi in Japanese section.

UH-60JA: Japanese Ground Self-Defence Force version; requirement for 50 to 70; procurement began in 1995; refer to Mitsubishi in Japanese section.

UH-60L: Replaced UH-60A in production for US Army from October 1989 (aircraft 89-26179 onwards); prototype (84-23953) first flight 22 March 1988; two preseries aircraft (89-26149 and 26154); first delivery 7 November 1989 to Texas ArNG. Powered by T700-GE-701C engines with uprated 2,535 kW (3,400 shp) transmission. Current production aircraft fitted with hover IR suppression system (HIRSS) to cool exhaust in hover as well as forward flight; older UH-60s retrofitted. Composites wide-chord main rotor blades of improved design flight tested at West Palm Beach, beginning 8 December 1993; new blade 16 per cent wider than current titanium rotor and has anhedral tip angled down at 20°; testing reveals much lower vibration plus anticipated benefits in payload, speed and manoeuvrability; projected retrofit from 1997 has still to be implemented. Underslung load capability increased to 4,082 kg (9,000 lb).

HH-60L: Alternative designation for medical evacuation version based on UH-60L airframe but incorporating UH-60Q specialised mission equipment. Sikorsky awarded US$11 million contract on 22 February 2000 for design definition and conversion of four UH-60Ls to this standard. Initial delivery expected in 2000.

MH-60L: Similar to MH-60A; for 160th SOAR, US Army; further modified as below.

AH-60L: 'Direct Action Penetrator'. Upgrade of MH-60L in 1990 with FLIR, radar and standard UH-60 external stores support system; two Black Hawk companies of 160th SOAR each have MH-60K platoon and AH-60L platoon. Armament mix includes multiple 30 mm Chain Gun, racks of four Hellfires and 2.75 in rocket pods, 40 mm grenade launcher or trainable 7.62 mm Gatling guns.

UH-60M: First use of designation was for proposed enhanced version for US Army. Cancelled early 1989 in favour of UH-60L. Designation subsequently re-used in 2000; see immediately below for details.

UH-60M: Improved Black Hawk version (originally known as the **UH-60L+**) intended to enter service with US Army in about 2002; initial stage of two-phase avionics and power plant modernisation effort that is intended to extend operational life by 25 to 30 years; second stage will be the UH-60X, described separately. Sikorsky awarded US$7.45 million contract in mid-2000 for risk reduction effort. In addition to purchasing some new-build aircraft under the next multiyear procurement programme (MYP6, FY02 onwards), US Army envisages procurement via major upgrade programme, whereby approximately 610 UH-60As will be brought to UH-60M standard, while another 357 UH-60As are expected to be similarly upgraded, before being further modified to the UH-60Q medical evacuation configuration. US Army also seeking information-technology enhancements to enable Black Hawk to operate on future digital battlefield, prompting Sikorsky to conceive and propose a 'growth plan' that includes the UH-60M and the still undefined UH-60X derivative, which is expected to be assigned to 'first-to-fight' units from about 2008.

Improvements envisaged for the UH-60M include a wide-chord, composite-spar main rotor, a digitised cockpit based on the MIL-STD-1553 databus and new avionics, an advanced flight control computer, new diagnostic monitoring systems, a strengthened fuselage, larger fuel tanks and advanced infra-red imaging capabilities.

Proposed upgrade will also allow entire US Army Black Hawk fleet to standardise on General Electric T700-GE-701C engine and improved durability main gearbox.

VH-60N: Nine for US Marine Corps Executive Flight Detachment of squadron HMX-1 at Quantico, Virginia, to replace UH-1Ns; deliveries started November 1988; known as VH-60A until redesignated 3 November 1989. Name **White Hawk** adopted by Marine Corps.

Additional equipment includes more durable gearbox, weather radar, SH-60B-type flight control system and ASI, -401 engines as in SH-60B, cabin soundproofing, VIP interior, cabin radio operator station, EMP hardening, 473 litre (125 US gallon; 104 Imp gallon) internal fuel tank and extensive avionics upgrading. SPAR (Special Progressive Aircraft Rework) being undertaken on VH-60N fleet from 1998 onwards.

UH-60P: South Korean Army version of UH-60L (S-70A-18) with minor avionics modifications to meet local requirements; first (KA-1602) of three UH-60Ls delivered by Sikorsky 10 December 1990; balance of 81 UH-60Ps on initial contract assembled locally by Korean Air Lines with increasing indigenous content, in US$500 million, five year programme. Deliveries from follow-on batch of 57 nearing completion in late 1998 (100th handed over in late 1996); acquisition of third batch to replace remaining UH-1 Iroquois still under consideration in third quarter of 1999. South Korea also has requirement for about 12 medical evacuation helicopters and could obtain UH-60Q or equivalent.

UH-60Q: 'Dustoff' (Dedicated Unhesitating Service To Our Fighting Forces) medical evacuation/search and rescue version for US Army. Development began in early 1990s, after Gulf War, when it was realised that a requirement existed for a longer-range medevac helicopter to replace the UH-1V Iroquois. A proof-of-principle conversion of a UH-60A (86-24560) was undertaken by Serv-Air Inc of Richmond, Kentucky, and flown for the first time as the **YUH-60A(Q)** on 31 January 1993. This aircraft was subsequently delivered to the Tennessee Army National Guard at Lovell Field, Chattanooga, on 12 March 1993, where it underwent a 12-month evaluation programme beginning in September 1993, with an organisation known as CECAT (Combat Enhancing Capable Aviation Team).

Sikorsky eventually selected as prime integrator for production and was awarded an initial contract on 9 February 1996 for two Phase 2 conversions in FY96, with second contract for further two in FY97; YUH-60A(Q) designation also applied to these aircraft, which participated in two year qualification programme, with initial flight qualification tests successfully completed in second quarter of 1997; formal operational test followed at Fort Campbell, Kentucky, between July and September 1998, using three YUH-60A(Q)s, with fourth delivered to CECAT in early 1999. Major subcontractors are Air Methods (medical interiors); Breeze-Eastern (HS-29900 external electric rescue hoist); BAE Systems Canada (mission management system); FLIR Systems Inc (SAFIRE thermal imaging system); Litton (LITOX onboard oxygen generating system); Simula (medical attendant seats) and Telephonics (intercom).

Definitive UH-60Q configuration will include a medical interior able to accommodate six stretcher patients, with

Sikorsky S-70A-28 of Turkish Army

integrated suction and oxygen systems plus defibrillation, ventilation and intubation equipment, as well as apparatus for monitoring of vital signs. It will also have a 'glass cockpit' incorporating Litton smart multifunction displays (SMFDs); Doppler 128C with embedded GPS; NVG-compatible lighting; AN/ARS-6(V)2 personnel locator system; HIRSS; chaff/flare dispensers; ESSS; -701C engines; plus an improved data modem and SINCGARS radios, which will allow it to transmit and receive digital data.

Following completion of Phase 2 qualification, work is due to begin on Phase 3 modification and procurement. Initial contract for six UH-60Q multimission medevac systems awarded in third quarter of 1999 to Air Methods, as first phase of LRIP; first two to be delivered in April 2000, with remainder following in June, and four are to be installed in new-build Black Hawks that are reportedly to be assigned to Army National Guard unit at Salem, Oregon. Further conversions will follow via UH-60M upgrade programme, with the US Army expected to receive 357 to equip first- and second-line medical evacuation units.

CH-60S KnightHawk: Potential key element in 1996 US Navy Helicopter Master Plan, which envisages retirement of CH/HH-46Ds, HH-60Hs and SH-3s by FY12 and their replacement by navalised CH-60S. Design is fundamentally baseline UH-60L Black Hawk with T700-GE-401C engines and dynamics of SH-60 Seahawk, plus automatic rotor blade folding system, folding tail pylon, improved durability gearbox, rotor brake, automatic flight control system (AFCS), HIFR capability and rescue hoist for SAR/CSAR missions. Will also feature 'glass cockpit', with active matrix liquid crystal displays (AMLCDs); common cockpit to be adopted by CH-60S and SH-60R, with Lockheed Martin securing US$61 million contract in August 1998 to develop and produce two prototypes for flight testing from late 1999. Equipment will include two integrated inertial navigation/GPS units, mass memory unit, mission computer, flight management computer, operational software and four Litton flat-panel displays replacing all but standby instruments; CH-60S cockpit will be 'scaled-down' version, with potential for upgrading if combat SAR mission is added at later date. CH-60S also to have provisions for external stores support system (ESSS), as available to US Army Black Hawks, allowing carriage of additional fuel and forward-firing weapons.

Design includes a convertible cargo handling system; when configured for pure cargo operations, CH-60 will carry two 1.02 × 1.22 × 1.02 m (40 × 48 × 40 in) tri-wall pallets with total weight of 1,588 to 1,814 kg (3,500 to 4,000 lb); as a personnel transport, will accommodate a crew of four, plus 13 passengers. Underslung loads up to 4,082 kg (9,000 lb) may also be carried, with total payload capacity about 4,536 kg (10,000 lb).

CH-60 potential first demonstrated by a modified UH-60L at Norfolk, Virginia, and San Diego, California, in June 1995, when internal and external cargo-carrying capability was studied by both Sikorsky and the US Navy. Detail design began October 1996, with subsequent developments including award in April 1997 of US$5.75 million contract to Sikorsky for CH-60 demonstrator. Demonstrator is hybrid vehicle, based around UH-60L (96-26673) borrowed from the US Army, married to components taken from SH-60F furnished by US Navy. Resulting YCH-60S (Navy identity 966673) made first flight on 6 October 1997 and was then used for joint Navy/Sikorsky 35 hour flight test programme that was completed on 10 January 1998. First shipboard demonstration accomplished on 19 November 1997, with YCH-60S completing 17 landings aboard combat store ship USS *Saturn*; initial trial included 12 vertical replenishment lifts with 680 kg (1,500 lb) slung load and three hot refuellings.

YCH-60S also being evaluated in 1999-2000 as potential airborne mine countermeasures (AMCM) platform to replace MH-53E; existing minesweeping sleds used by MH-53E are too heavy for CH-60, but new lightweight towed systems and laser imaging detection and ranging equipment are in prospect and could bestow adequate minehunting capability. Initial trials undertaken at Stratford during third quarter of 1999, followed by transfer to Patuxent River in fourth quarter for tow tests, plus carriage, winch deployment and recovery of AN/AQS-20/X mine detection sonar. Thicker frames for helicopter's rear-cabin structure will permit cable loads up to 2,722 kgf (6,000 lbf). Replacement of MH-53E by suitably modified CH-60S expected to begin in 2005, with ultimate total of 66 CH-60Ss planned to be configured for AMCM mission; full-scale development of this capability currently scheduled to begin in 2002.

Features not embodied in demonstrator include fuel dump vents, flotation gear, HIFR and navalised T700 engines. Avionics also unrepresentative, although featuring a databus and four 127 × 127 mm (5 × 5 in) Rockwell Collins AMLCDs for pilot and co-pilot vertical and horizontal situation data, plus two control display/navigation units and an LN-100G embedded GPS/INS.

Decision to proceed with low-rate initial production (LRIP) was taken in early 1998, although firm fixed-price contract for first lot not awarded to Sikorsky until September 1999; valued at US$67.4 million, this was for initial five aircraft, plus option for one more (taken up in

Trials YCH-60S carrying underslung load   1999/0051844

November 1999) and associated engineering and logistic services. Maiden flight of initial production CH-60S (165742) at Stratford, Connecticut on 27 January 2000; subsequently delivered to Patuxent River, Maryland, on 15 May to begin US Navy development testing and operational evaluation. At least four helicopters to be involved in trials, with initial technical evaluation to be completed by November 2000; three-month Opeval to follow, culminating in decision to launch full-rate production, which due to be taken in March 2001; other aspects of test programme include tow trials with AN/AQS-20/X mine detection sonar at Panama City, Florida, from September 2000.

Navy has eventual requirement for over 200 CH-60Ss, with initial quantity of 42 planned for fifth UH-60 multiyear procurement contract during FY97 to FY01, although it now appears that this figure will not be attained. First example funded in FY98, with five more (Lot 1) in FY99, 17 (Lot 2) in FY00 and 15 requested for FY01. Navy procurement in sixth multiyear purchase (FY02 to FY06) expected to total 117, beginning with 14 in FY02. Following trials at Patuxent River, first operating squadron to be HC-3 at North Island, California. Turkey also planning to acquire initial batch of four to six CH-60Ss.

UH-60X: Designation allocated to studies of potential upgrades to 255 existing UH-60L Black Hawks of US Army. Operational deployment scheduled for about 2008, with new variant being assigned to 'first-to-fight' units. Changes under consideration include those already earmarked for UH-60M, as well as incorporation of drive train and rotor systems from the S-92, plus more powerful engine currently under development as the Common Engine Program (CEP, formerly Advanced Common Engine). This is expected to result in a 2,237 kW (3,000 shp) engine, which will also be installed in Navy SH-60R as well as Army AH-64D Apache. UH-60X will also have a higher-capacity cargo hook.

Upgrade expected to offer significant benefits, compared with the original UH-60A version, including twice the payload and range capability and a 24 kt (44 km/h; 28 mph) speed advantage; maintenance costs also expected to fall to about half those of the UH-60A.

Operational requirements document for UH-60X drafted by US Army in last quarter of 1999 stipulates improved baseline performance with 25 per cent reduction in specific fuel consumption, a maximum TOW of 11,793 kg (26,000 lb) and increased fuel capacity of 1,560 litres (412 US gallons; 343 Imp gallons). Minimum mission requirement is to carry 4,100 kg (9,040 lb) external load at least 75 n miles (139 km; 86 miles) or an 11-man assault team at least 148 n miles (274 km; 170 miles).

Firehawk: Trials of specialist firefighting version began in July 1998, using modified UH-60L (96-26728) with extended landing gear and removable 3,785 litre (1,000 US gallon; 833 Imp gallon) ventral watertank manufactured by Aero Union of Chico, California. Replenishment of tank can be accomplished in two ways: by landing next to water source for water to be pumped into tank via side connector, or by hovering over source and using snorkel hose and pump assembly to suck water up. Subsequent three month demonstration of firefighting capability undertaken by Los Angeles County Fire Department proved validity of system and first customer was involved in negotiations with Sikorsky during fourth quarter of 1998. Identity of potential buyer not confirmed, but thought to be Sultan of Brunei. Demonstrator subsequently returned to US Army for further testing, before delivery to Oregon Army National Guard in 1999; Los Angeles County Fire Department later issued RFP for up to four dedicated firefighting helicopters in UH-60 size and weight category.

Maple Hawk: Unsuccessful contender in contest to supply Canadian Forces with new SAR helicopter; offer reportedly priced at C$300 million for 15 helicopters, but EHI EH 101 Cormorant selected in December 1997.

Battle Hawk: Offered to Australian Army for Project Air 87 armed reconnaissance requirement; based on MH-60K; two -701C engines and 2,535 kW (3,400 shp) transmission; up to 680 kg (1,500 lb) extra payload; improved electromagnetic and corrosion protection; belly turret with 20 mm Giat THL 20 cannon slaved to Elbit Toplite II FLIR/day TV/laser designator-ranger targeting sensor or Elbit Midash helmet sight; and full 'glass cockpit' with Rockwell Collins active matrix colour LCDs and HOCAS controls. Eliminated from bidding, but competition reopened in early 2000 and Battle Hawk reinstated along with Denel Red Hawk; other contenders include Bell AH-1Z, Agusta A 129, Boeing AH-64D and Eurocopter Tiger. Total of 25 to 30 helicopters required, with first squadron expected to become operational in 2007.

Exports comprise: S-70A-1: FMS deal for Royal Saudi Land Forces Army Aviation Command; 12 delivered January to April 1990 to squadron based at King

Khaled Military City; modified to **Desert Hawk** and one (delivered December 1990) fitted with VIP interior; Desert Hawk has 15 troop seats, blade erosion protection using polyurethane tape and spray-on coating, Racal Jaguar 5 frequency-hopping radio, provision for searchlights, internal auxiliary fuel tanks, and external hoist.

S-70A-1L: Medical evacuation version for Saudi Arabia; IR filtered searchlight, rescue hoist, improved AN/ARC-217 HF com, AN/ARN-147 VOR/ILS, AN/ARN-149 ADF, air conditioning and provision for six stretchers; eight delivered from December 1991; further eight required.

S-70A-5: Two for Philippine Air Force, delivered March 1984.

S-70A-9: Royal Australian Air Force; 39 replaced Bell UH-1s; deliveries from October 1987 to 1 February 1991; first completed by Sikorsky, remainder assembled by Hawker de Havilland in Australia; aircraft transferred to Australian Army in February 1989, but RAAF continues to maintain them.

S-70A-11: Three to Jordan in 1986-87.

S-70A-12: Japan Self-Defence Forces acquiring **UH-60J/JA** versions of Mitsubishi SH-60J for search and rescue. Sikorsky-built prototype (N7267D), plus two CKD kits, delivered late 1990. Further production by Mitsubishi (which see).

S-70A-16: Reserved for Westland Helicopters (see UK section of 1995-96 and earlier *Jane's*).

S-70A-17: Turkish Jandarma ordered six in September 1988; deliveries completed December 1988; further six (including two VIP) delivered from late 1990 to Turkish National Police. See also S-70A-28.

S-70A-18: Korea (see UH-60P).

S-70A-19: Reserved for GKN Westland of UK.

S-70A-21: Two VIP versions to Egypt, 1990. Further two VIP-configured UH-60L ordered by Egypt in third quarter of 1999; delivery due by end of 2003.

S-70A-22: Korean VIP version. Three aircraft built by Sikorsky.

S-70A-24: Two UH-60Ls for Mexico. Delivered 1991.

S-70A-25/26: Moroccan Gendarmerie ordered two Black Hawks with different seating arrangements in 1991; delivered October 1992; began operations 11 November 1992; fitted with colour weather radar.

S-70A-27: Hong Kong. Two delivered 16 December 1992 to Royal Hong Kong Auxiliary Air Force; unit became Government Flying Service 1 April 1993. Fitted with FLIR and searchlight. Requirement for further four reportedly existed in 1995, but only one additional aircraft delivered.

S-70A-28: Turkish follow-on batch; 90 ordered 8 December 1992, of which first five to Jandarma on 4 January 1993, followed by 40 to armed forces during 1993-94; balance of 45 to have been co-produced in Turkey by TAI (which see) but programme suspended. However, fresh negotiations for 50 additional Black Hawks concluded in latter half of 1998, with first five airlifted to Turkey by An-124 in mid-June 1999 and deliveries due to be completed in 2001. Final 30 produced with 'glass cockpit', which to be retrofitted to earlier machines; first flight of **S-70A-28D** in this guise on 29 March 2000. Installation includes four Rockwell Collins MFDs, dual flight management system and LN-100G INS/GPS; seven also to incorporate Tadiran Spectralink ASR-700 airborne search and rescue system (ASARS) for use by Turkish Army Special Forces in combat search and rescue and covert operations.

S-70A-30: One VIP transport ordered for Argentine Air Force, January 1994; delivered 4 September 1994.

S-70A-33: Four ordered by Brunei in 1995; delivered 1997-98. Equipment includes radar, AN/AAQ-21 FLIR and external stores support system.

S-70A-34: Malaysia ordered two S-70A Black Hawks in 1996 as replacements for AS 332 Super Puma in VIP transport role; first of pair delivered by end of 1997, with second following in February 1998.

S-70A-36: Brazil received four S-70As in August 1997 for use in Peru/Ecuador peacekeeping support mission; equipment includes GPS, HF radio, internal rescue hoist and weather radar.

S-70A-37: Version of Firehawk; two to Sultan of Brunei, 2000, replacing S-70Cs.

S-70A-39: Chilean order for one, announced in March 1998; delivered in July 1998, with further purchases expected to replace UH-1H Iroquois over next few years.

S-70A-41: Colombia. Believed to refer to 22 aircraft delivered during 1994 (two), 1995 (two), 1997 (seven) and 1998 (11); these are UH-60L derivative, unlike initial delivery of 1988-89 which was baseline UH-60A. Further helicopters requested in November 1999, with as many as 60 ultimately in prospect as part of US drive to combat increased narcotics trafficking in this region.

S-70A-50: Israel. Request for 15 UH-60Ls revealed by US DoD in April 1997; first so-called 'Peace Hawk' handed over at Stratford, Connecticut, on 23 March 1998, being airlifted (with four others) to Israel by C-5 on 27 May 1998. Deliveries completed by end of 1998. Additional 35 requested in September 2000 at estimated cost of US$525 million.

Direct transfers include one UH-60L to Bahrain, early 1991; five UH-60As delivered to Colombian Air Force in

US BLACK HAWK PROCUREMENT

FY	Lot	Army	USAF	USN	FMS	First aircraft*	Remarks
73	-	6				73-21650	RDT&E
77	1	15				77-22714	
78	2	56				78-22960	Includes YEH-60B (78-23013)
79	3	92				79-23265	
80	4	94				80-23416	
81	5	80	5			81-23547	
82	6	96	6			82-23660	
83	7	96				83-23837	} MYP1, 18 December 1981
84	8	96				84-23933	
85	9	104				85-24387	
86	10	96				86-24483	} MYP2, 31 October 1984
87	11	102	9			87-24579	
88	12	72	16		20[2]	88-26015	
89	13	72	18		5[3]	89-26123	} MYP3, 11 January 1988
90	14	76	22		1[4]	90-26218	
91	15	67[1]	15		8[5]	91-26318	
92	16	52	13			92-26408	
93	17	60				93-26473	
94	18	63				94-26533	} MYP4, 28 April 1992
95	19	68				95-26596	
96	20	65			2[3]	96-26664	
97	21	34	8		15[8]		
98	22	28		2		98-26795	} MYP5, 18 July 1997[6]
99	23	22		4			
00	24	19					
Totals		**1,612**	**112**	**6**	**51**		

Notes: [1] Includes 22 MH-60K for special operations
[2] For Saudi Arabia (13), Colombia (5) and Egypt (2)
[3] For Colombia
[4] For Bahrain
[5] For Saudi Arabia
[6] MYP5 (FY97-01) originally to consist of 108 aircraft (58 Army, 42 Navy and 8 USAF), with 36 in first year and 18 per year thereafter; however, additional aircraft have been purchased (or requested) in each of the first three years and further increases appear possible in remaining two years (FY00 and FY01)
[7] For Egypt
[8] For Israel
* Serial numbers not always consecutive; batches include transfers to USAF and Foreign Military Sales programme

July 1988 for anti-narcotics operations; five more sold February 1989. Israel received 10 former US Army UH-60As in August 1994, for 124 Squadron at Palmachim, under local name of Yanshuf (Owl).

Demonstration tour by US Army UH-60L in Norway and Sweden during February and March 1997; both countries have requirement for utility helicopter; collaborative programme involving all four Scandinavian countries and up to 80 helicopters is a possibility. Taiwan also requires new utility helicopter as replacement for existing UH-1H Iroquois; S-70A in contention and reportedly preferred option over competing Bell 412; up to 80 of chosen type to be obtained, with local manufacture expected to be a key factor in selection. Decision originally anticipated in 1998 but has been deferred, with initial contract for 20 to 25 helicopters likely. South Korea also due to select new attack helicopter, with AUH-60 Armed Black Hawk in contention; initial requirement exists for 18, with follow-on purchases likely to increase eventual total to 36.

Black Hawk version selected by Austria in third quarter of 2000 as new utility helicopter to begin replacement of Agusta-Bell 204/212 fleet. Initial purchase will involve nine helicopters at cost of US$186 million. Black Hawk also a contender for Greek Army requirement for up to 66 utility/combat search and rescue helicopters; comparative evaluation undertaken in July 2000, with Mil Mi-17 and Eurocopter AS 532 Cougar Mk II among candidates for co-production agreement.

S-70C: Commercial version, described separately and under SH-60B Seahawk.

CUSTOMERS: See tables. By 1999, over 2,400 H-60s of all variants had flown more than 3,600,000 hours. US Army total includes EH-60As and diversions to USAF, Bahrain, Colombia, Egypt and Saudi Arabia; 1,000th of S-70 series accepted 17 October 1988 and 2,000th May 1994; US Army Black Hawks in service in Germany, Hawaii and South Korea and with Army National Guard and Army Reserve.

US Army UH-60As loaned to the US Drug Enforcement Agency, augmenting five bought direct from Sikorsky. Fleet in 1999 totalled 13 (all with military serial numbers), plus three in storage and further two lost in accidents.

See Current Versions for export models and details.

COSTS: UH-60L US$8.6 million (1997) US Army unit cost; MH-60G US$10.2 million. Two VIP aircraft for Malaysia cost about US$20 million (1997). Two utility aircraft for Brunei, with firefighting kits, believed to cost US$25-30 million (1999).

DESIGN FEATURES: Represented new generation in technology for performance, survivability and ease of operation when introduced to replace UH-1 as US Army's main squad-

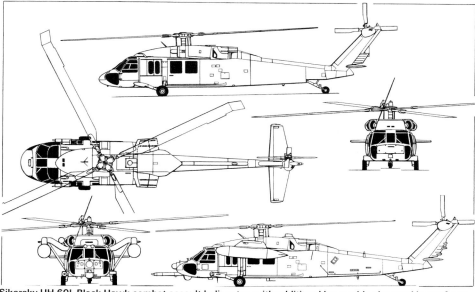

Sikorsky UH-60L Black Hawk combat assault helicopter, with additional lower side view and lower front view of MH-60K special operations variant *(Dennis Punnett/Jane's)*

BLACK HAWK DELIVERIES

	1978-83	1984	1985	1986	1987	1988	1989	1990	1991	1992	1993	1994	1995	1996	1997	1998	1999	Total
Argentina												1						1
Australia				1 (+13)	7 (+11)	13 (+14)	17	1										39
Bahrain							1											1
Brazil															4			4
Brunei			2											1	2	2		7
Chile																1		1
China		9	15															24
Colombia						5	5					2	2		7		11	32
Egypt								2							2			4
Hong Kong										2			1					3
Israel																15		15
Japan								1 (+2)										1
Jordan					1	2												3
Korea:																		
RoKAF										3								3
FMS										3								3
Co-prod								3 (+13)	4	(24)	(24)	(16)	(15)	(9)	(12)	(12)	(3)	7
Malaysia															1	1		2
Mexico										2								2
Morocco											2							2
Philippines			2															2
Saudi Arabia						2	10	1	4	4								21
Taiwan			11	3												4		18
Turkey						6			6			40	5			22		79
UK:																		
Westland					1													1
USA:																		
Air Force						11	16	18	26		7		5			8		91
Army	476	123	120	102	99	82	72	72	49	40	60	63	64	69	54	35	21	1,601
Navy																	2	2
DEA					1	4												5
SOA									1	10	8	4						23
Totals	476	134	146	110	117	118	118	129	67	65	108	80	67	70	70	66	56	1,971

Notes: Kits in parentheses; NOT to be included in totals (appear also under year of completion for Australia)
Yearly totals of US Army aircraft in period 1978 to 1983 were two, 36, 67, 119, 126 and 126 respectively
DEA: Drug Enforcement Agency (USA)
SOA: Special Operations Aircraft

carrying helicopter; adapted to wide variety of other roles, including several maritime applications. Four-blade main rotor; one-piece forged titanium rotor head with elastomeric blade retention bearings providing all movement and requiring no lubrication; hydraulic drag dampers; bifilar self-tuning vibration absorber above head; blades have 18° twist, and tips swept at 20°; thickness and camber vary over the length of blades, based on Sikorsky SC-1095 aerofoil; blades tolerant up to 23 mm hits and spar tubes pressurised with gauges to indicate loss of pressure following structural degradation.

Two pairs of tail rotor blades fastened in cross-beam arrangement, mounted to starboard; tail rotor pylon tilted to port to produce lift as well as anti-torque thrust and to extend permissible CG range; fixed fin large enough to allow controlled run-on landing following loss of tail rotor.

FLYING CONTROLS: Rotor pitch control powered by two independent hydraulic systems; Hamilton Sundstrand AFCS with digital three-axis autopilot provides speed and height control and coupled modes. Full-time autostabilisation includes feet-off heading hold cancelling torque-induced yaw at all airspeeds and during hover; positive fuselage attitude control provided by electrically driven variable incidence tailplane moving from +34° in hover to −6° during autorotation; angle is controlled by combined sensing of airspeed, collective-lever position, pitch attitude rate and lateral acceleration.

STRUCTURE: Main blade spar is formed and welded into oval titanium tube, with Nomex core, graphite trailing-edge, and covered by glass fibre/epoxy skin; titanium leading-edge abrasion strip and Kevlar tip. New main blades, with modified tips and 16 per cent increase in chord, under development for UH-60L; available for retrofit from 1997. Cross-beam composites tail rotor, eliminating all rotor head bearings. Light alloy airframe designed to retain 85 per cent of its flight deck and passenger space intact after vertical impact at 11.5 m (38 ft)/s, lateral impact at 9.1 m (30 ft)/s, and longitudinal impact at 12.2 m (40 ft)/s; also withstands simultaneous 20 g forward and 10 g downward impact; glass fibre and Kevlar used for cockpit doors, canopy, fairings and engine cowlings; glass fibre/Nomex floors; tailboom folds to starboard and main rotor mast can be lowered for transport/storage.

LANDING GEAR: Non-retractable tailwheel type with single wheel on each unit. Energy-absorbing main gear with a tailwheel which gives protection for the tail rotor in taxying over rough terrain or during a high-flare landing. Axle assembly and main gear oleo shock-absorbers by General Mechatronics. Mainwheel tyres size 26×10.00-11, pressure 8.96 to 9.65 bars (130 to 140 lb/sq in); tailwheel tyre size 15×6.00-6, pressure 6.21 to 6.55 bars (90 to 95 lb/sq in). Alaskan-based H-60s have Airglass Engineering ski landing gear.

POWER PLANT: Two 1,210 kW (1,622 shp) intermediate rating General Electric T700-GE-700 turboshafts initially. From late 1989 (UH-60L), two T700-GE-701C engines, each developing intermediate 1,342 kW (1,800 shp). (T700-GE-701A engines with maximum T-O rating of 1,285 kW; 1,723 shp optional in export models.) Transmission rating 2,109 kW (2,828 shp) in UH-60A, uprated to 2,535 kW (3,400 shp) in models with T700-GE-701C engines.

Two crashworthy, bulletproof fuel cells, with combined usable capacity of 1,361 litres (360 US gallons; 300 Imp gallons), aft of cabin. Single-point pressure refuelling, or gravity refuelling via point on each tank. Auxiliary fuel can be carried internally in one of several optional arrangements, or externally by the ESSS system. Two external tanks each of 871 litres (230 US gallons; 192 Imp gallons); up to two internal tanks, each of 700 litres (185 US gallons; 154 Imp gallons).

ACCOMMODATION: Two-man flight deck, with pilot and co-pilot on armour-protected seats. A third crew member is stationed in the cabin at the gunner's position adjacent forward cabin windows. Forward-hinged jettisonable door on each side for access to flight deck area. Main cabin open to cockpit to provide good communication with flight crew and forward view for squad commander. Accommodation for 11 fully equipped troops, or 14 in high-density configuration; 20 minimally armed personnel in optional configuration. Eight troop seats can be removed and replaced by four litters for medevac missions, or to make room for internal cargo. An optional layout is available to accommodate a maximum of six litter patients. Executive interiors for seven to 12 passengers available for the S-70A. Cabin heated and ventilated. Simula Safety Systems Inc received US$7.1 million contract in April 1999 covering supply of 290 cockpit airbag systems for installation in US Army UH-60A/L; this is low-rate initial production phase and total comprises 275 aircraft units and 15 spares.

External cargo hook, having a 3,630 kg (8,000 lb) lift capability, enables UH-60A to transport a 105 mm howitzer, its crew of five and 50 rounds of ammunition.

Australian Army Sikorsky S-70A-9 Black Hawk *(Australian Army)*

Rescue hoist of 272 kg (600 lb) capacity optional. Large rearward-sliding door on each side of fuselage for rapid entry and exit.

SYSTEMS: Solar 67 kW (90 hp) T-62T-40-1, Honeywell or Hamilton Sundstrand APU. An optional winterisation kit provides a second hydraulic accumulator installed in parallel with the APU hydraulic start accumulator, maintaining engine start capability at low ambient temperatures; Honeywell 30 to 40 kVA and 20 to 30 kVA electrical power generators; 17 Ah Ni/Cd battery. Engine fire extinguishing system. Rotor blade de-icing system standard on US Army aircraft, optional for export. Electric windscreen de-icing.

AVIONICS: Configurations vary between aircraft. Additional avionics and self-protection equipment installed in Enhanced Black Hawk, as described under Current Versions. Improvement options offered from 1996 for new-build and retrofit on S-70 series include 'glass cockpit' and digital avionics; equipment available includes EFIS and digital automated flight computer system (AFCS).

Comms: Raytheon AN/ARC-186 VHF-FM, Raytheon AN/ARC-115 VHF-AM, Raytheon AN/ARC-164 UHF-AM, Rockwell Collins AN/ARC-186(V) VHF-AM/FM, Honeywell AN/APX-100 IFF transponder, Raytheon TSEC/KT-28 voice security set, and intercom. HH-60G has AN/URC-108 satcom and is being upgraded with Rockwell Collins AN/ARC-210 integrated communications system; Rockwell Collins AN/ARC-220 nap of the earth digital radio and AN/ARC-222 installed on Block 152 Upgrade HH-60G.

Radar: MH-60K has Raytheon AN/APQ-147A terrain-following/terrain-avoidance radar, HH-60G has Honeywell AN/APN-239 (RDR-1400C) radar. AH-60L and some export S-70s also equipped with radar.

Flight: Hamilton Sundstrand AFCS with digital three-axis autopilot, Honeywell AN/ARN-123(V)1 VOR/marker beacon/glide slope receiver, Emerson AN/ARN-89 ADF, BAE Systems AN/ASN-128 Doppler, AN/ASN-43 gyrocompass, Honeywell AN/APN-209(V)2 radar altimeter. HH-60G has BAE Systems AN/ASN-137 Doppler, Rockwell Collins AN/ASN-149 GPS and Litton ring laser gyro INS (replacing Carousel IV).

Instrumentation: HH-60G has Teldix KG-10 map display. US Army MH-60 special operations versions to receive Elbit ANVIS 7 NVG/HUD system as retrofit; system already fitted in UH-60A and L. Elbit ANVIS/HUD system also ordered by Australia in August 1999; 12 units to be acquired for installation by Raytheon Australia on S-70A-9. NVG-compatible version of Lockheed Martin GH-3000 electronic standby instrument system to be installed on MH-60K.

Mission: HH-60G has Raytheon AN/AAQ-16 FLIR. UH-60Q to have FLIR Systems Inc AN/AAQ-21 SAFIRE thermal imaging system. Trials of Northrop Grumman Airborne Standoff Minefield Detection System (ASTAMIDS) undertaken over Bosnia in latter half of 1997; intended to be precursor of procurement of at least 85 production systems but poor test results, especially in areas of dense undergrowth, combined with budget cuts to force US Army to slow pace of development and delay deployment of this system.

Self-defence: Baseline UH-60 Black Hawk has Raytheon AN/APR-39(V)1 RWR, Sanders AN/ALQ-144 IR countermeasures set and BAE Systems M-130 chaff/flare dispenser. MH-60K has BAE Systems AN/AAR-47 missile warning system, Northrop Grumman AN/ALQ-136 pulse radio frequency jammer, Northrop Grumman AN/ALQ-162 CW radio jammer, Raytheon AN/APR-39A and AN/APR-44 pulse/CW warning receivers, Raytheon AN/AVR-2 laser detector, BAE Systems M-130 chaff/flare dispenser and Sanders AN/ALQ-144 IR countermeasures set. HH-60G has chaff/flare dispenser (BAE Systems M-130 being replaced by AN/ALE-40, but AN/ALE-47 to follow) and Sanders AN/ALQ-144 IR countermeasures set. Development testing of Sanders AN/ALQ-212 Advanced Threat IR Countermeasures (ATIRCM) system on EH/MH-60s of US Army to begin in 1999. AN/AAR-47 missile warning system, AN/ALR-69(V) RWR, AN/ALE-47 chaff/flare dispensers and AN/ALQ-213 EW management system installed on Block 152 Upgrade HH-60G.

EQUIPMENT: HH-60G has Lucas Aerospace internal rescue hoist with 76 m (250 ft) of cable. Lucas awarded contract in mid-2000 to supply initial batch of 21 hoists for installation on CH-60S and other S-70 variants, with potential options for additional 70 in 2001 and 2002 plus 34 in 2003. UH-60Q to have external Breeze-Eastern HS-29900 electric rescue hoist.

ARMAMENT: New production UH-60As and Ls from c/n 431 onward incorporate hardpoints for an external stores support system (ESSS). This consists of a combination of fixed provisions built into the airframe and four removable external pylons from which fuel tanks and a variety of weapons can be suspended. Able to carry more than 2,268 kg (5,000 lb) on each side of the helicopter, the ESSS can accommodate two 871 litre (230 US gallon; 192 Imp gallon) fuel tanks outboard, and two 1,703 litre (450 US gallon; 375 Imp gallon) tanks inboard. This allows the UH-60A to self-deploy 1,200 n miles (2,222 km; 1,381 miles) without refuelling. The ESSS also enables the Black Hawk to carry Hellfire laser-guided anti-armour missiles, gun or M56 mine dispensing pods, FIM-92 Stinger AAMs, ECM packs, rockets and motorcycles. Up to 16 Hellfires can be carried externally on the ESSS, with another 16 in the cabin to provide the capability to land and reload. (Laser designation provided by Bell OH-58 helicopter or ground troops.) Two pintle mounts in cabin, adjacent forward cabin windows on each side, can each accommodate a 0.50 in calibre General Electric GECAL 50 or 7.62 mm six-barrel Minigun.

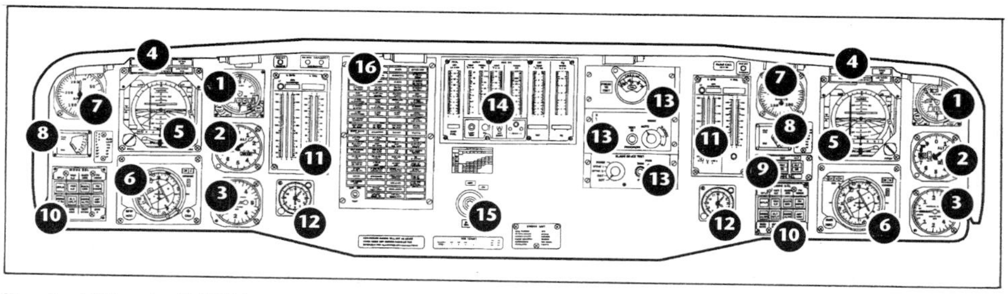

Standard Sikorsky S-70A instrument panel (international version)
1: radar altimeter, 2: barometric altimeter, 3: vertical speed indicator, 4: master warning panel, 5: vertical situation indicator (VSI), 6: horizontal situation indicator (HSI), 7: airspeed indicator (ASI), 8: stabilator position indicator, 9: command instrument system (CIS) mode select panel, 10: HSI/VSI mode select panel, 11: vertical instrument display system (VIDS) pilot display unit, 12: clock, 13: anti-ice system controls, 14: VIDS central display unit, 15: engine ignition switch, 16: caution advisory panel

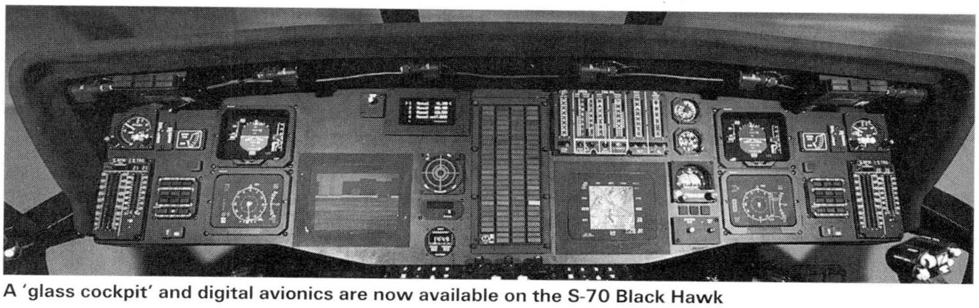

A 'glass cockpit' and digital avionics are now available on the S-70 Black Hawk

DIMENSIONS, EXTERNAL:
Main rotor diameter 16.36 m (53 ft 8 in)
Main rotor blade chord 0.53 m (1 ft 8¾ in)
Tail rotor diameter 3.35 m (11 ft 0 in)
Length overall: rotors turning 19.76 m (64 ft 10 in)
rotors and tail pylon folded 12.60 m (41 ft 4 in)
Length of fuselage:
UH-60A/HH-60G, excl flight refuelling probe 15.26 m (50 ft 0¾ in)
HH-60G, incl retracted refuelling probe 17.38 m (57 ft 0¼ in)
Fuselage max width: UH-60A 2.36 m (7 ft 9 in)
Max depth of fuselage 1.75 m (5 ft 9 in)
Height: overall, tail rotor turning 5.13 m (16 ft 10 in)
to top of rotor head 3.76 m (12 ft 4 in)
in air-transportable configuration 2.67 m (8 ft 9 in)
Tailplane span 4.38 m (14 ft 4½ in)
Tailplane chord 0.88 m (2 ft 10½ in)
Wheel track 2.705 m (8 ft 10½ in)
Wheelbase 8.83 m (28 ft 11¾ in)
Tail rotor ground clearance 1.98 m (6 ft 6 in)
Cabin doors (each): Height 1.37 m (4 ft 6 in)
Width 1.75 m (5 ft 9 in)

DIMENSIONS, INTERNAL:
Cabin: Volume 11.6 m^3 (410 cu ft)

AREAS:
Main rotor blades (each) 4.34 m^2 (46.70 sq ft)
Tail rotor blades (each) 0.41 m^2 (4.45 sq ft)
Main rotor disc 210.15 m^2 (2,262.0 sq ft)
Tail rotor disc 8.83 m^2 (95.03 sq ft)
Fin 3.00 m^2 (32.30 sq ft)
Tailplane 4.18 m^2 (45.00 sq ft)

WEIGHTS AND LOADINGS:
Weight empty: UH-60A 5,118 kg (11,284 lb)
UH-60L 5,224 kg (11,516 lb)
Payload: internal, UH-60A/L 1,197 kg (2,640 lb)
underslung, UH-60A 3,629 kg (8,000 lb)
underslung, UH-60L/Q and CH-60S 4,082 kg (9,000 lb)
Mission T-O weight: UH-60A 7,484 kg (16,500 lb)
UH-60L 7,907 kg (17,432 lb)
HH-60G 8,119 kg (17,900 lb)
MH-60K 11,113 kg (24,500 lb)
Max alternative T-O weight:
UH-60A 9,185 kg (20,250 lb)
UH-60L 11,113 kg (24,500 lb)
Max disc loading: UH-60L at mission T-O weight 36.7 kg/m^2 (7.52 lb/sq ft)
UH-60L at max alternative T-O weight 52.9 kg/m^2 (10.83 lb/sq ft)
Transmission loading: UH-60L at mission T-O weight and max power 3.04 kg/kW (5.00 lb/shp)
UH-60L at max alternative T-O weight and power 4.20 kg/kW (6.91 lb/shp)

PERFORMANCE (UH-60A at mission T-O weight, except where indicated):
Never-exceed speed (VNE):
UH-60L/Q 194 kt (359 km/h; 223 mph)
CH-60S 180 kt (333 km/h; 207 mph)
Max level speed at S/L 160 kt (296 km/h; 184 mph)
Max level speed at max T-O weight 158 kt (293 km/h; 182 mph)
Max cruising speed:
UH-60A 139 kt (257 km/h; 160 mph)
UH-60L 159 kt (294 km/h; 183 mph)
CH-60S 142 kt (263 km/h; 163 mph)
Single-engine cruising speed at 1,220 m (4,000 ft) and 35°C (95°F) 105 kt (195 km/h; 121 mph)
Vertical rate of climb at 1,220 m (4,000 ft) and 35°C (95°F): UH-60A 125 m (411 ft)/min
UH-60L 472 m (1,550 ft)/min
Service ceiling: UH-60A 5,790 m (19,000 ft)
UH-60L 5,835 m (19,140 ft)
Hovering ceiling: IGE at 35°C 2,895 m (9,500 ft)
OGE, ISA 3,170 m (10,400 ft)
OGE at 35°C: UH-60A 1,705 m (5,600 ft)
UH-60L 2,330 m (7,640 ft)
Range with max internal fuel at max T-O weight, 30 min reserves: UH-60A 319 n miles (592 km; 368 miles)
UH-60L 315 n miles (584 km; 363 miles)

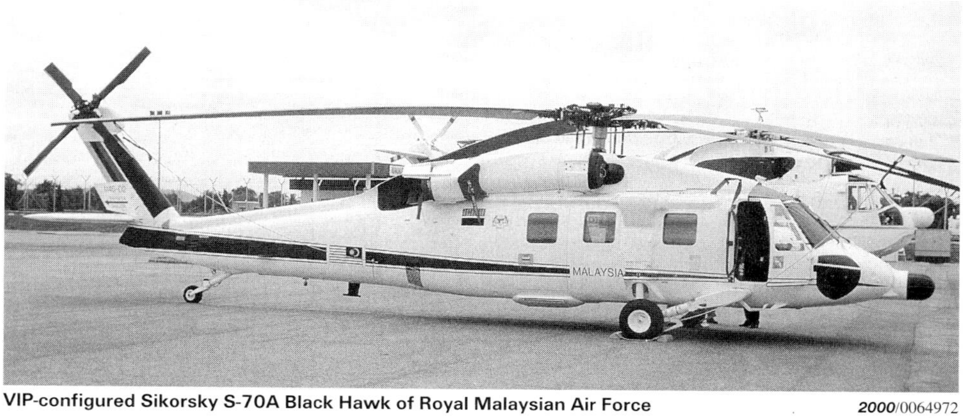

VIP-configured Sikorsky S-70A Black Hawk of Royal Malaysian Air Force *2000*/0064972

USA: AIRCRAFT—SIKORSKY

Prototype Sikorsky SH-60R on its maiden flight   2001/0034479

Range with external fuel tanks on ESSS pylons:
with two 870 litre (230 US gallon; 191.5 Imp gallon)
tanks   880 n miles (1,630 km; 1,012 miles)
with two 870 litre (230 US gallon; 191.5 Imp gallon)
and two 1,703 litre (450 US gallon; 375 Imp gallon)
tanks   1,200 n miles (2,222 km; 1,381 miles)
Endurance: UH-60A   2 h 18 min
UH-60L   2 h 6 min
UPDATED

SIKORSKY S-70B
US Navy designations: SH-60B and SH-60R Seahawk, SH-60F and HH-60H
US Coast Guard designation: HH-60J Jayhawk
Japan Maritime Self-Defence Force designation: SH-60J
Spanish Navy designation: HS.23
Republic of China Navy designation: S-70C(M)-1 and S-70C(M)-2 Thunderhawk
TYPE: Naval combat helicopter.
PROGRAMME: Naval development of Sikorsky UTTAS (UH-60A Black Hawk) utility helicopter; won US Navy LAMPS Mk III competition for shipboard helicopter in 1977; first flight of first of five YSH-60B prototypes (161169) 12 December 1979; development details in 1982-83 *Jane's*; first 18 SH-60Bs authorised FY82. Changed USN planning in 1993 resulted in premature end to SH-60B/F production; remanufacture planned of SH-60B/Fs and HH-60Hs as SH-60R.

Most recent development, revealed in model form at Asian Aerospace, Singapore, February 2000, involves incorporation of external stores support system (ESSS) of UH-60/S-70A series, as well as SH-60R sensor suite. According to Sikorsky, this proposed version evolved in response to requests from potential customers in and around Pacific Rim, such as Indonesia, Malaysia and Singapore, which all have unsatisfied shipborne helicopter requirements. If proceeded with, new version would have maximum take-off weight of around 11,340 kg (25,000 lb).
CURRENT VERSIONS: **SH-60B**: Initial production version for ASW/ASST; 181 built, excluding prototypes.
Detailed description applies to SH-60B, unless otherwise stated.
NSH-60B: Designation applied to two SH-60Bs (162337 and 162974) assigned to permanent test duties at Patuxent River, Maryland.

SH-60F: CV Inner Zone ASW helicopter, known as CV-Helo, for close-in ASW protection of aircraft carrier groups; US$50.9 million initial US Navy contract for full-scale development and production options placed 6 March 1985; replacing SH-3H Sea King; Seahawk prototype modified as SH-60F test aircraft; first flight 19 March 1987; initial fleet deployment by HS-2 aboard USS *Nimitz* in 1991. Currently assigned to 10 deployable squadrons (HS-2 to HS-8, HS-11, HS-14 and HS-15) plus two training units (HS-1 and HS-10). Production terminated with delivery of 82nd example 1 December 1994. To be phased out of carrier operations; conversions planned to SH-60R.

SH-60F has all LAMPS Mk III avionics, fairings and equipment removed, including cargo hook and RAST system main and tail probes, but installation provisions retained. Replaced by integrated ASW mission avionics including Honeywell AN/AQS-13F dipping sonar, MIL-STD-1553B databus, dual Litton AN/ASN-150 tactical navigation computers and AN/ASM-614 avionics support equipment, automatic flight control system with quicker automatic transition and both cable and Doppler autohover, tactical datalink with other aircraft, communications control system, multifunction keypads and displays for each of four crew members; internal/external fuel system and extra weapon station to port allowing carriage of three Mk 50 homing torpedoes; provision for surface search radar, FLIR, night vision equipment, passive ECM, MAD, air-to-surface missile capability, sonobuoy datalink, chaff/sonobuoy dispenser, attitude and heading reference system (AHRS), Navstar GPS, fatigue monitoring system and increase of maximum T-O weight to 10,659 kg (23,500 lb); secondary missions include SAR and plane guard.
YSH-60F: Designation applied to second production SH-60F (163283) which serves as 'prototype' on test duties at Patuxent River, Maryland. To be fitted with vectored thrust ducted propeller by Piasecki Aircraft Corporation by December 2004 for trials project as part of advanced technology demonstration programme. Cost of modification, testing and flight demonstration estimated at US$26.1 million.
HH-60H: US Navy procurement of 42 achieved in 1996 following resumption of deliveries on 1 December 1994, when first example of follow-on batch of 24 handed over at Stratford, Connecticut; used for strike-rescue/special warfare support (HCS); designated HH-60H in September 1986; first flight (163783) 17 August 1988; accepted by USN 30 March 1989; in service with HCS-4 at Norfolk, Virginia, January 1990; initial procurement ended with 18th delivery July 1991, completing HCS-5 at Point Mugu, California; both squadrons are Reserves. Regular SH-60F squadrons later added pairs of HH-60Hs for deployed duty when embarked aboard aircraft carriers; missions are to recover four-man crew at 250 n miles (463 km; 288 miles) from launch point or fly 200 n miles (371 km; 230 miles) and drop eight SEALs from 915 m (3,000 ft).

Close derivative of SH-60F, with same T700-GE-401C engines and HIRSS as SH-60B/F; equipment includes Litton AN/APR-39A(XE)2 RWR, Raytheon AN/AVR-2 laser warning receiver, Honeywell AN/AAR-47 missile plume detector, Lockheed Martin AN/ALE-47 chaff/flare dispenser, Sanders AN/ALQ-144 IR jammer, Elbit ANVIS 7 NVG/HUD system and two cabin-mounted M60D 7.62 mm machine guns; provision for weapon pylons; required to operate from decks of FFG-7, DD-963, CG-47 and larger vessels, as well as unprepared sites. Cubic AN/ARS-6 personnel locator system installed from FY91. Some equipped with Indal RAST (recovery assist, secure and traverse) equipment. Armament development authorised October 1991 for installation of Hellfire ASM, 70 mm (2.75 in) rockets and forward-firing guns.
HH-60J Jayhawk: Ordered in parallel with HH-60H; adapted for US Coast Guard medium-range recovery (MRR) role; last of 42 delivered in 1996. First flight (USCG 6001) 8 August 1989; first delivery to USCG (6002 at Elizabeth City CGAS) 16 June 1990; subsequently to Mobile, Traverse City, San Diego, Astoria, San Francisco, Cape Cod, Sitka, Kodiak and Clearwater CGAS. When carrying three 455 litre (120 US gallon; 100 Imp gallon) external tanks, HH-60J can fly out 300 n miles (556 km; 345 miles) and return with six survivors in addition to four-man crew, or loiter for 1 hour 30 minutes when investigating possible smugglers; other duties include law enforcement, drug interdiction, logistics, aids to navigation, environmental protection and military readiness; compatible with decks of 'Hamilton' and 'Bear' class USCG cutters. Equipment includes Honeywell RDR-1300C search/weather radar, AN/ARN-147 VOR/ILS, KDF 806 direction-finder, GPS, Tacan, VHF/UHF-DF, TacNav, dual U/VHF-FM radios, HF radio, IFF, V/U/HF IFF crypto computers, NVG-compatible cockpit, rescue hoist and external cargo hook.
XSH-60J: Japan Maritime Self-Defence Force (JMSDF) placed US$27 million order for two **S-70B-3**s for installation of Japanese avionics and mission equipment; first flights 31 August and early October 1987; 1,007 hour test programme by Japan Defence Agency Technical Research and Development Institute between 1 June 1989 and 7 April 1991 to evaluate largely Japanese avionics for SH-60J, but AN/APS-124 radar.
SH-60J: Mitsubishi (which see) is manufacturing SH-60J Seahawk for JMSDF.
SH-60R Strikehawk: Remanufactured Seahawk, also known as LAMPS Block II; combines SH-60B capabilities with dipping sonar of SH-60F; original plan anticipated first two conversions to be funded in FY98; 15 in FY99 and more thereafter; however, concerns over cost led to one year delay in launch of remanufacture programme, which began in FY00 with initial batch of four helicopters (ordered 25 April 2000), to be followed by four more in FY01 and eight in FY02. Thereafter, rate of conversion will increase to 24-27 per year. First conversion due for delivery to USN in 2001; HSL-41 at North Island, California, will be first squadron to receive SH-60R.

US Navy Helicopter Master Plan calls for all 170 existing SH-60Bs and 67 SH-60Fs to be converted by 2011, total of 188 planned, comprising entire SH-60B fleet and initial tranche of 18 SH-60Fs, leaving 59 SH-60Fs and 42 HH-60Hs in original configuration. Ultimately, remaining SH-60Fs to be upgraded and some HH-60Hs

SEAHAWK DELIVERIES

Year	SH-60B series							SH-60F series	HH-60H	HH-60J
	USN	Australia	Greece	Japan	Spain	Taiwan	Thailand	USN		
1983	9									
1984	27									
1985	24			2						
1986	24									
1987	22							2		
1988	16	(14)			6			10	8	
1989	8	4						19	4	6
1990	6	6				2		18	6	10
1991	6	6				8		17		12
1992	6							9		7
1993	9							7	2	4
1994	14		2						14	2
1995	8		3						8	1
1996	2									
1997			1			6				
1998			2							
Totals	181	16	8	2	6	10	6	82	42	42

Notes: Kits in parentheses; NOT to be included in totals (appear also under year of completion). No deliveries in 1999

SIKORSKY—AIRCRAFT: USA

Trials Sikorsky SH-60B deploying Raytheon/Thomson Marconi Sonar AN/ASQ-22 array destined for SH-60Rs *(L-3 Communications)*

may also be included in the remanufacture programme. Project will begin with SH-60B remanufacture which is expected to continue until FY09, with SH-60F upgrade starting in FY04 and running until FY12; in each case, Service Life Extension Program (SLEP) will double projected life to 20,000 flight hours. SH-60R systems to be orientated towards littoral warfare operations, with ability to process and prosecute large number of air and sea contacts in a comparatively confined space, the latter in relatively shallow water. New systems to enhance countermeasures and passive and active detection capability to be added; however, initial upgrade package abandoned on grounds of high cost in 1998, when less costly programme, making extensive use of COTS technology, was adopted.

Lockheed Martin secured US$61 million contract in third quarter of 1998 for development of common cockpit prototype applicable to SH-60R and CH-60S variants. Under terms of two-year contract, Lockheed Martin providing flight instrument displays, two MFDs, two operator keysets and digital communications suite as well as Litton integrated INS/GPS, mass memory unit, mission and flight management computers and applicable operational software for both versions. New 'glass cockpit' centred around Lockheed Martin-developed computer systems, but using use commercial PowerPC processors, with data presented to pilots via electronic flight instrument display and multifunction mission display.

Other changes to be made on SH-60R include deletion of MAD, addition of AGM-114 Hellfire anti-armour missile and two additional stores stations, databus, Telephonics AN/APS-147 multimode radar, Raytheon/Thomson-Marconi Sonar AN/AQS-22 (FLASH) advanced airborne low-frequency sonar, AN/AYK-14 mission processor and AN/UYS-2A enhanced modular signal processor, Lockheed Martin AN/ALQ-210 ESM, Raytheon AN/AAS-44 FLIR/laser ranger and NVG compatibility. MTOW expected to rise to 10,659 kg (23,500 lb). May also eventually be fitted with new, more powerful engine arising from Common Engine Program, currently in early stages of development.

Two SH-60Bs (162976 and 162977) allocated to serve as SH-60R prototypes; conversion undertaken at Owego, New York, where first was rolled out on 5 August 1999. First flight scheduled for October 1999, following electronic systems functional test and checkout on the ground, but was delayed until 11 December; first prototype half analogue/half 'glass cockpit' for initial testing, with full 'glass cockpit' installed after about three months. After initial trials at Owego, first prototype delivered to Patuxent River, Maryland, in early May 2000 for start of two-year Navy/contractor developmental test programme.

Further two development aircraft, to be remanufactured by Sikorsky, are expected to be delivered to Patuxent River in 2001.

Exports comprise: S-70B-1: Spanish Navy received six from December 1988 (designated **HS.23**) for operation from four FFG-7 frigates by Escuadrilla 010 at Rota; similar to USN SH-60B, but with Honeywell AN/AQS-13F dipping sonar. Spanish government approval to order additional six granted in December 1998, with order placed in third quarter 2000. Cost of deal is US$77.4 million, which includes funds to upgrade original six to same standard, including armament kits and compatibility with AGM-114 Hellfire missile.

S-70B-2: Royal Australian Navy selected Seahawk for role adaptable weapon system (RAWS) full-spectrum ASW helicopter with autonomous operating capability; order for eight confirmed 9 October 1984; eight more ordered May 1986; operates from RAN 'Adelaide' class (FFG-7) guided missile frigates; first flight (N7265H, now N24-001) at West Palm Beach 4 December 1987; 14 originally planned to be assembled in Australia, but announced late 1988 Sikorsky would assemble first eight; ASTA in Australia designated to assemble remainder in early 1989; first S-70B-2 handed over in USA 12 September 1989; formal acceptance at Nowra, New South Wales, 4 October 1989; Seahawk Introduction and Transition Unit became HS-816 at Nowra, 1 June 1991; formed first ship's flight, September 1991; squadron formally commissioned 23 July 1992. S-70B-2 has substantially different avionics from USN version: Racal Super Searcher radar (capable of tracking 32 surface targets) and Rockwell Collins advanced integrated avionics including cockpit controls and displays, navigation receivers, communications radios, airborne target handoff datalink and tactical data system (TDS); final delivery 11 September 1991. Upgrade of Australian Seahawks, known as Project Sea 1405, expected to include installation of Raytheon AN/AAQ-27 FLIR and an electronic warfare support measures package based on Elisra's AES-210 system; will also entail installation of Smiths NVG-compatible aircraft standby attitude indicators and Northrop Grumman AN/AAR-54(V) passive MAWS. Total of 16 Seahawks to be upgraded; first helicopter to undergo 32 month upgrade handed over to Tenix Defence Systems in first quarter 2000, with total cost expected to be US$63 million.

S-70B-6: Hybrid SH-60B/F for Greece, unofficially known as **Aegean Hawk**; selected December 1991 and initial quantity of five ordered 17 August 1992 for MEKO 200 frigates, with three options subsequently being converted to firm order and contract for further two being signed on 12 June 2000. Armament includes NFT Penguin Mk2 ASMs; avionics include AN/AQS-18(V)-3 dipping sonar, AN/APS-143(V3) radar and AN/ALR-66(V)-2 ESM; towed MAD and sonobuoy launcher omitted. First two delivered fourth quarter of 1994, with three more in 1995, one in 1997 and two in 1998.

S-70B-7: Six Seahawks ordered by Royal Thai Navy in October 1993; equipped for coastal surveillance, maritime patrol and SAR from aircraft carrier HTMS *Chakri Naruebet*; first handed over at Stratford on 6 March 1997, with all six delivered by June.

S-70B-28: Initial batch of four ordered by Turkish Navy on 14 February 1997, with option on another four subsequently converted to firm order, following negotiations that began in third quarter of 1997; at beginning of 2000, Turkish Navy announced plans to obtain another eight. L-3 Communications Ocean Systems HELRAS long-range dipping sonar to be installed. Original order includes supply of AGM-114 Hellfire ASM, with deliveries to begin in late 2000 for service aboard MEKO 200 frigates in ASW and surveillance roles. Turkey has ultimate requirement for up to 32 S-70Bs and is also considering acquisition of Exocet anti-ship missile.

S-70C(M)-1/2 Thunderhawk: Delivery began July 1990 to Taiwanese Navy of 10 SH-60B Seahawks; given S-70C(M)-1 designation; operated from Hualien by 701

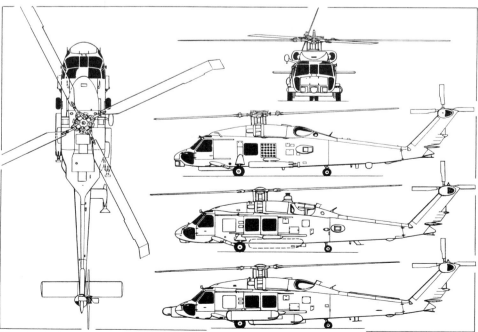

Sikorsky SH-60B Seahawk twin-turbine ASW/ASST helicopter, with additional side views of HH-60H (centre) and HH-60J Jayhawk (bottom) *(Dennis Punnett/Jane's)*

Squadron; shipboard deployment from 1993 aboard six FFG-7 frigates. Equipment includes Honeywell AN/AQS-18(V) dipping sonar, Telephonics AN/APS-128PC radar and Litton AN/ALR-606(V)-2 ESM integrated with radar antenna; no MAD. At least two of original batch modified for sigint gathering; large antenna array positioned on top of main rotor assembly. Sigint system used to intercept HF and VHF radio communications and may also possess jamming capability. Second order, for 11 S-70C(M)-2 version, placed on 26 June 1997.

CUSTOMERS: Total US Navy requirement originally 260 SH-60Bs; 186 on order, including five prototypes, when procurement prematurely terminated in FY94. First flight production Seahawk 11 February 1983; last SH-60B delivered to US Navy on 25 September 1996; first squadron was HSL-41 at NAS North Island, San Diego, California; operational deployment began 1984; 10 US Navy squadrons operating by March 1991 (HSLs 41, 43, 45, 47 and 49 at NAS North Island; 40, 42, 44, 46 and 48 at NAS Mayport, Florida); subsequently HSL-51 formed at Atsugi, Japan, 1 October 1991, and HSL-37 at NAS Barbers Point, Hawaii, began converting from SH-2Fs on 6 February 1992; SH-60Bs deployed in 'Oliver Hazard Perry' (FFG-7) class frigates, 'Spruance' class and Aegis equipped destroyers and 'Ticonderoga' class guided missile cruisers. US Navy originally required 150 SH-60Fs; total 82 completed, comprising seven preseries plus 18 each in FY88, 89 and 91, 12 in FY92 and nine in FY93; procurement then prematurely halted; two used for operational evaluation; in West Coast service with HS-2, 4, 6, 8, 10 and 14 squadrons at NAS North Island, California; HS-3 at Jacksonville, Florida, equipped from 27 August 1991 as first East Coast squadron, followed by HS-1, 5, 7, 11 and 15, of which training squadron HS-1 since disestablished, leaving HS-10 of Pacific Fleet to conduct all US Navy SH-60F instruction.

Exported to Australia, Greece, Japan, Spain, Taiwan, Thailand and Turkey (see Current Versions). S-70B-4 and -5 are derivatives of SH-60F and HH-60H, respectively; not taken up.

COSTS: US$20.25 million (1992) USN programme unit cost. Flyaway cost of final USN SH-60B about US$16 million; total SH-60R development programme costs expected to be around US$400 million, with unit flyaway cost US$16 million to US$18 million (FY96 dollars). Total cost of SH-60R programme stated to be US$2.5 billion (2000). Latest order for two aircraft from Greece valued at US$107 million.

DESIGN FEATURES: SH-60B Seahawk designed to provide all-weather detection, classification, localisation and interdiction of surface ships and submarines, either controlled through datalink from mother ship or operated independently; secondary missions include SAR, vertical replenishment, medevac, fleet support and communications relay.

Revised features, compared with UH-60A, include more powerful navalised GE T700-GE-401 engines, additional fuel, sensor operator's station, port-side internal launchers for 25 sonobuoys, pylon on starboard side of tailboom for MAD bird, lateral pylons for two torpedoes or external tanks, chin-mounted ESM pods, sliding cabin door, rescue hoist, electrically actuated blade folding, rotor brake, folding tail, short-wheelbase tailwheel landing gear with twin tailwheels stressed for lower crash impact, DAF Indal RAST recovery assist, secure and traversing for haul-down landings on small decks and moving into hangar, hovering in-flight refuelling system, and emergency flotation system; pilots' seats not armoured. SH-60B gives 57 minutes' more listening time on station and 45 minutes' more ship surveillance and targeting time than LAMPS Mk I.

Initial testing of new Fairey Hydraulics Decklock landing system for S-70B was completed in mid-1995; ensuing one year development programme was expected to lead to manufacture of prototype unit for operational trials. Decklock consists of a pair of steel jaws attached to a two-stage actuator which extends during approach to landing platform; jaws then automatically secure helicopter to deck-installed grid on landing, permitting operation without assistance of deck crew during storm-force weather conditions.

For operation in Gulf during mid-1980s Iran-Iraq war, 25 SH-60Bs fitted with upper and lower Sanders AN/ALQ-144 IR jammers, BAE Systems AN/ALE-39 chaff/flare dispensers, Honeywell AN/AAR-47 electro-optical missile warning, and a single 7.62 mm machine gun in door, for a weight penalty of 169 kg (369.5 lb); seven Seahawks fitted with Raytheon AN/AAS-38 FLIR on root weapon pylon with instantaneous relay to parent ship.

First Block I SH-60B update, introduced in production Lot 9, delivered from October 1991, includes provision for NFT AGM-119 Penguin anti-ship missile, Mk 50 advanced lightweight torpedo, Flightline AN/ARR-84 99-channel sonobuoy receiver (replacing ARR-75), Rockwell Collins AN/ARC-182 V/UHF FM radio and Rockwell Collins Class 3A Navstar GPS; before production cutbacks, 115 Penguin-capable Seahawks to come from retrofitting back to Lot 5, but only 28 launch kits (delivered 1997) so far ordered.

FLYING CONTROLS: As for UH-60.

STRUCTURE: Basically as for UH-60 plus marine corrosion protection; single cabin door, starboard side, narrower than on UH-60.

POWER PLANT: Two 1,260 kW (1,690 shp) intermediate rating General Electric T700-GE-401 turboshafts in early aircraft; 1,342 kW (1,800 shp) T700-GE-401C turboshafts introduced in 1988 and on HH-60H/J. Transmission rating 2,535 kW (3,400 shp). Internal fuel capacity 2,233 litres (590 US gallons; 491 Imp gallons). Hovering in-flight refuelling capability. Two 455 litre (120 US gallon; 100 Imp gallon) auxiliary fuel tanks on fuselage pylons optional (three on HH-60J). Hover IR suppressor subsystem (HIRSS) exhaust cowling fitted to HH-60H.

ACCOMMODATION: Pilot and airborne tactical officer/back-up pilot in cockpit, sensor operator in specially equipped station in cabin. Dual controls standard. Sliding door with jettisonable window on starboard side. Accommodation heated, ventilated and air conditioned.

SYSTEMS: Generally as for UH-60A.

AVIONICS: Refer also to Current Versions for variants other than SH-60B.

Comms: Rockwell Collins AN/ARC-159(V)2 UHF, Rockwell Collins AN/ARC-174(V)2 HF, Hazeltine AN/APX-76A(V) and Honeywell AN/APX-100(V)1 IFF transponders, TSEC/KG-45(E-1) communications security set, TSEC/KY-75 voice security set, Telephonics OK-374/ASC communications system control group. Satellite communications planned for SH-60R.

Radar: Raytheon AN/APS-124 search radar (Racal Super Searcher for Australia; Telephonics AN/APS-128PC for Taiwan).

Flight: Rockwell Collins AN/ARN-118(V) Tacan, Northrop Grumman AN/APN-127 Doppler, Rockwell Collins AN/ARA-50 UHF DF, Honeywell AN/APN-194(V) radar altimeter.

Mission: Sikorsky sonobuoy launcher, Flightline AN/ARR-75 and R-1651/ARA sonobuoy receiving sets (AN/ARR-84 receiver in Australian Seahawks and for USN Block 1 upgrade), Raytheon AN/ASQ-81(V)2 towed MAD (CAE AN/ASQ-504(V) internal MAD in Australian Seahawks), Raymond MU-670/ASQ magnetic tape memory unit, Astronautics IO-2177/ASQ altitude indicator, Fairchild AN/ASQ-164 control indicator set, Fairchild AN/ASQ-165 armament control indicator set, IBM AN/UYS-1(V)2 Proteus acoustic processor (Computing Devices UYS-503 for Australia) and CV-3252/A converter display, GD Information AN/AYK-14 (XN-1A) digital computer, Raytheon AN/ALQ-142 ESM, Sierra Research AN/ARQ-44 datalink and telemetry (Rockwell Collins DHS-901 in Australian Seahawks). SH-60F has Honeywell AN/AQS-13F dipping sonar (AN/AQS-18 in Taiwanese S-70s). During 1991 Gulf War, pod-mounted Hughes (now Raytheon) AN/AAQ-16 FLIR fitted to five SH-60Bs and Raytheon AN/AAQ-17 FLIR deployed on one SH-60B; BAE Systems Sea Owl IR turret evaluated later in 1991; Raytheon FLIR/laser designator tested on SH-60B in early 1994; option to procure 90 systems by 1997. Australian examples also acquired AN/AAQ-16 FLIR for 1991 Gulf War. Raytheon AN/AAQ-27 FLIR system to be adopted as part of planned upgrade to be undertaken by Tenix Defence Systems. US Navy to acquire airborne laser mine detection system for SH-60B/R; Northrop Grumman system selected in mid-2000, with 36 month engineering and manufacturing development (EMD) to begin in 2001 and production decision to be taken before the end of 2004.

Self-defence: ESM systems include Raytheon AN/APR-39 RWR on HH-60H; none on SH-60F. Australian Seahawks fitted with AN/ALE-47 chaff/flare dispensers and AN/AAR-47 missile detectors for 1991 Gulf War. Upgrade of Australian Seahawks expected to include adoption of ESM based on Elisra AES-210 system.

EQUIPMENT: External cargo hook (capacity 2,722 kg; 6,000 lb) and rescue hoist (272 kg; 600 lb) standard. SH-60B, F and R have provision for eight sonobuoys.

ARMAMENT: US Navy armament includes two Mk 46 torpedoes and (IOC 1993) NFT AGM-119B Penguin Mk 2 Mod 7 anti-shipping missiles. Block I upgrade integrates Penguin and Honeywell Mk 50 Advanced Lightweight Torpedo from 1993. HH-60H has two pintle-mounted M60D machine guns and cleared in 1996 to operate with AGM-114 Hellfire ASM; Navy to purchase sufficient Hellfire systems – comprising FLIR, laser designator and missile launcher – to equip four SH-60B/HH-60H helicopters on each of US east and west coasts. HH-60H also to receive 70 mm (2.75 in) rocket pods and forward-firing guns. Hellfire to be included in SH-60B/R armament for attacking small ships.

DIMENSIONS, EXTERNAL: As UH-60A except:

Length overall, rotors and tail pylon folded:
SH-60B 12.47 m (40 ft 11 in)
HH-60H 12.51 m (41 ft 0⅝ in)
HH-60J 13.13 m (43 ft 0⅞ in)
Length of fuselage: HH-60J 15.87 m (52 ft 1 in)
Width, rotors folded 3.26 m (10 ft 8½ in)
Height:
overall, tail rotor turning 5.18 m (17 ft 0 in)
overall, pylon folded 4.04 m (13 ft 3¼ in)
to top of rotor head 3.79 m (12 ft 5⅜ in)
Wheelbase 4.83 m (15 ft 10 in)
Tail rotor ground clearance 1.83 m (6 ft 0 in)
Main/tail rotor clearance 6.6 cm (2⅝ in)

AREAS: As UH-60A

WEIGHTS AND LOADINGS:
Weight empty: SH-60B ASW 6,191 kg (13,648 lb)
HH-60H 6,114 kg (13,480 lb)
HH-60J 6,086 kg (13,417 lb)
Useful load: HH-60J 3,551 kg (7,829 lb)
Internal payload: HH-60H 1,860 kg (4,100 lb)
Mission gross weight:
SH-60B ASW 9,182 kg (20,244 lb)
SH-60B ASST 8,334 kg (18,373 lb)
Max T-O weight:
SH-60B Utility, HH-60H 9,926 kg (21,884 lb)
SH-60F, SH-60R 10, 659 kg (23,500 lb)
HH-60J 9,637 kg (21,246 lb)
Max disc loading:
SH-60B Utility, HH-60H 47.2 kg/m² (9.67 lb/sq ft)
SH-60F, SH-60R 50.7 kg/m² (10.39 lb/sq ft)
HH-60J 45.9 kg/m² (9.39 lb/sq ft)
Transmission loading at max T-O weight and power:
SH-60B Utility, HH-60H 3.92 kg/kW (6.44 lb/shp)
SH-60F, SH-60R 4.21 kg/kW (6.91 lb/shp)
HH-60J 3.80 kg/kW (6.25 lb/shp)

PERFORMANCE:
Cruising speed at S/L:
HH-60H 147 kt (272 km/h; 169 mph)
HH-60J 146 kt (271 km/h; 168 mph)
Dash speed at 1,525 m (5,000 ft), tropical day:
SH-60B 126 kt (234 km/h; 145 mph)
Vertical rate of climb at S/L, 32.2°C (90°F):
SH-60B 213 m (700 ft)/min
Vertical rate of climb at S/L, 32.2°C (90°F), OEI:
SH-60B 137 m (450 ft)/min

UPDATED

One of four S-70C-6 Super Blue Hawks delivered to Taiwan in 1998 ***1999***/0051845

SIKORSKY S-70C

Republic of China Air Force name: Super Blue Hawk

TYPE: Multirole medium helicopter.

PROGRAMME: H-60 for customers not qualifying for FMS purchases. See below.

CUSTOMERS: Delivery of 24, designated S-70C-2, with nose radar to People's Republic of China completed December 1985 but offered for sale in 1992 (apparently without success) due to spares embargo; 14 to Republic of China Air Force, Taiwan in 1985-86; one each for GKN Westland (designated **WS-70L**) and Rolls-Royce in UK;

R-R aircraft used as testbed for R-R Turbomeca RTM 322 turboshafts; Westland demonstrator sold to Bahrain in 1996; two VIP-configured S-70C-14s for Brunei delivered 1986 but replaced by S-70A-37s in 2000; all based on Black Hawk. **S-70C(M)-1/2** designation and name Thunderhawk assigned to 10 Seahawks for Taiwan (see SH-60B entry) which ordered second Thunderhawk batch of 11 helicopters in 1997; first three of the follow-on batch delivered to Taiwan in mid-2000.

Further four S-70C-6 Super Blue Hawks handed over to Taiwan at Stratford, Connecticut, on 7 April 1998; for use in SAR role by Republic of China Air Force Air Rescue Group.

DESIGN FEATURES: Certified to FAR Pt 21.25; roles include utility transport, external lift, maritime and environmental survey, forestry and conservation, mineral exploration and heavy construction support; powered by General Electric CT7-2C or -2D engines; options include de-icing kit for main and tail rotors, cargo hook for 3,630 kg (8,000 lb) loads, rescue hoist, aeromedical evacuation kit and winterisation kit.

FLYING CONTROLS: As for UH-60.
STRUCTURE: As for UH-60.
POWER PLANT: Two 1,212 kW (1,625 shp) General Electric CT7-2C or 1,285 kW (1,723 shp) CT7-2D turboshafts, or equivalent military T700s. Rolls-Royce Turbomeca RTM 322 in demonstrator G-RRTM. Combined transmission rating (continuous) 2,334 kW (3,130 shp). Maximum fuel capacity 1,370 litres (362 US gallons; 301 Imp gallons).
ACCOMMODATION: Flight deck crew of two, with provision for 12 passengers in standard cabin configuration or up to 19 passengers in high-density layout. Forward-hinged door on each side of flight deck for access to cockpit area. Large rearward-sliding door on each side of main cabin.

DIMENSIONS, INTERNAL:
Cabin: Length	3.84 m (12 ft 7 in)
Max width	2.34 m (7 ft 8 in)
Max height	1.37 m (4 ft 6 in)
Floor area	8.18 m² (88 sq ft)
Volume	11.0 m³ (387 cu ft)
Baggage compartment volume	0.52 m³ (18.5 cu ft)

WEIGHTS AND LOADINGS:
Weight empty	4,607 kg (10,158 lb)
Max external load	3,630 kg (8,000 lb)
Max T-O weight	9,185 kg (20,250 lb)
Max disc loading	43.7 kg/m² (8.96 lb/sq ft)
Transmission loading at max T-O weight and power	3.94 kg/kW (6.47 lb/shp)

PERFORMANCE:
Never-exceed speed (V_{NE})	195 kt (361 km/h; 224 mph)
Max level speed at S/L	157 kt (290 km/h; 180 mph)
Cruising speed at S/L	145 kt (268 km/h; 167 mph)
Max rate of climb at S/L	615 m (2,020 ft)/min
Service ceiling	4,360 m (14,300 ft)
Service ceiling, OEI	1,095 m (3,600 ft)
Hovering ceiling OGE	1,204 m (3,950 ft)
Range at 135 kt (250 km/h; 155 mph) at 915 m (3,000 ft) with max standard fuel, 30 min reserves	255 n miles (473 km; 294 miles)
Range, max fuel, no reserves	297 n miles (550 km; 342 miles)

UPDATED

SIKORSKY S-76C+

TYPE: Multirole medium helicopter.
PROGRAMME: S-76A announced 19 January 1975; first flight (N762SA) 13 March 1977; certification to FAR Pt 29 in November 1978; deliveries started early 1979; delivery of Mk II began 1 March 1982; S-76B programme initiated October 1983; first flight (N3123U) 22 June 1984; final S-76B delivered December 1997; see 1997-98 and previous *Jane's* for details of these early models.

S-76C announced June 1989; replaced S-76B's P&WC PT6B turboshafts with Arriel engines; first flight 18 May 1990; FAA certification and first deliveries April 1991.

Following manufacture of c/n 760514 in mid-2000, production of S-76C+ fuselages progressively transferred to Aero Vodochody of the Czech Republic under a US$200 million contract, beginning with Sikorsky-built c/n 760515, which was shipped to the Czech Republic for final assembly and scheduled to return to the USA in March 2001. Three further shipsets to be sent from Sikorsky, with first wholly Czech-built fuselage expected to be completed in late 2001. Keystone Helicopter Corporation of West Chester, Pennsylvania, selected in April 2000 as principal completion centre for S-76C+s, following flight testing and certification at Sikorsky's Stratford, Connecticut, facility. Production rate 10 per year in 2000 and 2001, rising to 15 per year from 2002.

Improvements announced at Heli-Expo 2001 in February 2001 for introduction in 2004 include low-noise tail rotor and transmission, 6 per cent increase in take-off power, a fully integrated, 'all-glass cockpit', possibly based on the Honeywell Primus Epic system; Goodrich HUMS; a Differential GPS approach and landing system and passive and active noise/vibration control systems for the cabin.

CURRENT VERSIONS: **S-76C**: Initial production version, with Arriel 1S1 engines, now superseded by S-76C+.

S-76C+: FAA and CAA certification and first delivery, mid-1996; features Arriel 2S1 turboshafts with FADEC for improved single-engine performance and fuel efficiency. *As described.*

H-76 Eagle and S-76N Eagle: Military and naval developments of S-76B; last described in 1997-98 and 1998-99 editions, respectively.

Shadow: One-off modification (N765SA) of S-76; Sikorsky Helicopter Advanced Demonstrator and Operator Workload fitted with nose-mounted single-seat cockpit for US Army Rotorcraft Technology Integration (ARTI) programme; first flight 24 June 1985; used 1990 in Boeing Sikorsky LH First Team programme to flight test night vision pilotage system (NVPS) developed by Martin Marietta for RAH-66 Comanche; details in 1997-98 and earlier *Jane's*.

S-76D: Proposed utility variant to be developed jointly with Mil Helicopters of Russia. New features would include fixed landing gear, Mil-developed composites main rotor blades with electrothermal de-icing, X-configuration quiet tail rotor, aimed at reducing manufacturing costs and increasing payload and performance. New rotors would be offered as retrofit to existing S-76 models. If proceeded with, the S-76D would be produced in Russia for domestic and international markets.

CUSTOMERS: See table. The 500th S-76 was registered (N1WB) in May 1999. Deliveries in 1999 included at least one each to Japan and UK. Other recent customers include Palm Beach County Health Care Service, which has taken delivery of two in 'Trauma Hawks' EMS configuration; Copterline of Finland, which took delivery of two S-76C+s in March 2000 for scheduled services between Helsinki and Tallinn, Estonia; Toho Air Service of Japan, which has ordered one S-76C+ for delivery in second quarter 2001 for operations around six island destinations from its base on Hachijo Island; and Chinese Ministry of Communications, which ordered two SAR helicopters in December 2000 to be based at Shanghai. Additionally, UK Sikorsky distributor Air Hanson awarded a 10-year contract, beginning April 1998, to provide helicopter transportation for the British Royal Family, using S-76C+ G-XXEA, which replaces two Westland Wessex HCC. Mk 4s formerly operated by the RAF.

COSTS: Utility configuration US$6 million, Executive configuration US$7.7 million (1998). Direct operating cost US$69 per hour.

DESIGN FEATURES: Meets FAR Pt 29 with Category A IFR; intended for offshore support, business transport, medical evacuation and general utility use; technology and aerodynamics based on those of UH-60 Black Hawk.

Four-blade main rotor with high twist and varying section and camber based on Sikorsky SC-1095; tapered blade tip has 30° leading-edge sweep; fully articulated rotor head with single elastomeric bearings; hydraulic drag dampers; dual bifilar vibration absorber assemblies above rotor head; four-blade cross-beam tail rotor on port side; rotor brake optional. Main rotor blade life of 28,000 hours; tail rotor life unlimited. Optional manual blade folding.

Emergency medical service installation includes multiple-position pivoting primary patient litter, a second litter, and track-mounted seats for four attendants, forward and rear oxygen systems, and dual access to external power on ground; cabin volume 4.0 m³ (141 cu ft).

FLYING CONTROLS: Dual powered hydraulic controls with autostabilisation and autopilot; releasable spring-centring trim system for both cyclic and collective controls.

STRUCTURE: Main rotor blades have formed and welded titanium oval-section tubular main spar with Nomex honeycomb aerofoil core, glass fibre composites outer skin and titanium/nickel leading-edge abrasion strips. Fuselage has bonded aluminium/honeycomb skin panels to permit flush riveting; extensive use made of carbon fibre (replacing Kevlar and glass fibre in earlier S-76s). Aerospace Industrial Development Corporation (AIDC) of Taiwan signed five-year contract in December 1998 for supply of composites crew and passenger doors.

LANDING GEAR: Hydraulically retractable tricycle type, with single wheel on each unit. Nosewheel retracts rearward, main units inward into rear fuselage; all three wheels are enclosed by doors when retracted. Mainwheel tyres size 14.5×5.5-6 (14 ply) tubeless, pressure 11.38 bars (165 lb/sq in); nosewheel tyre size 5.00-4 (12 ply) tubeless, pressure 9.31 bars (135 lb/sq in). Hydraulic brakes; hydraulic mainwheel parking brake. Non-retractable tricycle gear, with low-pressure tyres and skid landing gear both optional.

POWER PLANT: Two Turbomeca Arriel 2S1 turboshafts with FADEC, each rated at 638 kW (856 shp) for take-off, 731 kW (980 shp) OEI for 30 seconds and 592 kW (794 shp) maximum continuous; transmission rating 1,193 kW (1,600 shp) for take-off and maximum continuous. Standard usable fuel capacity 1,064 litres (281 US gallons; 234 Imp gallons); optional auxiliary tank capacity 405 litres (107 US gallons; 89.1 Imp gallons). Self-sealing tanks optional.

S-76 DELIVERIES

Year	S-76A	S-76A+	S-76B	S-76C	S-76C+	Total	Cum
1979	35					35	35
1980	86					86	121
1981	50					50	171
1982	29					29	200
1983	36					36	236
1984	30					30	266
1985	8		11			19	285
1986	3		13			16	301
1987	3		7			10	311
1988	2		9			11	322
1989		3	14			17	339
1990	1	13	5			19	358
1991		1	8	10		19	377
1992			4	8		12	389
1993	1		3	7		11	400
1994			6	7		13	413
1995			6	10		16	429
1996			11	1	7	19	448
1997			4		14	18	466
1998					16	16	482
1999					7	7	489
Totals	**284**	**17**	**101**	**43**	**37**	**489**	

Note: Some S-76As retrofitted to A+

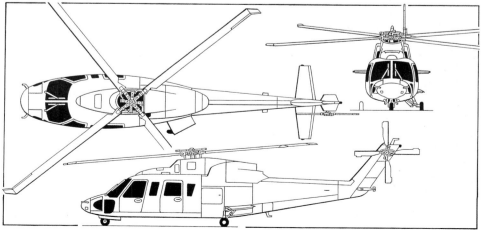

Sikorsky S-76C+ commercial transport helicopter (*Mike Keep/Jane's*)

USA: AIRCRAFT—SIKORSKY

ACCOMMODATION: One or two pilots and 12 to 13 passengers. Three four-abreast rows of seats, floor-mounted at a pitch of 79 cm (31 in). A number of executive layouts are available, including a four-passenger 'office in the sky' configuration. Executive versions have luxurious interior trim, full carpeting, special soundproofing, radiotelephone and co-ordinated furniture. Dual controls optional. Two large doors on each side of fuselage, hinged at forward edges; sliding doors available optionally. Baggage hold aft of cabin, with external door each side of fuselage. Cabin heated and ventilated; air conditioning optional. Windscreen demisting and dual windscreen wipers. Windscreen heating and external cargo hook optional.

SYSTEMS: Hydraulic system at pressure of 207 bars (3,000 lb/sq in) supplied by two pumps driven from main gearbox. Hydraulic system maximum flow rate 15.9 litres (4.2 US gallons; 3.5 Imp gallons)/min. Bootstrap reservoir. Pump head pressure 3.45 bars (50 lb/sq in). In VFR configuration, electrical system comprises two 200 A DC starter/generators and a 24 V 17 Ah Ni/Cd battery. In IFR configuration, system comprises gearbox-driven 7.5 kVA generator, and a 115 V 600 VA 400 Hz static inverter for AC power. 34 Ah battery optional. Engine fire detection and extinguishing system.

AVIONICS: *Comms:* Honeywell Primus II integrated radios.
Radar: Honeywell Primus 440 colour weather radar.
Flight: Honeywell SPZ-7600 dual digital AFCS. Optional Universal UNS-1D navigation management system, L-3 Com/Sextant TCAS and Argus moving map display.
Instrumentation: Four-tube Honeywell colour EFIS; Parker Hannifin Gull Electronics active colour matrix LCD three-screen integrated instrument display system (IIDS).

EQUIPMENT: Standard equipment includes cabin fire extinguishers; cockpit, cabin, instrument, navigation and anti-collision lights; landing light; external power socket; and utility soundproofing. Optional equipment includes 1,497 kg (3,000 lb) capacity cargo hook, 272 kg (600 lb) capacity rescue hoist, emergency flotation gear, engine air particle separators and litter installation.

DIMENSIONS, EXTERNAL:
Main rotor diameter	13.41 m (44 ft 0 in)
Main rotor blade chord	0.39 m (1 ft 3½ in)
Tail rotor diameter	2.44 m (8 ft 0 in)
Tail rotor blade chord	0.16 m (6½ in)
Length: overall, rotors turning	16.00 m (52 ft 6 in)
fuselage	13.21 m (43 ft 4 in)
Height overall, tail rotor turning	4.42 m (14 ft 6 in)
Tailplane span	3.05 m (10 ft 0 in)
Width of fuselage	2.13 m (7 ft 0 in)
Wheel track	2.44 m (8 ft 0 in)
Wheelbase	5.00 m (16 ft 5 in)
Tail rotor ground clearance	1.97 m (6 ft 5¾ in)

DIMENSIONS, INTERNAL:
Passenger cabin: Length	2.46 m (8 ft 1 in)
Max width	1.93 m (6 ft 4 in)
Max height	1.35 m (4 ft 5 in)
Floor area	4.18 m² (45.0 sq ft)
Volume	5.8 m³ (204 cu ft)
Baggage compartment volume	1.1 m³ (38 cu ft)

AREAS:
Main rotor disc	141.26 m² (1,520.5 sq ft)
Tail rotor disc	4.67 m² (50.27 sq ft)
Tailplane	2.00 m² (21.5 sq ft)

WEIGHTS AND LOADINGS:
Operating weight empty	2,545 kg (5,611 lb)
Weight empty, executive configuration	3,691 kg (8,138 lb)
Useful load	1,616 kg (3,562 lb)
Max fuel	3,030 kg (6,680 lb)
Max T-O weight	5,307 kg (11,700 lb)
Max disc loading	37.6 kg/m² (7.69 lb/sq ft)
Transmission loading at max T-O weight and power	4.45 kg/kW (7.31 lb/shp)

PERFORMANCE:
Max level speed	155 kt (287 km/h; 178 mph)
Long-range cruising speed	140 kt (259 km/h; 161 mph)
Max rate of climb at S/L	495 m (1,625 ft)/min
Service ceiling	3,871 m (12,700 ft)
Service ceiling, OEI	975 m (3,200 ft)
Hovering ceiling: IGE	1,722 m (5,650 ft)
OGE	549 m (1,800 ft)

Range at 140 kt (259 km/h; 161 mph) at 1,220 m (4,000 ft) with standard fuel:
no reserves	439 n miles (813 km; 505 miles)
30 min reserves	385 n miles (713 km; 443 miles)

UPDATED

Sikorsky S-76C+ twin-turbine helicopter in medical evacuation configuration *2001*/0103599

SIKORSKY S-92A HELIBUS

TYPE: Medium transport helicopter.

PROGRAMME: Announced March 1992; originally envisaged as S-92C 'Growth Hawk' development of S-70; market evaluation co-ordinated with Mitsubishi Corporation and Mitsubishi Heavy Industries. Launched at Paris Air Show, June 1995 as S-92A and military S-92IU (international utility). Risk-sharing partners Mitsubishi Heavy Industries (7.5 per cent), Jingdezhen Helicopter Group of China (2 per cent) and Gamesa of Spain (7 per cent), with Taiwan Aerospace (6.5 per cent) and Embraer (4 per cent) as additional fixed-price supplier/partners; Russia's Mil is associated with programme, but not yet a full partner; other suppliers include Aerazur (fuel cells), Goodrich (health and usage monitoring system), Dunlop (engine inlet), Hamilton Sundstrand (automatic flight control system and active vibration computers), Honeywell (radar and APU), Lucas-Western (rescue hoist), Martin-Baker (crew seats), Messier-Bugatti (wheels and brakes), Moog (active vibration controls), Parker Bertea (servos), Rockwell Collins (multifunctiion displays and nav/com suite); Universal (FMS) and Vickers (hydraulic pumps). Programme rationalisation resulted in discontinuation of S-92IU and adoption of standard civil/military configuration.

Design changes announced in July 2000 in response to customer requests include a 0.41 m (1 ft 4 in) increase in cabin length to permit installation of a 1.27 m (4 ft 2 in) wide cabin door to improve hoisting capability and to accommodate a Stokes litter during SAR operations; reduction in height of tail rotor pylon by about 1.02 m (3 ft 4 in) to offset additional weight of cabin extension; and relocation of the horizontal stabiliser. These changes will provide additional benefits in creating an improved fold configuration for shipboard operations, increased birdstrike protection deriving from the relocation of tail rotor drive shaft and controls aft of the tail spar, and a flatter hover attitude arising from a forward shift of the helicopter's centre of gravity, improving visibility for confined space and shipboard landing, and increasing aft fuselage ground clearance. The revised configuration will be incorporated from the third prototype and on all production aircraft.

Test programme comprises one ground test vehicle (PA1/GTV; first airframe built, first ground run 4 September 1998 and completed 200-hour FAA certification endurance run in September 1999) and four flying prototypes, of which first (PA2/N292SA) completed maiden flight at Sikorsky's Development Flight Center in Florida on 23 December 1998. GTV and first flying prototype have 1,305 kW (1,750 shp) CT7-6D engines; remaining three flying prototypes have production-standard CT7-8s and APUs. PA3/N392SA (second flying) joined the test programme in May 1999, flew in October, and is devoted principally to engine and AFCS development. PA4/N492SA incorporated airframe changes described above and final avionics updates and scheduled to join the test programme, in early 2001. PA5/N592SA, exhibited unflown at the Paris Air Show in June 1999, is in utility configuration with MIL-STD-1553 databus, rear loading ramp, folding seats for 22 troops, sliding windows, cargo hook and provision for 7.62 mm machine gun pintle mounts; was scheduled to fly in late 2000.

By October 2000, PA2 had flown 230 hours, and PA3 183 hours, including flights at maximum take-off weight of 13,998 kg (30,860 lb), demonstrating a range of 800 n miles (1,482 km; 921 miles) with mid-mission

Second flying prototype Sikorsky S-92A Helibus *2001*/0103595

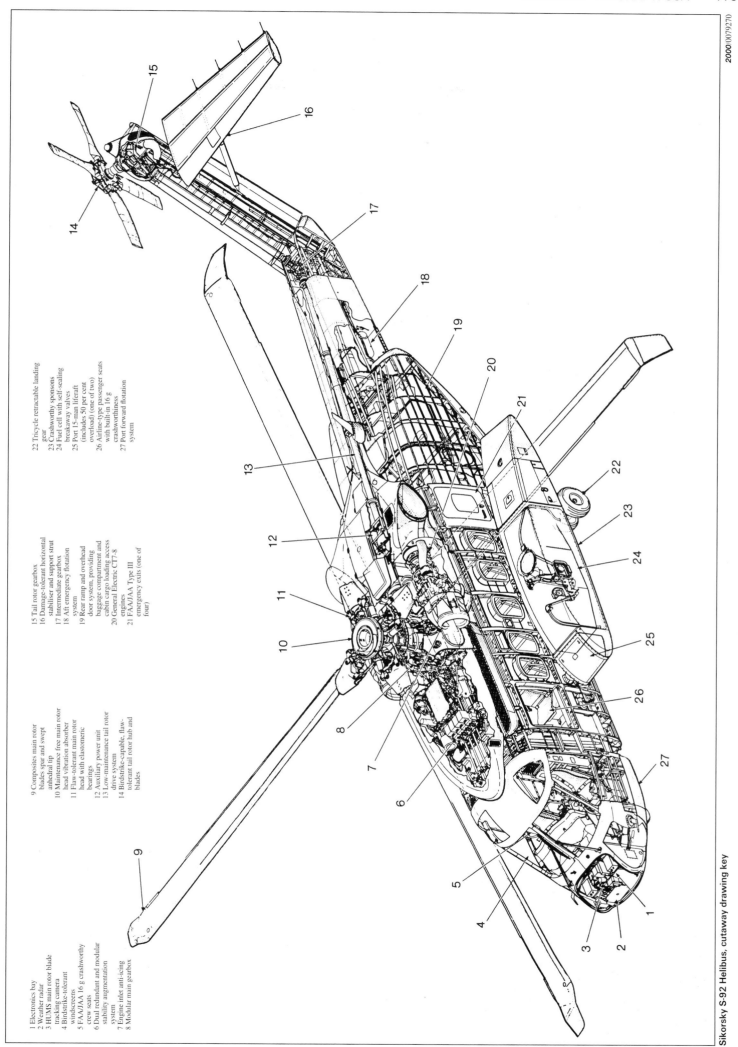

Sikorsky S-92 Helibus, cutaway drawing key

1. Electronics bay
2. Weather radar
3. HUMS main rotor blade tracking camera
4. Birdstrike-tolerant windscreens
5. FAA/JAA 16 g crashworthy crew seats
6. Dual redundant and modular stability augmentation system
7. Engine inlet anti-icing
8. Modular main gearbox
9. Composites main rotor blades spar and swept anhedral tip
10. Maintenance free main rotor head vibration absorber
11. Flaw-tolerant main rotor head with elastomeric bearings
12. Auxiliary power unit
13. Low-maintenance tail rotor drive system
14. Birdstrike-capable, flaw-tolerant tail rotor hub and blades
15. Tail rotor gearbox
16. Damage-tolerant horizontal stabiliser and support strut
17. Intermediate gearbox
18. Aft emergency flotation system
19. Rear ramp and overhead door system, providing baggage compartment and cabin cargo loading access
20. General Electric CT7-8 engines
21. FAA/JAA Type III emergency exits (one of four)
22. Tricycle retractable landing gear
23. Crashworthy sponsons
24. Fuel cell with self-sealing breakaway valves
25. Port 15-man liferaft (includes 50 per cent overload) (one of two)
26. Airline-type passenger seats with built-in 16 g crashworthiness
27. Port forward flotation system

USA: AIRCRAFT—SIKORSKY

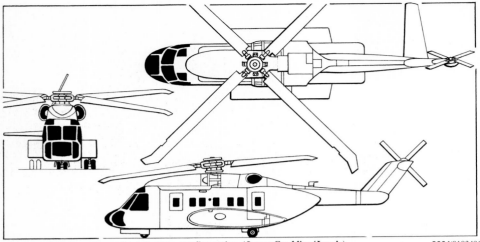

Sikorsky S-92 in envisaged production configuration *(James Goulding/Jane's)*

Interior of the military S-92 mockup

hover. FAR/JAR 29 certification expected in 2002 after 1,400-hour flight test programme, with first customer deliveries shortly thereafter. Target production: five in 2002, rising to 20 per year from 2003.

CUSTOMERS: Launch customer Cougar Helicopter of Eastern Canada ordered up to five for delivery in 2002. Other announced customers include Helijet of Vancouver, Aircontactgruppen AS of Norway (six for delivery between 2004 and 2007) and Copter Action of Finland. Selected in 2000 among final contenders for the Nordic Standard Helicopter Programme (NSHP) contract to provide up to 73 helicopters to Denmark, Finland, Norway and Sweden for SAR, transport and maritime missions. If selected, Patria Finavitec Oy of Finland would perform final assembly, test and some completion operations, and Saab AB of Sweden would provide NSHP systems design and integration.

Sikorsky is teamed with Bombardier Aerospace Defence Services to offer the S-92 as replacement for the Canadian Forces Sea King helicopters. If the S-92 is selected, Lockheed Martin Canada would be responsible for mission system integration, and Bombardier would perform interior completions, installation and checking of mission equipment, exterior painting and final acceptance and delivery, in addition to in-service support and fleet management. S-92 also proposed to Oman for its requirement for up to 50 SAR/troop transport helicopters.

COSTS: Programme US$600 million; unit cost US$13 million to US$14 million; target direct operating cost US$2,113 per hour (1999). Direct maintenance cost US$800 per hour (2000).

DESIGN FEATURES: Design, manufacture and assembly use CATIA database system; modular design simplifies customised configurations. Standard medium helicopter configuration, with engines above cabin; raised tailboom to permit rear loading via ramp; high-set, strut-braced tailplane to port. Prominent sponsons accommodate main landing gear and fuel.

Dynamic components based on proven Sikorsky technology to reduce development risks, including: titanium yoke-type infinite-life main rotor head with elastomeric bearings; four-blade, quick-release, all-composites main rotor with swept, tapered anhedral tips based on scaled-up version of blades tested on Black Hawk in 1995-96; blade chord increased 12 per cent, compared with Black Hawk; damping masses at mid-span of each blade; new transmission based on upgraded four-stage version of Black Hawk's three-stage main gearbox; rotor brake; new intermediate tail rotor gearbox; and new four-blade, fully articulated, birdstrike-tolerant tail rotor with individually removable blades to starboard, meeting FAR/JAR 29 birdstrike requirements. New-style vibration absorber, with Hamilton Sundstrand computer and Moog actuators, on top of rotor hub comprises metal drum enclosing five composites springs allowing absorber to move in opposition to in-plane forces.

FLYING CONTROLS: Similar, but not identical, to Black Hawk; dual digital AFCS with autopilot and dual independent, triple-axis stability augmentation features Hamilton Sundstrand primary processor based on that of RAH-66 Comanche, and is expected to have 8,000 hour MTBF.

STRUCTURE: Modular structure of aluminium and composites (about 40 per cent of structure is of composites, though mostly non-structural to reduce costs) designed to be highly crack-resistant, with extensive lightning/HIRF protection; composites main rotor blades (including spars); structure optimised for minimum parts count.

Sikorsky responsible for rotor and transmission systems, final assembly and flight test. Airframe largely designed and manufactured by 'Team S-92' partners, as follows: Mitsubishi (main cabin), AIDC of Taiwan (flight deck), Embraer (sponson front halves, landing gear and incorporation of Aerazur/Intertechnique fuel system), Gamesa (cabin interior, aft fuselage, tail boom and upper fuselage transmission housing), and Jingdezhen (vertical tail including horizontal stabiliser).

LANDING GEAR: Retractable tricycle; main units retract rearwards into sponsons; nosewheel retracts forwards under flight deck. Wheels and brakes supplied by Messier-Bugatti. Optional inflatable emergency flotation bags, usable up to Sea State 5.

POWER PLANT: Two GE CT7-8 turboshafts, each rated at 1,790 kW (2,400 shp) for T-O and 1,521 kW (2,040 shp) maximum continuous; 1,812 kW (2,430 shp) OEI 30 seconds; 1,790 kW (2,400 shp) OEI 2 minutes; and 1,755 kW (2,353 shp) OEI 30 minutes. Hamilton Sundstrand FADEC. Rolls-Royce Turbomeca RTM322 will be offered as alternative engine if demand is forthcoming. Transmission via modular compound planetary gearbox with 30 minute run-dry capability and 140 per cent overtorque certification, rating 3,117 kW (4,180 shp). Standard fuel totalling 1,136 litres (300 US gallons; 250 Imp gallons) in sponsons, each with gravity fuelling port. Auxiliary fuel options include two external tanks, each 871 litres (230 US gallons; 192 Imp gallons); single or dual 322 litre (85.0 US gallon; 70.8 Imp gallon) bench-type tanks in cabin, or two 700 litre (185 US gallon; 154 Imp gallon) tanks in cabin. Crash-resistant fuel system standard. In-flight refuelling probe optional.

ACCOMMODATION: Two-pilot crew on separate flight deck on FAA/JAA 16 *g* crashworthy seats, 19 passengers, with baggage on rear ramp, or up to three LD-3 cargo containers in civil version. Main door, starboard side, front in three options: (a) split horizontally in upward- and downward-hinged sections, latter including steps; (b) lower airstair section with inward/backward-sliding upper portion (allowing installation of external SAR winch) and (c) full-height sliding door incorporating Type IV emergency exit. Further three FAA/JAA Type IV cabin emergency exits, plus pop-out windows at each seat row, for rapid emergency evacuation. Total of 22 combat-ready troops in military version; both versions have rear-loading ramp. Military version has sliding cabin windows and weapons mounts. Martin-Baker crew and passenger seats. Accommodation is heated and ventilated with independent cockpit and cabin zones; air conditioning optional. Active noise suppression system under consideration.

SYSTEMS: Goodrich health and usage monitoring system standard, with cockpit displays and downloading facility to enable ground crew to access system via hand-held diagnostic equipment; active noise control system will reduce cabin noise by 3 to 4 dB; active vibration control may be employed in key airframe areas. BAE Systems CVR/FDR.

Honeywell 36-150 (S92) APU with in-flight start and continuous run capability. Electrical system similar to that of H-60 series, comprising two 45 kVA main gearbox-driven AC generators and one 35 kVA APU-driven back-up generator with improved generator control unit. Three hydraulic systems, supplied by main gearbox-driven pumps, two serving main and tail rotor and stability augmentation system, while third serves utilities and acts as back-up. Anti-icing system for engine inlets, windscreen and pitot head; main and tail rotor de-icing optional. Automatically deployed emergency flotation system meets or exceeds FAR/JAR stability requirements and has demonstrated Sea State 5 capability; externally mounted 14-person liferaft in forward end of each sponson, with 50 per cent overload capability.

AVIONICS: Open architecture avionics system accommodating ARINC 429 and MIL-STD-1553 interfaces with Rockwell Collins Pro Line 21 as core system.

Comms: Honeywell Primus II VHF.

Radar: Provision for radar in nose compartment.

Flight: UNS-1C FMS; Hamilton Sundstrand automatic flight control system featuring three-axis stability augmentation system and fully coupled dual digital autopilots with automatic approach-to-hover option.

Instrumentation: Rockwell Collins EFIS cockpit with four 203 × 152 mm (8 × 6 in) MFD-268EP active matrix liquid crystal displays (AMLCD) for PFD, EICAS, navigational and weather radar functions; fifth AMLCD, centre-mounted, optional for displaying sensor information such as moving map or FLIR data.

All avionics housed in removable mission equipment rack behind co-pilot station with wiring routed through conduits in fuselage frames for added protection.

EQUIPMENT: Optional hydraulically powered, electrically controlled SAR rescue hoist, maximum capacity 272 kg (600 lb), with cable viewing window and spotlight. Optional cargo handling system, which is compatible with 1.07 × 1.22 m (3 ft 6 in × 4 ft 0 in) pallets, includes 1,814 kg (4,000 lb) capacity centreline winch with 28 V DC winch motor, easy on/off cabin rollers, and cargo tiedowns at all major frames.

S-92 flight deck

S-92 rotor head

SIKORSKY to SINO SWEARINGEN—AIRCRAFT: USA

DIMENSIONS, EXTERNAL:
Main rotor diameter	17.71 m (56 ft 4 in)
Tail rotor diameter	3.35 m (11 ft 0 in)
Length overall, rotors turning	20.85 m (68 ft 5 in)
Fuselage: Length: airframe	17.73 m (58 ft 2 in)
incl tail rotor	18.03 m (59 ft 2 in)
Max width over sponsons	3.89 m (12 ft 9 in)
Height: overall, rotors turning	6.45 m (21 ft 2 in)
to top of main rotor head	4.37 m (14 ft 4 in)
Wheel track (c/l shock-absorbers)	3.17 m (10 ft 5 in)
Wheelbase	5.79 m (19 ft 0 in)
Forward door (stbd): Height	1.85 m (5 ft 11 in)
Width	1.02 m (3 ft 4 in)

DIMENSIONS, INTERNAL:
Cabin: Length	6.10 m (20 ft 0 in)
Max width	2.01 m (6 ft 7 in)
Max height	1.83 m (6 ft 0 in)
Volume	16.9 m³ (596 cu ft)
Baggage volume (civil)	3.1 m³ (110 cu ft)

AREAS:
Main rotor disc	231.55 m² (2,492.4 sq ft)
Tail rotor disc	8.83 m² (95.03 sq ft)

WEIGHTS AND LOADINGS (A: civil, B: military):
Weight empty: A	7,031 kg (15,500 lb)
B	6,895 kg (15,200 lb)
Max hook capacity: B	4,536 kg (10,000 lb)
Max T-O weight:	
internal load: A, B	11,430 kg (25,200 lb)
external load: A, B	12,020 kg (26,500 lb)
Max disc loading:	
internal load	49.4 kg/m² (10.11 lb/sq ft)
external load	51.9 kg/m² (10.63 lb/sq ft)
Transmission loading at max T-O weight and power:	
internal load	3.67 kg/kW (6.03 lb/shp)
external load	3.86 kg/kW (6.34 lb/shp)

PERFORMANCE (at max internal load T-O weight):
Never-exceed speed (V_{NE})	165 kt (305 km/h; 189 mph)
Max cruising speed	155 kt (287 km/h; 178 mph)
Econ cruising speed	140 kt (259 km/h; 161 mph)
Hovering ceiling, A: IGE	3,536 m (11,600 ft)
OGE	2,225 m (7,300 ft)
Range, normal fuel	490 n miles (907 km; 563 miles)

UPDATED

Non-flying first Sikorsky S-92, making a rare public appearance at Farnborough in July 2000 *(Paul Jackson/Jane's)* 2001/0103562

OTHER AIRCRAFT

Description of **RAH-66 Comanche** attack helicopter appears under Boeing Sikorsky in this section.

SINO SWEARINGEN

SINO SWEARINGEN AIRCRAFT COMPANY

1770 Skyplace Boulevard, San Antonio, Texas 78216
Tel: (+1 210) 258 86 99
Fax: (+1 210) 258 39 73
e-mail: ssacsj30@aol.com
Web: http://www.sj30jet.com
CHAIRMAN: Jack Sun
PRESIDENT AND CEO: Jack Braly
VICE-CHAIRMAN: Ed Swearingen
SENIOR VICE-PRESIDENT, OPERATIONS: Ronald D Neal
VICE-PRESIDENTS:
  Paul-David Bartles (Manufacturing)
  Robert E Homan (Quality Assurance)
  Chester J Schickling (Sales and Marketing)
  Alan Schwartz (Procurement)
  T H Tsiang (Programme Management)
  Roger Wilson (Engineering)
  Jim L Wooley (Human Resources)
DIRECTORS:
  John Karamanian (Product Support)
  Mike Potts (Corporate Communications)
  Donnie Rex (Management Information Systems)

Ed Swearingen is well known for designing Fairchild Merlin and Metro commuter and business aircraft, and for engineering such aircraft as Piper Twin Comanche and Lockheed JetStar II; also built prototype SA-32T for Jaffe (1991-92 *Jane's*). Current project is SJ30-2 business jet.

Swearingen Aircraft Inc re-formed on 5 January 1995 as joint venture company with Sino Aerospace of Taiwan, which provides financial support; will certify and produce SJ30-2 at new 8,130 m² (87,500 sq ft) facility at Martinsburg, West Virginia, where the workforce is expected to total 250 to 300 by the end of 2001.

UPDATED

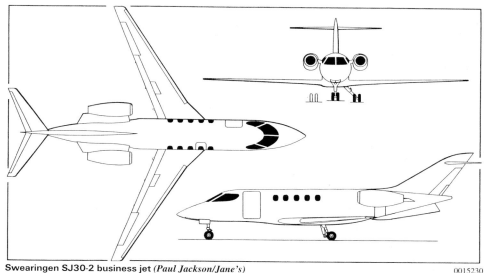

Swearingen SJ30-2 business jet *(Paul Jackson/Jane's)* 0015230

SINO SWEARINGEN SJ30-2

TYPE: Light business jet.
PROGRAMME: Announced 30 October 1986 as SA-30 Fanjet; Gulfstream Aerospace, Williams International and Rolls-Royce announced they were joining programme in October 1988 and aircraft renamed Gulfstream SA-30 Gulfjet; Gulfstream withdrew from programme 1 September 1989; place taken by Jaffe Group of San Antonio, Texas, and aircraft renamed Swearingen/Jaffe SJ30; now a Sino Swearingen project. First flight of prototype (N30SJ) 13 February 1991 (with FJ44-1 engines).

Certification originally intended in 1995 but delayed pending development of increased performance SJ30-2, *as described*, which features fuselage stretched by 1.32 m (4 ft 4 in), wing span increased by 1.83 m (6 ft 0 in), increased wing dihedral, revised wing/fuselage fairing, increased fuel capacity and more powerful FJ44-2A engines. Prototype (N30SJ, modified from SJ30 prototype) first flown (original engines) 8 November 1996.

Swearingen SJ30-2 prototype 2001/0100464

776　USA: AIRCRAFT—SINO SWEARINGEN

Instrument panel of the SJ30-2　　1999/0051851

SJ30-2 executive interior　　2001/0100463

FJ44-2A engines installed and flown for first time on 4 September 1997; total of 200 flight hours accumulated with these engines (or 300 hours overall in SJ30-2 configuration) by mid-March 1999, including operation at speeds up to M0.83 and maximum altitude of 13,105 m (43,000 ft), 1.8 g turns at 12,495 m (41,000 ft), verification of stall and engine-out minimum control speeds, stability and control tests, engine operating tests throughout the flight envelope, simulated icing tests, yaw/damper/rudder bias system testing and handling trials with simulated flight control failures including trim runaways and asymmetric speed brake deployment. N30SJ withdrawn from use in mid-1999.

Static and fatigue tests on a representative section of primary wing joint completed February 1999, to 150 per cent of limit load in 230,000 loading cycles. Risk-sharing partner Gamesa Aeronautica of Vitoria, Spain, manufacturing fuselages and wings of five certification test airframes; first of three production conforming flying prototypes (c/n 002/N138BF) rolled out at San Antonio 17 July 2000 and first flown 30 November 2000; c/n 003 and 004 will also be flying prototypes; c/n TF-2 and TF-3 are static test and fatigue airframes respectively, last-named being the final airframe to be assembled in Texas; assembly of first production airframe scheduled to begin at Martinsburg in fourth quarter 2000. The 1,400 hour flight test programme is scheduled to culminate in FAA certification in FAR Pt 23 Commuter category and first customer deliveries in fourth quarter 2001.

CUSTOMERS: Total of 175 orders placed by customers and distributors in Brazil, Canada, Germany, South Africa, Sweden, United Kingdom and USA by November 2000.

COSTS: US$4.3 million (1999).

DESIGN FEATURES: High-speed and high-altitude cruise with long-range capabilities, plus good slow speed handling; highly efficient swept wing planform.
Proprietary, computer-designed aerofoil sections. Wing sweep 32° 36′, dihedral 2° 30′.

FLYING CONTROLS: Conventional and manual. Electrically actuated variable incidence tailplane, trim tab on rudder, aileron force bias spring for trimming in roll axis; single ventral fin under tail for yaw damping; slotted Fowler trailing-edge flaps actuated electrically; hydraulically actuated full-span leading-edge slats; single electrohydraulically actuated spoiler/lift dumper panel on each wing ahead of flap.

STRUCTURE: All-metal with chemically milled skins on fuselage.

LANDING GEAR: Retractable tricycle type, manufactured by Derlan Inc of Santa Ana, California, with twin wheels on each unit. Trailing-link oleo-pneumatic suspension on main units. Hydraulic actuation, main units retracting inward and rearward into fuselage, nose unit forward. Hydraulically steerable (±60°) nose unit. All wheels 41 cm (16 in) diameter. Main unit tyre pressure 10.76 bars (155 lb/sq in), nosewheel tyre pressure 2.14 bars (31 lb/sq in). Standard air-cooled power braking; dual ABS anti-skid.

POWER PLANT: Two 10.23 kN (2,300 lb st) Williams FJ44-2A turbofans flat rated to 72°C, pod-mounted on pylons on sides of rear fuselage. Three fuel tanks, one in each wing, total 1,393 litres (368 US gallons; 306 Imp gallons), plus fuselage (torsion box) tank containing 1,317 litres (348 US gallons; 290 Imp gallons); combined capacity 2,710 litres (716 US gallons; 596 Imp gallons) under FAA FAR Pt 23 commuter category certification, 2,654 litres (701 US gallons; 584 Imp gallons) under JAA JAR23 certification. Single refuelling point. Oil capacity 3.4 litres (0.9 US gallon; 0.75 Imp gallon) per engine.

ACCOMMODATION: Pilot and one passenger (or co-pilot) on flight deck. Main cabin separated by a bulkhead; up to six passenger seats, each with adjustable reclining backs and retractable armrests, plus two foldaway tables and a computer power outlet; optional refreshment centre at front, and lavatory, washbasin and storage cabinets at rear. Airstair passenger door with light at front on port side. Emergency exit over starboard wing. Baggage compartment aft of main cabin, with external access via port side door under engine nacelle. Two-piece birdproof electrically heated wraparound windscreen.

SYSTEMS: Cabin pressurised to 0.83 bar (12.0 lb/sq in), maintaining sea level pressure to 12,500 m (41,000 ft); cabin heated by engine bleed air; cooled by a vapour-cycle system. Hydraulic system (207 bars; 3,000 lb/sq in) for actuation of leading-edge slats, airbrake/lift dumpers and landing gear extension/retraction. Two 300 A 28 V DC engine-driven starter/generators and static inverters. Redundant frequency-wild alternators provide power for windscreen heating. Wing and engine inlets anti-iced by engine bleed air; tailplane has electroexpulsive de-icing system. Oxygen system has 1,133 litre (40 cu ft) capacity.

AVIONICS: Honeywell Primus Epic CDS as core system.
Comms: Honeywell Primus II radio system; optional 8.33 kHz spacing.
Radar: Honeywell Primus 331 colour weather radar.
Flight: Dual IC-615 integrated avionics computers with combined flight director, autopilot and FMS/GPS functions; dual RVSM-compatible Honeywell AZ-850 micro air data computers; ADF, DME. Options include TCAS 2000, second FMS/GPS integrated with IC-615, second ADF, second DME and lightning sensor system.
Instrumentation: EFIS cockpit comprising three Honeywell DU-180 flat panel 203 × 254 mm (8 × 10 in) active matrix colour LCDs, including two PFDs and one MFD.

DIMENSIONS, EXTERNAL:
Wing span	12.90 m (42 ft 4 in)
Wing chord: at root	2.41 m (7 ft 11 in)
at tip	0.50 m (1 ft 7½ in)
Wing aspect ratio	9.4
Length overall	14.31 m (46 ft 11½ in)
Fuselage: Length	12.79 m (41 ft 11½ in)
Max diameter	1.52 m (5 ft 0 in)
Height overall	4.34 m (14 ft 3 in)
Tailplane span	4.44 m (14 ft 7 in)
Wheel track (mean)	2.09 m (6 ft 10¼ in)
Wheelbase	5.70 m (18 ft 8½ in)
Passenger door: Height	1.42 m (4 ft 8 in)
Width	0.81 m (2 ft 8 in)
Baggage door: Height	0.65 m (2 ft 1½ in)
Width	0.47 m (1 ft 6½ in)
Emergency door: Height	0.56 m (1 ft 10 in)
Width	0.48 m (1 ft 7 in)

DIMENSIONS, INTERNAL:
Cabin: Length:	
between pressure bulkheads	5.36 m (17 ft 7 in)
passenger section	3.81 m (12 ft 6 in)
Max width	1.43 m (4 ft 8½ in)
Max height	1.31 m (4 ft 3½ in)
Floor area	5.46 m² (58.8 sq ft)
Volume	5.4 m³ (190.6 cu ft)
Baggage compartment volume	1.7 m³ (60 cu ft)

AREAS:
Wings, gross	17.72 m² (190.7 sq ft)
Ailerons (total)	0.76 m² (8.20 sq ft)
Trailing-edge flaps (total)	3.31 m² (35.63 sq ft)
Leading-edge flaps (total)	2.83 m² (30.48 sq ft)
Fin	1.06 m² (11.40 sq ft)
Rudder, incl tab	0.69 m² (7.40 sq ft)
Tailplane	3.41 m² (36.70 sq ft)
Elevator, incl tabs	0.79 m² (8.53 sq ft)

WEIGHTS AND LOADINGS: (A: FAR Pt 23 commuter certification; B: JAR23 certification):
Weight empty, equipped: A, B	3,493 kg (7,700 lb)
Operating weight empty: A, B	3,583 kg (7,900 lb)
Max baggage weight: A, B	227 kg (500 lb)
Max payload: A, B	635 kg (1,400 lb)
Max fuel weight: A	2,177 kg (4,800 lb)
B	2,132 kg (4,700 lb)
Max T-O weight: A	5,987 kg (13,200 lb)
B	5,670 kg (12,500 lb)
Max ramp weight: A	6,033 kg (13,300 lb)
B	5,715 kg (12,600 lb)
Max landing weight: A	5,688 kg (12,540 lb)
B	5,670 kg (12,500 lb)

Roll-out of first production-conforming Sino Swearingen SJ30-2 on 17 July 2000　　2001/0100465

Max zero-fuel weight: A, B	4,218 kg (9,300 lb)
Max wing loading: A	338.0 kg/m² (69.22 lb/sq ft)
B	320.0 kg/m² (65.55 lb/sq ft)
Max power loading: A	293 kg/kN (2.87 lb/lb st)
B	277 kg/kN (2.72 lb/lb st)

PERFORMANCE (estimated):
Max operating Mach No. (M_{MO})	0.80
Long-range cruising speed	M0.78 (447 kt; 828 km/h; 514 mph)
Stalling speed, flaps down, flight idle power	91 kt (169 km/h; 105 mph)
Max rate of climb at S/L	1,390 m (4,560 ft)/min
Rate of climb at S/L, OEI	396 m (1,300 ft)/min
Service ceiling	14,935 m (49,000 ft)
Service ceiling, OEI	7,620 m (25,000 ft)
FAA T-O balanced field length: A	1,173 m (3,850 ft)
B	1,103 m (3,620 ft)
FAA landing distance at max landing weight: A, B	1,042 m (3,420 ft)

Range, with 45 min VFR reserves:
A	2,782 n miles (5,152 km; 3,201 miles)
B	2,530 n miles (4,685 km; 2,911 miles)

Range, pilot and three passengers, NBAA IFR reserves:
A	2,500 n miles (4,630 km; 2,877 miles)
B	2,230 n miles (4,130 km; 2,566 miles)

OPERATIONAL NOISE LEVELS (estimated):
T-O	74.7 EPNdB
Approach	93.6 EPNdB
Sideline	85.6 EPNdB

UPDATED

SKYCRAFT

SKYCRAFT INTERNATIONAL INC

6400 Homeworth Road, Homeworth, Ohio 44634
Tel: (+1 330) 862 28 68
Fax: (+1 330) 862 62 62
e-mail: PH6400@aol.com
PRESIDENT: John Hall

SkyCraft purchased rights to Super2 from Highlander Aircraft Corporation of Golden Valley, Minnesota, in 1999. Previously, Highlander had obtained the design from Aviation Scotland in UK and became sole source when co-licensee, ASL Hagfors Aero of Sweden, ceased trading in September 1995.

UPDATED

SKYCRAFT ARV SUPER2

TYPE: Side-by-side ultralight/kitbuilt.
PROGRAMME: Initial production in UK from assembly of Scottish-built kits. Original prototype Island Aircraft ARV-1 Super2 (G-OARV) first flew 11 March 1985; see 2000-01 and earlier *Jane's* for other non-US history.
Design rights acquired by Highlander in USA; company started construction of US prototype in September 1995 and this (N280KT) flew for first time in February 1996. By August 1996, Highlander had sold six kits, two of which were flying by year end. Programme transferred to SkyCraft in 1998.
Description applies to US-built version, unless otherwise stated.
CUSTOMERS: Total of 32 Super2s built by early 1994 (seven by Aviation Scotland). UK production then discontinued. Aircraft c/n 033 to Sweden in October 1994 as SE-KYP for demonstrations. Total of 36 aircraft believed flying by mid-2000.
COSTS: Quick-build kit US$17,950 (2000).
DESIGN FEATURES: Design objective low initial cost and low maintenance cost; engine with aluminium radiator in recessed duct in rear fuselage; superplastically formed aluminium alloy pressings to save weight; shoulder location of wing with forward sweep, low panel line and close cowling of small engine combine to provide optimum view from cabin, particularly in turns. Designed to BCAR Section K and FAR Pt 21.91; intended to be suitable for home building in kit form; quoted build time 900 hours.
Wing section NACA 2415 (modified); 5° forward sweep, 2° dihedral and 3° incidence.
FLYING CONTROLS: Conventional and manual. Mass-balanced ailerons operated by torque tube through leading-edge of manually operated, three-position, plain flaps; mass-balanced rudder and elevator with trim tab.
STRUCTURE: Single-spar aluminium alloy wing, cold bonded and flush riveted. Single uncompromised main bulkhead carries wings, bracing struts, controls, seats, fuel tank and main landing gear; conventional double beam forward of main bulkhead carries firewall, nose landing gear and engine; front fuselage skinning in four panels; rear fuselage of aluminium alloy with single curvature skinning. Cowlings, fairings and wingtips of glass fibre. Tail unit conventional aluminium alloy structure with three-spar fixed surfaces; single-spar rudder and elevator. Airframe corrosion-treated.
LANDING GEAR: Non-retractable tricycle type. Cantilever main legs of tapered steel leaf spring. All three wheels size 3.50×6.0, with tyres size 13×4.00-6, pressure 1.72 bars (25 lb/sq in). Hydraulic single disc brakes. Steerable nose leg has rubber shock rings and gas-filled damper.

ARV Super2 from the original production series *(Paul Jackson/Jane's)*   2000/0062382

POWER PLANT: One 59.6 kW (79.9 hp) Rotax 912 UL flat-four engine driving two-blade propeller. Mid-West rotary engine reinstated as option in 1998, along with 2SI three-cylinder two-stroke and Jabiru 2200 four-cylinder four-stroke engines. Fuel capacity 76 litres (20.0 US gallons; 16.7 Imp gallons) in single fuselage tank. Oil capacity 1.9 litres (0.5 US gallon; 0.4 Imp gallon).
ACCOMMODATION: Rearward-hinged one-piece Perspex canopy. Dual controls standard.
AVIONICS: To customers' requirements.
Comms: Optional Honeywell or Becker nav/com, Narco transponder.
Instrumentation: Three basic flight instruments plus engine instruments. Optional vacuum instruments driven by dual venturis mounted under front fuselage.

DIMENSIONS, EXTERNAL:
Wing span	8.99 m (29 ft 6 in)
Length overall	5.49 m (18 ft 0 in)
Height overall	2.29 m (7 ft 6 in)
Propeller diameter	1.07 m (3 ft 6 in)

DIMENSIONS, INTERNAL:
Cabin: Length	1.27 m (4 ft 2 in)
Max width	0.99 m (3 ft 3 in)
Max height	1.09 m (3 ft 7 in)

AREAS:
Wings, gross	8.59 m² (92.5 sq ft)

WEIGHTS AND LOADINGS:
Weight empty, equipped	313 kg (690 lb)
Max T-O and landing weight	529 kg (1,168 lb)
Max wing loading	61.7 kg/m² (12.63 lb/sq ft)
Max power loading	8.89 kg/kW (14.60 lb/hp)

PERFORMANCE (at max T-O weight):
Never-exceed speed (V_{NE})	131 kt (243 km/h; 151 mph)
Max level speed	119 kt (220 km/h; 137 mph)
Cruising speed at 75% power	100 kt (185 km/h; 115 mph)
Stalling speed, power off, 40° flap	40 kt (73 km/h; 45 mph)
Max rate of climb at S/L	244 m (800 ft)/min
T-O to 15 m (50 ft)	423 m (1,387 ft)
Landing from 15 m (50 ft)	442 m (1,450 ft)
Range	499 n miles (925 km; 575 miles)
g limits	+3.8/-2.0

UPDATED

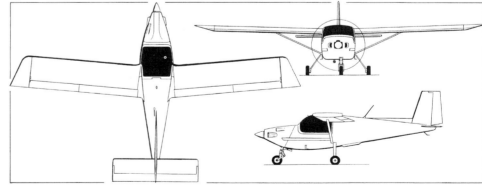

Skycraft ARV Super2 two-seat monoplane *(Dennis Punnett/Jane's)*

SKYSTAR

SKYSTAR AIRCRAFT CORPORATION

3901 Aviation Way, Caldwell, Idaho 83605
Tel: (+1 800) 554 83 69
Fax: (+1 800) 454 64 64
e-mail: info@skystar.com
Web: http://www.skystar.com
PRESIDENT AND CEO: Edward S Downs
VICE-PRESIDENT, ENGINEERING: Frank Miller

SkyStar Aircraft Corporation took over Kitfox programme from former Denney Aerocraft Company in 1993. Kitfox production continues. Local variants also produced by Skyfox in Australia (1999-2000 *Jane's*) and Aeropro in Slovak Republic (which see), but these companies are not affiliated with Skystar Aircraft Corporation and their products do not represent the latter's product line. By 2000, more than 4,000 Kitfox kits had been sold. In 1998, the company added the Kitfox Lite and in the following year streamlined its range. The Series 6 Kitfox and Kitfox Lite² were added in January 2000.

Rights to Aero Designs Pulsar purchased in late 1996, but in July 1999 the design was sold to the Pulsar Aircraft Corporation (which see).

UPDATED

SKYSTAR KITFOX

TYPE: Side-by-side kitbuilt.
PROGRAMME: Prototype Kitfox first flown 7 May 1984. Initial variants had MTOW below US 544 kg (1,200 lb) ultralight limit; these last described in 1999-2000 *Jane's*; kit production totalled 257 Model 1s, 491 Model 2s, and 467 Model 3s. Series IV (retrospectively renamed Classic IV) introduced 1991; 323 kits of 476 kg (1,050 lb) MTOW version produced before introduction of current version. First certified version was Series 5, introduced April 1994, but replaced by Series 6 in January 2000.
CURRENT VERSIONS: **Classic IV:** Ultralight version, introduced in 1992. Engines up to 74.6 kW (100 hp) can be fitted

under helmeted engine cowling. MTOW 544 kg (1,200 lb). **Speedster** option includes short-span wings and improved streamlining for up to 17 kt (32 km/h; 20 mph) additional speed.

Kitfox 6 Convertible: Introduced 2000; available in tailwheel and nosewheel subvariants. Considerably refined design, with only 5 per cent parts commonality with Kitfox V.

Detailed description applies to above version except where indicated.

Kitfox Lite: Described separately.
Kitfox Lite²: Described with Kitfox Lite.

CUSTOMERS: Some 4,000 Kitfox aircraft ordered, and over 2,500 built.
COSTS: Classic IV US$14,995; Series 6 US$19,495, excluding engine and avionics (2001).
DESIGN FEATURES: Designed to have good short-field performance. Constant-chord, high wing with V strut bracing each side; braced tailplane, mid-mounted. With Kitfox IV, introduced in 1991, new wing with laminar flow section and all-metal hinge brackets for full-span flaperons became standard; windscreen material thickened and given increased slope and, with other changes, resulted in increase in cruising speed; stalling and landing speeds decreased, former due to use of flaperons; wing folding standard. Optional agricultural spray pods. Quoted build time 1,000 hours.
FLYING CONTROLS: Manual. Junkers-type differential flaperons; rudder and elevators with mass balance.
STRUCTURE: Wings of aluminium alloy, plywood ribs and glass fibre tips, with fabric covering overall. Steel tube fuselage and tail unit, with fabric covering.
LANDING GEAR: Non-retractable type, with hydraulic disc brakes; tailwheel or tricycle configuration; can be converted after completion; optional speed fairings. Sprung aluminium main legs. Optional floats, amphibious floats and skis.
POWER PLANT: One 59.6 kW (79.9 hp) Rotax 912 UL flat-four. Options include 73.5 kW (98.6 hp) Rotax 912 ULS, 74.6 kW (100 hp) Teledyne Continental O-200, 93 kW (125 hp) Teledyne Continental IO-240B and 86.5 kW (116 hp) Textron Lycoming O-235. Fuel in two wing tanks, total capacity 102 litres (27.0 US gallons; 22.5 Imp gallons).
ACCOMMODATION: Two, side by side, can have 0.14 m³ (5 cu ft) storage space behind seat. Optional drop tank style cargo pod.

DIMENSIONS, EXTERNAL:
Wing span: normal	9.75 m (32 ft 0 in)
optional (Speedster)	8.85 m (29 ft 0½ in)
Wing aspect ratio, normal	7.8
Width, wings folded	2.44 m (8 ft 0 in)
Length: overall (nosewheel)	5.81 m (19 ft 1 in)
wings folded: tailwheel	6.68 m (21 ft 11 in)
nosewheel	6.71 m (22 ft 0 in)
Max width of fuselage	1.08 m (3 ft 6½ in)
Height overall: tailwheel	1.69 m (5 ft 6½ in)
nosewheel	2.54 m (8 ft 4 in)
Tailplane span	2.41 m (7 ft 11 in)
Wheel track	2.02 m (6 ft 7½ in)
Propeller diameter	1.83 m (6 ft 0 in)

AREAS:
Wings, gross (normal)	12.26 m² (132.0 sq ft)

WEIGHTS AND LOADINGS (Rotax 912 ULS):
Weight empty	340 kg (750 lb)
Baggage capacity	68 kg (150 lb)
Max T-O weight	703 kg (1,550 lb)
Max wing loading	57.3 kg/m² (11.74 lb/sq ft)
Max power loading	9.57 kg/kW (15.72 lb/hp)

PERFORMANCE (two crew, Rotax 912 ULS, unless stated otherwise):
Max level speed	111 kt (205 km/h; 125 mph)
Cruising speed	104 kt (193 km/h; 120 mph)
Stalling speed	38 kt (70 km/h; 43 mph)
Max rate of climb at S/L	427 m (1,400 ft)/min
T-O and landing run	92 m (300 ft)
g limits, pilot only	+3.8/−1.52

UPDATED

SkyStar Kitfox Speedster two-seat kitbuilt *(Paul Jackson/Jane's)*   2001/0092085

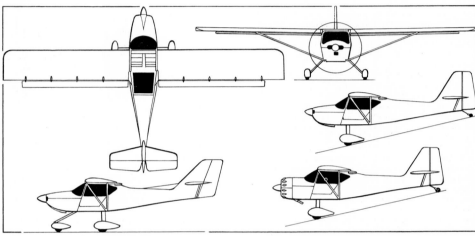

Continental/Lycoming-powered SkyStar Kitfox Series 6, with additional profiles of Rotax version with helmeted cowling and (bottom left) Series 5 Voyager, final examples of which are still being completed *(James Goulding/Jane's)*   2001/0092109

SKYSTAR KITFOX LITE and KITFOX LITE²

TYPE: Single-seat ultralight/kitbuilt and side-by-side ultralight/kitbuilt.
PROGRAMME: First displayed at Sun 'n' Fun in April 1998.
CURRENT VERSIONS: **Kitfox Lite:** Single-seat version.
Description applies to the above.
Kitfox Lite²: Two-seat training version introduced in 2000. Has 37.0 kW (49.6 hp) Rotax 503 UL-2V engine.
CUSTOMERS: 120 kits and 10 complete aircraft delivered by January 2001.
COSTS: Kitfox Lite US$14,995 including engine (2001). Kitfox Lite² US$21,995 including engine (2001).
DESIGN FEATURES: Generally as for Kitfox. Meets FAR Pt 103 standards. Wings fold for storage. Quoted build time 200 hours; wing pre-built by factory.
FLYING CONTROLS: Manual, including flaperons.
STRUCTURE: Welded 4130 steel fuselage; aluminium wing spar; aluminium and wood wing, all fabric covered.
LANDING GEAR: Tricycle or tailwheel configuration; free-castoring 6 cm (2½ in) tailwheel. 4.00-6 main tyres. Individual wheel brakes.
POWER PLANT: One 26.1 kW (35 hp) Skystar/2SI Series 460 engine driving Tennessee two-bladed propeller through 2.5:1 reduction gear. Fuel capacity 19 litres (5.0 US gallons; 4.2 Imp gallons).

DIMENSIONS, EXTERNAL:
Wing span	7.67 m (25 ft 2 in)
Length overall: Lite	5.00 m (16 ft 5 in)
Lite²	5.61 m (18 ft 5 in)
Height overall: Lite	1.57 m (5 ft 2 in)
Lite²	1.75 m (5 ft 9 in)

AREAS:
Wings, gross	9.38 m² (101.0 sq ft)

WEIGHTS AND LOADINGS:
Weight empty: Lite	115 kg (254 lb)
Lite²	223 kg (492 lb)
Max T-O weight: Lite	249 kg (550 lb)
Lite²	476 kg (1,050 lb)

PERFORMANCE:
Max level speed: Lite	55 kt (101 km/h; 63 mph)
Lite²	72 kt (133 km/h; 83 mph)
Normal cruising speed: Lite	48 kt (89 km/h; 55 mph)
Lite²	65 kt (121 km/h; 75 mph)
Stalling speed: Lite	24 kt (44 km/h; 27 mph)
Lite²	32 kt (58 km/h; 36 mph)
Max rate of climb at S/L	229 m (750 ft)/min
T-O run: Lite	31 m (100 ft)
Lite²	77 m (250 ft)
Landing run: Lite	23 m (75 ft)
Lite²	67 m (220 ft)

UPDATED

Late production Kitfox (Series V) Safari *(Paul Jackson/Jane's)*   2001/0092086

Single-seat Skyfox Lite   2000/0085833

SLIPSTREAM

SLIPSTREAM INDUSTRIES INC

W 8407 Cottonville Drive, Wautoma, Wisconsin 54982
Tel: (+1 920) 787 58 86 or (+1 800) 464 36 64
Fax: (+1 920) 787 57 15
e-mail: flygenesis@aol.com
Web: http://www.slipstreamind.com
PRESIDENT: Michael Puhl

SlipStream produces a range of ultralight aircraft, and also offers the Dragonfly, originally produced by Viking Aircraft.
NEW ENTRY

SLIPSTREAM DRAGONFLY

TYPE: Side-by-side ultralight kitbuilt.
PROGRAMME: Prototype first flight 16 June 1980. Plans available, together with preformed engine cowling and canopy. Kits of prefabricated component parts, requiring no complex jigging or tooling, also available; in this form aircraft known as 'Snap' Dragonfly. Kits estimated to save builder more than 700 working hours.
CURRENT VERSIONS: **Dragonfly Mark I**: Original configuration, with non-retractable mainwheels at tips of foreplane. EAA Oshkosh Outstanding Design, 1980.
 Dragonfly Mark II: In parallel production, for operation from unprepared strips and narrow taxiways. Main landing gear in short non-retractable cantilever units under wings, with individual hydraulic toe brakes, and increased foreplane and elevator areas; wheel track 2.44 m (8 ft).
 Dragonfly Mark III: Flight tested in 1985; non-retractable tricycle landing gear.
CUSTOMERS: More than 100 kits delivered, together with 2,000 sets of plans; some 70 kitbuilt and more than 430 plans-built aircraft flying.
COSTS: Kit: US$9,441 (2001). Plans: US$299.
DESIGN FEATURES: Unconventional shoulder-wing monoplane with equal-span foreplane; latter also mounting landing gear in one of two alternative locations.
 Wing section Eppler 1213. Foreplane GU25 section. Thickness/chord ratio 15 per cent; dihedral 3°; no incidence or sweepback. Foreplane t/c 17 per cent; anhedral 3° on Mk I or dihedral on Mk II; incidence −1°; no sweepback.
FLYING CONTROLS: Manual. Ailerons on inboard trailing-edge of wing; near full-span elevators on foreplane; horn-balanced rudder.
STRUCTURE: Composites wing; foreplane and tail unit structures of styrene foam, glass fibre, carbon fibre and epoxy. Semi-monocoque fuselage, formed (not carved) from 12.5 mm (½ in) thick urethane foam, with strips of 18 mm (¾ in) foam bonded along edges to allow large-radius external corners. Fuselage covered with glass fibre inside and out.
LANDING GEAR: See Current Versions. Brakes fitted.
POWER PLANT: One 44.5 kW (60 hp) 1,835 cc modified Volkswagen motorcar engine; 1,600 cc engine, rated at 33.5 kW (45 hp), optional. Fuel capacity 56.8 litres (15.0 US gallons; 12.5 Imp gallons).
DIMENSIONS, EXTERNAL:
 Wing span 6.71 m (22 ft 0 in)
 Foreplane span 6.71 m (22 ft 0 in)
 Length overall 5.79 m (19 ft 0 in)
 Height overall: Mk I 1.22 m (4 ft 0 in)
DIMENSIONS, INTERNAL:
 Cabin max width 1.09 m (3 ft 7 in)
AREAS:
 Wings, gross 4.24 m² (45.6 sq ft)
 Foreplane, net 4.33 m² (46.6 sq ft)
WEIGHTS AND LOADINGS:
 Weight empty 277 kg (610 lb)
 Max T-O weight 522 kg (1,150 lb)
PERFORMANCE:
 Max level speed at 75% power at 2,285 m (7,500 ft)
  156 kt (290 km/h; 180 mph)
 Cruising speed 143 kt (266 km/h; 165 mph)
 Stalling speed 42 kt (78 km/h; 48 mph)
 Max rate of climb at S/L 259 m (850 ft)/min
 T-O to 15 m (50 ft) 366 m (1,200 ft)
 Landing from 15 m (50 ft) 610 m (2,000 ft)
 Range 477 n miles (885 km; 550 miles)
UPDATED

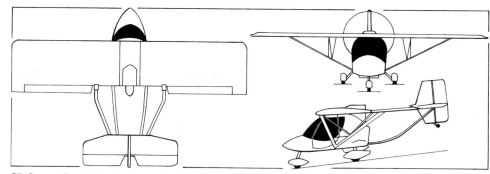

SlipStream Revelation two-seat ultralight *(Paul Jackson/Jane's)* 2001/0100466

SLIPSTREAM GENESIS and REVELATION

TYPE: Side-by-side ultralight kitbuilt.
PROGRAMME: Prototype Genesis first flew 1992; Revelation developed as a lower-powered 'recreational' version.
CURRENT VERSIONS: **Genesis**: *As described*.
 Revelation: Lower-powered version; can be upgraded to Genesis standard by replacement only of windscreen and engine.
CUSTOMERS: Total of 100 Genesis sold, and 70 flying, by mid-2000; 100 Revelations sold, and 75 flying, by same date.
COSTS: Kit including engine (2001): *Genesis* US$18,528; *Revelation* US$13,295.
DESIGN FEATURES: Quoted build time for Genesis 200 hours; for Revelation 150 hours (or 90 when opting for sailcloth-covered wing).
FLYING CONTROLS: Conventional and manual. Horn-balanced elevator.
STRUCTURE: Preformed glass fibre nose; fabric-covered strut-braced wings; tubular steel rear fuselage.
LANDING GEAR: Fixed tricycle layout with mechanical brakes; hydraulic brakes optional. Optional floats and speed fairings.
POWER PLANT: *Genesis*: One 47.8 kW (64.1 hp) Rotax 582 driving Tennessee two-blade wooden propeller. Fuel capacity 76 litres (20.0 US gallons; 16.7 Imp gallons). Other engines including Rotax 618, 912 UL and 912 ULS can be fitted.
 Revelation: One 39.3 kW (52.7 hp) Rotax 503B. Fuel capacity 38 litres (10.0 US gallons; 8.3 Imp gallons).
Data apply to both versions, except where otherwise stated.
DIMENSIONS, EXTERNAL:
 Wing span 9.35 m (30 ft 8 in)
 Length overall 5.89 m (19 ft 4 in)
 Height overall 1.92 m (6 ft 3½ in)
AREAS:
 Wings, gross 16.62 m² (179.0 sq ft)
WEIGHTS AND LOADINGS (A: Genesis, B: Revelation):
 Weight empty: A 256 kg (565 lb)
  B 211 kg (465 lb)
 Max T-O weight: A, B 544 kg (1,200 lb)
PERFORMANCE (A and B as above):
 Never-exceed speed (V_{NE}):
  A 104 kt (193 km/h; 120 mph)
 Normal cruising speed: A 65 kt (121 km/h; 75 mph)
  B 57 kt (106 km/h; 66 mph)
 Stalling speed: A, B 39 kt (72 km/h; 45 mph)
 Max rate of climb at S/L: A 244 m (800 ft)/min
  B 213 m (700 ft)/min
 T-O run: A, B 92 m (300 ft)
 Landing run: A 77 m (250 ft)
  B 46 m (150 ft)
 Range with max fuel: A 195 n miles (362 km; 225 miles)
  B 130 n miles (241 km; 150 miles)
NEW ENTRY

SLIPSTREAM SCEPTRE

TYPE: Single-seat ultralight kitbuilt.
PROGRAMME: Development of Genesis, introduced in 1999.
CUSTOMERS: One flying by mid-2000.
COSTS: US$10,995 (2001) including engine.
DESIGN FEATURES: Quoted build time 120 hours.
Data as Genesis, except that below.
POWER PLANT: One 39.3 kW (52.7 hp) Rotax 503B or 29.5 kW (39.6 hp) Rotax 447 UL-1V. Fuel capacity 19 litres (5.0 US gallons; 4.2 Imp gallons).
WEIGHTS AND LOADINGS (Rotax 447):
 Weight empty 163 kg (360 lb)
 Max T-O weight 408 kg (900 lb)
PERFORMANCE (Rotax 447):
 Normal cruising speed 48 kt (89 km/h; 55 mph)
 Stalling speed 24 kt (44 km/h; 27 mph)
 Max rate of climb at S/L 229 m (750 ft)/min
 T-O and landing run 31 m (100 ft)
 Range with max fuel 60 n miles (112 km; 70 miles)
NEW ENTRY

SLIPSTREAM SKYBLASTER

TYPE: Two-seat kitbuilt twin.
PROGRAMME: Introduced in 2000; based on Genesis/Revelation with 98 per cent commonality of parts.
CUSTOMERS: One flying by mid-2000.
COSTS: US$19,995 (2001).
DESIGN FEATURES: Two in-line engines, with counter-rotating propellers, mounted on front and back of wing; layout simplifies single-engine handling. Quoted build time 175 hours.
Data as Genesis, except that below.
POWER PLANT: Two 39.3 kW (52.7 hp) Rotax 503E engines.
WEIGHTS AND LOADINGS:
 Weight empty 318 kg (700 lb)
 Max T-O weight 635 kg (1,400 lb)
PERFORMANCE:
 Normal cruising speed 70 kt (129 km/h; 80 mph)
 Stalling speed 44 kt (81 km/h; 50 mph)
 Max rate of climb at S/L 244 m (800 ft)/min
 T-O run 61 m (200 ft)
NEW ENTRY

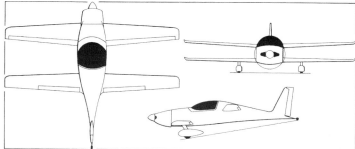

SlipStream Dragonfly Mark II *(Paul Jackson/Jane's)* 0016522

SlipStream Dragonfly Mark I

SONEX

SONEX LTD

PO Box 2521, Oshkosh, Wisconsin 54903-2521
Tel: (+1 920) 231 82 97
Fax: (+1 920) 426 83 33
e-mail: sonex@vbe.com
Web: http://www.sonex.ltd.com
PRESIDENT: John Monnett

EUROPEAN AGENT:
 ARES srl, Via Papiria 54a, I-00175 Roma, Italy

Tel: (+39 06) 76 90 01 32
Fax: (+39 06) 76 90 00 89
e-mail: arescomp@tin.it
Web: http://www.rce.it/sonex

NEW ENTRY

USA: AIRCRAFT—SONEX to STODDARD-HAMILTON

SONEX SONEX

TYPE: Side-by-side ultralight kitbuilt.
PROGRAMME: Announced March 1996 in response to request for ultralight to meet Italian requirements; first flight of prototype (N12SX) with Jabiru 2200 engine, 28 February 1998. Five prototypes. First customer-built aircraft flown 12 June 2000.
CUSTOMERS: At least four flying by mid-2000.
COSTS: Airframe kit US$11,120 tailwheel, US$11,355 tricycle; plans US$600 (including two-day workshop) (2000).
DESIGN FEATURES: Simple to assemble; 198 parts to easy-build kit. Wings removable for transport and storage. NACA 64-415 aerofoil.
FLYING CONTROLS: Conventional and manual. Flaps.
STRUCTURE: Of 6061 aluminium throughout.
LANDING GEAR: Choice of tailwheel or tricycle; both fixed. Titanium mainwheel legs. Azusa wheels and brakes.
POWER PLANT: Choice of power plants between 59.7 and 89.5 kW (80 and 120 hp), including Volkswagen and Jabiru 2200 and 3300. Two-blade, wooden, fixed-pitch propeller. Fuel capacity 61 litres (16.0 US gallons; 13.3 Imp gallons) in single fuel tank.

Prototype Sonex Sonex *(Paul Jackson/Jane's)*

DIMENSIONS, EXTERNAL:
Wing span	6.71 m (22 ft 0 in)
Length overall	5.36 m (17 ft 7 in)

AREAS:
Wings, gross	9.10 m² (98.0 sq ft)

WEIGHTS AND LOADINGS (59.7 kW; 80 hp Jabiru 2200):
Weight empty	249 kg (550 lb)
Max T-O weight	499 kg (1,100 lb)

PERFORMANCE::
Never-exceed speed (V$_{NE}$)	171 kt (317 km/h; 197 mph)
Max level speed	130 kt (241 km/h; 150 mph)
Normal cruising speed	113 kt (209 km/h; 130 mph)
Stalling speed: flaps up	40 kt (74 km/h; 46 mph)
flaps down	35 kt (65 km/h; 40 mph)
Max rate of climb at S/L	305 m (1,000 ft)/min
T-O run	92 m (300 ft)
Landing run	153 m (500 ft)
Range with max fuel	412 n miles (764 km; 475 miles)
g limits (Aerobatic)	+6/−3

NEW ENTRY

SPENCER

SPENCER AMPHIBIAN AIRCRAFT INC
PO Box 327, Kansas, Illinois 61933
Tel/Fax: (+1 217) 948 55 05
and
1629 Park Drive, Schaumburg, Illinois 60194
Tel/Fax: (+1 847) 882 01 23 and 56 78
e-mail: bobwan1@gateway.net
MANAGING DIRECTOR: Robert F Kerans

Percival Hopkins Spencer designed the Air Car S1 in 1939; the programme was sold to Republic Aircraft in 1941 and entered production in 1945 as the Seabee; the company built one C-1 version, 10 RC-2s and 1,050 RC-3s up to 1947. Spencer died in 1995, aged 97. Design rights were then offered for sale, but no purchaser had been found by mid-2000. Description last appeared in 2000-01 *Jane's*.

UPDATED

STODDARD-HAMILTON

STODDARD-HAMILTON AIRCRAFT INC
18701 58th Avenue Northeast, Arlington, Washington 98223
Tel: (+1 360) 435 85 33
Fax: (+1 360) 435 95 25
e-mail: airplanesales@stoddard-hamilton.com
Web: http://www.stoddard-hamilton.com
PRESIDENT: Bob Gavinsky
MARKETING DIRECTOR: Craig O'Neill

Company formed in 1979. Glasair I, II and II-S were superseded by Super II and III. Details of Glasair I in 1989-90 *Jane's*. Two-seat GlaStar added to product range in 1995; float production rights sold to Aero Marine Inc in 1996. Company had delivered more than 2,400 kits by April 1999.

Work began on a six-seater, named Aurora, in early 2000, but Stoddard-Hamilton ceased trading in May 2000 and filed for Chapter 11 bankruptcy on 17 July. At a bankruptcy hearing on 18 August 2000, the bulk of the assets was sold (subject to final agreement on some matters) to Anduril Private Investment Fund, run by Strider Capital Management. However at a court hearing on 27 November 2000 this agreement was set aside and bidding for the company was re-opened.

UPDATED

STODDARD-HAMILTON GLASAIR SUPER II

TYPE: Side-by-side kitbuilt.
PROGRAMME: Derived from Glasair TD, first flown in 1979. Jump-start kit options, to reduce quoted build time by around 750 hours, offered from 1999 onwards.
CURRENT VERSIONS: Glasair Super II (stretched) available in **RG** (retractable landing gear), **FT** (non-retractable tricycle gear) and **TD** (taildragger) forms introduced in 1992 to supersede earlier Glasair series.
CUSTOMERS: 200 kits sold, about 75 flying (by 1998).
COSTS: Kits: US$28,280 for RG; US$21,900 for FT; US$22,900 for TD, all without engine, propeller, instruments and avionics; engine US$27,800 for 180 hp version; fixed-pitch propeller US$1,500, constant-speed propeller US$5,700 (2000).
DESIGN FEATURES: Low-wing monoplane with moderately tapered wing and sweptback tail surfaces; optimised for high performance consistent with amateur assembly.
Wing section NASA LS(I)-0413; dihedral 3°; incidence 2° 20′. Upswept Hoerner style wingtips. Optional 61 cm (2 ft 0 in) extension on each wingtip.
FLYING CONTROLS: Conventional and manual. Horn-balanced rudder and elevators; ailerons with trim tab; plain flaps (slotted flaps optional).
STRUCTURE: Glass fibre and foam composites construction. One-piece wing spar.
LANDING GEAR: Three types available, as detailed above. RG version has inward-retracting mainwheels and rearward-retracting, free-castoring nosewheel. Brakes fitted. Wheel size 5.00-5 in. TD has full-swivel, locking tailwheel.
POWER PLANT: One 112 to 149 kW (150 to 200 hp) Textron Lycoming O-320 or IO-360 flat-four engine driving a Sensenich constant-speed metal propeller. Fuel capacity 182 litres (48.0 US gallons; 40.0 Imp gallons). Auxiliary tanks of 41.6 litres (11.0 US gallons; 9.2 Imp gallons) capacity in optional wingtip extensions.
ACCOMMODATION: Two persons, side by side. Enclosed cockpit with gull-wing doors. NACA air vents provide cabin ventilation.

DIMENSIONS, EXTERNAL (all versions):
Wing span: standard	6.81 m (22 ft 4 in)
optional	8.03 m (26 ft 4 in)
Wing aspect ratio: standard	6.1
optional	7.6
Length overall	6.31 m (20 ft 8½ in)
Height overall	2.07 m (6 ft 9½ in)
Propeller diameter	1.78 m (5 ft 10 in)

DIMENSIONS, INTERNAL:
Cabin max width	1.07 m (3 ft 6 in)
Baggage volume	0.34 m³ (12.0 cu ft)

AREAS:
Wings, gross: standard	7.55 m² (81.3 sq ft)
optional	8.50 m² (91.5 sq ft)

WEIGHTS AND LOADINGS:
Weight empty: RG	601 kg (1,325 lb)
FT	567 kg (1,250 lb)
TD	544 kg (1,200 lb)
Baggage capacity: all versions	45 kg (100 lb)
Max T-O weight: RG, FT	953 kg (2,100 lb)
TD	907 kg (2,000 lb)
Max wing loading: RG, FT	126.1 kg/m² (25.83 lb/sq ft)
TD	120.1 kg/m² (24.60 lb/sq ft)
Max power loading: 119 kW (160 hp) engine:	
FT	7.99 kg/kW (13.13 lb/hp)
134 kW (180 hp) engine:	
RG	7.10 kg/kW (11.67 lb/hp)
TD	6.76 kg/kW (11.11 lb/hp)

PERFORMANCE (180 hp engine):
Never-exceed speed (V$_{NE}$)	226 kt (418 km/h; 260 mph)
Max level speed at S/L:	
RG	207 kt (383 km/h; 238 mph)
FT, TD	198 kt (367 km/h; 228 mph)
Econ cruising speed at 2,440 m (8,000 ft):	
RG	192 kt (356 km/h; 221 mph)
FT, TD	182 kt (338 km/h; 210 mph)
Stalling speed, power off: plain flaps down:	
RG	57 kt (105 km/h; 65 mph)
FT, TD	55 kt (102 km/h; 63 mph)
slotted flaps down:	
RG	52 kt (95 km/h; 59 mph)
FT, TD	50 kt (92 km/h; 57 mph)
Max rate of climb at S/L: all	823 m (2,700 ft)/min
Service ceiling: all	approx 5,790 m (19,000 ft)
T-O and landing run: TD, FT	213 m (700 ft)
RG	244 m (800 ft)
Range: RG, standard fuel	1,080 n miles (2,000 km; 1,242 miles)
RG, with auxiliary fuel	1,330 n miles (2,463 km; 1,530 miles)
FT, TD, standard fuel	1,040 n miles (1,926 km; 1,196 miles)
FT, TD, with auxiliary fuel	1,278 n miles (2,366 km; 1,470 miles)
g limits at AUW of 708 kg (1,560 lb): all	+6/−4
	(+9/−6 ultimate)

UPDATED

Stoddard-Hamilton Glasair Super II-RG *(Paul Jackson/Jane's)*

STODDARD-HAMILTON GLASAIR III

TYPE: Side-by-side kitbuilt.
PROGRAMME: First flight July 1986.
CURRENT VERSIONS: **Glasair III:** *As described.*
  Glasair III Turbo: Powered by 224 kW (300 hp) Textron Lycoming TIO-540. Maximum level speed 284 kt (526 km/h; 327 mph); maximum rate of climb 1,143 m (3,750 ft)/min; withdrawn in 1998 and superseded by Glasair Super III.
  Glasair Super III: Prototype (N401KT) flying by third quarter 1998, when demonstrated 322 kt (596 km/h; 280 mph) at 10,060 m (33,000 ft), powered by IO-540-AA1B5 uprated with Honeywell turbocharger to 261 kW (350 hp) at 11,275 m (37,000 ft) and with Hartzell four-blade propeller. Predicted maximum speed of 350 kt (648 km/h; 403 mph). Has enlarged rudder, extra fuel capacity and highly modified cowling. Public debut at Sun 'n' Fun 1999.
CUSTOMERS: 330 kits delivered by 1998 (latest information), of which 125 then completed.
COSTS: Kit: US$39,980 without engine, propeller, instruments and avionics; engine US$38,000; constant-speed propeller US$5,700 (2000).
DESIGN FEATURES: Similar configuration to earlier Glasairs, but designed to offer better performance, constructional simplicity and economical kit price. Larger fuselage for increased baggage space, payload capacity and comfort, also improving the longitudinal and directional stability for better cross-country and IFR performance; windows to rear of cockpit doors. Thicker windscreen to improve protection against bird strikes at higher speeds, and additional glass fibre laminates, integral longerons, and lay-up schedule which provides structurally stronger and torsionally stiffer fuselage. Quoted build time 3,000 working hours.
  Wing section LS(I)-0413; wing incidence 2° 18′, dihedral 3°; strengthened; carries more fuel than previous models. Optional wingtip extensions for 8.33 m (27 ft 4 in) or 7.86 m (25 ft 9½ in) spans.
FLYING CONTROLS: As Glasair Super II. Slotted flaps optional.
LANDING GEAR: Retractable tricycle only, otherwise as Super II-RG. Mainwheels size 5.00-5; nosewheel 4.00-5. Differential Cleveland disc brakes on mainwheels.
POWER PLANT: One 224 kW (300 hp) Textron Lycoming IO-540-K1H5 flat-six piston engine driving a two-blade constant-speed propeller. Fuel capacity in wings 201 litres (53.0 US gallons; 44.1 Imp gallons). Fuselage header tank, capacity 30 litres (8.0 US gallons; 6.7 Imp gallons). Optional tanks in wingtip extensions, combined capacity 41.6 litres (11.0 US gallons; 9.2 Imp gallons).
ACCOMMODATION: As for Super II.

Data below refer to standard wing span Glasair III.
DIMENSIONS, EXTERNAL:
Wing span: standard	7.11 m (23 ft 4 in)
optional	8.33 m (27 ft 4 in)

Stoddard-Hamilton Glasair III *(Paul Jackson/Jane's)* 2000/0062937

Wing chord: at root	1.28 m (4 ft 2½ in)
at tip	0.84 m (2 ft 9 in)
Wing aspect ratio: standard	6.7
optional	8.2
Length overall	6.52 m (21 ft 4 in)
Height overall	2.29 m (7 ft 6 in)
Tailplane span	2.64 m (8 ft 8 in)
Wheel track	3.10 m (10 ft 2 in)

DIMENSIONS, INTERNAL: As for Glasair Super II
AREAS:
Wings, gross: standard	7.55 m² (81.3 sq ft)
optional	8.50 m² (91.5 sq ft)
Fin	1.06 m² (11.40 sq ft)
Rudder	0.57 m² (6.10 sq ft)
Tailplane	1.51 m² (16.30 sq ft)
Elevator	0.53 m² (5.70 sq ft)

WEIGHTS AND LOADINGS:
Weight empty	737 kg (1,625 lb)
Baggage capacity	45 kg (100 lb)
Max T-O weight: without wingtip extensions	1,089 kg (2,400 lb)
with wingtip extensions	1,134 kg (2,500 lb)
Max wing loading: standard	144.1 kg/m² (29.52 lb/sq ft)
optional	133.4 kg/m² (27.32 lb/sq ft)
Max power loading: standard	4.86 kg/kW (8.00 lb/hp)
optional	5.07 kg/kW (8.33 lb/hp)

PERFORMANCE (standard wings):
Never-exceed speed (VNE)	291 kt (539 km/h; 335 mph)
Max level speed at S/L	252 kt (467 km/h; 290 mph)
Cruising speed: at 75% power at 2,440 m (8,000 ft)	242 kt (448 km/h; 278 mph)
at 65% power at 2,440 m (8,000 ft)	230 kt (426 km/h; 265 mph)
Stalling speed: pilot only, plain flaps down	68 kt (126 km/h; 79 mph)
at max T-O weight, slotted flaps down	63 kt (117 km/h; 73 mph)
Max rate of climb at S/L	911 m (2,990 ft)/min
Service ceiling: Glasair III	approx 7,315 m (24,000 ft)
Glasair Super III	9,144 m (30,000 ft)
T-O run	274 m (900 ft)
Landing run	305 m (1,000 ft)
Range at 55% power:	
standard fuel	888 n miles (1,644 km; 1,021 miles)
with tip tanks	1,068 n miles (1,978 km; 1,229 miles)
g limits at AUW of 962 kg (2,120 lb): all	+6/−4
	+9/−6 ultimate

UPDATED

STODDARD-HAMILTON GLASTAR

TYPE: Side-by-side kitbuilt.
PROGRAMME: Designed by Arlington Aircraft Developments. Prototype (N824G) first flew 29 November 1994; deliveries of kits started 1995. Over 1,400 hours flown by December 1998. Prototype converted to tailwheel configuration and then back to tricycle. Chosen by EAA as flagship aircraft for Young Eagle junior pilot scheme; first customer-built aircraft completed in July 1996. Early examples had a 93 kW (125 hp) Teledyne Continental IO-240.
  Re-engineered version being built in Germany as OMF-160 Symphony (which see).
CUSTOMERS: Over 750 kits sold in 16 countries up to April 1999; about 100 then flying..
COSTS: US$24,750 per kit, excluding engine, instruments, upholstery and paint (2000).
DESIGN FEATURES: Conceived as inexpensive two-seat personal lightplane with wide performance. Convertible landing gear, float compatible. Wings fold and horizontal tail surfaces removable for compact hangarage or for trailer mounting. Quoted build time 1,000-1,500 hours. 'Jump-start' options introduced in November 1998 to reduce time by half; wing kits for this option are assembled in the Czech Republic.
  Braced, high wing of constant chord, unswept. Tailplane has root extensions; tall, sweptback fin. Wing section LS(I)-0413.
FLYING CONTROLS: Conventional and manual. Frise ailerons; Fowler flaps. Elevator tab for pitch trim. Ground-adjustable tab on ailerons. Three-position flaps: 0, 20 and 40°. Vortex generators added to wings in front of ailerons and at wingroots, and root strakes to horizontal stabiliser. Optional electric elevator trim.
STRUCTURE: Fuselage construction of GFRP with 4130 steel tube frame surrounding cockpit section. Wings and tail surfaces of 2024-T3 aluminium. Two-spar wings have six full ribs per side, plus stiffeners.
LANDING GEAR: Fixed nosewheel type with spats; steerable tailwheel or Aerocet 2200 float versions; no differences in airframe among landing gear alternatives. Tailwheel (15 cm; 6 in tyre) conversion effected in 3 hours; choice of three mainwheel sizes: 5.00-5 (with speed fairings), 6.00-6 and 8.00-6. Scott 3200 20 cm (8 in) tailwheel.
POWER PLANT: One 119 kW (160 hp) Textron Lycoming O-320-D1F flat-four, driving a Sensenich fixed-pitch or Hartzell constant-speed two-blade propeller; power plants in the range 93-168 kW (125-225 hp) are in use. Fuel capacity 114 litres (30.0 US gallons; 25.0 Imp gallons). Optional 76 litre (20.0 US gallon; 16.7 Imp gallon) auxiliary tanks.
ACCOMMODATION: Two persons side by side; seats adjust for pilot heights between 1.52 and 1.98 m (5 ft 0 in and 6 ft 6 in). Door each side. Baggage compartment behind crew with separate baggage door on port side.

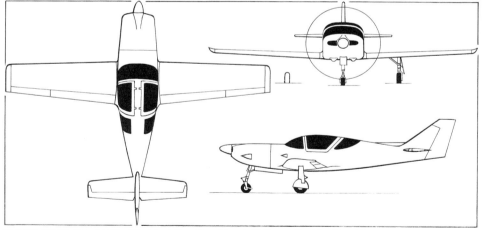

General arrangement of the Stoddard-Hamilton Glasair III *(James Goulding/Jane's)* 2001/0093599

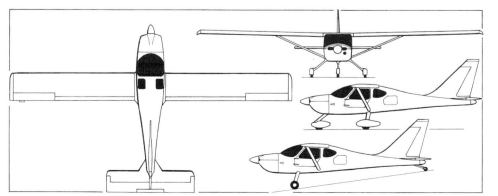

Tricycle version of GlaStar two-seat kitplane, with additional side view of tailwheel version *(Paul Jackson/Jane's)* 2001/0093597

782   USA: AIRCRAFT—STODDARD-HAMILTON to USN

SYSTEMS: Electrical system: 12 V 30 Ah battery; alternator fit appropriate to engine.
AVIONICS: To customer's specification.
DIMENSIONS, EXTERNAL:

Wing span	10.67 m (35 ft 0 in)
Wing chord, constant	1.12 m (3 ft 8 in)
Wing aspect ratio	9.6
Length: overall	6.78 m (22 ft 3 in)
wings folded	7.47 m (24 ft 6 in)
Height overall: tricycle	2.77 m (9 ft 1 in)
tailwheel	2.11 m (6 ft 11 in)
Tailplane span	3.28 m (10 ft 9 in)
Propeller diameter	1.83 m (6 ft 0 in)
Doors (each): Height	0.80 m (2 ft 7½ in)
Width	0.94 m (3 ft 1 in)
Height to sill	0.84 m (2 ft 9 in)

DIMENSIONS, INTERNAL:

Cabin max width	1.17 m (3 ft 10 in)
Baggage compartment volume	0.91 m³ (32.0 cu ft)

AREAS:

Wings, gross	11.89 m² (128.0 sq ft)
Ailerons (total)	0.37 m² (3.97 sq ft)
Trailing-edge flaps (total)	0.65 m² (6.95 sq ft)
Fin	1.29 m² (13.92 sq ft)
Rudder	0.47 m² (5.08 sq ft)
Tailplane	1.11 m² (12.00 sq ft)
Elevators (total)	0.99 m² (10.67 sq ft)

WEIGHTS AND LOADINGS:

Weight empty	544 kg (1,200 lb)
Fuel weight	92 kg (203 lb)
Baggage capacity	113 kg (250 lb)
Max T-O weight	889 kg (1,960 lb)
Max wing loading	74.8 kg/m² (15.31 lb/sq ft)
Max power loading	7.46 kg/kW (12.25 lb/hp)

Stoddard-Hamilton GlaStar assembled in the USA *(Paul Jackson/Jane's)*   2001/0093685

PERFORMANCE (119 kW, 160 hp, engine, constant-speed propeller):

Max level speed	145 kt (269 km/h; 167 mph)
Max cruising speed at 75% power	140 kt (259 km/h; 161 mph)
Stalling speed, power off:	
flaps up	49 kt (91 km/h; 57 mph)
flaps down	43 kt (79 km/h; 49 mph)
Max rate of climb at S/L	632 m (2,075 ft)/min
Service ceiling	6,096 m (20,000 ft)
T-O run	76 m (250 ft)
Landing run	91 m (300 ft)
Range with max fuel	828 n miles (1,533 km; 953 miles)
Endurance with max fuel	6 h
g limits	+3.8/–1.5

UPDATED

UNITED STATES AIR FORCE

AIR FORCE MATERIEL COMMAND

3753 Chidlaw Road, Wright-Patterson AFB, Ohio 45433-5006
Tel: (+1 513) 257 63 06
Fax: (+1 513) 257 25 58
e-mail: HQAFMC.PA@wpafb.af.mil
Web: http://www.afmc.wpafb.af.mil
COMMANDER, AFMC: General Lester Lyles
DIRECTOR, PUBLIC AFFAIRS: Col Donna Pastor

Formed 1 July 1992, incorporating former Air Force Logistics Command and Air Force Systems Command. Resulting Air Force Materiel Command (AFMC) is composed of four product centres, each of which is responsible for developing and acquiring specific weapon systems with the assistance of the Air Force Research Laboratory at Wright-Patterson AFB, Ohio and the Air Force Office of Scientific Research at Bolling AFB, Washington, DC. These comprise the Aeronautical Systems Center, Wright-Patterson AFB, Ohio; the Air Armament Center at Eglin AFB, Florida; the Electronic Systems Center, Hanscom AFB, Massachusetts; and the Space and Missile Systems Center, Los Angeles AFB, California.

AFMC also has two test establishments, namely Air Force Flight Test Center at Edwards AFB, California; and Arnold Engineering Development Center at Arnold AFB, Tennessee. These are responsible for evaluation of systems under development and in the field.

Five Air Logistics Centers (Ogden ALC at Hill AFB, Utah; Oklahoma City ALC at Tinker AFB, Oklahoma; Sacramento ALC at McClellan AFB, California; San Antonio ALC at Kelly AFB, Texas; and Warner Robins ALC at Robins AFB, Georgia) provide weapon system sustainment and maintenance services for just under 50 types of aircraft, with a total inventory of about 6,200 airframes and 20,000 engines; however, facilities at McClellan and Kelly both scheduled for closure by July 2001. Finally, two specialist organisations (Aerospace Maintenance and Regeneration Center at Davis-Monthan AFB, Arizona, and Air Force Security Assistance Center at Wright-Patterson AFB, Ohio provide for storage, regeneration, reclamation and disposal of aircraft and associated aerospace equipment.

At the end of 1999, AFMC manpower resources totalled 88,713 (27,942 military; 60,771 civilian), with the budget appropriation request for FY01 being US$39.5 billion.
UPDATED

REPLACEMENT INTERDICTOR AIRCRAFT (RIA)

Exploratory studies, for an aircraft to replace the Lockheed Martin F-117A Nighthawk and Boeing F-15E Eagle in the interdiction role and also augment the bomber force, have been launched by the US Air Force, with Boeing, Lockheed Martin and Northrop Grumman undertaking studies into a potential RIA. Service entry approximately 2015-20. No further news received by mid-2000.
UPDATED

UNITED STATES NAVY

NAVAL AIR SYSTEMS COMMAND

47123 Buse Road, Unit Moffett Building, Patuxent River, Maryland 20650-1547
Tel: (+1 301) 757 78 25
Web: http://www.navair.navy.mil

NAVAIR has over 31,500 military and civilian employees and provides total life cycle support of all naval aviation weapon systems. Subordinate elements include test organisations at China Lake, California; Patuxent River, Maryland and Point Mugu, California, as well as repair and overhaul depots at Cherry Point, North Carolina; Jacksonville, Florida; and North Island, California. It is prime authority for the USAF/USN Joint Strike Fighter, via the Program Executive Office (PEO); other PEOs have responsibility for tactical aircraft programmes; air ASW, assault and special mission programmes; and strike weapons and unmanned aviation.
UPDATED

JOINT STRIKE FIGHTER (JSF)

Note that specific details of the two competing JSF designs appear in the Boeing and Lockheed Martin entries earlier in the section.
TYPE: Multirole fighter.
PROGRAMME: Origins of Joint Strike Fighter (JSF) programme vested in separate USAF/USN Joint Advanced Strike Technology (JAST) and Defense Advanced Research Project Agency (DARPA) Common Affordable Lightweight Fighter (CALF) projects of early 1990s; designation **X-32** then assigned to planned CALF demonstrator (see US Navy entry in 1997-98 and previous editions of *Jane's* for more details of JAST and CALF programmes).

Projects merged in November 1994, as JAST, after Congressional directive in mid-1994; programme renamed JSF in latter half of 1995. Previously, formal request for proposals (RFP) for preliminary research contracts released on 2 September 1994, stipulating industry response by 4 November and issue of contract awards by 16 December.

Some elements of US industry joined forces to win JAST/JSF work, with international collaboration in evidence. McDonnell Douglas led one team after signing October 1994 memorandum of understanding (MoU) with Northrop Grumman and British Aerospace; each company submitted individual bids, but all three to participate in event of securing contract. Boeing also allied with Dassault of France on aspects of subsystem design effort.

Subsequent research contracts worth US$99.8 million were distributed between four companies: Boeing (US$27.6 million), Lockheed Martin (US$19.9 million), McDonnell Douglas (US$28.2 million) and Northrop Grumman (US$24.1 million). Further US$28 million allocated for associated avionics, propulsion systems, structures and materials, and modelling and simulation.

Merger of JAST and CALF resulted in expanded flight test programme, involving two finalists; each to build two demonstrators, one with ASTOVL capability and the other to use conventional take-off and landing (CTOL). Both variants to be built on common assembly line, with production for US and UK military expected to total around 3,000, including just over 750 ASTOVL-configured aircraft.

Draft RFP issued December 1995, with USA and UK signing MoU on 20 December 1995, which committed UK to participate in four year weapons system concept demonstration (WSCD) phase. MoU also stipulated that UK must contribute some 10 per cent (approximately US$200 million) of demonstration phase costs as full collaborative partner.

Formal release of the final RFP for JSF was expected on 7 March 1996, but was delayed to June 1996, with contract award date in November 1996. X-32 and **X-35** designations allocated to demonstration phase which was planned to conclude in February 2001, although it now appears likely to continue until mid-year; successful teams will each build two aircraft, with CTOL version to fly first. STOVL versions to fly second and participate in assessment and demonstration of hover and transition qualities.

Three candidates were in contention for WSCD, originating from Boeing, Lockheed Martin and McDonnell Douglas/British Aerospace/Northrop Grumman.

All three contenders chose Pratt & Whitney's F119 engine for their WSCD proposals, although a General Electric/Allison/Rolls-Royce team secured a US$7 million contract in March 1996 to examine alternative power plants. These were based on the General Electric F110 and YF120 engines, with the latter being chosen in May 1996 following Congressional directive aimed at fostering competition and also overcoming possible impact of developmental or operational problems with the F119. Further US$96 million multiyear contract awarded in February 1997 to cover technology maturation and core engine development of YF120-FX version over four year period; this likely to result in follow-on development programme starting in 2001, culminating in full-scale EMD from 2004. If necessary, it is planned that the F120 engine will be available from the 72nd production aircraft onwards.

On 16 November 1996, US Secretary of Defense, William J Perry announced that Boeing and Lockheed Martin had been chosen to participate in forthcoming WSCD and that the team headed by McDonnell Douglas had been eliminated. Simultaneously, Boeing was awarded a US$661.8 million contract for the next phase of the JSF programme, while Lockheed Martin received US$718.8 million; in addition, Pratt & Whitney secured a contract worth US$804 million for the associated Engine Ground and Flight Demonstration Program. Subsequently, Northrop Grumman and British Aerospace joined Lockheed Martin team.

Boeing's X-32 design incorporates a direct-lift concept using a two-dimensional thrust vectoring propulsion nozzle and swivelling retractable lift nozzles for the X-32B STOVL version; the X-32A CTOL derivative lacks STOVL features. Trials with a 94 per cent scale YF119-powered model of the STOVL configuration began at Tulalip, near Seattle, during 1995, as part of more than 5,300 hours of JAST/JSF-related testing which was completed in second quarter of 1996.

Lockheed Martin's X-35 design has a trapezoidal wing planform which initially featured foreplanes, although these since deleted; STOVL version embodies a lift fan, shaft-driven by a modified F119 with a vectoring lift/cruise nozzle developed by Rolls-Royce; lift fan replaced by extra fuel in the CTOL version. Lockheed Martin also turned to Russia for technical expertise, purchasing design data from Yakovlev; and used an 86 per cent subscale model (originally developed for the CALF project and fitted with a Pratt & Whitney F100-PW-220 engine plus an Allison shaft-driven lift fan) for testing. This was initially hover tested on an outdoor stand at NASA's Ames Research Center during July and August 1995, before being installed in the 24 × 36 m wind tunnel at Mountain View, California, in September for aerodynamic hover and transition trials which began in December. Wind-tunnel testing concluded on 5 March 1996, marking end of three-year effort to design, build and test large-scale powered model of VTOL version of JSF; this included 196 hours of propulsion system testing in 1995-96, representative of about 2,400 vertical take-offs and landings.

Further international interest resulted in three European nations joining JSF programme as limited collaborative partners at cost of US$10 million each, spread over five year period; Netherlands and Norway signed MoU committing them to participate in WSCD programme on 16 April 1997, with Denmark following suit later in year. In each case, JSF viewed as potential replacement for F-16 Fighting Falcon; Italy became 'informed partner' in April 1998 and announced intention in January 1999 of deeper involvement; Turkish partnership agreement signed 16 June 1999 as Foreign Military Sales customer (fourth-level participation). Other nations known to have expressed interest include Australia, Canada, France, Germany, Greece, Israel, Singapore, Spain and Sweden, which have all been briefed on JSF programme. For engineering and manufacturing development phase, four partnership options available. Most costly is Level 1, which entails responsibility for 10 per cent of cost; UK is only partner at this level. Italy, Netherlands and Turkey are Level 2 partners, each contributing 5 per cent of cost. Level 3 involves payment of 1 to 2 per cent, with Denmark and Norway having teamed up to share burden, while Canada meets the cost alone. Finally, Foreign Military Sales Level involves minimum contribution of US$75 million, but no subscribers by mid-2000.

Recent developments (1997-2000) include successful completion of initial and final design reviews. Lockheed Martin passed initial design review milestone in mid-June 1997, with Boeing following suit at beginning of September 1997, allowing both companies to begin the fabrication and assembly process. Final design review hurdle passed by Lockheed Martin in third quarter of 1998, with Boeing candidate completing this in fourth quarter; by then, both manufacturers reporting good progress with assembly of prototypes and Lockheed Martin had produced full-size mockup.

Notable power plant-related events and developments during same period include critical design review (CDR) of Pratt & Whitney F119 derivatives (SE611 for X-35; SE614 for X-32). CDR began in August 1997 and was completed successfully in third quarter of year, allowing work to start on assembly of both basic types of engine in October; ground testing of all four engine variants began during 1998, with SE611C for X-35A, SE611S for X-35B, SE614C for X-32A and SE614S for X-32B.

Multifunction integrated RF system/multifunction nose array (MIRFS/MFA) avionics package also specified for JSF during latter half of 1997; Northrop Grumman Electronic Systems and Sensors Division in competition with Raytheon Systems Company, with selection of winning equipment expected in 2001. Earlier, on 4 June 1997, Raytheon chosen to provide integrated core processor for Lockheed Martin X-35.

On armament front, consideration was given to development of advanced 25 mm calibre gun at cost of about US$60 million. This idea subsequently abandoned and Boeing/Mauser proposal for 27 mm gun adopted by both JSF teams in May 1999; integral carriage a possibility and has been endorsed by USAF, although US Navy has no requirement for a gun and is concerned about potential penalties associated with provision of integral armament; USMC known to prefer pod-mounted weapon and Royal Navy likely to follow suit.

Plans also revealed for various test projects allied to JSF programme. First announcement, in October 1997, concerned use of UK Defence Evaluation and Research Agency (DERA) vectored-thrust aircraft advanced control (VAAC) first-generation Harrier two-seater in testing flight control systems, which duly took place in 1998. Lockheed Martin revealed plans, in November 1997, to employ AFTI/F-16 for integrated subsystem technology demonstration of electrically operated flight control actuator system and modular 270 V DC electrical power system. Installation completed in 1998, with aircraft used for six month test and evaluation programme at Edwards AFB, California, during 1999.

JSF acquisition planning little affected by Quadrennial Defense Review of May 1997 and production run of successful design for US and UK still expected to number at least 3,000, following RAF revelation in May 1997 that it was considering CTOL version as replacement for Harrier GR. Mk 7. Subsequently, in June, USAF launched year-long study of plans to include some STOVL-configured aircraft in overall purchase as replacement for A-10 Thunderbolt II in CAS/BAI role; further announcement on 23 September 1997 revealed that USAF considering buying sufficient STOVL JSFs to equip two Fighter Wings (approximately 200 aircraft in total, including allowance for pilot training and maintenance requirements).

Lockheed Martin expected to fly demonstrators in first half of 2000, but had not done so by mid-October. Previously, in late 1999, Lockheed Martin modified original plan; initial aircraft will still be used to test CTOL as the X-35A, before being reconfigured as the X-35B for the STOVL demonstration; second aircraft, which was originally to be used as STOVL demonstrator, will now emerge as X-35C with carrier-compatible attributes of US Navy version. Boeing formally unveiled X-32A and X-32B variants at Palmdale on 14 December 1999; X-32A CTOL version then expected to make its first flight by April 2000 with the X-32B ASTOVL derivative following in mid-2000. However, maiden flight of X-32A delayed until 18 September 2000. This took place from Palmdale, with aircraft landing at Edwards AFB at end of 20-minute sortie; landing gear locked down throughout and not retracted until 10th flight on 10 October. At that time, Boeing expecting first flight of X-32B to take place in first quarter of 2001, this having completed initial series of engine runs on 21 September 2000. Following assessment of proposals and flight test results, selection of winning design currently expected to occur in May-June 2001, at which time engineering and manufacturing development (EMD) contract due to be awarded; however, EMD go-ahead may slip as consequence of possible budget cut imposed by Congress and could be delayed until October 2001, or even later.

Final Joint Operational Requirements Document (JORD) issued in March 2000, providing basis for proposals for the EMD phase that will follow the WSCD. The latter is expected to occupy some 48 to 51 months of combined flight and ground testing, although the two contractors involved in the WSCD are to be given considerable latitude in determining which aspects will be evaluated in flight and which by ground trials. EMD now due to start in June 2001 and extend to 2008, with the first flight of an EMD-dedicated JSF set for 2005, followed by the start of operational testing in 2008. Total of 10 aircraft to be produced for EMD, with all three planned variants represented.

Delivery of the first operational aircraft is scheduled for early in FY08 and the US production rate is expected to rise to a peak of 122 per year by 2011. Current plans envisage a start to be made on quantity production of the successful design in 2004, with procurement of long-lead items for phase one of low-rate initial production (LRIP 1); this will involve a total of 12 aircraft, with the first delivery in 2008, subject to release of full funding in 2005. Thereafter, three more LRIP batches are to be purchased, for delivery between the third quarter of 2007 and 2010, before the initial phase of full-rate production (FRP 1) is launched with long-lead items in 2007; this will be dependent upon full funding being approved in 2008, coincident with the JSF attaining IOC.

Affordability is a key consideration for JSF and is dictating some compromises on the part of the different elements of the US armed forces in order to meet unit flyaway cost targets; in first quarter of 1997, all versions comfortably within or below specified cost, with the latest target prices (1997) being US$28 million for the CTOL version, US$35 million for the STOVL version and US$38 million for the carrier-capable version.

Boeing's (X-32A) JSF contender was first to fly

JSF REQUIREMENTS
(1999 estimates)

Service	Qty	Remarks
US Air Force	1,763	Replaces A-10 and F-16; complements F-22. IOC 2011
US Navy	480	Complements F/A-18E/F. IOC 2012
US Marine Corps	609	Replaces AV-8B and F/A-18. IOC 2010
Royal Navy (UK)	60	Replaces Sea Harrier. IOC 2012
Royal Air Force (UK)	90	Replaces Harrier
Total	**3,002**	

JSF INITIAL PRODUCTION PLAN

Batch	Qty	FY	First Delivery
LRIP 1	13	05	2008
LRIP 2	30	06	2008
LRIP 3	54	07	
LRIP 4	86	08	
LRIP 5	115	09	
LRIP 6	166	10	
Total	**464**		

UPDATED

MULTIMISSION MARITIME AIRCRAFT (MMA)

In late 1997, US Navy initiated requirements study, lasting two years, into replacement for Lockheed Martin P-3C Orion maritime aircraft, as well as intelligence-gathering EP-3E, submarine-communications Boeing E-6 Mercury and tanker/transport Lockheed Martin KC-130 Hercules of US Marine Corps. Latter type to be replaced by new KC-130J aircraft from 2000 on, but maritime requirement remains. Further exploratory study contracts awarded in July 2000, when it was anticipated that service entry of MMA could occur as early as 2010 and will be no later than 2015.

Options being considered by US Navy include new development aircraft, off-the-shelf purchase of commercial airliners such as Boeing 737, or executive jets like Gulfstream V, acquisition of new-build Lockheed Martin P-3 Orion 21s with new engines, propellers and avionics, and remanufacturing existing P-3 Orions. Last-named is most affordable option, but least desirable. Analysis of alternatives expected to be completed in 2001, at which time US Navy will decide future course of MMA project. At present, development is likely to start in 2004, assuming continuation of MMA.

UPDATED

US LIGHT AIRCRAFT

US LIGHT AIRCRAFT CORPORATION
27080 Rancho Ballena Lane, Ramona, California 92065
Tel/Fax: (+1 619) 789 86 07
Web: http://www.flyhornet.com
PRESIDENT: James Millet

US Light Aircraft formed from partnership between Millett Engineering and Forespar Company in 1993.

NEW ENTRY

US LIGHT AIRCRAFT HORNET

TYPE: Tandem-seat ultralight kitbuilt.
PROGRAMME: Design began in 1991 and construction of prototype in 1992; first flight June 1993. Winner of 15 awards at Copperstate, Oshkosh and Sun 'n' Fun between 1993 and 1999. Qualifies under FAR Pt 21.191 as 51 per cent experimental kit.
CUSTOMERS: About 40 sold and 25 flying by mid-2000.
COSTS: Kit US$18,550, including engine; US$12,500 without engine (2001).
DESIGN FEATURES: 400 parts in kits include 7,500 factory-installed rivets. Quoted build time 300 hours.
FLYING CONTROLS: Conventional and manual. Electrically operated flaps with maximum deflection of 30°. Horizontal stabiliser trimmed electrically. Mass-balanced elevator.
STRUCTURE: Fabric-covered fuselage of 6063-T8 aluminium tube with tapered-base tubular longitudinal structure. Fabric-covered dual-spar wings with single bracing struts.
LANDING GEAR: Fixed tricycle type with 15 cm (6 in) Goodyear Airspring pneumatic suspension on mainwheels and 7.5 cm (3 in) Airspring on nosewheel; hydraulic brakes on mainwheels with aluminium Hegar wheels and 15×6-6 Cheng Shin tyres; nose tyre 13×5-6.
POWER PLANT: One 41.0 kW (55 hp) Hirth 2703 engine with electric start is standard, driving two-blade pusher propeller of diameter up to 1.52 m (5 ft 0 in); 48.5 kW (65 hp) Hirth 2706 optional. Fuel capacity 68 litres (18.0 US gallons; 15.0 Imp gallons) in two tanks between spars.

US Light Aircraft Hornet *(Geoffrey P Jones)*   2001/0100462

DIMENSIONS, EXTERNAL:
Wing span	8.38 m (27 ft 6 in)
Length overall	6.10 m (20 ft 0 in)

AREAS:
Wings, gross	12.82 m² (138.0 sq ft)

WEIGHTS AND LOADINGS:
Weight empty	215 kg (475 lb)
Max T-O weight	454 kg (1,000 lb)

PERFORMANCE (41.0 kW; 55 hp engine):
Never-exceed speed (V$_{NE}$)	108 kt (201 km/h; 125 mph)
Max level speed	87 kt (161 km/h; 100 mph)
Normal cruising speed at 75% power	61 kt (113 km/h 70 mph)
Stalling speed, flaps down	35 kt (65 km/h; 40 mph)
Max rate of climb at S/L	152 m (500 ft)/min
T-O run	122 m (400 ft)
Landing run	76 m (250 ft)
Range, 10% reserves	373 n miles (692 km; 430 miles)

NEW ENTRY

VAN'S

VAN'S AIRCRAFT INC
14401 NE Keil Road, Aurora, Oregon 97002
Tel: (+1 503) 678 65 45
e-mail: info@vansaircraft.com
Web: http://www.vansaircraft.com
PRESIDENT: Richard VanGrunsven

Company derives its name from its president's nickname and began in 1973 with the sale of plans (and, later, kits) for the RV-3 sporting homebuilt, details of which can be found in the 1992-93 *Jane's*. More than 9,000 Van's kits sold by September 2000, of which 2,488 were then flying. The company constructed a new 5,110 m² (55,000 sq ft) facility at Aurora, Oregon, to which it moved during 2000. All Van's kits conform to the FAA's 51 per cent rule.

UPDATED

VAN'S RV-4

TYPE: Tandem-seat kitbuilt.
PROGRAMME: First flight of prototype 21 August 1979. Plans and kits available to homebuilders.
CURRENT VERSIONS: **RV-4**: *As described.*
CUSTOMERS: Over 4,600 sets of plans and kits sold, with about 1,800 aircraft under construction and 950 RV-4s flying (2000 figures).
COSTS: Kit: US$12,116 (2000). Information pack: US$8.
DESIGN FEATURES: Cantilever low-wing monoplane with constant-chord, low aspect ratio wings, tapered horizontal tail surfaces and sweptback fin.
Wing section Van's Aircraft 135; dihedral 3° 30'; incidence 1°.
FLYING CONTROLS: Conventional and manual. Horn-balanced elevators; Frise ailerons; plain flaps. No tabs.
STRUCTURE: Generally of light alloy. I-beam main wing spar and pressed ribs; moulded glass fibre wingtips and engine cowling.
LANDING GEAR: Tailwheel type; fixed cantilever tapered steel spring main gear struts. Glass fibre speed fairings on mainwheels, tyre size 5.00-5; steerable tailwheel, tyre size 6 in.
POWER PLANT: One 112 kW (150 hp) Textron Lycoming O-320-E1F flat-four engine; two-blade fixed-pitch propeller. Options for engine up to 134 kW (180 hp). Fuel capacity 121 litres (32.0 US gallons; 26.6 Imp gallons).
ACCOMMODATION: Two persons in tandem under starboard-hinged canopy. Baggage compartments forward of instrument panel and aft of rear seat.

DIMENSIONS, EXTERNAL:
Wing span	7.01 m (23 ft 0 in)
Wing aspect ratio	4.8
Length overall	6.15 m (20 ft 2 in)
Height overall	1.65 m (5 ft 5 in)
Wheel track	1.88 m (6 ft 2 in)
Propeller diameter	1.73 m (5 ft 8 in)

AREAS:
Wings, gross	10.22 m² (110.0 sq ft)

WEIGHTS AND LOADINGS:
Weight empty	411 kg (905 lb)
Baggage capacity	13 kg (30 lb)
Max T-O weight	680 kg (1,500 lb)
Max wing loading	66.6 kg/m² (13.64 lb/sq ft)
Max power loading	6.09 kg/kW (10.00 lb/hp)

PERFORMANCE (112 kW; 150 hp engine; MTOW):
Max level speed at S/L	174 kt (322 km/h; 200 mph)
Cruising speed at 2,440 m (8,000 ft):	
at 75% power	163 kt (303 km/h; 188 mph)
at 55% power	148 kt (274 km/h; 170 mph)
Stalling speed	47 kt (87 km/h; 54 mph)
Max rate of climb at S/L	457 m (1,500 ft)/min
Service ceiling	5,485 m (18,000 ft)
T-O run	145 m (475 ft)
Landing run	130 m (425 ft)
Range with max fuel at 55% power	738 n miles (1,367 km; 850 miles)

UPDATED

VAN'S RV-6 and RV-6A

TYPE: Side-by-side kitbuilt.
PROGRAMME: Adaptation of RV-4 with side-by-side seats, initially demonstrated by private conversion (N6RV, modified by Art Chard) of RV-3 in 1977. First true RV-6 prototype (N66RV) flew 1985; RV-6A followed in July 1988.
CURRENT VERSIONS: **RV-6**: Tailwheel version.
RV-6A: Nosewheel version.
Air Beetle: See AIEP entry in Nigerian section of 1996-97 and earlier *Jane's*.
CUSTOMERS: More than 6,430 sets of RV-6/RV-6A plans and kits sold, with about 3,000 aircraft under construction and 700 RV-6s and 520 RV-6As flying by mid-2000.
COSTS: RV-6 kit US$13,316; quick-build kit US$20,816 (2000). Information pack: US$8. RV-6A kit US$14,143; quick-build kit US$21,643 (2000).
DESIGN FEATURES: Side-by-side, two-seat derivative of RV-4. *Data as for RV-4, except those below.*
LANDING GEAR: See Current Versions. Floatplane conversions offered by Copper Island Aviation of British Columbia.

Van's RV-4 kitbuilt (left) and an enhanced performance modification, the Harmon Rocket (Lycoming O-540 186 kW; 250 hp flat-six) *(Paul Jackson/Jane's)*   2001/0093686

VAN'S—AIRCRAFT: USA 785

POWER PLANT: One 112 to 134 kW (150 to 180 hp) Textron Lycoming flat-four engine. Fuel capacity 144 litres (38.0 US gallons; 31.6 Imp gallons). Powersport rotary-powered example flown 9 July 2000.
ACCOMMODATION: Side-by-side seating. Baggage compartments forward of instrument panel and behind seats. Hinged canopy.
DIMENSIONS, EXTERNAL:
Length overall: RV-6	6.15 m (20 ft 2 in)
RV-6A	6.02 m (19 ft 9 in)
Height overall: RV-6	1.60 m (5 ft 3 in)
RV-6A	2.03 m (6 ft 8 in)
Wheel track: RV-6, RV-6A	2.03 m (6 ft 8 in)

DIMENSIONS, INTERNAL:
Cabin max width	1.09 m (3 ft 7 in)

AREAS: As RV-4
WEIGHTS AND LOADINGS:
Weight empty: RV-6	438 kg (965 lb)
RV-6A	447 kg (985 lb)
Baggage capacity: RV-6, RV-6A	27 kg (60 lb)
Max T-O weight: RV-6	725 kg (1,600 lb)
RV-6A	748 kg (1,650 lb)
Max wing loading: RV-6	71.0 kg/m² (14.55 lb/sq ft)
RV-6A	73.2 kg/m² (15.00 lb/sq ft)
Max power loading: RV-6:	
150 hp	6.49 kg/kW (10.67 lb/hp)
180 hp	5.41 kg/kW (8.89 lb/hp)
RV-6A: 150 hp	6.70 kg/kW (11.00 lb/hp)
180 hp	5.58 kg/kW (9.17 lb/hp)

PERFORMANCE (112 kW; 150 hp engine; both variants, except where indicated):
Max level speed at S/L:	
RV-6	171 kt (317 km/h; 197 mph)
RV-6A	169 kt (313 km/h; 195 mph)
Cruising speed at 2,440 m (8,000 ft): at 75% power:	
RV-6	162 kt (299 km/h; 186 mph)
RV-6A	160 kt (296 km/h; 184 mph)
at 55% power: RV-6	146 kt (270 km/h; 168 mph)
RV-6A	145 kt (269 km/h; 167 mph)
Stalling speed	48 kt (89 km/h; 55 mph)
Max rate of climb at S/L: RV-6	413 m (1,355 ft)/min
RV-6A	398 m (1,305 ft)/min
Service ceiling: RV-6	4,815 m (15,800 ft)
RV-6A	4,495 m (14,750 ft)
T-O run: RV-6	168 m (550 ft)
RV-6A	171 m (560 ft)
Landing run	155 m (500 ft)
Range with max fuel at 55% power:	
RV-6	825 n miles (1,529 km; 950 miles)
RV-6A	760 n miles (1,408 km; 875 miles)

UPDATED

VAN'S RV-8 and RV-8A

TYPE: Tandem-seat kitbuilt.
PROGRAMME: First flown (N118RV) 22 July 1995; launched at Oshkosh five days later. Second prototype (N58RV) flew early 1997 and accumulated 325 hours in seven months.

Van's RV-6 tailwheel kitbuilt *(Paul Jackson/Jane's)*

Van's RV-6 lightplane, with additional side view of RV-6A *(Paul Jackson/Jane's)*

CURRENT VERSIONS: **RV-8**: Tailwheel version.
RV-8A: Tricycle version; first flown (N58VA) April 1998; public debut at Sun 'n' Fun, same month.
CUSTOMERS: By mid-2000, 1,300 sets of plans and kits had been sold and 100 were flying.
COSTS: RV-8 kit US$15,554, quick-build kit US$23,054; RV-8A kit US$15,908, quick-build kit US$23,408 (2000 prices).
DESIGN FEATURES: Similar in appearance to the RV-4, but with wider, longer fuselage; rounded engine cowling; non-sweptback main landing gear legs; larger canopy; however, very little parts commonality with RV-4.
Description generally as for RV-4/RV-6.
LANDING GEAR: Wittman leaf-spring main legs; nose- or tailwheel. Speed fairings on mainwheels and (when fitted) nosewheel.

POWER PLANT: One 149 kW (200 hp) Textron Lycoming IO-360-A16D four-cylinder piston engine driving two-blade Hartzell HC-C2YK-BF/F7666A4 constant-speed propeller. Fuel capacity 159 litres (42.0 US gallons; 35.0 Imp gallons). Oil capacity 7.6 litres (2.0 US gallons; 1.7 Imp gallons).
ACCOMMODATION: Tandem seating for pilot and passenger under sliding canopy. Separate windscreen.
Specification for RV-8/RV-8A, except where noted.
DIMENSIONS, EXTERNAL:
Wing span	7.01 m (23 ft 0 in)
Wing chord, constant	1.47 m (4 ft 10 in)
Wing aspect ratio	4.8
Length overall: RV-8	6.40 m (21 ft 0 in)
RV-8A	6.35 m (20 ft 10 in)
Height overall: RV-8	1.70 m (5 ft 7 in)
RV-8A	2.24 m (7 ft 4 in)
Wheel track	1.90 m (6 ft 3 in)
Propeller diameter	1.83 m (6 ft 0 in)

DIMENSIONS, INTERNAL:
Cockpit max width	0.91 m (3 ft 0 in)

AREAS:
Wings, gross	10.22 m² (110.0 sq ft)

WEIGHTS AND LOADINGS:
Weight empty: RV-8	494 kg (1,090 lb)
RV-8A	508 kg (1,120 lb)
Baggage capacity	45 kg (100 lb)
Max T-O weight	816 kg (1,800 lb)
Max wing loading	79.9 kg/m² (16.36 lb/sq ft)
Max power loading	5.48 kg/kW (9.00 lb/hp)

PERFORMANCE (two persons, 149 kW; 200 hp engine):
Max operating speed: RV-8	191 kt (354 km/h; 220 mph)
RV-8A	189 kt (351 km/h; 218 mph)
Cruising speed at 55% power at 2,440 m (8,000 ft):	
RV-8	162 kt (301 km/h; 187 mph)
RV-8A	161 kt (298 km/h; 185 mph)
Stalling speed, power off:	
one occupant: both	45 kt (83 km/h; 51 mph)
two occupants: both	51 kt (94 km/h; 58 mph)
Max rate of climb at S/L: RV-8	579 m (1,900 ft)/min
RV-8A	549 m (1,800 ft)/min
Service ceiling: RV-8	6,860 m (22,500 ft)
RV-8A	6,550 m (21,500 ft)
T-O run: both	155 m (500 ft)
Landing run: RV-8	152 m (500 ft)
RV-8A	168 m (550 ft)
Range at 55% power:	
RV-8	869 n miles (1,609 km; 1,000 miles)
RV-8A	773 n miles (1,432 km; 890 miles)

UPDATED

VAN'S RV-9A

TYPE: Side-by-side kitbuilt.
PROGRAMME: Prototype RV-9A (N96VA) first flown late 1997; was hybrid proof-of-concept aircraft using unique RV-6T fuselage plus new wings; had flown more than 200 hours when lost in accident 2 April 2000; second prototype, and first 'true' RV-9A (N129RV), flown 15 June 2000. Empennage kits became available in late 1999, to be followed by wing and fuselage/

Van's RV-8 tailwheel version *(Paul Jackson/Jane's)*

Van's RV-8 two-seat light aircraft, with additional side view of RV-8A *(Paul Jackson/Jane's)*

USA: AIRCRAFT—VAN'S to VIPER

finishing kits in mid-2000 and quick-build kits in late 2000.

DESIGN FEATURES: Based upon, and similar in appearance to, RV-6A, but with redesigned wing (Roncz aerofoil and sheared wingtips) and enlarged, constant-chord horizontal tail surfaces.

FLYING CONTROLS: Conventional and manual. Horn-balanced rudder. Slotted flaps extend approximately two-thirds of wing span, compared with one-half on other RV designs.

LANDING GEAR: Tricycle type; fixed. Brakes.

POWER PLANT: First prototype flown with 88.0 kW (118 hp) Textron Lycoming O-235-L2C flat-four driving two-blade Sensenich 72 × 66 fixed-pitch metal propeller. Second prototype has 119 kW (160 hp) Textron Lycoming O-320-D3G driving three-blade MT propeller. Engines between 80.5 and 119 kW (108 and 160 hp) can be fitted. Fuel capacity 129 litres (34.0 US gallons; 28.3 Imp gallons).

ACCOMMODATION: Pilot and passenger side by side under sliding canopy. Dual controls.

DIMENSIONS, EXTERNAL:
Wing span	8.53 m (28 ft 0 in)
Length overall	6.17 m (20 ft 3 in)
Height overall	2.26 m (7 ft 5 in)
Propeller diameter	1.83 m (6 ft 0 in)

AREAS:
Wings, gross	11.52 m² (124.0 sq ft)

WEIGHTS AND LOADINGS (119 kW; 160 hp engine):
Weight empty	460 kg (1,015 lb)
Baggage capacity	34 kg (75 lb)
Max T-O weight	793 kg (1,750 lb)
Max wing loading	68.9 kg/m² (14.11 lb/sq ft)
Max power loading	6.66 kg/kW (10.94 lb/hp)

PERFORMANCE (119 kW; 160 hp engine):
Max operating speed	168 kt (312 km/h; 194 mph)
Cruising speed: at 75% power	162 kt (299 km/h; 186 mph)
at 55% power	144 kt (267 km/h; 166 mph)
Stalling speed, flaps down	43 kt (79 km/h; 49 mph)
Max rate of climb at S/L (solo)	396 m (1,300 ft)/min
Service ceiling (solo)	5,640 m (18,500 ft)
T-O run	122 m (400 ft)
Landing run	137 m (450 ft)
Range at 75% power	634 n miles (1,174 km; 730 miles)
Endurance, no reserves	7 h 0 min

UPDATED

VAT

VERTICAL AVIATION TECHNOLOGIES INC

PO Box 2527, Sanford, Florida 32772-2527
Tel: (+1 407) 322 94 88
Fax: (+1 407) 330 26 47
PRESIDENT: Bradley G Clark

Vertical Aviation Technologies Inc is FAA approved repair facility for various Sikorsky helicopters. The company remanufactures Sikorsky S-55/H-19s and S-58/H-34s, as described in *Jane's Aircraft Upgrades*. In 1984, began to modify four-seat Sikorsky S-52-3 former production helicopter (of which 95 originally produced, mainly for US Marine Corps as HO5S) into kit for assembly by individuals, corporations or military; all tooling and fixtures completed by 1988.

UPDATED

VAT HUMMINGBIRD

TYPE: Four-seat helicopter kitbuilt.

PROGRAMME: Based on previously type-certified Sikorsky S-52-3 and first flown 1990. (S-52 prototype flew 12 February 1947; total 95 built.)

CUSTOMERS: Fourteen Hummingbird kits sold by early 1999, of which 11 flying.

COSTS: Kit US$140,500 (2000); plans US$375 (2000).

DESIGN FEATURES: Newly manufactured airframes and tailcones identical to S-52-3, except for nose section. Quoted build time 1,500 hours; kit includes original Sikorsky components.

FLYING CONTROLS: Conventional and manual. Two anhedral stabilisers.

STRUCTURE: Mainly aluminium, with GFRP nose and tailcone.

LANDING GEAR: Four-wheel type, fixed; hydraulic brakes.

POWER PLANT: One 194 kW (260 hp) Ford Taurus aluminium block V-8 water-cooled engine driving three-blade fully articulated main rotor and two-blade tail rotor. Fuel capacity 216 litres (57.0 US gallons; 47.5 Imp gallons). A Textron Lycoming power plant installation is currently under development.

ACCOMMODATION: Four persons, including pilot(s).

DIMENSIONS, EXTERNAL:
Main rotor diameter	10.06 m (33 ft 0 in)
Tail rotor diameter	1.75 m (5 ft 9 in)
Length: overall, rotors turning	12.11 m (39 ft 9 in)
fuselage	9.30 m (30 ft 6 in)
Height: to top of rotor head	2.62 m (8 ft 7 in)
overall	2.95 m (9 ft 8 in)
Fuselage max width	1.52 m (5 ft 0 in)
Wheel track: forward	1.83 m (6 ft 0 in)
rear	2.49 m (8 ft 2 in)
Wheelbase	1.88 m (6 ft 2 in)

DIMENSIONS, INTERNAL:
Cabin max width	1.83 m (6 ft 0 in)

AREAS:
Main rotor disc	79.46 m² (855.3 sq ft)
Tail rotor disc	2.41 m² (25.97 sq ft)

WEIGHTS AND LOADINGS:
Weight empty	816 kg (1,800 lb)
Max payload	408 kg (900 lb)
Max T-O weight	1,225 kg (2,700 lb)
Max disc loading	15.4 kg/m² (3.16 lb/sq ft)

PERFORMANCE (three-blade rotor):
Never-exceed speed (V_{NE})	95 kt (177 km/h; 110 mph)
Cruising speed	78 kt (145 km/h; 90 mph)
Max rate of climb at S/L	365 m (1,200 ft)/min
Service ceiling	3,350 m (11,000 ft)
Range	approx 347 n miles (643 km; 400 miles)

UPDATED

Vertical Aviation Technologies Hummingbird is based on the Sikorsky S-52

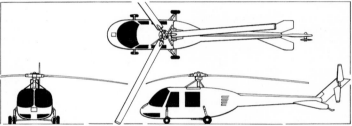

VAT Hummingbird four-seat helicopter *(James Goulding/Jane's)* 1999/0051865

VENTURE

VENTURE LIGHT AIRCRAFT RESOURCES LLC

Air Center West, Ryan Field, HCR 2, Box 270, Tucson, Arizona 85735
Tel/Fax: (+1 520) 883 83 95
e-mail: SpudSun@aol.com
Web: http://www.venture-thorpt-211.com

Venture builds kits of the Thorp T-211 under licence to AD Aerospace (which see in United Kingdom section). Kits are available to US constructors at US$9,995 (2000), less engine; kits sold in UK are sourced from Venture.

NEW ENTRY

VIKING

VIKING AIRCRAFT LIMITED

PO Box 646, Elkhorn, Wisconsin 53121-0646
Tel: (+1 414) 723 10 48
Fax: (+1 414) 723 10 49
e-mail: viking@pensys.com
Web: http://www.airsport.com/kits
OWNER: Patrick Taylor

Viking Aircraft Company formed 1979 to produce homebuilt aircraft. Dragonfly introduced 1980; Viking Aircraft Limited formed 1983 to produce prefabricated kits. In 2000, the Dragonfly was passed to SlipStream Industries (which see). Production of the Cygnet appears to have ceased.

UPDATED

VIKING DRAGONFLY

See SlipStream Dragonfly.

UPDATED

VIKING CYGNET SF-2A

Production of the Cygnet appears to have ceased. The aircraft last appeared in the 2000-01 edition.

UPDATED

VIPER

VIPER AIRCRAFT CORPORATION

3908 Stearman Avenue, Pasco, Washington 99301
Tel/Fax: (+1 509) 543 35 70
e-mail: info@viper-aircraft.com
Web: http://www.viper-aircraft.com
PRESIDENT: Scott Hanchette
VICE-PRESIDENT: Dan Hanchette

Viper Aircraft's first product is the ViperJet.

NEW ENTRY

VIPER VIPERJET

TYPE: Two-seat jet kitbuilt.

PROGRAMME: Development began in 1988; engineering by AirBoss Aerospace; tooling began in 1996; prototype was initially designed around a Teledyne Continental TSIO-520 and named ViperFan; in early 1999 a switch to jet propulsion was revealed and aircraft renamed ViperJet; development of ViperFan will continue as a lower priority. Prototype (N520VF) first flown 6 October 1999 and displayed publicly at Oshkosh in July 2000.

CURRENT VERSIONS: **ViperFan:** Initial piston-engined version.
ViperJet: Jet-powered version, *as described*.

CUSTOMERS: Three sold and one flying by end 2000.

COSTS: Kit US$134,900; engine extra (2001).

DESIGN FEATURES: High-performance personal jet at self-build price. Low wing with sweptback leading-edge and winglets; sweptback tail surfaces with fin fillet and ventral strakes.
Wing aerofoil section NACA 63-215 at root, NACA 63-212 at tip; horizontal tail NACA 63-012; vertical tail NACA 63-015.
Quoted build time 2,000 hours.

FLYING CONTROLS: Conventional and manual. Electric double-slotted flaps, span 1.65 m (5 ft 5 in) deflecting between 0

Views of Viper ViperJet during test flying

and 40°; aileron span 1.22 m (4 ft 0 in), deflection 27° up and 18° down; rudder deflection ±25°; elevator deflection ±30°. Dual controls.

STRUCTURE: Extensive use of composites. Preformed skins and wing spars.

LANDING GEAR: Retractable tricycle type; nosewheel retracts rearwards, mainwheels inwards.

POWER PLANT: Prototype initially powered by one refurbished Turbomeca Marboré 2 turbojet rated at 3.91 kN (880 lb st); later changed to converted General Electric T58-8F rated at 4.44 kN (1,000 lb st). Other engine options include versions of Marboré up to 4.89 kN (1,100 lb st). Fuel capacity 511 litres (135 US gallons; 112 Imp gallons).

ACCOMMODATION: Pilot and passenger in tandem in unpressurised cockpit.

DIMENSIONS, EXTERNAL:
Wing span	8.38 m (27 ft 6 in)
Wing chord: at root	1.52 m (5 ft 0 in)
at tip	0.63 m (2 ft 1 in)
Wing aspect ratio	7.7
Length overall	8.08 m (26 ft 6 in)
Height overall	3.53 m (11 ft 7 in)
Tailplane span	3.05 m (10 ft 0 in)

DIMENSIONS, INTERNAL:
Cockpit: Length	2.92 m (9 ft 7 in)
Max width	0.76 m (2 ft 6 in)
Max height	1.07 m (3 ft 6 in)

AREAS:
Wings, gross	9.10 m² (98.0 sq ft)
Trailing-edge flaps (total)	1.70 m² (18.25 sq ft)
Fin	1.92 m² (20.66 sq ft)
Rudder, incl tab	0.54 m² (5.80 sq ft)
Tailplane	2.42 m² (26.00 sq ft)

WEIGHTS AND LOADINGS (T58 engine):
Weight empty	907 kg (2,000 lb)
Max T-O weight	1,588 kg (3,500 lb)
Max wing loading	174.4 kg/m² (35.71 lb/sq ft)
Max power loading	357 kg/kN (3.50 lb/lb st)

PERFORMANCE:
Never-exceed speed (V_{NE})	356 kt (659 km/h; 410 mph)
Max level speed at 7,620 m (25,000 ft)	321 kt (595 km/h; 370 mph)
Stalling speed: flaps up	71 kt (131 km/h; 81 mph)
flaps down	66 kt (121 km/h; 75 mph)
Max rate of climb at S/L	1,370 m (4,500 ft)/min
Service ceiling	8,530 m (28,000 ft)
T-O run: flaps up	1,072 m (3,515 ft)
10° flap	738 m (2,420 ft)
Landing run	800 m (2,625 ft)
Range with max fuel	1,042 n miles (1,931 km/h; 1,200 miles)
g limits	±12

NEW ENTRY

VISIONAIRE

VISIONAIRE CORPORATION

Spirit of St Louis Airport, 595 Bell Avenue, Chesterfield, Missouri 63005
Tel: (+1 314) 530 10 07 and 530 64 00
Fax: (+1 314) 530 00 05
e-mail: info@visionaire.com
Web: http://www.visionaire.com
CHAIRMAN AND CEO: James O Rice Jr
PRESIDENT AND COO: Daniel D Mickelson
SENIOR VICE-PRESIDENT: Tom Stark
EXECUTIVE VICE-PRESIENT: Angelo Fiataruolo
VICE-PRESIDENT, MARKETING AND SALES: Barry Smith
VICE-PRESIDENT, STRATEGIC PLANNING: Jorge Perez
VICE-PRESIDENT, ENGINEERING: Joe Furnish
DIRECTOR, COMMUNICATIONS: Mark Jones
DIRECTOR, SALES: Jim Wickenhauser

EUROPEAN OFFICE:
VisionAire Europe SA, 7 Place du Théatre, L-2613 Luxembourg
Tel: (+352) 464 84 61
Fax: (+352) 46 48 45

VisionAire Corporation, formed in 1988 by James O Rice Jr, produces the Vantage all-composites business jet. Intended for mid-sized companies, the Vantage is specifically designed to offer cabin-class jet transportation at about half the cost of twin-jets currently on the market. Assembly of additional test aircraft was due to begin during first quarter of 1999 at 10,775 m² (116,000 sq ft) Ames, Iowa factory, opened 26 May 1998. An agreement of October 1997 provides for SimCom of Orlando, Florida, to provide pilot and maintenance training on behalf of Visionaire as part of the Vantage's purchase price. Visionaire employed 130 at the end of 1998, but laid off half of these in January 1999, pending a restructuring of the Vantage programme, which was completed in June 1999; by October 2000 workforce numbered 20.

FutureWorks, VisionAire's new product development division, introduced a full-size mockup, at EAA Oshkosh in 1998, of a two-seat tandem personal sport jet to be powered by a single turbofan. Designated VA-12, this is the first in a proposed line of aircraft to be marketed under the Spirit name. The introductory Spirit will be a twin-engined version of the initial prototype, optimised for good cross-country range, speed and carrying capacity. A third product, the VA-14 six-seat 'growth' model, was announced during 2000.

In mid-1998, VisionAire was in negotiation with Israeli government with a view to forming VisionArad company to build the Vantage at Arad, Israel. Nothing further has been heard. In late 2000, it announced the formation of a strategic partnership with Scaled Technology Works to design and develop the Vantage's wing, airframe and fuselage components. STW is also to build the prototypes and first two production examples.

UPDATED

VISIONAIRE VA-10 VANTAGE

TYPE: Business mono-jet.

PROGRAMME: Refinement of earlier design began on 15 January 1993; tested by two scale models; programme announced September 1995; design and construction of proof of concept (POC) aircraft started January 1996; built by Scaled Composites Inc under US$3.1 million contract; rolled out 8 November 1996; first flight (N247VA) 16 November 1996.

By 100th flight (180th hour), in February 1998, prototype had received 10 cm (4 in) stretch ahead of wings to accommodate aircraft systems; extra 2° of wing dihedral; further 27 per cent of horizontal tail area; and reduced rudder area. Changes then identified for following aircraft included redesigned landing gear; independently actuated mainwheel doors operating as brakes; reduced size windscreen posts; and a 1° reduction in jetpipe angle. Total of 193 sorties flown by August 2000; original plans for seven development aircraft: one POC, two flight development (VT1 and VT2), two preproduction and two static test.

In December 1998, work on Vantage was temporarily halted for a design review and restructuring as a consequence of control and weight problems, the aircraft being at least 363 kg (800 lb) over target. Modifications announced in June 1999 include one-piece fuselage, eliminating weight; revised, thinner wing for improved low-speed handling and lower drag; reduced forward sweep and whole wing moved aft; vertical tail increased in size and horizontal tail moved aft; engine moved back 51 cm (20 in) and up 13 cm (5 in) to reduce inlet curvature and decrease susceptibility to icing; main landing gear legs moved to wings and redesigned as trailing link units, as well as being reduced in height by 15 cm (6 in); baggage compartment repositioned in rear fuselage. Changes (to what is now known as Configuration 1A3) result in 181 kg (400 lb) MTOW increase.

Cost reduction plans announced in April 2000 to reduce required investment from US$130 million to US$90 million; involve modification of original prototype to exterior shape of final version and reduction in flight development and preproduction fleet from four to three.

VisionAire Vantage cabin interior

VisionAire Vantage instrumentation *(Paul Jackson/Jane's)*

First flight of VT-1 (initial development aircraft) due February 2002, followed by VT-2 in April 2002, and two preproduction aircraft (PT-1 and PT-2) afterwards (PT-1 due to make first flight May 2002). FAA certification to FAR Pt 23 Amdt 51 expected early 2003.

CUSTOMERS: Orders for over 110 aircraft by late 2000, including exports to nine countries; fractional ownership plan under consideration. Construction rate of 230 per year planned for fourth year of full production.

COSTS: US$2.195 million (2000). Direct operating cost (excluding finance and annual fixed expense; 456 hours per year utilisation) US$262 per hour.

DESIGN FEATURES: Low-cost high-performance corporate jet, meeting FAR Pt 23 Amendment 51 (single pilot operation) and making use of conventional aerodynamic techniques. Sweptforward wing minimises drag on mid-wing design and allows main spar to pass behind cabin. Bifurcated engine air intakes on fuselage shoulders, level with wing leading-edge.

Wing leading-edge sweep −6°; dihedral 5° 54′; aft-swept tips 1° washout; thickness/chord ratio 13.4 per cent.

FLYING CONTROLS: Conventional and manual. Aerodynamically balanced ailerons, elevators and rudder; three-axis electrical trim; yaw damper; autopilot. Electrically actuated Fowler 80-30 flaps; two-element spoilers on each wing.

STRUCTURE: Prepreg carbon-graphite construction. Monocoque fuselage structure with composites sandwich panels.

LANDING GEAR: Retractable tricycle type with hydraulic actuation. Single-wheel main units retract inward; single-wheel nose gear retracts forward. Forward main gear doors serve as independent airbrakes. ABSC wheels and hydraulic brakes.

POWER PLANT: One 13.55 kN (3,045 lb st) Pratt & Whitney Canada JT15D-5D turbofan. Two integral wing fuel tanks. Gravity refuelling point over each wing. Fuel capacity 908 litres (240 US gallons; 200 Imp gallons) in prototype; up to 1,261 litres (333 US gallons; 277 Imp gallons) in production version.

ACCOMMODATION: Pilot and co-pilot or passenger in cockpit, with two pairs of seats in cabin; seating by Fischer + Entwicklungen, crashworthy to 26 g. Forward baggage area; screened lavatory and baggage area at rear of cabin. Main door on port side, immediately behind cockpit. Cabin pressurised to equivalent altitude of 2,440 m (8,000 ft).

SYSTEMS: 28 V dual electrical power. Enviro systems electric vapour-cycle air conditioning. Cox mechanical de-icing.

AVIONICS: *Radar:* Honeywell RDR 2000 weather radar.
Flight: Integrated suite, comprising components from Garmin and Meggitt, includes Meggitt MAGIC integrated display and sensor system; Garmin GNS 530 comprising integrated IFR GPS/comm with VOR, LOC and glide slope, GTX327 digital transponder and GMA340 audio panels; and Meggitt autopilot.
Instrumentation: Dual IFR system with EFIS displays.

Proof of concept VisionAire VA-10 Vantage (N247VA)   2000/0075974

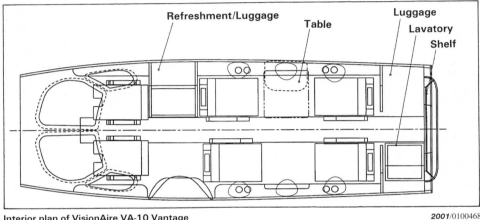

Interior plan of VisionAire VA-10 Vantage   2001/0100468

DIMENSIONS, EXTERNAL:
Wing span	14.48 m (47 ft 6 in)
Wing chord: at root	1.90 m (6 ft 3 in)
at tip	0.99 m (3 ft 3 in)
Wing aspect ratio	10.2
Length overall: prototype	12.53 m (41 ft 1½ in)
production	12.42 m (40 ft 9 in)
Fuselage: Length	11.32 m (37 ft 1½ in)
Max diameter	1.71 m (5 ft 7¼ in)
Height	4.37 m (14 ft 4 in)
Wheel track	3.61 m (11 ft 10 in)
Wheelbase	5.18 m (17 ft 0 in)
Passenger door: Width	0.75 m (2 ft 5½ in)

DIMENSIONS, INTERNAL:
Cabin (excl flight deck): Length	3.30 m (10 ft 10 in)
Max width	1.57 m (5 ft 2 in)
Max height	1.52 m (5 ft 0 in)
Floor area	5.48 m² (59 sq ft)
Volume	8.8 m³ (310 cu ft)
Baggage compartment volume:	
internal	0.90 m³ (31.8 cu ft)
external	0.42 m³ (15.0 cu ft)

AREAS:
Wings, gross	21.74 m² (234.0 sq ft)
Ailerons (total)	1.03 m² (11.10 sq ft)
Vertical tail surfaces (total)	5.72 m² (61.60 sq ft)
Tailplane	7.06 m² (76.00 sq ft)
Elevators (total)	1.35 m² (14.50 sq ft)

WEIGHTS AND LOADINGS:
Weight empty	2,236 kg (4,930 lb)
Max payload	849 kg (1,871 lb)
Max fuel weight	1,013 kg (2,234 lb)
Max T-O weight	3,719 kg (8,200 lb)
Max ramp weight	3,764 kg (8,300 lb)
Max landing weight	3,583 kg (7,900 lb)
Max zero-fuel weight	3,084 kg (6,800 lb)
Max wing loading	171.1 kg/m² (35.04 lb/sq ft)
Max power loading	274 kg/kN (2.69 lb/lb st)

PERFORMANCE (estimated):
Max operating Mach No.	0.65
Max cruising speed at 12,500 m (41,000 ft)	350 kt (648 km/h; 403 mph)
Econ cruising speed at 10,670 m (35,000 ft)	250 kt (463 km/h; 288 mph)
Stalling speed, power off, flaps down	70 kt (129 km/h; 80 mph)
Max rate of climb at S/L	1,220 m (4,000 ft)/min
Max certified altitude	12,500 m (41,000 ft)
T-O run	274 m (900 ft)
T-O to 15 m (50 ft)	610 m (2,000 ft)
Landing from 15 m (50 ft)	760 m (2,500 ft)
Landing run	363 m (1,190 ft)

Range with max fuel, IFR, 45 min reserves:
high-speed cruise with six occupants
1,000 n miles (1,852 km; 1,150 miles)
with two occupants
1,350 n miles (2,500 km; 1,554 miles)
g limits   +4.75/−2.5

UPDATED

VISIONAIRE VA-12C SPIRIT

TYPE: Two-seat jet sportplane.

PROGRAMME: Announced in single-engined form at Oshkosh in July 1998 and mockup displayed at NBAA Convention in October 1998; first flight then planned before end 2000. January 2001 now set as deadline for go-ahead decision; if approved, first deliveries planned for 2004.

CURRENT VERSIONS: Original **VA-12A** version to have been powered by one 4.00 kN (900 lb st) Williams FJX-2 engine; superseded by **VA-12B**, displayed at Oshkosh in July 1998; superseded in 1999 by revised **VA-12C**, *as described*. Other versions under consideration include a simplified single-engined, fixed-gear trainer and a four-seat cabin version.

CUSTOMERS: Orders for 40 received by mid-1999.

COSTS: US$750,000 (2000).

DESIGN FEATURES: Mid-mounted wing with moderate sweepback; twin engine air intakes on spine.

FLYING CONTROLS: Conventional and manual. Flaps.

STRUCTURE: Extensive use of composites.

LANDING GEAR: Retractable.

POWER PLANT: Two 3.11 kN (700 lb st) Williams FJX-2 turbofan engines.

ACCOMMODATION: Pilot and passenger in tandem under single-piece canopy; side-by-side seating under consideration.

WEIGHTS AND LOADINGS (provisional):
Weight empty	726 kg (1,600 lb)
Max T-O weight	1,587 kg (3,500 lb)
Max power loading	255 kg/kN (2.50 lb/lb st)

PERFORMANCE (provisional):
Max cruising speed at 9,145 m (30,000 ft)	400 kt (741 km/h; 460 mph)
Stalling speed: flaps up	61 kt (113 km/h; 71 mph)
flaps down	50 kt (93 km/h; 58 mph)
Max rate of climb at S/L	1,341 m (4,400 ft)/min
Range with max fuel	2,000 n miles (3,704 km; 2,301 miles)

UPDATED

Model of VisionAire VA-12C Spirit *(Paul Jackson/Jane's)*   2001/0100469

VORTEX

VORTEX AIRCRAFT COMPANY
500 Harbor West, Suite 1310, San Diego, California 92101
Tel: (+1 619) 234 53 87
Fax: (+1 619) 338 99 63
e-mail: vortexjet@aol.com
PRESIDENT AND CEO: Gary Kauffman

Vortex acquired design, manufacture and marketing rights to the civil version of Bede BD-10 following the bankruptcy of Bede Jet in February 1997. Production plans appear to have been abandoned; the aircraft was last described in the 2000-01 edition.

UPDATED

VSTOL

VSTOL AIRCRAFT CORPORATION
Suite 100, 8619 Northwest Brostrom Road, Kansas City, Missouri 64152
Tel/Fax: (+1 941) 936 12 61
e-mail: dick@vstolaircraft.com
Web: http://www.vstolaircraft.com
CEO: Dick Turner

VSTOL is responsible for marketing the SST 2000 Pairadigm and SS 2000 Super Solution series of kitbuilt aircraft. Kits are produced in Venezuela.

UPDATED

VSTOL SST 2000 PAIRADIGM
TYPE: Two-seat kitbuilt twin.
PROGRAMME: Developed from the Star Flight XC 280 Stiletto (see 1992-93 *Jane's*), first flown January 1997. A single-engined version is known as the **Super Solution 2000**.
CUSTOMERS: Over 400 flying by mid-2000. Some used for crop-spraying.
COSTS: Kit US$24,995 (1999).
DESIGN FEATURES: Two pusher engines mounted side by side at rear of fuselage; propellers partly overlap and rotate in same direction. Wings have full length leading-edge slats and are removable for storage. Quoted build time 250 hours.
FLYING CONTROLS: Conventional and manual. Actuation by stainless steel cables. Leading-edge slats, Frise ailerons, slotted flaps and horn-balanced rudder.
STRUCTURE: 4130 steel and aluminium tubing throughout; fabric-covered V-braced wings and composites fuselage skin; braced tailplane. Extended wings are optional. Cabin enclosure removable.
LANDING GEAR: Fixed; choice of nose- or tailwheel versions; no brakes; wheel size 152 mm (6 in). Steerable nosewheel on tricycle model.
POWER PLANT: Two engines, each between 37.3 and 74.6 kW (50 and 100 hp); 48.5 kW (65 hp) Hirth 2706 engines are recommended. Fuel capacity 75.7 litres (20 US gallons; 16.7 Imp gallons).

VSTOL Pairadigm *(Geoffrey P Jones)*   0016518

EQUIPMENT: Crop-spraying kit includes 189 litre (50.0 US gallon; 41.6 Imp gallon) chemical tank and 8.23 m (27 ft) sprayboom with 16 nozzles.
DIMENSIONS, EXTERNAL:
Wing span	10.52 m (34 ft 6 in)
Wing aspect ratio	6.3
Length overall	7.16 m (23 ft 6 in)
Height overall	2.59 m (8 ft 6 in)

AREAS:
Wings, gross	17.65 m² (190.0 sq ft)

WEIGHTS AND LOADINGS:
Weight empty	408 kg (900 lb)
Max T-O weight	771 kg (1,700 lb)
Max wing loading	43.7 kg/m² (8.95 lb/sq ft)

PERFORMANCE:
Never-exceed speed (V_{NE})	108 kt (201 km/h; 125 mph)
Max level speed	78 kt (144 km/h; 90 mph)
Normal cruising speed	61 kt (113 km/h; 70 mph)
Stalling speed	26 kt (47 km/h; 29 mph)
Max rate of climb at S/L	244 m (800 ft)/min
Rate of climb at S/L, OEI	61 m (200 ft)/min
Service ceiling	3,660 m (12,000 ft)
T-O run	46 m (150 ft)
Landing run	31 m (100 ft)
Range at cruising speed	208 n miles (386 km; 240 miles)
g limits	+4.4/–2.2

UPDATED

WACO CLASSIC

WACO CLASSIC AIRCRAFT CORPORATION
PO Box 1229, Battle Creek, Michigan 49016-1229
Tel: (+1 616) 565 10 00 or 565 32 00
Fax: (+1 616) 565 11 00 or 565 32 32
e-mail: flywaco@wacoclassic.com
Web: http://www.wacoclassic.com
PRESIDENT: Mitchell Lampert
GENERAL MANAGER: Patrick Horgan

Waco Classic (known until 1997 as Classic Aircraft) delivers some five or six aircraft per year. It originated in 1983 at Lansing, Michigan, and acquired the Waco YMF-5 type certificate from the FAA; manufacturing authority was assigned in February 1984. The company moved into a purpose-built, 2,800 m² (30,000 sq ft) factory at W K Kellogg Airport, Battle Creek, in April 2000; 90th aircraft handed over at official opening of factory on 20 May 2000.

UPDATED

WACO CLASSIC YMF SUPER
TYPE: Three-seat biplane.
PROGRAMME: Construction began March 1984 under type certificate of original Waco YMF-5; first flight of prototype (N1935B) 20 November 1985; FAA certification 11 March 1986; marketed initially as F-5, which was superseded by YMF Super, introduced in 1991.
CURRENT VERSIONS: **YMF Super:** Larger front cockpit than previous standard F-5 version (see 1992-93 *Jane's*) for commercial pleasure flying; enlarged forward door; front and rear cockpits 10 cm (4 in) longer; front cockpit 6.3 cm (2½ in) wider.
CUSTOMERS: Total 92 (including F-5s) delivered by July 2000. Exported to Australia, Canada, Germany, Kenya, Mexico, South Africa, Switzerland and the Caribbean.
COSTS: Standard equipped aircraft US$254,500 (1999).
DESIGN FEATURES: Single-bay biplane of 1930s vintage. Wings braced with N-type interplane struts, plus flying and landing wires, with double-N cabane. Elliptical wingtips. Cross-braced landing gear. Helmeted engine cowling. Wing section Clark Y; dihedral 2°, incidence 0° on upper and lower wings.

Waco Classic YMF Super three-seat biplane   *1999*/0085774

FLYING CONTROLS: Conventional and manual. Interconnected ailerons on upper and lower wings; no tabs. Horn-balanced tail surfaces. Tailplane trim by screw-jack actuator; ground-adjustable trim tab on rudder.
STRUCTURE: Modern construction techniques, tolerances and materials applied to original design. Steel interplane struts; streamlined stainless steel flying and landing wires; all-wood wing with Dacron covering; aluminium ailerons with external chordwise stiffening. Fuselage of 4130 welded steel tubes with internal oiling for corrosion protection; wooden bulkheads; Dacron covering. Braced welded steel tube tail surfaces with Dacron covering. Aluminium engine cowling.
LANDING GEAR: Tailwheel type; fixed. Shock-absorption by oil and spring shock-struts. Steerable tailwheel. Cleveland 30-67F hydraulic brakes on mainwheels only. Cleveland 40-101A mainwheels, tyre size 7.50-10; Cleveland 40-199A tailwheel, tyre size 3.50-10. Fairings standard on main legs and wheels; tailwheel fairing optional. Float and amphibious landing gear optional.
POWER PLANT: One 205 kW (275 hp) Jacobs R-755-B2 air-cooled radial engine (remanufactured), driving a Sensenich two-blade, fixed-pitch, wooden propeller. Constant-speed propeller with spinner optional. Fuel in two aluminium tanks in upper wing centre-section, total capacity 182 litres (48.0 US gallons; 40.0 Imp gallons).

Refuelling point for each tank in upper wing surface. Auxiliary tanks, capacity 45.4 litres (12.0 US gallons; 10.0 Imp gallons) each, optional in either or both inboard upper wing panels. Oil capacity 18.9 litres (5.0 US gallons; 4.2 Imp gallons).

ACCOMMODATION: Three seats in tandem open cockpits, two side by side in front position, single seat at rear. Controls in both cockpits; toe brakes in rear only. Seat belts with shoulder harness, and pilot's adjustable seat, standard. Lockable front and rear baggage compartments. Front windscreen removable.

SYSTEMS: 24 V electrical system with battery, alternator and starter for electrical supply to navigation, strobe and rear cockpit lights. Hydraulic system for brakes only.

AVIONICS: Customer specified. Honeywell KX 155 nav/com recommended.

Instrumentation: Compass, airspeed indicator, turn and bank indicator, rate of climb indicator, sensitive altimeter, recording tachometer, cylinder head temperature gauge and oil pressure and oil temperature gauges standard in rear cockpit. Front cockpit instruments optional. Other options include exhaust gas temperature gauge, carburettor temperature gauge, *g* meter, vacuum or electrically driven gyro system, Hobbs meter (engine-time recorder), outside air temperature gauge and manifold gauge.

EQUIPMENT: Instrument post lighting, heated pitot, three-colour paint scheme with choice of two designs, landing and taxying lights, front and rear cockpit heaters, also standard. Optional equipment includes ground service plug, flight-approved metal front cockpit cover, glider tow hook, deluxe interior with carpet, leather sidewalls and interior trim, and special exterior paint designs.

DIMENSIONS, EXTERNAL:
Wing span: upper	9.14 m (30 ft 0 in)
lower	8.18 m (26 ft 10 in)
Length overall	7.26 m (23 ft 10 in)
Height overall	2.59 m (8 ft 6 in)
Wheelbase	1.95 m (6 ft 5 in)
Propeller diameter	2.44 m (8 ft 0 in)

DIMENSIONS, INTERNAL:
Rear baggage compartment volume	0.21 m³ (7.5 cu ft)

AREAS:
Wings, gross	21.69 m² (233.5 sq ft)

WEIGHTS AND LOADINGS:
Basic weight empty	900 kg (1,985 lb)
Baggage capacity: front	11 kg (25 lb)
rear	34 kg (75 lb)
Max T-O and landing weight	1,338 kg (2,950 lb)
Max wing loading	61.7 kg/m² (12.63 lb/sq ft)
Max power loading	6.53 kg/kW (10.73 lb/hp)

PERFORMANCE:
Never-exceed speed (V_{NE})	186 kt (344 km/h; 214 mph)
Max level speed at S/L	117 kt (217 km/h; 135 mph)
Max cruising speed at S/L	104 kt (193 km/h; 120 mph)
Econ cruising speed at 2,440 m (8,000 ft)	95 kt (177 km/h; 110 mph)
Stalling speed, power off	51 kt (95 km/h; 59 mph)
Max rate of climb at S/L	235 m (770 ft)/min
T-O run	152 m (500 ft)
Range, standard fuel, 30 min reserves	286 n miles (531 km; 330 miles)

UPDATED

WAG-AERO

WAG-AERO INC
PO Box 181, 1216 North Road, Lyons, Wisconsin 53148
Tel: (+1 262) 763 95 86
Fax: (+1 262) 763 19 15
e-mail: wagaero-sales@wagaero.com
Web: http://www.wagaero.com
PRESIDENT: Mary M Myers
MARKETING SUPERVISOR: Mary Pat Henningfield

Company founded in early 1960s, initially to supply light aircraft spares. Three aircraft types currently offered, based on Piper designs. Founders Dick and Bobbie Wagner sold Wag-Aero to Bill Read and Mary Myers on 1 September 1995. Some 7,000 kits or sets of plans have been sold.

UPDATED

WAG-AERO SPORT TRAINER

TYPE: Tandem-seat kitbuilt.
PROGRAMME: Modernised kitbuilt version of 1937 Piper J-3. Sport Trainer first flew 12 March 1975.
CURRENT VERSIONS: **Sport Trainer:** Basic two-seat design. Quoted build time of 1,200 hours; conforms to FAA 51 per cent rule.

Description applies to Sport Trainer.

Acro Trainer: Differs from standard version by having strengthened fuselage, shortened wings (8.23 m; 27 ft 0 in), modified lift struts, improved wing fittings and rib spacing, and new leading-edge.

Observer: Replica of Piper L-4 military liaison aircraft.

Super Sport: Structural modifications to accept engines of up to 112 kW (150 hp), making it suitable for glider towing, bush operations, or as floatplane.

CUSTOMERS: Over 3,860 sets of plans and kits sold by end 1998.
COSTS: Plans: US$69.50; kit US$18,000.
DESIGN FEATURES: High-wing cabin monoplane with braced wings and tailplane.
FLYING CONTROLS: Conventional and manual. Horn-balanced rudder.
STRUCTURE: Fuselage and empennage of welded 4130 chrome molybdenum steel tube with fabric covering; wooden wing with light alloy leading-edge and fabric covering.
LANDING GEAR: Tailwheel type; fixed. Welded steel tube side Vs and half-axles. Bungee shock-absorbers. Speed fairings.
POWER PLANT: Can be powered by any flat-four Teledyne Continental, Franklin or Textron Lycoming engine between 48.5 and 93 kW (65 and 125 hp). Standard fuel capacity 45.4 litres (12.0 US gallons; 10.0 Imp gallons); auxiliary fuel capacity 98.4 litres (26.0 US gallons; 21.7 Imp gallons).
ACCOMMODATION: Two persons in tandem.

DIMENSIONS, EXTERNAL:
Wing span	10.73 m (35 ft 2½ in)
Wing aspect ratio	6.9
Length overall	6.82 m (22 ft 4½ in)
Height overall	2.03 m (6 ft 8 in)
Wheel track	1.80 m (5 ft 11 in)

AREAS:
Wings, gross	16.58 m² (178.5 sq ft)

WEIGHTS AND LOADINGS:
Weight empty	327 kg (720 lb)
Max T-O weight	635 kg (1,400 lb)
Max wing loading	38.3 kg/m² (7.84 lb/sq ft)
Max power loading	6.82 kg/kW (11.20 lb/hp)

PERFORMANCE:
Never-exceed speed (V_{NE})	104 kt (193 km/h; 120 mph)
Max level speed at S/L	89 kt (164 km/h; 102 mph)
Cruising speed	82 kt (151 km/h; 94 mph)
Stalling speed	34 kt (63 km/h; 39 mph)
Max rate of climb at S/L	149 m (490 ft)/min
Service ceiling	3,415 m (11,200 ft)
T-O run	122 m (400 ft)
Landing run	130 m (425 ft)
Range:	
at cruising speed with standard fuel	191 n miles (354 km; 220 miles)
with auxiliary fuel	395 n miles (732 km; 455 miles)

UPDATED

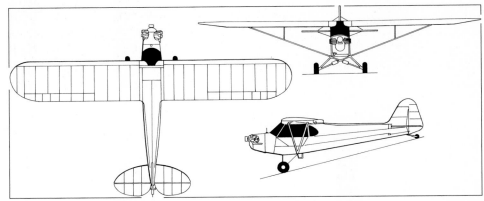

General arrangement of the Wag-Aero Sport Trainer *(Paul Jackson/Jane's)* 2000/0085835

WAG-AERO WAG-A-BOND

TYPE: Side-by-side kitbuilt.
PROGRAMME: Replica of 1948 Piper PA-15 Vagabond. Prototype Wag-A-Bond completed May 1978; plans and kits available. Conforms to FAA 51 per cent rule.
CUSTOMERS: More than 1,350 plans and kits sold by end 1998.
COSTS: Plans US$65 (2000).
DESIGN FEATURES: Two versions (**Classic** and **Traveler**); Traveler is modified and updated version of Vagabond with port and starboard doors, overhead skylight window, extended sleeping deck (conversion from aircraft to camper interior taking about 2 minutes and accommodating two persons), extended baggage area, engine of up to 85.7 kW (115 hp), and provision for full electrical system.
FLYING CONTROLS: Conventional and manual. Horn-balanced rudder. Tab on port elevator.
STRUCTURE: All-wood wing and aluminium alloy aileron structures. Welded 4130 steel tube and flat plate fuselage structure, and steel tube tail unit. Complete airframe is fabric covered.
LANDING GEAR: Non-retractable, tailwheel type; as Sport Trainer; mainwheel tyres 7.00-6; Cleveland brakes. Optional skis.
POWER PLANT: Traveler can be powered by a Textron Lycoming engine of 80.5 to 85.7 kW (108 to 115 hp). Classic can be powered by a Teledyne Continental engine of 48.5 to 74.5 kW (65 to 100 hp). Fuel capacity: Traveler 98.4 litres (26.0 US gallons; 21.6 Imp gallons), Classic 45.4 litres (12.0 US gallons; 10.0 Imp gallons).
ACCOMMODATION: Two persons side by side on adjustable bench seat; cabin includes baggage stowage.
SYSTEMS: Optional electrical system in Traveler.

DIMENSIONS, EXTERNAL:
Wing span	8.32 m (29 ft 3½ in)
Wing aspect ratio	5.0
Length overall	5.70 m (18 ft 8½ in)
Height overall	1.83 m (6 ft 0 in)
Wheel track	1.83 m (6 ft 0 in)

DIMENSIONS, INTERNAL:
Cabin max width	1.02 m (3 ft 4 in)

AREAS:
Wings, gross	13.75 m² (148.0 sq ft)

WEIGHTS AND LOADINGS:
Weight empty: Classic	318 kg (700 lb)
Traveler	340 kg (750 lb)
Baggage capacity: Classic	18 kg (40 lb)
Traveler	27 kg (60 lb)
Max T-O weight: Classic	567 kg (1,250 lb)
Traveler	658 kg (1,450 lb)

Wag-Aero Sport Trainer 0016517

Wag-Aero Wag-A-Bond 0016516

WAG-AERO to WARNER—AIRCRAFT: USA

Max wing loading: Classic	41.2 kg/m² (8.45 lb/sq ft)	Stalling speed:		
Traveler	47.8 kg/m² (9.80 lb/sq ft)	Classic, Traveler	48 kt (89 km/h; 55 mph)	
Max power loading: Classic	7.61 kg/kW (12.50 lb/hp)	Max rate of climb at S/L: Classic	190 m (625 ft)/min	
Traveler	7.67 kg/kW (12.61 lb/hp)	Traveler	259 m (850 ft)/min	

PERFORMANCE:
Max level speed: Classic 91 kt (169 km/h; 105 mph)
Traveler 118 kt (219 km/h; 136 mph)
Cruising speed: Classic 83 kt (153 km/h; 95 mph)
Traveler 91 kt (169 km/h; 105 mph)

Service ceiling: Traveler 4,260 m (14,000 ft)
T-O run: Traveler 119 m (390 ft)
Landing run: Traveler 232 m (760 ft)
Range: Traveler 521 n miles (965 km; 600 miles)

UPDATED

WAG-AERO 2+2 SPORTSMAN

TYPE: Four-seat kitbuilt.
PROGRAMME: Version of the Piper PA-14 Family Cruiser. Plans and material kits available. Conforms to FAA 51 per cent rule.
CUSTOMERS: Over 1,800 sets sold by end 1998.
COSTS: Plans: US$89; kit US$22,000 (2000).
DESIGN FEATURES: True four-seater, with option of hinged rear fuselage decking to provide access to baggage and rear seat areas. Conventional, high-wing configuration with V-type bracing struts, wire-braced empennage, mid-mounted tailplane and classic fin and rudder.
FLYING CONTROLS: Conventional and manual. Horn-balanced rudder. Upper and lower spoilers; trim tabs on both elevators.
STRUCTURE: Similar construction to Wag-A-Bond, with glass fibre wingtips. Alternatively, drawings and materials provided to modify standard PA-12, PA-14 or PA-18 wings. Prewelded fuselage structure also available.
LANDING GEAR: As Wag-A-Bond; main units use 8.00-6 or 7.00-6 tyres. Optional floats.
POWER PLANT: Engine of 93 to 149 kW (125 to 200 hp). Usable fuel capacity 148 litres (39.0 US gallons; 32.5 Imp gallons).
ACCOMMODATION: Four persons; two separate front seats, plus rear bench seat which is removable for stretcher or cargo carrying. Baggage area to rear of bench seat.

DIMENSIONS, EXTERNAL:
Wing span 10.90 m (35 ft 9 in)
Wing chord, constant 1.60 m (5 ft 3 in)
Wing aspect ratio 7.3
Length overall 7.12 m (23 ft 4½ in)
Height overall 2.02 m (6 ft 7½ in)
Wheel track 1.98 m (6 ft 6 in)

DIMENSIONS, INTERNAL:
Cabin max width 0.99 m (3 ft 3 in)

AREAS:
Wings, gross 16.18 m² (174.1 sq ft)

WEIGHTS AND LOADINGS:
Weight empty 490 kg (1,080 lb)
Max T-O weight 998 kg (2,200 lb)
Max wing loading 61.7 kg/m² (12.63 lb/sq ft)
Max power loading 6.70 kg/kW (11.00 lb/hp)

PERFORMANCE (typical; actual data depend on engine fitted):
Max level speed 112 kt (207 km/h; 129 mph)
Cruising speed 108 kt (200 km/h; 124 mph)
Stalling speed 44 kt (81 km/h; 50 mph)
Max rate of climb at S/L 244 m (800 ft)/min
Service ceiling 4,510 m (14,800 ft)
T-O run 70 m (230 ft)
Landing run 104 m (340 ft)
Range at cruising speed 582 n miles (1,078 km; 670 miles)

UPDATED

Wag-Aero Sportsman *(Paul Jackson/Jane's)*

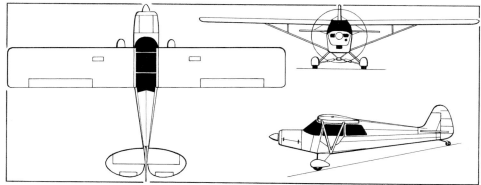
Wag-Aero 2+2 Sportsman *(James Goulding/Jane's)*

WARNER

WARNER AIRCRAFT COMPANY

9415 Laura Court, Seminole, Florida 33776-1625
Tel: (+1 727) 595 23 82
Fax: (+1 978) 887 05 55
e-mail: revolution@usa.net
Web: http://www.warnerair.com
PRESIDENT: Dana Axelrod

Warner Aerocraft markets the Warner Space Walker (see *Jane's* 1990-91), Revolution and Sportster.

UPDATED

WARNER SPORTSTER

TYPE: Tandem-seat sportplane kitbuilt.
PROGRAMME: Development of Country Air (later Warner) Space Walker, of which 46 built. Two-seat versions of Space Walker developed as Revolution I (48.5 kW; 65 hp Teledyne Continental) and Revolution II (74.6 kW; 100 hp Textron Lycoming); further power increase produced Sportster; prototype (N269U) first flown 18 April 1998.
CUSTOMERS: One flying by beginning 2000.
COSTS: Basic kit US$13,495; fast-build kit US$19,995 (2001).
DESIGN FEATURES: Quoted build time 700 hours. Roll rate 45°/s.
FLYING CONTROLS: Conventional and manual. Horn-balanced rudder and elevators; trim tab in port elevator.
STRUCTURE: 4130 steel tube and fabric fuselage with wooden turtle deck; fabric-covered two-spar wooden cantilever wings. Glass fibre cowling and wingtips. Tail unit identical to Revolution/Space Walker.
LANDING GEAR: Tailwheel type; fixed. Tubular welded V legs with tubular bracing and spring shock absorbers. 6.00-6 tyres, Cleveland wheels with disc brakes. Steerable 15 cm (6 in) Maule tailwheel. Speed fairings on mainwheels.
POWER PLANT: Engines from 63.4 to 119 kW (85 to 160 hp) can be fitted; prototype has 85.6 kW (115 hp) Textron Lycoming O-290-D. Fuel capacity 76 litres (20.0 US gallons; 16.7 Imp gallons).
ACCOMMODATION: Pilot and passenger in tandem in individual cockpits; choice of three styles of windscreen.

Warner Sportster, developed from Revolution *(Paul Jackson/Jane's)*

DIMENSIONS, EXTERNAL:
Wing span 7.92 m (26 ft 0 in)
Wing chord, constant 1.37 m (4 ft 6 in)
Wing aspect ratio 6.0
Length overall 6.20 m (20 ft 4 in)
Max diameter of fuselage 0.76 m (2 ft 6 in)
Height overall 1.68 m (5 ft 6 in)
Propeller ground clearance 0.46 m (1 ft 6 in)
Tailplane span 2.41 m (7 ft 11 in)
Wheel track 1.83 m (6 ft 0 in)
Wheelbase 4.41 m (14 ft 5½ in)

DIMENSIONS, INTERNAL:
Cockpit max width 0.76 m (2 ft 6 in)

AREAS:
Wings, gross 10.41 m² (112.0 sq ft)

WEIGHTS AND LOADINGS (O-290 engine):
Weight empty 386 kg (850 lb)
Baggage capacity 11 kg (25 lb)
Max T-O weight 680 kg (1,500 lb)
Max wing loading 65.4 kg/m² (13.39 lb/sq ft)
Max power loading 7.94 kg/kW (13.04 lb/hp)

PERFORMANCE (O-290 engine):
Never-exceed speed (V_{NE}) 130 kt (241 km/h; 150 mph)
Max level speed 115 kt (212 km/h; 132 mph)
Normal cruising speed 104 kt (193 km/h; 120 mph)
Stalling speed 40 kt (73 km/h; 45 mph)
Max rate of climb at S/L 358 m (1,175 ft)/min
Service ceiling 3,660 m (12,000 ft)
T-O run 130 m (425 ft)
Landing run 122 m (400 ft)
Range with max fuel 356 n miles (659 km; 410 miles)
g limits +6/−4

UPDATED

USA: AIRCRAFT—WEATHERLY to ZENITH

WEATHERLY

WEATHERLY AVIATION COMPANY
2100 Flightline Drive, PO Box 68, Lincoln, California 95648
Tel: (+1 916) 645 90 80
Fax: (+1 916) 645 18 16
PRESIDENT: Hal Weatherly

Weatherly developed conversion of Fairchild M-62 in the early 1960s for agricultural role, and later designed and built the Model 201 (see *Jane's* 1979-80 for details). Further refinement of the design resulted in the Weatherly 620, production of which continues at low rate.

UPDATED

WEATHERLY 620
TYPE: Agricultural sprayer.
PROGRAMME: Developed from Weatherly 201; prototype (N9256W) first flown 1979.
CURRENT VERSIONS: **620:** Initial version; no longer available. See *Jane's* 1982-83 for details.
620B: Current piston-engined version, powered by Pratt & Whitney radial engine. Length 8.15 m (26 ft 9 in); height 2.44 m (8 ft 0 in). Slightly different spray equipment but same sized hopper.
620TP: Initial turboprop version, with P&WC PT6A-11AG; no longer available. See *Jane's* 1982-83 for details.
620BTG: Current turboprop version; engine can be retrofitted to earlier versions.
Description applies to Weatherly 620BTG.
CUSTOMERS: By mid-2000, at least 155 Weatherly 620s of all versions had been built. Completion rate of one per year. Recent sales to Argentina (1997) and Australia (1998) as well as USA (2000).
COSTS: US$185,000, less engine (1998).
DESIGN FEATURES: High lift:drag ratio wings, with tip vanes for vortex diffusion and increase of swath width; vanes fold downwards for ground handling. Hinged wing leading-edges aid inspection.

Weatherly 620B (one Pratt & Whitney R-985) *(W J Taylor)* 2000/0084556

FLYING CONTROLS: Conventional and manual. Stainless steel rudder cables.
STRUCTURE: All-metal monocoque fuselage; metal wings and control surfaces; glass fibre tail fairings.
LANDING GEAR: Tailwheel type; fixed. Mainwheel oleos connected to spar to disperse stresses. Spring tailwheel assembly with fully castoring and locking wheel. Three-piston Cleveland brakes. Mainwheel tyre size 8.5-10; tailwheel 12.5-4.5.
POWER PLANT: *620B:* One Pratt & Whitney 336 kW (450 hp) R-985-AN radial engine driving a Hartzell three-blade propeller. Fuel capacity 369 litres (97.5 US gallons; 81.1 Imp gallons), of which 341 litres (90.1 US gallons; 75.1 Imp gallons) is usable.
620BTG: One Honeywell TPE331-1 turboprop, derated to 496 kW (665 shp), driving a McCauley 3GFR34C602/100LA-2 three-blade propeller. Fuel capacity 492 litres (131 US gallons; 108 Imp gallons) of which 439 litres (116 US gallons; 96.5 Imp gallons) usable.
ACCOMMODATION: Pilot only. Four-way adjustable mesh seat. Baggage compartment behind seat.
SYSTEMS: 24 V electrical system; 250 A starter/generator. Optional air conditioning and windshield washer/wiper.
EQUIPMENT: Agricultural dispersal system comprises a 1,344 litre (355 US gallon; 296 Imp gallon) hopper in forward fuselage, size 1.35 m³ (47.5 cu ft); Weath-Aero fan; 5 cm (2 in) SS bottom load pump; 5 cm (2 in) Agrinautics spray pump with three-way valve; 6.10 m (20 ft) pylon-mounted drop boom with nozzles every 15 cm (6 in); 63.5 cm (2 ft 1 in) Transland gate box. Main gear and cockpit wire deflectors, plus deflector cable from cabin to fin. Rinse-out system optional, as are stainless steel spray system and flow meters.

DIMENSIONS, EXTERNAL:
Wing span (with vanes extended)	14.22 m (46 ft 8 in)
Wing aspect ratio	7.9
Length overall	9.04 m (29 ft 8 in)
Height overall	2.90 m (9 ft 6 in)
Wheel track	3.20 m (10 ft 6 in)

AREAS:
Wings, gross	25.73 m² (277.0 sq ft)

WEIGHTS AND LOADINGS:
Weight empty	1,374 kg (3,030 lb)
Max T-O weight under CAM 8	2,721 kg (6,000 lb)
Max wing loading	105.8 kg/m² (21.66 lb/sq ft)
Max power loading	5.49 kg/kW (9.02 lb/shp)

PERFORMANCE:
Max operating speed	153 kt (283 km/h; 176 mph)
Econ cruising speed	122 kt (226 km/h; 140 mph)
Stalling speed	62 kt (115 km/h; 72 mph)
Max rate of climb at S/L	427 m (1,400 ft)/min
Service ceiling	4,572 m (15,000 ft)
T-O run	287 m (940 ft)

UPDATED

Weatherly 620BTG with retrofitted Honeywell TPE331-1 engine *(W J Taylor)* 2000/0084554

ZENITH

ZENITH AIRCRAFT COMPANY
Mexico Memorial Airport, PO Box 650, Mexico, Missouri 65265-0650
Tel: (+1 573) 581 90 00
Fax: (+1 573) 581 00 11
e-mail: info@zenithair.com
Web: http://www.zenithair.com
PRESIDENT: Sebastien C Heintz
VICE-PRESIDENT: Mathieu Heintz
MANAGER, ADMINISTRATION: Susan M McCullough
MANAGER, PRODUCTION: Nicholas M Heintz

Zenith Aircraft Company has acquired all rights to manufacture and sell the Canadian Zenair range of light aircraft, detailed below. For details of the CH 2000, see AMD in this section. By the end of 1998 (latest information supplied), over 1,800 kits had been produced.
In 1997, Zenith announced a joint venture with Czech Aircraft Works to assemble the CH 601 and CH 701 models; see CZAW entry in Czech Republic section.

UPDATED

ZENITH ZODIAC CH 601HD
TYPE: Side-by-side ultralight homebuilt.
PROGRAMME: First flight of CH 600 prototype June 1984. Plans and kits of parts (45 per cent premanufactured) for improved model followed. New **CH 601UL** developed to meet TP 10141 advanced ultralight (AULA) category in Canada. Construction of prototype CH 601 (59.7 kW; 80 hp Rotax 912) began September 1990 (first flight October that year). Kit production started January 1991. UL version (2001 cost US$12,380) remains available and was last described in the 1997-98 *Jane's*.
CH 601HD offers wider cockpit and extra power for increased performance and true cross-country capability;

Zenith CH 601 assembled in Malaysia *(Paul Jackson/Jane's)* 2001/0103548

CH 601HDS variant (first flown August 1991) has shorter, tapered 'speed' wing of 7.01 m (23 ft 0 in) span. HD and UL are available as kits and as 85 per cent pre-assembled kits requiring only quoted 120 working hours to complete, compared to quoted 400 hours for unassembled kits.
Description applies to CH 601HD.
CUSTOMERS: Over 300 of all versions flying by end 2000.
COSTS: Plans: US$315. Kit: US$12,620. Information pack: US$15 (2001). Materials and component parts available.
DESIGN FEATURES: Conventional low-wing monoplane with all-moving vertical tail. Conforms to Canadian TP 10141 airworthiness standards.
Wing section NACA 65018 (modified); dihedral 6°, incidence 8°.
FLYING CONTROLS: Manual. Long-span ailerons; elevators; and Zenith all-moving vertical fin.
STRUCTURE: All-metal monocoque. Single-piece, starboard-hinged canopy.
LANDING GEAR: Non-retractable tricycle or tailwheel type. Skis and floats optional.
POWER PLANT: One engine of 47.7 to 85.75 kW (64 to 115 hp), including 59.6 kW (79.9 hp) Rotax 912 UL, with reduction gear, driving three-blade ground adjustable propeller; 47.8 kW (64.1 hp) Rotax 582UL, 48.5 to 74.5 kW (65 to 100 hp) Teledyne Continental or 52 kW (70 hp) Volkswagen. 74.6 kW (100 hp) four-cylinder four-stroke

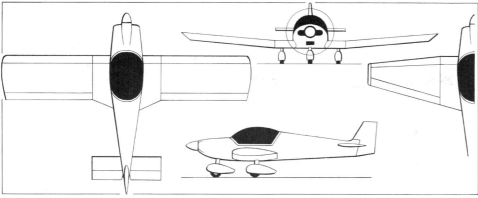

Zenith Zodiac CH 601HD, with half-plan view of tapered CH 601HDS wing *(James Goulding/Jane's)* 0016514

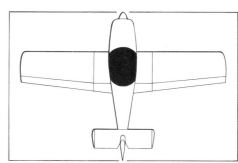

Plan view of Zenith Zodiac CH 601XL, showing revised wing shape of this version *(James Goulding/Jane's)* 2000/0062649

liquid-cooled Subaru EA81 is available from Stratos for the CH 601HD. Fuel capacity 60 litres (16.0 US gallons; 13.3 Imp gallons); wing baggage lockers may be fitted with two 26.5 litre (7.0 US gallon; 5.8 Imp gallon) auxiliary tanks.

DIMENSIONS, EXTERNAL:
Wing span	8.23 m (27 ft 0 in)
Wing chord (constant)	1.24 m (4 ft 1 in)
Wing aspect ratio	5.6
Length overall	5.79 m (19 ft 0 in)
Height overall: tailwheel	1.90 m (6 ft 2¾ in)
nosewheel	2.08 m (6 ft 10 in)
Tailplane span	2.29 m (7 ft 6 in)
Wheel track	1.88 m (6 ft 2 in)
Wheelbase	1.30 m (4 ft 3¼ in)
Propeller diameter	1.73 m (5 ft 8 in)

DIMENSIONS, INTERNAL:
Cockpit max width	1.12 m (3 ft 8 in)

AREAS:
Wings, gross	12.08 m² (130.0 sq ft)

WEIGHTS AND LOADINGS (Rotax 912 for power loading):
Weight empty	290 kg (640 lb)
Baggage capacity	36 kg (80 lb)
Max T-O weight	544 kg (1,200 lb)
Max wing loading	45.1 kg/m² (9.23 lb/sq ft)
Max power loading	9.11 kg/kW (15.00 lb/hp)

PERFORMANCE (Rotax 912):
Never-exceed speed (V_{NE})	130 kt (241 km/h; 150 mph)
Max level speed at S/L	117 kt (217 km/h; 135 mph)
Econ cruising speed	104 kt (193 km/h; 120 mph)
Stalling speed	41 kt (76 km/h; 47 mph)
Max rate of climb at S/L	351 m (1,150 ft)/min
Service ceiling	4,875 m (16,000 ft)
T-O run	131 m (430 ft)
Landing run	159 m (520 ft)
Range with max fuel	677 n miles (1,255 km; 780 miles)
Endurance	4 h
g limits	±6

UPDATED

ZENITH SUPER ZODIAC CH 601XL

TYPE: Side-by-side kitbuilt.
PROGRAMME: Certifiable version of CH 601 (which see), with new wing planform and 589 kg (1,300 lb) MTOW. Introduced at Oshkosh in July 1998.
Description generally as for CH 601, except that below.
CURRENT VERSIONS: **CH 601XL**: US version.
CH 601DX: German designation.
COSTS: Kit US$15,890 (2001).
DESIGN FEATURES: Wing section NACA 23018 at root, NACA 23015 at tip.
FLYING CONTROLS: Flaps added.
POWER PLANT: One 89 kW (120 hp) Jabiru 3300 flat-six or 86.5 kW (116 hp) Textron Lycoming O-235. Options include 73.5 kW (98.6 hp) Rotax 912 ULS and conversions of Subaru motorcar engines. Fuel capacity 91 litres (24.0 US gallons; 20.0 Imp gallons).

DIMENSIONS, EXTERNAL:
Wing span	8.23 m (27 ft 0 in)
Wing chord: at main panel root	1.60 m (5 ft 3 in)
at tip	1.33 m (4 ft 4¼ in)
Wing aspect ratio	5.5

AREAS:
Wings, gross	12.26 m² (132.0 sq ft)

WEIGHTS AND LOADINGS (O-235):
Weight empty	363 kg (800 lb)
Baggage capacity	18 kg (40 lb)
Max T-O weight	589 kg (1,300 lb)
Max wing loading	47.5 kg/m² (9.85 lb/sq ft)
Max power loading	6.82 kg/kW (11.21 lb/hp)

PERFORMANCE (O-235):
Never-exceed speed (V_{NE})	156 kt (289 km/h; 180 mph)
Max level speed	129 kt (238 km/h; 148 mph)
Normal cruising speed	120 kt (222 km/h; 138 mph)
Stalling speed: flaps down	39 kt (71 km/h; 44 mph)
Max rate of climb at S/L	283 m (930 ft)/min
Service ceiling	4,267 m (14,000 ft)
T-O and landing run	152 m (500 ft)
Max range	521 n miles (965 km; 600 miles)
Endurance, no reserves	4 h 30 min
g limits	±6

UPDATED

ZENITH GEMINI CH 620

TYPE: Two-seat kitbuilt twin.
PROGRAMME: Prototype construction began January 1996; first flight (N6265N) 23 July 1996; kit production and delivery yet to be confirmed.
CUSTOMERS: No aircraft registered in USA by January 2001.
COSTS: Airframe kit price US$19,950; power plant/instrument package US$20,205 (2000).
DESIGN FEATURES: Experimental category, low-wing design, employing many Zodiac components. Dihedral 0° 6′ 30″ outboard of engines. Slightly swept wing leading-edge. Starboard engine toed out 2° 30′. Quoted build time 750 hours.
FLYING CONTROLS: Manual. Zenair-type all-moving fin; mass-balanced ailerons and elevators; aileron movement ±3°.
STRUCTURE: Aluminium alloy, semi-monocoque, stressed skin construction; single-spar wing.
LANDING GEAR: Tailwheel or nosewheel type. 41 cm (16 in) mainwheels retract rearwards, protruding slightly from engine nacelles to minimise damage in a wheels-up landing. Tailwheel where fitted, is fixed, with optional spat.
POWER PLANT: Two 59.7 kW (80 hp) Jabiru 2200 flat-four piston engines driving two-blade fixed-pitch propellers. Standard fuel capacity 129 litres (34.0 US gallons; 28.3 Imp gallons).
ACCOMMODATION: Two persons side by side under single-piece canopy. Centrally mounted Y-shaped control stick.
SYSTEMS: Breakerless transistorised dual ignition; 12 V electric starter.

DIMENSIONS, EXTERNAL:
Wing span	8.53 m (28 ft 0 in)
Wing chord: at c/l	1.83 m (6 ft 0 in)
at tip	0.86 m (2 ft 9¾ in)
Wing aspect ratio	6.0
Length overall	6.27 m (20 ft 7 in)
Height overall	1.98 m (6 ft 6 in)
Wheel track	2.70 m (8 ft 10¼ in)
Propeller diameter	1.40 m (4 ft 7 in)

DIMENSIONS, INTERNAL:
Cabin max width	1.12 m (3 ft 8 in)

AREAS:
Wings, gross	12.63 m² (136.0 sq ft)

WEIGHTS AND LOADINGS:
Weight empty	354 kg (780 lb)
Max T-O weight	658 kg (1,450 lb)
Max wing loading	52.1 kg/m² (10.66 lb/sq ft)
Max power loading	5.51 kg/kW (9.06 lb/hp)

PERFORMANCE:
Max level speed	139 kt (257 km/h; 160 mph)

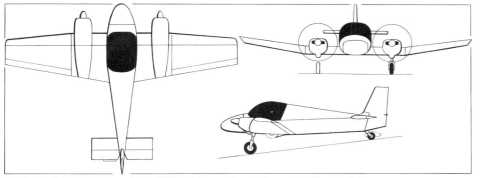

Zenith Gemini CH 620 (two Jabiru 2200 piston engines) *(James Goulding/Jane's)* 2000/0062648

Zenith Gemini CH 620 prototype

USA: AIRCRAFT—ZENITH

Cruising speed at 75% power	126 kt (233 km/h; 145 mph)
Stalling speed, flaps down	48 kt (89 km/h; 55 mph)
Max rate of climb at S/L	427 m (1,400 ft)/min
Rate of climb at S/L, OEI	91 m (300 ft)/min
Service ceiling	4,875 m (16,000 ft)
Service ceiling, OEI	915 m (3,000 ft)
T-O run	138 m (450 ft)
T-O to 15 m (50 ft)	259 m (850 ft)
Landing from 15 m (50 ft)	244 m (800 ft)
Landing run	152 m (500 ft)
Range, no reserves	650 n miles (1,203 km; 748 miles)

UPDATED

ZENITH STOL CH 701

TYPE: Side-by-side ultralight kitbuilt.
PROGRAMME: Experimental category prototype made first flight mid-1986. Plans, 49 or 85 per cent kits available. Meets Canadian TP 10141 standards for advanced ultralight trainers. German promotion as **CH-701D**. Also available in Czech Republic as **Kappa-1**, with slight performance differences. Ekoflug of Slovak Republic builds version designated MXP-740 (which see). CH-701-AG agricultural version ceased production after 55 completed. Quoted build time 400 hours.
CUSTOMERS: Over 500 flying by mid-2000.
COSTS: Plans: US$320. Kit: US$12,360. Information pack: US$35. Amphibian kit: US$17,730 (2001). Materials and component parts available.
DESIGN FEATURES: Strut-braced, high-wing, STOL cabin monoplane with foldable wings and fixed leading-edge slats. In most countries, CH 701 can be registered either as advanced ultralight (AUL) or as experimental homebuilt.
FLYING CONTROLS: Manual. Junkers-type flaperons; elevators; and Zenith all-moving fin. Full-span leading-edge slats for STOL performance.
STRUCTURE: All-metal semi-monocoque.
LANDING GEAR: Tricycle or tailwheel gear; optional floats, amphibious or ski gears.
POWER PLANT: One 37.0 kW (49.6 hp) Rotax 503 UL-2V engine in prototype, but 59.6 kW (79.9 hp) Rotax 912 UL, 47.8 kW (64.1 hp) Rotax 582 UL, 44.7 to 52.2 kW (60 to 70 hp) VW or 48.5 to 67.1 kW (65 to 90 hp) Teledyne Continental engine optional. Two- or three-blade propeller. Standard fuel capacity 42 litres (11.0 US gallons; 9.2 Imp gallons) in fuselage tank; optional 42 litres (11.0 US gallons; 9.2 Imp gallons) (total) in wing tanks.

DIMENSIONS, EXTERNAL:
Wing span	8.23 m (27 ft 0 in)
Length overall	6.38 m (20 ft 11 in)
Width, wings folded	2.06 m (6 ft 9 in)
Height overall	2.59 m (8 ft 6 in)

DIMENSIONS, INTERNAL:
Cabin max width	1.07 m (3 ft 6 in)

AREAS:
Wings, gross	11.33 m² (122.0 sq ft)

WEIGHTS AND LOADINGS:
Weight empty	208 kg (460 lb)
Max T-O weight	435 kg (960 lb)

PERFORMANCE (47.8 kW; 64.1 hp engine):
Max level speed at S/L	82 kt (152 km/h; 95 mph)
Max cruising speed	70 kt (129 km/h; 80 mph)
Stalling speed	27 kt (49 km/h; 30 mph)
Max rate of climb at S/L	427 m (1,400 ft)/min
T-O run	28 m (90 ft)
Landing run	24 m (80 ft)
Range: standard fuel	182 n miles (338 km; 210 miles)
with wing tanks	360 n miles (667 km; 415 miles)

UPDATED

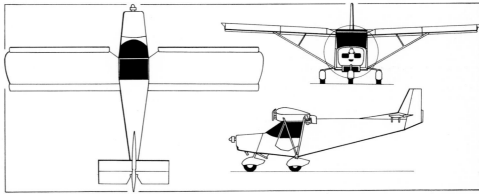

Zenith STOL CH 701 built in the UK *(Paul Jackson/Jane's)*

Zenith STOL CH 701 light aircraft *(Paul Jackson/Jane's)*

ZENITH STOL CH 801

TYPE: Four-seat kitbuilt.
PROGRAMME: Based on STOL CH 701 design; work started in 1988 for an undisclosed customer, but was suspended when order cancelled. Subsequently revived; prototype built by Flypass Ltd in Ontario, Canada and first flown March 1998. Launched at Oshkosh, July 1998, with kit deliveries scheduled for late 1998.
CUSTOMERS: Five sold and three flying by end 2000. European marketing by CZAW; UK PFA approval sought in 2001.
COSTS: Kit excluding engine US$20,950; plans US$420 (2001).
DESIGN FEATURES: High-wing, strut-braced monoplane. Although visually similar to the CH 701, no commonality of parts. Quoted build time 700 hours.
Wing aerofoil NACA65015.
FLYING CONTROLS: Manual. Full-span Junkers flaperons with outboard section offset 2° higher; deflection to 35° elevator; all-moving, horn-balanced fin. Fixed full-span leading-edge wing slats.
STRUCTURE: All-metal semi-monocoque construction, making extensive use of blind rivets. Single-spar 6061-T6 aluminium wing. Fuselage welded 4130 tubular steel frame.
LANDING GEAR: Tricycle type; cantilever mainwheel legs optimised for rough-field operations with Matco or Cleveland wheels and brakes. Nosewheel steerable ±15°. Tundra tyres on all wheels. Optional flotation gear.
POWER PLANT: One 134 kW (180 hp) Textron Lycoming O-360-A driving two-blade, fixed-pitch Sensenich 76-EMB-O-54 metal propeller in production prototype; engines in the range 112 to 179 kW (150 to 240 hp) can be fitted. Standard fuel capacity 151 litres (40.0 US gallons); optional fuel capacity 303 litres (80.0 US gallons; 66.6 Imp gallons).
ACCOMMODATION: Pilot and three passengers in enclosed cabin. Single door on port side. Underfuselage cargo pod optional.

DIMENSIONS, EXTERNAL:
Wing span	9.45 m (31 ft 0 in)
Wing chord, constant	1.60 m (5 ft 3 in)
Wing aspect ratio	5.8
Length overall	7.47 m (24 ft 6 in)
Height overall	2.67 m (8 ft 9 in)
Tailplane span	2.54 m (8 ft 4 in)
Tailplane chord	0.91 m (3 ft 0 in)
Wheel track	2.03 m (6 ft 8 in)
Wheelbase	1.65 m (5 ft 5 in)
Passenger door: Height	0.97 m (3 ft 2 in)
Width	0.91 m (3 ft 0 in)

DIMENSIONS, INTERNAL:
Cabin: Length	1.98 m (6 ft 6 in)
Max width	1.12 m (3 ft 8 in)
Max height	1.02 m (3 ft 4 in)

AREAS:
Wings, gross	15.51 m² (167.0 sq ft)
Tailplane	2.32 m² (25.00 sq ft)

WEIGHTS AND LOADINGS:
Weight empty	522 kg (1,150 lb)
Max T-O weight	975 kg (2,150 lb)
Max wing loading	62.9 kg/m² (12.87 lb/sq ft)
Max power loading	7.27 kg/kW (11.94 lb/hp)

PERFORMANCE:
Never-exceed speed (V_{NE})	130 kt (241 km/h; 150 mph)
Max level speed	96 kt (177 km/h; 110 mph)
Max cruising speed at 75% power	91 kt (169 km/h; 105 mph)
Stalling speed: flaps up	48 kt (89 km/h; 56 mph)
flaps down	34 kt (63 km/h; 39 mph)
Max rate of climb at S/L	219 m (720 ft)/min
Service ceiling	4,267 m (14,000 ft)

Zenith STOL CH 801 (Textron Lycoming O-360-A) *(Paul Jackson/Jane's)*

T-O run	119 m (390 ft)
T-O to 15 m (50 ft)	122 m (400 ft)
Landing from 15 m (50 ft)	91 m (300 ft)
Landing run	46 m (150 ft)
Range:	
with standard fuel	273 n miles (507 km; 315 miles)
with optional fuel	547 n miles (1,013 km; 630 miles)
Endurance: with standard fuel	3 h 0 min
with optional fuel	6 h 0 min
g limits	+6/−3

UPDATED

Zenith STOL CH 801 *(James Goulding/Jane's)*
1999/0051861

ZIVKO

ZIVKO AERONAUTICS INC

502 Airport Road, Building 11, Guthrie, Oklahoma 73044
Tel: (+1 405) 282 13 30
Fax: (+1 405) 282 13 39
e-mail: zivko@ionet.net
Web: http://www.zivko.com
PRESIDENT: Bill Zivko

Founded in 1987, Zivko manufactures the Edge 540 series of aerobatic monoplanes from a 1,040 m² (11,000 sq ft) factory on the airport at Guthrie and also provides wings for retrofitting to other aerobatic aircraft. In August 2000, it received FAA certification under Pt 21.191(g) for the Edge kits.

UPDATED

ZIVKO EDGE 540

TYPE: Aerobatic single-seat sportplane; aerobatic two-seat lightplane.
PROGRAMME: Initial version was Zivko Edge 360, later upgraded to Zivko Edge 540.
CURRENT VERSIONS: **540**: Single-seat version.
540T: Two-seat model introduced March 2000; differences listed below.
CUSTOMERS: At least 27 sold, including examples to Mexico, South Africa and Australia.
COSTS: Edge 540 US$224,908; Edge 540T US$238,477 (2001).
DESIGN FEATURES: Optimised for aerobatics; rate of roll 420°/s. Mid-positioned, tapered wing with large, horn-balanced ailerons. Transparencies in lower sides of cockpit.
FLYING CONTROLS: Conventional and manual. Ailerons and elevator actuated by pushrods; horn-balanced rudder by cables. Flight adjustable tabs in both elevators. Suspended, spade-type tab on each aileron.
STRUCTURE: 4130 Steel tube fuselage; upper fuselage and fairings of composites; lower aft fuselage fabric covered, stressed to ultimate ±15 g; all-composites fully cantilevered two-spar wing. Tail surfaces mutually wire-braced.
LANDING GEAR: Tailwheel type; fixed. Main legs aluminium sprung steel; Cleveland wheels and hydraulic brakes; optional speed fairings. Spring steel tail gear, with chrome-plated, solid tailwheel, choice of fixed or steerable.
POWER PLANT: One Textron Lycoming IO-540 modified to produce 254 kW (340 hp), driving Hartzell HC-C3YR-4AX three-blade, composites propeller. Edge 540 has single 72 litre (19.0 US gallon; 15.8 Imp gallon) forward fuselage fuel tank, plus optional total of 159 litres (42.0 US gallons; 35.0 Imp gallons) in wing tanks; Edge 540T has single 76 litre (20.0 US gallon; 16.6 Imp gallon) fuselage tank and two optional 83 litre (22.0 US gallon; 18.3 Imp gallon) wing tanks. Inverted oil and fuel systems.
ACCOMMODATION: Pilot (pilot and passenger in tandem in 540T) under one-piece perspex canopy; carbon fibre contoured seats.
SYSTEMS: 8 A alternator. Bendix fuel injection system.
DIMENSIONS, EXTERNAL:

Wing span: 540	7.42 m (24 ft 4 in)
540T	7.67 m (25 ft 2 in)
Length overall: 540	6.27 m (20 ft 7 in)
540T	6.91 m (22 ft 8 in)
Height overall	2.87 m (9 ft 5 in)

AREAS:

Wings, gross: 540	9.10 m² (98.0 sq ft)
540T	9.87 m² (106.2 sq ft)

WEIGHTS AND LOADINGS:

Weight empty: 540	531 kg (1,170 lb)
540T	572 kg (1,261 lb)
Max T-O weight: 540	816 kg (1,800 lb)
540T: Aerobatic	726 kg (1,600 lb)
Utility	885 kg (1,950 lb)
Max wing loading: 540	89.7 kg/m² (18.37 lb/sq ft)
540T: Aerobatic	73.6 kg/m² (15.07 lb/sq ft)
Utility	89.6 kg/m² (18.36 lb/sq ft)
Max power loading: 540	3.22 kg/kW (5.29 lb/hp)
540T: Aerobatic	2.86 kg/kW (4.71 lb/hp)
Utility	3.49 kg/kW (5.74 lb/hp)

PERFORMANCE:

Never-exceed speed (VNE)	230 kt (426 km/h; 265 mph)
Max cruising speed at 75% power:	
540	180 kt (333 km/h; 207 mph)
540T	195 kt (361 km/h; 224 mph)
Stalling speed: 540: Aerobatic	51 kt (95 km/h 59 mph)
Utility	61 kt (113 km/h; 71 mph)
540T: Aerobatic	53 kt (99 km/h; 61 mph)
Utility	61 kt (113 km/h; 71 mph)
Max rate of climb at S/L: 540	1,128 m (3,700 ft)/min
540T	1,067 m (3,500 ft)/min
g limits: Aerobatic	±10
Utility	±8

UPDATED

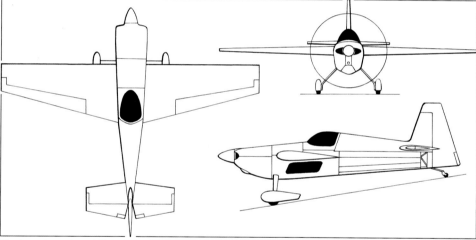

Single-seat version of Zivko Edge 540 *(James Goulding/Jane's)*
2001/0106842

Zivko Edge 540 N540KC, flown by Kirby Chambliss, gained 14th place in 2000 World Aerobatic Championships *(Paul Jackson/Jane's)*
2001/0103559

UZBEKISTAN

TAPO

TASHKENTSKOYE AVIATSIONNOYE PROIZDSTVENNOYE OBEDINENIE IMENI V P CHKALOVA (Tashkent Aircraft Production Enterprise named for V P Chkalov)

ulitsa Elbek 61, 700016 Tashkent
Tel: (+810 99871) 136 11 67
Fax: (+810 99871) 268 03 18
e-mail: oms@tapo.tashkent.ru
Web: http://www.uzbekistan.org/aircraft.html
GENERAL DIRECTOR: Vladim P Koucherov

Founded in 1932, this Enterprise is said to be the largest aircraft manufacturing centre on the Asian continent. It comprises five plants at Tashkent, Andizhan and Fergana. Products have included I-15s, I-16s, I-153s, Li-2s, Il-14s, An-8s, Ka-22s, An-12s, An-22s and Il-76s. It has manufactured wing centre-sections for the An-124 transport and two An-70 prototypes, and has repaired MiG and Sukhoi fighters and Tu-22M bombers. It developed the Il-78 tanker version and collaborated with Beriev on the A-50 AEW derivative of the Il-76. Low-rate production is under way of the Ilyushin Il-114 twin-turboprop transport and Il-76, for which TAPO is the sole source, although airframes are stored partly complete until orders are received.

TAPO was registered as a joint stock company in May 1996; the state holds 82.7 per cent, National Bank of Foreign Economic Activity 6.7 per cent and employees 10.6 per cent. An attempt in 1997 to sell 14 per cent of the company was unsuccessful; by mid-1998, offers were being invited for a 25 per cent share, but no acceptances have been reported. Company began airbrake manufacture for Avro RJ in 1997.

Subsidiary, TAPO-Avia operates commercial freight charter flights with fleet of four An-12s, one An-24 and five Il-76MDs.

UPDATED

YUGOSLAVIA, FEDERAL REPUBLIC

LOLA UTVA

LOLA UTVA FABRIKA AVIONA (LOLA Utva Aircraft Industry)

Jabučki put bb, YU-26000 Pančevo
Tel: (+381 13) 31 53 83 and 31 25 84
Fax: (+381 13) 31 98 59
COMMERCIAL AND MARKETING MANAGER: Cudic Zeljko

Utva (Sheldrake) Aircraft Industry formed at Zemun, 5 June 1937; to Pančevo 1939.

Main business now includes overhaul and maintenance of G-4 and Antonov An-2; production of machined parts and special tooling for Israel Aircraft Industries and Sogerma-SOCEA; and windscreens for G-4, Orao and Gazelle. The company planned to place the Lasta 2 primary/basic trainer in production and build a prototype of the new Utva-96 but, by January 2001, neither goal appeared to have been achieved. These aircraft were last detailed in the 1999-2000 *Jane's*.

UPDATED

LIGHTER THAN AIR

CANADA

21ST CENTURY

21ST CENTURY AIRSHIPS INC
110 Pony Drive, Newmarket, Ontario L3Y 4W2
Tel: (+1 905) 898 62 74
Fax: (+1 905) 898 72 45
e-mail: ballot@home.com
Web: http://www.21stCenturyAirships.com
CHAIRMAN AND CEO: Hokan Colting

Formed 1988 to focus on R&D aimed at improving airship low-speed manoeuvrability, a problem its chairman identified and has studied since 1978, particularly with spherical airships. He moved from Sweden to Canada 1981; has extensive experience in manufacturing and piloting lighter than air craft; achieved all objectives of his original 1978 research programme. Company has built and flight tested six spherical airship prototypes since 1991.

Details of the early prototypes can be found in the 1995-96 *Jane's*. Descriptions of current programmes follow.

UPDATED

SPAS-70 flight deck, with colour-coded engine instruments *2001*/0105831

21ST CENTURY AIRSHIPS SPAS-13

TYPE: Helium non-rigid special shape.
PROGRAMME: Demonstrator prototype: first flight (C-FRLM) 10 June 1994. Made 112 flights as a semi-rigid before being deflated on 20 November 1994. Reinflated in 1995 after a number of improvements was incorporated; 40 more flights completed; flew again in early 1998 as full-scale platform for a drag-reducing device which enables engine horsepower to be decreased while maintaining same speed. This invention being patented and will be incorporated in all the company's spherical airships.
DESIGN FEATURES: Spherical shape, with outer load-bearing envelope and separately sealed inner helium envelope. Only two ground crew required.
FLYING CONTROLS: Steering and altitude controlled through varied and deflected thrust technology developed and patented by 21st Century Airships Inc.
POWER PLANT: Two 37.0 kW (49.6 hp) Rotax 503 UL-2V piston engines.
ACCOMMODATION: Flight deck and passenger compartment inside load-bearing envelope.
DIMENSIONS, EXTERNAL:
  Envelope diameter   13.11 m (43 ft 0 in)
DIMENSIONS, INTERNAL:
  Envelope volume   1,167 m³ (41,200 cu ft)
WEIGHTS AND LOADINGS:
  Weight empty   772 kg (1,702 lb)
  Max T-O weight   1,100 kg (2,425 lb)
PERFORMANCE:
  Max level speed   30 kt (56 km/h; 35 mph)
  Max rate of climb at S/L   366 m (1,200 ft)/min
  Max tested pressure altitude   1,400 m (4,600 ft)
  Endurance, 30 min reserves   1 h 30 min

UPDATED

21ST CENTURY AIRSHIPS SPAS-70

TYPE: Helium non-rigid special shape.
PROGRAMME: Prototype (C-FYOK) made first flight 5 August 1997 with four 74.6 kW (100 hp) piston engines; being used for certification trials, which have been delayed to incorporate drag reducer (see SPAS-13 entry) and change to twin diesel engines. Latter installed in 1999, but funding problems still delaying flight testing in this form. Modified 1999-2000 to permit take-offs and landings on water.
DESIGN FEATURES: As described for SPAS-13.
FLYING CONTROLS: All controls are fly-by-wire; otherwise as described for SPAS-13.
POWER PLANT: Two 67.1 kW (90 hp) turbo-diesel engines planned for production version.
ACCOMMODATION: Four seats inside load-bearing envelope; only two fitted for certification flights; positioning of other two optional within large internal floor area.
AVIONICS: *Comms:* VHF transceivers; transponder.
  Flight: GPS nav receiver and radar altimeter. Goodrich Stormscope adverse weather warning system.
  Instrumentation: Regular engine monitoring, plus internal pressure gauges, outside/inside and helium temperature gauges, and fire warning lights for each engine and APU. All engine instruments and controls colour-coded to simplify management of four engines.
DIMENSIONS, EXTERNAL:
  Envelope diameter   17.07 m (56 ft 0 in)
DIMENSIONS, INTERNAL:
  Envelope volume   2,605 m³ (92,000 cu ft)
  Usable floor space inside load-bearing envelope
     65.0 m² (700 sq ft)
WEIGHTS AND LOADINGS:
  Weight empty   1,775 kg (3,913 lb)
  Max T-O weight   2,520 kg (5,555 lb)

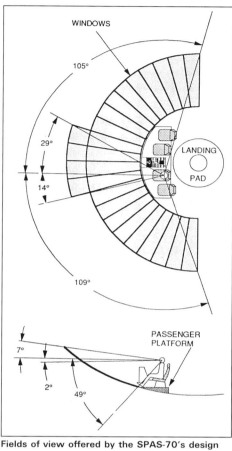

Fields of view offered by the SPAS-70's design
0016714

The four-seat, four-engined SPAS-70   *1999*/0051874

SPAS-13 in current non-rigid form, painted to resemble a baseball   *2001*/0105832

798　CANADA: LIGHTER THAN AIR—21ST CENTURY to PAN ATLANTIC

PERFORMANCE:
Max level speed	35 kt (64 km/h; 40 mph)
Max rate of climb at S/L	335 m (1,100 ft)/min
Max tested pressure altitude	3,155 m (10,350 ft)
Endurance, 30 min reserves	11 h

UPDATED

21ST CENTURY AIRSHIPS SPAS-75

TYPE: Helium non-rigid special shape.
PROGRAMME: Construction started 1997; to be completed once SPAS-70 is certified, and thus affected by delays in that programme. Apparently no further advanced by December 2000.
DESIGN FEATURES: As for SPAS-13 and SPAS-70; purpose-designed for aerial sightseeing tours.
FLYING CONTROLS: As for SPAS-70.
POWER PLANT: As for SPAS-70, except that engines are 112 kW (150 hp) each.
ACCOMMODATION: Pilot and up to 10 passengers inside load-bearing envelope.

DIMENSIONS, EXTERNAL:
Envelope diameter	22.86 m (75 ft 0 in)

DIMENSIONS, INTERNAL:
Envelope volume	6,230 m³ (220,000 cu ft)

WEIGHTS AND LOADINGS (estimated):
Weight empty	3,800 kg (8,378 lb)
Max T-O weight	6,600 kg (14,550 lb)

PERFORMANCE (estimated):
Max level speed	33 kt (61 km/h; 38 mph)

UPDATED

Artist's impression of the SPAS-75 interior
1999/0051876

PAN ATLANTIC

PAN ATLANTIC AEROSPACE CORPORATION (Subsidiary of Av-Intel Corporation)

PO Box 599, Station B, Ottawa, Ontario K1P 5P7
Tel: (+1 561) 776 13 51
Fax: (+1 561) 776 23 70
e-mail: ltadrone@aol.com
CHAIRMAN: Fredrick D Ferguson
VICE-PRESIDENT: Ian Stewart
INFORMATION OFFICER: Nick Baumberg

Pan Atlantic has been involved in lighter than air design and systems integration since 1978, including the Magnus Aerospace rotating sphere development project. Many of its designs have been awarded patents. Company's recent LTA-based projects have included a high-altitude, satellite-like platform in partnership with Skysat Communications Networks Corporation of New York; the Cargo Airship System (CAS); the LEAP surveillance drone; and the Magnus advertising vehicle. Descriptions of the Skysat and LEAP airships can be found in *Jane's Unmanned Aerial Vehicles and Targets*.

UPDATED

PAN ATLANTIC CARGO AIRSHIP SYSTEM (CAS)

TYPE: Helium non-rigid.
PROGRAMME: Conceived by F D Ferguson in 1988; two 15.24 m (50 ft) proof-of-concept prototypes test-flown between 1990 and 1992; US$30 million development programme, to build 183 m (600 ft) demonstration/production model to carry a 25 US ton payload, is under way; fabrication of demonstrator was scheduled to begin in June or July 1998 but had not been confirmed by mid-2000; however, programme reportedly still active at that time.
DESIGN FEATURES: New, patented technology based on modular, pressurised segments made of high-strength, lightweight materials and capable of absorbing gust and shock loads encountered by rigid airships. Intended for all-weather operation, carrying payloads from 22,680 to 453,590 kg (25 to 500 US tons). Design allows for effective load transfer throughout individual module envelopes. Airship is able to carry ISO containerised loads such as produce, livestock, manufactured goods and 'just in time' inventories at rates potentially comparable with road transport. Four tailfins, in X configuration.
FLYING CONTROLS: Conventional and manual. Ballonet in each segment (20 per cent of total volume); ruddervator on each tail surface.
POWER PLANT: Subscale (600 ft) prototype powered by six 224 kW (300 hp) Teledyne Continental flat-six piston engines. Full-size CAS to be powered by four 4,474 kW (6,000 shp) Rolls-Royce AE 2100 turboprops. See Weights and Loadings for fuel details.

DIMENSIONS, EXTERNAL (A: 600 ft prototype, B: full-size CAS):
Envelope: Length: A	183 m (600 ft)
B	457 m (1,500 ft)
Max diameter: A	24 m (80 ft)
B	61 m (200 ft)
Fineness ratio: A	7.6
B	7.5
Drag coefficient: A, B	0.026
Payload modules, each: Length: A	30 m (100 ft)
B	61 m (200 ft)
Max diameter: A	24 m (80 ft)
B	61 m (200 ft)
Front and rear modules, each:	
Length: A	46 m (150 ft)
B	107 m (350 ft)
Max diameter: A	24 m (80 ft)
B	61 m (200 ft)

DIMENSIONS, INTERNAL (A and B as above):
Envelope total volume: A	68,330 m³ (2,413,000 cu ft)
B	1.13 million m³ (39.8 million cu ft)
Ballonet total volume:	
A (at 20%)	13,670 m³ (482,600 cu ft)
B (at 20%)	230,000 m³ (8,000,000 cu ft)
Lifting volume: A	54,660 m³ (1,930,400 cu ft)
B	0.90 million m³ (31.8 million cu ft)

WEIGHTS AND LOADINGS (A and B as above):
Weight empty: A	25,855 kg (57,000 lb)
B	190,510 kg (420,000 lb)
Max fuel weight: A	10,614 kg (23,400 lb)
B	107,955 kg (238,000 lb)
Net available payload: A	19,325 kg (42,600 lb)
B	582,410 kg (1,284,000 lb)
Max T-O weight (gross lift at S/L):	
A	55,790 kg (123,000 lb)
B	891,760 kg (1,966,000 lb)

PERFORMANCE (estimated; A and B as above):
Max level speed: A	69 kt (128 km/h; 80 mph)
B	82 kt (153 km/h; 95 mph)
Cruising speed: A	64 kt (119 km/h; 74 mph)
B	66 kt (122 km/h; 76 mph)
Range:	
A at 61 kt (113 km/h; 70 mph)	
	2,607 n miles (4,828 km; 3,000 miles)
B at 66 kt (122 km/h; 76 mph)	
	3,476 n miles (6,437 km; 4,000 miles)
Endurance, conditions as for ranges above:	
A	40 h
B	52 h 36 min

UPDATED

Manufactured image of the planned Pan Atlantic Cargo Airship System prototype

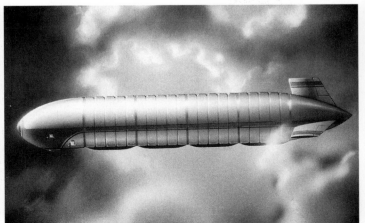

Production version of Pan Atlantic CAS will be 457 m (1,500 ft) long

CHINA, PEOPLE'S REPUBLIC

BKLAIC

BEIJING KEYUAN LIGHT AIRCRAFT INDUSTRIAL COMPANY LTD
7 Zhong Guan Cun South Road, Haidian, Beijing 100080
Tel: (+86 10) 62 57 28 22
Fax: (+86 10) 62 57 29 22

This company also produces the AD-200 lightplane described in the main Aircraft section.

UPDATED

BKLAIC ZHONGHUA

Promotion of this hot-air non-rigid continued until at least 1998, since which time no further reports have been received. An illustrated description last appeared in the 2000-01 *Jane's*.

UPDATED

CZECH REPUBLIC

KUBICEK

BALONY KUBICEK Spol. s r.o.
Francouzská 81, CR-602 00 Brno
Tel: (+420 5) 45 21 19 17 and 45 21 47 55
Fax: (+420 5) 45 21 30 74
e-mail: sales@kubicekballoons.cz
Web: http://www.kubicekballoons.cz
MANAGING DIRECTOR: Dipl Ing Ales Kubicek
SALES MANAGER: Michael Suchy

First Kubicek hot-air balloon was manufactured in 1983 and was followed by about 50 more of Kubicek design by various state-owned companies such as Aviatik and Aerotechnik. Present company was formed as private enterprise in 1991 and began new range of hot-air balloons under BB series designations (total of 126 produced by early 2000). Currently markets standard, ready-to-fly balloons in nine envelope sizes up to 6,000 m³ (211,900 cu ft); also special-shape balloons to order.

Kubicek's first hot-air airship design is the AV-1. A new design, the AV-2, made its maiden flight in 1999.

UPDATED

KUBICEK AV-1

TYPE: Hot-air non-rigid.
PROGRAMME: First flight 16 October 1993. Second example nearing completion in late 1998.
CUSTOMERS: First example (OK-3040) delivered to Czech company TICO in 1993-94; subsequently sold (1997) to Kubicek's Russian distributor Avgur (which see); now operated by Coca-Cola Moscow.
COSTS: US$85,000 (1997).
POWER PLANT: One 37.0 kW (49.6 hp) Rotax 503 UL-2V engine standard; also offered in Russia with optional 20.9 kW (28 hp) VAZ-1111 motorcar engine. Four-blade pusher propeller.
ACCOMMODATION: Pilot and co-pilot in side-by-side seats. Ground crew three to five persons.
DIMENSIONS, EXTERNAL:
  Envelope: Length   35.00 m (114 ft 10 in)
  Max diameter   12.00 m (39 ft 4½ in)
  Fineness ratio   2.9
  Height overall   14.55 m (47 ft 8¾ in)
DIMENSIONS, INTERNAL:
  Envelope volume   2,800 m³ (98,875 cu ft)

Kubicek AV-1 in the insignia of the Czech TICO company   0016726

WEIGHTS AND LOADINGS:
  Weight empty: envelope   195 kg (430 lb)
    gondola   240 kg (529 lb)
  Max T-O weight   750 kg (1,653 lb)
PERFORMANCE (Rotax 503):
  Max level speed   13.5 kt (25 km/h; 15.5 mph)
  Cruising speed   11 kt (20 km/h; 12 mph)
  Cruising altitude   800 m (2,620 ft)
  Endurance   1 h 30 min

UPDATED

KUBICEK AV-2

TYPE: Hot-air non-rigid.
PROGRAMME: First flight 26 May 1999; public debut (c/n 106/OK-8036) June 1999; then sold to ADS Kosice in Slovak Republic as OM-ADS.
POWER PLANT: One 34.0 kW (45.6 hp) Rotax 503 UL-1V engine standard; Rotax 912 optional. Fuel capacity 30 litres (7.9 US gallons; 6.6 Imp gallons).
ACCOMMODATION: Pilot and two passengers in all-metal (welded chromoly steel tube frame, duralumin skin) gondola.
DIMENSIONS, EXTERNAL:
  Envelope: Length   38.00 m (124 ft 8 in)
  Max diameter   14.20 m (46 ft 7 in)
  Gondola: Length   3.00 m (9 ft 10 in)
  Max width   1.70 m (5 ft 7 in)
  Max height   1.85 m (6 ft 0¾ in)
DIMENSIONS, INTERNAL:
  Envelope volume   3,500 m³ (123,601 cu ft)
WEIGHTS AND LOADINGS:
  Weight empty   440 kg (970 lb)
  Max T-O weight   870 kg (1,918 lb)
PERFORMANCE (Rotax 503):
  Max level speed   10.5 kt (20 km/h; 12 mph)
  Pressure ceiling (estimated)   2,440 m (8,000 ft)
  Endurance   1 h 30 min to 2 h 30 min

UPDATED

FRANCE

AERALL

ASSOCIATION POUR L'ETUDE ET LA RECHERCHE SUR LES AERONEFS ALLEGES
2 bis, avenue Odette, F-94130 Nogent-sur-Marne
Tel: (+33 1) 49 00 15 14
Fax: (+33 1) 47 75 09 56
PRESIDENT: Jean René Fontaine

NEW ENTRY

AERALL AVEA

TYPE: Helium non-rigid.
PROGRAMME: Originated in mid-1990s in balloon department of CNES space agency; has French Ministry of Transport, regional and private funding. Phase 1 (preliminary design evaluation) undertaken early 2000; Phase 2 (feasibility and market studies) was undertaken in late 2000 by Onera. Single MAP (see below) test-flown in September 2000; manufacture of full-size AVEA projected to start in 2004, with first flight 2006 and service entry in 2008.
DESIGN FEATURES: AVEA (*Aile Volante Epaisse Aérostatique:* thick aerostatic flying wing) is an unorthodox design comprising a variable number of tall, cylindrical, steerable modules (*modules aérostatiques pilotables,* or MAPs), attached in tandem and linked across their tops and bottoms by a trellis-like framework. Each MAP is provisionally 20 m (65.6 ft) in diameter and 100 m (328.1 ft) tall.

NEW ENTRY

GERMANY

CARGOLIFTER

CARGOLIFTER AG
Grüneburgweg 102, D-60323 Frankfurt-am-Main
Tel: (+49 69) 15 05 70
Fax: (+49 69) 150 58 18
e-mail: info@cargolifter.com
Web (1): http://www.cargolifter.de
Web (2): http://www.cargolifter-world.de
CHAIRMAN: Heinz Herrmann
PRESIDENT AND CEO: Dr Carl-Heinrich von Gablenz

HEAD OF DESIGN: Ralph Maurer
HEAD OF PRODUCTION: Christoph von Kessel
HEAD OF DEVELOPMENT OPERATIONS: Frank Bruno
MARKETING DIRECTOR: Hinrich Schliephack
MARKETING COMMUNICATIONS: Silke Rösser

Company formed on 1 September 1996 to continue development of CargoLifter large multipurpose airship for transportation of heavy and outsize cargoes. Nominal start-up capital of DM15.5 million; by July 2000, more than 30,000 shareholders in Germany and internationally and raised capital increased to €40.5 million. Full-size CL-160 preceded by subscale 'Joey' as proof-of-concept prototype; Skyship 600B, known as 'Charly', acquired in July 2000 for crew training.

Huge, 360 × 210 × 107 m (1,181 × 689 × 351 ft), 5.5 million m³ (194.2 million cu ft) hangar at Brand, south of Berlin, began construction March 1999 and was completed in late 2000; will house two CL-160s. Second hangar to be erected at new site in North Carolina, USA, for completion in about 2005.

UPDATED

CARGOLIFTER JOEY

TYPE: Helium semi-rigid.
PROGRAMME: One-eighth scale prototype to undertake materials, load suspension, flight behaviour and certification trials of CL-160; name Joey (as in baby kangaroo) chosen because dimensionally it could fit inside the loading bay 'pouch' of the CL-160. Designed by engineers from Stuttgart University; exhibited at Leipzig Trade Fair in May 1998; first flight (D-LJOE) 18 October 1999.
STRUCTURE: Envelope manufactured by Lindstrand Balloons (see under UK).
POWER PLANT: Two 17.2 kW (23 hp) Solo piston engines, each driving an MT 140R 60 propeller. Fuel capacity 40 litres (10.6 US gallons; 8.8 Imp gallons).
ACCOMMODATION: Small gondola for pilot and test-flight engineer.

DIMENSIONS, EXTERNAL:
Length overall	32.00 m (104 ft 11¾ in)
Max diameter	8.00 m (26 ft 3 in)
Fineness ratio	4.0

DIMENSIONS, INTERNAL:
Envelope volume	1,050 m³ (37,109 cu ft)

WEIGHTS AND LOADINGS:
Max payload	220 kg (485 lb)
Max T-O weight	830 kg (1,829 lb)

PERFORMANCE (estimated):
Max level speed	33 kt (61 km/h; 38 mph)
Cruising speed	20 kt (37 km/h; 23 mph)
Pressure ceiling	300 m (985 ft)
Endurance at cruising speed	4 h

UPDATED

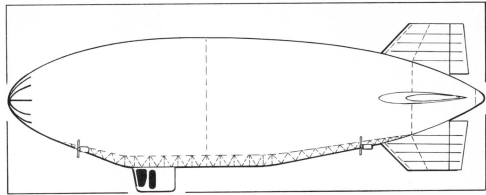

CargoLifter Joey one-eighth scale prototype for the CL-160 *(James Goulding/Jane's)* 1999/0051886

CARGOLIFTER CL-160

TYPE: Helium semi-rigid.
PROGRAMME: Began 1994; outcome of research project funded by German VDMA engineering plant construction organisation, indicating possibility of lower operating costs than conventional aircraft, higher speeds than a ship, and minimal need for ground handling infrastructures. Project has the support of Brandenburg state government. Design frozen 5 November 1998. Construction plans are for two prototypes and two preseries airships; first flight late 2002; first revenue service 2003; series production (up to four per year) 2004.
CUSTOMERS: Sixteen letters of intent (including one from United Nations) reportedly received by mid-2000, with others being negotiated.
COSTS: Estimates include US$134 million for development and first prototype, US$56 million for production CL-160 (1999).
DESIGN FEATURES: Semi-rigid structure with keel and catenary curtain; heart-shaped cross-section; large central cargo bay houses a specially designed 'multibox' container which can be loaded and unloaded from all sides; in-flight access to all main parts of the airship.
FLYING CONTROLS: Digital fly-by-wire control of aerodynamic control surfaces, all engines, and load exchange system.
STRUCTURE: Full-length carbon fibre structural keel supports cargo, load exchange installations, power plant and auxiliary systems; 20 cell envelope with central catenary curtain; internal ballonets. Nose and tail fuel/water tanks provide ballast in event of any gas cell failure (water replaces fuel as flight progresses). Subcontractors include MAN and Dornier Luftfahrt (keel) and Liebherr (loading crane).
POWER PLANT: Four MTU marine diesel engines, each with nominal rating of 1,425 kW (1,911 hp) and driving a fixed-axis tractor propeller, for main propulsion. Lower-powered turboshaft vertical thruster mounted on each main engine support, plus bow and stern thrusters for lateral manoeuvring. Total installed power 9,330 kW (12,512 hp). Turboshaft selection expected in mid-March 2001; choice between Rolls-Royce Turbomeca RTM322 (reportedly preferred choice), General Electric T700 and P&WC PT6.
ACCOMMODATION: Crew of 12. Payload is carried in an internal ventral cargo bay on a suspended load platform (multibox) that is winched up and down hydraulically for loading and unloading. Multibox can accommodate 36 standard 12.2 × 2.44 × 2.44 m (40 × 8 × 8 ft) freight containers. Winches are coupled electronically to keep load platform level regardless of airship movement.
Following data are provisional:

DIMENSIONS, EXTERNAL:
Envelope: Length, excl nose boom	260 m (853 ft)
Max diameter	65 m (213 ft)
Height overall	78 m (256 ft)
Tail unit span	74 m (243 ft)
Propeller diameter (main engines)	6.5 m (21.3 ft)

DIMENSIONS, INTERNAL:
Envelope volume	550,000 m³ (19.4 million cu ft)
Cargo multibox: Length	50 m (164 ft)
Width	8 m (26 ft)
Depth	8 m (26 ft)
Volume	3,200 m³ (113,000 cu ft)

WEIGHTS AND LOADINGS:
Max payload	120,000-160,000 kg (264,555-352,740 lb)
Max T-O weight	400,000 kg (881,850 lb)

PERFORMANCE (estimated):
Max level speed	72 kt (135 km/h; 83 mph)
Typical cruising speed at 2,000 m (6,560 ft)	49 kt (90 km/h; 56 mph)
Max flight altitude	2,000 m (6,560 ft)
Max range	5,400 n miles (10,000 km; 6,215 miles)

UPDATED

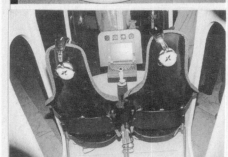

Crew gondola of the Joey prototype 1999/0051878

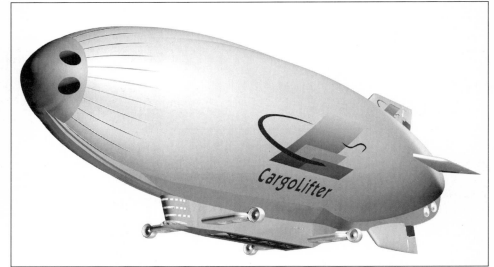

Computer-generated impression of the projected CargoLifter CL-160 2001/0106861

GEFA-FLUG

GESELLSCHAFT ZUR ENTWICKLUNG UND FÖRDERUNG AEROSTATISCHER FLUGSYSTEME GmbH (Aerostatic Flight Systems Development and Promotion Company)
Weststrasse 24C, D-52074 Aachen
Tel: (+49 241) 88 90 40
Fax: (+49 241) 889 04 20
e-mail: j.leisten@gefa-flug.de
MANAGING DIRECTOR: Karl Ludwig Busemeyer
SALES AND MARKETING DIRECTOR: Jürgen Leisten

GEFA-Flug was established 1975 to operate advertising and passenger-carrying hot-air balloons; has R&D department to develop airships for manned and unmanned civil and environmental research applications such as aerial photogrammetry and pollution monitoring.

Company currently operates more than six hot-air balloons, hot-air airships and a number of remotely controlled aerostats, and has sold about 100 systems (mostly unmanned and remotely controlled) worldwide. Its 20th hot-air airship was sold in mid-2000. Fleet includes three AS 105GD hot-air airships, including one four-seater; production of this model continues. Future market seen as survey projects for environmental research; GEFA-Flug also developing scientific measuring equipment in partnership with German Mining Technology, a partly government-funded body.

UPDATED

GEFA-FLUG AS 105 GD

TYPE: Hot-air non-rigid.
PROGRAMME: Modified version of AS 105 Mk II first flown 1990 and built until 1994 by Thunder & Colt in UK; German development began 1993. Certification (by UK CAA) and first production AS 105 GD completed in 1996; also certified by German LBA.
CUSTOMERS: Total of 19 built by end of 2000.
COSTS: Development programme DM1.4 million; standard airship approximately DM300,000 (2000).
DESIGN FEATURES: Longer and slimmer envelope than UK-built version; cruciform tail surfaces with rudder on each vertical fin; twin catenary load suspension system distributes loads of gondola weight and power plant forces evenly into envelope.

Ice-cold Warsteiner: Alpine setting for a GEFA-Flug AS 105 GD *2001/0098028*

FLYING CONTROLS: Burners located inside envelope and operated electrically, with manual override; pilot lights fitted with electric spark and piezoelectric ignition, specially modified to operate directly underneath inflation fan without blowing out; steering by rudder on each vertical fin, operated by cables connecting them to gondola.
STRUCTURE: Envelope made from high-tenacity and high temperature-resistant fabric; engine drives generator supplying electric fan used to pressurise envelope; additional air for envelope pressurisation via scoop located in propeller slipstream. Gondola is a chromoly aircraft-grade steel tube spaceframe with Macrolon skin panels; windscreen, of polycarbonate sheet, forms partially enclosed cockpit.
LANDING GEAR: Non-retractable; two front and two rear wheels; spring shock-absorbers.
POWER PLANT: One 47.8 kW (64.1 hp) Rotax 582 two-cylinder water-cooled engine, driving a Helix four-blade carbon fibre pusher propeller. Fuel capacity 28 litres (7.4 US gallons; 6.2 Imp gallons).
ACCOMMODATION: Pilot and up to three passengers in pairs.
AVIONICS: *Comms:* Dittel FSG 71 720-channel VHF transceiver; transponder optional.
Instrumentation: Standard VFR.
DIMENSIONS, EXTERNAL:
Envelope: Length overall 41.00 m (134 ft 6¼ in)
Max diameter 12.80 m (42 ft 0 in)
Fineness ratio 3.2
Tail unit span 18.00 m (59 ft 0¾ in)
Propeller diameter 1.60 m (5 ft 3 in)
Propeller ground clearance 0.10 m (4 in)
DIMENSIONS, INTERNAL:
Envelope volume 2,973 m³ (105,000 cu ft)
Gondola cabin: Length 4.00 m (13 ft 1½ in)
Max width 1.60 m (5 ft 3 in)
Max height 1.80 m (5 ft 10¾ in)
WEIGHTS AND LOADINGS:
Weight empty approx 450 kg (992 lb)
Max T-O weight: two-seat 850 kg (1,873 lb)
four-seat 900 kg (1,984 lb)
PERFORMANCE:
Max level speed 22 kt (40 km/h; 25 mph)
Max cruising speed 20 kt (37 km/h; 23 mph)
Econ cruising speed 15 kt (28 km/h; 17 mph)
Max rate of climb at S/L 183 m (600 ft)/min
Endurance: with max payload 1 h
with max fuel approx 5 h

UPDATED

WDL

WDL LUFTSCHIFFGESELLSCHAFT mbH
Flughafen Essen-Mülheim, D-45470 Mülheim/Ruhr
Tel: (+49 208) 37 80 80
Fax: (+49 208) 378 08 33 (Management)
(+49 208) 378 08 41 (Operations)
OWNER: Theodor Wüllenkemper
MARKETING, SALES AND OPERATIONS MANAGER: Arnold D Beier

WDL resumed airship construction in 1987 with first of new design known as WDL 1B; three of these built to date; described and illustrated in 1997-98 *Jane's*; further examples manufactured to firm order only.

Company plans include US$8 million project, launched in March 1997, for new factory and base at Ormond Beach Business Park and Airpark, Florida, USA. Contract for lease of 4.9 ha (12 acre) site signed in March 1998; construction due to begin March 2001; intends to build there, eventually, both the WDL 1B and a new, smaller airship.

UPDATED

ZEPPELIN

ZEPPELIN LUFTSCHIFFTECHNIK GmbH
Allmannsweilerstrasse 132, D-88046 Friedrichshafen
Tel: (+49 7541) 202 05
Fax: (+49 7541) 20 25 16
Web: http://www.zeppelin-nt.com
MANAGING DIRECTORS:
Max Mugler
Wolfgang von Zeppelin
SALES AND MARKETING MANAGER: Dietmar Blasius

Zeppelin Luftschifftechnik (ZLT) was formed in 1993 for the development and production of new technology (NT) airships. Luftschiffbau Zeppelin GmbH, of which ZLT is a subsidiary, was founded by Count Ferdinand von Zeppelin in 1908 and still exists under the auspices of the Zeppelin Foundation. The company is owned 51 per cent by Luftschiffbau Zeppelin; by Zeppelin GmbH, ZF Friedrichshafen AG and Lemförder Metallwaren AG, all of which have a 16.3 per cent share; and by Mrs Elizabeth Veil, granddaughter of Count Zeppelin.

VERIFIED

ZEPPELIN N 07
TYPE: Helium semi-rigid.
PROGRAMME: Study group for revival of modern semi-rigid airships formed 1989; concluded that combination of vectored thrust and new constructional approach offers best solution, resulting in NT programme; 10 m (32 ft 10 in) long, remotely controlled proof-of-concept model tested extensively in 1991. Design definition of N 07 series

Pitch and yaw control propellers in the N 07 tailcone *1999/0051881*

(originally designated N 05) completed late 1992; construction started October 1995, leading to completion in July 1997 and first flight (D-LZFN) on 18 September 1997 (69th anniversary of maiden flight (1928) of LZ 127 *Graf Zeppelin*) following public debut (April) at Friedrichshafen Air Show. Test flights undertaken from new airship base at Friedrichshafen Airport; damaged by gale in February 1999, but returned to test flying by following April; made 100th flight in August 1999; more than 400 hours (two-thirds of required total) flown by November 1999; D-LZFN named *Friedrichshafen* 2 July 2000; certification anticipated during 2000, but not announced by 1 March 2001. Second N 07 under construction; completion and first flight due in 2001.
CUSTOMERS: Five production airships sold to four pleasure cruise operators by early 1997; assembly of first began in 1998. No. 1 to be delivered to Skyship Cruise Ltd of Switzerland; No. 2/D-LION to RLBG Rheinische Luftschiffbetriebs GmbH at Bonn-Hangelar; No. 3 to K Ernsting's Zeppelin GmbH at Münster/Osnabrück; Nos. 4 and 5 to TransATLANTische Luftschiffahrt GmbH at Munich. No further orders by late 1999.
DESIGN FEATURES: Intended as precursor of product line of commercial airships for variety of applications including scientific research, environmental monitoring, TV missions and tourism. Internal 'prism' primary structure modified and optimised from Stuttgart University proposal of 1975 and British patent of 1924; single helium cell; placement of vectored thrust units near CG and also at rear enhances manoeuvrability and control at low speeds, further improving safety and reducing ground crew requirements to three persons, thereby also reducing operating costs. Tailfins in inverted Y configuration.
FLYING CONTROLS: Computer-assisted flight control, propulsion and landing. Dual fly-by-wire flight control system; no mechanical back-up. Elevator or rudder on each tail surface. Engine thrust control (see under Power Plant) in pitch and yaw.
STRUCTURE: Internal primary structure consists of three rows of aluminium tubular longerons, interconnected with triangular carbon fibre frames and cross-braced with Kevlar cables, providing internal prism-shaped structure accounting for only 12 per cent of gross weight; this frame and pressurised envelope are both load-carrying structures

GERMANY to NETHERLANDS: LIGHTER THAN AIR—ZEPPELIN to RIGIDAIR

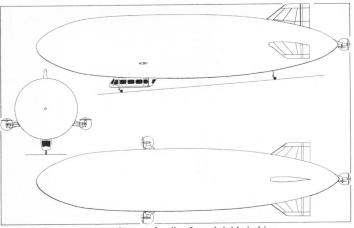

Zeppelin N 07 prototype of a new family of semi-rigid airships
(Paul Jackson/Jane's)

Zeppelin N 07 cabin interior
1999/0051880

to enhance safety factor (airship remains fully manoeuvrable even if internal pressure drops). Envelope, attached continuously to all longerons, is a high-strength, multilayer laminate of polyester and Tedlar; this material has low gas permeability and allows airship to be moored permanently in the open. Air ballonets in lower part of envelope afford protection to cabin and airframe in the event of a hard landing.

LANDING GEAR: Twin landing/ground handling wheels in tandem under gondola and rear of envelope. Airship can be moored to a fixed or mobile mast.

POWER PLANT: Three 149 kW (200 hp) Textron Lycoming IO-360 flat-four engines: one each side of hull above gondola in vectored-thrust propulsion unit, each driving a three-blade variable-pitch tractor propeller; engine in tailcone drives, via a belt-driven gearbox, a lateral fan (for precise yaw control during low-speed and hovering manoeuvres) and a vectored propeller used as a pusher in forward flight and for precise pitch control in landing configuration. Combination of thrust vectoring and fuel trim tanks virtually obviates need for ballast management.

ACCOMMODATION: Crew of two, on ergonomically designed flight deck; gondola main cabin accommodates up to 12 passengers on individual seats, with galley, lavatory and wardrobe provision, or equivalent cargo or other payload.

AVIONICS: *Radar:* Weather radar.
Flight: Dual digital fly-by-wire aerodynamic steering; triple thrust vector steering; modern LCD flight deck avionics, including EFIS, EICAS, radio nav and GPS.

Zeppelin *Friedrichshafen* after its naming ceremony in July 2000 *(Arnold Nayler)*
2001/0105837

DIMENSIONS, EXTERNAL:		
Envelope: Length overall		75.00 m (246 ft 0¾ in)
Max diameter		14.16 m (46 ft 5½ in)
Fineness ratio		5.3
Max width over propeller ducts		19.50 m (63 ft 11¾ in)
Height overall		17.20 m (56 ft 5¼ in)
DIMENSIONS, INTERNAL:		
Envelope volume		8,225 m³ (290,450 cu ft)
Ballonet volume		1,805 m³ (63,750 cu ft)
Gondola: Length		10.70 m (35 ft 1¼ in)
Max width		2.25 m (7 ft 4½ in)
WEIGHTS AND LOADINGS:		
Gondola: structural weight		440 kg (970 lb)
incl equipment and furniture		1,200 kg (2,646 lb)
Max payload		1,850 kg (4,078 lb)
Max T-O weight		7,900 kg (17,416 lb)
PERFORMANCE (estimated):		
Max level speed		70 kt (130 km/h; 80 mph)
Cruising speed		62 kt (115 km/h; 71 mph)
Cruising altitude		1,000 m (3,280 ft)
Pressure ceiling		2,500 m (8,200 ft)
Endurance at 38 kt (70 km/h; 43 mph):		
with max payload		14 h
with reduced payload		28 h

UPDATED

KOREA, SOUTH

KARI
KOREA AEROSPACE RESEARCH INSTITUTE
PO Box 113, Yu-Sung, 305600 Taejon

Tel: (+42 860) 23 52
Fax: (+42 860) 20 06
e-mail: hanch@viva.kari.re.kr
INFORMATION: Jong-Won Lee

The Aircraft Division of KARI is undertaking a feasibility study for a stratospheric airship. Details were originally expected to become available in mid-1998 on completion of the basic design, but none had been released by January 2001

UPDATED

NETHERLANDS

RIGIDAIR
RIGID AIRSHIP DESIGN NV
Franse Kampweg 7, NL-1243 JC 's-Graveland
Tel: (+31 35) 656 48 77
Fax: (+31 35) 656 30 07
e-mail: rigidair@worldonline.nl
Web: http://www.rigidair.com
MANAGING DIRECTOR: Nico Nanninga
NL-1 PROJECT LEADER: Ian Alexander
DESIGN: Arnout De Jong
OPERATIONS: Andrew Horler
AIRWORTHINESS: Koos Tromp
MARKETING: Jos Williams

Rigid Airship Design is a consortium formed on 26 May 1998 for the design, development and commercial manufacture of rigid airships, based on ideas put forward by Scottish airship expert Ian Alexander. Shareholders are RDM Aerospace NV (42.3 per cent); Airshot International NV (36.3 per cent); Stork NV (10.0 per cent); M Caransa BV (7.0 per cent); and Greenfield Capital Partners (4.4 per cent).

Initial (NL-1) project is for a 180 m (590 ft) long airship, provisionally designated RA-180. Subject to RLD design

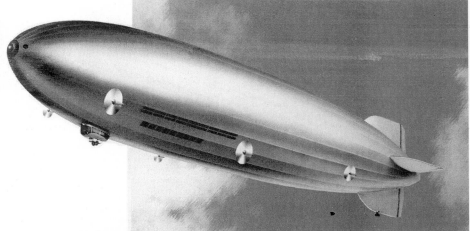

Artist's impression of Rigid Airship Design's projected RA-180 airship
1999/0041289

approval, Rigidair hopes to have the prototype ready for flight test by the end of 2001. Final assembly will take place in newly constructed hangar at Lelystad.

A US subsidiary, Rigid Airship USA Inc, was established in late 1998 and has ordered two RA-180 airships.

UPDATED

RIGIDAIR RA-180

TYPE: Helium rigid.

PROGRAMME: Design work continued in 1999; airship completion and first flight scheduled for 2001; registration PH-RAD reserved in September 1998.

CUSTOMERS: Two ordered by US subsidiary in early 1999.

DESIGN FEATURES: Incorporated technological advances are principally in the outer cover; gas cell material; power units, gearboxes and propellers; low-speed control; and water ballast recovery system. A mooring system, with a manpower requirement no greater than that for aeroplanes of similar capacity, is under development. Design has built-in stretch potential, capable of increasing useful lift to 50 tonnes.

First series will have multiple applications, from scheduled or tourist passenger flights to general freight transport and long-endurance operations.

FLYING CONTROLS: Vectored propeller thrust; water ballast recovery system to maintain equilibrium.

STRUCTURE: Robust skeleton of lightweight aluminium ring girders, longitudinal girders and bracing wires. Large, fireproof, laminated Mylar cell of helium between each set of ring girders (20 cells in all). Strong, synthetic fabric outer skin, inside which lie the engines, cargo compartment and passenger accommodation. Only the propellers, tailfins (four, cruciform configuration) and pilot's control cabin lie outside this skin.

POWER PLANT: Six 447 kW (600 hp) piston engines, four with vectoring propellers for T-O and landing, two with non-vectoring propellers for propulsion. Engines can be fitted with condensers which convert exhaust into water to keep weight and altitude stable.

ACCOMMODATION: External gondola for flight crew. Payload divided between two bays in forward part of hull, each with a capacity for up to 15 tonnes or 120 people. Modular payload system currently under consideration, offering choice of up to 10 different layouts for specific target markets.

Following data are provisional:

DIMENSIONS, EXTERNAL:

Hull: Length	180 m (590 ft)
Fineness ratio	6
Max diameter	30 m (98 ft)

DIMENSIONS, INTERNAL:

Hull helium volume	75,000 m³ (2.65 million cu ft)
Gross volume	83,100 m³ (2.93 million cu ft)

WEIGHTS AND LOADINGS:

Useful lift	35,000 kg (77,162 lb)
Total lifting capacity	more than 78,000 kg (171,960 lb)

PERFORMANCE (estimated):

Max level speed	80 kt (148 km/h; 92 mph)
Cruising speed	65 kt (120 km/h; 75 mph)
Range	10,800 n miles (20,000 km; 12,427 miles)

UPDATED

RUSSIAN FEDERATION

AEROSTATICA

AEROSTATIKA NAUCHNO-PROIZVODSTVENNAYA FIRMA (Aerostatica Scientific-Production Company)

ulitsa Krylatskaya 31-1-315, 121614 Moskva
Tel: (+7 095) 158 48 18
Fax: (+7 095) 415 26 30
PRESIDENT: Dr Alexander N Kirilin

Aerostatica is the English transliteration of Aerostatika, and is used for marketing. The company's first two airships, the 01 and 02, were described in the 1998-99 and earlier *Jane's*. It is currently developing an Aerostatica-200 twin-engined four-seater, to be followed by the larger Aerostatica-300. Former was planned to make maiden flight in 2000, but not reported by March 2001.

UPDATED

AEROSTATICA-200

TYPE: Helium non-rigid.

PROGRAMME: Financial backing from Mayor of Moscow; prototype under construction; targeted to fly mid-2000, but not yet reported (March 2001). Intended for environmental monitoring in Moscow Region.

POWER PLANT: Twin-engined.

ACCOMMODATION: Pilot and three passengers.

DIMENSIONS, INTERNAL:

Envelope volume	2,000 m³ (70,629 cu ft)

UPDATED

AEROSTATICA-300

TYPE: Helium non-rigid.

PROGRAMME: Collaborative design by Aerostatica and MAI; financial backing (1999) from Mayor of Moscow; originally targeted to fly in 2001.

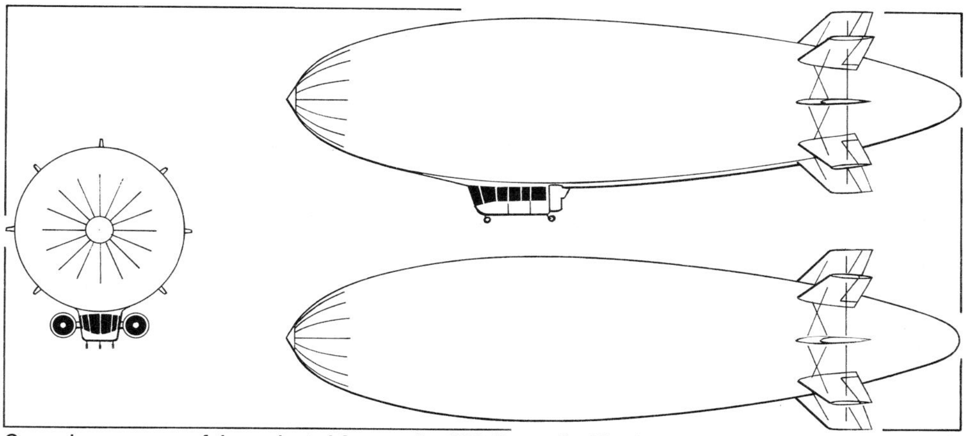

General appearance of the projected Aerostatica-300 *(James Goulding/Jane's)* **2000**/0079269

POWER PLANT: Two 73.5 kW (98.6 hp) Rotax 912 flat-four, thrust-vectoring engines mounted on gondola; single 157 kW (210 hp) LOM M337 flat-six engine at tail. Mogas fuel.

ACCOMMODATION: Pilot and up to eight passengers.

DIMENSIONS, EXTERNAL:

Length overall	50.51 m (165 ft 8½ in)
Max width over tailfins	13.985 m (45 ft 10½ in)
Fineness ratio	3.6
Height overall, incl gondola	16.195 m (53 ft 1½ in)
Width over propeller shrouds	7.79 m (25 ft 6¾ in)
Distance between propeller centres	5.085 m (16 ft 8¼ in)
Gondola: Length	7.15 m (23 ft 5½ in)
Max width	2.39 m (7 ft 10 in)

DIMENSIONS, INTERNAL:

Envelope volume	3,993 m³ (141,000 cu ft)
Ballonet volume	997 m³ (35,210 cu ft)

WEIGHTS AND LOADINGS:

Max payload	890 kg (1,962 lb)

PERFORMANCE (estimated):

Max level speed	62 kt (115 km/h; 71 mph)
Pressure ceiling	2,700 m (8,860 ft)
Range at max cruising speed	540 n miles (1,000 km; 621 miles)
Endurance at 27 kt (50 km/h; 31 mph) cruising speed	36 h

UPDATED

AVGUR

AVGUR VOZDUKHOPLAVATELNYI TSENTR (Augur Aeronautical Centre)

ulitsa Stepana Shutova 4, 109380 Moskva
Tel/Fax: (+7 095) 359 10 01 and 359 10 65
e-mail: augur@PBO.ru
Web: http://www.augur.pbo.ru

WORKS: St Petersburg
Tel/Fax: (+7 812) 173 61 75
GENERAL DIRECTOR: Smanislav Vladimirovich Fedorov

Formed 29 October 1991 by the Moscow Aviation Institute and a group of private shareholders, this company is successor to AIG Kovanko, established in November 1986 by Mr Fedorov. Early achievements included first hot-air balloon flight over Moscow (AIGK Aist) in September 1989 and first Russian advertising aerostat (unmanned AP-60) in December 1989. Became joint stock company on 22 January 1993; St Petersburg branch formed 4 November 1993.

Current range of inflatables comprises BB, Au-2250 and VB hot-air balloons; Au-8 captive manned aerostat; Au-900 manned aerostat; Au-2, Au-5, Au-6 and Au-10 captive advertising aerostats; a range of small geostats; and RD-1.5, RD-2 and RD-2.5 remotely piloted small airships. Other activities include rental and leasing of airships and manufacture of ground-based inflatables of all kinds. The company has a technical co-operation agreement with Worldwide Aeros Corporation of the USA (which see).

Avgur is the Russian distributor, under local designation AV-1R, for the Czech Kubicek AV-1 hot-air airship (which see). Its own airship programmes in 1999-2001 were the Au-11, Au-12, PD-3000 and MD-900, details of which appear below.

UPDATED

AVGUR Au-11

TYPE: Helium non-rigid.

PROGRAMME: Planned to fly in September 1998, but delayed; said to be in final assembly in February 2000, with projected first flight three months later.

COSTS: US$185,000 (1998).

DESIGN FEATURES: Single-seat version of Au-12; otherwise differs only in dimensions, power plant and range/endurance performance.

POWER PLANT: One 20.9 kW (28 hp) VAZ-1111 motorcar engine; no stern thruster engine.

DIMENSIONS, EXTERNAL:

Envelope: Length	27.50 m (90 ft 2¾ in)
Max diameter	6.88 m (22 ft 6¾ in)
Fineness ratio	4.0

DIMENSIONS, INTERNAL:

Envelope volume	669.0 m³ (23,626 cu ft)

PERFORMANCE (estimated): As for Au-12 except:

Max level speed	43 kt (80 km/h; 49 mph)
Cruising speed	32 kt (60 km/h; 37 mph)
Pressure ceiling	2,000 m (6,560 ft)
Range at cruising speed	162 n miles (300 km; 186 miles)
Endurance: at max speed	2 h 30 min
at cruising speed	5 h

UPDATED

AVGUR Au-12

TYPE: Helium non-rigid.

PROGRAMME: Two-seat enlarged version of single-seat Au-11. Prototype flown in 1998.

COSTS: US$275,000 (airship only); US$360,000 including ground equipment (1998).

DESIGN FEATURES: Intended for advertising, demonstration or patrol flights with purpose-designed onboard equipment, command uplink and information downlink. Space for two 20 × 6 m (65.6 × 19.7 ft) advertising displays on envelope sides. Conventional envelope shape; cruciform tailfins, each with elevator or rudder. Design is compliant with Russian Federation and US FAA airship design criteria.

FLYING CONTROLS: Engine thrust vectored in vertical plane by movable louvres; combined manual/electromechanical rudder control (cables); plus stern thruster. Two internal air ballonets.

STRUCTURE: Envelope of modern composites fabric on Lavsan base, with polyurethane skin and titanium oxide UV protection. Tail surfaces have tubular frames with high-strength, stretch-wrapped fabric skins. Reinforced nose with mooring cone.

LANDING GEAR: Single non-retractable and self-centring wheel under gondola. Provision for flotation bags.

POWER PLANT: One 64.9 kW (87 hp) Subaru 1.8 piston engine, with 30° up/105° down cascade-type thrust vectoring, mounted on rear of gondola; 2.2 kW stern thruster electric motor and propeller mounted on port side of lower tailfin.

ACCOMMODATION: Pilot and one passenger in gondola. Upward-opening door on each side. Ground crew of two or three.

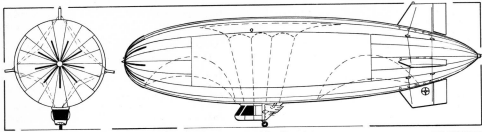
The Avgur Au-12 two-seat advertising airship *(James Goulding/Jane's)*

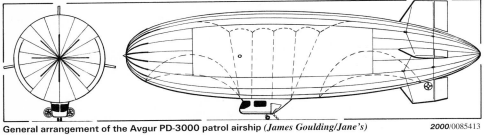
General arrangement of the Avgur PD-3000 patrol airship *(James Goulding/Jane's)*

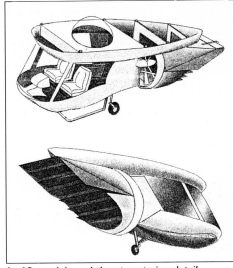
Au-12 gondola and thrust vectoring detail

AVIONICS: Basic VFR or IFR instrumentation. Radio modem or satcom datalink for patrol/surveillance and other special missions.
EQUIPMENT: Depending on mission, can include such items as radiation and chemical monitoring devices; electromagnetic and heat flow gauges; TV camera with video downlink; NP-1 scientific package including photon spectrometer, radiometer, polarimeter and space navigation set; and onboard computer.

DIMENSIONS, EXTERNAL:
Envelope: Length 34.50 m (113 ft 2¼ in)
Max diameter 9.875 m (32 ft 4¾ in)
Fineness ratio 3.5
Height overall 11.40 m (37 ft 4¾ in)
DIMENSIONS, INTERNAL:
Envelope volume 996.0 m³ (35,173 cu ft)
Ballonet volume (two, total) 250.0 m³ (8,829 cu ft)
WEIGHTS AND LOADINGS:
Weight empty 865 kg (1,907 lb)
Max payload approx 210 kg (463 lb)
PERFORMANCE (estimated):
Max level speed 43 kt (80 km/h; 50 mph)
Cruising speed 32 kt (60 km/h; 37 mph)
Cruising altitude 1,000 m (3,280 ft)
Pressure ceiling 2,000 m (6,560 ft)
Range at cruising speed above:
standard fuel 135 n miles (250 km; 155 miles)
with 20 litre (5.3 US gallon; 4.4 Imp gallon) auxiliary fuel tank 324 n miles (600 km; 372 miles)
Endurance: at max speed 2 h
at cruising speed above 5 h
UPDATED

AVGUR PD-3000

TYPE: Helium semi-rigid.
PROGRAMME: Development of PD-220 design (1999-2000 *Jane's*), which it apparently supersedes (*patrulnyi dirizhabl*: patrol airship). Revealed 1998 as PD-300, now apparently redesignated PD-3000. Reported by Novosti news agency in November 2000 as "built and tested" at Avialine Scientific and Technical Design Bureau in Moscow.
COSTS: Estimated US$650,000 for prototype, including spare set of engines (1998); production airship (assuming more than five ordered) approximately US$560,000.
DESIGN FEATURES: Designed for patrol, inspection, surveillance, people/payload transportation, search and rescue, photographic reconnaissance, agricultural roles, and other civil or paramilitary applications. Conventional shape envelope; cruciform tailfins, each with elevator or rudder. Compliant with Russian Federation and US FAA airship design criteria.
FLYING CONTROLS: Vectored thrust (+30/−105°) from cruise engines for fore-and-aft control; yaw control by stern thruster mounted on port side of lower vertical fin. Two internal air ballonets for pressure control; 200 litres (52.8 US gallons; 44.0 Imp gallons) water ballast for trim control.
STRUCTURE: Envelope of high-strength polyurethane-coated nylon with titaniaum dioxide UV protection; all-composites gondola.
LANDING GEAR: Single, non-retractable and self-centring wheel under rear of gondola.
POWER PLANT: Two 74.6 kW (100 hp) cruise engines, with similar cascade-type thrust vectoring to Au-12; 5 kW electric motor for directional control. Fuel capacity 285 litres (75.3 US gallons; 62.7 Imp gallons).
ACCOMMODATION: Pilot and co-pilot, with dual controls; detachable bulkhead, aft of which are seats for four passengers. Wardrobe, washstand and lavatory at rear. Access via let-down rear door with built-in steps. Accommodation heated and ventilated. Ground crew of four to six.

SYSTEMS: Include envelope anti-icing and APU.
AVIONICS: *Comms*: Dual nav/com radios; transponder.
Flight: Computer-aided flight controls; ADF; marker beacon receiver. Autopilot optional.
Instrumentation: IFR standard.
EQUIPMENT: Could include video cameras, loudhailers and other appropriate mission equipment.
DIMENSIONS, EXTERNAL:
Envelope: Length 45.40 m (148 ft 11½ in)
Max diameter 11.30 m (37 ft 1 in)
Fineness ratio 4.0
Gondola length 7.80 m (25 ft 7 in)
DIMENSIONS, INTERNAL:
Envelope volume 3,000 m³ (105,945 cu ft)
Ballonet volume (two, total) 750.0 m³ (26,486 cu ft)
AREAS:
Tail surface area (total) 65.90 m² (709.3 sq ft)
WEIGHTS AND LOADINGS:
Weight empty 1,200 kg (2,646 lb)
Max payload 500 kg (1,102 lb)
Max dynamic lift 400 kg (882 lb)
Max T-O weight 2,500 kg (5,511 lb)
PERFORMANCE (estimated):
Max level speed 70 kt (130 km/h; 80 mph)
Cruising speed 54 kt (100 km/h; 62 mph)
Cruising altitude 1,000 m (3,280 ft)
Pressure ceiling 2,000 m (6,560 ft)
Range at cruising speed above
518 n miles (960 km; 596 miles)
Endurance: at max speed 5 h
at cruising speed above 12 h
UPDATED

AVGUR MD-900

TYPE: Helium semi-rigid.
PROGRAMME: Details first made public August 1995, at which time company was seeking risk-sharing partners; in preliminary design by 1997 and still an active programme in 1999.
DESIGN FEATURES: *Modulnyi dirizhabl*: modular airship. Intended for variety of missions (see Equipment paragraph below); can be configured to transport passengers, cargo or specialised, removable mission modules. Conventional shape envelope with ventral gondola; four tailfins, each with rudder/elevator, indexed in X configuration; auxiliary fin between lower pair serves as mount for stern thruster engine.
FLYING CONTROLS: Rudders and elevators controlled hydraulically; stern thruster for yaw control. Two internal air ballonets for pressure control.
STRUCTURE: Envelope of domestic aerostatic fabric with an outer layer of Lavsan treated with titanium dioxide; fin truss of aluminium alloy; reinforced nose with mooring attachment. Semi-monocoque crew compartment.
LANDING GEAR: Non-retractable tricycle type.
POWER PLANT: Two MAN-D2866 thrust-vectoring turbo-diesel cruise engines (each 276 kW; 370 hp) plus a smaller turbo-diesel (37.3 kW; 50 hp) at rear for directional control.

Model of the Avgur MD-900 *(Paul Jackson/Jane's)*

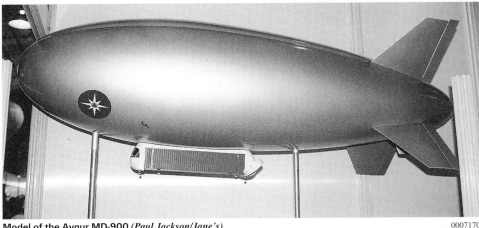
Avgur MD-900 fitted with open-sided cargo module, and showing alternative freight and passenger modules *(Paul Jackson/Jane's)*

ACCOMMODATION: Flight crew of three; dismountable payload module for up to 18 passengers or 24 casualties/medical staff. Ground crew four to six persons.
SYSTEMS: Single hydraulic system for tail control surface actuation.
AVIONICS: Flight envelope computer-controlled to reduce pilot workload; com/nav avionics standard.
EQUIPMENT: Standard gondola module is 10 × 3 × 2.5 m (32.8 × 9.8 × 8.2 ft), and can be outfitted as passenger transport or flying hospital, or with equipment for radar surveillance, ecological monitoring, electricity generation or other missions.
DIMENSIONS, EXTERNAL:

Envelope: Length	61.00 m (200 ft 1½ in)
Max diameter	17.10 m (56 ft 1¼ in)
Fineness ratio	3.6
Height overall	22.00 m (72 ft 2¼ in)
Payload module: Length	10.00 m (32 ft 9¾ in)
Max width	3.00 m (9 ft 10 in)
Max height	2.50 m (8 ft 2½ in)

DIMENSIONS, INTERNAL:

Envelope volume	9,050 m³ (319,600 cu ft)
Ballonet volume (two, total)	1,850 m³ (65,325 cu ft)

AREAS:

Tail surface area (total)	130.0 m² (1,399.3 sq ft)

WEIGHTS AND LOADINGS:

Weight empty	4,680 kg (10,318 lb)
Max payload	3,170 kg (6,989 lb)
Max dynamic lift	785 kg (1,730 lb)
Max T-O weight	8,630 kg (19,026 lb)

PERFORMANCE (estimated):

Max level speed	70 kt (130 km/h; 80 mph)
Cruising speed at 1,000 m (3,280 ft)	54 kt (100 km/h; 62 mph)
Typical operating altitude	up to 1,000 m (3,280 ft)
Pressure ceiling	2,000 m (6,560 ft)
Range at cruising speed above	1,620 n miles (3,000 km; 1,864 miles)
Endurance: at max speed	20 h
at cruising speed above	50 h

UPDATED

RVO

RUSSKOGO VOZDUZHOPLAVATEL'NOGO O B SHCHESTVA (RUSSIAN SOARING SOCIETY)
Web: http://www.pbo.ru

According to an announcement by the ITAR-TASS news agency in September 2000, the RVO is developing the design of a large, heavy-lift airship, designated **DTs-N1**, for transporting outsize cargo payloads. No details were known at the time of going to press.

NEW ENTRY

THERMOPLANE

THERMOPLANE DESIGN BUREAU, MOSCOW AVIATION INSTITUTE
bolshaya Volokolamskoe 4, 125871 Moskva
Tel: (+7 095) 158 41 27
Fax: (+7 095) 158 29 77
CHIEF DIRECTOR: Yuri G Ishkov

According to reports in early 1999, Thermoplane has submitted designs for an international patent for a modified version of the ALA-40 'two-gas' airship (see 1996-97 *Jane's*) capable of use as a shuttle-type air and space vehicle. Side-mounted jet/ramjet engines would accelerate the vehicle to Mach 5 before being jettisoned at 40 km (25 miles) altitude, when a liquid-propellant rocket engine would be ignited to propel the ALA-40 to orbital insertion speed. No further news of this project had emerged by the end of 2000.

Meanwhile, according to reports in June 2000, additional funding for the original Thermoplane project has been forthcoming from an Egypt-based, US-backed agricultural concern interested in a system to transport produce from Egypt to markets in Europe. A free-flying second prototype was then said to be nearing completion, with a modified third example (different propulsion and guidance systems) some 40 per cent complete.

UPDATED

VOZDUKh

VOZDUKh AERONAUTICAL ENTERPRISE
Moskva
CHIEF DESIGNER: Yuri S Boyko

NEW ENTRY

VOZDUKh B-150
TYPE: Helium semi-rigid hybrid.
PROGRAMME: Project first reported in early 2000; programme status not established at time of going to press.
DESIGN FEATURES: Envelope elliptical in profile, with heart-shaped cross-section; dorsal tailfin with inset rudder; anhedral tailplanes and elevators; large underfin with inset shrouded propeller. Powered by same engines as Mi-26 helicopter (which see).
Combines buoyant lift of helium airship with large-diameter, outrigged rotors which have reversible-pitch blades to provide assistance in ascent and descent and controlled acceleration during take-off and landing.
FLYING CONTROLS: Electromechanical and hydraulic.
POWER PLANT: Five 8,500 kW (11,399 shp) ZMKB Progress D-136 turboshafts, two driving each large-diameter rotor; rotors interconnected by synchronising shaft to maintain safe flight if both engines on one side fail. Fifth engine drives contrarotating pusher propellers aft of tailcone and lateral control shrouded propeller in ventral fin.

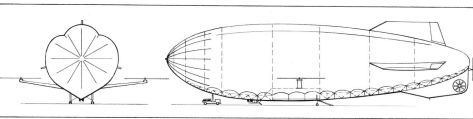

General arrangement of the semi-rigid B-150 *(James Goulding/Jane's)*

ACCOMMODATION: Crew of six.
Following data all provisional:
DIMENSIONS, EXTERNAL:

Length overall	187 m (614 ft)
Rotor diameter (each)	30 m (98 ft)
Width overall, incl rotors	100 m (328 ft)
Height overall	46 m (151 ft)
Propeller diameter: pusher pair, each	18 m (59.1 ft)
lateral controller	8 m (26.2 ft)

DIMENSIONS, INTERNAL:

Envelope volume	200,000 m³ (7,062,900 cu ft)
Payload compartment: Length	40 m (131.2 ft)
Width	6 m (19.7 ft)
Height	5 m (16.4 ft)

WEIGHTS AND LOADINGS:

Payload: typical	100,000 kg (220,460 lb)
max	150,000 kg (330,695 lb)

PERFORMANCE (design):

Max level speed	81 kt (150 km/h; 93 mph)
Cruising speed	70 kt (130 km/h; 81 mph)
Pressure ceiling	1,500 m (4,920 ft)
Range:	
with 100,000 kg (220,460 lb) payload	1,349 n miles (2,500 km; 1,553 miles)
with 150,000 kg (330,695 lb) payload	270 n miles (500 km; 310 miles)
Endurance with 100,000 kg (220,460 lb) payload	20 h

NEW ENTRY

UNITED KINGDOM

AIRSHIP TECHNOLOGIES — see ATG

ATG

ADVANCED TECHNOLOGIES GROUP
6th Floor, Town Hall, St Paul's Square, Bedford MK40 1SJ
Tel: (+44 1234) 22 18 00
Fax: (+44 1234) 22 18 01
e-mail: airshiptech@compuserve.com
Web: http://www.airship.com
CHAIRMAN AND COO: Air Marshal Sir John Walker
TECHNICAL DIRECTOR AND CEO: Roger Munk
DIRECTORS:
  John Wood (Sales)
  Gordon Taylor (Marketing)
  Peter Ward (Operations)
PUBLIC RELATIONS MANAGER: Joanna Amis

ATG is the trading name, from June 2000, of Airship Technologies Services Ltd, which was created in February 1996 to revive the activities of the former Airship Industries company (1991-92 *Jane's*), responsible for the British-designed Skyship 500/600 series. In current manufacture are the AT-04 and AT-10. Scale versions have also been tested of two other projects: the AT-08 unmanned airship for stratospheric communications use, and the ellipsoidal lifting-body SkyCat, a cargo carrier with roll-on/roll-off capability. A SkyKitten prototype for the SkyCat series was flown for the first time in June 2000.

NEW ENTRY

ATG AT-04
TYPE: Helium non-rigid.
PROGRAMME: Prototype under construction 1997 (envelope in USA, gondola at Cardington, Bedfordshire; final assembly at Weeksville, North Carolina). Targeted originally for 1998 first flight in USA and FAA certification six months later, but still awaiting engines in late 2000.
CUSTOMERS: Launch customer is World Airships Ltd, UK (one); two options by other customers by late 1999.
COSTS: Approximately US$15 million (1999).
DESIGN FEATURES: Constructional materials chosen for light weight and longer life than current standard; modular design and construction. Conventional shape envelope; four tailfins indexed in X configuration. Bow thruster permits ground handling and docking without need for ground crew. Computer-controlled LED display on each side of envelope.
FLYING CONTROLS: Dual-channel, optically signalled (fly-by-light) flight control system; rudder or elevator on each tailfin. Ballonets fore and aft for pressure and trim control.
STRUCTURE: Envelope of lightweight, long-life laminated fabric with built-in UV blocker and heat-sealed seams, manufactured by TCOM, USA. Composites tailfins and nose battens. Internally suspended gondola of Kevlar/epoxy fore and aft, with honeycomb centre-section.
LANDING GEAR: Semi-retractable tricycle type; twin wheels on each unit.

806   UK: LIGHTER THAN AIR—ATG

POWER PLANT: Three 447 kW (600 hp) (336 kW; 450 hp maximum continuous) Weslake-built Diesel Air horizontally opposed, two-stroke, direct injection diesel engines, one each side at rear of gondola and one within an envelope-mounted duct at the stern, plus bow thruster. Supercharged induction system. All three engines drive ducted propellers; main pair have vectored thrust (110° up/90° down). Economical cruise can be sustained by rear engine alone. Standard fuel capacity 3,000 litres (793 US gallons; 660 Imp gallons).

ACCOMMODATION: Two pilots, two flight attendants and 48 passengers. Flight deck module of gondola comprises two pilot stations with sidestick controls and large transparencies. Mid-body module contains passenger/payload compartment, four seat tracks for flexible location of equipment and seating, and two large passenger doors with built-in steps. Aft-body module accommodates galley and lavatory, plus either a panoramic viewing area for passenger operations or a full-width cargo door.

SYSTEMS: Electrical system employs three 50 kVA alternators for 115/200 V AC power at 400 Hz; starter/generator on each engine. Low-pressure pneumatic system for actuation of critical subsystems requiring low susceptibility to lightning strike and electromagnetic interference.

AVIONICS: *Flight:* Fly-by-light flight control system with autopilot; GPS navigation.
Instrumentation: Dual airline-standard VFR/IFR; full colour active matrix LCDs; four-tube EFIS displays. NVG-compatible flight deck.

Following data are provisional:

DIMENSIONS, EXTERNAL:
Overall: Length	82.1 m (269 ft 4 in)
Height	21.6 m (70 ft 10 in)
Envelope: Length	79.7 m (261 ft 6 in)
Max diameter	17.9 m (58 ft 9 in)
Fineness ratio	4.45
Gondola: Length	17.3 m (56 ft 9 in)
Max width	2.9 m (9 ft 6 in)

DIMENSIONS, INTERNAL:
Envelope volume	14,200 m³ (501,475 cu ft)
Ballonet volume (27% of gross)	3,834 m³ (135,400 cu ft)
Gondola: Max width	2.4 m (7 ft 10 in)

WEIGHTS AND LOADINGS:
Weight empty	7,031 kg (15,500 lb)
Max payload	7,000 kg (15,432 lb)
Max T-O weight	14,000 kg (30,865 lb)

PERFORMANCE (design):
Max level speed	87 kt (161 km/h; 100 mph)
Max sustained cruising speed (all engines)	75 kt (139 km/h; 86 mph)
Econ cruising speed (rear engine only)	45 kt (83 km/h; 52 mph)
Max rate of ascent and descent	762 m (2,500 ft)/min
Pressure ceiling	4,875 m (16,000 ft)
Min air turning radius	144 m (473 ft)
Max range	3,000 n miles (5,556 km; 3,452 miles)

UPDATED

ATG AT-10

TYPE: Helium non-rigid.
PROGRAMME: First flight planned by mid-2001.
COSTS: US$2.5 million (1999).
DESIGN FEATURES: Commercial airship with capability for advertising and camera platform use and for other applications such as pilot training. X configuration tailfins.
FLYING CONTROLS: Split-channel, optically signalled flight control system; provision for autopilot. Single ballonet centrally located over gondola for envelope pressure control.
STRUCTURE: Envelope of laminated translucent fabric with external catenary collar system supporting gondola. Provision for internal illumination system.
LANDING GEAR: Non-retractable bicycle type, coil-sprung with oleo damping; twin wheels on each unit.

Computer-graphic image of the ATG AT-10 four/five-passenger airship   2001/0103489

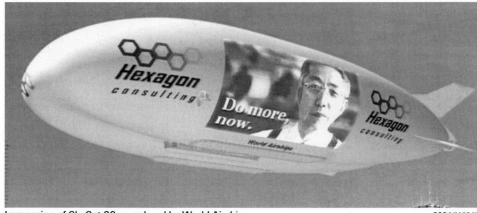
Impression of SkyCat 20 as ordered by World Airships   2001/0103490

POWER PLANT: Two 74.6 kW (100 hp) (59.7 kW; 80 hp maximum continuous) horizontally opposed, two-stroke, direct injection diesel engines with vectorable-thrust ducted propellers. Supercharged induction system.
ACCOMMODATION: One or two pilot stations with sidestick controls; four passenger seats on multiple tracks; large flight deck and passenger cabin transparencies; large passenger door on port side. Rear of gondola contains lavatory, ballonet air system, pneumatics, electrical services and cabin heater.
SYSTEMS: Low-pressure pneumatic system to power flight control actuators and provide low susceptibility to lightning strike and electromagnetic interference.

DIMENSIONS, EXTERNAL:
Length overall	41.40 m (135 ft 10 in)
Max diameter	10.70 m (35 ft 11¼ in)
Height overall	13.60 m (44 ft 7½ in)

DIMENSIONS, INTERNAL:
Envelope volume	2,500 m³ (88,290 cu ft)
Gondola: Length	4.40 m (14 ft 51/4 in)
Max width	1.52 m (4 ft 113/4 in)
Width at floor	0.83 m (2 ft 83/4 in)
Min height	1.86 m (6 ft 11/4 in)

WEIGHTS AND LOADINGS:
Max payload	740 kg (1,631 lb)

PERFORMANCE (design):
Max level speed	60 kt (111 km/h; 69 mph)
Cruising speed	50 kt (93 km/h; 58 mph)
Pressure ceiling	2,745 m (9,000 ft)
Max range at 30 kt (56 km/h; 35 mph)	1,020 n miles (1,889 km; 1,173 miles)
Endurance at 30 kt (56 km/h; 35 mph)	34 h

UPDATED

ATG SKYCAT

TYPE: Hybrid air vehicle.
PROGRAMME: Launched 29 June 2000, with three versions announced; numerals in designations indicate payload capacity in tonnes.
CURRENT VERSIONS: **SkyKitten:** 12.2 m (40 ft) long proof-of-concept prototypes; first flight (SkyKitten 1) 28 June 2000; more than 30 short flights, from the land and water, by early 2001; bow thruster added. SkyKitten 2 (10 per cent larger) then under construction; more powerful engines.
SkyCat 20: Initial operational version. Manufacture under way by mid-2000; first flight targeted for late 2001.
SkyCat 200: Second, and larger, operational version. Work already under way on payload module. Industrial partner IAR Brasov (which see under Romania) expected to begin deliveries of major subassemblies in early 2001.
SkyCat 1000: Largest in current range; go-ahead currently dependent upon additional financing.
CUSTOMERS: SkyCat 20 ordered by World Airships (UK) for delivery in 2002.
DESIGN FEATURES: Brings together lighter than air and hovercraft technologies, enabling airship to land on virtually any flat land or water surface without need for landing infrastructure or ground crew. Larger versions seen as heavy-lift military or cargo-carrying vehicles (SkyCat 1000 has payload capacity of 10 Boeing 747s) and can be constructed in the open, eliminating need for costly hangars.
Elliptical cross-section created by twin hulls, allied to cambered longitudinal shape, provides up to 40 per cent of lift.
FLYING CONTROLS: Flight capability derived from combination of aerodynamic lift and helium buoyancy. All versions have dual-channel, optically signalled flight control system, plus pressure control by multiple ballonets fore and aft in each of the outer hulls.
STRUCTURE: Envelope of laminated fabric with internal catenary system supporting payload module. Internal diaphragms supporting envelope shape allow for limited compartmentalisation, further enhancing vehicle's fail-safe properties. Construction based upon commonality of

Gondola of the AT-10 under construction (*Arnold Nayler*)   2001/0105838

Frontal view of SkyKitten 1, illustrating elliptical cross-section *(Arnold Nayler)*

Rear view of SkyKitten 1 prototype *(Arnold Nayler)*

parts with AT-04, allied to advanced hull form and unorthodox landing system. Payload module located on centreline, with roll-on/roll-off freight capability in SkyCat 200 and 1000.

LANDING GEAR: Air-cushion landing system. Hover skirts on underside of outer hulls provide amphibious capability, as well as enhancing ground handling compared with conventional airships. Hover skirts are 'sucked in' for clean in-flight profile and enhanced all-round field of view. System shares use of ballonet fans with hull pressure system.

POWER PLANT: *SkyCat 20:* Four 447 kW (600 hp) (358 kW; 480 hp maximum continuous) horizontally opposed, four-cylinder two-stroke, direct injection diesel engines, two mounted forward on hull and two on hull stern for cruise operation; all four in ducts with blown vanes to vector thrust for T-O, landing and ground handling. Supercharged induction system.

SkyCat 200: Four 5,966 kW (8,000 shp) (4,474 kW; 6,000 shp maximum continuous) turboshafts, in thrust-vectoring ducts positioned as for SkyCat 20.

SkyCat 1000: Six 11,185 kW (15,000 shp) (8,948 kW; 12,000 shp maximum continuous) turboshafts, two within each stern duct each driving a single propeller and one each side of hull forward.

ACCOMMODATION: Flight deck has side-by-side pilot stations, with sidestick controls and large transparencies for wide field of view; on larger models, also provides space (18.6 m^2; 200 sq ft in SkyCat 200, 83.6 m^2; 900 sq ft in 1000) for off-duty crew members.

SkyCat 20: Four passenger seats on individual tracks in payload compartment; large door each side with built-in steps; full-width door at rear for cargo use.

SkyCat 200 and 1000: Main deck provides clear load space for cargo/freight on a military-rated floor structure. Mezzanine decking can be provided to give multiple lower load area. Ramps (at rear on 200, fore and aft on 1000) provide roll-on/roll-off access to cargo area. Above door aperture is further accommodation space (37.2 m^2; 400 sq ft in 200, 195.1 m^2; 2,100 sq ft in 1000).

SYSTEMS: All versions have low-pressure pneumatic system as described for AT-04 and AT-10.

DIMENSIONS, EXTERNAL:
Length overall: 20	82.10 m (269 ft 4¼ in)
200	185.00 m (606 ft 11½ in)
1000	307.00 m (1,007 ft 2½ in)
Max width: 20	33.70 m (110 ft 6¾ in)
200	77.00 m (252 ft 7½ in)
1000	136.00 m (446 ft 2¼ in)
Height overall: 20	20.60 m (67 ft 7 in)
200	47.00 m (154 ft 2½ in)
1000	77.00 m (252 ft 7½ in)

DIMENSIONS, INTERNAL:
Envelope volume: 20	24,500 m^3 (865,210 cu ft)
200	457,500 m^3 (16.16 million cu ft)
1000	2 million m^3 (70.63 million cu ft)
Payload deck:	
Length: 20	24.60 m (80 ft 8½ in)
200	49.40 m (162 ft 1 in)
1000	80.80 m (265 ft 1 in)
Width: 20	3.10 m (10 ft 2 in)
200	7.50 m (24 ft 7¼ in)
1000	12.20 m (40 ft 0¼ in)
Height: 20	1.90 m (6 ft 2¾ in)
200	5.00 m (16 ft 4¾ in)
1000	8.00 m (26 ft 3 in)

WEIGHTS AND LOADINGS:
Max payload: 20	20,000 kg (44,092 lb)
200	200,000 kg (440,925 lb)
1000	1,000,000 kg (2,204,625 lb)

PERFORMANCE (estimated):
Max level speed: 20	87 kt (161 km/h; 100 mph)
200	90 kt (166 km/h; 103 mph)
1000	110 kt (203 km/h; 126 mph)
Cruising speed: 20, 200	75 kt (139 km/h; 86 mph)
1000	100 kt (185 km/h; 115 mph)
Pressure ceiling: 20, 200, 1000*	2,745 m (9,000 ft)
Range: 15	2,000 n miles (3,704 km; 2,301 miles)
200	3,225 n miles (5,972 km; 3,711 miles)
1000	4,000 n miles (7,408 km; 4,603 miles)

* 5,485 m (18,000 ft) for SkyCat 20 high-altitude version

NEW ENTRY

Computer-derived image comparing SkyCat 1000 with USAF C-5 Galaxy transports

---

CAMERON

CAMERON BALLOONS LTD

St John Street, Bedminster, Bristol BS3 4NH
Tel: (+44 117) 963 72 16
Fax: (+44 117) 966 11 68
e-mail: hcameron@cameronballoons.co.uk
Web (1): http://www.cameronballoons.co.uk
Web (2): http://www.thunderandcolt.co.uk
MANAGING DIRECTOR: Don Cameron
SALES DIRECTORS: Philip Dunnington
 Nick Purvis
MARKETING DIRECTOR: Alan Noble
TECHNICAL DIRECTORS: David Boxall
 Tom Sage
OPERATIONS AND PRODUCTION DIRECTOR: Jim Howard
PUBLIC RELATIONS DIRECTOR: Hannah Cameron

Cameron Balloons, formed in 1970, holds CAA, FAA, French CNT and German Musterzulassungsschein type certificates for its hot-air balloons. It is the world's largest manufacturer of conventional and special-shape balloons and, by 2000, had produced more than 4,800 from its main factory in Bristol. A sister company in Dexter, Michigan, USA, has produced more than 1,200.

Details of Cameron's conventional-shape balloon range are currently available on the company website. The company acquired the balloon assets of the former Thunder & Colt Ltd in late 1994. In June 2000, Cameron Balloons Ltd was offered and acquired preference and ordinary shares in Lindstrand Balloons Ltd (which see), making the Cameron Holdings Group by far the world's largest manufacturer of hot-air balloons.

Cameron also designs and produces hot-air airships, being the first company to develop a craft of this type. Current production hot-air airships are predominantly the DP-80, DP-90, AS-105 Mk 2 (1996-97 *Jane's*) and AS-120 Mk 2; no current production totals for these have been received, but one AS-120 was delivered to Russia in mid-2000.

UPDATED

CAMERON DP SERIES

TYPE: Hot-air non-rigid.
PROGRAMME: DP prototype (G-BMEZ) made first flight April 1986; DP series first/second/third in World Hot-Air Airship Championships 1990; DP 70 first in British and European Championships 1991; DP 80 holds current world endurance record and in 1994 received CAA approval for night flying. Cameron still building one or two a year, and series remains available to order; figures in designations indicate volume in thousands of cubic feet. Full description in 1996-97 and previous editions; shortened version follows.

CURRENT VERSIONS: **DP 60:** Single/two-place; details in 2000-01 and earlier editions.
 DP 70: As DP 60, but larger envelope; two-seater.
 DP 80: Two-seater, with better performance in hot/high conditions.
 DP 90: Larger-envelope version of DP 80.

CUSTOMERS: By late 2000 Cameron had completed, in addition to the DP 50 prototype, over 40 DP series airships; customer countries comprise Australia, Belgium, Brazil, Chile, Czech Republic, Germany, Hungary, Luxembourg, Switzerland, UK, USA and Venezuela.

DESIGN FEATURES: Single engine for both propulsion and pressure control; Cameron self-regulating pressure control system; can be flown unpressurised in event of engine failure and landed like hot-air balloon. Twin silencers minimise engine noise level; ergonomically designed cockpit.

Inflation by two 77.3, 86.4 or 109.1 litre (20.4, 22.8 or 28.8 US gallon; 17.0, 19.0 or 24.0 Imp gallon) propane burners with manual and remote piezoelectric/electronic ignition.

UK: LIGHTER THAN AIR—CAMERON to LINDSTRAND

Cameron DP 70 airship 0016724

POWER PLANT: Standard engine for DP 80/90 is 47.8 kW (64.1 hp) Rotax 582 UL driving a three-blade shrouded pusher propeller. Fuel capacity 22.7 litres (6.0 US gallons; 5.0 Imp gallons) of 40:1 two-stroke mixture.
SYSTEMS: 12 V 30 Ah battery.
ACCOMMODATION: Gondola seats two persons side by side with full-harness seat belts. Full-height polycarbonate windscreen.
DIMENSIONS, EXTERNAL:
Envelope: Length overall: DP 70 32.31 m (106 ft 0 in)
　DP 80　33.83 m (111 ft 0 in)
　DP 90　35.05 m (115 ft 0 in)
Height: DP 70　14.63 m (48 ft 0 in)
　DP 80　15.24 m (50 ft 0 in)
　DP 90　15.54 m (51 ft 0 in)
Max width: DP 70　11.89 m (39 ft 0 in)
　DP 80　12.19 m (40 ft 0 in)
　DP 90　12.80 m (42 ft 0 in)
Display area (each side): DP-80　92.0 m² (990 sq ft)
　DP-90　99.0 m² (1,066 sq ft)
DIMENSIONS, INTERNAL:
Envelope volume: DP 70　1,982 m³ (70,000 cu ft)
　DP 80　2,265 m³ (80,000 cu ft)
　DP 90　2,549 m³ (90,000 cu ft)
WEIGHTS AND LOADINGS:
Gondola (all models)　195 kg (430 lb)
Useful passenger load (S/L, 15°C):
　DP 70　222 kg (489 lb)
　DP 80　285 kg (628 lb)
　DP 90　359 kg (791 lb)
Max T-O weight: DP 70: one person　485 kg (1,069 lb)
　two persons　562 kg (1,239 lb)
PERFORMANCE:
No details provided

UPDATED

CAMERON ROZIERE BALLOONS

Rozières are combination helium/hot-air balloons in which a small propane burner maintains helium efficiency at night when solar heating is absent, providing greatly extended flight duration without expending helium or sand ballast. Cameron manufactures several sizes, from the R-15 to the R-900. Early achievements of this series were detailed in the 2000-01 and earlier editions.

Most recent of these was the 18,500 m³ (653,320 cu ft) R-650 *Breitling Orbiter 3* which took off from Chateau d'Oex, Switzerland, on 1 March 1999. The starting and finishing line in Mauritania (meridian 9° 27′ W) was crossed in the late evening of 4 March and re-crossed at 0954 GMT on 20 March, the balloon then continuing eastwards before landing near the Dakhla oasis in Egypt the following day. Three absolute world records were established by the flight: shortest time around the world (370 hours 24 minutes); duration (477 hours 47 minutes); and distance around the world (22,037.8 n miles; 40,814.0 km; 25,361.34 miles).

UPDATED

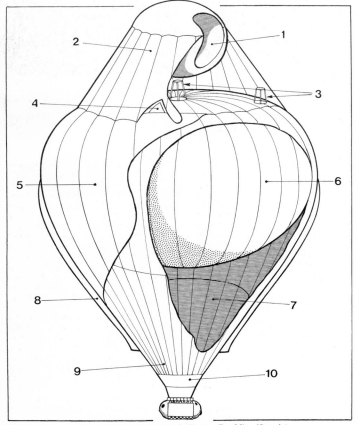

Main features of the *Breitling Orbiter 3* (James Goulding/Jane's)
1: Helium tent balloon, 2: Insulating tent, 3: Helium valves, 4: Chimney rip, 5: External insulating layer, 6: Helium cell, 7: Hot air cone, 8: Appendix, 9: Tear-out skirt, 10: Fireproofed layer
1999/0054436

Cameron *Breitling Orbiter 3* during its epic flight around the globe (Jean-François Luy/Breitling) 1999/0054437

LINDSTRAND

LINDSTRAND BALLOONS LTD
Maesbury Road, Oswestry, Shropshire SY10 8HA
Tel: (+44 1691) 67 18 88
Fax: (+44 1691) 67 99 91
e-mail: info@lindstrand.co.uk
Web: http://www.lindstrand.co.uk
MANAGING DIRECTOR: Per Lindstrand

Company formed November 1991 and had a 1997 workforce of more than 70. Product range includes hot-air and gas balloons and airships. The round-the-world Rozière balloon *ICO Global Challenger*, erroneously reported in earlier editions as a Cameron design, was in fact designed and built by Lindstrand Balloons. With a volume of 36,000 m³ (1,271,330 cu ft), it is the largest Rozière so far built.

Lindstrand Balloons Ltd also owns and services two GA 42 airships (c/n 01 and 04) (see Thunder & Colt entry in 1994-95 *Jane's*); these are available for sale. Other products include tethered aerostats supplied to the UK Ministry of Defence and for civilian passenger-carrying. Company is now part of Cameron (which see) Holdings Group, but Per Lindstrand retains a 35 per cent shareholding.

In February 1997, Lindstrand announced plans for a helium balloon to carry two passengers to a record height of 39,600 m (130,000 ft). This subsequently developed, in a 50/50 partnership with DaimlerChrysler Aerospace under a design study contract from the European Space Agency (ESA), into a solar-powered unmanned airship project, designed to loiter at 21,340 m (70,000 ft) and serve as a relay for mobile telephone communications. Further details can be found in *Jane's Unmanned Aerial Vehicles and Targets*.

UPDATED

LINDSTRAND HS 110

TYPE: Hot-air non-rigid.
PROGRAMME: Designed 1994, and several features tested that year on the AS 300 airship (see 1996-97 *Jane's*). First flown (G-LRBW) and received CAA certification in 1995. Version with internal illumination, for high visibility at night, is now certified.
CUSTOMERS: Total sales not revealed, but during 1999 was flying in Belgium, Canada, Mexico, Switzerland, UK and USA.
DESIGN FEATURES: Designed using latest computer-aided design technology. Conventional-shape envelope with cruciform tailfins and small ventral gondola; envelope can carry advertising artwork directly on fabric or on three banners (one each side and one under nose) which are attached by Velcro and can be changed in less than 1 hour.
FLYING CONTROLS: Burners activated from electrical switches on top of joystick located on control panel between seats. Rudder on each vertical fin.
STRUCTURE: Envelope is of high-tenacity polyamide fabric with heat-resistant polyurethane coating; double-fell, twin-needle seams reinforced with load tape; catenary load curtain and four gondola suspension cables each side. Tailfins, inflated from fan in gondola, have much higher

pressure than envelope and thus stay rigid even in full turns. Monocoque gondola built of glass fibre, Kevlar and carbon fibre. All metal components are of stainless steel or anodised aluminium.

LANDING GEAR: Non-retractable tricycle type.
POWER PLANT: One 48.5 kW (65 hp) Arrow GT 500 air-cooled flat-twin engine, driving a three-blade composites propeller. Fuel capacity 35 litres (9.25 US gallons; 7.7 Imp gallons). Modified Lindstrand Jetstream double burner for hot air supply (see under Weights and Loadings for fuel quantity). Burners can be fired singly or as a pair.
ACCOMMODATION: Side-by-side seats for pilot and one passenger in gondola, behind polycarbonate windscreen.
AVIONICS: *Instrumentation:* Altimeter, variometer, envelope temperature and pressure gauges, engine rpm indicator and engine cylinder head temperature gauge.

DIMENSIONS, EXTERNAL:
Length overall	34.00 m (111 ft 6½ in)
Envelope: Max diameter	13.20 m (43 ft 3¼ in)
Fineness ratio	2.6
Width over tailfins	14.80 m (48 ft 6¾ in)
Height overall	15.80 m (51 ft 10 in)
Wheel track	1.75 m (5 ft 9 in)
Gondola: Length	3.00 m (9 ft 10 in)
Max width	1.65 m (5 ft 5 in)
Height (incl landing gear)	1.80 m (5 ft 10¾ in)

DIMENSIONS, INTERNAL:
Envelope volume	3,125 m³ (110,350 cu ft)

AREAS:
Side banner displays (each)	300.0 m² (3,229 sq ft)
Nose banner display	14.0 m² (151 sq ft)

WEIGHTS AND LOADINGS:
Envelope	170 kg (375 lb)
Gondola, empty	184 kg (405 lb)
Burner fuel: standard	80 kg (176 lb)
max (solo pilot)	120 kg (264 lb)

PERFORMANCE:
Max level speed	20 kt (37 km/h; 23 mph)
Normal cruising speed	10-12 kt (18-22 km/h; 11-14 mph)
Endurance: standard burner fuel	2 h
max burner fuel	more than 3 h

UPDATED

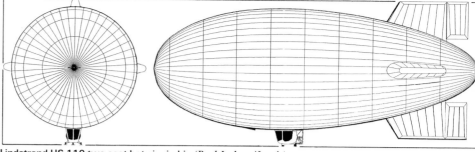
Lindstrand HS 110 two-seat hot-air airship *(Paul Jackson/Jane's)*

Lindstrand HS 110 (48.5 kW; 65 hp Arrow GT 500 engine) flying in Switzerland

UNITED STATES OF AMERICA

ABC

AMERICAN BLIMP CORPORATION
1900 North-East 25th Avenue, Suite 5, Hillsboro, Oregon 97124
Tel: (+1 503) 693 16 11
Fax: (+1 503) 681 09 06
Web: http://www.americanblimp.com
PRESIDENT AND CEO: James Thiele
EXECUTIVE VICE-PRESIDENT: Charles Ehrler
CHIEF ENGINEER: Rudy Bartel

MARKETING OFFICES:
302 Ritchie Highway, Severna Park, Maryland 21146-1910
Tel: (+1 410) 544 65 07
Fax: (+1 410) 544 65 09
e-mail: Jud4Blimps@aol.com
VICE-PRESIDENT, MARKETING: E Judson Brandreth Jr

In 1995 ABC announced the design of a larger (A-1-50) series of airships, a refinement of the earlier A-120 conceptual design with a new technology envelope. The civil version of the A-1-50 incorporates the ABC internal illumination system for advertising and is designated a Lightship; the non-illuminating surveillance version is designated as the Spector. Lightship envelopes remain translucent for internal illumination; Spector envelopes are opaque. Each is offered in three sizes to optimise costs with mission requirements. Other structural elements are identical.

The same envelope technology was introduced as an option for production on existing A-60+ Lightships from 1996 (first contract, for seven envelopes to be supplied by ILC Dover, announced 18 March 1996).

Addition to the range of the new, diesel-powered A-100 Lightship/Spector 31 was announced in June 1998. Company is currently developing airship designs for medium-altitude (2,000 to 3,000 m; 6,560 to 9,840 ft) operation with military and commercial payloads for surveillance and communications relay.

In 1999, ABC increased its manufacturing area by 30 per cent with a 410 m² (4,400 sq ft) extension at Hillsboro, which now covers almost 2,325 m² (25,000 sq ft). At the same time, a 280 m² (3,000 sq ft) fabrication shop was being readied at the Tillamook flight test centre. Workforce in 1999 was 45. On 12 February 2000, ABC's combined fleet of 17 A-60/A-1-50 airships achieved a cumulative total of 100,000 flight hours.

VERIFIED

ABC LIGHTSHIP A-60+
TYPE: Helium non-rigid.
PROGRAMME: A-50 prototype (see 1990-91 *Jane's*) made first flight 9 April 1988; first flight of production A-60 (1991-92 *Jane's*) June 1989; A-60 certified by FAA 18 May 1990 for day/night and VFR/IFR flight; first delivery (to Virgin Lightships Ltd) November 1990; A-60+ certified third quarter 1991 by FAA; APU upgrade certified by FAA, and FLIR Systems Inc gyrostabilised IR camera approved by FAA, in 1993; all A-60s converted to A-60+ in 1993. Certified also in Argentina, Australia, Brazil, Canada, Germany, Italy and South Africa; approved for operation in China, Japan and Turkey. Airship 05 reached 10,000 flight hours on 28 October 1999, and two others had achieved this total by February 2000. A Lightship Group two-man British crew in 016/N606LG set a new FAI Class BA absolute world speed record for gas airships on 19 January 2000, reaching 50.1 kt (92.8 km/h; 57.7 mph) over a 1 km course.

One A-60+ loaned by The Lightship Group was used in January 2000 for UK-based flight trials fitted with a DERA Mineseeker ultra-wideband, ground-penetrating radar for detection of undersurface metal and plastics mines. Subsequently, after the addition of a Wescam 16DS-M dual-sensor (daylight TV and FLIR) payload, this airship conducted a six-week programme (4 October to 14 November 2000) from bases in Italy and Kosovo, surveying 30 sites in the latter locality for mines and other unexploded ordnance on behalf of the United Nations Mine Action Co-ordination Centre (UN MACC).

CURRENT VERSIONS: **A-60:** Initial production version, described in 1991-92 *Jane's*. All since converted to A-60+.
A-60+: Current version from 1991; larger envelope; payload increased by 181 kg (400 lb) compared with A-60; Lightsign message board option from 1992.
Description applies to A-60+.
CUSTOMERS: Total of 23 registered by September 1998, of which 15 in contractor operation by early 2000: see table; 018 made its first flight 4 August 2000; 019 nearing completion at end of that year.
DESIGN FEATURES: Conventional-shape envelope; cruciform tail surfaces; gondola suspended by 12 catenary cables each attached to external patches, eliminating need for internal cables, gastight fittings and bellow sleeves. Can be inflated without a net by attaching ballasted gondola before adding helium; gondola can be removed from inflated Lightship without a net by ballasting catenary cables.
FLYING CONTROLS: Single internal ballonet; rudder or elevator on each of four tailfins; primary flight controls for left-hand front seat only.
STRUCTURE: Outer envelope skin of Dacron/Mylar with separate urethane film inner gastight bladder and single ballonet. All structural attachments to sewn outer bag, such as nose mooring, fin base and guy wires, car catenary and handling lines, made with webbing reinforcements sewn directly to envelope.

Lightship A-60+ five-seat airship *(Lindsay Peacock)*

A-60+ OPERATORS
(at January 2001)

c/n	Identity	Operator	Contractor	Country (base)
01	N2012P *Snoopy 1*	Icarus Aircraft	Met Life	USA
02	N2013K	American Blimp Corp	-	USA
03	N2017A/I-TIRE *Spirit of Europe 1*	Lightship Group	Goodyear	Italy
04	N2022B/ZS-ODY	Lightship Group	MTN	South Africa
05	N560VL *Skeeter*	Virgin Lightships	Monster.com	USA
06	N660VL	Virgin Lightships	CDW	USA
07	N760VL *Snoopy 2*	Icarus Aircraft	Met Life	USA
08	N860AB	Lightship Group	unknown	USA
09	N960AB	Beijing Orient	Beijing Orient	China
10	N3119W/VH-ZIC *Spirit of the South Pacific*	Lightship Group	Goodyear	Australia
11	N11ZP	Lightship Group	NatWest Bank	UK
12	N12ZP *Spirit of Europe 2*	Virgin Lightships	Goodyear	UK
13	N13ZP	[1]		USA
014	N604LG/PT-MKJ *Spirit of the Americas*	Lightship Group	Goodyear	Brazil
015	N605LG	Lightship Group	Premiere World	UK
016	N606LG	Lightship Group	Mazda	UK
017	N607LG	Lightship Group	Hood Dairies	USA
018	N618LG	Lightship Group	not known	Singapore
019-023	N619LG-N623LG[2]	Lightship Group	no allocation	USA

[1] Registration cancelled January 1999
[2] Reservations only; none completed by December 2000

LANDING GEAR: Single twin-wheel unit beneath gondola; small tailwheel at base of lower vertical fin.
POWER PLANT: Twin 59.7 kW (80 hp) Limbach L 2000 engines, pusher-mounted to enhance propulsive efficiency and reduce noise. Standard capacity of rear-mounted fuel tank is 280 litres (74.0 US gallons; 61.6 Imp gallons); can be refuelled in the field without mooring.
ACCOMMODATION: Two single seats and three-person rear bench seat in gondola.
SYSTEMS: APU capable of delivering 2.5 kW at 110 V certified December 1993 and now standard. Used mainly to power internal illumination lights, but can also be used for other airborne equipment.
EQUIPMENT: Certified for gyrostabilised TV or IR camera mount, complete with microwave downlink for live broadcast; can be operated simultaneously with APU.
DIMENSIONS, EXTERNAL:
Envelope: Length	39.01 m (128 ft 0 in)
Max diameter	10.01 m (32 ft 10 in)
Fineness ratio	3.9
Gondola: Length	3.96 m (13 ft 0 in)
Width	1.52 m (5 ft 0 in)
Height (incl landing gear)	2.90 m (9 ft 6 in)
Propeller diameter	1.52 m (5 ft 0 in)

DIMENSIONS, INTERNAL:
Envelope volume	1,926 m³ (68,000 cu ft)
Gondola: Cabin length	2.74 m (9 ft 0 in)
Cabin height	1.83 m (6 ft 0 in)

AREAS:
Tailfins (four, total)	42.74 m² (460.0 sq ft)

WEIGHTS AND LOADINGS:
Total weight empty	1,216 kg (2,680 lb)
Max buoyancy	1,814 kg (4,000 lb)
Max dynamic lift	113 kg (250 lb)
Max useful lift, ISA at 660 m (2,000 ft)	680 kg (1,500 lb)
Max gross weight	1,993 kg (4,394 lb)

PERFORMANCE:
Max level speed	46 kt (85 km/h; 53 mph)
Max rate of ascent	457 m (1,500 ft)/min
Service ceiling	2,225 m (7,300 ft)
Max rate of descent	396 m (1,300 ft)/min
Min T-O distance	112 m (366 ft)
Max range at 35 kt (65 km/h; 40 mph)	521 n miles (965 km; 600 miles)
Max endurance at 35 kt (65 km/h; 40 mph)	15 h

UPDATED

The Mineseeker A-60+ at Pristina, Kosovo, in November 2000 *(Mineseeker)* 2001/0083833

ABC LIGHTSHIP A-100
TYPE: Helium non-rigid.
PROGRAMME: Announced 26 June 1998; first flight planned for 1999, but delayed awaiting certification of SR305 engine. Prototype had not been completed or registered by December 2000.
CURRENT VERSIONS: To be available in **Lightship** and **Spector 31** variants.
CUSTOMERS: First delivery was originally planned in 1999 to customer in Mexico City.
DESIGN FEATURES: As A-60+/A-1-50 but improved payload and performance.
FLYING CONTROLS: Conventional and manual. Elevators and rudders controlled by cables from elevator wheel and rudder pedals. Single ballonet for trim, located at airship CG.
STRUCTURE: Envelope laminated from polyester, Mylar and other urethane films and adhesives, with outer film of Tedlar for UV protection; seams are heat-bonded. Control surfaces and fins of fabric-covered aluminium. Gondola of aluminium and fabric-covered steel tube.
POWER PLANT: Two 134 kW (180 hp) SMA SR305 diesel engines, each driving a Mühlbauer five-blade constant-speed reversible-pitch propeller. Standard fuel capacity 568 litres (150 US gallons; 125 Imp gallons).
ACCOMMODATION: Pilot and co-pilot or passenger in front of gondola, with three-person bench seat and optional sixth seat behind. Dual controls optional. Passenger door on port side; emergency exit on starboard side.
SYSTEMS: Electrical power from two 28 V, 90 A alternators; additional alternators optional.

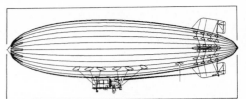

General appearance of the diesel-engined ABC Lightship A-100 1999/0051870

AVIONICS: Primarily from Honeywell Silver Crown range.
 Comms: Include KY 196 com radio, KT 76A transponder and KMA 24H audio panel.
 Flight: Include KX 155 nav/com and Apollo GX55 GPS.

DIMENSIONS, EXTERNAL:
Overall: Length	46.02 m (151 ft 0 in)
Width	12.19 m (40 ft 0 in)
Height	14.63 m (48 ft 0 in)
Envelope: Length	45.72 m (150 ft 0 in)
Max diameter	12.19 m (40 ft 0 in)
Fineness ratio	3.8
Gondola cabin: Length	3.05 m (10 ft 0 in)
Average width	1.52 m (5 ft 0 in)
Max height	1.93 m (6 ft 4 in)
Propeller diameter	1.68 m (5 ft 6 in)

DIMENSIONS, INTERNAL:
Envelope volume	3,256.4 m³ (115,000 cu ft)
Ballonet volume	846.7 m³ (29,900 cu ft)

WEIGHTS AND LOADINGS:
Operating weight empty	2,525 kg (5,567 lb)
Max payload	395 kg (870 lb)

PERFORMANCE (estimated):
Max operating speed	70 kt (129 km/h; 80 mph)
Max endurance	more than 20 h

UPDATED

ABC LIGHTSHIP/SPECTOR SERIES
TYPE: Helium non-rigid.
PROGRAMME: Three envelope sizes offered for both civil and surveillance versions. First to be built, a Lightship A-1-50 (N5132A, c/n 001), made its first flight on 8 January 1997. FAA type certificate issued 3 October 1997, at which time a total of 683 hours had been flown. Marketing designations are A-130, A-150 and A-170; designation on type certificate and registration documents of current version is A-1-50.
CURRENT VERSIONS: **Lightship A-130/Spector 36.**
 Lightship A-1-50/Spector 42. Only version so far built (early 2001).
 Lightship A-170/Spector 48.
CUSTOMERS: Five A-1-50s delivered by January 2000; see table. Further four (N156LG to N159LG) then under construction, but had not been completed by December 2000.
COSTS: Spector 42 approximately US$3.25 million (1996).
DESIGN FEATURES: As described for A-60+; larger gondola; also seat tracks for quick reconfiguration of cabin; water ballast trim system.
FLYING CONTROLS: Standard as for A-60+; dual controls and autopilot optional. Trimming with ballonet accomplished by transferring water between the nose and a tank in the rear portion of the gondola.
STRUCTURE: Single-walled ILC Dover envelope with single (26 per cent of volume) ballonet; Tedlar outer film plus inner ripstop fabric/helium barrier film laminated material with heat-sealed seams. Envelopes translucent on Lightship versions, opaque for Spector; those for Spector series have carbon black in the material for ease of maintenance.
LANDING GEAR: Non-retractable tricycle type. Spring-damped main units each with single wheel and tyre; slightly raised nose unit with twin wheels/tyres; small tailwheel at base of lower vertical fin.
POWER PLANT: Two 134 kW (180 hp) Textron Lycoming IO-360-B1G6 flat-four engines, each driving an MTV-25-B-C five-blade constant-speed, reversible-pitch tractor propeller. Rear-mounted fuel tank, standard capacity 571 litres (150.8 US gallons; 125.6 Imp gallons). Airship can be refuelled in the field without mooring.
ACCOMMODATION: Single seats for pilot and one passenger in cockpit; five single seats and a three-person bench seat in main cabin. Lavatory and galley optional.
SYSTEMS: Electrical system is 28 V DC, with an additional 2.5 kW/110 V at 60 Hz available from removable APU.
EQUIPMENT: Gondola hardpoint provisions for external installation of a variety of sensor equipment. Standard seat tracks in main cabin facilitate sensor/electronics rack installations.

DIMENSIONS, EXTERNAL (A: A-130/Spector 36, B: A-1-50/Spector 42, C: A-170/Spector 48):
Overall: Length: A	48.16 m (158 ft 0 in)
B	50.29 m (165 ft 0 in)
C	52.42 m (172 ft 0 in)
Width: A	13.72 m (45 ft 0 in)
B	14.02 m (46 ft 0 in)
C	14.63 m (48 ft 0 in)

ABC Lightship A-1-50 promoting an American beer *(Paul Jackson/Jane's)*

A-1-50 OPERATORS
(at January 2001)

c/n	Identity	Operator	Contractor	Country (base)
101	N151AB	Lightship Group	Sanyo	USA
102	N152LG *Bud 1*	Lightship Group	Budweiser	USA
103	N153LG/TC-AVK	Lightship Group	KOC Holdings	Turkey
104	N154ZP	Lightship Group	Vegas.com	USA
105	N155LG *Trumpasaurus*	Lightship Group	Monster.com	USA

Height: A	16.46 m (54 ft 0 in)	Main cabin length: A, B, C	3.35 m (11 ft 0 in)
B	16.76 m (55 ft 0 in)	Average cabin width: A, B, C	1.52 m (5 ft 0 in)
C	17.37 m (57 ft 0 in)	Cabin height: A, B, C	1.93 m (6 ft 4 in)
Envelope: Length: A	47.24 m (155 ft 0 in)	WEIGHTS AND LOADINGS:	
B	49.38 m (162 ft 0 in)	Total weight empty: A	2,797 kg (6,167 lb)
C	51.51 m (169 ft 0 in)	B	2,863 kg (6,311 lb)
Max diameter: A	12.50 m (41 ft 0 in)	C	2,925 kg (6,449 lb)
B	13.11 m (43 ft 0 in)	Max buoyancy: A	3,779 kg (8,332 lb)
C	13.72 m (45 ft 0 in)	B	4,055 kg (8,939 lb)
Fineness ratio: A, B, C	3.8	C	4,595 kg (10,131 lb)
Gondola: Length: A, B, C	8.08 m (26 ft 6 in)	Max dynamic lift: A	295 kg (650 lb)
Max width: A, B, C	6.25 m (20 ft 6 in)	B	340 kg (750 lb)
Max height: A, B, C	3.35 m (11 ft 0 in)	C	386 kg (850 lb)
Propeller diameter: A, B, C	1.68 m (5 ft 6 in)	Max useful lift at 656 m (2,000 ft), ISA:	
DIMENSIONS, INTERNAL:		A	1,143 kg (2,520 lb)
Envelope volume: A	3,681 m³ (130,000 cu ft)	B	1,708 kg (3,766 lb)
B	4,248 m³ (150,000 cu ft)	C	2,253 kg (4,967 lb)
C	4,814 m³ (170,000 cu ft)	Max T-O weight: A	3,808 kg (8,397 lb)
Gondola: Cabin overall length: A	4.67 m (15 ft 4 in)	B	4,394 kg (9,689 lb)
B, C	4.57 m (15 ft 0 in)	C	4,981 kg (10,981 lb)

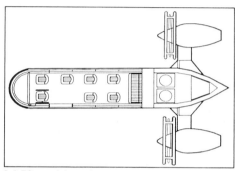

A-1-50 gondola and power plant arrangement

PERFORMANCE:
Max level speed: A	55 kt (101 km/h; 63 mph)
B	52 kt (96 km/h; 59 mph)
C	50 kt (92 km/h; 57 mph)
Max single-engine speed: A	35 kt (64 km/h; 40 mph)
B	33 kt (61 km/h; 38 mph)
C	31 kt (57 km/h; 35 mph)
Max rate of ascent: A, B, C	488 m (1,600 ft)/min
Service ceiling: A, B, C	3,050 m (10,000 ft)
Max rate of descent: A, B, C	427 m (1,400 ft)/min
Max range at 35 kt (64 km/h; 40 mph):	
A	564 n miles (1,046 km; 650 miles)
B	534 n miles (989 km; 615 miles)
C	504 n miles (933 km; 580 miles)
Max endurance at 35 kt (64 km/h; 40 mph): A	16 h
B	15 h
C	14 h

UPDATED

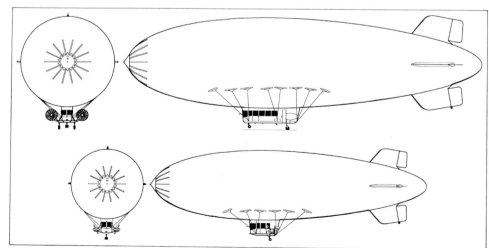

Comparative views, to scale, of the Lightship A-1-50 (top) and A-60+ *(Paul Jackson/Jane's)*

Inside the A-1-50's 10-seat gondola, looking forward

AEROS
WORLDWIDE AEROS CORPORATION
8411 Canoga Avenue, Chatsworth, California 91304
Tel: (+1 818) 993 55 33
Fax: (+1 818) 993 94 35
e-mail: aeros-us@worldnet.att.net
Web: http://www.aeros-airships.com

PRESIDENT: Igor Pasternak
COMMUNICATIONS: Eugene Slatkin

Formed in Ukraine in 1988, Aeros transferred its activities to the USA and was incorporated in Delaware in 1992, relocating to California in October 1993. Its current operation incorporates manufacturing and flight test facilities, R&D, management and marketing. Aeros Airship Company subsidiary formed September 1999 to lease and operate Aeros airships; operations began, with an Aeros 40-B, in November 1999. Aeros' technical capabilities are based on more than 20 years of research (including that initiated before 1988 by former Soviet design bureaux) into LTA technologies, especially those relating to cargo airships.

Details of the company's earlier D-1, D-4 and Aeros 50 designs have appeared in previous editions of *Jane's*. Its current products are the Aeros 40A, 40B and 40C family of passenger airships, all of which are designed to meet US FAA standards; 12 had been registered by the end of 2000.

UPDATED

AEROS 40A

TYPE: Helium semi-rigid.
PROGRAMME: Two sold to China Sky Corporation for government advertising in 1997.
DESIGN FEATURES: Include internal illumination, for night-time advertising, and easily changeable banners.
POWER PLANT: Two 50.7 kW (68 hp) Limbach L 1700 flat-four engines. Fuel capacity 83.3 litres (22.0 US gallons; 18.3 Imp gallons).
ACCOMMODATION: Pilot and one passenger. Single cockpit door.
DIMENSIONS, EXTERNAL:
Envelope: Length 37.25 m (122 ft 2½ in)
Max diameter 9.50 m (31 ft 2 in)
Fineness ratio 3.9
Height overall 11.86 m (38 ft 11 in)
DIMENSIONS, INTERNAL:
Envelope volume 1,700 m³ (60,035 cu ft)
Ballonet volume (two, total) 476 m³ (16,810 cu ft)

UPDATED

AEROS 40B SKY DRAGON

TYPE: Helium semi-rigid.
PROGRAMME: Prototype completed; first flight (N818AC) 11 September 1998; exported to China (Thakral Media Corporation) in November 1998; operating in Hong Kong 1999. Three more (one for Argos Medien AG of Germany and two more for China) under construction in 2000. FAA certification awarded 21 June 2000.
DESIGN FEATURES: Göttingen 409 root rib profile; X configuration tail unit, each with ruddervator. One ballonet forward and one aft (together, 30 per cent of total volume). Internal illumination, as in Aeros 40A.
FLYING CONTROLS: Fly-by-wire (pneumatic) with manual back-up.

Aeros 40B Sky Dragon twin-engined airship
2001/0104603

STRUCTURE: Envelope made up of 12 heat-sealed panels of a transparent, multilayer fabric (woven nylon, impregnated with polyurethane film, including two helium protection barriers) with double UV coating.
LANDING GEAR: Two mainwheels, with shock-absorbers.
POWER PLANT: Two 93 kW (125 hp) Teledyne Continental IO-240-B8 flat-four engines; MTV-7-D/LD170-12 reversible-pitch, three-blade pusher propellers. Fuel capacity 287 litres (75.9.0 US gallons; 63.2 Imp gallons), of which 265 litres (70.0 US gallons; 58.3 Imp gallons) are usable.
ACCOMMODATION: Pilot and four passengers (pilot and co-pilot only in prototype). Front two seats fully articulating and reclining; three bucket seats to rear. Full-depth windscreen; door, with upper and lower windows, in each side of gondola.
SYSTEMS: Automatic pressure maintenance system.
AVIONICS: Digital 'glass cockpit' package includes audio panel, voice annunciator, colour LEDs, GPS/com, second com, and transponder/encoder; readings displayed in both digital and analogue form. Provisions for full IFR package.
DIMENSIONS, EXTERNAL:
Envelope: Length 43.59 m (143 ft 0 in)
Max diameter 10.60 m (34 ft 9¼ in)
Fineness ratio 4.1
Height overall 13.35 m (43 ft 9½ in)

DIMENSIONS, INTERNAL:
Envelope volume 2,508 m³ (88,570 cu ft)
Ballonet volume (two, total) 752 m³ (26,557 cu ft)
WEIGHTS AND LOADINGS:
Useful load 907 kg (2,000 lb)
PERFORMANCE:
Max level speed 44 kt (82 km/h; 51 mph) IAS
Max certified altitude 2,885 m (7,500 ft)
Pressure ceiling 3,000 m (9,840 ft)
Max endurance 24 h

UPDATED

AEROS 40C

TYPE: Helium semi-rigid.
PROGRAMME: Projected for sightseeing tours to Grand Canyon; funding of US$5 million for two airships being sought in 1998. Prototype under construction 1999.
POWER PLANT: Two 224 kW (300 hp) Textron Lycoming IO-540-K2A5 flat-six engines. Fuel capacity 927 litres (245 US gallons; 204 Imp gallons).
ACCOMMODATION: Crew of two, plus up to 12 passengers. Cockpit configured for single-pilot operation.
DIMENSIONS, EXTERNAL:
Envelope: Length 60.00 m (196 ft 10¼ in)
Max diameter 14.90 m (48 ft 10½ in)
Fineness ratio 4.0
Height overall 18.70 m (61 ft 4¼ in)
DIMENSIONS, INTERNAL:
Envelope volume 7,000 m³ (247,205 cu ft)
Ballonet volume (two, total) 2,450 m³ (86,520 cu ft)
WEIGHTS AND LOADINGS:
Useful load 4,250 kg (9,370 lb)
PERFORMANCE:
Max level speed 64 kt (120 km/h; 74 mph)
Pressure ceiling 3,000 m (9,840 ft)
Max endurance 24 h

UPDATED

AHA

ADVANCED HYBRID AIRCRAFT INC

PO Box 144, Eugene, Oregon 97440
Tel/Fax: (+1 604) 541 80 64
e-mail: bnb@ahausa.com
Web: http://www.ahausa.com
MANAGER AND DIRECTOR: Bruce N Blake

Formerly based in Isle of Man, UK (see 1993-94 *Jane's*); relocated to Australia in 1993 and to USA in 1996. Venture capital for airship development, expected in early 1997, did not materialise, but reported that AHA signed letter of intent on 6 October 1998 with Xiao Bang Group (XBG) of China initiating 10 year joint venture to build and market new buoyant aircraft in China. Intention is to manufacture first Light Utility and a two-seat LV2 Hornet in USA, then to establish assembly plant in Zhuhai after training Chinese workforce. A similar joint venture in Morocco was in prospect at that time. In 1998, AHA was also reportedly seeking new investment to revive the projected Advanced Non-Rigid (ANR) airship last described in the 1992-93 *Jane's*.

Details of AHA's Hornet Hybrid RPV version and the Wasp remotely piloted blimp can be found in *Jane's Unmanned Aerial Vehicles and Targets*.

VERIFIED

AHA HORNET

TYPE: Helium non-rigid hybrid.
PROGRAMME: Design of Hornet LV (leisure variant) initiated October 1991; funding for LV located in 1995.

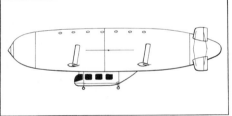

Hornet AW (right) and Light Utility projected versions *(James Goulding/Jane's)*

Manufacturing start-up and US promotional tour funded from sources within the USA. Manufacture is to FAA P-8110-2 airship design criteria standards.
CURRENT VERSIONS: **Hornet LV2:** Two-seat LV, available as kit. Meets FAA '49 per cent rule' to qualify for Experimental category operation. To be used for promotions at major world air shows and elsewhere.
Light Utility: More robust, four-seat version, intended to operate from rough fields and away from built-up areas. To be used for similar purposes to LV2.
Hornet AW: Aerial work version, planned to follow if sales of LV2 are successful; intended specifically to meet needs of electronic news-gathering organisations. Expected to provide capabilities similar to those of current twin-engined helicopters but at about half the acquisition and operating cost.
CUSTOMERS: First order for LV2, from a Florida-based group, was said to be imminent in November 1997 (latest

information received); qualified enquiries for LV2 and Light Utility then continued to be received.
COSTS: LV2, US$140,000 (uncertified) or US$250,000 (FAA certified); Light Utility (as four-seater), US$840,000; AW (basic configuration), US$1.26 million (1997).
DESIGN FEATURES: Two-seat gondola/cabin for LV2, with fixed tricycle landing gear. Hornet LV2 can be taxied like an ultralight, mast-moored by one person (mainly an automated operation), and remain on ground when vacated by pilot.

Hornet range designed to operate at heaviness of about 50 per cent, compared with conventional types with which operation at or near equilibrium is essential; one half of total lift is provided aerostatically by the buoyancy of the helium-filled envelope, the other half being generated aerodynamically by the stub-wings attached to the sides of the envelope. Hybrid buoyant aircraft do not require ballast and have flexibility to carry greater or lesser load, thus providing greater productivity.

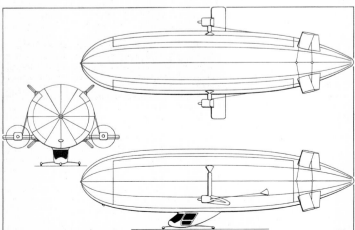

General arrangement of the AHA Hornet LV2 *(Mike Keep/Jane's)*

Scale model of the AHA Hornet LV2

Technology demonstration model of the AHA Light Utility airship 0016730

FLYING CONTROLS: Fly-by-wire; automatic envelope pressure control.
STRUCTURE: Major subcontractors expected to be ILC Dover (envelope) and Scaled Composites (gondola).
LANDING GEAR: Non-retractable tricycle type.
POWER PLANT: Two 67.1 kW (90 hp) MWE AE 100R rotary engines in LV2, mounted on stub-wings that can be tilted upward to vector thrust 30° downward for climb. Hornet AW would have four AE 100Rs and 45° thrust vectoring; Light Utility, two non-vectoring AE 100Rs.
ACCOMMODATION: Two seats in LV2; four seats standard, six or eight seats optional, in Light Utility.

DIMENSIONS, EXTERNAL:
Envelope: Length overall: LV2	19.51 m (64 ft 0 in)
AW Light Utility	30.48 m (100 ft 0 in)
Max diameter: LV2	4.88 m (16 ft 0 in)
AW	6.10 m (20 ft 0 in)
Light Utility	7.49 m (24 ft 7 in)
Fineness ratio: LV2, Light Utility	4.0
AW	5.0
Wing span: LV2	9.14 m (30 ft 0 in)
AW	9.21 m (30 ft 2½ in)
Light Utility	5.92 m (19 ft 5 in)
Wing chord, constant: LV2	1.52 m (5 ft 0 in)
AW	1.25 m (4 ft 1¼ in)
Light Utility	1.44 m (4 ft 8½ in)

DIMENSIONS, INTERNAL:
Envelope volume: LV2	250.0 m³ (8,830 cu ft)
AW	623.0 m³ (22,000 cu ft)
Light Utility	1,200.6 m³ (42,400 cu ft)
Max ballonet volume: LV2	37.5 m³ (1,325 cu ft)
AW	124.6 m³ (4,400 cu ft)
Light Utility	155.7 m³ (5,500 cu ft)

WEIGHTS AND LOADINGS:
Disposable load: LV2	209 kg (460 lb)
AW	427 kg (941 lb)
Light Utility	431 kg (950 lb)
Buoyancy: LV2	249 kg (550 lb)
AW	560 kg (1,235 lb)
Light Utility	1,080 kg (2,380 lb)
Max heaviness: LV2	249 kg (550 lb)
AW	720 kg (1,587 lb)
Light Utility	359 kg (791 lb)
Max T-O weight: LV2	499 kg (1,100 lb)
AW	1,280 kg (2,822 lb)
Light Utility	1,436 kg (3,166 lb)

PERFORMANCE (estimated):
Max level speed: LV2	61 kt (113 km/h; 70 mph)
AW	81 kt (150 km/h; 93 mph)
Light Utility	51 kt (94 km/h; 58 mph)
Ceiling: LV2	1,525 m (5,000 ft)
AW	2,135 m (7,000 ft)
Light Utility	2,745 m (9,000 ft)
Range: LV2, AW	200 n miles (370 km; 230 miles)
Light Utility	300 n miles (555 km; 345 miles)

UPDATED

AHA RIGID AIRSHIP

AHA has a 40-cabin, 110-passenger, pressure rigid airship in the design stage. Also of hybrid configuration, it is seen as having applications in the airborne leisure market or as competition on some sea ferry routes. No further details yet released.

VERIFIED

GSI

GLOBAL SKYSHIP INDUSTRIES INC

1001 Armstrong Boulevard, Unit A, Kissimmee, Florida 34741
Tel: (+1 407) 932 37 79
Fax: (+1 407) 932 29 16
e-mail: information@airshipoperations.com
Web (1): http://www.globalskyships.com
Web (2): http://www.airshipoperations.com
TECHNICAL DIRECTOR: Gary Burns
PUBLIC RELATIONS: Mary Kenny
e-mail: askmary@earthlink.net

Global Skyship Industries is the US subsidiary of Aviation Support Group Ltd of the UK which, in December 1996, acquired the assets of the former Westinghouse Airships Inc (WAI). As a result, GSI became holder of the type certificates for the UK-designed Skyship 500HL and Skyship 600 and the WAI Sentinel series. In late 1997, GSI announced that it had re-opened the production line and had these airships available for immediate delivery. One Skyship 600B was delivered to CargoLifter GmbH of Germany in July 2000.

See 1999-2000 *Jane's* for details of Sentinel 1000 and 1240, which remain available; no orders by end of 2000.

UPDATED

GSI SKYSHIP 500HL

TYPE: Helium non-rigid.
PROGRAMME: First flight (G-SKSB) (converted Skyship 500) 30 July 1987.
CUSTOMERS: Three conversions completed in UK; one (N501LP) in service in USA with Airship Operations Inc in 1998, but destroyed by severe storm on 7 September 1998. Type remains available from GSI which, in 2000, had three (including 07/N503LP and 09/N504LP) ready for assembly on receipt of orders.
DESIGN FEATURES: Combines modified Skyship 500 gondola with larger envelope of Skyship 600, providing ability to operate with greater payload in hotter climates and at higher altitudes than Skyship 500.
POWER PLANT: Two 152 kW (204 hp) Porsche 930/01 non-turbocharged engines; fuel capacity 545 litres (144 US gallons; 120 Imp gallons).

DIMENSIONS, EXTERNAL: As for Skyship 600 except:
Gondola: Length overall	9.24 m (30 ft 3½ in)
Max width	2.41 m (7 ft 10¾ in)

DIMENSIONS, INTERNAL: As for Skyship 600 except:
Gondola cabin: Length	4.20 m (13 ft 9½ in)
Height	1.96 m (6 ft 5 in)

WEIGHTS AND LOADINGS:
Max usable fuel	382 kg (842 lb)
Gross disposable load	2,190 kg (4,829 lb)

GSI Skyship 600B for Airship Operations Inc 1999/0098038

PERFORMANCE:
Max continuous speed	50 kt (93 km/h; 57 mph)
Cruising speed	up to 47 kt (87 km/h; 54 mph)
Pressure ceiling	3,050 m (10,000 ft)
Endurance at 35 kt (65 km/h; 40 mph)	17 h

UPDATED

GSI SKYSHIP 600

TYPE: Helium non-rigid.
PROGRAMME: First flight 6 March 1984; special category C of A awarded by UK CAA 1 September 1984; aerial work certification received second quarter 1986; full passenger-carrying C of A 8 January 1987, initiating Skycruise aerial sightseeing service over London, San Francisco, Munich and Sydney in 1987, and over Paris in 1988. First US FAA type certificate awarded to an airship for civil use was issued to Skyship 600 in May 1989. During 1990 Farnborough Air Show a Skyship 600 (G-SKSC) with two 227 litre (60 US gallon; 50 Imp gallon) long-endurance fuel tanks made unrefuelled flight of 50 hours 15 minutes; sufficient fuel for a further 20 hours remained at end of flight.
CURRENT VERSIONS: **Skyship 600**: Initital UK-built version (nine completed); two more due for completion by Global Skyships in 1999.
Description applies to Skyship 600 except where indicated.
Skyship 600B: With larger envelope, using Tedlar fabric, manufactured by TCOM LP. First example, converted from Skyship 600 c/n 07 (N602SK) and operated for Fuji Film by Airship Management Services, flew for first time 25 July 1998. Second example (c/n 10, N610SK) completed for Airship Operations Inc (Hilfiger contract) to replace destroyed Skyship 500HL.
CUSTOMERS: Total of nine built in UK, of which 01 (G-SKSC) purchased by Global Skyships from UK MoD and 07 (N602SK) converted to 600B in 1998; four others then still in operation with Interport Marine Agencies (03/G-SKSG), Airship International (02/N502LP and 04) and Wilmington Trust (06/N602SK). Two more (one 600 and one 600B) under construction in 1999, of which 02/B-04 (N602CL) delivered to CargoLifter, Germany in May 2000.
COSTS: Skyship 600 approximately US$5 million (1998).
DESIGN FEATURES: Non-rigid envelope, with one ballonet forward and one aft (together 26 per cent of total volume); ballonet air intake aft of each propulsor unit. Cruciform tail unit.
FLYING CONTROLS: Vectored thrust propulsion (see Power Plant); differential inflation of ballonets for static fore and aft trim; cable-operated rudders and elevators, each with spring tab; ballast in box below crew seats; disposable water ballast in tanks at rear.
STRUCTURE: Envelope manufactured by Aerazur (France) from single ply polyester fabric, coated with titanium dioxide-loaded polyurethane to reduce ultraviolet degradation; polyvinylidene chloride film bonded on to inner coating of polyurethane on inside of envelope minimises loss of helium gas. Four parabolic arch load curtains, carrying multiple Kevlar 29 gondola suspension cables. Nose structure is domed disc, moulded from GFRP and carrying fitting by which airship is moored to its mast. Each tail surface attached to envelope at root and braced by

Vectored-thrust power plant of the Skyship 600/600B

Skyship 600/600B cabin interior

wires on each side; all four surfaces constructed from interlocking ribs and spars of Fibrelam with GFRP skins.

One-piece moulded gondola of Kevlar-reinforced plastics, with flooring and bulkheads of Fibrelam panels; those forming engine compartment at rear are faced with titanium for fire protection.

LANDING GEAR: Single two-wheel assembly with double tyres, mounted beneath rear gondola.

POWER PLANT: Two 190 kW (255 hp) Porsche 930/67 six-cylinder air-cooled and turbocharged piston engines mounted in rear of gondola. Each drives a ducted propulsor consisting of a Hoffmann five-blade reversible-pitch propeller within an annular duct of carbon fibre-reinforced GFRP. Each propulsor can be rotated about its pylon attachment to gondola through an arc of 210°, 90° upward and 120° downward, vectored thrust thus providing both V/STOL and in-flight hovering ability.

Fuel tank, capacity 682 litres (180 US gallons; 150 Imp gallons), at rear of engine compartment. Auxiliary fuel tanks optional. Engine modifications include provision of automatic mixture control, fuel injection and electronic ignition.

ACCOMMODATION: Seats for pilot and co-pilot, with dual controls. Maximum capacity 13 passengers in addition to pilot.
SYSTEMS: 28 V electrical system, supplied by engine-driven alternators.
AVIONICS: Include Honeywell Silver Crown series dual nav/com, ADF, Omega, VOR/ILS and weather radar.
EQUIPMENT: Night signs (30.48 m; 100 ft long and 9.14 m; 30 ft high) are full-colour aerial boards comprising 8,400 primary-coloured lamps computerised to enable the display of signs, animation and graphic logos in 16 colours.

DIMENSIONS, EXTERNAL:
Envelope: Length overall: 600	59.00 m (193 ft 7 in)
600B	59.50 m (195 ft 2¾ in)
Max diameter	15.20 m (49 ft 10½ in)
Fineness ratio: 600	3.88
600B	3.91
Height overall, incl gondola and landing gear	20.30 m (66 ft 7¼ in)
Width over tailfins	19.20 m (63 ft 0 in)
Gondola: Length overall	11.67 m (38 ft 3½ in)
Max width	2.56 m (8 ft 4¾ in)
Propeller diameter	1.37 m (4 ft 6 in)

DIMENSIONS, INTERNAL:
Envelope volume: 600	6,666 m³ (235,400 cu ft)
600B	6,796 m³ (240,000 cu ft)
Ballonet volume (two, total)	1,733 m³ (61,200 cu ft)
Gondola cabin: Length	6.89 m (22 ft 7¼ in)
Height	1.92 m (6 ft 3½ in)
Floor area (usable)	12.00 m² (130.0 sq ft)

WEIGHTS AND LOADINGS:
Max usable fuel	484 kg (1,067 lb)
Gross disposable load	2,343 kg (5,165 lb)

PERFORMANCE:
Max continuous speed	58 kt (107 km/h; 67 mph)
Cruising speed at 70% power	52 kt (96 km/h; 60 mph)
Pressure ceiling	3,050 m (10,000 ft)
Still air range at 40 kt (74 km/h; 46 mph), without auxiliary tanks	550 n miles (1,019 km; 633 miles)

UPDATED

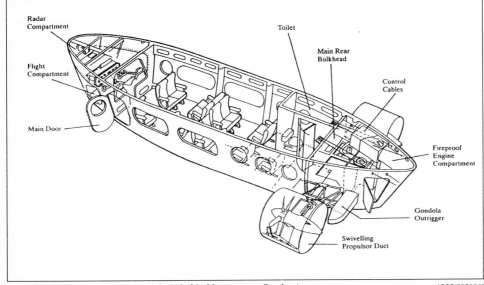
Skyship 600 13-passenger gondola *(Airship Management Services)*

INTERFACE

INTERFACE AIRSHIPS INC
PO Box 419, Terra Ceia, Florida 34250
e-mail: ecoblimp@usa.net
Web: http://www.ecoblimp.com

PRESIDENT: Brad Weigle
VICE-PRESIDENT: Joe Roman

Company formed 1994; built one example of Ecoblimp airship as pilot programme, using prototype for coastal oceanographic research in Florida. Plan was then to lease to research groups for similar work. See 1999-2000 *Jane's* for all known details.

UPDATED

OHIO

OHIO AIRSHIPS
Mantua, Ohio
Web: http://www.ohioairships.com

This company has a hybrid semi-rigid in the development stage, taxi tests of a subscale prototype having begun in March 2000. Design features include a semi-buoyant, aerofoil-shaped hull with an internal framework and five gas cells containing a combined 14.16 m³ (500 cu ft) of helium. Overall length is some 6.40 m (21 ft), dynamic lift is provided by 5.79 m (19 ft) span wings, and the prototype is powered by two wing-mounted glow-plug engines.

The full-size airship is projected as a heavy-lift cargo/freight carrier, with a possible size of up to 198,220 m³ (7 million cu ft) with a cargo container of 13,592 m³ (480,000 cu ft).

NEW ENTRY

ONAGO

ONAGO AIRCRAFT COMPANY
This company reports that it ceased trading in October 2000. A description of the Z-10 programme last appeared in the 2000-01 edition.

UPDATED

UPship

UPship CORPORATION
5198 Highway 84, Elba, Alabama 36323
Tel/Fax: (+1 334) 897 61 32
e-mail: airship@alaweb.com
Web: http://www.dirigible.com
PRESIDENT AND TECHNICAL DIRECTOR: Jesse Blenn

UPship's first airship to be built will be the 36 m, three-seat version described below. Design and cost projections for other, different sized versions have been made for craft with one, six, 16 and 50 seats; last-mentioned would be fitted with bow and stern thrusters for heavy lifting and precision hover.

VERIFIED

UPship 001

TYPE: Helium semi-rigid.
PROGRAMME: Design started 1989; original 750-001 (see 1992-93 *Jane's*) later enlarged as two-seat, 32 m 001 proof-of-concept vehicle; further enlarged in 1999 to 36 m, lower-drag envelope profile and three seats (or two occupants and heavier commercial or scientific payload). Progress by late 1998 (latest information received) had included land acquisitions, detail design work, granting of two US patents, and completion of residence and design studio at the operations site. Prototype construction and certification then planned to start in 2000 and estimated to take two years.
CUSTOMERS: Markets foreseen in tourism, advertising, scientific research, cargo, resource management, and minimum-impact logging operations, where UPships can be built and operated at competitive price.
COSTS: Up to US$2 million for set-up, construction and FAA certification of prototype; flyaway price up to US$500,000 (1999).
DESIGN FEATURES: Shape of minimum resistance, with propulsion engines in inverted V tailfins; nose-mounted thruster for control at all speeds. Designed for ease of maintenance, with reduced dependence on internal pressure. Greatly improved control, multiple helium cells and smoother ride increase usefulness and safety while minimising number of ground crew.
FLYING CONTROLS: Joystick regulates pitch and yaw by ruddervators in propeller slipstream, with aerodynamic assist; automatic pitch trim; foot pedals control bow thruster for enhanced and low-speed control (up/down, right/left and reverse) and for heavy lifting.
STRUCTURE: Aluminium and steel tubing hard structures; some carbon composites as appropriate; envelope of proprietary ripstop construction with internal divisions and three helium cells; tailfins deflect under excessive ground or air loads.
POWER PLANT: Three 17.9 kW (24 hp) König two-stroke radial piston engines with electric starting, two operating in tailfin openings and one as bow thruster. Fuel capacity 75.7 litres (20.0 US gallons; 16.7 Imp gallons).
ACCOMMODATION: Pilot and two passengers, seated in line; rear seat reclinable. Space aft of seats for rescuee(s), equipment or rest area. Cabin has side door and hatch for camera or rescue hoist; is electrically heated; noise- and vibration-free due to distance from engines.
SYSTEMS: Two alternators provide electrical power for night illumination system and cabin heating.

DIMENSIONS, EXTERNAL:
Envelope: Length overall	36.00 m (118 ft 1¼ in)
Max diameter	7.20 m (23 ft 7½ in)
Height overall	8.70 m (28 ft 6½ in)
Tail unit span	10.00 m (32 ft 9¾ in)
Propeller diameter	1.50 m (4 ft 11 in)

DIMENSIONS, INTERNAL:
Envelope volume	894.0 m³ (31,571 cu ft)
Ballonet volume (30%)	235.0 m³ (8,300 cu ft)
Helium fill	785.0 m³ (27,722 cu ft)
Gondola: Length	5.00 m (16 ft 4¾ in)
Max width	1.00 m (3 ft 3¼ in)

WEIGHTS AND LOADINGS (approx):
Weight empty	470 kg (1,036 lb)
Useful lift	315 kg (694 lb)

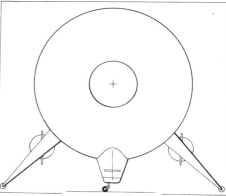

Nose view of UPship 001 as currently envisaged (*James Goulding/Jane's*)

Total static lift	785 kg (1,731 lb)
Additional thruster lift	50 kg (110 lb)

PERFORMANCE (estimated):
Max level speed	52 kt (98 km/h; 60 mph)
Cruising speed: at 50% power	42 kt (78 km/h; 48 mph)
at 25% power	34 kt (63 km/h; 39 mph)
Ceiling above T-O: normal	1,000 m (3,280 ft)
max	3,000 m (9,840 ft)

Range, two engines, 20% reserves:
at 50% power cruising speed	216 n miles (400 km; 248 miles)
at 25% power cruising speed	345 n miles (640 km; 397 miles)

Endurance, two engines, 20% reserves:
at 50% power cruising speed	5 h
at 25% power cruising speed	10 h

UPDATED

AIR-LAUNCHED MISSILES

Raytheon AIM-9 Sidewinder AAM being launched from a Lockheed Martin F-22 Raptor (USAF)

The following pages support the Armament paragraphs of aircraft descriptions in this book by explaining in brief how the potential of an individual aircraft is enhanced by its missile armament. Coverage is restricted to missiles carried by, or applicable to, aircraft in the current edition; for this reason, some older missiles are excluded, as are future projects still in the early stages of definition. Contents do include certain anti-tank and shoulder-launched anti-aircraft missiles which have airborne applications, mostly on helicopters.

To expedite retrieval of data, missiles are listed in alphabetical order of name or designation, with full cross-references to alternative epithets. In many instances, the 'manufacturer' of Chinese and Russian missiles is actually the export sales agency. More detailed information is contained in *Jane's Air-Launched Weapons*.

KEY

Roles		Guidance					
AAM	Air-to-air missile	AL	Active laser	LR	Laser radar		
ARM	Anti-radiation missile	AP	Autopilot	MMW	Millimetric-wave		
ASM	Air-to-surface missile	A/P	Active/passive radar	PR	Passive radar		
AshM	Anti-ship missile	ARH	Active radar homing	RC	Radio command		
ATM	Anti-tank missile	GPS	Global Positioning System	RF	Radio frequency		
LGB	Laser-guided bomb	I	Inertial	SAL	Semi-active laser		
SOM	Standoff missile	Im	Imaging	SARH	Semi-active radar homing		
		IR	Infra-red	T	Terrain reference		
		L	Laser	TV	Television		

Name/designation	Role	Manufacturer/country	Length m (in)	Diameter m (in)	Weight kg (lb)	Guidance	Range n miles (km)
3M55 see 'Yakhont'							
3M60 see 'Kayak'							
3M80 see Moskit							
9M14 see 'Sagger'							
9M17 see 'Swatter'							
9M32 see 'Grail'							
9M36 see 'Gremlin'							
9M39 see 'Grouse'							
9M114 see 'Spiral'							
9M120 see 'Spiral' and AT-X-16							
9M121 see AT-X-16							
9M313 see 'Gimlet'							
AA-6 see 'Acrid'							
AA-7 see 'Apex'							
AA-8 see 'Aphid'							
AA-9 see 'Amos'							
AA-10 see 'Alamo'							
AA-11 see 'Archer'							
AA-12 see 'Adder'							
AA-X-13 see R-37							
AAM-3, Type 90	AAM	Mitsubishi/Japan	3.00 (118)	0.13 (5)	91 (201)	IR	4.3 (8)
AAM-4	AAM	Mitsubishi/Japan				ARH	medium
AAM-5	AAM	not assigned/Japan				ImIR	
AAM-L, KS-172	AAM	Novator/Russia	7.40 (291)		750 (1,653)	I/ARH	216 (400)
AARGM	ARM	SAT & NAWC/USA				GPS/MMW	100 (185)
ACM, AGM-129	ASM	Raytheon/USA	6.35 (250)	0.70 (28)	1,250 (2,756)	I/LR	1,620 (3,000)
'Acrid', AA-6/R-46RD	AAM	Vympel/Russia	6.20 (244)	0.36 (14)	472 (1,041)	I/SARH	32 (60)
'Acrid', AA-6/R-46TD	AAM	Vympel/Russia	6.20 (244)	0.36 (14)	467 (1,030)	I/IR	27 (50)
'Adder', AA-12/R-77	AAM	Vympel/Russia	3.60 (142)	2.00 (79)	175 (386)	I/ARH	40 (75)
'Adder', R-77M-PD	AAM	Vympel/Russia	3.60 (142)	2.00 (79)	225 (496)	I/ARH	81 (150)
AFDS (unpowered)	SOM	LFK/Germany	3.47 (137)	0.63 (25)	660 (1,455)	GPS/I	
AGM-45 see Shrike							
AGM-65 see Maverick							
AGM-78 see Standard							
AGM-84 see Harpoon/SLAM							
AGM-88 see HARM							
AGM-114 see Hellfire							
AGM-119 see Penguin							
AGM-122 see Sidearm							
AGM-123 see Skipper							
AGM-129 see ACM							
AGM-130A	LGB	Boeing/USA	3.94 (155)	0.46 (18)	1,323 (2,917)	TV, ImIR	24 (45)
AGM-130C	LGB	Boeing/USA	3.95 (156)	0.46 (18)	1,353 (2,983)	TV/IIR	24 (45)
AGM-142 see Popeye							

AIR-LAUNCHED MISSILES

Name/designation	Role	Manufacturer/country	Length m (in)	Diameter m (in)	Weight kg (lb)	Guidance	Range n miles (km)
AGM-154C JSOW	SOM	Raytheon/USA	4.26 (168)		475 (1,047)/ 680 (1,499)	I/GPS	32 (60)
AGM-158 see JASSM							
AIM-7 see Sparrow							
AIM-9 see Sidewinder							
AIM-120 see AMRAAM							
AIM-132 see ASRAAM							
'Alamo', AA-10/R-27AE	AAM	Vympel/Russia	4.78 (188)	0.26 (10)	350 (772)	I/ARH	43 (80)
'Alamo', AA-10/R-27EM	AAM	Vympel/Russia	4.78 (188)	0.26 (10)	350 (772)	I/SARH	59 (110)
'Alamo', AA-10/R-27ER	AAM	Vympel/Russia	4.70 (185)	0.26 (10)	350 (772)	I/SARH	40 (75)
'Alamo', AA-10/R-27ET	AAM	Vympel/Russia	4.50 (177)	0.26 (10)	343 (756)	I/IR	38 (70)
'Alamo', AA-10/R-27R	AAM	Vympel/Russia	4.00 (157)	0.23 (9)	253 (558)	I/SARH	27 (50)
'Alamo', AA-10/R-27T	AAM	Vympel/Russia	3.70 (146)	0.23 (9)	254 (560)	I/IR	22 (40)

AA-10 'Alamo' AAM *(Paul Jackson/Jane's)*

AA-11 'Archer' short-range AAM *(Paul Jackson/Jane's)*

Name/designation	Role	Manufacturer/country	Length m (in)	Diameter m (in)	Weight kg (lb)	Guidance	Range n miles (km)
ALARM	ARM	Matra BAe/UK	4.30 (169)	0.22 (9)	265 (584)	PR	24 (45)
AM 39 see Exocet							
'Amos' AA-9/R-33	AAM	Vympel/Russia	4.15 (163)	0.38 (15)	490 (1,080)	I/SARH	65 (120)
AMRAAM, AIM-120A/B/C	AAM	Raytheon/USA	3.65 (144)	0.18 (7)	157 (346)	I/ARH	27 (50)
APACHE AP	SOM	Matra BAe/France	5.10 (201)	0.63 (25)	1,230 (2,712)	I/ARH	76 (140)
'Apex', AA-7/R-24R	AAM	Vympel/Russia	4.46 (176)	0.20 (8)	235 (518)	SARH	27 (50)
'Apex', AA-7/R-24T	AAM	Vympel/Russia	4.16 (164)	0.20 (8)	235 (518)	IR	27 (50)
'Aphid', AA-8/R-60	AAM	Vympel/Russia	2.08 (82)	0.13 (5)	65 (143)	IR	1.6 (3)
'Aphid', AA-8/R-60M	AAM	Vympel/Russia	2.08 (82)	0.13 (5)	65 (143)	AL	2.7 (5)
'Archer', AA-11/R-73M1	AAM	Vympel/Russia	2.90 (114)	0.17 (7)	105 (231)	I/IR	11 (20)
'Archer', AA-11/R-73M2	AAM	Vympel/Russia	2.90 (114)	0.17 (7)	110 (243)	I/IR	16 (30)
ARMAT	ARM	Matra BAe/France	4.15 (163)	0.40 (16)	550 (1,213)	I/PR	49 (90)
Armiger	ARM	BGT/Germany	3.90 (153)	0.20 (8)	230 (507)	PR/ImIR	108 (200)
AS-4 see 'Kitchen'							
AS-6 see 'Kingfish'							
AS-7 see 'Kerry'							
AS-9 see 'Kyle'							
AS-10 see 'Karen'							
AS-11 see 'Kilter'							
AS-12 see 'Kegler'							
AS-13 see 'Kingbolt'							
AS-14 see 'Kedge'							
AS-15 see 'Kent'							
AS 15TT	AShM	Aerospatiale Matra/France	2.30 (91)	0.18 (7)	96 (212)	Radio	8.1 (15)
AS-16 see 'Kickback'							
AS-17 see 'Krypton'							
AS-18 see 'Kazoo'							
AS-19 see 'Koala'							
AS-20 see 'Kayak'							
AS-30L	ASM	Aerospatiale Matra/France	3.65 (144)	0.34 (13)	520 (1,146)	I/SAL	5.4 (10)
AS 34 see Kormoran							
AS 37 see Martel							
ASM-1, Type 80	AShM	Mitsubishi/Japan	4.00 (157)	0.34 (13)	600 (1,323)	I/ARH	27 (50)
ASM-1C, Type 91	AShM	Mitsubishi/Japan	4.00 (157)	0.35 (14)	510 (1,124)	I/ARH	35 (65)
ASM-2, Type 93	AShM	Mitsubishi/Japan	4.10 (161)	0.35 (14)	520 (1,146)	I/ImIR	54 (100)
ASMP	SOM	Aerospatiale Matra/France	5.38 (212)	0.38 (15)	860 (1,896)	I/T	135 (250)
ASMP Plus	SOM	Aerospatiale Matra/France					270 (500)
Aspide 1	AAS	Alenia Marconi/Italy	3.70 (146)	0.20 (8)	220 (485)	SARH	19 (35)
Aspide 2	AAM	Alenia Marconi/Italy	3.65 (144)	0.21 (8)	225 (496)	ARH	22 (40)
ASRAAM (AIM-132)	AAM	Matra BAe/UK	2.90 (114)	0.17 (7)	87 (192)	ImIR	5.4 (10)
AT-2 see 'Swatter'							
AT-3 see 'Sagger'							
AT-6 see 'Spiral'							
AT-9 see 'Spiral 2'							
AT-12 see 'Swinger'							
AT-X-16, 9M120M/9M121 Vikhr M	ATM	Shipunov/Russia	2.80 (110)	0.13 (5)	45 (99)	SAL	5.4 (10)
Ataka see 'Swinger'							
ATAM see Mistral							
ATASK see Helstreak							
BGM-71 see TOW							
Black Shahine see SCALP EG							
Brimstone	ASM	Alenia Marconi/UK	1.63 (64)	0.18 (7)	50 (110)	MMW/I	4.3 (8)
Burya see 'Kitchen'							
C-101	AShM	CPMIEC/China	7.50 (295)	0.54 (21)	1,850 (4,079)	I/ARH	24 (45)
C-201, HY-4	AShM	CPMIEC/China	7.36 (290)	0.76 (30)	1,740 (3,836)	AP/ARH	73 (135)
C-601, CAS-1 'Kraken'/YJ-6	AShM	CPMIEC/China	7.36 (290)	0.76 (30)	2,440 (5,379)	AP/ARH	54 (100)
C-801, YJ-1	AShM	CPMIEC/China	4.65 (183)	0.36 (14)	655 (1,444)	I/ARH	27 (50)
C-802, YJ-2	AShM	CPMIEC/China	6.40 (252)	0.36 (14)	715 (1,576)	I/ARH	70 (130)
CAS-1 see C-601							
Darter, V-3C (see U-Darter)	AAM	Denel/South Africa	2.75 (108)	0.16 (6)	90 (198)	IR	2.7 (5)
DWS24/DWS39 (unpowered)	SOM	LFK/Germany	3.50 (138)		600 (1,323)	I	5.4 (10)
Exocet, AM 39	AShM	Aerospatiale Matra/France	4.70 (185)	0.35 (14)	670 (1,477)	I/ARH	27 (50)

AIR-LAUNCHED MISSILES

Name/designation	Role	Manufacturer/country	Length m (in)	Diameter m (in)	Weight kg (lb)	Guidance	Range n miles (km)
FIM-92 see Stinger							
Gabriel 3AS	AShM	IAI/Israel	3.85 (152)	0.34 (13)	560 (1,235)	I/ARH	19 (35)
Gabriel 4LR	AShM	IAI/Israel	4.70 (185)	0.44 (17)	960 (2,116)	I/ARH	108 (200)
'Gimlet', SA-16/9M313 Igla 1	AAM	Kolomna/Russia	1.69 (67)	0.07 (3)	11 (24)	IR	2.7 (5)
'Grail' SA-7/9M32 Strela 2	AAM	Turopov/Russia	1.22 (48)	0.07 (3)	10 (22)	IR	2.7 (5)
'Gremlin', SA-14/9M36 Strela 3	AAM	Turopov/Russia	1.47 (58)	0.07 (3)	11 (24)	IR	2.7 (5)
'Grouse', SA-18/9M39 Igla	AAM	Kolomna/Russia	1.69 (67)	0.07 (3)	11 (24)	IR	2.7 (5)

SA-18 Grouse AAM *(Paul Jackson/Jane's)*

Raytheon AGM-88 HARM *(Paul Jackson/Jane's)*

Name/designation	Role	Manufacturer/country	Length m (in)	Diameter m (in)	Weight kg (lb)	Guidance	Range n miles (km)
Hakim see PGM							
HARM, AGM-88A/B/B+/C/D	ARM	Raytheon/USA	4.17 (164)	0.25 (10)	361 (796)	PR	43 (80)
Harpoon, AGM-84A	AShM	Boeing/USA	3.90 (154)	0.34 (13)	530 (1,168)	I/ARH	65 (120)
Have Lite see Popeye 2							
Have Nap see Popeye 1							
Hellfire, AGM-114A	ATM	Hellfire Systems LLC/USA	1.63 (64)	0.18 (7)	46 (101)	SAL	4.3 (8)
Hellfire, AGM-114B/C	ATM	Hellfire Systems LLC/USA	1.73 (68)	0.18 (7)	48 (106)	SAL, ImIR or RF+IR	4.3 (8)
Hellfire, AGM-114F	ATM	Hellfire Systems LLC/USA	1.80 (71)	0.18 (7)	49 (108)	SAL	4.3 (8)
Hellfire 2, AGM-114K	ATM	Hellfire Systems LLC/USA	1.63 (64)	0.18 (7)	46 (101)	SAL	4.9 (9)
Hellfire 2, AGM-114 Longbow	ATM	Hellfire Systems LLC/USA	1.78 (70)	0.18 (7)	50 (110)	MMW/I	4.3 (8)
Helstreak/ATASK (Starstreak)	AAM	Shorts/UK	1.40 (55)	0.13 (5)	16 (35)	RC	3.2 (6)
HJ-8A	ATM	Norinco/China	0.88 (35)	0.12 (5)	11 (24)	Wire	1.6 (3)
HJ-8B	ATM	Norinco/China	1.00 (39)	0.12 (5)	13 (29)	Wire	2.2 (4)
HOT 1	ATM	Euromissile/Europe	1.27 (50)	0.14 (6)	24 (53)	Wire	2.2 (4)
HOT 2/3	ATM	Euromissile/Europe	1.30 (51)	0.15 (6)	24 (53)	Wire	2.2 (4)
Hsiung Feng 2	AShM	Chung Shan/Taiwan	3.90 (154)	0.34 (13)	520 (1,146)	I/ARH+ImIR	43 (80)
HY-4 see C-201							
Igla see 'Grouse' and 'Gimlet'							
IRIS-T	AAM	Bodensee/Germany	3.00 (118)	0.13 (5)	87 (192)	ImIR	6.5 (12)
JASSM, AGM-158	ASM	Lockheed Martin/USA	4.26 (168)		1,024 (2,257)	ImIR/GPS/INS	13.5 (250)
JSOW see AGM-154							
'Karen', AS-10/Kh-25MR	ASM	Zvezda/Russia	4.04 (159)	0.28 (11)	300 (661)	RC	5.4 (10)
'Karen', AS-10/Kh-25ML	ASM	Zvezda/Russia	4.04 (159)	0.28 (11)	300 (661)	SAL	11 (20)
'Kayak', AS-20/Kh-35/3M60 Uran	AShM	Zvezda/Russia	3.75 (148)	0.42 (17)	480 (1,058)	I/ARH	70 (130)
'Kazoo', AS-18/Kh-59M Ovod M	ASM	Raduga/Russia	5.37 (211)	0.38 (15)	930 (2,050)	I/TV	62 (115)
'Kedge', AS-14/Kh-29L	ASM	Vympel/Russia	3.87 (152)	0.38 (15)	657 (1,448)	SAL	5.4 (10)
'Kedge', AS-14/Kh-29T	ASM	Vympel/Russia	3.87 (152)	0.38 (15)	670 (1,477)	TV	6.5 (12)
'Kedge', AS-14/Kh-29TE	ASM	Vympel/Russia	3.87 (152)	0.38 (15)	700 (1,543)	TV	16 (30)
'Kegler', AS-12/Kh-25MP/Kh-27	ASM	Zvezda/Russia	4.36 (172)	0.28 (11)	310 (683)	I/PR	22 (40)
'Kent', AS-15A/Kh-55/RKV-500	SOM	Raduga/Russia	6.04 (238)	0.51 (20)	1,400 (3,086)	I/T	1,296 (2,400)
'Kent', AS-15B/Kh-55SM/RKV-500M	SOM	Raduga/Russia	7.10 (280)	0.77 (30)	1,700 (3,748)	I/T	1,620 (3,000)
KEPD 150 (PDWS 200)	SOM	Taurus Systems/International	4.50 (177)		1,060 (2,337)	I/GPS/T/ImIR	81 (150)
KEPD 350 (Taurus)	SOM	Taurus Systems/International	5.00 (197)		1,400 (3,086)	I/GPS/T/ImIR	189 (350)
'Kerry', AS-7/Kh-23	ASM	Zvezda/Russia	3.53 (139)	0.28 (11)	287 (633)	SAL or RC	2.7 (5)
Kh-15 see 'Kickback'							
Kh-22 see 'Kitchen'							
Kh-23 see 'Kerry'							
Kh-25 see 'Karen' and 'Kegler'							
Kh-27 see 'Kegler'							
Kh-28 see 'Kyle'							
Kh-29 see 'Kedge'							
Kh-31 see 'Krypton'							
Kh-35 see 'Kayak'							
Kh-41 see Moskit							
Kh-55/65 see 'Kent'							
Kh-58 see 'Kilter'							
Kh-59 see 'Kingbolt'							
Kh-101	SOM	Raduga/Russia	7.45 (293)		2,400 (5,291)	I/Im	2,700 (5,000)
Kh-102	SOM	Raduga/Russia	7.45 (293)		2,400 (5,291)	I/Im	2,700 (5,000)
'Kickback', AS-16/Kh-15/RKV-500B	ASM	Raduga/Russia	4.78 (188)	0.46 (18)	1,200 (2,646)	I/PR or I/ARH	81 (150)
'Kilter' AS-11/Kh-58E	ARM	Raduga/Russia	5.00 (197)	0.38 (15)	650 (1,433)	I/PR	86 (160)
'Kingbolt', AS-13/Kh-59 Ovod	ASM	Raduga/Russia	5.40 (213)	0.38 (15)	850 (1,874)	TV	86 (160)
'Kingfish', AS-6	SOM	Raduga/Russia	10.56 (416)	0.92 (36)	4,500 (9,921)	I/PR or I/ARH	216 (400)
'Kitchen', AS-4/Kh-22 Burya	SOM	Raduga/Russia	11.30 (445)	1.00 (39)	5,900 (13,007)	I/PR or I/ARH	216 (400)
'Koala', AS-19 (Kh-90/BL-10)	ASM	reportedly terminated					
Kokon see 'Spiral'							
Kormoran 1, AS 34	AShM	DASA/Germany	4.40 (173)	0.35 (14)	600 (1,323)	I/ARH	16 (30)

820 AIR-LAUNCHED MISSILES

Name/designation	Role	Manufacturer/country	Length m (in)	Diameter m (in)	Weight kg (lb)	Guidance	Range n miles (km)
Kormoran 2, AS 34	AShM	DASA/Germany	4.40 (173)	0.35 (14)	630 (1,389)	I/ARH	19 (35)
'Kraken' see C-601							

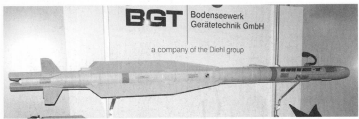

BGT Iris AAM *(Paul Jackson/Jane's)* 2000/0064935

AS-17 (Kh-31) 'Krypton' 2001/0105050

Name/designation	Role	Manufacturer/country	Length m (in)	Diameter m (in)	Weight kg (lb)	Guidance	Range n miles (km)
'Krypton', AS-17/Kh-31A-1	ASM	Zvezda/Russia	4.70 (185)	0.36 (14)	600 (1,323)	I/ARH	27 (50)
'Krypton', AS-17/Kh-31A-2	ASM	Zvezda/Russia	5.23 (206)	0.36 (14)	600 (1,323)	I/ARH	38 (70)
'Krypton', AS-17/Kh-31P-1	ASM	Zvezda/Russia	4.70 (185)	0.36 (14)	600 (1,323)	I/PR	54 (100)
'Krypton', AS-17/Kh-31P-2	ASM	Zvezda/Russia	5.23 (206)	0.36 (14)	600 (1,323)	I/PR	108 (200)
Kukri, V-3B	AAM	Denel/South Africa	2.94 (116)	0.13 (5)	73 (161)	IR	2.2 (4)
'Kyle', AS-9/Kh-28	ARM	Zvezda/Russia	6.00 (236)	0.43 (17)	715 (1,576)	PR	49 (90)
LY-60 see PL-11							
MAA-1 Piranha/Mol	AAM	Orbita/Brazil	2.82 (111)	0.15 (6)	90 (198)	IR	2.7 (5)
Magic 1, R 550	AAM	Matra BAe/France	2.72 (107)	0.16 (6)	89 (196)	IR	1.6 (3)
Magic 2, R 550	AAM	Matra BAe/France	2.75 (108)	0.16 (6)	89 (196)	IR	11 (20)
Marte 2	AShM	AOSM/Italy	4.80 (189)	0.32 (13)	340 (750)	I/ARH	11 (20)
Marte 2A	AShM	AOSM/Italy	3.90 (154)	0.32 (13)	269 (593)	I/AR	16 (30)
Marte 2B	ARM	AOSM/Italy	3.90 (154)	0.32 (13)	269 (593)	PR	32 (60)
Martel, AS 37	ARM	Matra BAe/France	4.20 (165)	0.40 (16)	535 (1,179)	PR	30 (55)
Maverick, AGM-65A	ASM	Raytheon/USA	2.49 (98)	0.31 (12)	210 (463)	TV	1.6 (3)
Maverick, AGM-65B	ASM	Raytheon/USA	2.49 (98)	0.31 (12)	210 (463)	TV	4.3 (8)
Maverick, AGM-65D	ASM	Raytheon/USA	2.49 (98)	0.31 (12)	220 (485)	ImIR	11 (20)
Maverick, AGM-65E	ASM	Raytheon/USA	2.49 (98)	0.31 (12)	293 (646)	SAL	11 (20)
Maverick, AGM-65F/G	ASM	Raytheon/USA	2.49 (98)	0.31 (12)	307 (677)	ImIR	13 (25)
Maverick, AGM-65H	ASM	Raytheon/USA	2.60 (102)	0.31 (12)	305 (672)	ARH	13 (25)
Meteor	AAM	Matra BAe/UK	3.65 (144)		160 (353)	ARH	81 (150)
MICA	AAM	Matra BAe/France	3.10 (122)	0.16 (6)	110 (243)	I/ARH or IR	27 (50)

Matra BAe MICA air-to-air missiles, showing IR (nearest) and active radar seekers *(Paul Jackson/Jane's)* 2000/0064975

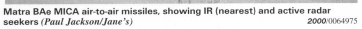

PGM-1A laser-guided version of Hakim ASM *(Paul Jackson/Jane's)* 2000/0064976

Name/designation	Role	Manufacturer/country	Length m (in)	Diameter m (in)	Weight kg (lb)	Guidance	Range n miles (km)
Mistral, ATAM	AAM	Matra BAe/France	1.80 (71)	0.09 (4)	18 (40)	IR	2.7 (5)
Mokopa, ZT6	ATM	Denel/South Africa	1.80 (71)	0.18 (7)	52 (115)	MMW or SAL	4.3 (8)
Moskit, Kh-41/3M80	ASM	Raduga/Russia	9.74 (383)	0.76 (30)	4,500 (9,921)	I/ARH or I/PR	135 (250)
Nag	ATM	DRDO/India			43 (95)	RC + ImIR or MMW	2.2 (4)
Nimrod	ASM	IAI/Israel	2.84 (112)	0.21 (8)	100 (220)	I/SAL	13 (25)
Ovod see 'Kazoo' and 'Kingbolt'							
Penguin 2, AGM-119B	AShM	Kongsberg/Norway	2.96 (117)	0.28 (11)	385 (849)	I/IR	19 (35)
Penguin 3, AGM-119A	AShM	Kongsberg/Norway	3.18 (125)	0.28 (11)	370 (816)	I/IR	30 (55)
PGM-A (1A, 2A, 3A) Hakim	ASM	Alenia Marconi/UK	3.60 (142)	0.35 (14)	300 (661)	I+SAL/IR/TV	11 (20)
PGM-B (1B, 2B, 3B) Hakim	ASM	Alenia Marconi/UK	4.00 (157)	0.43 (17)	1,115 (2,458)	I+SAL/IR/TV	11 (20)
Piranha see MAA-1							
PL-2/PL-3	AAM	CATIC/China	2.99 (118)	0.13 (5)	76 (168)	IR	1.6 (3)
PL-5	AAM	CATIC/China	2.89 (114)	0.13 (5)	85 (187)	IR	1.6 (3)
PL-7	AAM	CATIC/China	2.75 (108)	0.16 (6)	90 (198)	IR	1.6 (3)
PL-8	AAM	CATIC/China	3.00 (118)	0.16 (6)	120 (265)	IR	2.7 (5)
PL-9	AAM	Luoyang/China	2.99 (118)	0.16 (6)	115 (254)	IR	2.7 (5)
PL-10	AAM	CATIC/China	3.99 (157)	0.29 (11)	300 (661)	SARH	8.1 (15)
PL-11, LY-60	AAM	CATIC/China	3.89 (153)	0.20 (8)	220 (485)	SARH	13 (25)
Popeye 1, AGM-142A Have Nap	ASM	Rafael/Israel	4.82 (190)	0.53 (21)	1,360 (2,998)	I/TV or ImIR	43 (80)
Popeye 2, AGM-142B Have Lite	ASM	Rafael/Israel	4.00 (157)	0.53 (21)	1,115 (2,458)	I/TV or ImIR	40 (75)

Jane's All the World's Aircraft 2001-2002

Name/designation	Role	Manufacturer/country	Length m (in)	Diameter m (in)	Weight kg (lb)	Guidance	Range n miles (km)
Popeye, AGM-142C	ASM	Rafael/Israel		0.53 (21)		I/TV	
Popeye, AGM-142D	ASM	Rafael/Israel		0.53 (21)		I/ImIR	

Popeye 1/AGM-142A ASM *2001*/0105052

Israel's Python 4 AAM *2001*/0073040

Name/designation	Role	Manufacturer/country	Length m (in)	Diameter m (in)	Weight kg (lb)	Guidance	Range n miles (km)
Python 3	AAM	Rafael/Israel	3.00 (118)	0.16 (6)	120 (265)	IR	8.1 (15)
Python 4	AAM	Rafael/Israel	3.00 (118)	0.16 (6)	105 (231)	IR	8.1 (15)
QW-1 Vanguard	AAM	CPMIEC/China	1.53 (60)	0.07 (3)	16.5 (36)	IR	2.7 (5)
R-24 see 'Apex'							
R-27 see 'Alamo'							
R-33 see 'Amos'							
R-37, AA-X-13	AAM	Vympel/Russia	4.20 (165)	0.38 (15)	600 (1,323)	I/ARH	81 (150)
R-46 see 'Acrid'							
R-60 see 'Aphid'							
R-73 see 'Archer'							
R-77 see 'Adder'							
R 530 see Super 530							
R 550 see Magic							
RB 04	AShM	Saab/Sweden	4.25 (167)	0.50 (20)	600 (1,323)	I/AR	16 (30)
RB 05	ASM	Saab/Sweden	3.60 (142)	0.30 (12)	305 (672)	RC	4.3 (8)
RB 15F Mk 1	AShM	Saab/Sweden	4.35 (171)	0.50 (20)	598 (1,318)	I/AR	49 (90)
RB 15F Mk 2/3	AShM	Saab/Sweden	4.33 (170)	0.50 (20)	630 (1,389)	I/AR	81 (150)/108 (200)
RB 24J Swedish AIM-9J							
RB 71 Swedish Sky Flash							
RB 74 Swedish AIM-9L							
RB 75 Swedish Maverick							
SA-7 see 'Grail'							
SA-14 see 'Gremlin'							
SA-16 see 'Gimlet'							
SA-18 see 'Grouse'							
'Sagger', AT-3/9M14 Malyutka	ATM	Kolomna/Russia	0.86 (34)	0.13 (5)	113 (249)	Wire	1.6 (3)
SCALP EG (Storm Shadow)	SOM	Matra BAe/France (and UK)	5.10 (201)	0.63 (25)	1,300 (2,866)	I/MMW/ImIR	135 (250)
Sea Eagle	AShM	Matra BAe/UK	4.14 (163)	0.40 (16)	600 (1,323)	I/ARH	59 (110)
Sea Skua	AShM	Matra BAe/UK	2.50 (98)	0.25 (10)	147 (324)	SARH	8.1 (15)
Shafrir 2	AAM	Rafael/Israel	2.60 (102)	0.16 (6)	95 (209)	IR	1.6 (3)
Shrike, AGM-45	ARM	Raytheon/USA	3.05 (120)	0.20 (8)	177 (390)	PR	6.5 (12)
Shturm see 'Spiral 2'							
Sidearm, AGM-122	ARM	Motorola/USA	3.00 (118)	0.13 (5)	91 (200)	PR	4.3 (8)
Sidewinder, AIM-9L/M/S	AAM	several/USA and Europe	2.87 (113)	0.13 (5)	87 (192)	IR	4.3 (8)
Sidewinder, AIM-9P	AAM	several/USA	3.07 (121)	0.13 (5)	87 (192)	IR	4.3 (8)
Sidewinder, AIM-9R	AAM	several/USA	2.87 (113)	0.13 (5)	87 (192)	Visual band CCD	4.3 (8)
Sidewinder, AIM-9S	AAM	Raytheon/USA	2.87 (113)	0.13 (5)	86 (190)	IR	4.3 (8)
Sidewinder, AIM-9X	AAM	Raytheon/USA	2.90 (114)	0.13 (5)	85 (187)	ImIR	5.4 (10)

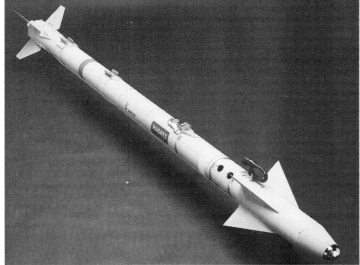

Black Shahine export version of Storm Shadow/SCALP EG
(Paul Jackson/Jane's) *2000*/0064977

Raytheon AIM-9X Sidewinder
2001/0022869

Name/designation	Role	Manufacturer/country	Length m (in)	Diameter m (in)	Weight kg (lb)	Guidance	Range n miles (km)
Skipper, AGM-123	LGB	ESC/USA	4.33 (170)	0.36 (14)	582 (1,283)	SAL	3.8 (7)
Sky Flash	AAM	Matra BAe/UK	3.66 (144)	0.20 (8)	195 (430)	SARH	22 (40)
Skyshark (unpowered)	ASM	CASMU/Italy	4.76 (187)		1,050 (2,315)	I	6.5 (12)
Skyshark (powered)	ASM	CASMU/Italy	4.76 (187)		1,170 (2,579)	I	13 (25)

AIR-LAUNCHED MISSILES

Name/designation	Role	Manufacturer/country	Length m (in)	Diameter m (in)	Weight kg (lb)	Guidance	Range n miles (km)
Sky Sword (Tien Chien) 1	AAM	Chung Shan/Taiwan	2.87 (113)	0.13 (5)	90 (198)	IR	2.7 (5)
Sky Sword (Tien Chien) 2	AAM	Chung Shan/Taiwan	3.60 (142)	0.20 (8)	190 (419)	SARH	22 (40)
SLAM, AGM-84E	ASM	Boeing/USA	4.50 (177)	0.34 (13)	630 (1,389)	I/GPS/ImIR	51 (95)
SLAM-ER, AGM-84H	ASM	Boeing/USA	4.37 (172)	0.34 (13)	726 (1,600)	I/GPS/ImIR	151 (280)
Sparrow, AIM-7F	AAM	Raytheon/USA	3.66 (144)	0.20 (8)	227 (500)	SARH	22 (40)
Sparrow, AIM-7M	AAM	Raytheon/USA	3.66 (144)	0.20 (8)	230 (507)	SARH	24 (45)
Sparrow, AIM-7P	AAM	Raytheon/USA	3.66 (144)	0.20 (8)	230 (507)	RC/SARH	24 (45)
Sparrow, AIM-7R	AAM	Raytheon/USA	3.66 (144)	0.20 (8)	230 (507)	RC/SARH/IR	24 (45)
'Spiral', AT-6/9M114 Kokon	ATM	Kolomna/Russia	1.83 (72)	0.13 (5)	34 (75)	RC	3.2 (6)
'Spiral 2', AT-9/9M/14 Shturm	ATM	Kolomna/Russia	1.83 (72)	0.13 (5)	40 (88)	RC	4.3 (8)
Standard, AGM-78	ARM	Raytheon/USA	4.57 (180)	0.34 (13)	615 (1,356)	PR	30 (55)
Starstreak see Helstreak							
Stinger, FIM-92	AAM	Raytheon/USA	1.52 (60)	0.07 (3)	16 (35)	IR	1.6 (3)
Storm Shadow see SCALP EG							
Strela see 'Grail' and 'Gremlin'							
Super 530D	AAM	Matra BAe/France	3.80 (150)	0.26 (10)	270 (595)	SARH	22 (40)
Super 530F1	AAM	Matra BAe/France	3.54 (139)	0.26 (10)	245 (540)	SARH	13 (25)
'Swatter', AT-2C/9M17 Skorpion	ATM	Nudelman/Russia	1.16 (46)	0.13 (5)	30 (66)	RC	2.2 (4)
Swift, ZT3/ZT35	ASM	Denel/South Africa	1.35 (53)/ 1.60 (63)	0.13 (5)	19 (42)	L	2.2 (4)
'Swinger', AT-12/9M120 Vikhr/Ataka	ATM	Shipunov/Russia	1.70 (67)	0.13 (5)	43 (95)	RC	4.3 (8)
Taurus see KEPD 350							
Tien Chien see Sky Sword							
Torgos	SOM	Kentron/South Africa					162 (300)
TOW, BGM-71A/B	ATM	Raytheon/USA	1.17 (46)	0.15 (6)	19 (42)	Wire	2.2 (4)
TOW, BGM-71C I-TOW	ATM	Raytheon/USA	1.45 (57)	0.15 (6)	19 (42)	Wire	2.2 (4)
TOW, BGM-71D/E TOW 2/2A	ATM	Raytheon/USA	1.55 (61)	0.15 (6)	22 (49)	Wire	2.2 (4)
TOW, BGM-71F TOW 2B	ATM	Raytheon/USA	1.17 (46)	0.15 (6)	23 (51)	Wire	2.2 (4)
TRIGAT, ATGW-3LR	ATM	consortium/Europe	1.57 (62)	0.15 (6)	48 (106)	ImIR	2.7 (5)
Type 80 see ASM-1							
Type 88 see ASM-2							
Type 90 see AAM-3							
U-Darter (see Darter)	AAM	Denel/South Africa	2.75 (108)	0.16 (6)	96 (212)	IR	4.3 (8)
Uran see 'Kayak'							
V-3B see Kukri							
V-3C see Darter							
Vikhr see 'Swinger' and AT-X-16							
X- see Kh-							
Yakhont (SS-N-26, 3M55)	AShM	Strela/Russia	8.30 (327)	0.67 (26)	2,550 (5,622)		
YJ-1 see C-801							
YJ-2 see C-802							
YJ-6 see C-601							
ZT3 and ZT35 see Swift							
ZT6 see Mokopa							

Yakhont air-launched anti-ship missile *(Paul Jackson/Jane's)*

AT-9 'Spiral 2' ATM *(Paul Jackson/Jane's)*

AGM-154 JSOW carriage trial on an F-16D Fighting Falcon *(Raytheon)*

AERO-ENGINES

Introduction

The following pages summarise the vital statistics of power plants mentioned in the main body of this book. They are divided into piston engines, turboprops/turboshafts and jet engines, and listed in alphabetical order of the manufacturer's name. Readers requiring further data on the two last-mentioned categories are referred to *Jane's Aero-Engines*. Some very small engines, employed by UAVs (and ultralights), are described in *Jane's Unmanned Aerial Vehicles and Targets*; those for large missiles and spacecraft, in *Jane's Space Directory*; and turboshafts, in *Jane's Helicopter Markets and Systems*. Space precludes an entry on each subvariant of more widely produced power plants, and therefore the aero-engine or helicopter publications should be consulted for these data.

Note that engine power ratings given below have been supplied, principally, by their manufacturers and are usually uninstalled output; data may thus vary from that quoted in the Aircraft section of this book, in which information is mainly supplied by the producer of the airframe. In most cases, the engine manufacturer's data reflect take-off power, generally under ISA sea-level conditions. Many gas-turbine engines, especially those for helicopters, are cleared to higher powers for brief periods in emergency.

UPDATED

Rolls-Royce jet engines, from smallest to largest
2001/0105047

Piston engines

Notes:
Arrangement: 4-O (2) means an engine with four cylinders, horizontally opposed, two-stroke; 8-IV (4) means eight cylinders in inverted-vee form, four-stroke; 4-X (2D) means four cylinders in X configuration, two-stroke diesel; 9-R is a nine-cylinder radial; L indicates in line. Wankel-type engines (W) are alternatively known as rotary (or, more accurately, rotating-piston) engines, re-using the term employed for early aero-engines based on the totally different principle of cylinders rotating about a stationary crankshaft.
Cooling: A = air cooling, L = liquid

In many cases the engines exist in numerous variants, for example with a geared drive or a supercharger (mechanically driven or turbo). Horsepower ratings to one place of decimals are based on metric (CV/PS) original data.

Engine type	Arrangement	Cooling	Cylinders			Weight, dry	Max power
			Bore	Stroke	Capacity		
Aero Prag (Czech Republic)							
AP-45							64.1 kW (86 hp)
Aerotechnik (Czech Republic)							
Mikron	4-L (4)	A	90.0 mm (3.54 in)	96.0 mm (3.78 in)	2,440 cc (149.0 cu in)	70.0 kg (154 lb)	48.5 kW (65 hp)
Arrow (Italy)							
AE 530AC	2-O (2)	A	74.6 mm (2.94 in)	61.0 mm (2.40 in)	533 cc (32.53 cu in)	50.0 kg (110 lb)	50.7 kW (68 hp)
AE 1070AC	4-O (2)	A	74.6 mm (2.94 in)	61.0 mm (2.40 in)	1,066 cc (65.1 cu in)	65.0 kg (143 lb)	90 kW (120 hp)
GP 1000	4-O (2)	A	74.6 mm (2.94 in)	57.0 mm (2.24 in)	996 cc (60.78 cu in)	65.0 kg (143 lb)	90 kW (120 hp)
GP 1500	6-O (2)	A	74.6 mm (2.94 in)	57.0 mm (2.24 in)	1,495 cc (91.2 cu in)	87.5 kg (193 lb)	134 kW (180 hp)
CAM (Canada)							
100	4-L (4)	L	74.0 mm (2.91 in)	86.5 mm (3.41 in)	1,488 cc (90.7 cu in)	92.1 kg (203 lb)	74.6 kW (100 hp)
Continental - see Teledyne Continental							
CRM (Italy)							
18D/SS	18-W (4D)	L	150.0 mm (5.91 in)	180.0 mm (7.09 in)	57,260 cc (3,495 cu in)	1,700 kg (3,745 lb)	1,380 kW (1,850 hp)
Diesel Air (UK)							
Dair 100	2 × 2-O	L	—	—	1,800 cc (109.8 cu in)	90 kg (198 lb)	74.6 kW (100 hp)
FAM (France)							
200	6-V (4)	L	—	—	3,000 cc (183.0 cu in)	179 kg (395 lb)	136 kW (182 hp)
HCI (USA)							
R180	5-R (4)	A	85.5 mm (3.365 in)	87.5 mm (3.445 in)	2,520 cc (154 cu in)	55.3 kg (122 lb)	55.9 kW (75 hp)
Hirth (Germany)							
F23A	2-O (2)	A	72.0 mm (2.835 in)	64.0 mm (2.52 in)	521 cc (31.79 cu in)	24.0 kg (52.9 lb)	29.8 kW (40 hp)
F30	4-O (2)	A	72.0 mm (2.835 in)	64.0 mm (2.52 in)	1,042 cc (63.6 cu in)	36.0 kg (79.4 lb)	70.8 kW (95 hp)
F30A	4-O (2)	A	72.0 mm (2.835 in)	64.0 mm (2.52 in)	1,042 cc (63.6 cu in)	39.0 kg (86.0 lb)	77.2 kW (104 hp)
F30A36	4-O (2)	A	72.0 mm (2.835 in)	64.0 mm (2.52 in)	1,042 cc (63.6 cu in)	39.0 kg (86.0 lb)	88.0 kW (118 hp)
F30ES	4-O (2)	A	72.0 mm (2.835 in)	64.0 mm (2.52 in)	1,042 cc (63.6 cu in)	42.0 kg (92.6 lb)	75.0 kW (101 hp)
F31	2-L (2)	A	76.0 mm (2.99 in)	69.0 mm (2.72 in)	625 cc (38.13 cu in)	26.5 kg (58.4 lb)	29.1 kW (39 hp)
F33A	1 (2)	A	76.0 mm (2.99 in)	69.0 mm (2.72 in)	313 cc (19.1 cu in)	12.7 kg (28.0 lb)	18.1 kW (24 hp)
F33B	1 (2)	A	76.0 mm (2.99 in)	69.0 mm (2.72 in)	313 cc (19.1 cu in)	13.0 kg (28.7 lb)	18.1 kW (24 hp)
2701	2-L (2)	A	70.0 mm (2.755 in)	64.0 mm (2.52 in)	493 cc (30.08 cu in)	32.8 kg (72.5 lb)	32.1 kW (43 hp)
2703	2-L (2)	A	72.0 mm (2.835 in)	64.0 mm (2.52 in)	521 cc (31.79 cu in)	32.8 kg (72.5 lb)	40.4 kW (55 hp)
2704	2-L (2)	A	76.0 mm (2.99 in)	69.0 mm (2.72 in)	625 cc (38.14 cu in)	31.0 kg (68.3 lb)	36.0 kW (48 hp)
2706	2-L (2)	A	76.0 mm (2.99 in)	69.0 mm (2.72 in)	625 cc (38.13 cu in)	30.2 kg (66.6 lb)	48.5 kW (65 hp)
3701	3-L (2)	A	76.0 mm (2.99 in)	69.0 mm (2.72 in)	939 cc (57.3 cu in)	37.0 kg (81.6 lb)	74.0 kW (99 hp)
HPower (USA)							
HKS 700E	(4)						44.7 kW (60 hp)
Jabiru (Australia)							
1600	4-O (4)	A	88.0 mm (3.465 in)	66.0 mm (2.60 in)	1,606 cc (98.0 cu in)	54.0 kg (119 lb)	44.7 kW (60 hp)
2200	4-O (4)	A	97.5 mm (3.84 in)	74.0 mm (2.91 in)	2,200 cc (134.3 cu in)	55.8 kg (123 lb)	59.7 kW (80 hp)
3300	6-O (4)	A	97.5 mm (3.84 in)	740 mm (2.91 in)	3,300 cc (201.4 cu in)	73.0 kg (161 lb)	89 kW (120 hp)
JPX (France)							
4T60/A	4-O (4)	A	93.0 mm (3.66 in)	75.4 mm (2.97 in)	2,050 cc (125.0 cu in)	73.0 kg (161 lb)	47.8 kW (65 hp)
4TX75/A	4-O (4)	A	95.0 mm (3.74 in)	82.0 mm (3.23 in)	2,325 cc (141.9 cu in)	78.0 kg (172 lb)	59.5 kW (79.8 hp)
4TX75/M	4-O (4)	A	95.0 mm (3.74 in)	82.0 mm (3.23 in)	2,325 cc (141.9 cu in)	71.0 kg (156.5 lb)	59.5 kW (79.8 hp)

824 AERO-ENGINES

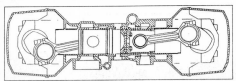

Cross-section of Diesel Air 100

LOM M337 *(Paul Jackson/Jane's)*

Mid West AE 50 rotary *(Paul Jackson/Jane's)*

Engine type	Arrangement	Cooling	Cylinders			Weight, dry	Max power
			Bore	Stroke	Capacity		
König (Germany)							
SD 750	4-R (2)	A	66.0 mm (2.60 in)	42.0 mm (1.655 in)	570 cc (34.78 cu in)	18.5 kg (41.0 lb)	20.8 kW (28 hp)
SF 930	4-R (2)	A	70.0 mm (2.755 in)	60.0 mm (2.36 in)	930 cc (56.75 cu in)	36.0 kg (79.4 lb)	35.8 kW (48 hp)
Limbach (Germany)							
L 550	4-O (2)	A	66.0 mm (2.60 in)	40.0 mm (1.57 in)	548 cc (33.44 cu in)	15.5 kg (34.0 lb)	32.0-33.6 kW (43-45 hp)
SL 1700	4-O (4)	A	88.0 mm (3.46 in)	69.0 mm (2.72 in)	1,680 cc (102.5 cu in)	73.0 kg (161 lb)	50.7 kW (68 hp)
L 1800	4-O (4)	A	90.0 mm (3.54 in)	69.0 mm (2.72 in)	1,756 cc (107.0 cu in)	70.0 kg (154 lb)	49.2 kW (66 hp)
L 2000	4-O (4)	A	90.0 mm (3.54 in)	78.4 mm (3.09 in)	1,994 cc (120.3 cu in)	70.0 kg (154 lb)	59.7 kW (80 hp)
L 2400EB	4-O (4)	A	97.0 mm (3.82 in)	82.0 mm (3.23 in)	2,424 cc (147.9 cu in)	82.0 kg (181 lb)	64.9 kW (87 hp)
L 2400EF	4-O (4)	A	97.0 mm (3.82 in)	82.0 mm (3.23 in)	2,424 cc (147.9 cu in)	82.0 kg (181 lb)	73.5 kW (99 hp)
L 2400EFI	4-O (4)	A+L	97.0 mm (3.82 in)	82.0 mm (3.23 in)	2,424 cc (147.9 cu in)	77.0 kg (170 lb)	74.6 kW (100 hp)
L 2400DWFIG	4-O (4)	A	97.0 mm (3.82 in)	82.0 mm (3.23 in)	2,424 cc (147.9 cu in)	100 kg (220 lb)	96 kW (128 hp)
L 2400EFI turbo	4-O (4)	A+L	97.0 mm (3.82 in)	82.0 mm (3.23 in)	2,424 cc (147.9 cu in)	105 kg (231 lb)	118 kW (158 hp)
LOM (Czech Republic)							
M132	4-L (4)	A	105.0 mm (4.13 in)	115.0 mm (4.53 in)	3,980 cc (242.9 cu in)	102 kg (225 lb)	90 kW (121 hp)
M137	6-L (4)	A	105.0 mm (4.13 in)	115.0 mm (4.53 in)	5,970 cc (364.3 cu in)	141 kg (311 lb)	134 kW (180 hp)
M332A	4-L (4)	A	105.0 mm (4.13 in)	115.0 mm (4.53 in)	3,980 cc (242.9 cu in)	102 kg (225 lb)	103 kW (138 hp)
M332B	4-L (4)	A	105.0 mm (4.13 in)	115.0 mm (4.53 in)	3,980 cc (242.9 cu in)	113 kg (249 lb)	118 kW (158 hp)
M337A	6-L (4)	A	105.0 mm (4.13 in)	115.0 mm (4.53 in)	5,970 cc (364.3 cu in)	153 kg (337 lb)	154 kW (207 hp)
M337B	6-L (4)	A	105.0 mm (4.13 in)	115.0 mm (4.53 in)	5,970 cc (364.3 cu in)	153 kg (337 lb)	173 kW (232 hp)
LPE (USA)							
IVG-600	8-IV (4)	L	—	—	9,832 cc (600.0 cu in)	255 kg (562 lb)	448 kW (600 hp)
Lycoming - see Textron Lycoming							
Mid-West (Germany)							
AE 50 Harrier	1-rotor (W)	A+L	—	—	294 cc (17.94 cu in)	33.0 kg (72.75 lb)	37.3 kW (50 hp)
AE75	3-L (2)	L	—	—	748 cc (45.50 cu in)	50.0 kg (110 lb)	57.4 kW (77 hp)
AE 100	2-rotor (W)	A+L	—	—	558 cc (35.90 cu in)	52.0 kg (115 lb)	70.8 kW (95 hp)
AE 110 Hawk	2-rotor (W)	A+L	—	—	588 cc (35.88 cu in)	53.0 kg (116.9 lb)	78.3 kW (105 hp)
Morane Renault – see Société de Motorisations Aéronautiques							
Nelson (USA)							
H-63CP	4-O (2)	A	68.3 mm (2.69 in)	70.0 mm (2.75 in)	1,030 cc (63.00 cu in)	30.8 kg (68 lb)	35.8 kW (48 hp)
Novikov (RKBM) (Russian Federation)							
DN-200	3 × 2-O (2D)	L	72.0 mm (2.835 in)	72.0 mm (2.835 in)	4,440 cc (270.9 cu in)	105 kg (231 lb)	110 kW (148 hp)
Orenda (Canada)							
OE600	8-V (4)	L	112.6 mm (4.433 in)	101.6 mm (4.00 in)	8,112 cc (495 cu in)	340 kg (750 lb)	447 kW (600 hp)
OE600 Turbo	8-V (4)	L	112.6 mm (4.433 in)	101.6 mm (4.00 in)	8,112 cc (495 cu in)		559 kW (750 hp)
PZL (Poland)							
PZL-3S	7-R (4)	A	155.5 mm (6.12 in)	155.0 mm (6.10 in)	20,600 cc (1,265 cu in)	411 kg (906 lb)	442 kW (592 hp)
PZL AI-14RA	9-R (4)	A	105.0 mm (4.125 in)	130.0 mm (5.118 in)	10,160 cc (620 cu in)	200 kg (441 lb)	191 kW (256 hp)
PZL ASz-62	9-R (4)	A	155.0 mm (6.10 in)	174.0 mm (6.85 in)	29,870 cc (1,823 cu in)	580 kg (1,279 lb)	735 kW (985 hp)
PZL K-9	9-R (4)	A	155.0 mm (6.10 in)	174.0 mm (6.85 in)	29,870 cc (1,823 cu in)	580 kg (1,279 lb)	860 kW (1,170 hp)
PZL-F 2A-120-C1	2-O (4)	A	117.48 mm (4.625 in)	88.9 mm (3.50 in)	1,916 cc (117.0 cu in)	58.5 kg (129 lb)	44.7 kW (60 hp)
PZL-F 4A-635-B31	4-O (4)	A	117.48 mm (4.625 in)	88.9 mm (3.50 in)	3,850 cc (235.0 cu in)	103 kg (226 lb)	86.5 kW (116 hp)
PZL-F 6A6350-C1	6-O (4)	A	117.48 mm (4.625 in)	88.9 mm (3.50 in)	5,735 cc (350.0 cu in)	150 kg (330 lb)	153 kW (205 hp)
Rotax (Austria)							
447 UL-1V	2-L (2)	A	67.5 mm (2.66 in)	61.0 mm (2.40 in)	436.5 cc (26.64 cu in)	26.8 kg (59.1 lb)	29.5 kW (39.6 hp)
447 UL-2V	2-L (2)	A	67.5 mm (2.66 in)	61.0 mm (2.40 in)	436.5 cc (26.64 cu in)	26.8 kg (59.1 lb)	31.0 kW (41.6 hp)
462	2-L (2)		—	—	—		38.8 kW (52 hp)
503 UL-1V	2-L (2)	A	72.0 mm (2.835 in)	61.0 mm (2.40 in)	496.7 cc (30.31 cu in)	31.4 kg (69.2 lb)	34.0 kW (45.6 hp)
503 UL-2V	2-L (2)	A	72.0 mm (2.835 in)	61.0 mm (2.40 in)	496.7 cc (30.31 cu in)	31.4 kg (69.2 lb)	37.0 kW (49.6 hp)
582 UL-2V	2-L (2)	L	76.0 mm (2.99 in)	64.0 mm (2.52 in)	580.7 cc (35.44 cu in)	27.4 kg (60.5 lb)	47.8 kW (64.1 hp)
618 UL-2V	2-L (2)	L	76.0 mm (2.99 in)	68.0 mm (2.68 in)	617.0 cc (37.65 cu in)	31.0 kg (68.3 lb)	55.0 kW (73.8 hp)
912 UL-2V	4-O (4)	A+L	79.5 mm (3.13 in)	61.0 mm (2.40 in)	1,211.2 cc (73.9 cu in)	59.0 kg (130 lb)	59.6 kW (79.9 hp)
912 ULS	4-O (4)	A+L	84.0 mm (3.31 in)	61.0 mm (2.40 in)	1,352 cc (82.50 cu in)	56.6 kg (125 lb)	73.5 kW (98.6 hp)
914 UL-2V/914F	4-O (4)	A+L	79.5 mm (3.13 in)	61.0 mm (2.40 in)	1,211.2 cc (73.9 cu in)	62.0 kg (137 lb)	84.6 kW (113.4 hp)
RotorWay (USA)							
RI 162	4-O (4)	L	—	—	2,660 cc (162.0 cu in)	77.1 kg (170 lb)	113 kW (152 hp)
SAEC (China)							
HS5: ASh-62IR made under licence, see PZL ASz-62 (Poland)							
Sauer (Germany)							
UL 1800	4-O (4)	A	90.0 mm (3.54 in)	69.0 mm (2.72 in)	1,745 cc (106.49 cu in)	56.5 kg (124.56 lb)	35.0 kW (46.9 hp)
SE 1800	4-O (4)	A	90.0 mm (3.54 in)	69.0 mm (2.72 in in)	1,745 cc (106.49 cu in)	63.3 kg (139.55 lb)	40.0 kW (53.6 hp)
SA 2100	4-O (4)	A	90.0 mm (3.54 in)	84.0 mm (3.31 in)	2,135 cc (130.3 cu in)	69.0 kg (152.1 lb)	59.0 kW (79.1 hp)
SD 2500	4-O (4)	A	97.0 mm (3.82 in)	84.0 mm (3.31 in)	2,481 cc (151.4 cu in)	79.0 kg (174.2 lb)	68.0 kW (91.2 hp)
SM 2700	4-O (4)	A	97.0 mm (3.82 in)	90 mm (3.54 in)	2,660 cc (162.3 cu in)	82.0 kg (180.8 lb)	75.0 kW (100.6 hp)

Rotax 503 *(Paul Jackson/Jane's)*

SMA SR305 *(Paul Jackson/Jane's)*

Verner SVS 1400 *(Paul Jackson/Jane's)*

PZL F4A-635 flat-four *(Paul Jackson/Jane's)*

Textron Lycoming TIO-540 six-cylinder air-cooled engine

Engine type	Arrangement	Cooling	Cylinders			Weight, dry	Max power
			Bore	Stroke	Capacity		
Société de Motorisations Aéronautiques (France)							
SR305	4-O (4D)	A	undisclosed	undisclosed	5,000 cc (305.1 cu in)	undisclosed	169 kW (227 hp)
Subaru (Japan)							
EJ22	4-L (4)	L	96.9 mm (3.81 in)	75.0 mm (2.95 in)	2,212 cc (136.0 cu in)	119 kg (264 lb)	119 kW (160 hp)
EA81-100	4-O (4)	L	92.0 mm (3.62 in)	67.0 mm (2.64 in)	1,781 cc (108.7 cu in)	97.1 kg (214 lb)	74.6 kW (100 hp)
EA81-140	4-O (4)	L	92.0 mm (3.62 in)	67.0 mm (2.64 in)	1,781 cc (108.7 cu in)	100 kg (222 lb)	104 kW (140 hp)
Teledyne Continental Motors (USA)							
CSD 283	4-O (2D)	—	—	—	—	—	149 kW (200 hp)
O-200-A	4-O (4)	A	103.2 mm (4.125 in)	98.4 mm (3.875 in)	3,280 cc (201.0 cu in)	99.8 kg (220 lb)	74.6 kW (100 hp)
IO-240	4-O (4)	A	112.7 mm (4.44 in)	98.4 mm (3.875 in)	3,940 cc (240.0 cu in)	113 kg (250 lb)	93 kW (125 hp)
IO-360-ES	6-O (4)	A	112.7 mm (4.44 in)	98.4 mm (3.875 in)	5,900 cc (360.0 cu in)	159 kg (350 lb)	157 kW (210 hp)
IO-360-KB	6-O (4)	A	112.7 mm (4.44 in)	98.4 mm (3.875 in)	5,900 cc (360.0 cu in)	148 kg (327 lb)	145.5 kW (195 hp)
TSIO-360-GB-C, D	6-O (4)	A	112.7 mm (4.44 in)	98.4 mm (3.875 in)	5,900 cc (360.0 cu in)	136 kg (300 lb)	168 kW (225 hp)
L/TSIO-360-EB, FB	6-O (4)	A	112.7 mm (4.44 in)	98.4 mm (3.875 in)	5,900 cc (360.0 cu in)	175 kg (385 lb)	149 kW (200 hp)
TSIO-360-GB, LB	6-O (4)	A	112.7 mm (4.44 in)	98.4 mm (3.875 in)	5,900 cc (360.0 cu in)	175 kg (386 lb)	157 kW (210 hp)
LTSIO-360-KB	6-O (4)	A	112.7 mm (4.44 in)	98.4 mm (3.875 in)	5,900 cc (360.0 cu in)	178 kg (392 lb)	164 kW (220 hp)
TSIO-360-MB	6-O (4)	A	112.7 mm (4.44 in)	98.4 mm (3.875 in)	5,900 cc (360.0 cu in)	187 kg (412 lb)	157 kW (210 hp)
IO-520-D	6-O (4)	A	133.0 mm (5.25 in)	101.6 mm (4.00 in)	8,500 cc (520.0 cu in)	208 kg (459 lb)	224 kW (300 hp)
IO-520-L	6-O (4)	A	133.0 mm (5.25 in)	101.6 mm (4.00 in)	8,500 cc (520.0 cu in)	212 kg (467 lb)	224 kW (300 hp)
IO-520-M, MB	6-O (4)	A	133.0 mm (5.25 in)	101.6 mm (4.00 in)	8,500 cc (520.0 cu in)	188 kg (415 lb)	213 kW (285 hp)
TSIO-520-AF	6-O (4)	A	133.0 mm (5.25 in)	101.6 mm (4.00 in)	8,500 cc (520.0 cu in)	198 kg (436 lb)	231 kW (310 hp)
TSIO-520-B, BB	6-O (4)	A	133.0 mm (5.25 in)	101.6 mm (4.00 in)	8,500 cc (520.0 cu in)	219 kg (483 lb)	213 kW (285 hp)
TSIO-520-BE	6-O (4)	A	133.0 mm (5.25 in)	101.6 mm (4.00 in)	8,500 cc (520.0 cu in)	223 kg (491 lb)	231 kW (310 hp)
TSIO-520-CE	6-O (4)	A	133.0 mm (5.25 in)	101.6 mm (4.00 in)	8,500 cc (520.0 cu in)	237 kg (527 lb)	242 kW (325 hp)
TSIO-520-C	6-O (4)	A	133.0 mm (5.25 in)	101.6 mm (4.00 in)	8,500 cc (520.0 cu in)	208 kg (458 lb)	213 kW (285 hp)
TSIO-520-E, EB	6-O (4)	A	133.0 mm (5.25 in)	101.6 mm (4.00 in)	8,500 cc (520.0 cu in)	219 kg (483 lb)	224 kW (300 hp)
TSIO-520-J, N, JB, NB	6-O (4)	A	133.0 mm (5.25 in)	101.6 mm (4.00 in)	8,500 cc (520.0 cu in)	221 kg (488 lb)	231 kW (310 hp)
TSIO-520-L, LB	6-O (4)	A	133.0 mm (5.25 in)	101.6 mm (4.00 in)	8,500 cc (520.0 cu in)	245 kg (539 lb)	231 kW (310 hp)
TSIO-520-M, P, R	6-O (4)	A	133.0 mm (5.25 in)	101.6 mm (4.00 in)	8,500 cc (520.0 cu in)	198 kg (436 lb)	231 kW (310 hp)
TSIO-520-T	6-O (4)	A	133.0 mm (5.25 in)	101.6 mm (4.00 in)	8,500 cc (520.0 cu in)	193 kg (426 lb)	231 kW (310 hp)
TSIO-520-UB	6-O (4)	A	133.0 mm (5.25 in)	101.6 mm (4.00 in)	8,500 cc (520.0 cu in)	192 kg (423 lb)	224 kW (300 hp)
TSIO-520-VB	6-O (4)	A	133.0 mm (5.25 in)	101.6 mm (4.00 in)	8,500 cc (520.0 cu in)	207 kg (457 lb)	242 kW (325 hp)
TSIO-520-WB	6-O (4)	A	133.0 mm (5.25 in)	101.6 mm (4.00 in)	8,500 cc (520.0 cu in)	189 kg (416 lb)	242 kW (325 hp)
GTSIO-520-D, H	6-O (4)	A	133.0 mm (5.25 in)	101.6 mm (4.00 in)	8,500 cc (520.0 cu in)	250 kg (550 lb)	280 kW (375 hp)
GTSIO-520-F, K	6-O (4)	A	133.0 mm (5.25 in)	101.6 mm (4.00 in)	8,500 cc (520.0 cu in)	272 kg (600 lb)	324 kW (435 hp)
GTSIO-520-L, M, N	6-O (4)	A	133.0 mm (5.25 in)	101.6 mm (4.00 in)	8,500 cc (520.0 cu in)	228 kg (502 lb)	280 kW (375 hp)
LTSIO-520-AE	6-O (4)	A	133.0 mm (5.25 in)	101.6 mm (4.00 in)	8,500 cc (520.0 cu in)	172 kg (380 lb)	187 kW (250 hp)
IO-550-B	6-O (4)	A	133.0 mm (5.25 in)	108.0 mm (4.25 in)	9,000 cc (550.0 cu in)	208 kg (462 lb)	224 kW (300 hp)
IO-550-C, E	6-O (4)	A	133.0 mm (5.25 in)	108.0 mm (4.25 in)	9,000 cc (550.0 cu in)	196 kg (433 lb)	224 kW (300 hp)
IO-550-F, L	6-O (4)	A	133.0 mm (5.25 in)	108.0 mm (4.25 in)	9,000 cc (550.0 cu in)	192 kg (423 lb)	224 kW (300 hp)
IO-550-G	6-O (4)	A	133.0 mm (5.25 in)	108.0 mm (4.25 in)	9,000 cc (550.0 cu in)	211 kg (465 lb)	209 kW (280 hp)
TSIO-550-B	6-O (4)	A	133.0 mm (5.25 in)	108.0 mm (4.25 in)	9,000 cc (550.0 cu in)	257 kg (566 lb)	261 kW (350 hp)
Voyager 550	6-O (4)	L	133.0 mm (5.25 in)	108.0 mm (4.25 in)	9,000 cc (550.0 cu in)	204 kg (450 lb)	261 kW (350 hp)
Voyager GT-550	6-O (4)	L	133.0 mm (5.25 in)	108.0 mm (4.25 in)	9,000 cc (550.0 cu in)	250 kg (550 lb)	298 kW (400 hp)
Textron Lycoming (USA)							
O-235-C	4-O (4)	A	111.0 mm (4.375 in)	98.4 mm (3.875 in)	3,850 cc (235.0 cu in)	97.5 kg (215 lb)	85.8 kW (115 hp)
O-235-I, M	4-O (4)	A	111.0 mm (4.375 in)	98.4 mm (3.875 in)	3,850 cc (235.0 cu in)	98.0 kg (218 lb)	88.0 kW (118 hp)
O-235N, P	4-O (4)	A	111.0 mm (4.375 in)	98.4 mm (3.875 in)	3,850 cc (235.0 cu in)	98.0 kg (218 lb)	86.5 kW (116 hp)
O-320-A, E	4-O (4)	A	130.0 mm (5.118 in)	98.4 mm (3.875 in)	5,200 cc (319.8 cu in)	110 kg (243 lb)	112 kW (150 hp)
(H)O-320-B2C	4-O (4)	A	130.0 mm (5.118 in)	98.4 mm (3.875 in)	5,200 cc (319.8 cu in)	115 kg (255 lb)	119 kW (160 hp)
O-320-D	4-O (4)	A	130.0 mm (5.118 in)	98.4 mm (3.875 in)	5,200 cc (319.8 cu in)	114 kg (253 lb)	119 kW (160 hp)
AEIO-320-D	4-O (4)	A	130.0 mm (5.118 in)	98.4 mm (3.875 in)	5,200 cc (319.8 cu in)	123 kg (271 lb)	119 kW (160 hp)
AEIO-320-E	4-O (4)	A	130.0 mm (5.118 in)	98.4 mm (3.875 in)	5,200 cc (319.8 cu in)	117 kg (258 lb)	112 kW (150 hp)
(L)IO-320-B, C	4-O (4)	A	130.0 mm (5.118 in)	98.4 mm (3.875 in)	5,200 cc (319.8 cu in)	117.5 kg (259 lb)	119 kW (160 hp)
(L)O-360-A	4-O (4)	A	130.0 mm (5.118 in)	111.0 mm (4.375 in)	5,920 cc (361.0 cu in)	120 kg (265 lb)	134 kW (180 hp)
O-360-F	4-O (4)	A	130.0 mm (5.118 in)	111.0 mm (4.375 in)	5,920 cc (361.0 cu in)	122 kg (269 lb)	134 kW (180 hp)
TO-360-C, F	4-O (4)	A	130.0 mm (5.118 in)	111.0 mm (4.375 in)	5,920 cc (361.0 cu in)	154 kg (343 lb)	157 kW (210 hp)
IO-360-A	4-O (4)	A	130.0 mm (5.118 in)	111.0 mm (4.375 in)	5,920 cc (361.0 cu in)	133 kg (293 lb)	149 kW (200 hp)
IO-360-B	4-O (4)	A	130.0 mm (5.118 in)	111.0 mm (4.375 in)	5,920 cc (361.0 cu in)	122 kg (268 lb)	134 kW (180 hp)
LIO-360-C	4-O (4)	A	130.0 mm (5.118 in)	111.0 mm (4.375 in)	5,920 cc (361.0 cu in)	134 kg (298 lb)	149 kW (200 hp)
TIO-360-C	4-O (4)	A	130.0 mm (5.118 in)	111.0 mm (4.375 in)	5,920 cc (361.0 cu in)	158 kg (348 lb)	157 kW (210 hp)
HIO-360-D1A	4-O (4)	A	130.0 mm (5.118 in)	111.0 mm (4.375 in)	5,920 cc (361.0 cu in)	146 kg (321 lb)	142 kW (190 hp)
(L)HIO-360-F1AD	4-O (4)	A	130.0 mm (5.118 in)	111.0 mm (4.375 in)	5,920 cc (361.0 cu in)	133 kg (293 lb)	142 kW (190 hp)
AEIO-360-A	4-O (4)	A	130.0 mm (5.118 in)	111.0 mm (4.375 in)	5,920 cc (361.0 cu in)	139 kg (307 lb)	149 kW (200 hp)
AEIO-360-B	4-O (4)	A	130.0 mm (5.118 in)	111.0 mm (4.375 in)	5,920 cc (361.0 cu in)	125 kg (277 lb)	134 kW (180 hp)
O-540-A	6-O (4)	A	130.0 mm (5.118 in)	111.0 mm (4.375 in)	8,860 cc (541.5 cu in)	161 kg (356 lb)	186 kW (250 hp)
IO-540-A1A5	6-O (4)	A	130.0 mm (5.118 in)	111.0 mm (4.375 in)	8,860 cc (541.5 cu in)	217 kg (479 lb)	201 kW (270 hp)
O-540-B	6-O (4)	A	130.0 mm (5.118 in)	111.0 mm (4.375 in)	8,860 cc (541.5 cu in)	166 kg (366 lb)	175 kW (235 hp)
IO-540-C	6-O (4)	A	130.0 mm (5.118 in)	111.0 mm (4.375 in)	8,860 cc (541.5 cu in)	170 kg (375 lb)	186 kW (250 hp)
IO-540-D	6-O (4)	A	130.0 mm (5.118 in)	111.0 mm (4.375 in)	8,860 cc (541.5 cu in)	173 kg (381 lb)	194 kW (260 hp)
O-540-E	6-O (4)	A	130.0 mm (5.118 in)	111.0 mm (4.375 in)	8,860 cc (541.5 cu in)	167 kg (368 lb)	194 kW (260 hp)
IO-540-F1B5	6-O (4)	A	130.0 mm (5.118 in)	111.0 mm (4.375 in)	8,860 cc (541.5 cu in)	167 kg (369 lb)	194 kW (260 hp)

Engine type	Arrangement	Cooling	Cylinders			Weight, dry	Max power
			Bore	Stroke	Capacity		
O-540-J	6-O (4)	A	130.0 mm (5.118 in)	111.0 mm (4.375 in)	8,860 cc (541.5 cu in)	162 kg (357 lb)	175 kW (235 hp)
IO-540-K	6-O (4)	A	130.0 mm (5.118 in)	111.0 mm (4.375 in)	8,860 cc (541.5 cu in)	201 kg (443 lb)	224 kW (300 hp)
IO-540-S	6-O (4)	A	130.0 mm (5.118 in)	111.0 mm (4.375 in)	8,860 cc (541.5 cu in)	200 kg (441 lb)	224 kW (300 hp)
IO-540-T4A5D	6-O (4)	A	130.0 mm (5.118 in)	111.0 mm (4.375 in)	8,860 cc (541.5 cu in)	187 kg (412 lb)	196 kW (260 hp)
IO-540-W1A5	6-O (4)	A	130.0 mm (5.118 in)	111.0 mm (4.375 in)	8,860 cc (541.5 cu in)	166 kg (367 lb)	175 kW (235 hp)
AEIO-540-D	6-O (4)	A	130.0 mm (5.118 in)	111.0 mm (4.375 in)	8,860 cc (541.5 cu in)	174 kg (386 lb)	194 kW (260 hp)
AEIO-540-L	6-O (4)	A	130.0 mm (5.118 in)	111.0 mm (4.375 in)	8,860 cc (541.5 cu in)	202 kg (445 lb)	224 kW (300 hp)
TIO-540-AF1A	6-O (4)	A	130.0 mm (5.118 in)	111.0 mm (4.375 in)	8,860 cc (541.5 cu in)	223 kg (491 lb)	201 kW (270 hp)
TIO-540-AE2A	6-O (4)	A	130.0 mm (5.118 in)	111.0 mm (4.375 in)	8,860 cc (541.5 cu in)	249 kg (549 lb)	261 kW (350 hp)
TIO-540-C	6-O (4)	A	130.0 mm (5.118 in)	111.0 mm (4.375 in)	8,860 cc (541.5 cu in)	205 kg (456 lb)	186 kW (250 hp)
(L)TIO-540-F	6-O (4)	A	130.0 mm (5.118 in)	111.0 mm (4.375 in)	8,860 cc (541.5 cu in)	233 kg (514 lb)	242 kW (325 hp)
(L)TIO-540-J	6-O (4)	A	130.0 mm (5.118 in)	111.0 mm (4.375 in)	8,860 cc (541.5 cu in)	235 kg (518 lb)	261 kW (350 hp)
TIO-540-S	6-O (4)	A	130.0 mm (5.118 in)	111.0 mm (4.375 in)	8,860 cc (541.5 cu in)	228 kg (502 lb)	224 kW (300 hp)
(L)TIO-540-U	6-O (4)	A	130.0 mm (5.118 in)	111.0 mm (4.375 in)	8,860 cc (541.5 cu in)	248 kg (547 lb)	261 kW (350 hp)
(L)TIO-540-V	6-O (4)	A	130.0 mm (5.118 in)	111.0 mm (4.375 in)	8,860 cc (541.5 cu in)	248 kg (547 lb)	269 kW (360 hp)
TIO-541-E	6-O (4)	A	130.0 mm (5.118 in)	111.0 mm (4.375 in)	8,860 cc (541.5 cu in)	270 kg (596 lb)	283 kW (380 hp)
TIGO-541-E	6-O (4)	A	130.0 mm (5.118 in)	111.0 mm (4.375 in)	8,860 cc (541.5 cu in)	319 kg (704 lb)	317 kW (425 hp)
AEIO-580	6-O (4)	A					246 kW (330 hp)
IO-720-A, B, D	8-O (4)	A	130.0 mm (5.118 in)	111.0 mm (4.375 in)	11,840 cc (722.0 cu in)	258 kg (568 lb)	298 kW (400 hp)
UFA (Russian Federation)							
Turbo-Diesel							331 kW (444 hp)
VAZ (Russian Federation)							
VAZ-416	2-rotor (W)	L	—	—	1,308 cc (79.8 cu in)	125 kg (275.6 lb)	134 kW (180 hp)
VAZ-4161	2-rotor (W)	L	—	—	1,308 cc (79.8 cu in)	125 kg (275.6 lb)	132 kW (178 hp)
VAZ-4162	2-rotor (W)	L	—	—	1,308 cc (79.8 cu in)	125 kg (275.6 lb)	147 kW (197 hp)
VAZ-426	3-rotor (W)	L	—	—	1,962 cc (119.7 cu in)	155 kg (341.7 lb)	201 kW (270 hp)
VAZ-4261	3-rotor (W)	L	—	—	1,962 cc (119.7 cu in)	155 kg (341.7 lb)	177 kW (237 hp)
VAZ-4262	3-rotor (W)	L	—	—	1,962 cc (119.7 cu in)	155 kg (341.7 lb)	198 kW (266 hp)
VAZ-4263	3-rotor (W)	L	—	—	1,962 cc (119.7 cu in)	145 kg (319.7 lb)	221 kW (296 hp)
VAZ-4265	3-rotor (W)	L	—	—	1,962 cc (119.7 cu in)	130 kg (287 lb)	201 kW (270 hp)
VAZ-526	4-rotor (W)	L	—	—	2,616 cc (159.6 cu in)	175 kg (386 lb)	298 kW (400 hp)
Verner (Germany)							
SVS 1400	2-O (4)	A	940 mm (3.70 in)	100.0 mm (3.94 in)	1,400 cc (85.4 cu in)	70.0 kg (154 lb)	58.8 kW (78.9 hp)
VM (Italy)							
1304HF	4-O (4D)	A	130.0 mm (5.118 in)	110.0 mm (4.33 in)	5,840 cc (356.4 cu in)	185 kg (408 lb)	154 kW (206 hp)
1306HF	6-O (4D)	A	130.0 mm (5.118 in)	110.0 mm (4.33 in)	8,760 cc (534.6 cu in)	243 kg (536 lb)	235 kW (315 hp)
1308HF	8-O (4D)	A	130.0 mm (5.118 in)	110.0 mm (4.33 in)	11,680 cc (713 cu in)	298 kg (657 lb)	316 kW (424 hp)
VOKBM (Russian Federation)							
M-3	3-R (4)	A	105.0 mm (4.125 in)	130.0 mm (5.118 in)	3,387 cc (206.5 cu in)	119 kg (262 lb)	77.2 kW (104 hp)
M-5	5-R (4)	A	105.0 mm (4.125 in)	130.0 mm (5.118 in)	5,644 cc (344.4 cu in)	115 kg (254 lb)	118 kW (158 hp)
M-7	7-R (4)	A	105.0 mm (4.125 in)	130.0 mm (5.118 in)	7,902 cc (482.2 cu in)	156 kg (343.9 lb)	191 kW (256 hp)
M-9F	9-R (4)	A	105.0 mm (4.125 in)	130.0 mm (5.118 in)	10,160 cc (620 cu in)	214 kg (472 lb)	294 kW (394 hp)
M-14NTK	9-R (4)	A	105.0 mm (4.125 in)	130.0 mm (5.118 in)	10,160 cc (620 cu in)		316 kW (424 hp)
M-14PF	9-R (4)	A	105.0 mm (4.125 in)	130.0 mm (5.118 in)	10,160 cc (620 cu in)	214 kg (472 lb)	294 kW (394 hp)
M-14PT	9-R (4)	A	105.0 mm (4.125 in)	130.0 mm (5.118 in)	10,160 cc (620 cu in)	217 kg (478 lb)	265 kW (355 hp)
Walter (Czech Republic)							
M202	2-O (2)	A	82.0 mm (3.23 in)	64.0 mm (2.52 in)	676 cc (41.25 cu in)	36.0 kg (79.4 lb)	48.5 kW (65 hp)
Wankel Rotary (Germany)							
LCR-407 SGti	1-rotor (W, D)	L	—	—	407 cc (24.8 cu in)	25.0 kg (55.1 lb)	27.2 kW (36.4 hp)
LOCR-407 SD	1-rotor (W, D)	L	—	—	407 cc (24.8 cu in)	38.0 kg (83.8 lb)	33.1 kW (44.4 hp)
LCR-814 TG ti	2-rotor (W, D)	L	—	—	814 cc (49.7 cu in)	35.0 kg (77.2 lb)	55.1 kW (73.9 hp)
LOCR-814 TD	2-rotor (W, D)	L	—	—	814 cc (49.7 cu in)	50.0 kg (110 lb)	66.2 kW (88.8 hp)
Twinpack	4-rotor (W, D)	L	—	—	—	119 kg (262 lb)	110 kW (148 hp)
Wilksch (UK)							
Diesel	—	—	—	—	—	—	89.5 kW (120 hp)
Zoche (Germany)							
ZO 01A	4-X (2D)	A	95.0 mm (3.74 in)	94.0 mm (3.70 in)	2,665 cc (162.6 cu in)	84.0 kg (185 lb)	110 kW (150 hp)
ZO 02A	8-X (2D)	A	95.0 mm (3.74 in)	94.0 mm (3.70 in)	5,330 cc (325.3 cu in)	118 kg (259 lb)	220 kW (300 hp)
Zöllner (Germany)							
DZ	4-O (4)	A	—	—	1,998 cc (121.9 cu in)	74.0 kg (163 lb)	86.8 kW (116 hp)

UPDATED

Wankel LCR-814 TG ti 2001/0105057

Zoche ZO 02 *(R J Malachowski)* 1999/0051897

Zöllner DZ *(Paul Jackson/Jane's)* 2000/0075980

Turboprop engines

Notes:
Arrangement: A = axial stages, C = centrifugal stages
Prop drive: FT = free turbine, SS = single shaft, 2S = two-shaft engine but no independent power turbine
For turboshaft engines, T-O rating is the maximum except contingency, usually a 30 minute, or maximum continuous power.

Engine type	Arrangement	Air flow	Prop drive	Length	Width	Dry weight	T-O rating
AlliedSignal - see Honeywell							
Allison – see Rolls-Royce (USA)							
DEMC (China)							
WJ5A I: AI-24A made under licence, details as AI-24T under Progress (ZMKB) (Ukraine)							2,080 kW (2,790 shp)
WJ5E	10A	14.6 kg (32.2 lb)/s	SS	2,381 mm (93.7 in)	770 mm (30.3 in)	720 kg (1,587 lb)	2,130 kW (2,856 shp) 720 kg (1,587 lb)*
General Electric (USA)							
CT7-5A	5A+C	4.41 kg (10.0 lb)/s	FT	2,438 mm (96 in)	737 mm (29 in)	355 kg (783 lb)	1,294 kW (1,735 shp)
CT7-9	5A+C	5.20 kg (11.5 lb)/s	FT	2,438 mm (96 in)	737 mm (29 in)	365 kg (805 lb)	<1,447 kW (1,940 shp)
Honeywell (USA)							
TPE331-3	C+C	3.54 kg (7.80 lb)/s	SS	1,092 mm (43 in)	533 mm (21 in)	161 kg (355 lb)	626 kW (840 shp)
TPE331-12	C+C	3.49 kg (7.7 lb)/s	SS	1,168 mm (46 in)	533 mm (21 in)	181 kg (400 lb)	834 kW (1,100 shp)
TPE331-14GR	C+C	5.26 kg (11.60 lb)/s	SS	1,333 mm (52.5 in)	533 mm (21 in)	281 kg (620 lb)	1,462 kW (1,960 shp)
LTP 101-700	A+C	2.31 kg (5.10 lb)/s	FT	949 mm (37.4 in)	592 mm (23.3 in)	147 kg (325 lb)	522 kW (700 shp)
KKBM (Russian Federation)							
NK-12MV	14A	65.0 kg (143 lb)/s	FT	4,785 mm (188 in)	1,150 mm (45.3 in)	2,900 kg (6,393 lb)	11,033 kW (14,795 shp)
Klimov (Russian Federation)							
TV3-117VMA-SB2	12A	9.0 kg (19.84 lb/s)	FT	2,055 mm (80.9 in)	650 mm (25.6 in)	560 kg (1,235 lb)	1,864 kW (2,500 shp)
TV7-117S	5A+C	7.95 kg (17.5 lb)/s	FT	2,143 mm (84.4 in)	886 mm (34.9 in)	520 kg (1,146 lb)	1,839 kW (2,466 shp)
LHTEC (USA)							
CTP800-4T Twin	2×(C+C)	2×3.27 kg (7.20 lb)/s	FT	1,769 mm (69.6 in)	1,276 mm (50.2 in)	546 kg (1,203 lb)	2,013 kW (2,700 shp)
OMKB (Russian Federation)							
TVD-20	7A+C	5.40 kg (11.9 lb)/s	FT	1,770 mm (69.7 in)	850 mm (33.5 in)	285 kg (628 lb)	1,029 kW (1,380 shp)
Pratt & Whitney Canada (Canada)							
PT6A-27	3A+C	3.08 kg (6.80 lb)/s	FT	1,575 mm (62 in)	483 mm (19 in)	149 kg (328 lb)	507 kW (680 shp)
PT6A-41	3A+C	3.40 kg (7.50 lb)/s	FT	1,701 mm (67 in)	483 mm (19 in)	183 kg (403 lb)	634 kW (850 shp)
PT6A-65R	4A+C	4.31 kg (9.50 lb)/s	FT	1,905 mm (75 in)	483 mm (19 in)	218 kg (481 lb)	1,026 kW (1,376 shp)
PW120	C, C	—	FT	2,134 mm (84 in)	635 mm (25 in)	418 kg (921 lb)	1,491 kW (2,100 shp)
PW121	C, C	—	FT	2,134 mm (84 in)	635 mm (25 in)	425 kg (936 lb)	1,603 kW (2,150 shp)
PW123 C/D	C, C	—	FT	2,134 mm (84 in)	635 mm (25 in)	450 kg (992 lb)	1,603 kW (2,150 shp)
PW127	C, C	—	FT	2,134 mm (84 in)	660 mm (26 in)	481 kg (1,060 lb)	2,051 kW (2,750 shp)
PW150A	2A+C	—	FT	2,423 mm (95.4 in)	767 mm (30.2 in)	690 kg (1,522 lb)	<5,593 kW (7,500 shp)
Progress (ZMKB) (Ukraine)							
AI-20M	10A	20.7 kg (45.6 lb)/s	SS	3,096 mm (121.9 in)	842 mm (33.15 in)	1,040 kg (2,293 lb)	2,940 kW (3,943 shp)
AI-24T	10A	14.4 kg (31.7 lb)/s	SS	2,346 mm (92.4 in)	677 mm (26.65 in)	600 kg (1,323 lb)	1,880 kW (2,520 shp)
PZL (Poland)							
TWD-10B	6A+C	4.58 kg (10.1 lb)/s	FT	2,060 mm (81.1 in)	555 mm (21.9 in)	230 kg (507 lb)	754 kW (1,011 shp)
RKBM (Russian Federation)							
TVD-1500S	3A+C	—	FT	1,965 mm (77.4 in)	620 mm (24.4 in)	245 kg (540 lb)	969 kW (1,300 shp)
Rolls-Royce (UK)							
Dart Mk 536	C+C	10.7 kg (23.5 lb)/s	SS	2,480 mm (97.6 in)	963 mm (37.9 in)	569 kg (1,257 lb)	1,700 kW (2,280 shp)
Tyne 21	6A, 9A	21.1 kg (46.5 lb)/s	2S	2,762 mm (108.7 in)	1,400 mm (55 in)	1,085 kg (2,391 lb)	3,357 kW (4,500 shp)
Rolls-Royce (USA)							
250-B17	6A+C	1.56 kg (3.45 lb)/s	FT	1,143 mm (45 in)	483 mm (19 in)	88.4 kg (195 lb)	313 kW (420 shp)
T56-15	14A	14.7 kg (32.4 lb)/s	SS	3,708 mm (146 in)	686 mm (27 in)	828 kg (1,825 lb)	3,424 kW (4,591 shp)
AE 2100A	14A	16.3 kg (36.0 lb)/s	FT	—	—	—	3,095 kW (4,150 shp)
AE 2100C	14A	—	FT	—	—	—	2,685 kW (3,600 shp)
AE 2100D2, D3	14A	16.3 kg (36.0 lb)/s	FT	2,743 mm (108 in)	1,151 mm (45.3 in)	702 kg (1,548 lb)	3,424 kW (4,591 shp)

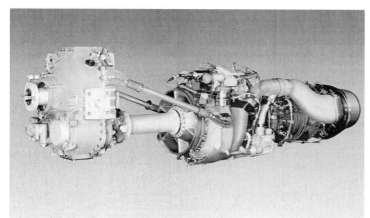

General Electric CT7 turboprop

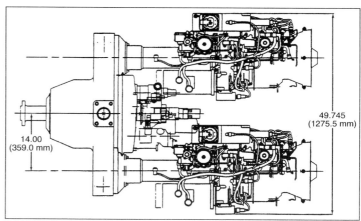

LHTEC CTP800-4T coupled turboprop

Walter M 601E turboprop

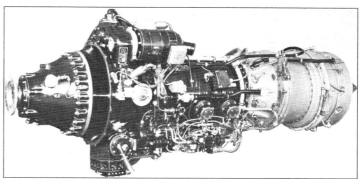

Progress (ZMKB) AI-24 single-shaft turboprop

Engine type	Arrangement	Air flow	Prop drive	Length	Width	Dry weight	T-O rating
SAEC (China)							
WJ6: AI-20M made under licence; details as under Progress (ZMKB) (Ukraine)							
Saturn (Russian Federation)							
AL-34-1	C	—	FT	1,700 mm (66.9 in)	640 mm (25.2 in)	178 kg (392 lb)	809 kW (1,085 shp)
Turbomeca (France)							
Arrius 1D	C	not disclosed	FT	826 mm (32.5 in)	476 mm (18.74 in)	111 kg (245 lb)	313 kW (420 shp)
Arrius 2F	C	not disclosed	FT	945 mm (37.5 in)	459 mm (18.07 in)	103 kg (227 lb)	376 kW (504 shp)
Walter (Czech Republic)							
M 601E	2A+C	3.60 kg (7.94 lb)/s	FT	1,675 mm (65.9 in)	590 mm (23.23 in)	200 kg (441 lb)	560 kW (751 shp)
M 601F	2A+C	3.60 kg (7.94 lb)/s	FT	1,675 mm (65.9 in)	590 mm (23.23 in)	202 kg (445 lb)	580 kW (778 shp)
M 602	C, C	7.33 kg (16.2 lb)/s	FT	2,669 mm (105 in)	753 mm (29.65 in)	570 kg (1,257 lb)	1,360 kW (1,824 shp)

*Residual thrust

UPDATED

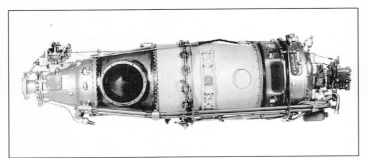

P&W Canada PT6A-65

Klimov TV3-117VMA-SB2 *(Paul Jackson/Jane's)*

AERO-ENGINES

Turboshaft engines

Notes:
See Turboprop engines

Engine type	Arrangement	Air flow	Drive	Length	Width	Dry weight	T-O rating
AlliedSignal – see Honeywell							
Allison – see Rolls-Royce							
Aviadvigatel (Russian Federation)							
D-25V	9A	26.2 kg (57.8 lb)/s	FT	2,737 mm (107.75 in)	1,086 mm (42.8 in)	1,325 kg (2,921 lb)	4,050 kW (5,430 shp)
CLXMW (China)							
WZ6		Turbomeca Turmo IIIC made under licence					1,145 kW (1,536 shp)
General Electric (USA)							
CT7-2	5A+C	4.5 kg (10.0 lb)/s	FT	1,194 mm (47.0 in)	635 mm (25.0 in)	212 kg (466 lb)	1,212 kW (1,625 shp)
CT7-6	5A+C	5.9 kg (13.0 lb)/s	FT	1,194 mm (47.0 in)	660 mm (26.0 in)	220 kg (485 lb)	1,491 kW (2,000 shp)
CT7-6D	5A+C	6.1 kg (13.5 lb)/s	FT	1,194 mm (47.0 in)	660 mm (26.0 in)	229 kg (504 lb)	1,514 kW (2,030 shp)
T64-419	14A	13.3 kg (29.4 lb)/s	FT	2,006 mm (79.0 in)	660 mm (26.0 in)	343 kg (755 lb)	3,542 kW (4,750 shp)
T700-401	5A+C	4.5 kg (10.0 lb)/s	FT	1,168 mm (46.0 in)	635 mm (25.0 in)	197 kg (434 lb)	1,260 kW (1,690 shp)
T700-700	5A+C	4.5 kg (10.0 lb)/s	FT	1,168 mm (46.0 in)	635 mm (25.0 in)	198 kg (437 lb)	1,210 kW (1,622 shp)
T700/T6	5A+C	5.9 kg (13.0 lb)/s	FT	1,205 mm (47.4 in)	660 mm (26.0 in)	220 kg (485 lb)	1,652 kW (2,215 shp)
Honeywell (USA) (formerly AlliedSignal)							
LTS 101-600A	1A+C	2.31 kg (5.10 lb)/s	FT	785 mm (30.9 in)	599 mm (23.6 in)	115 kg (253 lb)	459 kW (615 shp)
LTS 101-750B	1A+C	2.30 kg (5.10 lb)/s	FT	795 mm (31.3 in)	470 mm (18.5 in)	123 kg (271 lb)	515 kW (690 shp)
T53-703	5A+C	5.90 kg (13.0 lb)/s	FT	1,209 mm (47.6 in)	584 mm (23.0 in)	247 kg (545 lb)	1,343 kW (1,800 shp)
T5317A	5A+C	5.53 kg (12.2 lb)/s	FT	1,209 mm (47.6 in)	584 mm (23.0 in)	256 kg (564 lb)	1,119 kW (1,500 shp)
T55-714	7A+C	13.19 kg (29.1 lb)/s	FT	1,181 mm (46.5 in)	610 mm (24.0 in)	377 kg (832 lb)	3,630 kW (4,868 shp)
Klimov (Russian Federation)							
GTD-350 see under PZL (Poland)							
TV2-117	10A	8.4 kg (18.5 lb)/s	FT	2,842 mm (111.9 in)	547 mm (21.7 in)	338 kg (745 lb)	1,118 kW (1,500 shp)
TV3-117V	12A	8.7 kg (19.18 lb)/s	FT	2,085 mm (82.1 in)	640 mm (25.2 in)	285 kg (628 lb)	<1,633 kW (2,190 shp)
TV3-117VMA-SB3	12A	9.0 kg (19.84 lb)/s	FT	2,055 mm (80.9 in)	650 mm (25.6 in)	310 kg (683 lb)	1,864 kW (2,500 shp)
LHTEC (USA)							
T800-800	C+C	3.27 kg (7.20 lb)/s	FT	843.3 mm (33.2 in)	550.1 mm (21.7 in)	143 kg (315 lb)	995 kW (1,334 shp)
T800-801	C+C	4.43 kg (9.8 lb)/s	FT	843.3 mm (33.2 in)	550.1 mm (21.7 in)	149.7 kg (330 lb)	1,165 kW (1,563 shp)
CTS800-1G	C+C	4.01 kg (8.84 lb)/s	FT	856.0 mm (33.7 in)	550 mm (21.66 in)	149.7 kg (330 lb)	1,312 kW (1,760 shp)
CTS800-4	C+C	3.54 kg (7.8 lb)/s	FT	1,047.2 mm (41.23 in)	561.6 mm (22.11 in)	173.7 kg (383 lb)	1,016 kW (1,362 shp)
Mitsubishi (Japan)							
MG5-100	—	—	FT	—	—	—	597 kW (800 shp)
XTS1-10	—	—	FT	—	—	—	659 kW (884 shp)
MTR (International)							
390	C+C	3.20 kg (7.05 lb)/s	FT	1,078 mm (42.4 in)	442 mm (17.4 in)	169 kg (373 lb)	958 kW (1,285 shp)
OMKB (Russian Federation) GTD-3 see under Poland as PZL-10W							
Pratt & Whitney Canada (Canada)							
PT6B-36	3A+C	3.08 kg (6.80 lb)/s	FT	1,504 mm (59.2 in)	495 mm (19.5 in)	169 kg (373 lb)	732 kW (981 shp)
PT6T-3B	2×3A+C	2×3.08 kg (6.80 lb)/s	2×FT	1,702 mm (67.0 in)	1,118 mm (44.0 in)	299 kg (660 lb)	1,342 kW (1,800 shp)
PT6C-67	4A+C	4.3 kg (9.5 lb)	FT	1,930 mm (76.0 in)	483 mm (19.0 in)	—	1,252 kW (1,679 shp)
PW206A	C	—	FT	912 mm (35.9 in)	500 mm (19.7 in)	108 kg (237 lb)	477 kW (640 shp)
PW207D	C	—	FT	—	—	—	529 kW (710 shp)

Turbomeca Arriel 1C1 turboshaft

General Electric CT7 turboshaft

LHTEC T800 turboshaft

Rolls-Royce (formerly Allison) 250-C28 turboshaft

AERO-ENGINES

Engine type	Arrangement	Air flow	Drive	Length	Width	Dry weight	T-O rating
Progress (ZMKB) (Ukraine)							
D-136	6A, 7A	35.6 kg (78.4 lb)/s	FT	3,715 mm (146.3 in)	1,382 mm (4.4 in)	1,077 kg (2,374 lb)	7,457 kW (10,000 shp)
D-127	5A, 2A+C	27.4 kg (60.4 lb)/s	FT	3,950 mm (155.5 in)	1,400 mm (55.1 in)	—	10,700 kW (14,350 shp)
PZL (Poland)							
PZL-10W	6A+C	4.6 kg (10.1 lb)/s	FT	1,875 mm (73.8 in)	740 mm (29.0 in)	141 kg (310 lb)	662 kW (888 shp)
PZL GTD-350	7A+C	2.19 kg (4.83 lb)/s	FT	1,385 mm (54.53 in)	626 mm (24.9 in)	140 kg (307 lb)	294 kW (394 shp)
RKBM (Russian Federation)							
TVD-1500V (RD-600V)	3A+C	—	FT	1,250 mm (49.2 in)	620 mm (24.4 in)	220 kg (485 lb)	969 kW (1,300 shp)
Rolls-Royce (UK)							
Gem 42	4A, C	3.41 kg (7.50 lb)/s	FT	1,099 mm (43.2 in)	575 mm (22.6 in)	183 kg (404 lb)	746 kW (1,000 shp)
Gnome	10A	6.3 kg (13.8 lb)/s	FT	1,392 mm (54.8 in)	577 mm (22.7 in)	148 kg (326 lb)	1,145 kW (1,535 shp)
Rolls-Royce (USA) (formerly Allison)							
250-C20B	6A+C	1.56 kg (3.45 lb)/s	FT	985 mm (38.8 in)	483 mm (19.0 in)	71.5 kg (158 lb)	313 kW (420 shp)
250-C28	C	2.02 kg (4.45 lb)/s	FT	1,021 mm (40.2 in)	557 mm (21.9 in)	106 kg (233 lb)	373 kW (500 shp)
250-C30	C	2.54 kg (5.60 lb)/s	FT	1,041 mm (41.0 in)	557 mm (21.9 in)	114 kg (251 lb)	485 kW (650 shp)
250-C40B	C	2.77 kg (6.10 lb)/s	FT	1,041 mm (41 in)	557 mm (21.9 in)	127 kg (280 lb)	533 kW (715 shp)
T406	14A	16.1 kg (35.5 lb)/s	FT	1,958 mm (77.0 in)	671 mm (26.4 in)	440 kg (971 lb)	4,586 kW (6,150 shp)
RRTI (International)							
RTM 322-01/8	3A+C	5.75 kg (12.68 lb)/s	FT	1,171 mm (46.1 in)	604 mm (23.8 in)	240 kg (538 lb)	1,566 kW (2,100 shp)
RTM 322-01/9	3A+C	5.9 kg (13.0 lb)/s	FT	1,171 mm (46.1 in)	736 mm (28.98 in)	227 kg (500.4 lb)	1,735 kW (2,327 shp)
RTM322-01/12	3A+C	5.75 kg (12.68 lb)/s	FT	1,171 mm (46.1 in)	736 mm (28.98 in)	249 kg (549.0 lb)	1,566 kW (2,101 shp)
RTM322-02/8	3A+C	5.8 kg (12.79 lb)/s	FT	1,171 mm (46.1 in)	736 mm (28.98 in)	248 kg (546.7 lb)	1,670 kW (2,241 shp)
Soyuz (Russia)							
TV-O-100	2A+C	2.66 kg (5.86 lb)/s	FT	1,275 mm (50.2 in)	780 mm (30.7 in)	125 kg (276 lb)	537 kW (720 shp)
Turbomeca (France)							
Arriel 1	1A+C	not disclosed	SS	1,090 mm (42.9 in)	430 mm (16.9 in)	120 kg (265 lb)	551 kW (739 shp)
Arriel 2	1A+C	not disclosed	SS	—	—	—	632 kW (848 shp)
Arrius 1A	C	not disclosed	FT	782 mm (30.8 in)	360 mm (14.2 in)	87.0 kg (192 lb)	380 kW (509 shp)
Arrius 1A/1M	C	not disclosed	FT	793 mm (31.2 in)	367 mm (14.45 in)	101 kg (223 lb)	357 kW (479 shp)
Arrius 2B	C	not disclosed	FT	782 mm (30.8 in)	360 mm (14.2 in)	87.0 kg (192 lb)	519 kW (696 shp)
Arrius 2B1	C	not disclosed	FT	782 mm (30.8 in)	360 mm (14.2 in)	87.0 kg (192 lb)	559 kW (750 shp)
Arrius 2K1	C	not disclosed	—	968 mm (38.1 in)	470 mm (18.5 in)	115 kg (253.5 lb)	500 kW (670 shp)
Artouste III	A+C	4.31 kg (9.50 lb)/s	SS	1,815 mm (71.5 in)	507 mm (20.0 in)	178 kg (392 lb)	440 kW (590 shp)
Astazou XIVM	2A+C	2.49 kg (5.50 lb)/s	SS	1,470 mm (57.9 in)	460 mm (18.1 in)	160 kg (353 lb)	640 kW (858 shp)
Makila 1A1	3A+C	5.5 kg (12.1 lb)/s	FT	2,103 mm (82.8 in)	528 mm (20.8 in)	241 kg (531 lb)	1,357 kW (1,820 shp)
Makila 1A2	3A+C	5.5 kg (12.1 lb)/s	FT	2,117 mm (83.3 in)	498 mm (19.6 in)	247 kg (544 lb)	1,376 kW (1,845 shp)
TM 333	2A+C	—	FT	1,045 mm (41.1 in)	454 mm (17.9 in)	156 kg (345 lb)	747 kW (1,001 shp)
Turmo IIIC	1A+C	5.9 kg (13.0 lb)/s	FT	2,184 mm (85.5 in)	637 mm (25.1 in)	225 kg (496 lb)	1,104 kW (1,480 shp)
ZEF (China)							
WZ8: Turbomeca Arriel 1C made under licence							522 kW (700 shp)

UPDATED

Turbomeca Makila turboshaft

Turbomeca Arrius 2K turboshaft

AERO-ENGINES

Jet engines

Notes:
Arrangement: A = axial stages, C = centrifugal stages, F = fan stages, a/b = afterburner or augmentor
Air flow: Flow is through fan, where applicable, not core; BPR = bypass ratio; PF = propfan
<: Various ratings up to this maximum
[1] without reverser (other NK-8 family include it)
[2] measured over propfan blades
[3] has vectored nozzle(s)
W: Afterburning or augmented ('wet') thrust
D: Unaugmented ('dry') thrust of afterburning engine

Engine type	Arrangement	Air flow	BPR	Length	Diameter	Dry weight	T-O rating
Aviadvigatel (Russian Federation)							
D-20P	3F, 8A	113 kg (249 lb)/s	1.0	3,304 mm (130.0 in)	976 mm (38.3 in)	1,468 kg (3,236 lb)	53.0 kN (11,905 lb)
D-21A1	5F, 10A	153 kg (337 lb)/s	0.83	4,837 mm (190.4 in)	1,020 mm (40.2 in)	2,100 kg (4,630 lb)	52.3 kN (11,750 lb)
D-30 III	5F, 10A	127 kg (280 lb)/s	1.0	3,983 mm (156.8 in)	1,050 mm (41.3 in)	1,550 kg (3,417 lb)	66.7 kN (14,990 lb)
D-30F6	5F, 10A, a/b	150 kg (331 lb)/s	0.57	7,040 mm (277.2 in)	1,020 mm (40.2 in)	2,416 kg (5,326 lb)	186.1 kN (41,843 lb) W 93.2 kN (20,944 lb) D
D-30KU	3F, 11A	269 kg (593 lb)/s	2.42	5,700 mm (224.0 in)	1,560 mm (61.4 in)	2,668 kg (5,882 lb)	107.9 kN (24,250 lb)
D-30KU-90	3F, 13A	265 kg (584 lb)/s	2.44	5,700 mm (224.4 in)	1,560 mm (61.4 in)	2,400 kg (5,291 lb)	117.7 kN (26,455 lb)
PS-90A	2F+2A, 13A	470 kg (1,036 lb)/s	4.6	4,964 mm (195.4 in)	1,900 mm (74.8 in)	2,950 kg (6,503 lb)	156.9 kN (35,275 lb)
PS-90A10	1F, 12A	264 kg (582 lb)/s	3.76	4,280 mm (168.5 in)	1,400 mm (55.1 in)	1,900 kg (4,180 lb)	90.2 kN (20,283 lb)
PS-90A12	1F, 13A	370 kg (816 lb)/s	5.05	4,795 mm (188.8 in)	1,670 mm (65.8 in)	2,300 kg (5,071 lb)	117.7 kN (26,455 lb)
CFE (USA)							
738	1F, 5A+C	95.3 kg (210 lb)/s	5.3	2,514 mm (99.0 in)	1,092 mm (43.0 in)	601 kg (1,325 lb)	26.5 kN (5,957 lb)
CFM (International)							
CFM56-2B	1F+3A, 9A	370 kg (817 lb)/s	6.0	2,430 mm (95.7 in)	1,735 mm (68.3 in)	2,119 kg (4,671 lb)	97.9 kN (22,000 lb)
CFM56-3C	1F+3A, 9A	312 kg (688 lb)/s	5.0	2,360 mm (93.0 in)	1,524 mm (60.0 in)	1,951 kg (4,301 lb)	<104.5 kN (23,500 lb)
CFM56-5A	1F+3A, 9A	386 kg (852 lb)/s	6.0	2,422 mm (95.4 in)	1,735 mm (68.3 in)	2,257 kg (4,975 lb)	<117.9 kN (26,500 lb)
CFM56-5C	1F+4A, 9A	466 kg (1,027 lb)/s	6.6	2,616 mm (103.0 in)	1,836 mm (72.3 in)	2,492 kg (5,494 lb)	<151.3 kN (34,000 lb)
CFM56-7B	1F+4A, 9A	<355 kg (783 lb)/s	5.1	2,507 mm (98.7 in)	1,549 mm (61.0 in)	2,384 kg (5,256 lb)	<117.4 kN (26,400 lb)
CFM56-9	1F+2A, 9A	<282 kg (621 lb)/s	5.08	2,329 mm (91.7 in)	1,422 mm (56.0 in)	—	>82.3 kN (18,500 lb)
Engine Alliance (USA)							
GP7167	1F+3A, 9A	c907 kg (2,000 lb)/s	7	4,293 mm (169.0 in)	2,565 mm (101.0 in)	c5,216 kg (11,500 lb)	298 kN (67,000 lb)
GP7275	1F+4A, 9A	c1,179 kg (2,600 lb)/s	8	4,547 mm (179.0 in)	2,794 mm (110.0 in)	c6,033 kg (13,300 lb)	329 kN (74,000 lb)
Eurojet (International)							
EJ200	3F, 5A, a/b	<77.1 kg (170 lb)/s	c0.4	4,000 mm (157.0 in)	740 mm (29.0 in)	990 kg (2,183 lb)	c89.0 kN (20,000 lb) W c60 kN (13,500 lb) D
General Electric (USA)							
GE90-76B	1F+3A, 10A	1,361 kg (3,000 lb)/s	9.0	5,182 mm (204 in)	3,404 mm (134.0 in)	7,559 kg (16,664 lb)	340 kN (76,400 lb)
GE90-85B	1F+3A, 10A	1,415 kg (3,120 lb)/s	8.7	5,182 mm (204 in)	3,404 mm (134.0 in)	7,559 kg (16,664 lb)	377 kN (84,700 lb)
GE90-90B	1F+3A, 14A	1,449 kg (3,195 lb)/s	8.4	5,182 mm (204 in)	3,404 mm (134.0 in)	7,559 kg (16,664 lb)	401 kN (90,000 lb)
GE90-92B	1F+3A, 14A	1,461 kg (3,221 lb)/s	8.3	5,182 mm (204 in)	3,404 mm (134.0 in)	7,559 kg (16,664 lb)	409 kN (92,000 lb)
GE90-115B	—	—	—	—	—	c8,165 kg (18,000 lb)	c512 kN (115,000 lb)
CF6-50C2/E2	1F+3A, 14A	658 kg (1,450 lb)/s	4.3	4,394 mm (173.0 in)	2,195 mm (86.4 in)	3,956 kg (8,721 lb)	234 kN (52,500 lb)
CF6-80A2/A3	1F+3A, 14A	663 kg (1,460 lb)/s	4.6	3,998 mm (157.4 in)	2,195 mm (86.4 in)	3,819 kg (8,420 lb)	222.4 kN (50,000 lb)
CF6-80C2	1F+3A, 14A	802 kg (1,769 lb)/s	5.05	4,267 mm (168.0 in)	2,692 mm (106.0 in)	4,309 kg (9,499 lb)	<270 kN (60,690 lb)
CF6-80E1	1F+3A, 14A	874 kg (1,926 lb)/s	5.3	4,405 mm (173.5 in)	2,794 mm (110.0 in)	5,075 kg (11,189 lb)	<298 kN (66,870 lb)
CF34-3A	1F, 14A	151 kg (332 lb)/s	6.3	2,616 mm (103.0 in)	1,245 mm (49.0 in)	739 kg (1,625 lb)	<41.0 kN (9,220 lb)
CF34-8C1	1F, 10A	c181 kg (400 lb)/s	—	3,264 mm (128.5 in)	1,321 mm (52.0 in)	1,059 kg (2,335 lb)	56.38 kN (12,670 lb)
CF34-8D3	1F, 10A	c181 kg (400 lb)/s	—	3,264 mm (128.5 in)	1,321 mm (52.0 in)	1,120 kg (2,470 lb)	59.17 kN (13,300 lb)
CF34-8E	1F, 10A	—	—	—	—	1,120 kg (2,470 lb)	62.28 kN (14,000 lb)
CF34-10E	1F, 10A	—	5.3	—	—	—	82.26 kN (18,500 lb)
F101-102	2F, 9A, a/b	159.7 kg (352 lb)/s	2.01	4,590 mm (180.7 in)	1,402 mm (55.2 in)	2,023 kg (4,460 lb)	136.9 kN (30,780 lb) W
F110-100	3F, 9A, a/b	113 kg (250 lb)/s	0.87	4,630 mm (182.3 in)	1,181 mm (46.5 in)	1,492 kg (3,289 lb)	124.6 kN (28,000 lb) W 78.0 kN (17,530 lb) D
F110-129	3F, 9A, a/b	122 kg (270 lb)/s	0.76	4,620 mm (181.9 in)	1,181 mm (46.5 in)	1,791 kg (3,940 lb)	129.0 kN (29,000 lb) W 75.7 kN (17,000 lb) D
F110-400	3F, 9A, a/b	113 kg (250 lb)/s	0.87	4,620 mm (181.9 in)	1,181 mm (46.5 in)	1,599 kg (3,525 lb)	119.2 kN (26,080 lb) W 71.6 kN (16,800 lb) D
F118-100	3F, 9A	130+ kg (287+ lb)/s	—	2,553 mm (100.5 in)	1,181 mm (46.5 in)	1,452 kg (3,200 lb)	84.5 kN (19,000 lb)
F404-400	3F, 7A, a/b	64.4 kg (142 lb)/s	0.34	4,030 mm (158.8 in)	880 mm (34.8 in)	989 kg (2,180 lb)	71.2 kN (16,000 lb) W 48.9 kN (11,000 lb) D
F404-402	3F, 7A, a/b	66.2 kg (146 lb)/s	0.27	4,030 mm (158.8 in)	880 mm (34.8 in)	1,035 kg (2,282 lb)	78.7 kN (17,700 lb) W 53.2 kN (11,950 lb) D
F404-F1D2	3F, 7A	65 kg (143 lb)/s	—	—	880 mm (34.8 in)	—	46.8 kN (10,540 lb)
F414-400	3F, 7A, a/b	76.8 kg (169 lb)/s	—	—	—	—	97.9 kN (22,000 lb) W 55.6 kN (12,500 lb) D
GTRE (India)							
Kaveri	3F, 6A, a/b	—	—	—	—	—	81.0 kN (18,210 lb) W 52.0 kN (11,687 lb) D
Honeywell (USA)							
ALF502R-6	1F, 7A+C	87.0 kg (192 lb)/s	5.6	1,487 mm (58.6 in)	1,059 mm (41.7 in)	624 kg (1,375 lb)	33.4 kN (7,500 lb)
AS907	1F, 4A+C	—	—	—	—	—	28.9 kN (6,500 lb st)
AS977	1F, 4A+C	—	4.2	2,290 mm (90.0 in)	864 mm (34.0 in)	—	31.1 kN (7,000 lb)

CFM56-7 high-bypass turbofan

Saturn AL-31 *(Paul Jackson/Jane's)*

AERO-ENGINES

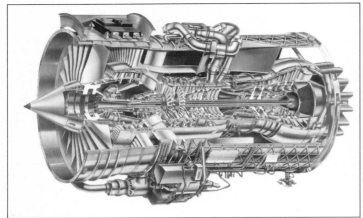

Cutaway of Rolls-Royce BR715

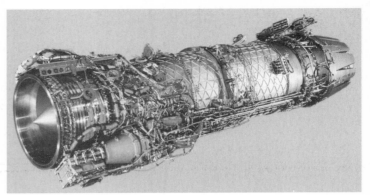

Eurojet EJ2000

Saturn AL-55 *(Paul Jackson/Jane's)*

Rolls-Royce Trent 8104 turbofan

Williams FJX-2 small turbofan *(Paul Jackson/Jane's)*

Engine type	Arrangement	Air flow	BPR	Length	Diameter	Dry weight	T-O rating
ATF3-6A	1F, 5A, C	73.5 kg (162 lb)/s	2.8	2,591 mm (102.0 in)	853 mm (33.6 in)	510 kg (1,125 lb)	24.2 kN (5,440 lb)
LF507	1F, 7A+C	87.0 kg (192 lb)/s	5.7	1,487 mm (58.6 in)	1,059 mm (41.7 in)	624 kg (1,375 lb)	31.1 kN (7,000 lb)
TFE731-2	1F, 4A+C	51.3 kg (113 lb)/s	2.66	1,520 mm (59.8 in)	716 mm (28.2 in)	337 kg (743 lb)	15.57 kN (3,500 lb)
TFE731-3	1F, 4A+C	53.7 kg (118 lb)/s	2.8	1,520 mm (59.8 in)	716 mm (28.2 in)	342 kg (754 lb)	16.46 kN (3,700 lb)
TFE731-5B	1F, 4A+C	64.9 kg (143 lb)/s	3.48	1,665 mm (65.5 in)	754 mm (29.7 in)	408 kg (899 lb)	21.13 kN (4,750 lb)
TFE731-20	1F, 4A+C	66.2 kg (146 lb)/s	3.1	1,547 mm (60.9 in)	716 mm (28.2 in)	406 kg (895 lb)	15.57 kN (3,500 lb)
TFE731-40	1F, 4A+C	65.8 kg (145 lb)/s	2.9	1,547 mm (60.9 in)	716 mm (28.2 in)	406 kg (895 lb)	18.91 kN (4,250 lb)
TFE731-60	1F, 4A+C	84.8 kg (187 lb)/s	3.9	2.083 mm (82.0 in)	781 mm (30.7 in)	448 kg (988 lb)	22.24 kN (5,000 lb)
TFE1042 (F124)	3F, 4A+C	42.7 kg (94.1 lb)/s	0.4	1,925 mm (75.8 in)	591 mm (23.25 in)	499 kg (1,100 lb)	28.0 kN (6,300 lb)
TFE1042-70 (F125)	3F, 4A+C, a/b	43.3 kg (95.4 lb)/s	0.4	3,561 mm (140.2 in)	591 mm (23.25 in)	617 kg (1,360 lb)	41.1 kN (9,250 lb) W 26.8 kN (6,025 lb) D
IAE (International)							
V2522-A5	1F+4A, 10A	384 kg (848 lb)/s	4.9	3,200 mm (126.0 in)	1,600 mm (63.0 in)	2,359 kg (5,200 lb)	97.9 kN (22,000 lb)
V2527-A5	1F+4A, 10A	384 kg (848 lb)/s	4.75	3,200 mm (126.0 in)	1,600 mm (63.0 in)	2,500 kg (5,511 lb)	117.9 kN (26,500 lb)
V2528-D5	1F+4A, 10A	384 kg (848 lb)/s	4.8	3,200 mm (126.0 in)	1,600 mm (63.0 in)	2,540 kg (5,600 lb)	124.6 kN (28,000 lb)
V2533-A5	1F+4A, 10A	384 kg (848 lb)/s	4.4	3,200 mm (126.0 in)	1,600 mm (63.0 in)	2,500 kg (5,511 lb)	146.8 kN (33,000 lb)
IHI (Japan)							
F3-30	2F, 5A	34.0 kg (75.0 lb)/s	0.9	1,340 mm (52.76 in)	560 mm (22.0 in)	340 kg (750 lb)	16.37 kN (3,680 lb)
IL (Poland)							
D-18A	2F, 5A	38.4 kg (84.66 lb)	0.7	1,940 mm (76.37 in)	750 mm (29.5 in)	380 kg (837.7 lb)	17.65 kN (3,968 lb)
K-15	6A	23.5 kg (51.8 lb)/s	0	1,560 mm (61.42 in)	725 mm (28.5 in)	320 kg (705.5 lb)	14.71 kN (3,307 lb)
SO-3	7A	—	0	2,151 mm (84.7 in)	707 mm (27.8 in)	321 kg (708 lb)	10.79 kN (2,425 lb)
KKBM (Russian Federation)							
NK-8-2U	2F+2A, 6A	228 kg (503 lb)/s	1.049	5,288 mm (206.2 in)	1,442 mm (56.8 in)	2,350 kg (5,180 lb)	103.0 kN (23,150 lb)
NK-8-4	2F+2A, 6A	222 kg (489 lb)/s	1.042	5,101 mm (201.0 in)	1,442 mm (56.8 in)	2,440 kg (5,379 lb)	103.0 kN (23,150 lb)
NK-86	2F+2A, 6A	292 kg (644 lb)/s	1.6	3,638 mm (143.0 in)¹	1,600 mm (63.0 in)	2,450 kg (5,401 lb)	127.5 kN (28,660 lb)
Klimov (Russian Federation)							
RD-33	4F, 9A, a/b	75.5 kg (166 lb)/s	0.49	4,229 mm (166.5 in)	1,040 mm (40.95 in)	1,055 kg (2,326 lb)	81.4 kN (18,300 lb) W 49.4 kN (11,110 lb) D
RD-35 – see PS/ZMK DV-2							
LM (China)							
WP6	9A, a/b	46.3 kg (102 lb)/s	0	5,483 mm (215.9 in)	668 mm (26.3 in)	725 kg (1,598 lb)	39.72 kN (8,929 lb) W 29.42 kN (6,614 lb) D
WP6A	9A, a/b	46.3 kg (102 lb)/s	0	5,483 mm (215.9 in)	668 mm (26.3 in)	725 kg (1,598 lb)	36.78 kN (8,267 lb) W
WS6	3F, 11A, a/b	155 kg (342 lb)/s	1.0	4,654 mm (183.2 in)	1,370 mm (53.94 in)	2,100 kg (4,630 lb)	122.1 kN (27,445 lb) W 71.1 kN (15,991 lb) D
LMC (China)							
WS6 uprated for JH-7							138.3 kN (31,085 lb) W 71.3 kN (16,027 lb) D

Jane's All the World's Aircraft 2001-2002

AERO-ENGINES

Engine type	Arrangement	Air flow	BPR	Length	Diameter	Dry weight	T-O rating
WP7B(BM)	3A+5A, a/b	65.0 kg (143 lb)/s	0	4,600 mm (181.1 in)	825 mm (32.5 in)	1,053 kg (2,321 lb)	59.8 kN (13,448 lb) W 43.2 kN (9,700 lb) D
WP7F	3A+5A, a/b	—	0	—	—	—	63.7 kN (14,330 lb) W 44.1 kN (9,921 lb) D
WP13	3A+5A, a/b	65.6 kg (145 lb)/s	0	4,600 mm (181.1 in)	907 mm (35.7 in)	1,211 kg (2,670 lb)	64.7 kN (14,550 lb) W 40.2 kN (9,039 lb) D
WP13A II	3A+5A, a/b	65.6 kg (145 lb)/s	0	5,150 mm (202.8 in)	907 mm (35.7 in)	1,201 kg (2,648 lb)	65.9 kN (14,815 lb) W 47.1 kN (10,582 lb) D
WP13B	3A+5A, a/b	—	0	—	—	—	68.7 kN (15,432 lb) W 47.1 kN (10,582 lb) D
WP13F	3A+5A, a/b	—	0	—	—	—	64.7 kN (14,550 lb) W 44.1 kN (9,921 lb) D
Pratt & Whitney (USA)							
F100-220	3F, 10A, a/b	—	0.7	5,280 mm (208.0 in)	1,181 mm (46.5 in)	1,451 kg (3,200 lb)	106.0 kN (23,830 lb) W 65.3 kN (14,670 lb) D
F100-220P	3F, 10A, a/b	—	0.6	5,280 mm (208.0 in)	1,181 mm (46.5 in)	1,526 kg (3,365 lb)	120.1 kN (27,000 lb) W 74.3 kN (16,700 lb) D
F100-229	3F, 10A, a/b	—	0.36	4,855 mm (191.15 in)	1,181 mm (46.5 in)	1,681 kg (3,705 lb)	129.5 kN (29,100 lb) W 79.2 kN (17,800 lb) D
F117-100	1F+4A, 12A	608 kg (1,340 lb)/s	6.0	3,729 mm (146.8 in)	2,154 mm (84.8 in)	3,220 kg (7,100 lb)	181.0 kN (40,700 lb)
F119-100	3F, 9A, a/b		0.45				c155.6 kN (35,000 lb) W
JSF-119	3F, 9A, a/b	—	—	—	—	—	c177.9 kN (40,000 lb)
J52-408	5A, 7A	64.9 kg (143 lb)/s	0	3,020 mm (118.9 in)	814 mm (32.1 in)	1,052 kg (2,318 lb)	49.8 kN (11,200 lb)
JT8D-219	1F+6A, 7A	221 kg (488 lb)/s	1.77	3,911 mm (154.0 in)	1,250 mm (49.2 in)	2,092 kg (4,612 lb)	93.4 kN (21,000 lb)
JT9D-7R4H	1F+4A, 11A	769 kg (1,695 lb)/s	4.8	3,371 mm (132.7 in)	2,463 mm (97.0 in)	4,029 kg (8,885 lb)	249 kN (56,000 lb)
PW2040	1F+4A, 12A	608 kg (1,340 lb)/s	6.0	3,729 mm (146.8 in)	2,154 mm (84.8 in)	3,311 kg (7,300 lb)	185.5 kN (41,700 lb)
PW4050	1F+4A, 11A	798 kg (1,759 lb)/s	5.1	3,371 mm (132.7 in)	2,463 mm (97.0 in)	4,179 kg (9,213 lb)	222 kN (50,000 lb)
PW4060	1F+4A, 11A	816 kg (1,800 lb)/s	4.8	3,371 mm (132.7 in)	2,463 mm (97.0 in)	4,179 kg (9,213 lb)	267 kN (60,000 lb)
PW4168	1F+5A, 11A	903 kg (1,990 lb)/s	5.1	4,143 mm (163.1 in)	2,715 mm (106.9 in)	6,396 kg (14,100 lb)	305 kN (68,600 lb)
PW4084	1F+6A, 11A	1,089 kg (2,400 lb)/s	6.4	4,869 mm (191.7 in)	3,010 mm (118.5 in)	6,768 kg (14,920 lb)	386 kN (86,760 lb)
PW4098	1F+7A, 11A	1,293 kg (2,850 lb)/s	5.8	4,945 mm (194.7 in)	3,035 mm (119.5 in)	7,484 kg (16,500 lb)	436 kN (98,000 lb)
PW6000	1F+4A, 5A	—	4.9	2,743 mm (108 in)	1,524 mm (60.0 in)	2,313 kg (5,100 lb)	<102.3 kN (23,000 lb)
PW8000	1F+3A, 5A	—	10.0	3,150 mm (124 in)	1,930 mm (76.0 in)	3,629 kg (8,000 lb)	<155.7 kN (35,000 lb)
PW7000 (XTE-66)	—	—	—	—	—	—	<155.7 kN (35,000 lb)
Pratt & Whitney Canada (Canada)							
JT15D-1B	1F+C	34.0 kg (75.0 lb)/s	3.3	1,506 mm (59.3 in)	691 mm (27.2 in)	235 kg (519 lb)	9.79 kN (2,200 lb)
JT15D-5	1F+1A, C	42.0 kg (92.6 lb)/s	—	1,549 mm (61.0 in)	711 mm (28.0 in)	288 kg (635 lb)	12.9 kN (2,900 lb)
PW305	1F, 4A+C	—	4.3	2,070 mm (81.5 in)	970 mm (38.2 in)	450 kg (993 lb)	23.4 kN (5,266 lb)
PW306	1F, 4A+C	—	4.5	1,920 mm (75.6 in)	970 mm (38.2 in)	473 kg (1,043 lb)	25.4 kN (5,700 lb)
PW308	1F, 4A+C	—	4.5	—	—	—	35.6 kN (8,000 lb)
PW308A	1F, 4A+C	—	3.8	—	—	—	28.9 kN (6,500 lb)
PW530A	1F, 2A+C	—	3.9	1,524 mm (60.0 in)	701 mm (27.6 in)	278 kg (613 lb)	13.34 kN (3,000 lb)
PW535A	1F, 2A+C	—	3.9	1,524 mm (60.0 in)	701 mm (27.6 in)	—	14.94 kN (3,360 lb)
PW545A	1F, 2A+C	—	4.0	1,727 mm (68.0 in)	813 mm (32.0 in)	347 kg (765 lb)	17.24 kN (3,876 lb)
Progress (ZMKB) (Ukraine)							
AI-22 (DV-22)	1F+2A, 7A	140 kg (309 lb)/s	c5	2,400 mm (94.5 in)	—	700 kg (1,543 lb)	37.95 kN (8,532 lb)
AI-222-25	1F, 7A	c40 kg (88 lb)/s	1.19	1,960 mm (77.2 in)	630 mm (24.8 in)	440 kg (970 lb)	24.51 kN (5,511 lb)
AI-25TL	3F, 7A	46.8 kg (103 lb)/s	2.0	1,993 mm (78.46 in)	820 mm (32.3 in)	350 kg (772 lb)	16.87 kN (3,792 lb)
D-18T	1F, 7A, 7A	765 kg (1,687 lb)/s	5.6	5,400 mm (212.6 in)	2,330 mm (91.7 in)	4,100 kg (9,039 lb)	230 kN (51,660 lb)
D-18TM	1F, 7A, 7A	770 kg (1,698 lb)/s	5.5	5,700 mm (224.5 in)	2,330 mm (91.7 in)	4,750 kg (10,472 lb)	248 kN (55,777 lb)
D-27 propfan	2PF, 5A, 2A+C	27.4 kg (60.4 lb)/s	29.95	—	—	1,650 kg (3,638 lb)	109.8 kN (24,690 lb)
D-36	1F, 6A, 7A	255 kg (562 lb)/s	6.3	3,470 mm (136.6 in)	1,373 mm (54.1 in)	1,109 kg (2,445 lb)	63.7 kN (14,330 lb)
D-436T	1F, 6A, 7A	265 kg (584 lb)/s	6.0	3,470 mm (136.6 in)	1,373 mm (54.1 in)	1,250 kg (2,756 lb)	73.5 kN (16,535 lb)
D-436T2	1F+A, 6A, 7A	265 kg (584 lb)/s	4.9	3,600 mm (141.7 in)	1,373 mm (54.1 in)	1,450 kg (3,197 lb)	80.4 kN (18,078 lb)
PS/ZMK (International)							
DV-2	1F+2A, 7A	49.5 kg (109 lb)/s	1.46	1,721 mm (67.75 in)	994 mm (39.1 in)	440 kg (970 lb)	21.58 kN (4,852 lb)
PW/MTU (International)							
MTFE	1F+2A, 6A	c370 kg (c815 lb)/s	5.2	2,515 mm (99.0 in)	1,397 mm (55.0 in)	1,769 kg (3,900 lb)	89.0 kN (20,000 lb)
RKBM (Russian Federation)							
RD-41	7A	53.5 kg (118 lb)/s	0	1,594 mm (62.75 in)	635 mm (25.0 in)	290 kg (639 lb)	40.2 kN (9,040 lb)
Rolls-Royce (Germany)							
BR710	1F, 10A	197 kg (435 lb)/s	4.2	3,409 mm (134.0 in)	1,321 mm (52.0 in)	1,633 kg (3,600 lb)	<68.9 kN (15,500 lb)
BR715	1F+2A, 10A	288 kg (636 lb)/s	4.5	3,599 mm (142.0 in)	1,575 mm (62.0 in)	2,114 kg (4,660 lb)	<102.3 kN (23,000 lb)
Rolls-Royce (UK)							
535C	1F, 6A, 6A	518 kg (1,142 lb)/s	4.4	3,010 mm (118.5 in)	1,877 mm (73.9 in)	3,309 kg (7,294 lb)	166.4 kN (37,400 lb)
535E4B	1F, 6A, 6A	522 kg (1,150 lb)/s	4.3	2,995 mm (117.9 in)	1,892 mm (74.5 in)	3,261 kg (7,189 lb)	<191.7 kN (43,100 lb)
Pegasus 11-61	3F, 8A[3]	208 kg (459 lb)/s	1.2	3,485 mm (137.2 in)	1,222 mm (48.1 in)	1,932 kg (4,260 lb)	105.9 kN (23,800 lb)
RB211-22B	1F, 7A, 6A	626 kg (1,380 lb)/s	5.0	3,033 mm (119.4 in)	2,154 mm (84.8 in)	4,171 kg (9,195 lb)	186.8 kN (42,000 lb)
RB211-524B	1F, 7A, 6A	671 kg (1,480 lb)/s	4.4	3,106 mm (122.3 in)	2,180 mm (85.8 in)	4,452 kg (9,814 lb)	222 kN (50,000 lb)
RB211-524G/H	1F, 7A, 6A	728 kg (1,604 lb)/s	4.3	3,175 mm (125.0 in)	2,192 mm (86.3 in)	4,387 kg (9,671 lb)	273 kN (60,600 lb)
Spey 512	5A, 12A	94.3 kg (208 lb)/s	0.71	2,911 mm (114.6 in)	942 mm (37.1 in)	1,168 kg (2,574 lb)	55.8 kN (12,550 lb)
Spey 807	5A, 12A	91.6 kg (202 lb)/s	0.96	2,456 mm (96.7 in)	825 mm (32.5 in)	1,096 kg (2,417 lb)	49.1 kN (11,030 lb)
Tay 611, 620	1F+3A, 12A	176 kg (388 lb)/s	3.18	2,405 mm (94.7 in)	1,118 mm (44.0 in)	1,422 kg (3,135 lb)	61.6 kN (13,850 lb)
Tay 651	1F+3A, 12A	193 kg (425.5 lb)/s	3.10	2,405 mm (94.7 in)	1,138 mm (44.8 in)	1,533 kg (3,380 lb)	68.5 kN (15,400 lb)
Trent 556	1F, 8A, 6A	858 kg (1,892 lb)/s	8.5	3,912 mm (154.0 in)	2,474 mm (97.4 in)	4,719 kg (10,400 lb)	252 kN (56,000 lb)
Trent 600	1F, 8A, 6A	922 kg (2,032 lb)/s	8.0	3,912 mm (154.0 in)	2,474 mm (97.4 in)	4,719 kg (10,400 lb)	306 kN (68,000 lb)
Trent 772	1F, 8A, 6A	897 kg (1,978 lb)/s	4.9	3,912 mm (154.0 in)	2,474 mm (97.4 in)	4,785 kg (10,550 lb)	320 kN (71,100 lb)
Trent 892	1F, 8A, 6A	1,200 kg (2,645 lb)/s	5.7	4,369 mm (172.0 in)	2,794 mm (110.0 in)	5,957 kg (13,133 lb)	411 kN (91,300 lb)
Trent 895	1F, 8A, 6A	1,217 kg (2,684 lb)/s	5.79	4,369 mm (172 in)	2,794 mm (110.0 in)	5,981 kg (13,186 lb)	425 kN (95,500 lb)
Trent 976	1F, 8A, 6A	1,179 kg (2,600 lb)/s	8.0	4,369 mm (172.0 in)	2,794 mm (110.0 in)	5,806 kg (12,800 lb)	338 kN (76,000 lb)
Viper 535	8A	23.9 kg (52.7 lb)/s	0	1,806 mm (71.1 in)	740 mm (29.1 in)	358 kg (790 lb)	14.94 kN (3,360 lb)
Viper 632	8A	26.5 kg (58.4 lb)/s	0	1,806 mm (71.1 in)	740 mm (29.1 in)	376.5 kg (830 lb)	17.66 kN (3,970 lb)
Viper 680	8A	27.2 kg (60.0 lb)/s	0	1,806 mm (71.1 in)	740 mm (29.1 in)	379 kg (836 lb)	19.39 kN (4,360 lb)
Rolls-Royce (USA)							
AE 3007A	1F, 14A	96.6 kg (213 lb)/s	5.0	2,705 mm (106.5 in)	1,105 mm (43.5 in)	717.1 kg (1,581 lb)	33.7 kN (7,580 lb)
AE 3007H	1F, 14A	118 kg (260 lb)/s	5.0	2,705 mm (106.5 in)	1,105 mm (43.5 in)	717.1 kg (1,581 lb)	36.9 kN (8,290 lb)
RRTI (International)							
Adour 811/815	2F, 5A, a/b	43.1 kg (95.0 lb)/s	0.75	2,970 mm (116.9 in)	559 mm (22.0 in)	738 kg (1,627 lb)	37.4 kN (8,400 lb) W 24.6 kN (5,520 lb) D
Adour 871	2F, 5A	44.0 kg (97.0 lb)/s	0.80	1,956 mm (77.0 in)	559 mm (22.0 in)	603 kg (1,330 lb)	26.6 kN (5,990 lb)
Samara (Russian Federation)							
NK-25	3F, 9A	339 kg (747 lb)/s	0.9	5,200 mm (205 in)	1,348 mm (53.1 in)	2,850 kg (6,283 lb)	245 kN (55,115 lb)
NK-44	—	—	6.0	—	3,100 mm (122.0 in)	—	392 kN (88,180 lb)
NK-93	2PF, 7A, 8A	1,000 kg (2,205 lb)/s	17.0	5,972 mm (235.1 in)	2,900 mm (114.2 in)[2]	3,650 kg (8,047 lb)	176.5 kN (39,683 lb)
NK-321	3F, 5A, 7A, a/b	365 kg (805 lb)/s	1.4	c6,000 mm (236.0 in)	1,460 mm (57.5 in)	3,442 kg (7,588 lb)	245 kN (55,077 lb) W 137.2 kN (30,843 lb) D

AERO-ENGINES

Engine type	Arrangement	Air flow	BPR	Length	Diameter	Dry weight	T-O rating
Saturn (Russian Federation)							
AL-21F3	14A, a/b	104 kg (229 lb)/s	—	5,340 mm (210.0 in)	1,030 mm (40.6 in)	1,800 kg (3,968 lb)	110.5 kN (24,800 lb) W 76.5 kN (17,200 lb) D
AL-31FM	4F, 9A, a/b	110 kg (243 lb)/s	0.57	4,950 mm (195.0 in)	1,277 mm (50.3 in)	1,520 kg (3,351 lb)	122.6 kN (27,560 lb) W 79.3 kN (17,857 lb) D
AL-37FU	4F, 9A, a/b	—	0.65	c5,000 mm (c197 in)	1,277 mm (50.3 in)	1,660 kg (3,660 lb)	142.2 kN (31,967 lb) W 83.4 kN (18,740 lb) D
AL-41 (SAT-41)	2F, 6A, a/b	c150 kg (331 lb)/s	—	—	—	c1,850 kg (c3,4078 lb)	c196.1 kN (c44,090 lb) W c113.8 kN (25,600 lb) D
AL-55	4F, 6A	29.8 kg (65.7 lb)/s	0.6	1,210 mm (47.6 in)	590 mm (23.2 in)	315 kg (694 lb)	21.58 kN (4,850 lb)
AL-55F	4F, 6A, a/b	29.8 kg (65.7 lb)/s	0.6	2,520 mm (99.2 in)	590 mm (23.2 in)	385 kg (849 lb)	34.34 kN (7,716 lb)
SNECMA (France)							
Atar 9K50	9A, a/b	73.0 kg (161 lb)/s	0	5,944 mm (234.0 in)	1,020 mm (40.2 in)	1,582 kg (3,487 lb)	70.6 kN (15,870 lb) W 49.2 kN (11,055 lb) D
M53 P2	3F, 5A, a/b	86.0 kg (190 lb)/s	0.35	5,070 mm (199.6 in)	1,055 mm (41.5 in)	1,500 kg (3,307 lb)	95.0 kN (21,355 lb) W 64.3 kN (14,455 lb) D
M88-2	3F, 6A, a/b	67.0 kg (148 lb)/s	0.25	3,540 mm (139.0 in)	660 mm (26.0 in)	909 kg (2,004 lb)	75.0 kN (16,872 lb) W 50.0 kN (11,250 lb) D
SNECMA-Pratt & Whitney Canada Joint Venture (International)							
SPW14	1F, 4A+C	—	c6	2,540 mm (100.0 in)	1,321 mm (50.0 in)	c1,361 kg (3,000 lb)	62.0 kN (13,940 lb)
Soyuz (Russian Federation)							
R-195	3A	67.0 kg (148 lb)/s	—	3,300 mm (130.0 in)	914 mm (36.0 in)	990 kg (2,183 lb)	44.13 kN (9,921 lb)
R-29-300	5A, 6A, a/b	110 kg (243 lb)/s	—	4,960 mm (195.0 in)	912 mm (35.9 in)	1,880 kg (4,145 lb)	122.3 kN (27,500 lb) W 78.5 kN (17,635 lb) D
R-35	5A, 6A, a/b	110 kg (243 lb)/s	—	4,950 mm (194.9 in)	912 mm (35.9 in)	1,765 kg (3,891 lb)	127.5 kN (28,660 lb) W 83.9 kN (18,850 lb) D
R-79	5F, 6A, a/b[3]	120 kg (265 lb)/s	1.0	5,229 mm (205.9 in)	1,100 mm (43.3 in)	2,750 kg (6,063 lb)	152.0 kN (34,170 lb) W 107.6 kN (24,200 lb) D
RD-1700	2F, 4A	—	0.78	—	—	298 kg (657 lb)	16.67 kN (3,748 lb)
Turbomeca-SNECMA (France)							
Larzac 04-20	2F, 4A	28.6 kg (63.0 lb)/s	1.04	1,179 mm (46.4 in)	602 mm (23.7 in)	302 kg (666 lb)	14.12 kN (3,175 lb)
Turbo-Union (International)							
RB199 Mk 105	3F, 3A, 6A, a/b	74.6 kg (164 lb)/s	c1	3,302 mm (130.0 in)	752 mm (29.6 in)	980 kg (2,160 lb)	74.7 kN (16,800 lb) W 42.5 kN (9,550 lb) D
Volvo/GE (Sweden)							
RM12	3F, 7A, a/b	68.0 kg (150 lb)/s	0.28	4,100 mm (161.4 in)	880 mm (34.8 in)	1,050 kg (2,315 lb)	80.5 kN (18,100 lb) W 54.0 kN (12,140 lb) D
Williams and Williams Rolls (USA)							
FJ33-1	1F+2A, C	—	—	1,216 mm (47.9 in)	439.4 mm (17.3 in)	<136 kg (300 lb)	5.34 kN (1,200 lb)
FJ44-1A	1F+1A, C	29.575 kg (65.2 lb)/s	3.28	1,356 mm (53.4 in)	530.9 mm (20.9 in)	205 kg (452 lb)	8.45 kN (1,900 lb)
FJ44-1C	1F+1A, C	26.49 kg (58.4 lb)/s	3.28	1,356 mm (53.4 in)	530.9 mm (20.9 in)	208 kg (459 lb)	6.67 kN (1,500 lb)
FJ44-2A	1F+3A, C	35.8 kg (79.0 lb)/s	4.1	1,507 mm (59.4 in)	553.7 mm (21.8 in)	—	10.23 kN (2,300 lb)
FJ44-2C	1F+3A, C	—	—	1,507 mm (59.4 in)	553.7 mm (21.8 in)	—	10.68 kN (2,400 lb)
FJX	1F+1A, C	—	—	1,041 mm (41.0 in)	368.3 mm (14.5 in)	<45 kg (100 lb)	3.11 kN (700 lb)
XAE (China)							
WS9: Rolls-Royce Spey Mk 202 made under licence; data presumed identical							
WP8	8A	150 kg (331 lb)/s	—	5,380 mm (211.8 in)	1,400 mm (55.1 in)	3,133 kg (6,906 lb)	93.2 kN (20,944 lb)

UPDATED

AERO-ENGINES

Auxiliary power units (APUs)

APU type	Start-up ceiling	Dry weight	S/L compression	Ground power	Rating
Aerosila (Russia)					
TA-4FE	<2,500 m (8,200 ft)	180 kg (397 lb)	2.4 bars (35 lb/sq in)	18 kVA	124 kW
TA-6A	<3,000 m (9,840 ft)	245 kg (540 lb)	4.4 bars (64 lb/sq in)	40 kVA	235 kW
TA-6V	<3,000 m (9,840 ft)	255 kg (562 lb)	4.4 bars (64 lb/sq in)	40 kVA	235 kW
TA-8	<3,000 m (9,840 ft)	165 kg (364 lb)	3.2 bars (46 lb/sq in)	—	107 kW
TA-8V	<5,000 m (16,400 ft)	185 kg (408 lb)	3.2 bars (46 lb/sq in)	40 kVA	107 kW
TA-8K	<3,500 m (11,480 ft)	178 kg (392 lb)	3.1 bars (45 lb/sq in)	60 kVA	55 kW
TA-12	<7,000 m (22,960 ft)	290 kg (639 lb)	4.8 bars (70 lb/sq in)	40 kVA	287 kW
TA-12-60	<7,000 m (22,960 ft)	335 kg (739 lb)	4.8 bars (70 lb/sq in)	60 kVA	287 kW
TA-14	<8,000 m (26,250 ft)	50 kg (110 lb)	3.6 bars (52 lb/sq in)	20 kVA	79 kW
TA-18-100	<11,000 m (36,090 ft)	140 kg (309 lb)	4.4 bars (64 lb/sq in)	60 kVA	214 kW
Honeywell (USA)					
36-100	6,095 m (20,000 ft)	53 kg (117 lb)	3.52 bars (51 lb/sq in)	30 kVA	127 kW
36-150W	6,095 m (20,000 ft)	52-55 kg (115-121 lb)	3.24 bars (47 lb/sq in)	10 kVA	142 kW
36-150CX/IAI	6,095 m (20,000 ft)	57 kg (126 lb)	3.38 bars (49 lb/sq in)	10 kVA	142 kW
36-150F2	6,095 m (20,000 ft)	55 kg (121 lb)	3.17 bars (46 lb/sq in)	10 kVA	142 kW
36-150F	6,095 m (20,000 ft)	53 kg (117 lb)	3.24 bars (47 lb/sq in)	10 kVA	142 kW
36-150DD	6,095 m (20,000 ft)	53 kg (117 lb)	3.24 bars (47 lb/sq in)	10 kVA	145 kW
36-150 M	6,095 m (20,000 ft)	65 kg (143 lb)	3.59 bars (52 lb/sq in)	23 kVA	145 kW
36-150RJ	11,280 m (37,000 ft)	73 kg (160 lb)	3.11 bars (45 lb/sq in)	30 kVA	179 kW
RE100	6,095 m (20,000 ft)	40 kg (88 lb)	3.45 bars (50 lb/sq in)	10 kVA	101 kW
RE220	13,105 m (43,000 ft)	109 kg (240 lb)	3.86 bars (56 lb/sq in)	45 kVA	261 kW
36-280	10,670 m (35,000 ft)	159 kg (350 lb)	3.24 bars (47 lb/sq in)	90 kVA	291 kW
36-300	10,670 m (35,000 ft)	136 kg (300 lb)	3.24 bars (47 lb/sq in)	90 kVA	291 kW
131-9A	12,495 m (41,000 ft)	159 kg (350 lb)	3.52 bars (51 lb/sq in)	90 kVA	343 kW
131-9B	12,495 m (41,000 ft)	186 kg (410 lb)	3.52 bars (51 lb/sq in)	96 kVA	90 kW
331-200	13,105 m (43,000 ft)	227 kg (500 lb)	3.45 bars (50 lb/sq in)	90 kVA	432 kW
331-250	13,105 m (43,000 ft)	227 kg (500 lb)	3.04 bars (44 lb/sq in)	90 kVA	447 kW
331-350	12,495 m (41,000 ft)	254 kg (560 lb)	3.45 bars (50 lb sq in)	115 kVA	618 kW
331-500	13,105 m (43,000 ft)	331 kg (730 lb)	3.65 bars (53 lb/sq in)	120 kVA	969 kW
700	7,620 m (25,000 ft)	295 kg (650 lb)	3.05 bars (44 lb/sq in)	100 kVA	648 kW
Microturbo (France)					
Saphir 20	6,095 m (20,000 ft)	38 kg (84 lb)		10 kVA	70 kW
Omsk (Russian Federation)					
VSU-10					
Pratt & Whitney (Canada)					
PW901	7,620 m (25,000 ft)	379 kg (835 lb)	3.72 bars (54 lb/sq in)	180 kVA	1,145 kW
ZMKB Progress (design), **Motor Sich** (manufacture) (Ukraine)					
AI-8	7,620 m (25,000 ft)	145 kg (320 lb)	—	—	60 kW
AI-9	—	45 kg (99 lb)	2.35 bars (34 lb/sq in)		
AI-9K	—	52 kg (115 lb)	2.35 bars (34 lb/sq in)		
AI-9V	—	70 kg (154 lb)	2.89 bars (41 lb/sq in)		
AI-9-3B	—	128 kg (282 lb)	3.92 bars (57 lb/sq in)	—	29 kW
Stupino (Russian Federation)					
TA-6					
TA-12					
Sundstrand (USA)					
APS500	<9,145 m (<30,000 ft)	46-52 kg (100-115 lb)	c3.6 bars (c52 lb/sq in)	19 kVA	200 kW
APS1000	<7,620 m (<25,000 ft)	70-82 kg (155-180 lb)	4.14 bars (60 lb/sq in)	45 kVA	261 kW
APS2000	11,280 m (37,000 ft)	132 kg (290 lb)	3.65 bars (53 lb/sq in)	60 kVA	376 kW
APS2100	11,280 m (37,000 ft)	122 kg (270 lb)	3.65 bars (53 lb/sq in)	60 kVA	376 kW
APS3200	11,885 m (39,000 ft)	138 kg (305 lb)	4.14 bars (60 lb/sq in)	90 kVA	450 kW
Ufa (Russian Federation)					
VD-100					

UPDATED

Honeywell 131-9 auxiliary power unit

INDEXES

AIRCRAFT

To help users of this title evaluate the published data, *Jane's Information Group* has divided entries into three categories.

- **N** NEW ENTRY — Information on new equipment and/or systems appearing for the first time in the title.
- **V** VERIFIED — The editor has made a detailed examination of the entry's content and checked its relevancy and accuracy for publication in the new edition to the best of his ability.
- **U** UPDATED — During the verification process, significant changes to content have been made to reflect the latest position known to *Jane's* at the time of publication.

Items in italics refer to entries which have been deleted from this edition with the relevant page numbers from last year.

AIRCRAFT APPEARING IN THIS EDITION ONLY. For those appearing in previous 10 editions, please see following index on tinted paper.

Index begins with aircraft generally identified only by numbers. Most were previously listed as 'Model 707' and so on under M

Entry	Page	Status
1-04 Baby Huey (Hillberg)	663	V
1S-2 (Il)	319	U
1-23 Manager (Il)	318	U
1-25 AS	319	U
1.42/44 MFI (MiG)	398	U
2+2 Sportsman (Wag-Aero)	790	U
3xtrim (Zaklady Lotnicze)	314	N
7ACA Champ (American Champion)	551	N
7ECA Citabria Aurora (American Champion)	551	U
7GCAA Citabria Adventure (American Champion)	552	U
7GCBC Citabria Explorer (American Champion)	552	U
8F (Renaissance)	749	U
8GCBC Scout (American Champion)	552	U
8KCAB Super Decathlon (American Champion)	553	U
11E Spartan (Luscombe)	698	U
19-25 Skyrocket III (Aviabellanca)	556	U
20XX (Boeing)	617	N
31A (Learjet)	671	U
45 (Learjet)	672	U
58 Baron (Raytheon)	738	U
60 (Learjet)	673	U
100 B Individual Lifting Vehicle (PAM)	718	U
114 (Boeing)	590	U
115 Commander (Commander)	644	N
172 Skyhawk (Cessna)	630	U
182S Skylane (Cessna)	632	U
200 (Extra)	148	U
200 King Air, Military versions (Raytheon)	741	U
206B-3 JetRanger III (Bell)	28	U
206H Stationair (Cessna)	633	U
206-L4 LongRanger IV (Bell)	29	U
208 Caravan I/U-27A (Cessna)	634	U
209 (Bell)	566	U
228 (HAL)	174	U
269, Model (Schweizer)	757	U
280 (Enstrom)	649	U
281 Proteus (Scaled Composites)	754	U
300 (Extra)	148	U
300C (Schweizer)	757	U
328JET (Fairchild Dornier)	150	U
330 (Extra)	148	U
330SP (Schweizer)	758	U
333 (Dragon Fly)	278	U
350 King Air (Raytheon)	742	U
358 (Gavilán)	85	U
382 U/V Hercules (Lockheed Martin)	682	U
400 (Extra)	149	U
400A Beechjet (Raytheon)	744	U
406 (Bell)	568	U
407 (Bell)	30	U
412 (Agusta)	275	U
412 (Bell)	31	U
414 (Boeing)	590	U
415 SuperScooper (Canadair)	45	U
427 (Bell)	30	U
430 (Bell)	32	U
480 (Enstrom)	649	U
508T (Gavilán)	85	U
525 Citation CJ1 (Cessna)	635	U
525A Citation CJ2 (Cessna)	636	U
528JET (Fairchild Dornier)	151	U
550 Citation Bravo (Cessna)	637	U
560 Citation Ultra (Cessna)	638	U
560XL Citation Excel (Cessna)	639	U
560 Citation Encore (Cessna)	638	U
650 Citation VII (Cessna)	640	U
660 Turbo-Thrush (Ayres)	563	U
665 Tiger/Tigre (Eurocopter)	223	U
680 Citation Sovereign (Cessna)	641	U
717-200 (Boeing)	619	U
728JET (Fairchild Dornier)	151	U
737 (Boeing)	599	U
737 (AEW&C) (Boeing)	587	U
737 MMMA (Boeing)	588	N
737-600 (Boeing)	599	U
737-700 (Boeing)	599	U
737-700 IGW (Boeing)	599	U
737-800 (Boeing)	599	U
737-900 (Boeing)	599	U
747-400 (Boeing)	602	U
747-400F (Boeing)	606	U
750 Citation X (Cessna)	641	U
757 (Boeing)	606	U
757-200 (Boeing)	606	U
757-300 (Boeing)	609	U
763 (Boeing)	610	U
767 (Boeing)	610	U
767-200 (Boeing)	610	U
767-300 (Boeing)	610	U
767-300 General Market Freighter (Boeing)	614	U
767-400 (Boeing)	611	U
767 AWACS (Boeing)	588	U
767 Military versions (Boeing)	588	U
777 (Boeing)	614	U
800XP Hawker (Raytheon)	745	U
901 Osprey (Bell Boeing)	570	U
928JET (Fairchild Dornier)	151	U
1125 Astra SPX (IAI)	262	U
1126 Galaxy (IAI)	264	U
1900D (Raytheon)	743	U
2000 (RAF)	58	U
2061 IHSRC (IAIO)	260	U
9000T (Avtekair)	561	U

A

Entry	Page	Status
A-1 (AMX)	207	U
A-1 Husky (Aviat)	557	U
A-1B (AMX)	207	U
A-5 (NAMC)	73	U
A-16, SKB-1 (SGAU)	425	U
A-29 (EMBRAER)	17	U
A-36 Bonanza (Raytheon)	736	U
A-50 Golden Eagle (KAI)	305	U
A-50 Mainstay (Beriev)	351	U
A 109 (Agusta)	273	U
A 119 Koala (Agusta)	274	U
A 129 International (Agusta)	272	U
A 129 Mangusta (Agusta)	271	U
A 200 King Air (Raytheon)	741	U
A 210 (Aquila)	144	U
A300-600 (Airbus)	182	U
A300-600 Convertible (Airbus)	184	U
A300-600 Freighter (Airbus)	184	U
A300-608ST Super Transporter (SATIC)	253	U
A310 (Airbus)	184	U
A310 AEW&C (Airbus)	203	U
A318 (Airbus)	189	U
A319 (Airbus)	190	U
A320 (Airbus)	186	U
A321 (Airbus)	191	U
A321 AGS (Airbus)	203	N
A330 (Airbus)	196	U
A340 (Airbus)	193	U
A380 (Airbus)	198	U
A400 Future Large Aircraft (Airbus)	200	U
AASI (see Advanced Aerodynamics and Structures Inc)	539	U
AAT (see Ameur Aviation Technologie)	108	U
AB 139 (Bell/Agusta)	209	U
ABS RF-9 (ABS)	481	U
AC-500 Aircar (NLA)	76	U
ACJ (Airbus)	190	U
AD-100 (NAI)	71	U
AD-200 (BKLAIC)	61	U
ADA (see Aeronautical Development Agency)	168	U
AD&D (see Aero-Design and Development)	260	U
Ae 270 Ibis (Ibis)	248	U
AERO-M (see Aktsionernoe Obshchestvo Aero-M)	342	U
AF-18A (Boeing)	577	U
AH-1W SuperCobra (Bell)	566	U
AH-1Z SuperCobra (Bell)	566	U
AH-1Z KingCobra (TAI)	494	N
AH-2A Rooivalk (Denel)	473	U
AH-6 (MD)	703	U
AH-60L (Sikorsky)	761	U
AH-64 Apache (Boeing)	594	U
AH-64 A and D Apache (Boeing)	594	U
AH-X (JGSDF)	293	N
AH-X (KAI)	306	N
AIDC (see Aerospace Industrial Development Corporation)	492	U
AII (see Aviation Industries of Iran)	257	U
AL-1A (Boeing)	589, 606	U
AIRTECH (see Aircraft Technology Industries)	203	U
ALBATROS (see Tsentr Nauchno-Tekhnicheskogo Tvorchestva Albatros)	346	V
AMD (see Aircraft Manufacturing & Development)	551	V
AMT-600 Guri (Aeromot)	16	N
AMX (AMX)	207	U
An-2 Colt (Antonov)	498	U
An-3 (Antonov)	498	U
An-12 Cub (Antonov)	498	U
An-24 Coke (Antonov)	498	U
An-26 Curl (Antonov)	498	U
An-28 (see PZL) (Antonov)	499	U
An-32 Cline (Antonov)	499	U
An-38 (Antonov)	500	U
An-70 (Antonov)	501	U
An-72 Coaler (Antonov)	502	U
An-72P Coaler (Antonov)	504	U
An-74 Coaler (Antonov)	502	U
An-74-300 (Antonov)	505	U
An-124 Condor (Antonov)	505	U
An-140 (Antonov)	507	U
An-174 (Antonov)	505	U
An-225 Mriya (Antonov)	509	U
AP.68TP-600 Viator (Vulcanair)	290	U
APM-20 Lionceau (Issoire)	134	U
APM-21 Lion (Issoire)	134	U
ARL-1 (CSIST)	493	U
AS-2 (Albatros)	346	U
AS-3 (Albatros)	346	U
AS 202 Bravo (FFA Bravo)	483	N
AS 332 Super Puma Mk I (Eurocopter)	226	U
AS 332L2 Super Puma Mk II (Eurocopter)	227	U
AS 350 Ecureuil/AStar (Eurocopter)	229	U
AS 350BA Ecureuil (IAR)	341	U
AS 355 Ecureuil 2/TwinStar (Eurocopter)	231	U
AS 355N Ecureuil 2 (IAR)	341	U
AS 365N Dauphin 2 (Eurocopter)	232	U
AS 532 Cougar Mk I (Eurocopter)	226	U
AS 532 Cougar Mk II (Eurocopter)	228	U
AS 550 Fennec (Eurocopter)	229	U
AS 555 Fennec (Eurocopter)	231	U
AS 565 Panther (army/air force) (Eurocopter)	234	U
AS 565 Panther (navy) (Eurocopter)	234	U
ASI (see American Sportscopter International Inc)	555	U
Astra, PC-7 Mk II M Turbo-Trainer (Pilatus)	486	U
AT-3 (Aero)	315	U
AT-29 (EMBRAER)	17	U
AT-92 (Pilatus)	486	U
AT-401B Air Tractor (Air Tractor)	546	U
AT-402 Turbo Air Tractor (Air Tractor)	546	U
AT-502 Turbo (Air Tractor)	547	U
AT-602 (Air Tractor)	548	U
AT-802 (Air Tractor)	548	U
ATF-18A (Boeing)	577	U
ATR 42 (ATR)	215	U
ATR 42 Surveyor (ATR)	216	U
ATR 72 (ATR)	217	U
ATT/SSTOL 'Super Frog' (Boeing)	590	U
ATX (CASA)	475	U
AUC (see American Utilicraft Corporation)	556	U
AVA-202 (AII)	257	U
AVA-404 (AII)	258	U
AVA-505 Thunder (AII)	258	U
AVIA (see Nauchno-Proizvodstvennoe Obedinenie Avia)	347	U
AVIC (see China Aviation Industry Corporation)	61	U
AVo 68-R Samburo (Nitsche)	161	U
AWT (see Advanced Wing Technologies)	28	N
ABS Aircraft AG (Switzerland)	481	V
Ace (Rusalen & Rusalen)	286	U
ACEAIR SA (Switzerland)	482	N
Accord-201 (Avia)	347	N
Acrobat, Aviatika-MAI-900 (Aviatika)	349	N

INDEXES: AIRCRAFT—A – B

AD AEROSPACE LTD (UK) ... 510 U
ADAM AIRCRAFT INDUSTRIES (USA) ... 540 N
ADVANCED AERODYNAMICS AND STRUCTURES INC (USA) ... 539 U
Advanced Aircraft Studies (BAE) ... 518 U
Advanced Light Helicopter (HAL) ... 170 U
Advanced Mobility Aircraft (Lockheed Martin) ... 685 U
Advanced Supersonic Airliner (International) ... 255 U
Advanced Theater Transport (Lockheed Martin) ... 686 N
Advanced Transport Studies (Lockheed Martin) ... 686 N
Advanced Turbo Trainer, PC-9 (M) (Pilatus) ... 487 U
ADVANCED WING TECHNOLOGIES (Canada) ... 28 U
Aeris 200 (Aceair) ... 482 N
AERMACCHI SpA (Italy) ... 266 U
AERO sp z.o.o. (Poland) ... 315 V
AEROCOMP (USA) ... 540 U
Aerocraft (Lockheed Martin) ... 687 U
AERO-DESIGN & DEVELOPMENT LTD (Israel) ... 260 U
AERO HOLDING AS (Czech Republic) ... 86 U
AEROKUHLMANN (France) ... 109 U
AEROLITES INC (USA) ... 543 U
Aeromaster (Aerolites) ... 543 N
AEROMOT INDUSTRIA MECHANICO-METALURGICA LTDA (Brazil) ... 15 U
AERONAUTICAL DEVELOPMENT AGENCY (India) ... 168 U
AERONAUTICA MACCHI SpA (Italy) ... 266 U
Aeropony (KNR) ... 667 U
Aeropract-21 M (Aeroprakt) ... 342 U
Aeropract-23 M (Aeroprakt) ... 342 U
Aeropract-25 (Aeroprakt) ... 343 U
Aeropract-27 (Aeroprakt) ... 343 U
AEROPRAKT OOO (Russian Federation) ... 342 U
AEROPRAKT FIRMA (Ukraine) ... 495 U
Aeroprakt-20 (Aeroprakt) ... 495 U
Aeroprakt-20R912 Sky Cruiser (Aeroprakt) ... 496 U
Aeroprakt-20SKh (Aeroprakt) ... 496 U
Aeroprakt-22 (Aeroprakt) ... 496 U
Aeroprakt-24 (Aeroprakt) ... 497 U
Aeroprakt-26 (Aeroprakt) ... 497 U
Aeroprakt-28 (Aeroprakt) ... 497 U
AEROPRO s.v.o (Slovak Republic) ... 472 N
AEROPROGRESS/ROKS-AERO (Russian Federation) ... 344 U
Aeroskiff (Aerolites) ... 543 N
AEROSPACE INDUSTRIAL DEVELOPMENT CORPORATION (Taiwan) ... 492 U
AEROSPATIALE MATRA (France) ... 130 U
AEROSPORT PTY LTD (Australia) ... 2 U
AEROSTAR SA, SC (Romania) ... 337 U
Aerostar 1 (Aerostar) ... 337 U
AEROSTAR AIRCRAFT CORPORATION LLC (USA) ... 543 U
AEROSTYLE ULTRALEICHT FLUGZEUGE GmbH (Germany) ... 142 U
AeroTiga, MD3-160 (SME) ... 308 U
AERO VODOCHODY AS (Czech Republic) ... 86 U
AGROLOT, FUNDACJA (Poland) ... 316 U
AGUSTAWESTLAND (International) ... 181 U
AIR & SPACE AMERICA INC (USA) ... 544 U
Airaile, S-12XL (Rans) ... 731 U
AIRBORNE INNOVATIONS LLC (USA) ... 544 U
AIRBUS INDUSTRIE (International) ... 181 U
AIRBUS INDUSTRIE ASIA (International) ... 200 U
AIRBUS MILITARY SAS (International) ... 200 U
Air Cam (Leza-Lockwood) ... 674 N
Aircar AC-500 (NLA) ... 76 U
AIR COMMAND INTERNATIONAL INC (USA) ... 544 U
AIRCRAFT DESIGNS INC (USA) ... 545 U
AIRCRAFT MANUFACTURING & DEVELOPMENT (USA) ... 551 V
AIRCRAFT TECHNOLOGY INDUSTRIES (International) ... 203 U
Aircraft Trials Programmes (NASA) ... 713 V
AirCruiser M9-200 (Millicer) ... 8 U
Airguard, F-7M (CAC) ... 63 U
AIR LIGHT GmbH (Germany) ... 143 U
Air Tractor AT-401B (Air Tractor) ... 546 U
AIR TRACTOR INC (USA) ... 546 U
Airtrainer CT4E (PAC) ... 312 U
AirTourer M10 (Millicer) ... 8 U
Airtruck C-5WA (IAI) ... 265 U
Airvan GA-8 (Gippsland) ... 5 U
Air Wolf M-26 (PZL) ... 325 U
Aist-2 T-411 (Khrunichev) ... 382 U
AKAFLIEG MÜNCHEN eV (Germany) ... 144 U
Akef (Boeing) ... 575 U
AKROTECH AVIATION INC (USA) ... 549 U
Aktai (Kazan) ... 383 U
AKTSIONERNOE OBSHCHESTVO AERO-M (Russian Federation) ... 342 U
Alaskan Bushmaster (Aviation Development) ... 559 N
Albatros L-39 (Aero) ... 87 U
Albatros L-139 (Aero) ... 87 U
ALBERTA AEROSPACE CORPORATION (Canada) ... 27 U
ALENIA AEROSPAZIO (Italy) ... 276 U
Alfa HB-207 (HB Flugtechnik) ... 12 U

Alize Association Zephyr ... 110 U
Allegro (Fantasy Air) ... 95 U
ALLIANCE AIRCRAFT CORPORATION (USA) ... 550 U
Alligator Ka-52 (Kamov) ... 377 U
Allround SF 40C (Scheibe) ... 164 U
ALP Aviation (Turkey) ... 494 U
AMERICAN CHAMPION AIRCRAFT CORPORATION (USA) ... 551 U
AMERICAN HOMEBUILTS' INC (USA) ... 553 V
AMERICAN SPORTSCOPTER INTERNATIONAL INC (USA) ... 555 U
AMERICAN UTILICRAFT CORPORATION (USA) ... 556 U
AMEUR AVIATION TECHNOLOGIE (France) ... 108 U
AMX INTERNATIONAL (International) ... 207 V
Angel CH-7 (Helisport) ... 281 N
Ansat (Kazan) ... 382 U
Ant PZL-126P (Agrolot) ... 316 U
ANTONOV (see AVIATSIONNY NAUCHNO-TEKHNICHESKY KOMPLEKS IMENI OK ANTONOVA) (Ukraine) ... 498 U
Apache AH-64 (Boeing) ... 594 U
Apache Longbow, WAH-64D (GKN Westland) ... 533 U
Apache Longbow (Boeing) ... 594 U
APEX INTERNATIONAL (France) ... 110 U
AQUILA TECHNISCHE ENTWICKLUNGEN GmbH (Germany) ... 144 U
ARCHEDYNE AEROSPACE (USA) ... 554 U
Archer III, PA-28-181 (Piper) ... 721 U
Arrow, PA-28R-201 (Piper) ... 722 U
ARSENYEVSKOYE AVIATSIONNOYE PROIZVODSTVENNOYE PREDPRIYATIE (Russian Federation) ... 346 U
Asso Vs (Rusalen & Rusalen) ... 286 U
ASSOCIATION ZEPHYR ... 110 U
ARV Super 2 (Skycraft) ... 777 U
AStar AS 350 (Eurocopter) ... 229 U
Astra, PC-7 Mk II M Turbo-Trainer (Pilatus) ... 486 U
Astra SPX, 1125 (IAI) ... 262 U
Atalef (Eurocopter) ... 234 U
ATEC v.o.s. (Czech Republic) ... 92 U
Atlantic 3 (Dassault) ... 122 U
Aurora (IAMI) ... 260 U
Avanti P.180 (Piaggio) ... 284 U
Avenger (Fisher) ... 653 U
AVIABELLANCA AIRCRAFT CORPORATION (USA) ... 556 U
AVIAKIT (France) ... 110 U
AVIAKOR MEZHDUNARODNAYA AVIATSIONNAYA KORPORATSIYA OAO (Russian Federation) ... 348 U
AVIANT, KIEVSKY GOSUDARSTVENNY AVIATSIONNY ZAVOD (Ukraine) ... 509 U
AVIASTAR ULYANOVSKY AVIATSIONNYI PROMSHLENNYI KOMPLEKS (Russian Federation) ... 348 U
AVIAT AIRCRAFT INC (USA) ... 557 U
Aviatika-MAI-217 (Aviatika) ... 350 N
Aviatika-MAI-890 (Aviatika) ... 348 U
Aviatika-MAI-890A (Aviatika) ... 349 U
Aviatika-MAI-900 Acrobat (Aviatika) ... 349 N
Aviatika-MAI-910 (Aviatika) ... 349 U
Aviatika-MAI-960 (Aviatika) ... 349 U
AVIATIKA KONTSERN (Russian Federation) ... 348 U
AVIATION DEVELOPMENT INTERNATIONAL LTD (USA) ... 559 N
AVIATION FARM LIMITED (Poland) ... 317 U
AVIATION INDUSTRIES OF IRAN (Iran) ... 257 U
AVIATON NAUCHNO-PROIZVODSTVENNAYA AVIATSIONNAYA FIRMA (Russian Federation) ... 350 U
AVIATSIONNY NAUCHNO-TEKHNICHESKY KOMPLEKS IMENI OK ANTONOVA ... 498 U
AVIATSIONNYI NAUCHNO-PROMYSHLENNYI KOMPLEKS MiG (Russian Federation) ... 389 U
AVIATSIONNY NAUCHNO-TEKHNISHESKY KOMPLEKS IMENI A N TUPOLEVA OAO (Russian Federation) ... 453 U
AVID AIRCRAFT INC (USA) ... 560 U
AVIOANE SA CRAIOVA, SC (Romania) ... 339 U
AVIONS DE TRANSPORT REGIONAL (International) ... 214 U
AVIONS FOURNIER (France) ... 134 U
AVIONS POTTIER (France) ... 136 V
Avro (see under BAE) ... 518 V
Avro RJ, Jumbolino (BAE) ... 518 U
AVTEKAIR INC ... 561 U
Ayres 7000 (Let) ... 99 U
AYRES CORPORATION (USA) ... 562 U
Azarakhsh (IAMI) ... 260 U
Azték ST-4 (Letov) ... 102 U

B

B-2B (Brantly) ... 627 U
B-6 (XAC) ... 82 U
B36TC, Turbo Bonanza (Raytheon) ... 737 U

B200 King Air (Raytheon) ... 740 U
B200 King Air, Military versions (Raytheon) ... 741 U
BA609 (Bell/Agusta) ... 211 U
BC-17X Globemaster (Boeing) ... 622 N
BD-12 (Bede) ... 565 U
BD-17 Nugget (Bede) ... 565 N
BD-100 Continental (Bombardier) ... 38 U
BD-700 Global Express (Bombardier) ... 34 U
Be-32 Cuff (Beriev) ... 352 U
Be-42 (Beriev) ... 353 U
Be-103 (Beriev) ... 353 U
Be-200 (Beriev) ... 354 U
Be-210 (Beriev) ... 356 U
Be-976, Il-76 SKIP (Beriev) ... 356 U
BF-16 (PAC) ... 312 U
BHEL (see Bharat Heavy Electricals Ltd) ... 169 U
BK 117 (Eurocopter/Kawasaki) ... 239 U
BKLAIC (see Beijing Keyuan Light Industrial Aircraft Company Ltd) ... 61 U
BN2A Mk III Trislander (B-N) ... 523 U
BN2B Islander (B-N) ... 521 U
BN2T Turbine Islander (B-N) ... 522 U
BN2T-4S Defender 4000 (B-N) ... 523 U
BO 105 (Eurocopter) ... 235 U
BO 105 CBS-5 (KAI) ... 306 U
BO 105 LS (Eurocopter) ... 54 U
BUAA (see Beijing University of Aeronautics and Astronautics) ... 62 U
Baaz MiG-29 (MiG) ... 389 U
Baby Huey 1-04 (Hillberg) ... 663 U
Backfire, Tu-22M (Tupolev) ... 453 U
Badger, M-203PW (Myasishchev) ... 422 U
BAE SYSTEMS AIRBUS (UK) ... 521 U
BAE SYSTEMS AUSTRALIA LIMITED (Australia) ... 2 U
BAE SYSTEMS plc (UK) ... 510 U
BAE SYSTEMS REGIONAL AIRCRAFT LTD (UK) ... 518 U
BAE SYSTEMS PROGRAMMES AND CUSTOMER SUPPORT GROUP (UK) ... 511 U
Baljims IA (AAT) ... 108 V
Barak, F-16C (Lockheed Martin) ... 689 U
Baron 58 (Raytheon) ... 738 U
BARR AIRCRAFT (USA) ... 564 U
Barrsix (Barr) ... 564 U
Barsuk M-203PW (Myasishchev) ... 422 U
Baz (Boeing) ... 575 U
Bear, Tu-95 (Tupolev) ... 454 U
Bear Tu-142 (Tupolev) ... 454 U
Bearcat (Aerolites) ... 543 N
BEDE AIRCRAFT CORPORATION (USA) ... 565 U
Bee M-16 (BUAA) ... 62 U
BEECH (see Raytheon Aircraft Company) ... 733 V
Beechjet 400A (Raytheon) ... 744 U
BEIJING KEYUAN LIGHT AIRCRAFT INDUSTRIAL COMPANY LTD (China) ... 61 U
BEIJING UNIVERSITY OF AERONAUTICS AND ASTRONAUTICS (China) ... 62 U
BELL/AGUSTA AEROSPACE COMPANY (International) ... 209 U
BEL-AIRE AVIATION (USA) ... 565 U
Bel-Aire 2000 ... 565 U
Bellaire (Liberty) ... 675 U
BELL BOEING (USA) ... 570 U
BELL HELICOPTER TEXTRON CANADA (Canada) ... 28 U
BELL HELICOPTER TEXTRON INC (USA) ... 566 U
Beluga (SATIC) ... 253 U
Belyi (Tupolev) ... 456 U
BERIEVA AVIATSIONNYI KOMPANIYA (Russian Federation) ... 350 U
Berkut (Rennaissance) ... 749 U
Berkut S-37 (Sukhoi) ... 444 U
Bestmann (KNR) ... 667 U
Beta II, R22 (Robinson) ... 751 U
B&F TECHNIK VERTRIEBS GmbH (Germany) ... 145 V
BHARAT HEAVY ELECTRICALS LTD (India) ... 169 U
Black Hawk (Sikorsky) ... 761 U
Blackjack, Tu-160 (Tupolev) ... 456 U
Black Shark (Kamov) ... 375 U
Blended Wing/Body studies (Boeing) ... 622 V
Blue Bird D139-PTI (Dorna) ... 258 U
Blue Eagle, AD-200 (BKLAIC) ... 61 U
BLUE YONDER AVIATION (Canada) ... 34 U
B-N GROUP (UK) ... 521 U
BOEING/BAe (International) ... 212 V
Boeing Business Jet (Boeing) ... 623 U
Boeing Business Jet 2 (Boeing) ... 624 U
Boeing Business Jet 3 (Boeing) ... 624 U
BOEING BUSINESS JETS (USA) ... 623 U
BOEING COMMERCIAL AIRPLANE GROUP (BCAG) (The Boeing Company) (USA) ... 598 U
Boeing Commercial Orders and Deliveries (Boeing) ... 598 U
BOEING COMPANY, THE (USA) ... 573 U
BOEING MILITARY AIRCRAFT AND MISSILE SYSTEMS GROUP (USA) ... 573 U
BOEING SIKORSKY (USA) ... 624 U

B – D—AIRCRAFT: INDEXES 841

BOEING PHANTOM WORKS (USA) 573 U
BOEING SPACE AND COMMUNICATIONS GROUP (USA) 622 U
BOMBARDIER AEROSPACE (Canada) 34 U
BOMBARDIER AEROSPACE CANADAIR OPERATIONS (Canada) 39 U
BOMBARDIER AEROSPACE DE HAVILLAND (Canada) 47 V
BOMBARDIER AEROSPACE LEARJET (USA) 671 U
Bonanza A36 (Raytheon) 736 U
Bongo NA40 (Unis) 104 U
Boomerang MB-300 (Morrow) 710 N
Boy, AMT-600 (Aeromot) 16 N
Brakeet F-16D (Lockheed Martin) 689 U
BRANTLY INTERNATIONAL INC (USA) 627 U
Brasilia, EMB-120 (EMBRAER) 19 U
Bravo AS202 (FFA Bravo) 483 N
Bravo 3000 (FFA Bravo) 484 N
Bravo M20M (Mooney) 709 U
Breeze A-25 (Aeroprakt) 343 U
Breeze M-28 (PZL) 320 U
Breezer (Aerostyle) 142 U
Bright Star (Custom Flight) 52 U
Bryza (PZL) 320 U
BUL AVIATION (France) 110 U
Bullfinch MIL Mi-52 (Mil) 419 U
Bumblebee, Dubna-1 (Dubna) 358 U
BURKHART GROB LUFT- UND RAUMFAHRT (Germany) 156 U
Bush Hawk FBA-2C (Found) 54 U
Business Jet (Boeing) 623 U
BUSSARD DESIGN GmbH (Germany) 146 N

C

C-5WA Airtruck (IAI) 265 U
C-12 (Raytheon) 741 U
C12J (Raytheon) 743 U
C12S (Raytheon) 742 U
C.15 (Boeing) 577 U
C.16 (Eurofighter) 242 U
C-17A Globemaster III (Boeing) 584 U
C-20F/G/H (Gulfstream Aerospace) 660 U
C-27J Spartan (Alenia) 276 U
C-29A (Raytheon) 748 U
C-37 (Gulfstream) 661 U
C-38A (IAI) 262 U
C-40A (Boeing) 599 U
C42 (Ikarus) 158 U
C90B King Air (Raytheon) 739 U
C90SE King Air (Raytheon) 739 U
C-98 (Cessna) 634 U
C-130J (Lockheed Martin) 682 U
C-212 Patrullero (CASA) 477 U
C-212 Series 300 (CASA) 475 U
C-212 Series 400 (CASA) 475 U
C-295M (CASA) 477 U
CA6 Criquet (Carlson) 628 U
CA-100 Century Jet 100 (Century) 629 U
CAC (see Chengdu Aircraft Industrial Group) 62 U
CAG (see Canadian Aerospace Group International) 51 U
CAP 10 B (CAP) 111 U
CAP 222 (CAP) 111 N
CAP 232 (CAP) 112 U
CCI CarterCopter (Carter) 628 U
CC-142 (de Havilland) 51 U
CC-144 (Canadair) 39 U
CC-144B (Canadair) 39 U
CC-150 Polaris (Airbus) 184 U
CE.15 (Boeing) 577 U
CE.16 (Eurofighter) 242 U
CE-144A (Canadair) 39 U
CF-188A/B (Boeing) 577 U
CH-7 Angel (Helisport) 281 N
CH-47 Chinook (Boeing) 590 U
CH-47 Chinook (Kawasaki) 297 U
CH-47J (Kawasaki) 297 U
CH-47JA (Kawasaki) 297 U
CH-50 (Eurocopter) 229 U
CH-53E Super Stallion (Sikorsky) 760 U
CH-55 (Eurocopter) 231 U
CH-60S (Sikorsky) 761 U
CH-146 Griffon (Bell) 31 U
CH 601HD Zodiac (Zenith) 792 U
CH 601XL Super Zodiac (Zenith) 793 U
CH 620 Gemini (Zenith) 793 U
CH 701 STOL (Zenith) 794 U
CH 701-AG STOL (Czaw) 93 V
CH 801 STOL (Zenith) 794 U
CH 2000, Zenith (AMD) 551 U
CHAIC (see Changhe Aircraft Industries Corporation) 66 U
Che-15 (Chernov) 356 U
Che-22 Corvette (Chernov) 356 U
Che-25 (Chernov) 357 U
CHERNOV (see Opytnyi Konstruktorskoye Byuro Chernov B & M OOO) 356 U
CHRDI (see Chinese Research and Development Institute) 67 N
CJ-6A (NAMC) 75 U
CKB-1 S-302 (SGAU) 426 U
CL-415 SuperScooper (Canadair) 45 U
CL-600 Challenger (Canadair) 39 U
CMC (see Chichester-Miles Consultants Ltd) 525 U
CN-235 (Airtech) 203 U
CN-235M (TAI) 494 U
CN-235 MPA (Airtech) 206 U
CN-235 MP Persuader (Airtech) 206 U
CN-235 Phoenix (Airtech) 204 U
CR100 (Dyn'Aero) 128 U
CR110 (Dyn'Aero) 128 U
CR120 (Dyn'Aero) 128 U
CRJ100 (Bombardier) 41 U
CRJ200 (Bombardier) 41 U
CRJ700 (Bombardier) 43 U
CRJ900 (Bombardier) 44 U
CSIST (see Chung Shan Institute of Science and Technology 493 U
CT (Flight Design) 155 U
CT4E Airtrainer (PAC) 312 U
CT-142 (de Havilland) 51 U
CT-155 (BAE) 511 U
CT-156 Harvard II (Raytheon) 734 U
CV-22 (Bell Boeing) 570 U
C-X (JASDF) 293 U
CZAW (see Czech Aircraft Works) 93 U

CAMERON & SONS AIRCRAFT (USA) 628 N
CANADAIR OPERATIONS, BOMBARDIER AEROSPACE (Canada) 39 U
CANADA AIR RV (Canada) 51 N
CANADIAN AEROSPACE GROUP INTERNATIONAL (Canada) 51 U
Candid, Il-76 (Ilyushin) 361 U
Canguro SF.600A (VulcanAir) 291 U
CAP AVIATION (France) 111 U
Captain SKB-1 S-400 (SGAU) 426 U
Carat TFK-2 (Technoflug) 166 U
Caravan IU-27A (Cessna) 634 U
Caravan II, F406 (Reims) 137 U
Careless, Tu-154M (Tupolev) 458 U
CARLSON AIRCRAFT INC (USA) 628 U
CARTERCOPTERS LLC (USA) 628 U
CarterCopter CCI (Carter) 628 U
Carthorse, GM-01 (Yalo/Aviata) 336 U
Cash, M-28 (PZL) 320 U
Centaur (Warrior) 532 U
CENTURY AEROSPACE CORPORATION (USA) 629 U
Century Jet 100, CA-100 (Century) 629 U
CENZIN FOREIGN TRADE ENTERPRISE CO LTD (Poland) 317 U
CESSNA AIRCRAFT COMPANY (USA) 630 U
CFM AIRCRAFT (UK) 524 U
Challenger CL-600 (Canadair) 39 U
Champ 7ACA (American Champion) 551 N
Champion (Avid) 560 U
CHANGHE AIRCRAFT INDUSTRIES CORPORATION (China) 66 U
CHENGDU AIRCRAFT INDUSTRIAL GROUP (China) 62 U
CHEREAU, ARNO, AERONAUTIQUE (France) 112 V
Chernaya Akula, Ka-50 (Kamov) 375 U
CHICHESTER-MILES CONSULTANTS LTD (UK) 525 U
CHINA AVIATION INDUSTRY CORPORATION (China) 61 U
CHINESE HELICOPTER RESEARCH AND DEVELOPMENT INSTITUTE (China) 67 N
Chinook 114 (Boeing) 590 U
Chinook 414 (Boeing) 590 U
Chinook CH-47 (Kawasaki/Boeing) 297 U
Chinook HC. Mk 2, 2A and 3 (Boeing) 590 U
Chinook production (Boeing) 593 U
Chiricahua UV-20A (Pilatus) 484 U
Chirok (L11/Pegas) 388 U
Choucas (Noin) 136 U
Chuji Jiaolianji-6A (NAMC) 75 U
Chrysalis (Stemme) 165 U
CHUNG SHAN INSTITUTE OF SCIENCE AND TECHNOLOGY (Taiwan) 493 U
CIRRUS DESIGN CORPORATION (USA) 642 U
Citabria Adventure, 7GCAA (American Champion) 552 U
Citabria Aurora 7ECA (American Champion) 551 U
Citabria Explorer 7GCBC (American Champion) 552 U
Citation VII 650 (Cessna) 640 U
Citation Bravo 550 (Cessna) 637 U
Citation CJ1, 525 (Cessna) 635 U
Citation CJ2, 525A (Cessna) 636 U
Citation Encore 560 (Cessna) 638 U
Citation Excel (Cessna) 639 U
Citation Sovereign, 680 (Cessna) 641 U
Citation Ultra 560 (Cessna) 638 U
Citation X 750 (Cessna) 641 U
CityHawk (AD&D) 261 N
CityStar, Yak-42 (Yakolev) 468 U
CLASSIC FIGHTER INDUSTRIES INCORPORATED (USA) 643 N
Cline, An-32 (Antonov) 499 U
Clobber, Yak-42 (Yakovlev) 468 U
Coaler, An-72 (Antonov) 502 U
Coaler, An-72P (Antonov) 504 U
Coaler, An-74 (Antonov) 502 U
Coke, An-24 (Antonov) 498 U
Colibri EC 120B (Eurocopter/CATIC/STAero) 238 U
Colt, An-2 (Antonov) 498 U
Columbia 300 (Lancair) 670 U
Comanche RAH-66 (Boeing Sikorsky) 624 U
Comet FK 12 (B&F) 145 U
Commander 115 (Commander) 644 N
Commander 147A (Air Command) 544 U
Commander 618 (Air Command) 545 U
COMMANDER AIRCRAFT COMPANY (USA) 644 U
Comp Air 3 (Aerocomp) 540 N
Comp Air 4 (Aerocomp) 540 U
Comp Air 6 (Aerocomp) 541 U
Comp Air 7 (Aerocomp) 541 U
Comp Air 8 (Aerocomp) 541 N
Comp Air 10 (Aerocomp) 542 U
Comp Monster (Aerocomp) 540 U
COMPOSITE TECHNOLOGY RESEARCH (Malaysia) 307 N
Condor, An-124 (Antonov) 505 U
Condor TL-132 (TL Ultralight) 103 U
Condor Plus TL-232 (TL Ultralight) 103 U
CONSTRUCCIONES AERONAUTICAS SA (Spain) 475 U
COSTRUZIONI AERONAUTICHE TECNAM SRL (Italy) 288 U
Continental BD-100 (Bombardier) 38 U
Coot (Ilyushin) 360 U
Coot-A Il-20M (Ilyushin) 360 U
Coot-B (Ilyushin) 360 U
Coot-C (Ilyushin) 360 U
Cormorant (EHI) 218 U
Corvette Che-22 (Chernov) 356 U
Cossack, An-225 Mriya (Antonov) 509 U
Cougar Mk I (TAI) 495 U
Cougar Mk I AS 532 (Eurocopter) 226 U
Cougar Mk II AS 532 (Eurocopter) 228 U
Courier S-7 (Rans) 731 U
Crane (NAL) 176 U
Creek TH-67 (Bell) 28 U
Cresco (PAC) 311 U
Criquet CA6 (Carlson) 628 N
Cruiser (Pulsar) 729 U
Cub, An-12 (Antonov) 498 U
Cuff, Be-32 (Beriev) 352 U
CULP'S SPECIALITIES (USA) 645 U
Curl, An-26 (Antonov) 498 U
CUSTOM FLIGHT COMPONENTS LTD (Canada) 52 V
CZECH AIRCRAFT WORKS s.r.o. (Czech Republic) 93 U

D

D4 BK Fascination (WD) 167 U
D139-PTI Blue Bird (Dorna) 258 U
DA 20-A1 Katana (Diamond) 52 U
DA 20-A2 Katana (Diamond) 53 U
DA 20-C1 Katana (Diamond) 53 U
DA 40-180 Diamond Star (Diamond) 11 U
DHC-8 Dash 8 Q100 (de Havilland) 47 U
DHC-8 Dash 8 Q200 (de Havilland) 47 U
DHC-8 Dash 8 Q300 (de Havilland) 48 U
DHC-8 Dash 8 Q400 (de Havilland) 49 U
DHC-8 Dash 8M (de Havilland) 51 U
DHC-8 Triton 300 (de Havilland) 51 U
DJ 2 (DJ) 53 V
DR 400 Dauphin (Robin) 140 U
DR 400/160 Major (Robin) 140 U
DR 400/180 Régent (Robin) 141 U
DR 500i President (Robin) 142 U
DUBNA (see Proizvodstvenno-Tekhnichesky Kompleks Dubnenskogo Mashinostroitelnogo Zavod) 358 U
DV 20 Katana (Diamond) 11 U

DAEWOO HEAVY INDUSTRIES CO LTD (Korea, South) 303 U
DAIMLERCHRYSLER AEROSPACE AIRBUS/ TUPOLEV (International) 214 U
Dakota Hawk (Fisher) 653 U
Darling M-12 (Aviacor) 348 U
Dash 8 Q100 DHC-8 (de Havilland) 47 U
Dash 8 Q200 DHC-8 (de Havilland) 47 U
Dash 8 Q300 DHC-8 (de Havilland) 48 U
Dash 8 Q400 DHC-8 (de Havilland) 49 U
Dash 8 8M, DHC-8 (de Havilland) 51 U
DASSAULT AVIATION (France) 113 U
Dauphin 2 AS 365N (Eurocopter) 232 U
Dauphin, DR 400 (Robin) 140 U
Dawn (FAJR) 259 N

INDEXES: AIRCRAFT—D – F

Entry	Page
Daya (Northrop Grumman)	715 U
Deer, M-202 PW (Myasishchev)	422 U
Defender (MD)	703 U
Defender (B-N)	522 U
Defender (Romaero)	341 U
Defender 4000 BN2T-4S (B-N)	523 U
DE HAVILLAND, BOMBARDIER AEROSPACE (Canada)	47 V
Delfin (Refly)	423 N
DELTA SYSTEM-AIR AS (Czech Republic)	93 U
DENEL AVIATION (South Africa)	473 U
DERRINGER AIRCRAFT COMPANY LLC (USA)	646 U
Devil (ENAER)	60 U
DFL Holdings Inc (USA)	647 U
DIAMOND AIRCRAFT CORPORATION (Canada)	52 U
DIAMOND AIRCRAFT INDUSTRIES GmbH (Austria)	10 U
Diamond Future Developments (Diamond)	12 N
Diamond Star DA 40-180 (Diamond)	11 U
DIRGANTARA INDONESIA, PT (Indonesia)	178 U
DJ AIRCRAFTS INTERNATIONAL (Canada}	53 U
Dolpheen (Eurocopter)	232 U
Dolphin DR 400 (Robin)	140 U
Dolphin, HH-65A (Eurocopter)	232 U
Dolphin, Z-9 (HAI)	68 U
DORNA, H.F. COMPANY (Iran)	258 U
Dornier 228 (HAL)	174 U
Dornier 328JET (Dornier Fairchild)	150 U
Dornier 528JET (Dornier Fairchild)	151 U
Dornier 728JET (Dornier Fairchild)	151 U
Dornier 928JET (Dornier Fairchild)	151 U
DORNIER LUFTFAHRT GmbH (Germany)	146 U
Dragon (Aeroprakt)	342 U
Dragon Fly 333 (Dragon Fly)	278 U
Dragonfly Canard Rotor/Wing (Boeing)	573 U
Dragonfly (IAIO)	260 U
Dragonfly (Slipstream)	779 N
DRAGON FLY SRL (Italy)	278 V
Dream, An-225 (Antonov)	509 U
DREAMWINGS LLC (USA)	647 U
Dromader M-18 (PZL)	322 U
Dromedary M-18 (PZL)	322 U
Dubna-1 Shmel (Dubna)	358 U
Dubna-2 Osa (Dubna)	358 U
DYN' AERO, SOCIETE (France)	128 U

E

E-2C (Northrop Grumman)	715 U
E2E Speedtwin (Speedtwin)	531 U
E-4 (Sat Saviat)	424 U
E-8 Joint STARS (Northrop Grumman)	716 U
E-9A (de Havilland)	51 U
E26 Tamiz (ENAER)	60 U
E-767 (Boeing)	588 U
EADS (see European Aeronautic Defence and Space Company)	214 U
EADS ATR (see Avions de Transport Regional)	214 U
EAL (see Ethiopian Airlines Enterprise)	108 U
EAL (see European Aerosystems Limited)	217 U
EC 120 B Colibri (Eurocopter/CATIC/STAero)	238 U
EC 130B (Eurocopter)	231 N
EC-130J Hercules (Lockheed Martin)	682 U
EC 135 (Eurocopter)	236 U
EC 145 (Eurocopter Kawasaki)	241 U
EC 155B (Eurocopter)	235 U
EC 165 (Eurocopter)	238 V
EC 635 (Eurocopter)	236 U
EE 10 Eaglet (Euro-ENAER)	309 U
EH 1-01 RotorMouse (Hillberg)	663 U
EH 1-02 TandemMouse (Hillberg)	663 U
EH-6 (MD)	703 U
EH-60A (Sikorsky)	761 U
EH 101 (EHI)	218 U
EHI (see EH Industries Limited)	218 U
EL-1 Gavilan 358 (AWT)	85 U
EMB-120 Brasilia (EMBRAER)	19 U
EMB-202 Ipanema (NEIVA)	26 U
EMB-314 Super Tucano (EMBRAER)	17 U
EMBRAER (see Empresa Brasileira de Aeronáutica SA)	17 U
ENAER (see Empresa Nacional de Aeronáutica de Chile)	60 U
ERJ-135 (EMBRAER)	22 U
ERJ-140 (EMBRAER)	23 U
ERJ-145 (EMBRAER)	20 U
ERJ-170 (EMBRAER)	24 U
ERJ-190 (EMBRAER)	24 U
EV97 Eurostar (EV-AT)	94 U
EV-AT (see Evektor-Aerotechnik AS)	94 U
E-Z Flyer (Blue Yonder)	34 U
EADS CASA (Spain)	475 U
EADS DEUTSCHLAND (Germany)	146 U
EADS DEUTSCHLAND AIRBUS (Germany)	148 U
EADS FRANCE (France)	130 U
EADS M (Germany)	147 U
EADS SOCATA (France)	130 U
Eagle II, Christen (Aviat)	558 U
Eagle 150 (Eagle Aircraft)	3 U
EAGLE AIRCRAFT PTY LTD (Australia)	3 U
Eagle, F-15C/D (Boeing)	575 U
Eagle, F-15E (Boeing)	575 U
Eagle M20S (Mooney)	710 U
Eagle, PC-12 (Pilatus)	491 U
Eaglet EE 10 (Euro-Enaer)	309 U
Easy Eagle (Grosso)	658 N
Echo P92 (Tecnam)	288 U
ECLIPSE AVIATION CORPORATION (USA)	648 U
Eclipse 500 (Eclipse)	648 U
Ecureuil 2/TwinStar AS 355 (Eurocopter)	231 U
Ecureuil AS 350BA (IAR)	341 U
Ecureuil AS 355N (IAR)	341 U
Ecureuil/AStar AS 350 (Eurocopter)	229 U
Edge 540 (Zivko)	795 U
EDRA HELICENTRO PECAS e MANUTENCAO LTDA (Brazil)	16 U
EH INDUSTRIES LIMITED (International)	218 U
E & K sp zoo (Poland)	317 U
EKOFLUG spol s r o (Slovak Republic)	473 U
EKSPERIMENTALNYI MASHINOSTROITELNYI ZAVOD IMENI VM MYASISHCHEVA (Russian Federation)	420 U
EL GAVILAN SA (Colombia)	85 U
Elitar IE-101 (Elitar)	358 U
ELITAR OOO (Russian Federation)	358 V
Ellipse (Explorer Aviation)	651 U
EMPRESA BRASILEIRA DE AERONAUTICA SA (Brazil)	17 U
EMPRESA NACIONAL DE AERONAUTICA DE CHILE (Chile)	60 U
Encore M20K (Mooney)	709 U
Enforcer, MH-90 (MD)	706 U
ENSTROM HELICOPTER CORPORATION, THE (USA)	649 U
Erica (Agusta)	275 U
Esquilo, HA-1 (Eurocopter)	229 U
Esquilo, TH-5 (Eurocopter)	229 U
Esquilo, UH-12 (Eurocopter)	229 U
Esquilo, UH-12B (Eurocopter)	231 U
Esquilo, VH-55 (Eurocopter)	231 U
Ethereal (Kazan)	382 U
ETHIOPIAN AIRLINES ENTERPRISE (Ethiopia)	108 U
ETRURIA TECHNOLOGY srl (Italy)	278 V
EURAVIAL (France)	134 U
EURO ALA (Italy)	279 V
EUROCOPTER SAS (International)	223 U
EUROCOPTER CANADA LTD (Canada)	53 U
EUROCOPTER/CATIC/STAero (International)	238 U
EUROCOPTER/KAWASAKI (International)	239 U
EURO-ENAER HOLDING BV (Netherlands)	309 U
EUROFAR (International)	247 U
EUROFIGHTER JAGDFLUGZEUG GmbH (International)	241 U
Eurofox (Aeropro)	472 U
EUROMIL (International)	247 U
Europa (Europa)	526 U
EUROPA AIRCRAFT COMPANY LTD (UK)	526 U
Europa Production Light Aircraft (Europa)	528 U
EUROPATROL (International)	247 U
EUROPEAN AERONAUTIC, DEFENCE AND SPACE COMPANY NV (International)	214 U
EUROPEAN AEROSYSTEMS LIMITED (International)	217 U
EUROPEAN MILITARY AIRCRAFT COMPANY (International)	233 U
European Tiltrotor Programme (International)	247 N
Eurostar, EV97 (EV-AT)	94 U
Eurostar VLA (EV-AT)	94 N
Exec 162F (Rotorway)	753 U
Experimental Crew Return Vehicle, X-38 (NASA)	712 U
EXPLORER AIRCRAFT CORPORATION PTY LTD (Australia)	4 U
EXPLORER AIRCRAFT INC (USA)	650 N
EXPLORER AVIATION (USA)	651 U
Explorer MD (MD)	706 U
Explorer (Explorer)	650 U
Explorer (Explorer Aircraft)	650 U
Express (Express)	652 U
EXPRESS AIRCRAFT COMPANY (USA)	652 U
Extra 200 (Extra)	148 U
Extra 300 (Extra)	148 U
Extra 330 (Extra)	148 U
Extra 400 (Extra)	149 U
EXTRA-FLUGZEUGBAU GmbH (Germany)	148 U

F

F1 (Farnborough)	528 U
F-2 (Mitsubishi)	298 U
F-7 (CAC)	62 U
F-8 (SAC)	78 U
F-8 IIM (SAC)	80 U
F.8L Falco (Sequoia)	759 U
F-10 (CAC)	66 U
F-11 Ferguson (Precision Tech)	727 N
F-11B Fergy (Precision Tech)	727 N
F-15 A/B Baz (Boeing)	575 U
F-15C Eagle (Boeing)	575 U
F-15C/D Akef (Boeing)	575 U
F-15D Eagle (Boeing)	575 U
F-15E Eagle (Boeing)	575 U
F-16 Customers, table of (Lockheed Martin)	694 U
F-16 Fighting Falcon (Lockheed Martin)	689 U
F-16A/B Netz (Lockheed Martin)	689 U
F-16A/B Production, table of (Lockheed Martin)	695 U
F-16C Barak (Lockheed Martin)	689 U
F-16D Brakeet (Lockheed Martin)	689 U
F-16C/D Fighting Falcon (KAI)	306 U
F-16C/D Fighting Falcon, (TAI)	494 U
F-16C/D Production, table of (Lockheed Martin)	696 U
F-22 Raptor (Lockheed Martin)	677 U
F.22 (General Avia)	280 U
F-28 (Enstrom)	649 U
F406 Caravan II (Reims)	137 U
F/A-18 Hornet (Boeing)	577 U
F/A-18A, B, C, D Hornet (Boeing)	577 U
F/A-18E/F Super Hornet (Boeing)	581 U
FBA-2C Bush Hawk (Found)	54 U
FBC-1 (XAC)	82 U
FC-1 (CAC)	65 U
FF-1080-200 Freight Feeder (AUC)	556 U
FI-X (JASDF)	293 U
FJ-100 (Aerostar)	543 U
FK 9 Mark 3 (B&F)	145 U
FK 12 Comet (B&F)	145 U
FK 14 Polaris (B&F)	146 U
FT-7 (GAIC)	67 U
FT-300 (NAI)	72 U
FTC-2000 (GAIC)	68 U
FU24-954 (PAC)	311 U
Fw 190A-8/N (Flug Werke)	156 U
F-X (KAI)	306 U
FAIRCHILD DORNIER (Germany)	150 N
FAIRCHILD DORNIER CORPORATION (USA)	653 U
FAJR AVIATION AND COMPOSITES INDUSTRIES (Iran)	259 N
Fajr-3 (FAJR)	259 U
Falco F.8L (Sequoia)	759 U
Falcon 50 EX (Dassault)	123 U
Falcon 900B/C (Dassault)	124 U
Falcon 900EX (Dassault)	124 U
Falcon 2000 (Dassault)	126 U
Falcon, M-201 (Myasishchev)	421 U
Falcon, W-3 (PZL Swidnik)	326 U
Falke 2000, SF 25C (Scheibe)	163 U
Fantan Q-5 (NAMC)	73 U
Farmer T-517 (Khrunichev)	387 U
FANTASY AIR GROUP–PROXIMEX s.r.o. (Czech Republic)	95 U
FARNBOROUGH-AIRCRAFT.COM (UK)	528 U
Fascination D4 BK (WD)	167 U
Fatman GA-200 (Gippsland)	4 U
Favorit SKB-1 (SGAU)	425 U
FEDERALNOYE GOSUDARSTVENNOYE UNITARNOYE PREDPRIYATIE, ROSSIYSKAYA (Russian Federation)	423 U
Fencer, Su-24 (Sukhoi)	427 U
Feniks GM-17 (Intracom)	250 N
Fennec AS 550 (Eurocopter)	229 U
Fennec AS 555 (Eurocopter)	231 U
Fennec (Aero)	89 U
Ferguson F11 (Precision Tech)	727 N
Fergy F11B (Precision Tech)	727 N
FFA BRAVO AG (Switzerland)	483 U
FFA FLUGZEUGWERKE ALTENRHEIN AG (Switzerland)	483 U
Fighting Falcon F-16 (Lockheed Martin)	689 U
Fighting Falcon F-16C/D (KAI)	306 U
Fighting Falcon F-16C/D (TAI)	494 U
FINAVITEC, PATRIA OY (Finland)	108 U
Finback-B (SAC)	78 U
Finist SM-92 (Technoavia)	451 U
Firefly (Kolb)	667 U
Firefly T67, T-3A (Slingsby)	530 U
Firefly T67M200 (Slingsby)	531 U
Firefly T67M260 (Slingsby)	531 U
Firnas 142 (ZLIN)	105 U
FISHER FLYING PRODUCTS (USA)	653 U
FLAMING AIR GmbH (Germany)	154 U
Flamingo T-433 (Khrunichev)	386 U
Flanker Su-27 (Sukhoi)	433 U
FLIGHT DESIGN GmbH (Germany)	155 U
Flightstar 115C (Flightstar)	654 N
Flightstar 115L (Flightstar)	654 U
FLIGHTSTAR SPORTSPLANES (USA)	654 U
FLIGHT TEAM ULTRALEICHTFLUGZEUGE und FLUGSCHULE (Germany)	155 U
FLUG WERKE GmbH (Germany)	156 U
Flyer Mark IV (Avid)	560 U
FLYING K ENTERPRISES INC (USA)	655 U
Flying Leopard (XAC)	82 U
FLY SYNTHESIS SRL (Italy)	279 U

F – I—AIRCRAFT: INDEXES 843

FOKKER AVIATION BV (Netherlands) ... 310 U
Formula (Flightstar) ... 654 N
FOUND AIRCRAFT CANADA (Canada) ... 54 U
FOURNIER, AVIONS (France) ... 134 N
Foxbat (Aeroprakt) ... 496 U
Foxhound, MiG-31 (MiG) ... 396 U
FRATI, STELIO (Italy) ... 280 U
Freebird II (Proshort) ... 728 U
FREEDOM LITE INC (Canada) ... 55 U
Fregata J6 (J&AS) ... 320 U
Freight Feeder FF-1080-200 (AUC) ... 556 U
Fresh TL-532 (TL Ultralight) ... 104 U
Frigate Bird (J&AS) ... 320 U
Frogfoot Su-25 (Sukhoi) ... 429 U
Frogfoot Su-28 (Sukhoi) ... 429 U
FUJI JUKOGYO KABUSHIKI KAISHA (Japan) ... 292 U
Fulcrum, MiG-29 (MiG) ... 389 U
FUNDACJA AGROLOT (Poland) ... 316 U
Fury, LP-1 (Lopresti) ... 698 U
Future Carrier-Borne Aircraft (RN) ... 529 U
Future Cargo Aircraft (Reims) ... 139 V
Future Large Aircraft A400 (Airbus) ... 200 U
Future Offensive Air System (RAF) ... 529 U
Future Organic Airborne Early Warning (FOAEW) (RN) ... 530 U
Future Strategic Tanker Aircraft (RAF) ... 529 U

G

G-3 Mirage (Remos) ... 163 U
G 115B, C and D (Grob) ... 156 U
G 115E, G and EG (Grob) ... 156 U
G 120A (Germany) ... 157 N
G-200 Giles (Akrotech) ... 549 U
G-202 Giles (Akrotech) ... 549 U
G-300 Giles (Akrotech) ... 550 U
G-750 Giles (Akrotech) ... 550 U
GA (see Gippsland Aeronautics) ... 4 U
GA-8 Airvan (Gippsland) ... 5 U
GA-200 Fatman (Gippsland) ... 4 U
GAIC (see Guizhou Aviation Industry Corporation) ... 67 U
GM-01 Gniady (Yalo/Aviata) ... 336 U
GM-11 (Intracom) ... 251 N
GM-12 (Intracom) ... 251 N
GM-16 (Intracom) ... 251 N
GM-17 Feniks (Intracom) ... 250 N
GM-19 (Intracom) ... 251 N
GM-28 (Intracom) ... 251 N
GR-582 Light Wing (Hughes) ... 6 U
GR-912 Light Wing (Hughes) ... 6 U
GROB (see Burkhart Grob Luft- und Raumfahrt) ... 156 U
GT-1 Trainer (Goair) ... 5 U

Gadfly (Vitek) ... 465 V
Gajaraj, Il-76 (Ilyushin) ... 361 U
Galaxy 1126 (IAI) ... 264 U
GALAXY AEROSPACE CORPORATION (USA) ... 655 U
Gavilan 358, EL-I (Gavilan) ... 85 U
Gavilan 508T (Gavilan) ... 85 U
GAVILÁN SA, EL (Colombia) ... 85 U
Gemini CH 620 (Zenith) ... 793 U
Gemini M-20 (PZL) ... 323 U
GENERAL AVIA COSTRUZIONI AERONAUTICHE SRL (Italy) ... 280 U
Genesis (Slipstream) ... 779 N
GIDROPLAN (Russian Federation) ... 359 N
Giles G-200 (Akrotech) ... 549 U
Giles G-202 (Akrotech) ... 549 U
Giles G-300 (Akrotech) ... 550 U
Giles G-750 (Akrotech) ... 550 U
GIPPSLAND AERONAUTICS PTY LTD (Australia) ... 4 U
GKN WESTLAND HELICOPTERS LIMITED (UK) ... 533 U
Glasair III (Stoddard-Hamilton) ... 781 U
Glasair Super II (Stoddard-Hamilton) ... 780 U
Glass Goose (Quikkit) ... 730 U
GlaStar (Stoddard Hamilton) ... 781 U
Global Express BD-700 (Bombardier) ... 34 U
Globemaster III C-17A (Boeing) ... 584 U
Globemaster BC-17X (Boeing) ... 622 N
Gniady GM-01 (Yalo/Aviata) ... 336 U
GOAIR PRODUCTS (Australia) ... 5 U
Golden Eagle (Sukhoi) ... 444 U
Golden Eagle A-50 (KAI) ... 305 U
Golden Eagle T-50 (KAI) ... 305 U
Golf P96 (Tecnam) ... 289 U
Goshawk T-45A (BAE) ... 511 U
Goshawk T-45 (Boeing/BAE) ... 212 U
GOSUDARSTVENNOYE UNITARNOYE PREDPRIYATIE AVIATSIONNYI VOYENNO-PROMYSHLENNYI KOMPLEKS SUKHOI (Russian Federation) ... 427 U
GOSUDARSTYENNYI KOSMICHESKII NAUCHNO-PROIZVODSTVENNYI TSENTR IMENI M.B. KHRUNICHEVA (Russian Federation) ... 383 U
GOSUDARSTVENNOYE UNITARNOYE PREDPRIYATIE KOMSOMOLSKOE-na-AMURE AVIATSIONNOYE PROIZVODSTEVENNOYE OBEDINENIE IMENI Yu A GAGARIN (Russian Federation) ... 387 U
Grach T-101 (Aeroprogress/ROKS-Aero) ... 344 U
Great Flight (KAI) ... 304 U
GREAT PLAINS AIRCRAFT SUPPLY CO INC (USA) ... 656 U
Griffin HT. Mk 1 (Bell) ... 31 U
Griffin (Saab) ... 479 U
Griffon 103 (Interplane) ... 95 U
Griffon (Agusta) ... 275 U
Griffon CH-146 (Bell) ... 31 U
GRIFFON AEROSPACE INC (USA) ... 656 U
Gripen, JAS 39 (Saab) ... 479 U
GROEN BROTHERS AVIATION INC (USA) ... 657 U
GROSSO AIRCRAFT INC (USA) ... 658 N
GUIZHOU AVIATION INDUSTRIAL CORPORATION (China) ... 67 U
GUIZHOU EVEKTOR AIRCRAFT CORPORATION (China) ... 68 N
Gulfstream IV-B (Gulfstream Aerospace) ... 659 U
Gulfstream IV-MPA (Gulfstream Aerospace) ... 659 U
Gulfstream IV-SP (Gulfstream Aerospace) ... 659 U
Gulfstream V (Gulfstream Aerospace) ... 661 U
Gulfstream SBJ (Gulfstream) ... 662 U
Gulfstream V-SP (Gulfstream) ... 661 N
GULFSTREAM AEROSPACE CORPORATION (USA) ... 658 U
Gull M-20 (PZL) ... 323 U
Guri AMT-600 (Aeromot) ... 16 N
Gyrfalcon, SKB-1 (SGAU) ... 426 U
Gzhel M-101T (Myasischev) ... 420 U

H

H-6 (XAC) ... 82 N
H-53E (Sikorsky) ... 760 U
HA-1 Esquilo (Eurocopter) ... 229 U
HAI (see Hafei Aviation Industry Company) ... 68 U
HAL (see Hindustan Aeronautics Limited) ... 170 V
HALO (Scaled Composites) ... 754 U
HAT (see Hellenic Aeronautical Technologies) ... 167 U
HB-2 (He-Ro's) ... 317 U
HB-204 Tornado (HB Flugtechnik) ... 13 U
HB-207 Alfa (HB Flugtechnik) ... 12 U
HD.21 (Eurocopter) ... 226 U
HELIBRAS (see Helicopteros do Brasil SA) ... 25 U
HGM-2 Ladybird (He-Ro's) ... 318 U
HH-60G Pave Hawk (Sikorsky) ... 761 U
HH-60H (Sikorsky) ... 768 U
HH-60J Jayhawk (Sikorsky) ... 768 U
HH-65A Dolphin (Eurocopter) ... 232 U
HJT-36 (HAL) ... 172 U
HK 36 (Diamond) ... 10 U
Hkp 9 (Eurocopter) ... 235 U
Hkp 10 (Eurocopter) ... 226 U
HM-1 (Eurocopter) ... 234 U
HOPE-X (NASDA) ... 302 U
HR 200/120B (Robin) ... 139 U
HR 200/160 (Robin) ... 139 U
HS.23 (Sikorsky) ... 768 U
HT.17 (Boeing) ... 590 U
HT.21 (Eurocopter) ... 226 U
HU.21L (Eurocopter) ... 228 U
HUGHES (see Howard Hughes Engineering) ... 6 U
HV-22 (Bell Boeing) ... 570 U

Haitun Z-9 (HAI) ... 68 U
HAFEI AVIATION INDUSTRY COMPANY (China) ... 68 U
HAL 100-seat Transport (HAL) ... 176 U
Halo, Mi-26 (MIL) ... 414 U
Hansa-3 (NAL) ... 177 U
HARBIN AIRCRAFT INDUSTRY GROUP (China) ... 71 U
Harvard II (Raytheon) ... 734 U
Havoc, Mi-28 (Mil) ... 416 U
Hawk 4 (Groen) ... 657 U
Hawk 6 (Groen) ... 658 U
Hawk 8 (Groen) ... 658 U
Hawk 50 series (BAE) ... 511 U
Hawk 60 series (BAE) ... 511 U
Hawk 100 series (BAE) ... 511 U
Hawk 200 series (BAE) ... 515 U
Hawk, IAR-99 (Avioane) ... 339 U
Hawk T. Mk 1 and 1A (BAE) ... 511 U
Hawker 450 (Raytheon) ... 748 N
Hawker 800 (KAC) ... 294 U
Hawker 800FI (Raytheon) ... 746 U
Hawker 800RA (Raytheon) ... 746 U
Hawker 800SIG (Raytheon) ... 746 U
Hawker 800SM (Raytheon) ... 746 U
Hawker 800XP (Raytheon) ... 745 U
Hawker Horizon (Raytheon) ... 747 U
Hawkeye E-2C (Northrop Grumman) ... 715 U
Haze, Mi-14 (Mil) ... 406 U
HB FLUGTECHNIK GmbH (Austria) ... 12 U
Helibus S-92A (Sikorsky) ... 772 U
Helicopter MK 2 (MK) ... 159 V
HELICOPTEROS DO BRASIL SA (Brazil) ... 25 U
HELISPORT, CH7, srl (Italy) ... 281 U
Helix-A, Ka-27 (Kamov) ... 371 U
Helix-B, Ka-29 (Kamov) ... 372 U
Helix-B, Ka-31 (Kamov) ... 372 U
Helix-C, Ka-32 (Kamov) ... 373 U
Helix-D, Ka-28 (Kamov) ... 371 U
HELLENIC AERONAUTICAL TECHNOLOGIES (Greece) ... 167 U
HENDERSON AERO SPECIALTIES INC (USA) ... 662 U
HEPP, AUGUST (Germany) ... 157 U
Hercules (Lockheed Martin) ... 682 U
Hercules 382U/V (Lockheed Martin) ... 682 U
Hercules C. Mk 4 and C. Mk 5 (Lockheed Martin) ... 682 U
Hermes (Aviakit) ... 110 U
Hermit Mi-34 (Mil) ... 417 U
Heron (AD&D) ... 261 N
Heron (Grob) ... 156 U
HE-RO'S DESIGN LUBLIN (Poland) ... 317 U
HILLBERG HELICOPTERS (USA) ... 663 V
Hind, Mi-24 (Mil) ... 410 U
Hind, Mi-25 (Mil) ... 410 U
Hind, Mi-35 (Mil) ... 410 U
HINDUSTAN AERONAUTICS LIMITED (India) ... 170 V
Hip, Mi-8 (V-8) (Mil) ... 404 U
Hip-G, Mi-9 (Mil) ... 406 U
Hip-H, Mi-17 (Mil) ... 407 U
Hip-H, Mi-171 (Mil) ... 407 U
Hip-H, Mi-172 (Mil) ... 407 U
HK AIRCRAFT TECHNOLOGIE AG (Germany) ... 157 V
Hokum Ka-50 (Kamov) ... 375 U
Hongdu AG Dragon (CAG) ... 51 U
HONGDU AVIATION INDUSTRY GROUP (China) ... 71 N
Hongzhaji-6 (XAC) ... 82 N
Hook, Mi-6 (Mil) ... 403 U
Hook-C Mi-22 (Mil) ... 410 U
Hornet (US LIGHT AIRCRAFT) ... 784 N
Hornet, F/A-18 (Boeing) ... 577 U
Hornet production, table of (Boeing) ... 578 U
HOWARD HUGHES ENGINEERING PTY LTD (Australia) ... 6 U
Hudournik, PC-9M Advanced Turbo Trainer (Pilatus) ... 487 U
HUMBERT AVIATION (France) ... 134 N
Hummingbird (AD&D) ... 260 U
Hummingbird (Eurocopter/CATIC/STAero) ... 238 U
Hummingbird (VAT) ... 786 U
Hummingbird PZL-110 (PZL Warszawa) ... 332 U
Hummingbird PZL-111 (PZL Warszawa) ... 333 U
Hummingbird T-311 (Aeroprogress/ROKS-Aero) ... 345 U
Husky A-1 (Aviat) ... 557 U
Hydroplane (Yalo/Aviata) ... 337 U
Hypersonic Flight Programmes (NASA) ... 713 U
HYUNDAI SPACE AND AIRCRAFT COMPANY (Korea, South) ... 304 U

I

I-1L (Interavia) ... 370 U
I-3 (Interavia) ... 370 U
I-23 Manager (IL) ... 318 U
I-25 AS (IL) ... 319 U
IA63 Pampa NG (LMAASA) ... 1 U
IAI (see Israel Aircraft Industries) ... 261 U
IAMI (see Iran Aircraft Manufacturing Industries Company) ... 259 U
IAIO (see Iran Aviation Industries Organisation) ... 259 U
IAPO (see IRKUTSKOYE AVIATSIONNOYE PROIZVODSTVENNOYE OBEDINENIE) ... 359 U
IAR-46 (IAR) ... 340 U
IAR-99 Şoim (Avioane) ... 339 U
IE-101 Eitar (Elitar) ... 358 U
III (see Iniziative Industriali Italiane) ... 282 V
IL (see Instytut Lotnictwa) ... 318 U
Il-18 Coot (Ilyushin) ... 360 U
Il-18D-36 (Ilyushin) ... 360 U
Il-20 Coot (Ilyushin) ... 360 U
Il-20M Coot A (Ilyushin) ... 360 U
Il-22M-11 (Ilyushin) ... 360 U
Il-24N (Ilyushin) ... 360 U
Il-38 May (Ilyushin) ... 360 U
Il-76 Candid (Ilyushin) ... 361 U
Il-76 SKIP, Be-976 (Beriev) ... 356 U
Il-76A (Beriev) ... 351 U
Il-76A-50 (Beriev) ... 351 U
Il-78M Midas (Ilyushin) ... 364 U
Il-96-300 (Ilyushin) ... 365 U
Il-96M (Ilyushin) ... 365 U
Il-96T (Ilyushin) ... 365 U
Il-98 (Ilyushin) ... 366 U
Il-100 (Ilyushin) ... 366 N
Il-103 (Ilyushin) ... 366 U
Il-112 (Ilyushin) ... 368 U

www.janes.com Jane's All the World's Aircraft 2001-2002

INDEXES: AIRCRAFT—I - L

Il-114 (Ilyushin)	368 U
Il-114FK (Ilyushin)	370 U
Il-114P (Ilyushin)	369 U
Il-140 (Ilyushin)	370 N
Il-214 (Ilushin)	370 N
ILYUSHIN (see Aviatsionnyi Kompleks Imeni S.V. Ilyushina)	359 U
IPTN (see Industri Pesawat Terbang Nusantara)	180 U
Iran 140 (IAMI)	260 U
IS-2 (IL)	319 U
IS-28M2/GR (IAR)	340 U
I/U-27A Caravan 208 (Cessna)	634 U
IAR SA, SC (Romania)	340 U
IAROM SA, SC (Romania)	337 U
Ibis Ae 270 (Ibis)	248 U
IBIS AEROSPACE LTD (International)	248 U
IE-101 Elitar (Elitar)	358 U
IKARUS DEUTSCHLAND (Germany)	158 V
ILYUSHIN (Russian Federation)	359 U
Impuls (SSVOBB)	310 U
Impulse (SSVOBB)	310 U
Individual Lifting Vehicle 100B (PAM)	718 U
INDUSTRIA AERONAUTICA NEIVA SA (Brazil)	26 U
INDUSTRI PESAWAT TERBANG NUSANTARA PT (Indonesia)	180 U
INIZIATIVE INDUSTRIALI ITALIANE SpA (Italy)	282 V
INSTYTUT LOTNICTWA (Poland)	318 U
INTERAVIA KONSTRUKTORSKOYE BURO AO (Russian Federation)	370 U
INTERPLANE spol s.r.o. (Czech Republic)	95 U
INTRACOM GENERAL MACHINERY SA (International)	250 N
Ipanema EMB-202 (NEIVA)	26 U
IRAN AIRCRAFT MANUFACTURING INDUSTRIES COMPANY (Iran)	259 U
IRAN AVIATION INDUSTRIES ORGANISATION (Iran)	259 U
Iridium M-93 (PZL Mielec)	324 U
Iridium M-96 (PZL Mielec)	324 U
IRKUTSKOYE AVIATSIONNOYE PROIZVODSTVENNOYE OBEDINENIE OAO (Russian Federation)	359 U
Iryda M-93 (PZL Mielec)	324 U
Iryda M-96 (PZL Mielec)	324 U
Iskierka, M-26 (PZL)	324 U
Islander (Romaero)	341 U
Islander AL. Mk 1 (B-N)	522 U
Islander BN2B (B-N)	521 U
Islander CC Mk 2/2A (B-N)	522 U
ISRAEL AIRCRAFT INDUSTRIES LTD (Israel)	261 U
ISSOIRE AVIATION (France)	134 U

J

J6 Fregata (J&AS)	320 U
J-7 (CAC)	62 U
J-8 (SAC)	78 U
J-10 (CAC)	66 U
J-11 (SAC)	80 U
J.300 Series 2 (Chereau)	112 U
JADC (see Japan Aircraft Development Corporation)	293 U
JAS 39 Gripen (Saab)	479 U
JASDF (see Nihon Koku Jieitai)	293 U
JGSDF (see Nihon Rikujyo Jieitai)	293 V
JH-7 (XAC)	82 U
JJ-7 (GAIC)	67 U
JMSDF (see Nihon Kaijyo Jieitai)	293 V
Jabiru (Jabiru)	7 U
JABIRU AIRCRAFT PTY LTD (Australia)	7 U
JABIRU ASIA (Sri Lanka)	478 U
Jackdaw (Noin)	136 U
Jaguar (SEPECAT)	254 V
Jaguar International (HAL)	173 U
JAPAN AIRCRAFT DEVELOPMENT CORPORATION (Japan)	293 U
J & AS AERO DESIGN sp z o o (Poland)	320 V
Jayhawk HH-60J (Sikorsky)	768 U
Jayhawk T-1A (Raytheon)	744 U
Jetcruzer 500 (AASI)	539 U
Jet Fox 97 (Euro ALA)	279 U
Jet Hawk 4T (Groen)	657 N
JetRanger III, 206B-3 (Bell)	28 U
Jianjiji-7 (CAC)	62 U
Jianjiji-8 (SAC)	78 U
Jianjiji-10 (CAC)	66 U
Jianjiji Hongzhaji-7 (XAC)	82 U
Jianjiji Jiaolianji-7 (GAIC)	67 U
John Doe (American Homebuilts')	553 U
Joint STARS E-8 (Northrop Grumman)	716 U
Joint Strike Fighter, X-32 (Boeing)	573 U
Joint Strike Fighter (Lockheed Martin)	687 U
Joint Strike Fighter (US Navy)	782 U
JOPLIN LIGHT AIRCRAFT (USA)	663 N
Jumbolino Avro RJ (BAE)	518 U
Jungmann T-131 (SSH)	335 U
JURCA, MARCEL (France)	135 U

K

K-8 Karakorum 8 (NAMC)	72 U
K-51 Peregrino (Kovacs)	25 U
K-1200 K-MAX (Kaman)	664 U
Ka-27 Helix (Kamov)	371 U
Ka-28 Helix (Kamov)	371 U
Ka-29 Helix-B (Kamov)	372 U
Ka-31 Helix-B (Kamov)	372 U
Ka-32 Helix-C (Kamov)	373 U
Ka-50 Chernaya Akula (Kamov)	375 U
Ka-52 Alligator (Kamov)	377 U
Ka-60 Kasatka (Kamov)	378 U
Ka-62 (Kamov)	378 U
Ka-64 Sky Horse (Kamov)	379 U
Ka-115 (Kamov)	379 U
Ka-215 (Kamov)	379 N
Ka-226A Sergei (Kamov)	379 U
KAC (see Kanematsu Aerospace Corporation)	294 U
KAI (see Korea Aerospace Industries)	304 U
KAL (see Korean Air Lines)	306 U
KAPO (see Kazanskoye Aviatsionnoye Proizvodstvennoye Obedinenie Imeni S.P. Gorbunova)	380 U
KAPP (see Kumertskoye Aviatsionnoye Proizvodstvennoye Predpriyatie)	381 U
KC-130J (Lockheed Martin)	682 U
KHRUNICHEV (see Gosudarstvennyi Kosmicheskii Nauchno-Proizvodstvennyi Tsentr Imeni M.B. Khrunicheva)	383 U
K-MAX K-1200 (Kaman)	664 U
KMH (KAI)	306 U
KMH (KAL)	307 U
KnAAPO (Russian Federation)	387 U
KP-2U Sova (Kappa)	96 U
KR-860 (Sukhoi)	450 U
KT-1 Woong-Bee (KAI)	304 U
Kai T-3 (Fuji)	293 U
KAISER FLUGZEUGBAU (Germany)	158 U
KAMAN AEROSPACE CORPORATION (USA)	664 U
KAMOV OAO (Russian Federation)	371 U
KANEMATSU AEROSPACE CORPORATION (Japan)	294 U
Kangaroo (VulcanAir)	291 U
Kania (PZL Swidnik)	326 U
Kapitan SKB-1 S-400 (SGAU)	426 U
KAPPA 77 AS (Czech Republic)	96 U
Karakorum 8 K-8 (NAMC)	72 U
Kasatka Ka-60 (Kamov)	378 U
Katana 100 (Diamond)	11 U
Katana DA 20-A1 (Diamond)	52 U
Katana DA 20-A2 (Diamond)	53 U
Katana DA 20-C1 (Diamond)	53 U
Katana Xtreme HK 36 (Diamond)	10 U
KAWASAKI JUKOGYO KABUSHIKI KAISHA (Japan)	294 U
KAZANSKOYE AVIATSIONNOYE PROIZVODSTVENNOYE OBEDINENIE IMENI S.P. GORBUNOVA (Russian Federation)	380 U
KAZANSKY VERTOLETNYI ZAVOD (Russian Federation)	381 U
KHARKOVSKOYE GOSUDARSTVENNOYE AVIATSIONNOYE PROIZVODSTVENYE PREDPRIYATIE (Ukraine)	509 U
KHRUNICHEV (Russian Federation)	383 U
Killer Whale Ka-60 (Kamov)	378 U
KIMBALL ENTERPRISES INC, JIM (USA)	666 U
KINETIC AVIATION (USA)	666 N
King Air 200 Military Versions (Raytheon)	741 U
King Air B200 (Raytheon)	740 U
King Air C90B (Raytheon)	739 U
King Air C90SE (Raytheon)	739 U
King Air 350 (Raytheon)	742 U
KingCobra AH-1Z (TAI)	494 N
Kiowa OH-58D (Bell)	568 U
Kiowa Warrior OH-58D(I) (Bell)	568 U
Kitfox (SkyStar)	777 U
Kitfox Lite (SkyStar)	778 U
Kitfox Lite² (SkyStar)	778 N
Kitty Hawk (PZL Swidnik)	326 U
KNR AIRCRAFT CO INC (USA)	667 U
Kittiwake (Kazan)	381 U
Koala A 119 (Agusta)	274 U
KOLB AIRCRAFT COMPANY, NEW INC (USA)	667 U
Kolbra (Kolb)	667 N
Koliber PZL-110 (PZL Warszawa)	333 U
Koliber Junior PZL-112 (PZL Warszawa)	334 U
Koliber 235 Senior PZL-111 (PZL Warszawa)	332 U
Kolibri T-311 (Aeroprogress/ROKS-Aero)	345 U
KOMPANIYA REFLAI (Russian Federation)	423 U
KOMPANIYA VITEK	465 U
KOREA AEROSPACE INDUSTRIES (Korea, South)	304 U
KOREAN AIR LINES CO LTD (Korea)	306 U
Korpen S 102B (Gulfstream)	660 U
KOVACS, JOSEPH (Brazil)	25 U
Krechet, SKB-1 (SGAU)	426 U
KUMERTSKOYE AVIATSIONNOYE PROIZVODSTVENNOYE PREDPRIYATIE (Russian Federation)	381 U

L

L-06 (SGAU)	425 N
L-39 Albatros (Aero)	87 U
L-39 MS (Aero)	89 U
L-59 (Aero)	89 U
L-139 Albatros (Aero)	87 U
L-159 (Aero)	89 U
L-410 UVP-E (Let)	97 U
L-420 (Let)	97 U
L-610G (Let)	99 U
LAVIASA (see Latin Americana de Aviacion SA)	1 U
LMAASA (see Lockheed Martin Aircraft Argentina)	1 U
Lak-52 (Aerostar)	338 U
LFS (Sukhoi)	447 V
LH-X (JGSDF)	293 U
LK-2M Sluka (Letov)	102 U
L11/Pegas (see Letno-ISSLEDOVATELSKY INSTITUT IMENI M.M. Gromova)	388 U
LM200 Loadmaster (Ayres)	563 U
LMZ (Russian Federation)	388 U
LOH (HAL)	172 U
LP-1 Fury (Lopresti)	698 U
LR-2 (Raytheon)	742 U
LS2 (HAT)	167 U
LT-11M Swati (BHEL)	169 U
LVM (Russian Federation)	388 U
Ladybird HGM-2 (He-Ro's)	318 U
Lambáda, UFM 11 (Urban)	104 U
Lambáda, UFM 13 (Urban)	104 U
LAMBERT AIRCRAFT ENGINEERING BVBA (Belgium)	13 U
Lancair IV (Lancair)	669 U
Lancair IVP (Lancair)	669 U
Lancair ES (Lancair)	668 U
LANCAIR GROUP INC (USA)	668 U
Lancair Super ES (Lancair)	669 U
Lancer (HAL)	174 U
Laser (Kolb)	668 N
LATIN AMERICANA de AVIACION SA (Argentina)	1 U
Learjet 31A (Learjet)	671 U
Learjet 45 (Learjet)	672 U
Learjet 60 (Learjet)	673 U
LEARJET (USA)	671 U
Legacy (EMBRAER)	23 N
Legacy 2000 (Lancair)	668 U
Legend (Performance)	720 U
Leopard (CMC)	525 V
LET AS, KUNOVICE (Czech Republic)	97 U
LETECKE OPRAVOVNE TRENCIN SP (Slovak Republic)	473 U
LETECKA TOVARNA s.r.o (Czech Republic)	101 U
LET-MONT s.r.o. (Czech Republic)	101 V
LETNO-ISSLEDOVATELSKY INSTITUT IMENI M.M. GROMOVA GOSUDARSTVENNYI NAUCHNYI TSENTR (Russian Federation)	388 U
LETOV AIR s.r.o (Czech Republic)	101 U
LMZ (Russian Federation)	388 U
LEZA-LOCKWOOD CORPORATION (USA)	674 U
LIBERTY AERONAUTICAL (USA)	675 U
LIBERTY AIRCRAFT PLC (UK)	529 N
Light Aircraft, Morane (EADS SOCATA)	132 U
Light Combat Aircraft (ADA)	168 U
Lightning (IAMI)	260 U
Lightning Bug (RFW)	750 U
LIGHT'S AMERICAN SPORTSCOPTER INC (Taiwan)	493 N
Light Wing GR-582 (Hughes)	6 U
Light Wing GR-912 (Hughes)	6 U
Light Wing Sport 2000 (Hughes)	6 U
Lion APM-21 (Issoire)	134 U
Lionceau APM-20 (Issoire)	134 U
Lion Cub (Issoire)	134 U
Lionheart (Griffon)	656 U
Little Bear (Henderson)	662 U
Little Spark M-26 (PZL)	324 U
LITTLE WING AUTOGYROS INC (USA)	675 V
Loadmaster, LM200 (Ayres)	563 U
LOCKHEED MARTIN AERONAUTICS COMPANY (USA)	676 V
LOCKHEED MARTIN AIRCRAFT ARGENTINA SA (Argentina)	1 U
LOCKHEED MARTIN CORPORATION (USA)	676 U
LOLA UTVA FABRIKA AVIONA (Yugoslavia)	796 U
LongRanger IV, 206L-4 (Bell)	29 U
LOPRESTI INC (USA)	698 V

L – O—AIRCRAFT: INDEXES

LUKHOVITSY MASHINOSTROITELNYI ZAVOD (Russian Federation)	388
LUSCOMBE AIRCRAFT CORPORATION (USA)	698
Lynx (GKN Westland)	534

M

M-7 Series (Maule)	699
M9-200 AirCruiser (Millicer)	8
M-10 AirTourer (Millicer)	8
M-12 (Aviacor)	348
M-16 Mifeng (BUAA)	62
M-18 Dromader (PZL)	322
M-19 Shark (Magni)	283
M-20 Mewa (PZL)	323
M-20 Talon (Magni)	283
M20K Encore (Mooney)	709
M20M Bravo (Mooney)	709
M20R Ovation (Mooney)	710
M20S Eagle (Mooney)	710
M-26 Iskierka (PZL)	324
M-28 Bryza (PZL)	320
M-28 Skytruck (PZL)	321
M80 (Masquito)	13
M-93 Iryda (PZL-Mielec)	324
M-96 Iryda (PZL-Mielec)	324
M-101T Gzhel (Myasishchev)	420
M-201 Sokol (Myasishchev)	421
M-202PW Olen (Myasishchev)	422
M-203PW Barsuk (Myasishchev)	422
M212, Mission (Lambert)	13
M-309 (Adam)	540
M-346 (Aermacchi)	270
M-500 (Myasishchev)	422
MA-60 (XAC)	83
MB-300 Boomerang (Morrow)	710
MB-339 (Aermacchi)	267
MCR01 (Dyn'Aero)	128
MCR4S (Dyn'Aero)	129
MD3-160 AeroTiga (SME)	308
MD-11 (Boeing)	618
MD 500 (MD)	702
MD 500 Defender (MD)	703
MD 520N (MD)	704
MD 530 (MD)	702
MD 530 Defender (MD)	703
MD 600N (MD)	705
MD Explorer (MD)	706
Me262 Messerschmit (Classic)	643
MEAA (see Malaysian Experimental Aircraft Association)	307
MFI 1.42, 1.44 (MiG)	398
MH-6 (MD)	703
MH-47 Chinook (Boeing)	590
MH-53E Sea Dragon (Sikorsky)	760
MH-60A (Sikorsky)	761
MH-60G Pave Hawk (Sikorsky)	761
MH-60K (Sikorsky)	761
MH-60L (Sikorsky)	761
MH-90 Enforcer (MD)	706
MH2000 (Mitsubishi)	300
Mi-6, Hook (Mil)	403
Mi-8, (V-8) Hip (Mil)	404
Mi-9, Hip-G (Mil)	406
Mi-14, Haze (Mil)	406
Mi-17, (Mi-8M) Hip-H (Mil)	407
Mi-17D (Kazan)	381
Mi-17KF Kittiwake (Kazan)	381
Mi-17N (Mil)	381
Mi-18 (see Mi-17) (Mil)	410
Mi-19 (see Mi-8MT/Mi-17) and Mi-9 (Mil)	410
Mi-22 Hook (Mil)	410
Mi-22, Hook-C (see Mi-6) (Mil)	410
Mi-24, Hind (Mil)	410
Mi-25, Hind (Mil)	410
Mi-26, Halo (Mil)	414
Mi-27 (Mil)	416
Mi-28, Havoc (Mil)	416
Mi-34, Hermit (Mil)	417
Mi-35 Hind (Mil)	410
Mi-38 (Euromil)	247
Mi-38 (Mil)	418
Mi-52 Snegir (Mil)	419
Mi-60 (Mil)	419
Mi-171, Hip (Mil)	407
Mi-172, Hip-H (Mil)	407
Mi-234 (Mil)	419
MiG (see Aviatsionnyi Nauchno-Promyshlennyi Kompleks MiG)	389
MiG 1.42 (1.44) MFI, (MiG)	398
MiG-29, Fulcrum (MiG)	389
MiG-31, Foxhound (MiG)	396
MiG-35 (MiG)	398
MiG 110 (MiG)	402
MiG 115 (MiG)	403
MiG-AT (MiG)	400
MIL (see Moskovsky Vertoletny Zavod Imeni M L Milya OAO)	403
MJ5 Sirocco (Jurca)	135
MK2 (MK Helicopter)	159
Model 12 (Kimball)	666
Model 269 (Schweizer)	757
MPA-X (JMSDF)	293
MPX (Diamond)	53
MRA-Mk4 Nimrod (BAE)	516
Mü30 Schlacro (AkaFlieg)	144
MV-22 (Bell Boeing)	570
MXP-740 (Ekoflug)	473
MY-103 (Mylius)	160
MY-104-200 (Mylius)	161
MYASISHCHEV (see Eksperimentalnyi Mashinostroitelnyi Zavod Imeni VM Myasishcheva)	420
Magic (Kaiser)	158
MAGNI GYRO (Italy)	283
Magnum (Avid)	560
Mainstay, A-50 (Beriev)	351
Major, DR 400/160 (Robin)	140
Mako EADS DEUTSCHLAND	146
MALAYSIAN EXPERIMENTAL AIRCRAFT ASSOCIATION (Malaysia)	307
Malibu Meridian PA-46-500TP (Piper)	726
Malibu Mirage PA-46-350P (Piper)	725
Manager I-23 (IL)	318
Mangusta, A 129 (Agusta)	271
MARCEL JURCA (France)	135
Masked Duck (Edra Helicentro)	16
MASQUITO AIRCRAFT n.v. (Belgium)	13
MAULE AIR INC (USA)	699
Maverick (Murphy)	57
MAVERICK AIR INC (USA)	701
May, Il-38 (Ilyushin)	360
MD HELICOPTERS INC (USA)	702
Medium Combat Aircraft (ADA)	169
Mercury (Khrunichev)	386
Merkury (Khrunichev)	386
Merlin GT (Aerocomp)	542
Merlin HC. Mk 3, EH 101 (EHI)	218
Merlin HM. Mks 1 and 2 (EHI)	218
Messerschmitt Me 262 (Classic)	643
Mewa, M-20 (PZL)	323
MEZHDUNARODNYYI AVIATSIONNYI KOMPANIYA ILUSHINA (Russian Federation)	359
MICCO AIRCRAFT COMPANY (USA)	708
Midas, Il-78 (Ilyushin)	364
Mifeng M-16 (BUAA)	62
MIL (Russian Federation)	403
Military Scout (Modus)	493
Millennium Swift (Aviat)	559
MILLICER AIRCRAFT INDUSTRIES (Australia)	7
Mirage, G-3 (Remos)	163
Mirage 2000 (Dassault)	113
Mriya, An-225 (Antonov)	509
Mission M212 (Lambert)	13
MITSUBISHI JUKOGYO KABUSHIKI KAISHA (Japan)	297
MK HELICOPTER GmbH (Germany)	159

Note: Aircraft previously identified as 'Model xxx' are now listed as numbers only at the beginning of the index.

MODUS VERTICRAFT CORPORATION (Taiwan)	493
Molniya-1 (Molniya)	419
MOLNIYA, NAUCHNO-PROIZVODSTVENNOYE OBEDINENIE MOLNIYA OAO (Russian Federation)	419
Mongoose A 129 (Agusta)	271
Monitor Jet (CAG)	51
Monoculp (Culp's)	645
MOONEY AIRCRAFT CORPORATION (USA)	709
Morane Light Aircraft (EADS SOCATA)	132
MORAVAN AEROPLANES INC (Czech Republic)	105
MOSKOVSKY VERTOLETNY ZAVOD IMIENI ML MILYA OAO (Russian Federation)	403
Moskvich Ka-115 (Kamov)	379
Mountain Goat (Kinetic)	666
Mrówka 2001, PZL-126P (Agrolot)	316
Multimission Maritime Aircraft (US Navy)	783
Multi Role Tanker Transport (Airbus)	202
Multirole Transport Aircraft (ADA)	169
MURPHY AIRCRAFT MANUFACTURING LTD (Canada)	55
Muscovite (Kamov)	379
Mushshak (PAC)	313
MYLIUS FLUGZEUGWERK GmbH & Co KG (Germany)	160

N

N-5A (NAMC)	75
N-250 (Dirgantara)	178
N-2130 (Dirgantara)	179
NA 40 Bongo (Unis)	104
NAI (see Nanjing Aeronautical Institute)	71
NAL (see National Aerospace Laboratories, India)	176
NAL (see National Aerospace Laboratory)	301
NAMC (see Nanchang Aircraft Manufacturing Company)	72
NAPO (see Gosuderstvennoye Uniternoye Predpriyatie Novosibirskoye Aviatsionnoye-Proizodstvennoye Obedinenie Imeni V P Chakalova)	422
NAS-332 Super Puma (Dirgantara)	180
NASA (see National Aeronautics and Space Administration)	711
NASA Space Programmes (NASA)	711
NASDA (see National Space Development Agency of Japan)	301
NBELL-412 (Dirgantara)	180
NEIVA (see Industria Aeronáutica Neiva S/A)	26
NH 90 (NH Industries)	251
NLA (see Nanjing Light Aircraft Incorporated Company)	76
NRJ (see New Regional Jet)	77
NANCHANG AIRCRAFT MANUFACTURING COMPANY (China)	72
NANJING AERONAUTICAL INSTITUTE (China)	71
NANJING GROEN AVIATION INDUSTRIAL LTD (China)	76
NANJING LIGHT AIRCRAFT INCORPORATED COMPANY (China)	76
NATIONAL AERONAUTICS AND SPACE ADMINISTRATION (USA)	711
NATIONAL AEROSPACE LABORATORIES (India)	176
NATIONAL AEROSPACE LABORATORY (Japan)	301
NATIONAL SPACE DEVELOPMENT AGENCY OF JAPAN (Japan)	301
NAUCHNO-KOMMERCHESKY FIRMA TECHNOAVIA (Russian Federation)	450
NAUCHNO-PROIZVODSTVENNOYE OBEDINENIE AVIA (Russian Federation)	347
NAUCHNO-PROIZVODSTVENNOYE OBEDINENIE MOLNIYA OAO (Russian Federation)	419
Nauticair 450 (Archedyne)	554
Netz F-16A/B (Lockheed Martin)	689
New Derringer (Derringer)	646
NEW KOLB AIRCRAFT COMPANY (USA)	667
NEW REGIONAL JET PROGRAMME MANAGEMENT COMPANY (China)	77
NH INDUSTRIES sarl (International)	251
Night Falcon (Lockheed Martin)	690
NIHON KAIJYO JIEITAI (Japan)	293
NIHON KOKU JIEITAI (Japan)	293
NIHON RIKUJYO JIEITAI (Japan)	293
Nimrod MRA.Mk 4 (BAE)	516
NITSCHE, FLUGZEUGBAU GmbH (Germany)	161
NOIN AERONAUTIQUE (France)	136
Nongye Feiji 5 (NAMC)	75
NORTHROP GRUMMAN CORPORATION (USA)	714
North Star (Custom Flight)	52
Nugget BD-17 (Bede)	565

O

OH-1 (Kawasaki)	295
OH-58D Kiowa (Bell)	568
OH-58D(I) Kiowa Warrior (Bell)	568
OMF (see Ostmechlenburgische Flugzeugbau)	162
OMF-160 Symphony (OMF)	162
OT-47B (Cessna)	638
Observer 2, P.68 (VulcanAir)	290
Observer 2TC, P.68 (VulcanAir)	290
Olen M-202PW (Myasishchev)	422
OPYTNO-KONSTRUKTORSKOYE BYURO IMENI AS YAKOVLEVA OAO (Russian Federation)	466
OPYTNYI KONSTRUKTORSKOYE BYURO SUKHOGO AOOT (Russian Federation)	427
OPYTNYI KONSTRUKTORSKOYE BYURO CHERNOV B& M OOO (Russian Federation)	356
Oriole, PZL-104M (PZL Warszawa)	331
Orlik PZL-130 (PZL Warszawa)	329
Orlik 2000 PZL-140 (PZL Warszawa)	334
Osa Dubna-2 (Dubna)	358
Osprey 901 (Bell Boeing)	570
OSTMECHLENBURGISCHE FLUGZEUGBAU GmbH (Germany)	162
OTKRITOYE AKTSIONERNOYE OBSHCHESTVO MOTOR-SICH (Ukraine)	509
Ovation M20R (Mooney)	710
Ovocl (Vitek)	465
Owl (IAIO)	260
Owl (Kappa)	96

P

P-51 G (Cameron)	628 U
P.68 Observer 2 (VulcanAir)	290 U
P.68 Observer 2TC (VulcanAir)	290 U
P.68C (VulcanAir)	290 U
P.68CTC (VulcanAir)	290 U
P92 Echo (Tecnam)	288 U
P96 Golf (Tecnam)	289 U
P.166 (Piaggio)	283 U
P.180 Avanti (Piaggio)	284 U
P200 series (Pottier)	136 U
P305 (Airbus)	200 U
PA-28-161 Warrior III (Piper)	721 U
PA-28-181 Archer III (Piper)	721 U
PA-28R-201 Arrow (Piper)	722 U
PA-32R-301 Saratoga II HP (Piper)	722 U
PA-32R-301T Saratoga II TC (Piper)	723 U
PA-34-220T Seneca V (Piper)	724 U
PA-44-180 Seminole (Piper)	725 U
PA-46-350P Malibu Mirage (Piper)	725 U
PA-46-500TP Malibu Meridian (Piper)	726 U
PAC (see Pacific Aerospace Corporation Limited)	311 U
PAC (see Pakistan Aeronautical Complex)	312 U
PADC (see Philippine Aerospace Development Corporation)	314 U
PAI (see Pacific Aeronautical Inc)	314 U
PAM (see Performance Aviation Manufacturing Group)	718 U
PC-6 Turbo-Porter (Pilatus)	484 U
PC-7 Turbo-Trainer (Pilatus)	486 U
PC-7 Mk II M Turbo-Trainer (Pilatus)	486 U
PC-7/CH Turbo-Trainer (Pilatus)	486 U
PC-9M Advanced Turbo Trainer (Pilatus)	487 U
PC-12/45 (Pilatus)	489 U
PC-12 Eagle (Pilatus)	491 U
PC-21 (Pilatus)	489 U
PT-6A (NAMC)	75
PT-X (JASDF)	293 N
PZL (see Polskie Zaklady Lotnicze)	320 U
PZL-104M Wilga 2000 (PZL Warszawa)	331 U
PZL-106BT Turbo-Kruk (PZL Warszawa)	332 U
PZL-110 Koliber (PZL Warszawa)	332 U
PZL-111 Koliber 235A Senior (PZL Warszawa)	333 U
PZL-112 Junior (PZL Warszawa)	334 U
PZL-126P Mrówka (Agrolot)	316 U
PZL-130 Orlik (PZL Warszawa)	329 U
PZL-140 Orlik 2000 (PZL Warszawa)	334 U
PZL-240 Pelikan (PZL Warszawa)	335 U
PZL MIELEC (see Zaklad Lotniczy PZL-Mielec Sp. z.o.o.)	324 U
PZL SWIDNIK SA (see Zygmunta Pulawskiego-PZL Swidnik)	326 U
PZL WARSZAWA-OKECIE (see Państwowe Zakłady Lotnicze Warszawa-Okecie)	329 U
PACIFIC AERONAUTICAL INC (Philippines)	314 U
PACIFIC AEROSPACE CORPORATION LIMITED (New Zealand)	311 U
Pairadigm, SST 2000 (VSTOL)	789 U
PAKISTAN AERONAUTICAL COMPLEX (Pakistan)	312 U
Pampa NG, 1A 63 (LMAASA)	1 U
PAŃSTWOWE ZAKŁADY LOTNICZE WARSZAWA-OKECIE (Poland)	329 U
Panther AS 565, (army/air force) (Eurocopter)	234 U
Panther AS 565, (navy) (Eurocopter)	234 U
PAPA 51 LTD (USA)	719 U
Parastu (IAIO)	260 U
Pashosh (EADS SOCATA)	131 U
PATRIA FINAVITEC OY (Finland)	108 U
Patrullero C-212 (CASA)	477 U
Paturi (Edra Helicentro)	16 U
Pave Hawk UH-60A (Sikorsky)	761 U
PAXMAN'S NORTHERN LITE AIRCRAFT (Canada)	58 U
PC-FLIGHT FLUGZEUGBAU GmbH (Germany)	162 V
Pegasus T-507 (Khrunichev)	386 U
Pelican Che-22 (Chevnov)	356 U
Pelican PZL-240 (PZL Warszawa)	335 U
Pelikan PZL-240 (PZL Warszawa)	335 U
Peregrine (Kovacs)	25 U
Peregrino K-51 (Kovacs)	25 U
PERFORMANCE AIRCRAFT (USA)	720 V
PERFORMANCE AVIATION MANUFACTURING GROUP (USA)	718 U
Persuader CN-235 MP (Airtech)	206 U
Pethen (Boeing)	594 U
PEZETEL (Poland)	320 U
PHANTOM WORKS, BOEING (USA)	573 U
PHILIPPINE AEROSPACE DEVELOPMENT CORPORATION (Philippines)	310 U
Phoenix CN-235 (Airtech)	204 U
Phoenix Fanjet (Alberta)	27 U
Phoenix GM-17 Feniks (Intracom)	250 N
Phoenix Kasatik M-12	348 U
PIAGGIO AERO INDUSTRIES SpA (Italy)	283 U
PIASECKI AIRCRAFT CORPORATION (USA)	720 U
PILATUS FLUGZEUGWERKE AG (Switzerland)	484 U
Pillan T-36 (ENAER)	60 U
PIPER AIRCRAFT INC, THE NEW (USA)	721 U
Piper New Projects (Piper)	727 N
Piper UL (Let-Mont)	101 U
Pocket Rocket (Hughes)	7
Polaris CC-150 (Airbus)	184 U
Polaris FK14 (B&F)	146 U
POLYOT PROIZVODSTVENNOYE OBEDINENIE POLET (Russian Federation)	423 U
POLSKIE ZAKLADY LOTNICZE Sp. z.o.o. (Poland)	320 U
POTTIER, AVIONS (France)	136 V
PRECISION TECH AIRCRAFT INC (USA)	727 N
Premier 1 (Raytheon)	735 U
Président, DR500i (Robin)	142 U
Pretty Flight (PC-Flight)	162 U
Private Explorer (Private Explorer)	727 U
PRIVATE EXPLORER INC (USA)	727 V
Proficient (PAC)	313 U
PROIZVODSTVENNO-TEKHNICHESKY KOMPLEKS DUBNENSKOGO MASHINOSTROITELNOGO ZAVOD AO (Russian Federation)	358 U
PROIZVODSTVENNOYE OBEDINENIE LEGKIE VERTOLETNYI MI OAO (Russian Federation)	388 U
PROSPORT AVIATION (USA)	728 N
Proteus 281 (Scaled Composites)	754 U
PT DIRGANTARA INDONESIA (Indonesia)	178 U
PT INDUSTRI PESAWAT TERBANG NUSANTARA (Indonesia)	180 U
Puelche (LAVIASA)	1 N
PULSAR AIRCRAFT CORPORATION (USA)	729 U
PZL-MIELEC DO-BRASIL FABRICA D'AVIOES LTDA (Brazil)	27 U

Q

Q-5 (NAMC)	73 U
Qiangjiji-5 (NAMC)	73 U
Quad Tiltrotor (Bell)	570 U
QUIKKIT (se Rainbow Flyers Inc) (USA)	730 U

R

R22 Beta II (Robinson)	751 U
R44 (Robinson)	752 U
R-80 Tiger Moth (Fisher)	653 U
R-99A (EMBRAER)	20 U
R-99B (EMBRAER)	20 U
R 2160 (Robin)	139 U
RAF (see Rotary Air Force)	58 U
RAF (see Royal Air Force)	529 U
RAH-66 Comanche (Boeing Sikorsky)	624 U
REFLY (see Kompaniya)	423 N
RF-9, ABS (ABS)	489 U
RFW (see Reflex Fibreglass Works)	750 U
RG-8A (Schweizer)	755 U
RJ (Avro)	518 U
RJ 70 (Avro)	518 U
RJ 85 (Avro)	518 U
RJ 100 (Avro)	518 U
RJX (Avro)	518 U
RMX-4 (MEAA)	307 U
RN (see Royal Navy)	529 U
RS-80 Tiger Moth (Fisher)	653 U
RSK MiG (Russian Federation)	423 N
RU-38A Twin Condor (Schweizer)	756 U
RV-4 (Van's)	784 U
RV-6 (Van's)	784 U
RV-6A (Van's)	784 U
RV-8 (Van's)	785 U
RV-8A (Van's)	785 U
RV-9A (Van's)	785 N
Ra'am (Boeing)	575 U
Rafale (Dassault)	117 U
RAINBOW FLYERS INC, QUIKKIT DIVISION (USA)	730 U
Rambler (Let-Mont)	101 U
RANS INC (USA)	731 U
Raptor (Bussard)	146 N
Raptor, F-22 (Lockheed Martin)	677 U
Raven 257 (Wolfsberg-Letov)	101 U
RAYTHEON AIRCRAFT COMPANY (USA)	733 U
Rebel (Murphy)	56 U
Red Kestrel, AH-2A Rooivalk (Denel)	473 U
Régent, DR 400/180 (Robin)	141 U
REFLEX FIBERGLASS WORKS INC (USA)	750 N
Regional Jet (Alliance)	550 U
REIMS AVIATION SA (France)	137 U
REMOS AIRCRAFT GmbH (Germany)	163 U
RENAISSANCE AIRCRAFT LLC (USA)	749 U
RENAISSANCE COMPOSITES INC (USA)	749 U
Renegade II (Murphy)	55 U
Renegade Spirit (Murphy)	56 U
Replacement Interdictor Aircraft (USAF)	782 U
Reusable Launch Vehicle X-33 (NASA)	711 U
Reusable Launch Vehicle X-34 (NASA)	711 U
Revelation (Slipstream)	779 N
Rhapsody (Dreamwings)	647 U
ROBIN AVIATION (France)	139 U
ROBINSON HELICOPTER COMPANY (USA)	751 U
ROMAERO SA, SC (Romania)	341 U
Rooivalk, AH-2A (Denel)	473 U
Rook T-101 (Aeroprogress/ROKS-Aero)	344 U
ROSTOVSKY VERTOLETNYI PROIZVODSTENNYI KOMPLEKS OAO ROSTVERTOL) (Russian Federation)	425 U
ROTARY AIR FORCE INC (Canada)	58 U
Rotax Falke, SF 25 C (Scheibe)	163 U
Roto-Pup (Little Wing)	675 U
RotorMouse EH 1-01 (Hillberg)	663 U
ROTORWAY INTERNATIONAL (USA)	753 U
ROYAL AIR FORCE (UK)	529 U
ROYAL NAVY (UK)	529 U
RUSALEN & RUSALEN SdF (Italy)	286 U
RV Griffin (Canada Air)	51 N

S

S-2B Pitts (Aviat)	558 U
S-2C Pitts (Aviat)	558 U
S2R Turbo-Thrush (Ayres)	562 U
S-7 Courier (Rans)	731 U
S8 (Stemme)	166 N
S10 (Stemme)	165 U
S-12S Super Airaile (Rans)	731 U
S-12XL Airaile (Rans)	731 U
S15 (Stemme)	165 U
S-16 Shekari (Rans)	732 U
S-17 Stinger (Rans)	732 U
S-18 Stinger II (Rans)	733 N
S-26 (Safire)	753 U
S-37 Berkut (Sukhoi)	444 U
S-49 (Sukhoi)	446 U
S-54 (Sukhoi)	446 U
S-55 (Sukhoi)	447 U
S-56 (Sukhoi)	447 U
S-70A (Sikorsky)	761 U
S-70B (Sikorsky)	768 U
S-70C (Sikorsky)	770 U
S-70C(M)-1 Thunderhawk (Sikorsky)	768 U
S-70C(M)-2 Thunderhawk (Sikorsky)	768 U
S-76C+ (Sikorsky)	771 U
S-80 (Sikorsky)	760 U
S-80 (Sukhoi)	449 U
S-92A Helibus (Sikorsky)	772 U
S 102B Kavpen (Gulfstream Aerospace)	660 U
S-302 (CKB-1) (SGAU)	426 U
S-400 SKB-1 (Kapitan) (SGAU)	426 U
SA 2-37A (Schweizer)	755 U
SA 2-38A (Schweizer)	756 U
SA-II Star Streak (CFM)	525 U
SAC (see Shaanxi Aircraft Company)	77 U
SAC (see Shenyang Aircraft Corporation)	78 U
SAIC (see Shanghai Aviation Industry Group)	81 U
SAMC (see Shijiazhuang Aircraft Manufacturing Corporation)	81 U
SATIC (see Special Aircraft Transport International Company)	253 U
SAT 'SAVIAT' (see Spetsialnoye Aviatsionnoye Tekhnology AO)	424 U
SB7L-360 Seeker (Seabird)	9 U
SB 427 (KAI)	306 U
SB-582 (SFLC)	82 U
SBJ (Gulfstream Aerospace)	662 U
SBM-03 Kos (E&K)	317 U
SEP (PZL Świdnik)	328 N
SEPECAT (see Société Européenne de Production de l'Avion E. C. A. T.)	254 V
SF 25 C Falke 2000 (Scheibe)	163 U
SF 25 C Rotax Falke (Scheibe)	163 U
SF 25 C Superschlepper (Scheibe)	164 N
SF 40 C Allround (Scheibe)	164 N
SF 45 SA Spirit (USA)	166 U
SF.260 (Aermacchi)	268 U
SF.260TP (Aermacchi)	270 U
SF.600A Canguro (VulcanAir)	291 U
SFLC (see Shanghai Feiteng LightPlane Company)	82 V
SGAU (see Samarsky Gosudartvennyi Aerokosmitsesky Universitet)	425 U
SH-2G Super Seasprite (Kaman)	664 U
SH-60B Seahawk (Sikorsky)	768 U
SH-60F (Sikorsky)	768 U
SH-60J (Sikorsky)	768 U
SH-60J (Mitsubishi/Sikorsky)	299 U
SH-60R Seahawk (Sikorsky)	768 U
SJ30-2 (Sino Swearingen)	775 U
SKB-1 Favorit (SGAU)	425 N
SL-39 (RSK MIG)	424 U
SM-92 Finist (Technoavia)	451 U
SM-92P (Technoavia)	452 U
SMG-92 (Technoavia)	452 U
SP-20 (MICCO)	708 U
SP-26 (MICCO)	708 U

S – T—AIRCRAFT: INDEXES

Entry	Page
SP-55M (Technoavia)	450
SR20 (Cirrus)	642
SR22 (Cirrus)	642
SR 2500 Super Rebel (Murphy)	57
SRA-4 (Gulfstream Aerospace)	660
SS-2A (Shinmaywa)	302
SS-11 Sky Watch (Freedom Lite)	55
SSH (see Serwis Samalstow Historycznich)	335
SSLF (see Shenyang Sailplane and Lightplane Factory)	82
SST 2000 Pairadigm (VSTOL)	789
SSVOBB (see Stichting Studenten Vliegtuigontwikkeling Bouw en Beheer)	310
ST-4 Azték (Letov)	102
ST Aero (see Singapore Technologies Aerospace)	472
STOL CH 701 (Zenith)	794
STOL CH801 (Zenith)	794
Su-24 Fencer (Sukhoi)	427
Su-25 Frogfoot (Sukhoi)	429
Su-25TM (Su-39) (Sukhoi)	431
Su-26 (Sukhoi)	447
Su-27 Flanker (Sukhoi)	433
Su-27 IB (Sukhoi)	436
Su-27K (Sukhoi)	438
Su-28 Frogfoot (Sukhoi)	429
Su-29 (Sukhoi)	447
Su-30 (Su-27PU) (Sukhoi)	438
Su-30M (Sukhoi)	439
Su-31 (Sukhoi)	448
Su-32FN and MF (Su-27 IB) (Sukhoi)	440
Su-33 (Su-27K)(Sukhoi)	440
Su-33UB (Su-27 KUB) (Sukhoi)	441
Su-34 (Su-27 IB) (Sukhoi)	441
Su-35 (Su-27M) (Sukhoi)	441
Su-37 (Su-27M) (Sukhoi)	442
Su-38L (Sukhoi)	449
Su-39 (Su-25TM) (Sukhoi)	444
SUKHOI (see Gosudarstvennoye Unitarnoye Predpriyatie Aviatsionnyi Voyenno-Promyshlennyi Kompleks Sukhoi)	427
SW-4 (PZL Swidnik)	328
SW-5 (PZL Swidnik)	328
SAAB AB (Sweden)	478
Safir 43 (ZLIN)	105
SAFIRE AIRCRAFT COMPANY (USA)	753
SAMARSKY GOSUDARTVENNYI AEROKOSMITSESKY UNIVERSITET (Russian Federation)	425
Samba UFM 10 (Urban)	105
Samburo AVo 68-R (Nitsche)	161
SAMSUNG AEROSPACE INDUSTRIES LTD (Korea, South)	307
Samurai Sword DA 20-AI (Diamond)	53
Samurai Sword DA 20-CI (Diamond)	53
Sanjaqak (IAIO)	260
Sāras (NAL)	176
Saratoga 11 HP PA-32R-301 (Piper)	722
Saratoga 11 TC, PA-32R-30IT (Piper)	723
SARATOVSKY AVIATSIONNY ZAVOD ZAO (Russian Federation)	425
SAUER, CHRIS (Switzerland)	492
Savaşan Şahin (TAI)	494, 689
Saviat E-4 (Sat Saviat)	424
SCALED COMPOSITES INC (USA)	754
Sceptre (Slipstream)	779
SCHEIBE FLUGZEUGBAU GmbH (Germany)	163
Schlacro Mü30 (AkaFlieg München)	144
SCHWEIZER AIRCRAFT CORPORATION (USA)	755
Scout 8GCBC (American Champion)	552
SCUB (AeroKuhlmann)	109
SEABIRD AVIATION AUSTRALIA PTY LTD (Australia)	9
Sea Dragon MH-53E (Sikorsky)	760
Seahawk SH-60B (Sikorsky)	768
Seahawk SH-60R (Sikorsky)	768
SEASTAR AIRCRAFT INC (USA)	759
SeaStar (Edra Helicentro)	16
Seastar (Seastar)	759
Sea Storm (SG Aviation)	287
SeaWing T-431 (Khrunichev)	385
Seeker SB7L-360 (Seabird)	9
Seminole PA-44-180 (Piper)	725
Senator (Elitar)	359
Seneca V PA-34-220T (Piper)	724
Sentinel, F28F-P (Enstrom)	649
SEQUOIA AIRCRAFT CORPORATION (USA)	759
Sergie Ka-226A (Kamov)	379
SERWIS SAMOLOTOW HISTORYCZNICH (Poland)	335
SG AVIATION (Italy)	286
SHAANXI AIRCRAFT COMPANY (China)	77
Shadow Series D (CFM)	524
Shabaviz 2-75 (IAIO)	260
Shahbaz (IAIO)	260
Shahed X-5 (IAIO)	260
Shamsher (HAL)	173
SHANGHAI AVIATION INDUSTRY GROUP (China)	81
SHANGHAI FEITENG LIGHTPLANE COMPANY (China)	82
Shark (EADS Deutschland)	146
Shark 280FX (Enstrom)	649
Shark M19 (Enstrom)	283
Shekari S-16 (Rans)	732
Shmel, Dubna-1 (Dubna)	358
SHENYANG AIRCRAFT CORPORATION (China)	78
SHENYANG SAILPLANE AND LIGHTPLANE FACTORY (China)	82
SHIJIAZHUANG AIRCRAFT MANUFACTURING CORPORATION (China)	81
SHINMAYWA KOGO KABUSHIKI KAISHA (Japan)	302
Sigma-4 (Albatros)	346
SIKORSKY AIRCRAFT (USA)	760
Silvercraft SH-4 (Sauer)	492
SINGAPORE TECHNOLOGIES AEROSPACE LTD (Singapore)	472
SINO SWEARINGEN AIRCRAFT COMPANY (USA)	775
Sinus (Flight Team)	155
Sirocco MJ5 (Jurca)	135
SKB-1 A-16 (SGAU)	425
SKB-1 S-400 Kapitan (SGAU)	426
SKB-1 Krechet (Gyrfalcon) (SGAU)	426
Sky Arrow (Rotax 912) (III)	282
Sky Arrow (Rotax 914) (III)	282
Skyboy (Interplane)	96
Skyblaster (Slipstream)	779
SKYCRAFT INTERNATIONAL INC (USA)	777
Sky Cruiser (Aeroprakt)	496
Skycycle (Carlson)	628
Skyhawk 172 (Cessna)	630
Sky Horse Ka-64 (Kamov)	379
Skylark T-420 (Khrunichev)	384
Skylane 182S (Cessna)	632
Skylifter (Yakovlev)	471
Sky Raider (Flying K)	655
Sky Raider II (Flying K)	655
Skyrocket III, 19-25 (Aviabellanca)	556
SKYSTAR AIRCRAFT CORPORATION (USA)	777
Skytruck M-28 (PZL)	321
Sky Watch SS-11 (Freedom Lite)	55
SLEPCEV, NESTOR (Australia)	9
SLINGSBY AVIATION LIMITED (UK)	530
Sling Shot (Kolb)	668
SLIPSTREAM INDUSTRIES (USA)	779
Sluka LK-2M (Letov)	102
SME AVIATION SDN BHD (Malaysia)	308
SMOLENSKY AVIATSIONNYI ZAVOD OAO (Russian Federation)	426
SOCATA GROUP AEROSPATIALE MATRA (France)	130
SOCIETE DYN'AERO (France)	128
SOCIETE EUROPEENNE DE PRODUCTION DE L'AVION E. C. A. T. (International)	254
Şoim, IAR-99 (Avioane)	339
Soko 2 (Soko)	14
SOKO AIR MOSTAR (Bosnia-Herzegovina)	14
Sokol M-201 (Myasishchev)	421
SOKOL NIZHEGORODSKY AVIASTROITELNYI ZAVOD AOOT (Russian Federation)	427
Sokól W-3 (PZL Swidnik)	326
Solo Aeropract-21M (Aeropract)	342
Sonerai I (Great Plains)	656
Sonerai II Series (Great Plains)	656
SONEX LTD (USA)	779
Sonex (Sonex)	780
Sopwith Pup (Culp's)	645
Sova, KP-2U (Kappa)	96
Space Shuttle (NAL)	301
Sparrowhawk (Gavilán)	85
Spartan, C-27J (Alenia)	276
Spartan 11E (Luscombe)	698
Special (Culp's)	645
SPECIAL AIRCRAFT TRANSPORT INTERNATIONAL COMPANY GIE (International)	253
SPEEDTWIN DEVELOPMENTS LTD (UK)	531
Speedtwin E2E (Speedtwin)	531
SPENCER AMPHIBIAN AIRCRAFT INC (USA)	780
SPETSIALNOYE AVIATSIONNOYE TEKHNOLOGY AO (Russian Federation)	424
Spirit SF 45 SA (USA)	166
Spirit VA-12C (VisionAire)	788
Sport 150 (Pulsar)	729
Sport 2000, Light Wing (Hughes)	6
Sportsman, 2+2 (Wag-Aero)	791
Sportster (Warner)	791
Sport Trainer (Wag-Aero)	790
Spotted Eaglet, PZL-130 (PZL Warszawa)	329
Spotted Eaglet, PZL-140 (PZL Warszawa)	334
Sprint, TB9 (EADS SOCATA)	130
Spyder (Flightstar)	654
Squall (Dassault)	117
Squirrel HT Mk 1 and HT Mk 2 (Eurocopter)	229
Starliner (Alliance)	550
Star Streak SA-II (CFM)	525
Star TL-96 (TL Ultralight)	103
Stationair 206H (Cessna)	633
STEMME GmbH & Co KG (Germany)	164
STICHTING STUDENTEN VLIEGTUIGONTWIKKELING BOUW EN BEHEER (Netherlands)	310
Stinger, S-17 (Rans)	732
Stinger II, S-18 (Rans)	733
ST JUST AVIATION INC (CANADA)	58
STODDARD-HAMILTON AIRCRAFT INC (USA)	780
Storch (Fly Synthesis)	279
STORCH AVIATION AUSTRALIA PTY LTD (Australia)	9
Storch SS Mk4 (Storch)	9
Stork T-411 (Khrunichev)	384
Storm (SG Aviation)	286
Streak (CFM)	525
STRELA PROIZVODSTVENNOYE OBEDINENIE (Russian Federation)	427
SUKHOI (Russian Federation)	427
Super Airale S-12S (Rans)	731
Super AMX (AMX)	208
Super Blue Hawk (Sikorsky)	770
SUPER-CHIPMUNK INC (Canada)	59
Super-Chipmunk (Super-Chipmunk)	59
SuperCobra (Bell)	566
Super Cruiser (Pulsar)	729
Super Cyclone (St Just)	59
Super Decathlon 8KCAB (American Champion)	553
Super Dimona HK-36 (Diamond)	10
Super ES (Lancair)	669
Super Frog, ATT/SSTOL (Boeing)	590
Super Hornet F/A-18E/F (Boeing)	581
Super Mushshak Agile (PAC)	313
Super Pulsar 100 (Pulsar)	729
Super Puma Mk I AS 332 (Eurocopter)	226
Super Puma Mk II AS 332 L2 (Eurocopter)	228
Super Puma, NAS-332 (Dirgantara)	180
SupaPup Mk 4 (Aerosport)	2
Super Rebel SR 2500 (Murphy)	57
Superschlepper SF 25C (Scheibe)	164
SuperScooper CL-415 (Canadair)	45
Super Seasprite (Kaman)	664
Supersonic Airliner Studies (International)	255
Super Stallion (Aircraft Designs)	545
Super Stallion CH-53E (Sikorsky)	760
Super Transporter A300-608ST (SATIC)	253
Super Toucan (Embraer)	17
Super Tucano EMB-314	17
Super Ximango AMT-200 (Aeromot)	15
Super Zodiac CH 601XL (Zenith)	793
Support, Amphibious and Battlefield Rotorcraft (RAF)	529
Surveyor, ATR 42 (ATR)	216
Sutlej, An-32 (Antonov)	499
Swallow (IAIO)	260
Swati LT-11M (BHEL)	169
S-WING spol s.r.o. (Czech Republic)	102
Swing (S-Wing)	102
SWISS AIRCRAFT AND SYSTEMS ENTERPRISE CORPORATION (SF) (Switzerland)	492
Symphony OMF-160 (OMF)	162

T

T-1A Jayhawk (Raytheon)	744
T-3 Kai (Fuji)	293
T-3A Firefly (Slingsby)	530
T-4 (Kawasaki)	294
T-5 (Fuji)	292
T-6A (Raytheon)	734
T-7 (Fuji)	293
T.16 (Dassault)	123
T.18 (Dassault)	124
T.19A (Airtech)	203
T.19B (Airtech)	203
T-21 (Khrunichev)	384
T-35 Pillan (ENAER)	60
T-45 Goshawk (BAE)	511
T-45 Goshawk (Boeing/BAE)	212
T-50 Golden Eagle (KAI)	305
T67 Firefly (Slingsby)	530
T67M200 Firefly (Slingsby)	531
T67M260 Firefly (Slingsby)	531
T-101 Grach (Aeroprogress/ROKS-Aero)	344
T-131 Jungmann (SSH)	335
T182T Turbo Skylane (Cessna)	632
T 206H Turbo Stationair (Cessna)	633
T-211, Thorp (AD)	510
T-260 (Aermacchi)	268
T-311 Kolibri (Aeroprogress/ROKS-Aero)	345
T-400 (Raytheon)	744
T-411 Aist (Khrunichev)	384
T-420 Skylark (Khrunichev)	384
T-421 Wolverine (Khrunichev)	385
T-431 Sea Wing (Khrunichev)	385

INDEXES: AIRCRAFT—T - X

Entry	Page
T-433 Flamingo (Khrunichev)	386 N
T-440 Merkury (Khrunichev)	386 U
T-507 (Khrunichev)	386 U
T-517 Farmer (Khrunichev)	387 U
TAAL (see Taneja Aerospace and Aviation)	177 U
TAC (see Taiwan Aerospace Corporation)	494 V
TAI (see Turkish Aerospace Industries Inc)	494 U
TAPO (Uzbekistan)	796 U
TASA (see Tbilisi Aviation State Association)	142 U
TB 9 Sprint (EADS SOCATA)	130 U
TB 9C Tampico Club (EADS SOCATA)	130 U
TB 10 Tobago (EADS SOCATA)	130 U
TB 10 Tobago GT (EADS SOCATA)	130 U
TB 20 Trinidad GT (EADS SOCATA)	131 U
TB 21 Trinidad GT Turbo (EADS SOCATA)	131 U
TB 200 Tobago (EADS SOCATA)	130 U
TB 200 Tobago XL (EADS SOCATA)	130 U
TBM 700 (EADS SOCATA)	133 U
TECHNOAVIA (see Nauchno-Kommerchesky Firma Technoavia)	450 U
TECNAM (see Costruzioni Aeronautiche Tecnam)	288 U
TFK-2 Carat (Technoflug)	166 U
TG-11A (Stemme)	165 U
TH-5 Esquilo (Eurocopter)	229 U
TH-28 (Enstrom)	649 U
TH-67 Creek (Bell)	28 U
TJ-1500 (Maverick Air)	701 U
TL-32 Typhoon (TL Ultralight)	103 U
TL-96 Star (TL Ultralight)	103 U
TL-132 Condor (TL Ultralight)	103 U
TL-232 Condor Plus (TL Ultralight)	103 U
TL-532 Fresh (TL Ultralight)	104 U
Tp 101 (Raytheon)	740 U
Tp 102 (Gulfstream Aerospace)	660 U
Tu-22M Backfire (Tupolev)	453 U
Tu-54 (Tupolev)	457 U
Tu-95 and Tu-142 Bear (Tupolev)	454 U
Tu-142 and Tu-95 Bear (Tupolev)	454 U
Tu-154M Careless (Tupolev)	458 U
Tu-156 (Tupolev)	459 U
Tu-160 Blackjack (Tupolev)	456 U
Tu-160SK (Tupolev)	457 V
Tu-204 (Tupolev)	459 U
Tu-204 Freighter Orel Avia (Tupolev)	461 U
Tu-206 (Tupolev)	462 U
Tu-214 (Tupolev)	459 U
Tu-216 (Tupolev)	462 U
Tu-224 (Tupolev)	459 U
Tu-234 (Tupolev)	459 U
Tu-244 (Tupolev)	462 U
Tu-245 (Tupolev)	462 U
Tu-324 (Tupolev)	462 U
Tu-330 (Tupolev)	462 U
Tu-334 (Tupolev)	463 U
Tu-336 (Tupolev)	463 U
Tu-338 (Tupolev)	462 U
Tu-344 (Tupolev)	464 U
Tu-354 (Tupolev)	463 U
Tu-414 (Tupolev)	464 N
TAIWAN AEROSPACE CORPORATION (Taiwan)	494 V
Tamiz E.26 (ENAER)	60 U
Tampico Club TB 9C (EADS SOCATA)	130 U
TandemMouse EH 1-02 (Hillberg)	663 U
TANEJA AEROSPACE AND AVIATION LTD (India)	177 U
TAGANROGSKY AVIATSIONNYI NAUCHNO-TEKHNICHESKY KOMPLEKS (Russian Federation)	480 U
Talon M20 (Magni)	283 U
Tango-2 (DFL)	647 U
TASHKENTSKOYE AVIATSIONNOYE PROIZDSTVENNOYE OBEDINENIE IMENI VP CHKALOVA (Uzbekistan)	796 U
TBILISI AVIATION STATE ASSOCIATION (Georgia)	142 U
Teal (L11/Pegas)	388 U
TECNAM, COSTRUZIONI AERONAUTICHE SRL (Italy)	288 U
TECHNOFLUG LEICHTFLUGZEUGBAU GmbH (Germany)	166 V
Tetras (Humbert)	134 N
Texan (Fly Synthesis)	280 U
Texan II, T-6A (Raytheon)	734 U
THE NEW PIPER AIRCRAFT INC (USA)	721 U
Thorp T-211 (AD)	510 U
Thrust-vectoring control programme (NASA)	713 U
Thunder AVA-505 (AII)	258 U
Thunderhawk S-70C (M-1) & (M-2) (Sikorsky)	768 U
Thunder Mustang (Papa 51)	719 U
Tiger 665 (Eurocopter)	223 U
Tiger Moth R-80 (Fisher)	653 U
Tiger Moth RS-80 (Fisher)	653 U
Tigre 665 (Eurocopter)	223 U
Tilt-Rotor Programme (Eurotilt)	247 U
TL ULTRALIGHT (Czech Republic)	103 V
Tobago, TB 10 (EADS SOCATA)	130 U
Tobago XL, TB 200 (EADS SOCATA)	130 U
Tornado HB-204 (HB Flugtechnik)	13 U
Tradewind (Association Zephyr)	110 U
Trainer GT-1 (Goair)	5 U
Trener Baby (Flaming Air)	154 N
Trinidad GT TB 20 (EADS SOCATA)	131 U
Trinidad GT Turbo, TB 21 (EADS SOCATA)	131 U
Trislander, BN2A Mk III (B-N)	523 U
Triton, DHC-8 Dash 8M (de Havilland)	51 U
TSENTR NAUCHNO-TEKNICHESKOGO TVORCHESTVA ALBATROS (Russian Federation)	346 U
Tulák (Let-Mont)	101 U
Tundra (Joplin)	663 U
TUPOLEV OAO (Russian Federation)	453 U
Turbine Islander BN2T (B-N)	522 U
Turbine Legend (Performance)	720 U
Turbo Air Tractor AT-402 (Air Tractor)	546 U
Turbo AT-502 (Air Tractor)	547 U
Turbo Bonanza, B36 TC (Raytheon)	737 U
Turbo-Kruk PZL-106BT (PZL Warszawa)	332 U
Turbo-Porter PC-6 (Pilatus)	484 U
Turbo-Raven PZL-106BT (PZL Warszawa)	332 U
Turbo Skylane T182T (Cessna)	632 U
Turbo Stationair T206H (Cessna)	633 U
Turbo-Thrush 660 (Ayres)	563 U
Turbo-Thrush S-2R (Ayres)	562 U
Turbo-Trainer PC-7 (Pilatus)	486 U
Turbo-Trainer PC-7 Mk II M (Pilatus)	486 U
Turbo Ximango AMT-300 (Aeromot)	15 U
TURKISH AEROSPACE INDUSTRIES INC (Turkey)	494 U
TUSHINSKY MASHINOSTROITELNYI ZAVOD AO (Russian Federation)	453 V
Tutor (Grob)	156 U
Twin Condor, Ru-38A (Schweizer)	756 U
Twinjet 1500 (Maverick Air)	701 U
Twin Panda, Y-12 (HAI)	69 U
Twin Squirrel (Eurocopter)	231 U
Twinstarr (Air & Space)	544 U
TwinStar AS 355 (Eurocopter)	231 U
Typhoon (Eurofighter)	242 U
Typhoon TL-32 (TL Ultralight)	103 U

U

U-4 (Gulfstream Aerospace)	660 U
U-27A Caravan (Cessna)	634 U
U-125 (KAC)	294 U
U-125 (Raytheon)	746 U
U-125A (KAC)	294 U
U-125A (Raytheon)	746 U
UC-35A (Cessna)	638 U
UC-35B (Cessna)	638 U
UC-35C (Cessna)	638
UC-35D (Cessna)	638 U
UFM 10 Samba (Urban)	105 U
UFM 11 Lambáda (Urban)	104 U
UFM 13 Lambáda (Urban)	104 U
UH-12 Esquilo (Eurocopter)	229 U
UH-12B Esquilo (Eurocopter)	231 U
UH-60A Black Hawk (Sikorsky)	761 U
UH-60A Pave Hawk (Sikorsky)	761 U
UH-60C Black Hawk (Sikorsky)	761 U
UH-60J (Mitsubishi/Sikorsky)	300 U
UH-60J (Sikorsky)	761 U
UH-60JA (Sikorsky)	761 U
UH-60L Black Hawk (Sikorsky)	761 U
UH-60M (Sikorsky)	761 U
UH-60P (KAL)	306 U
UH-60Q Black Hawk (Sikorsky)	761 U
UH-60X Black Hawk (Sikorsky)	761 U
UHCA/VLCT (see Ultra-High Capacity Airliner/Very Large Commercial Transport)	256 U
US-1A (ShinMaywa)	302 U
USA (see Ultraleichtflugzeugwerke Sachsen-Anhalt)	166 U
UV-20A Chiricahua (Pilatus)	484 U
ULAN-UDENSKY AVIATSIONNYI ZAVOD OAO (Russian Federation)	465 U
Ultra-High Capacity Airliner/Very Large Commercial Transport	256 U
ULTRALEICHTFLUGZEUGWERKE SACHSEN-ANHALT (Germany)	166 N
Ultrasport 254 (ASI)	555 U
Ultrasport 331 (ASI)	555 U
Ultrasport 496 (ASI)	555 U
Ultrasport family (LASI)	493 U
UNIS OBCHODNI spol. s.r.o. (Czech Republic)	104 U
UNITED STATES AIR FORCE (USA)	782 U
UNITED STATES NAVY (USA)	782 U
URBAN-AIR s.r.o. (Czech Republic)	104 U
US LIGHT AIRCRAFT CORPORATION (USA)	784 U

V

V-22 Osprey (Bell Boeing)	570 U
VA-10 Vantage (VisionAire)	787 U
VA-12C Spirit (VisionAire)	788 U
VA 300 (Vulcanair)	291 N
VAT (see Vertical Aviation Technologies)	786 U
VC-32A (Boeing)	660 U
VC-97 (EMBRAER)	19 U
VH-55 Esquilo (Eurocopter)	231 U
VH-60N (Sikorsky)	761 U
Vagabond (Etruria)	278 N
Vajra (Dassault)	113 U
Valkyrie (Dreamwings)	647 U
Valor (Aeroprakt)	496 U
VAN'S AIRCRAFT INC (USA)	784 U
Vantage VA-10 (VisionAire)	787 U
Vectored Thrust Ducted Propeller (Piasecki)	720 U
Vega (HK)	157 U
VENTURE LIGHT AIRCRAFT RESOURCES (USA)	786 N
Venture Star (NASA)	711 U
VERTICAL AVIATION TECHNOLOGIES INC (USA)	786 U
Verticraft (Modus)	493 U
Vertijet (Modus)	493 U
Very Large Commercial Transport (UHCA/VLCT)	256 U
Viator AP.68TP 600 (VulcanAir)	290 U
VIKING AIRCRAFT LIMITED (USA)	786 U
Viper (Paxman)	58 U
VIPER AIRCRAFT CORPORATION (USA)	786 U
Viperjet (Viper)	786 U
VISIONAIRE CORPORATION (USA)	787 U
Vista (Aeroprakt)	495 U
Vista Cruiser (Aeroprakt)	496 U
VITEK KOMPANIYA (Russian Federation)	465 U
VORONEZHSKOYE AVIATSIONNOYE PROIZVODSTVENNOYE OBEDINENIE OAO (Russian Federation)	465 U
VORTEX AIRCRAFT COMPANY (USA)	789 U
VSTOL AIRCRAFT CORPORATION (USA)	789 U
VTOL airliner studies (NAL)	301 U
Vulcan (Aeroprakt)	497 U
VULCANAIR SpA (Italy)	290 U

W

W-3 Sokól (PZL Swidnik)	326 U
WC-130J Hercules (Lockheed Martin)	682 U
WHGAC (see Wuhan Helicopter General Aviation Corporation)	82 U
WLAC-1 White Lightning (RFW)	750 U
WZL 3 (see Wojskowe Zaklady Lotnicze)	336 U
WACO CLASSIC AIRCRAFT CORPORATION (USA)	789 U
Wag-a-Bond (Wag-Aero)	790 U
WAG-AERO INC (USA)	790 U
WAH-64D Apache Longbow (GKN Westland)	533 U
WARNER AEROCRAFT COMPANY (USA)	791 U
Warrior III, PA-28-161 (Piper)	721 U
WARRIOR (AERO-MARINE) LTD (UK)	532 U
Wasp, Dubna-2 (Dubna)	358 U
Waverider Projects (NASA)	713 U
WEATHERLY AVIATION COMPANY (USA)	792 U
Weatherly 620 (Weatherly)	792 U
WD FLUGZEUGBAU GmbH (Germany)	167 U
Wega (HK)	157 U
WESTLAND HELICOPTERS LIMITED, GKN (UK)	533 U
Whisper (Kaiser)	159 U
White Colt (Kazan)	383 U
White Lightning WLAC-1 (RFW)	750 U
Wild Thing (Air Light)	143 U
Wilga 2000, PZL-104M (PZL Warszawa)	331 U
WINDEAGLE AIRCRAFT CORPORATION (Canada)	59 V
Windeagle (CAG)	51 U
WOJSKOWE ZAKLADY LOTNICZE Nr 3 (Poland)	336 U
WOLFSBERG AIRCRAFT CORPORATION NV (Belgium)	14 U
WOLFSBERG-EVEKTOR s.r.o. (Czech Republic)	105 U
Wolverine T-421 (Khrunichev)	385 U
Woong-Bee KT-1 (KAI)	304 U
WUHAN HELICOPTER GENERAL AVIATION CORPORATION (China)	82 U

X

X-32 Joint Strike Fighter (Boeing)	573 U
X-33 (NASA)	711 U
X-34 (NASA)	711 U
X-35 Joint Strike Fighter (Lockheed Martin)	687 U
X-38 (NASA)	712 U
X-44A (NASA)	713 V
XAC (see Xian Aircraft Company)	82 U
XL-2 (Liberty)	592 N

XIAN AIRCRAFT COMPANY (China)	82 U	
Ximango (Aeromot)	15 U	

Y

Y-5 (SAMC)	81 U
Y-7 (XAC)	84 U
Y-8 (SAC)	77 U
Y-12 (HAI)	69 U
Yak-3M (Yakovlev)	467 U
Yak-9U-M (Yakovlev)	467 U
Yak-18T (Yakovlev)	468 U
Yak-42 Clobber (Yakovlev)	468 U
Yak-52 (Yakovlev)	470 V
Yak-54, Yak-56 and Yak-57 (Yakovlev)	470 U
Yak-58 (Yakovlev)	471 U
Yak-112 (Yakovlev)	471 U
Yak-130 (Yakovlev)	466 U
Yak-131 (Yakovlev)	467 U
Yak-152 (Yakovlev)	470 N
Yak-SKh (Yakovlev)	471 N
YALO/AVIATA (see Zakład Naprawy I Budowy Sprzetu Latajacego Yalo)	336 U
YMF Super Waco (Classic)	789 U
YAKOVLEV AVIATSIONNOYE KORPORATSIYA OAO (Russian Federation)	466 V
YALO, ZAKŁAD NAPRAWY I BUDOWY SPRZETU LATAJACEGO YALO (Poland)	336 U
Yanshuf (Sikorsky)	761 U
Young Man T-131 (SSH)	335 N
Youngster (Fisher)	653 U
Youngster V (Fisher)	653 U
Yunshuji-5 (SAMC)	81 U
Yunshuji-7 (XAC)	84 U
Yunshuji-8 (SAC)	77 U
Yunshuji-12 (HAI)	69 U

Z

Z-9 Haitun (HAI)	68 U
Z-10 (CHRDI)	67 N
Z-11 (CHAIC)	66 U
Z-142C (Zlin)	106 V
Z-143 (Zlin)	107 U
Z-143 MAF (ZLIN)	107 U
Z-242 L (Zlin)	106 U
ZIU (TAI)	495 U
ZLIN (see Moravan Aeroplanes Inc)	105 U
'Z-X' (HAI)	69 U
ZAKŁAD NAPRAWY I BUDOWY SPRZETU LATAJACEGO YALO SC (Poland)	336 U
ZAKŁAD LOTNICZY PZL-MIELEC Sp. z.o.o. (Poland)	324 U
ZAKLADY LOTNICZE 3XTRIM Sp. z.o.o Poland	314 N
ZENAIR LTD (Canada)	59 U
ZENITH AIRCRAFT COMPANY (USA)	792 U
Zephyr (ATEC)	92 U
Zhishengji-9 (HAI)	68 U
Zhishengji-11 (CHAIC)	66 U
Ziu (TAI)	495 U
ZIVKO AERONAUTICS INC (USA)	795 U
Zodiac CH 601 HD (Zenith)	792 U
Zulu (BUL)	110 U
ZYGMUNTA PULAWSKIEGO-PZL SWIDNIK (Poland)	326 U

AIRCRAFT–PREVIOUS TEN EDITIONS

Index begins with aircraft generally identified only by numbers. Most were previously listed as 'Model 707' and so on under M

Aircraft	Edition	Page
2.5 ST Freighty (NIAT)	(1997-98)	428
6-650 Genesis (Gevers)	(1999-2000)	648
11 (Unikomtranso)	(1997-98)	469
16H-4 (Piasecki)	(1991-92)	475
18A (Air & Space)	(1997-98)	556
20 Tech 4 (New Technik)	(1991-92)	456
35A/36A (Learjet)	(1994-95)	547
35A/36A special missions versions (Learjet)	(1995-96)	550
44, Angel (King's)	(1994-95)	544
50 (Zlin)	(1994-95)	78
55B (Learjet)	(1991-92)	419
55C (Learjet)	(1991-92)	419
58 Baron (Ram/Beech)	(1991-92)	466
70, Model (Bombardier)	(1998-99)	35
082 Hercules (Lockheed Martin)	(1998-99)	665
90 (Zlin)	(1993-94)	71
107 (Boeing)	(1991-92)	366
110 Special (Aviat)	(2000-01)	576
114B Commander (Commander)	(2000-01)	660
115 (Aero Boero)	(1999-2000)	1
133-4.62 (Scaled Composites)	(1992-93)	438
137T Agro Turbo (Zlin)	(1995-96)	80
142 (Zlin)	(1995-96)	78
143 (Zlin)	(1995-96)	79
143 Triumph (Scaled Composites)	(1991-92)	473
145 (Myers)	(1991-92)	455
150 (Aero Boero)	(1997-98)	1
151 ARES (Scaled Composites/Rutan)	(1995-96)	643
152 (Cessna)	(1991-92)	385
172 (RAM/Cessna)	(1991-92)	465
180 AG (Aero Boero)	(1999-2000)	1
180 RVR (Aero Boero)	(1999-2000)	1
180 PSA (Aero Boero)	(1997-98)	2
182 Hercules (Lockheed Martin)	(1998-99)	665
184 Tempest (Jaran)	(1999-2000)	658
185 Orion (Lockheed Martin)	(2000-01)	695
185 Skywagon (Cessna)	(1991-92)	385
200 (Robin)	(1999-2000)	132
200 HISAR (Raytheon)	(2000-01)	754
200X (Moller)	(1993-94)	537
201AT Advanced Trainer (Mooney)	(1991-92)	453
201SE (Mooney)	(1991-92)	453
202 (Contender)	(1998-99)	634
202-11 Boomerang (Scaled Composites)	(1994-95)	626
205 (Bell)	(1991-92)	358
205A-1, Advanced (Fuji-Bell)	(1992-93)	157
205 MSE (Mooney)	(1994-95)	604
205B (Fuji Bell)	(1994-95)	228
205 UH-1P Huey II (Bell)	(1994-95)	492
206, Cessna, Turbine Pac (Soloy)	(1991-92)	490
206B Jet Ranger III (Bell)	(1992-93)	17
206L-3 Long Ranger III (Bell)	(1992-93)	18
206L-3ST Gemini ST (Tridair)	(1994-95)	647
206L-4ST Gemini ST (Tridair)	(1994-95)	647
206LT TwinRanger (Bell)	(1998-99)	25
207, Cessna, Turbine Pac (Soloy)	(1991-92)	490
208/208B Caravan Fire Fighter (Aero Union)	(1991-92)	338
209 HueyCobra (Modernised versions) (Bell)	(1993-94)	434
209 Improved SeaCobra (Bell)	(1994-95)	492
212 ASW (Agusta-Bell)	(1994-95)	213
212 Twin Two-Twelve (Bell)	(1998-99)	27
222 conversion (Heli-Air)	(1991-92)	540
222 SP (Heli-Air)	(1994-95)	540
228 Maritime Patrol Aircraft (HAL)	(1998-99)	162
228 Maritime Pollution Surveillance (HAL)	(1997-98)	160
228 Photogrammetry/Geo-survey (Dornier)	(1996-97)	123
228-212 (HAL)	(1997-98)	159
230 Bell	(1996-97)	28
234 (Boeing)	(1991-92)	369
252TSE (Mooney)	(1991-92)	454
257 TLS (Mooney)	(1991-92)	454
260 AG (Aero Boero)	(1997-98)	2
282 Hercules (Lockheed Martin)	(1998-99)	665
300 Sequoia (Sequoia)	(1995-96)	646
303 (Contender)	(1998-99)	634
305 (Brantley)	(1993-94)	462
328 (Dornier)	(2000-01)	153
340/340A (RAM/Cessna)	(1991-92)	466
340B (Saab)	(1999-2000)	477
360 (Boeing)	(1992-93)	353
382 Hercules (Lockheed Martin)	(1998-99)	665
385 TriStar conversions (Lockheed)	(1991-92)	430
400A (Avtek)	(1995-96)	486
401/402A/402B/402C (RAM/Cessna)	(1991-92)	466
402C (Cessna)	(1991-92)	385
406 AHIP (Bell)	(1994-95)	494
406 CS (Bell)	(1992-93)	345
407T (Bell)	(1995-96)	29
412 HP (Bell)	(1996-97)	26
412 SP (Agusta-Bell)	(1991-92)	157
414 (RAM/Cessna)	(1991-92)	466
414AW (RAM/Cessna)	(1991-92)	466
414AW Series V (Ram/Cessna)	(1991-92)	466
419 Express (Avtek)	(1999-2000)	557
421 Golden Eagle (Cessna)	(1991-92)	385
421C and 421CW (RAM/Cessna)	(1991-92)	466
425 Conquest I (Cessna)	(1991-92)	385
428JET (Dornier)	(2000-01)	155
441 Conquest II (Cessna)	(1991-92)	385
442 (Bell)	(1998-99)	31
500 Citation, Upgrade Programme (AlliedSignal)	(1994-95)	471
500/520/530 (KAL McDonnell Douglas)	(1994-95)	240
520 MK Black Tiger (KA)	(1994-95)	240
526 JPATS Citationjet (Cessna)	(1995-96)	519
528JET (Fairchild Dornier)	(1997-98)	137
550 Citation II (Cessna)	(1996-97)	591
560 Citation V (Cessna)	(1994-95)	525
560XL Citation Excel (Cessna)	(1997-98)	617
600 Thrush S2R-R1340 (Ayres)	(1996-97)	554
606 (Contender)	(1998-99)	634
645 (Lockheed Martin)	(1998-99)	661
650 Citation III (Cessna)	(1991-92)	390
650 Citation VI (Cessna)	(1996-97)	595
653 A/F-22X (Lockheed)	(1994-95)	553
660 Citation VII (Cessna)	(1991-92)	370
685 Orion (Lockheed Martin)	(2000-01)	695
700 Turbo Thrush (Ayres)	(1997-98)	570
707 (Boeing)	(1994-95)	499
707 Tanker/Transport (Boeing)	(1991-92)	372
707 TT (Alenia)	(1994-95)	218
707/720 conversions (IAI)	(1994-95)	205
720 AAT (Boeing)	(1991-92)	372
727 RE Quiet 727 (Valsan)	(1994-95)	495
727 update (Dee Howard)	(1991-92)	396
731 Falcon 20B Retrofit Programme (AlliedSignal)	(1993-94)	415
737 'Classic' (Boeing)	(2000-01)	614
737X (Boeing)	(1993-94)	443, 697
737-100 (Boeing)	(1995-96)	497
737-200 (Boeing)	(1995-96)	497
737-300 (Boeing)	(1999-2000)	591
737-300QC conversions (Pemco)	(1994-95)	615
737-400 (Boeing)	(1999-2000)	593
737-500 (Boeing)	(1999-2000)	594
747 Freighter conversions (Boeing)	(1991-92)	372
747-100 cargo conversion (Pemco)	(1994-95)	615
747-100 conversion (IAI)	(1994-95)	205
747-200 Freighter conversion (IAI)	(1994-95)	205
747-500X (Boeing)	(1996-97)	578
747-600X (Boeing)	(1996-97)	578
757QC (Pemco)	(1994-95)	615
767 Airborne Surveillance Testbed (Boeing)	(1991-92)	366
7J7-YXX (International)	(1992-93)	89, 354
1900 Exec-Liner (Beech)	(1991-92)	353
1900C Airliner (Beech)	(1991-92)	353
2000 (Load Ranger)	(1999-2000)	670
2000 (Saab)	(1999-2000)	480
2000 Starship 1 (Beech)	(1995-96)	637
2000A Starship (Raytheon)	(1995-96)	637
3000 (CASA)	(1995-96)	414
3140 Testbed (Robin)	(1994-95)	100
5800 (KFC)	(1992-93)	29

A

Aircraft	Edition	Page
A3XX (Airbus)	(2000-01)	205
A-4 Skyhawk upgrade (RNZAF)	(1991-92)	184
A-4S-1 Super Skyhawk (SA)	(1994-95)	392
A-4SU Super Skyhawk (SA)	(1994-95)	392
A-5, NAMC modernisation (Alenia)	(1994-95)	218
A-5M (NAMC)	(1997-98)	68
A-6E Intruder (Grumman)	(1993-94)	481
A-6E/TRAM (Grumman)	(1993-94)	481
A-7 Corsair II update programmes (LTV)	(1991-92)	430
A-11 (Hamilton)	(1994-95)	540
A-12 (Lockheed)	(1992-93)	402
A-12A (General Dynamics/McDonnell Douglas)	(1992-93)	382
A-22J Fanjet (Sadler)	(1994-95)	625
A-22 LASA (Sadler)	(1994-95)	624
A-27 (EMBRAER)	(2000-01)	17
A31X/AE-100 (AVIC/AIA/STPL)	(1997-98)	823
A-36 Halcón (CASA)	(1995-96)	411
A-36 Halcón (ENAER)	(1997-98)	54
A-40 Albatross (Beriev)	(1999-2000)	356
A-45 (Beriev)	(1998-99)	347
A 109A Mk II (Civil versions) (Agusta)	(1991-92)	154
A 109A Mk II (Military, Naval & Law Enforcement versions) (Agusta)	(1991-92)	154
A 109 EOA (Agusta)	(1991-92)	154
A 109HA (Agusta)	(1996-97)	244
A 109HO (Agusta)	(1996-97)	244
A 109KN (Agusta)	(1991-92)	155
A 139 Utility (Agusta)	(1996-97)	244
A-209 (Aero-M)	(1999-2000)	343
A-211 (Alfa-M)	(1999-2000)	349
A-230 (Aero-M)	(1999-2000)	344
A 300-600 Multi Role Tanker Transport (Airbus)	(1998-99)	190
A 316 (Airbus)	(1998-99)	177
A 317 (Airbus)	(1998-99)	177
A340-300 Advanced (Airbus)	(1995-96)	144
A340-300 Combi (Airbus)	(1992-93)	97
A340-300 Stretch (Airbus)	(1995-96)	144
A350 (Airbus)	(1991-92)	114
AA-2 Mamba (AAI)	(1995-96)	5
AA-23 STOL 180 (Taylorcraft)	(1991-92)	492
AA-23A1 Ranger (Taylorcraft)	(1991-92)	492
AA-23B Aerobat (Taylorcraft)	(1991-92)	492
AA-23B1 RS (Taylorcraft)	(1991-92)	492
AA200 Orion (CRSS)	(1993-94)	112
AA300 Rigel (CRSS)	(1993-94)	112
AA330 Theta (CRSS)	(1993-94)	112
AAA (See Advanced Amphibious Aircraft)	(1994-95)	126
AAA (see Aerostar Aircraft Corporation)	(1994-95)	446
AAC (see Aircraft Acquisition Corporation)	(1992-93)	323
AAI (see Australian Aircraft Industries)	(1996-97)	4
AC-4 Andorinha (Super Rotor)	(1992-93)	16
AC-05 Pijao (Aviones de Columbia)	(1994-95)	66
AC-130 (Lockheed Martin)	(1998-99)	665
AC-130U Spectre (Rockwell/Lockheed)	(1995-96)	641
ACT (see Aviation Composite Technology)	(1992-93)	174
ADC (CPCA)	(1998-99)	329
ADI (see Aero Designs Inc)	(1997-98)	554
AE-100 (AVIC/AIA/STPL)	(1997-98)	201
AE 206 Mistral (Aviasud)	(1998-99)	98
AE 207 Mistral Twin (Avaisud)	(1998-99)	98
AE 209 Albatros (Aviasud)	(1998-99)	99
AE 316 (AVIC/AIA/STPL)	(1998-99)	202
AE 317 (AVIC/AIA/STPL)	(1998-99)	202
AE 318 (AVIC/AIA/STPL)	(1998-99)	202
AEDECO (see Aeronautical Design Company)	(1996-97)	87
AERO (see Aero Vodochody)	(1995-96)	69
AEROTEK (see Aeronautical Systems Technology)	(1998-99)	460
A/F-117X (Lockheed Martin)	(1995-96)	568
AFI (see Aviation Franchising International)	(1997-98)	556
A/FX (USN)	(1994-95)	651
AG-5B Tiger (American General)	(1994-95)	472
AG-6 (Aerostar)	(1993-94)	249
AH-1 (4B)W Viper (Bell)	(1992-93)	343
AH-1E HueyCobra (Bell)	(1993-94)	434
AH-1F HueyCobra (Bell)	(1993-94)	434
AH-1F HueyCobra (IAR)	(1996-97)	330
AH-1J SeaCobra (American General)	(1994-95)	472
AH-1P HueyCobra (Bell)	(1993-94)	434
AHIP Army Helicopter Improvement Program (Bell)	(1994-95)	494
AH-1RO Dracula (IAR)	(2000-01)	348
AH-1RO SuperCobra (IAR)	(1997-98)	343, 826
AH-1S HueyCobra (Bell)	(1993-94)	434
AH-1S SuperCobra (Bell)	(2000-01)	297
AH-1T Improved SeaCobra (Bell)	(1994-95)	492
AH-1W Cobra Venom (Bell)	(1993-94)	436
AH-1W Venom (Bell)	(1994-95)	494
AIA (see Aviation Industries of Australia)	(1994-95)	4
AICSA (see Aero Industrial Colombian)	(1994-95)	65
AIEP (see Aeronautical Industrial Engineering and Project Management Company Ltd)	(1997-98)	315
AIPI (see Aerotech Industries Philippines)	(1994-95)	259
AI(R) (see Aero International (Regional))	(1999-2000)	177
AI(R) 58 (AI(R))	(1997-98)	171
AI(R) 70 (AI(R))	(1997-98)	171
AI(R) Jet (AI(R))	(1997-98)	171
AISA (see Aeronautica Industrial SA)	(1994-95)	398
AJS 37 Viggen (Saab)	(1994-95)	406
AK-21 (Delta-V)	(1999-2000)	308
ALCLH (IPTN)	(1997-98)	168
AMH (Fuji)	(1999-2000)	290
AMI (see Aero Modifications International)	(1992-93)	327
AMIT Fouga (AIA)	(1991-92)	149
AMTU IAR-707 (INAV)	(1995-96)	295

A—AIRCRAFT–PREVIOUS TEN EDITIONS: INDEXES 851

An-2 Colt (PZL) (2000-01) 331
An-3 (Antonov) (1992-93) 280
AN4 (Campana) (1998-99) 100
An-22 Antheus (Antonov) (1991-92) 231
An-24 Coke (Antonov) (1991-92) 232
An-30 Clank (Antonov) (1991-92) 233
An-71 Madcap (Antonov) (1997-98) 513
An-74 Variant Madcap (Antonov) (1991-92) 236
An-77 (Antonov) (1994-95) 425
An-180 (Antonov) (1997-98) 518
An-218 (Antonov) (1995-96) 443
An-225 Cossack (Antonov) (1995-96) 443
AOI (see Arab Organisation for Industrialisation) (1994-95) 79
AP (see Aeroplastika) (1995-96) 252
AP 68TP-600 Viator (TAAL) (1998-99) 166
ARA 3600 (Promavia) (1995-96) 14
ARCH 50 (DHI) (1994-95) 239
ARDC (see Air Force Research and Development Center) (2000-01) 320
ARTB (Lockheed) (1991-92) 429
ARV-1 Super2 (Aviation Scotland) (1994-95) 433
AS-2 (II) (1999-2000) 319
AS 18A Bravo (FFA) (1995-96) 425
AS-61 (Agusta-Sikorsky) (1994-95) 214
AS-61R Pelican (Agusta-Sikorsky) (1994-95) 214
AS100 (Aerospatiale) (1991-92) 58
AS 202 Bravo (FFA) (1996-97) 482
AS 202/32TP Turbine Bravo (FFA) (1997-98) 495
AS 330 Puma (Aerospatiale) (1991-92) 58
AS 330 Puma conversions (Atlas) (1994-95) 396
AS 332 Super Puma Mk II (Aerospatiale) (1991-92) 60
AS 332 Super Puma Mk I (Eurocopter) (1998-99) 213
AS 350 AllStar conversion (Rocky Mountain) (1993-94) 552
AS 352 Cougar Mk I (Eurocopter) (1998-99) 213
AS 366 Dauphin 2 (Aerospatiale) (1991-92) 66
AS 365X DGV Dauphin (Eurocopter) (1995-96) 180
AS 365X FBW Dauphin (Eurocopter) (1995-96) 180
AS 565 Panther (Helibras) (1993-94) 20
ASH-3H (Agusta-Sikorsky) (1994-95) 214
AST (see Aviaspetstrans) (2000-01) 356
AST (see Aviation Services Technologies) (1994-95) 474
ASTA (see AeroSpace Technologies of Australia Pty Ltd) (1995-96) 5
ASTOR Defender (Pilatus Britten-Norman) (1992-93) 310
AT³ (Scaled Composites) (1992-93) 438
AT-3 Tzu-Chung (AIDC) (1996-97) 492
AT-400 Turbo Air Tractor (Air Tractor) (1993-94) 413
AT-401A Air Tractor (Air Tractor) (1992-93) 324
AT-501 (Air Tractor) (1992-93) 325
AT-503 (Air Tractor) (1994-95) 470
AT-2000 (Aermacchi/DASA) (1994-95) 126
ATA (General Dynamics/McDonnell Douglas) (1992-93) 382
ATA (USAF) (1994-95) 649
ATF (Lockheed) (1993-94) 497
ATF (Northrop/McDonnell Douglas) (1991-92) 457
ATF (USAF) (1991-92) 495
ATL (Robin) (1991-92) 83
ATP (BAe) (1992-93) 295
ATP, Jetstream (BAe) (1993-94) 382
ATR (see Avions de Transport Regional) (2000-01) 215
ATR 52C (ATR) (1997-98) 199
ATR 82 (ATR) (1995-96) 159
ATT (Boeing) (1998-99) 574
ATTA 3000 (Promavia) (1995-96) 14
AU-24A (Helio) (1991-92) 413
AV-8B (McDonnell Douglas (Boeing)/BAe) (1997-98) 229
AV-8B II Plus (McDonnell Douglas (Boeing)/BAe) (1997-98) 234
AV-8S Harrier (BAe) (1991-92) 311
A-X (USN) (1993-94) 580

ABS AIRCRAFT (Germany) (1994-95) 103
ACA INDUSTRIES INC (USA) (1992-93) 323
Accord (Avia) (1998-99) 340
ACE (Denel) (1996-97) 465
Acrobat, Aviatika-900 (Aviatika) (1997-98) 364
Adnam 1 (IAF) (1994-95) 202
ADVANCED AIRCRAFT CORPORATION (USA) (1992-93) 323
ADVANCED AMPHIBIOUS AIRCRAFT (International) (1994-95) 126
Advanced Cargo Aircraft (US Army) (1995-96) 665
Advanced Military Programmes (MiG) (2000-01) 414
Advanced STOVL Strike Fighter (Lockheed) (1994-95) 563
Advanced Surveillance and Tracking Technology/ Airborne Radar Demonstrator (USAF) (1995-96) 664
Advanced Tactical Aircraft (General Dynamics/ McDonnell Douglas) (1991-92) 405, 495
Advanced Tactical Aircraft (USAF) (1994-95) 649
Advanced Tactical Fighter (McDonnell Douglas/ Northrop) (1991-92) 457
ADVENTURE AIR (USA) (1997-98) 555
Adventurer (Adventure Air) (1997-98) 555
AEA RESEARCH (Australia) (1998-99) 795
AERMACCHI/DASA (International) (1994-95) 126
Aerobat Model AA-23B (Taylorcraft) (1991-92) 492
AERO BOERO SA (Argentina) (1999-2000) 1
AERO CZECH AND SLOVAK AEROSPACE INDUSTRY LTD (Czech Republic) (1993-94) 63
AERO COMPOSITE TECHNOLOGY INC (USA) (1992-93) 497
AERO DESIGNS INC (USA) (1997-98) 554
AERODIS AMERICA INC (USA) (1992-93) 323
AERODYN (France) (1997-98) 97
AERO INDUSTRIAL COLOMBIANA (Colombia) (1994-95) 65
AERO INTERNATIONAL (REGIONAL) SAS (International) (1999-2000) 177
AERO MERCANTIL SA (Colombia) (1991-92) 48
AERO MODIFICATIONS INTERNATIONAL INC (USA) (1992-93) 327
AERONAUTICA INDUSTRIAL SA (Spain) (1994-95) 398
AERONAUTICAL INDUSTRIAL ENGINEERING AND PROJECT MANAGEMENT COMPANY LTD (Nigeria) (1997-98) 315
AERONAUTICAL DESIGN COMPANY (France) (1996-97) 87
AERONAUTICAL SYSTEMS TECHNOLOGY (South Africa) (1998-99) 460
AERONAUTIQUE SERVICE SA (France) (1999-2000) 104
AEROPLASTIKA (Lithuania) (1995-96) 252
Aeropony S-10 (Shadin) (1995-96) 646
AERORIC NAUCHNO-PROIZVODSTVENNOYE PREDPRIYATIE (Russian Federation) (2000-01) 355
AEROSPACE CONSORTIUM INC, THE (Canada) (1993-94) 20
Aero-Spaceplane studies (NAL) (2000-01) 306
Aero-Space Plane (NASA) (1993-94) 540
AEROSPACE TECHNOLOGIES OF AUSTRALIA PTY LTD (Australia) (1995-96) 5
Aerostar 3000 (AAC) (1994-95) 466
AEROSTAR AIRCRAFT CORPORATION (USA) (1994-95) 466
Aerostar P-3A (Aero Union) (1992-92) 338
AEROTECH INDUSTRIES PHILIPPINES (Philippines) (1994-95) 259
AERO UNION CORPORATION (USA) (1994-95) 468
Acro G115TA (Grob) (2000-01) 162
Acro-Husky (Aviat) (1996-97) 551
Aft-body strake system (Raisbeck) (1991-92) 465
AG-CAT CORPORATION (USA) (1997-98) 556
Ag-Cat Series (Schweizer) (1996-97) 722
Ag-Chuspi IAP-002 (Indaer Peru) (1992-93) 173
Agcraft (Taylorcraft) (1991-92) 492
Aggressor, Mi-24 (Orlando) (1994-95) 653
Ag Husky (Cessna) (1991-92) 385
Agile Falcon (General Dynamics) (1991-92) 403
AGRO-COPTERS LTDA (Colombia) (1994-95) 65
Agro Turbo Z 37T-2 (Zlin) (1996-97) 85
Agro Turbo Z 137T (Zlin) (1996-97) 85
Ag Trainer (Cessna/Avions de Colombia) (1994-95) 65
Ag Truck (Cessna) (1991-92) 385
AGUSTA (Italy) (2000-01) 273
Aiglon R1180 (Robin) (2000-01) 143
Aimak Il-87 (Ilyushin) (1999-2000) 365
Air Beetle (AIEP) (1996-97) 299
Airborne Standoff Radar (Astor) (RAF) (2000-01) 547
Aircar S-12 E (Spencer) (2000-01) 791
AIRCORP PTY LTD (Australia) (1994-95) 4
AIRCRAFT ACQUISITION CORPORATION (USA) (1992-93) 323
Air Force One VC-25A (Boeing) (1991-92) 372
AIR FORCE RESEARCH AND DEVELOPMENT CENTRE (Philippines) (2000-01) 320
Airfox, IAR-317 (IAR) (1991-92) 205
Airjet (ATR) (2000-01) 218
AIR-LIGHT FLUGZEUGVERTRIEB (Germany) (1997-98) 818
AIROD SENDIRIAN BERHAD (Malaysia) (1994-95) 242
Airone F.220 (General Avia) (2000-01) 285
Air Shark 1 (Freedom Master) (1993-94) 477
Airstar 2500 (Romaero) (1997-98) 345
AIRTECH CANADA (Canada) (1995-96) 24
Airtruk (PL-12) (Transavia) (1991-92) 6
Aist T-201 (Aeroprogress/ROKS-Aero) (1997-98) 350
AKROTECH EUROPE (France) (1998-99) 96
Albatross A-40 (Beriev) (1999-2000) 356
Albatros AE 209 (Aviasud) (1998-99) 99
Albatross Tanker conversion (Aero Union) (1991-92) 338
Alca L 159 (Aero) (1999-2000) 86
ALFA-M NAUCHNO-PROIZVODSVENNOYE PREDPRIYATIE (Russian Federation) (2000-01) 356
Alflex (NASDA) (1997-98) 298
Alizé (Dassault Aviation) (1991-92) 74
Allegro M20J (Mooney) (2000-01) 722
Alliance (SCTICSG) (1993-94) 688
ALLIEDSIGNAL AVIATION SERVICES (USA) (1994-95) 471
Allstar AS 350 conversion (Rocky Mountain) (1993-94) 552
ALPHA AIRLINES (Russia) (1994-95) 765
Alpha Jet (Dassault/Dornier) (1992-93) 102
AMECO-HAWK INTERNATIONAL (USA) (1992-93) 327
AMERICAN EUROCOPTER CORPORATION (USA) (1994-95) 472
AMERICAN GENERAL AIRCRAFT CORPORATION (USA) (1994-95) 472
AMERICAN REGIONAL AIRCRAFT INDUSTRIES (USA) (2000-01) 571
AMERICAN SPORTSCOPTER INTERNATIONAL (Taiwan) (1998-99) 486
AMF ENTERPRISES (UK) (2000-01) 525
Amphibian (Ross) (1998-99) 522
Anaconda T-710 (Aeroprogress/ ROKS-Aero) (1997-98) 357
Anafa (Bell) (1998-99) 27
Andorinha AC-4 (Super Rotor) (1992-93) 16
Angel 44 (King's) (1994-95) 544
Antek, An-2 (PZL Mielec) (1996-97) 314
Antheus, An-22 (Antonov) (1991-92) 231
Apache 1 (ACT) (1991-92) 186
APEC AEROSPACE (Singapore) (1998-99) 460
ARAB ORGANISATION FOR INDUSTRIALISATION (Egypt) (1994-95) 79
ARCTIC AIRCRAFT COMPANY (USA) (1997-98) 564
Arctic Tern (Arctic) (1994-95) 473
Arcturus CP-140A (IMP) (1993-94) 38
Arcturus CP-140A (Lockheed Martin) (1999-2000) 676
ARES (Scaled Composites/Rutan) (1995-96) 643
Ariel 212 (Zaitsev) (1999-2000) 465
ARROWTIME AVIATION SDN BHD (Malaysia) (1997-98) 304
ASL HAGFORS AERO B (Sweden) (1996-97) 472
Astra, 1125 (IAI) (1993-94) 182
Astra SP, 1125 (IAI) (1997-98) 254
Astra Galaxy (IAI) (1996-97) 235, 794
Asuka (NAL) (1991-92) 175
Atlant (Myasishchev) (1995-96) 363
Atlant VM-T (Myasishchev) (1991-92) 273
Atlantic 1 Modernisation (Alenia) (1991-92) 163
Atlantique 2 (Dassault) (1998-99) 112
ATLAS AIRCRAFT CORPORATION OF SOUTH AFRICA (PTY) LIMITED (South Africa) (1991-92) 210, 753
ATLAS AVIATION (PTY) LIMITED (South Africa) (1995-96) 407
Atlas Military Trainer (Atlas) (1991-92) 210
Aurora (USAF) (1995-96) 664
Aurora CP-140 (Lockheed Martin) (1999-2000) 676
AUSTRALIAN AIRCRAFT INDUSTRIES (Australia) (1996-97) 4
AUSTRALIAN AUTOGYRO CO, THE (Australia) (1993-94) 5
Avenger II A-12A (General Dynamics/McDonnell Douglas) (1992-93) 382
AVIA BALTIKA AVIATION (Lithuania) (2000-01) 313
AVIASPETSTRANS (Russian Federation) (2000-01) 356
AVIASUD INDUSTRIES (France) (1998-99) 98
AVIATEHNOLGIA (Moldova) (2000-01) 315
Aviatika-900 Acrobat (Aviatika) (1997-98) 364
AVIATION COMPOSITE TECHNOLOGY (Philippines) (1992-93) 174
AVIATION FRANCHISING INTERNATIONAL (USA) (1997-98) 556
AVIATION INDUSTRIES OF AUSTRALIA (Australia) (1994-95) 4
AVIATION SCOTLAND LTD (UK) (1994-95) 433
AVIATION SERVICES TECHNOLOGIES (USA) (1994-95) 474
AVIATSIONNYI KOMPLEKS IMENI S.V ILYUSHINA (Russian Federation) (2000-01) 369
AVIATSIONNYI NAUCHNO-PROMISHLENNYI KOMPLEKS-ANPK MiG IMIENI A. I. MIKOYANA (Russia) (1995-96) 340
AVIATSIONNYI NAUCHNO-TEKHNICHESKIY KOMPLEKS IMIENI A. N. TUPOLEVA (Russia) (1994-95) 370
AVIC/AIA/STPL (International) (1999-2000) 208
Aviocar, C-212P (CASA) (1991-92) 214
Aviocar, NC-121-200 (IPTN) (1999-2000) 176
Aviocar Special Mission Versions (CASA) (1993-94) 346
Aviojet, C-101 (CASA) (1995-96) 411
AVIOLIGHT (Italy) (1992-93) 153
AVIONES DE COLOMBIA (Colombia) (1996-97) 70
AVIONS FOURNIER (France) (1999-2000) 126
AVIONS MUDRY et CIE (France) (1997-98) 121, 817
AVIONS PHILIPPE MONIOT (France) (1999-2000) 128
AVIONS PIERRE ROBIN (France) (1998-99) 125
AVIOTECHNICA LTD (International) (1998-99) 203
Avroliner (Avro) (1995-96) 453
AVSTAR INC (USA) (1991-92) 343
Aya EC-130 (Lockheed Martin) (1998-99) 665
AYMET AEROSPACE INDUSTRIES (Turkey) (1994-95) 421
Aztec, amphibious (Wipaire) (1991-92) 496

852 INDEXES: AIRCRAFT-PREVIOUS TEN EDITIONS—**B – C**

B

B-2 Spirit (Northrop Grumman) .... *(1999-2000)* 705
B2-N Bushmaster (Aircorp) .... *(1991-92)* 3
B-6 (XAC) .... *(1992-93)* 44
B-7 (XAC) .... *(1998-99)* 73
B-52 Stratofortress (Boeing) .... *(1991-92)* 371
B200 SE King Air (Raytheon) .... *(1998-99)* 712
B200T Maritime Patrol (Raytheon) .... *(1998-99)* 713
BA-14B (MFI) .... *(1995-96)* 418
BAC (see Buchanan Aircraft Corporation) .... *(1995-96)* 5
BAC 1-11 (Dee Howard) .... *(1991-92)* 396
BAC-204 Ozzie Mozzie (BAC) .... *(1994-95)* 5
BAe Corporate 800 (BAe) .... *(1993-94)* 377
BAe Corporate 1000 (BAe) .... *(1993-94)* 378
BAe 125 Series 700-11 (BAe) .... *(1991-92)* 301
BAe 125-800 (BAe) .... *(1993-94)* 204
BAe 146 (BAe) .... *(1993-94)* 383
BAe 146 CC Mk 2 (BAe) .... *(1993-94)* 383
BAe 146 Military Variants (BAe) .... *(1992-93)* 298
BAe 146 NRA (BAe) .... *(1992-93)* 296
BAe 146 Series 100 (BAe) .... *(1993-94)* 383
BAe 146 Series 200 (BAe) .... *(1993-94)* 383
BAe 146 Series 300 (BAe) .... *(1993-94)* 383
BAe 146-200 cargo conversion (Pemco) .... *(1993-94)* 615
BAe 146-QC Convertible (BAe) .... *(1993-94)* 383
BAe 146-QT Quiet Trader (BAe) .... *(1993-94)* 383
BAe RJ70/RJ80 (BAe) .... *(1992-93)* 296
BAF (see British Air Ferries) .... *(1993-94)* 391
BD-4 (Bede) .... *(1992-93)* 504
BD-10 (Bede) .... *(1994-95)* 480
Be-12 Tchaika (Beriev) .... *(1991-92)* 239
Be-32K (IAR) .... *(1997-98)* 343
Be-42 Mermaid (Beriev) .... *(1998-99)* 347
BHK (See Bell Helicopter Korea) .... *(1994-95)* 238
BK 117 conversion (Heli-Air) .... *(1991-92)* 412
BN-2T Special Role Turbine Defenders (Pilatus Britten-Norman) .... *(1992-93)* 310
BN-2T Special Role Turbine Islander (Pilatus Britten-Norman) .... *(1992-93)* 310
BN2T-4R Defender (Pilatus Britten-Norman) .... *(1997-98)* 541
BO 105M/PAH-1/VBH/BSH-1 (Eurocopter) .... *(1994-95)* 169
BO 108 (Eurocopter) .... *(1992-93)* 117
BOR-4 (USSR) .... *(1991-92)* 298
BOR-5 (USSR) .... *(1991-92)* 298
BRJ-X (Bombardier) .... *(2000-01)* 41
BSKhS (Romashka) .... *(1996-97)* 418

Badger, Tu-16 (Tupolev) .... *(1994-95)* 370
Baghdad 1 (IAF) .... *(1993-94)* 181
Bahadur, MiG-27 (HAL) .... *(1997-98)* 161
Balbuzard L235 (AAT) .... *(1998-99)* 95
Bandeirante EMB-110 (EMBRAER) .... *(1994-95)* 14
Bandit (Avid) .... *(2000-01)* 577
Baron 58 (RAM/Beech) .... *(1991-92)* 466
BASLER TURBO CONVERSIONS (USA) .... *(1994-95)* 479
Bat (TAI) .... *(1998-99)* 487
Battlefield Lynx (Westland) .... *(1996-97)* 537
Beaver conversions (Wipaire) .... *(1991-92)* 496
Beaver, DHC-2/PZL-3S (Airtech Canada) .... *(1991-92)* 17
BEDE JET CORPORATION (USA) .... *(1994-95)* 480
Bee T-203 (Aeroprogress/ROKS-Aero) .... *(1997-98)* 350
Beech Mk II (Raytheon) .... *(1995-96)* 627
BEECH AIRCRAFT CORPORATION (USA) .... *(1995-96)* 489
BELL HELICOPTER KOREA (Korea, South) .... *(1994-95)* 238
BENGIS AIRCRAFT COMPANY (PTY) LTD (South Africa) .... *(1992-93)* 264
BERIEV DESIGN BUREAU (USSR) .... *(1991-92)* 239, 754
BILLIE AERO MARINE (France) .... *(2000-01)* 113
Bison M-3 (Myasishchev) .... *(1991-92)* 273
'Black' Projects (USAF) .... *(1996-97)* 744
Blinder, Tu-22 (Tupolev) .... *(1994-95)* 373
Blue Sky 91 (KAL) .... *(1997-98)* 302
BOEING CANADA LTD (Canada) .... *(1991-92)* 20
BOEING MILITARY AIRPLANES (USA) .... *(1992-93)* 348
Bohem 3 (Bohemia Air) .... *(1998-1999)* 83
BOHEMIA AIR (Czech Republic) .... *(1999-2000)* 88
Bonanza F33A (Raytheon) .... *(1997-98)* 722
Boomerang 202-11 (Scaled Composites) .... *(1994-95)* 626
BOWERS, PETER M. (USA) .... *(1994-95)* 520
BRANSON AIRCRAFT CORPORATION (USA) .... *(1993-94)* 461
BRANTLY HELICOPTER INDUSTRIES (USA) .... *(1994-95)* 521
Bravo 18A (FFA) .... *(1995-96)* 425
Bravo, AS 202 (FFA) .... *(1996-97)* 482
Brigantine, T-274 (Khrunichev) .... *(1997-98)* 393
BRISTOL AEROSPACE LTD (Canada) .... *(1995-96)* 31
BRITISH AEROSPACE AIRBUS LTD (UK) .... *(1993-94)* 435
BRITISH AEROSPACE ASSET MANAGEMENT (UK) .... *(1998-99)* 514
BRITISH AEROSPACE COMMERCIAL DIVISION (UK) .... *(1996-97)* 513
BRITISH AEROSPACE CORPORATE JETS LTD (UK) .... *(1993-94)* 377
BRITISH AEROSPACE REGIONAL AIRCRAFT LTD (UK) .... *(1995-96)* 452
BRITISH AIR FERRIES (UK) .... *(1993-94)* 391
BRITTEN-NORMAN (UK) .... *(2000-01)* 537
Bronco, OV-10D Plus (Rockwell) .... *(1991-92)* 468
Brushfire TG-10 (Venga/Baoshan) .... *(1995-96)* 210
BUCHANAN AIRCRAFT CORPORATION LTD (Australia) .... *(1995-96)* 5
BUSH CONVERSIONS INC (USA) .... *(1993-94)* 462
Bush Hawk FBA-2E (Found) .... *(1998-99)* 48
Bush Hawk-300, FBA-2C (Found) .... *(2000-01)* 58
Bushmaster B2N (Aircorp) .... *(1991-92)* 3
Business Jet (Piper) .... *(2000-01)* 741

C

C-5 Galaxy (Lockheed Martin) .... *(1995-96)* 566
C-12J (Beech) .... *(1991-92)* 353
C.14 (Dassault Aviation) .... *(1991-92)* 68
C-21A (Learjet) .... *(1994-95)* 547
C-23A Sherpa (Shorts) .... *(1992-93)* 312
C-23B Sherpa (Shorts) .... *(1992-93)* 312
C-26A/B (Fairchild Aerospace) .... *(1999-2000)* 644
C-27 (USAF) .... *(1991-92)* 495
C-27A Spartan (Alenia) .... *(2000-01)* 279
C-27A Spartan (Chrysler) .... *(1993-94)* 469
C-47TP Super Dakota (SAAF) .... *(1992-93)* 265
C90A King Air (Beech) .... *(1991-92)* 349
C-95 (EMBRAER) .... *(1991-92)* 10
C-101 Aviojet (CASA) .... *(1995-96)* 411
C-130 (Lockheed Martin) .... *(1998-99)* 665
C-130 Aerial spray system (Aero Union) .... *(1991-92)* 338
C-130 Auxiliary Fuel System (Aero Union) .... *(1991-92)* 338
C-130 Conversions (IAI) .... *(1994-95)* 205
C-130 Conversions (Lockheed) .... *(1991-92)* 430
C-130 Fireliner (Aero Union) .... *(1991-92)* 338
C-130 (Lockheed Martin) .... *(1998-99)* 665
C-130H (Lockheed Martin) .... *(1998-99)* 665
C-130H-30 (Lockheed Martin) .... *(1998-99)* 665
C-130H-MP (Lockheed Martin) .... *(1998-99)* 665
C-212P Aviocar (CASA) .... *(1991-92)* 214
CA-05 Christavia Mk 1 (Elmwood) .... *(1995-96)* 42
CA-22 Eland (Skyfox) .... *(1997-98)* 10
CA-25 Impala (Skyfox) .... *(1999-2000)* 9
CA-25N Gazelle (Skyfox) .... *(1999-2000)* 9
CAC (see Commercial Airplane Company) .... *(1994-95)* 227
CAF (see Changhe Aircraft Factory) .... *(1994-95)* 50
CAI (see Ciskei Aircraft Industries) .... *(1991-92)* 48
CALF (Boeing) .... *(1995-96)* 517
CALF (Lockheed Martin) .... *(1995-96)* 567
CALF (Northrop Grumman) .... *(1995-96)* 614
CAP 21 (Mudry) .... *(1993-94)* 87
CAP '92' (Mudry) .... *(1991-92)* 80
CAP 230 (Mudry) .... *(1991-92)* 79
CAP 231 (Mudry) .... *(1995-96)* 99, 786
CAP 231 EX (Mudry) .... *(1997-98)* 122
CAP X4 (Mudry) .... *(1991-92)* 80
CASA 3000 (CASA) .... *(1995-96)* 414
CAT (see Commuter Air Technology) .... *(1993-94)* 463
CATA (see Construction Aeronautique de Technologie Avancée .... *(2000-01)* 115
CAWAC (see Chengdu Asia Water Aircraft Company) .... *(1996-97)* 182
CB 206L-III (Metalnor) .... *(1992-93)* 32
CBA-123 Vector (EMBRAER/FMA) .... *(1994-95)* 157
CC-130 Hercules (Lockheed Martin) .... *(1998-99)* 665
CC-130T (Lockheed Martin) .... *(1998-99)* 665
CD 2 (Dornier Seastar) .... *(1996-97)* 286
CD 2 Seastar (Dornier/Composite) .... *(1991-92)* 94
CF-5A Upgrade programme (Bristol Aerospace) .... *(1991-92)* 23
CF-116 (Bristol Aerospace) .... *(1994-95)* 26
CH-8 Christavia Mk 4 (Elmwood) .... *(1995-96)* 42
CH-34 Super Puma (Aerospatiale) .... *(1991-92)* 58
CH-46 Sea Knight (Boeing) .... *(1991-92)* 366
CH-47C Chinook (Meridionali) .... *(1997-98)* 279
CH-135 (Bell) .... *(1998-99)* 27
CH-139 (Bell) .... *(1993-94)* 20
CH-147 (Boeing) .... *(1993-94)* 455
CH-148 Petrel (EHI) .... *(1993-94)* 140
CH-149 Chimo (EHI) .... *(1993-94)* 140
CH 200 (Zenair) .... *(1994-95)* 42
CH 250 (Zenair) .... *(1994-95)* 43
CH 300 (Zenair) .... *(1994-95)* 43
CH 601 (Zenair) .... *(1994-95)* 43
CH 601HD Zodiac (Derazona) .... *(1997-98)* 165
CH 601 HDS Super Zodiac (Zenith) .... *(1997-98)* 771
CH 601 (UL) Zodiac (Zenith) .... *(1997-98)* 770
CH 701-AG STOL .... *(1996-97)* 47
CH 2000 Zenith (Zenair) .... *(1999-2000)* 56
Che-15 (SGAU) .... *(1999-2000)* 421
Che-22 (Corvette) (SGAU) .... *(1999-2000)* 422
Che-25 (SGAU) .... *(1999-2000)* 422
Che-40 (SGAU) .... *(1999-2000)* 423
CHK-91 Chang-Gong 91 (KAL) .... *(1997-98)* 302
C/KC-130 Karnaf (Lockheed Martin) .... *(1998-99)* 665
CL-215 (Canadair) .... *(1991-92)* 26
CL-215T (Canadair) .... *(1994-95)* 33
CN-245 (Airtech) .... *(1998-99)* 195
CNIAR (see Centrul National al Industriei Aeronautice Române) .... *(1991-92)* 203
CP140 Aurora (Lockheed Martin) .... *(1999-2000)* 676
CP-140A Arcturus (IMP) .... *(1993-94)* 38
CP140A Arcturus (Lockheed Martin) .... *(1999-2000)* 676
CPCA (see Centrul de Proiectare SI Consulting Pentru Aviatie SA) .... *(1999-2000)* 341
C-Prop, cryogenic transport (Tupolev) .... *(1998-99)* 452
CR600 (Dyn' Aero) .... *(1998-99)* 120
CRJ-X (Canadair) .... *(1996-97)* 35
CRSS (see PT Cipta Restu Sarana Svaha) .... *(1994-95)* 125
CSH-2 Rooivalk (Denel) .... *(1998-99)* 461
CV 5800 (KFC) .... *(1994-95)* 40
C-XX (Cessna) .... *(1996-97)* 593

Cabri G2 (Guimbal) .... *(1994-95)* 93
Cadet (Robin) .... *(1992-93)* 69
CALIFORNIA HELICOPTER INTERNATIONAL (USA) .... *(1993-94)* 462
CALIFORNIA MICROWAVE INC (USA) .... *(1994-95)* 521
Camber, Il-86 (Ilyushin) .... *(1995-96)* 320
CAMPANA AVIATION (France) .... *(1998-99)* 100
Carajá NE-821 (Neiva) .... *(1992-93)* 16
Caravan, Fire Fighter (Aero Union) .... *(1991-92)* 338
CARDOEN, INDUSTRIAS, LTDA (Chile) .... *(1991-92)* 30
Cargo aircraft ground mobility system (Boeing) .... *(1991-92)* 372
Cargoliner (Cavenaugh) .... *(1991-92)* 385
Caribou DHC-4T (Newcal) .... *(1994-95)* 608
Careless, Tu-154C (Tupolev) .... *(1999-2000)* 452
Cava (Atlas) .... *(1991-92)* 210
CAVENAUGH AVIATION INC (USA) .... *(1993-94)* 463
C-CRAFT WOZLMSAYER (Austria) .... *(2000-01)* 9
CELAIR (PTY) LTD (South Africa) .... *(1992-93)* 264
CENTRUL NATIONAL AL INDUSTRIEI AERONAUTICE ROMANE (Romania) .... *(1991-92)* 203
CENTRUL DE PROIECTARE SI CONSULTING PENTRU AVIATIE SA (Romania) .... *(1999-2000)* 341
Centurion (Cessna) .... *(1991-92)* 385
Cessna 210 landing gear door modification (Sierra) .... *(1991-92)* 478
Cessna auxiliary fuel systems (Sierra) .... *(1991-92)* 478
Cessna suspended production (Cessna) .... *(1993-94)* 463
Challenger 601 (Canadair) .... *(1995-96)* 31
Challenger 604 (Canadair) .... *(1995-96)* 31
Chang-Gong 91, CHK-91 (KAL) .... *(1997-98)* 302
CHANGHE AIRCRAFT FACTORY (China) .... *(1994-95)* 50
Cheetah (Atlas) .... *(1994-95)* 394
Cheetah (HAL) .... *(1997-98)* 157
CHENGDU ASIA WATER AIRCRAFT COMPANY (International) .... *(1996-97)* 182
Chetak (HAL) .... *(1997-98)* 158
Chevvron 2-32 (AMF) .... *(2000-01)* 525
Cheyenne IIIA (Piper) .... *(1994-95)* 616
Cheyenne 400 (Piper) .... *(1994-95)* 617
Chimo Ch-149 (EHI) .... *(1993-94)* 140
CHINA NATIONAL AERO TECHNOLOGY IMPORT AND EXPORT CORPORATION (China) .... *(1999-2000)* 67
CHINCUL S. A. C. A. I. F. I. (Argentina) .... *(1994-95)* 2
Ching-Kuo, F-CK-1 (AIDC) .... *(2000-01)* 504
Chinook CH-47C (Meridionali) .... *(1997-98)* 279
Chinook Commercial (Boeing) .... *(1991-92)* 369
Christavia Mk 1 (Elmwood) .... *(1995-96)* 42
Christavia Mk 4 (Elmwood) .... *(1995-96)* 42
CHRYSLER TECHNOLOGIES AIRBORNE SYSTEMS (USA) .... *(1994-95)* 527
Chujiao-6A (NAMC) .... *(1995-96)* 59
Chuspi, IAP-001 (Indaer Peru) .... *(1992-93)* 173
CIPTA PT RESTU SARANA SYAHA (Indonesia) .... *(1994-95)* 125
Cirrus VK 30 (Cirrus) .... *(1996-97)* 597
Cirstel Alouette (Denel) .... *(1998-99)* 463
CISKEI AIRCRAFT INDUSTRIES (PTY) LTD (Ciskei) .... *(1991-92)* 48
Citation, Air ambulance equipment (Branson) .... *(1991-92)* 382
Citation, Extended range fuel system (Branson) .... *(1991-92)* 382
Citation, Extended width cargo door (Branson) .... *(1991-92)* 382
Citation II 550 (Cessna) .... *(1995-96)* 591
Citation II Weight increase (Branson) .... *(1991-92)* 382
Citation III (Cessna) .... *(1991-92)* 390
Citation V 560 (Cessna) .... *(1994-95)* 525
Citation VI 650 (Cessna) .... *(1996-97)* 595
Citation VII Model 660 (Cessna) .... *(1991-92)* 391
Citation 500 Upgrade Programme (AlliedSignal) .... *(1994-95)* 471
Citation S/II S550 (Cessna) .... *(1994-95)* 524
CitationJet JPATS 526 (Cessna) .... *(1995-96)* 519
Cityliner (Saab) .... *(1999-2000)* 477
CIVIL AVIATION DEPARTMENT (India) .... *(1991-92)* 100
Clank, An-30 (Antonov) .... *(1991-92)* 233
CLARK-NORMAN AIRCRAFT LTD (UK) .... *(1997-98)* 535
CLASSIC AIRCRAFT CORPORATION (USA) .... *(1997-98)* 621
Close Support Aircraft project (Sibnia) .... *(1998-99)* 416
Club Sprint (FLS) .... *(1994-95)* 447
Cobra Venom AH-1W (Bell) .... *(1993-94)* 436

C – E—AIRCRAFT–PREVIOUS TEN EDITIONS: INDEXES 853

Cock, An-22 (Antonov) .... *(1991-92)* 231
Coke (XAC) .... *(1991-92)* 46
Coke, An-24 (Antonov) .... *(1991-92)* 232
COLEMILL ENTERPRISES INC (USA) .... *(1994-95)* 528
Colt (SAP) .... *(1991-92)* 44
Colt AN-2 (PZL) .... *(2000-01)* 331
Comanchero (Schafer) .... *(1991-92)* 475
Comanchero 500 (Schafer) .... *(1991-92)* 475
Comanchero 750 (Schafer) .... *(1991-92)* 475
Combat Scout 406 CS (Bell) .... *(1992-93)* 345
Combat Support Helicopter Replacement (US Navy) .... *(1996-97)* 746
Commander 112 (Commander) .... *(1991-92)* 394
Commander 114B (Commander) .... *(2000-01)* 660
Commando (Westland) .... *(1994-95)* 462
COMMERCIAL AIRPLANE COMPANY (Japan) .... *(1994-95)* 227
Commercial Chinook (Boeing) .... *(1991-92)* 369
Commercial Hercules (Lockheed) .... *(1999-2000)* 680
Common Affordable Lightweight Fighter (Lockheed Martin) .... *(1995-96)* 567
Common Affordable Lightweight Fighter (Northrop Grumman) .... *(1995-96)* 614
Common Affordable Lightweight Fighter (US Navy) .... *(1995-96)* 666
Common Support Aircraft (Lockheed Martin) .... *(1999-2000)* 681
COMMUTER AIR TECHNOLOGY .... *(1993-94)* 463
COMTRAN LTD (USA) .... *(1993-94)* 472
CONAIR AVIATION LTD (Canada) .... *(1995-96)* 38
Concordino (SAAB) .... *(1999-2000)* 480
Condor Ru-38 (Schweizer) .... *(1994-95)* 627
CONDOR SA (Romania) .... *(1991-92)* 203
Condor T-204 (Aeroprogress) .... *(1996-97)* 751
Conquest I (Cessna) .... *(1991-92)* 385
Conquest II (Cessna) .... *(1991-92)* 385
CONSTRUCTION AERONAUTIQUE DE TECHNOLOGIE AVANCEE (France) .... *(2000-01)* 115
CONTENDER AIRCRAFT LTD (USA) .... *(1999-2000)* 640
COOK FLYING MACHINES (UK) .... *(1996-97)* 520
Coot-A, Il-20 DSR (Ilyushin) .... *(1996-97)* 358
Cora (Fantasy Air) .... *(2000-01)* 96
Corporate 800 (BAe) .... *(1993-94)* 377, 698
Corporate 1000 (BAe) .... *(1993-94)* 378
Corriedale SD27 (Sivel) .... *(1997-98)* 281
Corsair II Update Programmes (LTV) .... *(1991-92)* 430
Corvette Che-22 (SGAU) .... *(1999-2000)* 422
Corvette T-471 (Khrunichev) .... *(1997-98)* 398
Cossack An-225 (Antonov) .... *(1995-96)* 443
Cougar Mk I AS 352 (Eurocopter) .... *(1998-99)* 213
Courier 600, 700, 800, 900 (Helio) .... *(1991-92)* 412
Coyote (Montana) .... *(1995-96)* 609
CRANFIELD AERONAUTICAL SERVICES LIMITED (UK) .... *(1991-92)* 315
CRANFIELD INSTITUTE OF TECHNOLOGY, COLLEGE OF AERONAUTICS (UK) .... *(1991-92)* 315
CRO HOLDING (Czech Republic) .... *(2000-01)* 94
CROPLEASE PLC (UK) .... *(1993-94)* 393
Crosspointer TW-18 (Skyline) .... *(1997-98)* 482
Crusader (Dassault Aviation) .... *(1991-92)* 74
Cryoplane (DAA/Tupolev) .... *(2000-01)* 223
Cub (SAC) .... *(1991-92)* 41
Cuesta (EMBRAER) .... *(1994-95)* 20
Curl (XAC) .... *(1991-92)* 47
Curucaca (IPE) .... *(1992-93)* 16
Cutlass RG (Cessna) .... *(1991-92)* 385
Cygnet SF-2A (Viking) .... *(2000-01)* 798

D

D-500 (Egrett) .... *(1991-92)* 104
D-600 (Bell/Boeing) .... *(1996-97)* 565
DA 20-C1 Speed Katana (Diamond) .... *(1997-98)* 809
DA 40 (Diamond) .... *(1996-97)* 38
DAC (see Danubian Aircraft Company) .... *(1997-98)* 153
DASA (see DaimlerChrysler Aerospace) .... *(2000-01)* 152
DASA 80/130-passenger airliner (Deutsche Airbus) .... *(1991-92)* 88
DC-3 AMI Jet Prop (Professional Aviation) .... *(1994-95)* 398
DC-3 conversion, Turbo-67 (Basler) .... *(1991-92)* 346
DC-3-65TP Cargomaster (Schafer/AMI) .... *(1991-92)* 475
DC-3 65TPs (AMI) .... *(1992-93)* 327
DC-8 Series 71/72/73 (McDonnell Douglas) .... *(1991-92)* 435
DC-8 update (Dee Howard) .... *(1991-92)* 396
DC-9 cargo conversion (Pemco) .... *(1994-95)* 615
DC-9X (McDonnell Douglas) .... *(1994-95)* 575
DC-10 (McDonnell Douglas) .... *(1991-92)* 442
DC-130 Hercules (Lockheed Martin) .... *(1998-99)* 665
DHC-2/PZL-3S Beaver (Airtech Canada) .... *(1991-92)* 17
DHC-3/1000 Otter Airtech Canada .... *(1991-92)* 17
DHC-4T Caribou (Newcal) .... *(1994-95)* 608
DHC-7 ARL (California Microwave) .... *(1991-92)* 384
DHC-8 Dash 8 Series 100 (de Havilland) .... *(1997-98)* 45
DHI (see Daewoo Heavy Industries .... *(1999-2000)* 302
DK10 Dracula (CPCA) .... *(1998-99)* 329
DM3-1 Shmel (Dubna) .... *(1996-97)* 356
DM3-2 Osa (Dubna) .... *(1996-97)* 357
DMAV (see Dual Mode Air Vehicle) .... *(1992-93)* 373
DR 400 V6 (Robin) .... *(1997-98)* 128
DR 400/100 Cadet (Robin) .... *(1992-93)* 69
DR400/180R Remo (Robin) .... *(2000-01)* 142
DR 400/200R Remo 200 (Robin) .... *(1997-98)* 128
D-Series light aircraft (Daytona) .... *(1992-93)* 373
DV 20 Katana Xtreme (Diamond) .... *(1997-98)* 11
DV 22 Speed Katana (Diamond) .... *(1997-98)* 11

DAIMLERCHRYSLER AEROSPACE (Germany) .... *(2000-01)* 52
DAIMLERCHRYSLER AEROSPACE AIRBUS (Germany) .... *(2000-01)* 153
DALLACH, W.FLUGZEUG LEICHBAU (Germany) .... *(1999-2000)* 143
DANUBIAN AIRCRAFT COMPANY (Hungary) .... *(1997-98)* 153
Darling (Phoenix) .... *(1998-99)* 342
DASA/AEROSPATIALE/ALENIA (International) .... *(1991-92)* 120, 749
DASA/TUPOLEV (International) .... *(1991-92)* 750
Dash 8 Series 100 DHC-8 (de Havilland) .... *(1997-98)* 45
DASA/DENEL/HYUNDAI (International) *(1999-2000)* 212
DASSAULT AEROSPATIALE (France) .... *(1998-99)* 120
DASSAULT/DORNIER (International) .... *(1994-95)* 150
DATWYLER, MDC MAX, AG (Switzerland) .... *(1991-92)* 220
Dauphin (Helibras) .... *(1995-96)* 23
Dauphin AS 365X DGV (Eurocopter) .... *(1995-96)* 180
Dauphin AS 365X FBW (Eurocopter) .... *(1995-96)* 180
Dauphin X 380 DTP (Eurocopter) .... *(1995-96)* 180
Dauphin 2 AS 366 (Aerospatiale) .... *(1991-92)* 66
DAYTONA AIRCRAFT CONSTRUCTION INC (USA) .... *(1994-95)* 529
DE CHEVIGNY/WILSON (International) .... *(1994-95)* 150
DEE HOWARD COMPANY, THE (USA) .... *(1994-95)* 529
Deepak HPT-32 (HAL) .... *(1995-96)* 130
Defender, Battlefield Surveillance (Pilatus Britten-Norman) .... *(1994-95)* 453
Defender BN2T-4R (Pilatus Britten-Norman) .... *(1997-98)* 541
Defender, Border Patrol (Pilatus Britten-Norman) .... *(1994-95)* 454
Defender, ELINT (Pilatus Britten-Norman) .... *(1994-95)* 454
Defender, Maritime (Pilatus Britten-Norman) .... *(1994-95)* 454
Defiant 500 (UDRD) .... *(1992-93)* 175
Defiant 1000 (UDRD) .... *(1993-94)* 227
DELAERO BUSINESS AND COMMERCIAL AVIATION LTD (Russia) .... *(1992-93)* 198
Delphin (Myasishchev) .... *(1993-94)* 302
Delta, P.66D (Aviolight) .... *(1991-92)* 163
Delta Dart II (Richter) .... *(1998-99)* 153
DELTA-V (Moldova) .... *(2000-01)* 315
DERAZONA AVIATION INDUSTRY (Indonesia) .... *(1998-99)* 167
DEUTSCHE AEROSPACE AG (Germany) .... *(1994-95)* 104
DEUTSCHE AEROSPACE AIRBUS/TUPOLEV (International) .... *(1994-95)* 150
Diamond IA Long-range tank (Branson) .... *(1991-92)* 382
Dingo (Aero Ric) .... *(2000-01)* 355
Dino (Ganzavia) .... *(1991-92)* 100
Dino GAK-22 (GAK) .... *(1996-97)* 137
Dinosaur T-2402 (Khrunichev) .... *(1995-96)* 339
Discojet (Moller) .... *(1993-94)* 537
DORNA, H. F. (Iran) .... *(1991-92)* 144
Dorna Two (Dorna) .... *(1995-96)* 211
Dornier 228 (Dornier) .... *(1991-92)* 88
Dornier 228 Maritime Patrol (HAL) .... *(1998-99)* 162
Dornier 228 Maritime Pollution Surveillance (HAL) .... *(1997-98)* 160
Dornier 228 Photogrammetry/Geo-Survey (Dornier) .... *(1996-97)* 123
Dornier 228-212 (Dornier) .... *(1997-98)* 159
Dornier 328 (Dornier) .... *(2000-01)* 153
Dornier 328S, Stretched (Dornier) .... *(1993-94)* 99
Dornier 428JET (Dornier) .... *(2000-01)* 155
DORNIER COMPOSITE AIRCRAFT GmbH (Germany) .... *(1992-93)* 78
DORNIER LUFTFAHRT (Germany) .... *(1996-97)* 121
DORNIER SEASTAR MALAYSIA SDN BHD (Malaysia) .... *(1997-98)* 305
DOUGLAS AIRCRAFT COMPANY (USA) .... *(1997-98)* 691
DOUGLAS/CATIC (International) .... *(1993-94)* 138
Dracula AH-IRO (IAR) .... *(2000-01)* 348
Dracula DK-10 (CPCA) .... *(1998-99)* 329
Draken J35J (Saab-Scania) .... *(1991-92)* 220
Dromader Mini, M-21 (PZL Mielec) .... *(1991-92)* 193
Dromader Super, M-24 (PZL Mielec) .... *(1991-92)* 180
Dromander Water Bomber (Melex) .... *(1991-92)* 465
Dual aft-body stroke system (Raisbeck) .... *(1991-92)* 465
DUAL MODE AIR VEHICLE INC (USA) .... *(1992-93)* 373
Dual Pac conversions (Soloy) .... *(1993-94)* 572
Duet/Saras M-102 (Myasishchev/NAL) .... *(1997-98)* 238

E

E-1 (Sat Saviat) .... *(2000-01)* 436
E-2B (Grumman) .... *(1991-92)* 406
E-3B/C AWACS (Boeing) .... *(1993-94)* 452
E-3 Sentry (Boeing) .... *(1993-94)* 452
E-3D Sentry AEW Mk 1 (BAe/Boeing) .... *(1993-94)* 391, 452
E-3F (Boeing) .... *(1993-94)* 452
E-5 (Sat Saviat) .... *(1999-2000)* 420
E-6A Mercury/Tacamo II (Boeing) .... *(1993-94)* 454
E-8 (Boeing) .... *(1994-95)* 499
E.25 Mirlo (CASA) .... *(1995-96)* 411
E-86C J-STARS (Grumman) .... *(1993-94)* 485
EA-6B Prowler (Grumman) .... *(1993-94)* 482
EAA (see Eagle Aircraft Australia) .... *(1991-92)* 4
EAM (see Eagle Aircraft Malaysia) .... *(1998-99)* 297
EAPL (see Eagle Aircraft) .... *(1996-97)* 5
EC-18D (CTAS) .... *(1991-92)* 392
EC-95 (EMBRAER) .... *(1991-92)* 10
EC-130 Hercules/Aya/Sapeer (Lockheed Martin) .... *(1998-99)* 665
ECH-023 Namcu (ENAER) .... *(1999-2000)* 57
EDI (see Express Design Inc) .... *(1995-96)* 530
EF-111A update (Grumman) .... *(1991-92)* 410
EFA (see European Fighter Aircraft) .... *(1992-93)* 122
EFS (Mooney) .... *(1991-92)* 454
EFS (USAF) .... *(1992-93)* 456
EL-1 Model 358 (Gavilan) .... *(1994-95)* 67
EMB-110 Bandeirante (EMBRAER) .... *(1994-95)* 14
EMB-111 (EMBRAER) .... *(1993-94)* 14
EMB-135 (EMBRAER) .... *(1997-98)* 811
EMB-201A Ipanema (EMBRAER) .... *(1992-93)* 15
EMB-312 Tucano (EMBRAER) .... *(2000-01)* 17
EMB-312H Super Tucano (EMBRAER) .... *(1997-98)* 16
EMB-312HJ Super Tucano (Northrop Grumman) .... *(1995-96)* 617
EMB-720D Minuano (Neiva) .... *(1995-96)* 24
EMB-810D Seneca (Neiva) .... *(1995-96)* 24
E-Tan (IAC) .... *(1991-92)* 228
EX (Boeing) .... *(1994-95)* 517
EX (USN) .... *(1994-95)* 651

Eagle 300 (Celair) .... *(1992-93)* 264
Eagle series, Cessna, R/STOL mods (Sierra) .... *(1991-92)* 479
Eagle, F-15C/D (McDonnell Douglas) .... *(1993-94)* 521
Eagle, F-15DJ (Mitsubishi/McDonnell Douglas) .... *(1991-92)* 174
Eagle F-15J (Mitsubishi) .... *(2000-01)* 288
Eagle H-76 (Sikorsky) .... *(1997-98)* 755
Eagle M-99 Orkan (PZL-Mielec) .... *(1997-98)* 324
Eagle T208 (Khrunichev) .... *(1998-99)* 375
EAGLE AIRCRAFT (Australia) .... *(1996-97)* 5
EAGLE AIRCRAFT (MALAYSIA) .... *(1998-99)* 297
EAGLE AIRCRAFT AUSTRALIA (Australia) .... *(1991-92)* 4
Eastern Caravan, T-205 (Aeroprogress/ROKS-Aero) .... *(1997-98)* 351
ECOLE SUPERIEURE DES TECHNIQUES AERO TECHNIQUES ET CONSTRUCTION AUTOMOBILE (France) .... *(1999-2000)* 126
EGRETT (International) .... *(1996-97)* 182
Egrett II G-520 (Grob) .... *(1997-98)* 144
Egrett-1 (Egrett) .... *(1991-92)* 121
EIDGENOSSISCHES FLUGZEUGWERK (Switzerland) .... *(1995-96)* 431
EIS AIRCRAFT (Germany) .... *(1996-97)* 126
Eland CA-22 (Skyfox) .... *(1997-98)* 10
ELBIT LTD (Israel) .... *(1994-95)* 202
ELICOTTERI MERIDIONALI (Italy) .... *(1996-97)* 252
Elite (VAT) .... *(1994-95)* 652
ELMWOOD AVIATION (Canada) .... *(1995-96)* 42
EMBRAER/FMA (International) .... *(1995-96)* 168
EMIND (Yugoslavia) .... *(1991-92)* 497
Enhanced Flight Screener (USAF) .... *(1992-93)* 456
EPA AIRCRAFT COMPANY (UK) .... *(1994-95)* 455
Epervier (Epervier Aviation) .... *(1995-96)* 13
EPERVIER AVIATION (Belgium) .... *(1995-96)* 13
Eshet (EAL) .... *(1997-98)* 96
Esquilo (Helibras) .... *(2000-01)* 26
EUROCOPTER/MIL/KAZAN/KLIMOV (International) .... *(1994-95)* 174
Eurofighter 2000 (Eurofighter) .... *(1998-99)* 229, 803
EUROFLAG (International) .... *(1995-96)* 167
Eurofox (Evektor-Aerotechnik) .... *(2000-01)* 95
EURO-HERMESPACE SA (International) .... *(1993-94)* 163
European Fighter Aircraft (Eurofighter) .... *(1992-93)* 122
European Future Advanced Rotorcraft (International) .... *(2000-01)* 247
EUROSPACE COSTRUZIONI AERONAUTICHE (Italy) .... *(1999-2000)* 278
Eurotrainer 2000 (FFT) .... *(1991-92)* 97
Eurotrainer 2000A (FFT) .... *(1992-93)* 80
EVER (Egrett) .... *(1994-95)* 153
Excalibur F-15-F (Eurospace) .... *(1998-99)* 268
EXCALIBUR AVIATION COMPANY (USA) .... *(1993-94)* 474
EXCLUSIVE AVIATION (USA) .... *(2000-01)* 666
Exec 90 (Rotorway) .... *(1994-95)* 624
Executive 600 (Colemill) .... *(1991-92)* 393
Expediter 1 (Fairchild) .... *(1995-96)* 533
Expediter 23 (Fairchild Dornier) .... *(1999-2000)* 644
Experimental Aircraft 105 (USSR) .... *(1991-92)* 298
EXPERIMENTAL AVIATION INC (USA) .... *(1997-98)* 626

854 INDEXES: AIRCRAFT–PREVIOUS TEN EDITIONS—**E – H**

Explorer (De Chevigny/Wilson) (1994-95) 150
Explorer (Merlin) (1997-98) 698
Explorer MD 902 (McDonnell Douglas) (1997-98) 685
Express (Express Design) (1995-96) 530
EXPRESS DESIGN INC (USA) (1995-96) 530
Extender (McDonnell Douglas) (1991-92) 442
Extra 260 (Extra) (1991-92) 95

F

F1 Mirage (Dassault Aviation) (1992-93) 58
F-2 (Aerostar) (1997-98) 342
F-4 Phantom II (McDonnell Douglas) (1991-92) 435
F-4EJ Kai (Mitsubishi) (1994-95) 234
F-4F ICE Programme (DASA) (1994-95) 108
F-5 Plus upgrade (IAI) (1993-94) 205
F-5 Upgrade (SA) (1994-95) 392
F-5 Upgrade Programmes (Bristol Aerospace) (1994-95) 26
F-5 Waco Classic (Classic) (1992-93) 371
F-5E/RF-5E conversion (SA) (1993-94) 388
F-14 Tomcat (Grumman) (1993-94) 483
F-15C/D Eagle (McDonnell Douglas) (1993-94) 521
F-15DJ (Mitsubishi/McDonnell Douglas) (1991-92) 174
F-15E Eagle (McDonnell Douglas) (1994-95) 584
F-15-F 300M (Eurospace) (1998-99) 268
F-15F Eagle (McDonnell Douglas) (1994-95) 584
F-15-F Excalibur (Eurospace) (1998-99) 268
F-15H Eagle (McDonnell Douglas) (1992-93) 408
F-15J (Mitsubishi/Boeing) (2000-01) 304
F-15S/MTD (McDonnell Douglas) (1995-96) 581
F-15 upgrade (IAI) (1993-94) 184
F-22 Naval variant (Lockheed Martin) (1995-96) 557
F.22 Series (Taylorcraft) (1993-94) 574
F22A Classic 118 (Taylorcraft) (1991-92) 492
F-27 Firefighter (Conair) (1994-95) 35
F27 Friendship ARAT (SECA) (1991-92) 84
F-27 Large cargo door (Branson) (1991-92) 382
F33A Bonanza (Raytheon) (1997-98) 722
F-104S ASA (Alenia/Lockheed) (1991-92) 163
F-111 (BAe) (1993-94) 391
F-111 upgrade (Rockwell International) (1991-92) 468
F-111G (USAF/General Dynamics) (1991-92) 495
F-117A Nighthawk (Lockheed Martin) (1995-96) 567
F-117N Seahawk (Lockheed) (1994-95) 565
F.220 Airone (General Avia) (2000-01) 285
F1300 Jet Squalus (Promavia) (1996-97) 13
F/A-18 high-alpha research vehicle (NASA) (1991-92) 456
FA. MK2 Sea Harrier (BAE) (1999-2000) 515
FA-X (Fokker) (1996-97) 298
FBA-2C Bush Hawk-300 (Found) (2000-01) 58
F-CK-1 Ching-Kuo (AIDC) (2000-01) 504
FH-227 Large cargo door (Branson) (1991-92) 382
FK 11(B8F) (1999-2000) 142
FM-2 Air Shark 1 (Freedom Master) (1993-94) 477
FMA (see Fabrica Militar de Aviones) (1995-96) 3
FNJ (Fuji) (1998-99) 280
FR-06 Ranger 2000 (Rockwell/DASA) (1996-97) 229
FRWE (Australia) (1991-92) 745
FS-X (Mitsubishi) (1995-96) 245
FT-5 (CAC) (1994-95) 47
FTT Vector (DASA) (1999-2000) 144
F + W (see Swiss Federal Aircraft Factory) (1995-96) 431

FABRICA MILITAR DE AVIONES (Argentina) (1995-96) 3
FAIRCHILD AEROSPACE CORPORATION (USA) (2000-01) 668
FAIRCHILD DORNIER LUFTFAHRT BETEILIGUNGS (Germany) (1996-97) 791
FAIRCHILD DORNIER GERMANY DORNIER LUFTFAHRT (Germany) (1998-99) 138
Fajr (IRGC) (1991-92) 144
Falcon (PAC) (1997-98) 315
Falcon (Peregrine) (1996-97) 694
Falcon 20B Engine Retrofit Programme (AlliedSignal) (1994-95) 471
Falcon 20, 731 Engine retrofit programme (Garrett) (1991-92) 401
Falcon 900 (Dassault Aviation) (1992-93) 64
Falcon 9000 (Dassault) (1994-95) 92
Falcon PW300-F20 (Volpar) (1993-94) 584
Falcon SST (Dassault) (1999-2000) 124
Falcon T-401 (Aeroprogress/ROKS-Aero) (1997-98) 353
FAMILY AIR (Czech Republic) (1996-97) 78
Family Air (Family Air) (1996-97) 78
Fanjet A-22J (Sadler) (1994-95) 625
Fan Ranger (Rockwell International/ DASA) (1992-93) 136
Fantrainer 400/600 (RFB) (1991-92) 99
Fantrainer 400/600 (RTAF-RFB) (1991-92) 228
Fantrainer 800 (RFB) (1995-96) 119
FARRINGTON AIRCRAFT CORPORATION (USA) (1991-92) 401
Favorit (Aviatehnologia) (1999-2000) 308
FEDERALNOYE GOSUDARSTVENNOYE UNITARNOYE PREDPRIYATIE (Russian Federation) (2000-01) 400
Fennec (Helibras) (2000-01) 26
FFT GESELLSCHAFT FUR FLUGZEUG- UND FASERVERBUND-TECHNOLOGIE GmbH (Germany) (1992-93) 79
FFV AEROTECH (Sweden) (1991-92) 215
Fieldmaster NAC 6 (EPA) (1993-94) 393
Firebird (Intora) (2000-01) 546
Firecat (Conair) (1994-95) 35
Firefighter F27 (Conair) (1994-95) 35
Firefly T67M Mk 2 (Slingsby) (1999-2000) 532
Firemaster (Croplease) (1993-94) 393
Firestar tanker conversion (Aero Union) (1991-92) 338
Fishbed, MiG-21 (CAC) (1991-92) 33
Fishbed, MiG-21 (MiG) (1992-93) 214
Fitter-C, D, E, F, G, H, J and K (Sukhoi) (1995-96) 368
Flagon Su-15 (Sukhoi) (1992-93) 236
Flaming, PZL-105L (PZL Warszawa-Okecie) (1999-2000) 333
Flamingo, PZL-105L (PZL Warszawa) (1999-2000) 333
Flamingo, T-433 (Aeroprogress/ROKS-Aero) (1999-2000) 347
Flamingo III TM-20B (Bengis) (1992-93) 264
Flogger-A, B, C, E, F, G, H and K, MiG-23 (MiG) (1994-95) 324
Flogger D and J, MiG-27 (MiG) (1994-95) 326
Flogger-J (HAL) (1997-98) 161
FLS AEROSPACE (UK) (2000-01) 546
Fly Baby 1-A (Bowers) (1994-95) 520
Fly Baby 1-B (Bowers) (1994-95) 520
Fokker 50 (Fokker) (1997-98) 308
Fokker 50 Hot and High (Fokker) (1991-92) 181
Fokker 50 Maritime and Surveillance versions (Fokker) (1991-92) 181
Fokker 50 Special Mission Versions (Fokker) (1996-97) 292
Fokker 60 (Fokker) (1997-98) 310
Fokker 70 (Fokker) (1997-98) 312
Fokker 100 (Fokker) (1997-98) 310
Fokker 130 (Fokker) (1995-96) 262
Forger (Yakovlev) (1993-94) 331
FOURNIER, AVIONS (France) (1999-2000) 126
Foxbat MiG 25 (MiG) (1998-99) 380
Foxstar Baron (Colemill) (1991-92) 394
Freebird Mk 5 (Freewing) (1995-96) 536
FREEBIRD SPORT AIRCRAFT (USA) (2000-01) 669
Freedom Fighter (Northrop) (1992-93) 426
FREEDOM MASTER CORPORATION (USA) (1993-94) 477
Freestyle, Yak-141 (Yakovlev) (1993-94) 336
FREEWING AERIAL ROBOTICS CORPORATION (USA) (1995-96) 536
Freewing Concept (Freewing) (1995-96) 536
Fregat T-130 (Aeroprogress/ROKS/Aero) (1998-99) 335
Fregat T-230 (Khrunichev) (1998-99) 375
Freighty 2.5 (NIAT) (1997-98) 428
Friendship ARAT, F27 (SECA) (1991-92) 84
Frigate T-130 (Aeroprogress/ ROKS-Aero) (1997-98) 349
Frigate Bird T-230 (Khrunichev) (1998-99) 375
Fully enclosed landing gear doors (Raisbeck) (1991-92) 465
Future Advanced Small Airliner (DASA) (1994-95) 104
Future Aircraft Technology Enhancement (USAF) (2000-01) 794
Future Amphibious Support Helicopter (RN) (2000-01) 547
Future Light Aircraft (Diamond) (1999-2000) 12
Future SST (SCTICSG) (1991-92) 141

G

G2 Cabri (Guimbal) (1994-95) 93
G-4 Super Galeb (UTVA) (1994-95) 658
G-5 (UTVA) (1994-95) 660
G 109B (Grob) (1997-98) 141
G 115TA Acro (Grob) (2000-01) 162
G 164B Super-B Turbine (Ag-Cat) (1997-98) 556
G-164 C-T Turbo Cat (Marsh) (1991-92) 432
G222 (Alenia) (2000-01) 279
G 520 Egrett II (Grob) (1997-98) 144
G 850 Strato 2C (Grob) (1997-98) 146
GA-55 LightWing (Hughes) (1994-95) 8
GADC (see Global Aero Design Centre) (1998-99) 460
GAK (see Ganzavia Kft) (1996-97) 137
GAK-22 Dino (GAK) (1996-97) 137
GOHL (see Guangzhou Orlando Helicopters Ltd) (1992-93) 35
GF 200 (Grob) (1999-2000) 153
GF 250 (Grob) (1999-2000) 153
GF 300 (Grob) (1999-2000) 153
GF 350 (Grob) (1999-2000) 153
GP-60 (Myasishchev) (2000-01) 435
GR-582 Lightwing (Hughes) (1994-95) 7
GR-912 Lightwing (Hughes) (1994-95) 8

GafHawk 125-200 (Ameco-Hawk) (1991-92) 341
Galaxy (Lockheed Martin) (1995-96) 566
Galaxy 700 JetStar (Galaxy) (1993-94) 478
GALAXY GROUP (USA) (1993-94) 478
Galaxy Star (Galaxy) (1993-94) 478
GANZAVIA GT (Hungary) (1991-92) 100
GANZAVIA Kft (Hungary) (1996-97) 137
Gardian 2 (Dassault Aviation) (1991-92) 75
Gardian 50 (Dassault Aviation) (1991-92) 76
GARRETT GENERAL AVIATION SERVICES DIVISION (USA) (1992-93) 378
Gavião HB 315B (Helibras) (1993-94) 19
Gavilan (Aero Mercantil) (1991-92) 48
Gazelle (UTVA) (1994-95) 660
Gazelle CA-25N (Skyfox) (1999-2000) 9
Gazelle SA 342 (Eurocopter) (1994-95) 163
GEGASI INDUSTRIES SDN BHD (Malaysia) (1997-98) 306
Gemini CH620 (Zenair) (1996-97) 46
Gemini ST (Global) (1991-92) 406
Gemini ST (Tridair) (1994-95) 647
Gemini Ultra (Remus) (1999-2000) 160
Gemsbok (Atlas) (1991-92) 211
GENERAL DYNAMICS (USA) (1994-95) 535
GENERAL DYNAMICS/MCDONNELL DOUGLAS (USA) (1992-93) 382
Genesis 6-650 (Gevers) (1999-2000) 648
Geophysica M-55 (Myasishchev) (1996-97) 412
Geophysica-2 M-55 (Myasishchev) (1996-97) 413
GEPARD SENSOR TECHNOLOGIES SYSTEMS AG (Switzerland) (1991-92) 222
Gerfaut (Eurocopter) (1993-94) 156
GEVERS AIRCRAFT (USA) (2000-01) 670
G. F. GESTIONI INDUSTRIALI (Italy) (1993-94) 187
Glasair II-S (Stoddard-Hamilton) (1995-96) 660
GLOBAL AERO DESIGN CENTRE (Singapore) (1998-99) 460
GLOBAL HELICOPTER TECHNOLOGY (USA) (1994-95) 537
Goair Trainer (Goair Products) (1997-98) 6
Goblin (Aviotechnika) (1997-98) 202
Golden Eagle (Cessna) (1991-92) 385
GOMOLZIG INGENIEURBURO, HERBERT (Germany) (2000-01) 161
Grand 51 (Exclusive Aviation) (2000-01) 666
Greyhound 21 (Northrop Grumman) (2000-01) 731
Grif T-121 (Washington Aeroprogress) (1996-97) 751
Griffon EH 101 (EHI) (1996-97) 184
Griffon T-204 (Aeroprogress/ROKS-Aero) (1997-98) 351
Grillon 120 (Sellet-Pelletier) (1991-92) 85
GRUMMAN CORPORATION (USA) (1993-94) 480
Gyrfalcon (MAI-Avgur) (1999-2000) 387
GUANGZHOU ORLANDO HELICOPTERS LTD (China) (1992-93) 35
Guardian HU-25 (Dassault Aviation) (1991-92) 74
GUIMBAL, BRUNO (France) (1994-95) 93
GULFSTREAM AEROSPACE TECHNOLOGIES (USA) (1993-94) 487
Gulfstream IV (Gulfstream Aerospace) (1995-96) 538
Gulfstream XT (Gulfstream Aerospace) (1993-94) 487
GUNASEKERA AIRCRAFT (USA) (2000-01) 677
Gzhel, M-101T (Myasishchev) (1999-2000) 416

H

H-1 (Hamilton) (1997-98) 635
H2X Hawk III (Groen) (1999-2000) 650
H-6 (XAC) (1992-93) 44
H-40 (RFB) (1994-95) 113
H-42/45 (UTVA) (1994-95) 660
H-76 Eagle (Sikorsky) (1997-98) 755
H-76N (Sikorsky) (1991-92) 489
H-550A Stallion (Helio) (1991-92) 413
HAFFS (See Heliborne Aerial FireFighting System) (1991-92) 338
HAI (see Hellenic Aerospace Industry) (1994-95) 117
HAIG (see Hongdu Aviation Industry Group) (1999-2000) 64
HALE (SOCATA) (1996-97) 120
HB-23 Hobbyliner (HB-Aircraft) (1991-92) 7
HB-23 Scanliner (HB-Aircraft) (1991-92) 7
HB-202 (HB-Aircraft) (1991-92) 8
HB 315B Gaviao (Helibras) (1993-94) 19
HB 350 Esquilo (Helibras) (1993-94) 19
HB 355F2 Esquilo (Helibras) (1993-94) 19
HC-2 (He-Ro's) (1999-2000) 316
HC-130 Hercules (Lockheed Martin) (1998-99) 665
HD 19 (TAI) (1997-98) 505
HD.21 (Aerospatiale) (1991-92) 58
HD 21 (Eurocopter) (1998-99) 213
HDH (see Hawker de Havilland Ltd) (1995-96) 8
HD-XX (TAI) (1998-99) 487
HH-3F Pelican (Agusta/Sikorsky) (1991-92) 157
HK-36R Super Dimona (Diamond) (1996-97) 11
Hkp 10 (Aerospatiale) (1991-92) 58
Hkp 10 (Eurocopter) (1998-99) 213
HOTOL (BAe) (1992-93) 303
HP-32 Deepak (HAL) (1995-96) 130
HR 100/250 (Robin) (1997-98) 129
HS 748 ASP (HAL) (1998-99) 161
HS 748 ASWAC (HAL) (1995-96) 129
HT.21 (Aerospatiale) (1991-92) 58
HT 21 (Eurocopter) (1998-99) 213
HTT-34 (HAL) (1991-92) 104
HTT-35 (HAL) (1996-97) 141

H - K—AIRCRAFT-PREVIOUS TEN EDITIONS: INDEXES 855

HTT-38 (HAL) (2000-01) 179
HTTB (Lockheed) (1991-92) 429
HU-1 Seagull (SSLF) (1998-99) 72
HU-1H (Fuji-Bell) (1992-93) 157
HU-2B Petrel 650.B (SSLF) (1999-2000) 77
HU-2C Petrel 650C (SSLF) (2000-01) 83
HU-2D Petrel 650D (SSLF) (2000-01) 83
HU-16B Albatross Tanker conversion (Aero Union) (1991-92) 338
HU-25 Guardian (Dassault Aviation) (1991-92) 74
HX-1 (Hamilton) (1991-92) 412
HXT-2 (Hamilton) (1991-92) 412
HX-300 Surprise (Hepp) (2000-01) 163
HX-321 (Hamilton) (1997-98) 635

Haitun Z-9 (HAMC) (1994-95) 54
Haiyan (NAMC) (1991-92) 41
Halcón, A-36, T-36 (CASA) (1995-96) 411
Halcón, A-36 (ENAER) (1997-98) 54
Halcón T-36 (ENAER) (1996-97) 48
HAMILTON AEROSPACE (USA) (1997-98) 635
Hansa-2 NALLA (NAL) (1997-98) 162
Harke Mi-10 (MIL) (1991-92) 266
Harrier (BAe) (1993-94) 389
Harrier II, (McDonnell Douglas (Boeing)/BAe) (1997-98) 229
Harrier II Plus (McDonnell Douglas (Boeing)/BAe) (1997-98) 234
Harrier III (McDonnell Douglas/BAe) (1993-94) 167
Harrier GR. Mk 3 (BAe) (1991-92) 311
Harrier GR. Mk 5, 5A and 7, and T. Mk 10 (McDonnell Douglas (Boeing)/BAe) (1997-98) 229
Harrier T. Mk 4/4A (BAe) (1991-92) 311
Harrier T. Mk 4N/8 (BAe) (1993-94) 389
Harrier T. Mk 60 (BAe) (1993-94) 389
Hauler UH-12E (Rogerson Hiller) (1994-95) 623
Have Blue, XST (Lockheed) (1992-93) 402
Hawk T. Mk IFTS (BAe) (1994-95) 767
Hawk III, H2X (Groen) (1999-2000) 650
Hawk 72 (Hawk) (1993-94) 487
HAWK AIRCRAFT DEVELOPMENT CORPORATION (USA) (1993-94) 487
Hawker 800 (Raytheon) (1995-96) 464
Hawker 1000 (Raytheon) (1997-98) 733
Hawker PD1000Y (Raytheon) (1996-97) 716
HAWKER DE HAVILLAND LTD (Australia) (1995-96) 8
HB-AIRCRAFT INDUSTRIES LUFTFAHRZEUG AG (Austria) (1992-93) 9
HELI-AIR (USA) (1994-95) 540
Heliborne Aerial Fire Fighting System (Aero Union) (1991-92) 338
Heli-Camper (Orlando) (1993-94) 545
HELIO AIRCRAFT CORPORATION (USA) (1993-94) 389
Helitankers (Conair) (1994-95) 36
HELLENIC AEROSPACE INDUSTRY (Greece) (1994-95) 117
Heracles Molniya-1000 (Molniya) (1997-98) 425
Hercules C. Mk 1K, C. Mk 1W, Mk 2 and C, Mk 3 (Lockheed Martin) (1998-99) 665
Hercules, commercial (Lockheed Martin) (1999-2000) 680
Hercules conversion (IPTN) (1994-95) 126
Hercules conversions (IAI) (1994-95) 205
Hercules T.10, TK.10, TL.10 (Lockheed Martin) (1998-99) 665
Hermès (Euro-Hermespace) (1992-93) 124
Heron (General Avia) (2000-01) 285
HESA (Iran) (1998-99) 249
High-alpha research vehicle F/A-18 (NASA) (1991-92) 456
HIGHLANDER AIRCRAFT CORPORATION (USA) (1999-2000) 656
HighLander (HighLander) (1999-2000) 656
High Speed Commercial Transport (Boeing) (1999-2000) 618
HILLER AIRCRAFT CORPORATION (USA) (2000-01) 678
HIRT AIRCRAFT CORPORATION (USA) (1997-98) 637
HISAR 200 (Raytheon) (2000-01) 754
HOAC AUSTRIA FLUGZEUGWERK WIENER NEUSTADT GmbH (Austria) (1995-96) 11
Hobbyliner (HB-Aircraft) (1991-92) 7
Hobbyliner, HB-Aircraft (CAI) (1991-92) 48
HOFFMANN, WOLF, FLUGZEUGBAU AG (Germany) (1992-93) 82
Hokum, Ka-136 (Kamov) (1991-92) 254
HONDA MOTOR COMPANY (USA) (1993-94) 488
HONDA/MSU (International) (1995-96) 189
Hong-6 (XAC) (1992-93) 44
Hong-7 (XAC) (1995-96) 64
HONGDU AVIATION INDUSTRY GROUP (China) (1999-2000) 64
Hongzhaji-6 (XAC) (1992-93) 44
Hongzhaji-7 (XAC) (1995-96) 64
Hoodlum-A, Ka-26 (Kamov) (1993-94) 271
Hoodlum-B, Ka-126 (Kamov) (1997-98) 390
Hoplite, Mi-2 (PZL-Swidnik) (1994-95) 270
Hormone, Ka-25 (Kamov) (1995-96) 327
Horus (MBB) (1991-92) 94
HOWARD HUGHES ENGINEERING PTY LTD (Australia) (1995-96) 8
Huorpe (LAVIASA) (2000-01) 1
Huey II, UH-HIP (Bell) (1994-95) 492
Huey 800 (Global) (1994-95) 537
HueyCobra AH-IF (IAR) (1996-97) 330
HueyCobra, Modernised Versions (Bell) (1993-94) 434
Hummingbird (Aerotek) (1997-98) 477
Hummingbird RPX-Alpha (PADC) (1999-2000) 314
Hybrid Aircraft (Vivian Associates) (1992-93) 29
HYDRAVION LOURD (France) (1996-97) 119
Hydro 2000 (Hydro) (1997-98) 119
HYDRO 2000 INTERNATIONAL (France) (1999-2000) 127
HYDROPLANE COMPANY (Russian Federation) (1998-99) 350
Hyflex (NASDA) (1997-98) 298
Hytex (MBB) (1991-92) 94

I

I-1 (Interavia) (1994-95) 316
I-3 (Interavia) (1995-96) 326
IA 63 Pampa (LMAASA) (1998-99) 2
IAC (see International Aircraft Company) (1991-92) 228
IAF (see Iraqi Air Force) (1994-95) 202
IAR 330 Puma 2000 (IAR) (1994-95) 285
IAR-330L Puma (IAR) (1998-99) 329
IAI 1125 Astra (IAI) (1992-93) 141
IAP-001 Chuspi (Indaer Peru) (1992-93) 173
IAP-002 Ag-Chuspi (Indaer Peru) (1992-93) 173
IAP-003 Urpi (Indaer Peru) (1992-93) 174
IAR-93 (Soko/Avioane) (1997-98) 249
IAR-109 Swift (Avioane) (1995-96) 291
IAR-317 Airfox (IAR) (1991-92) 205
IAR-501 (INAV) (1999-2000) 342
IAR-503A (INAV) (1993-94) 248
IAR-705 (Avioane) (1993-94) 251
IAR-707 AMTU (INAV) (1995-96) 295
IAR-828 (IAR) (1993-94) 252
IAv CRAIOVA (see Intreprinderea de Avioane Craiova) (1991-92) 207
ICE F-4F Programme (MBB) (1994-95) 108
IG JAS (see Industrigruppen JAS AB) (1996-97) 472
Il-20 DSR Coot (Ilyushin) (1996-97) 358
Il-76 Command Post (Ilyushin) (1995-96) 319
Il-80 Maxdome (Ilyushin) (1996-97) 364
Il-82 (Ilyushin) (1999-2000) 365
Il-86 Camber (Ilyushin) (1995-96) 320
Il-86 Command Post; Maxdome (Ilyushin) (1995-96) 321
Il-87 Maxdome (Ilyushin) (1999-2000) 365
Il-90 (Ilyushin) (1991-92) 246
Il-102 (Ilyushin) (1993-94) 268
Il-106 (Ilyushin) (1998-99) 358
Il-108 (Ilyushin) (1993-94) 269
Il-X (Ilyushin) (1991-92) 249
INAV (see Institutul de Aviatie) (2000-01) 350
INDAER PERU (see Industria Aeronautica Del Peru SA) (1993-94) 226
IPE (see Industria Paranaense Estruturas) (1992-93) 16
IPE 06 Curucaca (IPE) (1992-93) 16
IR-02 (AII) (1997-98) 252
IR-12 (AII) (1998-99) 248
IR-H5 (AII) (1998-99) 247
IRGC (see Islamic Revolutionary Guards Corps) (1991-92) 144
ISAE (see Integrated Systems Aero Engineering) (1997-98) 637

IAR/AEDECO (International) (1997-98) 229
Idea PD.93 (Partenavia) (1996-97) 261
IKARUSFLUG LEICHTFLUGZEUGBAU (Germany) (2000-01) 165
ILYUSHIN DESIGN BUREAU (USSR) (1991-92) 241, 756
IMP AEROSPACE LTD (Canada) (1995-96) 43
Impala CA-25 (Skyfox) (1999-2000) 9
Improved SeaCobra (Bell) (1994-95) 492
INDUSTRIA AERONAUTICA DEL PERU SA (Peru) (1993-94) 226
INDUSTRIAL COMMERCIAL COMPANY 'OREL' (Russia) (1995-96) 367
INDUSTRIA METALURGICA DEL NORTE LTDA (Chile) (1992-93) 32
INDUSTRIA PARANAENSE DE ESTRUTURAS (Brazil) (1992-93) 16
INDUSTRIAS CARDOEN LTDA (Chile) (1991-92) 30
INDUSTRIGRUPPEN JAS AB (Sweden) (1996-97) 472
INSTITUTUL DE AVIATIE (Romania) (2000-01) 350
INTECO SRO (Czech Republic) (1995-96) 74
INTEGRATED SYSTEMS AERO ENGINEERING INC (USA) (1997-98) 637
INTERNATIONAL AIRCRAFT COMPANY (Thailand) (1991-92) 228
Interstate S1B2 Arctic Tern (Arctic) (1994-95) 473
Interstate S-4 Privateer (Arctic) (1996-97) 549
INTORA-FIREBIRD (USA) (2000-01) 546
INTREPRINDEREA DE AVIOANE CRAIOVA (Romania) (1991-92) 207
Intruder A-6E (Grumman) (1993-94) 481
In-wing weather radar (Sierra) (1991-92) 478
Ipanema EMB-201A (EMBRAER) (1992-93) 15
IRAQI AIR FORCE (Iraq) (1994-95) 202
Iridium I-22 (PZL Mielec) (1998-99) 308
Iridium M-95 (PZl Mielec) (1998-99) 309
Iridium M-97 (PZL Mielec) (1997-98) 324
Iroquois (Bell) (1991-92) 358
Iryda I-22 (PZL Mielec) (1998-99) 308
Iryda M-95 (PZL Mielec) (1998-99) 309
Iryda M-97 (PZL Mielec) (1997-98) 324
ISHIDA AEROSPACE RESEARCH INC (USA) (1995-96) 543
Iskra TS-11R (PZL Mielec) (1993-94) 231
ISLAMIC REVOLUTIONARY GUARDS CORPS (Iran) (1991-92) 144
ISOLAIR INC (USA) (1994-95) 542

J

J-5 Marco (Aviation Farm) (2000-01) 322
J-9 (CAC) (1994-95) 47
J-22 Orao (Soko/Avioane) (1997-98) 249
J35J Draken (Saab-Scania) (1991-92) 220
JA 37 Viggen (Saab-Scania) (1991-92) 219
JAST (McDonnell Douglas/BAe) (1995-96) 194
JC-130 Hercules (Lockheed Martin) (1998-99) 665
JDA (see Japan Defence Agency) (1996-97) 271
JEH (see Joint European Helicopter) (1991-92) 129
JJ-5 (CAC) (1994-95) 47
JPATS (see Joint Primary Aircraft Training System) (1994-95) 542
JW-1 (ACA) (1991-92) 335
JW-2 (ACA) (1991-92) 335
JW-3 (ACA) (1991-92) 335

Jabiru LSA (Jabiru) (1996-97) 7, 787
JAFFE AIRCRAFT INC (USA) (1992-93) 389
Jaguar International (SEPECAT) (1993-94) 179
JAMMAERO/STARCORP (USA) (2000-01) 678
JAPAN DEFENCE AGENCY (Japan) (1996-97) 271
JARAN AEROSPACE CORPORATION (USA) (2000-01) 678
JetCruzer 450 (AASI) (1995-96) 477
JetCruzer 620 (AASI) (1991-92) 335
JetCruzer 650 (AASI) (1997-98) 554
Jet Prop DC-3 AMI (Professional Aviation) (1994-95) 398
JetRanger III Model 206B (Bell) (1992-93) 17
Jet Squalus F1300 (Promavia) (1996-97) 13
Jetstream 41 (Jetstream) (1997-98) 532
Jetstream 51 (Jetstream) (1994-95) 451
Jetstream 61 (Jetstream) (1995-96) 459
Jetstream 71 (Jetstream) (1994-95) 451
Jetstream ATP (Jetstream) (1994-95) 450
JETSTREAM AIRCRAFT (UK) (1998-99) 514
JETSTREAM HOLDINGS (UK) (1993-94) 379
Jetstream Super 31 (Jetstream) (1994-95) 447
Jian-7 (CAC) (1995-96) 48
Jian-8 (SAC) (1995-96) 62
Jian Hong (XAC) (1995-96) 64
Jian Hong-7 (XAC) (1994-95) 62
Jianjiao-5 (CAC) (1994-95) 47
Jianjiao-7 (GAIC) (1995-96) 52
Jianjiji Hongzhai-7 (XAC) (1994-95) 62
Jianjiji Jiaolianji-5 (CAC) (1994-95) 47
Joint Advanced Strike Technology (McDonnell Douglas/BAe) (1995-96) 194
Joint Advanced Strike Technology (US Navy) (1995-96) 665
Joint Attack Fighter (USAF) (1994-95) 649
JOINT EUROPEAN HELICOPTER (International) (1991-92) 129
Joint Primary Aircraft Training System (USAF) (1994-95) 542
Joint Stealth Strike Aircraft (USAF) (1994-95) 649
JOINT-STOCK COMPANY AVIASPETSTRANS CONSORTIUM (Russian Federation) (1997-98) 360
Joint Strike Fighter (McDonnell Douglas/Northrop Grumman/BAe) (1997-98) 237
JORDAN AEROSPACE (Jordan) (1991-92) 177
JUNKERS FLUGZEUG BAU (Germany) (1999-2000) 156

K

K-250 (Kestrel) (2000-01) 680
KA-6D (Grumman) (1993-94) 481
Ka-25 Hormone (Kamov) (1995-96) 327
Ka-26 (Kamov) (1993-94) 271
Ka-26 Jet control testbed (Kamov) (1991-92) 250
Ka-34? (Kamov) (1991-92) 254
Ka-40 (Kamov) (2000-01) 385
Ka-118 (Kamov) (1992-93) 213
Ka-126 Hoodlum-B (Kamov) (1997-98) 390
Ka-128 (Kamov) (1996-97) 379
Ka-136 Hokum (Kamov) (1991-92) 254
KAIA (see Korea Aerospace Industries Association) (1996-97) 283

856 INDEXES: AIRCRAFT–PREVIOUS TEN EDITIONS—**K – M**

KBHC (see Korea Bell Helicopter Company) ... (1992-93) 165
KC-10A (McDonnell Douglas) ... (1991-92) 442
KC-130 (Lockheed Martin) ... (1998-99) 665
KC-130H (Lockheed Martin) ... (1998-99) 665
KC-135A Stratotanker (Boeing) ... (1991-92) 371
KCDC (see Korean Commercial Aircraft Development Consortium) ... (2000-01) 312
KDC-10 (McDonnell Douglas) ... (1994-95) 602
KE-3A (Boeing) ... (1991-92) 372
KFC (see Kelowna Flightcraft Group) ... (1995-96) 43
KF-X (SSA/General Dynamics) ... (1991-92) 179
KL-1 (Kestrel) ... (1996-97) 617
KM-2 Kai (Fuji) ... (1992-93) 156
KR-2PM Swift (Radwan) ... (2000-01) 343
KT-2 (SSA) ... (1999-2000) 305
KV-107IIA (Kawasaki) ... (1991-92) 173

Kaczor, PZL-107 (PZL Warszawa-Okecie) ... (1993-94) 244
KAMERTON-N LTD (Russian Federation) ... (1997-98) 383
KAMOV DESIGN BUREAU (USSR) ... (1991-92) 249
Kasatik M-12 (Phoenix) ... (1998-99) 342
Katana DA40 (Diamond) ... (1998-99) 9, 796
Katana T30 (Terzi) ... (1993-94) 203
Katran (GADC) ... (1997-98) 477
KATRAN (International) ... (1993-94) 163
Katran (Katran) ... (1993-94) 163
KCAD/AVIC/DASA/FOKKER (International) ... (1995-96) 189
KCDC/AVIC (International) ... (1996-97) 209
KELOWNA FLIGHTCRAFT GROUP (Canada) ... (1995-96) 43
KESTREL AIRCRAFT COMPANY (USA) ... (2000-01) 680
Kfir (IAI) ... (1994-95) 202
King Air C90/C90A update (Raisbeck) ... (1991-92) 456
King Air C90A (Beech) ... (1991-92) 349
King Air B200 Military Versions (Raytheon) ... (1998-99) 713
King Air B200 SE (Raytheon) ... (1998-99) 712
King Cat (Mid-Continent) ... (1991-92) 452
KING'S ENGINEERING FELLOWSHIP, THE (USA) ... (1994-95) 544
Kiowa (Bell) ... (1994-95) 494
Kiowa Warrior (Bell) ... (1994-95) 494
KIS (Tri-R) ... (1999-2000) 764
Kite T-620 (Aerospace/ROKS/Aero) ... (1999-2000) 347
Knight (Molniya) ... (1998-99) 406
Koala P220S (Evektor-Aerotechnik) ... (2000-01) 96
Kobra 2000 (IL) ... (1997-98) 319
KOC HOLDING (Turkey) ... (1991-92) 229
KOMSOMOLSK-ON-AMUR AIRCRAFT PRODUCTION ASSOCIATION (Russia) ... (1995-96) 339
Kong Yun (NAMC) ... (1991-92) 40
KOREA AEROSPACE INDUSTRIES ASSOCIATION (Korea, South) ... (1996-97) 283
KOREA AEROSPACE RESEARCH INSTITUTE (Korea, South) ... (1999-2000) 305
KOREA BELL HELICOPTER COMPANY (Korea) ... (1992-93) 165
KOREAN COMMERCIAL AIRCRAFT DEVELOPMENT CONSORTIUM (Korea) ... (2000-01) 312
Korean Indigenous Trainer (DHI) ... (1992-93) 164
Korint KTX-1 (DHI) ... (1993-94) 213
Korshun T-620 (Aerospace/ROKS/Aero) ... (1999-2000) 347
Krechet (MAI-Avgur) ... (1999-2000) 387
Kruk PZL-106B (PZL Warszawa-Okecie) ... (1998-99) 323
Korshun T-72DP (Washington Aeroprogress) ... (1995-96) 669
KUMERTAU AIRCRAFT PRODUCTION ENTERPRISE (Russian Federation) ... (1997-98) 398
Kuryer T-910 (Aeroprogress/ROKS-Aero) ... (1997-98) 357

L

L-2M Tech Two (New Technik) ... (1992-93) 456
L-13 Super Vivat (Aerotechnik) ... (1998-99) 82
L-13SDL Vivat (Aerotechnik) ... (1995-96) 73
L-13SDM Vivat (Aerotechnik) ... (1995-96) 73
L-13SE Vivat (Aerotechnik) ... (1994-95) 71
L-13SL Vivat (Aerotechnik) ... (1994-95) 71
L20J Allegro (Mooney) ... (2000-01) 722
L20T Predator (Mooney) ... (2000-01) 723
L-39 MS (Aero) ... (1991-92) 51
L-90 TP RediGo (Valmet) ... (1996-97) 86
L 100J Commercial Hercules (Lockheed) ... (1999-2000) 680
L 235 Balbuzard (AAT) ... (1998-99) 95
L-270 (Aero) ... (1991-92) 51
L 839 T-Bird II (Lockheed Martin) ... (1996-97) 239, 628
L-1011 conversions (Pemco) ... (1994-95) 615
L-1011 TriStar conversions (Lockheed) ... (1991-92) 430
LA-4-200 Amphibian (Lake) ... (1991-92) 415
LA-250 Renegade (Lake) ... (1996-97) 618
LAK-X (Avia Baltica) ... (1999-2000) 306
LASA A-22 (Sadler) ... (1994-95) 624
Lastal (UTVA) ... (1994-95) 656
Lasta 2 (UTVA) ... (1994-95) 656
LAVP-OZE (Aviacor) ... (2000-01) 358
LC-40 (Lancair) ... (1997-98) 642
LC-130 Hercules (Lockheed Martin) ... (1998-99) 665
LCPT (Rushjet/General Technologies/Aerosud) ... (1992-93) 136
LF 2 (HOAC) ... (1992-93) 9
LHX. (Boeing/Sikorsky) ... (1991-92) 381
LKhS-4 (Phöenix-Aviatechnica) ... (1994-95) 352
LMK.1 Oryx (CATA) ... (1999-2000) 107
LM 250 Loadmaster (Ayres) ... (1999-2000) 560
LOK (see Letecke Opravny Kbely) ... (1995-96) 77
LP-1 SwiftFury (LoPresti) ... (1992-93) 404
LRAACA (Lockheed) ... (1991-92) 426
LT-1 Swati (BHEL) ... (1994-95) 118

Ladoga-6 (Molniya) ... (1997-98) 425
Ladoga-9 (Molniya) ... (1997-98) 425
LAKE AIRCRAFT INC (USA) ... (1997-98) 641
Lama conversion (Rocky Mountain) ... (1993-94) 552
Lampyridae (DASA) ... (1995-96) 109
Lancair 235 (Lancair) ... (1993-94) 492
Lancair 320 Mark II (Lancair) ... (2000-01) 682
Lancair 360 Mk II (Lancair) ... (1999-2000) 682
Lancair 360 Turbo (Lancair) ... (1999-2000) 682
LARON AVIATION TECHNOLOGIES INC (USA) ... (1999-2000) 664
Lasta 2 (LOLA Utva) ... (1999-2000) 776
Lavi Technology Demonstrator (IAI) ... (1991-92) 146
Layang Project (ARDC) ... (1999-2000) 313
Learjet 35A/36A (Learjet) ... (1994-95) 547
Learjet 35A/36A Special missions versions (Learjet) ... (1995-96) 550
Learjet 55 Long-range tanks (Branson) ... (1991-92) 382
Learjet 55B (Learjet) ... (1991-92) 419
Learjet 55C (Learjet) ... (1991-92) 419
LEGKIE VERTOLETY MI NAUCHNO-PROIZVODSTVENNOYE OBEDINENIE (Russian Federation) ... (1999-2000) 387
Leshii SL-90 (Aviotechnica) ... (1997-98) 202
LETECKE OPRAVNY KBELY (Czech Republic) ... (1995-96) 77
LET KONCERNOVY PODNIK (Czechoslovakia) ... (1991-92) 51, 747
Light Aircraft (Toyota/Rutan) ... (1994-95) 626
Light Helicopter (Bell) ... (1994-95) 26
LIGHTNING BUG AIRCRAFT CORPORATION (USA) ... (1997-98) 647
Light Transport Aircraft (HAL) ... (1993-94) 111
LightWing GA-55 (Hughes) ... (1994-95) 8
LightWing GR-582 (Hughes) ... (1994-95) 7
LightWing GR-912 (Hughes) ... (1994-95) 8
Lincoln SV28 (Sivel) ... (1997-98) 282
Loadmaster, LM 250 (Ayres) ... (1999-2000) 560
LOAD RANGER (USA) ... (2000-01) 689
LOCK HAVEN AIRPLANE COMPANY (USA) ... (1993-94) 496
LOCKHEED FORT WORTH COMPANY (USA) ... (1994-95) 565
LOCKHEED MARTIN AIRCRAFT SERVICE (USA) ... (1995-96) 576
LongRanger III (Bell) ... (1992-93) 18
LongRanger IV, 206L-4 (Aymet) ... (1994-95) 421
LOPRESTI FLIGHT CONCEPTS (USA) ... (1993-94) 513
LOPRESTI PIPER AIRCRAFT ENGINEERING COMPANY (USA) ... (1991-92) 465
LOVAUX LIMITED (UK) ... (1991-92) 316
LTV AEROSPACE AND DEFENSE COMPANY (USA) ... (1992-93) 404
LUNDS TEKNISKE (Norway) ... (1995-96) 265

M

M-1 (Super Rotor) ... (1992-93) 17
M-4 (Myasishchev) ... (1991-92) 273
M-7 Starcraft (Maule) ... (1991-92) 434
M-12 Kasatik (Phoenix) ... (1998-99) 342
M-12 Tchaika (Beriev) ... (1991-92) 239
M-17 Mystic (Myasishchev) ... (1996-97) 412
M-21 Dromader Mini (PZL Mielec) ... (1991-92) 193
M-24 Dromader Super (PZL Mielec) ... (1991-92) 180
M-28 03/04 Skytruck Plus (PZL) ... (2000-01) 328
M-55 Geophysica (Myasishchev) ... (1996-97) 412
M-55 Geophysica-2 (Myasishchev) ... (1996-97) 413
M58 (Masquito) ... (1998-99) 11
M-90 Samson (Myasishchev) ... (1997-98) 428
M-95 Iryda (PZL Mielec) ... (1998-99) 309
M-97 Iryda (PZL Mielec) ... (1997-98) 324
M-99 Orkan (PZL Mielec) ... (1997-98) 324
M-101T Gzhel (Myasishchev) ... (1999-2000) 416
M-103 Skif (Myasishchev) ... (1993-94) 302
M-112 (Myasishchev) ... (1998-99) 407
M-150 (Myasishchev) ... (1999-2000) 417
M-200 (Myasishchev) ... (1995-96) 366
M200 Skycar (Moller) ... (1996-97) 682
M-290TP Redigo (Aermacchi) ... (1999-2000) 265
M-300 (Micco) ... (1998-99) 684
M400 Skycar (Moller) ... (1996-97) 682
M-500 (Myasishchev) ... (1998-99) 410
MA-2 Mamba (AIA) ... (1994-95) 4
MAC (see Melbourne Aircraft Corporation) ... (1991-92) 5
MAFFS (see Modular Aerial Fire Fighting System (Aero Union) ... (1991-92) 338
MAI-890 (Aviatika) ... (1993-94) 691
MAPO (see Federalnoye Gosudarstvennoye Unitarneoye Predpriyatie) ... (2000-01) 400
MAS (see Ministry of Aero-Space Industry) ... (1992-93) 32
MB-339A (Aermacchi) ... (1996-97) 239
MB-339CD (Aermacchi) ... (1995-96) 217
MB-339FD (Aermacchi) ... (1995-96) 217
MBB (see Messerschmitt-Bölkow-Blohm GmbH) ... (1992-93) 77
MC-130 Hercules (Lockheed Martin) ... (1998-99) 665
MD3 Swiss Trainer (MDB) ... (1993-94) 357
MD-10 (McDonnell Douglas) ... (1997-98) 695
MD-11 freight door (Marshall) ... (1991-92) 317
MD-12 (McDonnell Douglas) ... (1995-96) 607
MD-12X (McDonnell Douglas) ... (1991-92) 444
MD-17 (Boeing) ... (2000-01) 636
MD-20 (McDonnell Douglas) ... (1997-98) 698
MD-80 Cargo conversion (Pemco) ... (1994-95) 615
MD-80 Series (Boeing) ... (1999-2000) 613
MD-82 (SAMF/McDonnell Douglas) ... (1995-96) 63
MD-83 (SAMF/McDonnell Douglas) ... (1995-96) 63
MD-90 (Boeing) ... (2000-01) 633
MD-90-30T (SAMF/Boeing) ... (2000-01) 82
MD-95 (Douglas/CATIC) ... (1993-94) 138
MD-95 (McDonnell Douglas) ... (1997-98) 694
MD 500D (Kawasaki/Boeing) ... (1999-2000) 296
MD 902 Explorer (McDonnell Douglas) ... (1997-98) 685
MDX (McDonnell Douglas) ... (1991-92) 448
MD-XX (McDonnell Douglas) ... (1996-97) 677
METALNOR (see Industria Metalurgica del Norte) ... (1992-93) 32
MFI (see Malmo Forsknings & Innovations) ... (1996-97) 476
MFI-10C Vipan/Phönix (RFB) ... (1994-95) 113
MFI-11 (MFI) ... (1994-95) 405
MFT/LF (Embraer) ... (1991-92) 10
MGS-6 (Myasishchev) ... (1997-98) 428
MGS-8 (Myasishchev) ... (1997-98) 428
MH-02 (Honda/MSU) ... (1994-95) 179
MH-53J Pave Low III enhanced (Sikorsky) ... (1991-92) 480
Mi-2B (PZL Swidnik) ... (1991-92) 195
Mi-2 Hoplite (PZL-Swidnik) ... (1994-95) 270
Mi-10 (Mil) ... (1991-92) 266
Mi-10K (Mil) ... (1991-92) 266
Mi-34 VAZ (Mil) ... (1997-98) 421
Mi-40 (Mil) ... (1998-99) 405
Mi-46K (Mil) ... (1995-96) 360
Mi-46T (Mil) ... (1995-96) 359
Mi-52 (Mil) ... (1998-99) 405
Mi-54 (Mil) ... (1997-98) 423
Mi-58 (Mil) ... (1998-99) 405
MiG 1-2000 LFS (MiG) ... (2000-01) 411
MiG-18-50 (MiG) ... (1993-94) 288
MiG-21 (MiG) ... (1992-93) 214
MiG-21 upgrade (HAL) ... (1994-95) 124
MiG-21 upgrade (IAI) ... (1993-94) 184
MiG-21 2000 (IAI) ... (1994-95) 205
MiG-23 Flogger (MiG) ... (1994-95) 324
MiG-23 upgrade (AIA) ... (1993-94) 184
MiG-24 (MiG) ... (1991-92) 256
MiG-24 upgrade (HAL) ... (1994-95) 124
MiG-25 Foxbat (MiG) ... (1998-99) 380
MiG-27 (Flogger) (MiG) ... (1994-95) 326
MiG-27M (HAL) ... (1997-98) 161
MiG-33 (MiG) ... (1999-2000) 397
MiG 105, Experimental Aircraft (MiG) ... (1991-92) 298
MiG-110 (MAPO MiG) ... (1997-98) 407
MiG 125 (MiG) ... (2000-01) 415
MiG fighter, new (MiG) ... (1991-92) 263
MiG SVB (MiG) ... (1993-94) 288
MK-30 (DHI) ... (1994-95) 239
M.J.52 Zephyr (Jurca) ... (1993-94) 86
MM-1 (Myasishchev) ... (1994-95) 351
MPC75 (MPC Aircraft) ... (1991-92) 134
MSSA (Pilatus Britten-Norman) ... (1996-97) 526
MU-2 Express conversions (MU-2) ... (1991-92) 455
MX-7 Starcraft (Maule) ... (1991-92) 434
MY-102/200 (Mylius) ... (1999-2000) 157

MACAVIA INTERNATIONAL (USA) ... (1991-92) 431
Mach 10 Aircraft (Boeing) ... (1999-2000) 569
MACHEN INC (USA) ... (1993-94) 513
Madcap An-71 (Antonov) ... (1997-98) 513
Madcap, An-74 (Antonov) ... (1991-92) 236
MAGNUM AIRCRAFT (USA) ... (1991-92) 432
MAI-AVGUR VOZDUKHOPLAVATELNYITSENTR (Russian Federation) ... (2000-01) 399
Mail, Be-12 (Beriev) ... (1991-92) 239
MALMO FORSKNINGS & INNOVATIONS (Sweden) ... (1996-97) 476
Mamba AA-2 (AAI) ... (1995-96) 5
Mamba MA-2 (AIA) ... (1994-95) 4
MAPO 'MiG' (Russian Federation) ... (2000-01) 400
Marco J-5 (Aviation Farm) ... (2000-01) 322

M – P—AIRCRAFT-PREVIOUS TEN EDITIONS: INDEXES

Maritime Defender (Pilatus Britten-Norman) *(1994-95)* 454
MARSHALL OF CAMBRIDGE (ENGINEERING) LTD (UK) *(1992-93)* 307
MARSH AVIATION COMPANY (USA) *(1994-95)* 573
Martlet (C-Craft) *(2000-01)* 9
Martin T-420 (Khrunichev) *(1999-2000)* 385
Martin T-501 (Aeroprogress/ROKS-Aero) *(1997-98)* 355
Matador VA.1 (BAe) *(1991-92)* 311
Matador II, VA.2 (McDonnell Douglas (Boeing)/BAe) *(1997-98)* 229
Maxdome, Il-86 (Ilyushin) *(1996-97)* 364
Maxdome Il-87 (Ilyshin) *(1999-2000)* 365
MBB HELICOPTER CANADA LTD (Canada) *(1991-92)* 29
MBB HELICOPTER CORPORATION (USA) *(1991-92)* 434
MBB/KAWASAKI (International) *(1991-92)* 129
MCDONNELL AIRCRAFT AND MISSILE SYSTEMS (USA) *(1998-99)* 559
MCDONNELL DOUGLAS (Boeing)/BAE (International) *(1997-98)* 229
MCDONNELL DOUGLAS AEROSPACE (USA) *(1997-98)* 669
MCDONNELL DOUGLAS CORPORATION (USA) *(1998-99)* 683
MCDONNELL DOUGLAS HELICOPTER SYSTEMS (USA) *(1997-98)* 678
MCDONNELL DOUGLAS MILITARY TRANSPORT AIRCRAFT (USA) *(1997-98)* 687
MD HELICOPTERS (Netherlands) *(1999-2000)* 310
McGuire 11 (Gunasekera) *(1999-2000)* 656
MDB FLUGTECHNIK AG (Switzerland) *(1993-94)* 357
MELBOURNE AIRCRAFT CORPORATION PTY LTD (AUSTRALIA) *(1991-92)* 5
MELEX USA (USA) *(1996-97)* 680
Mercury (Aviation) *(1999-2000)* 353
Mercury E-6A (Boeing) *(1993-94)* 454
MERIDIONALI, ELICOTTERI (Italy) *(1997-98)* 278
Merkury (Aviation) *(1999-2000)* 353
Merlin (Merlin Aircraft) *(1995-96)* 608
Merlin IVC (SA227-AT) (Fairchild) *(1991-92)* 400
Merlin 23 (Fairchild Dornier) *(1999-2000)* 644
MERLIN AIRCRAFT (USA) *(1998-99)* 683
Merlin E-Z (Aerocomp) *(1998-99)* 533
Mermaid A-40 (Beriev) *(1999-2000)* 356
Mermaid B-42 (Beriev) *(1998-99)* 347
Messenger T-910 (Aeroprogress/ROKS-Aero) *(1997-98)* 357
MESSERSCHMITT-BÖLKOW-BLOHM GmbH (Germany) *(1992-93)* 77
Metallica LG2 (CRO) *(1998-1999)* 83
Metro III (Fairchild Aircraft) *(1991-92)* 398
Metro 23 (Fairchild Dornier) *(1999-2000)* 644
Metro 23 Special Mission Aircraft (Fairchild) *(1997-98)* 629
MID-CONTINENT AIRCRAFT CORPORATION (USA) *(1993-94)* 537
MIKOYAN DESIGN BUREAU (USSR) *(1991-92)* 255
MIL DESIGN BUREAU (Russia) *(1992-93)* 224
Mini-500 Bravo (Revolution) *(2000-01)* 763
Mini-Acro (Technoavia) *(1996-97)* 439
MINISTRY OF AERO-SPACE INDUSTRY (China) *(1992-93)* 32
Minuano (EMBRAER/Piper) *(1994-95)* 20
Minuano EMB-720D (NEIVA) *(1999-2000)* 26
Mirage III series (Dassault) *(1992-93)* 58
Mirage III/5 upgrade (IAI) *(1993-94)* 183, 184
Mirage F1 (Dassault Aviation) *(1992-93)* 58
Mirage advanced technology update programmes (Dassault Aviation) *(1991-92)* 67
Mirage Improvement Programme (F + W) *(1993-94)* 363
Mirage Safety Improvement Programme (SABCA) *(1994-95)* 13
Mirlo, E.25 (CASA) *(1995-96)* 411
Mistral AE-206 (Aviasud) *(1998-99)* 98
Mistral Twin AE-207 (Aviasud) *(1998-99)* 98

Note: Aircraft previously identified as 'Model xxx' are now listed as numbers only at the beginning of the index.

Modular Aerial Fire Fighting System (Aero Union) *(1991-92)* 338
Module (Aviastar) *(1999-2000)* 352
Moller 400 Volantor (Moller) *(1991-92)* 423
MOLLER INTERNATIONAL (USA) *(1997-98)* 699
Molniya-100 (Molniya) *(1997-98)* 424
Molniya-300 (Molniya) *(1997-98)* 424
Molniya-400 (Molniya) *(1997-98)* 425
Molniya-1000 Heracles (Molniya) *(1997-98)* 425
Mongol MiG-21 (MiG) *(1992-93)* 214
MONIOT, AVIONS PHILIPPE (France) *(1999-2000)* 128
Montalvá (Super Rotor) *(1992-93)* 17
MONTANA COYOTE INC (USA) *(1995-96)* 609
MORAVAN NARODNI PODNIK (Czechoslovakia) *(1991-92)* 54

MOSKOVSKII MASHINOSTROITELNYY ZAVOD SKOROST IMIENI A. S. YAKOVLEVA (Russia) *(1994-95)* 384
MOSKOVSKII MASHINOSTROITELNYY ZAVOD IMIENI A. I. MIKOYANA (Russia) *(1992-93)* 214
MPC AIRCRAFT GmbH (International) *(1992-93)* 128
Mriya, An-225 (Antonov) *(1995-96)* 443
Mrówka, PZL-126 (PZL Warszawa-Okecie) *(1993-94)* 245
MU-2 MODIFICATIONS INC (USA) *(1992-93)* 424
MUDRY et CIE, AVIONS (France) *(1997-98)* 121, 817
Multi-Role Fighter (USAF) *(1994-95)* 649
Multi-Role Tanker Transport (Airbus) *(1994-95)* 130
Mustang 11 M-11 (Mustang) *(1994-95)* 606
MUSTANG AERONAUTICS INC (USA) *(1994-95)* 606
MYASISHCHEV/NAL (International) *(1998-99)* 234
MYERS AVIATION (USA) *(1991-92)* 455
Mystère-Falcon 20 (Dassault Aviation) *(1991-92)* 74
Mystère-Falcon 100 (Dassault Aviation) *(1991-92)* 75
Mystère-Falcon 200 (Dassault Aviation) *(1991-92)* 74
Mystère-Falcon 900 (Dassault Aviation) *(1992-93)* 64
Mystic M-17 (Myasishchev) *(1996-97)* 412

N

N2XXM (Airtech) *(1998-99)* 195
N2XXM (IPTN) *(1998-99)* 169
N-62 (UTVA) *(1994-95)* 658
N-63 (LOLA Utva) *(1999-2000)* 776
N-63 (UTVA) *(1994-95)* 656
N-270 (IPTN/AMRAI) *(1999-2000)* 551
N-2130 (IPTN) *(1994-95)* 126
NAC 6 Fieldmaster (EPA) *(1993-94)* 393
NAC-100 (Nexus) *(1998-99)* 689
NAMC (see Nanchang Aircraft Manufacturing Company) *(1998-99)* 62
NAMC/PAC (International) *(1994-95)* 186
NASA Trials Programmes (NASA) *(1996-97)* 684
NATF, Naval Advanced Tactical Fighter (USN) *(1993-94)* 580
NBELL-407 (IPTN) *(1997-98)* 170
NBELL-430 (IPTN) *(1997-98)* 170
NBK-117 (IPTN) *(1991-92)* 105
NBO 105 (IPTN) *(1999-2000)* 176
NC-130 Hercules (Lockheed Martin) *(1998-99)* 665
NC-212-200 Aviocar (IPTN) *(1999-2000)* 176
NE-821 Carajá (NEIVA) *(1992-93)* 16
NH2 ALCLH (IPTN) *(1997-98)* 168
NH5 (IPTN) *(1997-98)* 168
NIAT (see JSC National Institute of Aviation Studies) *(1998-99)* 411
NIPPI (see Nihon Hikoki Kabushiki Kaisha) *(1994-95)* 236
NUAA/XALAC (see Nanjing University of Aeronautics and Astronautics) *(1998-99)* 66
NWI (see Northwest Industries Limited) *(1995-96)* 45

Nammer (IAI) *(1993-94)* 184
NANCHANG AIRCRAFT MANUFACTURING COMPANY (China) *(1998-99)* 62
NANJING UNIVERSITY OF AERONAUTICS AND ASTRONAUTICS (China) *(1998-99)* 66
NASH AIRCRAFT LTD (UK) *(1995-96)* 460
National Aero-Space Plane X-30 (NASA) *(1993-94)* 540, 698
NATIONAL INSTITUTE FOR AVIATION TECHNOLOGIES (Russian Federation) *(1998-99)* 411
Naval Advanced Tactical Fighter, NATF (USN) *(1993-94)* 580
NEW BRITISH AEROSPACE (UK) *(1999-2000)* 510
NEWCAL AVIATION INC (USA) *(1994-95)* 608
New Light Aircraft (Lockheed Martin) *(1998-99)* 624
New Light Aircraft (Socata) *(1997-98)* 818
NEW MYERS AIRCRAFT COMPANY, THE (USA) *(1997-98)* 704
New Small Airplane (Boeing) *(1997-98)* 588
New Strategic Aircraft (Lockheed Martin) *(1998-99)* 670
NEW TECHNIK INC (USA) *(1992-93)* 426
New Training Helicopter (US Army) *(1993-94)* 580
New Ultralight (BULAERO) *(1998-99)* 100
NEXUS (USA) *(1998-99)* 689
NGT Turboprop Trainer (Atlas) *(1993-94)* 339
Nightfox (McDonnell Douglas) *(1994-95)* 592
Nighthawk F-117A (Lockheed Martin) *(1995-96)* 567
NightRanger (Global) *(1991-92)* 405
NIHON HIKOKI KABUSHIKI KAISHA (Japan) *(1994-95)* 236
Nite-Writer (VAT) *(1994-95)* 652
Noga VI (IAR/AEDECO) *(1996-97)* 209
NOGATEC INTERNATIONAL (France) *(2000-01)* 136
Non-Development Airlift Aircraft (USAF) *(1996-97)* 744
Nong 5 (NAMC) *(1995-96)* 59
NORDHAM AIRCRAFT MODIFICATION DIVISON (USA) *(1994-95)* 609
NORTH AMERICAN AIRCRAFT MODIFICATION DIVISION, ROCKWELL (USA) *(1996-97)* 719
NORTHROP CORPORATION (USA) *(1993-94)* 542
Northrop TR-3A (USAF) *(1995-96)* 665

NORTHWEST INDUSTRIES LIMITED (Canada) *(1995-96)* 45
Novi Avion (VTI) *(1993-94)* 589
NOVOSIBIRSKOYE AVIATSIONNOYE-PROIZVODSTVENNOYE OBEDINEINE IMENI VP CHKALOVA GUP (Russian Federation) *(2000-01)* 435
Nuri upgrade, S-61A (Airod) *(1993-94)* 214

O

OA7-300 Optica (FLS) *(1994-95)* 446
OC-135B (USAF) *(1994-95)* 649
OGMA (see Oficinas Gerais de Material Aeronáutico) *(1995-96)* 288
OH-6D (Kawasaki) *(1999-2000)* 296
OH-6DA (Kawasaki) *(1999-2000)* 296
OH-58D Kiowa Warrior (Bell) *(1994-95)* 494
OH-X (Kawasaki) *(1996-97)* 275, 795
OV-10D Plus Bronco (Rockwell) *(1991-92)* 468

Observer P.68 (Partenavia) *(1991-92)* 165
Ocean Hawk SH-60F (Sikorsky) *(1992-93)* 447
OFICINAS GERAIS DE MATERIAL AERONAUTICO (Portugal) *(1995-96)* 288
Omega TB31 (SOCATA) *(1996-97)* 117
Omega 2 (ISAE) *(1997-98)* 637
OMEGA AEROSPACE CORPORATION (USA) *(1997-98)* 712
OMNIPOL COMPANY LIMITED (Czech Republic) *(1993-94)* 63
One-Eleven (Romaero) *(1997-98)* 345
One-Eleven 2400 (Dee Howard) *(1991-92)* 396
One-Eleven 2500 (Dee Howard) *(1991-92)* 396
OPERATION ABILITY LTD (UK) *(1991-92)* 318
Optica OA7-300 (FLS) *(1994-95)* 446
Optica OA7-300 (FLS) *(1997-98)* 538
Optica Scout Mk II (Lovaux) *(1991-92)* 316, 501
Opus 280 (ASL) *(1996-97)* 472, 612
Orao J22 (SOKO/Avioane) *(1997-98)* 249
OREL (see Industrial Commercial Company 'Orel') *(1995-96)* 367
Orel T-208 (Khrunichev) *(1998-99)* 375
Orel T-602 (Aeroprogress/ROKS-Aero) *(1995-96)* 305
Oriole PZL-104 (PZL Warszawa-Okecie) *(1999-2000)* 332
Orion (Lockheed Martin) *(2000-01)* 695
Orion, AA200 (CRSS) *(1993-94)* 112
Orion II P-3H (Lockheed) *(1993-94)* 697
Orion P-3C (Kawasaki/Lockheed) *(1993-94)* 207
Orkan M-99 (PZL Mielec) *(1997-98)* 324
ORLANDO HELICOPTER AIRWAYS INC (USA) *(1993-94)* 544
Orlik, PZL-130 (PZL Warszawa-Okecie) *(1991-92)* 201
Oryx (Denel) *(1999-2000)* 469
Oryx LMK.1 (CATA) *(1999-2000)* 107
Orzel PZL-140 (PZL Warszawa-Okecie) *(1995-96)* 287
Osprey V-22 (BAe) *(1997-98)* 528
Otter, DHC-3/1000 (Airtech Canada) *(1991-92)* 17
Ozzie Mozzie BAC-204 (BAC) *(1994-95)* 5

P

P-3 conversions (Lockheed) *(1991-92)* 430
P-3 Orion (Lockheed Martin) *(1999-2000)* 676
P-3A Aerostar (Aero Union) *(1991-92)* 338
P-3C (Kawasaki) *(1999-2000)* 282
P-3C update IV (Boeing) *(1991-92)* 366
P-3H Orion II (Lockheed) *(1993-94)* 697
P-7A (Lockheed) *(1991-92)* 426
P.66D Delta (Aviolight) *(1991-92)* 163
P.68 Observer (Partenavia) *(1991-92)* 165
P.68 Observer 2 (TAAL) *(1998-99)* 165
P.68 Observer 2TC (TAAL) *(1998-99)* 165
P.68 C (TAAL) *(1998-99)* 165
P.68 TC (TAAL) *(1998-99)* 165
P.95 (EMBRAER) *(1993-94)* 14
P120L (Eurocopter/CATIC/SA) *(1992-93)* 119
P132 (BAe) *(1992-93)* 296
P.134 (BAe) *(1993-94)* 378
P.135 (BAe) *(1993-94)* 378
P.210 Rocket (Riley) *(1991-92)* 467
P.210 Turbine (Advanced Aircraft) *(1991-92)* 336
P220S Koala (Evektor-Aerotechnik) *(2000-01)* 96
P-337 Sky Rocket (Cessna/Riley) *(1991-92)* 467
PA-18-150 Super Cub (Piper) *(1992-93)* 429
PA-28-140/-151 (RAM/Piper) *(1991-92)* 465
PA-32-301 Saratoga (Piper) *(1992-93)* 429
PA-34 220T Seneca IV (Piper) *(1996-97)* 697
PA-42-720 Cheyenne IIIA (Piper) *(1994-95)* 616
PA-42-1000 Cheyenne 400 (Piper) *(1994-95)* 617
PACI (see Philippine Aircraft Company Inc) *(1999-2000)* 314
PAH-2 Tiger (Eurocopter) *(1991-92)* 118
PC-9/A Pilatus (HDH) *(1992-93)* 7
PC-9 Mk II (Pilatus/Beech) *(1994-95)* 481
PD.90 Tapete Air Truck (Partenavia) *(1996-97)* 263
PD.93 Idea (Partenavia) *(1996-97)* 261
PD 374 (Raytheon) *(1995-96)* 639

858 INDEXES: AIRCRAFT–PREVIOUS TEN EDITIONS—**P – S**

PIC (see Promavia International Corporation) ... (1993-94) 39
PL-12-U (Transavia) ... (1991-92) 6
PPA-1, Agricultural aircraft (Emind) ... (1991-92) 497
PPA-2, Agricultural aircraft (Emind) ... (1991-92) 497
PR-01 Pond Racer (Scaled Composites) ... (1992-93) 439
PS-5 (HAMC) ... (1995-96) 53
PT CIPTA RESTU SARANA SVAHA (Indonesia) ... (1994-95) 125
PTS-2000 (Aermacchi/DASA) ... (1994-95) 126
PW (see Politechnika Warszawska) ... (1995-96) 270
PW-4 (PW) ... (1995-96) 270
PZL (see Zrzeszinie Wytworcow Sprzetu Lotniczego ll Silnikowego PZL) ... 1991-92) 188
PZL-104 Wilga 35 (PZL Warszawa-Okecie) ... (1999-2000) 332
PZL-104 Wilga 80 (PZL Warszawa-Okecie) ... (1999-2000) 332
PZL-105L Flaming (PZL Warszawa-Okecie) ... (1999-2000) 332
PZL-106B Kruk (PZL Warszawa-Okecie) ... (1998-99) 323
PZL-107 Kaczor (PZL Warszawa-Okecie) ... (1993-94) 244
PZL-130 Turbo Orlik (PZL Warszawa-Okecie) ... (1991-92) 201
PZL-140 Orzel (PZL Warszawa-Okecie) ... (1995-96) 287
PZL-230F Skorpion (PZL Warszawa-Okecie) ... (1997-98) 335

Pampa, IA 63 (LMASA) ... (1998-99) 2
Pampa 2000 International (LAASA) ... (1997-98) 3
PANAVIA AIRCRAFT (International) ... (2000-01) 257
Panda (GOHL) ... (1992-93) 35
Pantera 50C (ENAER) ... (1994-95) 46
Panther (Helibras) ... (1995-96) 23
Panther II (Colemill) ... (1991-92) 393
Panther 800 (Vought) ... (1994-95) 654
Panther Navajo (Colemill) ... (1991-92) 393
PARAGON AIRCRAFT CORPORATION (USA) ... (1996-97) 693
PARTENAVIA COSTRUZIONI AERONAUTICHE SpA (Italy) ... (1997-98) 279
Pathfinder III (Piasecki) ... (1991-92) 461
Pchel T-203 (Aeroprogress/ROKS-Aero) ... (1997-98) 350
Pegass (Delta) ... (1999-2000) 90
Pelican, AS-61R (Agusta-Bell) ... (1994-95) 214
PEMCO AEROPLEX INC (USA) ... (1994-95) 615
PEREGRINE FLIGHT INTERNATIONAL (USA) ... (1997-98) 713
Petrel (Nash) ... (1995-96) 460
Petrel 650C, HU-2C (SSLF) ... (2000-01) 83
Petrel 650D, HU-2D (SSLF) ... (2000-01) 83
Petrel CH-148 (EHI) ... (1993-94) 140
Phalcon 707 (AIA) ... (1998-99) 252
Phantom 2000 (IAI) ... (1993-94) 184
PHILEOAVIATION (MALAYSIA) SDN BHD (Malaysia) ... (1996-97) 288
PHILIPPINE AIRCRAFT COMPANY (Philippines) ... (1999-2000) 314
Phillicopter Mk 1 (Veetol Helicopters) ... (1997-98) 10
Phoenix Flyer (StarPAC) ... (1994-95) 644
PhoenixJet (Vortex) ... (2000-01) 800
PHOENIX OKB (Russian Federation) ... (1999-2000) 352
Phönix MFI-10C (RFB) ... (1994-95) 113
Pijao AC-05 (Aviones de Colombia) ... (1994-95) 66
PILATUS BRITTEN-NORMAN (UK) ... (1998-99) 518
Pillán, T35 (ENAER) ... (1994-95) 45
Pinguino F.22 (General Avia) ... (1997-98) 276
PIPER NORTH CORPORATION (USA) ... (1991-92) 464
Piranha (Sadler) ... (2000-01) 767
Pirania (IL) ... (1993-94) 228
Pole Star (HAMC) ... (1994-95) 54
POLITECHNIKA WARSZAWSKA (Poland) ... (1995-96) 270
Pond Racer (Scaled Composites) ... (1992-93) 439
PONY JSC (Russian Federation) ... (1999-2000) 419
Pony (Pony) ... (1998-99) 411
PRECEPTOR AIRCRAFT CORPORATION (USA) ... (2000-01) 741
Predator M20T (Mooney) ... (2000-01) 723
Premier Aztec SE (Lock Haven) ... (1993-94) 496
Prescott II (AFI) ... (1996-97) 541
President 600 (Colemill) ... (1991-92) 394
Président, DR (Robin) ... (1997-98) 818
Pressurised Centurion (Cessna) ... (1991-92) 385
Privateer S-4 Arctic (Arctic) ... (1996-97) 549
Priz (Riola) ... (1998-99) 412
Prize (Riola) ... (1998-99) 412
PROFESSIONAL AVIATION SERVICES (South Africa) ... (1995-96) 411
PROMAVIA (Belgium) ... (1998-99) 11
PROMAVIA INTERNATIONAL CORPORATION (Canada) ... (1993-94) 39
PROMAVIA/MiG (International) ... (1994-95) 195
Prop-Jet Bonanza (Tradewind) ... (1991-92) 493
Prowler EA-6B (Grumman) ... (1993-94) 482
Pucará IA 58A (FMA) ... (1994-95) 3
Puma, IAR-330L (IAR) ... (1998-99) 329
Puma, Persian Gulf War (Westland) ... (1991-92) 334
Puma AS330 conversion (Atlas) ... (1994-95) 396
Pursuit II, S-15 (Rans) ... (2000-01) 745
PZL-MIELEC DO BRAZIL FABRICA D'AVOIES (Brazil) ... (1998-99) 22

Q

Q-5K (NAMC) ... (1991-92) 40

Qiang-5 (NAMC) ... (1995-96) 56
Queenaire 800/8800 (Excalibur) ... (1991-92) 398
QUESTAIR INC (USA) ... (1997-98) 718
Quiet 727 (Valsan) ... (1991-92) 495
Quiet Turbofan propeller systems (Raisbeck) ... (1991-92) 465

R

R-02 Robert (SAU) ... (1998-99) 413
R-50 Robert (SAU) ... (2000-01) 437
R90-230 RG (Ruschmeyer) ... (1997-98) 149
R90-420 AT (Ruschmeyer) ... (1997-98) 149
R92 (Regioliner) ... (1992-93) 134
R-95 (EMBRAER) ... (1991-92) 10
R95 (Ruschmeyer) ... (1997-98) 149
R122 (Regioliner) ... (1992-93) 134
R1180 Aiglon (Robin) ... (2000-01) 143
R 3000 series (Robin) ... (1999-2000) 136
RC-130 Hercules (Lockheed Martin) ... (1998-99) 665
RF-4EJ (Mitsubishi) ... (1994-95) 234
RF-47 (Eurovial) ... (2000-01) 134
RFB (see Rhein-Flugzeugbau) ... (1996-97) 133
RG-8A (Schweizer) ... (1995-96) 643
RH-1100 (Rogerson Hiller) ... (1994-95) 623
RJ 115 (Avro) ... (1997-98) 528
RJ 120 (Avro) ... (1992-93) 710
RJX (Avro) ... (1994-95) 441
RNZAF (see Royal New Zealand Air Force) ... (1991-92) 184
RP-1 (Mitsubishi) ... (1997-98) 296
RPX-Alpha Hummingbird (PADC) ... (1999-2000) 314
RTAF (SWDC) (see Royal Thai Air Force) ... (1992-93) 279
RU-38 Condor (Schweizer) ... (1994-95) 627

RADWAN (Poland) ... (2000-01) 343
Rainbow Chaser (Jammaero/Starcorp) ... (2000-01) 678
RAISBECK ENGINEERING ... (1994-95) 619
RAM AIRCRAFT CORPORATION ... (1994-95) 619
Ram air recovery system (Raisbeck) ... (1991-92) 465
Ranger Model AA-23A1 (Taylorcraft) ... (1991-92) 482
Ranger 2000 FR 06 (Rockwell/DASA) ... (1996-97) 229
Ratnik (Kamerton-N) ... (1997-98) 383
Raven, EF-111A, update (Grumman) ... (1991-92) 410
Raven, PZL-106B (PZL Warszawa-Okecie) ... (1998-99) 323
RAYTHEON AIRCRAFT COMPANY (UK) ... (1995-96) 464
RAYTHEON CORPORATE JETS (UK) ... (1994-95) 454
Redigo M-290TP (Aermacchi) ... (1999-2000) 265
Regent 1500 (Advanced Aircraft) ... (1991-92) 336
REGIOLINER GmbH (International) ... (1992-93) 134
Regional Jetliner (Avro) ... (1993-94) 135
Regional Jet Series 100 & 200 (Canadair) ... (1999-2000) 41
Regional Jet Series 700 (Canadair) ... (1999-2000) 44
Regional Transport, 100/120-seat (KCAD/AVIC/DASA/FOKKER) ... (1995-96) 189
Remo 180, DR 400/180R (Robin) ... (2000-01) 142
Remo 200, DR 400/200R (Robin) ... (1997-98) 128
Renegade Turbo 270 (Lake) ... (1996-97) 618
Renegade LA-250 (Lake) ... (1996-97) 618
REVOLUTION HELICOPTER CORPORATION (USA) ... (2000-01) 763
RHEIN FLUGZEUGBAU (Germany) ... (1996-97) 133
RICHTER FLUGZEUGBAU (Germany) ... (1998-99) 153
RIDA-MDT LTD (Russian Federation) ... (1999-2000) 420
Rigel AA300 (CRSS) ... (1993-94) 112
RILEY INTERNATIONAL CORPORATION (USA) ... (1993-94) 549
Robert R-02 (SAU) ... (1998-99) 413
Robert R-50 (SAU) ... (2000-01) 437
Robin 200 (Robin) ... (1999-2000) 132
Robin 3140 Testbed (Robin) ... (1994-95) 100
Rocket P-210 (Riley) ... (1991-92) 467
ROCKWELL AEROSPACE AND DEFENCE (USA) ... (1996-97) 718, 799
ROCKWELL/DASA (International) ... (1997-98) 248
ROCKY MOUNTAIN HELICOPTERS INC (USA) ... (1993-94) 552
ROGERSON HILLER CORPORATION (USA) ... (1994-95) 623
ROKS-AERO CORPORATION (Russia) ... (1993-94) 303
ROMASHKA AGROTECHNICAL COMPANY (Russia) ... (1997-98) 430
Rombac 1-11 (Romaero) ... (1994-95) 286
ROSS AIRCRAFT COMPANY (UK) ... (1998-99) 522
ROYAL NEW ZEALAND AIR FORCE (New Zealand) ... (1991-92) 184
ROYAL THAI AIR FORCE (Thailand) ... (1992-93) 279
RUSCHMEYER LUFTFAHRTTECHNIK (Germany) ... (1998-99) 153
RUSJET/GENERAL TECHNOLOGIES/AEROSUD (International) ... (1992-93) 136

S

S1B2 Arctic Tern (Arctic Aircraft) ... (1994-95) 473
S-1-11B Super Stinker (Aviat) ... (1998-99) 544
S-1T Pitts Special (Aviat) ... (1999-2000) 554
S-2E Tracker upgrade (IMP) ... (1993-94) 38
S-2K (Sikorsky) ... (1993-94) 571
S2R-R1340 Thrush (Ayres) ... (1996-97) 554
S2R-R1820/510 Thrush (Ayres) ... (1996-97) 554
S2R-T Turbo Thrush (Marsh) ... (1991-92) 432
S-2S Pitts (Aviat) ... (2000-01) 575
S-2T Turbo Tracker (Grumman) ... (1991-92) 410
S-2T Turbo Tracker (Marsh) ... (1991-92) 432
S-2UP Tracker (IAI) ... (1994-95) 205
S-3 Viking (Lockheed) ... (1994-95) 557
S-4 Privateer (Arctic) ... (1996-97) 549
S-10 Aeropony (Shadin) ... (1995-96) 646
S10gsm (Gepard) ... (1991-92) 222
S-12 (Safire) ... (2000-01) 768
S-12-E Air Car (Spencer) ... (2000-01) 791
S-14 (Safire) ... (2000-01) 768
S-15 Pursuit (Rans) ... (2000-01) 745
S-18 (Sport Aircraft) ... (1994-95) 644
S-21 (Sukhoi) ... (1998-99) 435
S-21-G supersonic business jet (Sukhoi/Gulfstream) ... (1992-93) 139
S-32 (Sukhoi) ... (1997-98) 448
S-51 (Sukhoi) ... (1997-98) 448
S-55/H-19 (Orlando) ... (1991-92) 459
S-58/H-34 (Orlando) ... (1991-92) 460
S-58T (California Helicopter) ... (1991-92) 384
S-61 (Mitsubishi/Sikorsky) ... (1991-92) 174
S-61A Nuri Upgrade (Airod) ... (1993-94) 214
S-61/SH-3 (Sikorsky) ... (1991-92) 480
S-65/MH-53J Pave Low III Enhanced (Sikorsky) ... (1991-92) 480
S-76 (Sikorsky) ... (1996-97) 733
S-76 Shadow (Sikorsky) ... (1991-92) 489
S-76A+ (Sikorsky) ... (1991-92) 487
S-76B (Sikorsky) ... (1997-98) 753
S-76 Military versions (Sikorsky) ... (1998-99) 740
S-76N (Sikorsky) ... (1998-99) 740
S-80E (Sikorsky) ... (1997-98) 745
S-80M (Sikorsky) ... (1997-98) 745
S-84 (Sukhoi) ... (1997-98) 450
S-86 (Sukhoi) ... (1994-95) 369
S-86 (Sukhoi) ... (1998-99) 436
S-96 (Sukhoi) ... (1998-99) 436
S-99 (Sukhoi) ... (1992-93) 246
S 100B Argus (Saab) ... (1999-2000) 477
S-202 (SGAU) ... (1998-99) 415
S-302 (SGAU) ... (1998-99) 416
S.211 (Aermacchi) ... (1997-98) 263
S.211A (Northrop Grumman/Agusta) ... (1995-96) 617
S312 Tucano (Shorts) ... (1994-95) 457
S550 Citation S/11 (Cessna) ... (1994-95) 524
S-986 (Sukhoi) ... (1997-98) 450
SA (see Singapore Aerospace Ltd) ... (1995-96) 406
SA 2-37A (Schweizer) ... (1995-96) 643
SA 7 (STI) ... (1998-99) 8
SA-32T (Jaffe) ... (1991-92) 413
SA-204C (Snow Aviation) ... (1993-94) 571
SA-210AT (Snow Aviation) ... (1993-94) 571
SA227-AC Metro III (Fairchild Aircraft) ... (1991-92) 398
SA227-AT Merlin IVC (Fairchild Aircraft) ... (1991-92) 400
SA-227-CC Metro 23 (Fairchild Aerospace) ... (2000-01) 668
SA 227-DC Metro 23 (Fairchild Aerospace) ... (2000-01) 668
SA 342 Gazelle (Eurocopter) ... (1994-95) 163
SAAC (see Sammi Agusta Aerospace Corporation) ... (1992-93) 165
SAAF (see South African Air Force) ... (1992-93) 265
SABCA (see Société Anonyme Belge de Constructions Aéronautiques) ... (1995-96) 14
SAC (see Shenyang Aircraft Company) ... (1994-95) 59
SAGC (see Shenyang Aircraft Group Industry Corporation) ... (1995-96) 61
SAMF (see Shanghai Aircraft Manufacturing Company) ... (2000-01) 82
SAP (CSIST) ... (1998-99) 486
SAP (see Shijiazhuang Aircraft Plant) ... (1994-95) 61
SAR 42 (ATR) ... (1998-99) 201
SAU (see Nauchno-Proizvodstvennaya Korporatsiya Samoloty-Amfibyi Universalnyye (Russian Federation) ... (2000-01) 436
SB5 Sentinel (Seabird) ... (1991-92) 6
SB7L-360A Seeker (Seabird) ... (1997-98) 8
SC-95 (EMBRAER) ... (1991-92) 10
SC 01B Speed Canard (FFT) ... (1992-93) 79
SCTICSG (see Supersonic Commercial Transport International Co-operation Study Group) ... (1993-94) 178, 688

S—AIRCRAFT–PREVIOUS TEN EDITIONS: INDEXES 859

SD27 Corriedale (Sivel) ... (1997-98) 281
SECA (see Société d'Exploitation et de Constructions Aéronautiques) ... (1991-92) 84
SECC (see Shanghai Energy and Chemicals Corporation) ... (1998-99) 71
SEFA ASIA (see Société d'Etudes et de Fabrications Aéronautiques) ... (1994-95) 260
SF-2A Cygnet (Viking) ... (2000-01) 798
SF 25C Falke 1700 (Scheibe) ... (1999-2000) 161
SF 36 R (Scheibe) ... (1997-98) 150
SF.600A Canguro (SIAI-Marchetti) ... (1993-94) 167
SGAC (see Shenzen General Aviation Company) ... (1997-98) 75
SH-2 Seasprite (Kaman) ... (1994-95) 542
SH-2G Super Seasprite (Kaman) ... (1994-95) 542
SH-3D/TS (Agusta-Sikorsky) ... (1994-95) 214
SH-3H (Agusta-Sikorsky) ... (1994-95) 214
SH-5 (HAMC) ... (1997-98) 62
SH-5 (HAMC) ... (1995-96) 53
SJ30 (Swearingen) ... (1996-97) 737
SK60 (Saab) ... (1994-95) 406
SK-500 (StarKraft) ... (1998-99) 745
SK-700 (StarKraft) ... (1998-99) 745
SL (Myasishchev) ... (1992-93) 234
SL-90 Leshii (Aviotechnica) ... (1997-98) 202
SLAC (see Shenyang Light Aircraft Company) ... (1999-2000) 77
SM-95 Slava (Technoavia) ... (1996-97) 439
SMAN (see Société Morbihannaise d'Aéro Navigation) ... (1997-98) 129
SOGEPA (see Société de Gestion de Participations Aéronautiques ... (2000-01) 148
SOKO (see Vazduhoplovna Industrija Soko) ... (1993-94) 13
SONACA (see Société Nationale de Construction Aérospatiale) ... (1994-95) 13
SP-2H Firestar tanker conversion (Aero Union) ... (1991-92) 338
SP-91 (Technoavia) ... (1996-97) 436
SP-95 (Technoavia) ... (1999-2000) 445
SR-71 (Lockheed) ... (1992-93) 402
SSA-Kari (SSA) ... (1997-98) 303
SSBJ (Sukhoi/Gulfstream) ... (1992-93) 139
SSF (McDonnell Douglas/BAe) ... (1994-95) 184
SST-X001 (JADC) ... (1996-97) 271
ST-17 (CAT) ... (1991-92) 384
ST-50 (Israviation) ... (1998-99) 253
STI (see Sadleir Technology and Innovation) ... (1998-99) 7
STOL CH 701-AG (Zenair) ... (1996-97) 47
STOVL Prototype (Sadleir) ... (1994-95) 9
Su-15 (Sukhoi) ... (1992-93) 236
Su-17 Fitter (Sukhoi) ... (1995-96) 368
Su-20 Fitter (Sukhoi) ... (1995-96) 368
Su-22 Fitter (Sukhoi) ... (1995-96) 368
Su-25T (Su-34) (Sukhoi) ... (1993-94) 311
Su-26M (Sukhoi) ... (1996-97) 423
Su-31 (Sukhoi) ... (1992-93) 243
Su-32 (Sukhoi) ... (1994-95) 362
Su-33 (Sukhoi) ... (1999-2000) 435
Su-34 (Sukhoi) ... (1993-94) 363
Su-37 (Sukhoi) ... (1992-93) 244
Su-38 (Sukhoi) ... (1997-98) 445
Su-49 (Sukhoi) ... (1999-2000) 439
SV28 Lincoln (Sivel) ... (1997-98) 282

SABRELINER CORPORATION (USA) ... (1994-95) 624
Saab 35 Draken (Saab-Scania) ... (1991-92) 220
Saab 340 B (Saab) ... (1999-2000) 477
Saab 2000 (Saab) ... (1999-2000) 480
SADLEIR TECHNOLOGY AND INNOVATION (Australia) ... (1998-99) 7
SADLEIR VTOL INDUSTRIES AUSTRALIA (Australia) ... (1994-95) 9
SADLER AIRCRAFT CORPORATION (USA) ... (2000-01) 767
Safari An-28 (PZL Mielec) ... (1992-93) 178
Safety and performance conversions (Sierra) ... (1991-92) 478
SAMMI AGUSTA AEROSPACE CORPORATION LTD (South Korea) ... (1992-93) 165
Samson M-90 ... (1997-98) 428
Samurai Sword (Diamond) ... (1997-98) 11
SANGER (DASA) ... (1994-95) 108
Sapeer EC-130 Hercules (Lockheed Martin) ... (1998-99) 665
Sarohale TBM 60,000 (SOCATA) ... (1997-98) 134
Saviat E-1 (Sat Saviat) ... (2000-01) 436
Saviat E-5 (Sat Saviat) ... (1999-2000) 420
Scanliner (HB-Aircraft) ... (1991-92) 7
Scanliner, HB-Aircraft (CAI) ... (1991-92) 48
SCHAFER AIRCRAFT MODIFICATIONS INC (USA) ... (1993-94) 555
Scoutmaster (Lovaux) ... (1991-92) 316
Sea Bird (SEFA Asia) ... (1994-95) 260
SEABIRD AVIATION AUSTRALIA PTY LTD (Australia) ... (1997-98) 8
Seafire TA16 (Thurston) ... (1996-97) 742
Seagull, HU-1 (SSLF) ... (1998-99) 72
Sea Harrier FRS. Mk 1 (BAe) ... (1997-98) 524
Sea Harrier FRS. Mk 51 (BAe) ... (1997-98) 524
SEAIR PACIFIC PTY LTD (Australia) ... (1991-92) 6
Sea King (GKN Westland) ... (1997-98) 547
Sea King (Sikorsky) ... (1991-92) 480
Sea King conversion (IMP) ... (1994-95) 40
Sea Knight (Boeing) ... (1991-92) 366
Seamaster TA19 (Thurston) ... (1996-97) 742
Seaquest (Omega) ... (1997-98) 712
Seasprite (Kaman) ... (1994-95) 542
Seastar CD2 (Dornier Composite) ... (1991-92) 94
Seastar CD2 (Dornier Seastar Malaysia) ... (1996-97) 286
Sea Warrior (Agusta) ... (1994-95) 215
Seawolf (Lake) ... (1992-93) 393
Seeker SB7L-360A (Seabird) ... (1997-98) 8
SEFA ASIA INC (Philippines) ... (1993-94) 227
SELLET/PELLETIER HELICOPTERE (France) ... (1991-92) 85
Seneca III (EMBRAER/Piper) ... (1992-93) 16
Sentinel, SB5 (Seabird) ... (1991-92) 6
Sentry, E-3 (Boeing) ... (1993-94) 452
Sentry E-3D, AEW Mk I (BAe) ... (1992-93) 303
Sentry E-3D, AEW Mk I (Boeing) ... (1993-94) 452
Sequoia 300 (Sequoia) ... (1995-96) 646
SEYEDO SHOHADA PROJECT, DEFENCE INDUSTRIES (Iran) ... (1993-94) 181
SHADIN CO INC (USA) ... (1995-96) 646
Shadow (Sikorsky) ... (1991-92) 489
Shadow II (CFM) ... (1997-98) 535
Shadow Series C (CFM) ... (1998-99) 514
Shahal (IAI) ... (1993-94) 182
Shahbaz (PAC) ... (1997-98) 315
Shaman (Venga/Baoshan) ... (1995-96) 211
SHANGHAI AIRCRAFT MANUFACTURING FACTORY (China) ... (2000-01) 82
SHANGHAI ENERGY AND CHEMICALS CORPORATION (China) ... (1998-99) 71
SHENYANG AIRCRAFT CORPORATION (China) ... (1994-95) 59
SHENYANG AIRCRAFT GROUP INDUSTRY CORPORATION LTD (China) ... (1995-96) 61
SHENYANG LIGHT AIRCRAFT COMPANY (China) ... (1999-2000) 77
SHENZHEN GENERAL AVIATION COMPANY LTD (China) ... (1997-98) 75
Sherpa (Sherpa) ... (1994-95) 630
Sherpa (Shorts) ... (1993-94) 457
SHERPA AIRCRAFT MANUFACTURING (USA) ... (1997-98) 743
SHIJIAZHUANG AIRCRAFT PLANT (China) ... (1994-95) 61
SHIN MEIWA INDUSTRY CO LTD (Japan) ... (1992-93) 163
Shmel DM3-1 (Dubna) ... (1996-97) 356
SHORT BROTHERS PLC (UK) ... (1997-98) 544
Shorts 330 (Shorts) ... (1992-93) 311
Shorts 330-UTT (Shorts) ... (1992-93) 312
Shorts 360-300 (Shorts) ... (1991-92) 323
Shorts 360-300F (Shorts) ... (1991-92) 323
Short-field enhancement system (Raisbeck) ... (1991-92) 465
Shrike (Millicer) ... (1998-99) 796
Shuihong-5 (HAMC) ... (1995-96) 53
Shuishang Hongzhaji 5 (HAMC) ... (1995-96) 53
Shuttle Spacecraft (Rockwell International) ... (1993-94) 434
SIAI-MARCHETTI (Italy) ... (1996-97) 249
SIBERIAN AERONAUTICAL RESEARCH INSTITUTE SA CHAPLYGIN (Russian Federation) ... (1997-98) 432
SIERRA INDUSTRIES (USA) ... (1993-94) 559
Silhouette (Lunds Tekniske) ... (1995-96) 265
SINGAPORE AEROSPACE LTD (Singapore) ... (1995-96) 406
SIVEL (Italy) ... (1998-99) 274
Skif M-103 (Myasishchev) ... (1993-94) 302
Skorpion PZL-230F (PZL Warszawa-Okecie) ... (1997-98) 335
Skvorets T-407 (Aeroprogress/ROKS-Aero) ... (1998-99) 336
Skycar M-200 (Moller) ... (1996-97) 682
Skycar M-400 (Moller) ... (1996-97) 682
Skyfarmer (Transavia) ... (1992-93) 8
Skyfox (Skyfox) ... (1998-99) 304
SKYFOX AVIATION (Australia) ... (2000-01) 9
Skyhawk (Cessna) ... (1991-92) 385
Skyhawk upgrade (IAI) ... (1993-94) 184
Skyhawk upgrade (RNZAF) ... (1991-92) 184
Skyhook (The Australian Autogyro Co) ... (1993-94) 5
Skylane (Cessna) ... (1991-92) 385
Skylane RG (Cessna) ... (1991-92) 385
SKYLINE AVIATION TECHNOLOGIES (South Africa) ... (1998-99) 464
Skyranger (Aerodyn) ... (1997-98) 97
Sky Rocket P-337 (Riley) ... (1991-92) 467
Sky Rocket II (Flying K) ... (2000-01) 669
Skytruck Plus M-28 03/04 (PZL) ... (2000-01) 328
Skywagon (Cessna) ... (1991-92) 385
SNOW AVIATION INTERNATIONAL INC (USA) ... (1993-94) 571
SOCIÉTÉ ANONYME BELGE DE CONSTRUCTIONS AÉRONAUTIQUES (Belgium) ... (1995-96) 14
SOCIETE DE GESTION DE PARTICIPATIONS AERONAUTIQUES (France) ... (2000-01) 148
SOCIÉTÉ D'ÉTUDES ET DE FABRICATIONS AÉRONAUTIQUES (Philippines) ... (1994-95) 260
SOCIÉTÉ D'EXPLOITATION ET DE CONSTRUCTIONS AÉRONAUTIQUES (France) ... (1991-92) 84
SOCIETE MORBIHANNAISE D'AERO NAVIGATION (France) ... (1997-98) 129
SOCIÉTÉ NATIONALE DE CONSTRUCTION AEROSPATIALE (Belgium) ... (1994-95) 13
SOKO/AVIOANE (International) ... (1997-98) 249
SOKO/IAv CRAIOVA (International) ... (1991-92) 141
Sokol T-401 (Aeroprogress/ROKS-Aero) ... (1997-98) 353
SOLOY CORPORATION (USA) ... (1993-94) 572
Sootless exhaust stack fairings (Raisbeck) ... (1991-92) 465
SOUTH AFRICAN AIR FORCE (South Africa) ... (1992-93) 265
SOVIET AEROSPACE INDUSTRIES ASSOCIATION (Russia) ... (1992-93) 235
Space Shuttle (USSR) ... (1991-92) 297
Space transportation system (Rockwell) ... (1992-93) 434
Sparton C-27A (Alenia) ... (2000-01) 279
Special 110 (Aviat) ... (2000-01) 576
Special Mission Aircraft (Fairchild) ... (1997-98) 629
Special S-1T Pitts (Aviat) ... (1999-2000) 554
Spectre, AC-130U (Rockwell/Lockheed) ... (1995-96) 641
Speed Canard SC OIB (FFT) ... (1992-93) 79
Speed Katana, DA 20-C1 (Diamond) ... (1997-98) 809
Speed Katana, DV22 (Diamond) ... (1997-98) 11
Spiral project, 50-50 (USSR) ... (1991-92) 298
Spirit (Paragon) ... (1996-97) 693
Spirit (Questair) ... (1997-98) 718
Spirit B-2 (Northrop Grumman) ... (1999-2000) 705
SPORT AIRCRAFT INC (USA) ... (1995-96) 659
Sportsman VJ-22 (Volmer) ... (1994-95) 653
Sprint (FLS) ... (1994-95) 447
Sprinter T-430 (Khrunichev) ... (1997-98) 397
Stallion (Aircraft Designs) ... (2000-01) 544
Stallion (Helio) ... (1991-92) 413
Starboy 912 (Interplane) ... (2000-01) 98
Starcraft (Maule) ... (1991-92) 434
STARKRAFT (USA) ... (1998-99) 745
Starfire Bonanza (Colemill) ... (1991-92) 394
Starling T-407 (Aeroprogress/ROKS-Aero) ... (1998-99) 336
STAR OF PHOENIX AIRCRAFT (USA) ... (1994-95) 644
Starship 1, 2000 (Beech) ... (1995-96) 637
Star Streak (Airborne Innovations) ... (2000-01) 561
Statesman (BAe) ... (1992-93) 296
Sterkh T-201 (Khrunichev) ... (1999-2000) 384
Stiletto (MFI) ... (1993-94) 351
Stiletto T-9 (Terzi) ... (1998-99) 277
Stiletto, T-9R (Terzi Aerodine) ... (1992-93) 156
Storch Replica (Preceptor) ... (1999-2000) 718
Stork-2 T-411 (Aeroprogress/ROKS-Aero) ... (1997-98) 354
Stork T-201 (Aeroprogress/ROKS-Aero) ... (1997-98) 350
STOVL Strike Fighter (USN) ... (1994-95) 651
STRATEGIC BUSINESS UNIT HELICOPTERS (Germany) ... (1991-92) 91
STRATEGIC BUSINESS UNIT MILITARY AIRCRAFT (Germany) ... (1991-92) 93
STRATEGIC BUSINESS UNIT REGIONAL AIRCRAFT (Germany) ... (1991-92) 88
STRATEGIC BUSINESS UNIT SPACE TRANSPORTATION AND PROPULSION SYSTEMS (Germany) ... (1991-92) 94
Strategic Reconnaissance Aircraft (USAF) ... (1994-95) 649
Strato I (Grob) ... (1997-98) 144
Strato II (Egrett) ... (1991-92) 121
Strato 2C G 850 (Grob) ... (1997-98) 146
Stratocruzer 1250-ER (AASI) ... (1997-98) 554
Stratosfera M-17 (Myasishchev) ... (1995-96) 363
Stratofortress, B-52 (Boeing) ... (1991-92) 371
Stratotanker, KC-135A (Boeing) ... (1991-92) 371
Stretch 580 (KFC) ... (1991-92) 28
Strizh T-501 (Aeroprogress/ROKS-Aero) ... (1997-98) 355
SUKHOI DESIGN BUREAU (USSR) ... (1991-92) 275
SUKHOI/GULFSTREAM (International) ... (1992-93) 139
Super2 ARV-1 (Aviation Scotland) ... (1994-95) 433
Super-7 (CAC) ... (1995-96) 50
SUPER 580 AIRCRAFT COMPANY (USA) ... (1993-94) 573
Super 580 (Super 580 Aircraft) ... (1993-94) 573
Super Beaver (Wipaire) ... (1991-92) 496
Super Chevvron 2-45CS (AMF) ... (1998-99) 500
SuperCobra AH-1RO (IAR) ... (1997-98) 343
Super Courier (Helio) ... (1991-92) 413
Super Cub PA-18-150 (Piper) ... (1994-95) 616
Super Dakota C-47TP (SAAF) ... (1992-93) 265
Super Etendard (Dassault Aviation) ... (1991-92) 74
Super Galeb G-4 (UTVA) ... (1994-95) 658
Super Huey (Bell) ... (1991-92) 358
Super King Air 300 (Beech) ... (1991-92) 353
Super King Air 300 LW (Raytheon) ... (1996-97) 708
Super Lynx (Westland) ... (1996-97) 537
Super Phantom (IAI) ... (1994-95) 205
Super Puma Mk 1 AS 332 (Eurocopter) ... (1998-99) 213
Super Q (Comtran) ... (1991-92) 395
SUPER ROTOR INDUSTRIA AERONAUTICA LTDA, M. M. (Brazil) ... (1992-93) 16
Super Seasprite (Kaman) ... (1994-95) 542
Super Shadow Series D and D-D (CFM) ... (1997-98) 535

860 INDEXES: AIRCRAFT–PREVIOUS TEN EDITIONS—**S – U**

Super Skyhawk (SA) (1994-95) 392
Supersonic business jet (Sukhoi/Gulfstream) (1992-93) 139
SUPERSONIC COMMERCIAL TRANSPORT INTERNATIONAL CO-OPERATION STUDY GROUP (International) (1993-94) 178, 688
Supersonic STOVL fighter (USN) (1992-93) 457
Superstar 650 (Machen) (1991-92) 431
Superstar 680 (Machen) (1991-92) 431
Superstar 700 (Machen) (1991-92) 431
SuperStar F-22 (Lockheed) (1992-93) 396
Super Stinker S-1-11B (Aviat) (1998-99) 544
Super Tucano EMB-312H (EMBRAER) (1997-98) 16
Super Tucano EMB-312HJ (Northrop Grumman) (1995-96) 617
Super Vivat L-13 (Aerotechnik) (1998-99) 82
Super Zodiac CH 601 HDS (Zenith) (1997-98) 771
Surprise HX-300 (Hepp) (2000-01) 163
Swallow, Lasta 2 (LOLA Utva) (1999-2000) 776
Swati, LT-1 (BHEL) (1994-95) 118
SWEARINGEN AIRCRAFT INC (USA) (1995-96) 662
SwiftFury (LoPresti) (1992-93) 404
Swift, IAR-109 (Avioane) (1995-96) 291
Swift KR-2PM (Radwan) (2000-01) 343
SWISS FEDERAL AIRCRAFT FACTORY (F + W) (Switzerland) (1995-96) 431
Swiss Trainer MD3 (MDB) (1993-94) 357

T

'T-3A, Northrop' (USAF) (1995-96) 665
T-9 Stiletto (Terzi) (1998-99) 277
T-9R Stiletto (Terzi Aerodin) (1992-93) 156
T-10 (Lockheed Martin) (1998-99) 665
T.17 (Boeing) (1991-92) 372
T-27 (EMBRAER) (2000-01) 17
T30 Katana (Terzi) (1993-94) 203
T-33V (Volpar) (1993-94) 584
T-35 Pillán (ENAER) (1994-95) 45
T-35DT Turbo Pillán (ENAER) (1996-97) 48
T-36 Halcón (CASA) (1995-96) 411
T-36 Halcón (ENAER) (1996-97) 48
T-43A (Boeing) (1995-96) 497
T45 Turbine Dromader (Melex) (1991-92) 451
T-47A Citation S/11 (Cessna) (1994-95) 524
T-106 (Aeroprogress/ROKS-Aero) (1998-99) 335
T-60S (Sukhoi) (1997-98) 447
T67M Mk 2 Firefly (Slingsby) (1999-2000) 532
T-130 Fregat (Aeroprogress/ROKS-Aero) (1998-99) 335
T-121 Grif (Washington Aeroprogress) (1996-97) 751
T-201 Aist (Aeroprogress/ROKS-Aero) (1997-98) 350
T-201 Sterkh (Khrunichev) (1999-2000) 384
T-203 Pchela (Aeroprogress/ROKS-Aero) (1997-98) 350
T-204 Condor (Washington Aeroprogress) (1996-97) 751
T-204 Griffon (Aeroprogress/ROKS-Aero) (1997-98) 351
T-205 Eastern Caravan (Aeroprogress/ROKS-Aero) (1997-98) 351
T206/210 (RAM/Cessna) (1991-92) 465
T-208 Orel (Krunichev) (1998-99) 375
T-211 (Thorp) (1992-93) 454
T-230 Fregat (Khrunichev) (1998-99) 375
T-274 Brigantine (Krunichev) (1997-98) 393
T-274 Titan (Aerospace/ROKS-Aero) (1997-98) 351
T-300A Skyfarmer (Transavia) (1992-93) 8
T310 (RAM/Cessna) (1991-92) 465
T-400 Skyfarmer (Transavia) (1992-93) 8
T-401 Sokol (Aeroprogress/ROKS-Aero) (1997-98) 353
T-407 Skvorets (Aeroprogress/ROKS-Aero) (1998-99) 330
T-411 Wolverine (Khrunichev) (1997-98) 394
T-411 Wolverine (Washington Aeroprogress) (1997-98) 770
T-417 (Khrunichev) (2000-01) 396
T-422 Yastreb (Khrunichev) (1997-98) 396
T-430 Sprinter (Khrunichev) (1997-98) 397
T-433 Flamingo (Aeroprogress/ROKS-Aero) (1999-2000) 347
T-471 Corvette (Khrunichev) (1997-98) 398
T-501 Strizh (Aeroprogress/ROKS-Aero) (1997-98) 355
T-602 Ovel (Aeroprogress/ROKS-Aero) (1995-96) 305
T-610 Voyage (Aeroprogress/ROKS-Aero) (1997-98) 356
T-620 Korshun (Aeroprogress/ROKS-Aero) (1999-2000) 347
T-710 Anaconda (Aeroprogress/ROKS-Aero) (1997-98) 357
T-720 (Washington Aeroprogress) (1996-97) 751
T-910 Kuryer (Aeroprogress/ROKS-Aero) (1997-98) 357
T-2402 Dinosaur (Khrunichev) (1995-96) 339
TA-3T Triloader (Clark-Norman) (1996-97) 519
TA-16 Seafire (Thurston) (1996-97) 742
TA-19 Seamaster (Thurston) (1996-97) 742
TACAMO II, E-6 (Boeing) (1993-94) 454
TAH-1F HueyCobra (Bell) (1993-94) 434
TAV-8B (McDonnell Douglas (Boeing)/BAe) (1997-98) 229
TAV-8S Harrier (BAe) (1991-92) 311
TB 31 Omega (SOCATA) (1996-97) 117
TB 360 Tangara (SOCATA) (2000-01) 147
TBM 700S (SOCATA) (1995-96) 108
TBM 60,000 Sarohale (SOCATA) (1997-98) 134
TC-130 Hercules (Lockheed Martin) (1998-99) 665
TCM (see Teledyne Continental Motors) (1993-94) 575
TD Technology Demonstrator (IAI) (1994-95) 202
TG-10 Brushfire (Venga/Baosham) (1995-96) 210
TG-XI Yarasa (TAI) (1998-99) 487
TH-1S HueyCobra (Bell) (1993-94) 434
TK-10 (Lockheed Martin) (1998-99) 665
TL-10 (Lockheed Martin) (1998-99) 665
TKEF (see The King's Engineering Fellowship) (1993-94) 576
TM 20B Flamingo III (Bengis) (1992-93) 264
Tp 84 Hercules (Lockheed Martin) (1998-99) 665
Tp 89 (CASA) (1991-92) 213
Tp 100 (Saab) (1999-2000) 477
TR-3A Northrop (USAF) (1995-96) 665
TRANSAVIA (see Transfield Construction) (1992-93) 8
TS-2F Turbo Tracker (Marsh) (1994-95) 573
TS-11R Iskra (PZL Mielec) (1993-94) 231
TSC-1A3 Teal III (Thurston) (1996-97) 741
Tsu-Chiang, AT-3 (AIDC) (1991-92) 227
Tu-16 Badger (Tupolev) (1994-95) 370
Tu-22 Blinder (Tupolev) (1994-95) 373
Tu-24 (Tupolev) (1998-99) 443
Tu-34 (Tupolev) (1998-99) 444
Tu-130 (Tupolev) (1998-99) 444
Tu-134BSh (Tupolev) (1993-94) 327
Tu-134UBL (Tupolev) (1993-94) 327
Tu-136 (Tupolev) (2000-01) 469
Tu-144LL (Tupolev) (1998-99) 445
Tu-154 Careless (Tupolev) (1999-2000) 452
Tu-204C Freighter (Tupolev) (1999-2000) 455
Tu-230 (Tupolev) (1998-99) 449
Tu-244 (Tupolev) (1999-2000) 456
Tu-304 (Tupolev) (1999-2000) 456
Tu-306 (Tupolev) (1999-2000) 456
Tu-324 (Tupolev) (1999-2000) 456
Tu-414 (Tupolev) (1997-98) 468
Tupolev C-Prop (Tupolev) (1998-99) 452
TW-18 Crosspointer (Skyline) (1997-98) 482
TW-68 (Ishida) (1995-96) 543

TACAMO II, E-6A (Boeing) (1993-94) 454
Tacit Blue (USAF) (1996-97) 744
'Taildragger' conversions, Bolen (Bush) (1991-92) 384
Tail-less Aircraft Concepts (Lockheed Martin) (1998-99) 671
Tangara TB 360 (SOCATA) (2000-01) 147
Tapete Air Truck PD.90 (Partenavia) (1996-97) 263
TAYLORCRAFT AIRCRAFT (USA) (1993-94) 574
T-Bird II (Lockheed Martin/Aermacchi/Rolls-Royce) (1996-97) 239, 628
TBM SA (International) (1991-92) 143
Teal III TSC-1A3 (Thurston) (1996-97) 741
Tchaika (Beriev) (1991-92) 239
T-Craft (Vector) (1993-94) 582
Tech 2 (New Technik) (1991-92) 456
Tech 4 (New Technik) (1991-92) 456
TELEDYNE CONTINENTAL MOTORS (USA) (1993-94) 575
TELEDYNE RYAN AERONAUTICAL (USA) (1991-92) 493
Tempest 184 (Jaran) (1999-2000) 658
Terminator II (Isolair) (1994-95) 542
TERZI, PIETO (Italy) (1998-99) 277
THE KING'S ENGINEERING FELLOWSHIP (USA) (1994-95) 546
Theta AA330 (CRSS) (1993-94) 112
THORP AERO INC (USA) (1992-93) 454
Thrush S2R-R1340 (Ayres) (1996-97) 554
Thrush S2R-R1820/510 (Ayres) (1996-97) 554
THURSTON AEROMARINE CORPORATION (USA) (1997-98) 762
Tiger II (Northrop) (1992-93) 426
Tiger AG-5B (American General) (1994-95) 472
Tiger, AS332 (Aerospatiale) (1991-92) 59
Tigress (Lancair) (2000-01) 683
Tilt-rotor (Sikorsky) (1991-92) 452
Tiltrotor Concept (Bell Boeing) (2000-01) 586
Timberwolf (Venga/Baoshan) (1995-96) 211
Titan T-274 (Aeroprogress/ROKS-Aero) (1997-98) 351
Tomcat F-14 (Grumman) (1993-94) 483
Tornado (Mylius) (1999-2000) 157
Tornado 2000 (Panavia) (1992-93) 133
Tornado ADV (Panavia) (1999-2000) 250
Tornado ECR (Panavia) (1999-2000) 249
Tornado F Mk 2, 2A, 83 (Panavia) (1999-2000) 250
Tornado GR Mk 1, 1A, 1B, 484A (Panavia) (1999-2000) 246
Tornado IDS (Panavia) (1999-2000) 246
Toucan (EMBRAER) (2000-01) 17
Tracker S-2E upgrade (IMP) (1993-94) 38
Tracker S-2UP upgrading (IAI) (1994-95) 205
TRADEWIND TURBINES CORPORATION (USA) (1993-94) 577
Transall C-160 (International) (1991-92) 144
TRANSFIELD CONSTRUCTION PTY LTD, TRANSAVIA DIVISION (Australia) (1992-93) 8
Transregional 250 (CAT) (1991-92) 384
TRIDAIR HELICOPTERS INC (USA) (1994-95) 647
Triloader TA-3T (Clark-Norman) (1996-97) 519
TRILOADER AIRCRAFT CORPORATION (Belgium) (1996-97) 13
Trio (Hirt) (1996-97) 614
TRI-R TECHNOLOGIES (USA) (2000-01) 794
TriStar cargo conversions (Pemco) (1994-95) 615
TriStar conversions (Lockheed) (1991-92) 430
TriStar modification, civil (Marshall) (1991-92) 317
TriStar tanker conversion (Marshall) (1991-92) 317
Triumph (Scaled Composites/Rutan) (1991-92) 473
Tri-Z CH-300 (Zenair) (1994-95) 43
Tsu-Chiang AT-3 (AIDC) (1996-97) 492
Tucano EMB-312 (EMBRAER) (2000-01) 17
Tucano T. Mk I (Shorts) (1993-94) 399
Tundra (Laron) (1998-99) 655
TUPOLEV DESIGN BUREAU (USSR) (1991-92) 284, 756
Turbine 207 (MacAvia) (1991-92) 431
Turbine Bravo, AS 202/32TP (FFA) (1997-98) 495
Turbine Defender (Pilatus Britten-Norman) (1992-93) 309
Turbine Defender, Special Role (Pilatus Britten-Norman) (1992-93) 310
Turbine Dromader (Melex) (1991-92) 451
Turbine Exec (Hillberg) (2000-01) 678
Turbine Islander, Special Role (Pilatus Britten-Norman) (1992-93) 310
Turbine P-210 (Advanced Aircraft) (1991-92) 336
Turbine Pac conversions (Soloy) (1993-94) 572
Turbo-34 (Basler) (1994-95) 480
Turbo-67 DC-3 conversion (Basler) (1991-92) 346
Turbo 270 Renegade (Lake) (1996-97) 618
Turbo Cat, G-162 C-T (Marsh) (1991-92) 432
Turbo Centurion (Cessna) (1991-92) 385
Turbo Firecat (Conair) (1994-95) 35
Turbo Javelin PZL-106B (PZL Warszawa-Okecie) (1997-98) 338
Turbo Orlik (PZL Warszawa-Okecie) (1994-95) 278
Turbo Pillán, T-35DT (ENAER) (1996-97) 48
Turbo Sea Thrush, Terr-Mar (Ayres) (1991-92) 345
Turbo Skylane RG (Cessna) (1991-92) 385
TURBOTECH INC (USA) (1994-95) 648
Turbo Thrush, S2R-T (Marsh) (1991-92) 432
Turbo-Thrush 700 (Ayres) (1997-98) 570
Turbo Tracker S-2T (Grumman) (1991-92) 410
Turbo Tracker S-2T (Marsh) (1991-92) 432
Turbo Tracker TS-2F (Marsh) (1994-95) 573
Turbo Trainer, SA-32T (Jaffe) (1991-92) 413
TWIN COMMANDER AIRCRAFT CORPORATION (USA) (1993-94) 578
Twinbee (KARI) (1998-99) 294
Twinjet 1200 (Maverick Air) (1999-2000) 693
Twin Ranger, new (Bell) (1992-93) 344
Twin Ranger, 206LT (Bell) (1998-99) 25
Twin-T (Vector) (1993-94) 582
Twin Tech (New Technik) (1991-92) 456
Tzu-Chung AT-3 (AIDC) (1994-95) 420
Tzukit (IAI) (1991-92) 149

U

U-20 (Cessna) (1995-96) 522
UA-5 (Honda) (1993-94) 488
UC-26C (Fairchild Aerospace) (1999-2000) 644
UDRD (see Universal Dynamics Research and Development) (1993-94) 227
UH-1 (Bell) (1991-92) 358
UH-1H (Fuji-Bell) (1994-95) 228
UH-1HP (Global) (1992-93) 382
UH-1HP Huey II (Bell) (1994-95) 492
UH-1J (Fuji-Bell) (1994-95) 229
UH-1N (Bell) (1998-99) 27
UH-12E Hauler (Rogerson Hiller) (1994-95) 623
UH-12E3 (Hiller) (1998-99) 649
UH-12E3T (Hiller) (1998-99) 649
UH-12E5 (Hiller) (1997-98) 637
UH-46 Sea Knight (Boeing) (1991-92) 366
UHCA (Airbus) (1992-93) 97
UH-X (US Army) (1995-96) 665

Ultima (Junkers) (1999-2000) 156
Ultimate (Junkers) (1999-2000) 156
Ultra High Capacity Aircraft (Airbus) (1993-94) 126
Ultra Huey (UNC) (1994-95) 648
ULYANOVSK AVIATION INDUSTRIAL COMPLEX 'AVIASTAR' (Russia) (1995-96) 399
UNC HELICOPTER (USA) (1994-95) 648
UNIKOMTRANSO (Russian Federation) (1998-99) 453
UNITED STATES ARMY AVIATION AND TROOP COMMAND (USA) (1996-97) 744
UNIVERSAL DYNAMICS RESEARCH AND DEVELOPMENT (Philippines) (1993-94) 227
Urpi, IAP-003 (Indaer Peru) (1992-93) 174
Utility Transport (PADC) (1997-98) 317
Utva-75A (Utva) (1991-92) 500
Utva-75AG11 (Utva) (1991-92) 501

Entry	Edition	Page
Utva-95 (Utva)	(1997-98)	773
Utva-96 (LOLA Utva)	(1999-2000)	777

V

Entry	Edition	Page
V-1-A Vigilante (Ayres)	(1991-92)	345
V-2 Mi-2 (Mil)	(1994-95)	270
V-22 Osprey (BAe)	(1997-98)	528
VA.1 Matador (BAe)	(1991-92)	311
VA.2 Matador II (McDonnell Douglas (Boeing)/BAe)	(1997-98)	229
VC10 Tankers (BAe)	(1992-93)	303
VC-25A (Boeing)	(1991-92)	372
VC-130H Hercules (Lockheed Martin)	(1998-99)	665
VC-137 (Boeing)	(1993-94)	441
VC-X (USAF)	(1994-95)	649
VJ-22 Sportsman (Volmer)	(1994-95)	653
VK30 Cirrus (Cirrus)	(1996-97)	597
VM-23 Variant (Inteco)	(1995-96)	74
VM-T Atlant (Myasishchev)	(1995-96)	363
VM-T Atlant (Myasishchev)	(1991-92)	273
VSKhS (Romashka)	(1996-97)	418
VALMET AVIATION INDUSTRIES (Finland)	(1996-97)	86
VALSAN INC (USA)	(1993-94)	580
VARDAX CORPORATION (USA)	(1993-94)	581
Vazar Dash 3 (Vardex)	(1991-92)	496
VAZDUHOPLOVNA INDUSTRIJA SOKO DD (Yugoslavia)	(1993-94)	13
VAZDUHOPLOV TECHNIKI INSTITUT (Yugoslavia)	(1994-95)	660
Vector, CBA-123 (EMBRAER/FMA)	(1994-95)	157
Vector FTT (DASA)	(1999-2000)	144
VECTOR AIRCRAFT COMPANY (USA)	(1994-95)	653
VEETOL HELICOPTERS (Australia)	(1998-99)	8
VENGA/BAOSHAN (International)	(1995-96)	210
Venom AH-1W (Bell)	(1994-95)	494
Very Large Commercial Transport (Airbus)	(1993-94)	316
Viator AP 68TP 600 (TAAL)	(1998-99)	166
Viggen AJS 37 (Saab)	(1994-95)	406
Viggen, JA37 (Saab-Scania)	(1991-92)	219
Vigilant .Mk 1 (Grob)	(1997-98)	141
Vigilante V-1-A (Ayres)	(1991-92)	345
Viking (Lockheed)	(1994-95)	557
Vipan MFI-10C (RFB)	(1993-94)	687
Viper AH-1(4B)W (Bell)	(1992-93)	343
Viscount-life extension (BAF)	(1993-94)	391
Vistaplane (VAT)	(1994-95)	652
Vitaz (Molniya)	(1998-99)	406
Vivat L-13SDL (Aerotechnik)	(1995-96)	73
Vivat L-13SDM (Aerotechnik)	(1995-96)	73
Vivat L-13SE (Aerotechnik)	(1994-95)	71
Vivat L-13SL (Aerotechnik)	(1994-95)	71
VIVIAN ASSOCIATES LIMITED, LR (Canada)	(1992-93)	29
V-Jet-II (Scaled Composites)	(1999-2000)	742
Volantor (Moller)	(1991-92)	452
VOLMER AIRCRAFT (USA)	(1994-95)	653
VOLPAR AIRCRAFT CORPORATION (USA)	(1994-95)	654
VOUGHT AIRCRAFT COMPANY (USA)	(1994-95)	654
Voyage T-610 (Aeroprogress/ROKS-Aero)	(1997-98)	356
Voyager-500 (Revolution)	(2000-01)	763
VOYENNO-PROMYSHLENNYI KOMPLEKS MAPO (Russian Federation)	(2000-01)	400
VTOL AIRCRAFT PTY LTD (Australia)	(1991-92)	7
VTOL Concept (STI)	(1998-99)	8
VTOL INDUSTRIES AUSTRALIA (Australia)	(1994-95)	9

W

Entry	Edition	Page
WA-116 (Wallis)	(1992-93)	316
WA-116-T (Wallis)	(1992-93)	316
WA-116X (Wallis)	(1992-93)	316
WA-117/R-R MoD operations (Wallis)	(1991-92)	327
WA-121 (Wallis)	(1992-93)	316
WA-122/R-R (Wallis)	(1992-93)	316
WA-201 (Wallis)	(1992-93)	316
WAF 96 (Nogatec)	(1999-2000)	129
WC-130 Hercules (Lockheed Martin)	(1998-99)	665
WPAC (see Western Pacific Aviation Corporation)	(1996-97)	302
WS 70 (Westland)	(1991-92)	333
WS 70L (Westland)	(1995-96)	476
WSK-PZL MIELEC (see Wytwornia Sprzetu Komunikacyjnego-PZL Mielec)	(1993-94)	228
WSK-PZL SWIDNIK (see Wytwornia Sprzetu Komunikacyjnego Im.Zygmunta Pulawskiego-PZL Swidnik)	(1992-93)	181
Waco Classic F5 (Classic)	(1992-93)	371
WALLIS AUTOGYROS (UK)	(1994-95)	459
WASHINGTON AEROPROGRESS (USA)	(1997-98)	770
Werewolf, Ka 50 (Kamov)	(1994-95)	320
WESTERN PACIFIC AVIATION CORPORATION (Philippines)	(1996-97)	302
WHITE LIGHTNING AIRCRAFT CORPORATION (USA)	(1996-97)	752
Wilga 35 PZL-104 (PZL Warszawa-Okecie)	(1999-2000)	332
Wilga 80 PZL-104 (PZL Warszawa-Okecie)	(1999-2000)	332
Wilga 80-550 (Melex)	(1994-95)	465
Wing lockers (Raisbeck)	(1991-92)	465
WIPAIRE (USA)	(1994-95)	655
Wizard (Airborne Innovations)	(2000-01)	561
Wolverine T-411 (Khrunichev)	(1997-98)	394
Wolverine T-411 (Washington Aeroprogress)	(1997-98)	770
World Airlifter (Lockheed Martin)	(1997-98)	658
Wren AS 202/18A (FFA)	(1991-92)	221
WTA INC (USA)	(1991-92)	497
WYPOSAZEN AGROLOTNICZCH (Poland)	(1995-96)	268
WYTWORNIA SPRZETU KOMUNIKACY JNEGO (Poland)	(1999-2000)	319

X

Entry	Edition	Page
X4 (Robin)	(1997-98)	129
X-30 (NASA)	(1993-94)	540, 698
X-31A EFM (Rockwell/DASA)	(1996-97)	227
X-32 (Lockheed Martin)	(1999-2000)	681
X-36 (McDonnell Douglas)	(1997-98)	677
X 380 DTP Dauphin (Eurocopter)	(1995-96)	180
XJ-10 (CAC)	(1995-96)	51
XM-4 (Moller)	(1993-94)	537
XR Learjet (Dee Howard)	(1991-92)	396
XST (Northrop)	(1993-94)	542
XST Have Blue (Lockheed)	(1992-93)	402
XXJ (CAC)	(1998-99)	57
XIAMEN LIGHT AIRCRAFT COMPANY (China)	(1998-99)	66

Y

Entry	Edition	Page
Y7H (XAC)	(2000-01)	86
Yak-36M (Yakovlev)	(1993-94)	331
Yak-38 (Yakovlev)	(1993-94)	331
Yak-40TL (Yakovlev)	(1994-95)	385
Yak-41 (Yakovlev)	(1992-93)	257
Yak-42M (Yakovlev)	(1993-94)	333
Yak-44 (Yakovlev)	(1992-93)	259
Yak-44E (Yakovlev)	(1994-95)	387
Yak-46 propfan (Yakovlev)	(1993-94)	333
Yak-46 turbofan (Yakovlev)	(1993-94)	334
Yak-46-1 (Yakovlev)	(1997-98)	473
Yak-46-2 (Yakovlev)	(1996-97)	459
Yak-48 (Yakovlev)	(1996-97)	459
Yak-52 (Condor)	(1991-92)	203
Yak-55M (Yakovlev)	(1998-99)	456
Yak-56 (Yakovlev)	(1992-93)	260
Yak-77 (Yakovlev)	(1996-97)	460
Yak-141 Freestyle (Yakovlev)	(1993-94)	336
Yak-142 (Yakovlev)	(1997-98)	472
Yak-144 (Yakovlev)	(1993-94)	337
Yak-242 (Yakovlev)	(1997-98)	472
Yak-UTS (Yak 130) (Yakovlev)	(1993-94)	337
YF-23A (Northrop)	(1991-92)	457
YS-11EA (Nippon)	(1994-95)	237
YS-X (JADC)	(1995-96)	271
YSX-75B (JADC)	(1992-93)	158
YAKOVLEV DESIGN BUREAU (USSR)	(1991-92)	293, 756
Yamal (AST)	(2000-01)	356
Yavasa TG-XI (TAI)	(1998-99)	487
Yastreb T-422 (Khrunichev)	(1997-98)	396
Yasur 2000 (IAI)	(1994-95)	206
Yun-5 (SAMC)	(1995-96)	63
Yun-7 (XAC)	(1995-96)	65
Yun-8 (SAC)	(1995-96)	60
Yun-11B (HAMC)	(1995-96)	53
Yun-12 (II) (HAMC)	(1995-96)	54
Yunshuji-11B (I)(HAMC)	(1998-99)	60

Z

Entry	Edition	Page
Z-8 (CHAIC)	(1998-99)	57
Z-9 Haitun (HAMC/Eurocopter)	(1994-95)	54
Z 37T-2 Agro Turbo (Zlin)	(1996-97)	85
Z-50 (Zlin)	(1993-94)	70
Z 90 (Zlin)	(1993-94)	71
Z 137T Agro Turbo (Zlin)	(1996-97)	85
Z-142 (Zlin)	(1997-98)	93
ZLAC (see Zhongmeng Light Aircraft Corporation)	(1997-98)	80
Zafar 300 (Seyedo Shohada)	(1992-93)	139
ZAITSEV (Russian Federation)	(2000-01)	484
Zephyr M.J. 52 (Jurca)	(1993-94)	86
Zenith CH 200 (Zenair)	(1999-2000)	56
Zenith CH 250 (Zenair)	(1994-95)	43
Zhi-8 (CHAIC)	(1995-96)	51
Zhi-9 (HAMC)	(1994-95)	54
Zhi-9A Haitun (HAMC)	(1995-96)	53
Zhi-11 (CHAIC)	(1995-96)	52
Zhishengji-8 (CHAIC)	(1998-99)	57
Zhishengji-9 (HAMC)	(1994-95)	54
ZHONGMENG LIGHT AIRCRAFT CORPORATION (China)	(1997-98)	80
Zlin 50 (Zlin)	(1993-94)	70
Zlin 90 (Zlin)	(1993-94)	71
Zlin 137T Agro Turbo (Zlin)	(1996-97)	85
Zodiac CH 601 (Zenair)	(1994-95)	43
Zodiac CH 601 HD (Derazona)	(1997-98)	165
Zodiac CH 601 (UL) (Zenith)	(1997-98)	770
ZRZESZINIE WYTWORCOW SPRZETU LOTNICZEGO I SILNIKOWEGO PZL (Poland)	(1991-92)	188

PRIVATE AIRCRAFT

(Incorporating Sport, Homebuilt and Microlight Aircraft)

Private aircraft retained in the book are now listed in the main Aircraft index. Those listed below appeared in the 1992-93 and preceding edition, but aircraft and manufacturers now covered in the main Aircraft index are indicated thus*

A

A-1 (CCMI)	(1992-93)	554
A-2-S Barracuda (Hobbystica)	(1992-93)	559
A-2-2 Barracuda (Hobbystica)	(1992-93)	559
A-10 Silver Eagle (Mitchell Wing)	(1991-92)	599
AA-200 (Aerodis)	(1992-93)	498
ABC (Dimanchev/Valkanov)	(1992-93)	464
AG-38A Terrier (Mitchell Wing)	(1991-92)	560
AG-38-A War Eagle (Mitchell Wing)	(1991-92)	560
A. I. R.'s SA (see Air International Services)	(1992-93)	555
AMF-S14 Miranda (Falconar)	(1992-93)	467
AMF-S14H Miranda (Falconar)	(1992-93)	552
ARV-1K Golden Hawk (Falconar)	(1992-93)	552
ASL (see Club d'Avions Super Légers du Cher)	(1992-93)	555
ASL 18 La Guehe (ASL)	(1992-93)	555
ATI (see Advanced Technologies)	(1991-92)	540
ATOL (Martenko)	(1991-92)	511
ATTL-1 (Euronef)	(1992-93)	464
AUI (see Aviazione Ultraleggera Italiana)	(1992-93)	558
ACE AIRCRAFT COMPANY (USA)	(1992-93)	494
ACES HIGH LIGHT AIRCRAFT (Canada)	(1992-93)	551
Acroduster 1 SA-700 (Aerovant)	(1992-93)	499
Acroduster Too SA-750 (Stolp)	(1992-93)	539
Acro Sport I (Acro Sport)	(1992-93)	494
Acro Sport II (Acro Sport)	(1992-93)	494
ACRO SPORT INC (USA)	(1992-93)	494
Acro-Zenith CH 150 (Zenair)	(1992-93)	472
ADVANCED AVIATION INC (USA)	(1992-93) 495,	562
ADVANCED TECHNOLOGIES (USA)	(1991-92)	540
Adventurer (Adventure)	(1992-93)	496
Adventurer 2 + 2 (Adventure)	(1992-93)	496
Adventurer 4 Place (Adventure)	(1992-93)	496
AERMAS (see Automobiles Martini)	(1992-93)	554
Aermas 386 (Aermas)	(1992-93)	554
Aerobatic-Speedwings (Avid Aircraft)	(1992-93)	504
AEROCAR INC (USA)	(1992-93)	496
AERO COMPOSITE TECHNOLOGY INC (USA)	(1992-93)	497
AERO DESIGNS INC (USA)	(1992-93)	498
AERODIS AMERICA INC (USA)	(1992-93)	498
AERODIS, SARL (France)	(1991-92)	512
AEROLITES INC (USA)	(1992-93)	498
AERO MIRAGE (USA)	(1991-92)	535
AERONAUTIC 2000 (France)	(1991-92)	587
AEROSERVICE ITALIANA SRL (Italy)	(1992-93)	558
AERO-SOL (USA)	(1992-93)	562
AEROSPORT LTD (USA)	(1992-93)	499
AERO-TECH INDUSTRIES INC (USA)	(1992-93)	499
AEROTECH INTERNATIONAL LTD (UK)	(1991-92)	593
AEROVANT AIRCRAFT CORPORATION (USA)	(1992-93)	499
AERO VISIONS INTERNATIONAL (USA)	(1991-92)	536
Ag Bearcat (Aerolites)	(1992-93)	498
Ag-Hawk (CGS Aviation)	(1992-93)	563
AgStar (Pampa's Bull)	(1992-93)	550
*Airaile S-12 (Rans)		
*Airaile S-14 (Rans)		
AIR AND SPACE (USA)	(1992-93)	500
Air Arrow (Mountain Valley)	(1991-92)	506
Aircamper GN-1 (Grega)	(1992-93)	518
Airco DH2 (Redfern)	(1992-93)	532
AIR COMMAND INTERNATIONAL (USA)	(1992-93) 500,	562
AIR COMPOSITE (France)	(1992-93)	474
AIRCRAFT DESIGNS INC (USA)	(1992-93)	500
AIRCRAFT DEVELOPMENT INC (USA)	(1992-93)	562
AIRCRAFT DYNAMICS INC (USA)	(1992-93)	501
AIR CREATION (France)	(1992-93)	554
AIR INTERNATIONAL SERVICE SA (France)	(1992-93)	555
Air Master (Aerolights)	(1992-93)	498
AIR NOSTALGIA (USA)	(1992-93)	502
Airplume (Croses, Yves)	(1992-93)	555
AIR SERVICE (Hungary)	(1992-93)	557
AIRTECH CANADA (Canada)	(1991-92)	585
AKAFLIEG MÜNCHEN (see Flugtechnische Forschungsgruppe an der Technischen Universität München)	(1991-92)	522
Albatros (Vulcan)	(1992-93)	557
Albatros DVa (Ryder)	(1992-93)	534
ALIFERRARI (Italy)	(1992-93)	558
Allegro (A. I. R's SA)	(1992-93)	555
ALPHA (Poland)	(1992-93)	488
ALPHA AVIATION SUPPLY CO (USA)	(1992-93)	502
ALTURAIR (USA)	(1992-93)	502
American (American Air Jet)	(1991-92)	538
AMERICAN AIRCRAFT INC (USA)	(1992-93)	563
AMERICAN AIR JET INC (USA)	(1991-92)	538
AMF MICROFLIGHT LTD (UK)	(1992-93)	490
ANDREASSON, BJÖRN (Sweden)	(1992-93)	489
ANGLIN ENGINEERING (USA)	(1992-93)	502
Antares SB1 (Soyer/Barritault)	(1992-93)	484
ANTONOV (Ukraine)	(1992-93)	560
Arrow (Arrow Aircraft)	(1992-93)	551
ARROW AIRCRAFT CO (Canada)	(1992-93) 465,	551
Ascender II+ (Coldfire)	(1991-92)	596
Asterix 001 (Nickel & Foucard)	(1992-93)	480
Ausra VK-8 (Kensgaila)	(1992-93)	487
AUSTRALITE INC (USA)	(1992-93)	503
Autan M. J.53 (Jurca)	(1992-93)	478
Autogyro 18A (Air and space)	(1992-93)	500
AUTOMATED AIRCRAFT TOOLING (USA)	(1992-93)	503
AUTOMOBILES MARTINI SA (France)	(1992-93)	554
AviaStar (Pampa's Bull)	(1992-93)	550
AVIATECHNICA, CO-OPERATIVE DESIGN OFFICE (Russia)	(1992-93)	488
AVIATIKA LTD (Russia)	(1992-93)	489
*AVIAT INC (USA)		
AVIAZIONE ULTRALEGGERA ITALIANA (Italy)	(1992-93)	558
Avid Aerobat (Avid Aircraft)	(1992-93)	504
*AVID AIRCRAFT INC (USA)		
Avid Amphibian (Avid Aircraft)	(1992-93)	504
Avid Commuter (Avid Aircraft)	(1992-93)	504
Avid Flyer (Avid Aircraft)	(1991-92)	540
*Avid Flyer Mark IV (Avid Aircraft)		
Avid Landplane (Avid Aircraft)	(1992-93)	504
Avion (Ken Brock)	(1991-92)	596
AVIONS JODEL SA (France)	(1992-93)	476

B

B1-RD (Powers Bashforth)	(1992-93)	568
B1-RD Two-seater (Powers Bashforth)	(1992-93)	528
B-10 (Mitchell Wing)	(1991-92)	600
B-532 King Cobra B (Advanced Aviation)	(1992-93)	495
BA-4B (Andreasson)	(1992-93)	489
BA-12 Slandan (Andreasson, MFI)	(1992-93)	560
BAC (see Brutsche Aircraft Corporation)	(1992-93)	563
BD-4 (Bede)	(1992-93)	504
BD-5 (Alturair)	(1992-93)	502
BF-3 (CUP)	(1992-93)	558
BF-10 (Bebe)	(1992-93)	505
BLI-KEA (Hinz)	(1991-92)	522
BOAC (see Barney Oldfield Aircraft Company) (USA)	(1992-93)	505
BW-20 Giro Kopter (BW Rotor)	(1992-93)	563
BX-2 Cherry (Brändli)	(1992-93)	490
BX-200 (Bristol)	(1991-92)	543
Baby Ace Model D (Ace Aircraft)	(1992-93)	494
'Baby' Lakes (BOAC)	(1992-93)	505
Balade ST 80 (Stern)	(1992-93)	484
Baladin (Paumier)	(1992-93)	481
Baladin NP2 (Parent)	(1992-93)	481
Balerit (Mignet)	(1992-93)	556
*B & F TECHNIK VERTRIEBS GmbH (Germany)		
Bandit (Solar Wings)	(1992-93)	561
Banty (Butterfly Aero)	(1992-93)	563
BARNETT ROTORCRAFT (USA)	(1992-93)	504
BARNEY OLDFIELD AIRCRAFT COMPANY (USA)	(1992-93)	505
Baroudeur (Aéronautic 2000)	(1991-92)	587
Barracuda (Barracuda)	(1992-93)	504
BARRACUDA (see Bueth Enterprises Inc)	(1992-93)	504
Barracuda A-2-2 (Hobbystica)	(1992-93)	559
Barracuda A-2-S (Hobbystica)	(1992-93)	559
Bausalz (LO-Fluggeratebau)	(1992-93)	557
Bearcat (Aerolites)	(1992-93)	498
BEAVER RX ENTERPRISES LTD (Canada)	(1992-93)	552
Bébé Series (Jodel)	(1992-93)	476
BEDE-MICRO AVIATION INC (USA)	(1992-93)	505
BEIJING UNIVERSITY OF AERONAUTICS AND ASTRONAUTICS (China)	(1992-93)	553
BELIOVKIN (Latvia)	(1992-93)	559
*Berkut (Renaissance)		
BERNARD BROC ENTERPRISE (France)	(1992-93)	555
Beryl (Piel)	(1992-93)	482
Beta Bird (Hovey)	(1991-92)	598
Bifly (Lascaud)	(1992-93)	556
BIGUÁ (Argentina)	(1991-92)	502
Biguá amphibian (Biguá)	(1991-92)	502
Bird TA-2/3 (Taylor)	(1992-93)	541
Blue Max (Blue Max)	(1992-93)	550
BLUE MAX ULTRALIGHT (Australia)	(1992-93)	550
Blue Sky-3 (Korean Air)	(1992-93)	559
Bobcat (First Strike)	(1992-93)	514
Boondocker (For Two Enterprises)	(1992-93)	565
Boondocker 42 (For Two Enterprises)	(1992-93)	565
BRÄNDLI, MAX (Switzerland)	(1992-93)	490
BRISTOL, URIEL (Bristol)	(1991-92)	543
BROC ENTERPRISE, BERNARD (France)	(1992-93)	555
Brok-1M Garnys (LAK)	(1992-93)	559
BROKAW AVIATION INC (USA)	(1992-93)	507
BRÜGGER MAX (Switzerland)	(1992-93)	490
BRUTSCHE AIRCRAFT CORPORATION (USA)	(1992-93)	563
Buccaneer SX (Advanced Aviation)	(1992-93)	496
Buccaneer II, XA/650 (Advanced Aviation)	(1992-93)	496
Buckeye (Coldfire)	(1991-92)	597
Buckeye B (Coldfire)	(1992-93)	564
Buckeye 503B (Coldfire)	(1992-93)	564
'Buddy Baby' Lakes (BOAC)	(1992-93)	506
BUETH ENTERPRISES, WB, INC (USA)	(1992-93)	504
Bullet (Brokaw)	(1992-93)	507
Bumble Bee (Aircraft Designs)	(1992-93)	500
Bumble Camel (Smith, Ronald)	(1992-93)	553
BUSHBY AIRCRAFT INC (USA)	(1992-93)	507
Bushmaster I Mk2 (Snowbird)	(1992-93)	553
Butterfly (Butterfly Aero)	(1992-93)	563
BUTTERFLY AERO (USA)	(1992-93)	563
BW ROTOR COMPANY INC (USA)	(1992-93)	563

C

C22 (Ikarus)	(1992-93)	556
C22 HP (Ikarus)	(1992-93)	556
CA-61/-61R Mini Ace (Cvjetkovic)	(1992-93)	510
CA-65 (Cvjetkovic)	(1992-93)	510
CA-65A (Cvjetkovic)	(1992-93)	510
CB-1 Biplane (Hatz) (Kelly)	(1992-93)	520
CCMI (see Central China Mechanical Institute)	(1992-93)	554
*CFM (see Cook Flying|Machines)		
CH 150 (Zenair)	(1992-93)	472
CH 180 (Zenair)	(1992-93)	472
CH 600 (Zenair)	(1991-92)	510
*CH 601HD (Zenair)		
*CH 601HDS (Zenair)		
*CH 701 STOL (Zenair)		
*CH 701-AG STOL (Zenair)		
*CH 801 Super STOL (Zenair)	(1992-93)	474
CJ-3D Cracker Jack (Wood Wing)	(1992-93)	548
C. P. 70/750/751 Beryl (Piel)	(1992-93)	482
C. P. 90 Pinocchio (Piel)	(1992-93)	482
C. P. 150 Onyx (Piel)	(1992-93)	482
C. P. 300 series, Emeraude (Piel)	(1992-93)	481
C. P. 600 series, Super Diamant (Piel)	(1992-93)	482
C. P. 1320 (Piel)	(1992-93)	482
CUP (see Centro Ultraleggeri Partenopeo)	(1992-93)	558
Cabin Starduster (Stolp)	(1992-93)	540
CABRINHA ENGINEERING INC (USA)	(1992-93)	508
Cadi Z001 (Composite Aircraft)	(1992-93)	510
CALAIR CORPORATION (Australia)	(1991-92)	502
CANADIAN ULTRALIGHT MANUFACTURING LTD (Canada)	(1992-93)	552
Canelas Ranger (Martin Uhia)	(1992-93)	560
Capella (Flightworks)	(1992-93)	516
Capella XS (Flightworks)	(1992-93)	516
CARLSON AIRCRAFT INC (USA)	(1992-93) 508,	563
Carrera (Advanced Aviation)	(1992-93)	496
Carrera 182 (Advanced Aviation)	(1992-93)	496
CASSAGNERES E (USA)	(1992-93)	508
Cassutt IIIM (National)	(1992-93)	527
Cavalier SA 102.5 (MacFam)	(1992-93)	524
C_CON GmbH (Germany)	(1992-93)	556
Cekady (Cowan)	(1991-92)	504
Celebrity (Fisher Aero)	(1992-93)	516
Celerity (Mirage Aircraft)	(1992-93)	526
CENTRAIR SN (France)	(1992-93)	555

C – G—PRIVATE AIRCRAFT: INDEXES 863

CENTRAL CHINA MECHANICAL INSTITUTE (China) (1992-93) 554
CENTRO ULTRALEGGERI PARTENOPEO (Italy) (1992-93) 558
CESLOVAS (Lithuania) (1992-93) 559
CGS AVIATION INC (USA) (1992-93) 563
Challenger (Quad City) (1992-93) 568
Challenger II (Quad City) (1992-93) 568
Challenger II Special (Quad City) (1992-93) 530
Challenger Special (Quad City) (1992-93) 530
Chaos S-9 (Rans) (1992-93) 532
Charger MA-5 (Marquart) (1992-93) 526
Chaser S (Mc^2) (1992-93) 567
CHASLE, YVES (France) (1992-93) 555
Cherry BX-2 (Brändli) (1992-93) 490
*Chevvron 2-32C (AMF)
Chickinox (Dynali SA) (1992-93) 551
Chinook 1-S (Canadian Ultralight) (1992-93) 552
Chinook 2-S plus 2 (Canadian Ultralight) (1992-93) 552
CHUCK'S AIRCRAFT COMPANY (USA) (1992-93) 564
CICARÉ, AUGUSTO (Argentina) (1991-92) 502
CIRCA REPRODUCTIONS (Canada) (1992-93) 466
*CIRRUS DESIGN CORPORATION (USA)
Citabriette N3-2C (Mosler) (1991-92) 561
Classic (Cosy) (1992-93) 485
Classic (Fisher) (1992-93) 515
CLASSIC AIRCRAFT REPLICAS INC (USA) (1992-93) 509
Clipper II (Sunrise) (1992-93) 540
Clipper Super Sport (Sunrise) (1992-93) 571
Clipper Ultralight (Sunrise) (1992-93) 571
Cloud Dancer (US Aviation) (1992-93) 572
Club (Jodel) (1992-93) 476
CLUB D'AVIONS SUPERS LEGERS DU CHER (France) (1992-93) 555
CMI (Iraq) (1991-92) 522
Coach (Ehroflug) (1991-92) 589
Coach IIS (Ehroflug) (1992-93) 560
Cobra (Advanced Aviation) (1992-93) 495
Cobra (Cobra Helicopters) (1992-93) 509
Cobra B (Advanced Aviation) (1992-93) 562
COBRA HELICOPTERS (USA) (1992-93) 509
Cobra Mk 1 (Romain & Sons) (1992-93) 561
COLDFIRE SYSTEMS INC (USA) (1992-93) 564
Colibri 2 MB-2 (Brügger) (1992-93) 490
COLLINS AERO (USA) (1992-93) 509
COLOMBAN, MICHEL (France) (1992-93) 475
Commander 447 (Air Command) (1992-93) 562
Commander 500 (Air Command) (1992-93) 500
Commander 503 (Air Command) (1992-93) 500
Commander Elite 532 (Air Command) (1992-93) 500
Commander Elite 532 Two-seater (Air Command) (1992-93) 500
Commander Elite Sport 532 (Air Command) (1991-92) 536
Commander Elite Sport 582 (Air Command) (1992-93) 500
Commander Sport 447 (Air Command) (1992-93) 562
Commander Sport 503 (Air Command) (1992-93) 500
Commander Tandem 1000 (Air Command) (1992-93) 500
Commuter II-A (Cobra) (1992-93) 509
COMPOSITE AIRCRAFT DESIGN INC (USA) (1992-93) 510
COMPOSITE CONSTRUCTIONS GmbH (Germany) (1991-92) 589
CONDOR (UFM) (1992-93) 488
Convertible (Firebolt) (1992-93) 514
*COOK FLYING MACHINES (UK) (Now CFM)
CO-OPERATIVE DESIGN OFFICE AVIATECHNICA (Russia) (1992-93) 488
Cormorano (Aliferrari) (1992-93) 558
Corsair F4U (Air Nostalgia) (1992-93) 502
COSMOS SARL (France) (1992-93) 555
COSY EUROPE (Germany) (1992-93) 485
Cougar Model 1 (Acro Sport) (1992-93) 494
COUPE, AVIONS, JACQUES (France) (1992-93) 475
*Courier S-7 (Rans)
COWAN, PETER (Canada) (1991-92) 504
Coyote II S-6 (Rans) (1991-92) 601
Coyote II S-6ES (Rans) (1992-93) 569
Coyote S-4 (Rans) (1992-93) 569
Co Z DEVELOPMENT CORPORATION (USA) (1992-93) 510
Cozy Mk IV (Co Z) (1992-93) 510
Cracker Jack CJ-3D (Wood Wing) (1992-93) 548
Craft 200 (Craft Aerotech) (1992-93) 510
CRAFT AEROTECH (USA) (1992-93) 510
Cricri MC 15 (Colomban) (1992-93) 475
Criquet EC-6 (Croses) (1992-93) 476
Cropstar (Mainair Sport) (1992-93) 561
CROSES, EMILIEN (France) (1992-93) 476
CROSES, YVES (France) (1992-93) 555
Cross-Country (Polaris) (1992-93) 559
Cruiser (Falconar) (1992-93) 468
Crystall (Kuibyshev) (1992-93) 560
CSN (USA) (1992-93) 510
Cubmajor (Falconar) (1992-93) 467
Cuby I (Aces High) (1992-93) 551
Cuby I (Magal) (1991-92) 586
Cuby II (Aces High) (1992-93) 551
Cuby II (Magal) (1991-92) 586
Culex (Fisher Aero) (1992-93) 515
Culite (Aero Visions) (1991-92) 536
CUSTOM FLIGHT COMPONENTS (Canada) (1992-93) 466
CVJETKOVIC, ANTON (USA) (1992-93) 510
Cyclone V300 (Windryder) (1992-93) 548
Cygnus 21P (Gyro 2000) (1992-93) 518
Cygnus 21TX (Gyro 2000) (1992-93) 518

D

D 2 INC (USA) (1992-93) 510
D.9 Bébé Series (Jodel) (1992-93) 476
D.11 (Jodel) (1992-93) 476
D.18 (Jodel) (1992-93) 476
D.19 (Jodel) (1992-93) 477
D.112 Club (Jodel) (1992-93) 476
D.113 (Jodel) (1992-93) 476
D.119 (Jodel) (1992-93) 476
D.150 Mascaret (Jodel) (1992-93) 477
D-201 (D'Apuzzo) (1992-93) 511
DA-2A (Davis) (D2) (1992-93) 510
DA-2B (DZ) (1992-93) 510
DA-5A (Davis) (D2) (1992-93) 511
DA-7 (Davis) (1992-93) 512
DK-1 Der Kricket (FLSZ) (1992-93) 516

Daki 530Z/Ikenga (Gyro 2000) (1992-93) 518
D'APUZZO, NICHOLAS E (USA) (1992-93) 511
Dart M. J.3 (Jurca) (1992-93) 478
DAVIS, LEEON D (USA) (1992-93) 512
DAVIS WING LTD (USA) (1992-93) 512
DEDALUS Srl (Italy) (1992-93) 486
Delta (Nordpunkt) (1992-93) 490
Delta Agro (Beliovkin) (1992-93) 559
Delta Bird (Hovey) (1991-92) 598
Delta Dart II (C-Con) (1992-93) 556
Delta Hawk (Hovey) (1991-92) 598
Delta JD-2 (Dyke Aircraft) (1992-93) 512
Demoiselle (Ultra Efficient Products) (1992-93) 572
DENNEY AEROCRAFT COMPANY (USA) (1992-93) 512
Der Kricket, DK-1 (FLSZ) (1992-93) 516
DIEHL AERO-NAUTICAL (USA) (1991-92) 597
DIMANCHEV, GEORGI/VALKANOV, VESELIN (Bulgaria) (1992-93) 464
Dipper W-7 (Collins Aero) (1992-93) 509
Discovery (Progress) (1992-93) 529
Don Quixote J-1B (Alpha) (1992-93) 502
DORNA, H. F. CO (Iran) (1992-93) 486
Double Eagle TU-10 (Mitchell Aerospace) (1991-92) 600
*Dragonfly (Viking)
Drifter/ARV Series (Maxair) (1992-93) 526
DURAND ASSOCIATES INC (USA) (1992-93) 512
Durand Mk V (Durand) (1992-93) 512
DYKE AIRCRAFT (USA) (1992-93) 512
DYNALI SA (Belgium) (1992-93) 551

E

EAC-3 Pouplume (Croses) (1991-92) 513
EC-6 Criquet (Croses) (1992-93) 476
EC-8 Tourisme (Croses) (1991-92) 514
E-Racer (Sierra Delta) (1992-93) 535
EZ-1 (Aircraft Development) (1992-93) 562
E. Z Bird (Gyro 2000) (1992-93) 566

Eagle 2-PLC (Aeroservice Italiana) (1992-93) 558
Eagle II (Aviat) (1992-93) 503
EAGLE PERFORMANCE (USA) (1992-93) 564
Eagle XL (Aeroservice Italiana) (1992-93) 558
Eagle XL Country (Aeroservice Italiana) (1992-93) 558
EARLY BIRD AIRCRAFT COMPANY (USA) (1992-93) 564
EARTHSTAR AIRCRAFT INC (USA) (1992-93) 513, 564
EASTWOOD AIRCRAFT PTY LTD (Australia) (1992-93) 550
EICH, JAMES P (USA) (1991-92) 549
Emeraude (Piel) (1992-93) 481
EPA Costruzioni Aeronautiche (Italy) (1992-93) 558
Epervier (Epervier) (1992-93) 551
Epervier ARV (Epervier)
E Sporty (Falconar) (1992-93) 467
ETS NION AÉRONAUTIQUES (France) (1992-93) 556
EURONEF SA (Belgium) (1992-93) 464
Europlane (Stern) (1992-93) 484
EVANS AIRCRAFT (USA) (1992-93) 513
Excelsior (St Croix) (1992-93) 570
Explorer (Advanced Aviation) (1992-93) 496
*Express (Wheeler)
E. Z. Bird (Gyro 2000) (1992-93) 518

F

F-1B (Polaris) (1992-93) 559
F2-Foxcat (Aeroservice Italiana) (1992-93) 558
F2-Foxcat 2 PLC (Aeroservice Italiana) (1992-93) 558
F3 (CUP) (1992-93) 558
F4U Corsair (Air Nostalgia) (1992-93) 502
*F.8L Falco (Sequoia)
F9 (Falconar) (1992-93) 467
F10 (Falconar) (1992-93) 467
F11A (Falconar) (1992-93) 467
F12 (Falconar) (1992-93) 468
FA-1 (Fulton) (1991-92) 528
*FK9 (Funk)
FK9-503 (B & F Technik) (1992-93) 556
FLSZ (see Flight Level Six-Zero) (1992-93) 516
FP-101 (Fisher) (1992-93) 565
FP-202 Koala (Fisher) (1992-93) 565
FP-303 (Fisher) (1992-93) 565
FP-404 (Fisher) (1992-93) 515
FP-404 Classic (Fisher) (1992-93) 565
FP-404 Classic II (Fisher) (1992-93) 565
FP-505 Skeeter (Fisher) (1992-93) 565
FP-606 Sky Baby (Fisher) (1992-93) 565

*Falco F.8L (Sequoia)
Falcon (American Aircraft) (1992-93) 563
Falcon B.1 (SCWAL) (1992-93) 551
Falcon B.2 (SCWAL) (1992-93) 551
FALCONAR AVIATION (Canada) (1992-93) 467, 552
Farmate (SV Aircraft) (1992-93) 550
FIGHTER ESCORT WINGS (USA) (1992-93) 514
FIRE BOLT (USA) (1992-93) 514
Firebolt Convertible (Starfire) (1991-92) 572
Firestar KX (Kolb) (1992-93) 566
FIRST STRIKE AVIATION INC (USA) (1992-93) 514
FISHER AERO CORPORATION (USA) (1992-93) 515
FISHER FLYING PRODUCTS INC (USA) (1992-93) 514, 565
Flash-3 (Wolff) (1991-92) 524
FLIGHT LEVEL SIX-ZERO INC (USA) (1992-93) 516
FlightStar (Pampa's Bull) (1991-92) 583
FLIGHTWORKS CORPORATION (USA) (1992-93) 516
Flitzer Z-1 (Sky Craft) (1992-93) 492
Floater (Beaver) (1992-93) 552
FLOW STATE DESIGN (USA) (1992-93) 565
FLUGTECHNISCHE FORSCHUNGSGRUPPE an der TECHNISCHEN UNIVERSITÄT MÜNCHEN (Germany) (1991-92) 522
Flycycle (Kennedy Aircraft) (1991-92) 555
Flying Flea (Falconar Aviation) (1992-93) 468, 552
FOKKER DR 1 (Redfern) (1992-93) 532
FOR TWO ENTERPRISES (USA) (1992-93) 565
FOTON (USSR) (1991-92) 527
Fourtouna M. J.14 (Jurca) (1992-93) 478
Freedom 28 (BAC) (1992-93) 563
Freedom Machine KB-2 (Brock) (1992-93) 506
Free Spirit Mark II (Cabrinha) (1992-93) 508
FULTON AIRCRAFT LTD (UK) (1991-92) 528
Fun Fly (Swiss Aerolight) (1992-93) 560
Fun GT 447 (Air Creation) (1992-93) 554
FUNK, OTTO and PETER (Germany) (1991-92) 589
Fun Racer 447 (Air Creation) (1992-93) 555
Fury (Two Wings) (1992-93) 571

G

G-802 Orion (Aerodis) (1991-92) 512
GN-1 Aircamper (Grega) (1992-93) 518
GP3 Osprey II (Osprey Aircraft) (1992-93) 527
GP4 (Osprey Aircraft) (1992-93) 528
GT1000 GT (Rotary Air Force) (1992-93) 470
GW-1 Windwagon (Watson) (1992-93) 546

Gambit (SciCraft) (1992-93) 486
Gemini Flash 2 Alpha (Mainair Sports) (1992-93) 561
GEMINI INTERNATIONAL INC (USA) (1991-92) 597
GENERAL GLIDERS (Italy) (1992-93) 558
Gerfan (Air Composite) (1992-93) 474
Giro Kopter (BW Rotar) (1992-93) 563
*Glasair II-S (Stoddard-Hamilton)
*Glasair III (Stoddard-Hamilton)
*Glastar (Stoddard-Hamilton)
*Gnatsum M. J.7, M. J.77 (Jurca)
GOLDEN CIRCLE AIR (USA) (1992-93) 565
Golden Hawk (Falconar) (1992-93) 467, 552
Golden Interceptor (Eagle Performance) (1992-93) 564
Gotz 50 (Heinrich Zettle) (1992-93) 557
Graflite (Lundy) (1991-92) 558
*GREAT PLAINS AIRCRAFT SUPPLY CO INC (USA)
GREGA, JOHN W (USA) (1992-93) 518
GROEN BROTHERS AVIATION INC (USA) (1992-93) 518
Gryf ULMI (Tepelne) (1992-93) 554
Guppy SNS-2 (Sorrell) (1992-93) 536
Gypsy Trainer (R. D. Aircraft) (1992-93) 569
GYRO 2000 INC (USA) (1992-93) 518, 566
Gyrocopter (Wombat) (1992-93) 493
Gyroplane, JE-2 (Eich) (1991-92) 549

864 INDEXES: PRIVATE AIRCRAFT—**H - M**

H

H-3 Pegasus (Howland) .... (1992-93) 566
HA-2M Sportster (Aircraft Designs) .... (1992-93) 500
HM-290E Flying Flea (Falconar) .... (1992-93) 552
HM-293 (Falconar) .... (1992-93) 552
HM-360 (Falconar) .... (1992-93) 552
HM-380 (Falconar) .... (1992-93) 552
HM-1000 Balerit (Mignet) .... (1992-93) 556
HN 433 Menestral (Nicollier) .... (1992-93) 480
HN 434 Super Menestral (Nicollier) .... (1992-93) 480
HN 600 Week-End (Nicollier) .... (1992-93) 481
HN 700 Menestrel II (Nicollier) .... (1992-93) 480

HAPI ENGINES INC (USA) .... (1991-92) 553
Hawk (Kestrel) .... (1992-93) 468
Hawk Arrow (CGS Aviation) .... (1992-93) 564
Hawk II Arrow (CGS Aviation) .... (1992-93) 564
Hawk Classic (CGS Aviation) .... (1992-93) 564
Hawk II Classic (CGS Aviation) .... (1992-93) 564
Hawk Series (Groen) .... (1992-93) 518
HEADBERG AVIATION INC (USA) .... (1992-93) 519
Helicopter (BW Rotor) .... (1992-93) 563
HELICRAFT INC (USA) .... (1992-93) 519, 566
HFL-FLUGZEUGBAU GmbH (Germany) .... (1992-93) 556
High Gross STOL (Avid Aircraft) .... (1992-93) 504
Hi-Max (T. E. A. M.) .... (1992-93) 571
HINZ, LUCIA AND BERNHARD (Germany) .... (1991-92) 522
Hiperbipe SNS-7 (Sorrell) .... (1992-93) 536
Hiperlight EXB (Sorrell) .... (1992-93) 570
Hiperlight SNS-8 (Sorrell) .... (1992-93) 570
HIPP'S SUPERBIRDS INC (USA) .... (1992-93) 520, 566
Hirondelle (Chasle) .... (1992-93) 555
Hirondelle, PGK-1 (Western Aircraft) .... (1992-93) 472
HISTORICAL AIRCRAFT CORPORATION (USA) .... (1992-93) 520
Hitchiker (Labahan) .... (1991-92) 583
HLAMOT DESIGN (Hungary) .... (1992-93) 558
Hlamot-M (Hlamot Design) .... (1992-93) 558
HOBBYSTICA AVIO (Italy) .... (1992-93) 559
HOLCOMB, JERRY (USA) .... (1991-92) 554
Hollman Bumble Bee (Aircraft Designs) .... (1992-93) 500
Hollman Sportster (Aircraft Designs) .... (1992-93) 500
Honcho II (Magnum) .... (1992-93) 567
Horizon (Aero Visions) .... (1991-92) 536
Horizon 1 (Fisher Aero) .... (1992-93) 516
Horizon 2 (Fisher Aero) .... (1992-93) 516
HORNET MICROLIGHTS LTD (UK) .... (1991-92) 594
Hornet R (Hornet) .... (1991-92) 594
Hornet RS (Hornet) .... (1991-92) 594
Hornet RSE (Hornet) .... (1991-92) 594
HOVEY, ROBERT W. (USA) .... (1991-92) 598
HOWLAND AERO DESIGNS (USA) .... (1992-93) 566
Hummel Bird (Morry Hummel) .... (1992-93) 566
*Hummingbird (VAT)
Hummingbird 103 (Gemini) .... (1991-92) 597
Hummingbird Longwing (Gemini) .... (1991-92) 597
Hurricane (Windryder) .... (1992-93) 548
Hurricane 100 (Windryder) .... (1992-93) 548
HUSKY MANUFACTURING LTD (Canada) .... (1991-92) 505
HUTCHINSON AIRCRAFT COMPANY
Hybred R (Medway Microlights) .... (1992-93) 561
Hydroplum (Tisserand/Hydroplum) .... (1992-93) 476
Hydroplum II/Petrel (Tisserand/
HYDROPLUM SARL (France) .... (1992-93) 476

I

I-66L San Francesco (Iannotta) .... (1992-93) 486
IPE-06 (Industria Paranaense) .... (1991-92) 585

IANNOTTA DOTTING ORLANDO (Italy) .... (1992-93) 486
Ibis RJ-03 (Junqua) .... (1992-93) 477
Ikenga/Cygnus 21P (Gyro 2000) .... (1992-93) 518
Ikenga/Cygnus 21TX (Gyro 2000) .... (1992-93) 518
Intruder SNS-10 (Sorrell) .... (1992-93) 570
Invader Mk III-B (Ultra Efficient Products) .... (1992-93) 572

J

J-1B Don Quixote (Alpha) .... (1992-93) 502
J-2 Polonez (Janowski) .... (1992-93) 559
J-3 Kitten (Hipp's Superbirds) .... (1992-93) 566
J-4 Sportster (Hipp's Superbirds) .... (1992-93) 566
J 4B (Barnett Rotorcraft) .... (1992-93) 504
J 4B-2 (Barnett Rotorcraft) .... (1992-93) 504
J-5 Marco (Alpha) .... (1992-93) 488
J-6 Karatoo (Skyway Aircraft) .... (1991-92) 570
JB-1000 (Leader's International) .... (1992-93) 566
JC-01 (Coupé-Aviation) .... (1992-93) 475
JC-3 (Coupé-Aviation) .... (1992-93) 476
JC-200 (Coupé-Aviation) .... (1992-93) 476
JD-2 Delta (Dyke Aircraft) .... (1992-93) 512
JE-2 Gyroplane (Eich) .... (1991-92) 549
JN-1 (Jim Peris) .... (1992-93) 567
J. T.1 Monoplane (Taylor) .... (1992-93) 492
J. T.2 Titch (Taylor) .... (1992-93) 493

JANOWSKI, JAROSLOW (Poland) .... (1992-93) 559
JAVELIN AIRCRAFT COMPANY INC (USA) .... (1992-93) 520
Jenny (Early Bird) .... (1992-93) 564
JODEL, AVIONS SA (France) .... (1992-93) 476
Jr Ace, Pober (Acro Sport) .... (1992-93) 495
JULIAN, CHRIS (UK) .... (1991-92) 529
Junior Ace Model E (Ace Aircraft) .... (1992-93) 494
JUNQUA, ROGER (France) .... (1992-93) 477
Junster I (MacFam) .... (1992-93) 524
Junster II (MacFam) .... (1992-93) 524
*JURCA, MARCEL (France)

K

K-236 (Klampfl) .... (1992-93) 464
KB-2 Freedom Machine (Brock) .... (1992-93) 506
KB-3 (Ken Brock) .... (1992-93) 563
KR-1 (Rand Robinson) .... (1992-93) 531
KR-2 (Rand Robinson) .... (1992-93) 531
KR-100 (Rand Robinson) .... (1992-93) 531
KSML-IM Nyirseg-2 (Air Service) .... (1992-93) 558
KSML-P1 Nemere (Air Service) .... (1992-93) 557
KSML-P2 Nyirseg (Air Service) .... (1992-93) 557

Karatoo J-6 (Skyway Aircraft) .... (1991-92) 570
*Katana T30 (Terzi)
Katta (Yakovlev) .... (1991-92) 528
Kelly-D (Kelly) .... (1992-93) 520
KELLY, DUDLEY R (USA) .... (1992-93) 520
KEN BROCK MANUFACTURING INC (USA) .... (1992-93) 506, 563
KENNEDY AIRCRAFT COMPANY INC (USA) .... (1991-92) 555
KENSGAILA'S AIRCRAFT ENTERPRISE, V (Lithuania) .... (1992-93) 487
Kestrel (Helicraft) .... (1992-93) 566
KESTREL SPORT AVIATION (Canada) .... (1992-93) 468
KIMBERLEY, GARETH J. (Australia) .... (1992-93) 550
KIMBREL, MICHAEL G. (USA) .... (1991-92) 598
King Cobra B (Advanced Aviation) .... (1992-93) 495
Kis (Tri-R) .... (1992-93) 542
Kitfox III (Denney Aerocraft) .... (1991-92) 548
Kitfox IV (Denny Aerocraft) .... (1992-93) 512
Kit Hawk (Kestrel) .... (1992-93) 468
KLAMPFL (Austria) .... (1992-93) 464
Koala FP-202 (Fisher) .... (1992-93) 565
Kodiak (Sequoia) .... (1991-92) 569
KOLB AIRCRAFT INC (USA) .... (1992-93) 521, 566
KONSUPROD GmbH & Co KG (Germany) .... (1992-93) 557
Kopykat (Kopykat) .... (1992-93) 566
KOPYKAT (USA) .... (1992-93) 566
KOREAN AIR (South Korea) .... (1992-93) 559
KUIBYSHEV, PETER ALMURZIN (Russia) . (1992-93) 560

L

L5 (Lucas) .... (1992-93) 479
L6 (Lucas) .... (1992-93) 480
L7 (Lucas) .... (1992-93) 480
L10 (Lucas) .... (1992-93) 480
LA-4 Vortex (Langebro) .... (1992-93) 490
LACO-125 (Laven) .... (1991-92) 556
LACO-145 (Laven) .... (1991-92) 556
LACO-145A Special (Laven) .... (1992-93) 522
LAK (see Litovskaya Aviatsionnaya Konstruktsiya) .... (1991-92) 593
LK-2 Sluka (Letov) .... (1992-93) 554
LK-3 Nova (Letov) .... (1992-93) 554
LM-1 (Light Miniature) .... (1992-93) 522
LM-1U (Light Miniature) .... (1992-93) 522
LM-1X (Light Miniature) .... (1992-93) 522
LM-2P (Light Miniature) .... (1991-92) 556
LM-2U (Light Miniature) .... (1992-93) 523
LM-2X (Light Miniature) .... (1992-93) 523
LM-3U (Light Miniature) .... (1992-93) 523
LM-3X (Light Miniature) .... (1992-93) 523
LM-5X (Classic) .... (1992-93) 509
LO-120 (LO-Fluggeratebau) .... (1992-93) 557
LO-120 (Vogt) .... (1991-92) 590
LO-125 (LO-Fluggeratebau) .... (1992-93) 557
LO-150 (Vogt) .... (1991-92) 590
LP-01 Sibylle (Lendepergt) .... (1992-93) 479

LABAHAN, ROBERT (Australia) .... (1991-92) 583
La Guepe (ASL) .... (1992-93) 555
*Lancair IV (Lancair International)
Lancair 200 (Neico Aviation) .... (1991-92) 562
*Lancair 320 (Lancair International)
*Lancair 360 (Lancair International)
*LANCAIR INTERNATIONAL INC (USA)
Lancer MA-4 (Marquart) .... (1992-93) 525
LANGEBRO AVIATION (Sweden) .... (1992-93) 490
LASCAUD, ETS D (France) .... (1992-93) 556
Laser (Kolb) .... (1992-93) 522
LAVEN, JOE (USA) .... (1992-93) 522
LEADER'S INTERNATIONAL INC (USA) .... (1992-93) 566
LEADING EDGE AIR FOILS INC (USA) .... (1991-92) 567
Lederlin 380-L (Lederlin) .... (1992-93) 478
LEDERLIN, FRANÇOIS (France) .... (1992-93) 478
Legato (Automated Aircraft) .... (1992-93) 503
LEICHTFLUGZEUG GmbH & Co KG (Germany) .... (1992-93) 557
LENDEPERGT, PATRICK (France) .... (1992-93) 479
Leone T7 (AUI) .... (1992-93) 558
LETOV LTD (Czechoslovakia) .... (1992-93) 554
LIGETI AERONAUTICAL PTY LTD (Australia) .... (1991-92) 583
LIGHT MINIATURE AIRCRAFT INC (USA) .... (1992-93) 522
Lightning Bug (Lightning Bug Aircraft) .... (1992-93) 523
LIGHTNING BUG AIRCRAFT CORPORATION (USA) .... (1992-93) 523
Lightning P-38 (Mitchell Wing) .... (1991-92) 560
Lightning Sport Copter (Vancraft) .... (1992-93) 572
LITOVSKAYA AVIATSIONNAYA KONSTRUKTSIYA (USSR) .... (1991-92) 593
Little Dipper replica (R. D. Aircraft) .... (1991-92) 568
LOEHLE ENTERPRISES (USA) .... (1992-93) 524, 567
LO-FLUGGERATEBAU GmbH (Germany) .... (1992-93) 557
Lomac Trick (Polaris) .... (1991-92) 592
Lone Ranger (Striplin) .... (1992-93) 540
Lone Ranger Ultralight (Striplin) .... (1992-93) 571
LoneStar (Star Aviation) .... (1991-92) 572
Long Wing Super Pup (Preceptor) .... (1992-93) 529
LUCAS, EMILE (France) .... (1992-93) 479
Lucky (T. E. D. A.) .... (1991-92) 592
LUNDY, BRIAN (USA) .... (1991-92) 558

M

M-1 Midget Mustang (Bushby) .... (1992-93) 507
M-1 Slavutitch (Antonov) .... (1992-93) 560
M-1A Midget Mustang (Bushby) .... (1992-93) 507
M20 Venture (Questair) .... (1992-93) 530
M-80 (Morin) .... (1992-93) 480
MA-4 Lancer (Marquart) .... (1992-93) 525
MA-5 Charger (Marquart) .... (1992-93) 526
MAI-89 (Foton) .... (1991-92) 527
MAI-890 (Aviatika) .... (1992-93) 489
MB-2 Colibri 2 (Brügger) .... (1992-93) 490
MC 15 Cricri (Colomban) .... (1992-93) 475
MEG-2XH (Helicraft) .... (1992-93) 566
MFI (see Malmo Forsknings & Innovations AB) .... (1992-93) 560
MFI-9 HB (Andreasson) .... (1992-93) 490
MFS (see Microlight Flight Systems Pty
M. J.2 Tempete (Jurca) .... (1992-93) 477
M. J.3 Dart (Jurca) .... (1992-93) 478
M. J.4 Shadow (Jurca) .... (1992-93) 478
M. J.5 Sirocco Sport Wing (Jurca) .... (1992-93) 478
M. J.7 Gnatsum (Jurca) .... (1992-93) 478
M. J.8 (Jurca) .... (1992-93) 478
M. J.9 One-Oh-Nine (Jurca) .... (1992-93) 478
M. J.14 Fourtouna (Jurca) .... (1992-93) 478
M. J.22 Tempete (Jurca) .... (1992-93) 477
M. J.51 Sperocco (Jurca) .... (1992-93) 478
M. J.53 Autan (Jurca) .... (1992-93) 478
M. J.70 (Jurca) .... (1992-93) 478
M. J.77 Gnatsum (Jurca) .... (1992-93) 478
M. J.80 One-Nine-Oh (Jurca) .... (1992-93) 478
M. J.90 (Jurca) .... (1992-93) 478
MW5 Sorcerer (Whittaker) .... (1992-93) 562
MW5K Sorcerer (Aerotech) .... (1991-92) 593
MW6 Merlin (Whittaker) .... (1992-93) 562
MW7 (Whittaker) .... (1992-93) 562
MX Sport (Quicksilver) .... (1992-93) 569
MX Sprint (Quicksilver) .... (1992-93) 569

MACAIR INDUSTRIES INC (Canada) .... (1992-93) 468
MACFAM (USA) .... (1992-93) 524
Maestro (A. I. R's SA) .... (1992-93) 555
MAGAL HOLDINGS LTD (Canada) .... (1991-92) 586
Magnum V8 (O'Niél) .... (1992-93) 527
MAGNUM INDUSTRIES (USA) .... (1992-93) 525, 567
MAINAIR SPORTS LTD (UK) .... (1992-93) 561
Majorette (Falconar) .... (1992-93) 467
MALCOLM AIRPLANE COMPANY (USA) .... (1992-93) 567
MALMO FORSKNINGS & INNOVATIONS AB (Sweden) .... (1992-93) 560
Mangoos (Flow State Design) .... (1992-93) 565
Manx (Moshier Technologies) .... (1991-92) 561
Maranda (Falconar) .... (1992-93) 467, 552
Marco J-5 (Alpha) .... (1992-93) 488
MARKGRAFLICH BADISCHE VERWALTUNG (Germany) .... (1992-93) 557
Mariah, T-100D (Turner) .... (1992-93) 571
Mariner (Two Wings) .... (1992-93) 572
MARQUART, ED (USA) .... (1992-93) 525
MARTENKO FINLAND OY (Finland) .... (1992-93) 474
MARTIN UHIA (Spain) .... (1992-93) 560

M – S—PRIVATE AIRCRAFT: INDEXES 865

Mascaret D.150 (Jodel) .... *(1992-93)* 477
MAXAIR AIRCRAFT CORPORATION (USA) .... *(1992-93)* 526
Maya (Percy) .... *(1992-93)* 550
Mc² (MICROLIGHTS) (USA) .... *(1992-93)* 567
Medvegalis (Yuodinas) .... *(1992-93)* 559
MEDWAY MICROLIGHTS (UK) .... *(1992-93)* 561
MEM REPULOGEPES SZOLGALAT
Menestrel II HN 700 (Nicollier) .... *(1992-93)* 480
Menestrel HN 433 (Nicollier) .... *(1992-93)* 480
Merlin (Macair) .... *(1992-93)* 468
Merlin GT (Malcolm Airplane) .... *(1992-93)* 567
MICROFLIGHT AIRCRAFT LTD (UK) .... *(1991-92)* 594
Micro-Imp (Aerocar) .... *(1992-93)* 497
MICROLIGHT FLIGHT SYSTEMS PTY LTD (South Midget Mustang (Bushby) .... *(1992-93)* 507
Mifeng-II (Beijing University) .... *(1992-93)* 554
Mifeng-2 (Beijing University) .... *(1992-93)* 553
Mifeng-3 (Beijing University) .... *(1992-93)* 553
Mifeng-3C (Beijing University) .... *(1992-93)* 553
Mifeng-4 and 4A (Beijing University) .... *(1992-93)* 553
Mifeng-5 (Beijing University) .... *(1992-93)* 553
*Mini-500 (Revolution Helicopter)
Mini Ace (Cvjetkovic) .... *(1992-93)* 510
Minihawk (Falconar) .... *(1992-93)* 467
Mini-Imp (Aerocar) .... *(1992-93)* 496
Mini Master (Powers Bashforth) .... *(1992-93)* 528
MiniMAX (T. E. A. M.) .... *(1992-93)* 571
MIRAGE AIRCRAFT INC (USA) .... *(1992-93)* 526
Mistral 53 (Ikarusflug) .... *(1992-93)* 556
Mistral 462 (Aviasud) .... *(1992-93)* 474
Mistral 532 (Aviasud) .... *(1992-93)* 474
MITCHELL AEROSPACE COMPANY (USA) .... *(1992-93)* 526
MITCHELL WING (see Mitchell Aerospace Company) .... *(1992-93)* 526
Model 302 Kodiak (Sequoia) .... *(1991-92)* 569
Model 380 ATTL-1 (Euronef) .... *(1992-93)* 464
Model C (Cosmos Sarl) .... *(1992-93)* 555
Model D, Baby Ace (Ace Aircraft) .... *(1992-93)* 494
Model E, Junior Ace (Ace Aircraft) .... *(1992-93)* 494
Model S-12-E (Spencer) .... *(1992-93)* 536
Model 'S' Sidewinder (Smyth) .... *(1992-93)* 535
Model W-8 Tailwind (Wittman) .... *(1992-93)* 548
Model W-10 Tailwind (Wittman) .... *(1992-93)* 548
Monoplane, J. T.1 (Taylor) .... *(1992-93)* 492
Mono Z-CH 100 (Zenair) .... *(1992-93)* 472
MONTGOMERIE GYROCOPTERS, JIM (UK) .... *(1992-93)* 492
MORIN, A (France) .... *(1992-93)* 480
MORREY HUMMEL (USA) .... *(1992-93)* 566
Morrisey 2000 (Morrisey) .... *(1992-93)* 527
MORRISEY BILL (USA) .... *(1992-93)* 527
MOSHIER TECHNOLOGIES (USA) .... *(1991-92)* 561
MOSLER AIRFRAMES & POWER PLANTS (USA) .... *(1991-92)* 561, 600
MOUNTAIN VALLEY AIR (Canada) .... *(1991-92)* 506
MUP (Russia) .... *(1992-93)* 560
Mup-02M (Mup) .... *(1992-93)* 560
*MURPHY AVIATION LTD (Canada)
Mustang (Cobra Helicopters) .... *(1992-93)* 509
Mustang 5151 and 5151RG (Loehle) .... *(1992-93)* 524
Mustang P-51D (Falconar) .... *(1992-93)* 468
Mustang P-51D/TF (Fighter Escort) .... *(1992-93)* 514
MUSTANG HELICOPTER COMPANY (USA) .... *(1992-93)* 527

N

N3 Pup (Preceptor) .... *(1992-93)* 529, 568
N3-2 Pup (Preceptor) .... *(1992-93)* 529
N3-2C Citabriette .... *(1992-93)* 529
N3-C (Mosler) .... *(1991-92)* 561
N3-C (Preceptor) .... *(1992-93)* 529
NP2 Baladin (Parent) .... *(1992-93)* 481

NATIONAL AERONAUTICS & MANUFACTURING COMPANY (USA) .... *(1992-93)* 527
NEICO AVIATION INC (USA) .... *(1991-92)* 562
Nemere (Air Service) .... *(1992-93)* 557
NESSA AIRCRAFT COMPANY (Canada) .... *(1992-93)* 470
Nessall (Nessa Aircraft) .... *(1992-93)* 470
NICKEL & FOUCARD, RUDY, JOSEPH (France) .... *(1992-93)* 480
NICOLLIER, AVIONS H (France) .... *(1992-93)* 480
Nieuport 11 (Circa Reproductions) .... *(1992-93)* 466
Nieuport 11 (Leading Edge Air Foils) .... *(1992-93)* 567
Nieuport 12 (Circa Reproductions) .... *(1992-93)* 466
Nieuport 12 (Leading Edge Air Foils) .... *(1992-93)* 567
Nieuport 17/24 (Redfern) .... *(1992-93)* 532
Nifty (Morrisey) .... *(1992-93)* 527
NIKE AERONAUTICA SRL (Italy) .... *(1992-93)* 486
NION (see Ets Nion Aéronautiques) .... *(1992-93)* 556
NIPPER KITS AND COMPONENTS LTD (UK) .... *(1992-93)* 492
Nipper Mk IIIb (Nipper) .... *(1992-93)* 492
NOBLE HARDMAN AVIATION LTD
Nomad II (Magnum) .... *(1992-93)* 567
NORDPUNKT AG (Switzerland) .... *(1992-93)* 490
Norseman (Husky) .... *(1991-92)* 505
Nova (Aircraft Designs) .... *(1991-92)* 538
Nova (Letov) .... *(1992-93)* 554
Nuwaco T-10 (Aircraft Dynamics) .... *(1992-93)* 501
Nyirseg (Air Service) .... *(1992-93)* 557

O

Observer (Swiss Aerolight) .... *(1992-93)* 560
One-Nine-Oh (Jurca) .... *(1992-93)* 478
One-Oh-Nine M. J.9 (Jurca) .... *(1992-93)* 478
O'NIEL AIRPLANE COMPANY (USA) .... *(1992-93)* 527
Onyx C. P. 150 (Piel) .... *(1992-93)* 482
Optimum 88 (Optimum) .... *(1992-93)* 551
OPTIMUM AIRCRAFT GROUP (Bulgaria) .... *(1992-93)* 551
Orion G-802 (Aerodis) .... *(1991-92)* 512
Orlik RO-7 (Orlinski) .... *(1992-93)* 488
ORLINSKI, ROMAN (Poland) .... *(1992-93)* 488
Osprey II GP3 (Osprey Aircraft) .... *(1992-93)* 527
OSPREY AIRCRAFT (USA) .... *(1992-93)* 527

P

P-9 Pixie (Acro Sport) .... *(1992-93)* 494
P-38 Lightning (Mitchell Wing) .... *(1991-92)* 560
P-40 (Loehle) .... *(1992-93)* 524
P-51D Mustang (Falconar) .... *(1992-93)* 468
P.60 Minacro (Pottier) .... *(1992-93)* 482
P.70S (Pottier) .... *(1992-93)* 482
P.170S (Pottier) .... *(1992-93)* 483
P.180S (Pottier) .... *(1992-93)* 483
P.200 Series (Pottier) .... *(1992-93)* 483
PFA (see Popular Flying Association) .... *(1992-93)* 492
PGK-1 Hirondelle (Western Aircraft) .... *(1992-93)* 472
PL.2 (Pazmany) .... *(1992-93)* 528
PL.4A (Pazmany) .... *(1992-93)* 528
PL-9 Stork (Pazmany) .... *(1992-93)* 528
PM-2 (Paraplane) .... *(1991-92)* 567
PT-2 Sassy (ProTech Aircraft) .... *(1991-92)* 564
PT-2B Prostar (ProTech Aircraft) .... *(1992-93)* 530
PUL 9 (Nike) .... *(1992-93)* 486
PZL P.11c (Historical Aircraft) .... *(1992-93)* 520

Pago Jet (Pago Jet) .... *(1991-92)* 600
PAGO JET (USA) .... *(1991-92)* 600
PAMPA'S BULL SA (Argentina) .... *(1992-93)* 463, 550
Panther (Rotec) .... *(1992-93)* 569
Panther Plus (Rotec) .... *(1992-93)* 534
Panther 2 Plus (Rotec) .... *(1992-93)* 534
Papillon (Broc) .... *(1992-93)* 555
Parafan (Centrair) .... *(1991-92)* 555
Parakeet (Rose) .... *(1992-93)* 534
Paraplane (Markgraflich) .... *(1992-93)* 557
PARAPLANE CORPORATION (USA) .... *(1991-92)* 567
Paraplane PM-2 (Paraplane) .... *(1991-92)* 562
Parascender I (Powercraft) .... *(1992-93)* 568
Parascender II (Powercraft) .... *(1992-93)* 568
PARENT, NORBERT (France) .... *(1992-93)* 481
PARSONS, BILL (USA) .... *(1991-92)* 562
PAUMIER, NORBERT (France) .... *(1992-93)* 481
PAZMANY AIRCRAFT CORPORATION (USA) .... *(1992-93)* 528
Pegasus Q (Pegasus Transport) .... *(1991-92)* 594
Pegasus Q 462 (Solar Wings) .... *(1992-93)* 561
Pegasus Quastar (Solar Wings) .... *(1992-93)* 561
PEGASUS TRANSPORT SYSTEMS LTD (UK) .... *(1991-92)* 594
Pegasus XL (Solar Wings) .... *(1992-93)* 561
Pegasus XL-R (Pegasus Transport) .... *(1991-92)* 594
Pelican Club GS, PL and S (Ultravia) .... *(1992-93)* 471
Pelican Club UL (Ultravia) .... *(1992-93)* 553
Penetrater (Ultra Efficient Products) .... *(1992-93)* 572
PERCY, GRAHAM J (Australia) .... *(1992-93)* 550
Perigee (Holcomb) .... *(1991-92)* 554
PERIS, JIM (USA) .... *(1992-93)* 567
Phantom (Phantom Sport) .... *(1992-93)* 567
Phantom F11 (Phantom Sport) .... *(1992-93)* 568
PHANTOM SPORT PLANES INC (USA) .... *(1992-93)* 567
PHÖNIX-AVIATECHNICA (Russia) .... *(1992-93)* 489
PIEL, AVIONS CLAUDE (France) .... *(1992-93)* 481
PIETENPOL, BUCKEYE ASSOCIATION (USA) .... *(1992-93)* 528
PINAIRE ENGINEERING INC (USA) .... *(1992-93)* 568
Pinocchio C. P. 90 (Piel) .... *(1992-93)* 482
Pixie, P-9 Pober (Acro Sport) .... *(1992-93)* 494
POLARIS SRL (Italy) .... *(1992-93)* 559
Polonez J-2 (Janowski) .... *(1992-93)* 559
Poppy (Dedalus) .... *(1992-93)* 486
POPULAR FLYING ASSOCIATION, THE (UK) .... *(1992-93)* 492
POTTIER, JEAN (France) .... *(1992-93)* 482
Pouplume EAC-3 (Croses) .... *(1991-92)* 513
POWERCRAFT CORPORATION (USA) .... *(1992-93)* 568
POWERS BASHFORTH AIRCRAFT CORPORATION (USA) .... *(1992-93)* 528, 568
PRECEPTOR AIRCRAFT (USA) .... *(1992-93)* 529, 568
Predator (Mustang Helicopter) .... *(1992-93)* 527
PROGRESS AERO INC .... *(1992-93)* 529
Prop Kopter (BW Rotor) .... *(1992-93)* 563
Prostar PT-2B (ProTech Aircraft) .... *(1992-93)* 530
PROTECH AIRCRAFT INC (USA) .... *(1992-93)* 530
Prowler (Prowler Aviation) .... *(1992-93)* 530
PROWLER AVIATION LTD (USA) .... *(1992-93)* 530
Pulsar (Aero Designs) .... *(1992-93)* 498
Pup Series (Preceptor) .... *(1992-93)* 529
Pursuit S-11 (Rans) .... *(1992-93)* 532

Q

Q (Pegasus) .... *(1991-92)* 594

Qingting W-5 series (SAP) .... *(1992-93)* 554
Qingting W6 (SAP) .... *(1991-92)* 554
QUAD CITY ULTRALIGHT AIRCRAFT CORPORATION (USA) .... *(1992-93)* 530, 568
Quail (Aerosport) (D 2) .... *(1992-93)* 511
Quartz GT 462 (Air Creation) .... *(1991-92)* 588
Quartz GT 503 (Air Creation) .... *(1992-93)* 554
*QUESTAIR INC (USA)
QUICKSILVER ENTERPRISES INC (USA) .... *(1992-93)* 568
Quicksilver GT (Quicksilver) .... *(1992-93)* 568
Quicksilver GT 400 (Quicksilver) .... *(1991-92)* 601
Quicksilver GT 500 (Quicksilver) .... *(1992-93)* 569
Quicksilver MXL II Sport (Quicksilver) .... *(1992-93)* 569
Quicksilver MXL II Sprint (Quicksilver) .... *(1992-93)* 569

R

RB-2 (Roberts Ultralight) .... *(1991-92)* 602
RJ-03 Ibis (Junqua) .... *(1992-93)* 477
RO-7 Orlik (Orlinski) .... *(1992-93)* 488
RV-3 (Van's Aircraft) .... *(1992-93)* 543
*RV-4 (Van's Aircraft)
*RV-6 (Van's Aircraft)
*RV-6A (Van's Aircraft)
RX-35 (Beaver) .... *(1992-93)* 552
RX-550 (Beaver) .... *(1992-93)* 552
RX-650 (Beaver) .... *(1992-93)* 552

Rally 2B (Rotec) .... *(1992-93)* 570
Rally 3 (Rotec) .... *(1992-93)* 570
Rally Champ (Rotec) .... *(1992-93)* 570
Rally Sport (Rotec) .... *(1992-93)* 570
RAND ROBINSON ENGINEERING INC (USA) .... *(1992-93)* 531
Ranger (T. E. D. A) .... *(1991-92)* 592
Range Rider (Anglin) .... *(1992-93)* 502
*RANS INC (USA)
R. D. AIRCRAFT (USA) .... *(1992-93)* 532, 569
*Rebel (Murphy)
REDFERN WD (USA) .... *(1992-93)* 532
Reliant (Hipp's Superbirds) .... *(1992-93)* 566
Reliant II (Hipp's Superbirds) .... *(1992-93)* 520
Reliant II microlight (Hipp's Superbirds) .... *(1991-92)* 598
Reliant SX (Hipp's Superbirds) .... *(1992-93)* 520
Renegade II (Murphy)
Renegade Spirit (Murphy)
REPLICA PLANS (Canada) .... *(1992-93)* 470
REVOLUTION HELICOPTER CORPORATION INC (USA) .... *(1992-93)* 533
RICHARD WARNER AVIATION INC (USA) *(1992-93)* 546
Ritz Standard A1 (Falconar) .... *(1992-93)* 553
ROBERTS ULTRALIGHT AIRCRAFT (USA) .... *(1991-92)* 602
ROMAIN, J AND SONS (UK) .... *(1992-93)* 561
ROSE (USA) .... *(1992-93)* 534
ROTARY AIR FORCE INC (Canada) .... *(1992-93)* 470
ROTEC ENGINEERING INC (USA) .... *(1992-93)* 534, 570
RYDER INTERNATIONAL CORPORATION (USA) .... *(1992-93)* 534

S

S-4 Coyote (Rans) .... *(1992-93)* 569
S-6 Coyote II (Rans) .... *(1991-92)* 601
S-6ES Coyote II (Rans) .... *(1992-93)* 569
S-9 Chaos (Rans) .... *(1992-93)* 532
S-10 Sakota (Rans) .... *(1992-93)* 532
S-11 Pursuit (Rans) .... *(1992-93)* 532
S-12 Airaile (Rans) .... *(1992-93)* 532
S-14 Airaile (Rans) .... *(1992-93)* 532
SA 101 Super Starduster (Stolp) .... *(1992-93)* 540
SA 102.5 Cavalier (MacFam) .... *(1992-93)* 524
SA-300 Starduster Too (Stolp) .... *(1992-93)* 539
SA-500 Starlet (Stolp) .... *(1992-93)* 539
SA-700 Acroduster 1 (Aerovant) .... *(1992-93)* 499
SA-750 Acroduster Too (Stolp) .... *(1992-93)* 539
SA-900 V-Star (Stolp) .... *(1992-93)* 540
SAP (see Shijiazhuang Aircraft Plant) .... *(1991-92)* 554
SB1 Antares (Soyer/Barritault) .... *(1992-93)* 484
S. E.5A (Replica Plans) .... *(1992-93)* 470
SF-2A Cygnet (Hapi) .... *(1991-92)* 553
S. M. A. N. (see Société Morbihannaise d'Aéro Navigation) .... *(1992-93)* 484

866 INDEXES: PRIVATE AIRCRAFT—**S – V**

SNS-2 Guppy (Sorrell) (1992-93) 536
SNS-7 Hiperbipe (Sorrell) (1992-93) 536
SNS-8 Hiperlight (Sorrell) (1992-93) 570
SNS-8 Hiperlight EXP (Sorrell) (1992-93) 570
SNS-9 EXP II (Sorrell) (1992-93) 536
SNS-10 Intruder (Sorrell) (1992-93) 570
ST 80 Balade (Stern) (1992-93) 484
ST 87 Europlane (Stern) (1992-93) 484
*STOL CH 701 (Zenair)
SV-2 (Skywise) (1991-92) 583
SV 11B (SV Aircraft) (1992-93) 550
SX II Racer 503 (Air Creation) (1992-93) 555
SX GT 462 (Air Creation) (1991-92) 588
SX GT 503 (Air Creation) (1991-92) 588
SX GT 503S (Air Creation) (1992-93) 554
SX GT 582ES (Air Creation) (1992-93) 555
SX Racer (Air Creation) (1991-92) 588

Sadler Vampire (Skywise) (1991-92) 583
Saja (CMI) (1991-92) 522
Sakota S-10 (Rans) (1992-93) 532
Samolet (Aviatechnica) (1992-93) 488
San Francesco I-66L (Iannotta) (1992-93) 486
SAPHIR AMERICA (USA) (1991-92) 602
Sapphire (Winten) (1992-93) 551
Sassy, PT-2 (ProTech Aircraft) (1991-92) 564
SAURIER FLUG SERVICE (Germany) (1992-93) 557
Scamp A (Aerosport) (1992-93) 499
SCICRAFT LTD (Israel) (1992-93) 486
Scorcher (Mainair Sports) (1992-93) 561
Scout Mk III (Wheeler) (1992-93) 550
Scwal 100 (Scwal) (1992-93) 464
SCWAL SA (Belgium) (1992-93) 464, 551
SeaHawk (CGS Aviation) (1992-93) 564
Sea Hawker (Aero Composite) (1992-93) 497
Sea Panther (Rotec) (1992-93) 534
Seawind (SNA) (1992-93) 536
Sequoia Model 300 (Sequoia)
*SEQUOIA AIRCRAFT CORPORATION (USA)
Series 2000 autogyro (Rotary Air Force) (1992-93) 470
Shadow (Helicraft) (1992-93) 519
Shadow M. J.4. (Jurca) (1992-93) 478
*Shadow Series C and C-D (CFM)
*SHIJIAZHUANG AIRCRAFT PLANT (China)
Sibylle LP-01 (Lendepergt) (1992-93) 479
Sidewinder, Model S (Smyth) (1992-93) 535
SIERRA DELTA SYSTEMS (USA) (1992-93) 535
Silver Eagle II (Eagle Performance) (1992-93) 564
Silver Eagle A-10 (Mitchell Wing) (1991-92) 599
Silverwing Custom (Preceptor Aircraft) (1992-93) 568
Sirius (Nion Aeronautics) (1991-92) 589
Sirius C (Nion Aeronautiques) (1992-93) 556
Sirocco (Aviasud) (1992-93) 474
*Sirocco M. J. 5 (Jurca)
Sirocco Sport Wing M. J.5 (Jurca) (1992-93) 478
SIX-CHUTER INC (USA) (1992-93) 570
Sky Bird (BW Rotor) (1992-93) 563
Skybolt (Steen Aero Lab) (1992-93) 538
SKY CRAFT LTD (UK) (1992-93) 492
Sky Cycle (BW Rotor) (1992-93) 563
Skycycle (Kennedy Aircraft) (1991-92) 555
Skycycle (R. D. Aircraft) (1992-93) 569
Skycycle II (R. D. Aircraft) (1992-93) 569
Skycycle '87 (Carlson) (1992-93) 508
Skycycle Gipsy series (R. D. Aircraft) (1992-93) 569
Skye-Ryder (Six-Chuter) (1992-93) 570
Skyfly CA-65 (Cvjetkovic) (1992-93) 510
Skyfox (Calair) (1991-92) 502
Sky Kopter (BW Rotor) (1992-93) 563
Skylark (Airtech) (1991-92) 585
Skylark I (Helicraft) (1991-92) 566
Sky Pup (Sport Flight) (1992-93) 570
Sky Ranger (EPA) (1992-93) 558
Sky Ranger series (Striplin) (1992-93) 540
Sky-Rider (Kimberley) (1992-93) 550
Sky Scooter (Flaglor/Headberg) (1992-93) 519
Sky Sport (BW Rotor) (1992-93) 563
SKYTEK AUSTRALIA PTY LTD (Australia) (1991-92) 502
Skywalker II (Leichtflugzeug) (1992-93) 557
SKYWAY AIRCRAFT (USA) (1991-92) 570
SKYWISE ULTRAFLIGHT PTY LTD (Australia) (1991-92) 503, 583
Sky-Wolff (Wolff) (1991-92) 524
Slandan BA-12 (Andreasson/MFI) (1992-93) 560
Slavutitch M-1 (Antonov) (1992-93) 560
SMITH, RONALD M. B. (Canada) (1992-93) 553
SMITH, S & H, AIRCRAFT (USA) (1991-92) 570
SMYTH, JERRY (USA) (1992-93) 535
SNA INC (USA) (1991-92) 536
Sno-Bird (Ultralight Aircraft) (1991-92) 604
Sno Bird 503S (SnoBird) (1992-93) 570
Sno Bird 582S (SnoBird) (1992-93) 570
Sno Bird 636D (SnoBird) (1992-93) 570
Sno Bird Aircraft Inc (USA) (1992-93) 570
SNOWBIRD AEROPLANE CO. LTD, THE (UK) (1992-93) 561
SNOWBIRD AVIATION (Canada) (1992-93) 553
Snowbird Mk IV (Snowbird Aeroplane) (1992-93) 561
SOCIÉTÉ D'EXPLOITATION DES AÉRONEFS HENRI MIGNET (France) (1992-93) 556
SOCIÉTÉ MORBIHANNAISE D'AÉRO NAVIGATION (France) (1992-93) 484
SOLAR WINGS AVIATION LTD (UK) (1992-93) 561
SOLO WINGS (South Africa) (1992-93) 560
Sonerai I (Great Plains) (1992-93) 517
Sonic Spitfire (Sunrise) (1991-92) 603
Sooper-Coot Model A (Aerocar) (1992-93) 496
Sopwith Triplane (Circa Reproductions) (1992-93) 466
SORRELL AIRCRAFT COMPANY LTD (USA) (1992-93) 536
SOYER/BARRITAULT, CLAUDE/JEAN (France) (1992-93) 484
Spacewalker I (Anglin) (1992-93) 502
Spacewalker II (Anglin) (1992-93) 502
Sparrow (Carlson) (1992-93) 563
Sparrow II (Carlson) (1992-93) 508
Sparrow-ette (Carlson) (1992-93) 508
Sparrow Sport Special (Carlson) (1992-93) 508
Special (Laven) (1991-92) 556
Special (Pitts-Ultimate Aircraft) (1991-92) 508
Spectrum (Microflight Aircraft) (1991-92) 594
Spectrum (Spectrum Aircraft) (1992-93) 562
SPECTRUM AIRCRAFT (UK) (1992-93) 562
SPENCER AMPHIBIAN AIRCRAFT INC (USA) (1992-93) 536
Sperocco M. J.51 (Jurca) (1992-93) 478
Spitfire II Elite (Sunrise) (1992-93) 540, 571
Spitfire series (Sunrise) (1992-93) 571
SPORT FLIGHT ENGINEERING INC (USA) (1992-93) 570
Sport Helicopter (Star Aviation) (1992-93) 538
Sport Parasol (Loehle Aviation) (1992-93) 567
Sport Racer (Sport Racer) (1992-93) 537
SPORT RACER INC (USA) (1992-93) 537
Sportsman 2+2 (Wag-Aero) (1992-93) 546
Sportster HA-2M (Aircraft Designs) (1992-93) 500
*Sport Trainer (Wag-Aero)
Sportwing D-201 (D'Apuzzo) (1992-93) 511
Staaker Z-1 Flitzer (Sky Craft) (1992-93) 492
Stalker T-9 (Stoddard-Hamilton) (1991-92) 573
Stallion (Aircraft Designs) (1992-93) 501
STAR AVIATION INC (USA) (1992-93) 538
Starcruiser Gemini (Davis Wing) (1992-93) 512
Starduster Too SA-300 (Stolp) (1992-93) 539
STARFIRE AVIATION INC (USA) (1991-92) 572
STAR FLIGHT AIRCRAFT (USA) (1992-93) 570
Starlet (CSN) (1992-93) 510
Starlet SA-500 (Stolp) (1992-93) 539
ST CROIX AIRCRAFT (USA) (1992-93) 537, 570
STEEN AERO LAB INC (USA) (1992-93) 538
STERN, RÉNE (France) (1992-93) 484
STEWART 51 INC (USA) (1992-93) 538
STOLP STARDUSTER CORPORATION (USA) (1992-93) 539
Stork PL-9 (Pazmany) (1992-93) 528
Stratos (Ligeti Aeronautical) (1991-92) 583
Stratos 300 (HFL-Flugzeugbau) (1992-93) 556
STRIPLIN AIRCRAFT CORPORATION (USA) (1992-93) 540
Sunfun VJ-24W (Volmer) (1992-93) 572
Sunny (Tandem Aircraft) (1991-92) 590
Sunrise II (Dallach Flugzeuge) (1992-93) 556
SUNRISE ULTRALIGHT MANUFACTURING COMPANY (USA) (1992-93) 540, 571
Super Ace, Pober (Acro Sport) (1992-93) 495
Super Acro Sport (Acro Sport) (1992-93) 494
Super Acro-Zénith CH 180 (Zenair) (1992-93) 472
Super Adventurer (Adventure) (1992-93) 496
Super Adventurer 2 + 2 (Adventure) (1992-93) 496
'Super Baby' Lakes (BOAC) (1992-93) 506
Super Cat (First Strike) (1992-93) 514
Super Diamant (Piel) (1992-93) 482
Super Emeraude (Piel) (1992-93) 481
Super Hawk (Hovey) (1991-92) 598
Super Honcho (Magnum) (1992-93) 525
Super Kingfisher W-1 (Richard Warner) (1992-93) 546
Super Kitten (Hipp's Superbirds) (1992-93) 520
Super Koala (Fisher) (1992-93) 514
Super-Menestral HN 434 (Nicollier) (1992-93) 480
Super Pup (Preceptor) (1992-93) 529
Super Sportster (Hipp's Superbirds) (1992-93) 520
Super Starduster (Stolp) (1992-93) 540
Super STOL CH 801 (Zenair) (1992-93) 474
Super Zodiac CH 601HDS (Zenair) (1992-93) 473
SV AIRCRAFT PTY LTD (Australia) (1992-93) 550
SWICK AIRCRAFT (USA) (1992-93) 540
Swingwing VJ-23E (Volmer) (1992-93) 572
SWISS AEROLIGHT (Switzerland) (1992-93) 560

T

T-2 (Antonov) (1992-93) 560
T-4 (Antonov) (1992-93) 560
T7 Leone (AUI) (1992-93) 558
T-9 Stalker (Stoddard-Hamilton) (1991-92) 573
T-10 Nuwacot (Aircraft Dynamics) (1992-93) 501
T-100D Mariah (Turner) (1992-93) 571
T-110 (Turner) (1991-92) 576
T300 (Thruster Aircraft) (1992-93) 550
T500 (Thruster Aircraft) (1992-93) 550
TA-2/3 Bird (Taylor) (1992-93) 541
TA-3 (T. E. A. M.) (1991-92) 603
TA16 Trojan (Thurston) (1991-92) 576
TC-2 (Aero Mirage) (1991-92) 535
TR-200 (Tech-Aero) (1992-93) 542
TU-10 Double Eagle (Mitchell Wing) (1991-92) 600
T-Bird I (Golden Circle Air) (1992-93) 565
T-Bird II (Golden Circle Air) (1992-93) 565
T-Bird III (Golden Circle Air) (1992-93) 565
T-Bird Z2000 (Golden Circle Air) (1992-93) 566
T-Craft (Taylor Kits) (1992-93) 542
T. E. A. M. (see Tennessee Engineering and Manufacturing Inc) (1992-93) 571
TEDA (see Tecnologia Europea Divisione Aeronautica) (1991-92) 592

Tailwind W-8 (Wittman) (1992-93) 548
Tailwind W-10 (Wittman) (1992-93) 548
TAMIS MOTOR INDUSTRIES (Turkey) (1992-93) 560
TANDEM AIRCRAFT KG (Germany) (1991-92) 590
Tandem Trainer (Parsons) (1991-92) 562
Tandem Two-seater (Air Command) (1991-92) 537
TAYLOR AERO INC (USA) (1992-93) 541
TAYLOR KITS CORPORATION (USA) (1992-93) 541
TAYLOR, T (UK) (1992-93) 492
T Clip-Wing Taylorcraft (Swick) (1992-93) 540
TECH-AERO (USA) (1992-93) 542
TECNOLOGIA EUROPEA DIVISIONE AERONAUTICA Srl (Italy) (1991-92) 592
TEENIE COMPANY (USA) (1992-93) 542
Tempete (Jurca) (1992-93) 477
TENNESSEE ENGINEERING AND MANUFACTURING INC (USA) (1992-93) 571
TEPELNE ISOLACE BUDOV (Czechoslovakia) (1992-93) 554
Termite (Smith) (1991-92) 570
Terrier AG-38A (Mitchell Wing) (1991-92) 560
Texan Chuckbird (Chuck's Aircraft) (1992-93) 564
The Minimum (Saphir) (1991-92) 602
THRUSTER AIRCRAFT (AUSTRALIA) PTY LTD (Australia) (1992-93) 550
Thunder Gull J (Earthstar) (1992-93) 564
Thunder Gull JX (Earthstar) (1992-93) 513
THURSTON AEROMARINE CORPORATION (USA) (1992-93) 542
Titch J. T.2 (Taylor) (1992-93) 493
Tourisme EC-8 (Croses) (1991-92) 514
TRI-R TECHNOLOGIES (USA) (1992-93) 542
Trojan TA16 (Thurston) (1991-92) 576
Tukan (Aero-Sol) (1992-93) 562
TURNER AIRCRAFT INC (USA) (1992-93) 542, 571
TwinStar Mark III (Kolb) (1992-93) 521
Twin-T (Taylor Kits) (1992-93) 542
TWO WINGS AVIATION (USA) (1992-93) 542, 571
Tyro Mk 2 (Eastwood Aircraft) (1992-93) 550

U

U-2 Superwing (Mitchell Aerospace) (1991-92) 600
UFM (see Ultra Ligeros Franco Mexicanos) (1992-93) 488
ULM 1 Gryf (Tepelne) (1992-93) 554
ULM-1B Moskito (Konsuprod) (1992-93) 557

ULTIMATE AIRCRAFT (Canada) (1992-93) 470
Ultimate Aircraft 10 Dash 100/300 (Ultimate Aircraft) (1992-93) 470
Ultimate Aircraft 20 Dash 300 (Ultimate Aircraft) (1992-93) 471
Ultra-Aire 1 (Pinaire) (1992-93) 568
Ultrabat (Australite) (1992-93) 503
ULTRA EFFICIENT PRODUCTS INC (USA) (1992-93) 572
ULTRALEICHT-FLUGGERÄTEBAU GmbH (Germany) (1992-93) 557
ULTRA LIGEROS FRANCO MEXICANOS (Mexico) (1992-93) 488
ULTRALIGHT AIRCRAFT INC (USA) (1991-92) 604
Ultra Pup (Preceptor) (1992-93) 529, 568
ULTRAVIA AERO INC (Canada) (1992-93) 471, 553
US AVIATION (USA) (1992-93) 572

V

V-8 Special (V-8 Special) (1992-93) 542
V-8 Special SXS (V-8 Special) (1992-93) 542
VJ-23E Swingwing (Volmer) (1992-93) 572
VJ-24W Sunfun (Volmer) (1992-93) 572
VK-8 Ausra (Kensgaila) (1992-93) 487
VP-1 (Evans) (1992-93) 513
V-Star SA-900 (Stolp) (1992-93) 540

V-8 SPECIAL (USA) (1992-93) 542
Vagabond II (Saurier Flug Service) (1992-93) 557
Vancraft (Vancraft Copters) (1992-93) 543
VANCRAFT COPTERS (USA) (1992-93) 543, 572
Velocity (Velocity Aircraft) (1992-93) 544
VELOCITY AIRCRAFT (USA) (1992-93) 544
Venture M.20 (Questair) (1992-93) 530

V – Z—PRIVATE AIRCRAFT: INDEXES

*VERTICAL AVIATION TECHNOLOGIES INC (USA)
*VIKING AIRCRAFT LTD (USA)
VOGT, HELMUT (Germany) .... *(1991-92)* 590
Volksplane VP-1 (Evans) .... *(1992-93)* 513
Vortex LA-4 (Langebro) .... *(1992-93)* 490
VULCAN UL-AVIATION (Germany) .... *(1992-93)* 557

W

W-1 Super Kingfisher (Richard Warner) ...... *(1992-93)* 546
W-5 and W-5A (SAP) .... *(1992-93)* 554
W-5B (SAP) .... *(1992-93)* 554
W-6 Qingting (SAP) .... *(1991-92)* 554
W-7 Dipper (Collins Aero) .... *(1992-93)* 509
W8 Tailwind (Wittman) .... *(1992-93)* 548
W10 Tailwind (Wittman) .... *(1992-93)* 548

*Wag-a-Bond (Wag-Aero)
*WAG-AERO INC (USA)
War Eagle AG-38-A (Mitchell Wing) .... *(1991-92)* 560
WATSON WINDWAGON COMPANY
  (USA) .... *(1992-93)* 546
Weedhopper (Weedhopper) .... *(1992-93)* 572
WEEDHOPPER INC (USA) .... *(1992-93)* 572
Weedhopper Two-seater (Weedhopper) .... *(1992-93)* 572
Week-end HN 600 (Nicollier) .... *(1992-93)* 481
WESTERN AIRCRAFT SUPPLIES
  (Canada) .... *(1992-93)* 472
WEWYNE (USSR) .... *(1991-92)* 528

WHEELER AIRCRAFT (SALES) PTY LTD, RON
  (Australia) .... *(1992-93)* 550
Whing Ding II (WD-11) (Hovey) .... *(1991-92)* 598
*White Lightning (White Lightning)
*WHITE LIGHTNING AIRCRAFT CORPORATION
  (USA)
WHITTAKER, MICHAEL W. J. (UK) .... *(1992-93)* 562
Wichawk (Javelin) .... *(1992-93)* 520
WICKS AIRCRAFT SUPPLY (USA) .... *(1992-93)* 547
Wildente (Ultraleicht) .... *(1992-93)* 557
Wildente 2 (Ultraleicht) .... *(1992-93)* 557
Windlass Trike (Solo Wings) .... *(1992-93)* 560
WINDRYDER ENGINEERING INC (USA) .. *(1992-93)* 548
Windwagon (Watson) .... *(1992-93)* 546
WINTON, SCOTT (Australia) .... *(1992-93)* 551
WITTMAN, S. J. (USA) .... *(1992-93)* 548
WOLFF, ATELIERS PAUL
  (Luxembourg) .... *(1991-92)* 524
Wombat Gyrocopter (Julian) .... *(1991-92)* 529
WOMBAT GYROCOPTERS (UK) .... *(1992-93)* 493
WOOD WING SPECIALITY (USA) .... *(1992-93)* 582

X

XA/650 Buccaneer II (Advanced
  Aviation) .... *(1992-93)* 496
XC 280 Stiletto (Star Flight) .... *(1992-93)* 570
XC Series (Star Flight) .... *(1992-93)* 570

XP-Talon (Aero-Tech) .... *(1992-93)* 499
XTC Hydrolight (Diehl) .... *(1991-92)* 597

Y

YC-100 Series (Chasle) .... *(1992-93)* 555

YAKOVLEV (USSR) .... *(1991-92)* 528
Yarrow Arrow (Arrow Aircraft) .... *(1992-93)* 465
Yuca (Tamis Motor) .... *(1992-93)* 560
YUODINAS, KINTAUTAS (Lithuania) .... *(1992-93)* 559

Z

Z-1 Flitzer (Sky Craft) .... *(1992-93)* 492
Z-CH 100 Mono (Zenair) .... *(1992-93)* 472
Z-MAX (T. E. A. M.) .... *(1992-93)* 571
Zefiro 940 (General Gliders) .... *(1992-93)* 558
Zem-80 (Instytut Lotnictwa) .... *(1992-93)* 559
Zem-90 (Instytut Lotnictwa) .... *(1992-93)* 559
*Zenith CH 2000 (Zenair)
ZETTLE, HEINRICH (Germany) .... *(1992-93)* 557
Zodiac CH 600 (Zenair) .... *(1991-92)* 510
Zodiac CH 601 (Zenair)
Zodiac CH 601HD (Zenair) .... *(1992-93)* 473

SAILPLANES

Those listed below appeared in the 1992-93 and preceding edition, but powered gliders and manufacturers now covered in the main Aircraft index are indicated thus*

A

Entry	Edition	Page
AFH 24 (Akaflieg Hannover)	(1992-93)	582
AFH 26 (Akaflieg Hannover)	(1992-93)	582
AK-5 (Akaflieg Karlsruhe)	(1992-93)	582
AK-8 (Akaflieg Karlsruhe)	(1992-93)	582
ASH 25 (Schleicher)	(1992-93)	590
ASH 25 E (Schleicher)	(1992-93)	590
ASH 26 E (Schleicher)	(1992-93)	590
ASK 13 (Jubi)	(1992-93)	584
ASK 21 (Schleicher)	(1992-93)	588
ASK 23 (Schleicher)	(1992-93)	588
ASW 20 B (Schleicher)	(1992-93)	588
ASW 20 BL (Schleicher)	(1992-93)	588
ASW 20 C (Schleicher)	(1992-93)	588
ASW 20 CL (Schleicher)	(1992-93)	588
ASW 22 B (Schleicher)	(1992-93)	588
ASW 22 BE (Schleicher)	(1992-93)	588
ASW 24 (Schleicher)	(1992-93)	590
ASW 24 E (Schleicher)	(1992-93)	590
ASW 24 TOP (Schleicher)	(1992-93)	590
ASW 27 (Schleicher)	(1992-93)	590
ATS-1 Ardhra (Civil Aviation Department)	(1992-93)	594
AB-RADAB (Sweden)	(1992-93)	600
Acro IAR-35 (IAR)	(1992-93)	596
ADVANCED AVIATION INC (USA)	(1992-93)	600
AEROMOT-AERONAVES e MOTORES SA (Brazil)	(1992-93)	576
*AEROMOT-INDUSTRIA MECANICO- METALURGICA (Brazil)		
*AEROTECHNIK (Czech Republic)		
AKADEMISCHE FLIEGERGRUPPE BERLIN eV (Germany)	(1992-93)	578
AKADEMISCHE FLIEGERGRUPPE BRAUNSCHWEIG eV (Germany)	(1991-92)	614
AKADEMISCHE FLIEGERGRUPPE DARMSTADT eV (Germany)	(1991-92)	614
AKADEMISCHE FLIEGERGRUPPE HANNOVER eV (Germany)	(1992-93)	582
AKADEMISCHE FLIEGERGRUPPE KARLSRUHE eV (Germany)	(1992-93)	582
AKADEMISCHE FLIEGERGRUPPE STUTTGART eV (Germany)	(1992-93)	582
AKAFLIEG BERLIN (see Akademische Fliegergruppe Berlin eV)	(1992-93)	578
AKAFLIEG BRAUNSCHWEIG (see Akademische Fliegergruppe Braunschweig eV)	(1991-92)	614
AKAFLIEG DARMSTADT (see Akademische Fliegergruppe Darmstadt eV)	(1991-92)	614
AKAFLIEG HANNOVER (see Akademische Fliegergruppe Hannover eV)	(1992-93)	582
AKAFLIEG KARLSRUHE (see Akademische Fliegergruppe Karlsruhe eV)	(1992-93)	582
AKAFLIEG STUTTGART (see Akademische Fliegergruppe Stuttgart eV)	(1992-93)	582
ALEXANDER SCHLEICHER GmbH & CO (Germany)	(1992-93)	588
ANTONOV, OLEG K, DESIGN BUREAU (UKRAINE)	(1992-93)	600
Ardhra, ATS-1 (Civil Aviation Department)	(1992-93)	594
Atlas 21m (La Mouette)	(1992-93)	578
AUTO-AERO KÖZLEKEDÉSTECHNIKAI VALLALAT (Hungary)	(1992-93)	594
AVIONS RENÉ FOURNIER (France)	(1992-93)	578

B

Entry	Edition	Page
B-13 (Akaflieg Berlin)	(1992-93)	578
BRO-23KR Garnys (Oshkinis)	(1992-93)	600
Bakcyl PW-3 (PW)	(1992-93)	594
BURKHART GROB LUFT- UND RAUMFAHRT GmbH (Germany)	(1992-93)	582

C

Entry	Edition	Page
Carbon Dragon (Maupin)	(1992-93)	600
CELAIR MANUFACTURING AND EXPORT (PTY) LTD (South Africa)	(1992-93)	600
Celstar GA-1 (Celair)	(1992-93)	600
CENTRAIR, SA (France)	(1992-93)	578
CHENGDU AIRCRAFT CORPORATION (China)	(1991-92)	608
Chibis PPO-1 (Mikoyan)	(1992-93)	600
Chronos (Sarl La Mouette)	(1992-93)	578
CIVIL AVIATION DEPARTMENT (India)	(1992-93)	594
Club 101 (Centrair)	(1992-93)	578

D

Entry	Edition	Page
D 41 (Akaflieg Darmstadt)	(1991-92)	614
DG-300 Elan (Elan/Tovarna)	(1992-93)	602
DG-400 (Glaser-Dirks)	(1992-93)	582
DG-500 Elan (Glaser-Dirks)	(1992-93)	582
DG-600 (Glaser-Dirks)	(1992-93)	582
DG-800 (Glaser-Dirks)	(1992-93)	582
DWLKK (see Doswiadczalne) Warsztaty Lotni ze Konstrukcji Kompozytowych)	(1992-93)	594
Discus (Schempp-Hirth)	(1992-93)	584
DOKTOR FIBERGLAS (URSULA HÄNLE) (Germany)	(1992-93)	582
DOSWIADCZALNE WARSZTATY LOTNICZE KONSTRUKCJI KOMPOZYTOWYCH (Poland)	(1992-93)	594

E

Entry	Edition	Page
ESAG (see Eksprimentine Sportines Aviacijos Gamykla)	(1991-92)	632
EKAPRIMENTINE SPORTINES AVIACIJOS GAMYKLA (USSR)	(1991-92)	632
ELAN TOVARNA ŠPORTNEGAORODJA N. SOL. O (Yugoslavia)	(1992-93)	602
Elan, DG-300 (Elan Tovarna)	(1992-93)	602
Elan, DG-500 (Elan/Glaser-Dirks)	(1992-93)	582

F

Entry	Edition	Page
Fs-32 (Akaflieg Stuttgart)	(1992-93)	582
FACTORY SPORTINE (Lithuania)	(1992-93)	594
FALCONAR AVIATION LTD (Canada)	(1992-93)	576
Falke 90, SF-25C (Scheibe)	(1991-92)	620
Falke 92 SF-25C (Scheibe)	(1992-93)	584
Fauvel AV36 (Falconar)	(1992-93)	576
Fauvel AV361 (Falconar)	(1992-93)	576

G

Entry	Edition	Page
G 103C Twin III (Grob)	(1992-93)	582
G 109B (Grob)	(1992-93)	582
GA-1 Celstar (Celair)	(1992-93)	600
Gapa, PW-2 (PDWLKK)	(1992-93)	594
Gapa D PW-2D (DWLKK)	(1992-93)	594
Garnys, BRO-23KR, (Oshkinis)	(1992-93)	600
GENERAL GLIDERS (Italy)	(1992-93)	594
GLASER-DIRKS FLUGZEUGBAU GmbH (Germany)	(1992-93)	582
Gobé R-26SU (Auto-Aero)	(1992-93)	594
Gradient, VSO 10 (VSO)	(1991-92)	610
GROB (see Burkhart Grob Luft- Und Raumfahrt)	(1992-93)	582

H

Entry	Edition	Page
H 101 Salto 'Hänle' (Doktor Fiberglas)	(1992-93)	582
HK-36 Super Dimona (HOAC)	(1992-93)	576
*HK-36R Super Dimona (HOAC)		
HU-1, Seagull (Shenyang)	(1992-93)	576
*HU-2 Petrel 650B (Shenyang)		
*HOAC AUSTRIA FLUGZEUGWERK WIENER NEUSTADT (Austria)		
HAVRDA, HORST (Germany)	(1991-92)	616

I

Entry	Edition	Page
IAR-35 Acro (ICA)	(1992-93)	596
IPE 02b Nhapecam II (IPE)	(1992-93)	576
IPE-04 (IPE)	(1991-92)	608
IPE-05 Quero-Quero III (IPE)	(1992-93)	576
IPE-08 (IPE)	(1992-93)	576
IS-28B2 (IAR)	(1992-93)	596
*IS-28M2A (IAR)		
IS-29D2 (IAR)	(1992-93)	596
IS-30 (IAR)	(1991-92)	632
IS-32 (IAR)	(1991-92)	632
*IAR SA (Romania)		
IPE-INDUSTRIA PROJETOS E ESTRUTURAS AERONAUTICAS LTDA (Brazil)	(1992-93)	576
ISSOIRE-AVIATION SA (France)	(1991-92)	610

J

Entry	Edition	Page
Jantar Standard 3 (SZD)	(1992-93)	596
Janus (Schempp-Hirth)	(1992-93)	588
JASTREB FABRICA AVIONA I JEDRILICA (Yugoslavia)	(1992-93)	602
Jian Fan, X-7 (Chengdu)	(1991-92)	608
Jian Fan, X-9 (Shenyang)	(1992-93)	576
JUBI GmbH SPORTFLUGZEUGBAU (Germany)	(1992-93)	584
Junior (SZD)	(1992-93)	596
Junior 1 (Antonov)	(1992-93)	600

K

Entry	Edition	Page
KR-03A Puchatek (WSK)	(1992-93)	596
Kiwi (TWI)	(1992-93)	594
KLOTZ (Germany)	(1991-92)	616
Kuznechik PPO-2 (Mikoyan)	(1992-93)	600

L

Entry	Edition	Page
L-23 Super Blanik (LET)	(1992-93)	578
LAK-12 Lietuva (Factory Sportine)	(1992-93)	594
LAK-12 Lietuva 2R (Factory Sportine)	(1992-93)	594
LAK-16 (Esag)	(1991-92)	632
LAK-16M (Factory Sportine)	(1992-93)	594
LAK-17 (Factory Sportine)	(1992-93)	594
LS4 (Rolladen-Schneider)	(1992-93)	584
LS6 (Rolladen-Schneider)	(1992-93)	584
LS7 (Rolladen-Schneider)	(1992-93)	584
LET KONCERNOVY PODNIK (Czechoslovakia)	(1992-93)	578
Lietuva LAK-12 (Factory Sportine)	(1992-93)	594
Lietuva 2R LAK-12 (Factory Sportine)	(1992-93)	594
Lite-N-Nite (Michna)	(1991-92)	634

M

Entry	Edition	Page
Marianne (Centrair)	(1992-93)	578
MAUPIN, JIM (USA)	(1992-93)	600
MICHNA (USA)	(1991-92)	634
MIKOYAN (Russia)	(1992-93)	600
Moka 1 (Klotz)	(1991-92)	616
Monarch (Marske)	(1992-93)	600

N

Entry	Edition	Page
Nhapecan II, IPE-02b (IPE)	(1992-93)	676
Nimbus 3 (Schempp-Hirth)	(1992-93)	584
Nimbus 3D (Schempp-Hirth)	(1992-93)	584
Nimbus 4 (Schempp-Hirth)	(1992-93)	588

O

Entry	Edition	Page
OSKINIS (Russia)	(1992-93)	600

P

Entry	Edition	Page
PIK-20E2F (Issoire)	(1991-92)	610
PIK-30 (Issoire)	(1991-92)	610
PPO-1 Chibis (Mikoyan)	(1992-93)	600
PPO-2 Kuznechik (Mikoyan)	(1992-93)	600
PW-2 Gapa (PW)	(1991-92)	628
PW-2D Gapa (DWLKK)	(1992-93)	594
PW-3 Bakcyl (PW)	(1992-93)	594
PW-4 (PW)		
PW-5 (PW)	(1992-93)	596
Pégase A and B (Centrair)	(1992-93)	578
Petrel 650B (Shenyang)		
Piccolo (Technoflug)	(1991-92)	626

Piccolo B (Technoflug) .... (1992-93) 590
Pioneer II (Marske) .... (1992-93) 600
PRZEDSIEBIORSTWO DOSWIADCZALNO-
PRODUKCYJNE SZYBOWNICTWA
  (Poland) .... (1992-93) 596
Puchacz (SZD) .... (1992-93) 596
Puchatek KR-03A (WSK) .... (1992-93) 596

Q

Qian Jin, X-10 (Shenyang) .... (1992-93) 576
Quero-Quero III, IPE-05 (IPE) .... (1992-93) 576

R

R-26SU Gobé (Auto-Aero) .... (1992-93) 594
RF 5 (Fournier) .... (1992-93) 578
ROLLADEN-SCHNEIDER FLUGZEUGBAU GmbH
  (Germany) .... (1992-93) 584

S

S-1 Swift (Swift) .... (1992-93) 596
S-2A (Strojnik) .... (1992-93) 602
*S 10 (Stemme)
S10VC (Stemme) .... (1992-93) 590
SB-13 (Akaflieg Braunschweig) .... (1991-92) 614
SF-25C Falke 90 (Scheibe) .... (1991-92) 620
*SF-25C Falke 92 (Sheibe)
SF-25E Super-Falke (Scheibe) .... (1992-93) 584
SF-34B (Scheibe) .... (1992-93) 584
SGM 2-37 (Schweizer) .... (1992-93) 602
SGS 1-36 Sprite (Schweizer) .... (1992-93) 602
SGS 2-33A (Schweizer) .... (1992-93) 602
SH-2H (Havrda) .... (1991-92) 616
SSAKTB (see Specialus Sportines Avicijos Konstravimo -
  Technoginis Biuras) .... (1991-92) 632
ST-11 (Stralpes Aero) .... (1992-93) 578
ST-15 (Stralpes Aero) .... (1992-93) 578
SZD (see Przedsiebiorstow Doswiadczalno- rodukcyjne
  Szybownictwa) .... (1992-93) 596

SZD-48-3 Jantar Standard 3 (SZD) .... (1992-93) 596
SZD-50-3 Puchacz (SZD) .... (1992-93) 596
SZD-51-1 Junior (SZD) .... (1992-93) 596
SZD-55 (SZD) .... (1992-93) 596
SZD-56 (SZD) .... (1992-93) 596
SZD-59 (SZD) .... (1992-93) 596

*Salto H 101 'Hänle' (Doktor
  Fiberglas)* .... (1992-93) 582
SARL LA MOUETTE *(France)* .... (1992-93) 578
*SCHEIBE FLUGZEUGBAU GmbH (Germany)
SCHEMPP-HIRTH FLUGZEUGBAU GmbH & Co KG
  (Germany) .... (1992-93) 584
SCHLEICHER GmbH & Co, ALEXANDER
  (Germany) .... (1992-93) 588
SCHWEIZER AIRCRAFT CORPORATION
  (USA) .... (1992-93) 602
Sierra (Advanced Aviation) .... (1992-93) 600
SPECIALUS SPORTINES AVICIJOS KONSTRAVIMO -
  TECHNOLOGINIS BIURAS (USSR) .... (1991-92) 632
Sprite, SGS 1-36 (Schweizer) .... (1992-93) 602
Standard Cirrus 75-VTC (Jastreb) .... (1992-93) 602
Standard Cirrus G/81 (Jastreb) .... (1992-93) 602
*STEMME GmbH & Co KG (Germany)
STRALPES AÉRO SARL *(France)* .... (1992-93) 578
Stratus 500 (Stratus) .... (1992-93) 590
STRATUS UNTERNEHMENSBERATUNG GmbH
  (Germany) .... (1992-93) 590
STROJNIK, PROF. ALEX *(USA)* .... (1992-93) 602
Super Blanik L-23 (LET) .... (1992-93) 578
Super Dimona, HK-36 (HOAC) .... (1991-92) 608
Super Dimona HK-36R (HOAC)
Super Dimona SF-25E (Scheibe) .... (1992-93) 584
SWIFT LTD (Poland) .... (1992-93) 596
Swift S-1 (Swift) .... (1992-93) 596

T

Taifun 17E II (TWI) .... (1992-93) 590
TECHNICAL CENTRE, CIVIL AVIATION DEPARTMENT
  (India) .... (1992-93) 594
TECHNOFLUG LEICHTFLUGZEUGBAU GmbH
  (Germany) .... (1992-93) 590

TWI FLUGZEUGGESELLSCHAFT mbH
  (Germany) .... (1992-93) 590

V

VSO (see Vyvojová Skupina Orlican) .... (1991-92) 610
VSO 10 Gradient (VSO) .... (1991-92) 610
VUK-T (Jastreb) .... (1992-93) 602

Ventus (Schempp-Hirth) .... (1992-93) 588
VYVOJOVÁ SKUPINA ORLICAN
  (Czechoslovakia) .... (1991-92) 610

W

WSK (see Wytwornia Sprzetu
  Komunikacyjnego) .... (1992-93) 596

Windex 1200C (Radab) .... (1992-93) 600
Windrose (Maupin) .... (1992-93) 600
Woodstock One (Maupin) .... (1992-93) 600
WYTWORNIA SPRZETU KOMUNIKACYJNEGO
  (Poland) .... (1992-93) 596

X

X-7 Jian Fan (Chengdu) .... (1991-92) 608
X-9 Jian Fan (Shenyang) .... (1992-93) 576
X-10 Qian Jin (Shenyang) .... (1992-93) 576
XL-113 (Aerotechnik) .... (1992-93) 610

Ximango (Aeromot) .... (1992-93) 576

Z

Zefiro 940A (General Gliders) .... (1992-93) 594

HANG GLIDERS

Hang gliders are no longer listed in this book. Those listed below appeared in the 1992-93 and preceding edition

A

AOK (see Akademicki Osrodek Konstrukcyjny) .... *(1992-93)* 604

Ace/Ace RX (small) (Pegasus/Solar Wings) .... *(1992-93)* 604
Ace/Ace RX (medium) (Pegasus/Solar Wings) .... *(1992-93)* 604
Ace/Ace RX (large) (Pegasus/Solar Wings) .... *(1992-93)* 604
Ace Fun (Pegasus/Solar Wings) .... *(1992-93)* 604
Ace Sport (Pegasus/Solar Wings) .... *(1992-93)* 604
Ace Supersport (Pegasus/Solar Wings) .... *(1992-93)* 604
AEROTEC SARL (France) .... *(1992-93)* 604
Airfex (Finsterwalder) .... *(1992-93)* 604
AIRWAVE GLIDERS LTD (UK) .... *(1992-93)* 604
AKADEMICKI OSRODEK KONSTRUKCYJNY (Poland) .... *(1992-93)* 604
Alfa (Vega) .... *(1992-93)* 604
ANTONOV (see O. K. Antonov Design Bureau) .... *(1992-93)* 604
Atlas 14 (La Mouette) .... *(1992-93)* 604
Atlas 16 (La Mouette) .... *(1992-93)* 604
Atlas 18 (La Mouette) .... *(1992-93)* 604
Ayres (Polaris) .... *(1992-93)* 604

B

Bergfex (Finsterwalder) .... *(1992-93)* 604
Black Magic 22 (Airwave) .... *(1992-93)* 604
Black Magic 24 (Airwave) .... *(1992-93)* 604
Black Magic 27 (Airwave) .... *(1992-93)* 604

C

Calypso (Airwave) .... *(1992-93)* 604
Chiron 22 (Steinbach-Delta) .... *(1992-93)* 604
Chiron 25 (Steinbach-Delta) .... *(1992-93)* 604
Chiron 28 (Steinbach-Delta) .... *(1992-93)* 604
Classic (Firebird) .... *(1992-93)* 604
Cobra (La Mouette) .... *(1992-93)* 604
Comet 2B 165 (UP) .... *(1992-93)* 604
Comet 2B 185 (UP) .... *(1992-93)* 604
Compact (La Mouette) .... *(1992-93)* 604
Condor 5+ (Steinbach-Delta) .... *(1992-93)* 604

D

DELTA WING AVIATION (USA) .... *(1992-93)* 604
Divine 100 (Steinbach-Delta) .... *(1992-93)* 604
Divine 200 (Steinbach-Delta) .... *(1992-93)* 604
DRACHEN STUDIO KECUR GmbH (Germany) .... *(1992-93)* 604
Dream 165 (Delta Wing) .... *(1992-93)* 604
Dream 175 (Delta Wing) .... *(1992-93)* 604
Dream 185 (Delta Wing) .... *(1992-93)* 604

E

E Stratus Series (AOK) .... *(1992-93)* 604

Ephyr (Optimum) .... *(1992-93)* 604
Euro III (Steinbach-Delta) .... *(1992-93)* 604

F

F-14 (AOK) .... *(1992-93)* 604
FR 16 (Polaris) .... *(1992-93)* 604

Favorit (Optimum) .... *(1991-92)* 604
FINSTERWALDER GmbH (Germany) .... *(1992-93)* 604
FIREBIRD SCHWEIGER KG (Germany) .... *(1992-93)* 604
Flair (Rochelt) .... *(1992-93)* 604
Fun Air (Steinbach-Delta) .... *(1992-93)* 604
Funfex (Finsterwalder) .... *(1992-93)* 604

G

GT (Thalhofer) .... *(1992-93)* 604
GTR 148 (Moyes) .... *(1991-92)* 640
GTR 162 (Moyes) .... *(1991-92)* 640
GTR 175 (Moyes) .... *(1991-92)* 640
GTR 210 (Moyes) .... *(1991-92)* 640

Gemini M 134 (UP) .... *(1992-93)* 604
Gemini M 164 (UP) .... *(1992-93)* 604
Gemini M 184 (UP) .... *(1992-93)* 604
Genesis (Pacific Airwave) .... *(1992-93)* 604
Glidezilla 155 (UP) .... *(1992-93)* 604
Gryps (Polaris) .... *(1992-93)* 604
GUNTER ROCHELT (Germany) .... *(1992-93)* 604

H

HP AT 145 (Wills Wing) .... *(1992-93)* 604
HP AT 158 (Wills Wing) .... *(1992-93)* 604

I

INSTYTUT LOTNICTWA (Poland) .... *(1992-93)* 604

J

Joker (Thalhofer) .... *(1992-93)* 604

K

K2 (Airwave) .... *(1992-93)* 604
K3 (Airwave) .... *(1992-93)* 604

Komposit UT (Onpo Technologia) .... *(1992-93)* 604

L

LA MOUETTE, SARL (France) .... *(1992-93)* 604
Lightflex (Finsterwalder) .... *(1992-93)* 604

M

MX II (Vega) .... *(1992-93)* 604

Magic 6 (Airwave) .... *(1992-93)* 604
Magic Kiss 154 (Pacific Airwave) .... *(1992-93)* 604
Mars 150 (Moyes) .... *(1992-93)* 604
Mars 170 (Moyes) .... *(1991-92)* 604
Mars 190 (Moyes) .... *(1991-92)* 604
Minifex M2 (Finsterwalder) .... *(1992-93)* 604
Mission 170 (Moyes) .... *(1992-93)* 604
MOYES DELTA GLIDERS PTY LTD (Australia) .... *(1992-93)* 604
Mystic 155 (Delta Wing) .... *(1992-93)* 604
Mystic 166 (Delta Wing) .... *(1992-93)* 604
Mystic 177 (Delta Wing) .... *(1992-93)* 604

N

New Wave 15 (Firebird) .... *(1992-93)* 604
New Wave 16 (Firebird) .... *(1992-93)* 604
Nimbus (Swiss Aerolight) .... *(1992-93)* 604

O

O. K. ANTONOV DESIGN BUREAU (Ukraine) .... *(1992-93)* 604
ONPO TECHNOLOGIA (Russia) .... *(1992-93)* 604
Opit (Optimum) .... *(1992-93)* 604
OPTIMUM AIRCRAFT GROUP (Bulgaria) .... *(1992-93)* 604

P

PACIFIC AIRWAVE INC (USA) .... *(1992-93)* 604
Parafun (Steinbach) .... *(1992-93)* 604
Parafun 1 (Steinbach) .... *(1992-93)* 604
Para Safe (Steinbach) .... *(1992-93)* 604
PEGASUS TRANSPORT SYSTEMS LTD/ SOLAR WINGS LTD (UK) .... *(1992-93)* 604
Perfex (Finsterwalder) .... *(1992-93)* 604
POLARIS SRL (Italy) .... *(1992-93)* 604

Q

Quattro-S (Firebird) .... *(1992-93)* 604
Quattro-S Piccolo (Firebird) .... *(1992-93)* 604

R

Rapace 15 (Aerotec) .... *(1992-93)* 604
Rapace 16 (Aerotec) .... *(1992-93)* 604

ROCHELT, GÜNTER (Germany) .... *(1992-93)* 604
Rumour (Pegasus/Solar Wings) .... *(1992-93)* 604

S

SB-1 WARS (AOK) .... *(1992-93)* 604
SP (Steinbach-Delta) .... *(1992-93)* 604
SP Vario (Steinbach-Delta) .... *(1992-93)* 604

Schneid Air (Rochelt) .... *(1992-93)* 604
SEEDWINGS (USA) .... *(1992-93)* 604
Sensor 510-160 VG (Seedwings) .... *(1992-93)* 604
SKRAIDYKLES (Bulgaria) .... *(1991-92)* 639
Skyhawk 168 (Wills Wing) .... *(1991-92)* 642
Skyhawk 188 (Wills Wing) .... *(1991-92)* 642
Slavutitch Sport (Antonov) .... *(1992-93)* 604
Slavutitch-UT (Antonov) .... *(1992-93)* 604
Spectrum 144 (Wills Wing) .... *(1992-93)* 604
Spectrum 165 (Wills Wing) .... *(1992-93)* 604
Spit (Polaris) .... *(1992-93)* 604
Sport AT 150 (Wills Wing) .... *(1991-92)* 642
Sport AT 167 (Wills Wing) .... *(1991-92)* 642
Sport AT 180 (Wills Wing) .... *(1991-92)* 642
Sport American 167 (Wills Wing) .... *(1991-92)* 642
STEINBACH-DELTA FUN-AIR (Austria) .... *(1992-93)* 604
Stratus E Series (AOK) .... *(1992-93)* 604
Streak 130 (Delta Wing) .... *(1992-93)* 604
Streak 160 (Delta Wing) .... *(1992-93)* 604
Streak 180 (Delta Wing) .... *(1992-93)* 604
Sunfun, VJ-24 (Volmer) .... *(1992-93)* 604
Superfex (Finsterwalder) .... *(1992-93)* 604
Super Sport 143 (Wills Wing) .... *(1992-93)* 604
Super Sport 153 (Wills Wing) .... *(1992-93)* 604
Super Sport 163 (Wills Wing) .... *(1992-93)* 604
Swing (Thalhofer) .... *(1992-93)* 604
Swingwing, VJ-23 (Volmer) .... *(1992-93)* 604
SWISS AEROLIGHT (Switzerland) .... *(1992-93)* 604

T

THALHOFER TEAM (Germany) .... *(1992-93)* 604
Topfex (Finsterwalder) .... *(1992-93)* 604
Touring 13 (Polaris) .... *(1992-93)* 604
Touring 15 (Polaris) .... *(1992-93)* 604
Tropi 16 (Drachen) .... *(1992-93)* 604
Tropi 17 (Drachen) .... *(1992-93)* 604

U

Uno (Firebird) .... *(1992-93)* 604
Uno Jumbo (Firebird) .... *(1992-93)* 604
Uno Piccolo (Firebird) .... *(1992-93)* 604
UP INC (ULTRALITE PRODUCTS) (USA) .... *(1992-93)* 604

V

VJ-23 Swingwing (Volmer) .... *(1992-93)* 604
VJ-24 Sunfun (Volmer) .... *(1992-93)* 604

Vega 16/PR (Vega) .... *(1992-93)* 604
VEGA-DRACHENBAU PAWEL WIERZBOWSKI (Austria) .... *(1992-93)* 604
Vision Mk IV 17 (Pacific Airwave) .... *(1992-93)* 604
Vision Mk IV 19 (Pacific Airwave) .... *(1992-93)* 604
VOLMER AIRCRAFT (USA) .... *(1992-93)* 604

W

WILLS WING INC (USA) .... *(1992-93)* 604
WILLS WINGS DEUTSCHLAND (Germany) .... *(1992-93)* 604

X

XC-155 (Moyes) .... *(1992-93)* 604
XS-142 (Moyes) .... *(1992-93)* 604
XS-155 (Moyes) .... *(1992-93)* 604
XS-169 (Moyes) .... *(1992-93)* 604
XT-145 (Moyes) .... *(1992-93)* 604
XT-165 (Moyes) .... *(1992-93)* 604

Z

Z-87 (IL) .... *(1992-93)* 604
Z-90 (IL) .... *(1992-93)* 604

LIGHTER THAN AIR

To help users of this title evaluate the published data, *Jane's Information Group* has divided entries into three categories.

- **N** NEW ENTRY — Information on new equipment and/or systems appearing for the first time in the title.
- **V** VERIFIED — The editor has made a detailed examination of the entry's content and checked its relevancy and accuracy for publication in the new edition to the best of his ability.
- **U** UPDATED — During the vertification process, significant changes to content have been made to reflect the latest position known to *Jane's* at the time of publication.

Items in italics refer to entries which have been deleted from this edition with the relevant page numbers from last year.

(AIRSHIPS APPEARING IN THIS EDITION ONLY. For those in the previous ten editions, please see the following index on tinted paper)

21st CENTURY AIRSHIPS INC (Canada) .... 809 V

A

A-60 +, Lightship (ABC)	809	U
A-100, Lightship (ABC)	810	U
ABC (USA)	809	V
AERAL (France)	799	N
AEROS	811	V
AHA (USA)	812	V
AS-105 GD (GEFA-Flug)	800	U
AT-04 (ATG)	805	U
AT-10 (ATG)	806	U
ATG (UK)	805	N
Au-11 (Avgur)	803	U
Au-12 (Avgur)	803	U
AV-1 (Kubicek)	799	U
AV-2 (Kubicek)	799	U
ADVANCED HYBRID AIRCRAFT INC (USA)	812	V
ADVANCED TECHNOLOGIES GROUP (UK)	805	N
Aeros 40A (Aeros)	812	U
Aeros 40B Skydragon (Aeros)	812	U
Aeros 40C (Aeros)	812	U
Aerostatica-200 (Aerostatica)	803	U
Aerostatica-300 (Aerostatica)	803	U
AEROSTATIKA NAUCHNO-PROIZVODSTVENNAYA FIRMA (Russian Federation)	803	U
AIRSHIP TECHNOLOGIES LTD (UK)	805	V
AMERICAN BLIMP CORPORATION (USA)	809	V
ASSOCIATION POUR L'ETUDE ET LA RECHERCHE SUR LES AERONEFS ALLEGES (France)	799	N
AVGUR VOZDUKHOPLAVATELNYI TSENTR (Russian Federation)	803	U
Avea (AERAL)	799	N

B

B-150 (Vozdukh)	805	N
BKLAIC (China)	799	U
BEIJING KEYUAN LIGHT AIRCRAFT INDUSTRIAL COMPANY LTD (China)	799	U
Breitling Orbiter 3 (Cameron)	808	U

C

CL-160 (Cargolifter)	800	U
CAMERON BALLOONS LTD (UK)	807	U
Cargo Airship System (Pan Atlantic)	798	U
CARGOLIFTER AG (Germany)	799	U
CENTURY AIRSHIPS, 21st (Canada)	797	U

D

DP series (Cameron)	807	U

G

GEFA FLUG (Germany)	800	U
GSI (USA)	813	U
GESELLSCHAFT ZUR ENTWICKLUNG UND FÖRDERUNG AEROSTATISCHER FLUGSYSTEME mbH (Germany)	800	U
GLOBAL SKYSHIP INDUSTRIES INC (USA)	813	U

H

HS 110 (Lindstrand)	808	U
Hornet (AHA)	812	U

I

INTERFACE AIRSHIPS INC (USA)	814	U

J

Joey (Cargolifter)	800	U

K

KARI (Korea, South)	802	U
KOREA AEROSPACE RESEARCH INSTITUTE (Korea, South)	802	U
KUBICEK BALONY LTD (Czech Republic)	799	N

L

Lightship A-60+ (ABC)	809	U
Lightship A-100 (ABC)	810	U
Lightship/Spector Series (ABC)	810	U
LINDSTRAND BALLOONS LTD (UK)	808	U

M

MD-900 (Avgur)	804	U

N

N 07 (Zeppelin)	801	U

O

OHIO AIRSHIPS (USA)	814	N

P

PD-3000 (Avgur)	804	U

PAN ATLANTIC AEROSPACE CORPORATION (USA)	798	U

R

RA-180 (Rigidair)	803	U
RVO (Russian Federation)	805	N
RIGIDAIR (Netherlands)	802	U
Rigid Airship (AHA)	813	U
RIGID AIRSHIP DESIGN NV (Netherlands)	802	U
Rozière balloons (Cameron)	808	U
RUSSKOGO VOZDUZHOPLAVATEL 'NOGO OB SHCHESTVA (Russian Federation)	805	N

S

Skycat (ATG)	806	N
Skydragon, Aeros-40B (Aeros)	812	U
Skyship 500 HL (GSI)	813	U
Skyship 600 (GSI)	813	U
SPAS-13 (21st Century)	797	U
SPAS-70 (21st Century)	797	U
SPAS-75 (21st Century)	798	U

T

THERMOPLANE DESIGN BUREAU MOSCOW AVIATION INSTITUTE (Russian Federation)	805	U

U

UPship 001 (Upship)	815	U
UPship CORPORATION (USA)	815	V

V

VOZDUKH AERONAUTICAL ENTERPRISE (Russian Federation)	805	N

W

WDL LUFTSCHIFFGESELLSCHAFT mbH (Germany)	801	U
WORLDWIDE AEROS CORPORATION (USA)	811	U

Z

ZEPPELIN LUFTSCHIFFTECHNIK, GmbH (Germany)	801	V

LIGHTER THAN AIR – PREVIOUS TEN EDITIONS

Listing includes recreational balloons, which have not appeared after the 1992-93 edition

A

Entry	Edition	Page
A50 (Adams)	(1991-92)	658
A50S (Adams)	(1991-92)	659
A55 (Adams)	(1991-92)	659
A55S (Adams)	(1991-92)	659
A-60 Lightship (ABC)	(1991-92)	652
A60 (Adams)	(1991-92)	659
A60S (Adams)	(1991-92)	659
A-105 (Cameron)	(1991-92)	659
A-120 (Cameron)	(1991-92)	659
A-120 Lightship (ABC)	(1995-96)	683
A-140 (Cameron)	(1991-92)	659
A-160 (Cameron)	(1992-93)	611
A-180 (Cameron)	(1991-92)	660
A-210 (Cameron)	(1991-92)	660
A-250 (Cameron)	(1991-92)	660
A-300 (Cameron)	(1991-92)	660
A-375 (Cameron)	(1992-93)	611
A-530 (Cameron)	(1991-92)	660
AA-3 (Ballonfabrik)	(1992-93)	606
AA-4 (Ballonfabrik)	(1992-93)	606
AA-5 (Ballonfabrik)	(1992-93)	606
AA-6 (Ballonfabrik)	(1992-93)	606
AAC (see Advanced Airship Corporation)	(1994-95)	668
AB (Adams)	(1991-92)	659
AB-2 (Aerotechnik)	(1991-92)	658
AB-3 (Aerotechnik)	(1991-92)	658
AB-8 (Aerotechnik)	(1991-92)	659
AB-10 (Aerotechnik)	(1991-92)	660
ADA (see Airship Developments Australia)	(1992-93)	605
ADA-1200 (ADA)	(1991-92)	645
AEROS (see NPP Aerostatic Technics)	(1993-94)	594
ALA-40 (Thermoplane)	(1997-98)	782
ALA-100 (Thermoplane)	(1994-95)	667
ALA-300 (Thermoplane)	(1994-95)	667
ALA-500 (MAI)	(1992-93)	608
ALA-600 (Thermoplane)	(1995-96)	679
ANR (Advanced Airship)	(1992-93)	608
AS 56 (Thunder & Colt)	(1993-94)	598
AS 80 GD (Gefa-Flug)	(1991-92)	646
AS 80 Mk II (Cameron)	(1996-97)	763
AS 105 Mk II (Cameron)	(1996-97)	763
AS 261 (Thunder & Colt)	(1991-92)	651
AS 300 (Lindstrand)	(1996-97)	780
AV-1R (Avgur)	(1998-99)	767
AVIC (see No 605 Research Institue of Avic)	(1998-99)	767
AX-4 (Air Service)	(1992-93)	607
AX5-40M (Flamboyant)	(1991-92)	658
AX-6 (Fantasy)	(1992-93)	605
AX-6 (Kavanagh)	(1992-93)	605
AX6-56 (Flamboyant)	(1992-93)	658
AX6-56M (Flamboyant)	(1991-92)	658
AX-7 (Fantasy)	(1992-93)	605
AX-7 (Kavanagh)	(1992-93)	605
AX7-65 (Flamboyant)	(1991-92)	658
AX7-77M (Flamboyant)	(1991-92)	658
AX-8 (Fantasy)	(1992-93)	605
AX-8 (Kavanagh)	(1992-93)	605
AX8-85 (Flamboyant)	(1991-92)	659
AX8-105 (Flamboyant)	(1991-92)	659
AX9 (Galaxy)	(1992-93)	614
AX-9 (Kavanagh)	(1992-93)	605
AX-10 (Kavanagh)	(1992-93)	605
AX10-150 (Flamboyant)	(1991-92)	660
AX-11 (Kavanagh)	(1992-93)	605
AX 11-240 (Flamboyant)	(1991-92)	660
ADAMS' BALLOON LOFT INC (USA)	(1992-93)	613
ADVANCED AIRSHIP CORPORATION (UK)	(1994-95)	668
AERAZUR (France)	(1996-97)	757
AEROLIFT INC (USA)	(1991-92)	652
Aeros-50 (Aeros)	(1998-99)	774
AEROSTAR INTERNATIONAL INC (USA)	(1993-94)	599
Aerostatica-01 (Aerostatica)	(1998-99)	766
Aerostatica-02 (Aerostatica)	(1999-2000)	785
Aerostatica-300 (Aerostatica)	(1999-2000)	785
AEROTECHNIK (Czechoslovakia)	(1992-93)	606
AIR SERVICE REPULOGEPES SZOLGALAT (Hungary)	(1992-93)	607
Airship (Thompson)	(1992-93)	615
AIRSHIP DEVELOPMENTS AUSTRALIA PTY LTD (Australia)	(1992-93)	605
AIRSHIP INDUSTRIES (UK) LTD	(1991-92)	648
Airships, demonstrator (21st Century Airships)	(1994-95)	622
Airships, prototypes (21st Century Airships)	(1994-95)	661
Albatross (ADA)	(1991-92)	645
AVIAN BALLOON COMPANY (USA)	(1992-93)	614

B

Entry	Edition	Page
B 800/2-Ri (Ballonfabrik)	(1991-92)	658
BCI (see Buoyant-Copter)	(1998-99)	775
BUAA (see Beijing University of Aeronautics and Astronautics)	(1991-92)	645
BALLONFABRIK SEE- UND LUFTAUSRÜSTUNG GmbH und Co KG (Germany)	(1992-93)	606
BALLOONS SERVICE DI BONANNO LETTERIO (Italy)	(1991-92)	647
BALLOON WORKS, THE (USA)	(1992-93)	614
BALLONS CHAIZE (France)	(1992-93)	606
BARNES AIRSHIPS (USA)	(1993-94)	599
BEIJING UNIVERSITY OF AERONAUTICS AND ASTRONAUTICS (China)	(1991-92)	645
Breitling Orbiter 2 (Cameron)	(1998-99)	771
BUOYANT-COPTER (USA)	(1998-99)	771
Buoyant-Copter 18 (BCI)	(1997-98)	790

C

Entry	Edition	Page
C-56 (Kavanagh)	(1991-92)	658
C-65 (Kavanagh)	(1991-92)	658
C-77 (Kavanagh)	(1991-92)	658
CAS 1200 Cargo Airship System (Pan Atlantic)	(1995-96)	674
CS.1600 (Chaize)	(1991-92)	658
CS.1800 (Chaize)	(1991-92)	658
CS.2000 (Chaize)	(1991-92)	658
CS.2200 (Chaize)	(1991-92)	658
CS.3000 (Chaize)	(1991-92)	659
CS.4000 (Chaize)	(1991-92)	659
CHAIZE (see Ballons Chaize)	(1992-93)	606
Cloud Cruiser UM10-22 (Ulita)	(1992-93)	615
Cloud Cruiser UM10-23 (Ulita)	(1996-97)	771
Cloudhopper Midi (Thunder & Colt)	(1991-92)	658
Cloudhopper Super (Thunder & Colt)	(1991-92)	658
Concept AX-7 (Cameron)	(1992-93)	611

D

Entry	Edition	Page
D-1 (Aeros)	(1996-97)	769
D-4 (Aeros)	(1996-97)	769
D-4 (Flight Exploration)	(1997-98)	783
D-77 (Kavanagh)	(1991-92)	658
D-84 (Kavanagh)	(1991-92)	659
D-90 (Kavanagh)	(1991-92)	659
D-105 (Kavanagh)	(1991-92)	659
DEP-140 (Augur)	(1998-99)	768
DG-14 (Cameron)	(1994-95)	668
DKBA (see Dolgoprudnenskoe Kovarnyi Byuro Avtomatiki)	(1998-99)	769
DP-800 Ecologiya (DKBA)	(1997-98)	781
DP-6000 Vityaz (DKBA)	(1997-98)	782
DPD-5000	(1998-99)	768
DS-3 (DKBA)	(1993-94)	592
DEMENTYEV (Russia)	(1992-93)	607
DOLGOPRUDNENSKOE KOVARNYI BYURO AVTOMATIKI (Russian Federation)	(1998-99)	769
DragonFly 56 (Balloon Works)	(1992-93)	614
DragonFly 65 (Balloon Works)	(1992-93)	614
DragonFly 77 (Balloon Works)	(1992-93)	614
DragonFly 90 (Balloon Works)	(1992-93)	614
DragonFly 90DB (Balloon Works)	(1992-93)	614
DragonFly 105 (Balloon Works)	(1992-93)	614
DragonFly 560 (Balloon Works)	(1991-92)	658
DragonFly 650 (Balloon Works)	(1991-92)	659
DragonFly 770 (Balloon Works)	(1991-92)	659
DragonFly 900 (Balloon Works)	(1991-92)	659
DragonFly 1050 (Balloon Works)	(1991-92)	659

E

Entry	Edition	Page
E-120 (Kavanagh)	(1991-92)	659
E-140 (Kavanagh)	(1991-92)	659
E-160 (Kavanagh)	(1991-92)	660
E-200 (Kavanagh)	(1991-92)	660
E-210 (Kavanagh)	(1991-92)	660
E-240 (Kavanagh)	(1991-92)	660
E-260 (Kavanagh)	(1991-92)	660
Earthwinds Hilton (Raven)	(1995-96)	684
Ecoblimp (Interface)	(1999-2000)	795
Ecologiya DP-800 (DKBA)	(1997-98)	781

F

Entry	Edition	Page
FK-4 (HADG)	(1992-93)	605
FK-6 (HADG)	(1993-94)	591
FK-11 (HADG)	(1995-96)	674
FK-12 (HADG)	(1995-96)	674
FK-100 (Avic)	(1998-99)	762
Falcon II (Avian)	(1992-93)	614
Fantasy 6 (Fantasy)	(1991-92)	658
Fantasy 7 (Fantasy)	(1991-92)	658
Fantasy 8-90 (Fantasy)	(1991-92)	659
Fantasy 8-105 (Fantasy)	(1991-92)	659
FANTASY SKY PROMOTIONS INC (Canada)	(1992-93)	605
FireFly 8 (Balloon Works)	(1991-92)	659
FireFly 42 (Balloon Works)	(1992-93)	614
FireFly 65 (Balloon Works)	(1992-93)	614
FireFly 77 (Balloon Works)	(1992-93)	614
FireFly 90 (Balloon Works)	(1992-93)	614
FireFly 90DF (Balloon Works)	(1992-93)	614
FireFly 105 (Balloon Works)	(1992-93)	614
FireFly 140 (Balloon Works)	(1991-92)	659
FireFly 650 (Balloon Works)	(1991-92)	659
FireFly 770 (Balloon Works)	(1991-92)	659
FireFly 900 (Balloon Works)	(1991-92)	659
FireFly 1050 (Balloon Works)	(1991-92)	659
FLAMBOYANT BALLOONS (PTY) LTD (South Africa)	(1992-93)	608
FLIGHT EXPLORATION ASSOCIATION (UK)	(1998-99)	771

G

Entry	Edition	Page
GA-42 (Thunder & Colt)	(1994-95)	670
GB-1 (Thunder & Colt)	(1991-92)	658
GadFly 56 (Balloon Works)	(1992-93)	614
GadFly 560 (Balloon Works)	(1991-92)	658
Galaxy 7 (Galaxy)	(1992-93)	614
Galaxy 8 (Galaxy)	(1992-93)	614
GALAXY BALLOONS INC (USA)	(1992-93)	614
Global Challenger (Cameron)	(2000-01)	819
GOODYEAR AEROSPACE CORPORATION (USA)	(1991-92)	652

H

Entry	Edition	Page
H-34 Skyhopper (Cameron)	(1991-92)	658
HA-44 (Hamilton)	(2000-01)	817
HA 132 (Hamilton)	(1998-99)	769
HA 140 (Hamilton)	(1999-2000)	787
HADG (see Huahang Airship Development Group)	(1995-96)	674
HARI (see Hangzhou Airship Research Institute)	(1998-99)	763
HAMILTON AIRSHIP COMPANY, THE (USA)	(2000-01)	817
HANGZHOU AIRSHIP RESEARCH INSTITUTE (China)	(1998-99)	763
Holland Millennium Navigator (Rigid Airship)	(1998-99)	766
HUAHANG AIRSHIP DEVELOPMENT GROUP (China)	(1995-96)	674

K

Entry	Edition	Page
K 630/1-Ri (Ballonfabrik)	(1991-92)	658
K 780/2-Ri (Ballonfabrik)	(1991-92)	658
K 945/2-Ri (Ballonfabrik)	(1991-92)	658
K 1050/3-Ri (Ballonfabrik)	(1991-92)	658
K 1260/3-Ri (Ballonfabrik)	(1991-92)	658
K 1360/4-Ri (Ballonfabrik)	(1991-92)	658
K 1680/4-Ri (Ballonfabrik)	(1991-92)	658
KAVANAGH BALLOONS PTY LTD (Australia)	(1992-93)	605

L

Entry	Edition	Page
LASS (TCOM)	(1995-96)	686
LD (Adams)	(1991-92)	658
LD-S (Adams)	(1991-92)	658
LEMS (Pan Atlantic)	(1996-97)	756

L - Z–LIGHTER THAN AIR–PREVIOUS TEN EDITIONS: INDEXES 873

LZ NO5 (Zeppelin) .... (1994-95) 665
LZ N30 (Zeppelin) .... (1994-95) 666

Lightship A-60 (ABC) .... (1991-92) 652
Lightship A-120 (ABC) .... (1995-96) 683
Low Altitude Surveillance System (TCOM) .... (1995-96) 685

M

MA-55 (Avgur) .... (1998-99) 767
MAI (see Moscow Aviation Institute) .... (1992-93) 608
MATSS (TCOM) .... (1995-96) 685
MG-300 (Aerostar) .... (1991-92) 658
MG-1000 (Aerostar) .... (1991-92) 658
MLA-32-A (Spacial) .... (1991-92) 647

Magnum IX (Avian) .... (1992-93) 614
Magnum Super Sport (Avian) .... (1992-93) 614
Magnus 60 (Pan Atlantic) .... (1998-99) 762
Maritime Aerostat Tracking and Surveillance System (TCOM) .... (1995-96) 685
Mifeng-6 (BUAA) .... (1991-92) 645
Millennia (ABC) .... (1997-98) 788
Millennium Navigator (Rigid Airship) .... (1998-99) 766
Mirage 56 (Fantasy) .... (1992-93) 605
Mirage 65 (Fantasy) .... (1992-93) 605
Model 31Z (Thunder & Colt) .... (1991-92) 658
Model 42 Series 1 (Thunder & Colt) .... (1991-92) 658
Model 42A (Thunder & Colt) .... (1991-92) 658
Model 56 Series 1 (Thunder & Colt) .... (1991-92) 658
Model 56A (Thunder & Colt) .... (1991-92) 658
Model 65 Series 1 (Thunder & Colt) .... (1991-92) 658
Model 69A (Thunder & Colt) .... (1991-92) 659
Model 77 Series 1 (Thunder & Colt) .... (1991-92) 659
Model 77A (Thunder & Colt) .... (1991-92) 659
Model 84 Series 1 (Thunder & Colt) .... (1991-92) 659
Model 90 Series 1 (Thunder & Colt) .... (1991-92) 659
Model 90 Series 2 (Thunder & Colt) .... (1991-92) 659
Model 90A (Thunder & Colt) .... (1991-92) 659
Model 105 Series 1 (Thunder & Colt) .... (1991-92) 659
Model 105 Series 2 (Thunder & Colt) .... (1991-92) 659
Model 105A (Thunder & Colt) .... (1991-92) 659
Model 120A (Thunder & Colt) .... (1991-92) 659
Model 140 Series 2 (Thunder & Colt) .... (1991-92) 659
Model 160 Series 1 (Thunder & Colt) .... (1991-92) 660
Model 160A (Thunder & Colt) .... (1991-92) 660
Model 180 Series 1 (Thunder & Colt) .... (1991-92) 660
Model 180A (Thunder & Colt) .... (1991-92) 660
Model 240A (Thunder & Colt) .... (1991-92) 660
Model 300A (Thunder & Colt) .... (1991-92) 660
Model 400A (Thunder & Colt) .... (1991-92) 660
MOSCOW AVIATION INSTITUTE (Russia) .... (1992-93) 608

N

N17 (Zeppelin) .... (1996-97) 759
N30 (Zeppelin) .... (1996-97) 759
N-31 (Cameron) .... (1991-92) 658
N-42 (Cameron) .... (1991-92) 658
N-56 (Cameron) .... (1991-92) 658
N-65 (Cameron) .... (1991-92) 658
N-77 (Cameron) .... (1991-92) 658
N-90 (Cameron) .... (1991-92) 659
N-105 (Cameron) .... (1991-92) 659
N-120 (Cameron) .... (1992-93) 611
N-133 (Cameron) .... (1991-92) 659
N-145 (Cameron) .... (1991-92) 660
N-160 (Cameron) .... (1992-93) 611
N-180 (Cameron) .... (1991-92) 660
N-850 (Cameron) .... (1991-92) 660
NO5 (Zeppelin) .... (1994-95) 665

NPP AEROSTATIC TECHNICS (Ukraine) .... (1993-94) 594

O

O-31-O-77 (Cameron) .... (1991-92) 658
O-84 (Cameron) .... (1991-92) 659
O-90 (Cameron) .... (1992-93) 611
O-105 (Cameron) .... (1991-92) 659
O-120 (Cameron) .... (1991-92) 659
O-140 (Cameron) .... (1991-92) 659
O-160 (Cameron) .... (1991-92) 660
ONAGO AIRCRAFT COMPANY (USA) .... (2000-01) 825

P

PD-160 (Avgur) .... (1998-99) 768
PD-220 (Avgur) .... (2000-01) 816
PIG (see Promotional Ideas Group) .... (1998-99) 772
PIG 2 (PIG) .... (1995-96) 682
PIG 3 (PIG) .... (1997-98) 784

Pacific Flyer (Thunder & Colt) .... (1992-93) 611
PROMOTIONAL IDEAS GROUP (UK) .... (1998-99) 772
Puppy Chow (Boland) .... (1990-91) 678

R

RBX-7 (Bonanno) .... (1991-92) 647
RSZ (see Repulogepes Szologat) .... (1991-92) 657
RSZ-03/1 (RSZ) .... (1991-92) 659
RSZ-04/1 (RSZ) .... (1991-92) 659
RSZ-05 (RSZ) .... (1991-92) 658
RSZ-06/1 (Air Service) .... (1992-93) 607
RX-6 (Aerostar) .... (1991-92) 658
RX-7 (Aerostar) .... (1991-92) 659
RX-8 (Aerostar) .... (1991-92) 659
RX-100 (Cameron) .... (1992-93) 611

Rainbow-Stars (Boland) .... (1990-91) 678
RAVEN INDUSTRIES INC (USA) .... (1996-97) 770
RAVEN/SCALED (USA) .... (1996-97) 770
Red Baron (Boland) .... (1990-91) 680
REPULOGEPES SZOLOGAT (Hungary) .... (1991-92) 657
RESEARCH INSTITUTE, No 605 of AVIC (China) .... (1998-99) 762
Rover, A-2 (Boland) .... (1990-91) 672
Roziere R-15 (Cameron) .... (1992-93) 610
Roziere R-42 (Cameron) .... (1992-93) 610
Roziere R-60 (Cameron) .... (1992-93) 610
Roziere R-77 (Cameron) .... (1992-93) 610
Roziere R-225 (Cameron) .... (1992-93) 610

S

S-40A to S-50A (Aerostar) .... (1991-92) 658
S-52A to S-66A (Aerostar) .... (1991-92) 659
S-71A (Aerostar) .... (1991-92) 660
S-77A (Aerostar) .... (1991-92) 660
SARI (See Shanghai Aircraft Research Institute) .... (1997-98) 777
SPACIAL (see Servicios Publicitarios Aereos Construccion y Engenieria de Aeronaves Ligeras SA de CV) (Mexico) .... (1992-93) 607
SPAS-1 (21st Century) .... (1995-96) 672
SPAS-2 (21st Century) .... (1995-96) 673
SPAS-3 (21st Century) .... (1995-96) 673

SCALED COMPOSITES INC (USA) .... (1996-97) 770
Sentinel 1000 (GSI) .... (1999-2000) 795
Sentinel 1240 (GSI) .... (1999-2000) 795
Sentinel 5000 (Skyship) .... (1997-98) 785
Sentinel 8500 (Skyship) .... (1997-98) 786
SERVICIOS PUBLICITARIOS AEREOS CONSTRUCCION Y ENGENIERIA DE AERONAVES LIGERAS SA de CV (Mexico) .... (1992-93) 607
SHANGHAI AIRCRAFT RESEARCH INSTITUTE (China) .... (1997-98) 777
Shen Zhou-1 (SARI) .... (1995-96) 675
Shen Zhou-2 (SARI) .... (1995-96) 675
Skyhawk (Avian) .... (1992-93) 614
SKYRIDER AIRSHIPS INC (USA) .... (1991-92) 653
Skyship 500 (Slingsby) .... (1991-92) 649
Skyship 500 HL (WAI) .... (1994-95) 676
Skyship 600 (Skyship) .... (1997-98) 785
SKYSHIP/SENTINEL (UK) .... (1998-99) 772
SLINGSBY AVIATION LIMITED (UK) .... (1994-95) 669
Sparrow (Avian) .... (1992-93) 614

T

TCOM 32M (TCOM) .... (1995-96) 685
TCOM 71M (TCOM) .... (1995-96) 685

TCOM LIMITED PARTNERSHIP (USA) .... (1993-94) 615
TCOM LP (USA) .... (1996-97) 770
Tethered Aerostats (TCOM) .... (1993-94) 615
THE DIRIGIBLE AIRSHIP (USA) .... (1996-97) 771
Thermoplane (MAI) .... (1992-93) 608
THERMOPLANE DESIGN BUREAU, MOSCOW AVIATION INSTITUTE (Russia) .... (1997-98) 782
Thompson airship (Thompson) .... (1997-98) 790
THOMPSON, JAMES, AIAA (USA) .... (1997-98) 790
THUNDER & COLT LTD (UK) .... (1995-96) 682
Toluca (Spacial) .... (1991-92) 647
Turbo 8 (Avian) .... (1992-93) 614

U

UM10-22 Cloud Cruiser (Ulita) .... (1992-93) 615
UM10-23 Cloud Cruiser (Ulita) .... (1996-97) 771
UM30-71 (Ulita) .... (1994-95) 686
UM20-48 (Ulita) .... (1992-93) 616
UM25-64 (Ulita) .... (1992-93) 616
US-LTA (see US Lighter Than Air Corporation) .... (1998-99) 777
US-LTA 138-S (US-LTA) .... (1995-96) 687
US-LTA 185M (US-LTA) .... (1997-98) 791

Upship 750-001 (Upship) .... (1992-93) 616
UPSHIP PROJECT DIRIGIBLE AIRSHIPS .... (1993-94) 601
US LIGHTER THAN AIR CORPORATION (USA) .... (1998-99) 777

V

Viva-20 (Cameron) .... (1992-93) 610
Viva-31 (Cameron) .... (1992-93) 610
Viva-42 (Cameron) .... (1992-93) 610
Viva-56 (Cameron) .... (1992-93) 610
Viva-65 (Cameron) .... (1992-93) 610
Viva-77 (Cameron) .... (1992-93) 610
Viva-90 (Cameron) .... (1992-93) 610

W

WAI (see Westinghouse-Airship Industries) .... (1997-98) 791
WDL 1B (WDL) .... (1997-98) 778

WESTDEUTSCHE LUFTWERBUNG THEODOR WULLENKEMPER (Germany) .... (1996-97) 758
WESTINGHOUSE-AIRSHIPS INC (USA) .... (1997-98) 774
Whispership (Barnes) .... (1992-93) 614
WORLD QUEST (USA) .... (2000-01) 826

X

Xihu-5 (HARI) .... (1998-99) 763

Y

YEZ-2A (Skyship) .... (1997-98) 785

Z

Z-10 (Onago) .... (2000-01) 825
Zephyr 200 (Memphis) .... (1992-93) 614
Zhonghuai (BKLAIC) .... (2000-01) 811

AERO-ENGINES–PREVIOUS TEN EDITIONS

Detailed descriptions of aero-engines last appeared in the 1995-96 edition. For current data refer to table in this edition, or to full details in Jane's Aero-Engines

A

A650 (Alturdyne) (1992-93) 681
AE 301X (Allison) (1995-96) 756
AE 530 AC (Arrow) (1995-96) 715
AE 1070 AC (Arrow) (1995-96) 715
AE 1107 (Allison) (1995-96) 754
AE 2100 (Allison) (1995-96) 755
AE 3007 (Allison) (1995-96) 755
AE 3012 (Allison) (1995-96) 756
AI-14R (WSK-PZL Kalisz) (1995-96) 718
AI-20 (Progress) (1995-96) 739
AI-24 (Progress) (1995-96) 739
AI-25 (Progress) (1995-96) 740
AI-25TL (Progress) (1995-96) 740
AL-7 (Saturn) (1993-94) 642
AL-21 (Saturn) (1995-96) 732
AL-31 (Saturn) (1995-96) 732
AL-34 (Saturn) (1995-96) 733
AL-35 (Saturn) (1995-96) 733
ALF 502 (AlliedSignal) (1995-96) 751
AM-3 (Soyuz) (1993-94) 644
AM-5 (Soyuz) (1991-92) 705
AMI (see AeroMotion Inc) (1992-93) 681
AR.318 (Alfa Romeo Avio) (1992-93) 644
ASSAD (Russia) (1995-96) 720
ASTOVL (General Electric) (1995-96) 757
ASz-621R (WSK-PZL Kalisz) (1995-96) 718
ATF-3 (AlliedSignal) (1995-96) 749
AVIA-HSA (see Avia-Hamilton Standard Aviation Ltd) (1993-94) 612
AVIC (see Aviation Industries of China) (1995-96) 695

Adour (Rolls-Royce Turbomeca) (1995-96) 712
Advanced STOVL (Rolls-Royce) (1995-96) 748
AEROJET-GENERAL (USA) (1992-93) 678
AEROJET TECHSYSTEMS CO (USA) (1992-93) 678
AEROMOTION INC (USA) (1992-93) 681
AeroMotion Twin (AMI) (1992-93) 681
AEROSTAR SA (Romania) (1995-96) 720
AEROTECHNIK (Czech Republic) (1995-96) 700
Air breathing propulsion (CSD) (1992-93) 681
ALFA ROMEO AVIO SpA (Italy) (1995-96) 714
ALLIEDSIGNAL INC (USA) (1995-96) 749
ALLISON ENGINE COMPANY (1995-96) 753
ALTURDYNE (USA) (1992-93) 681
ALVIS UAV ENGINES (UK) (1994-95) 731
AO AVIADVIGATEL (Russia) (1993-94) 633
AOI ENGINE FACTORY (EF) (Egypt) (1995-96) 702
Arriel (Turbomeca) (1995-96) 704
Arrius (Turbomeca) (1995-96) 704
Arrius 1D (Turbomeca) (1995-96) 704
ARROW ENGINEERING SRL (Italy) (1995-96) 715
Artouste III (Turbomeca) (1995-96) 705
Astazou Turboprop (Turbomeca) (1991-92) 676
Astazou Turboshaft (Turbomeca) (1995-96) 704
Atar 9K50 (SNECMA) (1995-96) 702
ATELIERS JPX (France) (1991-92) 673
ATLAS AVIATION (South Africa) (1995-96) 738
AVIA-HAMILTON STANDARD AVIATION LTD (Czech Republic) (1993-94) 612
AVIA KUNCERNOVY PODNIK (Czech Republic) (1992-93) 628
AVIATION INDUSTRIES OF CHINA (China) (1995-96) 695
AVIO AR.318 (Alfa Romeo) (1995-96) 714

B

BAZ-21083 (Volszhsky) (1995-96) 737
BR 700 (BMW-RR) (1994-95) 694
BR 710 (BMW-RR) (1995-96) 706
BR 715 (BMW-RR) (1995-96) 706
BSEL (see Bet-Shemesh Engines) (1995-96) 714
BUAA (see Beijing University of Aeronautics and Astronautics) (1995-96) 695

BAKANOV (Russia) (1993-94) 635
BEIJING UNIVERSITY OF AERONAUTICS AND ASTRONAUTICS (China) (1995-96) 695
BET-SHEMESH ENGINES LTD (Israel) (1995-96) 714
BMW ROLLS-ROYCE GmbH (1995-96) 706
BOMBARDIER-ROTAX GmbH (Austria) (1995-96) 690

C

CAM (see Canadian Airmotive Inc) (1995-96) 691
CAM 100 (CAM) (1995-96) 691
CAREC (see China National Aero-Engine Corporation) (1995-96) 696
CATIC (see China National Aero-Technology Import and Export Corporation) (1992-93) 623
CEC (see Chengdu Engine Company) (1995-96) 696
CF 077A (Emdair) (1994-95) 731
CF 077B (Emdair) (1994-95) 731
CF 092A (Emdair) (1994-95) 731
CF 112 (Emdair) (1994-95) 731
CF6 (General Electric) (1995-96) 758
CF6-80C2 (General Electric) (1995-96) 759
CF6-80E1 (General Electric) (1995-96) 759
CF34 (General Electric) (1995-96) 757
CFE738 (CFE) (1995-96) 756
CFM56 (CFM International) (1995-96) 708
CFM88 (CFM International) (1995-96) 710
CLXMW (see Changzhou Lan Xiang Machinery Works) (1995-96) 696
CRM 18D/SS (CRM) (1995-96) 715
CSD (see Chemical Systems Division) (1992-93) 681
CT7 (General Electric) (1995-96) 761

CAMERA REYNAUD & CIE, SCS (France) (1992-93) 630
CANADIAN AIRMOTIVE INC (Canada) (1995-96) 691
CELMA-CIA ELECTROMECANICA (Brazil) (1995-96) 691
CFE COMPANY (USA) (1995-96) 756
CFM INTERNATIONAL SA (International) (1995-96) 708
CHANGZHOU LAN XIANG MACHINERY WORKS (China) (1995-96) 696
CHEMICAL SYSTEMS DIVISION (USA) (1992-93) 681
CHENGDU ENGINE COMPANY (China) (1995-96) 696
CHERNYSHOV (Russia) (1995-96) 724
CHINA NATIONAL AERO-ENGINE CORPORATION (China) (1995-96) 696
CHINA NATIONAL AERO-TECHNOLOGY IMPORT AND EXPORT CORPORATION (China) (1992-93) 623
Civil Spey RB163 (Rolls-Royce) (1993-94) 674
CRM (Italy) (1995-96) 715

D

D-15 (MKB) (1991-92) 697
D-18A (IL) (1995-96) 717
D-18T (Progress) (1995-96) 742
D-18TI (Progress) (1995-96) 742
D-18TM (Progress) (1995-96) 742
D-18TR (Progress) (1995-96) 742
D-21A1 (Aviadvigatel) (1995-96) 723
D-25V (JSC Aviadvigatel) (1994-95) 708
D-27 (Progress) (1995-96) 743
D-30 (JSC Aviadvigatel) (1995-96) 721
D-30F6 (JSC Aviadvigatel) (1995-96) 723
D-30K (JSC Aviadvigatel) (1995-96) 721
D-36 (Progress) (1995-96) 740
D-90A (Soloviev) (1991-92) 698
D-100/110/112 (JSC Aviadvigatel) (1995-96) 724
D-127 (Progress) (1995-96) 743
D-136 (Progress) (1995-96) 741
D-200 (JSC Aviadvigatel) (1995-96) 724
D-200/VAZ-430 (VAZ) (1995-96) 736
D-227 (ZMKB Progress) (1994-95) 730
D-236 (Progress) (1995-96) 741
D-336 (Progress) (1991-92) 702
D-436 (Progress) (1995-96) 741
D-727 (Progress) (1995-96) 743
DEMC (see Dongan Engine Manufacturing Company) (1995-96) 696
DN-200 (RKBM) (1995-96) 730
DV-2 (Progress) (1995-96) 740

DONGAN ENGINE MANUFACTURING COMPANY (China) (1995-96) 696
Dual Pac (Soloy) (1995-96) 767
DVADESETPRIVI MAJ BEOGRAD (Yugoslavia) (1995-96) 772
DYNA-CAM INDUSTRIES (USA) (1992-93) 681

E

EJ-200 (Eurojet) (1995-96) 710

EMDAIR LTD (UK) (1994-95) 731
EMG ENGINEERING COMPANY (USA) (1992-93) 682
EUROJET TURBO GmbH (International) (1995-96) 710

F

F3 (IHI) (1995-96) 716
F22 (Hirth) (1992-93) 636
F23A (Hirth) (1995-96) 706
F30 (Hirth) (1995-96) 707
F100 (Pratt & Whitney) (1995-96) 766
F103-GE-100 (General Electric) (1995-96) 758
F104-GA-100 (AlliedSignal) (1995-96) 749
F108 (CFM International) (1995-96) 708
F109-GA-100 (AlliedSignal) (1995-96) 749
F110 (General Electric) (1995-96) 759
F113 (Rolls-Royce) (1993-94) 674
F117 (Pratt & Whitney) (1995-96) 765
F118 (General Electric) (1995-96) 760
F119 (Pratt & Whitney) (1995-96) 766
F120 (General Electric) (1991-92) 726
F124-GA-100 (AlliedSignal) (1995-96) 750
F125-GA-100 (AlliedSignal) (1995-96) 750
F129 (Williams) (1995-96) 772
F263 R 53 (Hirth) (1992-93) 636
F402, Pegasus (Rolls-Royce) (1995-96) 748
F404 (General Electric) (1995-96) 756
F405 (Rolls-Royce Turbomeca) (1995-96) 712
F412 (General Electric) (1991-92) 724
F414 (General Electric) (1995-96) 757
FAM (see France Aéro Moteurs) (1992-93) 630
FAM (see France Aéro Moteurs) (1995-96) 702
F + E (see Fischer & Entwicklungen) (1992-93) 635
FJ44 (Williams) (1995-96) 772
FJR710 (NAL) (1991-92) 689
FNM 1 (see FN Moteurs SA) (1992-93) 619

FIATAVIO (Italy) (1995-96) 715
FISCHER + ENTWICKLUNGEN (Germany) (1992-93) 635
FISHER RESEARCH CORPORATION (USA) (1992-93) 682
FLYGMOTOR (Sweden) (1994-95) 725
FN MOTEURS SA (Belgium) (1992-93) 619
FRANCE AERO MOTEURS SARL (France) (1992-93) 630
FRANCE AERO MOTEURS SARL (France) (1995-96) 702

G

G8-2 (Gluhareff-EMG) (1992-93) 682
GADRI (see Guizhou Aero-Engine Design and Research Institute) (1995-96) 696
GE90 (General Electric) (1995-96) 757
GLC 38 (General Electric) (1994-95) 746
GMA 140 TK (Scoma) (1992-93) 63
GMA 501 (Allison) (1993-94) 661
GMA 1107 (Allison) (1993-94) 661
GMA 2100 (Allison) (1993-94) 662
GMA 3007 (Allison) (1993-94) 662
GP 1000 (Arrow) (1995-96) 715
GP 1500 (Arrow) (1995-96) 715
GT250 (Arrow) (1992-93) 644
GT500 (Arrow) (1992-93) 644
GT654 (Arrow) (1992-93) 644
GT1000 (Arrow) (1992-93) 644
GTD-3 (Mars) (1995-96) 728
GTD-350 (WSK-PZL-Rzeszów) (1995-96) 719
GTE-400 (Soyuz) (1995-96) 736
GTRE (see Gas Turbine Research Establishment) (1995-96) 708
GTX (GTRE) (1995-96) 708

Galigusov (Russia) (1995-96) 724
GARRETT (USA) (1994-95) 740
GAS TURBINE RESEARCH ESTABLISHMENT (India) (1995-96) 708
Gavrilov (Russia) (1995-96) 724
GE AIRCRAFT ENGINES (USA) (1995-96) 756
Gem (Rolls-Royce) (1995-96) 748
GENERAL ELECTRIC (USA) (1995-96) 756
Gnome (Rolls-Royce) (1995-96) 749
GOBLER-HIRTHMOTOREN GmbH (Germany) (1995-96) 706
GRETH, GERRY (USA) (1992-93) 690
GROUPEMENT TURBOMECA-SNECMA (GRTS) (France) (1995-96) 706
GUIZHOU AERO-ENGINE DESIGN AND RESEARCH INSTITUTE (China) (1995-96) 696

H

H-63 (Nelson) (1995-96) 762
HAL (see Hindustan Aeronautics Ltd.) (1995-96) 708
HDHV (see Hawker de Havilland Victoria Ltd) (1995-96) 690
HIRTH (see Gobler-Hirthmotoren) (1995-96) 706

H – P—AERO-ENGINES–PREVIOUS TEN EDITIONS: INDEXES

Entry	Edition	Page
HIRTH 2701 R 03 (Hirth)	(1995-96)	706
HIRTH 2702 R 03 (Hirth)	(1995-96)	706
HIRTH 2703 (Hirth)	(1995-96)	706
HIRTH 2704 (Hirth)	(1995-96)	707
HS7 (DEMC)	(1995-96)	696
HS8 (DEMC)	(1995-96)	696
HAWKER DE HAVILLAND VICTORIA LTD (Australia)	(1995-96)	690
HAWKER SIDDELEY CANADA INC (Canada)	(1991-92)	662
HINDUSTAN AERONAUTICS LTD (India)	(1995-96)	708
HIRTHMOTOREN, GÖBLER, GmbH (Germany)	(1992-93)	636

I

Entry	Edition	Page
IAE (see International Aero Engines AG)	(1995-96)	711
IAI (see Israel Aircraft Industries Ltd)	(1995-96)	714
IAME (see Ital-American Motor Engineering)	(1992-93)	645
IHI (see Ishikawajima-Harima Jukogyo Kabushiki Kaisha)	(1995-96)	716
IL (see Instytut Lotnictwa)	(1995-96)	717
IMAER (see Indústria Mecánica E Aeronautica Ltda)	(1992-93)	619
Imaer 1000 (Imaer)	(1992-93)	619
Imaer 2000 (Imaer)	(1992-93)	619
IO-240 (TCM)	(1995-96)	767
IO-320 Series (Textron Lycoming)	(1995-96)	770
IO-360 Series (Textron Lycoming)	(1995-96)	770
IO-520 Series (TCM)	(1995-96)	768
IO-540 Series (Textron Lycoming)	(1995-96)	770
IO-550 (TCM)	(1995-96)	768
IO-720 Series (Textron Lycoming)	(1995-96)	770
ITP (see Industria de Turbo Propulsores)	(1995-96)	738
INDUSTRIA DE TURBO PROPULSORES (Spain)	(1995-96)	738
INDÚSTRIA MECÁNICA E AERONAUTICA LTDA (Brazil)	(1992-93)	619
INSTYTUT LOTNICTWA (Poland)	(1995-96)	717
IN-TECH INTERNATIONAL, INC (USA)	(1992-93)	690
INTERNATIONAL AERO ENGINES AG (International)	(1995-96)	711
INTREPRINDEREA TURBOMECANICA BUCURESTI (Romania)	(1995-96)	720
ISHIKAWAJIMA-HARIMA JUKOGYO KABUSHIKI KAISHA (Japan)	(1995-96)	716
ISRAEL AIRCRAFT INDUSTRIES LTD (Israel)	(1995-96)	714
ITAL-AMERICAN MOTOR ENGINEERING (Italy)	(1992-93)	645
IVCHENKO PROGRESS ZMKB (Ukraine)	(1995-96)	739

J

Entry	Edition	Page
J52 (Pratt & Whitney)	(1995-96)	763
J69 (Teledyne CAE)	(1994-95)	756
J79-IAI-J1E (IAI)	(1992-93)	644
JPX 4T60/A (JPX)	(1995-96)	702
JPX 4TX 75 (JPX)	(1995-96)	702
JT8 (Pratt & Whitney)	(1995-96)	763
JT8D (Pratt & Whitney)	(1994-95)	749
JT8D-200 Series (Pratt & Whitney)	(1995-96)	763
JT9D (Pratt & Whitney)	(1995-96)	764
JT15D (P&WC)	(1995-96)	691
JTF10A (Pratt & Whitney)	(1991-92)	734
Jabiru 1600 (Jabiru Aircraft)	(1995-96)	690
JABIRU AIRCRAFT PTY (Australia)	(1995-96)	690
JAVELIN AIRCRAFT COMPANY INC (USA)	(1992-93)	690
JPX, SARL (France)	(1995-96)	702
JSC AVIADVIGATEL (Russia)	(1995-96)	721

K

Entry	Edition	Page
K-15 (IL)	(1995-96)	717
KFM 107 Maxi (IAME)	(1992-93)	645
KFM 112 (IAME)	(1992-93)	645
KJ12 (Kawasaki)	(1992-93)	647
KKBM (see Kuibyshev Engine Design Bureau)	(1995-96)	724
KAWASAKI HEAVY INDUSTRIES LTD (Japan)	(1995-96)	716
Khachaturov (Russia)	(1995-96)	724
KLIMOV CORPORATION (Russia)	(1995-96)	725
KONIG MOTORENBAU (Germany)	(1992-93)	636
KUIBYSHEV ENGINE DESIGN BUREAU (Russia)	(1995-96)	724
Kuznetsov (Russia)	(1995-96)	727

L

Entry	Edition	Page
L 90E (Limbach)	(1992-93)	636
L 275E (Limbach)	(1992-93)	636
L 550E (Limbach)	(1992-93)	636
L 1800 (Limbach)	(1995-96)	707
L 2000 (Limbach)	(1995-96)	707
L 2400 (Limbach)	(1995-96)	707
LEM 2fm 17 (Offmar)	(1992-93)	645
LF500 (AlliedSignal)	(1995-96)	751
LHTEC (see Light Helicopter Turbine Engine Company)	(1995-96)	762
LM (see Liming Engine Manufacturing Corporation)	(1995-96)	696
LMC (see Liyang|Machinery Corporation)	(1995-96)	698
LP512 (AlliedSignal)	(1995-96)	752
LPE (see Light Power Engine Corporation)	(1995-96)	762
LT101 (AlliedSignal)	(1995-96)	752
LTC1 (AlliedSignal)	(1995-96)	752
LTC4 (AlliedSignal)	(1995-96)	752
LTP 101 (AlliedSignal)	(1995-96)	753
LTS 101 (AlliedSignal)	(1995-96)	753
Larzac (Turbomeca/SNECMA)	(1995-96)	706
LIGHT HELICOPTER TURBINE ENGINE COMPANY (USA)	(1995-96)	762
LIGHT POWER ENGINE CORPORATION (USA)	(1995-96)	762
LIMBACH FLUGMOTOREN GmbH (Germany)	(1995-96)	707
LIMING ENGINE MANUFACTURING CORPORATION (China)	(1995-96)	696
LIYANG MACHINERY CORPORATION (China)	(1995-96)	698
LOM piston engines (LOM Prague)	(1995-96)	700
LOM PRAGUE (Czech Republic)	(1995-96)	700
LOTAREV/ZVL (International)	(1992-93)	642
LYCOMING TURBINE ENGINES (USA)	(1995-96)	751
LYULKA (USSR)	(1991-92)	697

M

Entry	Edition	Page
M-3 (VOKBM)	(1995-96)	737
M-14P (VOKBM)	(1995-96)	737
M-14V-26 (VMKB)	(1993-94)	648
M-16 (Bakanov)	(1993-94)	635
M-16 (VOKBM)	(1995-96)	737
M-17 (VOKBM)	(1995-96)	737
M-18 (VOKBM)	(1995-96)	737
M-19 (Bakanov)	(1993-94)	635
M-25 (VOKBM)	(1995-96)	737
M-29 (VOKBM)	(1995-96)	737
M53 (SNECMA)	(1995-96)	703
M88 (SNECMA)	(1995-96)	703
M 137A (Avia-HSA)	(1993-94)	612
M 202 (Walter)	(1995-96)	702
M 337A (Avia-HSA)	(1993-94)	613
M 601 (Walter)	(1995-96)	700
M 602 (Walter)	(1995-96)	701
MB-4-80 (Mudry)	(1992-93)	631
MKB (see Motorostroitelnoye Konstruktorskoye Buro)	(1991-92)	697
MS-1500 (Pieper)	(1992-93)	637
MTFE (see Mid Thrust Family Engine)	(1995-96)	711
MTFE engine (MTFE)	(1995-96)	711
MTR (see MTU TURBOMECA ROLLS-ROYCE GmbH)	(1995-96)	712
MTR 390 (MTR)	(1995-96)	712
MTU (see Motoren- und Turbinen-Union München GmbH)	(1995-96)	707
MUDRY (see Moteurs Mudry-Buchoux)	(1992-93)	631
MWAE (see Mid West Aero Engines)	(1995-96)	744
MWAE 75 (MWAE)	(1995-96)	744
MWAE 90 (MWAE)	(1992-93)	672
MWAE100R (MWAE)	(1995-96)	744
Makila (Turbomeca)	(1995-96)	705
MARQUARDT COMPANY, THE (USA)	(1992-93)	691
MARS, OMSK AIRCRAFT ENGINE DESIGN BUREAU (Russia)	(1995-96)	728
Merlyn (In-Tech)	(1992-93)	690
MICROTURBO INC (USA)	(1992-93)	692
MICROTURBO SA (France)	(1992-93)	630
MID THRUST FAMILY ENGINE (International)	(1995-96)	711
MID WEST AERO ENGINES (UK)	(1995-96)	744
Mikron (Aerotechnik)	(1995-96)	700
Mikulin (Russia)	(1995-96)	728
MITSUBISHI JUKOGYO KABUSHIKI KAISHA (Japan)	(1995-96)	717
Model 250 (Allison)	(1995-96)	753
Model 352 (Teledyne CAE)	(1992-93)	700
Model 501 (Allison)	(1995-96)	754
Model 535 (Rolls-Royce)	(1995-96)	746
Model 578DX (PW-Allison)	(1991-92)	736
Model 1000 (IMAER)	(1992-93)	619
Model 2000 (IMAER)	(1992-93)	619
MOLLER INTERNATIONAL (USA)	(1992-93)	691
MOMA STANOJLOVIC AIR DEPOT (Yugoslavia)	(1995-96)	772
MOTEURS MUDRY-BUCHOUX (France)	(1992-93)	631
MOTOREN- UND TURBINEN-UNION MÜNCHEN GmbH (Germany)	(1995-96)	707
MOTORLET A. S. (Czech Republic)	(1994-95)	689
MOTOROSTROITELNOYE KONSTRUKTORSKOYE BURO (USSR)	(1991-92)	697
MTU TURBOMECA ROLLS-ROYCE GmbH (International)	(1995-96)	712

N

Entry	Edition	Page
NAL (see National Aerospace Laboratory)	(1995-96)	717
NK-6 (KKBM)	(1993-94)	635
NK-8 (KKBM)	(1995-96)	725
NK-12M (KKBM)	(1995-96)	720
NK-16 (KKBM)	(1994-95)	712
NK-20 (KKBM)	(1993-94)	636
NK-22 (KKBM)	(1995-96)	725
NK-25 (KKBM)	(1995-96)	725
NK-44 (KKBM)	(1995-96)	725
NK-86 (KKBM)	(1995-96)	725
NK-88 (SAMARA)	(1995-96)	731
NK-92 (SAMARA)	(1995-96)	731
NK-93 (SAMARA)	(1995-96)	731
NK-144 (KKBM)	(1991-92)	695
NK-321 (SAMARA)	(1995-96)	731
NPT (see Noel Penny Turbines Ltd)	(1991-92)	710
NPT 301 (Noel Penny)	(1991-92)	710
NPT 754 (Noel Penny)	(1991-92)	710
NR 612 (Norton)	(1991-92)	710
NR 622 (Norton)	(1991-92)	709
NR 642 (Norton)	(1991-92)	709
NR 731 (Norton)	(1991-92)	709
NR 801 (Norton)	(1991-92)	709
NATIONAL AEROSPACE LABORATORY (Japan)	(1995-96)	717
NELSON AIRCRAFT CORPORATION (USA)	(1995-96)	762
NOEL PENNY TURBINES LTD (UK)	(1991-92)	710
NORTON MOTORS LTD (UK)	(1991-92)	709

O

Entry	Edition	Page
O-200 Series (TCM)	(1995-96)	767
O-235 Series (Textron Lycoming)	(1995-96)	770
O-320 Series (Textron Lycoming)	(1995-96)	770
O-360 Series (TCM)	(1995-96)	768
O-360 Series (Textron Lycoming)	(1995-96)	770
O-520 Series (TCM)	(1992-93)	697
O-540 Series (Textron Lycoming)	(1995-96)	770
O-550 (TCM)	(1992-93)	698
OMKB (see Omsk Aircraft Engine Design Bureau)	(1991-92)	699
OMS engine (Aerojet)	(1992-93)	678
OFFMAR AVIO srl (Italy)	(1992-93)	645
OMNIPOL AS (Czech Republic)	(1995-96)	700
OMSK AIRCRAFT ENGINE DESIGN BUREAU MARS (Russia)	(1995-96)	728
ORAO AIR DEPOT (Yugoslavia)	(1995-96)	772
ORAO AIR FORCE DEPOT (Yugoslavia)	(1991-92)	744
ORENDA (see Hawker Siddeley Canada Inc)	(1991-92)	662

P

Entry	Edition	Page
P&WC (see Pratt & Whitney Canada)	(1995-96)	691
P-020 (KKBM)	(1992-93)	652
P-065 (KKBM)	(1992-93)	652
PK6A (Klimov)	(1995-96)	727
PK100 (Klimov)	(1995-96)	727
PS (see Povazske Strojarne)	(1995-96)	738
PS-90A (JSC Aviadvigatel)	(1995-96)	721
PS-90A10 (Aviadvigatel)	(1995-96)	722
PS-90A12 (Aviadvigatel)	(1995-96)	722
PT6A (P&WC)	(1995-96)	693
PT6B/PT6C (P&WC)	(1995-96)	694
PT6T Twin-Pac (P&WC)	(1995-96)	695
PUL 212 (JPX)	(1992-93)	630
PUL 425 (JPX)	(1992-93)	630
PW100 (P&WC)	(1995-96)	692
PW200 (P&WC)	(1995-96)	695
PW300 (P&WC)	(1995-96)	692
PW300 Turboprop (P&WC)	(1994-95)	681
PW305 (P&WC)	(1993-94)	604
PW500 (P&WC)	(1995-96)	692
PW1120 (Pratt & Whitney)	(1991-92)	735
PW2000 (Pratt & Whitney)	(1995-96)	765
PW4000 (Pratt & Whitney)	(1995-96)	763
PW5000 (Pratt & Whitney)	(1991-92)	736
PZL (see Pezetel Foreign Trade Enterprise)	(1995-96)	717
PZL (see Polskie Zaklady Lotnicze)	(1991-92)	690
PZL-3S (WSK-PZL-Rzeszów)	(1995-96)	719
PZL-3SR (WSK-PZL-Rzeszów)	(1995-96)	719

INDEXES: AERO-ENGINES–PREVIOUS TEN EDITIONS—P - W

Entry	Edition	Page
PZL-10W (WSK-PZL-Rzeszów)	(1995-96)	719
PZL-Franklin engines (WSK-PZL Rzeszów)	(1995-96)	719
PZL KALISZ (see Wytwornia Sprzetu Komunikacy Jnego-PZL Kalize)	(1995-96)	718
PZL RZESZÓW (see Wytwornia Sprzetu Komunikacy Jnego-PZL Rzeszów)	(1995-96)	718
PARODI, MOTORSEGLERTECHNIK (Germany)	(1992-93)	637
Pegasus (Rolls-Royce)	(1995-96)	748
PEZETEL FOREIGN TRADE ENTERPRISE (Poland)	(1995-96)	717
PIAGGIO, INDUSTRIE AERONAUTICHE E MECCANICHE RINALDO SpA (Italy)	(1995-96)	715
PIEPER MOTORENBAU GmbH (Germany)	(1992-93)	637
PNPP AVIADVIGATEL (Russia)	(1992-93)	655
POLSKIE ZAKLADY LOTNICZE (Poland)	(1991-92)	690
PORSCHE, DR ING h c F, PORSCHE AG (Germany)	(1991-92)	680
POVAZSKE STROJARNE (Slovak Republic)	(1995-96)	738
PPO AVIADVIGATEL (USSR)	(1991-92)	699
PRATT & WHITNEY	(1995-96)	763
PRATT & WHITNEY CANADA (Canada)	(1995-96)	691
Product 25-11 (UMKPA)	(1995-96)	736
Product 55B (UMKPA)	(1995-96)	736
Product 95-1 (UMKPA)	(1995-96)	736
Product 95-111 (UMKPA)	(1995-96)	736
PROGRESS (Ukraine)	(1995-96)	739
PROGRESS/LOTAREV/ZVL (International)	(1992-93)	642
PS-ZMKB (International)	(1995-96)	712
PW-ALLISON (USA)	(1992-93)	696

R

Entry	Edition	Page
R-1E (Marquardt)	(1992-93)	691
R-11 (Soyuz)	(1993-94)	644
R-13 (Soyuz)	(1995-96)	733
R-15 (Soyuz)	(1993-94)	645
R-21 (Soyuz)	(1993-94)	645
R-25 (Soyuz)	(1995-96)	734
R-27 (Soyuz)	(1995-96)	734
R-28 (Soyuz)	(1995-96)	734
R-29 (Soyuz)	(1995-96)	734
R-35 (Soyuz)	(1995-96)	734
R-40A (Marquardt)	(1992-93)	691
R-79 (Soyuz)	(1995-96)	735
R-123-300 (Soyuz)	(1995-96)	736
R-126-300 (Soyuz)	(1995-96)	736
R-127-300 (Soyuz)	(1995-96)	736
R-195 (Soyuz)	(1995-96)	733
R-266/R-31 (Soyuz)	(1993-94)	664
RB 168 military Spey (Rolls-Royce)	(1995-96)	746
RB.163 civil Spey (Rolls-Royce)	(1992-93)	674
RB.183 (Rolls-Royce)	(1992-93)	675
RB199 (Turbo-Union)	(1995-96)	713
RB211 (Rolls-Royce)	(1995-96)	744
RD-9 (Soyuz)	(1993-94)	644
RD-33 (Klimov)	(1995-96)	727
RD-36-35 (RKBM)	(1995-96)	729
RD-36-51 (RKBM)	(1995-96)	729
RD-38 (RKBM)	(1995-96)	730
RD-41 (RKBM)	(1995-96)	730
RD-60 (RKBM)	(1995-96)	730
RD-60A (Klimov)	(1995-96)	727
RD-93 (Klimov)	(1995-96)	727
RD-500K (CEC)	(1995-96)	696
RI 162 (Rotorway)	(1995-96)	767
RM8 (Flygmotor)	(1994-95)	725
RKBM (see Rybinsk Engine-Building Design Office)	(1995-96)	729
RKBM Project (RKBM)	(1995-96)	730
RM12 (Volvo)	(1995-96)	739
RSRM (Thiokol)	(1992-93)	704
RTM 322 (Rolls-Royce Turbomeca)	(1995-96)	713
RTWD-14 (Saturn)	(1995-96)	733
RU-19 (Soyuz)	(1993-94)	645
Ramjets (Marquardt)	(1992-93)	691
RECTIMO AVIATION SA (France)	(1992-93)	631
REFRIGERATION EQUIPMENT WORKS (Poland)	(1992-93)	648
ROCKETDYNE (USA)	(1994-95)	753
ROLLS-ROYCE LYULKA (International)	(1991-92)	684
ROLLS-ROYCE plc (UK)	(1995-96)	744
ROLLS-ROYCE SATURN (International)	(1993-94)	625
ROLLS-ROYCE TURBOMECA LIMITED (International)	(1995-96)	712
ROTAX (see Bombardier-Rotax)	(1995-96)	690
Rotax 503 (Rotax)	(1995-96)	690
Rotax 532 (Rotax)	(1993-94)	603
Rotax 582 (Rotax)	(1995-96)	690
Rotax 618UL DCDI (Rotax)	(1995-96)	690
Rotax 912A (Rotax)	(1995-96)	690
Rotax 914 DCDI (Rotax)	(1995-96)	690
ROTORWAY INTERNATIONAL (USA)	(1995-96)	767
RYBINSK ENGINE-BUILDING DESIGN OFFICE (Russia)	(1995-96)	729

S

Entry	Edition	Page
SA (see Singapore Aerospace Ltd)	(1995-96)	737
SAC (see Shenyang Aero-Engine Company)	(1995-96)	699
SAMP (see Shanghai Aero-Engine Manufacturing Plant)	(1995-96)	699
SARI (see Shenyang Aviation Engine Research Institute)	(1995-96)	699
SAT-41 (Saturn)	(1995-96)	733
SC 430 (König)	(1992-93)	636
SD 570 (König)	(1992-93)	636
SE 1800 EIS (Sauer)	(1992-93)	638
SF 930 (König)	(1991-92)	678
SL 1700 (Limbach)	(1995-96)	707
SMPMC (see South Motive Power and Machinery Complex)	(1995-96)	699
SMR-95 (Klimov/Aerosud)	(1995-96)	727
SNECMA (see Société Nationale d'Étude et de Construction de Moteurs d'Aviation)	(1995-96)	702
SO-1 (IL)	(1991-92)	690
SO-3 (IL)	(1991-92)	690
SS 2100 H1S (Sauer)	(1992-93)	638
SSME (Rocketdyne)	(1994-95)	753
SSRM (Thiokol)	(1994-95)	760
ST 2500 H1S (Sauer)	(1992-93)	638
SALYUT (Russia)	(1995-96)	730
SAMARA STATE SCIENTIFIC & PRODUCTION ENTERPRISE (Russia)	(1995-96)	731
SARL JPX (France)	(1995-96)	702
SATURN NPO (Russia)	(1995-96)	732
SAUER MOTORENBAU GmbH (Germany)	(1992-93)	638
SCOMA-ÉNERGIE (France)	(1992-93)	631
SENER-INGENIERIA Y SISTEMAS (Spain)	(1994-95)	725
SHANGHAI AERO-ENGINE MANUFACTURING PLANT (China)	(1995-96)	699
SHENYANG AERO-ENGINE COMPANY (China)	(1995-96)	699
SHENYANG AVIATION ENGINE RESEARCH INSTITUTE (China)	(1995-96)	699
SHVETSOV (Russia)	(1995-96)	733
SINGAPORE AEROSPACE LTD (Singapore)	(1995-96)	737
SOCIETE NATIONALE D'ETUDE ET DE CONSTRUCTION DE MOTEURS D'AVIATION (France)	(1995-96)	702
SOCIETE TURBOMECA (France)	(1995-96)	703
SOLOY CORPORATION (USA)	(1995-96)	767
SOLOY DUAL PAC INC (USA)	(1995-96)	767
SOUTH MOTIVE POWER AND MACHINERY COMPLEX (China)	(1995-96)	699
SOYUZ (Russia)	(1995-96)	733
Space Shuttle OMS engine (Aerojet)	(1992-93)	678
Space Shuttle SSME (Rocketdyne)	(1994-95)	753
Space Shuttle SSRM (Thiokol)	(1994-95)	760
Spey, civil (Rolls-Royce)	(1992-93)	674
Spey, military (Rolls-Royce)	(1995-96)	746
Spey 807 (Fiat)	(1995-96)	715
Stamo MS 1500 (Pieper)	(1992-93)	637
SVERDLOV factory (Russia)	(1995-96)	736

T

Entry	Edition	Page
T53 (AlliedSignal)	(1995-96)	752
T55 (AlliedSignal)	(1995-96)	752
T56 (Allison)	(1995-96)	754
T63 (Allison)	(1995-96)	753
T64 (General Electric)	(1995-96)	760
T74 (P&WC)	(1995-96)	693, 694
T76 (AlliedSignal)	(1995-96)	750
T101 (P&WC)	(1995-96)	693
T400 (P&WC)	(1992-93)	623
T400 (P&WC)	(1995-96)	695
T406-AD-400 (Allison)	(1995-96)	754
T620 (JPX)	(1995-96)	702
T700 (General Electric)	(1995-96)	760
T703 (Allison)	(1995-96)	753
T800-LHT-800 (LHTEC)	(1995-96)	762
TCM (see Teledyne Continental Motors)	(1995-96)	767
TEI (see Tusas Engine Industries)	(1995-96)	739
TFA (Microturbo)	(1992-93)	631
TF30 (Pratt & Whitney)	(1991-92)	734
TF34 (General Electric)	(1995-96)	757
TFE109 (AlliedSignal)	(1995-96)	749
TFE731 (AlliedSignal)	(1995-96)	749
TFE1042-70 (AlliedSignal)	(1995-96)	750
TIO-540 Series (Textron Lycoming)	(1995-96)	770
TIO-541 Series (Textron Lycoming)	(1995-96)	770
TM 333 (Turbomeca)	(1995-96)	705
TOP (F + E)	(1992-93)	635
TP 319 Arrius (Turbomeca)	(1991-92)	675
TP-500 (TCM)	(1992-93)	698
TPE331 (AlliedSignal)	(1995-96)	750
TPF 351 (Garrett)	(1994-95)	742
TPJ 1304HF (VM Motori)	(1995-96)	716
TPJ 1306HF (VM Motori)	(1995-96)	716
TPJ 1308HF (VM Motori)	(1995-96)	716
TRB (Microturbo)	(1992-93)	630
TRD-3 (Soyuz)	(1993-94)	644
TRD-29 (Soyuz)	(1993-94)	645
TRD-31 (Saturn)	(1993-94)	661
TRD-37 (Soyuz)	(1993-94)	644
TRS 18 (Microturbo)	(1992-93)	631
TV-0-100 (Mars)	(1993-94)	638
TV-0-100 (Soyuz)	(1995-96)	735
TV116-300 (Soyuz)	(1995-96)	736
TV2-117 (Klimov)	(1995-96)	725
TV3-117 (Klimov)	(1995-96)	726
TV7-46 (Klimov)	(1995-96)	726
TV7-117 (Klimov)	(1995-96)	726
TVD-10 (Mars)	(1995-96)	728
TVD-20 (Mars)	(1995-96)	728
TVD-450 (Soyuz)	(1995-96)	736
TVD-1500 (RKBM)	(1995-96)	730
TWD-10B (WSK-PZL-Rzeszów)	(1995-96)	718
Tay (Rolls-Royce)	(1995-96)	747
TCAE TURBINE ENGINES (USA)	(1995-96)	767
TECHSPACE AERO SA (Belgium)	(1995-96)	691
TELEDYNE CAE DIVISION OF TELEDYNE INC (USA)	(1994-95)	756
TELEDYNE CONTINENTAL MOTORS (USA)	(1995-96)	767
TEXTRON LYCOMING (USA)	(1995-96)	770
THERMO-JET STANDARD INC (USA)	(1992-93)	703
Tornado (Fisher)	(1991-92)	721
Trent (Rolls-Royce)	(1995-96)	745
TRUD factory (Russia)	(1995-96)	736
TUMANSKY (RUSSIA)	(1995-96)	736
Turbine Pac (Soloy)	(1995-96)	767
Turbo 90 (CAM)	(1992-93)	619
TURBOMECA-SNECMA GROUPEMENT GRTS (France)	(1995-96)	706
TURBOMECA, SOCIETE (France)	(1995-96)	703
TURBOMECANICA (see Intreprinderea Turbomecanica Bucuresti)	(1995-96)	720
TURBO-UNION LTD (International)	(1995-96)	713
Turmo (Turbomeca)	(1995-96)	705
TUSAS ENGINE INDUSTRIES (Turkey)	(1995-96)	739
Tyne (Rolls-Royce)	(1995-96)	747
Type 4T60/A (JPX)	(1995-96)	702
Type 4TX 75 (JPX)	(1995-96)	702
Type 18D/SS (CRM)	(1995-96)	715
Type 301 (NPT)	(1991-92)	710
Type 352 (Teledyne CAE)	(1992-93)	700
Type 352 (Teledyne CAE)	(1994-95)	756
Type 535 (Rolls-Royce)	(1995-96)	746
Type 754 (NPT)	(1991-92)	710
Type 2701 R 03 (Hirth)	(1992-93)	636
Type 2702 R 03 (Hirth)	(1992-93)	636
Type 2703 (Hirth)	(1992-93)	636
Type 2704 (Hirth)	(1992-93)	636
Type R (MKB)	(1991-92)	698

U

Entry	Edition	Page
UDF (General Electric)	(1991-92)	726
UMKPA (Russia)	(1995-96)	736
UNITED TECHNOLOGIES PRATT & WHITNEY (USA)	(1995-96)	763
UNITED TURBINES (UK) LTD (UK)	(1992-93)	678
UP ARROW AIRCRAFT (USA)	(1992-93)	704

V

Entry	Edition	Page
V2500 (IAE)	(1995-96)	711
VAZ-413 (VAZ)	(1995-96)	736
VAZ-430/D-200 (VAZ)	(1995-96)	736
VD-7 (RKBM)	(1995-96)	730
VD-19 (RKBM)	(1995-96)	730
VOKBM (see Voronezhskoyeopytno-Konstruktorskoye Byuro Motorostroyeniya)	(1995-96)	737
VAZ car plant (Russia)	(1995-96)	736
VEDENEYEV (Russia)	(1994-95)	724
Viper (Rolls-Royce)	(1995-96)	747
Viper 600 (Fiat/Rolls-Royce)	(1995-96)	715
VM MOTORI SpA (Italy)	(1995-96)	716
VMKB Bureau (Russia)	(1993-94)	648
VOKBM Bureau (Russia)	(1994-95)	724
VOLSZHSKY bureau (Russia)	(1995-96)	737
VOLVO AERO CORPORATION (Sweden)	(1995-96)	738
VORONEZHSKOYEOPTYNO-KONSTRUKTORSKOYE BYURO MOTOROSTROYENIYA (Russia)	(1995-96)	737
Voyager 550 (TCM)	(1995-96)	768

W

Entry	Edition	Page
W 5/33 (Westmayer)	(1992-93)	618
WJ5 (DEMC)	(1995-96)	696

W – Z—AERO-ENGINES–PREVIOUS TEN EDITIONS: INDEXES

WJ6A (SMPMC) .... (1995-96) 699
WJ9 (ZAC) .... (1995-96) 699
WP5 (LM) .... (1992-93) 624
WP6 (LM) .... (1995-96) 696
WP7 (LM) .... (1995-96) 697
WP7B (LMC) .... (1995-96) 698
WP8 (XAE) .... (1993-94) 611
WP-11 (BUAA) .... (1995-96) 695
WP13 (LMC) .... (1995-96) 698
WP13G (CEC) .... (1995-96) 697
WP14 (CEC) .... (1995-96) 697
WS6 (LM) .... (1995-96) 697
WS8 (SAMP) .... (1995-96) 699
WS-9 (XAE) .... (1993-94) 612
WSK-PZL-KALISZ (see Wytwórnia Sprzetu
  Komunikacyjnego-PZL Kalisz) .... (1993-94) 630
WSK-PZL-RZESZOW (see Wytwórnia Sprzetu
  Komunikacyjnego-PZL Rzeszów) .... (1993-94) 630
WZ5 (ZARI) .... (1995-96) 700
WZ-6 (CLXMW) .... (1995-96) 696
WZ8 (SMPMC) .... (1995-96) 699

WESTERMAYER, OSKAR (Austria) .... (1992-93) 618
WILLIAMS INTERNATIONAL (USA) .... (1995-96) 772
WILLIAMSPORT (USA) .... (1992-93) 702

WYTWÓRNIA SPRZETU KOMUNIKACYJNEGO-PZL
  KALISZ (Poland) .... (1995-96) 718
WYTWÓRNIA SPRZETU KOMUNIKACYJNEGO-PZL
  RZESZÓW (Poland) .... (1995-96) 718

X

XAE (see Xian Aero-Engine
  Corporation) .... (1995-96) 699
XG-40 (Rolls-Royce) .... (1992-93) 676

XIAN AERO-ENGINE CORPORATION
  (China) .... (1995-96) 699

Y

YH40 (YMF) .... (1992-93) 627
YH280 (YMF) .... (1992-93) 627
YMF (see Yuhe Machine Factory) .... (1992-93) 627
YO-65 (Nelson) .... (1995-96) 762
YT702-LD-700 (AlliedSignal) .... (1995-96) 752

YUHE MACHINE FACTORY (China) .... (1992-93) 627

Z

ZAC (see Zhuzhou Aero-Engine
  Company) .... (1995-96) 699
ZARI (see Zhuzhou Aero-Engine Research
  Institute) .... (1995-96) 700
ZEF (see Zhuzhou Aero-Engine Factory) .... (1995-96) 700
ZMKB PROGRESS (Ukraine) .... (1995-96) 739
ZO 01A (Zoche) .... (1995-96) 708
ZO 02A (Zoche) .... (1995-96) 708
ZVL (see Zavody Na Vyrobu Lozisk) .... (1992-93) 630

ZHUZHOU AERO-ENGINE COMPANY
  (China) .... (1995-96) 699
ZHUZHOU AERO-ENGINE FACTORY
  (China) .... (1995-96) 700
ZHUZHOU AERO-ENGINE RESEARCH
  INSTITUTE (China) .... (1995-96) 700
ZOCHE, MICHAEL (Germany) .... (1995-96) 708

NOTES

NOTES

NOTES